CHILTON®

ASIAN
SERVICE MANUAL
2012 EDITION
VOLUME I
ACURA
HONDA

CENGAGE
Learning·

Australia • Brazil • Japan • Korea • Mexico • Singapore • Spain • United Kingdom • United States

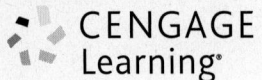
CENGAGE
Learning®

CHILTON®
Asian Service Manual
2012 Edition
Volume I
Acura and Honda

Vice President,
Technology & Trades Professional
Business Unit
Gregory L. Clayton

Publisher:
David Koontz

Director of Marketing:
Beth A. Lutz

Senior Production Director:
Wendy Troeger

Production Manager:
Sherondra Thedford

Senior Marketing Manager:
Jennifer Barbic

Associate Marketing Manager:
Rachael Torres

Chilton Content Specialist:
Paula Baillie

Graphical Designer:
Melinda Possinger

Art Director:
Benjamin Gleeksman

Sr. Content Project Manager:
William Tubbert

Senior Editors:
Christine L. Sheeky

Ryan Lee Price

Editors:
Dennis L. Bailey

Jim Bailey

For product information and technology assistance, contact us at
Professional & Career Group customer Support, 1-800-648-7450.
For permission to use material from this text or product,
submit all requests online at
www.cengage.com/permissions.
Further permissions questions can be e-mailed to
permissionrequest@cengage.com

ISBN-13: 978-1-2854-7105-1
ISBN-10: 1-2854-7105-9
ISSN: 2161-8755

Chilton
5 Maxwell Drive
Clifton Park, NY 12065-2919
USA

Chilton products are represented in Canada by Nelson Education, Ltd.

NOTICE TO THE READER

Printed in the United States of America
1 2 3 4 5 6 7 17 16 15 14 13

Contents

Sections

1   MDX

2   RDX

3   RL

4   TL

5   TSX

6   ZDX

    Acura Diagnostic Trouble Codes

7   Accord, Crosstour

8   Civic, Civic Hybrid

9   CR-V

10   CR-Z Hybrid

11   Element

12   Fit

13   Insight Hybrid

14   Odyssey

15   Pilot

16   Ridgeline

    Honda Diagnostic Trouble Codes

Model Index

Model	Section No.	Model	Section No.	Model	Section No.
A		**I**		**T**	
Accord	7-1	Insight Hybrid	13-1	TL	4-1
C		**M**		TSX	5-1
Civic	8-1	MDX	1-1	**Z**	
Civic Hybrid	8-1	**O**		ZDX	6-1
Crosstour	7-1	Odyssey	14-1		
CR-V	9-1	**P**			
CR-Z Hybrid		Pilot	15-1		
E		**R**			
Element	11-1	RDX	2-1		
F		Ridgeline	16-1		
Fit	12-1	RL	3-1		

USING THIS INFORMATION

Organization

To find where a particular model section or procedure is located, look in the Table of Contents. Main topics are listed with the page number on which they may be found. Following the main topics is an alphabetical listing of all of the procedures within the section and their page numbers.

Manufacturer and Model Coverage

This product covers 2011-2012 Asian models that are produced in sufficient quantities to warrant coverage, and which have technical content available from the vehicle manufacturers before our publication date. Although this information is as complete as possible at the time of publication, some manufacturers may make changes which cannot be included here. While striving for total accuracy, the publisher cannot assume responsibility for any errors, changes, or omissions that may occur in the compilation of this data.

Part Numbers and Special Tools

Part numbers and special tools are recommended by the publisher and vehicle manufacturer to perform specific jobs. Before substituting any part or tool for the one recommended, you must be completely satisfied that neither your personal safety, nor the performance of the vehicle will be endangered.

ACKNOWLEDGEMENT

Portions of materials contained herein have been reprinted under license from American Honda Corporation, License Agreement 11201AH.

No further reproduction or distribution of the material in this manual is allowed without the expressed written permission of the vehicle manufacturers and the publisher.

PRECAUTIONS

Before servicing any vehicle, please be sure to read all of the following precautions, which deal with personal safety, prevention of component damage, and important points to take into consideration when servicing a motor vehicle:

- Always wear safety glasses or goggles when drilling, cutting, grinding or prying.
- Steel-toed work shoes should be worn when working with heavy parts. Pockets should not be used for carrying tools. A slip or fall can drive a screwdriver into your body.
- Work surfaces, including tools and the floor should be kept clean of grease, oil or other slippery material.
- When working around moving parts, don't wear loose clothing. Long hair should be tied back under a hat or cap, or in a hair net.
- Always use tools only for the purpose for which they were designed. Never pry with a screwdriver.
- Keep a fire extinguisher and first aid kit handy.
- Always properly support the vehicle with approved stands or lift.
- Always have adequate ventilation when working with chemicals or hazardous material.
- Carbon monoxide is colorless, odorless and dangerous. If it is necessary to operate the engine with vehicle in a closed area such as a garage, always use an exhaust collector to vent the exhaust gases outside the closed area.

- When draining coolant, keep in mind that small children and some pets are attracted by ethylene glycol antifreeze, and are quite likely to drink any left in an open container, or in puddles on the ground. This will prove fatal in sufficient quantity. Always drain the coolant into a sealable container.
- To avoid personal injury, do not remove the coolant pressure relief cap while the engine is operating or hot. The cooling system is under pressure; steam and hot liquid can come out forcefully when the cap is loosened slightly. Failure to follow these instructions may result in personal injury. The coolant must be recovered in a suitable, clean container for reuse. If the coolant is contaminated it must be recycled or disposed of correctly.
- When carrying out maintenance on the starting system be aware that heavy gauge leads are connected directly to the battery. Make sure the protective caps are in place when maintenance is completed. Failure to follow these instructions may result in personal injury.
- Do not remove any part of the engine emission control system. Operating the engine without the engine emission control system will reduce fuel economy and engine ventilation. This will weaken engine performance and shorten engine life. It is also a violation of Federal law.
- Due to environmental concerns, when the air conditioning system is drained, the

refrigerant must be collected using refrigerant recovery/recycling equipment. Federal law requires that refrigerant be recovered into appropriate recovery equipment and the process be conducted by qualified technicians who have been certified by an approved organization, such as MACS, ASI, etc. Use of a recovery machine dedicated to the appropriate refrigerant is necessary to reduce the possibility of oil and refrigerant incompatibility concerns. Refer to the instructions provided by the equipment manufacturer when removing refrigerant from or charging the air conditioning system.

- Always disconnect the battery ground when working on or around the electrical system.
- Batteries contain sulfuric acid. Avoid contact with skin, eyes, or clothing. Also, shield your eyes when working near batteries to protect against possible splashing of the acid solution. In case of acid contact with skin or eyes, flush immediately with water for a minimum of 15 minutes and get prompt medical attention. If acid is swallowed, call a physician immediately. Failure to follow these instructions may result in personal injury.
- Batteries normally produce explosive gases. Therefore, do not allow flames, sparks or lighted substances to come near the battery. When charging or working near a battery, always shield your face and protect your eyes. Always provide ventilation. Failure to follow these instructions may result in personal injury.

- When lifting a battery, excessive pressure on the end walls could cause acid to spew through the vent caps, resulting in personal injury, damage to the vehicle or battery. Lift with a battery carrier or with your hands on opposite corners. Failure to follow these instructions may result in personal injury.

- Observe all applicable safety precautions when working around fuel. Whenever servicing the fuel system, always work in a well-ventilated area. Do not allow fuel spray or vapors to come in contact with a spark, open flame, or excessive heat (a hot drop light, for example). Keep a dry chemical fire extinguisher near the work area. Always keep fuel in a container specifically designed for fuel storage; also, always properly seal fuel containers to avoid the possibility of fire or explosion. Do not smoke or carry lighted tobacco or open flame of any type when working on or near any fuel-related components.

- Fuel injection systems often remain pressurized, even after the engine has been turned OFF. The fuel system pressure must be relieved before disconnecting any fuel lines. Failure to do so may result in fire and/or personal injury.

- The evaporative emissions system contains fuel vapor and condensed fuel vapor. Although not present in large quantities, it still presents the danger of explosion or fire. Disconnect the battery ground cable from the battery to minimize the possibility of an electrical spark occurring, possibly causing a fire or explosion if fuel vapor or liquid fuel is present in the area. Failure to follow these instructions can result in personal injury.

- The EPA warns that prolonged contact with used engine oil may cause a number of skin disorders, including cancer! You should make every effort to minimize your exposure to used engine oil. Protective gloves should be worn when changing oil. Wash your hands and any other exposed skin areas as soon as possible after exposure to used engine oil. Soap and water, or waterless hand cleaner should be used.

- Some vehicles are equipped with an air bag system, often referred to as a Supplemental Restraint System (SRS) or Supplemental Inflatable Restraint (SIR) system. The system must be disabled before performing service on or around system components, steering column, instrument panel components, wiring and sensors. Failure to follow safety and disabling procedures could result in accidental air bag deployment, possible personal injury and unnecessary system repairs.

- Always wear safety goggles when working with, or around, the air bag system. When carrying a non-deployed air bag, be sure the bag and trim cover are pointed away from your body. When placing a non-deployed air bag on a work surface, always face the bag and trim cover upward, away from the surface. This will reduce the motion of the module if it is accidentally deployed.

- Electronic modules are sensitive to electrical charges. The ABS module can be damaged if exposed to these charges.

- Brake pads and shoes may contain asbestos, which has been determined to be a cancer-causing agent. Never clean brake surfaces with compressed air. Avoid inhaling brake dust. Clean all brake surfaces with a commercially available brake cleaning fluid.

- When replacing brake pads, shoes, discs or drums, replace them as complete axle sets.

- When servicing drum brakes, disassemble and assemble one side at a time, leaving the remaining side intact for reference.

- Brake fluid often contains polyglycol ethers and polyglycols. Avoid contact with the eyes and wash your hands thoroughly after handling brake fluid. If you do get brake fluid in your eyes, flush your eyes with clean, running water for 15 minutes. If eye irritation persists, or if you have taken brake fluid internally, immediately seek medical assistance.

- Clean, high quality brake fluid from a sealed container is essential to the safe and proper operation of the brake system. You should always buy the correct type of brake fluid for your vehicle. If the brake fluid becomes contaminated, completely flush the system with new fluid. Never reuse any brake fluid. Any brake fluid that is removed from the system should be discarded. Also, do not allow any brake fluid to come in contact with a painted or plastic surface; it will damage the paint.

- Never operate the engine without the proper amount and type of engine oil; doing so will result in severe engine damage.

- Timing belt maintenance is extremely important! Many models utilize an interference- type, non freewheeling engine. If the timing belt breaks, the valves in the cylinder head may strike the pistons, causing potentially serious (also time-consuming and expensive) engine damage.

- Disconnecting the negative battery cable on some vehicles may interfere with the functions of the on-board computer system(s) and may require the computer to undergo a relearning process once the negative battery cable is reconnected.

- Steering and suspension fasteners are critical parts because they affect performance of vital components and systems and their failure can result in major service expense. They must be replaced with the same grade or part number or an equivalent part if replacement is necessary. Do not use a replacement part of lesser quality or substitute design. Torque values must be used as specified during reassembly.

ACURA

MDX

BRAKES 1-9

ANTI-LOCK BRAKE
 SYSTEM (ABS) 1-9
 General Information....................1-9
  Precautions...........................1-9
 Speed Sensors 1-9
  Removal & Installation............1-9
BLEEDING THE BRAKE
 SYSTEM......................1-10
 Bleeding Procedure 1-10
  Bleeding Procedure 1-10
FRONT DISC BRAKES 1-11
 Brake Caliper...........................1-11
  Removal & Installation............1-11
 Disc Brake Pads 1-11
  Removal & Installation............1-11
PARKING BRAKE 1-13
 Parking Brake Shoes 1-13
  Removal & Installation...........1-13
REAR DISC BRAKES 1-12
 Brake Caliper...........................1-12
  Removal & Installation............1-12
 Disc Brake Pads 1-12
  Removal & Installation............1-12

CHASSIS ELECTRICAL 1-14

AIR BAG (SUPPLEMENTAL
 RESTRAINT SYSTEM)........1-14
 General Information....................1-14
  Arming the System 1-15
  Clockspring Centering 1-15
  Disarming the
   System 1-15
  Removal & Installation..........1-15
  Service Precautions 1-14

DRIVE TRAIN..................1-16

 CV-Boots.................................1-16
  Inspection 1-16
 Differential Carrier 1-16
  Removal & Installation...........1-16
 Front Halfshaft..........................1-18
  Removal & Installation............1-18
 Propeller Shaft 1-19
  Removal & Installation............1-19

Rear Drive Axle.........................1-19
 Drain & Refill.............................1-19
 Fluid Recommendations 1-19
 Level Check 1-19
Rear Halfshaft 1-20
 Removal & Installation...........1-20

ENGINE COOLING 1-20

Engine Coolant.........................1-20
 Drain & Refill.........................1-20
Radiator 1-21
 Removal & Installation..........1-21
Thermostat 1-22
 Removal & Installation..........1-22
Water Pump 1-22
 Removal & Installation..........1-22

ENGINE ELECTRICAL.........1-23

BATTERY SYSTEM 1-23
 Battery.....................................1-23
  Battery Reconnect/Relearn
   Procedure 1-23
  Removal & Installation..........1-23
CHARGING SYSTEM 1-24
 Alternator 1-24
  Removal & Installation..........1-24
IGNITION SYSTEM 1-24
 Firing Orders 1-24
 Ignition Coil 1-24
  Removal & Installation..........1-24
 Ignition Timing.........................1-24
  Inspection & Adjustment 1-24
 Spark Plugs 1-24
  Removal & Installation..........1-24
STARTING SYSTEM 1-25
 Starter 1-25
  Removal & Installation..........1-25

ENGINE MECHANICAL 1-25

Accessory Drive Belt System.....1-25
 Adjustment 1-25
 Belt Routings 1-25
 Inspection 1-25
 Removal & Installation..........1-25
Air Intake System 1-26
 Removal & Installation..........1-26

Camshaft & Bearings 1-26
 Inspection 1-26
 Removal & Installation..........1-27
Catalytic Converter 1-27
 Removal & Installation..........1-27
Crankshaft Front Seal...............1-28
 Removal & Installation..........1-28
Crankshaft Pulley 1-29
 Removal & Installation..........1-29
Crankshaft Rear Cover &
 Seal..1-28
 Removal & Installation..........1-28
Cylinder Head 1-31
 Removal & Installation..........1-31
Cylinder Head Cover 1-29
 Removal & Installation..........1-29
Engine Cover............................1-32
 Removal & Installation..........1-32
Engine Oil & Filter 1-32
 OIL & Filter Change..............1-32
Engine Rear Cover....................1-32
 Removal & Installation..........1-32
Intake Manifold 1-33
 Removal & Installation..........1-33
Oil Pan 1-34
 Removal & Installation..........1-34
Oil Pump 1-34
 Removal & Installation..........1-34
Pistons & Rings 1-35
 Positioning 1-35
Rocker Arms.............................1-35
 Removal & Installation..........1-35
Timing Belt & Sprockets 1-36
 Removal & Installation..........1-36
Timing Belt Front Cover 1-36
 Removal & Installation..........1-36
Timing Belt Rear Cover 1-37
 Removal & Installation..........1-37
Valve Lash................................1-37
 Adjustment 1-37

ENGINE PERFORMANCE &
EMISSION CONTROLS 1-39

Accelerator Pedal Position
 (APP) Sensor 1-39
  Location...................................1-39
  Removal & Installation..........1-39

Air-Fuel Ratio (A/F) Sensor 1-39
   Location 1-39
   Removal & Installation 1-39
Camshaft Position (CMP)
   Sensor 1-39
   Location 1-39
   Removal & Installation 1-39
Crankshaft Position (CKP)
   Sensor 1-39
   Location 1-39
   Removal & Installation 1-39
Engine Coolant Temperature
   (ECT) Sensor 1-40
   Location 1-40
   Removal & Installation 1-40
Evaporative Emission
   Control System 1-40
   Removal & Installation 1-40
Exhaust Gas Recirculation
   (EGR) Valve 1-40
   Location 1-40
   Removal & Installation 1-42
Heated Oxygen (HO2S)
   Sensor 1-42
   Removal & Installation 1-42
Input Shaft Speed Sensor 1-42
   Removal & Installation 1-42
Intake Air Temperature (IAT)/
   Mass Airflow (MAF) Sensor 1-42
   Location 1-42
   Removal & Installation 1-42
Knock Sensor (KS) 1-43
   Location 1-43
   Removal & Installation 1-43
Manifold Absolute
   Pressure (MAP) Sensor 1-43
   Removal & Installation 1-43
Output Shaft Speed (OSS)
   Sensor 1-43
   Location 1-43
   Removal & Installation 1-43
Positive Crankcase
   Ventilation (PCV) Valve 1-44
   Location 1-44
   Removal & Installation 1-44
Powertrain Control
   Module (PCM) 1-44
   ECM/PCM Idle Learn
      Procedure 1-46
   Location 1-44
   Removal & Installation 1-44
Throttle Position Sensor
   (TPS) 1-46
   Location 1-46

Transaxle Fluid Temperature
   (TFT) Sensor 1-46
   Location 1-46
   Removal & Installation 1-46

FUEL 1-47

GASOLINE FUEL INJECTION SYSTEM 1-47
Fuel Filter 1-48
   Removal & Installation 1-48
Fuel Pump Control Module 1-49
   Removal & Installation 1-49
Fuel Rail & Injectors 1-49
   Removal & Installation 1-49
Fuel System Service
   Precautions 1-47
Fuel Tank 1-48
   Draining 1-48
   Removal & Installation 1-48
Relieving Fuel System
   Pressure 1-47
   With the HDS 1-47
   Without the HDS 1-47
Throttle Body 1-49
   Component Locations 1-49
   Removal & Installation 1-49

HEATING & AIR CONDITIONING SYSTEM 1-51

Blower Motor 1-51
   Removal & Installation ... 1-51
Heater Core 1-51
   Removal & Installation 1-51

PRECAUTIONS 1-9

SPECIFICATIONS AND MAINTENANCE CHARTS 1-3

Brake Specifications 1-7
Camshaft Specifications 1-5
Capacities 1-4
Crankshaft and Connecting
   Rod Specifications 1-5
Engine and Model Year
   Identification 1-3
Fluid Specifications 1-4
Engine Tune-Up Specifications ... 1-3
General Engine Specifications 1-3
Piston and Ring Specifications ... 1-5
Scheduled Maintenance
   Intervals 1-8

Tire Wheel and Ball Joint
   Specifications 1-7
Torque Specifications 1-6
Valve Specifications 1-4
Wheel Alignment 1-6

STEERING 000

Power Rack & Pinion
   Steering Gear 1-53
   Removal & Installation 1-53
Power Steering Pump 1-57
   Bleeding 1-57
   Fluid Replacement
      Procedure 1-57
   Removal & Installation 1-57

SUSPENSION 1-58

FRONT SUSPENSION 1-58
Knuckle 1-58
   Removal & Installation 1-58
Lower Ball Joints 1-58
   Removal & Installation 1-58
Lower Control Arms 1-59
   Removal & Installation 1-59
Stabilizer Bar & Links 1-59
   Removal & Installation 1-59
Struts 1-63
   Overhaul 1-64
   Removal & Installation 1-63
Wheel Hubs & Bearings 1-67
   Removal & Installation 1-67

REAR SUSPENSION 1-68
Coil Springs 1-68
   Removal & Installation 1-68
Knuckle 1-68
   Component Locations 1-68
   Removal & Installation 1-68
Lower Control Arms 1-70
   Removal & Installation 1-70
Shock Absorbers 1-71
   Removal & Installation 1-71
Stabilizer Bar & Links 1-72
   Removal & Installation 1-72
Trailing Arms 1-72
   Removal & Installation 1-72
Upper Ball Joints 1-73
   Removal & Installation 1-73
Upper Control Arms 1-74
   Removal & Installation 1-74
Wheel Hubs & Bearings 1-74
   Removal & Installation 1-74

SPECIFICATIONS AND MAINTENANCE CHART

ENGINE AND VEHICLE IDENTIFICATION

		Engine						Model Year	
Code	Liters (cc)	Cu. In.	Cyl.	Fuel Sys.	Engine Type	Eng. Mfg.	Code ①	Year
J37A1	3.7 (3664)	224	6	PGM-FI	SOHC	Honda	B	2011
							C	2012

PGM-FI: Programmed Fuel Injection

SOHC: Single Overhead Camshaft

① 10th digit of the Vehicle Identification Number (VIN)

71051_AMDX_C0001

GENERAL ENGINE SPECIFICATIONS

Year	Model	Engine Displacement Liters	Engine ID	Net Horsepower @ rpm	Net Torque @ rpm (ft. lbs.)	Bore x Stroke (in.)	Com-pression Ratio	Oil Pressure @ rpm
2011	MDX	3.7	J37A1	300@6000	275@5000	3.54x3.78	11.0:1	71@3000
2012	MDX	3.7	J37A1	300@6000	275@5000	3.54x3.78	11.0:1	71@3000

71051_AMDX_C0002

GASOLINE ENGINE TUNE-UP SPECIFICATIONS

Year	Engine Displacement Liters	Engine ID	Spark Plug Gap (in.)	Ignition Timing (deg.) MT	Ignition Timing (deg.) AT	Fuel Pump (psi)	Idle Speed (rpm) MT	Idle Speed (rpm) AT	Valve Clearance In.	Valve Clearance Ex.
2011	3.7	J37A1	0.039-0.043	NA	8-12 ②	57-64 ①	NA	660-760	0.008-0.009	0.011-0.013
2012	3.7	J37A1	0.039-0.043	NA	8-12 ②	57-64 ①	NA	660-760	0.008-0.009	0.011-0.013

NOTE: The Vehicle Emission Control Information label reflects specification changes during production and must be used if they differ from this chart.

① At idle, pressure regulator vacuum hose disconnected

② : Before Top Dead Center

71051_AMDX_C0003

CAPACITIES

| Year | Model | Engine Displacement Liters | Engine ID | Engine Oil with Filter (qts.) | Transmission (pts.) | | | Drive Axle | | Fuel Tank (gal.) | Cooling System (qts.) |
					5-Spd	6-Spd	Auto.	Front (pts.)	Rear (pts.)		
2011	MDX	3.7	J37A1	4.5	NA	NA	3.0	NA	5.34	21.0	7.7
2012	MDX	3.7	J37A1	4.5	NA	NA	3.0	NA	5.34	21.0	7.7

NOTE: All capacities are approximate. Add fluid gradually and ensure a proper fluid level is obtained.

NOTE: Capacities given are service capacities, not overhaul capacities

71051_AMDX_C0004

FLUID SPECIFICATIONS

| Year | Model | Engine Displ. Liters | Engine Oil | Man. Trans. | Auto. Trans. | Drive Axle | | Transfer Case | Power Steering Fluid | Brake Master Cylinder | Cooling System |
						Front	Rear				
2011	MDX	3.7	5W-20 Acura	Acura MTF	Acura ATF-Z1	NA	NA	①	Acura PS Fluid	Acura DOT 3	②
2012	MDX	3.7	5W-20 Acura	Acura MTF	Acura ATF-Z1	NA	NA	①	Acura PS Fluid	Acura DOT 3	②

① Hypoid gear oil SAE 90 or SAE 80W-90 viscosity, API classified GL4 or GL5 only

② Acura Long Life Antifreeze/Coolant-Type2

71051_AMDX_C0005

VALVE SPECIFICATIONS

| Year | Engine Displacement Liters | Engine ID | Seat Angle (deg.) | Face Angle (deg.) | Spring Test Pressure (lbs. @ in.) | Spring Installed Height (in.) | Stem-to-Guide Clearance (in.) | | Stem Diameter (in.) | |
							Intake	Exhaust	Intake	Exhaust
2011	3.7	J37A1	45	45	NA	NA	0.0008-0.0018	0.0022-0.0031	0.2159-0.2163	0.2146-0.2150
2012	3.7	J37A1	45	45	NA	NA	0.0008-0.0018	0.0022-0.0031	0.2159-0.2163	0.2146-0.2150

NA: Not Available

71051_AMDX_C0006

CAMSHAFT AND BEARING SPECIFICATIONS

All measurements are given in inches.

Year	Engine Displacement Liters	Engine ID	Journal Diameter	Brg. Oil Clearance	Shaft End-play	Runout	Journal Bore	Lobe Height Intake	Lobe Height Exhaust
2011	3.7	J37A1	NA	0.0020-0.0035	0.0020-0.0080	0.0010	NA	①	1.4326
2012	3.7	J37A1	NA	0.0020-0.0035	0.0020-0.0080	0.0010	NA	①	1.4326

NA: Information not available
① Primary: 1.3824 inches
  Mid: 1.4328 inches
  Secondary: 1.3824 inches

71051_AMDX_C0009

CRANKSHAFT AND CONNECTING ROD SPECIFICATIONS

All measurements are given in inches.

Year	Engine Displacement Liters	Engine ID	Crankshaft Main Brg. Journal Dia.	Main Brg. Oil Clearance	Shaft End-play	Thrust on No.	Connecting Rod Journal Diameter	Oil Clearance	Side Clearance
2011	3.7	J37A1	2.8337-2.8346	0.0007-0.0018	0.0040-0.0140	3	2.2431-2.2441	0.0008-0.0017	0.0060-0.0140
2012	3.7	J37A1	2.8337-2.8346	0.0007-0.0018	0.0040-0.0140	3	2.2431-2.2441	0.0008-0.0017	0.0060-0.0140

71051_AMDX_C0007

PISTON AND RING SPECIFICATIONS

All measurements are given in inches

Year	Engine Displacement Liters	Engine ID	Piston Clearance	Ring Gap Top Compression	Ring Gap Bottom Compression	Ring Gap Oil Control	Ring Side Clearance Top Compression	Ring Side Clearance Bottom Compression	Ring Side Clearance Oil Control
2011	3.7	J37A1	0.0002-0.0013	0.0120-0.0160	0.0160-0.0220	0.0080-0.0280	0.0022-0.0033	0.0012-0.0024	NA
2012	3.7	J37A1	0.0002-0.0013	0.0120-0.0160	0.0160-0.0220	0.0080-0.0280	0.0022-0.0033	0.0012-0.0024	NA

NA; Not Available

71051_AMDX_C0008

TORQUE SPECIFICATIONS
All readings in ft. lbs.

Year	Engine Displacement Liters	Engine ID	Cylinder Head Bolts	Main Bearing Bolts	Rod Bearing Bolts	Crankshaft Damper Bolts	Flywheel Bolts	Manifold		Spark Plugs	Oil Pan Drain Plug
								Intake	Exhaust		
2011	3.7	J37A1	①	②	③	④	54	16	40	13	29
2012	3.7	J37A1	①	②	③	④	54	16	40	13	29

① Step 1: 22 ft. lbs.
   Step 2: Rotate 90 degrees
   Step 3: Rotate an additional 90 degrees
   Step 4 (new bolts only): additional 90 degrees

② Step 1: Cap bolts 56 ft. lbs.
   Step 2: Side bolts 36 ft. lbs.

③ Step 1: 14 ft. lbs.
   Step 2: Rotate 90 degrees

④ Step 1: 47 ft. lbs.
   Step 2: Rotate 60 degrees

71051_AMDX_C0010

WHEEL ALIGNMENT

Year	Model		Caster		Camber		Toe-in
			Range (+/-Deg.)	Preferred Setting (Deg.)	Range (+/-Deg.)	Preferred Setting (Deg.)	(in.)
2011	MDX	F	±3.5	+4.12	1.00	-0.30	0 +/- 0.08
		R	—	—	0.45	-0.30	0.08 +/- 0.08
2012	MDX	F	±3.5	+4.12	1.00	-0.30	0 +/- 0.08
		R	—	—	0.45	-0.30	0.08 +/- 0.08

71051_AMDX_C0011

TIRE, WHEEL AND BALL JOINT SPECIFICATIONS

| Year | Model | OEM Tires | | Tire Pressures (psi) | | Wheel Size | Ball Joint Inspection | Lug Nut (ft. lbs.) |
		Standard	Optional	Front	Rear			
2011	MDX	P255/55R18	NA	NA	NA	NA	NS	80
2012	MDX	P255/55R18	NA	NA	NA	NA	NS	80

OEM: Original Equipment Manufacturer

PSI: Pounds Per Square Inch

NS: Not Specified by manufacturer

NA: Not available

71051_AMDX_C0012

BRAKE SPECIFICATIONS
All measurements in inches unless noted

| Year | Model | | Brake Disc | | | Brake Drum Diameter | | | Minimum Lining Thickness | | Brake Caliper | |
			Original Thickness	Minimum Thickness	Maximum Runout	Original Inside Diameter	Max. Wear Limit	Maximum Machine Diameter	Front	Rear	Bracket Bolts (ft. lbs.)	Mounting Bolts (ft. lbs.)
2011	MDX	F	1.100-1.111	1.020	0.004	NA	NA	NA	0.06	NA	101	53
		R	0.430-0.440	0.350	0.004	NA	NA	NA	NA	0.06	65	27
2012	MDX	F	1.100-1.111	1.020	0.004	NA	NA	NA	0.06	NA	101	53
		R	0.430-0.440	0.350	0.004	NA	NA	NA	NA	0.06	65	27

NA: Not Available

F: Front

R: Rear

71051_AMDX_C0013

SCHEDULED MAINTENANCE INTERVALS
ACURA—MDX

TO BE SERVICED	OF SERVIC	VEHICLE MILEAGE INTERVAL (x1000)														
		7.5	15	22.5	30	37.5	45	52.5	60	67.5	75	82.5	90	97.5	105	112.5	120
Accessory drive belts	I & A				✓				✓				✓				✓
Air cleaner element	R				✓				✓				✓				✓
Brake fluid	R	Every 3 years															
Brake hoses & lines (incl. ABS)	I		✓		✓		✓		✓		✓		✓		✓		✓
Cooling system hoses & connections	I		✓		✓		✓		✓		✓		✓		✓		✓
Engine coolant ①	R						✓						✓				
Engine oil	R	✓	✓	✓	✓	✓	✓	✓	✓	✓	✓	✓	✓	✓	✓	✓	✓
Engine oil and coolant levels	I	Inspect at each fuel stop															
Engine oil filter	R		✓		✓		✓		✓		✓		✓		✓		✓
Exhaust system	I		✓		✓		✓		✓		✓		✓		✓		✓
Fluid levels and condition	I		✓		✓		✓		✓		✓		✓		✓		✓
Front and rear brakes	I		✓		✓		✓		✓		✓		✓		✓		✓
Fuel lines & connection	I		✓		✓		✓		✓		✓		✓		✓		✓
Halfshaft boots	I		✓		✓		✓		✓		✓		✓		✓		✓
Idle speed	I & A														✓		
Parking brake system	I & A		✓		✓		✓		✓		✓		✓		✓		✓
Rear differential fluid	R	✓			✓		✓		✓				✓				✓
Rotate and inspect tires	I	✓	✓	✓	✓	✓	✓	✓	✓	✓	✓	✓	✓	✓	✓	✓	✓
Spark plugs	R														✓		
Supplemental Restraint System	I	Inspect the SRS 10 years after production															
Suspension components	I		✓		✓		✓		✓		✓		✓		✓		✓
Tie rod ends, steering gear box & boots	I		✓		✓		✓		✓		✓		✓		✓		✓
Timing belt	R														✓		
Transmission fluid	R						✓				✓				✓		
Valve clearance	I	Adjust if valves are noisy															
Water pump	S/I														✓		

R: Replace   I: Inspect   A: Adjust

① Every 12,000 miles or 10 years, then every 60,000 miles or 5 years

FREQUENT OPERATION MAINTENANCE (SEVERE SERVICE)

If a vehicle is operated under any of the following conditions it is considered severe service:

- Towing a trailer or using a camper or car-top carrier.
- Repeated short trips of less than 5 miles in temperatures below freezing, or trips of less than 10 miles in any temperature.
- Extensive idling or low-speed driving for long distances as in heavy commercial use, such as delivery, taxi or police cars.
- Operating on rough, muddy or salt-covered roads.
- Operating on unpaved or dusty roads.
- Driving in extremely hot (over 90°) conditions.

Air cleaner element: replace every 15,000 miles

Engine oil and filter: replace every 3750 miles or 6 months, whichever occurs first.

Timing belt: replace every 60,000 miles if the vehicle is regularly driven in temperatures above 110°F or below -20°F, or if frequently towing a trailer.

Transmission fluid: replace every 30,000 miles.

Rear differential fluid: replace every 60,000 miles.

Front and rear brakes: inspect every 7500 miles or 6 months, whichever occurs first.

Locks and hinges: lubricate every 15,000 miles.

Tie rods, steering gear box, boots: inspect every 7500 miles or 6 months, whichever occurs first.

Suspension components: inspect every 7500 miles or 6 months, whichever occurs first.

Halfshaft boots: inspect every 7500 miles or 6 months, whichever occurs first.

PRECAUTIONS

Before servicing any vehicle, please be sure to read all of the following precautions, which deal with personal safety, prevention of component damage, and important points to take into consideration when servicing a motor vehicle:

• Never open, service or drain the radiator or cooling system when the engine is hot; serious burns can occur from the steam and hot coolant.

• Observe all applicable safety precautions when working around fuel. Whenever servicing the fuel system, always work in a well-ventilated area. Do not allow fuel spray or vapors to come in contact with a spark, open flame, or excessive heat (a hot drop light, for example). Keep a dry chemical fire extinguisher near the work area. Always keep fuel in a container specifically designed for fuel storage; also, always properly seal fuel containers to avoid the possibility of fire or explosion. Refer to the additional fuel system precautions later in this section.

• Fuel injection systems often remain pressurized, even after the engine has been turned **OFF**. The fuel system pressure must be relieved before disconnecting any fuel lines. Failure to do so may result in fire and/or personal injury.

• Brake fluid often contains polyglycol ethers and polyglycols. Avoid contact with the eyes and wash your hands thoroughly after handling brake fluid. If you do get brake fluid in your eyes, flush your eyes with clean, running water for 15 minutes. If eye irritation persists, or if you have taken

brake fluid internally, IMMEDIATELY seek medical assistance.

• The EPA warns that prolonged contact with used engine oil may cause a number of skin disorders, including cancer. You should make every effort to minimize your exposure to used engine oil. Protective gloves should be worn when changing oil. Wash your hands and any other exposed skin areas as soon as possible after exposure to used engine oil. Soap and water, or waterless hand cleaner should be used.

• All new vehicles are now equipped with an air bag system, often referred to as a Supplemental Restraint System (SRS) or Supplemental Inflatable Restraint (SIR) system. The system must be disabled before performing service on or around system components, steering column, instrument panel components, wiring and sensors. Failure to follow safety and disabling procedures could result in accidental air bag deployment, possible personal injury and unnecessary system repairs.

• Always wear safety goggles when working with, or around, the air bag system. When carrying a non-deployed air bag, be sure the bag and trim cover are pointed away from your body. When placing a non-deployed air bag on a work surface, always face the bag and trim cover upward, away from the surface. This will reduce the motion of the module if it is accidentally deployed. Refer to the additional air bag system precautions later in this section.

• Clean, high quality brake fluid from a sealed container is essential to the safe and

proper operation of the brake system. You should always buy the correct type of brake fluid for your vehicle. If the brake fluid becomes contaminated, completely flush the system with new fluid. Never reuse any brake fluid. Any brake fluid that is removed from the system should be discarded. Also, do not allow any brake fluid to come in contact with a painted surface; it will damage the paint.

• Never operate the engine without the proper amount and type of engine oil; doing so WILL result in severe engine damage.

• Timing belt maintenance is extremely important. Many models utilize an interference-type, non-freewheeling engine. If the timing belt breaks, the valves in the cylinder head may strike the pistons, causing potentially serious (also time-consuming and expensive) engine damage. Refer to the maintenance interval charts for the recommended replacement interval for the timing belt, and to the timing belt section for belt replacement and inspection.

• Disconnecting the negative battery cable on some vehicles may interfere with the functions of the on-board computer system(s) and may require the computer to undergo a relearning process once the negative battery cable is reconnected.

• When servicing drum brakes, only disassemble and assemble one side at a time, leaving the remaining side intact for reference.

• Only an MVAC-trained, EPA-certified automotive technician should service the air conditioning system or its components.

BRAKES

GENERAL INFORMATION

PRECAUTIONS

• Certain components within the ABS system are not intended to be serviced or repaired individually.

• Do not use rubber hoses or other parts not specifically specified for and ABS system. When using repair kits, replace all parts included in the kit. Partial or incorrect repair may lead to functional problems and require the replacement of components.

• Lubricate rubber parts with clean, fresh brake fluid to ease assembly. Do not use shop air to clean parts; damage to rubber components may result.

• Use only DOT 3 brake fluid from an unopened container.

• If any hydraulic component or line is removed or replaced, it may be necessary to bleed the entire system.

• A clean repair area is essential. Always clean the reservoir and cap thoroughly before removing the cap. The slightest amount of dirt in the fluid may plug an orifice and impair the system function. Perform repairs after components have been thoroughly cleaned; use only denatured alcohol to clean components. Do not allow ABS components to come into contact with any substance containing mineral oil; this includes used shop rags.

• The Anti-Lock control unit is a microprocessor similar to other computer units in the vehicle. Ensure that the ignition switch is **OFF** before removing or installing controller harnesses. Avoid static electricity discharge at or near the controller.

ANTI-LOCK BRAKE SYSTEM (ABS)

• If any arc welding is to be done on the vehicle, the control unit should be unplugged before welding operations begin.

SPEED SENSORS

REMOVAL & INSTALLATION

Front

See Figure 1.

1. Turn the ignition switch to LOCK (0).
2. Remove the clip, then disconnect the wheel speed sensor connector.
3. Remove the bolts and the wheel speed sensor.

To install:

4. Install the wheel speed sensor in the reverse order of removal.

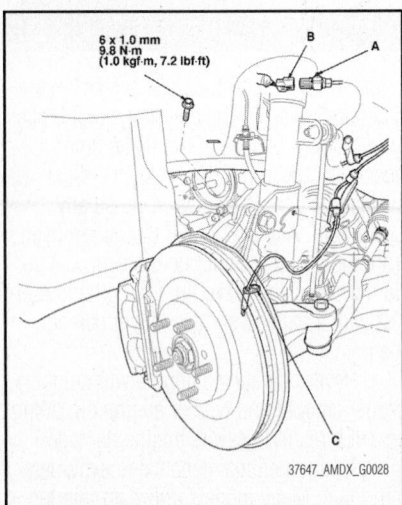

6 x 1.0 mm
9.8 N·m
(1.0 kgf·m, 7.2 lbf·ft)

37647_AMDX_G0028

Fig. 1 Remove the front wheel speed sensor

※※ **CAUTION**

Install the sensor carefully to avoid twisting the wires.

➡**If the wheel speed sensor comes in contact with the hub bearing unit, it is faulty.**

5. Start the engine, and check that the ABS, the VSA indicators, and the trailer stability assist warning goes off.

6. Test-drive the vehicle, and check that the ABS, the VSA indicators, and the trailer stability assist warning do not come on.

Rear

See Figure 2.

1. Turn the ignition switch to LOCK (0).

2. Disconnect the wheel speed sensor connector.

3. Remove the clips, the bolt, and the wheel speed sensor.

To install:

4. Install the wheel speed sensor in the reverse order of removal.

※※ **CAUTION**

Install the sensor carefully to avoid twisting the wires.

➡**If the wheel speed sensor comes in contact with the hub bearing unit, it is faulty.**

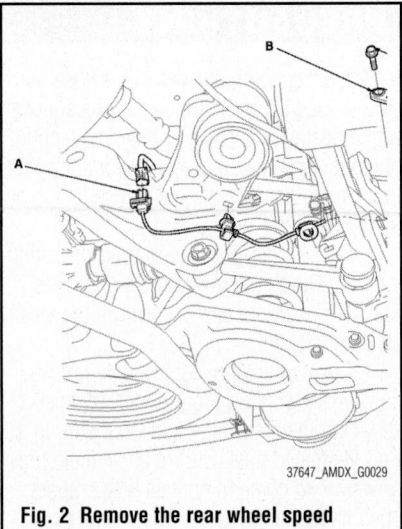

37647_AMDX_G0029

Fig. 2 Remove the rear wheel speed sensor

5. Start the engine, and check that the ABS, the VSA indicators, and the trailer stability assist warning goes off.

6. Test-drive the vehicle, and check that the ABS, the VSA indicators, and the trailer stability assist warning do not come on.

BRAKES   BLEEDING THE BRAKE SYSTEM

BLEEDING PROCEDURE

BLEEDING PROCEDURE

See Figure 3.

➡**Do not reuse the drained fluid. Use only clean Acura DOT 3 Brake Fluid from an unopened container. Using a non-Acura brake fluid can cause corrosion and shorten the life of the system.**

Make sure no dirt or other foreign matter is allowed to contaminate the brake fluid.

Do not spill brake fluid on the vehicle, it may damage the paint; if brake fluid does contact the paint, wash it off immediately with water.

The reservoir connected to the master cylinder must be at the MAX (upper) level mark at the start of the bleeding procedure

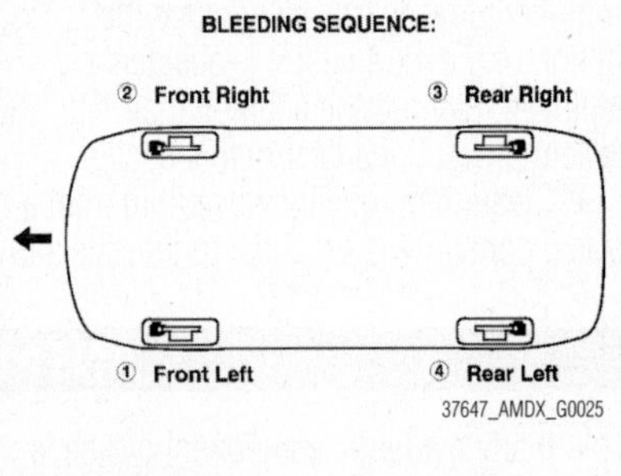

BLEEDING SEQUENCE:

② Front Right   ③ Rear Right

① Front Left   ④ Rear Left

37647_AMDX_G0025

Fig. 3 Bleed the calipers or the wheel cylinders in sequence

and checked after bleeding each wheel. Add fluid as required.

1. Make sure the brake fluid level in the reservoir is at the MAX (upper) level line.

2. Have someone slowly pump the brake pedal several times, then apply steady pressure.

3. Start the bleeding at the driver's side of the front brake system.

➡**Bleed the calipers or the wheel cylinders in the sequence shown.**

4. Attach a length of clear drain tube to the bleed screw, then, loosen the bleed screw to allow air to escape from the system. Then tighten the bleed screw securely.

5. Refill the master cylinder reservoir to the MAX (upper) level line.

6. Repeat the procedure for each brake circuit until there are no air bubbles are in the fluid.

BRAKES   **FRONT DISC BRAKES**

BRAKE CALIPER

REMOVAL & INSTALLATION

See Figure 4.

1. Raise and safely support the vehicle.
2. Remove the front wheel.

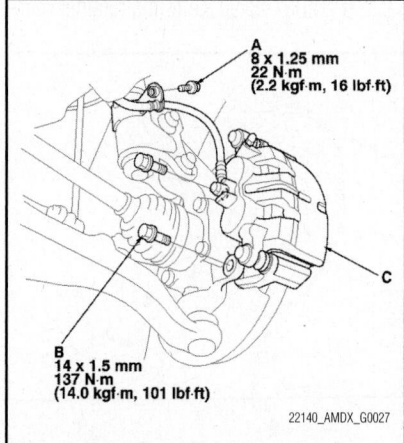

Fig. 4 Remove the bracket-to-knuckle bolts (B) to remove the caliper assembly (C) from the vehicle

3. Remove the brake hose bracket mounting bolt and banjo bolt to disconnect the brake hose from the caliper. Plug the brake hose to prevent excessive fluid loss.
4. Remove the brake bracket-to-knuckle bolts, then remove the caliper assembly from the knuckle.

To install:

5. Install the caliper assembly and tighten the mounting bolts to 101 ft. lbs. (137 Nm).
6. Connect the brake hose using new sealing washers and tighten the banjo bolt to 25 ft. lbs. (34 Nm).
7. The remainder of the installation is the reverse order of removal.

DISC BRAKE PADS

REMOVAL & INSTALLATION
See Figure 5.

1. Remove some brake fluid from the master cylinder.
2. Raise and safely support the vehicle
3. Remove the front wheel.
4. Remove the flange bolt, and pivot the caliper up out of the way.
5. Remove the pad shims and brake pads.
6. Remove the pad retainers.

➡**The upper and lower pad retainers are different. During installation, make sure the pad retainers are in proper positions.**

To install:

7. Install the pad retainers. Wipe excess assembly paste off the retainers. Keep the assembly paste off the discs and pads.
8. Mount the brake caliper piston compressor tool on the caliper body.
9. Press in the piston with the brake caliper piston compressor tool so the caliper will fit over the brake pads. Make sure the piston boot is in position to prevent damaging it when pivoting the caliper down.

➡**Be careful when pressing in the piston; brake fluid might overflow from the master cylinder's reservoir.**

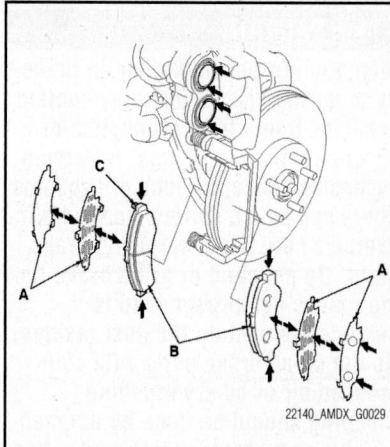

Fig. 5 Apply a thin coat assembly paste to the pad side of the shims (A) and back of the brake pads (B). Install the pad with the wear indicator (C) on the upper side

10. Remove the brake caliper piston compressor tool.
11. Apply a thin coat of M-77 assembly paste (P/N 08798-9010) to the pad side of the shims, the back of the brake pads and the other areas indicated by the arrows. Wipe excess assembly paste off the pad shims and brake pads. Contaminated brake discs or brake pads reduce stopping ability. Keep grease and assembly paste off the brake discs and brake pads.
12. Install the brake pads and pad shims correctly. Install the brake pad with the wear indicator on the upper inside. If you are reusing the brake pads, always reinstall the brake pads in their original positions to prevent a momentary loss of braking efficiency.
13. Pivot the caliper down into position. Install the flange bolt and tighten it to 53 ft. lbs. (72 Nm).
14. Clean the mating surfaces of the brake disc and the inside of the wheel, then install the front wheels.
15. Press the brake pedal several times to make sure the brakes work.
16. Add brake fluid as needed.
17. After installation, check for leaks at hose and line joints or connections, and retighten if necessary.
18. Test-drive the vehicle, then recheck for leaks.

❈❈ CAUTION

Dust and dirt accumulating on brake parts during normal use may contain asbestos fibers from production or aftermarket brake linings. Breathing excessive concentrations of asbestos fibers can cause serious bodily harm. Exercise care when servicing brake parts. Do not sand or grind brake lining unless equipment used is designed to contain the dust residue. Do not clean brake parts with compressed air or by dry brushing. Cleaning should be done by dampening the brake components with a fine mist of water, then wiping the brake components clean with a dampened cloth. Dispose of cloth and all residue containing asbestos fibers in an impermeable container with the appropriate label. Follow practices prescribed by the Occupational Safety and Health Administration (OSHA) and the Environmental Protection Agency (EPA) for the handling, processing, and disposing of dust or debris that may contain asbestos fibers.

BRAKE CALIPER

REMOVAL & INSTALLATION

See Figure 6.

1. Raise and safely support the vehicle.
2. Remove the rear wheel.
3. Remove the brake hose bracket mounting bolt and banjo bolt to disconnect the brake hose from the caliper. Plug the

brake hose to prevent excessive fluid loss.

4. Remove the brake caliper bracket mounting bolts, then remove the caliper assembly from the knuckle.

To install:

5. Install the caliper assembly to the knuckle. Tighten the mounting bolts to 65 ft. lbs. (88 Nm).
6. Connect the brake hose using new sealing washers and tighten the banjo bolt to 25 ft. lbs. (34 Nm).
7. The remainder of the installation is the reverse order of removal.

DISC BRAKE PADS

REMOVAL & INSTALLATION

See Figures 7 through 9.

Special Tools Required: Brake Caliper Piston Compressor 07AAE-SEPA101

❈❈ CAUTION

Frequent inhalation of brake pad dust, regardless of material composition, could be hazardous to your health. Avoid breathing dust particles. Never use an air hose or brush to clean brake assemblies. Use an OSHA-approved vacuum cleaner.

1. Remove some brake fluid from the master cylinder.
2. Raise and support the vehicle.
3. Remove the rear wheels.
4. Remove the flange bolt, and pivot the caliper up out of the way. Check the hose and pin boots for damage and deterioration.

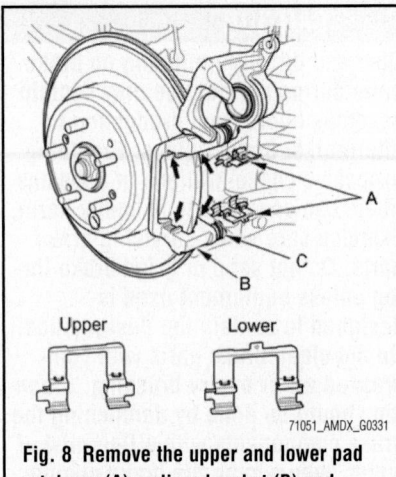

Fig. 8 Remove the upper and lower pad retainers (A); caliper bracket (B) and caliper pins (C)

5. Remove the pad shims and brake pads.

To install:

6. Remove the upper and lower pad retainers.

➡The upper and lower pad retainers are different. During installation, make sure the pad retainers are in their proper positions.

7. Clean the caliper bracket thoroughly; remove any rust, and check for grooves and cracks. Verify that the caliper pins move in and out smoothly. Clean and lube if needed.
8. Inspect the brake disc/drum, and check for damage and cracks.
9. Apply a thin coat of M-77 assembly paste (P/N 08798-9010) to the retainer mat-

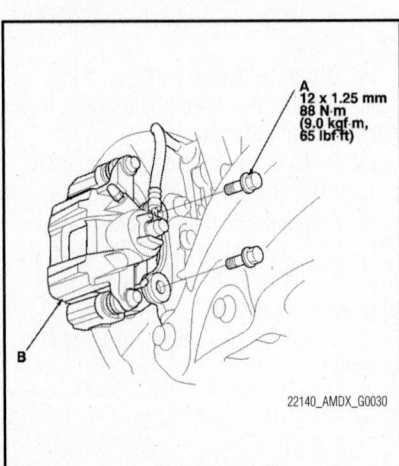

Fig. 6 Remove the brake caliper mounting bolts (A) to remove the caliper assembly

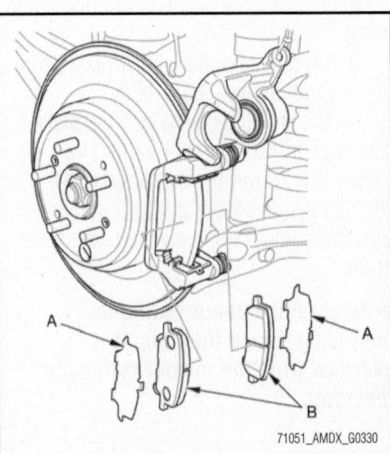

Fig. 7 Remove the pad shims (A) and brake pads (B)

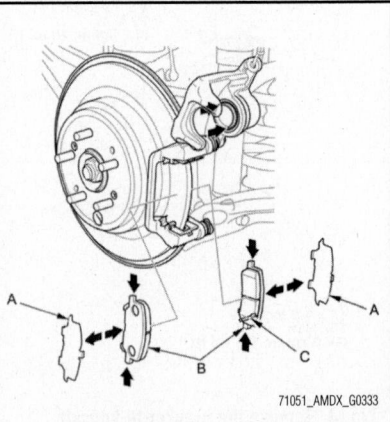

Fig. 9 Apply a thin coat of M-77 assembly paste to the pad side of the shims (A), the back of the brake pads (B), and the other areas indicated by the arrows; install the brake pad with the wear indicator (C) on the bottom inside

ing surface of the caliper bracket (indicated by the arrows and shaded area).

10. Install the upper and lower pad retainers. Wipe excess assembly paste off the retainers. Keep the assembly paste off the discs and pads.

11. Install the brake caliper piston compressor tool on the caliper body.

12. Press in the piston with the brake caliper piston compressor tool so the caliper will fit over the brake pads. Make sure the piston boot is in position to prevent damaging it when pivoting the caliper down.

➡ **Be careful when pressing in the piston; brake fluid might overflow from the master cylinder's reservoir. If brake fluid gets on any painted surface, wash it off immediately with water.**

13. Remove the brake caliper piston compressor tool.

14. Apply a thin coat of M-77 assembly paste (P/N 08798- 9010) to the pad side of the shims, the back of the brake pads, and the other areas indicated by the arrows. Wipe excess assembly paste off the pad shims and brake pads. Contaminated brake disc/drums or brake pads reduce stopping ability. Keep grease and assembly paste off the brake disc/drums and brake pads.

15. Install the brake pads and pad shims correctly. Install the brake pad with the wear indicator (C) on the bottom inside. If you are reusing the brake pads, always reinstall the brake pads in their original positions to prevent a temporary loss of braking efficiency.

16. Pivot the caliper down into position. Install the flange bolt, and tighten it to 27 ft. lbs. (37 Nm).

17. Clean the mating surfaces between the brake disc/drum and the inside of the wheel, then install the rear wheels.

18. Press the brake pedal several times to make sure the brakes work.

➡ **Engagement may require a greater pedal stroke immediately after the brake pads have been replaced as a set. Several applications of the brake pedal will restore the normal pedal stroke.**

19. Add brake fluid as needed.

20. After installation, check for leaks at hose and line joints or connections, and retighten if necessary.

21. Test-drive the vehicle, then check for leaks.

BRAKES **PARKING BRAKE**

PARKING BRAKE SHOES

REMOVAL & INSTALLATION
See Figures 10 through 14.

1. Raise and support the vehicle.
2. Remove the rear wheels.
3. Release the parking brake, and remove the rear brake disc/drum.
4. Disconnect and remove the upper return springs, then remove the shoe guide plate.
5. Remove the tension pins by pushing the respective retainer and turning the pin.
6. Remove the strut, and the rod spring.
7. Lower the parking brake shoe assembly.
8. Remove the forward brake shoe and the adjuster assembly by removing the lower return spring.
9. Remove the rearward brake shoe by

disconnecting the parking brake cable from the parking brake lever.

10. Remove the U-clip, wave washer, and parking brake lever from the brake shoe.

To install:

11. Apply Molykote® 44MA grease to the sliding surface of the pivot pin of the rearward brake shoe.

12. Install the parking brake lever and wave washer on the pivot pin, and secure with a new U-clip.

   a. Install the wave washer with its convex side facing out.

   b. Pinch the U-clip securely to prevent the parking brake lever from coming off the brake shoe.

13. Connect the parking brake cable to the parking brake lever. Apply Molykote® 44MA grease to the cable contact surface on the backing plate.

14. Apply a thin coat of Molykote® 44MA grease to the shoe ends and strut ends, sliding surfaces, and opposite edges of the parking brake shoe as shown. Wipe off any excess. Keep the grease off the brake linings.

15. Install the tension pin, the retainer spring, and the retainer on the rearward brake shoe. Make sure the tension pin does not contact the parking brake lever.

16. Install connecting rods A and B on the adjuster nut.

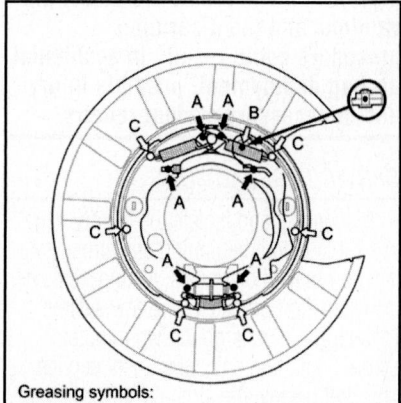

Greasing symbols:
➡● Brake shoe ends and strut ends
⇨○ Opposite edge of the shoe
⇨● Sliding surface

Fig. 12 Apply a thin coat of Molykote® 44MA grease to the shoe ends and strut ends (A), sliding surfaces (B), and opposite edges of the parking brake shoe (C) as shown

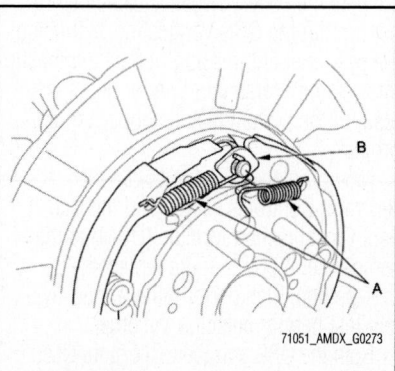

Fig. 10 Disconnect and remove the upper return springs (A), then remove the shoe guide plate (B)

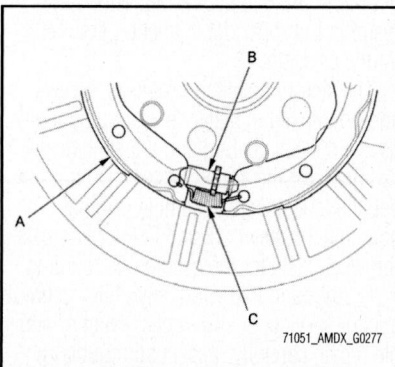

Fig. 11 Remove the forward brake shoe (A) and the adjuster assembly (B) by removing the lower return spring (C)

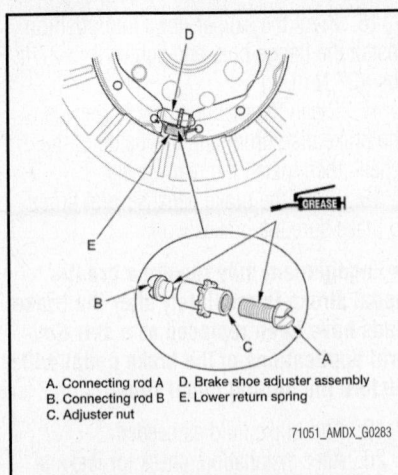

A. Connecting rod A   D. Brake shoe adjuster assembly
B. Connecting rod B   E. Lower return spring
C. Adjuster nut

71051_AMDX_G0283

Fig. 13 Install connecting rods A and B on the adjuster nut

Please note the following:
• Clean the threaded portions of connecting rod A and the sliding surface of

connecting rod B, then coat them with Molykote® 44MA grease.
• Shorten connecting rod A by fully turning the adjuster nut.

17. Position the brake shoe adjuster assembly on the parking brake shoes.

18. Hook the lower return spring on the parking brake shoes.

19. Install the rod spring to the strut first. Then install the strut on the parking brake shoes.

20. Install the tension pin, the retainer spring, and the retainer on the forward brake shoe.

21. Install the shoe guide plate.

22. Install the upper return springs as shown.

23. Install the rear brake disc/drum and the rear brake caliper bracket in the reverse order of removal.

24. Do the major parking brake adjustment.

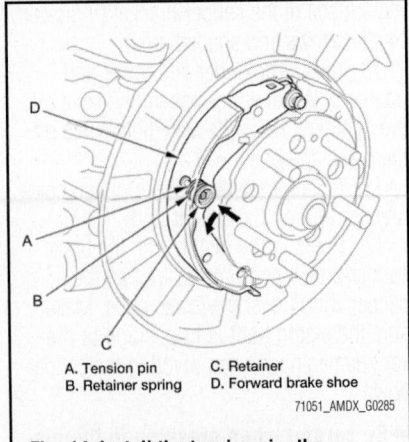

A. Tension pin   C. Retainer
B. Retainer spring   D. Forward brake shoe

71051_AMDX_G0285

Fig. 14 Install the tension pin, the retainer spring, and the retainer on the forward brake shoe

25. Clean the mating surface between the brake/drum and the inside of the wheel, then install the rear wheels.

CHASSIS ELECTRICAL

GENERAL INFORMATION

✳✳ CAUTION

These vehicles are equipped with an air bag system. The system must be disarmed before performing service on, or around, system components, the steering column, instrument panel components, wiring and sensors. Failure to follow the safety precautions and the disarming procedure could result in accidental air bag deployment, possible injury and unnecessary system repairs.

SERVICE PRECAUTIONS

Disconnect and isolate the battery negative cable before beginning any airbag system component diagnosis, testing, removal, or installation procedures. Allow system capacitor to discharge for two minutes before beginning any component service. This will disable the airbag system. Failure to disable the airbag system may result in accidental airbag deployment, personal injury, or death.

Do not place an intact undeployed airbag face down on a solid surface. The airbag will propel into the air if accidentally deployed and may result in personal injury or death.

When carrying or handling an undeployed airbag, the trim side (face) of the airbag should be pointing towards the body to minimize possibility of injury if accidental

AIR BAG (SUPPLEMENTAL RESTRAINT SYSTEM)

deployment occurs. Failure to do this may result in personal injury or death.

Replace airbag system components with OEM replacement parts. Substitute parts may appear interchangeable, but internal differences may result in inferior occupant protection. Failure to do so may result in occupant personal injury or death.

Wear safety glasses, rubber gloves, and long sleeved clothing when cleaning powder residue from vehicle after an airbag deployment. Powder residue emitted from a deployed airbag can cause skin irritation. Flush affected area with cool water if irritation is experienced. If nasal or throat irritation is experienced, exit the vehicle for fresh air until the irritation ceases. If irritation continues, see a physician.

Do not use a replacement airbag that is not in the original packaging. This may result in improper deployment, personal injury, or death.

The factory installed fasteners, screws and bolts used to fasten airbag components have a special coating and are specifically designed for the airbag system. Do not use substitute fasteners. Use only original equipment fasteners listed in the parts catalog when fastener replacement is required.

During, and following, any child restraint anchor service, due to impact event or vehicle repair, carefully inspect all mounting hardware, tether straps, and anchors for proper installation, operation, or damage. If a child restraint anchor is found damaged in any way, the anchor must be replaced. Fail-

ure to do this may result in personal injury or death.

Deployed and non-deployed airbags may or may not have live pyrotechnic material within the airbag inflator.

Do not dispose of driver/passenger/curtain airbags or seat belt tensioners unless you are sure of complete deployment. Refer to the Hazardous Substance Control System for proper disposal.

Dispose of deployed airbags and tensioners consistent with state, provincial, local, and federal regulations.

After any airbag component testing or service, do not connect the battery negative cable. Personal injury or death may result if the system test is not performed first.

If the vehicle is equipped with the Occupant Classification System (OCS), do not connect the battery negative cable before performing the OCS Verification Test using the scan tool and the appropriate diagnostic information. Personal injury or death may result if the system test is not performed properly.

Never replace both the Occupant Restraint Controller (ORC) and the Occupant Classification Module (OCM) at the same time. If both require replacement, replace one, then perform the Airbag System test before replacing the other.

Both the ORC and the OCM store Occupant Classification System (OCS) calibration data, which they transfer to one another when one of them is replaced. If both are replaced at the same time, an irreversible fault will be set in both modules and the OCS may malfunction and cause personal injury or death.

If equipped with OCS, the Seat Weight Sensor is a sensitive, calibrated unit and must be handled carefully. Do not drop or handle roughly. If dropped or damaged, replace with another sensor. Failure to do so may result in occupant injury or death.

If equipped with OCS, the front passenger seat must be handled carefully as well. When removing the seat, be careful when setting on floor not to drop. If dropped, the sensor may be inoperative, could result in occupant injury, or possibly death.

If equipped with OCS, when the passenger front seat is on the floor, no one should sit in the front passenger seat. This uneven force may damage the sensing ability of the seat weight sensors. If sat on and damaged, the sensor may be inoperative, could result in occupant injury, or possibly death.

DISARMING THE SYSTEM

Disconnect and isolate the negative battery cable. Wait 3 minutes for the system capacitor to discharge before performing any service.

ARMING THE SYSTEM

Connect the negative battery cable.

CLOCKSPRING CENTERING

See Figure 15.

1. Before installing the steering wheel, make sure the front wheels are pointing ahead, then center the cable reel. Do this by first rotating the cable reel clockwise until it stops. Then rotate it counterclockwise (about three turns) until the arrow mark on the cable reel label points straight up.

2. Position the two tabs of the turn signal canceling sleeve as shown, and install the steering wheel on to the steering column shaft, making sure the steering wheel hub engages the pins of the cable reel and tabs of the turn signal canceling sleeve. Do not tap on the steering wheel or steering column shaft when installing the steering wheel.

3. Install the steering wheel bolt and tighten it to 29 ft. lbs. (39 Nm), then connect the cable reel sub harness 20P connector.

REMOVAL & INSTALLATION

See Figures 16 through 19.

1. Make sure the front wheels are aligned straight ahead.

2. Do the battery terminal disconnection procedure, then wait at least 3 minutes before starting work.

3. Remove the driver's airbag.

4. Disconnect the cable reel sub harness 20P connector from the cable reel, then remove the steering wheel bolt.

5. Confirm that the front wheels point straight ahead, then remove the steering wheel with a steering wheel puller.

➡**Do not tap on the steering wheel or steering column shaft when removing the steering wheel.**

6. Remove the column cover screws, then remove the column covers.

7. Disconnect the dashboard wire harness 4P connector from the cable reel 4P connector, then disconnect the dashboard wire harness 20P connector from the cable reel.

8. Release the upper lock tab under the cable reel connector with a 90 degree hook-shaped tool. Slide the tool below the cable

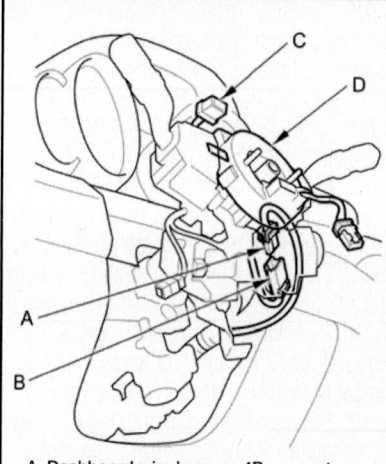

A. Dashboard wire harness 4P connector
B. Cable reel 4P connector
C. Dashboard wire harness 20P connector
D. Cable reel

71051_AMDX_G0315

Fig. 17 Disconnect the dashboard wire harness 4P connector from the cable reel 4P connector, then disconnect the dashboard wire harness 20P connector from the cable reel

reel connector just above the lock tab. Release the lower lock tab, and slide the cable reel off the column.

To install:

9. Before installing the steering wheel, align the front wheels straight ahead.

10. If not already done, do the battery terminal disconnection procedure, then wait at least 3 minutes before starting work.

11. Set the turn signal canceling sleeve so that the projections are aligned vertically.

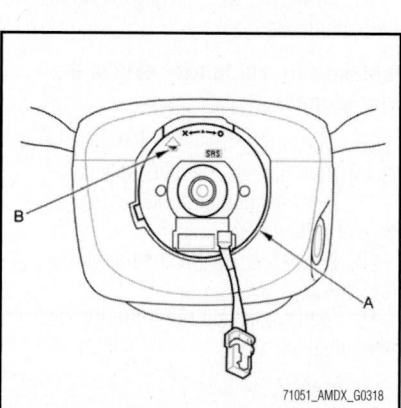

71051_AMDX_G0318

Fig. 15 Center the cable reel (A); do this by first rotating the cable reel clockwise until it stops, then rotate it counterclockwise (about three turns) until the arrow mark (B) on the cable reel label points straight up

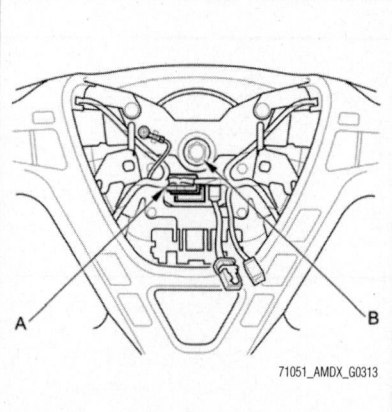

71051_AMDX_G0313

Fig. 16 Disconnect the cable reel sub harness 20P connector (A) from the cable reel, then remove the steering wheel bolt (B)

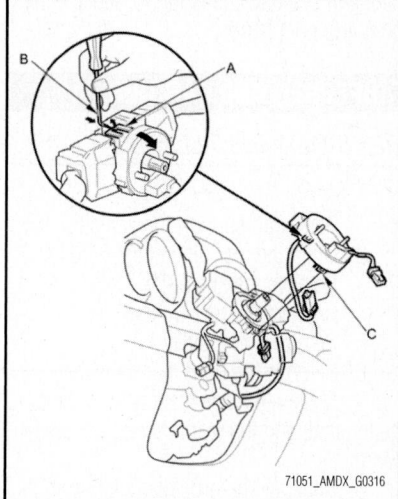

71051_AMDX_G0316

Fig. 18 Release the upper lock tab (A) under the cable reel connector with a 90 degree hook-shaped tool (B); release the lower lock tab (C)

12. Carefully install the cable reel on the steering column shaft. Then connect dashboard wire harness 20P connector to the cable reel, and connect the cable reel 4P connector to the dashboard wire harness 4P connector.

13. Install the steering column covers.

14. Before installing the steering wheel, make sure the front wheels are pointing ahead, then center the cable reel. Do this by first rotating the cable reel clockwise until it stops. Then rotate it counterclockwise (about three turns) until the arrow mark on the cable reel label points straight up.

15. Position the two tabs of the turn signal canceling sleeve as shown, and install the steering wheel on to the steering column shaft, making sure the steering wheel hub engages the pins of the cable reel and tabs of the turn signal canceling sleeve. Do not tap on the steering wheel or steering column shaft when installing the steering wheel.

16. Install the steering wheel bolt and tighten it to 29 ft. lbs. (39 Nm), then connect the cable reel sub harness 20P connector.

17. Install the driver's airbag.

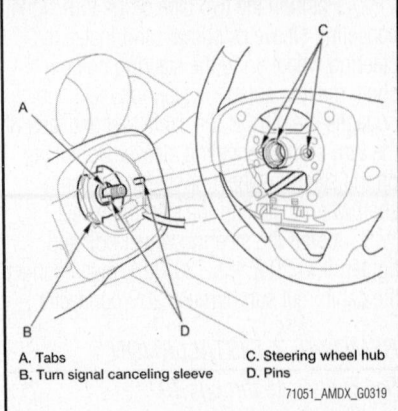

A. Tabs
B. Turn signal canceling sleeve
C. Steering wheel hub
D. Pins

71051_AMDX_G0319

Fig. 19 Position the two tabs of the turn signal canceling sleeve as shown, and install the steering wheel on to the steering column shaft, making sure the steering wheel hub engages the pins of the cable reel and tabs of the turn signal canceling sleeve

18. Do the battery terminal reconnection procedure.

19. Clear any DTCs with the HDS.

20. After installing the cable reel, confirm proper system operation:

a. Turn the ignition switch to ON (II); the SRS indicator should come on for about 6 seconds and then goes off.

b. After the SRS indicator has turned off, turn the steering wheel fully left and right to confirm the SRS indicator does not come on.

c. Make sure the horn and turn signal switches work properly.

d. Make sure the cruise control combination switch work.

e. Make sure the audio remote switch work.

f. Make sure the multi-information switch work.

g. Make sure the HFL-navigation voice control switch (with navigation) work.

h. Make sure the HFL switch (without navigation) work.

i. Make sure the distance switch work.

21. After installation, check the steering wheel spoke angle. If the steering spoke angles to the right and left are not equal (steering wheel is not centered), correct the engagement of the wheel/column shaft splines.

DRIVE TRAIN

CV-BOOTS

INSPECTION

1. Check the inboard boot and the outboard boot on the driveshaft for cracks, damage, leaking grease, and loose boot bands. If any damage is found, replace the boot and boot bands.

DIFFERENTIAL CARRIER

REMOVAL & INSTALLATION

See Figures 20 through 23.

1. Raise and support the vehicle.
2. Drain the rear differential fluid.
3. Remove the spare tire.
4. Remove the rear wheels.
5. Remove the muffler.
6. Remove the right rear driveshaft.
7. Make reference mark across the propeller shaft and the rear differential companion flange.
8. Separate the propeller shaft from the rear differential.

➡**Suspend the propeller shaft with an appropriate size nylon strap.**

9. Place a transmission jack under the rear differential.

10. Disconnect the left rear driveshaft inboard joint from the rear differential using the driveshaft remover and a hammer.

11. Remove the rear differential rear mounting bolts and the rear differential front mounting bracket bolts.

12. Lower the rear differential a little on the transmission jack, then remove the left rear driveshaft inboard joint from the rear differential.

71051_AMDX_G0416

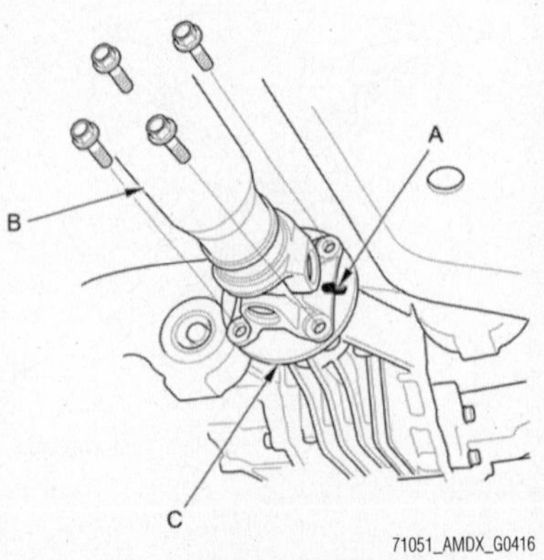

Fig. 20 Make reference mark (A) across the propeller shaft (B) and the rear differential companion flange (C)

➡**Make sure not to over extend the wire harness and the breather hose.**

13. Disconnect the right solenoid 4P connector and the rear differential fluid temperature sensor 2P connector, then remove the harness clips.

14. Disconnect the breather hose from the breather pipe.

15. Lower the rear differential slightly on the transmission jack.

➡**Make sure not to over extend the wire harness.**

16. Disconnect the left solenoid 4P connector, then remove the harness clips.

17. Disconnect the ground cable from the rear differential.

18. Lower the rear differential on the transmission jack.

19. Remove the set ring from the rear differential left side.

To install:

20. Place the rear differential on a transmission jack.

21. Install a new set ring into the groove of the rear differential left side.

22. If the original differential is being reinstalled, replace the set rings.

23. Raise the rear differential to the

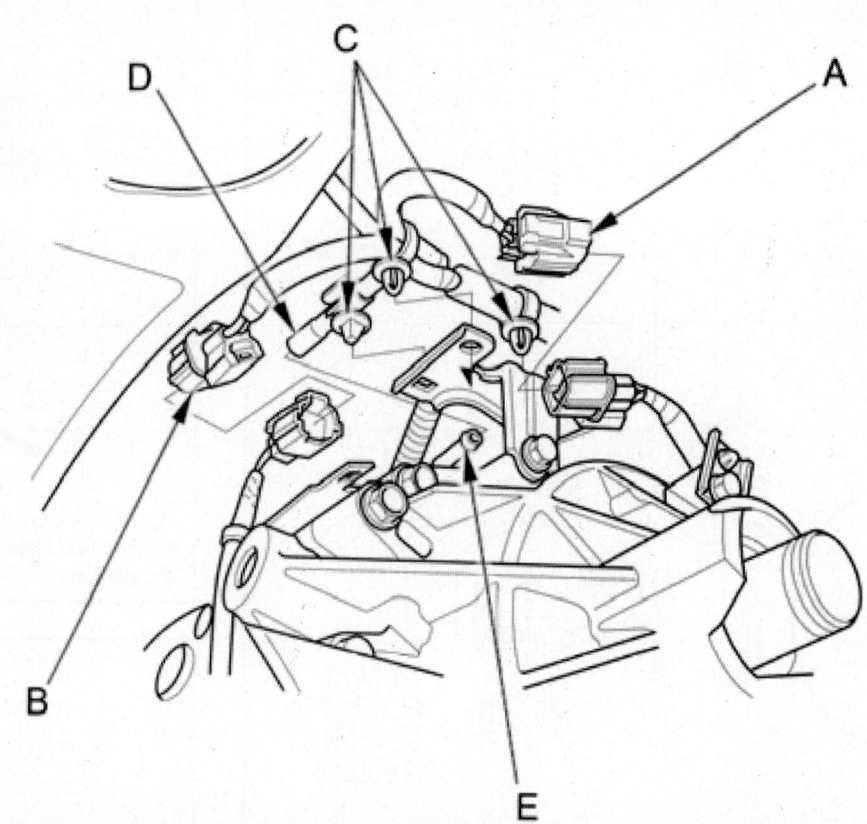

A. Right solenoid 4P connector
B. Rear differential fluid temperature
   sensor 2P connector
C. Harness clips

D. Breather hose
E. Breather pipe

71051_AMDX_G0418

Fig. 21 Disconnect the right solenoid 4P connector and the rear differential fluid temperature sensor 2P connector, then remove the harness clips

appropriate position on the transmission jack.

24. Connect the ground cable to the rear differential.

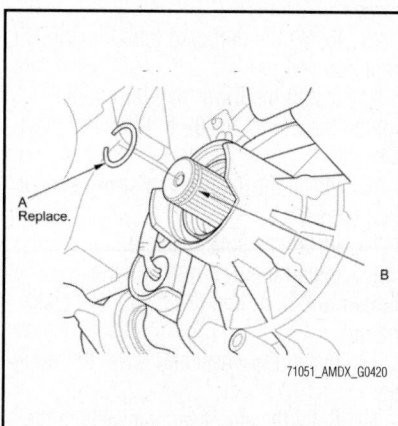

A
Replace.

B

71051_AMDX_G0420

Fig. 22 Install a new set ring (A) into the groove (B) of the rear differential left side

25. Connect the left solenoid 4P connector, then install the harness clips.

26. Raise the rear differential a little on the transmission jack.

27. Connect the breather hose to the breather pipe.

28. Connect the right solenoid 4P connector and the rear differential fluid temperature sensor 2P connector, then install the harness clips.

29. Raise the rear differential to the mounting level.

30. Apply appropriate grease to the left rear driveshaft inboard joint.

31. Insert the left rear driveshaft inboard joint into the rear differential.

32. Temporarily install the rear differential front mounting bracket bolts.

33. Loosen the rear differential front mounting bolts.

34. Install the new rear differential rear mounting bolts and tighten them to the specified torque in the sequence shown.

35. Replace the rear differential front mounting bolts, and tighten them to the specified torque.

36. Tighten the rear differential front mounting bracket bolts to the specified torque.

37. Attach the propeller shaft to the rear differential by aligning the reference mark you made during the removal procedure. Tighten the bolts to 53 ft. lbs. (72 Nm).

38. Install the right rear driveshaft.

39. Install a new gasket and the muffler with new nuts. Tighten the nuts to 40 ft. lbs. (54 Nm).

40. Install the rear wheels.

41. Install the spare tire.

42. Refill the rear differential with the recommended fluid.

43. Check the wheel alignment, and adjust it if necessary.

44. Test-drive the vehicle.

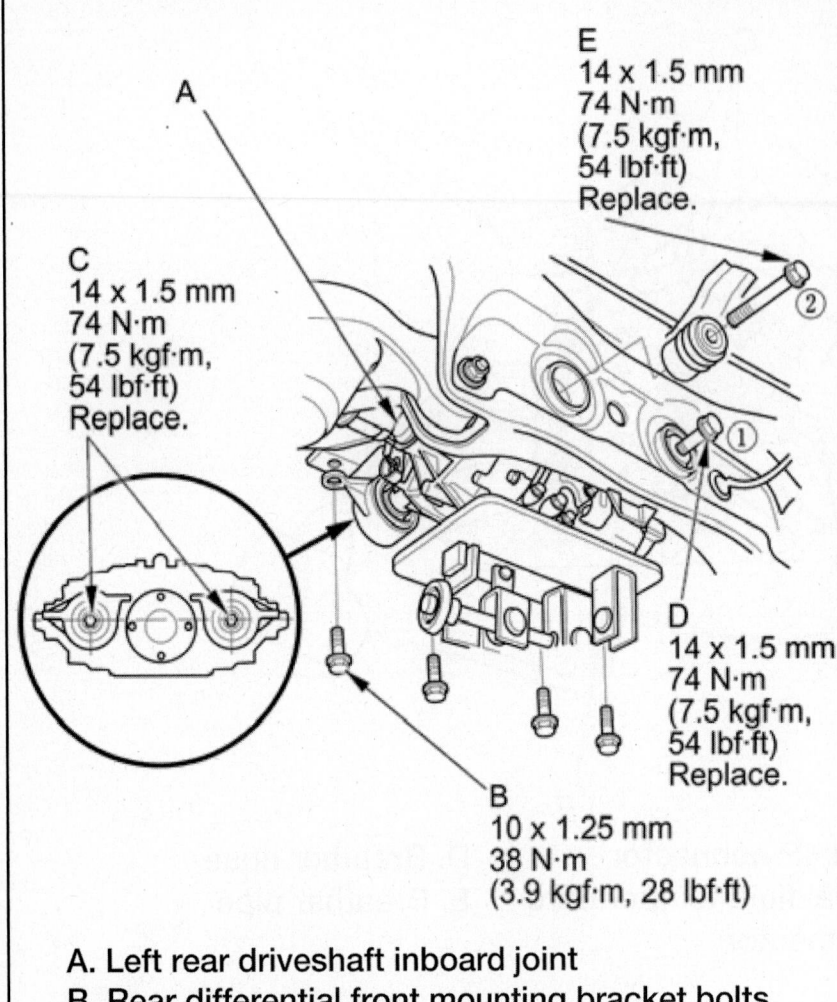

E
14 x 1.5 mm
74 N·m
(7.5 kgf·m,
54 lbf·ft)
Replace.

C
14 x 1.5 mm
74 N·m
(7.5 kgf·m,
54 lbf·ft)
Replace.

D
14 x 1.5 mm
74 N·m
(7.5 kgf·m,
54 lbf·ft)
Replace.

B
10 x 1.25 mm
38 N·m
(3.9 kgf·m, 28 lbf·ft)

A. Left rear driveshaft inboard joint
B. Rear differential front mounting bracket bolts
C. Rear differential front mounting bolts
D. Rear differential rear mounting bolt
E. Rear differential rear mounting bolt

71051_AMDX_G0421

Fig. 23 Insert the left rear driveshaft inboard joint into the rear differential

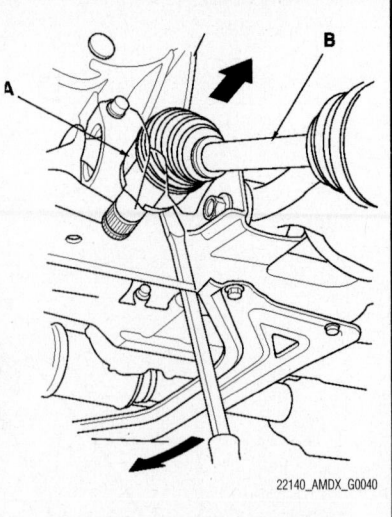

22140_AMDX_G0040

Fig. 24 To remove the left halfshaft (B), pry the inboard joint (A) from the differential with a prybar

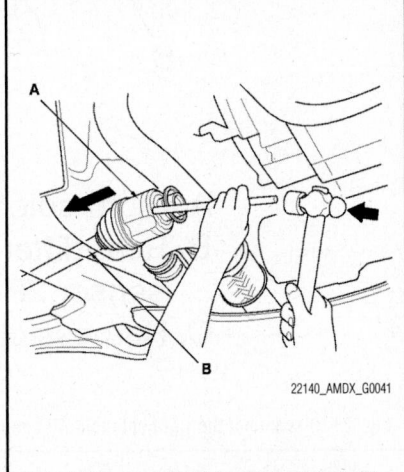

22140_AMDX_G0041

Fig. 25 If removing the right halfshaft (B), drive the inboard joint (A) off of the intermediate shaft using a drift and a hammer

FRONT HALFSHAFT

REMOVAL & INSTALLATION

See Figures 24 and 25.

1. Raise and safely support the vehicle.
2. Remove the front wheel
3. Pry up the locking tab on the spindle nut and remove the nut.
4. Remove the splash shield.
5. Drain the transaxle fluid.
6. Remove the suspension stroke sensor, if equipped.
7. Separate the lower ball joint from the wheel knuckle.
8. Loosen the halfshaft outboard joint from the front hub using a plastic hammer.

9. Pull the wheel knuckle outward and separate the outboard joint from the front hub.
10. If removing the left halfshaft, pry the inboard joint from the differential with a prybar. Remove the halfshaft as an assembly.
11. If removing the right halfshaft, drive the inboard joint off of the intermediate shaft using a drift and a hammer. Remove the halfshaft as an assembly.

To install:

➡**Use new circlips and cotter pins during installation.**

12. Insert the inboard end of the halfshaft into the differential (or intermedi-

ate shaft) until the set ring locks in the groove.
13. Install the outboard joint into the front hub.
14. Install the lower ball joint and tighten the nut to 76–83 ft. lbs. (103–113 Nm).
15. Install the suspension stroke sensor, if equipped.
16. Install the splash shield.
17. Install the new hub nut and tighten to 242 ft. lbs. (328 Nm) and stake the nut.
18. Install the wheel and lower the vehicle.
19. Refill the automatic transaxle to the correct level.
20. Test drive the vehicle and check for leaks.

PROPELLER SHAFT

REMOVAL & INSTALLATION

Rear

See Figure 26.

1. Raise and support the vehicle.
2. Remove the propeller shaft protector.
3. Make reference mark across the No. 1 propeller shaft and the transfer companion flange.
4. Remove the flange bolts.
5. Remove the center support bearing mounting bolts.
6. Make reference mark across the No. 2 propeller shaft and the rear differential companion flange.
7. Remove the flange bolts.

To install:

8. Set the No. 2 propeller shaft to the rear differential companion flange, by aligning the reference mark you made during the removal procedure. Then install new flange bolts to the specified torque.

➡**When replacing the propeller shaft or the rear differential, align the factory reference marks.**

9. Install the center support bearing mounting bolts. Tighten the bolts to 29 ft. lbs. (39 Nm).

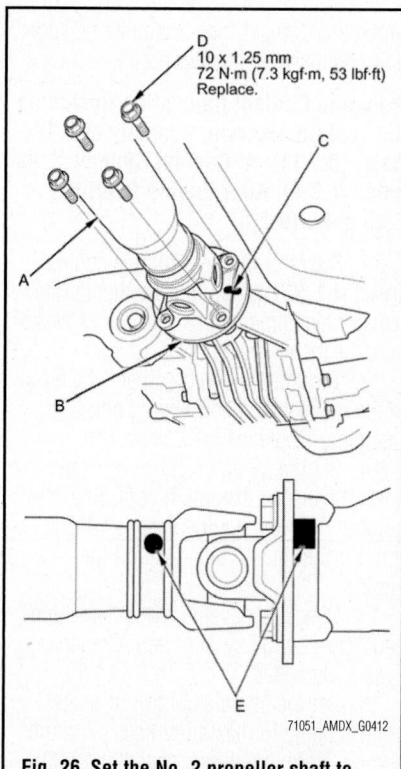

Fig. 26 Set the No. 2 propeller shaft to the rear differential companion flange

10. Set the No. 1 propeller shaft to the transfer companion flange by aligning the reference mark you made during the removal procedure. Then install new flange bolts and tighten to 53 ft. lbs. (72 Nm).
11. Install the propeller shaft protector. Tighten the bolts to 16 ft. lbs. (22 Nm).
12. If you installed a new propeller shaft, test-drive the vehicle at 55 mph (88 km/h), and check for noise or vibration. If there is a noise or vibration, rotate the propeller shaft 180 degrees from its current alignment with the rear differential companion flange, then recheck.

REAR DRIVE AXLE

FLUID RECOMMENDATIONS

Use Acura All-Wheel Drive Fluid (DPSF).

LEVEL CHECK

See Figure 27.

1. Park the vehicle on level ground, and turn the ignition switch to LOCK (0).
2. Use solvent and a brush to wash any oil and dirt off the differential fluid filler plugs, any dry them with compressed air.
3. Remove the filler plug and the sealing washer, then check the condition of the fluid, if the fluid is dirty or burnt, replace it.
4. Make sure the fluid is at the proper level. If the level is below the proper level, check for fluid leaks at the differential. If a problem is found, fix it before filling the differential with fluid.
5. The fluid level must reach up the bottom of the filler plug hole. If necessary, add the recommended fluid until it runs out, then reinstall the filler plug with a new sealing washer.

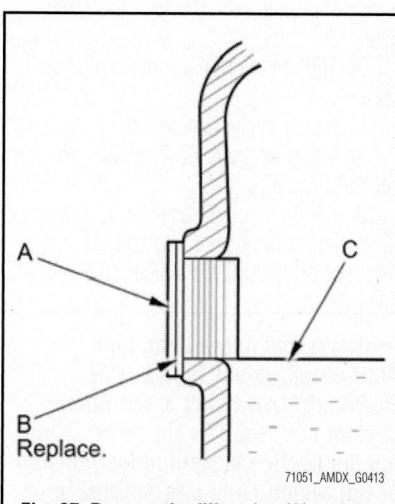

Fig. 27 Remove the filler plug (A) and the sealing washer (B), make sure the fluid is at the proper level (C)

DRAIN & REFILL

See Figure 28.

1. Park the vehicle on level ground, and turn the ignition switch to LOCK (0).
2. Remove the filler plug and the sealing washer.
3. Remove the drain plug and the sealing washer, and drain the fluid.
4. Reinstall the drain plug with a new sealing washer.
5. Refill the recommended fluid until it reaches the bottom of the filler plug hole.
6. Reinstall the filler plug with a new sealing washer.
7. If the Maintenance Minder required to replace the rear differential fluid, reset the Maintenance Minder, and this procedure is complete. If the Maintenance Minder did not require to replace the rear differential fluid, go to step 8.
8. With the ignition switch in LOCK (0), connect the HDS to the data link connector (DLC) behind the driver's dashboard lower cover.
9. Turn the ignition switch to ON (II).
10. Make sure the HDS communicates with the vehicle and the PCM. If it does not, go to the DLC circuit troubleshooting.
11. Select GAUGES in the BODY ELECTRICAL with the HDS.
12. Select ADJUSTMENT in the GAUGES with the HDS.
13. Select MAINTENANCE MINDER in the ADJUSTMENT with the HDS.
14. Select RESET in the MAINTENANCE MINDER with the HDS.
15. Select MAINTENANCE SUB ITEM 6 RESET with the HDS, and reset the rear differential fluid life.

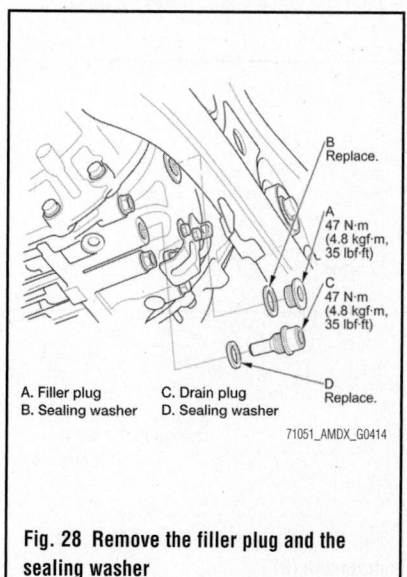

A. Filler plug   C. Drain plug
B. Sealing washer   D. Sealing washer

Fig. 28 Remove the filler plug and the sealing washer

REAR HALFSHAFT

REMOVAL & INSTALLATION

See Figure 29.

1. Raise and safely support the vehicle.
2. Remove the rear wheel.
3. Pry up the locking tab and remove the hub nut.
4. Remove the wheel sensor and harness clip.
5. Remove the lock pin from the upper arm ball joint castle nut, and remove the nut.
6. Separate the ball joint from the upper arm with the ball joint remover.
7. Remove the lower control arm.
8. Place a transmission jack under lower arm and remove the flange bolt.
9. Loosen the rear driveshaft outboard joint from the rear hub using a plastic hammer.
10. Pull the knuckle outward, then separate the rear driveshaft outboard joint.
11. Using the driveshaft remover and the hammer, pry out the inboard joint from the rear differential.
12. Remove the rear halfshaft, then remove the set ring.

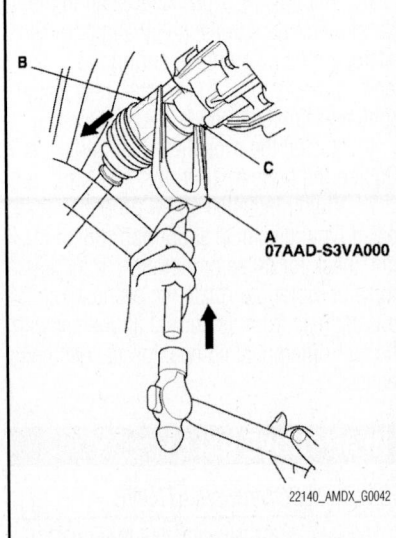
07AAD-S3VA000

22140_AMDX_G0042

Fig. 29 Using the driveshaft remover (A) and the hammer, pry out the inboard joint (B) from the rear differential (C)

To install:

13. Apply grease to the whole splined surface of the halfshaft.
14. Install a new set ring in the set ring groove of the differential.
15. Clean the areas where the driveshaft contacts the differential thoroughly with solvent or brake cleaner, and dry with compressed air. Insert the inboard end of the halfshaft into the differential until the set ring locks in the groove.
16. Pull the knuckle outward, and install the rear driveshaft outboard joint into the rear hub.
17. Install the wheel sensor bracket.
18. Install the lower arm flange bolt and tighten to 54 ft. lbs. (74 Nm).
19. Install the upper arm bolt and tighten to 47 ft. lbs. (64 Nm).
20. Install a new spindle nut and tighten to 181 ft. lbs. (245 Nm). Use a drift to stake the spindle nut shoulder once the nut is tightened to specification.
21. Install the wheel
22. Turn the wheel to make sure there is no binding between the driveshaft and wheel.
23. Refill the differential until the fluid level is at the bottom of the fill hole. A complete oil change would require 2.79 quarts of VTM–4 differential fluid. Install the plug and tighten to 35 ft. lbs. (47 Nm).
24. Check and adjust the wheel alignment.

ENGINE COOLING

ENGINE COOLANT

DRAIN & REFILL

See Figure 30.

1. Start the engine. Set the heater temperature control dial to maximum heat, then turn the ignition switch to LOCK (0). Make sure the engine and radiator are cool to the touch.

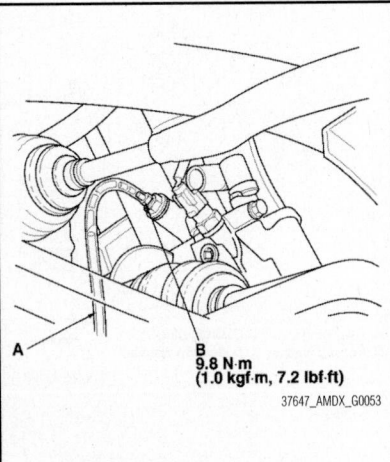
A
B 9.8 N·m
(1.0 kgf·m, 7.2 lbf·ft)
37647_AMDX_G0053

Fig. 30 Install a rubber hose (A) on the drain bolt (B)

2. Remove the radiator cap.
3. Remove the splash shield.
4. Loosen the drain plug and drain the coolant.
5. Install a rubber hose on the drain bolt located at the rear of the engine block, then loosen the drain bolt.
6. When the coolant stops draining, tighten the drain bolt. Remove the rubber hose.
7. Tighten the radiator drain plug securely.
8. Install the splash shield.
9. Remove, drain, and reinstall the coolant reservoir.
10. Fill the coolant reservoir to the MAX mark with Acura Long Life Antifreeze/Coolant Type 2 (P/N OL999-9011A).

➡**Always use Acura Long Life Antifreeze/Coolant Type 2 (P/N OL999-9011A). Using a non-Acura coolant can result in corrosion, causing the cooling system to malfunction or fail. Acura Long Life Antifreeze/ Coolant Type 2 is a mixture of 50 % antifreeze and 50 % water. Do not add water.**

11. Pour Acura Long Life Antifreeze/Coolant Type 2 into the radiator up to the base of the filler neck.

➡**Engine Coolant Capacities (Including the coolant reservoir capacity of 0.19 Gals. (0.7 L)): At Coolant Change: 1.93 gals. (7.3 L); After Engine Overhaul: 2.43 gals. (9.2 L)**

12. Start the engine. Hold the engine speed at 1,500 rpm until it warms up (the radiator fan comes on at least twice). Make sure the thermostat is open.
13. Turn off the engine. Check the level in the radiator, and add Acura Long Life Antifreeze/Coolant Type 2, if needed.
14. Set the climate control or heater control panel to maximum cool. Start the engine. Hold the engine speed at 1,500 rpm for 5 minutes, then turn off the engine.
15. Check the level in the radiator, and add Acura Long Life Antifreeze/Coolant Type 2, if needed.
16. Set the climate control or heater control panel to maximum heat. Start the engine. Hold the engine speed at 1,500 rpm for 5 minutes, then turn off the engine.

17. Check the level in the radiator, and add Acura Long Life Antifreeze/Coolant Type 2, if needed.

18. Set the climate control or heater control panel to maximum cool. Start the engine. Hold the engine speed at 1,500 rpm for 3 minutes, then turn off the engine.

19. Check the level in the radiator, and add Acura Long Life Antifreeze/Coolant Type 2, if needed.

20. Set the climate control or heater control panel to maximum heat. Start the engine. Hold the engine speed at 1,500 rpm for 3 minutes, then turn off the engine.

21. Check the level in the radiator, and add Acura Long Life Antifreeze/Coolant Type 2, if needed.

22. Repeat step 18 through 21 until the coolant level does not change in the radiator, then install the radiator cap loosely.

23. Set the climate control or heater control panel to maximum cool. Start the engine. Hold the engine speed at 2,500 rpm for 1 minute.

24. Set the climate control or heater control panel to maximum heat, and select high speed heat mode on the rear control panel. Measure the temperature of the air from the rear floor vent for 2 or 3 minutes.

25. With the engine idling, make sure you do not hear the sound of water flowing near the rear heater unit. If you hear water flowing, repeat step 18 through 21 several more times.

26. Install the radiator cap fully.

27. Fill the coolant reservoir with Acura Long Life Antifreeze/Coolant Type 2 up to the MAX mark on the coolant reservoir, then add an extra 0.11 gals. (0.4 L) coolant.

28. Clean up any spilled engine coolant.

29. If the maintenance minder required to replace the engine coolant, reset the maintenance minder, and this procedure is complete. If the maintenance minder did not require to replace the engine coolant, go to step 30.

30. Connect the Honda Diagnostic System (HDS) to the Data Link Connector (DLC).

31. Turn the ignition switch to ON (II).

32. Make sure the HDS communicates with the vehicle and the Powertrain Control Module (PCM). If it doesn't communicate, troubleshoot the DLC circuit.

33. Select BODY ELECTRICAL with the HDS.

34. Select ADJUSTMENT in the GAUGES MENU with the HDS.

35. Select RESET in the MAINTENANCE MINDER with the HDS.

36. Select MAINTENANCE SUB ITEM 5 RESET with the HDS.

37. Turn off the engine. Check the level in the radiator, and add Acura Long Life Antifreeze/Coolant Type 2, if needed.

➡**Removing the radiator cap while the engine is hot can cause the coolant to spray out. Always let the engine and radiator cool before removing the cap.**

RADIATOR

REMOVAL & INSTALLATION

See Figures 31 through 33.

1. Make sure you have the anti-theft code for the radio and the navigation system, then write down the frequencies for the radio's preset buttons.

2. Disconnect the negative cable from the battery first, then disconnect the positive cable. Wait at least 3 minutes before proceeding.

3. Remove the battery and battery tray.

4. Remove the air intake assembly.

5. Remove the front grille cover, then remove the front grille.

6. Disconnect the fan motor connectors, harness clamp, Engine Coolant Temperature (ECT) sensor 2 sub-harness connector and coolant reservoir hose.

7. Drain the engine coolant.

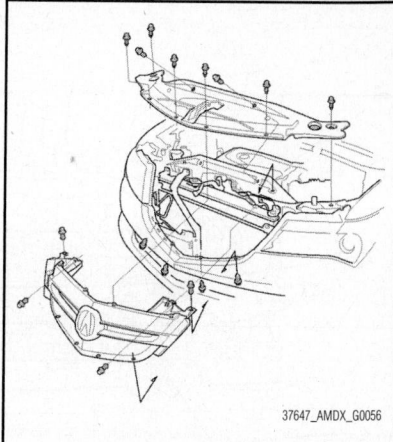

Fig. 31 Remove the front grille cover, then remove the front grille

8. Remove the upper radiator hose.

9. Unclamp the hose clamps and remove the ATF cooler hoses, then plug the line and hoses.

10. Remove the ATF cooler pipe mount bolts, then remove the ATF cooler pipe.

11. Remove the lower radiator hose and disconnect the ECT sensor 2 connector and loosen the two bolts.

12. Disconnect the hood latch switch connector, clamps, then remove the hood latch.

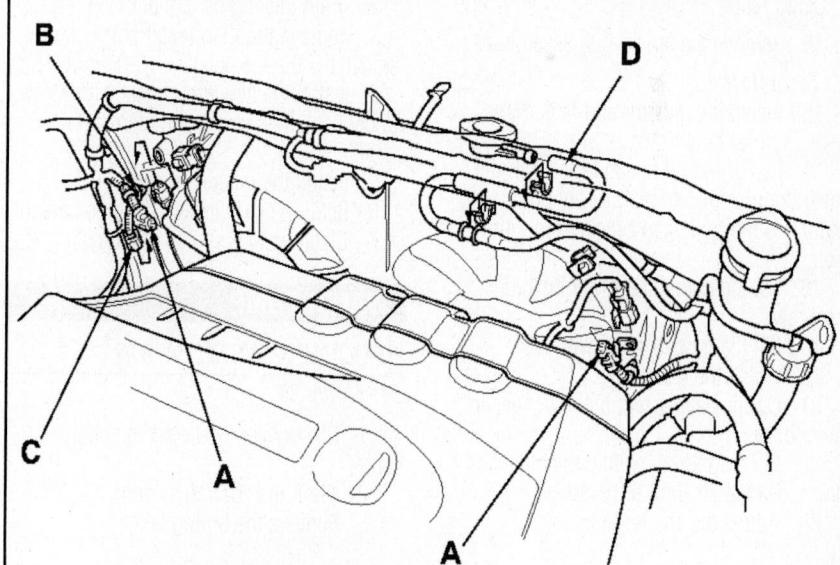

A. Fan motor connectors
B. Harness clamp
C. ECT sensor 2 sub-harness connector
D. Coolant reservoir hose

Fig. 32 Disconnect the fan motor connectors, harness clamp, Engine Coolant Temperature (ECT) sensor 2 sub-harness connector and coolant reservoir hose

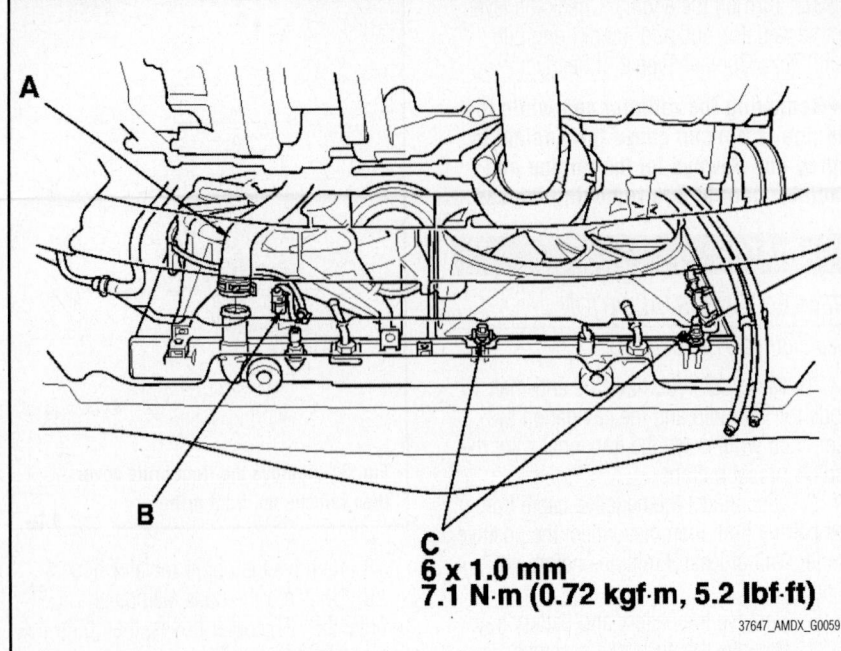

6 x 1.0 mm
7.1 N·m (0.72 kgf·m, 5.2 lbf·ft)

37647_AMDX_G0059

Fig. 33 Remove the lower radiator hose (A) and disconnect the ECT sensor 2 connector (B)

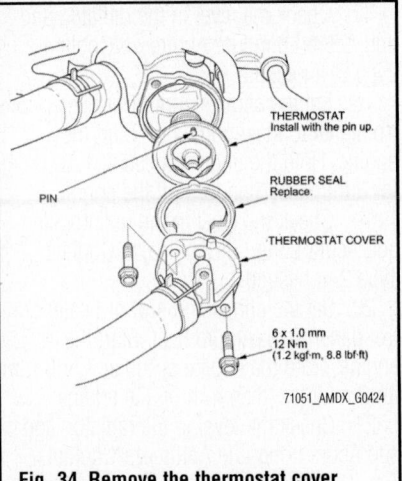

THERMOSTAT
Install with the pin up.

PIN

RUBBER SEAL
Replace.

THERMOSTAT COVER

6 x 1.0 mm
12 N·m
(1.2 kgf·m, 8.8 lbf·ft)

71051_AMDX_G0424

Fig. 34 Remove the thermostat cover, then remove the thermostat

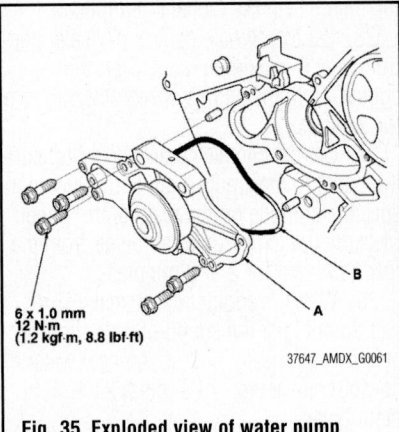

6 x 1.0 mm
12 N·m
(1.2 kgf·m, 8.8 lbf·ft)

B

A

37647_AMDX_G0061

Fig. 35 Exploded view of water pump mounting assembly

13. Remove the bulkhead bracket, and radiator and A/C condenser mount upper bracket/cushion, then remove the fan shroud mount bolts.

14. Move the fan shroud assemblies toward the engine, then pull up the radiator assembly.

15. Remove the fan shroud assemblies.

To install:

16. Install the radiator and fans in the reverse order of removal.

17. Install the bulkhead bracket in the reverse order of removal. Apply body paint to the bulkhead bracket mounting bolts.

18. Set the upper and lower cushions securely.

19. Fill the transmission with ATF.

20. Install the battery.

21. Do the battery terminal reconnection procedure.

22. Fill the radiator with engine coolant and bleed the air from the system.

23. Adjust the engine hood latch.

THERMOSTAT

REMOVAL & INSTALLATION

See Figure 34.

1. Remove the intake air duct.

2. Drain the engine coolant.

3. Remove the thermostat cover, then remove the thermostat.

4. Install the new thermostat with a new rubber seal, then install the thermostat cover.

5. Install the intake air duct.

6. Refill the radiator with engine coolant.

7. Clean up any spilled engine coolant.

WATER PUMP

REMOVAL & INSTALLATION

See Figure 35.

1. Disconnect the negative battery cable.

2. Drain the cooling system.

3. Remove the timing belt.

4. Remove the timing belt adjuster.

5. Remove the water pump mounting bolts.

6. Remove the water pump.

To install:

7. Install the water pump with a new O-ring. Tighten the mounting bolts to 105 inch lbs. (12 Nm).

8. The remainder of the installation is the reverse order of removal.

9. Refill the cooling system to the correct level.

10. Start the engine and check for leaks.

ENGINE ELECTRICAL

BATTERY

REMOVAL & INSTALLATION
See Figure 36.

➡ **The battery terminal disconnection/reconnection procedure must be done before and after doing this procedure. Some systems store data in memory that is lost when the battery is disconnected.**

1. Do the battery terminal disconnection procedure.

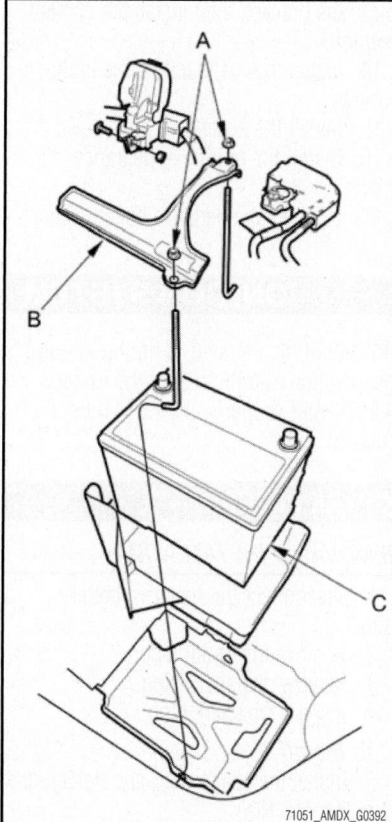

71051_AMDX_G0392

Fig. 36 Remove the two nuts (A) securing the battery hold down plate, then remove the battery setting plate (B) and the battery (C)

2. Remove the two nuts securing the battery hold down plate, then remove the battery setting plate and the battery.

To install:
3. Install the battery, then install the battery hold down plate.
4. Tighten the two nuts equally until the battery is stable.

➡ **Do not deform the battery hold down plate by over-tightening the nuts.**

5. Do the battery terminal reconnection procedure.

➡ **Make sure the battery is installed correctly, and the positive and negative terminals are not reverse connected.**

BATTERY RECONNECT/RELEARN PROCEDURE

✳✳ CAUTION

A battery can explode if you do not follow the proper procedure, causing serious injury to anyone nearby. Follow all procedures carefully and keep sparks and open flames away from the battery.

Disconnection

➡ **Some systems store data in memory that is lost when the battery is disconnected. Do the following procedures before disconnecting the battery.**

1. Make sure you have the anti-theft code for the audio or the audio-navigation unit.
2. Make sure the ignition switch is in LOCK (0).
3. Disconnect and isolate the negative cable from the battery.

➡ **Always disconnect the negative cable from the battery first.**

4. Disconnect the positive cable from the battery.

Reconnection
See Figure 37.

Please note the following:
• Some systems store data in memory that is lost when the battery is disconnected. Do the following procedures before disconnecting the battery.
• The battery sensor is initialized and recalibrated automatically after reconnecting.

1. Clean the battery terminals.
2. Test the battery.
3. Reconnect the positive cable to the battery first, then reconnect the negative cable to the battery.

➡ **Always connect the positive cable to the battery first.**

4. Apply multipurpose grease to the terminals to prevent corrosion.
5. Do the steering column position memorization procedure.
6. Enter the anti-theft code for the audio or the audio navigation unit.
7. Set the clock (for vehicles without navigation).
8. If equipped, initialize the navigation system.
9. If the power tailgate was opened while the battery was disconnected, the power tailgate control unit must be reset.

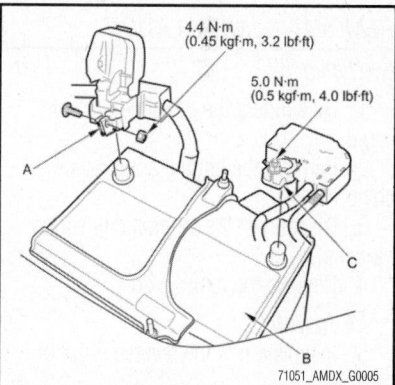

4.4 N·m
(0.45 kgf·m, 3.2 lbf·ft)

5.0 N·m
(0.5 kgf·m, 4.0 lbf·ft)

71051_AMDX_G0005

Fig. 37 Reconnect the positive cable (A) to the battery (B) first, then reconnect the negative cable (C) to the battery

ENGINE ELECTRICAL

ALTERNATOR

REMOVAL & INSTALLATION

See Figure 38.

1. Make sure you have the anti-theft code for the radio and the navigation system, then write down the frequencies for the radio's preset buttons.
2. Disconnect the negative cable from the battery first, then disconnect the positive cable. Wait at least 3 minutes before proceeding.
3. Remove the engine appearance cover.
4. Remove the accessory drive belt.
5. Remove the A/C suction line from the brackets.
6. Remove the coolant reservoir, then remove power steering fluid reservoir from the bracket.
7. Remove the harness bracket, then disconnect the A/C compressor clutch wiring harness.

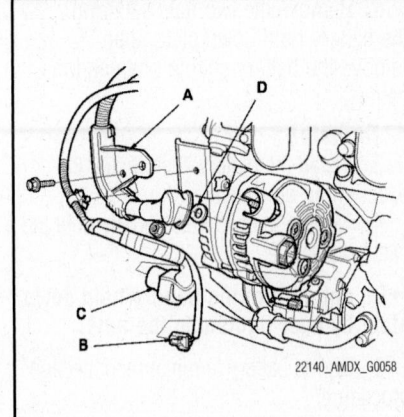

Fig. 38 Remove the following connectors to remove the alternator

8. Disconnect the alternator wiring harness and the BLK wire from the alternator.

CHARGING SYSTEM

9. Remove the mounting bolt, the bracket mounting bolt, then remove the alternator.

To install:

10. Install the alternator. Tighten the mounting bolt to 33 ft. lbs. (45 Nm) and bracket mounting bolt to 16 ft. lbs. (22 Nm).
11. Connect the alternator wiring harness and BLK wire to the alternator.
12. Install the harness bracket, then connect the A/C compressor clutch wiring harness
13. Install the power steering fluid reservoir to the bracket, then install the coolant reservoir.
14. Install the A/C suction line to the brackets.
15. Install the accessory drive belt.
16. Install the engine appearance cover.
17. Connect the negative battery cable.

ENGINE ELECTRICAL

FIRING ORDERS

The firing order is 1, 4, 2, 5, 3, 6.

IGNITION COIL

REMOVAL & INSTALLATION

See Figure 39.

1. Disconnect the negative battery cable.
2. Remove the engine appearance cover.
3. Disconnect the ignition coil wiring harness.
4. Remove the ignition coil.

To install:

5. Installation is the reverse order of removal.
6. Tighten the ignition coil mounting nut to 108 inch lbs. (12 Nm).

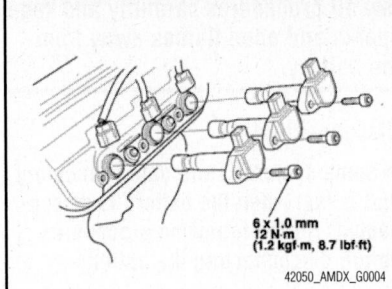

Fig. 39 Disconnect the ignition coil connector (A), then remove the ignition coils (B)

IGNITION TIMING

INSPECTION & ADJUSTMENT

Ignition timing is control by the Powertrain Control Module (PCM)

IGNITION SYSTEM

and cannot be adjusted. If timing is out of specification and the rest of the ignition system work properly, the PCM must be replaced.

SPARK PLUGS

REMOVAL & INSTALLATION

1. Disconnect the negative battery cable.
2. Remove the ignition coil.
3. Remove the spark plug.
4. Inspect the spark plug.

To install:

5. Install the spar k plug and tighten to 13 ft. lbs. (18 Nm).
6. Install the ignition coil.
7. Connect the negative battery cable.

ENGINE ELECTRICAL   **STARTING SYSTEM**

STARTER

REMOVAL & INSTALLATION

See Figures 40 and 41.

1. Make sure you have the anti-theft codes for the radio, and the navigation system, then write down the audio presets. Make sure the ignition switch is OFF.

2. Disconnect the negative cable from the battery first, then disconnect the positive cable. Wait at least 3 minutes before proceeding.

3. Remove the air intake assembly.

4. Remove the battery and battery tray.

5. Remove the starter wiring harness clamp.

6. Disconnect the positive starter cable from the B terminal. Disconnect the wiring harness from the S terminal.

7. Remove the starter mounting bolts and remove the starter.

To install:

8. Install the starter using a new gasket, then install the harness clamp, and positive starter cable and S terminal connector.

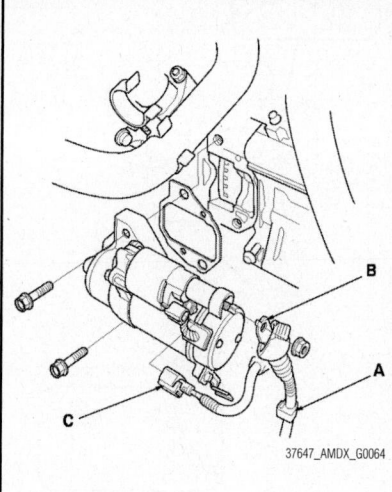

Fig. 40 Remove the starter wiring harness clamp (A)

Make sure the crimped side of the B terminal is facing out, away from the starter.

9. Install the battery base and battery.

10. Install the air cleaner assembly.

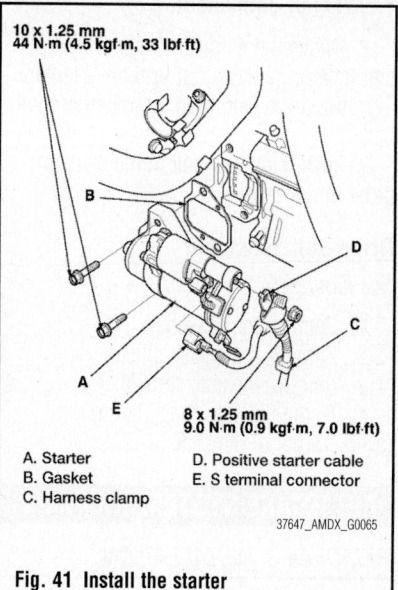

10 x 1.25 mm
44 N·m (4.5 kgf·m, 33 lbf·ft)

8 x 1.25 mm
9.0 N·m (0.9 kgf·m, 7.0 lbf·ft)

A. Starter
B. Gasket
C. Harness clamp
D. Positive starter cable
E. S terminal connector

Fig. 41 Install the starter

11. Do the battery terminal reconnection procedure.

12. Start the engine to make sure the starter works properly.

ENGINE MECHANICAL

➡ **Disconnecting the negative battery cable may interfere with the functions of the on board computer systems and may require the computer to undergo a relearning process, once the negative battery cable is reconnected.**

ACCESSORY DRIVE BELT SYSTEM

BELT ROUTINGS

See Figure 42.

INSPECTION

See Figure 43.

1. Inspect the belt for cracks and damage. If the belt is cracked or damaged, replace it.

2. Check that the auto-tensioner indica-

tor is within the standard range. If it is out of the standard range, replace the drive belt.

ADJUSTMENT

Belt tension is automatically maintained by a belt tensioner. No adjustments are necessary.

REMOVAL & INSTALLATION

Drive Belt

See Figure 44.

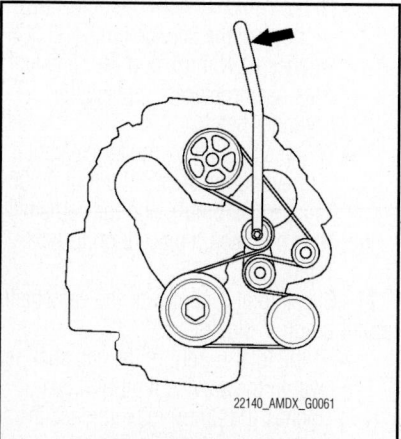

Fig. 42 Accessory drive belt routing

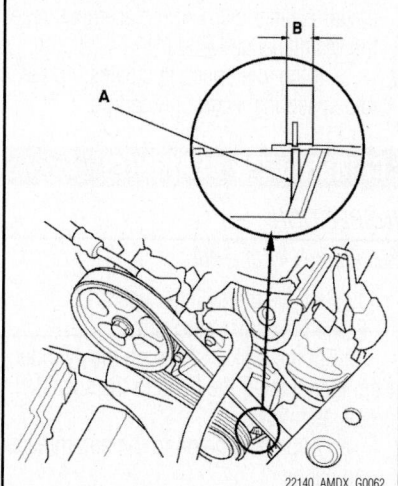

Fig. 43 Check that the auto-tensioner indicator (A) is within the standard range (B)

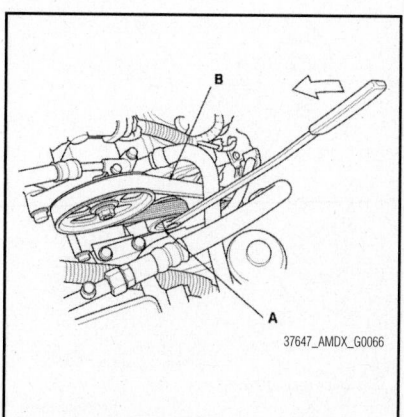

Fig. 44 Move the auto-tensioner with the belt tension release tool

➡ **For this procedure, you will need a Belt tension release tool (Snap-on YA9317) or equivalent.**

1. Move the auto-tensioner with the belt tension release tool to relieve tension from the drive belt, then remove the drive belt.

2. Install the new belt in the reverse order of removal.

Drive Belt Tensioner

See Figure 45.

1. Remove the drive belt.
2. Remove the splash shield.
3. Remove the auto-tensioner.
4. Install the auto-tensioner in the reverse order of removal.

AIR INTAKE SYSTEM

REMOVAL & INSTALLATION

Air Cleaner Assembly

See Figure 46.

1. Disconnect the MAF sensor/IAT sensor 5P connector.
2. Remove the bolts and, then remove harness clamps.
3. Remove the air cleaner.
4. Install the parts in the reverse order of removal.

Air Filter Element

1. Open the air cleaner housing cover.
2. Remove the air cleaner element from the air cleaner housing.

➡ **Be careful when removing the element, and do not turn upside down. Debris from the element may fall into the MAF sensor/IAT sensor.**

3. Check the air cleaner element for

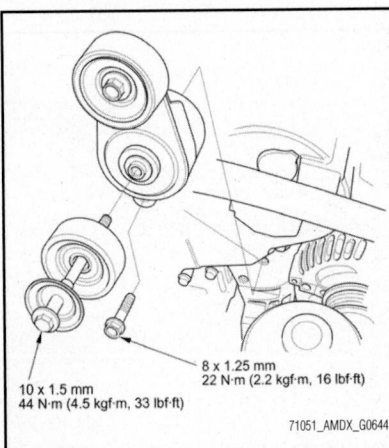

8 x 1.25 mm
22 N·m (2.2 kgf·m, 16 lbf·ft)

10 x 1.5 mm
44 N·m (4.5 kgf·m, 33 lbf·ft)

71051_AMDX_G0644

Fig. 45 Remove the auto-tensioner

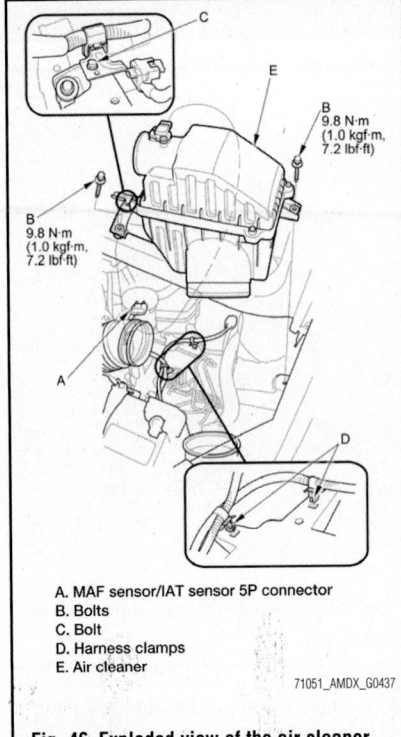

B
9.8 N·m
(1.0 kgf·m,
7.2 lbf·ft)

B
9.8 N·m
(1.0 kgf·m,
7.2 lbf·ft)

A. MAF sensor/IAT sensor 5P connector
B. Bolts
C. Bolt
D. Harness clamps
E. Air cleaner

71051_AMDX_G0437

Fig. 46 Exploded view of the air cleaner assembly

damage or clogging. If there is damage or clogging, replace the air cleaner element.

➡ **Do not use compressed air to clean the air cleaner element.**

4. Clean and remove any debris from inside the air cleaner housing.
5. Install the parts in the reverse order of removal.

   a. If you did not replace the air cleaner element, this procedure is complete.

   b. If the Maintenance Minder required the air cleaner element replacement, reset the Maintenance Minder.

   c. If the idle speed fluctuates, do the idle speed inspection procedure.

CAMSHAFT & BEARINGS

INSPECTION

See Figures 47 and 48.

1. Remove the cylinder head.
2. Remove the rocker arms.
3. Put the rocker shafts on the cylinder head, then tighten the bolts to the specified torque, as follows:

   a. Apply engine oil to the bolt threads and flange.

   b. Tighten the 8 x 1.255mm bolts to 17 ft. lbs. (24 Nm).

4. Seat the camshaft by pushing it toward the rear of the cylinder head.

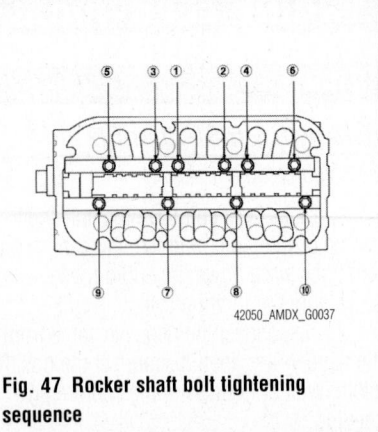

42050_AMDX_G0037

Fig. 47 Rocker shaft bolt tightening sequence

5. Zero the dial indicator against the end of the camshaft. Push the camshaft back and forth and read the end play. If the end play is beyond the service limit, replace the thrust cover and recheck. If it is still beyond the service limit, replace the camshaft:

   a. Standard (new): 0.002–0.008 inches (0.05–0.20mm)

   b. Service Limit: 0.008 inches (0.20mm)

6. Remove the camshaft thrust cover, then pull out the camshaft.

7. Wipe the camshaft clean, then inspect the lift ramps. Replace the camshaft if any lobes are pitted, scored or excessively worn.

8. Use a micrometer to measure the diameter of each camshaft journal.

9. Zero the gauge to the journal diameter.

10. Clean the camshaft bearing surfaces in the cylinder head. Measure the inside diameter of each camshaft bearing surface, and check for an out-of-round condition:

   - If the camshaft-to-holder clearance is within limits of 0.0020–0.0035 inches (0.050–0.089mm), go to step 11.

   - If the camshaft-to-holder clearance is beyond the service limit of 0.006 inches (0.15mm), and the camshaft has been replaced, replace the cylinder head.

   - If the camshaft-to-holder clearance is beyond the service limit of 0.006 inches (0.15mm), and the camshaft has not been replaced, go to next step.

11. Check total runout with the camshaft supported on V-blocks:

   - If the total runout of the camshaft is within the service limit of 0.001 inches (0.03mm) max, replace the cylinder head.

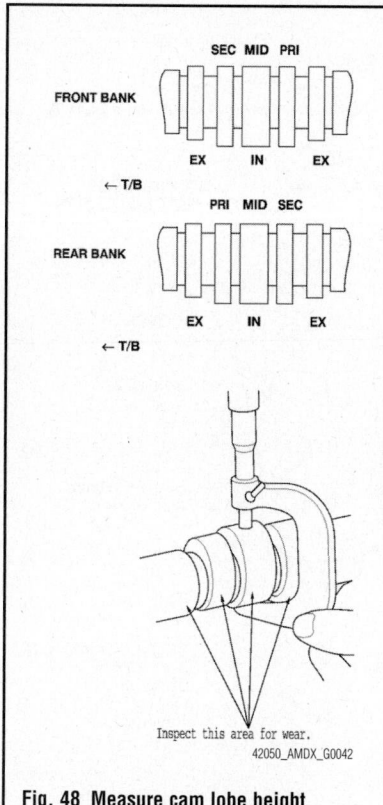

Fig. 48 Measure cam lobe height

- If the total runout is beyond the service limit 0.002 inches (0.04mm), replace the camshaft and recheck the oil clearance. If the oil clearance is still out of tolerance, replace the cylinder head.

12. Measure cam lobe height and compare with the following cam lobe height standard (new):

- PRI Intake: 1.3824 inches (35.112 mm)
- MID Intake: 1.4328 inches (36.394 mm)
- SEC Intake: 1.3824 inches (35.112 mm)
- Exhaust: 1.4326 inches (36.389 mm)

REMOVAL & INSTALLATION

Front Camshaft

See Figure 49.

1. Disconnect the negative and positive battery cable.
2. Remove the battery and battery box.
3. Drain the engine cooling system.
4. Disconnect the upper radiator hose.
5. Remove the exhaust gas recirculation valve.
6. Remove the timing belt.
7. Remove the intake manifold.
8. Remove the cylinder head cover

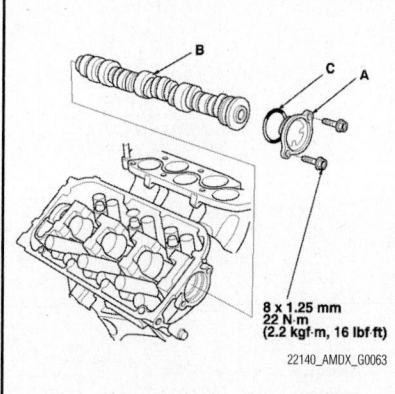

8 x 1.25 mm
22 N·m
(2.2 kgf·m, 16 lbf·ft)

22140_AMDX_G0063

Fig. 49 Remove the thrust cover (A) to remove the camshaft (B). Replace the O-ring (C) when reinstalling

9. Loosen the locknuts and adjusting screws.
10. Remove the bolts and the rocker arm assembly.

✳✳ CAUTION

Loosen the rocker shaft mounting bolts two turns at a time, to prevent damaging the valves or rocker arm assembly.

➡**When removing the rocker arm assembly, do not remove the rocker shaft mounting bolts. The bolts will keep the springs and the rocker arms on the shafts.**

11. Remove the front camshaft pulley.
12. Remove the thrust cover, the remove the camshaft.

To install:

13. Apply clean engine oil to the journals and cam lobes of the camshaft. Install the camshaft using a new O-ring. Tighten the thrust plate to 16 ft. lbs. (22 Nm).
14. Apply clean engine oil to the threads of the camshaft pulley mounting bolt, then install the front camshaft pulley.
15. Install the rocker arm assembly. Tighten the mounting bolts two turns at a time in the sequence shown to 17 ft. lbs. (24 Nm).
16. Install the timing belt.
17. Adjust the valve clearance.
18. The remainder of the installation is the reverse order of removal.
19. Refill the cooling system to the correct level.

Rear Camshaft

1. Disconnect the negative battery cable.

2. Drain the cooling system.
3. Remove the intake manifold.
4. Remove the purge joint.
5. Remove the brake lines from the master cylinder.
6. Remove the timing belt.
7. Remove the cylinder head cover.
8. Loosen the locknuts and adjusting screws.
9. Remove the bolts and the rocker arm assembly.

✳✳ CAUTION

Loosen the rocker shaft mounting bolts two turns at a time, to prevent damaging the valves or rocker arm assembly.

➡**When removing the rocker arm assembly, do not remove the rocker shaft mounting bolts. The bolts will keep the springs and the rocker arms on the shafts.**

10. Remove the rear camshaft pulley.
11. Remove the thrust cover, then remove the camshaft.

To install:

12. Apply clean engine oil to the journals and cam lobes of the camshaft. Install the camshaft using a new O-ring. Tighten the thrust plate to 16 ft. lbs. (22 Nm).
13. Apply clean engine oil to the threads of the camshaft pulley mounting bolt, then install the rear camshaft pulley.
14. Install the rocker arm assembly. Tighten the mounting bolts two turns at a time in the sequence shown to 17 ft. lbs. (24 Nm).
15. Install the timing belt.
16. Adjust the valve clearance.
17. Install the brake lines to the master cylinder and bleed the brake system.
18. The remainder of the installation is the reverse order of removal.
19. Refill the cooling system to the correct level.

CATALYTIC CONVERTER

REMOVAL & INSTALLATION

Under-Floor Three Way Catalytic Converter (TWC)

See Figure 50.

1. Raise the vehicle on a lift.
2. Remove the under-floor TWC.
3. Remove the converter cover.
4. Install the parts in the reverse order of removal with new gaskets and new self-locking nuts.

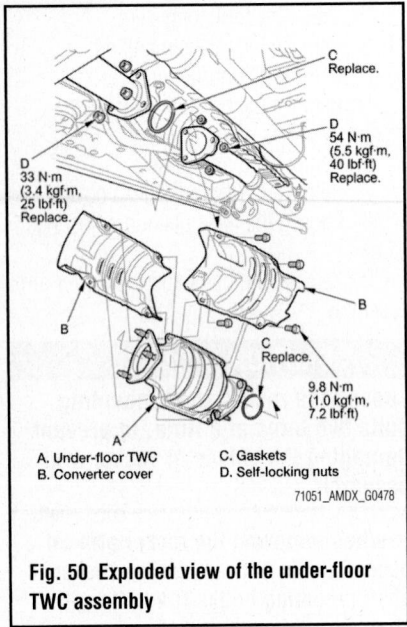

A. Under-floor TWC   C. Gaskets
B. Converter cover   D. Self-locking nuts

71051_AMDX_G0478

Fig. 50 Exploded view of the under-floor TWC assembly

Front Warm Up TWC (BANK 2)

See Figure 51.

➡️**If the warm up TWC is damaged internally, inspect the under floor TWC for debris.**

1. Remove the engine cover.
2. Remove the No. 5 ignition coil and the ignition coil heat insulator.
3. Remove the front A/F sensor (Sensor 1) and the front secondary HO2S (Sensor 2).
4. Remove the exhaust pipe A mounting nuts (front WU-TWC side).
5. Remove the EGR pipe.
6. Remove the A/C condenser fan assembly.

➡️**Place an appropriate size piece of cardboard in front of the radiator. The cardboard protects the radiator from damage when removing the warm up TWC.**

7. Remove the front WU-TWC bracket, and carefully remove the front WU-TWC.

To install:

8. Carefully install the front WU-TWC with a new gasket and new self-locking nuts. Tighten the nuts in a crisscross pattern in two or three steps.
9. Install the parts in the reverse order of removal.

Rear Warm Up TWC (BANK 1)

See Figure 52.

➡️**If the warm up TWC is damaged internally, inspect the under floor TWC for debris.**

1. Remove the rear A/F sensor

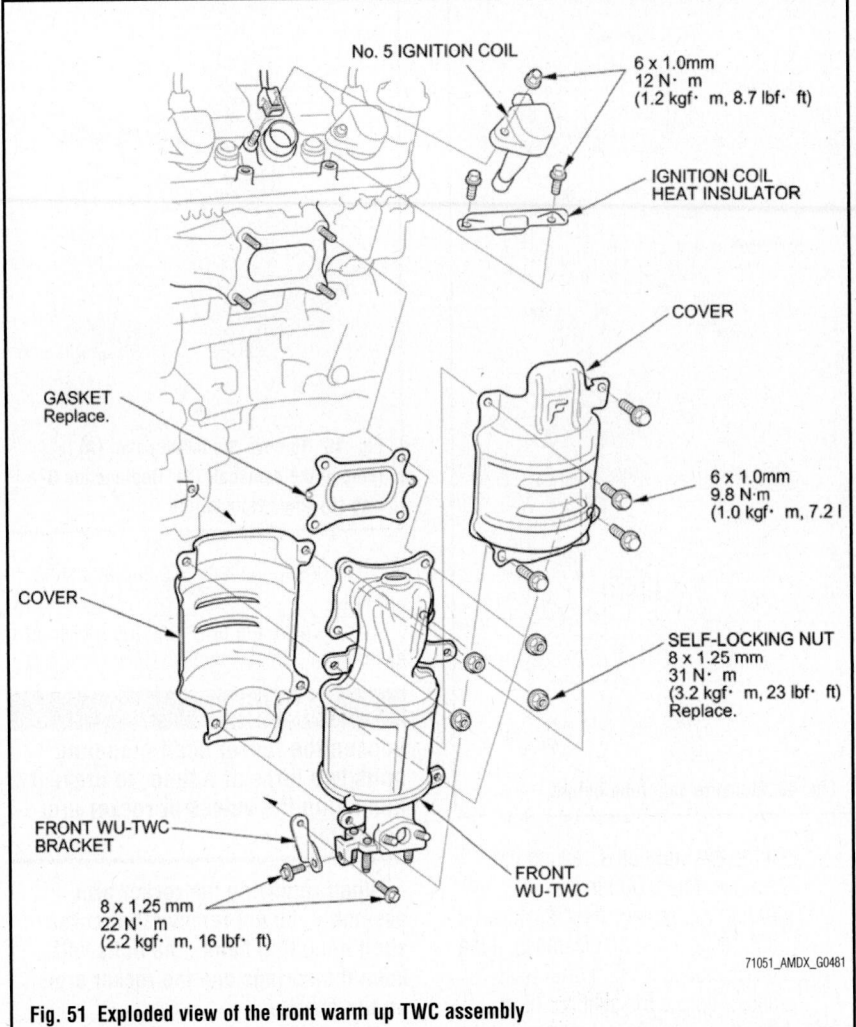

71051_AMDX_G0481

Fig. 51 Exploded view of the front warm up TWC assembly

(Sensor 1) and the rear secondary HO2S (Sensor 2).

2. Remove exhaust pipe A, then remove the rear WU-TWC bracket.
3. Remove the intermediate shaft.
4. Carefully remove the rear WU-TWC.

To install:

5. Carefully install the rear WU-TWC with a new gasket and new self-locking nuts. Tighten the nuts in a crisscross pattern in two or three steps.
6. Install the parts in the reverse order of removal.

CRANKSHAFT FRONT SEAL

REMOVAL & INSTALLATION

1. Remove the crankshaft position (CKP) sensor, the timing belt, and the timing belt drive pulley.
2. Remove the crankshaft front seal.

To install:

3. Clean and dry the crankshaft oil seal housing.

4. Apply a light coat of multipurpose grease to the crankshaft and to the lip of the seal.
5. Using the seal driver, drive in the crankshaft oil seal until the driver bottoms against the oil pump.
6. When the seal is in place, clean any excess grease off the crankshaft, and check that the oil seal lip is not distorted.
7. Install the timing belt drive pulley, the timing belt, and the CKP sensor.

CRANKSHAFT REAR COVER & SEAL

REMOVAL & INSTALLATION

1. Remove the transaxle assembly.
2. Remove the driveplate.
3. Remove the rear main seal.

To install:

4. Clean and dry the crankshaft oil seal housing.
5. Apply a light coat of multipurpose grease to the crankshaft and to the lip of the seal.

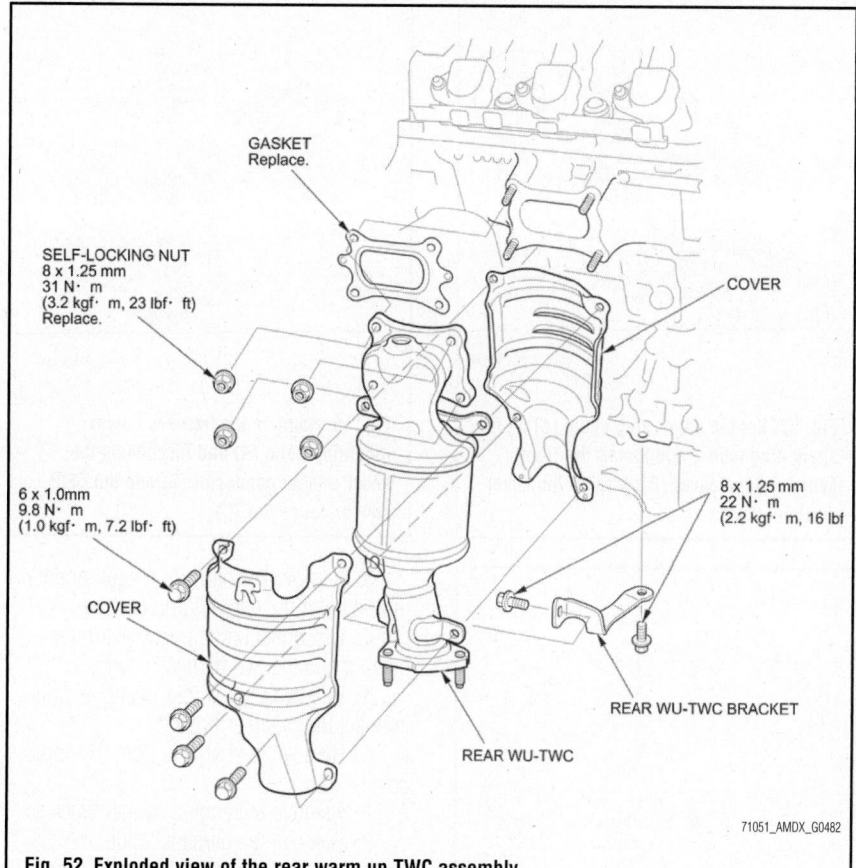

Fig. 52 Exploded view of the rear warm up TWC assembly

GASKET
Replace.

SELF-LOCKING NUT
8 x 1.25 mm
31 N· m
(3.2 kgf· m, 23 lbf· ft)
Replace.

COVER

6 x 1.0mm
9.8 N· m
(1.0 kgf· m, 7.2 lbf· ft)

COVER

8 x 1.25 mm
22 N· m
(2.2 kgf· m, 16 lbf

REAR WU-TWC BRACKET

REAR WU-TWC

71051_AMDX_G0482

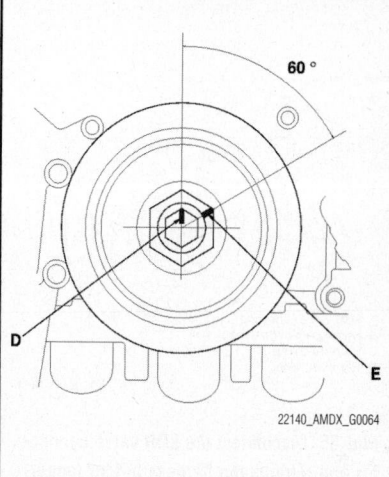

Fig. 54 Mark the bolt head and crankshaft pulley, then tighten the bolt an additional 60°. (The mark on the bolt head line up with the mark on the crankshaft pulley).

60°

D

E

22140_AMDX_G0064

6. Using a suitable seal driver tool, drive in the rear main seal until the driver attachment bottoms against the engine block end cover. Align the hole in the driver attachment with the pin on the crankshaft.

7. Clean any excess grease off the crankshaft, and check that the oil seal lip is not distorted.

8. Install the driveplate and transaxle assembly

CRANKSHAFT PULLEY

REMOVAL & INSTALLATION

See Figures 53 and 54.

➡**This procedure requires the use of the following special tools or their equivalents:**

- Holder handle 07JAB-001020A
- Holder attachment, 50mm, offset 07MAB-PY3010A
- Socket, 19mm 07JAA-001020A or a commercially available 19mm socket

1. Remove the right front wheel.
2. Remove the splash shield.
3. Remove the drive belt.
4. Hold the pulley with the holder handle and holder attachment.

5. Remove the bolt with a heavy duty 19mm socket and breaker bar, then remove the crankshaft pulley.

To install:

6. Remove any oil or clean the pulleys, crankshaft, bolt, and washer. Lubricate new engine oil.

7. Install the crankshaft pulley, and tighten the bolt as follows. Do not use an impact wrench.

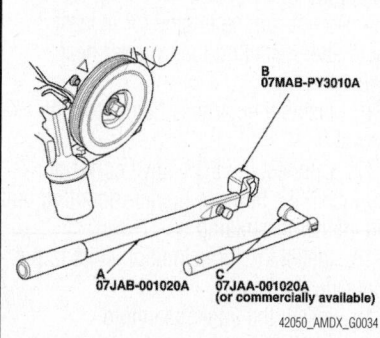

Fig. 53 Hold the pulley with the holder handle (A) and holder attachment (B). Remove the bolt with a heavy duty 19mm socket (C) and breaker bar, then remove the crankshaft pulley

B
07MAB-PY3010A

A
07JAB-001020A

C
07JAA-001020A
(or commercially available)

42050_AMDX_G0034

a. Hold the pulley with the handle and holder attachment, then tighten the bolt to 47 ft. lbs. (64 Nm) with a torque wrench and 19mm socket.

b. Mark the bolt head and crankshaft pulley as shown, then tighten the bolt an additional 60 degrees. (The mark on the bolt head line up with the mark on the crankshaft pulley).

8. Install the drive belt.
9. Install the splash shield.
10. Install the right front wheel.

CYLINDER HEAD COVER

REMOVAL & INSTALLATION

Front

See Figures 55 through 58.

1. Remove the intake manifold.
2. Remove the three ignition coils from the front cylinder head.
3. Disconnect the EGR valve connector and remove the harness holder mounting bolt and the harness clamp.
4. Remove the harness holder from the bracket.
5. Remove the front cylinder head cover.

To install:

6. Check the spark plug seals for damage. If any seals are damaged, replace it.
7. Thoroughly clean the head cover gasket and the groove of the cylinder head cover.

➡**Check and if necessary, replace the head cover gasket.**

8. Install the head cover gasket in the

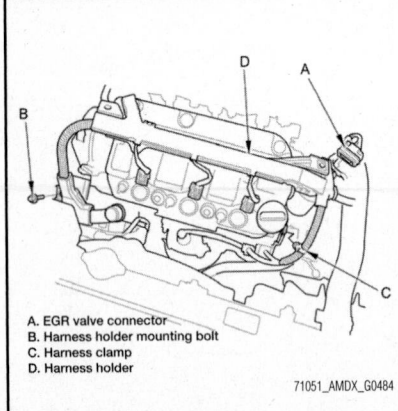

A. EGR valve connector
B. Harness holder mounting bolt
C. Harness clamp
D. Harness holder

71051_AMDX_G0484

Fig. 55 Disconnect the EGR valve connector and remove the harness holder mounting bolt and the harness clamp

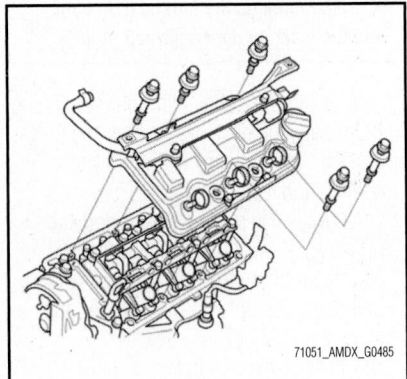

71051_AMDX_G0485

Fig. 56 Remove the front cylinder head cover

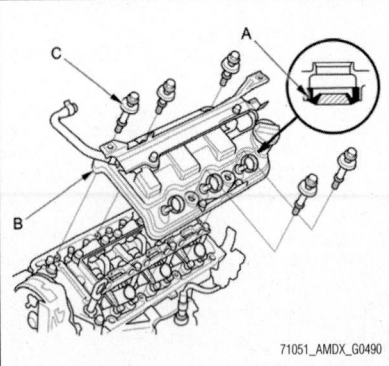

71051_AMDX_G0490

Fig. 57 Set the spark plug seals (A) on the spark plug tubes, and install the front cylinder head cover (B); inspect the cover washers (C)

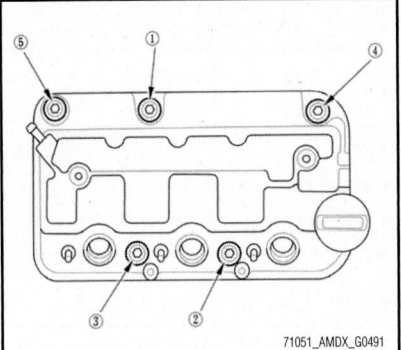

71051_AMDX_G0491

Fig. 58 Tighten the bolts in three steps. In the final step torque the bolts, in sequence

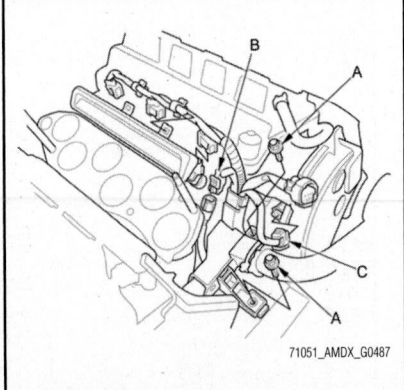

71051_AMDX_G0487

Fig. 59 Remove the harness holder mounting bolts (A) and disconnect the knock sensor connector (B) and the CMP sensor connector (C)

groove of the cylinder head cover. Make sure the head cover gasket is seated securely.

9. Remove all of the old liquid gasket from the rocker shaft holder and the cylinder head.

10. Clean the head cover contacting surfaces with a shop towel.

11. Apply liquid gasket (P/N 08717-0004, 08718-0003, 08718-0004, or 08718-0009) to the rocker shaft holder mating areas. Install the component within 5 minutes of applying the liquid gasket.

Please note the following:
• If you apply liquid gasket P/N 08718-0012, the component must be installed within 4 minutes.
• If too much time has passed after applying the liquid gasket, remove the old liquid gasket and residue, then reapply the new liquid gasket.

12. Set the spark plug seals on the spark plug tubes, and install the front cylinder head cover.

13. Inspect the spark plug seals for damage.

14. Inspect the cover washers. Replace any washer that is damaged or deteriorated.

15. Tighten the bolts in three steps. In the final step torque the bolts, in sequence, to 9 ft. lbs. (12 Nm).

Please note the following:
• Wait at least 30 minutes before filling the engine with oil.
• Do not run the engine for at least 3 hours after installing the cylinder head cover.

16. Install the harness holder to the bracket.

17. Connect the EGR valve connector and install the harness holder mounting bolt and the harness clamp.

18. Install the three ignition coils to the front cylinder head.

19. Install the intake manifold.

Rear

See Figures 59 through 61.

1. Remove the intake manifold.
2. Remove the three ignition coils from the rear cylinder head.

3. Remove the harness holder mounting bolt and the engine ground cable.

4. Disconnect the three injector connectors and the two harness clamps.

5. Remove the harness clamp and disconnect the breather hose.

6. Remove the harness from the upper cover.

7. Remove the engine hanger bracket.

8. Remove the harness holder mounting bolts and disconnect the knock sensor connector and the CMP sensor connector.

9. Remove the rear cylinder head cover.

To install:

10. Check the spark plug seals for damage. If any seals are damaged, replace it.

11. Thoroughly clean the head cover gasket and the groove of the cylinder head cover.

➡**Check and if necessary, replace the head cover gasket.**

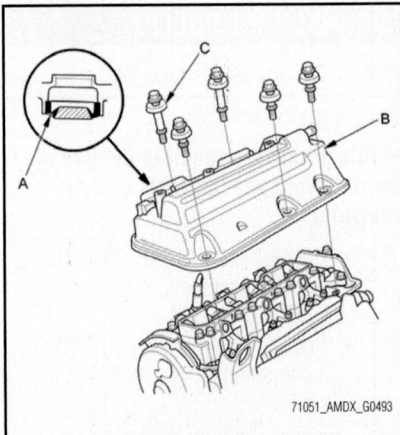

71051_AMDX_G0493

Fig. 60 Set the spark plug seals (A) on the spark plug tubes, and install the front cylinder head cover (B); inspect the cover washers (C)

12. Install the head cover gasket in the groove of the cylinder head cover. Make sure the head cover gasket is seated securely.

13. Remove all of the old liquid gasket from the rocker shaft holder and the cylinder head.

14. Clean the head cover contacting surfaces with a shop towel.

15. Apply liquid gasket (P/N 08717-0004, 08718-0003, 08718-0004, or 08718-0009) to the rocker shaft holder mating areas. Install the component within 5 minutes of applying the liquid gasket.

Please note the following:

• If you apply liquid gasket P/N 08718-0012, the component must be installed within 4 minutes.

• If too much time has passed after applying the liquid gasket, remove the old liquid gasket and residue, then reapply the new liquid gasket.

16. Set the spark plug seals on the spark plug tubes, and install the cylinder head cover.

17. Inspect the spark plug seals for damage.

18. Inspect the cover washers. Replace any washer that is damaged or deteriorated.

19. Tighten the bolts in three steps. In the final step torque the bolts, in sequence, to 9 ft. lbs. (12 Nm).

Please note the following:

• Wait at least 30 minutes before filling the engine with oil.

• Do not run the engine for at least 3 hours after installing the cylinder head cover.

20. Connect the knock sensor connector and the CMP sensor connector and install the harness holder mounting bolts.

21. Install the engine hanger bracket.

22. Install the harness to the upper cover.

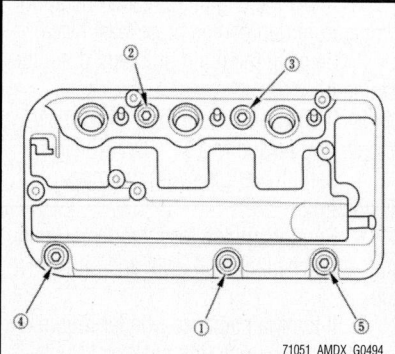

Fig. 61 Tighten the bolts in three steps. In the final step torque the bolts, in sequence

71051_AMDX_G0494

23. Connect the breather hose and install the harness clamp.

24. Connect the three injector connectors and the two harness clamps.

25. Install the harness holder mounting bolts and the engine ground cable.

26. Install the three ignition coils to the rear cylinder head.

27. Install the intake manifold.

CYLINDER HEAD

REMOVAL & INSTALLATION

See Figures 62 through 64.

1. Properly relieve the fuel system pressure.

2. Make sure you have the anti-theft code for the radio and the navigation system, then write down the frequencies for the radio's preset buttons.

3. Disconnect the negative cable from the battery first, then disconnect the positive cable. Wait at least 3 minutes before proceeding.

4. Drain the engine cooling system.

5. Remove the accessory drive belt.

6. Remove the power steering pump and power steering hose clamp.

7. Remove the alternator.

8. Remove the timing belt.

9. Remove the intake manifold.

10. Remove the ignition coils.

11. Disconnect the upper and lower radiator hoses from the engine assembly.

12. Remove the quick-connect cover, then disconnect the fuel supply hose.

13. Remove the purge joint.

14. Remove the following engine wiring harnesses:

• Six injector wiring harnesses
• Engine coolant temperature (ECT) sensor 1 wiring harness
• Crankshaft position (CKP) sensor wiring harness
• Exhaust gas recirculation (EGR) valve wiring harness
• Rocker arm oil control solenoid wiring harness
• Rocker arm oil pressure switch wiring harness
• Oil pressure switch wiring harness
• Two air fuel ratio (A/F) sensor wiring harnesses
• Two secondary heated oxygen sensor (secondary HO2S) wiring harnesses

15. Remove the front warm up three way catalytic converter (front WU-TWC) and rear warm up three way catalytic converter (rear WU-TWC).

16. Remove the connector bracket from the front cylinder head.

17. Remove the bracket from the rear cylinder head.

18. Remove the fuel rails.

19. Remove the water passage.

20. Remove the front and rear camshaft pulleys and back covers.

21. Remove the cylinder head cover as follows:

a. Remove the dipstick.

b. Remove the two bolts securing the harness holder, and disconnect the front air fuel ratio (A/F) sensor connector, front secondary heated oxygen sensor (secondary HO2S) connector, exhaust gas recirculation (EGR) valve connector and engine coolant temperature (ECT) sensor 1 connector.

c. Disconnect the three injector connectors from the injectors on the rear side cylinder head.

d. Remove the power steering hose bracket mounting bolt, the harness holder mounting bolts, and the engine ground cable bolt.

e. Remove the engine ground cable and disconnect the breather hose.

f. Remove the cylinder head covers.

22. Loosen the cylinder head bolts in sequence and ⅓ turns until all bolts are loose.

23. Remove the cylinder head.

To install:

24. Clean the cylinder head and engine block surface.

25. Clean and install the oil control orifices with new O-rings.

26. Install the dowel pins and new cylinder head gaskets.

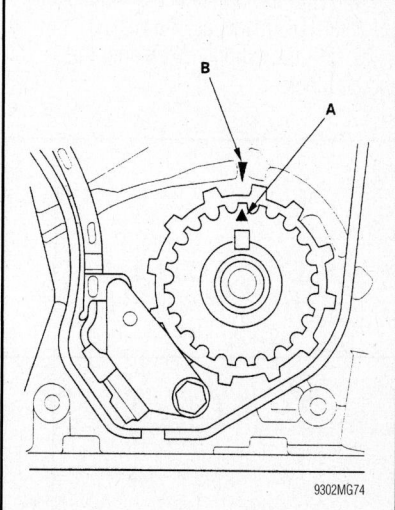

9302MG74

Fig. 62 Crankshaft timing belt sprocket TDC marks. Align sprocket mark (A) with pointer (B)

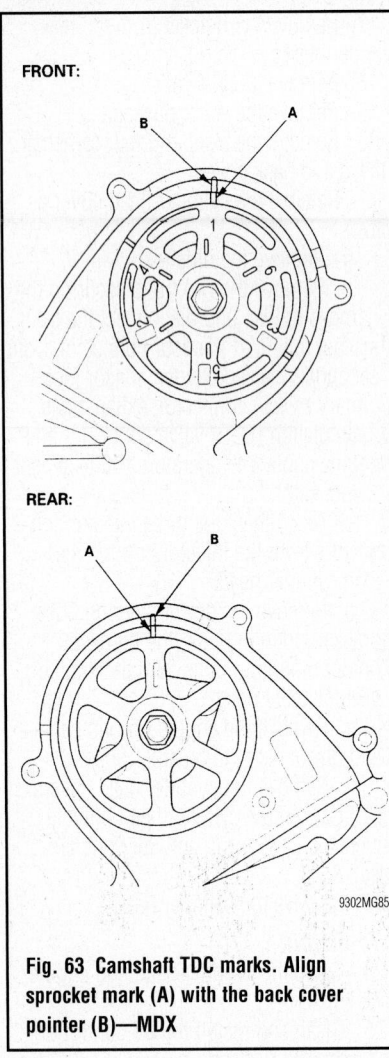

Fig. 63 Camshaft TDC marks. Align sprocket mark (A) with the back cover pointer (B)—MDX

27. Clean the timing belt pulleys, timing belt guide plate, and the upper and lower covers.

28. Align the crankshaft and camshaft sprocket TDC marks as shown.

29. Put the cylinder head onto the engine block.

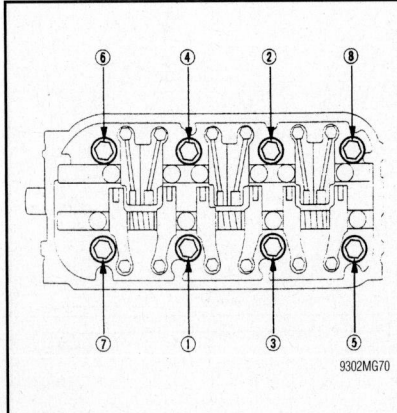

Fig. 64 Cylinder head bolt tightening sequence

30. Apply clean engine oil to the cylinder head bolt threads and flanges.

31. Tighten the cylinder head bolts in sequence as follows:
   a. Step 1: 22 ft. lbs. (29 Nm)
   b. Step 2: Plus 90 degrees
   c. Step 3: Plus an additional 90 degrees
   d. Step 4 (New bolts only): Plus an additional 90 degrees

32. Install the timing belt.

33. Adjust the valve clearance.

34. Install the cylinder head covers and tighten to 108 inch lbs. (12 Nm).

35. Install the water passage.

36. Install the front warm up three way catalytic converter (front WU-TWC) and rear warm up three way catalytic converter (rear WU-TWC).

37. Install the fuel rails.

38. Install the connector bracket to the front cylinder head.

39. Install the bracket to the rear cylinder head.

40. Install the purge joint and tighten the mounting nuts to 109 inch lbs. (12 Nm).

41. Connect the fuel supply hose and replace the quick-connect fitting cover.

42. Reconnect the engine wiring harnesses that were previously removed.

43. Connect the upper and lower radiator hoses.

44. Install the intake manifold.

45. Install the alternator.

46. Install the power steering pump and hose bracket.

47. Install the accessory drive belt.

48. Reconnect the battery cables.

49. After installation, check that all tubes, hoses and connectors are installed correctly.

50. Refill the engine cooling system to the correct level.

51. Start the engine and check for leaks.

ENGINE COVER

REMOVAL & INSTALLATION

See Figure 65.

1. Remove the engine appearance cover.

2. Installation is the reverse order of removal.

ENGINE REAR COVER

REMOVAL & INSTALLATION

1. Remove the transaxle assembly.
2. Remove the driveplate.
3. Remove the rear main seal.

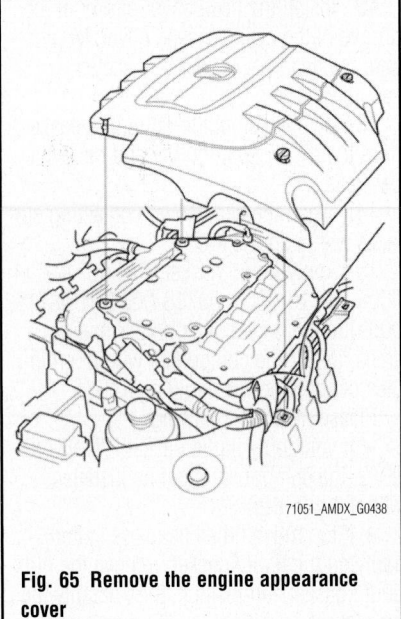

Fig. 65 Remove the engine appearance cover

To install:

4. Clean and dry the crankshaft oil seal housing.

5. Apply a light coat of multipurpose grease to the crankshaft and to the lip of the seal.

6. Using a suitable seal driver tool, drive in the rear main seal until the driver attachment bottoms against the engine block end cover. Align the hole in the driver attachment with the pin on the crankshaft.

7. Clean any excess grease off the crankshaft, and check that the oil seal lip is not distorted.

8. Install the driveplate and transaxle assembly.

ENGINE OIL & FILTER

OIL & FILTER CHANGE

Oil

1. Allow the engine to reach operating temperature (fan comes on at least twice).

2. Remove the drain bolt, and drain the engine oil.

3. Reinstall the drain bolt with a new washer and torque to 29 ft. lbs. (40 Nm).

4. Refill the engine with the recommended oil.

5. Run the engine for more than 3 minutes, then check the oil level and oil leakage.

6. If the Maintenance Minder required replacing the engine oil, reset the Maintenance Minder, and this procedure is complete. If the Maintenance Minder did not require engine oil replacement, go to step 7.

7. Turn the ignition switch to LOCK (0).

8. Connect the HDS to the DLC.

9. Turn the ignition switch to ON (II).

10. Make sure the HDS communicates with the vehicle and the PCM. If it does not communicate, troubleshoot the DLC circuit.

11. Select GAUGES in the BODY ELECTRICAL with the HDS.

12. Select ADJUSTMENT in the GAUGES with the HDS.

13. Select MAINTENANCE MINDER in the ADJUSTMENT with the HDS.

14. Select RESET in the MAINTENANCE MINDER with the HDS.

15. Select RESETTING THE ENGINE OIL LIFE with the HDS.

➡️**If you changed the ATF at the same time with the engine oil, select RESETTING THE ENGINE OIL LIFE AND ATF with the HDS instead.**

Oil Filter

See Figure 66.

➡️**Oil filter wrench 07AAA-PLCA100 should be used to remove the oil filter.**

1. Remove the oil filter with the oil filter wrench.

2. Inspect the filter to make sure the rubber seal is not stuck to the oil filter seating surface of the engine.

3. Inspect the threads and rubber seal on the new filter. Clean the seat on the engine block, then apply a light coat of new engine oil to the filter rubber seal. Use only filters with a built-in bypass system.

4. Install the oil filter by hand.

5. After the rubber seal seats, tighten the oil filter clockwise with the oil filter wrench ¾ turn clockwise

6. If four numbers or marks (1 to 4 or ▼ to ▼) are printed around the outside of the filter, you can use the following procedure to tighten the filter.

   a. Spin the filter on until its seal lightly seats against the oil filter base, and note which number or mark is at the bottom.

   b. Tighten the filter by turning it clockwise three numbers or marks from the one you noted. For example, if mark ▼ is at the bottom when the seal is lightly seated, tighten the filter until the mark ▼ comes around to the bottom.

7. After installation, fill the engine with engine oil to the specified level, run the engine for more than 3 minutes, then check for oil leakage.

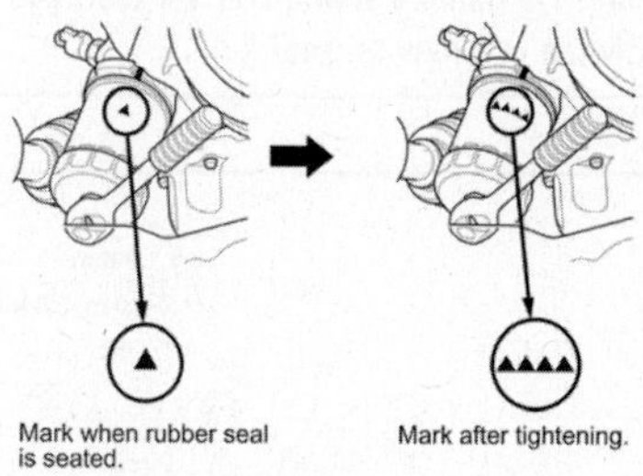

INTAKE MANIFOLD

REMOVAL & INSTALLATION

See Figures 67 through 70.

1. Disconnect the negative battery cable.

2. Remove the engine appearance cover.

3. Disconnect the breather pipe, then remove the intake air duct.

4. Disconnect the Positive Crankcase Ventilation (PCV) hose, the brake booster vacuum hose, vacuum hose, and the Intake Manifold Tuning (IMT) actuator connector.

5. Disconnect the Evaporative Emission (EVAP) canister hose, EVAP canister purge valve connector, throttle actuator connector, and Manifold Absolute Pressure (MAP) sensor connector.

6. Disconnect and plug the water bypass hoses.

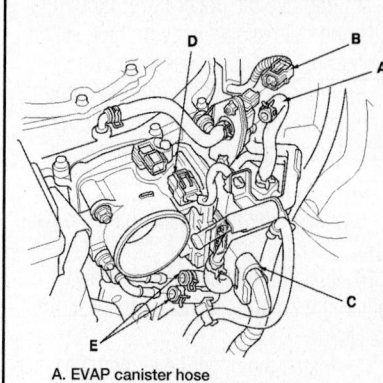

A. EVAP canister hose
B. EVAP canister purge valve connector
C. Throttle actuator connector
D. MAP sensor connector
E. Water bypass hoses

37647_AMDX_G0077

Fig. 68 Disconnect the EVAP canister hose, EVAP canister purge valve connector, throttle actuator connector, and MAP sensor connector

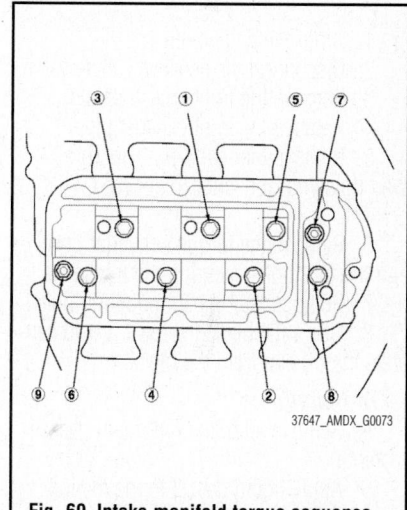

37647_AMDX_G0073

Fig. 69 Intake manifold torque sequence

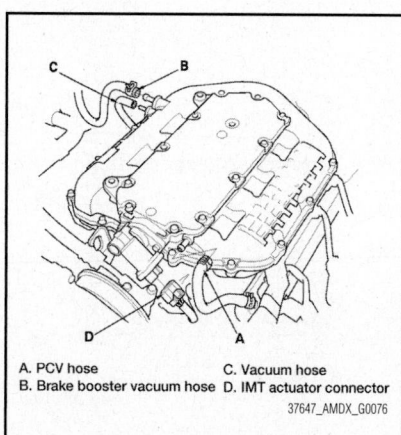

Number or Mark when rubber seal is seated	1 or ▼	2 or ▼▼	3 or ▼▼▼	4 or ▼▼▼▼
Number or Mark after tightening	4 or ▼▼▼▼	1 or ▼	2 or ▼▼	3 or ▼▼▼

Mark when rubber seal is seated. ▲

Mark after tightening. ▲▲▲▲

71051_AMDX_G0501

Fig. 66 Four numbers or marks are printed around the outside of the filter

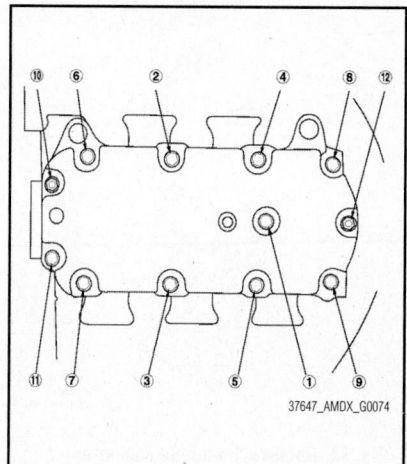

37647_AMDX_G0076

A. PCV hose   C. Vacuum hose
B. Brake booster vacuum hose   D. IMT actuator connector

Fig. 67 Disconnect the PCV hose, the brake booster vacuum hose, vacuum hose, and the IMT actuator connector

37647_AMDX_G0074

Fig. 70 Upper cover torque sequence

7. Remove the upper cover mounting bolts and nuts sequentially in three steps, then remove the upper cover.

8. Remove the intake manifold mounting bolts and nuts sequentially in three steps, then remove the intake manifold.

To install:

9. Install the new intake manifold gasket.

10. Install the intake manifold and tighten the bolts in sequence to 16 ft. lbs. (22 Nm).

11. Install the upper cover and new gasket. Tighten the bolts and nuts in sequence to 9 ft. lbs. (12 Nm).

12. The remainder of the installation is the reverse order of removal.

13. Start the engine and check for proper operation.

OIL PAN

REMOVAL & INSTALLATION

See Figures 71 through 73.

1. Drain the engine oil.
2. Disconnect the negative battery cable.
3. Remove the front splash shield.
4. Remove the front exhaust pipe.
5. Remove the rear Warm Up Three Way Catalytic Converter (rear WU-TWC) bracket.
6. Remove the torque converter cover and the four bolts securing the transaxle.
7. Remove the oil pan mounting bolts.
8. Use a flat bladed screwdriver to separate the oil pan from the block.

To install:

9. Remove all of the old liquid gasket material.

10. Apply liquid gasket to the mating surfaces of the oil pan as shown.

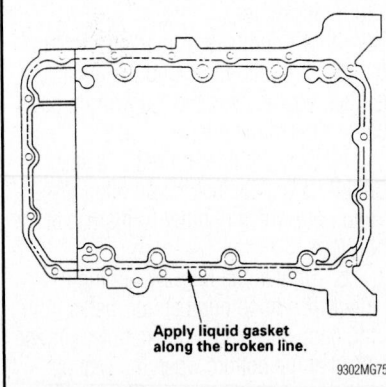

Apply liquid gasket along the broken line.

9302MG75

Fig. 72 Apply liquid gasket to the inner threads of the bolt holes and the engine block along the area indicated by the broken line

11. Install the oil pan on the engine block. Tighten the mounting bolts in sequence to 9 ft. lbs. (12 Nm).

12. Tighten the four bolts securing the transmission. Torque the bolts to 54 ft. lbs. (74 N m).

13. Install the torque converter cover.

14. Install the rear Warm Up Three Way Catalytic Converter (rear WU-TWC) bracket.

15. Install the front exhaust pipe using new gaskets and new self-locking nuts.

16. Install the splash shield.

17. Connect the negative battery cable.

➡**Wait at least 30 minutes before adding oil to the engine.**

18. Refill the engine with oil to the correct level.

OIL PUMP

REMOVAL & INSTALLATION

See Figures 74 and 75.

1. Remove the timing belt.
2. Remove the crankshaft position sensor.
3. Remove the rocker arm control solenoid/oil filter assembly.
4. Remove the oil pan.
5. Remove the oil screen before removing the oil pump.

To install:

6. Remove the old oil seal from the oil pump.

7. Gently tap in the new oil seal until the oil seal driver bottoms on the pump.

8. Remove all of the old liquid gasket from the oil pump mating surfaces, bolts, and bolt holes.

9. Clean and dry the oil pump mating surfaces.

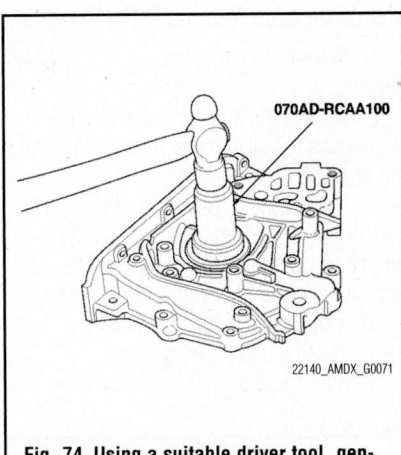

070AD-RCAA100

22140_AMDX_G0071

Fig. 74 Using a suitable driver tool, gently tap in a new oil seal

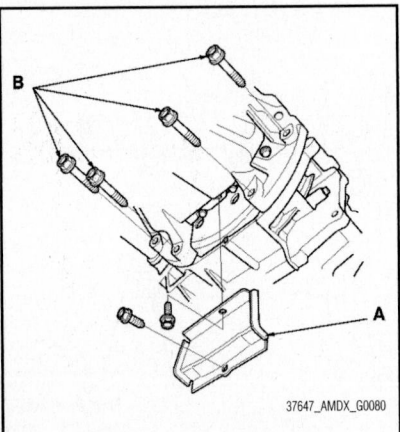

37647_AMDX_G0080

Fig. 71 Remove the torque converter cover (A) and the four bolts (B) securing the transaxle

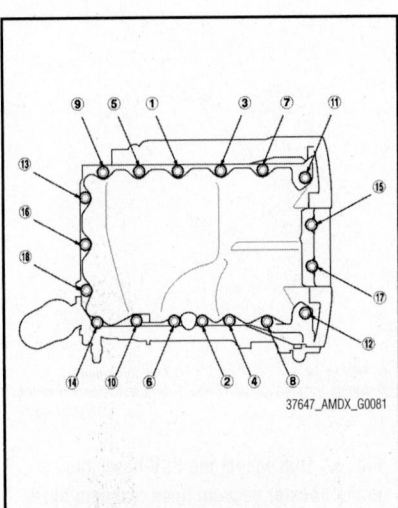

37647_AMDX_G0081

Fig. 73 Oil pan tightening sequence

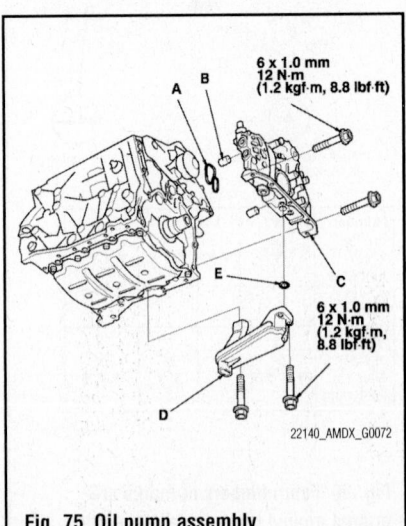

22140_AMDX_G0072

Fig. 75 Oil pump assembly

10. Apply liquid gasket evenly to the engine block mating surface of the oil pump. Install the component within 5 minutes of applying the liquid gasket.

11. Grease the lip of the oil seal, and apply oil to the new O-ring.

12. Install the dowel pins, then align the inner rotor with the crankshaft, and install the oil pump.

13. Clean the excess grease off the crankshaft, and check the seal for distortion.

14. Install the oil screen with new O-ring.

15. The remainder of the installation is the reverse order of removal.

16. Refill the engine with oil to the correct level.

PISTONS & RINGS

POSITIONING

See Figures 76 and 77.

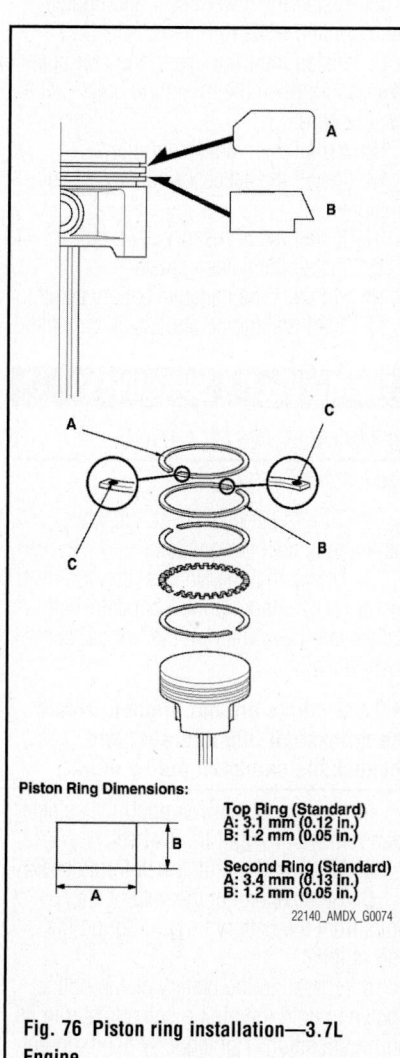

Piston Ring Dimensions:

Top Ring (Standard)
A: 3.1 mm (0.12 in.)
B: 1.2 mm (0.05 in.)

Second Ring (Standard)
A: 3.4 mm (0.13 in.)
B: 1.2 mm (0.05 in.)

22140_AMDX_G0074

Fig. 76 Piston ring installation—3.7L Engine

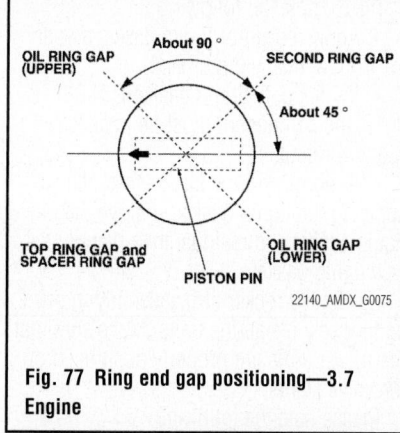

22140_AMDX_G0075

Fig. 77 Ring end gap positioning—3.7 Engine

ROCKER ARMS

REMOVAL & INSTALLATION

Front

See Figures 78 and 79.

1. Remove the cylinder head cover.
2. Loosen the locknuts and the adjusting screws.
3. Remove the rocker shaft bridge mounting bolts, the rocker shaft holder mounting bolts, and the rocker arm assembly.

  a. Loosen the rocker shaft bridge mounting bolts and the rocker shaft holder mounting bolts in sequence two turns at a time, to prevent damaging the valves or the rocker arm assembly.

  b. When removing the rocker arm assembly, do not remove the rocker shaft bridge mounting bolts and the rocker shaft holder mounting bolts. The bolts will keep the rocker arms on the shafts.

To install:

4. Remove all of the old liquid gasket from the rocker shaft holder and the cylinder head.

5. Apply liquid gasket (P/N 08717-0004, 08718-0003, 08718-0004, or 08718-

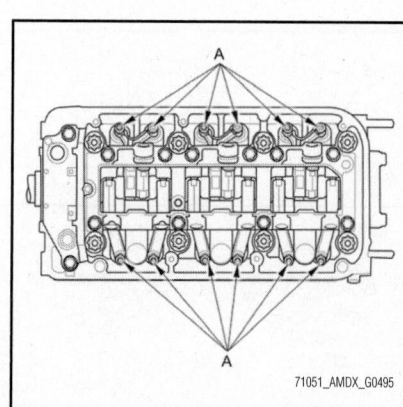

71051_AMDX_G0495

Fig. 78 Loosen the locknuts and the adjusting screws (A)

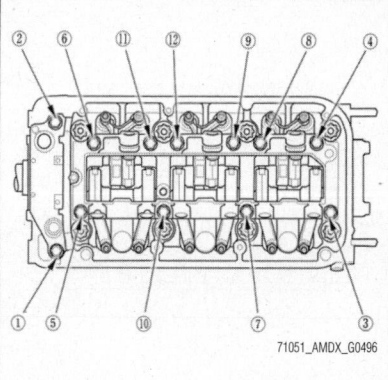

71051_AMDX_G0496

Fig. 79 Loosen the rocker shaft bridge mounting bolts and the rocker shaft holder mounting bolts in sequence two turns at a time, to prevent damaging the valves or the rocker arm assembly

0009) to the rocker shaft holder mating surface of the cylinder head. Install the component within 5 minutes of applying the liquid gasket.

Please note the following:

• Apply a bead of liquid gasket across the front of the cylinder head.

• If you apply liquid gasket P/N 08718-0012, the component must be installed within 4 minutes.

• If too much time has passed after applying the liquid gasket, remove the old liquid gasket and residue, then reapply the new liquid gasket.

6. Set the rocker arm assembly in place, and loosely install the bolts. Make sure that the rocker arms are properly positioned on the valve stems.

Please note the following:

• Wait at least 30 minutes before filling the engine with oil.

• Do not run the engine for at least 3 hours after installing the rocker arm assembly.

7. Tighten each bolt two turns at a time in the reverse sequence shown during the loosening of the bolts, to ensure that the rockers do not bind on the valves.

  a. When the rocker shaft bridge is reused: Specified Torque: 18 ft. lbs. (24.5 Nm).

  b. When a new rocker shaft bridge is installed: Specified Torque:

  • 1st pass: 21 ft. lbs. (28 Nm), then loosen the bolts.

  • 2nd pass: 18 ft. lbs. (24.5 Nm).

Rear

See Figures 80 and 81.

1. Remove the cylinder head cover.
2. Loosen the locknuts and the adjusting screws.

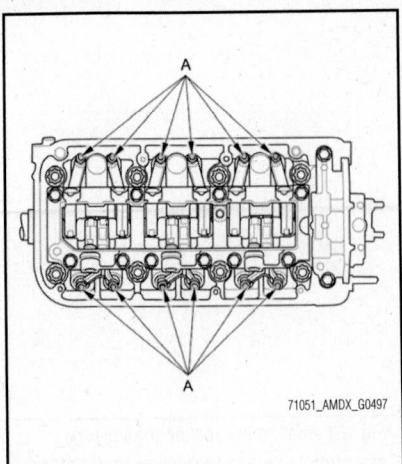

Fig. 80 Loosen the locknuts and the adjusting screws (A)

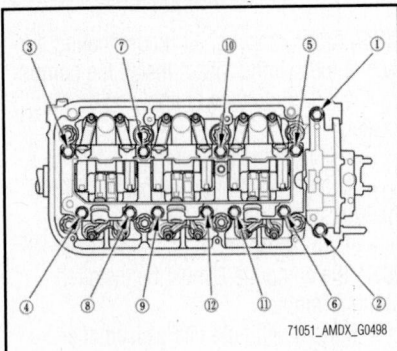

Fig. 81 Loosen the rocker shaft bridge mounting bolts and the rocker shaft holder mounting bolts in sequence two turns at a time, to prevent damaging the valves or rocker arm assembly

3. Remove the rocker shaft bridge mounting bolts, the rocker shaft holder mounting bolts, and the rocker arm assembly.

  a. Loosen the rocker shaft bridge mounting bolts and the rocker shaft holder mounting bolts in sequence two turns at a time, to prevent damaging the valves or rocker arm assembly.

  b. When removing the rocker arm assembly, do not remove the rocker shaft bridge mounting bolts and the rocker shaft holder mounting bolts. The bolts will keep the rocker arms on the shafts.

To install:

4. Remove all of the old liquid gasket from the rocker shaft holder and the cylinder head.

5. Apply liquid gasket (P/N 08717-0004, 08718-0003, 08718-0004, or 08718-0009) to the rocker shaft holder mating surface of the cylinder head. Install the component within 5 minutes of applying the liquid gasket.

Please note the following:
  • Apply a bead of liquid gasket across the front of the cylinder head.
  • If you apply liquid gasket P/N 08718-0012, the component must be installed within 4 minutes.
  • If too much time has passed after applying the liquid gasket, remove the old liquid gasket and residue, then reapply the new liquid gasket.

6. Set the rocker arm assembly in place, and loosely install the bolts. Make sure that the rocker arms are properly positioned on the valve stems.

Please note the following:
  • Wait at least 30 minutes before filling the engine with oil.
  • Do not run the engine for at least 3 hours after installing the rocker arm assembly.

7. Tighten each bolt two turns at a time in the reverse sequence shown during the loosening of the bolts, to ensure that the rockers do not bind on the valves.

  a. When the rocker shaft bridge is reused: Specified Torque: 18 ft. lbs. (24.5 Nm).

  b. When a new rocker shaft bridge is installed: Specified Torque:
  • 1st pass: 21 ft. lbs. (28 Nm), then loosen the bolts.
  • 2nd pass: 18 ft. lbs. (24.5 Nm).

TIMING BELT FRONT COVER

REMOVAL & INSTALLATION

See Figures 82 and 83.

1. Disconnect the negative battery cable.
2. Remove the splash shield.
3. Remove the accessory drive belt.
4. Remove the accessory drive belt auto-tensioner.
5. Support the engine assembly with a suitable jack.

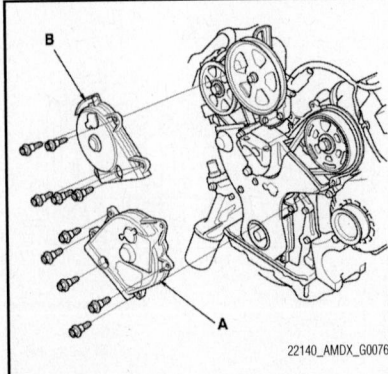

Fig. 82 Removing the front upper (A) and rear upper (B) cover

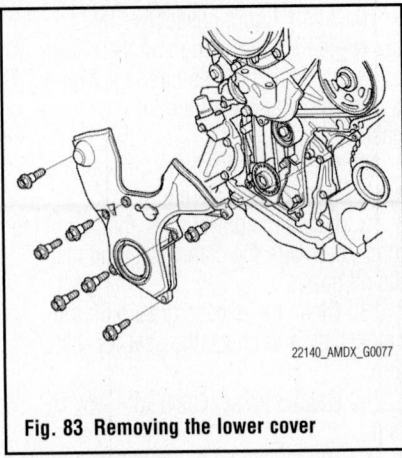

Fig. 83 Removing the lower cover

6. Remove the ground cable, then remove the upper half of the side engine mount bracket.
7. Remove the front upper and rear upper cover.
8. Remove the crankshaft pulley.
9. Remove the lower cover.

To install:

10. Install the lower cover and tighten the mounting bolts to 9 ft. lbs. (12 Nm).
11. Install the front upper and rear upper cover and tighten the mounting bolts to 9 ft. lbs. (12 Nm).
12. Install the crankshaft pulley.
13. Install the accessory drive belt auto-tensioner.
14. Install the accessory drive belt.
15. Install the splash shield.
16. Connect the negative battery cable.
17. Start the engine and check for leaks.

TIMING BELT & SPROCKETS

REMOVAL & INSTALLATION

See Figures 84 through 88.

1. Turn the crankshaft so the white mark aligns with the pointer.
2. Check that the No. 1 piston top dead center (TDC) mark on the front camshaft pulley and the pointer on the front upper cover are aligned.

➡ If the marks are not aligned, rotate the crankshaft 360 degrees, and recheck the camshaft pulley mark.

3. Raise and safely support the vehicle, then remove the right front wheel.
4. Remove the timing belt front covers.
5. Remove one of the battery clamp bolts from the battery tray, and grind the end of it flat.
6. Thread in the battery clamp bolt as shown to hold the timing belt adjuster in its current position. Tighten it by hand; do not use a wrench.

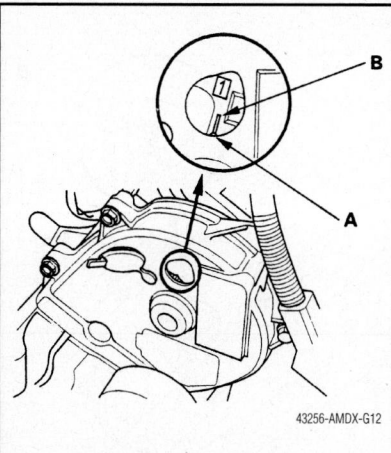

Fig. 84 Make sure the number 1 piston is at top dead center (A) on the front camshaft pulley and pointer (B)

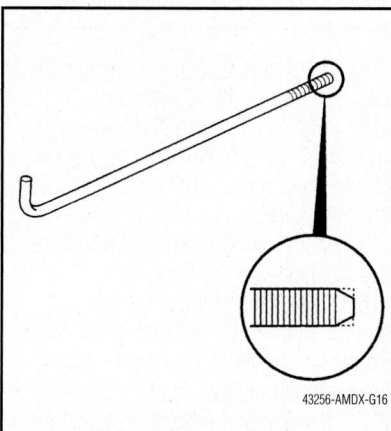

Fig. 85 Remove a battery clamp bolt and grind the end

7. Remove the timing belt guide plate.

8. Remove the lower half of the side engine mount bracket.

9. Remove the idler pulley bolt and idler pulley, then remove the timing belt. Discard the idler pulley bolt.

To install:

10. Clean the pulleys, belt guide plate and the upper and lower covers.

11. Set the timing belt drive pulley to TDC by aligning the TDC mark on the tooth of the belt drive pulley with the pointer on the oil pump.

12. Set the camshaft pulleys to TDC by aligning the TDC marks on the camshaft pulleys with the pointers on the back covers.

13. Loosely install the idler pulley with a new idler pulley bolt so the pulley can move but does not come off.

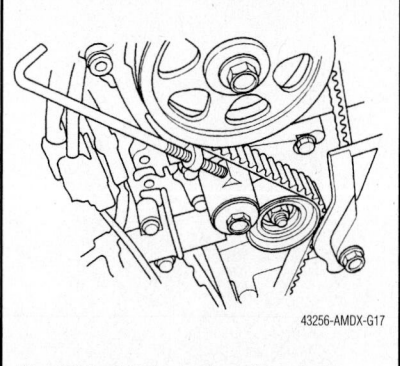

Fig. 86 Install the battery clamp bolt to hold the belt adjuster in position

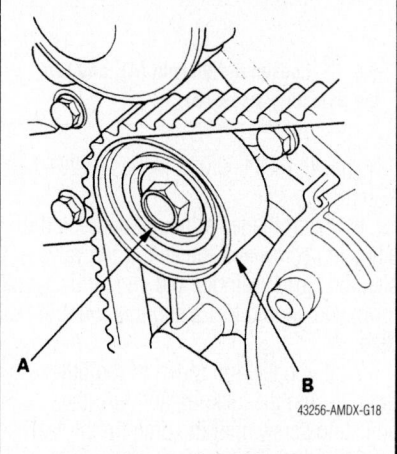

Fig. 87 Remove the idler pulley bolt (A), pulley (B) and the timing belt

➡️**If the auto-tensioner has extended and the timing belt cannot be installed, do the timing belt replacement procedure.**

14. Install the belt over the pulleys in this sequence; drive pulley, idler pulley, front camshaft pulley, water pump pulley, rear camshaft pulley and adjusting pulley.

15. Tighten the idler pulley bolt to 33 ft. lbs. (44 Nm).

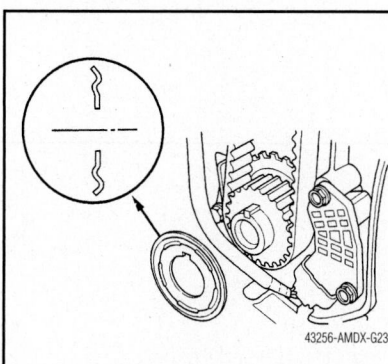

Fig. 88 Install the timing belt guide plate

16. Remove the battery clamp bolt from the back cover.

17. Install the lower half of the side engine mount bracket. Tighten the 3 long bolts to 33 ft. lbs. (44 Nm) and the one short bolt to 9 ft. lbs. (12 Nm).

18. Install the timing belt guide plate as shown.

19. Install the timing belt front covers.

20. Install the crankshaft pulley.

21. Rotate the crankshaft pulley about six turns clockwise so the timing belt positions itself on the pulleys.

22. Turn the crankshaft pulley so the white mark lines up with the pointer.

23. Check the camshaft pulley marks are aligned. If the marks are aligned, proceed to the next step. If the marks are not aligned, remove the timing belt and reinstall using the steps outlined before this step.

24. Install the upper half of the side engine mount bracket, and tighten the new mounting bolts to 33 ft. lbs. (44 Nm), then tighten the mass damper mounting bolt to 40 ft. lbs. (54 Nm).

25. Install the accessory drive belt and auto-tensioner.

26. Install the splash shield.

27. Install the right front wheel.

28. Start the engine and check for leaks.

TIMING BELT REAR COVER

REMOVAL & INSTALLATION

1. Remove the timing belt.
2. Remove the camshaft pulley.
3. Remove the rear cover.

To install:

4. Install the rear cover.
5. Install the camshaft pulley.
6. Install the timing belt.
7. Start the engine and check for leaks.

VALVE LASH

ADJUSTMENT

See Figures 89 through 92.

➡️**Adjust the valves only when the cylinder head temperature is less than 100°F (38°C).**

1. Remove the intake manifold.
2. Remove the cylinder head covers.
3. Set the No. 1 piston at top dead center (TDC).
4. Align the pointer on the front upper cover with the No. 1 piston TDC mark on the front camshaft pulley.

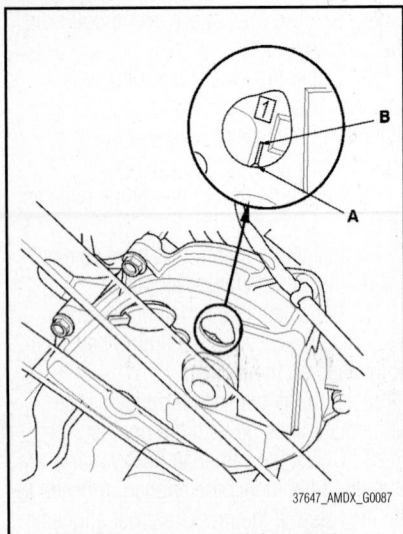

Fig. 89 Align the pointer (A) on the front upper cover with the No. 1 piston TDC mark (B) on the front camshaft pulley

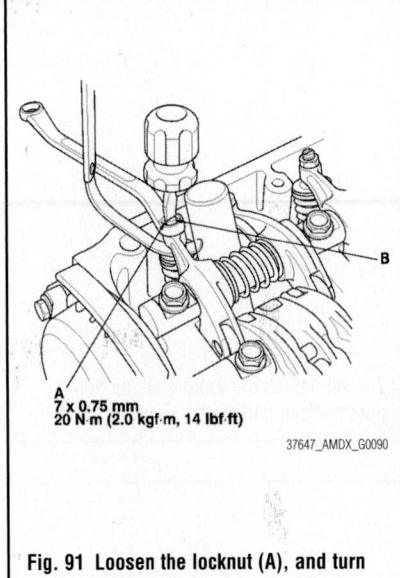

Fig. 91 Loosen the locknut (A), and turn the adjusting screw (B)

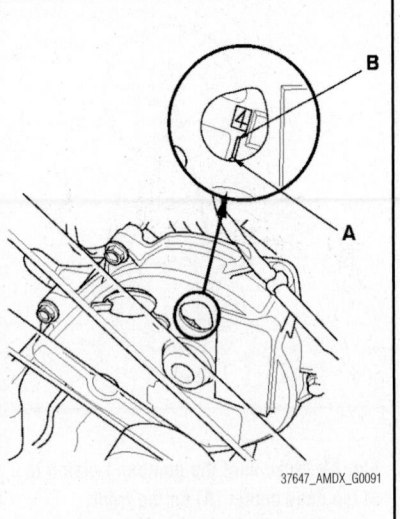

Fig. 92 Align the pointer (A) on the front upper cover with the No. 4 piston TDC mark (B) on the front camshaft pulley

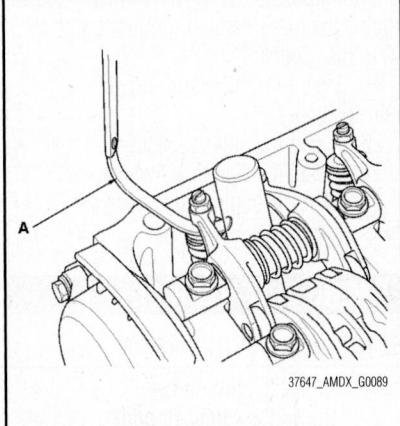

Fig. 90 Insert the feeler gauge (A) between the adjusting screw and the end of the valve stem on No. 1

5. Select the correct thickness feeler gauge for the valves you're going to check.
   a. Valve Clearance: Intake: 0.008–0.009 inches (0.20–0.24 mm);

Exhaust: 0.011–0.013 inches (0.28–0.32 mm)

6. Insert the feeler gauge between the adjusting screw and the end of the valve stem on No. 1 cylinder and slide it back and forth; you should feel a slight amount of drag.

7. If you feel too much or too little drag, loosen the locknut, and turn the adjusting screw until the drag on the feeler gauge is correct.

8. Tighten the locknut and recheck the clearance.

9. Repeat the adjustment, if necessary.

10. Rotate the crankshaft clockwise. Align the pointer on the front upper cover with the No. 4 piston TDC mark on the front camshaft pulley.

11. Check and, if necessary, adjust the valve clearance on No. 4 cylinder.

12. Rotate the crankshaft clockwise. Align the pointer on the front upper cover with the No. 2 piston TDC mark on the front camshaft pulley.

13. Check and, if necessary, adjust the valve clearance on No. 2 cylinder.

14. Rotate the crankshaft clockwise. Align the pointer on the front upper cover with the No. 5 piston TDC mark on the front camshaft pulley.

15. Check and, if necessary, adjust the valve clearance on No. 5 cylinder.

16. Rotate the crankshaft clockwise. Align the pointer on the front upper cover with the No. 3 piston TDC mark on the front camshaft pulley.

17. Check and, if necessary, adjust the valve clearance on No. 3 cylinder.

18. Rotate the crankshaft clockwise. Align the pointer on the front upper cover with the No. 6 piston TDC mark on the front camshaft pulley.

19. Check and, if necessary, adjust the valve clearance on No. 6 cylinder.

20. Install the cylinder head covers.

21. Install the intake manifold.

ENGINE PERFORMANCE & EMISSION CONTROLS

ACCELERATOR PEDAL POSITION (APP) SENSOR

LOCATION

The Accelerator Pedal Position (APP) sensor is located on the accelerator pedal module. The accelerator pedal position sensor is an integrated part of the accelerator pedal module. If the sensor is faulty, the entire module must be replaced.

REMOVAL & INSTALLATION

See Figure 93.

1. Disconnect the Accelerator Pedal Position (APP) sensor 6P connector.
2. Remove the accelerator pedal module.
3. Installation is the reverse order of removal.

AIR-FUEL RATIO (A/F) SENSOR

LOCATION

The Air Fuel Ratio (A/F) sensors are located on each bank of the engine just below the ignition coils.

REMOVAL & INSTALLATION

Front Bank (Bank 2)

See Figure 94.

1. Disconnect the front A/F sensor connector, then remove the A/F sensor.
2. Install the parts in the reverse order of removal.

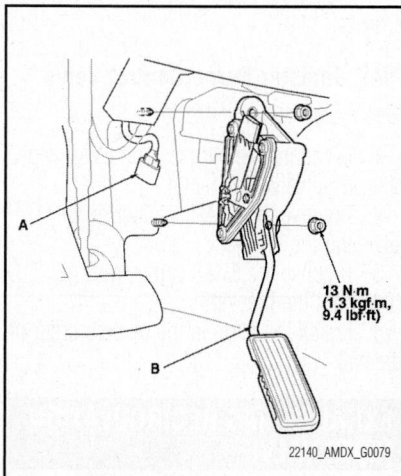

Fig. 93 Disconnect the APP sensor wiring harness (A) to remove the accelerator pedal module (B)

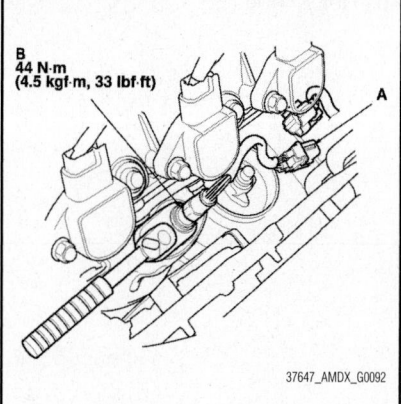

Fig. 94 Disconnect the front A/F sensor connector (A), then remove the A/F sensor (B)

Rear Bank (Bank 1)

See Figure 95.

1. Disconnect the rear A/F sensor connector, then remove the rear A/F sensor.
2. Install the parts in the reverse order of removal.

CAMSHAFT POSITION (CMP) SENSOR

LOCATION

The Camshaft Position (CMP) sensor is located on the timing belt cover.

REMOVAL & INSTALLATION

See Figures 96 and 97.

1. Remove the timing belt.
2. Remove the front camshaft pulley (CMP sensor pulse plate).
3. Disconnect the Camshaft Position Sensor (CMP), then remove the back cover.
4. Remove the CMP from the back cover.

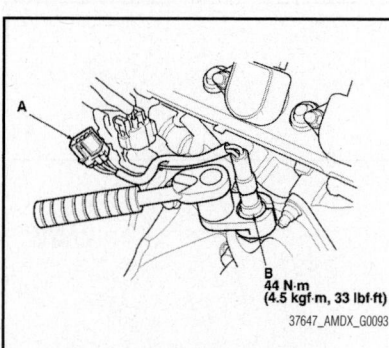

Fig. 95 Disconnect the rear A/F sensor connector (A), then remove the rear A/F sensor (B)

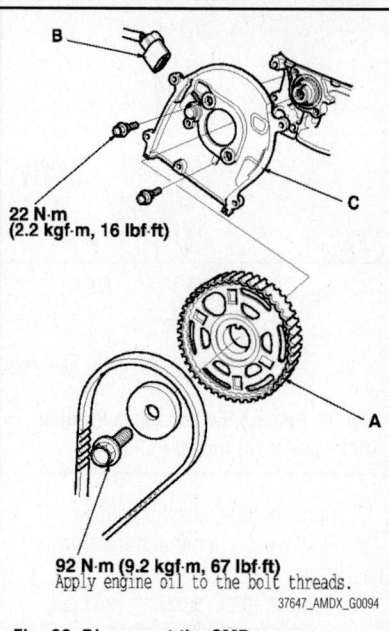

Fig. 96 Disconnect the CMP sensor connector (B), and remove the back cover (C)

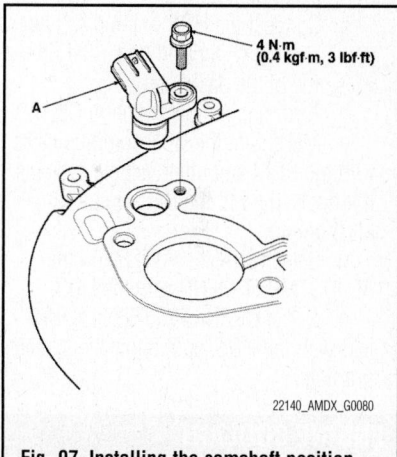

Fig. 97 Installing the camshaft position sensor into the back cover

5. Installation is the reverse order of removal.

CRANKSHAFT POSITION (CKP) SENSOR

LOCATION

The Crankshaft Position (CKP) sensor is located on the oil pump.

REMOVAL & INSTALLATION

See Figure 98.

1. Move the auto-tensioner to remove tension from the drive belt, then remove the accessory drive belt.

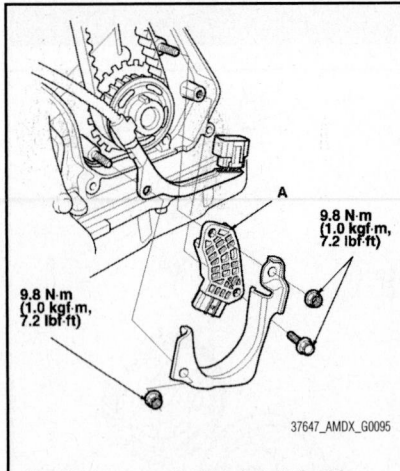

Fig. 98 Remove the Crankshaft Position (CKP) sensor (A) from the oil pump

2. Remove the crankshaft pulley.

3. Remove the upper and lower front covers from the engine.

4. Remove the Crankshaft Position (CKP) sensor from the oil pump.

5. Installation is the reverse order of removal.

6. After installation, perform the CKP learn procedure as follows:

   a. Connect the HDS to the data link connector (DLC) located under the driver's side of the dashboard.

   b. Turn the ignition switch to ON (II).

   c. Make sure the HDS communicates with the PCM and other vehicle systems. If it doesn't, go to the DLC circuit troubleshooting.

   d. Select CRANK PATTERN in the ADJUSTMENT MENU with the HDS.

   e. Select CRANK PATTERN LEARNING with the HDS, and follow the screen prompts.

ENGINE COOLANT TEMPERATURE (ECT) SENSOR

LOCATION

Sensor 1 is located in the engine compartment below the main underhood fuse box. Sensor 2 is located at the bottom of the radiator.

REMOVAL & INSTALLATION

ECT Sensor 1

See Figure 99.

1. Drain the engine cooling system.
2. Remove the engine appearance cover.
3. Remove the main under-hood fuse/relay box.
4. Disconnect the ECT sensor 1 connector.

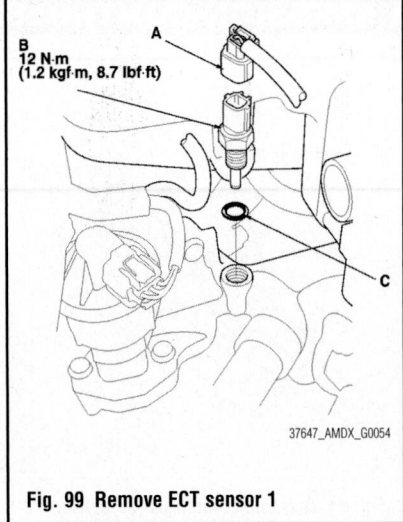

Fig. 99 Remove ECT sensor 1

5. Remove ECT sensor 1.
6. Install the parts in the reverse order of removal with a new O-ring, then refill the radiator with engine coolant.

ECT Sensor 2

See Figure 100.

1. Drain the engine cooling system.
2. Remove the splash shield.
3. Disconnect the ECT sensor 2 connector.
4. Remove ECT sensor 2.
5. Install the parts in the reverse order of removal with a new O-ring, then refill the radiator with engine coolant.

EVAPORATIVE EMISSION CONTROL SYSTEM

REMOVAL & INSTALLATION

EVAP Canister

See Figures 101 through 103.

1. Raise the vehicle on a lift.
2. Remove the EVAP canister cover.

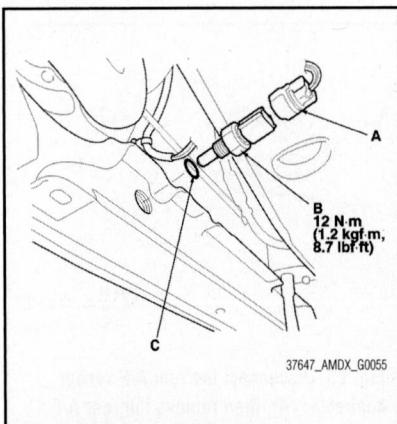

Fig. 100 Remove ECT sensor 2

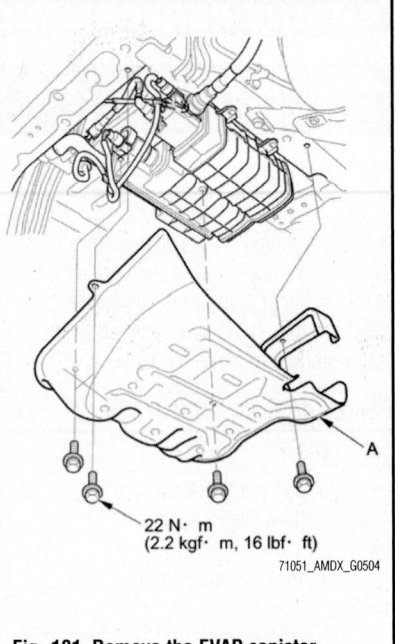

Fig. 101 Remove the EVAP canister cover (A)

3. Disconnect the quick-connect fitting, the hoses, the EVAP canister vent shut valve 2P connector, and the FTP sensor 3P connector, then remove the harness clamps.

4. Remove the bolts.

5. Remove the EVAP canister assembly.

6. Remove the EVAP canister vent shut valve and the FTP sensor from the canister.

7. Reassemble the EVAP canister vent shut valve and the FTP sensor onto the EVAP canister with new O-rings, a new cap, and a new retainer.

➡**Do not coat the O-rings with oil.**

8. Install the parts in the reverse order of removal.

EVAP Canister Purge Control Valve

See Figures 104 and 105.

1. Disconnect the hoses and EVAP canister purge valve connector.

2. Remove the EVAP canister purge valve and the bracket.

3. Remove the EVAP canister purge valve from the bracket.

4. Install the parts in the reverse order of removal

EXHAUST GAS RECIRCULATION (EGR) VALVE

LOCATION

The EGR valve is located on the left side of the engine.

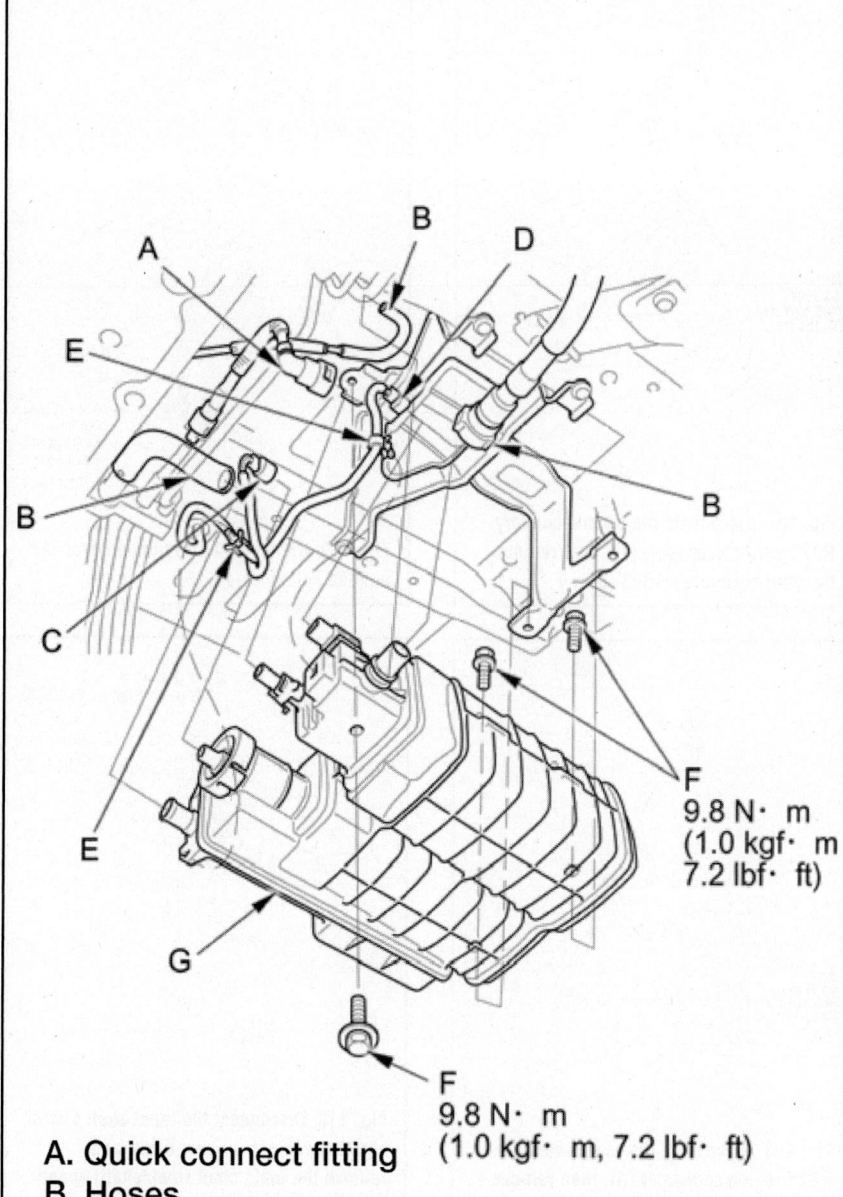

A. Quick connect fitting
B. Hoses
C. EVAP canister vent shut valve 2P connector
D. FTP sensor 3P connector
E. Harness clamps
F. Bolts
G. EVAP canister assembly

F
9.8 N· m
(1.0 kgf· m, 7.2 lbf· ft)

F
9.8 N· m
(1.0 kgf· m
7.2 lbf· ft)

71051_AMDX_G0505

Fig. 102 Disconnect the quick-connect fitting, the hoses, the EVAP canister vent shut valve 2P connector, and the FTP sensor 3P connector, then remove the harness clamps

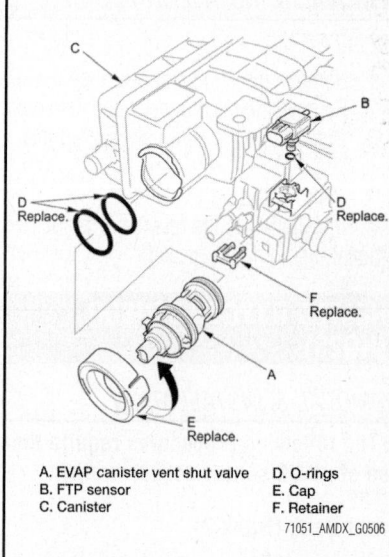

A. EVAP canister vent shut valve   D. O-rings
B. FTP sensor   E. Cap
C. Canister   F. Retainer

71051_AMDX_G0506

Fig. 103 Remove the EVAP canister vent shut valve and the FTP sensor from the canister

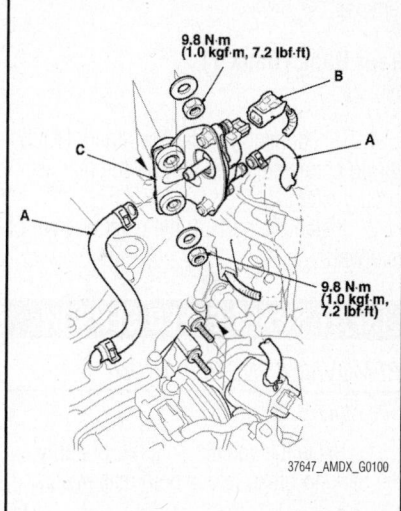

9.8 N·m
(1.0 kgf·m, 7.2 lbf·ft)

9.8 N·m
(1.0 kgf·m, 7.2 lbf·ft)

37647_AMDX_G0100

Fig. 104 Disconnect the hoses (A) and EVAP canister purge valve connector (B)

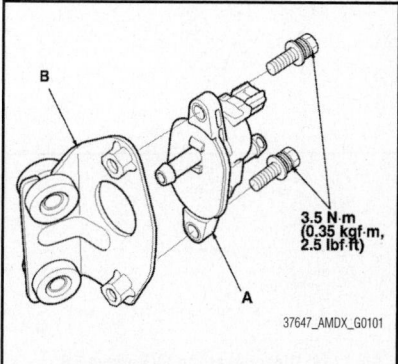

3.5 N·m
(0.35 kgf·m, 2.5 lbf·ft)

37647_AMDX_G0101

Fig. 105 Remove the EVAP canister purge valve (A) from the bracket (B)

REMOVAL & INSTALLATION

See Figure 106.

1. Remove the engine appearance cover.
2. Remove the main under-hood fuse box.
3. Disconnect the EGR valve 5P connector.
4. Remove the EGR valve.
5. Installation is the reverse order of removal, using a new gasket.

HEATED OXYGEN SENSOR (HO2S)

REMOVAL & INSTALLATION

➡**The following procedures require the use of an O2 sensor socket wrench.**

Front Bank (Bank 2)

See Figure 107.

1. Disconnect the front secondary HO2S wiring connector, then remove the front secondary HO2S.
2. Install the parts in the reverse order of removal.

Rear Bank (Bank 1)

See Figure 108.

1. Disconnect the rear secondary HO2S wiring connector, then remove the rear secondary HO2S.
2. Install the parts in the reverse order of removal.

INPUT SHAFT SPEED SENSOR

REMOVAL & INSTALLATION

See Figures 109 and 110.

1. Raise the vehicle on a lift, or apply the parking brake, block both rear wheels,

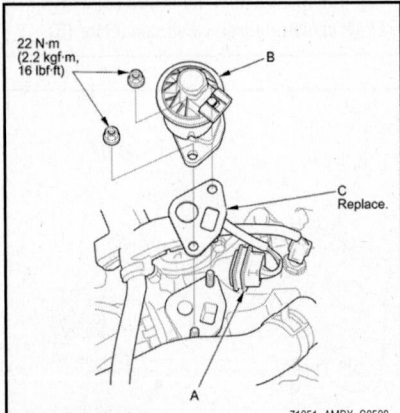

Fig. 106 Disconnect the EGR valve 5P connector (A), remove the EGR valve (B) and gasket (C)

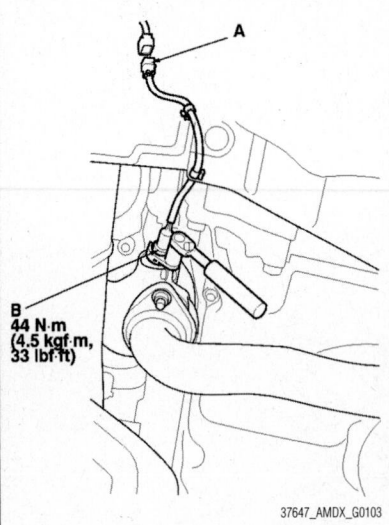

37647_AMDX_G0103

Fig. 107 Disconnect the front secondary HO2S wiring connector (A), then remove the front secondary HO2S (B)

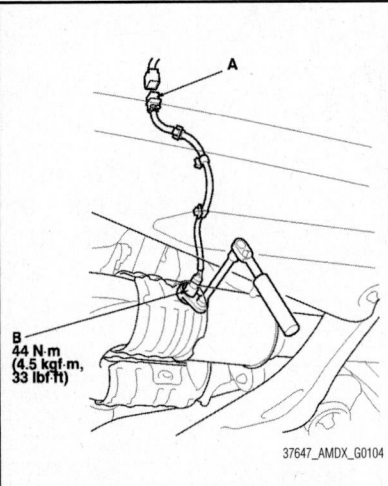

37647_AMDX_G0104

Fig. 108 Disconnect the rear secondary HO2S wiring connector (A), then remove the rear secondary HO2S (B)

and raise the front of the vehicle. Make sure it is securely supported.

2. Remove the front undercover and the splash shield.
3. Disconnect the input shaft (mainshaft) speed sensor connector, and remove the input shaft (mainshaft) speed sensor (A).

To install:

4. Apply ATF to a new O-ring, and install the O-ring over the input shaft (mainshaft) speed sensor. Then install the input shaft (mainshaft) speed sensor.
5. Check the connector for corrosion, dirt, or oil, and clean or repair if necessary, then connect the connector securely.

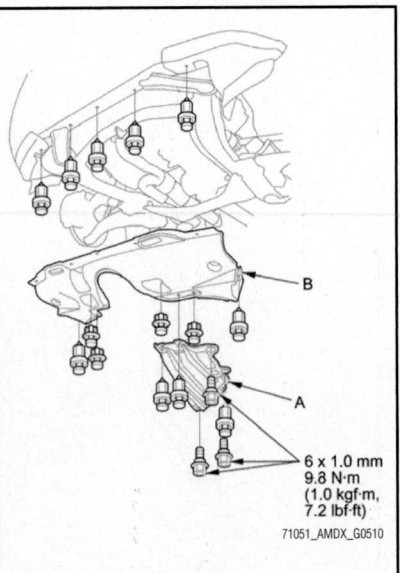

71051_AMDX_G0510

Fig. 109 Remove the front undercover (A) and the splash shield (B)

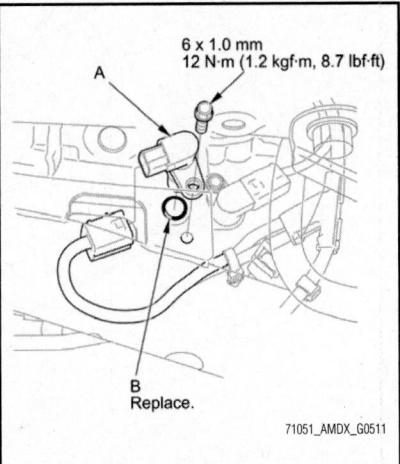

71051_AMDX_G0511

Fig. 110 Disconnect the input shaft (mainshaft) speed sensor connector, and remove the input shaft (mainshaft) speed sensor (A) and O-ring (B)

6. Install the splash shield and the front undercover.

INTAKE AIR TEMPERATURE (IAT)/MASS AIRFLOW (MAF) SENSOR

LOCATION

The IAT/MAF sensor is located in the air intake near the air cleaner.

REMOVAL & INSTALLATION

See Figure 111.

1. Disconnect the IAT/MAF sensor connector.
2. Remove the screw.

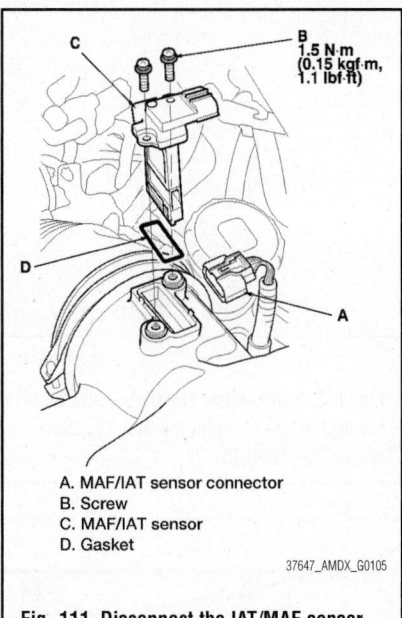

A. MAF/IAT sensor connector
B. Screw
C. MAF/IAT sensor
D. Gasket

37647_AMDX_G0105

Fig. 111 Disconnect the IAT/MAF sensor connector

3. Remove the IAT/MAF sensor.
4. Install the parts in the reverse order of removal with a new gasket.

KNOCK SENSOR (KS)

LOCATION

The Knock Sensor (KS) is located in the valley of the engine block.

REMOVAL & INSTALLATION

See Figure 112.

1. Remove the intake manifold.
2. Remove the injector rails and the injector base.

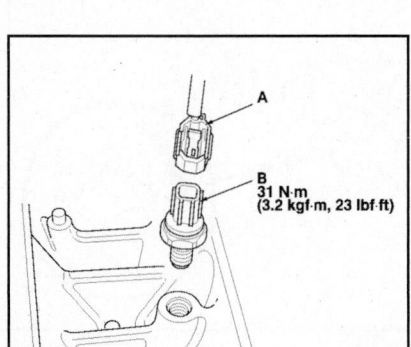

37647_AMDX_G0218

Fig. 112 Disconnect the Knock Sensor (KS) wiring connector (A), then remove the KS (B)

3. Disconnect the Knock Sensor (KS) wiring connector, then remove the KS.
4. Installation is the reverse order of removal. Tighten the knock sensor to 23 ft. lbs. (31 Nm).

MANIFOLD ABSOLUTE PRESSURE (MAP) SENSOR

REMOVAL & INSTALLATION

See Figure 113.

1. Remove the engine appearance cover.
2. Disconnect the MAP sensor connector.
3. Remove the screw.
4. Remove the MAP sensor.
5. Install the parts in the reverse order of removal with a new O-ring.

OUTPUT SHAFT SPEED (OSS) SENSOR

LOCATION

The Output Shaft Speed (OSS) sensor is located on the side of the transmission.

REMOVAL & INSTALLATION

See Figures 114 through 116.

1. Raise the vehicle up on a lift, or apply the parking brake, block both rear wheels, and raise the front of the vehicle. Make sure it is securely supported.
2. Remove the transmission undercover and splash shield.
3. Remove the damper from the front subframe.
4. Disconnect the output shaft (countershaft) speed sensor connector.
5. Remove the output shaft (countershaft) speed sensor and sensor washer.

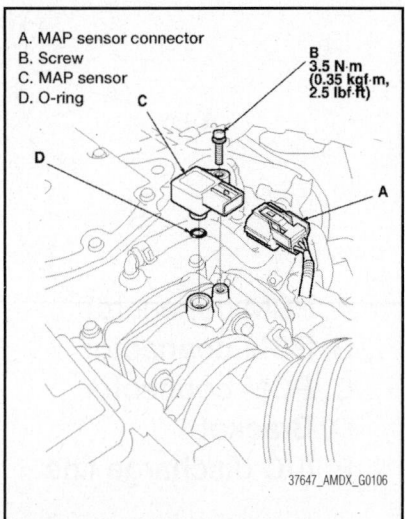

A. MAP sensor connector
B. Screw
C. MAP sensor
D. O-ring

37647_AMDX_G0106

Fig. 113 Disconnect the MAP sensor connector

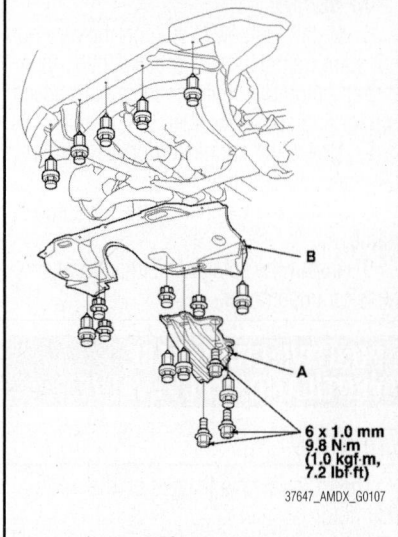

37647_AMDX_G0107

Fig. 114 Remove the transmission undercover (A) and splash shield (B)

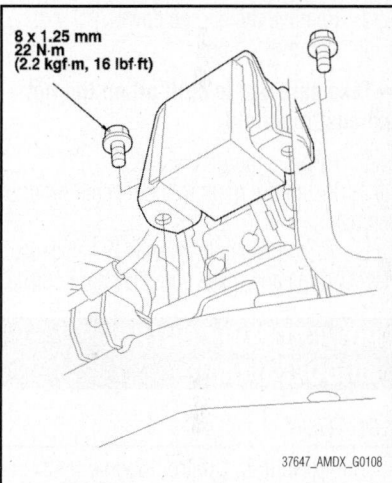

37647_AMDX_G0108

Fig. 115 Remove the damper from the front subframe

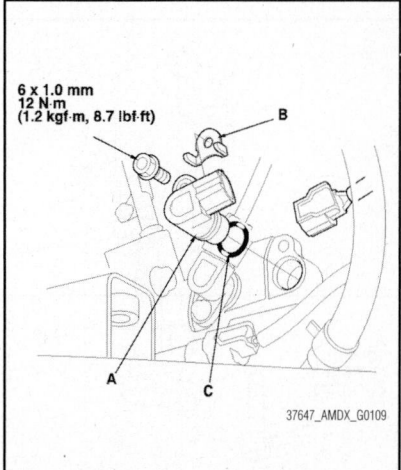

37647_AMDX_G0109

Fig. 116 Remove the output shaft (countershaft) speed sensor (A) and sensor washer (B)

To install:

6. Install the new O-ring on the new output shaft (countershaft) speed sensor, then install the output shaft (countershaft) speed sensor and sensor washer.

7. Check the connector for rust, dirt, or oil, then connect it securely.

8. Install the damper on the front subframe.

9. Install the splash shield and transmission undercover.

POSITIVE CRANKCASE VENTILATION (PCV) VALVE

LOCATION

The PCV valve is located on the intake manifold.

REMOVAL & INSTALLATION

See Figure 117.

1. Remove the engine cover.
2. Remove the bolt.

➡**Take care not to spill oil on the hot exhaust manifold.**

3. Remove the PCV valve.
4. Install the parts in the reverse order of removal.
5. When installing a new PCV valve, use new O-rings and make sure they are in place.

POWERTRAIN CONTROL MODULE (PCM)

LOCATION

The Powertrain Control Module (PCM) is located in the engine compartment on the passenger side fender in front of the underhood fuse relay box.

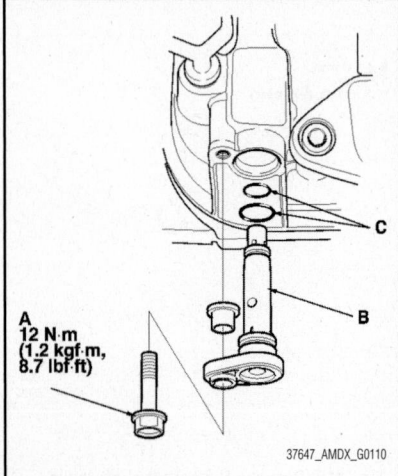

Fig. 117 Remove the bolt (A) and the PCV valve (B)

A
12 N·m
(1.2 kgf·m,
8.7 lbf·ft)

37647_AMDX_G0110

REMOVAL & INSTALLATION

See Figures 118 through 120.

Special Tools Required:
- Honda Diagnostic System (HDS) tablet tester
- Honda Interface Module (HIM) and an iN workstation with HDS and CM update software
- HDS pocket tester
- GNA600 and an iN workstation with HDS and CM update software

➡**Use any one of these update tools.**

- Make sure the HDS is loaded with the latest software version.
- If you are replacing the PCM after

71051_AMDX_G0512

Fig. 118 Remove the auxiliary under hood fuse/relay box (A) from the stay (B), then remove the bolt (C)

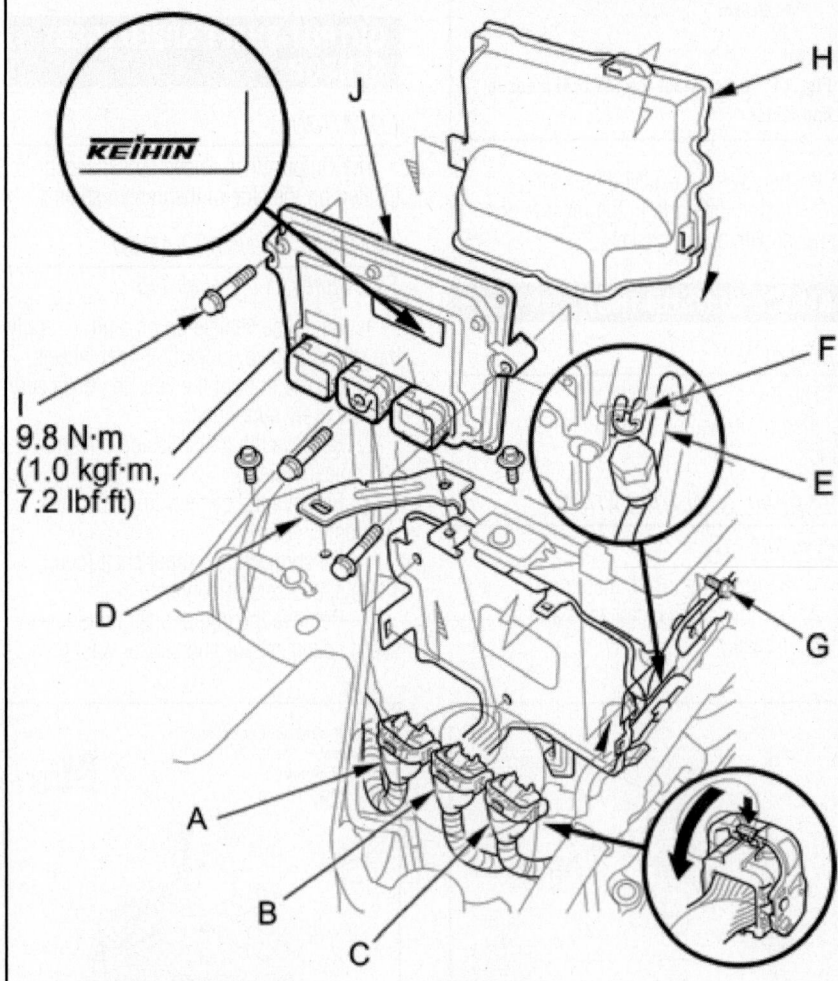

I
9.8 N·m
(1.0 kgf·m,
7.2 lbf·ft)

A. PCM connector
B. PCM connector
C. PCM connector
D. Bracket
E. A/C discharge line
F. Clip
G. A/C suction line
   mounting bracket bolt
H. Cover
I. Bolts
J. PCM

71051_AMDX_G0513

Fig. 119 Exploded view of Keihin PCM assembly

substituting a known-good PCM, reinstall the original PCM, then do this procedure.
- During the procedure, if any READ DATA, WRITE DATA, or other data checks fail, note the failure, then continue.

1. Connect the HDS to the Data Link Connector (DLC) located under the driver's side of the dashboard.

2. Turn the ignition switch to ON (II).

3. Make sure the HDS communicates with the PCM and other vehicle systems. If it doesn't, go to the DLC circuit trou-bleshooting. If you are returning from DLC circuit troubleshooting, skip steps 4 through 9, 20 through 26, and 29 through 31, and do this after replacing the PCM:
- Replace the engine oil and the engine oil filter.
- Replace the ATF.
- Clean the throttle body.

4. Select the PGM-FI system with the HDS.

5. Select the INSPECTION MENU with the HDS.

6. Select the ETCS TEST, then select the TP POSITION CHECK, and follow the screen prompts.

➡**If the TP POSITION CHECK indicates FAILED, continue with this procedure.**

7. Select the REPLACE PCM MENU, then select READ DATA and follow the screen prompts.

Please note the following:
- Doing this step copies (READS) the engine oil life data from the original PCM so you can later download (WRITES) it into the new PCM.
- If READ DATA indicates FAILED, continue with this procedure.

8. Select the A/T system with the HDS.

9. Select the REPLACE TCM/PCM MENU, then select READ DATA and follow the screen prompts.

Please note the following:
- Doing this step copies (READS) the ATF life data from the original PCM so you can later download (WRITES) it into the new PCM.
- If READ DATA indicates FAILED, continue with this procedure.

10. Turn the ignition switch to LOCK (0).

11. Jump the SCS line with the HDS.

12. Remove the auxiliary under hood fuse/relay box from the stay, then remove the bolt.

13. Remove the bracket, then free the A/C discharge line from the clip and remove the A/C suction line mounting bracket bolt.

14. Remove the cover, then disconnect PCM connectors A, B, and C.

➡**PCM connectors A, B, and C have symbols (A=□, B=△, C=○) embossed on them for identification.**

15. Keihin PCM: Remove the bolts, then remove the PCM.

16. Continental PCM: Remove the nuts, then remove the PCM.

17. Continental PCM: Remove the spacers.

To install:

18. Install the parts in the reverse order of removal.

19. Turn the ignition switch to ON (II).

20. Manually input the VIN to the PCM with the HDS.

➡**DTC P0630 "VIN Not Programmed or Mismatch" may be stored because the VIN has not been programmed into the PCM; ignore it, and continue this procedure.**

21. If the READ DATA (engine oil life) failed in step 7, go to step 24. Otherwise, go to step 22.

22. Select the PGM-FI system with the HDS.

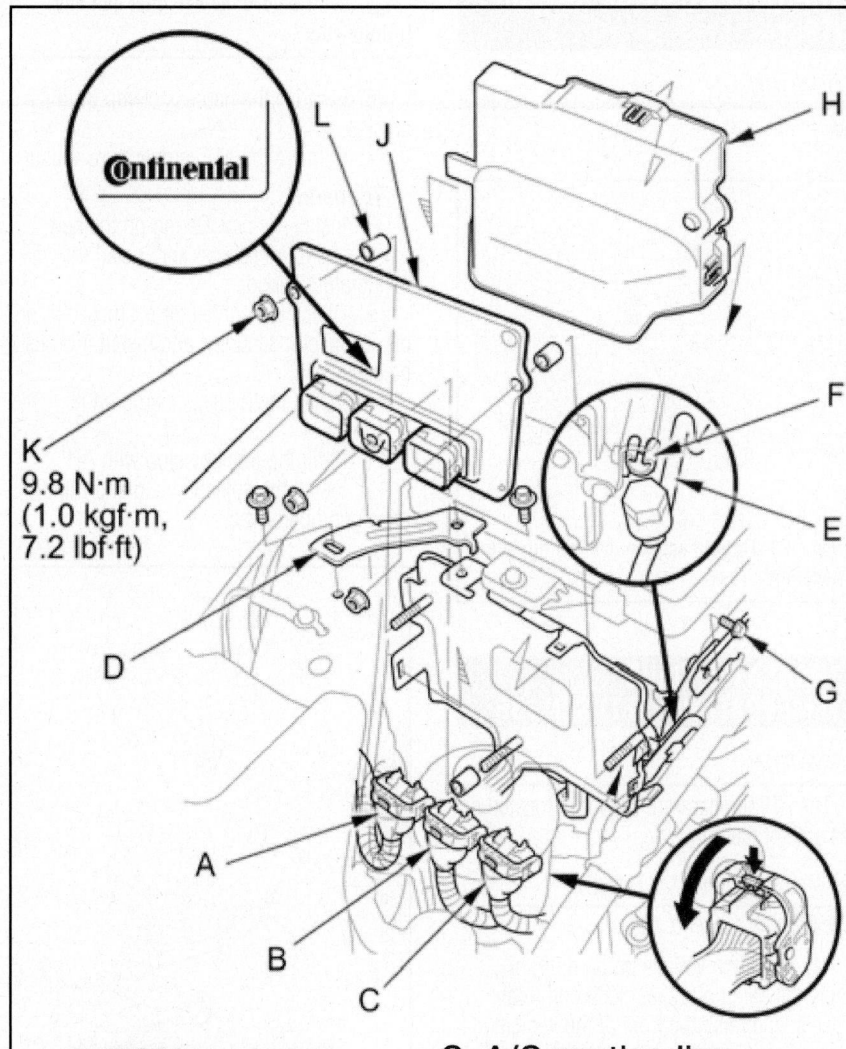

K
9.8 N·m
(1.0 kgf·m,
7.2 lbf·ft)

A. PCM connector
B. PCM connector
C. PCM connector
D. Bracket
E. A/C discharge line
F. Clip
G. A/C suction line mounting bracket bolt
H. Cover
J. PCM
K. Nuts
L. Spacers

71051_AMDX_G0514

Fig. 120 Exploded view of Continental PCM assembly

23. Select the REPLACE PCM MENU, then select WRITE DATA and follow the screen prompts.

➡**If the WRITE DATA indicates FAILED, continue with this procedure.**

24. If the READ DATA (ATF life) failed in step 9, go to step 27. Otherwise go to step 25.

25. Select the A/T SYSTEM with the HDS.

26. Select the REPLACE TCM/PCM MENU, then select WRITE DATA and follow the screen prompts.

➡**If the WRITE DATA indicates FAILED, continue with this procedure.**

27. Select IMMOBI system with the HDS.

28. Enter the immobilizer PCM code that you got from the iN, and use the PCM replacement procedure in the IMMOBI MENU of the HDS; it allows you to start the engine.

29. If the TP POSITION CHECK failed in step 6 clean the throttle body, then go to step 30.

30. If the READ DATA failed in step 7 or the WRITE DATA failed in step 23, replace the engine oil and engine oil filter, then go to step 31.

31. If the READ DATA failed in step 9 or the WRITE DATA failed in step 26, replace the ATF, then go to step 32.

32. Select PGM-FI system and reset the PCM with the HDS.

33. Update the PCM if it does not have the latest software.

34. Do the PCM idle learn procedure.

35. Do the CKP pattern learn procedure.

ECM/PCM IDLE LEARN PROCEDURE

The idle learn procedure must be done so the ECM/PCM can learn the engine idle characteristics.

Do the idle learn procedure whenever you do any of these actions:
• Replace the ECM/PCM.
• Reset the ECM/PCM.
• Update the ECM/PCM.
• Replace or clean the throttle body.
• Disassemble the engine or transmission.

➡**Erasing DTCs with the HDS does not require you to do the idle learn procedure.**

1. Make sure all electrical items (the A/C, the audio, the lights, etc.) are off.

2. Reset the ECM/PCM with the HDS.

3. Turn the ignition switch to ON (II), or press the engine start/stop button to select the ON mode, and wait 2 seconds.

4. Start the engine. Hold the engine speed at 3,000 rpm without load (A/T in P or N, M/T in neutral) until the radiator fan comes on, or until the engine coolant temperature reaches 194°F (90°C).

5. Let the engine idle for about 5 minutes with the throttle fully closed.

➡**If the radiator fan comes on, do not include its running time in the 5 minutes.**

THROTTLE POSITION SENSOR (TPS)

LOCATION
See Figure 121.

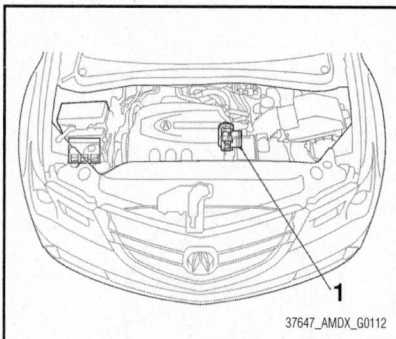

Fig. 121 Throttle actuator control module location

TRANSAXLE FLUID TEMPERATURE (TFT) SENSOR

LOCATION
The ATF temperature sensor is located on the side of the transaxle.

REMOVAL & INSTALLATION
See Figures 122 and 123.

1. Raise the vehicle up on a lift, or apply the parking brake, block both rear wheels, and raise the front of the vehicle. Make sure it is securely supported.

2. Remove the transmission undercover and splash shield.

3. Remove the drain plug, and drain the Automatic Transmission Fluid (ATF).

4. Reinstall the drain plug with a new sealing washer.

5. Disconnect the ATF temperature sensor connector, then remove the connector from the connector bracket.

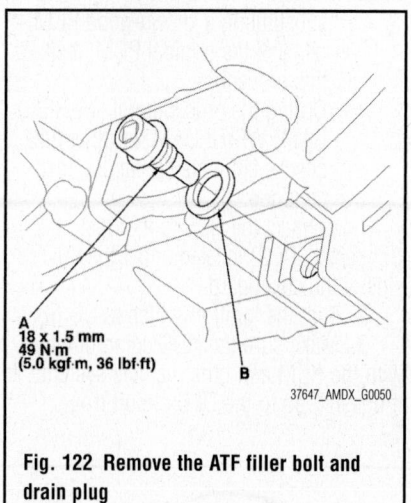

Fig. 122 Remove the ATF filler bolt and drain plug

6. Remove the harness clamp from its bracket.

7. Remove the ATF temperature sensor.

To install:
8. Install the new O-ring on the new ATF temperature sensor, and install the ATF temperature sensor.

9. Check the connector for rust, dirt, or oil. Connect it securely, and install it on its bracket.

10. Install the harness clamp on its bracket.

11. Refill the transmission with ATF.

12. Install the splash shield and transmission undercover.

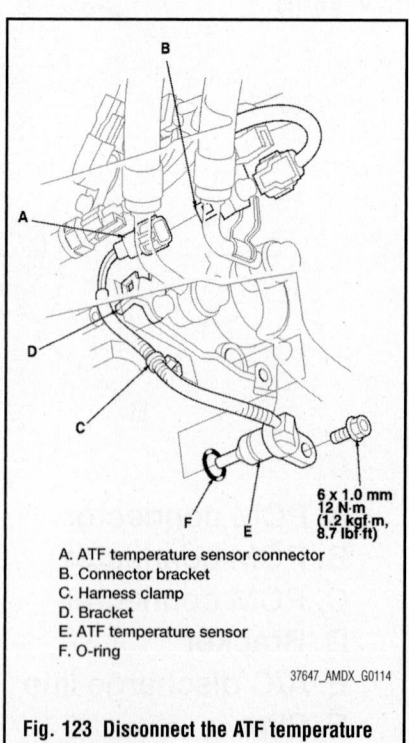

A. ATF temperature sensor connector
B. Connector bracket
C. Harness clamp
D. Bracket
E. ATF temperature sensor
F. O-ring

Fig. 123 Disconnect the ATF temperature sensor connector

FUEL

FUEL SYSTEM SERVICE PRECAUTIONS

Safety is the most important factor when performing not only fuel system maintenance but any type of maintenance. Failure to conduct maintenance and repairs in a safe manner may result in serious personal injury or death. Maintenance and testing of the vehicle's fuel system components can be accomplished safely and effectively by adhering to the following rules and guidelines.

• To avoid the possibility of fire and personal injury, always disconnect the negative battery cable unless the repair or test procedure requires that battery voltage be applied.

• Always relieve the fuel system pressure prior to disconnecting any fuel system component (injector, fuel rail, pressure regulator, etc.), fitting or fuel line connection. Exercise extreme caution whenever relieving fuel system pressure to avoid exposing skin, face and eyes to fuel spray. Please be advised that fuel under pressure may penetrate the skin or any part of the body that it contacts.

• Always place a shop towel or cloth around the fitting or connection prior to loosening to absorb any excess fuel due to spillage. Ensure that all fuel spillage (should it occur) is quickly removed from engine surfaces. Ensure that all fuel soaked cloths or towels are deposited into a suitable waste container.

• Always keep a dry chemical (Class B) fire extinguisher near the work area.

• Do not allow fuel spray or fuel vapors to come into contact with a spark or open flame.

• Always use a back-up wrench when loosening and tightening fuel line connection fittings. This will prevent unnecessary stress and torsion to fuel line piping.

• Always replace worn fuel fitting O-rings with new Do not substitute fuel hose or equivalent where fuel pipe is installed.

Before servicing the vehicle, make sure to also refer to the precautions in the beginning of this section as well.

RELIEVING FUEL SYSTEM PRESSURE

WITH THE HDS

See Figure 124.

✸✸ CAUTION

Before disconnecting fuel lines or hoses, relieve pressure from the system by disabling the fuel pump and

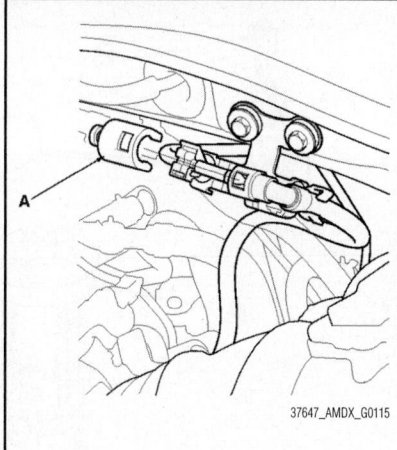

Fig. 124 Remove the quick-connect fitting cover (A)

37647_AMDX_G0115

then disconnecting the fuel tube/quick connect fitting in the engine compartment.

1. Turn the ignition switch to LOCK (0).
2. Connect the HDS to the data link connector (DLC) located under the driver's side of the dashboard.
3. Turn the ignition switch to ON (II).
4. Make sure the HDS communicates with the PCM. If it doesn't, go to the DLC circuit troubleshooting.
5. Turn the ignition switch to LOCK (0).
6. Remove the fuel fill cap to relieve the pressure in the fuel tank.
7. Turn the ignition switch to ON (II).
8. From the INSPECTION MENU of the HDS, select Fuel Pump OFF, then start the engine, and let it idle until it stalls.

➡**Do not allow the engine to idle above 1,000 rpm or the PCM will continue to operate the fuel pump. A DTC or a Temporary DTC may be set during this procedure. Check for DTCs, and clear them as needed.**

9. Turn the ignition switch to LOCK (0).
10. Do the battery terminal disconnection procedure.
11. Remove the quick-connect fitting cover.
12. Check the fuel quick-connect fitting for dirt, and clean it if needed.
13. Place a rag or shop towel over the quick-connect fitting.
14. Disconnect the quick-connect fitting: Hold the connector with one hand, and squeeze the retainer tabs with the other hand to release them from the locking tabs. Pull the connector off.

Please note the following:
• Be careful not to damage the line or other parts.
• Do not use tools.
• If the connector does not move, keep the retainer tabs pressed down, and alternately pull and push the connector until it comes off easily.
• Do not remove the retainer from the line; once removed, the retainer must be replaced with a new one.

15. After disconnecting the quick-connect fitting, check it for dirt or damage.
16. Do the battery terminal reconnection procedure.

WITHOUT THE HDS

See Figure 125.

✸✸ CAUTION

Before disconnecting fuel lines or hoses, relieve pressure from the system by disabling the fuel pump and then disconnecting the fuel tube/quick connect fitting in the engine compartment.

1. Remove PGM-FI main relay 2 from the under-dash fuse/relay box.
2. Start the engine, and let it idle until it stalls.

➡**If any DTCs are stored, clear and ignore them.**

3. Turn the ignition switch to LOCK (0).
4. Remove the fuel fill cap to relieve the pressure in the fuel tank.
5. Do the battery terminal disconnection procedure.
6. Remove the quick-connect fitting cover.
7. Check the fuel quick-connect fitting for dirt, and clean it if needed.
8. Place a rag or shop towel over the quick-connect fitting.
9. Disconnect the quick-connect fitting: Hold the connector with one hand, and squeeze the retainer tabs with the other hand to release them from the locking tabs. Pull the connector off.

Please note the following:
• Be careful not to damage the line or other parts.
• Do not use tools.
• If the connector does not move, keep the retainer tabs pressed down, and alternately pull and push the connector until it comes off easily.
• Do not remove the retainer from the line; once removed, the retainer must be replaced with a new one.

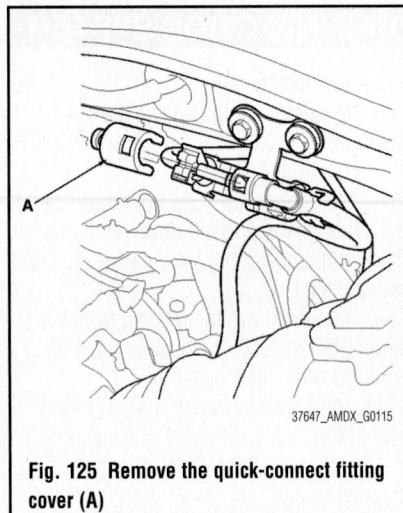

37647_AMDX_G0115

Fig. 125 Remove the quick-connect fitting cover (A)

10. After disconnecting the quick-connect fitting, check it for dirt or damage.

11. Do the battery terminal reconnection procedure.

FUEL FILTER

REMOVAL & INSTALLATION

See Figure 126.

The fuel filter should be replaced whenever the fuel pressure drops below the specified value, after making sure that the fuel pump and the fuel pressure regulator are OK.

1. Remove the fuel tank unit.

2. Remove the fuel filter set.

3. Check these items before installing the fuel tank unit:

   • When connecting the wire harness, make sure the connection is secure and the connectors are firmly locked into place.

   • When installing the fuel gauge sending unit, make sure the connection is secure and the connector is firmly locked into place. Be careful not to bend or twist it excessively.

4. Install the parts in the reverse order of removal with new O-rings. When installing the fuel tank unit, align the marks on the unit and the fuel tank.

➡**Coat the O-rings with clean engine oil.**

FUEL TANK

DRAINING

1. Remove the fuel tank unit.

2. Using a hand pump, a hose, and a container suitable for fuel, draw the fuel from the fuel tank.

3. Reinstall the fuel tank unit.

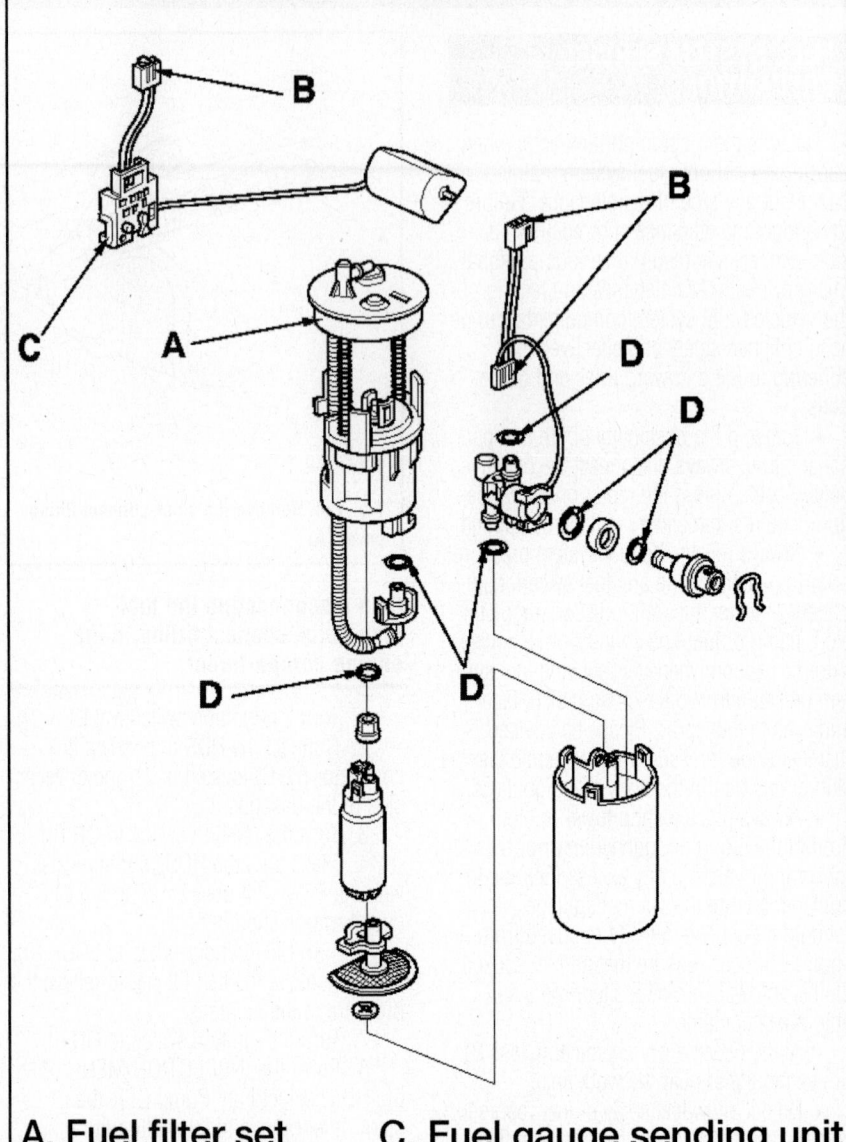

A. Fuel filter set
B. Connectors
C. Fuel gauge sending unit
D. O-rings

37647_AMDX_G0116

Fig. 126 Remove the fuel filter set

REMOVAL & INSTALLATION

See Figures 127 and 128.

1. Relieve the fuel pressure.

2. Drain the fuel tank, then reinstall the fuel tank unit without connecting the fuel tank unit 4P connector and the fuel tank unit quick-connect fitting.

3. Raise the vehicle on a lift.

4. Remove the exhaust pipe.

5. Remove the propeller shaft.

6. Remove the EVAP canister cover and fuel tank protector.

7. Loosen the clamp, and disconnect the tube. Slide back the clamps, then twist the hoses as you pull to avoid damaging them.

8. Disconnect the hoses.

9. Place a jack or other support under the fuel tank.

10. Remove the strap bolts and the straps.

11. Remove the fuel tank.

To install:

12. Install the parts in the reverse order of removal.

Please note the following:

• New fuel tanks have a ring pull at the fuel vapor hose connector. When you connect the hose and confirm that connection is

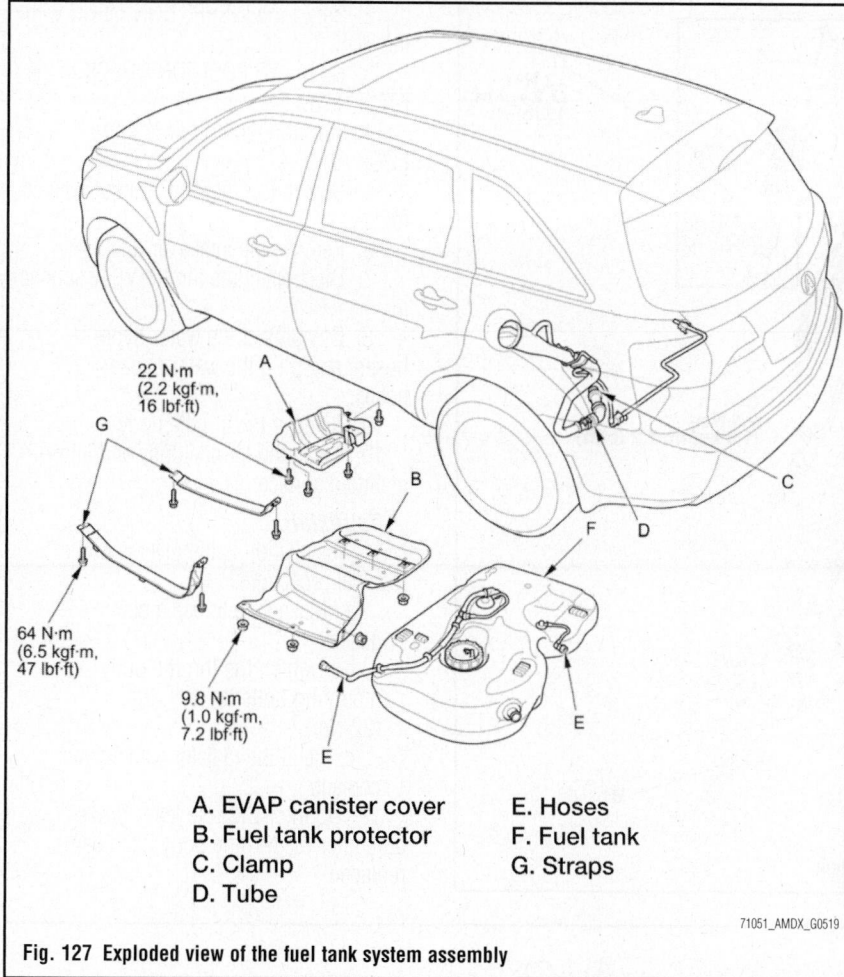

A. EVAP canister cover
B. Fuel tank protector
C. Clamp
D. Tube
E. Hoses
F. Fuel tank
G. Straps

71051_AMDX_G0519

Fig. 127 Exploded view of the fuel tank system assembly

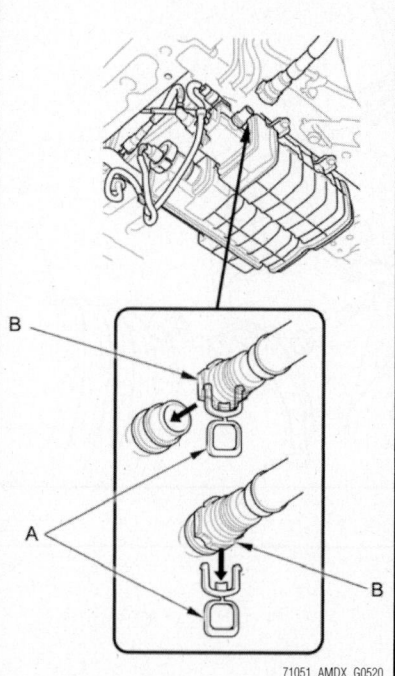

71051_AMDX_G0520

Fig. 128 New fuel tanks have a ring pull (A) at the fuel vapor hose connector (B)

secure, remove the ring pull by pulling it down.

- Before connecting the fuel fill pipe and the quick-connect fitting, check for dirt, and clean it if needed, taking care not to damage the fuel fill pipe and other parts.

FUEL PUMP CONTROL MODULE

REMOVAL & INSTALLATION

See Figure 129.

1. Remove the cargo floor lid.
2. On models with blind spot information (BSI), remove the BSI unit and/or the active damper control unit.
3. Disconnect the fuel pump control module connector.
4. Remove the fuel pump control module.
5. Install the parts in the reverse order of removal.

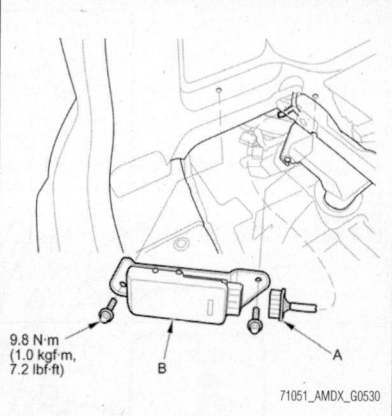

71051_AMDX_G0530

Fig. 129 Disconnect the fuel pump control module connector (A) and remove the fuel pump control module (B)

FUEL RAIL & INJECTORS

REMOVAL & INSTALLATION

See Figure 130.

1. Relieve the fuel pressure.
2. Remove the intake manifold.
3. Disconnect the connectors from the injectors.
4. Disconnect the quick-connect fitting.
5. Remove the fuel rail mounting bolts and nut from the fuel rails.
6. Remove the injector clips from the fuel rail.
7. Remove the injectors from the rails.

To install:

8. Coat the new O-ring with clean engine oil, and insert the injectors into the fuel rails.
9. Install the injector clips.
10. Coat the new injector O-ring with clean engine oil.
11. Install the injectors in the injector base.
12. Install the fuel rail mounting bolts and nut.
13. Install the injector connectors.
14. Connect the quick-connect fitting.
15. Turn the ignition switch to ON (II), but do not operate the starter. After the fuel pump runs for about 2 seconds, the fuel pressure in the fuel line rises. Repeat this two or three times, then check for fuel leaks.
16. Install the intake manifold.

THROTTLE BODY

COMPONENT LOCATIONS

The throttle body is mounted on the intake manifold.

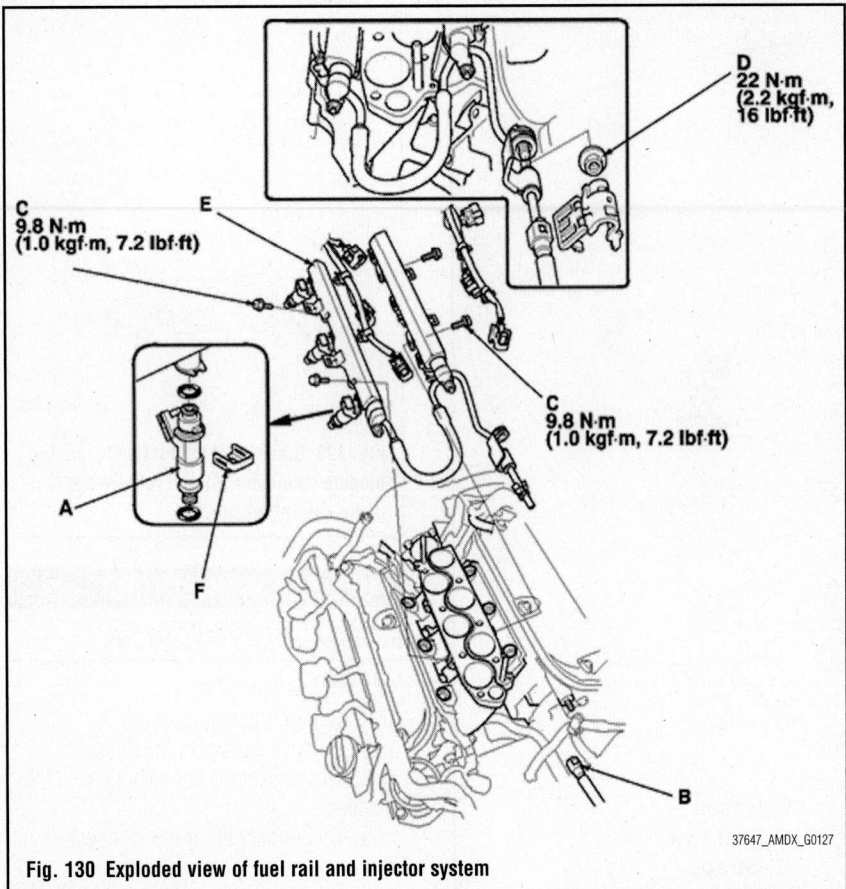

Fig. 130 Exploded view of fuel rail and injector system

C
9.8 N·m
(1.0 kgf·m, 7.2 lbf·ft)

D
22 N·m
(2.2 kgf·m, 16 lbf·ft)

C
9.8 N·m
(1.0 kgf·m, 7.2 lbf·ft)

37647_AMDX_G0127

2. Select the INSPECTION MENU with the HDS.

3. Do the TP POSITION CHECK in the ETCS TEST.

4. Turn the ignition switch to LOCK (0).

5. Disconnect the MAP sensor connector.

6. Remove the intake air duct.

7. Disconnect the throttle body connector.

8. Disconnect the water bypass hoses, and plug the water bypass hoses.

9. Remove the throttle body.

10. Clean the throttle body and intake manifold surface.

To install:

11. Installation is the reverse of the removal procedure.

   a. Make sure to use a new gasket.

   b. Tighten the throttle body mounting bolts to 16 ft. lbs. (22 Nm).

   c. Refill the radiator with engine coolant.

12. Do the PCM idle learn procedure after the throttle body has been replaced.

REMOVAL & INSTALLATION

See Figure 131.

➡Anytime you remove or replace the throttle body, you will need a suitable scan tool to perform an Idle Learn procedure.

✳✳ CAUTION

Do not insert your fingers into the installed throttle body when you turn the ignition switch ON (II) or while the ignition switch is ON (II). If you do, you will seriously injure your fingers if the throttle valve is activated.

➡If you are replacing the throttle body, begin at step 1. If you are removing the throttle body temporarily, begin at step 4.

1. Connect the HDS while the engine is stopped.

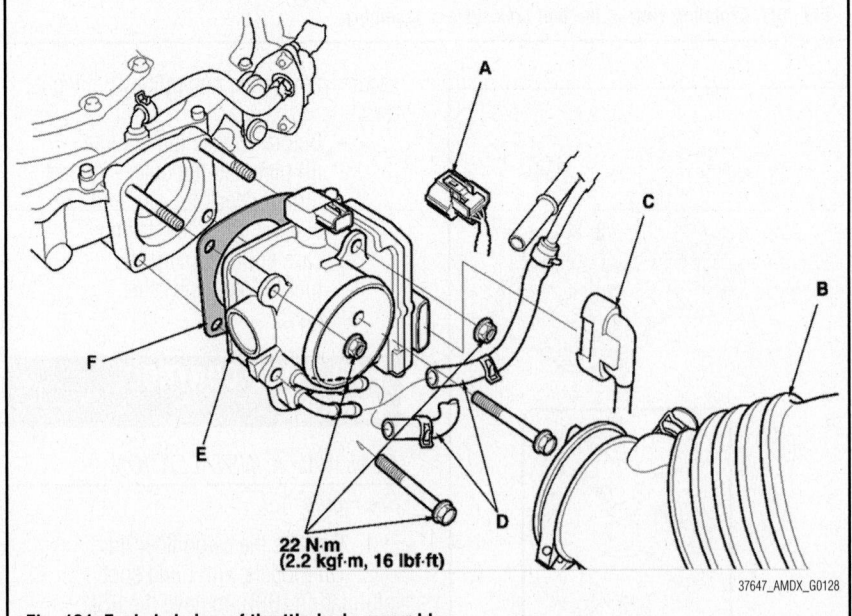

Fig. 131 Exploded view of throttle body assembly

22 N·m
(2.2 kgf·m, 16 lbf·ft)

37647_AMDX_G0128

HEATING & AIR CONDITIONING SYSTEM

BLOWER MOTOR

REMOVAL & INSTALLATION

See Figures 132 through 135.

1. Remove the glove box.
2. Detach the harness clips, then remove the bolts, and the glove box frame.
3. Cut the plastic cross brace in the glove box opening with diagonal cutters in the area shown, and discard it.
4. Detach the wire harness clip, then remove the self-tapping screws, and the passenger's heater duct.
5. Disconnect the connector from the front blower motor, then detach the wire harness clips.
6. Disconnect the connectors from the recirculation control motor and headlight leveling control unit, then detach the harness clip. Remove the self-tapping screws, the mounting nuts, and the blower unit.

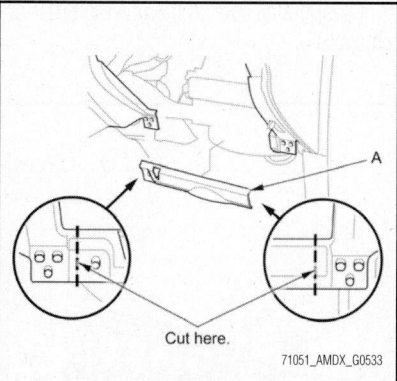

Cut here.

71051_AMDX_G0533

Fig. 132 Cut the plastic cross brace (A) in the glove box opening with diagonal cutters in the area shown, and discard it

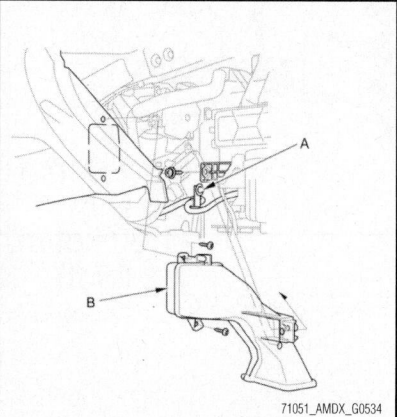

71051_AMDX_G0534

Fig. 133 Detach the wire harness clip (A), then remove the self-tapping screws, and the passenger's heater duct (B)

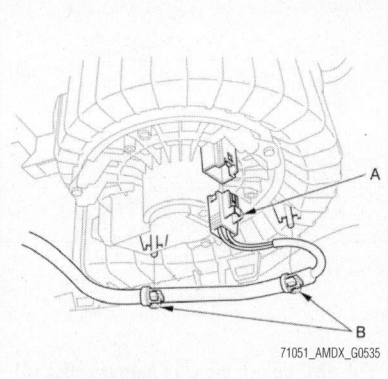

71051_AMDX_G0535

Fig. 134 Disconnect the connector (A) from the front blower motor, then detach the wire harness clips (B)

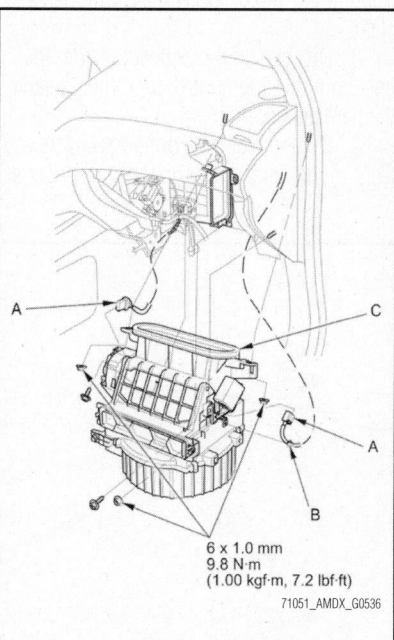

6 x 1.0 mm
9.8 N·m
(1.00 kgf·m, 7.2 lbf·ft)

71051_AMDX_G0536

Fig. 135 Disconnect the connectors (A), detach the harness clip (B) and remove the self-tapping screws, the mounting nuts, and the blower unit (C)

7. Install the unit in the reverse order of removal. Make sure that there is no air leakage.

HEATER CORE

REMOVAL & INSTALLATION

See Figures 136 through 142.

※※ CAUTION

SRS components are located in this area. Review the SRS component locations, and the precautions and procedures before doing repairs or service.

1. Do the battery terminal disconnection procedure.
2. Disconnect the front receiver line and front suction line from the front evaporator core.
3. When the engine is cool, drain the engine coolant from the radiator.
4. Front under the hood, remove the clamp. Slide the hose clamps back. Then disconnect the inlet heater hose and the outlet heater hose from the heater unit. Note the layout of the hose.

➡**Engine coolant will run out when the hoses are disconnected; drain it into a clean drip pan. Be sure not to let coolant spill on the electrical parts or the painted surfaces. If any coolant spills, rinse it off immediately.**

5. Remove the mounting nuts from the heater unit. Take care not to damage or bend the fuel lines or brake lines, etc.
6. Remove the dashboard.

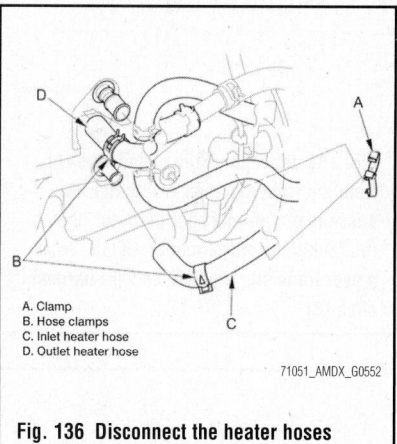

A. Clamp
B. Hose clamps
C. Inlet heater hose
D. Outlet heater hose

71051_AMDX_G0552

Fig. 136 Disconnect the heater hoses

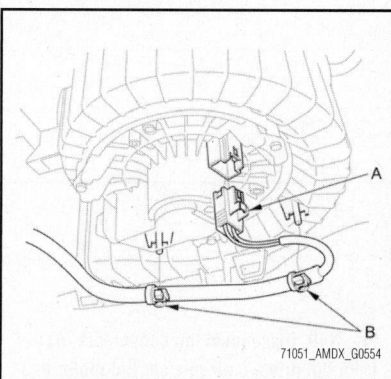

71051_AMDX_G0554

Fig. 137 Disconnect the connector (A) from the front blower motor and detach the wire harness clips (B)

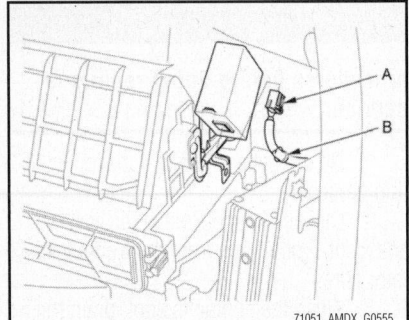

Fig. 138 Disconnect the connector (A) from the headlight leveling control unit, then detach the harness clip (B)

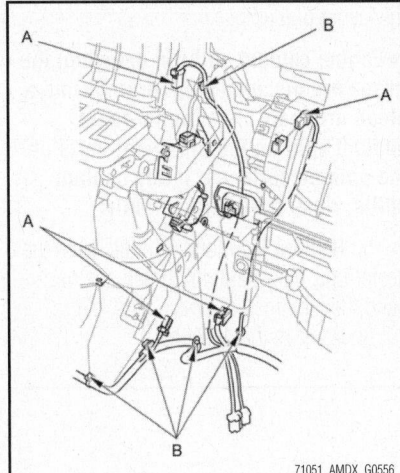

Fig. 139 Disconnect the connectors (A) from the front mode control motor, the passenger's air mix control motor, the recirculation control motor, and the front power transistor; detach the wire harness clips (B)

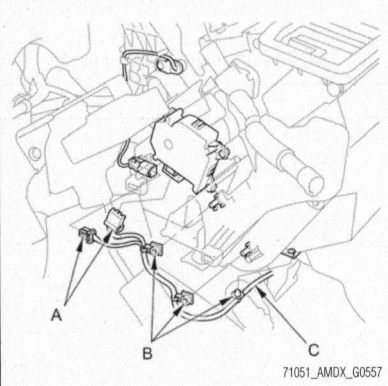

Fig. 140 Disconnect the connectors (A) from the driver's air mix control motor and the front evaporator temperature sensor; detach the wire harness clips (B), then remove the wire harness (C)

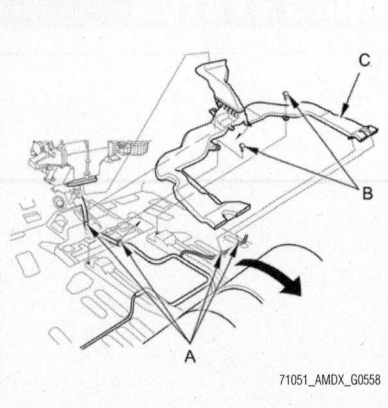

Fig. 141 Detach the wire harness clips (A) and the rear heater duct mounting clips (B), then remove the rear heater duct (C)

7. Disconnect the connector from the front blower motor. Detach the wire harness clips.

8. Disconnect the connector from the headlight leveling control unit, then detach the harness clip.

9. Disconnect the connectors from the front mode control motor, the passenger's air mix control motor, the recirculation control motor, and the front power transistor. Detach the wire harness clips.

10. Disconnect the connectors from the driver's air mix control motor and the front evaporator temperature sensor. Detach the wire harness clips, then remove the wire harness.

11. Turn over the carpet. Detach the wire harness clips and the rear heater duct mounting clips, then remove the rear heater duct.

12. Remove the mounting nuts. Slide the blower-heater unit, then remove the drain hose and blower-heater unit.

13. Remove the self-tapping screws and the passenger's heater duct.

14. Remove the self-tapping screws and the expansion valve cover.

15. Remove the self-tapping screw and the front heater core cover.

16. Remove the self-tapping screws, the heater pipe brackets, the grommets, and carefully pull out the front heater core.

To install:

17. Install the front heater core, and the front evaporator core in the reverse order of removal.

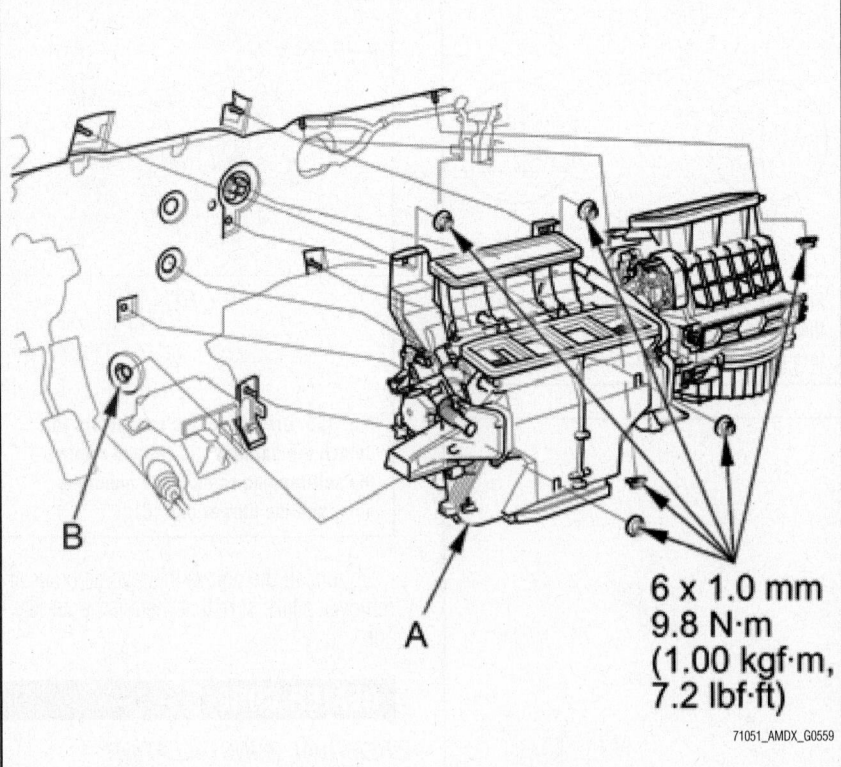

Fig. 142 Remove the mounting nuts, slide the blower-heater unit (A), then remove the drain hose (B) and blower-heater unit

6 x 1.0 mm
9.8 N·m
(1.00 kgf·m,
7.2 lbf·ft)

18. Install the heater unit in the reverse order of removal, and note these items:

a. Do not interchange the inlet and outlet heater hoses, and install the hose clamps securely.

b. Refill the cooling system with engine coolant.

c. Make sure that there is no coolant leakage.

d. Make sure that there is no air leakage.

19. Do the battery terminal reconnection procedure.

20. After installation, operate the heater unit in various functions to confirm that it works properly.

21. Run the self-diagnostic function to confirm that there are no problems in the system.

STEERING

POWER RACK & PINION STEERING GEAR

REMOVAL & INSTALLATION

See Figures 143 through 166.

Special Tools Required:
- Ball Joint Thread Protector, 12 mm 07AAF-SDAA100
- Ball Joint Remover, 28 mm 07MAC-SL0A202
- Engine Hanger Adapter Set VSB02C000031
- Engine Support Hanger, A and Reds AAR-T1256
- Engine Hanger Balance Bar VSB02C000019
- Front Subframe Adapter EQS02BMDXSB0

Note these items during removal:

a. Using clean solvent and a brush, wash any oil and dirt off the valve body unit, it's lines, and the end of the steering gearbox. Blow dry with compressed air.

b. Lower the front subframe from the body, and remove the steering gearbox through the gap produced by lowering the front subframe.

1. Make sure you have the anti-theft code for the radio and the navigation system, then write down the frequencies for the radio's preset buttons.

2. Remove the front bulkhead cover.

3. Remove the engine cover.

4. Drain the power steering fluid.

5. Disconnect the negative cable from the battery first, then disconnect the positive cable. Wait at least 3 minutes before proceeding.

6. Raise and safely support the vehicle.

7. Remove the front wheels.

8. Remove the driver's dashboard undercover.

9. Remove steering joint cover.

10. Remove the steering joint bolts, disconnect the steering joint by moving the steering joint toward the column.

11. Center the steering wheel spokes and install a commercially available steering wheel holder.

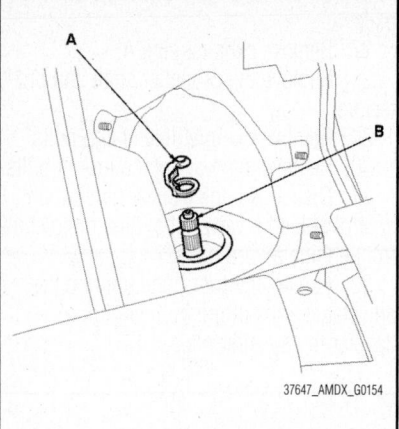

Fig. 143 Remove the center guide (A) (if equipped) from the top of the pinion shaft

12. Remove the center guide (if equipped) from the top of the pinion shaft, and discard it.

➡ **The center guide is for factory assembly use only.**

13. Apply vinyl tape to the splines on the pinion shaft.

14. Remove the air cleaner housing.

15. Remove the connector bracket from the front cylinder head. Use the bracket bolt

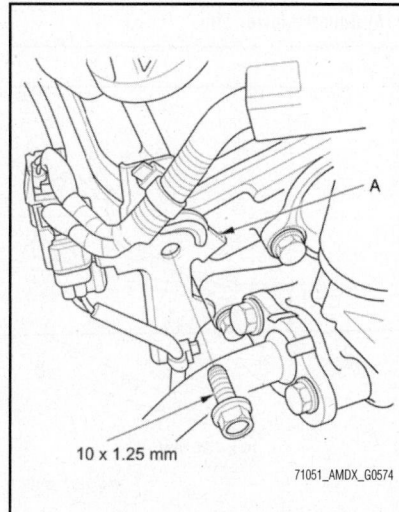

Fig. 144 Remove the connector bracket (A) from the front cylinder head

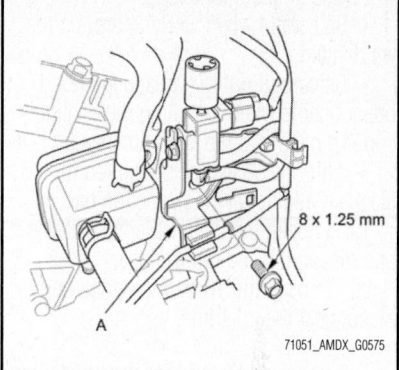

Fig. 145 Remove the engine mount control solenoid valve bracket (A) from the rear cylinder head

hole to attach the engine balance bar front arm.

16. Remove the engine mount control solenoid valve bracket from the rear cylinder head. Use the bracket bolt hole to attach the engine hanger balance bar rear arm.

17. Remove the service caps (A) for the front damper flange nuts from the cowl cover (B). Position the engine hanger

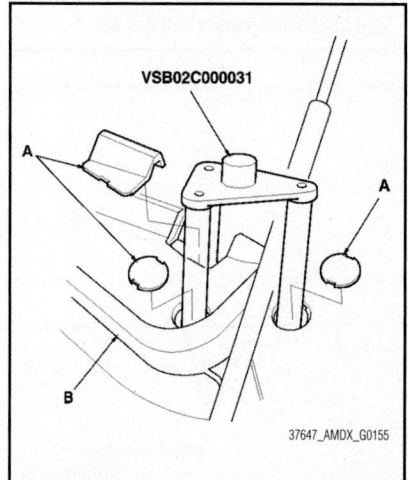

Fig. 146 Remove the service caps (A) for the front damper flange nuts from the cowl cover (B), then position the engine hanger adapter set (C)

adapter set (C) with the FRONT mark facing forward over the damper flange nuts.

18. Lift and support the engine with the engine support hanger (AAR-T1256) and the engine hanger balance bar (VSB02C000019). Attach the front arm to the front cylinder head with a spacer and a 10 mm bolt, and attach the rear arm to the rear cylinder head with an 8 mm bolt. Tighten the wing nut by hand to lift, and support the engine.

Please note the following:
• Be careful when working around the windshield.
• Be careful not to damage the hood opener cable when installing the engine support hanger at the front bulkhead.
• AAR-T1256 two sets required for stacking additional cross section bar.

19. Loosen the adjustable hose clamp, and disconnect the return hose.

20. Loosen the 16 mm flare nut, and disconnect the inlet line.

21. Remove the cotter pins from the tie-rod ball joints and remove the nuts.

22. Separate the tie-rod ball joints and knuckles using the ball joint remover.

23. Disconnect the lower arm ball joints from the knuckles.

24. Disconnect the stabilizer links from the lower arms.

25. Remove the front undercover and the front splash shield.

26. Remove the front subframe stiffener plate.

27. Remove exhaust pipe A.

28. Remove the propeller shaft and protectors.

29. Remove the inlet line clamp bolts.

30. Remove the rear engine mount bolts.

31. Disconnect the front engine mount control solenoid valve tube, then remove the front engine mount bolts.

32. Loosen the four bolts holding the adjustable arms of the front subframe adapter to its center plate.

33. Line up the slots in the arms with the bolt holes on the corner of the jack base, then attach the front subframe adapter to the jack base with the bolts that came with the jack. Tighten all bolts securely.

34. Raise the jack to vehicle height, then attach the front subframe adapter to the front subframe using the subframe stiffener mounting bolts and bolt holes.

35. Support the front subframe securely by raising the transmission jack.

36. Remove the transmission lower mount bolts.

37. Remove the stiffener mounting bolts from the front subframe front brackets.

38. Loosen the subframe mounting bolts on the front subframe front brackets so they are about 1 9/16 inches (40 mm) from the mounting surface. Do not loosen the special bolts more than necessary.

39. Remove the rear stiffener mounting bolts from the front subframe rear stiffener.

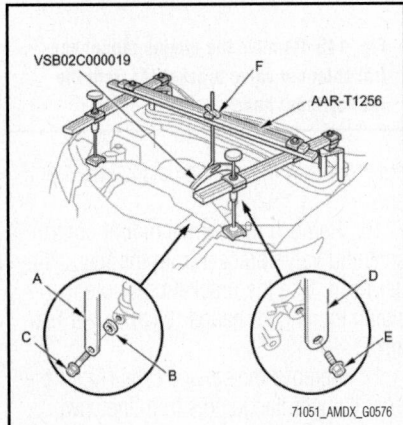

Fig. 147 Lift and support the engine with the engine support hanger (AAR-T1256) and the engine hanger balance bar

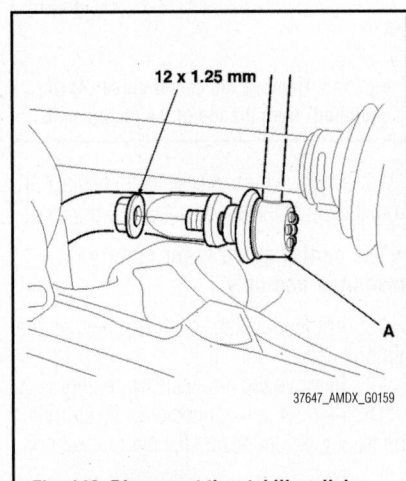

Fig. 149 Disconnect the stabilizer links (A) from the lower arms

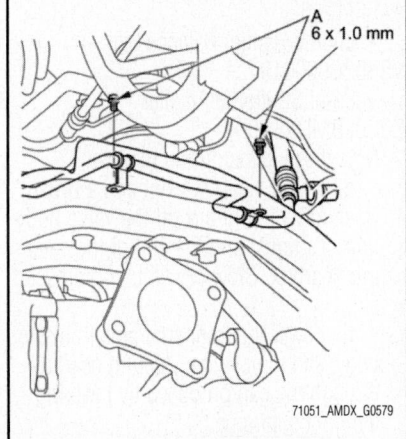

Fig. 151 Remove the inlet line clamp bolts (A)

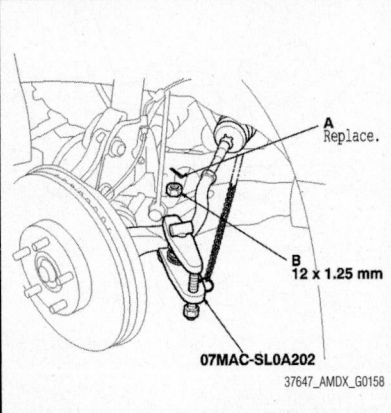

Fig. 148 Remove the cotter pins (A) from the tie-rod ball joints and remove the nuts (B)

Fig. 150 Remove the front subframe stiffener plate (A)

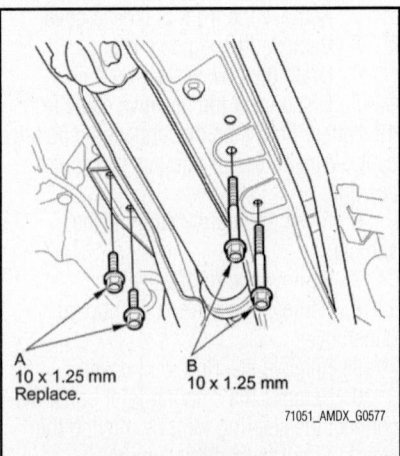

Fig. 152 Remove the rear engine mount bolts (A) and (B)

40. Loosen the special bolts on the front subframe rear brackets so they are about 1⁹⁄₁₆ inches (40 mm) from the mounting surface. Do not loosen the special bolts more than necessary.

41. Lower the transmission jack slowly until the front subframe has dropped about 1⁹⁄₁₆ inches (40 mm).

42. Remove the mounting bolts from the driver's side of the steering gearbox.

43. Remove the gearbox stiffener bracket from the driver's side of the front subframe.

44. Remove the mounting bolts from the passenger's side of the steering gearbox, then remove the mounting bracket and cushion.

45. Carefully move the steering gearbox toward the driver's side until the pinion shaft clears the wheel well opening on the frame.

46. Remove the steering gearbox through the wheel well opening on the driver's side.

To install:

> ❈❈ **CAUTION**
>
> Before installing the steering gearbox, make sure that no power steering fluid is on the mating surface of the steering gearbox and front subframe. To prevent the gearbox mounting bolts from loosening after the installation, remove any power steering fluid from the mount cushions and bolt holes.

47. Apply vinyl tape to the splines on the pinion shaft.

48. Install the pinion shaft grommet on the top of the valve housing.

49. Slide the steering gearbox between the front subframe and body from the driver's side.

50. Carefully move the steering gearbox toward the passenger's side until the pinion

shaft clears the wheel well opening on the body.

51. Continue moving the gearbox toward the passenger's side until the steering gearbox is in position.

52. Install the gearbox stiffener bracket on the left side of the front subframe, and tighten the bolts and nut to 28 ft. lbs. (38 Nm).

53. Loosely install the new mounting bolts on the left side of the steering gearbox.

54. Position the cutout on the mounting cushion as shown, and install it on the right side of the steering gearbox.

55. Install the gearbox mounting bracket over the mounting cushion, and loosely install the mounting bolts.

56. Tighten the mounting bolts on both sides of the steering gearbox to the specified torque alternately in two or more steps.

57. Loosely tighten the right rear subframe mounting bolt; insert the 15.7 mm

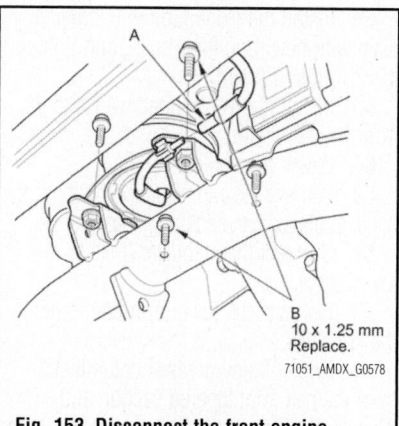

Fig. 153 Disconnect the front engine mount control solenoid valve tube (A), then remove the front engine mount bolts (B)

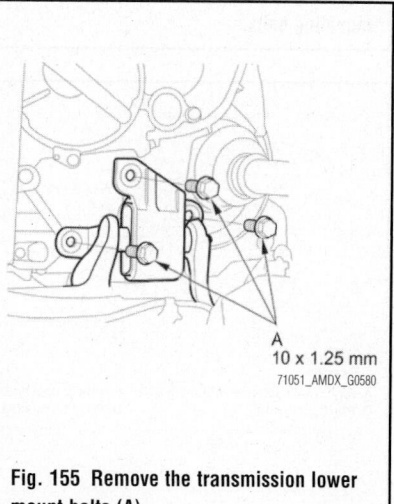

Fig. 155 Remove the transmission lower mount bolts (A)

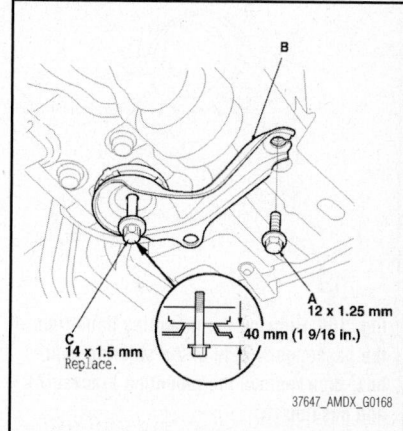

Fig. 157 Remove the mounting bolts (A) from the front subframe rear brackets (B)); subframe mounting bolt (C)

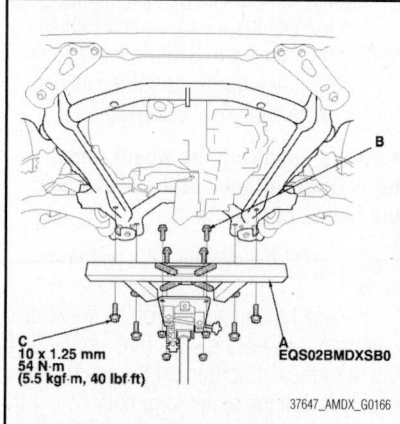

Fig. 154 Line up the slots in the arms with the bolt holes on the corner of the jack base

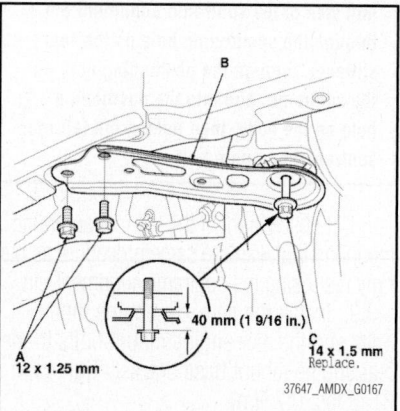

Fig. 156 Remove the stiffener mounting bolts (A) from the front subframe front brackets (B); subframe mounting bolt (C)

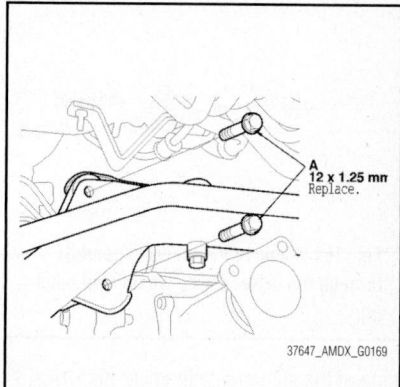

Fig. 158 Remove the mounting bolts (A) from the driver's side of the steering gear box

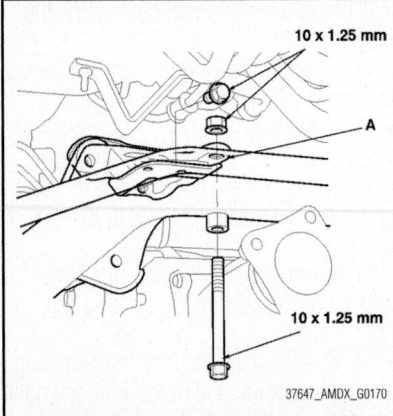

Fig. 159 Remove the gearbox stiffener bracket (A) from the driver's side of the front subframe

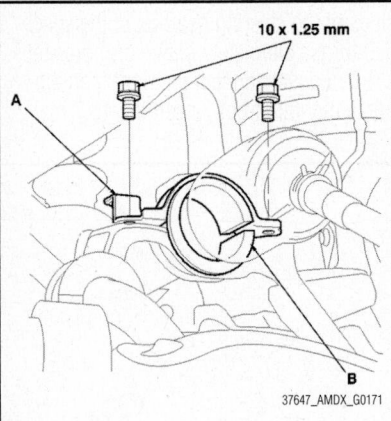

Fig. 160 Remove the mounting bolts from the passenger's side of the steering gearbox, then remove the mounting bracket (A) and cushion (B)

Fig. 161 Remove the steering gearbox through the driver's side wheel well opening

side of the subframe alignment pin (070AG-SJAA10S) through the positioning hole on the rear stiffener, through the positioning hole on the subframe, and into the positioning hole on the body, then tighten the left rear subframe mounting bolt.

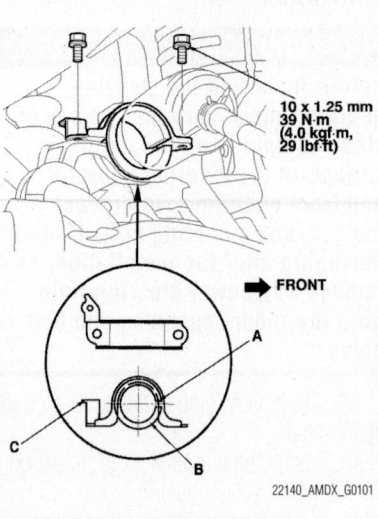

Fig. 162 Position the cutout (A) on the mounting cushion (B), and install it on the right side of the steering gearbox; install the gearbox mounting bracket (C) over the mounting cushion, and loosely install the mounting bolts.

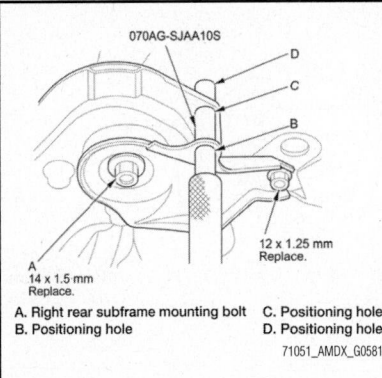

A. Right rear subframe mounting bolt
B. Positioning hole
C. Positioning hole
D. Positioning hole

Fig. 163 Loosely tighten the right rear subframe mounting bolt; insert the 15.7 mm side of the subframe alignment pin through the positioning hole on the rear stiffener, through the positioning hole on the subframe, and into the positioning hole on the body, then tighten the left rear subframe mounting bolt

58. Loosely tighten the left rear subframe mounting bolt with the same procedure as the right rear using the subframe alignment pin.

59. Install the new rear engine mount bolts and the rear engine mount bolts to the rear engine mount base bracket. Tighten to 31 ft. lbs. (42 Nm).

60. Loosely tighten transmission lower mount bolts.

61. Install the inlet line clamp bolts.

62. Tighten the front mount bolts, then

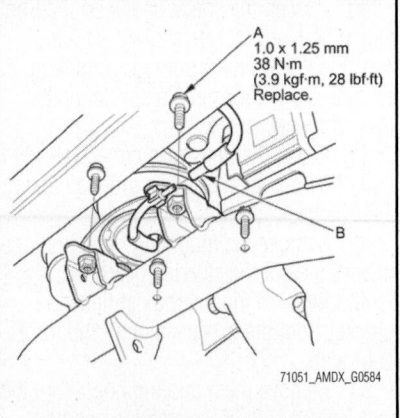

Fig. 164 Tighten the front mount bolts (A), then connect the front engine mount control solenoid valve tube (B)

connect the front engine mount control solenoid valve tube.

63. Install exhaust pipe A.

64. Install the propeller shaft and propeller shaft protectors.

65. Install the front subframe stiffener plate with new mounting bolts, and tighten to 43 ft. lbs. (59 Nm).

66. Install the front undercover and the front splash shield.

67. Lower the vehicle.

68. Remove the engine support hanger, the hanger balance bar, and hanger adapter set.

69. Connect the stabilizer links to the lower arms.

70. Connect the lower arm ball joints to the knuckles.

71. Wipe off any grease contamination from the ball joint tapered section and threads. Reconnect the tie-rod end ball joints to the knuckles. Install the new 12 mm nuts and tighten it to 44 ft. lbs. (60 Nm).

72. Install the new cotter pin and bend it.

73. Connect the return hose securely, and tighten the adjustable hose clamp.

74. Connect the inlet line, and tighten the 16 mm flare nut to 31 ft. lbs. (42 Nm).

75. Install the front wheel, then set the wheels in the straight ahead position.

➡**Before installing the wheel, clean the mating surfaces of the brake disc and inside of the wheel.**

76. Center the steering rack within its stroke.

77. Insert the upper end of the steering joint onto the steering shaft (line up the bolt hole with the flat portion on the shaft), and loosely install the upper joint bolt.

78. Slip the lower end of the steering joint onto the pinion shaft taking care to align the gap within the angle.

79. Remove the steering wheel holder tool.

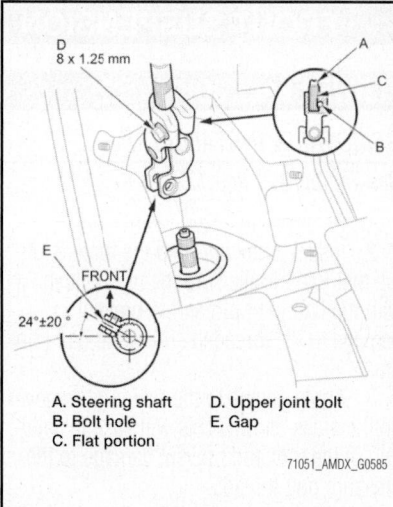

Fig. 165 Insert the upper end of the steering joint onto the steering shaft (line up the bolt hole with the flat portion on the shaft), and loosely install the upper joint bolt

A. Steering shaft   D. Upper joint bolt
B. Bolt hole   E. Gap
C. Flat portion

71051_AMDX_G0585

80. Align the bolt hole on the steering joint with the groove around the pinion shaft then loosely install the joint bolt.

81. Pull on the steering joint to make sure that the steering joint is fully seated, then tighten the lower joint bolt to the specified torque.

82. Tighten the upper joint bolt to the specified torque.

83. Install the steering joint cover.

84. Do the battery terminal reconnection procedure, and do these tasks:

a. Turn the ignition switch to ON (II); the SRS indicator should come on for about 6 seconds and then go off.

b. Make sure the horn and turn signal switches work properly.

c. Make sure the steering wheel switches work properly.

85. Fill the system with power steering fluid, and bleed air from the system.

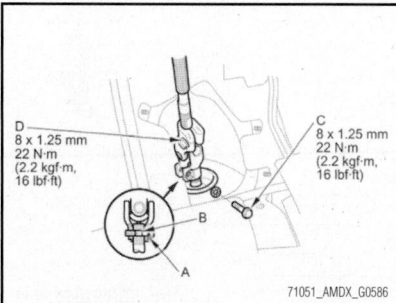

Fig. 166 Align the bolt hole on the steering joint with the groove around the pinion shaft then loosely install the joint bolt

71051_AMDX_G0586

86. After installation, do these checks:

a. Start the engine, allow it to idle, and turn the steering wheel from lock to lock several times to warm up the fluid. Check the gearbox for leaks.

b. Check the steering wheel spoke angle. If steering spoke angles to the right and left are not equal (steering wheel and rack are not centered), correct the engagement of the joint/pinion shaft serrations, then adjust the front toe by turning the tie-rod ends, if necessary.

c. Set the steering column to the center tilt position, then do the front toe inspection.

d. Do the VSA sensor neutral position memorization.

POWER STEERING PUMP

REMOVAL & INSTALLATION

See Figure 167.

1. Drain the power steering fluid from the reservoir.

2. Remove the engine appearance cover.

3. Remove the accessory drive belt from the pump pulley.

4. Cover the auto-tensioner, alternator, and A/C compressor with several shop towels to protect them from spilled power steering fluid. Disconnect the pump inlet hose and pump outlet hose from the pump, and plug them.

✱✱ WARNING

Take care not to spill the fluid on the body or parts. Wipe off any spilled fluid at once. Do not turn the steering wheel with the pump removed.

5. Remove the pump mounting bolts.

6. Cover the opening of the pump with a piece of tape to prevent foreign material from entering the pump.

To install:

7. Connect the pump inlet hose and pump outlet hose onto the new pump with the new O-ring.

8. Loosely install the pump in the pump bracket with the mounting bolts, then tighten the pump fittings securely.

9. Tighten the pump mounting bolts to 16 ft. lbs. (22 Nm).

10. Install the accessory drive belt.

11. Refill the power steering system to the correct level.

BLEEDING

1. Fill the reservoir to the upper level line.

2. Start the engine and run it at idle,

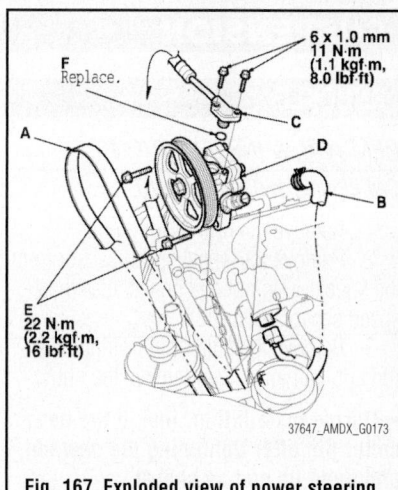

Fig. 167 Exploded view of power steering pump assembly

37647_AMDX_G0173

then turn the steering from lock-to-lock several times to bleed air from the system.

3. Recheck the fluid level and add some if necessary. Do not fill the reservoir beyond the upper level line.

4. If the fluid is contaminated, dark, or discolored, repeat the procedure as necessary.

FLUID REPLACEMENT PROCEDURE

1. Remove the reservoir from its holder. Raise the reservoir, then disconnect the return hose to drain the reservoir. Take care not to spill the fluid on the body and parts. Wipe off any spilled fluid at once.

➡**Inspect the reservoir screen for any debris. If the reservoir screen is clogged, replace the reservoir, and check for the source of the contamination.**

2. Connect a hose of suitable diameter to the disconnected return hose, and put the hose end in a suitable container.

3. Start the engine, let it run at idle, and turn the steering wheel from lock to lock several times. When fluid stops running out of the hose, shut off the engine. Discard the fluid.

➡**Stop the engine immediately once the fluid stops running out of the hose, shut off the engine. Discard the fluid.**

4. Reinstall the return hose on the reservoir.

5. Fill the reservoir to the upper level line.

6. Start the engine and run it at idle. Turn the steering from lock to lock several times to bleed air from the system.

7. Recheck the fluid level and add some if necessary. Do not fill the reservoir beyond the upper level line.

8. If the fluid is contaminated, dark, or discolored, repeat the procedure as necessary until the system is clean.

KNUCKLE

REMOVAL & INSTALLATION

See Figures 168 through 170.

1. Remove the hub bearing unit.
2. Remove the wheel speed sensor from the knuckle. Do not disconnect the wheel speed sensor connector.
3. Remove the cotter pin from the tie-rod end ball joint, then remove the nut.

➡**During installation, install the new cotter pin after tightening the new nut, and bend its end as shown.**

4. Disconnect the tie-rod end ball joint from the knuckle using the ball joint thread protector, and the ball joint remover.
5. Remove the lock pin from the lower arm ball joint, then remove the castle nut.

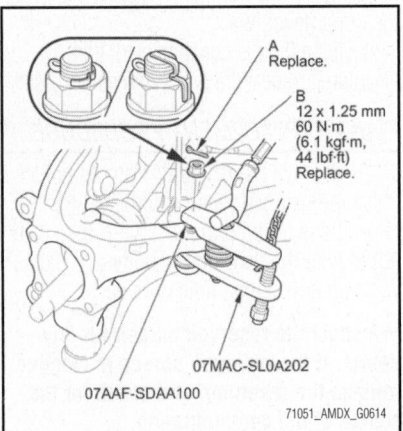

07MAC-SL0A202

07AAF-SDAA100

71051_AMDX_G0614

Fig. 168 Remove the cotter pin (A) from the tie-rod end ball joint, then remove the nut (B)

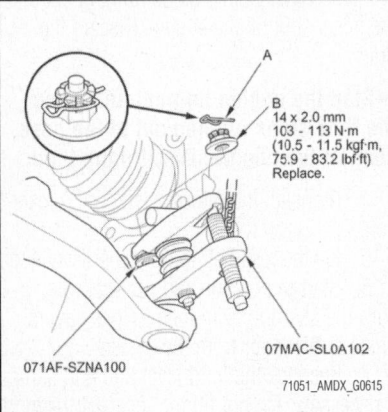

071AF-SZNA100

07MAC-SL0A102

71051_AMDX_G0615

Fig. 169 Remove the lock pin (A) from the lower arm ball joint, then remove the castle nut (B)

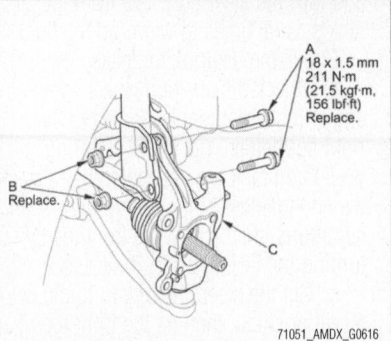

71051_AMDX_G0616

Fig. 170 Remove the damper pinch bolts (A) and flange nuts (B) from the damper, then remove the knuckle (C)

➡**During installation, install the lock pin as shown after tightening the new castle nut.**

6. Disconnect the lower arm ball joint from the knuckle using the ball joint thread protector, and the ball joint remover.

✳✳ WARNING

Be careful not to damage the ball joint boot when installing the remover. Do not force or hammer on the lower arm, or pry between the lower arm and the knuckle. You could damage the ball joint.

7. Remove the damper pinch bolts and flange nuts from the damper, then remove the knuckle.

To install:

➡**During installation, install new damper pinch bolts and new flange nuts.**

8. Install the knuckle in the reverse order of removal, and note these items:

   a. Be careful not to damage the ball joint boot when connecting the knuckle.

   b. Before connecting the lower arm ball joint to the knuckle, degrease the threaded section and tapered portion of the ball joint pin, the ball joint connecting hole, the threaded section and mating surfaces of the castle nut.

   c. Torque the castle nut to the lower torque specification, then tighten it only far enough to align the slot with the ball joint pin hole. Do not align the castle nut by loosening it.

9. Check the wheel alignment, and adjust it if necessary.

LOWER BALL JOINTS

REMOVAL & INSTALLATION

See Figures 171 and 172.

1. Remove the lower control arm.
2. Install a hex nut onto the threads of the ball joint. Make sure the nut is flush with the ball joint pin end to prevent damage to the threaded end of the ball joint pin.
3. Apply grease to the ball joint remover on the areas shown. This will ease installation of the tool and prevent damage to the pressure bolt threads.
4. Loosen the pressure bolt, and install the ball joint remover. Insert the jaws carefully, making sure not to damage the ball joint boot. Adjust the jaw spacing by turning the adjusting bolt.

✳✳ CAUTION

Fasten the safety chain securely to a suspension arm or the subframe. Do

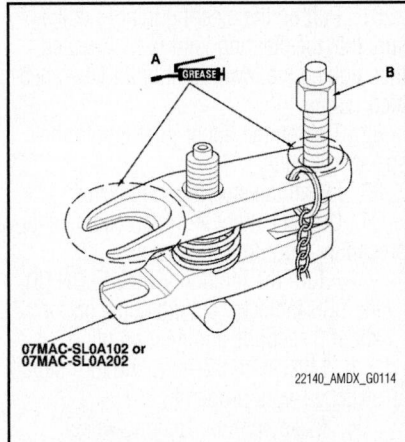

07MAC-SL0A102 or
07MAC-SL0A202

22140_AMDX_G0114

Fig. 171 Apply grease to the ball joint remover on the areas shown (A)

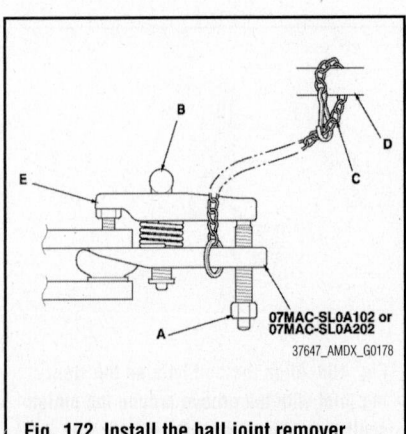

07MAC-SL0A102 or
07MAC-SL0A202

37647_AMDX_G0178

Fig. 172 Install the ball joint remover

not fasten it to a brake line or wire harness.

5. After adjusting the adjusting bolt, make sure the head of the adjusting bolt is in the position shown to allow the jaw to pivot.

6. With a wrench, tighten the pressure bolt until the ball joint pin pops loose from the ball joint connecting hole. If necessary, apply penetrating type lubricant to loosen the ball joint pin.

☀☀ CAUTION

Do not use pneumatic or electric tools on the pressure bolt.

7. Remove the ball joint remover, then remove the nut from the end of the ball joint pin, and pull the ball joint out of the ball joint connecting hole. Inspect the ball joint boot, and replace it if damaged.

LOWER CONTROL ARMS

REMOVAL & INSTALLATION

See Figures 173 through 175.

1. Raise the front of the vehicle, and support it with safety stands in the proper locations.

2. Remove the front wheel.

3. With active damper system: Remove the suspension stroke sensor from the lower arm.

➡**Use the new nut during reassembly.**

4. Remove the lock pin from the lower arm ball joint, then remove the castle nut.

➡**During installation, install the lock pin as shown after tightening the new castle nut.**

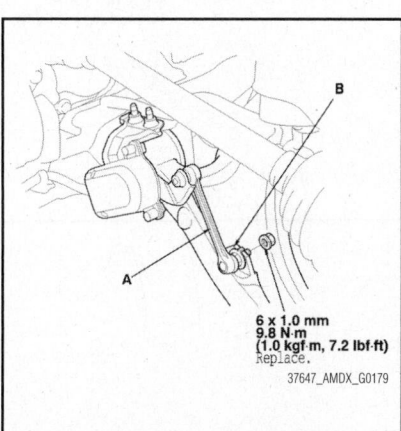

6 x 1.0 mm
9.8 N·m
(1.0 kgf·m, 7.2 lbf·ft)
Replace.

37647_AMDX_G0179

Fig. 173 Remove the suspension stroke sensor (A) from the lower arm (B)

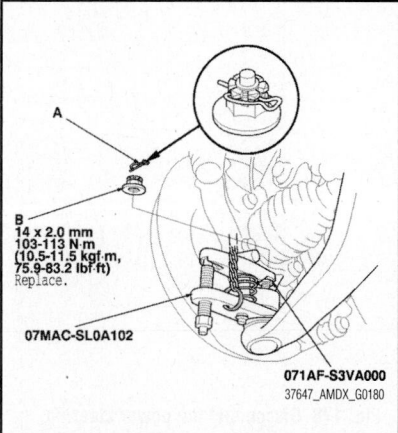

B
14 x 2.0 mm
103-113 N·m
(10.5-11.5 kgf·m,
75.9-83.2 lbf·ft)
Replace.

07MAC-SL0A102

071AF-S3VA000
37647_AMDX_G0180

Fig. 174 Remove the lock pin (A) from the lower arm ball joint, then remove the castle nut (B)

5. Disconnect the lower arm ball joint from the knuckle using the ball joint thread protector and the ball joint remover.

6. Remove the stabilizer bar bushing holder mounting bolt.

➡**During installation, install the mounting bolt after tightening the lower arm mounting 14 mm bolts to the specified torque value.**

7. Remove the lower arm mounting 14 mm and 16 mm bolts, then remove the lower arm from the front suspension subframe.

➡**Use the new mounting bolts during reassembly.**

8. Remove the lower arm stops.

➡**During installation, align the slot on the lower arm stop with the lug portion on the front side of lower arm bushing.**

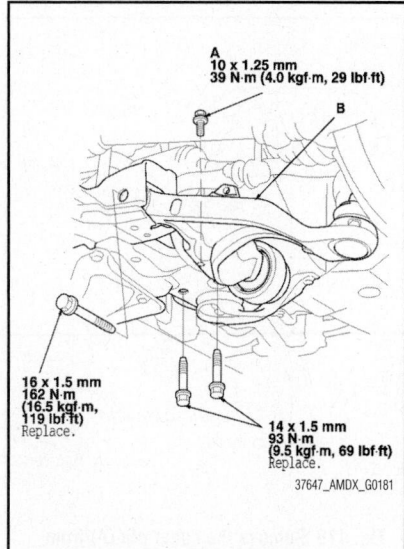

A
10 x 1.25 mm
39 N·m (4.0 kgf·m, 29 lbf·ft)

B

16 x 1.5 mm
162 N·m
(16.5 kgf·m,
119 lbf·ft)
Replace.

14 x 1.5 mm
93 N·m
(9.5 kgf·m, 69 lbf·ft)
Replace.

37647_AMDX_G0181

Fig. 175 Remove the stabilizer bar bushing holder mounting bolt (A)

To install:

9. Install the lower arm in the reverse order of removal, and note these items:

- First install all the components, and lightly tighten the bolts and nuts, then raise the suspension to load it with the vehicle's weight before fully tightening to the specified torque values.
- Be careful not to damage the ball joint boot when connecting the knuckle.
- Before connecting the lower arm ball joint to the knuckle, degrease the threaded section and tapered portion of the ball joint pin, the ball joint connecting hole, the threaded section and mating surfaces of the castle nut.
- Torque the castle nut to the lower torque specification, then tighten it only far enough to align the slot with the ball joint pin hole. Do not align the castle nut by loosening it.
- Before installing the wheel, clean the mating surfaces of the brake disc and the inside of the wheel.
- Check the wheel alignment, and adjust it if necessary.

10. With active damper system/Left side: Do the headlight initial position learning procedure.

STABILIZER BAR & LINKS

REMOVAL & INSTALLATION

Stabilizer Links

See Figure 176.

1. Raise the front of the vehicle, and support it with safety stands in the proper locations.

2. Remove the front wheel.

3. Remove the flange nuts while holding the respective joint pin with a hex wrench, then remove the stabilizer link.

To install:

4. Install the stabilizer link on the stabilizer bar and the damper with the joint pins set at the center of their range of movement.

5. Install the flange nuts, and lightly tighten them.

6. Place a jack under the lower arm, and raise the suspension to load it with the vehicle's weight.

7. Tighten the flange nuts to the specified torque value while holding the respective joint pin with a hex wrench.

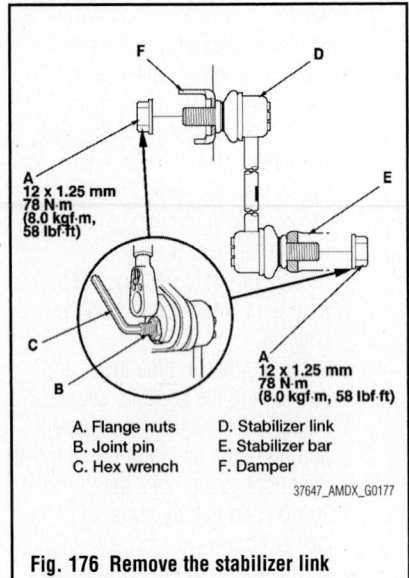

A. Flange nuts
B. Joint pin
C. Hex wrench
D. Stabilizer link
E. Stabilizer bar
F. Damper

37647_AMDX_G0177

Fig. 176 Remove the stabilizer link

8. Clean the mating surfaces of the brake disc and the inside of the wheel, then install the front wheel.

9. Test-drive the vehicle.

Stabilizer Bar

See Figures 177 through 194.

Special Tools Required:
• Ball Joint Thread Protector, 12 mm 07AAF-SDAA100
• Ball Joint Remover, 28 mm 07MAC-SL0A202
• Engine Hanger Adapter Set VSB02C000031
• Engine Support Hanger, A and Reds AAR-T1256
• Engine Hanger Balance Bar VSB02C000019

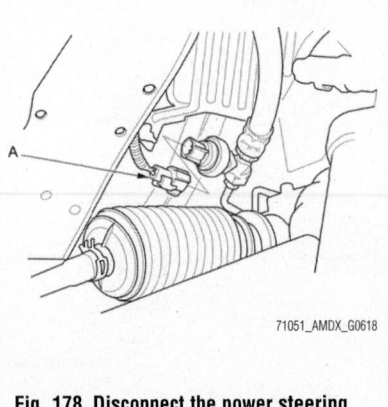

71051_AMDX_G0618

Fig. 178 Disconnect the power steering pressure (PSP) switch connector (A)

• Front Subframe Adapter EQS02BMDXSB0
• Subframe Alignment Pin 070AG-SJAA10S

➡**When you replace the stabilizer bar, lower the front subframe from the body, and replace the front stabilizer bar through the gap created by lowering the front subframe.**

1. Remove the engine cover.
2. Remove the front bulkhead cover.
3. Do the battery terminal disconnection procedure.
4. Raise and support the vehicle.
5. Remove the front wheels.
6. Remove the driver's dashboard undercover.
7. Remove the steering joint cover (A).
8. Loosen the upper steering joint bolt and remove the lower steering joint bolt. Disconnect the steering joint by sliding the steering joint into the column shaft. Tighten

the upper bolt to hold the steering joint in place.

Please note the following:
• If the center guide is in place and has not moved, leave it in place.
• If the center guide has moved or been removed, discard it.

9. Center the steering wheel spokes, and install a commercially available steering wheel holder tool.

10. Disconnect the power steering pressure (PSP) switch connector.

11. Remove the cotter pin from the tie-rod end ball joint, then remove the nut.

➡**During installation, install the new cotter pin after tightening the new nut, and bend its end as shown.**

12. Disconnect the tie-rod end ball joint from the knuckle using the ball joint thread protector and the ball joint remover.

13. Disconnect both sides of the stabilizer link from the stabilizer bar.

14. Remove the power steering pump outlet hose mounting bolt.

15. Remove the air cleaner housing.

16. Remove the connector bracket from the front cylinder head; use the bracket bolt hole to attach the engine balance bar front arm.

17. Remove the engine mount control solenoid valve bracket from the rear cylinder head; use the bracket bolt hole to attach the engine hanger balance bar rear arm.

18. Remove the service caps for the front damper flange nuts from the cowl cover. Position the engine hanger adapter set (VSB02C000031) with the FRONT mark facing forward over the damper flange nuts.

19. Lift and support the engine with the

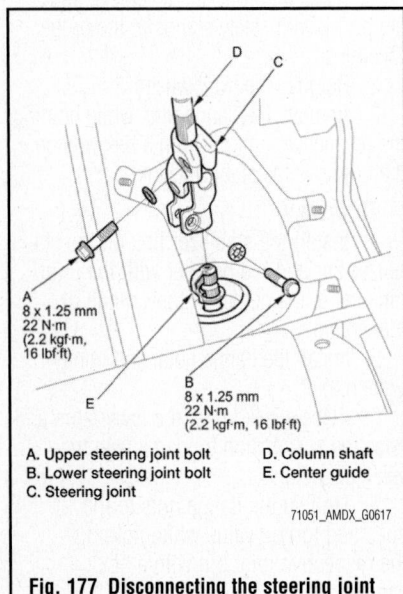

A. Upper steering joint bolt
B. Lower steering joint bolt
C. Steering joint
D. Column shaft
E. Center guide

71051_AMDX_G0617

Fig. 177 Disconnecting the steering joint

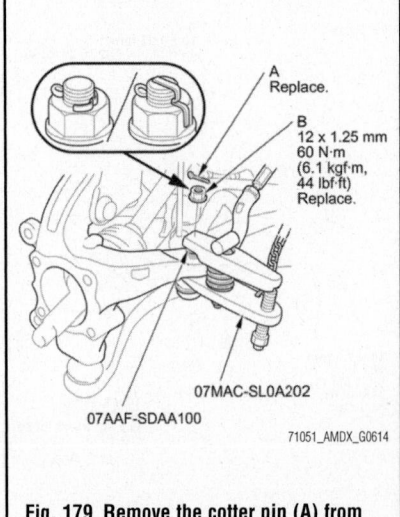

07MAC-SL0A202
07AAF-SDAA100

71051_AMDX_G0614

Fig. 179 Remove the cotter pin (A) from the tie-rod end ball joint, then remove the nut (B)

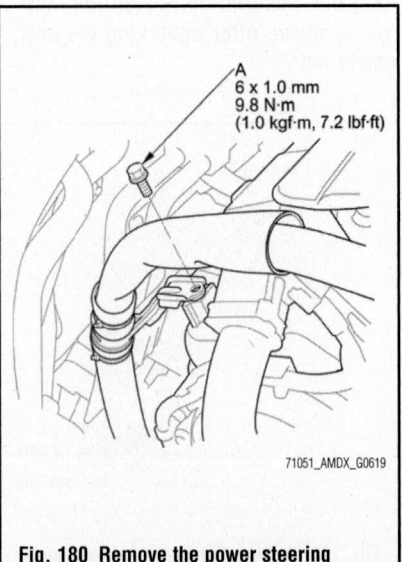

71051_AMDX_G0619

Fig. 180 Remove the power steering pump outlet hose mounting bolt (A)

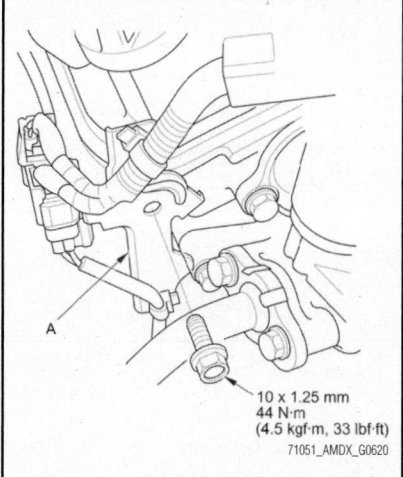

Fig. 181 Remove the connector bracket (A) from the front cylinder head; use the bracket bolt hole to attach the engine balance bar front arm

10 x 1.25 mm
44 N·m
(4.5 kgf·m, 33 lbf·ft)
71051_AMDX_G0620

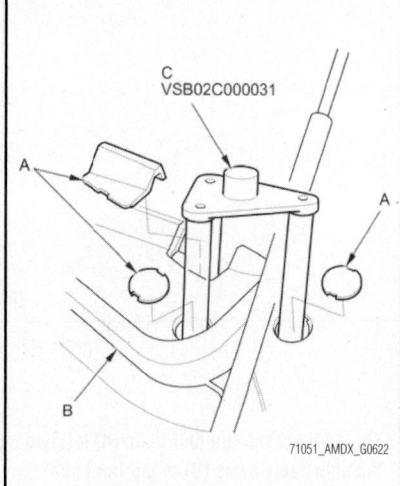

C
VSB02C000031

Fig. 183 Remove the service caps (A) for the front damper flange nuts from the cowl cover (B); engine hanger adapter set (C)

71051_AMDX_G0622

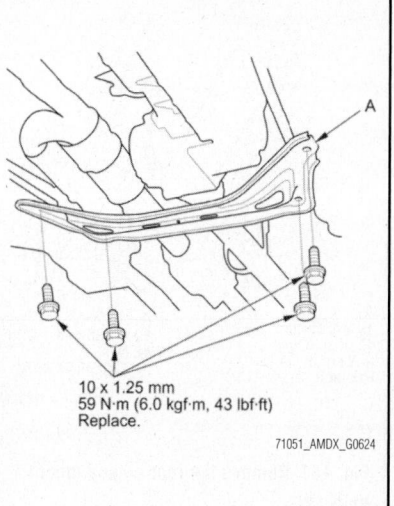

10 x 1.25 mm
59 N·m (6.0 kgf·m, 43 lbf·ft)
Replace.
71051_AMDX_G0624

Fig. 185 Remove the front subframe stiffener plate (A)

engine support hanger (AAR-T1256) and the engine hanger balance bar (VSB02C000019). Attach the front arm to the front cylinder head with a spacer and a 10 mm bolt, and attach the rear arm to the rear cylinder head with an 8 mm bolt. Tighten the wing nut by hand to lift, and support the engine.

Please note the following:
• Be careful when working around the windshield.
• Be careful not to damage the hood opener cable when installing the engine support hanger at the front bulkhead.
• AAR-T1256 two sets required for stacking additional cross section bar.

20. Raise and support the vehicle.

21. Remove the front undercover and the front splash shield.
22. Remove the front subframe stiffener plate.

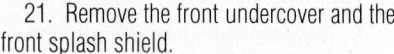

→Use new flange bolts during reassembly.

23. Remove exhaust pipe A.
24. Remove the propeller shaft and propeller shaft protectors.
25. Disconnect the front engine mount control solenoid valve tube, then remove the front engine mount bolts.
26. Remove the rear engine mount bolts.
27. Remove the transmission lower mount bolts.
28. Loosen the four bolts holding the

adjustable arms of the front subframe adapter to its center plate.

29. Line up the slots in the arms with the bolt holes on the corner of the jack base, then attach the front subframe adapter to the jack base with the bolts that came with the jack. Tighten all bolts securely.

30. Raise the jack to vehicle height, then attach the front subframe adapter to the front subframe using the subframe stiffener mounting bolts and bolt holes.

31. Support the front subframe securely by raising the transmission jack.

32. Remove the stiffener mounting bolts on both sides of the front subframe front bracket.

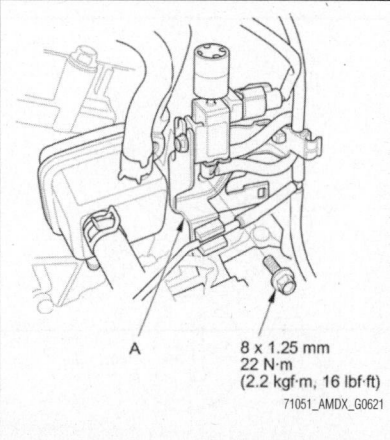

8 x 1.25 mm
22 N·m
(2.2 kgf·m, 16 lbf·ft)
71051_AMDX_G0621

Fig. 182 Remove the engine mount control solenoid valve bracket (A) from the rear cylinder head; use the bracket bolt hole to attach the engine hanger balance bar rear arm

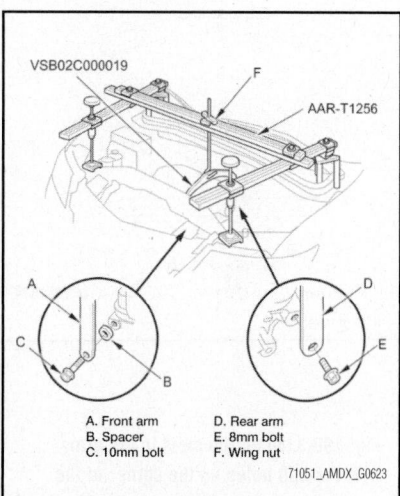

VSB02C000019
F
AAR-T1256

A
C
B
D
E

A. Front arm
B. Spacer
C. 10mm bolt
D. Rear arm
E. 8mm bolt
F. Wing nut

71051_AMDX_G0623

Fig. 184 Lift and support the engine with the engine support hanger and the engine hanger balance bar

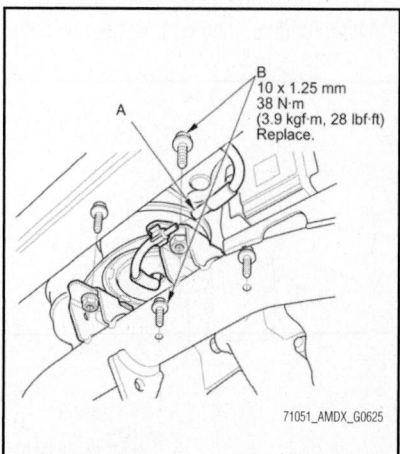

B
10 x 1.25 mm
38 N·m
(3.9 kgf·m, 28 lbf·ft)
Replace.

A

71051_AMDX_G0625

Fig. 186 Disconnect the front engine mount control solenoid valve tube (A), then remove the front engine mount bolts (B)

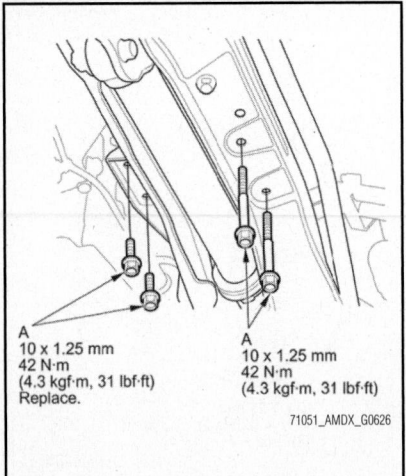

A
10 x 1.25 mm
42 N·m
(4.3 kgf·m, 31 lbf·ft)
Replace.

A
10 x 1.25 mm
42 N·m
(4.3 kgf·m, 31 lbf·ft)

71051_AMDX_G0626

Fig. 187 Remove the rear engine mount bolts (A)

33. Loosen the front side subframe mounting bolts until there is 1.6 inches (40 mm) distance between the bolt seat and the mounting surface. Do not loosen the mounting bolts more than necessary.

34. Remove the stiffener mounting bolts on both sides of the front subframe rear bracket.

35. Loosen the rear side subframe mounting bolts until there is 1.6 inches (40 mm) distance between the bolt seat and the mounting surface. Do not loosen the mounting bolts more than necessary.

36. Lower the transmission jack slowly with the front subframe adapter until the front subframe has dropped about 40 mm (1.6 in).

➡ **Do not try to lower the front sub-frame beyond the loosened subframe mounting bolts clearance.**

37. Remove the flange bolts and the bushing holders, then remove the bushings.

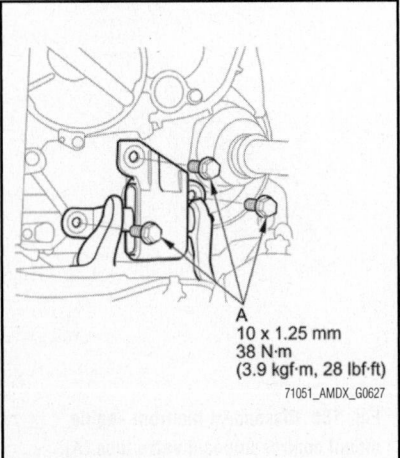

A
10 x 1.25 mm
38 N·m
(3.9 kgf·m, 28 lbf·ft)

71051_AMDX_G0627

Fig. 188 Remove the transmission lower mount bolts (A)

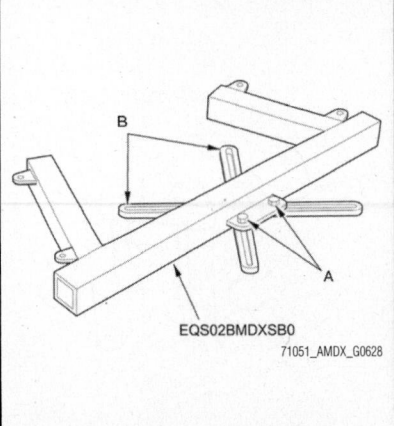

B

A

EQS02BMDXSB0

71051_AMDX_G0628

Fig. 189 Loosen the four bolts (A) holding the adjustable arms (B) of the front sub-frame adapter to its center plate

➡ **During installation, align the paint marks on the stabilizer bar with the side of the bushings.**

38. Move the stabilizer bar toward the passenger's side, and remove the stabilizer bar.

To install:

39. Install the stabilizer bar.
Please note the following:
• Note the right and left direction of the stabilizer bar.
• Note the direction of installation for the bushings.

40. Loosely tighten the new right-rear subframe mounting bolt and the new stiff-ener mounting bolt; insert the subframe

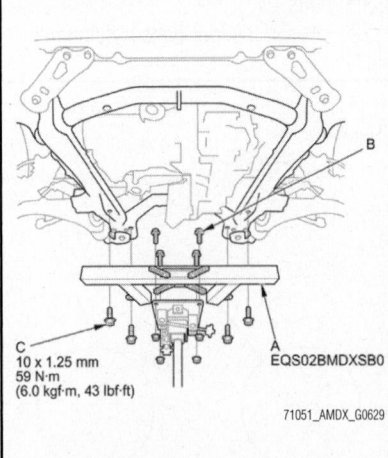

B

C
10 x 1.25 mm
59 N·m
(6.0 kgf·m, 43 lbf·ft)

A

EQS02BMDXSB0

71051_AMDX_G0629

Fig. 190 Line up the slots in the arms with the bolt holes on the corner of the jack base, then attach the front subframe adapter (A) to the jack base with the bolts (B) that came with the jack; subframe stiff-ener mounting bolts (C)

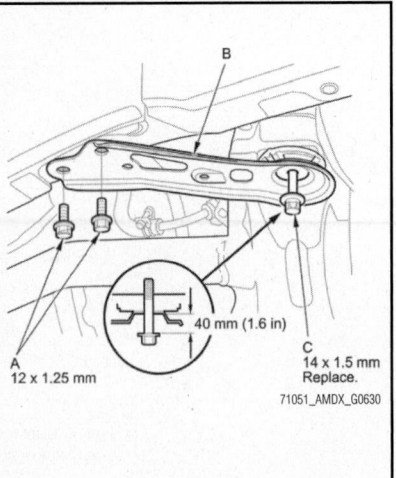

B

40 mm (1.6 in)

A
12 x 1.25 mm

C
14 x 1.5 mm
Replace.

71051_AMDX_G0630

Fig. 191 Remove the stiffener mounting bolts (A) on both sides of the front sub-frame front bracket (B); loosen the front side subframe mounting bolts (C)

alignment pin (070AG-SJAA10S) through the positioning hole on the front subframe rear bracket, through the positioning hole on the subframe, and into the positioning hole on the body, then tighten the subframe mounting bolt.

41. Loosely tighten the new left-rear subframe mounting bolt using the same procedure as the right-rear with the sub-frame alignment pin.

42. Loosely tighten the new subframe mounting bolt and the stiffener mounting bolts on both sides to the front subframe front bracket.

43. Tighten the subframe mounting bolts to the specified torque.

44. Tighten the front and rear stiffener mounting bolts to the specified torque.

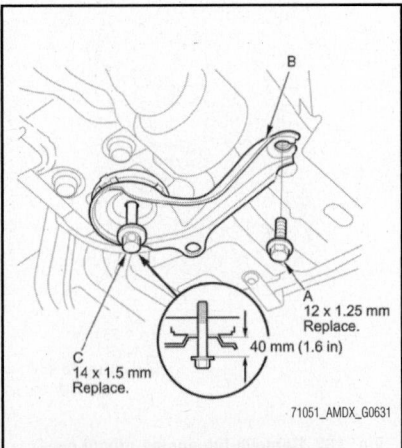

B

A
12 x 1.25 mm
Replace.

40 mm (1.6 in)

C
14 x 1.5 mm
Replace.

71051_AMDX_G0631

Fig. 192 Remove the stiffener mounting bolts (A) on both sides of the front sub-frame rear bracket (B); loosen the rear side subframe mounting bolts (C)

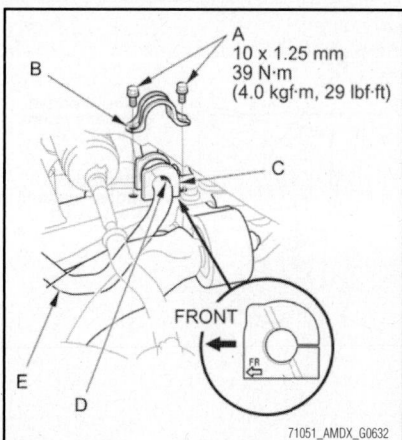

Fig. 193 Remove the flange bolts and the bushing holders, then remove the bushings

➡**Before tightening the stiffener mounting bolts, check that the positioning holes and slot are aligned using the subframe alignment pin.**

45. Install all of the removed parts in the reverse order of removal, and note these items:

   a. If the center guide is in place, use it to determine the steering joint installation angle.

   b. If the center guide is gone, check the steering joint installation angle.

   c. When connecting the front and rear engine mount, first lightly tighten the mounting bolts, supporting the engine/transmission with a floor jack, then remove the jack, and tighten the bolts to the specified torque.

   d. Before installing the wheel, clean the mating surfaces between the brake disc and inside of the wheel.

46. Do the battery terminal reconnection procedure, then turn the ignition switch to

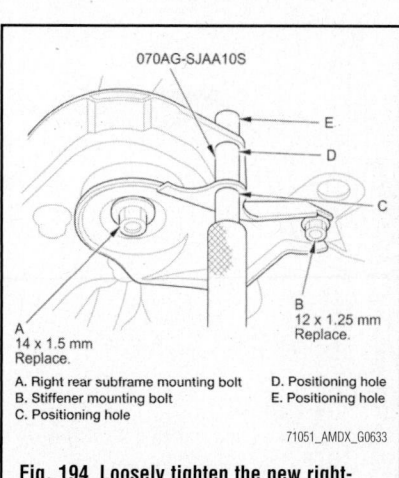

Fig. 194 Loosely tighten the new right-rear subframe mounting bolt and the new stiffener mounting bolt

ON (II) and check that the SRS indicator should come on for about 6 seconds and then go off.

47. Check the wheel alignment, and adjust it if necessary.

STRUTS

REMOVAL & INSTALLATION

See Figures 195 through 199.

➡**Except where called out, the illustrations show without active damper system.**

1. Raise the front of the vehicle, and support it with safety stands in the proper locations.

2. Remove the front wheel.

3. With active damper system: Disconnect the damper coil connector, and remove the flange bolt, and the harness clip.

➡**Be careful not to damage or contaminate the damper coil connector.**

4. Remove the wheel speed sensor harness and the brake hose from the damper. Do not disconnect the wheel speed sensor connector.

5. Remove the flange nut, while holding the joint pin with a hex wrench, and disconnect the stabilizer link from the damper.

6. Remove the damper pinch bolts and flange nuts from the damper.

7. Remove the cover and the service caps.

8. Remove the flange nuts from the top of the damper.

9. Remove the damper assembly.

➡**The damper springs are different, left and right. Mark the springs L and R before you continue.**

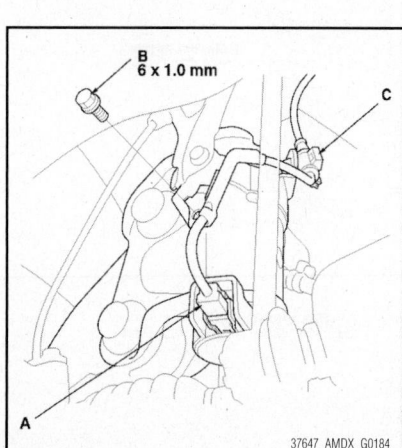

Fig. 195 Disconnect the damper coil connector (A), and remove the flange bolt (B), and the harness clip (C)

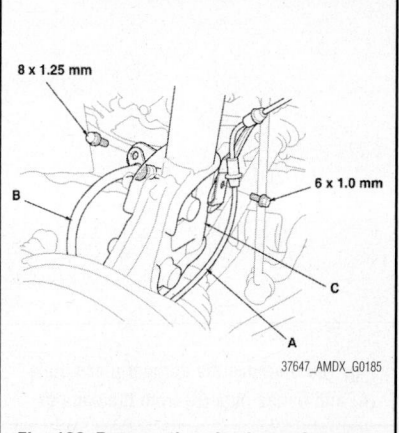

Fig. 196 Remove the wheel speed sensor harness (A) and the brake hose (B) from the damper (C)

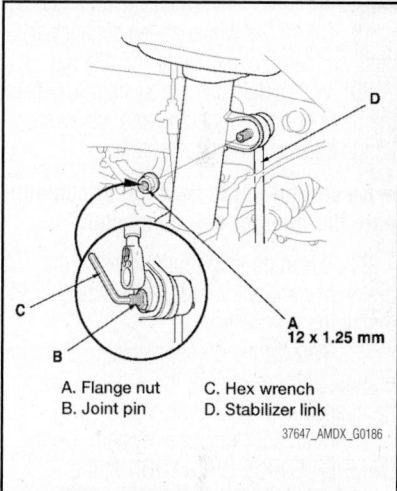

A. Flange nut   C. Hex wrench
B. Joint pin   D. Stabilizer link

Fig. 197 Remove the flange nut, while holding the joint pin with a hex wrench, and disconnect the stabilizer link from the damper

To install:

10. Install the damper assembly on to the frame. Note the direction of the damper mounting base.

➡**Be careful not to damage the body.**

11. Loosely install the new flange nuts to the top of the damper.

12. Loosely install new damper pinch bolts and new flange nuts to the damper.

13. Connect the stabilizer link to the damper, and loosely install a new flange nut.

14. Raise the front suspension with a floor jack to load the suspension with the vehicle's weight.

15. Tighten the flange nut to 58 ft. lbs. (78 N m), while holding the joint pin with the hex wrench

16. Tighten the flange nuts on top of the damper to 43 ft. lbs. (59 N m).

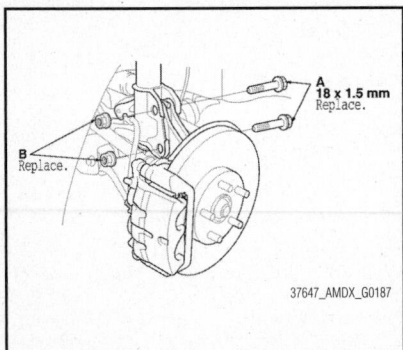

Fig. 198 Remove the damper pinch bolts (A) and flange nuts (B) from the damper

17. Tighten the damper pinch bolts and the flange nuts to 159 ft. lbs. (211 N m).
18. Install the cover and the service caps.
19. Install the wheel speed sensor harness and the brake hose to the damper.
20. With active damper system: Connect the damper coil connector, and install the flange bolt, and the harness clip.

➡**Be careful not to damage or contaminate the damper coil connecter.**

21. Clean the mating surfaces of the brake disc and the inside of the wheel, then install the front wheel.
22. With active damper system: Start the engine, then make sure there are no active damper system DTCs with the HDS.
23. With active damper system: Do the DAMPER FORCE OPERATION in the ACTIVE DAMPER SYSTEM INSPECTION MENU with the HDS, then make sure the all four damper units function normally.

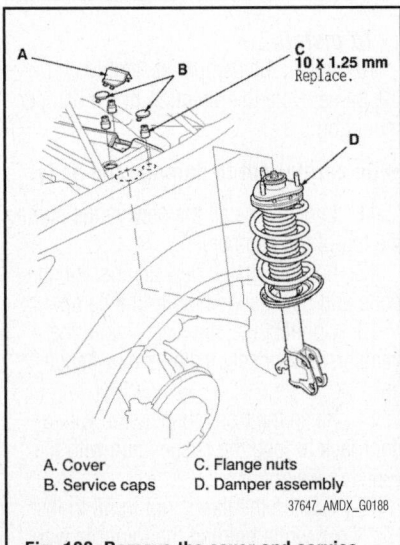

A. Cover
B. Service caps
C. Flange nuts
D. Damper assembly

Fig. 199 Remove the cover and service caps, the flange nuts and the damper assembly

24. Check the wheel alignment, and adjust it if necessary.

OVERHAUL

➡When compressing the damper spring, use a commercially available strut spring compressor (Branick MST-580A or Model 7200, or equivalent) according to the manufacturer's instructions.

➡Except where called out, the illustrations are shown without active damper system.

Disassembly

See Figures 200 through 202.

1. Compress the damper spring, then remove the self-locking nut while holding the damper shaft with a hex wrench or

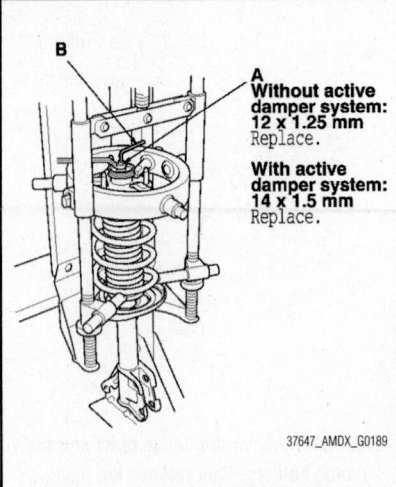

Fig. 200 Compress the damper spring, then remove the self-locking nut (A) while holding the damper shaft with a hex wrench or TORX® wrench (B)

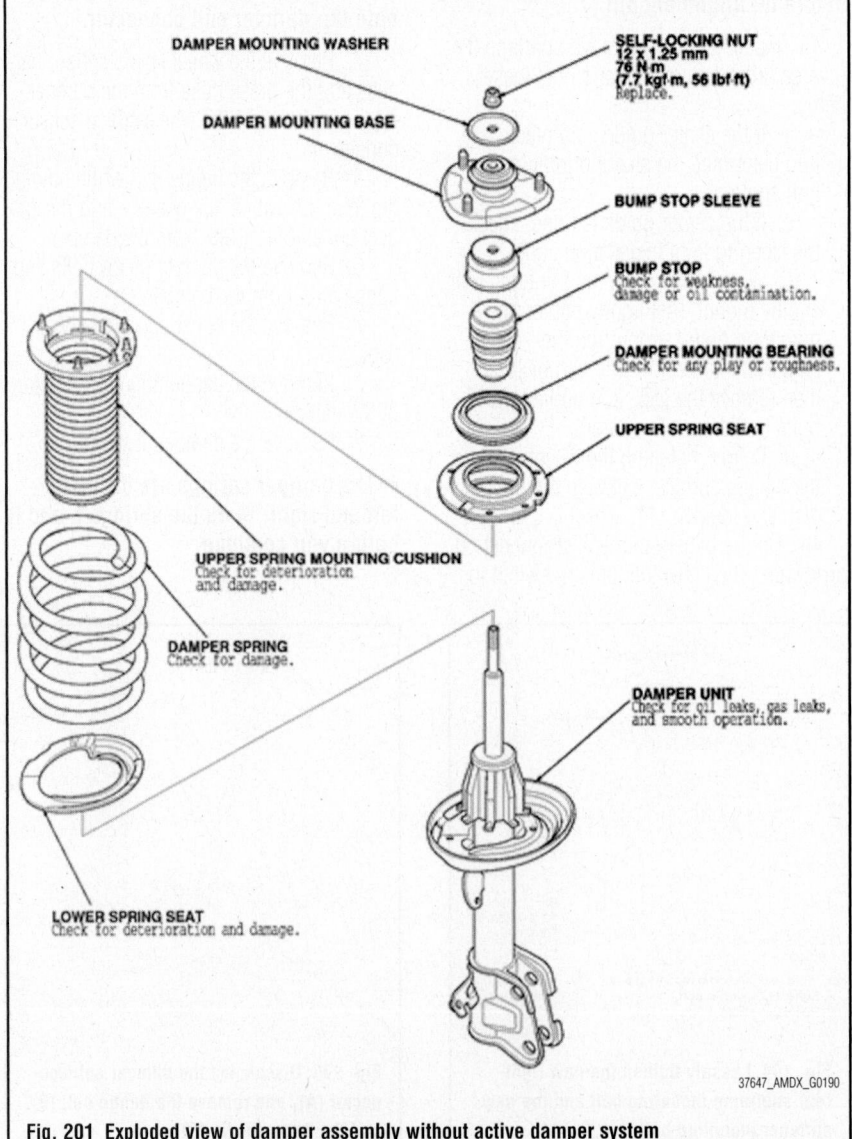

DAMPER MOUNTING WASHER

DAMPER MOUNTING BASE

SELF-LOCKING NUT
12 x 1.25 mm
76 N m
(7.7 kgf m, 56 lbf ft)
Replace.

BUMP STOP SLEEVE

BUMP STOP
Check for weakness,
damage or oil contamination.

DAMPER MOUNTING BEARING
Check for any play or roughness.

UPPER SPRING SEAT

UPPER SPRING MOUNTING CUSHION
Check for deterioration
and damage.

DAMPER SPRING
Check for damage.

DAMPER UNIT
Check for oil leaks, gas leaks,
and smooth operation.

LOWER SPRING SEAT
Check for deterioration and damage.

Fig. 201 Exploded view of damper assembly without active damper system

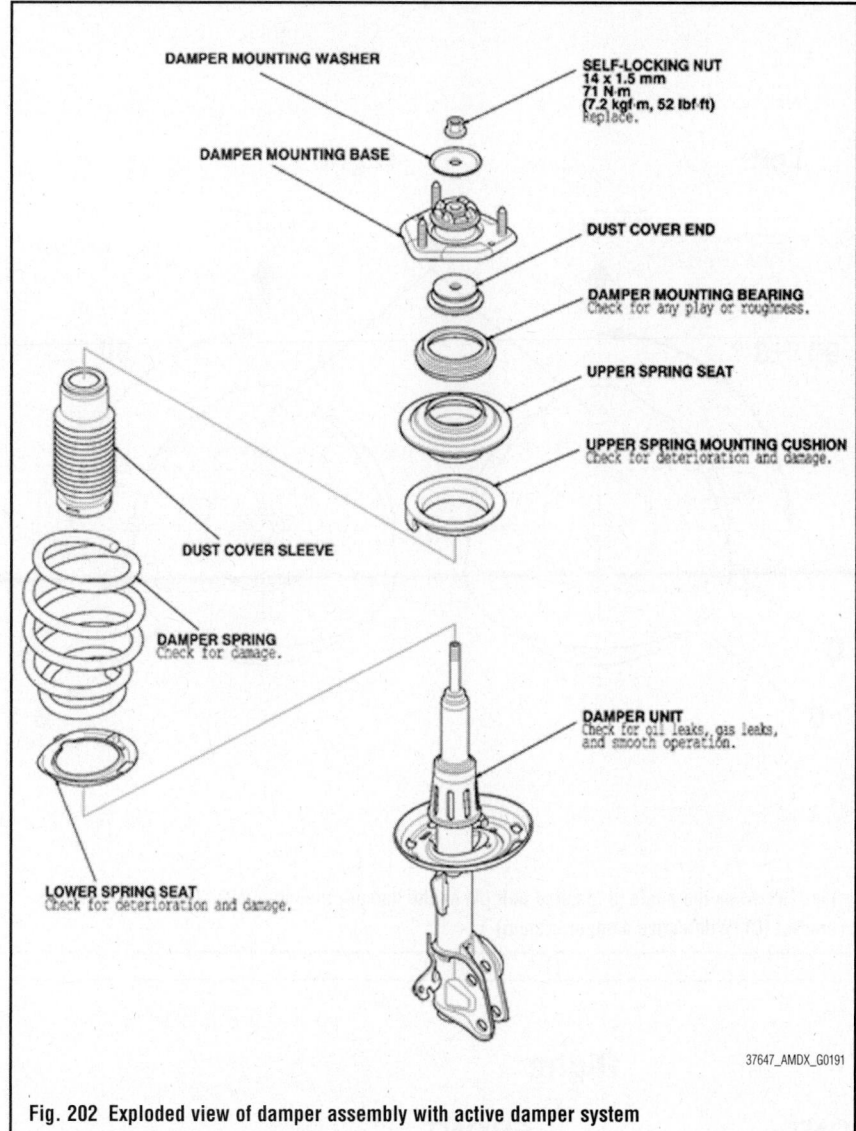

Fig. 202 Exploded view of damper assembly with active damper system

A. Damper spring   C. Upper end
B. Upper spring mounting cushion   D. Cushion stop

37647_AMDX_G0192

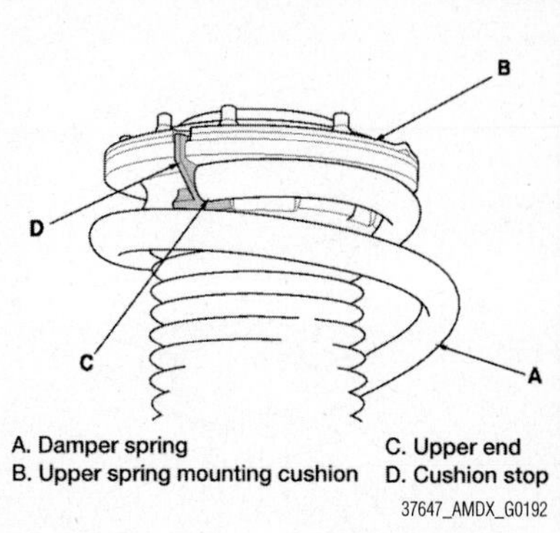

Fig. 203 Install the damper spring on the upper spring mounting cushion, then align the upper end of the damper spring with the cushion stop (Without active damper system)

A. Damper spring   C. Upper end
B. Upper spring mounting cushion   E. Cushion end

37647_AMDX_G0193

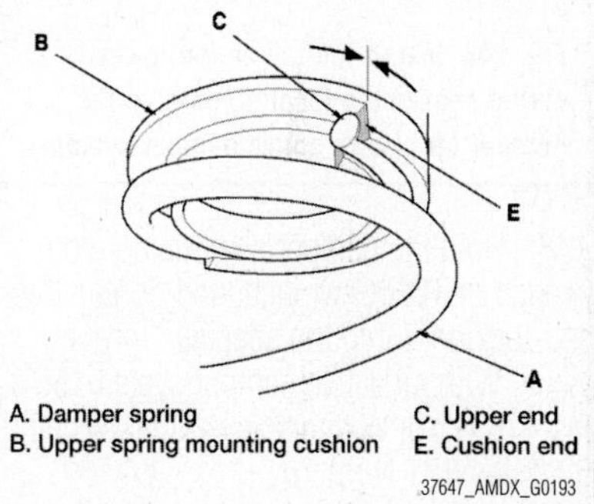

Fig. 204 Install the damper spring on the upper spring mounting cushion, then align the upper end of the damper spring with the cushion end (With active damper system)

TORX® wrench. Do not compress the damper spring more than necessary to remove the nut.

2. Release the pressure from the strut spring compressor, then disassemble the damper as shown in the Exploded View.

Inspection

1. Reassemble the damper mounting base, the damper mounting washer, and the self-locking nut.

2. Compress the damper assembly by hand, and check for smooth operation through a full stroke, both compression and extension. The damper should extend smoothly and constantly when compression is released. If it does not, the gas is leaking and the damper should be replaced.

3. Check for oil leaks, abnormal noises, and binding during these tests.

Reassembly

See Figures 203 through 209.

1. Install the damper spring on the upper spring mounting cushion, then align the upper end of the damper spring with the cushion stop or the cushion end.

2. Compress the damper spring.

3. Install the tab on the lower spring seat in the locating hole on the damper unit.

4. Align the lower end of the damper spring with the stepped part of the lower spring seat.

5. Install all the parts except the damper mounting washer and self-locking nut onto the damper unit by referring to the Exploded View.

6. Align the angle of the stud bolt on the damper mounting base and the damper bracket as shown.

7. Install the damper mounting washer and a new self-locking nut.

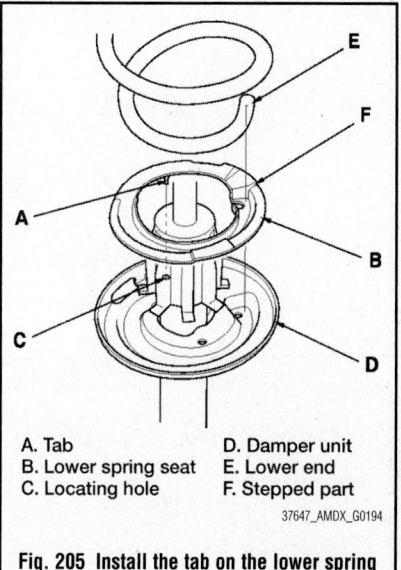

A. Tab   D. Damper unit
B. Lower spring seat   E. Lower end
C. Locating hole   F. Stepped part

37647_AMDX_G0194

Fig. 205 Install the tab on the lower spring seat in the locating hole on the damper unit (Without active damper system)

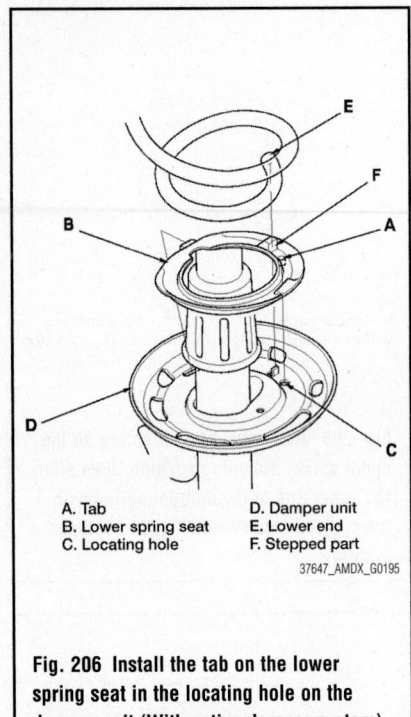

A. Tab
B. Lower spring seat
C. Locating hole
D. Damper unit
E. Lower end
F. Stepped part

37647_AMDX_G0195

Fig. 206 Install the tab on the lower spring seat in the locating hole on the damper unit (With active damper system)

8. Hold the damper shaft using a hex wrench or TORX® wrench, and tighten the self-locking nut to the specified torque value; Without active damper system: 56 ft. lbs. (76 N m); With active damper system: 52 ft. lbs. (71 N m).

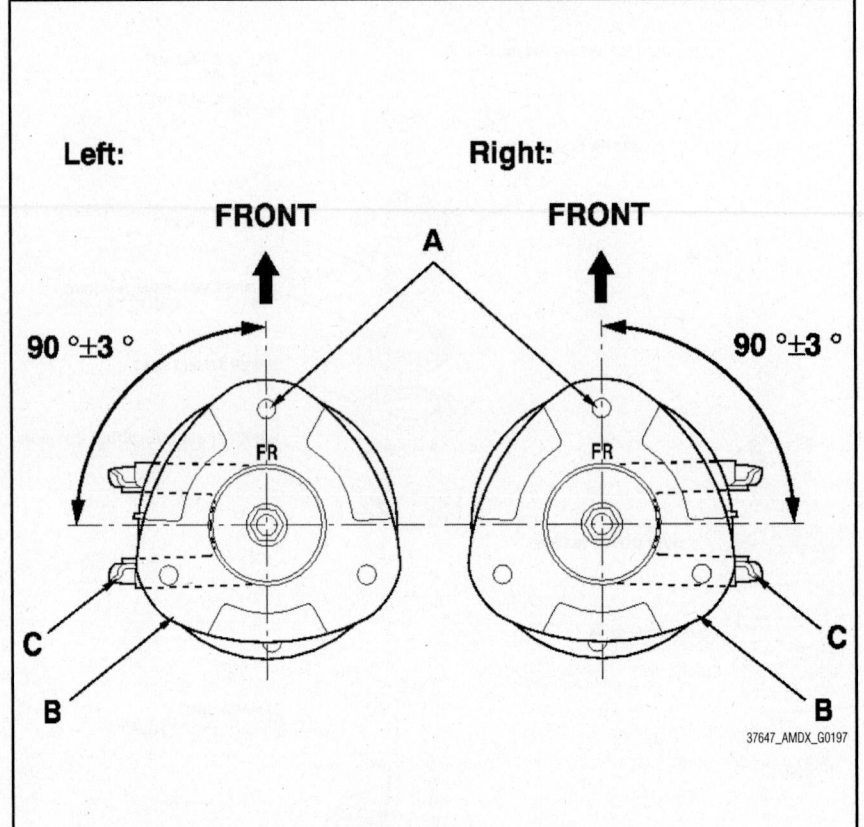

37647_AMDX_G0197

Fig. 208 Align the angle of the stud bolt (A) on the damper mounting base (B) and the damper bracket (C) (With active damper system)

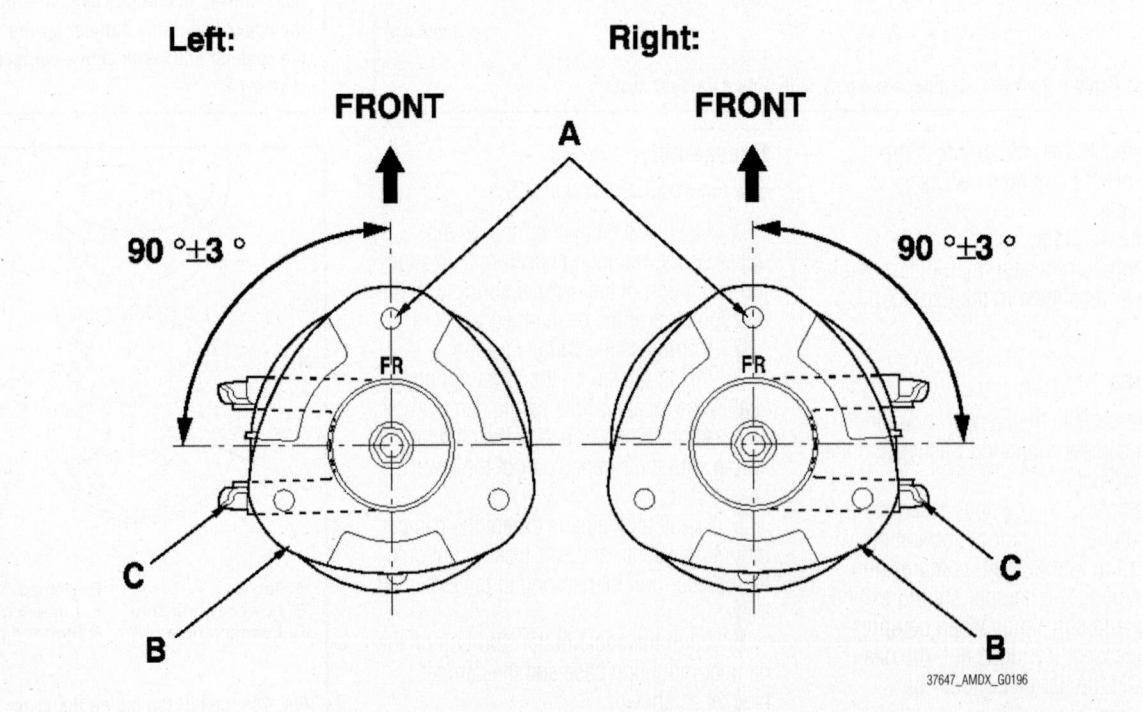

37647_AMDX_G0196

Fig. 207 Align the angle of the stud bolt (A) on the damper mounting base (B) and the damper bracket (C) (Without active damper system)

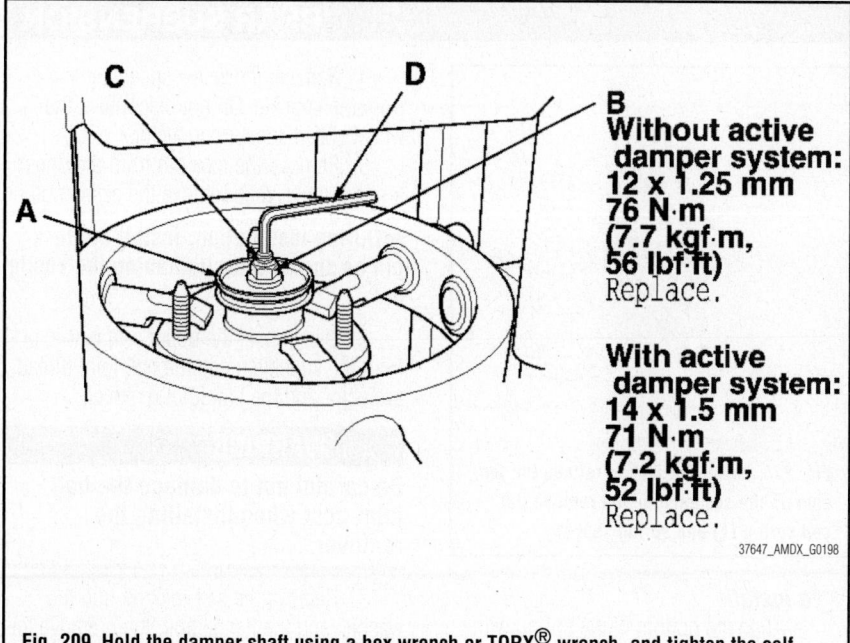

B
**Without active
damper system:
12 x 1.25 mm
76 N·m
(7.7 kgf·m,
56 lbf·ft)**
Replace.

**With active
damper system:
14 x 1.5 mm
71 N·m
(7.2 kgf·m,
52 lbf·ft)**
Replace.

37647_AMDX_G0198

Fig. 209 Hold the damper shaft using a hex wrench or TORX® wrench, and tighten the self-locking nut

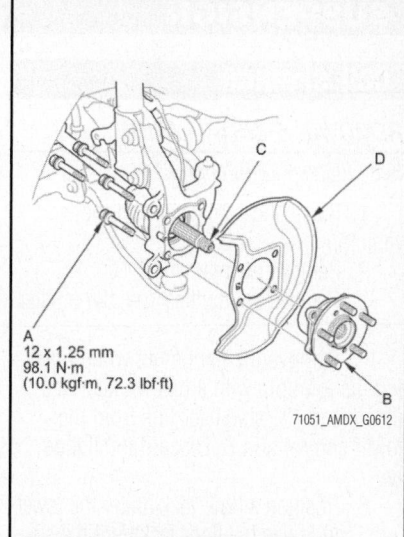

A
12 x 1.25 mm
98.1 N·m
(10.0 kgf·m, 72.3 lbf·ft)

71051_AMDX_G0612

Fig. 211 Remove the flange bolts, and remove the hub bearing unit by tapping the driveshaft end with a soft face hammer while drawing the hub bearing unit outward, then remove the splash guard

WHEEL HUBS & BEARINGS

REMOVAL & INSTALLATION

See Figures 210 and 211.

1. Raise and support the vehicle.
2. Remove the wheel nuts and the front wheel.
3. Remove the brake hose mounting bolt from the damper.
4. Remove the brake caliper bracket mounting bolts, then remove the caliper assembly from the knuckle. To prevent damage to the caliper assembly or brake hose, use a short piece of wire to hang the caliper assembly from the undercarriage. Do not twist the brake hose excessively.
5. Pry up the stake on the spindle nut, then remove the nut.
6. Remove the front brake disc.
7. Remove the flange bolts, and remove the hub bearing unit by tapping the driveshaft end with a soft face hammer while drawing the hub bearing unit outward, then remove the splash guard.

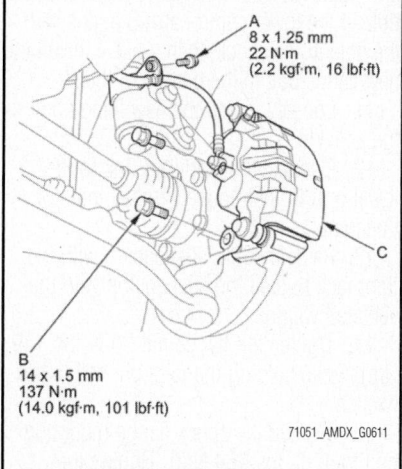

A
8 x 1.25 mm
22 N·m
(2.2 kgf·m, 16 lbf·ft)

C

B
14 x 1.5 mm
137 N·m
(14.0 kgf·m, 101 lbf·ft)

71051_AMDX_G0611

Fig. 210 Remove the brake hose mounting bolt (A) from the damper, the brake caliper bracket mounting bolts (B), then remove the caliper assembly (C)

Please note the following:
• Do not pull the driveshaft end outward. The driveshaft inboard joint may come apart.

• During installation, apply grease to the mating surfaces of the hub bearing unit and driveshaft outboard joint.
8. Check the hub bearing unit for damage and cracks.

To install:
9. Install the hub bearing unit in the reverse order of removal, and note these items:
   a. Use the new spindle nut during reassembly.
   b. Before installing the spindle nut, apply a small amount of engine oil to the seating surface of the nut. After tightening, use a drift to stake the spindle nut shoulder against the driveshaft.
   c. Before installing the brake disc, clean the mating surfaces between the hub bearing unit and the inside of the brake disc.
   d. Before installing the wheel, clean the mating surfaces between the brake disc and the inside of the wheel.
10. Check the wheel alignment, and adjust it if necessary.

COIL SPRINGS

REMOVAL & INSTALLATION

See Figures 212 through 214.

1. Raise and safely support the vehicle.
2. Remove the rear wheel.
3. Remove the muffler from the muffler hanger.
4. Remove the flange nut while holding the joint pin with a hex wrench, and disconnect the stabilizer link from the lower control arm B. Discard the flange nut.
5. Position a floor jack under the lower arm B and raise the floor jack until the suspension begins to compress.
6. Remove the flange bolt from the bottom of the damper.
7. Remove the flange bolt from the knuckle.
8. Lower the floor jack gradually and remove the spring and spring seat.

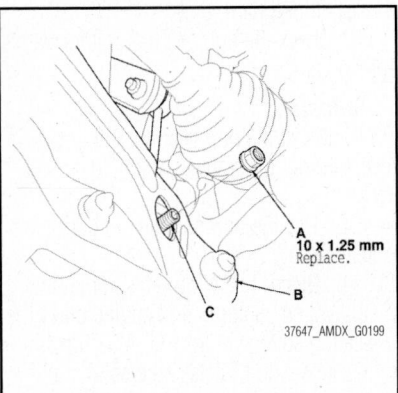

Fig. 212 Remove the flange nut (A) while holding the joint pin (C) with a hex wrench, and disconnect the stabilizer link from the lower arm B

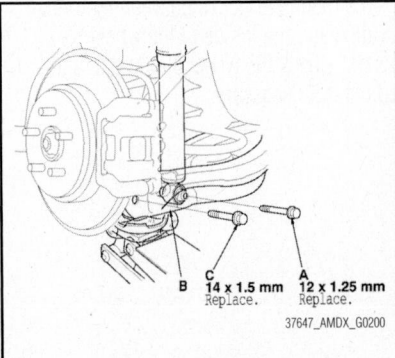

Fig. 213 Remove the flange bolt (A) from the bottom of the damper

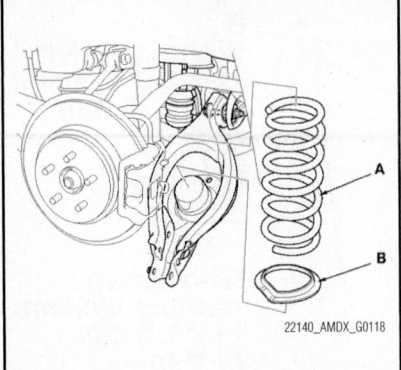

Fig. 214 Lower the jack to relieve the tension on the suspension and remove the coil spring (A) and spring seat (B)

To install:

9. Align the bottom of the spring with the stepped part of the spring seat, and install into the control arm.
10. With the jack under the control arm, raise the floor jack until the mounting hole in the lower control arm B aligns with the hole in the shock, then loosely install the new flange bolt to the bottom of the shock.
11. Loosely install the new flange bolt to the knuckle.
12. Connect the stabilizer link to the lower arm B, then loosely install the new flange nut.
13. Raise the rear suspension with the floor jack to load the suspension with the vehicle's weight.
14. Tighten the flange nut 36 ft. lbs. (49 Nm), while holding the joint pin with a hex wrench.
15. Tighten the shock flange mounting bolt to 47 ft. lbs. (64 Nm). Tighten the knuckle flange mounting bolt to 83 ft. lbs. (113 Nm).
16. Install the muffler to the muffler hanger.
17. Install the rear wheel.

KNUCKLE

COMPONENT LOCATIONS

See Figure 215.

REMOVAL & INSTALLATION

See Figures 216 through 221.

1. Remove the hub bearing unit.
2. Remove the brake hose bracket mounting bolt from the knuckle.
3. Remove the flange nuts, then remove the backing plate with the parking brake shoe assembly.

4. Remove the wheel speed sensor from the knuckle. Do not disconnect the wheel speed sensor connector.
5. Remove the lock pin from the upper arm ball joint, then remove the castle nut.

➡**During installation, install the lock pin as shown after tightening the castle nut.**

6. Disconnect the upper arm ball joint from the knuckle using the ball joint thread protector and the ball joint remover.

✳✳ WARNING

Be careful not to damage the ball joint boot when installing the remover.

7. Remove the self-locking nut, the washer, and the flange bolt, then remove the lower arm.
   Please note the following:
   • During installation, position the paint mark on the lower arm A toward outside of the vehicle.
   • Use a new self-locking nut and a new flange bolt, during reassembly.
8. Remove the flange nuts from the trailing arm.

➡**Use new flange nuts during reassembly.**

9. Place a floor jack under lower arm, and remove the flange bolt, then remove the knuckle.

➡**Use the new flange bolt during reassembly.**

To install:

10. Install the knuckle in the reverse order of removal, and note these items:
   a. First install all of the components, and lightly tighten the bolts and nuts, then raise the suspension to load it with the vehicle's weight before fully tightening to the specified torque.
   b. Be careful not to damage the ball joint boot when connecting the knuckle.
   c. Before connecting the upper arm ball joint to the knuckle, degrease the threaded section and tapered portion of the ball joint pin, the ball joint connecting hole, the threaded section and mating surfaces of the castle nut.
   d. Torque the castle nut to the lower torque specification, then tighten it only far enough to align the slot with the ball joint pin hole. Do not align the castle nut by loosening it.

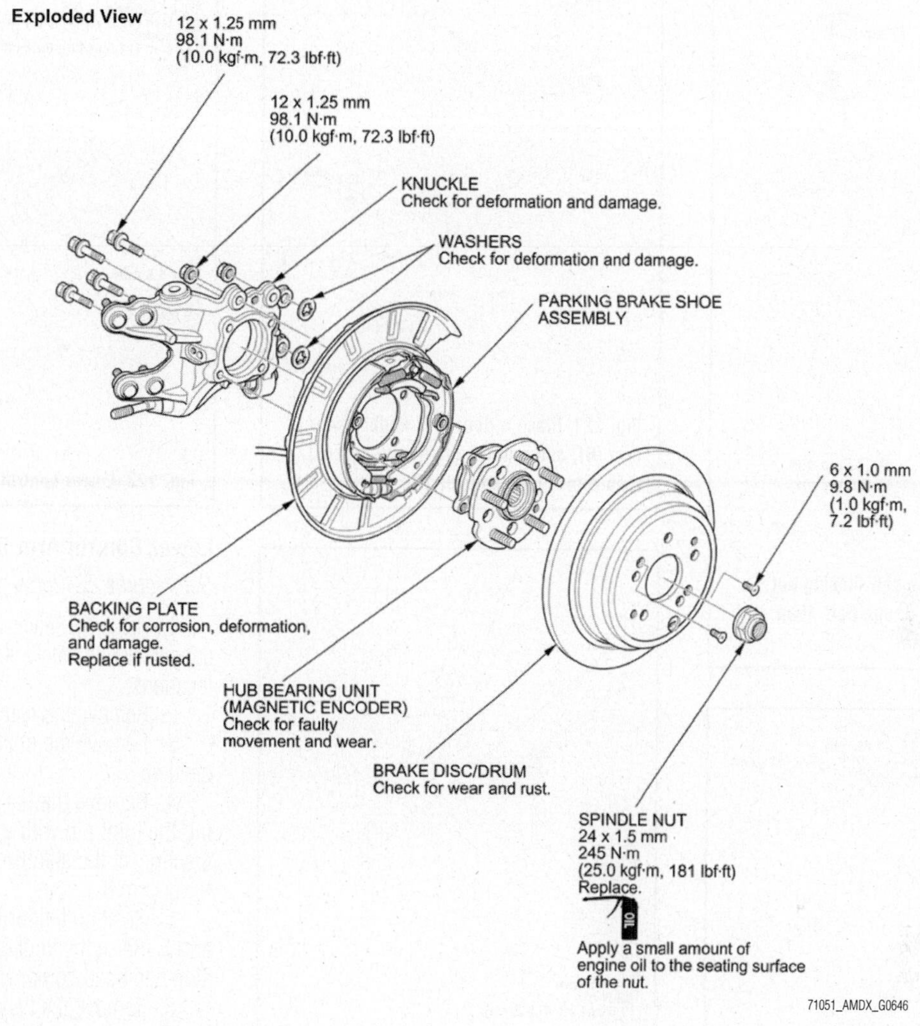

Exploded View

12 x 1.25 mm
98.1 N·m
(10.0 kgf·m, 72.3 lbf·ft)

12 x 1.25 mm
98.1 N·m
(10.0 kgf·m, 72.3 lbf·ft)

KNUCKLE
Check for deformation and damage.

WASHERS
Check for deformation and damage.

PARKING BRAKE SHOE
ASSEMBLY

6 x 1.0 mm
9.8 N·m
(1.0 kgf·m,
7.2 lbf·ft)

BACKING PLATE
Check for corrosion, deformation,
and damage.
Replace if rusted.

HUB BEARING UNIT
(MAGNETIC ENCODER)
Check for faulty
movement and wear.

BRAKE DISC/DRUM
Check for wear and rust.

SPINDLE NUT
24 x 1.5 mm
245 N·m
(25.0 kgf·m, 181 lbf·ft)
Replace.

Apply a small amount of
engine oil to the seating surface
of the nut.

71051_AMDX_G0646

Fig. 215 Exploded view of the rear knuckle and hub assembly

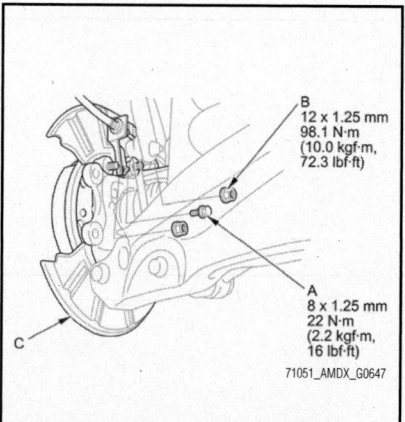

B
12 x 1.25 mm
98.1 N·m
(10.0 kgf·m,
72.3 lbf·ft)

A
8 x 1.25 mm
22 N·m
(2.2 kgf·m,
16 lbf·ft)

C

71051_AMDX_G0647

**Fig. 216 Remove the brake hose bracket
mounting bolt (A) from the knuckle,
remove the flange nuts (B), then remove
the backing plate (C)**

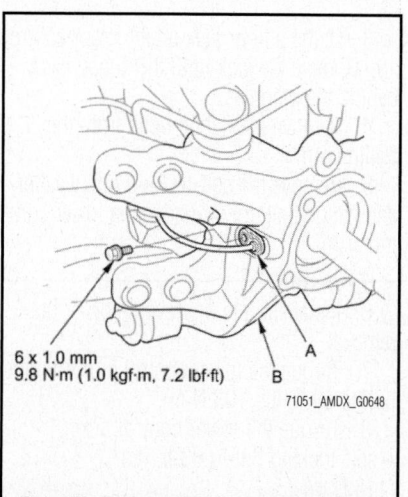

6 x 1.0 mm
9.8 N·m (1.0 kgf·m, 7.2 lbf·ft)

A

B

71051_AMDX_G0648

**Fig. 217 Remove the wheel speed sensor
(A) from the knuckle (B)**

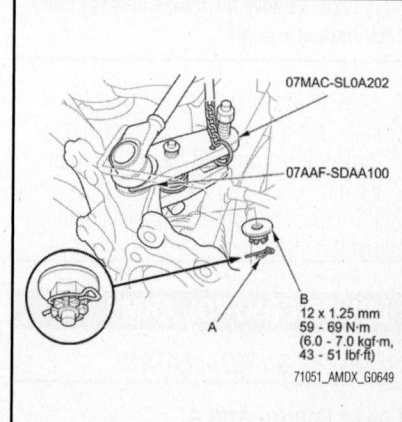

07MAC-SL0A202

07AAF-SDAA100

A

B
12 x 1.25 mm
59 - 69 N·m
(6.0 - 7.0 kgf·m,
43 - 51 lbf·ft)

71051_AMDX_G0649

**Fig. 218 Remove the lock pin (A) from the
upper arm ball joint, then remove the castle nut (B)**

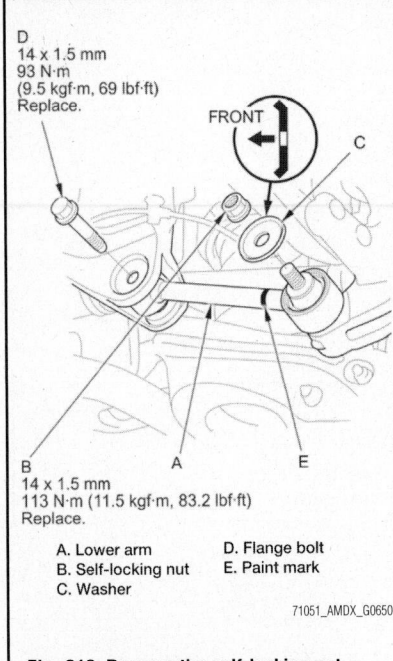

A. Lower arm
B. Self-locking nut
C. Washer
D. Flange bolt
E. Paint mark

71051_AMDX_G0650

Fig. 219 Remove the self-locking nut, the washer, and the flange bolt, then remove the lower arm

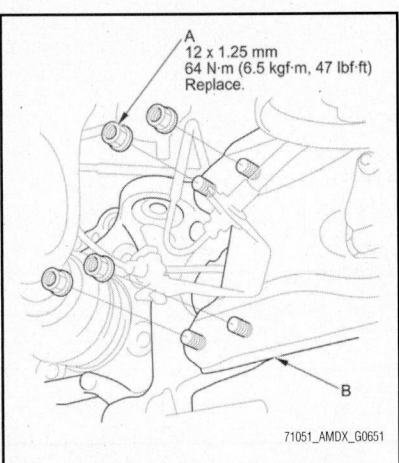

Fig. 220 Remove the flange nuts (A) from the trailing arm (B)

e. Before installing the wheel, clean the mating surfaces between the brake disc/drum and the inside of the wheel.

11. Check the wheel alignment, and adjust it if necessary.

LOWER CONTROL ARMS

REMOVAL & INSTALLATION

Lower Control Arm A

See Figures 222 and 223.

1. Raise and safely support the vehicle.
2. Remove the rear wheel.

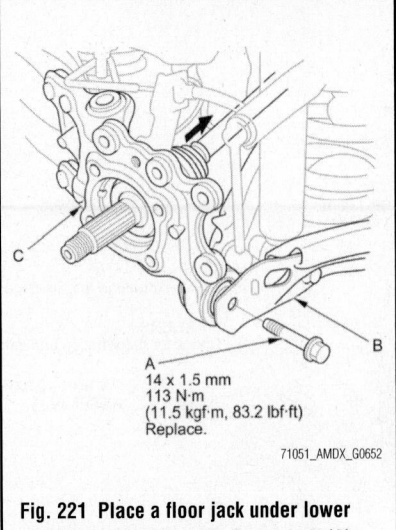

71051_AMDX_G0652

Fig. 221 Place a floor jack under lower arm (B), and remove the flange bolt (A), then remove the knuckle (C)

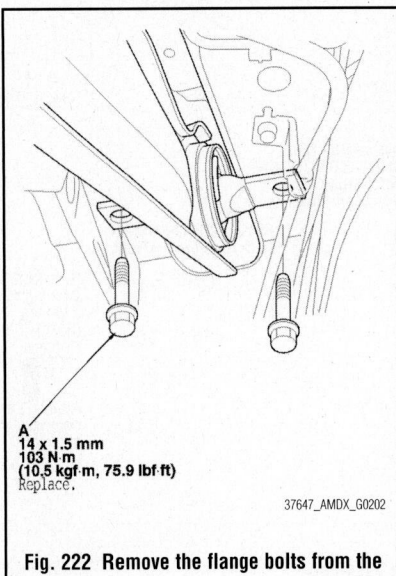

37647_AMDX_G0202

Fig. 222 Remove the flange bolts from the trailing arm

3. Using a floor jack under lower control arm B, raise the jack until the suspension begins to compress.
4. Remove the flange bolts from the trailing arm.
5. Remove the self-locking nut, washer, and flange bolt, then remove the lower control arm.

To install:

6. Installation is the reverse order of removal.
7. Torque the lower control arm A flange bolt to 69 ft. lbs. (93 N m).
8. Torque the lower control arm A self-locking nut to 83 ft. lbs (113 N m).
9. Torque the trailing arm flange bolts to 76 ft. lbs. (103 N m).

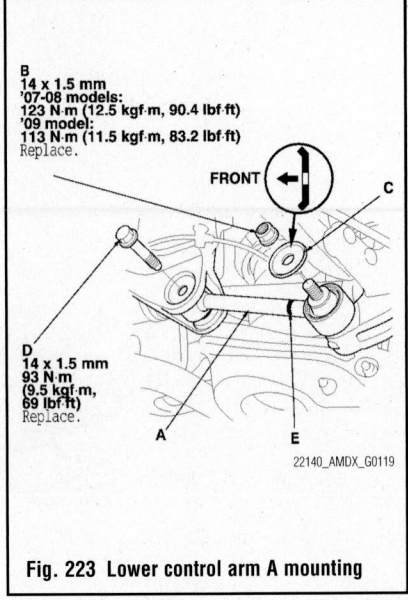

22140_AMDX_G0119

Fig. 223 Lower control arm A mounting

Lower Control Arm B

See Figures 224 through 226.

1. Raise the rear of the vehicle, and support it with safety stands in the proper locations.
2. Remove the rear wheel.
3. Remove the muffler from the muffler hangers.
4. Remove the flange nut while holding the joint pin with a hex wrench, and disconnect the stabilizer link from the lower arm B.
5. Position a floor jack under the lower arm B. Raise the floor jack until the suspension begins to compress.
6. Remove the flange bolt from the bottom of the damper.
7. Remove the flange bolt from the knuckle.
8. Lower the floor jack gradually.

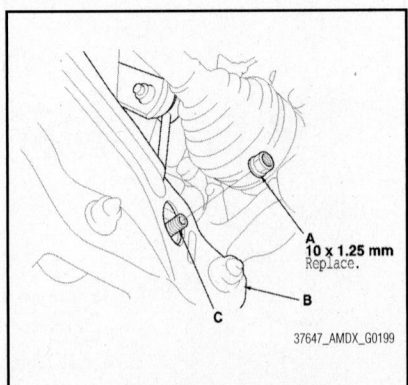

37647_AMDX_G0199

Fig. 224 Remove the flange nut (A) while holding the joint pin (C) with a hex wrench, and disconnect the stabilizer link from the lower arm B

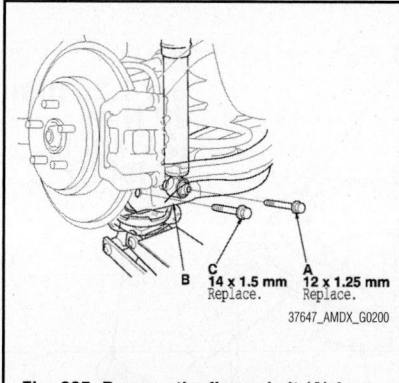

Fig. 225 Remove the flange bolt (A) from the bottom of the damper

9. Remove the spring and the lower spring seat.

➡**During installation, align the bottom of the spring with the stepped part of the lower spring seat and the lower arm B as shown.**

10. Mark the cam positions of the adjusting bolt and adjusting cam plate, then remove the self-locking nut, adjusting cam plate, and adjusting bolt.

11. Remove the lower arm B.

To install:

12. Install the lower arm B in the reverse order of removal, and note these items:
- Replace all bolts and nuts with new during installation.

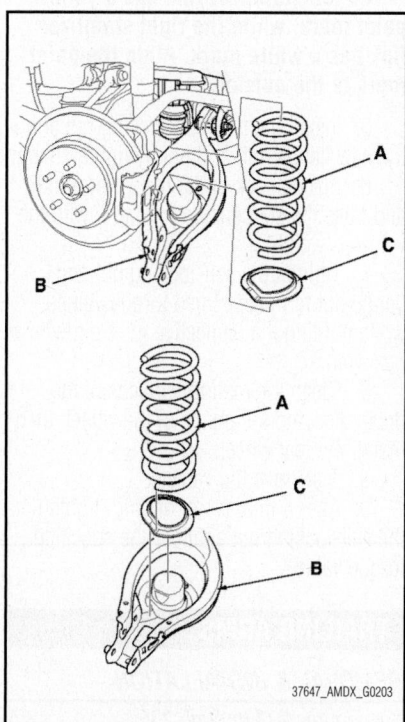

Fig. 226 Remove the spring (A) and the lower spring seat (C)

- First install all the suspension components, and lightly tighten the bolts and nuts, then raise the suspension to load it with the vehicle's weight before fully tightening to the specified torque values.
- Align the cam positions of the adjusting bolt and the adjusting cam plate with the marked positions when tightening.
- Before installing the wheel, clean the mating surfaces of the brake disc and the inside of the wheel.
- Check the wheel alignment, and adjust it if necessary.

13. Torque the self-locking nut to 98 ft. lbs. (132 N m).

14. Torque the flange bolt to knuckle to 83 ft. lbs. (113 N m).

15. Torque the lower damper flange bolt to 47 ft. lbs. (64 N m).

16. Torque the stabilizer link flange nut to 36 ft. lbs. (49 N m).

SHOCK ABSORBERS

REMOVAL & INSTALLATION

See Figures 227 through 229.

1. Raise the rear of the vehicle, and support it with safety stands in the proper locations.

2. Remove the rear wheel.

3. With active damper system: Disconnect the damper coil connector.

➡**Be careful not to damage or contaminate the damper coil connector.**

4. Remove the flange nut while holding the joint pin with a hex wrench, and disconnect the stabilizer link from the lower arm B.

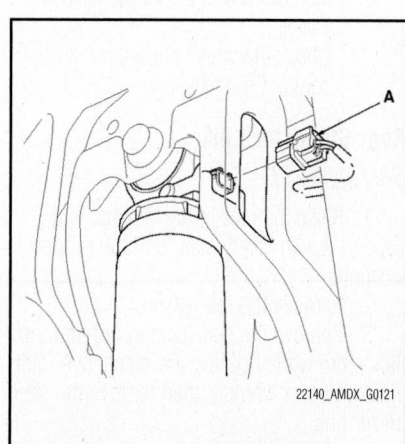

Fig. 227 Disconnect the wiring harness (A) of the active damper system, if equipped.

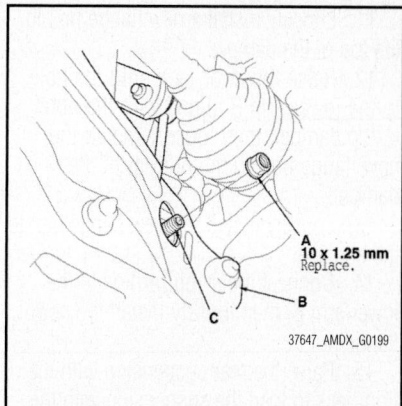

Fig. 228 Remove the flange nut (A) while holding the joint pin (C) with a hex wrench, and disconnect the stabilizer link from the lower arm B

5. Position a floor jack under the lower arm B. Raise the floor jack until the suspension begins to compress.

6. Remove the flange bolt from the bottom of the damper.

7. Remove the flange bolt from the knuckle.

8. Remove the flange bolt from the top of the damper.

9. Lower the floor jack gradually, then remove the damper from the vehicle.

To install:

10. Position the damper between the body and the lower arm B.

➡**With active damper system: Position the damper coil connector on the damper facing rearward.**

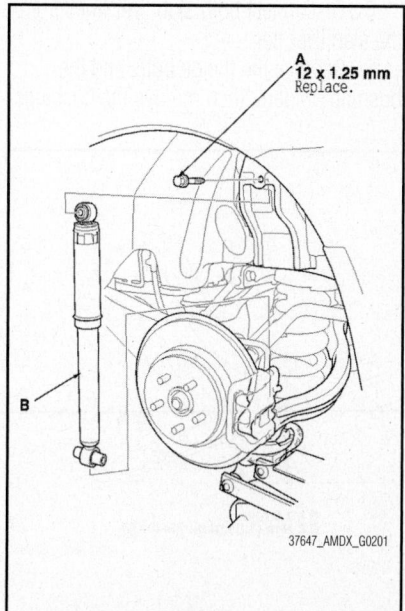

Fig. 229 Remove the flange bolt (A) from the top of the damper

11. Loosely install a new flange bolt to the top of the damper.

12. Raise the floor jack until the hole in the lower arm B aligns with the hole in the damper, then loosely install the new flange bolt to the bottom of the damper.

13. Loosely install the new flange bolt to the knuckle.

14. Connect the stabilizer link to the lower arm B, then loosely install the new flange nut.

15. Raise the rear suspension with the floor jack to load the suspension with the vehicle's weight.

16. Tighten the flange nut to 36 ft. lbs. (49 N m), while holding the joint pin with a hex wrench.

17. Tighten the flange bolts to the specified torque values
*   Upper and lower damper flange bolts: 47 ft. lbs. (64 N m)
*   Flange bolt to knuckle: 83 ft. lbs. (113 N m)

STABILIZER BAR & LINKS

REMOVAL & INSTALLATION

Stabilizer Bar

See Figures 230 and 231.

1. Raise the rear of the vehicle, and support it with safety stands in the proper locations.
2. Remove the rear wheels.
3. Remove the spare tire from the vehicle.
4. Remove the spare tire support bracket.
5. Disconnect both stabilizer links from the stabilizer bar.
6. Remove the flange bolts and the bushing holders, then remove the bushings

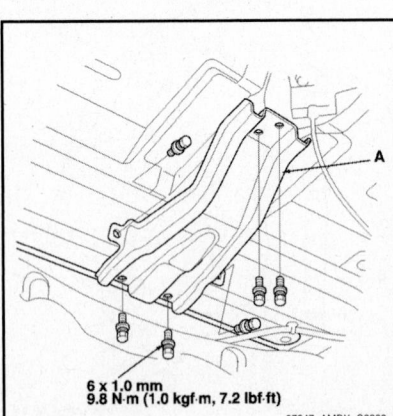

6 x 1.0 mm
9.8 N·m (1.0 kgf·m, 7.2 lbf·ft)

37647_AMDX_G0209

Fig. 230 Remove the spare tire support bracket (A)

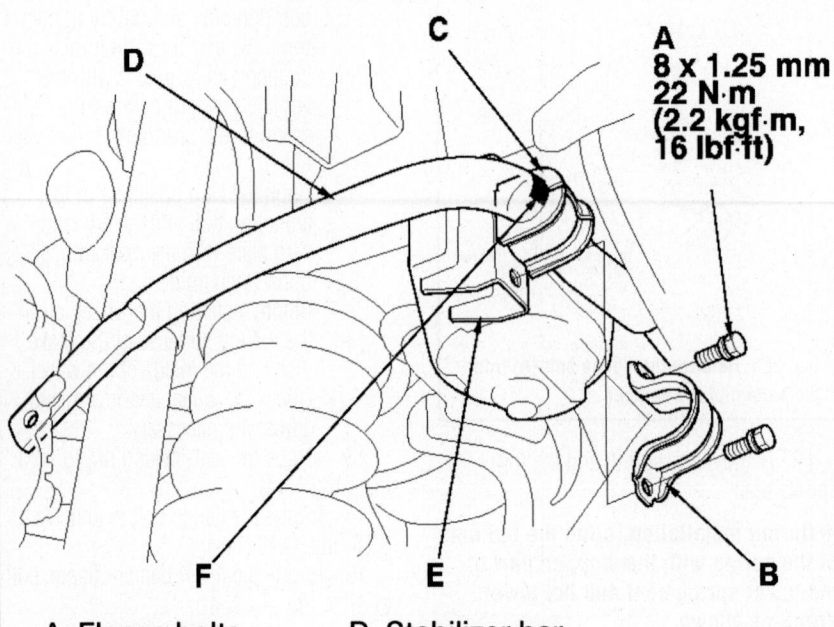

A
8 x 1.25 mm
22 N·m
(2.2 kgf·m, 16 lbf·ft)

A. Flange bolts
B. Bushing holders
C. Bushings
D. Stabilizer bar
E. Rear suspension subframe
F. Paint marks

37647_AMDX_G0210

Fig. 231 Remove the flange bolts and the bushing holders, then remove the bushings and the stabilizer bar from the rear suspension subframe

and the stabilizer bar from the rear suspension subframe.

To install:

7. Install the stabilizer bar in the reverse order of removal, and note these items:
*   Note the right and left direction of the stabilizer bar.
*   Align the paint marks on the stabilizer bar with the sides of the bushings.
*   Before installing the wheel, clean the mating surfaces of the brake disc/drum and the inside of the wheel.
*   Check the wheel alignment, and adjust it if necessary.

Rear Stabilizer Link

See Figure 233.

1. Raise the rear of the vehicle, and support it with safety stands in the proper locations.
2. Remove the rear wheel.
3. Remove the self-locking nut and the flange nut while holding the respective joint pin with a hex wrench, then remove the stabilizer link.

To install:

4. Install the stabilizer link on the stabilizer bar and lower arm B with the joint pins set at the center of their range of movement.

➡**The left stabilizer link has a yellow paint mark, while the right stabilizer link has a white mark. Align the paint mark to the outside.**

5. Install the new self-locking nut and the new flange nut, and lightly tighten them.

6. Place a jack under the lower arm B, and raise the suspension to load it with the vehicle's weight.

7. Tighten the self-locking nut and flange nut to the specified torque values while holding the respective joint pin with a hex wrench.

8. Clean the mating surfaces of the brake disc and the inside of the wheel, then install the rear wheel.

9. Test-drive the vehicle.

10. After 5 minutes of driving, tighten the self-locking nut again to the specified torque value.

TRAILING ARMS

REMOVAL & INSTALLATION

See Figures 233 through 235.

1. Raise the rear of the vehicle, and support it with safety stands in the proper locations.

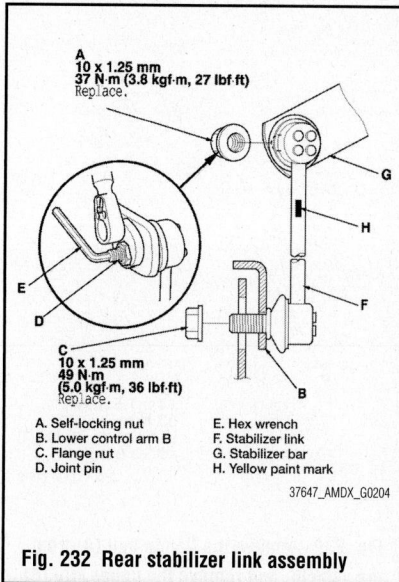

A. Self-locking nut
B. Lower control arm B
C. Flange nut
D. Joint pin
E. Hex wrench
F. Stabilizer link
G. Stabilizer bar
H. Yellow paint mark

37647_AMDX_G0204

Fig. 232 Rear stabilizer link assembly

2. Remove the rear wheel.

3. Remove the parking brake cable from the trailing arm.

4. Disconnect the brake line from the brake hose, then remove the brake hose retaining clip.

➡**To prevent the brake fluid from flowing, plug and cover the hose ends and joints with a shop towel or equivalent material.**

5. Disconnect the parking brake cable from the parking brake lever.

6. Remove the flange nuts from the trailing arm.

➡**Use the new flange nuts during reassembly.**

7. Remove the flange bolts from the trailing arm, then remove the trailing arm.

➡**Use the new flange bolts during reassembly.**

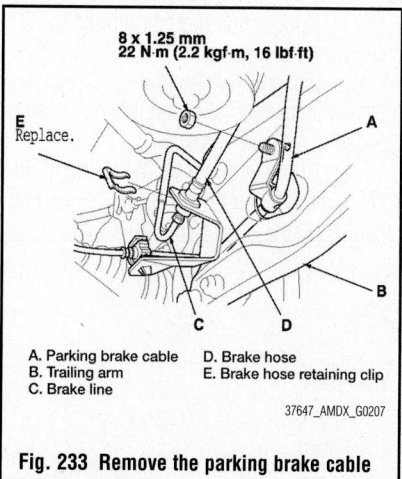

A. Parking brake cable
B. Trailing arm
C. Brake line
D. Brake hose
E. Brake hose retaining clip

37647_AMDX_G0207

Fig. 233 Remove the parking brake cable from the trailing arm

To install:

8. Install the trailing arm in the reverse order of removal, and note these items:

- First install all the suspension components, and lightly tighten the bolts and nuts, then raise the suspension to load it with the vehicle's weight before fully tightening to the specified torque values.
- Use the new brake hose retaining clip during reassembly.
- Before installing the wheel, clean the mating surfaces of the brake disc/drum and the inside of the wheel.
- Fill the master cylinder reservoir to the MAX (upper) level line, and bleed the brake system. Check for a leak at the brake hose/line joint, and retighten it if necessary.
- Check the wheel alignment, and adjust it if necessary.

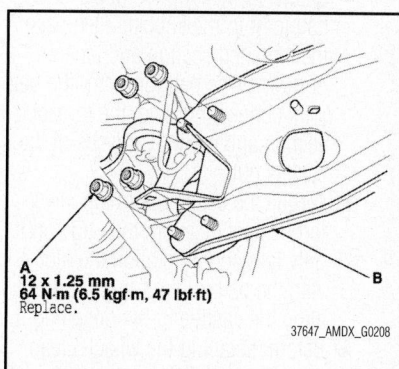

A
12 x 1.25 mm
64 N·m (6.5 kgf·m, 47 lbf·ft)
Replace.

37647_AMDX_G0208

Fig. 234 Remove the flange nuts (A) from the trailing arm (B)

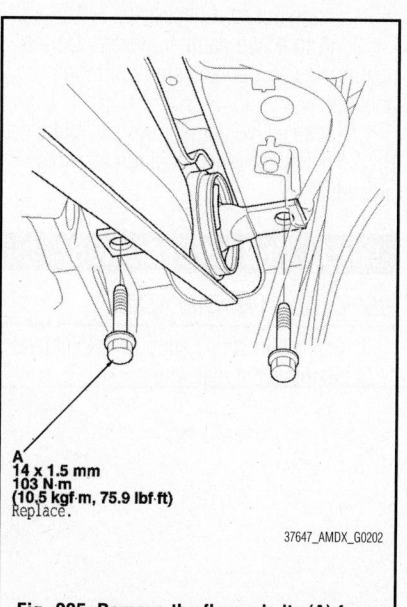

A
14 x 1.5 mm
103 N·m
(10.5 kgf·m, 75.9 lbf·ft)
Replace.

37647_AMDX_G0202

Fig. 235 Remove the flange bolts (A) from the trailing arm

UPPER BALL JOINTS

REMOVAL & INSTALLATION

See Figures 236 and 237.

1. Install a hex nut onto the threads of the ball joint. Make sure the nut is flush with the ball joint pin end to prevent damage to the threaded end of the ball joint pin.

2. Apply grease to the ball joint remover on the areas shown. This will ease installation of the tool and prevent damage to the pressure bolt threads.

3. Loosen the pressure bolt, and install the ball joint remover. Insert the jaws carefully, making sure not to damage the ball joint boot. Adjust the jaw spacing by turning the adjusting bolt.

❋❋ CAUTION

Fasten the safety chain securely to a suspension arm or the subframe. Do not fasten it to a brake line or wire harness.

4. After adjusting the adjusting bolt, make sure the head of the adjusting bolt is in the position shown to allow the jaw to pivot.

5. With a wrench, tighten the pressure bolt until the ball joint pin pops loose from the ball joint connecting hole. If necessary, apply penetrating type lubricant to loosen the ball joint pin.

❋❋ CAUTION

Do not use pneumatic or electric tools on the pressure bolt.

6. Remove the ball joint remover, then remove the nut from the end of the ball joint pin, and pull the ball joint out of the ball joint connecting hole. Inspect the ball joint boot, and replace it if damaged.

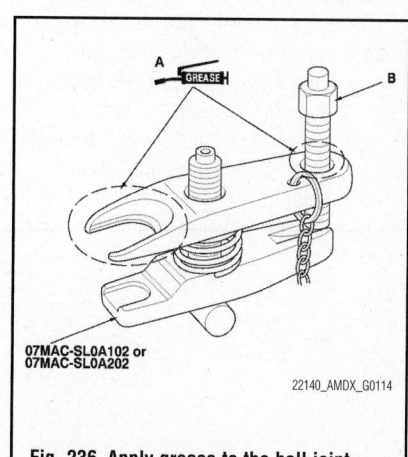

07MAC-SL0A102 or
07MAC-SL0A202

22140_AMDX_G0114

Fig. 236 Apply grease to the ball joint remover on the areas shown (A)

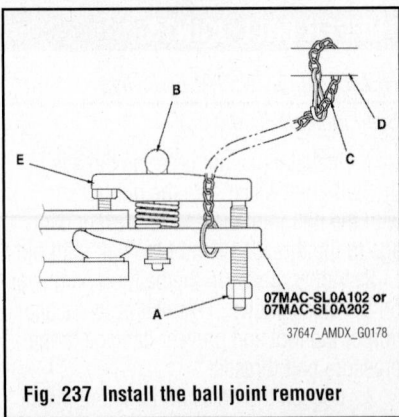

Fig. 237 Install the ball joint remover

UPPER CONTROL ARMS

REMOVAL & INSTALLATION

See Figures 238 and 239.

1. Raise the rear of the vehicle, and support it with safety stands in the proper locations.
2. Remove the rear wheel.
3. Position a floor jack under the lower arm B. Raise the floor jack until the suspension begins to compress.
4. With active damper system: Remove the suspension stroke sensor from the upper arm.
5. Remove the lock pin from the upper arm ball joint, then remove the castle nut.

➡**During installation, install the lock pin as shown after tightening the castle nut.**

6. Disconnect the upper arm ball joint from the knuckle using the ball joint remover.
7. Remove the flange bolt from the vehicle, and remove the upper arm.

➡**Use the new flange bolt during reassembly.**

To install:

8. Install the upper arm in the reverse order of removal, and note these items:
- First install all the components, and lightly tighten the bolts and nuts, then raise the suspension to load it with the vehicle's weight before fully tightening to the specified torque values.

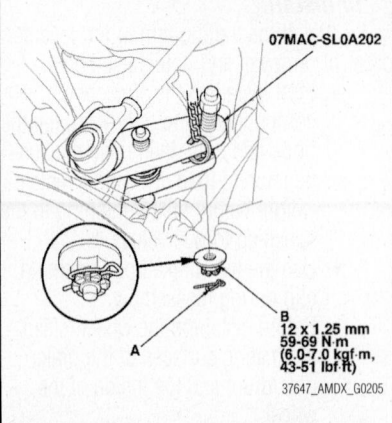

Fig. 238 Remove the lock pin (A) from the upper arm ball joint, then remove the castle nut (B)

- Be careful not to damage the ball joint boot when connecting the knuckle.
- Before connecting the upper arm ball joint to the knuckle, degrease the threaded section and tapered portion of the ball joint pin, the ball joint connecting hole, the threaded section and mating surfaces of the castles nut.
- Torque the castle nut to the lower torque specification, then tighten it only far enough to align the slot with the ball joint pin hole. Do not align the castle nut by loosening it.
- Before installing the wheel, clean the mating surfaces of the brake disc/drum and the inside of the wheel.
- Check the rear wheel alignment, and adjust if necessary.

9. With active damper system: Do the memorizing rear suspension full rebound position.
10. With active damper system/Left side: Do the headlight initial position learning procedure.

WHEEL HUBS & BEARINGS

REMOVAL & INSTALLATION

1. Raise and safely support the vehicle.
2. Remove the rear wheel.

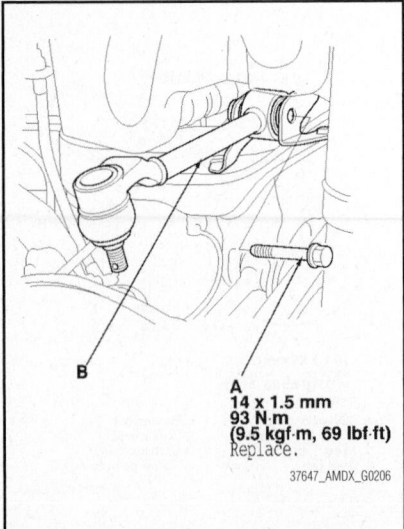

Fig. 239 Remove the flange bolt (A) from the vehicle, and remove the upper arm (B)

3. Remove the brake caliper bracket mounting bolts, then remove the caliper assembly from the knuckle.

✳✳ WARNING

To prevent damage to the caliper assembly or brake hose, use a short piece of wire to hang the caliper assembly from the undercarriage. Do not twist the brake hose excessively.

4. Remove the two mounting washers.
5. Raise the stake, then remove the spindle nut.
6. Release the parking brake, and remove the brake disc/drum.
7. Remove the mounting bolts, then remove the hub bearing unit.

To install:

8. Installation is the reverse order of removal. Note the following during installation:

a. Tighten hub mounting bolts to 72 ft. lbs. (98 Nm).

b. Using a new spindle nut, apply a small amount of clean engine oil to the seating surface of the nut.

c. Tighten the nut to 181 ft. lbs. (245 Nm), then use a drift to stake the spindle nut shoulder against the driveshaft.

ACURA

RDX

2

BRAKES 2-9

**ANTI-LOCK BRAKE SYSTEM
(ABS)............................ 2-9**
General Information.................. 2-9
  Precautions........................ 2-9
Speed Sensors...................... 2-9
  Removal & Installation........... 2-9
**BLEEDING THE BRAKE
SYSTEM....................... 2-10**
Bleeding Procedure................. 2-10
  Bleeding Procedure 2-10
  Bleeding the ABS
  System 2-11
FRONT DISC BRAKES 2-11
Brake Caliper........................ 2-11
  Removal & Installation......... 2-11
Disc Brake Pads 2-11
  Removal & Installation......... 2-11
PARKING BRAKE 2-13
Parking Brake Cables 2-13
  Adjustment 2-13
Parking Brake Shoes 2-13
  Removal & Installation......... 2-13
REAR DISC BRAKES 2-12
Brake Caliper........................ 2-12
  Removal & Installation......... 2-12
Disc Brake Pads 2-12
  Removal & Installation......... 2-12

CHASSIS ELECTRICAL 2-14

**AIR BAG (SUPPLEMENTAL
RESTRAINT SYSTEM)........ 2-14**
General Information.................. 2-14
  Arming the System 2-14
  Clockspring Centering 2-15
  Disarming the System........... 2-14
  Service Precautions 2-14

DRIVE TRAIN.................. 2-15

Front Driveshaft..................... 2-15
  Removal & Installation......... 2-15
Front Intermediate Driveshaft 2-17
  Removal & Installation......... 2-17
Pinion Oil Seal...................... 2-17
  Removal & Installation......... 2-17

Propeller Shaft 2-15
  Removal & Installation.......... 2-15
Rear Driveshaft 2-18
  Removal & Installation......... 2-18

ENGINE COOLING 2-19

Engine Coolant...................... 2-19
  Drain & Refill 2-19
Radiator............................. 2-20
  Removal & Installation......... 2-20
Thermostat 2-20
  Removal & Installation......... 2-20
Water Pump 2-21
  Removal & Installation......... 2-21

ENGINE ELECTRICAL......... 2-22

BATTERY SYSTEM 2-22
Battery 2-22
  Removal & Installation.......... 2-22
CHARGING SYSTEM 2-22
Alternator 2-22
  Removal & Installation......... 2-22
IGNITION SYSTEM 2-22
Firing Order......................... 2-22
Ignition Coil 2-22
  Removal & Installation......... 2-22
Ignition Timing...................... 2-23
  Adjustment 2-23
Spark Plugs.......................... 2-23
  Removal & Installation......... 2-23
STARTING SYSTEM 2-23
Starter 2-23
  Removal & Installation......... 2-23

ENGINE MECHANICAL 2-24

Accessory Drive Belts 2-24
  Adjustment 2-24
  Drive Belt Routing............... 2-24
  Inspection 2-24
  Removal & Installation......... 2-24
Air Intake System 2-24
  Removal & Installation......... 2-24
Camshaft & Valve Lifters........... 2-25
  Inspection 2-25
  Removal & Installation......... 2-27

Camshaft Chain &
  Sprockets 2-28
  Removal & Installation......... 2-28
Camshaft Chain Front Cover 2-29
  Removal & Installation......... 2-29
Crankshaft Front Seal............... 2-31
  Removal & Installation......... 2-31
Cylinder Head 2-31
  Removal & Installation......... 2-31
Exhaust Manifold 2-32
  Removal & Installation......... 2-32
Intake Manifold 2-32
  Removal & Installation......... 2-32
Oil Pan 2-33
  Removal & Installation......... 2-33
Oil Pump 2-34
  Inspection 2-35
  Removal & Installation......... 2-34
Pistons & Rings 2-35
  Positioning 2-35
Rear Main Seal 2-35
  Removal & Installation......... 2-35
Turbocharger System 2-36
  Removal & Installation......... 2-36
Valve Clearance 2-40
  Adjustment 2-40
Valve Covers 2-39
  Removal & Installation......... 2-39

**ENGINE PERFORMANCE &
EMISSION CONTROLS 2-41**

Air-Fuel Ratio (AFR) Sensor 2-41
  Location......................... 2-41
  Removal & Installation......... 2-41
Camshaft Position (CMP)
  Sensor 2-41
  Location......................... 2-41
  Removal & Installation......... 2-41
Component Locations 2-41
Crankshaft Position (CKP)
  Sensor 2-42
  Location......................... 2-42
  Removal & Installation......... 2-42
Engine Coolant Temperature
  (ECT) Sensor 2-42
  Location......................... 2-42
  Removal & Installation......... 2-42

Heated Oxygen (HO2S)
Sensor 2-43
Location................................2-43
Removal & Installation..........2-43
Input Shaft (Mainshaft) Speed
Sensor 2-43
Removal & Installation..........2-43
Intake Air Temperature (IAT)
Sensor & Mass Air
flow (MAF) Sensor..................2-44
Location................................2-44
Removal & Installation..........2-44
Knock Sensor (KS)....................2-44
Location................................2-44
Removal & Installation..........2-46
Manifold Absolute Pressure
(MAP) Sensor 2-46
Location................................2-46
Removal & Installation..........2-46
Output Shaft (Countershaft)
Speed (OSS) Sensor 2-47
Removal & Installation..........2-47
Positive Crankcase Ventilation
(PCV) Valve 2-47
Location................................2-47
Removal & Installation..........2-47
Powertrain Control Module
(PCM)....................................2-47
Location................................2-47
PCM Reset Procedure..........2-48
Removal & Installation..........2-47
Transmission Fluid Temperature
(TFT) Sensor 2-48
Removal & Installation..........2-48

FUEL............................2-49

GASOLINE FUEL INJECTION
SYSTEM......................2-49
Fuel Filter 2-50
Removal & Installation..........2-50
Fuel Pump Unit......................2-50
Removal & Installation..........2-50
Fuel Rail & Injectors 2-51
Removal & Installation..........2-51
Fuel System Service
Precautions 2-49

Fuel Tank................................2-53
Removal & Installation..........2-53
Relieving Fuel System
Pressure................................2-49
Throttle Body..........................2-53
Removal & Installation..........2-53

HEATING & AIR CONDITIONING
SYSTEM........................2-54

Blower Motor 2-54
Removal & Installation..........2-54
Heater Core 2-54
Removal & Installation..........2-54

PRECAUTIONS..................2-9

SPECIFICATIONS AND
MAINTENANCE CHARTS 2-3

Brake Specifications....................2-7
Camshaft Specifications..............2-5
Capacities 2-4
Crankshaft and Connecting Rod
Specifications 2-5
Engine and Model Year
Identification 2-3
Engine Tune-Up Specifications ...2-3
Fluid Specifications....................2-4
General Engine Specifications.....2-3
Piston and Ring
Specifications 2-5
Scheduled Maintenance
Intervals 2-8
Tire Wheel and Ball Joint
Specifications 2-7
Torque Specifications..................2-6
Valve Specifications 2-4
Wheel Alignment........................2-6

STEERING 2-55

Power Rack & Pinion Steering
Gear 2-55
Removal & Installation..........2-55
Power Steering Pump................2-57
Bleeding 2-57

Fluid Fill Procedure..............2-57
Removal & Installation..........2-57

SUSPENSION..................2-58

FRONT SUSPENSION 2-58
Knuckle, Spindle & Wheel
Bearing 2-58
Removal & Installation..........2-58
Lower Ball Joints......................2-60
Removal & Installation..........2-60
Lower Control Arms 2-60
Removal & Installation..........2-60
MacPherson Struts....................2-61
Overhaul 2-61
Removal & Installation..........2-61
Stabilizer Bar & Links 2-62
Removal & Installation..........2-62
Steering Knuckle 2-63
Removal & Installation..........2-63
Wheel Bearings 2-64
Adjustment 2-64
REAR SUSPENSION 2-64
Coil Springs..............................2-64
Removal & Installation..........2-64
Control Arms & Links 2-65
Removal & Installation..........2-65
Shock Absorbers......................2-66
Removal & Installation..........2-66
Stabilizer Bar 2-67
Removal & Installation..........2-67
Steering Knuckle 2-66
Removal & Installation..........2-66
Trailing Arm 2-68
Removal & Installation..........2-68
Wheel Bearings 2-68
Adjustment 2-68
Removal & Installation..........2-68

SPECIFICATIONS AND MAINTENANCE CHARTS

ENGINE AND VEHICLE IDENTIFICATION CHART

		Engine Code						Model Year	
Code	Liters (cc)	Cu. In.	Cyl.	Fuel Sys.	Engine Type	Eng. Mfg.		Code ①	Year
K23A1	2.3 (2350)	140	4	SMFI	DOHC	Honda		B	2011
								C.	2012

DOHC: Dual Overhead Cam

SMFI: Sequential Multi-port Fuel Injection

① 10th position of VIN

71051_ARDX_C0001

GENERAL ENGINE SPECIFICATIONS

Year	Model	Engine Displacement Liters (ID)	Net Horsepower @ rpm	Net Torque @ rpm (ft. lbs.)	Bore x Stroke (in.)	Compression Ratio	Oil Pressure @ rpm
2011	RDX	2.3 (K23A1)	240@6000	260@4500	3.39x3.90	8.8:1	44@3000
2012	RDX	2.3 (K23A1)	240@6000	260@4500	3.39x3.90	8.8:1	44@3000

SMFI: Sequential Multi-port Fuel Injection

71051_ARDX_C0002

ENGINE TUNE-UP SPECIFICATIONS

Year	Engine Displacement Liters (ID)	Spark Plug Gap (in.)	Ignition Timing (deg.)		Fuel Pump (psi)	Idle Speed (rpm)		Valve Clearance (in.)	
			MT	AT		MT	AT	In.	Ex.
2011	2.3 (K23A1)	0.028-0.031	—	12-14 ①	47-54	—	670-770	0.008-0.010	0.011-0.013
2012	2.3 (K23A1)	0.028-0.031	—	12-14 ①	47-54	—	670-770	0.008-0.010	0.011-0.013

NOTE: The Vehicle Emission Control Information label often reflects changes made during production and must be used if they differ from this chart.

NOTE: The fuel pressure readings are given with the vacuum hose disconnected

①: Before top dead center

71051_ARDX_C0003

CAPACITIES

Year	Model	Engine Displacement Liters (ID)	Engine Oil with Filter (qts.)	Transmission (pts.) 5-Spd	Transmission (pts.) Auto.	Transfer Case (pts.)	Drive Axle Front (pts.)	Drive Axle Rear (pts.)	Fuel Tank (gal.)	Cooling System (qts.)
2011	RDX	2.3 (K23A1)	4.7	—	①	②	—	③	18.0	④
2012	RDX	2.3 (K23A1)	4.7	—	①	②	—	③	18.0	④

NOTE: All capacities are approximate. Add fluid gradually and check to be sure a proper fluid level is obtained.

① Fluid change: 3.5 quarts

Overhaul: 8.2 quarts

② Fluid change: .45 quart

Overhaul: .48 quart

③ Fluid change: 2.67 quarts

Overhaul: 2.93 quarts

④ Fluid change: 1.85 gallons

Overhaul (engine): 2.22 gallons

71051_ARDX_C0004

FLUID SPECIFICATIONS

Year	Model	Engine Displacement Liters (ID)	Engine Oil	Auto. Trans.	Drive Axle	Power Steering Fluid	Brake Master Cylinder
2011	RDX	2.3 (K23A1)	①	②	③	④	DOT 3
2012	RDX	2.3 (K23A1)	①	②	③	④	DOT 3

DOT: Department Of Transportation

Note: If specification disagrees with specification in owners manual, use specification in owners manaual

① Mobil 1 (P/N 5w-30-MB1-000) or equivalent that meets Acure HTO-06 standard

② Acura ATF-Z1 fluid

③ Transfer case: API classified GL4 or GL5 only. SAE 90 or SAE 80W-90 viscosity.

Rear differential: Acura ATF-Z1 fluid

④ Acura power steering fluid

71051_ARDX_C0005

VALVE SPECIFICATIONS

Year	Engine Displacement Liters (ID)	Seat Angle (deg.)	Face Angle (deg.)	Spring Test Pressure (lbs. @ in.)	Spring Installed Height (in.)	Stem-to-Guide Clearance (in.) Intake	Stem-to-Guide Clearance (in.) Exhaust	Stem Diameter (in.) Intake	Stem Diameter (in.) Exhaust
2011	2.3 (K23A1)	45	45	NA	①	0.0012-0.0022	0.0022-0.0031	0.2156-0.2159	0.2146-0.2150
2012	2.3 (K23A1)	45	45	NA	①	0.0012-0.0022	0.0022-0.0031	0.2156-0.2159	0.2146-0.2150

NA: Not Available

① Valve spring free length:

Intake: 1.874 in.

Exhaust: 1.954 in.

71051_ARDX_C0006

CAMSHAFT SPECIFICATIONS

All measurements in inches unless noted

Year	Model	Engine Displacement Liters (ID)	Journal Dia.	Brg. Oil Clearance	Shaft End-play	Total Runout	Lobe Height Intake	Lobe Height Exhaust
2011	RDX	2.3 (K23A1)	NA	①	0.0020-0.0080	0.0012	②	1.3422
2012	RDX	2.3 (K23A1)	NA	①	0.0020-0.0080	0.0012	②	1.3422

NA: Not Available

① No. 1 Journal: 0.001-0.003 inch
All others: 0.002-0.004 inch

② Intake primary: 1.3356 inch. Intake secondary: 1.1668 inch.
All others: 0.002-0.004 inch

71051_ARDX_C0007

CRANKSHAFT AND CONNECTING ROD SPECIFICATIONS

All measurements are given in inches

Year	Engine Displacement Liters (ID)	Crankshaft Main Brg. Journal Dia.	Crankshaft Main Brg. Oil Clearance	Crankshaft Shaft End-play	Crankshaft Thrust on No.	Connecting Rod Journal Diameter	Connecting Rod Oil Clearance	Connecting Rod Side Clearance
2011	2.3 (K23A1)	①	②	0.0040-0.0140	NA	1.8888 1.8898	0.0013-0.0025	0.0060-0.0140
2012	2.3 (K23A1)	①	②	0.0040-0.0140	NA	1.8888 1.8898	0.0013-0.0025	0.0060-0.0140

NA: Not Available

① Except No. 3: 2.1648-2.1657 inches
No. 3: 2.1644-2.1654 inches

② Except No. 3: 0.0007-0.0016 inches
No. 3: 0.0010-0.0019 inches

71051_ARDX_C0008

PISTON AND RING SPECIFICATIONS

All measurements are given in inches

Year	Engine Displacement Liters (ID)	Piston Clearance	Ring Gap Top Compression	Ring Gap Bottom Compression	Ring Gap Oil Control	Ring Side Clearance Top Compression	Ring Side Clearance Bottom Compression	Ring Side Clearance Oil Control
2011	2.3 (K23A1)	0.0008-0.0016	0.0008-0.0012	0.016-0.0200	0.004-0.0120	0.0018-0.0035	0.0016-0.0026	0.0008-0.002
2012	2.3 (K23A1)	0.0008-0.0016	0.0008-0.0012	0.016-0.0200	0.004-0.0120	0.0018-0.0035	0.0016-0.0026	0.0008-0.002

71051_ARDX_C0009

TORQUE SPECIFICATIONS

All readings in ft. lbs.

Year	Engine Displacement Liters (ID)	Cylinder Head Bolts	Main Bearing Bolts	Rod Bearing Bolts	Crankshaft Damper Bolts	Flywheel Bolts	Manifold Intake	Manifold Exhaust	Spark Plugs	Oil Pan Drain Plug
2011	2.3 (K23A1)	①	②	③	④	NA	⑤	⑥	13	29
2012	2.3 (K23A1)	①	②	③	④	NA	⑤	⑥	13	29

NA: Not Available

① Step 1: 29 ft. lbs.
   Step 2: Tighten all bolts an additional 90 degrees. If using a new bolt tighten an extra 90 degrees

② Step 1: 22 ft. lbs.
   Step 2: additional 67 degrees

③ Step 1: 22 ft. lbs.
   Step 2: 90 degrees

④ Used bolt: 36 ft. lbs.
   New bolt: 130 ft. lbs., loosen and retighten to 36 ft. lbs.

⑤ Tighten bolts and nuts in a crisscross pattern beginning with the inner bolt, in three stages to 16 ft. lbs.

⑥ Tighten bolts and nuts in a crisscross pattern beginning with the inner bolt, in three stages to 33 ft. lbs.

71051_ARDX_C0010

WHEEL ALIGNMENT

Year	Model		Caster Range (+/-Deg.)	Caster Preferred Setting (Deg.)	Camber Range (+/-Deg.)	Camber Preferred Setting (Deg.)	Toe-in (in.)
2011	RDX	F	1.00	1.57	0.45	0	0+/-0.08
		R	—	—	0.45	-1.00	0.10+/-0.08
2012	RDX	F	1.00	1.57	0.45	0	0+/-0.08
		R	—	—	0.45	-1.00	0.10+/-0.08

71051_ARDX_C0011

TIRE, WHEEL AND BALL JOINT SPECIFICATIONS

Year	Model	OEM Tires		Tire Pressures (psi)		Wheel Size	Ball Joint Inspection	Lug Nut (ft. lbs.)
		Standard	Optional	Front	Rear			
2011	RDX	P235/65R18	①	32	32	R18 ②	NA	80
2012	RDX	P235/65R18	①	32	32	R18 ②	NA	80

OEM: Original Equipment Manufacturer

PSI: Pounds Per Square Inch

NA: Not Available

① Optional 18 inch tire sizes are: P225/60R18 and P245/55R18

② Optional 19 inch wheel available. Tire size is P245/55R19.

71051_ARDX_C0012

BRAKE SPECIFICATIONS

All measurements in inches unless noted

Year	Model		Brake Disc			Brake Drum Diameter			Minimum Lining Thickness		Brake Caliper	
			Original Thickness	Minimum Thickness	Maximum Runout	Original Inside Diameter	Max. Wear Limit	Maximum Machine Diameter	Front	Rear	Bracket Bolts (ft. lbs.)	Mounting Bolts (ft. lbs.)
2011	RDX	F	①	1.020	0.0016	—	—	—	0.060	—	101	37
		R	②	0.300	NA	—	—	—	—	0.060	80	17
2012	RDX	F	①	1.020	0.0016	—	—	—	0.060	—	101	37
		R	②	0.300	NA	—	—	—	—	0.060	80	17

NA: Not Available

F: Front

R: Rear

① 1.09-1.11 inch

② 0.35-0.36 inch

71051_ARDX_C0013

SCHEDULED MAINTENANCE INTERVALS
ACURA—RDX

TO BE SERVICED	OF SERVIC	VEHICLE MILEAGE INTERVAL (x1000)															
		7.5	15	22.5	30	37.5	45	52.5	60	67.5	75	82.5	90	97.5	105	112.5	120
Accessory drive belts	I & A				✓				✓				✓				✓
Air cleaner element	R				✓				✓				✓				✓
Brake fluid	R	Every 3 years															
Brake hoses & lines (incl. ABS)	I		✓		✓		✓		✓		✓		✓		✓		✓
Cooling system hoses & connections	I		✓		✓		✓		✓		✓		✓		✓		✓
Engine coolant ①	R						✓						✓				
Engine oil	R	✓	✓	✓	✓	✓	✓	✓	✓	✓	✓	✓	✓	✓	✓	✓	✓
Engine oil and coolant levels	I	Inspect at each fuel stop															
Engine oil filter	R		✓		✓		✓		✓		✓		✓		✓		✓
Exhaust system	I		✓		✓		✓		✓		✓		✓		✓		✓
Fluid levels and condition	I		✓		✓		✓		✓		✓		✓		✓		✓
Front and rear brakes	I		✓		✓		✓		✓		✓		✓		✓		✓
Fuel lines & connection	I		✓		✓		✓		✓		✓		✓		✓		✓
Halfshaft boots	I		✓		✓		✓		✓		✓		✓		✓		✓
Idle speed	I & A														✓		
Parking brake system	I & A		✓		✓		✓		✓		✓		✓		✓		✓
Rear differential fluid	R	✓			✓		✓		✓				✓				✓
Rotate and inspect tires	I	✓	✓	✓	✓	✓	✓	✓	✓	✓	✓	✓	✓	✓	✓	✓	✓
Spark plugs	R														✓		
Supplemental Restraint System	I	Inspect the SRS 10 years after production															
Suspension components	I		✓		✓		✓		✓		✓		✓		✓		✓
Tie rod ends, steering gear box & boots	I		✓		✓		✓		✓		✓		✓		✓		✓
Timing belt	R														✓		
Transmission fluid	R						✓				✓				✓		
Valve clearance	I	Adjust if valves are noisy															
Water pump	S/I														✓		

R: Replace   I: Inspect   A: Adjust

① Every 12,000 miles or 10 years, then every 60,000 miles or 5 years

FREQUENT OPERATION MAINTENANCE (SEVERE SERVICE)

If a vehicle is operated under any of the following conditions it is considered severe service:

- Towing a trailer or using a camper or car-top carrier.
- Repeated short trips of less than 5 miles in temperatures below freezing, or trips of less than 10 miles in any temperature.
- Extensive idling or low-speed driving for long distances as in heavy commercial use, such as delivery, taxi or police cars.
- Operating on rough, muddy or salt-covered roads.
- Operating on unpaved or dusty roads.
- Driving in extremely hot (over 90°) conditions.

Air cleaner element: replace every 15,000 miles.

Engine oil and filter: replace every 3750 miles or 6 months, whichever occurs first.

Timing belt: replace every 60,000 miles if the vehicle is regularly driven in temperatures above 110°F or below -20°F, or if frequently towing a trailer.

Transmission fluid: replace every 30,000 miles.

Rear differential fluid: replace every 60,000 miles.

Front and rear brakes: inspect every 7500 miles or 6 months, whichever occurs first.

Locks and hinges: lubricate every 15,000 miles.

Tie rods, steering gear box, boots: inspect every 7500 miles or 6 months, whichever occurs first.

Suspension components: inspect every 7500 miles or 6 months, whichever occurs first.

Halfshaft boots: inspect every 7500 miles or 6 months, whichever occurs first.

71051_ARDX_C0014

PRECAUTIONS

Before servicing any vehicle, please be sure to read all of the following precautions, which deal with personal safety, prevention of component damage, and important points to take into consideration when servicing a motor vehicle:

• Never open, service or drain the radiator or cooling system when the engine is hot; serious burns can occur from the steam and hot coolant.

• Observe all applicable safety precautions when working around fuel. Whenever servicing the fuel system, always work in a well-ventilated area. Do not allow fuel spray or vapors to come in contact with a spark, open flame, or excessive heat (a hot drop light, for example). Keep a dry chemical fire extinguisher near the work area. Always keep fuel in a container specifically designed for fuel storage; also, always properly seal fuel containers to avoid the possibility of fire or explosion. Refer to the additional fuel system precautions later in this section.

• Fuel injection systems often remain pressurized, even after the engine has been turned **OFF**. The fuel system pressure must be relieved before disconnecting any fuel lines. Failure to do so may result in fire and/or personal injury.

• Brake fluid often contains polyglycol ethers and polyglycols. Avoid contact with the eyes and wash your hands thoroughly after handling brake fluid. If you do get brake fluid in your eyes, flush your eyes with clean, running water for 15 minutes. If eye irritation persists, or if you have taken brake fluid internally, IMMEDIATELY seek medical assistance.

• The EPA warns that prolonged contact with used engine oil may cause a number of skin disorders, including cancer. You should make every effort to minimize your exposure to used engine oil. Protective gloves should be worn when changing oil. Wash your hands and any other exposed skin areas as soon as possible after exposure to used engine oil. Soap and water, or waterless hand cleaner should be used.

• All new vehicles are now equipped with an air bag system, often referred to as a Supplemental Restraint System (SRS) or Supplemental Inflatable Restraint (SIR) system. The system must be disabled before performing service on or around system components, steering column, instrument panel components, wiring and sensors. Failure to follow safety and disabling procedures could result in accidental air bag deployment, possible personal injury and unnecessary system repairs.

• Always wear safety goggles when working with, or around, the air bag system. When carrying a non-deployed air bag, be sure the bag and trim cover are pointed away from your body. When placing a non-deployed   face the bag and trim cover upward, away from the surface. This will reduce the motion of the module if it is accidentally deployed. Refer to the additional air bag system precautions later in this section.

• Clean, high quality brake fluid from a sealed container is essential to the safe and proper operation of the brake system. You should always buy the correct type of brake fluid for your vehicle. If the brake fluid becomes contaminated, completely flush the system with new fluid. Never reuse any brake fluid. Any brake fluid that is removed from the system should be discarded. Also, do not allow any brake fluid to come in contact with a painted surface; it will damage the paint.

• Never operate the engine without the proper amount and type of engine oil; doing so WILL result in severe engine damage.

• Timing belt maintenance is extremely important. Many models utilize an interference-type, non-freewheeling engine. If the timing belt breaks, the valves in the cylinder head may strike the pistons, causing potentially serious (also time-consuming and expensive) engine damage. Refer to the maintenance interval charts for the recommended replacement interval for the timing belt, and to the timing belt section for belt replacement and inspection.

• Disconnecting the negative battery cable on some vehicles may interfere with the functions of the on-board computer system(s) and may require the computer to undergo a relearning process once the negative battery cable is reconnected.

• When servicing drum brakes, only disassemble and assemble one side at a time, leaving the remaining side intact for reference.

• Only an MVAC-trained, EPA-certified automotive technician should service the air conditioning system or its components.

BRAKES

ANTI-LOCK BRAKE SYSTEM (ABS)

GENERAL INFORMATION

PRECAUTIONS

• Certain components within the ABS system are not intended to be serviced or repaired individually.

• Do not use rubber hoses or other parts not specifically specified for and ABS system. When using repair kits, replace all parts included in the kit. Partial or incorrect repair may lead to functional problems and require the replacement of components.

• Lubricate rubber parts with clean, fresh brake fluid to ease assembly. Do not use shop air to clean parts; damage to rubber components may result.

• Use only DOT 3 brake fluid from an unopened container.

• If any hydraulic component or line is removed or replaced, it may be necessary to bleed the entire system.

• A clean repair area is essential. Always clean the reservoir and cap thoroughly before removing the cap. The slightest amount of dirt in the fluid may plug an orifice and impair the system function. Perform repairs after components have been thoroughly cleaned; use only denatured alcohol to clean components. Do not allow ABS components to come into contact with any substance containing mineral oil; this includes used shop rags.

• The Anti-Lock control unit is a microprocessor similar to other computer units in the vehicle. Ensure that the ignition switch is **OFF** before removing or installing controller harnesses. Avoid static electricity discharge at or near the controller.

• If any arc welding is to be done on the vehicle, the control unit should be unplugged before welding operations begin.

SPEED SENSORS

REMOVAL & INSTALLATION

Front

See Figure 1.

1. Make sure that the ignition switch is in the LOCK (0) position.
2. Disconnect the negative, then the positive, battery cables.
3. Be sure that the ignition switch is in LOCK (0).
4. Raise and support the vehicle safely.

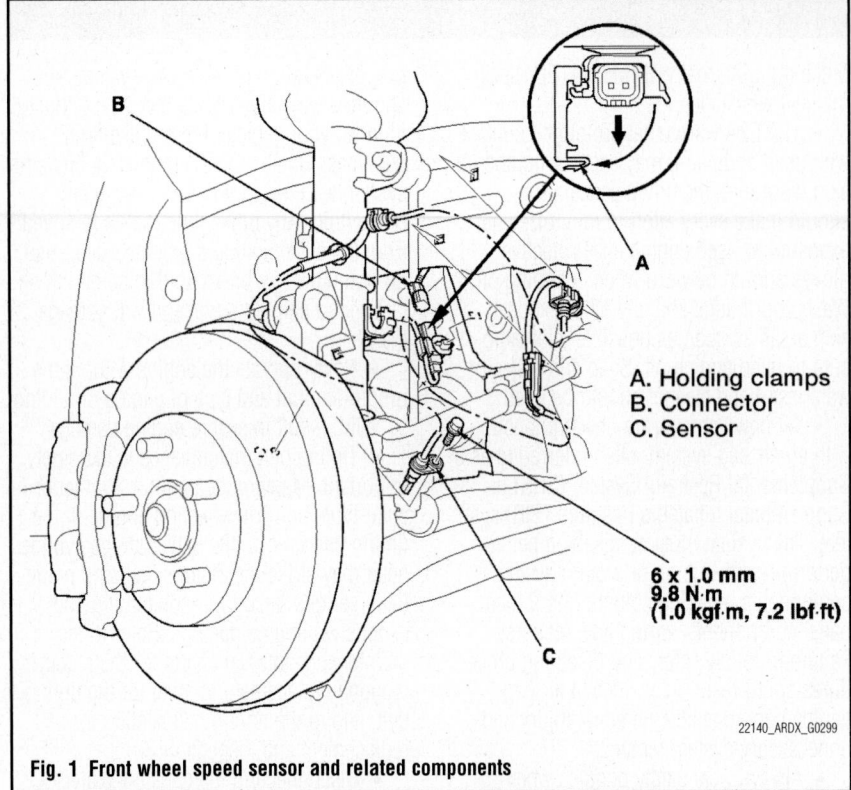

A. Holding clamps
B. Connector
C. Sensor

6 x 1.0 mm
9.8 N·m
(1.0 kgf·m, 7.2 lbf·ft)

22140_ARDX_G0299

Fig. 1 Front wheel speed sensor and related components

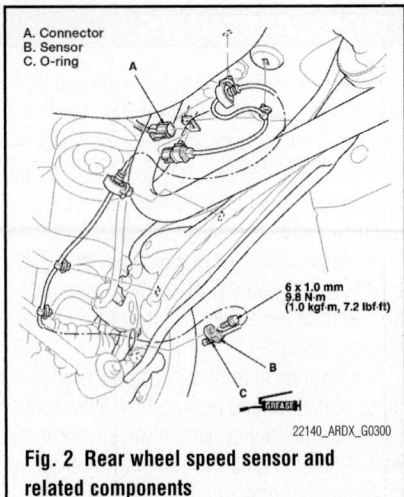

A. Connector
B. Sensor
C. O-ring

6 x 1.0 mm
9.8 N·m
(1.0 kgf·m, 7.2 lbf·ft)

22140_ARDX_G0300

Fig. 2 Rear wheel speed sensor and related components

5. Release the connector holding clamps. Disconnect the wheel speed sensor connector.

6. Remove the clips, bolt and wheel speed sensor.

To install:

7. Installation is the reverse of the removal procedure.

➡**Install the sensor carefully to avoid twisting the wires. If the sensor comes in contact with the wheel bearing, it is faulty and must be replaced. Do not drop the sensor.**

8. Connect the battery cables.

9. Turn the ignition switch ON (II), the SRS indicator should come on for about six seconds, then go off.

10. Start the engine and check that the ABS and VSA indicators do not stay on. Test drive the vehicle to ensure that the indicator lights do not come on.

11. Correct problems, as required.

Rear

See Figure 2.

1. Make sure that the ignition switch is in the LOCK (0) position.

2. Disconnect the negative, then the positive, battery cables.

3. Be sure that the ignition switch is in LOCK (0).

4. Raise and support the vehicle safely.

5. Disconnect the wheel speed sensor connector.

6. Remove the clips, bolt and wheel speed sensor.

To install:

7. Installation is the reverse of the removal procedure.

➡**Install the sensor carefully to avoid twisting the wires. If the sensor comes in contact with the wheel bearing, it is faulty and must be replaced. Do not drop the sensor.**

8. Connect the battery cables.

9. Turn the ignition switch ON (II), the SRS indicator should come on for about six seconds, then go off.

10. Start the engine and check that the ABS and VSA indicators do not stay on. Test drive the vehicle to ensure that the indicator lights do not come on.

11. Correct problems, as required.

BRAKES

BLEEDING THE BRAKE SYSTEM

BLEEDING PROCEDURE

BLEEDING PROCEDURE

See Figure 3.

✳✳ **WARNING**

Do not reuse the drained fluid. Use only clean Honda DOT 3 Brake Fluid from an unopened container. Using a non-Honda brake fluid can cause corrosion and shorten the life of the system.

✳✳ **WARNING**

Make sure no dirt or other foreign matter is allowed to contaminate the brake fluid.

✳✳ **WARNING**

Do not spill brake fluid on the vehicle, it may damage the paint; if brake fluid does contact the paint, wash it off immediately with water.

1. The reservoir on the master cylinder must be at the MAX (upper) level mark at the start of the bleeding procedure and checked after bleeding each brake caliper. Add fluid as required.

2. Make sure the brake fluid level in the reservoir is at the MAX (upper) level line.

3. Slide a piece of clear plastic hose over the first bleed screw, and submerge the other end in a container of new brake fluid.

4. Have someone slowly pump the brake pedal several times, and then apply steady pressure.

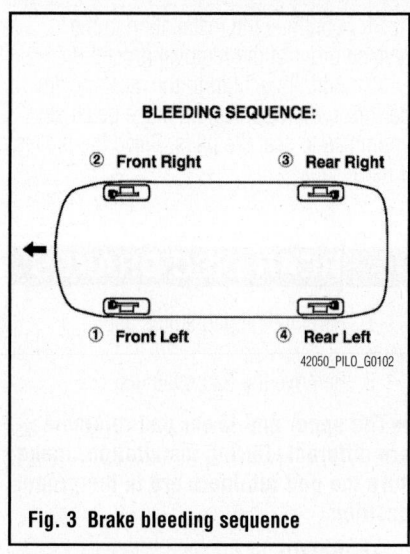

BLEEDING SEQUENCE:

② Front Right   ③ Rear Right

←

① Front Left   ④ Rear Left

42050_PILO_G0102

Fig. 3 Brake bleeding sequence

5. Starting at the left-front, loosen the brake bleed screw to allow air to escape from the system. Then tighten the bleed screw securely.

6. Repeat the procedure for each wheel in the sequence shown following until air bubbles no longer appear in the fluid.

7. Refill the master cylinder reservoir to the MAX (upper) level line.

BLEEDING THE ABS SYSTEM

> ❂❂ **WARNING**
>
> **Read and observe the Precautions in this section before starting work.**

1. The reservoir on the master cylinder must be at the MAX (upper) level mark at the start of the bleeding procedure and checked after bleeding each brake caliper. Add fluid as required.

2. Make sure the brake fluid level in the reservoir is at the MAX (upper) level line.

3. Slide a piece of clear plastic hose over the first bleed screw, and submerge the other end in a container of new brake fluid.

4. Have someone slowly pump the brake pedal several times, and then apply steady pressure.

5. Starting at the left-front, loosen the brake bleed screw to allow air to escape from the system. Then tighten the bleed screw securely.

6. Repeat the procedure for each wheel in the sequence shown following until air bubbles no longer appear in the fluid.

7. Refill the master cylinder reservoir to the MAX (upper) level line.

BRAKES

> ❂❂ **CAUTION**
>
> **Dust and dirt accumulating on brake parts during normal use may contain asbestos fibers from production or aftermarket brake linings. Breathing excessive concentrations of asbestos fibers can cause serious bodily harm. Exercise care when servicing brake parts. Do not sand or grind brake lining unless equipment used is designed to contain the dust residue. Do not clean brake parts with compressed air or by dry brushing. Cleaning should be done by dampening the brake components with a fine mist of water, then wiping the brake components clean with a dampened cloth. Dispose of cloth and all residue containing asbestos fibers in an impermeable container with the appropriate label. Follow practices prescribed by the Occupational Safety and Health Administration (OSHA) and the Environmental Protection Agency (EPA) for the handling, processing, and disposing of dust or debris that may contain asbestos fibers.**

BRAKE CALIPER

REMOVAL & INSTALLATION

1. Before servicing the vehicle, refer to the Precautions section.

2. As required, remove a small amount of brake fluid from the reservoir using a suction pump.

3. Raise and safely support the vehicle.

4. Remove the tire and wheel assembly.

5. Remove the brake hose mounting bolt. Disconnect and plug the brake line hose at the caliper.

6. Remove the brake caliper mounting bolts.

7. Remove the caliper assembly from its mounting.

> ❂❂ **CAUTION**
>
> **Do not allow the caliper to hang by the brake line hose, as damage to the hose may result.**

8. As required, remove the disc brake pads and shims from the caliper.

To install:

9. Clean the caliper thoroughly; remove any dirt or dust. Check the brake rotor for grooves or cracks and machine or replace, as necessary.

10. Installation is the reverse of the removal procedure.

11. Be sure to fill the brake system using the proper grade and type brake fluid.

12. Bleed the brake system.

13. Check for leaks and correct as required.

DISC BRAKE PADS

REMOVAL & INSTALLATION

See Figure 4.

FRONT DISC BRAKES

1. Before servicing the vehicle, refer to the Precautions section.

2. As required, remove a small amount of brake fluid from the reservoir using a suction pump.

3. Raise and safely support the vehicle.

4. Remove the tire and wheel assembly.

5. Remove the flange bolt while holding the caliper pin with a wrench.

➡**Be careful not to damage the pin boot, and pivot the caliper up and out of the way.**

6. Remove the pad shims and pads. Remove the pad retainers.

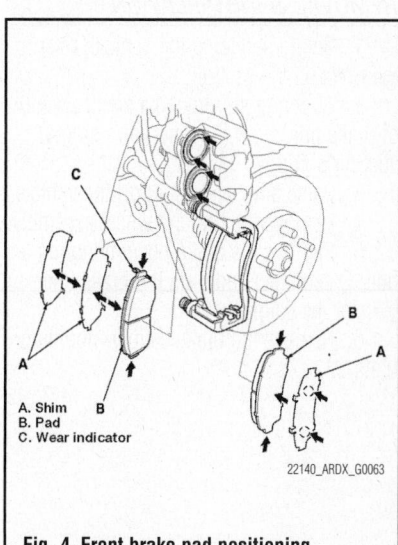

A. Shim
B. Pad
C. Wear indicator

22140_ARDX_G0063

Fig. 4 Front brake pad positioning

To install:

7. Install the pad retainers.

➡**Apply molybdenum brake grease to both surfaces of the shims and the back of the disc brake pads, prior to installation.**

8. Install the pads and shims.

9. Use a suitable tool to push caliper piston into its bore and enable the caliper to fit over the pads.

10. Lubricate the piston boot with silicon grease. Avoid twisting the boot.

11. Continue the installation in the reverse order of the removal procedure.

12. Add brake fluid to the master cylinder reservoir. Depress the brake pedal several times to seat the pads. Bleed the brakes if necessary.

BRAKES

⁜ CAUTION

Dust and dirt accumulating on brake parts during normal use may contain asbestos fibers from production or aftermarket brake linings. Breathing excessive concentrations of asbestos fibers can cause serious bodily harm. Exercise care when servicing brake parts. Do not sand or grind brake lining unless equipment used is designed to contain the dust residue. Do not clean brake parts with compressed air or by dry brushing. Cleaning should be done by dampening the brake components with a fine mist of water, then wiping the brake components clean with a dampened cloth. Dispose of cloth and all residue containing asbestos fibers in an impermeable container with the appropriate label. Follow practices prescribed by the Occupational Safety and Health Administration (OSHA) and the Environmental Protection Agency (EPA) for the handling, processing, and disposing of dust or debris that may contain asbestos fibers.

BRAKE CALIPER

REMOVAL & INSTALLATION

1. Before servicing the vehicle, refer to the Precautions section.

2. As required, remove a small amount of brake fluid from the reservoir using a suction pump.

3. Raise and safely support the vehicle.

4. Remove the tire and wheel assembly.

5. Remove the brake hose mounting bolt. Disconnect and plug the brake line hose at the caliper.

6. Remove the brake caliper mounting bolts.

7. Remove the caliper assembly from its mounting.

⁜ CAUTION

Do not allow the caliper to hang by the brake line hose, as damage to the hose may result.

8. As required, remove the disc brake pads and shims from the caliper.

To install:

9. Clean the caliper thoroughly; remove any dirt or dust. Check the brake rotor for grooves or cracks and machine or replace, as necessary.

10. Installation is the reverse of the removal procedure.

11. Be sure to fill the brake system using the proper grade and type brake fluid.

12. Bleed the brake system.

13. Check for leaks and correct as required.

DISC BRAKE PADS

REMOVAL & INSTALLATION

See Figure 5.

1. Before servicing the vehicle, refer to the Precautions section.

2. As required, remove a small amount of brake fluid from the reservoir using a suction pump.

3. Raise and safely support the vehicle.

4. Remove the tire and wheel assembly.

5. Remove the flange nuts and remove the brake hose mounting bracket.

6. Remove the flange bolt while holding the caliper pin with a wrench.

⁜ CAUTION

Be careful not to damage the pin boot, and pivot the caliper up and out of the way.

REAR DISC BRAKES

7. Remove the pad shims and pads.

8. Remove the pad retainers.

➡**The upper and lower pad retainers are different. During installation, make sure the pad retainers are in the proper position.**

To install:

9. Install the pad retainers.

10. Apply molybdenum brake grease to both surfaces of the shims and the back of the disc brake pads, prior to installation.

11. Install the pads and shims.

12. Use a suitable tool to push caliper piston into its bore and enable the caliper to fit over the pads.

13. Lubricate the piston boot with silicon grease. Avoid twisting the boot.

14. Continue the installation in the reverse order of the removal procedure.

15. Add brake fluid to the master cylinder reservoir. Depress the brake pedal several times to seat the pads. Bleed the brakes if necessary.

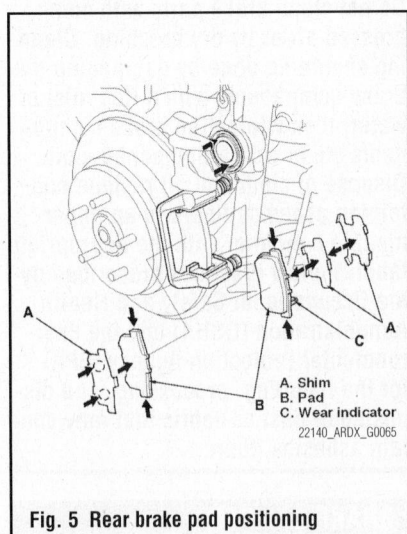

A. Shim
B. Pad
C. Wear indicator

22140_ARDX_G0065

Fig. 5 Rear brake pad positioning

BRAKES | **PARKING BRAKE**

PARKING BRAKE CABLES

ADJUSTMENT

1. Press the parking brake pedal with enough force to fully apply the parking brake. The parking brake pedal should be locked within 6–7 clicks.

2. Adjust the parking brake if the pedal clicks are not within the specification.

➡**Minor parking brake pedal adjustments (1 to 2 clicks) can be made with the adjusting nut. If a larger adjustment is required, follow the major adjustment procedure using the adjuster at the parking brake drum.**

3. After installing new parking brake shoes and/or new brake disc/drum, do the shoe lining break-in procedure.

Minor Adjustment

See Figure 6.

1. Raise the rear of the vehicle, and support it with safety stands in the proper locations.

2. Release the parking brake pedal fully.

3. Press the parking brake pedal 1 click.

4. Tighten the parking brake adjusting nut until the parking brakes drag slightly when the rear wheels are rotated.

5. Release the parking brake pedal fully, and check that the parking brakes do not drag when the rear wheels are rotated. Readjust if necessary.

6. Make sure the parking brakes are fully applied when the parking brake pedal is pressed all the way.

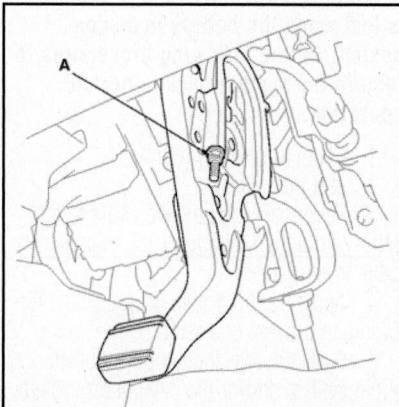

Fig. 6 Tighten the parking brake adjusting nut (A) until the parking brakes drag slightly when the rear wheels are rotated.

Major Adjustment

See Figure 7.

➡**This adjustment is to be done when replacing parking brake shoes.**

1. Before servicing the vehicle, refer to the Precautions section.

2. Raise and safely support the vehicle.

3. Fully release the parking brake lever.

4. Back off the parking brake adjusting nut on the parking brake pedal.

5. Remove the rear tire and wheel assemblies.

6. Remove the access plug.

7. Turn the ratchet teeth on the adjuster nut with a flat-tip screwdriver until the shoes lock against the parking brake drum. Then back off the adjuster 8 clicks, and install the access plug.

8. Clean the mating surface of the brake disc/drum and the inside of the wheel, then install the rear wheels.

9. Do the "Minor Adjustment" procedure.

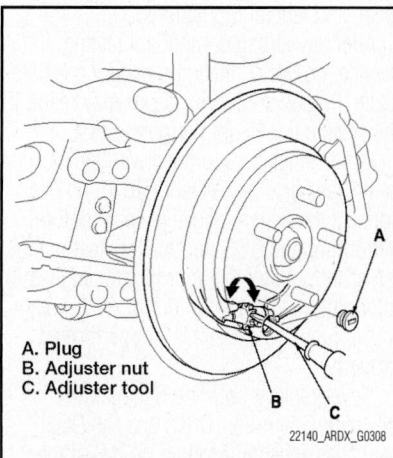

A. Plug
B. Adjuster nut
C. Adjuster tool

22140_ARDX_G0308

Fig. 7 Parking brake cable adjustment locating point

PARKING BRAKE SHOES

REMOVAL & INSTALLATION

See Figure 8.

1. Before servicing the vehicle, refer to the Precautions section.

2. As required, remove a small amount of brake fluid from the reservoir using a suction pump.

3. Raise and safely support the vehicle.

4. Remove the tire and wheel assembly.

5. Release the parking brake.

6. Remove the rear brake caliper.

7. Remove the rear disc brake rotor.

8. Disconnect and remove the brake spring and the upper return spring.

9. Disconnect and remove the lower return spring.

10. Remove the tension pins, by pushing on the appropriate retainer springs and turning the pin.

11. Remove the adjuster assembly, by moving the brake shoe forward.

12. Disconnect the parking brake cable from the parking brake lever.

13. Disconnect the rod spring and remove the strut.

14. Remove the parking brake shoes.

15. Remove the U-clip, wave washer, parking brake lever and the pivot pin from the shoe.

To install:

16. Apply molybdenum brake grease to the sliding surface of the pivot pin, prior to installation.

17. Installation is the reverse of the removal procedure.

18. Install the wave washer with its convex side facing out.

19. Pinch the U-clip securely to prevent the parking brake lever from coming out of the brake shoe.

20. Apply molybdenum brake grease.

21. Be sure that the adjuster pin is positioned correctly.

22. Fully release the parking brake pedal. Back off the pedal adjusting nut, on the parking brake pedal.

23. Remove the access plug.

24. Turn the ratchet teeth on the adjuster nut until the shoes lock against the parking brake drum.

25. Back off the adjuster eight clicks. Install the access plug.

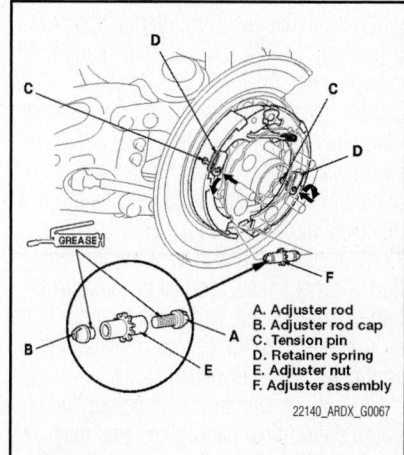

A. Adjuster rod
B. Adjuster rod cap
C. Tension pin
D. Retainer spring
E. Adjuster nut
F. Adjuster assembly

22140_ARDX_G0067

Fig. 8 Rear parking brake adjuster pin positioning

GENERAL INFORMATION

❊❊ CAUTION

These vehicles are equipped with an air bag system. The system must be disarmed before performing service on, or around, system components, the steering column, instrument panel components, wiring and sensors. Failure to follow the safety precautions and the disarming procedure could result in accidental air bag deployment, possible injury and unnecessary system repairs.

SERVICE PRECAUTIONS

Disconnect and isolate the battery negative cable before beginning any airbag system component diagnosis, testing, removal, or installation procedures. Allow system capacitor to discharge for two minutes before beginning any component service. This will disable the airbag system. Failure to disable the airbag system may result in accidental airbag deployment, personal injury, or death.

Do not place an intact undeployed airbag face down on a solid surface. The airbag will propel into the air if accidentally deployed and may result in personal injury or death.

When carrying or handling an undeployed airbag, the trim side (face) of the airbag should be pointing towards the body to minimize possibility of injury if accidental deployment occurs. Failure to do this may result in personal injury or death.

Replace airbag system components with OEM replacement parts. Substitute parts may appear interchangeable, but internal differences may result in inferior occupant protection. Failure to do so may result in occupant personal injury or death.

Wear safety glasses, rubber gloves, and long sleeved clothing when cleaning powder residue from vehicle after an airbag deployment. Powder residue emitted from a deployed airbag can cause skin irritation. Flush affected area with cool water if irritation is experienced. If nasal or throat irritation is experienced, exit the vehicle for fresh air until the irritation ceases. If irritation continues, see a physician.

Do not use a replacement airbag that is not in the original packaging. This may result in improper deployment, personal injury, or death.

The factory installed fasteners, screws and bolts used to fasten airbag components have a special coating and are specifically designed for the airbag system. Do not use substitute fasteners. Use only original equipment fasteners listed in the parts catalog when fastener replacement is required.

During, and following, any child restraint anchor service, due to impact event or vehicle repair, carefully inspect all mounting hardware, tether straps, and anchors for proper installation, operation, or damage. If a child restraint anchor is found damaged in any way, the anchor must be replaced. Failure to do this may result in personal injury or death.

Deployed and non-deployed airbags may or may not have live pyrotechnic material within the airbag inflator.

Do not dispose of driver/passenger/curtain airbags or seat belt tensioners unless you are sure of complete deployment.

Dispose of deployed airbags and tensioners consistent with state, provincial, local, and federal regulations.

After any airbag component testing or service, do not connect the battery negative cable. Personal injury or death may result if the system test is not performed first.

If the vehicle is equipped with the Occupant Classification System (OCS), do not connect the battery negative cable before performing the OCS Verification Test using the scan tool and the appropriate diagnostic information. Personal injury or death may result if the system test is not performed properly.

Never replace both the Occupant Restraint Controller (ORC) and the Occupant Classification Module (OCM) at the same time. If both require replacement, replace one, then perform the Airbag System test before replacing the other.

Both the ORC and the OCM store Occupant Classification System (OCS) calibration data, which they transfer to one another when one of them is replaced. If both are replaced at the same time, an irreversible fault will be set in both modules and the OCS may malfunction and cause personal injury or death.

If equipped with OCS, the Seat Weight Sensor is a sensitive, calibrated unit and must be handled carefully. Do not drop or handle roughly. If dropped or damaged, replace with another sensor. Failure to do so may result in occupant injury or death.

If equipped with OCS, the front passenger seat must be handled carefully as well. When removing the seat, be careful when setting on floor not to drop. If dropped, the sensor may be inoperative, could result in occupant injury, or possibly death.

If equipped with OCS, when the passenger front seat is on the floor, no one should sit in the front passenger seat. This uneven force may damage the sensing ability of the seat weight sensors. If sat on and damaged, the sensor may be inoperative, could result in occupant injury, or possibly death.

DISARMING THE SYSTEM

1. Before servicing the vehicle, refer to the Precautions section.
2. Some system store data in memory is lost when the battery is disconnected. Do the following procedures before disconnecting the battery:
   a. Make sure you have the anti-theft code(s) for the audio and/or the navigation system (if equipped).
   b. If you are replacing the audio unit, write down the audio presets (AM and FM), and the XM audio presets (if equipped), because the audio unit does not retain the presets after the battery is disconnected.
3. Make sure that the ignition switch is in the LOCK (0) position.
4. Disconnect the negative, then the positive, battery cables.

➡**Wait at least three minutes after disconnecting the battery cables before starting the repair procedure.**

ARMING THE SYSTEM

➡**Some system store data in memory is lost when the battery is disconnected. Do the following procedures to restore the system back to normal operation.**

1. Clean the battery terminals.
2. Test the battery.
3. Reconnect the positive cable to the battery first, then reconnect the negative cable to the battery.
4. Apply multipurpose grease to the terminals to prevent corrosion.
5. Enter the anti-theft code(s) for the audio system and/or the navigation system (if equipped).
6. Enter the audio presets (if applicable), and enter the XM audio presets (if equipped).

7. Set the clock (for vehicles without navigation).

CLOCKSPRING CENTERING

See Figure 9.

1. Before servicing the vehicle, refer to the Precautions section.

2. Make sure that the ignition switch is in the LOCK (0) position.

3. Disconnect the negative, then the positive, battery cables.

4. Be sure that the front wheels are in the straight ahead position.

5. Remove the steering wheel.

6. To center the cable reel, first rotate the cable reel clockwise until it stops.

7. Rotate it counterclockwise about three full turns.

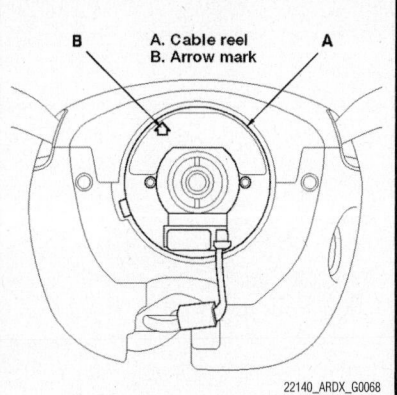

A. Cable reel
B. Arrow mark

22140_ARDX_G0068

Fig. 9 Cable reel arrow location and positioning: the arrow (mark B) on the cable reel label should point straight up.

8. The arrow on the cable reel label should point straight up.

9. Position the two tabs of the turn signal canceling sleeve, as shown.

10. Install the steering wheel onto the steering column shaft, making sure that the steering wheel hub engages the pins of the cable reel and the tabs of the turn signal canceling sleeve.

> ### ❊❊ CAUTION
> **Do not tap on the steering wheel of the steering column shaft when installing the steering wheel.**

11. Install the steering wheel. Tighten the retaining bolt to 29 ft. lbs. (39 Nm).

DRIVE TRAIN

PROPELLER SHAFT

REMOVAL & INSTALLATION

See Figures 10 and 11.

1. Raise and support the vehicle.

2. Remove the propeller shaft protector.

3. Make reference marks across the No. 1 propeller shaft and the transfer companion flange.

4. Remove the flange bolts.

5. Make reference marks across the No. 2 propeller shaft and the rear differential companion flange. Remove the flange bolts.

6. Remove the center support bearing mounting bolts, then remove the propeller shaft.

7. Install the components in reverse of removal procedure.

8. Tighten the flange bolts to 53 ft. lbs. (72 Nm).

9. Tighten the center bearing support bolts to 29 ft. lbs. (39 Nm).

FRONT DRIVESHAFT

REMOVAL & INSTALLATION

See Figures 12 through 17.

1. Raise and support the vehicle.

2. Remove the front wheel.

3. Pry up the stake on the spindle nut, then remove the nut.

4. Drain the transmission fluid, then reinstall the drain plug with a new sealing washer.

5. Remove the nuts and bolt, then separate the lower arm with a pry bar.

6. Pull the knuckle outward, and separate the outboard joint from the front hub using a soft face hammer.

7. Left driveshaft: Install the pry bar through the reference hole of the front subframe, and pry the inboard joint from the differential using a pry bar. Remove the driveshaft as an assembly.

Please note the following:

• Do not pull on the driveshaft or the inboard joint may come apart. Pull the inboard joint straight out to avoid damaging the oil seal.

• Be careful not to damage the oil seal or the end of the inboard joint with the pry bar.

8. Right driveshaft: Drive the inboard joint off of the intermediate shaft using a drift punch and a hammer. Remove the driveshaft as an assembly.

Please note the following:

• Do not pull on the driveshaft or the inboard joint may come apart.

• Be careful not to damage the end of the inboard joint with the drift.

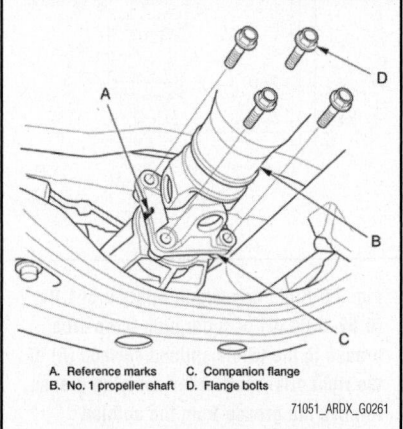

A. Reference marks   C. Companion flange
B. No. 1 propeller shaft   D. Flange bolts

71051_ARDX_G0261

Fig. 10 Make reference marks across the No. 1 propeller shaft and the transfer companion flange. Remove the flange bolts.

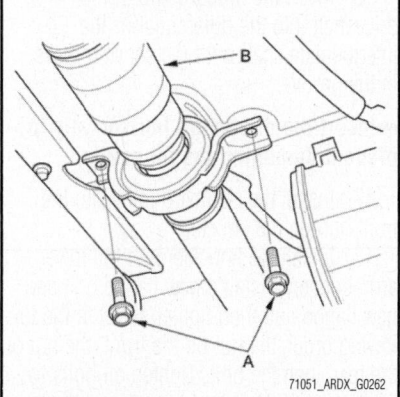

71051_ARDX_G0262

Fig. 11 Remove the center support bearing mounting bolts (A), then remove the propeller shaft (B).

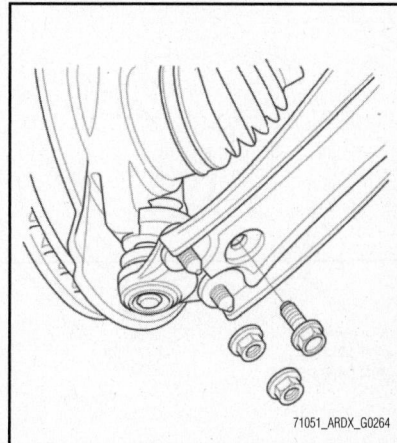

71051_ARDX_G0264

Fig. 12 Remove the nuts and bolt, then separate the lower arm with a pry bar.

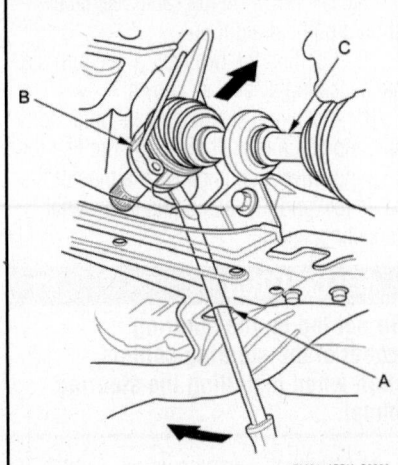

Fig. 13 Left driveshaft: Install the pry bar through the reference hole (A) of the front subframe, and pry the inboard joint (B) from the differential using a pry bar. Remove the driveshaft as an assembly. Do not pull on the driveshaft (C), or the inboard joint may come apart. Pull the inboard joint straight out to avoid damaging the oil seal.

9. Left driveshaft: Remove the set ring from the driveshaft inboard joint.

10. Right driveshaft: Remove the set ring from the intermediate shaft.

To install:

➡**Before starting installation, make sure the mating surfaces of the joint and the splined section are clean.**

11. Apply about 5 g (0.18 oz) moly 60 paste (P/N 08734-0001) to the contact area of the outboard joint and the front wheel bearing.

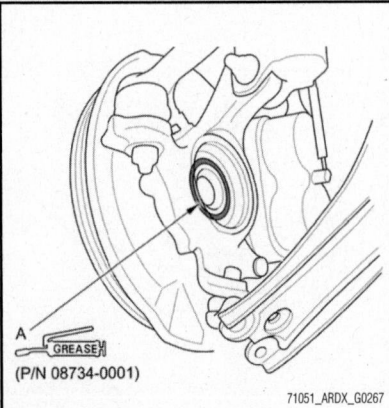

Fig. 14 Apply about 5 g (0.18 oz) moly 60 paste to the contact area (A) of the outboard joint and the front wheel bearing.

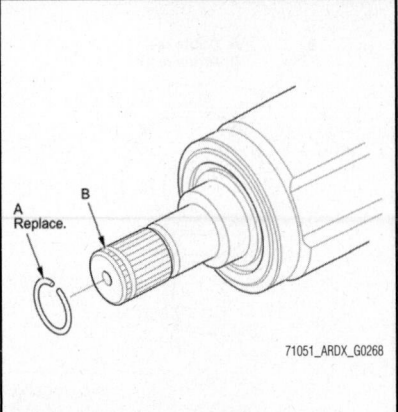

Fig. 15 Left driveshaft: Install a new set ring (A) into the set ring groove (B) of the left driveshaft inboard joint.

➡**The paste helps prevent noise and vibration.**

12. Clean the areas where the driveshaft contacts the differential thoroughly with solvent, and dry them with compressed air.

➡**Do not wash the rubber parts with solvent.**

13. Left driveshaft: Install a new set ring into the set ring groove of the left driveshaft inboard joint.

14. Right driveshaft: Install a new set ring into the set ring groove of the intermediate shaft.

15. Right driveshaft: Apply 0.5–1.0 g (0.02–0.04 oz) of super high temp urea grease (P/N 08798-9002) to the whole splined surface of the right driveshaft. After applying grease, remove the grease from the splined grooves at intervals of 2–3 splines and from the set ring groove so that air can bleed from the intermediate shaft.

16. Insert the inboard end of the driveshaft into the differential or the intermediate shaft until the set ring locks in the groove.

➡**Insert the driveshaft horizontally to prevent damaging the oil seal.**

17. Install the outboard joint into the front hub on the knuckle.

18. Install the knuckle onto the lower arm. Loosely install a new flange bolt and new flange nut, then tighten them in the following order, the nut on the front, the nut on the rear, then the bolt. Tighten all bolts to 38 ft. lbs. (52 Nm).

19. Apply a small amount of engine oil to the seating surface of a new spindle nut.

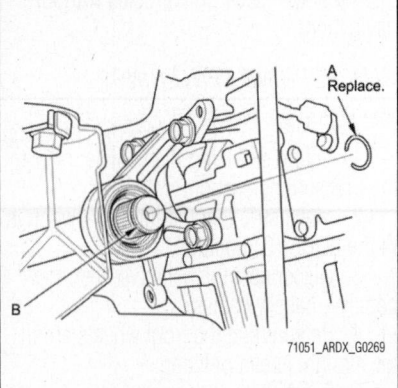

Fig. 16 Right driveshaft: Install a new set ring (A) into the set ring groove (B) of the intermediate shaft.

20. Install the spindle nut, then tighten it. After tightening, use a drift to stake the spindle nut shoulder against the driveshaft. Tighten the nut to 242 ft. lbs. (328 Nm).

21. Clean the mating surfaces of the brake disc and the wheel, then install the front wheel.

22. Turn the wheel by hand, and make sure there is no interference between the driveshaft and the surrounding parts.

23. Lower the vehicle.

24. Refill the transmission with recommended transmission fluid.

25. Check the wheel alignment, and adjust it if necessary.

26. Test-drive the vehicle.

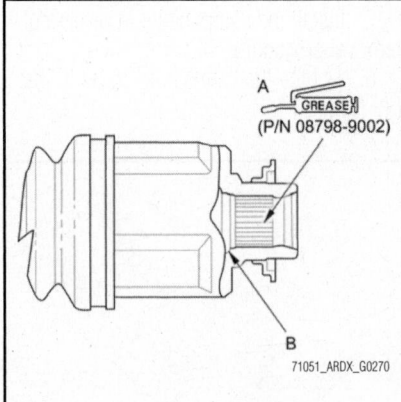

Fig. 17 Right driveshaft: Apply 0.5–1.0 g (0.02–0.04 oz) of super high temp urea grease to the whole splined surface (A) of the right driveshaft. After applying grease, remove the grease from the splined grooves at intervals of 2–3 splines and from the set ring groove (B) so that air can bleed from the intermediate shaft.

FRONT INTERMEDIATE DRIVESHAFT

REMOVAL & INSTALLATION

See Figure 18.

1. Drain the transmission fluid, then reinstall the drain plug with a new sealing washer.

2. Remove the right front driveshaft.

3. Remove the secondary HO2S (SENSOR 2) from the exhaust pipe.

4. Remove the heat shield.

5. Remove the flange bolt and the two dowel bolts.

6. Remove the intermediate shaft from the differential. Hold the intermediate shaft horizontal until it is clear of the differential to prevent damaging the oil seal.

To install:

7. Clean the areas where the intermediate shaft contacts the differential thoroughly with solvent, and dry them with compressed air.

➥**Do not wash the rubber parts with solvent.**

8. Install a new set ring into the set ring groove of the intermediate shaft.

9. Insert the intermediate shaft into the differential correctly.

10. Insert the intermediate shaft carefully to prevent damaging the oil seal.

11. Install the flange bolt and the two dowel bolts. Tighten the bolts to 29 ft. lbs. (39 Nm).

12. Install the heat shield and tighten the three bolts to 16 ft. lbs. (22 Nm).

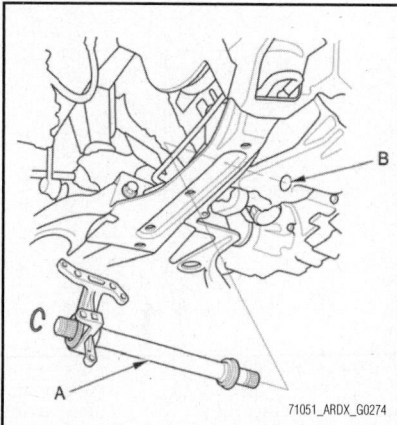

Fig. 18 Remove the intermediate shaft (A) from the differential. Hold the intermediate shaft horizontal until it is clear of the differential to prevent damaging the oil seal (B).

13. Install the secondary HO2S (sensor 2) in the exhaust pipe.

14. Install the right front driveshaft.

15. Refill the transmission with the recommended transmission fluid.

16. Check the wheel alignment, and adjust it if necessary.

17. Test-drive the vehicle.

PINION OIL SEAL

REMOVAL & INSTALLATION

A/T Differential

See Figures 19 and 20.

➥**Special Tools Required:**

- Driver Handle, 15 x 135L 07749-0010000
- Oil Seal Driver Attachment 07GAD-PG40100
- Oil Seal Driver Attachment, 58 mm 07JAD-PH80101

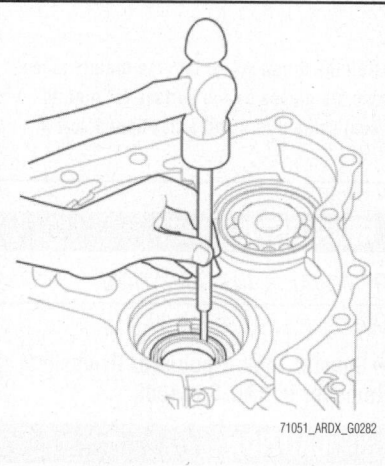

Fig. 19 Remove the oil seal from the transmission housing.

07749-0010000

07GAD-PG40100

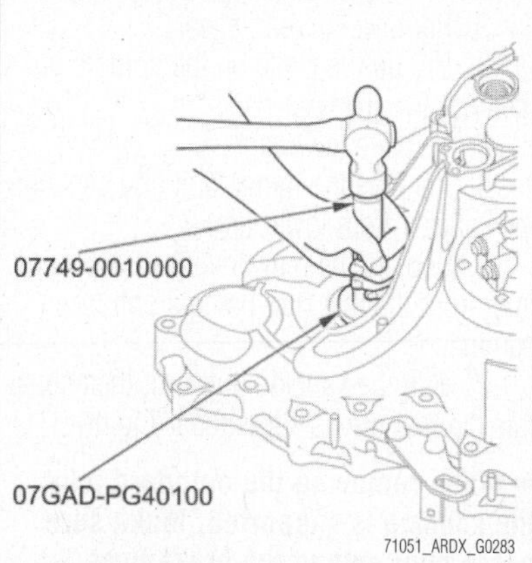

Fig. 20 Install a new oil seal flush with the transmission housing using the driver handle and the oil seal driver attachment.

1. Remove the oil seal from the transmission housing.

2. Remove the oil seal from the torque converter housing (similar as shown above).

To install:

3. Install a new oil seal flush with the transmission housing using the driver handle (15 x 135L) and the oil seal driver attachment.

4. Repeat this process for a new oil seal flush with the torque converter housing using the drive handle and the oil seal driver attachment (58 mm).

Rear Differential Side Case (with SH-AWD)

See Figures 21 through 23.

➥**Special Tools Required:**

- Driver Handle 07AAD-STKA100
- Oil Seal Driver 07AAD-STKA111
- Dust Seal Driver 07AAD-STKA120

1. Remove the rear differential.

2. Remove the 40 x 55 x 7.5 mm dust seal using a commercially available tool.

➥**Remove the dust seal with care so you do not damage the inside surface of the side case seal surface.**

3. Remove the thrust washer.

4. To remove the 34.7 x 54 x 9 mm oil seal, install the 6 mm screw into the maintenance hole of the oil seal as shown; the oil seal has four invisible places on the surface for maintenance (4 places, 90 degrees apart). Remove the oil seal using pliers and a flat-tipped screw driver as shown.

➥**Do not thread the 6 mm screws into the oil seal more than 5 mm (0.20 in).**

5. Install a new 34.7 x 54 x 9 mm oil

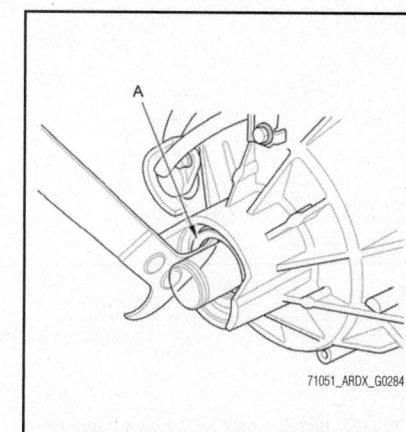

Fig. 21 Remove the 40 x 55 x 7.5 mm dust seal (A) using a commercially available tool.

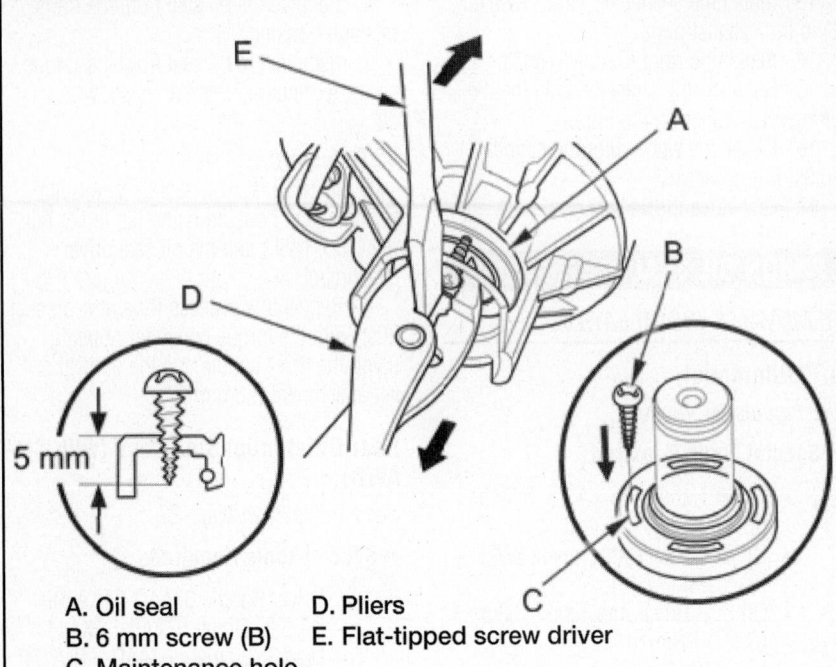

A. Oil seal
B. 6 mm screw (B)
C. Maintenance hole
D. Pliers
E. Flat-tipped screw driver

71051_ARDX_G0285

Fig. 22 To remove the 34.7 x 54 x 9 mm oil seal, install the 6 mm screw into the maintenance hole of the oil seal as shown; the oil seal has four invisible places on the surface for maintenance (4 places, 90 degrees apart). Remove the oil seal using pliers and a flat-tipped screw driver as shown.

seal into the rear differential side case using the oil seal driver and the driver handle.

6. Install the thrust washer, then install a new 40 x 55 x 7.5 mm dust seal using the dust seal driver and the driver handle.

7. Install the rear differential.

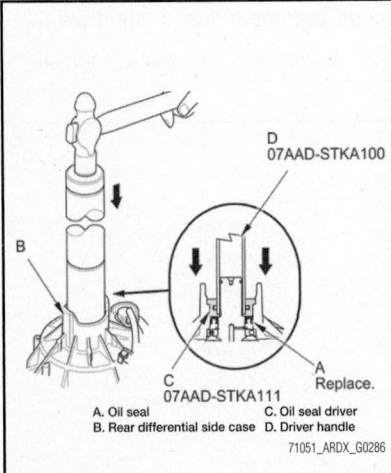

D
07AAD-STKA100

C
07AAD-STKA111

A Replace.

A. Oil seal   C. Oil seal driver
B. Rear differential side case   D. Driver handle

71051_ARDX_G0286

Fig. 23 Install a new 34.7 x 54 x 9 mm oil seal into the rear differential side case using the oil seal driver and the driver handle.

REAR DRIVESHAFT

REMOVAL & INSTALLATION

See Figures 24 through 29

➡**Special Tools Required: Driveshaft Remover 07AAD-S3VA000.**

❄❄ CAUTION

Be careful not to damage the brake hose, sensor, and harness.

1. Raise and support the vehicle.
2. Remove the rear wheel.
3. Pry up the stake on the spindle nut, then remove the nut.
4. Remove lower arm A.
5. Remove the flange bolt and separate the knuckle from lower arm B.
6. Loosen the rear driveshaft outboard joint from the rear hub using a soft face hammer.
7. Pull the knuckle outward, then separate the rear driveshaft outboard joint.

➡**When removing the outboard joint, the knuckle is supported. Make sure not to over extend the brake hose.**

8. Use a hammer to wedge the driveshaft remover between the inboard joint and the rear differential.

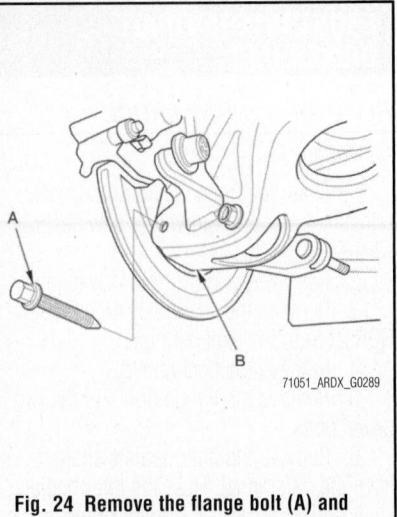

71051_ARDX_G0289

Fig. 24 Remove the flange bolt (A) and separate the knuckle from lower arm B.

9. Remove the rear driveshaft (A), then remove the set ring (B).

To install:

➡**Before starting installation, make sure the mating surfaces of the joint and the splined section are clean.**

10. Apply a small amount of super high temp urea grease (P/N 08798-9002) to the whole splined surface. After applying grease, remove the grease from the splined grooves at intervals of 2–3 splines and from the set ring groove so that air can bleed from the differential.

11. Clean the areas where the driveshaft contacts the differential thoroughly with solvent, and dry them with compressed air.

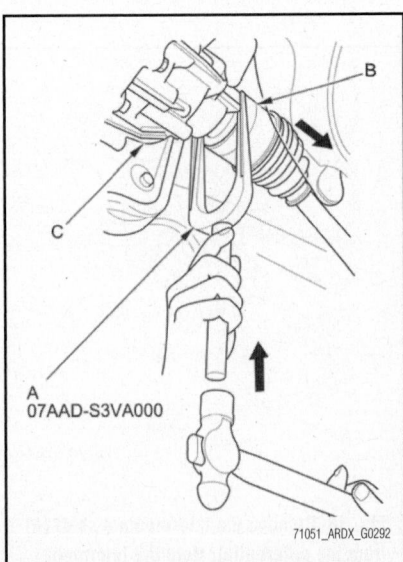

A
07AAD-S3VA000

71051_ARDX_G0292

Fig. 25 Use a hammer to wedge the driveshaft remover (A) between the inboard joint (B) and the rear differential (C).

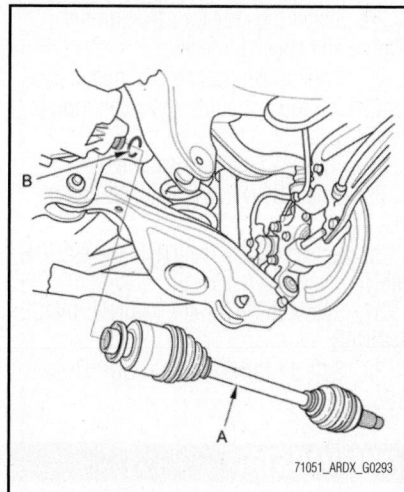

Fig. 26 Remove the rear driveshaft (A), then remove the set ring (B).

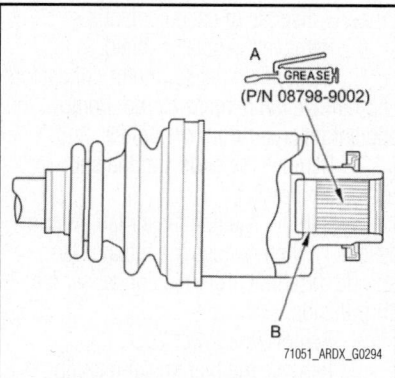

Fig. 27 Apply a small amount of super high temp urea grease (P/N 08798-9002) to the whole splined surface (A). After applying grease, remove the grease from the splined grooves at intervals of 2–3 splines and from the set ring groove (B) so that air can bleed from the differential.

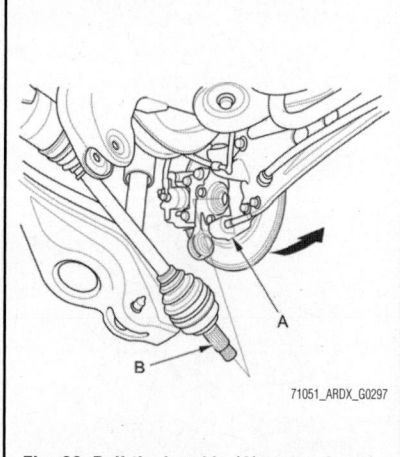

Fig. 28 Pull the knuckle (A) outward, and install the rear drive shaft outboard joint (B) into the rear hub.

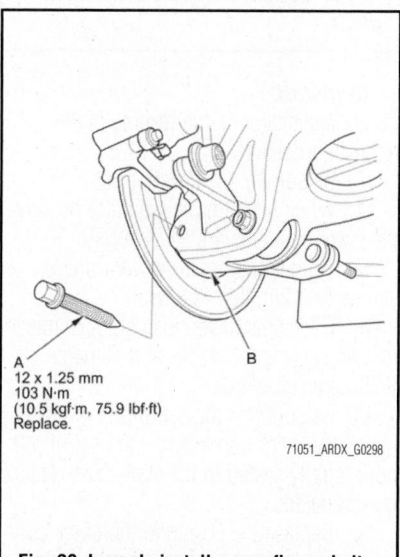

Fig. 29 Loosely install a new flange bolt (A) onto lower arm B.

➡Do not wash the rubber parts with solvent.

12. Install a new set ring into the set ring groove of the rear differential before installing the driveshaft.

13. Make sure the inboard joint is installed all the way into the rear differential and to ensure the stop ring is properly seated.

14. Pull the knuckle outward, and install the rear drive shaft outboard joint into the rear hub.

➡**When installing the outboard joint, support the knuckle. Make sure not to over extend the brake hose.**

15. Loosely install a new flange bolt onto lower arm B.

16. Install lower arm A, then loosely install a new flange bolt and a new self-locking nut with the washer. Position the "OUT" mark on lower arm A toward the outside of the vehicle.

17. Place a floor jack under the lower arm B, and raise the suspension to load it with the vehicle's weight.

18. Tighten the two flange bolts and the self-locking nut to the specified torque values, then remove the floor jack.

19. Apply a small amount of engine oil to the seating surface of a new spindle nut.

20. Install the spindle nut, then tighten it to 181 ft. lbs. (245 Nm). After tightening, use a drift to stake the spindle nut shoulder against the driveshaft.

21. Clean the mating surfaces of the brake disc and the wheel, then install the rear wheel.

22. Turn the wheel by hand, and make sure there is no interference between the driveshaft and surrounding parts.

23. Lower the vehicle.

24. Check the wheel alignment, and adjust it if necessary.

Test-drive the vehicle.

ENGINE COOLING

ENGINE COOLANT

DRAIN & REFILL

1. Start the engine. Set the heater temperature control dial to maximum heat, then turn off the ignition switch. Make sure the engine and radiator are cool to the touch.

2. Remove the radiator cap.

3. Loosen the drain plug and drain the coolant.

4. After the coolant has drained, tighten the radiator drain plug securely.

5. Remove the coolant reservoir and drain the coolant. Reinstall the coolant reservoir.

6. Fill the coolant reservoir to the MAX mark with Acura Long Life® Antifreeze/Coolant Type 2 (P/N OL999-9011A).

7. Pour Acura Long Life Antifreeze/Coolant Type 2 into the radiator up to the base of the filler neck.

➡**Always use the recommended coolant type. Using a non-Acura coolant can result in corrosion, causing the cooling system to malfunction or fail.**

➡**Acura Long Life Antifreeze/Coolant Type 2 is already a mixture of 50 % antifreeze and 50 % water. Do not add water. If the vehicles is regularly driven in very low temperatures (under -31°F, -35°C) a 60 % concentration of coolant should be used. To accomplish this, pour 0.4 US gal. (1.6 L) of Honda Extreme Cold Weather® Antifreeze/Coolant Type 2 into the radiator first, then add Acura Long Life Antifreeze/Coolant Type 2 until the radiator is full.**

8. Use the following specified amounts as a guide:

- Engine Coolant Capacities (Including the reserve tank capacity of 0.72 US qts. (0.7 L)

- At Coolant Change: 7.4 US qts. (7.0 L)
- After Engine Overhaul: 8.88 US qts. (8.4 L)

9. Loosely install the radiator cap.

10. Start the engine, and let it run until it warms up (the radiator fan comes on at least twice).

11. Turn off the engine. Check the level in the radiator and add Acura Long Life Antifreeze/Coolant Type 2 if needed.

❋❋ CAUTION

Removing the radiator cap while the engine is hot can cause the coolant to spray out. Always let the engine and radiator cool before removing the cap.

12. Put the radiator cap on tightly, then run the engine again and check for leaks.

13. Clean up any spilled engine coolant.

14. Connect the Honda Diagnostic System (HDS), or equivalent scan tool, to the data link connector (DLC).

15. Make sure the scan tool communicates with the vehicle and the powertrain control module (PCM). If it does not communicate, troubleshoot the DLC circuit.

16. Turn the ignition switch to ON (II).

17. Using the scan tool menu, select as follows:

- BODY ELECTRICAL
- GAUGE MENU, then ADJUSTMENT
- RESET in the MAINTENANCE MINDER
- MAINTENANCE SUB ITEM 5 RESET

RADIATOR

REMOVAL & INSTALLATION

See Figure 30.

1. Drain the cooling system. Properly dispose of used coolant.

2. Remove the condenser and radiator fan shroud assembly.

3. Remove the upper radiator hose.

4. Remove the lower radiator hose.

5. Disconnect and plug the transmission fluid cooler lines.

6. Disconnect the Engine Coolant Temperature (ECT) sensor 2 connector.

7. Remove the air intake duct cover and the four clips. Remove the upper radiator brackets.

8. Pull the radiator up and out of its mounting.

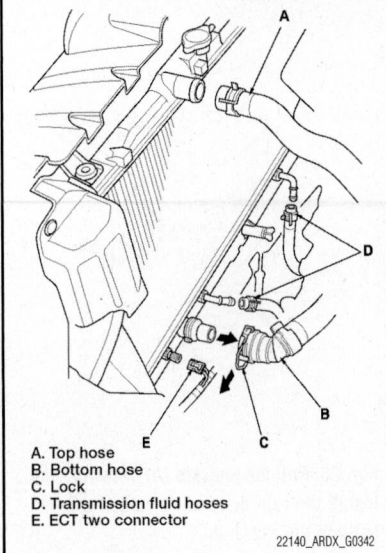

A. Top hose
B. Bottom hose
C. Lock
D. Transmission fluid hoses
E. ECT two connector

22140_ARDX_G0342

Fig. 30 Radiator removal disconnection points

To install:

9. Installation is the reverse of the removal procedure.

10. Be sure to use new O-rings.

11. When installing the radiator be sure the lower cushions are set securely.

12. Tighten the upper radiator bracket screws to 7.2 ft. lbs. (9.8 Nm).

13. Check the lower hose quick connector and set ring for cracks and damage. Replace as required.

14. Be sure that the set ring is in place inside the quick connector. If the set ring is not properly seated in the connector, replace the connector.

15. Replace the O-ring in the quick connector.

16. Check the lock. If it is damaged, replace it.

17. When installing a new lock to the connector, slide it straight down along the groove.

18. Install the lower hose on the quick connector. Install the clamp.

19. Push down the lock, then push the quick connector onto the radiator until you hear a click.

20. Continue the installation in the reverse order of the removal procedure.

21. Fill the cooling system with the proper grade and type coolant.

22. Bleed the air from the cooling system with the heater valve open.

23. Loosely install the radiator cap.

24. Start the engine and allow it to reach operating temperature until the fan comes on twice.

25. Turn the engine off. Check the coolant level. Correct as required.

26. Install the radiator cap. Run the engine and check for leaks.

27. Connect the Honda Diagnostic System (HDS) tool, or equivalent scan tool, to the DLC.

28. Turn the ignition switch ON (II).

29. Select BODY ELECTRICAL on the scan tool.

30. Select ADJUSTMENT in the GAUGE MENU.

31. Select RESET in the MAINTENANCE MINDER.

32. Select MAINTENANCE SUB ITEM 5 RESET.

THERMOSTAT

REMOVAL & INSTALLATION

See Figures 31 and 32.

1. Drain the cooling system. Be sure to properly dispose of used coolant.

2. Remove the splash shield.

3. Disconnect the fan motor connectors and remove the harness clamp. Remove the coolant reservoir from the holder.

4. Remove the clips and the support rod clamp bracket.

5. Remove the condenser fan shroud assembly, then remove the radiator fan shroud assembly from the condenser fan shroud side.

6. Remove the lower hose.

7. Remove the thermostat retaining bolts. Remove the thermostat from its mounting.

To install:

8. Installation is the reverse of the removal procedure.

9. Be sure to use a new O-ring.

10. Tighten the retaining bolts to 8.8 ft. lbs. (11.9 Nm).

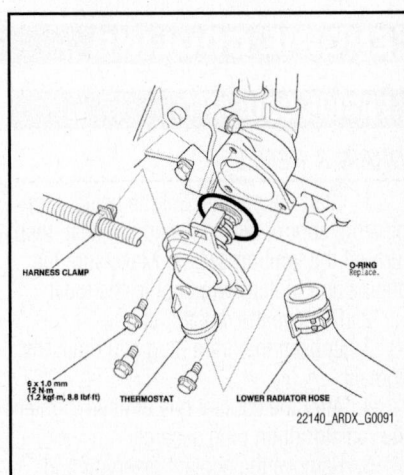

HARNESS CLAMP

O-RING
Replace.

6 x 1.0 mm
12 N·m
(1.2 kgf·m, 8.8 lbf·ft)

THERMOSTAT

LOWER RADIATOR HOSE

22140_ARDX_G0091

Fig. 31 Thermostat and related components

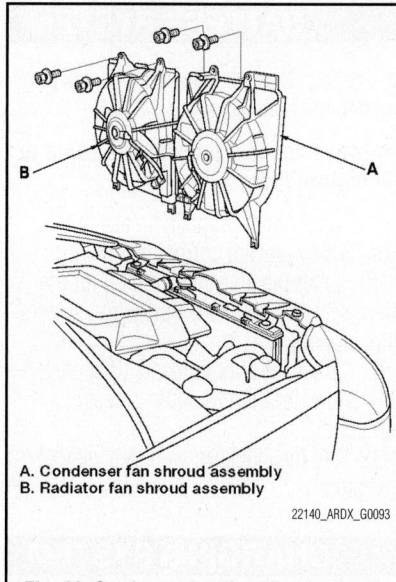

A. Condenser fan shroud assembly
B. Radiator fan shroud assembly

22140_ARDX_G0093

Fig. 32 Condenser fan shroud and related components

11. Fill the cooling system with the proper grade and type coolant.

12. Bleed the air from the cooling system with the heater valve open.

13. Loosely install the radiator cap.

14. Start the engine and allow it to reach operating temperature, until the radiator fan comes on twice.

15. Turn the engine off. Check the coolant level. Correct as required.

16. Install the radiator cap. Run the engine and check for leaks.

17. Connect the Honda Diagnostic System (HDS) tool, or equivalent scan tool, to the DLC.

18. Turn the ignition switch ON (II).

19. Select BODY ELECTRICAL on the scan tool.

20. Select ADJUSTMENT in the GAUGE MENU.

21. Select RESET in the MAINTENANCE MINDER.

22. Select MAINTENANCE SUB ITEM 5 RESET.

WATER PUMP

REMOVAL & INSTALLATION

See Figures 33 and 34.

1. Drain the cooling system. Be sure to properly dispose of used coolant.

2. Remove the splash shield.

3. Move the drive belt auto tensioner using a belt tension release tool, to relieve the tension on the drive belt. Remove the drive belt.

4. Hold the water pump pulley using an air conditioning clutch holder tool.

5. Remove the water pump pulley mounting bolts. Remove the water pump pulley.

6. Remove the water pump retaining bolts. Remove the water pump from its mounting.

To install:

7. Installation is the reverse of the removal procedure.

8. Be sure to use a new O-ring.

9. Tighten the retaining bolts to 8.8 ft. lbs. (11.9 Nm).

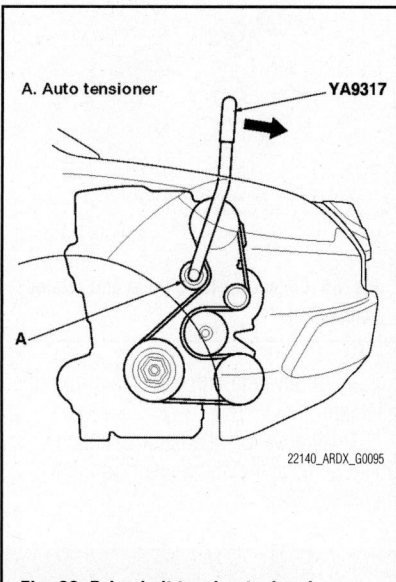

A. Auto tensioner   YA9317

22140_ARDX_G0095

Fig. 33 Drive belt tension tool and removal direction

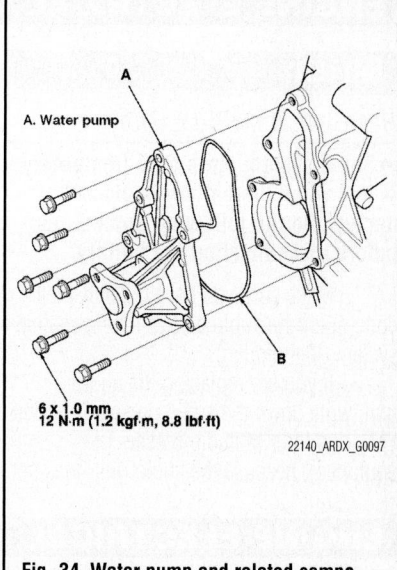

A. Water pump

6 x 1.0 mm
12 N·m (1.2 kgf·m, 8.8 lbf·ft)

22140_ARDX_G0097

Fig. 34 Water pump and related components

10. Fill the cooling system with the proper grade and type coolant.

11. Bleed the air from the cooling system with the heater valve open.

12. Loosely install the radiator cap.

13. Start the engine and allow it to reach operating temperature, until the radiator fan comes on twice.

14. Turn the engine off. Check the coolant level. Correct as required.

15. Install the radiator cap. Run the engine and check for leaks.

16. Connect the Honda Diagnostic System (HDS) tool, or equivalent scan tool, to the DLC.

17. Turn the ignition switch ON (II).

18. Select BODY ELECTRICAL on the scan tool.

19. Select ADJUSTMENT in the GAUGE MENU.

20. Select RESET in the MAINTENANCE MINDER.

21. Select MAINTENANCE SUB ITEM 5 RESET.

ENGINE ELECTRICAL   BATTERY SYSTEM

BATTERY

REMOVAL & INSTALLATION

➡**Some system store data in memory is lost when the battery is disconnected. Do the following procedures before disconnecting the battery.**

1. Make sure you have the anti-theft code(s) for the audio and/or the navigation system (if equipped).
2. If you are replacing the audio unit, write down the audio presets (AM and FM), and the XM audio presets (if equipped), because the audio unit does not retain the presets after the battery is disconnected.
3. Make sure the ignition switch is in LOCK (0).
4. Disconnect and isolate the negative cable from the battery.

➡**Always disconnect the negative cable from the battery first.**

5. Disconnect the positive cable from the battery.

To install:

6. Clean the battery terminals.
7. Test the battery.
8. Reconnect the positive cable to the battery first, then reconnect the negative cable to the battery.

➡**Always connect the positive cable to the battery first.**

9. Apply multipurpose grease to the terminals to prevent corrosion.
10. Enter the anti-theft code(s) for the audio system and/or the navigation system (if equipped).
11. Enter the audio presets (if applicable), and enter the XM audio presets (if equipped).
12. Set the clock (for vehicles without navigation).

ENGINE ELECTRICAL   CHARGING SYSTEM

ALTERNATOR

REMOVAL & INSTALLATION

See Figures 35 and 36.

1. Make sure you have the anti-theft code for the audio system and the navigation system (if equipped).
2. Make sure the ignition switch to LOCK (0).
3. Disconnect the negative cable from the battery.
4. Move the drive belt auto tensioner using a belt tension release tool, to relieve the tension on the drive belt. Remove the drive belt.
5. Remove the clip and remove the coolant reservoir from its holder.
6. Remove the clips and the support rod clamp bracket.
7. Disconnect the condenser fan motor connector. Remove the condenser fan shroud assembly.
8. Disconnect the alternator electrical connectors.

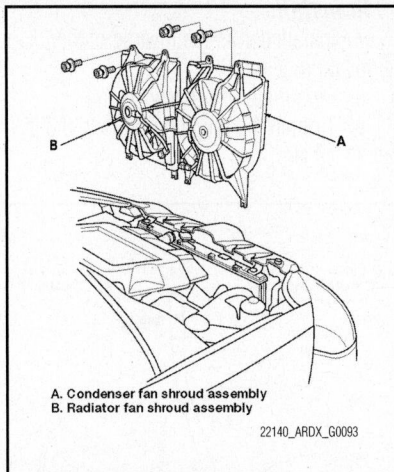

A. Condenser fan shroud assembly
B. Radiator fan shroud assembly

22140_ARDX_G0093

Fig. 35 Condenser fan shroud and related components

9. Remove the harness clamp from the alternator.
10. Remove the alternator retaining bolts. Remove the component from its mounting.

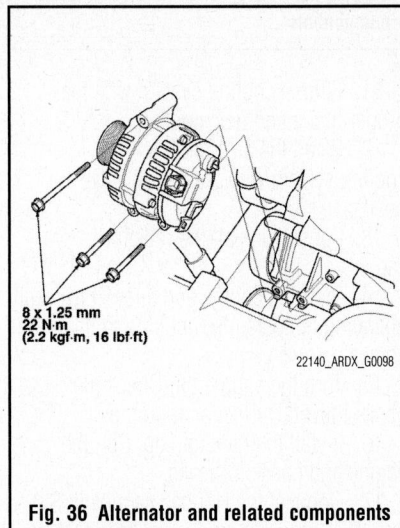

8 x 1.25 mm
22 N·m
(2.2 kgf·m, 16 lbf·ft)

22140_ARDX_G0098

Fig. 36 Alternator and related components

To install:

11. Installation is the reverse of the removal procedure.
12. Tighten the retaining bolts to 16 ft. lbs. (22 Nm).

ENGINE ELECTRICAL   IGNITION SYSTEM

FIRING ORDER

1–3–4–2

IGNITION COIL

REMOVAL & INSTALLATION

See Figure 37.

1. Before disconnecting the battery cables make sure you have the anti theft codes for the audio/navigation system.
2. Make sure that the ignition switch is in the LOCK (0) position.
3. Disconnect the negative battery cable.
4. Disconnect the positive battery cable.

➡**Wait at least three minutes after disconnecting the battery cables before starting the repair procedure.**

5. Remove the air charge cooler cover.
6. Disconnect the turbocharger boost sensor connector. Remove the vacuum hoses and the air bypass outlet connecting tube.

7. Remove the air charge cooler retaining bolts. Remove the air charge cooler.
8. Remove the ignition coil cover. Disconnect the ignition coil connectors.
9. Remove the ignition coils.

To install:

10. Tighten the coil retaining bolt to 8.8 ft. lbs. (11.9 Nm).
11. Install the air charge cooler and connect the intake air ducts.
12. Tighten the hose bands until edge (B) of the band aligns with the mark (C) on the band.

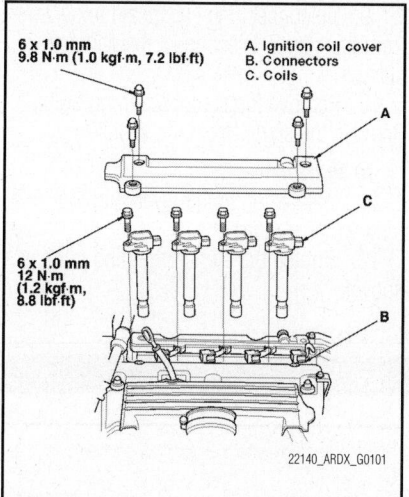

Fig. 37 Ignition coils and related components

→If the hose band edge exceeds the mark, replace the hose band.

13. Continue the installation in the reverse order of the removal procedure.

IGNITION TIMING

ADJUSTMENT

The ignition timing is not adjustable. It is controlled by the PCM.

SPARK PLUGS

REMOVAL & INSTALLATION

→Before disconnecting the battery cables make sure you have the anti theft codes for the audio/navigation system.

→Except when doing electrical inspections, always turn the ignition switch to LOCK (O), ground the SCS line with the Honda Diagnostic Service (HDS) tool, or equivalent scan tool, to take the PCM out of active status, disconnect the negative cable from the battery, then wait three minutes before starting the repair procedure. The SRS memory is not cleared even if the ignition switch is turned to LOCK (O) or the battery cables are disconnected from the battery.

1. Make sure that the ignition switch is in the LOCK (O) position.
2. Disconnect the negative battery cable.
3. Disconnect the positive battery cable.
4. Remove the ignition coils.
5. Remove the spark plugs.
6. Installation is the reverse of the removal procedure.

ENGINE ELECTRICAL

STARTING SYSTEM

STARTER

REMOVAL & INSTALLATION

See Figures 38 and 39.

1. Before servicing the vehicle, refer to the Precautions section.

→Before disconnecting the battery cables make sure you have the anti theft codes for the audio/navigation system.

→Except when doing electrical inspections, always turn the ignition switch to LOCK (O), ground the SCS line with the Honda Diagnostic Service (HDS) tool, or equivalent scan tool, to take the PCM out of active status, disconnect the negative cable from the battery,

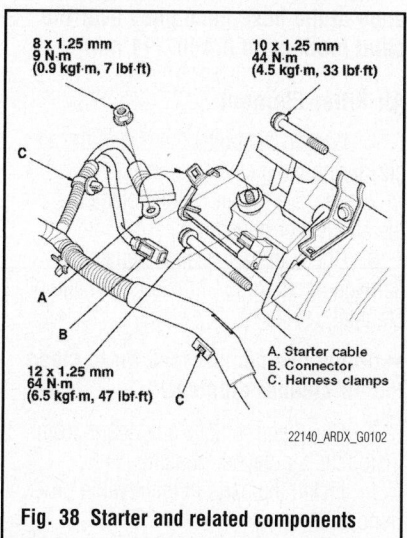

Fig. 38 Starter and related components

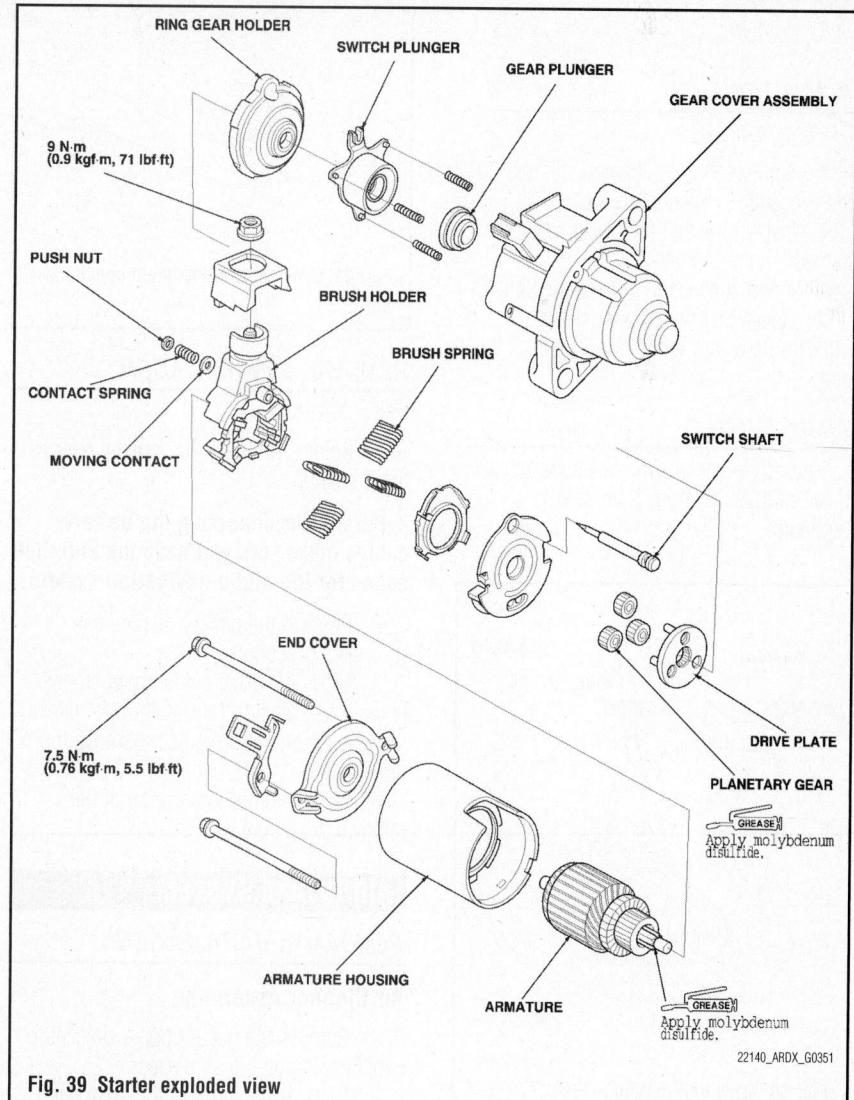

Fig. 39 Starter exploded view

then wait three minutes before starting the repair procedure. The SRS memory is not cleared even if the ignition switch is turned to LOCK (0) or the battery cables are disconnected from the battery.

2. Make sure that the ignition switch is in the LOCK (0) position.

3. Disconnect the negative battery cable.

4. Disconnect the positive battery cable.

5. Remove the clip and remove the coolant reservoir from its holder.

6. Remove the clips and the support rod clamp bracket.

7. Disconnect the condenser fan motor connector. Remove the condenser fan shroud assembly.

8. Remove the harness clamp and connector. Remove the intake manifold bracket.

9. Disconnect the electrical connectors from the starter. Disconnect the harness clamps.

10. Remove the starter retaining bolts. Remove the starter from the vehicle.

To install:

11. Installation is the reverse of the removal procedure.

12. Tighten the starter retaining bolts as shown.

ENGINE MECHANICAL

➡**Disconnecting the negative battery cable may interfere with the functions of the on board computer systems and may require the computer to undergo a relearning process, once the negative battery cable is reconnected.**

ACCESSORY DRIVE BELTS

DRIVE BELT ROUTING

See Figure 40.

INSPECTION

See Figure 41.

1. Before servicing the vehicle, refer to the Precautions section.

2. Inspect the belt for cracks and/or damage.

3. If damage exists, replace the belt.

4. Check that the auto tensioner indicator is within range. If not replace the belt.

ADJUSTMENT

The drive belt is adjusted automatically. If out of specification, it must be replaced.

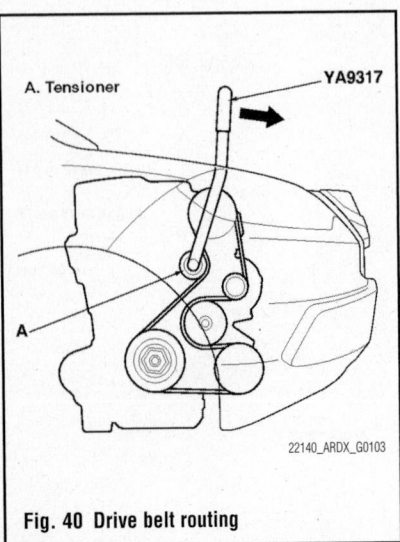

Fig. 40 Drive belt routing

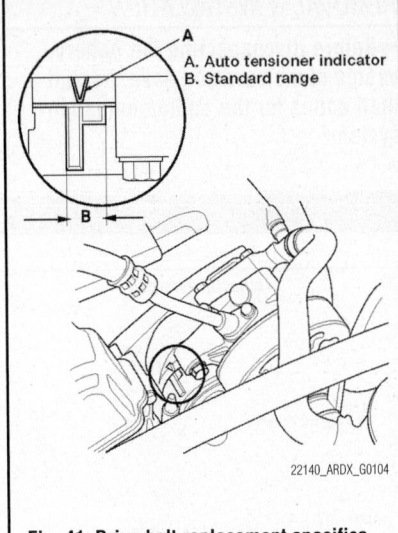

A. Auto tensioner indicator
B. Standard range

22140_ARDX_G0104

Fig. 41 Drive belt replacement specification

REMOVAL & INSTALLATION

See Figure 42.

1. Before servicing the vehicle, refer to the Precautions.

➡**Before disconnecting the battery cables make sure you have the anti theft codes for the audio/navigation system.**

2. Perform the battery disconnect/reconnect procedure.

3. Move the drive belt auto tensioner using a belt tension release tool, to relieve the tension on the drive belt. Remove the drive belt.

4. Installation is the reverse of the removal procedure.

AIR INTAKE SYSTEM

REMOVAL & INSTALLATION

Air Cleaner Assembly

1. Remove the battery upper bracket to avoid damaging the air cleaner.

2. Loosen the screw of the hose band.

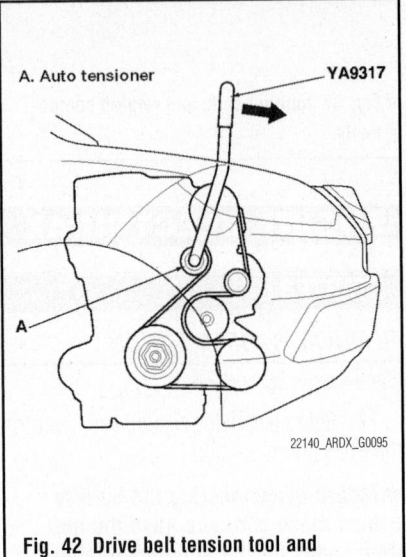

A. Auto tensioner   YA9317

22140_ARDX_G0095

Fig. 42 Drive belt tension tool and removal direction

3. Remove the bolts.

4. Remove the MAF/IAT sensor connector and harness clamps.

5. Remove the air cleaner.

6. Install the parts in the reverse order of removal.

➡**Tighten the hose band screw until the edge of the hose band goes over the paint mark up to 0.4 in. (11 mm).**

Air Filter Element

1. Loosen the bolts, then open the air cleaner housing cover.

2. Remove the air cleaner element from the air cleaner housing.

3. Check the air cleaner element for damage or clogging. If there is damage or clogged, replace it.

➡**Do not use compressed air to clean the air cleaner element.**

4. Clean and remove any debris from inside the air cleaner housing.

5. Install the parts in the reverse order of removal.

a. If you did not replace the air cleaner element, this procedure is complete.

b. If the maintenance minder required air cleaner replacement, reset the maintenance minder.

c. If you replace the air cleaner element, reset the PCM with the scan tool, and do the "PCM Idle Learn Procedure".

Air Intake Resonator

See Figure 43.

1. Do the battery removal procedure.
2. Remove the auxiliary under-hood fuse/relay box from the battery base.
3. Remove the air intake cover.
4. Remove the bolts, then remove the intake air resonator.
5. Install the parts in the reverse order of removal.
6. Do the battery installation procedure.

Charge Air Cooler Duct

See Figure 44.

1. With the help of an assistant, remove the bolts and the nut, then remove the charge air cooler air duct from the hood.
2. Install the air duct in the reverse order of removal.

Charge Air Cooler

See Figures 45 through 47.

1. Remove the charge air cooler cover.

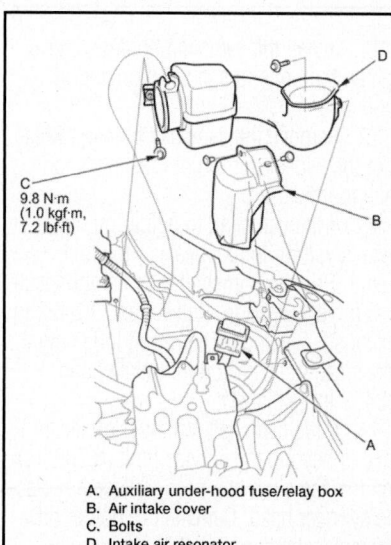

A. Auxiliary under-hood fuse/relay box
B. Air intake cover
C. Bolts
D. Intake air resonator

71051_ARDX_G0333

Fig. 43 Remove the auxiliary under-hood fuse/relay box from the battery base. Remove the air intake cover. Remove the bolts, then remove the intake air resonator.

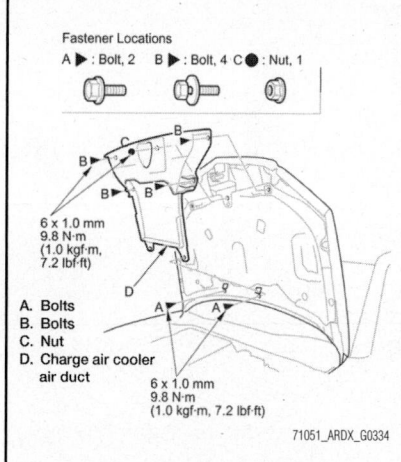

Fastener Locations
A ▶ : Bolt, 2   B ▶ : Bolt, 4 C ● : Nut, 1

6 x 1.0 mm
9.8 N·m
(1.0 kgf·m,
7.2 lbf·ft)

A. Bolts
B. Bolts
C. Nut
D. Charge air cooler air duct

6 x 1.0 mm
9.8 N·m
(1.0 kgf·m, 7.2 lbf·ft)

71051_ARDX_G0334

Fig. 44 With the help of an assistant, remove the bolts and the nut, then remove the charge air cooler air duct from the hood.

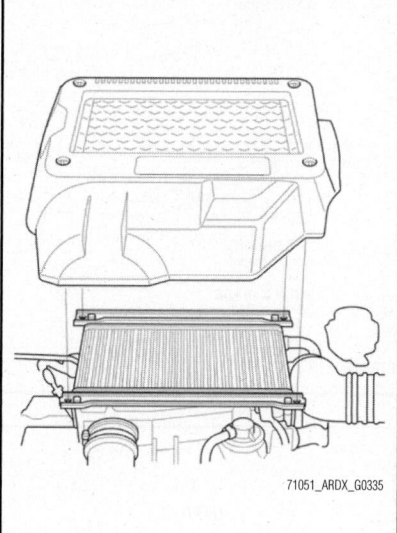

71051_ARDX_G0335

Fig. 45 Remove the charge air cooler cover.

2. Disconnect the turbocharger boost sensor connector.
3. Disconnect the vacuum hoses and the air bypass outlet connecting tube.
4. Remove the charge air cooler assembly.

To install:

5. Install the charge air cooler assembly and connect the intake air ducts (charge air cooler inlet hose and charge air cooler outlet tube). Tighten the hose bands until the edge of the band align with the mark on the band.

➡**If the hose band edge exceeds the mark, replace the hose band.**

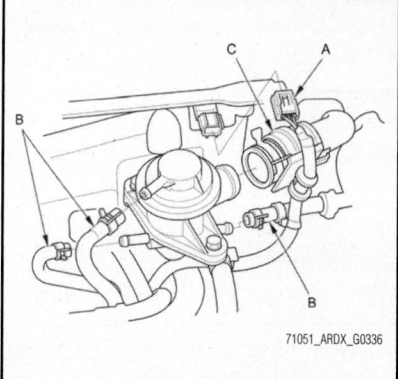

71051_ARDX_G0336

Fig. 46 Disconnect the turbocharger boost sensor connector (A). Disconnect the vacuum hoses (B) and the air bypass outlet connecting tube (C).

6. Connect the vacuum hoses and the air bypass outlet connecting tube.
7. Connect the turbocharger boost sensor connector.
8. Install the charge air cooler cover.

CAMSHAFT & VALVE LIFTERS

INSPECTION

See Figures 48 and 49.

✳✳ CAUTION

Do not rotate the camshaft during inspection.

1. Before servicing the vehicle, refer to the Precautions.

➡**Before disconnecting the battery cables make sure you have the anti theft codes for the audio/navigation system.**

2. Perform the battery disconnect/reconnect procedure.
3. Remove the rocker arm assembly.
4. Put the rocker shaft holders, the camshaft and the camshaft holders on the cylinder head. Tighten in sequence to specification.
5. Tighten 8x1.25mm bolts to 16 ft. lbs. (22 Nm), tighten 6x1.0mm bolts to 8.8 ft. lbs. (11.9 Nm) and tighten 6x1.0mm bolts 21, 22, 23.
6. Seat the camshaft by pushing it away from the camshaft pulley end of the cylinder head.
7. Zero the dial indicator gauge against the end of the camshaft, then push the camshaft back and forth, and read the end play.
8. If end play is beyond the service limit (0.02 in.) replace the cylinder head and

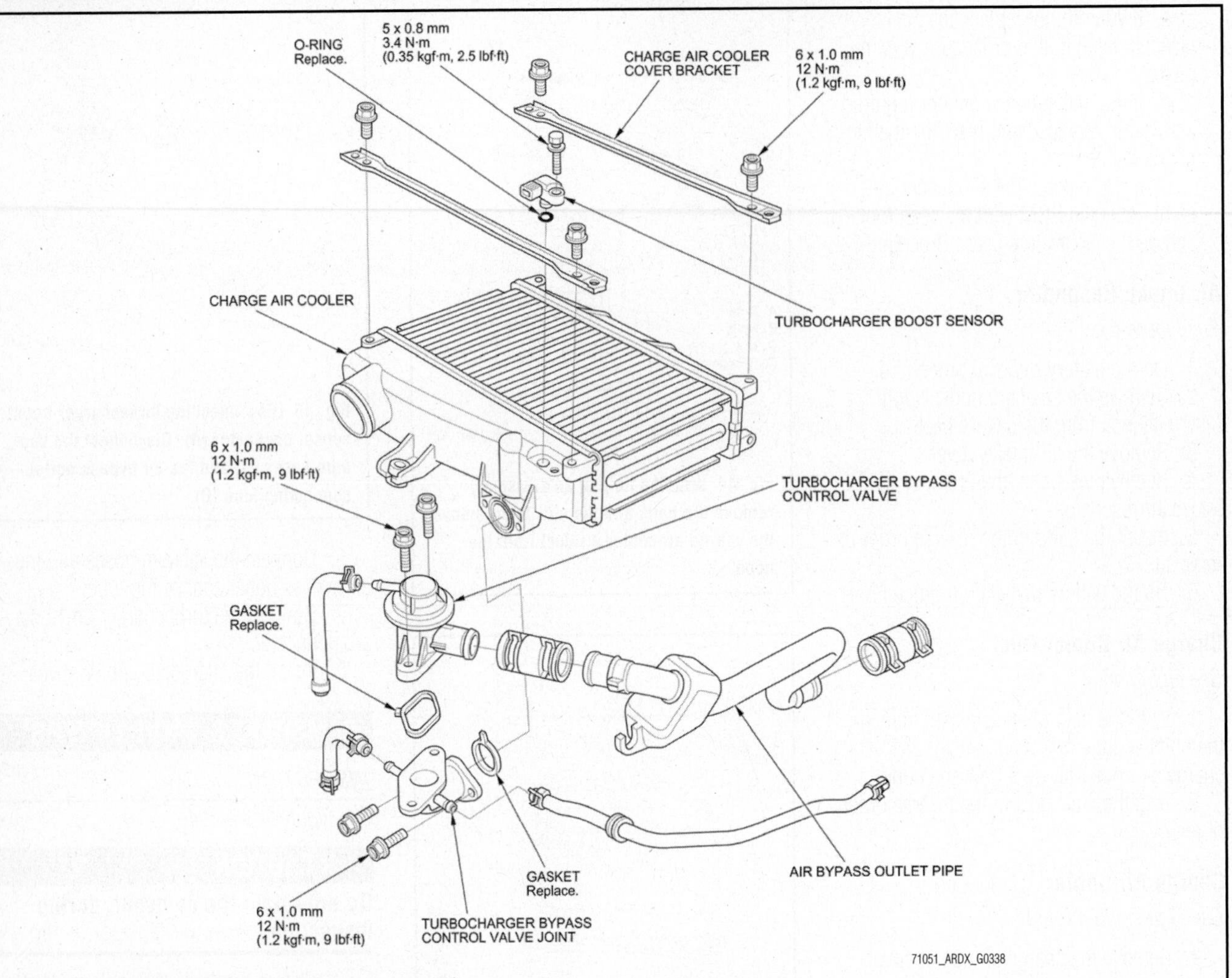

O-RING
Replace.

5 x 0.8 mm
3.4 N·m
(0.35 kgf·m, 2.5 lbf·ft)

CHARGE AIR COOLER
COVER BRACKET

6 x 1.0 mm
12 N·m
(1.2 kgf·m, 9 lbf·ft)

CHARGE AIR COOLER

TURBOCHARGER BOOST SENSOR

6 x 1.0 mm
12 N·m
(1.2 kgf·m, 9 lbf·ft)

TURBOCHARGER BYPASS
CONTROL VALVE

GASKET
Replace.

GASKET
Replace.

AIR BYPASS OUTLET PIPE

6 x 1.0 mm
12 N·m
(1.2 kgf·m, 9 lbf·ft)

TURBOCHARGER BYPASS
CONTROL VALVE JOINT

71051_ARDX_G0338

Fig. 47 Exploded view of the charge air cooler assembly

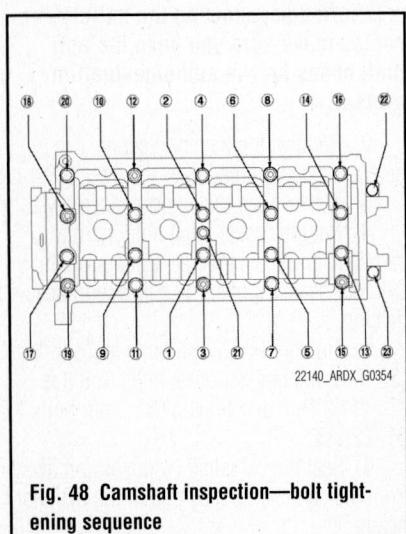

22140_ARDX_G0354

Fig. 48 Camshaft inspection—bolt tightening sequence

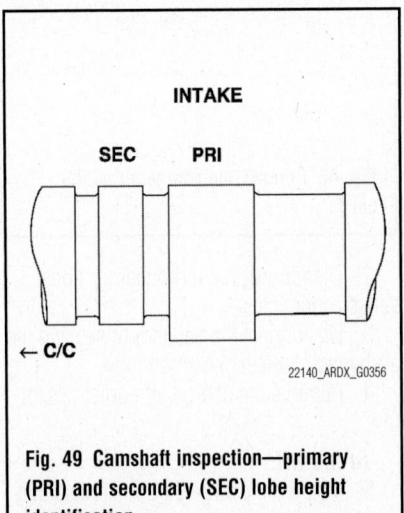

INTAKE

SEC   PRI

← C/C

22140_ARDX_G0356

Fig. 49 Camshaft inspection—primary (PRI) and secondary (SEC) lobe height identification

recheck. If specification is still not right, replace the camshaft. Camshaft end play standard (new) 0.002–0.008 in.

9. Unscrew the camshaft holder bolts two turns at a time, in a crisscross pattern.

Remove the camshaft holders from the cylinder head.

10. Lift the camshafts out of the cylinder head, wipe them clean and inspect the lift ramps. Replace the camshafts, as required.

11. Install the camshaft holders, then tighten the bolts to specification and in sequence.

12. Remove the camshaft holders. Measure the widest portion of Plastigage® on each journal.

13. If the camshaft to holder oil clearance is within limits, measure the lobe height. Primary camshaft lobe height specification is 1.3356 in. (intake) and 1.3422 in. (exhaust). Secondary camshaft lobe height specification is 1.1668 in. (intake) and 1.3422 in. (exhaust).

14. If the camshaft to holder oil clearance is beyond the service limit (0.006 in.) and the camshaft has been replaced, replace the cylinder head. Camshaft to holder oil clearance standard (new) No. 1 journal 0.001–0.003 in., all other journals 0.002–0.004 in.

15. If the camshaft to holder oil clearance is beyond the service limit and the camshaft has not been replaced, check the total runout with the camshaft supported in V-blocks.

16. If total runout is within specification, replace the cylinder head. Specification is standard (new) 0.001 in. max, service limit 0.002 in.

17. If total runout is beyond service limit, replace the camshaft and recheck the camshaft to holder oil clearance. If oil clearance is still beyond service limit, replace the cylinder head.

REMOVAL & INSTALLATION

See Figures 50 through 53.

1. Before servicing the vehicle, refer to the Precautions.

2. Perform the battery disconnect/reconnect procedure.

3. Remove the air charge cooler cover.

4. Disconnect the turbocharger boost sensor connector. Remove the vacuum hoses and the air bypass outlet connecting tube.

5. Remove the air charge cooler retaining bolts. Remove the air charge cooler.

6. Remove the fuel injector cover.

7. Remove the ignition coil cover. Disconnect the ignition coil connectors.

8. Remove the ignition coils.

9. Remove the power steering hose bracket, the breather hose and the dipstick.

10. Remove the bracket mounting bolts, from the valve cover.

11. Disconnect the rocker arm oil control solenoid connector, the rocker are oil pressure switch connector and the Variable Valve Timing Control (VTC) oil control solenoid valve connector. Remove the harness clamp.

12. Disconnect the Air/Fuel ratio (A/F) sensor connector, the oil pressure switch connector and the Crankshaft Position (CKP) sensor. Remove the harness clamps.

13. Remove the valve cover retaining bolts.

14. Remove the valve cover from the engine. Discard the gasket.

15. Remove the timing chain.

16. Loosen the rocker arm adjusting screws.

17. Remove the camshaft retaining bolts.

> **✳✳ WARNING**
>
> **To prevent damage to the camshafts, loosen the bolts in sequence, two turns at a time. Note that bolt one may not be used on all engines.**

18. Remove the camshaft chain guide B, the camshaft holders and the camshafts.

To install:

19. Make sure that the punch marks on the VTC actuator and the exhaust camshaft sprocket are facing UP. Position the camshafts in the holder.

20. Position the rocker arm caps and the chain guide B in place.

21. Tighten the camshaft retaining bolts to specification. Tighten 8x1.25 mm bolts to 16 ft. lbs. (22 Nm). Tighten 6x1.0 mm bolts to 8.8 ft. lbs. (11.9 Nm). (21, 22 and 23).

22. Continue the installation in the reverse order of the removal procedure.

23. Before installing the valve cover, clean the mating surfaces of the cylinder head and valve cover.

24. Install a new gasket in the groove of the valve cover.

25. Be sure that the mating surfaces are clean and dry.

26. Apply liquid gasket part number 08717-0004 or equivalent on the chain case and the number five rocker shaft holder mating surface areas.

> **✳✳ CAUTION**
>
> **Install the component within five minutes of applying the liquid gas-**

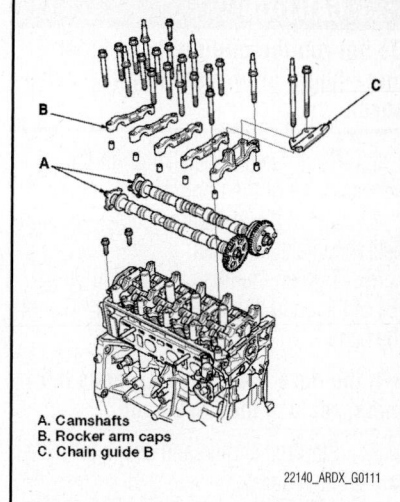

A. Camshafts
B. Rocker arm caps
C. Chain guide B

22140_ARDX_G0111

Fig. 52 Camshafts and related components

ket. Some liquid gasket materials require installation within four minutes. Be sure to read and follow the manufacturer's instruction. If too much time has passed before installing the component remove the liquid gasket and residue, then reapply new liquid gasket.

27. Position the spark plug seals on the spark plug tubes.

28. Position the valve cover on the cylinder head. Slide the cover back and forth to seat the gasket.

29. Inspect the valve cover washers, replace as required.

30. Tighten the retaining bolts in three steps to 8.8 ft. lbs. (11.9 Nm).

➡ **Wait at least 30 minutes to allow the gasket to cure before filling the engine with oil.**

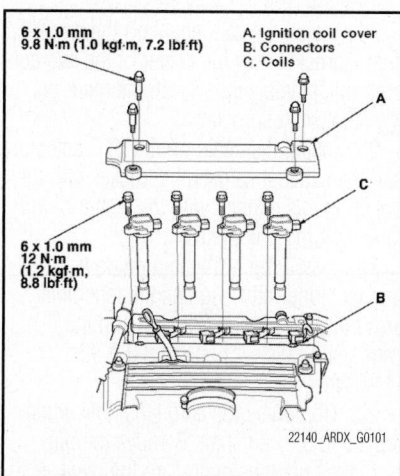

Fig. 50 Ignition coils and related components

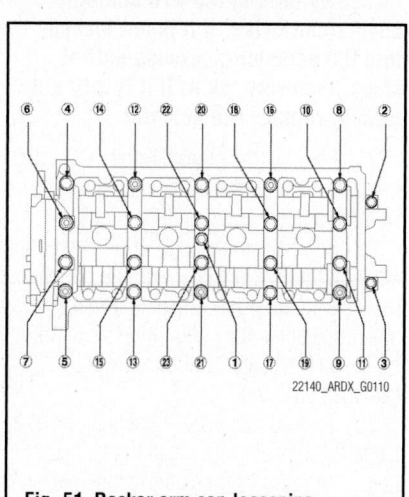

22140_ARDX_G0110

Fig. 51 Rocker arm cap loosening sequence

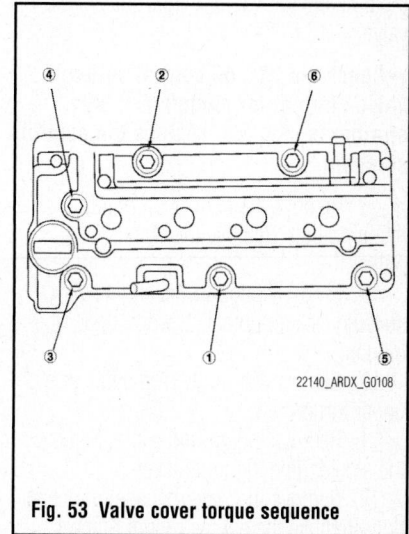

22140_ARDX_G0108

Fig. 53 Valve cover torque sequence

31. Continue the installation in the reverse order of the removal procedure.

32. Install the air charge cooler and connect the intake air ducts.

33. Tighten the hose bands until edge (B) of the band aligns with the mark (C) on the band.

➡**If the hose band edge exceeds the mark, replace the hose band.**

34. Start the engine and check for leaks, correct as required.

35. Adjust the valves.

36. Turn the ignition switch ON (II), the SRS indicator should come on for about six seconds, and then go off.

CAMSHAFT CHAIN & SPROCKETS

REMOVAL & INSTALLATION

See Figures 54 through 59.

1. Before servicing the vehicle, refer to the Precautions.

2. Perform the battery disconnect/reconnect procedure.

3. Remove the front splash shield.

4. Remove the drive belt.

5. Remove the valve cover.

6. Position the number one piston to TDC. The punch mark on the VTC actuator and the punch mark on the exhaust camshaft sprocket should be at the top. Align the TDC marks on the VTC actuator and the exhaust camshaft sprocket.

7. Disconnect the VTC oil control solenoid valve 2P connector. Remove the bolt and the VTC oil control solenoid valve.

➡**Check the VTC oil control solenoid valve strainer for clogging. If the strainer is clogged, replace the control solenoid.**

8. Remove the crankshaft pulley retaining bolt. Remove the crankshaft pulley.

9. Properly support the engine with a suitable jack and block of wood under the oil pan.

10. Remove the torque rod stiffener and upper torque rod.

11. Remove the ground cable. Remove the side engine mount bracket.

12. Remove the water bypass mounting bolt. Remove the engine mount bracket.

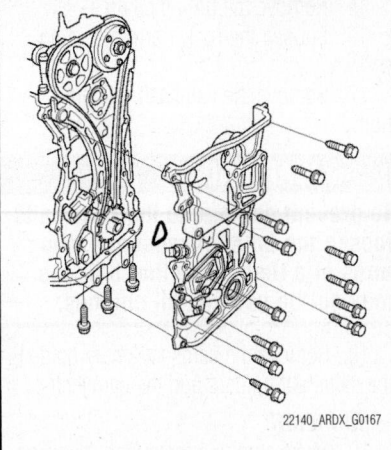

22140_ARDX_G0167

Fig. 54 Timing chain cover and retaining bolt locations

13. Remove the timing chain case cover retaining bolts. Remove the cover from its mounting on the engine.

14. Loosely install the crankshaft pulley.

15. Turn the crankshaft counterclockwise to compress the auto tensioner.

16. Align the holes on the lock and the auto tensioner. Insert a 0.06 in. diameter pin into the holes. Turn the crankshaft clockwise to secure the pin.

17. Remove the auto tensioner assembly retaining bolts. Remove the auto tensioner from its mounting.

18. Remove the cam chain guide B.

19. Remove the cam chain guide A and the tensioner arm.

20. Remove the cam chain.

To install:

➡**Keep the cam chain away from magnetic fields.**

➡**Be sure that the VTC actuator is locked by turning the VTC actuator counterclockwise. If it is not locked, turn the actuator clockwise until it stops, then recheck it. If it is still not locked, replace the actuator.**

21. Position the crankshaft at TDC. Align the TDC mark on the crankshaft sprocket with the pointer on the engine block.

22. Set the camshafts to TDC. The punch mark on the VTC actuator and the punch mark on the exhaust camshaft sprocket should be at the top. Align the TDC marks on the VTC actuator and the exhaust camshaft sprocket.

23. To hold the intake camshaft, insert a camshaft lock pin into the maintenance hole in the camshaft position pulse plate A and thru the number five rocker shaft holder.

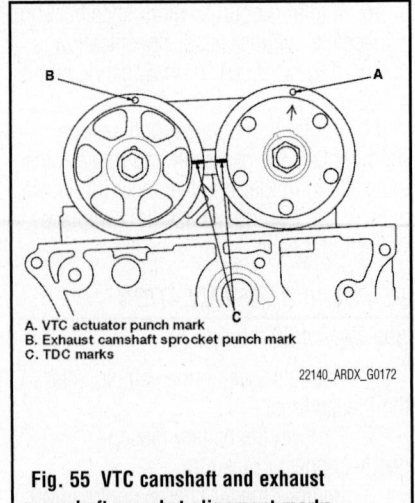

A. VTC actuator punch mark
B. Exhaust camshaft sprocket punch mark
C. TDC marks

22140_ARDX_G0172

Fig. 55 VTC camshaft and exhaust camshaft sprocket alignment marks

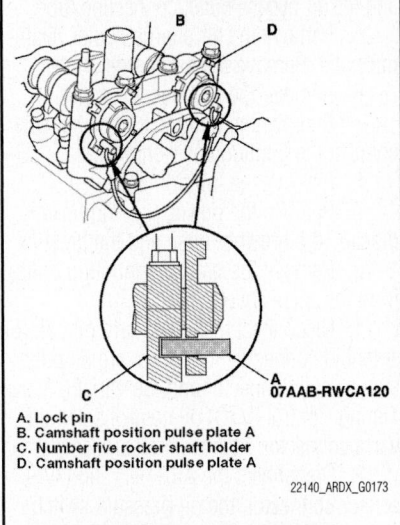

A. Lock pin
B. Camshaft position pulse plate A
C. Number five rocker shaft holder
D. Camshaft position pulse plate A

07AAB-RWCA120

22140_ARDX_G0173

Fig. 56 To hold the intake camshaft, insert a camshaft lock pin (A) into the maintenance hole in the camshaft position pulse plate A and thru the number five rocker shaft holder

24. Install the cam chain on the crankshaft sprocket with the center of the two colored link plates aligned with the mark on the crankshaft sprocket.

25. Install the chain on the VTC actuator and the exhaust camshaft sprocket with the punch marks aligned with the center of the three colored link plates.

26. Install the cam chain guide A and tensioner arm. Tighten the tensioner arm bolt to 16 ft. lbs (22 Nm) and the cam chain guide A bolts to 8.8 ft. lbs. (11.9 Nm).

27. Compress the auto tensioner when replacing the camshaft. Remove the pin from the auto tensioner. Turn the plate counterclockwise to release the lock. Press the rod and set the first cam to the edge of

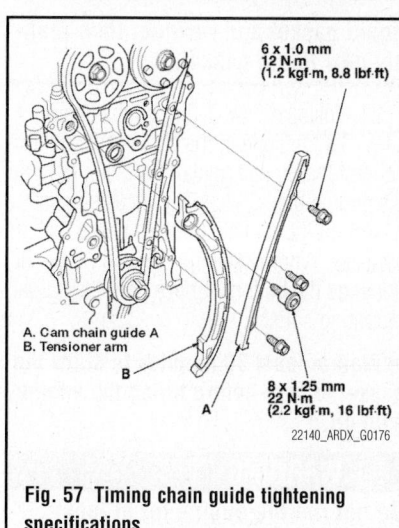

Fig. 57 Timing chain guide tightening specifications

A. Cam chain guide A
B. Tensioner arm

6 x 1.0 mm
12 N·m
(1.2 kgf·m, 8.8 lbf·ft)

8 x 1.25 mm
22 N·m
(2.2 kgf·m, 16 lbf·ft)

22140_ARDX_G0176

the rack. Insert a 0.05 in. diameter pin into the holes.

➡ **If the chain tensioner is not set up as described, the tensioner will be damaged.**

28. Install the auto tensioner. Tighten the retaining bolts to 8.8 ft. lbs. (11.9 Nm).

29. Install the cam chain guide B. Tighten the retaining bolts to 16 ft. lbs. (22 Nm).

30. Remove the lock pin from the auto tensioner.

31. Remove the camshaft lock pin set (tool).

32. Replace the timing cover oil seal. Be sure to use the proper seal installation tools.

➡ **Measure the distance between the chain case surface and the oil seal. Specification should be 1.30–1.33 in.**

33. Remove all the old liquid gasket material from the chain case mating surfaces, the bolts and the bolts holes of all mating surfaces. Clean and dry the chain case mating surfaces.

34. Apply liquid gasket to the engine block mating surface of the chain case. Apply a 0.12 in. diameter bead along the broken line as shown.

❊ CAUTION

Install the component within five minutes of applying the liquid gasket. Some liquid gasket materials require installation within four minutes. Be sure to read and follow the manufacturer's instruction. If too much time has passed before installing the component remove the liquid gasket and residue, then reapply new liquid gasket.

35. Apply liquid gasket to the engine block upper surface contact areas on the chain case and on the lower block upper surface contact areas of the chain case. Apply a 0.43 in. diameter and about 0.12 in. thickness of gasket to areas shown.

36. Apply liquid gasket to the oil pan mating surface of the chain case. Apply a 0.12 in. diameter bead along the broken line.

Install the component within five minutes of applying the liquid gasket. Some liquid gasket materials require installation within four minutes. Be sure to read and follow the manufacturer's instruction. If too much time has passed before installing the component remove the liquid gasket and residue, then reapply new liquid gasket.

37. Install a new O-ring on the chain case. Set the edge of the chain case to the edge of the oil pan. Install the chain case on the engine block. Wipe any excess liquid gasket on the oil pan and chain case mating

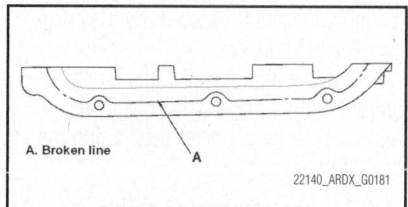

A. Broken line

22140_ARDX_G0181

Fig. 58 Apply liquid gasket part to the engine block upper surface contact areas (B) on the chain case and on the lower block upper surface contact areas (C) of the chain case, also, apply a 0.43 in. diameter and about 0.12 in. thickness of gasket to areas (B) and (C). Apply liquid gasket to the oil pan mating surface of the chain case. Apply a 0.12 in. diameter bead along the broken line (A) as shown.

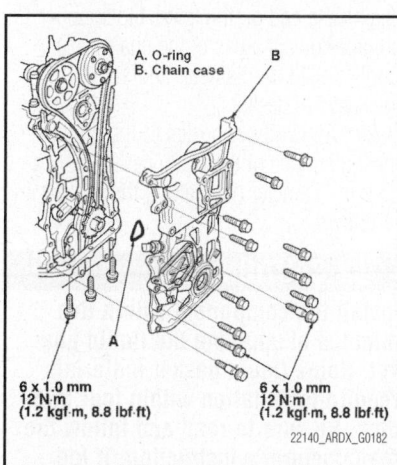

A. O-ring
B. Chain case

6 x 1.0 mm
12 N·m
(1.2 kgf·m, 8.8 lbf·ft)

6 x 1.0 mm
12 N·m
(1.2 kgf·m, 8.8 lbf·ft)

22140_ARDX_G0182

Fig. 59 Timing cover bolt installation locations

surfaces. When installing the chain case, do not slide the bottom surface onto the oil pan mounting surface.

➡ **Wait at least 30 minutes to allow the gasket to cure before filling the engine with oil.**

❊ CAUTION

Do not run the engine for at least three hours after installing the valve cover.

38. Tighten the cover retaining bolts to specification. Tightening specification is 8.8 ft. lbs. (11.9 Nm).

39. Continue the installation in the reverse order of the removal procedure.

40. When installing the VTC valve note the amount of valve opening by observing the position of the shoulder piston, thru the valve retard drain valve. If you see the shoulder of the piston, the valve is open and must be replaced.

41. Connect the battery positive terminal of the VTC terminal to the solenoid valve 2P connector terminal number two. Connect the battery negative terminal to the VTC solenoid valve 2P connector terminal number one.

42. Appearance of the inner valve in the port should be at least 0.05 in. If the inner valve does not open, replace it and repeat the procedure.

43. Perform the CKP pattern clear/relearn procedure as follows:

   a. Start the engine.

   b. Hold the engine speed at 3000 RPM, without a load in either PARK or NEUTRAL until the radiator fan comes on.

   c. Test drive the vehicle on a level road.

   d. Decelerate with the throttle fully closed from an engine speed of 2500 RPM down to 1000 RPM with the transmission in the S position 1st or 2nd gear.

   e. Repeat the previous step several times.

   f. Turn the ignition switch to LOCK (0).

   g. Turn the ignition switch to ON (II) and wait thirty seconds.

CAMSHAFT CHAIN FRONT COVER

REMOVAL & INSTALLATION

See Figures 60 and 61.

1. Before servicing the vehicle, refer to the Precautions.

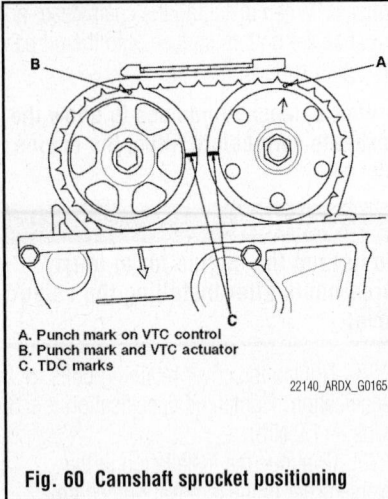

A. Punch mark on VTC control
B. Punch mark and VTC actuator
C. TDC marks

22140_ARDX_G0165

Fig. 60 Camshaft sprocket positioning at TDC

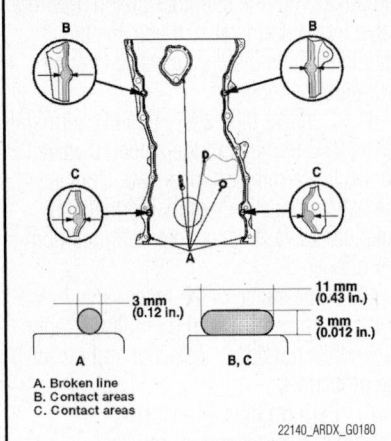

3 mm (0.12 in.)

11 mm (0.43 in.)

3 mm (0.012 in.)

A   B, C

A. Broken line
B. Contact areas
C. Contact areas

22140_ARDX_G0180

Fig. 61 Apply liquid gasket to the engine block mating surface of the chain case. Apply a 0.12 in. diameter bead along the broken line (A) as shown.

2. Perform the battery disconnect/reconnect procedure.

3. Remove the front splash shield.

4. Remove the drive belt.

5. Remove the valve cover.

6. Position the number one piston to TDC. The punch mark on the VTC actuator and the punch mark on the exhaust camshaft sprocket should be at the top. Align the TDC marks on the VTC actuator and the exhaust camshaft sprocket.

7. Disconnect the VTC oil control solenoid valve 2P connector. Remove the bolt and the VTC oil control solenoid valve.

➡**Check the VTC oil control solenoid valve strainer for clogging. If the strainer is clogged, replace the control solenoid.**

8. Remove the crankshaft pulley retaining bolt. Remove the crankshaft pulley.

9. Properly support the engine with a suitable jack and block of wood under the oil pan.

10. Remove the torque rod stiffener and upper torque rod.

11. Remove the ground cable. Remove the side engine mount bracket.

12. Remove the water bypass mounting bolt. Remove the engine mount bracket.

13. Remove the timing chain case cover retaining bolts. Remove the cover from its mounting on the engine.

To install:

14. Installation is the reverse of the removal procedure.

15. Replace the timing cover oil seal. Be sure to use the proper seal installation tools.

16. Measure the distance between the chain case surface and the oil seal. Specification should be 1.30–1.33 in.

17. Remove all the old liquid gasket material from the chain case mating surfaces, the bolts and the bolts holes of all mating surfaces. Clean and dry the chain case mating surfaces.

18. Apply liquid gasket to the engine block mating surface of the chain case. Apply a 0.12 in. diameter bead along the broken line, as shown.

➡**Install the component within five minutes of applying the liquid gasket. Some liquid gasket materials require installation within four minutes. Be sure to read and follow the manufacturer's instruction. If too much time has passed before installing the component remove the liquid gasket and residue, then reapply new liquid gasket.**

19. Apply liquid gasket to the engine block upper surface contact areas on the chain case and on the lower block upper surface contact areas of the chain case. Apply a 0.43 in. diameter and about 0.12 in. thickness of gasket.

20. Apply liquid gasket to the oil pan mating surface of the chain case. Apply a 0.12 in. diameter bead along the broken line as shown.

✳✳ CAUTION

Install the component within five minutes of applying the liquid gasket. Some liquid gasket materials require installation within four minutes. Be sure to read and follow the manufacturer's instruction. If too much time has passed before installing the component remove the

liquid gasket and residue, then reapply new liquid gasket.

21. Install a new O-ring on the chain case. Set the edge of the chain case to the edge of the oil pan. Install the chain case on the engine block. Wipe any excess liquid gasket on the oil pan and chain case mating surfaces. When installing the chain case, do not slide the bottom surface onto the oil pan mounting surface.

➡**Wait at least 30 minutes to allow the gasket to cure before filling the engine with oil.**

✳✳ WARNING

Do not run the engine for at least three hours after installing the valve cover.

22. Tighten the cover retaining bolts to specification. Tightening specification is 8.8 ft. lbs. (11.9 Nm).

23. Continue the installation in the reverse order of the removal procedure.

24. When installing the VTC valve note the amount of valve opening by observing the position of the shoulder piston, thru the valve retard drain valve. If you see the shoulder of the piston, the valve is open and must be replaced.

25. Connect the battery positive terminal of the VTC terminal to the solenoid valve 2P connector terminal number two. Connect the battery negative terminal to the VTC solenoid valve 2P connector terminal number one.

26. Appearance of the inner valve in the port should be at least 0.05 in. If the inner valve does not open, replace it and repeat the procedure.

27. Perform the CKP pattern clear/relearn procedure as follows:

   a. Start the engine.

   b. Hold the engine speed at 3000 RPM, without a load in either PARK or NEUTRAL until the radiator fan comes on.

   c. Test drive the vehicle on a level road.

   d. Decelerate with the throttle fully closed from an engine speed of 2500 RPM down to 1000 RPM with the transmission in the S position 1st or 2nd gear.

   e. Repeat the above step several times.

   f. Turn the ignition switch to LOCK (0).

   g. Turn the ignition switch to ON (II) and wait thirty seconds.

CRANKSHAFT FRONT SEAL

REMOVAL & INSTALLATION

See Figures 62 and 63.

1. Before servicing the vehicle, refer to the Precautions.
2. Perform the battery disconnect/reconnect procedure.
3. Remove the timing chain cover.
4. Carefully pry the old seal out of its mounting, using a suitable tool.

To install:

5. Position a new seal on the cover. Coat it with clean engine oil.
6. Use a seal driver tool to drive the new seal squarely into the chain case to the specified installed height.
7. Measure the distance between the chain case surface and the oil seal.
8. The oil seal installed height should be 1.30–1.33 in.

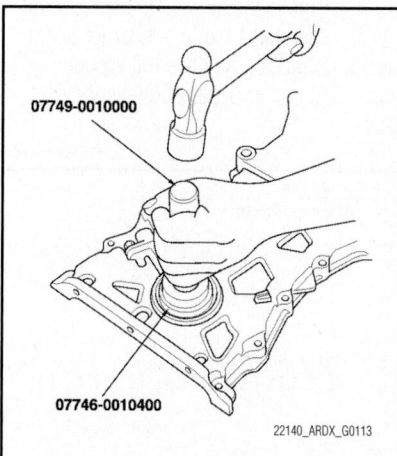

07749-0010000

07746-0010400

22140_ARDX_G0113

Fig. 62 Timing cover oil seal tool installation

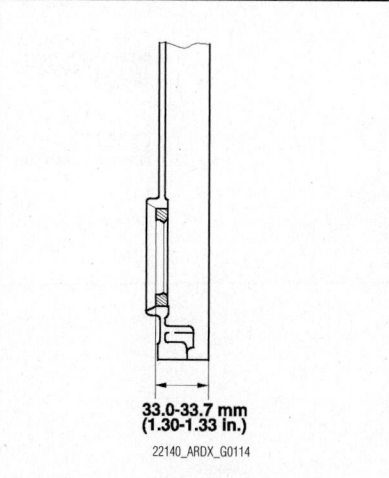

33.0-33.7 mm
(1.30-1.33 in.)

22140_ARDX_G0114

Fig. 63 Timing cover oil seal height specification

9. Continue the installation in the reverse order of the removal procedure.
10. Start the engine and check for leaks, correct as required.
11. Turn the ignition switch ON (II), the SRS indicator should come on for about six seconds, and then go off.

CYLINDER HEAD

REMOVAL & INSTALLATION

See Figures 64 through 67.

1. Before servicing the vehicle, refer to the Precautions.
2. Perform the battery disconnect/reconnect procedure.
3. Relieve the fuel system pressure.
4. Drain the engine coolant.
5. Remove the air cleaner housing assembly.
6. Remove the drive belt.
7. Remove the intake manifold.
8. Remove the exhaust manifold.
9. Remove the engine wire harness connectors and harness clamps from the cylinder head for the following:
   - Camshaft Position sensor A
   - Engine Coolant Temperature sensor
   - Camshaft Position sensor B
   - EVAP solenoid
10. Remove the quick connect fitting cover. Disconnect the fuel feed hose.
11. Disconnect the EVAP canister hose and the brake booster hoses.
12. Remove the two bolts retaining the EVAP canister purge control valve mounting bracket. Remove the harness clamps.
13. Disconnect the upper radiator hose, the heater hose and the water bypass hose.
14. Remove the timing chain.
15. Remove the camshafts.
16. Remove the cylinder head bolts.

✳✳ WARNING

To prevent damage to the head loosen the bolts a third turn at a time. Repeat the sequence until all bolts are loosened.

17. Remove the cylinder head from the engine.

To install:

18. Clean the cylinder head and block mating surfaces.
19. Install the new cylinder head gasket and dowel pins on the engine block.
20. Position the crankshaft to TDC. Align the TDC mark on the crankshaft sprocket with the pointer on the engine block.
21. Install the cylinder head.

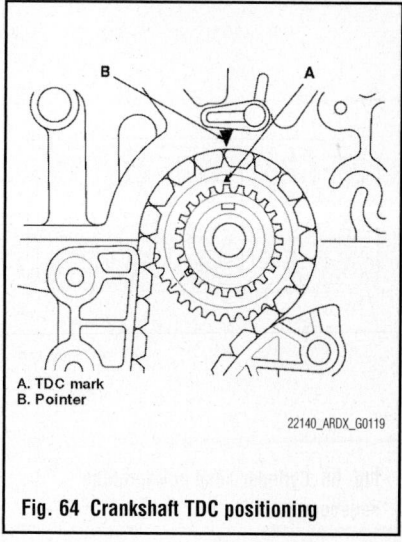

A. TDC mark
B. Pointer

22140_ARDX_G0119

Fig. 64 Crankshaft TDC positioning

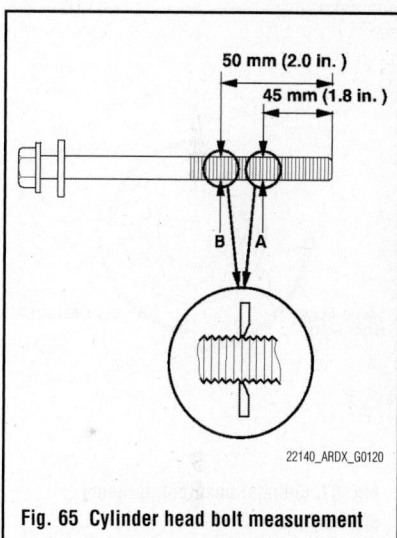

50 mm (2.0 in.)

45 mm (1.8 in.)

22140_ARDX_G0120

Fig. 65 Cylinder head bolt measurement

22. Measure the diameter of each cylinder head bolt at points "A" and "B", as shown. If either diameter is less than 0.42 in., replace the bolt.

➡**Apply engine oil to the threads and under the bolt heads, prior to installation.**

23. Tighten the cylinder head bolts to 29 ft. lbs. (39 Nm).

➡**Use a beam type torque wrench. If using a preset torque wench be sure to tighten slowly and do not over tighten. If the bolt makes any noise while tightening it, loosen it and retighten it.**

24. After tightening, tighten all bolts in two steps (90 degrees per step) in the same sequence used for tightening the cylinder head bolts in the step above.

➡**If you are using new bolts tighten an extra 90 degrees.**

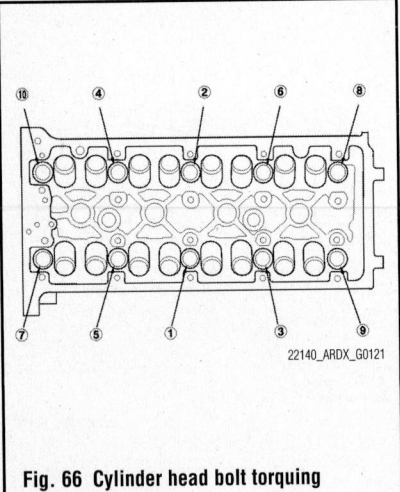

Fig. 66 Cylinder head bolt torquing sequence (1 of 2)

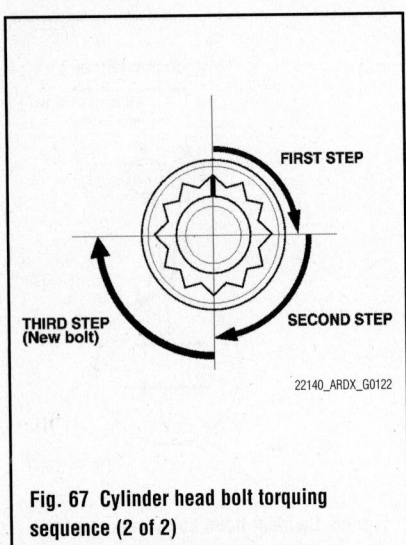

Fig. 67 Cylinder head bolt torquing sequence (2 of 2)

25. If the bolt you tightened is beyond the specified angle, re-measure the bolt. Do not loosen it back to the specified angle.

26. Continue the installation in the reverse order of the removal procedure.

27. Start the engine and check for leaks, correct as required.

28. Inspect the idle speed. Inspect the ignition timing.

29. Turn the ignition switch ON (II), the SRS indicator should come on for about six seconds, and then go off.

EXHAUST MANIFOLD

REMOVAL & INSTALLATION

See Figure 68.

1. Before servicing the vehicle, refer to the Precautions.

2. Perform the battery disconnect/reconnect procedure.

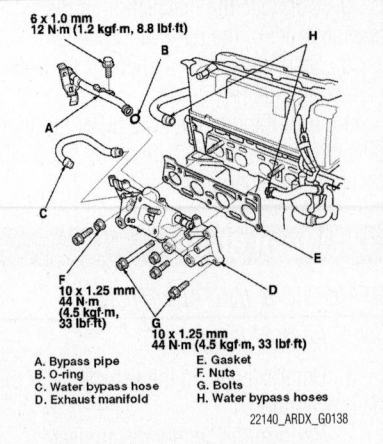

A. Bypass pipe
B. O-ring
C. Water bypass hose
D. Exhaust manifold
E. Gasket
F. Nuts
G. Bolts
H. Water bypass hoses

Fig. 68 Exhaust manifold and related components

3. Remove the turbocharger.

4. Remove the water bypass hoses.

5. Remove the exhaust manifold retaining bolts.

6. Remove the exhaust manifold from the engine.

To install:

7. Clean the mating surfaces of the engine and the exhaust manifold of dirt and debris.

8. Be sure to use a new gasket.

9. Tighten the retaining bolts and nuts to 33 ft. lbs. (45 Nm).

10. Continue the installation in the reverse order of the removal procedure.

11. Turn the ignition switch ON (II), the SRS indicator should come on for about six seconds, and then go off.

12. Road test the vehicle, correct problems as required.

INTAKE MANIFOLD

REMOVAL & INSTALLATION

See Figures 69 and 70.

1. Before servicing the vehicle, refer to the Precautions.

2. Perform the battery disconnect/reconnect procedure.

3. Relieve the fuel system pressure.

4. Remove the air charge cooler cover.

5. Disconnect the turbocharger boost sensor connector. Remove the vacuum hoses and the air bypass outlet connecting tube.

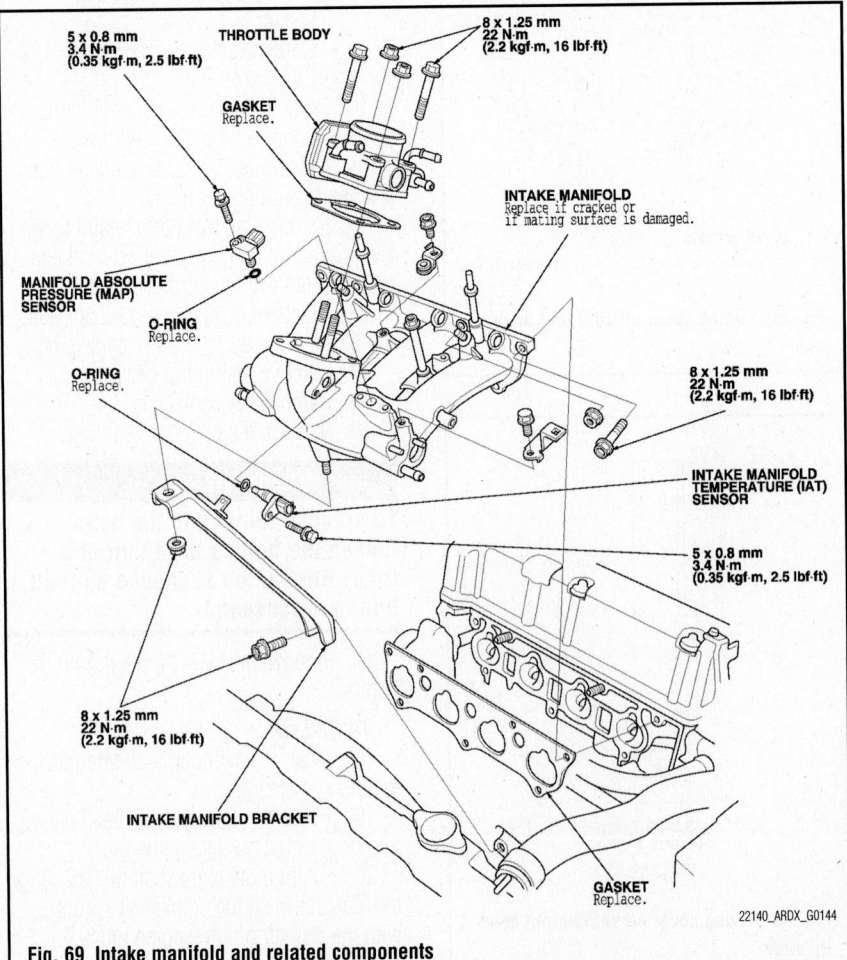

Fig. 69 Intake manifold and related components

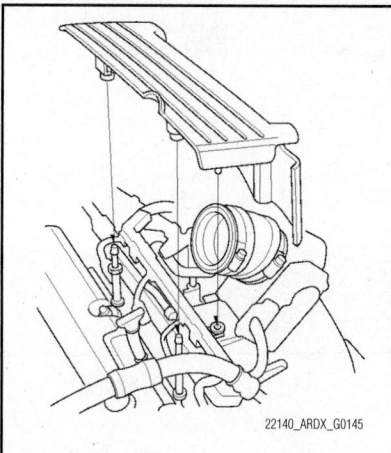

Fig. 70 Fuel injector cover and related components

6. Remove the air charge cooler retaining bolts. Remove the air charge cooler.

7. Remove the fuel injector cover. Remove the injector electrical connectors.

8. Remove the throttle actuator connector. Remove the Manifold Absolute Pressure (MAP) sensor connector. Remove the Intake Air Temperature (IAT) sensor connector. Remove the turbocharger bypass control solenoid valve connector.

9. Remove the quick connect fitting cover. Disconnect the fuel feed hose.

10. Remove the ground cable, the vacuum hoses, the brake booster vacuum hoses and the turbocharger bypass control solenoid valve bracket mounting bolt.

11. Remove the EVAP canister hose and the water bypass line bracket mounting bolt.

12. Remove the water bypass hoses, plug the lines.

13. Remove the PCV hose and the harness holder.

14. Remove the harness clamp and the connector, and then remove the intake manifold bracket.

15. Remove the intake manifold retaining bolts. Remove the intake manifold from the engine.

To install:

16. Clean the mating surfaces of the components to be sure they are free of dirt and debris.

17. Be sure to use a new gasket.

18. Tighten the retaining bolts in a crisscross pattern, beginning with the inner bolt to 16 ft. lbs. (22 Nm).

19. Continue the installation in the reverse order of the removal procedure.

20. Turn the ignition switch ON (II), the SRS indicator should come on for about six seconds, and then go off.

21. Road test the vehicle, correct problems as required.

OIL PAN

REMOVAL & INSTALLATION

See Figures 71 through 74.

1. Before servicing the vehicle, refer to the Precautions.

2. Perform the battery disconnect/reconnect procedure.

3. Raise and support the vehicle safely. Remove the front tire and wheel assemblies.

4. Drain the engine oil.

5. Remove the splash shield.

6. Remove the bolt securing the power steering line bracket, and unclamp the power steering line clamps on the front subframe.

7. Remove the bolts retaining the steering gearbox mounting brackets.

8. Remove the two bolts at the three way catalytic converter front joint.

9. Remove the lower torque rod.

10. Attach the front subframe adapter (VSB02C000016) to the subframe, by looping the strap over the front of the subframe. Secure the strap.

11. Raise the jack and line up the slots in the arms with the bolt holes on the corner of the jack base. Tighten the bolts.

12. Remove the mounting bolts. Loosen the mounting bolts so they are about ½ inch from the mounting surface.

13. Remove the subframe adapter.

14. Remove the lower torque rod bracket. Remove the torque converter cover inspection plate. Remove the transmission mounting bolts.

15. Remove the oil pan retaining bolts.

16. Using a suitable pry tool, separate the oil pan from the block.

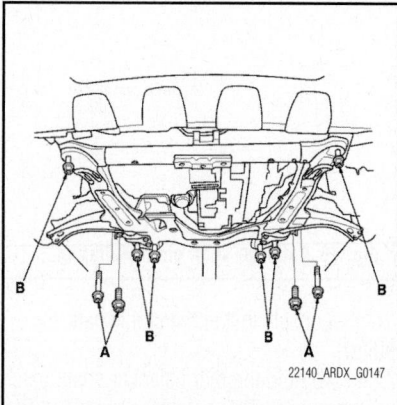

Fig. 71 Remove the mounting bolts (A). Loosen the mounting bolts (B) so they are about ½ inch from the mounting surface.

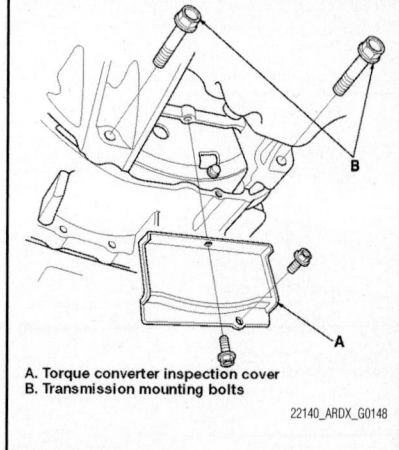

A. Torque converter inspection cover
B. Transmission mounting bolts

Fig. 72 Torque converter inspection cover and bolt locations

To install:

17. Clean the mating surfaces of the components to be sure they are free of dirt and debris.

18. Apply liquid gasket part number 08717-0004 or equivalent to the engine block mating surface of the oil pan and to the inside edge of the threaded bolt holes.

✴✴ CAUTION

Install the component within five minutes of applying the liquid gasket. Some liquid gasket materials require installation within four minutes. Be sure to read and follow the manufacturer's instruction. If too much time has passed before installing the component remove the liquid gasket and residue, then reapply new liquid gasket.

19. Apply liquid gasket about 0.12 in. diameter bead along the broken line.

20. Install the oil pan.

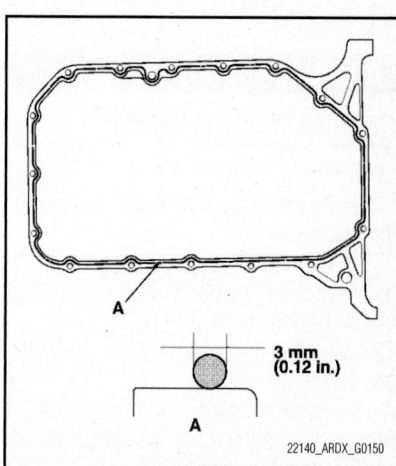

Fig. 73 Apply liquid gasket about 0.12 in. diameter bead along the broken line (A).

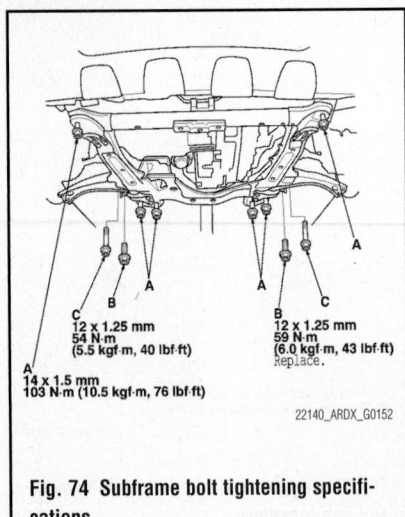

Fig. 74 Subframe bolt tightening specifications

21. Tighten the bolts in three steps. In the final step, tighten all bolts to 8.8 ft. lbs. (11.9 Nm). Wipe off the excess liquid gasket on each side of the crankshaft pulley and drive plate.

→Wait at least 30 minutes to allow the gasket to cure before filling the engine with oil.

※※ CAUTION
Do not run the engine for at least three hours after installing the oil pan.

22. Continue the installation in the reverse order of the removal procedure.

23. When installing the subframe be sure to tighten the bolts as shown.

24. Fill the engine with the proper grade and type engine oil.

25. Turn the ignition switch ON (II), the SRS indicator should come on for about six seconds, and then go off.

26. Road test the vehicle, correct problems as required.

OIL PUMP

REMOVAL & INSTALLATION

See Figures 75 through 78.

1. Before servicing the vehicle, refer to the Precautions.

2. Perform the battery disconnect/reconnect procedure.

3. Turn the crankshaft pulley so that TDC mark (on the pulley) lines up with the pointer (on the engine).

4. Raise and support the vehicle safely. Remove the front tire and wheel assemblies.

5. Drain the engine oil.

6. Remove the oil pan.

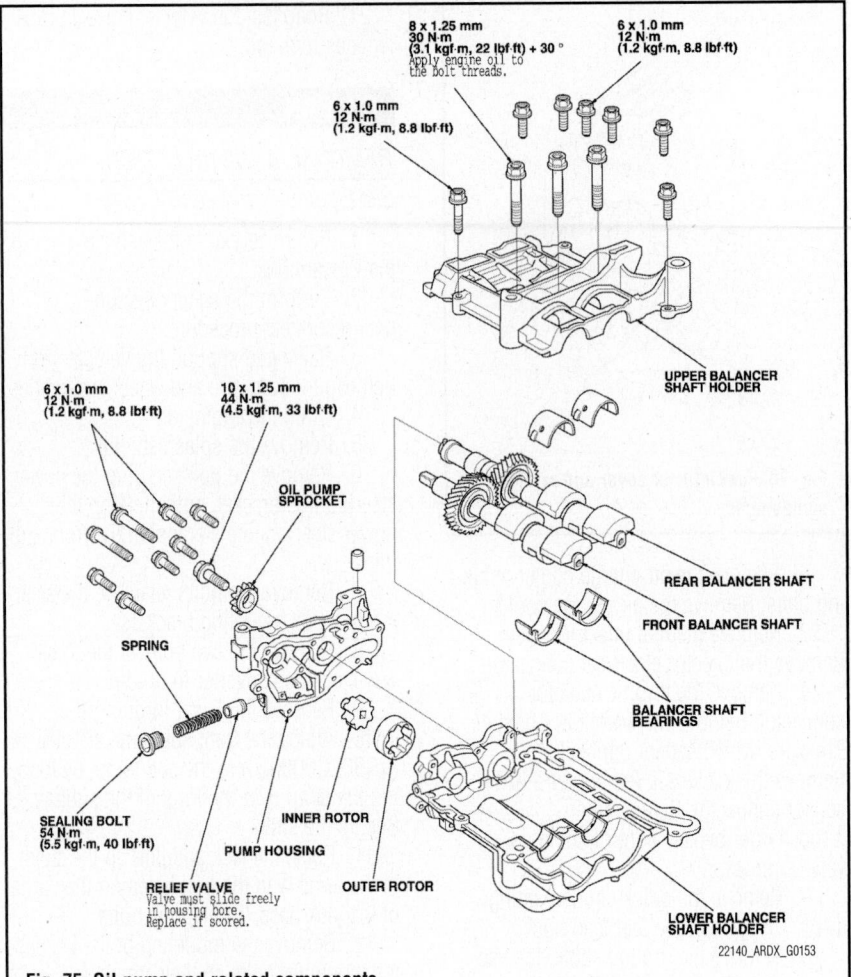

Fig. 75 Oil pump and related components

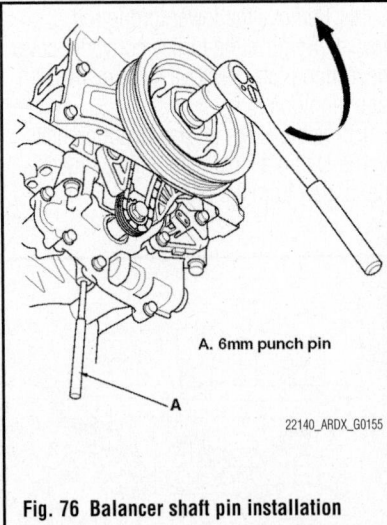

Fig. 76 Balancer shaft pin installation

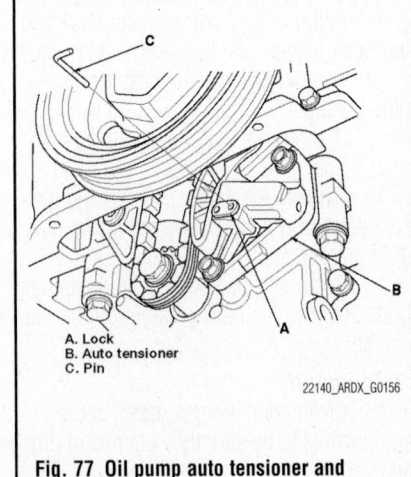

Fig. 77 Oil pump auto tensioner and related components

7. Loosely install the crankshaft pulley.

8. To hold the rear balancer shaft, insert a 6mm pin into the maintenance hole in the balancer shaft holder and through the rear balancer shaft.

9. Turn the crankshaft counterclockwise to compress the oil pump chain auto tensioner.

10. Align the holes on the lock and the oil pump chain auto tensioner. Insert a 0.06 in. diameter pin into the holes. Turn the crankshaft clockwise to secure the pin.

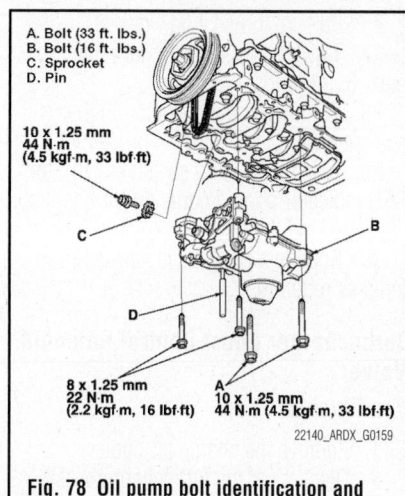

Fig. 78 Oil pump bolt identification and tightening specifications

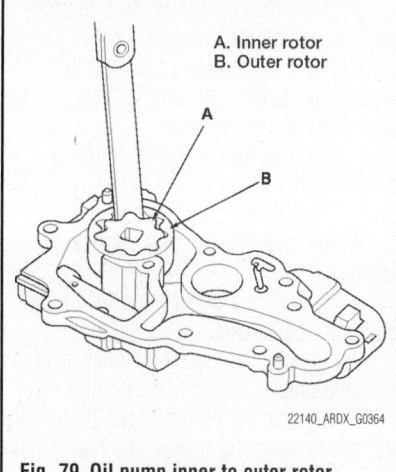

Fig. 79 Oil pump inner to outer rotor radial clearance check

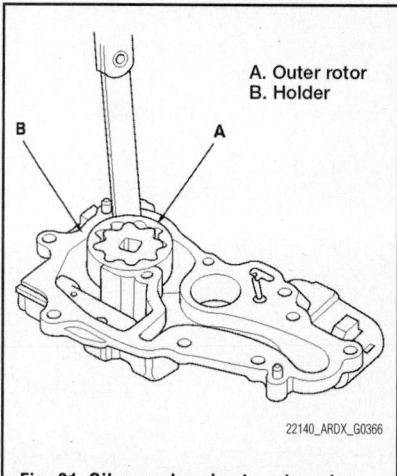

Fig. 81 Oil pump housing to outer rotor clearance check

11. Remove the oil pump chain auto tensioner.

12. Loosen the oil pump sprocket mounting bolt.

13. Remove the oil pump sprocket. Remove the oil pump retaining bolts. Remove the oil pump from its mounting.

To install:

14. Align the dowel pin on the rear balancer shaft with the mark on the oil pump.

15. To hold the balancer shaft, insert a 6mm pin into the maintenance hole in the lower balancer shaft holder and through the rear balancer shaft.

16. Apply clean engine oil to the bolt threads of the oil pump retaining bolts.

17. Loosely install the oil pump, then install the oil pump sprocket. Tighten the retaining bolt to 33 ft. lbs. (45 Nm).

18. Remove the pin driver.

19. Tighten the oil pump retaining bolts to specification. Tighten the 8x1.25mm bolts to 16 ft. lbs. (22 Nm). Tighten the 10x1.25mm bolts to 33 ft. lbs. (45 Nm).

20. Install the oil pump chain auto tensioner. Tighten the retaining bolts to 8.8 ft. lbs. (11.9 Nm). Remove the pin from the assembly.

21. Install the oil pan.

22. Continue the installation in the reverse order of the removal procedure.

23. Fill the engine with the proper grade and type engine oil.

24. Reconnect the battery. Perform the battery disconnect/reconnect procedure.

25. Road test the vehicle, correct problems as required.

INSPECTION

See Figures 79 through 81.

1. Remove the oil pump from the engine.

2. Remove the pump housing retaining bolts. Remove the pump housing.

3. Check the inner to outer radial clearance between the inner rotor and the outer rotor. It specification exceeds the service limit, replace the oil pump. Standard (new) service specification is 0.004–0.007 in. Service limit specification is 0.008 in.

4. Check the housing to rotor axial clearance between the rotor and the lower balancer shaft holder. It specification exceeds the service limit, replace the oil pump. Standard (new) service specification is 0.0012–0.0030 in. Service limit specification is 0.005 in.

5. Check the housing to outer rotor clearance between the outer rotor and the lower balancer shaft holder. It specification exceeds the service limit, replace the oil pump. Standard (new) service specification is 0.0014–0.0035 in. Service limit specification is 0.009 in.

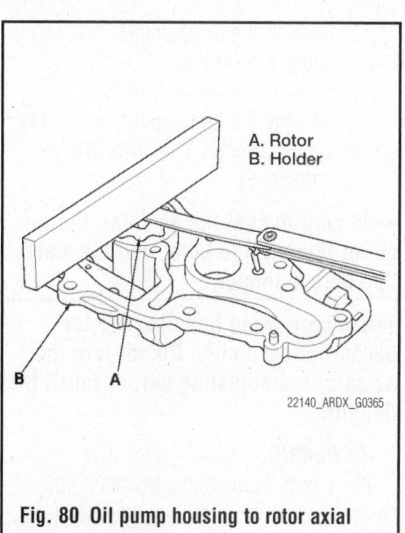

Fig. 80 Oil pump housing to rotor axial clearance check

PISTONS & RINGS

POSITIONING

See Figure 82.

REAR MAIN SEAL

REMOVAL & INSTALLATION

See Figures 83 and 84.

1. Remove the transmission.

2. Remove the drive plate.

3. Remove the seal from its mounting, using the proper seal removal tool.

To install:

4. Clean and dry the crankshaft oil seal housing.

5. Use the driver tool and the attachment to drive a new seal squarely into the block, to the specified installed height.

6. Measure the distance between the

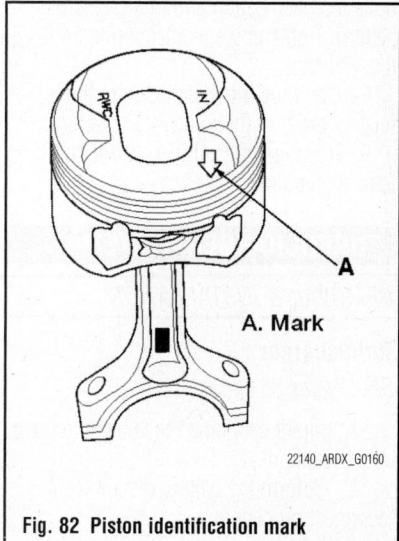

Fig. 82 Piston identification mark

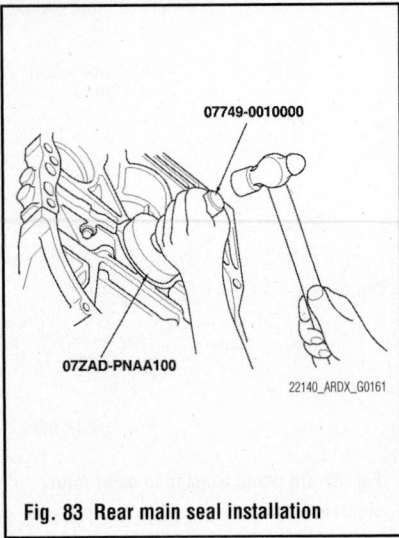

07749-0010000

07ZAD-PNAA100

22140_ARDX_G0161

Fig. 83 Rear main seal installation

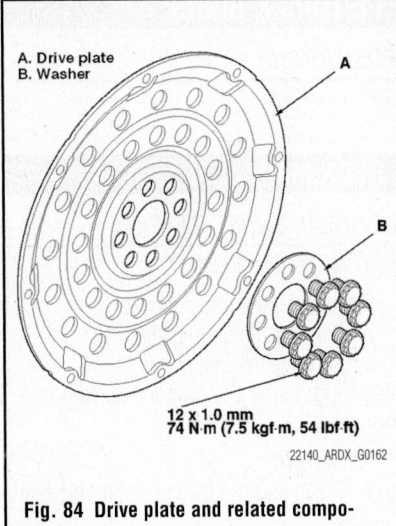

A. Drive plate
B. Washer

A

B

12 x 1.0 mm
74 N·m (7.5 kgf·m, 54 lbf·ft)

22140_ARDX_G0162

Fig. 84 Drive plate and related components

engine block and the oil seal. Specified height should be 0.001–0.047 in.

7. Install the drive plate. Tighten the bolts to specification and in a crisscross pattern. Tightening specification is 54 ft. lbs. (73 Nm).

8. Continue the installation in the reverse order of the removal procedure.

9. Road test the vehicle, correct problems as required.

TURBOCHARGER SYSTEM

REMOVAL & INSTALLATION

Turbocharger

See Figures 85 and 86.

1. Before servicing the vehicle, refer to the Precautions.

2. Perform the battery disconnect/reconnect procedure.

3. Drain the coolant. Be sure to properly dispose of used coolant.

4. Remove the cowl panel and the under cowl panel, removing fasteners on the passenger and driver sides.

5. Remove the air cleaner housing assembly.

6. Remove the charge air cooler.

7. Remove the warm up catalytic converter.

8. Remove the under hood fuse/relay box from the bracket.

9. Remove the master cylinder reservoir from the bracket. Remove the bracket.

10. Remove the heater hoses, then open the cable clamp and remove the heater valve cable.

11. Remove the heater valve mounting bolt. Remove the air bypass outlet pipe.

12. Remove the EVAP canister hose. Remove the intake air duct.

13. Remove the intake air ducts.

14. Remove the breather hose and the water bypass hose then remove the water bypass line mounting bolt.

15. Disconnect the turbocharger waste gate control solenoid valve connector and the turbocharger boost control solenoid valve connector.

16. Remove the vacuum hose, the vacuum line mounting bolt and the solenoid valve bracket mounting bolt, then loosen the solenoid valve bracket mounting bolt. Remove the solenoid valve bracket.

17. Remove the charge air cooler bracket. Remove the water bypass hose and the vacuum hoses. Remove the water bypass pipe from the turbocharger.

18. Remove the intake air duct.

19. Raise and safely support the vehicle.

20. Remove the oil return pipe, then remove the 16x1.5mm bolt at the oil feed pipe.

21. Remove the turbocharger bracket.

22. Lower the vehicle.

23. Remove the oil feed pipe.

24. Remove the turbocharger assembly retaining bolts. Remove the turbocharger from its mounting.

➡Be sure to seal the air inlet, the air outlet, the oil line hole, and the water line hole with tape.

➡Use care when handling the turbocharger assembly. Do not turn the actuator rod adjusting nut, or touch the impeller.

To install:

25. Clean the mating surfaces of the components to be sure they are free of dirt and debris.

26. Be sure to use a new gasket.

27. Tighten the turbocharger retaining bolts to 33 ft. lbs. (45 Nm).

28. Continue the installation in the reverse order of the removal procedure.

29. Turn the ignition switch ON (II), the SRS indicator should come on for about six seconds, and then go off.

30. Road test the vehicle, correct problems as required.

Turbocharger Boost Control Solenoid Valve

See Figure 87.

1. Remove the charge air cooler.

2. Disconnect the turbocharger wastegate control solenoid valve connector.

3. Disconnect the hoses and the turbocharger boost control valve connector.

4. Remove the turbocharger boost control solenoid valve.

5. Install the parts in the reverse order of removal.

Turbocharger Boost Sensor

See Figure 88.

1. Remove the charge air cooler cover.

2. Remove the charge air cooler cover stay.

3. Disconnect the turbocharger boost sensor connector.

4. Remove the turbocharger boost sensor.

5. Install the parts in the reverse order of removal with a new O-ring.

Turbocharger Bypass Control Solenoid

See Figure 89.

1. Remove the charge air cooler cover.

2. Disconnect the hoses.

3. Disconnect the turbocharger bypass control solenoid valve connector and the turbocharger bypass control solenoid valve.

4. Install the parts in the reverse order of removal.

Turbocharger Bypass Control Valve

See Figure 90.

1. Remove the charge air cooler cover.

2. Disconnect the turbocharger boost sensor connector.

3. Disconnect the vacuum hose and the air bypass outlet connecting tube, then remove the turbocharger bypass control valve.

4. Install the turbocharger bypass control valve with a new gasket.

5. Connect the vacuum hose and intake

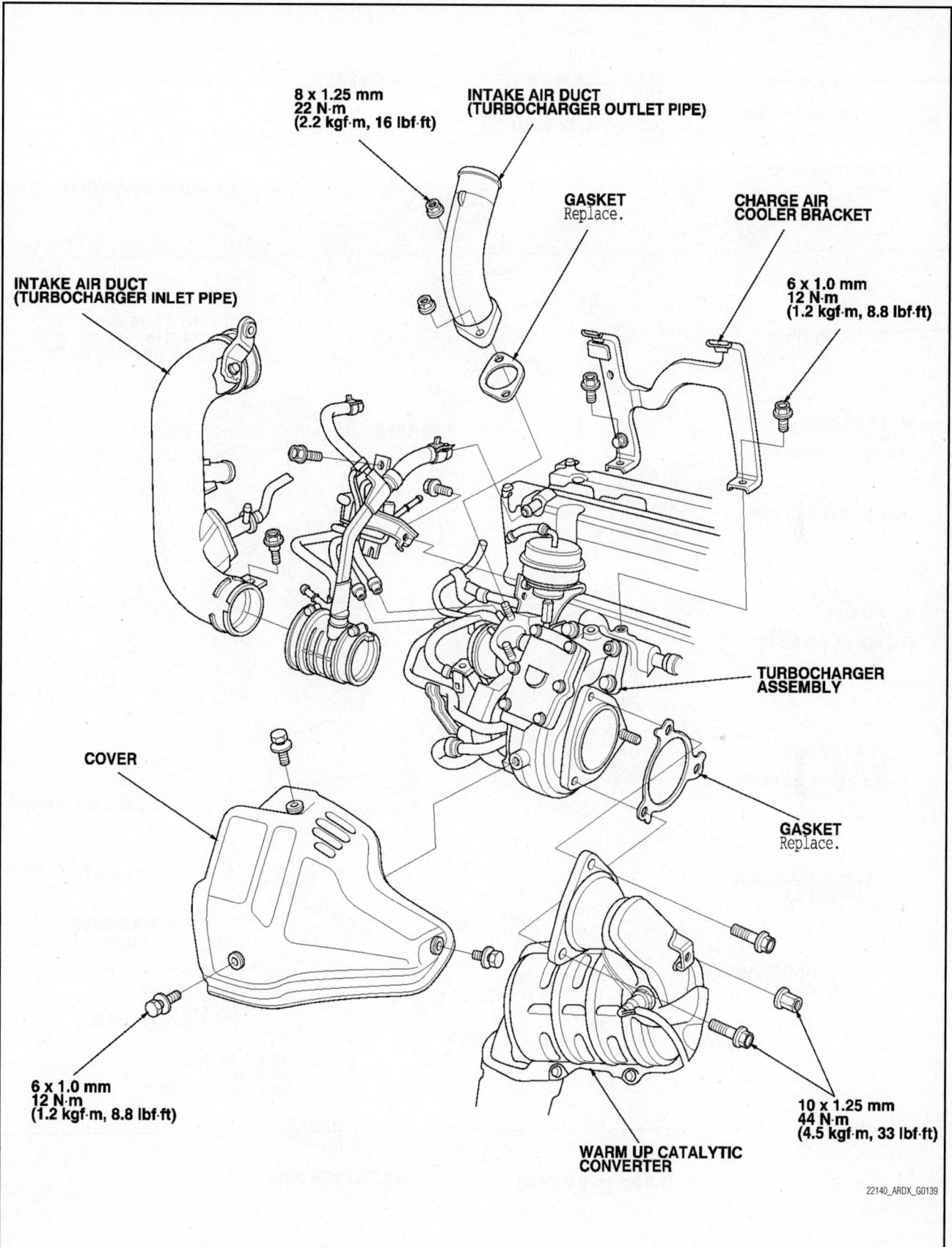

8 x 1.25 mm
22 N·m
(2.2 kgf·m, 16 lbf·ft)

INTAKE AIR DUCT
(TURBOCHARGER OUTLET PIPE)

GASKET
Replace.

CHARGE AIR
COOLER BRACKET

6 x 1.0 mm
12 N·m
(1.2 kgf·m, 8.8 lbf·ft)

INTAKE AIR DUCT
(TURBOCHARGER INLET PIPE)

TURBOCHARGER
ASSEMBLY

COVER

GASKET
Replace.

6 x 1.0 mm
12 N·m
(1.2 kgf·m, 8.8 lbf·ft)

10 x 1.25 mm
44 N·m
(4.5 kgf·m, 33 lbf·ft)

WARM UP CATALYTIC
CONVERTER

22140_ARDX_G0139

Fig. 85 Turbocharger and related components (1 of 2)

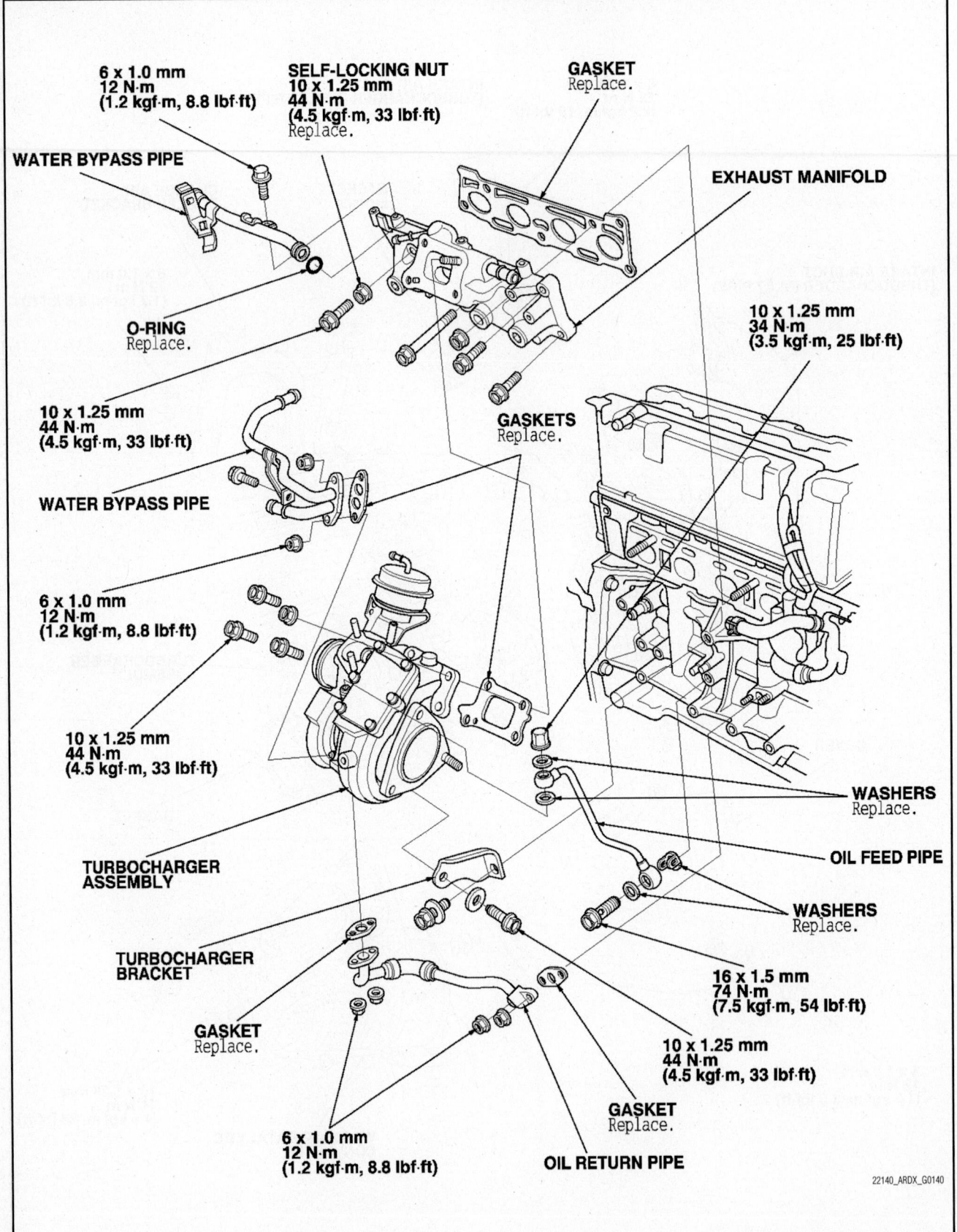

6 x 1.0 mm
12 N·m
(1.2 kgf·m, 8.8 lbf·ft)

SELF-LOCKING NUT
10 x 1.25 mm
44 N·m
(4.5 kgf·m, 33 lbf·ft)
Replace.

GASKET
Replace.

WATER BYPASS PIPE

EXHAUST MANIFOLD

O-RING
Replace.

10 x 1.25 mm
34 N·m
(3.5 kgf·m, 25 lbf·ft)

10 x 1.25 mm
44 N·m
(4.5 kgf·m, 33 lbf·ft)

WATER BYPASS PIPE

GASKETS
Replace.

6 x 1.0 mm
12 N·m
(1.2 kgf·m, 8.8 lbf·ft)

10 x 1.25 mm
44 N·m
(4.5 kgf·m, 33 lbf·ft)

WASHERS
Replace.

OIL FEED PIPE

TURBOCHARGER
ASSEMBLY

WASHERS
Replace.

TURBOCHARGER
BRACKET

16 x 1.5 mm
74 N·m
(7.5 kgf·m, 54 lbf·ft)

GASKET
Replace.

10 x 1.25 mm
44 N·m
(4.5 kgf·m, 33 lbf·ft)

6 x 1.0 mm
12 N·m
(1.2 kgf·m, 8.8 lbf·ft)

GASKET
Replace.

OIL RETURN PIPE

22140_ARDX_G0140

Fig. 86 Turbocharger and related components (2 of 2)

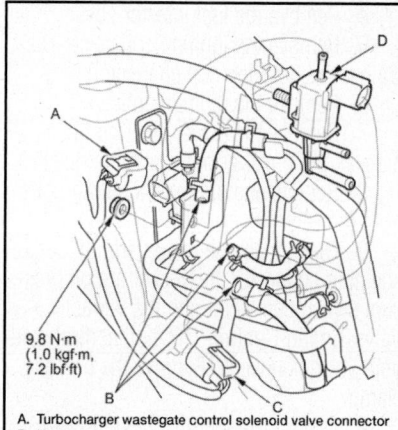

A. Turbocharger wastegate control solenoid valve connector
B. Hoses
C. Turbocharger boost control valve connector
D. Turbocharger boost control solenoid valve

71051_ARDX_G0367

Fig. 87 Disconnect the turbocharger wastegate control solenoid valve connector. Disconnect the hoses and the turbocharger boost control valve connector. Remove the turbocharger boost control solenoid valve.

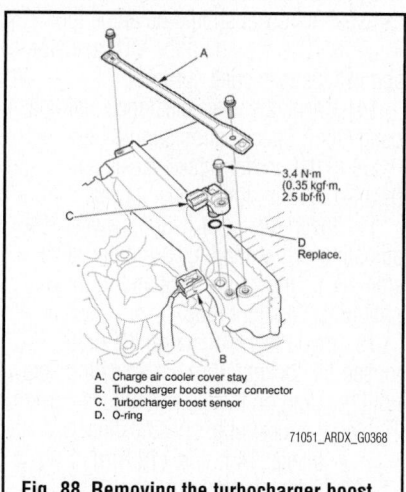

A. Charge air cooler cover stay
B. Turbocharger boost sensor connector
C. Turbocharger boost sensor
D. O-ring

71051_ARDX_G0368

Fig. 88 Removing the turbocharger boost sensor.

air duct, then connect the turbocharger boost sensor connector.

6. Install the charge air cooler cover.

Turbocharger Wastegate Control Valve

See Figure 91.

1. Remove the charge air cooler.
2. Disconnect the breather hose.
3. Loosen the nut and disconnect the turbocharger boost control solenoid valve connector.
4. Disconnect the hoses and the turbocharger wastegate control solenoid valve connector.
5. Remove the turbocharger wastegate control solenoid valve.

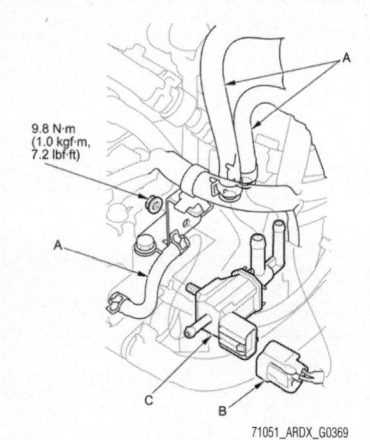

71051_ARDX_G0369

Fig. 89 Disconnect the hoses (A). Disconnect the turbocharger bypass control solenoid valve connector (B) and the turbocharger bypass control solenoid valve (C).

6. Install the parts in the reverse order of removal.

VALVE COVERS

REMOVAL & INSTALLATION

See Figures 91 through 93.

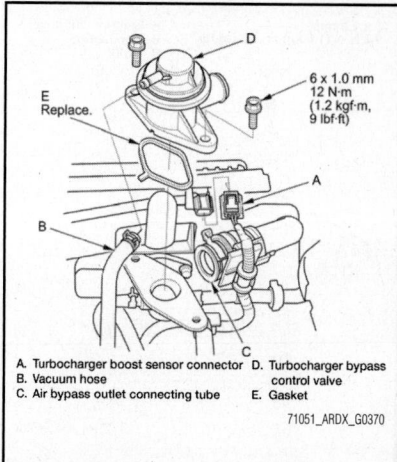

A. Turbocharger boost sensor connector
B. Vacuum hose
C. Air bypass outlet connecting tube
D. Turbocharger bypass control valve
E. Gasket

71051_ARDX_G0370

Fig. 90 Removing the turbocharger bypass control valve

1. Remove the air charge cooler cover.
2. Disconnect the turbocharger boost sensor connector. Remove the vacuum hoses and the air bypass outlet connecting tube.
3. Remove the air charge cooler retaining bolts. Remove the air charge cooler.
4. Remove the fuel injector cover.
5. Remove the ignition coil cover. Disconnect the ignition coil connectors.
6. Remove the ignition coils.
7. Remove the power steering hose

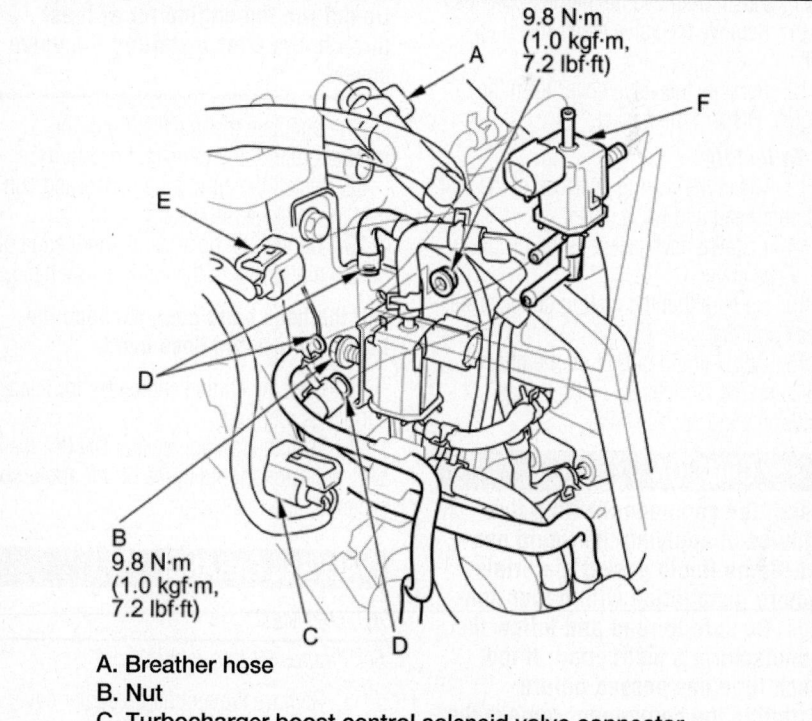

A. Breather hose
B. Nut
C. Turbocharger boost control solenoid valve connector
D. Hoses
E. Turbocharger wastegate control solenoid valve connector
F. Turbocharger wastegate control solenoid valve

71051_ARDX_G0483

Fig. 91 Removing the turbocharger wastegate control valve

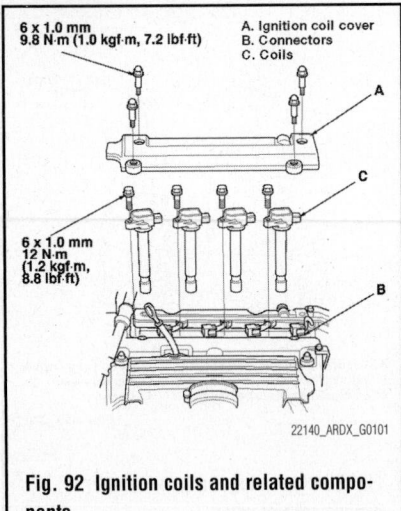

Fig. 92 Ignition coils and related components

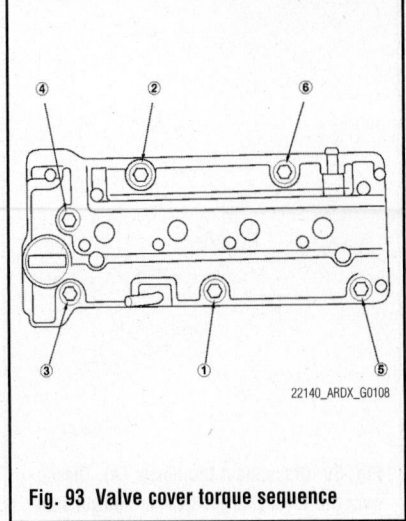

Fig. 93 Valve cover torque sequence

bracket, the breather hose and the dipstick.

8. Remove the bracket mounting bolts, from the valve cover.

9. Disconnect the rocker arm oil control solenoid connector, the rocker are oil pressure switch connector and the Variable Valve Timing Control (VTC) oil control solenoid valve connector. Remove the harness clamp.

10. Disconnect the Air/Fuel ratio (A/F) sensor connector, the oil pressure switch connector and the Crankshaft Position (CKP) sensor. Remove the harness clamps.

11. Remove the valve cover retaining bolts.

12. Remove the valve cover from the engine. Discard the gasket.

To install:

13. Clean the mating surfaces of the cylinder head and valve cover.

14. Install a new gasket in the groove of the valve cover.

15. Be sure that the mating surfaces are clean and dry.

16. Apply liquid gasket on the chain case and the number five rocker shaft holder mating surface areas.

❉❉ CAUTION

Install the component within five minutes of applying the liquid gasket. Some liquid gasket materials require installation within four minutes. Be sure to read and follow the manufacturer's instruction. If too much time has passed before installing the component remove the liquid gasket and residue, then reapply new liquid gasket.

17. Position the spark plug seals on the spark plug tubes.

18. Position the valve cover on the cylinder head. Slide the cover back and forth to seat the gasket.

19. Inspect the valve cover washers, replace as required.

20. Tighten the retaining bolts in three steps to 8.8 ft. lbs. (11.9 Nm).

➡**Wait at least 30 minutes to allow the gasket to cure before filling the engine with oil.**

❉❉ CAUTION

Do not run the engine for at least three hours after installing the valve cover.

21. Continue the installation in the reverse order of the removal procedure.

22. Install the air charge cooler and connect the intake air ducts.

23. Tighten the hose bands until edge of the band aligns with the mark on the band.

➡**If the hose band edge exceeds the mark, replace the hose band.**

24. Start the engine and check for leaks, correct as required.

25. Turn the ignition switch ON (II), the SRS indicator should come on for about six seconds, and then go off.

VALVE CLEARANCE

ADJUSTMENT

See Figures 94 through 97.

1. Remove the air charge cooler cover.

2. Disconnect the turbocharger boost sensor connector. Remove the vacuum hoses and the air bypass outlet connecting tube.

3. Remove the air charge cooler retaining bolts. Remove the air charge cooler.

4. Remove the fuel injector cover.

5. Remove the ignition coil cover. Disconnect the ignition coil connectors.

6. Remove the ignition coils.

7. Remove the power steering hose bracket, the breather hose, and the dipstick.

8. Remove the bracket mounting bolts from the valve cover.

9. Disconnect the rocker arm oil control solenoid connector, the rocker are oil pressure switch connector and the Variable Valve Timing Control (VTC) oil control solenoid valve connector. Remove the harness clamp.

10. Disconnect the Air/Fuel ratio (A/F) sensor connector, the oil pressure switch connector and the Crankshaft Position (CKP) sensor. Remove the harness clamps.

11. Remove the valve cover retaining bolts.

12. Remove the valve cover from the engine. Discard the gasket.

13. Position the number one piston at TDC. The punch mark on the VTC actuator and the punch mark on the exhaust camshaft sprocket should be at the top. Align the TDC marks on the VTC actuator and exhaust camshaft sprocket.

14. Check the valve clearance, using a gauge tool. Specification should be 0.008–0.010 in. for intake valves and 0.011–0.013 in. for exhaust valves.

15. Insert the feeler gauge between the adjusting screw and the end of the valve stem. Slide the gauge back and forth, you should feel a slight drag.

16. If you feel too much/little drag, loosen the locknut, turn the adjusting screw until the drag on the feeler gauge is correct. Tighten the locknut to specification:
- Intake: 14 ft. lbs. (19 Nm)
- Exhaust: 10 ft. lbs. (14 Nm)

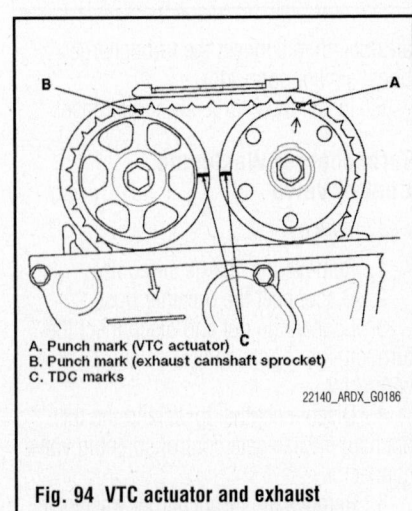

A. Punch mark (VTC actuator)
B. Punch mark (exhaust camshaft sprocket)
C. TDC marks

Fig. 94 VTC actuator and exhaust camshaft sprocket alignment marks

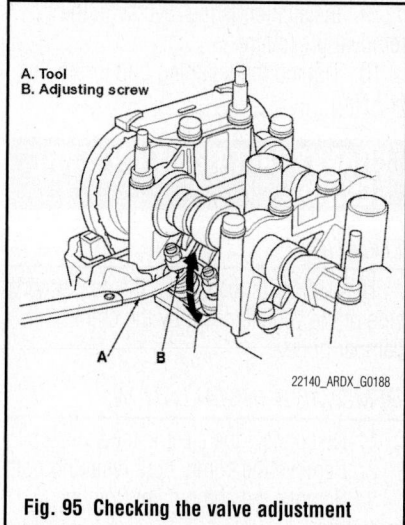

Fig. 95 Checking the valve adjustment

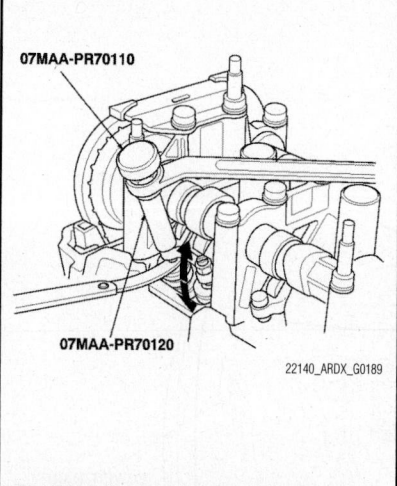

Fig. 96 Adjusting the valve adjustment

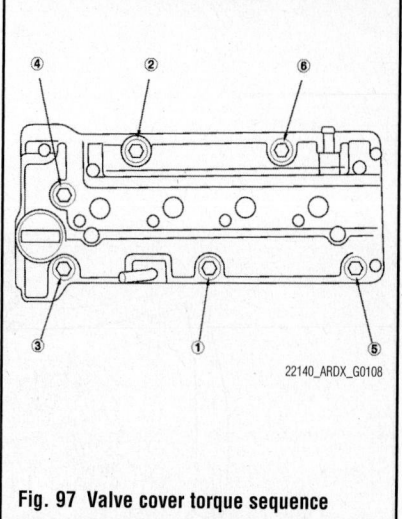

Fig. 97 Valve cover torque sequence

➡Be sure to apply clean engine oil to the nut threads.

17. Repeat the adjustment as required.

18. Rotate the crankshaft 180 degrees clockwise. Check and adjust the valve clearance on number three cylinder.

19. Rotate the crankshaft 180 degrees clockwise. Check and adjust the valve clearance on number four cylinder.

20. Rotate the crankshaft 180 degrees clockwise. Check and adjust the valve clearance on number two cylinder.

21. Clean the mating surfaces of the cylinder head and valve cover.

22. Install a new gasket in the groove of the valve cover.

23. Be sure that the mating surfaces are clean and dry.

24. Apply liquid gasket on the chain case and the number five rocker shaft holder mating surface areas.

✳✳ CAUTION

Install the component within five minutes of applying the liquid gasket. Some liquid gasket materials require installation within four minutes. Be sure to read and follow the manufacturer's instruction. If too much time has passed before installing the component remove the liquid gasket and residue, then reapply new liquid gasket.

25. Position the spark plug seals on the spark plug tubes.

26. Position the valve cover on the cylinder head. Slide the cover back and forth to seat the gasket.

27. Inspect the valve cover washers, replace as required.

28. Tighten the retaining bolts in three steps to 8.8 ft. lbs. (11.9 Nm).

➡Wait at least 30 minutes to allow the gasket to cure before filling the engine with oil. Do not run the engine for at least three hours after installing the valve cover.

29. Continue the installation in the reverse order of the removal procedure.

30. Install the air charge cooler and connect the intake air ducts.

31. Tighten the hose bands until edge of the band aligns with the mark on the band.

➡If the hose band edge exceeds the mark, replace the hose band.

32. Start the engine and check for leaks, correct as required.

33. Turn the ignition switch ON (II), the SRS indicator should come on for about six seconds, and then go off.

ENGINE PERFORMANCE & EMISSION CONTROLS

COMPONENT LOCATIONS

See Figures 98 through 105.

AIR-FUEL RATIO (AFR) SENSOR

LOCATION

See Figure 106.

The Air/Fuel (A/F) sensor is located just ahead of the warm-up three way catalytic converter.

REMOVAL & INSTALLATION

See Figure 107.

1. Disconnect the electrical connector.

2. Remove the component from its mounting, using a sensor removal tool.

To install:

3. Installation is the reverse of the removal procedure.

4. Tighten the sensor to 33 ft. lbs. (45 Nm).

CAMSHAFT POSITION (CMP) SENSOR

LOCATION

This sensor is located on the top of the engine mid way between the passenger's side of the vehicle and the driver's side of the vehicle.

REMOVAL & INSTALLATION

See Figures 108 and 109.

1. Remove the air cleaner, if removing sensor A.

2. Remove the air charge cooler, if removing sensor B.

3. Remove the air bypass outlet pipe.

4. Disconnect the electrical connector.

5. Remove the component retaining bolt.

6. Remove the component from its mounting.

7. Discard the O-ring.

To install:

8. Be sure to use a new O-ring.

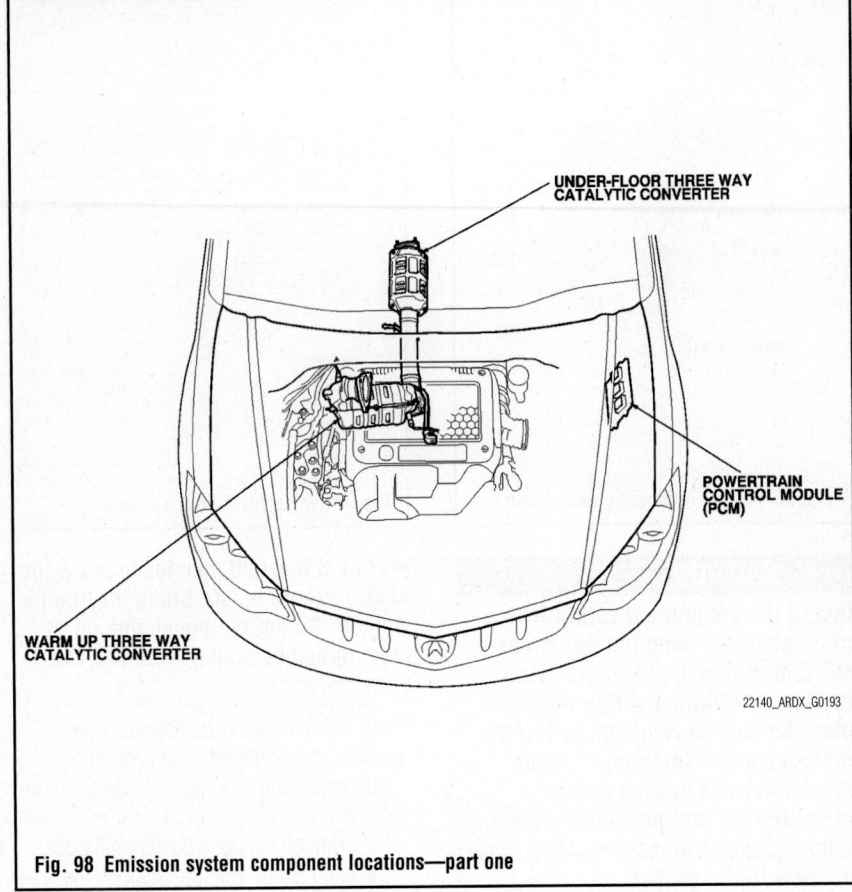

Fig. 98 Emission system component locations—part one

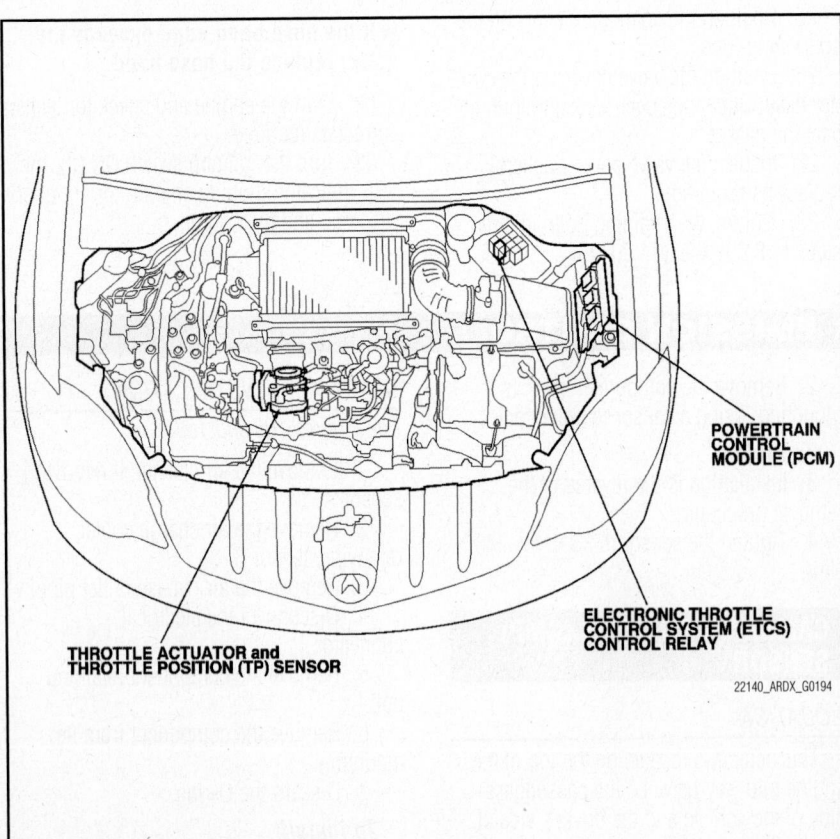

Fig. 99 Emission system component locations—part two

9. Installation is the reverse of the removal procedure.

10. Tighten the retaining bolt to 9 ft. lbs. (12 Nm).

CRANKSHAFT POSITION (CKP) SENSOR

LOCATION

This sensor is located on the passenger's side of the vehicle down by the crankshaft damper pulley.

REMOVAL & INSTALLATION

1. Disconnect the electrical connector.
2. Remove the component retaining bolt.
3. Remove the component from its mounting.
4. Discard the O-ring.

To install:

5. Be sure to use a new O-ring.
6. Installation is the reverse of the removal procedure.
7. Tighten the retaining bolt to 9 ft. lbs. (12 Nm).
8. Perform the "CKP Pattern Clear/Relearn Procedure".

ENGINE COOLANT TEMPERATURE (ECT) SENSOR

LOCATION

The Engine Coolant Temperature (ECT) Sensor 1 is located in the front of the engine block.

REMOVAL & INSTALLATION

ECT Sensor 1

See Figures 111 and 112.

1. Drain the engine coolant.
2. Remove the air cleaner.
3. Remove the charge air cooler.
4. Remove the ignition coil cover and bolt.
5. Disconnect the ignition coil connectors, CMP sensor A connector, CMP sensor B connector, and the EVAP canister purge valve connector.
6. Remove the harness holder.

➡**Be careful not to damage the ECT sensor 1 harness.**

7. Disconnect the ECT sensor 1 connector.
8. Remove ECT sensor 1.
9. Install the parts in the reverse order of removal with a new O-ring.
10. Refill the radiator with engine coolant.

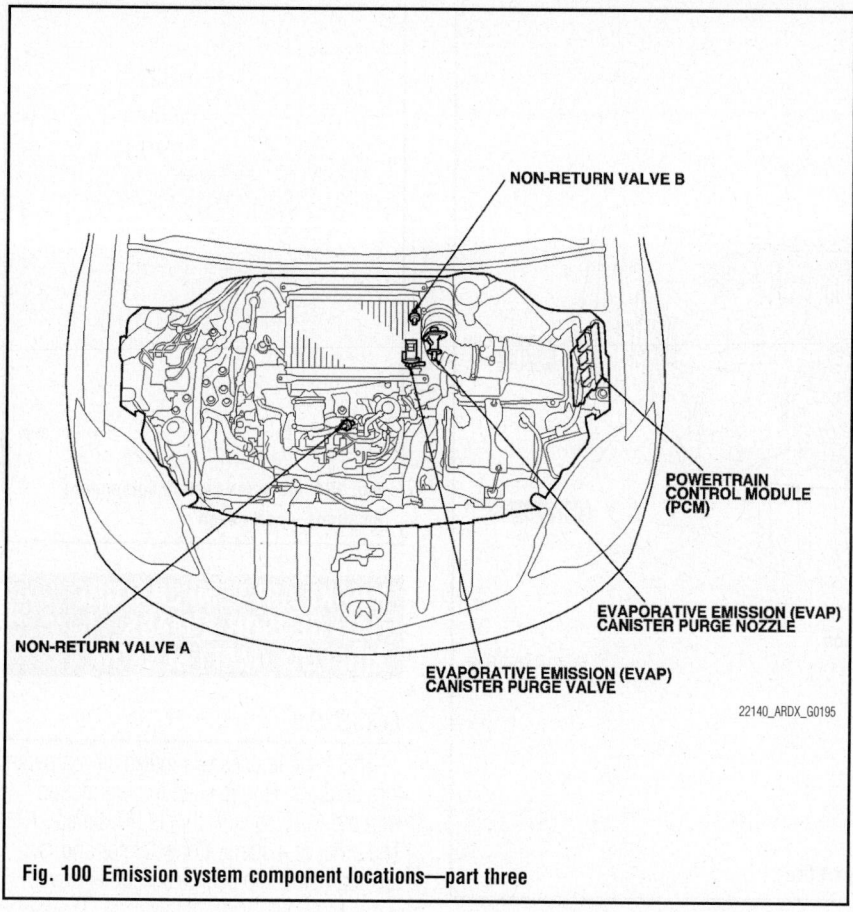

Fig. 100 Emission system component locations—part three

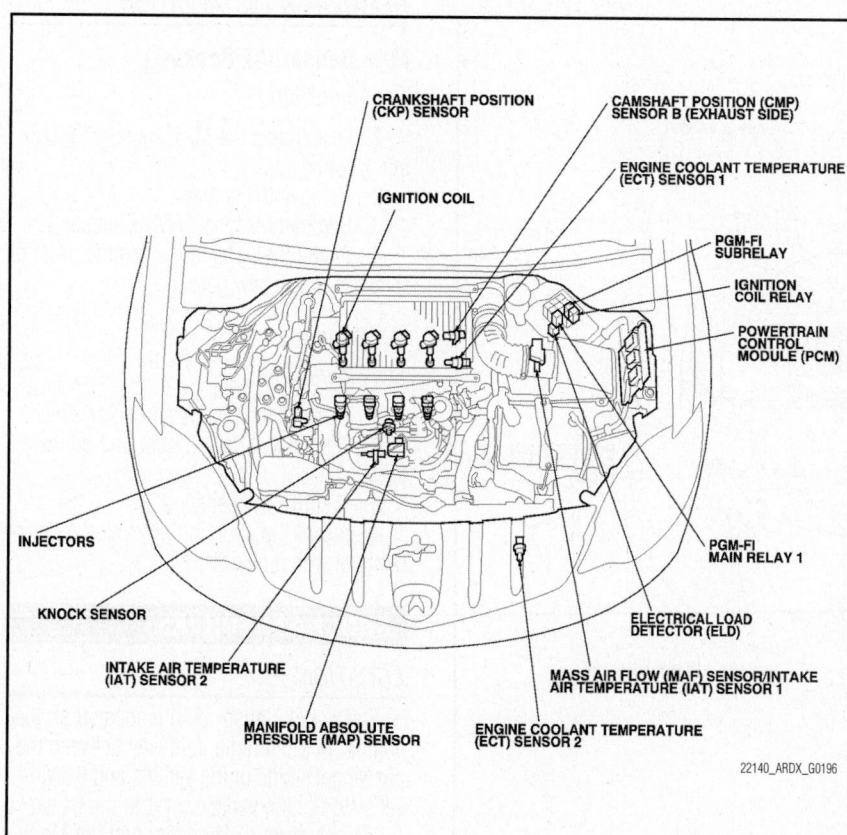

Fig. 101 Emission system component locations—part four

ECT Sensor 2

See Figure 113.

1. Drain the engine coolant.
2. Remove the splash shield.
3. Disconnect the ECT sensor 2 connector, then remove ECT sensor 2.
4. Install ECT sensor 2 with a new O-ring (C).
5. Install the splash shield.
6. Refill the radiator with engine coolant.

HEATED OXYGEN (HO2S) SENSOR

LOCATION

The Heated Oxygen (HO2S) sensor is located just behind the warm-up three way catalytic converter.

REMOVAL & INSTALLATION

See Figure 114.

1. Remove the WU-TWC stay.
2. Disconnect the electrical connector.
3. Remove the component from its mounting, using a sensor removal tool.

To install:

4. Installation is the reverse of the removal procedure.
5. Tighten the sensor to 33 ft. lbs. (45 Nm).

INPUT SHAFT (MAINSHAFT) SPEED SENSOR

REMOVAL & INSTALLATION

See Figure 115.

1. Raise the vehicle on a lift, or apply the parking brake, block the rear wheels, and raise the front of the vehicle. Make sure it is securely supported.
2. Remove the front splash shield.
3. Disconnect the input shaft (mainshaft) speed sensor connector, and remove the input shaft (mainshaft) speed sensor (A).

To install:

4. Install a new O-ring (B) on a new input shaft (mainshaft) speed sensor, then install the input shaft (mainshaft) speed sensor in the transmission housing.
5. Check the connector for rust, dirt, or oil, and clean or repair if necessary, then connect the connector securely.
6. Install the front splash shield.

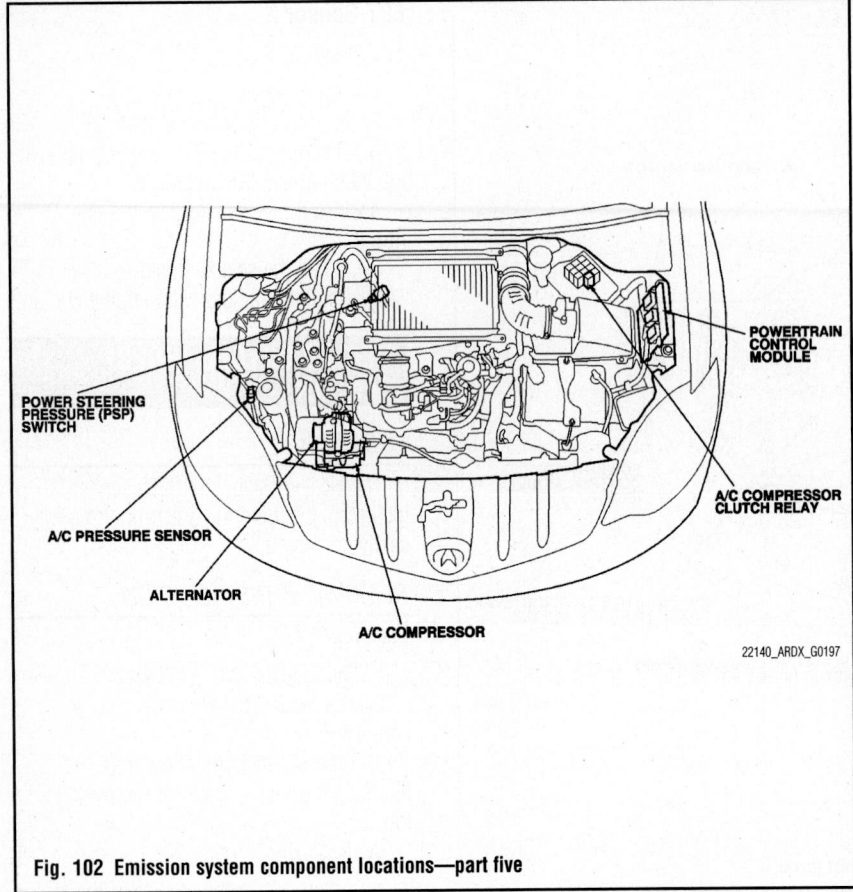

POWER STEERING PRESSURE (PSP) SWITCH

POWERTRAIN CONTROL MODULE

A/C PRESSURE SENSOR

A/C COMPRESSOR CLUTCH RELAY

ALTERNATOR

A/C COMPRESSOR

22140_ARDX_G0197

Fig. 102 Emission system component locations—part five

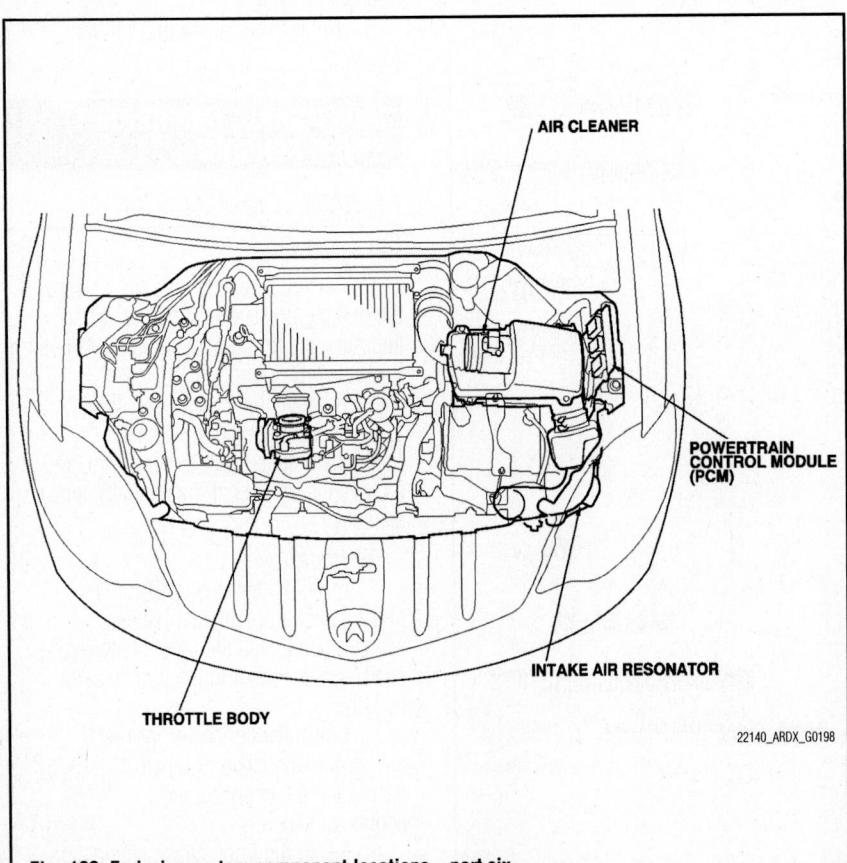

AIR CLEANER

POWERTRAIN CONTROL MODULE (PCM)

INTAKE AIR RESONATOR

THROTTLE BODY

22140_ARDX_G0198

Fig. 103 Emission system component locations—part six

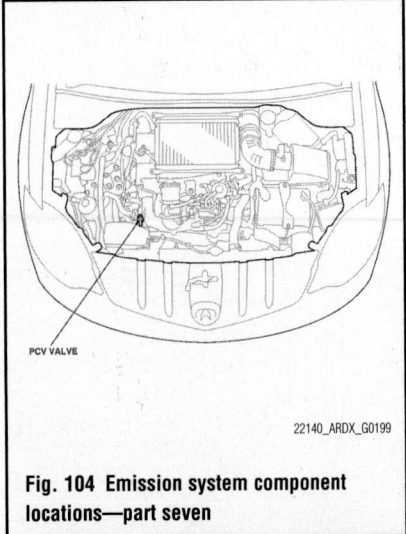

PCV VALVE

22140_ARDX_G0199

Fig. 104 Emission system component locations—part seven

INTAKE AIR TEMPERATURE (IAT) SENSOR & MASS AIRFLOW (MAF) SENSOR

LOCATION

This vehicle uses two intake air temperature sensors. One is used in conjunction with the MAF sensor; this is IAT sensor 1. The other is a stand-alone sensor and is located below the charge air cooler cover.

REMOVAL & INSTALLATION

MAF Sensor/IAT Sensor 1

See Figure 116.

1. Disconnect the MAF sensor/IAT sensor 1 connector.
2. Remove the screws.
3. Remove MAF sensor/IAT sensor 1.
4. Install the parts in the reverse order of removal with a new gasket.

IAT Sensor 2

See Figure 117.

1. Remove the charge air cooler cover.
2. Disconnect the IAT sensor 2 connector.
3. Remove IAT sensor 2.
4. Install the parts in the reverse order of removal with a new O-ring.

KNOCK SENSOR (KS)

LOCATION

The Knock Sensor (KS) is located on the engine, in the middle, mid way between the passenger's side of the vehicle and the driver's side of the vehicle, next to the Intake Air Temperature (IAT) sensor and the Manifold Absolute Pressure (MAP) sensor.

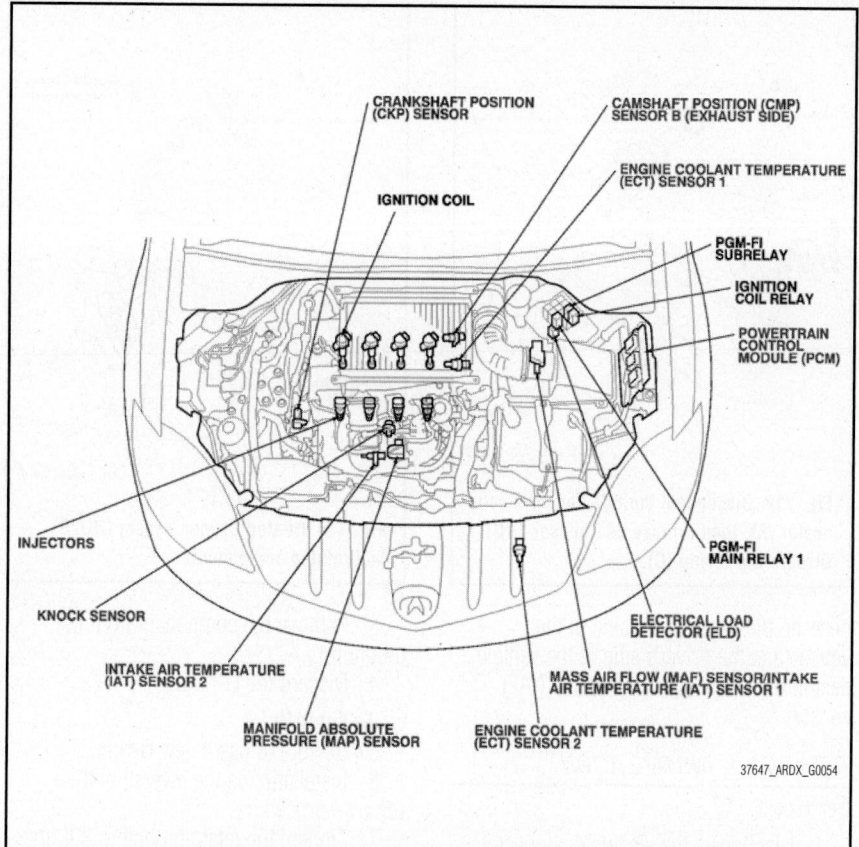

CRANKSHAFT POSITION
(CKP) SENSOR

CAMSHAFT POSITION (CMP)
SENSOR B (EXHAUST SIDE)

ENGINE COOLANT TEMPERATURE
(ECT) SENSOR 1

IGNITION COIL

PGM-FI
SUBRELAY

IGNITION
COIL RELAY

POWERTRAIN
CONTROL
MODULE (PCM)

INJECTORS

PGM-FI
MAIN RELAY 1

KNOCK SENSOR

ELECTRICAL LOAD
DETECTOR (ELD)

INTAKE AIR TEMPERATURE
(IAT) SENSOR 2

MASS AIR FLOW (MAF) SENSOR/INTAKE
AIR TEMPERATURE (IAT) SENSOR 1

MANIFOLD ABSOLUTE
PRESSURE (MAP) SENSOR

ENGINE COOLANT TEMPERATURE
(ECT) SENSOR 2

37647_ARDX_G0054

Fig. 105 Underhood locations of engine performance and emission control components

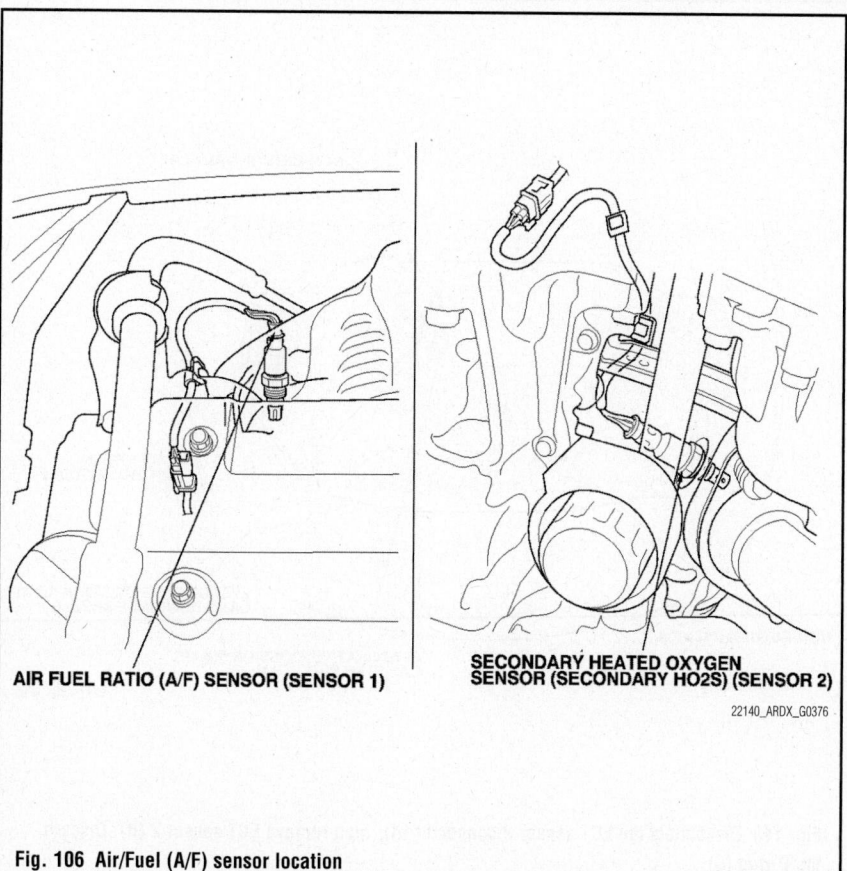

AIR FUEL RATIO (A/F) SENSOR (SENSOR 1)

SECONDARY HEATED OXYGEN
SENSOR (SECONDARY HO2S) (SENSOR 2)

22140_ARDX_G0376

Fig. 106 Air/Fuel (A/F) sensor location

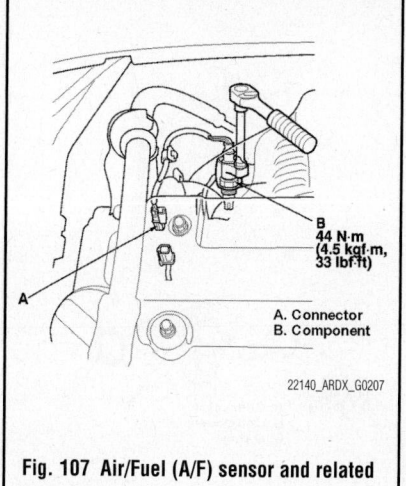

B
44 N·m
(4.5 kgf·m,
33 lbf·ft)

A

A. Connector
B. Component

22140_ARDX_G0207

Fig. 107 Air/Fuel (A/F) sensor and related components

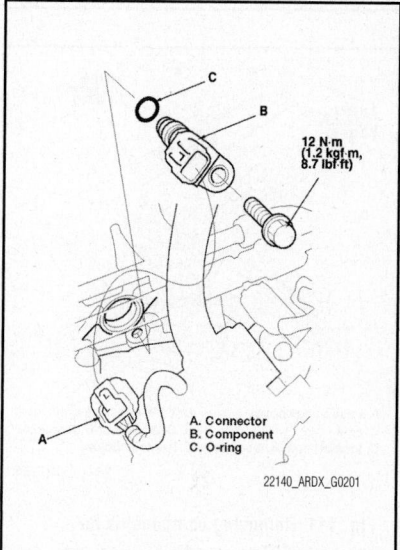

C

B

12 N·m
(1.2 kgf·m,
8.7 lbf·ft)

A

A. Connector
B. Component
C. O-ring

22140_ARDX_G0201

Fig. 108 Camshaft Position sensor A and related components

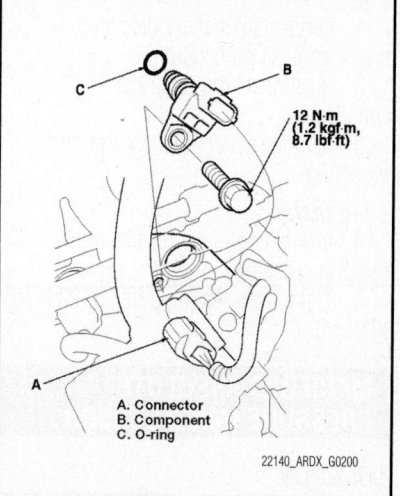

C

B

12 N·m
(1.2 kgf·m,
8.7 lbf·ft)

A

A. Connector
B. Component
C. O-ring

22140_ARDX_G0200

Fig. 109 Camshaft Position sensor B and related components

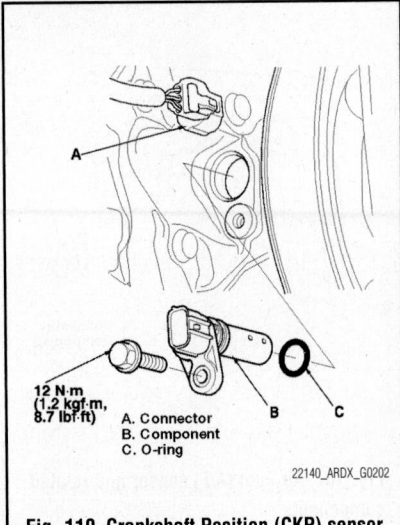

Fig. 110 Crankshaft Position (CKP) sensor and related components

12 N·m
(1.2 kgf·m,
8.7 lbf·ft)

A. Connector
B. Component
C. O-ring

22140_ARDX_G0202

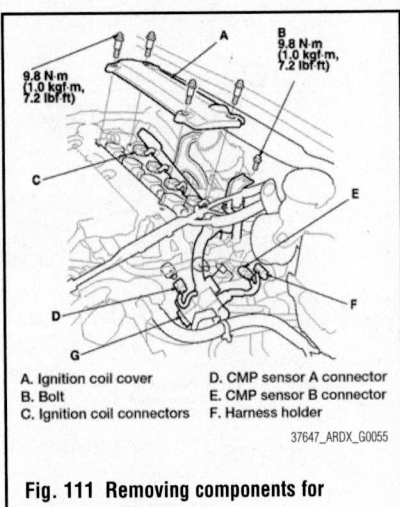

A. Ignition coil cover
B. Bolt
C. Ignition coil connectors
D. CMP sensor A connector
E. CMP sensor B connector
F. Harness holder

37647_ARDX_G0055

Fig. 111 Removing components for access to ECT sensor 1

B
9.8 N·m
(1.0 kgf·m,
7.2 lbf·ft)

9.8 N·m
(1.0 kgf·m,
7.2 lbf·ft)

REMOVAL & INSTALLATION

See Figure 118.

1. Perform the battery disconnect/reconnect procedure.
2. Disconnect the electrical connector.
3. Remove the component from its mounting.

To install:

4. Installation is the reverse of the removal procedure.
5. Tighten the retaining bolt to 23 ft. lbs. (31 Nm).

MANIFOLD ABSOLUTE PRESSURE (MAP) SENSOR

LOCATION

The Manifold Absolute Pressure (MAP) sensor is located on the engine mid way

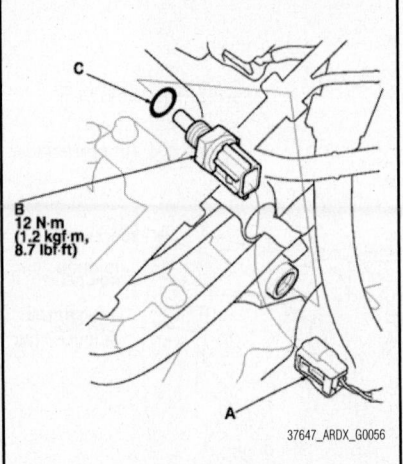

B
12 N·m
(1.2 kgf·m,
8.7 lbf·ft)

37647_ARDX_G0056

Fig. 112 Disconnect the ECT sensor 1 connector (A), then remove ECT sensor 1 (B). Discard the O-ring (C).

between the passenger's side of the vehicle and the driver's side of the vehicle, next to the Intake Air Temperature (IAT) sensor.

REMOVAL & INSTALLATION

See Figure 119.

1. Disconnect the electrical connector.
2. Remove the component retaining bolt.

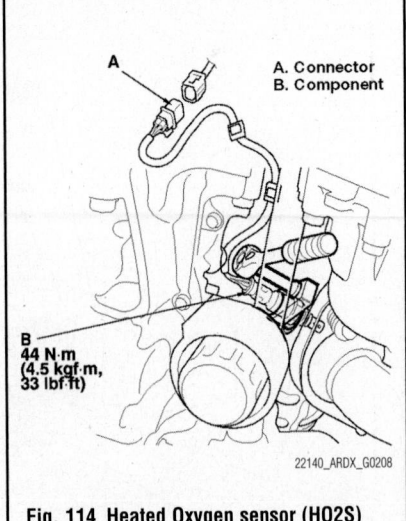

A. Connector
B. Component

B
44 N·m
(4.5 kgf·m,
33 lbf·ft)

22140_ARDX_G0208

Fig. 114 Heated Oxygen sensor (HO2S) and related components

3. Remove the component from its mounting.
4. Discard the O-ring.

To install:

5. Be sure to use a new O-ring.
6. Installation is the reverse of the removal procedure.
7. Tighten the retaining bolt to 3 ft. lbs. (4 Nm).

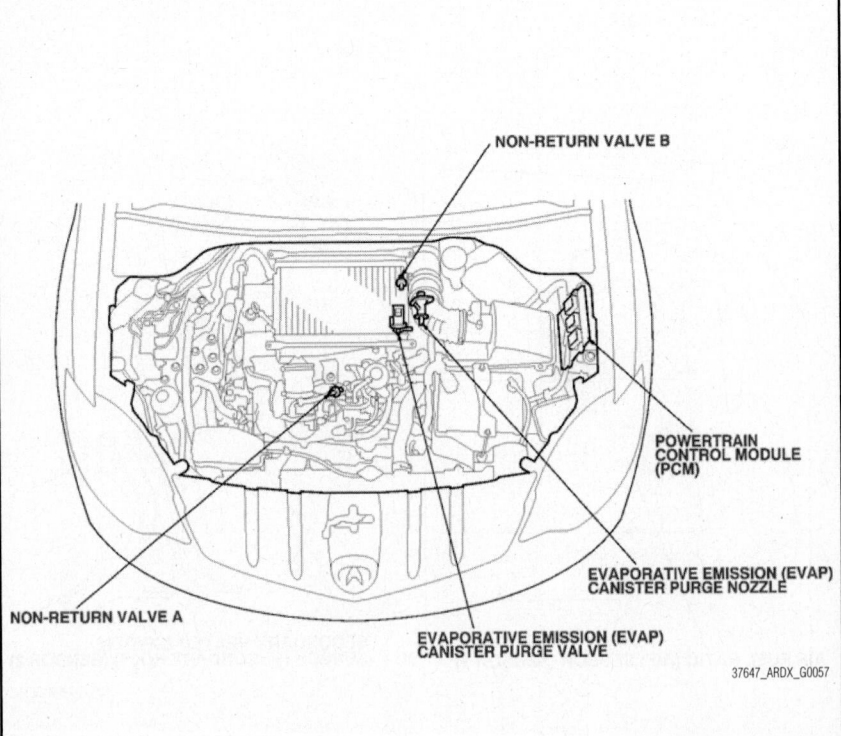

NON-RETURN VALVE B

POWERTRAIN
CONTROL MODULE
(PCM)

EVAPORATIVE EMISSION (EVAP)
CANISTER PURGE NOZZLE

EVAPORATIVE EMISSION (EVAP)
CANISTER PURGE VALVE

NON-RETURN VALVE A

37647_ARDX_G0057

Fig. 113 Disconnect the ECT sensor 2 connector (A), then remove ECT sensor 2 (B). Discard the O-ring (C).

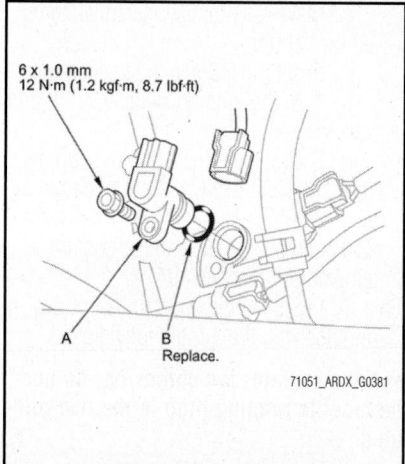

Fig. 115 Removing the input shaft (main-shaft) speed sensor (A) and O-ring (B)

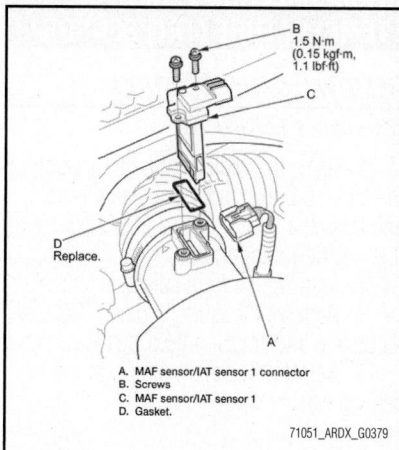

A. MAF sensor/IAT sensor 1 connector
B. Screws
C. MAF sensor/IAT sensor 1
D. Gasket.

Fig. 116 Removing the MAF sensor/IAT sensor 1

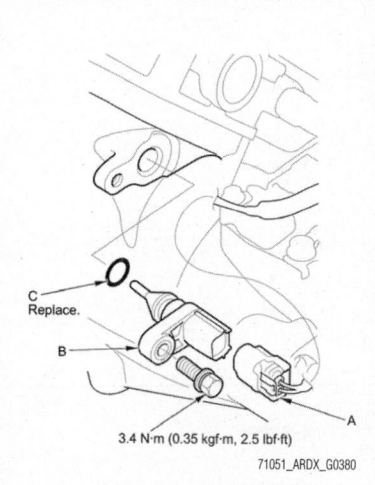

3.4 N·m (0.35 kgf·m, 2.5 lbf·ft)

Fig. 117 Disconnect the IAT sensor 2 connector (A). Remove IAT sensor 2 (B). Install the parts in the reverse order of removal with a new O-ring (C).

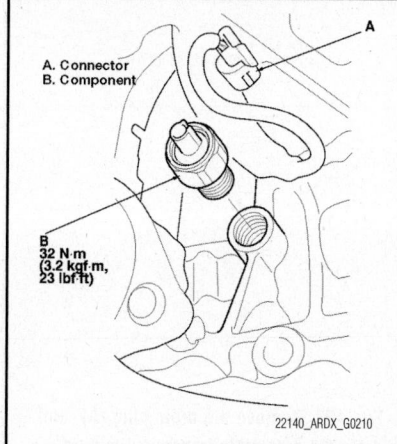

A. Connector
B. Component

Fig. 118 Knock sensor removal and installation

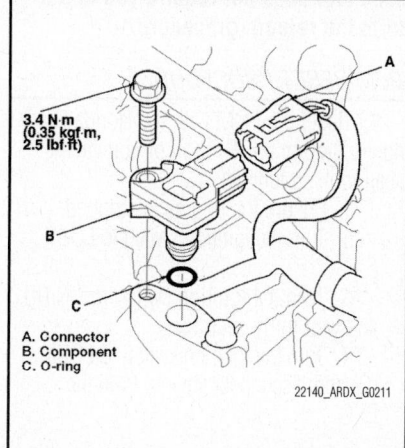

A. Connector
B. Component
C. O-ring

Fig. 119 MAP sensor and related components

OUTPUT SHAFT (COUNTERSHAFT) SPEED (OSS) SENSOR

REMOVAL & INSTALLATION
See Figure 120.

1. Raise the vehicle on a lift, or apply the parking brake, block the rear wheels, and raise the front of the vehicle. Make sure it is securely supported.
2. Remove the front splash shield.
3. Disconnect the output shaft (countershaft) speed sensor connector, and remove the output shaft (countershaft) speed sensor and the sensor washer (with SH-AWD).
4. Install a new O-ring on a new output shaft (countershaft) speed sensor, then install the output shaft (countershaft) speed sensor and the sensor washer (with SH-AWD) in the transmission housing.
5. Check the connector for rust, dirt, or

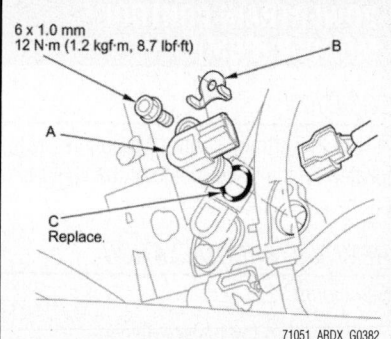

Fig. 120 Disconnect the output shaft (countershaft) speed sensor connector, and remove the output shaft (countershaft) speed sensor (A) and the sensor washer (with SH-AWD) (B). Install a new O-ring (C) on a new output shaft (countershaft) speed sensor, then install the output shaft (countershaft) speed sensor and the sensor washer (with SH-AWD) in the transmission housing.

oil, and clean or repair if necessary, then connect the connector securely.
6. Install the front splash shield.

POSITIVE CRANKCASE VENTILATION (PCV) VALVE

LOCATION

The Positive Crankcase Ventilation (PCV) valve is located in the front right side of the engine.

REMOVAL & INSTALLATION
See Figure 121.

1. Disconnect the PCV hose.
2. Remove the PCV valve.
3. Install the parts in the reverse order of removal with a new washer.

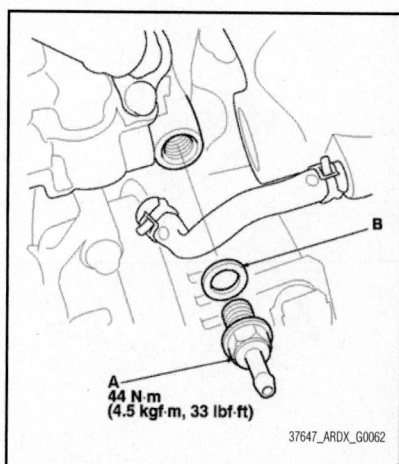

44 N·m (4.5 kgf·m, 33 lbf·ft)

Fig. 121 Showing the PCV valve (A) and the washer (B)

POWERTRAIN CONTROL MODULE (PCM)

LOCATION

The Powertrain Control Module (PCM) is located on the driver's side of the vehicle, on the inner fender well.

REMOVAL & INSTALLATION

See Figure 122.

1. Perform the battery disconnect/reconnect procedure.
2. Refer to the scan tool instructions for the particular model you are repairing. Use those procedures to position the PCM for replacement. Diagnostics for engine oil replacement, ATF fluid replacement, and throttle body cleaning will be addressed on both US and Canadian vehicles.
3. Disconnect the electrical connectors.

➡**These connectors have symbols embossed in them for identification.**

4. Remove the component retaining bolts.
5. Remove the component from its mounting.
6. Remove the cover and the bracket from the PCM

To install:

7. Installation is the reverse of the removal procedure.
8. Tighten the retaining bolt to 7 ft. lbs. (9 Nm).
9. Update the PCM if it does not have the latest software.

➡**The PCM relearn procedure must be performed whenever you replace the PCM, reset the PCM, update the PCM, remove the engine or replace/clean the throttle body. Erasing DTC's with the**

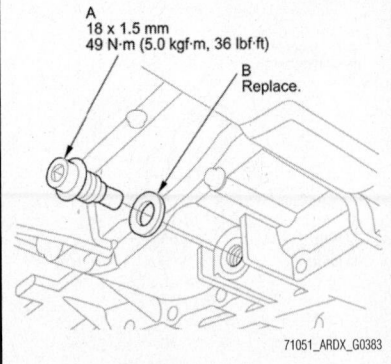

Fig. 123 Remove the drain plug (A), and drain the automatic transmission fluid (ATF). Reinstall the drain plug with a new sealing washer (B).

scan tool does not require you to perform the relearn procedure.

PCM RESET PROCEDURE

1. Reset the PCM with the Honda Diagnostic System (HDS) tool, or equivalent scan tool, as follows:
   a. Ensure the engine is stopped.
   b. Turn the ignition switch to LOCK (0).
   c. Turn the ignition switch to ON (II), and wait thirty seconds.
   d. Turn the ignition switch to LOCK (0) and disconnect the tool from the DLC.

e. Make sure that all electrical items are turned off.
   f. Reset the PCM with the scan tool.
   g. Turn the ignition switch ON (II) and wait two seconds.
   h. Start the engine. Hold the engine speed at 3000 RPM, without a load in either PARK or NEUTRAL until the radiator fan comes on or until the coolant temperature reaches 194° F.
   i. Let the engine idle for about five minutes with the throttle fully closed

➡**If the radiator fan comes on, do not include its running time in the five minutes.**

   j. Perform the "CKP Pattern Clear/Relearn Procedure".

TRANSMISSION FLUID TEMPERATURE (TFT) SENSOR

REMOVAL & INSTALLATION

See Figures 123 and 124.

1. Raise the vehicle on a lift, or apply the parking brake, block the rear wheels, and raise the front of the vehicle. Make sure it is securely supported.
2. Remove the front splash shield.
3. Remove the drain plug and drain the automatic transmission fluid (ATF).
4. Reinstall the drain plug with a new sealing washer.

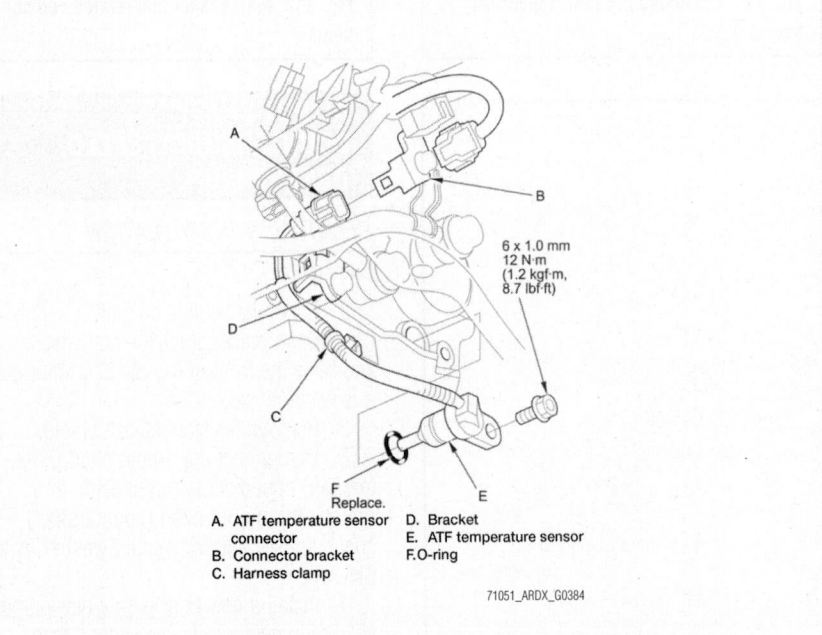

A. ATF temperature sensor connector
B. Connector bracket
C. Harness clamp
D. Bracket
E. ATF temperature sensor
F. O-ring

71051_ARDX_G0384

Fig. 124 Disconnect the ATF temperature sensor connector, then remove the connector from the connector bracket. Remove the harness clamp from its bracket. Remove the ATF temperature sensor and replace the sensor and the O-ring.

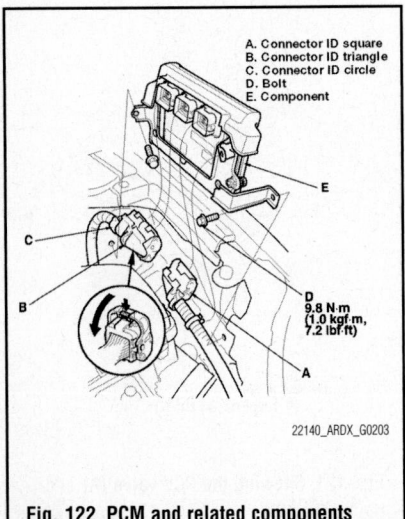

A. Connector ID square
B. Connector ID triangle
C. Connector ID circle
D. Bolt
E. Component

22140_ARDX_G0203

Fig. 122 PCM and related components

5. Disconnect the ATF temperature sensor connector, then remove the connector from the connector bracket.

6. Remove the harness clamp from its bracket.

7. Remove the ATF temperature sensor and replace the sensor and the O-ring.

To install:

8. Install a new O-ring on a new ATF temperature sensor, and install the ATF temperature sensor.

9. Check the connector for rust, dirt, or oil, and clean or repair if necessary, then connect the connector securely.

10. Install the harness clamp on its bracket.

11. Refill the transmission with ATF.

12. Install the front splash shield.

FUEL   GASOLINE FUEL INJECTION SYSTEM

FUEL SYSTEM SERVICE PRECAUTIONS

Safety is the most important factor when performing not only fuel system maintenance, but any type of maintenance. Failure to conduct maintenance and repairs in a safe manner may result in serious personal injury or death. Work on a vehicle's fuel system components can be accomplished safely and effectively by adhering to the following rules and guidelines.

• To avoid the possibility of fire and personal injury, always disconnect the negative battery cable unless the repair or test procedure requires that battery voltage be applied.

• Always relieve the fuel system pressure prior to disconnecting any fuel system component (injector, fuel rail, pressure regulator, etc.) fitting or fuel line connection. Exercise extreme caution whenever relieving fuel system pressure to avoid exposing skin, face and eyes to fuel spray. Please be advised that fuel under pressure may penetrate the skin or any part of the body that it contacts.

• Always place a shop towel or cloth around the fitting or connection prior to loosening to absorb any excess fuel due to spillage. Ensure that all fuel spillage is quickly removed from engine surfaces. Ensure that all fuel-soaked cloths or towels are deposited into a flame-proof waste container with a lid.

• Always keep a dry chemical (Class B) fire extinguisher near the work area.

• Do not allow fuel spray or fuel vapors to come into contact with a spark or open flame.

• Always use a second wrench when loosening or tightening fuel line connection fittings. This will prevent unnecessary stress and torsion on fuel piping. Always follow the proper torque specifications.

• Always replace worn fuel fitting O-rings with new ones. Do not substitute fuel hose where rigid pipe is installed.

RELIEVING FUEL SYSTEM PRESSURE

See Figures 125 and 126.

➡**Before doing this procedure, make sure the engine is cold.**

✳✳ CAUTION

Before disconnecting fuel lines or hoses, relieve pressure from the system by disabling the fuel pump, running the engine until it stalls, then and disconnecting the fuel line/quick connect fitting in the engine compartment.

1. Remove the driver's dashboard undercover.

2. Remove PGM-FI main relay 2 (FUEL PUMP) from the under-dash fuse/relay box.

3. Start the engine, and let it idle until it stalls.

➡**If any DTCs are stored, clear and ignore them.**

4. Turn the ignition switch to LOCK (0).

5. Remove the fuel fill cap to relieve the pressure in the fuel tank.

6. Perform the battery disconnect/reconnect procedure.

7. Remove the quick-connect fitting cover.

8. Check the fuel quick-connect fitting for dirt, and clean it if needed.

9. Place a rag or shop towel over the quick-connect fitting.

10. Disconnect the quick-connect fitting by holding the connector with one hand, and squeezing the retainer tabs with the other hand to release them from the locking tabs. Pull the connector off.

✳✳ CAUTION

Be careful not to damage the line or other parts. Do not use tools.

11. If the connector does not move, keep the retainer tabs pressed down, and alternately pull and push the connector until it comes off easily.

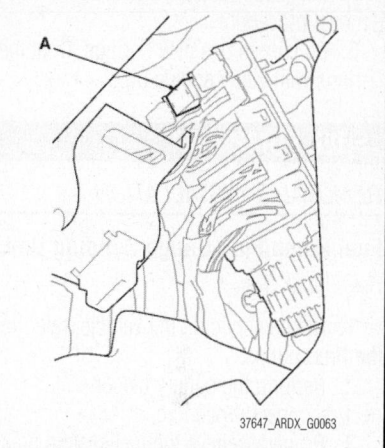

Fig. 125 Remove PGM-FI main relay 2 (FUEL PUMP) (A) from the under-dash fuse/relay box.

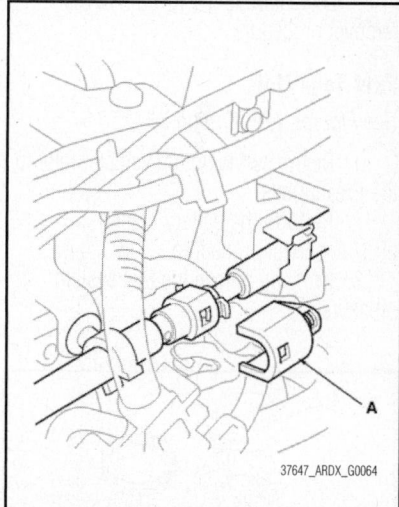

Fig. 126 Remove the quick-connect fitting cover (A).

➡**Do not remove the retainer from the line; once removed, the retainer must be replaced with a new one.**

12. After disconnecting the quick-connect fitting, check it for dirt or damage.

13. Perform the battery disconnect/reconnect procedure.

FUEL FILTER

REMOVAL & INSTALLATION

See Figure 127.

1. Properly relieve the fuel system pressure.
2. Remove the fuel tank.
3. Remove the fuel strainer set.

To install:

4. Installation is the reverse of the removal procedure.
5. Be sure to use new O-rings. Coat the O-rings with clean engine oil.

FUEL PUMP UNIT

REMOVAL & INSTALLATION

Fuel Pump/Fuel Gauge Sending Unit

See Figure 128.

1. Before servicing the vehicle, refer to the Precautions.
2. Perform the battery disconnect/reconnect procedure.
3. Properly relieve the fuel system pressure.
4. Remove the fuel tank.
5. Remove the fuel level sensor (fuel gauge sending unit) from the fuel tank unit.
6. Installation is the reverse of the removal procedure.

Fuel Tank Unit

See Figures 129 through 131.

1. Before servicing the vehicle, refer to the Precautions.
2. Perform the battery disconnect/reconnect procedure.
3. Properly relieve the fuel system pressure.

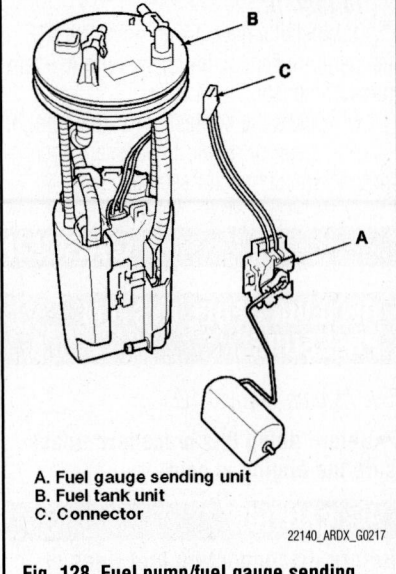

A. Fuel gauge sending unit
B. Fuel tank unit
C. Connector

22140_ARDX_G0217

Fig. 128 Fuel pump/fuel gauge sending unit related components

4. Remove the fuel cap.
5. Fold the rear seat down and remove the left rear door seal trim.
6. Pull back the carpet and expose the access panel.
7. Remove the access panel from the left side of the floor.

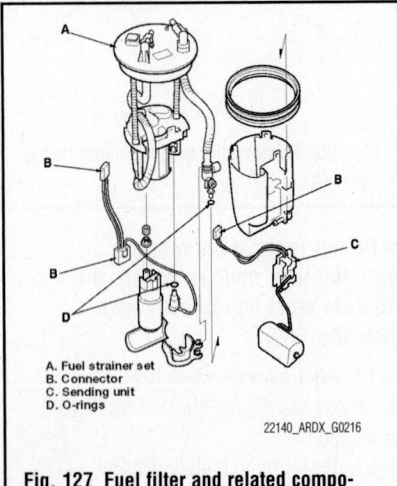

A. Fuel strainer set
B. Connector
C. Sending unit
D. O-rings

22140_ARDX_G0216

Fig. 127 Fuel filter and related components

8. Disconnect the fuel tank unit connector.
9. Disconnect the quick connect fittings from the fuel tank.
10. Using tool 07AAA-SOXA100 or equivalent, loosen the locknut.
11. Remove the locknut. Lift up the fuel tank unit from the fuel tank. Disconnect the transfer tube.

To install:

12. Temporarily attach a new base gasket to the assembly.
13. Position the mark on the transfer tube up.
14. Insert the assembly partially into the fuel tank.

➡ **Be careful not to damage the new base gasket. Be careful not to bend the sending unit. Do not coat the base gasket with oil.**

15. Transfer the base gasket from the sending unit to the fuel tank.
16. Align the marks on the fuel tank and the fuel gauge sending unit. Insert the sending unit and allow it to contact the base gasket.

❉❉❉ CAUTION

To avoid a fuel leak, check the base gasket visually or with your hand to

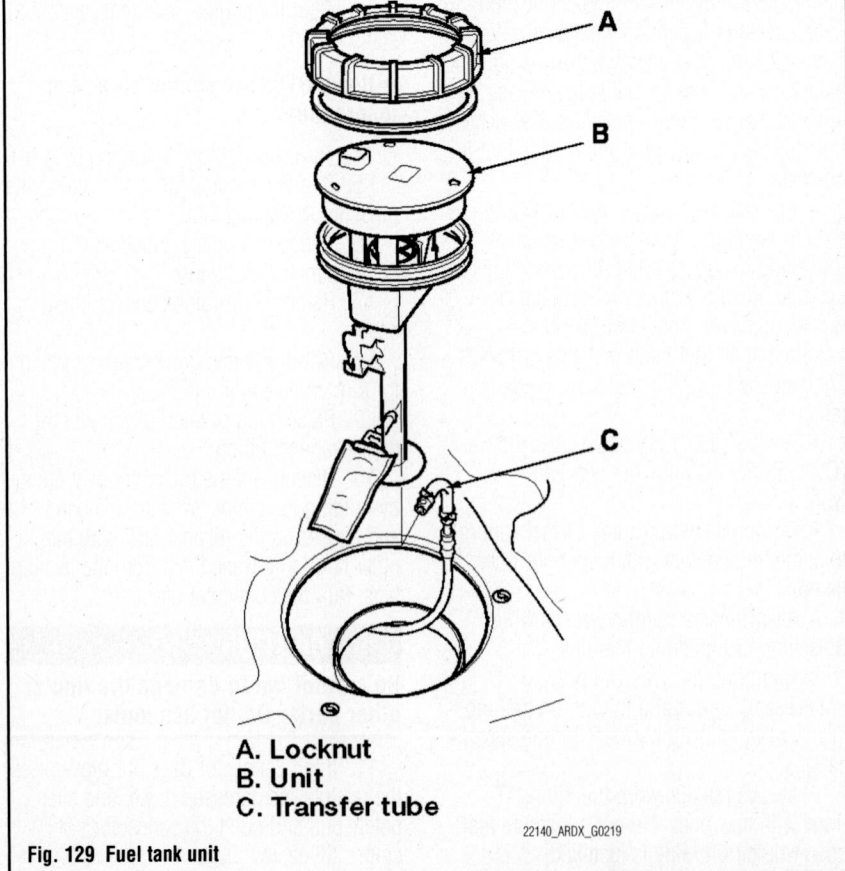

A. Locknut
B. Unit
C. Transfer tube

22140_ARDX_G0219

Fig. 129 Fuel tank unit

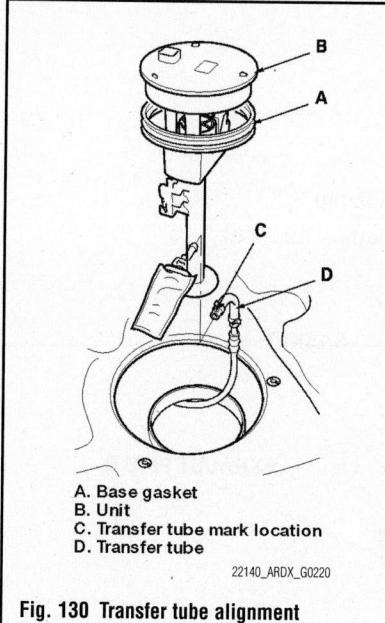

A. Base gasket
B. Unit
C. Transfer tube mark location
D. Transfer tube

22140_ARDX_G0220

Fig. 130 Transfer tube alignment

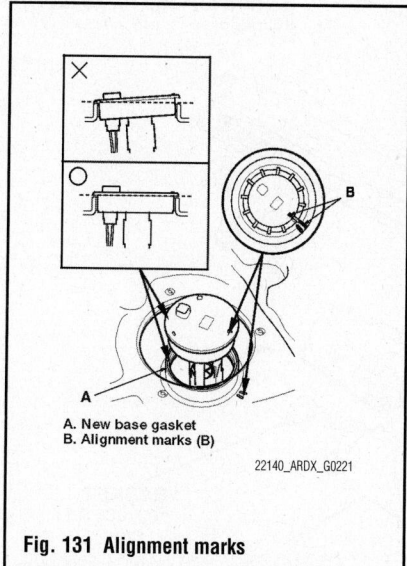

A. New base gasket
B. Alignment marks (B)

22140_ARDX_G0221

Fig. 131 Alignment marks

be sure it is not pinched. **Correct as required.**

17. Using the tool, tighten the locknut with the new locknut plate to specification. Specification is 52 ft. lbs. (70 Nm).

➡ **After tightening, be sure the marks are still aligned. To avoid a fuel leak, check the base gasket visually or with your hand to be sure it is not pinched. Correct as required.**

18. Connect the electrical connector.
19. Reconnect the negative battery cable and turn the ignition switch ON (II). The fuel pump will run for about two seconds and fuel pressure will rise.

➡ **Do not operate the starter motor.**

20. Repeat the above Step two or three times and check that there is no leakage in the fuel supply system.
21. Continue the installation in the reverse order of the removal procedure.

FUEL RAIL & INJECTORS

REMOVAL & INSTALLATION
See Figure 132.

1. Before servicing the vehicle, refer to the Precautions.
2. Perform the battery disconnect/reconnect procedure.
3. Properly relieve the fuel system pressure.
4. Remove the air charge cooler cover.
5. Disconnect the turbocharger boost sensor connector. Remove the vacuum hoses and the air bypass outlet connecting tube.
6. Remove the air charge cooler retaining bolts. Remove the air charge cooler.
7. Disconnect the throttle body connector. Remove the injector cover.
8. Disconnect the connectors from the injectors. Remove the harness holder from the fuel rail.
9. Disconnect the quick connectors.

10. Remove the fuel retaining nuts. Remove the fuel rail from its mounting.
11. Remove the injector clips from the injectors. Remove the injectors from the fuel rail.

To install:
12. Coat the new O-rings (black) with clean engine oil, and then insert the injectors into the fuel rail. Install the injector clips.
13. Coat the new O-rings (green) with clean engine oil, and then install the fuel rail and the injectors in the injector base.
14. Install the fuel rail retaining nuts. Tighten them to 16 ft. lbs. (22 Nm).
15. Connect the quick connect fitting.
16. Turn the ignition switch ON (II). The fuel pump will run for about two seconds and fuel pressure will rise.

➡ **Do not operate the starter motor.**

17. Repeat the above step two or three times and check that there is no leakage in the fuel supply system.
18. Continue the installation in the reverse order of the removal procedure.
19. Clear any stored DTC codes, as required.

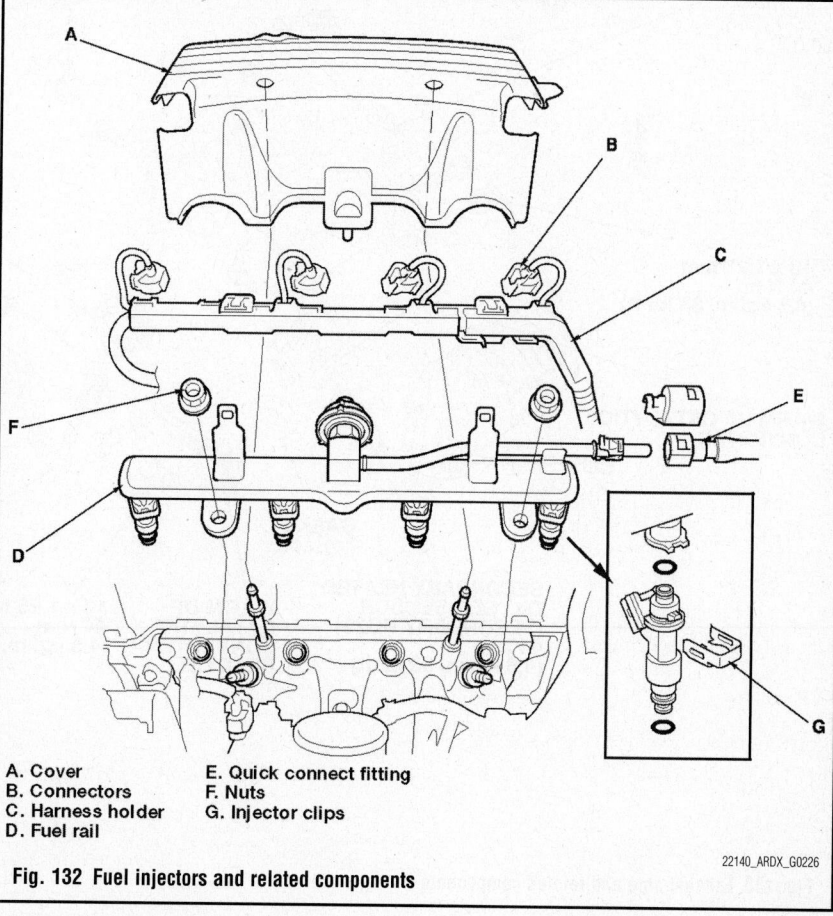

A. Cover
B. Connectors
C. Harness holder
D. Fuel rail
E. Quick connect fitting
F. Nuts
G. Injector clips

22140_ARDX_G0226

Fig. 132 Fuel injectors and related components

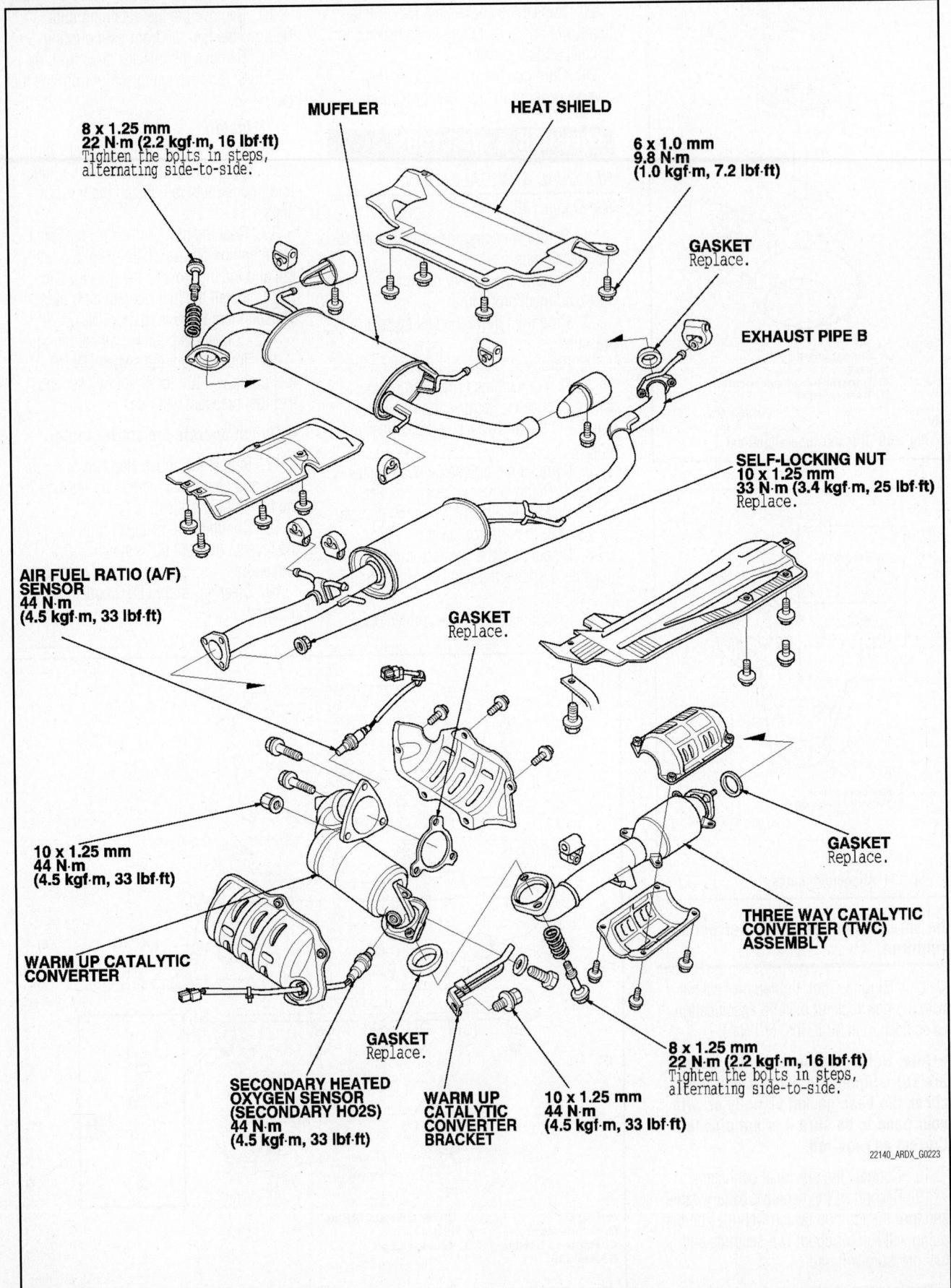

MUFFLER

HEAT SHIELD

8 x 1.25 mm
22 N·m (2.2 kgf·m, 16 lbf·ft)
Tighten the bolts in steps,
alternating side-to-side.

6 x 1.0 mm
9.8 N·m
(1.0 kgf·m, 7.2 lbf·ft)

GASKET
Replace.

EXHAUST PIPE B

SELF-LOCKING NUT
10 x 1.25 mm
33 N·m (3.4 kgf·m, 25 lbf·ft)
Replace.

AIR FUEL RATIO (A/F)
SENSOR
44 N·m
(4.5 kgf·m, 33 lbf·ft)

GASKET
Replace.

GASKET
Replace.

10 x 1.25 mm
44 N·m
(4.5 kgf·m, 33 lbf·ft)

THREE WAY CATALYTIC
CONVERTER (TWC)
ASSEMBLY

WARM UP CATALYTIC
CONVERTER

GASKET
Replace.

SECONDARY HEATED
OXYGEN SENSOR
(SECONDARY HO2S)
44 N·m
(4.5 kgf·ft, 33 lbf·ft)

WARM UP
CATALYTIC
CONVERTER
BRACKET

10 x 1.25 mm
44 N·m
(4.5 kgf·m, 33 lbf·ft)

8 x 1.25 mm
22 N·m (2.2 kgf·m, 16 lbf·ft)
Tighten the bolts in steps,
alternating side-to-side.

22140_ARDX_G0223

Fig. 133 Exhaust pipe and related components

FUEL TANK

REMOVAL & INSTALLATION

See Figures 133 and 134.

1. Before servicing the vehicle, refer to the Precautions.

2. Perform the battery disconnect/reconnect procedure.

3. Properly relieve the fuel system pressure.

4. Remove the fuel tank unit. Using a hand pump, hose and container drain the fuel from the tank.

5. Reinstall the fuel tank unit without connecting the quick connect fittings and the electrical connector.

6. Disconnect the secondary fuel gauge electrical connector.

7. Remove the fuel fill pipe cover. Disconnect the quick connect fitting.

8. Raise and support the vehicle safely.

9. Disconnect the fuel fill tube from the fuel tank. Slide back the clamp, then twist the tube as you pull to avoid damage. Disconnect the fuel vent tube.

10. Remove the exhaust pipe.

11. Remove the driveshaft.

12. Remove the fuel tank cover.

13. Properly support the fuel tank assembly by positioning a jack or support under the fuel tank.

14. Remove the fuel tank retaining straps.

15. Carefully lower the tank. Check to be sure nothing is attached. Remove the tank from the vehicle.

16. Installation is the reverse of the removal procedure.

17. Tighten the retaining strap bolts to 28 ft. lbs. (38 Nm).

THROTTLE BODY

REMOVAL & INSTALLATION

See Figure 135.

✳ CAUTION

Do not insert your fingers into the installed throttle body when you turn the ignition switch to ON (II) or while the ignition switch is in ON (II). If you do, you will seriously injure your fingers if the throttle valve is activated.

➡**If you are replacing or cleaning the throttle body, start at step 1. If you are removing the throttle body, start at step 4.**

1. USA, Canada models: Connect the HDS to the DLC while the engine is stopped.

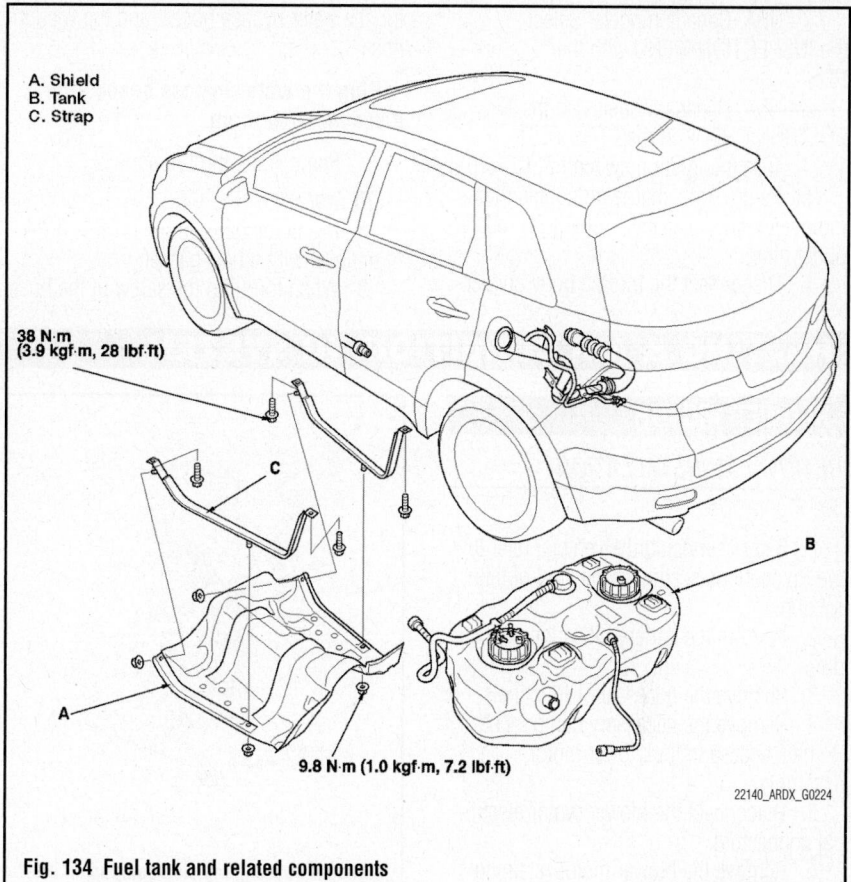

A. Shield
B. Tank
C. Strap

38 N·m
(3.9 kgf·m, 28 lbf·ft)

9.8 N·m (1.0 kgf·m, 7.2 lbf·ft)

22140_ARDX_G0224

Fig. 134 Fuel tank and related components

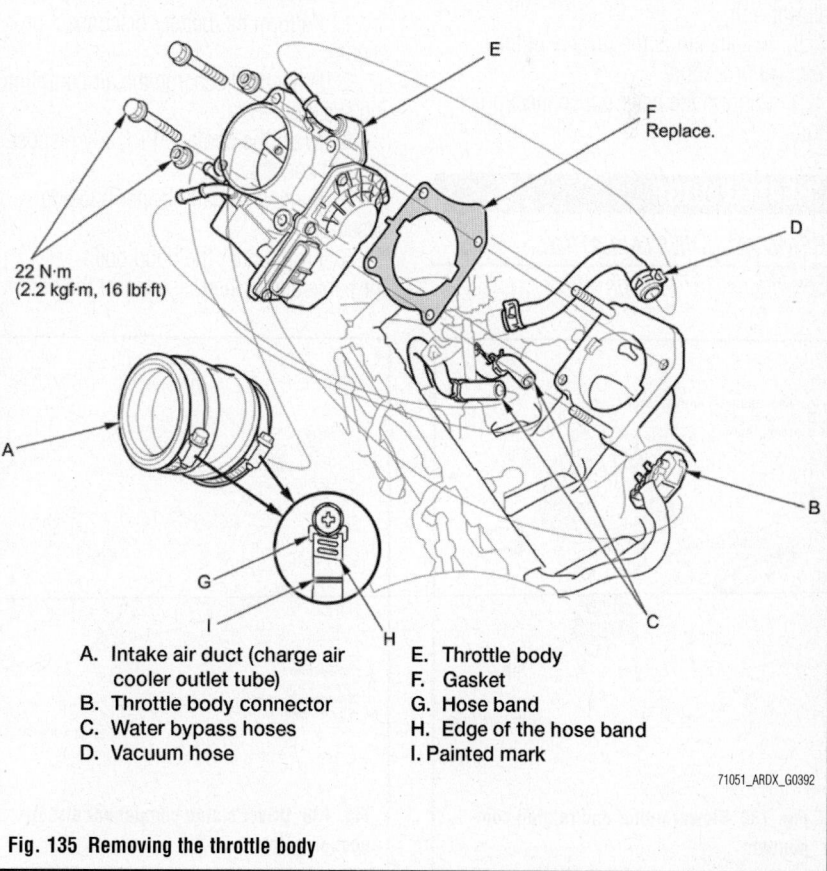

22 N·m
(2.2 kgf·m, 16 lbf·ft)

A. Intake air duct (charge air cooler outlet tube)
B. Throttle body connector
C. Water bypass hoses
D. Vacuum hose
E. Throttle body
F. Gasket
G. Hose band
H. Edge of the hose band
I. Painted mark

71051_ARDX_G0392

Fig. 135 Removing the throttle body

2. USA, Canada models: Select the INSPECTION MENU with the HDS.

3. USA, Canada models: Do the TP POSITION CHECK in the ETCS TEST.

4. Turn the ignition switch to LOCK (0).

5. Remove the charge air cooler, then remove the intake air duct (charge air cooler outlet tube).

6. Disconnect the throttle body connector, the water bypass hoses, and the vacuum hose.

➡**Plug the water bypass hoses after disconnecting them.**

7. Remove the throttle body.

To install:

8. Install the parts in the reverse order of removal with a new gasket.

9. When torquing the screw of the hose band, align the edge of the hose band with the mark painted on the hose band. If you tighten the screw over the mark, replace the hose band.

10. After installing the throttle body, do this:

a. Refill the radiator with engine coolant.

b. Do the PCM idle learn procedure.

HEATING & AIR CONDITIONING SYSTEM

BLOWER MOTOR

REMOVAL & INSTALLATION

See Figure 136.

1. Before servicing the vehicle, refer to the Precautions section at the start of this section.

2. Perform the battery disconnect procedure.

3. Remove the necessary trim panels.

4. Remove the necessary components to gain access to the blower motor assembly.

5. Disconnect the blower motor electrical connectors.

6. Remove the blower motor retaining screws.

7. Remove the blower motor from the blower unit.

8. Installation is the reverse of the removal procedure.

9. Perform the battery reconnect procedure.

HEATER CORE

REMOVAL & INSTALLATION

See Figures 137 through 140.

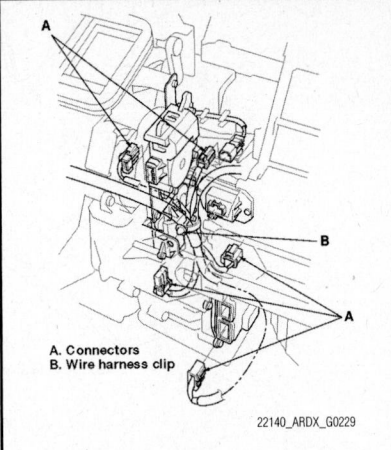

A. Connectors
B. Wire harness clip

22140_ARDX_G0229

Fig. 137 Passenger's side component disconnection points

1. Perform the battery disconnect procedure..

2. Properly discharge the air conditioning system.

3. Drain the coolant. Properly dispose of used coolant.

4. Remove the air cleaner housing assembly.

5. From under the hood open the cable clamp then disconnect the

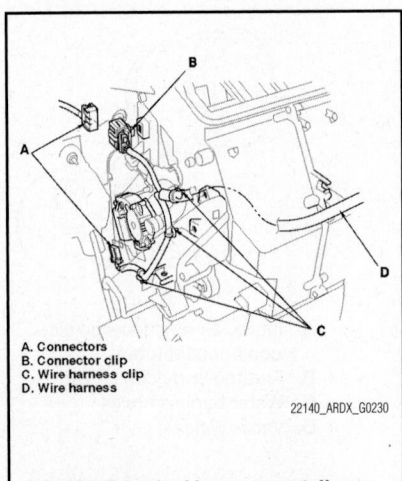

A. Connectors
B. Connector clip
C. Wire harness clip
D. Wire harness

22140_ARDX_G0230

Fig. 138 Driver's side component disconnection points

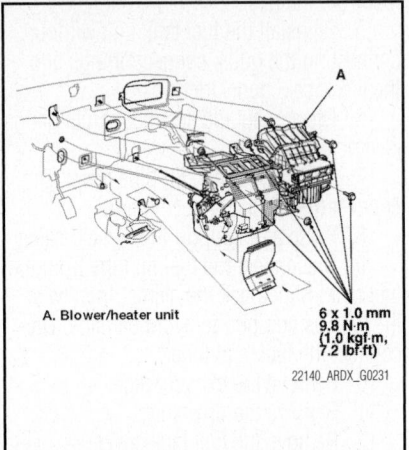

A. Blower/heater unit

6 x 1.0 mm
9.8 N·m
(1.0 kgf·m,
7.2 lbf·ft)

22140_ARDX_G0231

Fig. 139 Blower/heater unit and related components

heater valve cable from the heater valve arm. Turn the heater valve arm to the fully open position as shown.

6. Disconnect and plug the heater hoses at the heater core.

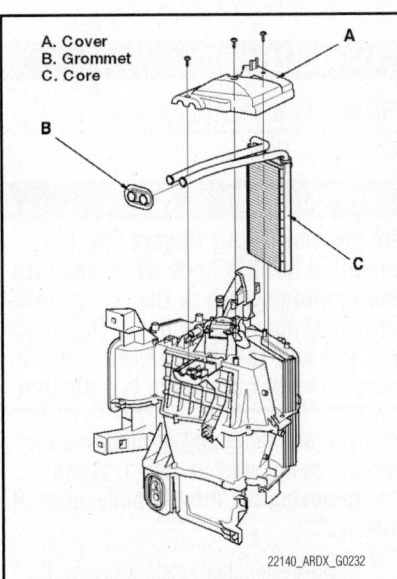

A. Cover
B. Grommet
C. Core

22140_ARDX_G0232

Fig. 140 Heater core and related components

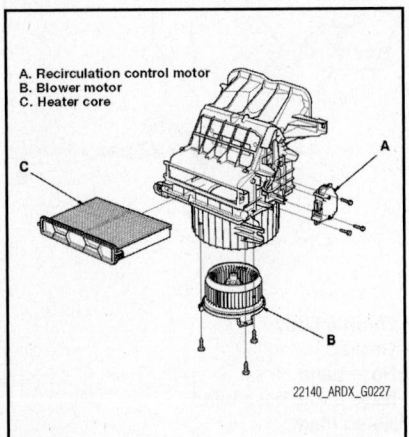

A. Recirculation control motor
B. Blower motor
C. Heater core

22140_ARDX_G0227

Fig. 136 Blower motor and related components

7. Remove the mounting nut from the heater unit.

8. Remove the lower instrument panel.

9. Disconnect the blower motor electrical connector. Remove the wire harness clip and ground terminal bolt.

10. Disconnect the connector from the recirculation control motor.

11. Disconnect the connectors from the climate control unit, mode control motor, passenger's air mix door motor, evaporator temperature sensor and the power transistor.

12. Remove the wire harness.

13. Disconnect the connectors from the driver's air mix door motor and air conditioning wire harness.

14. Remove the connector clip, the wire harness clips and the wire harness.

15. Remove the mounting bolt, mounting nuts, and remove the blower/heater unit from the vehicle.

16. Remove the self-taping screws. Remove the heater core cover.

17. Remove the grommet and carefully pull the heater core out of its mounting.

To install:

18. Installation is the reverse of the removal procedure.

19. Do not interchange the heater inlet and outlet hoses.

20. Be sure to use new O-rings, as required.

21. Perform the battery reconnect procedure.

22. Evacuate, recharge and leak test the system.

23. Fill the engine with the proper grade and type engine coolant.

24. Start the engine and check for leaks. Correct as required.

25. Be sure that the air conditioning system is functioning properly.

STEERING

POWER RACK & PINION STEERING GEAR

REMOVAL & INSTALLATION

See Figures 141 through 150.

1. Drain the power steering fluid.

2. Perform the battery disconnect procedure.

3. Raise the front of vehicle, and support it with safety stands in the proper locations.

4. Remove the front wheels.

5. Remove the steering wheel.

6. Remove the driver's dashboard undercover.

7. Remove the steering joint cover at the floorboard.

8. Remove the steering joint bolt, and disconnect the steering joint by moving the steering joint toward the column.

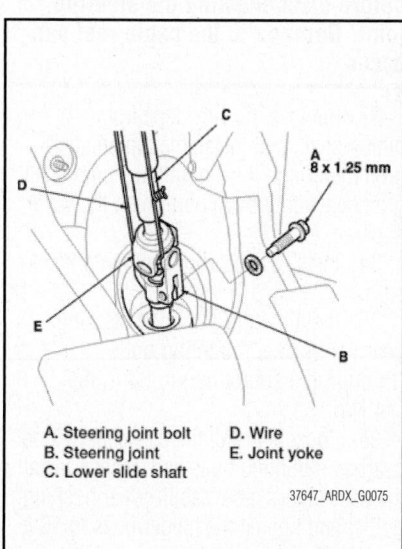

A. Steering joint bolt   D. Wire
B. Steering joint   E. Joint yoke
C. Lower slide shaft

37647_ARDX_G0075

Fig. 141 Removing the steering joint from the lower steering shaft

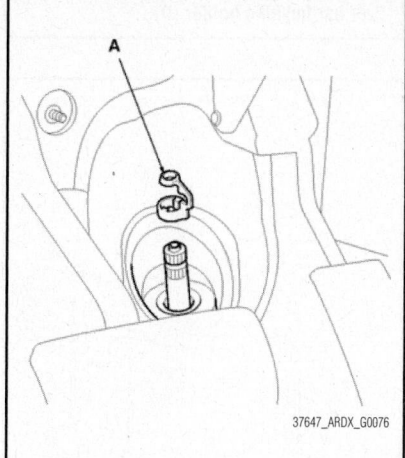

37647_ARDX_G0076

Fig. 142 Remove the center guide (A) (if equipped), and discard it.

9. Hold the lower slide shaft on the column with a piece of wire between the joint yoke on the lower slide shaft to the joint yoke on the upper shaft.

10. Remove the center guide (if equipped), and discard it.

➡**The center guide is for factory assembly only.**

11. Remove the tie rod ball joint nut cotter pin from and loosen the nut.

12. Separate the tie rod ball joint and knuckle using the ball joint remover.

13. Remove the air cleaner housing and the under-hood fuse/relay box.

14. Loosen the adjustable hose clamp, and disconnect the return hose. Remove the inlet line clamp bolt. Loosen the 18 mm flare nut, and disconnect the inlet line.

15. Remove the 10 mm flange bolt from the driver's side of the steering gearbox.

16. Disconnect the return line snap

holder clamp, and remove the return line clamp bolt.

17. Remove the pump outlet hose clamp from the steering gearbox.

18. Disconnect the return hose clips.

19. Remove the P/S heat shield.

20. Remove the 10 mm flange bolt and washer from the driver's side of the steering gearbox.

21. Remove the flange bolts, then remove the stabilizer bar bushing holder.

22. Repeat this process on the passenger's side of the steering gearbox.

23. Remove the steering gearbox bracket mounting bolts from both sides of the subframe.

24. Remove the gearbox mounting brackets.

25. Move the steering gearbox toward the front, and remove the pinion shaft grommet from the top of the valve housing. Apply vinyl tape to the splines on the pinion shaft.

26. Carefully move the steering gearbox and tie rods as an assembly toward the front side until the pinion shaft clears the wheel well opening on the frame.

27. Remove the steering gearbox through the wheel well opening on the driver's side.

✳✳ CAUTION

After removing the steering gearbox, make sure that no power steering fluid gets on the gearbox mount cushions, gearbox housing, or the surface of the front subframe. Wipe off any spilled fluid at once.

To install:

28. Using solvent and a brush, wash any oil and dirt off the valve body unit, its lines, and the end of the gearbox. Blow dry with compressed air.

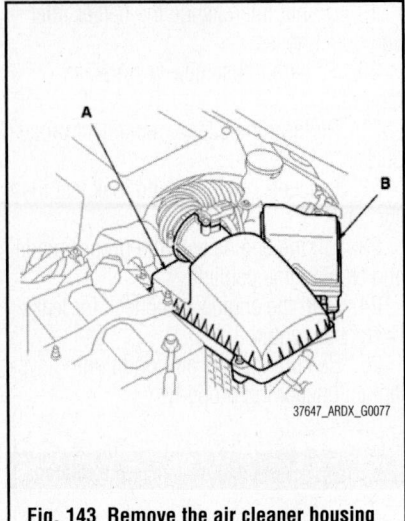

Fig. 143 Remove the air cleaner housing (A) and the under-hood fuse/relay box (B).

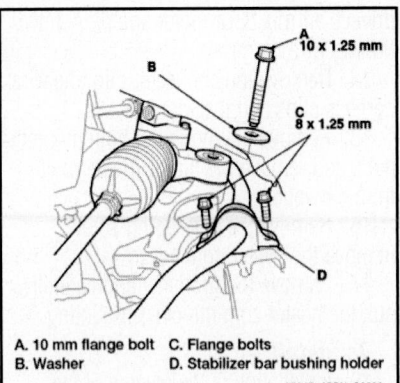

A. 10 mm flange bolt C. Flange bolts
B. Washer D. Stabilizer bar bushing holder

37647_ARDX_G0080

Fig. 146 Remove the 10 mm flange bolt (A) and washer (B) from the driver's side of the steering gearbox. Remove the flange bolts (C), then remove the stabilizer bar bushing holder (D).

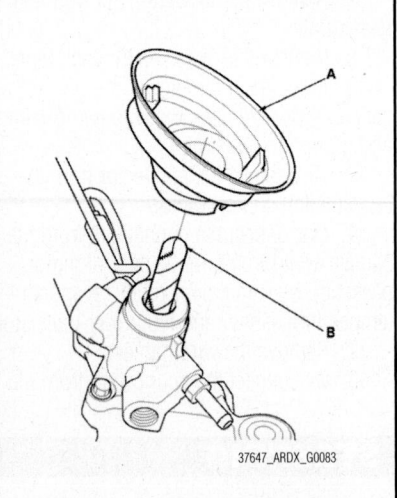

37647_ARDX_G0083

Fig. 149 Move the steering gearbox toward the front, and remove the pinion shaft grommet (A) from the top of the valve housing. Apply vinyl tape (B) to the splines on the pinion shaft.

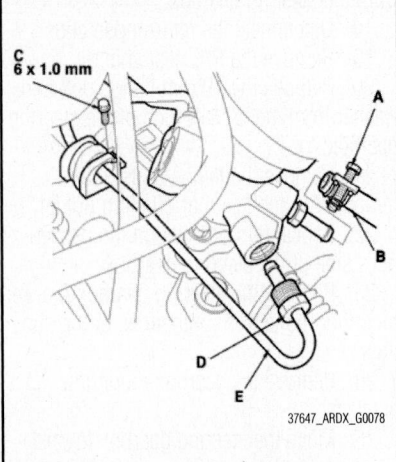

37647_ARDX_G0078

Fig. 144 Loosen the adjustable hose clamp (A), and disconnect the return hose (B). Remove the inlet line clamp bolt (C). Loosen the 18 mm flare nut (D), and disconnect the inlet line (E).

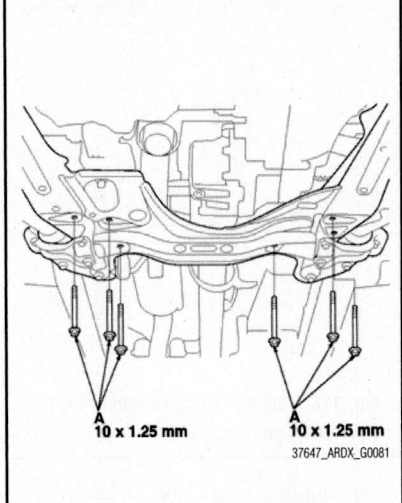

37647_ARDX_G0081

Fig. 147 Remove the steering gearbox bracket mounting bolts (A) from both sides of the subframe.

29. Slide the steering gearbox between the front subframe and the body from the driver's side. Place the gearbox in position on the front subframe.

30. Rotate the steering gearbox so the pinion shaft points downward.

31. Continue moving the gearbox toward the passenger's side until the steering gearbox is in position.

32. Make sure the power steering return line and feed line are routed above the gearbox.

✳✳ CAUTION

Be sure to remove the steering wheel before disconnecting the steering joint. Damage to the cable reel can occur.

33. Remove the vinyl tape from pinion shaft, and install the pinion shaft grommet. Align the slot in the pinion shaft grommet with the lug portion on the valve housing.

34. Install the gearbox mounting brackets.

35. Install and tighten the steering gearbox bracket mounting bolts (through the subframe) to 54 ft. lbs. (74 Nm).

36. Loosely install the passenger's side gearbox mounting bolts and washer. Install the passenger's side stabilizer bar bushing holder and tighten the holder bolts to 16 ft. lbs. (22 Nm).

37. Repeat the process on the driver's side.

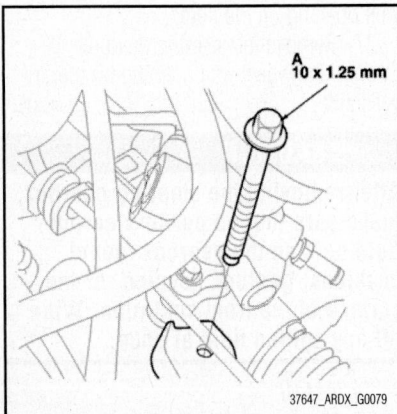

37647_ARDX_G0079

Fig. 145 Remove the 10 mm flange bolt (A) from the driver's side of the steering gearbox.

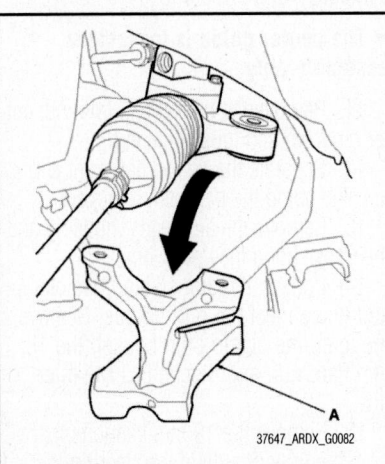

37647_ARDX_G0082

Fig. 148 Remove the gearbox mounting brackets (A).

38. Install the P/S heat shield.

39. Install the pump outlet hose clamp on the steering gearbox and connect the return hose clips

40. Connect the return line holder, and install the return line clamp bolt.

41. Loosely install the remaining 10 mm gearbox mounting bolt.

42. Tighten the gearbox mounting bolts, in an alternating pattern, to 43 ft. lbs. (59 Nm).

43. Connect the inlet line, and tighten the 18 mm flare nut to 35 ft. lbs. (47 Nm).

44. Connect the return hose securely, and tighten the adjustable hose clamp.

45. Install the inlet line clamp bolt

46. Install the under-hood fuse/relay box and the air cleaner housing.

47. Wipe off any grease contamination from the ball joint tapered section and threads. Reconnect the tie rod ends to the steering knuckles. Install the 12 mm nut, and tighten it to 40 ft. lbs. (54 Nm). Install a new cotter pin, and bend it as shown.

48. Center the steering rack within its stroke in steering joint connection.

49. With the rack in the straight ahead driving position, cut the wire and slip the lower end of the steering joint onto the pinion shaft in the range shown.

50. Align the bolt hole on the steering joint with the groove around the pinion shaft, then loosely install the joint bolt. Be sure that the joint bolt is securely in the groove in the pinion shaft. Pull on the

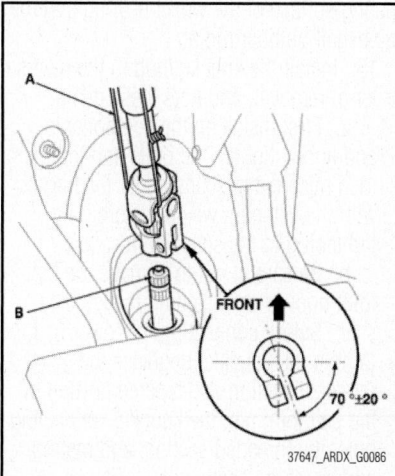

Fig. 150 With the rack in the straight ahead driving position, cut the wire (A) and slip the lower end of the steering joint onto the pinion shaft (B) in the range shown.

steering joint to make sure that the steering joint is fully seated. Tighten the steering joint bolt to 21 ft. lbs. (28 Nm).

51. Install the steering joint cover at the floorboard.

52. Install the driver's dashboard under-cover.

53. Install the front wheel, then set the wheels in the straight ahead position.

54. Install the steering wheel.

55. Perform the battery terminal reconnection procedure.

56. Perform the following tasks:

a. Verify cruise control, audio remote, navigation and HFL voice controls, horn, and turn signal switch operation.

b. Make sure the steering wheel is centered.

c. Fill the system with power steering fluid, and bleed air from the system.

57. After installation, do the following checks:

a. Start the engine, allow it to idle, and turn the steering wheel from lock-to-lock several times to warm up the fluid Check the gearbox for leaks..

b. Check the steering wheel spoke angle.

c. If steering spoke angles to the right and left are not equal (steering wheel and rack are not centered), correct the engagement of the joint/pinion shaft serrations.

d. Set the steering column to the center tilt position, and to the center telescopic position, then do the front toe inspection/adjustment.

POWER STEERING PUMP

REMOVAL & INSTALLATION

See Figure 151.

1. Drain the power steering reservoir. Be sure to properly dispose of used power steering fluid.

2. Remove the drive belt from the pump pulley.

3. Disconnect and plug the power steering pump fluid lines.

➥**Be sure to cover the auto tensioner, alternator and compressor with a shop towel in order to prevent fluid from spilling on these components.**

4. Remove the power steering pump retaining bolts. Remove the power steering pump from its mounting.

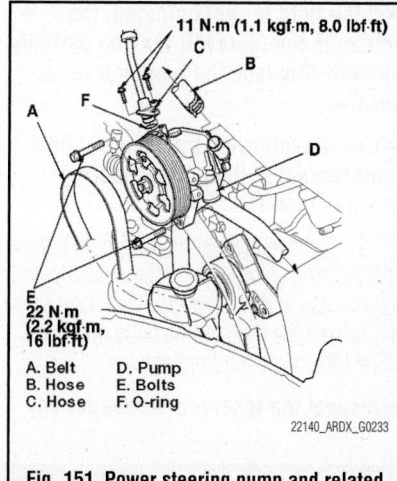

A. Belt   D. Pump
B. Hose   E. Bolts
C. Hose   F. O-ring

Fig. 151 Power steering pump and related components

➥**Cover the opening of the pump with a piece of tape to prevent dirt and debris from entering the pump.**

To install:

5. Connect the power steering pump hoes using new O-rings.

6. Loosely install the pump in the pump bracket with the mounting bolts. Tighten to 16 ft. lbs. (22 Nm).

7. Install the drive belt. Make sure that the belt is properly positioned on the pulleys.

8. Fill the reservoir with the proper grade and type power steering fluid.

BLEEDING

1. Before servicing the vehicle, refer to the Precautions section at the start of this vehicle menu.

2. Fill the power steering reservoir with the proper grade and type power steering fluid.

3. Start the engine and run it at fast idle.

4. Turn the steering wheel from lock-to-lock several times to bleed the air from the system.

5. Recheck the fluid level, correct as required.

6. Do not fill the reservoir past the upper level line.

FLUID FILL PROCEDURE

1. Check the reservoir at regular intervals, and add the recommended fluid as necessary.

⁂ CAUTION

Always use Acura Power Steering Fluid. Using any other type of power steering fluid or automatic transmission fluid can cause increased wear and poor steering in cold weather.

➡If the fluid is contaminated, the screen in the reservoir may be partially blocked. Replace the reservoir if necessary.

➡Use the following procedure if fluid replacement of the power steering pump is required:

2. Remove the reservoir from its holder. Raise the reservoir, then disconnect the return hose to drain the reservoir. Take care not to spill the fluid on the body and parts. Wipe off any spilled fluid at once.

➡Inspect the reservoir screen for any debris. If the reservoir screen is clogged, replace the reservoir.

3. Connect a hose of suitable diameter to the disconnected return hose, and put the hose end in a suitable container.
4. Start the engine, let it run at idle, and turn the steering wheel from lock-to-lock several times. When fluid stops running out of the hose, shut off the engine. Discard the fluid.
5. Reinstall the return hose on the reservoir.
6. Fill the reservoir to the upper level line.

7. Start the engine and run it at fast idle, then turn the steering from lock-to-lock several times to bleed air from the system.
8. Recheck the fluid level and add some if necessary. Do not fill the reservoir beyond the upper level line.
9. If the fluid is contaminated, dark, or discolored, repeat the procedure as necessary.
10. Note the system capacity:
   - 0.96 U.S. qt. (0.9 L) at disassembly
   - Reservoir capacity: 0.32 U.S. qt. (0.3 L)

SUSPENSION

FRONT SUSPENSION

KNUCKLE, SPINDLE & WHEEL BEARING

REMOVAL & INSTALLATION

Please note the following:
   - Special Tools Required: Ball Joint Thread Protector, 12 mm 07AAF-SDAA100.
   - Ball Joint Remover, 28 mm 07MAC-SL0A202
   - Ball Joint Thread Protector, 14 mm 07AAE-SJAA100
   - Ball Joint Remover, 32 mm 07MAC-SL0A102
   - Hub Dis/Assembly Pin, 42 mm 07GAF-SD4A100
   - Driver Handle, 15 x 135L 07749-0010000
   - Bearing Driver Attachment, 72 x 75 mm 07746-0010600
   - Support Base 07965-SD90100
   - Attachment, 96 mm 07948-SB00101

Knuckle/Hub Replacement

See Figures 152 through 154.

1. Raise and support the vehicle.
2. Remove the wheel nuts and the front wheel.
3. Remove the brake hose bracket mounting bolt.
4. Remove the brake caliper bracket mounting bolts, then remove the caliper assembly from the knuckle. To prevent damage to the caliper assembly or the brake hose, use a short piece of wire to hang the caliper assembly from the undercarriage. Do not twist the brake hose excessively.
5. Remove the wheel speed sensor from the knuckle. Do not disconnect the wheel speed sensor connector.
6. Pry up the stake on the spindle nut, then remove the nut.

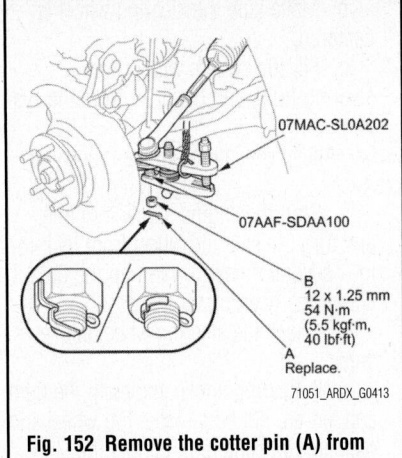

Fig. 152 Remove the cotter pin (A) from the tie-rod end ball joint, then remove the nut (B).

7. Remove the front brake disc.
8. Check the front hub for damage and cracks.
9. Remove the cotter pin from the tie-rod end ball joint, then remove the nut.
10. Disconnect the tie-rod end ball joint from the knuckle using the ball joint thread protector and the ball joint remover.
11. Remove the lock pin from the lower ball joint, then remove the castle nut.
12. Disconnect the lower ball joint from the knuckle using the ball joint thread protector and the ball joint remover.
13. Remove the damper pinch bolts and the flange nuts from the damper.
14. Remove the driveshaft outboard joint from the knuckle by tapping the driveshaft end with a soft face hammer while drawing the hub outward, then remove the knuckle.
Please note the following:
   - Do not pull the driveshaft end outward. The inner driveshaft joint may come apart.

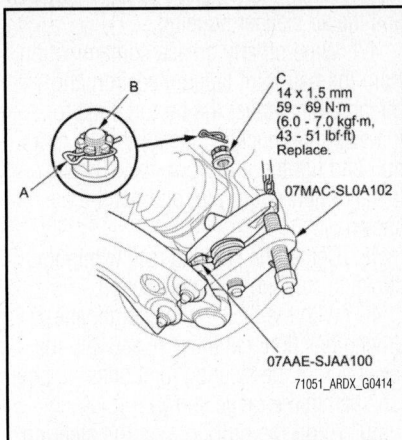

Fig. 153 Remove the lock pin (A) from the lower ball joint (B), then remove the castle nut (C).

   - During installation, apply grease to the mating surface of the wheel bearing and the driveshaft outboard joint.
15. Install the knuckle/hub in the reverse order of removal, and note these items:
   a. First install all the components, and lightly tighten the bolts and the nuts, then raise the suspension by loading it with the vehicle's weight before fully tightening to the specified torque.
   b. Be careful not to damage the ball joint boot when installing the knuckle.
   c. Before connecting the lower ball joint to the knuckle, degrease the threaded section and tapered portion of the ball joint pin, the knuckle connecting hole, the threaded section, and mating surface of the castle nut.
   d. Torque the castle nut to the lower torque specification, then tighten it only far enough to align the slot with the ball joint pin hole. Do not align the castle nut by loosening it.
   e. Use a new spindle nut during reassembly.

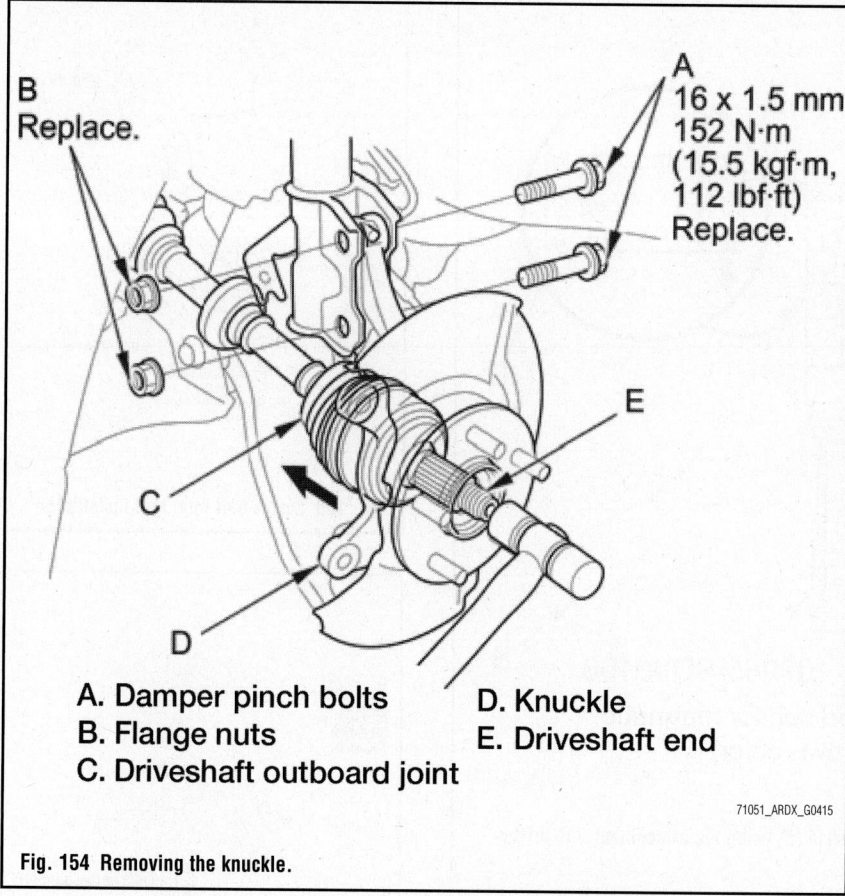

A. Damper pinch bolts
B. Flange nuts
C. Driveshaft outboard joint

D. Knuckle
E. Driveshaft end

B Replace.

A
16 x 1.5 mm
152 N·m
(15.5 kgf·m,
112 lbf·ft)
Replace.

71051_ARDX_G0415

Fig. 154 Removing the knuckle.

f. Before installing the spindle nut, apply a small amount of engine oil to the seating surface of the nut. After tightening, use a drift to stake the spindle nut shoulder against the driveshaft.

g. Before installing the brake disc, clean the mating surface of the front hub and the inside of the brake disc.

h. Before installing the wheel, clean the mating surface of the brake disc and the inside of the wheel.

16. Check the wheel alignment, and adjust it if necessary.

Wheel Bearing Replacement

See Figures 155 through 159.

1. Remove the knuckle/hub.

2. Separate the hub from the knuckle using the hub dis/assembly pin and a press. Hold the knuckle with the attachment of the press or equivalent tool. Be careful not to damage or deform the splash guard. Hold onto the hub to keep it from falling when pressed clear.

3. Press the wheel bearing inner race off of the hub using the hub dis/assembly pin, a commercially available bearing separator, and a press.

4. Remove the snap ring and the splash guard from the knuckle.

5. Press the wheel bearing out of the knuckle using the attachment, the driver handle, and a press.

6. Wash the knuckle and hub thoroughly in high flash point solvent before reassembly.

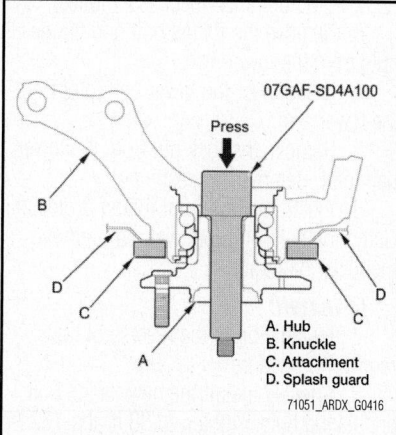

07GAF-SD4A100
Press

A. Hub
B. Knuckle
C. Attachment
D. Splash guard

71051_ARDX_G0416

Fig. 155 Separate the hub from the knuckle using the hub dis/assembly pin and a press. Hold the knuckle with the attachment of the press or equivalent tool. Be careful not to damage or deform the splash guard. Hold onto the hub to keep it from falling when pressed clear.

7. Press a new wheel bearing (A) into the knuckle (B) using the old bearing (C), a steel plate (D), the attachment, the support base, and a press.

8. Install the wheel bearing with the wheel speed sensor magnetic encoder (E) (brown color), toward the inside of the knuckle.

9. Remove any oil, grease, dust, metal debris, and other foreign material from the encoder surface.

10. Keep all magnetic tools away from the encoder surface.

11. Be careful not to damage the magnetic encoder surface when you insert the wheel bearing.

12. Install a new snap ring securely in the knuckle.

13. Install the splash guard and tighten the screws to the specified torque.

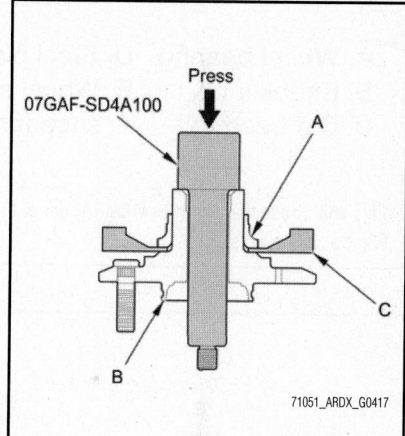

07GAF-SD4A100
Press

A
C
B

71051_ARDX_G0417

Fig. 156 Press the wheel bearing inner race (A) off of the hub (B) using the hub dis/assembly pin, a commercially available bearing separator (C), and a press.

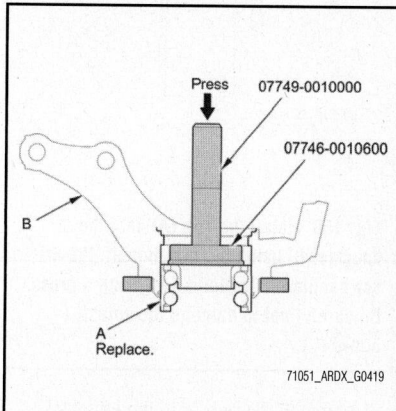

Press
07749-0010000
07746-0010600

A
Replace.

71051_ARDX_G0419

Fig. 157 Press the wheel bearing (A) out of the knuckle (B) using the attachment, the driver handle, and a press.

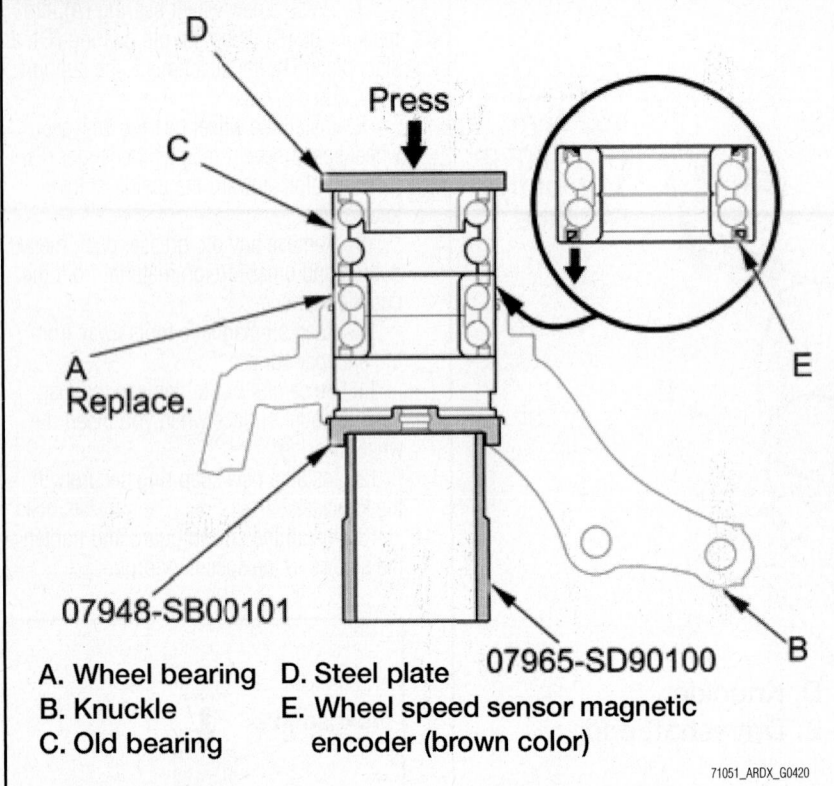

A. Wheel bearing D. Steel plate
B. Knuckle E. Wheel speed sensor magnetic
C. Old bearing encoder (brown color)

71051_ARDX_G0420

Fig. 158 Press the wheel bearing (A) out of the knuckle (B) using the attachment, the driver handle, and a press.

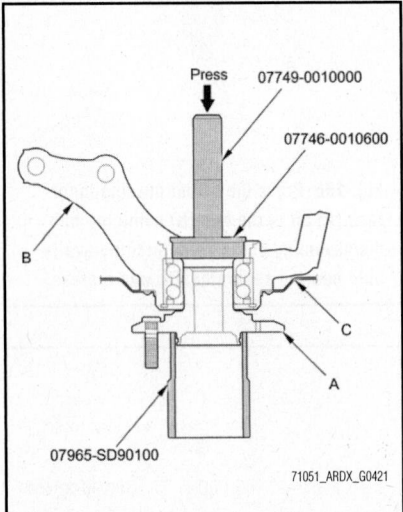

Fig. 159 Install the hub (A) onto the knuckle (B) using the attachment, the driver handle, the support base, and a press. Be careful not to damage the splash guard (C).

14. Install the hub onto the knuckle using the attachment, the driver handle, the support base, and a press. Be careful not to damage the splash guard.

15. Install the knuckle/hub.

LOWER BALL JOINTS

REMOVAL & INSTALLATION

See Figures 160 and 161.

1. Disconnect the positive battery cable.
2. Raise and support the vehicle safely. Remove the tire and wheel assemblies.
3. Remove the flange bolt and flange nuts from the lower arm.
4. Disconnect the lower ball joint from the lower arm.
5. Remove the lock pin from the lower ball joint. Remove the castle nut.
6. Install the ball joint thread protector. Using the ball joint removal tool, remove the lower ball joint.

To install:

7. Installation is the reverse of the removal procedure.
8. Loosely install the new flange bolt and flange nuts. Tighten to 38 ft. lbs. (52 Nm) in the following order:
   - Front nut
   - Rear nut
   - Rear bolt
9. Tighten the castle nut to 43–54 ft. lbs. (58–73 Nm), and then tighten it only enough to install the cotter pin. Do not align the castle nut by loosening it.

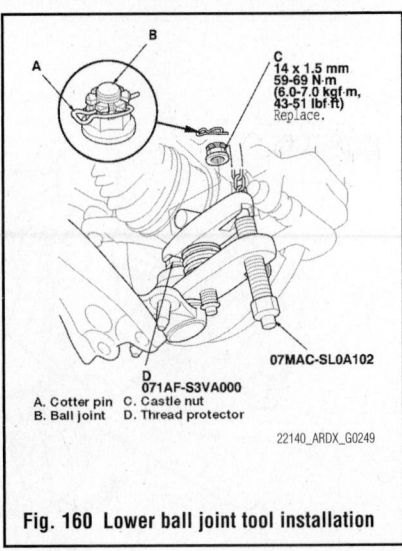

A. Cotter pin C. Castle nut
B. Ball joint D. Thread protector

22140_ARDX_G0249

Fig. 160 Lower ball joint tool installation

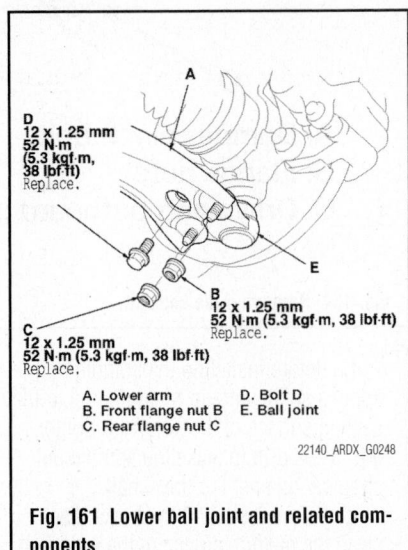

A. Lower arm D. Bolt D
B. Front flange nut B E. Ball joint
C. Rear flange nut C

22140_ARDX_G0248

Fig. 161 Lower ball joint and related components

➤First install all the components, and lightly tighten the bolts and nuts, then tighten the lower ball joint to the lower arm to specification. Raise the suspension to load it with the vehicles weight before fully tightening the lower ball joint to the knuckle to specification.

10. Check and adjust the wheel alignment, as required.

LOWER CONTROL ARMS

REMOVAL & INSTALLATION

See Figures 162 and 163.

1. Raise and support the vehicle safely. Remove the tire and wheel assemblies.
2. Remove the flange bolts and the bushing holder bolts from the front stabilizer bar.
3. Remove the flange bolt and flange nuts from the lower arm.

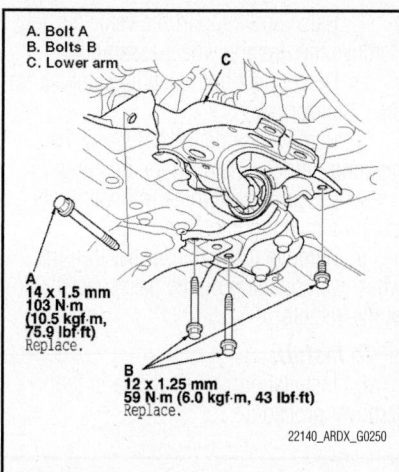

A. Bolt A
B. Bolts B
C. Lower arm

A
14 x 1.5 mm
103 N·m
(10.5 kgf·m,
75.9 lbf·ft)
Replace.

B
12 x 1.25 mm
59 N·m (6.0 kgf·m, 43 lbf·ft)
Replace.

22140_ARDX_G0250

Fig. 162 Lower control arm and related components

4. Disconnect the lower ball joint from the lower arm.

5. Remove the lower arm mounting bolt.

6. Remove the lower arm mounting bolts. Remove the lower arm from the front suspension subframe.

To install:

7. Installation is the reverse of the removal procedure.

8. Be sure to use new bolts when installing the lower control arm.

9. Tighten the stabilizer bushing bolts to 16 ft. lbs. (22 Nm).

10. Loosely install the new flange bolt and flange nuts. Tighten to 38 ft. lbs. (52 Nm) in the following order:
- Front flange nut
- Rear flange nut
- Bolt

11. Tighten the castle nut to 43–54 ft. lbs. (58–73 Nm), and then tighten it only enough to install the cotter pin. Do not align the castle nut by loosening it.

➡**First install all the components, and lightly tighten the bolts and nuts, then tighten the lower ball joint to the lower arm to specification. Raise the suspension to load it with the vehicles weight before fully tightening the lower ball joint to the knuckle to specification.**

12. Check and adjust the wheel alignment, as required.

MACPHERSON STRUTS

REMOVAL & INSTALLATION

See Figures 163 through 165.

1. Raise and support the vehicle safely. Remove the tire and wheel assemblies.

2. Remove the wheel sensor harness clip, the wire guide and the brake hose bracket from the strut. Do not disconnect the wheel sensor connector.

3. Disconnect the stabilizer link from the strut.

4. Remove the strut pinch bolts and self locking nuts from the strut.

5. Remove the lid from the cowl cover.

6. Remove the flange nuts from the top of the strut.

➡**Strut springs are different, left and right. Mark the springs "L" and "R" before removing them.**

7. Remove the strut from the vehicle.

To install:

8. Installation is the reverse of the removal procedure.

9. Be sure to use new bolts and nuts, as required.

10. Loosely install the new upper strut retaining flange nuts.

11. Loosely install the new front strut retaining bolts and new self locking nuts.

12. Connect the stabilizer link to the strut. Tighten the new nut to 58 ft. lbs. (78 Nm).

13. Install the wheel speed sensor harness clip, wire guide and brake hose bracket to the strut.

14. Raise the front suspension with a floor jack to load the suspension with the vehicles weight.

15. Tighten the strut bolts and self locking nuts to 122 ft. lbs. (165 Nm).

16. Tighten the upper strut nuts to specification. Specification is 33 ft. lbs. (45 Nm).

17. Install the cowl cover.

18. Check and adjust the wheel alignment, as required.

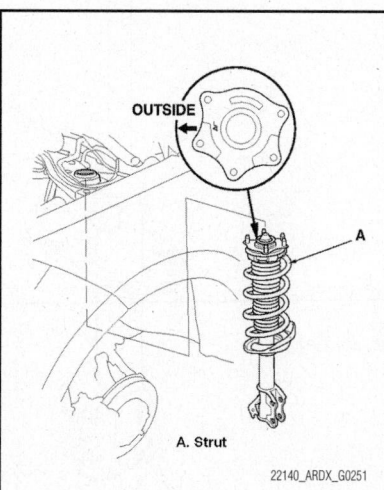

OUTSIDE

A. Strut

22140_ARDX_G0251

Fig. 163 Front strut installation and positioning

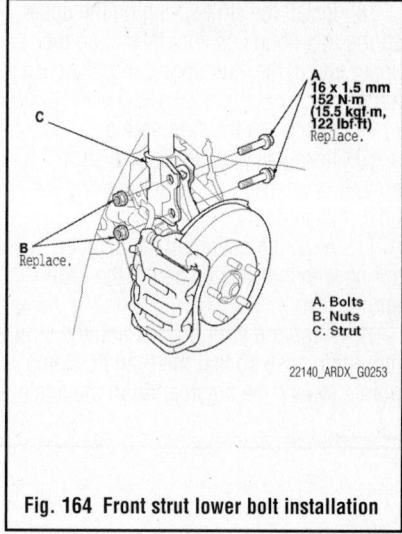

A
16 x 1.5 mm
152 N·m
(15.5 kgf·m,
122 lbf·ft)
Replace.

C

B
Replace.

A. Bolts
B. Nuts
C. Strut

22140_ARDX_G0253

Fig. 164 Front strut lower bolt installation

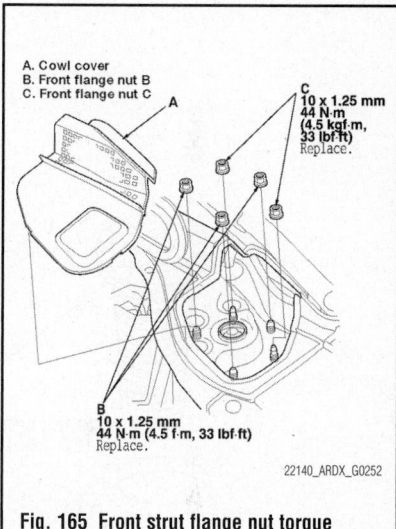

A. Cowl cover
B. Front flange nut B
C. Front flange nut C

C
10 x 1.25 mm
44 N·m
(4.5 kgf·m,
33 lbf·ft)
Replace.

B
10 x 1.25 mm
44 N·m (4.5 f·m, 33 lbf·ft)
Replace.

22140_ARDX_G0252

Fig. 165 Front strut flange nut torque sequence

OVERHAUL

See Figures 166 through 169.

1. Raise and support the vehicle safely. Remove the tire and wheel assemblies.

2. Remove the strut.

3. Position the strut in a suitable holding fixture.

4. Remove the cap from the top of the strut assembly.

5. Compress the strut spring and remove the self locking nut, using the strut nut adapter and a 17mm deep socket while holding the strut shaft with a wrench.

➡**Do not compress the spring more than necessary to remove the nut.**

6. Release the pressure from the strut spring compressor tool.

7. Disassemble the strut. Inspect and replace defective components, as required.

8. Install the strut spring on the upper spring mounting cushion, then align the upper end of the strut spring with the cushion stop.

9. Compress the strut spring.

10. Install all parts except the strut mounting washer and the self locking nut on to the strut unit.

11. Align the bottom of the damper spring with the stepped part of the lower spring seat.

12. Align the strut bracket and the strut mounting base so that the triangle stamp points toward the outside. Align the angle

of the stud bolt on the strut bracket as shown.

13. Install a new self locking nut. Tighten the nut to 33 ft. lbs. (45 Nm). Install the cap.

14. Continue the installation in the reverse order of the removal procedure.

STABILIZER BAR & LINKS

REMOVAL & INSTALLATION

Stabilizer Bar
See Figure 170.

1. Raise and support the vehicle safely. Remove the tire and wheel assemblies.

2. Disconnect both stabilizer links from the stabilizer bar.

3. Remove the flange bolts and the bushing holders. Remove the bushings.

4. Disconnect both tie rod ball joints from the steering knuckles.

5. Remove the stabilizer bar thru the wheel well opening on the passenger's side of the vehicle.

To install:

6. Installation is the reverse of the removal procedure.

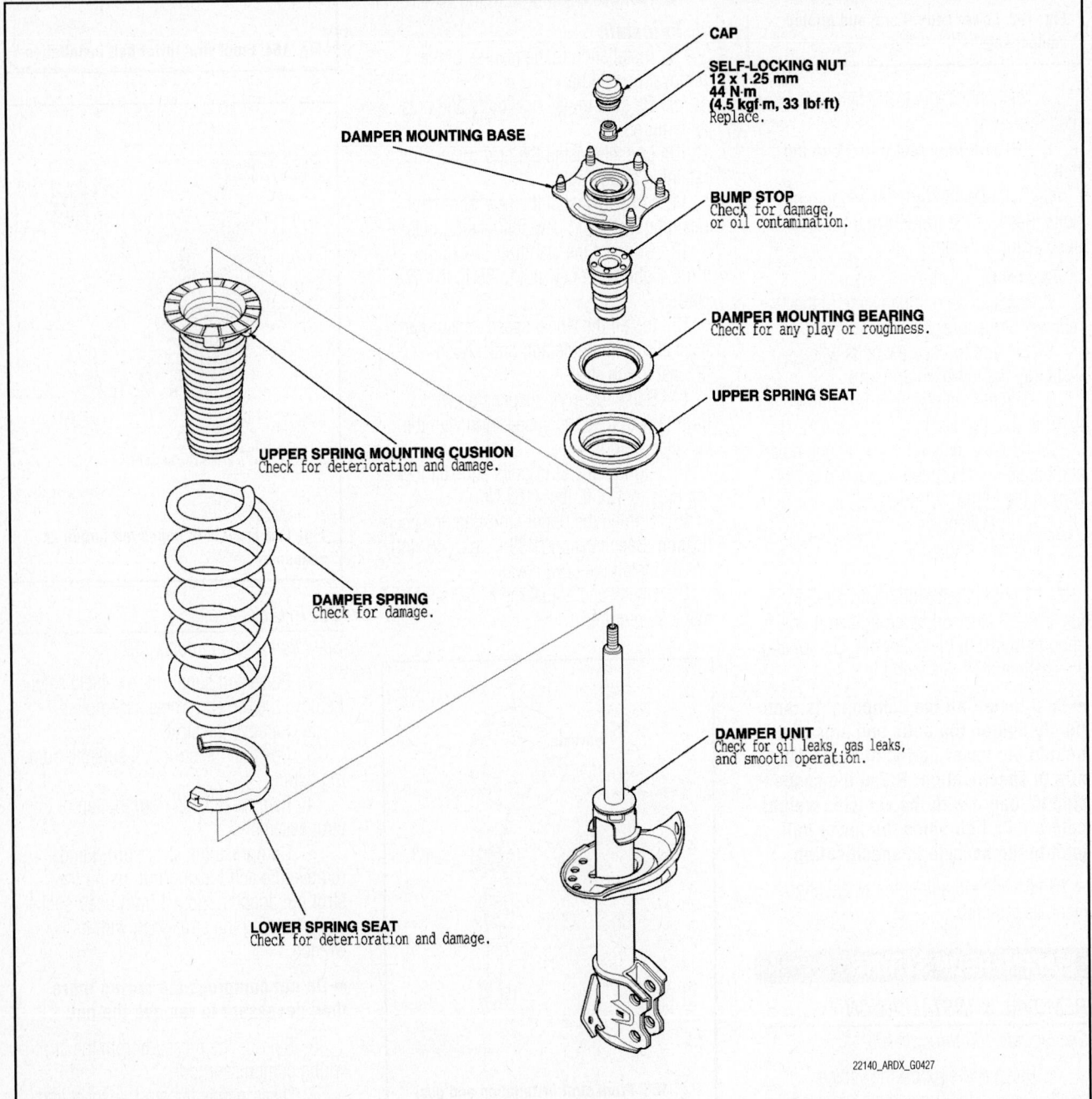

CAP

SELF-LOCKING NUT
12 x 1.25 mm
44 N·m
(4.5 kgf·m, 33 lbf·ft)
Replace.

DAMPER MOUNTING BASE

BUMP STOP
Check for damage,
or oil contamination.

DAMPER MOUNTING BEARING
Check for any play or roughness.

UPPER SPRING SEAT

UPPER SPRING MOUNTING CUSHION
Check for deterioration and damage.

DAMPER SPRING
Check for damage.

DAMPER UNIT
Check for oil leaks, gas leaks,
and smooth operation.

LOWER SPRING SEAT
Check for deterioration and damage.

22140_ARDX_G0427

Fig. 166 Front strut and related components

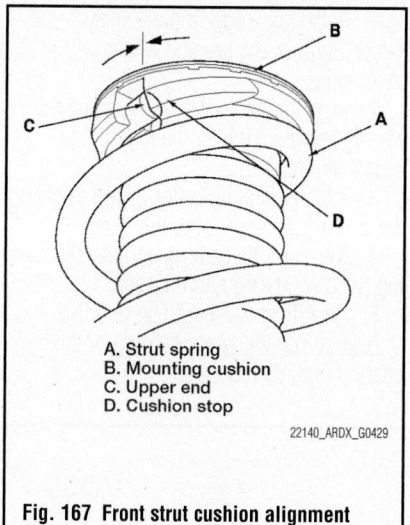

Fig. 167 Front strut cushion alignment

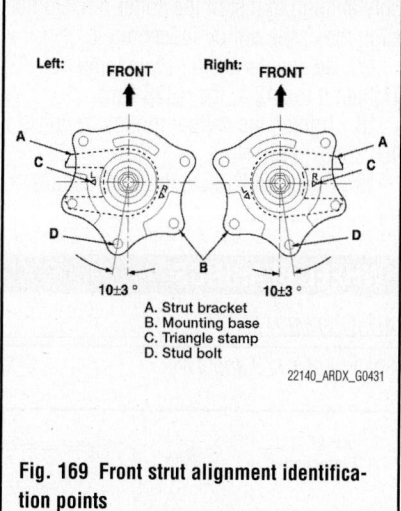

Fig. 169 Front strut alignment identification points

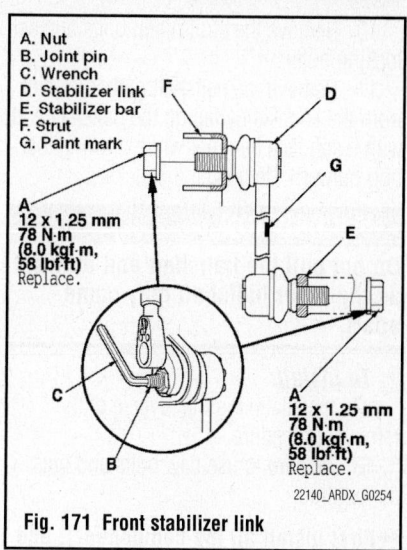

Fig. 171 Front stabilizer link

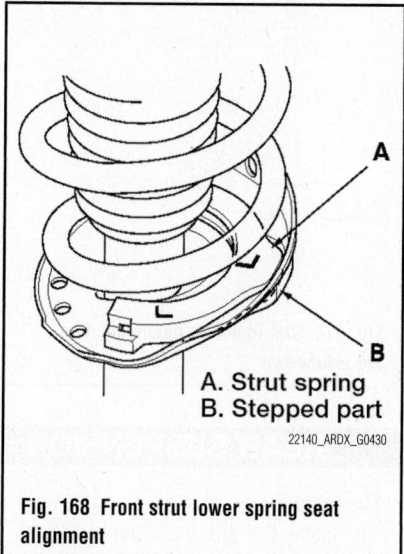

Fig. 168 Front strut lower spring seat alignment

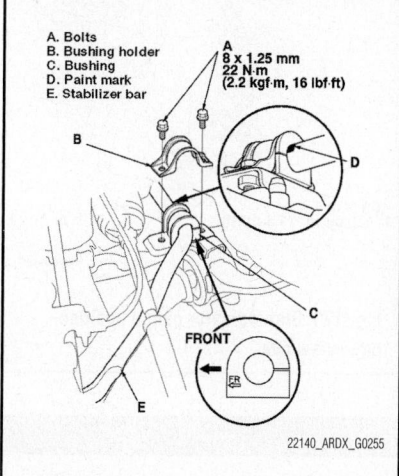

Fig. 170 Front stabilizer and related components

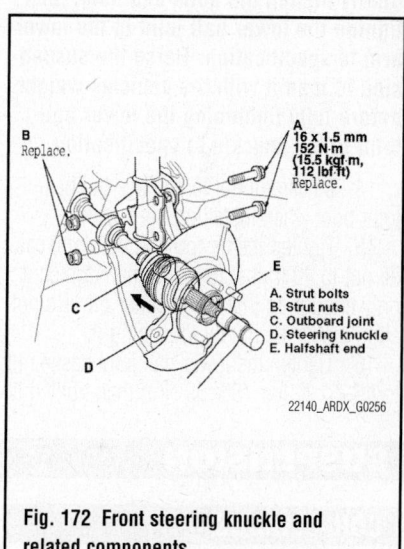

Fig. 172 Front steering knuckle and related components

7. Be sure to use new bolts and nuts, as required.

8. Note the right and left direction of the stabilizer bar.

9. Align the paint marks on the stabilizer bar with the sides of the bushings.

10. Note the fore/aft direction of the bushing.

11. Check and adjust the wheel alignment, as required.

Stabilizer Link

See Figure 171.

1. Raise and support the vehicle safely. Remove the tire and wheel assemblies.

2. Remove the flange nuts while holding the respective joint pin, with the hex wrench. Remove the stabilizer link.

To install:

3. Installation is the reverse of the removal procedure.

4. Be sure to use new bolts and nuts, as required.

5. Install the stabilizer link on the stabilizer bar and the strut with the joint pins set at the center of their range movement.

➡**The stabilizer link has a paint mark. Align the paint mark on the stabilizer link facing inward.**

6. Install the new flange nuts and lightly tighten them. Tighten them to specification while holding the respective joint pin with the hex wrench. Specification is 58 ft. lbs. (79 Nm).

7. Check and adjust the wheel alignment, as required.

STEERING KNUCKLE

REMOVAL & INSTALLATION

See Figure 172.

1. Raise and support the vehicle safely. Remove the tire and wheel assemblies.

2. Remove the brake hose mounting bolt. Remove the caliper and position it to the side. Do not allow the caliper to hang by the brake hose.

3. Remove the wheel speed sensor from the knuckle. Do not disconnect the wheel speed sensor connector.

4. Remove the spindle nut. Remove the rotor.

5. Check the hub for damage and cracks.

6. Remove the cotter pin from the tie rod end ball joint. Remove the nut.

7. Disconnect the tie rod ball joint from the steering knuckle using a ball joint removal tool.

8. Remove the lock pin from the lower ball joint. Remove the castle nut.

9. Disconnect the lower ball joint from the knuckle.

10. Remove the strut pinch bolts and self locking nuts.

11. Remove the halfshaft outboard joint from the knuckle by taping the halfshaft end with a soft face hammer while drawing the hub outward. Remove the knuckle.

✳✳ CAUTION

Do not pull the halfshaft end outward as the inner halfshaft may come apart.

To install:

12. Installation is the reverse of the removal procedure.

13. Be sure to use new bolts and nuts, as required.

➡**First install all the components, and lightly tighten the bolts and nuts, then tighten the lower ball joint to the lower arm to specification. Raise the suspension to load it with the vehicles weight before fully tightening the lower ball joint to the knuckle to specification.**

14. Be careful not to damage the ball joint boot when installing the knuckle.

15. Tighten the tie rod end ball joint castle nut to 40 ft. lbs. (54 Nm) then tighten it only enough to install the cotter pin. Do not align the castle nut by loosening it.

16. Tighten the lower ball joint castle nut to 43–51 ft. lbs. (58–69 Nm) then tighten it only enough to install the cotter pin. Do not align the castle nut by loosening it.

17. Be sure to use a new spindle nut. Tighten it to 242 ft. lbs. (328 Nm).

18. Tighten the caliper mounting bolts to 101 ft. lbs. (137 Nm).

19. Check and adjust the wheel alignment, as required.

WHEEL BEARINGS

ADJUSTMENT

See Figures 173 and 174.

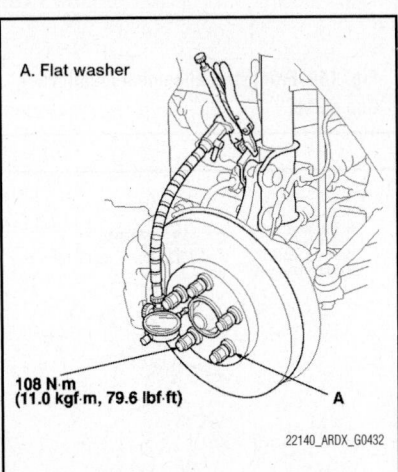

Fig. 173 Dial indicator gauge positioning—view one

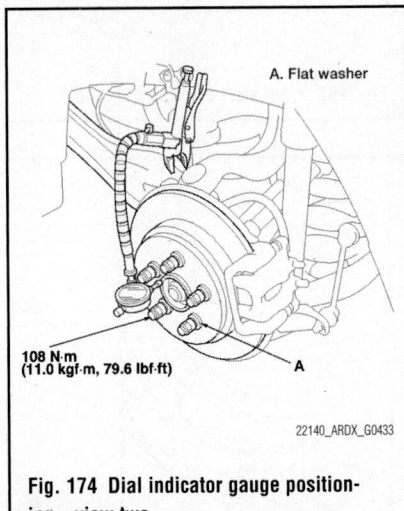

Fig. 174 Dial indicator gauge positioning—view two

1. Raise and support the vehicle safely. Remove the tire and wheel assemblies.

2. Install suitable flat washers and the wheel lug nuts. Tighten the lug nuts to specification.

3. Install a dial indicator gauge. Position the dial gauge against the hub flange.

4. Measure the bearing endplay by moving the disc inward and outward.

5. Specification should be 0–0.002 in.

6. Replace the wheel bearing or the hub bearing unit, as required.

SUSPENSION

COIL SPRINGS

REMOVAL & INSTALLATION

See Figures 175 and 176.

1. Raise and support the vehicle safely. Remove the tire and wheel assemblies.

2. Position the floor jack at the connecting point of the lower control arm and the lower strut mount.

3. Disconnect the stabilizer link from the lower arm.

4. Remove the flange bolt that connects the lower arm and the knuckle.

5. Remove the flange bolt that connects the lower arm and the bottom of the strut.

6. Lower the floor jack, slowly.

7. Remove the spring and the lower spring cushion.

8. Remove the flange bolt that connects to the body. Remove the spring mounting collar and the spring mounting cushion, as required.

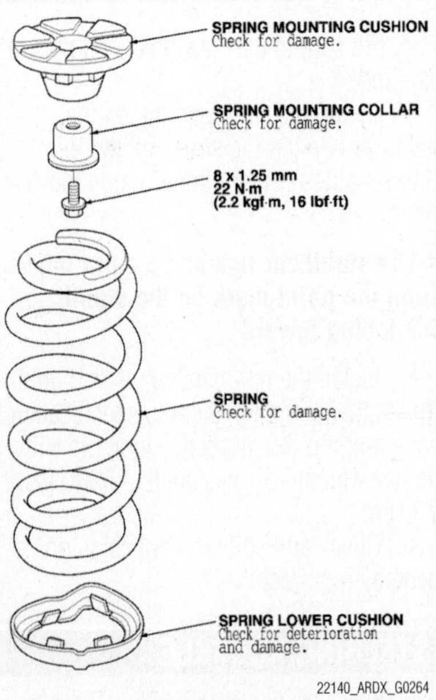

SPRING MOUNTING CUSHION
Check for damage.

SPRING MOUNTING COLLAR
Check for damage.

8 x 1.25 mm
22 N·m
(2.2 kgf·m, 16 lbf·ft)

SPRING
Check for damage.

SPRING LOWER CUSHION
Check for deterioration and damage.

22140_ARDX_G0264

Fig. 175 Rear spring and related components

REAR SUSPENSION

To install:

9. Install the spring and the lower spring cushion. Align the bottom of the spring and the lower spring cushion, as shown.

10. Position the floor jack at the connecting point of the lower control arm.

11. Slowly raise the jack until you align the bolt hole with the holes in the lower arm and the knuckle.

12. Loosely install a new flange bolt.

13. Compress the strut by hand until you can align the bolt hole with the holes in the lower arm and the strut.

14. Loosely install a new flange bolt.

15. Connect the stabilizer link to the lower arm.

16. Raise the rear suspension with the floor jack until the vehicle just lifts off the safety stands.

17. Tighten the flange bolts and the self locking nut to specification. Bolt specification is 76 ft. lbs. (103 Nm). Self locking nut specification is 29 ft. lbs. (39 Nm).

18. Continue the installation in the reverse order of the removal procedure.

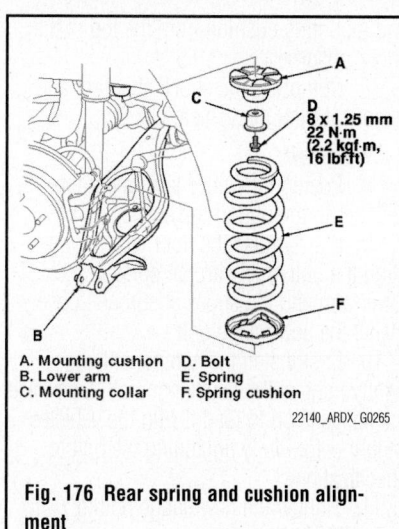

Fig. 176 Rear spring and cushion alignment

A. Mounting cushion   D. Bolt
B. Lower arm   E. Spring
C. Mounting collar   F. Spring cushion

22140_ARDX_G0265

19. Check and adjust the rear alignment, as required.

CONTROL ARMS & LINKS

REMOVAL & INSTALLATION

Lower Control Arm

See Figures 177 through 179.

1. Raise and support the vehicle safely. Remove the tire and wheel assemblies.
2. Position the floor jack at the connecting point of the lower control arm and the lower strut mount.
3. Disconnect the stabilizer link from the lower arm.
4. Remove the flange bolt that connects the lower arm and the knuckle.
5. Remove the flange bolt that connects the lower arm and the bottom of the strut.
6. Lower the floor jack, slowly.
7. Remove the spring and the lower spring cushion.
8. Remove the flange bolt that connects to the body. Remove the spring mounting collar and the spring mounting cushion, as required.
9. Mark the cam positions of the adjusting bolt and the adjusting cam. Remove the nut, adjusting cam and adjusting bolt.
10. Set aside the nut and control arm mounting nut. Remove the lower arm.

To install:

11. Position the lower arm. Loosely install a new adjusting bolt, adjusting cam and self locking nut.

➡**At final tightening the torque the nut to 43 ft. lbs. (58 Nm).**

12. Install the spring and the lower spring cushion. Align the bottom of the

spring and the lower spring cushion, as shown.

13. Position the floor jack at the connecting point of the lower control arm.
14. Slowly raise the jack until you align the bolt hole with the holes in the lower arm and the knuckle.
15. Loosely install a new flange bolt.
16. Compress the strut by hand until you can align the bolt hole with the holes in the lower arm and the strut.
17. Loosely install a new flange bolt.
18. Connect the stabilizer link to the lower arm.
19. Raise the rear suspension with the floor jack until the vehicle just lifts off the safety stands.
20. Tighten the flange bolts and the self locking nut to specification. Bolt specification is 76 ft. lbs. (103 Nm). Self locking nut specification is 29 ft. lbs. (39 Nm).
21. Continue the installation in the reverse order of the removal procedure.

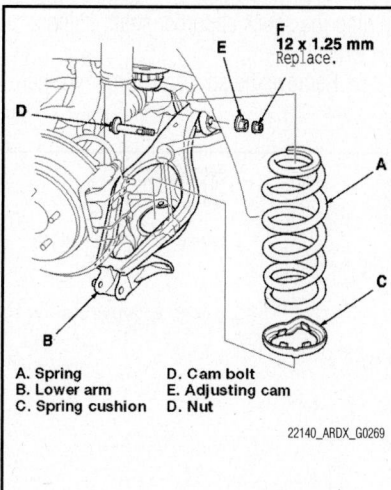

A. Spring   D. Cam bolt
B. Lower arm   E. Adjusting cam
C. Spring cushion D. Nut

22140_ARDX_G0269

Fig. 177 Adjusting cam bolt location

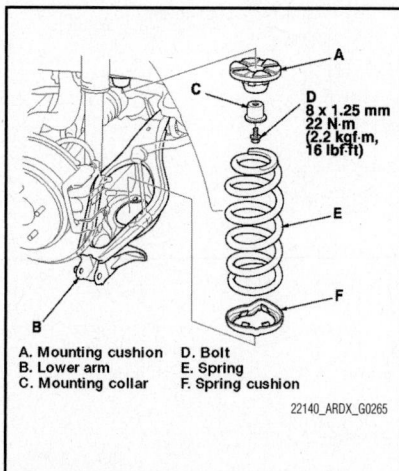

A. Mounting cushion   D. Bolt
B. Lower arm   E. Spring
C. Mounting collar   F. Spring cushion

22140_ARDX_G0265

Fig. 178 Rear spring and cushion alignment

22. Check and adjust the rear alignment, as required.

Lower Control Rod

See Figure 179.

1. Raise and support the vehicle safely. Remove the tire and wheel assemblies.
2. Position the floor jack at the connecting point of the lower control arm and the knuckle.
3. Remove the self locking nut, the washer and the flange bolt.
4. Remove the lower control rod from its mounting.

To install:

5. Installation is the reverse of the removal procedure.
6. Tighten the new flange bolt to 76 ft. lbs. (103 Nm). Tighten the new nut to 98 ft. lbs. (133 Nm).

➡**First install all the components and lightly tighten the bolts and nuts, then raise the suspension to load it with the vehicles weight before fully tightening the bolts and nuts to specification.**

7. Check and adjust the rear alignment, as required.

Upper Control Arm

See Figures 180 and 181.

1. Raise and support the vehicle safely. Remove the tire and wheel assemblies.
2. Position a floor jack at the connecting point of the lower arm and knuckle.
3. Remove the lock pin from the upper arm ball joint and loosen the nut.
4. Using the proper tool disconnect the upper arm ball joint from the knuckle.
5. Remove the flange bolt. Remove the upper arm from the vehicle.

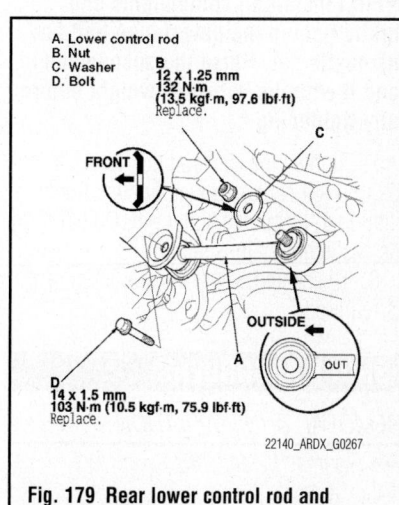

A. Lower control rod
B. Nut
C. Washer
D. Bolt

22140_ARDX_G0267

Fig. 179 Rear lower control rod and related components

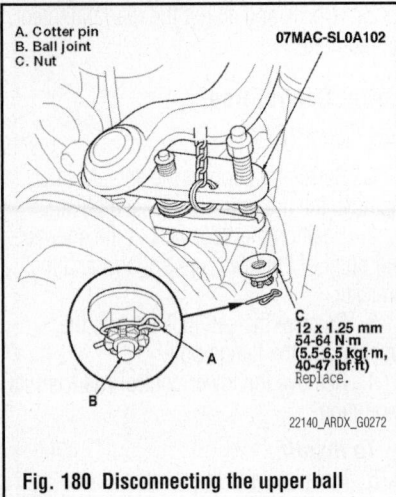

A. Cotter pin
B. Ball joint
C. Nut

07MAC-SL0A102

C
12 x 1.25 mm
54-64 N·m
(5.5-6.5 kgf·m,
40-47 lbf·ft)
Replace.

Fig. 180 Disconnecting the upper ball joint

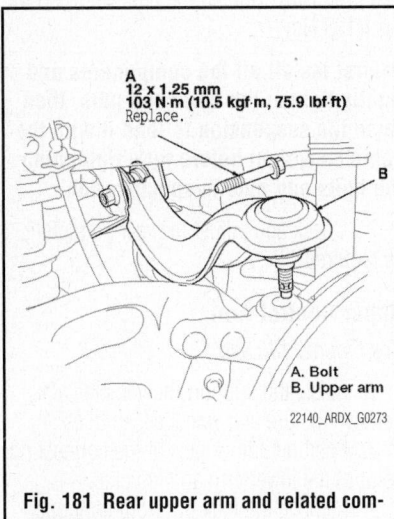

A
12 x 1.25 mm
103 N·m (10.5 kgf·m, 75.9 lbf·ft)
Replace.

B

A. Bolt
B. Upper arm

22140_ARDX_G0273

Fig. 181 Rear upper arm and related components

To install:

6. Installation is the reverse of the removal procedure.

➡ First install all components and lightly tighten the new flange bolt and the castle nut. Raise the suspension to load it with the vehicle's weight before fully tightening.

7. Tighten the castle nut to 43–47 ft. lbs. (58–64 Nm), and then tighten it only enough to install the cotter pin. Do not align the castle nut by loosening it.

8. Check and adjust the rear alignment, as required.

SHOCK ABSORBERS

REMOVAL & INSTALLATION

See Figures 182 and 183.

1. Raise and support the vehicle safely. Remove the tire and wheel assemblies.

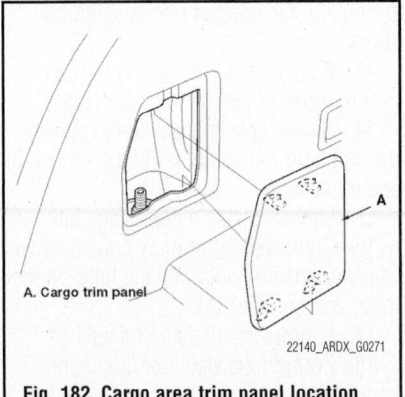

A. Cargo trim panel

22140_ARDX_G0271

Fig. 182 Cargo area trim panel location

2. Position a floor jack at the connecting point of the lower arm and the knuckle. Raise the floor jack until the suspension begins to compress.

3. Remove the flange bolt from the bottom of the strut.

4. Remove the lid from the cargo area side trim panel.

5. Remove the self locking nut, while holding the shock absorber shaft, using a hex wrench.

6. Remove the shock absorber washer

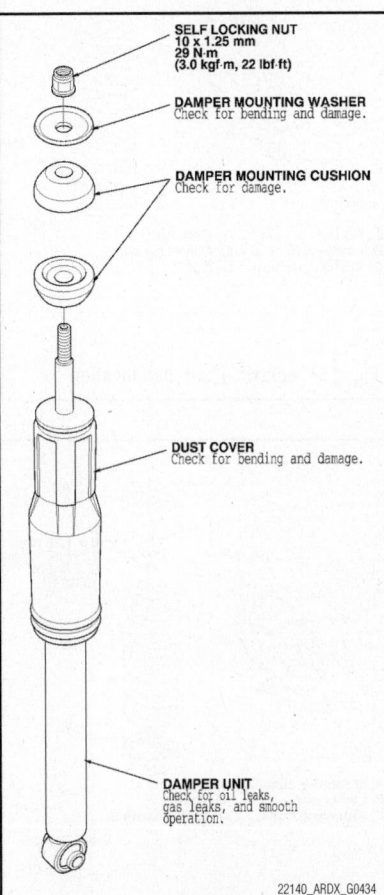

SELF LOCKING NUT
10 x 1.25 mm
29 N·m
(3.0 kgf·m, 22 lbf·ft)

DAMPER MOUNTING WASHER
Check for bending and damage.

DAMPER MOUNTING CUSHION
Check for damage.

DUST COVER
Check for bending and damage.

DAMPER UNIT
Check for oil leaks,
gas leaks, and smooth
operation.

22140_ARDX_G0434

Fig. 183 Rear shock absorber and related components

and mounting cushion from the top of the shock absorber.

7. Compress the shock absorber and remove it from the vehicle.

To install:

8. Position the floor jack under the lower arm, to support the suspension.

9. Slowly raise the floor jack until you align the bolt hole with the holes in the lower arm and the shock absorber. Loosely install the new flange bolt.

10. First install the component and lightly tighten the new flange bolt, then raise the suspension to load it with the vehicles weight before fully tightening the bolt to specification.

11. Tighten the new flange bolt to 76 ft. lbs. (103 Nm).

12. Continue the installation in the reverse order of the removal procedure.

13. Tighten the new self locking nut to 22 ft. lbs. (30 Nm).

14. Check and adjust the rear alignment, as required.

STEERING KNUCKLE

REMOVAL & INSTALLATION

See Figures 184 through 188.

1. Raise and support the vehicle safely. Remove the tire and wheel assemblies.

2. Remove the hub bearing unit.

3. Remove the parking brake cable from the trailing arm.

4. Remove the flange nuts and remove the backing plate.

➡ Hang the backing plate, using mechanics wire, to prevent damage to the parking brake cable and the backing plate.

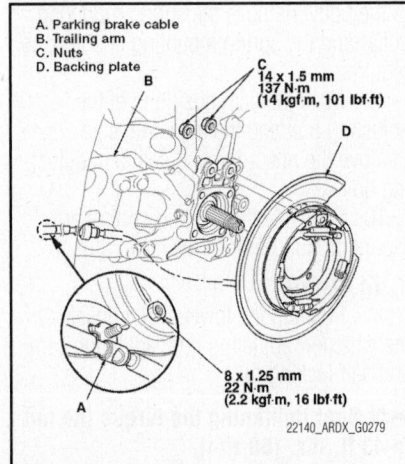

A. Parking brake cable
B. Trailing arm
C. Nuts
D. Backing plate

C
14 x 1.5 mm
137 N·m
(14 kgf·m, 101 lbf·ft)

D

8 x 1.25 mm
22 N·m
(2.2 kgf·m, 16 lbf·ft)

22140_ARDX_G0279

Fig. 184 Rear knuckle and related components

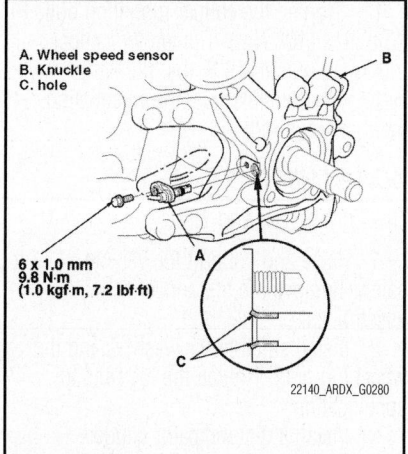

Fig. 185 Rear wheel speed sensor location

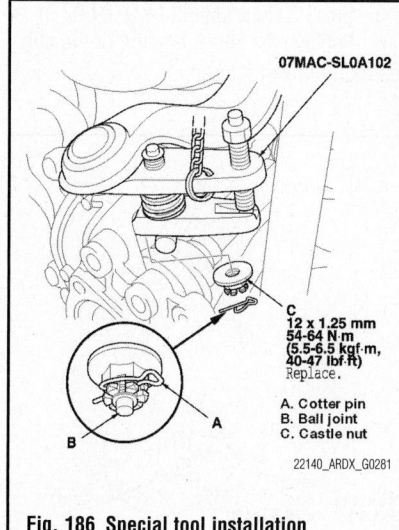

Fig. 186 Special tool installation

5. Remove the wheel speed sensor from the knuckle. Do not disconnect the sensor at the connector.

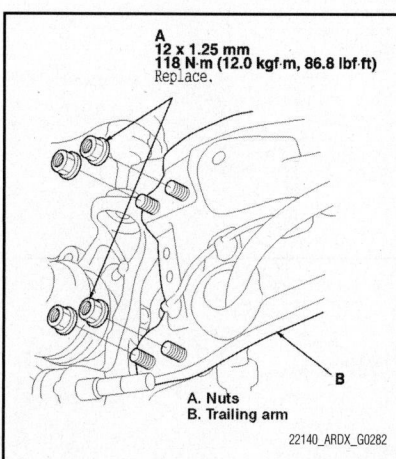

Fig. 187 Trailing arm to knuckle retaining nut locations

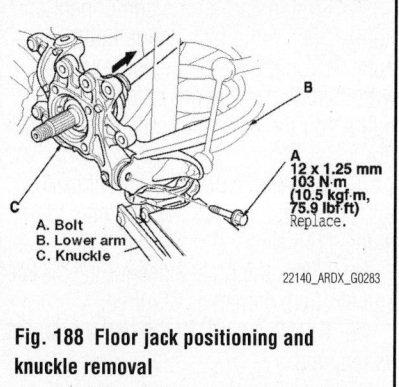

Fig. 188 Floor jack positioning and knuckle removal

6. Remove the lock pin from the upper ball joint. Loosen the castle nut.

7. Using the proper removal tool disconnect the upper arm ball joint from the knuckle.

8. Remove the lower control arm link.

9. Remove the locking nut and separate the knuckle from the trailing arm.

10. Position a floor jack under the lower arm. Remove the flange bolt. Remove the knuckle from the vehicle.

To install:

11. Installation is the reverse of the removal procedure.

➡**Before installation apply multi grease to the inside hole on the knuckle.**

12. Be sure to use new bolts and nuts, as required.

13. Tighten the new knuckle flange bolt to 76 ft. lbs. (103 Nm). Tighten the trailing arm to knuckle retaining nuts to 87 ft. lbs. (118 Nm).

➡**First install all the components and lightly tighten the bolts and nuts, then raise the suspension to load it with the vehicles weight before fully tightening the bolts and nuts to specification.**

14. Check and adjust the rear alignment, as required.

STABILIZER BAR

REMOVAL & INSTALLATION

Stabilizer Bar

See Figure 189.

1. Raise and support the vehicle safely. Remove the tire and wheel assemblies.

2. Disconnect both stabilizer links from the stabilizer bar.

3. Remove the flange bolts and the bushing holders.

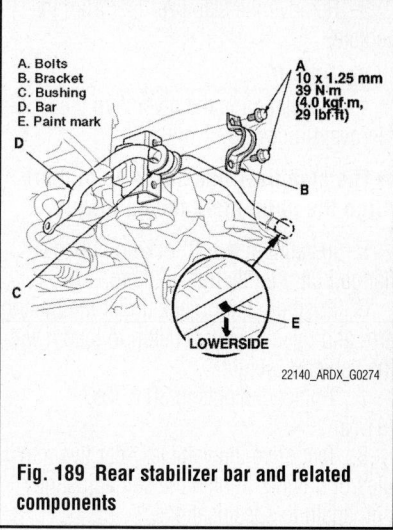

Fig. 189 Rear stabilizer bar and related components

4. Remove the stabilizer bar from its mounting.

To install:

5. Installation is the reverse of the removal procedure.

6. Tighten the stabilizer bar bushing bolts to 29 ft. lbs. (30 Nm).

➡**Note the right and left direction of the stabilizer bar. The bar has a paint mark on the center facing downward.**

7. Check and adjust the rear alignment, as required.

Stabilizer Link

See Figure 190.

1. Raise and support the vehicle safely. Remove the tire and wheel assemblies.

2. Remove the self locking nut and the flange nut while holding the respective joint pin with the hex wrench.

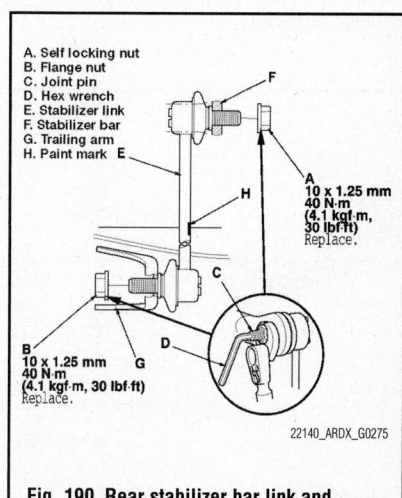

Fig. 190 Rear stabilizer bar link and related components

3. Remove the stabilizer link from the vehicle.

To install:

4. Installation is the reverse of the removal procedure.

➡**The stabilizer link has a paint mark. Align the paint mark facing inward.**

5. Install a new self-locking nut and flange bolt. Lightly tighten them.

6. Position a floor jack under the trailing arm and raise the suspension to load it with the vehicle's weight.

7. Tighten the nuts to 30 ft. lbs. (41 Nm).

8. Test drive the vehicle. After five minutes of driving, tighten the self adjusting nut, again to specification.

9. Check and adjust the rear alignment, as required.

TRAILING ARM

REMOVAL & INSTALLATION

See Figure 191.

1. Raise and support the vehicle safely. Remove the tire and wheel assemblies.

2. Remove the parking brake shoe. Remove the parking brake cable from the backing plate.

3. Remove the parking brake cable from the trailing arm.

4. Remove the brake hose mounting bracket from the trailing arm.

5. Position a floor jack at the connecting point of the lower arm and knuckle.

6. Remove the self locking nuts and flange bolts from the trailing arm.

7. Remove the trailing arm from its mounting.

To install:

8. Installation is the reverse of the removal procedure.

9. First install all the component and lightly tighten the new flange bolts and nuts, then raise the suspension to load it with the vehicles weight before fully tightening the bolts and nuts to specification.

10. Tighten the new flange bolt to 76 ft. lbs. (103 Nm). Tighten the new nuts to 87 ft. lbs. (118 Nm).

11. Check the brake hose for interference and twisting, correct as required.

12. Check and adjust the rear alignment, as required.

WHEEL BEARINGS

REMOVAL & INSTALLATION

See Figure 192.

1. Raise and support the vehicle safely. Remove the tire and wheel assemblies.

2. Remove the caliper and position it to the side. Do not allow the caliper to hang by the brake hose. Remove the two washers.

3. Remove the spindle nut.

4. Release the parking brake, if necessary. Remove the rotor.

5. Remove the hub bearing unit retaining bolts. Remove the component from the vehicle.

To install:

6. Installation is the reverse of the removal procedure.

7. Be sure to use new bolts and nuts, as required.

8. First install all the components and lightly tighten the bolts and nuts, then raise the suspension to load it with the vehicles weight before fully tightening the bolts and nuts to specification.

9. Check and adjust the rear alignment, as required.

10. Be sure to use a new spindle nut. Tighten it to 181 ft. lbs. (245 Nm).

11. Tighten the caliper mounting bolts to 80 ft. lbs. (108 Nm). Tighten the caliper mounting nuts to 16 ft. lbs. (22 Nm).

12. Check and adjust the wheel alignment, as required.

ADJUSTMENT

See Figures 193 and 194.

1. Raise and support the vehicle safely. Remove the tire and wheel assemblies.

2. Install suitable flat washers and the wheel lug nuts. Tighten the lug nuts to specification.

3. Install a dial indicator gauge. Position the dial gauge against the hub flange.

4. Measure the bearing endplay by moving the disc inward and outward.

5. Specification should be 0–0.002 in.

6. Replace the wheel bearing or the hub bearing unit, as required.

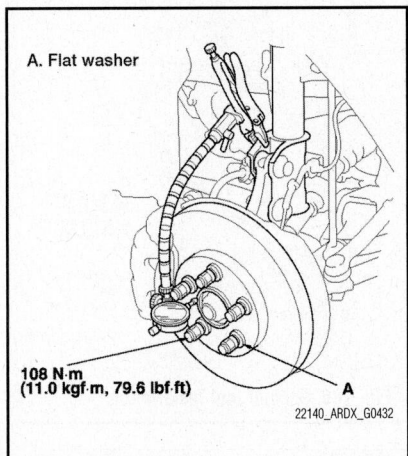

A. Flat washer

108 N·m
(11.0 kgf·m, 79.6 lbf·ft)   A

22140_ARDX_G0432

Fig. 193 Dial indicator gauge positioning—view one

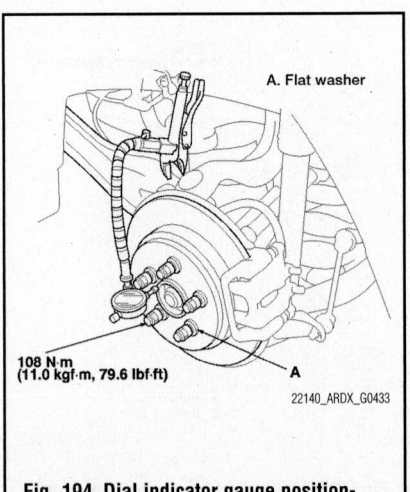

A. Flat washer

108 N·m
(11.0 kgf·m, 79.6 lbf·ft)   A

22140_ARDX_G0433

Fig. 194 Dial indicator gauge positioning—view two

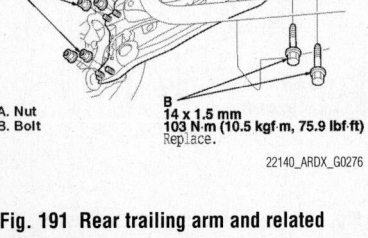

A
12 x 1.25 mm
118 N·m
(12.0 kgf·m,
86.8 lbf·ft)
Replace.

A. Nut
B. Bolt

B
14 x 1.5 mm
103 N·m (10.5 kgf·m, 75.9 lbf·ft)
Replace.

22140_ARDX_G0276

Fig. 191 Rear trailing arm and related components

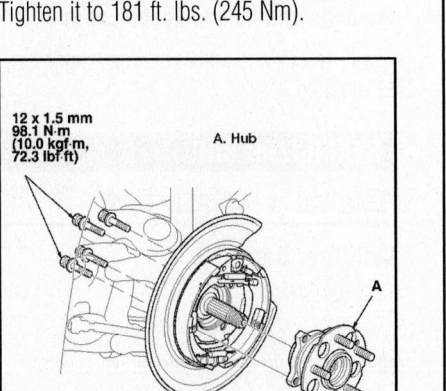

12 x 1.5 mm
98.1 N·m
(10.0 kgf·m,
72.3 lbf·ft)

A. Hub

A

22140_ARDX_G0278

Fig. 192 Rear hub and related components

ACURA

RL

BRAKES 3-3

ANTI-LOCK BRAKE SYSTEM (ABS)............................3-9
General Information....................3-9
  Precautions..........................3-9
Speed Sensors 3-10
  Removal & Installation........3-10
BLEEDING THE BRAKE SYSTEM............................3-10
Bleeding Procedure...................3-10
  Bleeding the ABS
  System 3-10
FRONT DISC BRAKES 3-10
Brake Caliper............................3-10
  Removal & Installation........3-10
Brake Pads 3-10
  Removal & Installation.........3-10
PARKING BRAKE 3-13
Parking Brake Shoes 3-13
  Removal & Installation........3-13
REAR DISC BRAKES 3-11
Brake Caliper............................3-11
  Removal & Installation.........3-11
Brake Pads 3-12
  Removal & Installation.........3-12

CHASSIS ELECTRICAL 3-14

AIR BAG (SUPPLEMENTAL RESTRAINT SYSTEM).......3-14
General Information....................3-14
  Arming the System 3-14
  Disarming the System..........3-14
  Service Precautions 3-14

DRIVE TRAIN..................3-15

CV-Joint....................................3-15
  Overhaul 3-15
Differential Carrier 3-16
  Removal & Installation.........3-16
Front Axle Shaft (Driveshaft) 3-16
  Removal & Installation.........3-16
Intermediate Shaft 3-17
  Removal & Installation.........3-17
Propeller Shaft 3-17
  Removal & Installation.........3-17

Rear Halfshafts 3-18
  Removal & Installation..........3-18

ENGINE COOLING 3-19

Engine Coolant.........................3-19
  Drain & Refill.......................3-19
Engine Fan & Shroud...............3-19
  Removal & Installation.........3-19
Radiator....................................3-20
  Removal & Installation.........3-20
Thermostat 3-20
  Removal & Installation.........3-20
Water Pump 3-21
  Removal & Installation.........3-21

ENGINE ELECTRICAL.........3-21

BATTERY SYSTEM 3-21
Battery 3-21
  Battery Disconnect
  Procedure 3-21
  Battery Reconnect
  Procedure 3-21
  Removal & Installation.........3-21
CHARGING SYSTEM 3-22
Alternator 3-22
  Removal & Installation.........3-22
IGNITION SYSTEM 3-22
Firing Order..............................3-22
Ignition Coil Pack.....................3-22
  Removal & Installation.........3-22
Ignition Timing 3-22
  Adjustment 3-22
STARTING SYSTEM 3-22
Starter 3-22
  Removal & Installation.........3-22

ENGINE MECHANICAL 3-23

Accessory Drive Belt System.....3-23
  Adjustment 3-23
  Inspection 3-23
  Removal & Installation.........3-23
Air Intake System 3-23
  Removal & Installation.........3-23
Camshaft & Bearings 3-23
  Inspection 3-23
  Removal & Installation.........3-24

Crankshaft Front Seal...............3-24
  Removal & Installation.........3-24
Crankshaft Rear Seal 3-25
  Removal & Installation.........3-25
Cylinder Head 3-25
  Removal & Installation.........3-25
Driveplate 3-26
  Removal & Installation.........3-26
Exhaust System 3-26
  Removal & Installation.........3-26
Intake Manifold 3-26
  Removal & Installation.........3-26
Oil Pan 3-27
  Removal & Installation.........3-27
Oil Pump 3-28
  Inspection 3-28
  Removal & Installation.........3-28
Pistons & Rings 3-28
  Positioning 3-28
Rocker Arm (Valve) Cover 3-28
  Removal & Installation.........3-28
Rocker Arms.............................3-29
  Removal & Installation.........3-29
Timing Belt & Sprockets 3-30
  Removal & Installation.........3-30
Valve Lash................................3-31
  Adjustment 3-31

ENGINE PERFORMANCE & EMISSION CONTROLS 3-32

Accelerator Pedal Position (APP) Sensor 3-32
  Location...............................3-32
  Removal & Installation.........3-32
Air-Fuel Ratio Sensor (A/F) 3-32
  Location...............................3-32
  Removal & Installation.........3-32
Camshaft Position (CMP) Sensor 3-32
  Location...............................3-32
  Removal & Installation.........3-32
Component Locations 3-32
Crankshaft Position (CKP) Sensor 3-32
  CKP Pattern Clear/CKP Pattern
  Learn Procedure 3-32

Location 3-32
Removal & Installation..........3-32
Engine Coolant Temperature
(ECT) Sensor 3-33
Location 3-33
Removal & Installation..........3-33
Evaporative Emission Control
System 3-34
EVAP Canister......................3-34
EVAP Canister Purge Control
Valve 3-34
Exhaust Gas Recirculation
(EGR) Valve...........................3-34
Location 3-34
Removal & Installation..........3-35
Heated Oxygen (HO2S) Sensor ..3-35
Removal & Installation..........3-35
Input Speed Sensor...................3-35
Removal & Installation..........3-35
Knock Sensor (KS)....................3-36
Location 3-36
Removal & Installation..........3-36
Manifold Absolute Pressure
(MAP) Sensor 3-36
Location 3-36
Removal & Installation..........3-36
Mass Air Flow/Intake Air
Temperature (MAF/IAT)
Sensor 3-36
Location 3-36
Removal & Installation..........3-36
Output Shaft Speed (OSS)
Sensor 3-36
Location 3-36
Removal & Installation..........3-36
Positive Crankcase Ventilation
(PCV Valve............................3-36
Location 3-36
Removal & Installation..........3-36
Powertrain Control Module
(PCM) 3-37
Location 3-37
PCM Idle Learn
Procedure 3-38
Removal & Installation..........3-37
Throttle Actuator Control
(TAC) 3-38
Location 3-38
Removal & Installation..........3-38
Throttle Position Sensor
(TPS) 3-38

Location 3-38
Removal & Installation..........3-38
Transmission Fluid Temperature
(TFT) Sensor 3-38
Location 3-38
Removal & Installation..........3-38

FUEL............................3-39

**GASOLINE FUEL INJECTION
SYSTEM........................3-39**
Fuel Filter 3-39
Removal & Installation..........3-39
Fuel Rail & Injectors 3-39
Removal & Installation..........3-39
Fuel System Service
Precautions 3-39
Fuel Tank 3-40
Draining 3-40
Removal & Installation..........3-40
Relieving Fuel System
Pressure 3-39
Throttle Body........................3-40
Removal & Installation..........3-40

**HEATING & AIR CONDITIONING
SYSTEM........................3-41**
Blower Motor 3-41
Removal & Installation..........3-41
Heater Core 3-41
Removal & Installation..........3-41

PRECAUTIONS.................3-9

**SPECIFICATIONS AND
MAINTENANCE CHARTS 3-3**
Brake Specifications...................3-7
Camshaft Specifications..............3-5
Capacities 3-4
Crankshaft and Connecting Rod
Specifications 3-5
Engine and Model Year
Identification 3-3
Engine Tune-Up
Specifications 3-3
Fluid Specifications...................3-4
General Engine Specifications.....3-3
Piston and Ring
Specifications 3-5

Scheduled Maintenance
Intervals 3-8
Tire Wheel and Ball Joint
Specifications 3-7
Torque Specifications.................3-6
Valve Specifications 3-4
Wheel Alignment......................3-6

STEERING 3-44
Power Rack & Pinion Steering
Gear 3-44
Removal & Installation..........3-44
Power Steering Pump.................3-45
Fluid Fill Procedure 3-46
Removal & Installation..........3-45

SUSPENSION..................3-46

FRONT SUSPENSION 3-46
Knuckle, Hub & Bearing............3-46
Removal & Installation..........3-46
Lower Ball Joints......................3-48
Removal & Installation..........3-48
Lower Control Arms 3-48
Removal & Installation..........3-48
Stabilizer Bar & Links 3-49
Removal & Installation..........3-49
Struts (Front Damper &
Spring) 3-50
Overhaul 3-51
Removal & Installation..........3-50
Upper Control Arms 3-51
Removal & Installation..........3-51
REAR SUSPENSION 3-52
Coil Springs..........................3-52
Removal & Installation..........3-52
Control Arms 3-53
Removal & Installation..........3-53
Knuckle/Hub Bearing Unit.........3-54
Removal & Installation..........3-54
Shock Absorbers......................3-56
Removal & Installation..........3-56
Stabilizer Bar & Links 3-57
Removal & Installation..........3-57
Wheel Bearings 3-57
Inspection 3-57

SPECIFICATIONS AND MAINTENANCE CHARTS

ENGINE AND VEHICLE IDENTIFICATION

	Engine						Model Year	
Code	Liters (cc)	Cu. In.	Cyl.	Fuel Sys.	Engine Type	Eng. Mfg.	Code ①	Year
J37A2	3.7 (3700)	226	6	SMFI	SOHC	Honda	B	2011
							C	2012

SMFI: Sequential Multi-Port Fuel Injection

SOHC: Single Overhead Camshaft

① 10th digit of the Vehicle Identification Number (VIN)

71051_ACRL_C0001

GENERAL ENGINE SPECIFICATIONS

Year	Model	Engine Displacement Liters	Engine ID	Net Horsepower @ rpm	Net Torque @ rpm (ft. lbs.)	Bore x Stroke (in.)	Compression Ratio	Oil Pressure @ rpm
2011	RL	3.7	J37A2	300@6300	271@5000	3.42X3.89	10.5:1	44@3000
2012	RL	3.7	J37A2	300@6300	271@5000	3.42X3.89	10.5:1	44@3000

71051_ACRL_C0002

GASOLINE ENGINE TUNE-UP SPECIFICATIONS

Year	Engine Displacement Liters	Spark Plug Gap (in.)	Ignition Timing (deg.) MT	Ignition Timing (deg.) AT	Fuel Pump (psi)	Idle Speed (rpm) MT	Idle Speed (rpm) AT	Valve Clearance In.	Valve Clearance Ex.
2011	3.7	0.039-0.043	NA	8-12 ①	57-64 ②	NA	630-730	0.008-0.009	0.011-0.013
2012	3.7	0.039-0.043	NA	8-12 ①	57-64 ②	NA	630-730	0.008-0.009	0.011-0.013

NOTE: The Vehicle Emission Control Information label reflects specification changes during production and must be used if they differ from this chart.

NA: Not Applicable

① Before Top Dead Center

② At idle, pressure regulator vacuum hose disconnected

71051_ACRL_C0003

CAPACITIES

Year	Model	Engine Displacement Liters	Engine ID	Engine Oil with Filter (qts.)	Transmission (qts.)		Drive Axle		Fuel Tank (gal.)	Cooling System (qts.)
					Service	Overhaul	Front (pts.)	Rear (pts.)		
2011	RL	3.7	J37A2	4.5 ①	3.3	8.2	0.9 ②	0.77	19.3	6.4 ③
2012	RL	3.7	J37A2	4.5 ①	3.3	8.2	0.9 ②	0.77	19.3	6.4 ③

NOTE: All capacities are approximate. Add fluid gradually and ensure a proper fluid level is obtained.

① Engine oil w/filter at change.
  Engine oil capacity at overhaul: 5.3 qts.

② Transfer assembly fluid capacity at change.
  Transfer assembly fluid capacity at overhaul: 0.96 pts.

③ Cooling capacity at change.
  Cooling capacity at engine overhaul: 9.1 qts.

71051_ACRL_C0004

FLUID SPECIFICATIONS

Year	Model	Engine Displ. Liters	Engine Oil	Man. Trans.	Auto. Trans.	Drive Axle		Transfer Assembly	Power Steering Fluid	Brake Master Cylinder	Cooling System
						Front	Rear				
2011	RL	3.7	5W-20 Honda	NA	Acura ATF-DW-1	NA	Acura ATF-Z1	GL4, GL5 SAE 90	Acura PS Fluid	Acura DOT 3	①
2012	RL	3.7	5W-20 Honda	NA	Acura ATF-DW-1	NA	Acura ATF-Z1	GL4, GL5 SAE 90	Acura PS Fluid	Acura DOT 3	①

DOT: Department Of Transportation

NA: Not Applicable

① Acura Long Life Antifreeze/Coolant-Type2

71051_ACRL_C0005

VALVE SPECIFICATIONS

Year	Engine Displacement Liters	Engine ID	Seat Angle (deg.)	Face Angle (deg.)	Spring Test Pressure (lbs. @ in.)	Spring Installed Height (in.)	Stem-to-Guide Clearance (in.)		Stem Diameter (in.)	
							Intake	Exhaust	Intake	Exhaust
2011	3.7	J37A2	45	45	NA	NA	0.0008-0.0017	0.0022-0.0032	0.2159-0.2163	0.2146-0.2150
2012	3.7	J37A2	45	45	NA	NA	0.0008-0.0017	0.0022-0.0032	0.2159-0.2163	0.2146-0.2150

NA: Not Available

71051_ACRL_C0006

CAMSHAFT AND BEARING SPECIFICATIONS

All measurements are given in inches.

Year	Engine Displacement Liters	Engine ID	Journal Diameter	Brg. Oil Clearance	Shaft End-play	Runout	Journal Bore	Lobe Height Intake	Lobe Height Exhaust
2011	3.7	J37A2	NA	0.0020-0.0035	0.0020-0.0080	0.0010	NA	①	②
2012	3.7	J37A2	NA	0.0020-0.0035	0.0020-0.0080	0.0010	NA	①	②

NA: Not Available

① Primary: 1.3504 in.
  Secondary: 1.4146 in.
② Primary: 1.4462 in.
  Secondary: 1.4713 in.

71051_ACRL_C0009

CRANKSHAFT AND CONNECTING ROD SPECIFICATIONS

All measurements are given in inches.

Year	Engine Displacement Liters	Engine ID	Crankshaft Main Brg. Journal Dia.	Crankshaft Main Brg. Oil Clearance	Crankshaft Shaft End-play	Crankshaft Thrust on No.	Connecting Rod Journal Diameter	Connecting Rod Oil Clearance	Connecting Rod Side Clearance
2011	3.7	J37A2	2.8337-2.8346	0.0008-0.0018	0.004-0.014	NA	2.2431-2.2441	0.0008-0.0017	0.006-0.014
2012	3.7	J37A2	2.8337-2.8346	0.0008-0.0018	0.004-0.014	NA	2.2431-2.2441	0.0008-0.0017	0.006-0.014

NA: Not available.

71051_ACRL_C0007

PISTON AND RING SPECIFICATIONS

All measurements are given in inches

Year	Engine Displacement Liters	Engine ID	Piston Clearance	Ring Gap Top Compression	Ring Gap Bottom Compression	Ring Gap Oil Control	Ring Side Clearance Top Compression	Ring Side Clearance Bottom Compression	Ring Side Clearance Oil Control
2011	3.7	J37A2	0.0002-0.0013	0.012-0.015	0.016-0.021	0.008-0.027	0.0022-0.0033	0.0012-0.0024	NA
2012	3.7	J37A2	0.0002-0.0013	0.012-0.015	0.016-0.021	0.008-0.027	0.0022-0.0033	0.0012-0.0024	NA

NA: Not Applicable

71051_ACRL_C0008

TORQUE SPECIFICATIONS
All readings in ft. lbs.

Year	Engine Displacement Liters	Engine ID	Cylinder Head Bolts	Main Bearing Bolts	Rod Bearing Bolts	Crankshaft Damper Bolts	Flywheel Bolts	Manifold Intake	Manifold Exhaust	Spark Plugs	Oil Pan Drain Plug
2011	3.7	J37A2	①	②	③	④	54	16	NA	16	29
2012	3.7	J37A2	①	②	③	④	54	16	NA	16	29

NA: Not applicable.

① Step 1: 22 ft. lbs.

Step 2: Rotate 90 degrees

Step 3: Rotate an additional 90 degrees

② Main bearing cap bolts: 54 ft. lbs.

Bearing cap side bolts: 36 ft. lbs.

③ Step 1: 14 ft. lbs.

Step 2: Rotate 90 degrees

④ Step 1: 47 ft. lbs.

Step 2: Rotate 90 degrees

71051_ACRL_C0010

WHEEL ALIGNMENT

Year	Model		Caster Range (+/-Deg.)	Caster Preferred Setting (Deg.)	Camber Range (+/-Deg.)	Camber Preferred Setting (Deg.)	Toe-in (in.)
2011	RL	F	0.30	+2.10	0.30	-0.08	0 +/- 0.08
		R	—	—	0.30	-1.15	0.08 +/- 0.08
2012	RL	F	0.30	+2.10	0.30	0.00	0 +/- 0.08
		R	—	—	0.30	0.08	0.08 +/- 0.08

71051_ACRL_C0011

TIRE, WHEEL AND BALL JOINT SPECIFICATIONS

Year	Model	OEM Tires Standard	OEM Tires Optional	Tire Pressures (psi)② Front	Tire Pressures (psi)② Rear	Wheel Size	Ball Joint Inspection	Lug Nut (ft. lbs.)
2011	RL	P245/45/R18	①	32	30	18	NS	94
2012	RL	P245/45/R18	①	32	30	18	NS	94

OEM: Original Equipment Manufacturer

PSI: Pounds Per Square Inch

NS: Not Specified by manufacturer

① Optional tire sizes available in 18 in. 19 in., and 20 in. wheel applications.

② Normal driving; for high-speed driving: 36 front; 35 rear.

71051_ACRL_C0012

BRAKE SPECIFICATIONS
All measurements in inches unless noted

Year	Model		Brake Disc Original Thickness	Brake Disc Minimum Thickness	Brake Disc Maximum Runout	Minimum Lining Thickness Front	Minimum Lining Thickness Rear	Brake Caliper Bracket Bolts (ft. lbs.)	Brake Caliper Mounting Bolts (ft. lbs.)
2011	RL	F	1.10-1.11	1.024	0.002	0.063	—	58	—
		R	0.626-0.634	0.551	0.002	—	0.063	80	17
2012	RL	F	1.10-1.11	1.024	0.002	0.063	—	58	—
		R	0.626-0.634	0.551	0.002	—	0.063	80	17

F: Front

R: Rear

71051_ACRL_C0013

SCHEDULED MAINTENANCE INTERVALS
Acura—RL

TO BE SERVICED	TYPE OF SERVICE	SERVICE INTERVALS						
		Symbol A	Symbol B	1	2	3	4	5
Air cleaner element	Replace				✓			
Brake fluid	Inspect		✓					
Brake hoses and lines	Inspect		✓					
Clutch fluid (if equipped)	Inspect		✓					
Drive belt	Inspect				✓			
Driveshaft boots	Inspect		✓					
Dust & pollen filter	Replace				✓			
Engine coolant	Inspect		✓					
Engine coolant	Replace							✓
Engine oil	Replace	✓						
Engine oil & filter	Replace		✓					
Exhaust system	Inspect		✓					
Front & rear brakes	Inspect		✓					
Fuel lines	Inspect		✓					
Parking brake adjustment	Inspect		✓					
Power steering fluid	Inspect		✓					
Spark plugs	Replace						✓	
Suspension components	Inspect		✓					
Tie-rod ends, steering gearbox and boots	Inspect		✓					
Timing belt	Replace						✓	
Tires	Rotate			✓				
Transmission fluid	Inspect		✓					
Transmission fluid	Replace					✓		
Valve clearance	Inspect						✓	
Windshield washer fluid	Inspect		✓					

NOTES:

Refer to the Owner's Manual for complete explanation of the service maintenance reminder system.

The maintenance minder is an important feature of the multi-information display. Based on engine operating conditions, the RL's onboard computer (PCM) remaining engine oil life. The system also displays the remaining engine oil life along with the code(s) for other scheduled maintenance items needing service.

If the message "SERVICE DUE NOW" does not appear more than 12 months after the display is reset, change the engine oil every year.

Independent of the maintenance messages in the multi-information display, replace the brake fluid every 3 years.

Inspect idle speed every 160,000 miles (256,000 km).

Adjust the valves during services A, B, 1, 2 or 3 only if they are noisy.

71051_ACRL_C0014

PRECAUTIONS

Before servicing any vehicle, please be sure to read all of the following precautions, which deal with personal safety, prevention of component damage, and important points to take into consideration when servicing a motor vehicle:

• Never open, service or drain the radiator or cooling system when the engine is hot; serious burns can occur from the steam and hot coolant.

• Observe all applicable safety precautions when working around fuel. Whenever servicing the fuel system, always work in a well-ventilated area. Do not allow fuel spray or vapors to come in contact with a spark, open flame, or excessive heat (a hot drop light, for example). Keep a dry chemical fire extinguisher near the work area. Always keep fuel in a container specifically designed for fuel storage; also, always properly seal fuel containers to avoid the possibility of fire or explosion. Refer to the additional fuel system precautions later in this section.

• Fuel injection systems often remain pressurized, even after the engine has been turned **OFF**. The fuel system pressure must be relieved before disconnecting any fuel lines. Failure to do so may result in fire and/or personal injury.

• Brake fluid often contains polyglycol ethers and polyglycols. Avoid contact with the eyes and wash your hands thoroughly after handling brake fluid. If you do get brake fluid in your eyes, flush your eyes with clean, running water for 15 minutes. If eye irritation persists, or if you have taken brake fluid internally, IMMEDIATELY seek medical assistance.

• The EPA warns that prolonged contact with used engine oil may cause a number of skin disorders, including cancer. You should make every effort to minimize your exposure to used engine oil. Protective gloves should be worn when changing oil. Wash your hands and any other exposed skin areas as soon as possible after exposure to used engine oil. Soap and water, or waterless hand cleaner should be used.

• All new vehicles are now equipped with an air bag system, often referred to as a Supplemental Restraint System (SRS) or Supplemental Inflatable Restraint (SIR) system. The system must be disabled before performing service on or around system components, steering column, instrument panel components, wiring and sensors. Failure to follow safety and disabling procedures could result in accidental air bag deployment, possible personal injury and unnecessary system repairs.

• Always wear safety goggles when working with, or around, the air bag system. When carrying a non-deployed air bag, be sure the bag and trim cover are pointed away from your body. When placing a non-deployed air bag on a work surface, always face the bag and trim cover upward, away from the surface. This will reduce the motion of the module if it is accidentally deployed. Refer to the additional air bag system precautions later in this section.

• Clean, high quality brake fluid from a sealed container is essential to the safe and proper operation of the brake system. You

should always buy the correct type of brake fluid for your vehicle. If the brake fluid becomes contaminated, completely flush the system with new fluid. Never reuse any brake fluid. Any brake fluid that is removed from the system should be discarded. Also, do not allow any brake fluid to come in contact with a painted surface; it will damage the paint.

• Never operate the engine without the proper amount and type of engine oil; doing so WILL result in severe engine damage.

• Timing belt maintenance is extremely important. Many models utilize an interference-type, non-freewheeling engine. If the timing belt breaks, the valves in the cylinder head may strike the pistons, causing potentially serious (also time-consuming and expensive) engine damage. Refer to the maintenance interval charts for the recommended replacement interval for the timing belt, and to the timing belt section for belt replacement and inspection.

• Disconnecting the negative battery cable on some vehicles may interfere with the functions of the on-board computer system(s) and may require the computer to undergo a relearning process once the negative battery cable is reconnected.

• When servicing drum brakes, only disassemble and assemble one side at a time, leaving the remaining side intact for reference.

• Only an MVAC-trained, EPA-certified automotive technician should service the air conditioning system or its components.

BRAKES

ANTI-LOCK BRAKE SYSTEM (ABS)

GENERAL INFORMATION

PRECAUTIONS

• Certain components within the ABS system are not intended to be serviced or repaired individually.

• Do not use rubber hoses or other parts not specifically specified for and ABS system. When using repair kits, replace all parts included in the kit. Partial or incorrect repair may lead to functional problems and require the replacement of components.

• Lubricate rubber parts with clean, fresh brake fluid to ease assembly. Do not use shop air to clean parts; damage to rubber components may result.

• Use only DOT 3 brake fluid from an unopened container.

• If any hydraulic component or line is removed or replaced, it may be necessary to bleed the entire system.

• A clean repair area is essential. Always clean the reservoir and cap thoroughly before removing the cap. The slightest amount of dirt in the fluid may plug an orifice and impair the system function. Perform repairs after components have been thoroughly cleaned; use only denatured alcohol to clean components. Do not allow ABS components to come into contact with any substance containing mineral oil; this includes used shop rags.

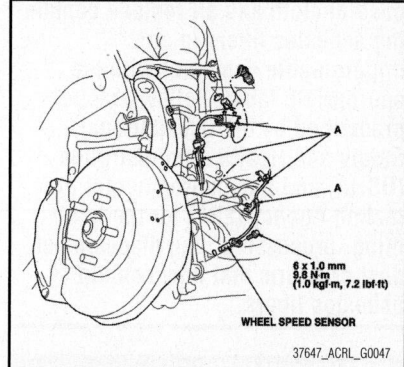

6 x 1.0 mm
9.8 N·m
(1.0 kgf·m, 7.2 lbf·ft)

WHEEL SPEED SENSOR

37647_ACRL_G0047

Fig. 1 Replace the clamps (A) when the wheel speed sensors are removed or replaced—front wheel sensors

- The Anti-Lock control unit is a microprocessor similar to other computer units in the vehicle. Ensure that the ignition switch is **OFF** before removing or installing controller harnesses. Avoid static electricity discharge at or near the controller.
- If any arc welding is to be done on the vehicle, the control unit should be unplugged before welding operations begin.

SPEED SENSORS

REMOVAL & INSTALLATION

See Figure 1.

✳✳ WARNING

Read and observe the Precautions in this section before starting work.

checked after bleeding each brake caliper. Add fluid as required.

2. Make sure the brake fluid level in the reservoir is at the MAX (upper) level line.

3. Slide a piece of clear plastic hose over the first bleed screw, and submerge the other end in a container of new brake fluid.

4. Have someone slowly pump the brake pedal several times, and then apply steady pressure.

➡ **Install the sensor carefully to avoid twisting the wires.**

1. Replace the clamps when the wheel speed sensors are removed or replaced.

2. Use the figures for reference when removing and installing the wheel speed sensors.

BLEEDING THE BRAKE SYSTEM

5. Starting at the left-front, loosen the brake bleed screw to allow air to escape from the system. Then tighten the bleed screw securely.

6. Repeat the procedure for each wheel in the sequence shown following until air bubbles no longer appear in the fluid.

7. Refill the master cylinder reservoir to the MAX (upper) level line.

BRAKES

BLEEDING PROCEDURE

BLEEDING THE ABS SYSTEM

Read and observe the Precautions in this section before starting work.

1. The reservoir on the master cylinder must be at the MAX (upper) level mark at the start of the bleeding procedure and

BRAKES

✳✳ CAUTION

Dust and dirt accumulating on brake parts during normal use may contain asbestos fibers from production or aftermarket brake linings. Breathing excessive concentrations of asbestos fibers can cause serious bodily harm. Exercise care when servicing brake parts. Do not sand or grind brake lining unless equipment used is designed to contain the dust residue. Do not clean brake parts with compressed air or by dry brushing. Cleaning should be done by dampening the brake components with a fine mist of water, then wiping the brake components clean with a dampened cloth. Dispose of cloth and all residue containing asbestos fibers in an impermeable container with the appropriate label. Follow practices prescribed by the Occupational Safety and Health Administration (OSHA) and the Environmental Protection Agency (EPA) for the handling, processing, and disposing of dust or debris that may contain asbestos fibers.

BRAKE CALIPER

REMOVAL & INSTALLATION

See Figure 2.

✳✳ WARNING

Read and observe the Precautions in this section before starting work.

1. Raise and support the vehicle.
2. Remove the wheel nuts, and the front wheel, taking care not to scratch the caliper.
3. Remove the brake hose mounting bracket.
4. Remove the brake caliper bracket mounting bolts, and remove the caliper assembly from the knuckle.

✳✳ CAUTION

To prevent damage to the caliper assembly or brake hose, use a short piece of wire to hang the caliper assembly from the undercarriage. Do not twist the brake hose with force.

5. Installation is the reverse of the removal procedure.
6. Tighten the caliper mounting bolts to 58 ft. lbs. (78 Nm).

BRAKE PADS

REMOVAL & INSTALLATION

See Figures 3 and 4.

✳✳ WARNING

Read and observe the Precautions in this section before starting work.

FRONT DISC BRAKES

➡ **Due to the high performance nature of the brake system, the rotors and pads may wear faster. Be sure to inspect the front rotor thickness anytime the front pads are replaced.**

1. Raise and support the vehicle.
2. Remove the front wheels.
3. Check the thickness of the inner brake pad and outer pad. Do not include the thickness of the brake pad backing plate.
4. Brake pad thickness should be:
   - Standard: 0.41-0.45 in. (10.5-11.5 mm)
   - Service limit: 0.06 in. (1.6 mm)
5. If any part of the brake pad thickness

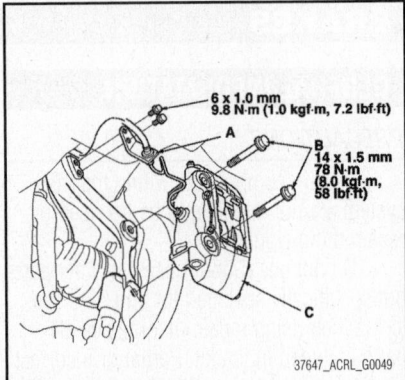

37647_ACRL_G0049

Fig. 2 Remove the brake hose mounting bracket (A). Remove the brake caliper bracket mounting bolts (B), and remove the caliper assembly (C) from the knuckle.

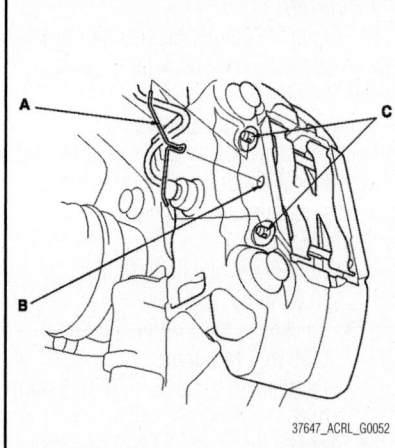

Fig. 3 Turn and twist out the clip (A) from the caliper hole (B), and pull the clip out from the pad pins (C).

is less than the service limit, replace all of the front brake pads as a set.

6. To replace the pads, first remove some brake fluid from the master cylinder.

7. Turn and twist out the clip from the caliper hole and pull the clip out from the pad pins.

8. Remove the pad pins and the pad spring.

9. Remove the pads.

To install:

10. Mount the brake caliper piston compressor tool on the caliper.

11. Press in the piston with the brake caliper piston compressor tool so that the caliper will fit over the pads. Make sure the piston boot is in position to prevent damaging it.

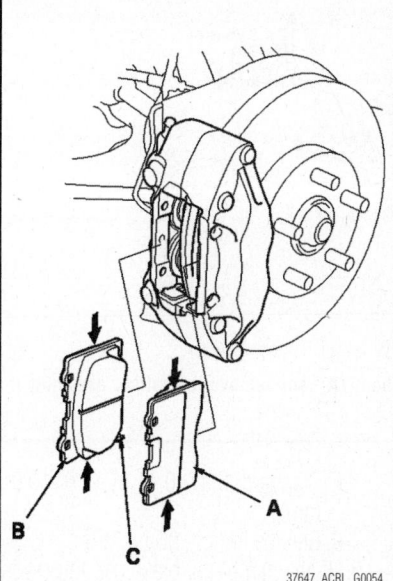

Fig. 4 Apply a thin coat of M-77, or equivalent, assembly paste to the pad sides of the pad shim (A), the back of the brake pads (B), and the other areas indicated by the arrows. Wipe excess assembly paste off the pads. Install the brake pad with the wear indicator (C) on the inside.

✳✳ CAUTION

Be careful when pressing in the piston; brake fluid might overflow from the master cylinder's reservoir. If brake fluid gets on any painted surface, wash it off immediately with water.

12. Remove the brake caliper piston compressor tool.

13. Apply a thin coat of M-77, or equivalent, assembly paste to the pad sides of the pad shim, the back of the brake pads, and the other areas indicated by the arrows. Wipe excess assembly paste off the pads.

➡**Contaminated brake discs or pads reduce stopping ability. Keep paste off the brake discs and pads.**

14. Install the brake pads correctly. Install the brake pad with the wear indicator on the inside.

15. If reusing the brake pads, always reinstall the brake pads in their original positions to prevent a temporary loss of braking efficiency.

16. Install the pad spring. Hold the pad spring and install the pad pins into the caliper from the outside to the inside of vehicle.

17. First, insert the clip ends to the pad pins, and then twist the clip into the caliper hole to stabilize.

18. Clean the mating surface between the brake disc and the inside of the wheel, then install the front wheels.

19. Press the brake pedal several times to make sure the brakes work.

➡**Engagement may require a greater pedal stroke immediately after the brake pads have been replaced as a set. Several applications of the brake pedal will restore the normal pedal stroke.**

20. Add brake fluid as needed.

21. After installation, check for leaks at the brake hose and line joints or connections, and retighten if necessary.

22. Reinstall the front wheels, then test-drive the vehicle.

23. Check for leaks.

BRAKES

✳✳ CAUTION

Dust and dirt accumulating on brake parts during normal use may contain asbestos fibers from production or aftermarket brake linings. Breathing excessive concentrations of asbestos fibers can cause serious bodily harm. Exercise care when servicing brake parts. Do not sand or grind brake lining unless equipment used is designed to contain the dust residue. Do not clean brake parts with compressed air or by dry brushing. Cleaning should be done by dampening the brake components with a fine mist of water, then wiping the brake components clean with a dampened

cloth. Dispose of cloth and all residue containing asbestos fibers in an impermeable container with the appropriate label. Follow practices prescribed by the Occupational Safety and Health Administration (OSHA) and the Environmental Protection Agency (EPA) for the handling, processing, and disposing of dust or debris that may contain asbestos fibers.

BRAKE CALIPER

REMOVAL & INSTALLATION

See Figure 5.

1. Raise and support the vehicle.

REAR DISC BRAKES

2. Remove the wheel nuts, and the rear wheel.

3. Remove the brake caliper bracket mounting bolts and remove the caliper assembly from the knuckle.

✳✳ CAUTION

To prevent damage to the caliper assembly or brake hose, use a short piece of wire to hang the caliper assembly from the undercarriage. Do not twist the brake hose excessively.

4. Installation is the reverse of the removal procedure.

5. Tighten the caliper mounting bolts to 80 ft. lbs. (108 Nm).

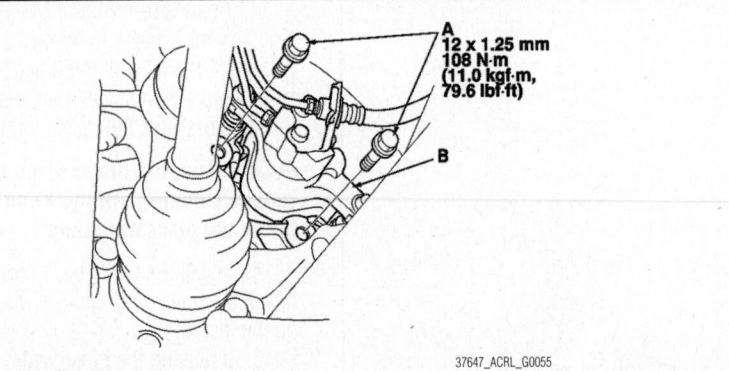

Fig. 5 Remove the brake caliper bracket mounting bolts (A), and remove the caliper assembly (B) from the knuckle.

BRAKE PADS

REMOVAL & INSTALLATION

See Figures 6 and 7.

✳✳ WARNING

Read and observe the Precautions in this section before starting work.

➡ **Due to the high performance nature of the brake system, the rotors and pads may wear faster. Be sure to inspect the front rotor thickness anytime the front pads are replaced.**

1. Raise and support the vehicle.
2. Remove the rear wheels.
3. Check the thickness of the inner brake pad and outer pad. Do not include the thickness of the brake pad backing plate.
4. Brake pad thickness should be:

- Standard: 0.37–0.41 in. (9.5–10.5 mm)
- Service limit: 0.06 in. (1.6 mm)

5. If any part of the brake pad thickness is less than the service limit, replace all of the rear brake pads as a set.
6. To replace the rear pads, first remove some brake fluid from the master cylinder.
7. Release the parking brake.
8. Remove the caliper bracket mounting bolts while holding the caliper pins with a wrench being careful not to damage the pin boot, and remove the caliper. Check the hose and pin boots for damage and deterioration. Thoroughly clean the outside of the caliper to prevent dust and dirt from entering inside. Support the caliper with a piece of wire so it does not hang from the brake hose.
9. Remove the pad shims and brake pads.
10. Remove the pad retainers.

To install:

11. Clean the caliper bracket thoroughly; remove any rust, and check for grooves and cracks.
12. Verify that the caliper pins move in and out smoothly. Clean and lube if needed.
13. Inspect the brake disc for runout, thickness, parallelism, and check for damage and cracks.

- Runout service limit: 0.0016 in. (0.04 mm)
- Thickness max. refinishing limit: 0.55 in. (14.0 mm)
- Parallelism: 0.0006 in. (0.015 mm) max.

14. Apply a thin coat of M-77 assembly paste to the retainers mating surfaces of the caliper bracket.
15. Install the pad retainers. Wipe excess assembly paste off the retainers.

✳✳ CAUTION

Contaminated brake discs and pads reduce stopping ability. Keep assembly paste off the discs and pads.

16. Install the pad retainers.
17. Apply M-77 assembly paste to the pad side of the shims and the back of the brake pads and the other areas indicated by the arrows. Wipe excess assembly paste off the pad shims and brake pads. Contaminated brake discs or pads reduce stopping ability. Keep assembly paste off the discs and pads.
18. Install the brake pads and pad shims on the caliper bracket. Install the inner brake pad with its wear indicator facing on top.
19. If reusing the brake pads, always reinstall the brake pads in their original positions to prevent a temporary loss of braking efficiency.
20. Push in the piston so the caliper will fit over the brake pads. Make sure the piston boot is in position to prevent damaging it when installing the caliper.

✳✳ CAUTION

Be careful when pushing in the caliper, brake fluid might overflow from the master cylinder's reservoir.

21. Apply a thin coat of M-77 assembly paste to the piston edges on their mating surfaces against the inner pad shim.
22. Install the brake caliper.
23. Install the caliper bolts, and torque them to 17 ft. lbs. (23 Nm) while holding the caliper pins with a wrench, being careful not to damage the pin boot.
24. Clean the mating surface between

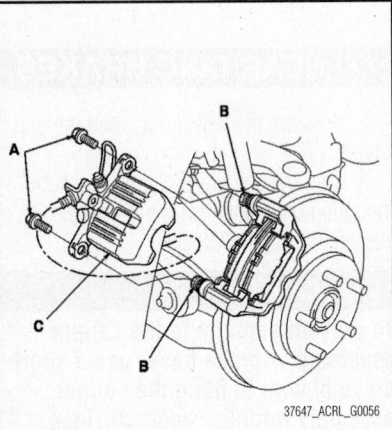

Fig. 6 Remove the caliper bracket mounting bolts (A) while holding the caliper pins (B) with a wrench being careful not to damage the pin boot, and remove the caliper (C).

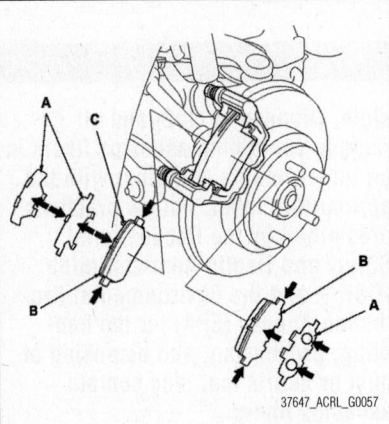

Fig. 7 Apply M-77 assembly paste to the pad side of the shims (A) and the back of the brake pads (B) and the other areas indicated by the arrows. Install the inner brake pad with its wear indicator (C) facing on top.

the brake disc/drum and the inside of the wheel, then install the rear wheels.

25. Press the brake pedal several times to make sure the brakes work, then road-test the vehicle.

➡**Engagement may require a greater pedal stroke immediately after the brake pads have been replaced as a set. Several applications of the brake pedal will restore the normal pedal stroke.**

26. Add brake fluid as needed.
27. After installation, check for leaks at the hose and line joints and connections, and retighten if necessary. Test-drive the vehicle, then recheck for leaks.

BRAKES

PARKING BRAKE SHOES

REMOVAL & INSTALLATION

See Figures 8 and 9.

✳✳ WARNING

Read and observe the Precautions in this section before starting work.

1. Raise and support the vehicle.

2. Remove the rear wheels.
3. Release the parking brake and remove the rear brake caliper and the brake disc/drum.
4. Disconnect and remove the upper brake spring.
5. Remove the adjuster assembly.
6. Disconnect and remove the lower brake spring.
7. Remove the parking brake lever

PARKING BRAKE

assembly and disconnect it from the parking brake cable end.

8. Remove the tension pins by pushing the retainer springs and turning the pins.
9. Remove the parking brake shoes.

To install:

10. Apply Molykote 44MA grease, or equivalent, to the shoe ends and connecting rod ends, sliding surfaces, and opposite edges of the parking brake shoe, as shown. Wipe off any excess. Keep grease off the brake linings.

11. Reinstall the tension pins and retainer springs.
12. Connect the parking brake cable end to the parking brake lever assembly, and install the parking brake lever assembly.
13. Reinstall the lower brake spring.
14. Clean the threaded portions of the longer clevis pin and coat the threads of the longer clevis pin with grease.
15. Clean the sliding surface of the shorter clevis pin, and coat the sliding surface of the shorter clevis pin with grease.
16. Install the longer clevis pin and pin on the adjuster and shorten the longer clevis pin by turning the adjuster.
17. Position the brake shoe adjuster assembly on the parking brake shoes.
18. Reinstall the upper brake spring.
19. Install the rear brake disc/drum and the rear brake caliper.

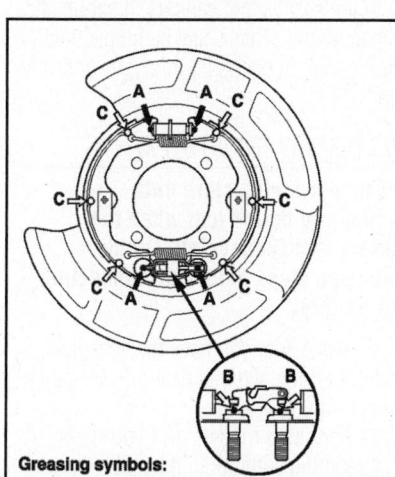

Greasing symbols:
➡● **Brake shoe ends and connecting rod ends**
⇨● **Sliding surface**
⟿○ **Opposite edge of the shoe**

37647_ACRL_G0063

Fig. 8 Apply Molykote 44MA grease, or equivalent, to the shoe ends and connecting rod ends (A), sliding surfaces (B), and opposite edges of the parking brake shoe (C) as shown. Wipe off any excess. Keep grease off the brake linings.

A. Longer clevis pin   D. Adjuster assembly
B. Shorter clevis pin   E. Upper brake spring
C. Adjuster

37647_ACRL_G0064

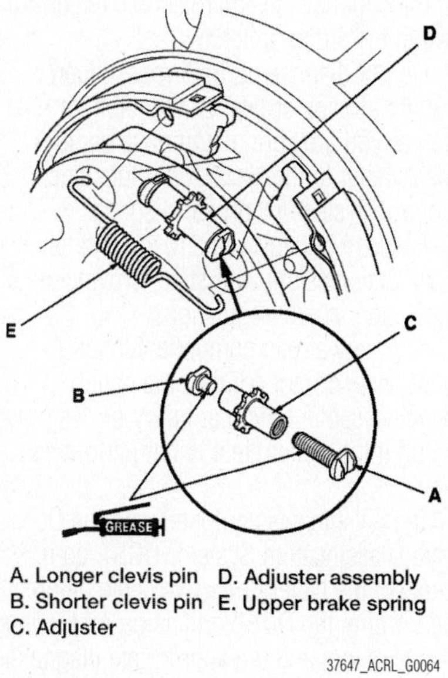

Fig. 9 Reassembling the parking brake shoe adjuster

CHASSIS ELECTRICAL AIR BAG (SUPPLEMENTAL RESTRAINT SYSTEM)

GENERAL INFORMATION

✳✳ CAUTION

These vehicles are equipped with an air bag system. The system must be disarmed before performing service on, or around, system components, the steering column, instrument panel components, wiring and sensors. Failure to follow the safety precautions and the disarming procedure could result in accidental air bag deployment, possible injury and unnecessary system repairs.

SERVICE PRECAUTIONS

Disconnect and isolate the battery negative cable before beginning any airbag system component diagnosis, testing, removal, or installation procedures. Allow system capacitor to discharge for two minutes before beginning any component service. This will disable the airbag system. Failure to disable the airbag system may result in accidental airbag deployment, personal injury, or death.

Do not place an intact undeployed airbag face down on a solid surface. The airbag will propel into the air if accidentally deployed and may result in personal injury or death.

When carrying or handling an undeployed airbag, the trim side (face) of the airbag should be pointing towards the body to minimize possibility of injury if accidental deployment occurs. Failure to do this may result in personal injury or death.

Replace airbag system components with OEM replacement parts. Substitute parts may appear interchangeable, but internal differences may result in inferior occupant protection. Failure to do so may result in occupant personal injury or death.

Wear safety glasses, rubber gloves, and long sleeved clothing when cleaning powder residue from vehicle after an airbag deployment. Powder residue emitted from a deployed airbag can cause skin irritation. Flush affected area with cool water if irritation is experienced. If nasal or throat irritation is experienced, exit the vehicle for fresh air until the irritation ceases. If irritation continues, see a physician.

Do not use a replacement airbag that is not in the original packaging. This may result in improper deployment, personal injury, or death.

The factory installed fasteners, screws and bolts used to fasten airbag components have a special coating and are specifically designed for the airbag system. Do not use substitute fasteners. Use only original equipment fasteners listed in the parts catalog when fastener replacement is required.

During, and following, any child restraint anchor service, due to impact event or vehicle repair, carefully inspect all mounting hardware, tether straps, and anchors for proper installation, operation, or damage. If a child restraint anchor is found damaged in any way, the anchor must be replaced. Failure to do this may result in personal injury or death.

Deployed and non-deployed airbags may or may not have live pyrotechnic material within the airbag inflator.

Do not dispose of driver/passenger/curtain airbags or seat belt tensioners unless you are sure of complete deployment. Refer to the Hazardous Substance Control System for proper disposal.

Dispose of deployed airbags and tensioners consistent with state, provincial, local, and federal regulations.

After any airbag component testing or service, do not connect the battery negative cable. Personal injury or death may result if the system test is not performed first.

If the vehicle is equipped with the Occupant Classification System (OCS), do not connect the battery negative cable before performing the OCS Verification Test using the scan tool and the appropriate diagnostic information. Personal injury or death may result if the system test is not performed properly.

Never replace both the Occupant Restraint Controller (ORC) and the Occupant Classification Module (OCM) at the same time. If both require replacement, replace one, then perform the Airbag System test before replacing the other.

Both the ORC and the OCM store Occupant Classification System (OCS) calibration data, which they transfer to one another when one of them is replaced. If both are replaced at the same time, an irreversible fault will be set in both modules and the OCS may malfunction and cause personal injury or death.

If equipped with OCS, the Seat Weight Sensor is a sensitive, calibrated unit and must be handled carefully. Do not drop or handle roughly. If dropped or damaged, replace with another sensor. Failure to do so may result in occupant injury or death.

If equipped with OCS, the front passenger seat must be handled carefully as well. When removing the seat, be careful when setting on floor not to drop. If dropped, the sensor may be inoperative, could result in occupant injury, or possibly death.

If equipped with OCS, when the passenger front seat is on the floor, no one should sit in the front passenger seat. This uneven force may damage the sensing ability of the seat weight sensors. If sat on and damaged, the sensor may be inoperative, could result in occupant injury, or possibly death.

DISARMING THE SYSTEM

➡**Some systems store data in memory that is lost when the battery is disconnected. Do the following steps before disconnecting the battery.**

1. Make sure you have the anti-theft code(s) for the audio and/or the navigation system (if equipped).
2. For some models or if you're replacing the audio unit, it may be necessary to write down the audio presets (AM and FM), and the XM radio presets (if equipped), because the audio unit does not retain the presets after the battery is disconnected.
3. Make sure the ignition switch is in LOCK (0).
4. Disconnect and isolate the negative cable from the battery.

✳✳ CAUTION

Always disconnect the negative cable from the battery first.

5. Disconnect the positive cable from the battery.
6. Wait at least 3 minutes for the battery and SRS system to discharge before performing any repairs.

ARMING THE SYSTEM

➡**Some systems store data in memory that is lost when the battery is disconnected. Do the following steps to restore the**

systems back to normal operation.

1. Clean the battery terminals.
2. Reconnect the positive cable to the battery first, then reconnect the negative cable to the battery.

✳✳ CAUTION

Always connect the positive cable to the battery first.

3. Apply multipurpose grease to the terminals to prevent corrosion.
4. Enter the anti-theft code(s) for the audio system and/or the navigation system (if equipped).
5. Enter the audio presets (if applicable), and enter the XM radio presets (if equipped).
6. Set the clock (for vehicles without navigation).

DRIVE TRAIN

CV-JOINT

OVERHAUL

Inboard Joint Side

See Figure 10.

1. Remove the boot bands. Be careful not to damage the boot.
  a. If the boot band is a welded type, cut the boot band.
  b. If the boot band is a double loop type, lift up the band end, then push it into the clip.
  c. If the boot band is a low profile type, pinch the boot band, using commercially available boot band pliers.
2. Make marks on each roller and the inboard joint to identify the locations of the rollers to the grooves on the inboard joint.

➡**Do not engrave or scribe any marks on the rolling surface.**

3. Remove the inboard joint onto a clean shop towel. Be careful not to drop the rollers when separating them from the inboard joint.
4. Make marks on the spider that match the marks on the rollers, then remove the rollers.

➡**Do not engrave or scribe any marks on the rolling surface.**

5. Remove the circlip.
6. Make marks on the spider and the driveshaft to identify the position of the spider on the shaft.
7. Remove the spider, using a puller, if necessary.
8. Wrap the splines on the driveshaft with vinyl tape to prevent damaging the boot.
9. Remove the inboard boot. Be careful not to damage the boot.
10. Remove the vinyl tape.

11. Reassemble in reverse order, making note to align all reference marks.

Outboard Joint Side

See Figure 11.

1. Remove the boot bands. Lift up the three tabs, using a screwdriver, then release the band. Be careful not to damage the boot.
2. Slide the outboard boot partially toward the inboard joint side. Be careful not to damage the boot.
3. Wipe off the grease to expose the driveshaft and the outboard joint inner race.
4. Make a mark on the driveshaft at the same level as the outboard joint end.
5. Securely clamp the driveshaft in a bench vise with a shop towel.
6. Remove the outboard joint, using the 26 x 1.5 mm threaded adapter and a commercially available 5/8" x 18 UNF slide hammer.
7. Remove the driveshaft from the bench vise.
8. Remove the stop ring from the driveshaft.
9. Wrap the splines on the driveshaft with vinyl tape to prevent damaging the boot.

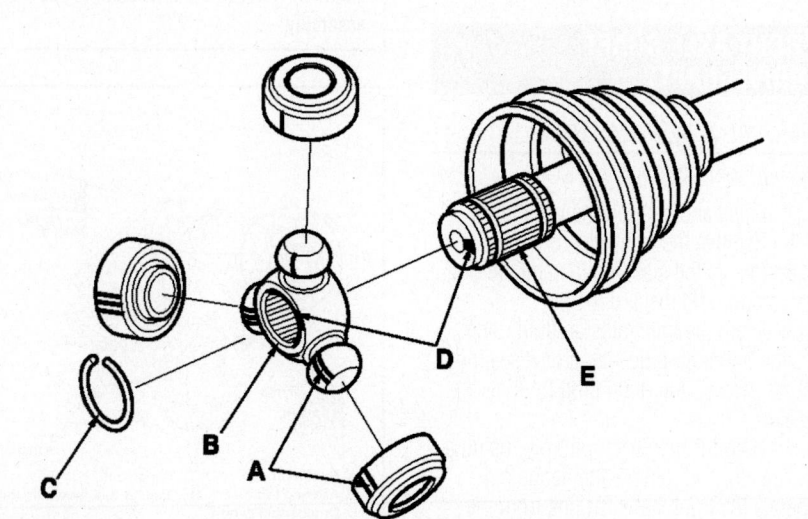

A. Marks on spider to roller
B. Spider
C. Circlip
D. Marks on spider to driveshaft
E. Driveshaft

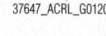
37647_ACRL_G0120

Fig. 10 Marking spider during CV joint disassembly

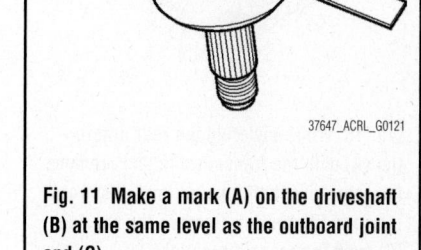
37647_ACRL_G0121

Fig. 11 Make a mark (A) on the driveshaft (B) at the same level as the outboard joint end (C).

10. Remove the outboard boot. Be careful not to damage the boot.

11. Remove the vinyl tape.

12. Reassemble in reverse order, making note to align all reference marks.

DIFFERENTIAL CARRIER

REMOVAL & INSTALLATION

See Figures 12 and 13.

1. Raise the vehicle on a lift.

2. Drain the rear differential fluid.

3. Remove the propeller shaft.

4. Disconnect the rear differential harness connectors.

5. Disconnect the breather hose.

6. Disconnect the inboard driveshaft joints from the rear differential using a pair of screwdrivers.

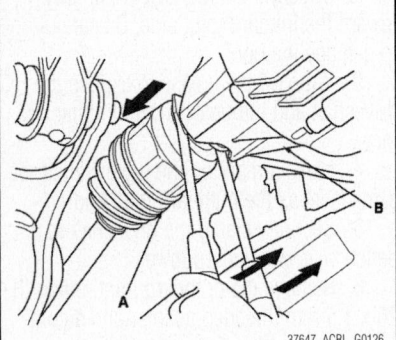

Fig. 12 Disconnect the inboard driveshaft joints (A) from the rear differential (B) using a pair of screwdrivers.

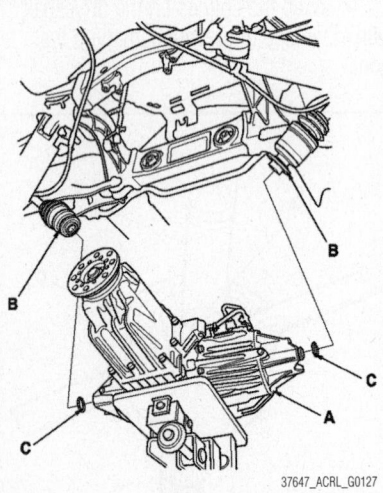

Fig. 13 While lowering the rear differential (A) with the transmission jack, remove the rear driveshaft inboard joints (B) from the rear differential. Remove the set rings (C) from the rear differential.

7. Place the transmission jack under the rear differential.

8. Remove the two rear differential rear mounting bolts (horizontal through the crossmember).

9. Remove the four rear differential front mounting bolts (vertical at the front housing).

10. While lowering the rear differential with the transmission jack, remove the rear driveshaft inboard joints from the rear differential. Remove the set rings from the rear differential.

To install:

11. Place the rear differential on a transmission jack.

12. Install new set rings into the side case grooves of the rear differential (where the driveshafts will join).

13. Apply appropriate grease to the splines of the rear driveshaft inboard joints.

14. Raise the rear differential to the mounting level, then install the rear driveshaft inboard joints.

15. Install four new rear differential front mounting bolts. Torque to 22 ft. lbs. (30 Nm).

16. Install two new rear differential rear mounting bolts (through the crossmember). Torque to 54 ft. lbs. (74 Nm).

17. Connect the breather hose.

18. Connect the rear differential harness connectors and.

19. Install the propeller shaft.

20. Refill the rear differential fluid.

21. Test-drive the vehicle.

FRONT AXLE SHAFT (DRIVESHAFT)

REMOVAL & INSTALLATION

See Figures 14 through 16.

1. Raise and support the vehicle.

2. Remove the front wheels.

3. Pry up the stake on the spindle nut, then remove and discard the nut.

4. Drain the transmission fluid, then reinstall the drain plug with a new sealing washer. Tighten the drain plug to 36 ft. lbs. (49 Nm).

5. Hold the stabilizer joint pin, using a hex wrench, and remove the flange nut. Separate the front stabilizer link from the lower arm.

6. Remove the damper fork mounting nut, the flange bolt, and the damper pinch bolt, then remove the damper fork.

7. Remove the steering knuckle holder bolt and nut.

8. Pull the knuckle outward, and separate the outboard joint from the front hub using a plastic hammer.

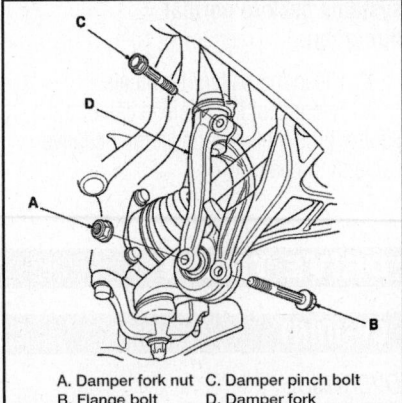

A. Damper fork nut　C. Damper pinch bolt
B. Flange bolt　　　D. Damper fork

37647_ACRL_G0115

Fig. 14 Remove the damper fork mounting nut, the flange bolt, and the damper pinch bolt, then remove the damper fork.

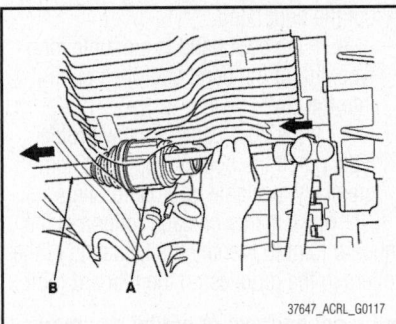

37647_ACRL_G0117

Fig. 15 Remove the right driveshaft by driving the inboard joint (A) off of the intermediate shaft, using a drift and a hammer. Remove the driveshaft (B) as an assembly.

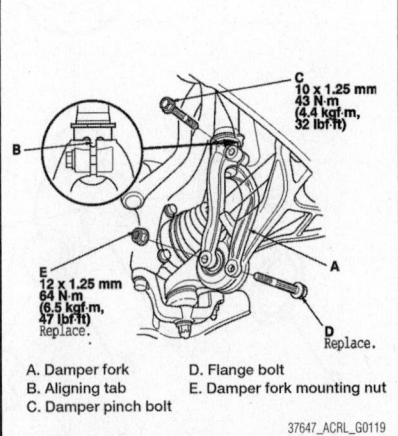

A. Damper fork　　　　D. Flange bolt
B. Aligning tab　　　　E. Damper fork mounting nut
C. Damper pinch bolt

37647_ACRL_G0119

Fig. 16 Install the damper fork over the driveshaft and onto the lower arm. Install the damper in the damper fork so the aligning tab is aligned with the slot in the damper fork. Loosely install the damper pinch bolt. Loosely install a new flange bolt and a new damper fork mounting nut.

9. Remove exhaust pipe A.

10. Remove the left driveshaft by prying the inboard joint from the differential using a prybar. Remove the driveshaft as an assembly.

➡**Do not pull on the driveshaft or the inboard joint may come apart. Pull the inboard joint straight out to avoid damaging the oil seal. Be careful not to damage the oil seal or the end of the inboard joint with the prybar.**

11. Remove the right driveshaft by driving the inboard joint off of the intermediate shaft, using a drift and a hammer. Remove the driveshaft as an assembly.

➡**Do not pull on the driveshaft, or the inboard joint may come apart.**

12. If necessary, remove the set ring from the left driveshaft inboard joint and the set ring from the intermediate shaft.

To install:

➡**Before starting installation, make sure the mating surfaces of the joint and the splined section are clean.**

13. Apply about 0.18 oz. of moly 60 paste (P/N 08734-0001) to the contact area of the outboard joint and front wheel bearing.

➡**The paste helps prevent noise and vibration.**

14. Install a new set ring into the set ring groove of the left driveshaft inboard joint.

15. Install a new set ring into the set ring groove of the intermediate shaft.

16. Apply 0.02-0.04 oz. of super high temp urea grease (P/N 08798-9002) to the whole splined surface of the right driveshaft.

17. After applying grease, remove the grease from the splined grooves at intervals of 2-3 splines and from the set ring groove so that air can bleed from the intermediate shaft.

18. Clean the areas where the driveshaft contacts the differential thoroughly with solvent and dry them with compressed air.

➡**Do not wash the rubber parts with solvent.**

19. Insert the inboard end of the driveshaft into the differential or intermediate shaft until the set ring locks in the groove.

➡**Insert the driveshaft horizontally to prevent damaging the oil seal.**

20. Install the outboard joint into the front hub on the knuckle.

21. Install exhaust pipe A.

22. Install the knuckle holder to the knuckle, and then tighten a new knuckle holder bolt and a new nut. Tighten the bolt and nut to 54 ft. lbs. (74 Nm).

23. Install the damper fork over the driveshaft and onto the lower arm. Install the damper in the damper fork so the aligning tab is aligned with the slot in the damper fork. Loosely install the damper pinch bolt.

24. Loosely install a new flange bolt and a new damper fork mounting nut.

25. Tighten the pinch bolt to 32 ft. lbs. (43 Nm). Tighten the flange bolt and nut to 47 ft. lbs. (64 Nm).

26. Connect the front stabilizer link to the lower arm and loosely install a new flange nut. Hold the stabilizer link ball joint pin with a hex wrench, and tighten the new flange nut to 40 ft. lbs. (54 Nm).

27. Place a floor jack under the lower arm, and raise the suspension to load it with the vehicle's weight.

➡**Do not put the floor jack under the ball joint.**

28. Tighten the damper pinch bolt and the damper fork mounting nut while holding the flange bolt to the specified torque values, then remove the floor jack.

29. Apply a small amount of engine oil to the seating surface of a new spindle nut.

30. Install the spindle nut, then tighten it to 242 ft. lbs. (329 Nm). After tightening, use a drift to stake the spindle nut shoulder against the driveshaft.

31. Clean the mating surfaces of the brake discs and the wheels, then install the front wheels.

32. Turn the front wheel by hand, and make sure there is no interference between the driveshaft and the surrounding parts.

33. Lower the vehicle.

34. Refill the transmission with recommended transmission fluid.

35. Check the wheel alignment, and adjust it if necessary.

36. Test-drive the vehicle.

INTERMEDIATE SHAFT

REMOVAL & INSTALLATION
See Figure 17.

1. Drain the transmission fluid, then reinstall the drain plug with a new sealing washer.

2. Remove exhaust pipe A.

3. Remove the right front driveshaft.

4. Remove the rear warm up three way catalytic converter (WU-TWC) bracket.

5. Remove the flange bolt and the two dowel bolts.

6. Remove the intermediate shaft from

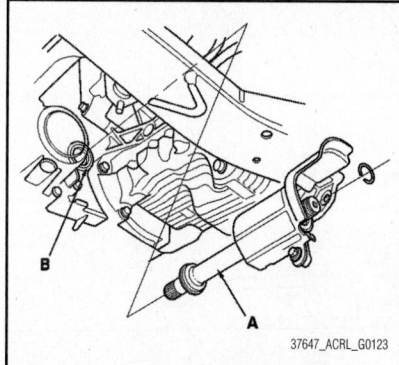

Fig. 17 Remove the intermediate shaft (A) from the front differential. Hold the intermediate shaft horizontal until it is clear of the differential to prevent damaging the oil seal (B).

the front differential. Hold the intermediate shaft horizontal until it is clear of the differential to prevent damaging the oil seal.

To install:

7. Clean the areas where the driveshaft contacts the differential thoroughly with solvent, and dry them with compressed air.

➡**Do not wash the rubber parts with solvent.**

8. Install a new set ring into the set ring groove of the intermediate shaft.

9. Insert the intermediate shaft into the differential correctly.

➡**Insert the intermediate shaft carefully to prevent damaging the oil seal.**

10. Install the flange bolt and the two dowel bolts. Torque to 29 ft. lbs. (39 Nm).

11. Install the rear warm up three way catalytic converter (WU-TWC) bracket. Torque the bolts to 16 ft. lbs. (22 Nm).

12. Install the right front driveshaft.

13. Install exhaust pipe A.

14. Refill the transmission with the recommended transmission fluid.

15. Check the wheel alignment, and adjust it if necessary.

16. Test-drive the vehicle.

PROPELLER SHAFT

REMOVAL & INSTALLATION
See Figures 18 and 19.

1. Raise and support the vehicle.

2. Remove the muffler with the exhaust pipe B.

3. Remove the No. 1 propeller shaft protector.

4. Make reference marks across the No. 1 propeller shaft and the transfer companion flange.

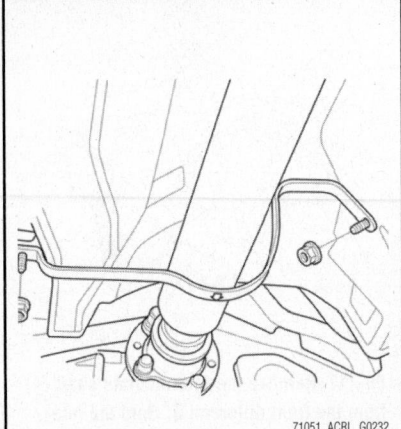

71051_ACRL_G0232

Fig. 18 Make reference marks across the No. 1 propeller shaft and the transfer companion flange. Wrap the length of each propeller shaft with cardboard, at least 5 mm thick, before removing the propeller shaft. Remove the mounting nuts.

5. Wrap the length of each propeller shaft with cardboard, at least 5 mm thick, before removing the propeller shaft. Remove the mounting nuts.

6. Make reference marks across the No. 2 propeller shaft and the rear differential companion flange. Remove the flange bolts.

7. Remove the center support bearing mounting bolts.

✳✳ CAUTION

Do not exceed the angle of center support bearing joint.

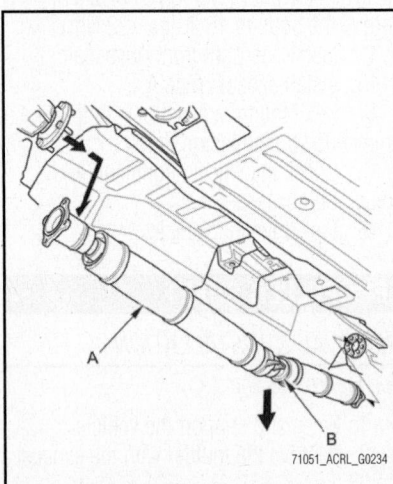

71051_ACRL_G0234

Fig. 19 Slide the propeller shaft (A) toward the rear differential, then flex the No. 2 joint (B), and remove the No. 1 propeller shaft from the transfer.

8. Slide the propeller shaft toward the rear differential, then flex the No. 2 joint, and remove the No. 1 propeller shaft from the transfer.

➡**Keep the protector on the propeller shaft after removing it from the vehicle. Be careful not to damage the propeller shaft.**

➡**Wrap the length of each propeller shaft with cardboard, at least 5 mm thick to protect the carbon fiber shaft during removal and installation, before removing the propeller shaft.**

9. Remove the propeller shaft from the vehicle.

To install:

10. Set the No. 1 propeller shaft to the transfer companion flange by flexing the No. 2 joint, then install the propeller shaft to the vehicle.

11. Align the reference marks made during the removal procedure. Then install the new mounting nuts finger tight.

12. Install the center support bearing mounting bolts. Make sure to use new bolts. Tighten to 29 ft. lbs. (39 Nm).

13. Set the No. 2 propeller shaft to the rear differential companion flange by aligning the reference marks made during the removal procedure. Then install the new flange bolts to 54 ft. lbs. (74 Nm).

➡**When replacing the propeller shaft or the rear differential, align the factory reference marks (E).**

14. Tighten the No. 1 propeller shaft mounting nuts to 54 ft. lbs. (74 Nm).

15. Remove the cardboard wrapping from the propeller shaft.

16. Turn the shaft protector hanger arrow toward the front of the vehicle and install the No. 1 propeller shaft protector. Tighten protector bolts to 16 ft. lbs. (22 Nm).

17. Install a new gasket, then install the muffler with exhaust pipe B. Tighten nuts to 40 ft. lbs. (54 Nm).

18. If you installed a new propeller shaft, test-drive the vehicle at 55 mph (88 km/h), and check for noise or vibration. If there is a noise or vibration, rotate the propeller shaft 180 degrees from its current alignment with the rear differential companion flange, then recheck.

REAR HALFSHAFTS

REMOVAL & INSTALLATION

See Figures 20 and 21.

1. Raise and support the vehicle.
2. Remove the rear wheels.
3. Pry up the stake on the spindle nut, then remove the nut.
4. Remove the rear differential, and disconnect the inboard joint from the rear differential.
5. Tapping on the outer end of the driveshaft, separate the rear driveshaft outboard joint from the rear hub, using a plastic hammer.
6. Remove the rear driveshaft. Be careful not to damage the wheel speed sensor.
7. Pull on the outer joint. Do not pull on the driveshaft because the joint may come apart.
8. Remove the set ring from the rear differential.

To install:

9. Before starting installation, make sure the mating surfaces of the joint and the splined section are clean.

10. Apply super-high temp urea grease to the whole splined surface end of the driveshaft.

11. After applying grease, remove the grease from the splined grooves at intervals of 2-3 splines and from the set ring groove so that air can bleed from the differential.

12. Install the outboard joint of the driveshaft into the rear hub. Be careful not to damage the wheel speed sensor.

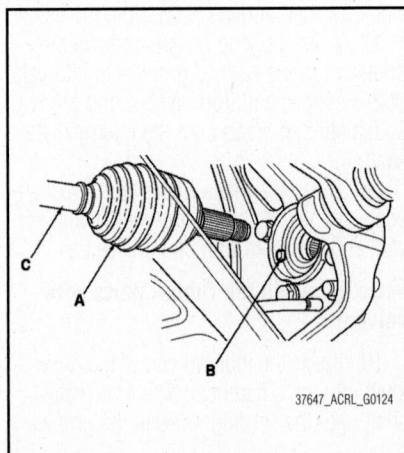

37647_ACRL_G0124

Fig. 20 Remove the rear driveshaft (A). Be careful not to damage the wheel speed sensor (B). Pull on the outer joint. Do not pull on the driveshaft (C) because the joint may come apart.

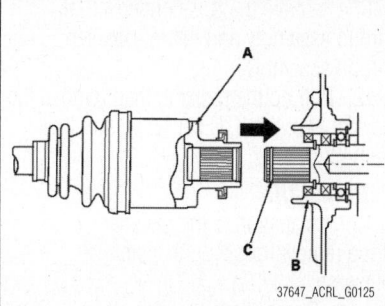

Fig. 21 Make sure the inboard joint (A) is installed all the way into the rear differential (B) and to ensure the set ring (C) is properly seated.

13. Install new set rings in the set ring groove of the rear differential.

14. Clean the areas where the driveshaft contacts the rear differential thoroughly with solvent, and dry with compressed air.

➡**Do not wash the rubber parts with solvent.**

15. Make sure the inboard joint is installed all the way into the rear differential and to ensure the set ring is properly seated.

16. Install the rear differential.

17. Apply a small amount of engine oil to the seating surface of a new spindle nut.

18. Install the spindle nut, then tighten it to 181 ft. lbs. (245 Nm). After tightening, use a drift to stake the spindle nut shoulder against the driveshaft.

19. Clean the mating surfaces of the brake discs and the wheels, then install the rear wheels.

20. Turn the rear wheel by hand, and make sure there is no interference between the driveshaft and the surrounding parts.

21. Lower the vehicle.

22. Check the wheel alignment, and adjust it if necessary.

23. Test-drive the vehicle.

ENGINE COOLING

ENGINE COOLANT

DRAIN & REFILL

1. Start the engine. Set the heater temperature control dial to maximum heat, then turn the ignition switch to LOCK (0). Make sure the engine and the radiator are cool to the touch.

2. Remove the radiator cap.

3. Loosen the drain plug and drain the coolant.

4. Install a rubber hose on the drain bolt located at the rear of the cylinder block, then loosen the drain bolt.

5. When the coolant stops draining, tighten the drain bolt. Remove the rubber hose.

6. Tighten the radiator drain plug securely.

7. Remove the coolant reservoir and drain the coolant, then reinstall the coolant reservoir.

8. Fill the coolant reservoir to the MAX mark with Acura Long Life Antifreeze/Coolant Type 2, or equivalent.

9. Pour the antifreeze/coolant into the radiator up to the base of the filler neck.

➡**Always use Acura Long Life Antifreeze/Coolant Type 2 (P/N OL999-9011A). Using a non-Acura coolant can result in corrosion, causing the cooling system to malfunction or fail.**

➡**Acura Long Life Antifreeze/Coolant Type 2 is a mixture of 50 % antifreeze and 50 % water. Do not add water.**

10. Note these engine coolant capacities (including the reserve tank capacity of 0.153 gal. (0.58 L):
- After Coolant Change: 1.59 gal. (6.0 L)
- After Engine Overhaul: 2.27 gal. (8.6 L)

11. Install the radiator cap loosely.

12. Start the engine and let it run until it warms up (the radiator fan comes on at least twice).

13. Turn off the engine. Check the level in the radiator and add antifreeze/coolant if needed.

✳✳ CAUTION

Removing the radiator cap while the engine is hot can cause the coolant to spray out. Always let the engine and the radiator cool before removing the cap.

14. Put the radiator cap on tightly, then run the engine again and check for leaks.

15. Clean up any spilled engine coolant.

16. Connect the Honda Diagnostic System (HDS), or equivalent scan tool, to the data link connector (DLC):

a. Make sure the HDS communicates with the vehicle and the powertrain control module (PCM). If it doesn't communicate, troubleshoot the DLC circuit.

b. Turn the ignition switch to ON (II).

c. Select BODY ELECTRICAL with the scan tool.

d. Select ADJUSTMENT in the GAUGE MENU with the scan tool.

e. Select RESET in the MAINTENANCE MINDER with the scan tool.

f. Select MAINTENANCE SUB ITEM 5 RESET with the scan tool.

ENGINE FAN & SHROUD

REMOVAL & INSTALLATION
See Figures 22 and 23.

1. Remove the right side fender trim.

2. Raise the vehicle on the lift to full height.

3. Remove the splash shield.

4. Remove the harness connectors, the ground cable, and the harness clamps. Loosen the A/C condenser fan shroud mounting bolts.

5. Lower the vehicle on the lift.

6. Remove the air intake duct cover and the coolant reservoir.

7. Disconnect the radiator fan control (RFC) connectors and the A/C condenser

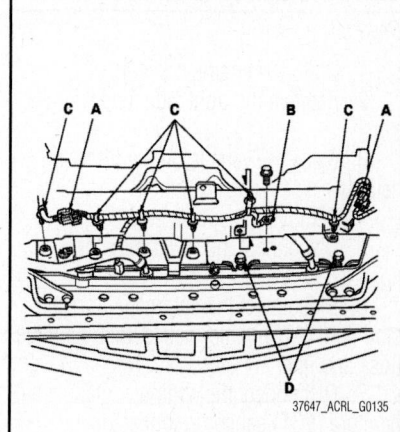

Fig. 22 Remove the harness connectors (A), the ground cable (B), and the harness clamps (C). Loosen the A/C condenser fan shroud mounting bolts (D).

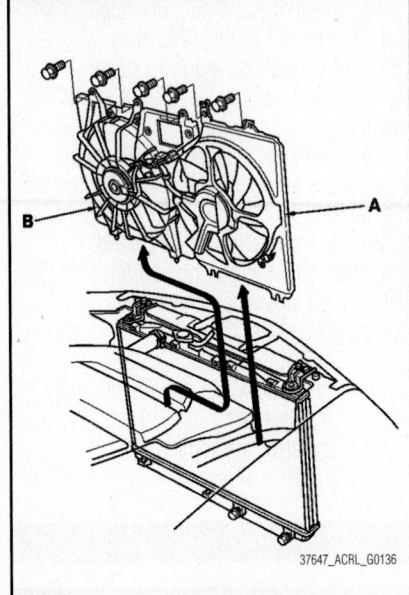

Fig. 23 Remove the condenser fan shroud assembly (A) and the radiator fan shroud assembly (B) from right side of the engine compartment.

fan motor connector, then remove the harness clamps.

8. Remove the Radiator Fan Control (RFC) unit.

9. Remove the condenser fan shroud assembly and the radiator fan shroud assembly from right side of the engine compartment.

10. Installation is the reverse of the removal procedure.

RADIATOR

REMOVAL & INSTALLATION

See Figure 24.

1. Drain the engine coolant.
2. Remove the right side fender trim.
3. Raise the vehicle on the lift to full height.
4. Remove the splash shield.
5. Disconnect the lower radiator hose.
6. Disconnect the automatic transmission fluid (ATF) cooler hoses, then plug the hose and line.
7. Disconnect the engine coolant temperature (ECT) sensor 2 connector.
8. Lower the vehicle on the lift.
9. Disconnect the upper radiator hose and the coolant reservoir hose.
10. Remove the air intake duct cover, the coolant reservoir, and the radiator upper brackets.

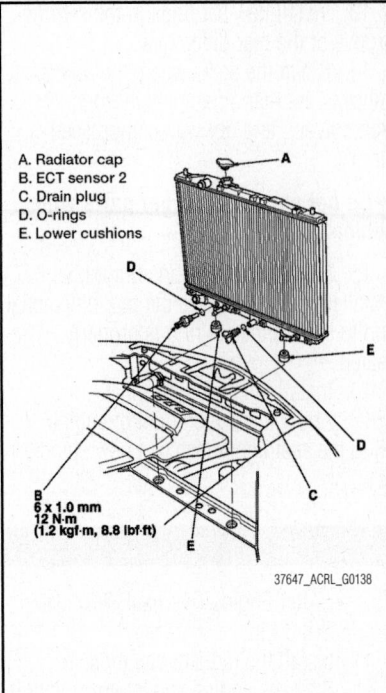

A. Radiator cap
B. ECT sensor 2
C. Drain plug
D. O-rings
E. Lower cushions

B
6 x 1.0 mm
12 N·m
(1.2 kgf·m, 8.8 lbf·ft)

Fig. 24 Removing/installing the radiator

11. Remove the A/C condenser fan shroud assembly and the radiator fan shroud assembly.

12. Pull up the radiator, then remove the radiator cap, the ETC sensor 2, and the drain plug.

To install:

13. Installation is the reverse of the removal procedure. Note the following:

   a. Reassemble the radiator with new O-rings.

   b. Install the radiator. Make sure the lower cushions are set securely.

   c. Install the radiator fan shroud assembly and the A/C condenser fan shroud assembly.

THERMOSTAT

REMOVAL & INSTALLATION

See Figure 25.

1. Do the battery removal procedure.
2. Drain the engine coolant.

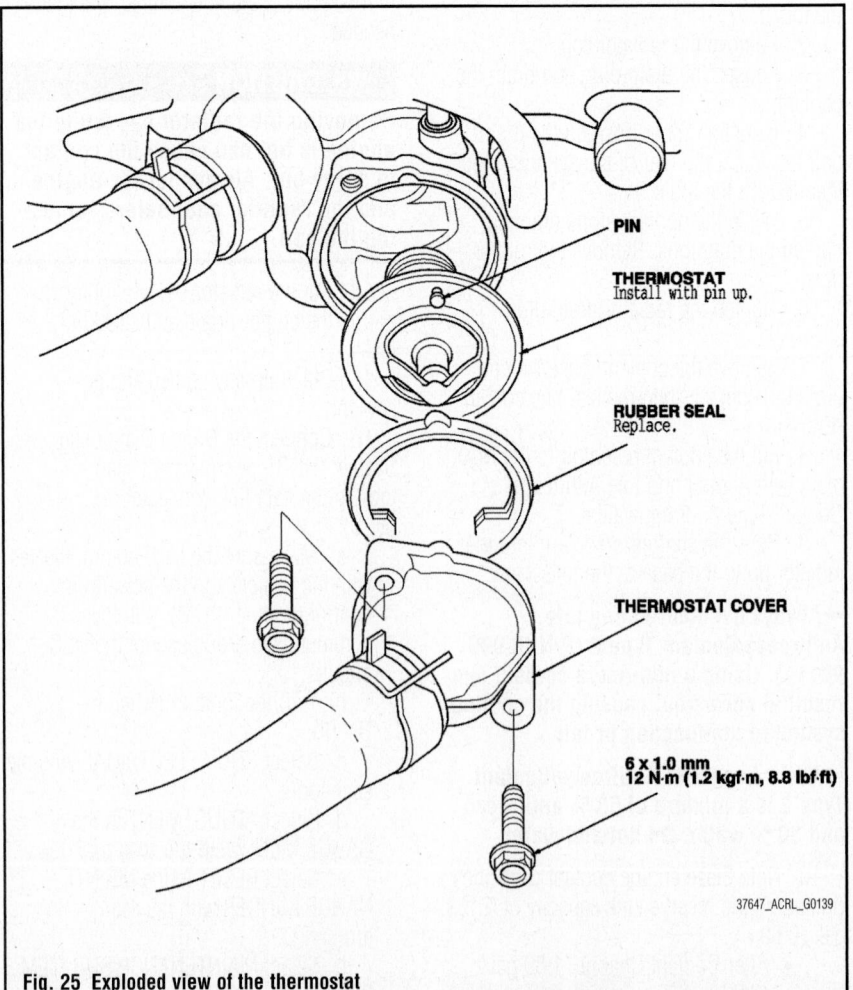

PIN

THERMOSTAT
Install with pin up.

RUBBER SEAL
Replace.

THERMOSTAT COVER

6 x 1.0 mm
12 N·m (1.2 kgf·m, 8.8 lbf·ft)

Fig. 25 Exploded view of the thermostat

3. Remove the thermostat cover, then remove the thermostat.

4. Install the thermostat with a new rubber seal and with the pin on the top side.

5. Do the battery installation procedure.

6. Refill the radiator with engine coolant, then bleed the air from the cooling system.

7. Clean up any spilled engine coolant.

WATER PUMP

REMOVAL & INSTALLATION
See Figure 26.

1. Drain the engine coolant.

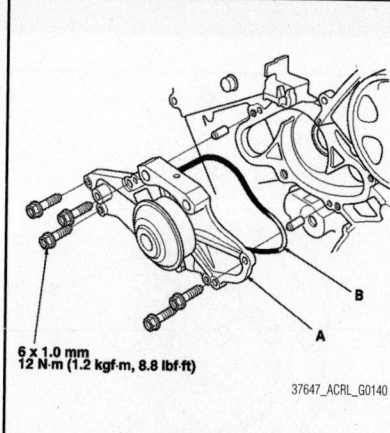

6 x 1.0 mm
12 N·m (1.2 kgf·m, 8.8 lbf·ft)

37647_ACRL_G0140

Fig. 26 Exploded view of the water pump (A), showing the O-ring (B) and related components

2. Remove the timing belt and belt adjuster.

3. Remove the water pump by removing the five bolts.

To install:
4. Inspect and clean the O-ring groove and the mating surface of the engine block.

5. Install the water pump with a new O-ring in the reverse order of removal.

6. Clean up any spilled engine coolant.

7. Install the timing belt adjuster and timing belt.

8. Refill the radiator with engine coolant, then bleed the air from the cooling system.

ENGINE ELECTRICAL

BATTERY SYSTEM

BATTERY

REMOVAL & INSTALLATION

➡The battery terminal disconnection/reconnection procedure must be done before and after doing this procedure. Some systems store data in memory that is lost when the battery is disconnected.

1. Do the "Battery Disconnect Procedure".

2. Remove the two nuts securing the battery setting plate, then remove the battery setting plate and the battery.

3. Install the battery, then install the battery setting plate.

4. Tighten the two nuts equally until the battery is stable.

➡**Do not deform the battery setting plate by over-tightening the nuts.**

5. Do the "Battery Reconnect Procedure".

➡**Make sure the battery is installed correctly, and the positive terminal and the negative terminal are not reversed.**

BATTERY DISCONNECT PROCEDURE

➡**Some systems store data in memory that is lost when the battery is disconnected. Do the following steps before disconnecting the battery.**

1. Make sure you have the anti-theft code(s) for the audio and/or the navigation system (if equipped).

➡**For some models, or if replacing the audio unit, it may be necessary to write down the audio presets (AM and FM), and the XM radio presets (if equipped), because the audio unit does not retain the presets after the battery is disconnected.**

2. Make sure the ignition switch is in LOCK.

3. Disconnect and isolate the negative cable from the battery.

➡**Always disconnect the negative cable from the battery first.**

4. Disconnect the positive cable from the battery.

BATTERY RECONNECT PROCEDURE

➡**Some systems store data in memory that is lost when the battery is disconnected. Do the following steps to restore the systems back to normal operation.**

1. Clean the battery terminals.
2. Test the battery.
3. Reconnect the positive cable to the battery first, then reconnect the negative cable to the battery.

➡**Always connect the positive cable to the battery first.**

4. Apply multipurpose grease to the terminals to prevent corrosion.

5. Enter the anti-theft code(s) for the audio system and/or the navigation system (if equipped).

6. Enter the audio presets (if applicable), and enter the XM radio presets (if equipped).

7. Set the clock (for vehicles without navigation).

8. Do the steering column position memorization.

ENGINE ELECTRICAL

ALTERNATOR

REMOVAL & INSTALLATION

See Figure 27.

1. Do the battery terminal disconnect procedure, then wait at least 3 minutes before beginning work.
2. Remove the upper grille cover.
3. Raise the vehicle on the lift to full height.
4. Remove the splash shield.
5. Remove the ground cable, the harness clamps, and the connector from the A/C condenser fan shroud.
6. Loosen the two bolts securing the A/C condenser fan shroud.
7. Lower the vehicle on the lift.
8. Remove the radiator fan control unit.

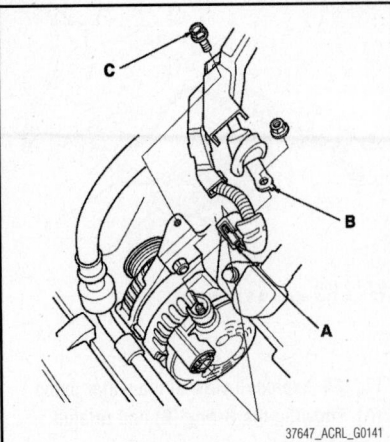

37647_ACRL_G0141

Fig. 27 Disconnect the alternator connector (A) and BLK wire (B) from the alternator. Remove the bolt (C) securing the harness holder.

CHARGING SYSTEM

9. Remove the coolant reservoir.
10. Remove the two bolts and remove the A/C condenser fan shroud.
11. Remove the drive belt.
12. Disconnect the alternator connector and BLK wire from the alternator. Remove the bolt securing the harness holder.
13. Remove the mounting bolts for the alternator bracket and alternator, then remove the alternator.
14. Installation is the reverse of the removal procedure.
15. Torque the long alternator mounting bolt to 33 ft. lbs. (44 Nm) and the alternator bracket bolt to 16 ft. lbs. (22 Nm).

ENGINE ELECTRICAL

FIRING ORDER

Firing order for this engine is:
1–4–2–5–3–6

IGNITION COIL PACK

REMOVAL & INSTALLATION

Front

1. Remove the engine cover.
2. Remove the upper grille cover.

3. Remove the coolant reservoir from the bracket.
4. Disconnect the ignition coil connectors, then remove the front ignition coils.
5. Install the front ignition coils in the reverse order of removal.

Rear

1. Disconnect the ignition coil connectors, then remove the rear ignition coils.

IGNITION SYSTEM

2. Install the rear ignition coils in the reverse order of removal.

IGNITION TIMING

ADJUSTMENT

➡**No adjustment is possible**

ENGINE ELECTRICAL

STARTER

REMOVAL & INSTALLATION

See Figure 28.

1. Do the battery terminal disconnect procedure, then wait at least 3 minutes before beginning work.
2. Remove the engine cover.
3. Remove the vacuum hose, transmission dipstick, and vacuum line mounting bolt (all located above the starter).
4. Remove the harness clamp. Disconnect the starter cable from the B terminal, then disconnect the BLK/WHT wire from the S terminal.

5. Remove the two bolts holding the starter, then remove the starter.

To install:

6. Install the starter, using a new gasket, then install the harness clamp and connect the B terminal and BLK/WHT wire. Make sure the crimped side of the B terminal faces away from the starter when you connect it. Tighten the starter mounting bolts to 33 ft. lbs. (44 Nm).
7. Install the vacuum hose, transmission dipstick, and vacuum line mounting bolt.
8. Install the engine cover.
9. Do the battery terminal reconnect procedure.

STARTING SYSTEM

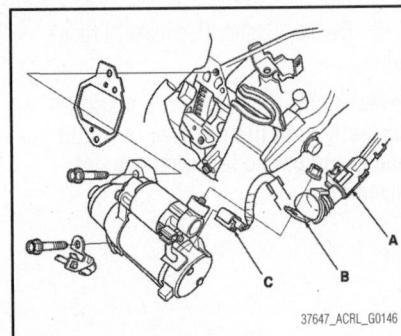

37647_ACRL_G0146

Fig. 28 Remove the harness clamp (A) and disconnect the starter cable (B) from the B terminal, then disconnect the BLK/WHT wire (C) from the S terminal.

ENGINE MECHANICAL

➡Disconnecting the negative battery cable may interfere with the functions of the on board computer systems and may require the computer to undergo a relearning process, once the negative battery cable is reconnected.

ACCESSORY DRIVE BELT SYSTEM

INSPECTION

See Figure 29.

1. Remove the right upper fender trim.
2. Inspect the belt for cracks or damage. If the belt is cracked or damaged, replace it.
3. Check that the auto-tensioner indicator is within the standard range. If it is out of the standard range, replace the drive belt.

ADJUSTMENT

➡Accessory drive belt is automatically adjusted by the tensioner.

REMOVAL & INSTALLATION

See Figure 30.

1. Remove the right upper fender trim.
2. Remove the engine cover.
3. Move the auto-tensioner (A) using the belt tension release tool in the direction

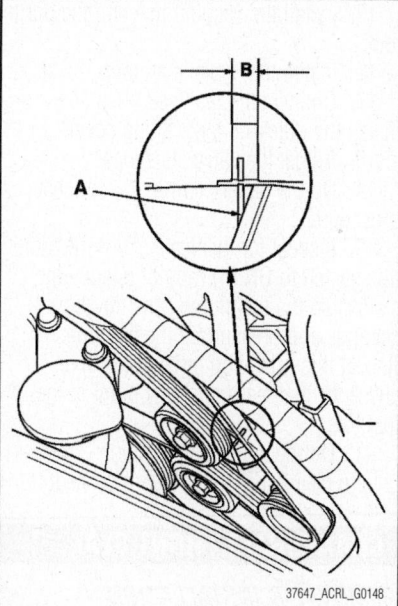

37647_ACRL_G0148

Fig. 29 Check that the auto-tensioner indicator (A) is within the standard range (B) as shown. If it is out of the standard range, replace the drive belt.

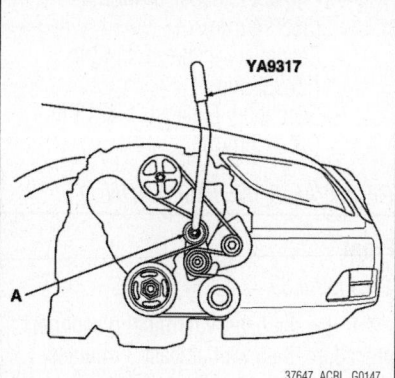

37647_ACRL_G0147

Fig. 30 Move the auto-tensioner (A) using the belt tension release tool in the direction shown to relieve tension from the drive belt, then remove the drive belt.

shown to relieve tension from the drive belt, then remove the drive belt.

4. Install the new belt in the reverse order of removal.

AIR INTAKE SYSTEM

REMOVAL & INSTALLATION

Air Cleaner Assembly

1. Remove the battery trim and the left upper fender trim.
2. Disconnect the MAF sensor/IAT sensor connector.
3. Remove the bolts and the air cleaner.
4. Install the parts in the reverse order of removal.

Air Filter Element

1. Remove the battery trim, and the left upper fender trim.
2. Open the air cleaner housing cover.
3. Remove the air cleaner element from the air cleaner housing cover.
4. Check the air cleaner element for damage or clogging. If it is damage or clogged, replace it.

➡Do not use compressed air to clean the air cleaner element.

5. Clean and remove any debris from inside the air cleaner housing.
6. Install the parts in the reverse order of removal.
   a. If you did not replace the air cleaner element, this procedure is complete.

b. If the maintenance minder required air cleaner element replacement, reset the maintenance minder.

c. If you replaced the air cleaner element, reset the PCM, and do the PCM idle learn procedure.

d. If the idle speed fluctuates, do the idle speed inspection.

CAMSHAFT & BEARINGS

INSPECTION

See Figures 31 and 32.

1. Remove the cylinder head.
2. Remove the rocker arm assembly.
3. On the front camshaft, put the rocker shafts bridge and the rocker shaft holder on the front cylinder head, then tighten the bolts to 16 ft. lbs. (22 Nm), in the sequence shown.
4. On the rear camshaft, put the rocker shaft bridge and the rocker shaft holder on the rear cylinder head, then tighten the bolts to 16 ft. lbs. (22 Nm), in the sequence shown.
5. Seat the camshaft by pushing it toward the rear of the cylinder head.
6. Zero a dial indicator against the end of the camshaft. Push the camshaft back and forth and read the end play.
   a. If the end play is beyond the service limit, replace the thrust cover and recheck.
   b. If it is still beyond the service limit, replace the camshaft.
7. Camshaft End Play:
   - Standard (New): 0.002-0.008 in. (0.05-0.20 mm)
   - Service Limit: 0.008 in. (0.20 mm)
8. Remove the camshaft thrust cover, then pull out the camshaft.
9. Wipe the camshaft clean, then inspect the lift ramps. Replace the camshaft if any lobes are pitted, scored, or excessively worn.
10. Measure the diameter of each camshaft journal.
11. Zero the gauge to the journal diameter.
12. Clean the camshaft bearing surfaces in the cylinder head. Measure the inside diameter of each camshaft bearing surface, and check for an out-of-round condition.
    a. If the camshaft-to-holder clearance is within limits, skip the next step. If the camshaft-to-holder clearance is beyond the service limit and the camshaft has been replaced, replace the cylinder head.

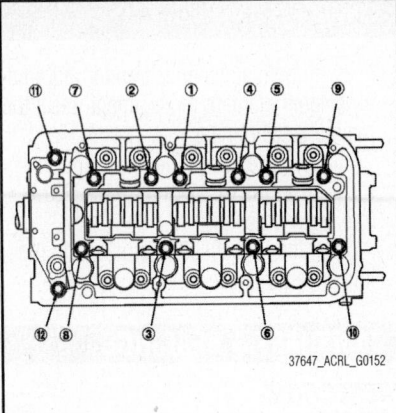

Fig. 31 Showing the camshaft bolt tightening sequence

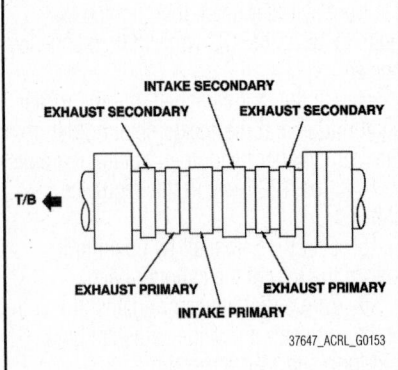

Fig. 32 Showing the camshaft cam lobes

b. If the camshaft-to-holder clearance is beyond the service limit and the camshaft has not been replaced, go to the next step.

13. Camshaft-to-Holder Oil Clearance:
- Standard (New): 0.0020-0.0035 in. (0.050-0.089 mm)
- Service Limit: 0.006 in. (0.15 mm)

14. Check total runout with the camshaft supported on V-blocks.

a. If the total runout of the camshaft is within the service limit, replace the cylinder head.

b. If the total runout is beyond the service limit, replace the camshaft and recheck the oil clearance. If the oil clearance is still out of tolerance, replace the cylinder head.

15. Camshaft Total Runout:
- Standard (New): 0.001 in. (0.03 mm) max.
- Service Limit: 0.002 in. (0.04 mm)

16. Measure the cam lobe height.

17. Cam Lobe Height Standard (New):

- Primary Intake: 1.3504 in. (34.299 mm)
- Primary Exhaust: 1.4462 in. (36.734 mm)
- Secondary Intake: 1.4146 in. (35.932 mm)
- Secondary Exhaust: 1.4713 in. (37.370 mm)

REMOVAL & INSTALLATION

Front

See Figure 33.

1. Do the battery terminal disconnect procedure, then wait at least 3 minutes before beginning work.

2. Remove the battery and battery box.

3. Drain the engine cooling system.

4. Disconnect the upper radiator hose.

5. Remove the exhaust gas recirculation valve and the stud bolts.

6. Remove the timing belt.

7. Remove the rocker arm assembly.

8. Remove the front camshaft pulley.

9. Remove the thrust cover, then remove the front camshaft.

To install:

10. Install the front camshaft and thrust cover. Tighten the bolts in the sequence shown.

➡**Always use a new O-ring.**

11. Apply new engine oil to the journals and the cam lobes.

12. Install the rocker arm assembly.

13. Install the timing belt.

14. Install the stud bolts, then install the EGR valve.

15. Install the upper radiator hose and the lower radiator hose.

16. Adjust the valve clearance.

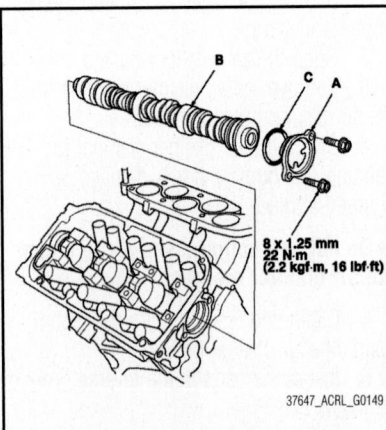

8 x 1.25 mm
22 N·m
(2.2 kgf·m, 16 lbf·ft)

Fig. 33 Remove the thrust cover (A), then remove the front camshaft (B).

17. Do the battery terminal reconnect procedure.

18. Fill the radiator with engine coolant and bleed the air out.

19. Do the crankshaft position (CKP) pattern clear/CKP pattern learn procedure.

Rear

See Figures 34 and 35.

1. Do the battery terminal disconnect procedure, then wait at least 3 minutes before beginning work.

2. Relieve the fuel pressure.

3. Remove the quick-connect fitting cover, then disconnect the fuel feed hose.

4. Remove the intake manifold.

5. Remove the purge joint.

6. Remove the two bolts securing the vacuum line.

7. Remove the timing belt.

8. Remove the rocker arm assembly.

9. Remove the rear camshaft pulley.

10. Remove the thrust cover, then remove the rear camshaft.

To install:

11. Install the rear camshaft in the reverse order of removal. Tighten the bolts in the sequence shown.

➡**Always use a new O-ring.**

12. Apply new engine oil to the journals and cam lobes.

13. Install the rocker arm assembly.

14. Install the timing belt.

15. Install the vacuum line and the purge joint.

16. Install the intake manifold.

17. Connect the fuel feed hose, then install the quick-connect fitting cover.

18. Adjust the valve clearance.

19. Do the battery terminal reconnect procedure.

20. Inspect for fuel leaks. Turn the ignition switch to ON (II) (do not operate the starter) so the fuel pump runs for about 2 seconds and pressurizes the fuel line. Repeat this operation three times, then check for fuel leakage at any point in the fuel line.

21. Do the crankshaft position (CKP) pattern clear/CKP pattern learn procedure.

CRANKSHAFT FRONT SEAL

REMOVAL & INSTALLATION

1. Remove the crankshaft position (CKP) sensor.

2. Remove the timing belt, and the timing belt drive pulley.

3. Remove the crankshaft front seal.

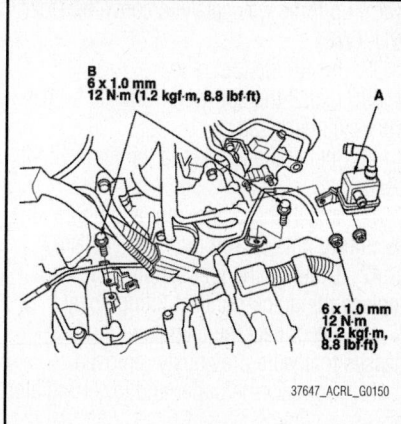

Fig. 34 Remove the purge joint (A). Remove the two bolts (B) securing the vacuum line.

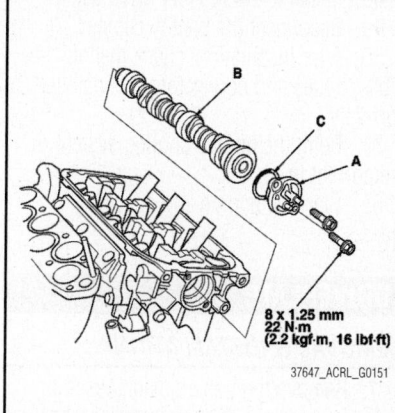

Fig. 35 Remove the thrust cover (A), then remove the rear camshaft (B).

To install:

4. Clean and dry the crankshaft oil seal housing.

5. Apply a light coat of multipurpose grease to the crankshaft and to the lip of the seal.

6. Using a proper seal driver, squarely drive in the crankshaft oil seal until the driver bottoms against the oil pump.

7. When the seal is in place, clean any excess grease off the crankshaft, and check that the oil seal lip is not distorted.

8. Install the timing belt drive pulley, the timing belt, and the CKP sensor.

CRANKSHAFT REAR SEAL

REMOVAL & INSTALLATION

See Figure 36.

Special Tools Required:
- Driver Handle, 15 x 135L 07749-0010000

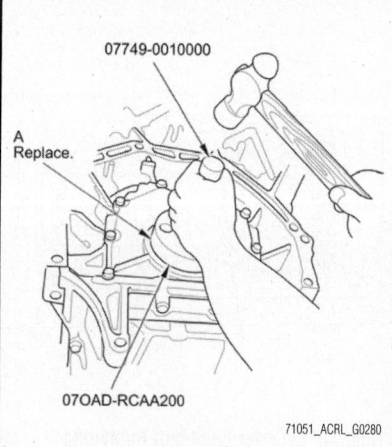

Fig. 36 Using the driver handle, 15 x 135L and the driver attachment, 106 mm, drive in the new crankshaft oil seal (A) until the driver attachment bottoms on the engine block end cover. Align the hole in the driver attachment with the pin on the crankshaft.

- Driver Attachment, 106 mm 07OAD-RCAA200

1. Remove the transmission and the drive plate.

2. Remove the transmission end crankshaft oil seal.

3. Clean and dry the crankshaft oil seal housing.

4. Apply a light coat of new engine oil to the lip of the crankshaft oil seal.

To install:

5. Using the driver handle, 15 x 135L and the driver attachment, 106 mm, drive in the new crankshaft oil seal until the driver attachment bottoms on the engine block end cover. Align the hole in the driver attachment with the pin on the crankshaft.

6. Clean the excess oil off the crankshaft, and check that the oil seal lip is not distorted.

7. Install the drive plate, and the transmission.

CYLINDER HEAD

REMOVAL & INSTALLATION

See Figures 37 through 40.

1. Properly relieve the fuel system pressure.

2. Make sure you have the anti-theft code for the radio and the navigation system, then write down the frequencies for the radio's preset buttons.

3. Disconnect the negative cable from the battery first, then disconnect the positive cable. Wait at least 3 minutes before proceeding.

4. Drain the engine cooling system.

5. Remove the accessory drive belt.

6. Remove the power steering pump and power steering hose clamp.

7. Remove the alternator.

8. Remove the timing belt.

9. Remove the intake manifold.

10. Remove the ignition coils.

11. Disconnect the upper and lower radiator hoses from the engine assembly.

12. Remove the quick-connect cover, then disconnect the fuel supply hose.

13. Remove the purge joint.

14. Remove the following engine wiring harnesses:

- Six injector wiring harnesses
- Engine coolant temperature (ECT) sensor 1 wiring harness
- Crankshaft position (CKP) sensor wiring harness
- Exhaust gas recirculation (EGR) valve wiring harness
- Rocker arm oil control solenoid wiring harness
- Rocker arm oil pressure switch wiring harness
- Oil pressure switch wiring harness
- Two air fuel ratio (A/F) sensor wiring harnesses
- Two secondary heated oxygen sensor (secondary HO2S) wiring harnesses

15. Remove the front warm up three way catalytic converter (front WU-TWC) and rear warm up three way catalytic converter (rear WU-TWC).

16. Remove the connector bracket from the front cylinder head.

17. Remove the bracket from the rear cylinder head.

18. Remove the fuel rails.

19. Remove the water passage.

20. Remove the front and rear camshaft pulleys and back covers.

21. Remove the cylinder head cover as follows:

   a. Remove the dipstick.

   b. Remove the two bolts securing the harness holder, and disconnect the front air fuel ratio (A/F) sensor connector, front secondary heated oxygen sensor (secondary HO2S) connector, exhaust gas recirculation (EGR) valve connector and engine coolant temperature (ECT) sensor 1 connector.

   c. Disconnect the three injector connectors from the injectors on the rear side cylinder head.

   d. Remove the power steering hose bracket mounting bolt, the harness

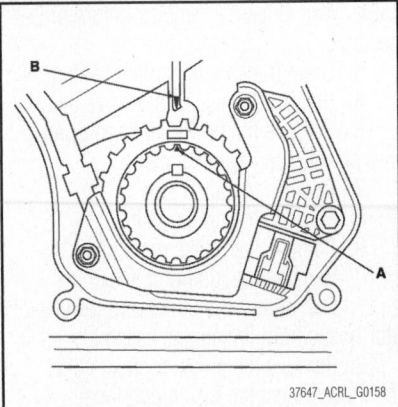

Fig. 37 Crankshaft timing belt sprocket TDC marks. Align sprocket mark (A) with pointer (B)

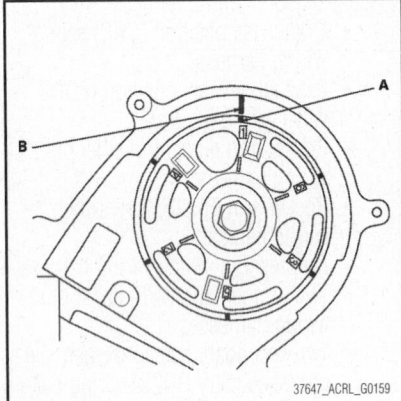

Fig. 38 Camshaft TDC marks. Align sprocket mark (A) with the back cover pointer (B)—front head shown; rear aligns the same

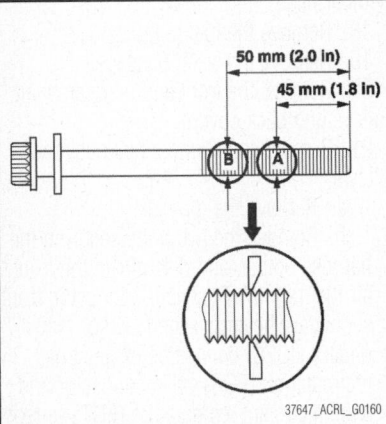

Fig. 39 Measure the diameter of each cylinder head bolt at point A and point B. If either diameter is less than 10.6 mm (0.42 in), replace the cylinder head bolt.

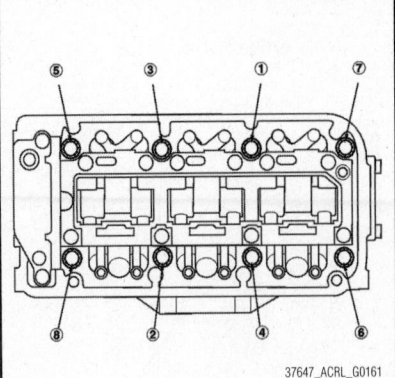

Fig. 40 Cylinder head bolt tightening sequence—front head shown; rear is the same sequence

holder mounting bolts, and the engine ground cable bolt.

  e. Remove the engine ground cable and disconnect the breather hose.

  f. Remove the cylinder head covers.

22. Loosen the cylinder head bolts in sequence and ⅓ turns until all bolts are loose.

23. Remove the cylinder head.

To install:

24. Clean the cylinder head and engine block surface.

25. Clean and install the oil control orifices with new O-rings.

26. Install the dowel pins and new cylinder head gaskets.

27. Clean the timing belt pulleys, timing belt guide plate, and the upper and lower covers.

28. Align the crankshaft and camshaft sprocket TDC marks as shown.

29. Put the cylinder head onto the engine block.

30. Measure the diameter of each cylinder head bolt at point A and point B. If either diameter is less than 10.6 mm (0.42 in), replace the cylinder head bolt.

31. Apply clean engine oil to the cylinder head bolt threads and flanges.

32. Tighten the cylinder head bolts in sequence as follows:

  a. Step 1: 22 ft. lbs. (29 Nm)

  b. Step 2: Plus 90 degrees

  c. Step 3: Plus an additional 90 degrees, if using new bolts

33. Install the timing belt.

34. Adjust the valve clearance.

35. Install the cylinder head covers and tighten to 108 inch lbs. (12 Nm).

36. Install the water passage.

37. Install the front warm up three way catalytic converter (front WU-TWC) and rear

warm up three way catalytic converter (rear WU-TWC).

38. Install the fuel rails.

39. Install the connector bracket to the front cylinder head.

40. Install the bracket to the rear cylinder head.

41. Install the purge joint and tighten the mounting nuts to 109 inch lbs. (12 Nm).

42. Connect the fuel supply hose and replace the quick-connect fitting cover.

43. Reconnect the engine wiring harnesses that were previously removed.

44. Connect the upper and lower radiator hoses.

45. Install the intake manifold.

46. Install the alternator.

47. Install the power steering pump and hose bracket.

48. Install the accessory drive belt.

49. Reconnect the battery cables.

50. After installation, check that all tubes, hoses and connectors are installed correctly.

51. Refill the engine cooling system to the correct level.

52. Start the engine and check for leaks.

DRIVEPLATE

REMOVAL & INSTALLATION

1. Remove the transmission assembly.

2. Remove the driveplate and washer from the engine.

To install:

3. Install the driveplate and washer on the crankshaft. Tighten the bolts in a criss-cross pattern to 54 ft. lbs. (74 Nm).

4. Install the transmission assembly into the vehicle.

EXHAUST SYSTEM

REMOVAL & INSTALLATION

See Figure 41.

➡**Manufacturer does not provide specific instructions for these components. Use the illustration as reference when servicing this system.**

INTAKE MANIFOLD

REMOVAL & INSTALLATION

See Figures 42 and 43.

1. Remove the battery trim.

2. Remove the engine cover.

3. Drain the engine coolant.

4. Disconnect the manifold absolute pressure (MAP) sensor connector, then

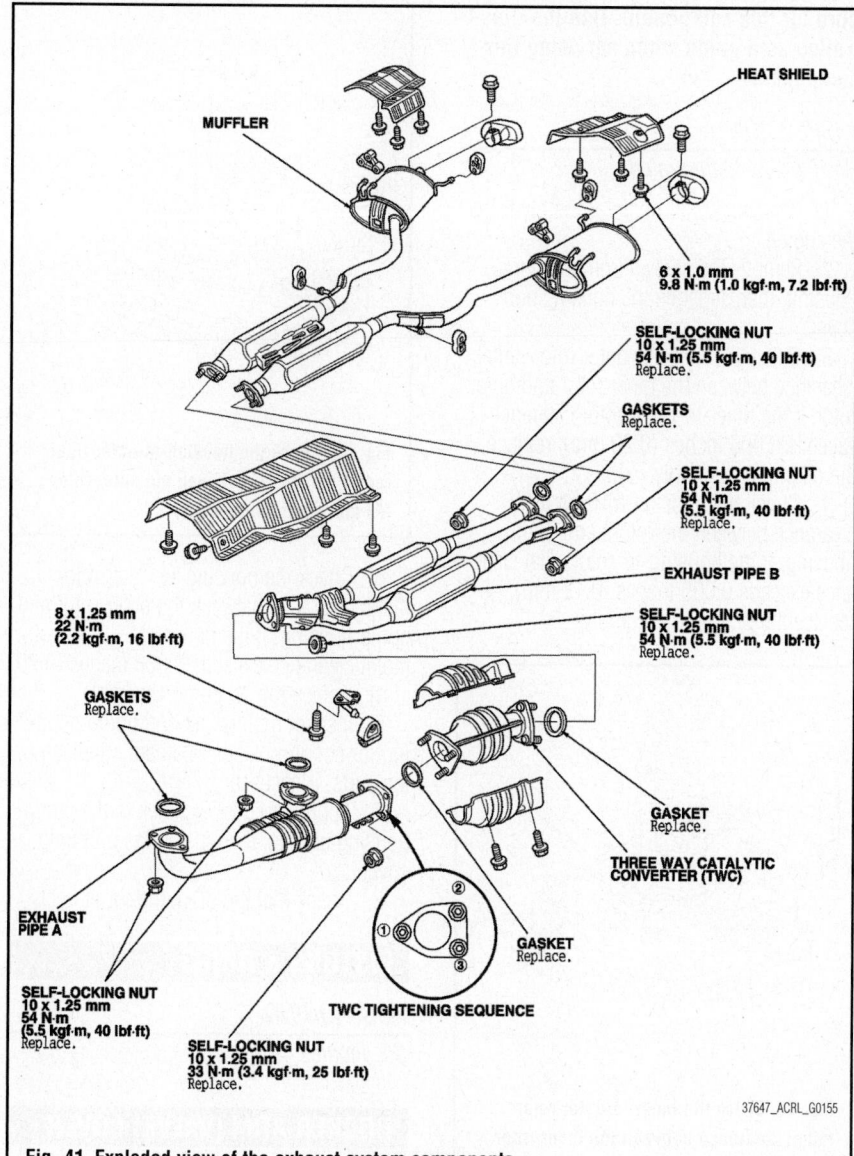

Fig. 41 Exploded view of the exhaust system components

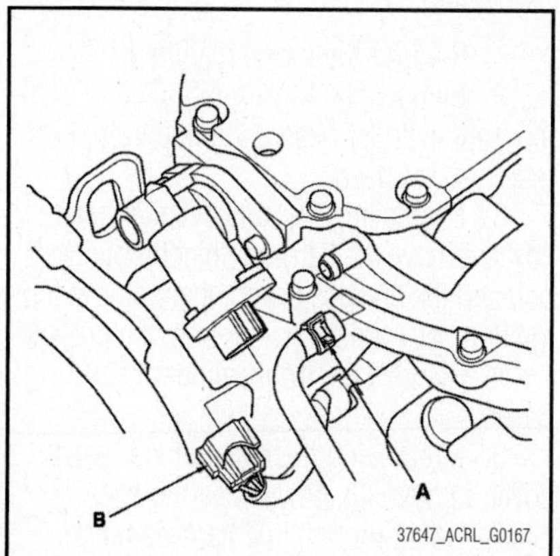

MUFFLER

HEAT SHIELD

6 x 1.0 mm
9.8 N·m (1.0 kgf·m, 7.2 lbf·ft)

SELF-LOCKING NUT
10 x 1.25 mm
54 N·m (5.5 kgf·m, 40 lbf·ft)
Replace.

GASKETS
Replace.

SELF-LOCKING NUT
10 x 1.25 mm
54 N·m
(5.5 kgf·m, 40 lbf·ft)
Replace.

EXHAUST PIPE B

8 x 1.25 mm
22 N·m
(2.2 kgf·m, 16 lbf·ft)

SELF-LOCKING NUT
10 x 1.25 mm
54 N·m (5.5 kgf·m, 40 lbf·ft)
Replace.

GASKETS
Replace.

EXHAUST
PIPE A

SELF-LOCKING NUT
10 x 1.25 mm
54 N·m
(5.5 kgf·m, 40 lbf·ft)
Replace.

SELF-LOCKING NUT
10 x 1.25 mm
33 N·m (3.4 kgf·m, 25 lbf·ft)
Replace.

TWC TIGHTENING SEQUENCE

GASKET
Replace.

THREE WAY CATALYTIC
CONVERTER (TWC)

37647_ACRL_G0155

remove the breather pipe and the intake air duct.

5. Disconnect the throttle actuator connector and the evaporative emission (EVAP) canister purge valve connector, then remove the water bypass hoses, the EVAP hose, the brake booster vacuum hose, and the vacuum hose.

6. Remove the positive crankcase ventilation (PCV) hose and disconnect the intake manifold tuning (IMT) actuator connector.

7. Remove the upper cover mounting bolts and nuts in an alternating pattern, starting from the corner bolt, in three steps, then remove the upper cover.

8. Remove the intake manifold mounting bolts and nuts sequentially in three steps, then remove the intake manifold.

To install:

9. Install the intake manifold. Tighten the bolts and nuts sequentially in three steps to a final torque of 16 ft. lbs. (22 Nm). Always use a new intake manifold gasket.

10. Install the upper cover. Starting from a corner bolt, tighten the bolts and nuts sequentially in three steps, to a final torque of 9 ft. lbs. (12 Nm). Always use a new gasket.

11. Install the remaining components in reverse of the removal procedure.

12. Refill the cooling system.

OIL PAN

REMOVAL & INSTALLATION

See Figure 44.

1. Drain the engine oil.

2. Do the battery terminal disconnect procedure, then wait at least 3 minutes before beginning work.

3. Remove the front splash shield.

4. Remove the front exhaust pipe.

5. Remove the rear Warm Up Three Way Catalytic Converter (rear WU-TWC) bracket.

6. Remove the torque converter cover and the four bolts securing the transaxle.

7. Remove the oil pan mounting bolts.

8. Use a flat bladed screwdriver to separate the oil pan from the block.

To install:

9. Remove all of the old liquid gasket material.

10. Apply liquid gasket to the mating surfaces of the oil pan as shown.

11. Install the oil pan on the engine block. Tighten the mounting bolts in sequence to 9 ft. lbs. (12 Nm).

12. Tighten the four bolts securing the transmission. Torque the bolts to 54 ft. lbs. (74 Nm).

37647_ACRL_G0167

Fig. 42 Remove the positive crankcase ventilation (PCV) hose (A), and disconnect the intake manifold tuning (IMT) actuator connector (B).

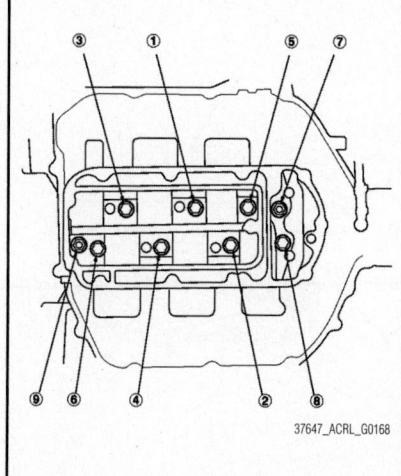

37647_ACRL_G0168

Fig. 43 Intake manifold bolt tightening sequence

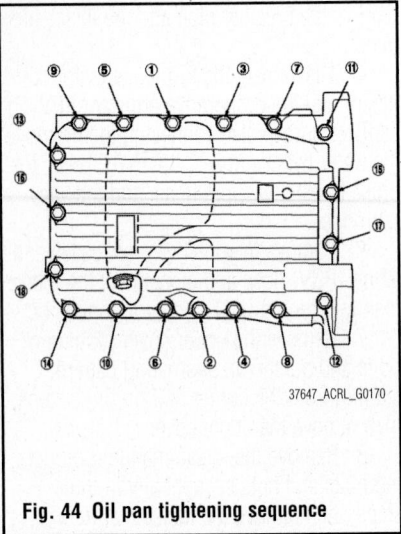

Fig. 44 Oil pan tightening sequence

13. Install the torque converter cover.

14. Install the rear Warm Up Three Way Catalytic Converter (rear WU-TWC) bracket.

15. Install the front exhaust pipe using new gaskets and new self-locking nuts.

16. Install the splash shield.

17. Do the battery terminal reconnect procedure.

➡**Wait at least 30 minutes before adding oil to the engine.**

18. Refill the engine with oil to the correct level.

OIL PUMP

REMOVAL & INSTALLATION

See Figure 45.

➡**Manufacturer does not provide a specific removal and installation proce-**

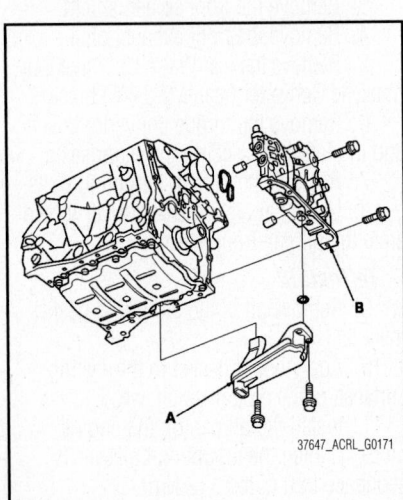

Fig. 45 Showing the location of the oil pump screen (A) and oil pump (B)

dure for this component. Use the illustration as a guide when servicing this component.

INSPECTION

See Figures 46 through 48.

1. Remove the oil pump assembly from the engine.

2. Remove the screws from the pump housing, then separate the housing and cover.

3. Check the inner-to-outer rotor radial clearance between the inner rotor and outer rotor. If the inner-to-outer rotor clearance exceeds 0.008 inches (0.20 mm), replace the oil pump assembly.

4. Check the housing-to-rotor axial clearance between the rotors and pump housing. If the housing-to-rotor axial clearance exceeds 0.005 inches (0.12 mm), replace the oil pump assembly.

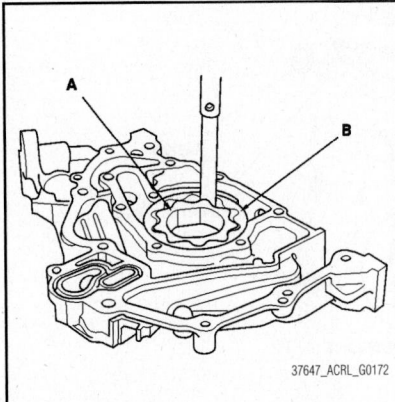

Fig. 46 Check the inner-to-outer rotor radial clearance between the inner rotor (A) and outer rotor (B)

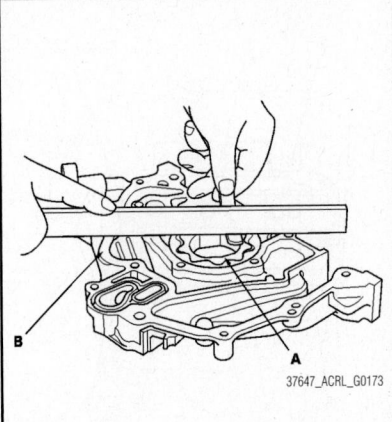

Fig. 47 Check the housing-to-rotor axial clearance between the rotors (A) and pump housing (B)

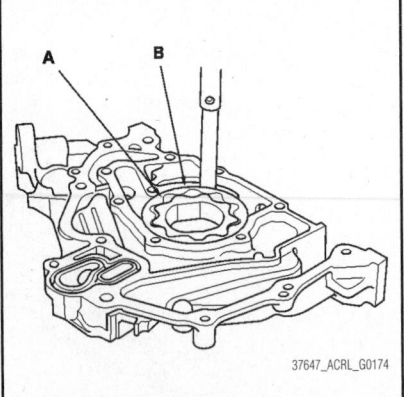

Fig. 48 Check the housing-to-outer rotor radial clearance between the outer rotor (A) and pump housing (B)

5. Check the housing-to-outer rotor radial clearance between the outer rotor and pump housing. If the housing-to-outer rotor radial clearance exceeds 0.008 inches (0.20 mm), replace the oil pump assembly.

6. Inspect both rotors and pump housing for scoring or other damage. Replace the parts, if necessary.

7. Apply liquid thread lock to the pump housing screws, then install the oil pump cover.

8. Check that the oil pump turns freely.

PISTONS & RINGS

POSITIONING

See Figures 49 and 50.

ROCKER ARM (VALVE) COVER

REMOVAL & INSTALLATION

See Figure 51.

1. Remove the intake manifold.

2. Remove the three ignition coils from the front cylinder head and three from the rear cylinder head.

3. If removing the front cylinder head cover, remove the harness holder mounting bolt and the harness clamp, then remove the ignition coil harness holder from the bracket.

4. Remove the front cylinder head cover.

5. If removing the rear cylinder head cover, remove the power steering hose clamp mounting bolt, the harness holder mounting bolts, the harness clamp, and the breather hose.

6. Remove the wire harness from the rear upper cover.

7. Remove the rear cylinder head cover.

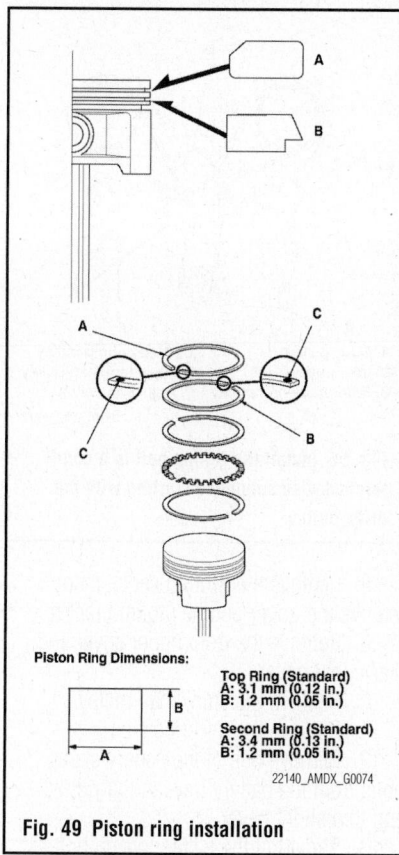

Piston Ring Dimensions:

Top Ring (Standard)
A: 3.1 mm (0.12 in.)
B: 1.2 mm (0.05 in.)

Second Ring (Standard)
A: 3.4 mm (0.13 in.)
B: 1.2 mm (0.05 in.)

22140_AMDX_G0074

Fig. 49 Piston ring installation

To install:

8. Thoroughly clean the head cover gasket and the groove.

9. Install the head cover gasket in the groove of the cylinder head cover. Make sure the head cover gasket is seated securely.

10. Remove all of the old liquid gasket from the rocker shaft holder and the cylinder head.

11. Clean the head cover contacting surfaces with a shop towel.

12. Apply liquid gasket (P/N 08717-0004 or equivalent) evenly to the rocker shaft holder mating areas.

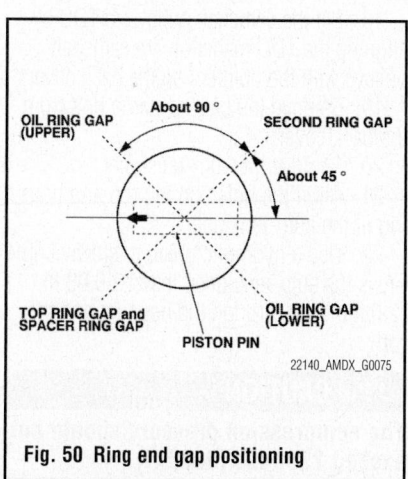

22140_AMDX_G0075

Fig. 50 Ring end gap positioning

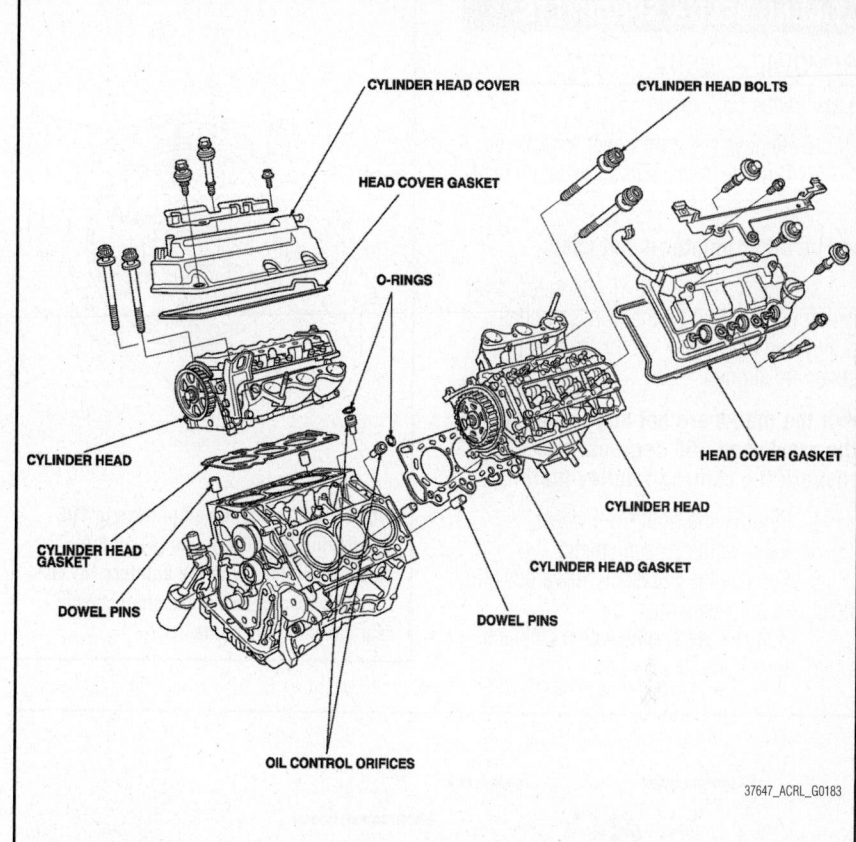

Fig. 51 Exploded view of the cylinder head components, showing the valve covers (cylinder head covers)

13. Install the component within 5 minutes of applying the liquid gasket.

14. Install all remaining components in reverse of the removal procedure.

15. Tighten the valve cover (cylinder head cover) bolts to 9 ft. lbs. (12 Nm).

ROCKER ARMS

REMOVAL & INSTALLATION

See Figure 52.

1. Remove the cylinder head cover.

2. Loosen the rocker arm locknuts and adjusting screws.

3. Remove the rocker arm assembly as follows:

   a. Unscrew the rocker shaft bolts 2 turns at a time in a criss-cross pattern to avoid damaging the vales or rocker assembly.

   b. Do not remove the rocker shaft bolts. These bolts keep the springs and rocker arms on the shafts.

4. Remove the rocker arms and shafts from the vehicle as an assembly.

➡**Keep all valve train components in order for assembly.**

5. Remove the rocker arms and springs from the rocker arm shafts.

To install:

6. Assemble the rocker arms and springs to the rocker arm shafts in their original positions.

7. Install the rocker arm assemblies. Tighten the bolts in sequence and in multiple passes to 17 ft. lbs. (24 Nm).

8. Adjust the valve clearance.

9. Install the cylinder head cover.

10. Install the intake manifold.

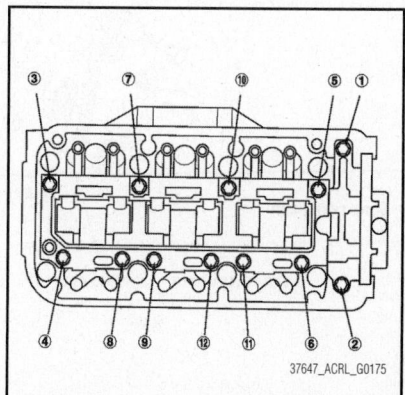

37647_ACRL_G0175

Fig. 52 Rocker arm shaft tightening sequence

TIMING BELT & SPROCKETS

REMOVAL & INSTALLATION

See Figures 53 through 55

1. Remove the right upper fender trim.
2. Turn the crankshaft so its white mark lines up with the pointer.

➡The other pointer is not used.

3. Check that the No. 1 piston top dead center (TDC) mark on the front camshaft pulley and the pointer on the front upper cover are aligned.

➡If the marks are not aligned, rotate the crankshaft 360 degrees, and recheck the camshaft pulley mark.

4. Remove the right front wheel.
5. Remove the splash shield.
6. Remove the accessory drive belt and drive belt auto-tensioner.
7. Support the engine with a jack and wood block under the oil pan.

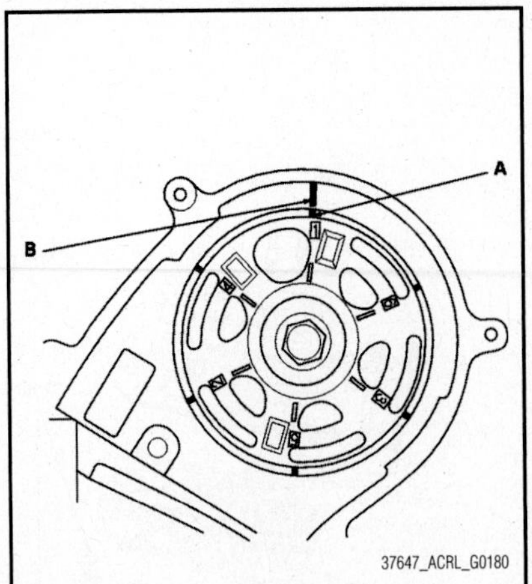
A
B
37647_ACRL_G0180

Fig. 54 Set the camshaft pulleys to TDC by aligning the TDC marks (A) on the camshaft pulleys with the pointers (B) on the back covers—front camshaft shown; rear camshaft the same

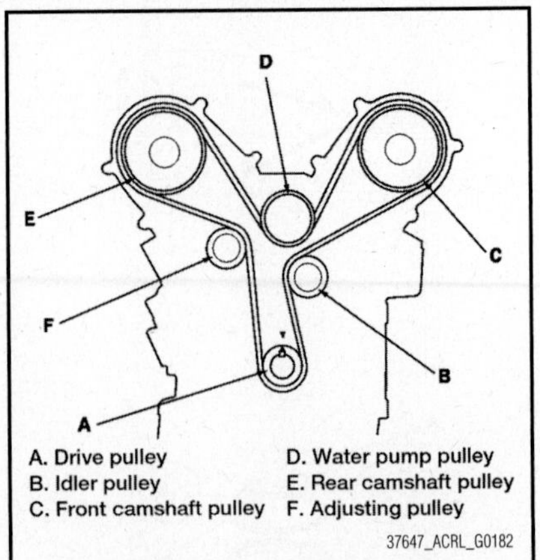
D
E
C
F
B
A
A. Drive pulley
B. Idler pulley
C. Front camshaft pulley
D. Water pump pulley
E. Rear camshaft pulley
F. Adjusting pulley
37647_ACRL_G0182

Fig. 55 Install the timing belt in a counterclockwise sequence starting with the drive pulley.

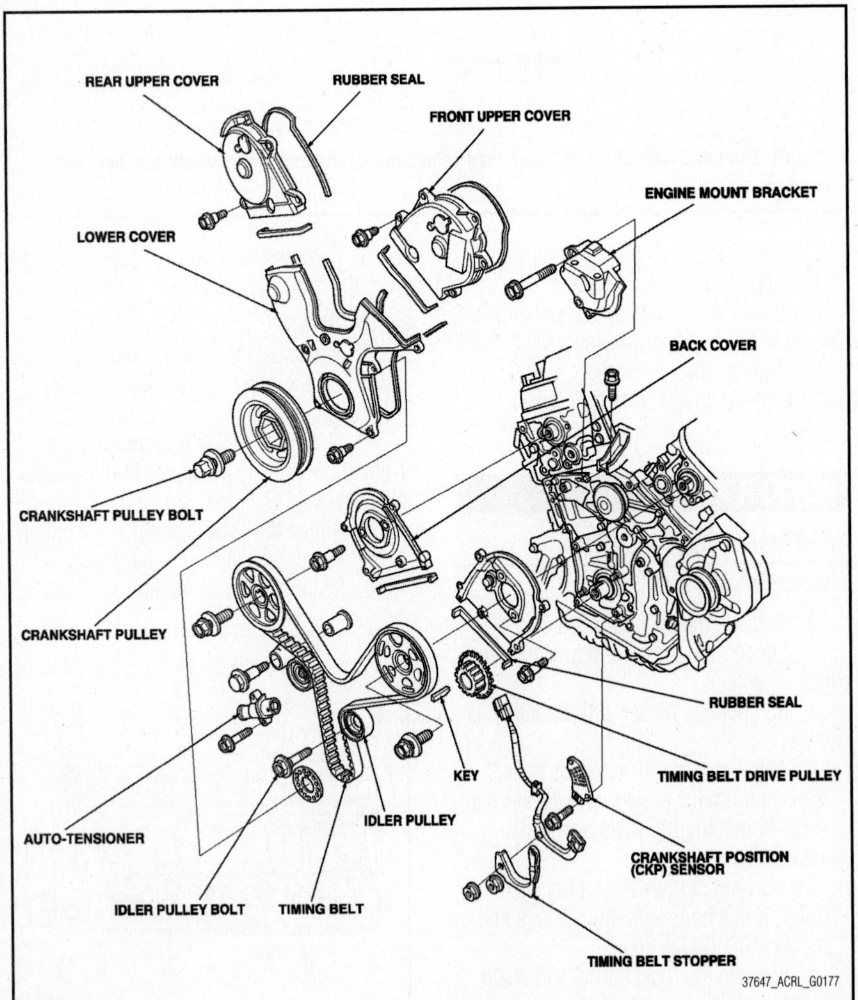
REAR UPPER COVER
RUBBER SEAL
FRONT UPPER COVER
ENGINE MOUNT BRACKET
LOWER COVER
BACK COVER
CRANKSHAFT PULLEY BOLT
CRANKSHAFT PULLEY
RUBBER SEAL
KEY
TIMING BELT DRIVE PULLEY
IDLER PULLEY
AUTO-TENSIONER
CRANKSHAFT POSITION (CKP) SENSOR
IDLER PULLEY BOLT
TIMING BELT
TIMING BELT STOPPER
37647_ACRL_G0177

Fig. 53 Exploded view of the front covers and timing belt components

8. Remove the ground cable, then remove the upper engine mount bracket
9. Remove the front upper cover and rear upper cover.
10. Remove the crankshaft pulley.
11. Remove the lower cover.
12. Remove one of the battery clamp bolts from the battery tray, and grind the end to a slight bevel.
13. Thread in the battery clamp bolt as shown to hold the timing belt adjuster in its current position. Tighten it by hand; do not use a wrench.
14. Remove the side engine mount bracket.
15. Remove the idler pulley bolt and the idler pulley, then remove the timing belt. Discard the idler pulley bolt.

To install:

16. Clean the timing belt pulleys, the timing belt guide plate, and the upper and lower covers.
17. Set the timing belt drive pulley to top dead center (TDC) by aligning the TDC mark on the tooth of the timing belt drive pulley with the pointer on the oil pump.
18. Set the camshaft pulleys to TDC by aligning the TDC marks on the camshaft pulleys with the pointers on the back covers.
19. Remove the battery clamp bolt from the back cover.
20. Remove the auto-tensioner.
21. Align the holes on the rod and housing of the auto-tensioner.
22. Use a hydraulic press to slowly compress the auto-tensioner. Insert a 0.08 in. (2.0 mm) pin through the housing and the rod.

⁂ CAUTION

The compression pressure should not exceed 2200 lbs (9800 N).

23. Install the auto-tensioner. Make sure the pin stays in place.

24. With the battery clamp bolt in place to hold the timing belt adjuster, loosely install the idler pulley with a new idler pulley bolt so the pulley can move but does not come off.

25. Install the timing belt in a counterclockwise sequence starting with the drive pulley.

26. Tighten the idler pulley bolt to 33 ft. lbs. (44 Nm).

27. Remove the pin from the auto-tensioner.

28. Remove the battery clamp bolt from the back cover.

29. Install the side engine mount bracket. Tighten the three horizontal bolts to 33 ft. lbs. (44 Nm) and the one smaller vertical bolt to 9 ft. lbs. (12 Nm).

30. Install the timing belt guide plate on the crankshaft, with the beveled edge facing the engine block.

31. Install the lower cover.

32. Install the crankshaft pulley.

33. Install the front upper cover and the rear upper cover.

34. Rotate the crankshaft pulley about six turns clockwise so the timing belt positions itself on the pulleys.

35. Turn the crankshaft pulley so the white mark lines up with the pointer.

36. Through the hole in the rear cover, check the camshaft pulley marks are aligned for TDC.

37. Install the upper engine mount bracket, then tighten the two vertical bolts to 40 ft. lbs. (54 Nm), and then the one smaller horizontal bolt to 47 ft. lbs. (64 Nm).

38. Install the ground cable.

39. Install the drive belt auto-tensioner.

40. Install the drive belt.

41. Install the splash shield.

42. Install the right front wheel.

43. Do the crankshaft position (CKP) pattern clear/CKP pattern learn procedure.

VALVE LASH

ADJUSTMENT

See Figures 56 through 58.

➡**Adjust the valves only when the cylinder head temperature is less than 100°F (38°C).**

1. Remove the intake manifold.

2. Remove the cylinder head covers.

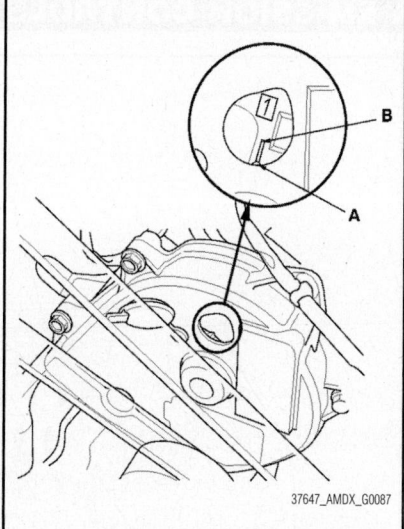

37647_AMDX_G0087

Fig. 56 Align the pointer (A) on the front upper cover with the No. 1 piston TDC mark (B) on the front camshaft pulley

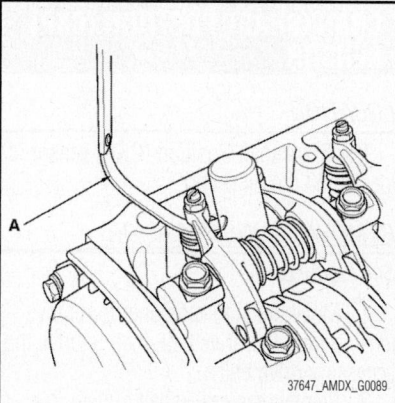

37647_AMDX_G0089

Fig. 57 Insert the feeler gauge (A) between the adjusting screw and the end of the valve stem on No. 1

3. Set the No. 1 piston at top dead center (TDC).

4. Align the pointer on the front upper cover with the No. 1 piston TDC mark on the front camshaft pulley.

5. Select the correct thickness feeler gauge for the valves you're going to check.

　a. Valve Clearance: Intake: 0.008–0.009 inches (0.20–0.24 mm); Exhaust: 0.011–0.013 inches (0.28–0.32 mm)

6. Insert the feeler gauge between the adjusting screw and the end of the valve stem on No. 1 cylinder and slide it back and forth; you should feel a slight amount of drag.

7. If you feel too much or too little drag, loosen the locknut, and turn the

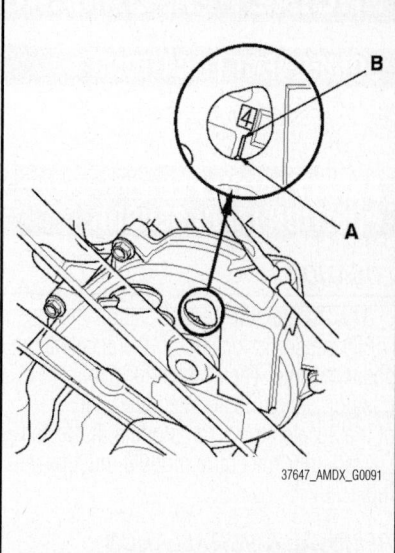

37647_AMDX_G0091

Fig. 58 Align the pointer (A) on the front upper cover with the No. 4 piston TDC mark (B) on the front camshaft pulley

adjusting screw until the drag on the feeler gauge is correct.

8. Tighten the locknut and recheck the clearance.

9. Repeat the adjustment, if necessary.

10. Rotate the crankshaft clockwise. Align the pointer on the front upper cover with the No. 4 piston TDC mark on the front camshaft pulley.

11. Check and, if necessary, adjust the valve clearance on No. 4 cylinder.

12. Rotate the crankshaft clockwise. Align the pointer on the front upper cover with the No. 2 piston TDC mark on the front camshaft pulley.

13. Check and, if necessary, adjust the valve clearance on No. 2 cylinder.

14. Rotate the crankshaft clockwise. Align the pointer on the front upper cover with the No. 5 piston TDC mark on the front camshaft pulley.

15. Check and, if necessary, adjust the valve clearance on No. 5 cylinder.

16. Rotate the crankshaft clockwise. Align the pointer on the front upper cover with the No. 3 piston TDC mark on the front camshaft pulley.

17. Check and, if necessary, adjust the valve clearance on No. 3 cylinder.

18. Rotate the crankshaft clockwise. Align the pointer on the front upper cover with the No. 6 piston TDC mark on the front camshaft pulley.

19. Check and, if necessary, adjust the valve clearance on No. 6 cylinder.

20. Install the cylinder head covers.

21. Install the intake manifold.

ENGINE PERFORMANCE & EMISSION CONTROLS

COMPONENT LOCATIONS

See Figures 59 through 61.

ACCELERATOR PEDAL POSITION (APP) SENSOR

LOCATION

The Accelerator Pedal Position (APP) sensor is located on the accelerator pedal module. The accelerator pedal position sensor is an integrated part of the accelerator pedal module. If the sensor is faulty, the entire module must be replaced.

REMOVAL & INSTALLATION

See Figure 62.

1. Disconnect the APP sensor connector.
2. Release the lock tab on the base of the accelerator pedal, then remove the accelerator pedal module.
3. Install the parts in the reverse order of removal.

AIR-FUEL RATIO SENSOR (A/F)

LOCATION

The Air Fuel Ratio (A/F) sensors are located on each bank of the engine just below the ignition coils.

REMOVAL & INSTALLATION

See Figure 63.

1. Disconnect the front A/F sensor connector, then remove the A/F sensor.
2. Install the parts in the reverse order of removal.
3. Tighten the sensor to 33 ft. lbs. (44 Nm).

CAMSHAFT POSITION (CMP) SENSOR

LOCATION

The Camshaft Position (CMP) sensor is located on the timing belt back cover.

REMOVAL & INSTALLATION

See Figure 64.

1. Remove the timing belt.
2. Remove the front camshaft pulley (CMP sensor pulse plate).
3. Disconnect the Camshaft Position Sensor (CMP), then remove the back cover.

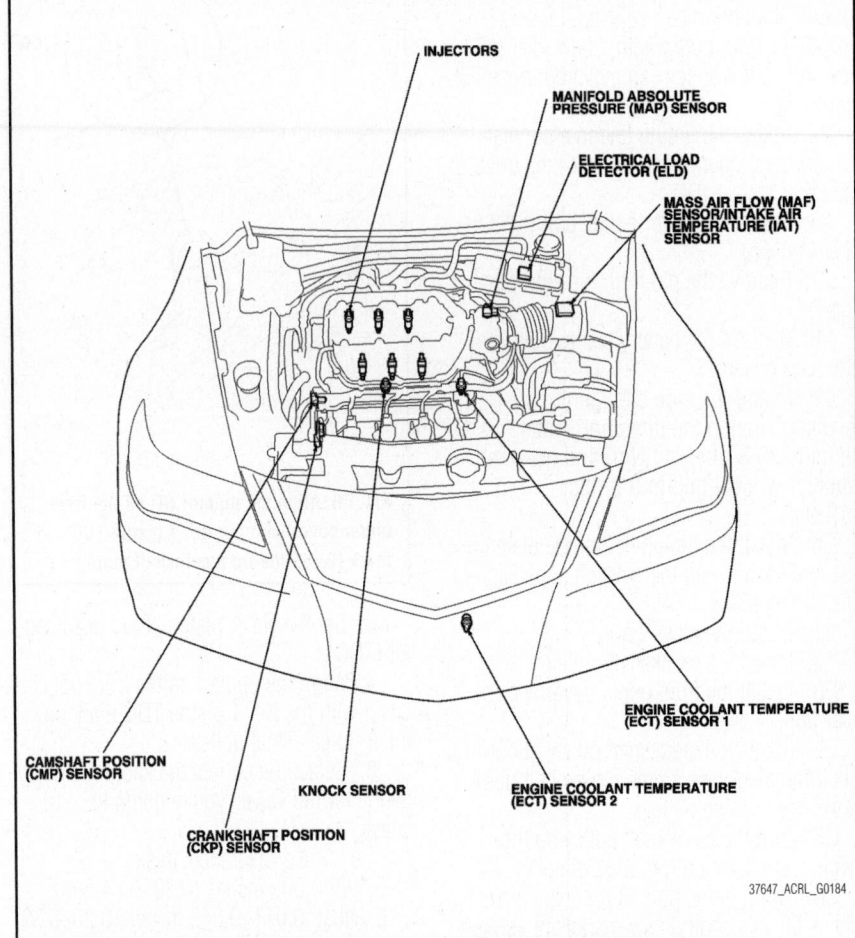

Fig. 59 Underhood engine performance & emission control component locations

37647_ACRL_G0184

4. Remove the CMP from the back cover.
5. Installation is the reverse order of removal.
6. Do the CKP pattern clear/CKP pattern learn procedure.

CRANKSHAFT POSITION (CKP) SENSOR

LOCATION

The Crankshaft Position (CKP) sensor is located on the oil pump.

REMOVAL & INSTALLATION

See Figure 65.

1. Move the auto-tensioner to remove tension from the drive belt, then remove the accessory drive belt.
2. Remove the crankshaft pulley.
3. Remove the upper and lower front covers from the engine.
4. Remove the Crankshaft Position (CKP) sensor from the oil pump.

5. Installation is the reverse order of removal.
6. Do the CKP Pattern Clear/CKP Pattern Learn Procedure.

CKP PATTERN CLEAR/CKP PATTERN LEARN PROCEDURE

➡ **This procedure allows the clear and learn process without a scan tool.**

1. Start the engine. Hold the engine speed at 3,000 RPM without load (in P or N) until the radiator fan comes on.
2. Test-drive the vehicle on a level road: Decelerate (with the throttle fully closed) from an engine speed of 2,500 RPM down to 1,000 RPM with the transmission in 2.
3. Test-drive the vehicle on a level road: Decelerate (with the throttle fully closed) from an engine speed of 5,000 RPM down to 3,000 RPM with the transmission in 2.
4. Repeat step 2 and 3 several times.
5. Turn the ignition switch to LOCK (0).
6. Turn the ignition switch to ON (II), and wait 30 seconds.

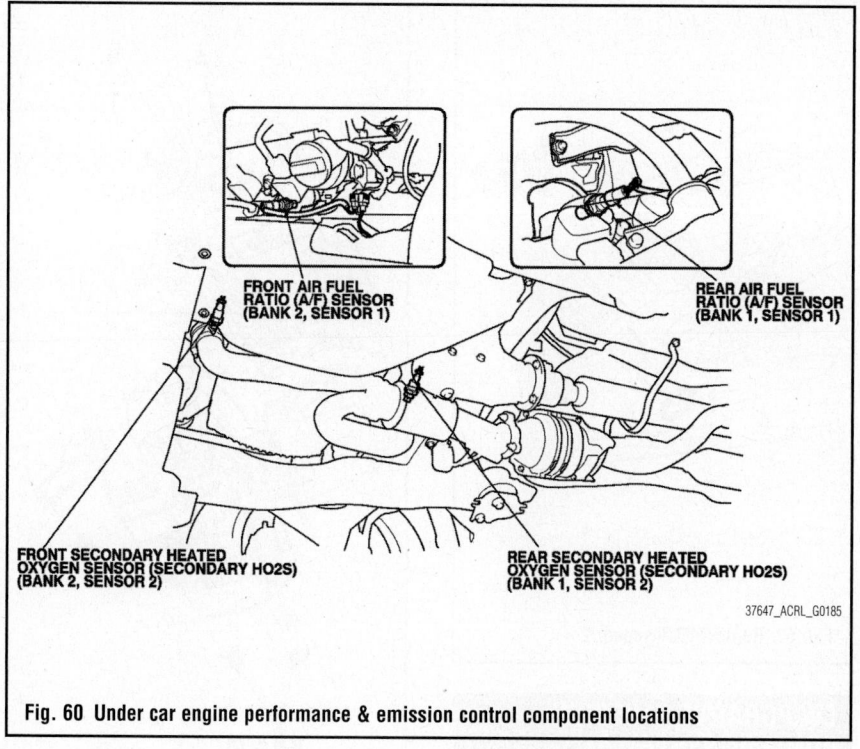

Fig. 60 Under car engine performance & emission control component locations

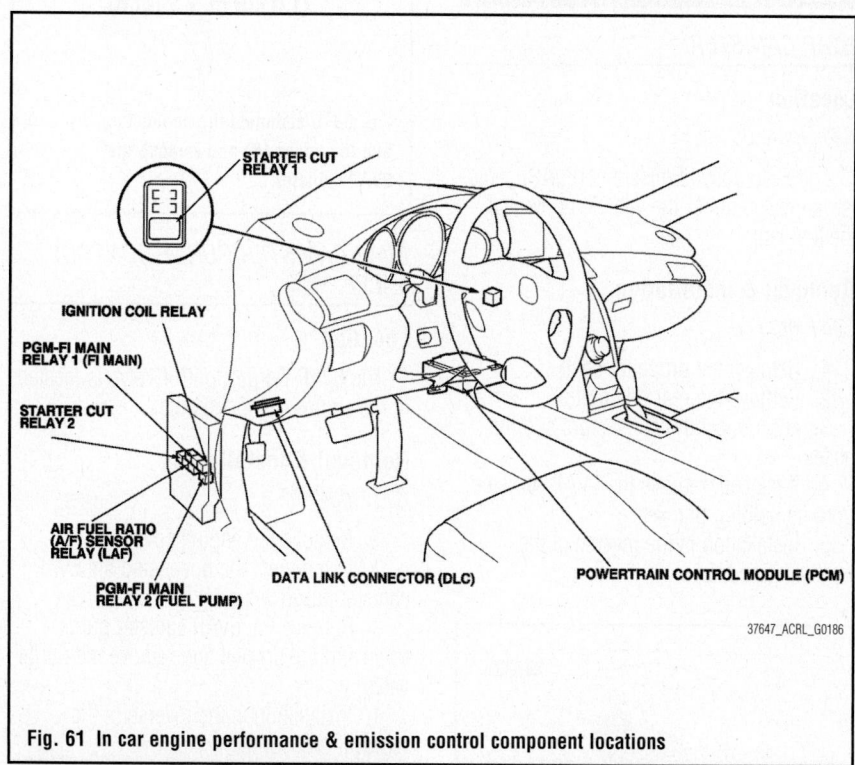

Fig. 61 In car engine performance & emission control component locations

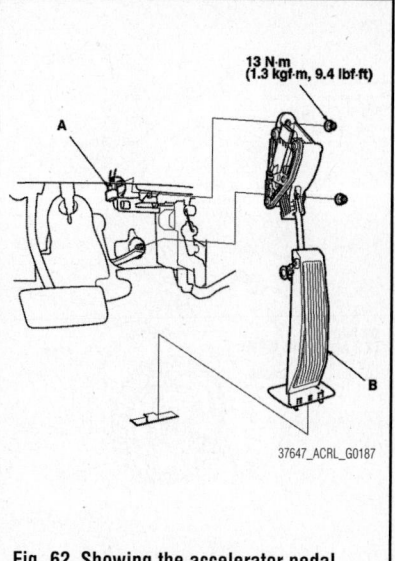

Fig. 62 Showing the accelerator pedal base (A) and module (B)

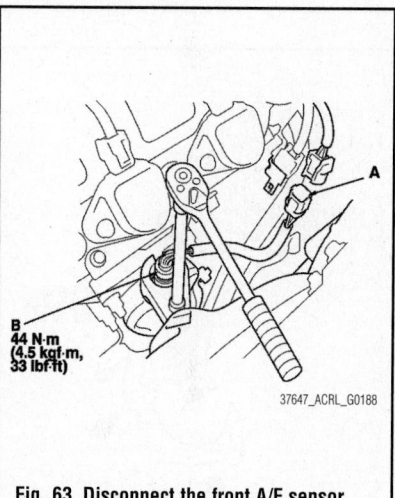

Fig. 63 Disconnect the front A/F sensor connector (A), then remove the A/F sensor (B)—bank 2 shown; bank 1 is the same

ENGINE COOLANT TEMPERATURE (ECT) SENSOR

LOCATION

Sensor 1 is located in the engine compartment below the main underhood fuse box. Sensor 2 is located at the bottom of the radiator.

REMOVAL & INSTALLATION

ECT Sensor 1

See Figure 66.

1. Drain the engine cooling system.
2. Remove the engine appearance cover.
3. Remove the main under-hood fuse/relay box.
4. Disconnect the ECT sensor 1 connector.
5. Remove ECT sensor 1.
6. Install the parts in the reverse order of removal with a new O-ring, then refill the radiator with engine coolant.

ECT Sensor 2

See Figure 67.

1. Drain the engine cooling system.
2. Remove the splash shield.
3. Disconnect the ECT sensor 2 connector.
4. Remove ECT sensor 2.
5. Install the parts in the reverse order of removal with a new O-ring, then refill the radiator with engine coolant.

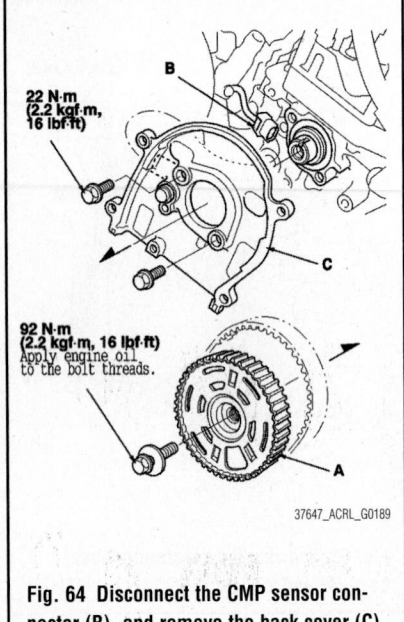

Fig. 64 Disconnect the CMP sensor connector (B), and remove the back cover (C)

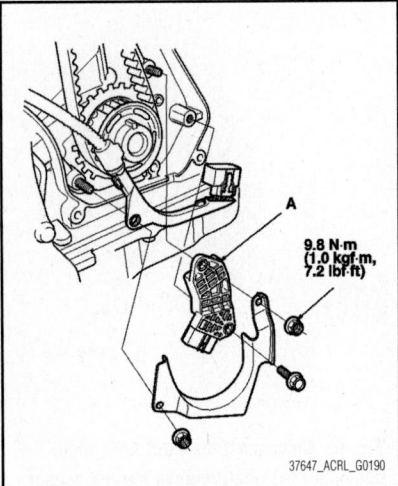

Fig. 65 Remove the Crankshaft Position (CKP) sensor (A) from the oil pump

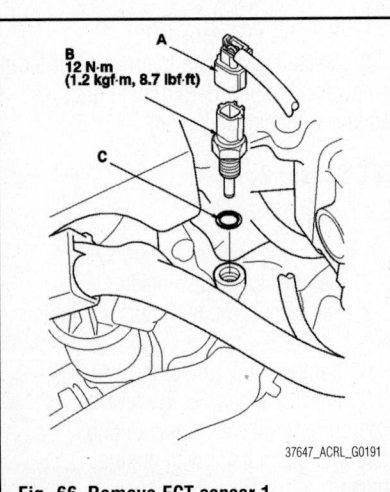

Fig. 66 Remove ECT sensor 1

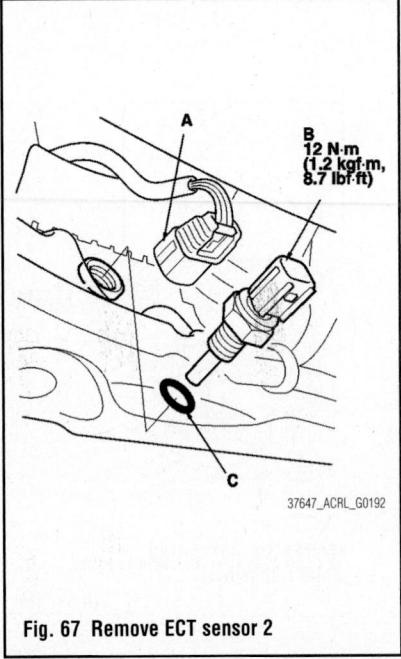

Fig. 67 Remove ECT sensor 2

EVAPORATIVE EMISSION CONTROL SYSTEM

EVAP CANISTER

Location

See Figure 68.

The Evaporative Emissions (EVAP) canister is located under the vehicle body near the fuel tank.

Removal & Installation

See Figure 69.

1. Remove the propeller shaft.
2. Remove the rear differential.
3. Disconnect the connectors and the hoses.
4. If needed, remove the EVAP canister from its holding bracket.
5. Installation is the reverse of the removal procedure.

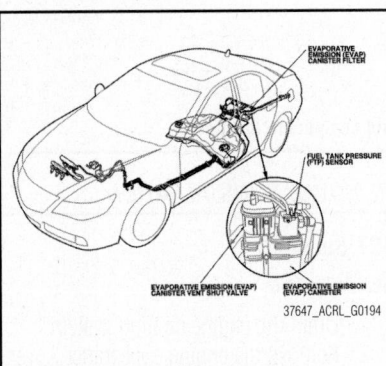

Fig. 68 Showing the under car locations of the EVAP system components

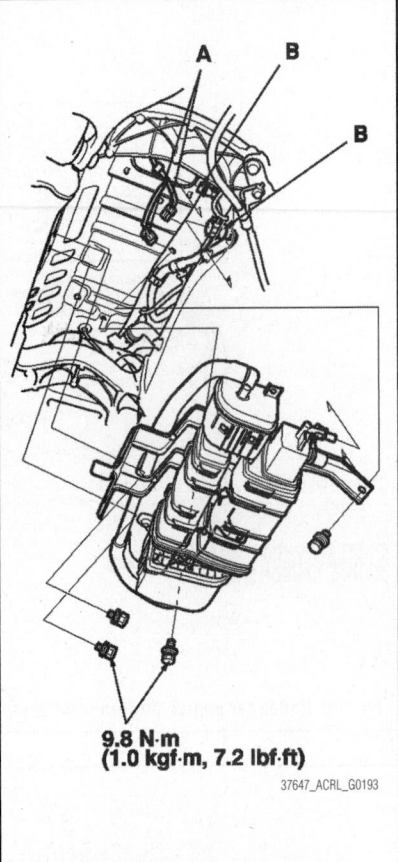

Fig. 69 Disconnect the connectors (A) and the hoses (B) and remove the EVAP canister.

EVAP CANISTER PURGE CONTROL VALVE

Location

The EVAP Purge Control Valve is located in the engine compartment.

Removal & Installation

See Figure 70.

1. Remove the engine cover.
2. Disconnect the hoses and the EVAP canister purge valve 2P connector.
3. Remove the EVAP canister purge valve and the bracket and remove the purge valve.
4. Installation is the reverse of the removal procedure.

EXHAUST GAS RECIRCULATION (EGR) VALVE

LOCATION

See Figure 71.

The Exhaust Gas Recirculation (EGR) valve is located near the front left side of the engine.

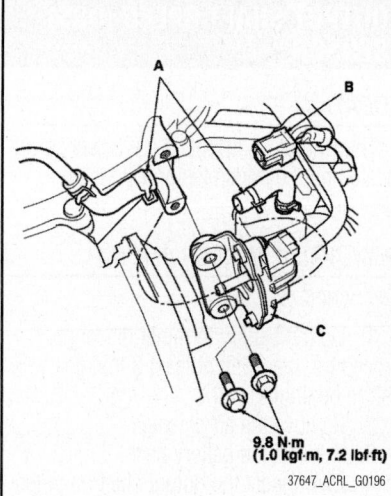

Fig. 70 Disconnect the hoses (A) and the EVAP canister purge valve 2P connector (B). Remove the EVAP canister purge valve and the bracket (C).

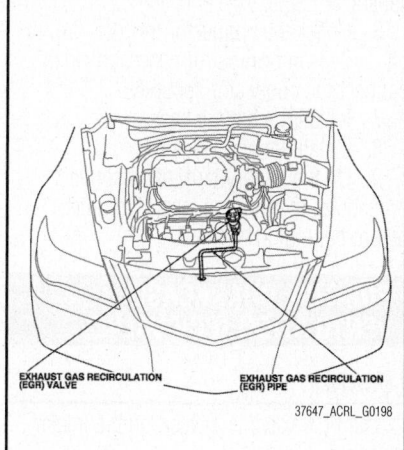

Fig. 71 Showing the underhood location of the EGR valve

REMOVAL & INSTALLATION

See Figure 72.

1. Remove the engine cover.
2. Disconnect the EGR valve 5P connector.
3. Remove the EGR valve and gasket.
4. Install the parts in the reverse order of removal with a new gasket.

HEATED OXYGEN (HO2S) SENSOR

REMOVAL & INSTALLATION

Front Bank (Bank 2)

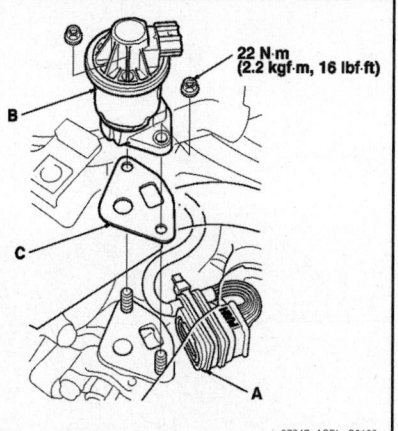

Fig. 72 Disconnect the EGR valve 5P connector (A), then remove the EGR valve (B) and gasket (C).

See Figure 73.

1. Disconnect the front secondary HO2S wiring connector, then remove the front secondary HO2S.
2. Install the parts in the reverse order of removal.

Rear Bank (Bank 1)

See Figure 74.

1. Disconnect the rear secondary HO2S wiring connector, then remove the rear secondary HO2S.
2. Install the parts in the reverse order of removal.

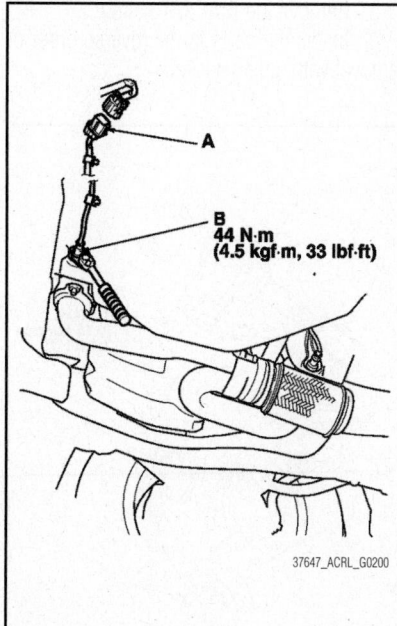

Fig. 73 Disconnect the front secondary HO2S wiring connector (A), then remove the front secondary HO2S (B)

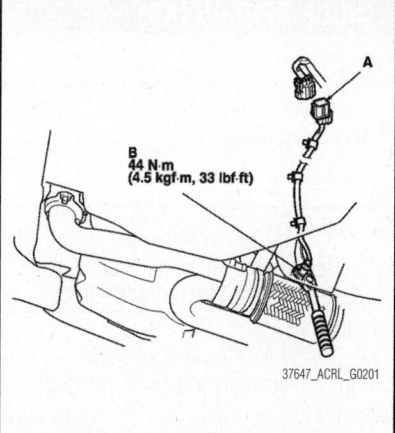

Fig. 74 Disconnect the rear secondary HO2S wiring connector (A), then remove the rear secondary HO2S (B)

INPUT SPEED SENSOR

REMOVAL & INSTALLATION

See Figure 75.

1. Raise the vehicle on a lift, or apply the parking brake, block both rear wheels, and raise the front of the vehicle.
2. Make sure it is securely supported.
3. Remove the splash shield.
4. Disconnect the input shaft (mainshaft) speed sensor connector, and remove the input shaft (mainshaft) speed sensor.

To install:

5. Install a new O-ring on a new input shaft (mainshaft) speed sensor, then install the input shaft (mainshaft) speed sensor.
6. Check the connector for corrosion, dirt, or oil, and clean or repair if necessary, then connect the connector securely.
7. Install the splash shield.

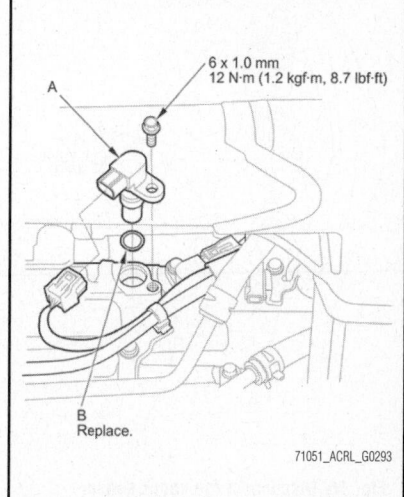

Fig. 75 Showing input shaft speed sensor (A) and O-ring (B)

KNOCK SENSOR (KS)

LOCATION

The knock sensor is located in the valley of the engine block.

REMOVAL & INSTALLATION

See Figure 76.

1. Remove the intake manifold.
2. Remove the injector rails and the injector base.
3. Disconnect the knock sensor wiring connector, then remove the knock sensor.
4. Installation is the reverse order of removal. Tighten the knock sensor to 23 ft. lbs. (31 Nm).

MANIFOLD ABSOLUTE PRESSURE (MAP) SENSOR

LOCATION

The Manifold Absolute Pressure (MAP) sensor is located on the left rear side of the engine, near the throttle body connection.

REMOVAL & INSTALLATION

See Figure 77.

1. Remove the engine appearance cover.
2. Disconnect the MAP sensor connector.
3. Remove the screw and the MAP sensor and O-ring.
4. Install the parts in the reverse order of removal with a new O-ring.

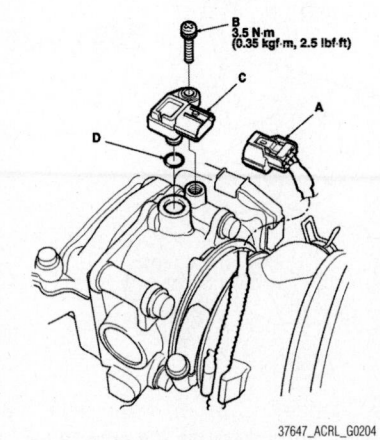

Fig. 77 Disconnect the MAP sensor connector. Remove the screw and the MAP sensor and O-ring.

MASS AIR FLOW/INTAKE AIR TEMPERATURE (MAF/IAT) SENSOR

LOCATION

The MAF/IAT sensor is located in the air intake near the air cleaner.

REMOVAL & INSTALLATION

See Figure 78.

1. Disconnect the MAF/IAT sensor connector.
2. Remove the screw.
3. Remove the MAF/IAT sensor.
4. Install the parts in the reverse order of removal with a new gasket.

OUTPUT SHAFT SPEED (OSS) SENSOR

LOCATION

The Output Shaft Speed (OSS) sensor is located on the transmission extension housing.

REMOVAL & INSTALLATION

See Figure 79.

1. Do the battery terminal disconnect procedure, then wait at least 3 minutes before beginning work.
2. Remove the air cleaner.
3. Remove the battery base.
4. Disconnect the output shaft (countershaft) speed sensor connector, and remove the output shaft (countershaft) speed sensor and O-ring.

To install:

5. Install the new O-ring on the new output shaft (countershaft) speed sensor, then install the output shaft (countershaft) speed sensor.
6. Check the connector for rust, dirt, or oil, and clean or repair if necessary, then connect the connector securely.
7. Install the battery base.
8. Install the air cleaner.
9. Do the battery terminal reconnect procedure, then wait at least 3 minutes before beginning work.

POSITIVE CRANKCASE VENTILATION (PCV VALVE

LOCATION

The PCV valve is located on the intake manifold.

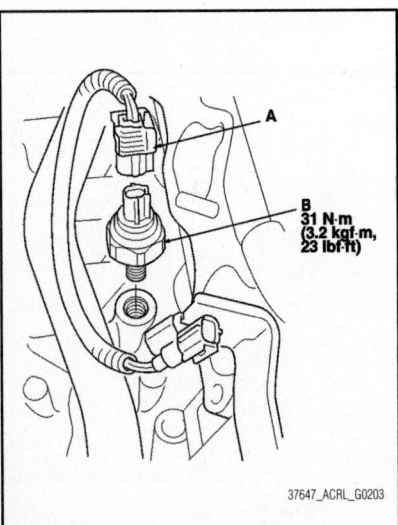

Fig. 76 Disconnect the knock sensor wiring connector (A), then remove the knock sensor (B)

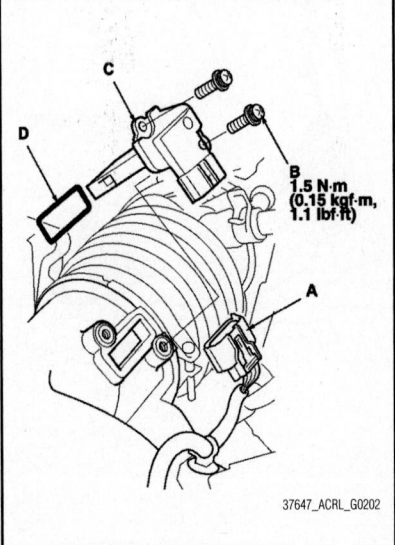

Fig. 78 Disconnect the MAF/IAT sensor connector

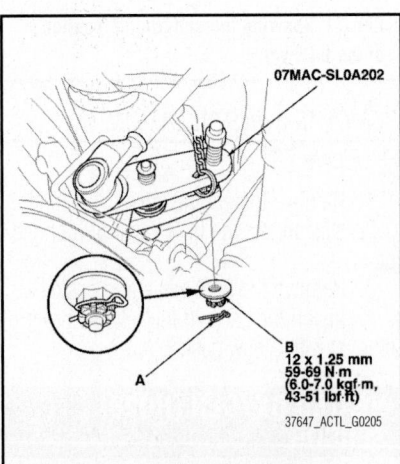

Fig. 79 Disconnect the output shaft (countershaft) speed sensor connector, and remove the output shaft (countershaft) speed sensor (A) and O-ring (B).

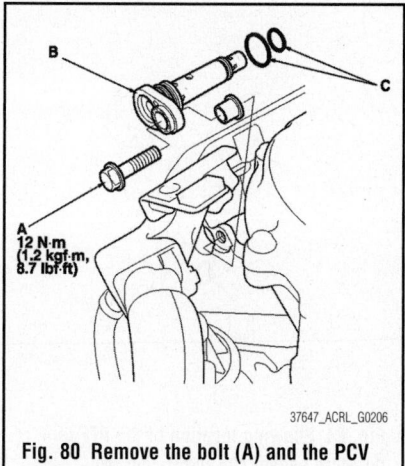

Fig. 80 Remove the bolt (A) and the PCV valve (B)

REMOVAL & INSTALLATION

See Figure 80.

1. Remove the engine cover.
2. Remove the bolt.

➥**Take care not to spill oil on the hot exhaust manifold.**

3. Remove the PCV valve.
4. Install the parts in the reverse order of removal.
5. When installing a new PCV valve, use new O-rings and make sure they are in place.

POWERTRAIN CONTROL MODULE (PCM)

LOCATION

See Figure 81.

The Powertrain Control Module (PCM) is located under the center of the instrument panel, behind the center console.

REMOVAL & INSTALLATION

➥**Make sure any scan tool being used has the latest software version.**

➥**If replacing the PCM after substituting a known-good PCM, reinstall the original PCM, then do this procedure.**

➥**During the procedure, if any READ DATA, WRITE DATA, or other data checks fail, note the failure, then continue.**

1. Connect the HDS, or equivalent scan tool, to the data link connector (DLC) located under the driver's side of the dashboard.
2. Turn the ignition switch to ON (II).
3. Make sure the scan tool communicates with the PCM and other vehicle systems.
   a. If it doesn't, go to the DLC circuit troubleshooting.

   b. If you are returning from DLC circuit troubleshooting, skip steps 4 through 9, 17 through 19, 22, and 23, then do this after replacing the PCM:
4. Replace the engine oil and the engine oil filter.
5. Clean the throttle body.
6. Select the PGM-FI system with the scan tool.
7. Select the INSPECTION MENU with the scan tool.
8. Select the ETCS TEST, then select the TP POSITION CHECK, and follow the screen prompts.

➥**If the TP POSITION CHECK indicates FAILED, continue with this procedure.**

9. Select the REPLACE PCM MENU, then select READ DATA, and follow the screen prompts.

➥**Doing this step copies (READS) the engine oil life data from the original PCM so you can later download (WRITE) it into the new PCM.**

10. If the READ DATA indicates FAILED, continue with this procedure.
11. Turn the ignition switch to LOCK (0).
12. Do the battery terminal disconnect procedure, then wait at least 3 minutes before beginning work.
13. Remove the front console covers, and pull back the carpet.
14. Remove the ducts.
15. Disconnect the PCM connectors.
16. Remove the bolts, then remove the PCM.

To install:

17. Install the parts in the reverse order of removal.
18. Do the battery terminal reconnect procedure.
19. Turn the ignition switch to ON (II).
20. Manually input the VIN to the PCM with the scan tool.

➥**DTC P0630 VIN Not Programmed or Mismatch may be stored because the VIN has not been programmed into the PCM, ignore it, and continue this procedure.**

21. If the READ DATA (engine oil life) failed in step 7, go to step 21. Otherwise, go to step 19.
22. Select the PGM-FI system with the HDS.
23. Select the REPLACE PCM MENU, then select WRITE DATA, and follow the screen prompts.

➥**If the WRITE DATA indicates FAILED, continue this procedure.**

24. Select IMMOBI system with the scan tool.
25. Enter the immobilizer PCM code that you got from the software, and use the PCM replacement procedure in the scan tool; it allows you to start the engine.
26. If the TP POSITION CHECK failed in step 6, clean the throttle body, then go to step 24.
27. If the READ DATA failed in step 7 or the WRITE DATA failed in step 20, replace the engine oil and engine oil filter, then go to step 25.

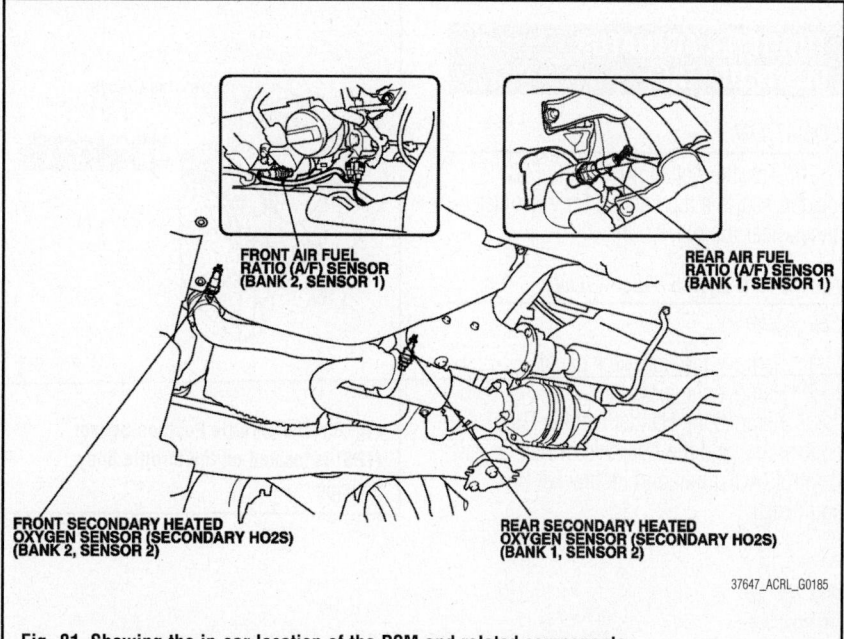

Fig. 81 Showing the in-car location of the PCM and related components

28. Select PGM-FI system, and reset the PCM with the scan tool.

29. Update the PCM if it does not have the latest software.

30. Do the PCM idle learn procedure.

31. Do the CKP Clear/Learn procedure.

PCM IDLE LEARN PROCEDURE

The idle learn procedure must be done so the PCM can learn the engine idle characteristics.

1. Do the idle learn procedure whenever you do any of these actions:

- Replace the PCM.
- Reset the PCM.
- Update the PCM.
- Replace or clean the throttle body.
- Disassemble the engine or the transmission.

➡**Clearing the DTCs with the scan tool does not require you to do the idle learn procedure.**

2. Make sure all electrical items (the A/C, the audio, the rear window defogger, the lights, etc.) are off.

3. Reset the PCM with the scan tool.

4. Turn the ignition switch to ON (II), and wait 2 seconds.

5. Start the engine. Hold the engine speed at 3,000 RPM without load (in P or N) until the radiator fan comes on, or until the engine coolant temperature reaches 194°F (90°C).

6. Let the engine idle for about 5 minutes with the throttle fully closed.

7. If the radiator fan comes on, do not include its running time in the 5 minutes.

THROTTLE ACTUATOR CONTROL (TAC)

LOCATION

The Throttle Actuator Control (TAC) is located near the top edge of the right kick panel, near the blower unit.

REMOVAL & INSTALLATION

See Figure 82.

1. Remove the right kick panel.
2. Remove the glove box.
3. Remove the HandsFreeLink (HFL) control unit and the Adaptive Cruise Control (ACC) unit and its bracket (if equipped).

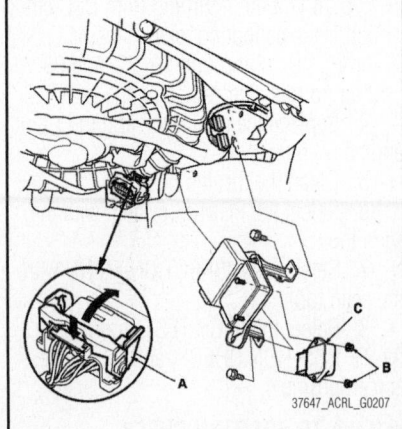

Fig. 82 Disconnect the throttle actuator control module 16P connector (A). Remove the nuts (B) and the throttle actuator control module (C).—with ACC shown; without ACC similar

4. Disconnect the throttle actuator control module 16P connector.

5. Remove the nuts and the throttle actuator control module.

6. Install the parts in the reverse order of removal.

THROTTLE POSITION SENSOR (TPS)

LOCATION

See Figure 83.

The Throttle Position Sensor (TPS) is located on the throttle body housing.

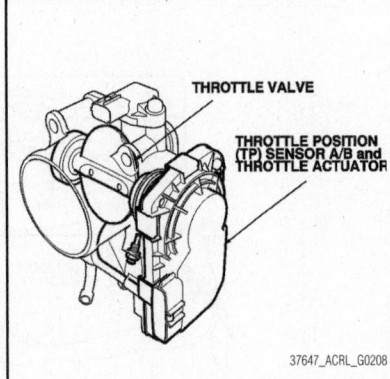

Fig. 83 The Throttle Position Sensor (TPS) is located on the throttle body housing

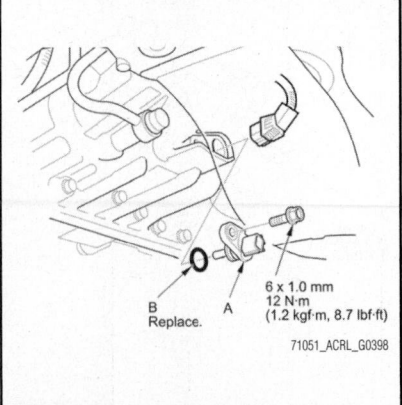

Fig. 84 Showing location of the ATF temperature sensor (A) and O-ring (B)

REMOVAL & INSTALLATION

➡**Manufacturer does not provide a specific removal and installation procedure for this component. Use the location illustration as a guide when servicing this component.**

TRANSMISSION FLUID TEMPERATURE (TFT) SENSOR

LOCATION

See Figure 84.

REMOVAL & INSTALLATION

1. Raise the vehicle on a lift, or apply the parking brake, block both rear wheels, and raise the front of the vehicle.

2. Make sure it is securely supported.

3. Remove the splash shield.

4. Remove the drain plug and sealing washer and drain the automatic transmission fluid (ATF).

5. Reinstall the drain plug with a new sealing washer.

6. Disconnect the ATF temperature sensor connector, then remove the ATF temperature sensor.

To install:

7. Install a new O-ring on a new ATF temperature sensor, and install the ATF temperature sensor.

8. Check the connector for corrosion, dirt, or oil, and clean or repair if necessary, then connect the connector securely.

9. Install the splash shield.

10. Refill the transmission with ATF.

FUEL   **GASOLINE FUEL INJECTION SYSTEM**

FUEL SYSTEM SERVICE PRECAUTIONS

Safety is the most important factor when performing not only fuel system maintenance but any type of maintenance. Failure to conduct maintenance and repairs in a safe manner may result in serious personal injury or death. Maintenance and testing of the vehicle's fuel system components can be accomplished safely and effectively by adhering to the following rules and guidelines.

• To avoid the possibility of fire and personal injury, always disconnect the negative battery cable unless the repair or test procedure requires that battery voltage be applied.

• Always relieve the fuel system pressure prior to disconnecting any fuel system component (injector, fuel rail, pressure regulator, etc.), fitting or fuel line connection. Exercise extreme caution whenever relieving fuel system pressure to avoid exposing skin, face and eyes to fuel spray. Please be advised that fuel under pressure may penetrate the skin or any part of the body that it contacts.

• Always place a shop towel or cloth around the fitting or connection prior to loosening to absorb any excess fuel due to spillage. Ensure that all fuel spillage (should it occur) is quickly removed from engine surfaces. Ensure that all fuel soaked cloths or towels are deposited into a suitable waste container.

• Always keep a dry chemical (Class B) fire extinguisher near the work area.

• Do not allow fuel spray or fuel vapors to come into contact with a spark or open flame.

• Always use a back-up wrench when loosening and tightening fuel line connection fittings. This will prevent unnecessary stress and torsion to fuel line piping.

• Always replace worn fuel fitting O-rings with new Do not substitute fuel hose or equivalent where fuel pipe is installed.

Before servicing the vehicle, make sure to also refer to the precautions in the beginning of this section as well.

RELIEVING FUEL SYSTEM PRESSURE

See Figure 85.

1. Remove the left kick panel.
2. Remove PGM-FI main relay 2 (FUEL PUMP) from the driver's under-dash fuse/relay box.

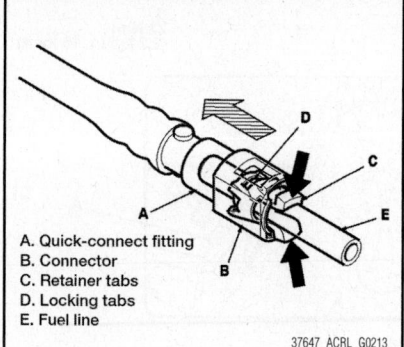

A. Quick-connect fitting
B. Connector
C. Retainer tabs
D. Locking tabs
E. Fuel line

37647_ACRL_G0213

Fig. 85 Disconnect the quick-connect fitting (A): Hold the connector (B) with one hand, and squeeze the retainer tabs (C) with the other hand to release them from the locking tabs (D). Pull the connector off. Be careful not to damage the line (E) or other parts.

3. Start the engine, and let it idle until it stalls.

➡ **If any DTCs are stored, clear and ignore them.**

4. Turn the ignition switch to LOCK (0).
5. Remove the fuel fill cap to relieve the pressure in the fuel tank.
6. Do the battery terminal disconnect procedure, then wait at least 3 minutes before beginning work.
7. Remove the quick-connect fitting cover.
8. Check the fuel quick-connect fitting for dirt, and clean it if needed.
9. Place a rag or a shop towel over the quick-connect fitting.
10. Disconnect the quick-connect fitting as follows:

a. Hold the connector with one hand, and squeeze the retainer tabs with the other hand to release them from the locking tabs.

b. Pull the connector off. Be careful not to damage the line or other parts.

➡ **Do not use tools.**

11. If the connector does not move, keep the retainer tabs pressed down, and alternately pull and push the connector until it comes off easily.
12. Do not remove the retainer from the line; once removed, the retainer must be replaced with a new one. After disconnecting the quick-connect fitting, check it for dirt or damage.

FUEL FILTER

REMOVAL & INSTALLATION
See Figure 86.

The fuel filter should be replaced whenever the fuel pressure drops below the specified value, after making sure that the fuel pump and the fuel pressure regulator are OK.

1. If the fuel tank is full, drain the fuel.
2. Remove the fuel tank unit.
3. Remove the fuel filter set.
4. Check these items before installing the fuel tank unit:

a. When connecting the wire harness, make sure the connection is secure and the connectors are firmly locked into place.

b. When installing the fuel gauge sending unit, make sure the connection is secure and the connector is firmly locked into place. Be careful not to bend or twist it excessively.

To install:
5. Install the parts in the reverse order of removal with a new base gasket and a new O-ring.
6. Coat the O-rings with clean engine oil; do not use any other oils or fluids.
7. Do not pinch the O-rings during installation.
8. Use all the new parts supplied in the fuel filter replacement kit.

FUEL RAIL & INJECTORS

REMOVAL & INSTALLATION
See Figure 87.

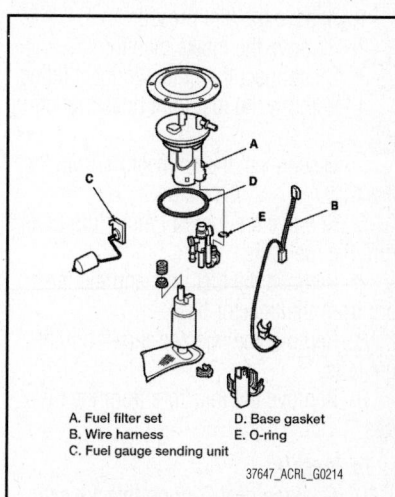

A. Fuel filter set
B. Wire harness
C. Fuel gauge sending unit
D. Base gasket
E. O-ring

37647_ACRL_G0214

Fig. 86 Showing the component of the fuel tank unit, with the fuel filter

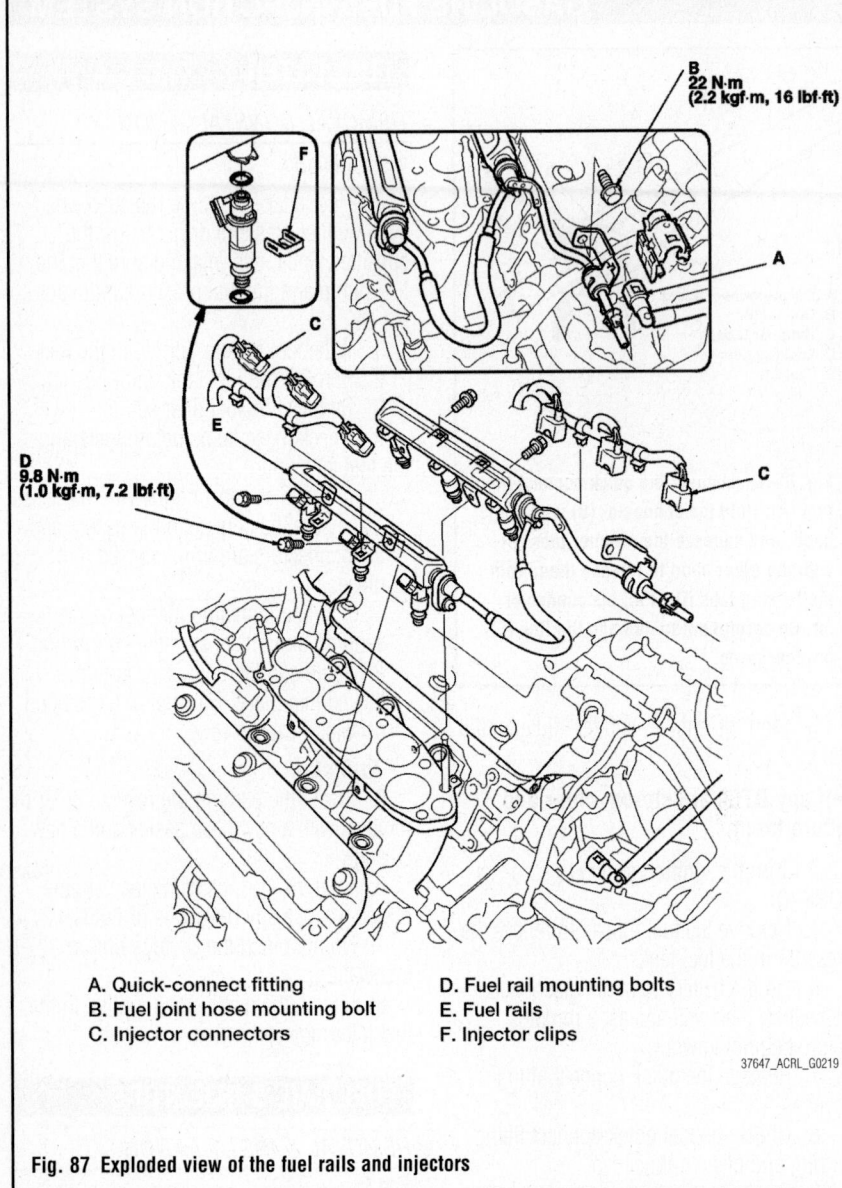

B
22 N·m
(2.2 kgf·m, 16 lbf·ft)

A

D
9.8 N·m
(1.0 kgf·m, 7.2 lbf·ft)

C

A. Quick-connect fitting
B. Fuel joint hose mounting bolt
C. Injector connectors
D. Fuel rail mounting bolts
E. Fuel rails
F. Injector clips

37647_ACRL_G0219

Fig. 87 Exploded view of the fuel rails and injectors

1. Relieve the fuel pressure.
2. Remove the intake manifold.
3. Disconnect the quick-connect fitting.
4. Remove the fuel joint hose mounting bolt.
5. Disconnect the connectors from the injectors.
6. Remove the fuel rail mounting bolts from the fuel rails.
7. Remove the fuel rails and the injectors from the injector base.
8. Remove the injector clips from the fuel rails.
9. Remove the injectors from the fuel rails.

To install:

10. Coat the new O-rings (black) with clean engine oil, and insert the injectors into the fuel rails.

11. Install the injector clips.
12. Coat the new injector O-rings (green) with clean engine oil.
13. Install the fuel rails and the injectors in the injector base.
14. Install the fuel rail mounting bolts, and connect the injector connectors.
15. Install the fuel joint hose mounting bolt.
16. Connect the quick-connect fitting.
17. Turn the ignition switch to ON (II), but do not operate the starter. After the fuel pump runs for about 2 seconds, the fuel rail will be pressurized. Repeat this two or three times, then check for fuel leaks.
18. Install the intake manifold with a new gasket.

FUEL TANK

DRAINING

1. Remove the secondary fuel gauge sending unit.
2. Using a hand pump, a hose, and a container suitable for fuel, draw the fuel from the secondary fuel gauge sending unit side of the tank.
3. Remove the fuel tank unit.
4. Using a hand pump, a hose, and a container suitable for fuel, draw the fuel from the fuel tank unit side of the tank.

REMOVAL & INSTALLATION

1. Drain the fuel tank.
2. Reinstall the fuel tank unit and the secondary fuel gauge sending unit without connecting the fuel tank unit 5P connector, the secondary fuel gauge sending unit 5P connector and the quick-connect fitting.
3. Remove the propeller shaft.
4. Remove the rear differential.
5. Raise the vehicle on a lift.
6. Remove the fuel tank covers and parking brake cable bolts.
7. Remove the clip from the EVAP canister bracket.
8. Disconnect the hoses. Slide back the clamps, then twist the hoses as you pull to avoid damaging them.
9. Place a jack or other support under the tank.
10. Remove the strap bolts and the straps.
11. Remove the fuel tank. If it sticks to the undercoat on its mount, carefully pry it off the mount.

To install:

12. Install the parts in the reverse order of removal.

➡ **The new fuel vent hose and the fuel vent return hose have ring pull at the connectors. When you connect the hoses and confirm that the connections are secure, remove the ring pulls by pulling them down.**

13. Before connecting the fuel fill pipe and the quick-connect fitting, check for dirt, and clean it if needed, taking care not to damage the fuel fill pipe and other parts.

THROTTLE BODY

REMOVAL & INSTALLATION

See Figure 88.

❊❊ WARNING

Do not insert your fingers into the installed throttle body when you turn

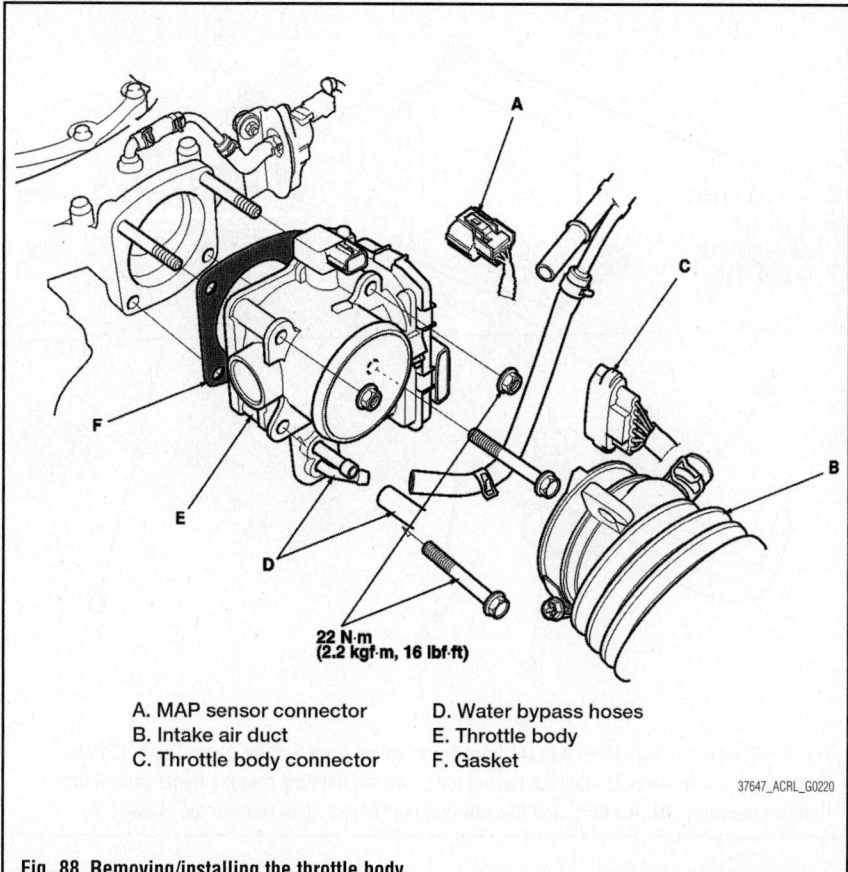

22 N·m
(2.2 kgf·m, 16 lbf·ft)

A. MAP sensor connector
B. Intake air duct
C. Throttle body connector
D. Water bypass hoses
E. Throttle body
F. Gasket

37647_ACRL_G0220

Fig. 88 Removing/installing the throttle body

the ignition switch to ON (II) or while the ignition switch is ON (II). If you do, you will seriously injure your fingers if the throttle valve is activated.

➡If you are replacing the throttle body, begin at step 1. If you are removing the throttle body temporarily, begin at step 4.

1. Connect the scan tool while the engine is stopped.
2. Select the INSPECTION MENU with the scan tool.
3. Do the TP POSITION CHECK in the ETCS TEST.
4. Turn the ignition switch to ON (II).
5. Disconnect the MAP sensor connector.
6. Remove the intake air duct.
7. Disconnect the throttle body connector.
8. Disconnect and plug the water bypass hoses.
9. Remove the throttle body.

To install:
10. Install the parts in the reverse order of removal with a new gasket, then do this:
   a. Refill the radiator with engine coolant.
   b. Do the PCM idle learn procedure.

HEATING & AIR CONDITIONING SYSTEM

BLOWER MOTOR

REMOVAL & INSTALLATION

See Figures 89 and 90.

1. Read the Precautions information in this section before beginning work.
2. Remove the glove box.
3. Remove the right kick panel.
4. Remove the USB adapter unit.
5. Cut the plastic cross brace in the glove box opening with diagonal cutters in the area shown. Retain the plastic cross brace to be reinstalled later.

➡Use the grommets and the self-tapping screws to reinstall the plastic cross brace.

6. Remove the wire harness clips and the connector clips and then remove the bolts and the glove box frame.
7. Disconnect the connectors from the blower motor, the throttle actuator control module subharness, and the dashboard wire harnesses.
8. Disconnect the connector from the recirculation control motor. Remove the

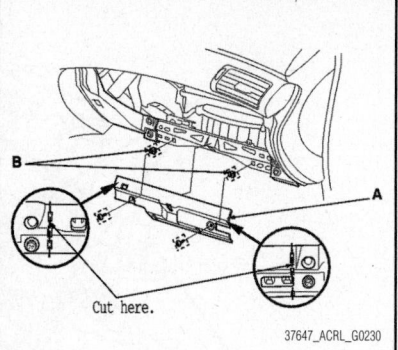

Cut here.

37647_ACRL_G0230

Fig. 89 Cut the plastic cross brace (A) in the glove box opening with diagonal cutters in the area shown. Retain the plastic cross brace to be reinstalled later. Use the grommets (B) and the self-tapping screws to reinstall the plastic cross brace.

mounting nuts, the bolts, and the blower unit.
9. Install the unit in the reverse order of removal. Make sure that there is no air leakage.

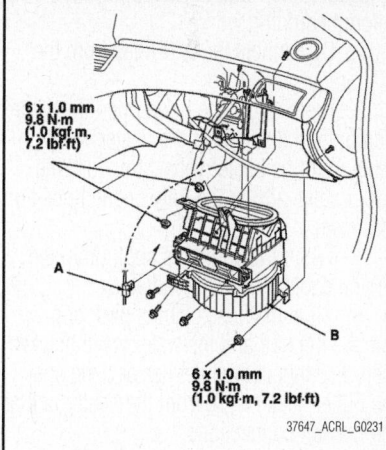

6 x 1.0 mm
9.8 N·m
(1.0 kgf·m,
7.2 lbf·ft)

6 x 1.0 mm
9.8 N·m
(1.0 kgf·m, 7.2 lbf·ft)

37647_ACRL_G0231

Fig. 90 Disconnect the connector (A) from the recirculation control motor. Remove the mounting nuts, the bolts, and the blower unit (B).

HEATER CORE

REMOVAL & INSTALLATION

See Figures 91 through 95.

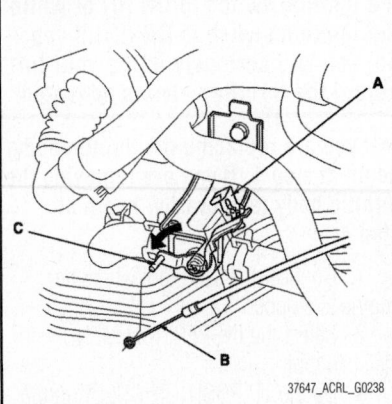

Fig. 91 From under the hood open the cable clamp (A), then disconnect the heater valve cable (B) from the heater valve arm (C). Turn the heater valve arm to the fully opened position as shown.

➡The heater unit, in this application, is the heater and A/C unit housing. It also contains the evaporator core.

❊❊ WARNING

SRS components are located in this area. Review the SRS component locations and the precautions and procedures before doing repairs or service.

  1. Do the battery terminal disconnect procedure, then wait at least 3 minutes before beginning work.
  2. Disconnect the A/C lines from the evaporator core.
  3. From under the hood open the cable clamp, then disconnect the heater valve cable from the heater valve arm. Turn the heater valve arm to the fully opened position as shown.
  4. When the engine is cool, drain the engine coolant from the radiator.
  5. Slide the hose clamps back and remove the bolt and the water valve bracket, then disconnect the inlet heater hose and the outlet heater hose from the heater unit at the firewall connections.

❊❊ CAUTION

Engine coolant will run out when the hoses are disconnected; drain it into a clean drip pan. Be sure not to let coolant spill on the electrical parts or the painted surfaces. If any coolant spills, rinse it off immediately.

  6. Remove the firewall mounting nut that retains the heater unit. Take care not to damage or bend the fuel lines or the brake lines.

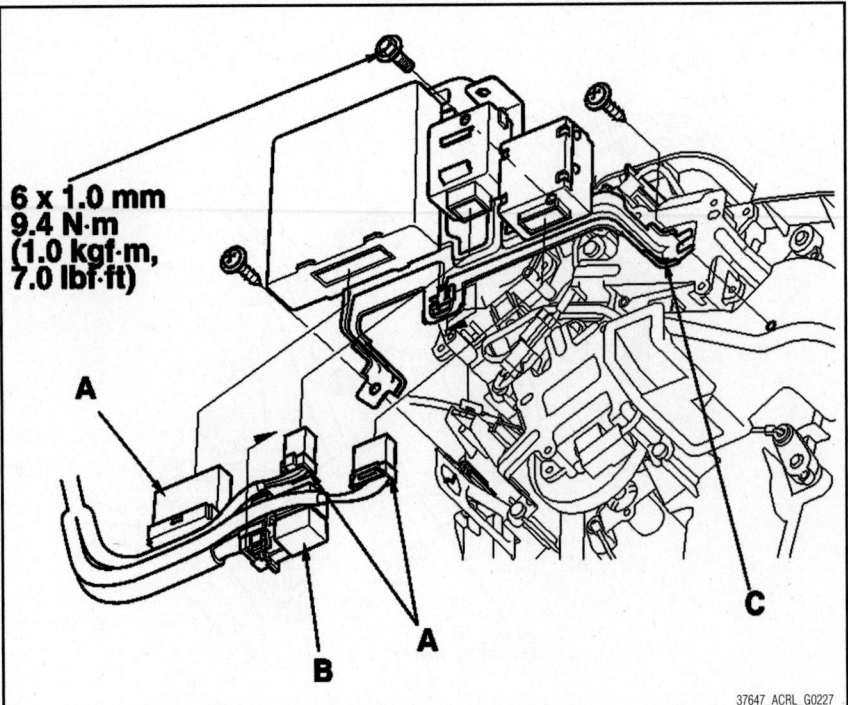

6 x 1.0 mm 9.4 N·m (1.0 kgf·m, 7.0 lbf·ft)

Fig. 92 Disconnect the connectors (A) from the adaptive front lighting control unit, the electronically controlled power steering control unit, and the daytime running lights control unit. Remove the relay (B), the bolt, and the self-tapping screws, then remove the bracket (C).

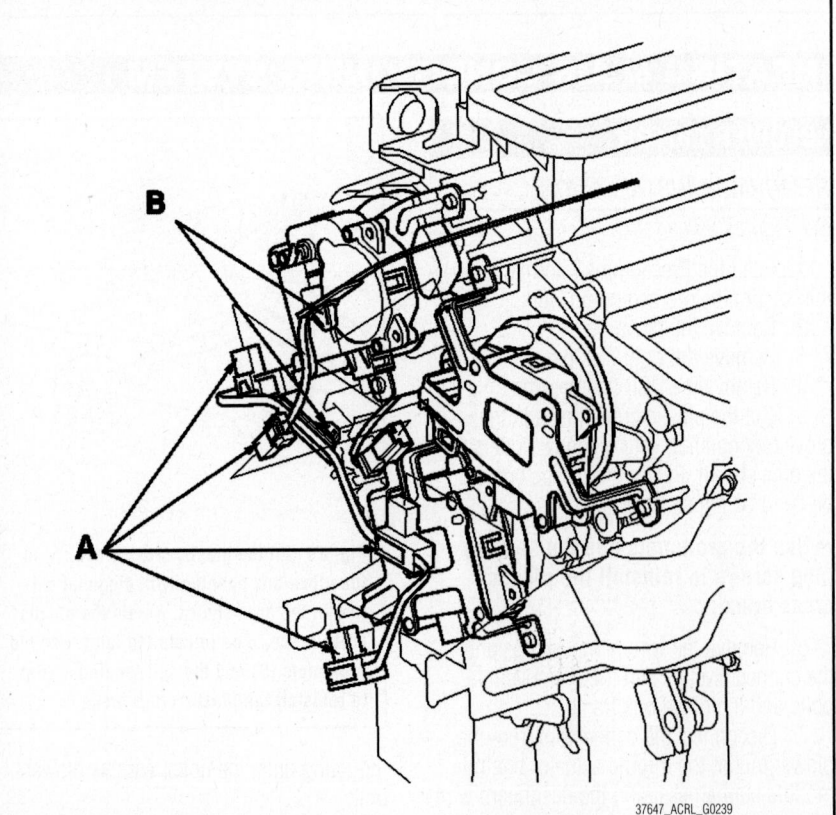

Fig. 93 Disconnect the connectors (A) from the driver's cool vent control motor, the driver's air mix control motor, the mode control motor, and the evaporator temperature sensor, then remove the wire harness clips (B).

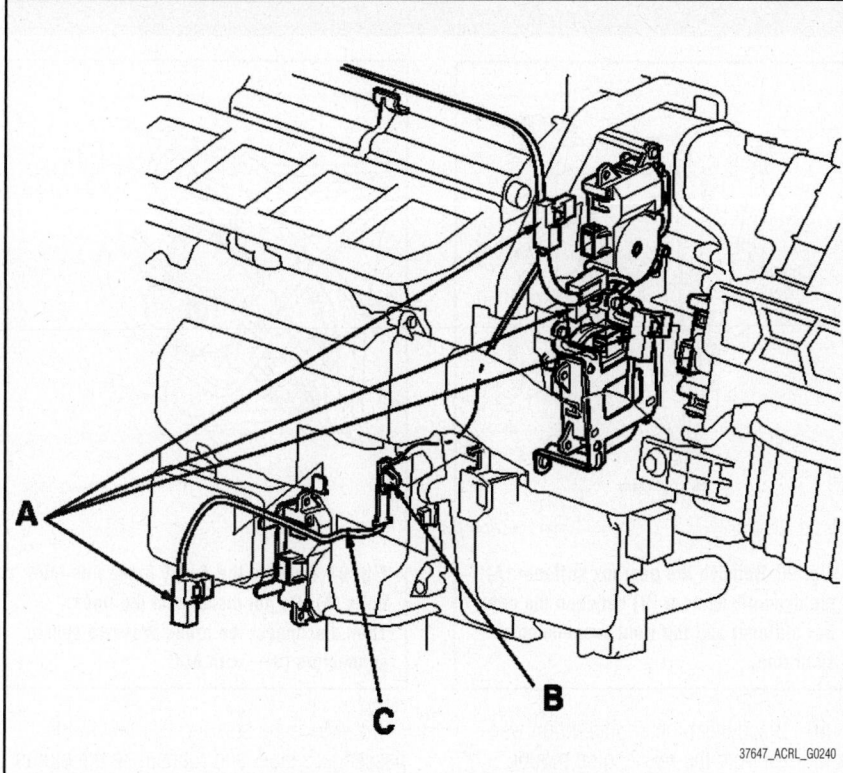

37647_ACRL_G0240

Fig. 94 Disconnect the connectors (A) from the recirculation control motor, the passenger's cool vent control motor, the passenger's air mix control motor, and the rear vent control motor, and then remove the wire harness clip (B) and the wire harness (C).

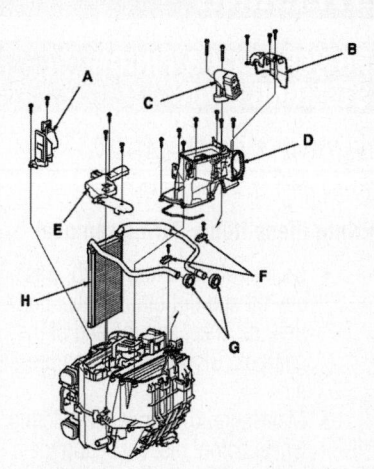

A. Heater air duct
B. Expansion valve cover
C. Expansion valve
D. Joint duct
E. Heater core cover
F. Heater pip brackets
G. Grommets
H. Heater core

37647_ACRL_G0242

Fig. 95 Remove the self-tapping screws and the passenger's heater duct, then remove the expansion valve cover, the bolts, and the expansion valve. Remove the self-tapping screws, the joint duct and the heater core cover . Remove the self-tapping screws, the heater pipe brackets and the grommets, then carefully pull out the heater core.

7. Remove the dashboard.

8. Disconnect the connectors from the adaptive front lighting control unit, the electronically controlled power steering control unit, and the daytime running lights control unit. Remove the relay, the bolt, and the self-tapping screws, then remove the bracket.

9. Disconnect the connectors from the driver's cool vent control motor, the driver's air mix control motor, the mode control motor, and the evaporator temperature sensor, then remove the wire harness clips.

10. Disconnect the connectors from the blower motor, the power transistor, the control motor relay, the throttle actuator control module subharness, and the dashboard wire harnesses. Remove the self-tapping screw, the connector clip and the wire harness clips.

11. Disconnect the connectors from the recirculation control motor, the passenger's cool vent control motor, the passenger's air mix control motor, and the rear vent control motor, and then remove the wire harness clip and the wire harness.

12. Remove the clips, the mounting nuts, and the blower-heater unit.

13. Remove the self-tapping screws and the passenger's heater duct, then remove the expansion valve cover, the bolts, and the expansion valve. Remove the self-tapping screws, the joint duct, and the heater core cover. Remove the self-tapping screws, the heater pipe brackets, and the grommets, then carefully pull out the heater core.

To install:

14. Install the heater core and the evaporator core in the reverse order of removal.

15. Install the heater unit in the reverse order of removal, and note these items:

a. Do not interchange the inlet and the outlet heater hoses, and install the hose clamps securely.

b. Refill the cooling system with engine coolant.

c. Adjust the heater valve cable.

d. Make sure that there is no coolant leakage.

e. Make sure that there is no air leakage.

16. Do the battery terminal reconnect procedure.

STEERING

POWER RACK & PINION STEERING GEAR

REMOVAL & INSTALLATION

See Figures 96 through 98.

➡Note these items during removal:

• Using solvent and a brush, wash any oil and dirt off the valve body unit, its lines, and the end of the gearbox. Blow dry with compressed air.
• Make sure to remove the steering wheel before disconnecting the steering joint. Damage to the cable reel can occur.

1. Do the battery terminal disconnect procedure, then wait at least 3 minutes before beginning work.
2. Remove the left and right upper fender trim and left upper trim.
3. Drain the power steering fluid.
4. Remove the underhood relay box, then release the wire harness clips.
5. Remove the front strut brace.
6. Raise the front of the vehicle, and support it with safety stands in the proper locations.
7. Remove the driver's air bag, and the steering wheel.
8. Remove the steering joint cover at the floorboard.
9. Remove the steering joint bolts, then disconnect the steering joint from the pinion shaft and the steering column shaft.

➡Do not turn the steering column shaft after disconnecting the steering joint.

10. Remove the steering shaft center guide. Remove steering joint cover. Be careful not to damage the mating surface on joint cover and the pinion shaft grommet. Replace the cover seal if necessary.
11. Remove the pinion shaft grommet from the top of the valve body unit. Apply vinyl tape to the splines on the pinion shaft.
12. Remove and discard the cotter pin from the tie-rod ball joint nut, and loosen the nut.
13. Separate the tie-rod ball joint from the steering knuckle.
14. Remove the nuts from the rear engine mount.
15. Remove the rear engine mount and the engine mount vacuum hose holder mounting bolt on the gearbox stiffener.

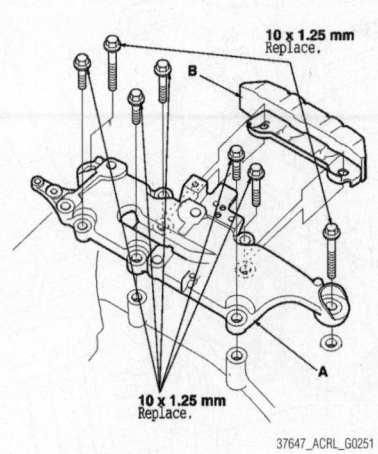

Fig. 96 Remove the gearbox stiffener (A) the dynamic damper (B) between the gearbox stiffener and the front suspension subframe.

16. Disconnect the engine mount vacuum hose from the rear engine mount.
17. Remove the feed line holder mounting bolt and A/T shift cable holder mounting bolt on the gearbox stiffener. Disconnect the return hose from the clamp.
18. Remove the p/s fluid feed line holder mounting bolt and return line holder mounting bolt on the gearbox.
19. Disconnect the Electronically Controlled Power Steering (ECPS) valve motor connector.
20. Place several shop towels under the line connections, and cover the gearbox mounting cushion to protect it from the power steering fluid.
21. Loosen the adjustable hose clamp and disconnect the return hose from the steering gearbox.
22. Loosen the 14 mm flare nut and disconnect the feed line from the steering gearbox.
23. Remove the return line connector nipple from the steering gearbox.
24. Remove the P/S gearbox heat shield.
25. Remove the gearbox stiffener, the dynamic damper between the gearbox stiffener and the front suspension subframe.
26. Remove the steering gearbox mounting bolt and washer on the right gearbox mount. Repeat for the other side.
27. Remove the 4-way brake line joint bolts. Do not disconnect the lines. Then disconnect the brake pressure sensor connectors if equipped with Adaptive Cruise Control (ACC).

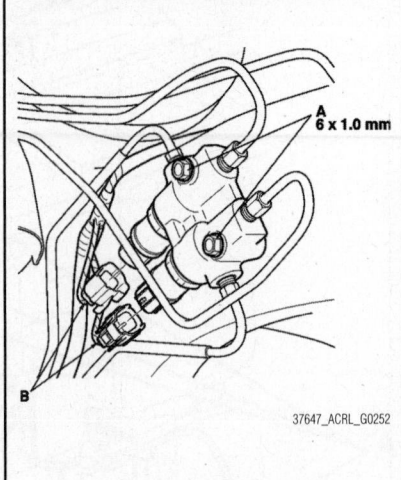

Fig. 97 Remove the 4-way brake line joint bolts (A). Do not disconnect the lines. Then disconnect the brake pressure sensor connectors (B)—with ACC

28. Move the steering gearbox to the passenger's side, and rotate it so the pinion shaft points toward the front of the vehicle.
29. Carefully move the steering gearbox as an assembly toward the left side of the vehicle until the pinion shaft clears the wheel well opening. Be careful not to damage the brake lines with the pinion shaft.
30. Lift the passenger's side of the gearbox, and remove the steering gearbox through the wheel well opening on the driver's side as shown by the arrow.
31. After removing the steering gearbox, make sure that no power steering fluid gets on the gearbox mount cushions, the gearbox housing, the surface of the front suspension the subframe, and the stiffener. Wipe off any spilled fluid at once.

To install:

✳✳ CAUTION

Do not turn the steering column shaft. If the column shaft is turned, do the "VSA Sensor Neutral Position Memorization" with the scan tool after installing the steering gearbox.

32. Before installing the steering gearbox, make sure that no power steering fluid is on the mating surface of the gearbox and front suspension subframe. To prevent the gearbox mounting bolts from loosening after the installation, remove any power steering fluid from the mount cushions and bolt holes.

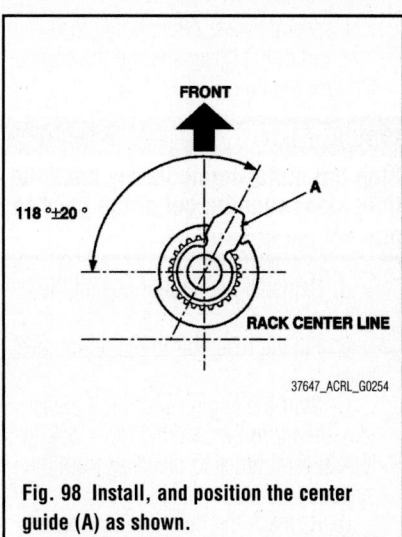

FRONT

118 °±20 °

A

RACK CENTER LINE

37647_ACRL_G0254

Fig. 98 Install, and position the center guide (A) as shown.

33. Apply vinyl tape to the splines on the pinion shaft for protection.

34. Apply a mild soap and water solution to both sides of the mount cushion mating surfaces.

35. Lift and pass the passenger's side of the steering gearbox through the wheel well opening on the driver's side.

36. Carefully move the steering gearbox toward the passenger's side until the pinion shaft clears the wheel well opening on the body.

37. Rotate the steering gearbox so the pinion shaft points upward, then move the steering gearbox to the driver's side.

38. Install the 4-way brake line joint. If equipped with ACC, reconnect the brake pressure sensor connectors.

39. Install the new steering gearbox mounting bolt and washer on the left front gearbox mount, then tighten the steering gearbox mounting bolt to 58 ft. lbs. (78 Nm). Repeat on the right side.

40. Install the gearbox stiffener and the dynamic damper, then tighten the new mounting bolts to 25 ft. lbs. (34 Nm).

41. Install the new steering gearbox mounting bolts, then tighten the steering gearbox mounting bolts to 58 ft. lbs. (78 Nm).

42. Install the P/S heat shield.

43. Install the return line connector nipple in the steering gearbox, then tighten it to 21 ft. lbs. (28 Nm).

44. Connect the feed line and tighten the 14 mm flare nut to 27 ft. lbs. (37 Nm).

45. Connect the return hose securely, and tighten the adjustable hose clamp.

46. Connect the ECPS valve motor connector.

47. Install the feed line holder mounting bolt and return line holder mounting bolt on the gearbox.

48. Install the feed line holder mounting bolt and A/T shift cable holder mounting bolt on the gearbox stiffener.

49. Install the return hose in the clamp.

50. Connect the engine mount vacuum hose in the rear engine mount.

51. Install the rear engine mount, then tighten the new mounting bolts and the engine mount vacuum hose holder mounting bolt on the gearbox stiffener to 40 ft. lbs. (54 Nm).

52. Install the nuts on the rear engine mount, then tighten them to 47 ft. lbs. (64 Nm).

53. Wipe off any grease from the ball joint tapered section and threads, then reconnect the tie-rod end to the knuckle arms.

54. Install the tie-rod end ball joint nut, and tighten it to 43 ft. lbs. (59 Nm), then install the new cotter pin.

55. Install the front wheels, and set them in the straight ahead position.

56. Remove the vinyl tape from the pinion shaft, and install the pinion shaft grommet. Align the slot in the pinion shaft grommet with the lug portion on the valve housing.

57. Install the new cover seal all the way around in the steering joint cover. Make sure there are no wrinkles in the seal, then install the steering joint cover.

58. Install, and position the center guide as shown.

59. Install the steering joint, aligning the slit of the steering joint with the tab of the center guide.

60. Align the bolt hole on the steering joint with the groove around the pinion shaft, and loosely install the joint bolts. Be sure that the joint bolt is securely in the groove in the pinion shaft. Pull on the steering joint to make sure that the steering joint is fully seated. Tighten the steering joint bolts to 21 ft. lbs. (28 Nm).

61. Install the steering joint cover at the floorboard.

62. Install the relay box, then connect the wire harness clips.

63. Install the front strut brace.

64. Install the steering wheel and the driver's air bag.

65. Do the battery terminal reconnect procedure.

66. Do these tasks:
   a. Turn the ignition switch to ON (II) and check that the SRS indicator comes on for about 6 seconds and then goes off.

   b. Make sure the horn and the turn signal switches work properly.

   c. Make sure the steering wheel switches work properly.

   d. Fill the system with power steering fluid, and bleed air from the system.

67. After installation, do the following checks.

   a. Start the engine, allow it to idle, and turn the steering wheel from lock-to-lock several times to warm up the fluid. Check the gearbox for leaks.

   b. Check that the ECPS indicator does not come on.

   c. Check the steering wheel spoke angle. If the steering spoke angles to the right and left are not equal (steering wheel and rack are not centered), correct the engagement of the joint/pinion shaft serrations.

68. Set the steering column to the center tilt position, then do the front toe inspection.

69. Install the left and right upper fender trim and left upper trim.

POWER STEERING PUMP

REMOVAL & INSTALLATION

1. Place a suitable container under the vehicle.

2. Remove the right upper fender trim.

3. Drain the power steering fluid from the reservoir.

4. Remove the drive belt from the pump pulley.

5. Cover the auto-tensioner, the alternator, and the A/C compressor with several shop towels to protect them from spilled power steering fluid. Disconnect the pump inlet hose and the pump outlet hose from the pump, and plug them.

❊❊ CAUTION

Take care not to spill the fluid on the body or parts. Wipe off any spilled fluid at once. Do not turn the steering wheel with the pump removed.

6. Cover the opening of the pump with a piece of tape to prevent foreign material from entering the pump.

7. Remove the pump mounting bolts, then remove the power steering pump.

To install:

8. Connect the pump inlet hose and the pump outlet hose onto the new pump with a new O-ring.

9. Loosely install the pump in the pump bracket with the mounting bolts, then tighten the pump fittings securely. Tighten the pump mounting bolts to 16 ft. lbs. (22 Nm).

10. Install the drive belt. Note these items during belt installation:

a. Inspect the belt for wear and cracks. Replace the belt if necessary.

b. Make sure that the belt is properly positioned on the pulleys.

✳✳ CAUTION

Do not get power steering fluid or grease on the auto-tensioner, alternator, A/C compressor, and drive belt or pulley faces. Clean off any fluid or grease before installation.

11. Fill the reservoir to the upper level line.

12. Start the engine, and check for fluid leaks.

13. Install the right upper fender trim.

FLUID FILL PROCEDURE

1. Remove the right upper fender trim.

2. Check the reservoir at regular intervals, and add the recommended fluid as necessary to fill the reservoir to the upper level line.

➡**Manufacturers recommends to always use Acura-brand Power Steering Fluid. Using any other type of power steering fluid or automatic transmission fluid can cause increased wear, and poor steering in cold weather.**

3. If drain and refill is required, proceed as follows:

a. Raise the reservoir, then disconnect the return hose to drain the reservoir. Take care not to spill the fluid on the body and parts. Wipe off any spilled fluid at once.

➡ **Inspect the reservoir screen for any debris. If the reservoir screen is clogged, replace the reservoir.**

b. Connect a hose of suitable diameter to the disconnected return hose, and put the hose end in a suitable container.

c. Start the engine, let it run at idle, and turn the steering wheel from lock-to-lock several times. When fluid stops running out of the hose, shut off the engine. Discard the fluid.

✳✳ CAUTION

Stop the motor immediately once the fluid stops running out of the hose to prevent pump damage.

d. Reinstall the return hose on the reservoir.

e. Fill the reservoir to the upper level line.

f. Start the engine and run it at fast idle, then turn the steering from lock-to-lock several times to bleed air from the system.

g. Recheck the fluid level and add some if necessary. Do not fill the reservoir beyond the upper level line.

h. If the fluid is contaminated, dark, or discolored, repeat the procedure as necessary until the system is clean.

4. Install the right upper fender trim.

SUSPENSION

FRONT SUSPENSION

KNUCKLE, HUB & BEARING

REMOVAL & INSTALLATION

Special Tools Required:
- Ball Joint Thread Protector, 12 mm: 07AAF-SDAA100
- Ball Joint Remover, 32 mm: 07MAC-SL0A102

➡**Refer to the exploded view illustration as needed during the following procedure, including tightening specifications.**

Hub Bearing Unit Replacement

See Figures 99 through 101.

1. Raise and support the vehicle.

2. Remove the wheel nuts, and the front wheel, taking care not to scratch the caliper.

3. Remove the brake hose mounting bracket.

4. Remove the brake caliper bracket mounting bolts and remove the caliper assembly from the knuckle.

➡**To prevent damage to the caliper assembly or the brake hose, use a short piece of wire to hang the caliper assembly from the undercarriage. Do not twist the brake hose excessively.**

5. Remove the wheel speed sensor and the O-ring from the knuckle. Do not disconnect the wheel sensor connector.

6. Pry up the stake, then remove the spindle nut.

➡**To avoid damage, do not strike aluminum parts with a metal hammer. If necessary, tap gently with a soft face hammer.**

7. Remove the front brake disc.

8. Remove the hub bearing unit mounting bolts.

9. Remove the hub bearing unit by tapping the driveshaft end with a soft face hammer while drawing the hub bearing unit outward.

➡**Do not pull the driveshaft end outward. The driveshaft inboard joint may come apart.**

To install:

➡**During installation, apply grease to the mating surfaces of the hub bearing unit and driveshaft outboard joint.**

10. Check the hub bearing unit for damage and cracks.

11. Install the hub bearing unit in the reverse order of removal, and note these items:

a. Use a new spindle nut during reassembly.

b. Before installing the spindle nut, apply a small amount of engine oil to the seating surface of the nut. After tightening, use a drift to stake the spindle nut shoulder against the driveshaft.

c. Before installing the brake disc, clean the mating surfaces of the front hub bearing unit and the inside of the brake disc.

d. Before installing the wheel, clean the mating surfaces of the brake disc and the inside of the wheel. Tighten wheel nuts to 94 ft. lbs. (124 Nm).

12. Check the wheel alignment, and adjust it if necessary.

Knuckle Replacement

See Figures 102 and 103.

1. Remove the hub bearing unit.

2. Remove the splash guard.

3. Remove the wheel speed sensor from the knuckle. Do not disconnect the wheel speed sensor connector.

➡**Use a new wheel speed sensor bracket and new clips during reassembly.**

4. Remove the cotter pin from the tie-rod end ball joint, then loosen the nut.

➡**During installation, install a new cotter pin after tightening the nut, and bend its end as shown.**

5. Disconnect the tie-rod end ball joint from the knuckle using the ball joint thread protector and the ball joint remover.

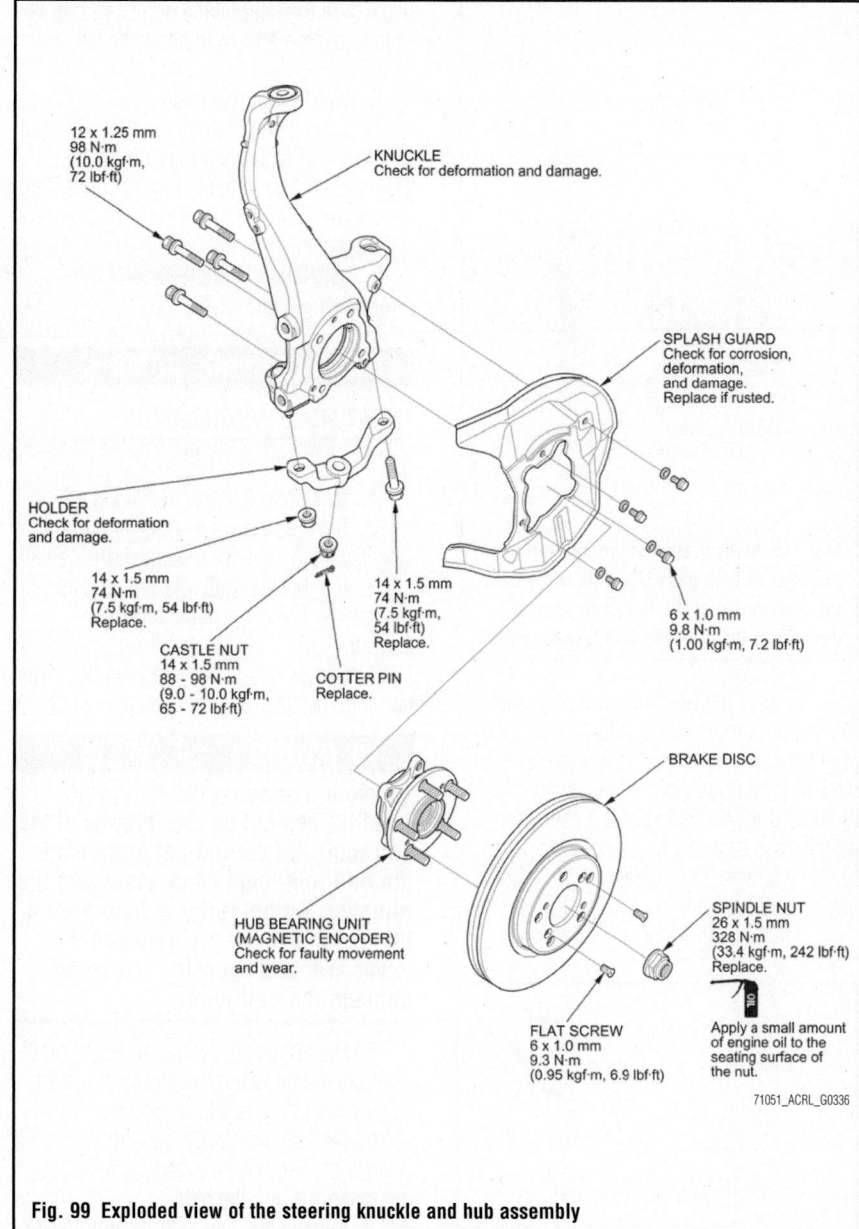

Fig. 99 Exploded view of the steering knuckle and hub assembly

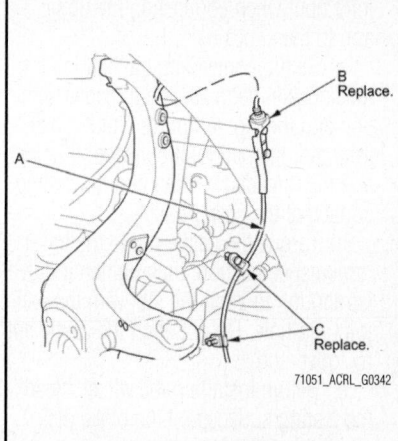

Fig. 102 Remove the wheel speed sensor (A) from the knuckle. Do not disconnect the wheel speed sensor connector. Use a new wheel speed sensor bracket (B) and new clips (C) during reassembly.

6. Remove the holder from the knuckle.

➡**Use a new flange bolt and a self-locking nut during reassembly.**

7. Remove the lock pin from the upper arm ball joint, then remove the castle nut.

8. Disconnect the upper arm ball joint from the knuckle using the ball joint thread protector and the ball joint remover, then remove the knuckle.

To install:

9. Install the knuckle in the reverse order of removal, and note these items:

a. First, install all the components, and lightly tighten the bolts and nuts, then raise the suspension to load it with the vehicle's weight before fully tightening to the specified torque.

b. Be careful not to damage the ball

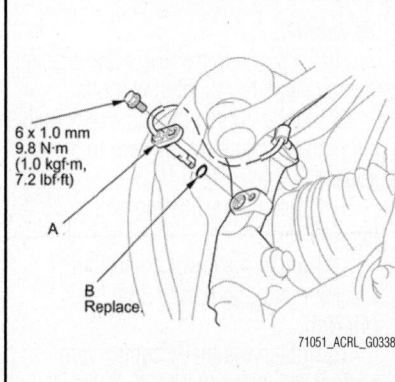

Fig. 100 Remove the wheel speed sensor (A) and the O-ring (B) from the knuckle. Do not disconnect the wheel sensor connector.

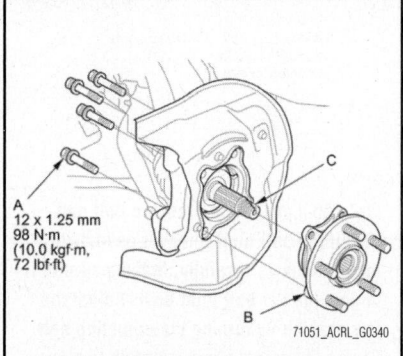

Fig. 101 Remove the hub bearing unit mounting bolts (A). Remove the hub bearing unit (B) by tapping the driveshaft end (C) with a soft face hammer while drawing the hub bearing unit outward.

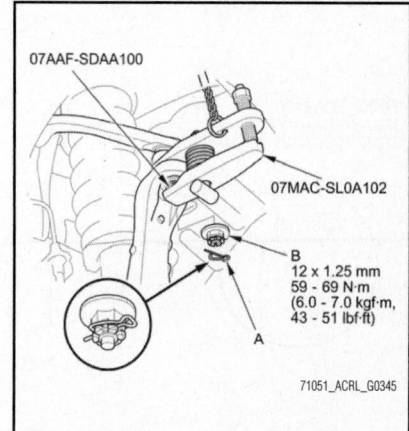

Fig. 103 Remove the lock pin (A) from the upper arm ball joint, then remove the castle nut (B).

joint boot when connecting the upper arm to the knuckle.

c. Before connecting the ball joint to the knuckle, degrease the threaded section and the tapered portion of the ball joint pin, the ball joint connecting hole, and the threaded section and the mating surface of the castle nut.

d. Torque the castle nut to the lower torque specification, then tighten it only far enough to align the slot with the ball joint pin hole. Do not align the castle nut by loosening it.

e. Before installing the wheel, clean the mating surfaces on the brake disk and the inside of the wheel.

f. Check the wheel alignment, and adjust it if necessary.

LOWER BALL JOINTS

REMOVAL & INSTALLATION

See Figures 104 through 106.

➡**Always use a ball joint remover tool and a thread protector tool to disconnect a ball joint. Do not strike the housing or any other part of the ball joint connection to disconnect it.**

1. Install a hex nut or the ball joint thread protector onto the threads of the ball joint.

➡**Using a hex nut, make sure the nut is flush with the ball joint pin end to prevent damage to the threaded end of the ball joint pin.**

2. Apply grease to the ball joint remover on both ends. This will ease the installation of the tool, and prevent damage to the pressure bolt threads.

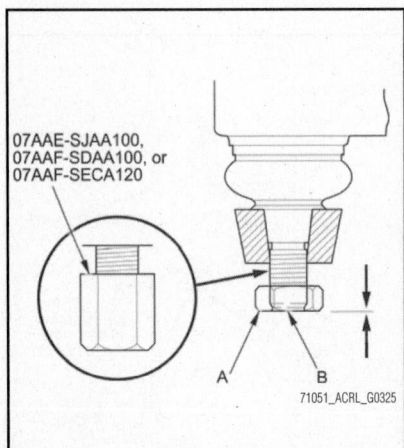

Fig. 104 Install a hex nut (A) or the ball joint thread protector onto the threads of the ball joint (B).

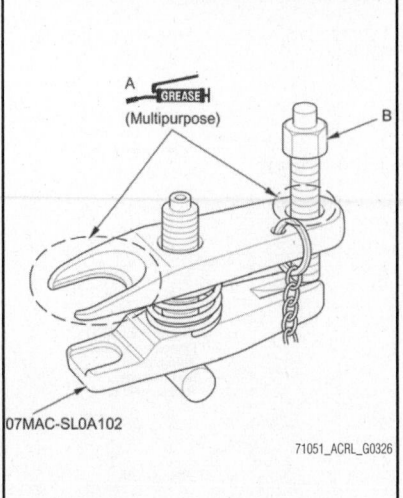

Fig. 105 Apply grease to the ball joint remover on both ends (A). This will ease the installation of the tool, and prevent damage to the pressure bolt threads (B).

3. Loosen the pressure bolt, and install the ball joint remover as shown. Insert the jaws carefully, making sure not to damage the ball joint boot. Adjust the jaw spacing by turning the adjusting bolt. Fasten the safety chain securely to a suspension arm or the subframe. Do not fasten it to a brake line or wire harness.

4. After adjusting the adjusting bolt,

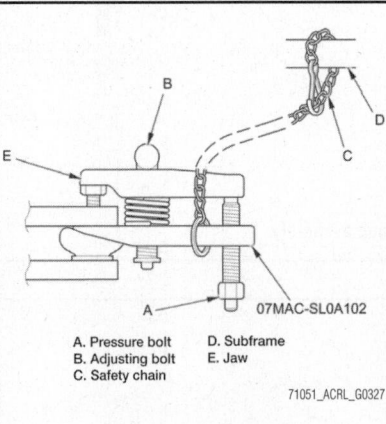

Fig. 106 Loosen the pressure bolt and install the ball joint remover as shown. Insert the jaws carefully, making sure not to damage the ball joint boot. Adjust the jaw spacing by turning the adjusting bolt. Fasten the safety chain securely to a suspension arm or the subframe. Do not fasten it to a brake line or wire harness. After adjusting the adjusting bolt, make sure the head of the adjusting bolt is in the position shown to allow the jaw to pivot.

make sure the head of the adjusting bolt is in the position shown to allow the jaw to pivot.

5. Remove the ball joint.

6. Remove the tool, then remove the nut from the end of the ball joint pin, and pull the ball joint out of the ball joint connecting hole. Inspect the ball joint boot, and replace it if damaged.

7. Installation is the reverse of the removal procedure.

LOWER CONTROL ARMS

REMOVAL & INSTALLATION

See Figures 107 and 108.

1. Raise and support the vehicle.

2. Remove the wheel nuts, and the front wheels taking care not to scratch the caliper.

3. Remove the bolt and the nut and remove the holder from the steering knuckle.

4. Remove the cotter pin from the lower ball joint castle nut, then remove the nut.

❋❋ CAUTION

To avoid damaging the ball joint, install a hex nut on the threads of the ball joint. Be careful not to damage the ball joint boot when installing the remover. Do not force or hammer on the lower arm, or pry between the lower arm and knuckle. You could damage the ball joint.

5. Disconnect the lower arm ball joint from the holder using the ball joint thread protector and the ball joint remover.

6. Remove the flange nut and the washer, then remove the damper fork mounting nut and the bolt.

7. Remove the flange bolts and the self-locking nut and washer, then remove the lower control arm.

To install:

8. In case of a loose lower arm bracket, follow the steps to retighten its bolts:

a. Lightly tighten the three new bolts, aligning the center of the cam to the edge.

b. Tighten the adjusting bolt to 36 ft. lbs. (49 Nm).

c. Tighten the lower control arm bracket mounting bolt to 36 ft. lbs. (49 Nm).

d. Tighten the lower control arm bracket mounting bolt to 36 ft. lbs. (49 Nm).

9. Install the lower arm in the reverse order of removal, and note these items:

a. First install all the components,

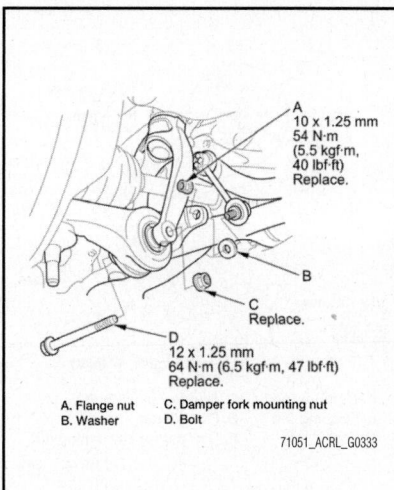

Fig. 107 Remove the flange nut and the washer, then remove the damper fork mounting nut and the bolt.

and lightly tighten the bolts and the nuts, then raise the suspension to load it with the vehicle's weight before fully tightening it to the specified torque. Do not place the jack against the ball joint pin of the knuckle.

b. Be careful not to damage the ball joint boot when installing the knuckle.

c. Before connecting the lower ball joint to the holder, degrease the threaded section and the tapered portion of the ball joint pin, the holder connecting hole, the threaded section and the mating surface of the castle nut.

d. Torque the castle nut to the lower torque specification, then tighten it only far enough to align the slot with the ball joint pin hole. Do not align the castle nut by loosening it.

e. Install a new cotter pin on the castle nut after torquing.

f. Before installing the wheel, clean the mating surfaces of the brake disc and the inside of the wheel.

g. Note the tightening specifications:
- Vertical flange bolt: 69 ft. lbs. (93 Nm)
- Horizontal flange bolt: 65 ft. lbs. (88 Nm)
- Damper fork bolt: 47 ft. lbs. (64 Nm)
- Stabilizer link locking nut: 40 ft. lbs. (54 Nm)

10. Check the wheel alignment, and adjust it if necessary.

STABILIZER BAR & LINKS

REMOVAL & INSTALLATION

Stabilizer Bar

1. Raise and support the vehicle.
2. Remove the wheel nuts, and the front wheels, taking care not to scratch the caliper.
3. Note the location and alignment of paint marks on the end of the stabilizer bar and bushings.
4. Disconnect the stabilizer links from the stabilizer bar on the right and left sides.
5. Remove the front suspension subframe from the body.
6. Remove the flange bolts and the stabilizer bar bushing holders, then remove the bushings and the stabilizer bar.

To install:

7. Install the stabilizer bar in the reverse order of removal, and note these items:

a. Note the right and left direction of the stabilizer bar.

b. Align the ends of the paint marks on the stabilizer bar with each end of the bushings.

c. Note the fore/aft direction of the bushing holders.

8. Clean the mating surface of the brake disc and the inside of wheel, then install the front wheel.

9. Check the wheel alignment, and adjust it if necessary.

Stabilizer Links

See Figures 109 and 110.

1. Raise and support the vehicle.
2. Remove the wheel nuts, and the front wheels, taking care not to scratch the caliper.
3. Remove the self-locking nuts while holding the respective joint pin with a hex wrench and remove the washer and the stabilizer link.
4. Place the floor jack under the lower arm, and raise the suspension.

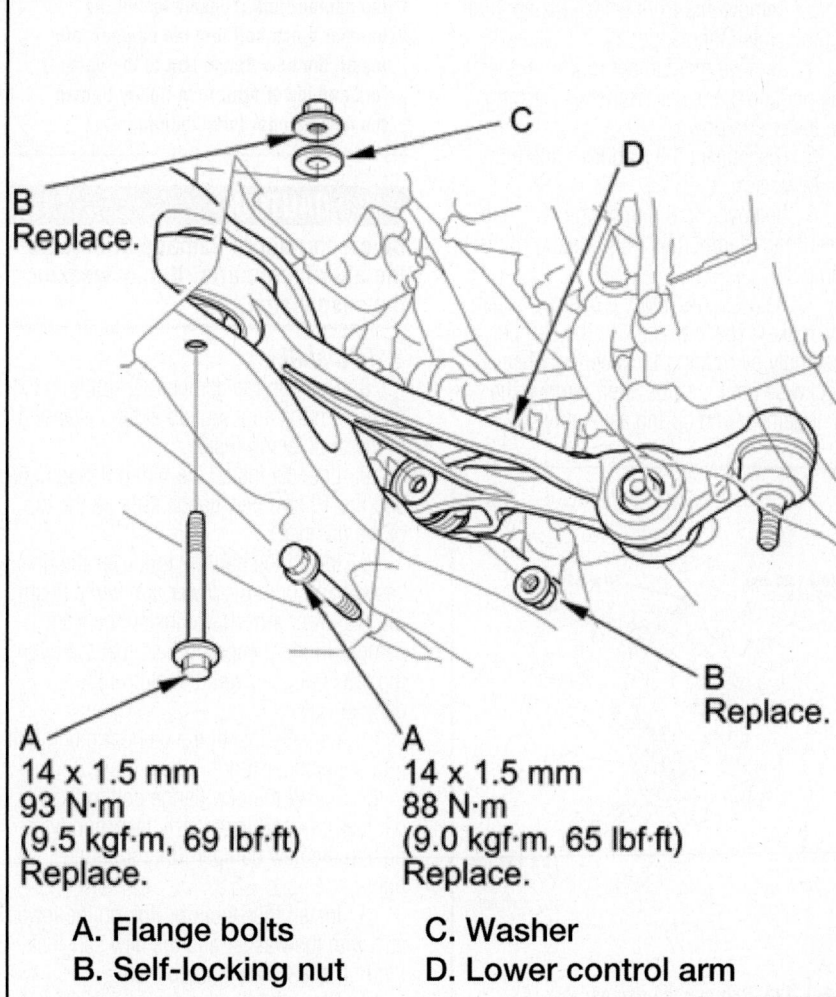

A
14 x 1.5 mm
93 N·m
(9.5 kgf·m, 69 lbf·ft)
Replace.

A
14 x 1.5 mm
88 N·m
(9.0 kgf·m, 65 lbf·ft)
Replace.

A. Flange bolts
B. Self-locking nut
C. Washer
D. Lower control arm

71051_ACRL_G0334

Fig. 108 Remove the flange bolts and the self-locking nut and washer, then remove the lower control arm.

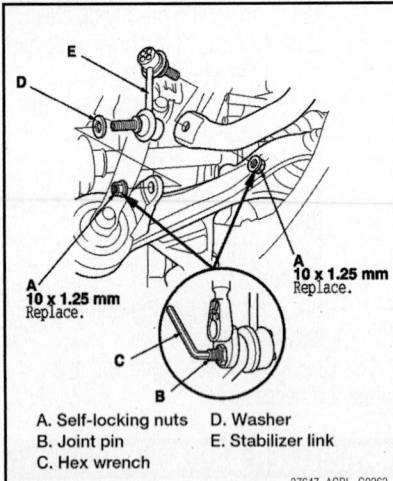

A. Self-locking nuts   D. Washer
B. Joint pin   E. Stabilizer link
C. Hex wrench

37647_ACRL_G0263

**Fig. 109 Remove the self-locking nuts (A)
while holding the respective joint pin (B)
with a hex wrench (C), and remove the
washer (D) and the stabilizer link (E).**

➡ **Do not place the jack against the
knuckle.**

To install:

5. Install the stabilizer link on the stabilizer bar and the lower arm with the joint
pins set at the center of their range of
movement. The paint mark on the stabilizer
link is near the upper end of the stabilizer
link.

6. Install the washer and the new self-
locking nuts and lightly tighten them.

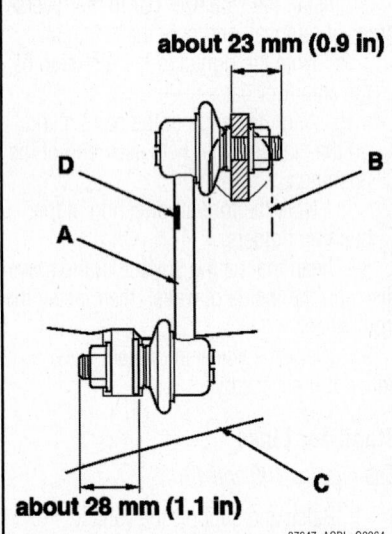

about 23 mm (0.9 in)

about 28 mm (1.1 in)

37647_ACRL_G0264

**Fig. 110 Install the stabilizer link (A) on
the stabilizer bar (B) and the lower arm
(C) with the joint pins set at the center of
their range of movement. The paint mark
(D) on the stabilizer link is near the upper
end of the stabilizer link.**

7. Tighten the new self-locking nuts
to the specified torque while holding
the respective joint pin with a hex
wrench:
   • Upper nut: 32 ft. lbs. (43 Nm)
   • Lower nut: 40 ft. lbs. (54 Nm)

8. Clean the mating surface of the brake
disc and the inside of the wheel, then install
the front wheel.

9. Test-drive the vehicle.

10. After 5 minutes of driving, tighten
the self-locking nuts again to the specified
torque.

STRUTS (FRONT DAMPER &
SPRING)

REMOVAL & INSTALLATION

See Figures 111 and 112.

1. Remove the upper fender trims and
left upper trim.

2. Raise and support the vehicle.

3. Remove the front wheel, taking care
not to scratch the caliper.

4. Remove the damper fork from the
damper and the lower arm while pushing
the lower arm down.

5. Disconnect the stabilizer link from
the lower arm.

6. Remove the 8 mm flange nuts and
the 10 mm flange nuts from the top of the
damper.

7. Before removing the damper assembly, make a space to remove the damper
assembly by rotating the steering wheel
clockwise until it stops, then remove the
front damper and spring assembly.

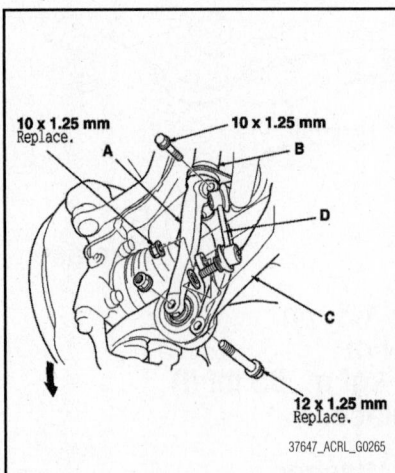

37647_ACRL_G0265

**Fig. 111 Remove the damper fork (A)
from the damper (B) and the lower arm (C)
while pushing the lower arm down. Disconnect the stabilizer link (D) from the
lower arm.**

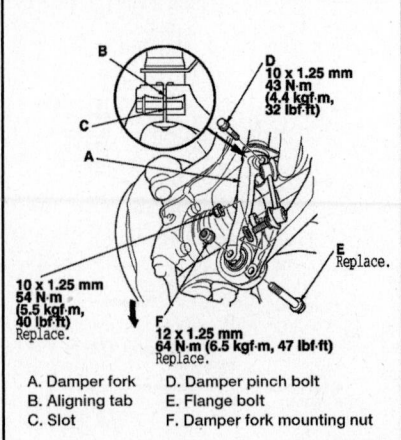

A. Damper fork   D. Damper pinch bolt
B. Aligning tab   E. Flange bolt
C. Slot   F. Damper fork mounting nut

37647_ACRL_G0266

**Fig. 112 Install the damper fork over the
driveshaft and onto the lower arm while
pushing the lower arm down. Install the
front damper in the damper fork so that
the aligning tab is aligned with the slot in
the damper fork. Loosely install the
damper pinch bolt into the damper fork.
Install the new flange bolt to the damper
fork and lower arm, then lightly tighten
the new damper fork mounting nut.**

✳✳ CAUTION

**Be careful not to damage or scratch
the aluminum parts. It may weaken
the suspension.**

To install:

8. Position the damper assembly in the
body, with the aligning tab at the lower end
of the strut facing inside.

9. Loosely install the 8 mm flange nuts
and the 10 mm new flange nuts on the top
of the damper.

10. Install the damper fork over the driveshaft and onto the lower arm while pushing the lower arm down. Install the front
damper in the damper fork so that the aligning tab is aligned with the slot in the
damper fork.

11. Loosely install the damper pinch bolt
into the damper fork.

12. Install the new flange bolt to the
damper fork and lower arm, then lightly
tighten the new damper fork mounting
nut.

13. Install the stabilizer link on the lower
arm with the washer and the new self-locking nut, and lightly tighten them.

14. Place the floor jack under the lower
arm, and raise the suspension to load it
with the vehicle's weight.

15. Tighten the flange nuts on the top of
the damper to the specified torque.

16. Tighten the damper pinch bolt, the flange nut on the damper fork, and the self-locking nut to the specified torque.

17. Tighten torques are as follows:
- Damper pinch bolt: 32 ft. lbs. (43 Nm)
- Lower fork nut: 47 ft. lbs. (64 Nm)
- Lower link nut: 40 ft. lbs. (54 Nm)
- 8 mm upper strut nuts: 25 ft. lbs. (34 Nm)
- 10 mm upper strut nuts: 39 ft. lbs. (55 Nm)

18. Clean the mating surfaces of the brake disc and the inside of the wheel, then install the front wheel.

19. Install the upper fender trims.

OVERHAUL

Disassembly & Inspection

See Figure 113.

➡**When compressing the damper spring, use a commercially available strut spring compressor (Branick MST-580A or Model 7200, or equivalent).**

1. Compress the damper spring, then remove the self-locking nut while holding the damper shaft with a hex wrench. Do not compress the damper spring more than is necessary to remove the self-locking nut.

2. Release the pressure from the strut spring compressor, then disassemble the damper as shown.

3. Reassemble all the parts, except for the spring mounting cushion and spring.

4. Compress the damper assembly by hand, and check for smooth operation through a full stroke, both compression and extension. The damper should extend smoothly and constantly when compression is released. If it does not, the gas is leaking and the damper should be replaced.

5. Check for oil leaks, abnormal noises, and binding during these tests.

Reassembly

See Figure 114.

1. Align the damper mounting base and spring mounting cushion, then reassemble the damper except for the washer and self-locking nut.

2. Install the damper assembly on the strut spring compressor and compress the spring lightly.

3. Align the bottom of the spring and the stepped part of the lower spring seat.

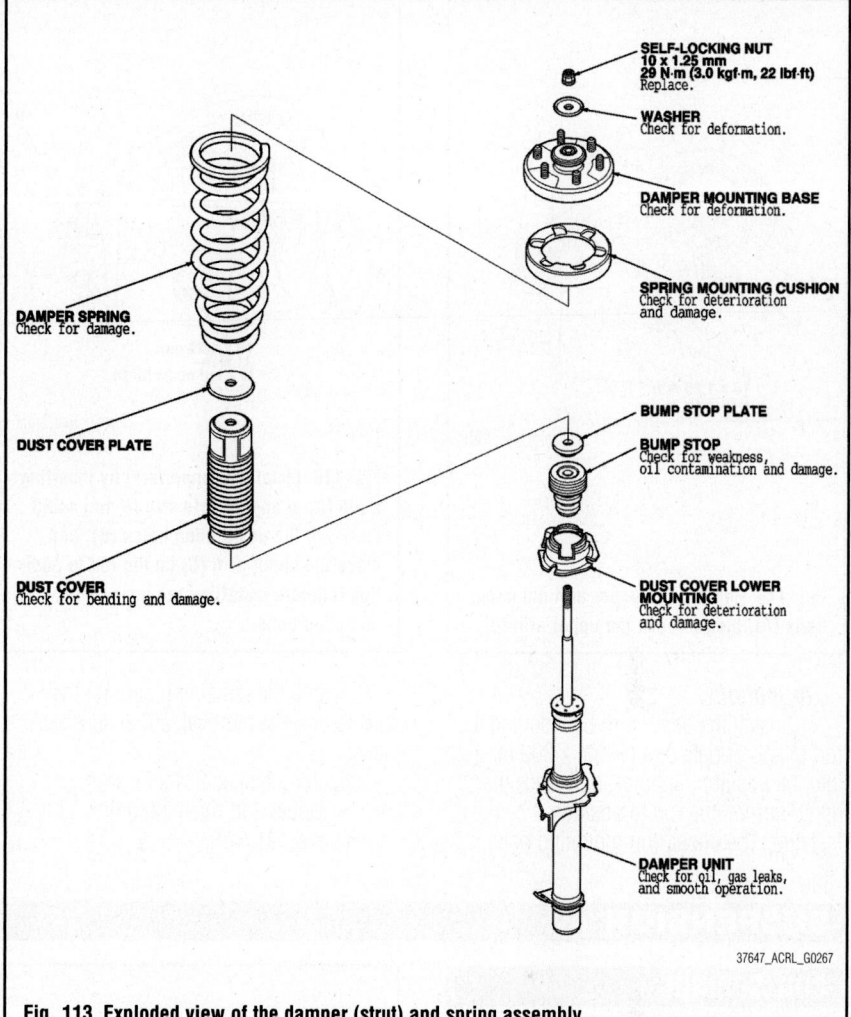

DAMPER SPRING
Check for damage.

DUST COVER PLATE

DUST COVER
Check for bending and damage.

SELF-LOCKING NUT
10 x 1.25 mm
29 N·m (3.0 kgf·m, 22 lbf·ft)
Replace.

WASHER
Check for deformation.

DAMPER MOUNTING BASE
Check for deformation.

SPRING MOUNTING CUSHION
Check for deterioration and damage.

BUMP STOP PLATE

BUMP STOP
Check for weakness, oil contamination and damage.

DUST COVER LOWER MOUNTING
Check for deterioration and damage.

DAMPER UNIT
Check for oil, gas leaks, and smooth operation.

37647_ACRL_G0267

Fig. 113 Exploded view of the damper (strut) and spring assembly

4. Position the damper mounting base so the stud bolt is aligned with the aligning tab in the damper unit.

5. Compress the damper spring. Do not compress the spring excessively.

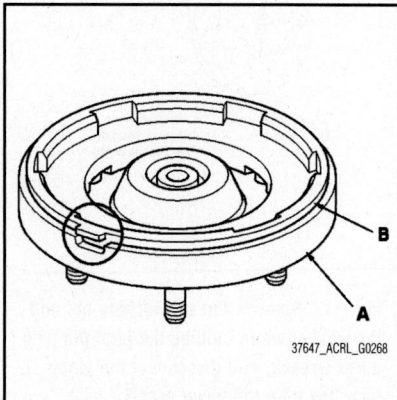

37647_ACRL_G0268

Fig. 114 Align the damper mounting base (A) and spring mounting cushion (B) as shown, then reassemble the damper except for the washer and self-locking nut.

6. Install the washer and a new 10 mm self-locking nut. Hold the damper shaft with a hex wrench and tighten the 10 mm self-locking nut to the specified torque of 22 ft. lbs. (29 Nm).

7. Remove the damper assembly from the strut spring compressor.

UPPER CONTROL ARMS

REMOVAL & INSTALLATION

See Figures 115 and 116.

1. Remove the front damper.

2. Remove the wheel speed sensor bracket from the upper arm.

3. Remove the cotter pin from the upper arm ball joint, then loosen the ball joint nut.

4. Disconnect the upper arm ball joint from the knuckle using the ball joint thread protector and the ball joint remover.

5. Remove the upper arm mounting bolts, then remove the upper arm.

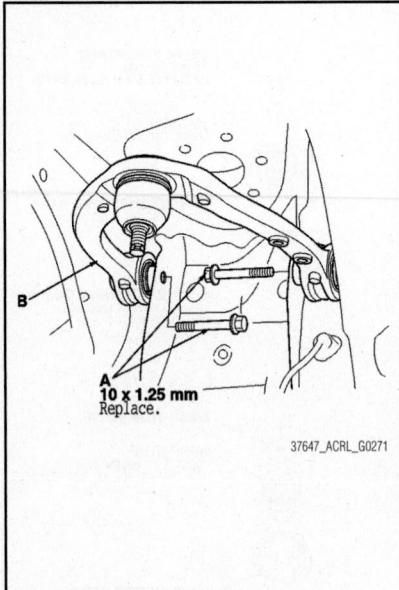

Fig. 115 Remove the upper arm mounting bolts (A), then remove the upper arm (B).

To install:

6. Install the upper arm by inserting a rod of appropriate size (6 mm x 300 mm) into the positioning holes, and place the upper arm on the rod to position it before installing the upper arm mounting bolts.

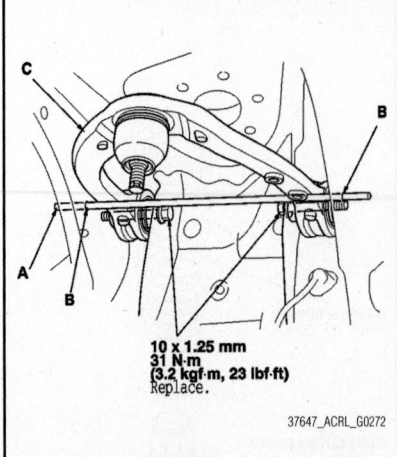

Fig. 116 Install the upper arm by inserting a rod (A) of appropriate size (6 mm x 300 mm) into the positioning holes (B), and place the upper arm (C) on the rod to position it before installing the upper arm mounting bolts.

7. Install the remaining parts in the reverse order of removal, and note these items:

  a. Torque specifications are:
  • Upper arm retaining bolts: 23 ft. lbs. (31 Nm)

• Upper ball joint castle nut (new): 43–51 ft. lbs. (59–69 Nm)

  b. First install all the components, and lightly tighten the bolts and the nuts, then raise the suspension to load it with the vehicle's weight before fully tightening it to the specified torque. Do not place the jack against the ball joint pin of the knuckle.

  c. Be careful not to damage the ball joint boot when installing the knuckle.

  d. Before connecting the ball joint to the knuckle, degrease the threaded section and the tapered portion of the ball joint pin, the knuckle connecting hole, and the threaded section and the mating surface of the castle nut.

  e. Torque the castle nut to the lower torque specification, then tighten it only far enough to align the slot with the ball joint pin hole. Do not align the castle nut by loosening it.

➡️**Use a new wheel speed sensor bracket on reassembly.**

  f. Before installing the wheel, clean the mating surfaces on the brake disc and the inside of the wheel.

  g. Check the front wheel alignment, and adjust it if necessary.

SUSPENSION

REAR SUSPENSION

COIL SPRINGS

REMOVAL & INSTALLATION

See Figures 117 and 118.

1. Raise and support the vehicle.
2. Remove the wheel nuts, and the rear wheel.
3. Remove the self-locking nut and the washer while holding the joint pin with a hex wrench, and disconnect the stabilizer link from the lower arm B.
4. Place a floor jack at the connecting point of the lower arm B and the stabilizer link.
5. Remove the cotter pin from the lower arm ball joint, and loosen the castle nut.
6. Disconnect the lower arm ball joint from the knuckle using the ball joint thread protector and the ball joint remover.
7. Lower the floor jack gradually.
8. Remove the spring, the spring mounting cushion, and the lower spring seat.

To install:

9. Install the spring mounting cushion, the spring mount cushion, then align the bottom of the spring, the stepped part of the lower spring seat and lower arm B.

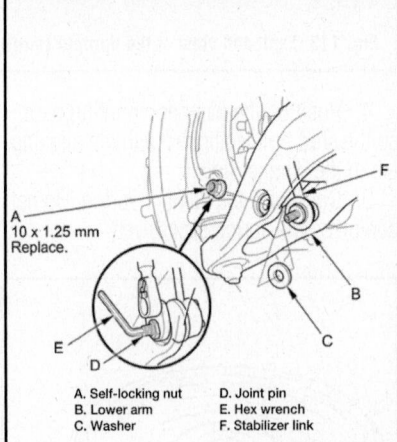

A. Self-locking nut D. Joint pin
B. Lower arm   E. Hex wrench
C. Washer   F. Stabilizer link

Fig. 117 Remove the self-locking nut and the washer while holding the joint pin with a hex wrench, and disconnect the stabilizer link from the lower arm B.

10. Place the floor jack at the connecting point of the lower arm and the stabilizer link.
11. Align the bottom of the spring and

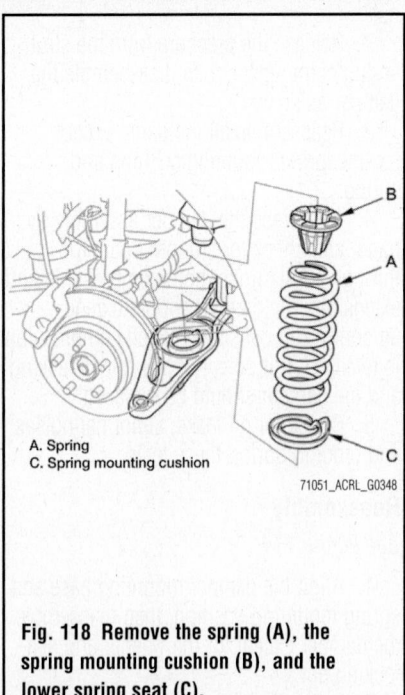

A. Spring
C. Spring mounting cushion

Fig. 118 Remove the spring (A), the spring mounting cushion (B), and the lower spring seat (C).

the stepped part of the lower spring seat, and the lower arm B.

12. Place the floor jack at the connecting point of lower arm and the stabilizer link.

13. Raise the jack slowly until you can align the bolt hole of the lower arm B and the knuckle ball joint pin, then loosely install a new castle nut.

14. Install the stabilizer link on the lower arm B with the washer and a new self-locking nut, and lightly tighten them.

15. Raise the rear suspension with a floor jack to load it with the vehicle's weight.

16. Tighten the castle nut and self-locking nut to the specified torque of 54–61 ft. lbs. (74–83 Nm).

➡Torque the castle nut to the lower torque specification, then tighten it only far enough to align the slot with the ball joint pin hole. Do not align the castle nut by loosening it.

17. Insert a new cotter pin into the ball joint pin from the rear to the front of the vehicle, and bend its end.

18. Clean the mating surface of the brake disc/drum and the inside of the wheel, then install the rear wheel.

19. Check the rear wheel alignment, and adjust it if necessary.

CONTROL ARMS

REMOVAL & INSTALLATION

Rear Control Arm

See Figure 119.

1. Raise and support the vehicle.
2. Remove the wheel nuts, and the rear wheel.
3. Remove the control arm mounting nut and the washer from the knuckle side.
4. Mark the cam positions of the adjusting bolt and the adjusting cam, then remove the self-locking nut, the adjusting cam, and the adjusting bolt.
5. Remove the control arm.

To install:

6. Install the control arm in the reverse order of removal, and note these items:

➡Use the figure to identify nuts when tightening to specification.

a. First install all of the components, and lightly tighten the bolts and nuts, then raise the suspension to load it with the vehicle's weight before fully tightening to the specified torque.

b. Align the cam positions of the adjusting bolt and the adjusting cam with the marked positions when tightening.

c. Tighten all mounting hardware to the specified torque.

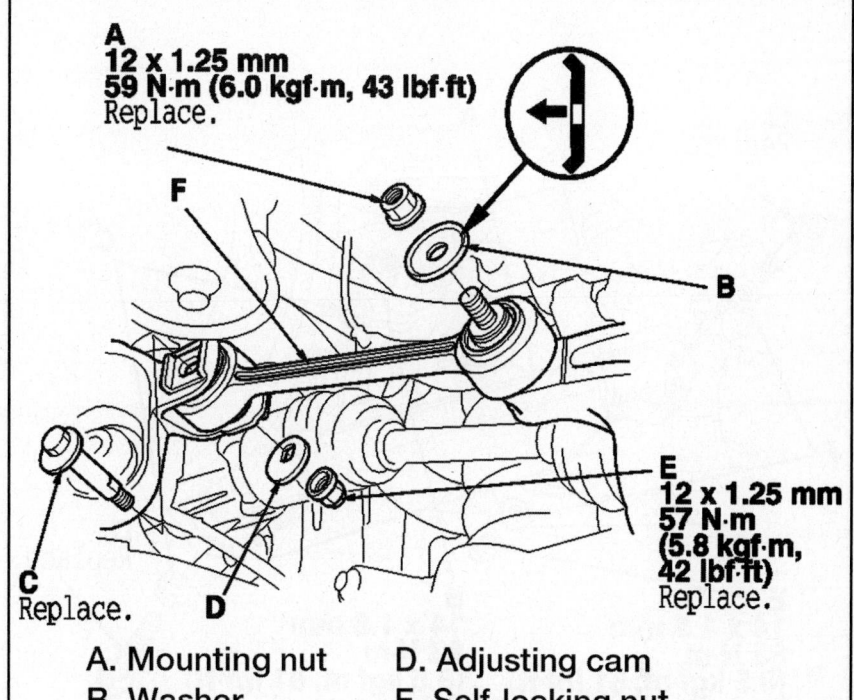

A. Mounting nut
B. Washer
C. Adjusting bolt
D. Adjusting cam
E. Self-locking nut
F. Rear control arm

Fig. 119 Removing/installing the rear control arm

d. Before installing the wheel, clean the mating surfaces on the brake disc/drum and inside of the wheel.
7. Check the wheel alignment, and adjust it if necessary.

Lower Arm A

See Figure 120.

1. Raise and support the vehicle.
2. Remove the wheel nuts, and the rear wheel.
3. Remove the lower arm mounting nut, the washers and the mounting bolt from the knuckle side.
4. Remove the self-locking nut, the washers, and the flange bolt, then remove the lower arm A.

To install:

5. Install the lower arm A in the reverse order of removal, and note these items:

➡Use the figure to identify nuts when tightening to specification.

a. First install all of the components, and lightly tighten the bolts and nuts, then raise the suspension to load it with the vehicle's weight before fully tightening to the specified torque.

b. Before installing the wheel, clean the mating surfaces on the brake

disc/drum and the inside of the wheel.
6. Check the wheel alignment, and adjust it if necessary.

Lower Arm B

See Figures 121 and 122.

1. Raise and support the vehicle.
2. Remove the rear wheel.
3. Remove the self-locking nut and the washer while holding the joint pin with a hex wrench, and disconnect the stabilizer link from lower arm B.
4. Place a floor jack at the connecting point of lower arm B and the stabilizer link.
5. Remove the cotter pin from the lower arm B ball joint, and loosen the castle nut.
6. Disconnect the lower arm ball joint from the knuckle using the ball joint thread protector and the ball joint remover.

➡Be careful not to damage the ball joint boot when installing the remover.

7. Lower the floor jack gradually.
8. Remove the spring, the spring mounting cushion, and the lower spring seat.
9. Remove the self-locking nut and the flange bolt, then remove lower arm B.

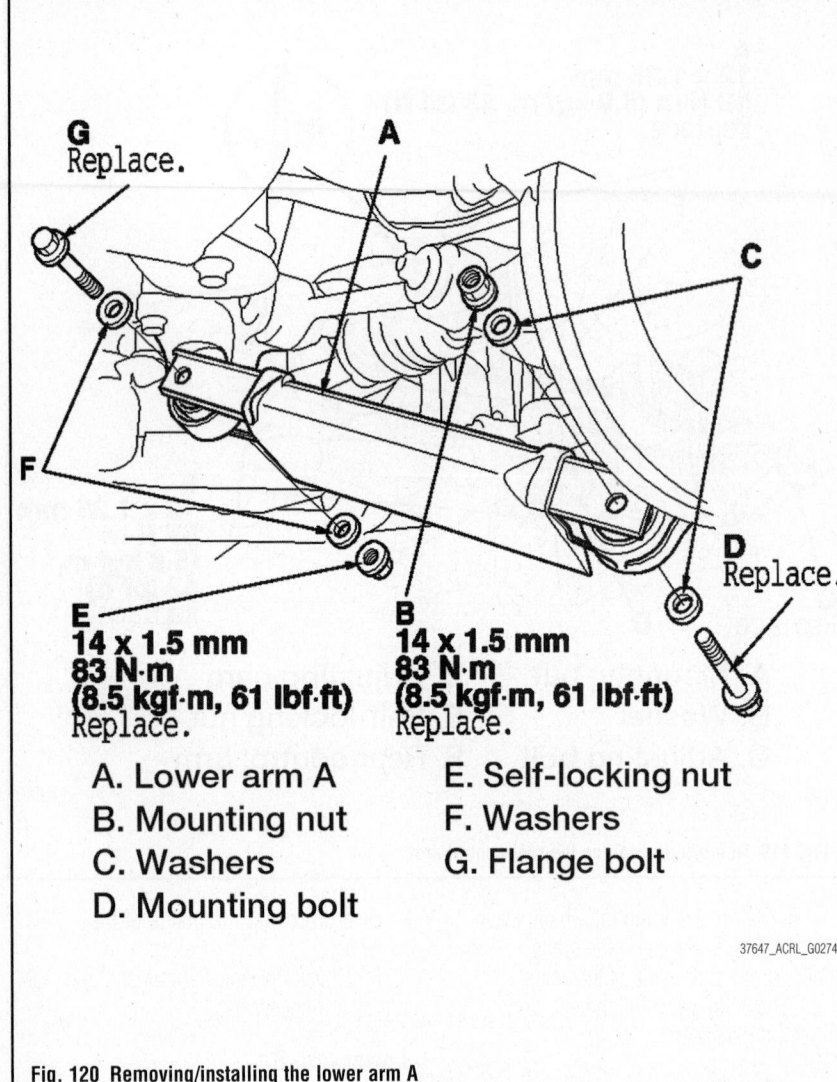

Fig. 120 Removing/installing the lower arm A

G Replace.

A

C

F

E
**14 x 1.5 mm
83 N·m
(8.5 kgf·m, 61 lbf·ft)
Replace.**

B
**14 x 1.5 mm
83 N·m
(8.5 kgf·m, 61 lbf·ft)
Replace.**

D Replace.

A. Lower arm A
B. Mounting nut
C. Washers
D. Mounting bolt

E. Self-locking nut
F. Washers
G. Flange bolt

37647_ACRL_G0274

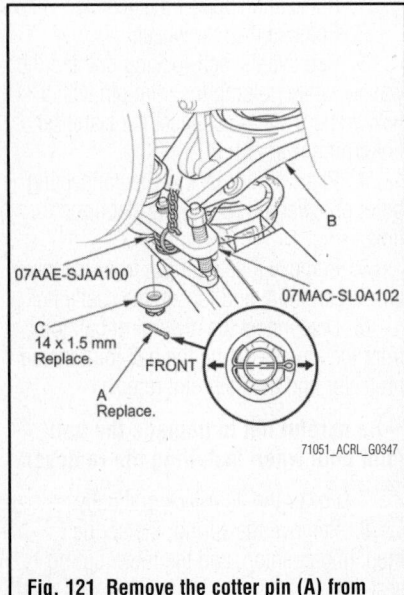

Fig. 121 Remove the cotter pin (A) from the lower arm B ball joint (B), and loosen the castle nut (C).

07AAE-SJAA100

B

07MAC-SL0A102

C
14 x 1.5 mm
Replace.

FRONT

A
Replace.

71051_ACRL_G0347

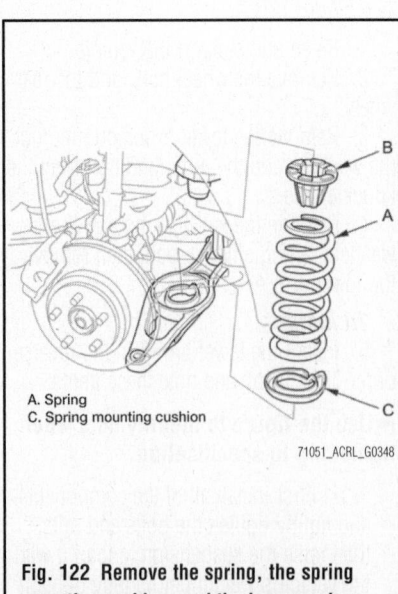

Fig. 122 Remove the spring, the spring mounting cushion, and the lower spring seat. Remove the self-locking nut and the flange bolt, then remove lower arm B.

B

A

C

A. Spring
C. Spring mounting cushion

71051_ACRL_G0348

To install:

10. Position lower arm B, install a new flange bolt, and loosely install a new self-locking nut.

11. Install the spring mounting cushion and the spring.

12. Align the bottom of the spring, the stepped part of the lower spring seat and lower arm B.

13. Place a floor jack at the connecting point of lower arm B and the stabilizer link.

14. Raise the jack slowly until you can align the bolt hole of lower arm B and the knuckle ball joint pin, then loosely install a new castle nut.

15. Raise the rear suspension with a floor jack to load it with the vehicle's weight.

16. Tighten the castle nut to the specified torque.

Please note the following:

• Torque the castle nut to the lower torque specification, then tighten it only far enough to align the slot with the ball joint pin hole. Do not align the castle nut by loosening it.

• Insert a new cotter pin into the ball joint pin from the rear to the front of the vehicle, and bend its end over properly.

17. Install the stabilizer link on lower arm B with the washer and a new self-locking nut, and lightly tighten them.

18. Tighten the self-locking nut to the specified torque while holding the respective joint pin with a hex wrench.

19. Clean the mating surface of the brake disc/drum the inside of the wheel, then install the rear wheel.

20. Check the rear wheel alignment, and adjust it if necessary.

KNUCKLE/HUB BEARING UNIT

REMOVAL & INSTALLATION

Special Tools Required:
• Ball Joint Remover, 32 mm 07MAC-SL0A102
• Ball Joint Thread Protector, 12 mm 07AAF-SDAA100
• Ball Joint Thread Protector, 14 mm 07AAE-SJAA100

Hub Bearing Unit Replacement

See Figures 123 through 125.

1. Raise and support the vehicle.

2. Remove the wheel nuts, and the rear wheel.

3. Remove the brake caliper bracket mounting bolts and remove the caliper assembly from the knuckle.

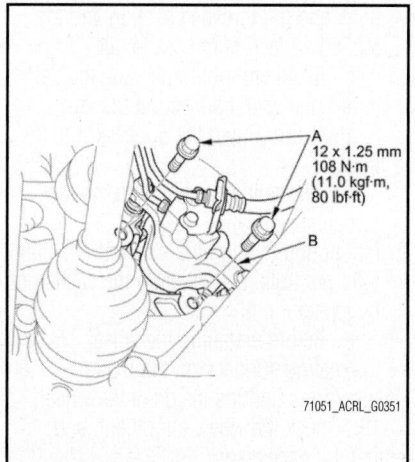

Fig. 123 Remove the brake caliper bracket mounting bolts (A), and remove the caliper assembly (B) from the knuckle.

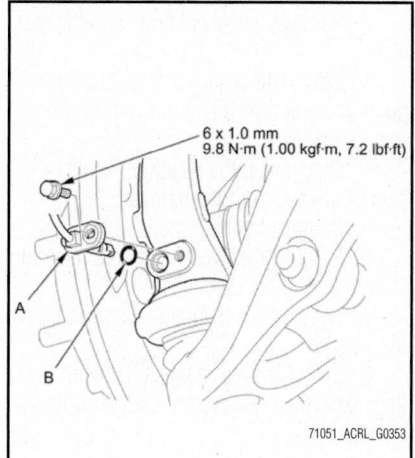

Fig. 124 Remove the wheel speed sensor (A) and the O-ring (B) from the knuckle. Do not disconnect the wheel speed sensor connector.

➡**To prevent damage to the caliper assembly or brake hose, use a short piece of wire to hang the caliper assembly from the undercarriage. Do not twist the brake hose excessively.**

4. Remove the two washers.

➡**During installation, make sure the washers are installed between the brake caliper bracket and the knuckle.**

5. Remove the wheel speed sensor and the O-ring from the knuckle. Do not disconnect the wheel speed sensor connector.

6. Pry up the stake and remove the spindle nut.

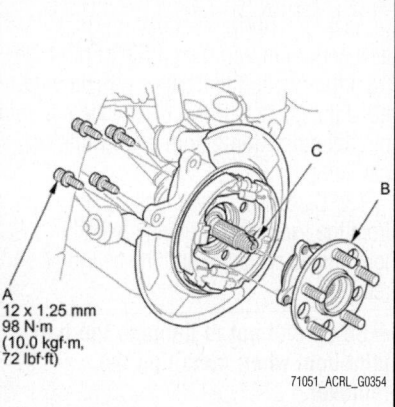

Fig. 125 Remove the flange bolts (A), and remove the hub bearing unit (B) by tapping the driveshaft end (C) with a soft face hammer while drawing the hub bearing unit outward.

7. Remove the rear brake disc/drum.

8. Remove the flange bolts and remove the hub bearing unit by tapping the driveshaft end with a soft face hammer while drawing the hub bearing unit outward.

➡**Do not pull the driveshaft end outward. The driveshaft inboard joint may come apart.**

To install:

9. Install the hub bearing unit in the reverse order of removal, and note these items:

   a. Use a new spindle nut on reassembly.

   b. Before installing the spindle nut, apply a small amount of engine oil to the seating surface of the nut. After tightening, use a drift to stake the spindle nut shoulder against the driveshaft.

   c. Before installing the brake disc/drum, clean the mating surfaces of the hub bearing unit and the inside of the brake disc/drum.

   d. Before installing the wheel, clean the mating surfaces of the brake disc/drum and the inside of the wheel.

10. Check the wheel alignment, and adjust it if necessary.

Knuckle Replacement

See Figures 126 through 130.

1. Remove the hub bearing unit.

2. Remove lower arm A mounting nut, the washers, and the mounting bolt from the knuckle side.

➡**Use new lower arm A mounting nuts and new bolts during reassembly.**

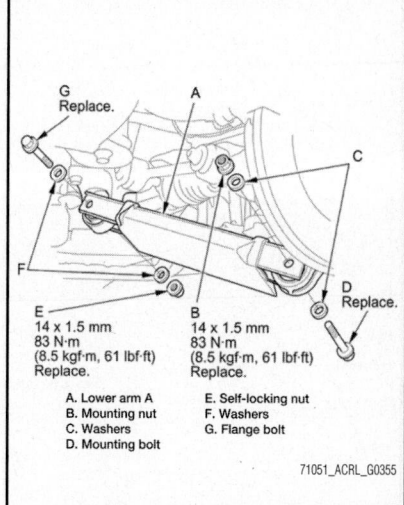

Fig. 126 Remove lower arm A mounting nut, the washers, and the mounting bolt from the knuckle side. Remove the self-locking nut, the washers, and the flange bolt, then remove the lower arm A.

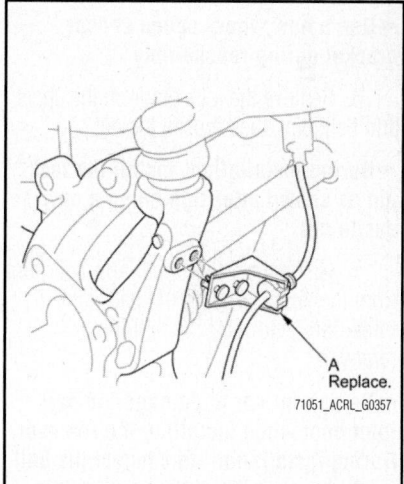

Fig. 127 Remove the wheel speed sensor bracket (A) from the knuckle. Do not disconnect the wheel speed sensor connector.

3. Remove the self-locking nut, the washers, and the flange bolt, then remove the lower arm A.

4. Remove the backing plate mounting nuts, then remove the backing plate.

✳✳ CAUTION

To prevent damage to the backing plate or parking brake shoes assembly and cable, use a short piece of wire to hang the backing plate from the undercarriage. Do not twist the parking brake cable excessively.

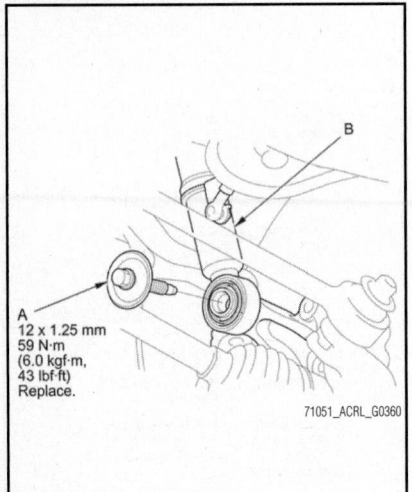

A
12 x 1.25 mm
59 N·m
(6.0 kgf·m,
43 lbf·ft)
Replace.

71051_ACRL_G0360

Fig. 128 Remove the damper (shock absorber) lower mounting bolt (A), and disconnect the damper (B) from the knuckle.

5. Remove the wheel speed sensor bracket from the knuckle. Do not disconnect the wheel speed sensor connector.

➡️ Use a new wheel speed sensor bracket during reassembly.

6. Remove the lock pin from the upper arm ball joint, then loosen the nut.

➡️ During installation, install the lock pin as shown after tightening a new castle nut.

7. Disconnect the upper arm ball joint from the knuckle using the ball joint thread protector and the ball joint remover.

➡️ Be careful not to damage the ball joint boot when installing the remover. During installation, to connect the ball joint, raise up the suspension with a jack .

8. Remove the self-locking nut and the washer, then disconnect the stabilize link from lower arm B.

➡️ Use a new self-locking nut during reassembly.

9. Remove the damper (shock absorber) lower mounting bolt and disconnect the damper from the knuckle.

➡️ Use a new damper lower mounting bolt during reassembly.

10. Place a floor jack at the connecting point of lower arm B and the stabilizer link.

11. Remove the cotter pin from the lower arm B ball joint, and loosen the nut.

12. Use the new castle nut during reassembly. During installation, insert a new cotter pin into the ball joint pin hole from the rear to the front of vehicle, and bend its end as shown. Check the ball joint pin hole direction before connecting the ball joint.

13. Disconnect the knuckle ball joint from the lower arm B using the ball joint thread protector and the ball joint remover.

➡️ Be careful not to damage the ball joint boot when installing the remover.

14. Remove the control arm mounting nut and the washer from the knuckle, then remove the knuckle.

➡️ Use a new control arm mounting nut during reassembly.

To install:

15. Install the knuckle in the reverse order of removal, and note these items:

a. First install all the components and lightly tighten the bolts and nuts, then raise the suspension to load it with the vehicle's weight before fully tightening to the specified torque.

b. Before connecting the ball joint to the knuckle and lower arm B, degrease the threaded section and the tapered portion of the ball joint pin, the connecting

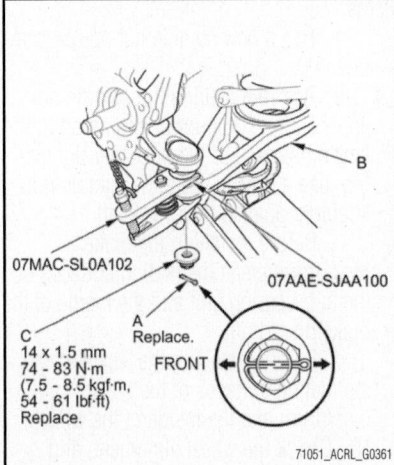

07MAC-SL0A102

07AAE-SJAA100

C
14 x 1.5 mm
74 - 83 N·m
(7.5 - 8.5 kgf·m,
54 - 61 lbf·ft)
Replace.

A
Replace.
FRONT

71051_ACRL_G0361

Fig. 129 Remove the cotter pin (A) from the lower arm B ball joint (B), and loosen the nut (C). Use the new castle nut during reassembly. During installation, insert a new cotter pin into the ball joint pin hole from the rear to the front of vehicle, and bend its end as shown. Check the ball joint pin hole direction before connecting the ball joint.

hole, and the threaded section and the mating surface of the castle nut.

c. Be careful not to damage the ball joint boot when connecting the knuckle to lower arm B and the upper arm to the knuckle.

d. Torque the castle nut to the lower torque specification, then tighten it only far enough to align the slot with the ball joint pin hole. Do not align the castle nut by loosening it.

e. Before installing the wheel, clean the mating surfaces on the brake disc/drum and the inside of the wheel.

16. Check the wheel alignment, and adjust it if necessary.

SHOCK ABSORBERS

REMOVAL & INSTALLATION

See Figure 131.

1. Raise and support the vehicle.

2. Remove the wheel nuts, and the rear wheel.

3. Place a floor jack at the connecting point of lower arm B and the stabilizer link to support them.

4. Disconnect the headlight leveling sensor linkage from the damper.

5. Remove the rear shelf.

6. Remove the two flange nuts retaining the top end of the damper.

7. Remove the damper lower bolt from the knuckle.

8. Lower the rear suspension with the floor jack, then remove the damper from the vehicle.

To install:

9. Place a floor jack at the connecting point of lower arm B and the stabilizer link.

10. Compress the damper by hand, and move it into position.

11. Loosely tighten a new damper lower mounting bolt.

12. Loosely install new flange nuts.

13. Raise the rear suspension with a floor jack to load it with the vehicle's weight.

14. Tighten the flange nuts and the damper lower mounting bolt to the specified torque:

- Upper flange nuts: 28 ft. lbs. (38 Nm)
- Lower mounting bolt: 43 ft. lbs. (59 Nm)

15. Connect the headlight leveling sensor linkage to the damper with the self-locking nut.

16. Clean the mating surface of the brake disc/drum and the inside of the wheel, then install the rear wheel.

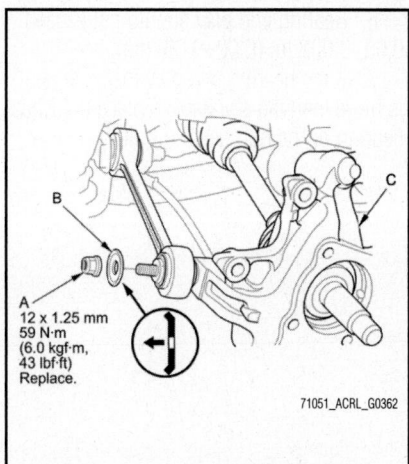

Fig. 130 Remove the control arm mounting nut (A) and the washer (B) from the knuckle (C), then remove the knuckle.

17. Check the rear wheel alignment, and adjust it if necessary.

18. Do the "Adaptive Front Lighting Control Unit Learning Procedure".

STABILIZER BAR & LINKS

REMOVAL & INSTALLATION

Stabilizer Bar

1. Raise and support the vehicle.

2. Remove the wheel nuts, and the rear wheels.

3. Disconnect the stabilizer links from the stabilizer bar on the right and left sides.

4. Remove the flange bolts and the bushing holders, then remove the bushing and the stabilizer bar.

To install:

5. Install the stabilizer bar in the reverse order of removal, and note these items:

   a. Note the right and left direction of the stabilizer bar.

   b. Align the ends of the paint marks on the stabilizer bar with each end of the bushings.

6. Tighten the stabilizer bar mounting bolts to 36 ft. lbs. (49 Nm).

   a. Clean the mating surface of the brake disc/drum and the inside of the wheel, then install the rear wheel.

7. Check the wheel alignment, and adjust it if necessary.

Stabilizer Links

1. Raise and support the vehicle.

2. Remove the wheel nuts, and the rear wheel.

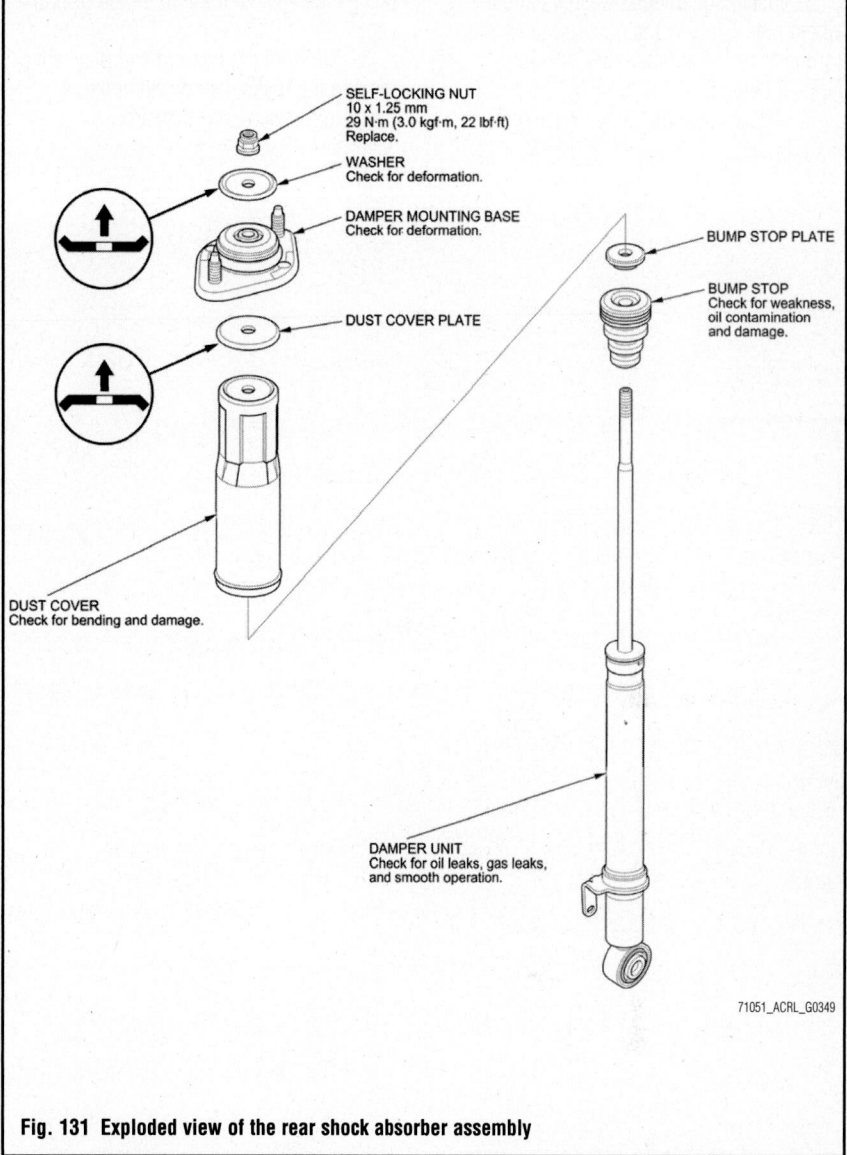

Fig. 131 Exploded view of the rear shock absorber assembly

3. Remove the self-locking nuts while holding the respective joint pin with a hex wrench, then remove the washer and the stabilizer link.

To install:

4. Place a floor jack at the connecting point of lower arm B and the stabilizer link, and raise the suspension.

5. Install the stabilizer link on the stabilizer bar and lower arm B with the joint pins set at the center of their range of the movement.

➡**The stabilizer link has a paint mark in the middle of the link. The paint mark indicates the difference between the left and right stabilizer link.**

6. Install the washer and new self-locking nuts, and lightly tighten them.

7. Tighten the self-locking nuts to the

specified torque while holding the respective joint pin with a hex wrench.

8. Torque specifications are:
- Lower link nut: 40 ft. lbs. (54 Nm)
- Upper link nut: 32 ft. lbs. (43 Nm)

9. Clean the mating surface of the brake disc/drum and the inside of the wheel, then install the rear wheel.

10. Test-drive the vehicle.

11. After 5 minutes of driving, tighten the self-locking nuts again to the specified torque.

WHEEL BEARINGS

INSPECTION

1. Raise and support the vehicle.

2. Remove the wheel nuts, and the wheels, taking care not to scratch the caliper.

3. Install suitable flat washers and the wheel nuts. Tighten the nuts to the specified torque to hold the brake disc securely against the hub.

4. Attach the dial gauge. Place the dial gauge against the hub flange or hub cap.

5. Measure the bearing end play while moving the brake disc or the brake disc/drum inward and outward.

6. Bearing end play should not exceed 0.00–0.002 in. (0.00–0.05 mm).

7. If the bearing end play measurement is more than the standard, replace the wheel bearing or the hub bearing unit.

ACURA

TL

4

BRAKES **4-10**

ANTI-LOCK BRAKE SYSTEM (ABS)...................**4-8**
General Information...................4-8
  Precautions...................4-8
Speed Sensors...................4-8
  Removal & Installation...........4-8
BLEEDING THE BRAKE SYSTEM **4-9**
Bleeding Procedure...................4-9
  Bleeding the ABS System 4-9
FRONT DISC BRAKES **4-10**
Brake Caliper...................4-10
  Removal & Installation...........4-10
Brake Pads 4-10
  Removal & Installation 10
PARKING BRAKE...................**12**
Parking Brake Cables...................12
  Removal & Installation...........12
Parking Brake Shoes...................12
  Removal & Installation...........12
REAR DISC BRAKES **11**
Brake Caliper 11
  Removal & Installation...........11
Disc Brake Pads...................11
  Removal & Installation 11

CHASSIS ELECTRICAL **13**

AIR BAG (SUPPLEMENTAL RESTRAINT SYSTEM) **13**
General Information 13
  Arming the System...................14
  Disarming the System.............14
  Service Precautions...............13

DRIVE TRAIN **14**

Differential Carrier 14
  Removal & Installation 14
Driveshaft 15
  Removal & Installation 15
Front Halfshafts 16
  Removal & Installation 16
Front Intermediate Shaft...................17
  Removal & Installation 17

Rear Driveshaft 18
  Removal & Installation.............18

ENGINE COOLING **18**

Cooling Fans 18
  Removal & Installation 18
Engine Coolant 19
  Drain & Refill Procedure 19
Radiator 19
  Removal & Installation 19
Thermostat 20
  Removal & Installation 20
Water Pump 20
  Removal & Installation 20

ENGINE ELECTRICAL **21**

BATTERY SYSTEM...............**21**
Battery 21
  Removal & Installation 21
CHARGING SYSTEM **21**
Alternator 21
  Removal & Installation 21
IGNITION SYSTEM **22**
Firing Order 22
Ignition Coils 22
  Removal & Installation 22
Ignition Timing 23
  Inspection & Adjustment...........23
STARTING SYSTEM **23**
Starter 23
  Removal & Installation 23

ENGINE MECHANICAL...........**24**

Accessory Drive Belts 24
  Adjustment 24
  Belt Routings 24
  Inspection 24
  Removal & Installation 24
Air Cleaner 24
  Removal & Installation 24
Camshaft & Bearings 24
  Inspection 24
  Removal & Installation 25
Crankshaft Front Seal 26
  Removal & Installation 26

Cylinder Head 26
  Removal & Installation 26
Cylinder Head Covers 28
  Removal & Installation 28
Driveplate 30
  Removal & Installation 30
Engine Cover 30
  Removal & Installation 30
Exhaust System 31
  Removal & Installation 31
Flywheel 31
  Removal & Installation 31
Intake Manifold 31
  Removal & Installation 31
Oil Pan 32
  Removal & Installation 32
Oil Pump 33
  Removal & Installation 33
Pistons & Rings...................34
  Positioning...................34
Rocker Arms 34
  Removal & Installation 34
Timing Belt & Sprockets...................36
  Removal & Installation 36
Timing Belt Front Cover...................36
  Removal & Installation 36
Valve Lash 38
  Adjustment 38

ENGINE PERFORMANCE & EMISSION CONTROLS **39**

Accelerator Pedal Position (APP) Sensor...................39
  Location 39
  Removal & Installation 39
Air-Fuel Ratio (AFR) Sensor 40
  Location 40
  Removal & Installation 40
Camshaft Position Sensor 40
  Removal & Installation 40
Clutch Pedal Position Switch 40
  Location 40
  Removal & Installation 40
Component Locations...................39
Crankshaft Position Sensor 40

CKP Pattern Clear/CKP
Pattern Learn..................40
Removal & Installation 40
Engine Control Module/
Powertrain Control Module
(ECM/PCM) 41
Ecm/pcm IDLE Learn
Procedure 42
Location 41
Removal & Installation 41
Engine Coolant Temperature
(ECT) Sensor 42
Removal & Installation 42
Evaporative Emission Control
System 43
EVAP Canister 43
EVAP Canister Purge
Control Valve 44
Exhaust Gas Recirculation
(EGR) Valve 45
Removal & Installation 45
Fuel Tank Pressure (FTP)
Sensor 45
Location 45
Removal & Installation 45
Heated Oxygen (HO2S)
Sensor 46
Location 46
Removal & Installation 46
Input Speed Sensor (ISS) 46
Removal & Installation 46
Knock Sensor (KS) 46
Location 46
Removal & Installation 46
Manifold Absolute Pressure
(MAP) Sensor....................47
Removal & Installation 47
Mass Airflow (MAF)/Intake Air
Temperature (IAT) Sensor 47
Location 47
Removal & Installation 47
Output Shaft Speed (OSS)
Sensor 47
Location 47
Removal & Installation 47
Positive Crankcase Ventilation
(PCV) Valve 47
Location 47
Removal & Installation 47

Transmission Fluid Temperature
(TFT) Sensor 48
Location 48
Removal & Installation 48

FUEL 48

**GASOLINE FUEL INJECTION
SYSTEM 48**
Fuel Filter 50
Removal & Installation 50
Fuel Rail & Injectors 52
Removal & Installation 52
Fuel System Service
Precautions 48
Fuel Tank 49
Draining 49
Removal & Installation 50
Relieving Fuel System
Pressure 48
With the HDS 48
Without the HDS 49
Throttle Body 53
Removal & Installation 53

**HEATING & AIR CONDITIONING
SYSTEM 54**

Blower Motor 54
Removal & Installation 54
Heater Core....................55
Removal & Installation 55
Heater Unit....................56
Removal & Installation 56

PRECAUTIONS 8

**SPECIFICATIONS AND
MAINTENANCE CHARTS 3**

Brake Specifications 7
Camshaft Specifications 5
Capacities 4
Crankshaft and Connecting
Rod Specifications....................5
Engine and Model Year
Identification....................3
Engine Tune-Up Specifications.......3
Fluid Specifications 4

General Engine Specifications 3
Piston and Ring Specifications.......5
Scheduled Maintenance
Intervals....................7
Tire Wheel and Ball Joint
Specifications 6
Torque Specifications 6
Valve Specifications....................4
Wheel Alignment 6

STEERING 57

Power Steering Gear 57
Removal & Installation 57

SUSPENSION 61

FRONT SUSPENSION 61
Lower Ball Joints 61
Removal & Installation 61
Lower Control Arms....................61
Removal & Installation 61
Stabilizer Bar & Links....................63
Removal & Installation 63
Steering Knuckle....................65
Removal & Installation 65
Struts (Damper/Spring) 66
Overhaul....................67
Removal & Installation 66
Upper Ball Joints 68
Removal & Installation 68
Upper Control Arms....................68
Removal & Installation 68
Wheel Hubs & Bearings 69
Removal & Installation 69
REAR SUSPENSION 70
Control Arms (Lower &
Upper) 70
Removal & Installation 70
Control Rods (Control Arm) 71
Removal & Installation 71
Stabilizer Bar & Links....................72
Removal & Installation 72
Struts (Damper/Spring) 73
Overhaul....................73
Removal & Installation 73
Wheel Hubs & Bearings 75
Removal, Repacking, &
Installation 75

SPECIFICATIONS AND MAINTENANCE CHARTS

ENGINE AND VEHICLE IDENTIFICATION

			Engine				Model Year	
Code	Liters (cc)	Cu. In.	Cyl.	Fuel Sys.	Engine Type	Eng. Mfg.	Code ①	Year
J35Z6	3.5 (3490)	213	6	PGM-FI	SOHC	Honda	B	2011
J37A4	3.7 (3704)	226	6	PGM-FI	SOHC	Honda	C	2012

PGM-FI: Programmed Fuel Injection

SOHC: Single Overhead Camshaft

① 10th digit of the Vehicle Identification Number (VIN)

71051_ACTL_C0001

GENERAL ENGINE SPECIFICATIONS

Year	Model	Engine Displacement Liters	Engine ID	Net Horsepower @ rpm	Net Torque @ rpm (ft. lbs.)	Bore x Stroke (in.)	Com-pression Ratio	Oil Pressure @ rpm
2011	TL	3.5	J35Z6	280@6200	254@5000	3.50x3.66	11.2:1	71@3000
	TL	3.7	J37A4	305@6300	275@5000	3.54x3.78	11.2:1	71@3000
2012	TL	3.5	J35Z6	280@6200	254@5000	3.50x3.66	11.2:1	71@3000
	TL	3.7	J37A4	305@6300	275@5000	3.54x3.78	11.2:1	71@3000

71051_ACTL_C0002

GASOLINE ENGINE TUNE-UP SPECIFICATIONS

Year	Engine Displacement Liters	Engine ID	Spark Plug Gap (in.)	Ignition Timing (deg.) MT	AT	Fuel Pump (psi)	Idle Speed (rpm) MT	AT	Valve Clearance In.	Ex.
2011	3.5	J35Z6	0.039-0.043	NA	8-12 ①	57-64 ②	NA	630-730	0.008-0.009	0.011-0.013
	3.7	J37A4	0.039-0.043	NA	8-12 ①	57-64 ②	700-800	630-730	0.008-0.009	0.011-0.013
2012	3.5	J35Z6	0.039-0.043	NA	8-12 ①	57-64 ②	NA	630-730	0.008-0.009	0.011-0.013
	3.7	J37A4	0.039-0.043	NA	8-12 ①	57-64 ②	700-800	630-730	0.008-0.009	0.011-0.013

NA: Not Available

NOTE: The Vehicle Emission Control Information label reflects specification changes during production and must be used if they differ from this chart.

① Before Top Dead Center

② Pressure with fuel pressure gauge connected

71051_ACTL_C0003

CAPACITIES

Year	Model	Engine Displacement Liters	Engine ID	Engine Oil with Filter (qts.)	Transmission (pts.)		Fuel Tank (gal.)	Cooling System (qts.)
					6-Spd	Auto.		
2011	TL	3.5	J35Z6	4.5	—	①	18.5	6.6
		3.7	J37A4	4.5	4.4	①	18.5	6.6
2012	TL	3.5	J35Z6	4.5	—	①	18.5	6.6
		3.7	J37A4	4.5	4.4	①	18.5	6.6

NOTE: All capacities are approximate. Add fluid gradually and ensure a proper fluid level is obtained.

NOTE: Capacities given are service capacities, not overhaul capacities

① Fluid change with SH-AWD: 6.0 pts.

　Fluid change without SH-AWD: 7.0 pts.

71051_ACTL_C0004

FLUID SPECIFICATIONS

Year	Model	Engine Displacement Liters	Engine Oil	Man. Trans.	Auto. Trans.	Transfer Case	Power Steering Fluid	Brake Master Cylinder	Cooling System
2011	TL	3.5	5W-20 Acura	—	Acura ATF-Z1	SAE 90 GL4 /GL5	Acura PS Fluid	Acura DOT 3	①
	TL	3.7	5W-20 Acura	Acura MTF	Acura ATF-Z1	SAE 90 GL4 /GL5	Acura PS Fluid	Acura DOT 3	①
2012	TL	3.5	5W-20 Acura	—	Acura ATF-Z1	SAE 90 GL4 /GL5	Acura PS Fluid	Acura DOT 3	①
	TL	3.7	5W-20 Acura	Acura MTF	Acura ATF-Z1	SAE 90 GL4 /GL5	Acura PS Fluid	Acura DOT 3	①

DOT: Department of Transportation

① Acura Long Life Antifreeze/Coolant-Type2

71051_ACTL_C0005

VALVE SPECIFICATIONS

Year	Engine Displacement Liters	Engine ID/VIN	Seat Angle (deg.)	Face Angle (deg.)	Spring Test Pressure (lbs. @ in.)	Spring Installed Height (in.)	Stem-to-Guide Clearance (in.)		Stem Diameter (in.)	
							Intake	Exhaust	Intake	Exhaust
2011	3.5	J35Z6	45	45	NA	NA	0.0008-0.0018	0.0022-0.0031	0.2159-0.2163	0.2146-0.2150
	3.7	J37A4	45	45	NA	NA	0.0008-0.0018	0.0022-0.0031	0.2159-0.2163	0.2146-0.2150
2012	3.5	J35Z6	45	45	NA	NA	0.0008-0.0018	0.0022-0.0031	0.2159-0.2163	0.2146-0.2150
	3.7	J37A4	45	45	NA	NA	0.0008-0.0018	0.0022-0.0031	0.2159-0.2163	0.2146-0.2150

NA: Not Available

71051_ACTL_C0006

CAMSHAFT AND BEARING SPECIFICATIONS

All measurements are given in inches.

Year	Engine Displacement Liters	Engine ID	Journal Diameter	Brg. Oil Clearance	Shaft End-play	Runout	Journal Bore	Lobe Height Intake	Lobe Height Exhaust
2011	3.5	J35Z6	NA	0.0020-0.0035	0.0020-0.0080	0.0010	NA	①	1.4472
	3.7	J37A4	NA	0.0020-0.0035	0.0020-0.0080	0.0010	NA	②	③
2012	3.5	J35Z6	NA	0.0020-0.0035	0.0020-0.0080	0.0010	NA	①	1.4472
	3.7	J37A4	NA	0.0020-0.0035	0.0020-0.0080	0.0010	NA	②	③

NA: Not Available

① Primary: 1.3504 inches
   Secondary: 1.4024 inches
② Primary: 1.3504 innches
   Secondary: 1.4146 inches
③ Primary: 1.4462 inches
   Secondary: 1.4713 inches

71051_ACTL_C0009

CRANKSHAFT AND CONNECTING ROD SPECIFICATIONS

All measurements are given in inches.

Year	Engine Displacement Liters	Engine ID	Crankshaft Main Brg. Journal Dia.	Crankshaft Main Brg. Oil Clearance	Crankshaft Shaft End-play	Thrust on No.	Connecting Rod Journal Diameter	Connecting Rod Oil Clearance	Connecting Rod Side Clearance
2011	3.5	J35Z6	2.8337-2.8346	0.0007-0.0018	0.0040-0.0140	3	2.1644-2.1654	0.0008-0.0017	0.0060-0.0140
	3.7	J37A4	2.8337-2.8346	0.0007-0.0018	0.0040-0.0140	3	2.2431-2.2441	0.0008-0.0017	0.0060-0.0140
2012	3.5	J35Z6	2.8337-2.8346	0.0007-0.0018	0.0040-0.0140	3	2.1644-2.1654	0.0008-0.0017	0.0060-0.0140
	3.7	J37A4	2.8337-2.8346	0.0007-0.0018	0.0040-0.0140	3	2.2431-2.2441	0.0008-0.0017	0.0060-0.0140

71051_ACTL_C0007

PISTON AND RING SPECIFICATIONS

All measurements are given in inches

Year	Engine Displacement Liters	Engine ID	Piston Clearance	Ring Gap Top Compression	Ring Gap Bottom Compression	Ring Gap Oil Control	Ring Side Clearance Top Compression	Ring Side Clearance Bottom Compression	Ring Side Clearance Oil Control
2011	3.5	J35Z6	0.0006-0.0016	0.0080-0.0014	0.0160-0.0220	0.0080-0.0280	0.0022-0.0031	0.0012-0.0022	NA
	3.7	J37A4	0.0006-0.00160	0.0120-0.0160	0.0160-0.0220	0.0080-0.0140	0.0022-0.0033	0.0012-0.0024	NA
2012	3.5	J35Z6	0.0006-0.0016	0.0080-0.0014	0.0160-0.0220	0.0080-0.0280	0.0022-0.0031	0.0012-0.0022	NA
	3.7	J37A4	0.0006-0.00160	0.0120-0.0160	0.0160-0.0220	0.0080-0.0140	0.0022-0.0033	0.0012-0.0024	NA

NA: Not Applicable

71051_ACTL_C0008

TORQUE SPECIFICATIONS
All readings in ft. lbs.

Year	Engine Displacement Liters	Engine ID/VIN	Cylinder Head Bolts	Main Bearing Bolts	Rod Bearing Bolts	Crankshaft Damper Bolts	Flexplate/ Flywheel Bolts	Manifold Intake	Manifold Exhaust	Spark Plugs	Oil Pan Drain Plug
2011	3.5	J35Z6	①	②	③	④	54/76	16	23	13	29
	3.7	J37A4	①	②	③	④	54/76	16	23	13	29
2012	3.5	J35Z6	①	②	③	④	54/76	16	23	13	29
	3.7	J37A4	①	②	③	④	54/76	16	23	13	29

① Step 1: 22 ft. lbs.

  Step 2: Rotate 90 degrees

  Step 3: Rotate an additional 90 degrees

  Step 4 (new bolts only): additional 90 degrees

② Step 1: 54 ft. lbs.

  Step 2: Repeat Step 1 to ensure 54 ft. lbs.

  Step 3: Side bolts to 36 ft. lbs.

  Step 4: Repeat Step 3 to ensure 36 ft. lbs.

③ Step 1: 14 ft. lbs.

  Step 2: Additional 90 degrees

④ Step 1: 47 ft. lbs.

  Step 2: Additional 60 degrees

71051_ACTL_C0010

WHEEL ALIGNMENT

Year	Model		Caster Range (+/-Deg.)	Caster Preferred Setting (Deg.)	Camber Range (+/-Deg.)	Camber Preferred Setting (Deg.)	Toe-in (in.)
2011	TL	F	0.45	+3.28	+0.30, -0.45	-0.30	0 +/- 0.08
		R	—	—	+0.30, -0.45	-1.00	0 +/- 0.08
2012	TL	F	0.45	+3.28	+0.30, -0.45	-0.30	0 +/- 0.08
		R	—	—	+0.30, -0.45	-1.00	0 +/- 0.08

71051_ACTL_C0011

TIRE, WHEEL AND BALL JOINT SPECIFICATIONS

Year	Model	OEM Tires Standard	OEM Tires Optional	Tire Pressures (psi) Front	Tire Pressures (psi) Rear	Wheel Size	Ball Joint Inspection	Lug Nut (ft. lbs.)
2011	TL	P245/45VR18	NS	32	32	18 x 8	NS	94
2012	TL	P245/45VR18	NS	32	32	18 x 8	NS	94

OEM: Original Equipment Manufacturer

PSI: Pounds Per Square Inch

NS: Not Specified by manufacturer

71051_ACTL_C0012

BRAKE SPECIFICATIONS
All measurements in inches unless noted

| Year | Model | | Brake Disc | | | Brake Drum Diameter | | | Minimum Lining Thickness | | Brake Caliper | |
			Original Thickness	Minimum Thickness	Maximum Runout	Original Inside Diameter	Max. Wear Limit	Maximum Machine Diameter	Front	Rear	Bracket Bolts (ft. lbs.)	Mounting Bolts (ft. lbs.)
2011	TL	F	1.100-1.110	1.020	0.002	—	—	—	0.06	—	101	53
		R	0.430-0.440	0.350	0.002	—	—	—	—	0.06	65	16
2012	TL	F	1.100-1.110	1.020	0.002	—	—	—	0.06	—	101	53
		R	0.430-0.440	0.350	0.002	—	—	—	—	0.06	65	16

F: Front

R: Rear

71051_ACTL_C0013

SCHEDULED MAINTENANCE INTERVALS
Acura—TL

| TO BE SERVICED | TYPE OF SERVICE | SERVICE INTERVALS | | | | | | |
		Symbol A	Symbol B	1	2	3	4	5
Air cleaner element	Replace				✓			
Brake fluid	Inspect		✓					
Brake hoses and lines	Inspect		✓					
Clutch fluid (if equipped)	Inspect		✓					
Drive belt	Inspect				✓			
Driveshaft boots	Inspect		✓					
Dust & pollen filter	Replace				✓			
Engine coolant	Inspect		✓					
Engine coolant	Replace							✓
Engine oil	Replace	✓						
Engine oil & filter	Replace		✓					
Exhaust system	Inspect		✓					
Front & rear brakes	Inspect		✓					
Fuel lines	Inspect		✓					
Parking brake adjustment	Inspect		✓					
Power steering fluid	Inspect		✓					
Spark plugs	Replace						✓	
Suspension components	Inspect		✓					
Tie-rod ends, steering gearbox and boots	Inspect		✓					
Timing belt	Replace						✓	
Tires	Rotate			✓				
Transmission fluid	Inspect		✓					
Transmission fluid	Replace					✓		
Valve clearance	Inspect						✓	
Windshield washer fluid	Inspect		✓					

71051_ACTL_C0014

PRECAUTIONS

Before servicing any vehicle, please be sure to read all of the following precautions, which deal with personal safety, prevention of component damage, and important points to take into consideration when servicing a motor vehicle:

• Never open, service or drain the radiator or cooling system when the engine is hot; serious burns can occur from the steam and hot coolant.

• Observe all applicable safety precautions when working around fuel. Whenever servicing the fuel system, always work in a well-ventilated area. Do not allow fuel spray or vapors to come in contact with a spark, open flame, or excessive heat (a hot drop light, for example). Keep a dry chemical fire extinguisher near the work area. Always keep fuel in a container specifically designed for fuel storage; also, always properly seal fuel containers to avoid the possibility of fire or explosion. Refer to the additional fuel system precautions later in this section.

• Fuel injection systems often remain pressurized, even after the engine has been turned OFF. The fuel system pressure must be relieved before disconnecting any fuel lines. Failure to do so may result in fire and/or personal injury.

• Brake fluid often contains polyglycol ethers and polyglycols. Avoid contact with the eyes and wash your hands thoroughly after handling brake fluid. If you do get brake fluid in your eyes, flush your eyes with clean, running water for 15 minutes. If eye irritation persists, or if you have taken

brake fluid internally, IMMEDIATELY seek medical assistance.

• The EPA warns that prolonged contact with used engine oil may cause a number of skin disorders, including cancer. You should make every effort to minimize your exposure to used engine oil. Protective gloves should be worn when changing oil. Wash your hands and any other exposed skin areas as soon as possible after exposure to used engine oil. Soap and water, or waterless hand cleaner should be used.

• All new vehicles are now equipped with an air bag system, often referred to as a Supplemental Restraint System (SRS) or Supplemental Inflatable Restraint (SIR) system. The system must be disabled before performing service on or around system components, steering column, instrument panel components, wiring and sensors. Failure to follow safety and disabling procedures could result in accidental air bag deployment, possible personal injury and unnecessary system repairs.

• Always wear safety goggles when working with, or around, the air bag system. When carrying a non-deployed air bag, be sure the bag and trim cover are pointed away from your body. When placing a non-deployed air bag on a work surface, always face the bag and trim cover upward, away from the surface. This will reduce the motion of the module if it is accidentally deployed. Refer to the additional air bag system precautions later in this section.

• Clean, high quality brake fluid from a sealed container is essential to the safe and

proper operation of the brake system. You should always buy the correct type of brake fluid for your vehicle. If the brake fluid becomes contaminated, completely flush the system with new fluid. Never reuse any brake fluid. Any brake fluid that is removed from the system should be discarded. Also, do not allow any brake fluid to come in contact with a painted surface; it will damage the paint.

• Never operate the engine without the proper amount and type of engine oil; doing so WILL result in severe engine damage.

• Timing belt maintenance is extremely important. Many models utilize an interference-type, non-freewheeling engine. If the timing belt breaks, the valves in the cylinder head may strike the pistons, causing potentially serious (also time-consuming and expensive) engine damage. Refer to the maintenance interval charts for the recommended replacement interval for the timing belt, and to the timing belt section for belt replacement and inspection.

• Disconnecting the negative battery cable on some vehicles may interfere with the functions of the on-board computer system(s) and may require the computer to undergo a relearning process once the negative battery cable is reconnected.

• When servicing drum brakes, only disassemble and assemble one side at a time, leaving the remaining side intact for reference.

• Only an MVAC-trained, EPA-certified automotive technician should service the air conditioning system or its components.

BRAKES

GENERAL INFORMATION

PRECAUTIONS

• Certain components within the ABS system are not intended to be serviced or repaired individually.

• Do not use rubber hoses or other parts not specifically specified for and ABS system. When using repair kits, replace all parts included in the kit. Partial or incorrect repair may lead to functional problems and require the replacement of components.

• Lubricate rubber parts with clean, fresh brake fluid to ease assembly. Do not use shop air to clean parts; damage to rubber components may result.

• Use only DOT 3 brake fluid from an unopened container.

• If any hydraulic component or line is removed or replaced, it may be necessary to bleed the entire system.

• A clean repair area is essential. Always clean the reservoir and cap thoroughly before removing the cap. The slightest amount of dirt in the fluid may plug an orifice and impair the system function. Perform repairs after components have been thoroughly cleaned; use only denatured alcohol to clean components. Do not allow ABS components to come into contact with any substance containing mineral oil; this includes used shop rags.

• The Anti-Lock control unit is a microprocessor similar to other computer units in the vehicle. Ensure that the ignition switch is OFF before removing or installing con-

ANTI-LOCK BRAKE SYSTEM (ABS)

troller harnesses. Avoid static electricity discharge at or near the controller.

• If any arc welding is to be done on the vehicle, the control unit should be unplugged before welding operations begin.

SPEED SENSORS

REMOVAL & INSTALLATION

Front

See Figure 1.

1. Turn the ignition switch to LOCK (0), or press the engine start/stop button to select the OFF mode.
2. Remove the front wheel.
3. Release the clamp, then disconnect the wheel speed sensor connector.

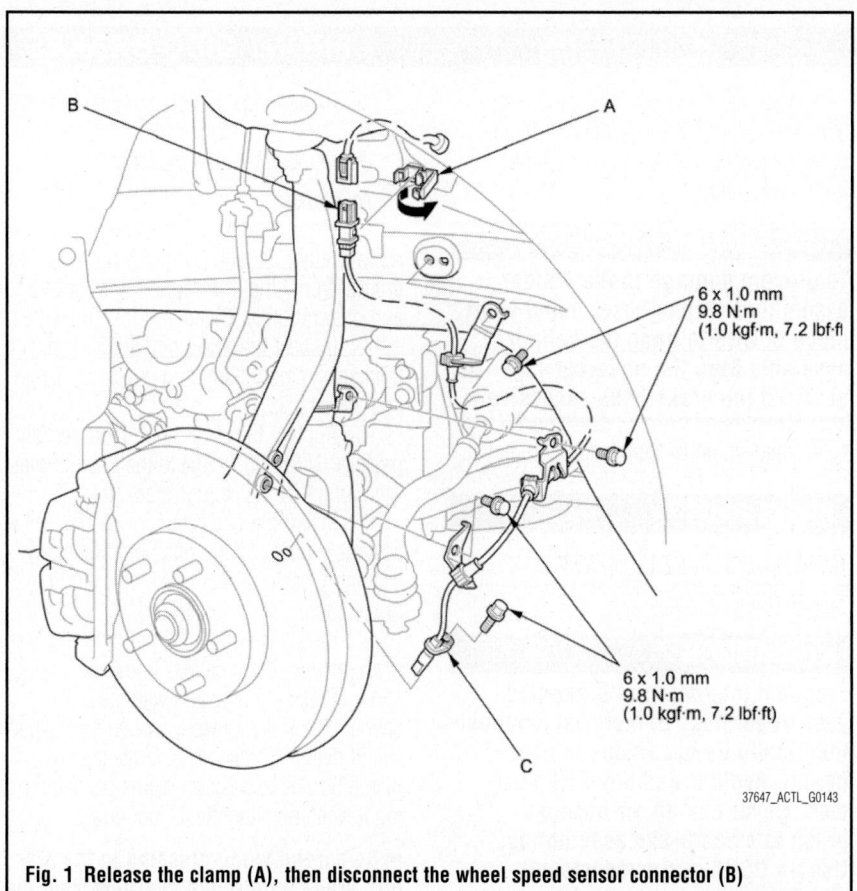

6 x 1.0 mm
9.8 N·m
(1.0 kgf·m, 7.2 lbf·ft)

6 x 1.0 mm
9.8 N·m
(1.0 kgf·m, 7.2 lbf·ft)

37647_ACTL_G0143

Fig. 1 Release the clamp (A), then disconnect the wheel speed sensor connector (B)

4. Remove the bolts and the wheel speed sensor.

To install:

5. Install the wheel speed sensor in the reverse order of removal, and note these items:

- Do not twist the sensor wires.
- If the wheel speed sensor comes in contact with the wheel bearing, it is faulty.
- Make sure there is no debris in the sensor mounting hole.

6. Start the engine, and make sure the ABS and the VSA indicators go off.

7. Test-drive the vehicle, and make sure the ABS and the VSA indicators do not come on.

Rear

1. Turn the ignition switch to LOCK (0), or press the engine start/stop button to select the OFF mode.

2. Remove the rear wheel.

3. Release the clamp, then disconnect the wheel speed sensor connector.

4. Remove the clamps, the bolt, and the wheel speed sensor.

To install:

5. Install the wheel speed sensor in the reverse order of removal, and note these items:

- Do not twist the sensor wires.
- If the wheel speed sensor comes in contact with the hub bearing unit, it is faulty.
- Make sure there is no debris in the sensor mounting hole.
- Lubricate the O-ring on the wheel speed sensor.

6. Start the engine, and make sure the ABS and the VSA indicators go off.

7. Test-drive the vehicle, and make sure the ABS and the VSA indicators do not come on.

BRAKES   BLEEDING THE BRAKE SYSTEM

BLEEDING PROCEDURE

BLEEDING THE ABS SYSTEM

✳✳ WARNING

Read and observe the "PRECAU-TIONS" in this section before starting work.

1. The reservoir on the master cylinder must be at the MAX (upper) level mark at the start of the bleeding procedure and checked after bleeding each brake caliper. Add fluid as required.

2. Make sure the brake fluid level in the reservoir is at the MAX (upper) level line.

3. Slide a piece of clear plastic hose over the first bleed screw, and submerge the other end in a container of new brake fluid.

4. Have someone slowly pump the brake pedal several times, and then apply steady pressure.

5. Starting at the left-front, loosen the brake bleed screw to allow air to escape from the system. Then tighten the bleed screw securely.

6. Repeat the procedure for each wheel in the sequence shown following until air bubbles no longer appear in the fluid.

7. Refill the master cylinder reservoir to the MAX (upper) level line.

BRAKES   **FRONT DISC BRAKES**

Dust and dirt accumulating on brake parts during normal use may contain asbestos fibers from production or aftermarket brake linings. Breathing excessive concentrations of asbestos fibers can cause serious bodily harm. Exercise care when servicing brake parts. Do not sand or grind brake lining unless equipment used is designed to contain the dust residue. Do not clean brake parts with compressed air or by dry brushing. Cleaning should be done by dampening the brake components with a fine mist of water, then wiping the brake components clean with a dampened cloth. Dispose of cloth and all residue containing asbestos fibers in an impermeable container with the appropriate label. Follow practices prescribed by the Occupational Safety and Health Administration (OSHA) and the Environmental Protection Agency (EPA) for the handling, processing, and disposing of dust or debris that may contain asbestos fibers.

BRAKE CALIPER

REMOVAL & INSTALLATION

See Figure 2.

➡ **Keep any grease off the brake disc and the brake pads.**

1. Raise and support the vehicle.
2. Remove the front wheel.

3. Remove the brake hose mounting bolt, the brake caliper bracket mounting bolts, then remove the caliper assembly from the knuckle.

❊❊ **CAUTION**

To prevent damage to the caliper assembly or brake hose, use a short piece of wire to hang the caliper assembly from the undercarriage. Do not twist the brake hose excessively.

4. Installation is the reverse of removal.

BRAKE PADS

REMOVAL & INSTALLATION
See Figures 3 and 4.

❊❊ **CAUTION**

Frequent inhalation of brake pad dust, regardless of material composition, could be hazardous to your health. Avoid breathing dust particles. Never use an air hose or brush to clean brake assemblies. Use an OSHA-approved vacuum cleaner.

1. Remove some brake fluid from the master cylinder.
2. Raise and support the vehicle.
3. Remove the front wheels.
4. Remove the brake hose mounting bolt, the flange bolt, and pivot the caliper up out of the way.
5. Check the hose and the pin boots for damage and deterioration.

6. Remove the pad shims and the brake pads.
7. Remove the pad retainers.

To install:

8. Clean the caliper bracket thoroughly; remove any rust, and check for grooves and cracks. Verify that the caliper pins move in and out smoothly. Clean and lube if needed.
9. Inspect the brake disc for runout, thickness, parallelism, and check for damage and cracks.
10. Apply a thin coat of M-77 assembly paste (P/N 08798-9010) to the retainer mating surface of the caliper bracket.
11. Install the pad retainers. Wipe excess assembly paste off the retainers. Keep any assembly paste off the brake disc and the brake pads.
12. Install the brake caliper piston compressor tool on the caliper body.
13. Press in the piston with the brake caliper piston compressor tool so the caliper will fit over the brake pads. Make sure the piston boot is in position to prevent damaging it when pivoting the caliper down.

➡ **Be careful when pressing in the piston; brake fluid might overflow from the master cylinder's reservoir. If brake fluid gets on any painted surface, wash it off immediately with water.**

14. Remove the brake caliper piston compressor tool.
15. Apply a thin coat of M-77 assembly paste (P/N 08798-9010) to the pad side of the shims, the back of the brake pads. Wipe excess assembly paste off the pad shims and the brake pads friction material.

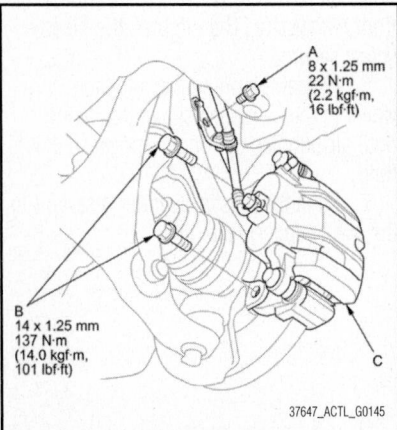

Fig. 2 Remove the brake hose mounting bolt (A), the brake caliper bracket mounting bolts (B), then remove the caliper assembly (C) from the knuckle

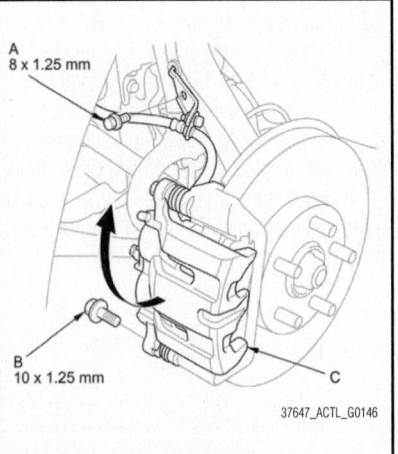

Fig. 3 Remove the brake hose mounting bolt (A), the flange bolt (B), and pivot the caliper (C) up out of the way

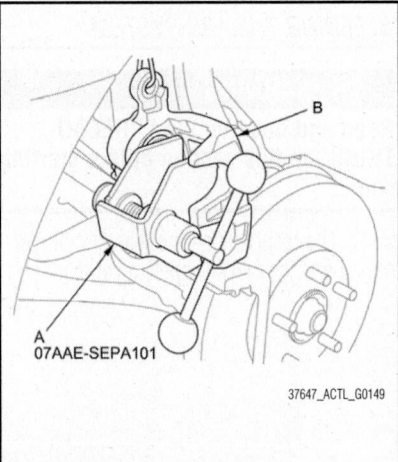

Fig. 4 Install the brake caliper piston compressor tool (A) on the caliper body (B)

> **⁂ CAUTION**
>
> **Keep grease and assembly paste off the brake disc and the brake pads. Contaminated brake disc or brake pads reduce stopping ability.**

16. Install the brake pads and pad shims correctly. Install the brake pad with the wear indicator on the upper inside. If you are reusing the brake pads, always reinstall the brake pads in their original positions to prevent a temporary loss of braking efficiency.

17. Pivot the caliper down into position. Install the flange bolt, and tighten it to 53 ft. lbs. (72 Nm).

18. Install the brake hose mounting bolt.

19. Clean the mating surface between the brake disc and the inside of the wheel, then install the front wheels.

20. Press the brake pedal several times to make sure the brakes work.

➡Engagement may require a greater pedal stroke immediately after the brake pads have been replaced as a set. Several applications of the brake pedal will restore the normal pedal stroke.

21. Add brake fluid as needed.

22. After installation, check for leaks at hose and line joints or connections, and retighten if necessary.

23. Test-drive the vehicle, then check for leaks.

BRAKES

> **⁂ CAUTION**
>
> **Dust and dirt accumulating on brake parts during normal use may contain asbestos fibers from production or aftermarket brake linings. Breathing excessive concentrations of asbestos fibers can cause serious bodily harm. Exercise care when servicing brake parts. Do not sand or grind brake lining unless equipment used is designed to contain the dust residue. Do not clean brake parts with compressed air or by dry brushing. Cleaning should be done by dampening the brake components with a fine mist of water, then wiping the brake components clean with a dampened cloth. Dispose of cloth and all residue containing asbestos fibers in an impermeable container with the appropriate label. Follow practices prescribed by the Occupational Safety and Health Administration (OSHA) and the Environmental Protection Agency (EPA) for the handling, processing, and disposing of dust or debris that may contain asbestos fibers.**

BRAKE CALIPER

REMOVAL & INSTALLATION
See Figure 5.

➡Keep any grease off the brake disc/drum and the brake pads.

1. Raise and support the vehicle.
2. Remove the rear wheel.
3. Release the parking brake lever fully.
4. Remove the brake hose mounting bolt and the brake caliper bracket mounting bolts, then remove the caliper assembly from the knuckle.

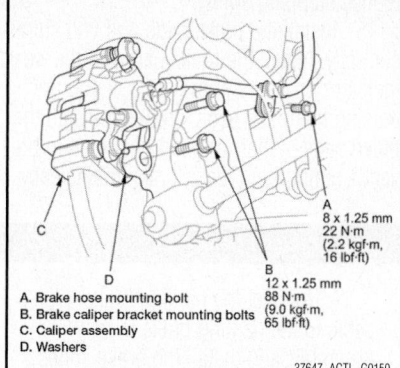

A. Brake hose mounting bolt
B. Brake caliper bracket mounting bolts
C. Caliper assembly
D. Washers

A 8 x 1.25 mm
22 N·m
(2.2 kgf·m,
16 lbf·ft)

B 12 x 1.25 mm
88 N·m
(9.0 kgf·m,
65 lbf·ft)

37647_ACTL_G0150

Fig. 5 Remove the brake hose mounting bolt and the brake caliper bracket mounting bolts, then remove the caliper assembly from the knuckle

> **⁂ CAUTION**
>
> **To prevent damage to the caliper assembly or brake hose, use a short piece of wire to hang the caliper assembly from the undercarriage. Do not twist the brake hose excessively.**

5. Installation is the reverse of removal.

6. Tighten the caliper bracket mounting bolts to 65 ft. lbs. (88 Nm).

DISC BRAKE PADS

REMOVAL & INSTALLATION
See Figures 5 and 6.

> **⁂ CAUTION**
>
> **Frequent inhalation of brake pad dust, regardless of material composition, could be hazardous to your health. Avoid breathing dust particles. Never use an air hose or brush to clean brake assemblies. Use an OSHA-approved vacuum cleaner.**

REAR DISC BRAKES

1. Remove some brake fluid from the master cylinder.
2. Raise and support the vehicle.
3. Remove the rear wheels.
4. Remove the brake hose mounting bolt and the brake caliper bracket mounting bolts, then remove the caliper assembly from the knuckle.

> **⁂ CAUTION**
>
> **To prevent damage to the caliper assembly or brake hose, use a short piece of wire to hang the caliper assembly from the undercarriage. Do not twist the brake hose excessively.**

5. Remove the pad shims and the brake pads.
6. Remove the upper and lower pad retainers.

➡The upper and lower pad retainers are different. During installation, make sure the pad retainers are in their proper positions.

To install:

7. Clean the caliper bracket thoroughly; remove any rust, and check for grooves and

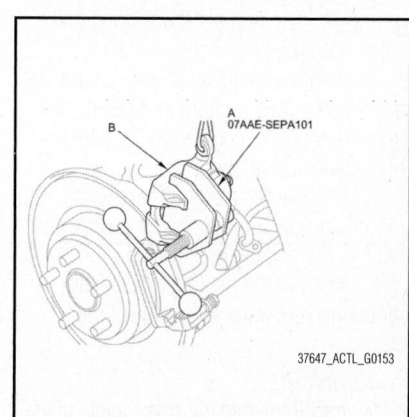

A 07AAE-SEPA101

B

37647_ACTL_G0153

Fig. 6 Install the brake caliper piston compressor tool (A) on the caliper body (B)

cracks. Verify that the caliper pins move in and out smoothly. Clean and lube if needed.

8. Inspect the brake disc/drum for runout, thickness, parallelism, and check for damage and cracks.

9. Apply a thin coat of M-77 assembly paste (P/N 08798-9010) to the retainer mating surface of the caliper bracket (indicated by the arrows).

10. Install the upper and lower pad retainers. Wipe excess assembly paste off the retainers. Keep the assembly paste off the brake disc/drum and the brake pads.

11. Install the brake caliper piston compressor tool on the caliper body.

12. Press in the piston with the brake caliper piston compressor tool so the caliper will fit over the brake pads. Make sure the piston boot is in position to prevent damaging it when pivoting the caliper down.

➡Be careful when pressing in the piston; brake fluid might overflow from the master cylinder's reservoir. If brake fluid gets on any painted surface, wash it off immediately with water.

13. Remove the brake caliper piston compressor tool.

14. Apply a thin coat of M-77 assembly paste (P/N 08798-9010) to the pad side of the shims and the back of the brake pads. Wipe excess assembly paste off the pad shims and the brake pads friction material. Keep grease and assembly paste off the brake disc/drum and the brake pads. Contaminated brake disc/drum or brake pads reduce stopping ability.

15. Install the brake pads and pad shims correctly. Install the brake pad with the wear indicator on the bottom inside. If you are reusing the brake pads, always reinstall the brake pads in their original positions to prevent a temporary loss of braking efficiency.

16. Pivot the caliper down into position. Install the flange bolt, and tighten it to 16 ft. lbs. (22 Nm).

17. Install the brake hose mounting bolt.

18. Clean the mating surfaces between the brake disc/drum and the inside of the wheel, then install the rear wheels.

19. Press the brake pedal several times to make sure the brakes work.

➡Engagement may require a greater pedal stroke immediately after the brake pads have been replaced as a set. Several applications of the brake pedal will restore the normal pedal stroke.

20. Add brake fluid as needed.

21. After installation, check for leaks at hose and line joints or connections, and retighten if necessary.

22. Test-drive the vehicle, then recheck for leaks.

BRAKES

PARKING BRAKE CABLES

REMOVAL & INSTALLATION

See Figure 7.

➡The parking brake cables must not be bent or distorted. This will lead to stiff operation and premature cable failure.

1. Release the parking brake lever fully.
2. Remove the center console.
3. Remove the wire guide base, then disconnect the parking brake cables from the equalizer.
4. Without SH-AWD: Remove the heat shield from over the catalytic converter.
5. Remove the parking brake cable grommets, and pull out the parking brake cable from the body floor.
6. Remove the parking brake shoes (as described in this section) and disconnect the parking brake cable from the parking brake lever assembly.
7. Remove the parking brake cable mounting nuts from the backing plate and remove the parking brake cable.
8. Remove the parking brake cable mounting hardware, then remove the cable.

To install:

9. Install the parking brake cable in the reverse order of removal, and note these items:

   a. Be careful not to bend or distort the cable.

   b. Connect the parking brake cable to the parking brake lever assembly, and install the brake shoes and drum.

   c. Adjust the parking brake. Apply the parking brake firmly 10 times then adjust it again.

PARKING BRAKE

PARKING BRAKE SHOES

REMOVAL & INSTALLATION

See Figures 8 and 9.

❋❋ CAUTION

Frequent inhalation of brake pad dust, regardless of material composition,

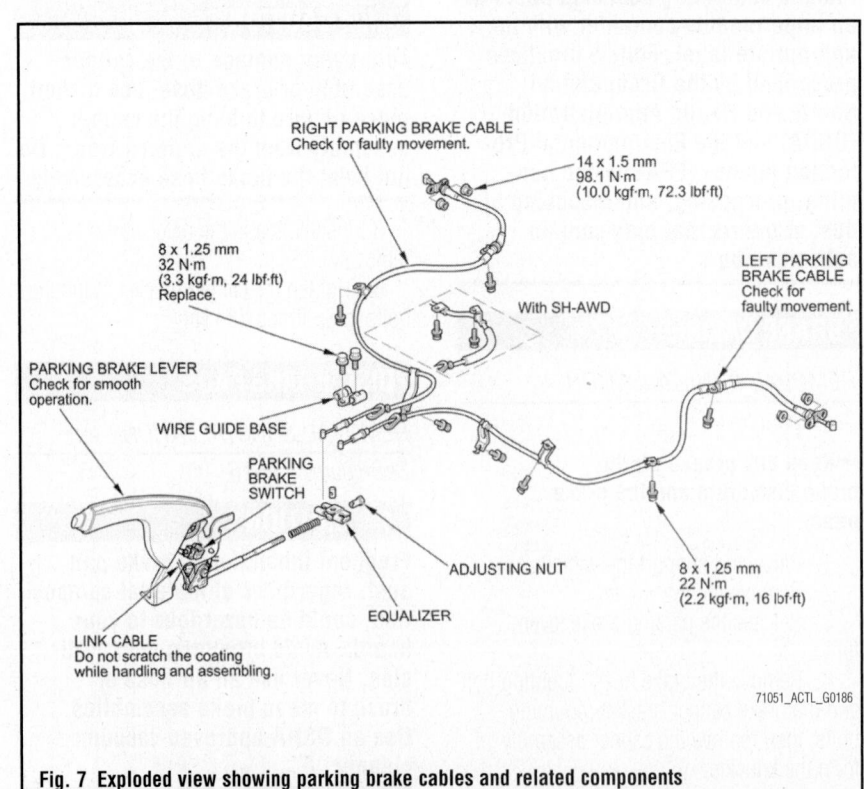

Fig. 7 Exploded view showing parking brake cables and related components

71051_ACTL_G0186

could be hazardous to your health. Avoid breathing dust particles. Never use an air hose or brush to clean brake assemblies. Use an OSHA-approved vacuum cleaner.

1. Raise and support the vehicle.
2. Remove the rear wheels.
3. Release the parking brake, and remove the rear brake disc/drum.
4. Disconnect and remove the upper return spring.
5. Remove the adjuster assembly.
6. Disconnect and remove the lower return spring.
7. Remove the parking brake lever assembly, and disconnect it from the parking brake cable end.
8. Remove the tension pins by pushing respective retainer springs, and turning the pin.
9. Remove the parking brake shoes.

To install:
10. Apply Molykote 44MA grease to the shoe ends, sliding surfaces, and opposite edges of the parking brake shoe as shown. Wipe off any excess. Keep grease off the brake linings.
11. Reinstall the tension pins and retainer springs.
12. Connect the parking brake cable end

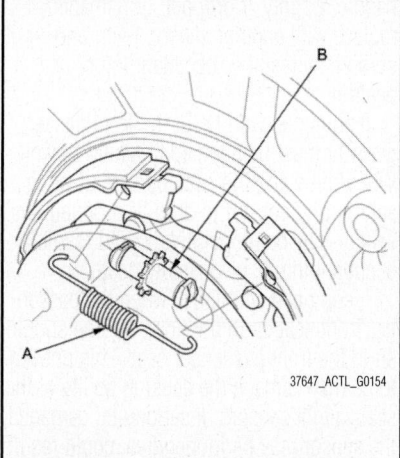

Fig. 8 Disconnect and remove the upper return spring (A)and adjuster assembly (B)

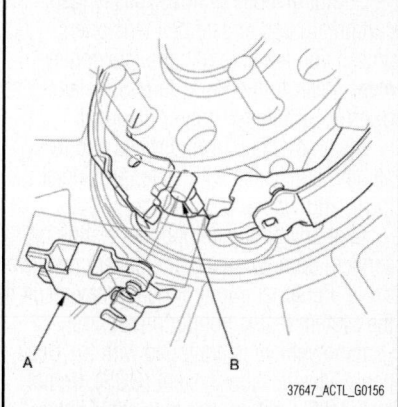

Fig. 9 Remove the parking brake lever assembly (A), and disconnect it from the parking brake cable end (B)

to the parking brake lever assembly, and install the parking brake lever assembly.
13. Reinstall the lower return spring.
14. Clean the threaded portions of the clevis pin A, and coat the threads of the clevis pin A with grease. Clean the sliding surface of the clevis pin B, and coat the sliding surface of the clevis pin B with grease. Install the clevis pin A and B on the adjuster and shorten the clevis pin A by turning the adjuster.

15. Position the brake shoe adjuster assembly on the parking brake shoes.
16. Reinstall the upper return spring.
17. Install the rear brake disc/drum and the rear brake caliper.
18. Do the major parking brake adjustment.
19. Clean the mating surfaces between the brake disc/drum and the inside of the wheel, then install the rear wheels.

CHASSIS ELECTRICAL AIR BAG (SUPPLEMENTAL RESTRAINT SYSTEM)

GENERAL INFORMATION

✳✳ CAUTION

These vehicles are equipped with an air bag system. The system must be disarmed before performing service on, or around, system components, the steering column, instrument panel components, wiring and sensors. Failure to follow the safety precautions and the disarming procedure could result in accidental air bag deployment, possible injury and unnecessary system repairs.

SERVICE PRECAUTIONS

Disconnect and isolate the battery negative cable before beginning any airbag system component diagnosis, testing, removal, or installation procedures. Allow system capacitor to discharge for two minutes before beginning any component service. This will disable the airbag system. Failure to disable the airbag system may result in

accidental airbag deployment, personal injury, or death.

Do not place an intact undeployed airbag face down on a solid surface. The airbag will propel into the air if accidentally deployed and may result in personal injury or death.

When carrying or handling an undeployed airbag, the trim side (face) of the airbag should be pointing towards the body to minimize possibility of injury if accidental deployment occurs. Failure to do this may result in personal injury or death.

Replace airbag system components with OEM replacement parts. Substitute parts may appear interchangeable, but internal differences may result in inferior occupant protection. Failure to do so may result in occupant personal injury or death.

Wear safety glasses, rubber gloves, and long sleeved clothing when cleaning powder residue from vehicle after an airbag deployment. Powder residue emitted from a deployed airbag can cause skin irritation. Flush affected area with cool water if irritation is experienced. If nasal or throat irrita-

tion is experienced, exit the vehicle for fresh air until the irritation ceases. If irritation continues, see a physician.

Do not use a replacement airbag that is not in the original packaging. This may result in improper deployment, personal injury, or death.

The factory installed fasteners, screws and bolts used to fasten airbag components have a special coating and are specifically designed for the airbag system. Do not use substitute fasteners. Use only original equipment fasteners listed in the parts catalog when fastener replacement is required.

During, and following, any child restraint anchor service, due to impact event or vehicle repair, carefully inspect all mounting hardware, tether straps, and anchors for proper installation, operation, or damage. If a child restraint anchor is found damaged in any way, the anchor must be replaced. Failure to do this may result in personal injury or death.

Deployed and non-deployed airbags may or may not have live pyrotechnic material within the airbag inflator.

Do not dispose of driver/passenger/curtain airbags or seat belt tensioners unless you are sure of complete deployment. Refer to the Hazardous Substance Control System for proper disposal.

Dispose of deployed airbags and tensioners consistent with state, provincial, local, and federal regulations.

After any airbag component testing or service, do not connect the battery negative cable. Personal injury or death may result if the system test is not performed first.

If the vehicle is equipped with the Occupant Classification System (OCS), do not connect the battery negative cable before performing the OCS Verification Test using the scan tool and the appropriate diagnostic information. Personal injury or death may result if the system test is not performed properly.

Never replace both the Occupant Restraint Controller (ORC) and the Occupant Classification Module (OCM) at the same time. If both require replacement, replace one, then perform the Airbag System test before replacing the other.

Both the ORC and the OCM store Occupant Classification System (OCS) calibration data, which they transfer to one another when one of them is replaced. If both are replaced at the same time, an irreversible fault will be set in both modules and the OCS may malfunction and cause personal injury or death.

If equipped with OCS, the Seat Weight Sensor is a sensitive, calibrated unit and must be handled carefully. Do not drop or handle roughly. If dropped or damaged, replace with another sensor. Failure to do so may result in occupant injury or death.

If equipped with OCS, the front passenger seat must be handled carefully as well. When removing the seat, be careful when setting on floor not to drop. If dropped, the sensor may be inoperative, could result in occupant injury, or possibly death.

If equipped with OCS, when the passenger front seat is on the floor, no one should sit in the front passenger seat. This uneven force may damage the sensing ability of the seat weight sensors. If sat on and damaged, the sensor may be inoperative, could result in occupant injury, or possibly death.

DISARMING THE SYSTEM

➡**Some systems store data in memory that is lost when the battery is disconnected. Do the following steps before disconnecting the battery.**

1. Make sure you have the anti-theft code(s) for the audio and/or the navigation system (if equipped).
2. For some models or if you're replacing the audio unit, it may be necessary to write down the audio presets (AM and FM), and the XM radio presets (if equipped), because the audio unit does not retain the presets after the battery is disconnected.
3. Make sure the ignition switch is in LOCK (0).
4. Disconnect and isolate the negative cable from the battery.

✳✳ CAUTION

Always disconnect the negative cable from the battery first.

5. Disconnect the positive cable from the battery.
6. Wait at least 3 minutes for the battery and SRS system to discharge before performing any repairs.

ARMING THE SYSTEM

➡**Some systems store data in memory that is lost when the battery is disconnected. Do the following steps to restore the systems back to normal operation.**

1. Clean the battery terminals.
2. Reconnect the positive cable to the battery first, then reconnect the negative cable to the battery.

✳✳ CAUTION

Always connect the positive cable to the battery first.

3. Apply multipurpose grease to the terminals to prevent corrosion.
4. Enter the anti-theft code(s) for the audio system and/or the navigation system (if equipped).
5. Enter the audio presets (if applicable), and enter the XM radio presets (if equipped).
6. Set the clock (for vehicles without navigation).
7. Do the steering column position memorization.

DRIVE TRAIN

DIFFERENTIAL CARRIER

REMOVAL & INSTALLATION

See Figures 10 through 12.

➡**Special Tools Required: Driveshaft Remover 07AAD-S3VA000.**

1. Raise and support the vehicle.
2. Drain the rear differential fluid.
3. Remove the rear wheels.
4. Remove exhaust pipe B and the muffler.
5. Remove the lower arm A mounting bolts and the lower arm B mounting bolts.
6. Disconnect the inboard joints from the rear differential using the driveshaft remover and a hammer.
7. Pull the knuckle outward, and separate the rear driveshaft inboard joints from the rear differential.
8. Make reference mark across the No. 3 propeller shaft and the rear differential companion flange. Separate the No. 3 propeller shaft from the rear differential.

➡**Suspend the propeller shaft with an appropriate size nylon strap.**

9. Disconnect the rear differential subharness connectors and remove the harness clips.
10. Remove the rear subframe rear bracket bolts.
11. Place a transmission jack under the rear differential.
12. Remove the rear differential rear mount bolts and the rear differential front mounting bracket nuts.
13. Slightly lower the rear differential on the transmission jack, then disconnect the breather hose.
14. Lower the rear differential on the transmission jack.
15. Remove the set ring from the rear differential.

To install:

16. Place the rear differential on a transmission jack.
17. Install a new set ring into the groove of the rear differential.
18. Raise the rear differential to the appropriate position, then connect the breather hose.
19. Raise the rear differential to the mounting level.
20. Loosely install new rear differential rear mount bolts and new rear differential front mounting bracket nuts.
21. Loosen the rear differential front mounting bracket bolts.
22. Tighten the rear differential rear mount bolt and the rear differential front mounting bracket nuts to the specified torque (shown in the illustration).

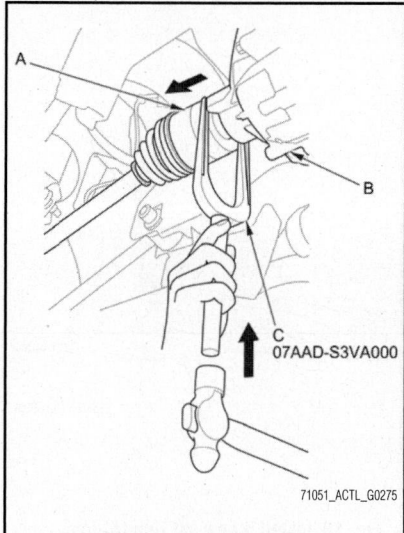

Fig. 10 Disconnect the inboard joints (A) from the rear differential (B) using the driveshaft remover (C) and a hammer.

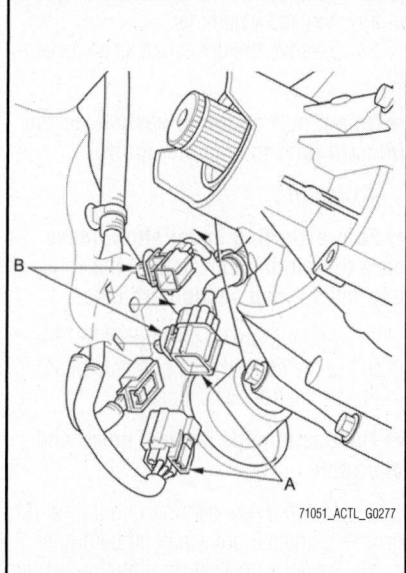

Fig. 11 Disconnect the rear differential subharness connectors (A), and remove the harness clips (B).

23. Tighten the rear differential rear mount bolt to the specified torque (shown in the illustration).

24. Tighten the rear differential front mounting bracket bolts to the specified torque (shown in the illustration).

25. Install the rear subframe rear bracket bolts to 20 ft. lbs. (27 Nm).

26. Connect the rear differential subharness connectors and install the harness clips.

27. Attach the No. 3 propeller shaft to the rear differential by aligning the reference mark you made during the removal proce-

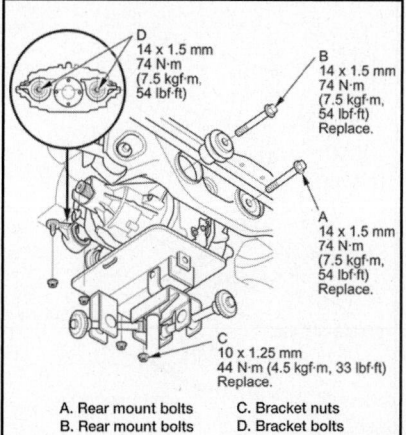

Fig. 12 Loosely install new rear differential rear mount bolts and new rear differential front mounting bracket nuts. Loosen the rear differential front mounting bracket bolts.

dure. Tighten the bolts to 53 ft. lbs. (72 Nm).

28. Apply appropriate grease to the rear driveshaft inboard joints.

29. Pull the knuckle outward, and insert the rear driveshaft inboard joints into the rear differential.

30. Loosely install the new lower arm A mounting bolts and the new lower arm B mounting bolts.

31. Place a floor jack under the lower arm, and raise the suspension to load it with the vehicle's weight.

32. Tighten the lower arm A mounting bolt and the lower arm B mounting bolt to 43 ft. lbs. (59 Nm), then remove the floor jack.

33. Install exhaust pipe B and the muffler.

34. Install the rear wheels.

35. Refill the rear differential with the recommended fluid.

36. Test-drive the vehicle.

DRIVESHAFT

REMOVAL & INSTALLATION

See Figures 13 and 14.

1. Raise and support the vehicle.

2. Remove exhaust pipe B and the muffler.

3. Place a transmission jack under the No. 2 propeller shaft.

4. Make reference mark across the No. 1 propeller shaft and the transfer companion flange.

5. Remove the No. 1 propeller shaft flange nuts.

6. Make reference mark across the No. 3 propeller shaft and the rear differential companion flange.

7. Remove the No. 3 propeller shaft flange bolts.

8. Remove the No. 1 and the No. 2 center support bearings mounting bolts.

9. Move the propeller shaft toward the rear differential, then lower the propeller shaft assembly on the jack.

➡**Secure the shaft to the transmission jack with a tie down to prevent it from rolling off.**

To install:

10. Install in the reverse order of removal, noting the following:

a. Align reference marks made during removal.

b. Tighten center bearing support bolts to 29 ft. lbs. (39 Nm).

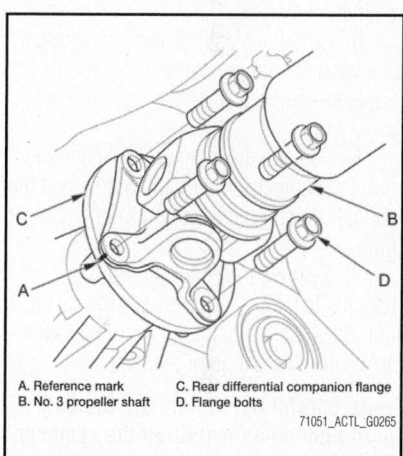

Fig. 13 Make reference mark across the No. 3 propeller shaft and the rear differential companion flange. Remove the No. 3 propeller shaft flange bolts.

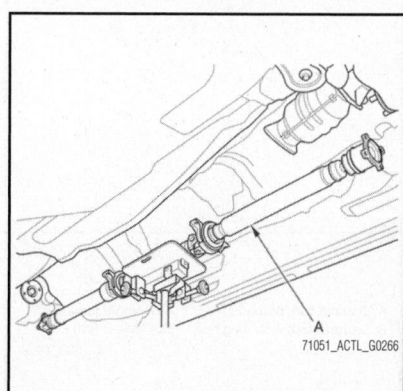

Fig. 14 Move the propeller shaft (A) toward the rear differential, then lower the propeller shaft assembly on the jack.

c. Tighten the No. 3 flange bolts to 53 ft. lbs. (72 Nm).

d. Tighten the No. 1 flange nuts to 53 ft. lbs. (72 Nm).

e. If you installed a new propeller shaft, test-drive the vehicle at 55 mph (88 km/h), and check for noise or vibration. If there is a noise or vibration, rotate the propeller shaft 180 degrees from its current alignment with the rear differential companion flange, then recheck.

FRONT HALFSHAFTS

REMOVAL & INSTALLATION

See Figures 15 through 18.

1. Raise and support the vehicle.
2. Remove the front wheels.
3. Pry up the stake on the spindle nut, then remove the nut.
4. Drain the transmission fluid, then reinstall the drain plug with a new sealing washer:
5. Hold the stabilizer joint pin using a hex wrench, and remove the flange nut. Separate the front stabilizer link from the lower arm.
6. Remove the damper fork mounting nut, the damper fork mounting bolt, and the damper pinch bolt, then remove the damper fork.
7. Remove the cotter pin from the knuckle ball joint, then remove the castle nut. Separate the ball joint from the lower arm using the ball joint remover.

➡**Be careful not to damage the ball joint boot when installing the remover. Do not force or hammer on the lower**

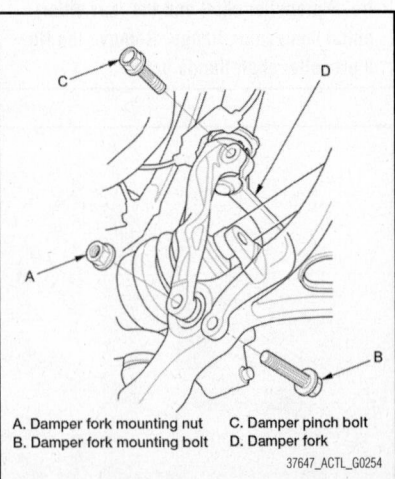

A. Damper fork mounting nut C. Damper pinch bolt
B. Damper fork mounting bolt D. Damper fork

37647_ACTL_G0254

Fig. 15 Remove the damper fork mounting nut, the damper fork mounting bolt, and the damper pinch bolt, then remove the damper fork

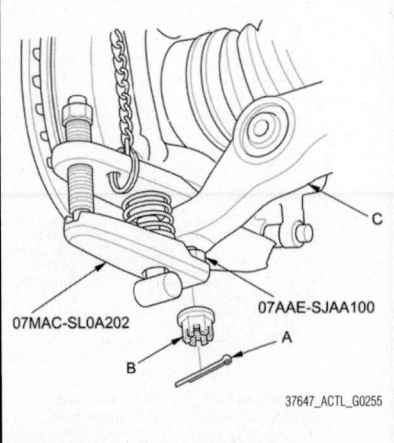

07MAC-SL0A202   07AAE-SJAA100

37647_ACTL_G0255

Fig. 16 Remove the cotter pin (A) from the knuckle ball joint, then remove the castle nut (B). Separate the ball joint from the lower arm (C) using the ball joint remover.

arm, or pry between the lower arm and the knuckle. You could damage the ball joint.

8. Pull the knuckle outward, and separate the outboard joint from the front hub using a soft face hammer.
9. Remove exhaust pipe A.
10. Left driveshaft: Pry the inboard joint from the differential using a prybar.
11. Remove the driveshaft as an assembly.

➡**Do not pull on the driveshaft, or the inboard joint may come apart. Pull the inboard joint straight out to avoid damaging the oil seal. Be careful not to damage the oil seal or the end of the inboard joint using the prybar.**

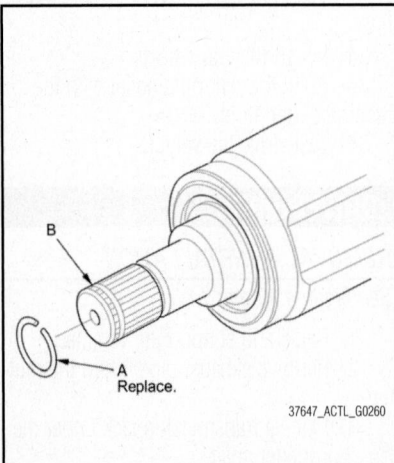

Replace.

37647_ACTL_G0260

Fig. 17 Install a new set ring (A) into the set ring groove (B) of the left driveshaft inboard joint

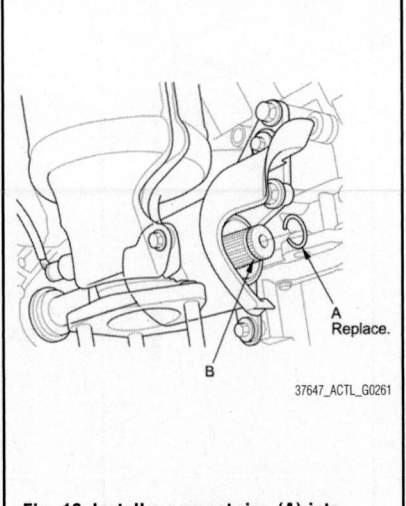

Replace.

37647_ACTL_G0261

Fig. 18 Install a new set ring (A) into the set ring groove (B) of the intermediate shaft

12. Right driveshaft: Drive the inboard joint off of the intermediate shaft using a drift punch and a hammer.
13. Remove the driveshaft as an assembly.

➡**Do not pull on the driveshaft, or the inboard joint may come apart.**

To install:

➡**Before starting installation, make sure the mating surfaces of the joint and the splined section are clean.**

14. Apply Moly 60 paste (P/N 08734-0001) to the contact area of the outboard joint and the front wheel bearing.

➡**The paste helps prevent noise and vibration.**

15. Install a new set ring into the set ring groove of the left driveshaft inboard joint.
16. Install a new set ring into the set ring groove of the intermediate shaft.
17. Apply small quantity of super high temperature grease (P/N 08798-9002) to the whole splined surface of the right driveshaft.
18. After applying grease, remove the grease from the splined grooves at intervals of 2–3 splines and from the set ring groove so that air can bleed from the intermediate shaft.
19. Clean the areas where the driveshaft contacts the differential thoroughly with solvent, and dry them with compressed air.

➡**Do not wash the rubber parts with solvent.**

20. Insert the inboard end of the driveshaft into the differential or the intermediate

shaft until the set ring locks in the groove.

➡Insert the driveshaft horizontally to prevent damaging the oil seal.

21. Install the outboard joint into the front hub on the knuckle.
22. Install exhaust pipe A.
23. Wipe off any grease contamination from the ball joint tapered section and threads, then install the knuckle onto the lower arm. Be careful not to damage the ball joint boot. Wipe off the grease before tightening the nut at the ball joint. Torque the castle nut to 69–76 ft. lbs. (93–103 Nm), then tighten it only far enough to align the slot with the ball joint pin hole.

➡Make sure the ball joint boot is not damaged or cracked. Do not align the nut by loosening it.

24. Install a new cotter pin into the ball joint pin hole, and bend the cotter pin.
25. Install the damper fork over the driveshaft and onto the lower arm. Install the damper in the damper fork so the aligning tab is aligned with the slot in the damper fork. Loosely install the damper pinch bolt.
26. Loosely install a new flange bolt and a new damper fork mounting nut.
27. Connect the front stabilizer link to the lower arm, and loosely install the flange nut. Hold the stabilizer link ball joint pin using a hex wrench, and tighten the flange nut to 58 ft. lbs. (78 Nm).
28. Place a floor jack under the lower arm, and raise the suspension to load it with the vehicle's weight.

➡Do not put the floor jack under the ball joint.

29. Tighten the damper pinch bolt to 36 ft. lbs. (49 Nm) and the damper fork mounting nut to 47 ft. lbs. (64 Nm)while holding the flange bolt, then remove the floor jack.
30. Apply a small amount of engine oil to the seating surface of a new spindle nut.
31. Install the spindle nut, then tighten it to 242 ft. lbs. (329 Nm). After tightening, use a drift to stake the spindle nut shoulder against the driveshaft.
32. Clean the mating surfaces of the brake disc and the wheel, then install the front wheels.
33. Turn the front wheel by hand, and make sure there is no interference between the driveshaft and the surrounding parts.
34. Lower the vehicle.
35. Refill the transmission with recommended transmission fluid:
36. Check the wheel alignment, and adjust it if necessary.

FRONT INTERMEDIATE SHAFT

REMOVAL & INSTALLATION

See Figures 19 through 21.

1. Drain the transmission fluid, then reinstall the drain plug with a new sealing washer.
2. Remove exhaust pipe A.
3. Remove the right front driveshaft..
4. Remove the rear warm up three way catalytic converter (rear WU-TWC) bracket.

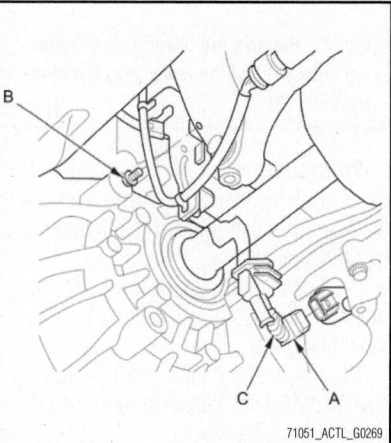

Fig. 19 Disconnect the CKP sensor connector (A), and remove the sensor harness clamp bracket bolt (B). Swing the sensor harness (C) out of the way to prevent damaging it.

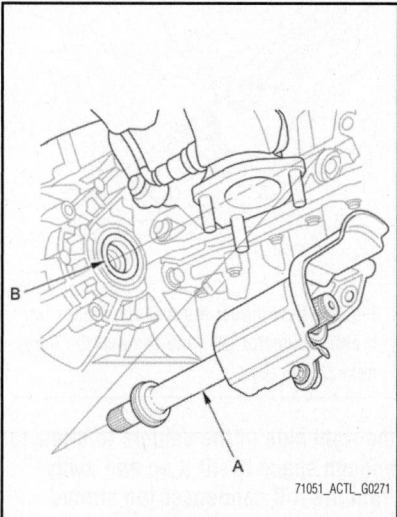

Fig. 20 Remove the intermediate shaft (A) from the differential. Hold the intermediate shaft horizontal until it is clear of the differential to prevent damaging the oil seal (B).

5. Remove the crankshaft position (CKP) sensor cover.
6. Disconnect the CKP sensor connector and remove the sensor harness clamp bracket bolt.
7. Swing the sensor harness out of the way to prevent damaging it.
8. Remove the flange bolt and the two dowel bolts.
9. Remove the intermediate shaft from the differential. Hold the intermediate shaft horizontal until it is clear of the differential to prevent damaging the oil seal.

To install:

10. Clean the areas where the intermediate shaft contacts the differential thoroughly with solvent, and dry them with compressed air.

➡Do not wash the rubber parts with solvent.

11. Install a new set ring into the set ring groove of the intermediate shaft.
12. Insert the intermediate shaft into the differential correctly.

➡Insert the intermediate shaft carefully to prevent damaging the oil seal.

13. Install the flange bolt and the two dowel bolts. Tighten bolts to 29 ft. lbs. (39 Nm).
14. Install the rear warm up three way catalytic converter (rear WU-TWC) bracket. Tighten bracket bolts to 16 ft. lbs. (22 Nm).
15. Install the right front driveshaft.
16. Connect the crankshaft position

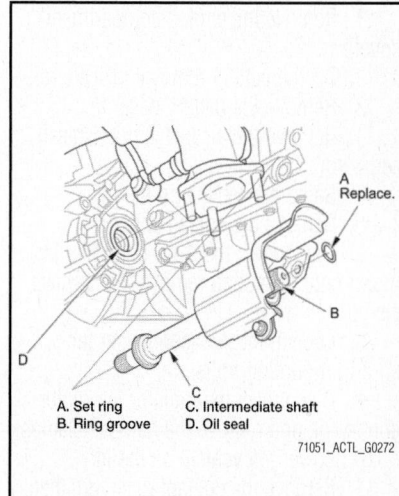

A. Set ring
B. Ring groove
C. Intermediate shaft
D. Oil seal

Fig. 21 Install a new set ring into the set ring groove of the intermediate shaft. Insert the intermediate shaft into the differential correctly. Insert the intermediate shaft carefully to prevent damaging the oil seal.

(CKP) sensor connector (A), and install the sensor harness clamp bracket.

17. Install the CKP sensor cover.

18. Install exhaust pipe A.

19. Refill the transmission with the recommended transmission fluid:

20. Check the wheel alignment, and adjust it if necessary.

21. Test-drive the vehicle.

REAR DRIVESHAFT

REMOVAL & INSTALLATION

See Figures 22 and 23.

1. Raise the vehicle on a lift.
2. Drain the rear differential fluid.
3. Remove the rear wheels.
4. Remove exhaust pipe B and the muffler.
5. Remove the lower arm A mounting bolts and the lower arm B mounting bolts.
6. Disconnect the inboard joints from the rear differential using the driveshaft remover and a hammer.
7. Pull the knuckle outward, and separate the rear driveshaft inboard joints from the rear differential.

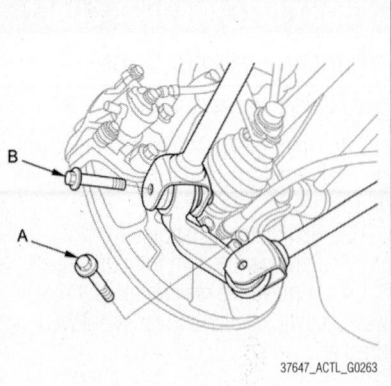

Fig. 22 Remove the lower arm A mounting bolts (A) and the lower arm B mounting bolts (B)

To install:

8. Pull the knuckle outward, and insert the rear driveshaft inboard joints into the rear differential.

9. Loosely install new lower arm A mounting bolts and new lower arm B mounting bolts.

10. Place a floor jack under the lower arm, and raise the suspension to load it with the vehicle's weight.

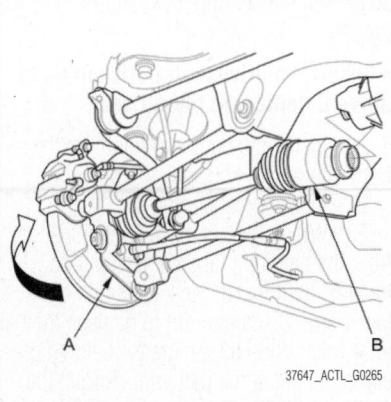

Fig. 23 Pull the knuckle (A) outward, and separate the rear driveshaft inboard joints (B) from the rear differential

11. Tighten the lower arm A mounting bolt and the lower arm B mounting bolt to 43 ft. lbs. (59 Nm).

12. Install exhaust pipe B and the muffler.

13. Install the rear wheels.

14. Refill the rear differential with the recommended fluid.

15. Test-drive the vehicle.

ENGINE COOLING

COOLING FANS

REMOVAL & INSTALLATION

See Figures 24 through 26.

1. Remove the engine compartment covers.
2. Do the battery removal procedure.
3. Remove the battery base.
4. Remove the air intake duct splash separator.
5. Raise the vehicle on the lift.
6. Remove the splash shield.
7. Disconnect the A/C condenser fan motor connector, and remove the harness clamp.
8. Loosen the A/C condenser fan shroud mounting bolts.
9. Disconnect the radiator fan motor connector, and remove the harness clamps.
10. Lower the vehicle on the lift.
11. Remove the coolant reservoir, then remove the A/C condenser fan shroud assembly.
12. Remove the upper brackets.
13. Remove the radiator fan shroud assembly.

➡**Pull-up the radiator, then move the radiator fan shroud assembly toward**

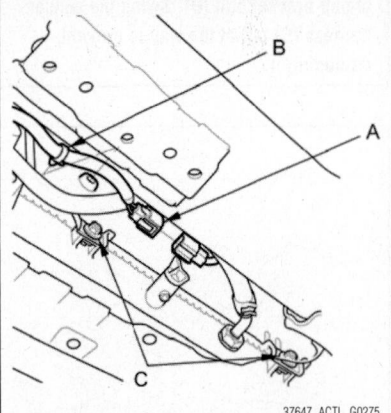

Fig. 24 Disconnect the A/C condenser fan motor connector (A), and remove the harness clamp (B)

the right side of the vehicle to allow for enough space to lift it up and away from the A/C condenser fan shroud assembly.

14. Disassemble the fan shrouds.

To install:

15. Assemble the fan shrouds.

16. Install the left radiator fan shroud assembly.

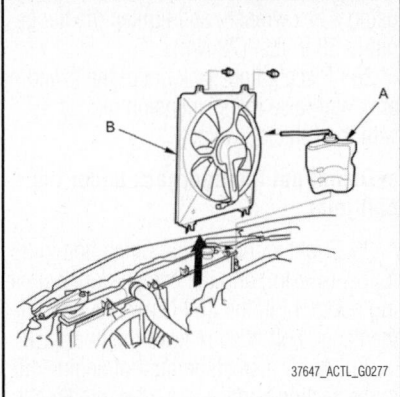

Fig. 25 Remove the coolant reservoir (A), then remove the A/C condenser fan shroud assembly (B)

17. Install the upper brackets.

18. Install the A/C condenser fan shroud assembly, then install the coolant reservoir.

19. Raise the vehicle on the lift.

20. Connector the radiator fan motor connector, and install the harness clamps.

21. Tighten the A/C condenser fan shroud mounting bolts.

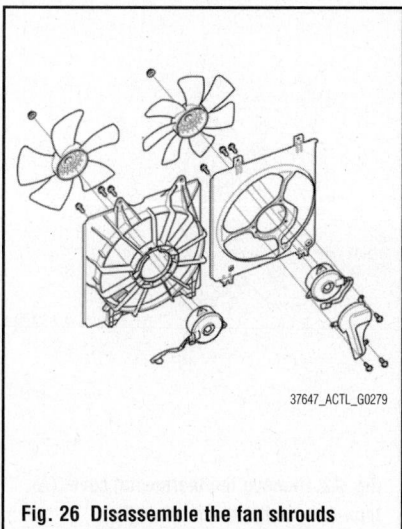

Fig. 26 Disassemble the fan shrouds

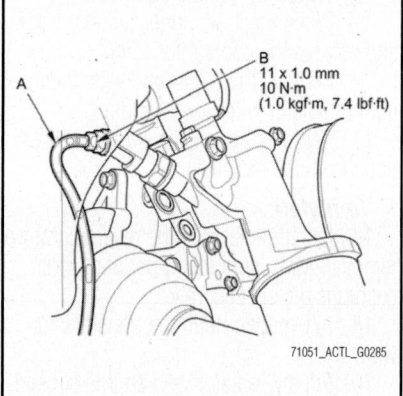

Fig. 27 Install a rubber hose (A) on the drain bolt (B) located at the rear of the engine block, then loosen the drain bolt.

22. Connect the A/C condenser fan motor connector, and install the harness clamp.

23. Install the splash shield.

24. Lower the vehicle on the lift.

25. Install the air intake duct splash separator.

26. Install the battery base.

27. Do the battery installation procedure.

28. Install the engine compartment covers.

ENGINE COOLANT

DRAIN & REFILL PROCEDURE

See Figure 27.

1. Remove the engine compartment covers.

2. Wait until the engine is cool, then carefully remove the radiator cap.

3. Loosen the drain plug, and drain the coolant.

4. Install a rubber hose on the drain bolt located at the rear of the engine block, then loosen the drain bolt.

5. When the coolant stops draining, tighten the drain bolt. Remove the rubber hose.

6. Tighten the radiator drain plug securely.

7. Remove, drain, and reinstall the coolant reservoir.

8. Fill the coolant reservoir to the MAX mark with Acura Long Life Antifreeze/Coolant Type 2.

9. Pour Acura Long Life Antifreeze/Coolant Type 2 into the radiator up to the base of the filler neck.

➡**Always use Acura Long Life Antifreeze/Coolant Type 2. Using a non-Acura coolant can result in corrosion,**
causing the cooling system to malfunction or fail. Acura Long Life Antifreeze/Coolant Type 2 is a mixture of 50 % antifreeze and 50 % water. Do not add water.

10. Loosely install the radiator cap.

11. Set the heater temperature control to maximum heat.

12. Start the engine, and let it run until it warms up (the radiator fan comes on at least twice).

13. Turn off the engine. Check the level in the radiator, and add Acura Long Life Antifreeze/Coolant Type 2, if needed.

14. Put the radiator cap on securely, then run the engine again, and check for leaks.

15. Clean up any spilled engine coolant.

16. Install the engine compartment covers.

17. If the maintenance minder required engine coolant replacement, reset the maintenance minder, and this procedure is complete. If the maintenance minder did not require engine coolant replacement, go to the next step.

18. Turn the ignition switch to LOCK (0), or press the engine start/stop button to select the OFF mode.

19. Connect the Honda Diagnostic System (HDS) to the data link connector (DLC).

20. Turn the ignition switch to ON (II), or press the engine start/stop button to select the ON mode.

21. Make sure the HDS communicates with the vehicle and the engine control module (ECM)/powertrain control module (PCM). If it does not communicate, troubleshoot the DLC circuit.

22. Select GAUGES in the BODY ELECTRICAL with the HDS.

23. Select ADJUSTMENT in the GAUGES with the HDS.

24. Select MAINTENANCE MINDER in the ADJUSTMENT with the HDS.

25. Select RESET in the MAINTENANCE MINDER with the HDS.

26. Select MAINTENANCE SUB ITEM 5 RESET with the HDS.

RADIATOR

REMOVAL & INSTALLATION

See Figures 28 through 31.

1. Remove the engine compartment covers.

2. Drain the engine coolant.

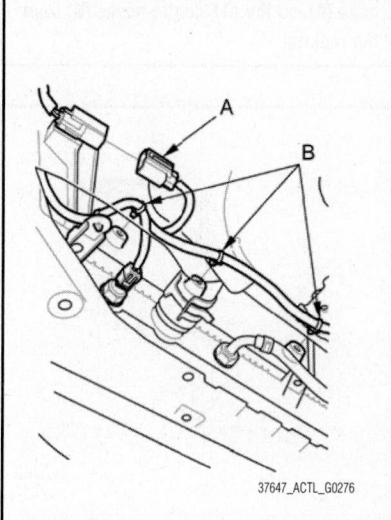

Fig. 28 Disconnect the radiator fan motor connector (A), and remove the harness clamps (B)

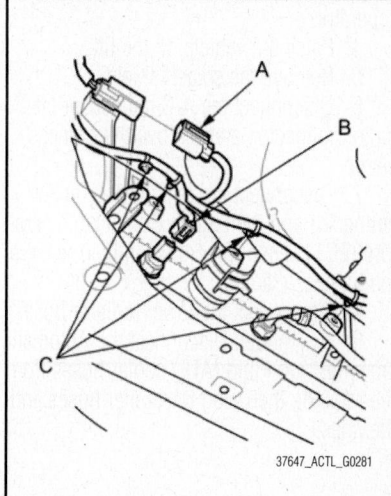

Fig. 29 Disconnect the radiator fan motor connector (A) and the ECT sensor 2 connector (B), then remove the harness clamps (C)

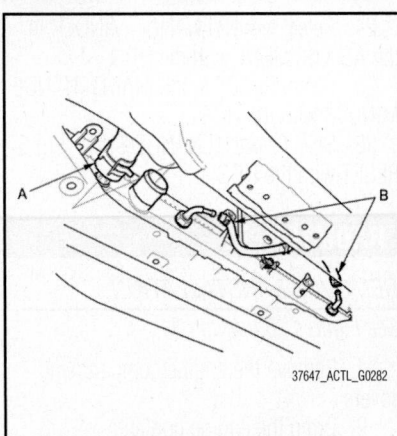

Fig. 30 Disconnect the lower radiator hose (A)and the ATF cooler hoses (B) from the radiator

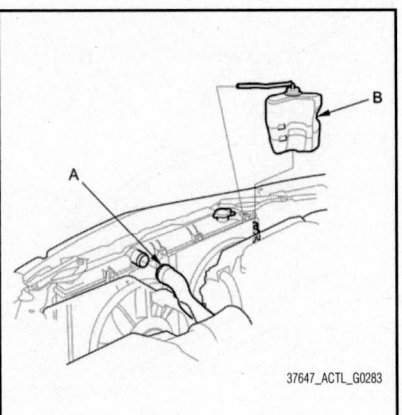

Fig. 31 Disconnect the upper radiator hose (A) and remove the coolant reservoir (B)

3. Remove the air intake duct splash separator.

4. Raise the vehicle on the lift.

5. Remove the splash shield.

6. Disconnect the A/C condenser fan motor connector, and remove the harness clamp.

7. Disconnect the radiator fan motor connector and the Engine Coolant Temperature (ECT) sensor 2 connector, then remove the harness clamps.

8. Disconnect the lower radiator hose.

9. A/T model: Disconnect the Automatic Transmission Fluid (ATF) cooler hoses from the radiator, then plug the cooler hoses and the lines.

10. Lower the vehicle on the lift.

11. Disconnect the upper radiator hose and remove the coolant reservoir.

12. Remove the upper brackets, then pull up the radiator with fan shrouds.

13. Remove the fan shrouds and other parts from the radiator.

To install:

14. Install radiator in the reverse order of removal. Make sure the upper and lower cushions are set securely.

15. A/T model: Refill the transmission with ATF.

16. Fill the radiator with engine coolant, and bleed the air from the cooling system.

17. Clean up any spilled engine coolant.

18. Install the engine compartment covers.

THERMOSTAT

REMOVAL & INSTALLATION

See Figure 32.

1. Remove the engine compartment covers.

2. Drain the engine coolant.

3. Do the battery removal procedure.

4. A/T model: Remove harness clamps and disconnect the positive starter cable.

5. A/T model: Disconnect the A/T clutch pressure control solenoid valve C connector, and remove the bolt and the harness holder.

6. Remove the thermostat cover, then remove the thermostat.

To install:

7. Install the new thermostat with a new rubber seal in the reverse order of the removal.

8. Do the battery installation procedure.

9. Refill the radiator with engine coolant, and bleed the air from the cooling system.

10. Clean up any spilled engine coolant.

11. Install the engine compartment covers.

WATER PUMP

REMOVAL & INSTALLATION

See Figure 33.

1. Drain the engine coolant.

2. Remove the timing belt.

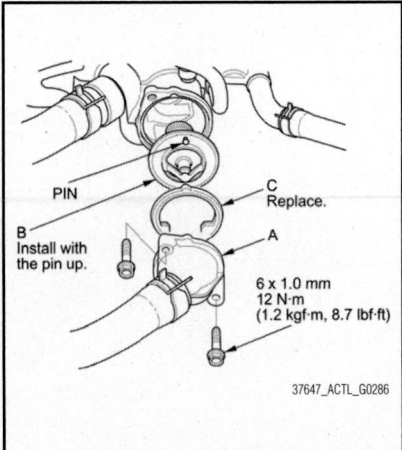

Fig. 32 Remove the thermostat cover (A), then remove the thermostat (B)

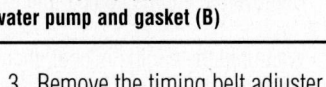

Fig. 33 Remove the five bolts securing the water pump (A), then remove the water pump and gasket (B)

3. Remove the timing belt adjuster.

4. Remove the five bolts securing the water pump, then remove the water pump and gasket.

To install:

5. Inspect and clean the O-ring groove and the mating surface of the engine block.

6. Install the water pump with a new gasket.

7. Clean up any spilled engine coolant.

8. Install the timing belt adjuster:

9. Install the timing belt:

10. Refill the radiator with engine coolant, and bleed the air from the cooling system.

ENGINE ELECTRICAL   BATTERY SYSTEM

BATTERY

REMOVAL & INSTALLATION

See Figure 34.

➡**The battery terminal disconnection and reconnection procedure must be done before and after doing this procedure. Some systems store data in the memory that is lost when the battery is disconnected.**

1. Do the battery terminal disconnection procedure.
2. Remove the three clips and air intake duct splash separator.
3. Remove the two nuts securing the battery setting plate, then remove the battery setting plate and the battery.

To install:

4. Install the battery, then install the battery setting plate.
5. Tighten the two nuts equally until the battery is stable.

➡**Do not deform the battery setting plate by over-tightening the nuts.**

6. Install the air intake duct splash separator and three clips.
7. Do the battery terminal reconnection procedure.

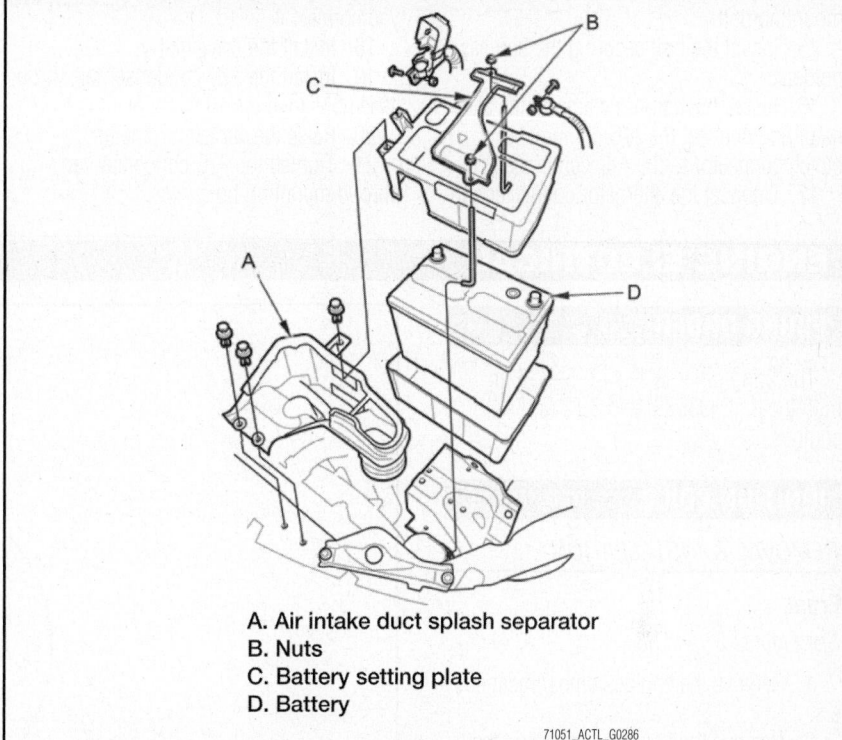

A. Air intake duct splash separator
B. Nuts
C. Battery setting plate
D. Battery

71051_ACTL_G0286

Fig. 34 Exploded view of battery and components

➡**Make sure the battery is correctly installed, and that the positive terminal** and the negative terminal are not connected in reverse.

ENGINE ELECTRICAL   CHARGING SYSTEM

ALTERNATOR

REMOVAL & INSTALLATION

See Figures 35 and 36.

1. Remove the engine compartment covers.
2. Do the battery terminal disconnection procedure.
3. Raise the vehicle on the lift.
4. Remove the splash shield.
5. Disconnect the A/C condenser fan motor connector and remove the harness clamp.
6. Loosen the A/C condenser fan shroud mounting bolts.
7. Lower the vehicle on the lift.
8. Remove the coolant reservoir, then remove the A/C condenser fan shroud assembly.
9. Remove the drive belt.
10. Disconnect the alternator connector and the positive alternator cable from the alternator.
11. Remove the harness clamp from the alternator and disconnect the A/C compres-

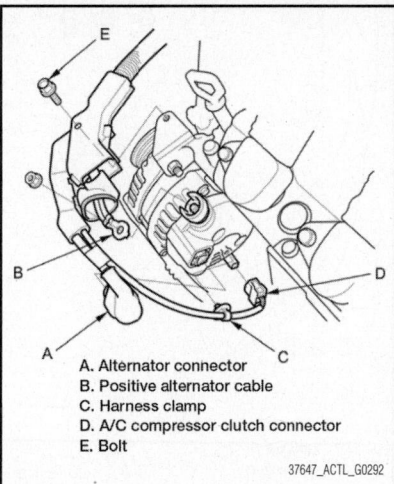

A. Alternator connector
B. Positive alternator cable
C. Harness clamp
D. A/C compressor clutch connector
E. Bolt

37647_ACTL_G0292

Fig. 35 Disconnect the alternator connector and the positive alternator cable from the alternator

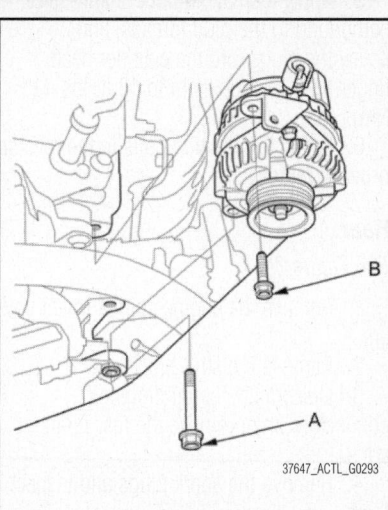

37647_ACTL_G0293

Fig. 36 Remove the mounting bolt (A) and the alternator bracket mounting bolt (B), then remove the alternator

sor clutch connector from the A/C compressor.
12. Remove the bolt securing the harness holder.

13. Remove the mounting bolt and the alternator bracket mounting bolt, then remove the alternator.

To install:

14. Install the alternator, then tighten the mounting bolt and the alternator bracket mounting bolt.

15. Install the bolt securing the harness holder.

16. Install the harness clamp to the alternator and connect the A/C compressor clutch connector to the A/C compressor.

17. Connect the alternator connecter and the positive alternator cable to the alternator. Make sure the crimped side of the ring terminal faces away from the alternator when you connect it.

18. Install the drive belt.

19. Install the A/C condenser fan shroud assembly, then install the coolant reservoir.

20. Raise the vehicle on the lift.

21. Tighten the A/C condenser fan shroud mounting bolts.

22. Connect the A/C condenser fan motor connector and install the harness clamp.

23. Install the splash shield.

24. Lower the vehicle on the lift.

25. Do the battery terminal reconnection procedure.

26. Install the engine compartment covers.

ENGINE ELECTRICAL

IGNITION SYSTEM

FIRING ORDER

The firing order is 1–4–2–5–3–6 for both the 3.5L (J35Z6) and 3.7L (J37A4) engines.

IGNITION COILS

REMOVAL & INSTALLATION

Front

See Figures 37 and 38.

1. Remove the engine compartment covers.

2. Remove the harness clamp and the bolt, then remove the harness holder from the bracket.

3. Disconnect the ignition coil connectors, then remove the front ignition coils.

4. Remove the spark plugs and inspect them.

5. Apply a small amount of anti-seize compound to the plug threads, and screw the plugs into the cylinder head, finger tight. Torque them to 13 ft. lbs. (18 Nm).

6. Install the ignition coils in the reverse order of removal.

Rear

See Figure 39.

1. Remove the engine compartment covers.

2. Remove the strut brace.

3. Disconnect the ignition coil connectors, then remove the rear ignition coils.

4. Remove the spark plugs and inspect them.

5. Apply a small amount of anti-seize compound to the plug threads, and screw the plugs into the cylinder head, finger tight. Torque them to 13 ft. lbs. (18 Nm).

6. Install the ignition coils in the reverse order of removal.

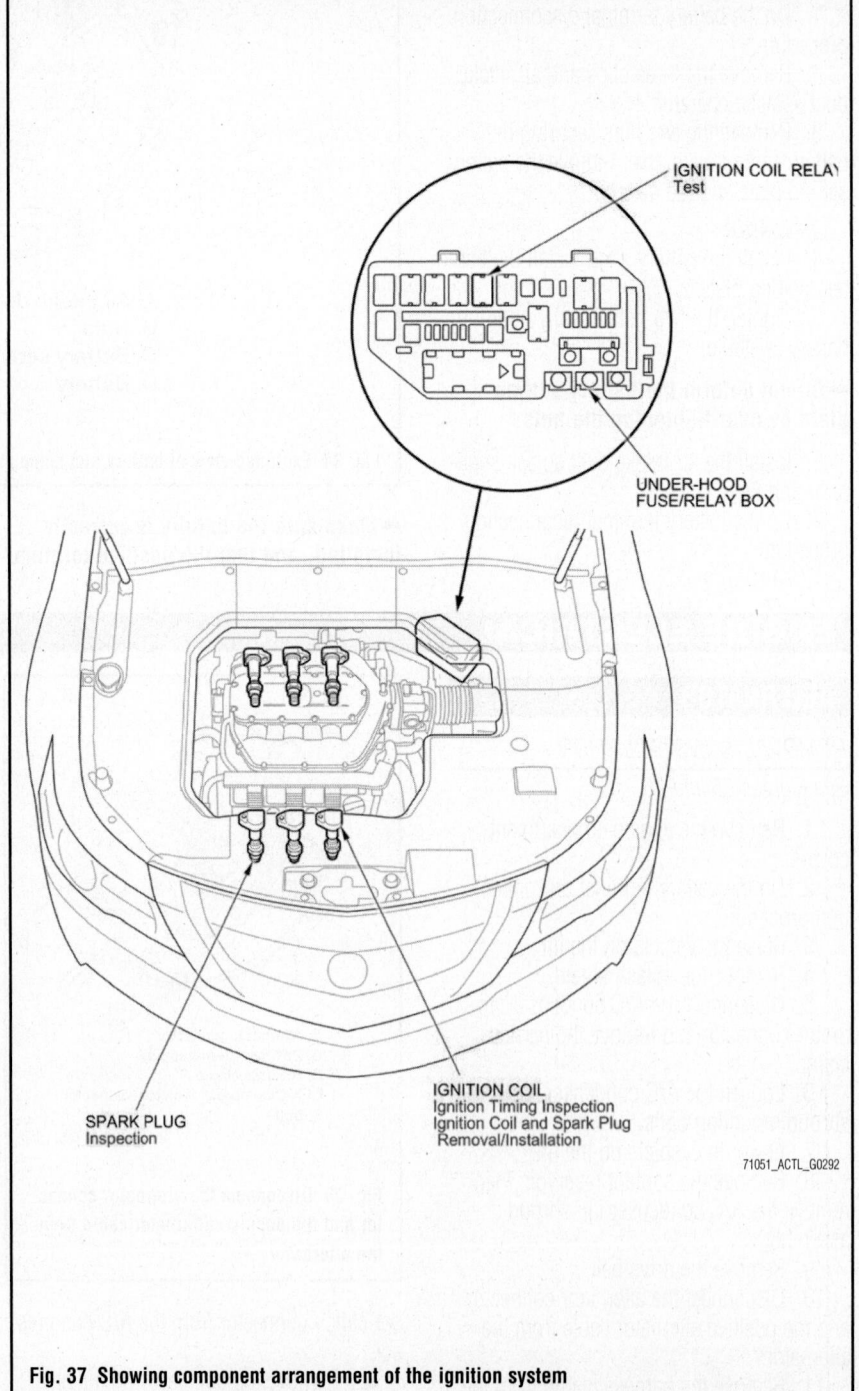

Fig. 37 Showing component arrangement of the ignition system

71051_ACTL_G0292

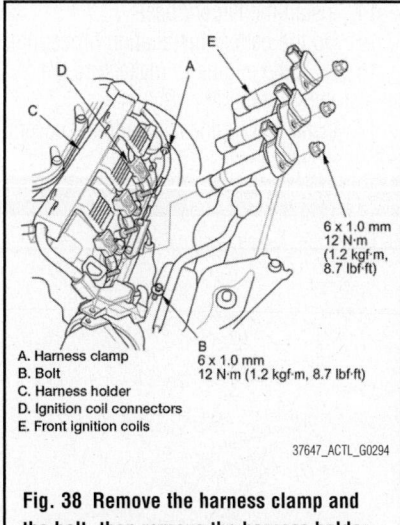

A. Harness clamp
B. Bolt
C. Harness holder
D. Ignition coil connectors
E. Front ignition coils

6 x 1.0 mm
12 N·m
(1.2 kgf·m,
8.7 lbf·ft)

B
6 x 1.0 mm
12 N·m (1.2 kgf·m, 8.7 lbf·ft)

37647_ACTL_G0294

Fig. 38 Remove the harness clamp and the bolt, then remove the harness holder from the bracket

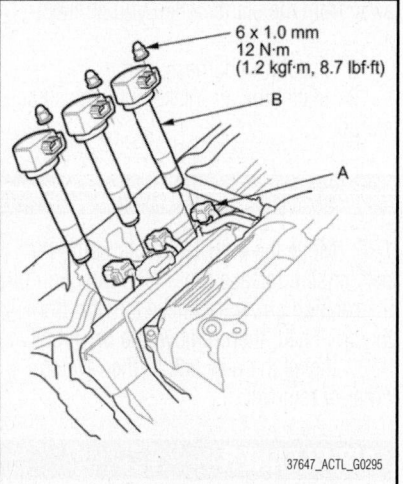

6 x 1.0 mm
12 N·m
(1.2 kgf·m, 8.7 lbf·ft)

37647_ACTL_G0295

Fig. 39 Disconnect the ignition coil connectors (A), then remove the rear ignition coils (B)

IGNITION TIMING

INSPECTION & ADJUSTMENT

1. Remove the engine compartment covers.

2. Connect the Honda Diagnostic System (HDS) to the Data Link Connector (DLC).

3. Turn the ignition switch to ON (II), or press the engine start/stop button to select the ON mode.

4. Make sure the HDS communicates with the vehicle and the Engine Control Module (ECM)/Powertrain Control Module (PCM). If it does not communicate, troubleshoot the DLC circuit.

5. Check for DTCs. If a DTC is present, diagnose and repair the cause before continuing with this test.

6. Start the engine. Hold the engine speed at 3,000 rpm with no load (in neutral (M/T model) or N or P (A/T model)) until the radiator fan comes on, then let it idle.

7. Check the idle speed.

8. Jump the SCS line with the HDS.

9. Connect the timing light to the No. 1 ignition coil harness.

10. Aim the light toward the pointer on the timing belt cover. Check the ignition timing under a no load condition (headlights, blower fan, rear window defogger, and air conditioner are turned off).

➡**The other pointer is not used.**

Ignition timing specification: 10°±2° BTDC (RED mark) at idle in N or P

11. If the ignition timing differs from the specification, check the cam timing. If the cam timing is OK, update the ECM/PCM if it does not have the latest software, or substitute a known-good ECM/PCM, then recheck. If the system works properly, and the ECM/PCM was substituted, replace the original ECM/PCM.

12. Disconnect the HDS and the timing light.

13. Install the engine compartment covers.

ENGINE ELECTRICAL

STARTER

REMOVAL & INSTALLATION

See Figure 40.

1. Remove the engine compartment covers.

2. Do the battery removal procedure.

3. Remove the battery base.

4. Remove the air intake duct splash separator.

5. Remove the harness clamp.

6. Disconnect the positive starter cable and the S terminal connector.

7. A/T model: Remove the upper radiator hose bracket and the dipstick.

8. Remove the two bolts holding the starter, then remove the starter.

To install:

9. Install the starter, then tighten the mounting bolts. Tighten to 54 ft. lbs. (74 Nm).

➡**Always use a new gasket (A/T model).**

STARTING SYSTEM

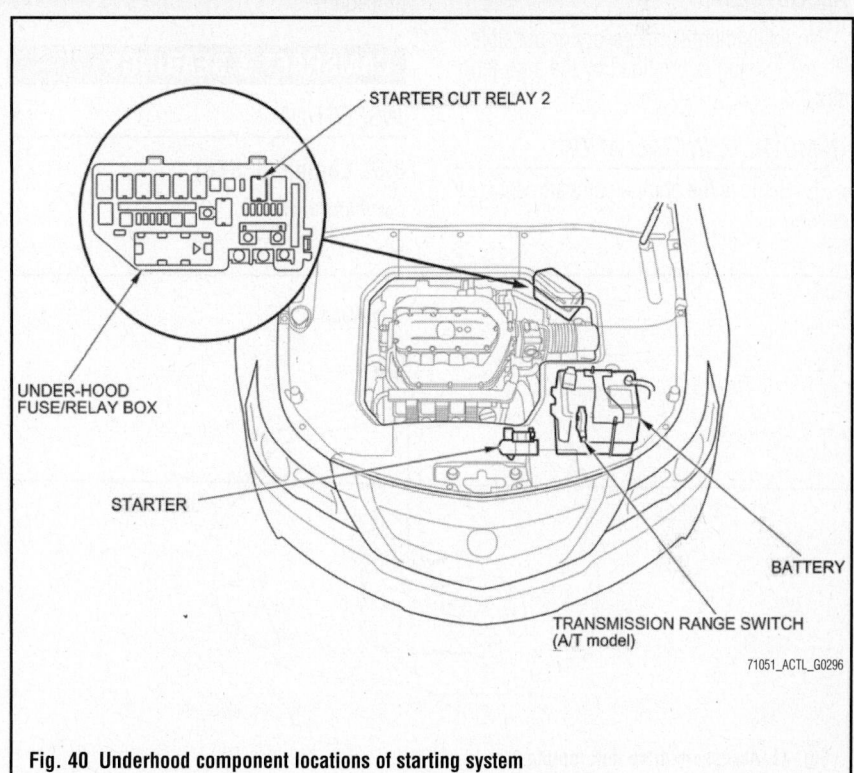

STARTER CUT RELAY 2

UNDER-HOOD FUSE/RELAY BOX

STARTER

BATTERY

TRANSMISSION RANGE SWITCH (A/T model)

71051_ACTL_G0296

Fig. 40 Underhood component locations of starting system

10. A/T model: Install the upper radiator hose bracket and the dipstick.

11. Connect the positive starter cable and the S terminal connector. Make sure the crimped side of the ring terminal faces away from the starter when you connect it.

12. Install the harness clamp.

13. Install the air intake duct splash separator.

14. Install the battery base.

15. Do the battery installation procedure.

16. Start the engine to make sure the starter works properly.

17. Install the engine compartment covers.

ENGINE MECHANICAL

➡ **Disconnecting the negative battery cable may interfere with the functions of the on board computer systems and may require the computer to undergo a relearning process, once the negative battery cable is reconnected.**

ACCESSORY DRIVE BELTS

BELT ROUTINGS

See Figure 41.

Refer to the accompanying illustration for belt routing.

INSPECTION

1. Remove the engine compartment covers.

2. Inspect the belt for cracks or damage. If the belt is cracked or damaged, replace it.

3. Check that the position of the auto-tensioner indicator is within the standard range. If it is out of the standard range, replace the drive belt.

4. Install the engine compartment covers.

ADJUSTMENT

No adjustment is necessary or possible. Proper tension is provided by the auto-tensioner.

REMOVAL & INSTALLATION

1. Remove the engine compartment covers.

2. Move the auto-tensioner using the belt tension release tool in the direction of the rotation arrow to relieve tension from the drive belt, then remove the drive belt.

3. Install the new belt in the reverse order of removal.

AIR CLEANER

REMOVAL & INSTALLATION

See Figure 43.

1. Do the battery removal procedure.

2. Disconnect the MAF sensor/IAT sensor 5P connector.

3. Remove the harness clamps and the bolt.

4. Loosen the band, then remove the air cleaner housing.

5. Install the parts in the reverse order of removal.

➡ **When tightening the screw of the hose band, align the edge of the hose band between the marks (or within the mark) painted on the hose band.**

6. Do the battery installation procedure.

CAMSHAFT & BEARINGS

INSPECTION

3.5L Engine (J35Z6)

See Figure 44.

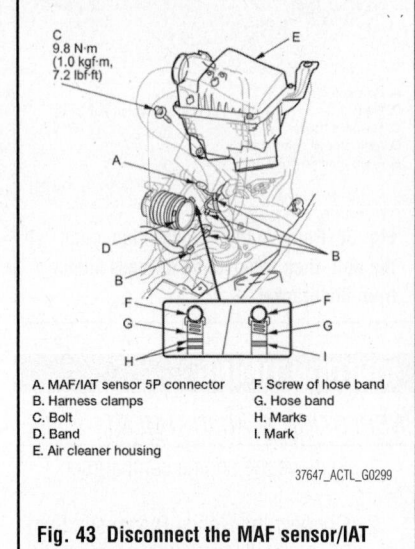

A. MAF/IAT sensor 5P connector
B. Harness clamps
C. Bolt
D. Band
E. Air cleaner housing
F. Screw of hose band
G. Hose band
H. Marks
I. Mark

37647_ACTL_G0299

Fig. 43 Disconnect the MAF sensor/IAT sensor 5P connector

1. Remove the cylinder head.

2. Remove the rocker arm assembly.

3. Disassemble the rocker arm assembly.

4. Front: Put the rocker shafts bridge and the rocker shaft holder on the front cylinder head, then tighten the bolts to 16 ft. lbs. (22 Nm).

5. Rear: Put the rocker shaft bridge and the rocker shaft holder on the rear cylinder head, then tighten the bolts to 16 ft. lbs. (22 Nm).

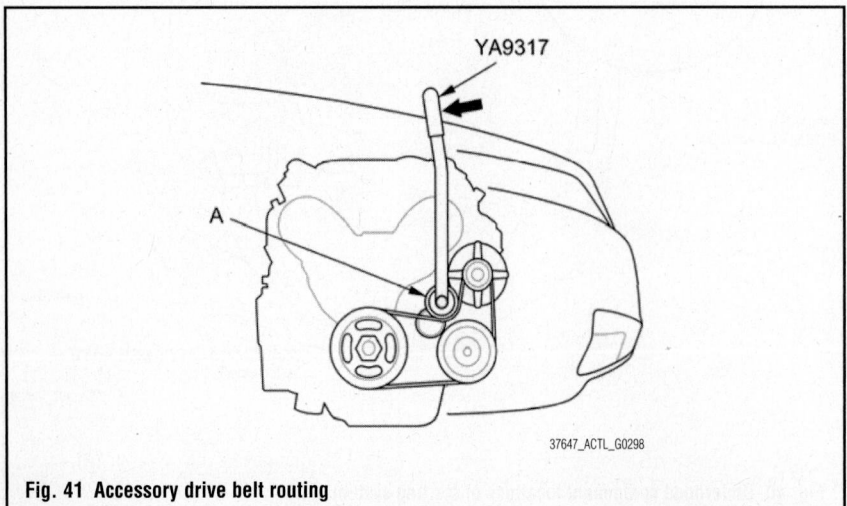

Fig. 41 Accessory drive belt routing

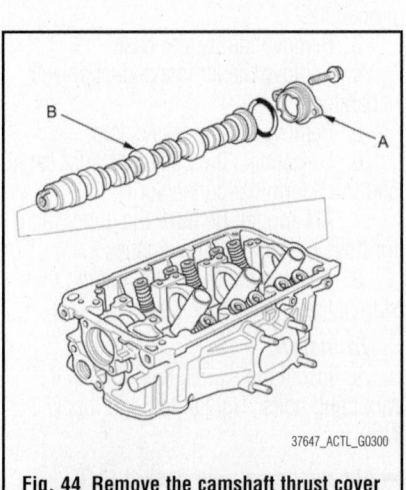

Fig. 44 Remove the camshaft thrust cover (A), then pull out the camshaft (B)

6. Seat the camshaft by pushing it toward the rear of the cylinder head.

7. Zero the dial indicator against the end of the camshaft. Push the camshaft back and forth and read the end play. If the end play is beyond the service limit, replace the thrust cover and recheck. If it is still beyond the service limit, replace the camshaft.

8. Remove the camshaft thrust cover, then pull out the camshaft.

9. Wipe the camshaft clean, then inspect the lift ramps. Replace the camshaft if any lobes are pitted, scored, or excessively worn.

10. Measure the diameter of each camshaft journal.

11. Zero the gauge to the journal diameter.

12. Clean the camshaft bearing surfaces in the cylinder head. Measure the inside diameter of each camshaft bearing surface, and check for an out-of-round condition.

   a. If the camshaft-to-holder clearance is within limits, go to step 14.

   b. If the camshaft-to-holder clearance is beyond the service limit and the camshaft has been replaced, replace the cylinder head.

   c. If the camshaft-to-holder clearance is beyond the service limit and the camshaft has not been replaced, go to step 13.

➡Camshaft-to-Holder Oil Clearance specifications:

- Standard (New): 0.0020–0.0035 inches (0.050–0.089 mm)
- Service Limit: 0.006 inches (0.15 mm)

13. Check total runout with the camshaft supported on V-blocks.

   a. If the total runout of the camshaft is within the service limit, replace the cylinder head.

   b. If the total runout is beyond the service limit, replace the camshaft and recheck the oil clearance. If the oil clearance is still out of tolerance, replace the cylinder head.

➡Camshaft Total Runout

- Standard (New): 0.001 inches (0.03 mm) max.
- Service Limit: 0.002 inches (0.04 mm)

14. Measure the cam lobe height.

➡Cam Lobe Height Standard (New):

- Intake Primary: 1.3504 inches
- Intake Secondary: 1.4024 inches
- Exhaust: 1.4472 inches

3.7L Engine (J37A4)

1. Remove the cylinder head.
2. Remove the rocker arm assembly.
3. Disassemble the rocker arm assembly.
4. Front: Put the rocker shafts bridge and the rocker shaft holder on the front cylinder head, then tighten the bolts to 16 ft. lbs. (22 Nm).
5. Rear: Put the rocker shaft bridge and the rocker shaft holder on the rear cylinder head, then tighten the bolts to 16 ft. lbs. (22 Nm).
6. Seat the camshaft by pushing it toward the rear of the cylinder head.
7. Zero the dial indicator against the end of the camshaft. Push the camshaft back and forth and read the end play. If the end play is beyond the service limit, replace the thrust cover and recheck. If it is still beyond the service limit, replace the camshaft.
8. Remove the camshaft thrust cover, then pull out the camshaft.
9. Wipe the camshaft clean, then inspect the lift ramps. Replace the camshaft if any lobes are pitted, scored, or excessively worn.
10. Measure the diameter of each camshaft journal.
11. Zero the gauge to the journal diameter.
12. Clean the camshaft bearing surfaces in the cylinder head. Measure the inside diameter of each camshaft bearing surface, and check for an out-of-round condition.

   a. If the camshaft-to-holder clearance is within limits, go to step 14.

   b. If the camshaft-to-holder clearance is beyond the service limit and the camshaft has been replaced, replace the cylinder head.

   c. If the camshaft-to-holder clearance is beyond the service limit and the camshaft has not been replaced, go to step 13.

Camshaft-to-Holder Oil Clearance specifications:
- Standard (New): 0.0020–0.0035 inches (0.050–0.089 mm)
- Service Limit: 0.006 inches (0.15 mm)

13. Check total runout with the camshaft supported on V-blocks.

   a. If the total runout of the camshaft is within the service limit, replace the cylinder head.

   b. If the total runout is beyond the service limit, replace the camshaft and recheck the oil clearance. If the oil clearance is still out of tolerance, replace the cylinder head.

Camshaft Total Runout
- Standard (New): 0.001 inches (0.03 mm) max.
- Service Limit: 0.002 inches (0.04 mm)

14. Measure the cam lobe height.

Cam Lobe Height Standard (New):
- Intake Primary: 1.3504 inches
- Intake Secondary: 1.4146 inches
- Exhaust Primary: 1.4462 inches
- Exhaust Secondary: 1.4713 inches

REMOVAL & INSTALLATION

3.5L Engine (J35Z6) & J37A4 Engines

Front

See Figures 45 and 46.

1. Remove the engine compartment covers.
2. Do the battery removal procedure.
3. Drain the engine coolant.
4. Disconnect the radiator hoses.

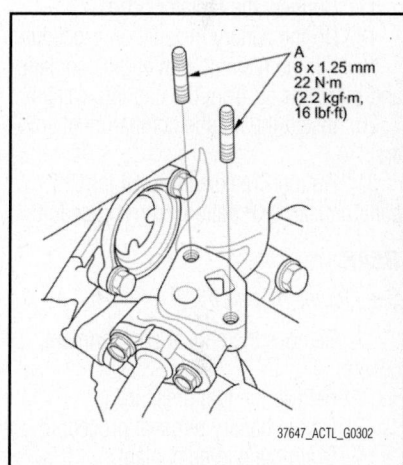

Fig. 45 Remove the EGR valve stud bolts (A)

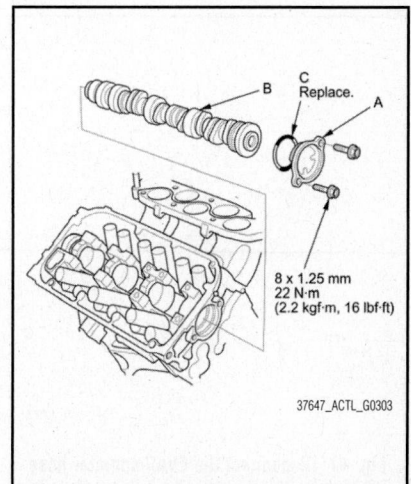

Fig. 46 Remove the thrust cover (A), then remove the front camshaft (B)

5. Remove the exhaust gas recirculation (EGR) valve.

6. Remove the EGR valve stud bolts.

7. Remove the timing belt.

8. Remove the front rocker arm assembly.

9. Remove the front camshaft pulley.

10. Remove the thrust cover, then remove the front camshaft.

To install:

11. Install the front camshaft in the reverse order of removal. Always use a new O-ring. Apply new engine oil to the journals and the cam lobes.

12. Apply new engine oil to the threads of the camshaft pulley mounting bolt, then install the front camshaft pulley.

13. Install the front rocker arm assembly, then tighten the mounting bolts.

14. Install the timing belt.

15. Adjust the valve clearance.

16. Install the EGR valve stud bolts, then install the EGR valve.

17. Connect the radiator hoses.

18. Do the battery installation procedure.

19. Fill the radiator with engine coolant, and bleed the air from the cooling system.

20. Install the engine compartment covers.

21. Do the Crankshaft Position (CKP) pattern clear/CKP pattern learn procedure.

REAR

See Figures 47 and 48.

1. Remove the engine compartment covers.

2. Relieve the fuel pressure.

3. Do the battery removal procedure.

4. Drain the engine coolant.

5. Remove the under-hood fuse/relay box from the bracket.

6. Remove the air cleaner assembly.

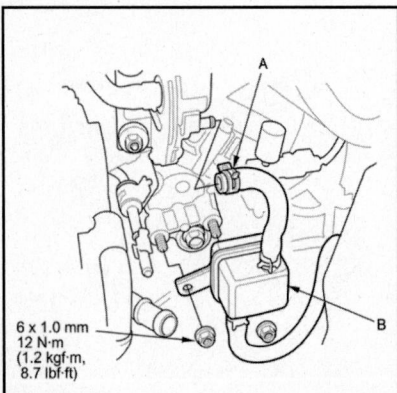

6 x 1.0 mm
12 N·m
(1.2 kgf·m, 8.7 lbf·ft)

37647_ACTL_G0305

Fig. 47 Disconnect the EVAP canister hose (A), then remove the EVAP canister joint (B) with the bracket

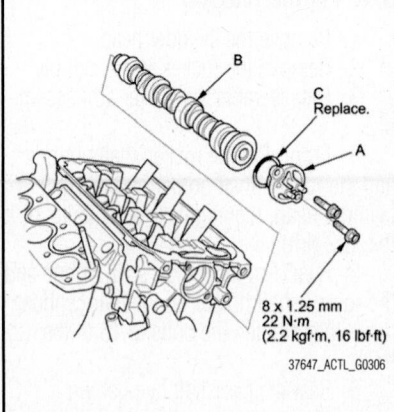

8 x 1.25 mm
22 N·m
(2.2 kgf·m, 16 lbf·ft)

37647_ACTL_G0306

Fig. 48 Remove the thrust cover (A), then remove the rear camshaft (B)

7. Remove the quick-connect fitting cover, then disconnect the fuel feed hose.

8. Disconnect the heater hoses.

9. Disconnect the Evaporative Emission (EVAP) canister hose, then remove the EVAP canister joint with the bracket.

10. Remove the timing belt.

11. Remove the rear rocker arm assembly.

12. Remove the rear camshaft pulley.

13. Remove the thrust cover, then remove the rear camshaft.

To install:

14. Install the rear camshaft in the reverse order of removal. Always use a new O-ring. Apply new engine oil to the journals and cam lobes.

15. Apply new engine oil to the threads of the camshaft pulley mounting bolt, then install the rear camshaft pulley.

16. Install the rear rocker arm assembly, then tighten the mounting bolts.

17. Install the timing belt.

18. Adjust the valve clearance.

19. Install the EVAP canister joint with the bracket, then connect the EVAP canister hose.

20. Connect the heater hoses.

21. Connect the fuel feed hose, then install the quick-connect fitting cover.

22. Install the air cleaner assembly.

23. Install the under-hood fuse/relay box to the bracket.

24. Do the battery installation procedure.

25. Inspect for fuel leaks. Turn the ignition switch to ON (II), or press the engine start/stop button to select the ON mode (do not operate the starter) so the fuel pump runs for about 2 seconds and pressurizes the fuel line. Repeat this operation three times, then check for fuel leakage at any point in the fuel line.

26. Fill the radiator with engine coolant, and bleed the air from the cooling system.

27. Install the engine compartment covers.

28. Do the Crankshaft Position (CKP) pattern clear/CKP pattern learn procedure.

CRANKSHAFT FRONT SEAL

REMOVAL & INSTALLATION

See Figure 49.

1. Remove the timing belt drive pulley.
   a. Remove the timing belt.
   b. Remove the timing belt stopper, then remove the timing belt drive pulley and the key.

2. Remove the pulley end crankshaft oil seal.

To install:

3. Clean and dry the crankshaft oil seal housing.

4. Apply a light coat of new engine oil to the lip of the crankshaft oil seal.

5. Using the oil seal driver, 64 mm, drive in the new crankshaft oil seal until the oil seal driver bottoms against the oil pump. When the seal is in place, clean any excess grease off the crankshaft, and check that the oil seal lip is not distorted.

6. Install the timing belt drive pulley.

CYLINDER HEAD

REMOVAL & INSTALLATION

See Figures 50 through 54.

1. Remove the engine compartment covers.

2. Relieve the fuel pressure.

3. Do the battery terminal disconnection procedure.

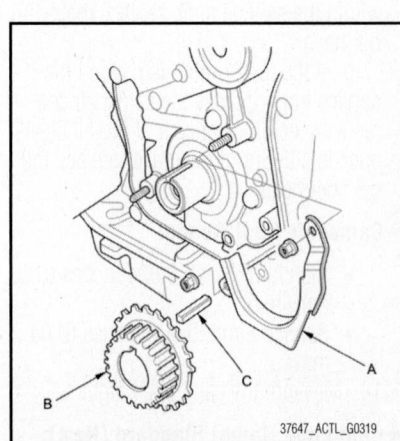

37647_ACTL_G0319

Fig. 49 Remove the timing belt stopper (A), then remove the timing belt drive pulley (B) and the key (C)

4. Drain the engine coolant.
5. Remove the alternator.
6. Remove the intake manifold.
7. Remove the six ignition coils.
8. Remove the timing belt.
9. Disconnect the following engine wire harness connectors, and remove the wire harness clamps from the cylinder head:

- Six injector connectors
- Knock sensor connector
- Engine Coolant Temperature (ECT) sensor 1 connector
- Engine mount control solenoid valve connector
- Camshaft Position (CMP) sensor connector
- Rocker arm oil control solenoid connector
- Rocker arm oil pressure switch connector
- Two Air Fuel Ratio (A/F) sensor connectors

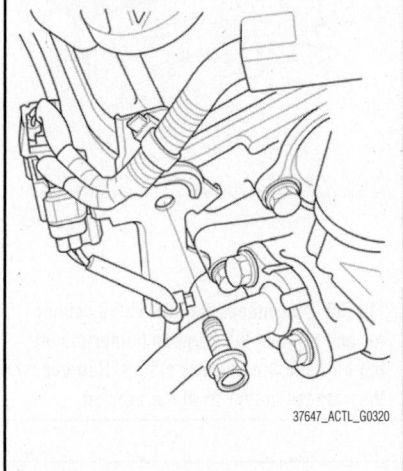

Fig. 50 Remove the connector bracket from the front cylinder head

- Two secondary Heated Oxygen Sensor (secondary HO2S) connectors

10. Remove the front Warm Up Three Way Catalytic Converter (front WU-TWC) and the rear Warm Up Three Way Catalytic Converter (rear WU-TWC).
11. Remove the quick-connect fitting cover, then disconnect the fuel feed hose.
12. Remove the connector bracket from the front cylinder head.
13. Remove the engine mount control solenoid valve bracket from the rear cylinder head.
14. Remove the Evaporative Emission (EVAP) canister joint with the bracket.
15. Remove the injector bases.
16. Remove the water passage.
17. Remove the camshaft pulleys and the back covers.
18. Remove the cylinder head covers.
19. Remove the cylinder head bolts. To prevent warpage, loosen the bolts in sequence ⅓ turn at a time; repeat the sequence until all bolts are loosened.
20. Remove the cylinder heads.

To install:

21. Clean the cylinder head and the engine block surface.
22. Clean and install the oil control orifices with new O-rings.
23. Install the dowel pins and the new cylinder head gaskets.
24. Clean the timing belt pulleys, the timing belt guide plate, and the upper and lower covers.
25. Set the timing belt drive pulley to Top Dead Center (TDC) by aligning the TDC mark on the tooth of the timing belt drive pulley with the pointer on the oil pump.
26. Set the camshaft pulleys to TDC by aligning the TDC marks on the camshaft

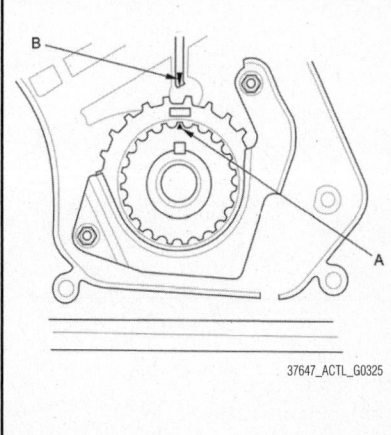

Fig. 52 Set the timing belt drive pulley to TDC by aligning the TDC mark (A) on the tooth of the timing belt drive pulley with the pointer (B) on the oil pump

pulleys with the pointers on the back covers.

27. Install the cylinder heads on the engine block.
28. Measure the diameter of each cylinder head bolt at point A and point B.
29. If either diameter is less than 0.44 inches (11.3 mm), replace the cylinder head bolt.
30. Apply new engine oil to the threads and under the bolt heads of all cylinder head bolts.
31. Torque the cylinder head bolts in sequence to 22 ft. lbs. (29 Nm) using a beam-type torque wrench. When using a preset click-type torque wrench, be sure to tighten slowly and do not overtighten. If a bolt makes any noise while you are torquing it, loosen the bolt and retighten it from the first step.
32. After torquing, tighten all cylinder head bolts in two steps (90° per step) using the sequence shown in step 11. If you are using a new cylinder head bolt, tighten the bolt an extra 90°.

➡️**Remove the cylinder head bolt if you tightened it beyond the specified angle, and go back to step 8 of the procedure. Do not loosen it back to the specified angle.**

33. Install the timing belt.
34. Adjust the valve clearance.
35. Install the cylinder head covers.
36. Install the water passage.
37. Install the injector bases.
38. Install the connector bracket to the front cylinder head.
39. Install the Evaporative Emission (EVAP) canister joint with the bracket.

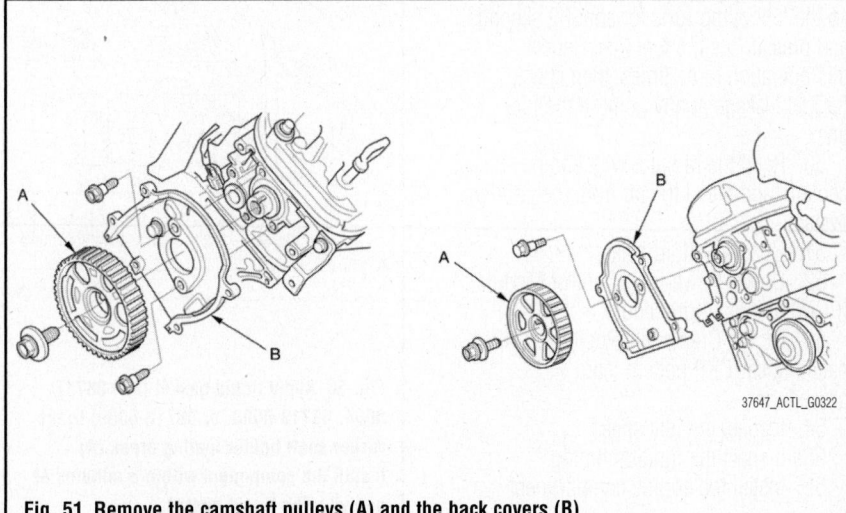

Fig. 51 Remove the camshaft pulleys (A) and the back covers (B)

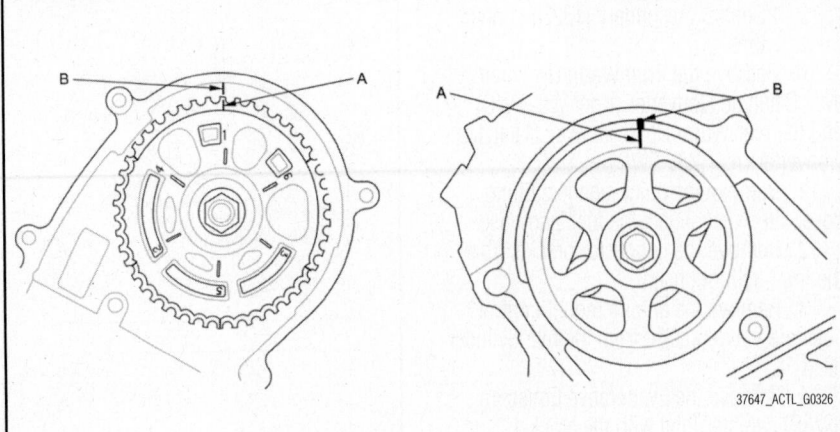

Fig. 53 Set the camshaft pulleys to TDC by aligning the TDC marks (A) on the camshaft pulleys with the pointers (B) on the back covers

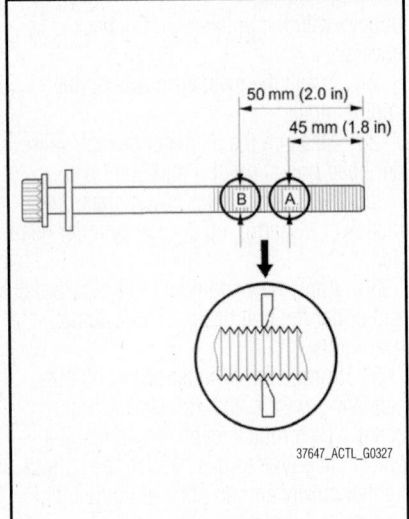

Fig. 54 Measure the diameter of each cylinder head bolt at point A and point B

40. Install the engine mount control solenoid valve bracket to the rear cylinder head.

41. Connect the fuel feed hose, then install the quick-connect fitting cover.

42. Install the front Warm Up Three Way Catalytic Converter (front WU-TWC) and the rear Warm Up Three Way Catalytic Converter (rear WU-TWC).

43. Connect the following engine wire harness connectors, and install the wire harness clamps to the cylinder head:
- Six injector connectors
- Knock sensor connector
- Engine Coolant Temperature (ECT) sensor 1 connector
- Engine mount control solenoid valve connector

- Camshaft Position (CMP) sensor connector
- Rocker arm oil control solenoid connector
- Rocker arm oil pressure switch connector
- Two Air Fuel Ratio (A/F) sensor connectors
- Two secondary Heated Oxygen Sensor (secondary HO2S) connectors

44. Install the six ignition coils.

45. Install the intake manifold.

46. Install the alternator.

47. Do the battery terminal reconnection procedure.

48. After installation, check that all tubes, hoses, and connectors are installed correctly.

49. Inspect for fuel leaks. Turn the ignition switch to ON (II), or press the engine start/stop button to select the ON mode (do not operate the starter) so the fuel pump runs for about 2 seconds and pressurizes the fuel line. Repeat this operation three times, then check for fuel leakage at any point in the fuel line.

50. Refill the radiator with engine coolant, and bleed the air from the cooling system.

51. Check for fluid leaks.

52. Do the Powertrain Control Module (PCM) idle learn procedure.

53. Do the Crankshaft Position (CKP) pattern clear/CKP pattern learn procedure.

54. Inspect the idle speed.

55. Inspect the ignition timing.

56. Install the engine compartment covers.

CYLINDER HEAD COVERS

REMOVAL & INSTALLATION

➡This procedure is valid for both 3.5L and 3.7L. All illustrations are from 3.5L, but 3.7L will be very similar.

Front

See Figures 55 through 58.

1. Remove the intake manifold.

2. Remove the three ignition coils from the front cylinder head.

3. Disconnect the EGR valve connector and remove the harness holder mounting bolt and the harness clamp.

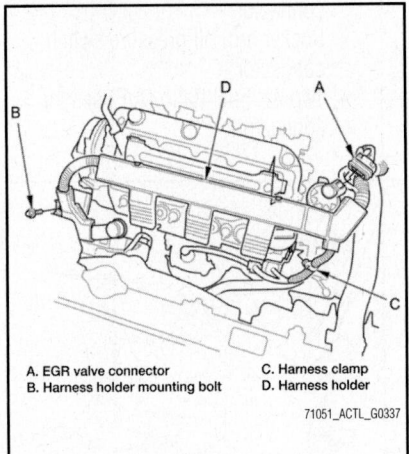

A. EGR valve connector
B. Harness holder mounting bolt
C. Harness clamp
D. Harness holder

Fig. 55 Disconnect the EGR valve connector and remove the harness holder mounting bolt and the harness clamp. Remove the harness holder from the bracket.

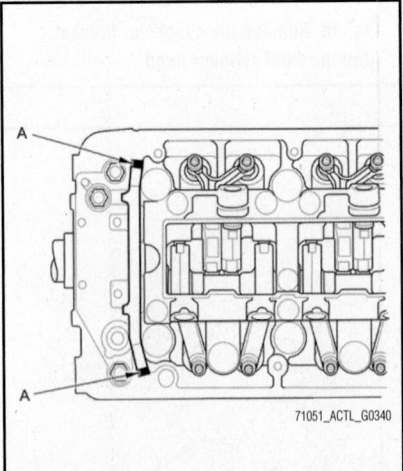

Fig. 56 Apply liquid gasket (P/N 08717-0004, 08718-0003, or 08718-0009) to the rocker shaft holder mating areas (A). Install the component within 5 minutes of applying the liquid gasket.

4. Remove the harness holder from the bracket.

5. Remove the 5 bolts and the front cylinder head cover.

To install:

6. Check the spark plug seals for damage. If any seal are damaged, replace it.

7. Thoroughly clean the head cover gasket and the groove.

8. Install the head cover gasket in the groove of the cylinder head cover. Make sure the head cover gasket is seated securely.

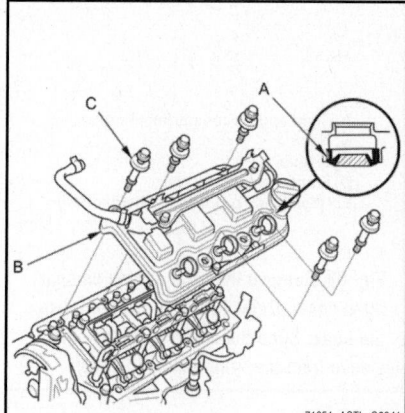

Fig. 57 Set the spark plug seals (A) on the spark plug tubes, and install the front cylinder head cover (B). Inspect the spark plug seals for damage. Inspect the cover washers (C). Replace any washer that is damaged or deteriorated.

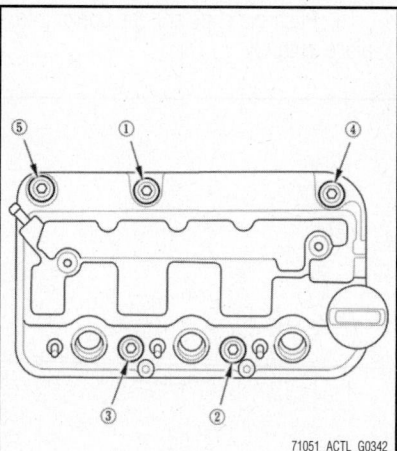

Fig. 58 Tighten the bolts in three steps. In the final step torque all bolts, in sequence, 9 ft. lbs. (12 Nm).

9. Remove all of the old liquid gasket from the rocker shaft holder and the cylinder head.

10. Clean the head cover contacting surfaces with a shop towel.

11. Apply liquid gasket (P/N 08717-0004, 08718-0003, or 08718-0009) to the rocker shaft holder mating areas. Install the component within 5 minutes of applying the liquid gasket.

Please note the following:

• If you apply liquid gasket P/N 08718-0012, the component must be installed within 4 minutes.

• If too much time has passed after applying the liquid gasket, remove the old liquid gasket and residue, then reapply the new liquid gasket.

12. Set the spark plug seals (A) on the spark plug tubes, and install the front cylinder head cover (B).

13. Inspect the spark plug seals for damage.

14. Inspect the cover washers (C). Replace any washer that is damaged or deteriorated.

15. Tighten the bolts in three steps. In the final step torque all bolts, in sequence, 9 ft. lbs. (12 Nm).

Please note the following:

• Wait at least 30 minutes before filling the engine with oil.

• Do not run the engine for at least 3 hours after installing the cylinder head cover.

16. Install the harness holder to the bracket.

17. Connect the EGR valve connector and tighten the harness holder mounting bolt and the harness clamp.

18. Install the three ignition coils to the front cylinder head.

19. Install the intake manifold.

Rear

See Figures 59 through 60.

1. Remove the strut brace.

2. Remove the intake manifold.

3. Remove the three ignition coils from the rear cylinder head.

4. Remove the harness holder mounting bolts and the engine ground cable.

5. Disconnect the three injector connectors and the two harness clamps.

6. Remove the harness clamp and disconnect the breather hose.

7. Remove the harness from the upper cover.

8. Remove the engine hanger bracket.

9. Remove the harness holder mounting bolts and disconnect the knock sensor connector and the CMP sensor connector.

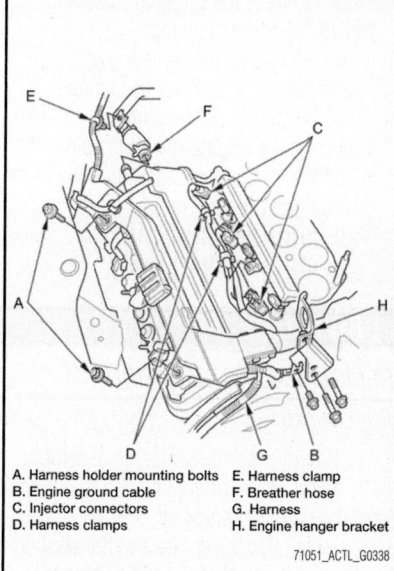

A. Harness holder mounting bolts
B. Engine ground cable
C. Injector connectors
D. Harness clamps
E. Harness clamp
F. Breather hose
G. Harness
H. Engine hanger bracket

Fig. 59 Remove the external components in order to remove the rear cylinder head cover

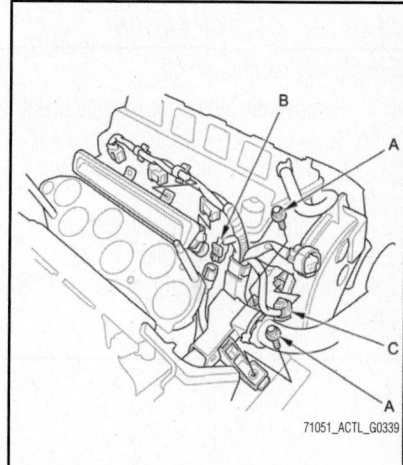

Fig. 60 Remove the harness holder mounting bolts (A) and disconnect the knock sensor connector (B) and the CMP sensor connector (C).

10. Remove the 5 bolts and the rear cylinder head cover.

To install:

11. Install valve cover and gasket using same instructions as for Front cover, then continue as follows:

a. Connect the knock sensor connector and the CMP sensor connector and install the harness holder mounting bolts.

b. Install the engine hanger bracket.

c. Install the harness to the upper cover.

d. Connect the breather hose and install the harness clamp.

e. Reconnect the three injector connectors and the two harness clamps.

f. Install the harness holder mounting bolts and the engine ground cable.

g. Install the three ignition coils to the rear cylinder head.

h. Install the intake manifold.

i. Install the strut brace.

DRIVEPLATE

REMOVAL & INSTALLATION

See Figure 61.

1. Remove the transmission assembly.

2. Remove the driveplate and the washer from the engine crankshaft.

3. Install the driveplate and the washer on the engine crankshaft, and tighten the eight bolts in a crisscross pattern in at least two steps.

4. Install the transmission assembly.

ENGINE COVER

REMOVAL & INSTALLATION

See Figures 62 through 65.

1. Remove the left engine compartment cover:

a. Detach the clips and release the hook.

b. Slide the cover to release the sleeve from the front bulkhead cover.

2. Remove the right engine compartment cover.

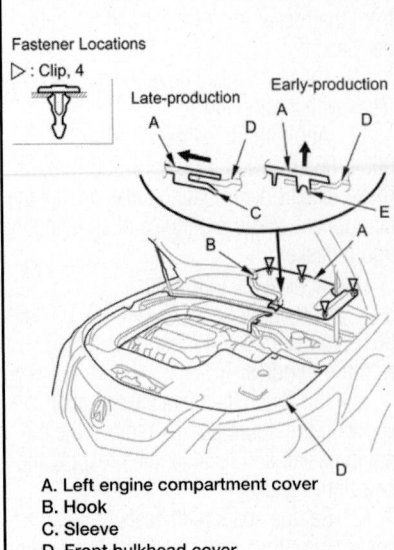

A. Left engine compartment cover
B. Hook
C. Sleeve
D. Front bulkhead cover
E. Hook (early production, if equipped)

71051_ACTL_G0313

Fig. 62 Remove the left engine compartment cover. Detach the clips and release the hook. Slide the cover to release the sleeve from the front bulkhead cover.

a. Detach the clips and release the hook.

b. Slide the cover to release the sleeve from the front bulkhead cover.

➡**For some early-production models: Release the hook before removing the cover.**

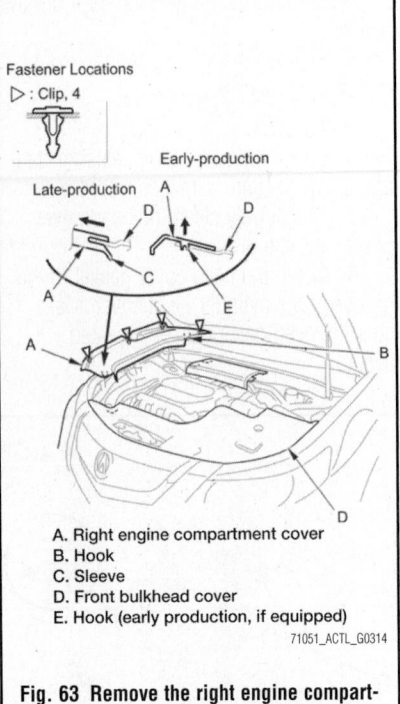

A. Right engine compartment cover
B. Hook
C. Sleeve
D. Front bulkhead cover
E. Hook (early production, if equipped)

71051_ACTL_G0314

Fig. 63 Remove the right engine compartment cover. Detach the clips and release the hook. Slide the cover to release the sleeve from the front bulkhead cover.

3. Detach the clips, and release the projection, then remove the front bulkhead cover.

4. Detach the clip, and release the hook, then remove the strut brace cover.

To install:

5. Install the covers in the reverse order of removal, and note these items:

a. If the clips are damaged or stress-whitened, replace them with new ones.

b. Push the clips and the hooks into place securely.

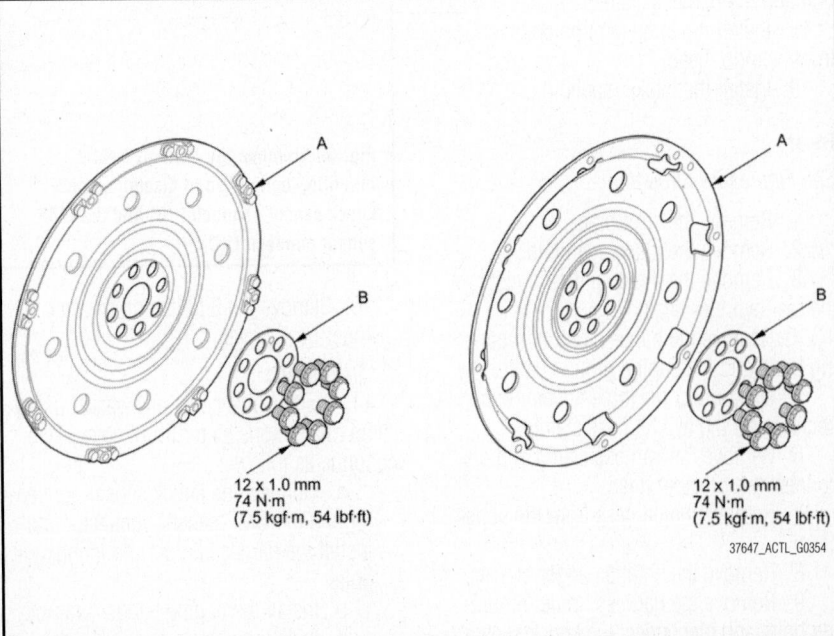

Fig. 61 Remove the driveplate (A) and the washer (B) from the engine crankshaft

12 x 1.0 mm
74 N·m
(7.5 kgf·m, 54 lbf·ft)

37647_ACTL_G0354

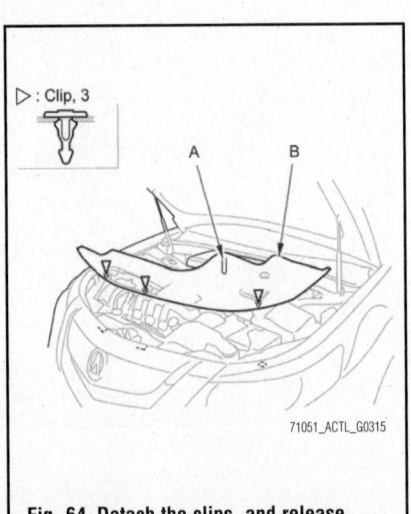

71051_ACTL_G0315

Fig. 64 Detach the clips, and release the projection (A), then remove the front bulkhead cover (B).

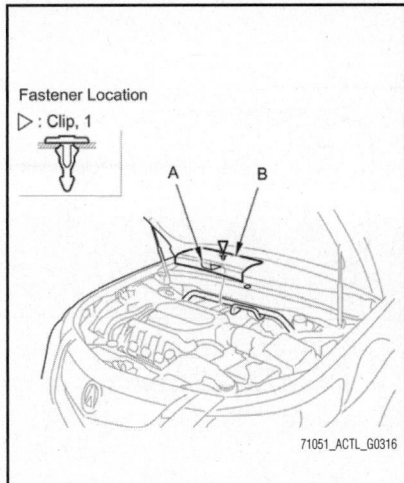

Fastener Location
▷ : Clip, 1

71051_ACTL_G0316

Fig. 65 Detach the clip, and release the hook (A), then remove the strut brace cover (B).

EXHAUST SYSTEM

REMOVAL & INSTALLATION

See Figures 66 and 67.

➡**Manufacturer does not provide a specific removal and installation procedure. Use the referenced illustrations as reference when servicing.**

FLYWHEEL

REMOVAL & INSTALLATION

See Figure 68.

1. Install the ring gear holder.
2. Loosen the flywheel mounting bolts in a crisscross pattern in several steps. Remove the bolts, then remove the flywheel and the ring gear holder.

3. Install the flywheel on the crankshaft, and install the mounting bolts finger-tight.
4. Install the ring gear holder, then torque the flywheel mounting bolts in a crisscross pattern in several steps to 76 ft. lbs. (103 Nm).

INTAKE MANIFOLD

REMOVAL & INSTALLATION

See Figures 69 through 71.

1. Remove the engine compartment covers.
2. Disconnect the Manifold Absolute Pressure (MAP) sensor connector and the breather pipe, then remove the intake air duct.
3. Disconnect the throttle actuator connector, the Evaporative Emission (EVAP) canister purge valve connector, the water bypass hoses, the EVAP canister hose, the brake booster vacuum hose, and the vacuum hose.
4. Disconnect the Positive Crankcase Ventilation (PCV) hose and the Intake Manifold Tuning (IMT) actuator connector.
5. Remove the upper cover mounting bolts and nuts sequentially in three steps, then remove the upper cover.
6. Remove the intake manifold mounting bolts and nuts sequentially in three steps, then remove the intake manifold.

To install:
7. Install the intake manifold. Tighten the mounting bolts and nuts sequentially in three steps to 16 ft. lbs. (22 Nm). Always use a new intake manifold gasket.
8. Install the upper cover. Tighten the mounting bolts and nuts sequentially in three steps to 9 ft. lbs. (12 Nm). Always use a new gasket.
9. Connect the PCV hose and the IMT actuator connector.
10. Connect the throttle actuator connector, the EVAP canister purge valve connector, the water bypass hoses, the EVAP canister hose, the brake booster vacuum hose, and the vacuum hose.
11. Install the intake air duct, then connect the breather pipe and the MAP sensor connector.
12. Clean up any spilled engine coolant.
13. After installation, check that all tubes, hoses, and connectors are installed correctly.
14. Refill the radiator with engine coolant, and bleed the air from the cooling system.
15. Install the engine compartment covers.

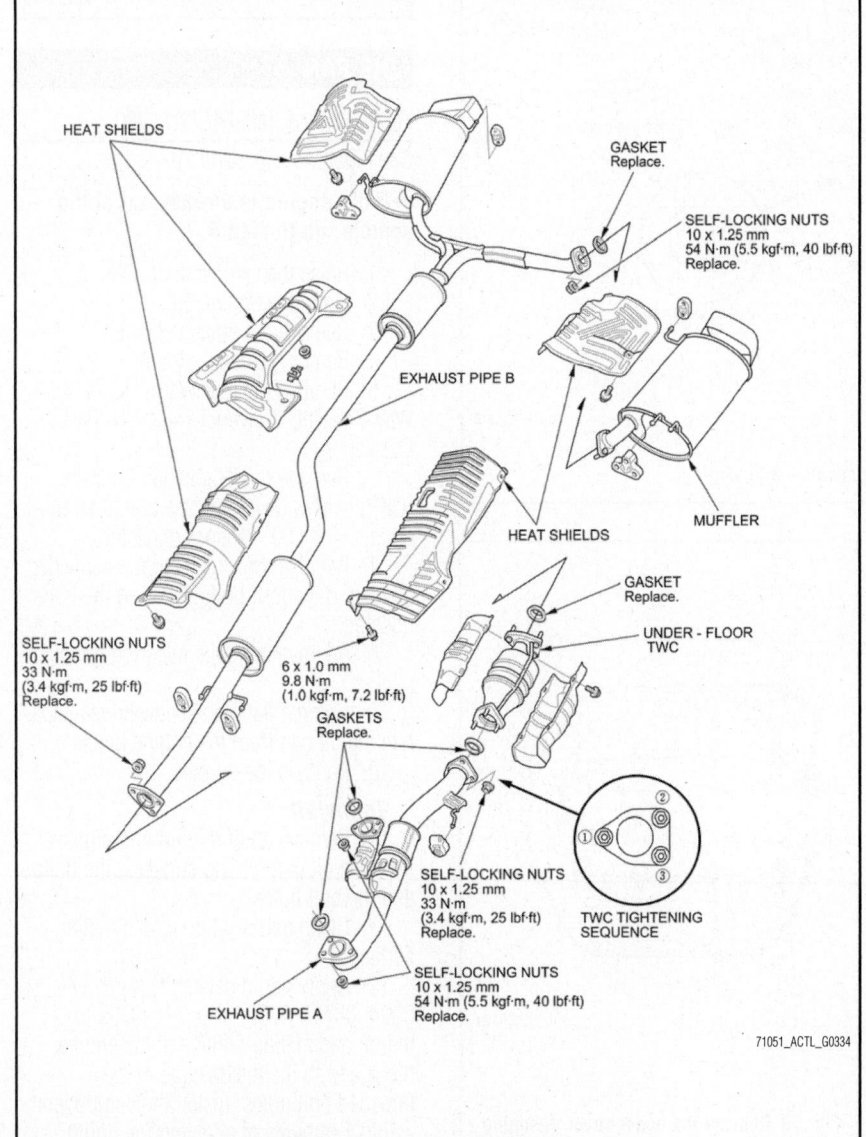

HEAT SHIELDS

GASKET
Replace.

SELF-LOCKING NUTS
10 x 1.25 mm
54 N·m (5.5 kgf·m, 40 lbf·ft)
Replace.

EXHAUST PIPE B

MUFFLER

HEAT SHIELDS

GASKET
Replace.

UNDER - FLOOR
TWC

SELF-LOCKING NUTS
10 x 1.25 mm
33 N·m
(3.4 kgf·m, 25 lbf·ft)
Replace.

6 x 1.0 mm
9.8 N·m
(1.0 kgf·m, 7.2 lbf·ft)

GASKETS
Replace.

SELF-LOCKING NUTS
10 x 1.25 mm
33 N·m
(3.4 kgf·m, 25 lbf·ft)
Replace.

TWC TIGHTENING
SEQUENCE

SELF-LOCKING NUTS
10 x 1.25 mm
54 N·m (5.5 kgf·m, 40 lbf·ft)
Replace.

EXHAUST PIPE A

71051_ACTL_G0334

Fig. 66 Exploded view of the exhaust system components—3.5L engine

NOTE: Use new gaskets and new self-locking nuts when reassembling.

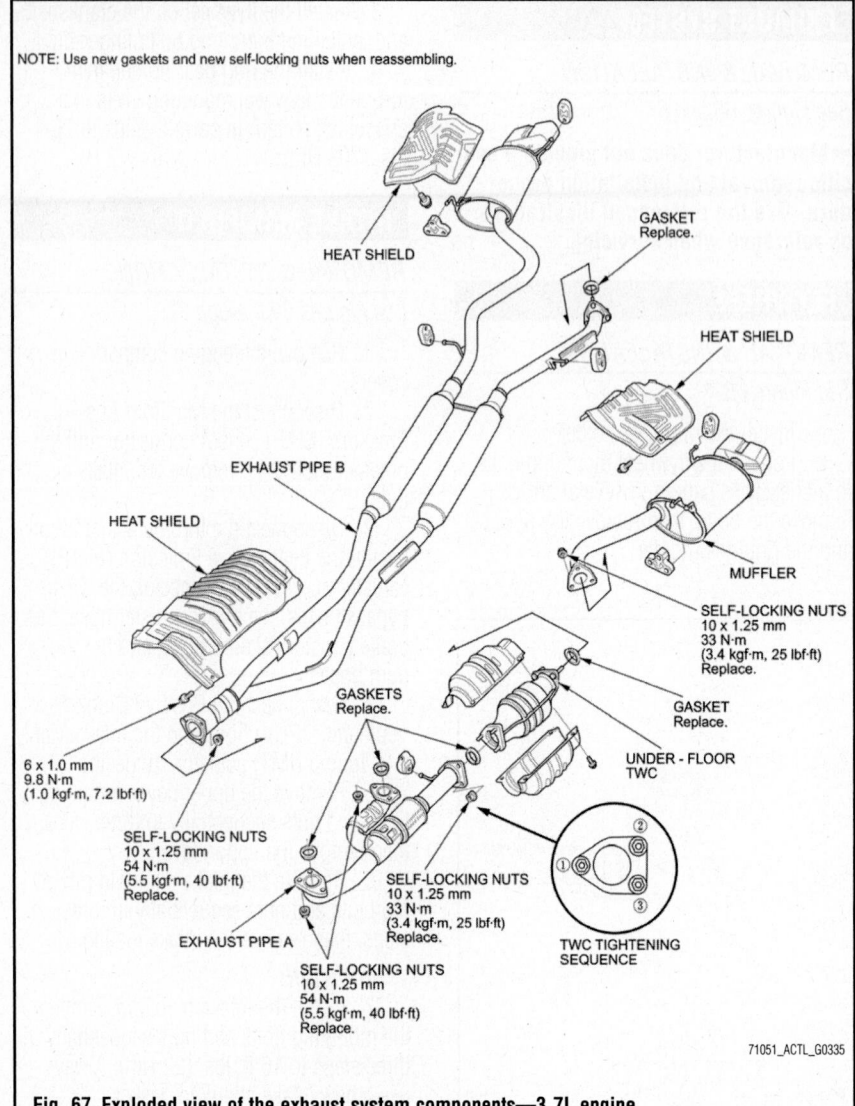

HEAT SHIELD

GASKET
Replace.

HEAT SHIELD

EXHAUST PIPE B

HEAT SHIELD

MUFFLER

SELF-LOCKING NUTS
10 x 1.25 mm
33 N·m
(3.4 kgf·m, 25 lbf·ft)
Replace.

GASKETS
Replace.

GASKET
Replace.

6 x 1.0 mm
9.8 N·m
(1.0 kgf·m, 7.2 lbf·ft)

UNDER - FLOOR
TWC

SELF-LOCKING NUTS
10 x 1.25 mm
54 N·m
(5.5 kgf·m, 40 lbf·ft)
Replace.

SELF-LOCKING NUTS
10 x 1.25 mm
33 N·m
(3.4 kgf·m, 25 lbf·ft)
Replace.

EXHAUST PIPE A

TWC TIGHTENING
SEQUENCE

SELF-LOCKING NUTS
10 x 1.25 mm
54 N·m
(5.5 kgf·m, 40 lbf·ft)
Replace.

71051_ACTL_G0335

Fig. 67 Exploded view of the exhaust system components—3.7L engine

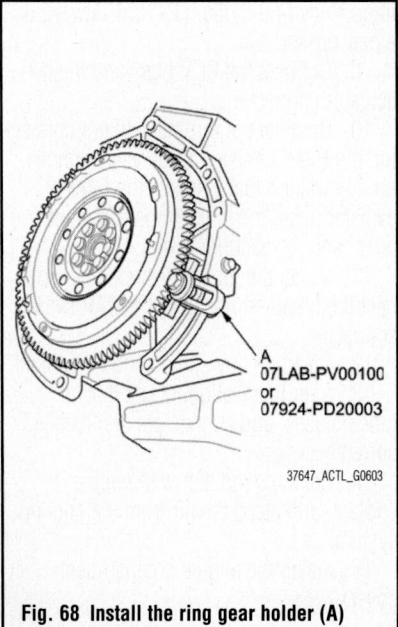

A
07LAB-PV00100
or
07924-PD20003

37647_ACTL_G0603

Fig. 68 Install the ring gear holder (A)

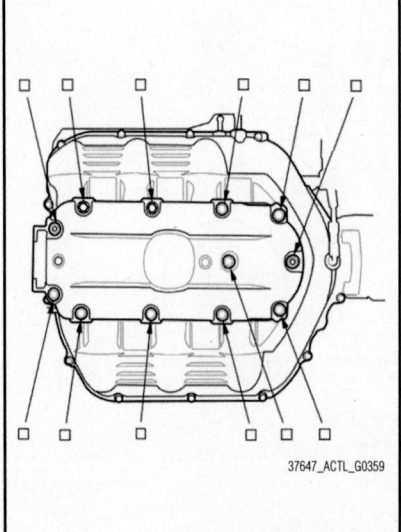

37647_ACTL_G0359

Fig. 69 Remove the upper cover mounting bolts and nuts sequentially in three steps

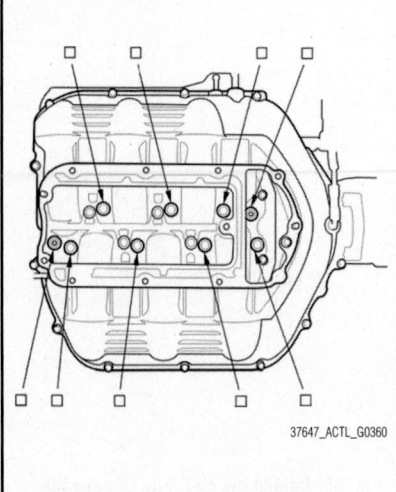

37647_ACTL_G0360

Fig. 70 Remove the intake manifold mounting bolts and nuts sequentially in three steps, then remove the intake manifold

OIL PAN

REMOVAL & INSTALLATION

See Figures 72 through 74.

➡️**If the engine is already out of the vehicle, go to step 6.**

1. Raise the vehicle on the lift.
2. Drain the engine oil.
3. Remove the splash shield.
4. Remove exhaust pipe A.
5. Remove the rear Warm Up Three Way Catalytic Converter (rear WU-TWC) bracket.
6. Remove the Crankshaft Position (CKP) sensor cover and the bolt, then disconnect the CKP sensor connector.
7. Remove the clutch/torque converter cover and the four bolts securing the transmission.
8. Remove the bolts securing the oil pan.
9. Using a flat blade screwdriver, separate the oil pan from the engine block.
10. Remove the oil pan.

To install:

11. Remove all of the old liquid gasket from the oil pan mating surfaces, the bolts, and the bolt holes.
12. Clean and dry the oil pan mating surfaces.
13. Apply liquid gasket, P/N 08717-0004, 08718-0003, or 08718-0009 to the oil pan mating surface of the engine block and to the inside edge of the threaded bolt holes. Install the component within 5 minutes of applying the liquid gasket.

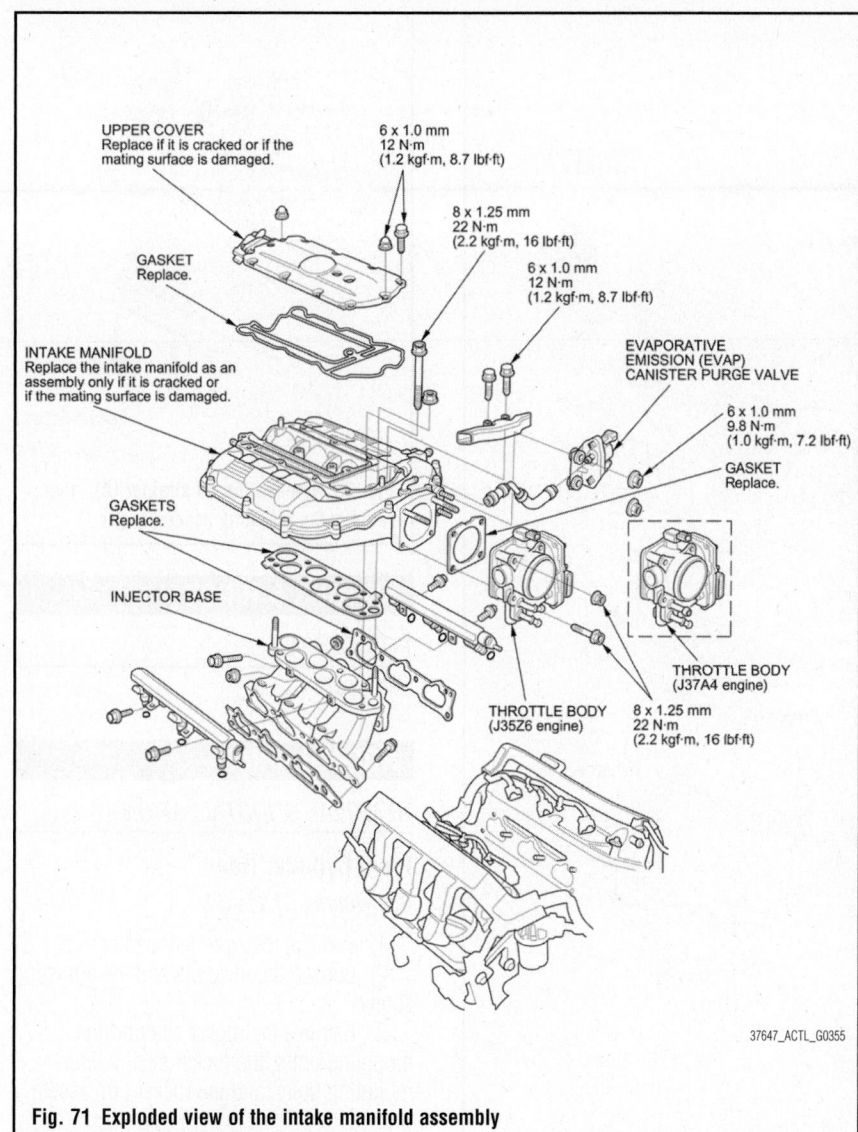

Fig. 71 Exploded view of the intake manifold assembly

Labels in figure:

UPPER COVER
Replace if it is cracked or if the
mating surface is damaged.

GASKET
Replace.

INTAKE MANIFOLD
Replace the intake manifold as an
assembly only if it is cracked or
if the mating surface is damaged.

GASKETS
Replace.

INJECTOR BASE

6 x 1.0 mm
12 N·m
(1.2 kgf·m, 8.7 lbf·ft)

8 x 1.25 mm
22 N·m
(2.2 kgf·m, 16 lbf·ft)

6 x 1.0 mm
12 N·m
(1.2 kgf·m, 8.7 lbf·ft)

EVAPORATIVE
EMISSION (EVAP)
CANISTER PURGE VALVE

6 x 1.0 mm
9.8 N·m
(1.0 kgf·m, 7.2 lbf·ft)

GASKET
Replace.

THROTTLE BODY
(J37A4 engine)

THROTTLE BODY
(J35Z6 engine)

8 x 1.25 mm
22 N·m
(2.2 kgf·m, 16 lbf·ft)

37647_ACTL_G0355

➡**Apply a bead of liquid gasket along
the mating edge of the oil pan. If you
apply liquid gasket P/N 08718-0012,
the component must be installed within
4 minutes. If too much time has passed
after applying the liquid gasket,
remove the old liquid gasket and
residue, then reapply the new liquid
gasket.**

14. Install the oil pan on the engine
block.

15. Tighten the bolts in three steps. In
the final step, torque all bolts, in sequence,
to 9 ft. lbs. (12 Nm).

➡**Wait at least 30 minutes before filling the engine with oil. Do not run the
engine for at least 3 hours after
installing the oil pan.**

16. Tighten the four bolts securing the
transmission, then install the clutch/torque
converter cover.

17. Connect the Crankshaft Position
(CKP) sensor connector, then install the
CKP sensor cover and the bolt.

18. Install the rear Warm Up Three Way
Catalytic Converter (rear WU-TWC) bracket.

19. If the engine is still in the vehicle, do
the following steps.

20. Install exhaust pipe A using new
gaskets and new self-locking nuts.

21. Install the splash shield.

22. Refill the engine with the recommended engine oil.

OIL PUMP

REMOVAL & INSTALLATION

See Figures 75 through 77.

1. Drain the engine oil.
2. Remove the timing belt:
3. Remove the timing belt drive pulley
from the crankshaft:

4. Attach the chain hoist to the engine
hook on the engine hanger bracket.

5. Remove the rocker arm oil control
solenoid/oil filter assembly.

6. Remove the oil pan.

7. Remove the oil strainer.

8. Remove the mounting bolts, then
remove the oil pump assembly.

To install:

9. Remove the old oil seal from the oil
pump.

10. Clean and dry the crankshaft oil seal
housing.

11. Using the oil seal driver, 64 mm,
drive in the new crankshaft oil seal until the
oil seal driver bottoms on the pump.

12. Remove all of the old liquid gasket
from the oil pump mating surfaces, the
bolts, and the bolt holes.

13. Clean and dry the oil pump mating
surfaces.

14. Apply liquid gasket, P/N 08717-
0004, 08718-0003, or 08718-0009,
to the engine block mating surface of the
oil pump and to the inside edge of the
threaded bolt holes. Install the component
within 5 minutes of applying the liquid
gasket.

➡**Apply a bead of liquid gasket along
the mating surface. If you apply liquid
gasket P/N 08718-0012, the component
must be installed within 4 minutes. If
too much time has passed after applying the liquid gasket, remove all of the
old liquid gasket and residue, then
reapply the new liquid gasket.**

15. Apply a light coat of new engine
oil to the lip of the crankshaft oil seal,
and apply new engine oil to the new
O-ring.

16. Install the dowel pins, then align the
inner rotor with the crankshaft, and install
the oil pump.

➡**Wait at least 30 minutes before filling the engine with oil. Do not run the
engine for at least 3 hours after
installing the oil pump.**

17. Clean the excess grease off the
crankshaft, and check the seal for distortion.

18. Install the oil strainer with a new O-
ring.

19. Install the oil pan.

20. Install the rocker arm oil control
solenoid/oil filter assembly with a new
rocker arm oil control solenoid filter.

21. Install the timing belt drive pulley to
the crankshaft:

22. Install the timing belt:

23. Remove the chain hoist.

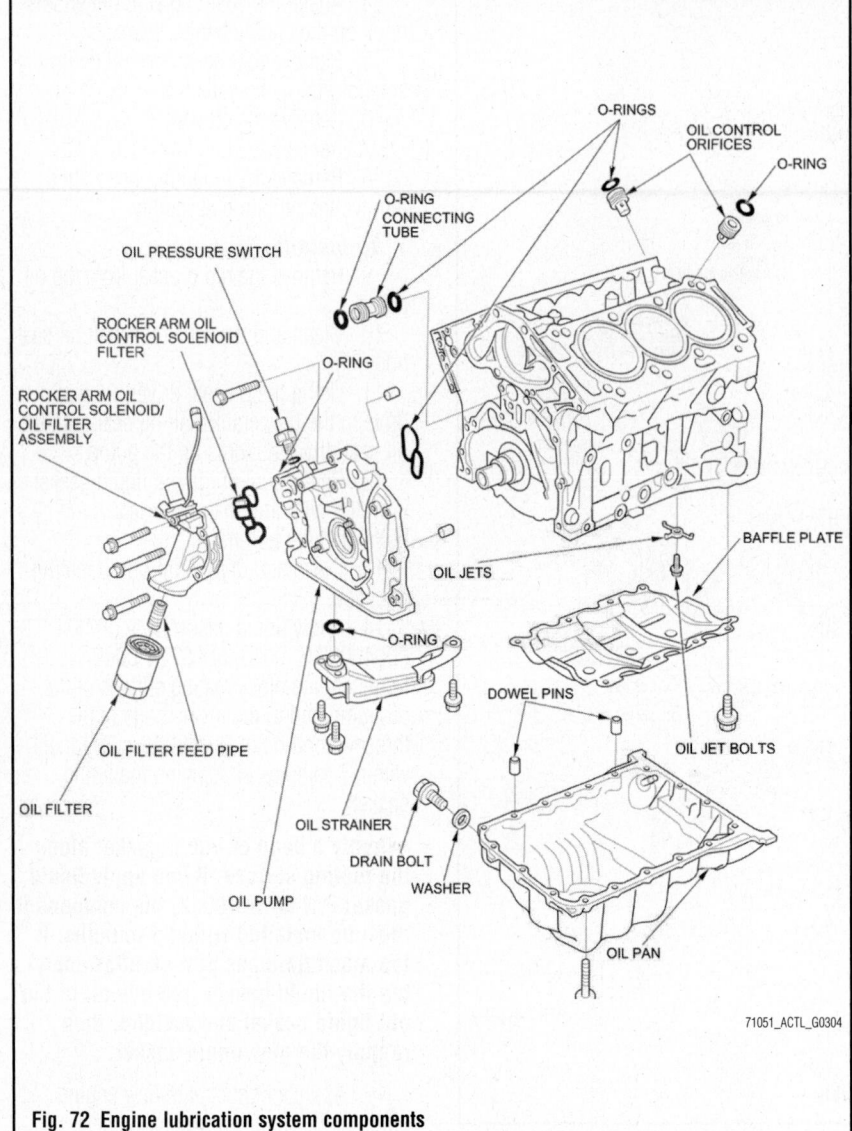

Fig. 72 Engine lubrication system components

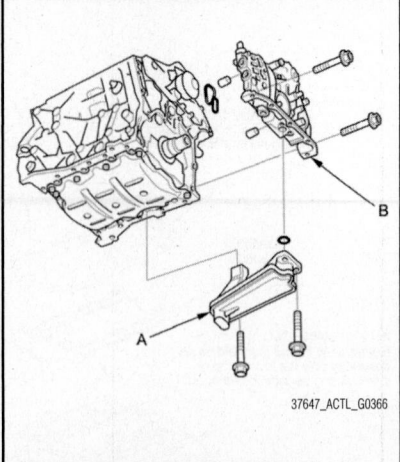

Fig. 75 Remove the oil strainer (A) , then remove the oil pump assembly (B)

PISTONS & RINGS

POSITIONING

See Figure 78.

ROCKER ARMS

REMOVAL & INSTALLATION

Front Cylinder Head

See Figures 79 and 80.

1. Remove the cylinder head cover.
2. Loosen the locknuts and the adjusting screws
3. Remove the rocker shaft bridge mounting bolts, the rocker shaft holder mounting bolts, and the rocker arm assembly.

   a. Loosen the rocker shaft bridge mounting bolts and the rocker shaft holder mounting bolts in sequence two turns at a time, to prevent damaging the valves or the rocker arm assembly.

   b. When removing the rocker arm assembly, do not remove the rocker shaft bridge mounting bolts and the rocker shaft holder mounting bolts. The bolts will keep the rocker arms on the shafts.

   To install:

4. Apply liquid gasket, P/N 08717-0004, 08718-0003, or 08718-0009 to the rocker shaft holder mating surface of the cylinder head. Install the component within 5 minutes of applying the liquid gasket.

➡Apply a bead of liquid gasket along the mating surfaces. If you apply liquid gasket P/N 08718-0012, the component must be installed within 4 minutes. If too much time has passed after apply-

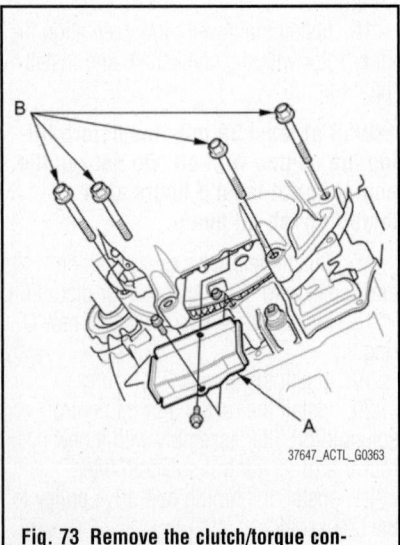

Fig. 73 Remove the clutch/torque converter cover (A) and the four bolts (B) securing the transmission

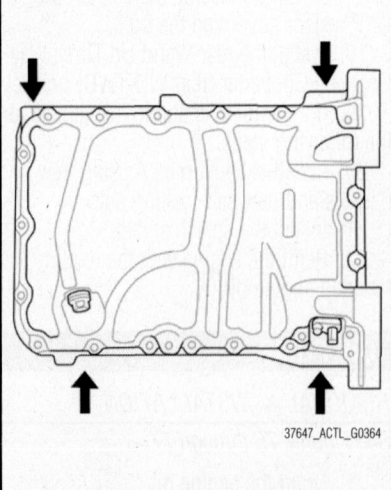

Fig. 75 Using a flat blade screwdriver, separate the oil pan from the engine block in the places shown

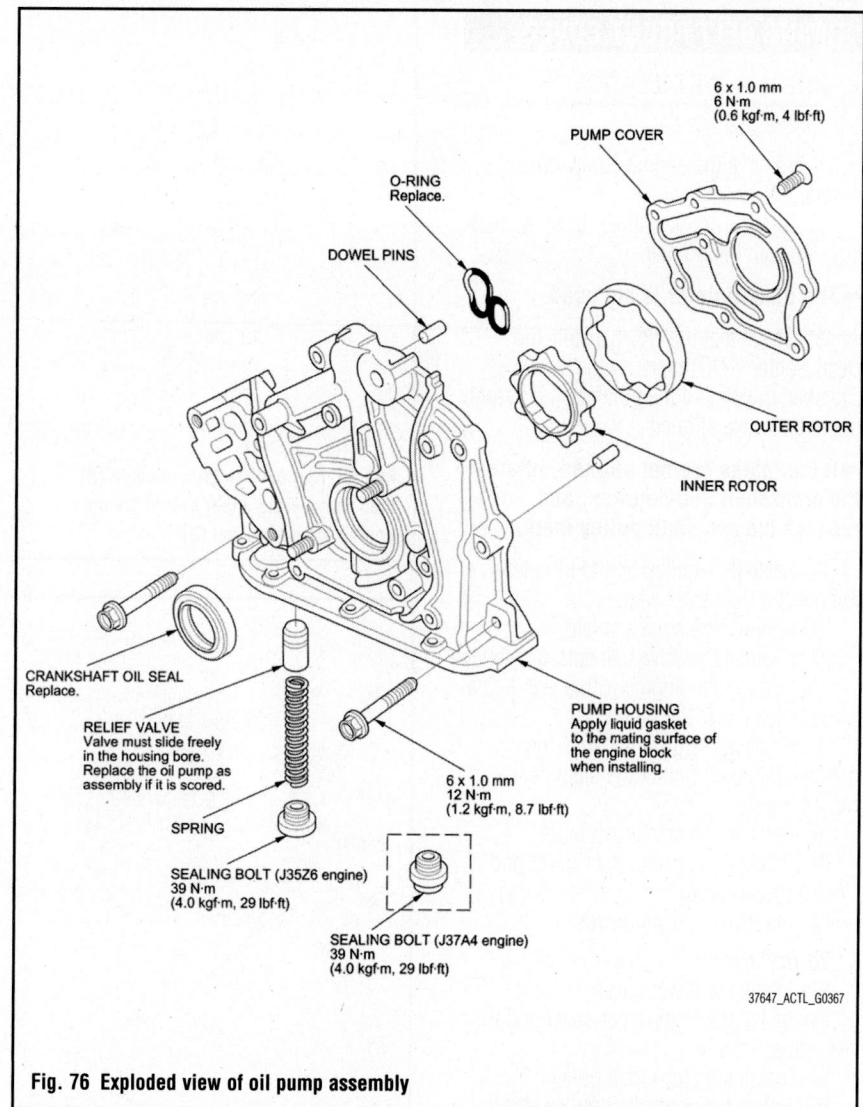

Fig. 76 Exploded view of oil pump assembly

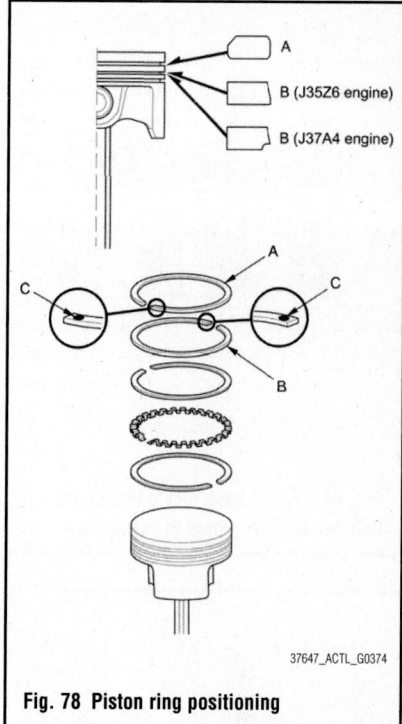

Fig. 78 Piston ring positioning

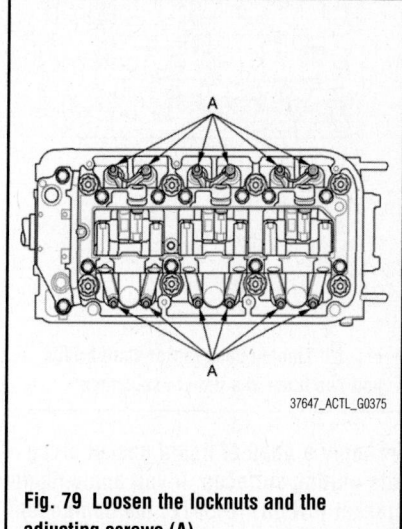

Fig. 79 Loosen the locknuts and the adjusting screws (A)

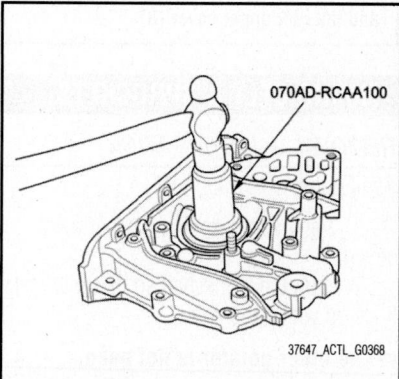

Fig. 77 Using the oil seal driver, 64 mm, drive in the new crankshaft oil seal until the oil seal driver bottoms on the pump

ing the liquid gasket, remove the old liquid gasket and residue, then reapply the new liquid gasket.

5. Set the rocker arm assembly in place, and loosely install the bolts. Make sure that

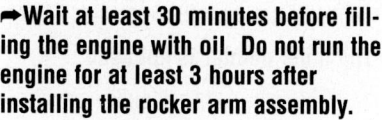

the rocker arms are properly positioned on the valve stems.

➡**Wait at least 30 minutes before filling the engine with oil. Do not run the engine for at least 3 hours after installing the rocker arm assembly.**

6. Tighten each bolt two turns at a time in the sequence shown to ensure that the rockers do not bind on the valves.

Rear Cylinder Head

See Figure 81.

1. Remove the cylinder head cover.
2. Loosen the locknuts and the adjusting screws.
3. Remove the rocker shaft bridge mounting bolts, the rocker shaft holder mounting bolts, and the rocker arm assembly:

    a. Loosen the rocker shaft bridge mounting bolts and the rocker shaft holder mounting bolts in sequence two

turns at a time, to prevent damaging the valves or the rocker arm assembly.

    b. When removing the rocker arm assembly, do not remove the rocker shaft bridge mounting bolts and the rocker shaft holder mounting bolts. The bolts will keep the rocker arms on the shafts.

To install:

4. Apply liquid gasket, P/N 08717-0004, 08718-0003, or 08718-0009 to the rocker shaft holder mating surface of the cylinder head. Install the component within 5 minutes of applying the liquid gasket.

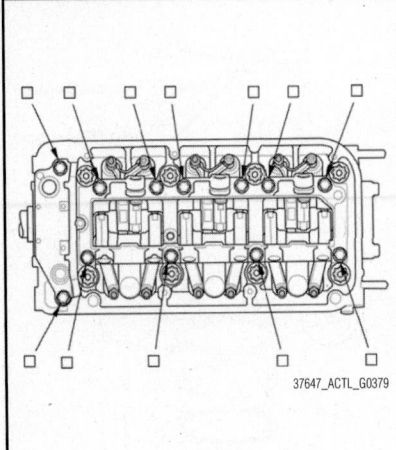

Fig. 80 Tighten each rocker shaft bridge bolt two turns at a time in sequence

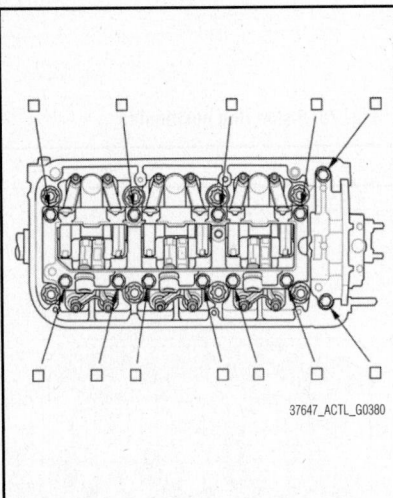

Fig. 81 Tighten each rocker shaft bridge bolt two turns at a time in sequence

➡**Apply a bead of liquid gasket along the mating surfaces. If you apply liquid gasket P/N 08718-0012, the component must be installed within 4 minutes. If too much time has passed after applying the liquid gasket, remove the old liquid gasket and residue, then reapply the new liquid gasket.**

5. Set the rocker arm assembly in place, and loosely install the bolts. Make sure that the rocker arms are properly positioned on the valve stems.

➡**Wait at least 30 minutes before filling the engine with oil. Do not run the engine for at least 3 hours after installing the rocker arm assembly.**

6. Tighten each bolt two turns at a time in the sequence shown to ensure that the rockers do not bind on the valves.

TIMING BELT FRONT COVER

REMOVAL & INSTALLATION

See Figures 82 and 83.

1. Remove the engine compartment covers.
2. Turn the crankshaft so its white mark lines up with the pointer.

➡**The other pointer is not used.**

3. Check that the No. 1 piston Top Dead Center (TDC) mark on the front camshaft pulley and the pointer on the front upper cover are aligned.

➡**If the marks are not aligned, rotate the crankshaft 360 degrees, and recheck the camshaft pulley mark.**

4. Raise the vehicle on the lift, then remove the right front wheel.
5. Remove the splash shield.
6. Remove the drive belt auto-tensioner.
7. Support the engine with a jack and a wood block under the oil pan.
8. Remove the ground cable, then remove the upper half of the side engine mount bracket.
9. Remove the crankshaft pulley.
10. Remove the front upper cover and the rear upper cover.
11. Remove the lower cover.

To install:
12. Install the lower cover.
13. Install the front upper cover and the rear upper cover.
14. Install the crankshaft pulley.
15. Rotate the crankshaft pulley about six turns clockwise so the timing belt positions itself on the pulleys.
16. Turn the crankshaft pulley so its white mark lines up with the pointer.

➡**The other pointer is not used.**

17. Check the camshaft pulley marks.

➡**If the marks are not aligned, rotate the crankshaft 360 degrees, and recheck the camshaft pulley mark. If the camshaft pulley marks are at TDC, go to step 17. If the camshaft pulley marks are not at TDC, remove the timing belt and repeat steps 2 through 16.**

18. Install the upper half of the side engine mount bracket, then tighten the mounting bolts.
19. Install the ground cable.
20. Install the drive belt auto-tensioner.
21. Install the splash shield.
22. Install the right front wheel.
23. Install the engine compartment covers.

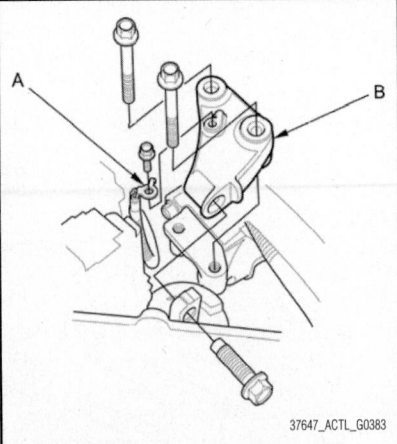

Fig. 82 Remove the ground cable (A), then remove the upper half of the side engine mount bracket (B)

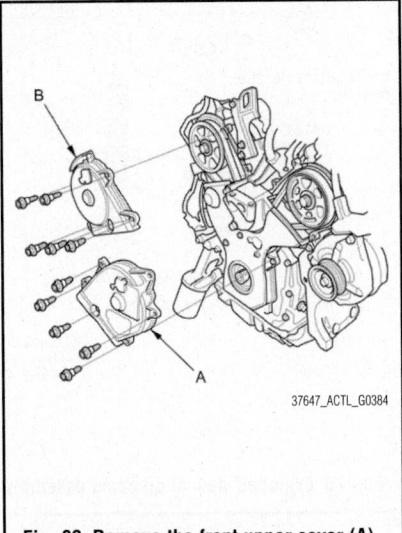

Fig. 83 Remove the front upper cover (A) and the rear upper cover (B)

TIMING BELT & SPROCKETS

REMOVAL & INSTALLATION

See Figures 84 through 87.

1. Remove the engine compartment covers.
2. Turn the crankshaft so its white mark lines up with the pointer.

➡**The other pointer is not used.**

3. Check that the No. 1 piston Top Dead Center (TDC) mark on the front camshaft pulley and the pointer on the front upper cover are aligned.

➡**If the marks are not aligned, rotate the crankshaft 360 degrees, and recheck the camshaft pulley mark.**

4. Raise the vehicle on the lift, then remove the right front wheel.

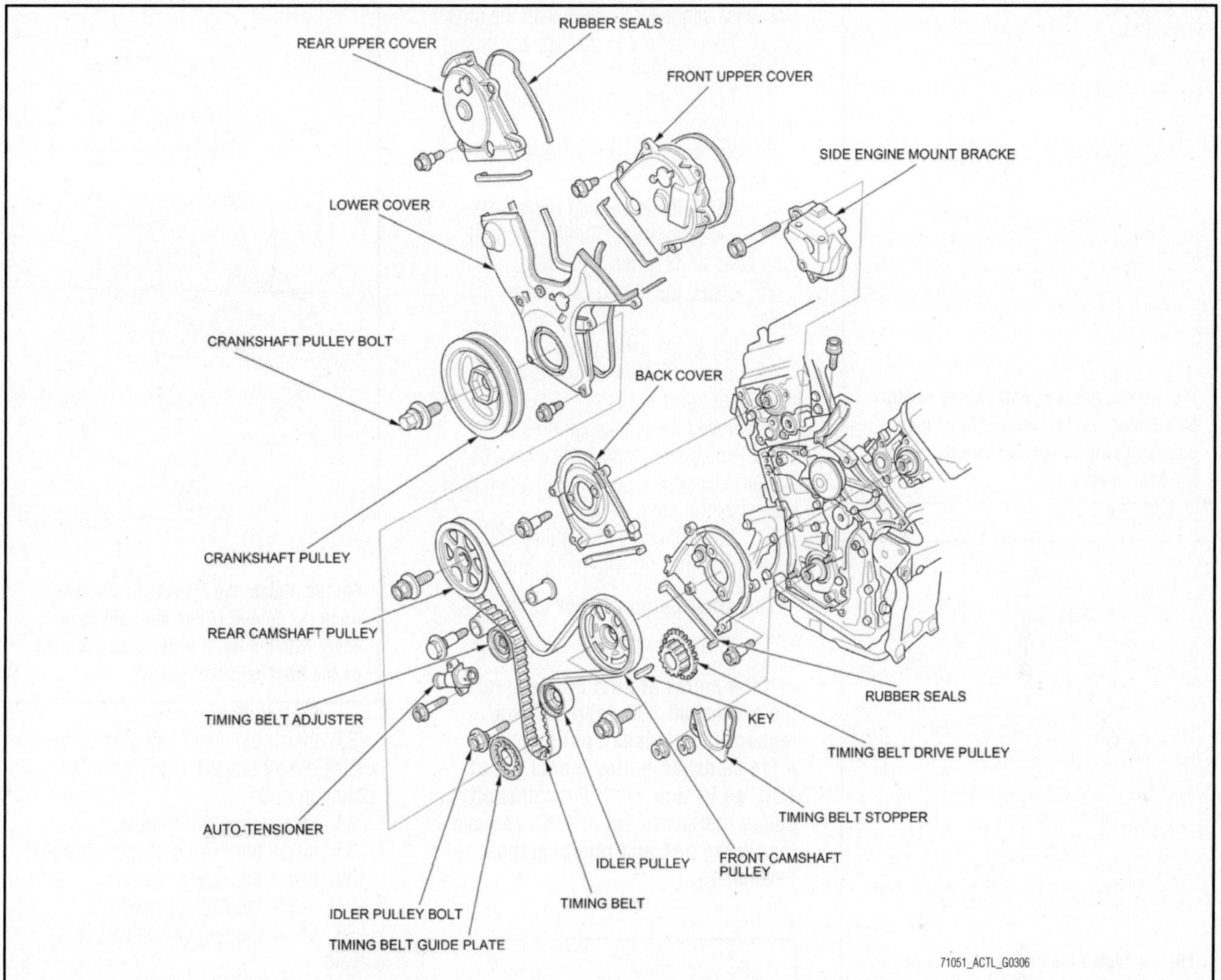

RUBBER SEALS

REAR UPPER COVER

FRONT UPPER COVER

SIDE ENGINE MOUNT BRACKE

LOWER COVER

CRANKSHAFT PULLEY BOLT

BACK COVER

CRANKSHAFT PULLEY

REAR CAMSHAFT PULLEY

RUBBER SEALS

TIMING BELT ADJUSTER

KEY

TIMING BELT DRIVE PULLEY

TIMING BELT STOPPER

AUTO-TENSIONER

FRONT CAMSHAFT
PULLEY

IDLER PULLEY

IDLER PULLEY BOLT

TIMING BELT

TIMING BELT GUIDE PLATE

71051_ACTL_G0306

Fig. 84 Showing engine timing belt front cover, timing belt, sprockets and related components—3.5L engine (J35Z6) & 3.7L engine (J37A4)

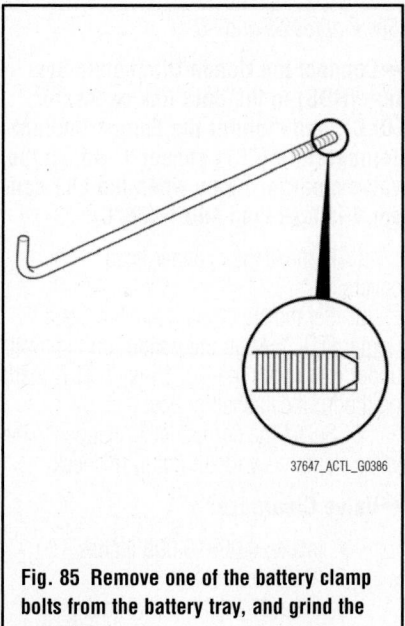

37647_ACTL_G0386

Fig. 85 Remove one of the battery clamp bolts from the battery tray, and grind the end of it

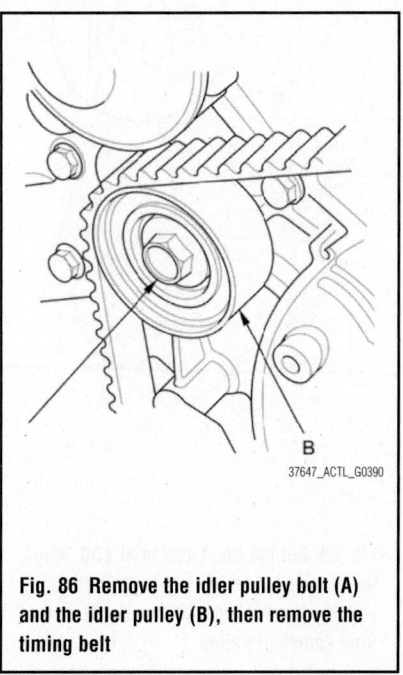

B

37647_ACTL_G0390

Fig. 86 Remove the idler pulley bolt (A) and the idler pulley (B), then remove the timing belt

5. Remove the splash shield.

6. Remove the drive belt auto-tensioner.

7. Support the engine with a jack and a wood block under the oil pan.

8. Remove the ground cable, then remove the upper half of the side engine mount bracket.

9. Remove the crankshaft pulley.

10. Remove the front upper cover and the rear upper cover.

11. Remove the lower cover.

12. Remove one of the battery clamp bolts from the battery tray, and grind the end of it.

13. Thread the battery clamp bolt into hold the timing belt adjuster in its current position. Tighten it by hand, do not use a wrench.

14. Remove the timing belt guide plate.

15. Remove the lower half of the side engine mount bracket.

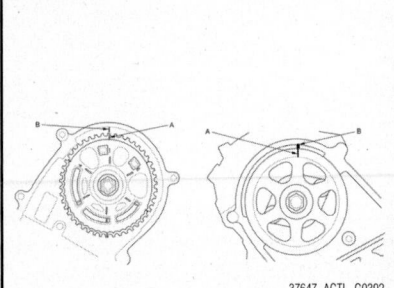

Fig. 87 Set the camshaft pulleys to TDC by aligning the TDC marks (A) on the camshaft pulleys with the pointers (B) on the back covers

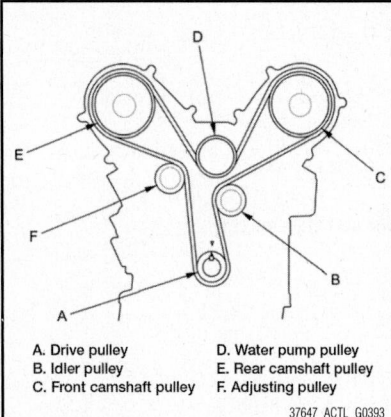

A. Drive pulley
B. Idler pulley
C. Front camshaft pulley
D. Water pump pulley
E. Rear camshaft pulley
F. Adjusting pulley

37647_ACTL_G0393

Fig. 88 Install the timing belt in a counterclockwise sequence starting with the drive pulley

16. Remove the idler pulley bolt and the idler pulley, then remove the timing belt. Discard the idler pulley bolt.

To install:

17. Clean the timing belt pulleys, the timing belt guide plate, and the upper and lower covers.

18. Set the timing belt drive pulley to Top Dead Center (TDC) by aligning the TDC mark on the tooth of the timing belt drive pulley with the pointer on the oil pump.

19. Set the camshaft pulleys to TDC by aligning the TDC marks on the camshaft pulleys with the pointers on the back covers.

20. Loosely install the idler pulley with a new idler pulley bolt so the pulley can move but does not come off.

21. If the auto-tensioner has extended and the timing belt cannot be installed, do the timing belt replacement procedure.

22. Install the timing belt in a counter-clockwise sequence starting with the drive pulley. Take care not to damage the timing belt during installation.

23. Tighten the idler pulley bolt to 33 ft. lbs. (44 Nm).

24. Remove the battery clamp bolt from the back cover.

25. Install the lower half of the side engine mount bracket. Tighten the three long bolts to 33 ft. lbs. (44 Nm).

26. Install the timing belt guide plate.

27. Install the lower cover.

28. Install the front upper cover and the rear upper cover.

29. Install the crankshaft pulley.

30. Rotate the crankshaft pulley about six turns clockwise so the timing belt positions itself on the pulleys.

31. Turn the crankshaft pulley so its white mark lines up with the pointer.

➡The other pointer is not used.

32. Check the camshaft pulley marks.

➡**If the marks are not aligned, rotate the crankshaft 360 degrees, and recheck the camshaft pulley mark. If the camshaft pulley marks are at TDC, go to step 17. If the camshaft pulley marks are not at TDC, remove the timing belt and repeat steps 2 through 16.**

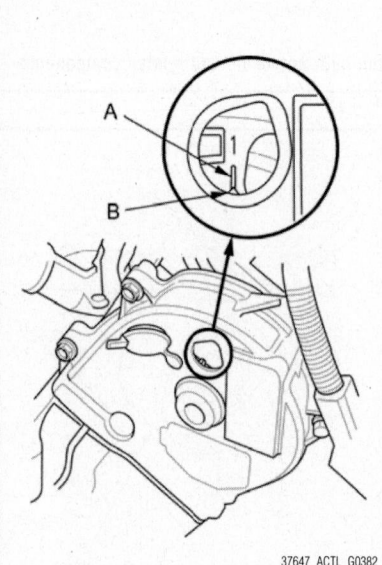

Fig. 89 Set the No. 1 piston at TDC. Align the pointer (A) on the front upper cover with the No. 1 piston TDC mark (B) on the front camshaft pulley

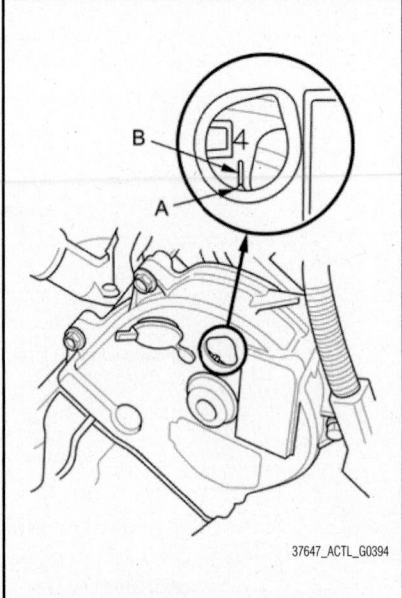

Fig. 90 Rotate the crankshaft clockwise. Align the pointer (A) on the front upper cover with the No. 4 piston TDC mark (B) on the front camshaft pulley

33. Install the upper half of the side engine mount bracket, then tighten the mounting bolts.

34. Install the ground cable.

35. Install the drive belt auto-tensioner.

36. Install the splash shield.

37. Install the right front wheel.

38. Install the engine compartment covers.

VALVE LASH

ADJUSTMENT

See Figures 89 and 90.

➡**Connect the Honda Diagnostic System (HDS) to the data link connector (DLC), and monitor the Engine Coolant Temperature (ECT) sensor 1. Adjust the valve clearance only when the ECT sensor 1 is less than 100°F (38°C).**

1. Remove the cylinder head covers.

2. Set the No. 1 piston at Top Dead Center (TDC). Align the pointer on the front upper cover with the No. 1 piston TDC mark on the front camshaft pulley.

3. Select the correct feeler gauge for the valve clearance you are going to check.

➡**Valve Clearance:**

- Intake: 0.008–0.009 inches (0.20–0.24 mm)
- Exhaust: 0.011–0.013 inches (0.28–0.32 mm)

4. Insert the feeler gauge between the adjusting screw and the end of the valve stem on the No. 1 cylinder, and slide it back and forth; you should feel a slight amount of drag.

5. If you feel too much or too little drag, loosen the locknut, and turn the adjusting screw until the drag on the feeler gauge is correct.

6. While holding the adjusting screw with the screw driver, tighten the locknut, then recheck the clearance. Repeat the adjustment, if necessary.

➡**Specified Torque:**

- Intake: 14 ft. lbs. (20 Nm)
- Exhaust: 10 ft. lbs. (14 Nm)

➡**Apply new engine oil to the nut threads.**

7. Rotate the crankshaft clockwise. Align the pointer on the front upper cover with the No. 4 piston TDC mark on the front camshaft pulley.

8. Check and, if necessary, adjust the valve clearance on the No. 4 cylinder.

9. Rotate the crankshaft clockwise. Align the pointer on the front upper cover with the No. 2 piston TDC mark on the front camshaft pulley.

Check and, if necessary, adjust the valve clearance on the No. 2 cylinder.

10. Rotate the crankshaft clockwise. Align the pointer on the front upper cover

with the No. 5 piston TDC mark on the front camshaft pulley.

11. Check and, if necessary, adjust the valve clearance on the No. 5 cylinder.

12. Rotate the crankshaft clockwise. Align the pointer on the front upper cover with the No. 3 piston TDC mark on the front camshaft pulley.

13. Check and, if necessary, adjust the valve clearance on the No. 3 cylinder.

14. Rotate the crankshaft clockwise. Align the pointer on the front upper cover with the No. 6 piston TDC mark on the front camshaft pulley.

15. Check and, if necessary, adjust the valve clearance on the No. 6 cylinder.

16. Install the cylinder head covers.

ENGINE PERFORMANCE & EMISSION CONTROLS

COMPONENT LOCATIONS

See Figure 91.

ACCELERATOR PEDAL POSITION (APP) SENSOR

LOCATION

The Accelerator Pedal Position (APP)

sensor is located at the top of the accelerator pedal assembly.

REMOVAL & INSTALLATION

See Figures 92 and 93.

1. Remove the driver's dashboard undercover.

2. Disconnect the APP sensor 6P connector.

3. Remove the two mounting nuts, then remove the clip. Using a flat-tip screwdriver, push the lock, then remove the accelerator pedal pad from the pedal stop.

4. Remove the accelerator pedal module.

➡**The APP sensor is not available separately. Do not disassemble the accelerator pedal module. If the pedal stop is damaged, replace it by pulling back the carpet, and removing the nuts, the bolt, and the pedal stop.**

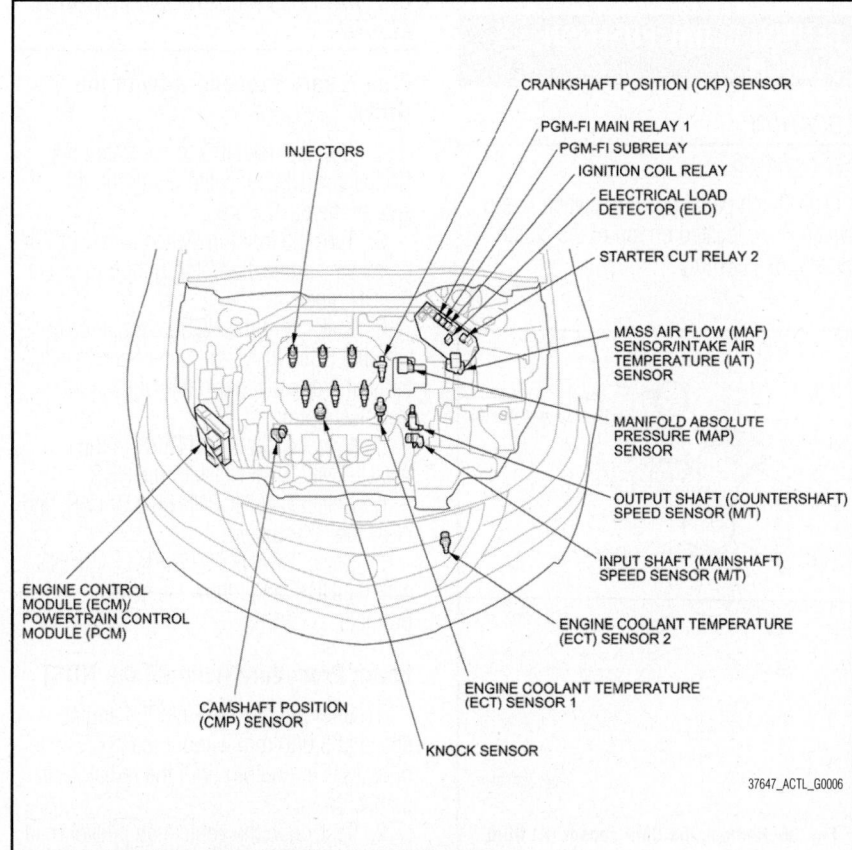

Fig. 91 Underhood engine control component locations

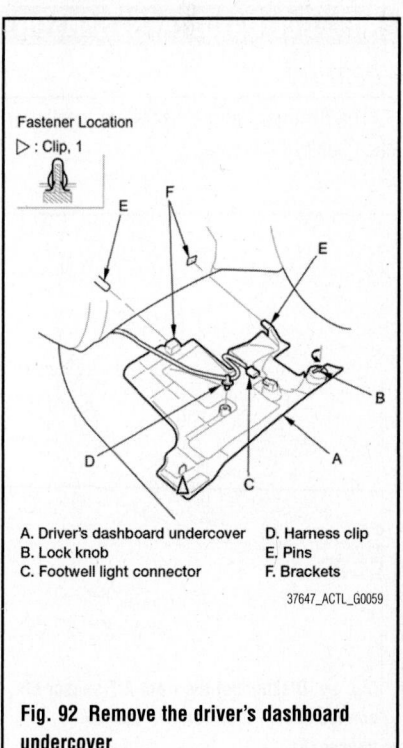

A. Driver's dashboard undercover
B. Lock knob
C. Footwell light connector
D. Harness clip
E. Pins
F. Brackets

37647_ACTL_G0059

Fig. 92 Remove the driver's dashboard undercover

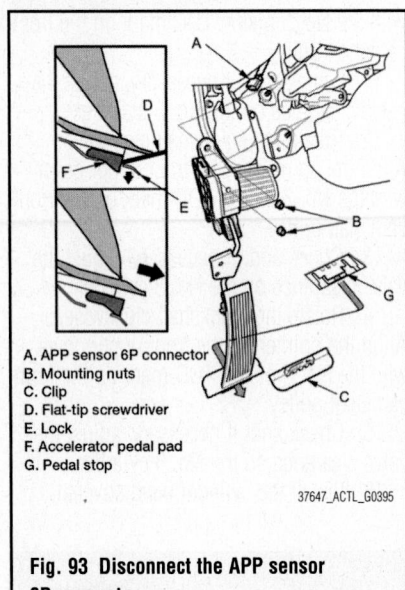

A. APP sensor 6P connector
B. Mounting nuts
C. Clip
D. Flat-tip screwdriver
E. Lock
F. Accelerator pedal pad
G. Pedal stop

37647_ACTL_G0395

Fig. 93 Disconnect the APP sensor 6P connector

To install:

5. Install the accelerator pedal pad to the pedal stop. Slide the pad until it locks with a clicking sound.

6. Install a new clip.

➡**The clip must be replaced whenever the accelerator pedal module is installed. Make sure that the accelerator pedal module and the clip are secure.**

7. Install the nuts, and connect the APP sensor 6P connector.

AIR-FUEL RATIO (AFR) SENSOR

LOCATION

The Air Fuel Ratio sensors are located on each bank of the engine.

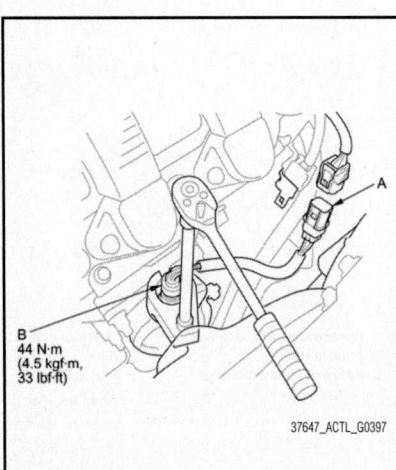

B
44 N·m
(4.5 kgf·m, 33 lbf·ft)

A

37647_ACTL_G0397

Fig. 94 Disconnect the front A/F sensor 6P connector (A), then remove the A/F sensor (B)

REMOVAL & INSTALLATION

Front Bank (Bank 2)

See Figure 94.

1. Disconnect the front A/F sensor 6P connector, then remove the A/F sensor.

2. Install the parts in the reverse order of removal.

Rear Bank (Bank 1)

1. Disconnect the rear A/F sensor 6P connector, then remove the A/F sensor.

2. Install the parts in the reverse order of removal.

CAMSHAFT POSITION SENSOR

REMOVAL & INSTALLATION

See Figure 95.

1. Remove the timing belt.
2. Remove the front camshaft pulley (CMP sensor pulse plate).
3. Disconnect the CMP sensor connector, then remove the back cover.
4. Remove the CMP sensor from the back cover.
5. Install the parts in the reverse order of removal. Install the timing belt.
6. Do the CKP pattern clear/CKP pattern learn procedure.

CLUTCH PEDAL POSITION SWITCH

LOCATION

See Figure 96.

The clutch pedal position switch A and switch B are located on top of the clutch pedal arm assembly.

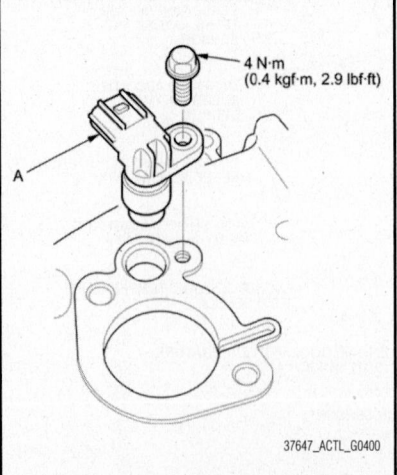

4 N·m
(0.4 kgf·m, 2.9 lbf·ft)

A

37647_ACTL_G0400

Fig. 95 Remove the CMP sensor (A) from the back cover

REMOVAL & INSTALLATION

See Figure 97.

1. Disconnect the applicable clutch pedal position switch A or B connector.

2. Loosen the locknut, then remove clutch pedal position switch.

3. Install clutch pedal position switch and adjust the pedal height.

CRANKSHAFT POSITION SENSOR

REMOVAL & INSTALLATION

See Figure 98.

1. Raise the vehicle on a lift.

➡**Make the vehicle is level, because engine oil will drip out when you remove the sensor.**

2. Remove the CKP sensor cover.
3. Disconnect the CKP sensor connector.
4. Remove the CKP sensor.
5. Install the parts in the reverse order of removal with a new O-ring.
6. Do the CKP pattern clear/CKP pattern learn procedure.
7. Check for engine oil level, and add more oil if needed.

CKP PATTERN CLEAR/CKP PATTERN LEARN

Clear/Learn Procedure (with the HDS)

1. Connect the HDS to the Data Link Connector (DLC) located under the driver's side of the dashboard.

2. Turn the ignition switch to ON (II), or press the engine start/stop button to select the ON mode.

3. Make sure the HDS communicates with the ECM/PCM and other vehicle systems. If it doesn't, go to the DLC circuit troubleshooting.

4. Select CRANK PATTERN in the ADJUSTMENT MENU with the HDS.

5. Select CRANK PATTERN CLEAR, and clear the CKP pattern.

6. Select CRANK PATTERN LEARNING with the HDS, and follow the screen prompts.

Learn Procedure (without the HDS)

1. Start the engine. Hold the engine speed at 3,000 rpm without load (A/T in P or N, M/T in neutral) until the radiator fan comes on.

2. Test-drive the vehicle on a level road: Decelerate (with the throttle fully closed)

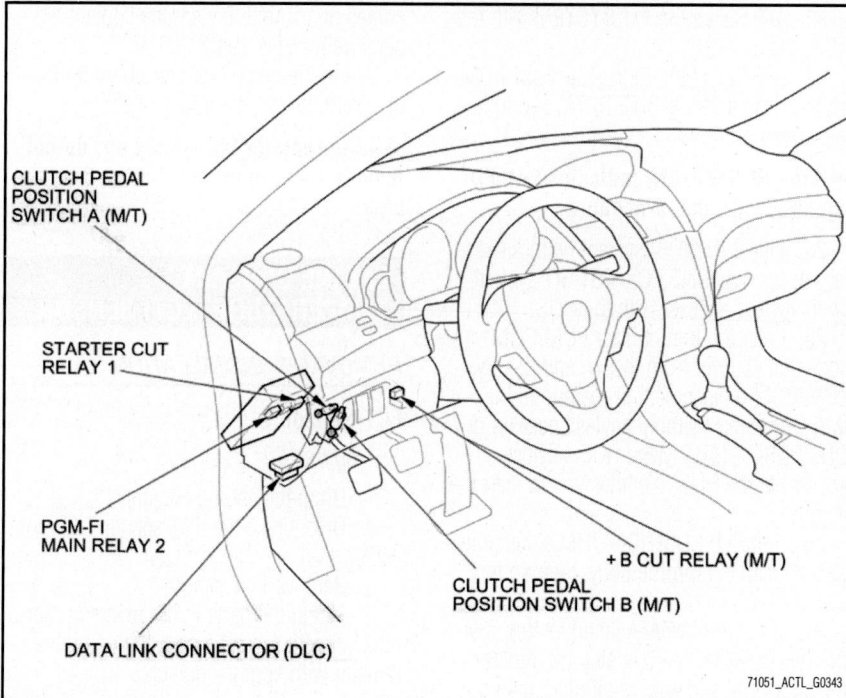

CLUTCH PEDAL
POSITION
SWITCH A (M/T)

STARTER CUT
RELAY 1

PGM-FI
MAIN RELAY 2

+ B CUT RELAY (M/T)

CLUTCH PEDAL
POSITION SWITCH B (M/T)

DATA LINK CONNECTOR (DLC)

71051_ACTL_G0343

Fig. 96 The clutch pedal position switch A and switch B are located on top of the clutch pedal arm assembly.

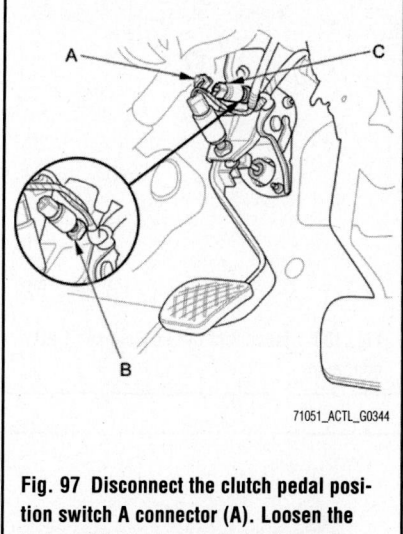

71051_ACTL_G0344

Fig. 97 Disconnect the clutch pedal position switch A connector (A). Loosen the locknut (B), then remove clutch pedal position switch A (C)—clutch pedal position switch A indicated; switch B is similar

from an engine speed of 2,500 rpm down to 1,000 rpm with the Transmission in 2nd.

3. Repeat step 2 several times.

4. Turn the ignition switch to LOCK (0), or press the engine start/stop button to select the OFF mode.

5. Turn the ignition switch to ON (II), or press the engine start/stop button to select the ON mode, and wait 30 seconds.

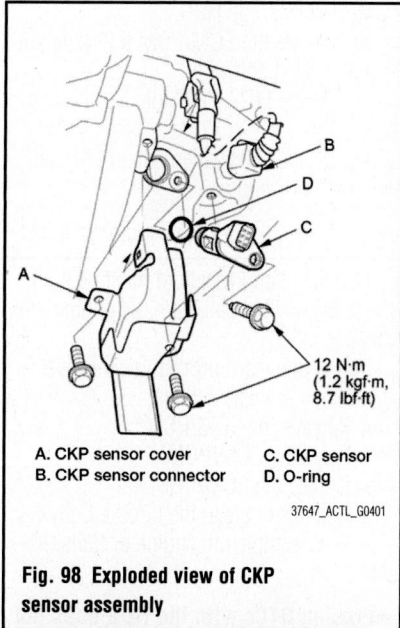

A. CKP sensor cover
B. CKP sensor connector
C. CKP sensor
D. O-ring

12 N·m
(1.2 kgf·m,
8.7 lbf·ft)

37647_ACTL_G0401

Fig. 98 Exploded view of CKP sensor assembly

ENGINE CONTROL MODULE/POWERTRAIN CONTROL MODULE (ECM/PCM)

LOCATION

The Engine control Module (ECM)/Powertrain Control Module (PCM) is located in the front of the engine compartment on the passenger's side.

REMOVAL & INSTALLATION

See Figure 99.

1. Connect HDS to the Data Link Connector (DLC) located under the driver's side of the dashboard.

2. Turn the ignition switch to ON (II), or press the engine start/stop button to select the ON mode.

3. Make sure the HDS communicates with the ECM/PCM and other vehicle systems. If it doesn't, go to the DLC circuit troubleshooting. If you are returning from the DLC circuit troubleshooting, skip steps 4 through 9, 18 through 23, and 26 through 28, and do this after replacing the ECM/PCM:
   - Replace the engine oil and the engine oil filter.
   - Replace the ATF (A/T).
   - Clean the throttle body.

4. Select the PGM-FI system with the HDS.

5. Select the INSPECTION MENU with the HDS.

6. Select the ETCS TEST, then select the TP POSITION CHECK, and follow the screen prompts.

➡**If the TP POSITION CHECK indicates FAILED, continue with this procedure.**

7. Select the REPLACE PCM MENU, then select READ DATA, and follow the screen prompts.

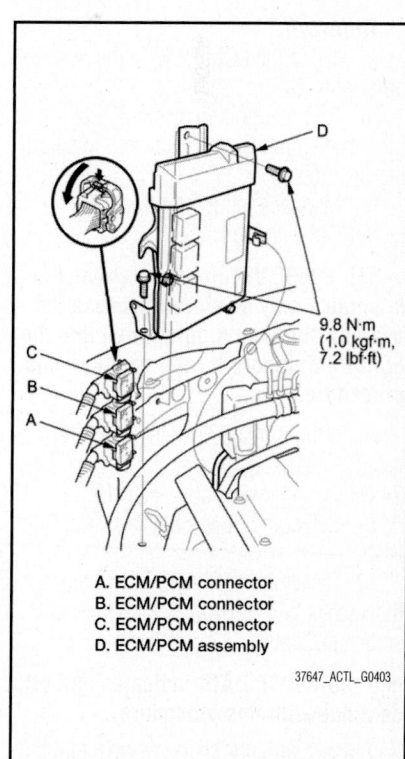

9.8 N·m
(1.0 kgf·m,
7.2 lbf·ft)

A. ECM/PCM connector
B. ECM/PCM connector
C. ECM/PCM connector
D. ECM/PCM assembly

37647_ACTL_G0403

Fig. 99 Disconnect ECM/PCM connectors, then remove the ECM/PCM assembly

➡Doing this step copies (READS) the engine oil life data from the original ECM/PCM so you can later download (WRITES) it into the new ECM/PCM. If READ DATA indicates FAILED, continue with this procedure.

8. A/T: Select the A/T system with the HDS.

9. A/T: Select the REPLACE TCM/PCM MENU, then select READ DATA, and follow the screen prompts.

➡Doing this step copies (READS) the ATF life data from the original PCM so you can later download (WRITES) it into the new PCM. If READ DATA indicates FAILED, continue with this procedure.

10. Turn the ignition switch to LOCK (0), or press the engine start/stop button to select the OFF mode.

11. Jump the SCS line with the HDS.

12. Canada model: If equipped, remove the ECM/PCM connector guard.

13. Disconnect ECM/PCM connectors, then remove the ECM/PCM assembly.

➡ECM/PCM connectors A, B, and C have symbols (A: square, B: triangle, C: circle) embossed on them for identification.

14. Remove the cover and the bracket from the ECM/PCM.

To install:

15. Install the ECM/PCM in the reverse order of removal.

16. Turn the ignition switch to ON (II), or press the engine start/stop button to select the ON mode.

17. Manually input the VIN to the ECM/PCM with the HDS.

➡DTC P0630 VIN Not Programmed or Mismatch may be stored because the VIN has not been programmed into the ECM/PCM; ignore it, and continue this procedure.

18. If the READ DATA (engine oil life) failed in step 7, go to step 21 (A/T) or step 24 (M/T). Otherwise, go to step 19.

19. Select the PGM-FI system with the HDS.

20. Select the REPLACE PCM MENU, then select WRITE DATA, and follow the screen prompts.

➡If the WRITE DATA indicates FAILED, continue with this procedure.

21. A/T: If the READ DATA (ATF life) failed in step 9, go to step 24. Otherwise go to step 22.

22. A/T: Select the A/T SYSTEM with the HDS.

23. A/T: Select the REPLACE TCM/PCM MENU, then select WRITE DATA, and follow the screen prompts.

➡If the WRITE DATA indicates FAILED, continue with this procedure.

24. Select IMMOBI system (without keyless access) or ONE PUSH START system (with keyless access) with the HDS.

25. Enter the immobilizer ECM/PCM code that you got from the iN, and use the ECM/PCM replacement procedure in the IMMOBI Menu (without keyless access) or ONE PUSH START Menu (with keyless access) of the HDS; it allows you to start the engine.

26. If the TP POSITION CHECK failed in step 6 clean the throttle body, then go to step 27.

27. If the READ DATA failed in step 7 or the WRITE DATA failed in step 20, replace the engine oil and engine oil filter, then go to step 28 (A/T) or step 29 (M/T).

28. If the READ DATA failed in step 9 or the WRITE DATA failed in step 23, replace the ATF, then go to step 29.

29. Select PGM-FI system, and reset the ECM/PCM with the HDS.

30. Update the ECM/PCM if it does not have the latest software.

31. Do the ECM/PCM idle learn procedure.

32. Do the CKP pattern clear/CKP pattern learn procedure.

ECM/PCM IDLE LEARN PROCEDURE

The idle learn procedure must be done so the ECM/PCM can learn the engine idle characteristics.

Do the idle learn procedure whenever you do any of these actions:

• Replace the ECM/PCM.
• Reset the ECM/PCM.
• Update the ECM/PCM.
• Replace or clean the throttle body.
• Disassemble the engine or transmission.

➡Erasing DTCs with the HDS does not require you to do the idle learn procedure.

1. Make sure all electrical items (the A/C, the audio, the lights, etc.) are off.

2. Reset the ECM/PCM with the HDS.

3. Turn the ignition switch to ON (II), or press the engine start/stop button to select the ON mode, and wait 2 seconds.

4. Start the engine. Hold the engine speed at 3,000 rpm without load (A/T in P or N, M/T in neutral) until the radiator fan

comes on, or until the engine coolant temperature reaches 194°F (90°C).

5. Let the engine idle for about 5 minutes with the throttle fully closed.

➡If the radiator fan comes on, do not include its running time in the 5 minutes.

ENGINE COOLANT TEMPERATURE (ECT) SENSOR

REMOVAL & INSTALLATION

ECT Sensor 1

See Figure 100.

1. Drain the engine coolant.

2. Disconnect the ECT sensor 1 2P connector.

3. Remove ECT sensor 1.

4. Install the parts in the reverse order of removal with a new O-ring, then refill the radiator with engine coolant.

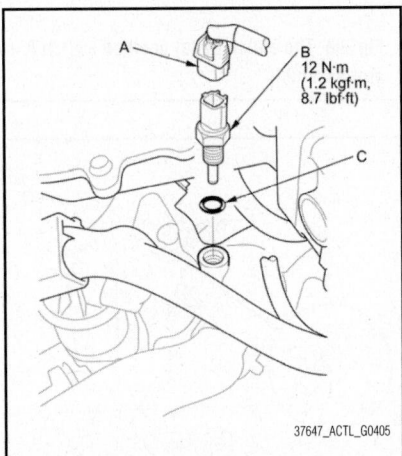

Fig. 100 Disconnect the ECT sensor 1 2P connector

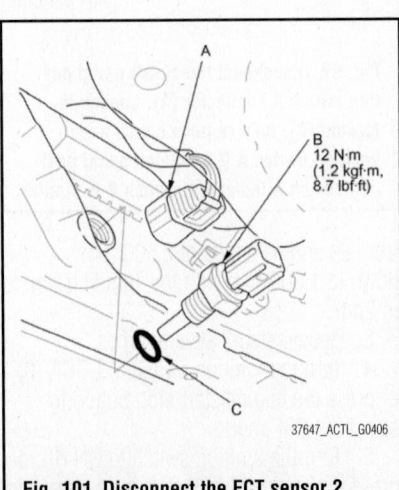

Fig. 101 Disconnect the ECT sensor 2 2P connector

ECT Sensor 2

See Figure 101.

1. Remove the splash shield.
2. Drain the engine coolant.
3. Disconnect the ECT sensor 2 2P connector.
4. Remove ECT sensor 2.
5. Install the parts in the reverse order of removal with a new O-ring, then refill the radiator with engine coolant.

EVAPORATIVE EMISSION CONTROL SYSTEM

EVAP CANISTER

Location

See Figure 102.

Removal & Installation

With SH-AWD

See Figures 103 and 104.

1. Remove the rear differential.
2. Remove the fuel tank.
3. Remove the EVAP canister filter.
4. Disconnect the quick-connect fitting, the hose, the FTP sensor 3P connector, and the EVAP canister vent shut valve 2P connector.

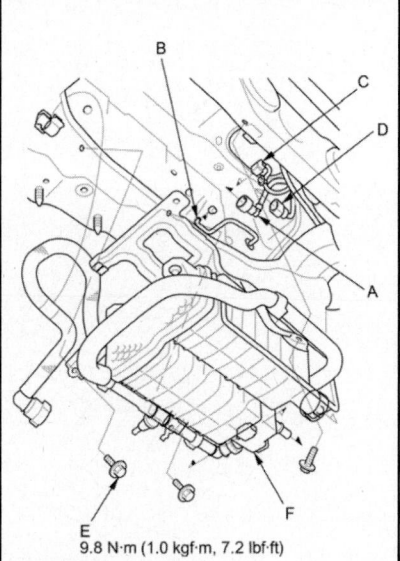

E
9.8 N·m (1.0 kgf·m, 7.2 lbf·ft)

A. Quick-connect fitting
B. Hose
C. FTP sensor 3P connector
D. EVAP canister vent shut valve 2P connector
E. Bolts
F. EVAP canister assembly

37647_ACTL_G0409

Fig. 103 Disconnect the quick-connect fitting, the hose, the FTP sensor 3P connector, and the EVAP canister vent shut valve 2P connector

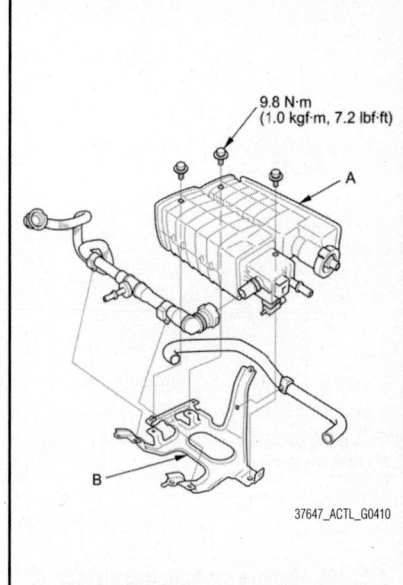

9.8 N·m
(1.0 kgf·m, 7.2 lbf·ft)

37647_ACTL_G0410

Fig. 104 Remove the EVAP canister (A) from the bracket (B)

5. Remove the bolts and the EVAP canister assembly.
6. Remove the EVAP canister from the bracket.
7. Remove the EVAP canister vent shut valve and the FTP sensor from the EVAP canister.

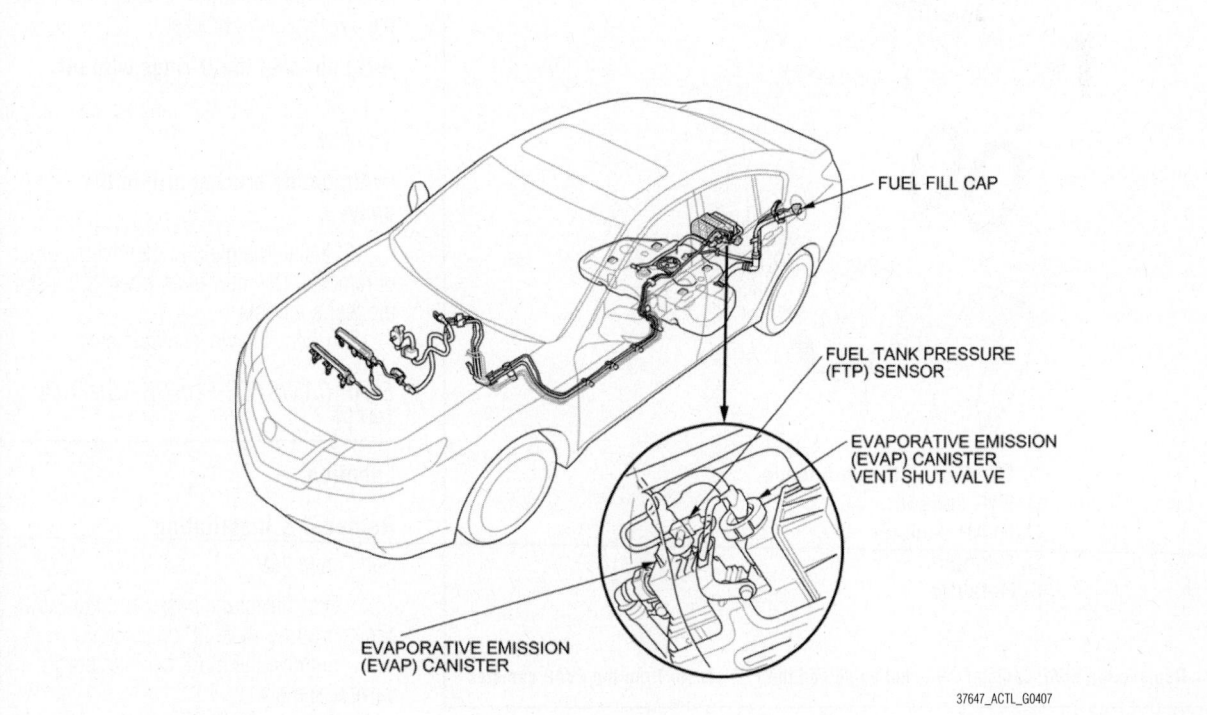

FUEL FILL CAP

FUEL TANK PRESSURE (FTP) SENSOR

EVAPORATIVE EMISSION (EVAP) CANISTER VENT SHUT VALVE

EVAPORATIVE EMISSION (EVAP) CANISTER

37647_ACTL_G0407

Fig. 102 EVAP canister system component locations

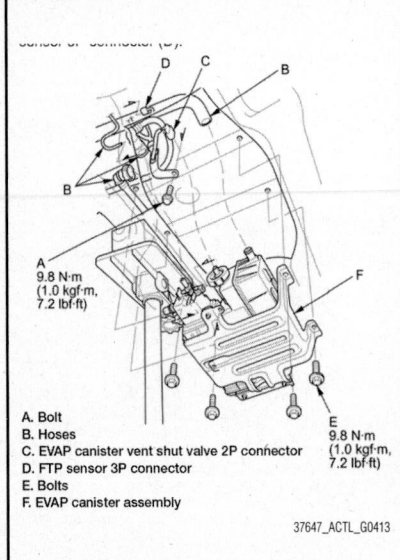

A. Bolt
B. Hoses
C. EVAP canister vent shut valve 2P connector
D. FTP sensor 3P connector
E. Bolts
F. EVAP canister assembly

A
9.8 N·m
(1.0 kgf·m,
7.2 lbf·ft)

E
9.8 N·m
(1.0 kgf·m,
7.2 lbf·ft)

37647_ACTL_G0413

Fig. 105 Remove the bolt, and disconnect the hoses, the EVAP canister vent shut valve 2P connector, and the FTP sensor 3P connector

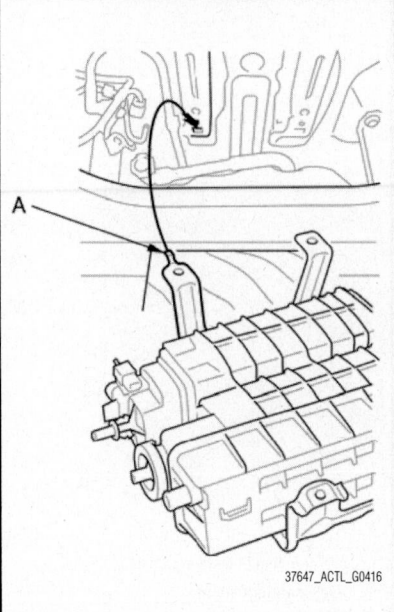

A

37647_ACTL_G0416

Fig. 107 Attach the bracket arm (A) to the body

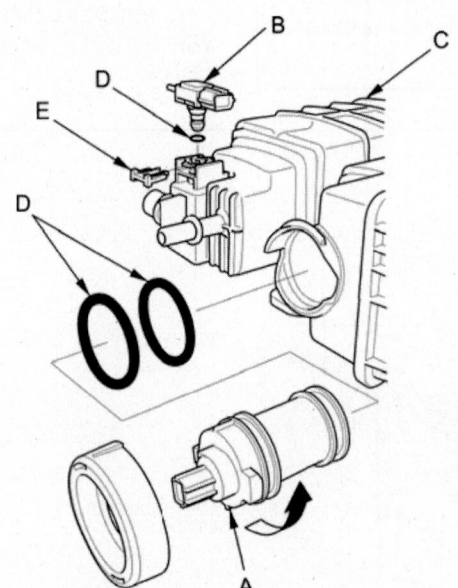

A. EVAP canister shut valve
B. FTP sensor
C. EVAP canister
D. O-rings
E. Retainer

37647_ACTL_G0415

Fig. 106 Remove the EVAP canister vent shut valve and the FTP sensor from the EVAP canister

8. Reassemble the EVAP canister with new O-rings and a new retainer, then install the EVAP canister bracket.

➡**Do not coat the O-rings with oil.**

9. Install the parts in the reverse order of removal.

➡The new fuel vapor hose has a ring pull at the connector. When you connect the connector and confirm that the connection is secure, remove the ring pull by pulling it down.

Without SH-AWD

See Figures 105 through 107.

1. Raise the vehicle on a lift.
2. Remove the wheel sensor harness clamps.
3. Support the rear subframe with a transmission jack and a wooden block.
4. Remove the rear subframe mounting bolts.
5. Lower the transmission jack and the rear subframe about 50 mm.

➡**Be careful not to damage the connecting parts.**

6. Remove the bolt, and disconnect the hoses, the EVAP canister vent shut valve 2P connector, and the FTP sensor 3P connector.
7. Remove the bolts, then remove the EVAP canister assembly.
8. Remove the EVAP canister from the EVAP canister bracket.
9. Remove the EVAP canister vent shut valve and the FTP sensor from the EVAP canister.
10. Reassemble the EVAP canister with new O-rings and a new retainer, then install the EVAP canister bracket.

➡**Do not coat the O-rings with oil.**

11. Install the EVAP canister assembly to the body.

➡**Attach the bracket arm to the body.**

12. Install the parts in the reverse order of removal. Use new bolts when you install the rear subframe.
13. Check the wheel alignment.

EVAP CANISTER PURGE CONTROL VALVE

Location

Removal & Installation

See Figure 109.

1. Disconnect the hoses and the EVAP canister purge valve 2P connector.
2. Remove the EVAP canister purge valve assembly.
3. Remove the EVAP canister purge valve from the bracket.
4. Install the parts in the reverse order of removal.

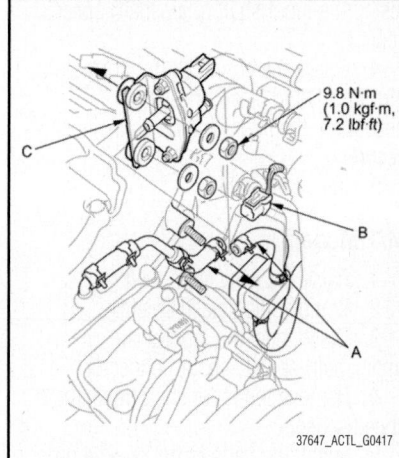

Fig. 109 Disconnect the hoses (A) and the EVAP canister purge valve 2P connector (B)

FUEL TANK PRESSURE (FTP) SENSOR

LOCATION

See Figure 109.

REMOVAL & INSTALLATION

With SH-AWD

See Figures 110 through 112.

1. Remove the trunk front trim panel.
   a. Remove the trunk floor.
   b. Remove the tool box spacer.
   c. Release the clips and the Velcro fasteners, then remove the trunk front trim panel.
2. Remove the access panel from the floor.
3. Disconnect the hose and the FTP sensor 3P connector.
4. Remove the FTP sensor.
5. Install the parts in the reverse order of removal with a new O-ring and a new retainer.

Without SH-AWD

See Figures 113 and 114.

1. Remove the bolt, and disconnect the EVAP canister vent shut valve 2P connector, the FTP sensor 3P connector, and the hoses.
2. Remove the bolts, and move the EVAP canister assembly to the rear.
3. Remove the FTP sensor.
4. Install the parts in the reverse order of removal with a new O-ring and a new retainer.

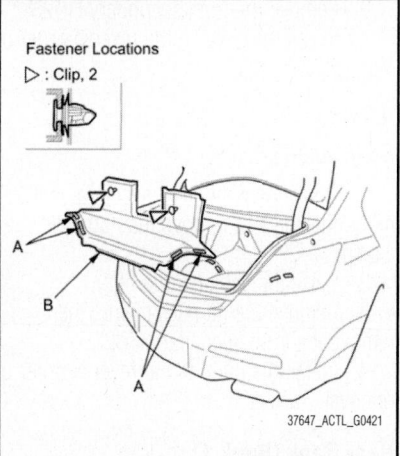

Fig. 110 Release the clips and the Velcro fasteners (A), then remove the trunk front trim panel (B)

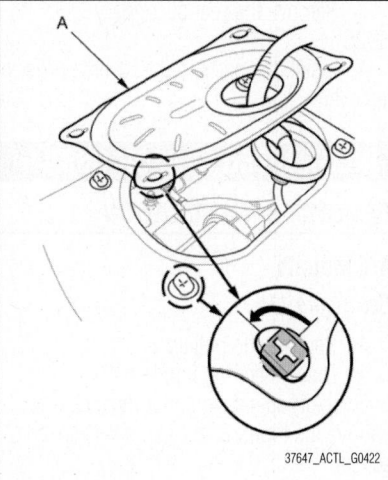

Fig. 111 Remove the access panel (A) from the floor

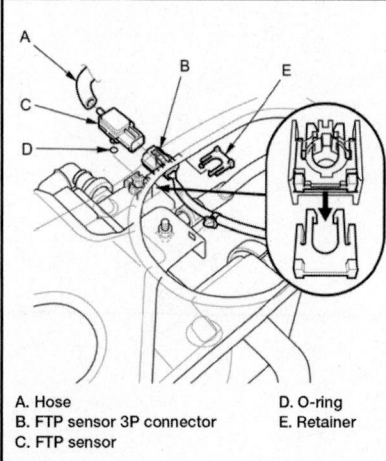

A. Hose
B. FTP sensor 3P connector
C. FTP sensor
D. O-ring
E. Retainer

Fig. 112 Disconnect the hose and the FTP sensor 3P connector

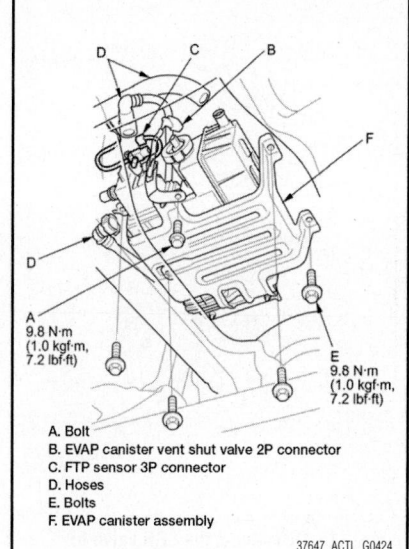

A. Bolt
B. EVAP canister vent shut valve 2P connector
C. FTP sensor 3P connector
D. Hoses
E. Bolts
F. EVAP canister assembly

Fig. 113 Remove the bolt, and disconnect the EVAP canister vent shut valve 2P connector, the FTP sensor 3P connector, and the hoses

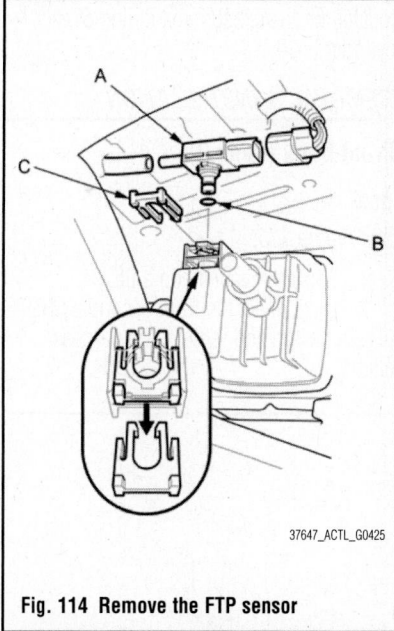

Fig. 114 Remove the FTP sensor

EXHAUST GAS RECIRCULATION (EGR) VALVE

REMOVAL & INSTALLATION

See Figure 115.

1. Disconnect the EGR valve 5P connector.
2. Remove the EGR valve.
3. Install the parts in the reverse order of removal with a new gasket.

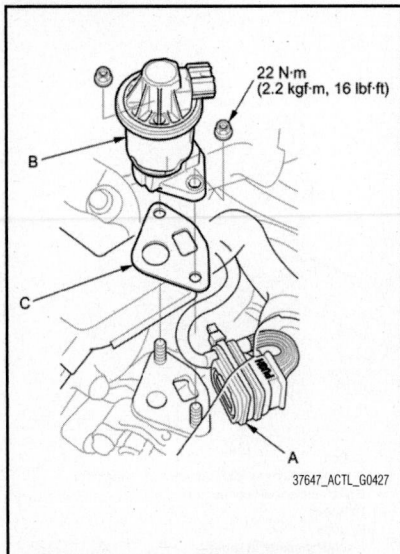

Fig. 115 Disconnect the EGR valve 5P connector

HEATED OXYGEN SENSOR (HO2S)

LOCATION

The Heated Oxygen Sensors (HO2S) are located on the exhaust manifolds of each cylinder head.

REMOVAL & INSTALLATION

Front Bank (Bank 2)

M/T

See Figure 116.

1. Raise the vehicle on a lift.
2. Disconnect the front secondary HO2S 4P connector, and remove the harness clamp.

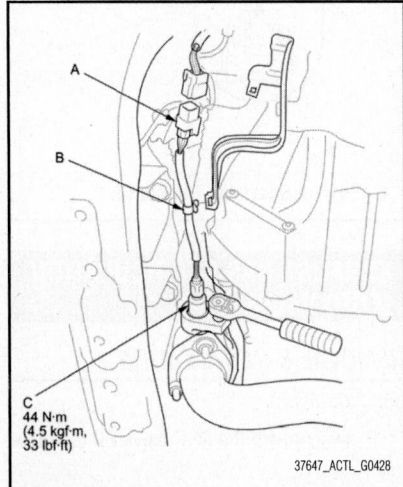

Fig. 116 Disconnect the front secondary HO2S 4P connector, and remove the harness clamp

3. Remove the front secondary HO2S.
4. Install the parts in the reverse order of removal.

A/T

See Figure 117.

1. Disconnect the front secondary HO2S 4P connector, and remove the harness clamps.
2. Raise the vehicle on a lift.
3. Remove the harness clamp, then remove the front secondary HO2S.
4. Install the parts in the reverse order of removal.

Rear Bank (Bank 1)

1. Raise the vehicle on a lift.
2. Disconnect the rear secondary HO2S 4P connector, and remove the harness clamps.
3. Remove the rear secondary HO2S.
4. Install the parts in the reverse order of removal.

INPUT SPEED SENSOR (ISS)

REMOVAL & INSTALLATION

A/T Models

See Figure 118.

1. Remove the battery.
2. Disconnect the input shaft (mainshaft) speed sensor connector, and remove the input shaft (mainshaft) speed sensor.
3. Install a new O-ring on a new input shaft (mainshaft) speed sensor, then

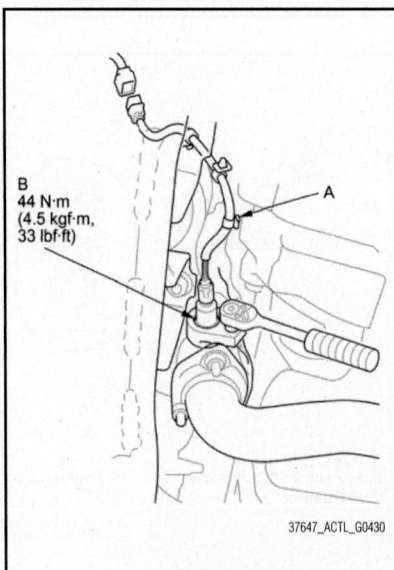

Fig. 117 Remove the harness clamp, then remove the front secondary HO2S

install the input shaft (mainshaft) speed sensor.
4. Check the connector for corrosion, dirt, or oil, and clean or repair if necessary, then connect the connector securely.
5. Install the battery.

M/T Models

1. Raise the vehicle on a lift.
2. Remove the splash shield.
3. Disconnect the input shaft (mainshaft) speed sensor connector.
4. Remove the input shaft (mainshaft) speed sensor.
5. Install the parts in the reverse order of removal with a new O-ring.
6. If needed, refill the manual transmission fluid.

KNOCK SENSOR (KS)

LOCATION

The Knock Sensor (KS) is located in the valley of the engine block between the two cylinder heads.

REMOVAL & INSTALLATION

See Figure 119.

1. Remove the intake manifold.
2. Remove the injector base.
3. Disconnect the knock sensor connector.
4. Remove the knock sensor.
5. Install the parts in the reverse order of removal.

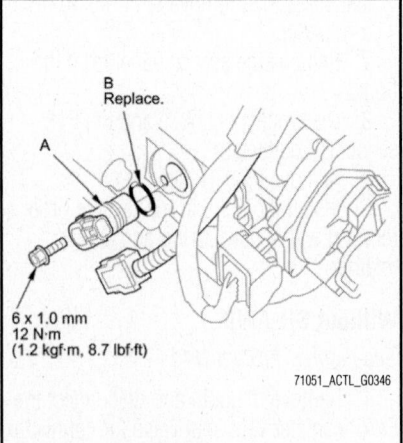

Fig. 118 Disconnect the input shaft (mainshaft) speed sensor connector, and remove the input shaft (mainshaft) speed sensor (A). Install a new O-ring (B) on a new input shaft (mainshaft) speed sensor, then install the input shaft (mainshaft) speed sensor.

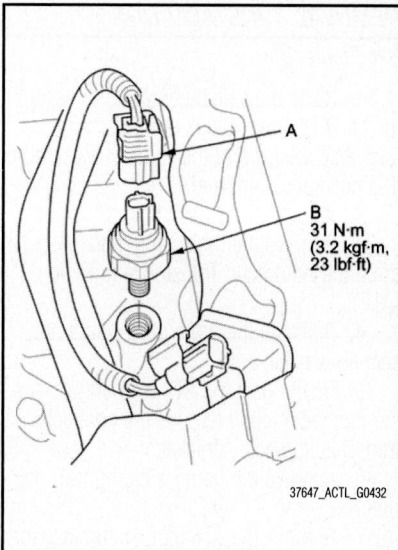

Fig. 119 Disconnect the knock sensor connector

MANIFOLD ABSOLUTE PRESSURE (MAP) SENSOR

REMOVAL & INSTALLATION

See Figure 120.

1. Disconnect the MAP sensor connector.
2. Remove the screw.
3. Remove the MAP sensor.
4. Install the parts in the reverse order of removal with a new O-ring.

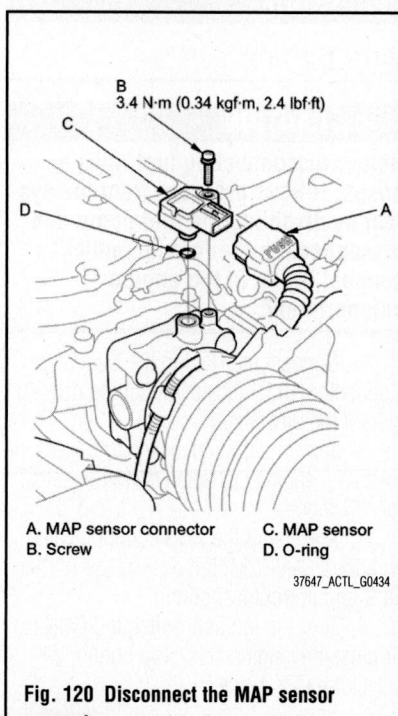

A. MAP sensor connector
B. Screw
C. MAP sensor
D. O-ring

37647_ACTL_G0434

Fig. 120 Disconnect the MAP sensor connector

MASS AIRFLOW (MAF)/INTAKE AIR TEMPERATURE (IAT) SENSOR

LOCATION

The MAF/IAT sensor is located on the engine air intake.

REMOVAL & INSTALLATION

See Figure 121.

1. Disconnect the MAF/IAT sensor 5P connector.
2. Remove the screws.
3. Remove the MAF/IAT sensor.
4. Install the parts in the reverse order of removal with a new O-ring.

OUTPUT SHAFT SPEED (OSS) SENSOR

LOCATION

The Output Shaft Speed (OSS) sensor is located on the side of the A/T transaxle.

REMOVAL & INSTALLATION

See Figure 122.

1. Do the battery removal procedure.
2. Disconnect the output shaft (countershaft) speed sensor connector, and remove the output shaft (countershaft) speed sensor.

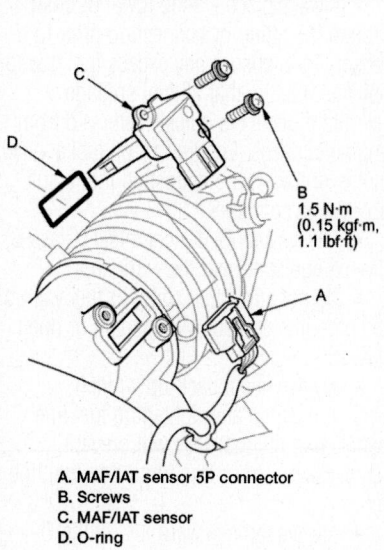

A. MAF/IAT sensor 5P connector
B. Screws
C. MAF/IAT sensor
D. O-ring

37647_ACTL_G0433

Fig. 121 Disconnect the MAF/IAT sensor 5P connector

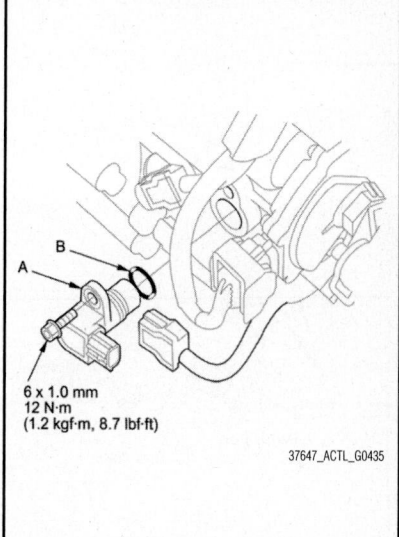

Fig. 122 Disconnect the output shaft (countershaft) speed sensor connector, and remove the output shaft (countershaft) speed sensor

3. Install a new O-ring on a new output shaft (countershaft) speed sensor, then install the output shaft (countershaft) speed sensor and the sensor washer.
4. Check the connector for rust, dirt, or oil, and clean or repair if necessary, then connect the connector securely.
5. Do the battery installation procedure.

POSITIVE CRANKCASE VENTILATION (PCV) VALVE

LOCATION

The Positive Crankcase Ventilation PCV) valve is located on the front right area of the engine below the harness cover.

REMOVAL & INSTALLATION

See Figure 123.

1. Lift up the harness cover.
2. Remove the bolt.
3. Remove the PCV valve.

➡**Do not to spill oil on the hot exhaust manifold.**

4. Install the parts in the reverse order of removal.

➡**When installing a new PCV valve, make sure the O-rings are in place. When installing a used PCV valve, use new O-rings.**

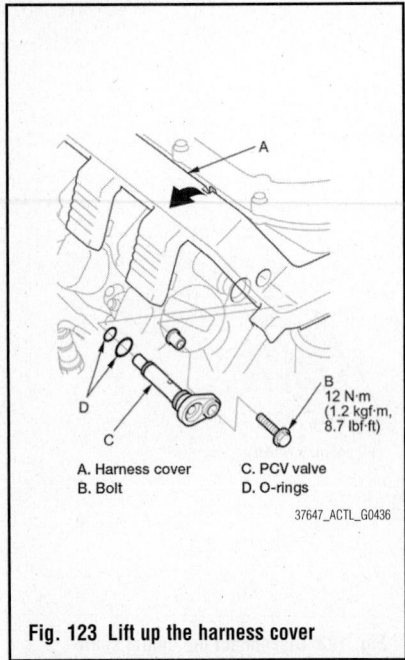

A. Harness cover
B. Bolt
C. PCV valve
D. O-rings

37647_ACTL_G0436

Fig. 123 Lift up the harness cover

TRANSMISSION FLUID TEMPERATURE (TFT) SENSOR

LOCATION

The ATF (automatic transmission fluid) temperature sensor is located on the transaxle case.

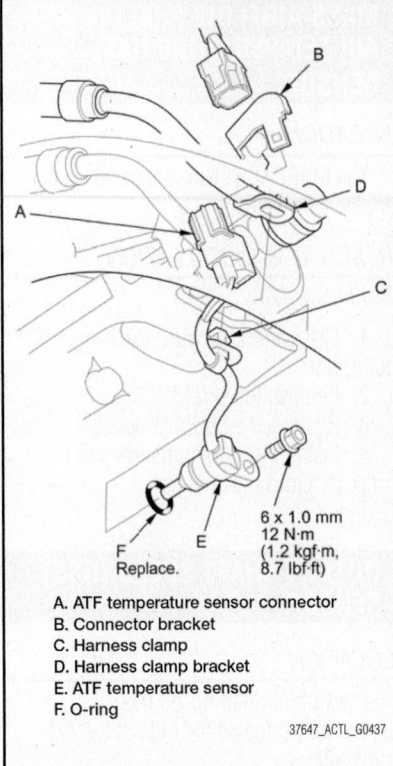

6 x 1.0 mm
12 N·m
(1.2 kgf·m,
8.7 lbf·ft)

Replace.

A. ATF temperature sensor connector
B. Connector bracket
C. Harness clamp
D. Harness clamp bracket
E. ATF temperature sensor
F. O-ring

37647_ACTL_G0437

Fig. 124 Disconnect the ATF temperature sensor connector, then remove the connector from the connector bracket

REMOVAL & INSTALLATION

See Figure 124.

1. Raise the vehicle on a lift, or apply the parking brake, block both rear wheels, and raise the front of the vehicle. Make sure it is securely supported.

2. Remove the splash shield.

3. Remove the drain plug, and drain the Automatic Transmission Fluid (ATF).

4. Reinstall the drain plug with a new sealing washer.

5. Disconnect the ATF temperature sensor connector, then remove the connector from the connector bracket.

6. Remove the harness clamp from its bracket.

7. Remove the ATF temperature sensor, and replace the sensor.

8. Install a new O-ring on a new ATF temperature sensor, and install the ATF temperature sensor.

9. Check the connector for rust, dirt, or oil, and clean or repair if necessary, then connect the connector securely, and install it on its bracket.

10. Install the harness clamp on its bracket.

11. Refill the transmission with ATF.

12. Install the splash shield.

FUEL   GASOLINE FUEL INJECTION SYSTEM

FUEL SYSTEM SERVICE PRECAUTIONS

Safety is the most important factor when performing not only fuel system maintenance but any type of maintenance. Failure to conduct maintenance and repairs in a safe manner may result in serious personal injury or death. Maintenance and testing of the vehicle's fuel system components can be accomplished safely and effectively by adhering to the following rules and guidelines.

• To avoid the possibility of fire and personal injury, always disconnect the negative battery cable unless the repair or test procedure requires that battery voltage be applied.

• Always relieve the fuel system pressure prior to disconnecting any fuel system component (injector, fuel rail, pressure regulator, etc.), fitting or fuel line connection. Exercise extreme caution whenever relieving fuel system pressure to avoid exposing skin, face and eyes to fuel spray. Please be advised that fuel under pressure may penetrate the skin or any part of the body that it contacts.

• Always place a shop towel or cloth around the fitting or connection prior to loosening to absorb any excess fuel due to spillage. Ensure that all fuel spillage (should it occur) is quickly removed from engine surfaces. Ensure that all fuel soaked cloths or towels are deposited into a suitable waste container.

• Always keep a dry chemical (Class B) fire extinguisher near the work area.

• Do not allow fuel spray or fuel vapors to come into contact with a spark or open flame.

• Always use a back-up wrench when loosening and tightening fuel line connection fittings. This will prevent unnecessary stress and torsion to fuel line piping.

• Always replace worn fuel fitting O-rings with new Do not substitute fuel hose or equivalent where fuel pipe is installed.

Before servicing the vehicle, make sure to also refer to the precautions in the beginning of this section as well.

RELIEVING FUEL SYSTEM PRESSURE

WITH THE HDS

✳✳ CAUTION

Before disconnecting fuel lines or hoses, relieve pressure from the system by disabling the fuel pump and disconnecting the fuel line/quick connect fitting in the engine compartment.

1. Connect the HDS to the Data Link Connector (DLC) located under the driver's side of the dashboard.

2. Turn the ignition switch to ON (II), or press the engine start/stop button to select the ON mode.

3. Make sure the HDS communicates with the ECM/PCM. If it doesn't, go to the DLC circuit troubleshooting.

4. Turn the ignition switch to LOCK (0), or press the engine start/stop button to select the OFF mode.

5. Remove the fuel fill cap to relieve the pressure in the fuel tank.

6. Turn the ignition switch to ON (II), or press the engine start/stop button to select the ON mode.

7. From the INSPECTION MENU of the HDS, select Fuel Pump OFF, then start the engine, and let it idle until it stalls.

➡**Do not allow the engine to idle above 1,000 rpm or the ECM/PCM will continue to operate the fuel pump. Pending or Confirmed DTC may be set during this procedure. Check for DTCs, and clear them as needed.**

8. Turn the ignition switch to LOCK (0), or press the engine start/stop button to select the OFF mode.

9. Do the battery terminal disconnection procedure.

10. Remove the cover and the quick-connect fitting cover.

11. Check the quick-connect fitting for dirt, and clean it if needed.

12. Place a rag or a shop towel over the quick-connect fitting.

13. Disconnect the quick-connect fitting.

  a. Hold the connector with one hand, and squeeze the retainer tabs with the other hand to release them from the locking tabs.

  b. Pull the connector off.

➡**Be careful not to damage the line or other parts. Do not use tools. If the connector does not move, keep the retainer tabs pressed down, and alternately pull and push the connector until it comes off easily. Do not remove the retainer from the line; once removed, the retainer must be replaced with a new one.**

14. After disconnecting the quick-connect fitting, check it for dirt or damage; without SH-AWD system, with SH-AWD system.

15. Do the battery terminal reconnection procedure.

WITHOUT THE HDS

See Figure 125.

1. Remove the driver's dashboard undercover.

2. Remove PGM-FI main relay 2 from the driver's under-dash fuse/relay box.

3. Start the engine, and let it idle until it stalls.

➡**If any DTCs are stored, clear and ignore them.**

4. Turn the ignition switch to LOCK (0), or press the engine start/stop button to select the OFF mode.

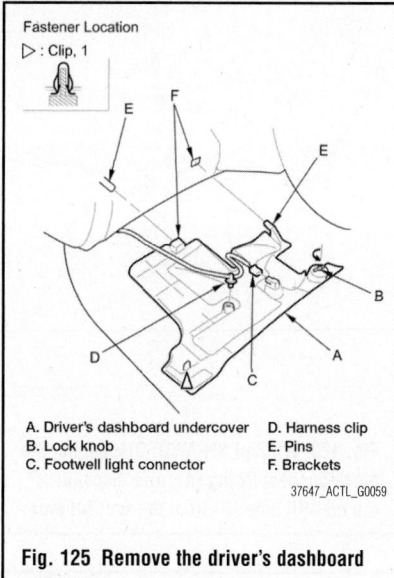

Fig. 125 Remove the driver's dashboard undercover

A. Driver's dashboard undercover
B. Lock knob
C. Footwell light connector
D. Harness clip
E. Pins
F. Brackets

37647_ACTL_G0059

5. Remove the fuel fill cap to relieve the pressure in the fuel tank.

6. Do the battery terminal disconnection procedure.

7. Remove the cover and the quick-connect fitting cover.

8. Check the quick-connect fitting for dirt, and clean it if needed.

9. Place a rag or a shop towel over the quick-connect fitting.

Disconnect the quick-connect fitting.

  a. Hold the connector with one hand, and squeeze the retainer tabs with the other hand to release them from the locking tabs.

  b. Pull the connector off.

➡**Be careful not to damage the line or other parts. Do not use tools. If the connector does not move, keep the retainer tabs pressed down, and alternately pull and push the connector until it comes off easily. Do not remove the retainer from the line; once removed, the retainer must be replaced with a new one.**

10. After disconnecting the quick-connect fitting, check it for dirt or damage; without SH-AWD system, with SH-AWD system.

11. Do the battery terminal reconnection procedure.

FUEL TANK

DRAINING

Without SH-AWD

See Figure 126.

1. Remove the fuel tank unit.

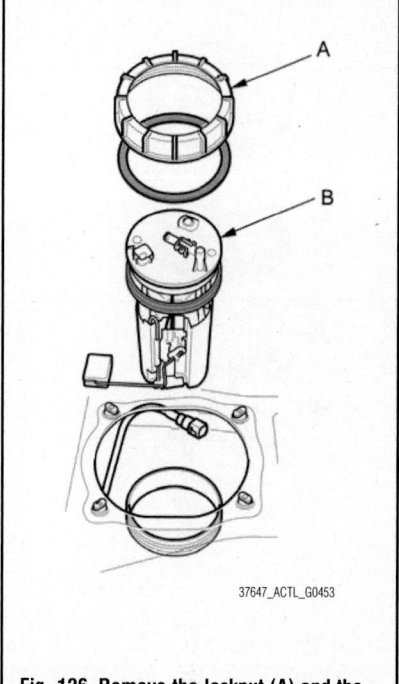

Fig. 126 Remove the locknut (A) and the fuel tank unit (B)

37647_ACTL_G0453

  a. Relieve the fuel pressure.
  b. Remove the rear seat cushion.
  c. Remove the access panel from the floor.
  d. Disconnect the fuel tank unit 4P connector.
  e. Disconnect the quick-connect fitting from the fuel tank unit.
  f. Using the special tool, loosen the locknut.
  g. Remove the locknut and the fuel tank unit.

2. Using a hand pump, draw the fuel from the fuel tank.

3. Install the fuel tank unit, as described in this section.

With SH-AWD

See Figure 127.

1. Remove the fuel tank unit.
  a. Relieve the fuel pressure.
  b. Remove the rear seat cushion.
  c. Remove the access panel from the left side of the floor.
  d. Disconnect the fuel tank unit 4P connector.
  e. Disconnect the quick-connect fitting from the fuel tank unit.
  f. Using the special tool, loosen the locknut.
  g. Remove the locknut, and lift up the fuel tank unit.
  h. Disconnect the transfer tube, and remove the fuel tank unit.

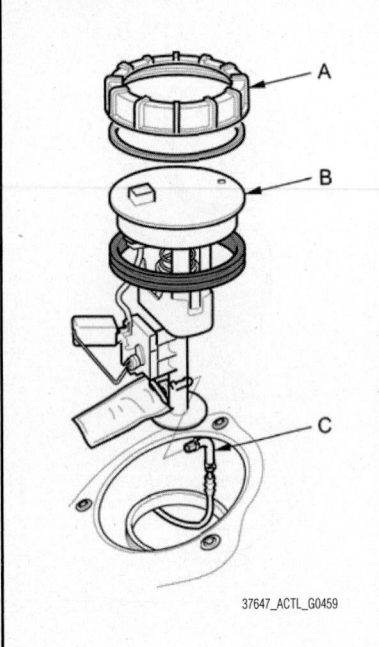

Fig. 127 Remove the locknut (A), and lift up the secondary fuel gauge sending unit (B) from the fuel tank

2. Remove the secondary fuel gauge sending unit.

   a. Remove the access panel from the right side of the floor.

   b. Disconnect the secondary fuel gauge sending unit 4P connector.

   c. Using the special tool, loosen the locknut.

   d. Remove the locknut, and lift up the secondary fuel gauge sending unit from the fuel tank.

   e. Disconnect the transfer tube, and remove the secondary fuel gauge sending unit.

3. Using a hand pump, a hose, and a container suitable for fuel, draw the fuel from the fuel tank.

4. Install the fuel tank unit, as described in this section.

REMOVAL & INSTALLATION

See Figures 128 through 131.

1. Drain the fuel tank.

2. Reinstall the fuel tank unit without connecting the fuel tank unit 4P connector and the quick-connect fitting.

3. Remove the fuel fill pipe cover.

   a. Remove the left rear wheel.

   b. Remove the fuel fill pipe cover.

   c. Disconnect the fuel fill tube:

   • Without SH-AWD: Disconnect the quick-connect fitting, and discon-

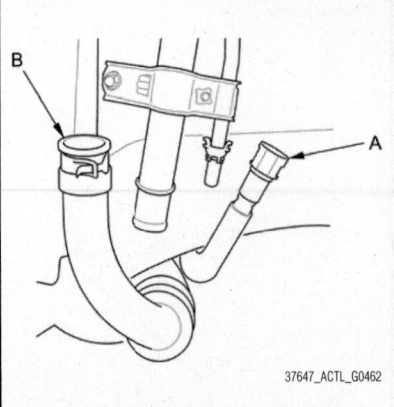

Fig. 128 Without SH-AWD: Disconnect the quick-connect fitting (A), and disconnect the fuel fill tube (B) from the fuel fill pipe

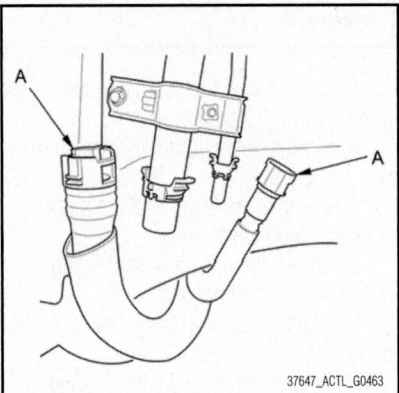

Fig. 129 With SH-AWD: Disconnect the quick-connect fittings (A)

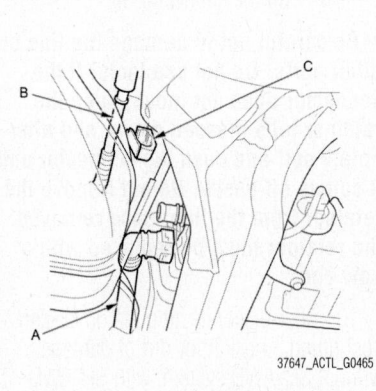

Fig. 130 Disconnect the hose (A) from the EVAP canister

nect the fuel fill tube from the fuel fill pipe by sliding back the clamp on the fuel fill tube, then twist the hose as you pull to avoid damaging the hose or pipe.

   • With SH-AWD: Disconnect the quick-connect fittings.

   d. Remove the fuel fill pipe.

4. Raise the vehicle on a lift.

5. Disconnect the hose from the EVAP canister.

6. Remove the hose from the clamp.

➡**Be careful not to damage the hose.**

7. Remove the exhaust pipe.

8. Remove the middle floor undercover.

➡**With SH-AWD: The left side is shown; the right side is similar.**

   a. Remove the left middle floor undercover and the right middle floor undercover.

   b. Remove the bolts.

   c. Right side (without SH-AWD): Remove the nut.

   d. Release the hook(s).

9. Remove the fuel tank protector.

10. Remove the right parking brake cable mounting bolts.

11. Place a jack or other support under the fuel tank.

12. Remove the strap bolts and the straps.

13. Remove the fuel tank.

To install:

14. Install the parts in the reverse order of removal.

   a. New fuel tanks have ring pull at the fuel vapor hose connector. When you connect the hose and confirm that the connection is secure, remove the ring pull by pulling it down.

   b. Before connecting the fuel fill pipe and the quick-connect fitting, check for dirt, and clean it if needed, taking care not to damage the fuel fill pipe and other parts.

   c. When installing the fuel tank protector, make sure to insert it into the clip in the direction shown.

FUEL FILTER

REMOVAL & INSTALLATION

With SH-AWD

➡**The fuel filter should be replaced whenever the fuel pressure drops below the specified value, after making sure that the fuel pump and fuel pressure regulator are OK.**

1. Remove the fuel tank unit.

2. Remove the fuel filter set.

3. Check these items before installing the fuel tank unit:

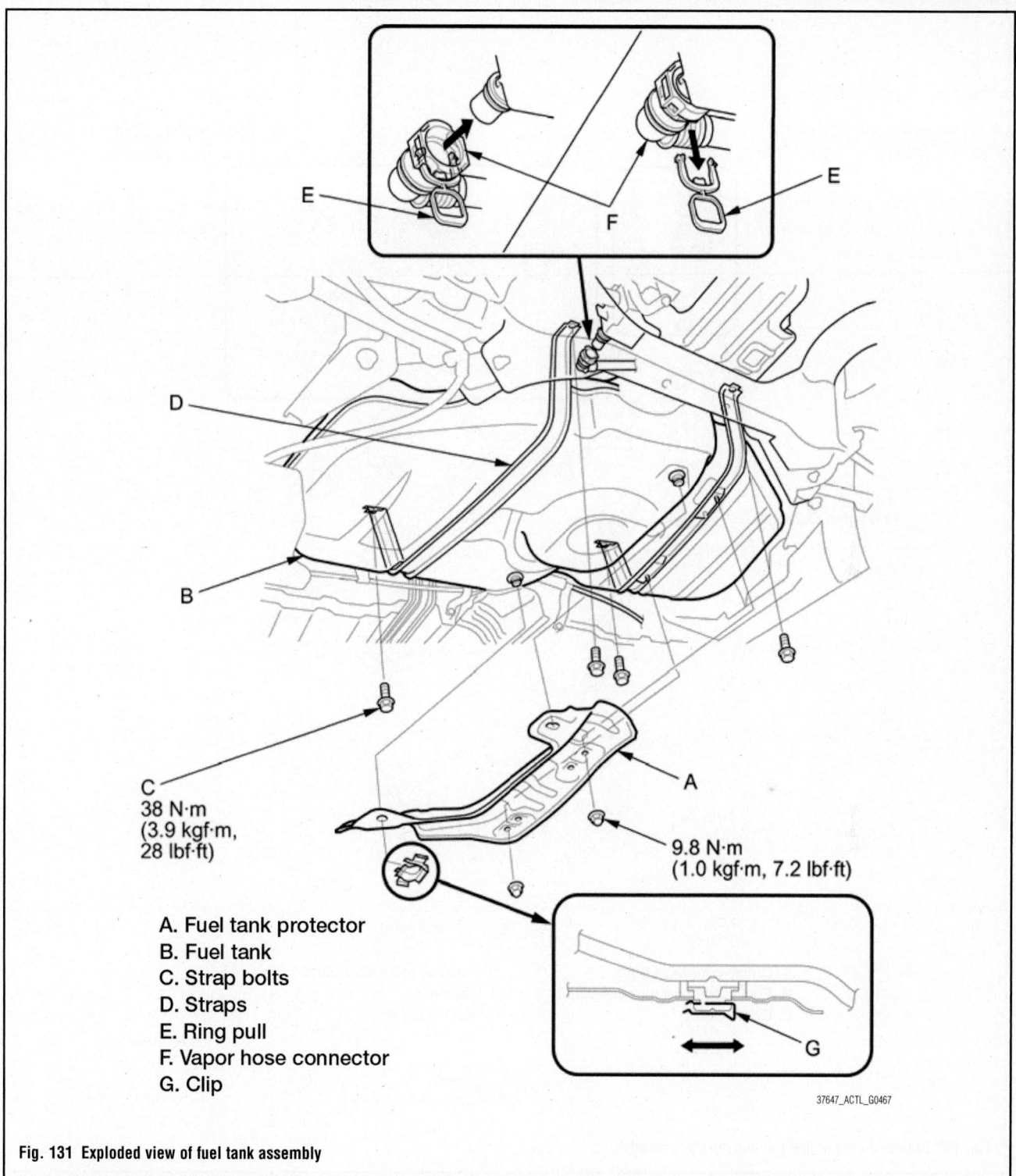

A. Fuel tank protector
B. Fuel tank
C. Strap bolts
D. Straps
E. Ring pull
F. Vapor hose connector
G. Clip

C
38 N·m
(3.9 kgf·m,
28 lbf·ft)

9.8 N·m
(1.0 kgf·m, 7.2 lbf·ft)

37647_ACTL_G0467

Fig. 131 Exploded view of fuel tank assembly

- When connecting the wire harness, make sure the connection is secure and the connectors are firmly locked into place.
- When installing the fuel gauge sending unit, make sure the connection is secure and the connector is firmly locked into place. Be careful not to bend or twist it excessively.

4. Install the parts in the reverse order of removal with new O-rings. When installing the fuel tank unit, align the marks on the unit and the fuel tank.

➡Coat the O-rings with clean engine oil; do not use any other oil or fluid. Do not pinch the O-rings during installation. Use all the new parts supplied in the fuel filter replacement kit.

Without SH-AWD

➡The fuel filter should be replaced whenever the fuel pressure drops below the specified value, after making sure that the fuel pump and the fuel pressure regulator are OK.

1. Remove the fuel tank unit.
2. Remove the fuel filter set.

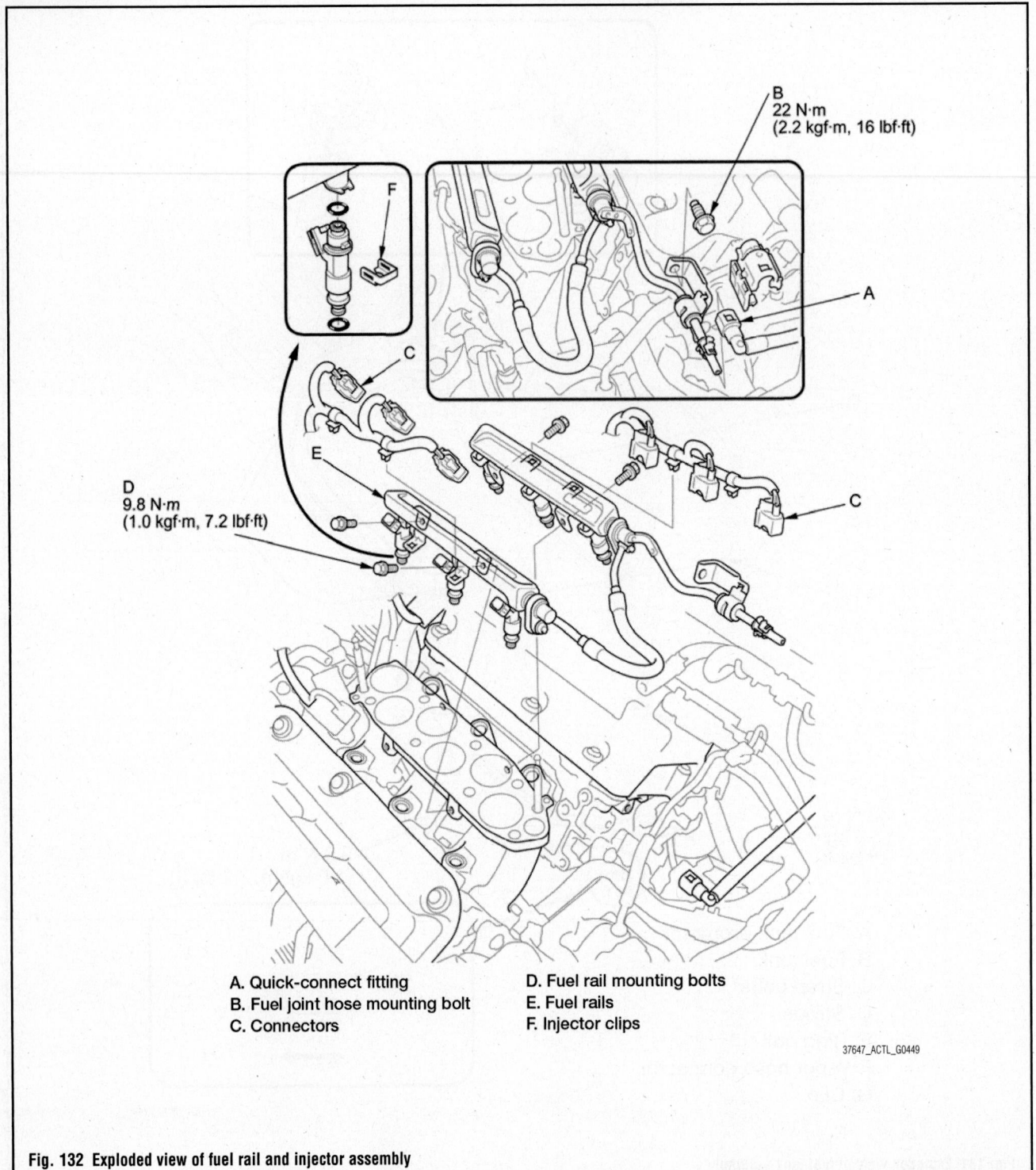

B
22 N·m
(2.2 kgf·m, 16 lbf·ft)

A

C

D
9.8 N·m
(1.0 kgf·m, 7.2 lbf·ft)

E

C

F

A. Quick-connect fitting
B. Fuel joint hose mounting bolt
C. Connectors

D. Fuel rail mounting bolts
E. Fuel rails
F. Injector clips

37647_ACTL_G0449

Fig. 132 Exploded view of fuel rail and injector assembly

3. Check these items before installing the fuel tank unit:

- When connecting the wire harness, make sure the connection is secure and the connectors are firmly locked into place.
- When installing the fuel gauge sending unit, make sure the connection is secure and the connector is firmly locked into place. Be care-

ful not to bend or twist it excessively.

4. Install the parts in the reverse order of removal with new O-rings and a new bracket. When installing the fuel tank unit, align the marks on the unit and the fuel tank.

➡**Coat the O-rings with clean engine oil; do not use any other oil or fluid. Do not pinch the O-rings during installa-**

tion. Use all the new parts supplied in the fuel filter replacement kit.

FUEL RAIL & INJECTORS

REMOVAL & INSTALLATION

See Figures 132 and 133.

1. Relieve the fuel pressure.
2. Remove the intake manifold.

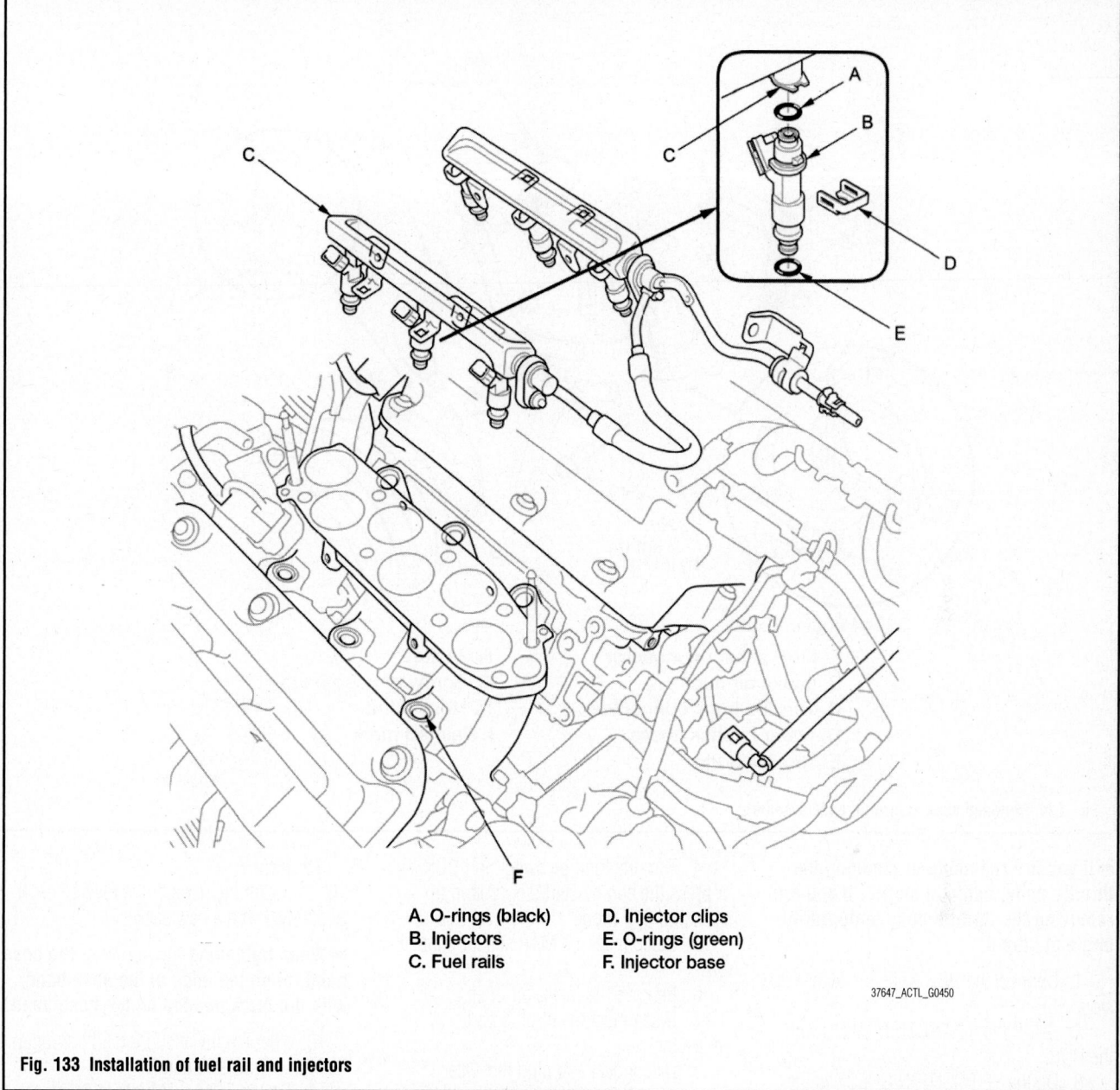

A. O-rings (black)   D. Injector clips
B. Injectors   E. O-rings (green)
C. Fuel rails   F. Injector base

37647_ACTL_G0450

Fig. 133 Installation of fuel rail and injectors

3. Disconnect the quick-connect fitting.

4. Remove the fuel joint hose mounting bolt.

5. Disconnect the connectors from the injectors.

6. Remove the fuel rail mounting bolts from the fuel rails.

7. Remove the fuel rails and the injectors from the injector base.

8. Remove the injector clips from the fuel rails.

9. Remove the injectors from the fuel rails.

To install:

10. Coat the new O-rings (black) with clean engine oil, and insert the injectors into the fuel rails.

11. Install the injector clips.

12. Coat the new injector O-rings (green) with clean engine oil.

13. Install the fuel rails and the injectors in the injector base.

14. Install the fuel rail mounting bolts, and connect the injector connectors.

15. Install the fuel joint hose mounting bolt.

16. Connect the quick-connect fitting.

17. Turn the ignition switch to ON (II), or press the engine start/stop button to select the ON mode, but do not operate the starter. After the fuel pump runs for about 2 seconds, the fuel rail will be pressurized. Repeat this two or three times, then make sure there are no fuel leaks.

18. Install the intake manifold with a new gasket.

THROTTLE BODY

REMOVAL & INSTALLATION

See Figure 134.

✳✳ CAUTION

Do not insert your fingers into the installed throttle body when you turn the ignition switch to ON (II)/press the engine start/stop button, or while the ignition switch is in ON (II)/vehicle in ON mode. If you do, you will seriously injure your fingers if the throttle valve is activated.

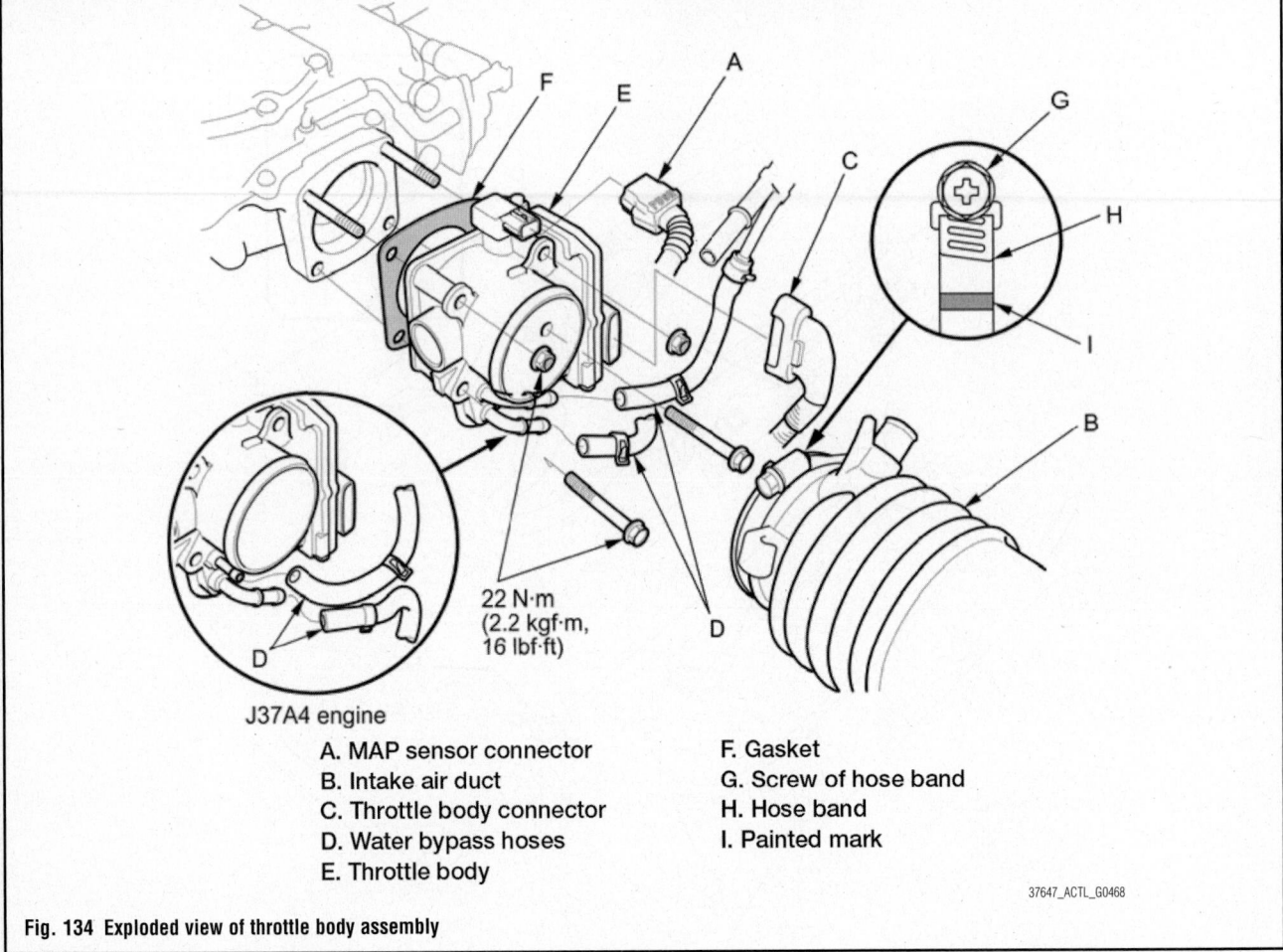

J37A4 engine

A. MAP sensor connector
B. Intake air duct
C. Throttle body connector
D. Water bypass hoses
E. Throttle body

F. Gasket
G. Screw of hose band
H. Hose band
I. Painted mark

22 N·m
(2.2 kgf·m,
16 lbf·ft)

37647_ACTL_G0468

Fig. 134 Exploded view of throttle body assembly

➡**If you are replacing or cleaning the throttle body, begin at step 1. If you are removing the throttle body temporarily, begin at step 4.**

1. Connect the HDS while the engine is stopped.
2. Select the INSPECTION MENU with the HDS.
3. Do the TP POSITION CHECK in the ETCS TEST.

4. Turn the ignition switch to LOCK (0), or press the engine start/stop button to select the OFF mode.
5. Disconnect the MAP sensor connector.
6. Remove the intake air duct.
7. Disconnect the throttle body connector.
8. Disconnect and plug the water bypass hoses, then remove the throttle body.

To install:

9. Install the parts in the reverse order of removal with a new gasket.

➡**When tightening the screw of the hose band, align the edge of the hose band with the mark painted on the hose band.**

10. After the installation, refill the radiator with engine coolant.
11. Do the ECM/PCM idle learn procedure.

HEATING & AIR CONDITIONING SYSTEM

BLOWER MOTOR

REMOVAL & INSTALLATION

See Figures 135 through 139.

1. Remove the passenger's dashboard undercover.
2. Remove the passenger's dashboard trim.
3. Remove the glove box.
   a. Remove the bolts and the screws.
   b. Pull the glove box by hand to detach the clips.
   c. While holding the glove box, dis-

connect the glove box light bulb, the trunk lid opener main switch connector, and the keyless access system mode switch connector (for some models).
4. Remove the harness clips, the bolts, and the glove box frame.
5. Cut the plastic cross brace in the glove box opening with diagonal cutters in the area shown, and discard it.
6. Release the wire from its harness clip, then remove the self-tapping screws, and the passenger's heater duct.

7. Disconnect the connector from the blower motor, and detach the wire harness clips.
8. Remove the self-tapping screws and disconnect the climate control unit connector, and recirculation control motor connector.
9. Disconnect the right engine compartment harness connector and harness clips.
10. Remove the mounting nuts, and the blower unit.
11. Install the unit in the reverse order of removal. Make sure that there are no air leaks.

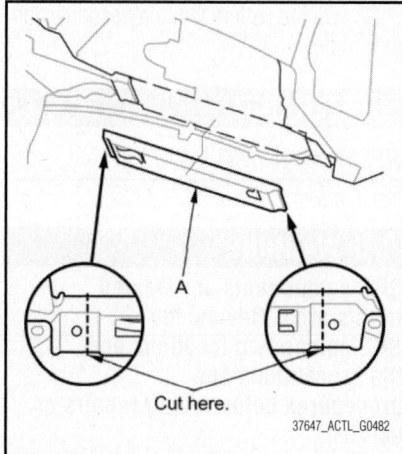

Fig. 135 Cut the plastic cross brace (A) in the glove box opening

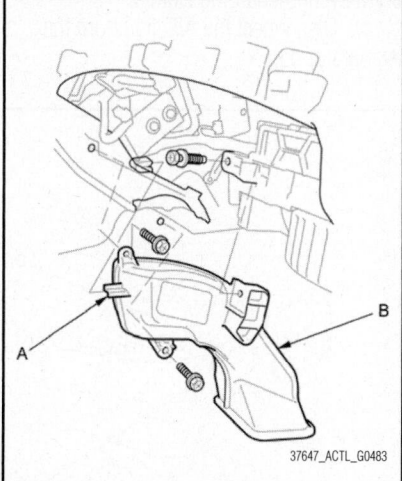

Fig. 136 Release the wire from its harness clip (A), remove the self-tapping screws, and the passenger's heater duct (B)

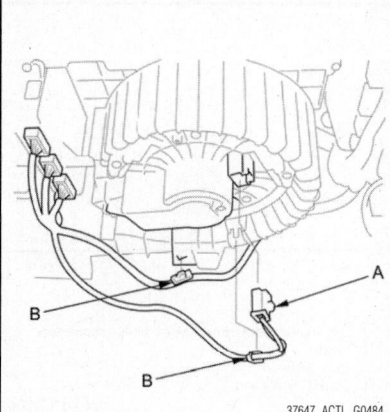

Fig. 137 Disconnect the connector (A) from the blower motor, and detach the wire harness clips (B)

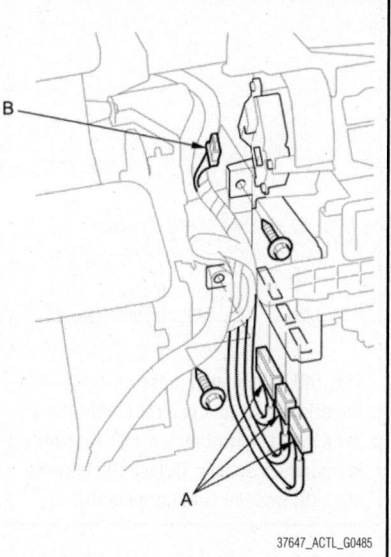

Fig. 138 Disconnect the climate control unit connector (A), and recirculation control motor connector (B)

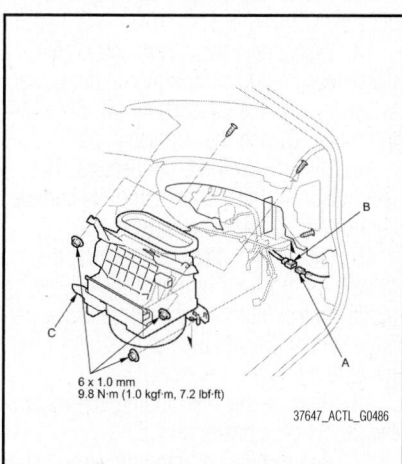

Fig. 139 Disconnect the right engine compartment harness connector (A) and harness clips (B)

HEATER CORE

REMOVAL & INSTALLATION
See Figures 140 through 145.

❉❉ WARNING

SRS components are located in this area. Review the SRS component locations and the precautions and procedures before doing repairs or service.

1. Do the battery terminal disconnection procedure.
2. Recover the refrigerant with a recovery/recycling/charging station.

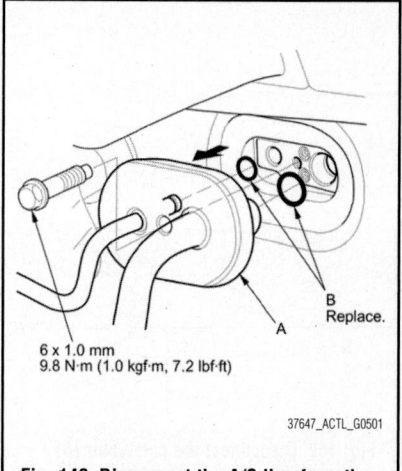

Fig. 140 Disconnect the A/C line from the evaporator core

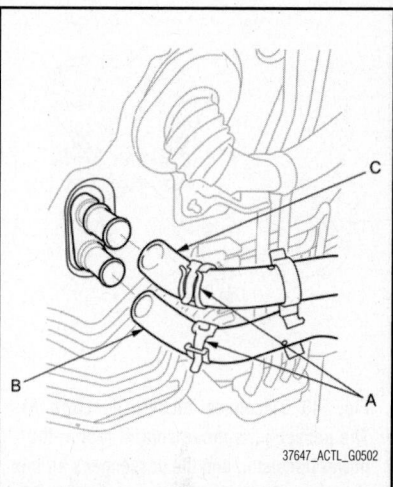

Fig. 141 Disconnect the inlet heater hose (B) and the outlet heater hose (C) from the heater unit

3. Disconnect the A/C line from the evaporator core.
4. When the engine is cool, drain the engine coolant from the radiator.
5. From under the hood, slide the heater hose clamps back.
6. Disconnect the inlet heater hose and the outlet heater hose from the heater unit. Note and/or mark the layout of the hoses.

➡**Engine coolant will run out when the hoses are disconnected; drain it into a clean drip pan. Be sure not to let coolant spill on the electrical parts or the painted surfaces. If any coolant spills, rinse it off immediately.**

7. Remove the mounting nut from the heater unit. Take care not to damage or bend the fuel lines or the brake lines.
8. Remove the dashboard.
9. Disconnect the left engine compartment wire harness connectors.

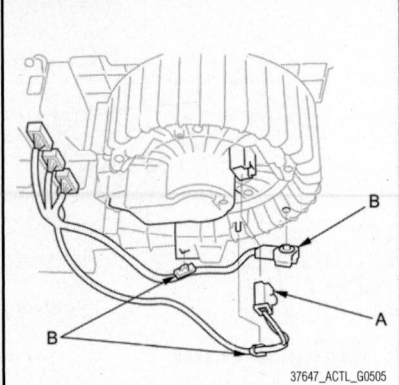

Fig. 142 Disconnect the connector (A) from the blower motor. Remove the wire harness clips (B).

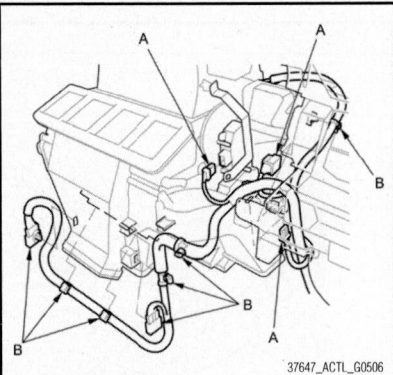

Fig. 143 Disconnect these connectors (A): The passenger's mode control motor, the power transistor, and the passenger's air mix control motor. Detach the harness clips (B)

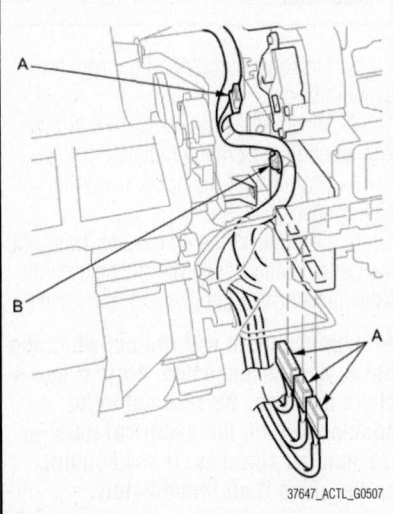

Fig. 144 Disconnect these connectors (A): The recirculation control motor, and the climate control unit. Detach the harness clip (B).

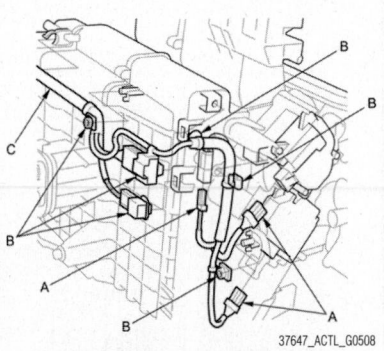

Fig. 145 Disconnect these connectors (A): the driver's air mix control motor, the driver's mode control motor, and evaporator temperature sensor. Detach the harness clips (B) and the wire harness (C).

10. Remove the ducts, and the drain hose.

11. Remove the mounting nuts, and the blower-heater unit.

12. Disconnect the connector from the blower motor. Remove the wire harness clips.

13. Disconnect these connectors: The passenger's mode control motor, the power transistor, and the passenger's air mix control motor. Detach the harness clips.

14. Disconnect these connectors: The recirculation control motor, and the climate control unit. Detach the harness clip.

15. Disconnect these connectors: the driver's air mix control motor, the driver's mode control motor, and evaporator temperature sensor. Detach the harness clips and the wire harness.

16. Remove the self-tapping screws and the passenger's heater duct.

17. Remove the self-tapping screws and the joint duct.

18. Remove the self-tapping screw and the heater core cover.

19. Remove the self-tapping screw, the heater pipe bracket, and the grommet, and carefully pull out the heater core.

To install:

20. Install the heater core in the reverse order of removal.

21. Install the heater unit in the reverse order of removal, and note these items:

- Do not interchange the inlet and outlet heater hoses, and install the hose clamps securely.
- Refill the cooling system with engine coolant.
- Make sure that there is no coolant leakage.
- Make sure that there is no air leakage.
- Charge the system.

22. Do the battery terminal reconnection procedure.

HEATER UNIT

REMOVAL & INSTALLATION

See Figures 146 and 147.

❊❊ WARNING

SRS components are located in this area. Review the SRS component locations and the precautions and procedures before doing repairs or service.

1. Do the battery terminal disconnection procedure.

2. Recover the refrigerant with a recovery/recycling/charging station.

3. Disconnect the A/C line from the evaporator core.

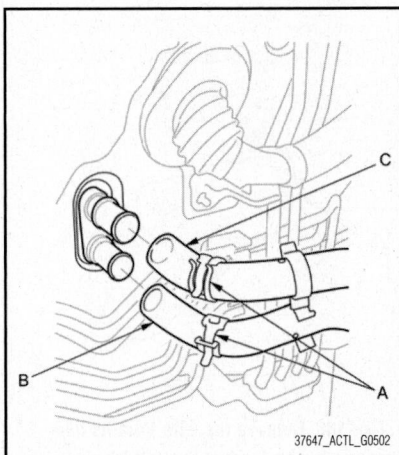

Fig. 146 Disconnect the inlet heater hose (B) and the outlet heater hose (C) from the heater unit

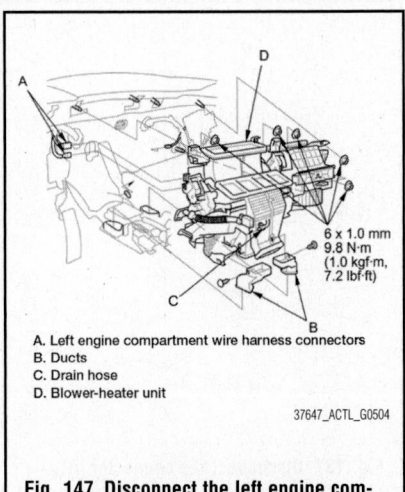

A. Left engine compartment wire harness connectors
B. Ducts
C. Drain hose
D. Blower-heater unit

6 x 1.0 mm
9.8 N·m
(1.0 kgf-m,
7.2 lbf·ft)

Fig. 147 Disconnect the left engine compartment wire harness connectors

4. When the engine is cool, drain the engine coolant from the radiator.

5. From under the hood, slide the heater hose clamps back.

6. Disconnect the inlet heater hose and the outlet heater hose from the heater unit. Note and/or mark the layout of the hoses.

➡**Engine coolant will run out when the hoses are disconnected; drain it into a clean drip pan. Be sure not to let coolant spill on the electrical parts or the painted surfaces. If any coolant spills, rinse it off immediately.**

7. Remove the mounting nut from the heater unit. Take care not to damage or bend the fuel lines or the brake lines.

8. Remove the dashboard.

9. Disconnect the left engine compartment wire harness connectors.

10. Remove the ducts, and the drain hose.

11. Remove the mounting nuts, and the blower-heater unit.

To install:

12. Install the heater unit in the reverse order of removal, and note these items:

- Do not interchange the inlet and outlet heater hoses, and install the hose clamps securely.
- Refill the cooling system with engine coolant.
- Make sure that there is no coolant leakage.
- Make sure that there is no air leakage.
- Charge the system.

13. Do the battery terminal reconnection procedure.

STEERING

POWER STEERING GEAR

REMOVAL & INSTALLATION

See Figures 148 through 158.

✳✳ WARNING

SRS components are located in this area. Review the SRS component locations and the precautions and procedures before doing repairs or service.

➡**Note these items during removal:**

- Use solvent and a brush, wash any oil and dirt off the end of the steering gearbox. Avoid any electrical parts. Blow dry with compressed air.
- Be sure to remove the steering wheel before disconnecting the steering joint, or damage to the cable reel can occur.

1. Do the battery terminal disconnection procedure.

2. Raise and support the vehicle.

3. Remove the front wheels.

4. Remove the driver's airbag, and the steering wheel.

5. Remove the steering joint cover.

6. Loosen the upper steering joint bolt, and remove the lower steering joint bolt.

7. Disconnect the steering joint by sliding the steering joint into the column shaft. Tighten the upper steering joint bolt to hold the steering joint in place.

➡**Do not disconnect the steering joint from the column shaft.**

8. Remove the center guide (if equipped) from the top of the pinion shaft, and discard it.

9. Apply vinyl tape to the splines on the pinion shaft.

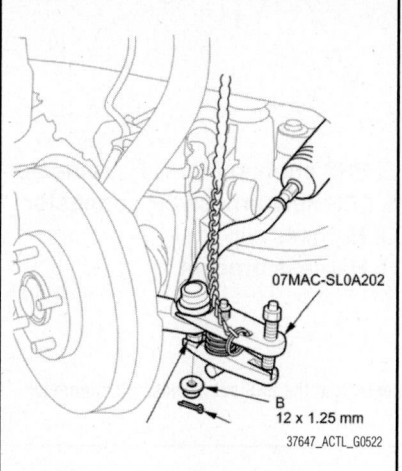

07MAC-SL0A202

B
12 x 1.25 mm

37647_ACTL_G0522

Fig. 148 Remove the cotter pin (A) from the tie-rod end ball joint, then remove the nut (B)

10. Disconnect the hood support struts from both sides of the pivot ball (bolted to the hood). Secure the hood in a vertical position. Remove the right side pivot ball and install it into the lower threaded hole, then reattach the support strut.

➡**Do not attempt to close the hood with the support strut in the vertical position, as it will damage the support strut and hood.**

11. Remove the left and right engine compartment covers, the strut brace cover, and the front bulkhead cover, then remove the cowl cover.

12. Remove the strut brace.

13. Remove the cotter pin from the tie-rod end ball joint, then remove the nut on both sides.

14. Disconnect the tie-rod end ball joint from the knuckle using the ball joint thread protector and the ball joint remover.

➡**Be careful not to damage the ball joint boot when installing the remover.**

15. Remove the P/S heat shield.

16. Disconnect the EPS subharness 12P connector and the EPS subharness 3P connector by pushing the lock and pulling up the lever. Wrap the connectors with vinyl tape to avoid contamination from grease or water.

17. Remove the EPS subharness 12P connector, the EPS subharness 3P connector, and the harness clip from the harness bracket.

18. Remove the harness clips, then remove the heat shield bracket from the steering gearbox.

19. Remove the steering gearbox mounting bolts and washers from the steering gearbox.

20. Remove the connector bracket from the front cylinder head; use the bracket bolt hole to attach the engine hanger balance bar front arm.

21. Disconnect the vacuum hose, and remove the harness clamp, then remove the engine mount control solenoid valve bracket from the rear cylinder head; use the bracket bolt hole to attach the 2008 V6 attachment arm.

22. Lift and support the engine with the engine support hanger and the engine hanger balance bar. Attach the front arm to the front cylinder head with a spacer and a 10 x 1.25 mm bolt. Remove the rear arm from the engine hanger balance bar, and install the 2008 V6 attachment arm, then attach the 2008 V6 attachment arm to the rear cylinder head with the 8 x 1.25 mm bolt.

23. M/T: Remove the rear engine mount.

24. A/T: Remove the rear engine mount stop and the heat shield, then remove the rear engine mount.

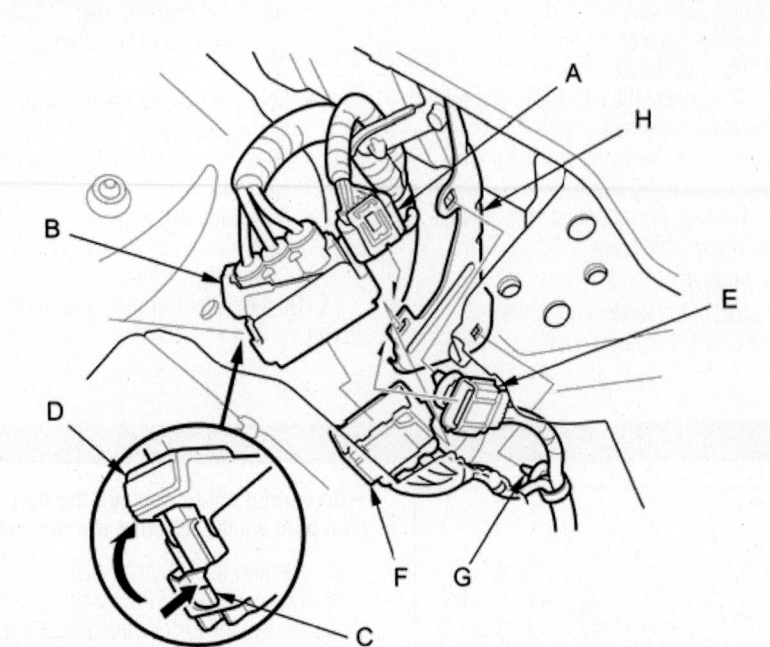

A. EPS subharness 12P connector
B. EPS subharness 3P connector
C. Lock
D. Lever

E. EPS subharness 12P connector
F. EPS subharness 3P connector
G. Harness clip
H. Harness bracket

37647_ACTL_G0523

Fig. 149 Disconnect the EPS subharness 12P connector and the EPS subharness 3P connector by pushing the lock and pulling up the lever

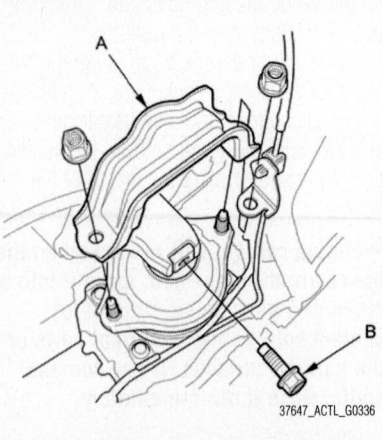

37647_ACTL_G0336

Fig. 152 Remove the front engine mount stop (A) and the vacuum hose bracket (B), then remove the front engine mount bolt (C)

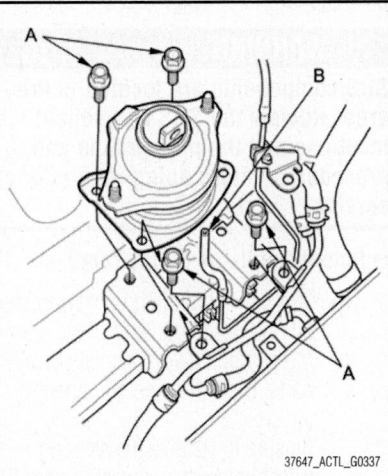

37647_ACTL_G0337

Fig. 153 M/T: Remove the front engine mount mounting bolts (A), then remove the front engine mount (B), and disconnect the vacuum hose (C)

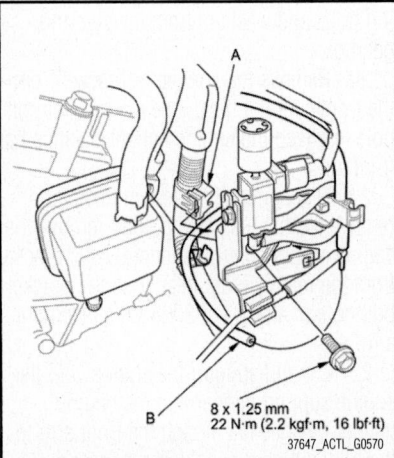

8 x 1.25 mm
22 N·m (2.2 kgf·m, 16 lbf·ft)
37647_ACTL_G0570

Fig. 150 Disconnect the vacuum hose, and remove the harness clamp, then remove the engine mount control solenoid valve bracket from the rear cylinder head

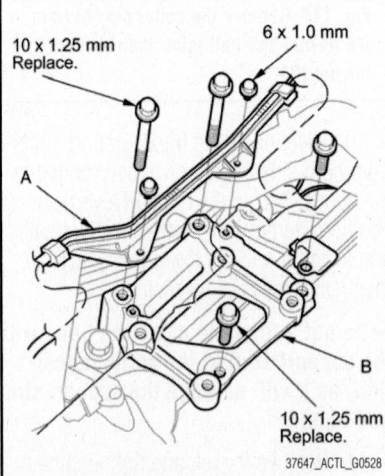

10 x 1.25 mm
Replace.

6 x 1.0 mm

10 x 1.25 mm
Replace.
37647_ACTL_G0528

Fig. 151 Remove the EPS subharness holder (A), then remove the rear engine mount base (B)

25. A/T: Disconnect the vacuum hose.
26. Remove the EPS subharness holder, then remove the rear engine mount base.
27. Remove the front engine mount stop and the vacuum hose bracket, then remove the front engine mount bolt.

28. M/T: Remove the front engine mount mounting bolts, then remove the front engine mount, and disconnect the vacuum hose.
29. Raise the vehicle on a lift.
30. Remove the front splash shield.

31. Remove the front subframe stiffener.
32. Remove exhaust pipe A.
33. With SH-AWD: Remove the propeller shaft.
34. M/T: Remove the transfer assembly.
35. Remove the cotter pin from the knuckle ball joint, then remove the castle nut.
36. Disconnect the knuckle ball joint from the lower arm using the ball joint thread protector and the ball joint remover.

➡ Be careful not to damage the ball joint boot when installing the remover. Do not force or hammer on the lower arm, or pry between the lower arm and the knuckle. You could damage the ball joint.

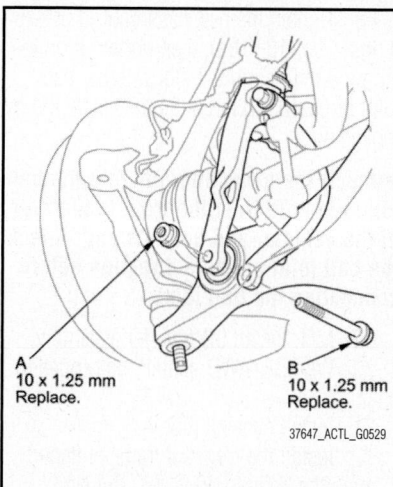

Fig. 154 Remove the damper fork mounting nut (A) while holding the mounting bolt (B), then remove the mounting bolt

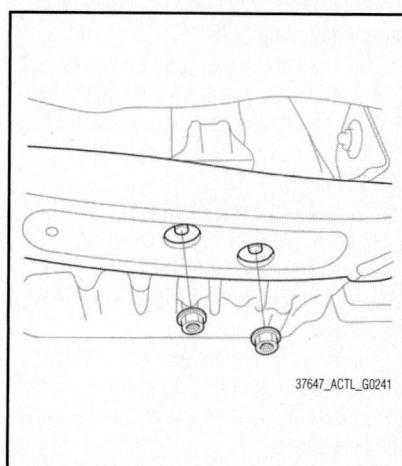

Fig. 155 Remove the transmission lower mount nuts

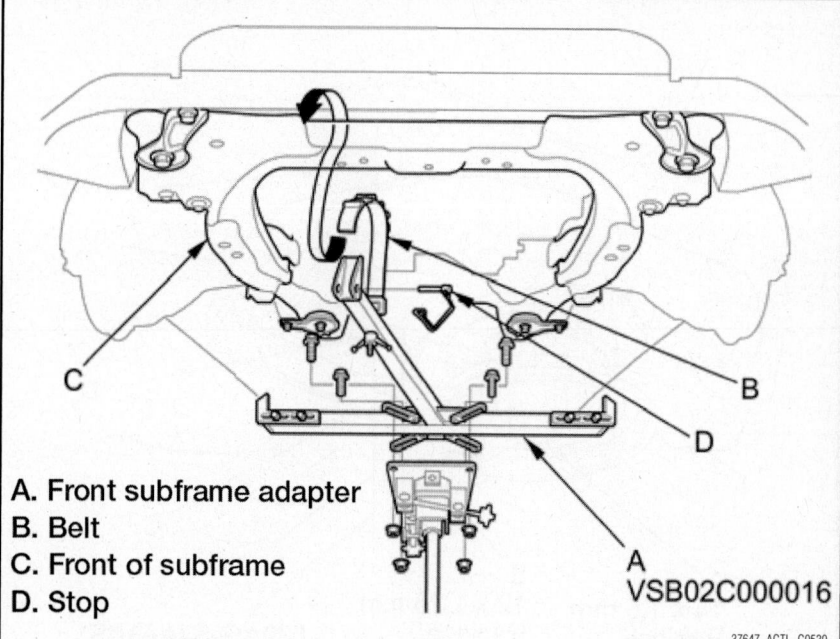

A. Front subframe adapter
B. Belt
C. Front of subframe
D. Stop

VSB02C000016

Fig. 156 Attach the front subframe adapter to the subframe by hanging the belt over the front of the subframe, then secure the belt with its stop

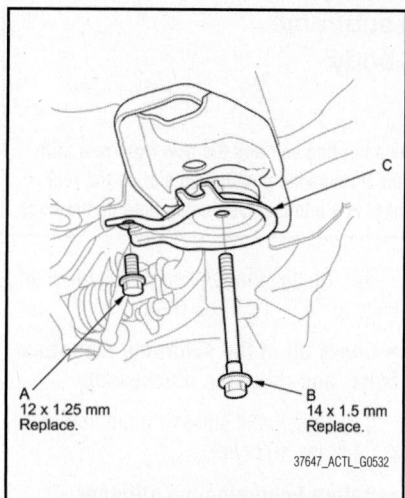

Fig. 157 Remove the stiffener mounting bolt (A) and subframe mounting bolts (B) from the rear stiffeners (C)

37. Remove the damper fork mounting nut while holding the mounting bolt, then remove the mounting bolt.

38. Remove the subframe middle mounts on both sides.

39. Remove the transmission lower mount nuts.

40. Attach the front subframe adapter to the subframe by hanging the belt over the front of the subframe, then secure the belt with its stop.

41. Raise the jack and line up the slots in the front subframe adapter arms with the bolt holes on the jack base, then securely attach them with four bolts.

42. Remove the stiffener mounting bolts on both sides.

43. Loosen the subframe mounting bolts so they are about 0.75 inches (20 mm) from the mounting surface. Do not loosen the

front subframe mounting bolts more than necessary.

44. Remove the stiffener mounting bolt and subframe mounting bolts from the rear stiffeners on both sides.

45. Lower the jack slowly until the front subframe has dropped about 2.75 inches (70 mm).

46. Remove the EPS subharness from the steering gearbox.

➡Disconnect the EPS subharness 3P connector by pushing the lock and pulling down the lever.

47. Carefully move the steering gearbox toward the driver's side until the pinion shaft clears the fenderwell opening on the body.

48. Remove the steering gearbox through the fenderwell opening on the driver's side.

49. Remove the pinion shaft grommet from the top of the torque sensor.

To install:

50. Before installing the steering gearbox, make sure that no grease is on the mating surface of the steering gearbox and the front subframe. To prevent the gearbox mounting bolts from loosening after the installation, remove any grease from the bolt holes.

51. Wrap vinyl tape around the splines on the pinion shaft.

52. Install the pinion shaft grommet. Align the cutout in the pinion shaft grommet with the lug portion on the torque sensor. The grommet must not have a gap at the mating surface of the grommet and torque sensor.

53. Slide the steering gearbox between the front subframe and the body from the driver's side.

54. Carefully move the steering gearbox toward the passenger's side until the pinion shaft clears the wheel well opening on the body.

55. Continue moving the steering gearbox toward the passenger's side until the steering gearbox is in position.

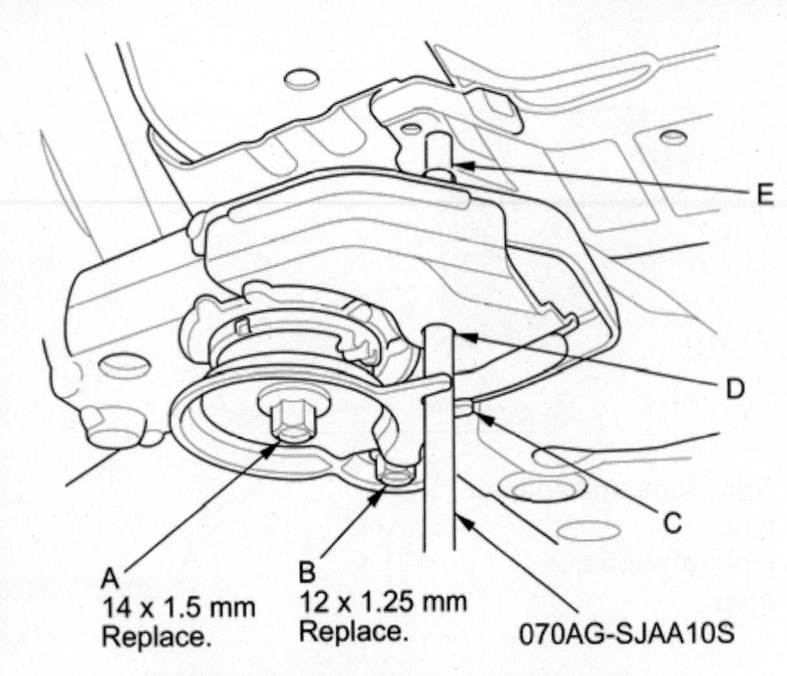

A
14 x 1.5 mm
Replace.

B
12 x 1.25 mm
Replace.

070AG-SJAA10S

A. Right rear subframe mounting bolt
B. Right rear stiffener mounting bolt
C. Positioning slot
D. Positioning hole on subframe
E. Positioning hole on body

37647_ACTL_G0352

Fig. 158 Loosely install the new right-rear subframe mounting bolt and the new right-rear stiff-ener mounting bolt; insert the subframe alignment pin through the positioning slot on the rear stiffener, through the positioning hole on the subframe, and into the positioning hole on the body

56. Install the EPS subharness to the steering gearbox.

➡**Pull up the lever of the EPS motor 3P connector, then confirm the connector is fully seated.**

57. Loosely tighten the new right-rear subframe mounting bolt and the new stiff-ener mounting bolt; insert the 15.7 mm side of the subframe alignment pin through the positioning slot on the rear stiffener, through the positioning hole on the sub-frame, and into the positioning hole on the body.

58. Loosely tighten the new left-rear subframe mounting bolt and the new stiff-ener mounting bolt with the same procedure as the right rear using the subframe align-ment pin.

59. Loosely tighten the new front side subframe mounting bolt and the stiffener mounting bolt on both sides.

60. Tighten the right-rear subframe mounting bolt to 76 ft. lbs. (103 Nm).

61. Tighten the left-rear subframe mounting bolt to 76 ft. lbs. (103 Nm).

62. Tighten the front subframe mounting bolts to 76 ft. lbs. (103 Nm).

➡**Check all of the subframe mounting bolts, and retighten if necessary.**

63. Tighten the stiffener mounting bolts to 69 ft. lbs. (93 Nm).

➡**Before tightening the stiffener mounting bolts, check that the position-ing holes and slot are aligned using the subframe alignment pin.**

64. Install the transmission lower mount nuts), and tighten them to 33 ft. lbs. (44 Nm).

65. Install the subframe middle mounts on both sides with new mounting bolts, and tighten the long bolt to 33 ft. lbs. (44 Nm). Tighten the two short bolts to 36 ft. lbs. (49 Nm).

66. Remove the transmission jack sup-porting the front subframe.

67. Install the new damper fork mount-ing bolt and new mounting nut, and loosely tighten the nut while holding the mounting bolt.

68. Tighten the new castle nut to 69–76 ft. lbs. (93–103 Nm), then tighten it only far enough to align the slot with the ball joint pin hole. Do not align the castle nut by loosening it.

➡**Insert the new cotter pin into the ball joint pin hole from the front to the rear of the vehicle, and bend its end. Check the ball joint pin hole direction before connecting the ball joint.**

69. M/T: Install the transfer assembly.

70. With SH-AWD: Install the propeller shaft.

71. Install exhaust pipe A.

72. Install the front subframe stiffener. Tighten the bolts to 40 ft. lbs. (54 Nm).

73. Install the front splash shield.

74. Lower the vehicle.

75. M/T: Connect the vacuum hose, then align the front engine mount with its mount-ing holes, and install new front engine mount mounting bolts.

76. Tighten the new front engine mount bolt to 40 ft. lbs. (54 Nm), then install the front engine mount stop and the vacuum hose bracket tightening the nuts to 47 ft. lbs. (64 Nm).

77. Install the rear engine mount base, then install the EPS subharness.

78. M/T: Tighten the new rear engine mount mounting bolts to 40 ft. lbs. (54 Nm) in a crisscross sequence.

79. A/T: Tighten the new rear engine mount mounting bolt to 40 ft. lbs. (54 Nm), then install the rear engine mount stop and the heat shield.

80. A/T: Connect the vacuum hose.

81. Remove the engine support hanger and the engine hanger balance bar.

82. Install the connector bracket to the front cylinder head.

83. Install the engine mount control solenoid valve bracket to the rear cylinder head, then install the harness clamp and connect the vacuum hose.

84. Install the washers and new steering gearbox mounting bolts to the steering gearbox, then loosely tighten them.

➡**Make sure that the washers are in the correct position.**

85. Tighten the steering gearbox mount-ing bolts to 54 ft. lbs. (74 Nm).

➡**Check all of the steering gearbox mounting bolts, and retighten if neces-sary.**

86. Install the heat shield bracket to the steering gearbox, then install the harness clips.

87. Install the EPS subharness 12P con-

nector, the EPS subharness 3P connector, and the harness clip to the harness bracket.

88. Connect the EPS subharness 12P connector and EPS subharness 3P connector, and pull down the lever, then confirm the connector is fully seated.

89. Install the P/S heat shield with the flange bolts, and tighten them to 7 ft. lbs. (10 Nm).

90. Install the strut brace.

91. Install the cowl cover and engine compartment cover.

92. Install the hood support struts to the upper location on both sides of the hood.

93. Wipe off any grease contamination from the ball joint tapered section and threads. Connect the tie-rod end ball joint to the knuckle. Install the new nut, and tighten it to 44 ft. lbs. (60 Nm).

94. Install a new cotter pin, and bend it.

95. Place a floor jack under the lower arm, and raise the suspension to load it with the vehicle's weight. Do not place the jack against the ball joint pin of the knuckle.

96. Tighten the damper fork mounting nut while holding the mounting bolt to the specified torque.

97. Install the front wheels, then set the wheels in the straight ahead position.

➡**Before installing the wheel, clean the mating surface between the brake disc and the inside of the wheel.**

98. Remove the vinyl tape around the splines on the pinion shaft.

99. Loosen the upper steering joint bolt, and slip the lower end of the steering joint onto the pinion shaft taking care to align the gap within the angle.

100. Align the bolt hole on the steering joint with the groove around the pinion shaft, then loosely install the lower steering joint bolt. Be sure that the joint bolt is securely in the groove in the pinion shaft.

101. Pull on the steering joint to make sure that the steering joint is fully seated, then tighten the lower joint bolt to 21 ft. lbs (28 Nm).

102. Tighten the upper steering joint bolt to 21 ft. lbs (28 Nm).

103. Install the steering joint cover.

104. Install the steering wheel, and the driver's airbag.

105. With the tires raised off the ground (vehicle on a lift), check for the following symptoms by turning the steering wheel fully to the right and left several times:

- Rubbing sound coming from the lower steering column area: Probable cause is the steering column joint is contacting the cover.
- Grating sound from the lower steering column area or a rough feeling during steering: Probable cause is poor engagement of the pinion shaft splines.

- Noise from around the steering wheel during steering: Probable cause is poor engagement of the SRS cable reel with the steering wheel or a damaged cable reel.

106. Do the battery terminal reconnection procedure, and check these items:

- Turn the ignition switch to ON (II), or press the engine start/stop button to select the ON mode, and check that the SRS indicator comes on for about 6 seconds, and then goes off.
- Make sure the horn and turn signal switches work properly.
- Make sure the steering wheel switches work properly.

107. After installation, check these items:

- Start the engine, allow it to idle, and turn the steering wheel from lock to lock several times.
- Check that the EPS indicator does not come on.
- Check the steering wheel spoke angle. If steering spoke angles to the right and left are not equal (steering wheel and rack are not centered), correct the engagement of the joint/pinion shaft splines.
- Do the torque sensor neutral position memorization.
- With SH-AWD: Do the VSA sensor neutral position memorization.
- Check the wheel alignment, and adjust it if necessary.

SUSPENSION

FRONT SUSPENSION

LOWER BALL JOINTS

REMOVAL & INSTALLATION

See Figures 159 and 160.

➡**Always use a ball joint remover to disconnect a ball joint. Do not strike the housing or any other part of the ball joint connection to disconnect it.**

1. Install a hex nut or the ball joint thread protector onto the threads of the ball joint.

➡**Using a hex nut, make sure the nut is flush with the ball joint pin end to prevent damage to the threaded end of the ball joint pin.**

2. Apply grease to the ball joint remover. This will ease the installation of the tool, and prevent damage to the pressure bolt threads.

3. Loosen the pressure bolt, and install the ball joint remover as shown. Insert the jaws carefully, making sure not to damage

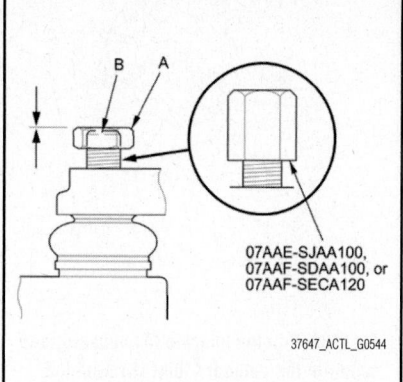

07AAE-SJAA100,
07AAF-SDAA100, or
07AAF-SECA120

37647_ACTL_G0544

Fig. 159 Install a hex nut (A) or the ball joint thread protector onto the threads of the ball joint (B)

the ball joint boot. Adjust the jaw spacing by turning the adjusting bolt.

➡**Fasten the safety chain securely to a suspension arm or the subframe. Do not fasten it to a brake line or wire harness.**

4. After adjusting the adjusting bolt, make sure the head of the adjusting bolt is in position to allow the jaw to pivot.

5. With a wrench, tighten the pressure bolt until the ball joint pin pops loose from the ball joint connecting hole. If necessary, apply penetrating type lubricant to loosen the ball joint pin.

➡**Do not use pneumatic or electric tools on the pressure bolt.**

6. Remove the ball joint remover, then remove the nut or the ball joint thread protector from the end of the ball joint pin, and pull the ball joint out of the ball joint connecting hole. Inspect the ball joint boot, and replace it if damaged.

LOWER CONTROL ARMS

REMOVAL & INSTALLATION

See Figures 161 through 163.

1. Raise and support the vehicle.
2. Remove the front wheel.

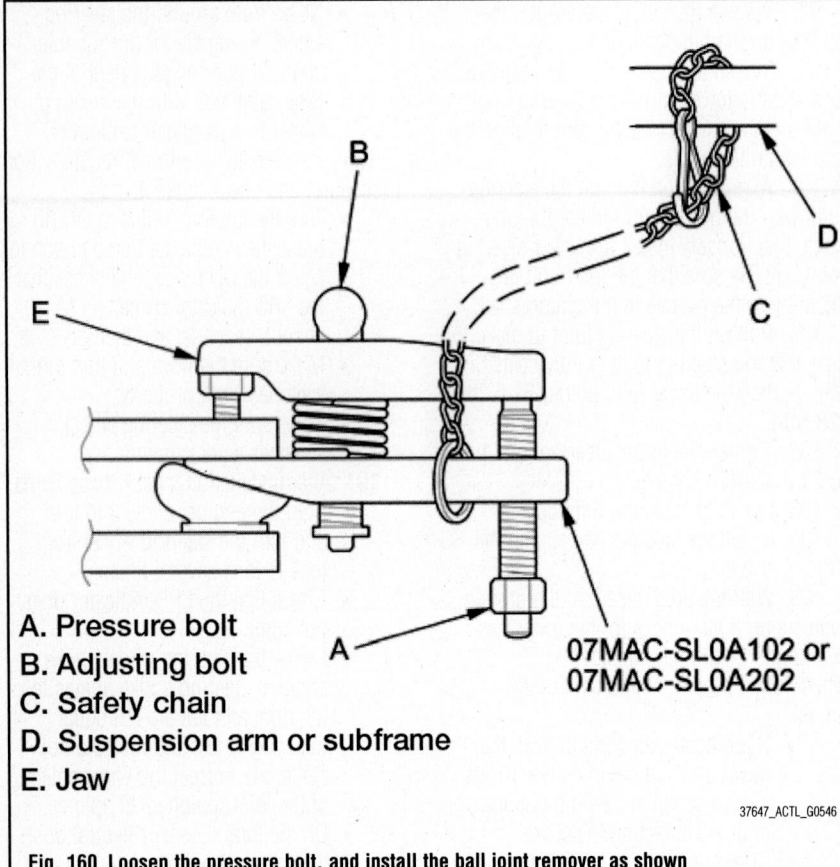

A. Pressure bolt
B. Adjusting bolt
C. Safety chain
D. Suspension arm or subframe
E. Jaw

07MAC-SL0A102 or
07MAC-SL0A202

37647_ACTL_G0546

Fig. 160 Loosen the pressure bolt, and install the ball joint remover as shown

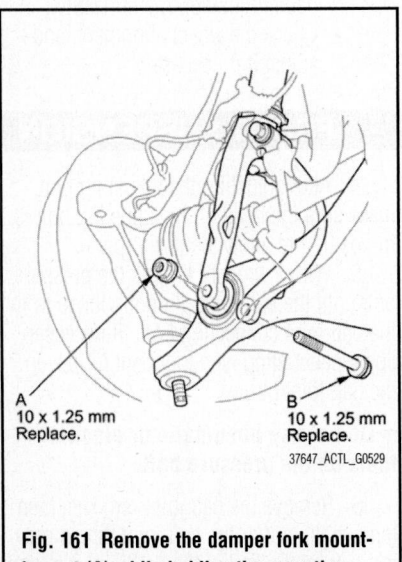

37647_ACTL_G0529

Fig. 161 Remove the damper fork mounting nut (A) while holding the mounting bolt (B), then remove the mounting bolt

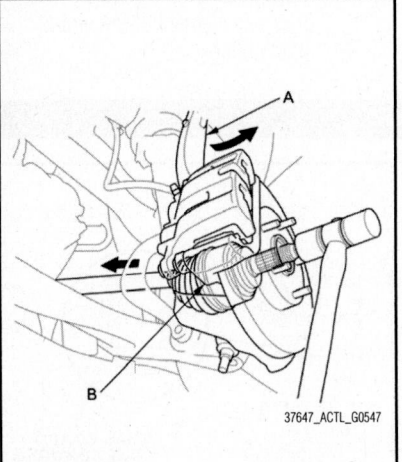

37647_ACTL_G0547

Fig. 162 Pull the knuckle (A) outward, and separate the outboard joint (B) from the front hub using a soft face hammer

3. Remove the spindle nut.
4. Remove the front splash shield.
5. Remove the damper fork mounting nut while holding the mounting bolt, then remove the mounting bolt.

➡Use the new damper fork mounting bolt and the new mounting nut, and

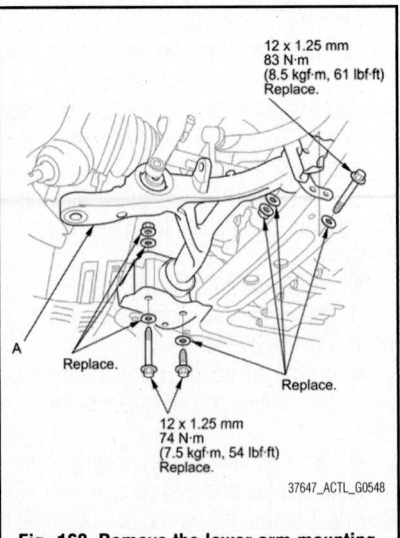

37647_ACTL_G0548

Fig. 163 Remove the lower arm mounting bolts, mounting nuts, and the washers, then remove the lower arm (A)

torque the nut while holding the bolt during reassembly.

6. Disconnect the stabilizer link from the lower arm.
7. Remove the cotter pin from the knuckle ball joint, then remove the castle nut.

➡Use the new castle nut during reassembly. During installation, insert the new cotter pin into the ball joint pin hole from the front to the rear of the vehicle, and bend it. Check the ball joint pin hole direction before connecting the ball joint.

8. Disconnect the knuckle ball joint from the lower arm using the ball joint thread protector and the ball joint remover.

➡Be careful not to damage the ball joint boot when installing the remover. Do not force or hammer on the lower arm, or pry between the lower arm and the knuckle. You could damage the ball joint.

9. Pull the knuckle outward, and separate the outboard joint from the front hub using a soft face hammer.

➡Do not pull the driveshaft end outward. The driveshaft inboard joint may come apart.

10. Remove the lower arm mounting bolts, mounting nuts, and the washers, then remove the lower arm.

To install:

➡Use new lower arm mounting bolts, new mounting nuts, and new washers during reassembly.

11. Install the lower arm in the reverse order of removal, and note these items:
  • First install all of the components, and lightly tighten the bolts and the nuts, then raise the suspension to load it with the vehicle's weight before fully tightening it to the

specified torque. Do not place the jack against the ball joint pin of the knuckle.

- Be careful not to damage the ball joint boot when connecting the knuckle.
- Before connecting the ball joint, degrease the threaded section and the tapered portion of the ball joint pin, the ball joint connecting hole, the threaded section, and the mating surfaces of the castle nut.
- Torque the castle nut to the lower torque specification, then tighten it only far enough to align the slot with the ball joint pin hole. Do not align the castle nut by loosening it.
- Use a new spindle nut on reassembly.
- Before installing the spindle nut, apply a small amount of engine oil to the seating surface of the nut. After tightening, use a drift to stake the spindle nut shoulder against the driveshaft.
- Before installing the wheel, clean the mating surfaces of the brake disc and the inside of the wheel.

12. Check the wheel alignment, and adjust it if necessary.

STABILIZER BAR & LINKS

REMOVAL & INSTALLATION

Front Stabilizer Link

See Figure 164.

1. Raise and support the vehicle.
2. Remove the front wheel.
3. Remove the flange nuts while holding the respective joint pin with a hex wrench, then remove the stabilizer link.
4. Install the stabilizer link on the stabilizer bar and the lower arm with the joint pins set at the center of their range of movement.

➡**The stabilizer link has a paint mark. The left stabilizer link is marked with yellow paint, and the right stabilizer link is marked with white paint.**

5. Install the new flange nuts, and tighten them to 58 ft. lbs. (78 Nm) while holding the respective joint pin with a hex wrench.
6. Clean the mating surfaces of the brake disc and the inside of the wheel, then install the front wheel.

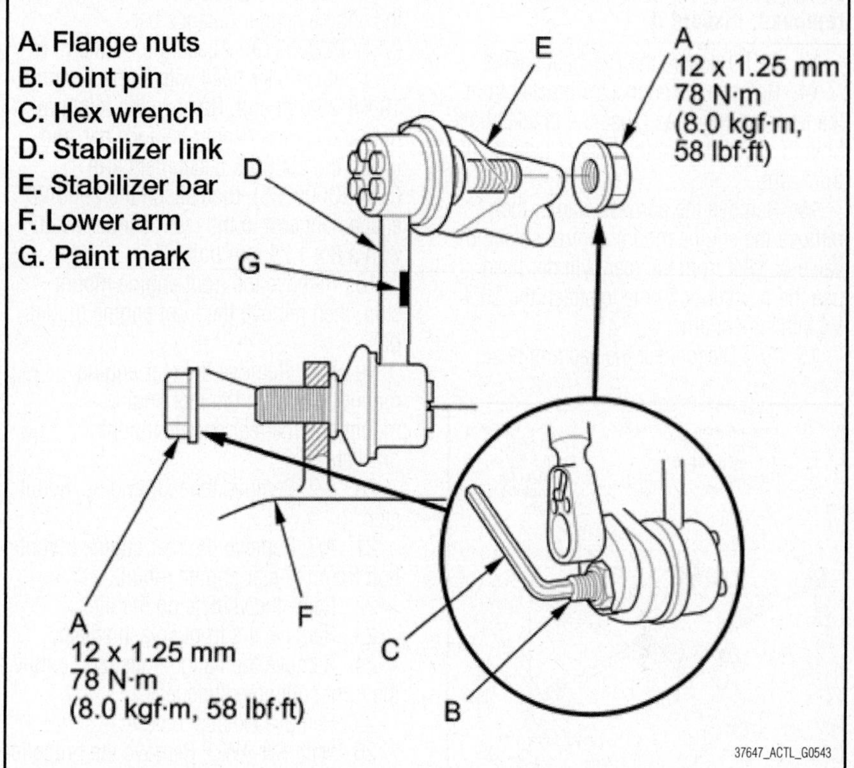

Stabilizer Bar

See Figures 165 through 173.

1. Note these items during replacement:

A. Flange nuts
B. Joint pin
C. Hex wrench
D. Stabilizer link
E. Stabilizer bar
F. Lower arm
G. Paint mark

A
12 x 1.25 mm
78 N·m
(8.0 kgf·m, 58 lbf·ft)

A
12 x 1.25 mm
78 N·m
(8.0 kgf·m, 58 lbf·ft)

37647_ACTL_G0543

Fig. 164 Remove the front stabilizer link

- Be sure to remove the steering wheel before disconnecting the steering joint. Damage to the cable reel can occur.
- Lower the front subframe from the body, and replace the front stabilizer bar through the gap created by lowering the front subframe.

2. Disconnect the support struts from both sides of the pivot ball (bolted to the hood). Secure the hood in a vertical position. Install the right side pivot ball into the lower threaded hole.

➡**Do not attempt to close the hood with the support strut in the vertical position, as it will damage the support strut and hood.**

3. Remove the engine compartment covers
4. Do the battery terminal disconnection procedure.
5. Remove the front strut brace.
6. Raise and support the vehicle.
7. Remove the front wheels.
8. Disconnect both sides of the stabilizer link from the stabilizer bar.
9. Disconnect both sides of the tie-rod end ball joint from the knuckle.

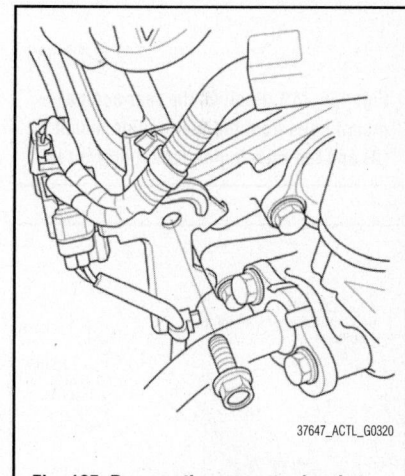

37647_ACTL_G0320

Fig. 165 Remove the connector bracket from the front cylinder head

10. Remove the driver's airbag and the steering wheel.
11. Remove the steering joint cover.
12. Loosen the upper steering joint bolt, and remove the lower steering joint bolt. Disconnect the steering joint by sliding the steering joint into the column shaft. Tighten the steering joint upper bolt to hold the column shaft.

➡**If the center guide is in place and**

has not moved, leave it in place. If the center guide has moved or been removed, discard it.

13. M/T: Remove the P/S heat shield.

14. Remove the connector bracket from the front cylinder head; use the bracket bolt hole to attach the engine hanger balance bar front arm.

15. Remove the harness clamp, then remove the engine mount control solenoid valve bracket from the rear cylinder head; use the bracket bolt hole to attach the 2008 V6 attachment arm.

16. A/T: Disconnect the vacuum hose.

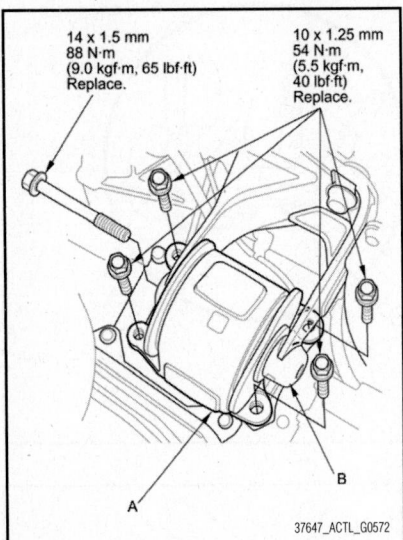

Fig. 166 M/T: Remove the rear engine mount bolts from the rear engine mount (A) and the rear engine mount bracket (B)

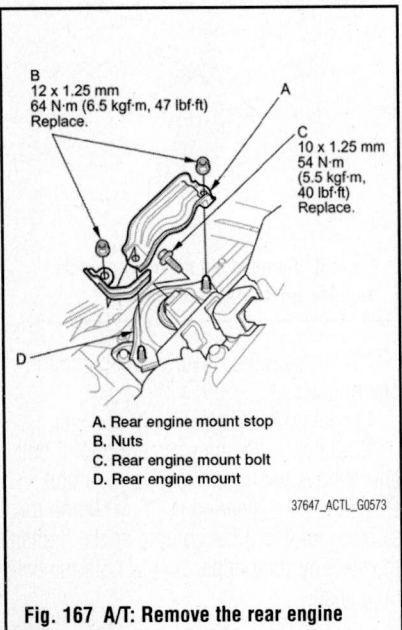

A. Rear engine mount stop
B. Nuts
C. Rear engine mount bolt
D. Rear engine mount

Fig. 167 A/T: Remove the rear engine mount stop (A)

17. Lift and support the engine with the engine support hanger (AAR-T1256) and the engine hanger balance bar (VSB02C000019). Attach the front arm to the front cylinder head with a spacer and a 10 x 1.25 mm bolt. Remove the rear arm from the engine hanger balance bar, and install the 2008 V6 attachment arm (SIL02C000033), then attach the 2008 V6 attachment arm to the rear cylinder head with a 8 x 1.25 mm bolt.

18. Remove the front engine mount stop, then remove the front engine mount bolt.

19. M/T: Remove the rear engine mount bolts from the rear engine mount and the rear engine mount bracket.

20. A/T: Remove the rear engine mount stop.

21. A/T: Remove the rear engine mount bolt from the rear engine mount.

22. Raise the vehicle on the lift.

23. Remove the front splash shield.

24. Remove the flange bolts, and remove the front subframe stiffener.

25. Remove exhaust pipe A:

26. With SH-AWD: Remove the propeller shaft.

27. Remove the front subframe mounting bolts, and the front subframe middle mounts.

28. Remove the nuts securing the lower transmission mount.

29. Attach the subframe adapter (VSB02C000016) to a transmission jack, then raise up the jack to the subframe. Hang the belt over the front of the subframe, then secure the belt by inserting the stop and then tighten the wing nut.

30. Remove the stiffener mounting bolt on both sides of the front stiffener.

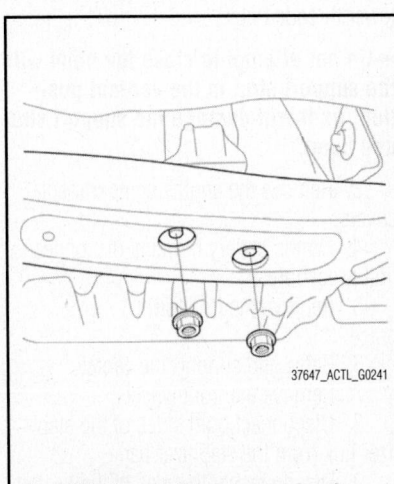

Fig. 168 Remove the nuts securing the lower transmission mount

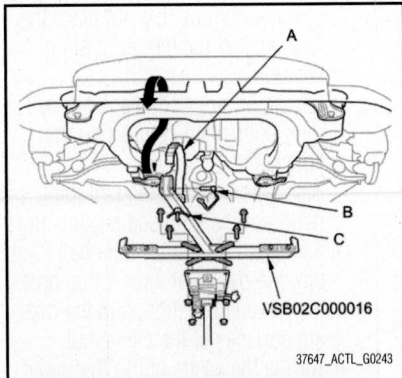

Fig. 169 Attach the subframe adapter to the front subframe by looping the strap (A) over the front of the front subframe, then secure the strap with the stop (B), then tighten the wing nut (C)

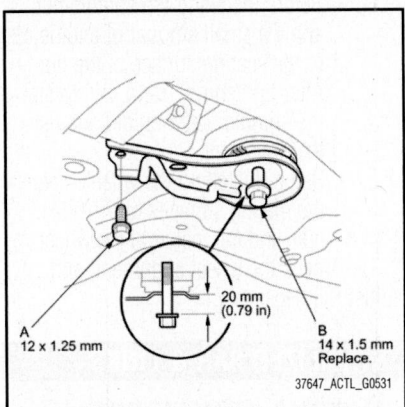

Fig. 170 Remove the stiffener mounting bolt (A) on both sides of the front stiffener (B)

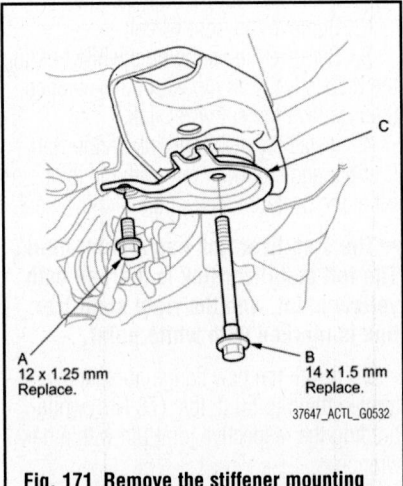

Fig. 171 Remove the stiffener mounting bolt on both sides of the rear stiffener

31. Loosen the subframe mounting bolts to obtain a 0.39 inches (10 mm) distance between the bolt seat and the mounting surface. Do not loosen the mounting bolts more than necessary.

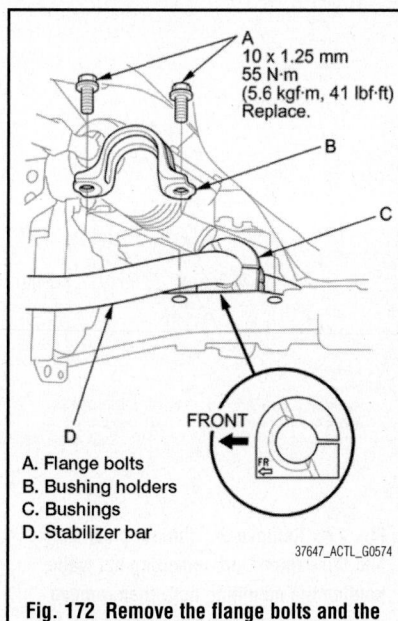

A. Flange bolts
B. Bushing holders
C. Bushings
D. Stabilizer bar

37647_ACTL_G0574

Fig. 172 Remove the flange bolts and the bushing holders, then remove the bushings

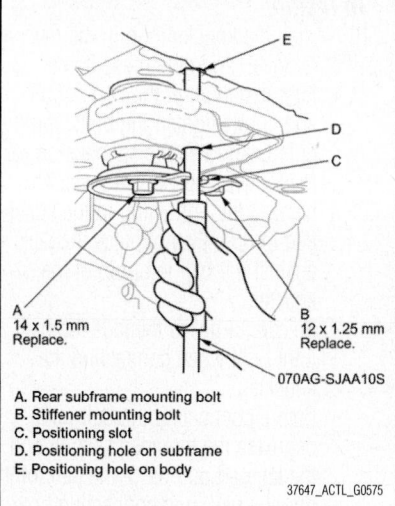

A. Rear subframe mounting bolt
B. Stiffener mounting bolt
C. Positioning slot
D. Positioning hole on subframe
E. Positioning hole on body

37647_ACTL_G0575

Fig. 173 Insert the 15.7 mm side of the subframe alignment pin through the positioning slot on the rear stiffener, through the positioning hole on the front subframe, and into the positioning hole on the body

32. Remove the stiffener mounting bolt on both sides of the rear stiffener.

33. Loosen the subframe mounting bolts to obtain a 1.18 inches (30 mm) distance between the bolt seat and the mounting surface. Do not loosen the mounting bolts more than necessary.

34. Lower the transmission jack slowly with the front subframe adapter until the front subframe has dropped about 1.18 inches (30 mm).

➡**Do not lower the front subframe beyond the loosened subframe mounting bolts clearance.**

35. Remove the flange bolts and the bushing holders, then remove the bushings.

36. Move the stabilizer bar toward the passenger's side, and remove the stabilizer bar.

To install:

37. Install the stabilizer bar noting the following:
  • Note the right and left direction of the stabilizer bar.
  • Note the direction of installation for the bushings.
  • Do not set the bushings on the bent or curved part of the stabilizer bar.

38. Loosely install the new right rear subframe mounting bolt and the new stiffener mounting bolt; insert the 15.7 mm side of the subframe alignment pin (070AG-SJAA10S) through the positioning slot on the rear stiffener, through the posi-

tioning hole on the front subframe, and into the positioning hole on the body, then tighten the right rear subframe mounting bolt.

39. Loosely install the new left rear subframe mounting bolt and the new stiffener mounting bolt with the same procedure as the right rear using the subframe alignment pin.

40. Loosely install the new subframe mounting bolt and the stiffener mounting bolt to both front stiffeners.

41. Tighten the right rear subframe mounting bolt to 76 ft. lbs. (103 Nm).

42. Tighten the left rear subframe mounting bolt to 76 ft. lbs. (103 Nm).

43. Tighten the front side subframe mounting bolts to 76 ft. lbs. (103 Nm).

➡**Check all of the subframe mounting bolts, and retighten if necessary.**

44. Tighten the stiffener mounting bolts to 69 ft. lbs. (93 Nm).

➡**Before tightening the stiffener mounting bolts, check that the positioning holes and slot are aligned using the subframe alignment pin.**

45. Install all of the removed parts in the reverse order of removal, and note these items:
  • If the center guide is in place, use it to determine the steering joint installation angle.
  • If the center guide is gone, check the steering joint installation angle.

• Check the steering wheel installation.
• When connecting the front and rear engine mount, first lightly tighten the mounting bolts, supporting the engine/transmission with a floor jack, then remove the jack, and tighten the bolts to the specified torque.
• Before installing the wheel, clean the mating surfaces of the brake disc and inside of the wheel.

46. Do the battery terminal reconnection procedure, then turn the ignition switch to ON (II) and check that the SRS indicator should come on for about 6 seconds and then go off.

47. Check the wheel alignment, and adjust it if necessary.

STEERING KNUCKLE

REMOVAL & INSTALLATION

See Figures 174 and 175.

1. Raise and support the vehicle.
2. Remove the wheel nuts and the front wheel.
3. Remove the brake hose bracket mounting bolt.
4. Remove the brake caliper bracket mounting bolts, then remove the caliper assembly from the knuckle. To prevent damage to the caliper assembly or the brake hose, use a short piece of wire to hang the caliper assembly from the undercarriage. Do not twist the brake hose excessively.
5. Remove the wheel speed sensor harness bracket and the wheel speed sensor from the knuckle.

➡**Do not disconnect the wheel speed sensor connector.**

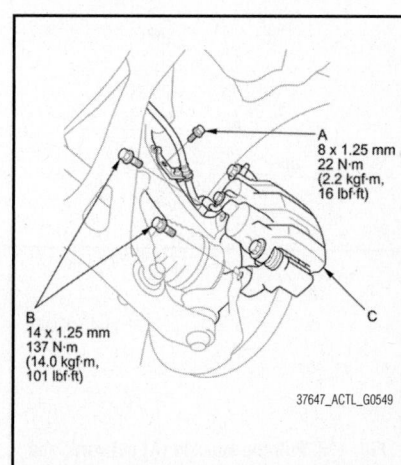

37647_ACTL_G0549

Fig. 174 Remove the brake hose bracket mounting bolt (A)

6. Pry up the stake on the spindle nut, then remove the nut.

7. Remove the front brake disc.

8. Check the front hub for damage and cracks.

9. Remove the cotter pin from the tie-rod end ball joint, then remove the nut.

10. Disconnect the tie-rod end ball joint from the knuckle using the ball joint thread protector and the ball joint remover.

11. Remove the cotter pin from the knuckle ball joint, then remove the castle nut.

12. Disconnect the knuckle ball joint from the lower arm using the ball joint thread protector and the ball joint remover.

➡Be careful not to damage the ball joint boot when installing the remover. Do not force or hammer on the lower arm, or pry between the lower arm and the knuckle. You could damage the ball joint.

13. Remove the cotter pin from the upper arm ball joint, then remove the castle nut.

14. Disconnect the upper arm ball joint from the knuckle using the ball joint thread protector and the ball joint remover.

15. Pull the knuckle outward, and separate the outboard joint from the front hub using a soft face hammer, then remove the knuckle/hub.

➡Do not pull the driveshaft end outward. The driveshaft inboard joint may come apart. During installation, apply grease to the mating surfaces of the wheel bearing and the driveshaft outboard joint.

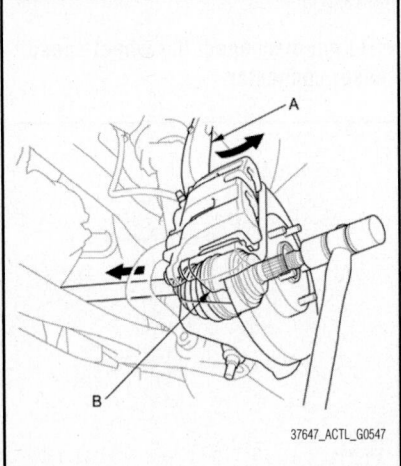

Fig. 175 Pull the knuckle (A) outward, and separate the outboard joint (B) from the front hub using a soft face hammer

To install:

16. Install the knuckle/hub in the reverse order of removal, and note these items:

• First install all of the components, and lightly tighten the bolts and the nuts, then raise the suspension to load it with the vehicle's weight before fully tightening to the specified torque. Do not place the jack against the ball joint pin of the knuckle.

• Be careful not to damage the ball joint boot when connecting the knuckle.

• Before connecting the ball joint, degrease the threaded section and the tapered portion of the ball joint pin, the ball joint connecting hole, the threaded section, and the mating surfaces of the castle nut.

• Torque the castle nut to the lower torque specification, then tighten it only far enough to align the slot with the ball joint pin hole. Do not align the castle nut by loosening it.

• Use a new spindle nut on reassembly.

• Before installing the spindle nut, apply a small amount of engine oil to the seating surface of the nut. After tightening, use a drift to stake the spindle nut shoulder against the driveshaft.

• Before installing the brake disc, clean the mating surfaces of the front hub and the inside of the brake disc.

• Before installing the wheel, clean the mating surfaces of the brake disc and the inside of the wheel.

17. Check the wheel alignment, and adjust it if necessary.

STRUTS (DAMPER/SPRING)

REMOVAL & INSTALLATION

See Figures 176 through 179.

1. Raise and support the vehicle.

2. Remove the front wheel.

3. Remove the wheel speed sensor harness bracket mounting bolt.

4. Remove the damper pinch bolt and the damper fork mounting nut while holding the mounting bolt, then remove the damper fork from the damper and the lower arm.

5. Remove the front strut brace mounting nuts.

6. Remove the damper mounting nuts from the top of the damper. Do not let the damper/spring drop down under its own weight.

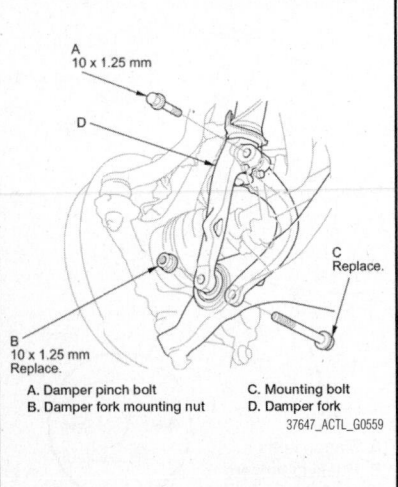

A. Damper pinch bolt
B. Damper fork mounting nut
C. Mounting bolt
D. Damper fork

Fig. 176 Remove the damper pinch bolt and the damper fork mounting nut while holding the mounting bolt, then remove the damper fork from the damper and the lower arm

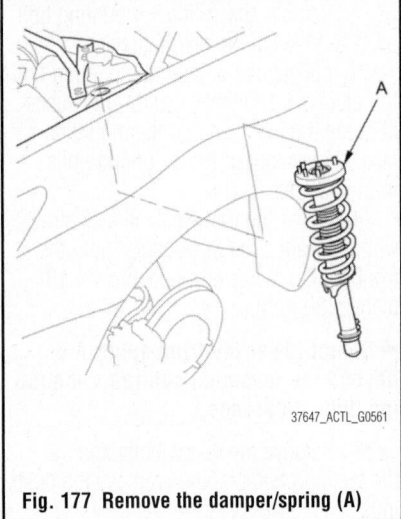

Fig. 177 Remove the damper/spring (A)

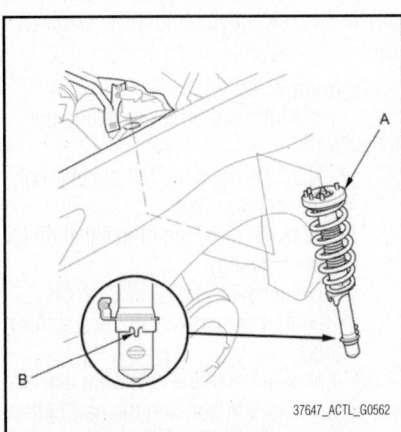

Fig. 178 Position the damper/spring (A) in the body with the aligning tab (B) facing inside

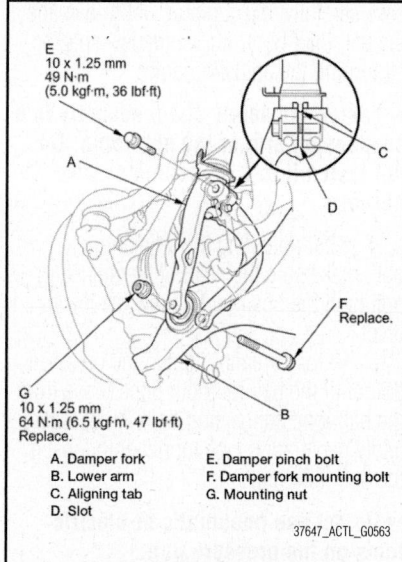

Fig. 179 Install the damper fork over the driveshaft and onto the lower arm. Install the aligning tab on the damper unit into the slot of the damper fork

7. Remove the damper/spring.

To install:

8. Position the damper/spring in the body with the aligning tab facing inside.

9. Loosely install the new damper mounting nuts to the top of the damper.

10. Loosely install the front strut brace mounting nuts.

11. Install the damper fork over the driveshaft and onto the lower arm. Install the aligning tab on the damper unit into the slot of the damper fork.

12. Loosely install the damper pinch bolt into the damper fork.

13. Connect the damper fork and the lower arm with the new damper fork mounting bolt, then lightly tighten the new mounting nut.

14. Place a floor jack under the lower arm, and raise the suspension to load it with the vehicle's weight.

15. Tighten the damper pinch bolt to 36 ft. lbs. (49 Nm) and the damper fork mounting nut while holding the mounting bolt to 47 ft. lbs. (64 Nm).

16. Tighten the damper mounting nuts to 41 ft. lbs. (55 Nm) and front strut brace mounting nuts on top of the damper to 16 ft. lbs. (22 Nm).

17. Install the wheel speed sensor harness bracket.

18. Clean the mating surfaces of the brake disc and the inside of the wheel, then install the front wheel.

19. Check the wheel alignment, and adjust it if necessary.

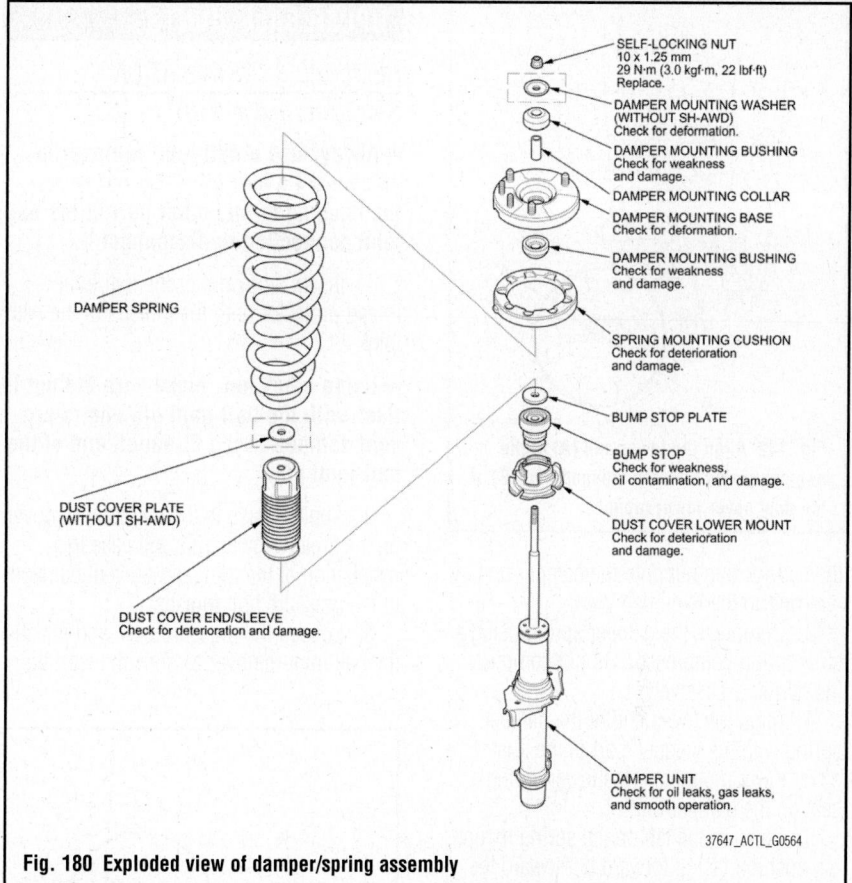

Fig. 180 Exploded view of damper/spring assembly

OVERHAUL

Disassembly

See Figure 181.

1. Compress the damper spring, then remove the self-locking nut while holding the damper shaft with a hex wrench. Do not compress the damper spring more than is necessary to remove the self-locking nut.

2. Release the pressure from the strut spring compressor, then disassemble the damper as shown in the Exploded View.

Inspection

1. Reassemble all the parts, except for the damper spring.

2. Compress the damper assembly by hand, and check for smooth operation through a full stroke, both compression and extension. The damper should extend smoothly and constantly when compression is released. If it does not, the gas is leaking and the damper should be replaced.

3. Check for oil leaks, abnormal noises, and binding during these tests.

Reassembly

See Figures 181 through 183.

1. Install the spring mounting cushion on the damper mounting base by aligning the tab and notch.

2. Install all the parts except the damper mounting washer (Without SH-AWD) and

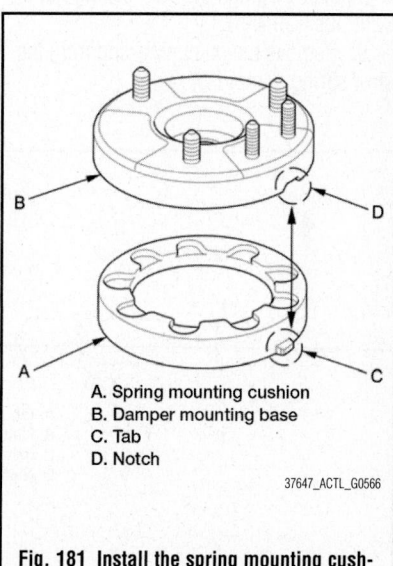

A. Spring mounting cushion
B. Damper mounting base
C. Tab
D. Notch

Fig. 181 Install the spring mounting cushion on the damper mounting base by aligning the tab and notch

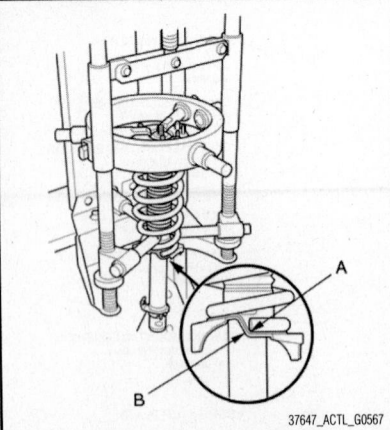

Fig. 182 Align the lower end (A) of the damper spring with the stepped part (B) of the dust cover lower mount

the self-locking nut onto the damper unit by referring to the Exploded View.

3. Compress the damper spring using a strut spring compressor. Do not compress the spring excessively.

4. Align the lower end of the damper spring with the stepped part of the dust cover lower mount and the lower spring seat on the damper unit.

5. Position the tab on the spring mounting cushion facing forward but toward the inside of the vehicle.

6. Align the angle of the stud on the damper mounting base with the aligning tab on the bottom of the damper unit.

7. Install the damper mounting washer (Without SH-AWD) and the new self-locking nut.

8. Hold the damper shaft with a hex wrench, and tighten the self-locking nut to 22 ft. lbs. (29 Nm).

9. Remove the damper/spring from the strut spring compressor.

UPPER BALL JOINTS

REMOVAL & INSTALLATION

See Figures 184 and 185.

➡**Always use a ball joint remover to disconnect a ball joint. Do not strike the housing or any other part of the ball joint connection to disconnect it.**

1. Install a hex nut or the ball joint thread protector onto the threads of the ball joint.

➡**Using a hex nut, make sure the nut is flush with the ball joint pin end to prevent damage to the threaded end of the ball joint pin.**

2. Apply grease to the ball joint remover on the areas shown. This will ease the installation of the tool, and prevent damage to the pressure bolt threads.

3. Loosen the pressure bolt, and install the ball joint remover as shown. Insert the

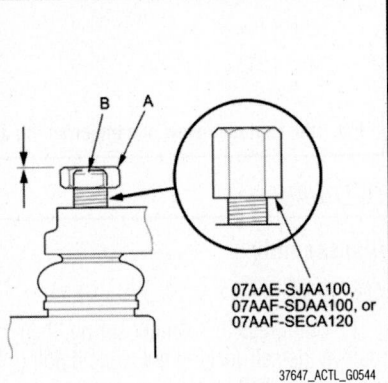

07AAE-SJAA100,
07AAF-SDAA100, or
07AAF-SECA120

Fig. 184 Install a hex nut (A) or the ball joint thread protector onto the threads of the ball joint (B)

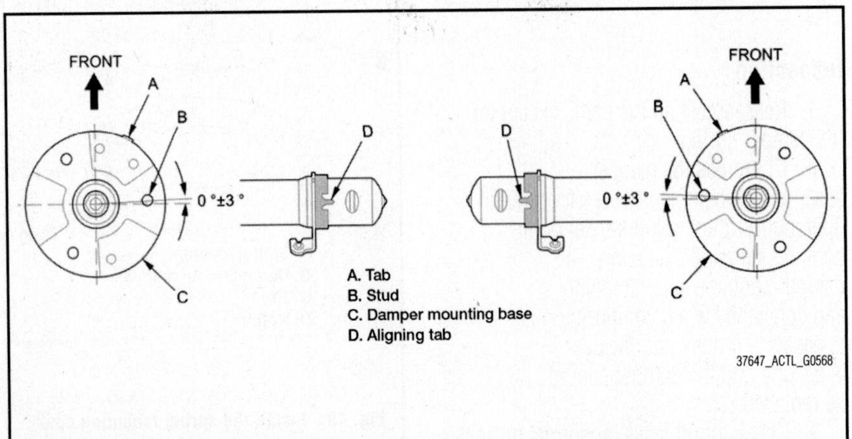

jaws carefully, making sure not to damage the ball joint boot. Adjust the jaw spacing by turning the adjusting bolt.

➡**Fasten the safety chain securely to a suspension arm or the subframe. Do not fasten it to a brake line or wire harness.**

4. After adjusting the adjusting bolt, make sure the head of the adjusting bolt is in the position to allow the jaw to pivot.

5. With a wrench, tighten the pressure bolt until the ball joint pin pops loose from the ball joint connecting hole. If necessary, apply penetrating type lubricant to loosen the ball joint pin.

➡**Do not use pneumatic or electric tools on the pressure bolt.**

6. Remove the ball joint remover, then remove the nut or the ball joint thread protector from the end of the ball joint pin, and pull the ball joint out of the ball joint connecting hole. Inspect the ball joint boot, and replace it if damaged.

UPPER CONTROL ARMS

REMOVAL & INSTALLATION

See Figure 186.

1. Raise and support the vehicle.
2. Remove the front wheel.
3. Remove the front damper/spring.
4. Remove the cotter pin from the upper arm ball joint, then remove the castle nut.
5. Disconnect the upper arm ball joint from the knuckle using the ball joint thread protector and the ball joint remover.
6. Remove the upper arm mounting bolts, then remove the upper arm.
7. Install the upper arm in the reverse order of removal, and note these items:

- First install all of the components, and lightly tighten the bolts and the nuts, then raise the suspension to load it with the vehicle's weight before fully tightening it to the specified torque. Tighten the upper arm mounting bolts to 23 ft. lbs. (31 Nm).

✳✳ CAUTION

Do not place the jack against the ball joint pin of the knuckle.

- Be careful not to damage the ball joint boot when connecting the knuckle.

FRONT

FRONT

0°±3°

0°±3°

A. Tab
B. Stud
C. Damper mounting base
D. Aligning tab

37647_ACTL_G0568

Fig. 183 Position the tab on the spring mounting cushion facing forward but toward the inside of the vehicle

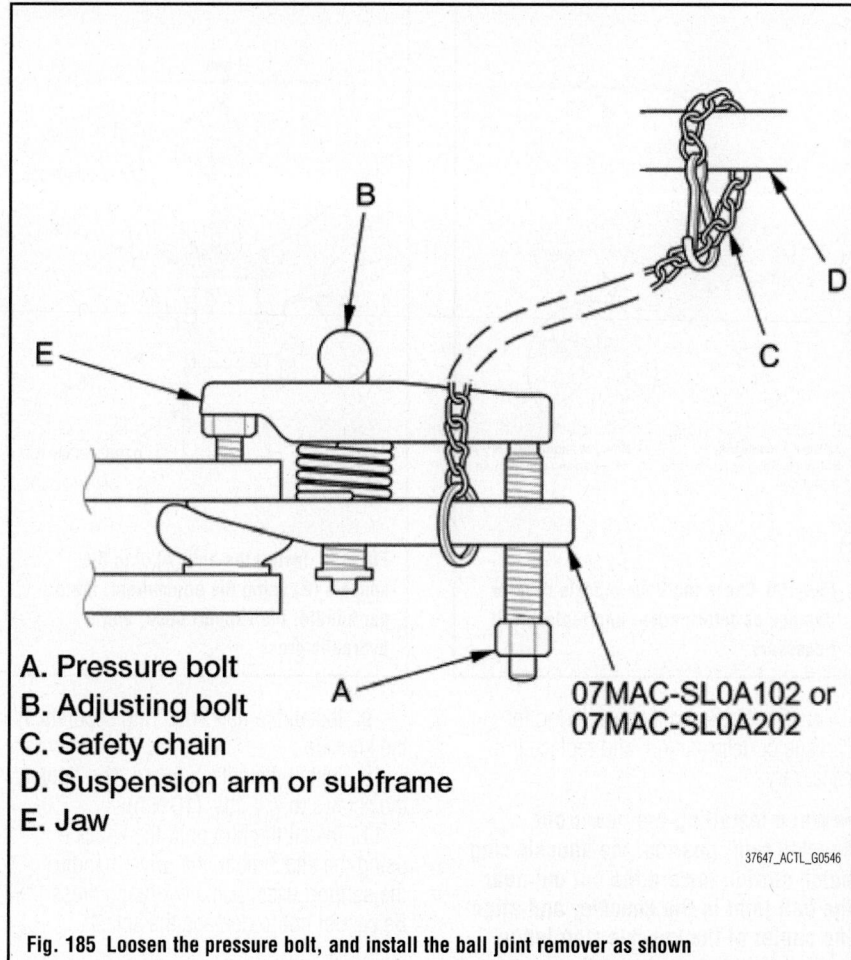

A. Pressure bolt
B. Adjusting bolt
C. Safety chain
D. Suspension arm or subframe
E. Jaw

07MAC-SL0A102 or
07MAC-SL0A202

37647_ACTL_G0546

Fig. 185 Loosen the pressure bolt, and install the ball joint remover as shown

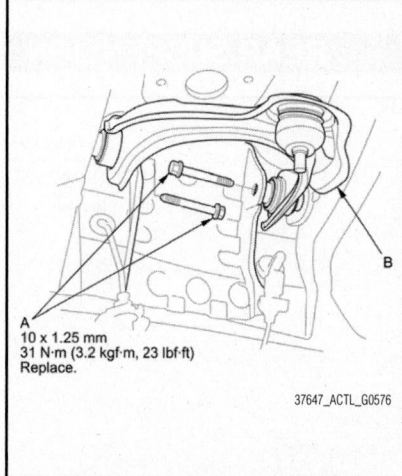

A
10 x 1.25 mm
31 N·m (3.2 kgf·m, 23 lbf·ft)
Replace.

37647_ACTL_G0576

Fig. 186 Remove the upper arm mounting bolts (A), then remove the upper arm (B)

- Before connecting the ball joint, degrease the threaded section and the tapered portion of the ball joint pin, the ball joint connecting hole, the threaded section, and the mating surfaces of the castle nut.
- Torque the castle nut to 33–38 ft. lbs. (44–52 Nm), then tighten it

only far enough to align the slot with the ball joint pin hole. Do not align the castle nut by loosening it.
- Before installing the wheel, clean the mating surfaces of the brake disc and the inside of the wheel.

8. Check the wheel alignment, and adjust it if necessary.

WHEEL HUBS & BEARINGS

REMOVAL & INSTALLATION

See Figures 187 through 191.

1. Remove the knuckle/hub.
2. Separate the hub from the knuckle using the hub disassembly/assembly pin and a hydraulic press. Hold the knuckle with the attachment of the hydraulic press or equivalent tool. Be careful not to damage or deform the splash guard. Hold onto the hub to keep it from falling when pressed clear.
3. Press the wheel bearing inner race off of the hub using the hub disassembly/assembly pin, a commercially available bearing separator, and a press.

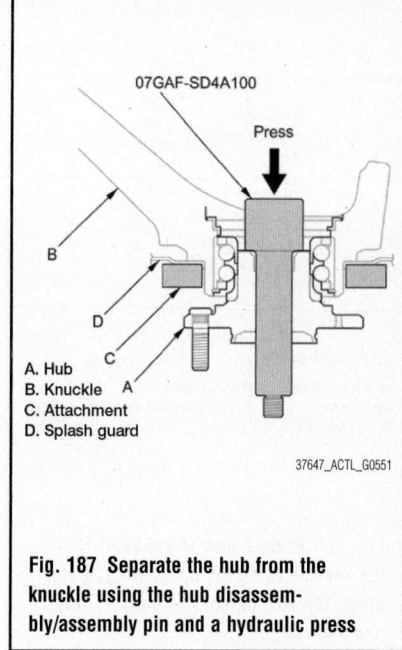

A. Hub
B. Knuckle
C. Attachment
D. Splash guard

07GAF-SD4A100
Press

37647_ACTL_G0551

Fig. 187 Separate the hub from the knuckle using the hub disassembly/assembly pin and a hydraulic press

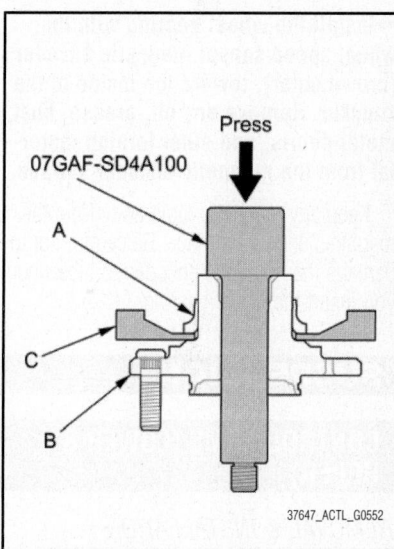

07GAF-SD4A100
Press

37647_ACTL_G0552

Fig. 188 Press the wheel bearing inner race (A) off of the hub (B) using the hub disassembly/assembly pin, a commercially available bearing separator (C), and a press

4. Remove the snap ring and the splash guard from the knuckle.
5. Press the wheel bearing out of the knuckle using the attachment, the driver handle, and a press.
6. Wash the knuckle and the hub thoroughly in high flash point solvent before reassembly.
7. Press a new wheel bearing into the knuckle using the old bearing, a steel plate, the attachment, the support base, and a press.

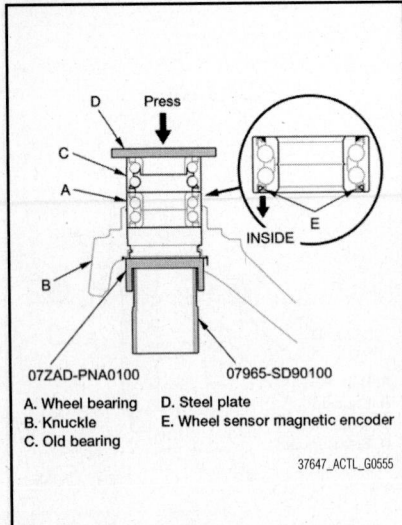

Fig. 189 Press a new wheel bearing into the knuckle using the old bearing, a steel plate, the attachment, the support base, and a press

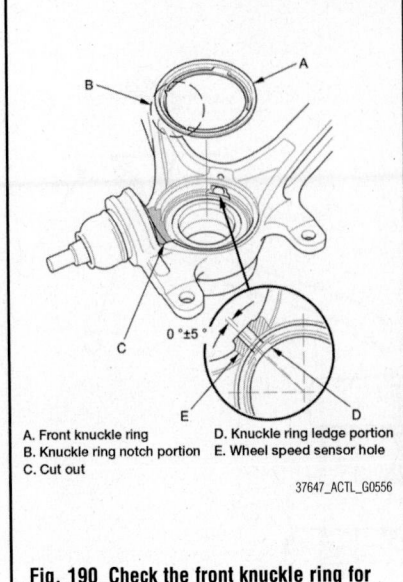

Fig. 190 Check the front knuckle ring for damage or deformation, and replace it if necessary

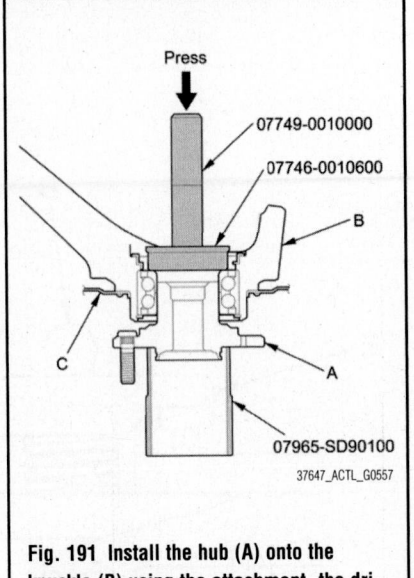

Fig. 191 Install the hub (A) onto the knuckle (B) using the attachment, the driver handle, the support base, and a hydraulic press

➡️**Install the wheel bearing with the wheel speed sensor magnetic encoder (brown color), toward the inside of the knuckle. Remove any oil, grease, dust, metal debris, and other foreign material from the magnetic encoder surface.**

Keep any magnetic tools away from the magnetic encoder surface. Be careful not to damage the magnetic encoder surface when you insert the wheel bearing.

8. Check the front knuckle ring for damage or deformation, and replace it if necessary.

➡️**When installing the new front knuckle ring, position the knuckle ring notch portion toward the cut out near the ball joint in the knuckle, and align the center of the knuckle ring ledge portion with the center of the wheel speed sensor hole on the knuckle.**

9. Install the new snap ring securely in the knuckle.
10. Install the splash guard, and tighten the screws to 7 ft. lbs. (10 Nm).
11. Install the hub onto the knuckle using the attachment, the driver handle, the support base, and a hydraulic press. Be careful not to damage the splash guard.
12. Install the knuckle/hub.

SUSPENSION

CONTROL ARMS (LOWER & UPPER)

REMOVAL & INSTALLATION

Rear Lower Arm A

1. Raise and support the vehicle.
2. Remove the rear wheel.
3. Remove the parking brake cable mounting bolt.
4. Remove the lower arm A mounting bolts, then remove lower arm A.

To install:

➡️**Use new mounting bolts during reassembly.**

5. Install lower arm A in the reverse order of removal, and note these items:
   - First install all of the components, and lightly tighten the bolts, then raise the suspension to load it with the vehicle's weight before fully tightening to the specified torque.

   - Before installing the wheel, clean the mating surfaces on the brake disc/drum and the inside of the wheel.
6. Check the wheel alignment, and adjust it if necessary.

Rear Lower Arm B

See Figure 192.

1. Raise and support the vehicle.
2. Remove the rear wheel.
3. Remove the lower arm B mounting bolts, then remove lower arm B.

To install:

➡️**Use new mounting bolts during reassembly.**

4. Install lower arm B in the reverse order of removal, and note these items:
   - First install all of the components, and lightly tighten the bolts, then raise the suspension to load it with

REAR SUSPENSION

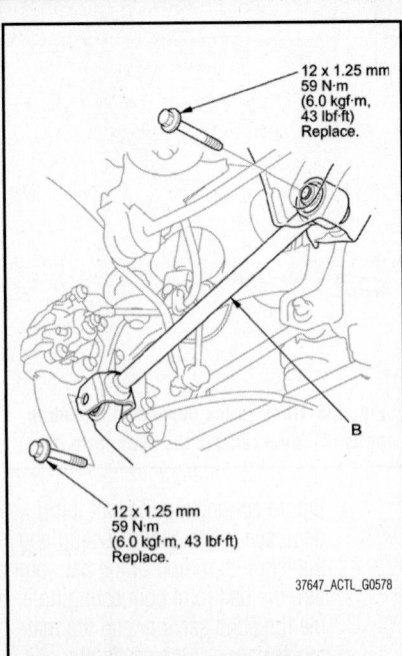

Fig. 192 Remove the lower arm B mounting bolts, then remove lower arm B

the vehicle's weight before fully tightening to the specified torque.

- Before installing the wheel, clean the mating surfaces on the brake disc/drum and the inside of the wheel.

5. Check the wheel alignment, and adjust it if necessary.

Rear Upper Arm

See Figure 193.

➡**Special Tools Required:**

- Ball Joint Thread Protector, 14 mm 07AAE-SJAA100
- Ball Joint Remover, 32 mm 07MAC-SL0A102

1. Raise and support the vehicle.
2. Remove the rear wheel.
3. Remove the rear damper/spring.
4. Release the parking brake lever fully.
5. Remove the brake caliper bracket mounting bolts, then remove the caliper assembly from the knuckle.

➡**To prevent damage to the caliper assembly or the brake hose, use a short piece of wire to hang the caliper assembly from the undercarriage. Do not twist the brake hose excessively.**

6. Remove the lock pin from the upper arm ball joint, then remove the castle nut.

7. Disconnect the upper arm ball joint from the knuckle using the ball joint thread protector and the ball joint remover.

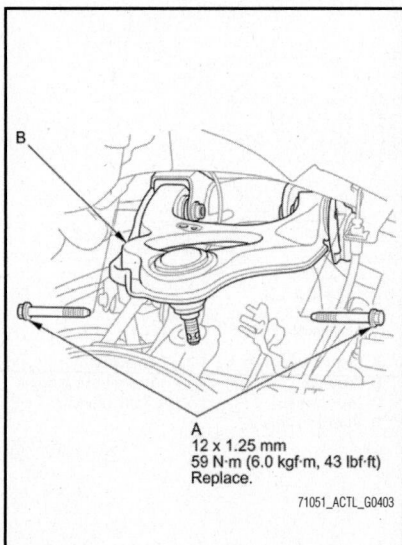

A
12 x 1.25 mm
59 N·m (6.0 kgf·m, 43 lbf·ft)
Replace.

71051_ACTL_G0403

Fig. 193 Remove the upper arm mounting bolts (A), then remove the upper arm (B).

8. Remove the wheel speed sensor harness bracket.
9. Remove the upper arm mounting bolts, then remove the upper arm.

To install:

10. Install the upper arm in the reverse order of removal, and note these items:

a. Use new mounting bolts during reassembly.

b. First install all of the components, and lightly tighten the bolts and the nuts, then raise the suspension to load it with the vehicle's weight before fully tightening to the specified torque.

c. Be careful not to damage the ball joint boot when connecting the knuckle.

d. Make sure the are reinstalled between the brake caliper bracket and the knuckle, if they are removed.

e. Before connecting the ball joint, degrease the threaded section and the tapered portion of the ball joint pin, the ball joint connecting hole, the threaded

section, and the mating surfaces of the castle nut.

f. During installation, to connect the ball joint, position a floor jack under the connecting point of the knuckle and lower arm A, and raise the suspension with the jack.

g. Torque the castle nut to the lower torque specification, then tighten it only far enough to align the slot with the ball joint pin hole. Do not align the castle nut by loosening it.

h. Before installing the wheel, clean the mating surfaces between the brake disc/drum and the inside of the wheel.

11. Check the wheel alignment, and adjust it if necessary.

CONTROL RODS (CONTROL ARM)

REMOVAL & INSTALLATION

See Figure 194.

1. Raise and support the vehicle.
2. Remove the rear wheel.

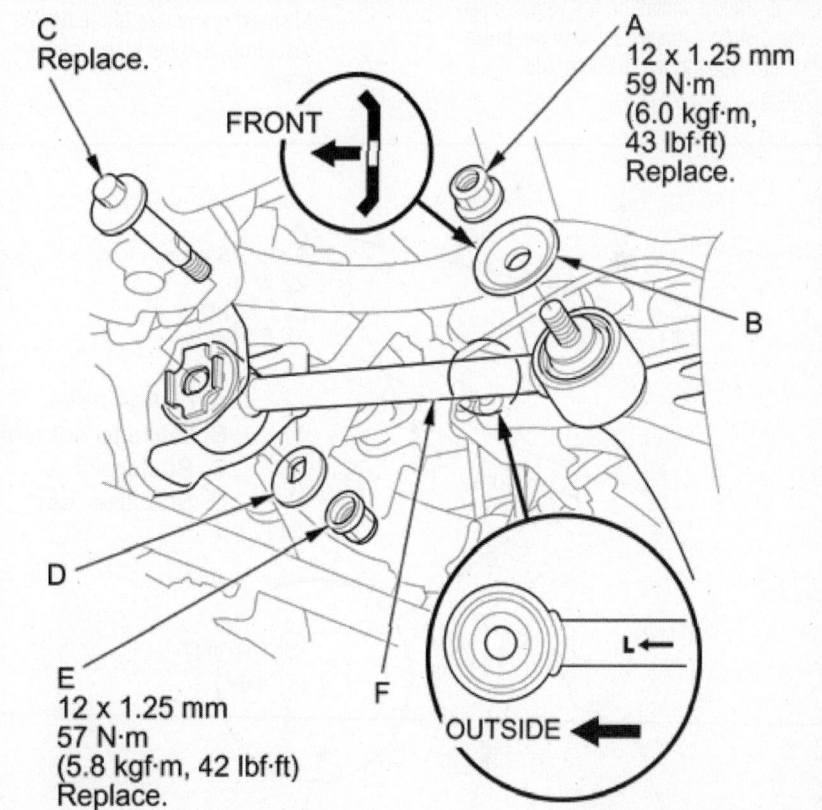

A. Control arm mounting self-locking nut
B. Washer
C. Cam adjusting bolt
D. Adjusting cam plate
E. Self-locking nut
F. Control arm

71051_ACTL_G0404

Fig. 194 Removing the control rod (control arm)

3. Remove the control arm mounting self-locking nut and the washer from the knuckle side.

4. Mark the cam positions of the adjusting bolt and the adjusting cam plate with the frame.

5. Remove the self-locking nut while holding the adjusting bolt, then remove the adjusting cam plate, the adjusting bolt, and the control arm.

To install:

6. Install the control arm in the reverse order of removal, and note these items:

a. Use a new self-locking nuts and new bolts during reassembly.

b. First install all of the components, and lightly tighten the bolts and the nuts, then raise the suspension to load it with the vehicle's weight before fully tightening to the specified torque.

c. Position the extended surfaces of the cam on the adjusting bolt and the adjusting cam plate facing down.

d. Align the cam positions of the adjusting bolt and the adjusting cam plate with the marked positions on the frame when tightening the self-locking nut.

e. Before installing the wheel, clean the mating surfaces between the brake disc/drum and the inside of the wheel.

7. Check the wheel alignment, and adjust it if necessary.

STABILIZER BAR & LINKS

REMOVAL & INSTALLATION

Stabilizer Bar

See Figure 195.

1. Raise and support the vehicle.
2. Remove the rear wheels.
3. Disconnect both stabilizer links from the stabilizer bar.
4. Remove the flange bolts and the bushing holders, then remove the bushings and the stabilizer bar.

To install:

5. Install the stabilizer bar in the reverse order of removal, and note these items:

- Note the right and left direction of the stabilizer bar.
- Note the direction of installation for the bushing.
- Do not set the bushings on the bent or curved part of the stabilizer bar.
- Before installing the wheel, clean the mating surfaces of the brake disc/drum and the inside of the wheel.

Stabilizer Link

See Figure 196.

1. Raise and support the vehicle.
2. Remove the rear wheel.
3. Remove the self-locking nuts while holding the respective joint pin with a hex wrench, then remove the stabilizer link.

To install:

4. Install the stabilizer link on the stabilizer bar and the brake hose bracket with the joint pins set at the center of their range of movement.

➡**The stabilizer link has a paint mark. The paint mark indicates the difference between the left and right stabilizer links. Install the end of the stabilizer link with the paint mark in the upper position.**

5. Install the new self-locking nuts, and tighten them to the specified torque while holding the respective joint pin with a hex wrench.

6. Clean the mating surfaces of the brake disc/drum and the inside of the wheel, then install the rear wheel.

7. Test-drive the vehicle.

8. After 5 minutes of driving, tighten the self-locking nuts again to the specified torque.

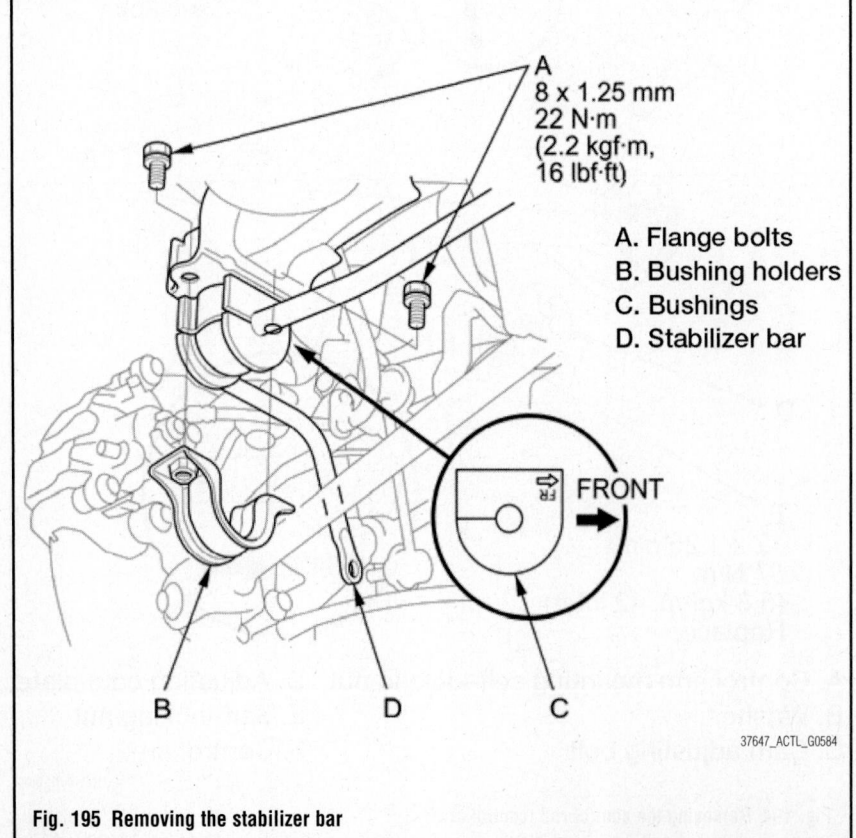

A
8 x 1.25 mm
22 N·m
(2.2 kgf·m,
16 lbf·ft)

A. Flange bolts
B. Bushing holders
C. Bushings
D. Stabilizer bar

FRONT

37647_ACTL_G0584

Fig. 195 Removing the stabilizer bar

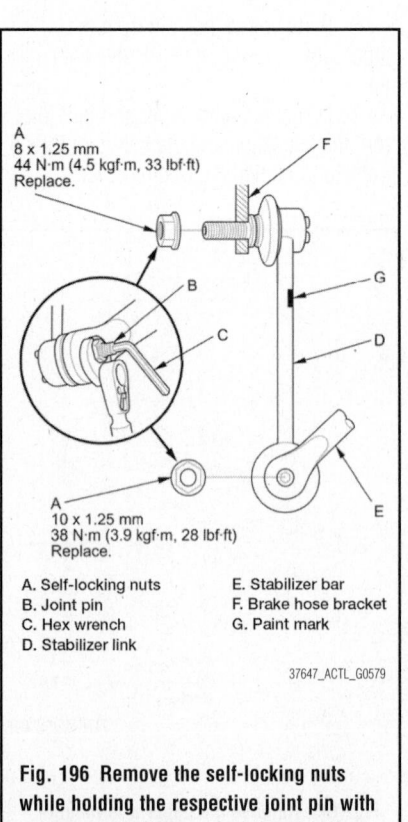

A
8 x 1.25 mm
44 N·m (4.5 kgf·m, 33 lbf·ft)
Replace.

A
10 x 1.25 mm
38 N·m (3.9 kgf·m, 28 lbf·ft)
Replace.

A. Self-locking nuts   E. Stabilizer bar
B. Joint pin   F. Brake hose bracket
C. Hex wrench   G. Paint mark
D. Stabilizer link

37647_ACTL_G0579

Fig. 196 Remove the self-locking nuts while holding the respective joint pin with a hex wrench, then remove the stabilizer link

STRUTS (DAMPER/SPRING)

REMOVAL & INSTALLATION

See Figures 197 through 200.

1. Raise and support the vehicle.
2. Remove the rear wheel.
3. Remove the brake hose mounting bolt.
4. Remove the rear seat.
5. Remove the damper mounting nuts from the top of the damper.
6. Remove the brake hose bracket mounting nuts, then disconnect the brake hose bracket from the knuckle.
7. Remove the damper lower mounting bolt.
8. Remove the damper/spring by lowering the rear suspension.

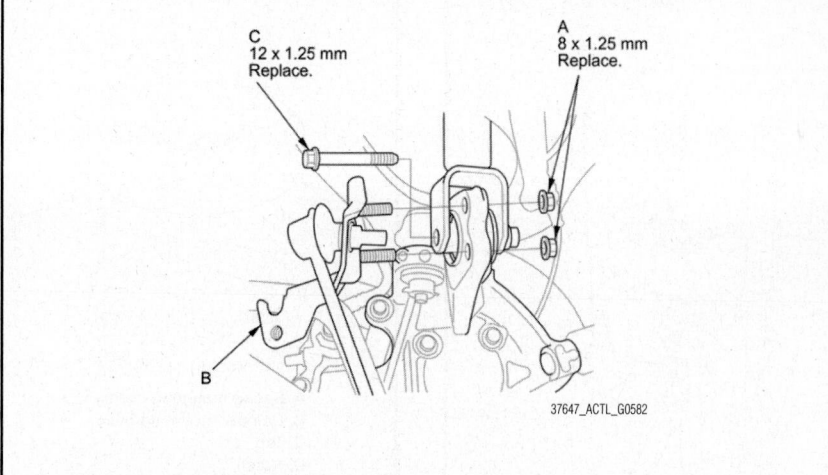

Fig. 199 Remove the brake hose bracket mounting nuts (A), then disconnect the brake hose bracket (B) from the knuckle

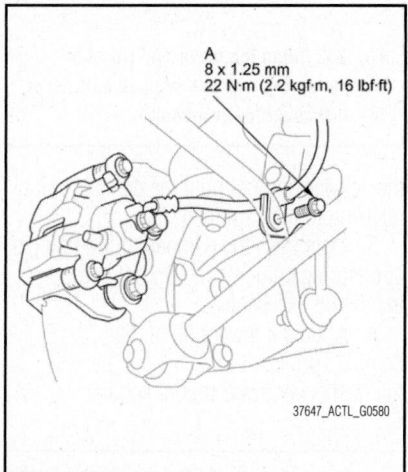

Fig. 197 Remove the brake hose mounting bolt (A)

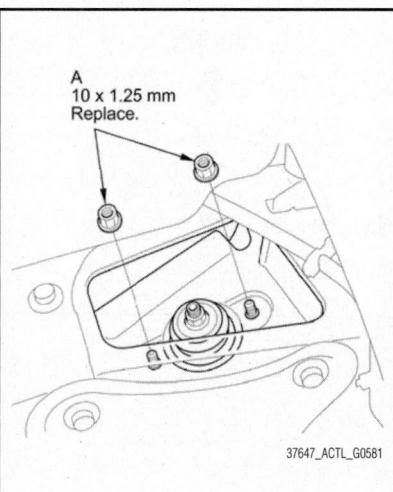

Fig. 198 Remove the damper mounting nuts (A) from the top of the damper

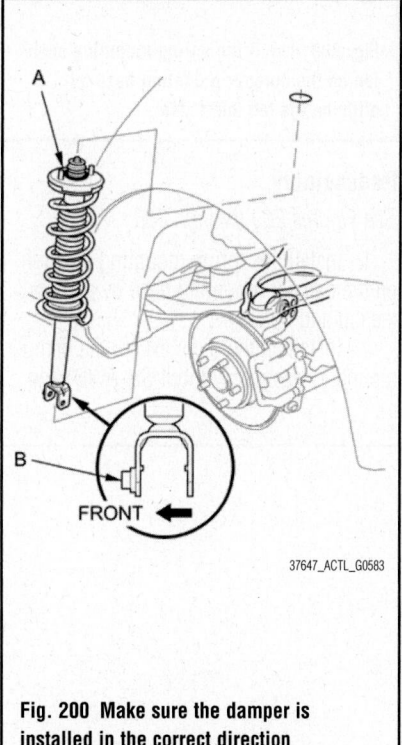

Fig. 200 Make sure the damper is installed in the correct direction

To install:

9. Lower the rear suspension, and position the damper/spring in the body with the welded nut on the bottom of the damper facing forward.

➡**Make sure the damper is installed in the correct direction.**

10. Loosely install the new damper mounting nuts to the top of the damper.
11. Loosely install the new damper lower mounting bolt on the bottom of the damper. Connect the brake hose bracket to the

knuckle, and loosely install the new brake hose bracket mounting nuts.

12. Place a floor jack under the connecting point of the knuckle and lower arm A, and raise the suspension to load it with the vehicle's weight.
13. Tighten the damper lower mounting bolt to 47 ft. lbs. (64 Nm) and the brake hose bracket mounting nuts to 13 ft. lbs. (18 Nm).
14. Tighten the damper mounting nuts on top of the damper to 41 ft. lbs. (55 Nm).
15. Install the brake hose mounting bolt, and tighten the mounting bolt to 16 ft. lbs. (22 Nm).
16. Install the rear seat.
17. Clean the mating surfaces of the brake disc/drum and the inside of the wheel, then install the rear wheel.
18. Check the wheel alignment, and adjust it if necessary.

OVERHAUL

Disassembly

See Figure 201.

1. Compress the damper spring, then remove the self-locking nut while holding the damper shaft with a hex wrench. Do not compress the damper spring more than is necessary to remove the self-locking nut.
2. Release the pressure from the strut spring compressor, then disassemble the damper as shown in the Exploded View.

Inspection

1. Reassemble all the parts, except for the damper spring.
2. Compress the damper assembly by

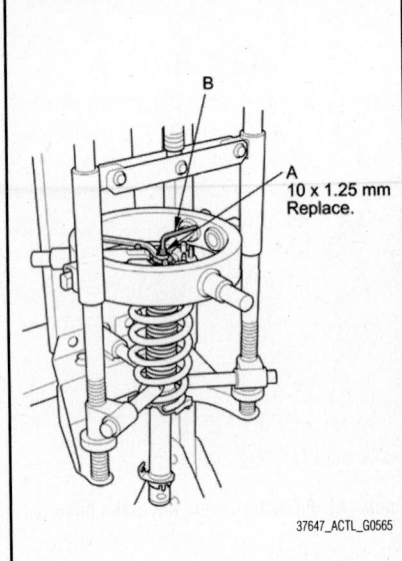

Fig. 201 Compress the damper spring, then remove the self-locking nut (A) while holding the damper shaft with a hex wrench (B)

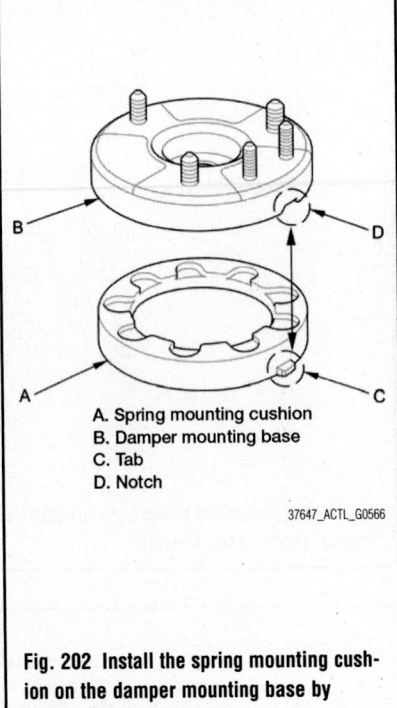

A. Spring mounting cushion
B. Damper mounting base
C. Tab
D. Notch

Fig. 202 Install the spring mounting cushion on the damper mounting base by aligning the tab and notch

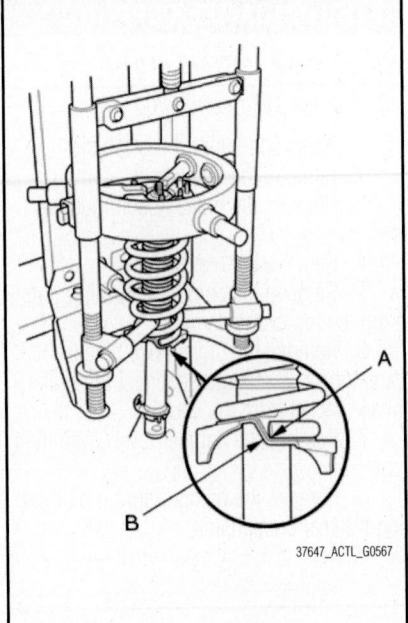

Fig. 203 Align the lower end (A) of the damper spring with the stepped part (B) of the dust cover lower mount

hand, and check for smooth operation through a full stroke, both compression and extension. The damper should extend smoothly and constantly when compression is released. If it does not, the gas is leaking and the damper should be replaced.

3. Check for oil leaks, abnormal noises, and binding during these tests.

Reassembly

See Figures 202 through 206.

1. Install the spring mounting cushion on the damper mounting base by aligning the tab and notch.

2. Install all the parts except the damper mounting washer (Without SH-AWD) and

the self-locking nut onto the damper unit by referring to the Exploded View.

3. Compress the damper spring using a strut spring compressor. Do not compress the spring excessively.

4. Align the lower end of the damper spring with the stepped part of the dust cover lower mount and the

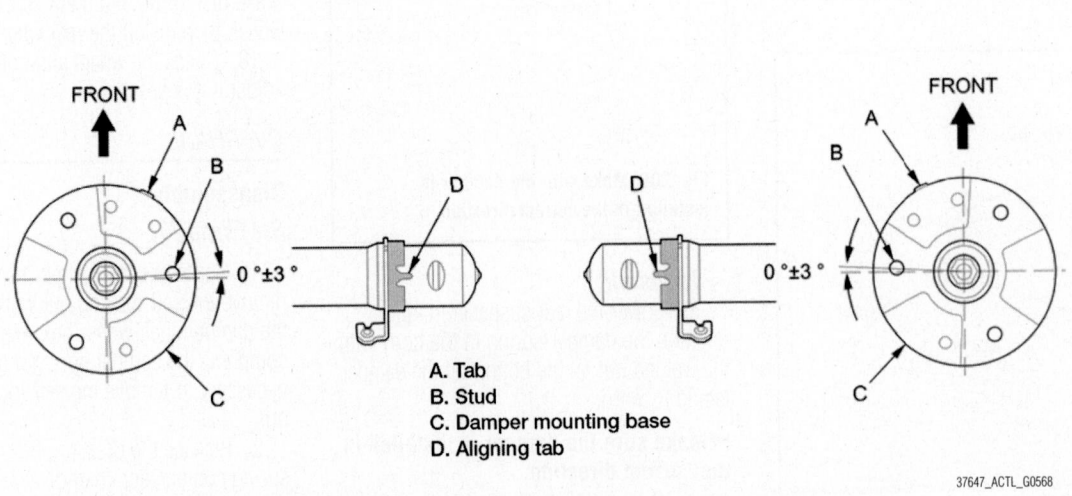

A. Tab
B. Stud
C. Damper mounting base
D. Aligning tab

Fig. 204 Position the tab on the spring mounting cushion facing forward but toward the inside of the vehicle

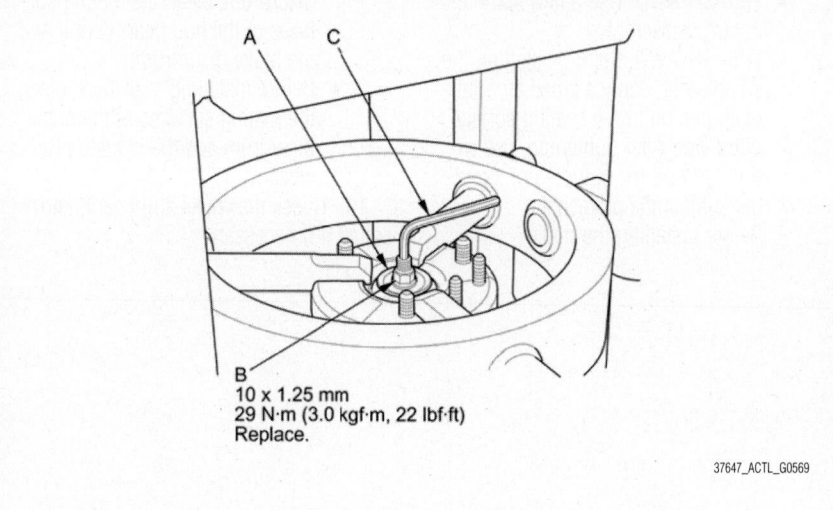

Fig. 205 Install the damper mounting washer (Without SH-AWD) (A) and the new self-locking nut (B)

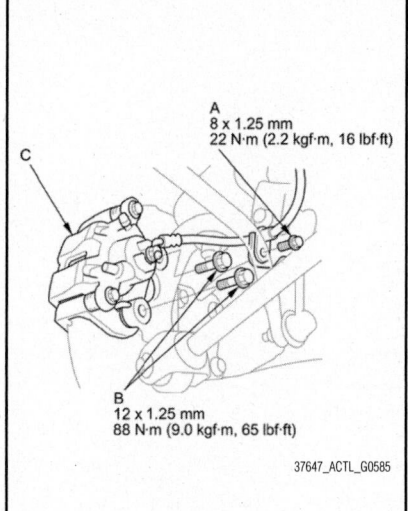

Fig. 206 Remove the brake hose mounting bolt (A) and the brake caliper bracket mounting bolts (B), then remove the caliper assembly (C)

lower spring seat on the damper unit.

5. Position the tab on the spring mounting cushion facing forward but toward the inside of the vehicle.

6. Align the angle of the stud on the damper mounting base with the aligning tab on the bottom of the damper unit.

7. Install the damper mounting washer (Without SH-AWD) and the new self-locking nut.

8. Hold the damper shaft with a hex wrench, and tighten the self-locking nut to 22 ft. lbs. (29 Nm).

caliper assembly from the undercarriage. Do not twist the brake hose excessively.

6. Remove the two washers.

➡**During installation, make sure the washers are installed between the brake caliper bracket and the knuckle.**

7. With SH-AWD: Pry up the stake on the spindle nut, then remove the spindle nut.

8. Remove the rear brake disc/drum.

9. Without SH-AWD: Remove the hub bearing unit and the O-ring.

10. With SH-AWD: Remove the flange bolts, and remove the hub bearing unit by tapping the driveshaft end with a soft face hammer while drawing the hub bearing unit outward.

➡**Do not pull the driveshaft end outward. The driveshaft inboard joint may come apart.**

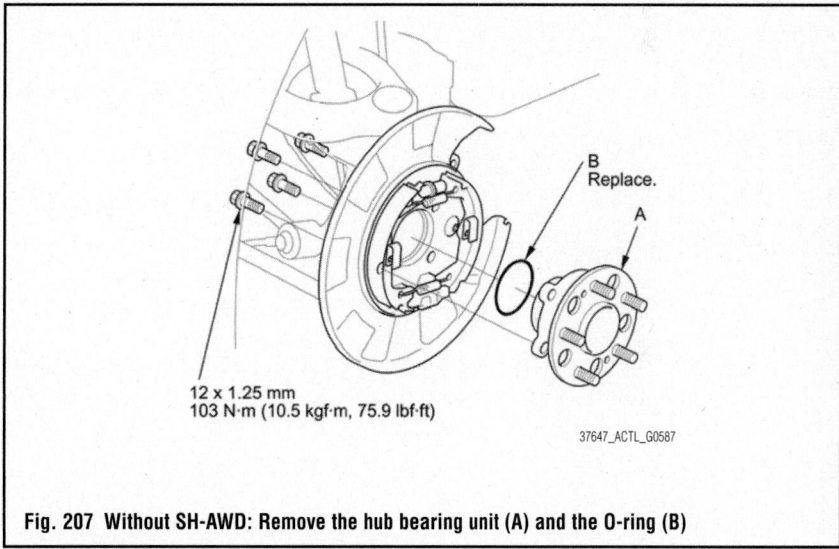

Fig. 207 Without SH-AWD: Remove the hub bearing unit (A) and the O-ring (B)

9. Remove the damper/spring from the strut spring compressor.

WHEEL HUBS & BEARINGS

REMOVAL, REPACKING, & INSTALLATION

See Figures 207 through 209.

1. Raise and support the vehicle.
2. Remove the wheel nuts, and the rear wheel.
3. Release the parking brake lever fully.
4. Remove the brake hose mounting bolt.
5. Remove the brake caliper bracket mounting bolts, then remove the caliper assembly from the knuckle. To prevent damage to the caliper assembly or the brake hose, use a short piece of wire to hang the

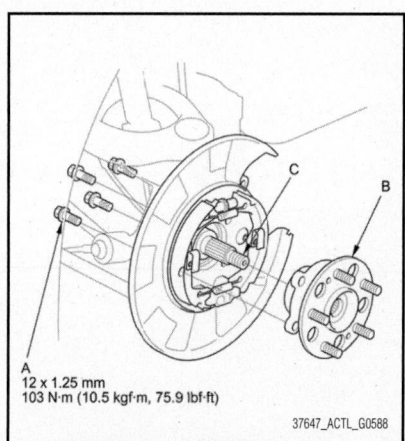

Fig. 208 With SH-AWD: Remove the flange bolts (A), and remove the hub bearing unit (B) by tapping the driveshaft end (C) with a soft face hammer

11. Check the hub bearing unit for damage and cracks.

To install:

12. Install the hub bearing unit in the reverse order of removal, and note these items:

- Without SH-AWD: Use a new O-ring on reassembly.

- With SH-AWD: Use a new spindle nut on reassembly.
- With SH-AWD: Before installing the spindle nut, apply a small amount of engine oil to the seating surface of the nut. After tightening, use a drift to stake the spindle nut shoulder against the driveshaft.
- Before installing the brake disc/drum, clean the mating surfaces of the hub bearing unit and the brake disc/drum.
- Before installing the wheel, clean the mating surfaces of the brake disc/drum and the inside of the wheel.

13. Check the wheel alignment, and adjust it if necessary.

ACURA

TSX

5

BRAKES	5-10

ANTI-LOCK BRAKE SYSTEM (ABS).............5-10
General Information...................5-10
  Precautions..........................5-10
Speed Sensors 5-10
  Removal & Installation..........5-10
BLEEDING THE BRAKE SYSTEM.....................5-11
Bleeding Procedure..................5-11
  Bleeding Procedure 5-11
  Bleeding the ABS System 5-11
FRONT DISC BRAKES 5-12
Brake Caliper...........................5-12
  Removal & Installation..........5-12
Brake Pads 5-12
  Removal & Installation..........5-12
REAR DISC BRAKES 5-13
Brake Caliper...........................5-13
  Removal & Installation..........5-13
Disc Brake Pads 5-13
  Removal & Installation..........5-13

CHASSIS ELECTRICAL	5-14

AIR BAG (SUPPLEMENTAL RESTRAINT SYSTEM)........5-14
General Information...................5-14
  Arming the System 5-15
  Clockspring Centering..........5-15
  Disarming the System...........5-15
  Service Precautions.............5-14

DRIVE TRAIN	5-16

Front Halfshafts.......................5-16
  Removal & Installation..........5-16

ENGINE COOLING	5-20

Electric Engine Fan..................5-21
  Removal & Installation..........5-21
Engine Coolant........................5-20
  Drain & Refill........................5-20
Radiator...................................5-22
  Removal & Installation..........5-22
Thermostat 5-24
  Removal & Installation..........5-24

Water Pump 5-26
  Removal & Installation..........5-26

ENGINE ELECTRICAL	5-26

BATTERY SYSTEM 5-26
Battery....................................5-26
  Battery Reconnect/Relearn Procedure 5-27
  Removal & Installation..........5-26
CHARGING SYSTEM 5-27
Alternator 5-27
  Removal & Installation..........5-27
IGNITION SYSTEM 5-29
Firing Order.............................5-29
Ignition Coils 5-29
  Removal & Installation..........5-29
Ignition Timing.........................5-29
  Inspection & Adjustment 5-29
STARTING SYSTEM 5-30
Starter 5-30
  Removal & Installation..........5-30

ENGINE MECHANICAL	5-31

Accessory Drive Belt System.....5-31
  Adjustment 5-31
  Inspection 5-31
  Removal & Installation..........5-31
Air Cleaner 5-32
  Removal & Installation..........5-32
Camshaft & Bearings 5-32
  Inspection 5-32
  Removal & Installation..........5-35
Cylinder Head 5-36
  Removal & Installation..........5-36
Exhaust System 5-39
  Removal & Installation..........5-39
Intake Manifold 5-41
  Removal & Installation..........5-41
Oil Pan 5-43
  Removal & Installation..........5-43
Oil Pump 5-45
  Removal & Installation..........5-45
Pistons & Rings 5-47
  Positioning 5-47
Rear Main Seal.........................5-47
  Removal & Installation..........5-47

Timing Belt & Sprockets 5-48
  Removal & Installation..........5-48
Timing Belt Front Cover............5-48
  Removal & Installation..........5-48
Valve Lash...............................5-50
  Adjustment 5-50

ENGINE PERFORMANCE & EMISSION CONTROLS	5-51

Accelerator Pedal Position (AAP) Sensor 5-51
  Location...............................5-51
  Removal & Installation..........5-51
Air-Fuel Ratio (AFR) Sensor......5-51
  Location...............................5-51
  Removal & Installation..........5-51
Camshaft Position (CMP) Sensor 5-51
  Removal & Installation..........5-51
Clutch Pedal Position (CPP) Switch 5-52
  Location...............................5-52
  Removal & Installation..........5-54
Component Locations 5-51
Crankshaft Position (CKP) Sensor 5-54
  CKP Pattern Clear/CKP Pattern Learn.....................000
  Removal & Installation..........5-54
Engine Control Module (ECM) 5-55
  Location...............................5-55
  Removal & Installation..........5-55
Engine Coolant Temperature (ECT) Sensor 5-56
  Removal & Installation..........5-56
EVAP Canister 5-57
  Location...............................5-57
  Removal & Installation..........5-57
EVAP Canister Purge Control Valve 5-59
  Location...............................5-59
  Removal & Installation..........5-59
Exhaust Gas Recirculation (EGR) Valve 5-59
  Location...............................5-59
  Removal & Installation..........5-59

Heated Oxygen (HO2S) Sensor .5-60
    Removal & Installation..........5-60
Input Speed Sensor (ISS)..........5-60
    Removal & Installation..........5-60
Knock Sensor (KS)....................5-61
    Operation..............................5-61
    Removal & Installation..........5-61
Manifold Absolute Pressure
    (MAP) Sensor.......................5-61
    Removal & Installation..........5-61
Mass Airflow (MAF)/Intake Air
    Temperature (IAT) Sensor.......5-61
    Location...............................5-61
    Removal & Installation..........5-61
Output Shaft Speed (OSS)
    Sensor.................................5-62
    Location...............................5-62
    Removal & Installation..........5-62
Positive Crankcase
    Ventilation (PCV) Valve..........5-62
    Removal & Installation..........5-62
Powertrain Control
    Module (PCM)......................5-63
    Location...............................5-63
    Removal & Installation..........5-63
Throttle Control Actuator..........5-64
    Location...............................5-64
    Removal & Installation..........5-64
Throttle Position Sensor............5-64
    Location...............................5-64
    Removal & Installation..........5-64
Transmission Fluid
    Temperature (TFT) Sensor.......5-64
    Location...............................5-64
    Removal & Installation..........5-64
Variable Timing Camshaft
    (VTC) Oil Control
    Solenoid Valve......................5-65
    Location...............................5-65
    Removal & Installation..........5-65

FUEL.............................5-65

**GASOLINE FUEL INJECTION
SYSTEM.......................5-65**
    Fuel Filter............................5-66
       Removal & Installation..........5-66
    Fuel Rail & Injectors 5-67
       Removal & Installation..........5-67
    Fuel System Service
       Precautions 5-65
    Relieving Fuel System
       Pressure............................5-65
    Throttle Body.........................5-69
       Removal & Installation..........5-69

**HEATING & AIR CONDITIONING
SYSTEM.......................5-72**
    Blower Motor 5-72
       Removal & Installation..........5-72
    Heater Core 5-72
       Removal & Installation..........5-72

PRECAUTIONS 5-10

**SPECIFICATIONS AND
MAINTENANCE CHARTS......000**
    Brake Specifications................5-8
    Camshaft Specifications..............5-5
    Capacities 5-4
    Crankshaft and Connecting
       Rod Specifications 5-6
    Engine and Model Year
       Identification 5-3
    Engine Tune-Up Specifications ...5-3
    Fluid Specifications.................5-4
    General Engine Specifications.....5-3
    Piston and Ring Specifications ...5-6
    Scheduled Maintenance
       Intervals 5-9

Tire Wheel and Ball Joint
    Specifications 5-8
Torque Specifications 5-7
Valve Specifications 5-5
Wheel Alignment.......................5-7

STEERING 000

Power Steering Gearbox............5-74
    Removal & Installation..........5-74

SUSPENSION...................5-78

FRONT SUSPENSION 5-78
Knuckle & Spindle 5-78
    Removal & Installation..........5-78
Lower Ball Joints......................5-79
    Removal & Installation..........5-79
Lower Control Arms 5-80
    Removal & Installation..........5-80
Stabilizer Bar& Links................5-81
    Removal & Installation..........5-81
Struts......................................5-82
    Overhaul 5-84
    Removal & Installation..........5-82
Upper Ball Joints......................5-85
    Removal & Installation..........5-85
Upper Control Arms 5-86
    Removal & Installation..........5-86
Wheel Hubs & Bearings............5-87
    Removal, Packing, &
       Installation........................5-87
REAR SUSPENSION 5-87
Control Links 5-87
    Removal & Installation..........5-87
Knuckle/Hub Bearing Unit.........5-90
    Removal & Installation..........5-90
Stabilizer Bar & Links 5-92
    Removal & Installation..........5-92
Strut 5-93
    Removal & Installation..........5-93

SPECIFICATIONS AND MAINTENANCE CHARTS

ENGINE AND VEHICLE IDENTIFICATION

		Engine					Model Year	
Code	Liters (cc)	Cu. In.	Cyl.	Fuel Sys.	Engine Type	Eng. Mfg.	Code ①	Year
K24Z3	2.4 (2354)	144	4	MPFI	DOHC	Honda	B	2011
J35Z6	3.5 (3490)	213	6	PGM-FI	SOHC	Honda	C	2012

MPFI: Multi-Port Fuel Injection

PGM-FI: Programmed fuel injection

DOHC: Double Overhead Camshaft

SOHC: Single Overhead Camshaft

① 10th digit of the Vehicle Identification Number (VIN)

71051_ATSX_C0001

GENERAL ENGINE SPECIFICATIONS

Year	Model	Engine Displacement Liters	Engine ID	Net Horsepower @ rpm	Net Torque @ rpm (ft. lbs.)	Bore x Stroke (in.)	Com-pression Ratio	Oil Pressure @ rpm
2011	TSX	2.4	K24Z3	201@7000	172@3400	3.43X3.90	11.0:1	44@3000
		3.5	J35Z6	280@6200	254@5000	3.50X3.66	11.2:1	71@3000
2012	TSX	2.4	K24Z3	201@7000	172@3400	3.43X3.90	11.0:1	44@3000
		3.5	J35Z6	280@6200	254@5000	3.50X3.66	11.2:1	71@3000

71051_ATSX_C0002

GASOLINE ENGINE TUNE-UP SPECIFICATIONS

Year	Engine Displacement Liters	Engine ID	Spark Plug Gap (in.)	Ignition Timing (deg.) MT	AT	Fuel Pump (psi)	Idle Speed (rpm) MT	AT	Valve Clearance In.	Ex.
2011	2.4	K24Z3	0.039-0.043	6-10 ①	6-10 ①	48-55	700-800	750-850	0.008-0.010	0.010-0.011
	3.5	J35Z6	0.039-0.043	—	8-12 ①	57-64	—	630-730	0.008-0.009	0.011-0.013
2012	2.4	K24Z3	0.039-0.043	6-10 ①	6-10 ①	48-55	700-800	750-850	0.008-0.010	0.010-0.011
	3.5	J35Z6	0.039-0.043	—	8-12 ①	57-64	—	630-730	0.008-0.009	0.011-0.013

NOTE: The Vehicle Emission Control Information label reflects specification changes during production and must be used if they differ from this chart.

①: Before Top Dead Center

71051_ATSX_C0003

CAPACITIES

Year	Model	Engine Displacement Liters	Engine ID	Engine Oil with Filter (qts.)	Transmission (pts.)		Fuel Tank (gal.)	Cooling System (qts.)
					6-Spd	Auto.		
2011	TSX	2.4	K24Z3	4.2	4.2	5.2	18.5	6.5
		3.5	J35Z6	4.5	—	6.0	18.5	7.0
2012	TSX	2.4	K24Z3	4.2	4.2	5.2	18.5	6.5
		3.5	J35Z6	4.5	—	6.0	18.5	7.0

NOTE: All capacities are approximate. Add fluid gradually and ensure a proper fluid level is obtained.

NOTE: Capacities given are service capacities, not overhaul capacities

71051_ATSX_C0004

FLUID SPECIFICATIONS

Year	Model	Engine Displacement Liters	Engine Oil	Man. Trans.	Auto. Trans.	Power Steering Fluid	Brake Master Cylinder	Cooling System
2011	TSX	2.4	5W-20 Acura	Acura MTF	Acura ATF-Z1	Acura PS Fluid	Acura DOT 3	①
		3.5	5W-20 Honda	—	Acura ATF-Z1	Acura PS Fluid	Acura DOT 3	①
2012	TSX	2.4	5W-20 Acura	Acura MTF	Acura ATF-Z1	Acura PS Fluid	Acura DOT 3	①
		3.5	5W-20 Honda	—	Acura ATF-Z1	Acura PS Fluid	Acura DOT 3	①

DOT: Department Of Transportation

① Acura Long Life Antifreeze/Coolant-Type2

71051_ATSX_C0005

GENERAL ENGINE SPECIFICATIONS

Year	Model	Engine Displacement Liters	Engine ID	Net Horsepower @ rpm	Net Torque @ rpm (ft. lbs.)	Bore x Stroke (in.)	Com- pression Ratio	Oil Pressure @ rpm
2011	TSX	2.4	K24Z3	201@7000	172@3400	3.43X3.90	11.0:1	44@3000
		3.5	J35Z6	280@6200	254@5000	3.50X3.66	11.2:1	71@3000
2012	TSX	2.4	K24Z3	201@7000	172@3400	3.43X3.90	11.0:1	44@3000
		3.5	J35Z6	280@6200	254@5000	3.50X3.66	11.2:1	71@3000

71051_ATSX_C0006

CAMSHAFT AND BEARING SPECIFICATIONS
All measurements are given in inches.

Year	Engine Displacement Liters	Engine ID	Journal Diameter	Brg. Oil Clearance	Shaft End-play	Runout	Journal Bore	Lobe Height Intake	Lobe Height Exhaust
2011	2.4	K24Z3	NA	①	0.0020-0.0080	0.0010	NA	②	1.3500
	3.5	J35Z6	NA	0.0020-0.0035	0.0020-0.0080	0.0010	NA	③	1.4472
2012	2.4	K24Z3	NA	①	0.0020-0.0080	0.0010	NA	②	1.3500
	3.5	J35Z6	NA	0.0020-0.0035	0.0020-0.0080	0.0010	NA	③	1.4472

NA: Information not available

① No. 1 Journal: 0.0010-0.0030 inches

   Other Journals: 0.0020-0.0040 inches

② Primary: 1.3285 inches

   Mid: 1.359 inches

   Secondary: 1.3285 inches

③ Primary: 1.3504 inches

   Secondary: 1.4024 inches

71051_ATSX_C0009

CRANKSHAFT AND CONNECTING ROD SPECIFICATIONS

All measurements are given in inches.

Year	Engine Displacement Liters	Engine ID	Crankshaft				Connecting Rod		
			Main Brg. Journal Dia.	Main Brg. Oil Clearance	Shaft End-play	Thrust on No.	Journal Diameter	Oil Clearance	Side Clearance
2011	2.4	K24Z3	①	②	0.0040-0.0140	4	1.8888-1.8898	0.0013-0.0026	0.0060-0.0140
	3.5	J35Z6	2.8337-2.8346	0.0007-0.0017	0.0040-0.0140	3	2.1644-2.1654	0.0008-0.0017	0.0060-0.0140
2012	2.4	K24Z3	①	②	0.0040-0.0140	4	1.8888-1.8898	0.0013-0.0026	0.0060-0.0140
	3.5	J35Z6	2.8337-2.8346	0.0007-0.0017	0.0040-0.0140	3	2.1644-2.1654	0.0008-0.0017	0.0060-0.0140

① Nos. 1, 2, 4 and 5: 2.1647-2.1657 inches

No. 3: 2.1644-2.1654 inches

② Nos. 1, 2, 4 and 5: 0.0007-0.0016 inches

No. 3: 0.0010-0.0019 inches

71051_ATSX_C0007

PISTON AND RING SPECIFICATIONS

All measurements are given in inches

Year	Engine Displacement Liters	Engine ID	Piston Clearance	Ring Gap			Ring Side Clearance		
				Top Compression	Bottom Compression	Oil Control	Top Compression	Bottom Compression	Oil Control
2011	2.4	K24Z3	0.0008-0.0016	0.0080-0.0014	0.0200-0.0260	0.0080-0.0280	0.0024-0.0033	0.0016-0.0026	NA
	3.5	J35Z6	0.0006-0.0016	0.0080-0.0140	0.0160-0.0220	0.0080-0.0280	0.0022-0.0031	0.0012-0.0022	NA
2012	2.4	K24Z3	0.0008-0.0016	0.0080-0.0014	0.0200-0.0260	0.0080-0.0280	0.0024-0.0033	0.0016-0.0026	NA
	3.5	J35Z6	0.0006-0.0016	0.0080-0.0140	0.0160-0.0220	0.0080-0.0280	0.0022-0.0031	0.0012-0.0022	NA

NA: Not Available

71051_ATSX_C0008

TORQUE SPECIFICATIONS
All readings in ft. lbs.

Year	Engine Displacement Liters	Engine ID/VIN	Cylinder Head Bolts	Main Bearing Bolts	Rod Bearing Bolts	Crankshaft Damper Bolts	Flexplate/ Flywheel Bolts	Manifold		Spark Plugs	Oil Pan Drain Plug
								Intake	Exhaust		
2011	2.4	K24Z3	①	②	③	④	5490	16	33	13	30
	3.5	J35Z6	⑤	⑥	⑦	⑧	54/90	16	23	13	29
2012	2.4	K24Z3	①	②	③	④	5490	16	33	13	30
	3.5	J35Z6	⑤	⑥	⑦	⑧	54/90	16	23	13	29

① Step 1: 29 ft. lbs.

   Step 2: Rotate 90 degrees

   Step 3: Rotate an additional 90 degrees

   Step 4 (new bolts only): additional 90 degrees

② Step 1: 22 ft. lbs.

   Step 2: Rotate 48 degrees

③ Step 1: 30 ft. lbs.

   Step 2: Rotate 120 degrees

④ Step 1: 36 ft. lbs.

   Step 2: For a new bolt or a new crankshaft only, tighten to 130 ft.lbs.,

   then remove the bolt and tighten it to 36 ft.lbs.

   Step 3: Rotate an additional 90 degrees

⑤ Step 1: 22 ft. lbs.

   Step 2: Rotate 90 degrees

   Step 3: Rotate an additional 90 degrees

   Step 4 (new bolts only): additional 90 degrees

⑥ Step 1: 54 ft. lbs.

   Step 2: Repeat Step 1 to ensure 54 ft. lbs.

   Step 3: Side bolts to 36 ft. lbs.

   Step 4: Repeat Step 3 to ensure 36 ft. lbs.

⑦ Step 1: 14 ft. lbs.Cap bolts: 29 ft. lbs.

   Step 2: Rotate 90 degrees

⑧ Step 1: 47 ft. lbs.

   Step 2: Rotate 60 degrees

71051_ATSX_C0010

WHEEL ALIGNMENT

Year	Model		Caster Range (+/-Deg.)	Caster Preferred Setting (Deg.)	Camber Range (+/-Deg.)	Camber Preferred Setting (Deg.)	Toe-in (in.)
2011	TSX (4 cyl)	F	0.10/0.50	+3.47	0.30	0.00	0 +/- 0.08
		R	—	—	0.30	-1.00	0.08 +/- 0.08
	TSX (6cyl)	F	0.10/0.50	+3.52	0.30	-0.03	0 +/- 0.08
		R	—	—	0.30	-1.12	0.08 +/- 0.08
2012	TSX (4 cyl)	F	0.10/0.50	+3.47	0.30	0.00	0 +/- 0.08
		R	—	—	0.30	-1.00	0.08 +/- 0.08
	TSX (6cyl)	F	0.10/0.50	+3.52	0.30	-0.03	0 +/- 0.08
		R	—	—	0.30	-1.12	0.08 +/- 0.08

71051_ATSX_C0011

TIRE, WHEEL AND BALL JOINT SPECIFICATIONS

| Year | Model | OEM Tires | | Tire Pressures (psi) | | Wheel Size | Ball Joint Inspection | Lug Nut (ft. lbs.) |
		Standard	Optional	Front	Rear			
2011	TSX (4 cyl)	P225/50R17	NS	33	33	17x7.5	NS	80
	TSX (6cyl)	P235/45R18	NS	33	33	18x8	NS	80
2012	TSX (4 cyl)	P225/50R17	NS	33	33	17x7.5	NS	80
	TSX (6cyl)	P235/45R18	NS	33	33	18x8	NS	80

OEM: Original Equipment Manufacturer

PSI: Pounds Per Square Inch

NS: Not Specified by manufacturer

71051_ATSX_C0012

BRAKE SPECIFICATIONS
All measurements in inches unless noted

| Year | Model | | Brake Disc | | | Brake Drum Diameter | | | Minimum Lining Thickness | | Brake Caliper | |
			Original Thickness	Minimum Thickness	Maximum Runout	Original Inside Diameter	Max. Wear Limit	Maximum Machine Diameter	Front	Rear	Bracket Bolts (ft. lbs.)	Mounting Bolts (ft. lbs.)
2011	TSX	F	1.100-1.100	1.020	0.002	NA	NA	NA	0.06	NA	80	37
		R	0.350-0.360	0.310	0.002	NA	NA	NA	NA	0.04	80	17
2012	TSX	F	1.100-1.100	1.020	0.002	NA	NA	NA	0.06	NA	80	37
		R	0.350-0.360	0.310	0.002	NA	NA	NA	NA	0.04	80	17

NA: Not Available

F: Front

R: Rear

71051_ATSX_C0013

SCHEDULED MAINTENANCE INTERVALS
ACURA—TSX

TO BE SERVICED	OF SERVIC	VEHICLE MILEAGE INTERVAL (x1000)															
		7.5	15	22.5	30	37.5	45	52.5	60	67.5	75	82.5	90	97.5	105	112.5	120
Accessory drive belts	I & A				✓				✓				✓				✓
Air cleaner element	R				✓				✓				✓				✓
Brake fluid	R	Every 3 years															
Brake hoses & lines (incl. ABS)	I		✓		✓		✓		✓		✓		✓		✓		✓
Cooling system hoses & connections	I		✓		✓		✓		✓		✓		✓		✓		✓
Engine coolant ①	R					✓							✓				
Engine oil	R	✓	✓	✓	✓	✓	✓	✓	✓	✓	✓	✓	✓	✓	✓	✓	✓
Engine oil and coolant levels	I	Inspect at each fuel stop															
Engine oil filter	R		✓		✓		✓		✓		✓		✓				✓
Exhaust system	I		✓		✓		✓		✓		✓		✓				✓
Fluid levels and condition	I		✓		✓		✓		✓		✓		✓				✓
Front and rear brakes	I		✓		✓		✓		✓		✓		✓				✓
Fuel lines & connection	I		✓		✓		✓		✓		✓		✓				✓
Halfshaft boots	I		✓		✓		✓		✓		✓		✓				✓
Idle speed	I & A														✓		
Parking brake system	I & A		✓		✓		✓		✓		✓		✓		✓		✓
Rear differential fluid	R	✓			✓		✓		✓				✓				✓
Rotate and inspect tires	I	✓	✓	✓	✓	✓	✓	✓	✓	✓	✓	✓	✓	✓	✓	✓	✓
Spark plugs	R														✓		
Supplemental Restraint System	I	Inspect the SRS 10 years after production															
Suspension components	I		✓		✓		✓		✓		✓		✓		✓		✓
Tie rod ends, steering gear box & boots	I		✓		✓		✓		✓		✓		✓		✓		✓
Timing belt	R														✓		
Transmission fluid	R						✓				✓				✓		
Valve clearance	I	Adjust if valves are noisy															
Water pump	S/I														✓		

R: Replace   I: Inspect   A: Adjust

① Every 12,000 miles or 10 years, then every 60,000 miles or 5 years

FREQUENT OPERATION MAINTENANCE (SEVERE SERVICE)

If a vehicle is operated under any of the following conditions it is considered severe service:

- Towing a trailer or using a camper or car-top carrier.
- Repeated short trips of less than 5 miles in temperatures below freezing, or trips of less than 10 miles in any temperature.
- Extensive idling or low-speed driving for long distances as in heavy commercial use, such as delivery, taxi or police cars.
- Operating on rough, muddy or salt-covered roads.
- Operating on unpaved or dusty roads.
- Driving in extremely hot (over 90°) conditions.

Air cleaner element: replace every 15,000 miles

Engine oil and filter: replace every 3750 miles or 6 months, whichever occurs first.

Timing belt: replace every 60,000 miles if the vehicle is regularly driven in temperatures above 110°F or below -20°F, or if frequently towing a trailer.

Transmission fluid: replace every 30,000 miles.

Rear differential fluid: replace every 60,000 miles.

Front and rear brakes: inspect every 7500 miles or 6 months, whichever occurs first.

Locks and hinges: lubricate every 15,000 miles.

Tie rods, steering gear box, boots: inspect every 7500 miles or 6 months, whichever occurs first.

Suspension components: inspect every 7500 miles or 6 months, whichever occurs first.

Halfshaft boots: inspect every 7500 miles or 6 months, whichever occurs first.

PRECAUTIONS

Before servicing any vehicle, please be sure to read all of the following precautions, which deal with personal safety, prevention of component damage, and important points to take into consideration when servicing a motor vehicle:

• Never open, service or drain the radiator or cooling system when the engine is hot; serious burns can occur from the steam and hot coolant.

• Observe all applicable safety precautions when working around fuel. Whenever servicing the fuel system, always work in a well-ventilated area. Do not allow fuel spray or vapors to come in contact with a spark, open flame, or excessive heat (a hot drop light, for example). Keep a dry chemical fire extinguisher near the work area. Always keep fuel in a container specifically designed for fuel storage; also, always properly seal fuel containers to avoid the possibility of fire or explosion. Refer to the additional fuel system precautions later in this section.

• Fuel injection systems often remain pressurized, even after the engine has been turned **OFF**. The fuel system pressure must be relieved before disconnecting any fuel lines. Failure to do so may result in fire and/or personal injury.

• Brake fluid often contains polyglycol ethers and polyglycols. Avoid contact with the eyes and wash your hands thoroughly after handling brake fluid. If you do get brake fluid in your eyes, flush your eyes with clean, running water for 15 minutes. If eye irritation persists, or if you have taken brake fluid internally, IMMEDIATELY seek medical assistance.

• The EPA warns that prolonged contact with used engine oil may cause a number of skin disorders, including cancer. You should make every effort to minimize your exposure to used engine oil. Protective gloves should be worn when changing oil. Wash your hands and any other exposed skin areas as soon as possible after exposure to used engine oil. Soap and water, or waterless hand cleaner should be used.

• All new vehicles are now equipped with an air bag system, often referred to as a Supplemental Restraint System (SRS) or Supplemental Inflatable Restraint (SIR) system. The system must be disabled before performing service on or around system components, steering column, instrument panel components, wiring and sensors. Failure to follow safety and disabling procedures could result in accidental air bag deployment, possible personal injury and unnecessary system repairs.

• Always wear safety goggles when working with, or around, the air bag system. When carrying a non-deployed air bag, be sure the bag and trim cover are pointed away from your body. When placing a non-deployed air bag on a work surface, always face the bag and trim cover upward, away from the surface. This will reduce the motion of the module if it is accidentally deployed. Refer to the additional air bag system precautions later in this section.

• Clean, high quality brake fluid from a sealed container is essential to the safe and proper operation of the brake system. You should always buy the correct type of brake fluid for your vehicle. If the brake fluid becomes contaminated, completely flush the system with new fluid. Never reuse any brake fluid. Any brake fluid that is removed from the system should be discarded. Also, do not allow any brake fluid to come in contact with a painted surface; it will damage the paint.

• Never operate the engine without the proper amount and type of engine oil; doing so WILL result in severe engine damage.

• Timing belt maintenance is extremely important. Many models utilize an interference-type, non-freewheeling engine. If the timing belt breaks, the valves in the cylinder head may strike the pistons, causing potentially serious (also time-consuming and expensive) engine damage. Refer to the maintenance interval charts for the recommended replacement interval for the timing belt, and to the timing belt section for belt replacement and inspection.

• Disconnecting the negative battery cable on some vehicles may interfere with the functions of the on-board computer system(s) and may require the computer to undergo a relearning process once the negative battery cable is reconnected.

• When servicing drum brakes, only disassemble and assemble one side at a time, leaving the remaining side intact for reference.

• Only an MVAC-trained, EPA-certified automotive technician should service the air conditioning system or its components.

BRAKES

ANTI-LOCK BRAKE SYSTEM (ABS)

GENERAL INFORMATION

PRECAUTIONS

• Certain components within the ABS system are not intended to be serviced or repaired individually.

• Do not use rubber hoses or other parts not specifically specified for and ABS system. When using repair kits, replace all parts included in the kit. Partial or incorrect repair may lead to functional problems and require the replacement of components.

• Lubricate rubber parts with clean, fresh brake fluid to ease assembly. Do not use shop air to clean parts; damage to rubber components may result.

• Use only DOT 3 brake fluid from an unopened container.

• If any hydraulic component or line is removed or replaced, it may be necessary to bleed the entire system.

• A clean repair area is essential. Always clean the reservoir and cap thoroughly before removing the cap. The slightest amount of dirt in the fluid may plug an orifice and impair the system function. Perform repairs after components have been thoroughly cleaned; use only denatured alcohol to clean components. Do not allow ABS components to come into contact with any substance containing mineral oil; this includes used shop rags.

• The Anti-Lock control unit is a microprocessor similar to other computer units in the vehicle. Ensure that the ignition switch is **OFF** before removing or installing controller harnesses. Avoid static electricity discharge at or near the controller.

• If any arc welding is to be done on the vehicle, the control unit should be unplugged before welding operations begin.

SPEED SENSORS

REMOVAL & INSTALLATION

Front

See Figure 1.

1. Turn the ignition switch to LOCK (0).
2. Release the clamp, then disconnect the wheel speed sensor connector.
3. Remove the bolts and the wheel speed sensor.

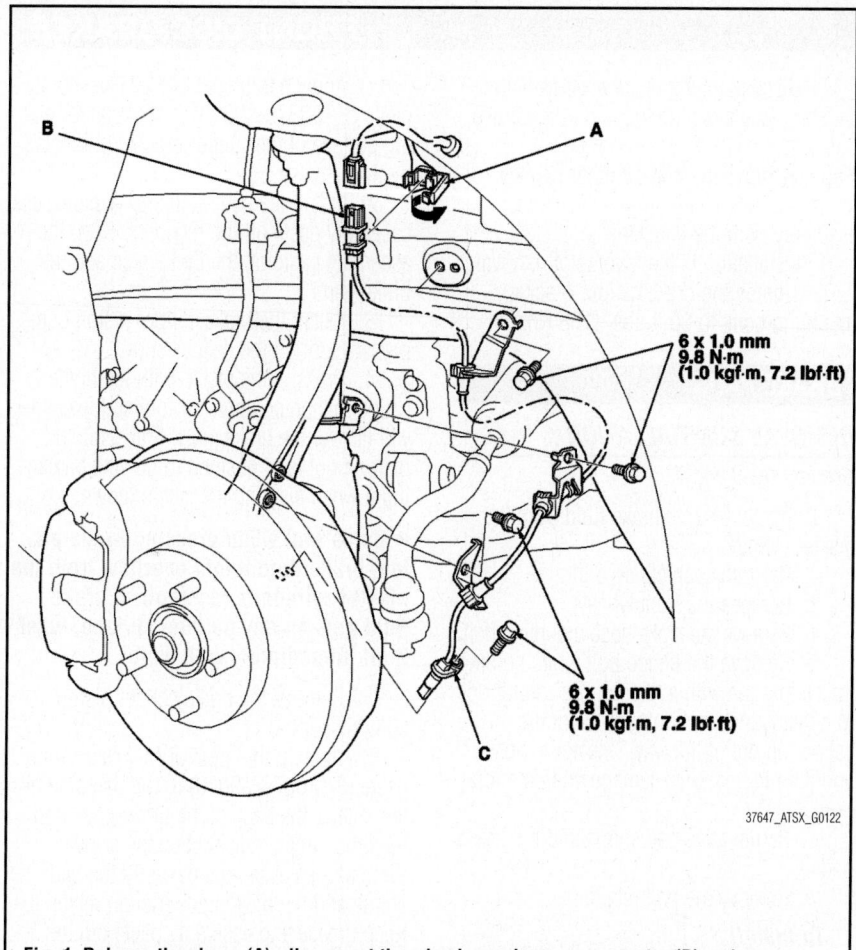

6 x 1.0 mm
9.8 N·m
(1.0 kgf·m, 7.2 lbf·ft)

6 x 1.0 mm
9.8 N·m
(1.0 kgf·m, 7.2 lbf·ft)

37647_ATSX_G0122

Fig. 1 Release the clamp (A), disconnect the wheel speed sensor connector (B) and remove the wheel speed sensor (C)

4. Install the wheel speed sensor in the reverse order of removal, and note these items:
- Do not twist the sensor wires.
- If the wheel speed sensor comes in contact with the wheel bearing, it is faulty.
- Make sure there is no debris in the sensor mounting hole.

5. Start the engine, and test-drive the vehicle. Make sure that the ABS and VSA indicators turn off during the test-drive and do not come back on.

Rear

1. Turn the ignition switch to LOCK (0).
2. Release the clamp, then disconnect the wheel speed sensor connector.
3. Remove the clamps, the bolt, and the wheel speed sensor.
4. Install the wheel speed sensor in the reverse order of removal, and note these items:
- Do not twist the sensor wires.
- If the wheel speed sensor comes in contact with the hub bearing unit, it is faulty.
- Make sure there is no debris in the sensor mounting hole.
- Lubricate the O-ring on the wheel speed sensor.

5. Start the engine, and test-drive the vehicle. Make sure that the ABS and VSA indicators turn off during the test-drive and do not come back on.

BRAKES

BLEEDING THE BRAKE SYSTEM

BLEEDING PROCEDURE

BLEEDING PROCEDURE

1. Make sure the brake fluid level in the reservoir is at the MAX (upper) level line.
2. Have someone slowly pump the brake pedal several times, then apply steady pressure.
3. Start the bleeding at the driver's side of the front brake system.

➡ **Bleed the calipers in the sequence shown.**

4. Attach a length of clear drain tube to the bleed screw, then loosen the bleed screw to allow air to escape from the system. Then tighten the bleed screw securely.
5. Refill the master cylinder reservoir to the MAX (upper) level line.
6. Repeat the procedure for each brake circuit until there are no air bubbles in the fluid.

BLEEDING THE ABS SYSTEM

The ABS brake system is bled in the usual fashion with no special procedures required. Refer to the bleeding procedure located in this section. Make certain the master cylinder reservoir is filled before the bleeding is begun and check the level frequently.

✴✴ CAUTION

Dust and dirt accumulating on brake parts during normal use may contain asbestos fibers from production or aftermarket brake linings. Breathing excessive concentrations of asbestos fibers can cause serious bodily harm. Exercise care when servicing brake parts. Do not sand or grind brake lining unless equipment used is designed to contain the dust residue. Do not clean brake parts with compressed air or by dry brushing. Cleaning should be done by dampening the brake components with a fine mist of water, then wiping the brake components clean with a dampened cloth. Dispose of cloth and all residue containing asbestos fibers in an impermeable container with the appropriate label. Follow practices prescribed by the Occupational Safety and Health Administration (OSHA) and the Environmental Protection Agency (EPA) for the handling, processing, and disposing of dust or debris that may contain asbestos fibers.

BRAKE CALIPER

REMOVAL & INSTALLATION

See Figure 2.

1. Raise the vehicle on a lift.
2. Remove the front wheels.
3. Remove the brake hose mounting bolt.

4. Disconnect the brake hose from the caliper. Place a cap on the end of the brake hose.
5. Remove the brake caliper bracket mounting bolts, then remove the caliper assembly from the knuckle.
6. Installation is the reverse of removal.
7. Tighten the brake caliper bracket mounting bolts to 80 ft. lbs. (108 Nm).

BRAKE PADS

REMOVAL & INSTALLATION

See Figures 3 and 4.

1. Remove some brake fluid from the master cylinder.
2. Raise the vehicle on a lift.
3. Remove the front wheels.
4. Remove the brake hose mounting bolt.
5. Remove the flange bolt while holding the caliper pin with a wrench. Be careful not to damage the pin boot, and pivot the caliper up out of the way. Check the hose and the pin boots for damage and deterioration.
6. Remove the pad shims and the brake pads.
7. Remove the pad retainers.

To install:

8. Clean the caliper bracket thoroughly; remove any rust, and check for grooves and cracks.
9. Verify that the caliper pins move in and out smoothly. Clean and lube if needed.
10. Inspect the brake disc for runout, thickness, parallelism, and check for damage and cracks.

11. Apply a thin coat of M-77 assembly paste (P/N 08798-9010) to the retainer mating surface of the caliper bracket (indicated by the arrows).
12. Install the pad retainers. Wipe excess assembly paste off the retainers. Keep the assembly paste off the brake disc and the brake pads.
13. Install the brake caliper piston compressor tool on the caliper body.
14. Press in the piston with the brake caliper piston compressor tool so the caliper will fit over the brake pads. Make sure the piston boot is in position to prevent damaging it when pivoting the caliper down.

➡ Be careful when pressing in the piston; brake fluid might overflow from the master cylinder's reservoir. If brake fluid gets on any painted surface, wash it off immediately with water.

15. Remove the brake caliper piston compressor tool.
16. Apply a thin coat of M-77 assembly paste (P/N 08798-9010) to the pad side of the shims, the back of the brake pads and the other areas indicated by the arrows. Wipe excess assembly paste off the pad shims and the brake pads' friction material. Keep grease and assembly paste off the brake disc and the brake pads. Contaminated brake disc or brake pads reduce stopping ability.
17. Install the brake pads and the pad shims correctly. Install the brake pad with the wear indicator on the upper inside. If you are reusing the brake pads, always reinstall the brake pads in their original posi-

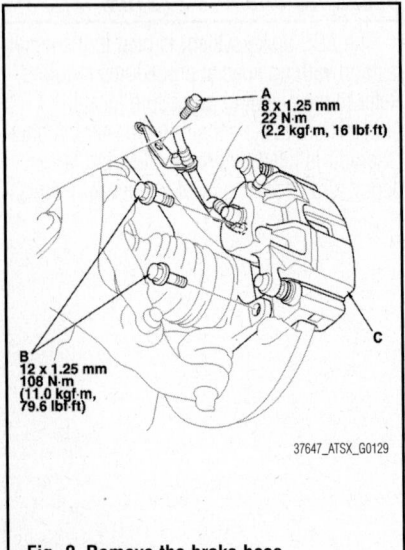

A
8 x 1.25 mm
22 N·m
(2.2 kgf·m, 16 lbf·ft)

B
12 x 1.25 mm
108 N·m
(11.0 kgf·m,
79.6 lbf·ft)

C

37647_ATSX_G0129

Fig. 2 Remove the brake hose mounting bolt

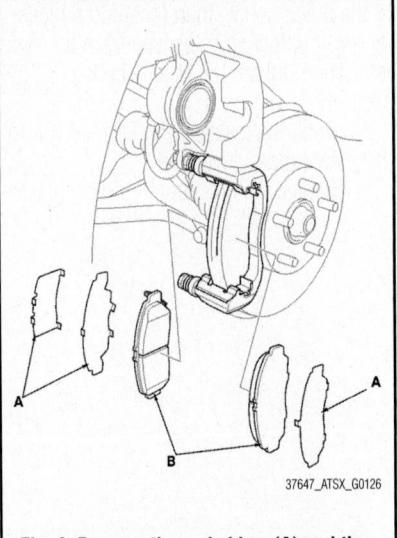

37647_ATSX_G0126

Fig. 3 Remove the pad shims (A) and the brake pads (B)

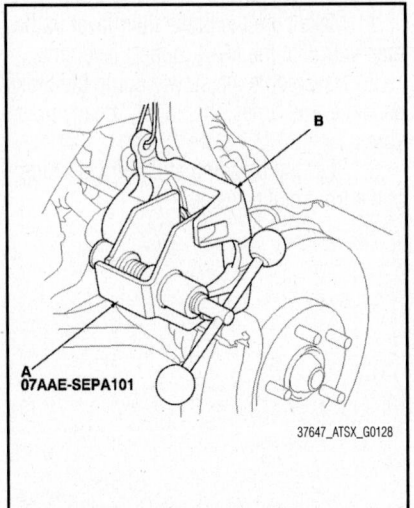

07AAE-SEPA101

B

A

37647_ATSX_G0128

Fig. 4 Install the brake caliper piston compressor tool (A) on the caliper body (B)

tions to prevent a temporary loss of braking efficiency.

18. Pivot the caliper down into position. Install the flange bolt, and tighten it to 37 ft. lbs. (50 Nm) while holding the caliper pin with a wrench being careful not to damage the pin boot.

19. Install the brake hose mounting bolt. Tighten the bolt to 16 ft. lbs. (22 Nm).

20. Clean the mating surfaces between the brake disc and the inside of the wheel, then install the front wheels.

21. Press the brake pedal several times to make sure the brakes work.

➡**Engagement may require a greater pedal stroke immediately after the brake pads have been replaced as a set. Several applications of the brake**

pedal will restore the normal pedal stroke.

22. Add brake fluid as needed.

23. After installation, check for leaks at hose and line joints or connections, and retighten if necessary.

24. Test-drive the vehicle, then recheck for leaks.

BRAKES

✳✳ CAUTION

Dust and dirt accumulating on brake parts during normal use may contain asbestos fibers from production or aftermarket brake linings. Breathing excessive concentrations of asbestos fibers can cause serious bodily harm. Exercise care when servicing brake parts. Do not sand or grind brake lining unless equipment used is designed to contain the dust residue. Do not clean brake parts with compressed air or by dry brushing. Cleaning should be done by dampening the brake components with a fine mist of water, then wiping the brake components clean with a dampened cloth. Dispose of cloth and all residue containing asbestos fibers in an impermeable container with the appropriate label. Follow practices prescribed by the Occupational Safety and Health Administration (OSHA) and the Environmental Protection Agency (EPA) for the handling, processing, and disposing of dust or debris that may contain asbestos fibers.

BRAKE CALIPER

REMOVAL & INSTALLATION

See Figures 5 and 6.

1. Raise the vehicle on a lift.
2. Remove the rear wheel.
3. Release the parking brake lever fully.
4. Loosen the parking brake cable adjusting nut.
5. Remove the flange bolt from the arm.
6. Disconnect the parking brake cable from the lever.
7. Remove the brake hose mounting bolt.
8. Disconnect the brake hose from the caliper. Place a cap on the end of the brake hose.

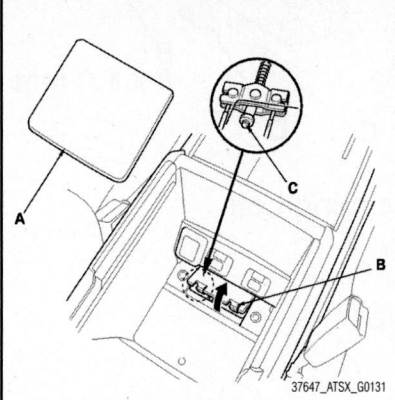

Fig. 5 Remove the console box mat (A), open the lid (B) then loosen the parking brake cable adjusting nut (C)

9. Remove the brake caliper bracket mounting bolts, and remove the caliper assembly from the knuckle.

➡**Make sure the washers are in position on reassembly, if they are removed.**

10. Installation is the reverse of removal.

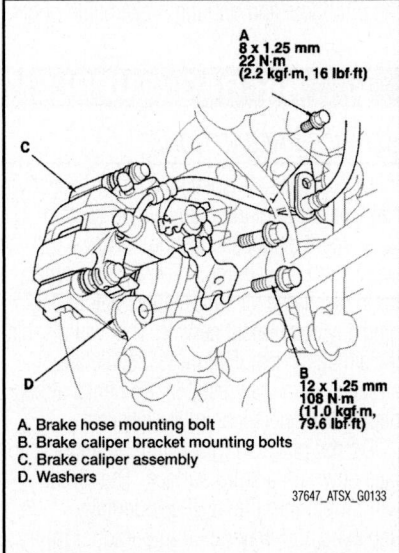

A. Brake hose mounting bolt
B. Brake caliper bracket mounting bolts
C. Brake caliper assembly
D. Washers

37647_ATSX_G0133

Fig. 6 Remove the brake caliper assembly

REAR DISC BRAKES

DISC BRAKE PADS

REMOVAL & INSTALLATION

See Figures 7 and 8.

1. Remove some brake fluid from the master cylinder.
2. Raise the vehicle on a lift.
3. Remove the rear wheels.
4. Remove the brake hose mounting bolt.
5. Remove the flange bolts while holding respective caliper pin with a wrench. Be careful not to damage the pin boot, and remove the caliper. Check the hose, the pin boots, and the parking brake cable boots for damage and deterioration.

➡**Do not twist the brake hose and the parking brake cable to prevent damage.**

6. Remove the pad shim and the brake pads.
7. Remove the pad retainers (A).
8. Clean the caliper bracket thoroughly; remove any rust, and check for grooves and cracks.
9. Verify that the caliper pins move in and out smoothly. Clean and lube if needed.
10. Inspect the brake disc for runout, thickness, parallelism, and check for damage and cracks.
11. Apply a thin coat of M-77 assembly paste (P/N 08798-9010) to the retainer mating surface of the caliper bracket (indicated by the arrows).
12. Install the pad retainers. Wipe excess assembly paste off the retainers. Keep the assembly paste off the brake disc and the brake pads.
13. Apply a thin coat of M-77 assembly paste (P/N 08798-9010) to the pad side of the shim, the back of the brake pads, and the other areas indicated by the arrows. Wipe excess assembly paste off the pad shim and the brake pads friction material. Keep grease and assembly paste off the brake disc and the brake pads. Contami-

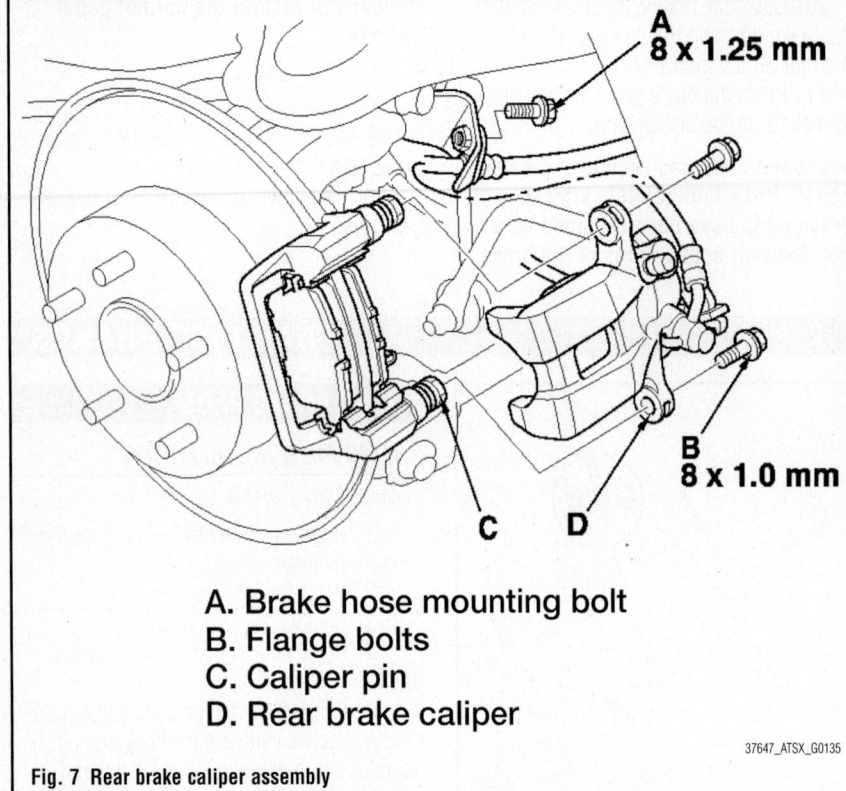

A. Brake hose mounting bolt
B. Flange bolts
C. Caliper pin
D. Rear brake caliper

37647_ATSX_G0135

Fig. 7 Rear brake caliper assembly

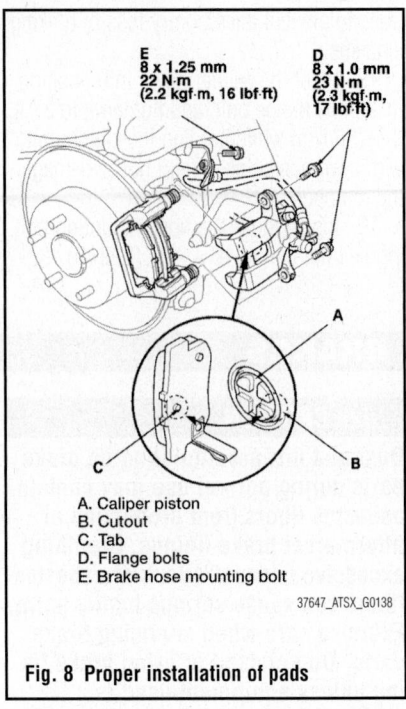

A. Caliper piston
B. Cutout
C. Tab
D. Flange bolts
E. Brake hose mounting bolt

37647_ATSX_G0138

Fig. 8 Proper installation of pads

nated brake disc or brake pads reduce stopping ability.

14. Install the brake pads and pad shim correctly. Install the brake pad with the wear indicator on the bottom inside. If you are reusing the brake pads, always reinstall the brake pads in their original positions to prevent a temporary loss of braking efficiency.

15. Rotate the caliper piston clockwise into the cylinder, then align the cutout in the piston with the tab on the inner pad by turning the piston back. Lubricate the boot with rubber grease to avoid twisting the piston boot. If the piston boot is twisted, back it out so it is positioned properly.

➡Be careful when moving the piston back in the caliper; brake fluid might overflow from the master cylinder's reservoir. If brake fluid gets on any painted surface, wash it off immediately with water.

16. Install the caliper. Install the flange bolts, and tighten it to the specified torque while holding the respective caliper pins with a wrench being careful not to damage the pin boots and parking brake cable boots.

17. Install the brake hose mounting bolt.

18. Clean the mating surfaces between the brake disc and the inside of the wheel, then install the rear wheels.

19. Press the brake pedal several times to make sure the brakes work.

➡Engagement may require a greater pedal stroke immediately after the brake pads have been replaced as a set. Several applications of the brake pedal will restore the normal pedal stroke.

20. Add brake fluid as needed.

21. After installation, check for leaks at hose and line joints or connections, and retighten if necessary.

22. Test-drive the vehicle, then recheck for leaks.

CHASSIS ELECTRICAL   AIR BAG (SUPPLEMENTAL RESTRAINT SYSTEM)

GENERAL INFORMATION

❋❋ CAUTION

These vehicles are equipped with an air bag system. The system must be disarmed before performing service on, or around, system components, the steering column, instrument panel components, wiring and sensors. Failure to follow the safety precautions and the disarming procedure could result in accidental air bag deployment, possible injury and unnecessary system repairs.

SERVICE PRECAUTIONS

Disconnect and isolate the battery negative cable before beginning any airbag system component diagnosis, testing, removal, or installation procedures. Allow system capacitor to discharge for two minutes before beginning any component service. This will disable the airbag system. Failure to disable the airbag system may result in accidental airbag deployment, personal injury, or death.

Do not place an intact undeployed airbag face down on a solid surface. The airbag will propel into the air if accidentally deployed and may result in personal injury or death.

When carrying or handling an undeployed airbag, the trim side (face) of the airbag should be pointing towards the body to minimize possibility of injury if accidental deployment occurs. Failure to do this may result in personal injury or death.

Replace airbag system components with OEM replacement parts. Substitute parts may appear interchangeable, but internal differences may result in inferior occupant protection. Failure to do so may result in occupant personal injury or death.

Wear safety glasses, rubber gloves, and long sleeved clothing when cleaning powder residue from vehicle after an airbag deployment. Powder residue emitted from a deployed airbag can cause skin irritation. Flush affected area with cool water if irritation is experienced. If nasal or throat irritation is experienced, exit the vehicle for fresh air until the irritation ceases. If irritation continues, see a physician.

Do not use a replacement airbag that is not in the original packaging. This may result in improper deployment, personal injury, or death.

The factory installed fasteners, screws and bolts used to fasten airbag components have a special coating and are specifically designed for the airbag system. Do not use substitute fasteners. Use only original equipment fasteners listed in the parts catalog when fastener replacement is required.

During, and following, any child restraint anchor service, due to impact event or vehicle repair, carefully inspect all mounting hardware, tether straps, and anchors for proper installation, operation, or damage. If a child restraint anchor is found damaged in any way, the anchor must be replaced. Failure to do this may result in personal injury or death.

Deployed and non-deployed airbags may or may not have live pyrotechnic material within the airbag inflator.

Do not dispose of driver/passenger/curtain airbags or seat belt tensioners unless you are sure of complete deployment. Refer to the Hazardous Substance Control System for proper disposal.

Dispose of deployed airbags and tensioners consistent with state, provincial, local, and federal regulations.

After any airbag component testing or service, do not connect the battery negative cable. Personal injury or death may result if the system test is not performed first.

If the vehicle is equipped with the Occupant Classification System (OCS), do not connect the battery negative cable before performing the OCS Verification Test using the scan tool and the appropriate diagnostic information. Personal injury or death may result if the system test is not performed properly.

Never replace both the Occupant Restraint Controller (ORC) and the Occupant Classification Module (OCM) at the same time. If both require replacement,

replace one, then perform the Airbag System test before replacing the other.

Both the ORC and the OCM store Occupant Classification System (OCS) calibration data, which they transfer to one another when one of them is replaced. If both are replaced at the same time, an irreversible fault will be set in both modules and the OCS may malfunction and cause personal injury or death.

If equipped with OCS, the Seat Weight Sensor is a sensitive, calibrated unit and must be handled carefully. Do not drop or handle roughly. If dropped or damaged, replace with another sensor. Failure to do so may result in occupant injury or death.

If equipped with OCS, the front passenger seat must be handled carefully as well. When removing the seat, be careful when setting on floor not to drop. If dropped, the sensor may be inoperative, could result in occupant injury, or possibly death.

If equipped with OCS, when the passenger front seat is on the floor, no one should sit in the front passenger seat. This uneven force may damage the sensing ability of the seat weight sensors. If sat on and damaged, the sensor may be inoperative, could result in occupant injury, or possibly death.

DISARMING THE SYSTEM

➡**Some systems store data in memory that is lost when the battery is disconnected. Do the following steps before disconnecting the battery.**

1. Make sure you have the anti-theft code(s) for the audio and/or the navigation system (if equipped).
2. For some models or if you're replacing the audio unit, it may be necessary to write down the audio presets (AM and FM), and the XM radio presets (if equipped), because the audio unit does not retain the presets after the battery is disconnected.
3. Make sure the ignition switch is in LOCK (0).
4. Disconnect and isolate the negative cable from the battery.

✲✲ CAUTION

Always disconnect the negative cable from the battery first.

5. Disconnect the positive cable from the battery.
6. Wait at least 3 minutes for the battery and SRS system to discharge before performing any repairs.

ARMING THE SYSTEM

➡**Some systems store data in memory that is lost when the battery is disconnected. Do the following steps to restore the systems back to normal operation.**

1. Clean the battery terminals.
2. Reconnect the positive cable to the battery first, then reconnect the negative cable to the battery.

✲✲ CAUTION

Always connect the positive cable to the battery first.

3. Apply multipurpose grease to the terminals to prevent corrosion.
4. Enter the anti-theft code(s) for the audio system and/or the navigation system (if equipped).
5. Enter the audio presets (if applicable), and enter the XM radio presets (if equipped).
6. Set the clock (for vehicles without navigation).
7. Do the steering column position memorization.

CLOCKSPRING CENTERING

1. First rotate the cable reel clockwise until it stops.
2. Then rotate it counterclockwise (three full turns) until the arrow mark on the cable reel label points straight up.
3. Position the two tabs of the turn signal canceling sleeve as shown in the installation, and install the steering wheel on to the steering column shaft, making sure the steering wheel hub engages the pins of the cable reel and tabs of the turn signal canceling sleeve. Do not tap on the steering wheel or steering column shaft when installing the steering wheel.

DRIVE TRAIN

FRONT HALFSHAFTS

REMOVAL & INSTALLATION

Front Halfshaft

4-Cylinder Engine

See Figures 9 through 16.

Special Tools Required:
- Ball Joint Thread Protector, 14 mm 07AAE-SJAA100
- Ball Joint Remover, 28 mm 07MAC-SL0A202

1. Raise and support the vehicle.
2. Remove the front wheel.
3. Pry up the stake on the spindle nut, then remove the nut.
4. Drain the transmission fluid, then reinstall the drain plug with a new sealing washer.
5. Hold the stabilizer link joint pin using a hex wrench, and remove the flange nut. Separate the front stabilizer link from the lower arm.

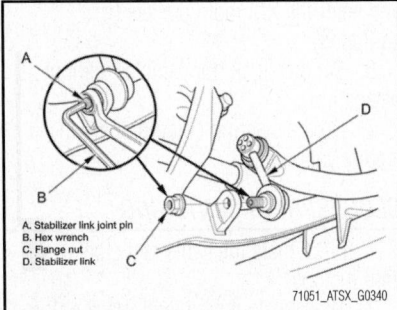

A. Stabilizer link joint pin
B. Hex wrench
C. Flange nut
D. Stabilizer link

71051_ATSX_G0340

Fig. 9 Hold the stabilizer link joint pin using a hex wrench, and remove the flange nut

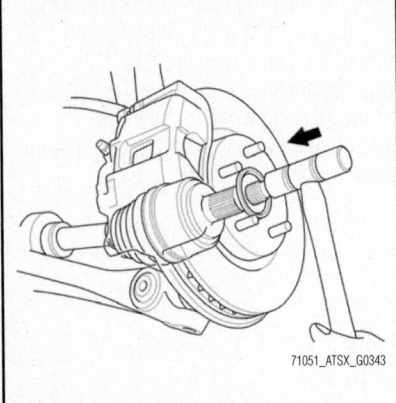

71051_ATSX_G0343

Fig. 10 Pull the knuckle outward, and separate the outboard joint from the front hub using a soft face hammer

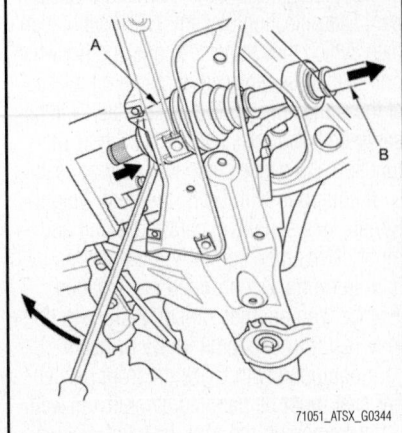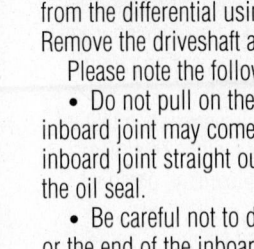

71051_ATSX_G0344

Fig. 11 Left driveshaft: Pry the inboard joint (A) from the differential using a pry bar, remove the driveshaft (B) as an assembly

6. Remove the damper pinch bolt and the damper fork mounting nut while holding the mounting bolt, then remove the damper fork from the damper and the lower arm.
7. Remove the cotter pin from the knuckle ball joint, and remove the castle nut. Separate the ball joint from the lower arm using the 14 mm ball joint thread protect and the 28 mm ball joint remover.

Please note the following
- Be careful not to damage the ball joint boot when installing the remover.
- Do not force or hammer on the lower arm, or pry between the lower arm and the knuckle. You could damage the ball joint.

8. Pull the knuckle outward, and separate the outboard joint from the front hub using a soft face hammer.

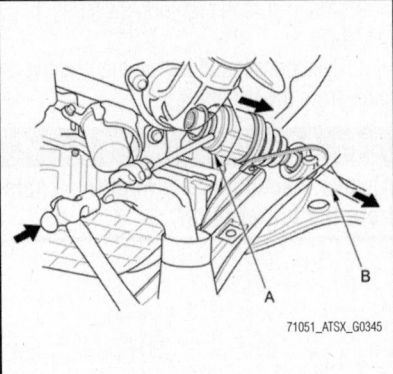

71051_ATSX_G0345

Fig. 12 Right driveshaft: Drive the inboard joint (A) off of the intermediate shaft using a drift punch and a hammer, remove the driveshaft (B) as an assembly

9. Left driveshaft: Pry the inboard joint from the differential using a pry bar. Remove the driveshaft as an assembly.
   Please note the following
   - Do not pull on the driveshaft, or the inboard joint may come apart. Pull the inboard joint straight out to avoid damaging the oil seal.
   - Be careful not to damage the oil seal or the end of the inboard joint with the pry bar.

10. Right driveshaft: Drive the inboard joint off of the intermediate shaft using a drift punch and a hammer. Remove the driveshaft as an assembly.

➡**Do not pull on the driveshaft, or the inboard joint may come apart.**

11. Remove the set ring from the left driveshaft inboard joint.
12. Remove the set ring from the intermediate shaft.

To install:

➡**Before starting installation, make sure the mating surfaces of the joint and the splined section are clean.**

13. Apply about 5 g (0.18 oz) of Moly 60 paste (P/N 08734-0001) to the contact area (A) of the outboard joint and the front wheel bearing.

➡**The paste helps prevent noise and vibration.**

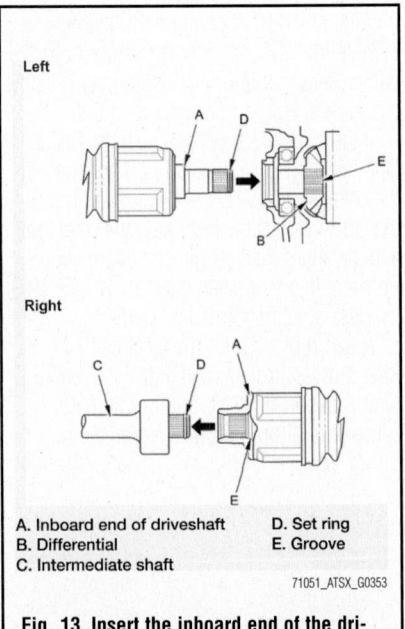

A. Inboard end of driveshaft
B. Differential
C. Intermediate shaft
D. Set ring
E. Groove

71051_ATSX_G0353

Fig. 13 Insert the inboard end of the driveshaft into the differential or the intermediate shaft until the set ring locks in the groove

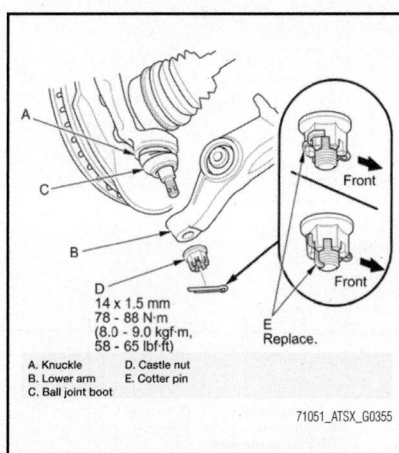

Fig. 14 Wipe off any grease contamination from the ball joint tapered section and threads, then install the knuckle onto the lower arm

14. Install a new set ring into the set ring groove of the left driveshaft inboard joint.

15. Install a new set ring into the set ring groove of the intermediate shaft.

16. Apply 0.02–0.04 oz (0.5–1.0 g) of Super High Temp Urea Grease (P/N 08798-9002) to the whole splined surface of the right driveshaft. After applying grease, remove the grease from the splined grooves at intervals of 2–3 splines and from the set ring groove so that air can bleed from the intermediate shaft.

17. Clean the areas where the driveshaft contacts the differential thoroughly with solvent, and dry them with compressed air.

➡ **Do not wash the rubber parts with solvent.**

18. Insert the inboard end of the driveshaft into the differential or the intermediate shaft until the set ring locks in the groove.

➡ **Insert the driveshaft horizontally to prevent damaging the oil seal.**

19. Install the outboard joint into the front hub on the knuckle.

20. Wipe off any grease contamination from the ball joint tapered section and threads, then install the knuckle onto the lower arm. Be careful not to damage the ball joint boot. Wipe off the grease before tightening the nut at the ball joint. Torque the castle nut to the lower torque specification, then tighten it only far enough to align the slot with the ball joint pin hole.

Please note the following
• Make sure the ball joint boot is not damaged or cracked.
• Do not align the nut by loosening it.

21. Install a new cotter pin into the ball joint pin hole, and bend the cotter pin as shown.

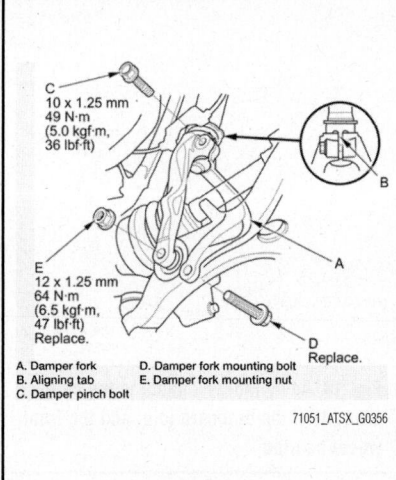

Fig. 15 Install the damper fork over the driveshaft and onto the lower arm

22. Install the damper fork over the driveshaft and onto the lower arm. Install the damper in the damper fork so the aligning tab is aligned with the slot in the damper fork. Loosely install the damper pinch bolt.

23. Loosely install a new damper fork mounting bolt and a new damper fork mounting nut.

24. Connect the front stabilizer link to the lower arm, and loosely install a new flange nut. Hold the stabilizer link joint pin using a hex wrench, and tighten the flange nut.

25. Place a floor jack under the lower arm, and raise the suspension to load it with the vehicle's weight.

➡ **Do not put the floor jack under the ball joint.**

26. Tighten the damper pinch bolt and the damper fork mounting nut while holding the damper fork mounting bolt to the speci-

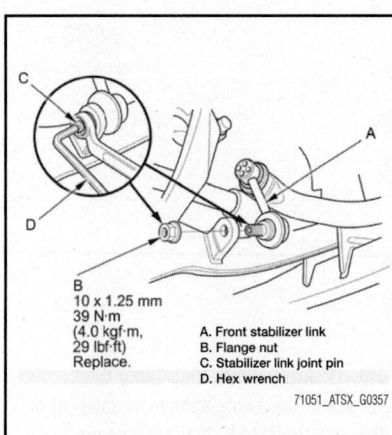

Fig. 16 Connect the front stabilizer link to the lower arm, and loosely install a new flange nut

fied torque values, then remove the floor jack.

27. Apply a small amount of engine oil to the seating surface of a new spindle nut.

28. Install the spindle nut, then tighten it. After tightening to 242 ft. lbs. (328 Nm), use a drift to stake the spindle nut shoulder against the driveshaft.

29. Clean the mating surfaces between the brake disc and the inside of the wheel, then install the front wheel.

30. Turn the wheel by hand, and make sure there is no interference between the driveshaft and surrounding parts.

31. Refill the transmission with the recommended transmission fluid.

32. Lower the vehicle.

33. Check the wheel alignment, and adjust it if necessary.

34. Test-drive the vehicle.

6-Cylinder Engine

See Figures 17 through 21.

Special Tools Required:
• Ball Joint Thread Protector, 14 mm 07AAE-SJAA100
• Ball Joint Remover, 28 mm 07MAC-SL0A202

1. Raise and support the vehicle.
2. Remove the front wheel.
3. Pry up the stake on the spindle nut, then remove the nut.
4. Drain the transmission fluid, then reinstall the drain plug with a new sealing washer.
5. Hold the stabilizer link joint pin using a hex wrench, and remove the flange nut. Separate the front stabilizer link from the lower arm.
6. Remove the damper pinch bolt and the damper fork mounting nut while holding the mounting bolt, then remove the damper fork from the damper and the lower arm.
7. Remove the cotter pin from the knuckle ball joint, then remove the castle nut. Separate the ball joint from the lower arm using the 14 mm ball joint thread protector and the 28 mm ball joint remover.

Please note the following
• Be careful not to damage the ball joint boot when installing the remover.
• Do not force or hammer on the lower arm, or pry between the lower arm and the knuckle. You could damage the ball joint.

8. Pull the knuckle outward, and separate the outboard joint from the front hub using a soft face hammer.
9. Remove exhaust pipe A.
10. Left driveshaft: Pry the inboard joint from the differential using a pry bar. Remove the driveshaft as an assembly.

Please note the following

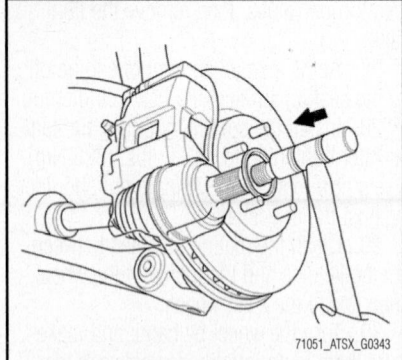

Fig. 17 Pull the knuckle outward, and separate the outboard joint from the front hub using a soft face hammer

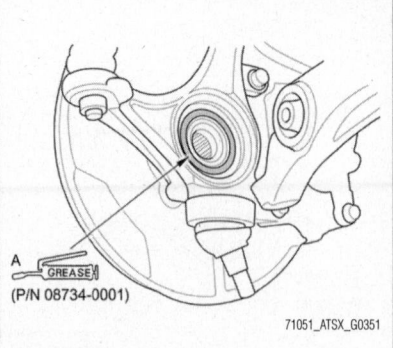

Fig. 18 Apply Moly 60 paste to the contact area (A) of the outboard joint and the front wheel bearing

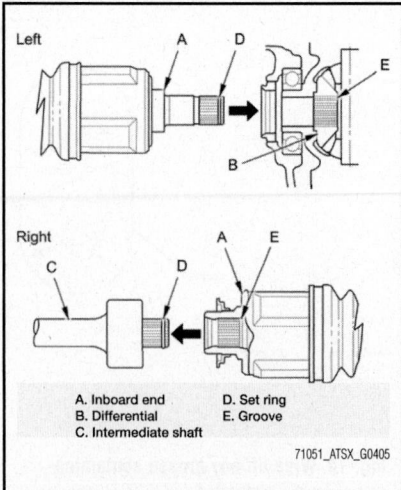

Fig. 20 Insert the inboard end of the driveshaft into the differential or the intermediate shaft until the set ring locks in the groove

• Do not pull on the driveshaft, or the inboard joint may come apart. Pull the inboard joint straight out to avoid damaging the oil seal.

• Be careful not to damage the oil seal or the end of the inboard joint with the pry bar.

11. Right driveshaft: Drive the inboard joint off of the intermediate shaft using a drift punch and a hammer. Remove the driveshaft as an assembly.

➡**Do not pull on the driveshaft, or the inboard joint may come apart.**

12. Remove the set ring from the left driveshaft inboard joint.

13. Remove the set ring from the intermediate shaft.

To install:

➡**Before starting installation, make sure the mating surfaces of the joint and the splined section are clean.**

14. Apply about 0.18 oz (5 g) of Moly 60 paste (P/N 08734-0001) to the contact area (A) of the outboard joint and the front wheel bearing.

➡**The paste helps prevent noise and vibration.**

15. Install a new set ring into the set ring groove of the left driveshaft inboard joint.

16. Install a new set ring into the set ring groove of the intermediate shaft.

17. Apply 0.02–0.04 oz (0.5–1.0 g) of Super High Temp Urea Grease (P/N 08798-9002) to the whole splined surface (A) of the right driveshaft. After applying grease, remove the grease from the splined grooves at intervals of 2–3 splines and from the set ring groove (B) so that air can bleed from the intermediate shaft.

18. Clean the areas where the driveshaft contacts the differential thoroughly with solvent, and dry them with compressed air.

➡**Do not wash the rubber parts with solvent.**

19. Insert the inboard end of the driveshaft into the differential or the intermediate shaft until the set ring locks in the groove.

➡**Insert the driveshaft horizontally to prevent damaging the oil seal.**

20. Install the outboard joint into the front hub on the knuckle.

21. Install exhaust pipe A.

22. Wipe off any grease contamination from the ball joint tapered section and threads, then install the knuckle onto the lower arm. Be careful not to damage the ball joint boot. Wipe off the grease before tightening the nut at the ball joint. Torque the castle nut to the lower torque specification, then tighten it only far enough to align the slot with the ball joint pin hole.

Please note the following

• Make sure the ball joint boot is not damaged or cracked.

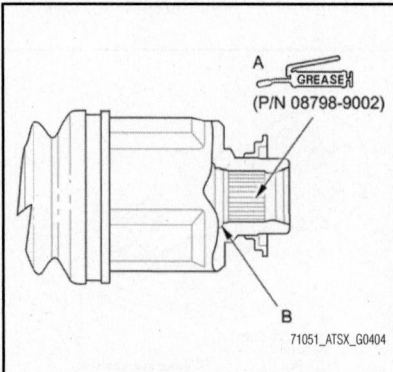

Fig. 19 Apply Super High Temp Urea Grease to the whole splined surface (A) of the right driveshaft, then remove the grease from the splined grooves at intervals of 2–3 splines and from the set ring groove (B)

• Do not align the nut by loosening it.

23. Install a new cotter pin into the ball joint pin hole, and bend the cotter pin as shown.

24. Install the damper fork over the driveshaft and onto the lower arm. Install the damper in the damper fork so the aligning tab is aligned with the slot in the damper fork. Loosely install the damper pinch bolt.

25. Loosely install a new damper fork mounting bolt and a new damper fork mounting nut.

26. Connect the front stabilizer link to the lower arm, and loosely install a new flange nut. Hold the stabilizer link joint pin using a hex wrench, and tighten the flange nut.

27. Place a floor jack under the lower arm, and raise the suspension to load it with the vehicle's weight.

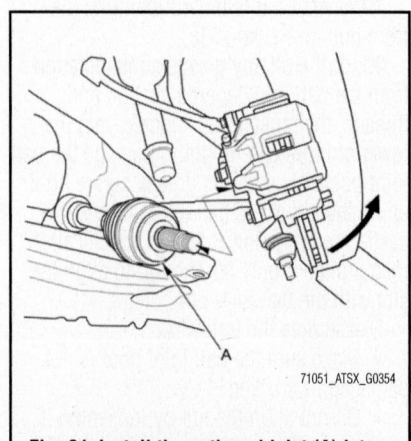

Fig. 21 Install the outboard joint (A) into the front hub on the knuckle

→**Do not put the floor jack under the ball joint.**

28. Tighten the damper pinch bolt and the damper fork mounting nut while holding the damper fork mounting bolt to the specified torque values, then remove the floor jack.

29. Apply a small amount of engine oil to the seating surface of a new spindle nut.

30. Install the spindle nut, then tighten it. After tightening to 242 ft. lbs. (328 Nm), use a drift to stake the spindle nut shoulder against the driveshaft.

31. Clean the mating surfaces between the brake disc and the inside of the wheel, then install the front wheel.

32. Turn the wheel by hand, and make sure there is no interference between the driveshaft and surrounding parts.

33. Refill the transmission with the recommended transmission fluid.

34. Lower the vehicle.

35. Check the wheel alignment, and adjust it if necessary.

36. Test-drive the vehicle.

Intermediate Shaft

4-Cylinder Engine

See Figure 22.

1. Drain the transmission fluid, then reinstall the drain plug with a new sealing washer:

2. Remove the right driveshaft.

3. M/T model: Remove the CKP sensor cover from the engine block.

4. Remove the flange bolt and the two dowel bolts.

5. Remove the intermediate shaft from the differential. Hold the intermediate shaft horizontal until it is clear of the differential to prevent damaging the oil seal.

To install:

6. Clean the areas where the intermediate shaft contacts the differential thoroughly with solvent, and dry them with compressed air.

→**Do not wash the rubber parts with solvent.**

7. Install a new set ring onto the set ring groove of the intermediate shaft.

8. Insert the intermediate shaft into the differential correctly.

→**Insert the intermediate shaft carefully to prevent damaging the oil seal.**

9. Install the flange bolt and the two dowel bolts.

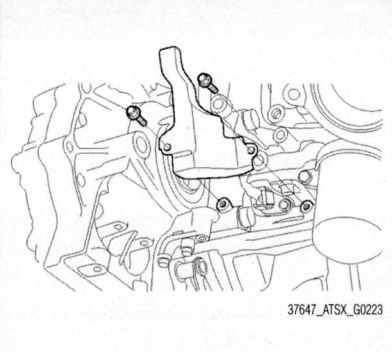

Fig. 22 M/T model: Remove the CKP sensor cover (A) from the engine block

10. Install the right driveshaft.

11. M/T model: Install the CKP sensor cover to the engine block.

12. Refill the transmission with the recommended transmission fluid:

13. Check the wheel alignment, and adjust it if necessary.

14. Test-drive the vehicle.

6-Cylinder Engine

See Figures 23 and 24.

1. Drain the transmission fluid, then reinstall the drain plug with a new sealing washer.

2. Remove exhaust pipe A.

3. Remove the right driveshaft.

4. Remove the rear Warm Up Three Way Catalytic Converter (WU-TWC) bracket.

5. Remove the Crankshaft Position (CKP) sensor cover.

6. Disconnect the CKP sensor connector, and remove the sensor harness clamp bracket bolt.

7. Swing the sensor harness out of the way to prevent damaging it.

8. Remove the flange bolt and the two dowel bolts.

9. Remove the intermediate shaft from the differential. Hold the intermediate shaft horizontal until it is clear of the differential to prevent damaging the oil seal.

To install:

10. Clean the areas where the intermediate shaft contacts the differential thoroughly with solvent, and dry them with compressed air.

→**Do not wash the rubber parts with solvent.**

11. Install a new set ring into the set ring groove of the intermediate shaft.

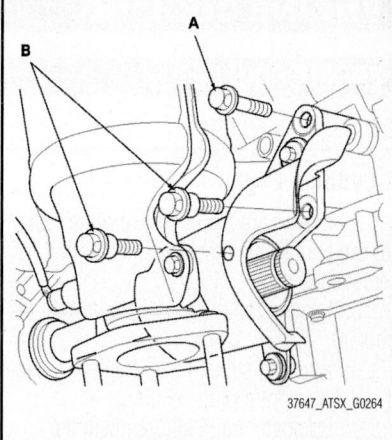

Fig. 23 Remove the flange bolt (A) and the two dowel bolts (B)

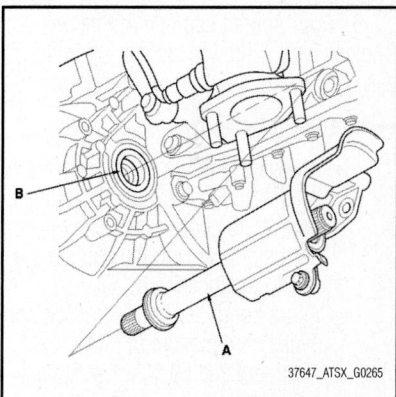

Fig. 24 Remove the intermediate shaft (A) from the differential; prevent damaging the oil seal (B)

12. Insert the intermediate shaft into the differential correctly.

→**Insert the intermediate shaft carefully to prevent damaging the oil seal.**

13. Install the flange bolt and the two dowel bolts.

14. Install the rear WU-TWC bracket.

15. Install the right driveshaft.

16. Connect the CKP sensor connector, and install the sensor harness clamp bracket.

17. Install the CKP sensor cover.

18. Install exhaust pipe A.

19. Refill the transmission with the recommended transmission fluid.

20. Check the wheel alignment, and adjust it if necessary.

21. Test-drive the vehicle.

ENGINE COOLING

ENGINE COOLANT

DRAIN & REFILL

4-Cylinder Engine

1. Make sure the engine and the radiator are cool to the touch, then carefully remove the cap.

2. Loosen the drain plug, and drain the coolant.

3. After the coolant has drained, tighten the radiator drain plug securely.

4. Remove, drain, and reinstall the coolant reservoir.

5. Fill the coolant reservoir to the MAX mark with Acura Long Life Antifreeze/Coolant Type 2.

6. Pour Acura Long Life Antifreeze/Coolant Type 2 into the radiator up to the base of the filler neck.

➡Note the following:

- Always use Acura Long Life Antifreeze/Coolant Type 2. Using a non-Acura coolant can result in corrosion, causing the cooling system to malfunction or fail.
- Acura Long Life Antifreeze/Coolant Type 2 is a mixture of 50 % antifreeze and 50 % water. Do not add water.
- If the vehicle is regularly driven in very low temperatures (under -31°F, -35°C) a 60 % concentration of coolant should be used. To accomplish this, pour 0.4 gal. (1.6 L) of Acura Extreme Cold Weather Antifreeze/Coolant Type 2 into the radiator first, then add Acura Long Life antifreeze/Coolant Type 2 until the radiator is full.

7. Loosely install the radiator cap.

8. Start the engine, and let it run until it warms up (the radiator fan comes on at least twice).

9. Turn off the engine. Check the level in the radiator, and add Acura Long Life Antifreeze/Coolant Type 2, if needed.

10. Put the radiator cap on securely, then start the engine again, and check for leaks.

11. Clean up any spilled engine coolant.

12. If the maintenance minder required engine coolant replacement, reset the maintenance minder, and this procedure is complete. If the maintenance minder did not require engine coolant replacement, go to step 15.

13. Turn the ignition switch to LOCK (0).

14. Connect the Honda Diagnostic System (HDS) to the Data Link Connector (DLC).

15. Turn the ignition switch to ON (II).

16. Make sure the HDS communicates with the vehicle and the Engine Control Module (ECM)/Powertrain Control Module (PCM). If it does not communicate, troubleshoot the DLC circuit.

17. Select GAUGES in the BODY ELECTRICAL with the HDS.

18. Select ADJUSTMENT in the GAUGES with the HDS.

19. Select MAINTENANCE MINDER in the ADJUSTMENT with the HDS.

20. Select RESET in the MAINTENANCE MINDER with the HDS.

21. Select MAINTENANCE SUB ITEM 5 RESET with the HDS.

6-Cylinder Engine

See Figures 25 and 26.

1. Remove the engine compartment covers.

 a. Remove the left engine compartment cover.

 b. Detach the clip, and release the hooks.

 c. Pull up the left engine compartment cover to release the pins.

 d. Remove the right engine compartment cover.

 e. Detach the clip, and release the hooks.

 f. Pull up the right engine compartment cover to release the pins.

 g. Detach the clips, and release the hooks, then remove the front grille cover.

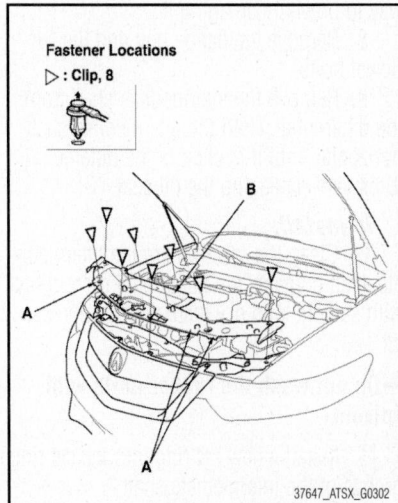

Fig. 25 Detach the clips, and release the hooks (A), then remove the front grille cover (B)

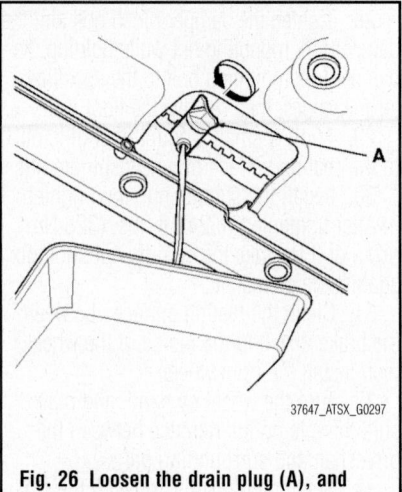

Fig. 26 Loosen the drain plug (A), and drain the coolant

 h. If necessary, detach the clips, then remove the front fender trim.

➡The left front fender trim is shown; the right front fender trim is similar.

2. Make sure the engine and the radiator are cool to the touch, then carefully remove the radiator cap.

3. Loosen the drain plug, and drain the coolant.

4. Install a rubber hose on the drain bolt located at the rear of the engine block, then loosen the drain bolt.

5. When the coolant stops draining, tighten the drain bolt. Remove the rubber hose.

6. Tighten the radiator drain plug securely.

7. Remove, drain, and reinstall the coolant reservoir.

8. Fill the coolant reservoir to the MAX mark with Acura Long Life Antifreeze/Coolant Type 2 (P/N OL999-9011A).

9. Pour Acura Long Life Antifreeze/Coolant Type 2 into the radiator up to the base of the filler neck.

➡Note the following:

- Always use Acura Long Life Antifreeze/Coolant Type 2. Using a non-Acura coolant can result in corrosion, causing the cooling system to malfunction or fail.
- Acura Long Life Antifreeze/Coolant Type 2 is a mixture of 50 % antifreeze and 50 % water. Do not add water.

10. Loosely install the radiator cap.

11. Start the engine, and let it run until it warms up (the radiator fan comes on at least twice).

12. Turn off the engine. Check the level in the radiator, and add Acura Long Life Antifreeze/Coolant Type 2, if needed.

13. Put the radiator cap on securely, then run the engine again, and check for leaks.

14. Clean up any spilled engine coolant.

15. Install the engine compartment covers.

16. If the maintenance minder required engine coolant replacement, reset the maintenance minder, and this procedure is complete. If the maintenance minder did not require engine coolant replacement, go to step 17.

17. Turn the ignition switch to LOCK (0).

18. Connect the Honda Diagnostic System (HDS) to the Data Link Connector (DLC).

19. Turn the ignition switch to ON (II).

20. Make sure the HDS communicates with the vehicle and the Powertrain Control Module (PCM). If it does not communicate, troubleshoot the DLC circuit.

21. Select GAUGES in the BODY ELECTRICAL with the HDS.

22. Select ADJUSTMENT in the GAUGES with the HDS.

23. Select MAINTENANCE MINDER in the ADJUSTMENT with the HDS.

24. Select RESET in the MAINTENANCE MINDER with the HDS.

25. Select MAINTENANCE SUB ITEM 5 RESET with the HDS.

ELECTRIC ENGINE FAN

REMOVAL & INSTALLATION

4-Cylinder Engine

See Figures 27 and 28.

1. Disconnect the fan motor connectors, and remove the harness clamp.

2. Remove the A/C condenser fan shroud assembly.

3. Remove the radiator fan shroud assembly.

➡**Move the radiator fan shroud assembly toward the right side of the vehicle to allow enough space to lift it up and away from the A/C condenser fan shroud assembly.**

4. Disassemble the fan shrouds.

To install:

5. Assemble the fan shrouds.

6. Install the radiator fan shroud assembly.

7. Install the A/C condenser fan shroud assembly.

8. Connect the fan motor connectors, and install the harness clamp.

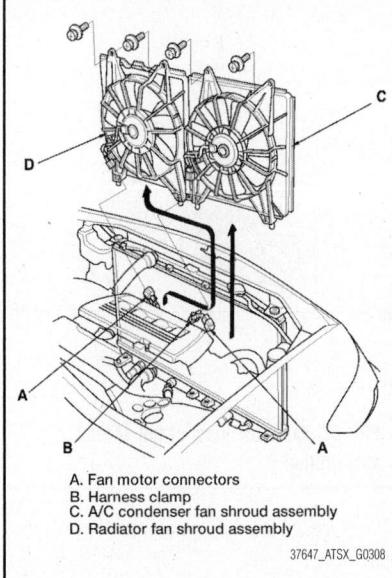

A. Fan motor connectors
B. Harness clamp
C. A/C condenser fan shroud assembly
D. Radiator fan shroud assembly

37647_ATSX_G0308

Fig. 27 Remove the radiator fan shroud assembly

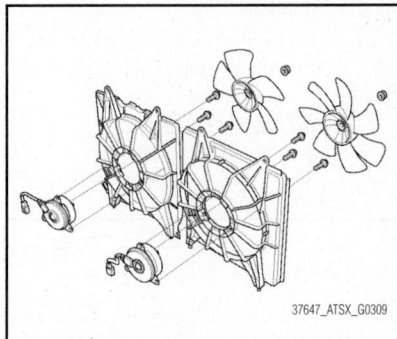

37647_ATSX_G0309

Fig. 28 Exploded view of fan shroud assemblies

6-Cylinder Engine

See Figures 29 through 32.

1. Remove the engine compartment covers.

    a. Remove the left engine compartment cover.

    b. Detach the clip, and release the hooks.

    c. Pull up the left engine compartment cover to release the pins.

    d. Remove the right engine compartment cover.

    e. Detach the clip, and release the hooks.

    f. Pull up the right engine compartment cover to release the pins.

    g. Detach the clips, and release the hooks, then remove the front grille cover.

    h. If necessary, detach the clips, then remove the front fender trim.

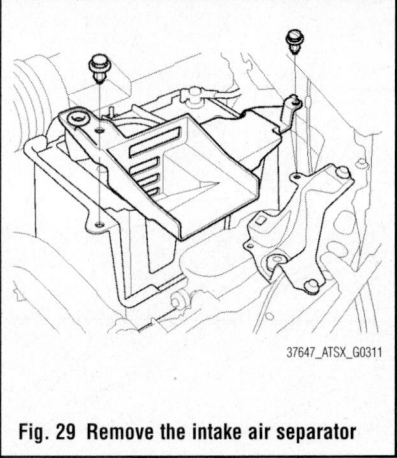

37647_ATSX_G0311

Fig. 29 Remove the intake air separator

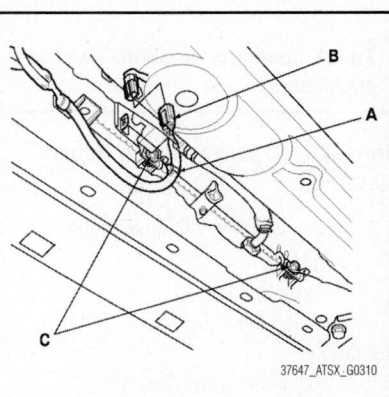

37647_ATSX_G0310

Fig. 30 Disconnect the A/C condenser fan motor connector (A), remove the harness clamp (B) and loosen the A/C condenser fan shroud mounting bolts (C)

➡**The left front fender trim is shown; the right front fender trim is similar.**

2. Do the battery removal procedure.

3. Remove the battery base.

4. Remove the intake air separator.

5. Remove the intake air ducts.

6. Raise the vehicle on the lift.

7. Remove the splash shield.

8. Disconnect the A/C condenser fan motor connector, and remove the harness clamp.

9. Loosen the A/C condenser fan shroud mounting bolts.

10. Disconnect the radiator fan motor connector.

11. Lower the vehicle on the lift.

12. Remove the A/C condenser fan shroud assembly.

13. Remove the upper brackets.

14. Remove the radiator fan shroud assembly.

➡**Pull-up the radiator, then move the radiator fan shroud assembly toward the right side of the vehicle to allow for enough space to lift it up and away**

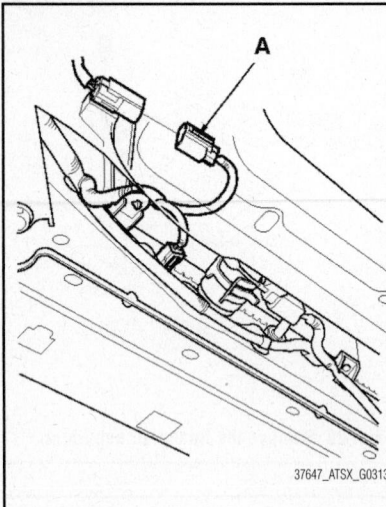

Fig. 31 Disconnect the radiator fan motor connector (A)

from the A/C condenser fan shroud assembly.

15. Disassemble the fan shrouds.

To install:

16. Assemble the fan shrouds.
17. Install the radiator fan shroud assembly.
18. Install the upper brackets.

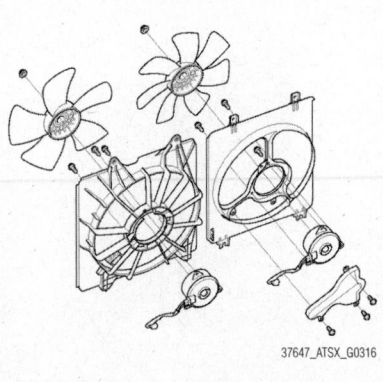

Fig. 32 Exploded view of fan shroud assemblies

19. Install the A/C condenser fan shroud assembly.
20. Raise the vehicle on the lift.
21. Connector the radiator fan motor connector.
22. Tighten the A/C condenser fan shroud mounting bolts.
23. Connect the A/C condenser fan motor connector, and install the harness clamp.
24. Install the splash shield.
25. Lower the vehicle on the lift.

26. Install the intake air duct.
27. Install the intake air separator.
28. Install the battery base.
29. Do the battery installation procedure.
30. Install the engine compartment covers.

RADIATOR

REMOVAL & INSTALLATION

4-Cylinder Engine

See Figures 33 through 35.

1. Drain the engine coolant.
2. Raise the vehicle on the lift to full height.
3. Remove the splash shield.
4. Disconnect the lower radiator hose.
5. A/T model: Disconnect the Automatic Transmission Fluid (ATF) cooler hoses, then plug the hoses and the lines.
6. Disconnect the fan motor connectors and Engine Coolant Temperature (ECT) sensor 2 connector, then remove the harness clamps.
7. Lower the vehicle on the lift.
8. Remove the grille cover.
9. Remove the clips securing the intake air duct cover.
10. Disconnect the upper radiator hose.
11. Disconnect the coolant reservoir

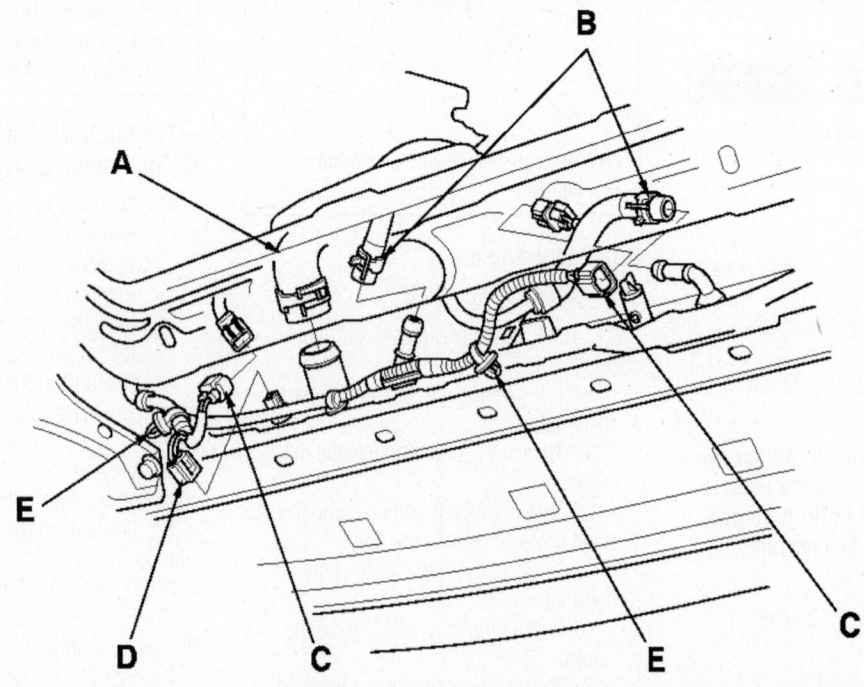

A. Lower radiator hose
B. ATF cooler hoses
C. Fan motor connectors
D. ECT sensor 2 connector
E. Harness clamps

Fig. 33 Disconnect the lower radiator hose

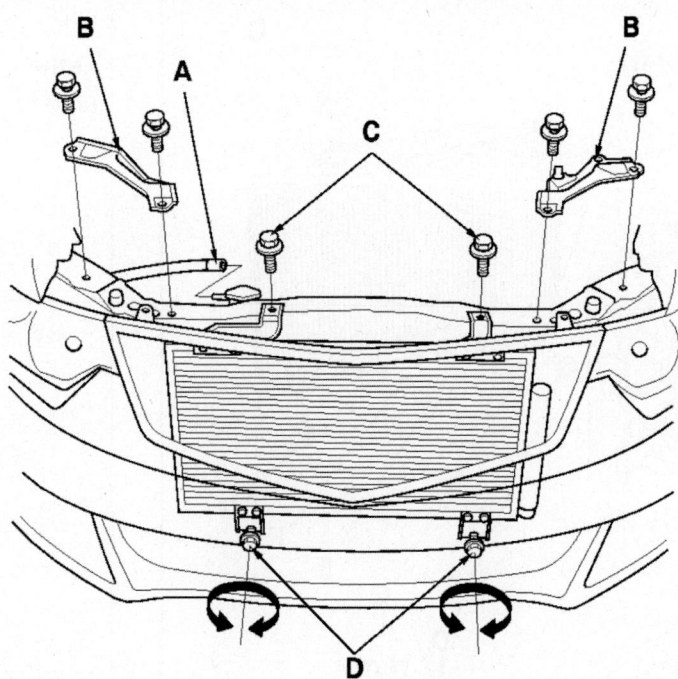

A. Coolant reservoir hose
B. Radiator upper brackets
C. A/C condenser upper mounting bolts
D. A/C condenser lower mounting bolts

37647_ATSX_G0322

Fig. 34 Disconnect the coolant reservoir hose and remove the radiator upper brackets

hose and remove the radiator upper brackets.

12. Remove the A/C condenser upper mounting bolts and loosen the A/C condenser lower mounting bolts.

13. Pull up the radiator, then remove the radiator fan shroud assembly, the A/C condenser fan shroud assembly, the radiator cap, ECT sensor 2, and the drain plug.

To install:

14. Reassemble the radiator with new O-rings.

15. Install the radiator. Make sure the lower cushions are set securely.

16. Set the A/C condenser in place, then tighten the upper mounting bolts and the A/C condenser lower mounting bolts.

17. Connect the coolant reservoir hose and install the radiator upper brackets.

18. Connect the upper radiator hose.

19. Install the clips securing the intake air duct cover.

20. Install the grille cover.

21. Raise the vehicle on the lift to full height.

22. Connect the fan motor connectors and ECT sensor 2 connector, then install the harness clamps.

23. Connect the lower radiator hose.

24. A/T model: Connect the Automatic Transmission Fluid (ATF) cooler hoses.

25. Install the splash shield.

26. Lower the vehicle on the lift.

27. Refill the radiator with engine coolant, and bleed the air from the cooling system with the heater valve open.

28. Clean up any spilled engine coolant.

6-Cylinder Engine

See Figures 36 through 40.

1. Remove the engine compartment covers.

   a. Remove the left engine compartment cover.

   b. Detach the clip, and release the hooks.

   c. Pull up the left engine compartment cover to release the pins.

   d. Remove the right engine compartment cover.

   e. Detach the clip, and release the hooks.

   f. Pull up the right engine compartment cover to release the pins.

   g. Detach the clips, and release the hooks, then remove the front grille cover.

   h. If necessary, detach the clips, then remove the front fender trim.

➡ **The left front fender trim is shown; the right front fender trim is similar.**

2. Drain the engine coolant.

3. Remove the intake air separator.

4. Remove the intake air ducts.

5. Raise the vehicle on the lift.

6. Remove the splash shield.

7. Disconnect the A/C condenser fan motor connector, and remove the harness clamp.

8. Loosen the A/C condenser fan shroud mounting bolts.

9. Disconnect the radiator fan motor connector and the Engine Coolant Temperature (ECT) sensor 2 connector, then remove the harness clamp.

10. Lower the vehicle on the lift.

11. Disconnect the upper radiator hose, then remove the radiator fan shroud assembly and the A/C condenser fan shroud assembly.

12. Raise the vehicle on the lift.

13. Disconnect the lower radiator hose.

14. Disconnect the Automatic Transmission Fluid (ATF) cooler hoses from the radiator, then plug the cooler hoses and the lines.

15. Lower the vehicle on the lift.

16. Remove the upper brackets, then pull up the radiator.

17. Remove the other parts from the radiator.

To install:

18. Install radiator in the reverse order of removal.

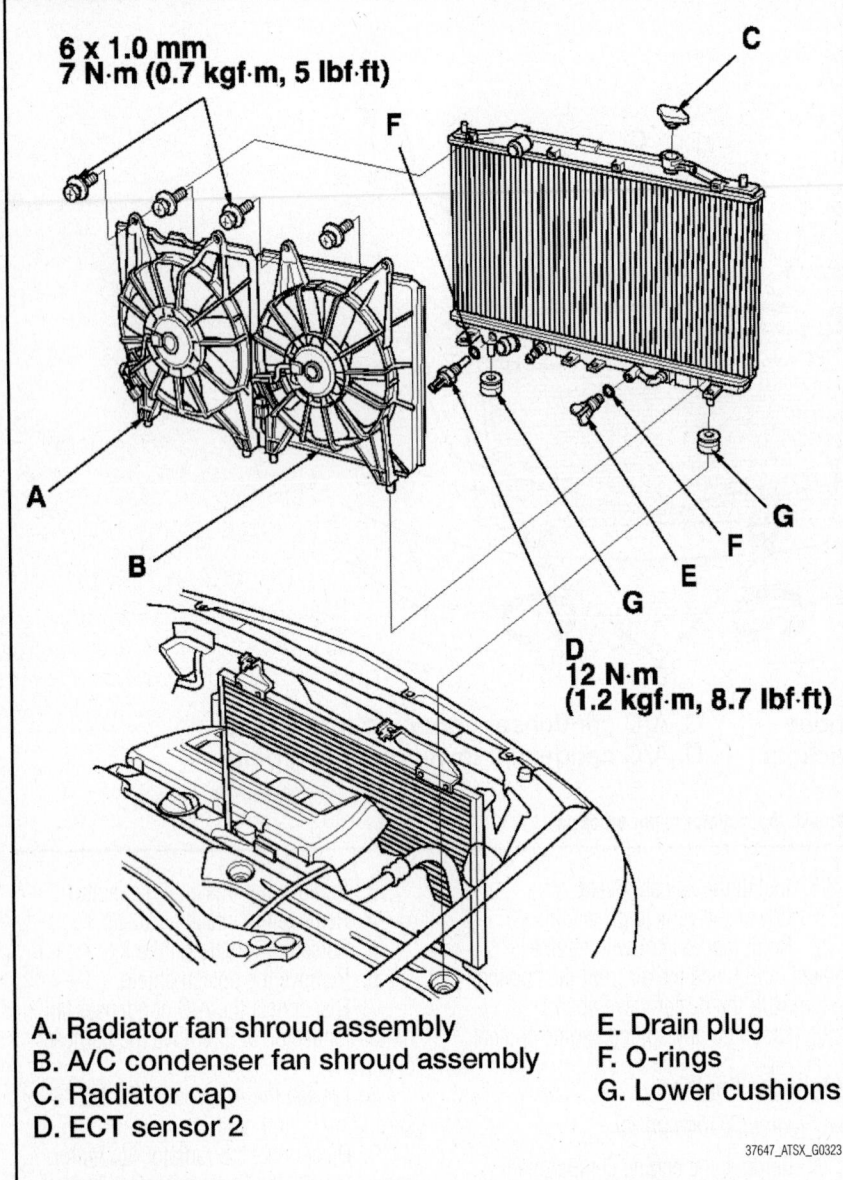

A. Radiator fan shroud assembly
B. A/C condenser fan shroud assembly
C. Radiator cap
D. ECT sensor 2

E. Drain plug
F. O-rings
G. Lower cushions

37647_ATSX_G0323

Fig. 35 Remove the radiator

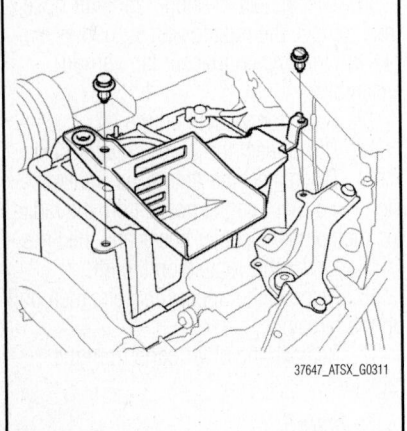

37647_ATSX_G0311

Fig. 36 Remove the intake air separator

➡**Make sure the upper and lower cushions are set securely.**

19. Fill the radiator with engine coolant, and bleed the air from the cooling system with the heater valve open.
20. Clean up any spilled engine coolant.
21. Install the engine compartment covers.

THERMOSTAT

REMOVAL & INSTALLATION

4-Cylinder Engine

See Figures 41 and 42.

1. Drain the engine coolant.
2. Clean any dirt off the quick connector, the thermostat cover, and the lower radiator hose.

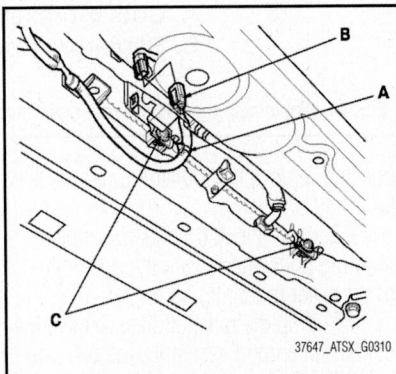

37647_ATSX_G0312

Fig. 37 Remove the intake air ducts

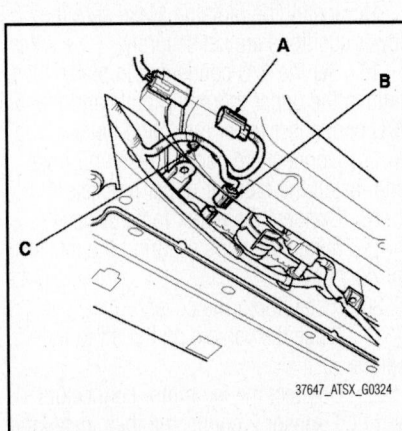

37647_ATSX_G0310

Fig. 38 Disconnect the A/C condenser fan motor connector (A), remove the harness clamp (B) and loosen the A/C condenser fan shroud mounting bolts (C)

37647_ATSX_G0324

Fig. 39 Disconnect the radiator fan motor connector (A) and the ECT sensor 2 connector (B), then remove the harness clamp (C)

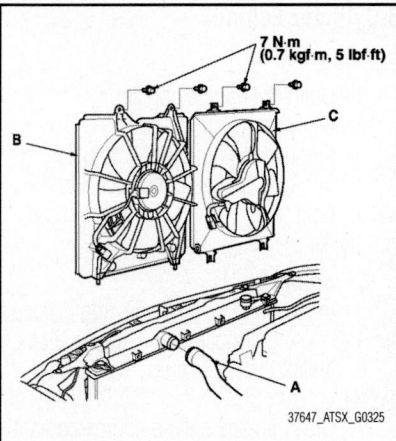

Fig. 40 Disconnect the upper radiator hose (A), then remove the radiator fan shroud assembly (B) and the A/C condenser fan shroud assembly (C)

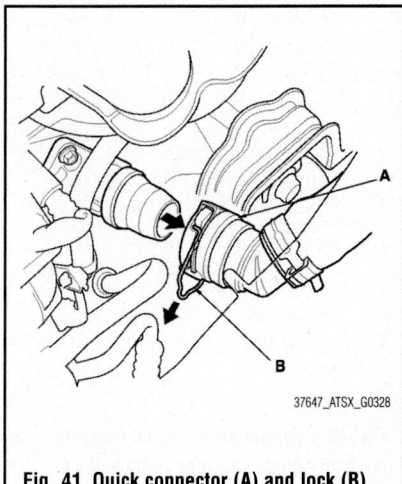

Fig. 41 Quick connector (A) and lock (B)

3. Pull out the lock by hand, then wiggle the quick connector loose, and remove it from the thermostat cover. Do not use any tools to remove the quick connector.

4. Remove the thermostat.

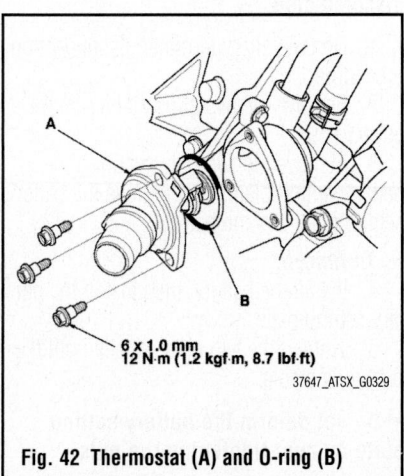

Fig. 42 Thermostat (A) and O-ring (B)

5. Install a new thermostat with a new O-ring.

6. Check the quick connector and the set ring for cracks or damage. If the connector and/or the set ring are cracked or damaged, replace the connector.

7. Make sure the set ring is in place inside the quick connector. If the set ring is off the connector, replace the quick connector.

8. Replace a new O-ring in the quick connector.

9. Check the lock. If the lock is damaged or deformed, replace it. When installing the new lock to the connector, push it straight down along the groove.

10. Clean the connecting surface of the thermostat cover, then apply clean engine coolant around the connecting surface.

11. Push the lock down, then push the quick connector onto the thermostat cover until you hear it click.

12. Refill the radiator with engine coolant, and bleed the air from the cooling system with the heater valve open.

6-Cylinder Engine

See Figures 43 and 44.

1. Remove the engine compartment covers.

a. Remove the left engine compartment cover.

b. Detach the clip, and release the hooks.

c. Pull up the left engine compartment cover to release the pins.

d. Remove the right engine compartment cover.

e. Detach the clip, and release the hooks.

f. Pull up the right engine compartment cover to release the pins.

g. Detach the clips, and release the hooks, then remove the front grille cover.

h. If necessary, detach the clips, then remove the front fender trim.

➡**The left front fender trim is shown; the right front fender trim is similar.**

2. Drain the engine coolant.

3. Do the battery removal procedure.

4. Remove harness clamps and disconnect the positive starter cable.

5. Disconnect the A/T clutch pressure control solenoid valve C connector, and remove the bolt and the harness holder.

6. Remove the thermostat cover, then remove the thermostat.

7. Install the new thermostat with a new

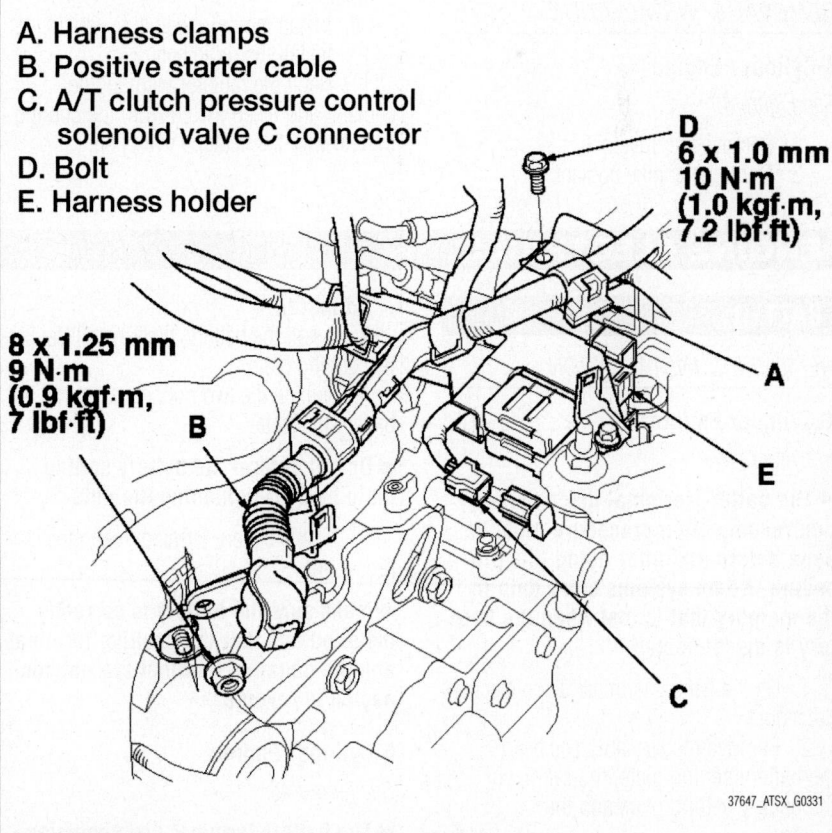

A. Harness clamps
B. Positive starter cable
C. A/T clutch pressure control solenoid valve C connector
D. Bolt
E. Harness holder

Fig. 43 Remove harness clamps and disconnect the positive starter cable

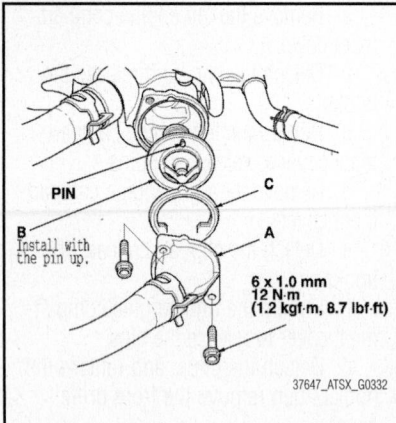

Fig. 44 Thermostat cover (A), thermostat (B), rubber seal (C)

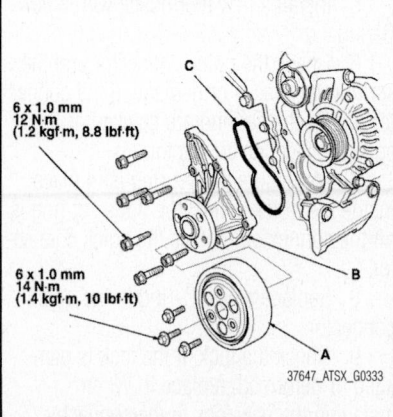

Fig. 45 Exploded view of water pump assembly showing the water pump pulley (A), water pump (B) and O-ring (C)

rubber seal in the reverse order of the removal.

8. Do the battery installation procedure.

9. Refill the radiator with engine coolant, and bleed the air from the cooling system with the heater valve open.

10. Clean up any spilled engine coolant.

11. Install the engine compartment covers.

WATER PUMP

REMOVAL & INSTALLATION

4-Cylinder Engine

See Figure 45.

1. Remove the drive belt.
2. Drain the engine coolant.

3. Remove the drive belt auto-tensioner.

4. Remove the water pump pulley.

5. Remove the six bolts securing the water pump, then remove the water pump.

To install:

6. Inspect and clean the O-ring groove and the mating surface of the water passage.

7. Install the water pump with a new O-ring and the water pump pulley.

8. Clean up any spilled engine coolant.

9. Install the drive belt auto-tensioner.

10. Install the drive belt.

11. Refill the radiator with engine coolant, and bleed the air from the cooling system with the heater valve open.

6-Cylinder Engine

See Figure 46

1. Drain the engine coolant.
2. Remove the timing belt.
3. Remove the timing belt adjuster.
4. Remove the five bolts securing the water pump, then remove the water pump and gasket.

To install:

5. Inspect and clean the O-ring groove and the mating surface of the engine block.

6. Install the water pump with a new gasket.

7. Clean up any spilled engine coolant.

8. Install the timing belt adjuster:

9. Install the timing belt:

10. Refill the radiator with engine coolant, and bleed the air from the cooling system.

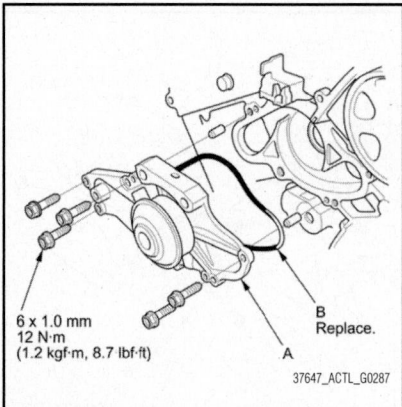

Fig. 46 Remove the five bolts securing the water pump (A), then remove the water pump and gasket (B)

ENGINE ELECTRICAL

BATTERY

REMOVAL & INSTALLATION

4-Cylinder Engine

See Figure 47.

➡The battery terminal disconnection/reconnection procedure must be done before and after doing this procedure. Some systems store data in the memory that is lost when the battery is disconnected.

1. Do the battery terminal disconnection procedure.

2. Remove the two nuts securing the battery setting plate, then remove the battery setting plate and the battery.

To install:

3. Install the battery, then install the battery setting plate.

4. Tighten the two nuts equally until the battery is stable.

➡Do not deform the battery setting plate by over-tightening the nuts.

5. Do the battery terminal reconnection procedure.

➡Make sure the battery is correctly installed and that the positive terminal and the negative terminal are not connected in reverse.

6-Cylinder Engine

See Figure 48.

➡The battery terminal disconnection and reconnection procedure must be

BATTERY SYSTEM

done before and after doing this procedure. Some systems store data in the memory that is lost when the battery is disconnected.

1. Do the battery terminal disconnection procedure.

2. Remove the two clips and air intake cover.

3. Remove the two nuts securing the battery setting plate, then remove the battery setting plate and the battery.

To install:

4. Install the battery, then install the battery setting plate.

5. Tighten the two nuts equally until the battery is stable.

➡Do not deform the battery setting plate by over-tightening the nuts.

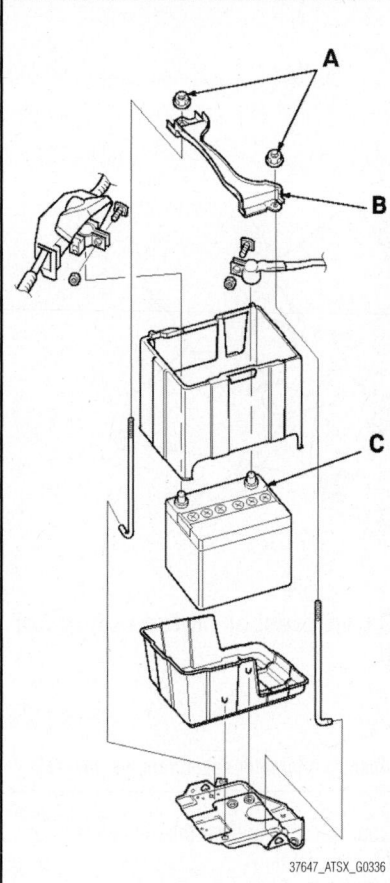

Fig. 47 Remove the two nuts (A) securing the battery setting plate, then remove the battery setting plate (B) and the battery (C)

6. Install the air intake cover and two clips.

7. Do the battery terminal reconnection procedure.

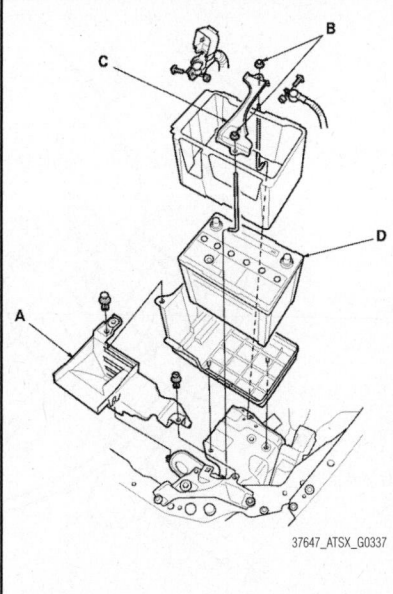

Fig. 48 Exploded view of battery assembly

➡**Make sure the battery is correctly installed, and that the positive terminal and the negative terminal are not connected in reverse.**

BATTERY RECONNECT/RELEARN PROCEDURE

Disconnection

➡**Some systems store data in memory (including seat position, mirror position, etc) that is lost when the battery is disconnected. Do the following procedures before disconnecting the battery.**

1. Make sure you have the anti-theft code(s) for the audio and/or the navigation system (if equipped).

2. If you are replacing the audio unit, write down the audio presets (AM and FM), and the XM audio presets (if equipped), because the audio unit does not retain the presets after the battery is disconnected.

3. Make sure the ignition switch is in LOCK (0).

4. Disconnect and isolate the negative cable from the battery.

➡**Always disconnect the negative cable from the battery first.**

5. Disconnect the positive cable from the battery.

Reconnection

➡**Some systems store data in memory (including seat position, mirror position, etc) that is lost when the battery is disconnected. Do the following procedures to restore the systems back to normal operation.**

1. Clean the battery terminals.
2. Test the battery.
3. Reconnect the positive cable to the battery first, then reconnect the negative cable to the battery.

➡**Always connect the positive cable to the battery first.**

4. Apply multipurpose grease to the terminals to prevent corrosion.

5. Enter the anti-theft code(s) for the audio system and/or the navigation system (if equipped).

6. Enter the audio presets (if applicable), and enter the XM audio presets (if equipped).

7. Set the clock (for vehicles without navigation).

ENGINE ELECTRICAL

ALTERNATOR

REMOVAL & INSTALLATION

4-Cylinder Engine

See Figures 49 and 50.

1. Do the battery terminal disconnection procedure.

2. Remove the accessory drive belt.

3. Remove the two bolts securing the alternator.

4. Disconnect the alternator connector and the positive alternator cable, and remove the harness clamp, then remove the alternator.

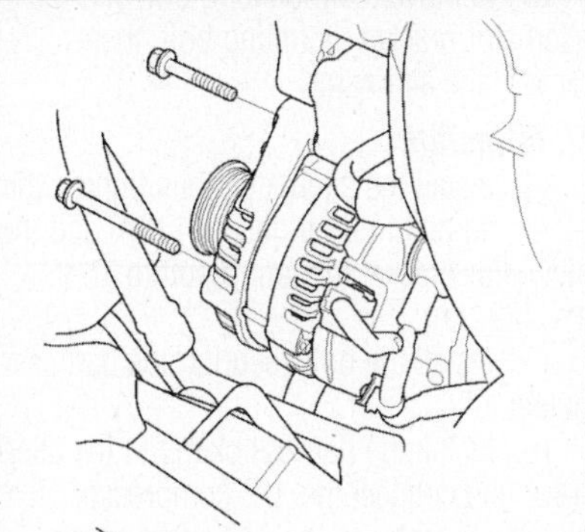

CHARGING SYSTEM

Fig. 49 Remove the two bolts securing the alternator

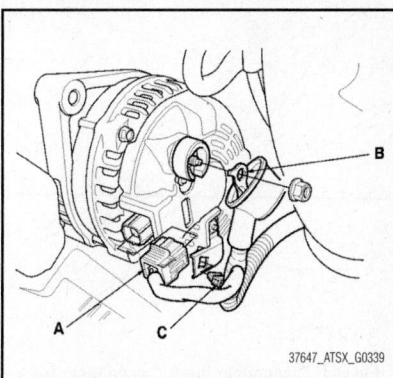

Fig. 50 Disconnect the alternator connector (A) and the positive alternator cable (B), and remove the harness clamp (C)

To install:

5. Install the alternator, then connect the alternator connector and the positive alternator cable, and install the harness clamp. Make sure the crimped side of the ring terminal faces away from the alternator when you connect it.

6. Tighten the two bolts securing the alternator. Tighten the short bolt to 16 ft. lbs. (22 Nm). Tighten the long bolt to 33 ft. lbs. 44 Nm).

7. Install the drive belt.

8. Do the battery terminal reconnection procedure.

6-Cylinder Engine

See Figures 51 through 53.

1. Remove the engine compartment covers.

   a. Remove the left engine compartment cover.

   b. Detach the clip, and release the hooks.

   c. Pull up the left engine compartment cover to release the pins.

   d. Remove the right engine compartment cover.

   e. Detach the clip, and release the hooks.

   f. Pull up the right engine compartment cover to release the pins.

   g. Detach the clips, and release the hooks, then remove the front grille cover.

   h. If necessary, detach the clips, then remove the front fender trim.

➡**The left front fender trim is shown; the right front fender trim is similar.**

2. Do the battery terminal disconnection procedure.

3. Raise the vehicle on the lift.

4. Remove the splash shield.

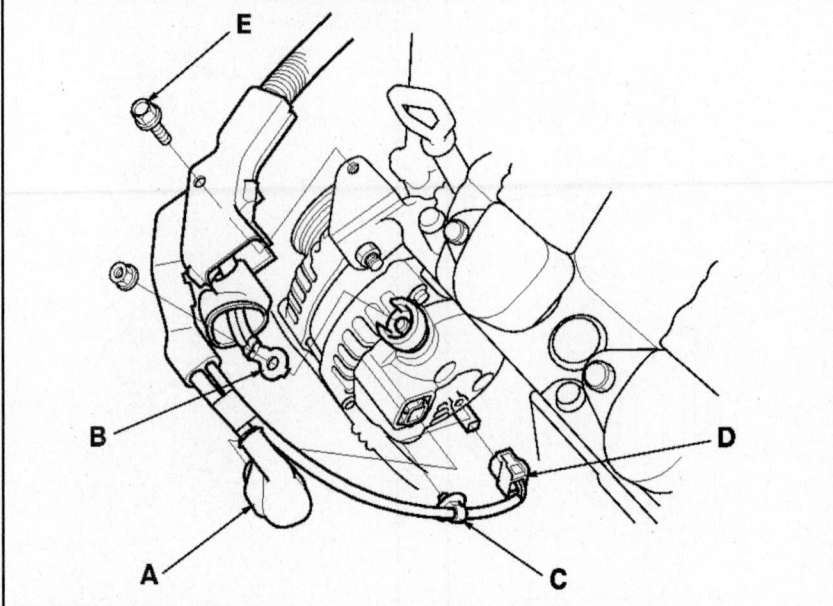

A. Alternator connector
B. Positive alternator cable
C. Harness clamp
D. A/C compressor clutch connector
E. Bolt

37647_ATSX_G0340

Fig. 52 Disconnect the alternator connector and the positive alternator cable from the alternator

5. Disconnect the A/C condenser fan motor connector and remove the harness clamp.

6. Loosen the A/C condenser fan shroud mounting bolts.

7. Lower the vehicle on the lift.

8. Remove the A/C condenser fan shroud assembly.

9. Remove the drive belt.

10. Disconnect the alternator connector and the positive alternator cable from the alternator.

11. Remove the harness clamp from the alternator and disconnect the A/C compressor clutch connector from the A/C compressor.

12. Remove the bolt securing the harness holder.

13. Remove the mounting bolt and the alternator bracket mounting bolt, then remove the alternator.

To install:

14. Install the alternator, then tighten the mounting bolt to 33 ft. lbs. (44 Nm) and the alternator bracket mounting bolt to 16 ft. lbs. (22 Nm).

15. Install the bolt securing the harness holder.

16. Install the harness clamp to the alternator and connect the A/C compressor clutch connector to the A/C compressor.

17. Connect the alternator connecter and the positive alternator cable to the alternator. Make sure the crimped side of the ring ter-

minal faces away from the alternator when you connect it.

18. Install the drive belt.

19. Install the A/C condenser fan shroud assembly.

20. Raise the vehicle on the lift.

21. Tighten the A/C condenser fan shroud mounting bolts.

22. Connect the A/C condenser fan motor connector and install the harness clamp.

23. Install the splash shield.

24. Lower the vehicle on the lift.

25. Do the battery terminal reconnection procedure.

26. Install the engine compartment covers.

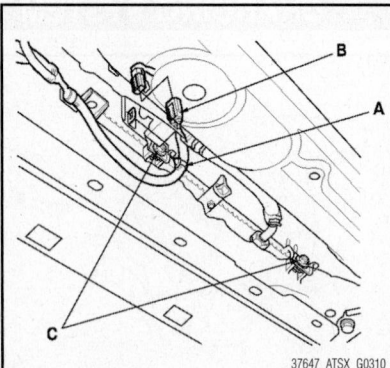

37647_ATSX_G0310

Fig. 51 Disconnect the A/C condenser fan motor connector (A), remove the harness clamp (B) and loosen the A/C condenser fan shroud mounting bolts (C)

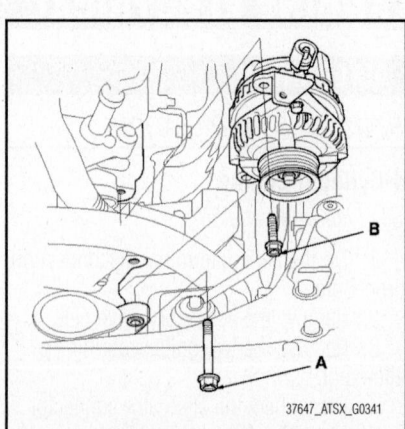

37647_ATSX_G0341

Fig. 53 Remove the mounting bolt (A) and the alternator bracket mounting bolt (B), then remove the alternator

ENGINE ELECTRICAL

IGNITION SYSTEM

FIRING ORDER

4-Cylinder Engine

The firing order is 1–3–4–2.

6-Cylinder Engine

The firing order is 1–4–2–5–3–6.

IGNITION COILS

REMOVAL & INSTALLATION

4-Cylinder Engine

See Figure 54.

1. Remove the ignition coil cover.
2. Disconnect the ignition coil connectors, then remove the ignition coils.
3. Install the ignition coils in the reverse order of removal.

6-Cylinder Engine

Front

See Figure 55.

1. Remove the engine compartment covers.
2. Remove the harness clamp and the bolt, then remove the harness holder from the bracket.
3. Disconnect the ignition coil connectors, then remove the front ignition coils.
4. Install the front ignition coils in the reverse order of removal.

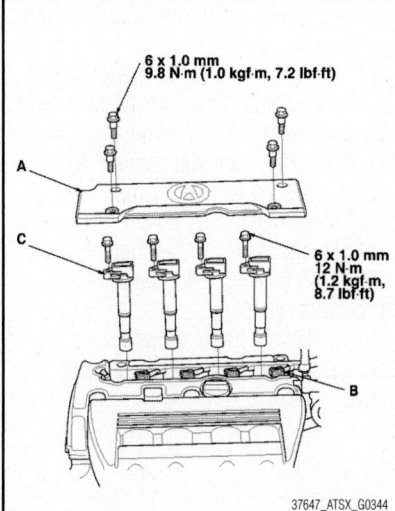

Fig. 54 Remove the ignition coil cover (A), disconnect the ignition coil connectors (B), then remove the ignition coils (C)

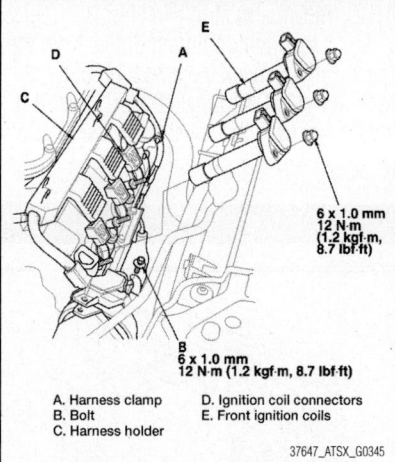

A. Harness clamp
B. Bolt
C. Harness holder
D. Ignition coil connectors
E. Front ignition coils

37647_ATSX_G0345

Fig. 55 Remove the harness clamp and the bolt, then remove the harness holder from the bracket

Rear

See Figure 56.

1. Remove the engine compartment covers.
2. Remove the strut brace.
3. Disconnect the ignition coil connectors, then remove the rear ignition coils.
4. Install the rear ignition coils in the reverse order of removal.

IGNITION TIMING

INSPECTION & ADJUSTMENT

4-Cylinder Engine

1. Connect the Honda Diagnostic System (HDS) to the Data Link Connector (DLC).

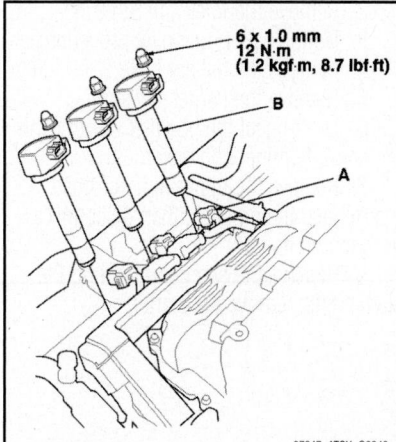

37647_ATSX_G0346

Fig. 56 Disconnect the ignition coil connectors (A), then remove the rear ignition coils (B)

2. Turn the ignition switch to ON (II).
3. Make sure the HDS communicates with the vehicle and the Engine Control Module (ECM)/Powertrain Control Module (PCM) If it does not communicate, troubleshoot the DLC circuit..
4. Check for DTCs. If a DTC is present, diagnose and repair the cause before continuing with this test.
5. Start the engine. Hold the engine speed at 3,000 rpm with no load (in N or P (A/T model) or neutral (M/T model)) until the radiator fan comes on, then let it idle.
6. Check the idle speed.
7. Jump the SCS line with the HDS.
8. Connect the timing light to the service loop (white tape).
9. Aim the light toward the pointer on the cam chain case. Check the ignition timing under a no load condition (headlights, blower fan, rear window defogger, and air conditioner are turned off).

➡**The other pointer is not used.**

Ignition Timing specifications:
• M/T model: 8°±2° BTDC (RED mark) at idle in neutral
• A/T model: 8°±2° BTDC (RED mark) at idle in N or P

10. If the ignition timing differs from the specification, check the cam timing. If the cam timing is OK, update the ECM/PCM if it does not have the latest software, or substitute a known-good ECM/PCM, then recheck. If the system works properly, and the ECM/PCM was substituted, replace the original ECM/PCM.
11. Disconnect the HDS and the timing light.

6-Cylinder Engine

1. Remove the engine compartment covers.
2. Connect the Honda Diagnostic System (HDS) to the Data Link Connector (DLC).
3. Turn the ignition switch to ON (II).
4. Make sure the HDS communicates with the vehicle and the Engine Control Module (ECM)/Powertrain Control Module (PCM). If it does not communicate, troubleshoot the DLC circuit.
5. Check for DTCs. If a DTC is present, diagnose and repair the cause before continuing with this test.
6. Start the engine. Hold the engine speed at 3,000 rpm with no load (in neutral (M/T model) or N or P (A/T model)) until the radiator fan comes on, then let it idle.
7. Check the idle speed.

8. Jump the SCS line with the HDS.

9. Connect the timing light to the No. 1 ignition coil harness.

10. Aim the light toward the pointer on the timing belt cover. Check the ignition timing under a no load condition (headlights, blower fan, rear window defogger, and air conditioner are turned off).

➡**The other pointer is not used.**

Ignition timing specification: 10 °±2 ° BTDC (RED mark) at idle in N or P

11. If the ignition timing differs from the specification, check the cam timing. If the cam timing is OK, update the ECM/PCM if it does not have the latest software, or sub-

stitute a known-good ECM/PCM, then recheck. If the system works properly, and the ECM/PCM was substituted, replace the original ECM/PCM.

12. Disconnect the HDS and the timing light.

13. Install the engine compartment covers.

ENGINE ELECTRICAL

STARTER

REMOVAL & INSTALLATION

4-Cylinder Engine

See Figures 57 and 58.

1. Do the battery terminal disconnection procedure.

2. Remove the intake manifold.

3. Disconnect the positive starter cable from the B terminal, and the S terminal connector.

4. Remove the harness clamp.

5. Remove the two bolts securing the starter, then remove the starter.

To install:

6. Install the starter, then tighten the two bolts. Tighten the 12 x 1.25 mm bolt to 47 ft. lbs. (64 Nm). Tighten the 10 x 1.25 mm bolt to 33 ft. lbs. (44 Nm).

7. Install the harness clamp.

8. Connect the positive starter cable to

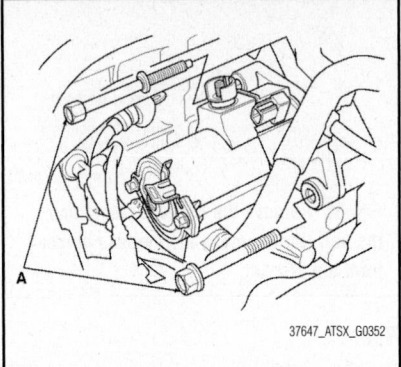

37647_ATSX_G0352

Fig. 58 Remove the two bolts (A) securing the starter, then remove the starter

the B terminal, and the S terminal connector. Make sure the crimped side of the ring terminal faces away from the starter when you connect it.

9. Install the intake manifold.

10. Do the battery terminal reconnection procedure.

11. Start the engine to make sure the starter works properly.

6-Cylinder Engine

See Figure 59.

1. Remove the engine compartment covers.

2. Remove the intake air separator.

3. Do the battery removal procedure.

4. Remove the battery base.

5. Remove the harness clamp.

6. Disconnect the positive starter cable and the S terminal connector.

7. Remove the upper radiator hose bracket and the dipstick, then disconnect the vacuum hose.

8. Remove the two bolts holding the starter, then remove the starter.

STARTING SYSTEM

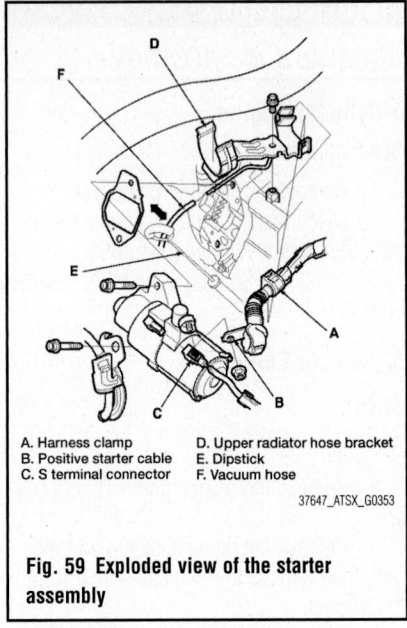

A. Harness clamp
B. Positive starter cable
C. S terminal connector
D. Upper radiator hose bracket
E. Dipstick
F. Vacuum hose

37647_ATSX_G0353

Fig. 59 Exploded view of the starter assembly

To install:

9. Install the starter, then tighten the mounting bolts to 33 ft. lbs. (44 Nm).

➡**Always use a new gasket.**

10. Connect the vacuum hose, then install the upper radiator hose bracket and the dipstick.

11. Connect the positive starter cable and the S terminal connector. Make sure the crimped side of the ring terminal faces away from the starter when you connect it.

12. Install the harness clamp.

13. Install the battery base.

14. Do the battery installation procedure.

15. Start the engine to make sure the starter works properly.

16. Install the intake air separator.

17. Install the engine compartment covers.

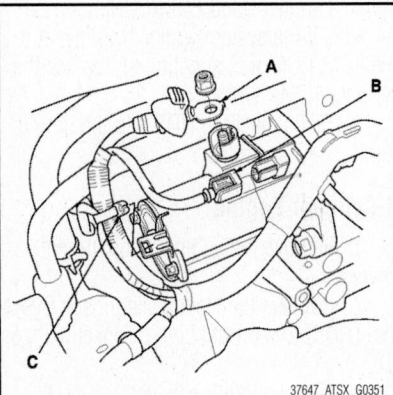

37647_ATSX_G0351

Fig. 57 Disconnect the positive starter cable (A) from the B terminal, and the S terminal connector (B)

ENGINE MECHANICAL

→Disconnecting the negative battery cable may interfere with the functions of the on board computer systems and may require the computer to undergo a relearning process, once the negative battery cable is reconnected.

ACCESSORY DRIVE BELT SYSTEM

ADJUSTMENT

No adjustment is necessary or possible. Proper tension is provided by the auto-tensioner.

INSPECTION

1. Inspect the belt for cracks or damage. If the belt is cracked or damaged, replace it.
2. Check that the position of the auto-tensioner indicator is within the standard range. If it is out of the standard range, replace the drive belt.

REMOVAL & INSTALLATION

4-Cylinder Engine

See Figure 60.

Special Tools Required: Belt tension release tool Snap-On YA9317 or equivalent, commercially available
1. Move the auto-tensioner with the belt tension release tool in the direction shown to relieve tension from the drive belt, then remove the drive belt.
2. Install the new belt in the reverse order of removal.

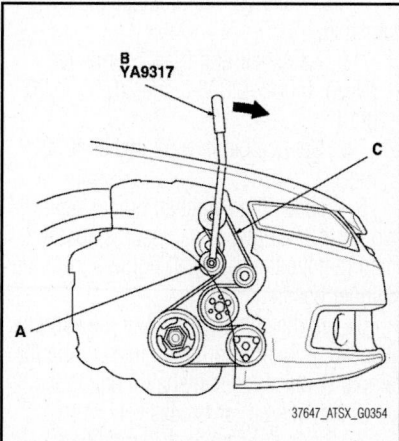

Fig. 60 Move the auto-tensioner (A) with the belt tension release tool (B) to relieve tension from the drive belt (C), then remove the drive belt

6-Cylinder Engine

See Figure 61.

1. Move the auto-tensioner with the belt tension release tool in the direction shown to relieve tension from the drive belt, then remove the drive belt.
2. Install the new belt in the reverse order of removal.

Drive Belt Idler Pulley

4-Cylinder Engine

See Figure 62.

1. Remove the drive belt.
2. Remove the idler pulley.
3. Install the idler pulley in the reverse order of removal.

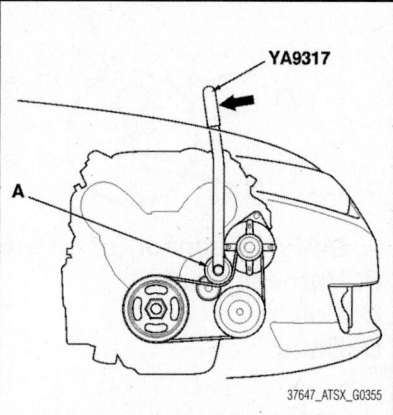

Fig. 61 Move the auto-tensioner (A) with the belt tension release tool to relieve tension from the drive belt, then remove the drive belt

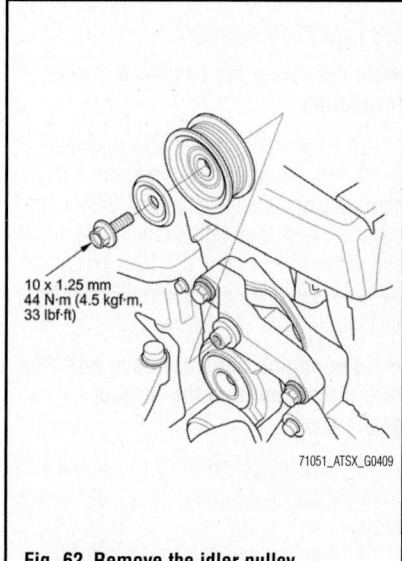

Fig. 62 Remove the idler pulley

Drive Belt Tensioner

4-Cylinder Engine

See Figures 63 and 64.

1. Remove the drive belt.
2. Remove the idler pulley base.
3. Remove the auto-tensioner.
4. Install the auto-tensioner, then tighten the bolt/nut in the numbered sequence shown.
5. Install the idler pulley base.
6. Install the drive belt.

6-Cylinder Engine

See Figure 65.

1. Remove the drive belt.
2. Remove the front splash shield.
3. Remove the auto-tensioner.
4. Install the auto-tensioner in the reverse order of removal.

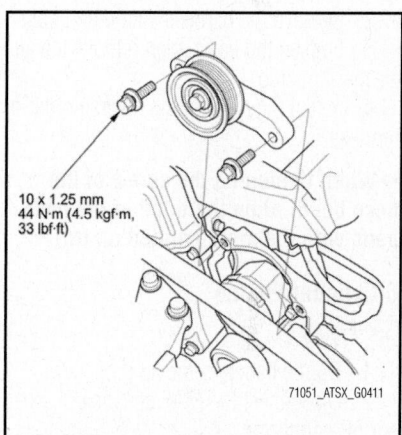

Fig. 63 Remove the idler pulley base

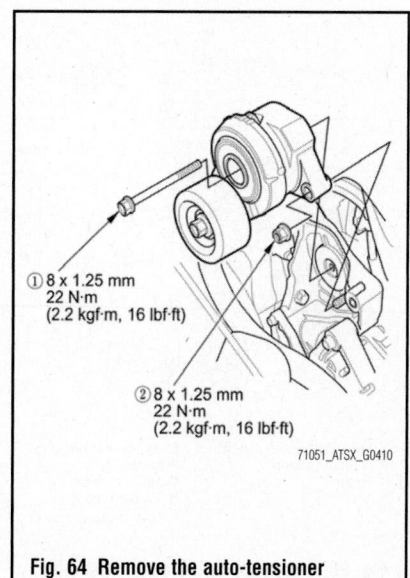

Fig. 64 Remove the auto-tensioner

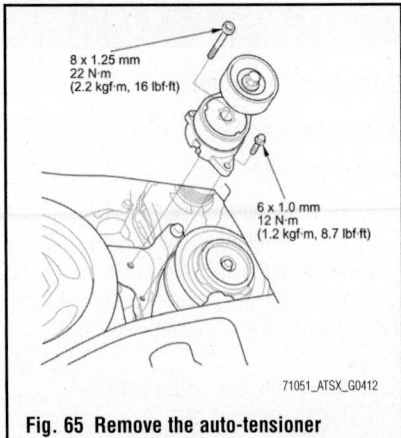

Fig. 65 Remove the auto-tensioner

AIR CLEANER

REMOVAL & INSTALLATION

4-Cylinder Engine

See Figure 66.

1. Disconnect the MAF sensor/IAT sensor connector.
2. Remove the harness clamp and bolts.
3. Loosen the band, then remove the air cleaner housing.
4. Install the parts in the reverse order of removal.

➡**When tightening the screw of the hose band, align the edge of the hose band with the mark painted on it.**

6-Cylinder Engine

See Figure 67.

1. Do the battery removal procedure.
2. Disconnect the MAF sensor/IAT sensor 5P connector.

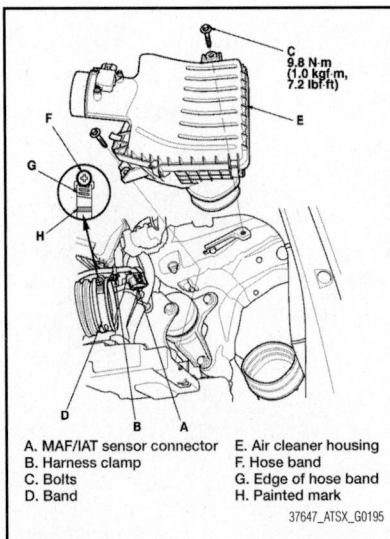

A. MAF/IAT sensor connector E. Air cleaner housing
B. Harness clamp   F. Hose band
C. Bolts   G. Edge of hose band
D. Band   H. Painted mark

37647_ATSX_G0195

Fig. 66 Remove the air cleaner assembly

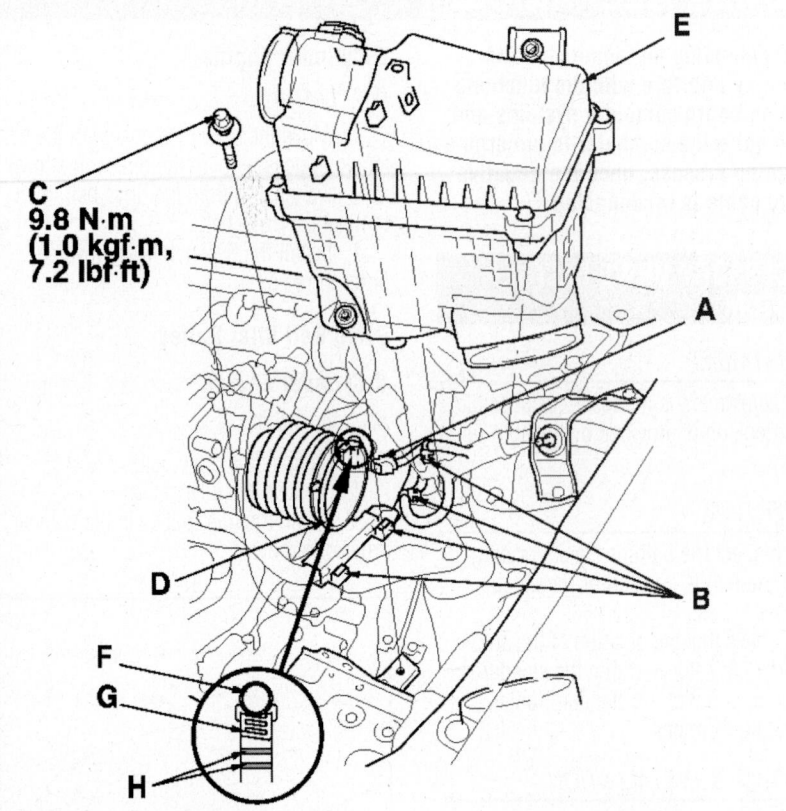

A. MAF/IAT sensor 5P connector
B. Harness clamp
C. Bolt
D. Band
E. Air cleaner housing
F. Screw of the hose band
G. Edge of the hose band
H. Paint marks

37647_ATSX_G0356

Fig. 67 Remove the air cleaner assembly

CAMSHAFT & BEARINGS

INSPECTION

4-Cylinder Engine

See Figures 68 through 72.

➡**Do not rotate the camshaft during inspection.**

1. Remove the rocker arm assembly.
2. Put the rocker shaft holders, the camshaft, and the camshaft holders on the cylinder head, then tighten the bolts, in sequence, to 16 ft. lbs. (22 Nm) except for bolts 21, 22, and 23. Tighten these three bolts to 9 ft. lbs. (12 Nm).

➡**If the engine does not have bolt 21, skip it and continue the torque sequence.**

3. Seat the camshaft by pushing it away from the camshaft pulley end of the cylinder head.
4. Zero the dial indicator against the end of the camshaft, then push the camshaft back and forth, and read the end play. If the end play is beyond the service limit, replace the cylinder head and recheck. If it is still beyond the service limit, replace the camshaft.

   a. Camshaft End Play: Standard (New): 0.002–0.008 inches (0.05–0.20 mm)
   b. Service Limit: 0.02 inches (0.4 mm)

5. Loosen the camshaft holder bolts two turns at a time, in a crisscross pattern. Then remove the camshaft holders from the cylinder head.
6. Lift the camshafts out of the cylinder head, wipe them clean, then inspect the lift ramps. Replace the camshaft if any lobes are pitted, scored, or excessively worn.
7. Clean the camshaft journal surfaces in the cylinder head, then set the camshafts back in place. Place a Plastigage strip across each journal.
8. Install the camshaft holders, then

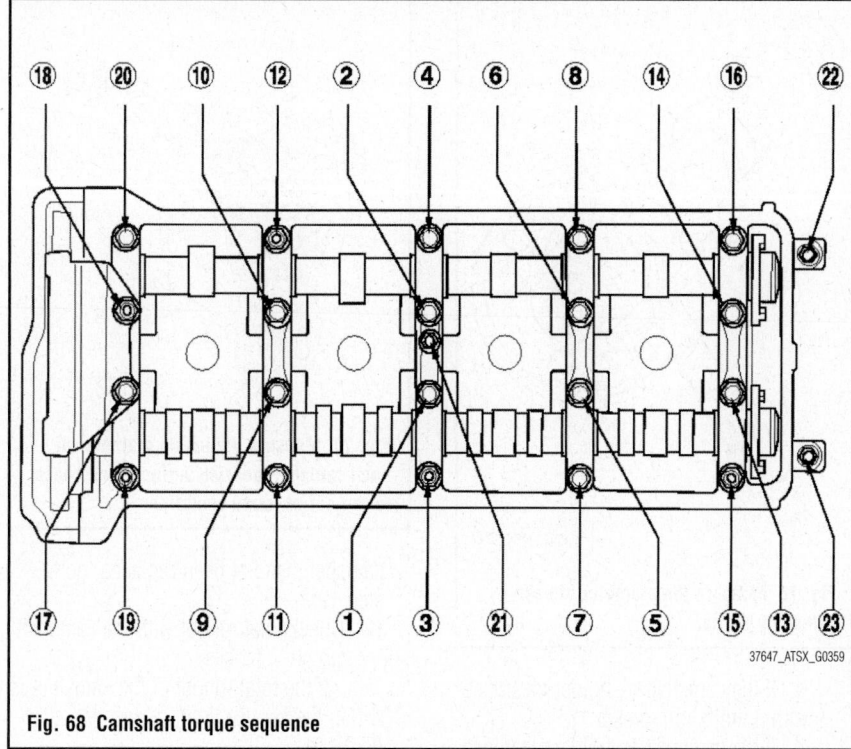

Fig. 68 Camshaft torque sequence

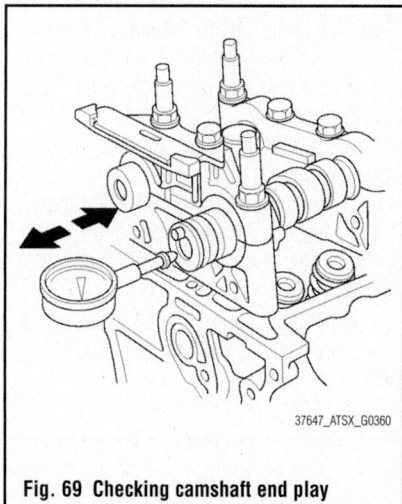

Fig. 69 Checking camshaft end play

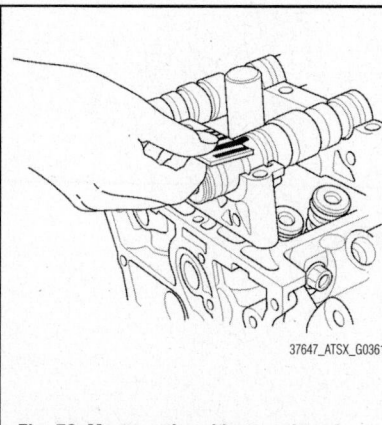

Fig. 70 Measure the widest portion of Plastigage on each journal

tighten the bolts to the specified torque as shown in step 2.

9. Remove the camshaft holders. Measure the widest portion of Plastigage on each journal.

a. If the camshaft-to-holder clearance is within limits, go to step 11.

b. If the camshaft-to-holder clearance is beyond the service limit, and the camshaft has been replaced, replace the cylinder head.

c. If the camshaft-to-holder clearance is beyond the service limit, and the camshaft has not been replaced, go to step 10.

➡**Camshaft-to-Holder Oil Clearance: Standard (New):**

- No. 1 Journal: 0.001–0.003 inches (0.030–0.069 mm)
- No. 2, 3, 4, 5 Journals: 0.002–0.004 inches (0.060–0.099 mm)
- Service Limit: 0.006 inches (0.15 mm)

10. Check the total runout with the camshaft supported on V-blocks.

a. If the total runout of the camshaft is within the service limit, replace the cylinder head.

b. If the total runout is beyond the service limit, replace the camshaft and recheck the camshaft-to-holder oil clearance. If the oil clearance is still beyond the service limit, replace the cylinder head.

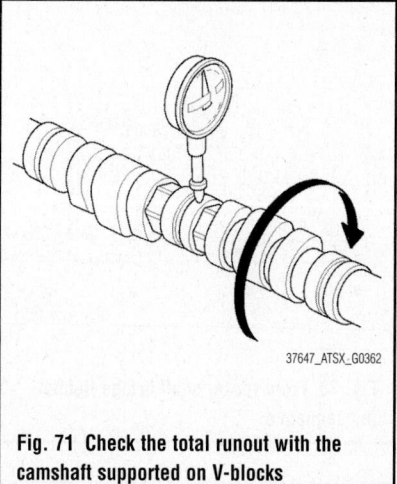

Fig. 71 Check the total runout with the camshaft supported on V-blocks

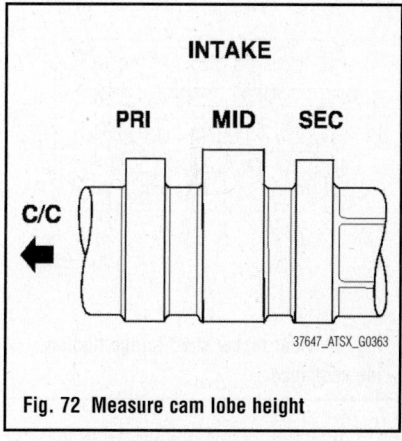

Fig. 72 Measure cam lobe height

c. Camshaft Total Runout: Standard (New): 0.001 inches (0.03 mm) max.

d. Service Limit: 0.002 inches (0.04 mm)

11. Measure cam lobe height.

➡**Cam Lobe Height Standard (New):**

- INTAKE
- Primary: 1.3285 inches (33.744 mm)
- Mid: 1.3959 inches (35.456 mm)
- Secondary: 1.3285 inches (33.744 mm)
- EXHAUST
- 1.3500 inches (34.291 mm)

6-Cylinder Engine

See Figures 73 through 79.

1. Remove the cylinder head.
2. Remove the rocker arm assembly.
3. Front: Put the rocker shafts bridge and the rocker shaft holder on the front cylinder head, then tighten the bolts to 16 ft. lbs. (22 Nm).
4. Rear: Put the rocker shaft bridge and the rocker shaft holder on the rear cylinder

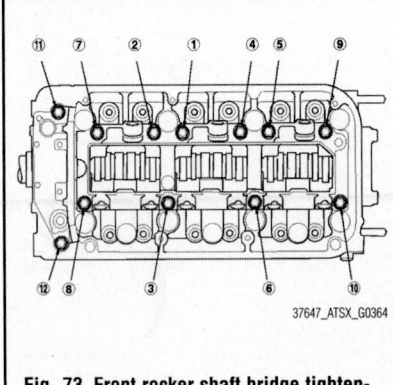

Fig. 73 Front rocker shaft bridge tightening sequence

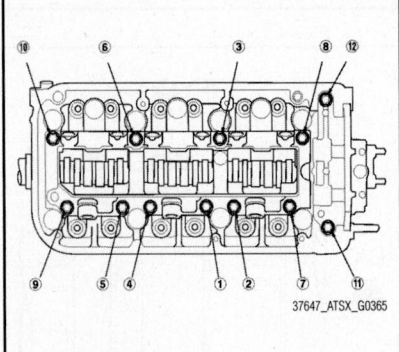

Fig. 74 Rear rocker shaft bridge tightening sequence

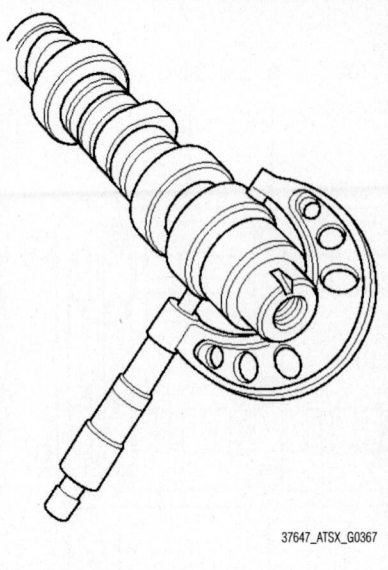

Fig. 75 Measure the diameter of each camshaft journal

a. If the camshaft-to-holder clearance is within limits, go to step 13.

b. If the camshaft-to-holder clearance is beyond the service limit and the camshaft has been replaced, replace the cylinder head.

Camshaft-to-Holder Oil Clearance: Standard (New): 0.0020–0.0035 inches (0.050–0.089 mm)

Service Limit: 0.0060 inches (0.15 mm)

a. If the camshaft-to-holder clearance is beyond the service limit and the

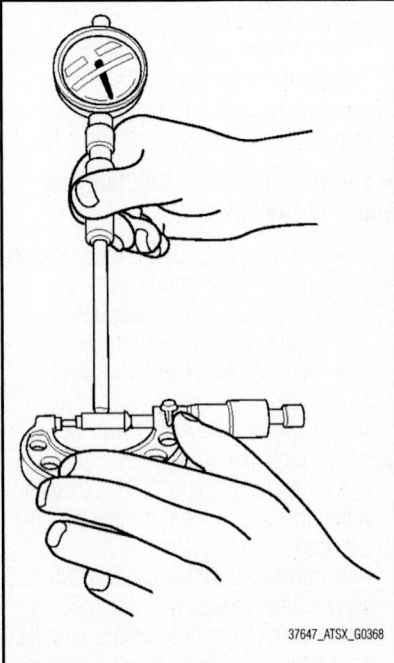

Fig. 76 Zero the gauge to the journal diameter

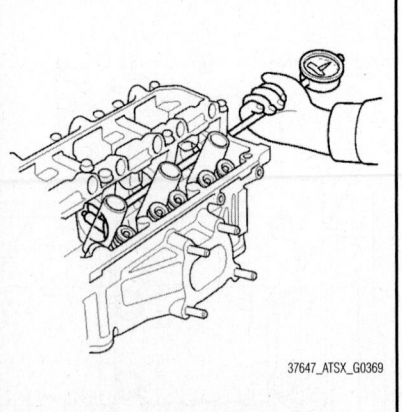

Fig. 77 Measure the inside diameter of each camshaft bearing surface, and check for an out-of-round condition

camshaft has not been replaced, go to step 12.

12. Check total runout with the camshaft supported on V-blocks.

a. If the total runout of the camshaft is within the service limit, replace the cylinder head.

b. If the total runout is beyond the service limit, replace the camshaft and recheck the oil clearance. If the oil clearance is still out of tolerance, replace the cylinder head.

Camshaft Total Runout: Standard (New): 0.0012 inches (0.03 mm) max.

Service Limit: 0.002 inches (0.04 mm)

13. Measure the cam lobe height.

Cam Lobe Height Standard (New):
- INTAKE
- Primary: 1.3504 inches (34.299 mm)
- Secondary: 31.4024 inches (5.621 mm)

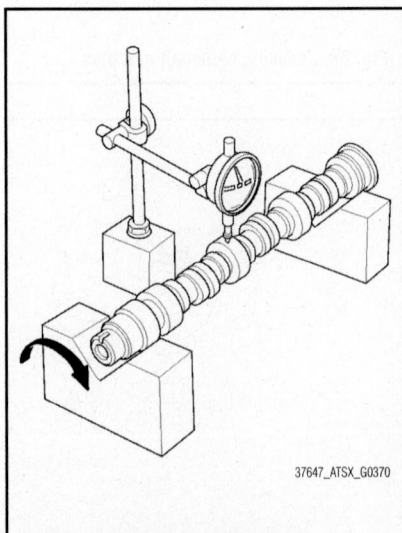

Fig. 78 Check total runout with the camshaft supported on V-blocks

head, then tighten the bolts to 16 ft. lbs. (22 Nm).

5. Seat the camshaft by pushing it toward the rear of the cylinder head.

6. Zero the dial indicator against the end of the camshaft. Push the camshaft back and forth and read the end play. If the end play is beyond the service limit, replace the thrust cover and recheck. If it is still beyond the service limit, replace the camshaft.

a. Camshaft End Play: Standard (New): 0.002–0.008 inches (0.05-0.20 mm)

b. Service Limit: 0.008 inches (0.20 mm)

7. Remove the camshaft thrust cover, then pull out the camshaft.

8. Wipe the camshaft clean, then inspect the lift ramps. Replace the camshaft if any lobes are pitted, scored, or excessively worn.

9. Measure the diameter of each camshaft journal.

10. Zero the gauge to the journal diameter.

11. Clean the camshaft bearing surfaces in the cylinder head. Measure the inside diameter of each camshaft bearing surface, and check for an out-of-round condition.

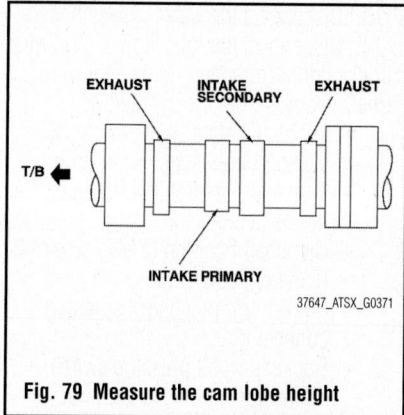

Fig. 79 Measure the cam lobe height

• EXHAUST: 1.4472 inches (36.760 mm)

REMOVAL & INSTALLATION

6-Cylinder Engine

Front

See Figures 80 and 81.

1. Remove the engine compartment covers.
2. Do the battery removal procedure.
3. Drain the engine coolant.
4. Disconnect the radiator hoses.
5. Remove the exhaust gas recirculation (EGR) valve.
6. Remove the EGR valve stud bolts.
7. Remove the timing belt.
8. Remove the front rocker arm assembly.
9. Remove the front camshaft pulley.
10. Remove the thrust cover, then remove the front camshaft.

To install:

11. Install the front camshaft in the reverse order of removal. Always use a new

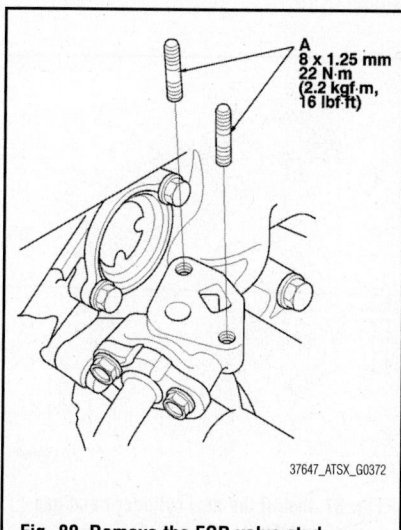

Fig. 80 Remove the EGR valve stud bolts (A)

O-ring. Apply new engine oil to the journals and the cam lobes.

12. Apply new engine oil to the threads of the camshaft pulley mounting bolt, then install the front camshaft pulley.
13. Install the front rocker arm assembly, then tighten the mounting bolts.
14. Install the timing belt.
15. Adjust the valve clearance.
16. Install the EGR valve stud bolts, then install the EGR valve.
17. Connect the radiator hoses.
18. Do the battery installation procedure.
19. Fill the radiator with engine coolant, and bleed the air from the cooling system with the heater valve open.
20. Install the engine compartment covers.
21. Do the Crankshaft Position (CKP) pattern clear/CKP pattern learn procedure.

Rear

See Figures 82 and 83.

1. Remove the engine compartment covers.
2. Relieve the fuel pressure.
3. Do the battery removal procedure.
4. Drain the engine coolant.
5. Remove the under-hood fuse/relay box from the bracket.
6. Remove the air cleaner assembly.
7. Remove the quick-connect fitting cover, then disconnect the fuel feed hose.
8. Disconnect the heater hoses.
9. Disconnect the Evaporative Emission (EVAP) canister hose, then remove the EVAP canister joint.
10. Remove the upper transmission mount.
11. Remove the timing belt.
12. Remove the rear rocker arm assembly.
13. Remove the rear camshaft pulley.

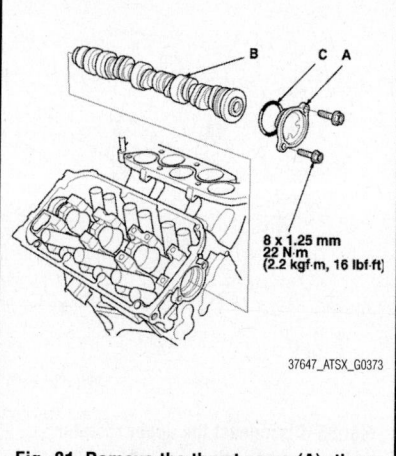

Fig. 81 Remove the thrust cover (A), then remove the front camshaft (B)

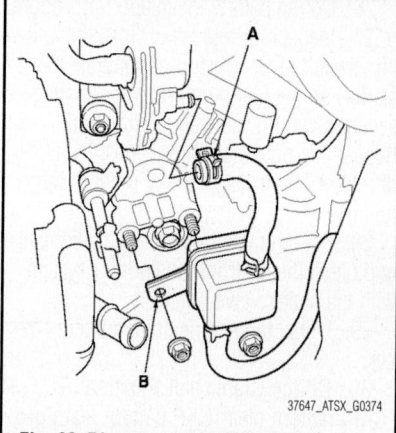

Fig. 82 Disconnect the EVAP canister hose (A), then remove the EVAP canister joint (B)

14. Remove the thrust cover, then remove the rear camshaft.

To install:

15. Install the rear camshaft in the reverse order of removal. Always use a new O-ring. Apply new engine oil to the journals and cam lobes.
16. Apply new engine oil to the threads of the camshaft pulley mounting bolt, then install the rear camshaft pulley.
17. Install the rear rocker arm assembly, then tighten the mounting bolts.
18. Install the timing belt.
19. Install the upper transmission mount.
20. Adjust the valve clearance.
21. Install the EVAP canister joint, then connect the EVAP canister hose.
22. Connect the heater hoses.
23. Connect the fuel feed hose, then install the quick-connect fitting cover.
24. Install the air cleaner assembly.
25. Install the under-hood fuse/relay box to the bracket.

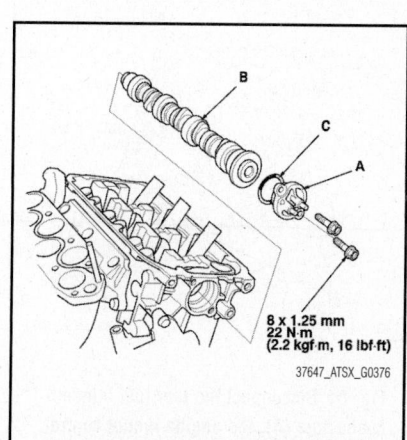

Fig. 83 Remove the thrust cover (A), then remove the rear camshaft (B)

26. Do the battery installation procedure.

27. Inspect for fuel leaks. Turn the ignition switch to ON (II) (do not operate the starter) so the fuel pump runs for about 2 seconds and pressurizes the fuel line. Repeat this operation three times, then check for fuel leakage at any point in the fuel line.

28. Fill the radiator with engine coolant, and bleed the air from the cooling system with the heater valve open.

29. Install the engine compartment covers.

30. Do the Crankshaft Position (CKP) pattern clear/CKP pattern learn procedure.

CYLINDER HEAD

REMOVAL & INSTALLATION

4-Cylinder Engine

See Figures 84 through 90.

➡ **Prior to starting the removal of the cylinder head, note the following:**

- Use fender covers to avoid damaging painted surfaces.
- To avoid damage, unplug the wiring connectors carefully while holding the connector portion.
- Connect the Honda Diagnostic System (HDS) to the Data Link Connector (DLC), and monitor the Engine Coolant Temperature (ECT) sensor 1. To avoid damaging the cylinder head, wait until the ECT sensor 1 temperature drops below 100°F (38°C) before loosening the cylinder head bolts.
- Mark all wiring and hoses to avoid misconnection. Also, be sure that

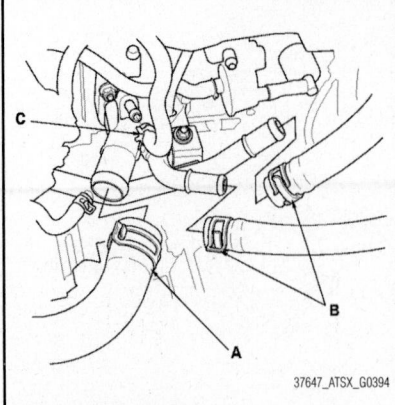

Fig. 85 Disconnect the upper radiator hose (A), the heater hoses (B), and the water bypass hose (C)

they do not contact other wiring or hoses, or interfere with other parts.

1. Remove the strut brace.
2. Relieve the fuel pressure.
3. Drain the engine coolant.
4. Remove the drive belt.
5. Remove the intake manifold.
6. Remove the catalytic converter.
7. Remove the Evaporative Emission (EVAP) canister hose.
8. Remove the quick-connect fitting cover, then disconnect the fuel feed hose.
9. Disconnect the four fuel injector connectors, the engine mount control solenoid connector, and remove the ground cables.
10. Remove the four bolts securing the EVAP canister purge valve bracket.
11. Disconnect the upper radiator hose, the heater hoses, and the water bypass hose.
12. Remove the two bolts securing the connecting pipe.

13. Disconnect the water bypass hose.
14. Disconnect the following engine wire harness connectors, and remove the wire harness clamps from the cylinder head:
    - Engine Coolant Temperature (ECT) sensor 1 connector
    - Camshaft Position (CMP) sensor A (Intake) connector
    - Camshaft Position (CMP) sensor B (Exhaust) connector
    - Rocker arm oil control solenoid connector
    - Rocker arm oil pressure switch connector
    - EVAP canister purge valve connector
    - Variable Valve Timing Control (VTC) oil control solenoid valve connector
    - Engine oil pressure switch connector
15. Remove the cam chain.
16. Remove the rocker arm assembly.
17. Remove the cylinder head bolts. To prevent warpage, loosen the bolts in sequence ⅓ turn at a time; repeat the sequence until all bolts are loosened.
18. Remove the cylinder head.

To install:

19. Install a new coolant separator in the engine block whenever the engine block is replaced.
20. Clean the cylinder head and the engine block surface.
21. Install the new cylinder head gasket and the dowel pins on the engine block. Always use a new cylinder head gasket.
22. Set the crankshaft to Top Dead Center (TDC). Align the TDC mark on the crankshaft sprocket with the pointer on the engine block.
23. Install the cylinder head on the engine block.

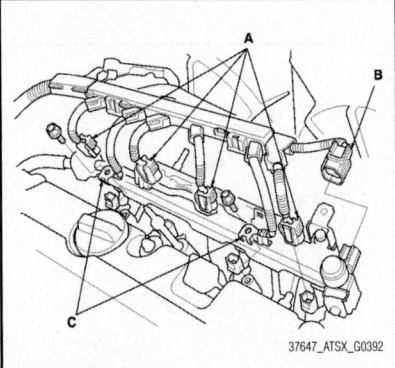

Fig. 84 Disconnect the four fuel injector connectors (A), the engine mount control solenoid connector (B), and remove the ground cables (C)

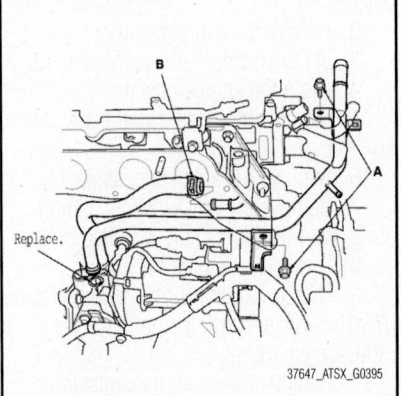

Fig. 86 Remove the two bolts (A) securing the connecting pipe and disconnect the water bypass hose (B)

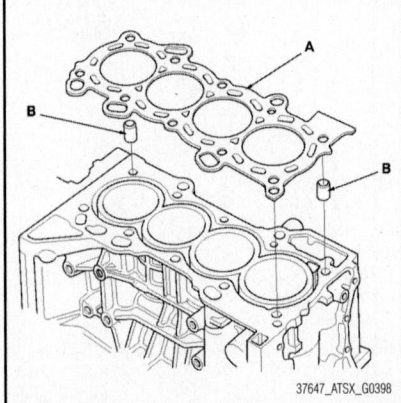

Fig. 87 Install the new cylinder head gasket (A) and the dowel pins (B) on the engine block

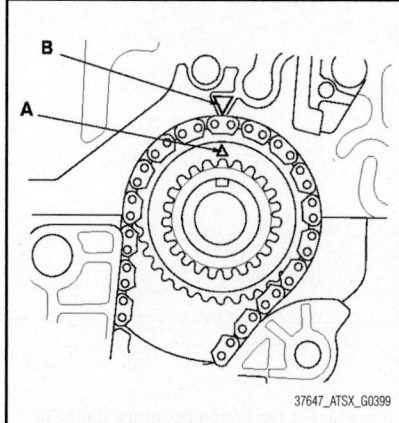

Fig. 88 Align the TDC mark (A) on the crankshaft sprocket with the pointer (B) on the engine block

24. Measure the diameter of each cylinder head bolt at point A and point B.

25. If either diameter is less than 0.42 inches (10.6 mm), replace the cylinder head bolt.

26. Apply new engine oil to the threads and under the bolt heads of all cylinder head bolts.

27. Torque the cylinder head bolts in sequence to 29 ft. lbs. (39 Nm). Use a beam-type torque wrench. When using a preset click-type torque wrench, be sure to tighten slowly and do not overtighten. If a bolt makes any noise while you are torquing it, loosen the bolt and retighten it from the first step.

28. After torquing, tighten all cylinder head bolts in two steps (90° per step) using the sequence shown in step 9. If you are using a new cylinder head bolt, tighten the bolt an extra 90°.

➡**Remove the cylinder head bolt if you tightened it beyond the specified angle,**

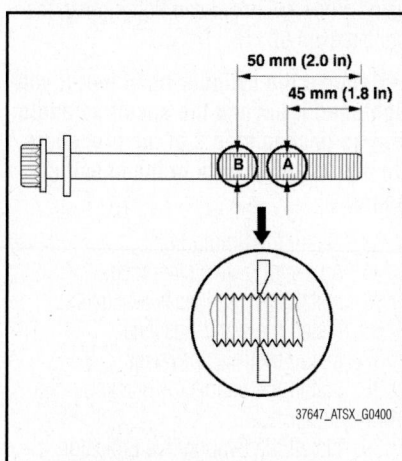

Fig. 89 Measure the diameter of each cylinder head bolt at point A and point B

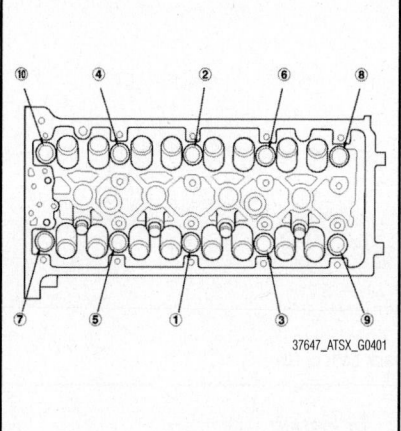

Fig. 90 Torque the cylinder head bolts in sequence

and go back to step 6 of the procedure. Do not loosen it back to the specified angle.

29. Install the rocker arm assembly.
30. Install the cam chain.
31. Connect the following engine wire harness connectors, and install the wire harness clamps to the cylinder head:

- Engine Coolant Temperature (ECT) sensor 1 connector
- Camshaft Position (CMP) sensor A (Intake) connector
- Camshaft Position (CMP) sensor B (Exhaust) connector
- Rocker arm oil control solenoid connector
- Rocker arm oil pressure switch connector
- Evaporative Emission (EVAP) canister purge valve connector
- Variable Valve Timing Control (VTC) oil control solenoid valve connector
- Engine oil pressure switch connector

32. Install the two bolts securing the connecting pipe.
33. Install the water bypass hose.
34. Connect the upper radiator hose, the heater hoses, and the water bypass hose.
35. Install the four bolts securing the EVAP canister purge valve bracket.
36. Connect the four fuel injector connectors, the engine mount control solenoid connector, and install the ground cables.
37. Connect the fuel feed hose, then install the quick-connect fitting cover.
38. Connect the EVAP canister hose.
39. Install the catalytic converter.
40. Install the intake manifold.
41. Install the drive belt.
42. Install the strut brace.

43. After installation, check that all tubes, hoses, and connectors are installed correctly.

44. Inspect for fuel leaks. Turn the ignition switch to ON (II) (do not operate the starter) so the fuel pump runs for about 2 seconds and pressurizes the fuel line. Repeat this operation three times, then check for fuel leakage at any point in the fuel line.

45. Refill the radiator with engine coolant, and bleed the air from the cooling system with the heater valve open.

46. Check for fluid leaks.

47. Do the Engine Control Module (ECM)/Powertrain Control Module (PCM) idle lean procedure.

48. Do the Crankshaft Position (CKP) pattern clear/CKP pattern lean procedure.

49. Inspect the idle speed.

50. Inspect the ignition timing.

6-Cylinder Engine

See Figures 91 through 97.

1. Remove the engine compartment covers.
2. Relieve the fuel pressure.
3. Do the battery terminal disconnection procedure.
4. Drain the engine coolant.
5. Remove the alternator.
6. Remove the intake manifold.
7. Remove the six ignition coils.
8. Remove the timing belt.
9. Disconnect the following engine wire harness connectors, and remove the wire harness clamps from the cylinder head:

- Six injector connectors
- Knock sensor connector
- Engine Coolant Temperature (ECT) sensor 1 connector
- Engine mount control solenoid valve connector
- Camshaft Position (CMP) sensor connector
- Rocker arm oil control solenoid connector
- Rocker arm oil pressure switch connector
- Two Air Fuel Ratio (A/F) sensor connectors
- Two secondary Heated Oxygen Sensor (secondary HO2S) connectors

10. Remove the front Warm Up Three Way Catalytic Converter (front WU-TWC) and the rear Warm Up Three Way Catalytic Converter (rear WU-TWC).

11. Remove the quick-connect fitting cover, then disconnect the fuel feed hose.

12. Remove the connector bracket from the front cylinder head.

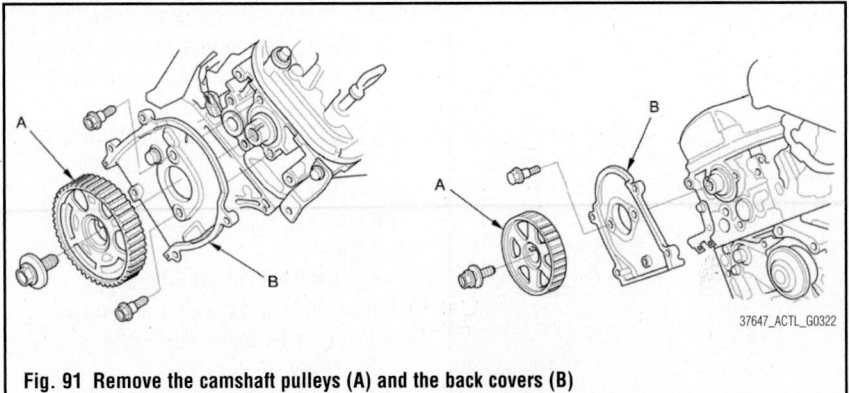

37647_ACTL_G0322

Fig. 91 Remove the camshaft pulleys (A) and the back covers (B)

13. Remove the engine mount control solenoid valve bracket from the rear cylinder head.

14. Remove the Evaporative Emission (EVAP) canister joint with the bracket.

15. Remove the injector bases.

16. Remove the water passage.

17. Remove the camshaft pulleys and the back covers.

18. Remove the cylinder head covers.

19. Remove the cylinder head bolts. To prevent warpage, loosen the bolts in sequence ⅓ turn at a time; repeat the sequence until all bolts are loosened.

20. Remove the cylinder heads.

To install:

21. Clean the cylinder head and the engine block surface.

22. Clean and install the oil control orifices with new O-rings.

23. Install the dowel pins and the new cylinder head gaskets.

24. Clean the timing belt pulleys, the timing belt guide plate, and the upper and lower covers.

25. Set the timing belt drive pulley to Top Dead Center (TDC) by aligning the TDC mark on the tooth of the timing belt drive pulley with the pointer on the oil pump.

26. Set the camshaft pulleys to TDC by

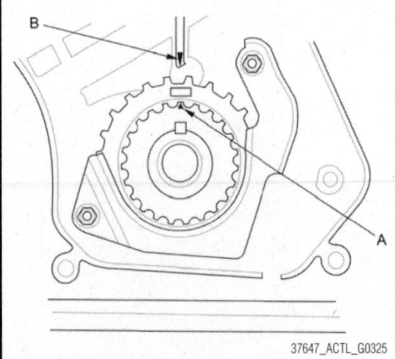

37647_ACTL_G0325

Fig. 93 Set the timing belt drive pulley to TDC by aligning the TDC mark (A) on the tooth of the timing belt drive pulley with the pointer (B) on the oil pump

aligning the TDC marks on the camshaft pulleys with the pointers on the back covers.

27. Install the cylinder heads on the engine block.

28. Measure the diameter of each cylinder head bolt at point A and point B.

29. If either diameter is less than 0.44 inches (11.3 mm), replace the cylinder head bolt.

30. Apply new engine oil to the threads and under the bolt heads of all cylinder head bolts.

31. Torque the cylinder head bolts in sequence to 22 ft. lbs. (29 Nm) using a beam-type torque wrench. When using a preset click-type torque wrench, be sure to tighten slowly and do not overtighten. If a bolt makes any noise while you are torquing it, loosen the bolt and retighten it from the first step.

32. After torquing, tighten all cylinder head bolts in two steps (90° per step) using the sequence shown in step 11. If you are using a new cylinder head bolt, tighten the bolt an extra 90°.

➡Remove the cylinder head bolt if you tightened it beyond the specified angle, and go back to step 8 of the procedure. Do not loosen it back to the specified angle.

33. Install the timing belt.

34. Adjust the valve clearance.

35. Install the cylinder head covers.

36. Install the water passage.

37. Install the injector bases.

38. Install the connector bracket to the front cylinder head.

39. Install the Evaporative Emission (EVAP) canister joint with the bracket.

40. Install the engine mount control

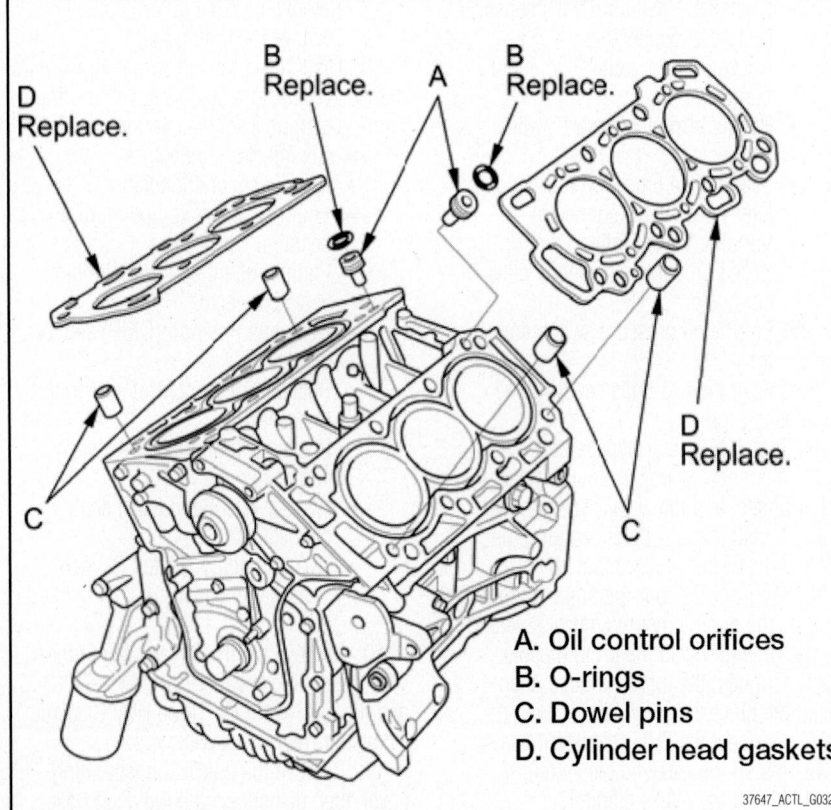

A. Oil control orifices
B. O-rings
C. Dowel pins
D. Cylinder head gaskets

37647_ACTL_G0324

Fig. 92 Clean and install the oil control orifices with new O-rings

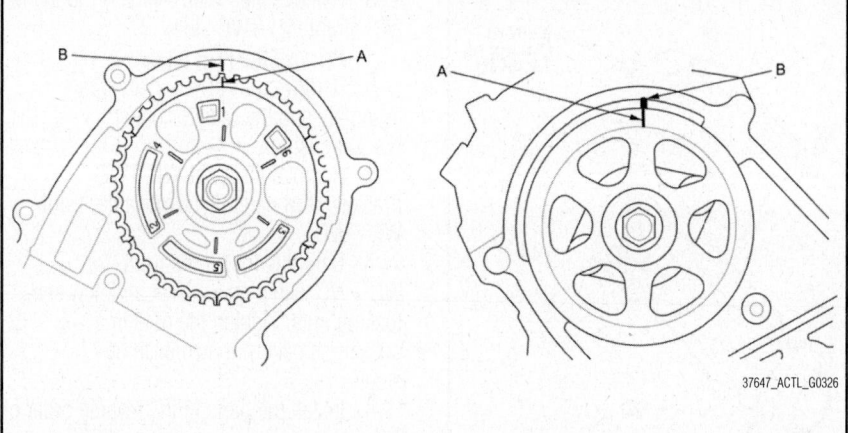

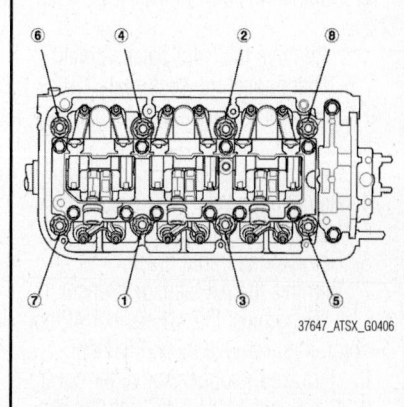

Fig. 94 Set the camshaft pulleys to TDC by aligning the TDC marks (A) on the camshaft pulleys with the pointers (B) on the back covers

Fig. 97 Torque the rear cylinder head bolts in sequence

solenoid valve bracket to the rear cylinder head.

41. Connect the fuel feed hose, then install the quick-connect fitting cover.

42. Install the front Warm Up Three Way Catalytic Converter (front WU-TWC) and the rear Warm Up Three Way Catalytic Converter (rear WU-TWC).

43. Connect the following engine wire harness connectors, and install the wire harness clamps to the cylinder head:

- Six injector connectors
- Knock sensor connector
- Engine Coolant Temperature (ECT) sensor 1 connector
- Engine mount control solenoid valve connector
- Camshaft Position (CMP) sensor connector
- Rocker arm oil control solenoid connector
- Rocker arm oil pressure switch connector

- Two Air Fuel Ratio (A/F) sensor connectors
- Two secondary Heated Oxygen Sensor (secondary HO2S) connectors

44. Install the six ignition coils.

45. Install the intake manifold.

46. Install the alternator.

47. Do the battery terminal reconnection procedure.

48. After installation, check that all tubes, hoses, and connectors are installed correctly.

49. Inspect for fuel leaks. Turn the ignition switch to ON (II), (do not operate the starter) so the fuel pump runs for about 2 seconds and pressurizes the fuel line. Repeat this operation three times, then check for fuel leakage at any point in the fuel line.

50. Refill the radiator with engine coolant, and bleed the air from the cooling system.

51. Check for fluid leaks.

52. Do the Powertrain Control Module (PCM) idle learn procedure.

53. Do the Crankshaft Position (CKP) pattern clear/CKP pattern learn procedure.

54. Inspect the idle speed.

55. Inspect the ignition timing.

56. Install the engine compartment covers.

EXHAUST SYSTEM

REMOVAL & INSTALLATION

4-Cylinder Engine Catalytic Converter

WARM UP TWC

See Figures 98 through 101.

➡ **Special Tools Required: O2 sensor wrench, Snap-On YA8875, or SWR2, or equivalent, commercially available**

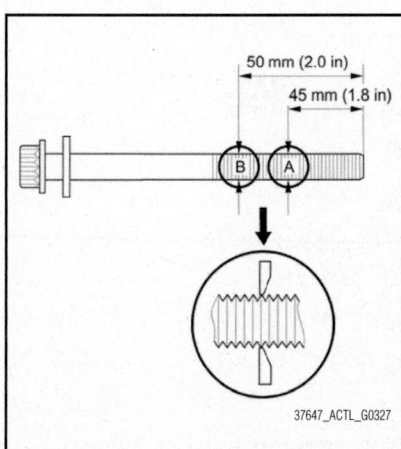

Fig. 95 Measure the diameter of each cylinder head bolt at point A and point B

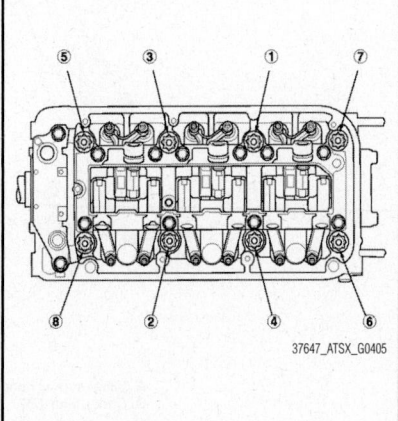

Fig. 96 Torque the front cylinder head bolts in sequence

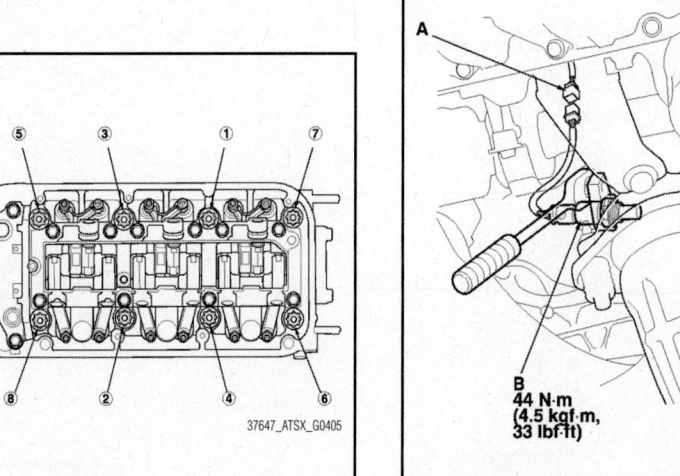

Fig. 98 Disconnect the secondary HO2S 4P connector (A), then remove secondary HO2S (B)

1. Raise the vehicle on a lift.
2. Remove the secondary HO2S (Sensor 2).
    a. Remove the front splash shield.
    b. Disconnect the secondary HO2S 4P connector, then remove secondary HO2S.
3. Remove the bolts.
4. Remove the WU-TWC bracket.
5. Lower the vehicle.
6. Remove the frame brace.
7. Remove the A/F sensor (Sensor 1).
    a. Disconnect the A/F sensor 4P connector, then remove the A/F sensor.
8. Remove the upper converter cover.
9. Remove the WU-TWC and the gaskets.
10. Remove the converter cover.
11. Install the parts in the reverse order of removal with new gaskets.

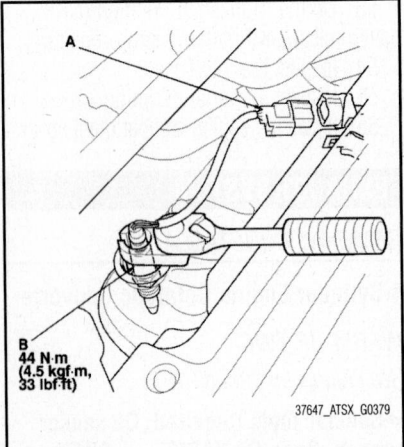

Fig. 99 Disconnect the A/F sensor 4P connector (A), then remove the A/F sensor (B)

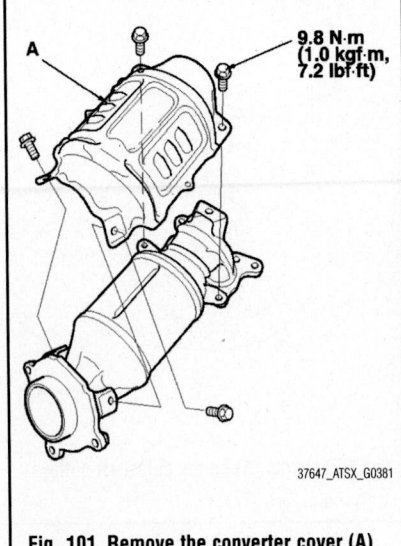

Fig. 101 Remove the converter cover (A)

Under-Floor TWC

See Figure 102.

1. Raise the vehicle on a lift.
2. Remove the exhaust pipe hangers.
3. Remove the under-floor TWC.
4. Install the parts in the reverse order of removal with new gaskets and new self-locking nuts.

6-Cylinder Engine Catalytic Converter

WARM UP TWC FRONT (BANK 2)

See Figure 103.

1. Remove the No. 2 the ignition coil and the ignition coil heat insulator.
2. Remove the front A/F sensor (Sensor 1) and the front secondary HO2S (Sensor 2).

3. Remove the exhaust pipe A mounting nuts (front WU-TWC side).
4. Remove the EGR pipe.
5. Remove the A/C condenser fan assembly and the radiator upper bracket/cushion.
6. Remove the front WU-TWC bracket, then carefully remove the front WU-TWC.
7. Carefully install the front WU-TWC with a new gasket and new self-locking nuts. Tighten the nuts in a crisscross pattern in two or three steps.
8. Install the parts in the reverse order of removal.

WARM UP TWC FRONT (BANK 1)

See Figure 104.

1. Remove the strut brace.
2. Remove the cowl cover.
3. Remove the engine wire harness holder mounting bolts.
4. Remove the No. 4 ignition coil and the ignition coil heat insulator.
5. Remove the P/S heat shield.
6. Remove the rear A/F sensor (Sensor 1) and the rear secondary HO2S (Sensor 2).
7. Remove the exhaust pipe A mounting nuts (rear WU-TWC side).
8. Carefully remove the rear WU-TWC.
9. Carefully install the rear WU-TWC with a new gasket and new self-locking nuts. Tighten the nuts in a crisscross pattern in two or three steps.
10. Install the parts in the reverse order of removal.

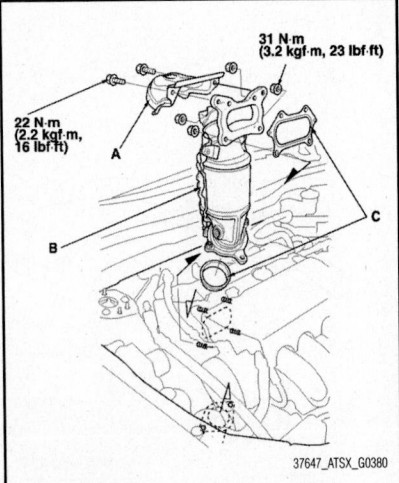

Fig. 100 Remove the upper converter cover (A), the WU-TWC (B) and the gaskets (C)

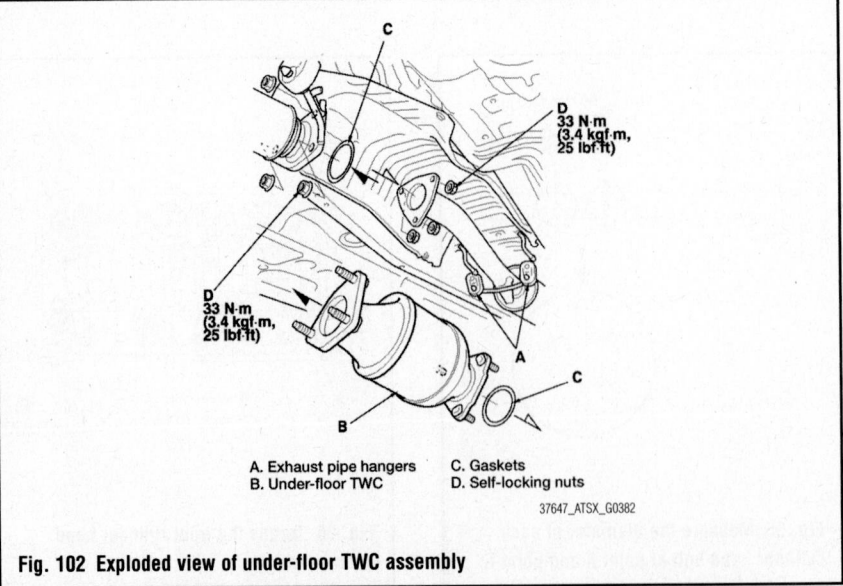

A. Exhaust pipe hangers   C. Gaskets
B. Under-floor TWC   D. Self-locking nuts

Fig. 102 Exploded view of under-floor TWC assembly

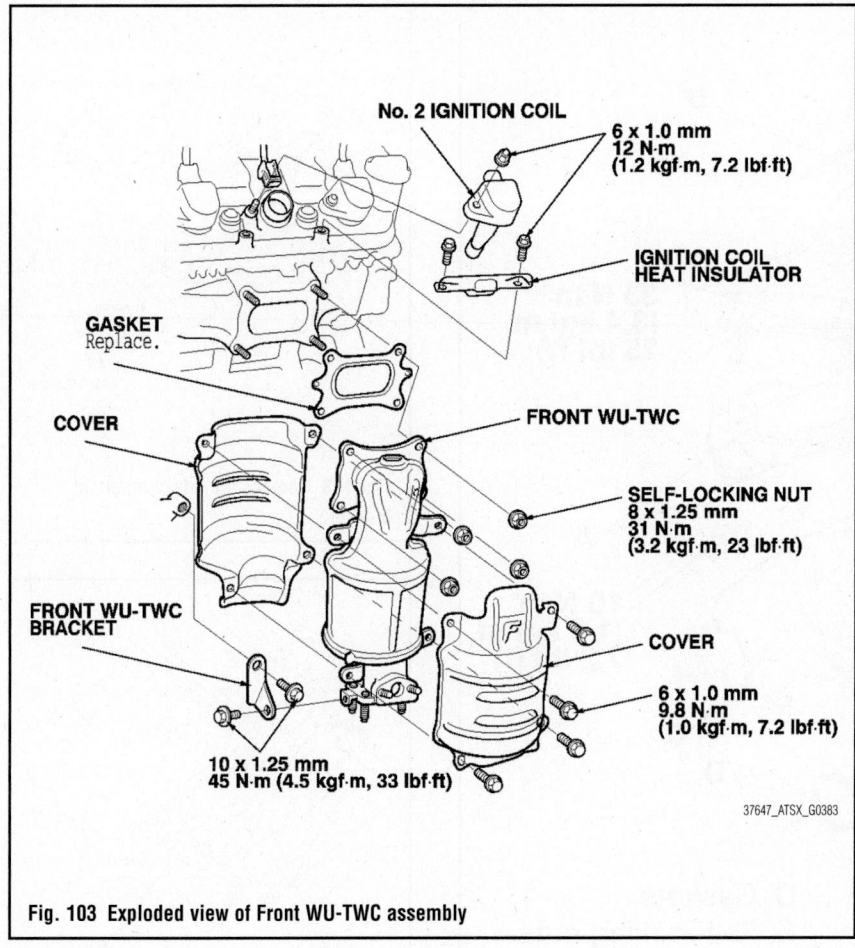

Fig. 103 Exploded view of Front WU-TWC assembly

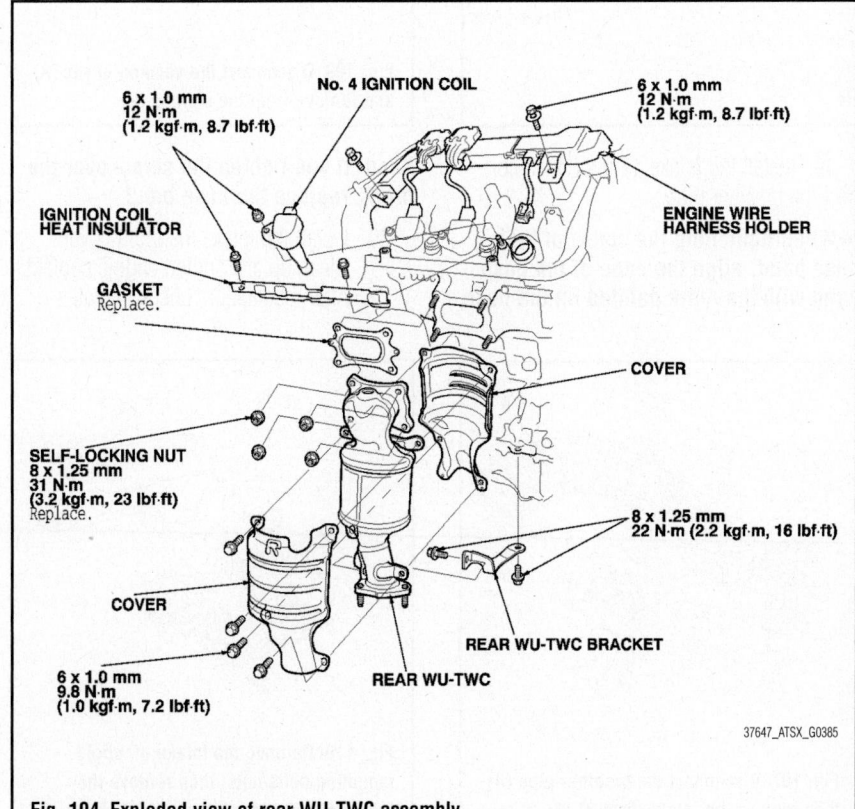

Fig. 104 Exploded view of rear WU-TWC assembly

Under-Floor Twc

See Figure 105.

1. Raise the vehicle on a lift.
2. Remove the exhaust pipe hangers.
3. Remove the under-floor TWC.
4. Remove the converter cover.
5. Install the parts in the reverse order of removal with new gaskets and new self-locking nuts.

INTAKE MANIFOLD

REMOVAL & INSTALLATION

4-Cylinder Engine

See Figures 106 through 110.

1. Remove the intake manifold cover.
2. Disconnect the breather pipe, then remove the intake air duct.
3. Disconnect the Evaporative Emission (EVAP) canister hose and the brake booster vacuum hose.
4. Disconnect the water bypass hoses, then plug the water bypass hoses.
5. Disconnect the Manifold Absolute Pressure (MAP) sensor connector (A) and the throttle actuator connector (B), then remove the wire harness clamps (C).
6. Remove the intake manifold bracket.
7. Disconnect the vacuum hoses from the intake manifold.
8. Remove the vacuum hoses from the clamps.
9. Remove the intake manifold mounting bolts/nuts, then remove the intake manifold from the cylinder head.
10. Disconnect the Positive Crankcase Ventilation (PCV) hose (A) from the intake manifold (B).

To install:

11. Connect the PCV hose to the intake manifold.
12. Install the intake manifold with new gaskets, and tighten the bolts and nuts in a crisscross pattern in three steps, beginning with the inner bolt. Tighten the bolts and nuts to 16 ft. lbs. (22 Nm).
13. Connect the vacuum hoses to the intake manifold.
14. Install the vacuum hoses to the clamps.
15. Install the intake manifold bracket. Tighten the bolt and nuts to 16 ft. lbs. (22 Nm).
16. Connect the MAP sensor connector and the throttle actuator connector, then install the wire harness clamps.
17. Connect the water bypass hoses.
18. Connect the EVAP canister hose and the brake booster vacuum hose.

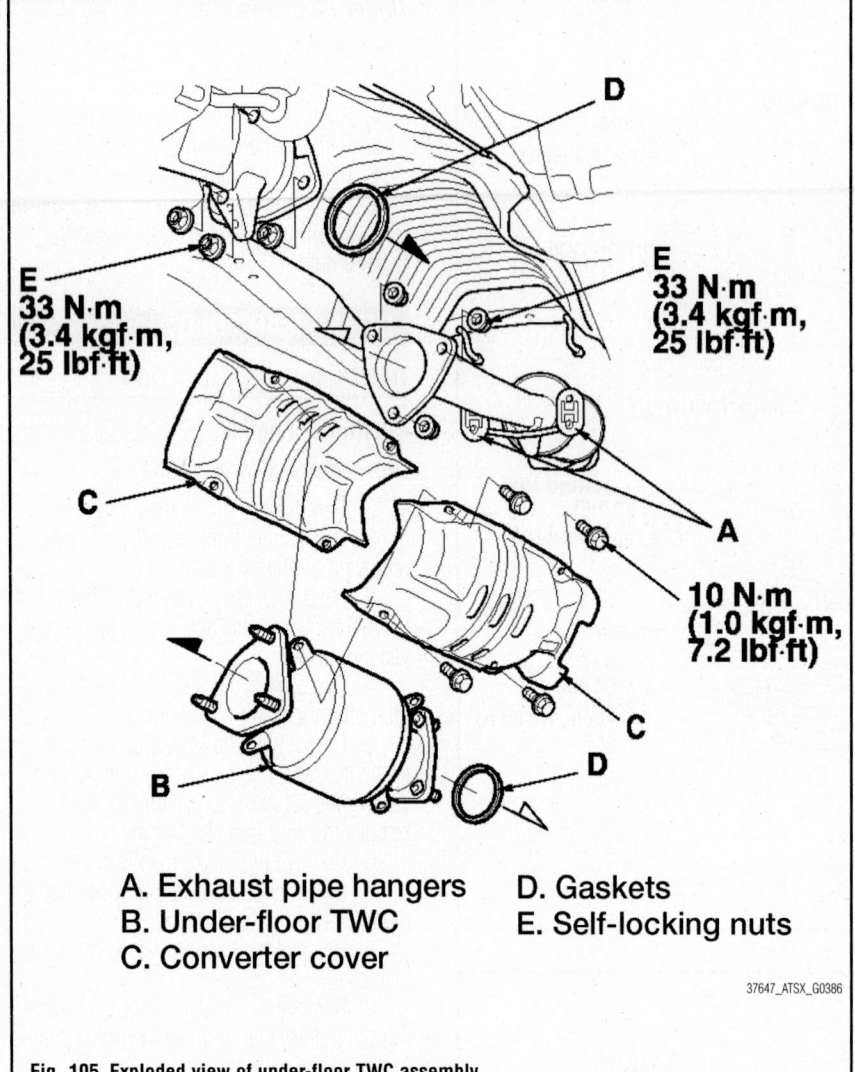

E
**33 N·m
(3.4 kgf·m,
25 lbf·ft)**

E
**33 N·m
(3.4 kgf·m,
25 lbf·ft)**

C

A

**10 N·m
(1.0 kgf·m,
7.2 lbf·ft)**

C

B

D

A. Exhaust pipe hangers
B. Under-floor TWC
C. Converter cover
D. Gaskets
E. Self-locking nuts

37647_ATSX_G0386

Fig. 105 Exploded view of under-floor TWC assembly

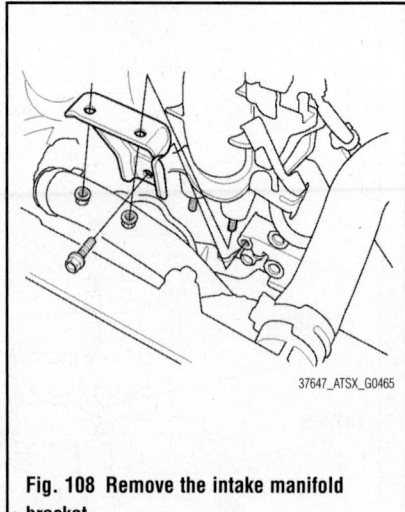

37647_ATSX_G0465

Fig. 108 Remove the intake manifold bracket

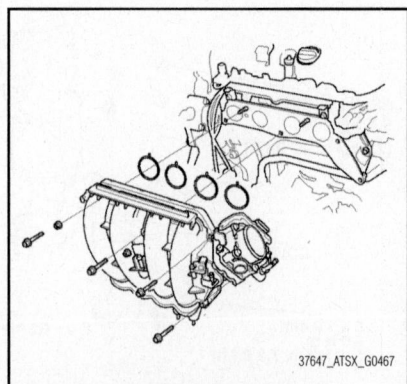

B

A

37647_ATSX_G0466

Fig. 109 Disconnect the vacuum hoses (A) and remove from the clamps (B)

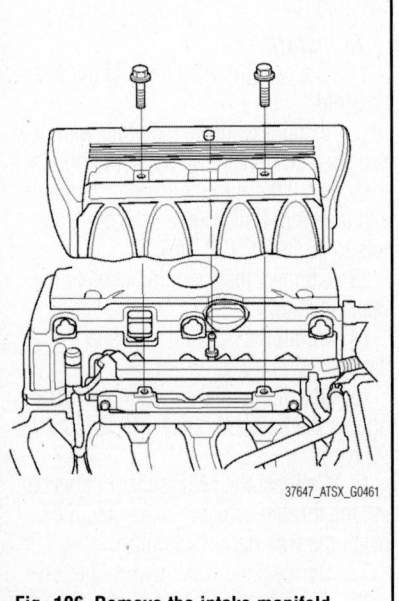

37647_ATSX_G0461

Fig. 106 Remove the intake manifold cover

19. Install the intake air duct, then connect the breather pipe.

➡**When tightening the screw of the hose band, align the edge of the hose band with the mark painted on the hose**

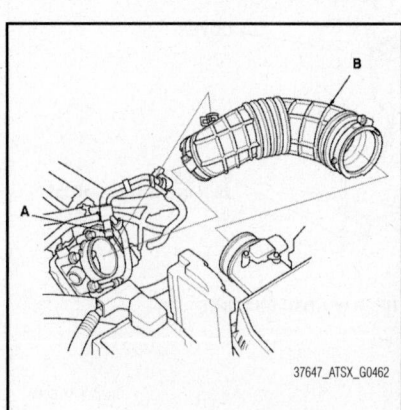

B

A

37647_ATSX_G0462

Fig. 107 Disconnect the breather pipe (A), then remove the intake air duct (B)

band. If you tighten the screw over the mark, replace the hose band.

20. Install the intake manifold cover.
21. Clean up any spilled engine coolant.
22. After installation, check that all

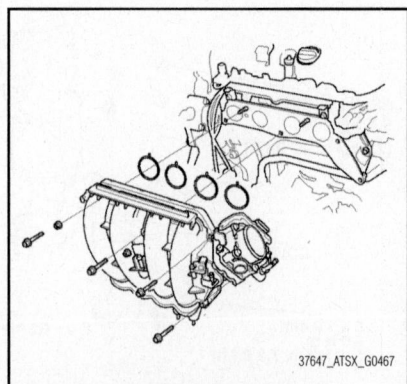

37647_ATSX_G0467

Fig. 110 Remove the intake manifold mounting bolts/nuts, then remove the intake manifold from the cylinder head

tubes, hoses, and connectors are installed correctly.

23. Refill the radiator with engine coolant, and bleed the air from the cooling system with the heater valve open.

6-Cylinder Engine

See Figures 111 through 114.

1. Remove the engine compartment covers.

2. Disconnect the Manifold Absolute Pressure (MAP) sensor connector and the breather pipe, then remove the intake air duct.

3. Disconnect the throttle actuator connector, the Evaporative Emission (EVAP) canister purge valve connector, the water bypass hoses, the EVAP canister hose, the brake booster vacuum hose, and the vacuum hose.

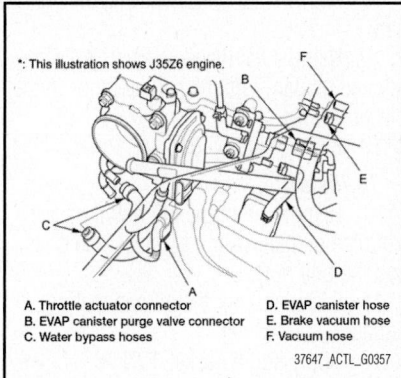

A. Throttle actuator connector
B. EVAP canister purge valve connector
C. Water bypass hoses
D. EVAP canister hose
E. Brake vacuum hose
F. Vacuum hose

37647_ACTL_G0357

Fig. 111 Disconnect the throttle actuator connector, the EVAP canister purge valve connector, the water bypass hoses, the EVAP canister hose, the brake booster vacuum hose, and the vacuum hose

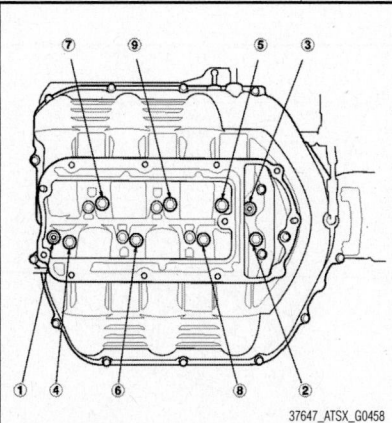

37647_ATSX_G0458

Fig. 112 Remove the intake manifold mounting bolts and nuts sequentially in three steps, then remove the intake manifold

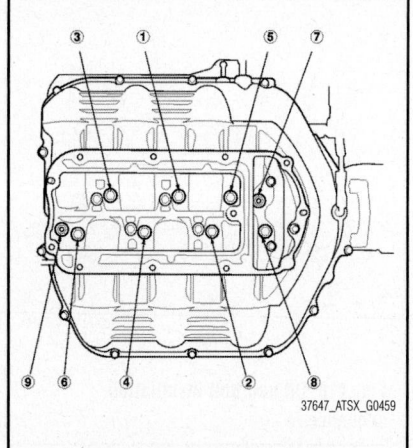

37647_ATSX_G0459

Fig. 113 Install the intake manifold

4. Disconnect the Positive Crankcase Ventilation (PCV) hose and the Intake Manifold Tuning (IMT) actuator connector.

5. Remove the upper cover mounting bolts and nuts sequentially in three steps, then remove the upper cover.

6. Remove the intake manifold mounting bolts and nuts sequentially in three steps, then remove the intake manifold.

To install:

7. Install the intake manifold. Tighten the mounting bolts and nuts sequentially in three steps to 16 ft. lbs. (22 Nm). Always use a new intake manifold gasket.

8. Install the upper cover. Tighten the mounting bolts and nuts sequentially in three steps to 9 ft. lbs. (12 Nm). Always use a new gasket.

9. Connect the PCV hose and the IMT actuator connector.

10. Connect the throttle actuator connector, the EVAP canister purge valve connec-

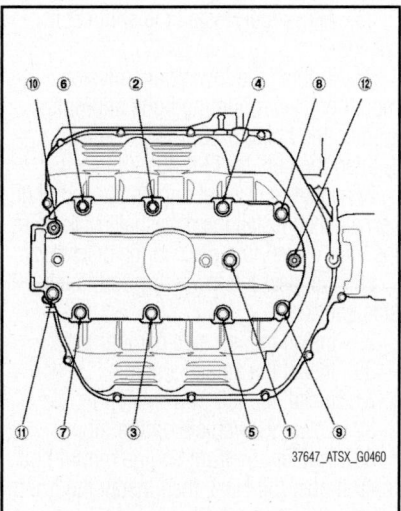

37647_ATSX_G0460

Fig. 114 Install the upper cover

tor, the water bypass hoses, the EVAP canister hose, the brake booster vacuum hose, and the vacuum hose.

11. Install the intake air duct, then connect the breather pipe and the MAP sensor connector.

12. Clean up any spilled engine coolant.

13. After installation, check that all tubes, hoses, and connectors are installed correctly.

14. Refill the radiator with engine coolant, and bleed the air from the cooling system..

15. Install the engine compartment covers.

OIL PAN

REMOVAL & INSTALLATION

4-Cylinder Engine

See Figures 115 through 118.

1. If the engine is already out of the vehicle, go to step 13.

2. Disconnect the vacuum hose and remove the, front engine mount stop, then remove the front engine mount bolt.

3. Raise the vehicle on the lift.

4. Remove the left front wheel.

5. Remove the splash shield.

6. Drain the engine oil.

7. Remove the left side damper fork.

8. Separate the left side knuckle from the lower arm.

9. Remove the left side driveshaft. Coat all precision-finished surface with new engine oil. Tie a plastic bag over the driveshaft end.

10. Remove the lower transmission mount bracket mounting bolts and nuts.

11. A/T model: Remove the shift cable cover.

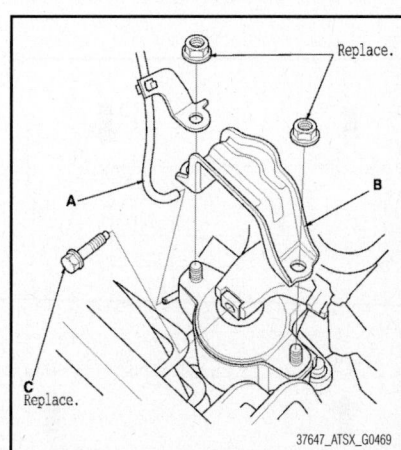

37647_ATSX_G0469

Fig. 115 Disconnect the vacuum hose (A) and remove the, front engine mount stop (B), then remove the front engine mount bolt (C)

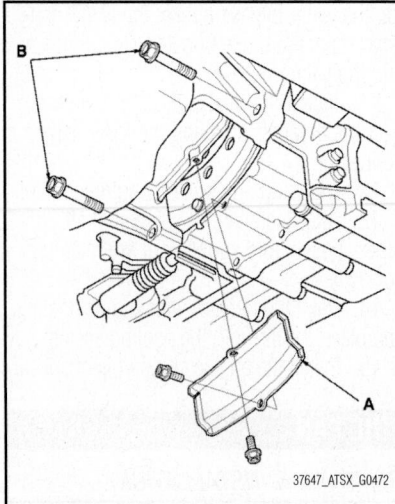

Fig. 116 Remove the clutch cover/torque converter cover (A) and the transmission mounting bolts (B)

12. Use a transmission jack to lift the transmission 1.2–1.6 inches (30–40 mm).

13. Remove the clutch cover/torque converter cover and the transmission mounting bolts.

14. Remove the bolts securing the oil pan.

15. Using a flat blade screwdriver, separate the oil pan from the engine block in the places shown.

16. Remove the oil pan.

To install:

17. Remove all of the old liquid gasket from the oil pan mating surfaces, the bolts, and the bolt holes.

18. Clean and dry the oil pan mating surfaces.

19. Apply liquid gasket (P/N 08717-0004, 08718-0003, or 08718-0009) to the engine block mating surface of the oil pan

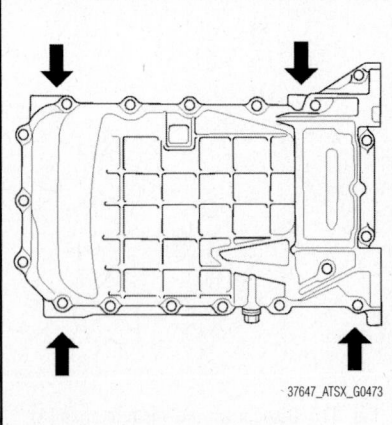

Fig. 117 Separate the oil pan from the engine block in the places shown

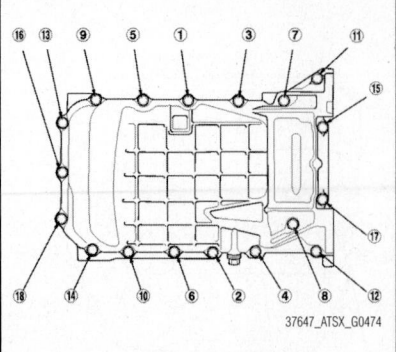

Fig. 118 Oil pan bolt installation sequence

and to the inside edge of the threaded bolt holes. Install the component within 5 minutes of applying the liquid gasket.

➡**If too much time has passed after applying the liquid gasket, remove the old liquid gasket and residue, then reapply new liquid gasket.**

20. Install the oil pan.

21. Tighten the bolts in three steps. In the final step, torque all bolts, in sequence, to 9 ft. lbs. (12 Nm). Wipe off the excess liquid gasket on the each side of crankshaft pulley and the flywheel/drive plate.

➡**Wait at least 30 minutes before filling the engine with oil. Do not run the engine for at least 3 hours after installing the oil pan.**

22. Install the clutch cover/torque converter cover and the transmission mounting bolts.

23. If the engine is still in the vehicle, do steps 8 through 18.

24. Lower the transmission jack from the transmission.

25. A/T model: Install the shift cable cover.

26. Tighten the lower transmission mount bracket mounting bolts and nuts. Tighten the bolts to 40 ft. lbs. (54 Nm). Tighten the nuts to 33 ft. lbs. (44 Nm).

27. Install a new set ring on the end of driveshaft, then install the driveshaft. Make sure the ring "clicks" into place in the differential.

28. Connect the lower arm to the left side knuckle.

29. Install the left side damper fork.

30. Install the splash shield.

31. Install the left front wheel.

32. Lower the vehicle on the lift.

33. Tighten the front engine mount bolt to 40 ft. lbs. (54 Nm), then install the front engine mount stop and connect the vacuum hose.

34. Refill the engine with recommended engine oil.

6-Cylinder Engine

See Figures 119 through 121.

➡**If the engine is already out of the vehicle, go to step 6.**

1. Raise the vehicle on the lift.

2. Drain the engine oil.

3. Remove the splash shield.

4. Remove exhaust pipe A.

5. Remove the rear Warm Up Three Way Catalytic Converter (rear WU-TWC) bracket.

6. Remove the Crankshaft Position (CKP) sensor cover and the bolt, then disconnect the CKP sensor connector.

7. Remove the clutch/torque converter cover and the four bolts securing the transmission.

8. Remove the bolts securing the oil pan.

9. Using a flat blade screwdriver, separate the oil pan from the engine block in the places shown.

10. Remove the oil pan.

To install:

11. Remove all of the old liquid gasket from the oil pan mating surfaces, the bolts, and the bolt holes.

12. Clean and dry the oil pan mating surfaces.

13. Apply liquid gasket, P/N 08717-0004, 08718-0003, or 08718-0009 to the oil pan mating surface of the engine block and to the inside edge of the threaded bolt holes. Install the component within 5 minutes of applying the liquid gasket.

➡**Apply a bead of liquid gasket along the mating edge of the oil pan. If you**

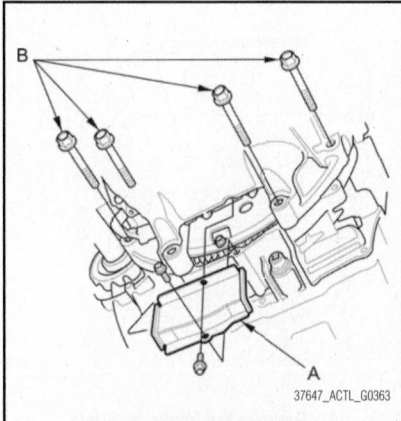

Fig. 119 Remove the clutch/torque converter cover (A) and the four bolts (B) securing the transmission

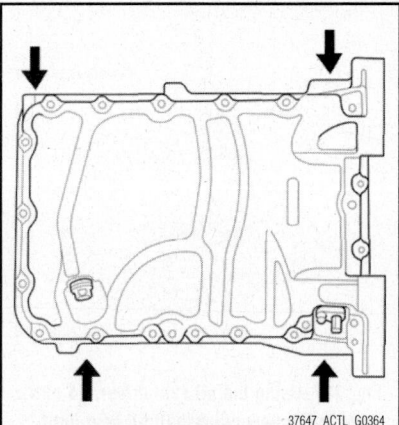

Fig. 120 Using a flat blade screwdriver, separate the oil pan from the engine block in the places shown

apply liquid gasket P/N 08718-0012, the component must be installed within 4 minutes. If too much time has passed after applying the liquid gasket, remove the old liquid gasket and residue, then reapply the new liquid gasket.

14. Install the oil pan on the engine block.

15. Tighten the bolts in three steps. In the final step, torque all bolts, in sequence, to 9 ft. lbs. (12 Nm).

➡**Wait at least 30 minutes before filling the engine with oil. Do not run the engine for at least 3 hours after installing the oil pan.**

16. Tighten the four bolts securing the transmission to 54 ft. lbs. (74 Nm), then install the clutch/torque converter cover.

17. Connect the Crankshaft Position (CKP) sensor connector, then install the CKP sensor cover and the bolt.

18. Install the rear Warm Up Three Way Catalytic Converter (rear WU-TWC) bracket.

19. If the engine is still in the vehicle, do the following steps.

20. Install exhaust pipe A using new gaskets and new self-locking nuts.

21. Install the splash shield.

22. Refill the engine with the recommended engine oil.

OIL PUMP

REMOVAL & INSTALLATION

4-Cylinder Engine

See Figures 122 through 125.

1. Turn the crankshaft pulley so its Top Dead Center (TDC) mark lines up with the pointer.

➡**The other pointer is not used.**

2. Remove the oil pan.

3. To hold the rear balancer shaft, insert a 6 mm long pin punch (A) (Snap-On PPC108LA or equivalent) into the maintenance hole in the balancer shaft holder and through the rear balancer shaft.

4. Turn the crankshaft counterclockwise to compress the oil pump chain auto-tensioner.

5. Align the holes on the lock and the oil pump chain auto-tensioner, then insert a 0.12 inches (3.0 mm) diameter pin into the holes. Turn the crankshaft clockwise to secure the pin.

6. Remove the oil pump chain auto-tensioner.

7. Loosen the oil pump sprocket mounting bolt.

8. Remove the oil pump sprocket, then remove the oil pump.

To install:

9. Make sure the No. 1 piston Top Dead Center (TDC) mark lines up with the pointer.

10. Align the dowel pin on the rear balancer shaft with the mark on the oil pump.

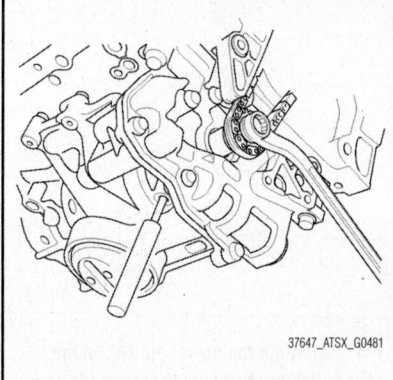

Fig. 123 Loosen the oil pump sprocket mounting bolt.

11. To hold the rear balancer shaft, insert a 6 mm long pin punch (Snap-O n PPC108LA or equivalent) into the maintenance hole in the balancer shaft holder and through the rear balancer shaft.

12. Apply new engine oil to the threads of the oil pump mounting bolts and oil pump sprocket mounting bolt, then loosely install the oil pump with a new O-ring, then install the oil pump sprocket.

13. Tighten the oil pump mounting bolts and the oil pump sprocket mounting bolt.

14. Remove the 6 mm pin punch.

15. Install the oil pump chain auto-tensioner.

➡**If the 1.2 inches (3.0 mm) diameter pin come off the auto-tensioner, compress the auto-tensioner.**

16. Remove the pin from the oil pump chain auto-tensioner.

17. Install the oil pan.

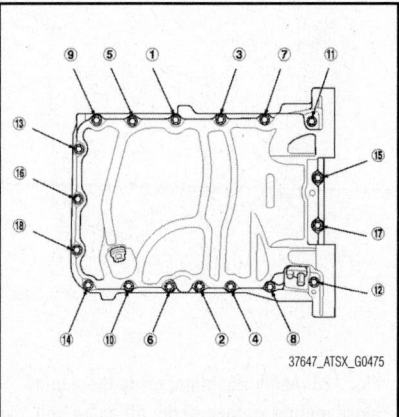

Fig. 121 Oil pan bolt installation sequence

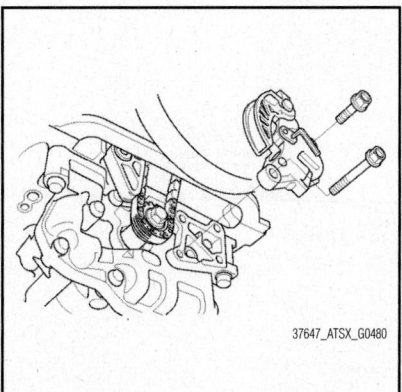

Fig. 122 Remove the oil pump chain auto-tensioner

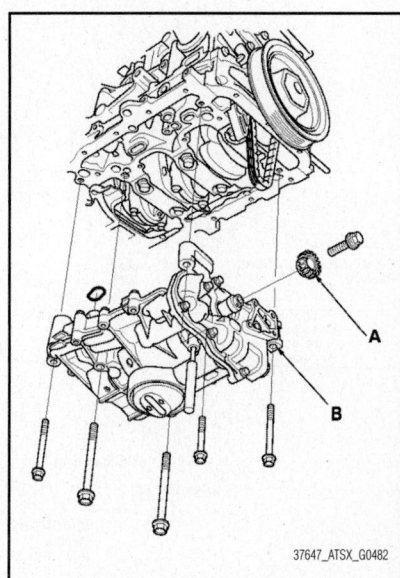

Fig. 124 Remove the oil pump sprocket (A), then remove the oil pump (B).

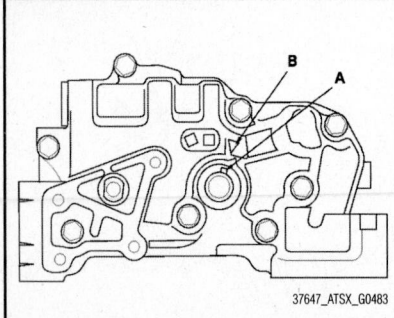

Fig. 125 Align the dowel pin (A) on the rear balancer shaft with the mark (B) on the oil pump.

6-Cylinder Engine

See Figures 126 through 128.

1. Drain the engine oil.
2. Remove the timing belt:
3. Remove the timing belt drive pulley from the crankshaft:
4. Attach the chain hoist to the engine hook on the engine hanger bracket.

5. Remove the rocker arm oil control solenoid/oil filter assembly.
6. Remove the oil pan.
7. Remove the oil strainer.
8. Remove the mounting bolts, then remove the oil pump assembly.

To install:

9. Remove the old oil seal from the oil pump.
10. Clean and dry the crankshaft oil seal housing.
11. Using the oil seal driver, 64 mm, drive in the new crankshaft oil seal until the oil seal driver bottoms on the pump.
12. Remove all of the old liquid gasket from the oil pump mating surfaces, the bolts, and the bolt holes.
13. Clean and dry the oil pump mating surfaces.
14. Apply liquid gasket, P/N 08717-0004, 08718-0003, or 08718-0009, to the engine block mating surface of the oil pump and to the inside edge of the threaded bolt holes. Install the component within 5 minutes of applying the liquid gasket.

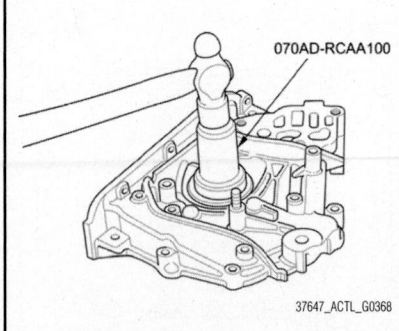

Fig. 127 Using the oil seal driver, 64 mm, drive in the new crankshaft oil seal until the oil seal driver bottoms on the pump

➡️**Apply a bead of liquid gasket along the mating surface. If you apply liquid gasket P/N 08718-0012, the component must be installed within 5 minutes. If too much time has passed after applying the liquid gasket, remove all of the old liquid gasket and residue, then reapply the new liquid gasket.**

15. Apply a light coat of new engine oil to the lip of the crankshaft oil seal, and apply new engine oil to the new O-ring.
16. Install the dowel pins, then align the inner rotor with the crankshaft, and install the oil pump.

➡️**Wait at least 30 minutes before filling the engine with oil. Do not run the engine for at least 3 hours after installing the oil pump.**

17. Clean the excess grease off the crankshaft, and check the seal for distortion.
18. Install the oil strainer with a new O-ring.
19. Install the oil pan.

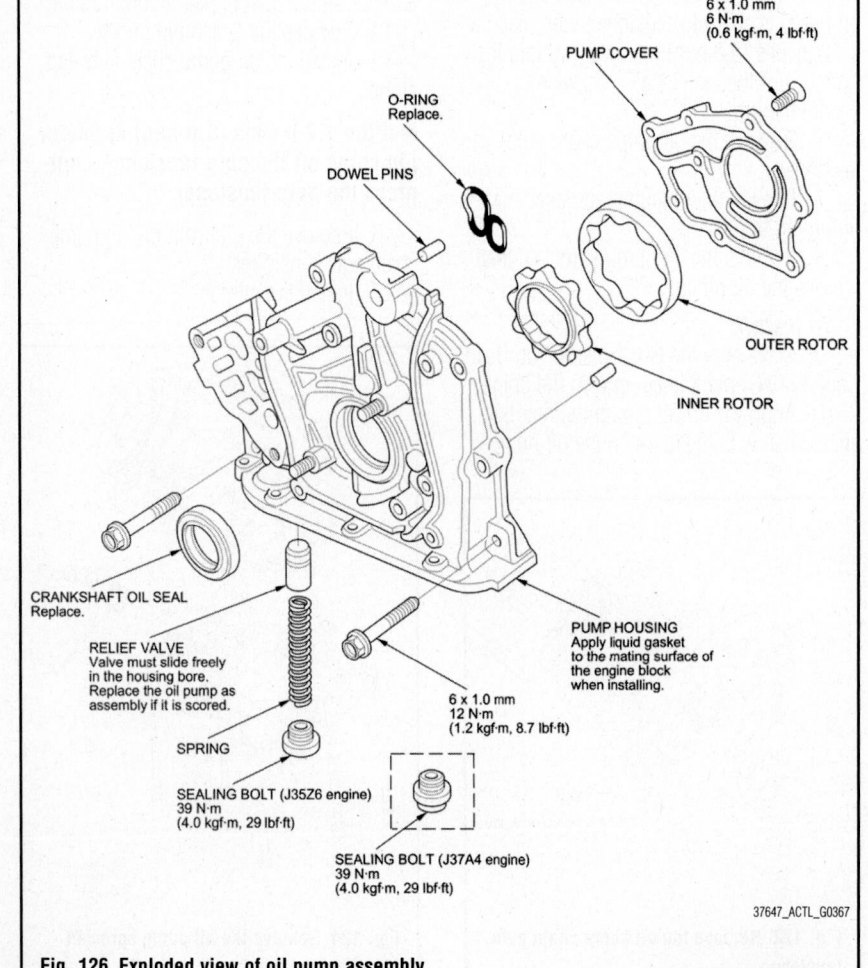

Fig. 126 Exploded view of oil pump assembly

Fig. 128 Apply liquid gasket to the engine block mating surface of the oil pump and to the inside edge of the threaded bolt holes

20. Install the rocker arm oil control solenoid/oil filter assembly with a new rocker arm oil control solenoid filter.

21. Install the timing belt drive pulley to the crankshaft:

22. Install the timing belt:

23. Remove the chain hoist.

PISTONS & RINGS

POSITIONING

4-Cylinder Engine

See Figure 129.

6-Cylinder Engine

See Figure 130.

REAR MAIN SEAL

REMOVAL & INSTALLATION

4-Cylinder Engine

See Figures 131 and 132.

Special Tools Required:
- Driver Handle, 15 x 135L 07749-0010000
- Oil Seal Driver Attachment, 96 mm 07ZAD-PNAA100

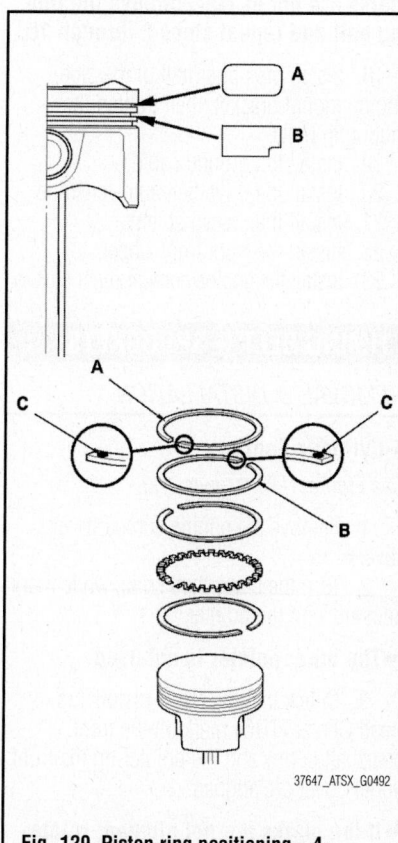

Fig. 129 Piston ring positioning—4-Cylinder Engine

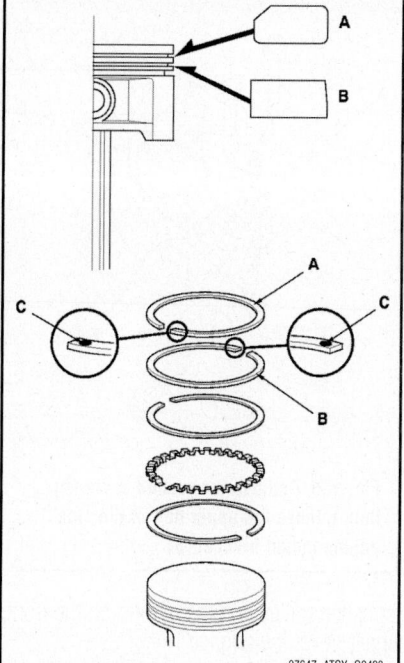

Fig. 130 Piston ring positioning—6-Cylinder Engine

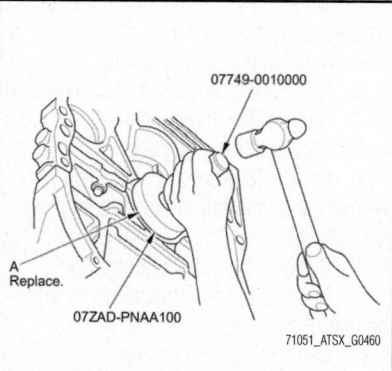

Fig. 131 Drive a new crankshaft oil seal (A) squarely into the engine block to the specified installed height

1. Remove the transmission:

2. M/T: Remove the pressure plate, the clutch disk, and the flywheel.

3. A/T: Remove the drive plate.

4. Remove the crankshaft oil seal.

To install:

5. Clean and dry the crankshaft oil seal housing.

6. Apply a light coat of new engine oil to the lip of the crankshaft oil seal.

7. Use the driver handle, 15 x 135L and the oil seal driver attachment, 96 mm to drive a new crankshaft oil seal (A) squarely into the engine block to the specified installed height.

8. Clean the excess oil off the crank-

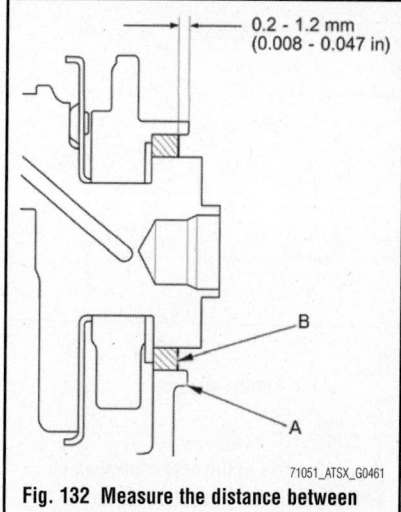

Fig. 132 Measure the distance between the engine block (A) and the crankshaft oil seal (B)

shaft, and check that the oil seal lip is not distorted.

9. Measure the distance between the engine block and the crankshaft oil seal.

➡ Oil Seal Installed Height: 0.0085–0.047 inches (0.2–1.2 mm)

10. M/T: Install the flywheel, the clutch disk, and the pressure plate.

11. A/T: Install the drive plate.

12. Install the transmission:

6-Cylinder Engine

See Figure 133.

Special Tools Required:
- Driver Handle, 15 x 135L 07749-0010000
- Oil Seal Driver Attachment, 106 mm 070AD-RCA0200

1. Remove the transmission and the drive plate.

2. Remove the transmission end crankshaft oil seal.

3. Clean and dry the crankshaft oil seal housing.

4. Apply a light coat of new engine oil to the lip of the crankshaft oil seal.

5. Using the driver handle, 15 x 135L and the oil seal driver attachment, 106 mm, drive in the new crankshaft oil seal until the oil seal driver attachment bottoms on the engine block end cover. Align the hole in the oil seal driver attachment with the pin on the crankshaft.

6. Clean any excess oil off the crankshaft, and check that the oil seal lip is not distorted.

7. Install the drive plate and the transmission.

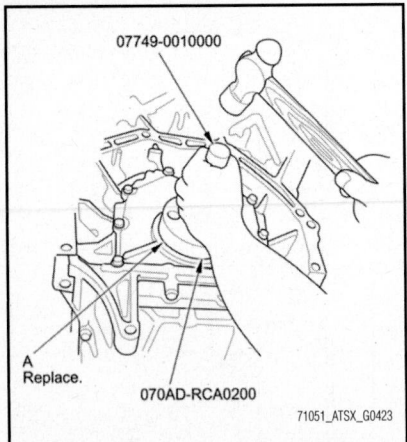

Fig. 133 Drive in the new crankshaft oil seal (A) until the oil seal driver attachment bottoms on the engine block end cover

TIMING BELT FRONT COVER

REMOVAL & INSTALLATION

6-Cylinder Engine

See Figures 134 through 137.

1. Remove the engine compartment covers.

2. Turn the crankshaft so its white mark lines up with the pointer.

➡**The other pointer is not used.**

3. Check that the No. 1 piston Top Dead Center (TDC) mark on the front

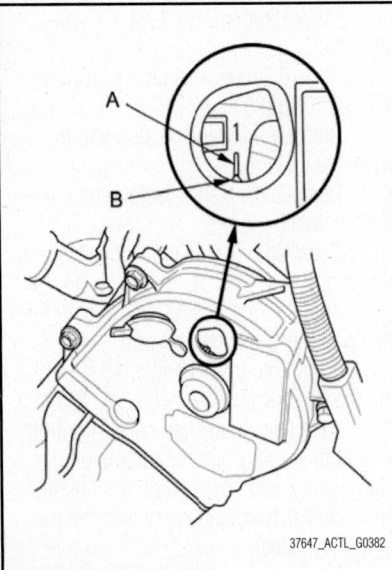

37647_ACTL_G0382

Fig. 134 Check that the No. 1 piston TDC mark (A) on the front camshaft pulley and the pointer (B) on the front upper cover are aligned

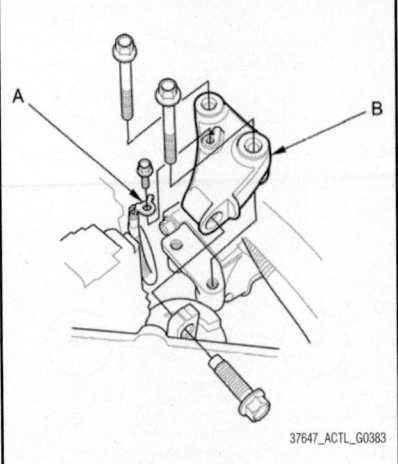

37647_ACTL_G0383

Fig. 135 Remove the ground cable (A), then remove the upper half of the side engine mount bracket (B)

camshaft pulley and the pointer on the front upper cover are aligned.

➡**If the marks are not aligned, rotate the crankshaft 360 degrees, and recheck the camshaft pulley mark.**

4. Raise the vehicle on the lift, then remove the right front wheel.

5. Remove the splash shield.

6. Remove the drive belt auto-tensioner.

7. Support the engine with a jack and a wood block under the oil pan.

8. Remove the ground cable, then remove the upper half of the side engine mount bracket.

9. Remove the crankshaft pulley.

10. Remove the front upper cover and the rear upper cover.

11. Remove the lower cover.

To install:

12. Install the lower cover.

13. Install the front upper cover and the rear upper cover.

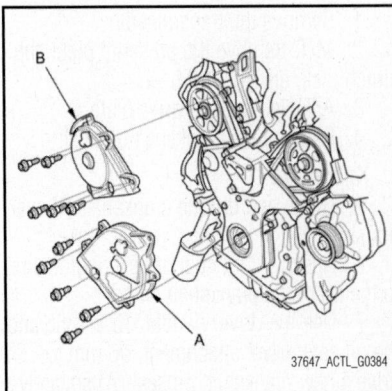

B

A

37647_ACTL_G0384

Fig. 136 Remove the front upper cover (A) and the rear upper cover (B)

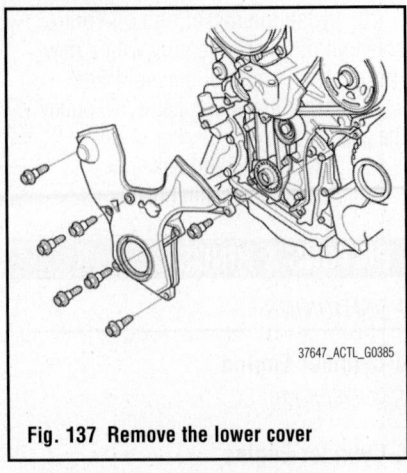

37647_ACTL_G0385

Fig. 137 Remove the lower cover

14. Install the crankshaft pulley.

15. Rotate the crankshaft pulley about six turns clockwise so the timing belt positions itself on the pulleys.

16. Turn the crankshaft pulley so its white mark lines up with the pointer.

➡**The other pointer is not used.**

17. Check the camshaft pulley marks.

➡**If the marks are not aligned, rotate the crankshaft 360 degrees, and recheck the camshaft pulley mark. If the camshaft pulley marks are at TDC, go to step 17. If the camshaft pulley marks are not at TDC, remove the timing belt and repeat steps 2 through 16.**

18. Install the upper half of the side engine mount bracket, then tighten the mounting bolts.

19. Install the ground cable.

20. Install the drive belt auto-tensioner.

21. Install the splash shield.

22. Install the right front wheel.

23. Install the engine compartment covers.

TIMING BELT & SPROCKETS

REMOVAL & INSTALLATION

6-Cylinder Engine

See Figures 138 through 143.

1. Remove the engine compartment covers.

2. Turn the crankshaft so its white mark lines up with the pointer.

➡**The other pointer is not used.**

3. Check that the No. 1 piston Top Dead Center (TDC) mark on the front camshaft pulley and the pointer on the front upper cover are aligned.

➡**If the marks are not aligned, rotate the crankshaft 360 degrees, and recheck the camshaft pulley mark.**

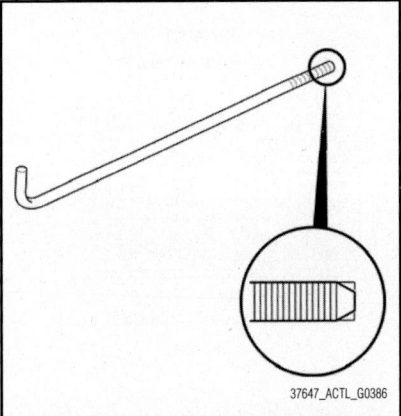

37647_ACTL_G0386

Fig. 138 Remove one of the battery clamp bolts from the battery tray, and grind the end of it

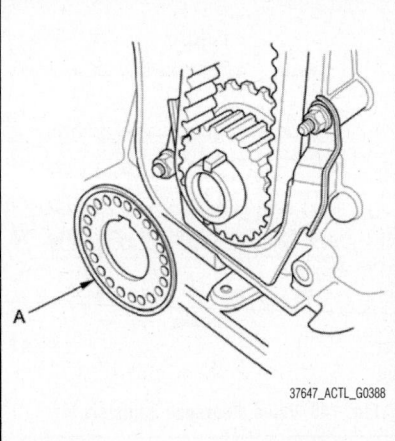

37647_ACTL_G0388

Fig. 140 Remove the timing belt guide plate (A)

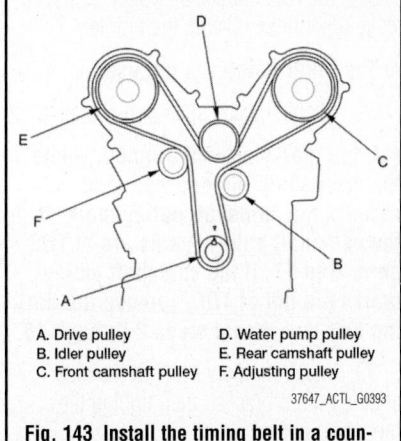

A. Drive pulley　　　　D. Water pump pulley
B. Idler pulley　　　　E. Rear camshaft pulley
C. Front camshaft pulley　　F. Adjusting pulley

37647_ACTL_G0393

Fig. 143 Install the timing belt in a counterclockwise sequence starting with the drive pulley

4. Raise the vehicle on the lift, then remove the right front wheel.

5. Remove the splash shield.

6. Remove the drive belt auto-tensioner.

7. Support the engine with a jack and a wood block under the oil pan.

8. Remove the ground cable, then remove the upper half of the side engine mount bracket.

9. Remove the crankshaft pulley.

10. Remove the front upper cover and the rear upper cover.

11. Remove the lower cover.

12. Remove one of the battery clamp bolts from the battery tray, and grind the end of it.

13. Thread the battery clamp bolt into hold the timing belt adjuster in its current position. Tighten it by hand, do not use a wrench.

14. Remove the timing belt guide plate.

15. Remove the lower half of the side engine mount bracket.

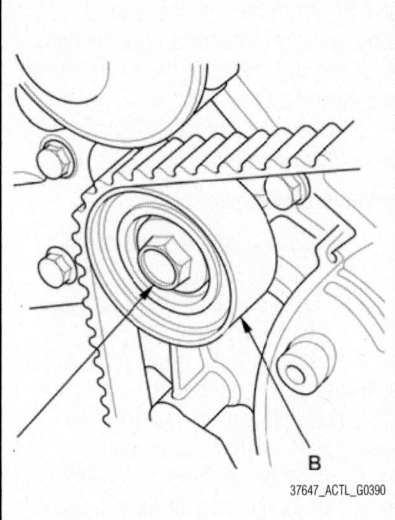

37647_ACTL_G0390

Fig. 141 Remove the idler pulley bolt (A) and the idler pulley (B), then remove the timing belt

16. Remove the idler pulley bolt and the idler pulley, then remove the timing belt. Discard the idler pulley bolt.

To install:

17. Clean the timing belt pulleys, the timing belt guide plate, and the upper and lower covers.

18. Set the timing belt drive pulley to Top Dead Center (TDC) by aligning the TDC mark on the tooth of the timing belt drive pulley with the pointer on the oil pump.

19. Set the camshaft pulleys to TDC by aligning the TDC marks on the camshaft pulleys with the pointers on the back covers.

20. Loosely install the idler pulley with a new idler pulley bolt so the pulley can move but does not come off.

21. If the auto-tensioner has extended and the timing belt cannot be installed, do the timing belt replacement procedure.

22. Install the timing belt in a counterclockwise sequence starting with the drive pulley. Take care not to damage the timing belt during installation.

23. Tighten the idler pulley bolt to 33 ft. lbs. (44 Nm).

24. Remove the battery clamp bolt from the back cover.

25. Install the lower half of the side engine mount bracket. Tighten the three long bolts to 33 ft. lbs. (44 Nm).

26. Install the timing belt guide plate.

27. Install the lower cover.

28. Install the front upper cover and the rear upper cover.

29. Install the crankshaft pulley.

30. Rotate the crankshaft pulley about six turns clockwise so the timing belt positions itself on the pulleys.

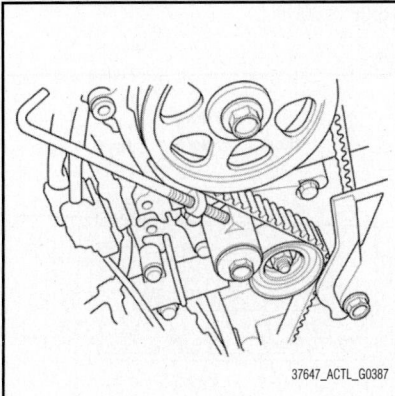

37647_ACTL_G0387

Fig. 139 Thread the battery clamp bolt into hold the timing belt adjuster in its current position

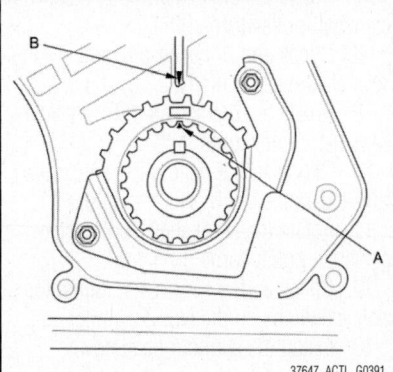

37647_ACTL_G0391

Fig. 142 Set the timing belt drive pulley to TDC by aligning the TDC mark (A) on the tooth of the timing belt drive pulley with the pointer (B) on the oil pump

31. Turn the crankshaft pulley so its white mark lines up with the pointer.

➡**The other pointer is not used.**

32. Check the camshaft pulley marks.

➡**If the marks are not aligned, rotate the crankshaft 360 degrees, and recheck the camshaft pulley mark. If the camshaft pulley marks are at TDC, go to step 17. If the camshaft pulley marks are not at TDC, remove the timing belt and repeat steps 2 through 16.**

33. Install the upper half of the side engine mount bracket, then tighten the mounting bolts.
34. Install the ground cable.
35. Install the drive belt auto-tensioner.
36. Install the splash shield.
37. Install the right front wheel.
38. Install the engine compartment covers.

VALVE LASH

ADJUSTMENT

4-Cylinder Engine

See Figures 144 and 145.

Special Tools Required:
- Adjuster 07MAA-PR70110
- Locknut wrench 07MAA-PR70120

➡**Connect the Honda Diagnostic System (HDS) to the Data Link Connector (DLC) and monitor the Engine Coolant Temperature (ECT) sensor 1 with the HDS. Adjust the valve clearance only when the ECT sensor 1 temperature is less than 100°F (38°C).**

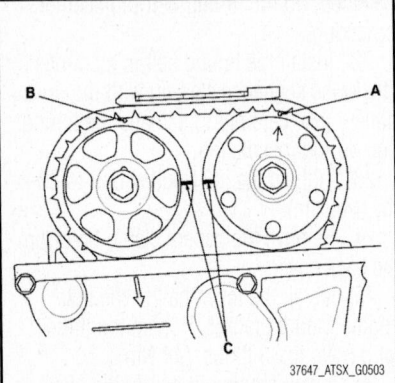

Fig. 144 The punch mark (A) on the VTC actuator and the punch mark (B) on the exhaust camshaft sprocket should be at the top. Align the TDC marks (C) on the VTC actuator and the exhaust camshaft sprocket

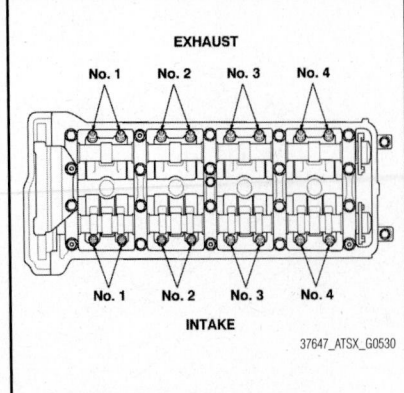

Fig. 145 Valve clearance adjusters

1. Remove the cylinder head cover.
2. Set the No. 1 piston at Top Dead Center (TDC). The punch mark on the Variable Valve Timing Control (VTC) actuator and the punch mark on the exhaust camshaft sprocket should be at the top. Align the TDC marks on the VTC actuator and the exhaust camshaft sprocket.
3. Select the correct feeler gauge for the valve clearance you are going to check.

➡**Valve Clearance:**

- Intake: 0.008–0.010 inches (0.21–0.25 mm)
- Exhaust: 0.010–0.011 inches (0.25–0.29 mm)

4. Insert the feeler gauge between the adjusting screw and the end of the valve stem, and slide it back and forth; you should feel a slight amount of drag.
5. If you feel too much or too little drag, loosen the locknut with the locknut wrench and the adjuster, and turn the adjusting screw until the drag on the feeler gauge is correct.
6. Tighten the locknut to 10 ft. lbs. (14 Nm), and recheck the clearance. Repeat the adjustment if necessary.
7. Rotate the crankshaft 180° clockwise (camshaft pulley turns 90°).
8. Check and, if necessary, adjust the valve clearance on the No. 3 cylinder.
9. Rotate the crankshaft 180° clockwise (camshaft pulley turns 90°).
10. Check and, if necessary, adjust the valve clearance on the No. 4 cylinder.
11. Rotate the crankshaft 180° clockwise (camshaft pulley turns 90°).
12. Check and, if necessary, adjust the valve clearance on the No. 2 cylinder.
13. Install the cylinder head cover.

6-Cylinder Engine

See Figures 146 through 148.

➡**Connect the Honda Diagnostic System (HDS) to the data link connector**

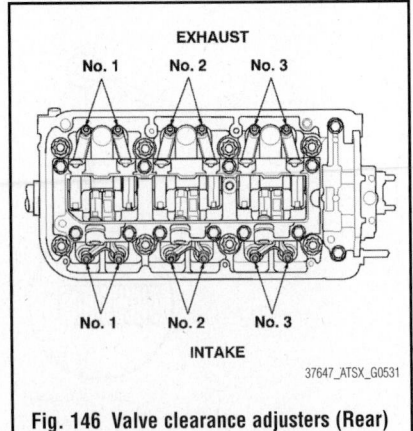

Fig. 146 Valve clearance adjusters (Rear)

(DLC), and monitor the Engine Coolant Temperature (ECT) sensor 1. Adjust the valve clearance only when the ECT sensor 1 is less than 100°F (38°C).

1. Remove the cylinder head covers.
2. Set the No. 1 piston at Top Dead Center (TDC). Align the pointer on the front upper cover with the No. 1 piston TDC mark on the front camshaft pulley.
3. Select the correct feeler gauge for the valve clearance you are going to check.

➡**Valve Clearance:**

- Intake: 0.008–0.009 inches (0.20–0.24 mm)
- Exhaust: 0.011–0.013 inches (0.28–0.32 mm)

➡**Apply new engine oil to the nut threads.**

4. Insert the feeler gauge between the adjusting screw and the end of the valve stem on the No. 1 cylinder, and slide it back and forth; you should feel a slight amount of drag.
5. If you feel too much or too little drag, loosen the locknut, and turn the adjusting screw until the drag on the feeler gauge is correct.

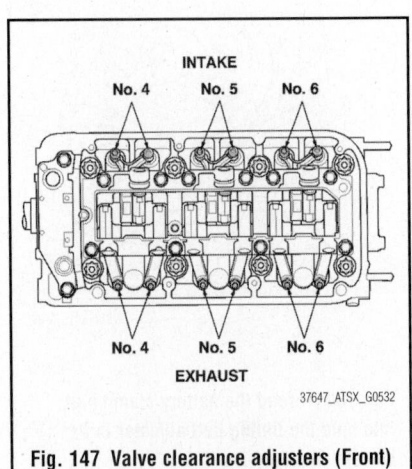

Fig. 147 Valve clearance adjusters (Front)

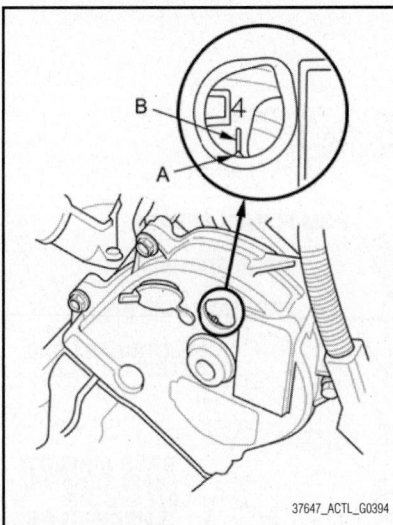

Fig. 148 Rotate the crankshaft clockwise. Align the pointer (A) on the front upper cover with the No. 4 piston TDC mark (B) on the front camshaft pulley

37647_ACTL_G0394

6. While holding the adjusting screw with the screw driver, tighten the locknut, then recheck the clearance. Repeat the adjustment, if necessary.

➡**Specified Torque:**

- Intake: 14 ft. lbs. (20 Nm)
- Exhaust: 10 ft. lbs. (14 Nm)

➡**Apply new engine oil to the nut threads.**

7. Rotate the crankshaft clockwise. Align the pointer on the front upper cover with the No. 4 piston TDC mark on the front camshaft pulley.

8. Check and, if necessary, adjust the valve clearance on the No. 4 cylinder.

9. Rotate the crankshaft clockwise. Align the pointer on the front upper cover with the No. 2 piston TDC mark on the front camshaft pulley.

Check and, if necessary, adjust the valve clearance on the No. 2 cylinder.

10. Rotate the crankshaft clockwise. Align the pointer on the front upper cover with the No. 5 piston TDC mark on the front camshaft pulley.

11. Check and, if necessary, adjust the valve clearance on the No. 5 cylinder.

12. Rotate the crankshaft clockwise. Align the pointer on the front upper cover with the No. 3 piston TDC mark on the front camshaft pulley.

13. Check and, if necessary, adjust the valve clearance on the No. 3 cylinder.

14. Rotate the crankshaft clockwise. Align the pointer on the front upper cover with the No. 6 piston TDC mark on the front camshaft pulley.

15. Check and, if necessary, adjust the valve clearance on the No. 6 cylinder.

16. Install the cylinder head covers.

ENGINE PERFORMANCE & EMISSION CONTROLS

COMPONENT LOCATIONS

4-Cylinder Engine

See Figure 149.

6-Cylinder Engine

See Figure 150.

ACCELERATOR PEDAL POSITION (AAP) SENSOR

LOCATION

The Accelerator Pedal Position (APP) sensor is located at the top of the accelerator pedal assembly.

REMOVAL & INSTALLATION

See Figure 86 and 151.

1. Disconnect the APP sensor 6P connector.

2. Remove the two mounting nuts, then remove the clip. Using a flat-tip screwdriver, push the lock, then remove the accelerator pedal pad from the pedal stop.

3. Remove the accelerator pedal module.

➡**The APP sensor is not available separately. Do not disassemble the accelerator pedal module. If the pedal stop is damaged, replace it by pulling back the carpet, and removing the nuts, the bolt, and the pedal stop.**

To install:

4. Install the accelerator pedal pad to the pedal stop. Slide the pad until it locks with a clicking sound.

5. Install a new clip.

➡**The clip must be replaced whenever the accelerator pedal module is installed. Make sure that the accelerator pedal module and the clip are secure.**

6. Install the nuts, and connect the APP sensor 6P connector.

AIR-FUEL RATIO (AFR) SENSOR

LOCATION

See Figures 152 through 154.

REMOVAL & INSTALLATION

4-Cylinder Engine

See Figure 155.

Special Tools Required: O2 sensor wrench, Snap-On YA8875, SWR2, or equivalent, commercially available

1. Disconnect the A/F sensor 4P connector, then remove the A/F sensor.

2. Install the parts in the reverse order of removal.

6-Cylinder Engine

Front Bank (Bank 2)

See Figure 156.

1. Disconnect the front A/F sensor 6P connector, then remove the A/F sensor.

2. Install the parts in the reverse order of removal.

Rear Bank (Bank 1)

See Figure 157.

1. Disconnect the rear A/F sensor 6P connector, then remove the A/F sensor.

2. Install the parts in the reverse order of removal.

CAMSHAFT POSITION (CMP) SENSOR

REMOVAL & INSTALLATION

4-Cylinder Engine

CMP Sensor A

See Figure 158.

1. Disconnect the CMP sensor A connector.

2. Remove CMP sensor A from the intake camshaft side of the cylinder head.

3. Install the parts in the reverse order of removal with a new O-ring.

CMP Sensor B

See Figures 159 and 160.

1. Disconnect the connector and hoses from the EVAP canister purge valve, then remove the EVAP canister purge valve assembly.

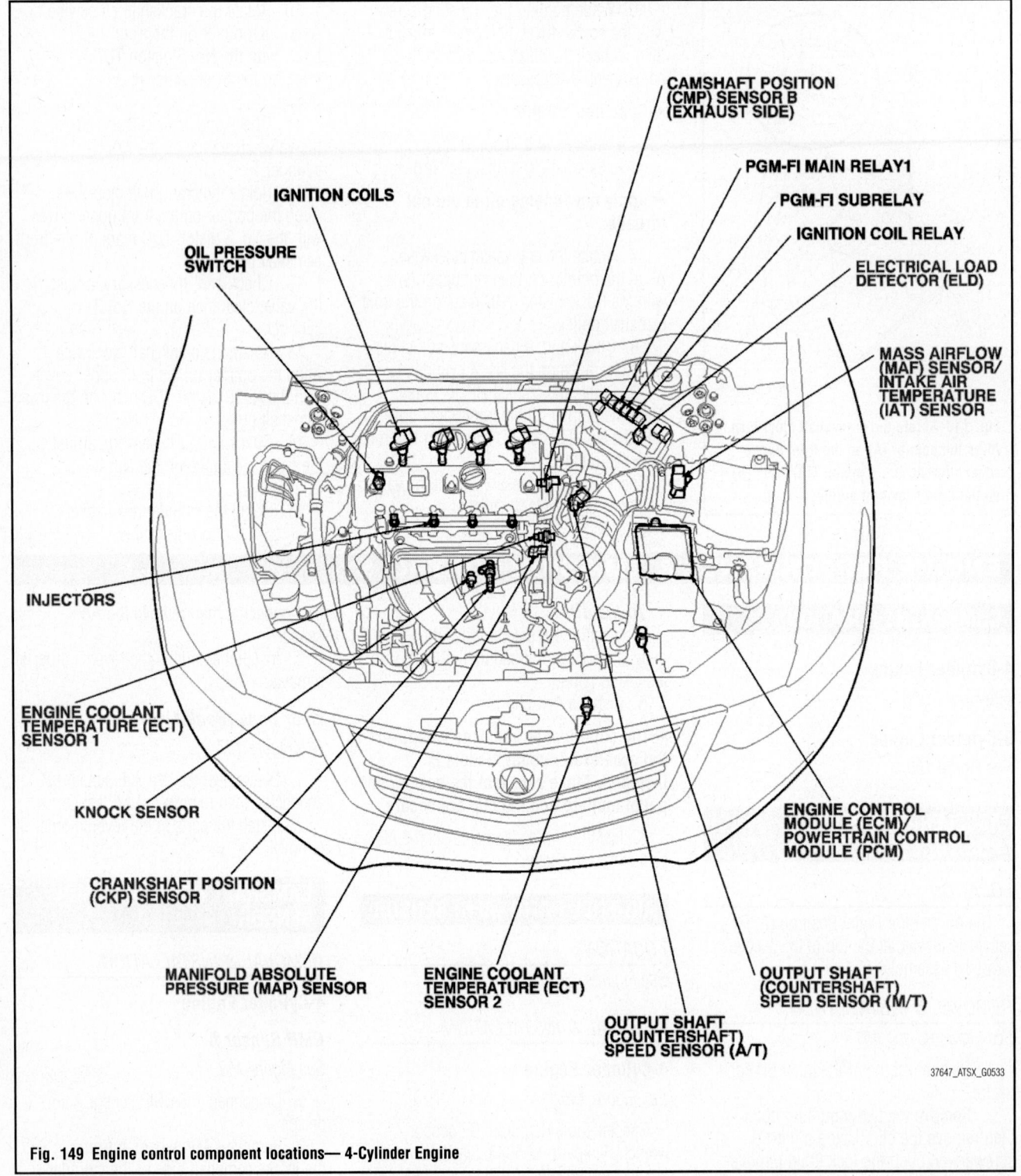

Fig. 149 Engine control component locations— 4-Cylinder Engine

2. Disconnect the CMP sensor B connector.

3. Remove CMP sensor B.

4. Install the parts in the reverse order of removal with a new O-ring.

6-Cylinder Engine

See Figures 161 and 162.

1. Remove the timing belt.

2. Remove the front camshaft pulley (CMP sensor pulse plate).

3. Disconnect the CMP sensor connector, then remove the back cover.

4. Remove the CMP sensor from the back cover.

5. Install the parts in the reverse order of removal.

6. Install the timing belt.

7. Do the CKP pattern clear/CKP pattern learn procedure.

CLUTCH PEDAL POSITION (CPP) SWITCH

LOCATION

The clutch pedal position switch is located at the top of the clutch pedal assembly.

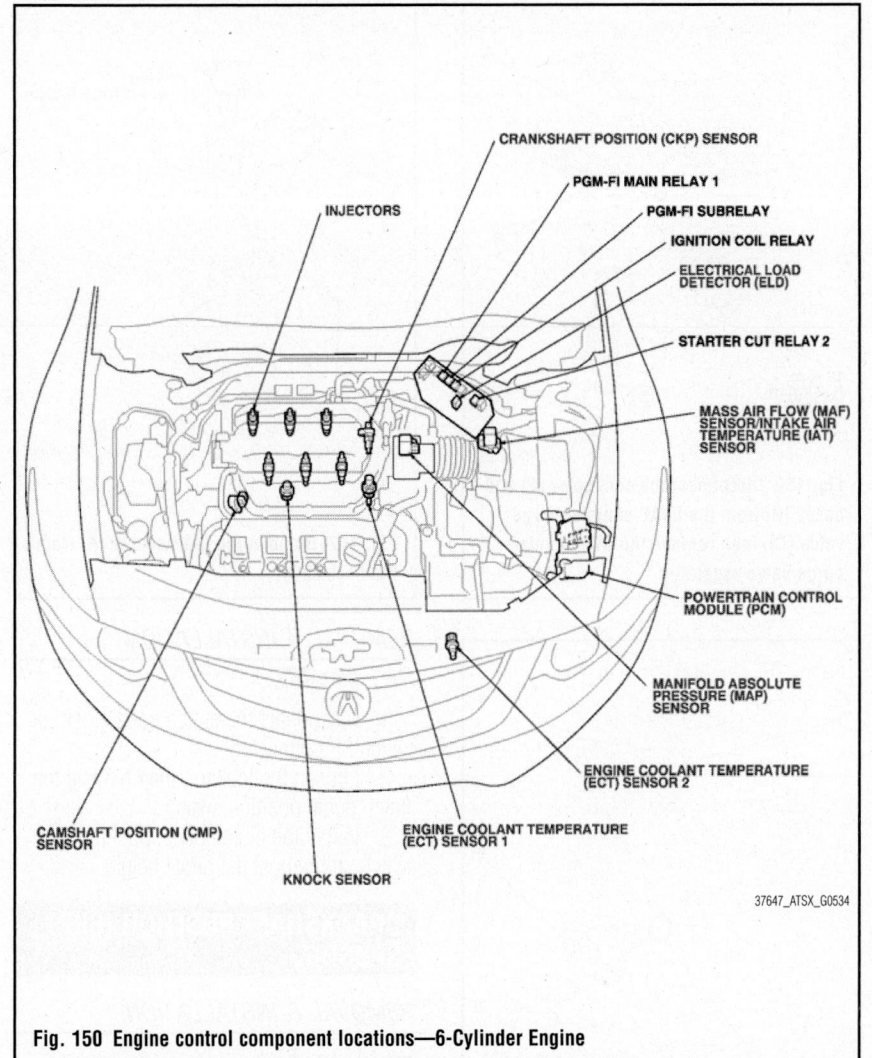

Fig. 150 Engine control component locations—6-Cylinder Engine

CRANKSHAFT POSITION (CKP) SENSOR
PGM-FI MAIN RELAY 1
PGM-FI SUBRELAY
IGNITION COIL RELAY
ELECTRICAL LOAD DETECTOR (ELD)
STARTER CUT RELAY 2
MASS AIR FLOW (MAF) SENSOR/INTAKE AIR TEMPERATURE (IAT) SENSOR
POWERTRAIN CONTROL MODULE (PCM)
MANIFOLD ABSOLUTE PRESSURE (MAP) SENSOR
ENGINE COOLANT TEMPERATURE (ECT) SENSOR 2
ENGINE COOLANT TEMPERATURE (ECT) SENSOR 1
KNOCK SENSOR
CAMSHAFT POSITION (CMP) SENSOR
INJECTORS

37647_ATSX_G0534

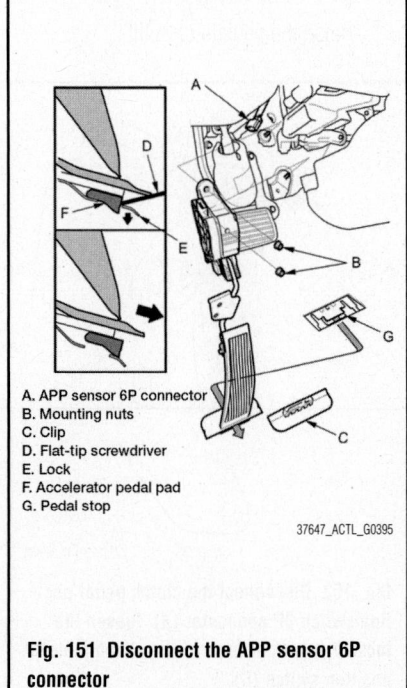

A. APP sensor 6P connector
B. Mounting nuts
C. Clip
D. Flat-tip screwdriver
E. Lock
F. Accelerator pedal pad
G. Pedal stop

37647_ACTL_G0395

Fig. 151 Disconnect the APP sensor 6P connector

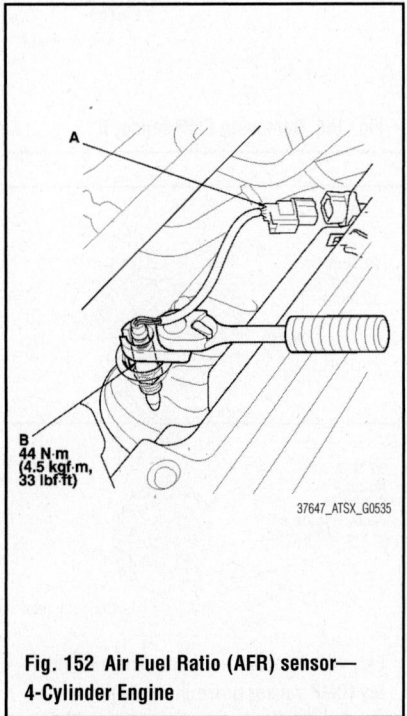

B
44 N·m
(4.5 kgf·m, 33 lbf·ft)

37647_ATSX_G0535

Fig. 152 Air Fuel Ratio (AFR) sensor—4-Cylinder Engine

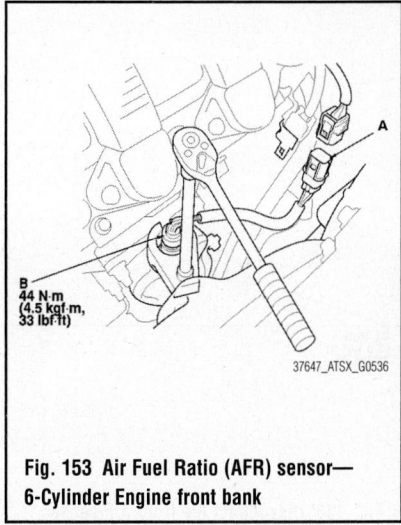

A
B
44 N·m
(4.5 kgf·m, 33 lbf·ft)

37647_ATSX_G0536

Fig. 153 Air Fuel Ratio (AFR) sensor—6-Cylinder Engine front bank

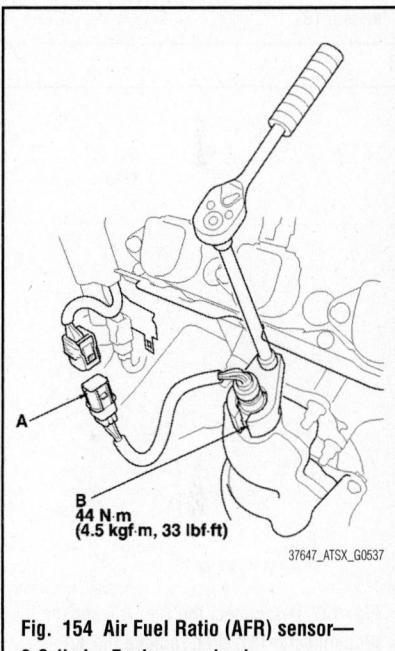

A
B
44 N·m
(4.5 kgf·m, 33 lbf·ft)

37647_ATSX_G0537

Fig. 154 Air Fuel Ratio (AFR) sensor—6-Cylinder Engine rear bank

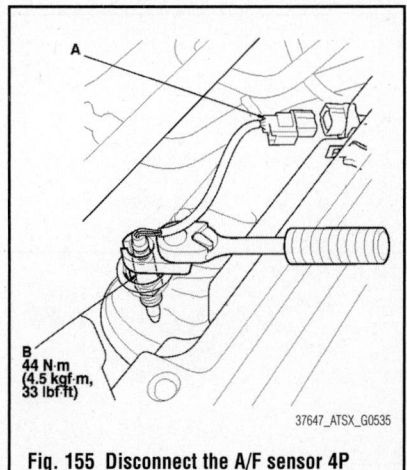

A
B
44 N·m
(4.5 kgf·m, 33 lbf·ft)

37647_ATSX_G0535

Fig. 155 Disconnect the A/F sensor 4P connector (A), then remove the A/F sensor (B)

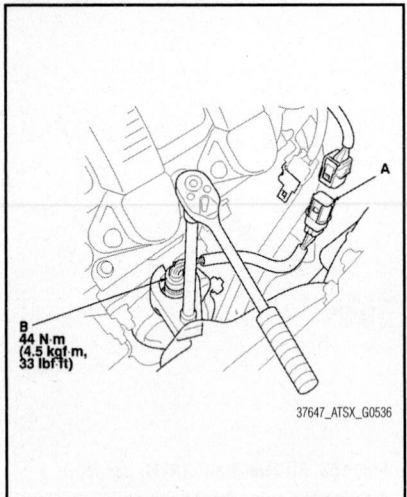

Fig. 156 Disconnect the front A/F sensor 6P connector (A), then remove the A/F sensor (B)

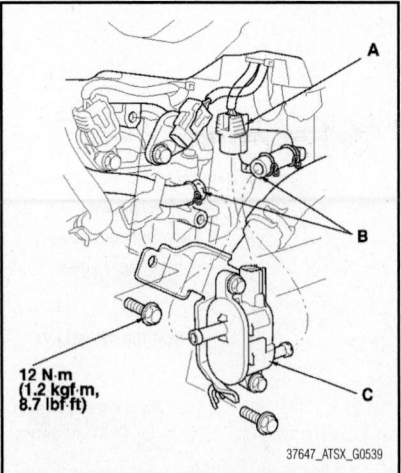

Fig. 159 Disconnect the connector (A) and hoses (B) from the EVAP canister purge valve (C), then remove the EVAP canister purge valve assembly.

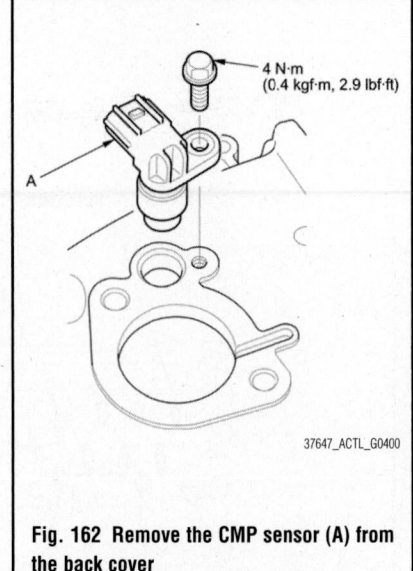

Fig. 162 Remove the CMP sensor (A) from the back cover

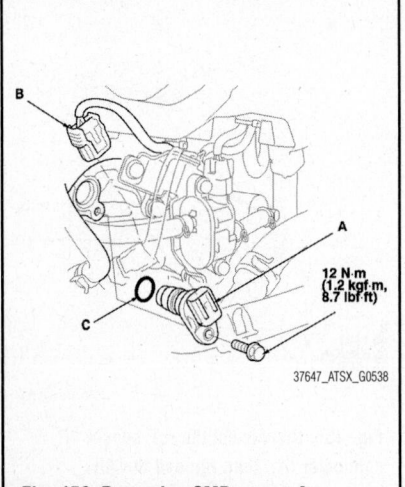

Fig. 157 Disconnect the rear A/F sensor 6P connector (A), then remove the A/F sensor (B)

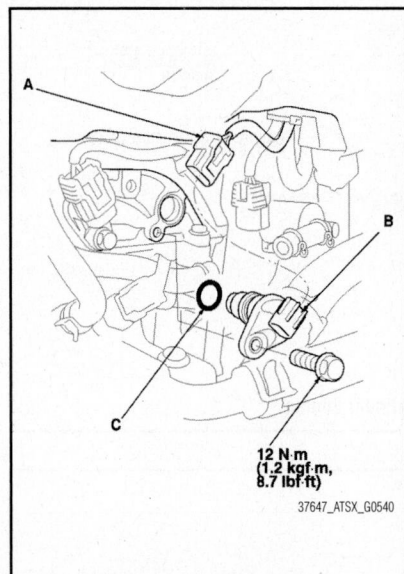

Fig. 160 Removing CMP sensor B

REMOVAL & INSTALLATION

See Figure 163.

1. Disconnect the clutch pedal position switch 3P connector.
2. Loosen the locknut, then remove the clutch pedal position switch.
3. Install the clutch pedal position switch, and adjust the pedal height.

CRANKSHAFT POSITION (CKP) SENSOR

REMOVAL & INSTALLATION

4-Cylinder Engine

See Figure 164.

1. Raise the vehicle on a lift.

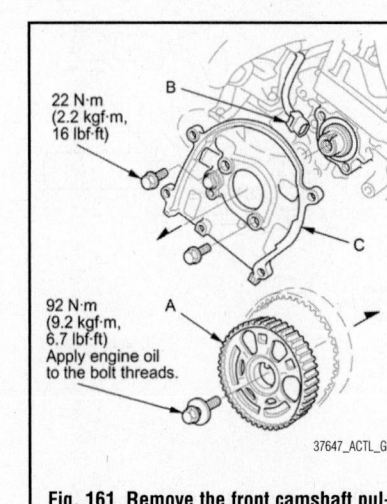

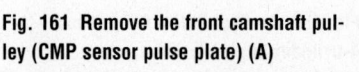

Fig. 158 Removing CMP sensor A

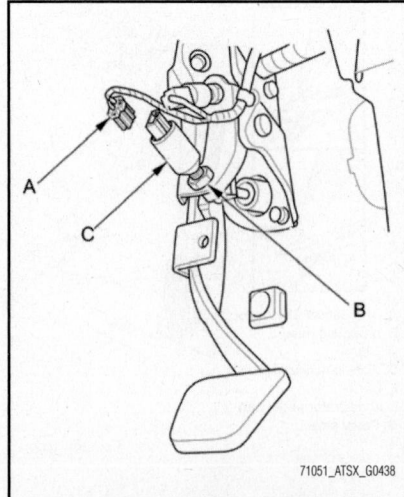

Fig. 161 Remove the front camshaft pulley (CMP sensor pulse plate) (A)

Fig. 163 Disconnect the clutch pedal position switch 3P connector (A), loosen the locknut (B), then remove the clutch pedal position switch (C)

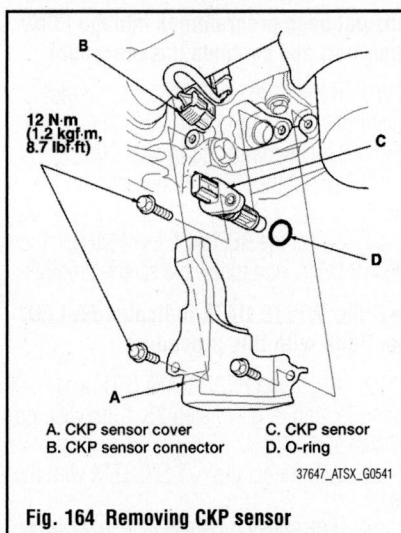

A. CKP sensor cover
B. CKP sensor connector
C. CKP sensor
D. O-ring

37647_ATSX_G0541

Fig. 164 Removing CKP sensor

➡**Make sure the vehicle is level, because engine oil will drip out when you remove the sensor.**

2. Remove the CKP sensor cover.
3. Disconnect the CKP sensor connector.
4. Remove the CKP sensor.
5. Install the parts in the reverse order of removal with a new O-ring.
6. Do the CKP pattern clear/CKP pattern learn procedure.
7. Check the engine oil level, and add more oil if needed.

6-Cylinder Engine

See Figure 165.

1. Raise the vehicle on a lift.

➡**Make sure the vehicle is level, because engine oil will drip out when you remove the sensor.**

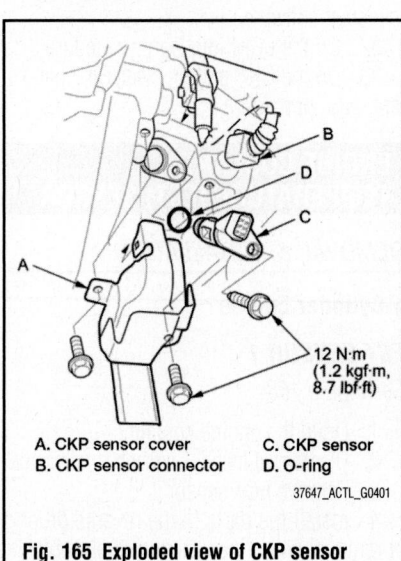

A. CKP sensor cover
B. CKP sensor connector
C. CKP sensor
D. O-ring

12 N·m
(1.2 kgf·m,
8.7 lbf·ft)

37647_ACTL_G0401

Fig. 165 Exploded view of CKP sensor assembly

2. Remove the CKP sensor cover.
3. Disconnect the CKP sensor connector.
4. Remove the CKP sensor.
5. Install the parts in the reverse order of removal with a new O-ring.
6. Do the CKP pattern clear/CKP pattern learn procedure.
7. Check for engine oil level, and add more oil if needed.

CKP PATTERN CLEAR/CKP PATTERN LEARN

Clear/Learn Procedure (with the HDS)

1. Connect the HDS to the Data Link Connector (DLC) located under the driver's side of the dashboard.
2. Turn the ignition switch to ON (II), or press the engine start/stop button to select the ON mode.
3. Make sure the HDS communicates with the ECM/PCM and other vehicle systems. If it doesn't, go to the DLC circuit troubleshooting.
4. Select CRANK PATTERN in the ADJUSTMENT MENU with the HDS.
5. Select CRANK PATTERN CLEAR, and clear the CKP pattern.
6. Select CRANK PATTERN LEARNING with the HDS, and follow the screen prompts.

Learn Procedure (without the HDS)

1. Start the engine. Hold the engine speed at 3,000 rpm without load (A/T in P or N, M/T in neutral) until the radiator fan comes on.
2. Test-drive the vehicle on a level road: Decelerate (with the throttle fully closed) from an engine speed of 2,500 rpm down to 1,000 rpm with the Transmission in 2nd.
3. Repeat step 2 several times.
4. Turn the ignition switch to LOCK (0), or press the engine start/stop button to select the OFF mode.
5. Turn the ignition switch to ON (II), or press the engine start/stop button to select the ON mode, and wait 30 seconds.

ENGINE CONTROL MODULE (ECM)

LOCATION

The ECM is used in the 4-cylinder engine models. It is located next to the battery in the engine compartment.

REMOVAL & INSTALLATION

4-Cylinder Engine

See Figures 166 and 167.

Special Tools Required:
• Honda Diagnostic System (HDS) tablet tester
• Honda Interface Module (HIM) and an iN workstation with the latest HDS software version
• HDS pocket tester
• GNA600 and an iN workstation with the latest HDS software version

➡**Any one of the above updating tools can be used.**

1. Connect the HDS to the Data Link Connector (DLC) located under the driver's side of the dashboard.
2. Turn the ignition switch to ON (II).
3. Make sure the HDS communicates with the ECM and other vehicle systems. If it doesn't, go to the DLC circuit troubleshooting. If you are returning from the DLC circuit troubleshooting, skip steps 4 through 9, 19 through 24, and 27 through 29, and do these procedures after replacing the ECM:
   a. Replace the engine oil and the engine oil filter.
   b. Replace the ATF (A/T).
   c. Clean the throttle body.
4. Select the PGM-FI system with the HDS.
5. Select the INSPECTIONMENU with the HDS.
6. Select the ETCS TEST, then select the TP POSITION CHECK, and follow the screen prompts.

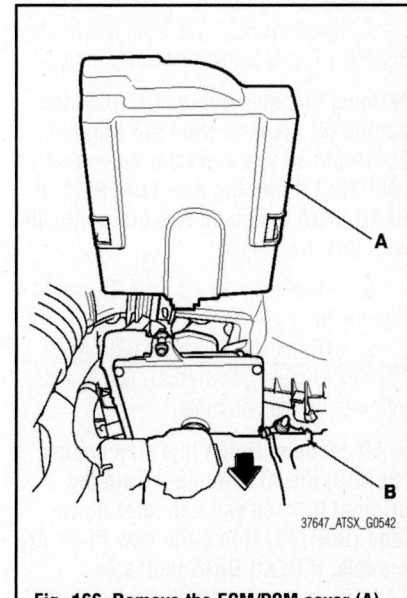

37647_ATSX_G0542

Fig. 166 Remove the ECM/PCM cover (A) and the battery setting plate (B)

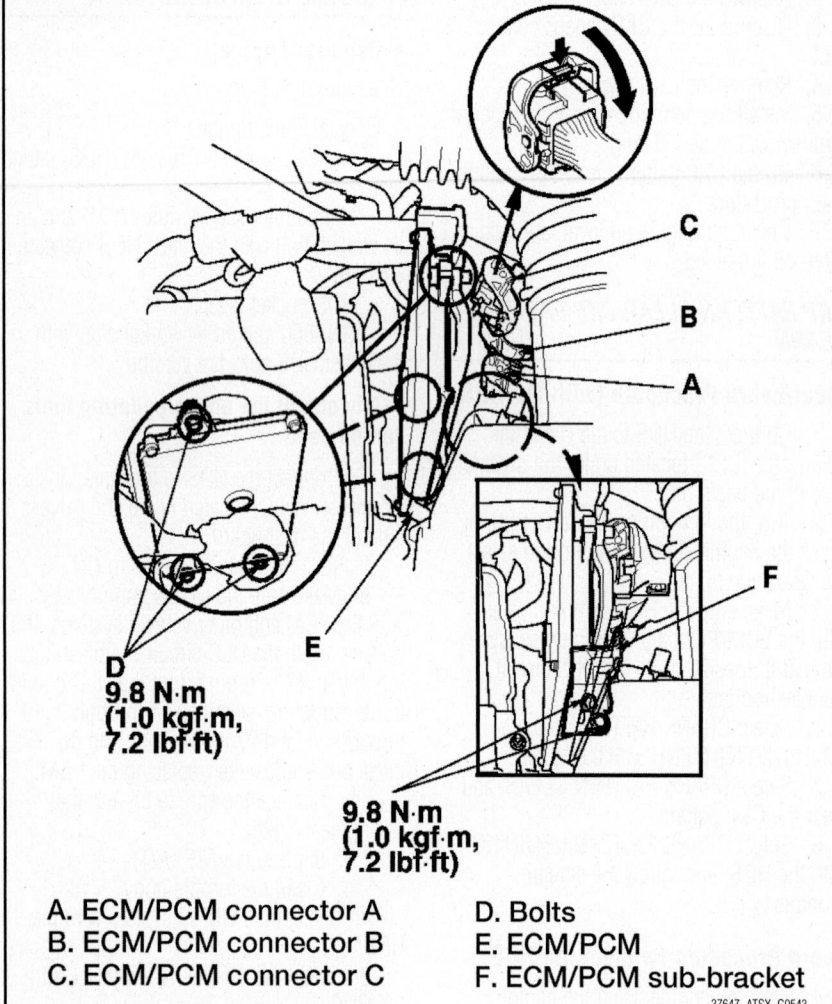

D.
9.8 N·m
(1.0 kgf·m,
7.2 lbf·ft)

9.8 N·m
(1.0 kgf·m,
7.2 lbf·ft)

A. ECM/PCM connector A
B. ECM/PCM connector B
C. ECM/PCM connector C
D. Bolts
E. ECM/PCM
F. ECM/PCM sub-bracket

37647_ATSX_G0543

Fig. 167 ECM/PCM disconnect and removal

➡**If the TP POSITION CHECK indicates FAILED, continue with this procedure.**

7. Select the REPLACE ECM MENU, then READ DATA, and follow the screen prompts.

➡**Doing this step copies (READS) the engine oil life data from the original ECM/PCM so you can later download (WRITES) it into the new ECM/PCM. If READ DATA indicates FAILED, continue with this procedure.**

8. A/T models: Select the A/T system with the HDS.
9. A/T models: Select the REPLACE TCM/PCM MENU, then READ DATA, and follow the screen prompts.

➡**A/T models: Doing this step copies (READS) the ATF life data from the original PCM so you can later download (WRITES) it into the new PCM. A/T models: If READ DATA indicates FAILED, continue with this procedure.**

10. Turn the ignition switch to LOCK (0).

11. Jump the SCS line with the HDS.
12. Remove the ECM/PCM cover.
13. Remove the battery setting plate, and reposition the battery away from the ECM.

➡**Do not disconnect the battery terminals.**

14. Remove the bolts.
15. Disconnect ECM/PCM connectors A, B, and C, then remove the ECM.

➡**ECM/PCM connectors A, B, and C have symbols (A=□, B=△, C=○) embossed on them for identification. Canada model: When disconnecting the ECM connectors, remove the ECM/PCM sub-bracket.**

16. Install the parts in the reverse order of removal.
17. Turn the ignition switch to ON (II).
18. Manually input the VIN to the ECM/PCM with the HDS.

➡**DTC P0630 VIN Not Programmed or Mismatch may be stored because the VIN**

has not been programmed into the ECM; ignore it, and continue this procedure.

19. If the READ DATA (engine oil life) failed in step 7, go to step 22 (A/T) or step 25 (M/T). Otherwise, go to step 20.
20. Select the PGM-FI system with the HDS.
21. Select the REPLACE ECM MENU, then WRITE DATA, and follow the screen prompts.

➡**If the WRITE DATA indicates FAILED, continue with this procedure.**

22. A/T: If the READ DATA (ATF life) failed in step 8, go to step 25. Otherwise go to step 23.
23. A/T: Select the A/T SYSTEM with the HDS.
24. A/T: Select the REPLACE TCM/PCM MENU, then WRITE DATA, and follow the screen prompts.

➡**A/T: If the WRITE DATA indicates FAILED, continue with this procedure.**

25. Select IMMOBI system with the HDS.
26. Enter the immobilizer ECM code that you got from the iN, and use the ECM/PCM replacement procedure in the IMMOBI MENU of the HDS; it allows you to start the engine.
27. If the TP POSITION CHECK failed in step 6, clean the throttle body, then go to step 28.
28. If the READ DATA failed in step 7 or the WRITE DATA failed in step 21, replace the engine oil and engine oil filter, then go to step 29 (A/T) or step 30 (M/T).
29. A/T: If the READ DATA failed in step 9 or the WRITE DATA failed in step 24, replace the ATF, then go to step 30.
30. Select PGM-FI system, and reset the ECM with the HDS.
31. Update the ECM if it does not have the latest software.
32. Do the ECM idle learn procedure.
33. Do the CKP pattern clear/CKP pattern learn procedure.

ENGINE COOLANT TEMPERATURE (ECT) SENSOR

REMOVAL & INSTALLATION

4-Cylinder Engine

ECT SENSOR 1

See Figure 168.

1. Drain the engine coolant.
2. Disconnect the ECT sensor 1 connector.
3. Remove ECT sensor 1.
4. Install the parts in the reverse order of removal with a new O-ring, then refill the radiator with engine coolant.

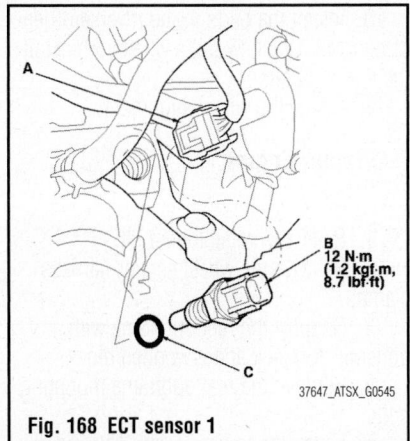

Fig. 168 ECT sensor 1

ECT Sensor 2

See Figure 169.

1. Remove the front splash shield.
2. Drain the engine coolant.
3. Disconnect the ECT sensor 2 connector, then remove ECT sensor 2.
4. Install ECT sensor 2 with a new O-ring.
5. Install the front splash shield.
6. Refill the radiator with engine coolant.

6-Cylinder Engine

ECT Sensor 1

See Figure 170.

1. Drain the engine coolant.
2. Disconnect the ECT sensor 1 2P connector.
3. Remove ECT sensor 1.
4. Install the parts in the reverse order of removal with a new O-ring, then refill the radiator with engine coolant.

ECT Sensor 2

See Figure 171.

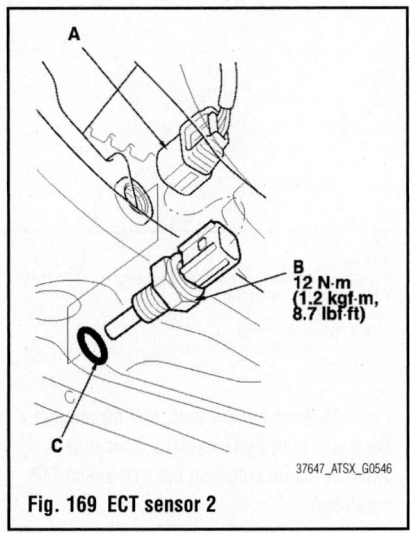

Fig. 169 ECT sensor 2

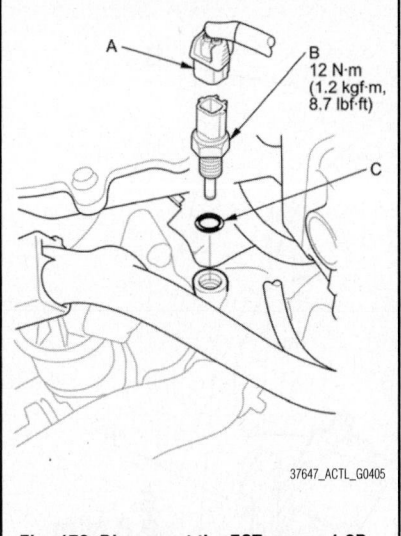

Fig. 170 Disconnect the ECT sensor 1 2P connector

1. Remove the splash shield.
2. Drain the engine coolant.
3. Disconnect the ECT sensor 2 2P connector.
4. Remove ECT sensor 2.
5. Install the parts in the reverse order of removal with a new O-ring, then refill the radiator with engine coolant.

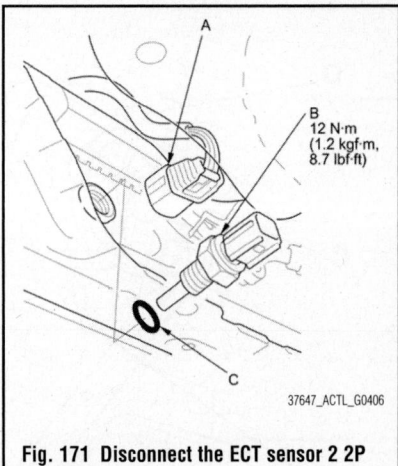

Fig. 171 Disconnect the ECT sensor 2 2P connector

EVAP CANISTER

LOCATION

The EVAP canister is located under the rear of the vehicle near the fuel tank.

REMOVAL & INSTALLATION

4-Cylinder Engine

See Figures 172 through 174.

1. Raise the vehicle on a lift.

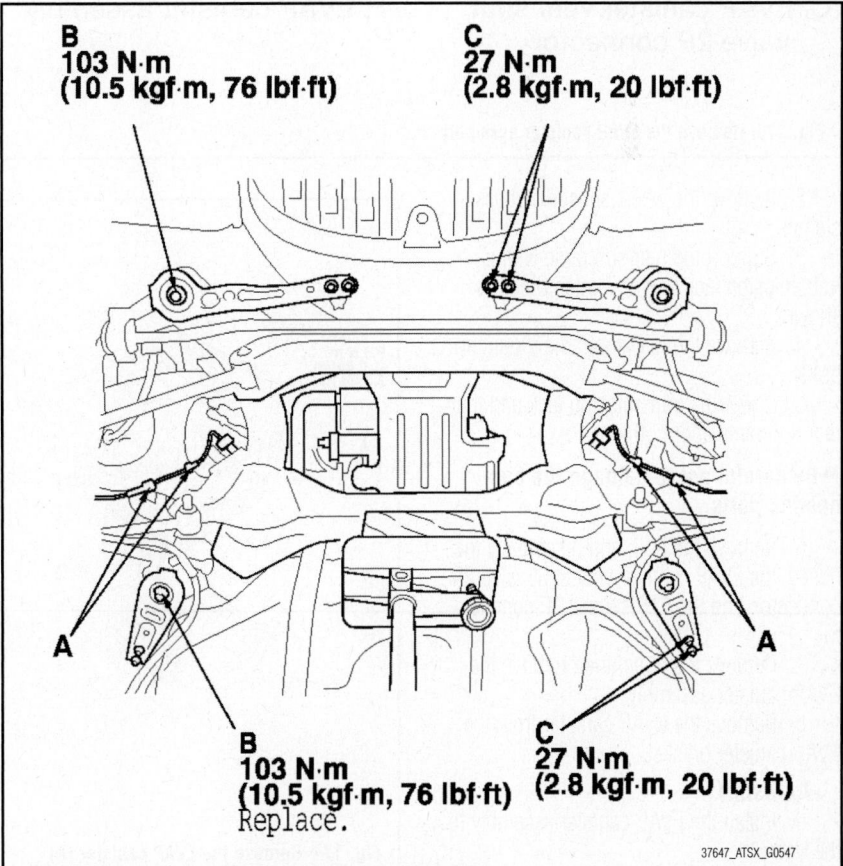

Fig. 172 Remove the wheel sensor harness clamps (A)

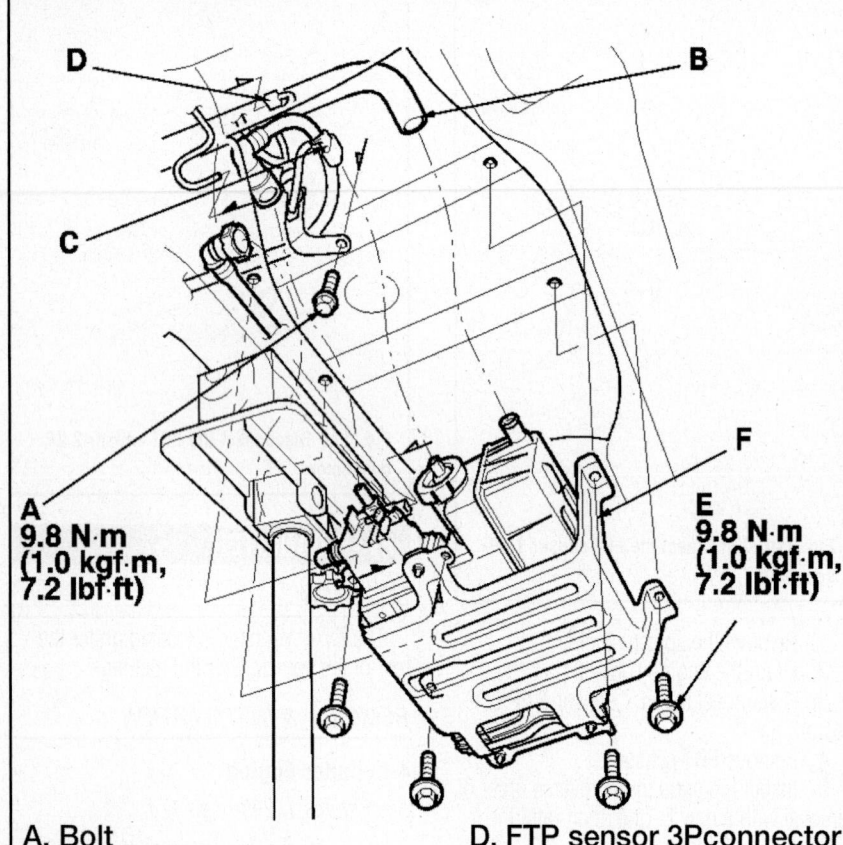

A.
**9.8 N·m
(1.0 kgf·m,
7.2 lbf·ft)**

E.
**9.8 N·m
(1.0 kgf·m,
7.2 lbf·ft)**

A. Bolt
B. Hoses
C. EVAP canister vent shut
   valve 2P connector

D. FTP sensor 3Pconnector
E. Bolts
F. EVAP canister assembly

37647_ATSX_G0548

Fig. 173 Remove the EVAP canister assembly

2. Remove the wheel sensor harness clamps.

3. Support the rear subframe with a transmission jack and a wooden block as shown.

4. Remove the rear subframe mounting bolts.

5. Lower the transmission jack and the rear subframe about 50 mm.

➡Be careful not to damage the connecting parts.

6. Remove the bolt, and disconnect the hoses, the EVAP canister vent shut valve 2P connector, and the FTP sensor 3P connector.

7. Remove the bolts, then remove the EVAP canister assembly.

8. Remove the EVAP canister from the EVAP canister bracket.

To install:

9. Install the EVAP canister assembly to the body.

➡Attach the bracket arm to the body.

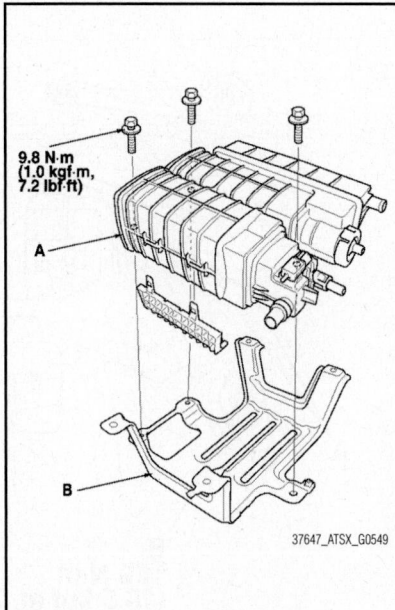

**9.8 N·m
(1.0 kgf·m,
7.2 lbf·ft)**

A

B

37647_ATSX_G0549

Fig. 174 Remove the EVAP canister (A) from the EVAP canister bracket (B)

10. Install the parts in the reverse order of removal. Use new bolts when you install the rear subframe.

11. Check the wheel alignment.

6-Cylinder Engine

See Figures 175 through 177.

1. Raise the vehicle on a lift.

2. Remove the wheel sensor harness clamps.

3. Support the rear subframe with a transmission jack and a wooden block.

4. Remove the rear subframe mounting bolts.

5. Lower the transmission jack and the rear subframe about 50 mm.

➡Be careful not to damage the connecting parts.

6. Remove the bolt, and disconnect the hoses, the EVAP canister vent shut valve 2P connector, and the FTP sensor 3P connector.

7. Remove the bolts, then remove the EVAP canister assembly.

8. Remove the EVAP canister from the EVAP canister bracket.

To install:

9. Install the EVAP canister assembly to the body.

➡Attach the bracket arm to the body.

10. Install the parts in the reverse order of removal. Use new bolts when you install the rear subframe.

11. Check the wheel alignment.

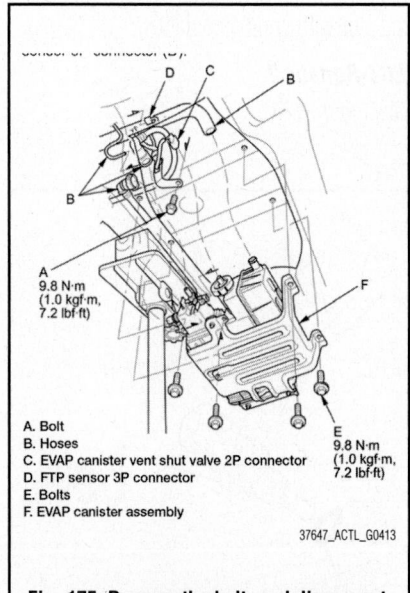

A.
**9.8 N·m
(1.0 kgf·m,
7.2 lbf·ft)**

E.
**9.8 N·m
(1.0 kgf·m,
7.2 lbf·ft)**

A. Bolt
B. Hoses
C. EVAP canister vent shut valve 2P connector
D. FTP sensor 3P connector
E. Bolts
F. EVAP canister assembly

37647_ACTL_G0413

Fig. 175 Remove the bolt, and disconnect the hoses, the EVAP canister vent shut valve 2P connector, and the FTP sensor 3P connector

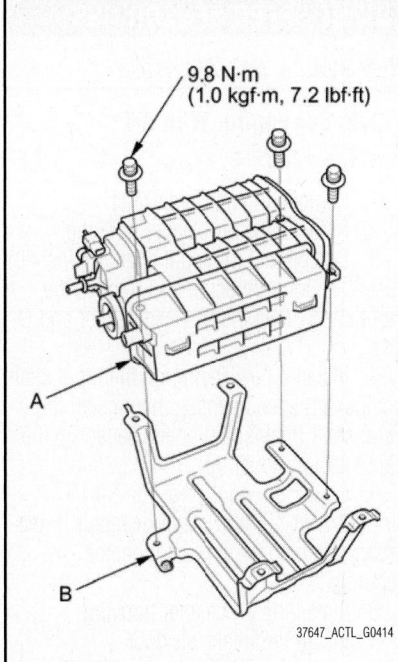

Fig. 176 Remove the EVAP canister (A) from the EVAP canister bracket (B)

EVAP CANISTER PURGE CONTROL VALVE

LOCATION

The EVAP canister purge valve is located on the end of the intake manifold near the air intake duct.

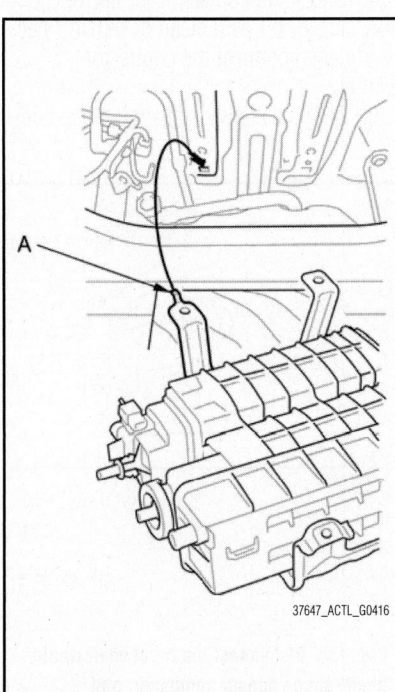

Fig. 177 Attach the bracket arm (A) to the body

REMOVAL & INSTALLATION

4-Cylinder Engine

See Figure 178.

1. Disconnect the EVAP canister purge valve 2P connector.
2. Disconnect the hoses, then remove the EVAP canister purge valve.
3. Install the parts in the reverse order of removal.

6-Cylinder Engine

See Figures 179 and 180.

1. Disconnect the hoses and the EVAP canister purge valve 2P connector.
2. Remove the EVAP canister purge valve assembly.
3. Remove the EVAP canister purge valve from the bracket.

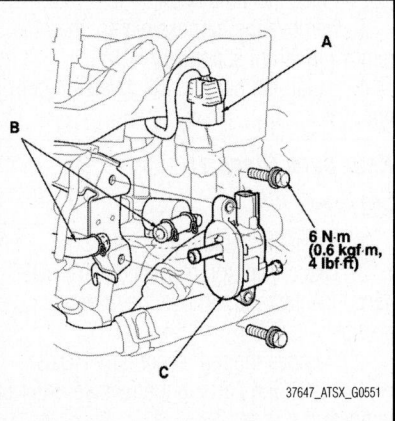

Fig. 178 Disconnect the EVAP canister purge valve 2P connector (A), the hoses (B), then remove the EVAP canister purge valve (C)

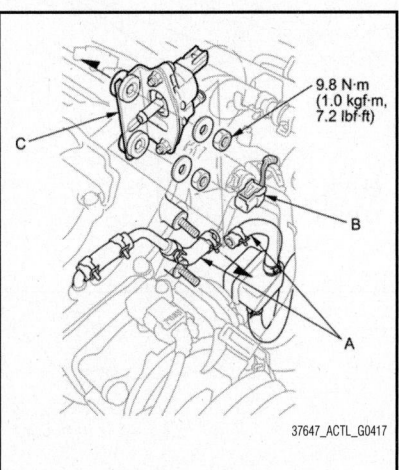

Fig. 179 Disconnect the hoses (A) and the EVAP canister purge valve 2P connector (B)

4. Install the parts in the reverse order of removal.

EXHAUST GAS RECIRCULATION (EGR) VALVE

LOCATION

The EGR valve is located at the front left of the engine.

REMOVAL & INSTALLATION

6-Cylinder Engine

See Figure 181.

1. Disconnect the EGR valve 5P connector.
2. Remove the EGR valve.
3. Install the parts in the reverse order of removal with a new gasket.

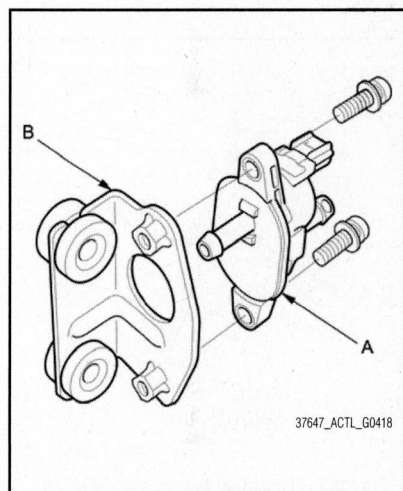

Fig. 180 Remove the EVAP canister purge valve (A) from the bracket (B)

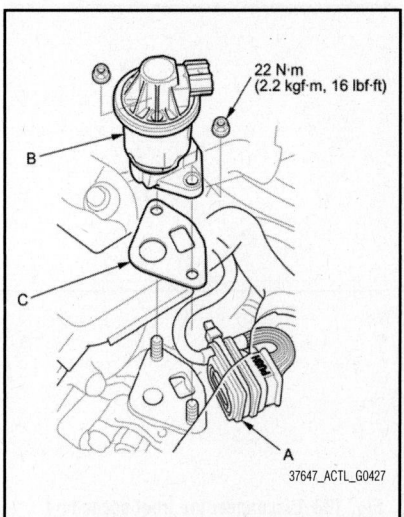

Fig. 181 Disconnect the EGR valve 5P connector

REMOVAL & INSTALLATION

Special Tools Required: O2 sensor wrench, Snap-On YA8875, SWR2, or equivalent, commercially available

4-Cylinder Engine

See Figure 182.

1. Remove the front splash shield.
2. Disconnect the secondary HO2S 4P connector, then remove secondary HO2S.
3. Install the parts in the reverse order of removal.

6-Cylinder Engine

Front Bank (Bank 2)

See Figures 183 and 184.

1. Disconnect the front secondary HO2S 4P connector, and remove the harness clamps.

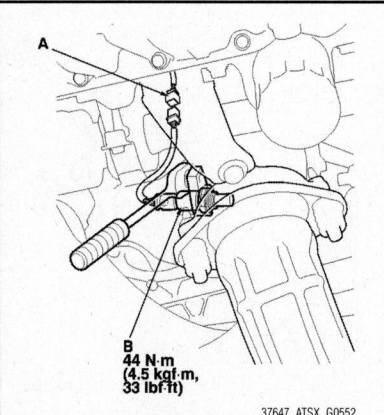

Fig. 182 Disconnect the secondary HO2S 4P connector (A), then remove secondary HO2S (B)

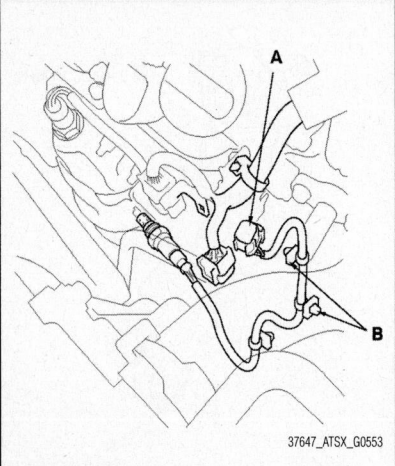

Fig. 183 Disconnect the front secondary HO2S 4P connector (A), and remove the harness clamps (B)

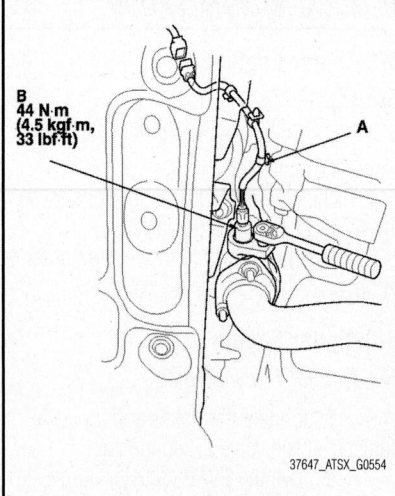

Fig. 184 Remove the harness clamp (A), then remove the front secondary HO2S (B)

2. Raise the vehicle on a lift.
3. Remove the harness clamp, then remove the front secondary HO2S.
4. Install the parts in the reverse order of removal.

Rear Bank (Bank 1)

See Figure 185.

1. Raise the vehicle on a lift.
2. Disconnect the rear secondary HO2S 4P connector, and remove the harness clamps.
3. Remove the rear secondary HO2S.
4. Install the parts in the reverse order of removal.

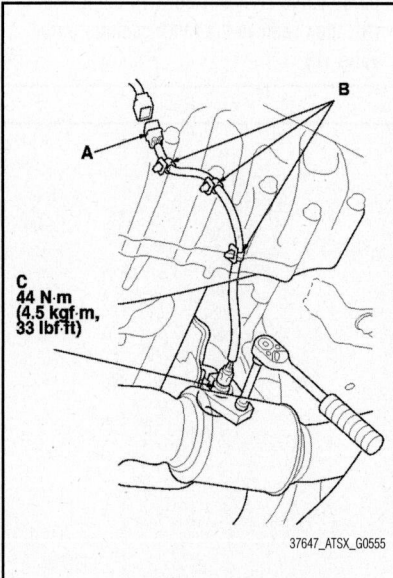

Fig. 185 Disconnect the rear secondary HO2S 4P connector (A), and remove the harness clamps (B)

INPUT SPEED SENSOR (ISS)

REMOVAL & INSTALLATION

4-Cylinder Engine With A/T

See Figure 186.

1. Remove the intake air duct.
2. Remove the air cleaner housing.
3. Disconnect the input shaft (mainshaft) speed sensor connector, and remove the input shaft (mainshaft) speed sensor and O-ring.
4. Install a new O-ring on the input shaft (mainshaft) speed sensor, then install the input shaft (mainshaft) speed sensor in the transmission housing.
5. Check the connector for corrosion, dirt, or oil, and clean or repair if necessary, then connect the connector securely.
6. Install the air cleaner housing.
7. Install the intake air duct.

6-Cylinder Engine

See Figure 187.

1. Do the battery removal procedure.
2. Disconnect the input shaft (mainshaft) speed sensor connector, and remove the input shaft (mainshaft) speed sensor and O-ring.
3. Install a new O-ring on a new input shaft (mainshaft) speed sensor, then install the input shaft (mainshaft) speed sensor.
4. Check the connector for corrosion, dirt, or oil, and clean or repair if necessary, then connect the connector securely.
5. Do the battery installation procedure.

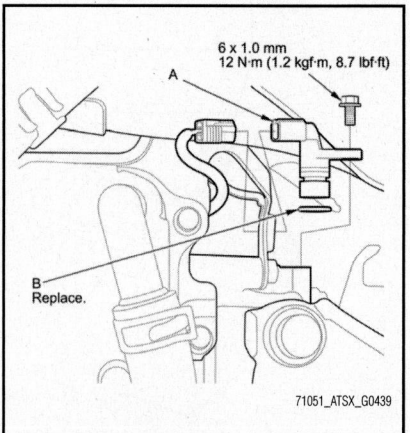

Fig. 186 Disconnect the input shaft (mainshaft) speed sensor connector, and remove the input shaft (mainshaft) speed sensor (A) and O-ring (B)

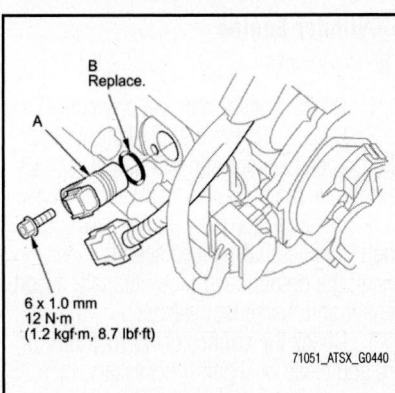

Fig. 187 Disconnect the input shaft (mainshaft) speed sensor connector, and remove the input shaft (mainshaft) speed sensor (A) and O-ring (B)

KNOCK SENSOR (KS)

OPERATION

The Knock Sensor (KS) is a piezo-electric (pressure-sensitive) electronic device that enables the Powertrain Control Module (PSM) to control the ignition timing for optimum performance, and to protect the engine from potentially damaging levels of detonation. The PCM uses the KS system to test for abnormal engine noise that may indicate detonation, which is also known as spark knock.

REMOVAL & INSTALLATION

4-Cylinder Engine

See Figure 188.

1. Remove the intake manifold.
2. Disconnect the knock sensor connector.
3. Remove the knock sensor.

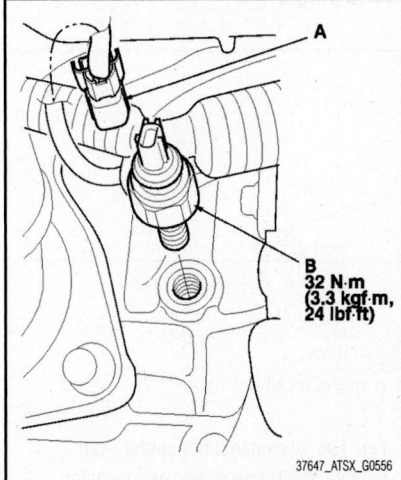

Fig. 188 Disconnect the knock sensor connector (A) and remove the knock sensor (B)

4. Install the parts in the reverse order of removal.

6-Cylinder Engine

See Figure 189.

1. Remove the intake manifold.
2. Remove the injector base.
3. Disconnect the knock sensor connector.
4. Remove the knock sensor.
5. Install the parts in the reverse order of removal.

MANIFOLD ABSOLUTE PRESSURE (MAP) SENSOR

REMOVAL & INSTALLATION

4-Cylinder Engine

See Figure 190.

1. Disconnect the MAP sensor connector.
2. Remove the MAP sensor.

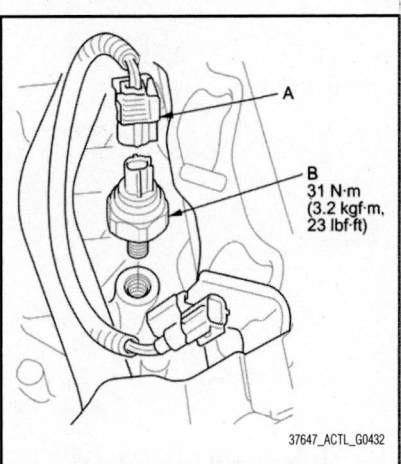

Fig. 189 Disconnect the knock sensor connector

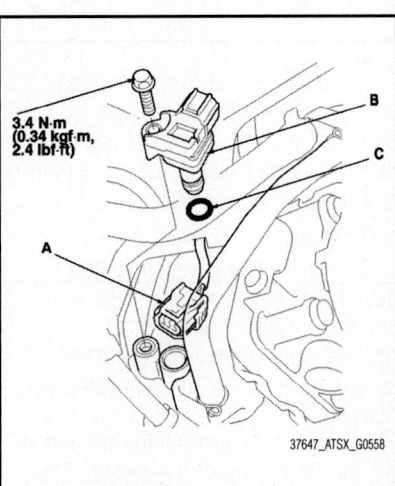

Fig. 190 Disconnect the MAP sensor connector (A) and remove the MAP sensor (B)

3. Install the parts in the reverse order of removal with a new O-ring.

6-Cylinder Engine

See Figure 191.

1. Disconnect the MAP sensor connector.
2. Remove the screw.
3. Remove the MAP sensor.
4. Install the parts in the reverse order of removal with a new O-ring.

MASS AIRFLOW (MAF)/INTAKE AIR TEMPERATURE (IAT) SENSOR

LOCATION

The MAF/IAT sensor is located on the engine air intake.

REMOVAL & INSTALLATION

4-Cylinder Engine

See Figure 192.

1. Disconnect the MAF sensor/IAT sensor connector.
2. Remove the screws.
3. Remove the MAF sensor/IAT sensor.
4. Install the parts in the reverse order of removal with a new O-ring.

6-Cylinder Engine

See Figure 193.

1. Disconnect the MAF/IAT sensor 5P connector.

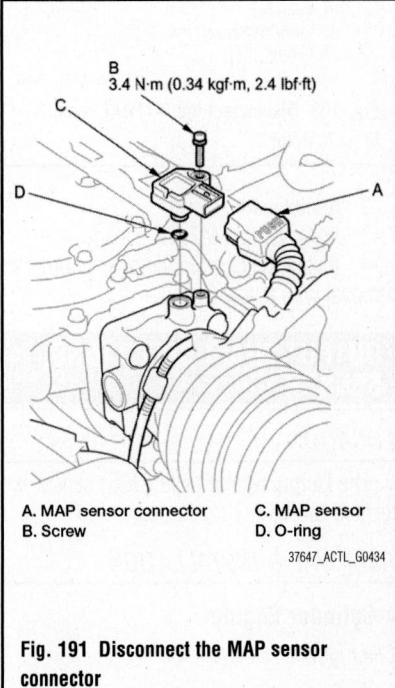

A. MAP sensor connector	C. MAP sensor
B. Screw	D. O-ring

Fig. 191 Disconnect the MAP sensor connector

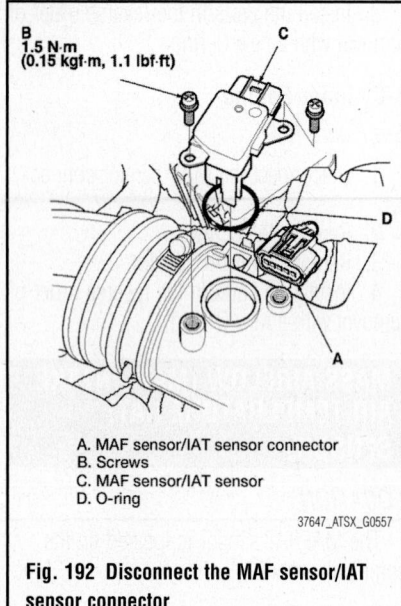

A. MAF sensor/IAT sensor connector
B. Screws
C. MAF sensor/IAT sensor
D. O-ring

37647_ATSX_G0557

Fig. 192 Disconnect the MAF sensor/IAT sensor connector

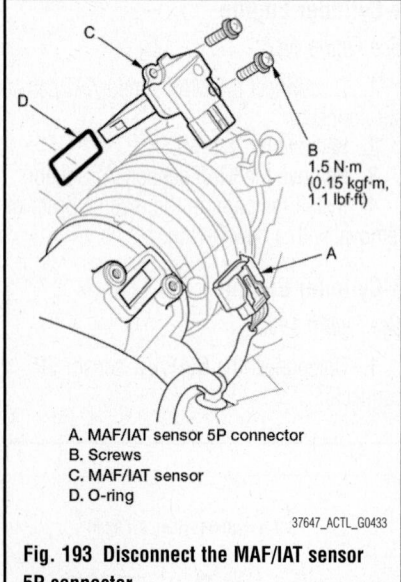

A. MAF/IAT sensor 5P connector
B. Screws
C. MAF/IAT sensor
D. O-ring

37647_ACTL_G0433

Fig. 193 Disconnect the MAF/IAT sensor 5P connector

2. Remove the screws.
3. Remove the MAF/IAT sensor.
4. Install the parts in the reverse order of removal with a new O-ring.

OUTPUT SHAFT SPEED (OSS) SENSOR

LOCATION

The Output Shaft Speed (OSS) sensor is located on the A/T transaxle.

REMOVAL & INSTALLATION

4-Cylinder Engine

See Figures 194 and 195.

1. Remove the intake air duct.

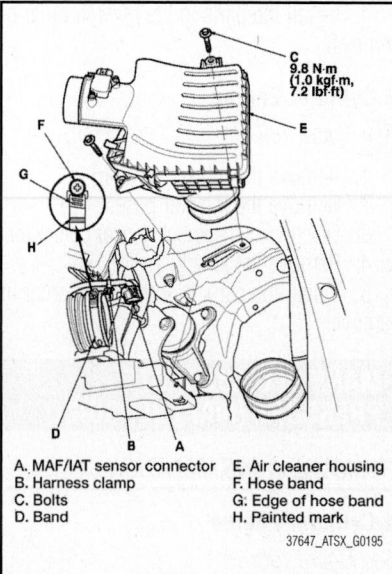

A. MAF/IAT sensor connector   E. Air cleaner housing
B. Harness clamp   F. Hose band
C. Bolts   G. Edge of hose band
D. Band   H. Painted mark

37647_ATSX_G0195

Fig. 194 Remove the air cleaner housing

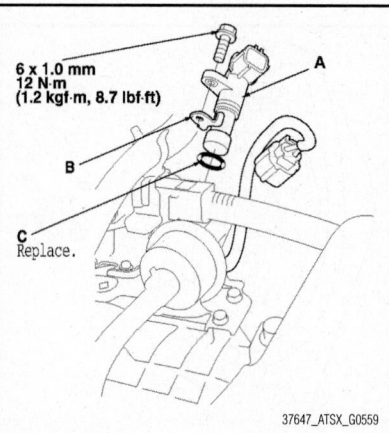

Fig. 195 Disconnect the output shaft (countershaft) speed sensor connector, and remove the output shaft (countershaft) speed sensor (A) and the speed sensor washer (B)

2. Remove the air cleaner housing.
3. Disconnect the output shaft (countershaft) speed sensor connector, and remove the output shaft (countershaft) speed sensor and the speed sensor washer.

To install:

4. Install a new O-ring on a new output shaft (countershaft) speed sensor, then install the output shaft (countershaft) speed sensor with the speed sensor washer in the transmission housing.
5. Check the connector for rust, dirt, or oil, and clean or repair if necessary. Then connect the connector securely.
6. Install the air cleaner housing.
7. Install the intake air duct.

6-Cylinder Engine

See Figure 196.

1. Do the battery removal procedure.
2. Disconnect the output shaft (countershaft) speed sensor connector, and remove the output shaft (countershaft) speed sensor.
3. Install a new O-ring on a new output shaft (countershaft) speed sensor, then install the output shaft (countershaft) speed sensor and the sensor washer.
4. Check the connector for rust, dirt, or oil, and clean or repair if necessary, then connect the connector securely.
5. Do the battery installation procedure.

POSITIVE CRANKCASE VENTILATION (PCV) VALVE

REMOVAL & INSTALLATION

4-Cylinder Engine

See Figure 197.

1. Disconnect the PCV hose.
2. Remove the PCV valve.
3. Install the parts in the reverse order of removal with a new washer.

6-Cylinder Engine

See Figure 198.

1. Lift up the harness cover.
2. Remove the bolt.
3. Remove the PCV valve.

➡**Do not to spill oil on the hot exhaust manifold.**

4. Install the parts in the reverse order of removal.

➡**When installing a new PCV valve, make sure the O-rings are in place. When installing a used PCV valve, use new O-rings.**

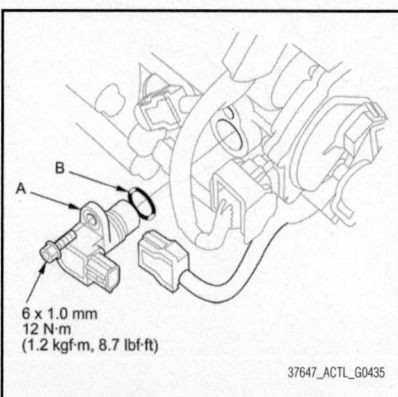

37647_ACTL_G0435

Fig. 196 Disconnect the output shaft (countershaft) speed sensor connector, and remove the output shaft (countershaft) speed sensor

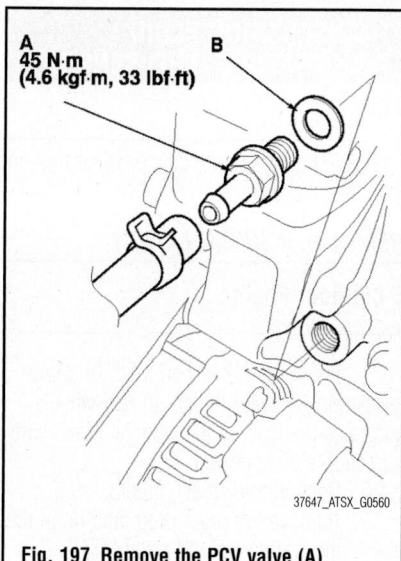

Fig. 197 Remove the PCV valve (A)

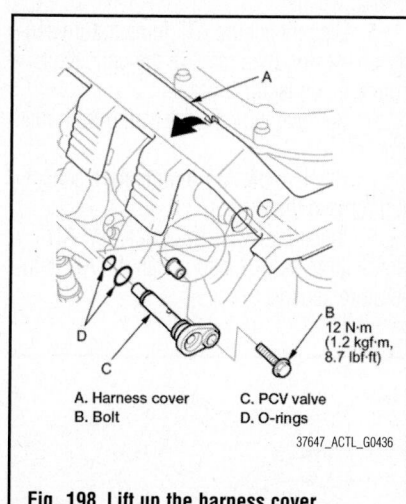

A. Harness cover
B. Bolt
C. PCV valve
D. O-rings

Fig. 198 Lift up the harness cover

POWERTRAIN CONTROL MODULE (PCM)

LOCATION

The PCM is used in the 6-cylinder engine and is located in the engine compartment behind the left headlight.

REMOVAL & INSTALLATION

See Figure 199.

Special Tools Required:
• Honda Diagnostic System (HDS) tablet tester
• Honda Interface Module (HIM) and an iN workstation with the latest HDS software version
• HDS pocket tester
• GNA600 and an iN workstation with the latest HDS software version

➡**Any one of the above updating tools can be used.**

1. Connect HDS to the Data Link Connector (DLC) located under the driver's side of the dashboard.
2. Turn the ignition switch to ON (II).
3. Make sure the HDS communicates with the PCM and other vehicle systems. If it doesn't, go to the DLC circuit troubleshooting. If you are returning from the DLC circuit troubleshooting, skip steps 4 through 9, 20 through 25, and 28 through 30, and do this after replacing the PCM:
   • Replace the engine oil and the engine oil filter.
   • Replace the ATF.
   • Clean the throttle body.
4. Select the PGM-FI system with the HDS.
5. Select the INSPECTIONMENU with the HDS.
6. Select the ETCS TEST, then select the TP POSITION CHECK, and follow the screen prompts.

➡**If the TP POSITION CHECK indicates FAILED, continue with this procedure.**

7. Select the REPLACE PCM MENU, then select READ DATA, and follow the screen prompts.

➡**Doing this step copies (READS) the engine oil life data from the original PCM so you can later download (WRITES) it into the new PCM. If READ DATA indicates FAILED, continue with this procedure.**

8. Select the A/T system with the HDS.
9. Select the REPLACE TCM/PCM MENU, then select READ DATA, and follow the screen prompts.

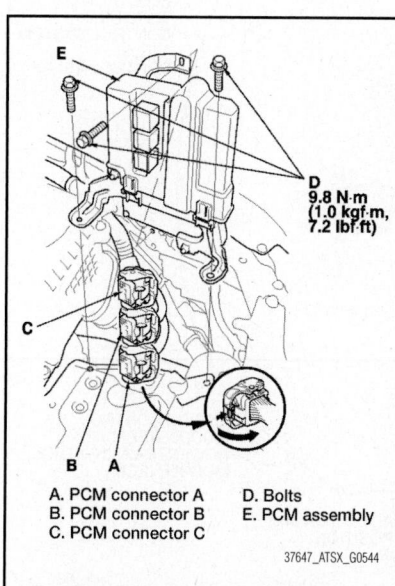

A. PCM connector A
B. PCM connector B
C. PCM connector C
D. Bolts
E. PCM assembly

Fig. 199 Disconnecting and removing the PCM

➡**Doing this step copies (READS) the ATF life data from the original PCM so you can later download (WRITES) it into the new PCM. If READ DATA indicates FAILED, continue with this procedure.**

10. Turn the ignition switch to LOCK (0).
11. Jump the SCS line with the HDS.
12. Do the battery removal procedure.
13. Remove the bolts.
14. Disconnect PCM connectors A, B, and C, then remove the PCM assembly.

➡**PCM connectors A, B, and C have symbols (A=□, B=△, C=○) embossed on them for identification.**

15. Disconnect the cover and the bracket from the PCM.
16. Install a known-good PCM in the reverse order of removal.
17. Do the battery installation procedure.
18. Turn the ignition switch to ON (II).

➡**DTC P0630 VIN Not Programmed or Mismatch may be stored because the VIN has not been programmed into the PCM; ignore it, and continue this procedure.**

19. Manually input the VIN to the PCM with the HDS.
20. If the READ DATA (engine oil life) failed in step 7, go to step 23. Otherwise, go to step 21.
21. Select the PGM-FI system with the HDS.
22. Select the REPLACE PCM MENU, then select WRITE DATA, and follow the screen prompts.

➡**If the WRITE DATA indicates FAILED, continue with this procedure.**

23. If the READ DATA (ATF life) failed in step 9, go to step 26. Otherwise go to step 24.
24. Select the A/T SYSTEM with the HDS.
25. Select the REPLACE TCM/PCM MENU, then select WRITE DATA, and follow the screen prompts.

➡**If the WRITE DATA indicates FAILED, continue with this procedure.**

26. Select IMMOBI system with the HDS.
27. Enter the immobilizer PCM code that you got from the iN, and use the PCM replacement procedure in the IMMOBI MENU of the HDS; it allows you to start the engine.
28. If the TP POSITION CHECK failed in step 6 clean the throttle body, then go to step 26.

29. If the READ DATA failed in step 7 or the WRITE DATA failed in step 22, replace the engine oil and engine oil filter, then go to step 30.

30. If the READ DATA failed in step 9 or the WRITE DATA failed in step 25, replace the ATF, then go to step 31.

31. Select PGM-FI system, and reset the PCM with the HDS.

32. Update the PCM if it does not have the latest software.

33. Do the PCM idle learn procedure.

34. Do the CKP pattern clear/CKP pattern learn procedure.

THROTTLE CONTROL ACTUATOR

LOCATION

The throttle actuator is located between the air intake duct and the intake manifold.

4-Cylinder Engine

See Figure 200.

6-Cylinder Engine

See Figure 201.

REMOVAL & INSTALLATION

➡The manufacturer does not provide a specific Removal and Installation procedure for this component. Refer to the graphic(s) when servicing this component.

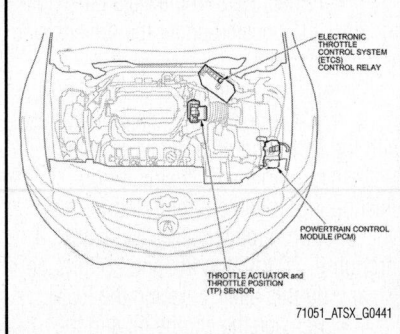

Fig. 201 Throttle actuator and throttle position sensor locations

THROTTLE POSITION SENSOR

LOCATION

The throttle position sensor is located between the air intake duct and the intake manifold.

4-Cylinder Engine

See Figure 200.

6-Cylinder Engine

See Figure 201.

REMOVAL & INSTALLATION

➡The manufacturer does not provide a specific Removal and Installation procedure for this component. Refer to the graphic(s) when servicing this component.

TRANSMISSION FLUID TEMPERATURE (TFT) SENSOR

LOCATION

The ATF temperature sensor is located on the transaxle case.

REMOVAL & INSTALLATION

6-Cylinder Engine

See Figure 202.

1. Raise the vehicle on a lift, or apply the parking brake, block both rear wheels, and raise the front of the vehicle. Make sure it is securely supported.

2. Remove the splash shield.

3. Remove the drain plug, and drain the Automatic Transmission Fluid (ATF).

4. Reinstall the drain plug with a new sealing washer.

5. Disconnect the ATF temperature sensor connector, then remove the connector from the connector bracket.

6. Remove the harness clamp from its bracket.

7. Remove the ATF temperature sensor, and replace the sensor.

8. Install a new O-ring on a new ATF temperature sensor, and install the ATF temperature sensor.

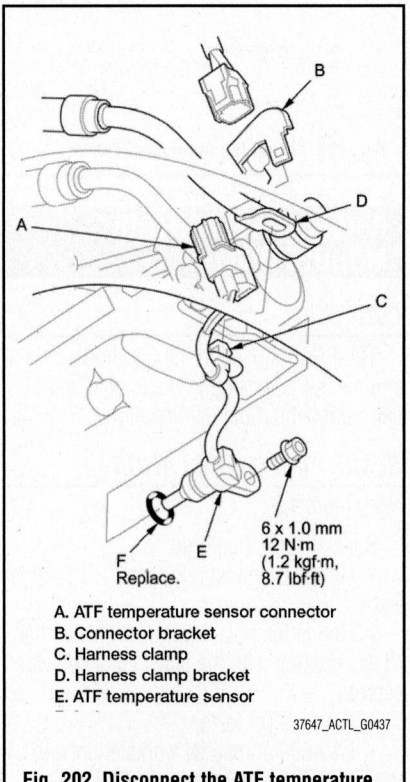

A. ATF temperature sensor connector
B. Connector bracket
C. Harness clamp
D. Harness clamp bracket
E. ATF temperature sensor

37647_ACTL_G0437

Fig. 202 Disconnect the ATF temperature sensor connector, then remove the connector from the connector bracket

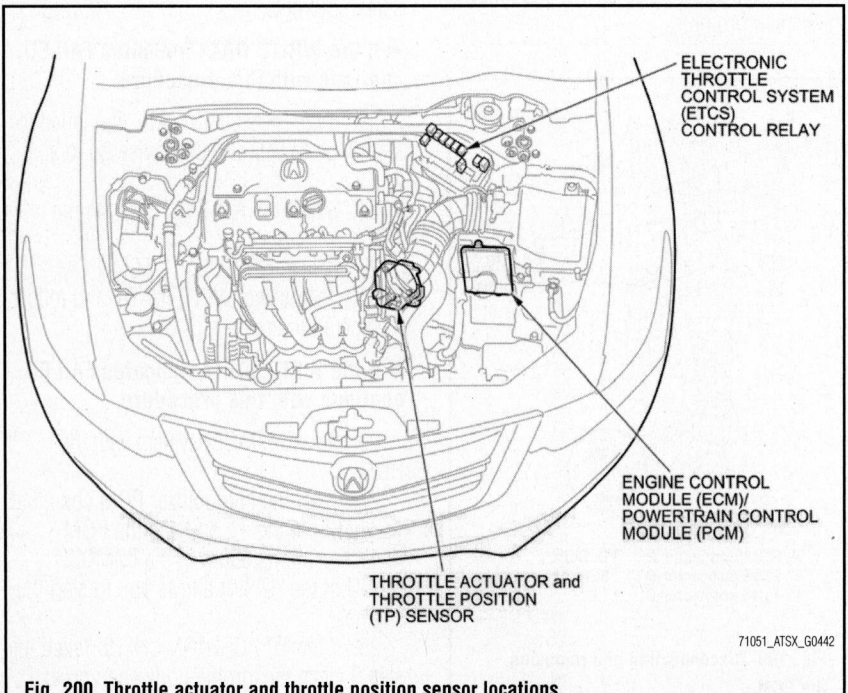

Fig. 200 Throttle actuator and throttle position sensor locations

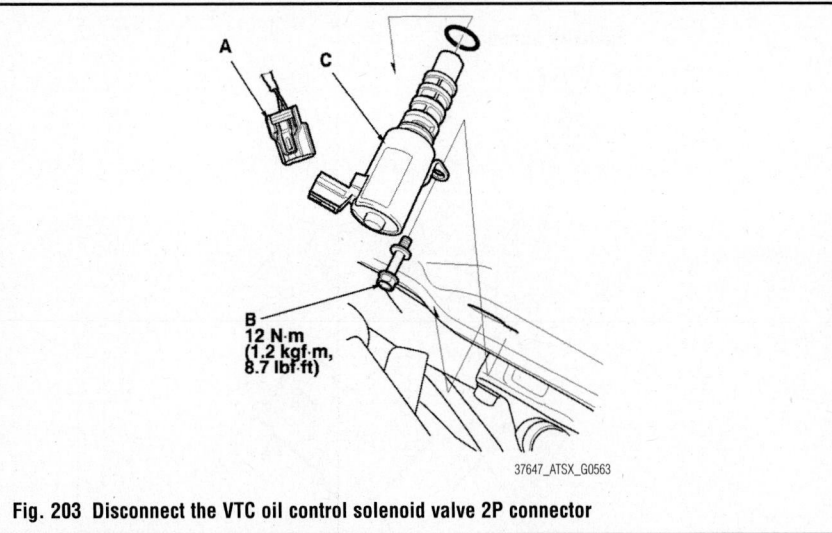

Fig. 203 Disconnect the VTC oil control solenoid valve 2P connector

37647_ATSX_G0563

VARIABLE TIMING CAMSHAFT (VTC) OIL CONTROL SOLENOID VALVE

LOCATION

4-Cylinder Engine

The VTC oil control solenoid valve is located on the cylinder head.

REMOVAL & INSTALLATION

4-Cylinder Engine

See Figure 203.

1. Disconnect the VTC oil control solenoid valve 2P connector.
2. Remove the bolt and the VTC oil control solenoid valve.
3. Installation is the reverse of removal.

---

9. Check the connector for rust, dirt, or oil, and clean or repair if necessary, then connect the connector securely, and install it on its bracket.

10. Install the harness clamp on its bracket.
11. Refill the transmission with ATF.
12. Install the splash shield.

FUEL

GASOLINE FUEL INJECTION SYSTEM

FUEL SYSTEM SERVICE PRECAUTIONS

Safety is the most important factor when performing not only fuel system maintenance but any type of maintenance. Failure to conduct maintenance and repairs in a safe manner may result in serious personal injury or death. Maintenance and testing of the vehicle's fuel system components can be accomplished safely and effectively by adhering to the following rules and guidelines.

• To avoid the possibility of fire and personal injury, always disconnect the negative battery cable unless the repair or test procedure requires that battery voltage be applied.

• Always relieve the fuel system pressure prior to disconnecting any fuel system component (injector, fuel rail, pressure regulator, etc.), fitting or fuel line connection. Exercise extreme caution whenever relieving fuel system pressure to avoid exposing skin, face and eyes to fuel spray. Please be advised that fuel under pressure may penetrate the skin or any part of the body that it contacts.

• Always place a shop towel or cloth around the fitting or connection prior to loosening to absorb any excess fuel due to spillage. Ensure that all fuel spillage (should it occur) is quickly removed from engine surfaces. Ensure that all fuel soaked cloths or towels are deposited into a suitable waste container.

• Always keep a dry chemical (Class B) fire extinguisher near the work area.

• Do not allow fuel spray or fuel vapors to come into contact with a spark or open flame.

• Always use a back-up wrench when loosening and tightening fuel line connection fittings. This will prevent unnecessary stress and torsion to fuel line piping.

• Always replace worn fuel fitting O-rings with new Do not substitute fuel hose or equivalent where fuel pipe is installed.

Before servicing the vehicle, make sure to also refer to the precautions in the beginning of this section as well.

RELIEVING FUEL SYSTEM PRESSURE

✳✳ CAUTION

Before disconnecting fuel lines or hoses, relieve pressure from the system by disabling the fuel pump and disconnecting the fuel line/quick connect fitting in the engine compartment.

With the HDS

1. Connect the HDS to the Data Link Connector (DLC) located under the driver's side of the dashboard.
2. Turn the ignition switch to ON (II).

3. Make sure the HDS communicates with the ECM/PCM. If it doesn't, go to the DLC circuit troubleshooting.
4. Turn the ignition switch to LOCK (0).
5. Remove the fuel fill cap to relieve pressure in the fuel tank.
6. Turn the ignition switch to ON (II).
7. From the INSPECTIONMENU of the HDS, select Fuel Pump OFF, then start the engine, and let it idle until it stalls.

➡**Do not allow the engine to idle above 1,000 rpm or the ECM/PCM will continue to operate the fuel pump. Pending or Confirmed DTC may be set during this procedure. Check for DTCs, and clear them as needed.**

8. Turn the ignition switch to LOCK (0).
9. Do the battery terminal disconnection procedure.
10. Remove the cover and the quick-connect fitting cover.
11. Check the quick-connect fitting for dirt, and clean it if needed.
12. Place a rag or a shop towel over the quick-connect fitting.
13. Disconnect the quick-connect fitting.
   a. Hold the connector with one hand, and squeeze the retainer tabs with the other hand to release them from the locking tabs.
   b. Pull the connector off.

➡**Be careful not to damage the line or other parts. Do not use tools. If the** connector does not move, keep the retainer tabs pressed down, and alternately pull and push the connector until it comes off easily. Do not remove the retainer from the line; once removed, the retainer must be replaced with a new one.

14. After disconnecting the quick-connect fitting, check it for dirt or damage.

15. Do the battery terminal reconnection procedure.

Without the HDS

See Figure 204.

1. Remove the driver's dashboard undercover.

2. Remove PGM-FI main relay 2 from the driver's under-dash fuse/relay box.

3. Start the engine, and let it idle until it stalls.

➡ **If any DTCs are stored, clear and ignore them.**

4. Turn the ignition switch to LOCK (0).

5. Remove the fuel fill cap to relieve the pressure in the fuel tank.

6. Do the battery terminal disconnection procedure.

7. Remove the cover and the quick-connect fitting cover.

8. Check the quick-connect fitting for dirt, and clean it if needed.

9. Place a rag or a shop towel over the quick-connect fitting.

10. Disconnect the quick-connect fitting.

　a. Hold the connector with one hand, and squeeze the retainer tabs with the other hand to release them from the locking tabs.

　b. Pull the connector off.

➡ **Be careful not to damage the line or other parts. Do not use tools. If the connector does not move, keep the retainer tabs pressed down, and alternately pull and push the connector until it comes off easily. Do not remove the retainer from the line; once removed, the retainer must be replaced with a new one.**

11. After disconnecting the quick-connect fitting, check it for dirt or damage.

12. Do the battery terminal reconnection procedure.

FUEL FILTER

REMOVAL & INSTALLATION

See Figure 205.

➡ **The fuel filter should be replaced whenever the fuel pressure**

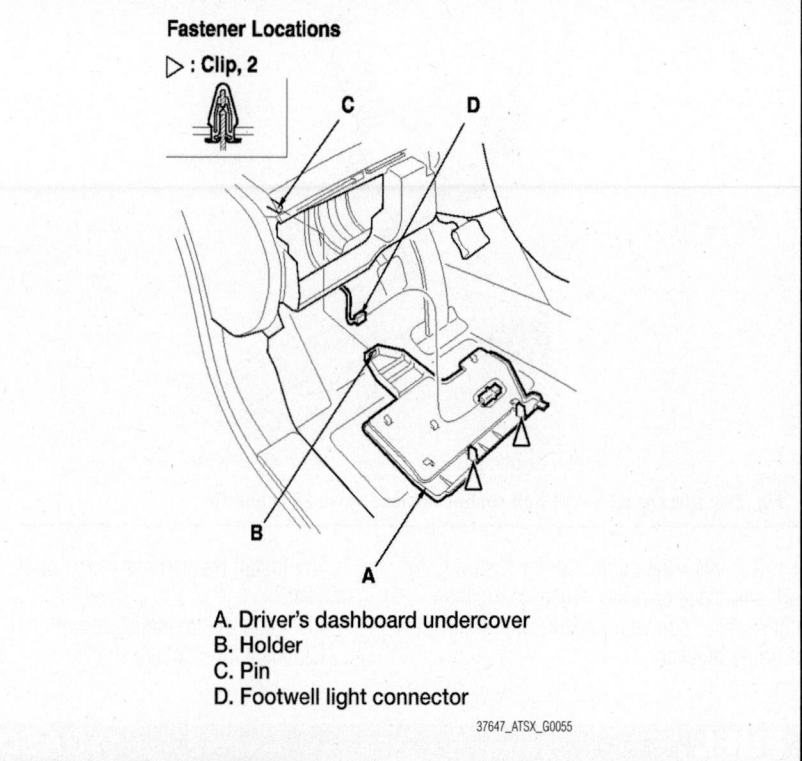

Fastener Locations

▷ : Clip, 2

A. Driver's dashboard undercover
B. Holder
C. Pin
D. Footwell light connector

37647_ATSX_G0055

Fig. 204 Remove driver's dashboard undercover

drops below the specified value, after making sure that the fuel pump and the fuel pressure regulator are OK.

1. Remove the fuel tank unit.
2. Remove the fuel filter set.

To install:

3. Check these items before installing the fuel tank unit:

　• When connecting the wire harness, make sure the connection is secure and the con-

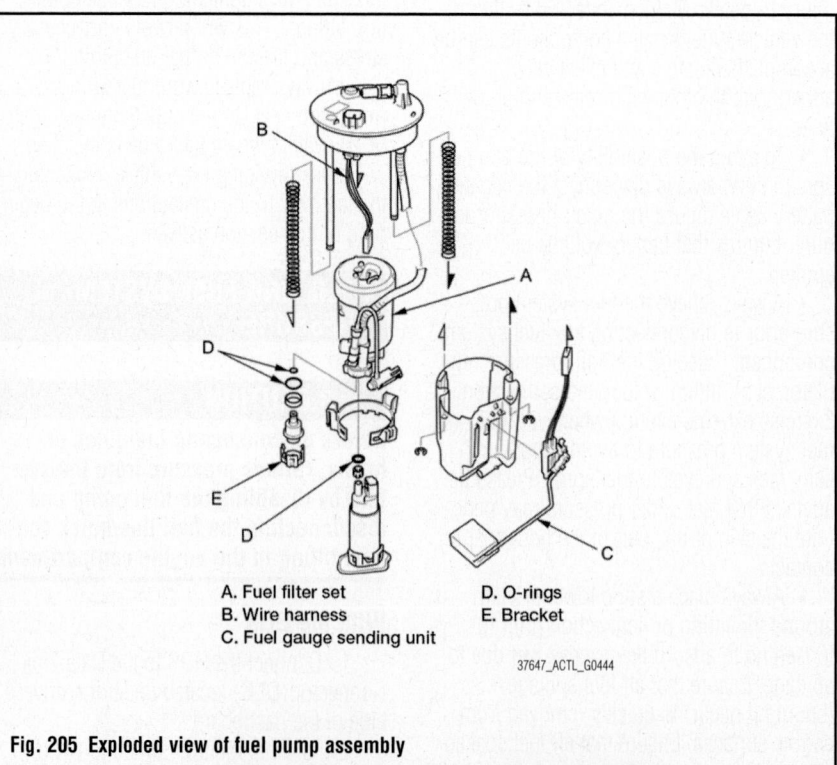

A. Fuel filter set
B. Wire harness
C. Fuel gauge sending unit
D. O-rings
E. Bracket

37647_ACTL_G0444

Fig. 205 Exploded view of fuel pump assembly

nectors are firmly locked into place.

- When installing the fuel gauge sending unit, make sure the connection is secure and the connector is firmly locked into place. Be careful not to bend or twist it excessively.

4. Install the parts in the reverse order of removal with new O-rings and a new bracket. When installing the fuel tank unit, align the marks on the unit and the fuel tank.

➡**Coat the O-rings with clean engine oil; do not use any other oil or fluid. Do not pinch the O-rings during installation. Use all the new parts supplied in the fuel filter replacement kit.**

FUEL RAIL & INJECTORS

REMOVAL & INSTALLATION

4-Cylinder Engine

See Figures 206 and 207.

1. Relieve the fuel pressure.
2. Remove the engine cover.
3. Remove the quick-connect fitting cover, then disconnect the quick-connect fitting.
4. Disconnect the injector connectors and the engine mount control solenoid valve connector.
5. Remove the ground cable bolts.
6. Remove the fuel rail mounting nuts from the fuel rail.
7. Remove the fuel rail and the injectors from the injector base.

8. Remove the injector clips from the fuel rail.
9. Remove the injectors from the fuel rail.

To install:

10. Coat the new O-rings (black) with clean engine oil, and insert the injectors into the fuel rail.
11. Install the injectors clips.
12. Coat the new injector O-rings (brown) with clean engine oil.
13. Install the fuel rail and the injectors in the injector base.
14. Install the fuel rail mounting nuts and the ground cable bolts.
15. Connect the injector connectors and the engine mount control solenoid valve connector.

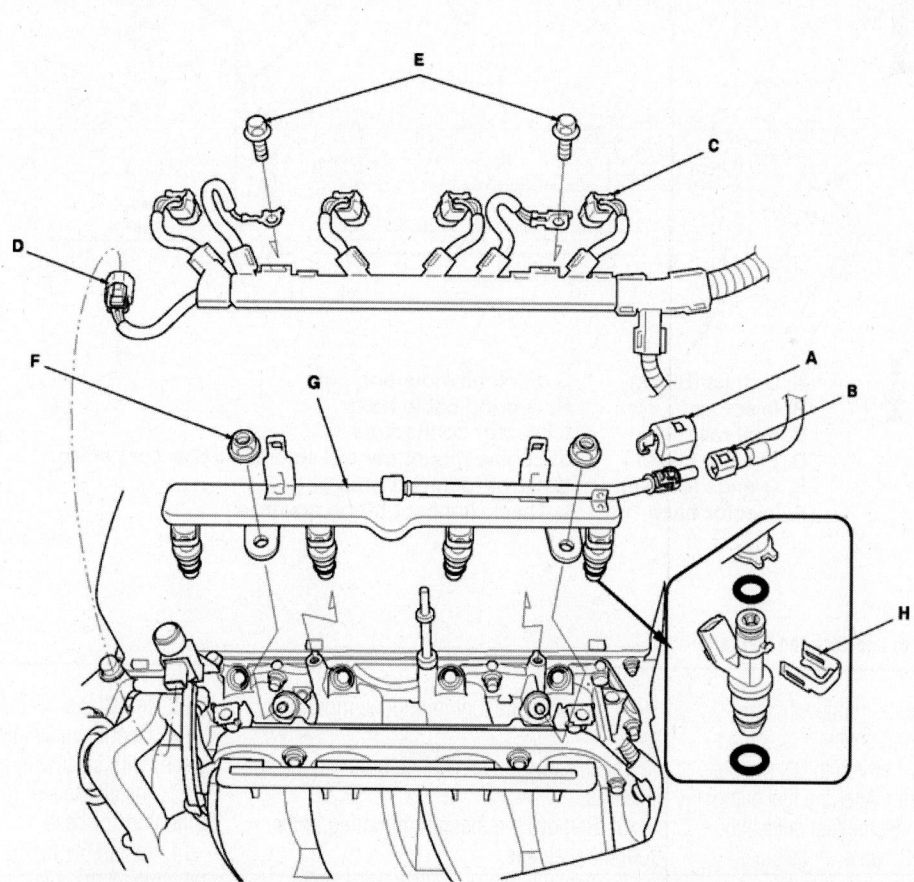

A. Quick-disconnect fitting cover
B. Quick-disconnect fitting
C. Injector connectors
D. Engine mount control solenoid valve connector
E. Ground cable bolts
F. Fuel rail mounting nuts
G. Fuel rail
H. Injector clips

37647_ATSX_G0568

Fig. 206 Fuel rail assembly

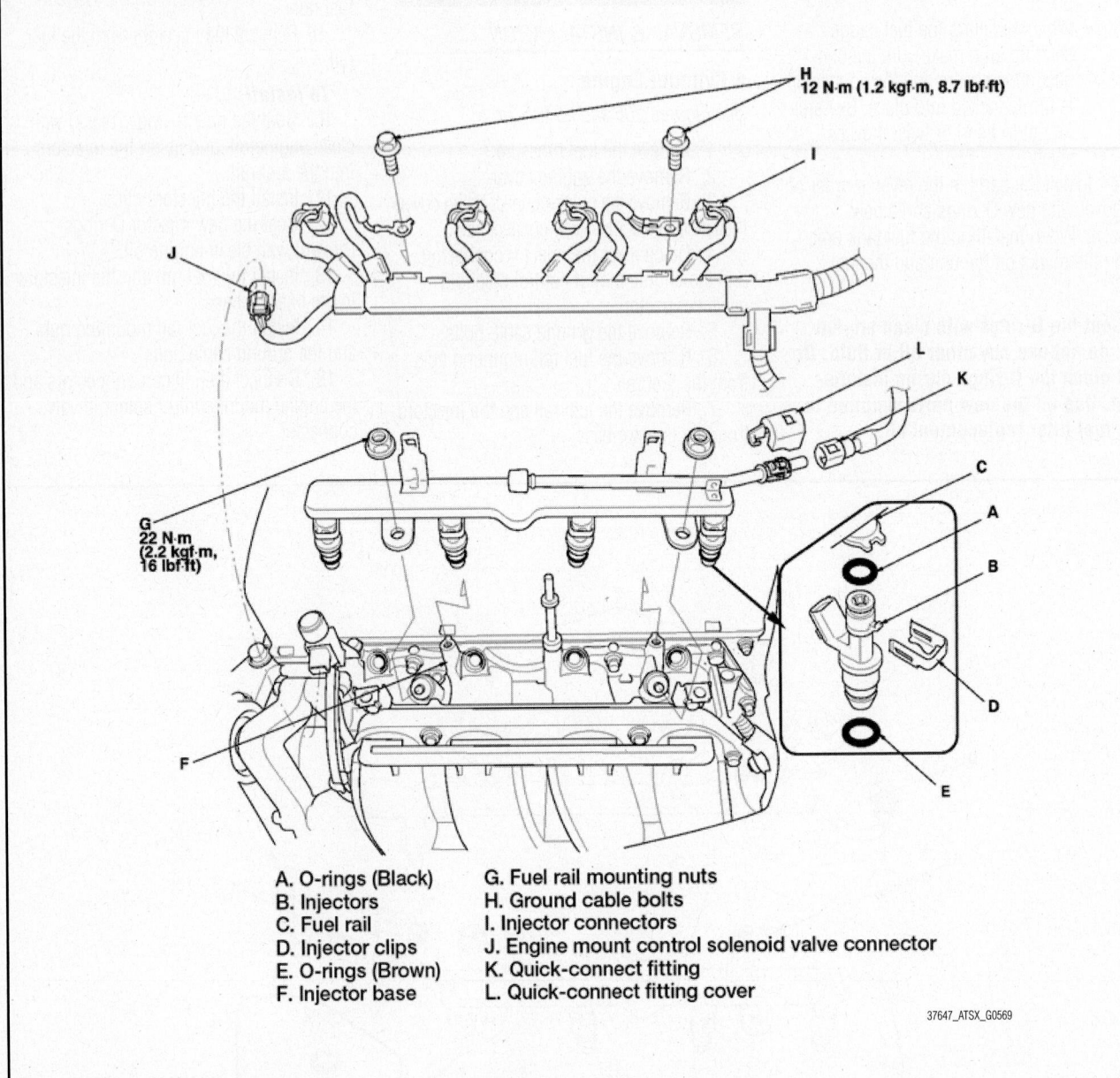

A. O-rings (Black)
B. Injectors
C. Fuel rail
D. Injector clips
E. O-rings (Brown)
F. Injector base
G. Fuel rail mounting nuts
H. Ground cable bolts
I. Injector connectors
J. Engine mount control solenoid valve connector
K. Quick-connect fitting
L. Quick-connect fitting cover

37647_ATSX_G0569

Fig. 207 Installation of injectors and fuel rail

16. Connect the quick-connect fitting, and quick-connect fitting cover.

17. Turn the ignition switch to ON (II), but do not operate the starter. After the fuel pump runs for about 2 seconds, the fuel rail will be pressurized. Repeat this two or three times, then make sure there are no fuel leaks.

18. Reinstall the engine cover.

6-Cylinder Engine

See Figures 208 and 209.

1. Relieve the fuel pressure.
2. Remove the intake manifold.
3. Disconnect the quick-connect fitting.

4. Remove the fuel joint hose mounting bolt.

5. Disconnect the connectors from the injectors.

6. Remove the fuel rail mounting bolts from the fuel rails.

7. Remove the fuel rails and the injectors from the injector base.

8. Remove the injector clips from the fuel rails.

9. Remove the injectors from the fuel rails.

To install:

10. Coat the new O-rings (black) with clean engine oil, and insert the injectors into the fuel rails.

11. Install the injector clips.

12. Coat the new injector O-rings (green) with clean engine oil.

13. Install the fuel rails and the injectors in the injector base.

14. Install the fuel rail mounting bolts, and connect the injector connectors.

15. Install the fuel joint hose mounting bolt.

16. Connect the quick-connect fitting.

17. Turn the ignition switch to ON (II), but do not operate the starter. After the fuel pump runs for about 2 seconds, the fuel rail will be pressurized. Repeat this two or three times, then make sure there are no fuel leaks.

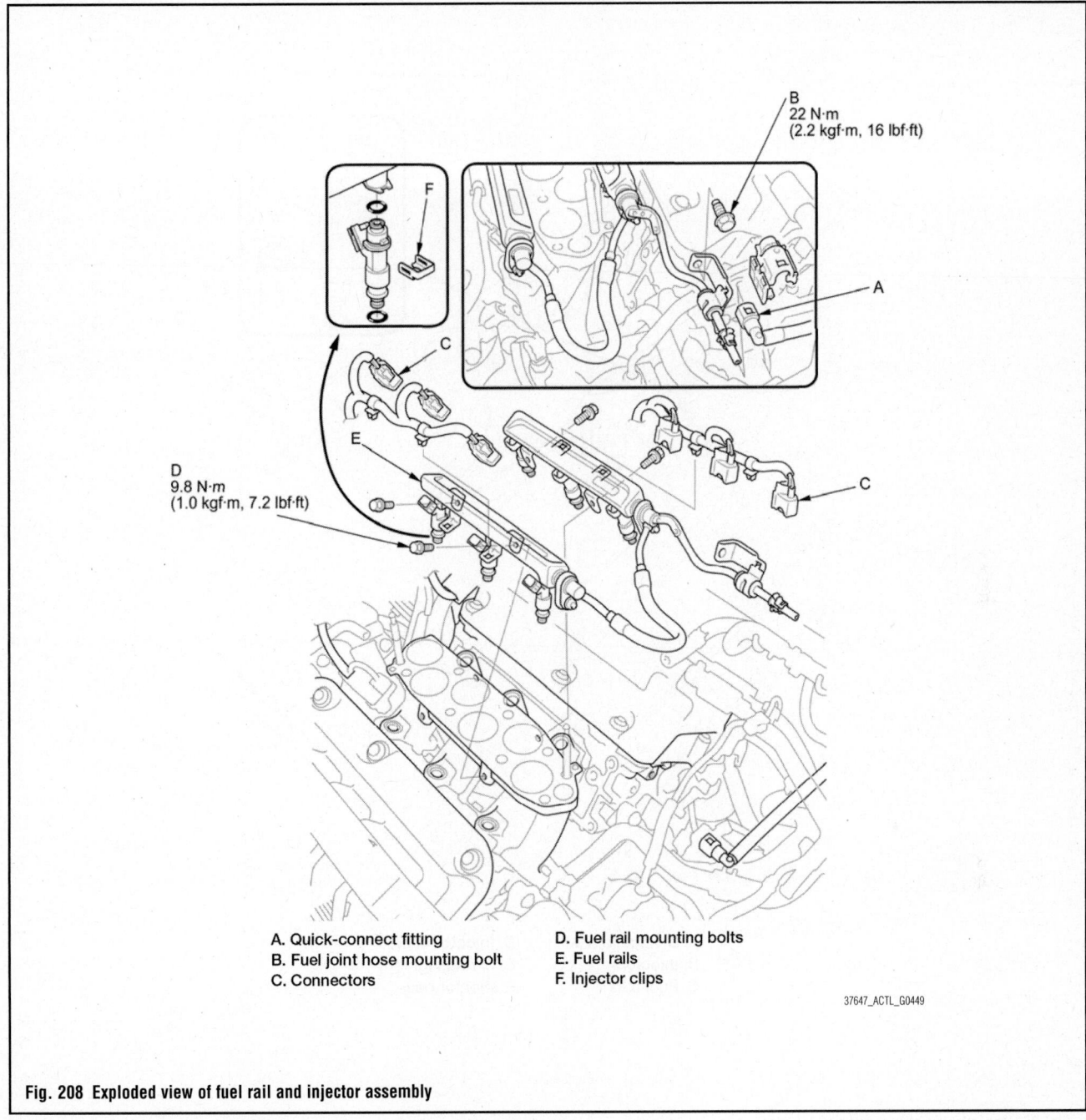

A. Quick-connect fitting
B. Fuel joint hose mounting bolt
C. Connectors
D. Fuel rail mounting bolts
E. Fuel rails
F. Injector clips

37647_ACTL_G0449

Fig. 208 Exploded view of fuel rail and injector assembly

18. Install the intake manifold with a new gasket.

THROTTLE BODY

REMOVAL & INSTALLATION

4-Cylinder Engine

See Figures 210 and 211.

1. Turn the ignition switch to LOCK (0).
2. Remove the intake air duct.
3. Disconnect the throttle body connector.
4. Disconnect and plug the water bypass hoses.

5. Remove the throttle body.
6. Install the parts in the reverse order of removal with a new gasket.

➡**When tightening the screw of the hose band, align the edge of the hose band with the mark painted on it.**

7. After the installation, refill the radiator with engine coolant.
8. Connect the HDS to the data link connector (DLC) located under the driver's side of the dashboard.
9. Turn the ignition switch to ON (II).

10. Reset the ECM/PCM with the HDS.
11. Select the ETCS TEST in the INSPECTION MENU with the HDS.
12. Select TP POSITION CHECK and clear the throttle position (TP) learned value.
13. Turn the ignition switch to LOCK (0).
14. Turn the ignition switch to ON (II), and wait 2 seconds without pressing the accelerator pedal.
15. Do the ECM/PCM idle learn procedure.

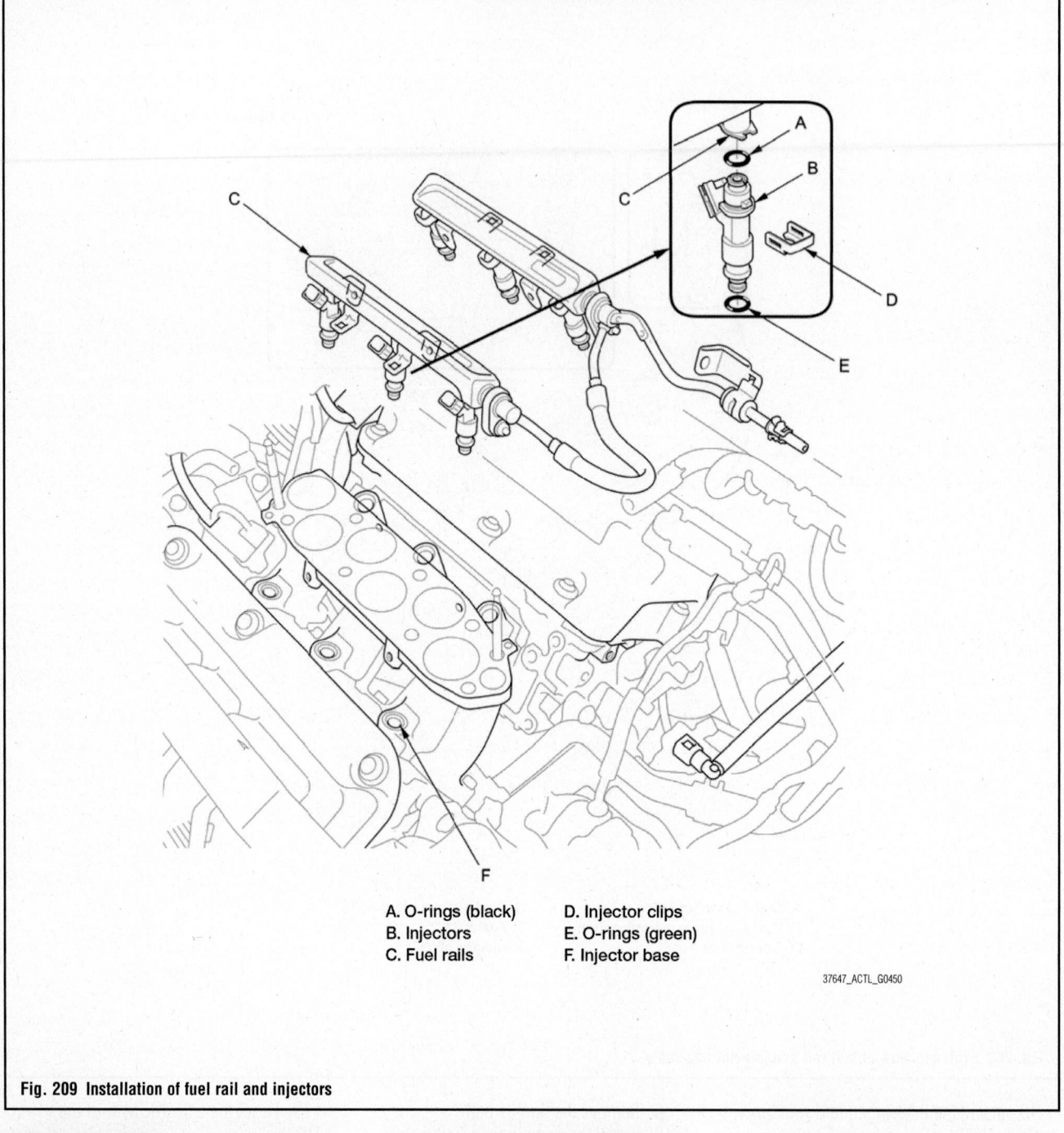

A. O-rings (black) D. Injector clips
B. Injectors E. O-rings (green)
C. Fuel rails F. Injector base

37647_ACTL_G0450

Fig. 209 Installation of fuel rail and injectors

6-Cylinder Engine

❈❈ CAUTION

Do not insert your fingers into the installed throttle body when you turn the ignition switch to ON (II), or while the ignition switch is in ON (II). If you do, you will seriously injure your fingers if the throttle valve is activated.

1. Make sure the ignition switch turned LOCK (0).
2. Disconnect the MAP sensor connector.
3. Remove the intake air duct.

4. Disconnect the throttle body connector.
5. Disconnect and plug the water bypass hoses, then remove the throttle body.
6. Install the parts in the reverse order of removal with a new gasket. After you torque the hose clamp screw, make sure the clearance is less than 0.04 inches (1.0 mm).
7. Refill the radiator with engine coolant.

➡**If it is in case of replacing or cleaning the throttle body, go to step 8. If it isn't, this procedure is complete.**

8. Connect the HDS to the data link connector (DLC) located under the driver's side of the dashboard.
9. Turn the ignition switch to ON (II).
10. Reset the PCM with the HDS.
11. Select the ETCS TEST in the INSPECTION MENU with the HDS.
12. Select TP POSITION CHECK and clear the throttle position (TP) learned value.
13. Turn the ignition switch to LOCK (0).
14. Turn the ignition switch to ON (II), and wait 2 seconds without pressing the accelerator pedal.
15. Do the PCM idle learn procedure.

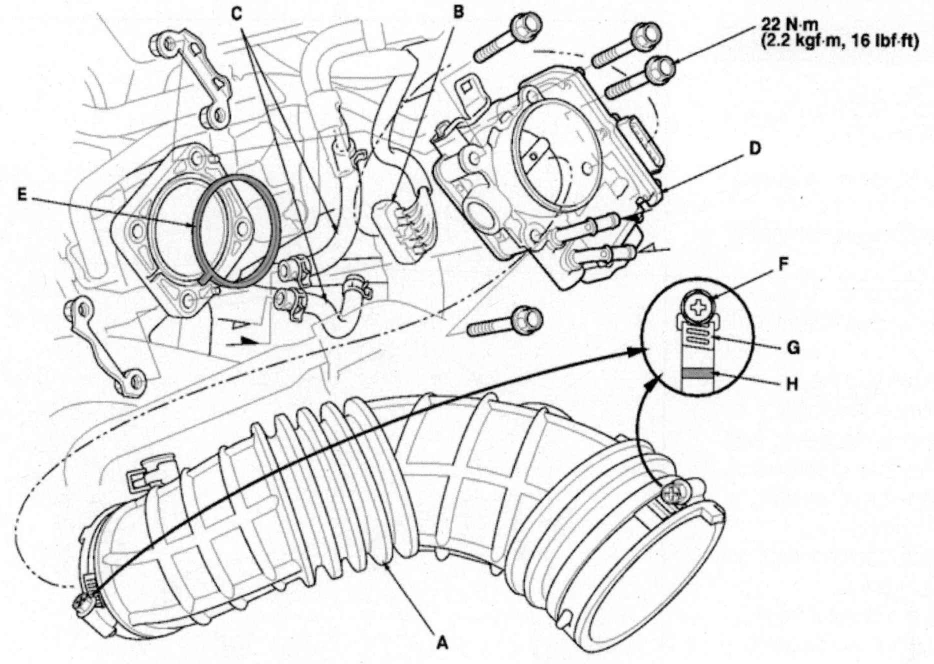
C
B
22 N·m
(2.2 kgf·m, 16 lbf·ft)
D
E
F
G
H
A

A. Intake air duct
B. Throttle body connector
C. Water bypass hoses
D. Throttle body
E. Gasket
F. Hose band
G. Edge of hose band
H. Painted mark

37647_ATSX_G0578

Fig. 210 Exploded view of throttle body removal

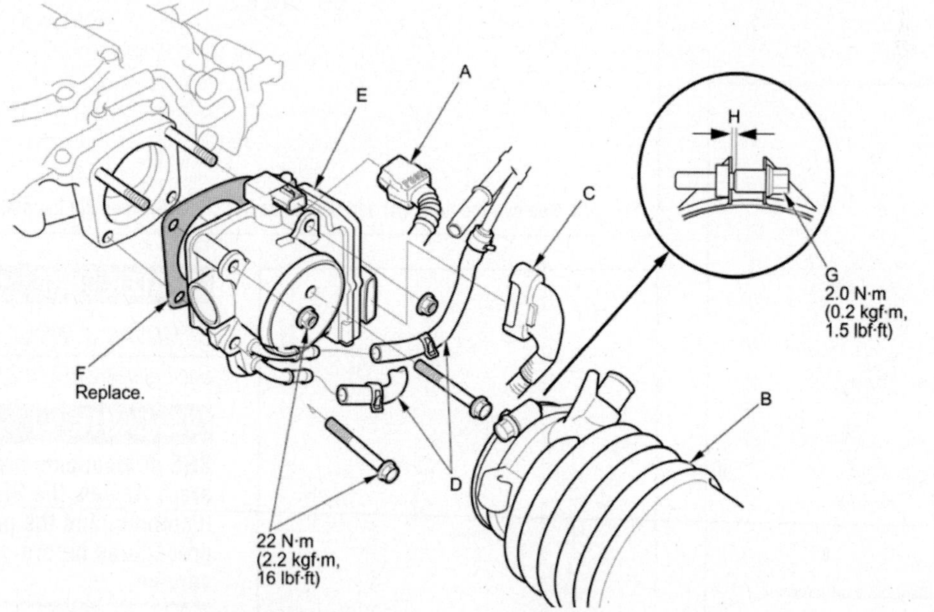
A
E
H
C
G
2.0 N·m
(0.2 kgf·m,
1.5 lbf·ft)
F
Replace.
B
D
22 N·m
(2.2 kgf·m,
16 lbf·ft)

A. MAP sensor connector
B. Intake air duct
C. Throttle body connector
D. Water bypass hoses
E. Throttle body
F. Gasket
G. Hose screw clamp
H. Clearance

71051_ATSX_G0482

Fig. 211 Exploded view of the throttle body assembly

HEATING & AIR CONDITIONING SYSTEM

BLOWER MOTOR

REMOVAL & INSTALLATION

See Figures 212 through 215.

1. Remove the passenger's dashboard undercover.

2. Remove the passenger's dashboard trim.

3. Remove the glove box.

4. Remove the passenger's dashboard center lower cover.

5. Remove the bolts and the connector clips. Then cut the plastic cross brace in the glove box opening with diagonal cutters in the area shown. Retain the plastic cross brace it will be reinstalled.

6. Remove the self-tapping screws, and the passenger's heater duct.

7. Disconnect the connector from the blower motor and the wire harness clip.

8. Disconnect the left engine compartment subharness connector from the bracket and harness clips.

9. Disconnect the connector from the recirculation control motor.

10. Remove the self-tapping screw, the mounting nuts, and the blower unit.

11. Install the unit in the reverse order of removal. Make sure that there are no air leaks.

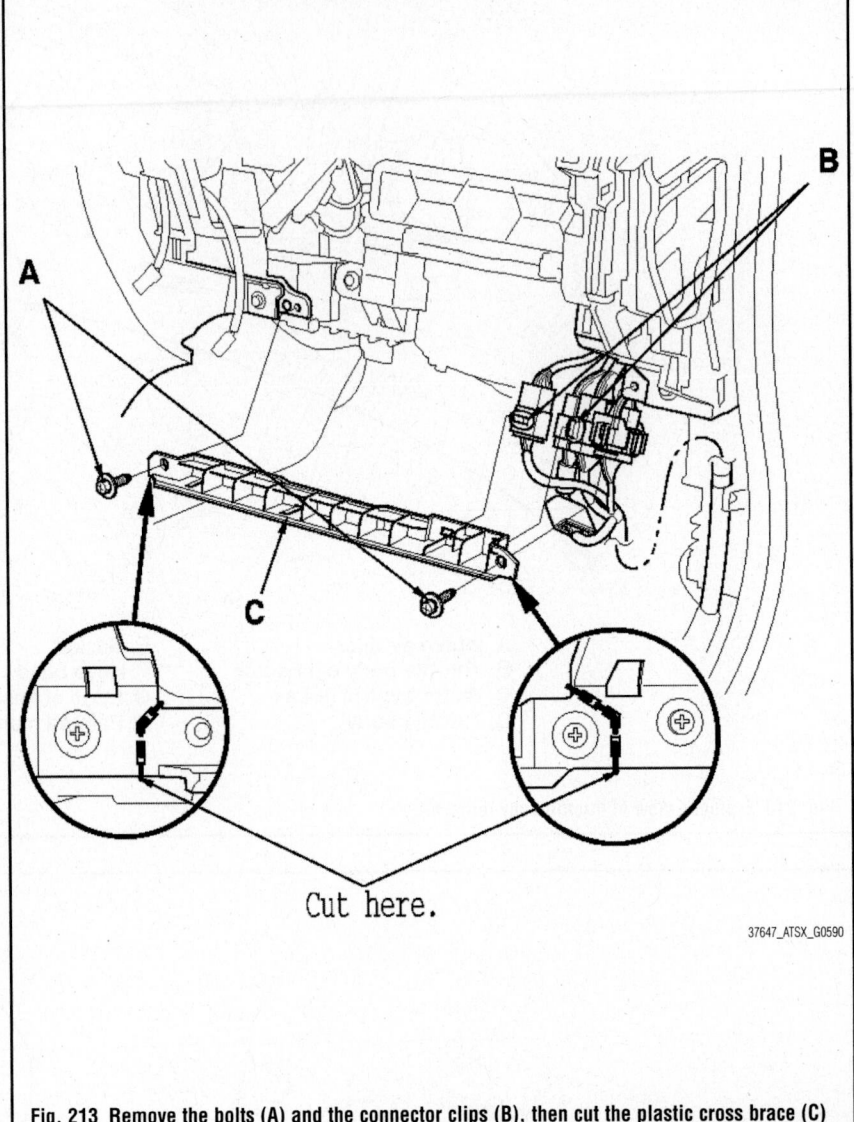

Fig. 213 Remove the bolts (A) and the connector clips (B), then cut the plastic cross brace (C)

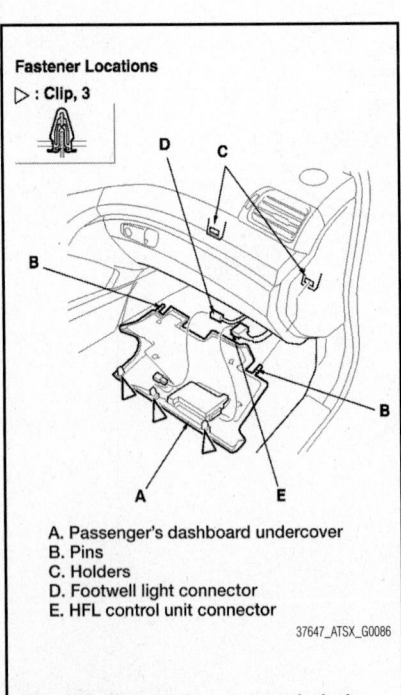

A. Passenger's dashboard undercover
B. Pins
C. Holders
D. Footwell light connector
E. HFL control unit connector

37647_ATSX_G0086

Fig. 212 Remove the passenger's dashboard undercover

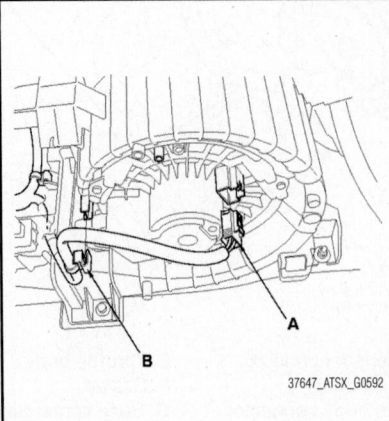

37647_ATSX_G0592

Fig. 214 Disconnect the connector (A) from the blower motor and the wire harness clip (B)

HEATER CORE

REMOVAL & INSTALLATION

See Figures 216 and 217.

❋❋ CAUTION

SRS components are located in this area. Review the SRS component locations, and the precautions and procedures before doing repairs or service.

1. Do the battery terminal disconnection procedure.

2. Recover the refrigerant with a recovery/recycling/charging station.

3. Disconnect the A/C line from the evaporator core.

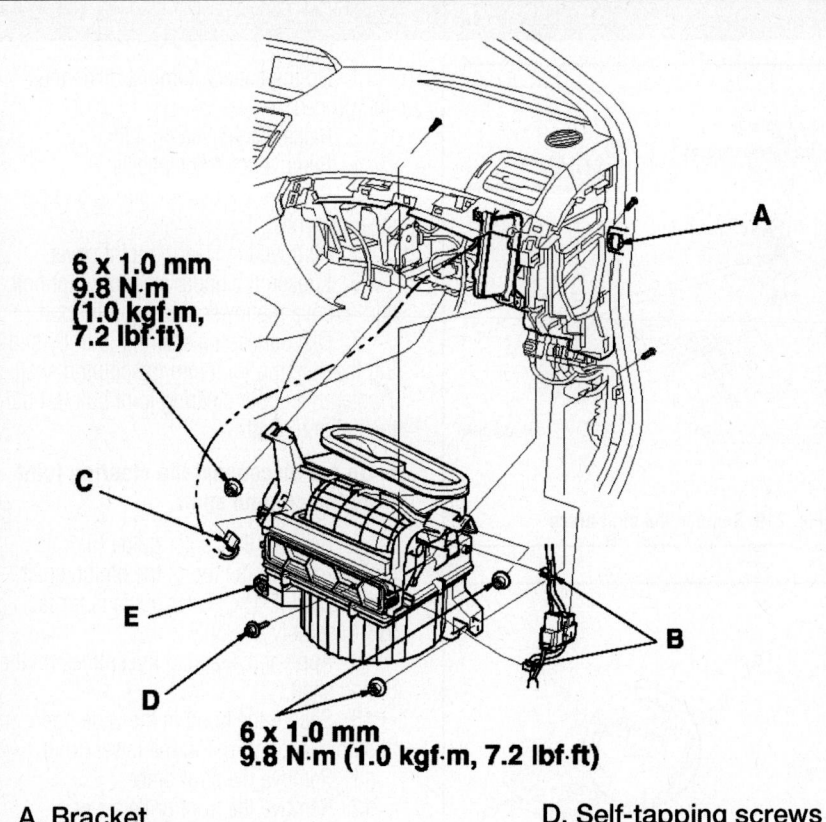

**6 x 1.0 mm
9.8 N·m
(1.0 kgf·m,
7.2 lbf·ft)**

**6 x 1.0 mm
9.8 N·m (1.0 kgf·m, 7.2 lbf·ft)**

A. Bracket
B. Harness clips
C. Recirculation control motor connector
D. Self-tapping screws
E. Blower unit

37647_ATSX_G0593

Fig. 215 Disconnect and remove the blower unit

4. When the engine is cool, drain the engine coolant from the radiator.

5. From under the hood, slide the hose clamps back. Disconnect the inlet heater

hose and the outlet heater hose from the heater unit. Note the layout of the hoses.

➡**Engine coolant will run out when the hoses are disconnected; drain it into a**

clean drip pan. Be sure not to let coolant spill on the electrical parts or the painted surfaces. If any coolant spills, rinse it off immediately.

6. Remove the mounting nut from the heater unit. Take care not to damage or bend the fuel lines or the brake lines, etc..

7. Remove the dashboard.

8. Disconnect the connector. Detach the connector clip, the harness clips, the clips, remove the ducts and the drain hose. Then remove the mounting nuts and the blower-heater unit.

9. Disconnect the connector from the blower motor. Detach the harness clip.

10. Disconnect the connectors: The mode control motor, the power transistor, the evaporator temperature sensor, the passenger's air mix control motor, and the recirculation control motor. Detach the harness clips.

11. Disconnect the connector from the driver's air mix control motor. Detach the harness clips and the connector clip, remove the wire harness.

12. Remove the self-tapping screws and the passenger's heater duct. Remove the self-tapping screws, the joint duct, and seal. Remove the self-tapping screw and the heater core cover. Remove the self-tapping screw, the heater pipe bracket, and the grommet, and carefully pull out the heater core.

To install:

13. Install the heater core, and the evaporator core in the reverse order of removal.

14. Install the heater unit in the reverse order of removal, and note these items:

   a. Do not interchange the inlet and outlet heater hoses, and install the hose clamps securely.

   b. Refill the cooling system with engine coolant.

   c. Make sure that there is no coolant leakage.

   d. Make sure that there is no air leakage.

   e. Make sure that there is no refrigerant leakage.

15. Do the battery terminal reconnection procedure.

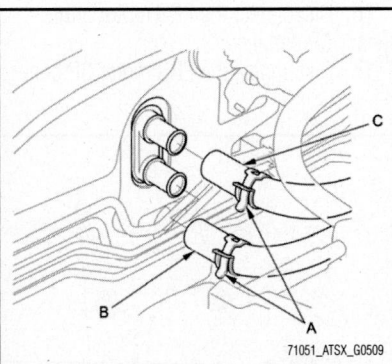

71051_ATSX_G0509

Fig. 216 From under the hood, slide the hose clamps (A) back, disconnect the inlet heater hose (B) and the outlet heater hose (C) from the heater unit

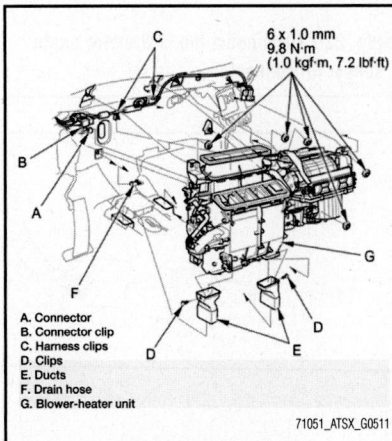

**6 x 1.0 mm
9.8 N·m
(1.0 kgf·m, 7.2 lbf·ft)**

A. Connector
B. Connector clip
C. Harness clips
D. Clips
E. Ducts
F. Drain hose
G. Blower-heater unit

71051_ATSX_G0511

Fig. 217 Exploded view of the blower-heater unit assembly

STEERING

POWER STEERING GEARBOX

REMOVAL & INSTALLATION

4-Cylinder Engine

See Figures 218 through 233.

Special Tools Required:
- Ball joint thread protector, 12 mm 07AAF-SDAA100
- Ball joint remover, 28 mm 07MAC-SL0A202
- Subframe alignment pin 070AG-SJAA10S
- Engine hanger adapter VSB02C000015*
- Front subframe adapter VSB02C000016*
- Engine support hanger, A and Reds AAR-T1256*

➡*These special tools are available through the Acura Tool and Equipment Program, 888-424-6857.

➡Note these items during removal:

- Use solvent and a brush, wash any oil and dirt off the end of the steering gearbox. Avoid any electrical parts. Blow dry with compressed air.
- Be sure to remove the steering wheel before disconnecting the steering joint, or damage to the cable reel can occur.

✳✳ WARNING

SRS components are located in this area. Review the SRS components locations and the precautions and procedures before doing repairs or service.

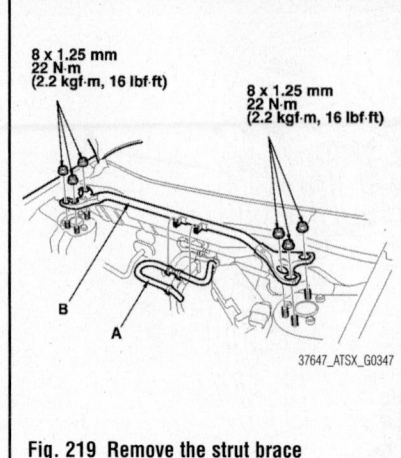

Fig. 219 Remove the strut brace

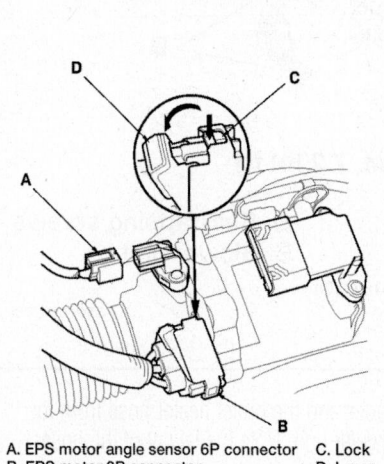

A. EPS motor angle sensor 6P connector   C. Lock
B. EPS motor 3P connector   D. Lever

37647_ATSX_G0630

Fig. 220 Disconnect the EPS motor angle sensor 6P connector

1. Do the battery terminal disconnection procedure.
2. Raise the vehicle on a lift.
3. Remove the front wheels.
4. Remove the drivers' airbag and the steering wheel.
5. Remove the steering joint cover.
6. Loosen the upper steering joint bolt, and remove the lower steering joint bolt.
7. Disconnect the steering joint by sliding the steering joint into the column shaft. Tighten the upper steering joint bolt to hold the column shaft.

➡**Do not disconnect the steering joint from the column shaft.**

8. Remove the center guide (if equipped) from the top of the pinion shaft, and discard it. The center guide is for factory assembly use only.
9. Apply vinyl tape to the splines on the pinion shaft.
10. Secure the hood in the wide open position (support rod in the lower hole).
11. Remove the strut brace.
12. Remove the front grille cover.
13. Remove the cotter pin from the tie-rod end ball joint, then remove the nut on both sides.
14. Disconnect the tie-rod end ball joint from the knuckle using the ball joint thread protector and the ball joint remover on both sides.

➡**Be careful not to damage the ball joint boot when installing the remover.**

15. Remove the P/S heat shield.
16. Disconnect the EPS motor angle sensor 6P connector.
17. Disconnect the EPS motor 3P con-

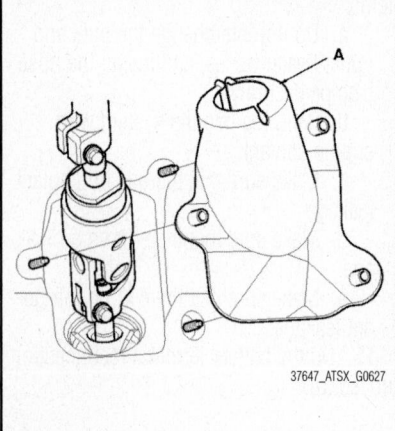

37647_ATSX_G0627

Fig. 218 Remove the steering joint cover (A)

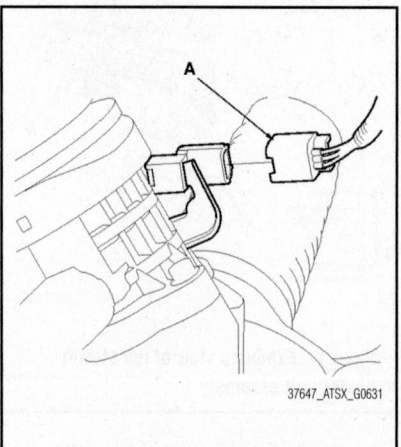

37647_ATSX_G0631

Fig. 221 Disconnect the EPS subharness 6P connector (A)

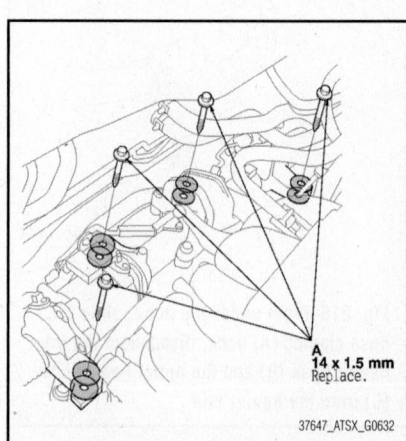

37647_ATSX_G0632

Fig. 222 Remove the steering gearbox mounting bolts (A) and washers (B) from the steering gearbox

nector by pushing the lock and pulling down the lever. Wrap the connectors with vinyl tape to avoid contamination from grease or water.

18. Disconnect the EPS subharness 6P connector. Wrap the connectors with vinyl tape to avoid contamination from grease or water.

19. Remove the steering gearbox mounting bolts and washers from the steering gearbox.

20. Remove the ignition coil cover.

21. Remove the harness cover bracket bolts.

22. Attach the engine hanger adapter (VSB02C000015) to the threaded hole located on the rear side of the cylinder head.

23. Install the engine support hanger (AAR-T1256) to the vehicle, and attach the hook to the slotted hole in the engine hanger adapter (VSB02C000015). Tighten the wing nut by hand to lift and support the engine.

➡**Be careful when working around the windshield. Be careful not to damage the hood opener cable when installing the engine support hanger at the front bulkhead.**

24. Remove the rear engine mount.

25. A/T: Remove the upper base bracket from the base bracket.

26. Remove the front engine mount stop, then remove the front engine mount bolt.

27. Raise the vehicle on a lift.

28. Remove the front splash shield.

29. Remove exhaust pipe A.

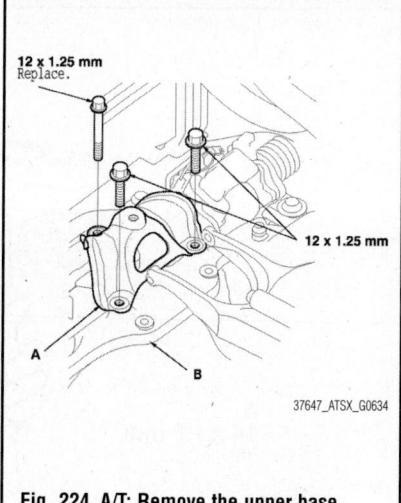

Fig. 224 A/T: Remove the upper base bracket (A) from the base bracket (B)

30. A/T: Disconnect the shift cable from the selector control lever.

31. Attach the front subframe adapter (VSB02C000016) to the subframe by hanging the belt over the front of the subframe, then secure the belt with its stop.

32. Raise the jack and line up the slots in the front subframe adapter arms with the bolt holes on the jack base, then securely attach them with four bolts.

33. Remove the transmission lower mount bolts.

34. Remove the subframe middle mount on both sides.

35. Remove the stiffeners mounting bolt on both sides.

36. Loosen the subframe mounting bolts so they are about 13/16 inches (20 mm)

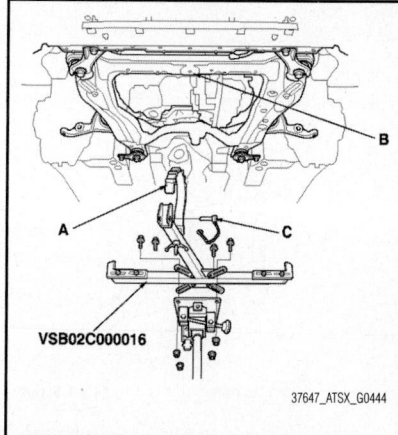

Fig. 226 Attach the front subframe adapter to the subframe by hanging the belt (A) over the front of the subframe (B), then secure the belt with its stop (C)

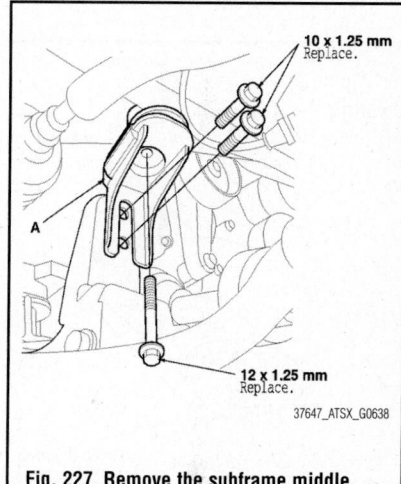

Fig. 227 Remove the subframe middle mount (A) on both sides

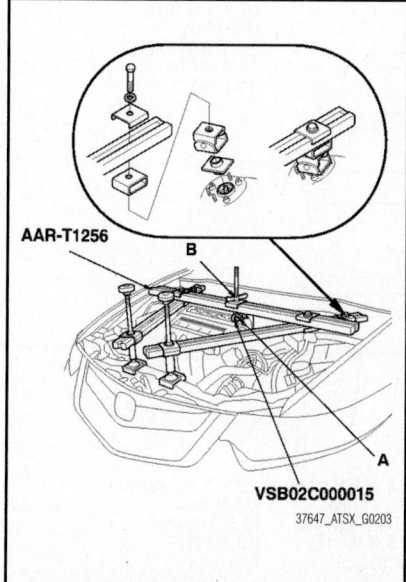

Fig. 223 Install the engine support hanger (AAR-T1256) to the vehicle

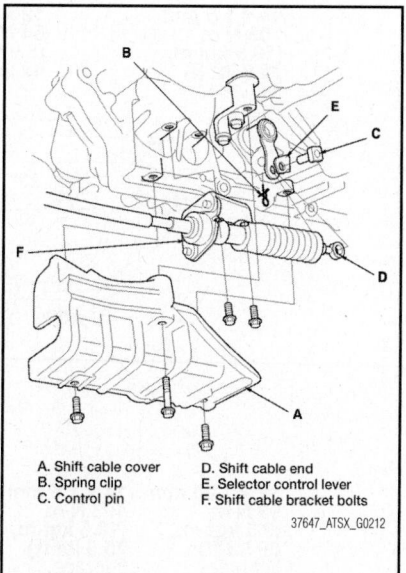

A. Shift cable cover
B. Spring clip
C. Control pin
D. Shift cable end
E. Selector control lever
F. Shift cable bracket bolts

Fig. 225 A/T: Disconnect the shift cable from the selector control lever

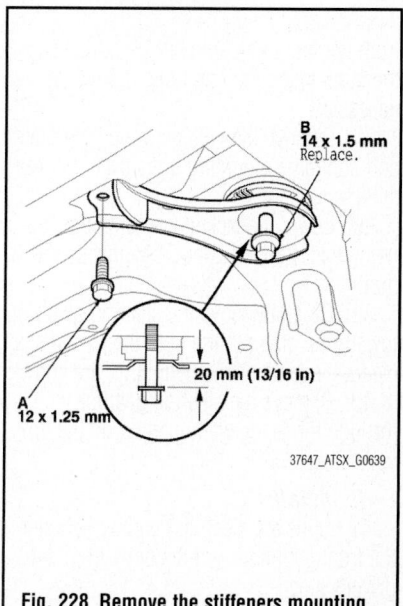

Fig. 228 Remove the stiffeners mounting bolt (A) on both sides

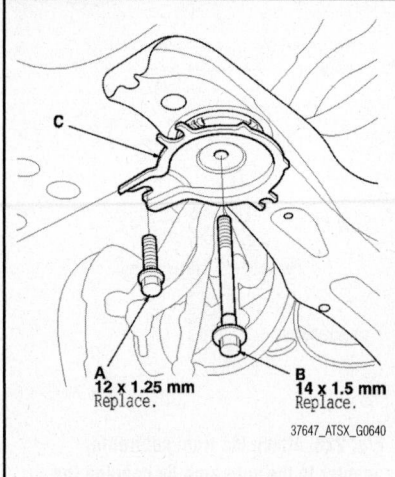

Fig. 229 Remove the stiffener mounting bolts (A) and subframe mounting bolts (B) from the rear stiffeners (C)

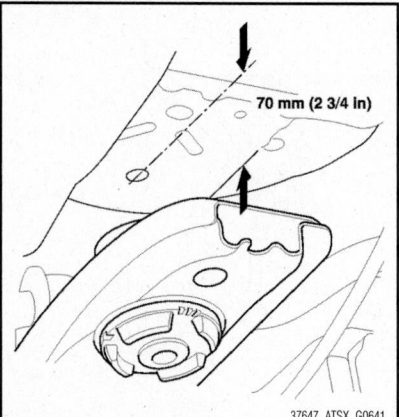

Fig. 230 Lower the jack slowly until the subframe has dropped about 2 ¾ inches (70 mm)

from the mounting surface. Do not loosen the subframe mounting bolts more than necessary.

37. Remove the stiffener mounting bolts and subframe mounting bolts from the rear stiffeners.

38. Lower the jack slowly until the subframe has dropped about 2 ¾ inches (70 mm).

39. Carefully move the steering gearbox toward the driver's side until the pinion shaft clears the fenderwell opening on the body.

40. Remove the steering gearbox through the fenderwell opening on the driver's side.

To install:

41. Slide the steering gearbox between the front subframe and the body from the driver's side.

42. Carefully move the steering gearbox toward the passenger's side until the pinion

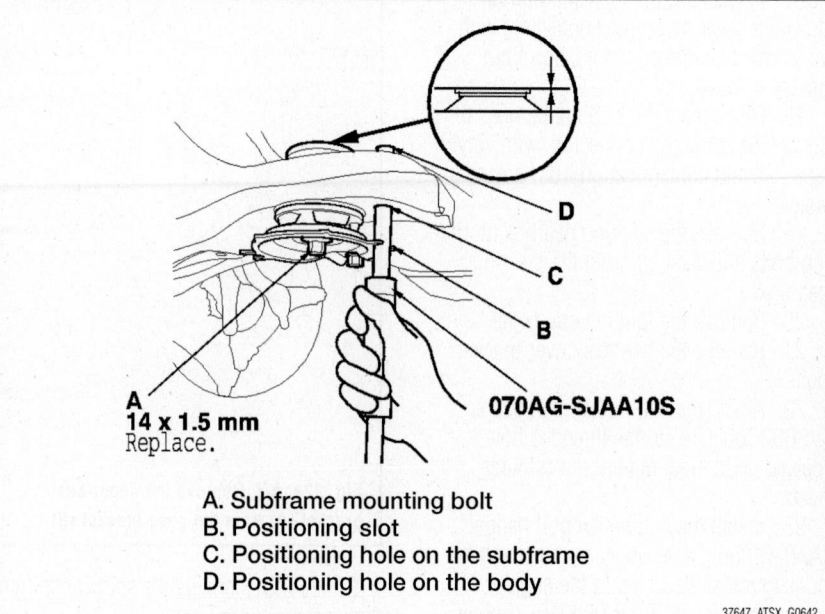

A. Subframe mounting bolt
B. Positioning slot
C. Positioning hole on the subframe
D. Positioning hole on the body

Fig. 231 Insert the subframe alignment pin through the positioning slot on the right rear stiffener, through the positioning hole on the subframe, and into the positioning hole on the body

shaft clears the wheel well opening on the body.

43. Continue moving the steering gearbox toward the passenger's side until the steering gearbox is in position.

44. Install the subframe rear stiffeners, and loosely install the new subframe mounting bolts, new rear stiffener mounting bolts, and front stiffener mounting bolts.

45. Loosely tighten the subframe mounting bolt in the right rear stiffener until the subframe insulator contacts the body; insert

the subframe alignment pin (070AG-SJAA10S) through the positioning slot on the right rear stiffener, through the positioning hole on the subframe, and into the positioning hole on the body.

46. Loosely tighten the left rear subframe mounting bolt with the same procedure as the right rear using the subframe alignment pin.

47. Loosely tighten the subframe mounting bolt and the stiffener mounting bolt on both sides to the front stiffener.

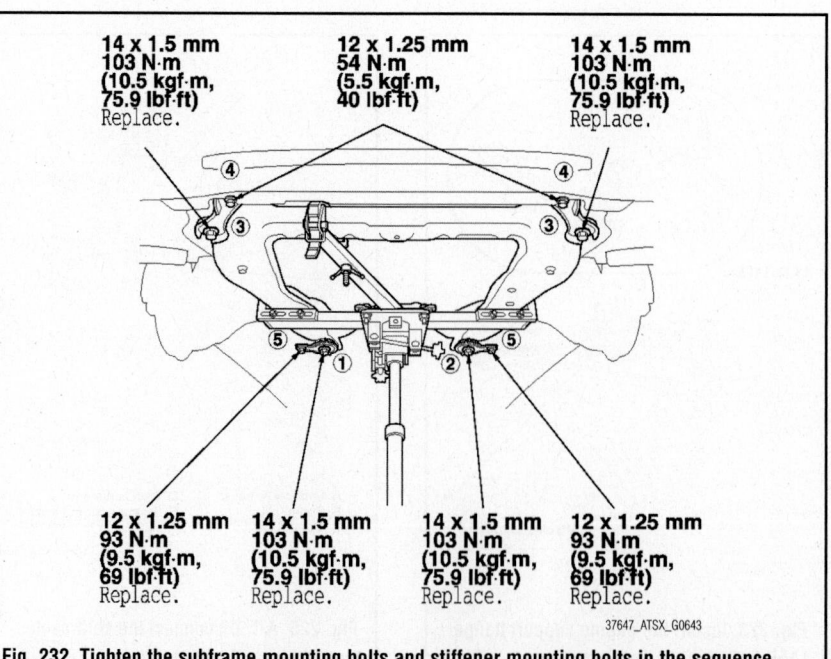

Fig. 232 Tighten the subframe mounting bolts and stiffener mounting bolts in the sequence

48. Tighten the subframe mounting bolts and stiffener mounting bolts to the specified torque in the sequence shown.

➡**Tighten ① and ② in the same procedure as before using the subframe alignment pin. Check all of the subframe mounting bolts, and retighten if necessary. Before tightening the stiffener mounting bolts, check that the positioning holes and slot are aligned using the subframe alignment pin.**

49. Install the transmission lower mount bolts, and tighten them to 40 ft. lbs. (54 Nm).

50. Install the subframe middle mount with new mounting bolts, and tighten the 10 x 1.25 mm bolts to 36 ft. lbs. (49 Nm) and the 12 x 1.25 mm bolts to 33 ft. lbs. (44 Nm) on both sides.

51. Remove the transmission jack supporting the subframe.

52. A/T: Connect the shift cable to the selector control lever.

53. Install the exhaust pipe A.

54. Install the front splash shield.

55. Lower the vehicle.

56. Tighten the new front engine mount bolt to 40 ft. lbs. (54 Nm), then install the front engine mount stop with new mount nuts. Tighten the nuts to 58 ft. lbs (78 Nm).

57. A/T: Install the upper base bracket to the base bracket with a new mounting bolt and mounting bolts, then tighten them to 40 ft. lbs. (54 Nm).

58. Install the rear engine mount with new mounting bolts, and tighten the 10 x 1.25 mm bolts to 40 ft. lbs. (54 Nm) and the 12 x 1.25 mm bolt to 58 ft. lbs. (78 Nm).

59. Remove the engine support hanger.

60. Remove the engine hanger adapter.

61. Install the harness cover bracket.

62. Install the ignition coil cover.

63. Install the washers and new steering gearbox mounting bolts to the steering gearbox, then loosely tighten them.

➡**Make sure that the washers are in the correct position.**

64. Tighten the steering gearbox mounting bolts to 54 ft. lbs. (74 Nm) in the sequence shown.

➡**Check all of the steering gearbox mounting bolts, and retighten if necessary.**

65. Connect the EPS subharness 6P connector to the steering gearbox.

66. Remove the vinyl tape, then connect the EPS motor angle sensor 6P connector to the steering gearbox.

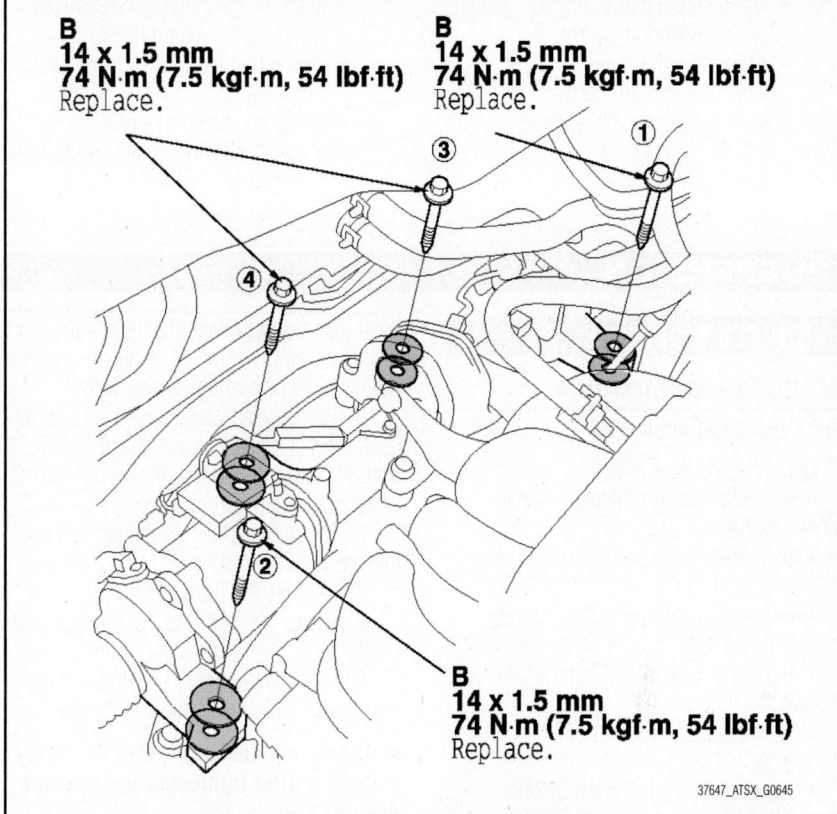

B
14 x 1.5 mm
74 N·m (7.5 kgf·m, 54 lbf·ft)
Replace.

B
14 x 1.5 mm
74 N·m (7.5 kgf·m, 54 lbf·ft)
Replace.

B
14 x 1.5 mm
74 N·m (7.5 kgf·m, 54 lbf·ft)
Replace.

37647_ATSX_G0645

Fig. 233 Tighten the steering gearbox mounting bolts (B) in sequence

67. Connect the EPS motor 3P connector, then pull up the lever, and confirm the connector is fully seated.

68. Install the P/S heat shield with the flange bolts.

69. Install the strut brace.

70. Install the front grille cover.

71. Reinstall the hood support rod to the upper location of the hood.

72. Wipe off any grease contamination from the ball joint tapered section and threads. Reconnect the tie-rod end ball joint to the knuckle. Install the nut, and tighten it to 40 ft. lbs. (54 Nm).

73. Install a new cotter pin, and bend it.

74. Install the front wheels, then set the wheels in the straight ahead position.

➡**Before installing the wheel, clean the mating surface between the brake disc and the inside of the wheel.**

75. Remove the vinyl tape around the splines on the pinion shaft.

76. Loosen the upper steering joint bolt and slip the lower end of the steering joint onto the pinion shaft taking care to align the gap within the angle of 35 degrees +/- 20 degrees.

➡**Pick up the tabs of the pinion shaft grommet, and turn up the lip of the**

pinion shaft grommet securely in place. Make sure that light does not enter from the space between the pinion shaft grommet and the body.

77. Align the bolt hole on the steering joint with the groove around the pinion shaft, then loosely install the lower steering joint bolt. Be sure that the joint bolt is securely in the groove in the pinion shaft.

78. Pull on the steering joint to make sure that the steering joint is fully seated, then tighten the lower joint bolt to the specified torque.

79. Tighten the upper steering joint bolt to 21 ft. lbs. (28 Nm).

80. Install the steering joint cover.

81. Install the steering wheel, and the driver's airbag.

82. With the tires raised off the ground (vehicle on a lift), check for rubbing or grating noises by turning the steering wheel fully to the right and left several times.

83. Do the battery terminal reconnection procedure, and check these items:

- Turn the ignition switch to ON (II), and check that the SRS indicator comes on for about 6 seconds, and then goes off.

- Make sure the horn and turn signal switches work properly.
- Make sure the steering wheel switches work properly.

84. After installation, check these items:
- Start the engine, allow it to idle, and turn the steering wheel from lock to lock several times.
- Check that the EPS indicator does not come on.
- Check the steering wheel spoke angle. If steering spoke angles to the right and left are not equal (steering wheel and rack are not centered), correct the engagement of the joint/pinion shaft serrations.

85. Do the torque sensor neutral position memorization.

86. Check the wheel alignment, and adjust it if necessary.

SUSPENSION

FRONT SUSPENSION

KNUCKLE & SPINDLE

REMOVAL & INSTALLATION

See Figures 234 through 237.

Special Tools Required:
- Ball joint thread protector, 14 mm 07AAE-SJAA100
- Ball joint thread protector, 12 mm 07AAF-SDAA100
- Ball joint thread protector, 10 mm 07AAF-SECA120
- Ball joint remover, 28 mm 07MAC-SL0A202
- Hub dis/assembly tool 07GAF-SD4A100
- Attachment, 72 x 75 mm 07746-0010600
- Driver handle 07749-0010000
- Attachment, 80 x 96 mm 07ZAD-PNA0100
- Support base 07965-SD90100

1. Raise the vehicle on a lift.
2. Remove the wheel nuts and the front wheel.
3. Remove the brake hose bracket mounting bolt.
4. Remove the brake caliper bracket mounting bolts, then remove the caliper assembly from the knuckle. To prevent damage to the caliper assembly or the brake hose, use a short piece of wire to hang the caliper assembly from the undercarriage. Do not twist the brake hose excessively.

5. Remove the wheel speed sensor harness bracket and the wheel speed sensor from the knuckle. Do not disconnect the wheel speed sensor connector.
6. Pry up the stake on the spindle nut, then remove the nut.
7. Remove the front brake disc.
8. Check the front hub for damage and cracks.
9. Remove the cotter pin from the tie-rod end ball joint, then remove the nut.

➡**During installation, install the new cotter pin after tightening the nut, and bend its end.**

10. Disconnect the tie-rod end ball joint from the knuckle using the ball joint thread protector and the ball joint remover.
11. Remove the cotter pin from the knuckle ball joint, then remove the castle nut.

➡**During installation, insert the new cotter pin into the ball joint pin hole from the front to the rear of the vehicle, and bend its end as shown. Check the ball joint pin hole direction before connecting the ball joint.**

12. Disconnect the knuckle ball joint

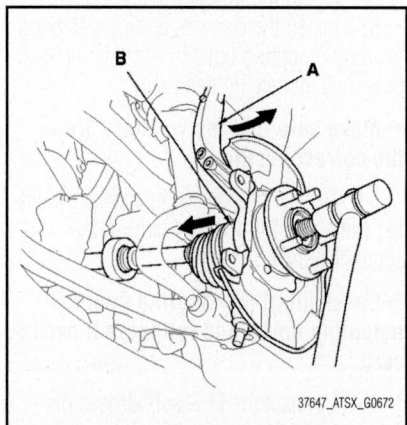

Fig. 236 Remove the cotter pin (A) from the upper arm ball joint, then remove the castle nut (B)

from the lower arm using the ball joint thread protector and the ball joint remover.

➡**Be careful not to damage the ball joint boot when installing the remover. Do not force or hammer on the lower arm, or pry between the lower arm and the knuckle. You could damage the ball joint.**

13. Remove the cotter pin from the upper arm ball joint, then remove the castle nut.

➡**During installation, insert the new cotter pin into the ball joint pin hole from the front to the rear of the vehicle,**

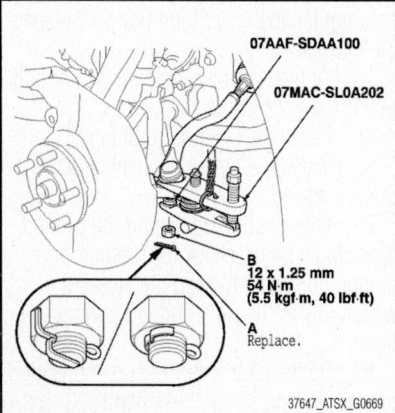

Fig. 234 Remove the cotter pin (A) from the tie-rod end ball joint, then remove the nut (B)

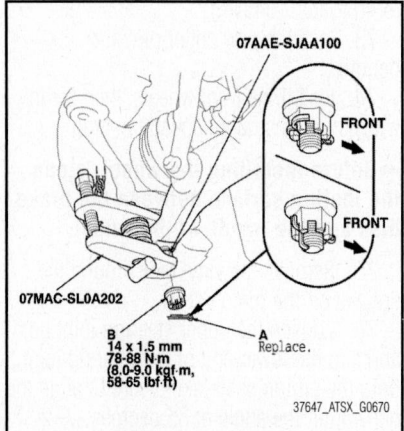

Fig. 235 Remove the cotter pin (A) from the knuckle ball joint, then remove the castle nut (B)

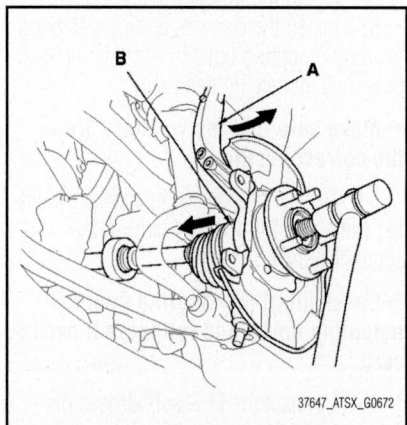

Fig. 237 Pull the knuckle (A) outward, and separate the outboard joint (B) from the front hub using a plastic hammer

and bend its end as shown. Check the ball joint pin hole direction before connecting the ball joint.

14. Disconnect the upper arm ball joint from the knuckle using the ball joint thread protector and the ball joint remover.

15. Pull the knuckle outward, and separate the outboard joint from the front hub using a plastic hammer.

➡**Do not pull the driveshaft end outward. The driveshaft inboard joint may come apart. During installation, apply grease to the mating surfaces of the wheel bearing and the driveshaft outboard joint.**

To install:

16. Install the knuckle/hub in the reverse order of removal, and note these items:
- First install all of the components, and lightly tighten the bolts and the nuts, then raise the suspension to load it with the vehicle's weight before fully tightening to the specified torque. Do not place the jack against the ball joint pin of the knuckle.
- Be careful not to damage the ball joint boot when connecting the knuckle.
- Before connecting the ball joint, degrease the threaded section and the tapered portion of the ball joint pin, the ball joint connecting hole, the threaded section, and the mating surfaces of the castle nut.
- Torque the castle nut to the lower torque specification, then tighten it only far enough to align the slot with the ball joint pin hole. Do not align the castle nut by loosening it.
- Use a new spindle nut on reassembly.
- Before installing the spindle nut, apply a small amount of engine oil to the seating surface of the nut. After tightening, use a drift to stake the spindle nut shoulder against the driveshaft.
- Before installing the brake disc, clean the mating surfaces of the front hub and the inside of the brake disc.
- Before installing the wheel, clean the mating surfaces of the brake disc and the inside of the wheel.

17. Check the wheel alignment, and adjust it if necessary.

LOWER BALL JOINTS

REMOVAL & INSTALLATION

See Figures 238 and 239.

Special Tools Required:
- Ball joint thread protector, 14 mm 07AAE-SJAA100

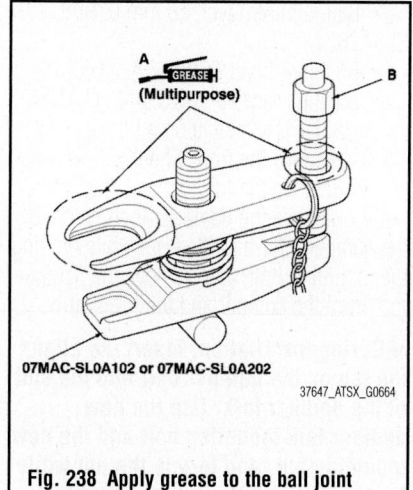

Fig. 238 Apply grease to the ball joint remover on the areas shown (A)

- Ball joint thread protector, 12 mm 07AAF-SDAA100
- Ball joint thread protector, 10 mm 07AAF-SECA120
- Ball joint remover, 32 mm 07MAC-SL0A102
- Ball joint remover, 28 mm 07MAC-SL0A202

➡**Always use a ball joint remover to disconnect a ball joint. Do not strike the housing or any other part of the ball joint connection to disconnect it.**

1. Install a hex nut or the ball joint thread protector onto the threads of the ball joint.

➡**Using a hex nut, make sure the nut is flush with the ball joint pin end to prevent damage to the threaded end of the ball joint pin.**

2. Apply grease to the ball joint remover on the areas shown. This will ease the installation of the tool, and prevent damage to the pressure bolt threads.

3. Loosen the pressure bolt, and install the ball joint remover as shown. Insert the jaws carefully, making sure not to damage the ball joint boot. Adjust the jaw spacing by turning the adjusting bolt.

➡**Fasten the safety chain securely to a suspension arm or the subframe. Do not fasten it to a brake line or wire harness.**

4. After adjusting the adjusting bolt, make sure the head of the adjusting bolt is in the position shown to allow the jaw to pivot.

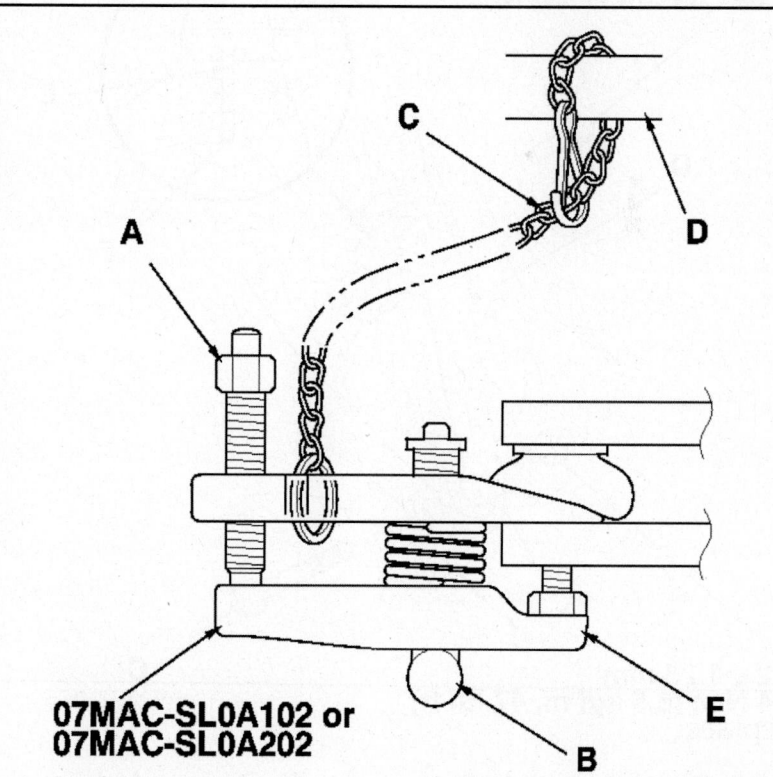

07MAC-SL0A102 or 07MAC-SL0A202

A. Pressure bolt
B. Adjusting bolt
C. Safety chain
D. Suspension arm or subframe
E. Jaw

Fig. 239 Loosen the pressure bolt, and install the ball joint remover as shown

5. With a wrench, tighten the pressure bolt until the ball joint pin pops loose from the ball joint connecting hole. If necessary, apply penetrating type lubricant to loosen the ball joint pin.

➡ **Do not use pneumatic or electric tools on the pressure bolt.**

6. Remove the ball joint remover, then remove the nut or the ball joint thread protector from the end of the ball joint pin, and pull the ball joint out of the ball joint connecting hole. Inspect the ball joint boot, and replace it if damaged.

LOWER CONTROL ARMS

REMOVAL & INSTALLATION

See Figures 240 and 241.

Special Tools Required:
- Ball joint thread protector, 14 mm 07AAE-SJAA100

- Ball joint remover, 28 mm 07MAC-SL0A202
- Bushing driver 070AF-TA0A100
- Bushing receiver set 070AF-TA0A220

1. Raise the vehicle on a lift.
2. Remove the front wheel.
3. Remove the front splash shield.
4. Remove the damper pinch bolt and the damper fork mounting nut while holding the mounting bolt, then remove the damper fork from the damper and the lower arm.

➡ **During installation, insert the aligning tab on the damper unit into the slot of the damper fork. Use the new damper fork mounting bolt and the new mounting nut, and torque the nut while holding the bolt during reassembly.**

5. Disconnect the stabilizer link from the lower arm.
6. Remove the cotter pin from the knuckle ball joint, then remove the castle nut.

➡ **During installation, insert the new cotter pin into the ball joint pin hole from the front to the rear of the vehicle, and bend it. Check the ball joint pin hole direction before connecting the ball joint.**

7. Disconnect the knuckle ball joint from the lower arm using the ball joint thread protector and the ball joint remover.

➡ **Be careful not to damage the ball joint boot when installing the remover. Do not force or hammer on the lower arm, or pry between the lower arm and the knuckle. You could damage the ball joint.**

8. Remove the lower arm mounting bolts, and remove the lower arm.

➡ **Use new lower arm mounting bolts during reassembly.**

To install:

9. Install the lower arm in the reverse order of removal, and note these items:
- First install all of the components, and lightly tighten the bolts and the nuts, then raise the suspension to load it with the vehicle's weight before fully tightening it to the specified torque. Do not place the jack against the ball joint pin of the knuckle.
- Be careful not to damage the ball joint boot when connecting the knuckle.
- Before connecting the ball joint, degrease the threaded section and the tapered portion of the ball joint pin, the ball joint connecting hole, the threaded section, and the mating surfaces of the castle nut.
- Torque the castle nut to the lower torque specification, 58–65 ft. lbs. (78–88 Nm), then tighten it only far

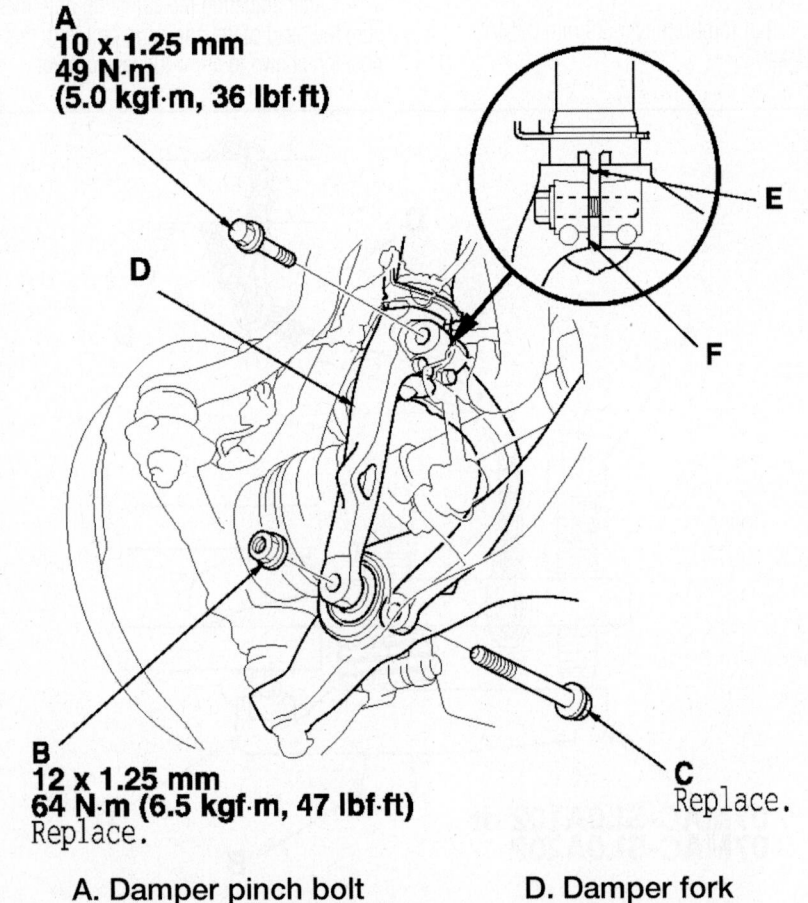

A
10 x 1.25 mm
49 N·m
(5.0 kgf·m, 36 lbf·ft)

B
12 x 1.25 mm
64 N·m (6.5 kgf·m, 47 lbf·ft)
Replace.

C
Replace.

A. Damper pinch bolt
B. Damper fork mounting nut
C. Mounting bolt
D. Damper fork
E. Aligning tab
F. Slot

37647_ATSX_G0666

Fig. 240 Remove the damper pinch bolt and the damper fork mounting nut while holding the mounting bolt

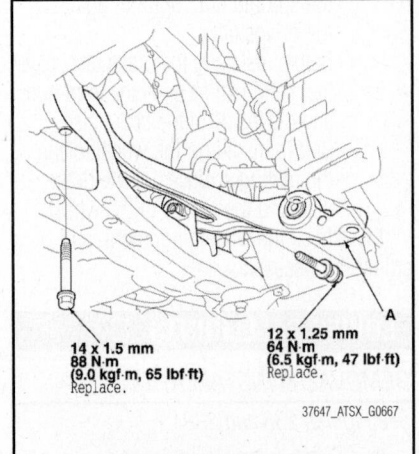

14 x 1.5 mm
88 N·m
(9.0 kgf·m, 65 lbf·ft)
Replace.

12 x 1.25 mm
64 N·m
(6.5 kgf·m, 47 lbf·ft)
Replace.

37647_ATSX_G0667

Fig. 241 Remove the lower arm mounting bolts, and remove the lower arm (A)

enough to align the slot with the ball joint pin hole. Do not align the castle nut by loosening it.

- Before installing the wheel, clean the mating surfaces of the brake disc and the inside of the wheel.

10. Check the wheel alignment, and adjust it if necessary.

STABILIZER BAR & LINKS

REMOVAL & INSTALLATION

Stabilizer Links

See Figure 242.

1. Raise the vehicle on a lift.
2. Remove the front wheel.
3. Remove the self-locking nut and the flange nut while holding the respective joint pin with a hex wrench, then remove the stabilizer link.

To install:

4. Install the stabilizer link on the stabilizer bar and the lower arm with the joint pins set at the center of their range of movement.

➡**The stabilizer link has a paint mark. The left stabilizer link is marked with yellow paint, and the right stabilizer link is marked with white paint.**

5. Install the new self-locking nut and the new flange nut, and tighten them to the specified torque while holding the respective joint pin with a hex wrench.
6. Clean the mating surfaces of the brake disc and the inside of the wheel, then install the front wheel.
7. Test-drive the vehicle.
8. After 5 minutes of driving, tighten the self-locking nut again to the specified torque.

Stabilizer Bar

See Figure 243.

1. Raise the vehicle on a lift.
2. Remove the front wheels.

3. Disconnect both tie-rod ball joints from the knuckle.
4. Disconnect both stabilizer links from the stabilizer bar.
5. Remove the flange bolts and the bushing holders, then remove the bushings and the stabilizer bar.

To install:

➡**During installation, align the paint marks on the stabilizer bar with the side of the bushings.**

6. Install the stabilizer bar in the reverse order of removal, and note these items:

- Note the right and left direction of the stabilizer bar.
- Note the direction of installation for the bushings.
- Before installing the wheel, clean the mating surfaces of the brake disc and the inside of the wheel.

7. Check the wheel alignment, and adjust it if necessary.

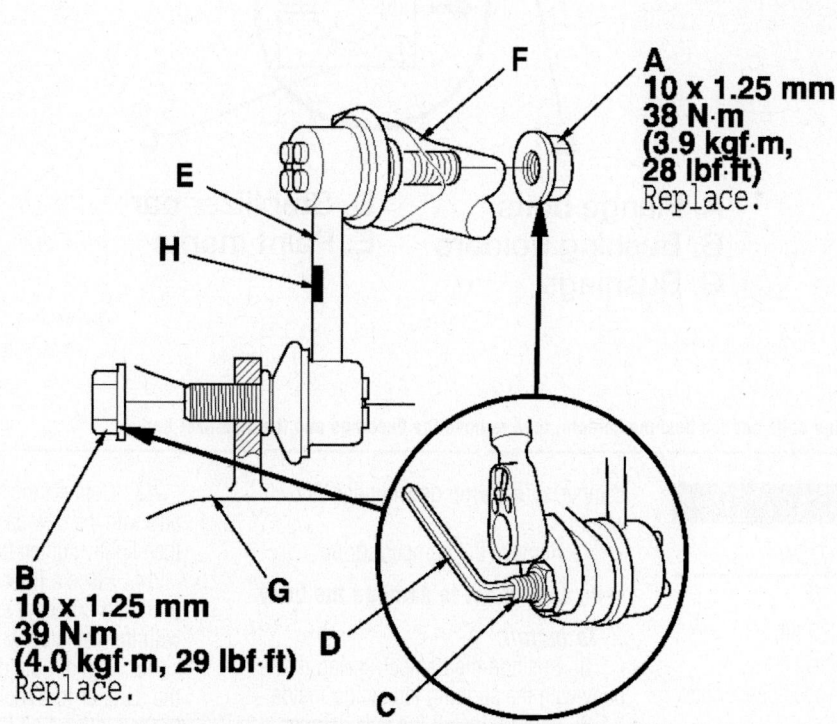

A **10 x 1.25 mm 38 N·m (3.9 kgf·m, 28 lbf·ft) Replace.**

B **10 x 1.25 mm 39 N·m (4.0 kgf·m, 29 lbf·ft) Replace.**

A. Self-locking nut
B. Flange nut
C. Joint pin
D. Hex wrench
E. Stabilizer link
F. Stabilizer bar
G. Lower arm
H. Paint mark

37647_ATSX_G0662

Fig. 242 Front stabilizer link

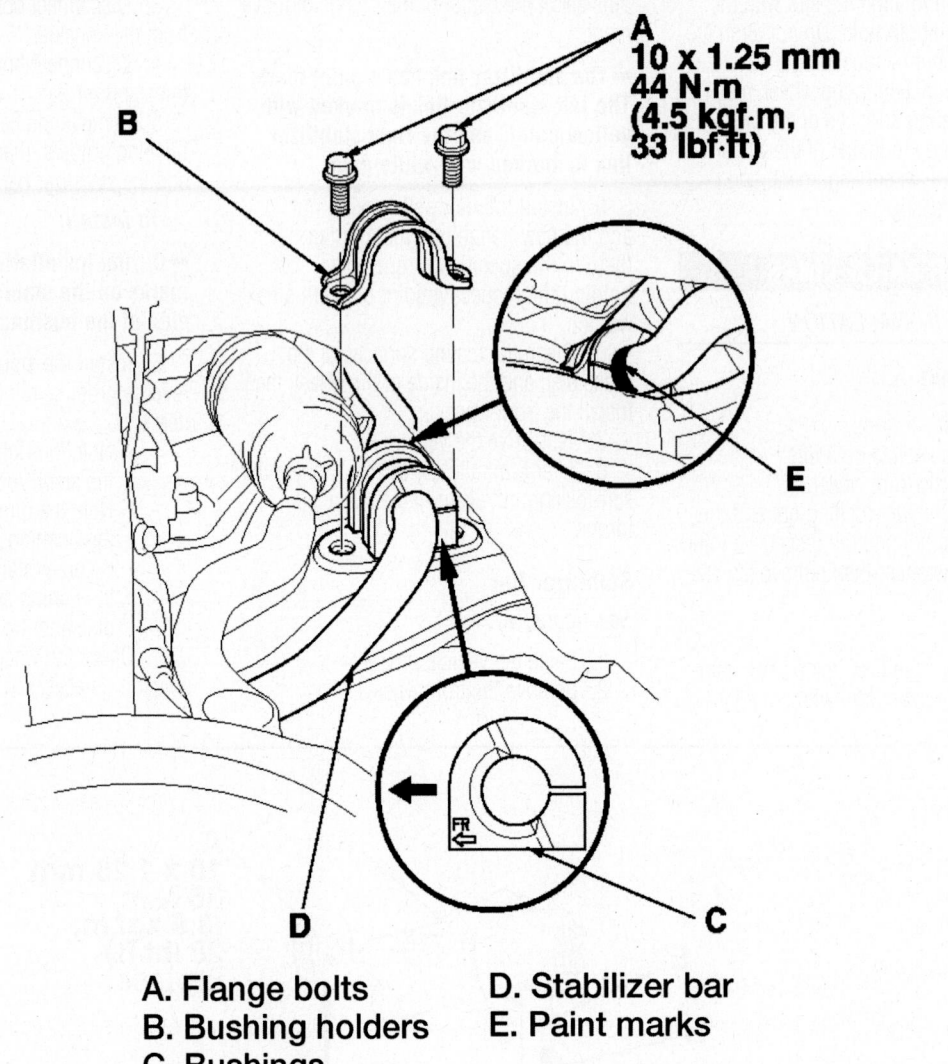

A. Flange bolts
B. Bushing holders
C. Bushings

D. Stabilizer bar
E. Paint marks

37647_ATSX_G0679

Fig. 243 Remove the flange bolts and the bushing holders, then remove the bushings and the stabilizer bar

STRUTS

REMOVAL & INSTALLATION

See Figures 244 through 248.

1. Raise the vehicle on a lift.
2. Remove the front wheel.
3. Remove the wheel speed sensor harness bracket mounting bolt.
4. Remove the damper pinch bolt and the damper fork mounting nut while holding the mounting bolt, then remove the damper fork from the damper and the lower arm.
5. Remove the front strut brace mounting nuts.
6. Remove the damper mounting nuts from the top of the damper. Do not let the

damper/spring drop down under its own weight.
7. Remove the damper/spring.

➡**Be careful not to damage the body.**

To install:

8. Position the damper/spring in the body with the aligning tab facing inside.
9. Loosely install the new damper mounting nuts to the top of the damper.
10. Loosely install the front strut brace mounting nuts.
11. Install the damper fork over the driveshaft and onto the lower arm. Install the aligning tab on the damper unit into the slot of the damper fork.
12. Loosely install the damper pinch bolt into the damper fork.

13. Connect the damper fork and the lower arm with the new damper fork mounting bolt, then lightly tighten the new mounting nut.
14. Place a floor jack under the lower arm, and raise the suspension to load it with the vehicle's weight.
15. Tighten the damper pinch bolt and the damper fork mounting nut while holding the mounting bolt to the specified torque. Damper pinch bolt: tighten to 36 ft. lbs. (49 Nm); damper fork mounting nut: tighten to 47 ft. lbs. (64 Nm).
16. Tighten the damper mounting nuts and front strut brace mounting nuts on top of the damper to the specified torque. Damper mounting nuts: tighten to 41 ft. lbs. (55 Nm); strut brace mounting nuts: tighten to 16 ft. lbs. (22 Nm).

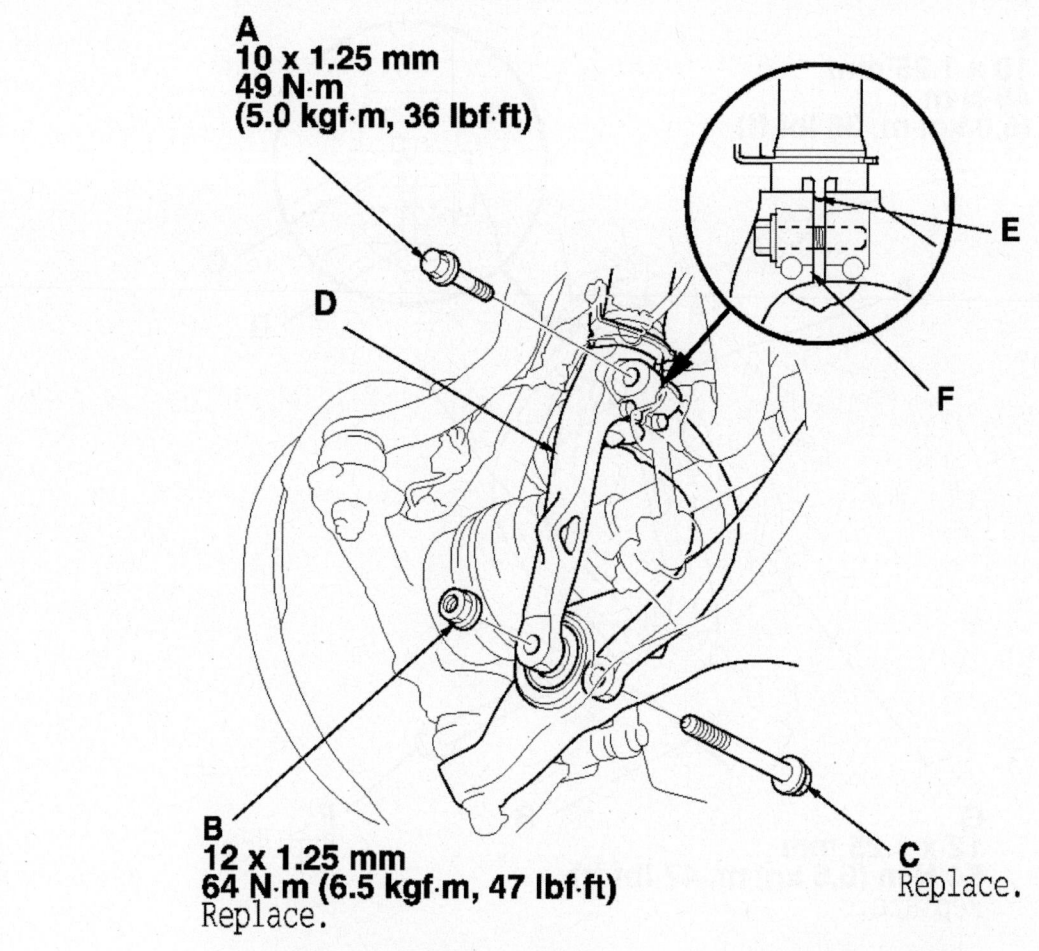

A
10 x 1.25 mm
49 N·m
(5.0 kgf·m, 36 lbf·ft)

E

D

F

B
12 x 1.25 mm
64 N·m (6.5 kgf·m, 47 lbf·ft)
Replace.

C
Replace.

A. Damper pinch bolt
B. Damper fork mounting nut
C. Mounting bolt
D. Damper fork
E. Aligning tab
F. Slot

37647_ATSX_G0666

Fig. 244 Remove the damper pinch bolt and the damper fork mounting nut while holding the mounting bolt

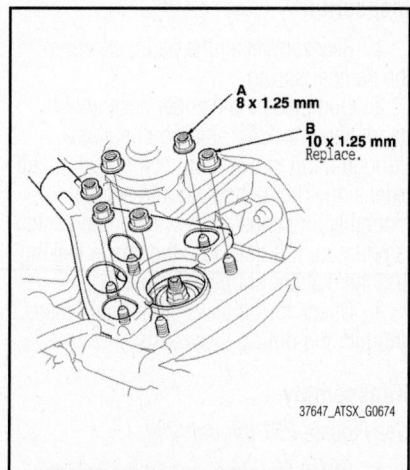

A
8 x 1.25 mm

B
10 x 1.25 mm
Replace.

37647_ATSX_G0674

Fig. 245 Remove the front strut brace mounting nuts (A) and the damper mounting nuts (B)

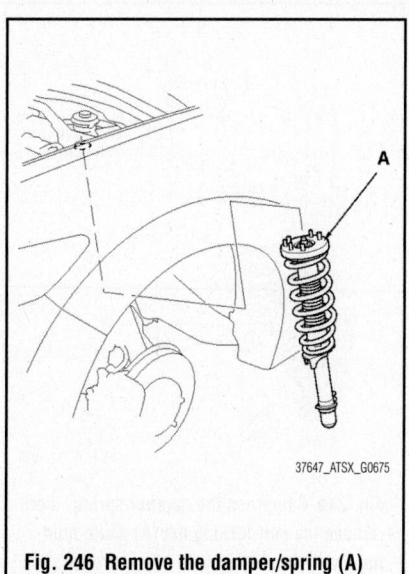

A

37647_ATSX_G0675

Fig. 246 Remove the damper/spring (A)

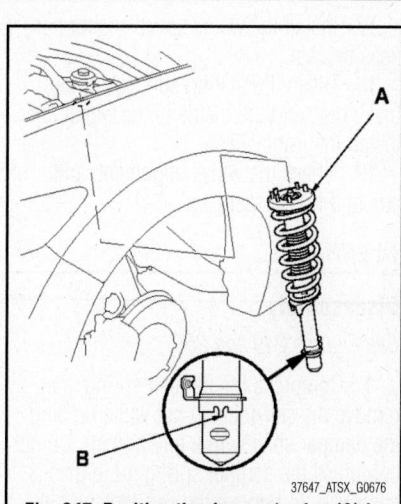

A

B

37647_ATSX_G0676

Fig. 247 Position the damper/spring (A) in the body with the aligning tab (B) facing inside

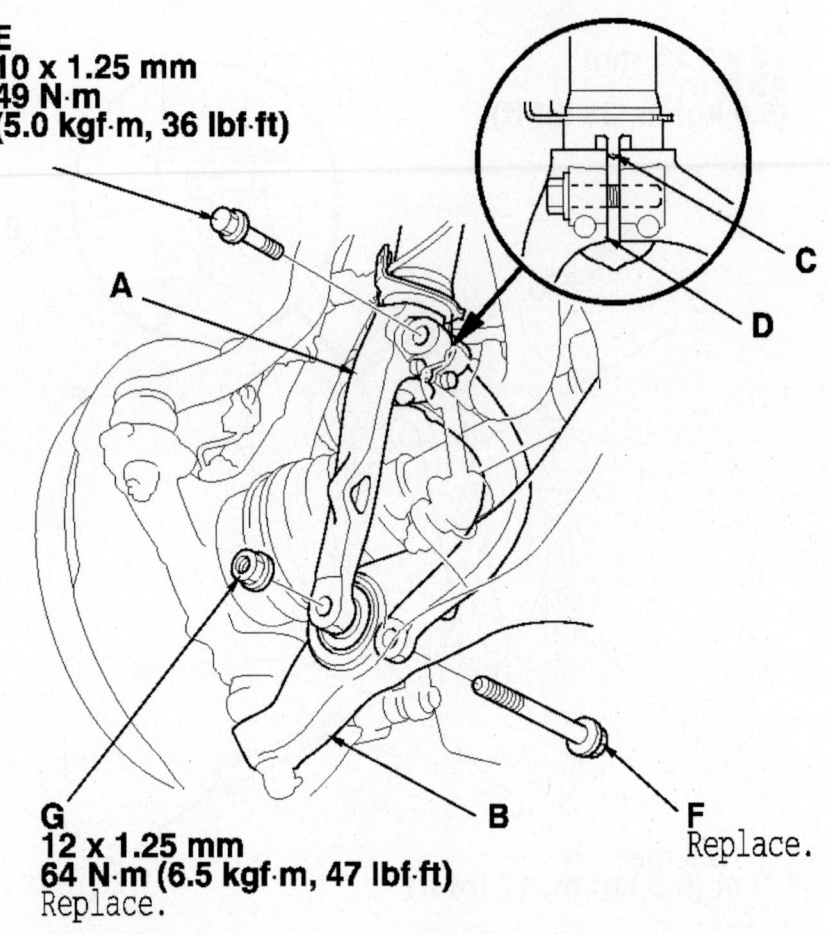

E
10 x 1.25 mm
49 N·m
(5.0 kgf·m, 36 lbf·ft)

C

D

A

G
12 x 1.25 mm
64 N·m (6.5 kgf·m, 47 lbf·ft)
Replace.

B

F
Replace.

A. Damper fork
B. Lower arm
C. Aligning tab
D. Slot

E. Damper pinch bolt
F. Damper fork mounting bolt
G. Mounting nut

37647_ATSX_G0677

Fig. 248 Installing damper fork

17. Install the wheel speed sensor harness bracket.

18. Clean the mating surfaces of the brake disc and the inside of the wheel, then install the front wheel.

19. Check the wheel alignment, and adjust it if necessary.

OVERHAUL

Disassembly

See Figures 249 and 250.

1. Compress the damper spring, then remove the self-locking nut while holding the damper shaft with a hex wrench. Do not compress the damper spring more than is necessary to remove the self-locking nut.

2. Release the pressure from the strut spring compressor, then disassemble the damper as shown in the Exploded View.

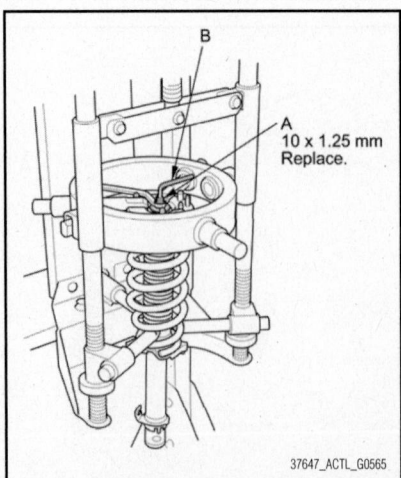

B

A
10 x 1.25 mm
Replace.

37647_ACTL_G0565

Fig. 249 Compress the damper spring, then remove the self-locking nut (A) while holding the damper shaft with a hex wrench (B)

Inspection

1. Reassemble all the parts, except for the damper spring.

2. Compress the damper assembly by hand, and check for smooth operation through a full stroke, both compression and extension. The damper should extend smoothly and constantly when compression is released. If it does not, the gas is leaking and the damper should be replaced.

3. Check for oil leaks, abnormal noises, and binding during these tests.

Reassembly

See Figures 251 through 254.

1. Install the spring mounting cushion on the damper mounting base by aligning the tab and notch.

2. Install all the parts except the damper

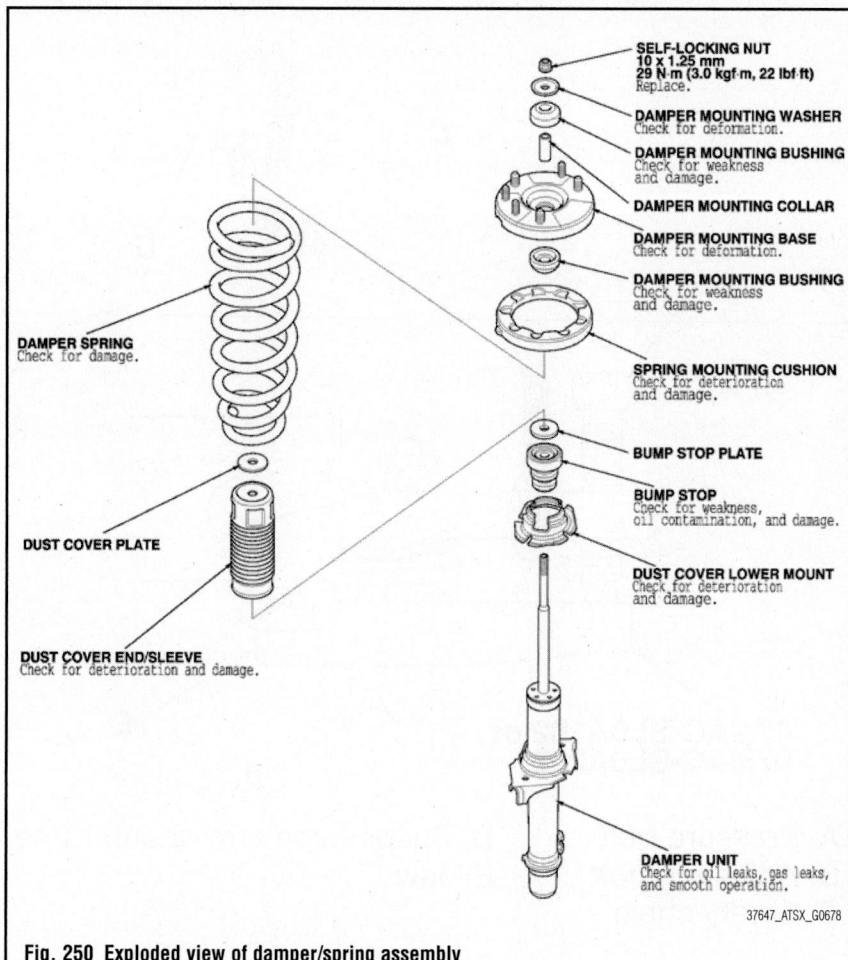

Fig. 250 Exploded view of damper/spring assembly

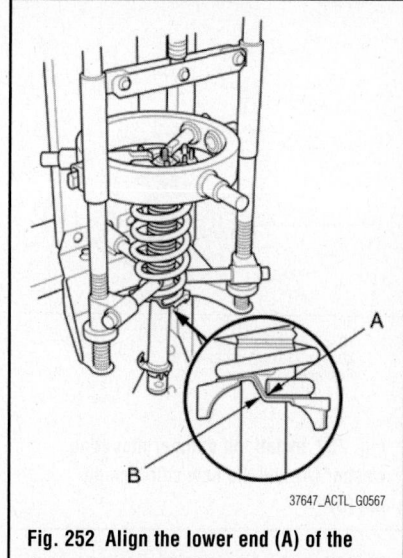

Fig. 252 Align the lower end (A) of the damper spring with the stepped part (B) of the dust cover lower mount

7. Install the damper mounting washer and the new self-locking nut.

8. Hold the damper shaft with a hex wrench, and tighten the self-locking nut to 22 ft. lbs. (29 Nm).

9. Remove the damper/spring from the strut spring compressor.

UPPER BALL JOINTS

REMOVAL & INSTALLATION

See Figures 255 and 256.

Special Tools Required:
• Ball joint thread protector, 14 mm 07AAE-SJAA100
• Ball joint thread protector, 12 mm 07AAF-SDAA100
• Ball joint thread protector, 10 mm 07AAF-SECA120

mounting washer and the self-locking nut onto the damper unit by referring to the Exploded View.

3. Compress the damper spring using a strut spring compressor. Do not compress the spring excessively.

4. Align the lower end of the damper spring with the stepped part of the dust cover lower mount and the lower spring seat on the damper unit.

5. Position the tab on the spring mounting cushion facing forward but toward the inside of the vehicle.

6. Align the angle of the stud on the damper mounting base with the aligning tab on the bottom of the damper unit.

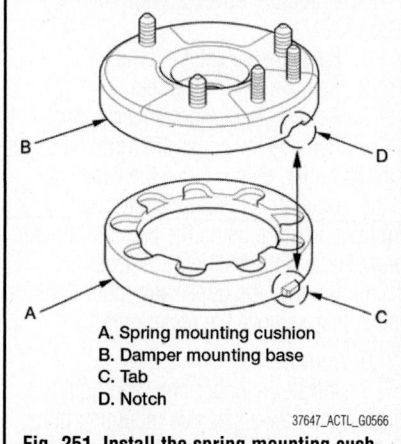

A. Spring mounting cushion
B. Damper mounting base
C. Tab
D. Notch

Fig. 251 Install the spring mounting cushion on the damper mounting base by aligning the tab and notch

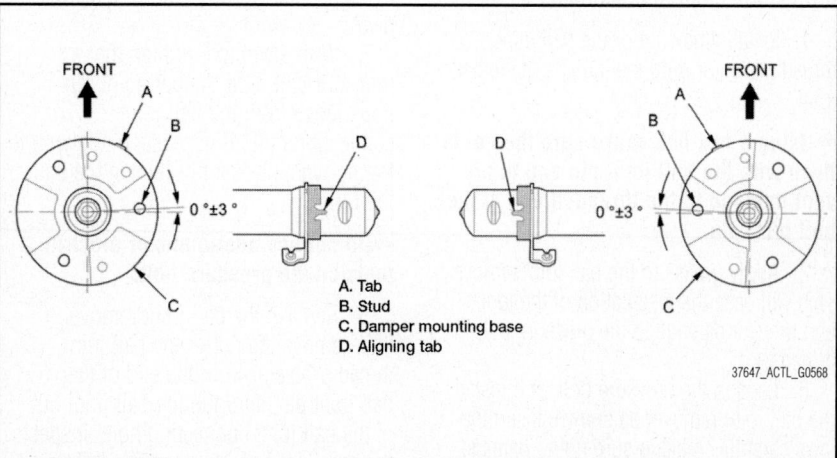

A. Tab
B. Stud
C. Damper mounting base
D. Aligning tab

Fig. 253 Position the tab on the spring mounting cushion facing forward but toward the inside of the vehicle

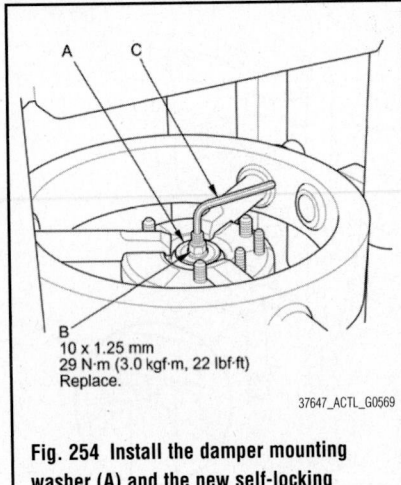

Fig. 254 Install the damper mounting washer (A) and the new self-locking nut (B)

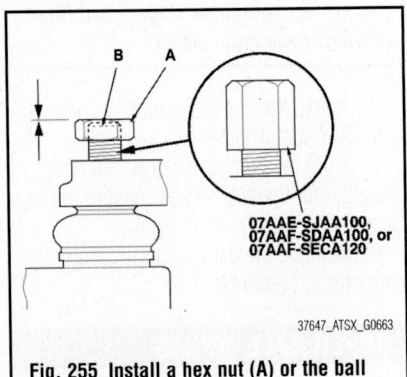

Fig. 255 Install a hex nut (A) or the ball joint thread protector onto the threads of the ball joint (B)

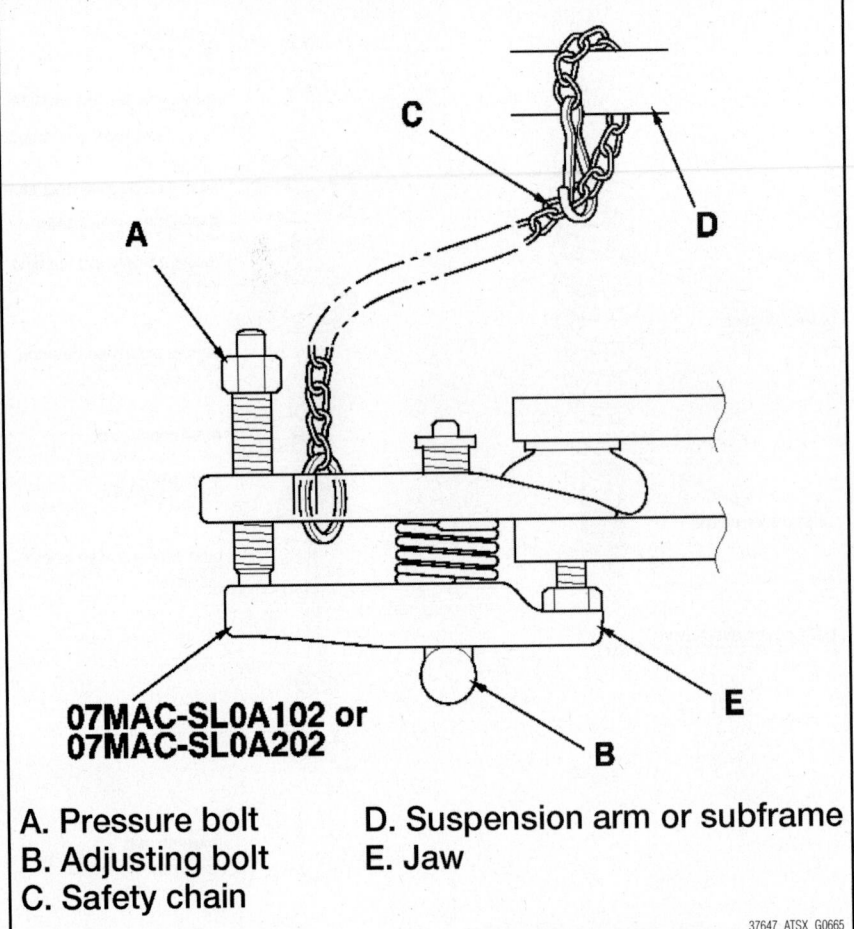

07MAC-SL0A102 or
07MAC-SL0A202

A. Pressure bolt   D. Suspension arm or subframe
B. Adjusting bolt   E. Jaw
C. Safety chain

Fig. 256 Loosen the pressure bolt, and install the ball joint remover as shown

• Ball joint remover, 32 mm 07MAC-SL0A102
• Ball joint remover, 28 mm 07MAC-SL0A202

➡**Always use a ball joint remover to disconnect a ball joint. Do not strike the housing or any other part of the ball joint connection to disconnect it.**

1. Install a hex nut or the ball joint thread protector onto the threads of the ball joint.

➡**Using a hex nut, make sure the nut is flush with the ball joint pin end to prevent damage to the threaded end of the ball joint pin.**

2. Apply grease to the ball joint remover. This will ease the installation of the tool, and prevent damage to the pressure bolt threads.

3. Loosen the pressure bolt, and install the ball joint remover as shown. Insert the jaws carefully, making sure not to damage the ball joint boot. Adjust the jaw spacing by turning the adjusting bolt.

➡**Fasten the safety chain securely to a suspension arm or the subframe. Do not fasten it to a brake line or wire harness.**

4. After adjusting the adjusting bolt, make sure the head of the adjusting bolt is in the position shown to allow the jaw to pivot.

5. With a wrench, tighten the pressure bolt until the ball joint pin pops loose from the ball joint connecting hole. If necessary, apply penetrating type lubricant to loosen the ball joint pin.

➡**Do not use pneumatic or electric tools on the pressure bolt.**

6. Remove the ball joint remover, then remove the nut or the ball joint thread protector from the end of the ball joint pin, and pull the ball joint out of the ball joint connecting hole. Inspect the ball joint boot, and replace it if damaged.

UPPER CONTROL ARMS

REMOVAL & INSTALLATION

See Figures 257 through 259.

Special Tools Required:
• Ball joint thread protector, 10 mm 07AAF-SECA120
• Ball joint remover, 28 mm 07MAC-SL0A202

1. Raise the vehicle on a lift.
2. Remove the front wheel.
3. Remove the front damper/spring.
4. Remove the cotter pin from the upper arm ball joint, then remove the castle nut.
5. Disconnect the upper arm ball joint from the knuckle using the ball joint thread protector and the ball joint remover.
6. Remove the upper arm mounting bolts, then remove the upper arm.

To install:

7. Install the upper arm, and lightly tighten the new upper arm mounting bolts, then connect the knuckle, and lightly tighten the castle nut.

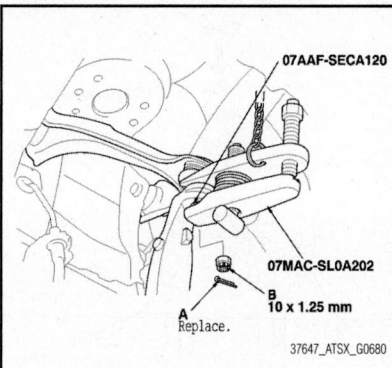

Fig. 257 Remove the cotter pin (A) from the upper arm ball joint, then remove the castle nut (B)

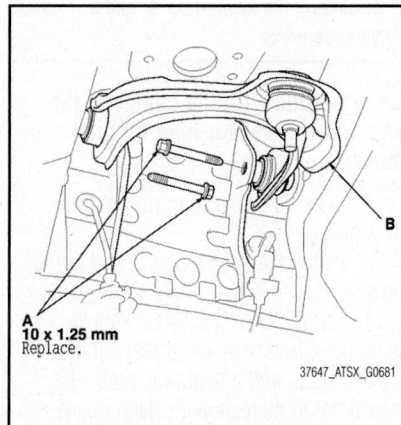

Fig. 258 Remove the upper arm mounting bolts (A), then remove the upper arm (B)

➡ **Be careful not to damage the ball joint boot when connecting the knuckle. Before connecting the ball joint, degrease the threaded section and the tapered portion of the ball joint pin, the ball joint connecting hole, the threaded section**

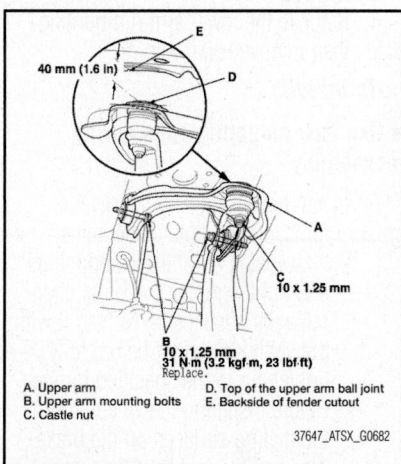

A. Upper arm
B. Upper arm mounting bolts
C. Castle nut
D. Top of the upper arm ball joint
E. Backside of fender cutout

Fig. 259 Install the upper arm, and lightly tighten the new upper arm mounting bolts

and the mating surfaces of the castle nut.

8. Place a floor jack under the lower arm, and raise the suspension until the clearance between the top of the upper arm ball joint and the backside of the fender cut out point is 1.6 inches (40 mm), then tighten the upper arm mounting bolts to 23 ft. lbs. (31 Nm).

➡ **To measure the specified clearance, temporarily remove the front inner fender.**

9. Lower the floor jack
10. Install the front damper/spring.
11. Place the floor jack under the lower arm, and raise the suspension to load it with the vehicle's weight.
12. Tighten the castle nut on the upper arm ball joint to 33–38 ft. lbs. (44–52 Nm).

➡ **Torque the castle nut to the lower torque specification, then tighten it only far enough to align the slot with the ball joint pin hole. Do not align the castle nut by loosening it. Insert the new cotter pin into the ball joint pin hole from the front to the rear of the vehicle, and bend its end. Check the ball joint pin hole direction before connecting the ball joint.**

13. Clean the mating surfaces of the brake disc and the inside of the wheel, then install the front wheel.
14. Check the wheel alignment, and adjust it if necessary.

WHEEL HUBS & BEARINGS

REMOVAL, PACKING, & INSTALLATION

See Figures 260 through 265.

1. Remove the knuckle/hub.
2. Separate the hub from the knuckle using the hub dis/assembly pin and a hydraulic press. Hold the knuckle with the attachment of the hydraulic press or equivalent tool. Be careful not to damage or deform the splash guard. Hold onto the hub to keep it from falling when pressed clear.
3. Press the wheel bearing inner race off of the hub using the hub dis/assembly pin, a commercially available bearing separator, and a press.
4. Remove the snap ring and the splash guard from the knuckle.
5. Press the wheel bearing out of the knuckle using the attachment, the driver handle, and a press.

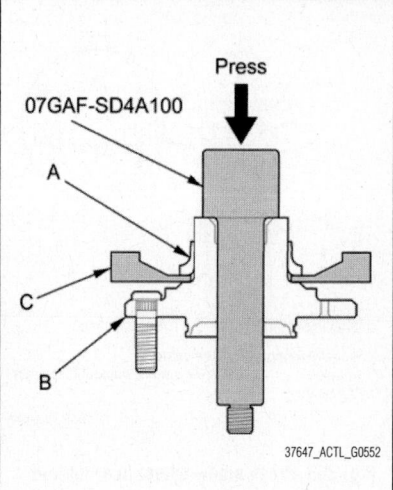

Fig. 260 Press the wheel bearing inner race (A) off of the hub (B) using the hub dis/assembly pin, a commercially available bearing separator (C), and a press

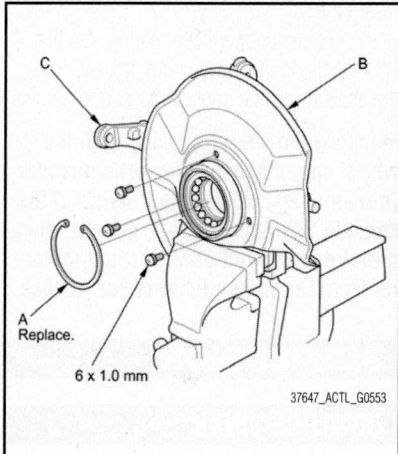

Fig. 261 Remove the snap ring (A) and the splash guard (B) from the knuckle (C)

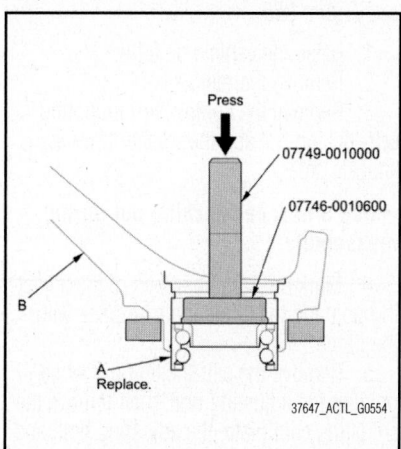

Fig. 262 Press the wheel bearing (A) out of the knuckle (B) using the attachment, the driver handle, and a press

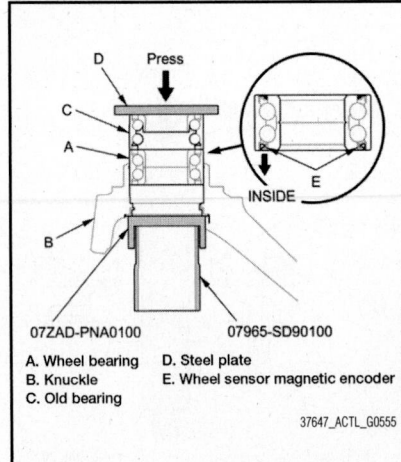

Fig. 263 Press a new wheel bearing into the knuckle using the old bearing, a steel plate, the attachment, the support base, and a press

A. Wheel bearing D. Steel plate
B. Knuckle   E. Wheel sensor magnetic encoder
C. Old bearing

07ZAD-PNA0100   07965-SD90100

37647_ACTL_G0555

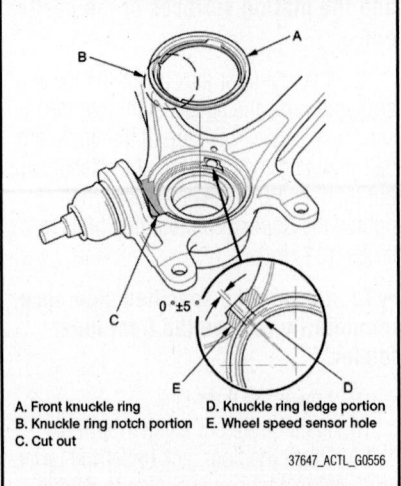

Fig. 264 Check the front knuckle ring for damage or deformation, and replace it if necessary

A. Front knuckle ring   D. Knuckle ring ledge portion
B. Knuckle ring notch portion E. Wheel speed sensor hole
C. Cut out

37647_ACTL_G0556

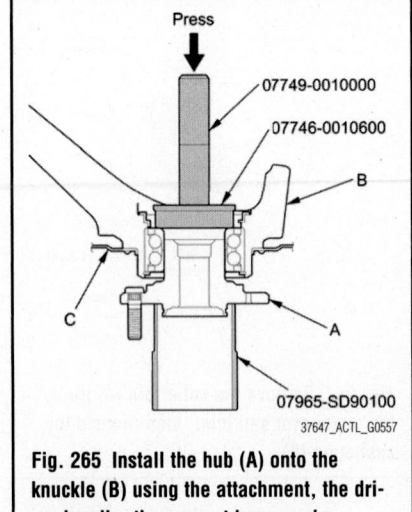

Fig. 265 Install the hub (A) onto the knuckle (B) using the attachment, the driver handle, the support base, and a hydraulic press

07749-0010000
07746-0010600

07965-SD90100

37647_ACTL_G0557

6. Wash the knuckle and the hub thoroughly in high flash point solvent before reassembly.

7. Press a new wheel bearing into the knuckle using the old bearing, a steel plate, the attachment, the support base, and a press.

➡**Install the wheel bearing with the wheel speed sensor magnetic encoder (brown color), toward the inside of the knuckle. Remove any oil, grease, dust, metal debris, and other foreign material from the magnetic encoder surface.**

Keep any magnetic tools away from the magnetic encoder surface. Be careful not to damage the magnetic encoder surface when you insert the wheel bearing.

8. Check the front knuckle ring for damage or deformation, and replace it if necessary.

➡**When installing the new front knuckle ring, position the knuckle ring notch portion toward the cut out near the ball joint in the knuckle, and align the center of the knuckle ring**

ledge portion with the center of the wheel speed sensor hole on the knuckle.

9. Install the new snap ring securely in the knuckle.

10. Install the splash guard, and tighten the screws to 7 ft. lbs. (10 Nm).

11. Install the hub onto the knuckle using the attachment, the driver handle, the support base, and a hydraulic press. Be careful not to damage the splash guard.

12. Install the knuckle/hub.

SUSPENSION

CONTROL LINKS

REMOVAL & INSTALLATION

Rear Control Arm

See Figure 266.

1. Raise the vehicle on a lift.
2. Remove the rear wheel.
3. Remove the control arm mounting self-locking nut and the washer from the knuckle side.

➡**Use a new self-locking nut during reassembly.**

4. Mark the cam positions of the adjusting bolt and the adjusting cam plate with the frame.

5. Remove the self-locking nut while holding the adjusting bolt, then remove the adjusting cam plate, the adjusting bolt, and the control arm.

To install:

➡**Use a new adjusting bolt and a new self-locking nut during reassembly.**

6. Install the control arm in the reverse order of removal, and note these items:

- First install all of the components, and lightly tighten the bolts and the nuts, then raise the suspension to load it with the vehicle's weight before fully tightening to the specified torque.
- Position the extended surfaces of the cam on the adjusting bolt and the adjusting cam plate facing down.
- Align the cam positions of the adjusting bolt and the adjusting cam plate with the marked positions on the frame when tightening the self-locking nut.
- Before installing the wheel, clean the mating surfaces on the brake disc and the inside of the wheel.

7. Check the wheel alignment, and adjust it if necessary.

REAR SUSPENSION

Rear Lower Arm A

See Figure 267.

1. Raise the vehicle on a lift.
2. Remove the rear wheel.
3. Remove the parking brake cable mounting bolt.
4. Remove the lower arm A mounting bolts, then remove lower arm A.

To install:

➡**Use new mounting bolts during reassembly.**

5. Install lower arm A in the reverse order of removal, and note these items:

- First install all of the components, and lightly tighten the bolts, then raise the suspension to load it with the vehicle's weight before fully tightening to the specified torque.
- Before installing the wheel, clean the mating surfaces on the brake disc and the inside of the wheel.

6. Check the wheel alignment, and adjust it if necessary.

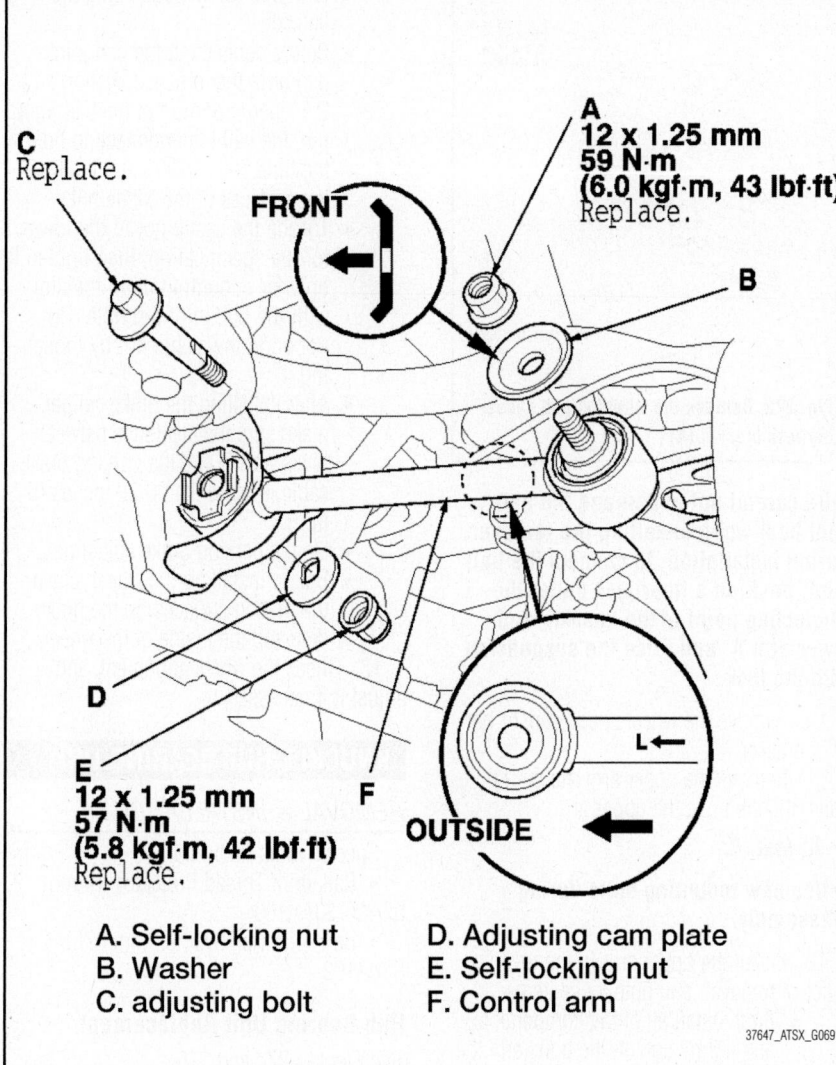

A. Self-locking nut
B. Washer
C. adjusting bolt
D. Adjusting cam plate
E. Self-locking nut
F. Control arm

37647_ATSX_G0691

Fig. 266 Remove the self-locking nut while holding the adjusting bolt, then remove the adjusting cam plate, the adjusting bolt, and the control arm

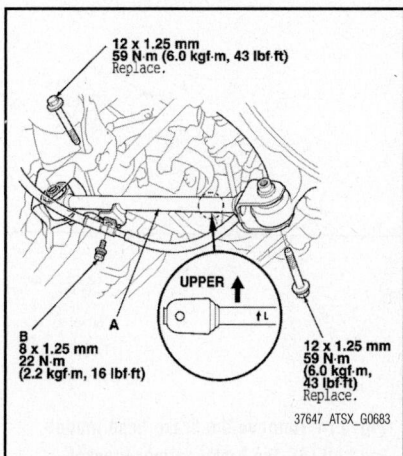

37647_ATSX_G0683

Fig. 267 Remove the lower arm A mounting bolts, then remove lower arm A

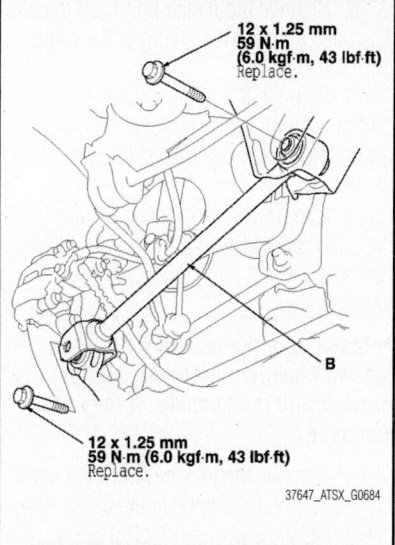

37647_ATSX_G0684

Fig. 268 Remove the lower arm B mounting bolts, then remove lower arm B

Rear Lower Arm B

See Figure 268.

1. Raise the vehicle on a lift.
2. Remove the rear wheel.
3. Remove the lower arm B mounting bolts, then remove lower arm B.

To install:

➡**Use new mounting bolts during reassembly.**

4. Install lower arm B in the reverse order of removal, and note these items:
- First install all of the components, and lightly tighten the bolts, then raise the suspension to load it with the vehicle's weight before fully tightening to the specified torque.
- Make sure the clearance between

lower arm B and the parking brake cable is more than 0.20 inches (5 mm).
- Before installing the wheel, clean the mating surfaces on the brake disc and the inside of the wheel.

5. Check the wheel alignment, and adjust it if necessary.

Rear Upper Arm

See Figures 269 through 273.

Special Tools Required:
- Ball joint thread protector, 14 mm 07AAE-SJAA100
- Ball joint remover, 32 mm 07MAC-SL0A102

1. Raise the vehicle on a lift.
2. Remove the rear wheel.
3. Remove the rear damper/spring.
4. Release the parking brake lever fully.
5. Loosen the parking brake cable adjusting nut.

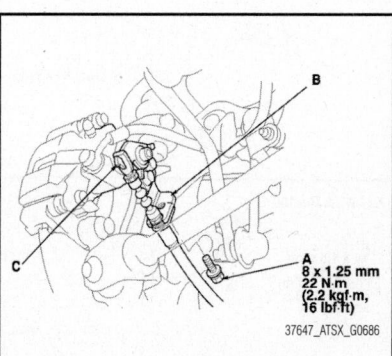

37647_ATSX_G0686

Fig. 269 Remove the flange bolt (A) from the arm (B), then disconnect the parking brake cable from the lever (C)

6. Remove the flange bolt from the arm. Then disconnect the parking brake cable from the lever.

7. Remove the brake caliper bracket mounting bolts, then remove the caliper assembly from the knuckle. To prevent damage to the caliper assembly or the brake hose, use a short piece of wire to hang the caliper assembly from the undercarriage. Do not twist the brake hose excessively.

➡ **Make sure the washers are reinstalled between the brake caliper bracket and the knuckle, if they are removed.**

8. Remove the lock pin from the upper arm ball joint, then remove the castle nut.

➡ **During installation, install the lock pin as shown after tightening the new castle nut.**

9. Disconnect the upper arm ball joint from the knuckle using the ball joint thread protector and the ball joint remover.

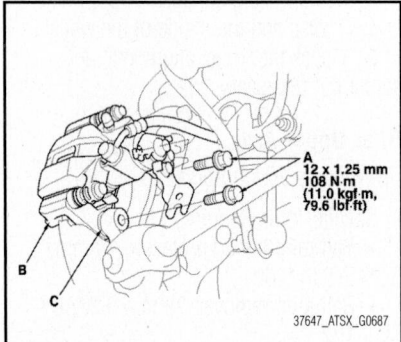

Fig. 270 Remove the brake caliper bracket mounting bolts (A), then remove the caliper assembly (B) from the knuckle

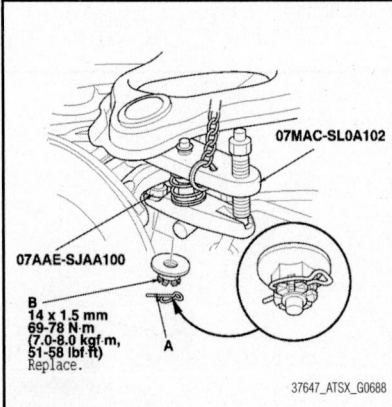

Fig. 271 Remove the lock pin (A) from the upper arm ball joint, then remove the castle nut (B)

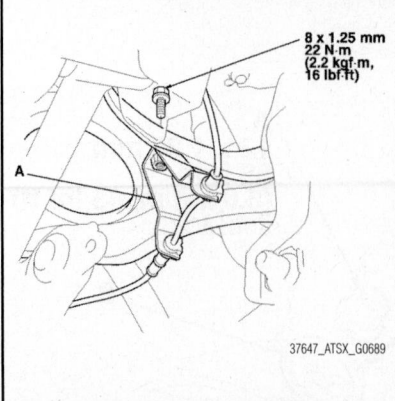

Fig. 272 Remove the wheel speed sensor harness bracket (A)

➡ **Be careful not to damage the ball joint boot when installing the remover. During installation, to connect the ball joint, position a floor jack under the connecting point of the knuckle and lower arm A, and raise the suspension with the jack.**

10. Remove the wheel speed sensor harness bracket.

11. Remove the upper arm mounting bolts, then remove the upper arm.

To install:

➡ **Use new mounting bolts during reassembly.**

12. Install the upper arm in the reverse order of removal, and note these items:
   - First install all of the components, and lightly tighten the bolts and the nuts, then raise the suspension to load it with the vehicle's weight before fully tightening to the specified torque.
   - Be careful not to damage the ball

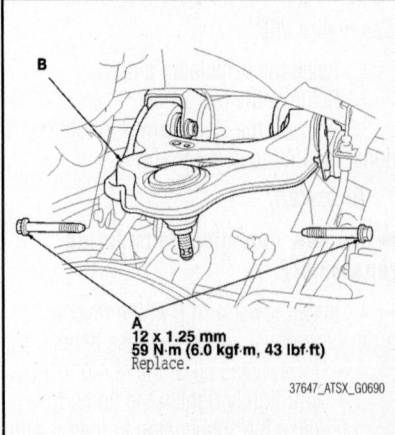

Fig. 273 Remove the upper arm mounting bolts (A), then remove the upper arm (B)

joint boot when connecting the knuckle.
   - Before connecting the ball joint, degrease the threaded section and the tapered portion of the ball joint pin, the ball joint connecting hole, the threaded section, and the mating surfaces of the castle nut.
   - Torque the castle nut to the lower torque specification, then tighten it only far enough to align the slot with the ball joint pin hole. Do not align the castle nut by loosening it.
   - After installing the brake caliper, make sure the clearance between lower arm B and the parking brake cable is more than 0.20 inches (5 mm).
   - Do the parking brake adjustment.
   - Before installing the wheel, clean the mating surfaces on the brake disc and the inside of the wheel.

13. Check the wheel alignment, and adjust it if necessary.

KNUCKLE/HUB BEARING UNIT

REMOVAL & INSTALLATION

Special Tools Required:
   - Ball Joint Thread Protector, 14 mm 07AAE-SJAA100
   - Ball Joint Remover, 32 mm 07MAC-SL0A102

Hub Bearing Unit Replacement

See Figures 274 and 275.

1. Raise and support the vehicle.
2. Remove the wheel nuts and the rear wheel.
3. Release the parking brake lever fully.

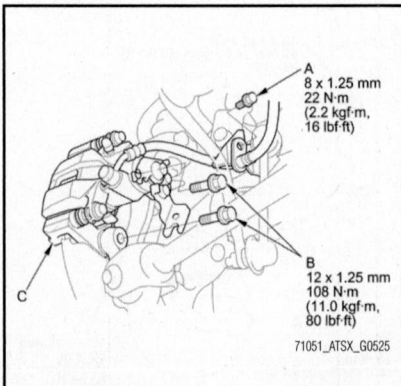

Fig. 274 Remove the brake hose mounting bolt (A), the brake caliper bracket mounting bolts (B), then remove the caliper assembly (C) from the knuckle

4. Loosen the parking brake cable adjusting nut.

5. Remove the flange bolt from the arm. Then disconnect the parking brake cable from the lever.

6. Remove the brake hose mounting bolt.

7. Remove the brake caliper bracket mounting bolts, then remove the caliper assembly from the knuckle. To prevent damage to the caliper assembly or the brake hose, use a short piece of wire to hang the caliper assembly from the undercarriage. Do not twist the brake hose excessively.

8. Remove the two washers.

➡**During installation, make sure the washers are installed between the brake caliper bracket and the knuckle.**

9. Remove the rear brake disc.

10. Remove the hub bearing unit and the O-ring.

11. Check the hub bearing unit for damage and cracks.

To install:

12. Install the hub bearing unit in the reverse order of removal, and note these items:

   a. Use a new O-ring on reassembly.

   b. After installing the brake caliper, make sure the clearance between lower arm B and the parking brake cable is more than 5 mm (0.20 in).

   c. Before installing the brake disc, clean the mating surfaces between the hub bearing unit and the brake disc.

   d. Do the parking brake adjustment.

   e. Before installing the wheel, clean the mating surfaces between the brake disc and the inside of the wheel.

13. Check the wheel alignment, and adjust it if necessary.

Knuckle Replacement

See Figures 276 through 279.

1. Remove the hub bearing unit.
2. Remove the splash guard.
3. Remove the lock pin from the upper arm ball joint, then remove the castle nut.

➡**During installation, install the lock pin as shown after tightening the new castle nut.**

4. Disconnect the upper arm ball joint from the knuckle using the ball joint thread protector and the ball joint remover.

   Please note the following
   • Be careful not to damage the ball joint boot when installing the remover.
   • During installation, to connect the ball joint, raise up the suspension with a jack.

5. Remove the wheel speed sensor from the knuckle. Do not disconnect the wheel speed sensor connector.

6. Remove the flange nut while holding the joint pin with a hex wrench, then disconnect the stabilizer link from the knuckle, and remove the brake hose bracket.

➡**Use the new flange nut during reassembly.**

7. Remove the damper lower mounting bolt.

➡**Use the new mounting bolt during reassembly.**

8. Remove the control arm mounting self-locking nut and the washer.

➡**Use a new self-locking nut during reassembly.**

9. Remove the lower arm A mounting bolt, and the lower arm B mounting bolt, then remove the knuckle.

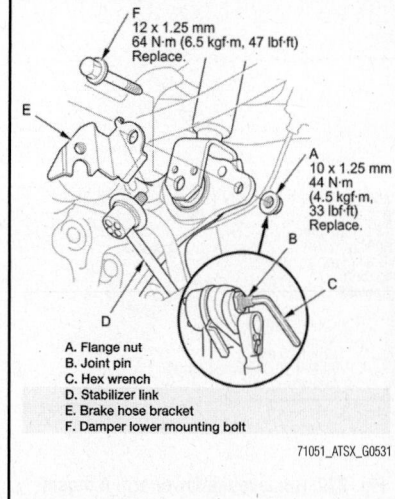

A. Flange nut
B. Joint pin
C. Hex wrench
D. Stabilizer link
E. Brake hose bracket
F. Damper lower mounting bolt

71051_ATSX_G0531

Fig. 277 Remove the flange nut while holding the joint pin with a hex wrench, then disconnect the stabilizer link from the knuckle, and remove the brake hose bracket

➡**Use new mounting bolts during reassembly.**

To install:

10. Install the knuckle in the reverse order of removal, and note these items:

   a. First install all of the components, and lightly tighten the bolts and the nuts, then raise the suspension to load it with the vehicle's weight before fully tightening to the specified torque.

   b. Be careful not to damage the ball joint boot when connecting the knuckle.

   c. Before connecting the ball joint, degrease the threaded section and the tapered portion of the ball joint pin, the ball joint connecting hole, the threaded

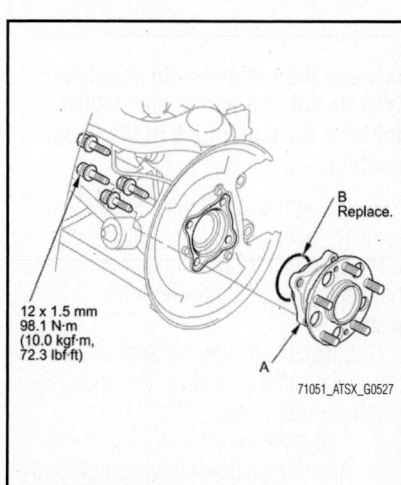

Fig. 275 Remove the hub bearing unit (A) and the O-ring (B)

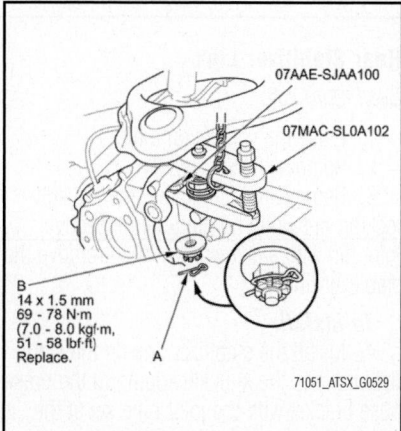

Fig. 276 Remove the lock pin (A) from the upper arm ball joint, then remove the castle nut (B)

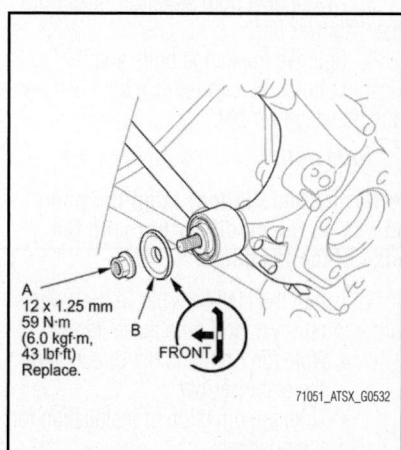

Fig. 278 Remove the control arm mounting self-locking nut (A) and the washer (B)

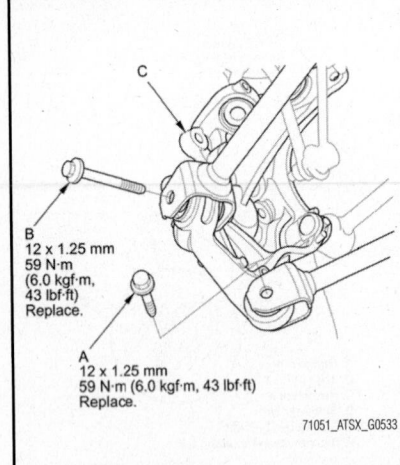

Fig. 279 Remove the lower arm A mounting bolt (A), and the lower arm B mounting bolt (B), then remove the knuckle (C)

section, and the mating surfaces of the castle nut.

d. Torque the castle nut to the lower torque specification, then tighten it only far enough to align the slot with the ball joint pin hole. Do not align the castle nut by loosening it.

e. Before installing the wheel, clean the mating surfaces between the brake disc and the inside of the wheel.

11. Check the wheel alignment, and adjust it if necessary.

STABILIZER BAR & LINKS

REMOVAL & INSTALLATION

Rear Stabilizer Bar

See Figure 280.

1. Raise the vehicle on a lift.
2. Remove the rear wheels.
3. Disconnect both stabilizer links from the stabilizer bar.
4. Remove the flange bolts and the bushing holders, then remove the bushings and the stabilizer bar.

To install:

➡**During installation, align the paint marks on the stabilizer bar with the side of the bushings.**

5. Install the stabilizer bar in the reverse order of removal, and note these items:
  - Note the right and left direction of the stabilizer bar.
  - Note the direction of installation for the bushing.
  - Before installing the wheel, clean the mating surfaces of the brake disc and the inside of the wheel.

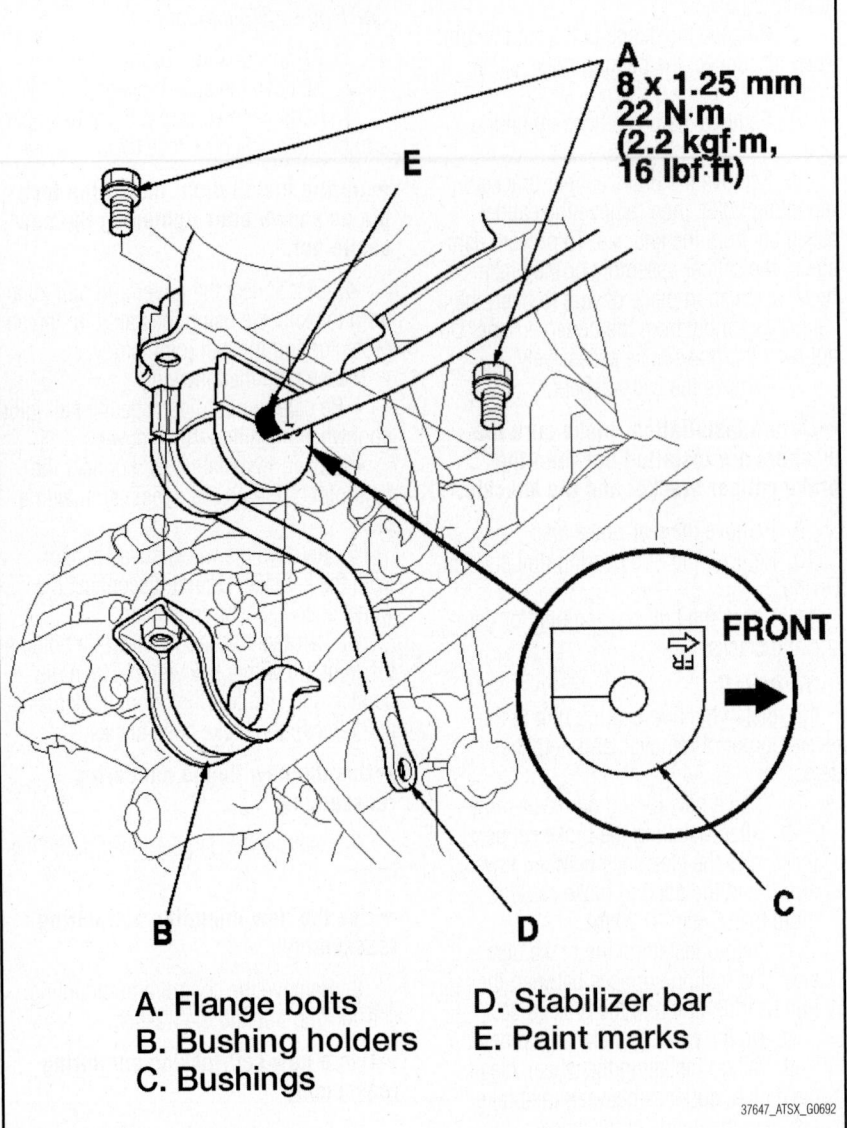

A. Flange bolts
B. Bushing holders
C. Bushings
D. Stabilizer bar
E. Paint marks

Fig. 280 Remove the flange bolts and the bushing holders, then remove the bushings and the stabilizer bar

Rear Stabilizer Link

See Figure 281.

1. Raise the vehicle on a lift.
2. Remove the rear wheel.
3. Remove the flange nut and the self-locking nut while holding the respective joint pin with a hex wrench, then remove the stabilizer link.

To install:

4. Install the stabilizer link on the stabilizer bar and the knuckle adding in the brake hose bracket with the joint pins set at the center of their range of movement.

➡**The stabilizer link has a paint mark. The paint mark indicates the difference** between the left and right stabilizer links. Install the end of the stabilizer link with the paint mark in the upper position.

5. Install the new flange nut and the new self-locking nut, and tighten them to the specified torque while holding the respective joint pin with a hex wrench.

6. Clean the mating surfaces of the brake disc and the inside of the wheel, then install the rear wheel.

7. Test-drive the vehicle.

8. After 5 minutes of driving, tighten the self-locking nuts again to the specified torque.

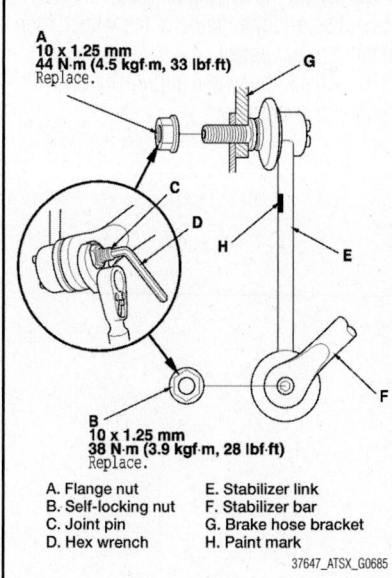

Fig. 281 Remove the flange nut and the self-locking nut while holding the respective joint pin with a hex wrench, then remove the stabilizer link

STRUT

REMOVAL & INSTALLATION

See Figures 282 through 286.

1. Raise the vehicle on a lift.
2. Remove the rear wheel.
3. Fold down the rear seat-back, then remove the lid.

➡**Lift up the tab inside underneath the lid first using a flat-tipped screwdriver, then release the hooks.**

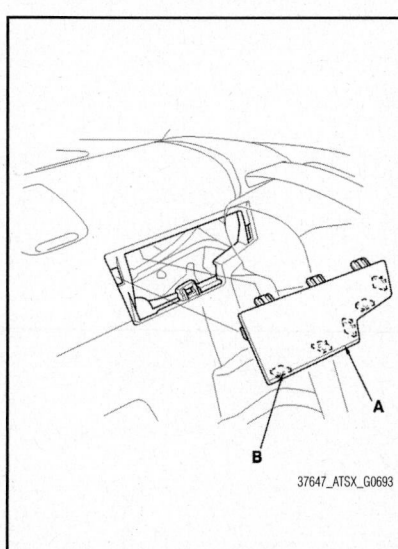

Fig. 282 Remove the lid (A) and lift up the tab (B)

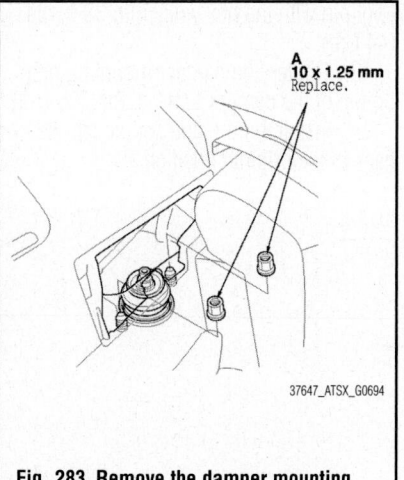

Fig. 283 Remove the damper mounting nuts (A) from the top of the damper

4. Remove the damper mounting nuts from the top of the damper.
5. Remove the flange nut while holding the joint pin with a hex wrench, then disconnect the stabilizer link from the knuckle, and remove the brake hose bracket.
6. Remove the damper lower mounting bolt.
7. Remove the damper/spring by lowering the rear suspension.

➡**Be careful not to damage the body.**

To install:

8. Lower the rear suspension, and

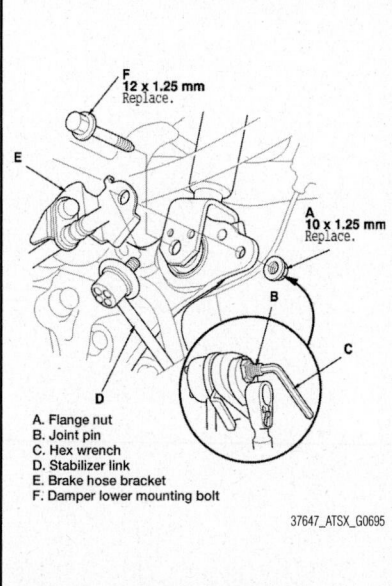

Fig. 284 Remove the flange nut while holding the joint pin with a hex wrench, then disconnect the stabilizer link from the knuckle, and remove the brake hose bracket

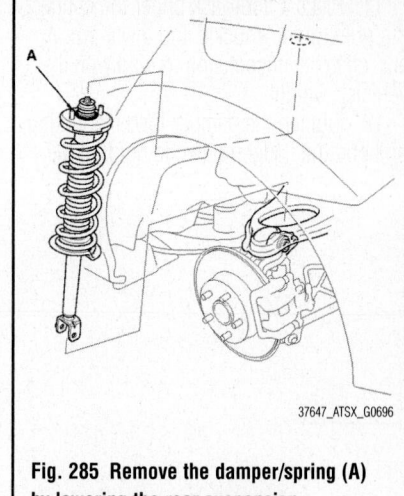

Fig. 285 Remove the damper/spring (A) by lowering the rear suspension

position the damper/spring in the body with the welded nut on the bottom of the damper facing forward.

➡**Be careful not to damage the body. Make sure the damper is installed in the correct direction.**

9. Loosely install the new damper mounting nuts to the top of the damper.
10. Loosely install the new damper lower mounting bolt on the bottom of the damper. Connect the stabilizer link to the brake hose bracket to the knuckle, and loosely install the new flange nut.

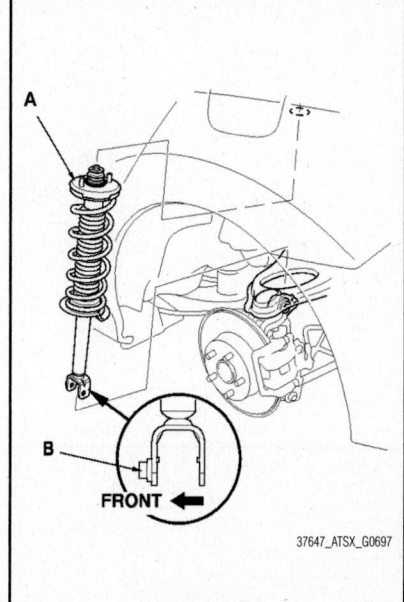

Fig. 286 Lower the rear suspension, and position the damper/spring (A) in the body with the welded nut (B) on the bottom of the damper facing forward

11. Place a floor jack under the connecting point of the knuckle and lower arm A, and raise the suspension to load with the vehicle's weight.

12. Tighten the damper lower mounting bolt and the flange nut while holding the joint pin with the hex wrench to 33 ft. lbs. (44 Nm).

13. Tighten the damper mounting nuts on top of the damper to 41 ft. lbs. (55 Nm).

14. Install the lid, and set the rear seatback to the original position.

15. Clean the mating surfaces of the brake disc and the inside of the wheel, then install the rear wheel.

16. Check the wheel alignment, and adjust it if necessary.

ACURA

ZDX

6

BRAKES 6-9

**ANTI-LOCK BRAKE
SYSTEM (ABS)** 6-9
General Information.................... 6-9
Precautions.......................... 6-9
Speed Sensors........................ 6-9
Removal & Installation........... 6-9
**BLEEDING THE BRAKE
SYSTEM** 6-11
Bleeding Procedure................. 6-11
Bleeding Procedure 6-11
FRONT DISC BRAKES 6-11
Brake Calipers 6-11
Removal & Installation.......... 6-11
Brake Pads 6-11
Removal & Installation.......... 6-11
PARKING BRAKE 6-14
Parking Brake Shoes 6-14
Removal & Installation.......... 6-14
REAR DISC BRAKES 6-12
Brake Caliper....................... 6-12
Removal & Installation.......... 6-12
Disc Brake Pads 6-13
Removal & Installation.......... 6-13

CHASSIS ELECTRICAL 6-15

**AIR BAG (SUPPLEMENTAL
RESTRAINT SYSTEM)** 6-15
General Information.................. 6-15
Arming the System 6-15
Clockspring Centering.......... 6-15
Disarming the System.......... 6-15
Service Precautions 6-15

DRIVE TRAIN 6-16

Differential Carrier 6-16
Removal & Installation.......... 6-16
Driveshaft........................... 6-17
Removal & Installation.......... 6-17
Front Halfshaft...................... 6-18
Removal & Installation.......... 6-18
Rear Drive Axle..................... 6-19
Drain & Refill 6-19
Rear Halfshaft....................... 6-19
Removal & Installation.......... 6-19

ENGINE COOLING 6-20

Electric Engine Fan.................. 6-21
Removal & Installation.......... 6-21
Engine Coolant...................... 6-20
Bleeding 6-20
Drain & Refill 6-20
Radiator.............................. 6-21
Removal & Installation.......... 6-21
Thermostat.......................... 6-21
Removal & Installation.......... 6-21
Water Pump 6-21
Removal & Installation.......... 6-21

ENGINE ELECTRICAL 6-23

BATTERY SYSTEM 6-23
Battery............................... 6-23
Battery Reconnect/Relearn
Procedure 6-23
Removal & Installation.......... 6-23
CHARGING SYSTEM 6-23
Alternator 6-23
Removal & Installation.......... 6-23
IGNITION SYSTEM 6-24
Firing Order......................... 6-24
Ignition Coils and Spark Plugs ... 6-24
Removal & Installation.......... 6-24
Ignition Timing...................... 6-24
Inspection & Adjustment 6-24
STARTING SYSTEM 6-25
Starter 6-25
Removal & Installation.......... 6-25

ENGINE MECHANICAL 6-25

Accessory Drive Belt System..... 6-25
Adjustment 6-25
Belt Routings 6-25
Inspection 6-25
Removal & Installation.......... 6-25
Air Cleaner 6-25
Removal & Installation.......... 6-25
Camshaft & Bearings 6-26
Inspection 6-26
Removal & Installation.......... 6-27
Catalytic Converter 6-28
Removal & Installation.......... 6-28

Crankshaft Front Seal............... 6-30
Removal & Installation.......... 6-30
Crankshaft Pulley 6-29
Removal & Installation.......... 6-29
Crankshaft Rear Cover
& Seal 6-30
Removal & Installation.......... 6-30
Cylinder Head 6-30
Removal & Installation.......... 6-30
Driveplate........................... 6-32
Removal & Installation.......... 6-32
Front Cover & Seal 6-33
Removal & Installation.......... 6-33
Intake Manifold 6-33
Removal & Installation.......... 6-33
Main Bearing Torque
Sequence 6-35
Oil Pan 6-35
Removal & Installation.......... 6-35
Oil Pump 6-36
Removal & Installation.......... 6-36
Pistons & Rings 6-37
Positioning 6-37
Rocker Arms......................... 6-37
Removal & Installation.......... 6-37
Timing Belt & Sprockets 6-38
Removal & Installation.......... 6-38
Timing Belt Front Cover 6-38
Removal & Installation.......... 6-38
Timing Belt Rear Cover 6-39
Removal & Installation.......... 6-39
Valve Lash........................... 6-39
Adjustment 6-39

ENGINE PERFORMANCE &
EMISSION CONTROLS 6-40

Accelerator Pedal Position
(APP) Sensor 6-40
Location.............................. 6-40
Removal & Installation.......... 6-40
Air-Fuel Ratio (AFR) Sensor...... 6-40
Location.............................. 6-40
Removal & Installation.......... 6-40
Camshaft Position (CMP)
Sensor 6-40
Location.............................. 6-40
Removal & Installation.......... 6-40

Component Locations 6-40
Crankshaft Position (CKP)
Sensor 6-41
    Location 6-41
    Removal & Installation..........6-42
Engine Coolant Temperature
(ECT) Sensor 6-42
    Location 6-42
    Removal & Installation..........6-42
EVAP Canister 6-42
    Location 6-42
    Removal & Installation..........6-42
Exhaust Gas Recirculation
(EGR) Valve 6-43
    Location 6-43
    Removal & Installation..........6-43
Heated Oxygen (HO2S)
Sensor 6-43
    Removal & Installation..........6-43
Input Shaft Speed Sensor.........6-44
    Removal & Installation..........6-44
Knock Sensor (KS)....................6-44
    Location 6-44
    Removal & Installation..........6-44
Manifold Absolute Pressure
(MAP) Sensor 6-44
    Location 6-44
    Removal & Installation..........6-45
Mass Airflow (MAF)/Intake
Air Temperature (IAT)
Sensor 6-45
    Location 6-45
    Removal & Installation..........6-45
Output Shaft Speed (OSS)
Sensor 6-45
    Removal & Installation..........6-45
Positive Crankcase
Ventilation (PCV) Valve...........6-45
    Removal & Installation..........6-45
Powertrain Control
Module (PCM)........................6-46
    Location................................6-46
    PCM Idle Learn
    Procedure 6-46
    Removal & Installation..........6-46
Transmission Fluid
Temperature (TFT) Sensor.......6-47
    Location................................6-47
    Removal & Installation..........6-47

FUEL............................6-47

**GASOLINE FUEL
INJECTION SYSTEM..........6-47**
Fuel Filter....................................6-48
    Removal & Installation..........6-48
Fuel Rail & Injectors 6-49
    Removal & Installation..........6-49
Fuel System Service
Precautions 6-47
Fuel Tank....................................6-49
    Draining 6-49
    Removal & Installation..........6-50
Relieving Fuel System
Pressure.................................6-47
Throttle Body 6-50
    Removal & Installation..........6-50

**HEATING & AIR
CONDITIONING SYSTEM 6-51**

Blower Motor 6-51
    Removal & Installation..........6-51
Heater Core 6-51
    Removal & Installation..........6-51

PRECAUTIONS.................6-9

**SPECIFICATIONS AND
MAINTENANCE CHARTS 6-3**

Brake Specifications...................6-7
Camshaft Specifications.............6-5
Capacities 6-4
Crankshaft and Connecting
Rod Specifications 6-5
Engine and Model Year
Identification 6-3
Engine Tune-Up Specifications ..6-3
Fluid Specifications....................6-4
General Engine Specifications.....6-3
Piston and Ring Specifications ...6-5
Scheduled Maintenance
Intervals 6-8
Tire Wheel and Ball Joint
Specifications 6-7
Torque Specifications 6-6
Valve Specifications 6-4
Wheel Alignment........................6-6

STEERING 6-53

Power Rack & Pinion
Steering Gear 6-53
    Removal & Installation..........6-53
Power Steering Pump.................6-56
    Bleeding................................6-56
    Fluid Fill Procedure 6-56
    Removal & Installation..........6-56

SUSPENSION.................6-56

FRONT SUSPENSION..........6-56
Knuckle 6-56
    Removal & Installation..........6-56
Lower Ball Joints.......................6-57
    Removal & Installation..........6-57
Lower Control Arms 6-58
    Control Arm Bushing
    Replacement 6-59
    Removal & Installation..........6-58
Stabilizer Bar & Links 6-59
    Removal & Installation..........6-59
Struts 6-61
    Overhaul 6-62
    Removal & Installation..........6-61
Wheel Hubs & Bearings............6-63
    Removal, Packing, &
    Installation............................6-63
REAR SUSPENSION............6-64
Coil Springs...............................6-64
    Removal & Installation..........6-64
Knuckle/Hub Bearing Unit.........6-65
    Removal & Installation..........6-65
Lower Control Arms 6-66
    Removal & Installation..........6-66
Shock Absorbers.......................6-67
    Removal & Installation..........6-67
Stabilizer Bar & Links 6-67
    Removal & Installation..........6-67
Trailing Arms.............................6-68
    Removal & Installation..........6-68
Upper Ball Joints.......................6-68
    Removal & Installation..........6-68
Upper Control Arms 6-69
    Removal & Installation..........6-69
Wheel Hubs & Bearings............6-70
    Removal, Repacking, &
    Installation............................6-70

SPECIFICATIONS AND MAINTENANCE CHARTS

ENGINE AND VEHICLE IDENTIFICATION CHART

Engine Code							Model Year	
Code	Liters (cc)	Cu. In.	Cyl.	Fuel Sys.	Engine Type	Eng. Mfg.	Code ①	Year
J37A5	3.7 (3664)	226	6	PGM-FI ③	SOHC	Honda	B	2011
							C	2012

SOHC: Single Overhead Cam

SMFI: Sequential Multi-port Fuel Injection

① 10th position of VIN

② VTEC: variable timing electronically controlled

③ PGM-FI: multi-point fuel injection

71051_AZDX_C0001

GENERAL ENGINE SPECIFICATIONS

Year	Model	Engine Displacement Liters	Engine ID	Net Horsepower @ rpm	Net Torque @ rpm (ft. lbs.)	Bore x Stroke (in.)	Compression Ratio	Oil Pressure @ rpm
2011	ZDX	3.7	J37A5	300@6300	270@4500	3.54x3.78	11.2:1	44@3000
2012	ZDX	3.7	J37A5	300@6300	270@4500	3.54x3.78	11.2:1	44@3000

71051_AZDX_C0002

ENGINE TUNE-UP SPECIFICATIONS

Year	Engine Displacement Liters	Engine ID	Spark Plug Gap (in.)	Ignition Timing (deg.) MT	AT	Fuel Pump (psi)	Idle Speed (rpm) MT	AT	Valve Clearance (in.) In.	Ex.
2011	3.7	J37A5	0.039-0.043	—	8-12 ②	55-63	—	660-760	0.008-0.009	0.011-0.013
2012	3.7	J37A5	0.039-0.043	—	8-12 ②	55-63	—	660-760	0.008-0.009	0.011-0.013

NOTE: The Vehicle Emission Control Information label often reflects changes made during production and must be used if they differ from this chart.

NOTE: The fuel pressure readings are given with the vacuum hose connected to the regulator and the engine running

①: Before top dead center

71051_AZDX_C0003

CAPACITIES

Year	Model	Engine Displacement Liters	Engine ID	Engine Oil with Filter (qts.)	Transmission (qts.) 5-Spd	Transmission (qts.) Auto.	Fuel Tank (gal.)	Cooling System (qts.)
2011	ZDX	3.7	J37A5	4.5	—	3.3 ①	21.0	7.1 ②
2012	ZDX	3.7	J37A5	4.5	—	3.3 ①	21.0	7.1 ②

NOTE: All capacities are approximate. Add fluid gradually and check to be sure a proper fluid level is obtained.

① 3.3 qts for service; 8.5 qts for overhaul.

② 7.1 qts for service; 9.1 qts for engine overhaul.

71051_AZDX_C0004

FLUID SPECIFICATIONS

Year	Model	Engine Displ. Liters	Engine Oil	Man. Trans.	Auto. Trans.	Power Steering Fluid	Brake Master Cylinder	Cooling System
2011	ZDX	3.7	5W-20 Acura	N/A	Acura ATF-Z1	Acura PS Fluid	Acura DOT 3	①
2012	ZDX	3.7	5W-20 Acura	N/A	Acura ATF-Z1	Acura PS Fluid	Acura DOT 3	①

DOT: Department Of Transpotation

N/A: Not Applicable

① Acura Long Life Antifreeze/Coolant-Type2

71051_AZDX_C0005

VALVE SPECIFICATIONS

Year	Engine Displacement Liters	Engine ID	Seat Angle (deg.)	Face Angle (deg.)	Spring Test Pressure (lbs. @ in.)	Spring Installed Height (in.)	Stem-to-Guide Clearance (in.) Intake	Stem-to-Guide Clearance (in.) Exhaust	Stem Diameter (in.) Intake	Stem Diameter (in.) Exhaust
2011	3.7	J37A5	NA	NA	NA	NA	0.00079-0.0018	0.00217-0.0032	0.2159-0.2163	0.2146-0.2150
2012	3.7	J37A5	NA	NA	NA	NA	0.00079-0.0018	0.00217-0.0032	0.2159-0.2163	0.2146-0.2150

NA: Not Available

71051_AZDX_C0007

CAMSHAFT AND BEARING SPECIFICATIONS

All measurements are given in inches.

Year	Engine Displacement Liters	Engine VIN	Journal Diameter	Brg. Oil Clearance	Shaft End-play	Runout	Journal Bore	Lobe Height Intake	Lobe Height Exhaust
2011	3.7	J37A5	NA	0.0020-0.0035	0.0020-0.0080	0.0012	NA	①	1.4472
2012	3.7	J37A5	NA	0.0020-0.0035	0.0020-0.0080	0.0012	NA	①	1.4472

NA: Information not available

① Primary: 1.35035 in.

  Secondary: 1.41464 in.

71051_AZDX_C0006

CRANKSHAFT AND CONNECTING ROD SPECIFICATIONS

All measurements are given in inches

Year	Engine Displacement Liters	Engine ID	Crankshaft Main Brg. Journal Dia.	Main Brg. Oil Clearance	Shaft End-play	Thrust on No.	Connecting Rod Journal Diameter	Oil Clearance	Side Clearance
2011	3.7	J37A5	2.8337-2.8346	0.00075-0.0018	0.0039-0.0138	3	2.24315-2.24409	0.0002-0.00055	0.0059-0.0138
2012	3.7	J37A5	2.8337-2.8346	0.00075-0.0018	0.0039-0.0138	3	2.24315-2.24409	0.0002-0.00055	0.0059-0.0138

71051_AZDX_C0008

PISTON AND RING SPECIFICATIONS

All measurements are given in inches

Year	Engine Disp. Liters	Engine ID	Piston Clearance	Ring Gap Top Compression	Ring Gap Bottom Compression	Ring Gap Oil Control	Ring Side Clearance Top Compression	Ring Side Clearance Bottom Compression	Ring Side Clearance Oil Control
2011	3.7	J37A5	0.00016-0.00126	0.0118-0.0157	0.0157-0.0217	0.0079-0.0138	0.00217-0.00335	0.00118-0.0024	NA
2012	3.7	J37A5	0.00016-0.00126	0.0118-0.0157	0.0157-0.0217	0.0079-0.0138	0.00217-0.00335	0.00118-0.0024	NA

NA: Not Available

71051_AZDX_C0009

TORQUE SPECIFICATIONS

All readings in ft. lbs.

Year	Engine Displacement Liters	Engine ID	Cylinder Head Bolts	Main Bearing Bolts	Rod Bearing Bolts	Crankshaft Damper Bolts	Flywheel Bolts	Manifold Intake	Manifold Exhaust	Spark Plugs	Oil Pan Drain Plug
2011	3.7	J37A5	①	②	③	④	55	16	40 ⑤	18	30
2012	3.7	J37A5	①	②	③	④	55	16	40 ⑤	18	30

NOTE: Dip main bearing bolts and crankshaft damper bolt in clean engine oil prior to tightening.

① 12-point head bolts

Step 1: Torque all bolts to 22 ft. lbs.

Step 2: Torque all bolts an additional 90 deg.

Step 3: Torque all bolts an additional 90 deg.

New Bolt Only: an additional 90 deg.

② Cap bolts: 54 ft. lbs.

Side bolts: 36 ft. lbs.

③ Step 1: 14 ft. lbs.

Step 2: 90 degrees

④ Step 1: 48 ft. lbs.

Step 2: 65 degrees

⑤ Exhaust pipe-to-engine: 40 ft. lbs.; no exhaust manifold.

71051_AZDX_C0010

WHEEL ALIGNMENT

Year	Model		Caster Range (+/-Deg.)	Caster Preferred Setting (Deg.)	Camber Range (+/-Deg.)	Camber Preferred Setting (Deg.)	Toe-in (in.)
2011	ZDX	F	0.36	+4.38	0.45	-0.30	0+/-0.08
		R	—	—	0.45	-0.30	0+/-0.08
2012	ZDX	F	0.36	+4.38	0.45	-0.30	0+/-0.08
		R	—	—	0.45	-0.30	0+/-0.08

71051_AZDX_C0011

TIRE, WHEEL AND BALL JOINT SPECIFICATIONS

Year	Model	OEM Tires		Tire Pressures (psi)		Wheel Size	Ball Joint Inspection	Lug Nut Torque (ft. lbs.)
		Standard	Optional	Front	Rear			
2011	ZDX	P255/50R19	None	①	①	8.5	NS	94
2012	ZDX	P255/50R19	None	①	①	8.5	NS	94

OEM: Original Equipment Manufacturer

PSI: Pounds Per Square Inch

NS: Not specified by manufacturer

① See placard on vehicle

71051_AZDX_C0012

BRAKE SPECIFICATIONS
All measurements in inches unless noted

Year	Model		Brake Disc			Minimum Lining Thickness		Brake Caliper	
			Original Thickness	Minimum Thickness	Maximum Runout	Front	Rear	Bracket Bolts (ft. lbs.)	Mounting Bolts (ft. lbs.)
2011	ZDX	F	1.102	1.024	0.002	0.063	—	101	53
		R	0.433	0.354	0.002	—	0.030	65	27
2012	ZDX	F	1.102	1.024	0.002	0.063	—	101	53
		R	0.433	0.354	0.002	—	0.030	65	27

F: Front

R: Rear

71051_AZDX_C0013

SCHEDULED MAINTENANCE INTERVALS
ACURA—ZDX

TO BE SERVICED	OF SERVIC	VEHICLE MILEAGE INTERVAL (x1000)															
		7.5	15	22.5	30	37.5	45	52.5	60	67.5	75	82.5	90	97.5	105	112.5	120
Accessory drive belts	I & A				✓				✓				✓				✓
Air cleaner element	R				✓				✓				✓				✓
Brake fluid	R	Every 3 years															
Brake hoses & lines (incl. ABS)	I		✓		✓		✓		✓		✓		✓		✓		✓
Cooling system hoses & connections	I		✓		✓		✓		✓		✓		✓		✓		✓
Engine coolant A	R						✓						✓				
Engine oil	R	✓	✓	✓	✓	✓	✓	✓	✓	✓	✓	✓	✓	✓	✓	✓	✓
Engine oil and coolant levels	I	Inspect at each fuel stop															
Engine oil filter	R		✓		✓		✓		✓		✓		✓		✓		✓
Exhaust system	I		✓		✓		✓		✓		✓		✓		✓		✓
Fluid levels and condition	I		✓		✓		✓		✓		✓		✓		✓		✓
Front and rear brakes	I		✓		✓		✓		✓		✓		✓		✓		✓
Fuel lines & connection	I		✓		✓		✓		✓		✓		✓		✓		✓
Halfshaft boots	I		✓		✓		✓		✓		✓		✓		✓		✓
Idle speed	I & A														✓		
Parking brake system	I & A		✓		✓		✓		✓		✓		✓		✓		✓
Rear differential fluid	R	✓			✓		✓		✓				✓				✓
Rotate and inspect tires	I	✓	✓	✓	✓	✓	✓	✓	✓	✓	✓	✓	✓	✓	✓	✓	✓
Spark plugs	R														✓		
Supplemental Restraint System	I	Inspect the SRS 10 years after production															
Suspension components	I		✓		✓		✓		✓		✓		✓		✓		✓
Tie rod ends, steering gear box & boots	I		✓		✓		✓		✓		✓		✓		✓		✓
Timing belt	R														✓		
Transmission fluid	R						✓				✓				✓		
Valve clearance	I	Adjust if valves are noisy															
Water pump	S/I														✓		

R: Replace   I: Inspect   A: Adjust

A Every 12,000 miles or 10 years, then every 60,000 miles or 5 years

FREQUENT OPERATION MAINTENANCE (SEVERE SERVICE)

If a vehicle is operated under any of the following conditions it is considered severe service:

- Towing a trailer or using a camper or car-top carrier.
- Repeated short trips of less than 5 miles in temperatures below freezing, or trips of less than 10 miles in any temperature.
- Extensive idling or low-speed driving for long distances as in heavy commercial use, such as delivery, taxi or police cars.
- Operating on rough, muddy or salt-covered roads.
- Operating on unpaved or dusty roads.
- Driving in extremely hot (over 90°) conditions.

Air cleaner element: replace every 15,000 miles

Engine oil and filter: replace every 3750 miles or 6 months, whichever occurs first.

Timing belt: replace every 60,000 miles if the vehicle is regularly driven in temperatures above 110°F or below -20°F, or if frequently towing a trailer.

Transmission fluid: replace every 30,000 miles.

Rear differential fluid: replace every 60,000 miles.

Front and rear brakes: inspect every 7500 miles or 6 months, whichever occurs first.

Locks and hinges: lubricate every 15,000 miles.

Tie rods, steering gear box, boots: inspect every 7500 miles or 6 months, whichever occurs first.

Suspension components: inspect every 7500 miles or 6 months, whichever occurs first.

Halfshaft boots: inspect every 7500 miles or 6 months, whichever occurs first.

71051_AZDX_C0014

PRECAUTIONS

Before servicing any vehicle, please be sure to read all of the following precautions, which deal with personal safety, prevention of component damage, and important points to take into consideration when servicing a motor vehicle:

• Never open, service or drain the radiator or cooling system when the engine is hot; serious burns can occur from the steam and hot coolant.

• Observe all applicable safety precautions when working around fuel. Whenever servicing the fuel system, always work in a well-ventilated area. Do not allow fuel spray or vapors to come in contact with a spark, open flame, or excessive heat (a hot drop light, for example). Keep a dry chemical fire extinguisher near the work area. Always keep fuel in a container specifically designed for fuel storage; also, always properly seal fuel containers to avoid the possibility of fire or explosion. Refer to the additional fuel system precautions later in this section.

• Fuel injection systems often remain pressurized, even after the engine has been turned **OFF**. The fuel system pressure must be relieved before disconnecting any fuel lines. Failure to do so may result in fire and/or personal injury.

• Brake fluid often contains polyglycol ethers and polyglycols. Avoid contact with the eyes and wash your hands thoroughly after handling brake fluid. If you do get brake fluid in your eyes, flush your eyes with clean, running water for 15 minutes. If eye irritation persists, or if you have taken

brake fluid internally, IMMEDIATELY seek medical assistance.

• The EPA warns that prolonged contact with used engine oil may cause a number of skin disorders, including cancer. You should make every effort to minimize your exposure to used engine oil. Protective gloves should be worn when changing oil. Wash your hands and any other exposed skin areas as soon as possible after exposure to used engine oil. Soap and water, or waterless hand cleaner should be used.

• All new vehicles are now equipped with an air bag system, often referred to as a Supplemental Restraint System (SRS) or Supplemental Inflatable Restraint (SIR) system. The system must be disabled before performing service on or around system components, steering column, instrument panel components, wiring and sensors. Failure to follow safety and disabling procedures could result in accidental air bag deployment, possible personal injury and unnecessary system repairs.

• Always wear safety goggles when working with, or around, the air bag system. When carrying a non-deployed air bag, be sure the bag and trim cover are pointed away from your body. When placing a non-deployed air bag on a work surface, always face the bag and trim cover upward, away from the surface. This will reduce the motion of the module if it is accidentally deployed. Refer to the additional air bag system precautions later in this section.

• Clean, high quality brake fluid from a sealed container is essential to the safe and

proper operation of the brake system. You should always buy the correct type of brake fluid for your vehicle. If the brake fluid becomes contaminated, completely flush the system with new fluid. Never reuse any brake fluid. Any brake fluid that is removed from the system should be discarded. Also, do not allow any brake fluid to come in contact with a painted surface; it will damage the paint.

• Never operate the engine without the proper amount and type of engine oil; doing so WILL result in severe engine damage.

• Timing belt maintenance is extremely important. Many models utilize an interference-type, non-freewheeling engine. If the timing belt breaks, the valves in the cylinder head may strike the pistons, causing potentially serious (also time-consuming and expensive) engine damage. Refer to the maintenance interval charts for the recommended replacement interval for the timing belt, and to the timing belt section for belt replacement and inspection.

• Disconnecting the negative battery cable on some vehicles may interfere with the functions of the on-board computer system(s) and may require the computer to undergo a relearning process once the negative battery cable is reconnected.

• When servicing drum brakes, only disassemble and assemble one side at a time, leaving the remaining side intact for reference.

• Only an MVAC-trained, EPA-certified automotive technician should service the air conditioning system or its components.

BRAKES

ANTI-LOCK BRAKE SYSTEM (ABS)

GENERAL INFORMATION

PRECAUTIONS

• Certain components within the ABS system are not intended to be serviced or repaired individually.

• Do not use rubber hoses or other parts not specifically specified for and ABS system. When using repair kits, replace all parts included in the kit. Partial or incorrect repair may lead to functional problems and require the replacement of components.

• Lubricate rubber parts with clean, fresh brake fluid to ease assembly. Do not use shop air to clean parts; damage to rubber components may result.

• Use only DOT 3 brake fluid from an unopened container.

• If any hydraulic component or line is removed or replaced, it may be necessary to bleed the entire system.

• A clean repair area is essential. Always clean the reservoir and cap thoroughly before removing the cap. The slightest amount of dirt in the fluid may plug an orifice and impair the system function. Perform repairs after components have been thoroughly cleaned; use only denatured alcohol to clean components. Do not allow ABS components to come into contact with any substance containing mineral oil; this includes used shop rags.

• The Anti-Lock control unit is a microprocessor similar to other computer units in the vehicle. Ensure that the ignition switch is **OFF** before removing or installing con-

troller harnesses. Avoid static electricity discharge at or near the controller.

• If any arc welding is to be done on the vehicle, the control unit should be unplugged before welding operations begin.

SPEED SENSORS

REMOVAL & INSTALLATION

Front

See Figure 1.

1. Turn the ignition switch to LOCK (0), or press the engine start/stop button to select the OFF mode.

2. Remove the clip, then disconnect the wheel speed sensor connector.

3. Remove the bolts and the wheel speed sensor.

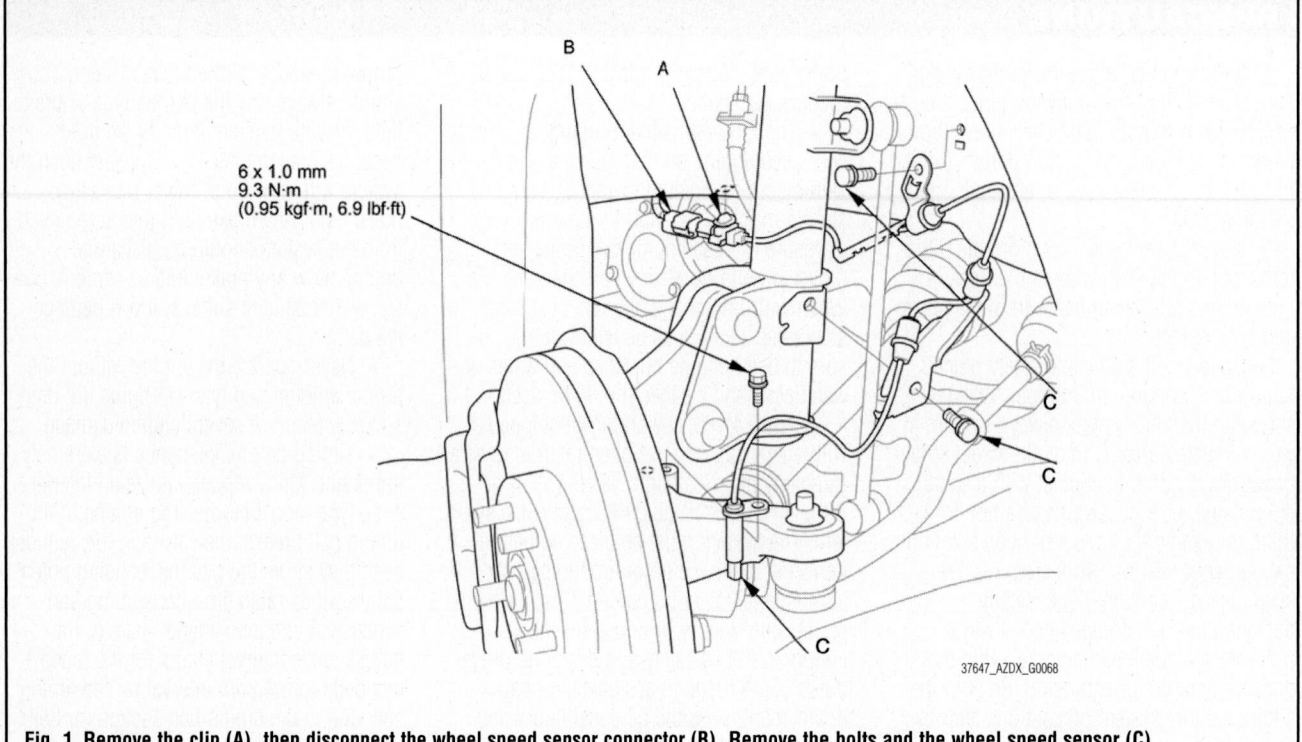

Fig. 1 Remove the clip (A), then disconnect the wheel speed sensor connector (B). Remove the bolts and the wheel speed sensor (C).

To install:

4. Install the wheel speed sensor in the reverse order of removal, and note these items:

   a. Do not twist the sensor wires.

   b. If the wheel speed sensor comes in contact with the hub bearing unit, it is faulty.

   c. Make sure there is no debris in the sensor mounting hole.

5. Start the engine, and make sure the ABS and the VSA indicators go off.

6. Test-drive the vehicle, and make sure the ABS and the VSA indicators do not come on.

Rear

See Figure 2.

1. Turn the ignition switch to LOCK (0), or press the engine start/stop button to select the OFF mode.

2. Remove the clip, then disconnect the wheel speed sensor connector.

3. Remove the clip (left side), the bolts, and the wheel speed sensor.

To install:

4. Install the wheel speed sensor in the reverse order of removal, and note these items:

   a. Do not twist the sensor wires.

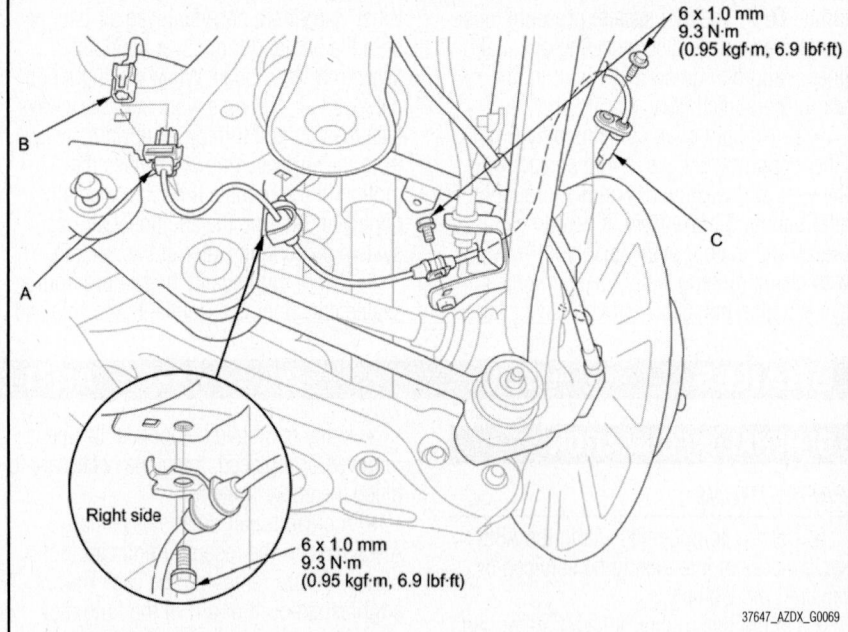

Fig. 2 Remove the clip (A), then disconnect the wheel speed sensor connector (B). Remove the bolts and the wheel speed sensor (C).

   b. If the wheel speed sensor comes in contact with the hub bearing unit, it is faulty.

   c. Make sure there is no debris in the sensor mounting hole.

5. Start the engine, and make sure the ABS and the VSA indicators go off.

6. Test-drive the vehicle, and make sure the ABS and the VSA indicators

BRAKES

BLEEDING THE BRAKE SYSTEM

BLEEDING PROCEDURE

BLEEDING PROCEDURE

See Figure 3.

➡Do not reuse the drained fluid. Use only clean Honda DOT 3 Brake Fluid from an unopened container. Using a non-Honda brake fluid can cause corrosion and shorten the life of the system.

Do not mix different brands of brake fluid; they may not be compatible.

Make sure no dirt or other foreign matter is allowed to contaminate the brake fluid.

✳✳ WARNING

Do not spill brake fluid on the vehicle, it may damage the paint; if brake

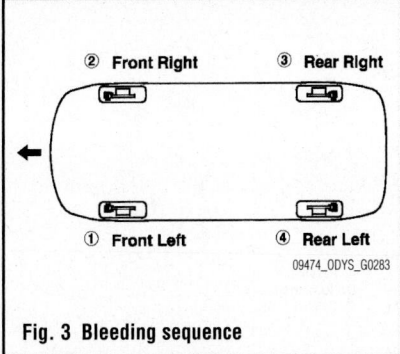

Fig. 3 Bleeding sequence

fluid does contact the paint, wash it off immediately with water.

The reservoir on the master cylinder must be at the MAX (upper) level mark at

the start of the bleeding procedure and checked after bleeding each brake caliper. Add fluid as required.

1. Make sure the brake fluid level in the reservoir is at the MAX (upper) level line.
2. Attach a length of clear drain tube to the bleed screw.
3. Have someone slowly pump the brake pedal several times, then apply steady pressure.
4. Starting at the left-front, loosen the brake bleed screw to allow air to escape from the system. Then tighten the bleed screw securely.
5. Repeat the procedure for each caliper until no air bubbles are in the fluid. Bleed the calipers in the sequence shown.
6. Refill the master cylinder reservoir to the MAX (upper) level line.

BRAKES

FRONT DISC BRAKES

✳✳ CAUTION

Dust and dirt accumulating on brake parts during normal use may contain asbestos fibers from production or aftermarket brake linings. Breathing excessive concentrations of asbestos fibers can cause serious bodily harm. Exercise care when servicing brake parts. Do not sand or grind brake lining unless equipment used is designed to contain the dust residue. Do not clean brake parts with compressed air or by dry brushing. Cleaning should be done by dampening the brake components with a fine mist of water, then wiping the brake components clean with a dampened cloth. Dispose of cloth and all residue containing asbestos fibers in an impermeable container with the appropriate label. Follow practices prescribed by the Occupational Safety and Health Administration (OSHA) and the Environmental Protection Agency (EPA) for the handling, processing, and disposing of dust or debris that may contain asbestos fibers.

BRAKE CALIPERS

REMOVAL & INSTALLATION

See Figure 4.

➡Keep grease away from the brake disc and the brake pads.

1. Raise and support the vehicle.
2. Remove the wheel.
3. Remove the brake hose bracket mounting bolt (A).
4. Remove the brake caliper bracket mounting bolts (B), then remove the caliper assembly (C) from the knuckle.

➡To prevent damage to the caliper assembly or brake hose, use a short piece of wire to hang the caliper assembly from the undercarriage. Do not twist the brake hose excessively.

5. Installation is the reverse of the removal procedure.

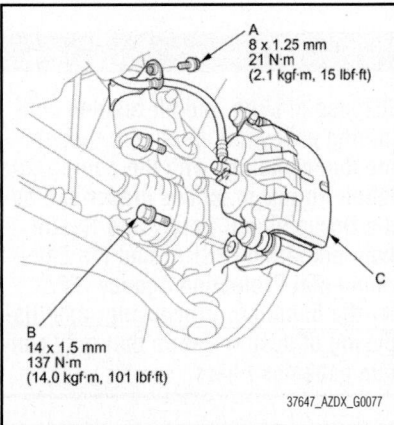

Fig. 4 Remove the brake hose bracket mounting bolt (A). Remove the brake caliper bracket mounting bolts (B), then remove the caliper assembly (C) from the knuckle.

BRAKE PADS

REMOVAL & INSTALLATION

See Figures 5 and 6.

1. Raise and support the vehicle.
2. Remove the front wheels.
3. Check the thickness of the inner pad and the outer pad. Do not include the thickness of the backing plate.
4. If any part of the brake pad thickness is less than the service limit of 0.63 inches (1.6 mm), replace the front brake pads as a set.
5. Clean the mating surfaces between the brake disc and the inside of the wheel, then install the front wheels.
6. Remove some brake fluid from the master cylinder.

✳✳ WARNING

Frequent inhalation of brake pad dust, regardless of material composition, could be hazardous to your health. Avoid breathing dust particles. Never use an air hose or brush to clean brake assemblies. Use an OSHA-approved vacuum cleaner.

7. Remove the front wheels.
8. Remove the lower caliper flange bolt and pivot the caliper up out of the way. Check the hose and the pin boots for damage and deterioration.
9. Remove the pad shims and the brake pads.
10. Remove the upper and lower pad retainers.

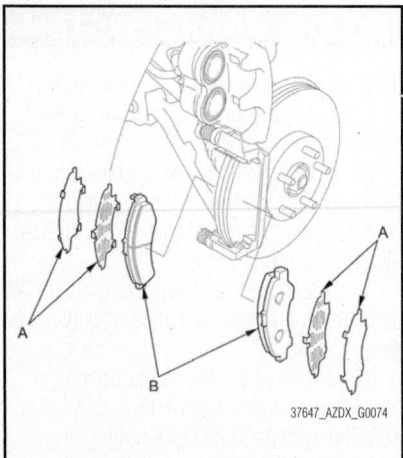

37647_AZDX_G0074

Fig. 5 Remove the pad shims (A) and the brake pads (B).

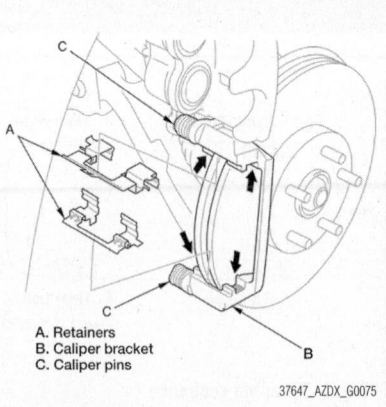

A. Retainers
B. Caliper bracket
C. Caliper pins

37647_AZDX_G0075

Fig. 6 Removing the upper and lower pad retainers and checking caliper bracket and pins

➡The upper and lower pad retainers are different. During installation, make sure the pad retainers are in the proper positions.

11. Clean the caliper bracket thoroughly; remove any rust, and check for grooves and cracks.

12. Verify that the caliper pins move in and out smoothly. Clean and lube if needed.

13. Inspect the brake disc for runout, thickness, parallelism, and check for damage and cracks. See "Specifications" in this section.

To install:

14. Apply a thin coat of M-77 assembly paste to the retainer mating surface of the caliper bracket (indicated by the arrows).

15. Install the upper and lower pad retainers. Wipe off the excess assembly paste from the retainers.

※※ CAUTION

Keep the assembly paste away from the brake disc and the brake pads.

16. Install the brake caliper piston compressor tool on the caliper body.

17. Press in the piston with the brake caliper piston compressor tool so the caliper will fit over the brake pads. Make sure the piston boot is in position to prevent damaging it when pivoting the caliper down.

➡Be careful when pressing in the piston; brake fluid might overflow from the master cylinder's reservoir. If brake fluid gets on any painted surface, wash it off immediately with water.

18. Remove the brake caliper piston compressor tool.

19. Apply a thin coat of M-77 assembly paste to the pad side of the shims, the back of the brake pads (B), and the other areas indicated by the arrows. Wipe off the excess assembly paste from the pad shims and brake pads friction material.

※※ CAUTION

Keep grease and assembly paste away from the brake disc and brake pads. Contaminated brake disc or brake pads reduce stopping ability.

20. Install the brake pads and pad shims correctly. Install the brake pad with the wear indicator on the upper inside. If you are reusing the brake pads, always reinstall the brake pads in their original positions to prevent a temporary loss of braking efficiency.

21. Pivot the caliper down into position. Install the flange bolt and tighten it to 53 ft. lbs. (72 Nm).

22. Clean the mating surfaces between the brake disc and the inside of the wheel, then install the front wheels.

23. Press the brake pedal several times to make sure the brakes work.

➡Engagement may require a greater pedal stroke immediately after the brake pads have been replaced as a set.

24. Several applications of the brake pedal will restore the normal pedal stroke.

25. Add brake fluid as needed.

26. After installation, check for leaks at hose and line joints or connections, and retighten if necessary.

Test-drive the vehicle, then recheck for leaks.

BRAKES

※※ CAUTION

Dust and dirt accumulating on brake parts during normal use may contain asbestos fibers from production or aftermarket brake linings. Breathing excessive concentrations of asbestos fibers can cause serious bodily harm. Exercise care when servicing brake parts. Do not sand or grind brake lining unless equipment used is designed to contain the dust residue. Do not clean brake parts with compressed air or by dry brushing. Cleaning should be done by dampening the brake components with a fine mist of water, then wiping the brake components clean with a dampened cloth.

Dispose of cloth and all residue containing asbestos fibers in an impermeable container with the appropriate label. Follow practices prescribed by the Occupational Safety and Health Administration (OSHA) and the Environmental Protection Agency (EPA) for the handling, processing, and disposing of dust or debris that may contain asbestos fibers.

BRAKE CALIPER

REMOVAL & INSTALLATION

See Figure 7.

1. Raise and support the vehicle.
2. Remove the rear wheel.

REAR DISC BRAKES

3. Release the parking brake.
4. Remove the brake caliper bracket mounting bolts, then remove the caliper assembly from the knuckle.

※※ CAUTION

To prevent damage to the caliper assembly or brake hose, use a short piece of wire to hang the caliper assembly from the undercarriage. Do not twist the brake hose excessively.

5. Installation is the reverse of the removal procedure.
6. Torque the caliper bracket mounting bolts to 65 ft. lbs. (88 Nm).

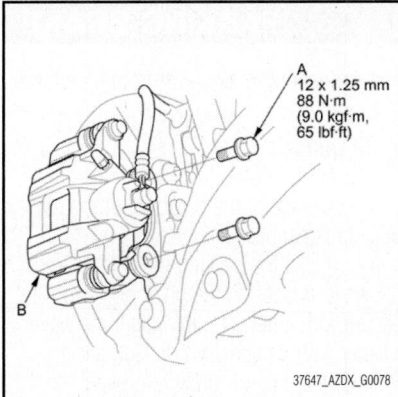

Fig. 7 Remove the brake caliper bracket mounting bolts (A), then remove the caliper assembly (B) from the knuckle.

DISC BRAKE PADS

REMOVAL & INSTALLATION
See Figures 8 and 9.

✻✻ CAUTION

Frequent inhalation of brake pad dust, regardless of material composition, could be hazardous to your health. Avoid breathing dust particles. Never use an air hose or brush to clean brake assemblies. Use an OSHA-approved vacuum cleaner.

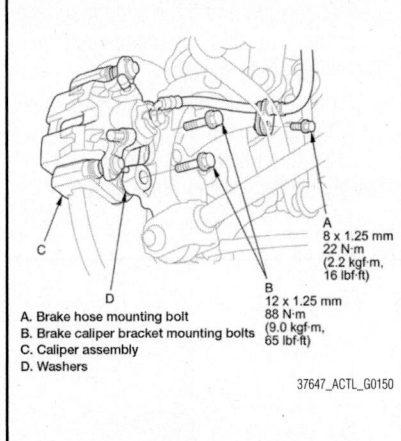

A. Brake hose mounting bolt
B. Brake caliper bracket mounting bolts
C. Caliper assembly
D. Washers

Fig. 8 Remove the brake hose mounting bolt and the brake caliper bracket mounting bolts, then remove the caliper assembly from the knuckle

1. Remove some brake fluid from the master cylinder.
2. Raise and support the vehicle.
3. Remove the rear wheels.
4. Remove the brake hose mounting bolt and the brake caliper bracket mounting bolts, then remove the caliper assembly from the knuckle.

✻✻ CAUTION

To prevent damage to the caliper assembly or brake hose, use a short piece of wire to hang the caliper assembly from the undercarriage. Do not twist the brake hose excessively.

5. Remove the pad shims and the brake pads.
6. Remove the upper and lower pad retainers.

➡**The upper and lower pad retainers are different. During installation, make sure the pad retainers are in their proper positions.**

To install:

7. Clean the caliper bracket thoroughly; remove any rust, and check for grooves and cracks. Verify that the caliper pins move in and out smoothly. Clean and lube if needed.
8. Inspect the brake disc/drum for runout, thickness, parallelism, and check for damage and cracks.
9. Apply a thin coat of M-77 assembly paste (P/N 08798-9010) to the retainer mating surface of the caliper bracket (indicated by the arrows).
10. Install the upper and lower pad retainers. Wipe excess assembly paste off

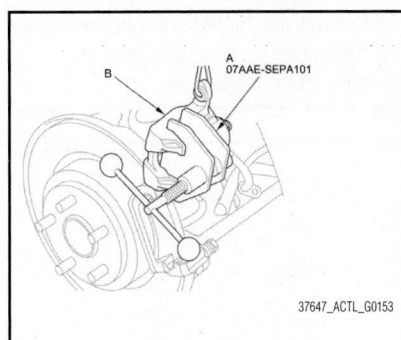

Fig. 9 Install the brake caliper piston compressor tool (A) on the caliper body (B)

the retainers. Keep the assembly paste off the brake disc/drum and the brake pads.

11. Install the brake caliper piston compressor tool on the caliper body.
12. Press in the piston with the brake caliper piston compressor tool so the caliper will fit over the brake pads. Make sure the piston boot is in position to prevent damaging it when pivoting the caliper down.

➡**Be careful when pressing in the piston; brake fluid might overflow from the master cylinder's reservoir. If brake fluid gets on any painted surface, wash it off immediately with water.**

13. Remove the brake caliper piston compressor tool.
14. Apply a thin coat of M-77 assembly paste (P/N 08798-9010) to the pad side of the shims and the back of the brake pads. Wipe excess assembly paste off the pad shims and the brake pads friction material. Keep grease and assembly paste off the brake disc/drum and the brake pads. Contaminated brake disc/drum or brake pads reduce stopping ability.
15. Install the brake pads and pad shims correctly. Install the brake pad with the wear indicator on the bottom inside. If you are reusing the brake pads, always reinstall the brake pads in their original positions to prevent a temporary loss of braking efficiency.
16. Pivot the caliper down into position. Install the flange bolt, and tighten it to 16 ft. lbs. (22 N m).
17. Install the brake hose mounting bolt.
18. Clean the mating surfaces between the brake disc/drum and the inside of the wheel, then install the rear wheels.
19. Press the brake pedal several times to make sure the brakes work.

➡**Engagement may require a greater pedal stroke immediately after the brake pads have been replaced as a set. Several applications of the brake pedal will restore the normal pedal stroke.**

20. Add brake fluid as needed.
21. After installation, check for leaks at hose and line joints or connections, and retighten if necessary.
22. Test-drive the vehicle, then recheck for leaks.

PARKING BRAKE SHOES

REMOVAL & INSTALLATION

See Figures 10 through 15.

1. Raise and support the vehicle.
2. Remove the rear wheels.
3. Release the parking brake and remove the rear brake disc/drum.
4. Disconnect and remove the upper return springs, then remove the shoe guide plate.
5. Remove the tension pins by pushing the respective retainers and turning the pin.
6. Remove the strut and the rod spring.
7. Lower the parking brake shoe assembly.
8. Remove the forward brake shoe and the adjuster assembly by removing the lower return spring.
9. Remove the rearward brake shoe by disconnecting the parking brake cable from the parking brake lever.
10. Remove the U-clip, the wave washer,

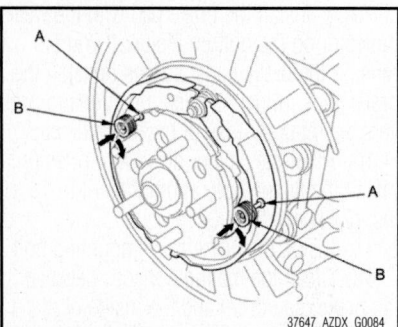

Fig. 10 Remove the tension pins (A) by pushing the respective retainers (B) and turning the pin.

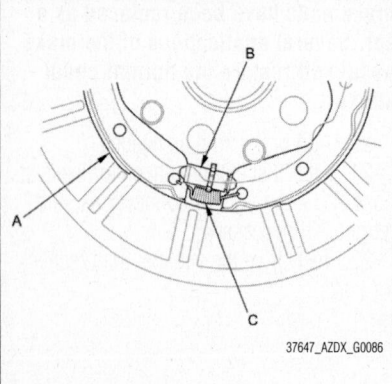

Fig. 11 Remove the forward brake shoe (A) and the adjuster assembly (B) by removing the lower return spring (C).

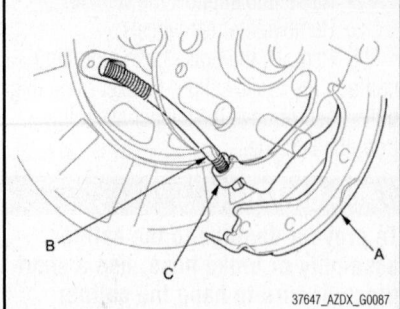

Fig. 12 Remove the rearward brake shoe (A) by disconnecting the parking brake cable (B) from the parking brake lever (C).

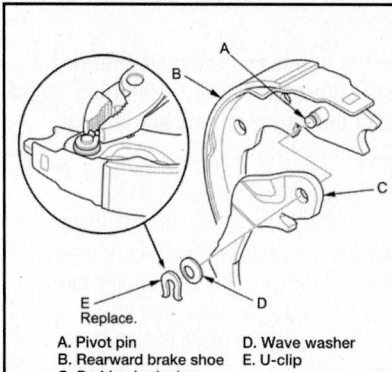

A. Pivot pin   D. Wave washer
B. Rearward brake shoe   E. U-clip
C. Parking brake lever

Fig. 13 Apply Molykote 44MA grease to the sliding surface of the pivot pin of the rearward brake shoe. Install the parking brake lever and the wave washer on the pivot pin, and secure with a new U-clip. Install the wave washer with its convex side facing out. Pinch the U-clip securely to prevent the parking brake lever from coming off the brake shoe.

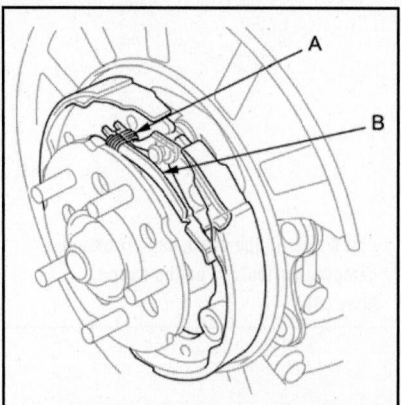

Fig. 14 Install the rod spring (A) to the strut (B) first. Then install the strut on the parking brake shoes.

and the parking brake lever from the brake shoe.

To install:

11. Apply Molykote 44MA grease to the sliding surface of the pivot pin of the rearward brake shoe.
12. Install the parking brake lever and the wave washer on the pivot pin, and secure with a new U-clip. Install the wave washer with its convex side facing out. Pinch the U-clip securely to prevent the parking brake lever from coming off the brake shoe.
13. Connect the parking brake cable to the parking brake lever. Apply Molykote 44MA grease to the cable contact surface on the backing plate.
14. Apply a thin coat of Molykote 44MA grease to the shoe ends and strut ends, sliding surfaces, and opposite edges of the parking brake shoe.

➡ **Wipe off any excess. Keep the grease off the brake linings.**

15. Install the tension pin, the retainer spring, and the retainer on the rearward brake shoe.

➡ **Make sure the tension pin does not contact the parking brake lever.**

16. Install connecting rods on the adjuster nut and note the following:
    a. Clean the threaded portions of connecting rod A and the sliding surface of connecting rod B, then coat them with Molykote 44MA grease.
    b. Shorten connecting rod A by fully turning the adjuster nut.
17. Position the brake shoe adjuster assembly on the parking brake shoes.

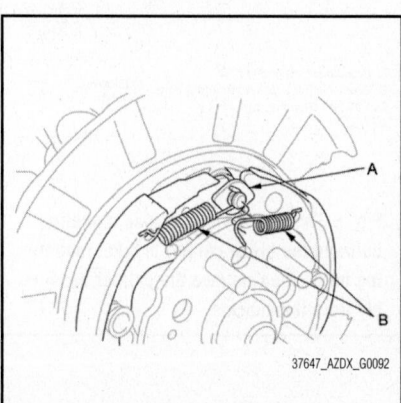

Fig. 15 Install the shoe guide plate (A), then install the upper return springs (B) as shown.

18. Hook the lower return spring on the parking brake shoes.

19. Install the rod spring to the strut first. Then install the strut on the parking brake shoes.

20. Install the tension pin, the retainer spring, and the retainer on the forward brake shoe.

21. Install the shoe guide plate, then install the upper return springs.

22. Install the rear brake disc/drum and the rear brake caliper bracket.

23. Do the parking brake "Major Adjustment".

24. Clean the mating surfaces between the brake/drum and the inside of the wheel, then install the rear wheels.

CHASSIS ELECTRICAL / AIR BAG (SUPPLEMENTAL RESTRAINT SYSTEM)

GENERAL INFORMATION

❋❋ CAUTION

These vehicles are equipped with an air bag system. The system must be disarmed before performing service on, or around, system components, the steering column, instrument panel components, wiring and sensors. Failure to follow the safety precautions and the disarming procedure could result in accidental air bag deployment, possible injury and unnecessary system repairs.

SERVICE PRECAUTIONS

Disconnect and isolate the battery negative cable before beginning any airbag system component diagnosis, testing, removal, or installation procedures. Allow system capacitor to discharge for two minutes before beginning any component service. This will disable the airbag system. Failure to disable the airbag system may result in accidental airbag deployment, personal injury, or death.

Do not place an intact undeployed airbag face down on a solid surface. The airbag will propel into the air if accidentally deployed and may result in personal injury or death.

When carrying or handling an undeployed airbag, the trim side (face) of the airbag should be pointing towards the body to minimize possibility of injury if accidental deployment occurs. Failure to do this may result in personal injury or death.

Replace airbag system components with OEM replacement parts. Substitute parts may appear interchangeable, but internal differences may result in inferior occupant protection. Failure to do so may result in occupant personal injury or death.

Wear safety glasses, rubber gloves, and long sleeved clothing when cleaning powder residue from vehicle after an airbag deployment. Powder residue emitted from a deployed airbag can cause skin irritation. Flush affected area with cool water if irritation is experienced. If nasal or throat irritation is experienced, exit the vehicle for fresh air until the irritation ceases. If irritation continues, see a physician.

Do not use a replacement airbag that is not in the original packaging. This may result in improper deployment, personal injury, or death.

The factory installed fasteners, screws and bolts used to fasten airbag components have a special coating and are specifically designed for the airbag system. Do not use substitute fasteners. Use only original equipment fasteners listed in the parts catalog when fastener replacement is required.

During, and following, any child restraint anchor service, due to impact event or vehicle repair, carefully inspect all mounting hardware, tether straps, and anchors for proper installation, operation, or damage. If a child restraint anchor is found damaged in any way, the anchor must be replaced. Failure to do this may result in personal injury or death.

Deployed and non-deployed airbags may or may not have live pyrotechnic material within the airbag inflator.

Do not dispose of driver/passenger/ curtain airbags or seat belt tensioners unless you are sure of complete deployment. Refer to the Hazardous Substance Control System for proper disposal.

Dispose of deployed airbags and tensioners consistent with state, provincial, local, and federal regulations.

After any airbag component testing or service, do not connect the battery negative cable. Personal injury or death may result if the system test is not performed first.

If the vehicle is equipped with the Occupant Classification System (OCS), do not connect the battery negative cable before performing the OCS Verification Test using the scan tool and the appropriate diagnostic information. Personal injury or death may result if the system test is not performed properly.

Never replace both the Occupant Restraint Controller (ORC) and the Occupant Classification Module (OCM) at the same time. If both require replacement, replace one, then perform the Airbag System test before replacing the other.

Both the ORC and the OCM store Occupant Classification System (OCS) calibration data, which they transfer to one another when one of them is replaced. If both are replaced at the same time, an irreversible fault will be set in both modules and the OCS may malfunction and cause personal injury or death.

If equipped with OCS, the Seat Weight Sensor is a sensitive, calibrated unit and must be handled carefully. Do not drop or handle roughly. If dropped or damaged, replace with another sensor. Failure to do so may result in occupant injury or death.

If equipped with OCS, the front passenger seat must be handled carefully as well. When removing the seat, be careful when setting on floor not to drop. If dropped, the sensor may be inoperative, could result in occupant injury, or possibly death.

If equipped with OCS, when the passenger front seat is on the floor, no one should sit in the front passenger seat. This uneven force may damage the sensing ability of the seat weight sensors. If sat on and damaged, the sensor may be inoperative, could result in occupant injury, or possibly death.

DISARMING THE SYSTEM

Disconnect and isolate the negative battery cable. Wait 3 minutes for the system capacitor to discharge before performing any service.

ARMING THE SYSTEM

After performing service, connect the negative battery cable to re-arm the SRS system.

CLOCKSPRING CENTERING

1. First rotate the cable reel clockwise until it stops.

2. Then rotate it counterclockwise (three full turns) until the arrow mark on the cable reel label points straight up.

3. Position the two tabs of the turn signal canceling sleeve as shown, and install the steering wheel on to the steering column shaft, making sure the steering wheel hub engages the pins of the cable reel and tabs of the turn signal canceling sleeve. Do not tap on the steering wheel or steering column shaft when installing the steering wheel.

DRIVE TRAIN

DIFFERENTIAL CARRIER

REMOVAL & INSTALLATION

See Figures 16 through 20.

Special Tools Required: Driveshaft
Remover 07AAD-S3VA000
1. Raise and support the vehicle.
2. Drain the rear differential fluid.
3. Remove the rear wheels.
4. Remove the lower arm A
mounting bolts and the lower arm B
mounting bolts.
5. Disconnect the inboard joints from
the rear differential using the driveshaft
remover and a hammer.
6. Pull the knuckle outward, and separate the rear driveshaft inboard joints from
the rear differential.

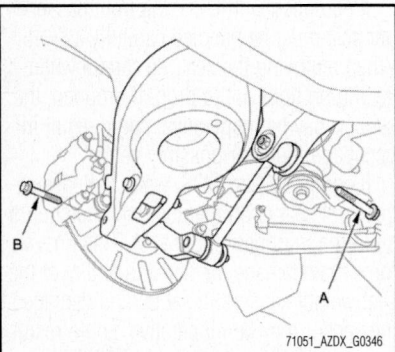

**Fig. 16 Remove the lower arm A mount-
ing bolts (A) and the lower arm B mounting bolts (B)**

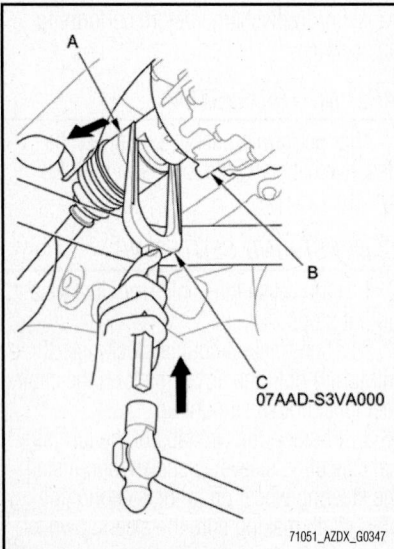

**Fig. 17 Disconnect the inboard joints (A)
from the rear differential (B) using the driveshaft remover (C) and a hammer**

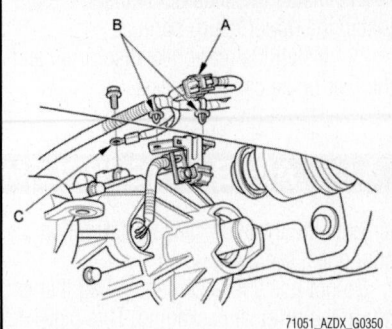

**Fig. 18 Disconnect the left solenoid 4P
connector (A), then remove the harness
clips (B); disconnect the ground cable (C)
from the rear differential**

7. Make reference mark across the No.
2 propeller shaft and the rear differential
companion flange. Separate the No. 2 propeller shaft from the rear differential.

➡**Suspend the propeller shaft with an
appropriate size nylon strap.**

8. Remove the spare tire.
9. Place a transmission jack under the
rear differential.
10. Remove the rear differential rear
mount bolts and the rear differential front
mounting bracket bolts.
11. Slightly lower the rear differential on
the transmission jack.

➡**Make sure you do not over extend the
wire harness and the breather hose.**

12. Disconnect the left solenoid 4P connector, then remove the harness clips.
13. Disconnect the ground cable from
the rear differential.
14. Slightly lower the rear differential on
the transmission jack.
15. Disconnect the right solenoid 4P
connector and the rear differential fluid

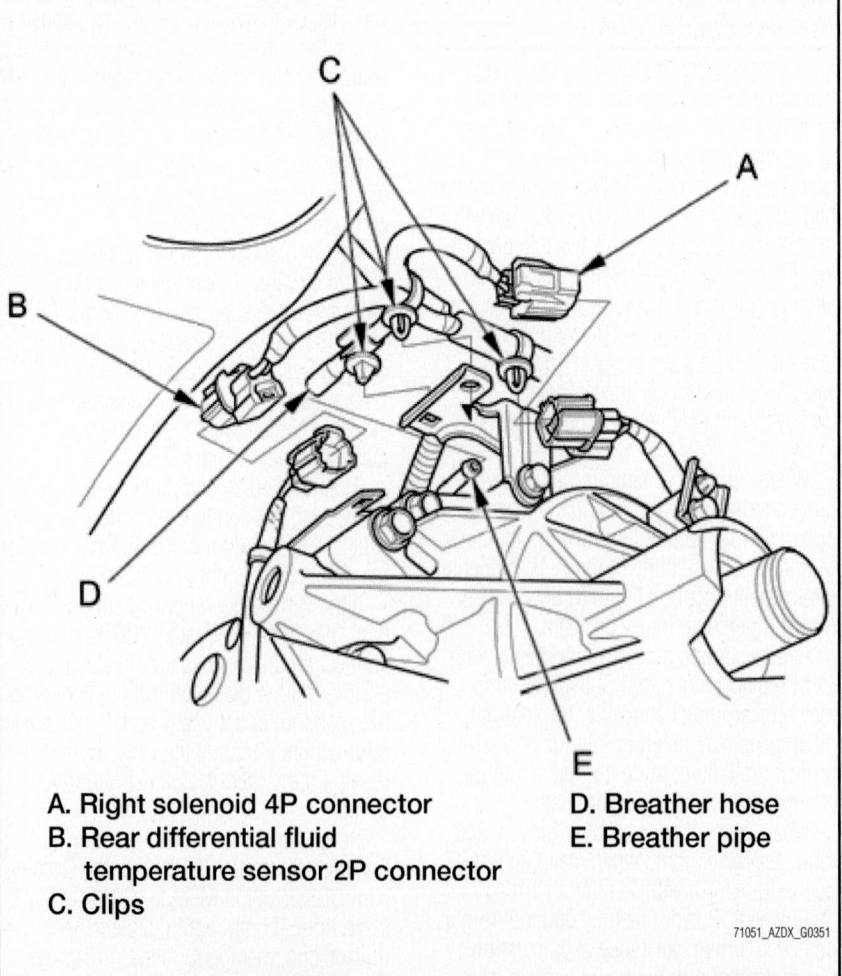

A. Right solenoid 4P connector
B. Rear differential fluid
  temperature sensor 2P connector
C. Clips
D. Breather hose
E. Breather pipe

**Fig. 19 Disconnect the right solenoid 4P connector and the rear differential fluid temperature
sensor 2P connector, then remove the clips**

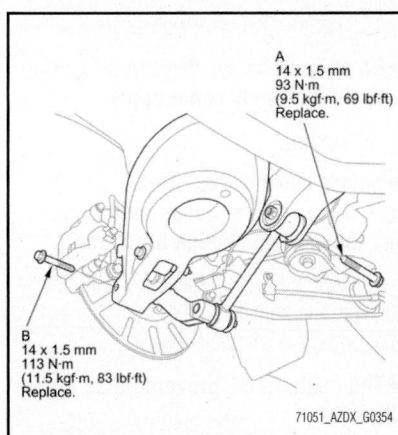

A
14 x 1.5 mm
93 N·m
(9.5 kgf·m, 69 lbf·ft)
Replace.

B
14 x 1.5 mm
113 N·m
(11.5 kgf·m, 83 lbf·ft)
Replace.

71051_AZDX_G0354

Fig. 20 Loosely install the new lower arm A mounting bolts (A) and the new lower arm B mounting bolts (B)

temperature sensor 2P connector, then remove the clips.

16. Disconnect the breather hose from the breather pipe.

17. Lower the rear differential on the transmission jack.

18. Remove the set rings from the rear differential.

To install:

19. Place the rear differential on a transmission jack.

20. Install new set rings into the groove of the rear differential.

21. Raise the rear differential to the appropriate position on the transmission jack.

22. Connect the ground cable to the rear differential.

23. Connect the left solenoid 4P connector, then install the harness clips.

24. Raise the rear differential a little on the transmission jack.

25. Connect the breather hose to the breather pipe.

26. Connect the right solenoid 4P connector and the rear differential fluid temperature sensor 2P connector, then install the clips.

27. Raise the rear differential to the mounting level.

28. Temporarily install the rear differential front mounting bracket bolts.

29. Loosen the rear differential front mounting bolts.

30. Install the new rear differential rear mount bolts and tighten them to the specified torque in the sequence shown.

31. Replace the rear differential front mounting bolts and tighten them to the specified torque.

32. Tighten the rear differential front mounting bracket bolts to the specified torque.

33. Attach the No. 2 propeller shaft to the rear differential companion flange by aligning the reference marks made during the removal procedure.

34. Apply appropriate grease to the rear driveshaft inboard joints.

35. Pull the knuckle outward, and insert the rear driveshaft inboard joints into the rear differential.

36. Loosely install the new lower arm A mounting bolts and the new lower arm B mounting bolts.

37. Place a floor jack under the lower arm, and raise the suspension to load it with the vehicle's weight.

38. Tighten the lower arm A mounting bolts and the lower arm B mounting bolts to the specified torque, then remove the floor jack.

39. Install the rear wheels.

40. Install the spare tire.

41. Refill the rear differential with the recommended fluid.

42. Check the wheel alignment, and adjust it if necessary.

43. Test-drive the vehicle.

DRIVESHAFT

REMOVAL & INSTALLATION

Rear

See Figures 21 through 23.

1. Raise and support the vehicle.

2. Remove the propeller shaft protector.

3. Make reference mark across the No. 1 propeller shaft and the transfer companion flange.

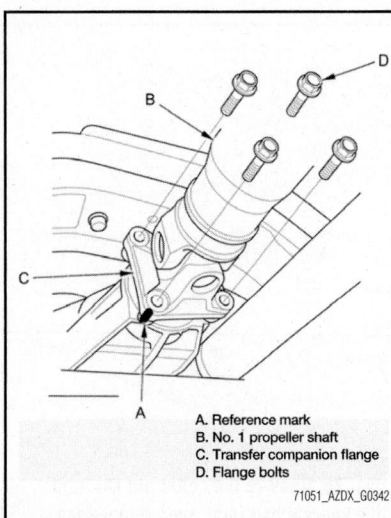

A. Reference mark
B. No. 1 propeller shaft
C. Transfer companion flange
D. Flange bolts

71051_AZDX_G0342

Fig. 21 Make reference mark across the No. 1 propeller shaft and the transfer companion flange

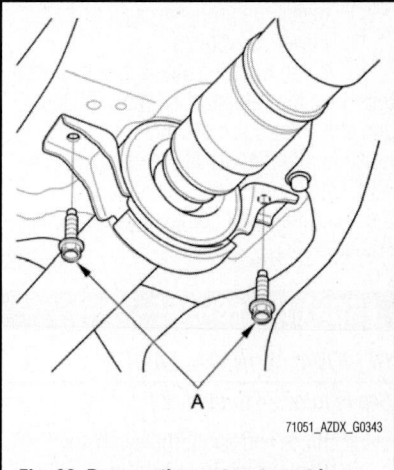

A

71051_AZDX_G0343

Fig. 22 Remove the center support bearing mounting bolts (A)

4. Remove the No. 1 propeller shaft flange bolts.

5. Remove the center support bearing mounting bolts.

6. Make reference mark across the No. 2 propeller shaft and the rear differential companion flange.

7. Remove the No. 2 propeller shaft flange bolts.

To install:

8. Set the No. 2 propeller shaft to the rear differential companion flange by aligning the reference mark you made during the removal procedure. Then install new No. 2 propeller shaft flange bolts to 53 ft. lbs. (72 Nm).

➡ **When replacing the propeller shaft or the rear differential, align the factory reference marks.**

9. Install new center support bearing mounting bolts. Tighten to 29 ft. lbs. (39 Nm).

10. Set the No. 1 propeller shaft to the transfer companion flange by aligning the reference mark you made during the removal procedure. Then install new No. 1 propeller shaft flange bolts to 53 ft. lbs. (72 Nm).

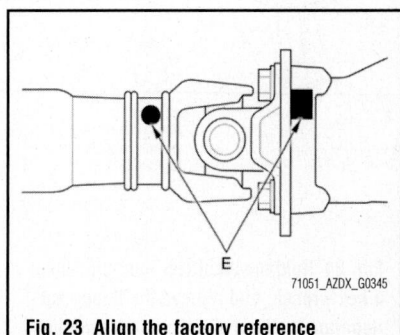

E

71051_AZDX_G0345

Fig. 23 Align the factory reference marks (E)

11. Install the propeller shaft protector.

12. Lower the vehicle.

13. If you installed a new propeller shaft, test-drive the vehicle at 55 mph (88 km/h), and check for noise or vibration. If there is a noise or vibration, rotate the propeller shaft 180 degrees from its current alignment with the rear differential companion flange, then recheck.

FRONT HALFSHAFT

REMOVAL & INSTALLATION

See Figures 24 through 29.

1. Raise and support the vehicle.

2. Remove the front wheels.

3. Pry up the stake on the spindle nut, then remove the nut.

4. Drain the transmission fluid, then reinstall the drain plug with a new sealing washer:

5. Hold the stabilizer joint pin using a hex wrench, and remove the flange nut. Separate the front stabilizer link from the lower arm.

6. Remove the damper fork mounting nut, the damper fork mounting bolt, and the damper pinch bolt, then remove the damper fork.

7. Remove the cotter pin from the knuckle ball joint, then remove the castle nut. Separate the ball joint from the lower arm using the ball joint remover.

➡ **Be careful not to damage the ball joint boot when installing the remover. Do not force or hammer on the lower arm, or pry between the lower arm and the knuckle. You could damage the ball joint.**

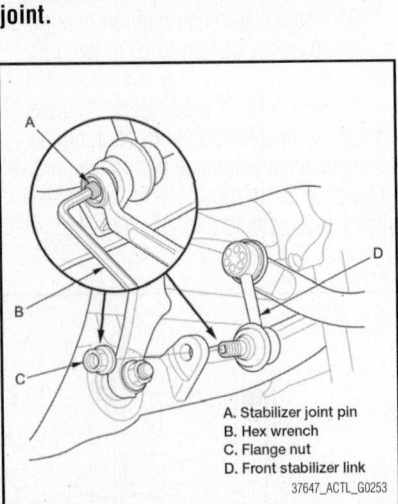

A. Stabilizer joint pin
B. Hex wrench
C. Flange nut
D. Front stabilizer link

37647_ACTL_G0253

Fig. 24 Hold the stabilizer joint pin using a hex wrench, and remove the flange nut. Separate the front stabilizer link from the lower arm.

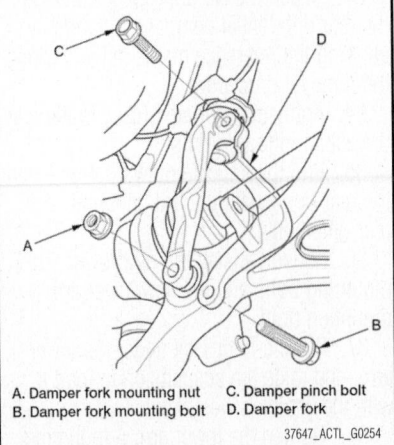

A. Damper fork mounting nut   **C. Damper pinch bolt**
B. Damper fork mounting bolt   **D. Damper fork**

37647_ACTL_G0254

Fig. 25 Remove the damper fork mounting nut, the damper fork mounting bolt, and the damper pinch bolt, then remove the damper fork

8. Pull the knuckle outward, and separate the outboard joint from the front hub using a soft face hammer.

9. Remove exhaust pipe A.

10. Left driveshaft: Pry the inboard joint from the differential using a prybar.

11. Remove the driveshaft as an assembly.

➡ **Do not pull on the driveshaft, or the inboard joint may come apart. Pull the inboard joint straight out to avoid damaging the oil seal. Be careful not to damage the oil seal or the end of the inboard joint using the prybar.**

12. Right driveshaft: Drive the inboard joint off of the intermediate shaft using a drift punch and a hammer.

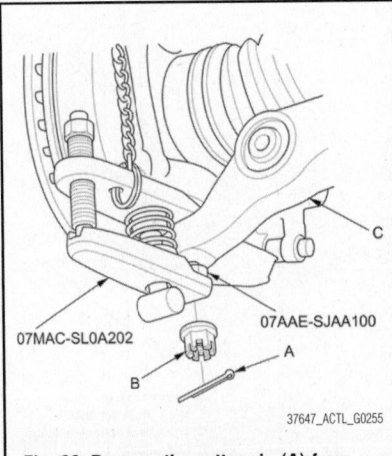

07MAC-SL0A202   07AAE-SJAA100

37647_ACTL_G0255

Fig. 26 Remove the cotter pin (A) from the knuckle ball joint, then remove the castle nut (B). Separate the ball joint from the lower arm (C) using the ball joint remover.

13. Remove the driveshaft as an assembly.

➡ **Do not pull on the driveshaft, or the inboard joint may come apart.**

To install:

➡ **Before starting installation, make sure the mating surfaces of the joint and the splined section are clean.**

14. Apply Moly 60 paste (P/N 08734-0001) to the contact area of the outboard joint and the front wheel bearing.

➡ **The paste helps prevent noise and vibration.**

15. Install a new set ring into the set ring groove of the left driveshaft inboard joint.

16. Install a new set ring into the set ring groove of the intermediate shaft.

17. Apply small quantity of super high temperature grease (P/N 08798-9002) to the whole splined surface of the right driveshaft.

18. After applying grease, remove the grease from the splined grooves at intervals of 2–3 splines and from the set ring groove so that air can bleed from the intermediate shaft.

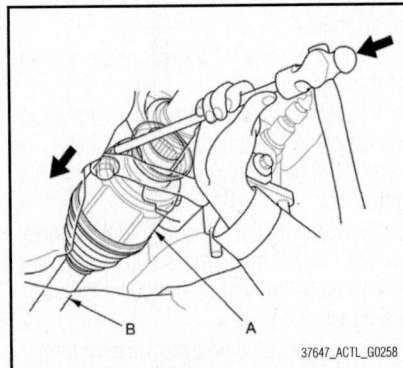

37647_ACTL_G0258

Fig. 27 Drive the inboard joint (A) off of the intermediate shaft using a drift punch and a hammer

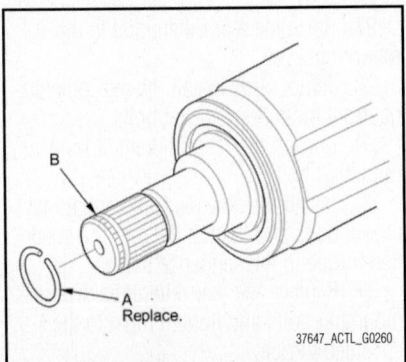

A. Replace.

37647_ACTL_G0260

Fig. 28 Install a new set ring (A) into the set ring groove (B) of the left driveshaft inboard joint

19. Clean the areas where the driveshaft contacts the differential thoroughly with solvent, and dry them with compressed air.

➡**Do not wash the rubber parts with solvent.**

20. Insert the inboard end of the driveshaft into the differential or the intermediate shaft until the set ring locks in the groove.

➡**Insert the driveshaft horizontally to prevent damaging the oil seal.**

21. Install the outboard joint into the front hub on the knuckle.

22. Install exhaust pipe A.

23. Wipe off any grease contamination from the ball joint tapered section and threads, then install the knuckle onto the lower arm. Be careful not to damage the ball joint boot. Wipe off the grease before tightening the nut at the ball joint. Torque the castle nut to 69–76 ft. lbs. (93–103 N m), then tighten it only far enough to align the slot with the ball joint pin hole.

➡**Make sure the ball joint boot is not damaged or cracked. Do not align the nut by loosening it.**

24. Install a new cotter pin into the ball joint pin hole, and bend the cotter pin.

25. Install the damper fork over the driveshaft and onto the lower arm. Install the damper in the damper fork so the aligning tab is aligned with the slot in the damper fork. Loosely install the damper pinch bolt.

26. Loosely install a new flange bolt and a new damper fork mounting nut.

27. Connect the front stabilizer link to the lower arm, and loosely install the flange nut. Hold the stabilizer link ball joint pin using a hex wrench, and tighten the flange nut to 58 ft. lbs. (78 N m).

28. Place a floor jack under the lower arm, and raise the suspension to load it with the vehicle's weight.

➡**Do not put the floor jack under the ball joint.**

29. Tighten the damper pinch bolt to 36 ft. lbs. (49 N m) and the damper fork mounting nut to 47 ft. lbs. (64 N m)while holding the flange bolt, then remove the floor jack.

30. Apply a small amount of engine oil to the seating surface of a new spindle nut.

31. Install the spindle nut, then tighten it to 242 ft. lbs. (329 N m). After tightening, use a drift to stake the spindle nut shoulder against the driveshaft.

32. Clean the mating surfaces of the brake disc and the wheel, then install the front wheels.

33. Turn the front wheel by hand, and make sure there is no interference between the driveshaft and the surrounding parts.

34. Lower the vehicle .

35. Refill the transmission with recommended transmission fluid:

36. Check the wheel alignment, and adjust it if necessary.

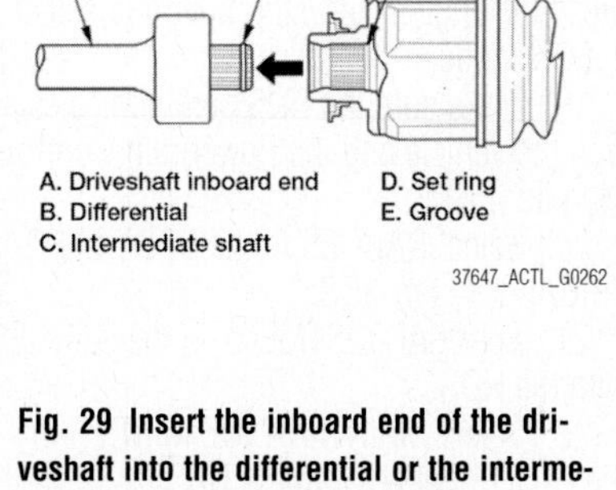

REAR DRIVE AXLE

DRAIN & REFILL

See Figure 30.

1. Park the vehicle on level ground, and turn the ignition switch to LOCK (0), or press the engine start/stop button to select the OFF mode.

2. Remove the filler plug and the sealing washer.

3. Remove the drain plug and the sealing washer, and drain the fluid.

4. Reinstall the drain plug with a new sealing washer.

5. Refill the recommended fluid until it reaches the bottom of the filler plug hole.

6. Reinstall the filler plug with a new sealing washer.

7. If the Maintenance Minder required to replace the rear differential fluid, reset the Maintenance Minder, and this procedure is complete. If the Maintenance Minder did not require to replace the rear differential fluid, go to step 8.

8. With the ignition switch in LOCK (0), connect the HDS to the data link connector (DLC) behind the driver's dashboard lower cover.

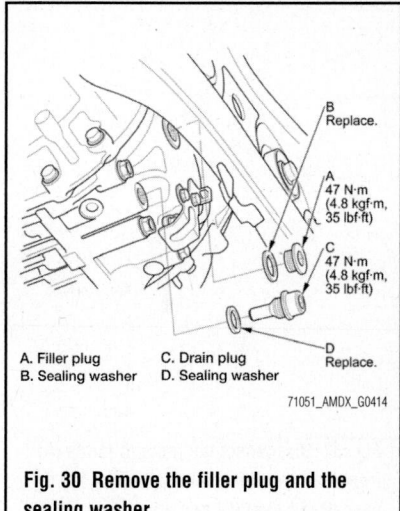

A. Filler plug   C. Drain plug
B. Sealing washer   D. Sealing washer

71051_AMDX_G0414

Fig. 30 Remove the filler plug and the sealing washer

9. Turn the ignition switch to ON (II), or press the engine stop/start button to select the ON mode.

10. Make sure the HDS communicates with the vehicle and the PCM. If it does not, go to the DLC circuit troubleshooting.

11. Select GAUGES in the BODY ELECTRICAL with the HDS.

12. Select ADJUSTMENT in the GAUGES with the HDS.

13. Select MAINTENANCE MINDER in the ADJUSTMENT with the HDS.

14. Select RESET in the MAINTENANCE MINDER with the HDS.

15. Select MAINTENANCE SUB ITEM 6 RESET with the HDS, and reset the rear differential fluid life.

REAR HALFSHAFT

REMOVAL & INSTALLATION

See Figures 31 through 33.

1. Raise the vehicle on a lift.
2. Drain the rear differential fluid.

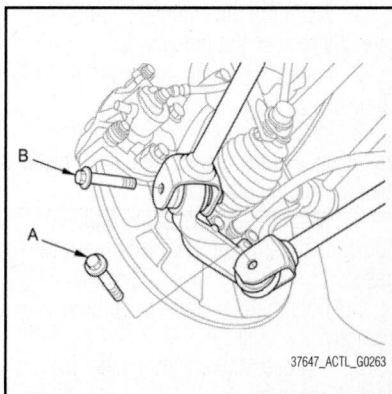

37647_ACTL_G0263

Fig. 31 Remove the lower arm A mounting bolts (A) and the lower arm B mounting bolts (B)

Left

Right

A. Driveshaft inboard end   D. Set ring
B. Differential   E. Groove
C. Intermediate shaft

37647_ACTL_G0262

Fig. 29 Insert the inboard end of the driveshaft into the differential or the intermediate shaft until the set ring locks in the groove

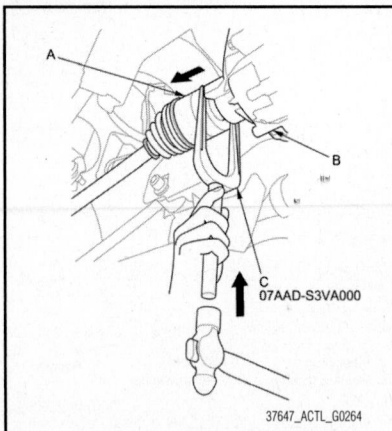

Fig. 32 Disconnect the inboard joints (A) from the rear differential (B) using the driveshaft remover (C) and a hammer

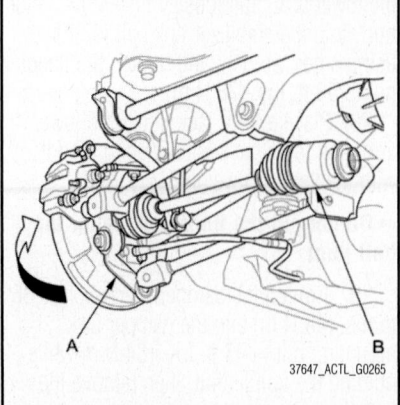

Fig. 33 Pull the knuckle (A) outward, and separate the rear driveshaft inboard joints (B) from the rear differential

3. Remove the rear wheels.

4. Remove exhaust pipe B and the muffler.

5. Remove the lower arm A mounting bolts and the lower arm B mounting bolts.

6. Disconnect the inboard joints from the rear differential using the driveshaft remover and a hammer.

7. Pull the knuckle outward, and separate the rear driveshaft inboard joints from the rear differential.

To install:

8. Pull the knuckle outward, and insert the rear driveshaft inboard joints into the rear differential.

9. Loosely install new lower arm A mounting bolts and new lower arm B mounting bolts.

10. Place a floor jack under the lower arm, and raise the suspension to load it with the vehicle's weight.

11. Tighten the lower arm A mounting bolt and the lower arm B mounting bolt to 43 ft. lbs. (59 N m).

12. Install exhaust pipe B and the muffler.

13. Install the rear wheels.

14. Refill the rear differential with the recommended fluid.

15. Test-drive the vehicle.

ENGINE COOLING

ENGINE COOLANT

BLEEDING

1. Loosely install the radiator cap.

2. Start the engine, and let it run until it warms up (the radiator fan comes on at least twice).

3. Turn off the engine. Check the level in the radiator, and add Acura Long Life Antifreeze/Coolant Type 2, if needed.

4. Put the radiator cap on securely, then run the engine again, and check for leaks.

5. Clean up any spilled engine coolant.

DRAIN & REFILL

See Figure 34.

1. Wait until the engine is cool, then carefully remove the radiator cap.

2. Loosen the drain plug and drain the coolant.

3. Install a rubber hose on the drain bolt located at the rear of the engine block, then loosen the drain bolt.

4. When the coolant stops draining, tighten the drain bolt. Remove the rubber hose.

5. Tighten the radiator drain plug securely.

6. Remove, drain, and reinstall the coolant reservoir.

7. Fill the coolant reservoir to the MAX mark with Acura Long Life Antifreeze/Coolant Type 2.

8. Pour coolant into the radiator up to the base of the filler neck.

9. Engine Coolant Capacities:
   - At Coolant Change: 1.77 gal. (6.7 L)
   - After Engine Overhaul: 2.27 gal. (8.6 L)

➡**Manufacturer recommends to always use Acura Long Life Antifreeze/Coolant Type 2. Using a non-Acura coolant can result in corrosion, causing the cooling system to malfunction or fail.**

➡**Acura Long Life Antifreeze/Coolant Type 2 is a mixture of 50% antifreeze and 50% water. Do not add water.**

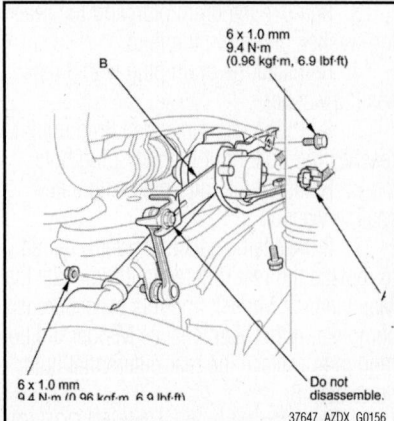

Fig. 34 Install a rubber hose (A) on the drain bolt (B) located at the rear of the engine block, then loosen the drain bolt.

10. Loosely install the radiator cap.

11. Start the engine, and let it run until it warms up (the radiator fan comes on at least twice).

12. Turn off the engine. Check the level in the radiator, and add coolant, if needed.

13. Put the radiator cap on securely, then run the engine again, and check for leaks.

14. Clean up any spilled engine coolant.

15. If the maintenance minder required engine coolant replacement, reset the maintenance minder, and this procedure is complete. If the maintenance minder did not require engine coolant replacement, go to step 15.

16. Turn the ignition switch to LOCK (0), or press the engine start/stop button to select the OFF mode.

17. Connect the Honda Diagnostic System (HDS), or equivalent scan tool, to the data link connector (DLC).

18. Turn the ignition switch to ON (II), or press the engine start/stop button to select the ON mode.

19. Make sure the HDS communicates with the vehicle and the Powertrain Control Module (PCM).

20. Select GAUGES in the BODY ELECTRICAL with the HDS.

21. Select ADJUSTMENT in the GAUGES with the HDS.

22. Select MAINTENANCE MINDER in the ADJUSTMENT with the HDS.

23. Select RESET in the MAINTENANCE MINDER with the HDS.

24. Select MAINTENANCE SUB ITEM 5 RESET with the HDS.

ELECTRIC ENGINE FAN

REMOVAL & INSTALLATION

See Figures 35 and 36.

1. Drain the engine coolant.
2. Remove the bulkhead cover.
3. Remove the battery.
4. Remove the battery base.
5. Disconnect the radiator fan motor connector, the engine coolant temperature (ECT) sensor 2 sub harness connector, and the ECT sensor 2 connector, then remove the harness clamp.
6. Disconnect the upper radiator hose.
7. Remove the radiator fan shroud assembly.
8. Remove the harness clamp and the wire harness.
9. Disassemble the radiator fan shroud assembly.

To install:

10. Reassemble the radiator fan shroud assembly.
11. Install the wire harness and the harness clamp.
12. Install the radiator fan shroud assembly, then tighten the bolts.
13. Connect the upper radiator hose.
14. Install the harness clamp, then connect the ECT sensor 2 connector, the ECT sensor 2 sub harness connector, and the radiator fan motor connector.
15. Install the battery base.
16. Do the battery installation procedure.
17. Install the bulkhead cover.
18. Fill the radiator with engine coolant, and bleed the air from the cooling system.
19. Clean up any spilled engine coolant.

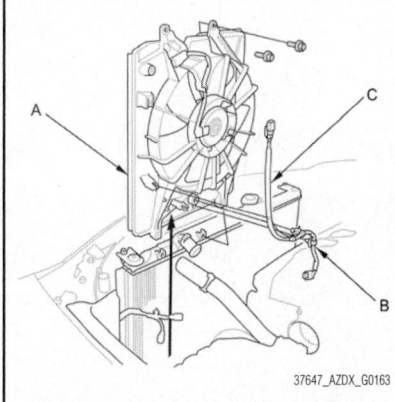

Fig. 36 Remove the radiator fan shroud assembly (A). Remove the harness clamp (B) and the wire harness (C).

RADIATOR

REMOVAL & INSTALLATION

See Figures 37 and 38.

1. Drain the engine coolant.
2. Remove the bulkhead cover.
3. Remove the battery.
4. Remove the battery base.
5. Remove the A/C condenser fan shroud assembly..
6. Remove the radiator fan shroud assembly.
7. Disconnect the lower radiator hose.
8. Remove the power steering (P/S) hose clamp.
9. Disconnect the ATF cooler hoses from the transaxle, then plug the ATF cooler hoses and the lines.
10. Disconnect the ATF cooler hoses, then plug the ATF cooler hoses and the lines.
11. Remove the radiator upper brackets, then pull up the radiator.

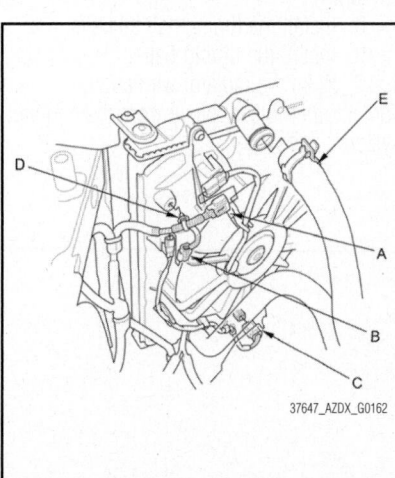

Fig. 35 Detaching connectors, clamps and radiator hose

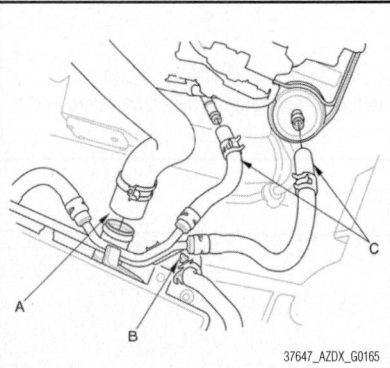

Fig. 37 Disconnect the lower radiator hose (A), remove the power steering (P/S) hose clamp (B), and disconnect the ATF cooler hoses (C) from the transaxle.

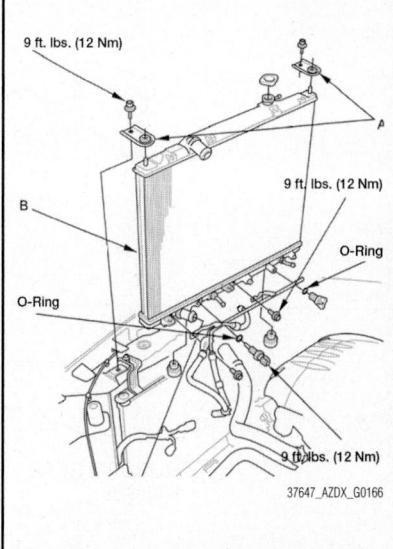

Fig. 38 Remove the radiator upper brackets (A), then pull up the radiator (B).

12. Remove the other parts from the radiator.
13. Install the radiator in the reverse order of removal. Make sure the upper and lower cushions are set securely.
14. Install the battery base.
15. Do the battery installation procedure.
16. Refill the transaxle with ATF.
17. Fill the radiator with engine coolant, and bleed the air from the cooling system.
18. Clean up any spilled engine coolant.

THERMOSTAT

REMOVAL & INSTALLATION

See Figure 39.

1. Drain the engine coolant.
2. Remove the intake air duct.
3. Remove the thermostat cover, then remove the thermostat.

To install:

4. Install the new thermostat with a new rubber seal, then install the thermostat cover.
5. Install the intake air duct.
6. Refill the radiator with engine coolant, and bleed the air from the cooling system.
7. Clean up any spilled engine coolant.

WATER PUMP

REMOVAL & INSTALLATION

See Figure 40.

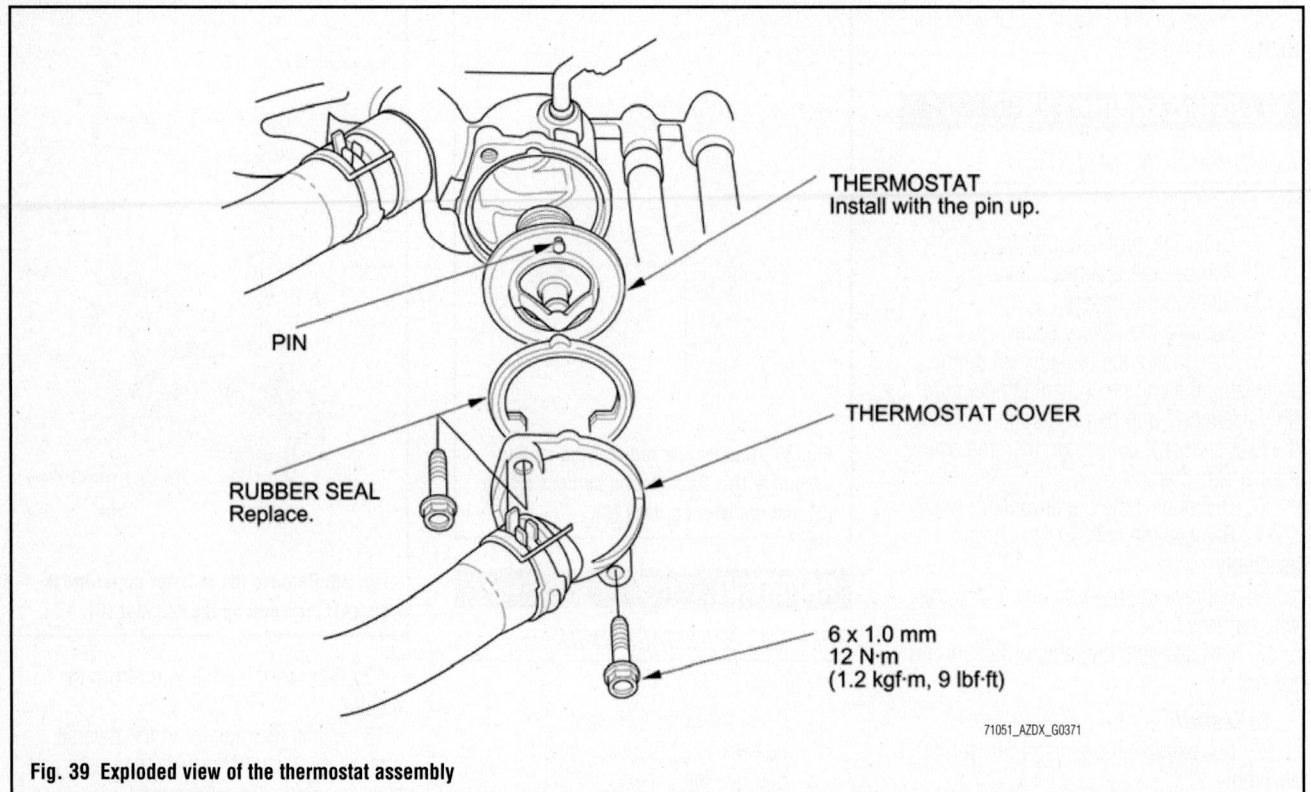

THERMOSTAT
Install with the pin up.

PIN

THERMOSTAT COVER

RUBBER SEAL
Replace.

6 x 1.0 mm
12 N·m
(1.2 kgf·m, 9 lbf·ft)

71051_AZDX_G0371

Fig. 39 Exploded view of the thermostat assembly

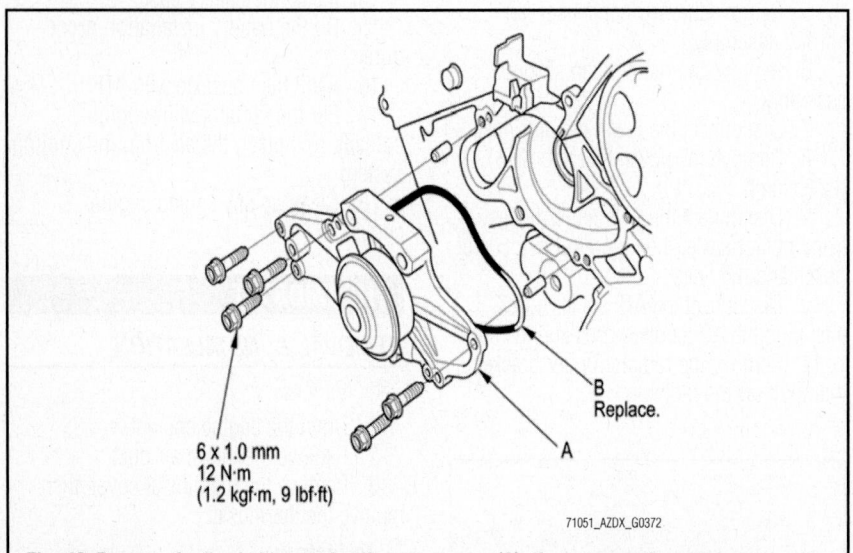

6 x 1.0 mm
12 N·m
(1.2 kgf·m, 9 lbf·ft)

B
Replace.

A

71051_AZDX_G0372

Fig. 40 Remove the five bolts securing the water pump (A), then remove the water pump and the O-ring (B)

1. Drain the engine coolant.
2. Remove the timing belt.
3. Remove the timing belt adjuster.
4. Remove the five bolts securing the water pump, then remove the water pump.
5. Inspect and clean the O-ring groove and the mating surface of the engine block.

To install:

6. Install the water pump with a new O-ring.
7. Clean up any spilled engine coolant.
8. Install the timing belt adjuster.
9. Install the timing belt.
10. Refill the radiator with engine coolant, then bleed the air from the cooling system.

ENGINE ELECTRICAL | BATTERY SYSTEM

BATTERY

REMOVAL & INSTALLATION

See Figure 41.

➡**The battery terminal disconnection/reconnection procedure must be done before and after doing this procedure. Some systems store data in memory that is lost when the battery is disconnected.**

1. Do the battery terminal disconnection procedure.
2. Remove the two nuts securing the battery hold down plate, then remove the battery setting plate and the battery.

To install:
3. Install the battery, then install the battery hold down plate.
4. Tighten the two nuts equally until the battery is stable.

➡**Do not deform the battery hold down plate by over-tightening the nuts.**

5. Do the battery terminal reconnection procedure.

➡**Make sure the battery is installed correctly, and the positive and negative terminals are not reverse connected.**

BATTERY RECONNECT/RELEARN PROCEDURE

Disconnection

➡**Some systems store data in memory (including seat position, mirror position, etc) that is lost when the battery is disconnected. Do the following procedures before disconnecting the battery.**

1. Make sure to have the anti-theft code for the audio system or the audio-navigation unit.
2. Make sure the ignition switch is in LOCK (0), or the vehicle ignition in the OFF mode.
3. Disconnect and isolate the negative cable from the battery.

➡**Always disconnect the negative cable from the battery first.**

4. Disconnect the positive cable from the battery.

Reconnection

➡**Some systems store data in memory (including seat position, mirror position, etc) that is lost when the battery is disconnected. Do the following procedures to restore the systems back to normal operation.**

1. Clean the battery terminals.
2. Test the battery.
3. Reconnect the positive cable to the battery first, then reconnect the negative cable to the battery.

➡**Always connect the positive cable to the battery first.**

4. Apply multipurpose grease to the terminals to prevent corrosion.
5. Do the steering column position memorization.
6. Enter the anti-theft code for the audio system or the audio-navigation unit.

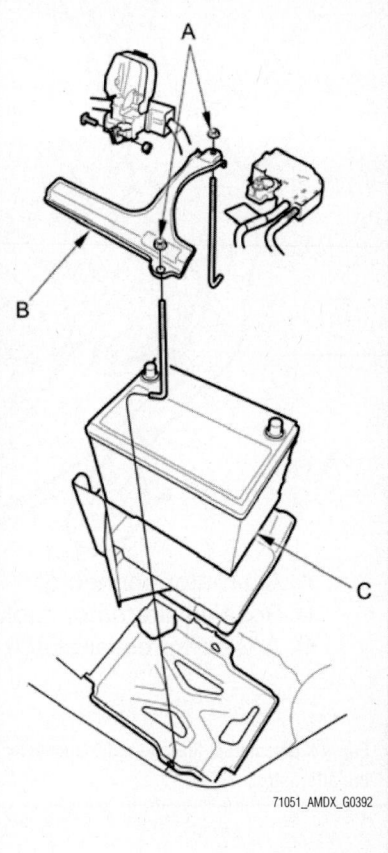

71051_AMDX_G0392

Fig. 41 Remove the two nuts (A) securing the battery hold down plate, then remove the battery setting plate (B) and the battery (C)

7. Set the clock (for vehicles without navigation).

ENGINE ELECTRICAL | CHARGING SYSTEM

ALTERNATOR

REMOVAL & INSTALLATION

See Figures 42 and 43.

1. Do the battery terminal disconnection procedure.
2. Remove the A/C condenser fan shroud assembly.
3. Remove the drive belt.
4. Disconnect the alternator connector and the positive alternator cable from the alternator.
5. Disconnect the A/C compressor clutch connector from the A/C compressor.
6. Remove the bolt securing the harness holder.
7. Remove the mounting bolt and the alternator bracket mounting bolt, then remove the alternator.

To install:
8. Install the alternator, then tighten the mounting bolt to 33 ft. lbs. (45 Nm) and tighten the alternator bracket mounting bolt to 16 ft. lbs. (22 Nm).
9. Install the bolt securing the harness holder.
10. Connect the A/C compressor clutch connector to the A/C compressor.
11. Connect the alternator connecter and the positive alternator cable to the alternator. Make sure the crimped side of the ring terminal faces away from the alternator when you connect it.
12. Install the drive belt.
13. Install the A/C condenser fan shroud assembly.
14. Do the battery terminal reconnection procedure.

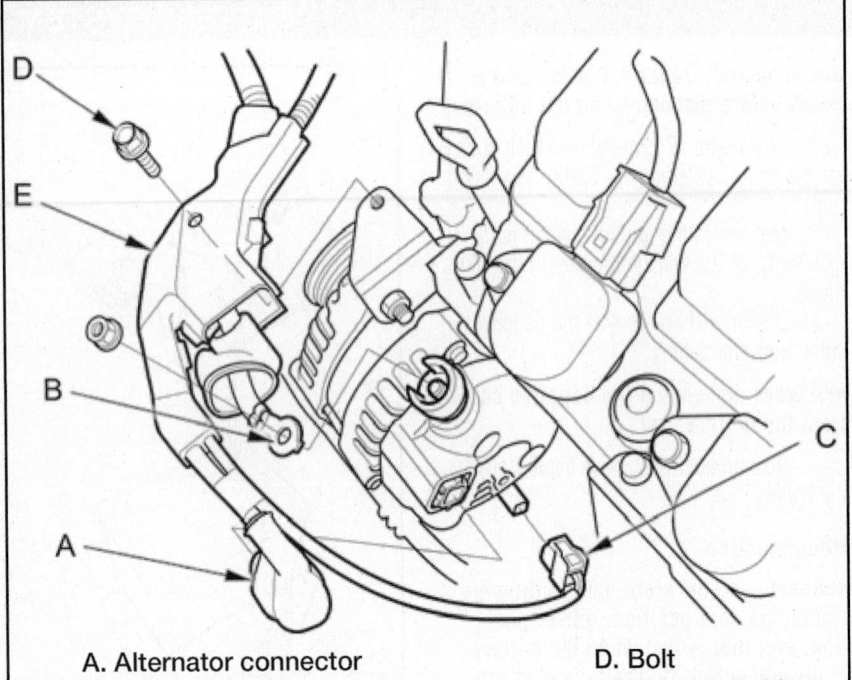

A. Alternator connector
B. Positive alternator cable
C. A/C compressor clutch connector
D. Bolt
E. Harness holder

71051_AZDX_G0374

Fig. 42 Disconnect the alternator connector and the positive alternator cable from the alternator

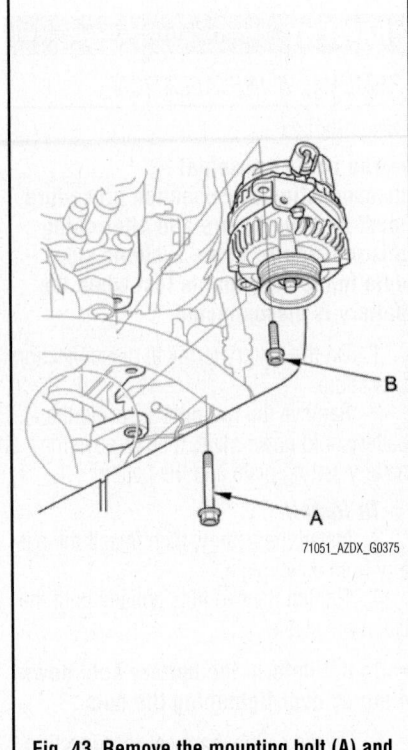

71051_AZDX_G0375

Fig. 43 Remove the mounting bolt (A) and the alternator bracket mounting bolt (B), then remove the alternator

ENGINE ELECTRICAL

FIRING ORDER

Engine firing order is 1-4-2-5-3-6.

IGNITION COILS AND SPARK PLUGS

REMOVAL & INSTALLATION

1. Remove the engine cover.
2. Disconnect the ignition coil connectors, then remove the ignition coils.
3. Remove the spark plugs and inspect them.
4. Apply a small amount of anti-seize compound to the plug threads, and screw the plugs into the cylinder head, finger tight.
5. Torque them to 16 ft. lbs. (22 Nm).
6. Install the ignition coils in the reverse order of removal.

IGNITION TIMING

INSPECTION & ADJUSTMENT

See Figure 44.

1. Connect the HDS to the DLC.
2. Turn the ignition switch to ON (II), or

press the engine start/stop button to select the ON mode.
3. Make sure the HDS communicates with the vehicle and the PCM. If it does not communicate, troubleshoot the DLC circuit.
4. Check for DTCs. If a DTC is present, diagnose and repair the cause before continuing with this test.
5. Start the engine. Hold the engine speed at 3,000 rpm with no load (in P or N) until the radiator fan comes on, then let it idle.
6. Check the idle speed.
7. Jump the SCS line with the HDS.
8. Connect the timing light to the No. 1 ignition coil harness.
9. Aim the light toward the pointer on the timing belt lower cover. Check the ignition timing under a no load condition (headlights, blower fan, rear window defogger, and air conditioner are turned off).

➡The other pointer is not used.

➡Ignition Timing: 10±2°BTDC (RED mark) at idle in P or N

10. If the ignition timing differs from the specification, check the camshaft timing. If the camshaft timing is OK, update the PCM

IGNITION SYSTEM

if it does not have the latest software, or substitute a known-good PCM, then recheck. If the system works properly, and the PCM was substituted, replace the original PCM.
11. Disconnect the HDS and the timing light.

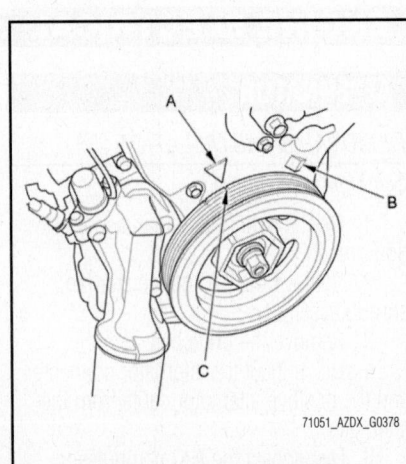

71051_AZDX_G0378

Fig. 44 Aim the light toward the pointer (A) on the timing belt lower cover; other pointer (B), red mark (C)

ENGINE ELECTRICAL

STARTING SYSTEM

STARTER

REMOVAL & INSTALLATION

See Figure 45.

1. Remove the battery and the battery base.

2. Disconnect the positive starter cable and the S terminal connector.

3. Remove the lower radiator hose bracket from the starter and remove the dipstick.

4. Remove the two bolts holding the starter, then remove the starter.

To install:

5. Install the starter, then tighten the mounting bolts to 33 ft. lbs. (45 Nm).

➡**Always use a new gasket.**

6. Install the lower radiator hose bracket to the starter and install the dipstick.

7. Connect the positive starter cable

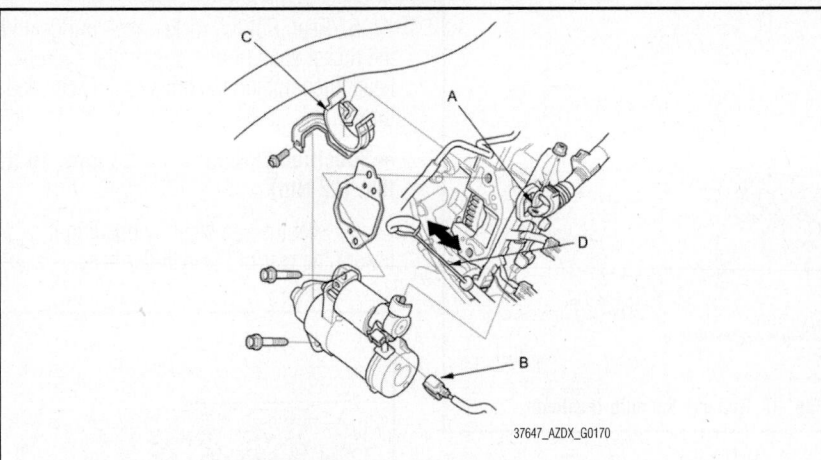

Fig. 45 Disconnect the positive starter cable and the S terminal connector. Remove the lower radiator hose bracket from the starter and remove the dipstick.

and the S terminal connector. Make sure the crimped side of the ring terminal faces away from the starter when you connect it.

8. Install the battery base and install the battery.

9. Start the engine to make sure the starter works properly.

ENGINE MECHANICAL

➡**Disconnecting the negative battery cable may interfere with the functions of the on board computer systems and may require the computer to undergo a relearning process, once the negative battery cable is reconnected.**

ACCESSORY DRIVE BELT SYSTEM

ADJUSTMENT

No adjustment is necessary or possible. Proper tension is provided by the auto-tensioner.

BELT ROUTINGS

See Figure 46.

INSPECTION

1. Remove the engine compartment covers.

2. Inspect the belt for cracks or damage. If the belt is cracked or damaged, replace it.

3. Check that the position of the auto-tensioner indicator is within the standard range. If it is out of the standard range, replace the drive belt.

4. Install the engine compartment covers.

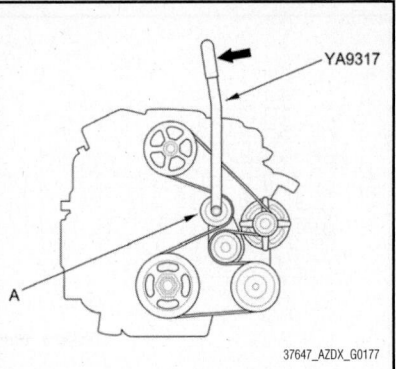

Fig. 46 Showing the accessory belt routing, the location of the auto tensioner (A) and using a lever to remove tension during belt removal

REMOVAL & INSTALLATION

Drive Belt

See Figure 47.

➡**For this procedure, you will need a belt tension release tool (Snap-On® YA9317) or equivalent.**

1. Move the auto-tensioner with the belt tension release tool to relieve tension from the drive belt, then remove the drive belt.

2. Install the new belt in the reverse order of removal.

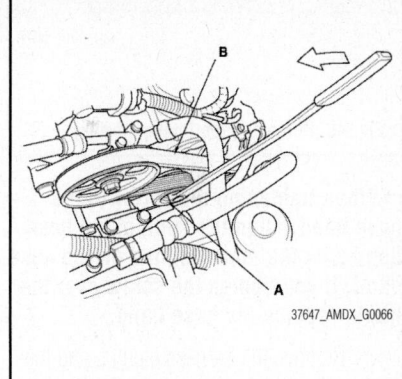

Fig. 47 Move the auto-tensioner with the belt tension release tool

Drive Belt Tensioner

See Figure 48.

1. Remove the drive belt.
2. Remove the splash shield.
3. Remove the auto-tensioner.
4. Install the auto-tensioner in the reverse order of removal.

AIR CLEANER

REMOVAL & INSTALLATION

See Figure 49.

1. Disconnect the breather pipe.
2. Disconnect the MAF/IAT sensor connector, then remove the intake air duct.

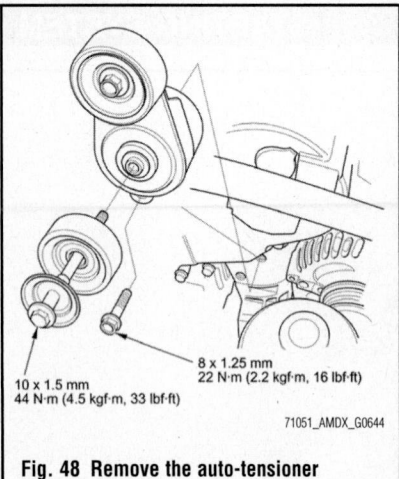

Fig. 48 Remove the auto-tensioner

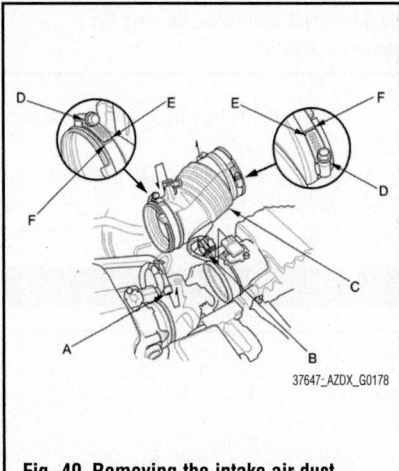

Fig. 49 Removing the intake air duct

➡When tightening the screw of the hose band, align the edge of the hose band with the mark painted on the hose band. If you tighten the screw over the mark, replace the hose band.

3. Remove the harness clamps and the harness bracket to clear access to the air cleaner.

4. Remove the air cleaner.

5. Install the parts in the reverse order of removal.

CAMSHAFT & BEARINGS

INSPECTION

See Figures 50 through 56.

1. Remove the cylinder head.
2. Remove the rocker arm assembly.
3. Disassemble the rocker arm assembly.
4. Front: Put the rocker shafts bridge and the rocker shaft holder on the front cylinder head, then tighten the bolts to the specified torque.

➡Specified Torque: 8 x 1.25 mm: 16 ft. lbs. (22 Nm)

5. Rear: Put the rocker shaft bridge and the rocker shaft holder on the rear cylinder head, then tighten the bolts to the specified torque.

➡Specified Torque: 8 x 1.25 mm: 16 ft. lbs. (22 Nm)

6. Seat the camshaft by pushing it toward the rear of the cylinder head.

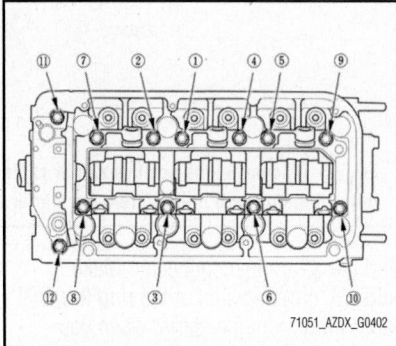

Fig. 50 Front: Put the rocker shafts bridge and the rocker shaft holder on the front cylinder head, then tighten the bolts

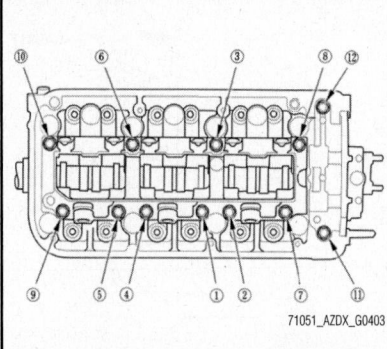

Fig. 51 Rear: Put the rocker shaft bridge and the rocker shaft holder on the rear cylinder head, then tighten the bolts

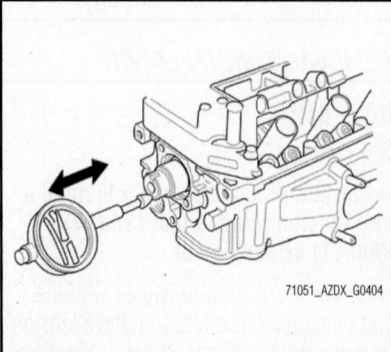

Fig. 52 Zero the dial indicator against the end of the camshaft

7. Zero the dial indicator against the end of the camshaft. Push the camshaft back and forth and read the end play. If the end play is beyond the service limit, replace the thrust cover and recheck. If it is still beyond the service limit, replace the cylinder head. If it is still beyond the service limit, replace the camshaft.

➡Camshaft End Play:

- Standard (New): 0.0020–0.0079 inches (0.05–0.20 mm)
- Service Limit: 0.0079 inches (0.20 mm)

8. Remove the camshaft thrust cover, then pull out the camshaft.

9. Wipe the camshaft clean, then inspect the lift ramps. Replace the camshaft if any lobes are pitted, scored, or excessively worn.

10. Measure the diameter of each camshaft journal.

11. Zero the gauge to the journal diameter.

12. Clean the camshaft bearing surfaces in the cylinder head. Measure the inside diameter of each camshaft bearing surface, and check for an out-of-round condition.

a. If the camshaft-to-holder clearance is within limits, go to step 14.

b. If the camshaft-to-holder clearance is beyond the service limit and the camshaft has been replaced, replace the cylinder head.

c. If the camshaft-to-holder clearance is beyond the service limit and the camshaft has not been replaced, go to step 13.

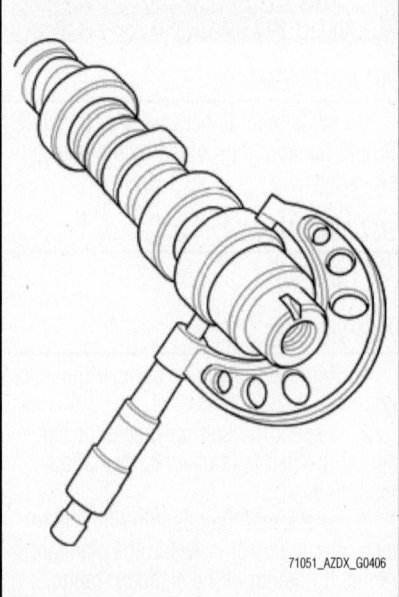

Fig. 53 Measure the diameter of each camshaft journal

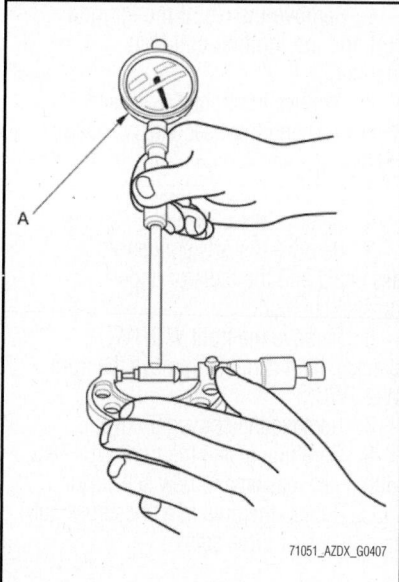

Fig. 54 Zero the gauge (A) to the journal diameter

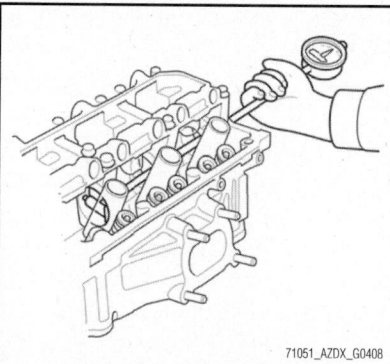

Fig. 55 Measure the inside diameter of each camshaft bearing surface, and check for an out-of-round condition

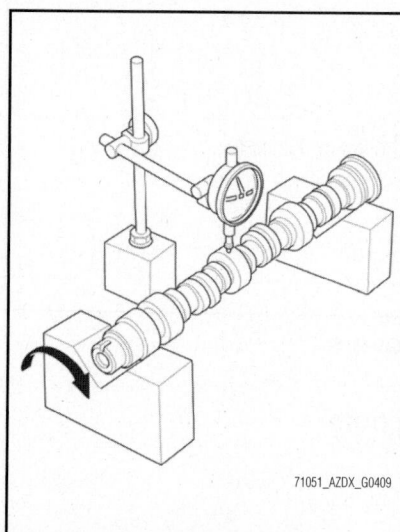

Fig. 56 Check total runout with the camshaft supported on V-blocks

➡**Camshaft-to-Holder Oil Clearance:**

- Standard (New): 0.00197–0.00350 inches (0.050–0.089 mm)
- Service Limit: 0.0059 inches (0.15 mm)

13. Check total runout with the camshaft supported on V-blocks.

    a. If the total runout of the camshaft is within the service limit, replace the cylinder head.

    b. If the total runout is beyond the service limit, replace the camshaft and recheck the oil clearance. If the oil clearance is still out of tolerance, replace the cylinder head.

➡**Camshaft Total Runout:**

- Standard (New): 0.0012 inches (0.03 mm) maximum.
- Service Limit: 0.0016inches (0.04 mm)

REMOVAL & INSTALLATION

Front

See Figure 57.

1. Do the battery removal procedure.
2. Remove the battery base.
3. Drain the engine coolant.
4. Disconnect the radiator hoses.
5. Remove the EGR valve.
6. Remove the EGR valve stud bolts.
7. Remove the timing belt. See "Timing Belt & Sprockets" in this section.
8. Remove the front rocker arm assembly. See "Rocker Arms" in this section.
9. Remove the front camshaft pulley.
10. Remove the thrust cover, then remove the front camshaft and O-ring.

To install:

11. Install the front camshaft in reverse order of removal.

    a. Always use a new O-ring.

    b. Apply new engine oil to the journals and the cam lobes.

    c. Apply new engine oil to the threads of the camshaft pulley mounting bolt, then install the front camshaft pulley.

12. Install the front rocker arm assembly, then tighten the mounting bolts, in an alternating sequence, to 16 ft. lbs. (22 Nm).
13. Install the timing belt.
14. Adjust the valve clearance.
15. Install the EGR valve stud bolts, then install the EGR valve.
16. Connect the radiator hoses.
17. Install the battery base.
18. Install the battery.
19. Fill the radiator with engine coolant, and bleed the air from the cooling system.

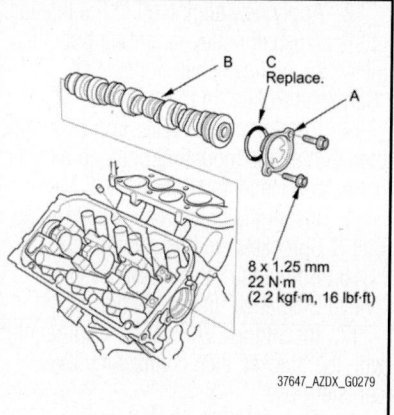

Fig. 57 Removing/installing the thrust cover (A), front camshaft (B) and O-ring (C)

Rear

See Figure 58.

1. Relieve the fuel pressure.
2. Remove the battery.
3. Drain the engine coolant.
4. Remove the air cleaner.
5. Remove the quick-connect fitting cover, then disconnect the fuel feed hose.
6. Disconnect the heater hoses.
7. Disconnect the EVAP canister hose, then remove the EVAP canister purge joint with the bracket.
8. Remove the timing belt.
9. Remove the rear rocker arm assembly.
10. Remove the rear camshaft pulley.
11. Remove the thrust cover, then remove the rear camshaft.

To install:

12. Install the rear camshaft in reverse order of removal.

    a. Always use a new O-ring.

    b. Apply new engine oil to the journals and cam lobes.

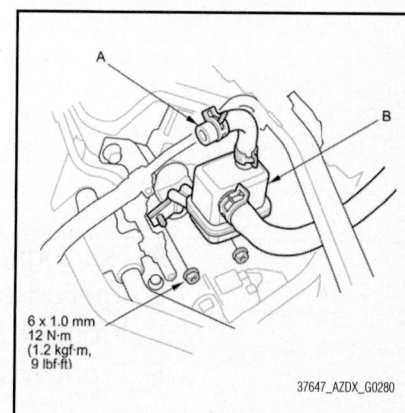

Fig. 58 Disconnect the EVAP canister hose (A), then remove the EVAP canister purge joint (B) with the

13. Apply new engine oil to the threads of the camshaft pulley mounting bolt, then install the rear camshaft pulley. See "Camshaft Pulley" in this section.

14. Install the rear rocker arm assembly, then tighten the mounting bolts, in an alternating sequence, to 16 ft. lbs. (22 Nm).

15. Install the timing belt. See "Timing Belt & Sprockets" in this section.

16. Adjust the valve clearance. See "Valve Clearance" in this section.

17. Install the EVAP canister purge joint with the bracket, then connect the EVAP canister hose.

18. Connect the heater hoses.

19. Connect the fuel feed hose, then install the quick-connect fitting cover.

20. Install the air cleaner.

21. Install the battery.

22. Inspect for fuel leaks. Turn the ignition switch to ON (II), or press the engine start/stop button to select the ON mode (do not operate the starter) so the fuel pump runs for about 2 seconds and pressurizes the fuel line. Repeat this operation three times, then check for fuel leakage at any point in the fuel line.

23. Fill the radiator with engine coolant, and bleed the air from the cooling system.

CATALYTIC CONVERTER

REMOVAL & INSTALLATION

Front Warm-Up Three Way Catalytic Converter (WU-TWC) (Bank 2)
See Figure 59.

1. Remove the No. 2 the ignition coil and the ignition coil heat insulator.

2. Remove the front A/F sensor (Sensor 1) and front secondary HO2S (Sensor 2).

3. Remove exhaust pipe A.

4. Remove the EGR pipe.

5. Remove the A/C condenser fan assembly and the radiator upper bracket/cushion.

6. Remove the front WU-TWC bracket, then carefully remove the front WU-TWC.

7. Remove the converter covers.

8. Carefully install the front WU-TWC with a new gasket and new self-locking nuts. Tighten the nuts in a crisscross pattern in two or three steps.

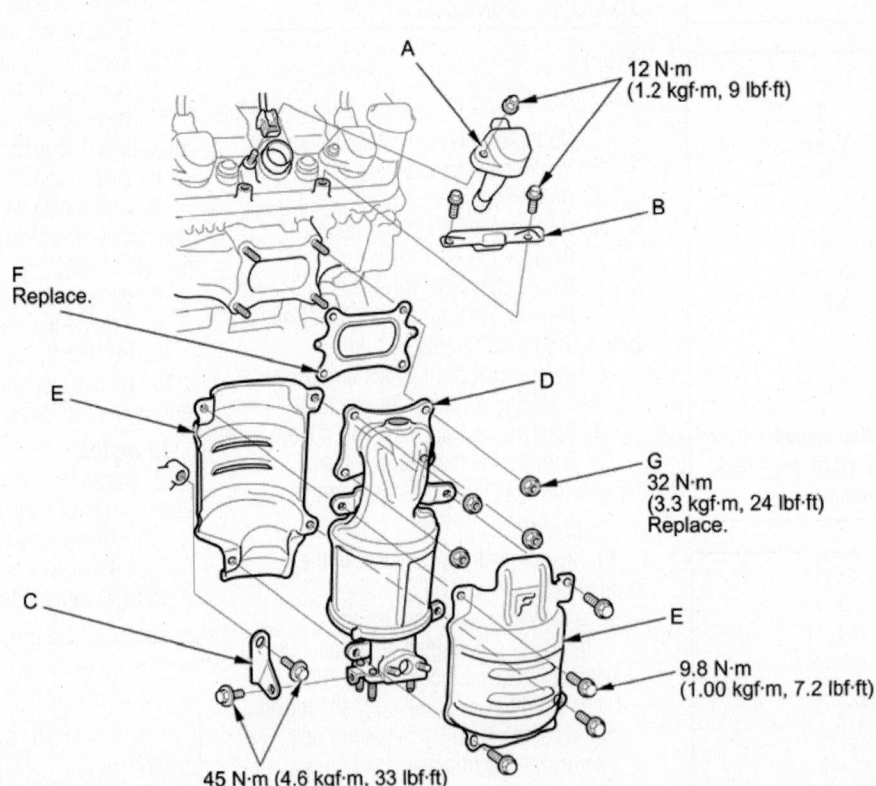

A. No. 2 ignition coil
B. Ignition coil heat insulator
C. Front WU-TWC bracket
D. Front WU-TWC
E. Converter covers
F. Gasket
G. Self-locking nuts

71051_AZDX_G0436

Fig. 59 Exploded view of the front WU-TWC (Bank 2) assembly

9. Install the parts in the reverse order of removal.

Rear WU-TWC (Bank 1)

See Figure 60.

1. Remove the rear A/F sensor (Sensor 1) and rear secondary HO2S (Sensor 2).
2. Remove exhaust pipe A.
3. Remove the intermediate shaft.
4. Remove the rear WU-TWC bracket, then carefully remove the rear WU-TWC.
5. Remove the converter covers.
6. Carefully install the rear WU-TWC with a new gasket and new self-locking nuts. Tighten the nuts in a crisscross pattern in two or three steps.
7. Install the parts in the reverse order of removal.

Under Floor Three Way Catalytic Converter (TWC)

See Figure 61.

1. Raise the vehicle on a lift.
2. Remove the exhaust pipe hanger.
3. Remove the under-floor TWC.
4. Remove the converter covers.
5. Install the parts in the reverse order of removal with new gaskets and new self-locking nuts.

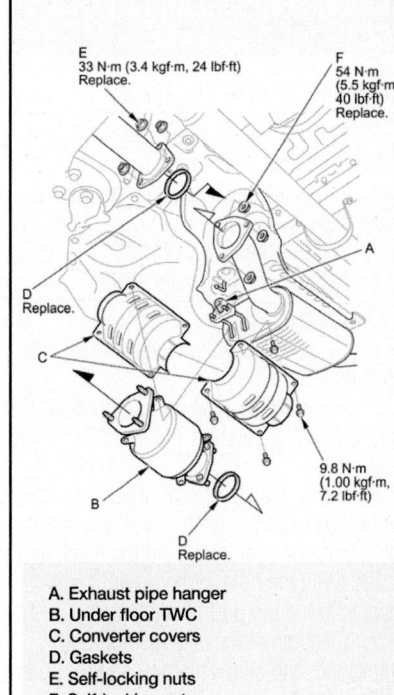

E
33 N·m (3.4 kgf·m, 24 lbf·ft)
Replace.

F
54 N·m
(5.5 kgf·m,
40 lbf·ft)
Replace.

D
Replace.

9.8 N·m
(1.00 kgf·m,
7.2 lbf·ft)

D
Replace.

A. Exhaust pipe hanger
B. Under floor TWC
C. Converter covers
D. Gaskets
E. Self-locking nuts
F. Self-locking nuts

71051_AZDX_G0418

Fig. 61 Exploded view of the under floor TWC assembly

CRANKSHAFT PULLEY

REMOVAL & INSTALLATION

See Figures 62 through 64.

Special Tools Required:
- Holder Handle 07JAB-001020B
- Holder Attachment, 50 mm, Offset 07MAB-PY3010A
- Socket, 19 mm 07JAA-001020A

1. Raise the vehicle on the lift.
2. Remove the right front wheel.
3. Remove the splash shield and the engine undercover.
4. Remove the drive belt.
5. Hold the pulley with the holder handle and the holder attachment, 50 mm, offset.
6. Remove the bolt with a heavy duty socket, 19 mm and a breaker bar, then remove the crankshaft pulley.

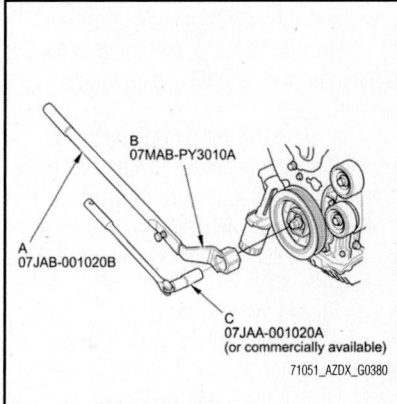

B
07MAB-PY3010A

A
07JAB-001020B

C
07JAA-001020A
(or commercially available)

71051_AZDX_G0380

Fig. 62 Hold the pulley with the holder handle (A) and the holder attachment, 50 mm, offset (B); remove the bolt with a heavy duty socket, 19 mm (C) and a breaker bar, then remove the crankshaft pulley

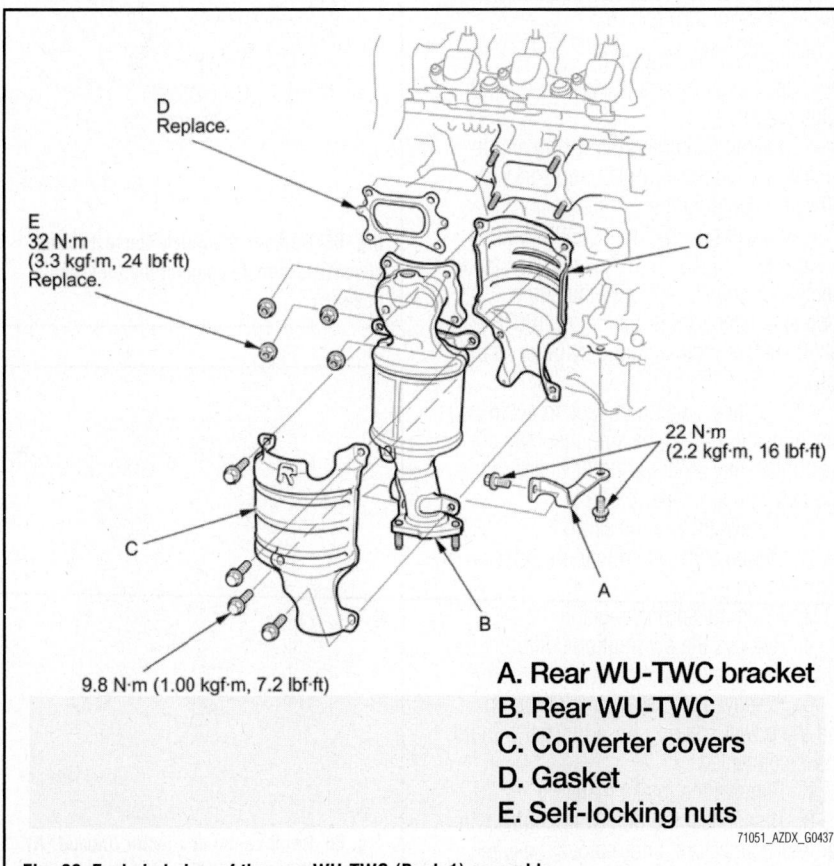

D
Replace.

E
32 N·m
(3.3 kgf·m, 24 lbf·ft)
Replace.

C

22 N·m
(2.2 kgf·m, 16 lbf·ft)

C

A

B

9.8 N·m (1.00 kgf·m, 7.2 lbf·ft)

A. **Rear WU-TWC bracket**
B. **Rear WU-TWC**
C. **Converter covers**
D. **Gasket**
E. **Self-locking nuts**

71051_AZDX_G0437

Fig. 60 Exploded view of the rear WU-TWC (Bank 1) assembly

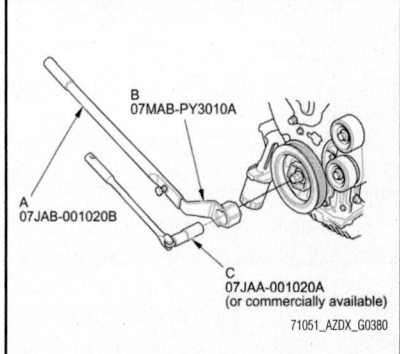

B
07MAB-PY3010A

A
07JAB-001020B

C
07JAA-001020A
(or commercially available)

71051_AZDX_G0380

Fig. 63 Hold the pulley with the holder handle (A) and the holder attachment (B); torque wrench and a heavy duty socket (C)

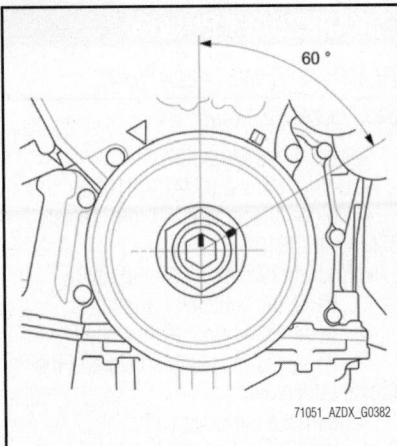

Fig. 64 Tighten the bolt an additional 60 degrees

To install:

7. Remove any oil and clean the pulley, the crankshaft, the bolt, and the washer. Lubricate with new engine oil as shown.

8. Install the crankshaft pulley, and tighten the bolt. Do not use an impact wrench.

a. Hold the pulley with the holder handle and the holder attachment. Torque the bolt to 48 ft. lbs. (65 Nm) with a torque wrench and a heavy duty socket.

b. Tighten the bolt an additional 60 degrees.

9. Install the drive belt.

10. Install the splash shield and the engine undercover.

11. Install the right front wheel.

CRANKSHAFT FRONT SEAL

REMOVAL & INSTALLATION

1. Remove the crankshaft position (CKP) sensor, the timing belt, and the timing belt drive pulley. For additional information, refer to the following section, "Timing Belt, Removal & Installation."

2. Remove the crankshaft front seal.

To install:

3. Clean and dry the crankshaft oil seal housing.

4. Apply a light coat of multipurpose grease to the crankshaft and to the lip of the seal.

5. Using the seal driver, drive in the crankshaft oil seal until the driver bottoms against the oil pump.

6. When the seal is in place, clean any excess grease off the crankshaft, and check that the oil seal lip is not distorted.

7. Install the timing belt drive pulley, the timing belt, and the CKP sensor.

CRANKSHAFT REAR COVER & SEAL

REMOVAL & INSTALLATION

1. Remove the transaxle assembly. For additional information, refer to the following section, "Automatic Transaxle, Removal & Installation."

2. Remove the driveplate.

3. Remove the rear main seal.

To install:

4. Clean and dry the crankshaft oil seal housing.

5. Apply a light coat of multipurpose grease to the crankshaft and to the lip of the seal.

6. Using a suitable seal driver tool, drive in the rear main seal until the driver attachment bottoms against the engine block end cover. Align the hole in the driver attachment with the pin on the crankshaft.

7. Clean any excess grease off the crankshaft, and check that the oil seal lip is not distorted.

8. Install the driveplate and transaxle assembly

CYLINDER HEAD

REMOVAL & INSTALLATION

See Figures 65 through 78.

Please note the following:
• Use fender covers to avoid damaging painted surfaces.
• To avoid damaging the wiring and terminals, unplug the wiring connectors carefully while holding the connector portion.
• Connect the HDS to the DLC, and monitor ECT SENSOR 1. To avoid damaging the cylinder head, wait until the engine coolant temperature drops below 100°F (38°C) before loosening the cylinder head bolts.
• Mark all wiring and hoses to avoid misconnection. Also, be sure that they do not contact any other wiring or hoses, or interfere with any other parts.

1. Relieve the fuel pressure.

2. Do the battery terminal disconnection procedure.

3. Drain the engine coolant.

4. Remove the six ignition coils.

5. Remove the alternator.

6. Remove the power steering pump and the power steering hose bracket with its hoses connected.

7. Remove the intake manifold.

8. Disconnect the following engine wire harness connectors, and remove the wire harness clamps from the cylinder head:

• Six injector connectors
• Knock sensor connector
• ECT sensor 1 connector
• Engine mount control solenoid valve connector
• EGR valve connector
• CMP sensor connector
• Rocker arm oil control solenoid connector
• Rocker arm oil pressure switch connector
• Two A/F sensor connectors
• Two secondary HO2S connectors

9. Remove the front warm up TWC and the rear warm up TWC.

10. Remove the quick-connect fitting cover, then disconnect the fuel feed hose.

11. Remove the EVAP canister purge joint with the bracket.

12. Remove the connector bracket from the front cylinder head.

13. Remove the engine mount control solenoid valve bracket from the rear cylinder head.

14. Remove the injector bases.

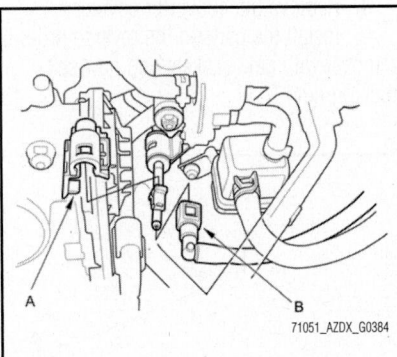

Fig. 65 Remove the quick-connect fitting cover (A), then disconnect the fuel feed hose (B)

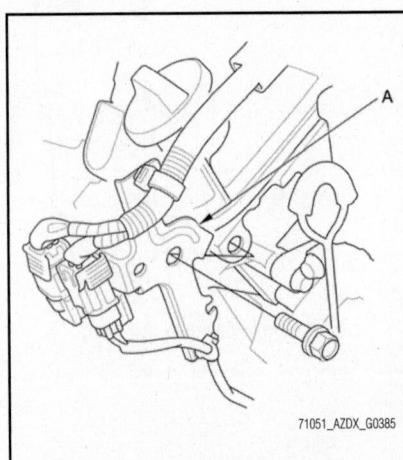

Fig. 66 Remove the connector bracket (A) from the front cylinder head

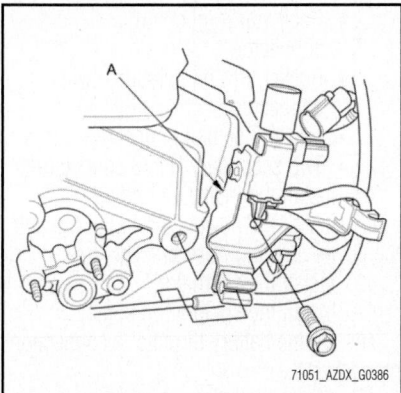

Fig. 67 Remove the engine mount control solenoid valve bracket (A) from the rear cylinder head

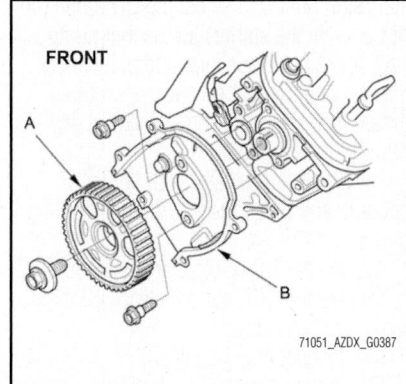

Fig. 68 Remove the camshaft pulley (A) and the back cover (B)—Front

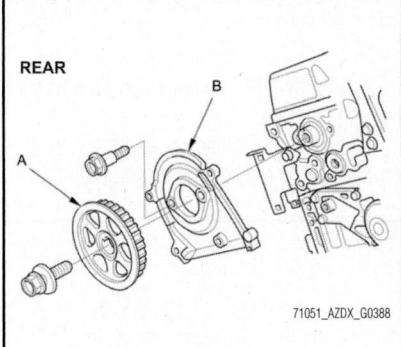

Fig. 69 Remove the camshaft pulley (A) and the back cover (B)—Rear

15. Remove the water passage.
16. Remove the timing belt.
17. Remove the camshaft pulleys and the back covers.
18. Remove the cylinder head covers.
19. Remove the cylinder head bolts. To prevent warpage, loosen the bolts in sequence ⅓ turn at a time; repeat the sequence until all bolts are loosened.

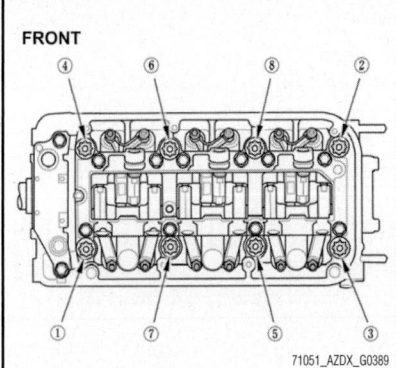

Fig. 70 Remove the cylinder head bolts, to prevent warpage, loosen the bolts in sequence—Front

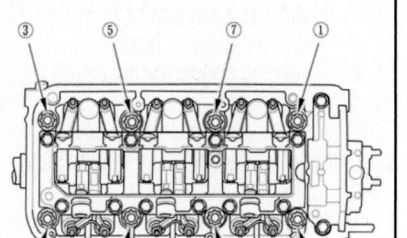

Fig. 71 Remove the cylinder head bolts, to prevent warpage, loosen the bolts in sequence—Rear

20. Remove the cylinder heads.

To install:
21. Clean the cylinder head and the engine block surface.
22. Clean and install the oil control orifices with new O-rings.

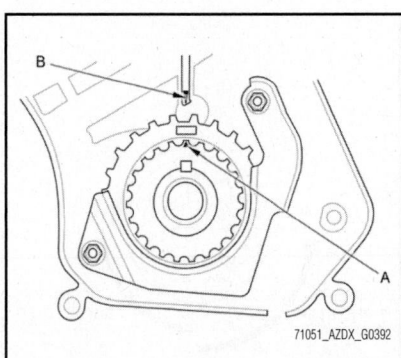

Fig. 72 Set the timing belt drive pulley to top dead center (TDC) by aligning the TDC mark (A) on the tooth of the timing belt drive pulley with the pointer (B) on the oil pump

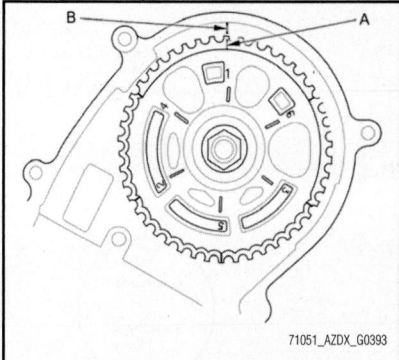

Fig. 73 Set the camshaft pulleys to TDC by aligning the TDC marks (A) on the camshaft pulleys with the pointers (B) on the back covers—Front

23. Install the dowel pins and the new cylinder head gaskets.
24. Clean the timing belt pulleys, the timing belt guide plate, and the upper and lower covers.
25. Set the timing belt drive pulley to top dead center (TDC) by aligning the TDC mark on the tooth of the timing belt drive pulley with the pointer on the oil pump.
26. Set the camshaft pulleys to TDC by aligning the TDC marks (A) on the camshaft pulleys with the pointers (B) on the back covers.
27. Install the cylinder heads on the engine block.
28. Measure the diameter of each cylinder head bolt at point A and point B.
29. If either diameter is less than 0.445 inches (11.3 mm), replace the cylinder head bolt.
30. Apply new engine oil to the threads and under the bolt heads of all cylinder head bolts.

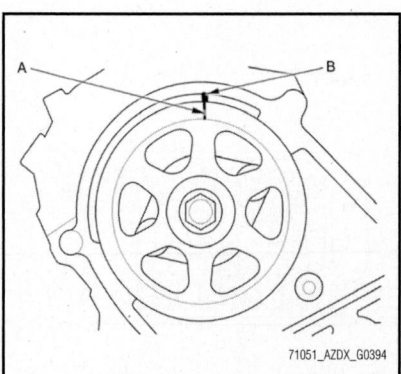

Fig. 74 Set the camshaft pulleys to TDC by aligning the TDC marks (A) on the camshaft pulleys with the pointers (B) on the back covers—Rear

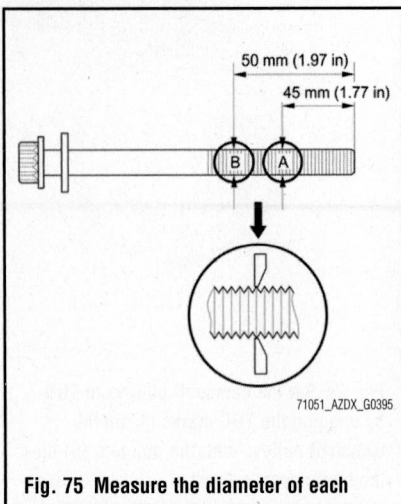

Fig. 75 Measure the diameter of each cylinder head bolt at point A and point B

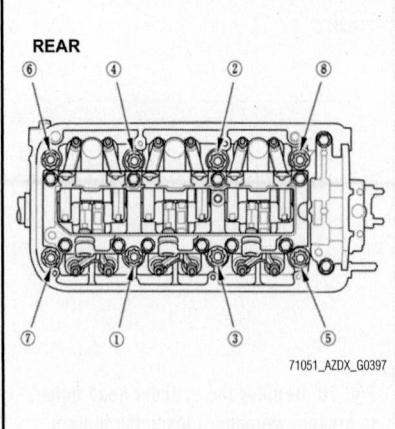

Fig. 77 Torque the cylinder head bolts in sequence—Rear

- Rocker arm oil control solenoid connector
- Rocker arm oil pressure switch connector
- Two A/F sensor connectors
- Two secondary HO2S connectors

44. Install the intake manifold.

45. Install the power steering pump and the power steering hose bracket.

46. Install the alternator.

47. Install the six ignition coils.

48. Do the battery terminal reconnection procedure.

49. After installation, check that all tubes, hoses, and connectors are installed correctly.

50. Inspect for fuel leaks. Turn the ignition switch to ON (II), or press the engine start/stop button to select the ON mode (do not operate the starter) so the fuel pump runs for about 2 seconds and pressurizes the fuel line. Repeat this operation three times, then check for fuel leakage at any point in the fuel line.

51. Refill the radiator with engine coolant, and bleed the air from the cooling system.

52. Do the PCM idle learn procedure.

53. Do the CKP pattern clear/CKP pattern learn procedure.

54. Inspect the idle speed.

55. Inspect the ignition timing.

31. Torque the cylinder head bolts in sequence to 22 ft. lbs. (30 Nm) with a beam-type torque wrench if possible. When using a preset click-type torque wrench, be sure to tighten slowly and do not over-tighten. If a bolt makes any noise while you are torquing it, loosen the bolt and retighten it from the first step.

32. After torquing, tighten all cylinder head bolts in two steps (90 degrees per step) using the sequence shown in the previous step. If you are using a new cylinder head bolt, tighten the bolt an extra 90 degrees.

➡**Remove the cylinder head bolt if you tightened it beyond the specified angle, and go back to step 8 of the procedure. Do not loosen it back to the specified angle.**

33. Install the timing belt.

34. Adjust the valve clearance.

35. Install the cylinder head covers.

36. Install the water passage.

37. Install the injector bases.

38. Install the connector bracket to the front cylinder head.

39. Install the engine mount control solenoid valve bracket to the rear cylinder head.

40. Install the EVAP canister purge joint with the bracket.

41. Connect the fuel feed hose, then install the quick-connect fitting cover.

42. Install the front warm up TWC and the rear warm up TWC.

43. Connect the following engine wire harness connectors, and install the wire harness clamps to the cylinder head:

- Six injector connectors
- Knock sensor connector
- ECT sensor 1 connector
- Engine mount control solenoid valve connector
- EGR valve connector
- CMP sensor connector

DRIVEPLATE

REMOVAL & INSTALLATION

See Figure 79.

1. Remove the transmission assembly.

2. Remove the driveplate and the washer from the engine.

3. Install the driveplate and the washer on the engine crankshaft, and tighten the

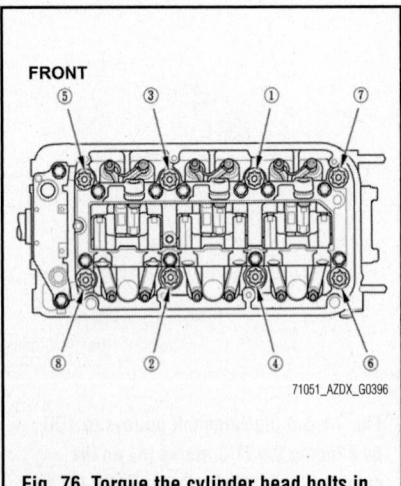

Fig. 76 Torque the cylinder head bolts in sequence—Front

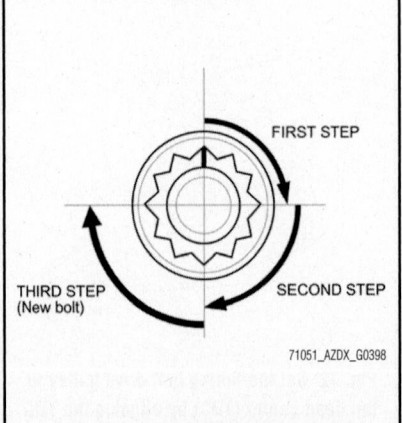

Fig. 78 After torquing, tighten all cylinder head bolts in two steps (90 degrees per step) using the sequence shown

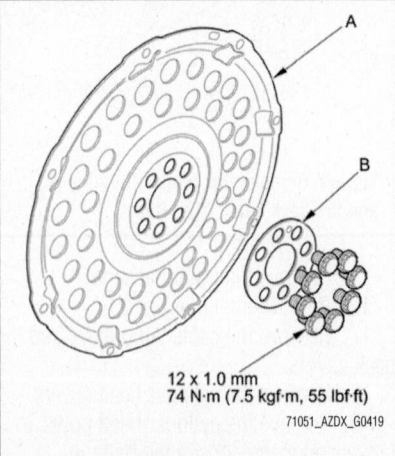

12 x 1.0 mm
74 N·m (7.5 kgf·m, 55 lbf·ft)

Fig. 79 Remove the driveplate (A) and the washer (B) from the engine

eight bolts in a crisscross pattern in at least two steps.

4. Install the transmission assembly.

FRONT COVER & SEAL

REMOVAL & INSTALLATION

See Figures 80 through 82.

1. Remove the engine compartment covers. Refer to Engine Mechanical section for Engine Compartment Covers Removal and Installation.

2. Turn the crankshaft so its white mark lines up with the pointer.

➡**The other pointer is not used.**

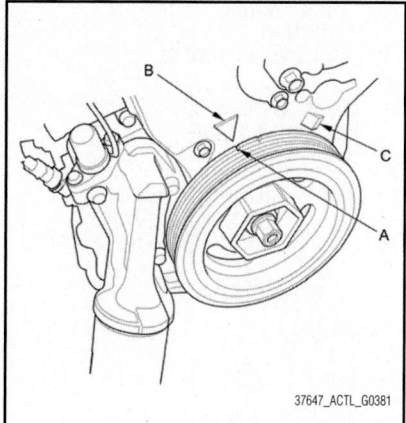

37647_ACTL_G0381

Fig. 80 Turn the crankshaft so its white mark (A) lines up with the pointer (B)

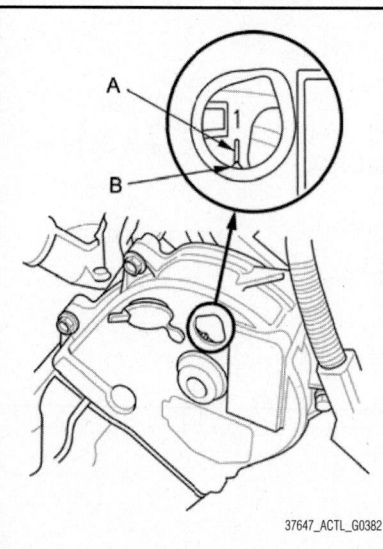

37647_ACTL_G0382

Fig. 81 Check that the No. 1 piston TDC mark (A) on the front camshaft pulley and the pointer (B) on the front upper cover are aligned

3. Check that the No. 1 piston Top Dead Center (TDC) mark on the front camshaft pulley and the pointer on the front upper cover are aligned.

➡**If the marks are not aligned, rotate the crankshaft 360 degrees, and recheck the camshaft pulley mark.**

4. Raise the vehicle on the lift, then remove the right front wheel.

5. Remove the splash shield.

6. Remove the drive belt auto-tensioner.

7. Support the engine with a jack and a wood block under the oil pan.

8. Remove the ground cable, then remove the upper half of the side engine mount bracket.

9. Remove the crankshaft pulley.

10. Remove the front upper cover and the rear upper cover.

11. Remove the lower cover.

To install:

12. Install the lower cover.

13. Install the front upper cover and the rear upper cover.

14. Install the crankshaft pulley.

15. Rotate the crankshaft pulley about six turns clockwise so the timing belt positions itself on the pulleys.

16. Turn the crankshaft pulley so its white mark lines up with the pointer.

➡**The other pointer is not used.**

17. Check the camshaft pulley marks.

➡**If the marks are not aligned, rotate the crankshaft 360 degrees, and recheck the camshaft pulley mark. If the camshaft pulley marks are at TDC, go to step 17. If the camshaft pulley marks are not at TDC, remove the timing belt and repeat steps 2 through 16.**

18. Install the upper half of the side engine mount bracket, then tighten the mounting bolts.

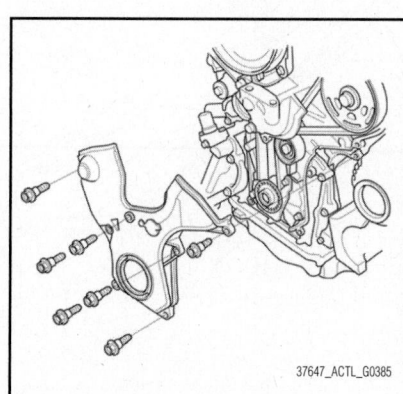

37647_ACTL_G0385

Fig. 82 Remove the lower cover

19. Install the ground cable.

20. Install the drive belt auto-tensioner.

21. Install the splash shield.

22. Install the right front wheel.

23. Install the engine compartment covers.

INTAKE MANIFOLD

REMOVAL & INSTALLATION

See Figures 83 and 84.

1. Remove the engine cover.

2. Disconnect the breather pipe.

3. Remove the intake air duct.

4. Disconnect the PCV hose, the brake booster vacuum hose, the vacuum hose, and the IMT actuator connector from the intake manifold.

5. Disconnect the EVAP canister purge hose (A), the EVAP canister purge valve connector (B), the throttle actuator connector (C), and the manifold absolute pressure (MAP) sensor connector (D). Disconnect and plug the water bypass hoses (E).

6. Remove the upper cover mounting bolts and nuts, using an alternating sequence, in three steps, then remove the upper cover.

7. Remove the intake manifold mounting bolts and nuts, using an alternating sequence, in three steps.

8. Remove the intake manifold.

To install:

9. Install the intake manifold. Tighten the bolts and nuts in an alternating sequence, in three steps. Always use a new intake manifold gasket. Tighten the bolts and nuts to 16 ft. lbs. (22 Nm).

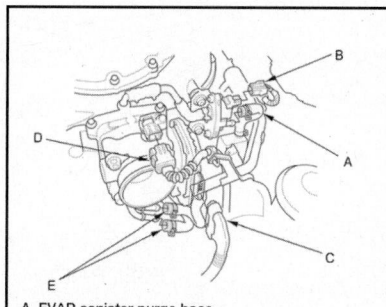

A. EVAP canister purge hose
B. EVAP canister purge valve connector
C. Throttle actuator connector
D. Manifold absolute pressure (MAP) sensor connector
E. Water bypass hoses

37647_AZDX_G0290

Fig. 83 Disconnect the EVAP canister purge hose, the EVAP canister purge valve connector, the throttle actuator connector, and the manifold absolute pressure (MAP) sensor connector. Disconnect and plug the water bypass hoses.

10. Install the upper cover. Tighten the bolts and nuts sequentially in three steps to 9 ft. lbs. (12 Nm). Always use a new gasket.

11. Connect the water bypass hoses. Connect the MAP sensor connector, the throttle actuator connector, the EVAP canister purge valve connector, and the EVAP canister purge hose.

12. Connect the IMT actuator connector, the vacuum hose, the brake booster vacuum hose, and the PCV hose.

13. Install the intake air duct.

➡️**When tightening the screw of the hose band, align the edge of the hose band with the mark painted on the hose band. If you tighten the screw over the mark, replace the hose band.**

14. Connect the breather pipe.
15. Install the engine cover.
16. Clean up any spilled engine coolant.
17. After installation, check that all

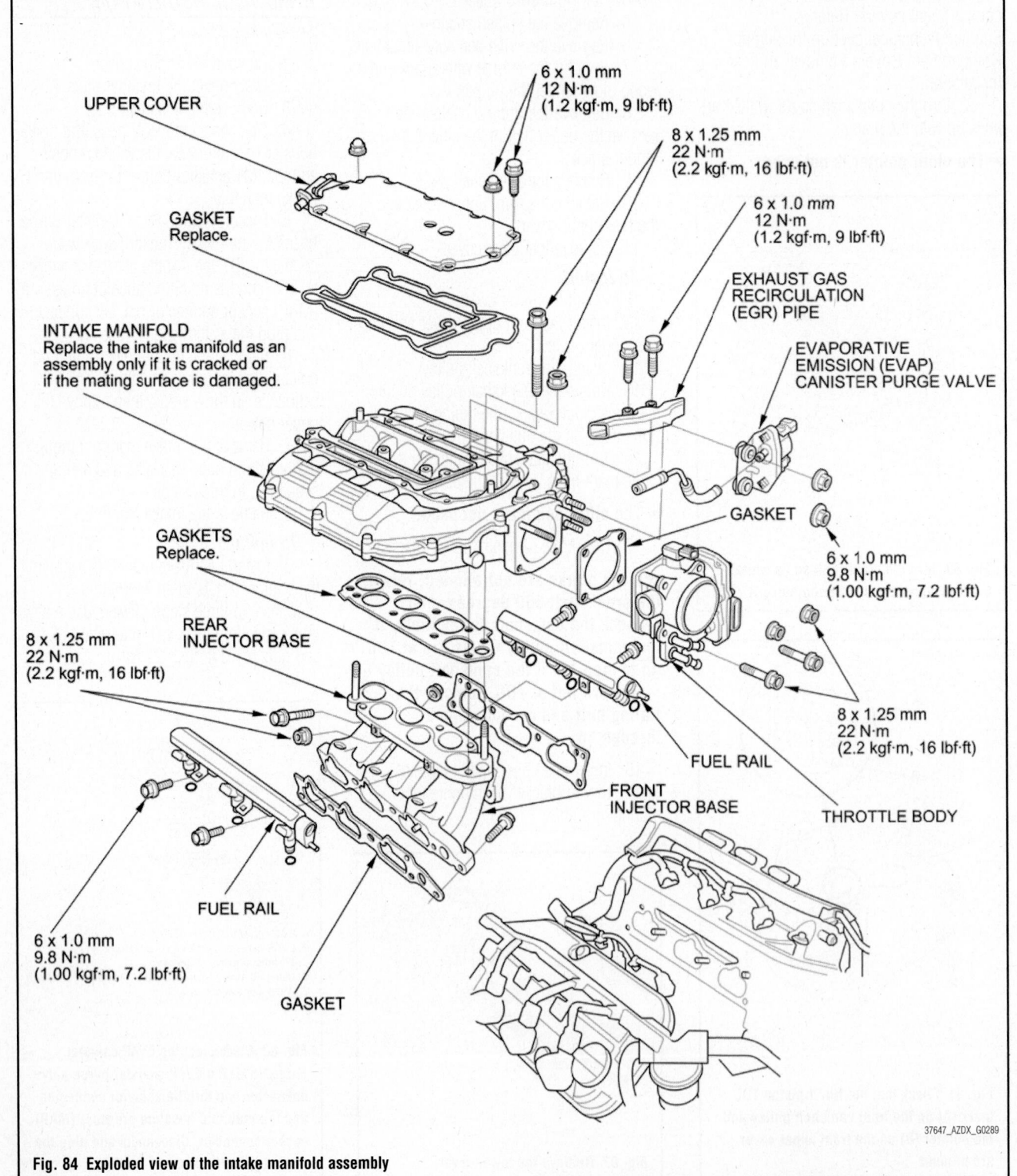

Fig. 84 Exploded view of the intake manifold assembly

UPPER COVER

GASKET
Replace.

INTAKE MANIFOLD
Replace the intake manifold as an assembly only if it is cracked or if the mating surface is damaged.

GASKETS
Replace.

8 x 1.25 mm
22 N·m
(2.2 kgf·m, 16 lbf·ft)

REAR INJECTOR BASE

FUEL RAIL

6 x 1.0 mm
9.8 N·m
(1.00 kgf·m, 7.2 lbf·ft)

GASKET

6 x 1.0 mm
12 N·m
(1.2 kgf·m, 9 lbf·ft)

8 x 1.25 mm
22 N·m
(2.2 kgf·m, 16 lbf·ft)

6 x 1.0 mm
12 N·m
(1.2 kgf·m, 9 lbf·ft)

EXHAUST GAS RECIRCULATION (EGR) PIPE

EVAPORATIVE EMISSION (EVAP) CANISTER PURGE VALVE

GASKET

6 x 1.0 mm
9.8 N·m
(1.00 kgf·m, 7.2 lbf·ft)

8 x 1.25 mm
22 N·m
(2.2 kgf·m, 16 lbf·ft)

FUEL RAIL

FRONT INJECTOR BASE

THROTTLE BODY

37647_AZDX_G0289

tubes, hoses, and connectors are installed correctly.

18. Refill the radiator with engine coolant, and bleed the air from the cooling system.

MAIN BEARING TORQUE SEQUENCE

See Figure 85.

OIL PAN

REMOVAL & INSTALLATION

See Figures 86 through 89.

➡ **If the engine is already out of the vehicle, go to step 7.**

1. Raise the vehicle on the lift.

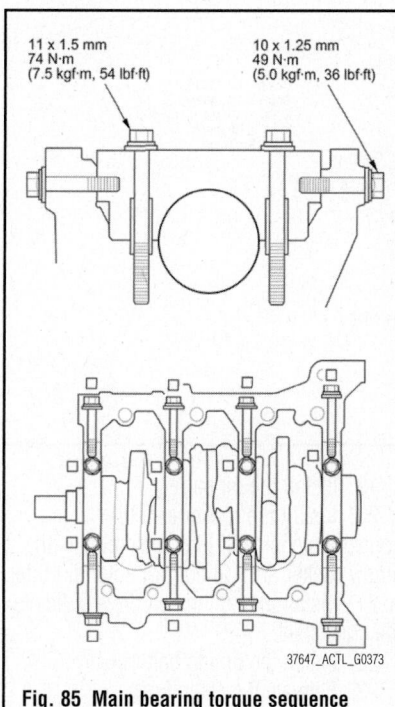

Fig. 85 Main bearing torque sequence

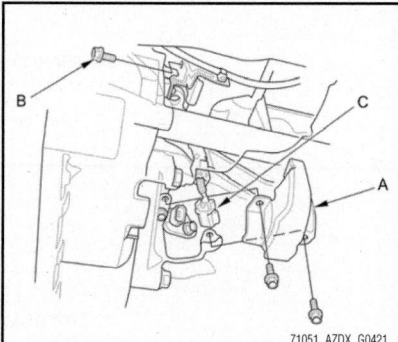

Fig. 86 Remove the CKP sensor cover (A) and the bolt (B), then disconnect the CKP sensor connector (C)

2. Drain the engine oil.
3. Remove the splash shield.
4. Remove the front subframe stiffener.
5. Remove exhaust pipe A.
6. Remove the rear Warm Up Three Way Catalytic Converter (rear WU-TWC) bracket.
7. Remove the Crankshaft Position (CKP) sensor cover and the bolt, then disconnect the CKP sensor connector.
8. Remove the torque converter cover and the four bolts securing the transaxle.
9. Remove the bolts securing the oil pan.
10. Using a flat blade screwdriver, separate the oil pan from the engine block in the places shown.
11. Remove the oil pan.

To install:

12. Remove all of the old liquid gasket from the oil pan mating surfaces, the bolts, and the bolt holes.

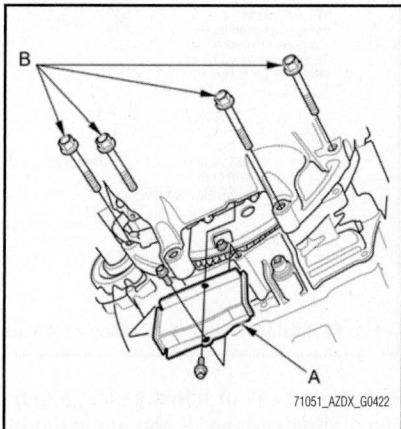

Fig. 87 Remove the torque converter cover (A) and the four bolts (B) securing the transaxle

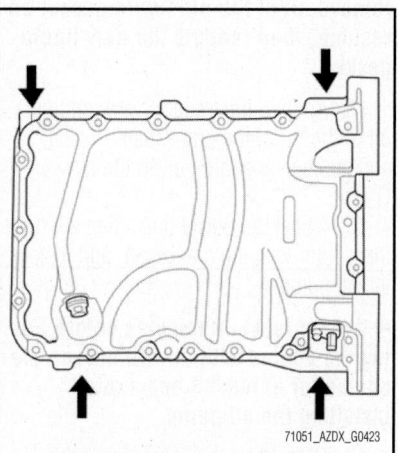

Fig. 88 Using a flat blade screwdriver, separate the oil pan from the engine block in the places shown

13. Clean and dry the oil pan mating surfaces.
14. Apply liquid gasket, P/N 08717-0004, 08718-0003, or 08718-0009 to the oil pan mating surface of the engine block and to the inside edge of the threaded bolt holes. Install the component within 5 minutes of applying the liquid gasket.

➡ **Apply a bead of liquid gasket along the mating edge of the oil pan. If you apply liquid gasket P/N 08718-0012, the component must be installed within 4 minutes. If too much time has passed after applying the liquid gasket, remove the old liquid gasket and residue, then reapply the new liquid gasket.**

15. Install the oil pan on the engine block.
16. Tighten the bolts in three steps. In the final step, torque all bolts, in sequence, to 9 ft. lbs. (12 Nm).

➡ **Wait at least 30 minutes before filling the engine with oil. Do not run the engine for at least 3 hours after installing the oil pan.**

17. Tighten the four bolts securing the transaxle to 55 ft. lbs. (75 Nm), then install the torque converter cover.
18. Connect the Crankshaft Position (CKP) sensor connector, then install the CKP sensor cover and the bolt.
19. Install the rear Warm Up Three Way Catalytic Converter (rear WU-TWC) bracket. Tighten the bolts to 16 ft. lbs. (22 Nm).
20. If the engine is still in the vehicle, do the following steps.
21. Install exhaust pipe A using new gaskets and new self-locking nuts.

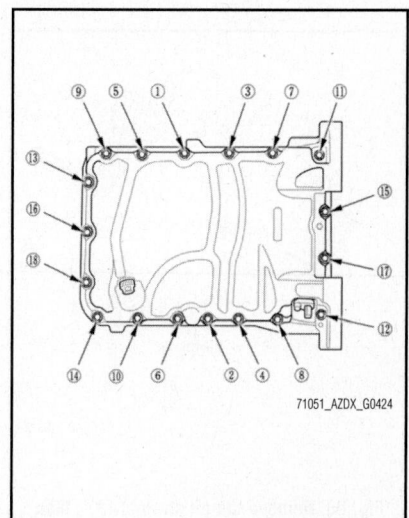

Fig. 89 Tighten the bolts in three steps

22. Install the front subframe stiffener.
23. Install the splash shield.
24. Refill the engine with the recommended engine oil.

OIL PUMP

REMOVAL & INSTALLATION

See Figures 90 through 92.

1. Drain the engine oil.
2. Remove the timing belt:
3. Remove the timing belt drive pulley from the crankshaft:
4. Attach the chain hoist to the engine hook on the engine hanger bracket.
5. Remove the rocker arm oil control solenoid/oil filter assembly.
6. Remove the oil pan.
7. Remove the oil strainer.
8. Remove the mounting bolts, then remove the oil pump assembly.

To install:

9. Remove the old oil seal from the oil pump.
10. Clean and dry the crankshaft oil seal housing.
11. Using the oil seal driver, 64 mm, drive in the new crankshaft oil seal until the oil seal driver bottoms on the pump.
12. Remove all of the old liquid gasket from the oil pump mating surfaces, the bolts, and the bolt holes.
13. Clean and dry the oil pump mating surfaces.
14. Apply liquid gasket, P/N 08717-0004, 08718-0003, or 08718-0009, to the engine block mating surface of the oil pump and to the inside edge of the threaded bolt holes. Install the component within 5 minutes of applying the liquid gasket.

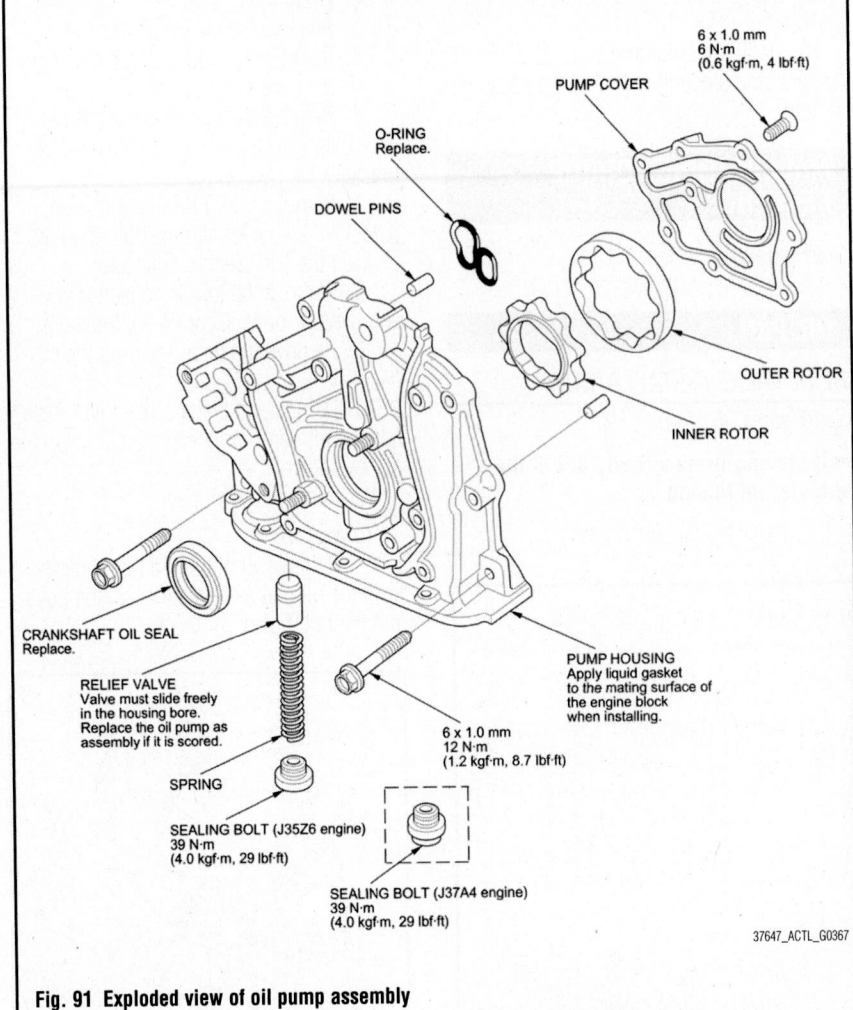

Fig. 91 Exploded view of oil pump assembly

➡ **Apply a bead of liquid gasket along the mating surface. If you apply liquid gasket P/N 08718-0012, the component must be installed within 4 minutes. If too much time has passed after applying the liquid gasket, remove all of the old liquid gasket and residue, then reapply the new liquid gasket.**

15. Apply a light coat of new engine oil to the lip of the crankshaft oil seal, and apply new engine oil to the new O-ring.
16. Install the dowel pins, then align the inner rotor with the crankshaft, and install the oil pump.

➡ **Wait at least 30 minutes before filling the engine with oil. Do not run the engine for at least 3 hours after installing the oil pump.**

17. Clean the excess grease off the crankshaft, and check the seal for distortion.
18. Install the oil strainer with a new O-ring.

19. Install the oil pan.
20. Install the rocker arm oil control solenoid/oil filter assembly with a new rocker arm oil control solenoid filter.
21. Install the timing belt drive pulley to the crankshaft:
22. Install the timing belt:
23. Remove the chain hoist.

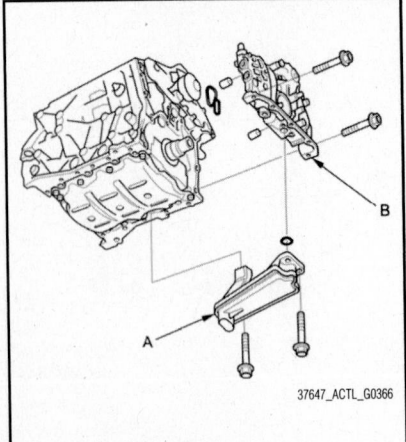

Fig. 90 Remove the oil strainer (A), then remove the oil pump assembly (B)

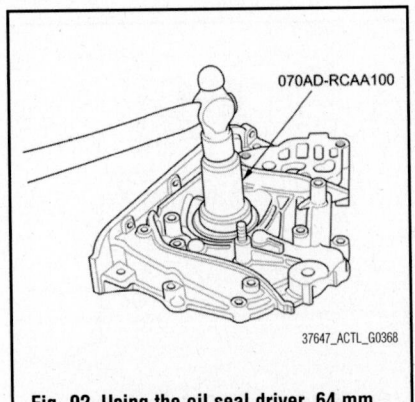

Fig. 92 Using the oil seal driver, 64 mm, drive in the new crankshaft oil seal until the oil seal driver bottoms on the pump

PISTONS & RINGS

POSITIONING

See Figure 93.

ROCKER ARMS

REMOVAL & INSTALLATION

Front Cylinder Head

See Figures 94 through 96.

1. Remove the cylinder head cover.
2. Loosen the locknuts and the adjusting screws
3. Remove the rocker shaft bridge mounting bolts, the rocker shaft holder mounting bolts, and the rocker arm assembly.

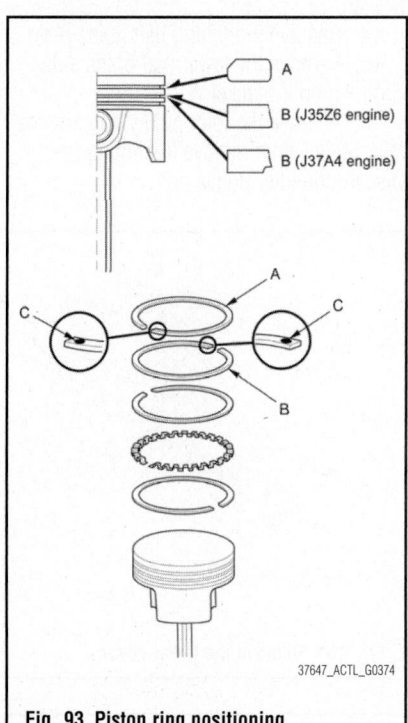

Fig. 93 Piston ring positioning

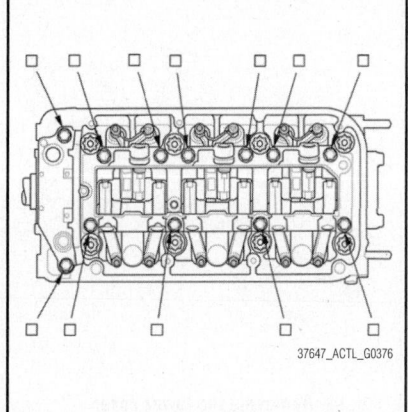

Fig. 95 Front rocker shaft bridge bolt loosening sequence

a. Loosen the rocker shaft bridge mounting bolts and the rocker shaft holder mounting bolts in sequence two turns at a time, to prevent damaging the valves or the rocker arm assembly.

b. When removing the rocker arm assembly, do not remove the rocker shaft bridge mounting bolts and the rocker shaft holder mounting bolts. The bolts will keep the rocker arms on the shafts.

To install:

4. Apply liquid gasket, P/N 08717-0004, 08718-0003, or 08718-0009 to the rocker shaft holder mating surface of the cylinder head. Install the component within 5 minutes of applying the liquid gasket.

➡Apply a bead of liquid gasket along the mating surfaces. If you apply liquid gasket P/N 08718-0012, the component must be installed within 4 minutes. If too much time has passed after applying the liquid gasket, remove the old liquid gasket and residue, then reapply the new liquid gasket.

5. Set the rocker arm assembly in place, and loosely install the bolts. Make sure that the rocker arms are properly positioned on the valve stems.

➡Wait at least 30 minutes before filling the engine with oil. Do not run the engine for at least 3 hours after installing the rocker arm assembly.

6. Tighten each bolt two turns at a time in the sequence shown to ensure that the rockers do not bind on the valves.

Rear Cylinder Head

See Figure 97.

1. Remove the cylinder head cover.
2. Loosen the locknuts and the adjusting screws.
3. Remove the rocker shaft bridge mounting bolts, the rocker shaft holder mounting bolts, and the rocker arm assembly.

a. Loosen the rocker shaft bridge mounting bolts and the rocker shaft holder mounting bolts in sequence two turns at a time, to prevent damaging the valves or the rocker arm assembly.

b. When removing the rocker arm assembly, do not remove the rocker shaft bridge mounting bolts and the rocker shaft holder mounting bolts. The bolts will keep the rocker arms on the shafts.

To install:

4. Apply liquid gasket, P/N 08717-0004, 08718-0003, or 08718-0009 to the rocker shaft holder mating surface of the cylinder head. Install the component within 5 minutes of applying the liquid gasket.

➡Apply a bead of liquid gasket along the mating surfaces. If you apply liquid gasket P/N 08718-0012, the component must be installed within 4 minutes. If too much time has passed after applying the liquid gasket, remove the old

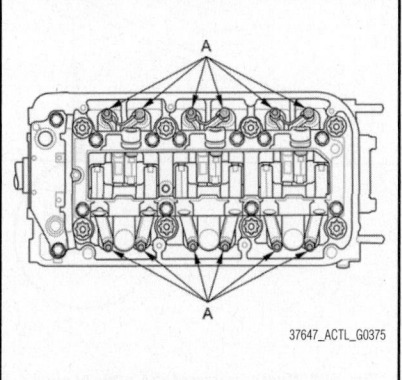

Fig. 94 Loosen the locknuts and the adjusting screws (A)

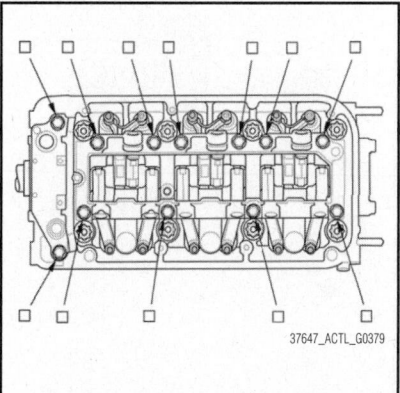

Fig. 96 Tighten each rocker shaft bridge bolt two turns at a time in sequence

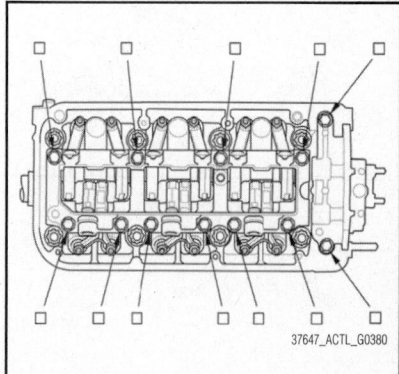

Fig. 97 Tighten each rocker shaft bridge bolt two turns at a time in sequence

liquid gasket and residue, then reapply the new liquid gasket.

5. Set the rocker arm assembly in place, and loosely install the bolts. Make sure that the rocker arms are properly positioned on the valve stems.

➡**Wait at least 30 minutes before filling the engine with oil. Do not run the engine for at least 3 hours after installing the rocker arm assembly.**

6. Tighten each bolt two turns at a time in the sequence shown to ensure that the rockers do not bind on the valves.

TIMING BELT FRONT COVER

REMOVAL & INSTALLATION

See Figures 98 and 99.

1. Disconnect the negative battery cable.
2. Remove the splash shield.
3. Remove the accessory drive belt.
4. Remove the accessory drive belt auto-tensioner.
5. Support the engine assembly with a suitable jack.
6. Remove the ground cable, then remove the upper half of the side engine mount bracket.
7. Remove the front upper and rear upper cover.
8. Remove the crankshaft pulley.
9. Remove the lower cover.

To install:

10. Install the lower cover and tighten the mounting bolts to 9 ft. lbs. (12 Nm).
11. Install the front upper and rear upper cover and tighten the mounting bolts to 9 ft. lbs. (12 Nm).
12. Install the crankshaft pulley.

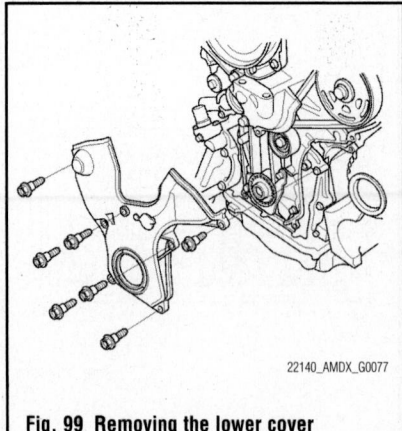

Fig. 99 Removing the lower cover

13. Install the accessory drive belt auto-tensioner.
14. Install the accessory drive belt.
15. Install the splash shield.
16. Connect the negative battery cable.
17. Start the engine and check for leaks.

TIMING BELT & SPROCKETS

REMOVAL & INSTALLATION

See Figures 100 through 107.

1. Remove the engine compartment covers.
2. Turn the crankshaft so its white mark lines up with the pointer.

➡**The other pointer is not used.**

3. Check that the No. 1 piston Top Dead Center (TDC) mark on the front camshaft pulley and the pointer on the front upper cover are aligned.

➡**If the marks are not aligned, rotate the crankshaft 360 degrees, and recheck the camshaft pulley mark.**

4. Raise the vehicle on the lift, then remove the right front wheel.
5. Remove the splash shield.
6. Remove the drive belt auto-tensioner.
7. Support the engine with a jack and a wood block under the oil pan.
8. Remove the ground cable, then remove the upper half of the side engine mount bracket.
9. Remove the crankshaft pulley.
10. Remove the front upper cover and the rear upper cover.
11. Remove the lower cover.
12. Remove one of the battery clamp bolts from the battery tray, and grind the end of it.
13. Thread the battery clamp bolt into hold the timing belt adjuster in its current position. Tighten it by hand, do not use a wrench.
14. Remove the timing belt guide plate.
15. Remove the lower half of the side engine mount bracket.
16. Remove the idler pulley bolt and the idler pulley, then remove the timing belt. Discard the idler pulley bolt.

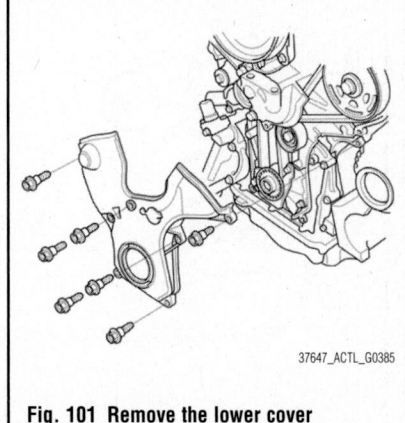

Fig. 101 Remove the lower cover

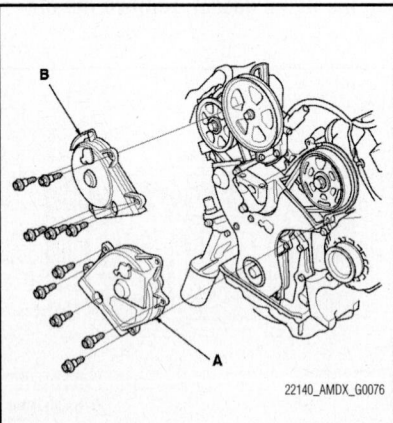

Fig. 98 Removing the front upper (A) and rear upper (B) cover

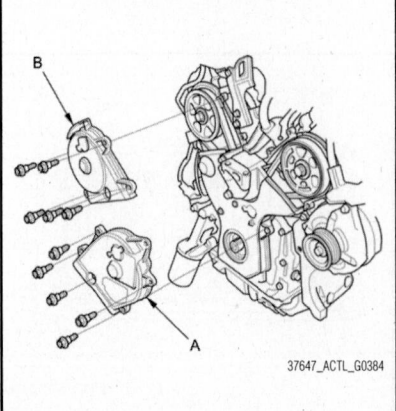

Fig. 100 Remove the front upper cover (A) and the rear upper cover (B)

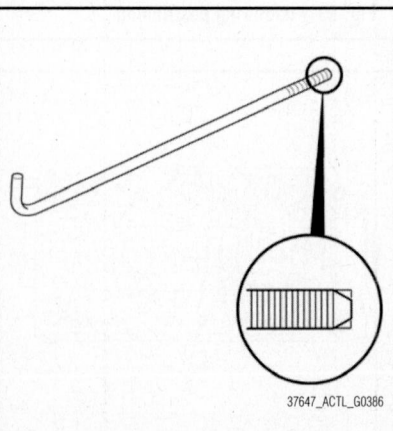

Fig. 102 Remove one of the battery clamp bolts from the battery tray, and grind the end of it

Fig. 103 Thread the battery clamp bolt into hold the timing belt adjuster in its current position

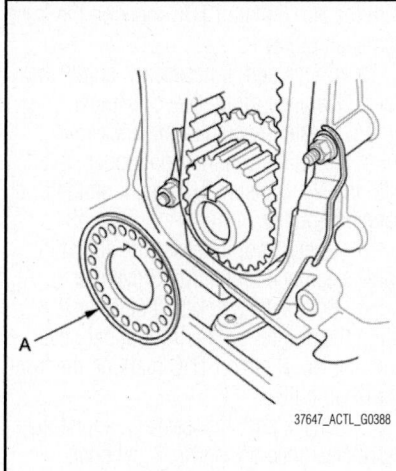

Fig. 104 Remove the timing belt guide plate (A)

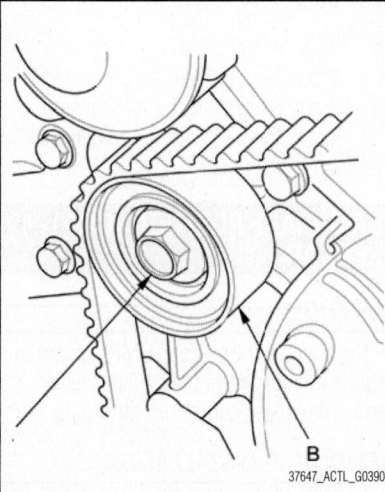

Fig. 105 Remove the idler pulley bolt (A) and the idler pulley (B), then remove the timing belt

To install:

17. Clean the timing belt pulleys, the timing belt guide plate, and the upper and lower covers.

18. Set the timing belt drive pulley to Top Dead Center (TDC) by aligning the TDC mark on the tooth of the timing belt drive pulley with the pointer on the oil pump.

19. Set the camshaft pulleys to TDC by aligning the TDC marks on the camshaft pulleys with the pointers on the back covers.

20. Loosely install the idler pulley with a new idler pulley bolt so the pulley can move but does not come off.

21. If the auto-tensioner has extended and the timing belt cannot be installed, do the timing belt replacement procedure.

22. Install the timing belt in a counter-clockwise sequence starting with the drive pulley. Take care not to damage the timing belt during installation.

23. Tighten the idler pulley bolt to 33 ft. lbs. (44 N m).

24. Remove the battery clamp bolt from the back cover.

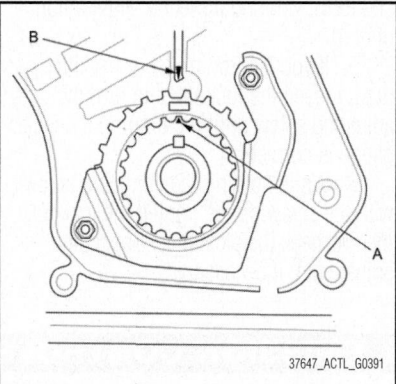

Fig. 106 Set the timing belt drive pulley to TDC by aligning the TDC mark (A) on the tooth of the timing belt drive pulley with the pointer (B) on the oil pump

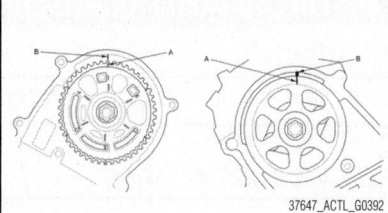

Fig. 107 Set the camshaft pulleys to TDC by aligning the TDC marks (A) on the camshaft pulleys with the pointers (B) on the back covers

25. Install the lower half of the side engine mount bracket. Tighten the three long bolts to 33 ft. lbs. (44 N m).

26. Install the timing belt guide plate.

27. Install the lower cover.

28. Install the front upper cover and the rear upper cover.

29. Install the crankshaft pulley.

30. Rotate the crankshaft pulley about six turns clockwise so the timing belt positions itself on the pulleys.

31. Turn the crankshaft pulley so its white mark lines up with the pointer.

➡**The other pointer is not used.**

32. Check the camshaft pulley marks.

➡**If the marks are not aligned, rotate the crankshaft 360 degrees, and recheck the camshaft pulley mark. If the camshaft pulley marks are at TDC, go to step 17. If the camshaft pulley marks are not at TDC, remove the timing belt and repeat steps 2 through 16.**

33. Install the upper half of the side engine mount bracket, then tighten the mounting bolts.

34. Install the ground cable.

35. Install the drive belt auto-tensioner.

36. Install the splash shield.

37. Install the right front wheel.

38. Install the engine compartment covers.

TIMING BELT REAR COVER

REMOVAL & INSTALLATION

1. Remove the timing belt. For additional information, refer to the following section, "Timing Belt, Removal & Installation."

2. Remove the camshaft pulley.

3. Remove the rear cover.

To install:

4. Install the rear cover.

5. Install the camshaft pulley.

6. Install the timing belt.

7. Start the engine and check for leaks.

VALVE LASH

ADJUSTMENT

See Figures 108.

➡**Connect the Honda Diagnostic System (HDS) to the data link connector (DLC), and monitor the Engine Coolant Temperature (ECT) sensor 1. Adjust the valve clearance only when the ECT sensor 1 is less than 100°F (38°C).**

1. Remove the cylinder head covers.

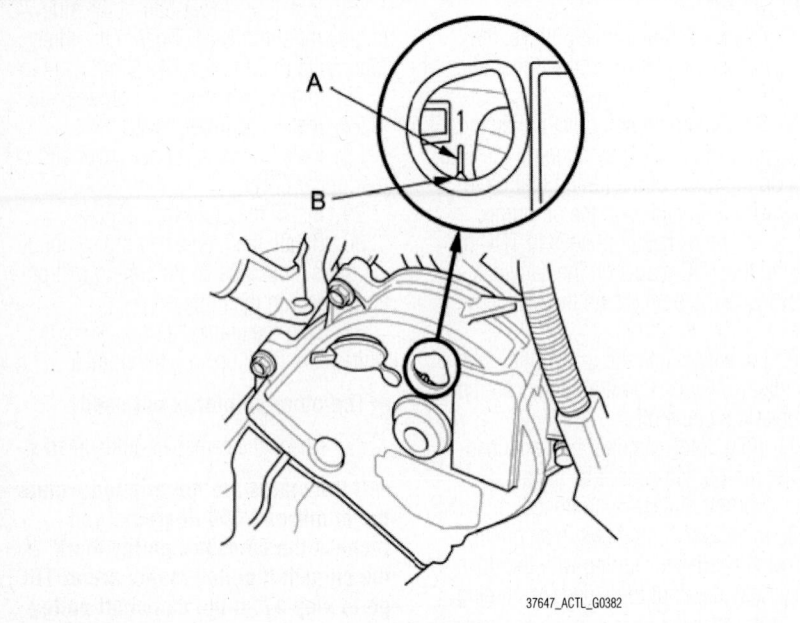

37647_ACTL_G0382

Fig. 108 Set the No. 1 piston at TDC. Align the pointer (A) on the front upper cover with the No. 1 piston TDC mark (B) on the front camshaft pulley

2. Set the No. 1 piston at Top Dead Center (TDC). Align the pointer on the front upper cover with the No. 1 piston TDC mark on the front camshaft pulley.

3. Select the correct feeler gauge for the valve clearance you are going to check.

Valve Clearance:
- Intake: 0.008–0.009 inches (0.20–0.24 mm)
- Exhaust: 0.011–0.013 inches (0.28–0.32 mm)

4. Insert the feeler gauge between the adjusting screw and the end of the valve stem on the No. 1 cylinder, and slide it back and forth; you should feel a slight amount of drag.

5. If you feel too much or too little drag, loosen the locknut, and turn the adjusting screw until the drag on the feeler gauge is correct.

6. While holding the adjusting screw with the screw driver, tighten the locknut, then recheck the clearance. Repeat the adjustment, if necessary.

Specified Torque:
- Intake: 14 ft. lbs. (20 Nm)
- Exhaust: 10 ft. lbs. (14 Nm)

➡ **Apply new engine oil to the nut threads.**

7. Rotate the crankshaft clockwise. Align the pointer on the front upper cover with the No. 4 piston TDC mark on the front camshaft pulley.

8. Check and, if necessary, adjust the valve clearance on the No. 4 cylinder.

9. Rotate the crankshaft clockwise. Align the pointer on the front upper cover with the No. 2 piston TDC mark on the front camshaft pulley.

Check and, if necessary, adjust the valve clearance on the No. 2 cylinder.

10. Rotate the crankshaft clockwise. Align the pointer on the front upper cover with the No. 5 piston TDC mark on the front camshaft pulley.

11. Check and, if necessary, adjust the valve clearance on the No. 5 cylinder.

12. Rotate the crankshaft clockwise. Align the pointer on the front upper cover with the No. 3 piston TDC mark on the front camshaft pulley.

13. Check and, if necessary, adjust the valve clearance on the No. 3 cylinder.

14. Rotate the crankshaft clockwise. Align the pointer on the front upper cover with the No. 6 piston TDC mark on the front camshaft pulley.

15. Check and, if necessary, adjust the valve clearance on the No. 6 cylinder.

16. Install the cylinder head covers.

ENGINE PERFORMANCE & EMISSION CONTROLS

COMPONENT LOCATIONS

See Figure 110.

ACCELERATOR PEDAL POSITION (APP) SENSOR

LOCATION

The Accelerator Pedal Position (APP) Sensor is on the top end of the accelerator pedal assembly. The APP sensor is not serviceable as a separate component. If the sensor is tested faulty, the pedal assembly must be replaced.

REMOVAL & INSTALLATION

See Figure 111.

1. Disconnect the accelerator pedal position (APP) sensor 6P connector.
2. Remove the accelerator pedal module.

➡ **The APP sensor is not available separately. Do not disassemble the accelerator pedal module.**

3. Install the parts in the reverse order of removal.

AIR-FUEL RATIO (AFR) SENSOR

LOCATION

The Air Fuel Ratio sensors are located on each bank of the engine.

REMOVAL & INSTALLATION

Front Bank

See Figure 112.

1. Disconnect the front A/F sensor 6P connector, then remove the A/F sensor.
2. Install the parts in the reverse order of removal.

Rear Bank

See Figure 113.

1. Disconnect the rear A/F sensor 6P connector, then remove the A/F sensor.
2. Install the parts in the reverse order of removal.

CAMSHAFT POSITION (CMP) SENSOR

LOCATION

The Camshaft Position (CMP) sensor is located in the front of the cylinder head, behind the upper camshaft pulley plate.

REMOVAL & INSTALLATION

See Figures 114 and 115.

1. Remove the timing belt.
2. Remove the front camshaft pulley (CMP sensor pulse plate).

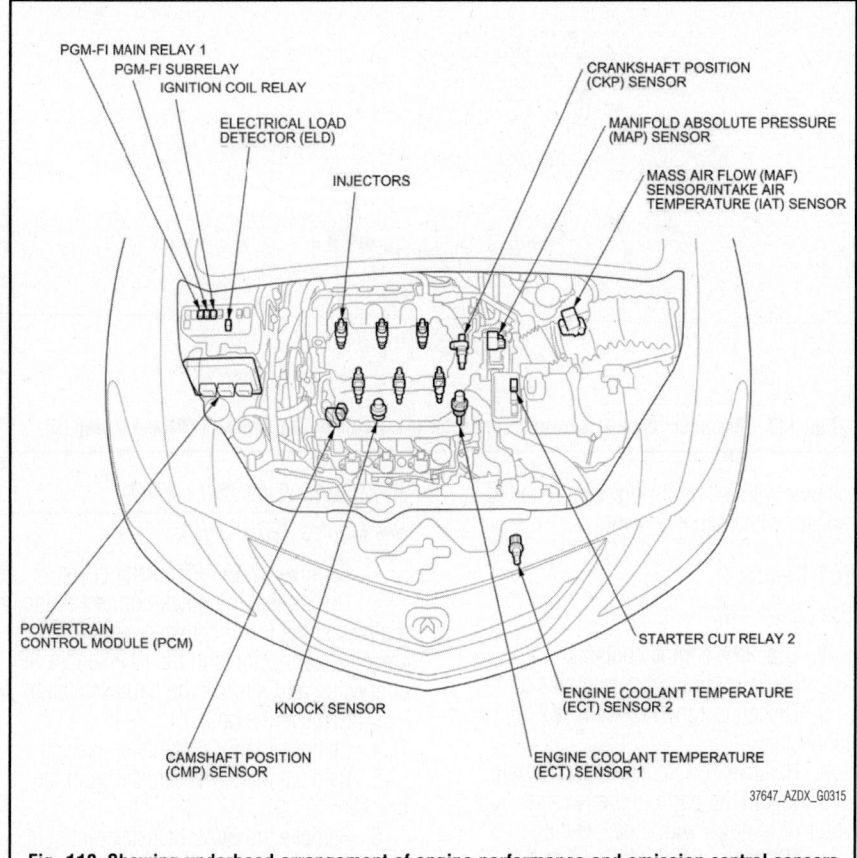

PGM-FI MAIN RELAY 1
PGM-FI SUBRELAY
IGNITION COIL RELAY
ELECTRICAL LOAD
DETECTOR (ELD)
INJECTORS
CRANKSHAFT POSITION
(CKP) SENSOR
MANIFOLD ABSOLUTE PRESSURE
(MAP) SENSOR
MASS AIR FLOW (MAF)
SENSOR/INTAKE AIR
TEMPERATURE (IAT) SENSOR
POWERTRAIN
CONTROL MODULE (PCM)
STARTER CUT RELAY 2
KNOCK SENSOR
ENGINE COOLANT TEMPERATURE
(ECT) SENSOR 2
CAMSHAFT POSITION
(CMP) SENSOR
ENGINE COOLANT TEMPERATURE
(ECT) SENSOR 1

37647_AZDX_G0315

Fig. 110 Showing underhood arrangement of engine performance and emission control sensors

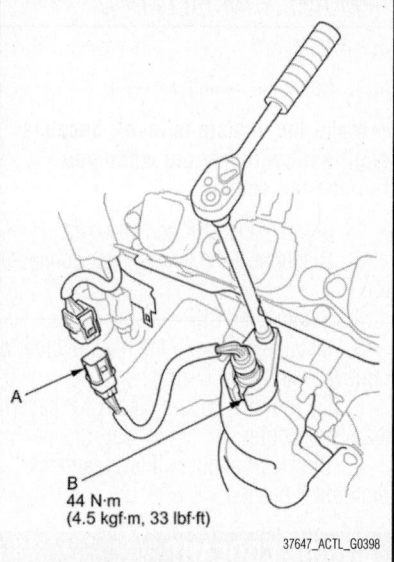

B
44 N·m
(4.5 kgf·m, 33 lbf·ft)

37647_ACTL_G0398

Fig. 113 Disconnect the rear A/F sensor 6P connector (A), then remove the A/F sensor (B)

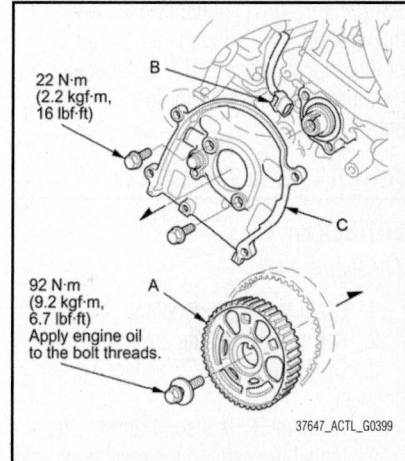

22 N·m
(2.2 kgf·m,
16 lbf·ft)

92 N·m
(9.2 kgf·m,
6.7 lbf·ft)
Apply engine oil
to the bolt threads.

37647_ACTL_G0399

Fig. 114 Remove the front camshaft pulley (CMP sensor pulse plate) (A)

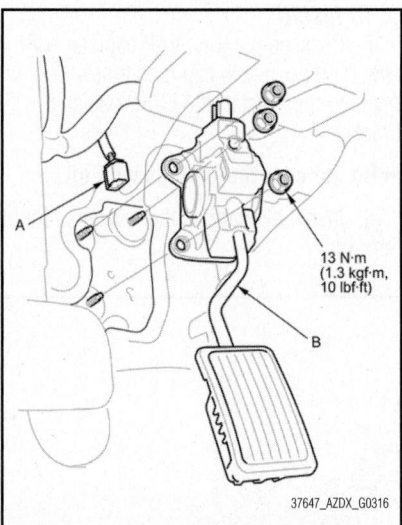

13 N·m
(1.3 kgf·m,
10 lbf·ft)

37647_AZDX_G0316

Fig. 111 Disconnect the accelerator pedal position (APP) sensor 6P connector (A). Remove the accelerator pedal module (B).

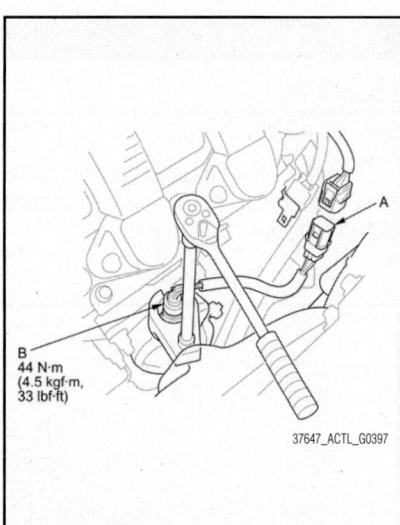

B
44 N·m
(4.5 kgf·m,
33 lbf·ft)

37647_ACTL_G0397

Fig. 112 Disconnect the front A/F sensor 6P connector (A), then remove the A/F sensor (B)

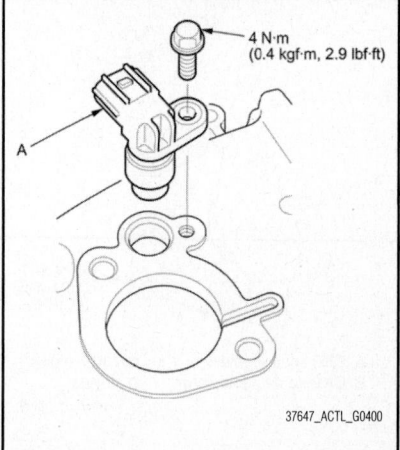

4 N·m
(0.4 kgf·m, 2.9 lbf·ft)

37647_ACTL_G0400

Fig. 115 Remove the CMP sensor (A) from the back cover

3. Disconnect the CMP sensor connector, then remove the back cover.

4. Remove the CMP sensor from the back cover.

5. Install the parts in the reverse order of removal. Install the timing belt.

6. Do the CKP pattern clear/CKP pattern learn procedure.

CRANKSHAFT POSITION (CKP) SENSOR

LOCATION

The Crankshaft Position (CKP) Sensor is located on the forward side of the engine block, near the crank pulley.

REMOVAL & INSTALLATION

See Figure 116.

1. Raise the vehicle on a lift.

➡ **Make the vehicle is level, because engine oil will drip out when you remove the sensor.**

2. Remove the CKP sensor cover.
3. Disconnect the CKP sensor connector.
4. Remove the CKP sensor.
5. Install the parts in the reverse order of removal with a new O-ring.
6. Do the CKP pattern clear/CKP pattern learn procedure.
7. Check for engine oil level, and add more oil if needed.

ENGINE COOLANT TEMPERATURE (ECT) SENSOR

LOCATION

There are two Engine Coolant Temperature (ECT) sensors used on this engine. ECT 1 is located on top of the engine, while ECT 2 is located on the lower front edge of the engine.

REMOVAL & INSTALLATION

ECT Sensor 1

See Figure 117.

1. Drain the engine coolant.
2. Remove the engine cover.
3. Disconnect the ECT sensor 1 connector.
4. Remove ECT sensor 1 and O-ring.
5. Install the parts in the reverse order of

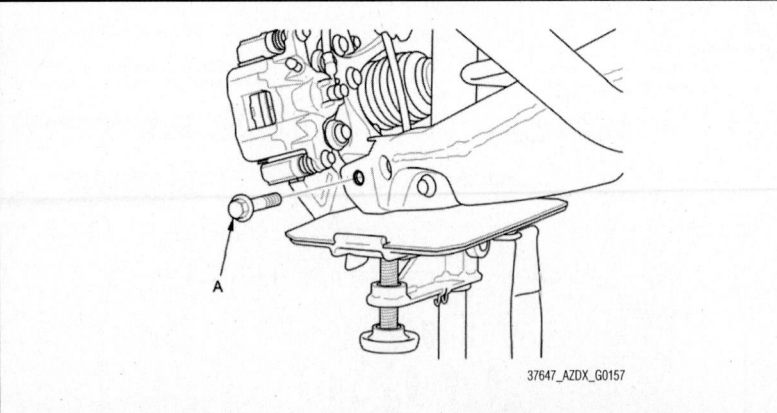

removal with a new O-ring, then refill the radiator with engine coolant.

ECT Sensor 2

See Figure 118.

1. Drain the engine coolant.
2. Remove the front splash shield.
3. Disconnect the ECT sensor 2 connector.
4. Remove ECT sensor 2 and O-ring.
5. Install the parts in the reverse order of removal with a new O-ring, then refill the radiator with engine coolant.

EVAP CANISTER

LOCATION

See Figure 119.

The Evaporative Emissions (EVAP) canister is located under the vehicle, just in front of the fuel tank.

Fig. 117 Disconnect the ECT sensor 1 connector (A). Remove ECT sensor 1 (B) and O-ring (C).

REMOVAL & INSTALLATION

See Figures 120 and 121.

1. Remove the EVAP canister cover.
2. Disconnect the quick-connect fitting, the hoses, the EVAP canister vent shut valve 2P connector, and the FTP sensor 3P connector, and remove the harness clamps.
3. Remove the bolts.
4. Remove the EVAP canister assembly.
5. Remove the EVAP canister from the bracket.
6. Remove the EVAP canister vent shut valve and the FTP sensor from the EVAP canister.

To install:

7. Reassemble the EVAP canister with new O-rings, a new cap, and a new retainer, then install the EVAP canister bracket.

➡ **Do not coat the O-rings with oil.**

8. Install the parts in the reverse order of removal.

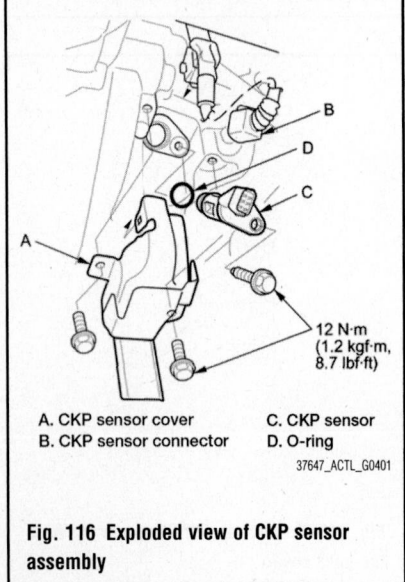

A. CKP sensor cover
B. CKP sensor connector
C. CKP sensor
D. O-ring

12 N·m
(1.2 kgf·m,
8.7 lbf·ft)

37647_ACTL_G0401

Fig. 116 Exploded view of CKP sensor assembly

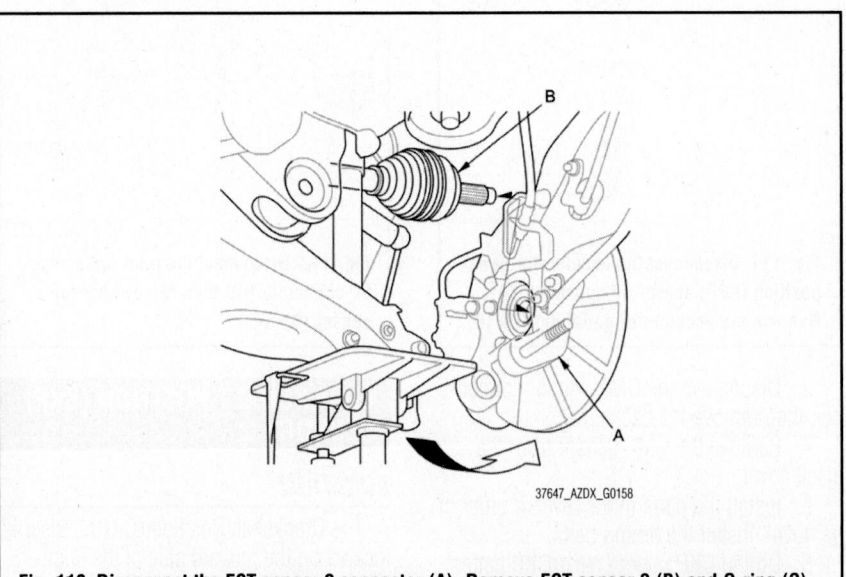

37647_AZDX_G0158

Fig. 118 Disconnect the ECT sensor 2 connector (A). Remove ECT sensor 2 (B) and O-ring (C).

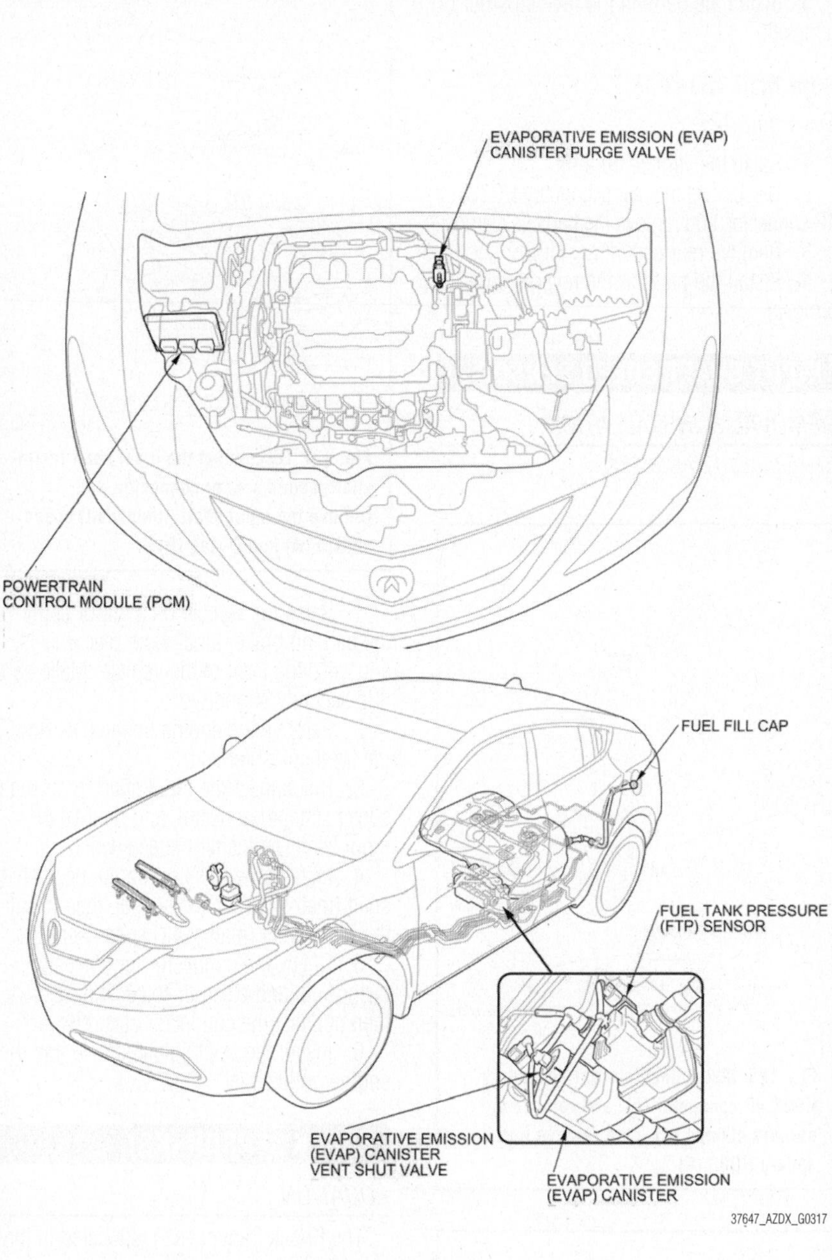

EVAPORATIVE EMISSION (EVAP) CANISTER PURGE VALVE
POWERTRAIN CONTROL MODULE (PCM)
FUEL FILL CAP
FUEL TANK PRESSURE (FTP) SENSOR
EVAPORATIVE EMISSION (EVAP) CANISTER VENT SHUT VALVE
EVAPORATIVE EMISSION (EVAP) CANISTER
37647_AZDX_G0317

Fig. 119 Showing the location of the EVAP system components

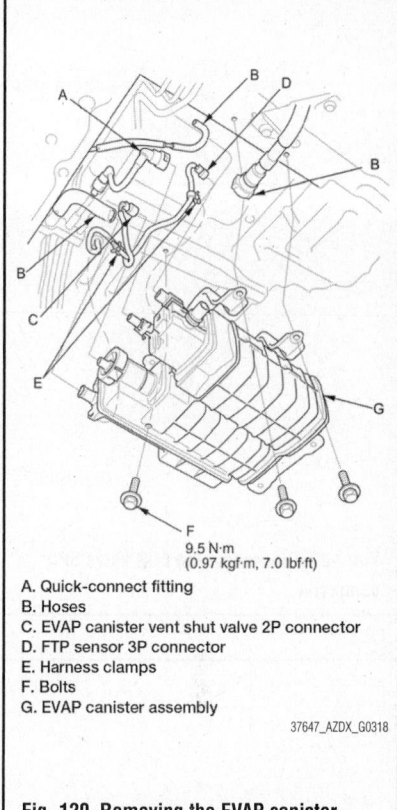

A
B
D
B
B
C
E
G
F
9.5 N·m (0.97 kgf·m, 7.0 lbf·ft)
A. Quick-connect fitting
B. Hoses
C. EVAP canister vent shut valve 2P connector
D. FTP sensor 3P connector
E. Harness clamps
F. Bolts
G. EVAP canister assembly
37647_AZDX_G0318

Fig. 120 Removing the EVAP canister and related components

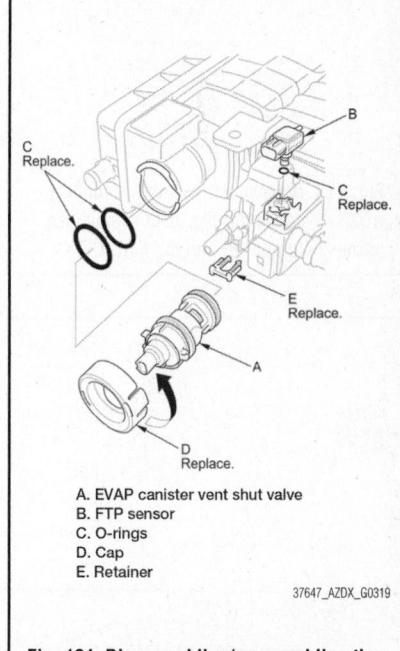

B
C Replace.
C Replace.
E Replace.
A
D Replace.
A. EVAP canister vent shut valve
B. FTP sensor
C. O-rings
D. Cap
E. Retainer
37647_AZDX_G0319

Fig. 121 Disassembling/reassembling the EVAP canister shut valve and FTP sensor

EXHAUST GAS RECIRCULATION (EGR) VALVE

LOCATION

The EGR valve is located on the left side of the engine.

REMOVAL & INSTALLATION

See Figure 122.

1. Disconnect the EGR valve 5P connector.
2. Remove the EGR valve.
3. Install the parts in the reverse order of removal with a new gasket.

REMOVAL & INSTALLATION

➡The following procedures require the use of an O2 sensor socket wrench.

Front Bank (Bank 2)

See Figures 123 and 124.

1. Disconnect the front secondary HO2S 4P connector, then remove the harness clamps and the bolt.
2. Raise the vehicle on a lift.
3. Remove the front splash shield.
4. Remove front secondary HO2S.

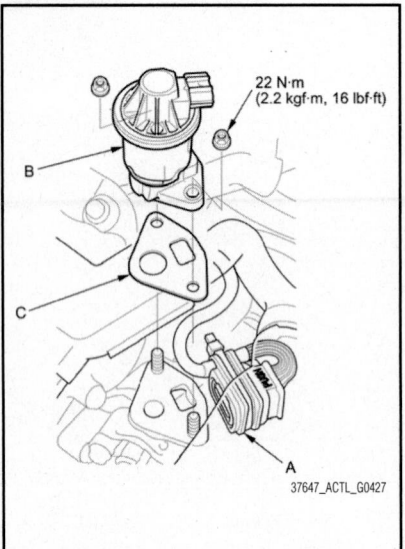

Fig. 122 Disconnect the EGR valve 5P connector

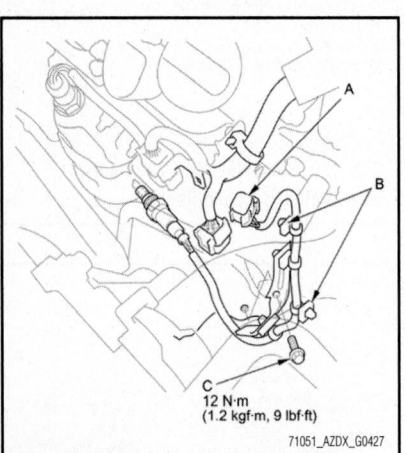

Fig. 123 Disconnect the front secondary HO2S 4P connector (A), then remove the harness clamps (B) and the bolt (C)

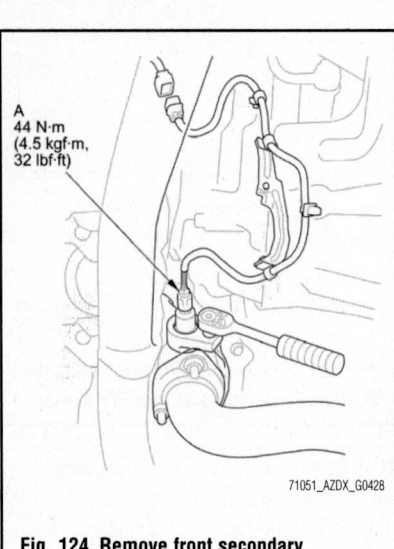

Fig. 124 Remove front secondary HO2S (A)

5. Install the parts in the reverse order of removal.

Rear Bank (Bank 1)

See Figure 125.

1. Raise the vehicle on a lift.
2. Disconnect the rear secondary HO2S 4P connector, and remove the harness clamps.
3. Remove rear secondary HO2S.
4. Install the parts in the reverse order of removal.

INPUT SHAFT SPEED SENSOR

REMOVAL & INSTALLATION

See Figures 126 and 127.

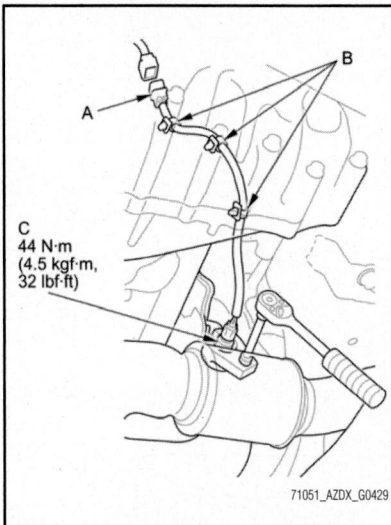

Fig. 125 Disconnect the rear secondary HO2S 4P connector (A), and remove the harness clamps (B), then remove rear secondary HO2S (C)

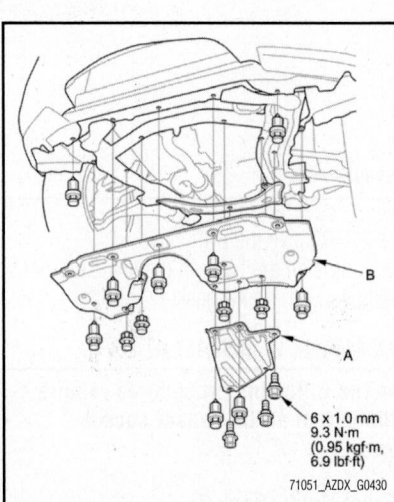

Fig. 126 Remove the engine undercover (A) and the front splash shield (B)

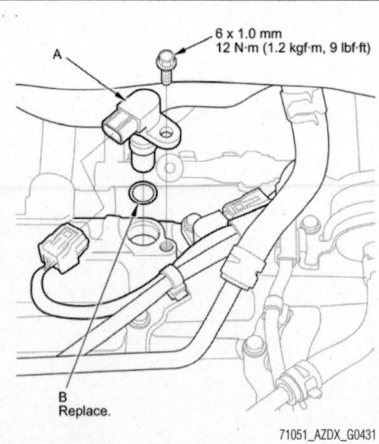

Fig. 127 Disconnect the input shaft (mainshaft) speed sensor connector, and remove the input shaft (mainshaft) speed sensor (A) and O-ring (B)

1. Raise the vehicle on a lift, or apply the parking brake, block both rear wheels, and raise the front of the vehicle. Make sure it is securely supported.
2. Remove the engine undercover and the front splash shield.
3. Disconnect the input shaft (mainshaft) speed sensor connector, and remove the input shaft (mainshaft) speed sensor.
4. Install a new O-ring on the new input shaft (mainshaft) speed sensor, then install the input shaft (mainshaft) speed sensor.
5. Check the connector for corrosion, dirt, or oil, and clean or repair if necessary, then connect the connector securely.
6. Install the front splash shield and the engine undercover.

KNOCK SENSOR (KS)

LOCATION

The Knock Sensor (KS) is located in the valley of the engine block.

REMOVAL & INSTALLATION

See Figure 128.

1. Remove the intake manifold.
2. Remove the injector rails and the injector base.
3. Disconnect the knock sensor 1P connector, then remove the knock sensor.
4. Install the parts in the reverse order of removal.

MANIFOLD ABSOLUTE PRESSURE (MAP) SENSOR

LOCATION

The Manifold Absolute Pressure (MAP) sensor is located near the throttle body.

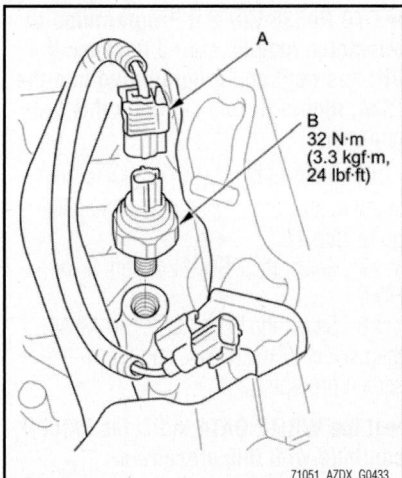

Fig. 128 Disconnect the knock sensor 1P connector (A), then remove the knock sensor (B)

REMOVAL & INSTALLATION

See Figure 129.

1. Remove the engine appearance cover.
2. Disconnect the MAP sensor connector.
3. Remove the screw.
4. Remove the MAP sensor.
5. Install the parts in the reverse order of removal with a new O-ring.

MASS AIRFLOW (MAF)/INTAKE AIR TEMPERATURE (IAT) SENSOR

LOCATION

The MAF/IAT sensor is located in the air intake near the air cleaner.

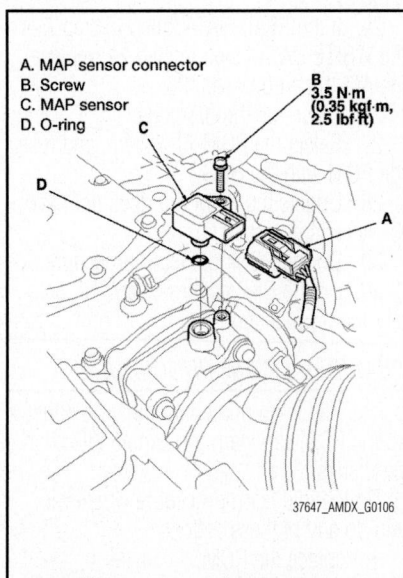

A. MAP sensor connector
B. Screw
C. MAP sensor
D. O-ring

B
3.5 N·m
(0.35 kgf·m, 2.5 lbf·ft)

37647_AMDX_G0106

Fig. 129 Disconnect the MAP sensor connector

REMOVAL & INSTALLATION

See Figure 130.

1. Disconnect the MAF/IAT senor connector.
2. Remove the screw.
3. Remove the MAF/IAT senor.
4. Install the parts in the reverse order of removal with a new gasket.

OUTPUT SHAFT SPEED (OSS) SENSOR

REMOVAL & INSTALLATION

See Figure 131.

1. Raise the vehicle on a lift, or apply the parking brake, block both rear wheels, and raise the front of the vehicle. Make sure it is securely supported.
2. Remove the engine undercover and the front splash shield.
3. Disconnect the output shaft (countershaft) speed sensor connector, and remove the output shaft (countershaft) speed sensor and O-ring.

To install:

4. Install a new O-ring on the new output shaft (countershaft) speed sensor, then install the output shaft (countershaft) speed sensor.
5. Check the connector for rust, dirt, or oil, and clean or repair if necessary, then connect the connector securely.
6. Install the front splash shield and the engine undercover.

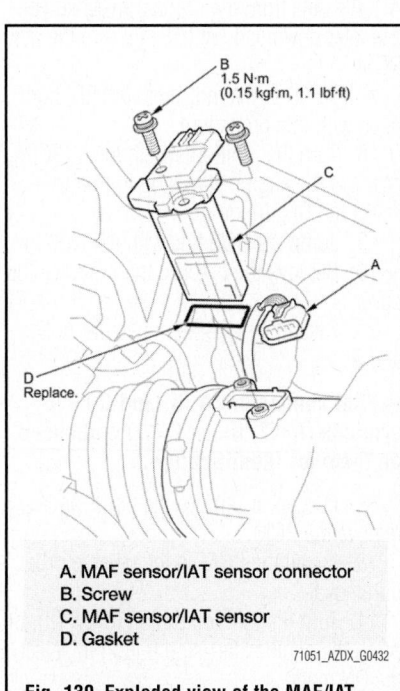

A. MAF sensor/IAT sensor connector
B. Screw
C. MAF sensor/IAT sensor
D. Gasket

71051_AZDX_G0432

Fig. 130 Exploded view of the MAF/IAT sensor assembly

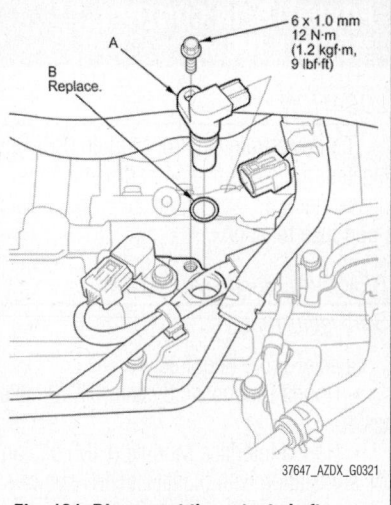

37647_AZDX_G0321

Fig. 131 Disconnect the output shaft (countershaft) speed sensor connector, and remove the output shaft (countershaft) speed sensor (A) and O-ring (B).

POSITIVE CRANKCASE VENTILATION (PCV) VALVE

REMOVAL & INSTALLATION

See Figure 132.

1. Remove the engine cover.
2. Lift up the engine harness holder.
3. Remove the bolt.
4. Remove the PCV valve.

➡**Take care not to spill oil on the hot exhaust manifold.**

5. Install the parts in the reverse order of removal.
   Please note the following:
   • When installing a new PCV valve, make sure the O-rings are in place.
   • When installing a used PCV valve, use new O-rings.

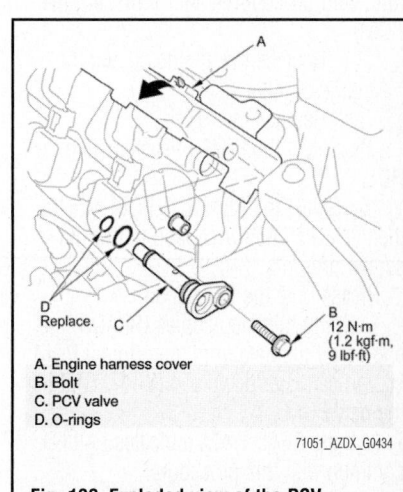

A. Engine harness cover
B. Bolt
C. PCV valve
D. O-rings

71051_AZDX_G0434

Fig. 132 Exploded view of the PCV valve assembly

POWERTRAIN CONTROL MODULE (PCM)

LOCATION

The Powertrain Control Module (PCM) is located in the engine compartment on the passenger side fender in front of the under-hood fuse relay box.

REMOVAL & INSTALLATION

See Figure 133.

Special Tools Required:
• Honda Diagnostic System (HDS) tablet tester
• Honda Interface Module (HIM) and an iN workstation with the latest HDS software version
• HDS pocket tester
• GNA600 and an iN workstation with the latest HDS software version
• MVCI unit with the latest control module (CM) update software installed

➡Any one of the above updating tools can be used.

➡Make sure the HDS/iN workstation or the MVCI unit has the latest HDS software version.

1. Connect HDS to the data link connector (DLC) located under the driver's side of the dashboard.
2. Turn the ignition switch to ON (II), or press the engine start/stop button to select the ON mode.
3. Make sure the HDS communicates with the PCM and all other vehicle systems. If it doesn't, go to the DLC circuit troubleshooting. If the result of the DLC circuit troubleshooting is to replace the PCM, skip steps 4 through 7, 16 through 21, and 24 through 25, and do the following procedures after replacing the PCM:
   a. Replace the engine oil and the engine oil filter.
   b. Replace the ATF.
4. Select the PGM-FI system with the HDS.
5. Select the REPLACE PCM MENU, then select READ DATA, and follow the screen prompts.
Please note the following:
• Doing this step copies (READS) the engine oil life data from the original PCM so you can later download (WRITES) it into the new PCM.
• If the READ DATA indicates FAILED, continue with this procedure.
6. Select the A/T system with the HDS.
7. Select the REPLACE TCM/PCM

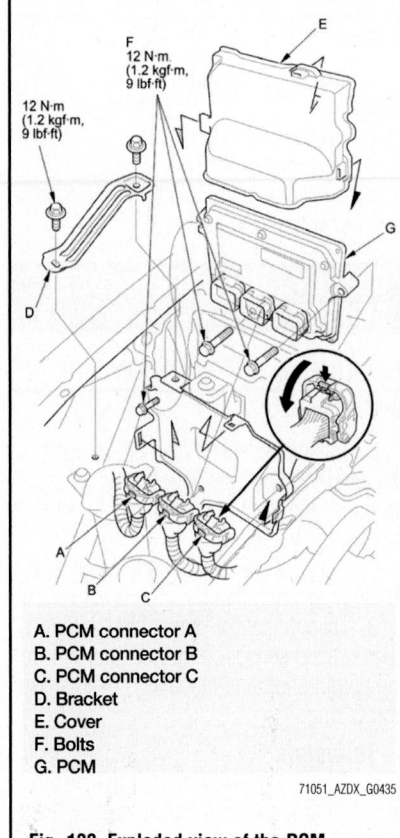

A. PCM connector A
B. PCM connector B
C. PCM connector C
D. Bracket
E. Cover
F. Bolts
G. PCM

71051_AZDX_G0435

Fig. 133 Exploded view of the PCM assembly

MENU, then select READ DATA, and follow the screen prompts.
Please note the following:
• Doing this step copies (READS) the ATF life data from the original PCM so you can later download (WRITES) it into the new PCM.
• If READ DATA indicates FAILED, continue with this procedure.
8. Turn the ignition switch to LOCK (0), or press the engine start/stop button to select the OFF mode.
9. Jump the SCS line with the HDS.
10. Remove the bracket, then remove the cover.
11. Disconnect PCM connectors A, B, and C.

➡PCM connectors A, B, and C have symbols (A=□, B=△, C=○) embossed on them for identification.

12. Loosen or remove the bolts, and remove the PCM.
13. Install the PCM in the reverse order of removal.
14. Turn the ignition switch to ON (II), or press the engine start/stop button to select the ON mode.
15. Manually input the VIN to the PCM with the HDS.

➡DTC P0630 VIN Not Programmed or Mismatch may be stored because the VIN has not been programmed into the PCM; ignore it, and continue this procedure.

16. If the READ DATA (engine oil life) failed in step 5, go to step 19. Otherwise, go to step 17.
17. Select the PGM-FI system with the HDS.
18. Select the REPLACE PCM MENU, then select WRITE DATA, and follow the screen prompts.

➡If the WRITE DATA indicates FAILED, continue with this procedure.

19. If the READ DATA (ATF life) failed in step 7, go to step 22. Otherwise go to step 20.
20. Select the A/T SYSTEM with the HDS.
21. Select the REPLACE TCM/PCM MENU, then select WRITE DATA, and follow the screen prompts.

➡If the WRITE DATA indicates FAILED, continue with this procedure.

22. Select the IMMOBI system (without keyless access) or ONE PUSH START system (with keyless access) with the HDS.
23. Enter the immobilizer PCM code that you got from the iN, and use the PCM replacement procedure in the IMMOBI Menu (without keyless access) or ONE PUSH START Menu (with keyless access) of the HDS; it allows you to start the engine.
24. If the READ DATA failed in step 5 or the WRITE DATA failed in step 18, replace the engine oil and the engine oil filter, then go to step 25.
25. If the READ DATA failed in step 7 or the WRITE DATA failed in step 21, replace the ATF, then go to step 26.
26. Clean the throttle body.
27. Select the PGM-FI system, and reset the PCM with the HDS.
28. Update the PCM if it does not have the latest software.
29. Do the PCM idle learn procedure.
30. Do the CKP pattern clear/CKP pattern learn procedure.

PCM IDLE LEARN PROCEDURE

The idle learn procedure must be done so the PCM can learn the engine idle characteristics.

Do the idle learn procedure whenever you do any of these actions:
• Replace the PCM.
• Reset the PCM.
• Update the PCM.

- Replace or clean the throttle body.
- Disassemble the engine or the transmission.

➡**Clearing the DTCs with the HDS does not require you to do the idle learn procedure.**

1. Make sure all electrical items (the A/C, the audio, the lights, etc.) are off.
2. Reset the PCM with the HDS.
3. Turn the ignition switch to ON (II), or press the engine start/stop button to select the ON mode, and wait 2 seconds.
4. Start the engine. Hold the engine speed at 3,000 rpm without load (in P or N) until the radiator fan comes on, or until the engine coolant temperature reaches 194°F (90°C).
5. Let the engine idle for about 5 minutes with the throttle fully closed.

➡**If the radiator fan comes on, do not include its running time in the 5 minutes.**

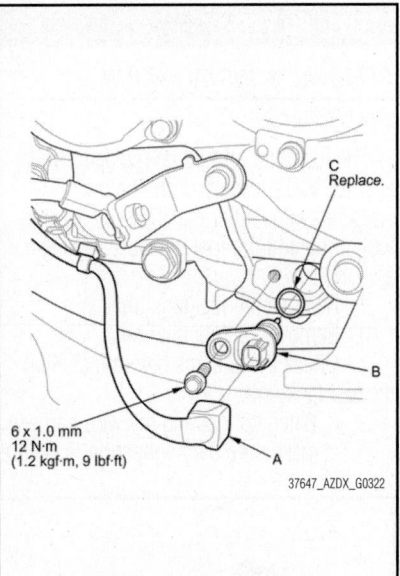

6 x 1.0 mm
12 N·m
(1.2 kgf·m, 9 lbf·ft)

37647_AZDX_G0322

Fig. 134 Disconnect the ATF temperature sensor connector (A), then remove the ATF temperature sensor (B).

TRANSMISSION FLUID TEMPERATURE (TFT) SENSOR

LOCATION

The ATF Temperature Sensor is located on the side of the transmission housing.

REMOVAL & INSTALLATION

See Figure 134.

1. Raise the vehicle on a lift, or apply the parking brake, block both rear wheels, and raise the front of the vehicle. Make sure it is securely supported.
2. Remove the engine undercover and the front splash shield.
3. Remove the drain plug and drain the automatic transmission fluid (ATF).
4. Reinstall the drain plug with a new sealing washer. Tighten to 36 ft. lbs. (49 Nm).
5. Disconnect the ATF temperature sensor connector, then remove the ATF temperature sensor.

FUEL

GASOLINE FUEL INJECTION SYSTEM

FUEL SYSTEM SERVICE PRECAUTIONS

Safety is the most important factor when performing not only fuel system maintenance but any type of maintenance. Failure to conduct maintenance and repairs in a safe manner may result in serious personal injury or death. Maintenance and testing of the vehicle's fuel system components can be accomplished safely and effectively by adhering to the following rules and guidelines.

- To avoid the possibility of fire and personal injury, always disconnect the negative battery cable unless the repair or test procedure requires that battery voltage be applied.
- Always relieve the fuel system pressure prior to disconnecting any fuel system component (injector, fuel rail, pressure regulator, etc.), fitting or fuel line connection. Exercise extreme caution whenever relieving fuel system pressure to avoid exposing skin, face and eyes to fuel spray. Please be advised that fuel under pressure may penetrate the skin or any part of the body that it contacts.
- Always place a shop towel or cloth around the fitting or connection prior to loosening to absorb any excess fuel due to spillage. Ensure that all fuel spillage (should it occur) is quickly removed from engine surfaces. Ensure that all fuel soaked

cloths or towels are deposited into a suitable waste container.

- Always keep a dry chemical (Class B) fire extinguisher near the work area.
- Do not allow fuel spray or fuel vapors to come into contact with a spark or open flame.
- Always use a back-up wrench when loosening and tightening fuel line connection fittings. This will prevent unnecessary stress and torsion to fuel line piping.
- Always replace worn fuel fitting O-rings with new Do not substitute fuel hose or equivalent where fuel pipe is installed.

Before servicing the vehicle, make sure to also refer to the precautions in the beginning of this section as well.

RELIEVING FUEL SYSTEM PRESSURE

✱✱ CAUTION

Before disconnecting fuel lines or hoses, relieve pressure from the system by disabling the fuel pump and then disconnecting the fuel line/quick-connect fitting in the engine compartment.

With the HDS

1. Connect the HDS to the data link connector (DLC) located under the driver's side of the dashboard.

2. Turn the ignition switch to ON (II), or press the engine start/stop button to select the ON mode.
3. Make sure the HDS communicates with the PCM. If it doesn't, go to the DLC circuit troubleshooting.
4. Turn the ignition switch to LOCK (0), or press the engine start/stop button to select the OFF mode.
5. Remove the fuel fill cap to relieve the pressure in the fuel tank.
6. Turn the ignition switch to ON (II), or press the engine start/stop button to select the ON mode.
7. From the INSPECTION MENU of the HDS, select Fuel Pump OFF, then start the engine, and let it idle until it stalls.
Please note the following:

- Do not allow the engine to idle above 1,000 rpm or the PCM will continue to operate the fuel pump.
- A Pending or Confirmed DTC may be set during this procedure. Check for DTCs, and clear them as needed.

8. Turn the ignition switch to LOCK (0), or press the engine start/stop button to select the OFF mode.
9. Do the battery terminal disconnection procedure.
10. Remove the quick-connect fitting cover.
11. Check the fuel quick-connect fitting for dirt, and clean it if needed.
12. Place a rag or shop towel over the quick-connect fitting.

13. Disconnect the quick-connect fitting: Hold the connector with one hand, and squeeze the retainer tabs with the other hand to release them from the locking tabs. Pull the connector off.

Please note the following:

• Be careful not to damage the line or other parts.

• Do not use tools.

• If the connector does not move, keep the retainer tabs pressed down, and alternately pull and push the connector until it comes off easily.

• Do not remove the retainer from the line; once removed, the retainer must be replaced with a new one.

14. After disconnecting the quick-connect fitting, check it for dirt or damage.

15. Do the battery terminal reconnection procedure.

Without the HDS

1. Remove PGM-FI main relay 2 from the driver's under-dash fuse/relay box.

2. Start the engine, and let it idle until it stalls.

➡ **If any DTCs are stored, clear and ignore them.**

3. Turn the ignition switch to LOCK (0), or press the engine start/stop button to select the OFF mode.

4. Remove the fuel fill cap to relieve the pressure in the fuel tank.

5. Do the battery terminal disconnection procedure.

6. Remove the quick-connect fitting cover.

7. Check the fuel quick-connect fitting for dirt, and clean it if needed.

8. Place a rag or shop towel over the quick-connect fitting.

9. Disconnect the quick-connect fitting: Hold the connector with one hand, and squeeze the retainer tabs with the other hand to release them from the locking tabs. Pull the connector off.

Please note the following:

• Be careful not to damage the line or other parts.

• Do not use tools.

• If the connector does not move, keep the retainer tabs pressed down, and alternately pull and push the connector until it comes off easily.

• Do not remove the retainer from the line; once removed, the retainer must be replaced with a new one.

10. After disconnecting the quick-connect fitting, check it for dirt or damage.

11. Do the battery terminal reconnection procedure.

FUEL FILTER

REMOVAL & INSTALLATION

See Figure 135.

The fuel filter should be replaced whenever the fuel pressure drops below the specified value, after making sure that the fuel pump and the fuel pressure regulator are OK.

1. Remove the fuel tank unit.

2. Remove the fuel filter set.

3. Check these items before installing the fuel tank unit:

• When connecting the wire harness, make sure the connection is secure and the connectors are firmly locked into place.

• When installing the fuel gauge sending unit, make sure the connection is secure and the connector is firmly locked into place. Be careful not to bend or twist it excessively.

4. Install the parts in the reverse order of removal with new O-rings. When installing the fuel tank unit, align the marks on the unit and the fuel tank.

➡ **Coat the O-rings with clean engine oil.**

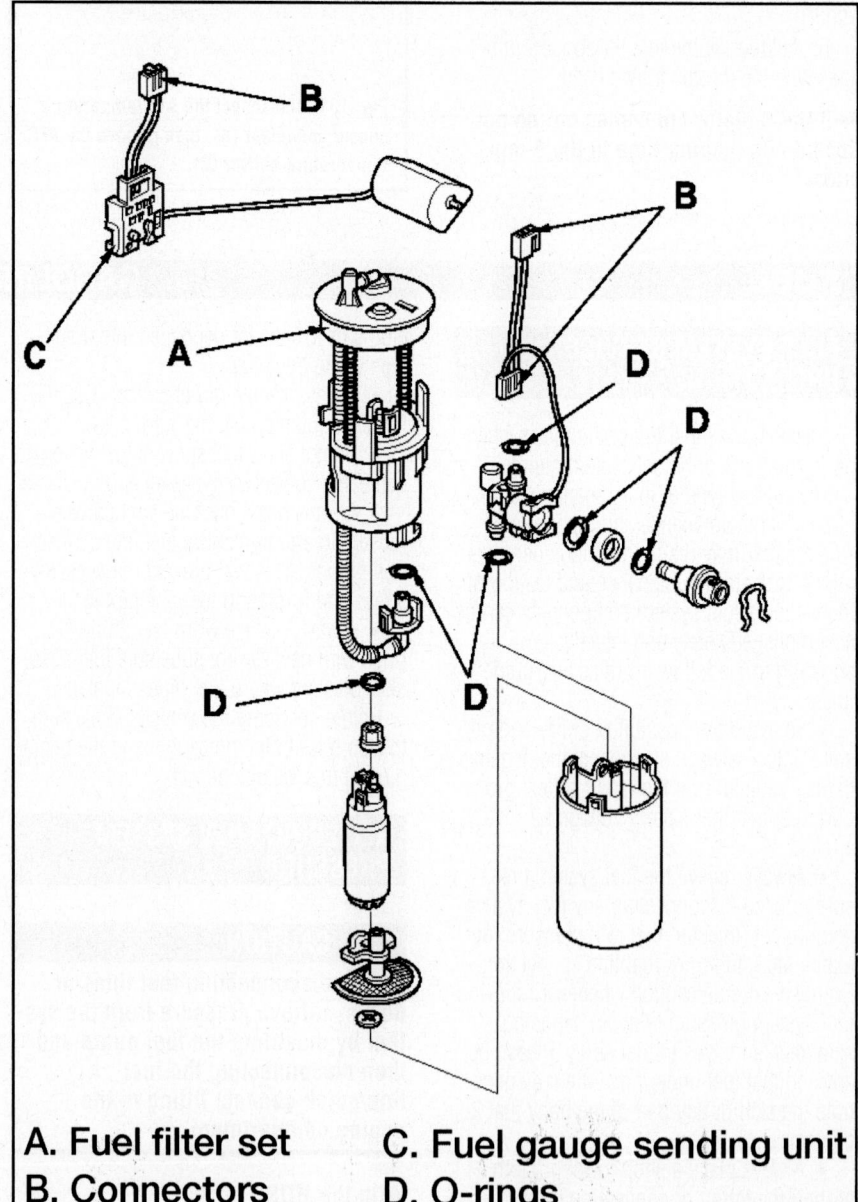

A. **Fuel filter set**
B. **Connectors**
C. **Fuel gauge sending unit**
D. **O-rings**

37647_AMDX_G0116

Fig. 135 Remove the fuel filter set

FUEL RAIL & INJECTORS

REMOVAL & INSTALLATION

See Figure 136.

1. Relieve the fuel pressure. See "Fuel System Pressure" in this section.
2. Remove the intake manifold. See "Intake Manifold" in "ENGINE MECHANICAL" section.
3. Disconnect the quick-connect fitting.
4. Remove the fuel joint hose mounting bolt.
5. Disconnect the connectors from the injectors.
6. Remove the fuel rail mounting bolts from the fuel rails.
7. Remove the fuel rails and the injectors from the injector base.
8. Remove the injector clips from the fuel rails.
9. Remove the injectors from the fuel rails.

To install:

10. Coat the new O-rings (black) with clean engine oil, and insert the injectors into the fuel rails.
11. Install the injector clips.
12. Coat the new injector O-rings (green) with clean engine oil.
13. Install the fuel rails and the injectors in the injector base.
14. Reinstall the fuel rail mounting bolts, and connect the injector connectors.
15. Reinstall the fuel joint hose mounting bolt.
16. Connect the quick-connect fitting.
17. Turn the ignition switch to ON (II), or press the engine start/stop button to select the ON mode, but do not operate the starter. After the fuel pump runs for about 2 seconds, the fuel rail will be pressurized.
18. Repeat this two or three times, then make sure there are no fuel leaks.
19. Reinstall the intake manifold with a new gasket.

FUEL TANK

DRAINING

1. Remove the fuel tank unit.
2. Using a hand pump, a hose, and a container suitable for fuel, draw the fuel from the fuel tank.
3. Reinstall the fuel tank unit.

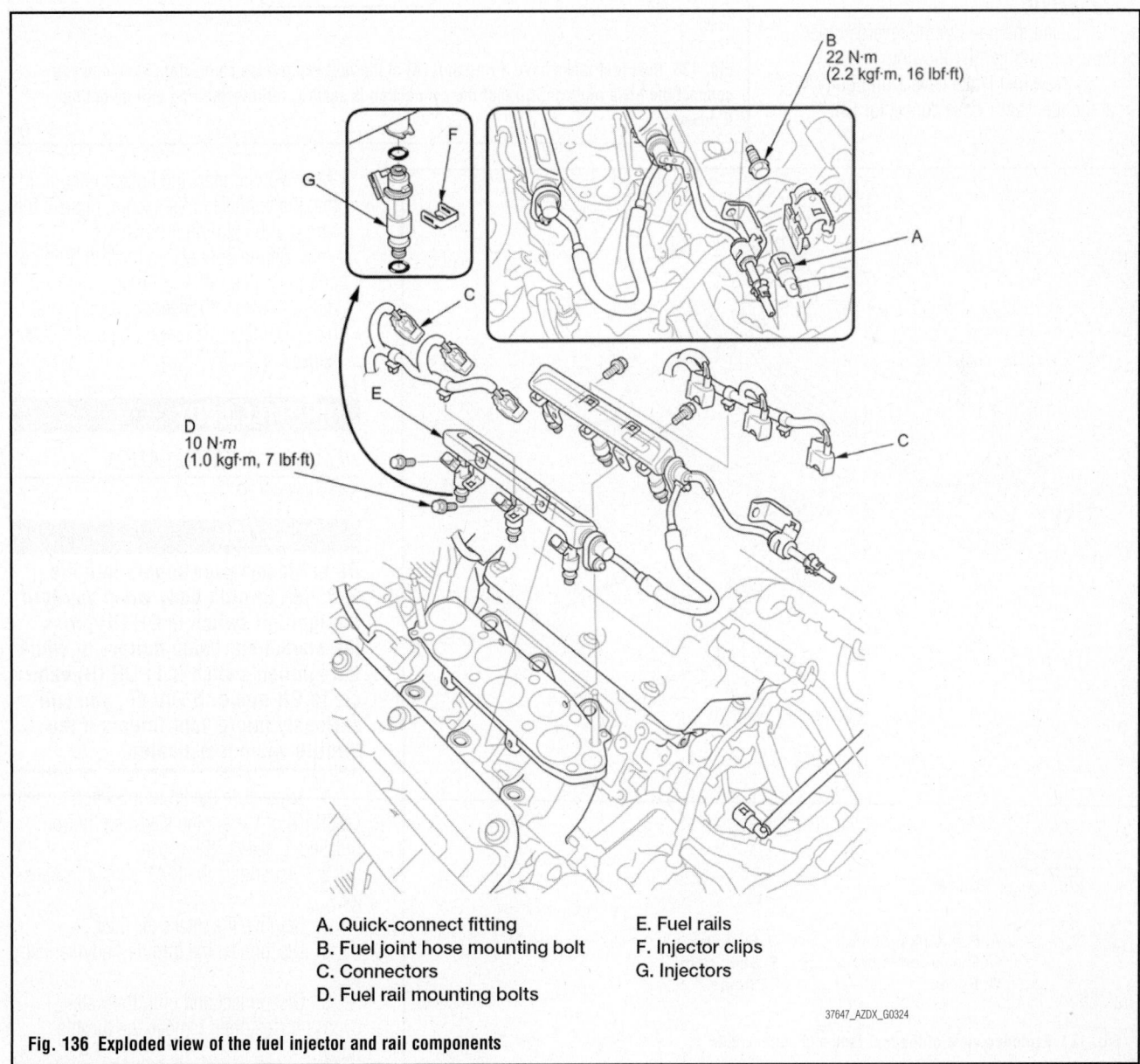

A. Quick-connect fitting
B. Fuel joint hose mounting bolt
C. Connectors
D. Fuel rail mounting bolts
E. Fuel rails
F. Injector clips
G. Injectors

37647_AZDX_G0324

Fig. 136 Exploded view of the fuel injector and rail components

REMOVAL & INSTALLATION

See Figures 137 and 138.

1. Drain the fuel tank, then reinstall the fuel tank unit without connecting the fuel tank unit 4P connector and the fuel tank unit quick-connect fitting.

2. Raise the vehicle on a lift.

3. Remove the muffler.

4. Remove the propeller shaft.

5. Remove the EVAP canister cover.

6. Remove the fuel tank protector.

7. Disconnect the quick-connector fitting and the hoses. Slide back the clamps, then twist the hoses as you pull to avoid damaging them.

8. Place a jack or other support under the fuel tank, then remove the strap bolts and the straps.

9. Remove the fuel tank.

To install:

10. Install the parts in the reverse order of removal, noting the following:

   a. New fuel tanks have a ring pull (A) at the fuel vapor hose connector (B).

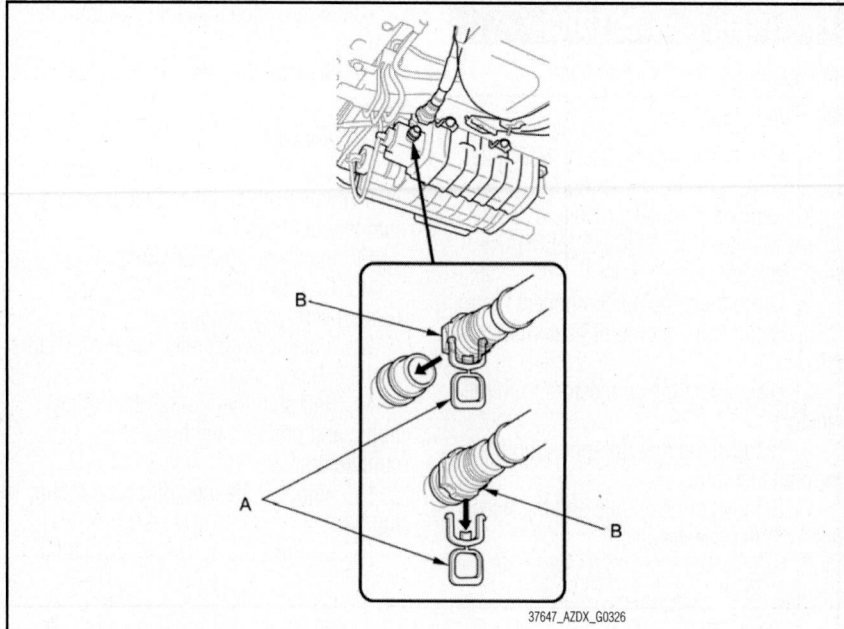

Fig. 138 New fuel tanks have a ring pull (A) at the fuel vapor hose connector (B). When you connect the hose and confirm that the connection is secure, remove the ring pull by pulling it down.

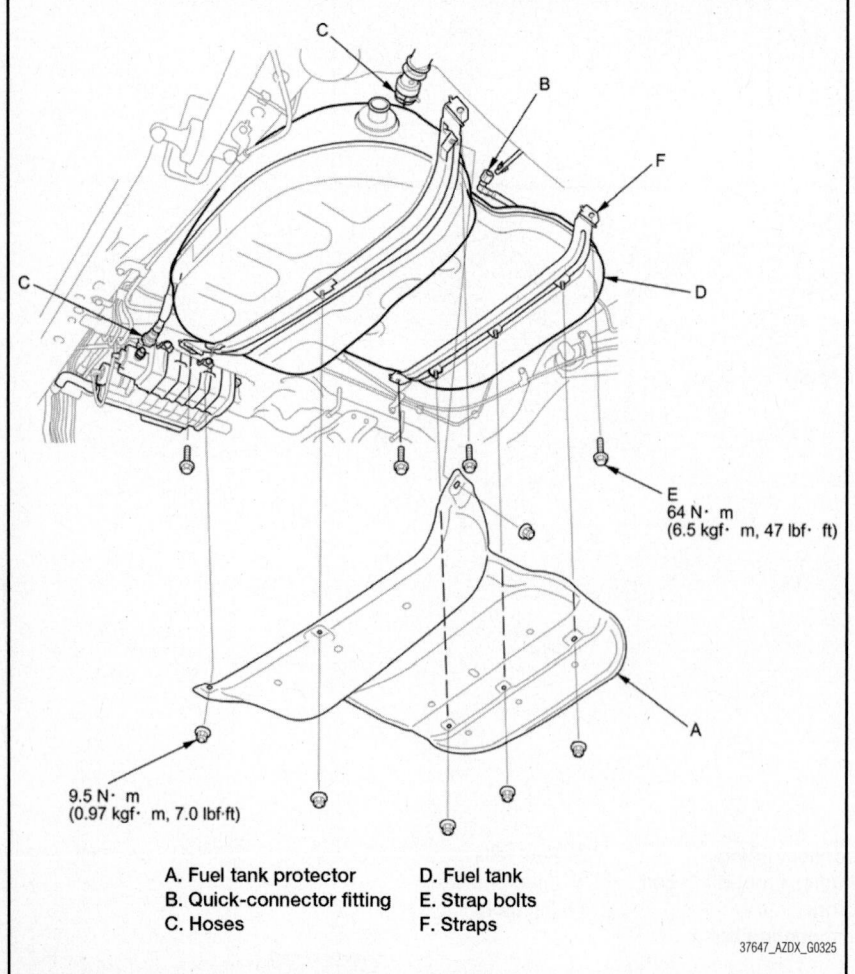

A. Fuel tank protector   D. Fuel tank
B. Quick-connector fitting   E. Strap bolts
C. Hoses   F. Straps

Fig. 137 Exploded view of the fuel tank and connections

When you connect the hose and confirm that the connection is secure, remove the ring pull by pulling it down.

   b. Before connecting the fuel fill pipe and the quick-connect fitting, check for dirt, and clean it if needed, taking care not to damage the fuel fill pipe and other parts.

THROTTLE BODY

REMOVAL & INSTALLATION

See Figure 139.

❋❋ CAUTION

Do not insert your fingers into the installed throttle body when you turn the ignition switch to ON (II)/press the engine start/stop button, or while the ignition switch is in ON (II)/vehicle in ON mode. If you do, you will seriously injure your fingers if the throttle valve is activated.

1. Make sure the ignition switch turned LOCK (0) or the engine start/stop button pressed to select OFF mode.

2. Disconnect the MAP sensor connector.

3. Remove the intake air duct.

4. Disconnect the throttle body connector.

5. Disconnect and plug the water bypass hoses, then remove the throttle body.

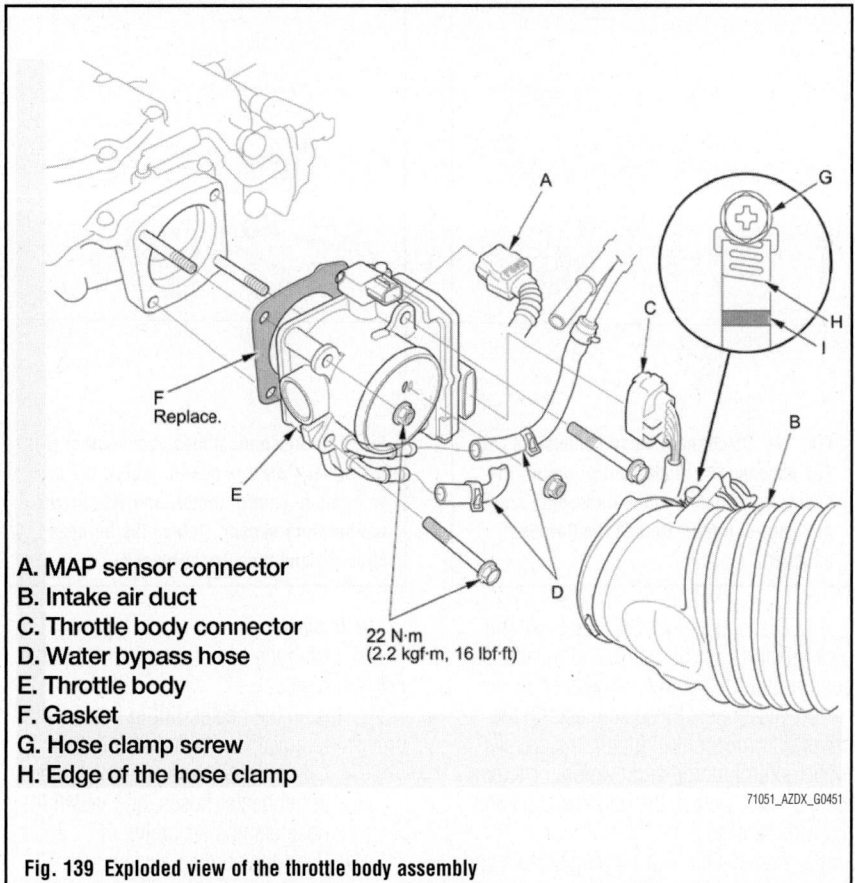

A. MAP sensor connector
B. Intake air duct
C. Throttle body connector
D. Water bypass hose
E. Throttle body
F. Gasket
G. Hose clamp screw
H. Edge of the hose clamp

22 N·m
(2.2 kgf·m, 16 lbf·ft)

71051_AZDX_G0451

Fig. 139 Exploded view of the throttle body assembly

6. Install the parts in the reverse order of removal with a new gasket.
   Please note the following:
   • When tightening the screw of the hose clamp, align the edge of the hose clamp with the painted mark.
   • After installation, refill the radiator with engine coolant.
   • If it is in case of replacing or cleaning the throttle body, go to step 7. If it isn't, this procedure is complete.
7. Connect the HDS to the data link connector (DLC) located under the driver's side of the dashboard.
8. Press the engine start/stop button to select the ON mode.
9. Reset the PCM with the HDS.
10. Select the ETCS TEST in the INSPECTION MENU with the HDS.
11. Select TP POSITION CHECK and clear the throttle position (TP) learned value.
12. Turn the ignition switch to LOCK (0) or press the engine start/stop button to select the OFF mode.
13. Turn the ignition switch to ON (II), or press the engine start/stop button to select the ON mode, and wait 2 seconds without pressing the accelerator pedal.
14. Do the PCM idle learn procedure.

HEATING & AIR CONDITIONING SYSTEM

BLOWER MOTOR

REMOVAL & INSTALLATION
See Figure 140.

1. The recirculation control motor, the blower motor, and the dust and pollen filter can be replaced without removing the blower unit.

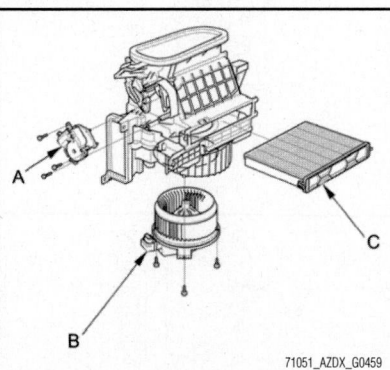

71051_AZDX_G0459

Fig. 140 The recirculation control motor (A), the blower motor (B), and the dust and pollen filter (C) can be replaced without removing the blower unit

2. Before reassembly, make sure that the recirculation control linkage and door move smoothly without binding.
3. After reassembly, make sure the recirculation control motor runs smoothly.

HEATER CORE

REMOVAL & INSTALLATION
See Figures 141 through 147.

✳✳ WARNING

SRS components are located in this area. Review the SRS component locations and the precautions and procedures before doing repairs or service.

1. Do the battery terminal disconnection procedure.
2. Recover the refrigerant with a recovery/recycling/charging station.
3. Disconnect the A/C line from the evaporator core.
4. When the engine is cool, drain the engine coolant from the radiator.

5. From under the hood, slide the heater hose clamps back.
6. Disconnect the inlet heater hose and the outlet heater hose from the heater unit. Note and/or mark the layout of the hoses.

➡**Engine coolant will run out when the hoses are disconnected; drain it into a**

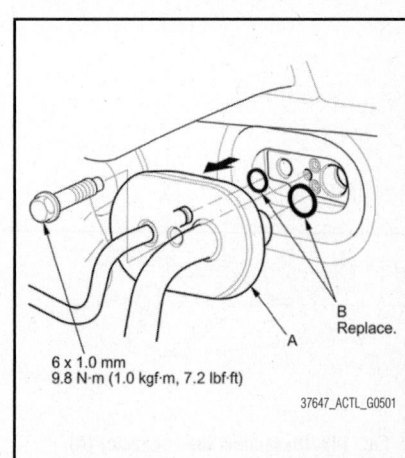

6 x 1.0 mm
9.8 N·m (1.0 kgf·m, 7.2 lbf·ft)

37647_ACTL_G0501

Fig. 141 Disconnect the A/C line from the evaporator core

clean drip pan. Be sure not to let coolant spill on the electrical parts or the painted surfaces. If any coolant spills, rinse it off immediately.

7. Remove the mounting nut from the heater unit. Take care not to damage or bend the fuel lines or the brake lines.

8. Remove the dashboard. Refer to Instrument Panel (Dashboard) Removal and Installation in the Body section.

9. Disconnect the left engine compartment wire harness connectors.

10. Remove the ducts, and the drain hose.

11. Remove the mounting nuts, and the blower-heater unit.

12. Disconnect the connector from the blower motor. Remove the wire harness clips.

13. Disconnect these connectors: The passenger's mode control motor, the power transistor, and the passenger's air mix control motor. Detach the harness clips.

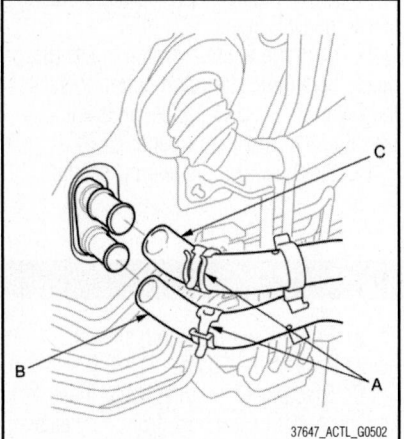

Fig. 142 Disconnect the inlet heater hose (B) and the outlet heater hose (C) from the heater unit

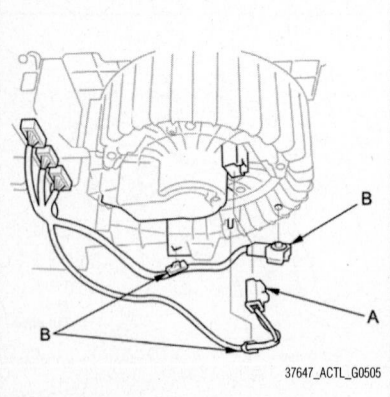

Fig. 143 Disconnect the connector (A) from the blower motor. Remove the wire harness clips (B).

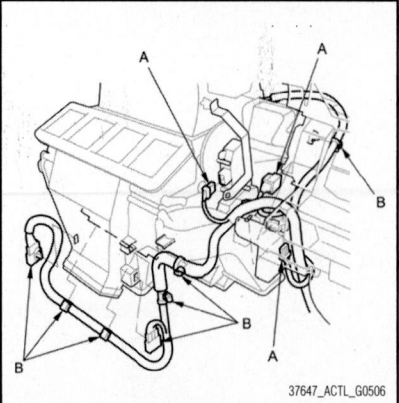

Fig. 144 Disconnect these connectors (A): The passenger's mode control motor, the power transistor, and the passenger's air mix control motor. Detach the harness clips (B)

14. Disconnect these connectors: The recirculation control motor, and the climate control unit. Detach the harness clip.

15. Disconnect these connectors: the driver's air mix control motor, the driver's mode control motor, and evaporator temperature sensor. Detach the harness clips and the wire harness.

16. Remove the self-tapping screws and the passenger's heater duct.

17. Remove the self-tapping screws and the joint duct.

18. Remove the self-tapping screw and the heater core cover.

19. Remove the self-tapping screw, the heater pipe bracket, and the grommet, and carefully pull out the heater core.

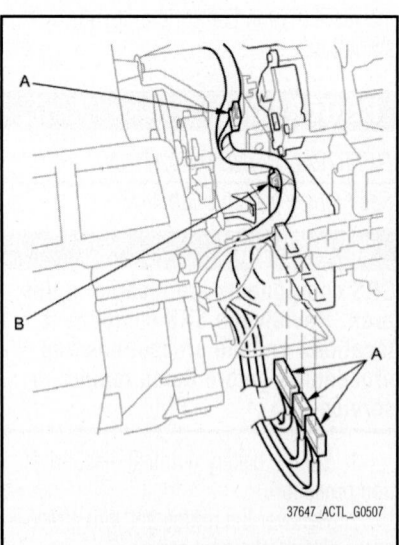

Fig. 145 Disconnect these connectors (A): The recirculation control motor, and the climate control unit. Detach the harness clip (B).

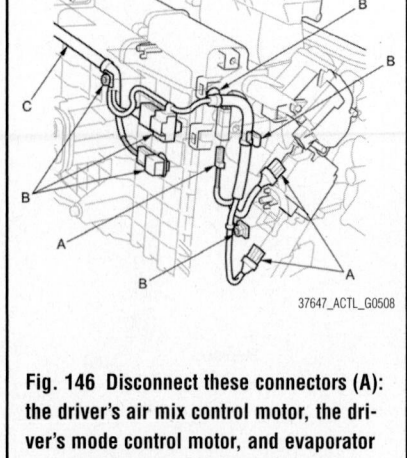

Fig. 146 Disconnect these connectors (A): the driver's air mix control motor, the driver's mode control motor, and evaporator temperature sensor. Detach the harness clips (B) and the wire harness (C).

To install:

20. Install the heater core in the reverse order of removal.

21. Install the heater unit in the reverse order of removal, and note these items:
- Do not interchange the inlet and outlet heater hoses, and install the hose clamps securely.
- Refill the cooling system with engine coolant.
- Make sure that there is no coolant leakage.
- Make sure that there is no air leakage.
- Charge the system.

22. Do the battery terminal reconnection procedure.

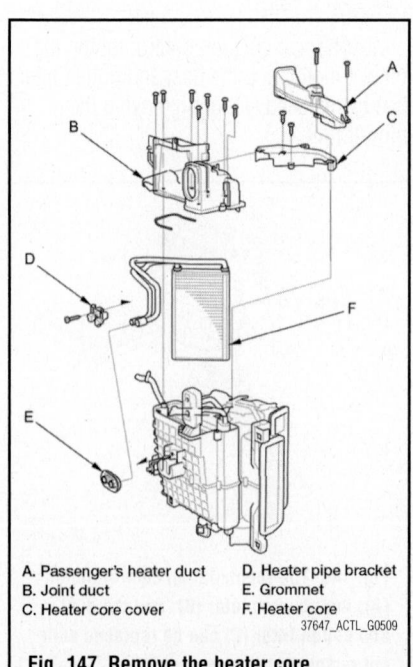

A. Passenger's heater duct
B. Joint duct
C. Heater core cover
D. Heater pipe bracket
E. Grommet
F. Heater core

Fig. 147 Remove the heater core

STEERING

POWER RACK & PINION STEERING GEAR

REMOVAL & INSTALLATION

See Figures 148 through 158.

➡This procedure covers both standard hydraulic power steering system vehicles and vehicles with Electronically Controlled Power Steering (ECPS).

1. Using clean solvent and a brush, wash any oil and dirt off the valve body unit, it's lines, and the end of the steering gearbox. Blow dry with compressed air.

2. Remove the passenger's side front fender trim.

3. Pry the hood support strut clip, and release the hood support strut, then move the hood to a vertical position and install a hood prop bar (07AAZ-SZNA100).

4. Remove the bulkhead cover.

5. Remove the engine cover.

6. Drain the power steering fluid.

7. Do the battery terminal disconnection procedure. See "BATTERY SYSTEM" section. Wait at least 3 minutes before proceeding.

8. Raise and support the vehicle.

9. Remove the front wheels.

10. Pull back the carpet, then remove the steering joint cover.

11. Loosen the upper steering joint bolt, and remove the lower steering joint bolt. Disconnect the steering joint by sliding the steering joint into the column shaft. Tighten the upper steering joint bolt to hold the steering joint in place.

12. Center the steering wheel spokes, and install a commercially available steering wheel holder tool.

13. Remove the center guide (if equipped) from the top of the pinion shaft.

14. Wrap vinyl tape over the splines on the steering gear pinion shaft.

15. With ECPS: Disconnect the ECPS valve motor connector.

16. With active damper system: Remove the suspension stroke sensor harness clips on both sides.

17. Loosen the adjustable hose clamp and disconnect the return hose.

18. Loosen the 16 mm flare nut and disconnect the inlet line.

19. Remove the cotter pin from the tie-rod end ball joint, then remove the nut.

20. Disconnect the tie-rod end ball joint from the knuckle using the ball joint thread protector and the ball joint remover on both sides.

21. Disconnect the stabilizer link from the stabilizer bar on both sides.

22. Install an appropriate engine support assembly and apply tension to support the engine.

23. Remove the front engine mount stop and the front engine mount control solenoid valve line bracket, then remove the front engine mount mounting bolt.

24. Raise the vehicle.

25. Remove the engine undercover and the front splash shield.

26. Disconnect the front engine mount control solenoid valve tube and remove the clip. Remove the front engine mount mounting bolts.

27. Remove the front subframe stiffener plate.

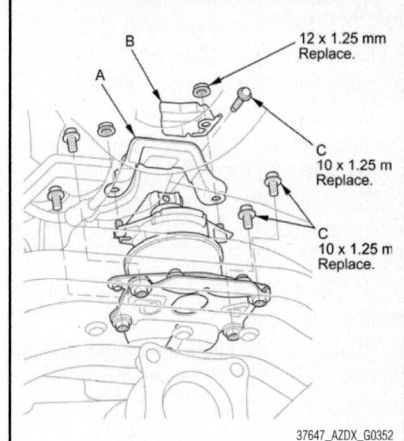

Fig. 150 Remove the rear mount stop (A), heat shield (B), and the rear engine mount mounting bolts (C), then move the rear engine mount aside.

28. Remove exhaust pipe A.

29. Remove the propeller shaft from the transfer shaft flange, after making a reference mark across the connection.

30. Remove the inlet line clamp bolts.

31. Remove the rear mount stop, heat shield, and the rear engine mount mounting bolts, then move the rear engine mount aside.

32. Remove the rear engine mount base bracket.

33. Remove the steering gearbox mounting bolts from the driver's side of the steering gearbox.

34. Remove the gearbox stiffener bracket from the driver's side of the front subframe.

35. Remove the steering gearbox mounting bolts from the passenger's side of the

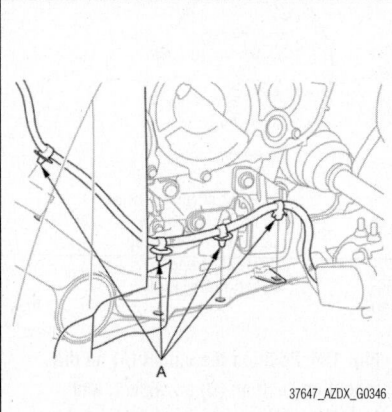

Fig. 148 With active damper system: Remove the suspension stroke sensor harness clips (A) on both sides.

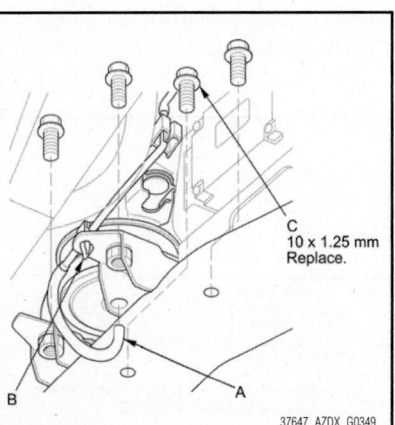

Fig. 149 Disconnect the front engine mount control solenoid valve tube (A), and remove the clip (B). Remove the front engine mount mounting bolts (C).

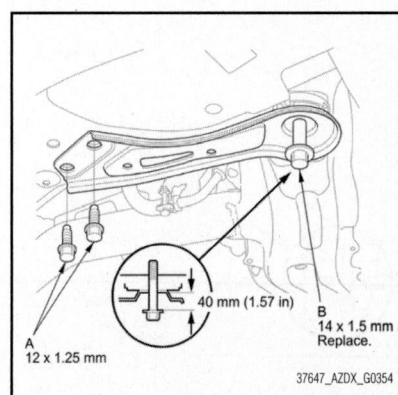

Fig. 151 Remove the front stiffener mounting bolts (A) on both sides. Loosen the subframe mounting bolts (B) so they are about 1.57 inches (40 mm) from the mounting surface.

steering gearbox, then remove the mounting bracket and cushion.

36. Attach a subframe adapter (VSB02C000016 or equivalent) to the subframe by looping a strap over the front of the subframe, then secure the strap.

37. Raise a jack and secure it with the bolt holes on the jack base, then securely attach them with four bolts.

38. Remove the lower transmission mount mounting bolts and the ground cable bolt.

39. Remove the front stiffener mounting bolts on both sides.

40. Loosen the subframe mounting bolts so they are about 1.57 inches (40 mm) from the mounting surface.

➡ **Do not loosen the subframe mounting bolts more than necessary.**

41. Remove the rear stiffener mounting bolts on both sides.

42. Loosen the subframe mounting bolts so they are about 1.57 inches (40 mm) from the mounting surface.

➡ **Do not loosen the subframe mounting bolts more than necessary.**

43. Lower the transmission jack slowly until the front subframe has dropped about 1.57 inches (40 mm).

44. Carefully move the steering gearbox toward the driver's side until the pinion shaft clears the wheel well opening on the frame.

45. Remove the steering gearbox through the wheel well opening on the driver's side.

46. Remove the pinion shaft grommet from the top of the valve housing.

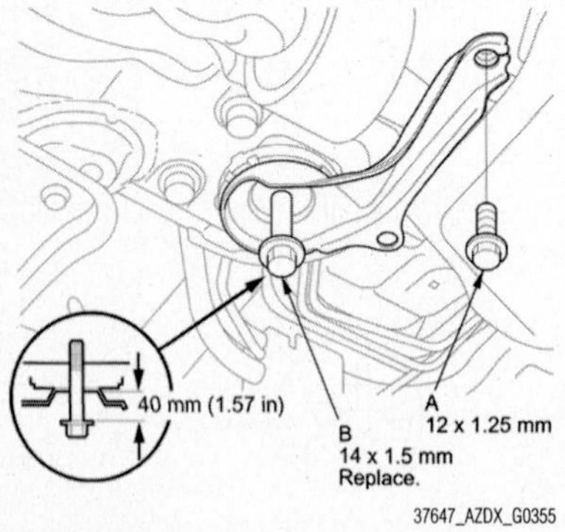

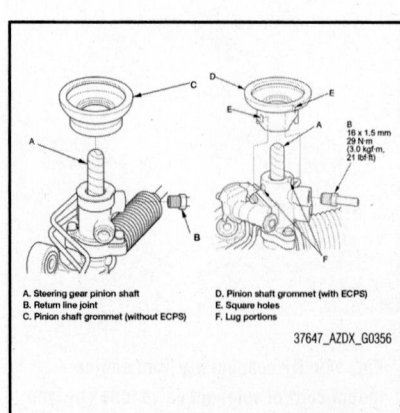

Fig. 152 Remove the rear stiffener mounting bolts (A) on both sides. Loosen the subframe mounting bolts (B) so they are about 1.57 inches (40 mm) from the mounting surface.

47. Remove the return line joint if necessary.

48. After removing the steering gearbox, make sure that no power steering fluid gets on the gearbox mount cushions, gearbox housing, surface of the front subframe and stiffener. Wipe off any spilled fluid at once.

To install:

49. Before installing the steering gearbox, make sure that no power steering fluid is on the mating surface of the steering gearbox and front subframe. To prevent the gearbox mounting bolts from loosening after the installation, remove any power steering fluid from the mount cushions and bolt holes.

50. Wrap vinyl tape over the splines on the steering gear pinion shaft.

51. Install the return line joint if removed.

52. Without ECPS: Install the pinion shaft grommet on the top of the valve housing.

53. With ECPS: Install the pinion shaft grommet. Align the square holes in the pinion shaft grommet with the lug portions on the top of the valve housing. The grommet must not have a gap at the mating surface of the grommet and valve housing.

54. Slide the steering gearbox between the front subframe and body from the driver's side.

55. Carefully move the steering gearbox toward the passenger's side until the pinion shaft clears the wheel well opening on the body.

56. Continue moving the steering gearbox toward the passenger's side until the steering gearbox is in position.

57. Raise the front subframe up to the body. Loosely install the new subframe mounting bolts and the front stiffener mounting bolts and the new rear stiffener mounting bolts.

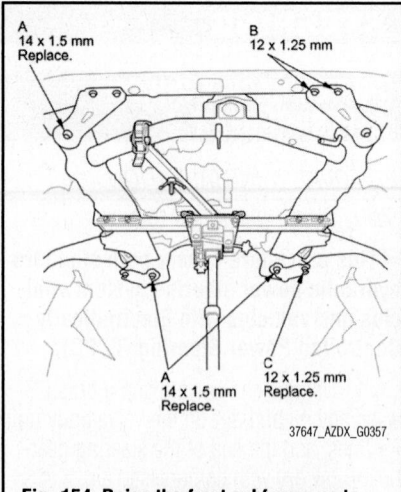

Fig. 154 Raise the front subframe up to the body. Loosely install the new subframe mounting bolts (A) and the front stiffener mounting bolts (B), and the new rear stiffener mounting bolts (C).

58. Align the front subframe with the subframe alignment pin.

59. Tighten the lower transmission mount mounting bolts and the ground cable bolt.

60. Lower the transmission jack supporting the subframe.

61. Position the cutout on the mounting

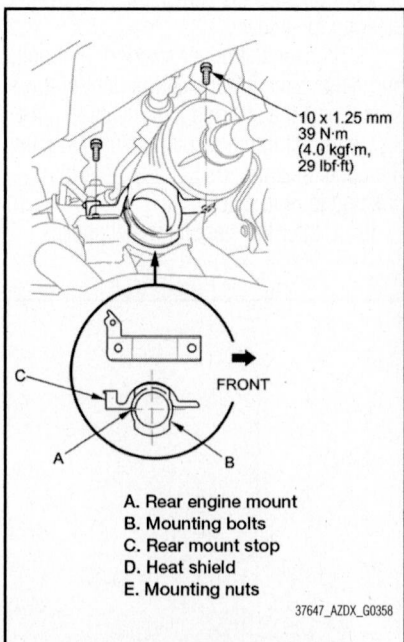

A. Rear engine mount
B. Mounting bolts
C. Rear mount stop
D. Heat shield
E. Mounting nuts

Fig. 155 Position the cutout (A) on the mounting cushion (B) as shown, and install it on the passenger's side of the steering gearbox. Install the gearbox mounting bracket (C) over the mounting cushion, and loosely install the mounting bolts.

Fig. 153 Reassembling the components on the pinion shaft

cushion, as shown, and install it on the passenger's side of the steering gearbox.

62. Install the gearbox mounting bracket over the mounting cushion, and loosely install the mounting bolts.

63. Install the gearbox stiffener bracket (A) on the driver's side of the front subframe, and tighten the bolts and nut to the specified torque of 28 ft. lbs. (38 Nm).

64. Loosely install the new steering gearbox mounting bolts on the driver's side of the steering gearbox.

65. Tighten the steering gearbox mounting bolts on both sides to 37 ft. lbs. (50 Nm), alternately in two or more steps.

66. Install the rear engine mount base bracket with new mounting bolts and tighten to 31 ft. lbs. (42 Nm).

67. Install the rear engine mount with new mounting bolts and tighten to the respective specified torques shown in the figure.

68. Install the rear mount stop and the heat shield with new mounting nuts and tighten to the specified torque in the figure.

69. Install the power steering inlet line clamp bolts, and tighten to 7 ft. lbs. (10 Nm).

70. Install exhaust pipe A and tighten the new flange-to-converter nuts to 24 ft. lbs. (33 Nm), and the new flange-to-engine nuts to 40 ft. lbs. (54 Nm).

71. Install the propeller shaft onto the transfer shaft flange and tighten the new nuts to 53 ft. lbs. (72 Nm).

72. Install the front subframe stiffener plate with new mounting bolts and tighten to the specified torque of 44 ft. lbs. (59 Nm).

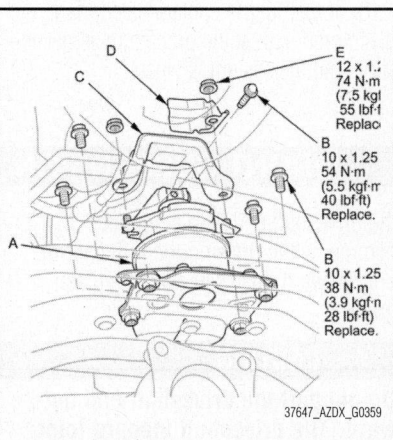

Fig. 156 Install the rear engine mount with new mounting bolts and tighten to the respective specified torques shown in the figure. Install the rear mount stop and the heat shield with new mounting nuts and tighten to the specified torque as shown.

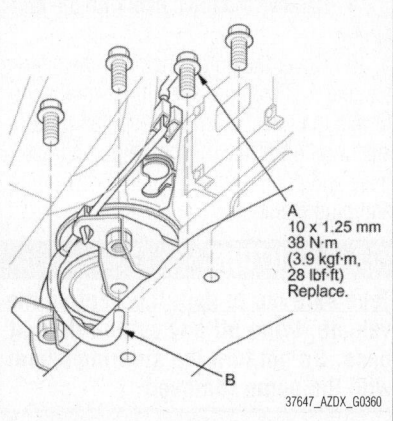

Fig. 157 Install the new front engine mount mounting bolts (A) and tighten to 28 ft. lbs. (38 Nm), then connect the front engine mount control solenoid valve tube (B).

73. Install the engine undercover and the front splash shield.

74. Install the new front engine mount mounting bolts and tighten to 28 ft. lbs. (38 Nm), then connect the front engine mount control solenoid valve tube.

75. Lower the vehicle.

76. Install the new front engine mount mounting bolt and install the front engine mount stop and the vacuum hose bracket with new mounting nuts.

77. Remove the engine support hanger, the engine hanger balance bar, and engine hanger adapter set, then install service caps to the cowl cover.

78. Install the engine mount control

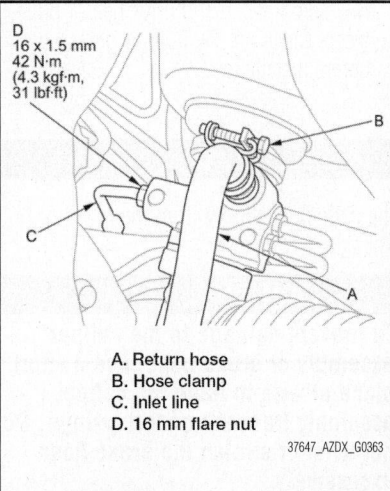

A. Return hose
B. Hose clamp
C. Inlet line
D. 16 mm flare nut

Fig. 158 Connect the return hose securely and tighten the adjustable hose clamp. Connect the inlet line and tighten the 16 mm flare nut to the specified torque of 31 ft. lbs. (42 Nm).

solenoid valve bracket to the rear cylinder head.

79. Install the connector bracket to the front cylinder head.

80. Connect the stabilizer link to the stabilizer bar on both sides. Tighten the bolts to 28 ft. lbs. (38 Nm).

81. On both sides, wipe off any grease contamination from the ball joint tapered section and threads. Reconnect the tie-rod end ball joint to the knuckle. Install the new nut and tighten it to the specified torque of 44 ft. lbs. (60 Nm). Install a new cotter pin.

82. Connect the return hose securely and tighten the adjustable hose clamp. Connect the inlet line and tighten the 16 mm flare nut to the specified torque of 31 ft. lbs. (42 Nm).

83. With active damper system: Install the suspension stroke sensor harness clips on both sides.

84. With ECPS: Connect the ECPS valve motor connector.

85. Remove the hood prop bar, then reattach the hood support strut on the pivot bolt.

86. Install the passenger's side front fender trim.

87. Install the front wheels, then set the wheels in the straight ahead position.

➡**Before installing the wheel, clean the mating surfaces between the brake disc and the inside of the wheel.**

88. Center the steering rack within its stroke.

89. Remove the vinyl tape around the splines on the pinion shaft.

90. Loosen the upper steering joint bolt, then slip the lower end of the steering joint onto the pinion shaft taking care to align the gap within the angle.

91. Remove the steering wheel holder tool.

92. Align the bolt hole on the steering joint with the groove around the pinion shaft then loosely install the lower steering joint bolt.

93. Pull on the steering joint to make sure that the steering joint is fully seated, then tighten the lower joint bolt, then the upper steering joint bolt, to the specified torque of 16 ft. lbs. (22 Nm).

94. Install the steering joint cover.

95. Set the carpet in place.

96. Do the battery terminal reconnection procedure, and check these items:

   a. Make sure the horn and turn signal switches work properly.

   b. Make sure the steering wheel switches work properly.

c. Fill the system with power steering fluid, and bleed air from the system.

d. Install the engine cover.

97. After installation, check these items:

a. Start the engine, allow it to idle, and turn the steering wheel from lock to lock several times to warm up the fluid.

b. Check the gearbox for leaks.

c. Check the steering wheel spoke angle. If steering spoke angles to the right and left are not equal (steering wheel and rack are not centered), correct the engagement of the joint/pinion shaft serrations.

d. Set the steering column to the center tilt position and to the center telescopic position, then do the front toe inspection.

e. Do the "VSA Sensor Neutral Position Memorization".

POWER STEERING PUMP

BLEEDING

1. Fill the reservoir to the upper level line.

2. Start the engine and run it at idle, then turn the steering from lock-to-lock several times to bleed air from the system.

3. Recheck the fluid level and add some if necessary. Do not fill the reservoir beyond the upper level line.

4. If the fluid is contaminated, dark, or discolored, repeat the procedure as necessary.

REMOVAL & INSTALLATION

1. Remove the engine cover.

2. Place a suitable container under the vehicle to catch any spilled fluid.

3. Drain the power steering fluid from the reservoir.

4. Remove the drive belt from the pump pulley.

5. Cover the auto-tensioner, the alternator, and A/C compressor with several shop towels to protect them from spilled power steering fluid. Disconnect the pump inlet hose and pump outlet hose from the pump, and plug them.

✳✳ CAUTION

Take care not to spill the fluid on the vehicle. Wipe off any spilled fluid at once. Do not turn the steering wheel with the pump removed.

6. Remove the pump mounting bolts, then remove the pump.

7. Cover the opening of the pump to prevent foreign material from entering the pump.

To install:

8. Transfer the pump inlet hose and the pump outlet hose from the original pump onto the new pump with the new O-ring.

9. Loosely install the pump in the pump bracket with the mounting bolts, then tighten the pump fittings to the specified torque of 8 ft. lbs. (11 Nm).

10. Tighten the pump mounting bolts to the specified torque of 16 ft. lbs. (22 Nm).

11. Install the drive belt.

12. Note these items during drive belt installation:

a. Inspect the belt for wear and cracks. Replace the belt if necessary.

b. Make sure that the belt is properly positioned on the pulleys.

c. Do not get power steering fluid or grease on the auto-tensioner, alternator, A/C compressor, and drive belt or pulley faces. Clean off any fluid or grease before installation.

13. Fill the reservoir to the upper level line.

14. Install the engine cover.

15. Start the engine, and check for leaks.

FLUID FILL PROCEDURE

1. Remove the reservoir from its holder. Raise the reservoir, then disconnect the return hose to drain the reservoir. Take care not to spill the fluid on the body and parts. Wipe off any spilled fluid at once.

➡Inspect the reservoir screen for any debris. If the reservoir screen is clogged, replace the reservoir, and check for the source of the contamination.

2. Connect a hose of suitable diameter to the disconnected return hose, and put the hose end in a suitable container.

3. Start the engine, let it run at idle, and turn the steering wheel from lock to lock several times. When fluid stops running out of the hose, shut off the engine. Discard the fluid.

➡Stop the engine immediately once the fluid stops running out of the hose, shut off the engine. Discard the fluid.

4. Reinstall the return hose on the reservoir.

5. Fill the reservoir to the upper level line.

6. Start the engine and run it at idle. Turn the steering from lock to lock several times to bleed air from the system.

7. Recheck the fluid level and add some if necessary. Do not fill the reservoir beyond the upper level line.

8. If the fluid is contaminated, dark, or discolored, repeat the procedure as necessary until the system is clean.

SUSPENSION

FRONT SUSPENSION

KNUCKLE

REMOVAL & INSTALLATION

Hub Bearing Unit

See Figure 159.

➡Refer to the exploded view figure for reference.

1. Raise and support the vehicle.

2. Remove the wheel nuts and the front wheel.

3. Remove the brake hose mounting bolt from the damper.

4. Remove the brake caliper bracket mounting bolts, then remove the caliper assembly from the knuckle.

✳✳ CAUTION

To prevent damage to the caliper assembly or brake hose, use a short piece of wire to hang the caliper assembly from the undercarriage. Do not twist or stretch the brake hose excessively.

5. Pry up the stake on the spindle nut, then remove the nut.

6. Remove the front brake disc.

7. Remove the flange bolts, and remove the hub bearing unit by tapping the driveshaft end with a soft face hammer while drawing the hub bearing unit outward, then remove the splash guard.

✳✳ CAUTION

Do not pull the driveshaft end outward. The driveshaft inboard joint may come apart.

8. Check the hub bearing unit for damage and cracks.

To install:

➡Torque settings are shown in the exploded view figure.

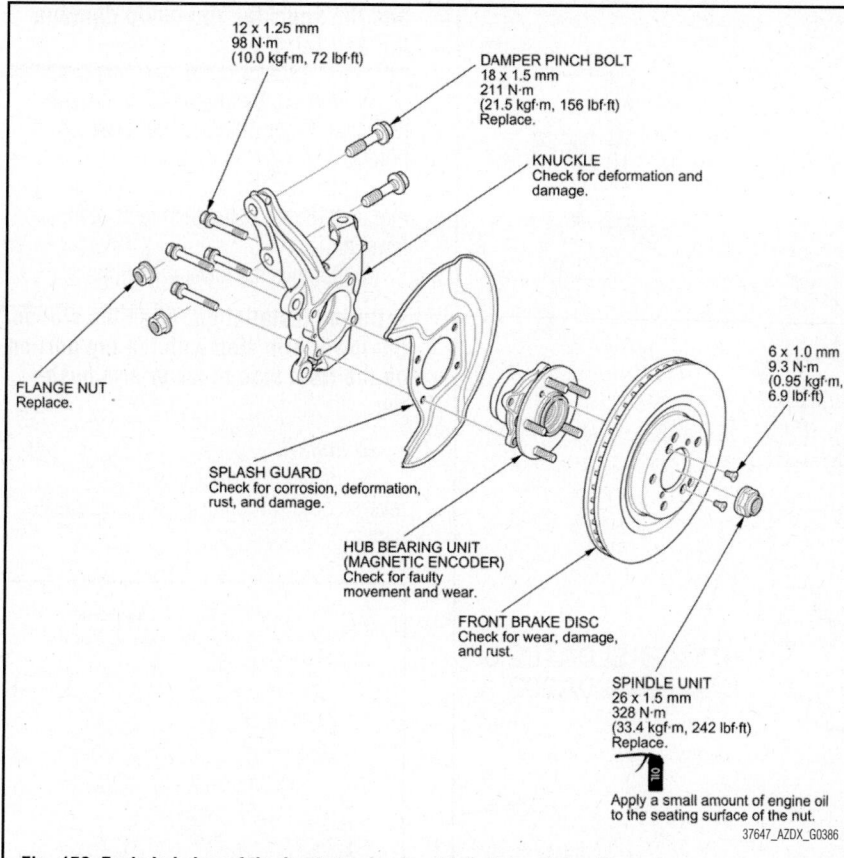

Fig. 159 Exploded view of the front steering knuckle/hub bearing unit

12 x 1.25 mm
98 N·m
(10.0 kgf·m, 72 lbf·ft)

DAMPER PINCH BOLT
18 x 1.5 mm
211 N·m
(21.5 kgf·m, 156 lbf·ft)
Replace.

KNUCKLE
Check for deformation and damage.

FLANGE NUT
Replace.

SPLASH GUARD
Check for corrosion, deformation, rust, and damage.

HUB BEARING UNIT
(MAGNETIC ENCODER)
Check for faulty movement and wear.

FRONT BRAKE DISC
Check for wear, damage, and rust.

6 x 1.0 mm
9.3 N·m
(0.95 kgf·m, 6.9 lbf·ft)

SPINDLE UNIT
26 x 1.5 mm
328 N·m
(33.4 kgf·m, 242 lbf·ft)
Replace.

Apply a small amount of engine oil to the seating surface of the nut.

37647_AZDX_G0386

a. During installation, apply grease to the mating surfaces of the hub bearing unit and driveshaft outboard joint.

b. Install the hub bearing unit in the reverse order of removal, and note these items:

c. Use the new spindle nut during reassembly.

d. Before installing the spindle nut, apply a small amount of engine oil to the seating surface of the nut.

e. After tightening, use a drift to stake the spindle nut shoulder against the driveshaft.

f. Before installing the brake disc, clean the mating surfaces of the hub bearing unit and the inside of the brake disc.

g. Before installing the wheel, clean the mating surfaces of the brake disc and the inside of the wheel.

h. Check the wheel alignment, and adjust it if necessary.

Knuckle Replacement

➡**Refer to the exploded view.**

1. Remove the hub bearing unit as above.

2. Remove the wheel speed sensor from the knuckle, but do not disconnect the wheel speed sensor connector.

3. Remove the cotter pin from the tie-rod end ball joint, then remove the nut.

4. Disconnect the tie-rod end ball joint from the knuckle using the ball joint thread protector and the ball joint remover.

5. Remove the lock pin from the lower arm ball joint, then remove the castle nut.

6. Disconnect the lower arm ball joint from the knuckle using the ball joint thread protector and the ball joint remover, using care not to damage the ball joint boot.

✳✳ CAUTION

Do not force or hammer on the lower arm, or pry between the lower arm and the knuckle. You could damage the ball joint.

7. Remove the damper pinch bolts and flange nuts from the damper, then remove the knuckle.

To install:

8. Install the knuckle in the reverse order of removal, and note these items:

a. During installation, install new damper pinch bolts and new flange nuts.

b. Be careful not to damage the ball joint boot when connecting the knuckle.

c. During installation, install the new

cotter pin after tightening the new nut, and properly bend its end.

d. Before connecting the ball joint to the knuckle, degrease the threaded section and the tapered portion of the ball joint pin, the ball joint connecting hole, the threaded section and the mating surfaces of the castle nut.

e. Torque the castle nut to the lower torque specification, then tighten it only far enough to align the slot with the ball joint pin hole. Do not align the castle nut by loosening it.

f. Check the wheel alignment, and adjust it if necessary.

LOWER BALL JOINTS

REMOVAL & INSTALLATION

See Figures 160 and 161.

➡**Always use a ball joint remover to disconnect a ball joint. Do not strike the housing or any other part of the ball joint connection to disconnect it.**

1. Install a hex nut or the ball joint thread protector onto the threads of the ball joint.

➡**Using a hex nut, make sure the nut is flush with the ball joint pin end to prevent damage to the threaded end of the ball joint pin.**

2. Apply grease to the ball joint remover. This will ease the installation of the tool, and prevent damage to the pressure bolt threads.

3. Loosen the pressure bolt, and install the ball joint remover as shown. Insert the jaws carefully, making sure not to damage the ball joint boot. Adjust the jaw spacing by turning the adjusting bolt.

➡**Fasten the safety chain securely to a suspension arm or the subframe. Do**

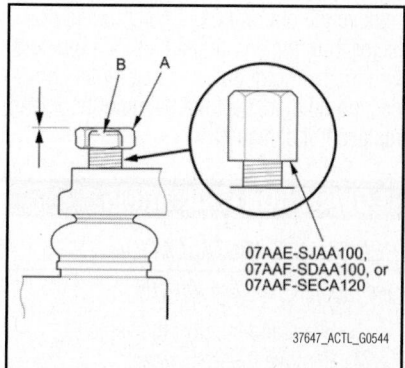

07AAE-SJAA100, 07AAF-SDAA100, or 07AAF-SECA120

37647_ACTL_G0544

Fig. 160 Install a hex nut (A) or the ball joint thread protector onto the threads of the ball joint (B)

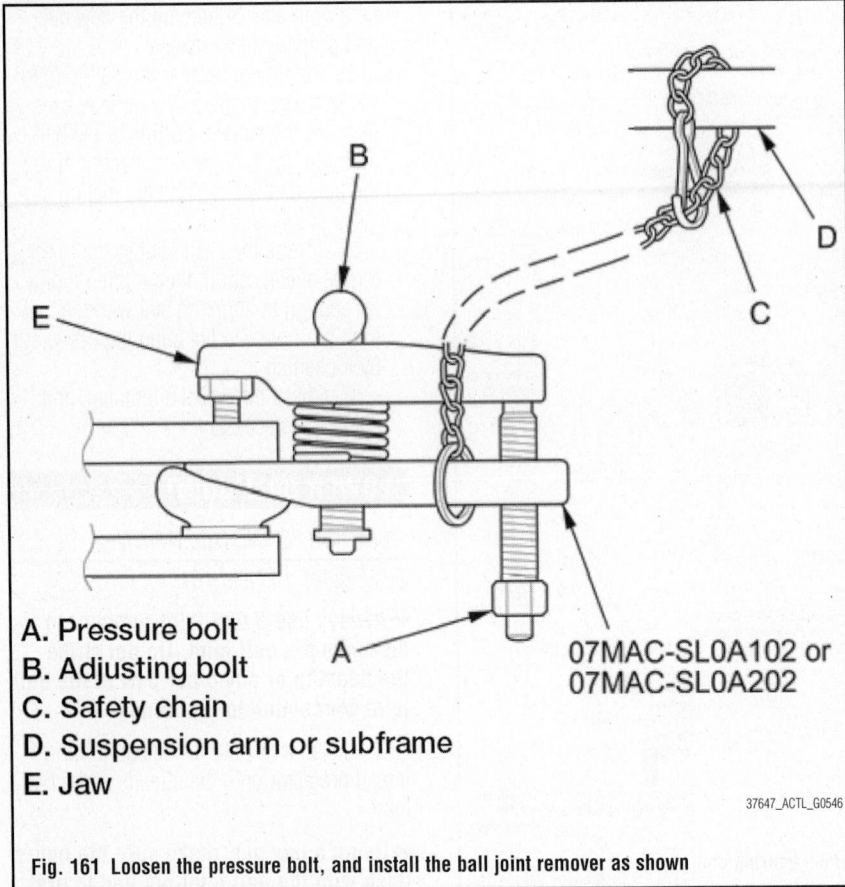

A. Pressure bolt
B. Adjusting bolt
C. Safety chain
D. Suspension arm or subframe
E. Jaw

07MAC-SL0A102 or
07MAC-SL0A202

37647_ACTL_G0546

Fig. 161 Loosen the pressure bolt, and install the ball joint remover as shown

not fasten it to a brake line or wire harness.

4. After adjusting the adjusting bolt, make sure the head of the adjusting bolt is in position to allow the jaw to pivot.

5. With a wrench, tighten the pressure bolt until the ball joint pin pops loose from the ball joint connecting hole. If necessary, apply penetrating type lubricant to loosen the ball joint pin.

➡ Do not use pneumatic or electric tools on the pressure bolt.

6. Remove the ball joint remover, then remove the nut or the ball joint thread protector from the end of the ball joint pin, and pull the ball joint out of the ball joint connecting hole. Inspect the ball joint boot, and replace it if damaged.

LOWER CONTROL ARMS

REMOVAL & INSTALLATION

See Figures 162 through 164.

1. Raise and support the vehicle.
2. Remove the front wheel.
3. With active damper system: Remove the suspension stroke sensor from the lower arm and detach the harness clip.

4. Remove the lock pin from the lower arm ball joint, then remove the castle nut.

5. Disconnect the lower arm ball joint from the knuckle using the ball joint thread protector and the ball joint remover.

✳ CAUTION

Do not force or hammer on the lower arm, or pry between the lower arm

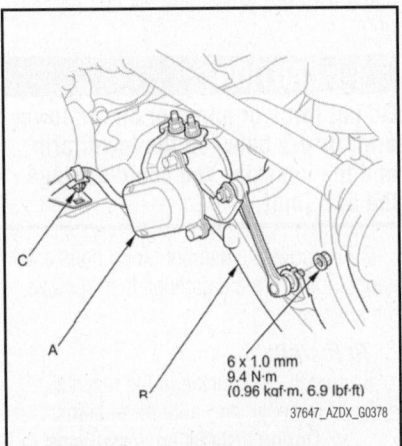

6 x 1.0 mm
9.4 N·m
(0.96 kgf·m, 6.9 lbf·ft)

37647_AZDX_G0378

Fig. 162 With active damper system: Remove the suspension stroke sensor (A) from the lower arm (B) and detach the harness clip (C).

and the knuckle. You could damage the ball joint.

6. Remove the mounting bolt of the rear side on the stabilizer bar bushing holder.

7. Remove the lower arm mounting 14 mm and 16 mm bolts, then remove the lower arm.

8. Remove the lower arm stops.

➡ During installation, align the slot on the lower arm stop with the lug portion on the front side of lower arm bushing.

To install:

9. Install the lower arm in the reverse order of removal, and note these items:

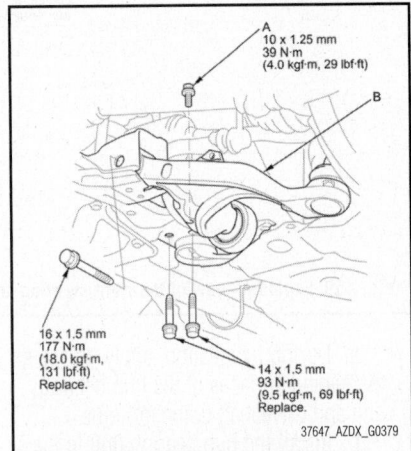

A
10 x 1.25 mm
39 N·m
(4.0 kgf·m, 29 lbf·ft)

B

16 x 1.5 mm
177 N·m
(18.0 kgf·m, 131 lbf·ft)
Replace.

14 x 1.5 mm
93 N·m
(9.5 kgf·m, 69 lbf·ft)
Replace.

37647_AZDX_G0379

Fig. 163 Remove the mounting bolt (A) of the rear side on the stabilizer bar bushing holder. Remove the lower arm mounting 14 mm and 16 mm bolts, then remove the lower arm (B).

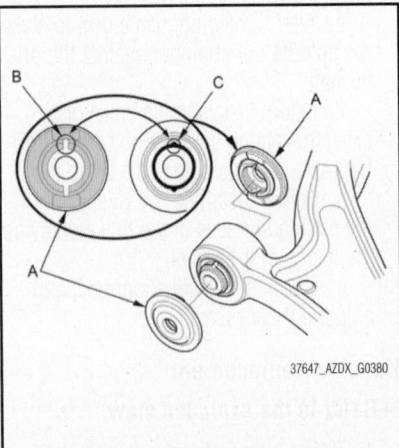

37647_AZDX_G0380

Fig. 164 Remove the lower arm stops (A). During installation, align the slot (B) on the lower arm stop with the lug portion (C) on the front side of lower arm bushing.

a. First install all of the components, and lightly tighten the bolts and the nuts, then raise the suspension to load it with the vehicle's weight before fully tightening to the specified torque as shown in the figure.

b. Do not place the jack against the ball joint of the lower arm.

c. Be careful not to damage the ball joint boot when connecting the knuckle.

d. Before connecting the ball joint to the knuckle, degrease the threaded section and the tapered portion of the ball joint pin, the ball joint connecting hole, the threaded section and the mating surfaces of the castle nut.

e. Torque the castle nut to the lower torque specification, then tighten it only far enough to align the slot with the ball joint pin hole. Do not align the castle nut by loosening it.

f. Before installing the wheel, clean the mating surfaces of the brake disc and the inside of the wheel.

10. Check the wheel alignment, and adjust it if necessary.

CONTROL ARM BUSHING REPLACEMENT

See Figure 165.

1. Remove the self-locking nut, then remove the rear side lower arm bushing with bracket.

2. Align the angle of the lower arm center line and the under edge line of the bushing bracket as shown.

3. Install a new self-locking nut, then tighten the nut to the specified torque of 119 ft. lbs.(162 Nm).

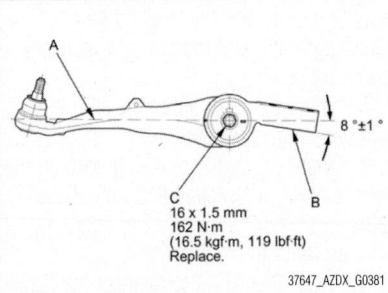

16 x 1.5 mm
162 N·m
(16.5 kgf·m, 119 lbf·ft)
Replace.

37647_AZDX_G0381

Fig. 165 Align the angle of the lower arm center line (A) and the under edge line (B) of the bushing bracket as shown. Install a new self-locking nut (C), then tighten the nut to the specified torque of 119 ft. lbs.(162 Nm).

STABILIZER BAR & LINKS

REMOVAL & INSTALLATION

Stabilizer Links

See Figure 166.

1. Raise and support the vehicle.
2. Remove the front wheel.
3. Remove the flange nuts while holding the respective joint pin with a hex wrench, then remove the stabilizer link.

To install:

4. Install the stabilizer link on the stabilizer bar and the damper with the joint pins set at the center of their range of movement. Note the stabilizer link has a white paint mark.

5. Install the flange nuts, and tighten them to the specified torque while holding the respective joint pin with a hex wrench.

6. Clean the mating surfaces of the brake disc and the inside of the wheel, then install the front wheel.

Stabilizer Bar

See Figures 167 through 170.

These special tools, or equivalent, are required during this procedure:
- Hood Prop Bar 07AAZ-SZNA100
- Ball Joint Thread Protector, 12 mm 07AAF-SDAA100

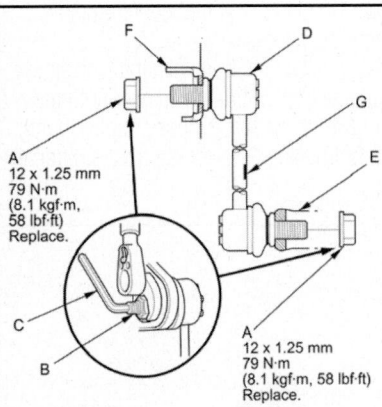

A
12 x 1.25 mm
79 N·m
(8.1 kgf·m,
58 lbf·ft)
Replace.

A
12 x 1.25 mm
79 N·m
(8.1 kgf·m, 58 lbf·ft)
Replace.

A. Flange nuts   E. Stabilizer bar
B. Joint pin   F. Damper
C. Hex wrench   G. White paint mark
D. Stabilizer link

37647_AZDX_G0393

Fig. 166 Remove the flange nuts while holding the respective joint pin with a hex wrench, then remove the stabilizer link. Install the stabilizer link on the stabilizer bar and the damper with the joint pins set at the center of their range of movement. Note the stabilizer link has a white paint mark.

- Ball Joint Remover, 28 mm 07MAC-SL0A202
- Engine Hanger Adapter Set VSB02C000031
- Subframe Adapter VSB02C000016

➡**When you replace the stabilizer bar, lower the front subframe from the body, and replace the stabilizer bar through the gap created by lowering the front subframe.**

1. Remove the passenger's side front fender trim.

2. Pry the hood support strut clip, and release the hood support strut, then move the hood to a vertical position and install a hood prop bar (o7AAZ-SZNA100, or equivalent).

3. Do the battery terminal disconnection procedure. See "BATTERY SYSTEM" section. Wait at least 3 minutes before proceeding.

4. Remove the engine cover.

5. Remove the bulkhead cover.

6. Raise and support the vehicle.

7. Remove the front wheels.

8. Pull back the carpet, then remove the steering joint cover.

9. Loosen the upper steering joint bolt, and remove the lower steering joint bolt. Disconnect the steering joint by sliding the steering joint into the column shaft. Tighten the upper bolt to hold the steering joint in place.

➡**If the center guide is in place and has not moved, leave it in place. If the center guide has moved or been removed, discard it.**

10. Center the steering wheel spokes, and install a commercially available steering wheel holder tool.

11. Remove the cotter pin from the tie-rod end ball joint, then remove the nut.

12. Disconnect the tie-rod end ball joint from the knuckle using the ball joint thread protector and the ball joint remover.

13. Disconnect both sides of the stabilizer link from the stabilizer bar.

14. Remove the power steering pump outlet hose mounting bolt.

15. Remove the connector bracket from the front cylinder head; use the bracket bolt hole to attach the engine hanger balance bar front arm (from Engine Hanger Adapter Set VSB02C000031).

16. Remove the engine mount control solenoid valve bracket from the rear cylinder head; use the bracket bolt hole to attach the engine hanger balance bar rear arm (from Engine Hanger Adapter Set VSB02C000031).

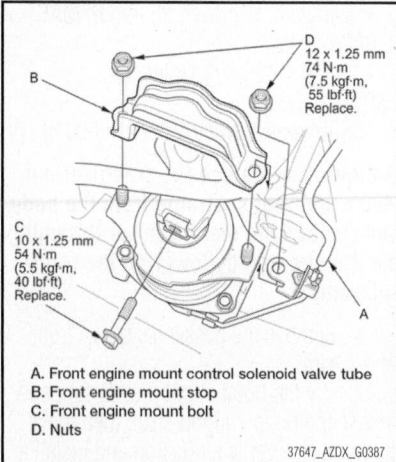

A. Front engine mount control solenoid valve tube
B. Front engine mount stop
C. Front engine mount bolt
D. Nuts

37647_AZDX_G0387

Fig. 167 Disconnect the front engine mount control solenoid valve tube, and remove the front engine mount stop, then remove the front engine mount bolt. Discard the nuts and bolt.

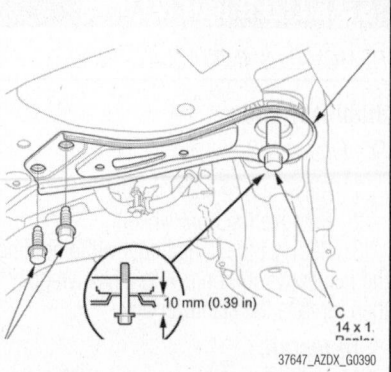

37647_AZDX_G0390

Fig. 168 Remove the stiffener mounting bolts (A) on both sides of the front stiffener (B). Loosen the subframe mounting bolts (C) to obtain a 10 mm (0.39 in) distance between the bolt seat and the mounting surface. Do not loosen the mounting bolts more than necessary.

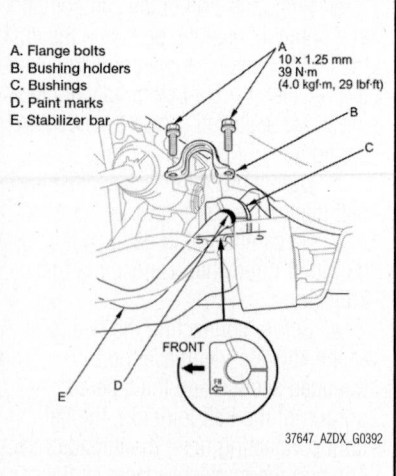

A. Flange bolts
B. Bushing holders
C. Bushings
D. Paint marks
E. Stabilizer bar

37647_AZDX_G0392

Fig. 170 Remove the flange bolts and the bushing holders, then remove the bushings. Note the paint marks on the stabilizer bar which will be aligned with the side of the bushings during installation.

17. Remove the service caps for the front damper flange nuts from the cowl cover. Position the engine hanger adapter set (VSB02C000031) with the "FRONT" mark facing forward over the damper flange nuts.

18. Support the engine with the engine support hanger, and the engine hanger balance bar.

19. Disconnect the front engine mount control solenoid valve tube, and remove the front engine mount stop, then remove the front engine mount bolt. Discard the nuts and bolt.

20. Raise the vehicle on a lift.

21. Remove the engine undercover.

22. Remove the front splash shield.

23. Remove the front subframe stiffener plate.

24. Remove exhaust pipe A.

25. Remove the P/S line bracket, then remove the rear engine mount base mounting bolts.

26. Remove the ground cable bolt and remove the lower transmission mount mounting bolts.

27. Attach the subframe adapter (VSB02C000016) to a transmission jack, then raise up the jack to the subframe. Loop the belt over the front of the subframe, then secure the belt by inserting the stop and then tightening the wing nut.

28. Raise the jack and line up the slots in the subframe adapter arms with the bolt holes on the jack base, then securely attach them with four bolts.

29. Remove the stiffener mounting bolts on both sides of the front stiffener.

30. Loosen the subframe mounting bolts to obtain a 10 mm (0.39 in) distance between the bolt seat and the mounting surface. Do not loosen the mounting bolts more than necessary.

31. Remove the stiffener mounting bolts on both sides of the rear stiffener.

32. Loosen the subframe mounting bolts to obtain a 30 mm (1.18 in) distance between the bolt seat and the mounting surface. Do not loosen the mounting bolts more than necessary.

33. Lower the transmission jack slowly with the front subframe adapter until the front subframe has dropped about 30 mm (1.18 in). Do not lower the front subframe

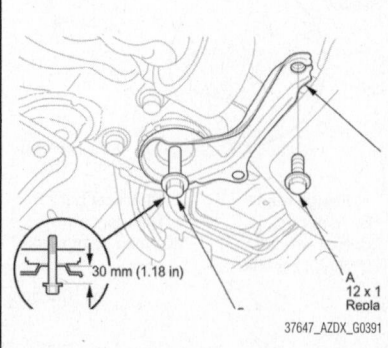

37647_AZDX_G0391

Fig. 169 Remove the stiffener mounting bolts (A) on both sides of the rear stiffener (B). Loosen the subframe mounting bolts (C) to obtain a 30 mm (1.18 in) distance between the bolt seat and the mounting surface. Do not loosen the mounting bolts more than necessary.

beyond the loosened subframe mounting bolts clearance.

34. Remove the flange bolts and the bushing holders, then remove the bushings. Note the paint marks on the stabilizer bar which will be aligned with the side of the bushings during installation.

35. Move the stabilizer bar toward the passenger's side, and remove the stabilizer bar.

To install:

➡**Refer to the appropriate figure in the removal steps for specified torque settings.**

36. Install the stabilizer bar.

37. Note the direction of installation for the bushings.

38. Raise the front subframe up to the body, loosely install the new subframe mounting bolts and the front stiffener mounting bolts, and the new rear stiffener mounting bolts.

39. Align the front subframe with the subframe alignment pin.

40. Install all of the removed parts in the reverse order of removal, and note these items:

   a. Refer to stabilizer link removal/installation to connect the stabilizer bar to the links.

   b. During installation, align the paint marks on the stabilizer bar with the side of the bushings.

   c. Use new mounting bolts for the bolt of the front side during reassembly.

   d. Use new front subframe stiffener flange bolts during reassembly.

e. Use new nuts and a front engine mount bolt during reassembly.

f. If the center guide is in place, use it to determine the steering joint installation angle.

g. If the center guide is gone, check the steering joint installation angle.

h. During installation, install the new cotter pin after tightening the new nut, and properly bend its end.

i. Before installing the wheel, clean the mating surfaces on the brake disc and the inside of the wheel.

41. Do the battery terminal reconnection procedure.

42. Check the wheel alignment, and adjust it if necessary.

STRUTS

REMOVAL & INSTALLATION

See Figures 171 through 173.

1. Raise and support the vehicle.
2. Remove the front wheel.
3. With active damper system: Disconnect the damper coil 2P connector, and remove the flange bolt, and detach the harness clip.

➡ **Be careful not to damage or contaminate the connector.**

4. Remove the wheel speed sensor harness and the brake hose from the damper. Do not disconnect the wheel speed sensor connector.

5. Remove the flange nut, while holding the joint pin with a hex wrench, and disconnect the stabilizer link from the damper.

6. Remove the damper pinch bolts and flange nuts from the damper.

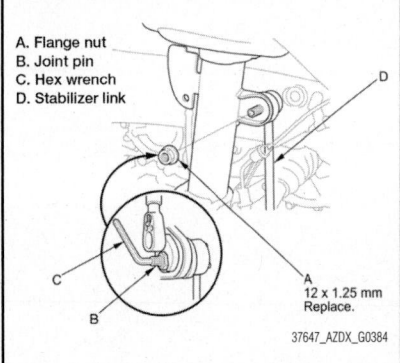

A. Flange nut
B. Joint pin
C. Hex wrench
D. Stabilizer link

12 x 1.25 mm
Replace.

37647_AZDX_G0384

Fig. 172 Remove the flange nut, while holding the joint pin with a hex wrench, and disconnect the stabilizer link from the damper.

✳✳ CAUTION

Do not allow the knuckle to rotate too far outward. This may allow the driveshaft inboard joint to come apart.

7. Remove the service caps from the shock tower and remove the flange nuts from the top of the damper.

8. Remove the damper/spring assembly.

➡ **The left and right damper springs are different. Mark the springs L and R before you continue.**

To install:

➡ **Do not fully tighten nuts or bolts until instructed to do so.**

9. Install the damper/spring on to the frame. Note the direction of the damper mounting base as shown.

10. Loosely install the new flange nuts to the top of the damper.

11. Loosely install new damper pinch bolts and new flange nuts to the lower end of the damper.

12. Connect the stabilizer link to the damper, and loosely install a new flange nut.

13. Tighten the lower flange nut to 58 ft. lbs. (79 Nm), while holding the joint pin with the hex wrench.

14. Tighten the flange nuts on top of the damper to 44 ft. lbs. (59 Nm).

15. Tighten the damper pinch bolts while holding the flange nuts to 156 ft. lbs. (211 Nm).

16. Install the service caps.

17. Install the wheel speed sensor harness and the brake hose to the damper.

18. With active damper system: Connect the damper 2P coil connector and install the flange bolt and attach the harness clip.

✳✳ CAUTION

Be careful not to damage or contaminate the connecter.

19. Clean the mating surfaces of the brake disc and the inside of the wheel, then install the front wheel.

20. With active damper system: Start the engine, then make sure there are no active damper system DTCs shown, using the HDS, or equivalent scan tool.

21. With active damper system: Do the DAMPER FORCE OPERATION TEST in the FUNCTIONAL TESTS MENU with the HDS, then make sure all four damper units function normally.

22. Check the wheel alignment, and adjust it if necessary.

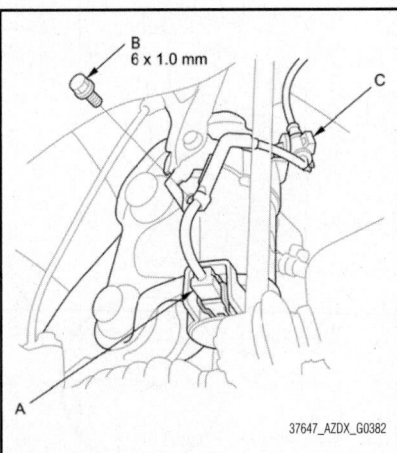

B
6 x 1.0 mm

37647_AZDX_G0382

Fig. 171 With active damper system: Disconnect the damper coil 2P connector (A), and remove the flange bolt (B), and detach the harness clip (C).

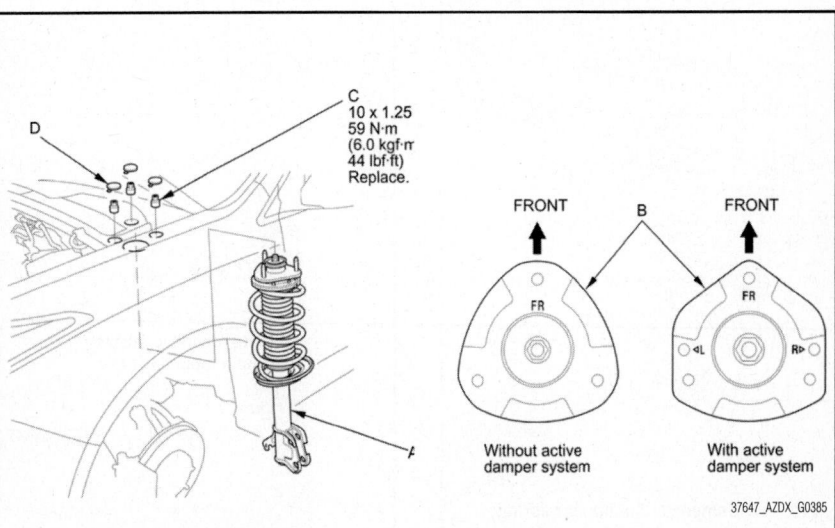

C
10 x 1.25
59 N·m
(6.0 kgf·m
44 lbf·ft)
Replace.

FRONT   B   FRONT

Without active   With active
damper system   damper system

37647_AZDX_G0385

Fig. 173 Install the damper/spring assembly (A) on to the frame. Note the direction of the damper mounting base (B) as shown.

OVERHAUL

Disassembly

See Figures 174 and 175.

1. Compress the damper spring, then remove the self-locking nut while holding the damper shaft with a hex wrench. Do not compress the damper spring more than is necessary to remove the self-locking nut.

2. Release the pressure from the strut spring compressor, then disassemble the damper as shown in the Exploded View.

Inspection

1. Reassemble all the parts, except for the damper spring.

2. Compress the damper assembly by hand, and check for smooth operation through a full stroke, both compression and extension. The damper should extend smoothly and constantly when compression is released. If it does not, the gas is leaking and the damper should be replaced.

3. Check for oil leaks, abnormal noises, and binding during these tests.

Reassembly

See Figures 176 through 179.

1. Install the spring mounting cushion on the damper mounting base by aligning the tab and notch.

2. Install all the parts except the damper mounting washer (Without SH-AWD) and the self-locking nut onto the damper unit by referring to the Exploded View.

3. Compress the damper spring using a

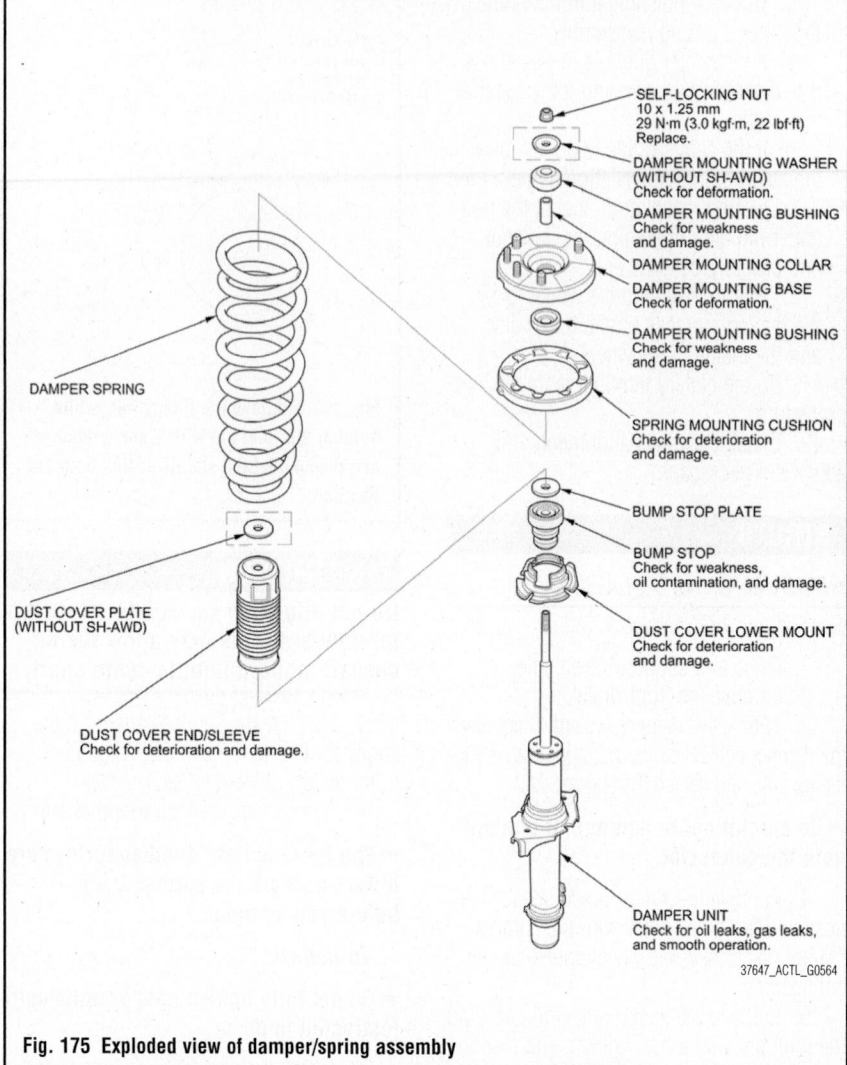

Fig. 175 Exploded view of damper/spring assembly

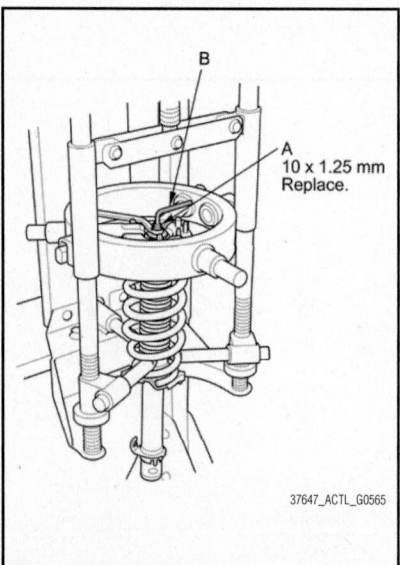

Fig. 174 Compress the damper spring, then remove the self-locking nut (A) while holding the damper shaft with a hex wrench (B)

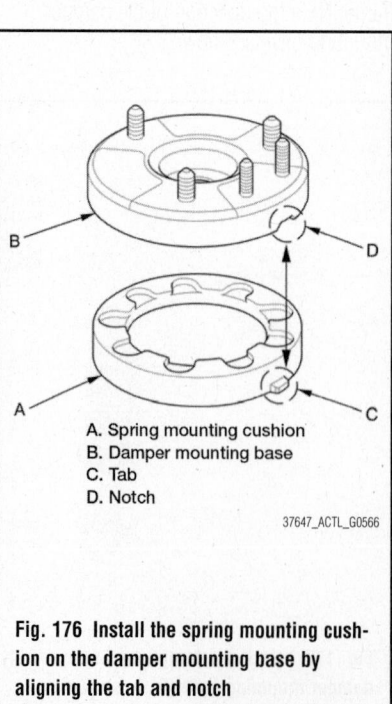

A. Spring mounting cushion
B. Damper mounting base
C. Tab
D. Notch

Fig. 176 Install the spring mounting cushion on the damper mounting base by aligning the tab and notch

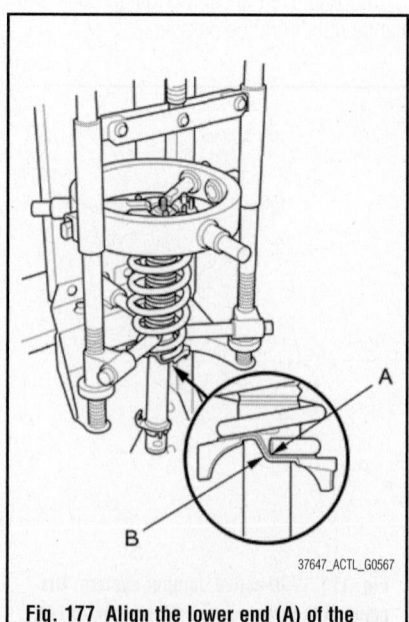

Fig. 177 Align the lower end (A) of the damper spring with the stepped part (B) of the dust cover lower mount

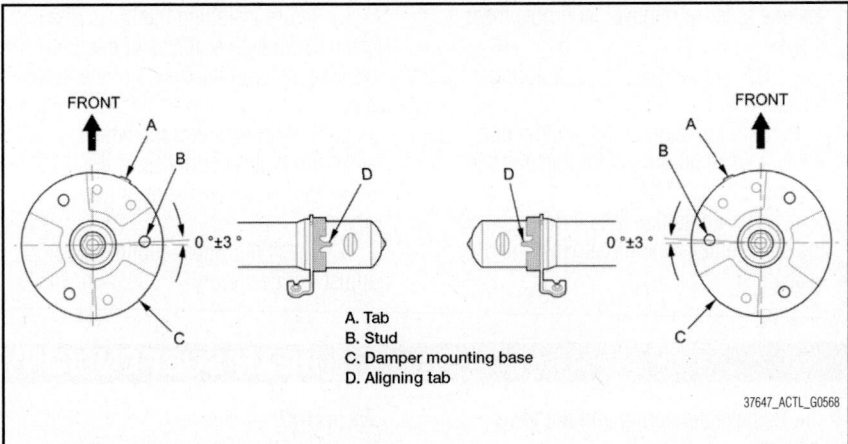

Fig. 178 Position the tab on the spring mounting cushion facing forward but toward the inside of the vehicle

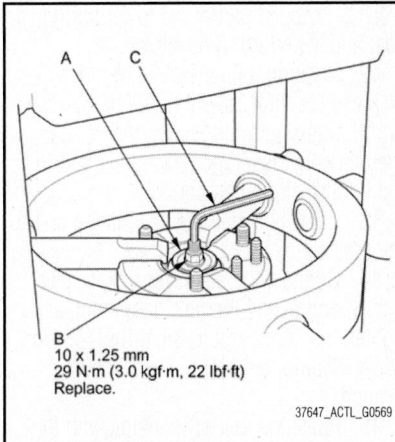

10 x 1.25 mm
29 N·m (3.0 kgf·m, 22 lbf·ft)
Replace.

Fig. 179 Install the damper mounting washer (Without SH-AWD) (A) and the new self-locking nut (B)

strut spring compressor. Do not compress the spring excessively.

4. Align the lower end of the damper spring with the stepped part of the dust cover lower mount and the lower spring seat on the damper unit.

5. Position the tab on the spring mounting cushion facing forward but toward the inside of the vehicle.

6. Align the angle of the stud on the damper mounting base with the aligning tab on the bottom of the damper unit.

7. Install the damper mounting washer (Without SH-AWD) and the new self-locking nut.

8. Hold the damper shaft with a hex wrench, and tighten the self-locking nut to 22 ft. lbs. (29 N·m).

9. Remove the damper/spring from the strut spring compressor.

WHEEL HUBS & BEARINGS

REMOVAL, PACKING, & INSTALLATION

Hub Bearing Unit

See Figure 180.

➡**Refer to the exploded view figure for reference.**

1. Raise and support the vehicle.
2. Remove the wheel nuts and the front wheel.
3. Remove the brake hose mounting bolt from the damper.
4. Remove the brake caliper bracket mounting bolts, then remove the caliper assembly from the knuckle.

✳✳ CAUTION

To prevent damage to the caliper assembly or brake hose, use a short piece of wire to hang the caliper assembly from the undercarriage. Do not twist or stretch the brake hose excessively.

5. Pry up the stake on the spindle nut, then remove the nut.
6. Remove the front brake disc.
7. Remove the flange bolts, and remove the hub bearing unit by tapping the driveshaft end with a soft face hammer while drawing the hub bearing unit outward, then remove the splash guard.

✳✳ CAUTION

Do not pull the driveshaft end outward. The driveshaft inboard joint may come apart.

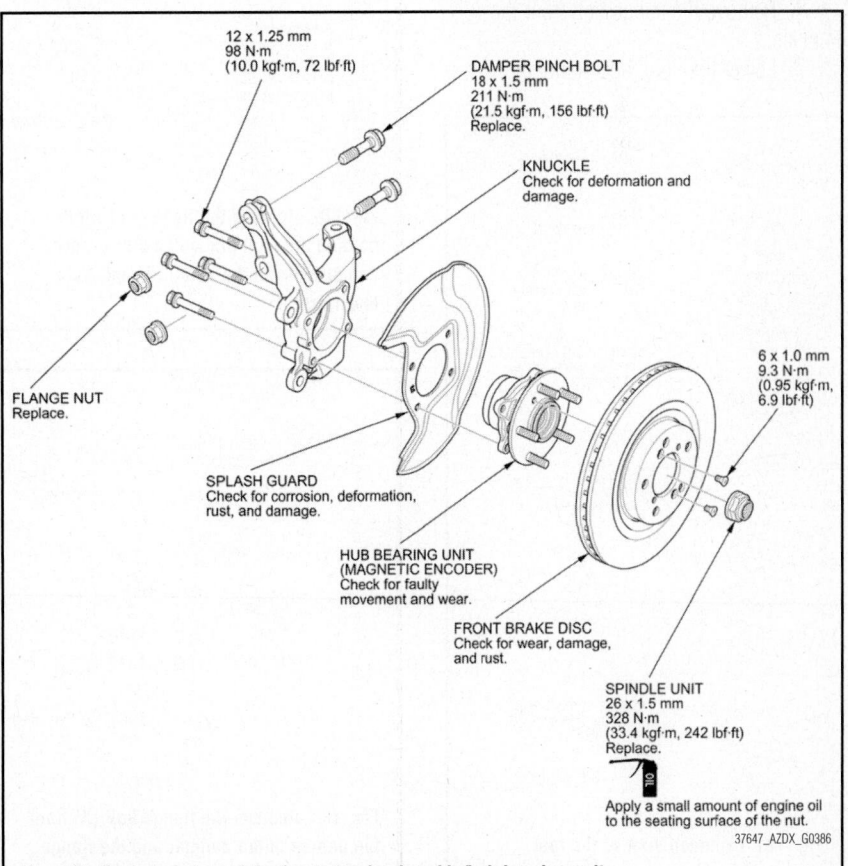

Fig. 180 Exploded view of the front steering knuckle/hub bearing unit

8. Check the hub bearing unit for damage and cracks.

To install:

➡️ **Torque settings are shown in the exploded view figure.**

a. During installation, apply grease to the mating surfaces of the hub bearing unit and driveshaft outboard joint.

b. Install the hub bearing unit in the reverse order of removal, and note these items:

c. Use the new spindle nut during reassembly.

d. Before installing the spindle nut, apply a small amount of engine oil to the seating surface of the nut.

e. After tightening, use a drift to stake the spindle nut shoulder against the driveshaft.

f. Before installing the brake disc, clean the mating surfaces of the hub bearing unit and the inside of the brake disc.

g. Before installing the wheel, clean the mating surfaces of the brake disc and the inside of the wheel.

h. Check the wheel alignment, and adjust it if necessary.

SUSPENSION

REAR SUSPENSION

COIL SPRINGS

REMOVAL & INSTALLATION

See Figures 181 through 185.

➡️ **Refer to the exploded view as needed during the following procedure.**

1. Raise and support the vehicle.
2. Remove the rear wheel.
3. Remove the flange nut while holding the joint pin with a hex wrench, and disconnect the stabilizer link from lower arm B.
4. Position a floor jack under lower arm B. Raise the floor jack until the suspension begins to compress.
5. Remove the flange bolt from the bottom of the damper.
6. Remove the flange bolt from the knuckle.
7. Lower the floor jack gradually.

8. Remove the spring and the lower spring seat.
9. Remove the flange bolt that connects the body, and remove the bump stop, the spring guide, and the spring mounting cushion if necessary.

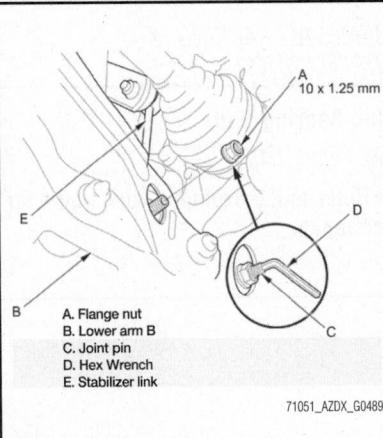
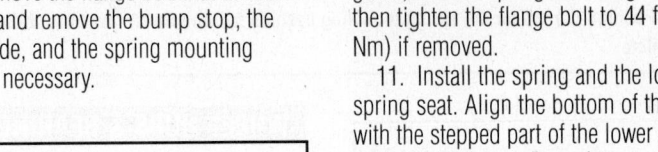

Fig. 182 Remove the flange nut while holding the joint pin with a hex wrench, and disconnect the stabilizer link from lower arm B

A. Flange nut
B. Lower arm B
C. Joint pin
D. Hex Wrench
E. Stabilizer link

71051_AZDX_G0489

To install:

10. Install the bump stop, the spring guide, and the spring mounting cushion, then tighten the flange bolt to 44 ft. lbs. (59 Nm) if removed.

11. Install the spring and the lower spring seat. Align the bottom of the spring with the stepped part of the lower spring seat and lower arm B as shown.

12. Position a floor jack under lower arm B. Raise the floor jack until the hole in lower arm B aligns with the hole in the damper, then loosely install the new flange bolt to the bottom of the damper.

13. Loosely install the new flange bolt to the knuckle.

14. Connect the stabilizer link to lower arm B, and install the new flange nut, and tighten the flange nut to 36 ft. lbs. (49 Nm), while holding the joint pin with the hex wrench.

15. Raise the rear suspension with the floor jack to load the suspension with the vehicle's weight.

16. Tighten the flange bolts to the specified torque.

a. Tighten the bolt to the knuckle to 83 ft. lbs. (113 Nm).

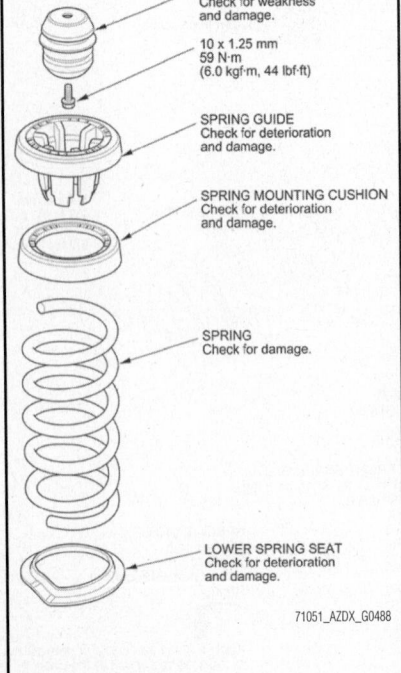

BUMP STOP
Check for weakness and damage.

10 x 1.25 mm
59 N·m
(6.0 kgf·m, 44 lbf·ft)

SPRING GUIDE
Check for deterioration and damage.

SPRING MOUNTING CUSHION
Check for deterioration and damage.

SPRING
Check for damage.

LOWER SPRING SEAT
Check for deterioration and damage.

71051_AZDX_G0488

Fig. 181 Exploded view of the rear spring assembly

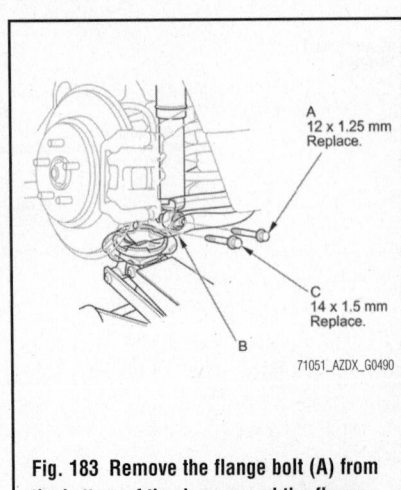

A
12 x 1.25 mm
Replace.

C
14 x 1.5 mm
Replace.

B

71051_AZDX_G0490

Fig. 183 Remove the flange bolt (A) from the bottom of the damper and the flange bolt (C) from the knuckle; lower arm B (B)

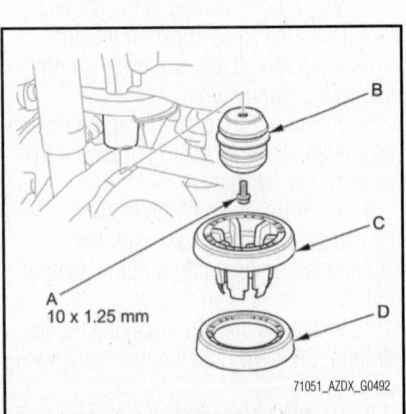

B

C

A
10 x 1.25 mm

D

71051_AZDX_G0492

Fig. 184 Remove the flange bolt that connects the body, and remove the bump stop, the spring guide, and the spring mounting cushion if necessary

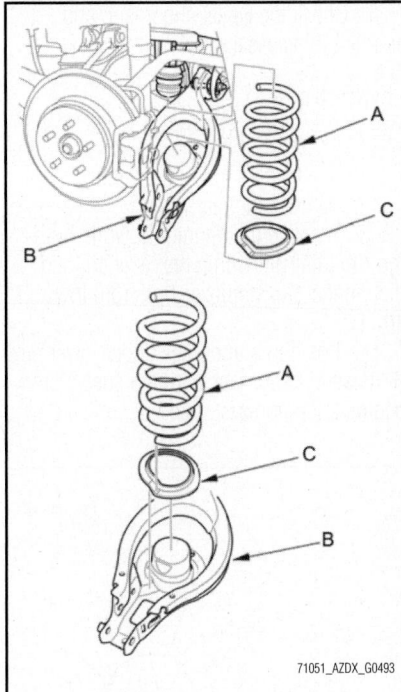

Fig. 185 Install the spring (A) and the lower spring seat (C), align the bottom of the spring with the stepped part of the lower spring seat and lower arm B (B) as shown

b. Tighten the bolt to the damper to 47 ft. lbs. (64 Nm).

17. Clean the mating surfaces between the brake disc/drum and the inside of the wheel, then install the rear wheel.

18. Check the wheel alignment, and adjust it if necessary.

KNUCKLE/HUB BEARING UNIT

REMOVAL & INSTALLATION

Hub Bearing Unit Replacement

See Figure 186.

➡**Refer to the exploded view illustration for reference during this procedure.**

1. Raise and support the vehicle.
2. Remove the wheel nuts and rear wheel.
3. Remove the brake caliper bracket mounting bolts, then remove the caliper assembly from the knuckle.

✳✳ CAUTION

To prevent damage to the caliper assembly or the brake hose, use a short piece of wire to hang the caliper assembly from the undercarriage. Do not twist the brake hose excessively.

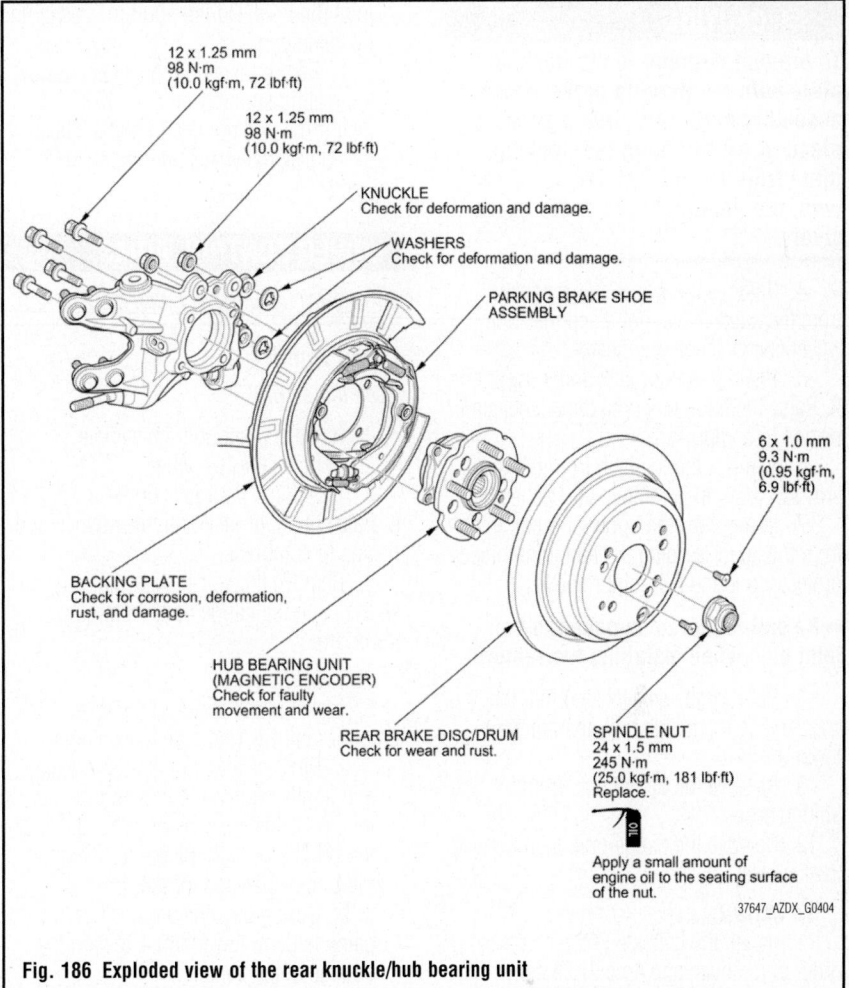

Fig. 186 Exploded view of the rear knuckle/hub bearing unit

4. Remove the two washers from the caliper-knuckle mounting tangs.

5. Pry up the stake on the spindle nut, then remove the nut.

6. Release the parking brake lever fully, and remove the rear brake disc/drum.

7. Remove the hub bearing unit.

✳✳ CAUTION

Do not pull the driveshaft end outward. The driveshaft inboard joint may come apart.

8. Check the hub bearing unit for damage and cracks.

To install:

9. Install the hub bearing unit in the reverse order of removal, and note these items:

a. Use a new spindle nut during reassembly.

b. Before installing the spindle nut, apply a small amount of engine oil to the seating surface of the nut.

c. After tightening, use a drift to stake the spindle nut shoulder against the driveshaft.

d. Before installing the brake disc/drum, clean the mating surfaces of the hub bearing unit and the inside of the brake disc/drum.

e. During installation, make sure the washers are installed between the brake caliper bracket and the knuckle.

f. Before installing the wheel, clean the mating surfaces of the brake disc/drum and the inside of the wheel.

10. Check the wheel alignment, and adjust it if necessary.

Knuckle Replacement

➡**Refer to the exploded view illustration for reference during this procedure.**

1. Remove the hub bearing unit.
2. Remove the brake hose bracket mounting bolt from the knuckle.
3. Remove the flange nuts, then remove the backing plate with the parking brake shoe assembly.

⁂ CAUTION

To prevent damage to the backing plate with the parking brake shoes assembly and cable, use a short piece of wire to hang the backing plate from the undercarriage. Do not twist the parking brake cable excessively.

4. Remove the wheel speed sensor from the knuckle. Do not disconnect the wheel speed sensor connector.

5. Position a floor jack under lower arm B. Raise the floor jack until the suspension begins to compress.

6. Remove the lock pin from the upper arm ball joint, then remove the castle nut.

7. Disconnect the upper arm ball joint from the knuckle using the ball joint thread protector and the ball joint remover.

➡ **Be careful not to damage the ball joint boot when installing the remover.**

8. Remove the self-locking nut, the washer, and the flange bolt, then remove lower arm A.

9. Remove the flange nuts from the trailing arm.

10. Remove the flange bolt, then remove the knuckle.

To install:

11. Install the knuckle in the reverse order of removal, and note these items:

a. First install all of the components, and lightly tighten the bolts and the nuts, then raise the suspension to load it with the vehicle's weight before fully tightening to the specified torque.

b. Use new flange nuts during reassembly.

c. During installation, position the paint mark on lower arm A toward the outside of the vehicle.

d. Use a new self-locking nut and the new flange bolt, during reassembly.

e. Be careful not to damage the ball joint boot when connecting the knuckle.

f. Before connecting the ball joint, degrease the threaded section and the tapered portion of the ball joint pin, the ball joint connecting hole, and the threaded section and the mating surfaces of the castle nut.

g. During installation, install the lock pin by properly bending it over the edge of the castle nut after tightening the castle nut.

h. Torque the castle nut to the lower torque specification, then tighten it only far enough to align the slot with the ball

joint pin hole. Do not align the castle nut by loosening it.

i. Before installing the wheel, clean the mating surfaces on the brake disc/drum and the inside of the wheel.

12. Check the wheel alignment, and adjust it if necessary.

LOWER CONTROL ARMS

REMOVAL & INSTALLATION

Lower Arm A

See Figure 187.

1. Raise and support the vehicle.
2. Remove the rear wheel.
3. Position a floor jack under lower arm B. Raise the floor jack until the suspension begins to compress.
4. Remove the self-locking nut, the washer, and the flange bolt, then remove lower arm A.

To install:

5. Install lower arm A in the reverse order of removal, and note these items:

a. First install all of the components, and lightly tighten the bolt and the nut, then raise the suspension to load it with the vehicle's weight before fully tightening to the specified torque.

b. During installation, position the paint mark on lower arm A toward the outside of the vehicle.

c. Use a new self-locking nut and the new flange bolt, during assembly.

d. Before installing the wheel, clean the mating surfaces of the brake disc/drum and the inside of the wheel.

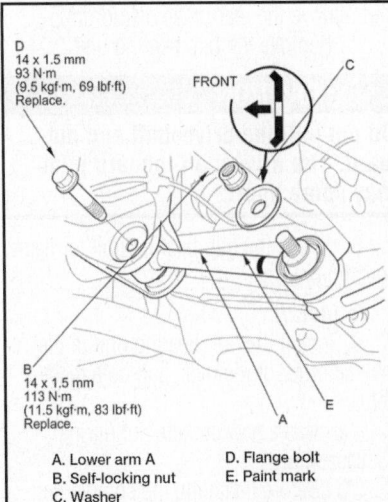

Fig. 187 Remove the self-locking nut, the washer, and the flange bolt, then remove lower arm A.

6. Check the wheel alignment, and adjust it if necessary.

Lower Arm B

See Figures 188 and 189.

1. Raise and support the vehicle.
2. Remove the rear wheel.
3. Remove the flange nut while holding the joint pin with a hex wrench, and disconnect the stabilizer link from lower arm B.
4. Position a floor jack under lower arm B. Raise the floor jack until the suspension begins to compress.

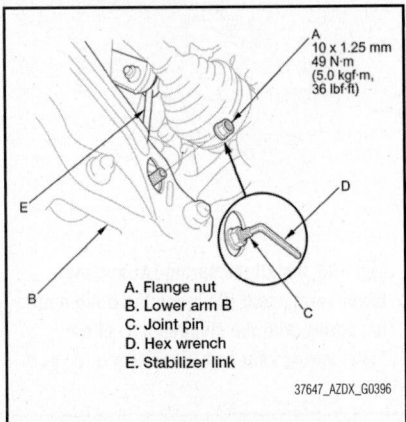

A. Flange nut
B. Lower arm B
C. Joint pin
D. Hex wrench
E. Stabilizer link

37647_AZDX_G0396

Fig. 188 Remove the flange nut while holding the joint pin with a hex wrench, and disconnect the stabilizer link from lower arm B.

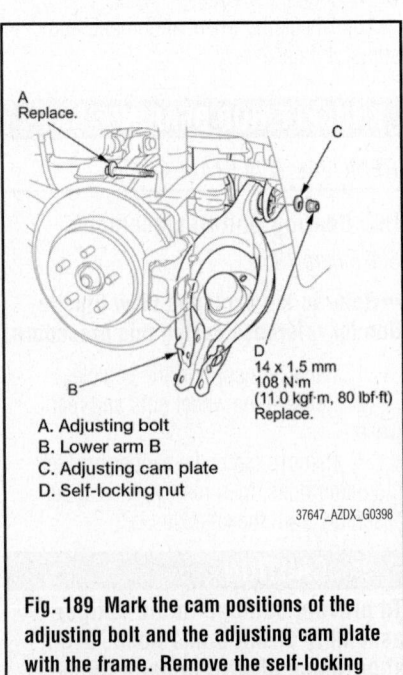

A. Adjusting bolt
B. Lower arm B
C. Adjusting cam plate
D. Self-locking nut

37647_AZDX_G0398

Fig. 189 Mark the cam positions of the adjusting bolt and the adjusting cam plate with the frame. Remove the self-locking nut while holding the adjusting bolt, then remove the adjusting cam plate, the adjusting bolt, and lower arm B.

5. Remove the flange bolt from the bottom of the damper.

6. Remove the flange bolt from the knuckle.

7. Lower the floor jack gradually.

8. Remove the spring and the lower spring seat from the lower arm B.

9. Mark the cam positions of the adjusting bolt and the adjusting cam plate with the frame. Remove the self-locking nut while holding the adjusting bolt, then remove the adjusting cam plate, the adjusting bolt, and lower arm B.

To install:

10. Install lower arm B in the reverse order of removal, and note these items:

a. First install all of the suspension components, and lightly tighten the bolts and the nuts, then raise the suspension to load it with the vehicle's weight before fully tightening to the specified torques:

- Self-locking nut on cam adjusting bolt: 80 ft. lbs. (108 Nm)
- Knuckle flange bolt: 83 ft. lbs. (113 Nm)
- Lower damper flange bolt: 47 ft. lbs. (64 Nm)
- Stabilizer link flange nut: 36 ft. lbs. (49 Nm)

b. Use a new adjusting bolt and a new self-locking nut during reassembly.

c. Use the new flange bolts during reassembly.

d. Align the bottom of the spring with the stepped part of the lower spring seat and lower arm B as shown in the figure under removal steps.

e. Align the cam positions of the adjusting bolt and the adjusting cam plate with the marked positions on the frame when tightening the self-locking nut.

f. Before installing the wheel, clean the mating surfaces of the brake disc/drum and the inside of the wheel.

11. Check the wheel alignment, and adjust it if necessary.

SHOCK ABSORBERS

REMOVAL & INSTALLATION

See Figure 190.

1. Raise and support the vehicle.

2. Remove the rear wheel.

3. With active damper system: Disconnect the damper coil 2P connector, and detach the harness clip. Use care not to damage or contaminate the connector.

4. Remove the flange nut while holding the joint pin with a hex wrench, and disconnect the stabilizer link from lower arm B.

5. Position a floor jack under lower arm

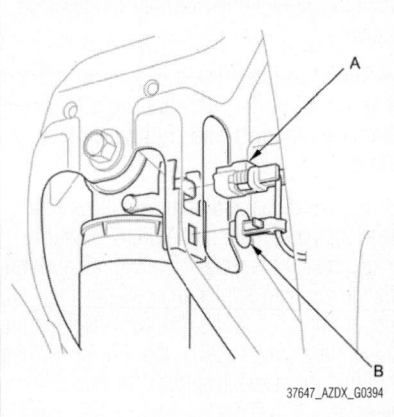

37647_AZDX_G0394

Fig. 190 With active damper system: Disconnect the damper coil 2P connector (A), and detach the harness clip (B). Use care not to damage or contaminate the connector.

B. Raise the floor jack until the suspension begins to compress.

6. Remove the flange bolt from the bottom of the damper.

7. Remove the flange bolt from the knuckle.

8. Remove the flange bolt from the top of the damper.

9. Lower the floor jack gradually, then remove the damper from the vehicle.

Inspection:

10. Compress the damper assembly by hand, and check for smooth operation through a full stroke, both compression and extension. The damper should extend smoothly and constantly when compression is released. If it does not, the gas is leaking and the damper should be replaced.

11. Check for oil leaks, abnormal noises, and binding during these tests.

To install:

12. Position the damper between the body and lower arm B.

13. With active damper system: Position the damper coil 2P connector on the damper facing rearward.

14. Loosely install a new flange bolt to the top of the damper.

15. Raise the floor jack until the hole in lower arm B aligns with the hole in the damper, then loosely install the new flange bolt to the bottom of the damper.

16. Loosely install the new flange bolt to the knuckle.

17. Connect the stabilizer link to lower arm B, and install the new flange nut, and tighten the flange nut to 36 ft. lbs. (49 Nm), while holding the joint pin with the hex wrench.

18. Raise the rear suspension with the floor jack to load the suspension with the vehicle's weight.

19. Tighten the flange bolts on the top of the damper to 47 ft. lbs. (64 Nm) and bottom of the damper to 47 ft. lbs. (64 Nm).

20. Tighten the flange bolt to the knuckle to 83 ft. lbs. (113 Nm).

21. With active damper system: Connect the damper coil 2P connector, and attach the harness clip. Use care not to damage or contaminate the connector.

22. Clean the mating surfaces of the brake disc/drum and the inside of the wheel, then install the rear wheel.

23. With active damper system: Do the memorizing rear suspension full rebound position.

24. With active damper system: Start the engine, then make sure there are no active damper system DTCs with the HDS.

25. With active damper system: Do the DAMPER FORCE OPERATION TEST in the FUNCTIONAL TESTS MENU with the HDS, or equivalent scan tool, then make sure all four damper units function normally.

26. Check the wheel alignment, and adjust it if necessary.

STABILIZER BAR & LINKS

REMOVAL & INSTALLATION

Stabilizer Bar

See Figure 191.

1. Raise and support the vehicle.

2. Remove the rear wheels.

3. Remove the spare tire from the vehicle.

4. Remove the spare tire support bracket.

5. Disconnect both stabilizer links from the stabilizer bar.

6. Remove the flange bolts and the bushing holders, then remove the bushings and the stabilizer bar. Note the paint marks on the stabilizer bar which will be aligned with the side of the bushings during installation.

To install:

7. Install the stabilizer bar in the reverse order of removal, and note these items:

a. Note the right and left direction of the stabilizer bar.

b. Align the paint marks on the stabilizer bar with the side of the bushings.

c. See "Stabilizer Link" to connect the stabilizer bar to the links.

d. Before installing the wheel, clean the mating surfaces of the brake disc/drum and the inside of the wheel.

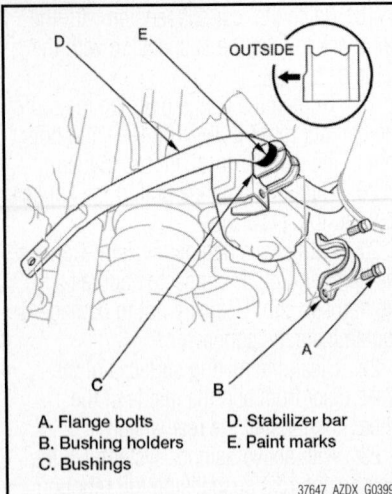

A. Flange bolts
B. Bushing holders
C. Bushings
D. Stabilizer bar
E. Paint marks

37647_AZDX_G0399

Fig. 191 Remove the flange bolts and the bushing holders, then remove the bushings and the stabilizer bar. Note the paint marks on the stabilizer bar which will be aligned with the side of the bushings during installation.

Stabilizer Link

See Figure 192.

1. Raise and support the vehicle.
2. Remove the rear wheel.
3. Remove the self-locking nut and the flange nut while holding the respective joint pin with a hex wrench, then remove the stabilizer link.

To install:

4. Install the stabilizer link on the stabilizer bar and lower arm B with the joint

A
10 x 1.25 mm
37 N·m (3.8 kgf·m, 27 lbf·ft)
Replace.

C
10 x 1.25 mm
49 N·m
(5.0 kgf·m, 36 lbf·ft)

A. Self-locking nut
B. Lower arm B
C. Flange nut
D. Joint pin
E. Hex wrench
F. Stabilizer link
G. Stabilizer bar
H. Paint mark

37647_AZDX_G0400

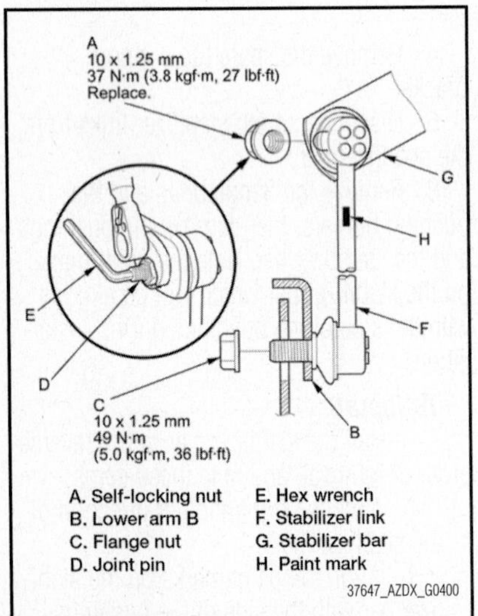

Fig. 192 Remove the self-locking nut and the flange nut while holding the respective joint pin with a hex wrench, then remove the stabilizer link.

pins set at the center of their range of movement.

➡**The stabilizer link has a paint mark. The paint mark indicates the difference between the left and right stabilizer links.**

5. Install the end of the stabilizer link with the paint mark in the upper position.
6. Install a new self-locking nut and the flange nut, and tighten the self-locking upper nut to 27 ft. lbs. (37 Nm). Tighten the lower flange nut to 36 ft. lbs. (49 Nm) while holding the respective joint pin with a hex wrench.
7. Clean the mating surfaces of the brake disc/drum and the inside of the wheel, then install the rear wheel.
8. Test-drive the vehicle.
9. After 5 minutes of driving, tighten the self-locking nut again to 27 ft. lbs. (37 Nm).

TRAILING ARMS

REMOVAL & INSTALLATION

See Figures 193 and 194.

1. Raise and support the vehicle.
2. Remove the rear wheel.
3. Disconnect the parking brake cable from the parking brake lever.
4. Remove the parking brake cable mounting bolts from the backing plate.
5. Remove the parking brake cable nut from the trailing arm.
6. Disconnect the brake line from the brake hose, then remove the brake hose clip.

➡**Plug the end of a hose and joint to prevent spilling brake fluid.**

7. Remove the wheel speed sensor harness bolt.
8. Remove the flange nuts from the trailing arm.

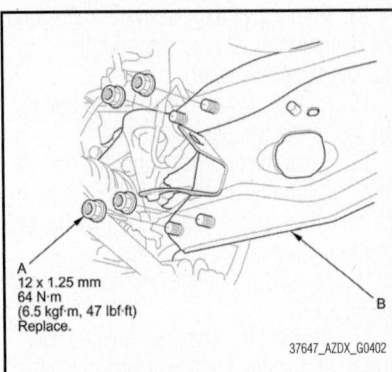

A
12 x 1.25 mm
64 N·m
(6.5 kgf·m, 47 lbf·ft)
Replace.

37647_AZDX_G0402

Fig. 193 Remove the flange nuts (A) from the trailing arm (B).

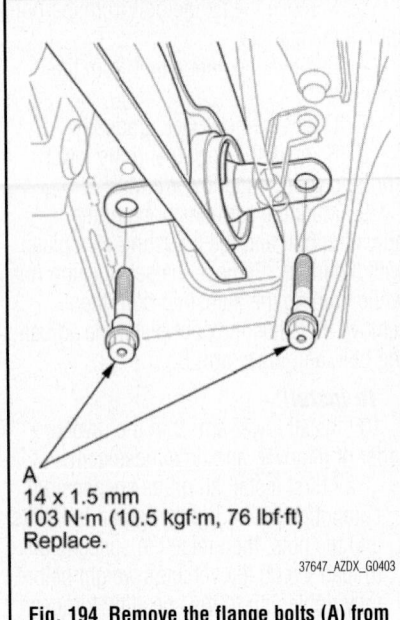

A
14 x 1.5 mm
103 N·m (10.5 kgf·m, 76 lbf·ft)
Replace.

37647_AZDX_G0403

Fig. 194 Remove the flange bolts (A) from the trailing arm (B), then remove the trailing arm.

9. Remove the flange bolts from the trailing arm, then remove the trailing arm.

To install:

10. Install the trailing arm in the reverse order of removal, and note these items:

a. First install all of the components, and lightly tighten the bolts and the nuts, then raise the suspension to load it with the vehicle's weight before fully tightening to the specified torque:

- Flange bolts: 76 ft. lbs. (103 Nm)
- Flange nuts: 47 ft. lbs. (64 Nm)

b. Use new flange nuts during reassembly.
c. Use new brake hose clip during reassembly.
d. Do the parking brake adjustment.
e. Fill the master cylinder reservoir to the MAX (upper) level line, and bleed the brake system. Check for a leak at the brake hose/line joint, and retighten it if necessary.
f. Before installing the wheel, clean the mating surfaces on the brake disc/drum and the inside of the wheel.

11. Check the wheel alignment, and adjust it if necessary.

UPPER BALL JOINTS

REMOVAL & INSTALLATION

See Figures 195 and 196.

➡**Always use a ball joint remover to disconnect a ball joint. Do not strike the housing or any other part of the ball joint connection to disconnect it.**

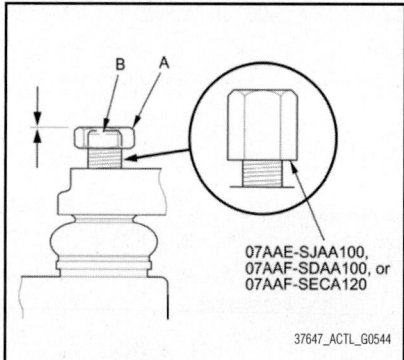

Fig. 195 Install a hex nut (A) or the ball joint thread protector onto the threads of the ball joint (B)

1. Install a hex nut or the ball joint thread protector onto the threads of the ball joint.

➡**Using a hex nut, make sure the nut is flush with the ball joint pin end to prevent damage to the threaded end of the ball joint pin.**

2. Apply grease to the ball joint remover. This will ease the installation of the tool, and prevent damage to the pressure bolt threads.

3. Loosen the pressure bolt, and install the ball joint remover as shown. Insert the jaws carefully, making sure not to damage

the ball joint boot. Adjust the jaw spacing by turning the adjusting bolt.

➡**Fasten the safety chain securely to a suspension arm or the subframe. Do not fasten it to a brake line or wire harness.**

4. After adjusting the adjusting bolt, make sure the head of the adjusting bolt is in position to allow the jaw to pivot.

5. With a wrench, tighten the pressure bolt until the ball joint pin pops loose from the ball joint connecting hole. If necessary, apply penetrating type lubricant to loosen the ball joint pin.

➡**Do not use pneumatic or electric tools on the pressure bolt.**

6. Remove the ball joint remover, then remove the nut or the ball joint thread protector from the end of the ball joint pin, and pull the ball joint out of the ball joint connecting hole. Inspect the ball joint boot, and replace it if damaged.

UPPER CONTROL ARMS

REMOVAL & INSTALLATION
See Figure 197.

1. Raise and support the vehicle.
2. Remove the rear wheel.

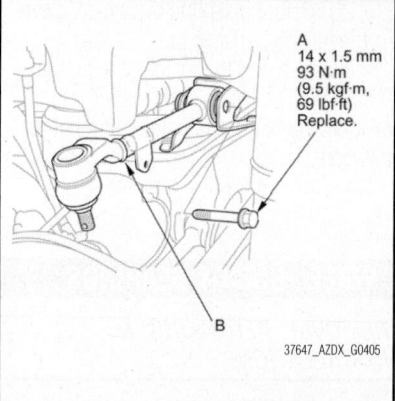

Fig. 197 Remove the flange bolt (A) and remove the upper arm (B).

3. Position a floor jack under lower arm B. Raise the floor jack until the suspension begins to compress.

4. With active damper system: Remove the suspension stroke sensor from the upper arm.

5. Remove the lock pin from the upper arm ball joint, then remove the castle nut.

6. Disconnect the upper arm ball joint from the knuckle using the ball joint thread protector and the ball joint remover. Use care not to damage the ball joint boot when installing the remover.

7. Remove the flange bolt and remove the upper arm.

To install:

8. Install the upper arm in the reverse order of removal, and note these items:

a. Use the new flange bolt during reassembly.

b. First install all of the components, and lightly tighten the bolt and the nut, then raise the suspension to load it with the vehicle's weight before fully tightening to the specified torque:
- Upper arm flange bolt: 69 ft. lbs. (93 Nm)
- Upper arm castle nut: 44–51 ft. lbs. (59–69 Nm)

c. Be careful not to damage the ball joint boot when connecting the knuckle.

d. Before connecting the ball joint to the knuckle, degrease the threaded section and the tapered portion of the ball joint pin, the ball joint connecting hole, the threaded section and the mating surfaces of the castle nut.

e. Torque the castle nut to the lower torque specification, then tighten it only far enough to align the slot with the ball joint pin hole. Do not align the castle nut by loosening it.

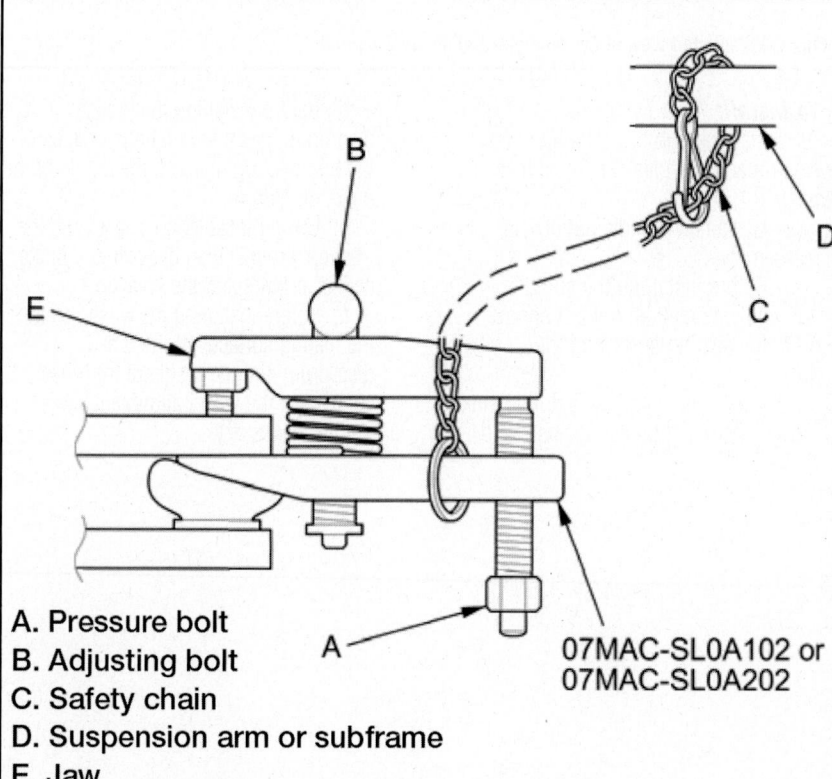

A. Pressure bolt
B. Adjusting bolt
C. Safety chain
D. Suspension arm or subframe
E. Jaw

Fig. 196 Loosen the pressure bolt, and install the ball joint remover as shown

f. Before installing the wheel, clean the mating surfaces of the brake disc/drum and the inside of the wheel.

9. With active damper system: Do the memorizing rear suspension full rebound position.

10. Check the wheel alignment, and adjust it if necessary.

WHEEL HUBS & BEARINGS

REMOVAL, REPACKING, & INSTALLATION

See Figure 198.

➡**Refer to the exploded view illustration for reference during this procedure.**

1. Raise and support the vehicle.

2. Remove the wheel nuts and rear wheel.

3. Remove the brake caliper bracket mounting bolts, then remove the caliper assembly from the knuckle.

✳✳ CAUTION

To prevent damage to the caliper assembly or the brake hose, use a short piece of wire to hang the caliper assembly from the undercarriage. Do not twist the brake hose excessively.

4. Remove the two washers from the caliper-knuckle mounting tangs.

5. Pry up the stake on the spindle nut, then remove the nut.

6. Release the parking brake lever fully, and remove the rear brake disc/drum.

7. Remove the hub bearing unit.

✳✳ CAUTION

Do not pull the driveshaft end outward. The driveshaft inboard joint may come apart.

8. Check the hub bearing unit for damage and cracks.

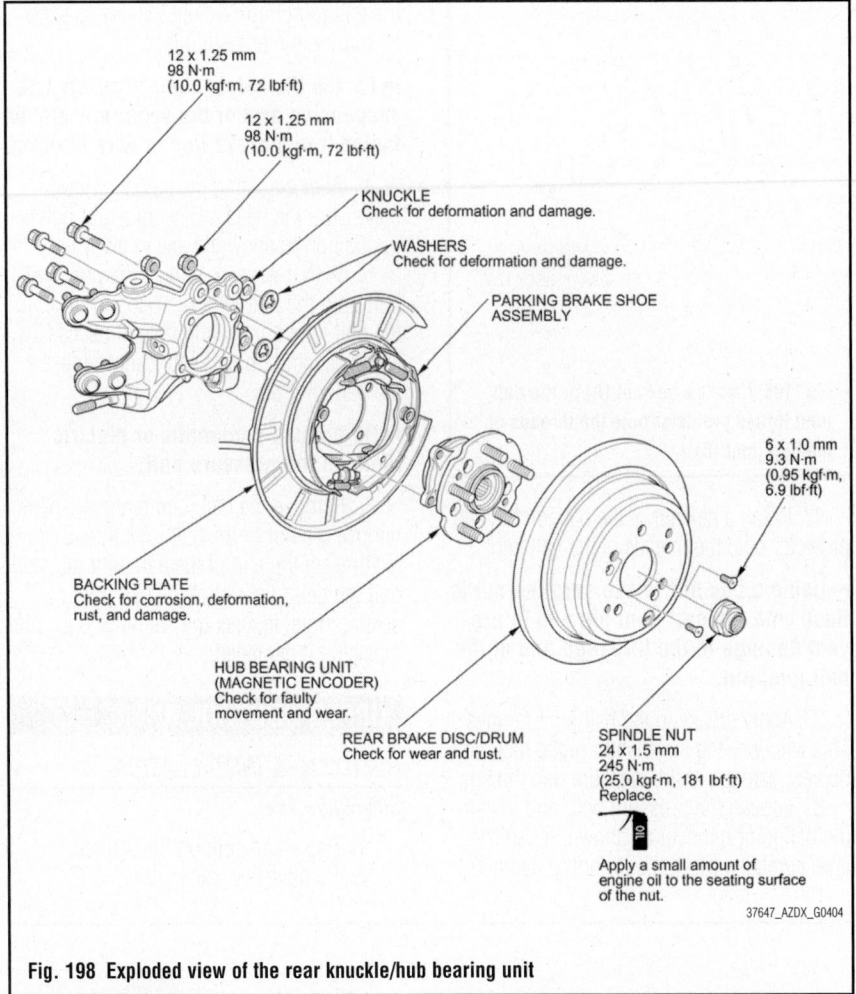

12 x 1.25 mm
98 N·m
(10.0 kgf·m, 72 lbf·ft)

12 x 1.25 mm
98 N·m
(10.0 kgf·m, 72 lbf·ft)

KNUCKLE
Check for deformation and damage.

WASHERS
Check for deformation and damage.

PARKING BRAKE SHOE
ASSEMBLY

6 x 1.0 mm
9.3 N·m
(0.95 kgf·m,
6.9 lbf·ft)

BACKING PLATE
Check for corrosion, deformation,
rust, and damage.

HUB BEARING UNIT
(MAGNETIC ENCODER)
Check for faulty
movement and wear.

REAR BRAKE DISC/DRUM
Check for wear and rust.

SPINDLE NUT
24 x 1.5 mm
245 N·m
(25.0 kgf·m, 181 lbf·ft)
Replace.

Apply a small amount of
engine oil to the seating surface
of the nut.

37647_AZDX_G0404

Fig. 198 Exploded view of the rear knuckle/hub bearing unit

To install:

9. Install the hub bearing unit in the reverse order of removal, and note these items:

a. Use a new spindle nut during reassembly.

b. Before installing the spindle nut, apply a small amount of engine oil to the seating surface of the nut.

c. After tightening, use a drift to stake the spindle nut shoulder against the driveshaft.

d. Before installing the brake disc/drum, clean the mating surfaces of the hub bearing unit and the inside of the brake disc/drum.

e. During installation, make sure the washers are installed between the brake caliper bracket and the knuckle.

f. Before installing the wheel, clean the mating surfaces of the brake disc/drum and the inside of the wheel.

10. Check the wheel alignment, and adjust it if necessary.

ACURA

Diagnostic Trouble Codes

DIAGNOSTIC TROUBLE CODES **DTC-1**

OBD II Trouble Code List (P0XXX Codes) DTC-2
OBD II Trouble Code List (P1XXX Codes) DTC-28
OBD II Trouble Code List (P2XXX Codes) DTC-34
OBD II Trouble Code List (P3XXX Codes) DTC-42
OBD II Vehicle Applications DTC-1
    Acura DTC-1

DIAGNOSTIC TROUBLE CODES

OBD II VEHICLE APPLICATIONS

ACURA

MDX
2011–2012
- 3.7L V6 SMFI . . . . Engine Code: J37A1

RDX
2011–2012
- 2.3L I4 SMFI . . . .Engine Code: K23A1

RL
2011–2012
- 3.7L V6 SMFI. . . . Engine Code: J37A2

TL
2011–2012
- 3.5L V6 PGM-FI. . . Engine Code: J35Z6
- 3.7L V6 PGM-FI. . . Engine Code: J37A4

TSX
2011–2012
- 2.4L I4 MP-FI. . . Engine Code: K24Z3
- 3.5L V6 PGM-FI. . . Engine Code: J37Z6

TSX
2011–2012
- 3.7L V6 PGM-FI. . . Engine Code: J37A5

OBD II Trouble Code List (P0XXX Codes)

DTC	Trouble Code Title and Conditions
DTC: P0010 **1T PCM, MIL: Yes** **Year:** 2011, 2012 **Model:** RDX **Engine:** 2.3L L4	**Variable Valve Timing Control (VTC) Oil Control Solenoid Valve Malfunction :** With the engine running one of the following conditions must be met: Condition 1: Output duty is 40% or more and the VTC current is 200 mA or less for at least 5 seconds. Condition 2: Output duty is 5% or less and the VTC current is 800 mA or more for at least 5 seconds. **NOTE: Before you troubleshoot, record all freeze data and any on-board snapshot, and review the general troubleshooting information.**
DTC: P0011 **2T PCM, MIL: Yes** **Year:** 2011, 2012 **Model:** RDX, TSX **Engine:** 2.3L L4, 2.4L L4, 3.5L V6	**Variable Valve Timing Control (VTC) System Malfunction :** With the engine running and at operating temperature. The difference between the timing control command and the actual timing of the camshaft is $+/-$ 50 degrees or more for at least 15 seconds. **NOTE: Before you troubleshoot, record all freeze data and any on-board snapshot, and review the general troubleshooting information.**
DTC: P0034 **1T PCM, MIL: Yes** **Year:** 2011, 2012 **Model:** RDX **Engine:** 2.3L L4	**Turbocharger Bypass Control Solenoid Valve Circuit Low Voltage:** With the engine running, battery voltage a minimum od 10.5 V and DTC P0035 is not active. The return signal is OPEN and this condition continues at least 5 seconds, though the close signal is output to the turbo-charger bypass control solenoid valve. **NOTE: Before you troubleshoot, record all freeze data and any on-board snapshot, and review the general troubleshooting information.**
DTC: P0035 **1T PCM, MIL: Yes** **Year:** 2011, 2012 **Model:** RDX **Engine:** 2.3L L4	**Turbocharger Bypass Control Solenoid Valve Circuit High Voltage:** With the engine running, battery voltage a minimum of 10.5V and DTC P0034 is not active. The return signal is closed and this condition continues at least 5 seconds, though the open signal is output to the turbo-charger bypass control solenoid valve. **NOTE: Before you troubleshoot, record all freeze data and any on-board snapshot, and review the general troubleshooting information.**
DTC: P0045 **1T PCM, MIL: Yes** **Year:** 2011, 2012 **Model:** RDX **Engine:** 2.3L L4	**Turbocharger Boost Control Solenoid Valve Circuit Malfunction:** Engine coolant temperature (ECT SENSOR 1) above 158 ºF (70 ºC), Transmission in D, Engine speed above 3,800 rpm, REL TP SENSOR above 59 deg, Drive at least 2 seconds. The return signal does not change as set by the output of the turbo-charger boost control solenoid valve, and this condition continues for at least 2 seconds. When duty is between 98-2% and there is no return signal for 2 seconds or more continuously. **NOTE: Before you troubleshoot, record all freeze data and any on-board snapshot, and review the general troubleshooting information.**
DTC: P0096 **2T PCM, MIL: Yes** **Year:** 2011, 2012 **Model:** RDX **Engine:** 2.3L L4	**Intake Air Temperature (IAT) Sensor 2 Circuit Range/Performance Problem:** With the engine OFF for a minimum of 6 hours, and DTcs P0112, P0113, P0116, P0117, P0118, P0125, P1116, P2183, P2184, P2185 and P2610 are not active. A malfunction is detected if these following 3 conditions are present after the engine and ignition switch have been turned OFF for at least 6 hours before restarting the engine. 1). When the temperature difference between the IAT2 and the ECT1 is 48 ºF (27 ºC) or more. 2). When the temperature difference between the IAT2 and the ECT2 is 52 ºF (29 ºC) or more. 1). When the temperature difference between the ECT2 and the ECT1 is 75 ºF (42 ºC) or more.
DTC: P0097 **1T PCM, MIL: Yes** **Year:** 2011, 2012 **Model:** RDX **Engine:** 2.3L L4	**Intake Air Temperature (IAT) Sensor 2 Circuit Low Voltage :** With the ignition ON the IAT sensor 2 voltage is 0.07 V or less for at least 5 seconds. **NOTE: Before you troubleshoot, record all freeze data and any on-board snapshot, and review the general troubleshooting information.**
DTC: P0098 **1T PCM, MIL: Yes** **Year:** 2011, 2012 **Model:** RDX **Engine:** 2.3L L4	**Intake Air Temperature (IAT) Sensor 2 Circuit High Voltage:** With the ignition ON the IAT sensor output voltage is 4.93 V or more for at least 5 seconds. **NOTE: Before you troubleshoot, record all freeze data and any on-board snapshot, and review the general troubleshooting information.**
DTC: P0101 **2T PCM, MIL: Yes** **Year:** 2011, 2012 **Model:** MDX, RDX, RL, TL, TSX, ZDX **Engine:** 2.3L L4, 2.4L L4, 3.5L V6, 3.7L V6	**Mass Airflow (MAF) Sensor Circuit Range/Performance Problem:** Elapsed time after engine start 5 seconds, engine coolant temperature 156 ºF (69 ºC), engine speed 650-2,100 rpm. No active DTCs P0102, P0103, P0107, P0108, P0112, P0113, P0117, P0118, P0171, P0172, P0174, P0175, P0300, P0301, P0302, P0303, P0304, P0305, P0306, P0335, P0339, P0340, P0341, P0344, P0401, P0404, P0443, P0496, P0497, P0506, P0507, P1128, P1129, P145C, P2413, P2646, P2647, P2648, P2649. The difference between the amount of intake air measured by the MAF sensor and the amount of intake air calculated from the MAP sensor output is out of the normal area for at least 10 seconds.
DTC: P0102 **1T PCM, MIL: Yes** **Year:** 2011, 2012 **Model:** MDX, RDX, RL, TL, TSX, ZDX **Engine:** 2.3L L4, 2.4L L4, 3.5L V6, 3.7L V6	**Mass Air Flow (MAF) Sensor Circuit Low Voltage:** The lower limit of the MAF sensor output is specified. If the output is below that limit, the PCM detects a malfunction and stores a DTC. Execution is continuous and the duration time is 2 seconds or more. The MAF sensor input voltage is 0.1 volt or less for at least 2 seconds. **NOTE: Before you troubleshoot, record all freeze data and any on-board snapshot.**

DTC	Trouble Code Title and Conditions
DTC: P0103 **1T PCM, MIL: Yes** **Year:** 2011, 2012 **Model:** MDX, RDX, RL, TL, TSX, ZDX **Engine:** 2.3L L4, 2.4L L4, 3.5L V6, 3.7L V6	**Mass Airflow (MAF) Sensor Circuit High Voltage:** The upper limit of the MAF sensor output is specified. If the output is above that limit, the PCM detects a malfunction and stores a DTC. Execution is continuous and the duration time is 2 seconds or more. The MAF sensor input voltage is 4.89 V or more for at least 2 seconds. P0102 is not active. **NOTE: Before you troubleshoot, record all freeze data and any on-board snapshot.**
DTC: P0107 **1T PCM, MIL: Yes** **Year:** 2011, 2012 **Model:** RDX, TSX **Engine:** 2.3L L4, 2.4L L4, 3.5L V6	**Manifold Absolute Pressure (MAP) Sensor Circuit Low Voltage:** With the engine running the Map sensor output voltage is 0.24 volts or less for at least 2 seconds.
DTC: P0107 **1T PCM, MIL: Yes** **Year:** 2011, 2012 **Model:** MDX, RL, TL, ZDX **Engine:** 3.5L V6, 3.7L V6	**Manifold Absolute Pressure (MAP) Sensor Circuit Low Voltage (PGM-FI System):** The MAP sensor outputs low signal voltage at high vacuum (throttle valve closed) and high signal voltage at low vacuum (throttle valve wide open). If a signal voltage from the MAP sensor is a set value or less, the Powertrain Control Module (PCM) detects a malfunction and a DTC is stored. The execution time is continuous and the duration time is 2 seconds or more. The MAP sensor output voltage is 0.23 V or less for at least 2 seconds. DTC P0108 is not active. **NOTE: Before you troubleshoot, record all freeze data and any on-board snapshot.**
DTC: P0108 **1T PCM, MIL: Yes** **Year:** 2011, 2012 **Model:** MDX, RL, TL, ZDX **Engine:** 3.5L V6, 3.7L V6	**MAP Sensor Circuit High Voltage (A/T/System) (With Navigation):** The MAP sensor outputs low signal voltage at high vacuum (throttle valve closed) and high signal voltage at low vacuum (throttle valve wide open). If a signal voltage from the MAP sensor is a set value or more, the Powertrain Control Module (PCM) detects a malfunction and a DTC is stored. The execution time is continuous and the duration time is 2 seconds or more. DTC P0107 is not active. The MAP sensor output voltage is 4.49 V or more for at least 2 seconds. **NOTE: Before you troubleshoot, record all freeze data and any on-board snapshot.**
DTC: P0108 **1T PCM, MIL: Yes** **Year:** 2011, 2012 **Model:** RDX, TSX **Engine:** 2.3L L4, 2.4L L4, 3.5L V6	**Manifold Absolute Pressure (MAP) Sensor Circuit High Voltage:** With the engine running the Map sensor output voltage value is 4.49 V for at least 2 seconds. **NOTE: Before you troubleshoot, record all freeze data and any on-board snapshot, and review the general troubleshooting information.**
DTC: P0111 **2T PCM, MIL: Yes** **Year:** 2011, 2012 **Model:** MDX, RL, TL, TSX, ZDX **Engine:** 2.4L L4, 3.5L V6, 3.7L V6	**Intake Air Temperature (IAT) Sensor Circuit Range/Performance:** The execution is once per driving cycle and the duration time is 10 seconds or more. Engine OFF time is 6 hours. DTCs P0112, P0113, P0116, P0117, P0118, P0125, P1116, P2183, P2184, P2185, P2610 are not active. A malfunction is detected if the following three conditions are present after the engine has stopped and the ignition switch has been turned to LOCK (0) for at least 6 hours before restarting the engine: * When the temperature difference between the IAT and ECT1 is 57 ºF (32 ºC) or more. * When the temperature difference between the IAT and ECT2 is 30 ºF (17 ºC) or more. * When the temperature difference between the ECT2 and ECT1 is 73 ºF (41 ºC) or more. **NOTE: Before you troubleshoot, record all freeze data and any on-board snapshot.**
DTC: P0112 **1T PCM, MIL: Yes** **Year:** 2011, 2012 **Model:** TL, TSX **Engine:** 2.4L L4, 3.5L V6	**Intake Air Temperature (IAT) Sensor Circuit Low Voltage:** With the ignition ON the IAT sensor output voltage is 0.08 V or less for at least 2 seconds.
DTC: P0112 **1T PCM, MIL: Yes** **Year:** 2011, 2012 **Model:** RDX **Engine:** 2.3L L4	**Intake Air Temperature (IAT) Sensor 1 Circuit Low Voltage:** With the ignition ON the IAT sensor 1 output voltage is 0.07 V or less for at least 5 seconds. **NOTE: Before you troubleshoot, record all freeze data and any on-board snapshot, and review the general troubleshooting information.**
DTC: P0112 **1T PCM, MIL: Yes** **Year:** 2011, 2012 **Model:** MDX, RL, TL, ZDX **Engine:** 3.5L V6, 3.7L V6	**Intake Air Temperature (IAT) Sensor Circuit Low Voltage:** If the IAT sensor output voltage is excessively low, the Powertrain Control Module (PCM) detects a malfunction and a DTC is stored. The execution time is continuous and the duration time is 2 seconds or more. The IAT sensor output voltage is 0.08 V or less for at least 2 seconds. **NOTE: Before you troubleshoot, record all freeze data and any on-board snapshot.**
DTC: P0113 **1T PCM, MIL: Yes** **Year:** 2011, 2012 **Model:** RDX **Engine:** 2.3L L4	**Intake Air Temperature (IAT) Sensor 1 Circuit High Voltage:** With the ignition ON the IAT sensor 1 output voltage is 4.93 V or less for at least 5 seconds. **NOTE: Before you troubleshoot, record all freeze data and any on-board snapshot, and review the general troubleshooting information.**

DTC	Trouble Code Title and Conditions
DTC: P0113 **1T PCM, MIL: Yes** **Year:** 2011, 2012 **Model:** TSX **Engine:** 2.4L L4, 3.5L V6	**Intake Air Temperature (IAT) Sensor Circuit High Voltage:** With ignition ON, the IAT sensor output voltage is 4.92 V or more for at least 2 seconds.
DTC: P0113 **1T PCM, MIL: Yes** **Year:** 2011, 2012 **Model:** MDX, RL, TL, ZDX **Engine:** 3.5L V6, 3.7L V6	**Intake Air Temperature (IAT) Sensor Circuit High Voltage:** The IAT sensor resistance varies depending on temperature. The output voltage and the sensor resistance increase as the intake air temperature decreases. Conversely, the output voltage and the sensor resistance decrease as the intake air temperature increases. If the IAT sensor output voltage is excessively high, the Powertrain Control Module (PCM) detects a malfunction and a DTC is stored. The execution time is continuous and the duration time is 2 seconds or more. P0112 is not active. The IAT sensor output voltage is 4.92 V or more for at least 2 seconds. **NOTE: Before you troubleshoot, record all freeze data and any on-board snapshot.**
DTC: P0116 **2T PCM, MIL: Yes** **Year:** 2011, 2012 **Model:** RL, TL **Engine:** 3.5L V6, 3.7L V6	**Engine Coolant Temperature (ECT) Sensor Range/Performance Problem :** Malfunction 1 (Slow Response): The engine coolant temperature does not reach (15°C) 86°F within 1,200 seconds. Malfunction 2 (Stuck): The ECT sensor output voltage does not vary by 60 mV or more within 1,200 seconds. **NOTE: If DTC P0117 and/or P0118 are stored at the same time as DTC P0116, troubleshoot those DTCs first, then recheck for DTC P0116.**
DTC: P0116 **2T PCM, MIL: Yes** **Year:** 2011, 2012 **Model:** MDX, RL, TL, TSX, ZDX **Engine:** 2.4L L4, 3.5L V6, 3.7L V6	**Engine Coolant Temperature (ECT) Sensor 1 Circuit Range/Performance Problem :** The execution time is once per driving cycle and the duration time is 10 seconds or more. The following DTCs are not active P0101, P0102, P0103, P0107, P0108, P0117, P0118, P0134, P0135, P0154, P0155, P0171, P0172, P0174, P0175, P0300, P0301, P0302, P0303, P0304, P0305, P0306, P0335, P0339, P0340, P0344, P0401, P0404, P0443, P0496, P0497, P0506, P0507, P0627, P1077, P1078, P1128, P1129, P1172, P1174, P145C, P2195, P2197, P2227, P2228, P2229, P2237, P2238, P2240, P2241, P2243, P2245, P2247, P2249, P2251, P2252, P2254, P2255, P2413, P2610, P2646, P2647, P2648, P2649. Malfunction determination 1: With a completely cooled engine (one that has been off for at least 6 hours): When the change in coolant temperature after 42 minutes or more of running time or drive 5 miles (7 km) or more of driving time is 50 °F (10 °C) or less, a malfunction is detected. Malfunction determination 2: With a partially cooled engine (one that has been off for less than 6 hours): When the difference between the coolant temperature after 42 minutes or more of running time or drive 5 miles (7 km) or more of driving time the coolant temperature after the engine has been off for 150 minutes and then run for 10 seconds is 50°F (10 °C) or less, a malfunction is detected. **NOTE: Before you troubleshoot, record all freeze data and any on-board snapshot.**
DTC: P0116 **2T PCM, MIL: Yes** **Year:** 2011, 2012 **Model:** RDX **Engine:** 2.3L L4	**Engine Coolant Temperature (ECT) Sensor 1 Circuit Range/Performance Problem:** A malfunction is detected if the following three conditions are present after the engine has stopped and the ignition switch has been turned to LOCK (0) for at least 6 hours before restarting the engine: (1) When the temperature difference between the IAT and ECT1 is 57 °F (32 °C) or more. (2) When the temperature difference between the IAT and ECT2 is 30 °F (17 °C) or more. (3) When the temperature difference between the ECT2 and ECT1 is 73 °F (41 °C) or more. **NOTE: If DTC P0111 is stored at the same time as DTC P1116, troubleshoot DTC P0111 first, then recheck for DTC P1116.**
DTC: P0117 **1T PCM, MIL: Yes** **Year:** 2011, 2012 **Model:** MDX, RDX, RL, TL, TSX, ZDX **Engine:** 2.3L L4, 2.4L L4, 3.5L V6, 3.7L V6	**Engine Coolant Temperature (ECT) Sensor 1 Circuit Low Voltage:** If the ECT sensor 1 output voltage is less than a set value when the engine coolant temperature is high, the PCM detects a malfunction and a DTC is stored. The execution time is continuous and the duration time is 2 seconds or more. P0118 is not active. The ECT sensor 1 output voltage is 0.08 V or less for at least 2 seconds. **NOTE: Before you troubleshoot, record all freeze data and any on-board snapshot.**
DTC: P0118 **1T PCM, MIL: Yes** **Year:** 2011, 2012 **Model:** MDX, RDX, RL, TL, TSX, ZDX **Engine:** 2.3L L4, 2.4L L4, 3.5L V6, 3.7L V6	**Engine Coolant Temperature (ECT) Sensor 1 Circuit High Voltage:** If the ECT sensor 1 output voltage is less than the set value when the engine coolant temperature is low, the PCM detects a malfunction and a DTC is stored. The execution time is continuous and the duration time is 2 seconds or more. DTC P0117 is not active. The ECT sensor 1 output voltage is 4.92 V or more for at least 2 seconds. **NOTE: Before you troubleshoot, record all freeze data and any on-board snapshot.**
DTC: P0122 **1T PCM, MIL: Yes** **Year:** 2011, 2012 **Model:** MDX, ZDX **Engine:** 3.7L V6	**Throttle Position (TP) Sensor A Circuit Low Voltage:** If the signal from TP sensor A is less than a fixed value for a set time, the PCM detects a TP sensor A malfunction and stores a DTC. The execution time is continuous and the duration time is 200 milliseconds or more. With the engine running and DTCs P0123, P2101, P2118, P2135, P2176 are not active. **NOTE: Before you troubleshoot, record all freeze data and any on-board snapshot.**

DTC	Trouble Code Title and Conditions
DTC: P0122 **1T PCM, MIL: Yes** **Year:** 2012 **Model:** TL **Engine:** 3.5L V6	**Throttle Position (TP) Sensor A Circuit Low Voltage:** With the ignition switch ON (II), the TP sensor output voltage is 0.3 V or less for at least 200 milliseconds. **NOTE: Before you troubleshoot, record all freeze data and any on-board snapshot, and review the general troubleshooting information.**
DTC: P0122 **1T PCM, MIL: Yes** **Year:** 2011, 2012 **Model:** RDX, RL, TL, TSX **Engine:** 2.3L L4, 2.4L L4, 3.5L V6, 3.7L V6	**Throttle Position (TP) Sensor A Circuit Low Voltage :** With the ignition ON the TP sensor 1 output voltage value is 0.20 V or less for at least 0.1 second. **NOTE: Before you troubleshoot, record all freeze data and any on-board snapshot, and review the general troubleshooting information.**
DTC: P0123 **1T PCM, MIL: Yes** **Year:** 2011, 2012 **Model:** MDX, RDX, RL, TL, TSX, ZDX **Engine:** 2.3L L4, 2.4L L4, 3.5L V6, 3.7L V6	**Throttle Position (TP) Sensor A Circuit High Voltage:** If the signal from TP sensor A is more than a fixed value for a set time, the PCM detects a TP sensor A malfunction and stores a DTC. The execution time is continuous and the duration time is 200 milliseconds or more. Ignition is running and DTCs P0122, P2101, P2118, P2135, P2176 are not active. The TP sensor A output voltage is 4.8 V or more for at least 200 milliseconds. **NOTE: Before you troubleshoot, record all freeze data and any on-board snapshot.**
DTC: P0125 **2T PCM, MIL: Yes** **Year:** 2011, 2012 **Model:** MDX, RDX, RL, TL, TSX, ZDX **Engine:** 2.3L L4, 2.4L L4, 3.5L V6, 3.7L V6	**Engine Coolant Temperature (ECT) Sensor 1 Malfunction/Slow Response:** As the engine coolant warms, the ECT sensor 1 resistance decreases, and the PCM detects a low signal voltage. If the ECT sensor 1 output voltage does not reach a specified temperature at which closed-loop control for stoichiometric air/fuel ratio starts within a set time, depending on the initial coolant temperature after starting the engine, the PCM detects a malfunction. **NOTE: Before you troubleshoot, record all freeze data and any on-board snapshot.**
DTC: P0128 **2T PCM, MIL: Yes** **Year:** 2012 **Model:** TL **Engine:** 3.5L V6	**Cooling System Malfunction :** With the engine OFF for at least 6 hours, coolant temperature between 20 to 123 ºF (-7 to 50 ºC), the malfunction threshold are as follows. Malfunction determination 1: If the difference between the current measured coolant temperature at the radiator (ECT 2) and the initial coolant temperature at the radiator is at least 13 ºF (7 ºC) when the estimated coolant temperature at the engine (ECT 1) reaches 164 ºF (73 ºC), a malfunction is detected (Thermostat Stuck Open); or if the coolant temperature at the radiator (ECT 2) only reaches 68 ºF (20 ºC), a malfunction is detected. Malfunction determination 2: When the calculated engine coolant temperature (ECT 1) reaches 158 ºF (70 ºC) before the measured engine coolant temperature (ECT 1) reaches 158 ºF (70 ºC), a malfunction is detected.
DTC: P0128 **2T PCM, MIL: Yes** **Year:** 2011, 2012 **Model:** MDX, RL, ZDX **Engine:** 3.7L V6	**Cooling System Malfunction:** The execution time is once per drive cycle and the duration time is dependent on driving conditions. Malfunction determination 1: If the difference between the current measured coolant temperature at the radiator (ECT 2) and the initial coolant temperature at the radiator (ECT 2) is at least 21 ºF (12 ºC) when the calculated coolant temperature at the engine (ECT 1) reaches 159 ºF (71 ºC), a malfunction is detected (thermostat stuck open); or if the coolant temperature at the radiator (ECT 2) only reaches 68 ºF (20 ºC), a malfunction is detected (thermostat malfunction). Malfunction determination 2: When the calculated engine coolant temperature (ECT 1) reaches 158 ºF (70 ºC) before the measured engine coolant temperature (ECT 1) reaches 158 ºF (70 ºC), a malfunction is detected. **NOTE: Before you troubleshoot, record all freeze data and any on-board snapshot.**
DTC: P0128 **2T PCM, MIL: Yes** **Year:** 2011, 2012 **Model:** RDX, RL, TL, TSX **Engine:** 2.3L L4, 2.4L L4, 3.5L V6, 3.7L V6	**Cooling System Malfunction:** With engine running the ECT sensor output (70 C) 158 F or less, an estimated engine coolant temperature is (75 C) 168 F or more. The difference between the estimated engine coolant temperature and the ECT sensor output is (15 C) 27 F or more. **NOTE: If the DTCs listed below are stored at the same time as DTC P0128, troubleshoot those DTCs first:** P0107, P0108, P1128, P1129, P1106, P1107, P1108, P1259, P0401, P0116, P0117, P0118, P0112, P0113, P0335, P0336, P0300-P0306, P0505, P1519, then recheck for P0128.
DTC: P0133 **2T O2S1, MIL: Yes** **Year:** 2011, 2012 **Model:** MDX, RL, TL, TSX, ZDX **Engine:** 3.5L V6, 3.7L V6	**Rear Air/Fuel Ratio (A/F) Sensor (Bank 1, Sensor 1) Malfunction/Slow Response:** The execution time is once per drive cycle and the duration time is 6.8 seconds or more. Engine coolant temperature 156 ºF (69 ºC), Intake air temperature, Fuel trim, 0.73-1.47, Engine speed 1,250-2,200 rpm, Vehicle speed 33 mph or more. The total rear A/F sensor output value is 11 or less in 6.5 seconds. **NOTE: Before you troubleshoot, record all freeze data and any on-board snapshot.** Conditions for illuminating the MIL When a malfunction is detected during the first drive cycle, a Temporary DTC is stored in the PCM memory. If the malfunction recurs during the next (second) drive cycle, the MIL comes on and the DTC and the freeze frame data are stored.

DTC	Trouble Code Title and Conditions
DTC: P0133 **2T PCM, MIL: Yes** **Year:** 2011, 2012 **Model:** RDX, TSX **Engine:** 2.3L L4, 2.4L L4	**Air/Fuel Ratio (A/F) Sensor (Sensor 1) Malfunction/Slow Response:** With the engine running and in closed loop, and the vehicle speed at least 33 mph. The total A/F sensor (SENSOR 1) output value for A/T models is 20 and for M/T models 38 or less in 0.8 seconds. **NOTE: Before you troubleshoot, record all freeze data and any on-board snapshot, and review the general troubleshooting information. If DTC P0139 is stored at the same time as DTC P0133, troubleshoot DTC P0139 first, then recheck for DTC P0133.**
DTC: P0134 **2T PCM, MIL: Yes** **Year:** 2011, 2012 **Model:** MDX, RL, TL, TSX, ZDX **Engine:** 3.5L V6, 3.7L V6	**Rear Air/Fuel Ratio (A/F) Sensor (Bank 1, Sensor 1) Heater System Malfunction:** The execution time is continuous and the duration time is 40 seconds or more. Battery voltage minimum 10.5 volts. Malfunction determination 1: The rear A/F sensor (bank 1, sensor 1) internal resistance value is 110 ohms or more for at least 40 seconds right after the engine starts. Malfunction determination 2: The rear A/F sensor (bank 1, sensor 1) internal resistance value is 110 ohms or more for at least 15 seconds. **NOTE: Before you troubleshoot, record all freeze data and any on-board snapshot.** The rear A/F sensor (bank 1, sensor 1) internal resistance value is 200 ohms or more for at least 1 second.
DTC: P0134 **2T PCM, MIL: Yes** **Year:** 2011, 2012 **Model:** RDX, TSX **Engine:** 2.3L L4, 2.4L L4	**Air/Fuel Ratio (A/F) Sensor (Sensor 1) Heater System Malfunction :** With the engine running for 40 seconds or more and battery voltage above 10.5 V. Malfunction 1: From start up the A/F sensor (SENSOR 1) internal resistance value is 90 ohms or more for at least 40 seconds right after the engine starts. Malfunction 2: With the engine hot and at operating temperature the A/F sensor (SENSOR 1) internal resistance value is 250 ohms or more for at least 1.0 second. **NOTE: Before you troubleshoot, record all freeze data and any on-board snapshot, and review the general troubleshooting information. If DTC P0135 is stored at the same time as DTC P0134, troubleshoot DTC P0135 first, then recheck for DTC P0134.**
DTC: P0135 **1T PCM, MIL: Yes** **Year:** 2011, 2012 **Model:** MDX, RL, TL, TSX, ZDX **Engine:** 3.5L V6, 3.7L V6	**Rear Air/Fuel Ratio (A/F) Sensor (Bank 1, Sensor 1) Heater Circuit Malfunction:** The execution time is continuous and the duration time is 2 seconds or more. Engine running and battery at 10 volts or more. No return signal HIGH is detected when the PCM output duty is less than 2%. Return signal does not change when the PCM output duty is more than 20% and less than 80%. No return signal LOW is detected when the PCM output duty is more than 8%. **NOTE: Before you troubleshoot, record all freeze data and any on-board snapshot.**
DTC: P0135 **2T PCM, MIL: Yes** **Year:** 2011, 2012 **Model:** RDX, TSX **Engine:** 2.3L L4, 2.4L L4	**Air/Fuel Ratio (A/F) Sensor (Sensor 1) Heater Circuit Malfunction :** With the engine running and the engine temperature at least 69 degrees the following malfunctions are detected. Malfunction 1: The heater current is 0.8 A or less for at least 4 seconds while the heater is active, and the heater current is 0.8 A or more for at least 4 seconds while the heater is not active. Malfunction 2: The heater current is 15.2 A or more for at least 0.6 second. **NOTE: Before you troubleshoot, record all freeze data and any on-board snapshot, and review the general troubleshooting information.**
DTC: P0137 **1T PCM, MIL: Yes** **Year:** 2011, 2012 **Model:** MDX, RL, TL, TSX, ZDX **Engine:** 3.5L V6, 3.7L V6	**Rear Secondary Heated Oxygen Sensor (Secondary HO2S (Bank 1, Sensor 2) Circuit Low Voltage:** The execution time is continuous and the duration time is 40 seconds or more. The system is in closed loop with the engine coolant temperature at 156 ºF (69 ºC) and the intake air temperature at -13 ºF (-25 ºC). The rear secondary HO2S output voltage is 0.05 V or less for at least 40 seconds. After current is applied to the rear secondary HO2S heater, if the rear secondary HO2S output continues low (lean) during feedback control, a malfunction is detected and a DTC is stored. **NOTE: Before you troubleshoot, record all freeze data and any on-board snapshot.**
DTC: P0137 **1T PCM, MIL: Yes** **Year:** 2011, 2012 **Model:** RDX, TSX **Engine:** 2.3L L4, 2.4L L4	**Secondary HO2S (Sensor 2) Circuit Low Voltage :** With engine running the secondary HO2S (Sensor 2) output voltage remains in Zone 1 (0.29 V or below) for at least 72 seconds.
DTC: P0138 **2T PCM, MIL: Yes** **Year:** 2011, 2012 **Model:** TSX **Engine:** 2.4L L4	**Secondary HO2S (Sensor 2) Circuit High Voltage :** With engine running the Secondary HO2S (Sensor 2) remains in Zone 4 (0.80 V or above) for at least 72 seconds.
DTC: P0138 **1T PCM, MIL: Yes** **Year:** 2011, 2012 **Model:** MDX, RL, TL, TSX, ZDX **Engine:** 3.5L V6, 3.7L V6	**Rear Secondary Heated Oxygen Sensor (Secondary HO2S (Bank 1, Sensor 2) Circuit High Voltage:** The execution time is continuous and the duration time is 2 seconds or more. The rear secondary HO2S output voltage is 1.270 V or more for at least 5 seconds. After current is applied to the rear secondary HO2S heater, if the rear secondary HO2S output continues to exceed the upper limit used during feedback control, a malfunction is detected and a DTC is stored. **NOTE: Before you troubleshoot, record all freeze data and any on-board snapshot.**

DTC	Trouble Code Title and Conditions
DTC: P0139 **2T PCM, MIL: Yes** **Year:** 2011, 2012 **Model:** MDX, RL, TL, TSX, ZDX **Engine:** 3.5L V6, 3.7L V6	**Rear Secondary Heated Oxygen Sensor (Secondary HO2S (Bank 1, Sensor 2) Circuit Slow Response:** If the response time of the rear secondary HO2S becomes longer than the specified time after current to the secondary HO2S heater is applied, a malfunction is detected and a DTC is stored. The execution time is once per driving cycle and the duration time is 20.7 seconds or less. When a malfunction is detected during the first drive cycle, a Temporary DTC is stored in the PCM memory. If the malfunction recurs during the next (second) drive cycle, the MIL comes on and the DTC and the freeze frame data are stored. **NOTE: Before you troubleshoot, record all freeze data and any on-board snapshot.**
DTC: P0139 **2T PCM, MIL: Yes** **Year:** 2011, 2012 **Model:** RDX, TSX **Engine:** 2.3L L4, 2.4L L4	**Secondary HO2S (Sensor 2) Slow Response :** Engine started, vehicle driven in closed loop in 4th or 6th gear at over 55 mph at steady speed, and the PCM detected the HO2S response time to switch between 300-600 mv was too slow, or that the rich to lean or lean to rich switch time was too slow.
DTC: P0141 **1T PCM, MIL: Yes** **Year:** 2011, 2012 **Model:** RDX, TSX **Engine:** 2.3L L4, 2.4L L4	**Secondary HO2S (Sensor 2) Heater Circuit Malfunction :** With engine running the current is 0.5 A or less, or 3.6 A or more for at least 5 seconds when the heater is on.
DTC: P0141 **1T PCM, MIL: Yes** **Year:** 2011, 2012 **Model:** MDX, RL, TL, TSX, ZDX **Engine:** 3.5L V6, 3.7L V6	**Rear Secondary Heated Oxygen Sensor (Bank 1 Sensor 2) Heater Circuit Malfunction:** The rear secondary HO2S heater output is 0.38 A or less, or 3.33 A or more, for at least 5 seconds when the heater is on. Engine running the battery voltage (IGP terminal of PCM) is 10.5-16 V, DTCs P0117, P0118 are not active. If the rear secondary HO2S heater draws more or less than a specified amperage, the PCM detects a malfunction and a DTC is stored. **NOTE: Before you troubleshoot, record all freeze data and any on-board snapshot.**
DTC: P0153 **2T PCM, MIL: Yes** **Year:** 2011, 2012 **Model:** MDX, RL, TL, TSX, ZDX **Engine:** 3.5L V6, 3.7L V6	**Front Air/Fuel Ratio (A/F) Sensor (Bank 2, Sensor 1) Malfunction/Slow Response:** The execution time is once per drive cycle and the duration time is 6.8 seconds or more. Engine coolant temperature 156 ºF (69 ºC), Intake air temperature, Fuel trim, 0.73-1.47, Engine speed 1,250-2,200 rpm, Vehicle speed 33 mph or more. The total front A/F sensor (bank 2, sensor 1) output value is 34 or less in 6.8 seconds.
DTC: P0154 **2T PCM, MIL: Yes** **Year:** 2011, 2012 **Model:** MDX, RL, TL, TSX, ZDX **Engine:** 3.5L V6, 3.7L V6	**Front Air Fuel Ratio (A/F) Sensor (Bank 2, Sensor 1) Heater System Malfunction :** The execution time is continuous and the duration time is 40 seconds or more. Battery voltage minimum 10.5 volts. Malfunction determination 1: The front A/F sensor (bank 2, sensor 1) internal resistance value is 200 or more for at least 40 seconds right after the engine starts. Malfunction determination 2: The front A/F sensor (bank 2, sensor 1) internal resistance value is 270 or more for at least 1.0 second. **NOTE: Before you troubleshoot, record all freeze data and any on-board snapshot.**
DTC: P0155 **2T PCM, MIL: Yes** **Year:** 2011, 2012 **Model:** MDX, RL, TL, TSX, ZDX **Engine:** 3.5L V6, 3.7L V6	**Front Air/Fuel Ratio (A/F) Sensor (Bank 2, Sensor 1) Heater Circuit Malfunction:** The execution time is continuous and the duration time is 2 seconds or more. Engine running and battery voltage at 10 volts or more. One of these 2 conditions must be met for at least 2 seconds: (1) No return signal HIGH is detected when the PCM output duty is less than 20%. (2) Return signal does not change when the PCM output duty is more than 20% and less than 80%. **NOTE: Before you troubleshoot, record all freeze data and any on-board snapshot.**
DTC: P0157 **1T O2S HTR2, MIL: Yes** **Year:** 2011, 2012 **Model:** MDX, RL, TL, TSX, ZDX **Engine:** 3.5L V6, 3.7L V6	**Front Secondary HO2S (Bank 2, Sensor 2) Circuit Low Voltage :** The execution time is continuous and the duration time is 40 seconds or more. The system is in closed loop with the engine coolant temperature at 156 ºF (69 ºC) and the intake air temperature at -13 ºF (-25 ºC). The rear secondary HO2S output voltage is 0.05 V or less for at least 40 seconds. After current is applied to the rear secondary HO2S heater, if the rear secondary HO2S output continues low (lean) during feedback control, a malfunction is detected and a DTC is stored. **NOTE: Before you troubleshoot, record all freeze data and any on-board snapshot.**
DTC: P0158 **2T PCM, MIL: Yes** **Year:** 2011, 2012 **Model:** MDX, RL, TL, TSX, ZDX **Engine:** 3.5L V6, 3.7L V6	**Front Secondary HO2S (Bank 2, Sensor 2) Circuit High Voltage :** With the vehicle at operating temperature and driven between 1,000-3,000 rpm for at least 1 minute, the rear secondary HO2S output voltage is 1.270 V or more for at least 5 seconds. After current is applied to the rear secondary HO2S heater, if the rear secondary HO2S output continues to exceed the upper limit used during feedback control, a malfunction is detected and a DTC is stored. **NOTE: Before you troubleshoot, record all freeze data and any on-board snapshot.**
DTC: P0159 **2T PCM, MIL: Yes** **Year:** 2011, 2012 **Model:** MDX, RL, TL, TSX, ZDX **Engine:** 3.5L V6, 3.7L V6	**Front Secondary HO2S (Bank 2, Sensor 2) Slow Response :** The execution time is once per driving cycle and the duration time is 20.7 seconds or less. Engine coolant temperature is 156 ºF (69 ºC), intake air temperature -13 ºF (-25 ºC), engine speed 1,180-2,000 rpms and vehicle speed is 30 mph. When the front secondary HO2S output drops to the response deterioration judgment threshold value and the response characteristics measurement is finished.(0.77-4.58 seconds)

DTC	Trouble Code Title and Conditions
DTC: P0161 **1T PCM, MIL: Yes** **Year:** 2011, 2012 **Model:** MDX, RL, TL, TSX, ZDX **Engine:** 3.5L V6, 3.7L V6	**Front Secondary HO2S (Bank 2, Sensor 2) Heater Circuit Malfunction :** The front secondary HO2S heater output is 0.38 A or less, or 3.33 A or more, for at least 5 seconds when the heater is on. Engine running the battery voltage (IGP terminal of PCM) is 10.5-16 V, DTCs P0117, P0118 are not active. If the front secondary HO2S heater draws more or less than a specified amperage, the PCM detects a malfunction and a DTC is stored. **NOTE: Before you troubleshoot, record all freeze data and any on-board snapshot.**
DTC: P0171 **2T PCM, MIL: Yes** **Year:** 2011, 2012 **Model:** RDX, TSX **Engine:** 2.3L L4, 2.4L L4	**Fuel System Too Lean:** With the the engine at operating temperature in closed loop, and the engine speed between 550-650 rpm. The Long term fuel trim is higher than 1.33 (+33%). If any related DTCs are listed first troubleshoot those DTCs first, then recheck for P0171. **NOTE: Before you troubleshoot, record all freeze data and any on-board snapshot, and review the general troubleshooting information.**
DTC: P0171 **2T PCM, MIL: Yes** **Year:** 2011, 2012 **Model:** MDX, RL, TL, TSX, ZDX **Engine:** 3.5L V6, 3.7L V6	**Rear Bank (Bank 1) Fuel System Too Lean:** If long term fuel trim is higher than normal (too lean), a malfunction in the fuel metering components is detected and a DTC is stored. Long term fuel trim is higher than 1.188 (+18.8 %). Engine is at running temperature, vehicle speed at 550-4,000 rpms and the system is in closed loop. **NOTE: Before you troubleshoot, record all freeze data and any on-board snapshot.**
DTC: P0172 **2T PCM, MIL: Yes** **Year:** 2011, 2012 **Model:** RDX, TSX **Engine:** 2.3L L4, 2.4L L4	**Fuel System Too Rich:** With the vehicle running, engine rpm between 550-650 and in closed loop. The long term fuel trim is lower than 0.785 (-215%). If any related DTCs are present, troubleshoot those DTCs first, then recheck P0172. **NOTE: Before you troubleshoot, record all freeze data and any on-board snapshot, and review the general troubleshooting information.**
DTC: P0172 **2T PCM, MIL: Yes** **Year:** 2011, 2012 **Model:** MDX, RL, TL, TSX, ZDX **Engine:** 3.5L V6, 3.7L V6	**Rear Bank (Bank 1) Fuel System Too Rich:** The execution time is once per driving cycle and the duration time is 13 seconds or more. If long term fuel trim is lower than normal (too rich), a malfunction in the fuel metering components is detected and a DTC is stored. Engine is at running temperature, vehicle speed at 550-4,000 rpm and the system is in closed loop the long term fuel trim is lower than 0.820 (-18.0 %). **NOTE: Before you troubleshoot, record all freeze data and any on-board snapshot.**
DTC: P0174 **1T PCM, MIL: Yes** **Year:** 2011, 2012 **Model:** MDX, RL, TL, TSX, ZDX **Engine:** 3.5L V6, 3.7L V6	**Front Bank (Bank 2) Fuel System Too Lean :** If long term fuel trim is higher than normal (too lean), a malfunction in the fuel metering components is detected and a DTC is stored. Long term fuel trim is higher than 1.188 (+18.8 %). Engine is at running temperature, vehicle speed at 550-4,000 rpm and the system is in closed loop. **NOTE: Before you troubleshoot, record all freeze data and any on-board snapshot**
DTC: P0175 **2T PCM, MIL: Yes** **Year:** 2011, 2012 **Model:** MDX, RL, TL, TSX, ZDX **Engine:** 3.5L V6, 3.7L V6	**Front Bank (Bank 2) Fuel System Too Rich :** The execution time is once per driving cycle and the duration time is 13 seconds or more. If long term fuel trim is lower than normal (too rich), a malfunction in the fuel metering components is detected and a DTC is stored. Engine is at running temperature, vehicle speed at 550-4,000 rpm and the system is in closed loop the long term fuel trim is lower than 0.820 (-18.0 %). **NOTE: Before you troubleshoot, record all freeze data and any on-board snapshot.**
DTC: P0201 **1T CCM, MIL: Yes** **Year:** 2011, 2012 **Model:** MDX, ZDX **Engine:** 3.7L V6	**Fuel Injector 1 Circuit Malfunction:** Engine started, system voltage over 9v, and PCM detected the injector voltage for Cylinder 1 did not equal the system voltage with the injector commanded "off", or that the injector voltage did not equal zero (0) volts with the injector commanded "on".
DTC: P0202 **1T CCM, MIL: Yes** **Year:** 2011, 2012 **Model:** MDX, ZDX **Engine:** 3.7L V6	**Fuel Injector 2 Circuit Malfunction:** Engine started, system voltage over 9v, and PCM detected the injector voltage for Cylinder 2 did not equal the system voltage with the injector commanded "off", or that the injector voltage did not equal zero (0) volts with the injector commanded "on".
DTC: P0203 **1T CCM, MIL: Yes** **Year:** 2011, 2012 **Model:** MDX, ZDX **Engine:** 3.7L V6	**Fuel Injector 3 Circuit Malfunction:** Engine started, system voltage over 9v, and PCM detected the injector voltage for Cylinder 3 did not equal the system voltage with the injector commanded "off", or that the injector voltage did not equal zero (0) volts with the injector commanded "on".
DTC: P0204 **1T CCM, MIL: Yes** **Year:** 2011, 2012 **Model:** MDX, ZDX **Engine:** 3.7L V6	**Fuel Injector 4 Circuit Malfunction:** Engine started, system voltage over 9v, and PCM detected the injector voltage for Cylinder 4 did not equal the system voltage with the injector commanded "off", or that the injector voltage did not equal zero (0) volts with the injector commanded "on".

DTC	Trouble Code Title and Conditions
DTC: P0205 **1T CCM, MIL: Yes** **Year:** 2011, 2012 **Model:** MDX, ZDX **Engine:** 3.7L V6	**Fuel Injector 5 Circuit Malfunction:** Engine started, system voltage over 9v, and PCM detected the injector voltage for Cylinder 5 did not equal the system voltage with the injector commanded "off", or that the injector voltage did not equal zero (0) volts with the injector commanded "on".
DTC: P0206 **1T CCM, MIL: Yes** **Year:** 2011, 2012 **Model:** MDX, ZDX **Engine:** 3.7L V6	**Fuel Injector 6 Circuit Malfunction:** Engine started, system voltage over 9v, and PCM detected the injector voltage for Cylinder 6 did not equal the system voltage with the injector commanded "off", or that the injector voltage did not equal zero (0) volts with the injector commanded "on".
DTC: P0222 **1T PCM, MIL: Yes** **Year:** 2011, 2012 **Model:** MDX, RDX, RL, TL, TSX, ZDX **Engine:** 2.3L L4, 2.4L L4, 3.5L V6, 3.7L V6	**Throttle Position (TP) Sensor B Circuit Low Voltage:** The execution time is continuous and the duration time is 200 milliseconds or more. Ignition on and DTCs P0223, P2101, P2118, P2135, P2176 are not active. The TP sensor B output voltage is 0.3 V or less for at least 200 milliseconds. If the signal from TP sensor B is less than a fixed value for a set time, the PCM detects a TP sensor B malfunction and stores a DTC. **NOTE: Before you troubleshoot, record all freeze data and any on-board snapshot.**
DTC: P0223 **1T PCM, MIL: Yes** **Year:** 2011, 2012 **Model:** MDX, RDX, RL, TL, TSX, ZDX **Engine:** 2.3L L4, 2.4L L4, 3.5L V6, 3.7L V6	**Throttle Position (TP) Sensor B Circuit High Voltage :** The execution time is continuous and the duration time is 200 milliseconds or more. Ignition ON and DTCs P0222, P2101, P2118, P2135, P2176 are not active. The TP sensor B output voltage is 4.8 V or more for at least 200 milliseconds. If the signal from TP sensor B is more than a fixed value for a set time, the PCM detects a TP sensor B malfunction and stores a DTC. **NOTE: Before you troubleshoot, record all freeze data and any on-board snapshot.**
DTC: P0234 **2T PCM, MIL: Yes** **Year:** 2011, 2012 **Model:** RDX **Engine:** 2.3L L4	**Turbocharger Overboost Problem:** With the engine running at a speed 3,800 rpm or more without a load and at operating temperature. The boost pressure is at a specified pressure or more, and the fuel cut for engine protection operates for at least 2 seconds. **NOTE: If any of the DTCs listed below are indicated at the same time as DTC P0234, troubleshoot those DTCs first, then recheck for P0234. P0107, P0108, P1128, P1129: Manifold Absolute Pressure (MAP) sensor. P0236, P0237, P0238: Turbocharger boost sensor.**
DTC: P0236 **2T PCM, MIL: Yes** **Year:** 2011, 2012 **Model:** RDX **Engine:** 2.3L L4	**Turbocharger Boost Sensor Circuit Range/Performance Problem:** With the engine running at a speed of 1,750 or more, and the throttle position 22º or more for at least 5 seconds. When the fluctuation value of the tubocharger boost sensor measured at each detection of negative pressure area determination and high throttle area determination, a STUCK failure is detected. **NOTE: Before you troubleshoot, record all freeze data and any on-board snapshot, and review the general troubleshooting information.**
DTC: P0237 **1T PCM, MIL: Yes** **Year:** 2011, 2012 **Model:** RDX **Engine:** 2.3L L4	**Turbocharger Boost Sensor Circuit Low Voltage:** With the engin running the turbocharger boost sensor output voltage value is 0.23V or less for at least 2 seconds. **NOTE: Before you troubleshoot, record all freeze data and any on-board snapshot, and review the general troubleshooting information.**
DTC: P0238 **1T PCM, MIL: Yes** **Year:** 2011, 2012 **Model:** RDX **Engine:** 2.3L L4	**Turbocharger Boost Sensor Circuit High Voltage:** With the engine running, and DTC P0237 not active the tubocharger boost sensor output voltage value is 4.49V or more for at least 2 seconds. **NOTE: Before you troubleshoot, record all freeze data and any on-board snapshot, and review the general troubleshooting information.**
DTC: P0243 **1T PCM, MIL: Yes** **Year:** 2011, 2012 **Model:** RDX **Engine:** 2.3L L4	**Turbocharger Wastegate Control Solenoid Valve Circuit Malfunction:** With the engine running, battery voltage a minimum of 10.5V, waste gate solenoid valve output duty between 2-98%. The return signal does not change as set by the output of the turbocharger wastgate control solenoid valve and this condition continues at least 2 seconds. **NOTE: Before you troubleshoot, record all freeze data and any on-board snapshot, and review the general troubleshooting information.**
DTC: P0299 **2T PCM, MIL: Yes** **Year:** 2011, 2012 **Model:** RDX **Engine:** 2.3L L4	**Turbocharger Underboost Problem:** Wit the the engine speed at 3,800 rpm without a load and at operating temperature, the difference between the upper limit boost pressure output and the actual boost pressure output is 26 kPa (7.8 in.HG, 200 mmHg) or more for at least 4 seconds. **NOTE: Before you troubleshoot, record all freeze data and any on-board snapshot, and review the general troubleshooting information.**

DTC	Trouble Code Title and Conditions
DTC: P0300 **2T PCM, MIL: Yes** **Year:** 2011, 2012 **Model:** MDX, RDX, RL, TL, TSX, ZDX **Engine:** 2.3L L4, 2.4L L4, 3.5L V6, 3.7L V6	**Random Misfire Detected:** The execution time is once per driving cycle and the duration time is 200 revolutions for Type 1 and 1,000 revolutions for Type 2. Conditions for illuminating the MIL Type 1 and Type 2. * Misfire Type 1 (Severe) Per 200 revolutions 27-90 times. If a type 1 misfire (catalyst damaging) occurs once, the MIL blinks once per second, a Temporary DTC is stored, and the high rpm fuel injection stop system activates. The fuel injection stops, at high rpm only, on the cylinder that has the highest misfire counts. the MIL continues to blink, and the fuel injection stays off at high rpm, until the drive is completed. If a type 1 misfire occurs during a second drive cycle, the MIL and fuel injection behave the same and a DTC is stored. After a type 1 misfire has been detected during two drive cycles, the MIL comes on and stays on beginning with the third drive cycle, unless the Temporary DTC has been cleared by the PCM. Even if the MIL is on, it will start blinking if a type 1 misfire occurs. If the malfunction recurs during the next (second) drive cycle, the MIL comes on and the DTC and the freeze frame data are stored. * Misfire Type 2 (Light) Per 1,000 revolutions 55 times. If a type 2 misfire (emission-related but not severe enough to immediately damage the catalyst) occurs, a Temporary DTC is stored, but the MIL does come on or blink. If a type 2 misfire occurs during a second drive cycle, the MIL comes on and stays on unless the Temporary DTC has been cleared by the PCM. **NOTE: Before you troubleshoot, record all freeze data and any on-board snapshot.**
DTC: P0301 **2T PCM, MIL: Yes** **Year:** 2011, 2012 **Model:** MDX, RDX, RL, TL, TSX, ZDX **Engine:** 2.3L L4, 2.4L L4, 3.5L V6, 3.7L V6	**Cylinder 1 Misfire Detected :** The execution time is once per driving cycle and the duration time is 200 revolutions for Type 1 and 1,000 revolutions for Type 2. Conditions for illuminating the MIL Type 1 and Type 2. * Misfire Type 1 (Severe) Per 200 revolutions 27-90 times. If a type 1 misfire (catalyst damaging) occurs once, the MIL blinks once per second, a Temporary DTC is stored, and the high rpm fuel injection stop system activates. The fuel injection stops, at high rpm only, on the cylinder that has the highest misfire counts. the MIL continues to blink, and the fuel injection stays off at high rpm, until the drive is completed. If a type 1 misfire occurs during a second drive cycle, the MIL and fuel injection behave the same and a DTC is stored. After a type 1 misfire has been detected during two drive cycles, the MIL comes on and stays on beginning with the third drive cycle, unless the Temporary DTC has been cleared by the PCM. Even if the MIL is on, it will start blinking if a type 1 misfire occurs. If the malfunction recurs during the next (second) drive cycle, the MIL comes on and the DTC and the freeze frame data are stored. * Misfire Type 2 (Light) Per 1,000 revolutions 55 times. If a type 2 misfire (emission-related but not severe enough to immediately damage the catalyst) occurs, a Temporary DTC is stored, but the MIL does come on or blink. If a type 2 misfire occurs during a second drive cycle, the MIL comes on and stays on unless the Temporary DTC has been cleared by the PCM. **NOTE: Before you troubleshoot, record all freeze data and any on-board snapshot.**
DTC: P0302 **2T PCM, MIL: Yes** **Year:** 2011, 2012 **Model:** MDX, RDX, RL, TL, TSX, ZDX **Engine:** 2.3L L4, 2.4L L4, 3.5L V6, 3.7L V6	**Cylinder 2 Misfire Detected :** The execution time is once per driving cycle and the duration time is 200 revolutions for Type 1 and 1,000 revolutions for Type 2. Conditions for illuminating the MIL Type 1 and Type 2. * Misfire Type 1 (Severe) Per 200 revolutions 27-90 times. If a type 1 misfire (catalyst damaging) occurs once, the MIL blinks once per second, a Temporary DTC is stored, and the high rpm fuel injection stop system activates. The fuel injection stops, at high rpm only, on the cylinder that has the highest misfire counts. the MIL continues to blink, and the fuel injection stays off at high rpm, until the drive is completed. If a type 1 misfire occurs during a second drive cycle, the MIL and fuel injection behave the same and a DTC is stored. After a type 1 misfire has been detected during two drive cycles, the MIL comes on and stays on beginning with the third drive cycle, unless the Temporary DTC has been cleared by the PCM. Even if the MIL is on, it will start blinking if a type 1 misfire occurs. If the malfunction recurs during the next (second) drive cycle, the MIL comes on and the DTC and the freeze frame data are stored. * Misfire Type 2 (Light) Per 1,000 revolutions 55 times. If a type 2 misfire (emission-related but not severe enough to immediately damage the catalyst) occurs, a Temporary DTC is stored, but the MIL does come on or blink. If a type 2 misfire occurs during a second drive cycle, the MIL comes on and stays on unless the Temporary DTC has been cleared by the PCM. **NOTE: Before you troubleshoot, record all freeze data and any on-board snapshot.**

DTC	Trouble Code Title and Conditions
DTC: P0303 **2T PCM, MIL: Yes** **Year:** 2011, 2012 **Model:** MDX, RDX, RL, TL, TSX, ZDX **Engine:** 2.3L L4, 2.4L L4, 3.5L V6, 3.7L V6	**Cylinder 3 Misfire Detected :** The execution time is once per driving cycle and the duration time is 200 revolutions for Type 1 and 1,000 revolutions for Type 2. Conditions for illuminating the MIL Type 1 and Type 2. * Misfire Type 1 (Severe) Per 200 revolutions 27-90 times. If a type 1 misfire (catalyst damaging) occurs once, the MIL blinks once per second, a Temporary DTC is stored, and the high rpm fuel injection stop system activates. The fuel injection stops, at high rpm only, on the cylinder that has the highest misfire counts. the MIL continues to blink, and the fuel injection stays off at high rpm, until the drive is completed. If a type 1 misfire occurs during a second drive cycle, the MIL and fuel injection behave the same and a DTC is stored. After a type 1 misfire has been detected during two drive cycles, the MIL comes on and stays on beginning with the third drive cycle, unless the Temporary DTC has been cleared by the PCM. Even if the MIL is on, it will start blinking if a type 1 misfire occurs. If the malfunction recurs during the next (second) drive cycle, the MIL comes on and the DTC and the freeze frame data are stored. * Misfire Type 2 (Light) Per 1,000 revolutions 55 times. If a type 2 misfire (emission-related but not severe enough to immediately damage the catalyst) occurs, a Temporary DTC is stored, but the MIL does come on or blink. If a type 2 misfire occurs during a second drive cycle, the MIL comes on and stays on unless the Temporary DTC has been cleared by the PCM. **NOTE: Before you troubleshoot, record all freeze data and any on-board snapshot.**
DTC: P0304 **2T PCM, MIL: Yes** **Year:** 2011, 2012 **Model:** MDX, RDX, RL, TL, TSX, ZDX **Engine:** 2.3L L4, 2.4L L4, 3.5L V6, 3.7L V6	**Cylinder 4 Misfire Detected :** The execution time is once per driving cycle and the duration time is 200 revolutions for Type 1 and 1,000 revolutions for Type 2. Conditions for illuminating the MIL Type 1 and Type 2. * Misfire Type 1 (Severe) Per 200 revolutions 27-90 times. If a type 1 misfire (catalyst damaging) occurs once, the MIL blinks once per second, a Temporary DTC is stored, and the high rpm fuel injection stop system activates. The fuel injection stops, at high rpm only, on the cylinder that has the highest misfire counts. the MIL continues to blink, and the fuel injection stays off at high rpm, until the drive is completed. If a type 1 misfire occurs during a second drive cycle, the MIL and fuel injection behave the same and a DTC is stored. After a type 1 misfire has been detected during two drive cycles, the MIL comes on and stays on beginning with the third drive cycle, unless the Temporary DTC has been cleared by the PCM. Even if the MIL is on, it will start blinking if a type 1 misfire occurs. If the malfunction recurs during the next (second) drive cycle, the MIL comes on and the DTC and the freeze frame data are stored. * Misfire Type 2 (Light) Per 1,000 revolutions 55 times. If a type 2 misfire (emission-related but not severe enough to immediately damage the catalyst) occurs, a Temporary DTC is stored, but the MIL does come on or blink. If a type 2 misfire occurs during a second drive cycle, the MIL comes on and stays on unless the Temporary DTC has been cleared by the PCM. **NOTE: Before you troubleshoot, record all freeze data and any on-board snapshot.**
DTC: P0305 **2T PCM, MIL: Yes** **Year:** 2011, 2012 **Model:** MDX, RL, TL, ZDX **Engine:** 3.5L V6, 3.7L V6	**Cylinder 5 Misfire Detected :** The execution time is once per driving cycle and the duration time is 200 revolutions for Type 1 and 1,000 revolutions for Type 2. Conditions for illuminating the MIL Type 1 and Type 2. * Misfire Type 1 (Severe) Per 200 revolutions 27-90 times. If a type 1 misfire (catalyst damaging) occurs once, the MIL blinks once per second, a Temporary DTC is stored, and the high rpm fuel injection stop system activates. The fuel injection stops, at high rpm only, on the cylinder that has the highest misfire counts. the MIL continues to blink, and the fuel injection stays off at high rpm, until the drive is completed. If a type 1 misfire occurs during a second drive cycle, the MIL and fuel injection behave the same and a DTC is stored. After a type 1 misfire has been detected during two drive cycles, the MIL comes on and stays on beginning with the third drive cycle, unless the Temporary DTC has been cleared by the PCM. Even if the MIL is on, it will start blinking if a type 1 misfire occurs. If the malfunction recurs during the next (second) drive cycle, the MIL comes on and the DTC and the freeze frame data are stored. * Misfire Type 2 (Light) Per 1,000 revolutions 55 times. If a type 2 misfire (emission-related but not severe enough to immediately damage the catalyst) occurs, a Temporary DTC is stored, but the MIL does come on or blink. If a type 2 misfire occurs during a second drive cycle, the MIL comes on and stays on unless the Temporary DTC has been cleared by the PCM. **NOTE: Before you troubleshoot, record all freeze data and any on-board snapshot.**

DTC	Trouble Code Title and Conditions
DTC: P0306 **1T PCM, MIL: Yes** **Year:** 2011, 2012 **Model:** MDX, RL, TL, TSX, ZDX **Engine:** 3.5L V6, 3.7L V6	**Cylinder 6 Misfire Detected:** The execution time is once per driving cycle and the duration time is 200 revolutions for Type 1 and 1,000 revolutions for Type 2. Conditions for illuminating the MIL Type 1 and Type 2. * Misfire Type 1 (Severe) Per 200 revolutions 27-90 times. If a type 1 misfire (catalyst damaging) occurs once, the MIL blinks once per second, a Temporary DTC is stored, and the high rpm fuel injection stop system activates. The fuel injection stops, at high rpm only, on the cylinder that has the highest misfire counts. the MIL continues to blink, and the fuel injection stays off at high rpm, until the drive is completed. If a type 1 misfire occurs during a second drive cycle, the MIL and fuel injection behave the same and a DTC is stored. After a type 1 misfire has been detected during two drive cycles, the MIL comes on and stays on beginning with the third drive cycle, unless the Temporary DTC has been cleared by the PCM. Even if the MIL is on, it will start blinking if a type 1 misfire occurs. If the malfunction recurs during the next (second) drive cycle, the MIL comes on and the DTC and the freeze frame data are stored. * Misfire Type 2 (Light) Per 1,000 revolutions 55 times. If a type 2 misfire (emission-related but not severe enough to immediately damage the catalyst) occurs, a Temporary DTC is stored, but the MIL does come on or blink. If a type 2 misfire occurs during a second drive cycle, the MIL comes on and stays on unless the Temporary DTC has been cleared by the PCM. **NOTE: Before you troubleshoot, record all freeze data and any on-board snapshot**
DTC: P0325 **1T PCM, MIL: Yes** **Year:** 2011, 2012 **Model:** MDX, RDX, RL, TL, TSX, ZDX **Engine:** 2.3L L4, 2.4L L4, 3.5L V6, 3.7L V6	**Knock Sensor Circuit Malfunction :** The execution time is continuous and the duration time is 5 seconds or more. Engine running, speed 2,000 rpm or more, coolant temerature at 140 ºF (60 ºC) or more. No signals from the knock sensor are detected for at least 5 seconds. If the signals from the knock sensor do not vary for a set time, the PCM detects a malfunction and stores a DTC. **NOTE: Before you troubleshoot, record all freeze data and any on-board snapshot.**
DTC: P0335 **1T PCM, MIL: Yes** **Year:** 2011, 2012 **Model:** RDX, TSX **Engine:** 2.3L L4, 2.4L L4, 3.5L V6	**Crankshaft Position (CKP) Sensor No Signal:** With the engine running the PCM has detected no CKP pulses over 75 times.
DTC: P0335 **1T PCM, MIL: Yes** **Year:** 2012 **Model:** TL **Engine:** 3.5L V6	**Crankshaft Position (CKP) Sensor No Signal :** With the engine running with the speed over 750 rpm, and DTCs P0365, P0369 are not active. No input signals from the CKP sensor are detected while the signals from the CMP sensor are detected at least 125 times. **NOTE: Before you troubleshoot, record all freeze data and any on-board snapshot, and review the general troubleshooting information.**
DTC: P0335 **1T PCM, TCIL: Yes** **Year:** 2011, 2012 **Model:** MDX, RL, ZDX **Engine:** 3.7L V6	**Crankshaft Position (CKP) Sensor A No Signal :** The execution is continuous and the duration is for 2 seconds or more (when the engine speed is 750 rpm). MIL OFF, D indicator blinks. No signals from the CKP sensor are input at least 63 times. If no pulsing signals from the CKP sensor are detected, a malfunction is detected and a DTC is stored.
DTC: P0339 **1T PCM, MIL: Yes** **Year:** 2011, 2012 **Model:** RDX, TL, TSX **Engine:** 2.3L L4, 2.4L L4, 3.5L V6	**Crankshaft Position (CKP) Sensor Circuit Intermittent Interruption :** With the engine running othet than 58 pulses are detected during intervals between reference pulses for each crank revolution. This condition has been detected 30 times. **NOTE: Before you troubleshoot, record all freeze data and any on-board snapshot, and review the general troubleshooting information.**
DTC: P0339 **1T PCM, MIL: Yes** **Year:** 2011, 2012 **Model:** MDX, RL, ZDX **Engine:** 3.7L V6	**Crankshaft Position (CKP) Sensor A Circuit Intermittent Interruption :** Other than 22 pulses are detected during intervals between reference pulses for each crankshaft revolution. This condition has been detected at least 30 times. An abnormal amount of pulsing signals are detected from CKP sensor A. **NOTE: Before you troubleshoot, record all freeze data and any on-board snapshot.**
DTC: P0340 **1T PCM, MIL: Yes** **Year:** 2011, 2012 **Model:** RDX **Engine:** 2.3L L4	**Camshaft Position (CMP) Sensor No Signal :** With the engine running no signals from the CMP sensor are detected while signals from the CKP sensor are detected 352 times in succession. DTcs P0335, P0339 and P0344 are not active. **NOTE: Before you troubleshoot, record all freeze data and any on-board snapshot, and review the general troubleshooting information.**
DTC: P0340 **1T PCM, MIL: Yes** **Year:** 2011, 2012 **Model:** RL, TL **Engine:** 3.5L V6, 3.7L V6	**Camshaft Position (CMP) Sensor No Signal:** With the engine running no signals from CMP sensor A are detected while signals from the CKP sensor are detected 352 times in succession. The execution time is continuous and DTCs P0335, P0339, P0344 are not active. **NOTE: Before you troubleshoot, record all freeze data and any on-board snapshot.**

DTC	Trouble Code Title and Conditions
DTC: P0341 **2T PCM, MIL: Yes** **Year:** 2011, 2012 **Model:** RDX, TSX **Engine:** 2.3L L4, 2.4L L4	**Camshaft Position (CMP) Sensor A and CKP Sensor Incorrect Phase Detected :** With the engine running and at operating temperature, and the vehicle speed between 19-38 mph (30-60 km/h) for at least 5 seconds. Malfuntions are as follows: Malfunction with VTC OFF: The gap between the CMP sensor A pulse and the middle of the CMP sensor A assembly is 10 degees or more for at least 5 seconds. Malfunction with VTC ACTIVE: The timing of the camshaft is out of specified range (Other than when BTDC is between 10-100 degrees) for at least 5 seconds. **NOTE: Before you troubleshoot, record all freeze data and any on-board snapshot, and review the general troubleshooting information.**
DTC: P0344 **1T PCM, MIL: Yes** **Year:** 2011, 2012 **Model:** RL, TL **Engine:** 3.5L V6, 3.7L V6	**Camshaft Position (CMP) Sensor Intermittent Interruption:** With the engine running more or less than two CMP sensor pulses are detected during intervals between the CKP standard pulses. This condition occurs at least 30 times. The execution time is continuous and DTCs P0335, P0339, P0340 are not active. **NOTE: Before you troubleshoot, record all freeze data and any on-board snapshot.**
DTC: P0344 **2T PCM, MIL: Yes** **Year:** 2011, 2012 **Model:** RDX, TSX **Engine:** 2.3L L4, 2.4L L4	**Camshaft Position (CMP) Sensor A Circuit Intermittent Interruption:** With the engine running, more or less than two CMP sensor A pulses are detected during intervals between the CKP standard pulses. This condition occurs at least 30 times. **NOTE: Before you troubleshoot, record all freeze data and any on-board snapshot, and review the general troubleshooting information.**
DTC: P0351 **1T PCM, MIL: Yes** **Year:** 2011, 2012 **Model:** MDX, RL, TL, TSX, ZDX **Engine:** 2.4L L4, 3.5L V6, 3.7L V6	**Cylinder 1 Ignition Coil Circuit Malfunction:** With the engine running, the execution time is continuous and the duration time is 5 seconds or more. (when the engine speed is 700 rpm) The return signal does not change for at least 5 seconds when the ignition coil is triggered. **NOTE: Before you troubleshoot, record all freeze data and any on-board snapshot.**
DTC: P0352 **1T PCM** **Year:** 2011, 2012 **Model:** MDX, RL, TL, TSX, ZDX **Engine:** 2.4L L4, 3.5L V6, 3.7L V6	**Cylinder 2 Ignition Coil Circuit Malfunction:** With the engine running, the execution time is continuous and the duration time is 5 seconds or more. (when the engine speed is 700 rpm) The return signal does not change for at least 5 seconds when the ignition coil is triggered.
DTC: P0353 **1T PCM, MIL: Yes** **Year:** 2011, 2012 **Model:** MDX, RL, TL, TSX, ZDX **Engine:** 3.5L V6, 3.7L V6	**Cylinder 3 Ignition Coil Circuit Malfunction:** With the engine running, the execution time is continuous and the duration time is 5 seconds or more. (when the engine speed is 700 rpm) The return signal does not change for at least 5 seconds when the ignition coil is triggered. **NOTE: Before you troubleshoot, record all freeze data and any on-board snapshot**
DTC: P0354 **1T PCM, MIL: Yes** **Year:** 2011, 2012 **Model:** MDX, RL, TL, TSX, ZDX **Engine:** 3.5L V6, 3.7L V6	**Cylinder 4 Ignition Coil Circuit Malfunction:** With the engine running, the execution time is continuous and the duration time is 5 seconds or more. (when the engine speed is 700 rpm) The return signal does not change for at least 5 seconds when the ignition coil is triggered. **NOTE: Before you troubleshoot, record all freeze data and any on-board snapshot**
DTC: P0355 **1T PCM, MIL: Yes** **Year:** 2011, 2012 **Model:** MDX, RL, TL, TSX, ZDX **Engine:** 3.5L V6, 3.7L V6	**Cylinder 5 Ignition Coil Circuit Malfunction:** With the engine running, the execution time is continuous and the duration time is 5 seconds or more. (when the engine speed is 700 rpm) The return signal does not change for at least 5 seconds when the ignition coil is triggered. **NOTE: Before you troubleshoot, record all freeze data and any on-board snapshot**
DTC: P0356 **1T PCM, MIL: Yes** **Year:** 2011, 2012 **Model:** MDX, RL, TL, TSX, ZDX **Engine:** 3.5L V6, 3.7L V6	**Cylinder 6 Ignition Coil Circuit Malfunction:** With the engine running, the execution time is continuous and the duration time is 5 seconds or more. (when the engine speed is 700 rpm) The return signal does not change for at least 5 seconds when the ignition coil is triggered. **NOTE: Before you troubleshoot, record all freeze data and any on-board snapshot**
DTC: P0365 **1T PCM, MIL: Yes** **Year:** 2011, 2012 **Model:** RDX, TSX **Engine:** 2.3L L4, 2.4L L4	**Camshaft Position (CMP) Sensor B No Signal:** With the engine running, NO CMP sensor B pulsing signals are detected for at least 50 times in succession.
DTC: P0365 **1T PCM, MIL: Yes** **Year:** 2011, 2012 **Model:** TL, TSX, ZDX **Engine:** 3.5L V6, 3.7L V6	**Camshaft Position (CMP) Sensor Circuit No Signal:** With the engine running at 750 rpm, DTCs P0335, P0339 and P0369 are not active. No input signals from the CMP sensor are detected while the signals from the CKP sensor are detected at least 300 times. **NOTE: Before you troubleshoot, record all freeze data and any on-board snapshot**

DTC	Trouble Code Title and Conditions
DTC: P0369 **1T PCM, MIL: Yes** **Year:** 2011, 2012 **Model:** TL, TSX, ZDX **Engine:** 3.5L V6, 3.7L V6	**Camshaft Position (CMP) Sensor Circuit Intermittent Interruption:** Other than 30 times of CMP pulses are detected during two cycles of CKP. This condition has been detected at least 10 times. **NOTE: Before you troubleshoot, record all freeze data and any on-board snapshot.**
DTC: P0369 **1T PCM, MIL: Yes** **Year:** 2011, 2012 **Model:** RDX, TSX **Engine:** 2.3L L4, 2.4L L4	**Camshaft Position (CMP) Sensor B Intermittent Interruption:** With the engine running, more or less than four CMP B pulsing signals are detected during intervals between standard pulses at least 30 times in succession. **NOTE: Before you troubleshoot, record all freeze data and any on-board snapshot, and review the general troubleshooting information.**
DTC: P0385 **1T PCM, MIL: Yes** **Year:** 2011, 2012 **Model:** MDX, RL, TL, ZDX **Engine:** 3.5L V6, 3.7L V6	**Crankshaft Position (CKP) Sensor B No Signal :** With the engine running the execution is continuous and P0385, P0389 are not active. No signals from CKP sensor A are detected while signals from CKP sensor B are detected 127 times in succession. If no pulsing signals are detected from CKP sensor A, a malfunction is detected and a DTC is stored. **NOTE: Before you troubleshoot, record all freeze data and any on-board snapshot.**
DTC: P0389 **1T PCM, MIL: Yes** **Year:** 2011, 2012 **Model:** MDX, RL, TL, ZDX **Engine:** 3.5L V6, 3.7L V6	**Crankshaft Position (CKP) Sensor B Intermittent Interruption :** Engine running, other than 22 pulses are detected during intervals between reference pulses for each crankshaft revolution. This condition has been detected at least 30 times. An abnormal amount of pulsing signals are detected from CKP sensor B. **NOTE: Before you troubleshoot, record all freeze data and any on-board snapshot.**
DTC: P0400 **2T PCM, MIL: Yes** **Year:** 2011, 2012 **Model:** RL, TL, TSX, ZDX **Engine:** 3.5L V6, 3.7L V6	**Exhaust Gas Recirculation (EGR) System Leak Detected:** With engine running, rpm 1,500-3,500 and system in closed loop the A/F sensor response parameter according to the opening and closing of the EGR is 14,746 (LT80A) or less. **NOTE: Before you troubleshoot, record all freeze data and any on-board snapshot.**
DTC: P0401 **2T PCM, MIL: Yes** **Year:** 2011, 2012 **Model:** MDX, RL, TL, TSX, ZDX **Engine:** 3.5L V6, 3.7L V6	**Exhaust Gas Recirculation (EGR) Insufficient Flow:** Engine temperature is at a minimum of 156 ºF (69 ºC), engine speed is between 1,100-2,400 rpm with the map value of 14 kPa (4.0 in.Hg, 100 mmHg), throttle fully closed and battery 10.5 volt minimum. The execution time is once per driving cycle and the duration time is 3 seconds or more. The ratio of the current EGR flow to the normal EGR flow is 15% or less for at least 3 seconds. **NOTE: Before you troubleshoot, record all freeze data and any on-board snapshot.**
DTC: P0404 **2T PCM, MIL: Yes** **Year:** 2011, 2012 **Model:** MDX, RL, TL, TSX, ZDX **Engine:** 3.5L V6, 3.7L V6	**Exhaust Gas Recirculation (EGR) Control Circuit Range/Performance Problem :** Vehicle at a speed between 15-75 mph, actual valve lift command at 0.012 in. (0.3 mm). The execution time is once per driving cycle and the duration time is 5 seconds or more. The difference between the command value of the amount of EGR valve lift in the PCM and the actual amount of valve lift is 0.041 in. (1.020 mm) or more for at least 5 seconds. **NOTE: Before you troubleshoot, record all freeze data and any on-board snapshot.**
DTC: P0406 **1T PCM, MIL: Yes** **Year:** 2011, 2012 **Model:** MDX, RL, TL, TSX, ZDX **Engine:** 3.5L V6, 3.7L V6	**Exhaust Gas Recirculation (EGR) Valve Position Sensor Circuit High Voltage:** Wth the engine running the execution time is continuous and the duration time is 2 seconds or more. The EGR valve position sensor output voltage is 4.88 V or more for at least 2 seconds. **NOTE: Before you troubleshoot, record all freeze data and any on-board snapshot.**
DTC: P0420 **2T PCM, MIL: Yes** **Year:** 2011, 2012 **Model:** RDX, TSX **Engine:** 2.3L L4, 2.4L L4	**Catalytic System Efficiency Below Threshold :** With the vehicle driven to a speed of 16-75 mph at less than 3500 rpm in closed loop for 2-5 minutes, ECT sensor more than 140ºF, MAF sensor from 8-50 g/sec, engine load less than 99% (±8%), predicted Catalyst temperature over 750ºF, and the PCM detected the catalyst oxygen storage capacity was below an acceptable threshold during the test. **NOTE: If some of the DTCs listed below are stored at the same time as DTC P0420, troubleshoot those DTCs first, then recheck for DTC P0420.**
DTC: P0420 **2T PCM, MIL: Yes** **Year:** 2011, 2012 **Model:** MDX, RL, TL, TSX, ZDX **Engine:** 3.5L V6, 3.7L V6	**Rear Bank Catalyst System Efficiency Below Threshold (Bank 1):** The execution time is once per driving cycle and the duration time is 102 seconds or more. The number of detections is 752 (CTAGLT67) or more. **NOTE: Before you troubleshoot, record all freeze data and any on-board snapshot.**
DTC: P0430 **2T PCM, MIL: Yes** **Year:** 2011, 2012 **Model:** MDX, TL, TSX, ZDX **Engine:** 3.5L V6, 3.7L V6	**Front Bank Catalyst System Efficiency Below Threshold (Bank 2) :** The execution time is once per driving cycle and the duration time is 102 seconds or more. The number of detections is 720 (CTAGLT68) or more. **NOTE: Before you troubleshoot, record all freeze data and any on-board snapshot.**

DTC	Trouble Code Title and Conditions
DTC: P0443 **1T PCM, MIL: Yes** **Year:** 2011, 2012 **Model:** MDX, RDX, RL, TL, TSX, ZDX **Engine:** 2.3L L4, 2.4L L4, 3.5L V6, 3.7L V6	**Evaporative Emission (EVAP) Canister Purge Valve Circuit Malfunction:** With the engine running, EVAP canister purge valve output duty at 2% to 98% the execution time is continuous and the duration time is 5 seconds or more. When the return signal does not change according to the EVAP canister purge valve duty cycle for a set time, the PCM detects a malfunction. The return signal does not change according to the EVAP canister purge valve output for at least 5 seconds. **NOTE: Before you troubleshoot, record all freeze data and any on-board snapshot.**
DTC: P0451 **2T PCM, MIL: Yes** **Year:** 2011, 2012 **Model:** MDX, RDX, RL, TL, TSX, ZDX **Engine:** 2.3L L4, 2.4L L4, 3.5L V6, 3.7L V6	**Fuel Tank Pressure (FTP) Sensor Circuit Range/Performance Problem:** Elapsed time after starting the engine is 2 seconds with the throttle fully closed. The execution time is once per driving cycle and the duration time is 20 seconds or more. The FTP sensor output fluctuates by 0.3 kPa (0.1 in.Hg, 2 mmHg) or more at least five times within 3 seconds. **NOTE: Before you troubleshoot, record all freeze data and any on-board snapshot.**
DTC: P0452 **2T PCM, MIL: Yes** **Year:** 2011, 2012 **Model:** MDX, RDX, RL, TL, TSX, ZDX **Engine:** 2.3L L4, 2.4L L4, 3.5L V6, 3.7L V6	**Fuel Tank Pressure (FTP) Sensor Circuit Low Voltage:** Elapsed time after starting the engine is 2 seconds at idle. The execution time is once per driving cycle and the duration time is 3 seconds or more. The output from the fuel tank pressure sensor is less than -7 kPa (-2.1 in.Hg, -55 mmHg) for at least 3 seconds. **NOTE: Before you troubleshoot, record all freeze data and any on-board snapshot.**
DTC: P0453 **2T PCM, MIL: Yes** **Year:** 2011, 2012 **Model:** MDX, RDX, RL, TL, TSX, ZDX **Engine:** 2.3L L4, 2.4L L4, 3.5L V6, 3.7L V6	**Fuel Tank Pressure (FTP) Sensor Circuit High Voltage:** Elapsed time after starting the engine is 2 seconds at idle. The execution time is once per driving cycle and the duration time is 3 seconds or more. If the FTP sensor output voltage is higher than a target value within a set time after starting the engine in a cold condition. The output from the fuel tank pressure sensor is more than 8 kPa (2.2 in.Hg, 55 mmHg) for at least 3 seconds. **NOTE: Before you troubleshoot, record all freeze data and any on-board snapshot.**
DTC: P0455 **2T PCM, MIL: Yes** **Year:** 2011, 2012 **Model:** MDX, RDX, TL, TSX, ZDX **Engine:** 2.3L L4, 2.4L L4, 3.5L V6, 3.7L V6	**Evaporative Emission (EVAP) System Large Leak Detected:** The execution time is once per driving cycle and the duration time is 36 minutes and 37 seconds, or less. Here is an overview of the malfunction detection for the EONV method: 1: Judgement of detection of 0.09 inch leak as normal operation 2: Judgement of detection of 0.02 inch leak as normal operation 3: Detection of 0.02 inch leak 4: Detection of atmospheric pressure failure 5: Flickering of the FTP sensor The execution time is once per driving cycle and the duration time is 36 minutes and 37 seconds, or less. The variation of pressure inside the fuel tank is 0.03 kPa (0.009 in.Hg, 0.24 mmHg) or more. **NOTE: Before you troubleshoot, record all freeze data and any on-board snapshot.**
DTC: P0456 **2T PCM, MIL: Yes** **Year:** 2011, 2012 **Model:** MDX, RDX, RL, TL, TSX, ZDX **Engine:** 2.3L L4, 2.4L L4, 3.5L V6, 3.7L V6	**EVAP System Very Small Leak Detected :** The execution time is once per driving cycle and the duration time is at least 16 minutes and 37 seconds, but not more than 36 minutes and 37 seconds. Here is an overview of the malfunction detection for the EONV method: 1: Judgement of detection of 0.09 inch leak as normal operation 2: Judgement of detection of 0.02 inch leak as normal operation 3: Detection of 0.02 inch leak 4: Detection of atmospheric pressure failure 5: Flickering of the FTP sensor The execution time is once per driving cycle and the duration time is 36 minutes and 37 seconds, or less. The variation of pressure inside the fuel tank is 0.03 kPa (0.009 in.Hg, 0.24 mmHg) or more. **NOTE: Before you troubleshoot, record all freeze data and any on-board snapshot.**
DTC: P0457 **3T PCM, MIL: Yes** **Year:** 2011, 2012 **Model:** MDX, RDX, TL, TSX, ZDX **Engine:** 2.3L L4, 2.4L L4, 3.5L V6, 3.7L V6	**Evaporative Emission (EVAP) System Leak Detected/Fuel Fill Cap Loose or Missing:** Engine coolant temperature before EVAP purge control starts 140 ºF (60 ºC), 2 mph (2 km/h), Barometric pressure, 76 kPa (22.5 in.Hg, 569 mmHg), Battery voltage 10.5 volts or more, fuel trim 0.73-1.47, system is in close loop. The execution time is continuous and the duration time is 12 seconds or more. The output from the fuel cap monitor is 0.053 or less for at least 12 seconds (when there is no NG judgment history in this drive cycle). P0455 or P0456 are judged as NG. **NOTE: Before you troubleshoot, record all freeze data and any on-board snapshot.**

DTC	Trouble Code Title and Conditions
DTC: P0461 **1T PCM** **Year:** 2011, 2012 **Model:** RL, ZDX **Engine:** 3.7L V6	**Fuel Level Sensor (Fuel Gauge Sending Unit) Circuit Range/Performance Problem :** The execution time is every 125 miles (200 km).The change in the fuel level sensor output is 3.5 % or less. **NOTE: Because it requires 162 miles (260 km) of driving without refueling to complete this diagnosis, DTC P0461 cannot be duplicated during this troubleshooting.**
DTC: P0461 **1T PCM** **Year:** 2011, 2012 **Model:** MDX, RDX, TL, TSX, ZDX **Engine:** 2.3L L4, 2.4L L4, 3.5L V6, 3.7L V6	**Fuel Level Sensor (Fuel Gauge Sending Unit) Range/Performance Problem :** The execution time is every 125 miles (200 km), DTCs P0462, P0463, U0028, U0155 are not active. If the powertrain control module (PCM) receives no change in the fuel level sensor output after driving for a specified number of miles, it detects a malfunction. The change in the fuel level sensor output is 3.5 % or less.
DTC: P0462 **1T PCM** **Year:** 2011, 2012 **Model:** MDX, RDX, RL, TL, TSX, ZDX **Engine:** 2.3L L4, 2.4L L4, 3.5L V6, 3.7L V6	**Fuel Level Sensor (Fuel Gauge Sending Unit) Circuit Low Voltage:** The fuel level sensor (fuel gauge sending unit) output voltage is 0.10 V or less for at least 5 seconds. The execution time is continuous DTCs P0463, U0155 are not active. **NOTE: Before you troubleshoot, record all freeze data and any on-board snapshot.**
DTC: P0463 **1T PCM** **Year:** 2011, 2012 **Model:** MDX, RDX, RL, TL, TSX, ZDX **Engine:** 2.3L L4, 2.4L L4, 3.5L V6, 3.7L V6	**Fuel Level Sensor (Fuel Gauge Sending Unit) Circuit High Voltage :** The execution time is every 125 miles (200 km), DTCs P0462, P0463, U0028, U0155 are not active. If the Powertrain Control Module (PCM) receives no change in the fuel level sensor output after driving for a specified number of miles, it detects a malfunction. The fuel level sensor output voltage is 4.92 V or more for at least 5 seconds.
DTC: P0480 **1T PCM** **Year:** 2011, 2012 **Model:** RL **Engine:** 3.7L V6	**Radiator Fan Control (RFC) System Malfunction:** With ignition switch ON (II) the RFC terminal voltage is low for at least 20 seconds, and the same condition continues even when the power reset is repeated 5 times. (Power is supplied for about 5 seconds after the power reset.) **NOTE: Before you troubleshoot, record all freeze data and any on-board snapshot.**
DTC: P0496 **2T PCM, MIL: Yes** **Year:** 2011, 2012 **Model:** MDX, RDX, RL, TL, TSX, ZDX **Engine:** 2.3L L4, 2.4L L4, 3.5L V6, 3.7L V6	**Evaporative Emission (EVAP) System High Purge Flow Detected:** Monitor execution is once per driving cycle. The output from the EVAP canister purge valve is 0.2 kPa (0.07 in.Hg, 2 mmHg) or more for at least 10 seconds. **NOTE: Before you troubleshoot, record all freeze data and any on-board snapshot.**
DTC: P0497 **2T PCM, MIL: Yes** **Year:** 2011, 2012 **Model:** MDX, RDX, RL, TL, TSX, ZDX **Engine:** 2.3L L4, 2.4L L4, 3.5L V6, 3.7L V6	**Evaporative Emission (EVAP) System Low Purge Flow Detected:** The execution time is once per driving cycle and P145C is judged as NG. Enable conditions are low load duration time 10 seconds, Wait for 10 seconds after the ignition switch is turned to LOCK (0) and Engine coolant temperature before EVAP purge control starts is 140 ºF (60 ºC). The output from the fuel tank pressure sensor is 0.2 kPa (0.07 in.Hg, 2 mmHg) or less for at least 10 seconds. **NOTE: Before you troubleshoot, record all freeze data and any on-board snapshot.**
DTC: P0498 **1T PCM, MIL: Yes** **Year:** 2011, 2012 **Model:** MDX, RL, ZDX **Engine:** 3.7L V6	**Evaporative Emission (EVAP) Canister Vent Shut Valve Control Circuit Low Voltage:** The execution time is continuous and the duration time is 5 seconds or more. If the return signal is OFF when the powertrain control module (PCM) outputs the ON signal to the EVAP canister vent shut valve, the PCM detects a malfunction. DTC P0499 is not active. The return signal is Low for at least 5 seconds when the PCM outputs the ON signal to the EVAP canister vent shut valve.
DTC: P0498 **1T PCM, MIL: Yes** **Year:** 2011, 2012 **Model:** RDX, TL, TSX **Engine:** 2.3L L4, 2.4L L4, 3.5L V6	**Evaporative Emission (EVAP) Canister Vent Shut Valve Control Circuit Low Voltage:** The execution time is continuous and the duration time is 5 seconds or more. If the return signal is OFF (Low) when the Powertrain Control Module (PCM) outputs the ON signal to the EVAP canister vent shut valve, the PCM detects a malfunction. The return signal is Low for at least 5 seconds when the PCM outputs the ON signal to the EVAP canister vent shut valve.
DTC: P0499 **1T PCM, MIL: Yes** **Year:** 2011, 2012 **Model:** MDX, RDX, RL, TL, TSX, ZDX **Engine:** 2.3L L4, 2.4L L4, 3.5L V6, 3.7L V6	**Evaporative Emission (EVAP) Canister Vent Shut Valve Control Circuit High Voltage:** If the return signal is ON when the Powertrain Control Module (PCM) outputs the OFF signal to the EVAP canister vent shut valve, the PCM detects a malfunction.The execution time is continuous and the duration time is 5 seconds or more. DTC P0498 is not active. The return signal is ON for at least 5 seconds when the PCM outputs the Low signal to the EVAP canister vent shut valve. **NOTE: Before you troubleshoot, record all freeze data and any on-board snapshot.**

DTC	Trouble Code Title and Conditions
DTC: P0506 **2T PCM, MIL: Yes** **Year:** 2011, 2012 **Model:** MDX, RDX, TL, TSX, ZDX **Engine:** 2.3L L4, 2.4L L4, 3.5L V6, 3.7L V6	**Idle Control System RPM Lower Than Expected:** The execution time is once per driving cycle and the duration time is 20 seconds or more. Throttle fully closed, fuel trim 0.75-1.47, intake air temperature 19 ºF (-7 ºC) and battery voltage 10.5 or more. If the actual idle speed varies beyond a specified value from the target speed over a certain period of time, the PCM detects a malfunction. The actual idle speed is at least 100 rpm less than the target idle speed for at least 20 seconds. **NOTE: Before you troubleshoot, record all freeze data and any on-board snapshot.**
DTC: P0507 **2T PCM, MIL: Yes** **Year:** 2011, 2012 **Model:** MDX, RDX, RL, TL, TSX, ZDX **Engine:** 2.3L L4, 2.4L L4, 3.5L V6, 3.7L V6	**Idle Control System RPM Higher Than Expected:** Enable conditions are as follows: coolant temperature 156 ºF (69 ºC) minimum, intake air temperature 19 ºF (-7 ºC), fuel trim 0.73-1.47, throttle closed and battery voltage at least 10.5 volts. The execution time is once per driving cycle and the duration time is 20 seconds or more. If the actual idle speed varies beyond a specified value from the target speed over a certain period of time, the PCM detects a malfunction. The actual idle speed is at least 200 rpm greater than the target idle speed for at least 20 seconds. **NOTE: Before you troubleshoot, record all freeze data and any on-board snapshot.**
DTC: P050A **2T PCM, MIL: Yes** **Year:** 2011, 2012 **Model:** MDX, RDX, RL, TL, TSX, ZDX **Engine:** 2.3L L4, 2.4L L4, 3.5L V6, 3.7L V6	**Cold Start Idle Air Control System Performance Problem:** The execution time is once per driving cycle and the duration time is 10 seconds or more. When the actual amount of air is less than the target amount, a malfunction is detected. The total airflow is decreased by a factor of 0.693 for at least 10 seconds. **NOTE: Before you troubleshoot, record all freeze data and any on-board snapshot.**
DTC: P050B **2T PCM, MIL: Yes** **Year:** 2011, 2012 **Model:** MDX, RL, TL, TSX, ZDX **Engine:** 2.4L L4, 3.5L V6, 3.7L V6	**Cold Start Ignition Timing Performance Problem:** The execution time is once per driving cycle and the duration time is 3.5 seconds or more and the engine speed is 2,100 rpm or more for at least 3.5 seconds. When the actual engine speed is a specified value or more, and it continues for a specified time, a malfunction is detected. **NOTE: Before you troubleshoot, record all freeze data and any on-board snapshot.**
DTC: P050B **2T PCM, MIL: Yes** **Year:** 2011, 2012 **Model:** RDX **Engine:** 2.3L L4	**Cold Start Ignition Timing Control System Performance Problem:** With the vehicle at idle and throttle position fully closed, the engine speed is 2,150 rpm or more for at least 3.5 seconds. **NOTE: Before you troubleshoot, record all freeze data and any on-board snapshot, and review the general troubleshooting information.**
DTC: P0522 **1T PCM, MIL: Yes** **Year:** 2012 **Model:** TL **Engine:** 3.5L V6	**Rocker Arm Oil Pressure Sensor Circuit Low Voltage :** With the engine running and the barometric pressure a minimum of 69 Kpa (20.5 in. Hg. 520 mmHg) or more, the rocker arm oil pressure sensor output voltage is 0.18 V or less for at least 2 seconds.
DTC: P0523 **1T PCM, MIL: Yes** **Year:** 2012 **Model:** TL **Engine:** 3.5L V6	**Rocker Arm Oil Pressure Sensor Circuit High Voltage:** With the engine speed at a maximum of 6,000 rpm, engine coolant temperature a minimum of 104 ºF (40 ºC), the EOP sensor output voltage is 4.79 V or more for at least 2 seconds.
DTC: P0532 **2T PCM** **Year:** 2011, 2012 **Model:** RDX, TSX **Engine:** 2.3L L4, 2.4L L4, 3.5L V6	**A/C Pressure Sensor Circuit Low Voltage:** Ignition switch ON (II), DTC P0533 is not active. The A/C pressure sensor output voltage is 0.24 V for at least 10 seconds. **NOTE: Before you troubleshoot, record all freeze data and any on-board snapshot, and review the general troubleshooting information.**
DTC: P0533 **1T PCM** **Year:** 2011, 2012 **Model:** RDX **Engine:** 2.3L L4	**A/C Pressure Sensor Circuit High Voltage:** With the ignition switch on, the A/C pressure sensor output voltage is 4.74 V or more for at least 10 seconds. DTC P0532 is not active. **NOTE: Before you troubleshoot, record all freeze data and any on-board snapshot, and review the general troubleshooting information.**
DTC: P0562 **1T PCM** **Year:** 2011, 2012 **Model:** MDX, RDX, RL, TL, TSX, ZDX **Engine:** 2.3L L4, 2.4L L4, 3.5L V6, 3.7L V6	**Charging System Low Voltage:** The execution time is continuous and the duration time is 60 seconds or more, engine speed 550 rpm. When the IGP (power source) terminal voltage is a set value or less for a set time, the PCM detects a malfunction. The IGP terminal voltage is 11.0 V or less for at least 60 seconds. **NOTE: Before you troubleshoot, record all freeze data and any on-board snapshot. If any high current load accessories are installed, this DTC can be set.**

DTC	Trouble Code Title and Conditions
DTC: P0563 **1T PCM, MIL: Yes** **Year:** 2011, 2012 **Model:** MDX, RL, TL, ZDX **Engine:** 3.5L V6, 3.7L V6	**Powertrain Control Module (PCM) Power Source Circuit Unexpected Voltage:** The PCM operates for at least 5 seconds after the ignition switch is turned to LOCK (0). Battery voltage 10.1 volts. (IGP terminal of PCM) **NOTE: Before you troubleshoot, record all freeze data and any on-board snapshot.**
DTC: P0563 **1T PCM, MIL: Yes** **Year:** 2011, 2012 **Model:** RDX, TSX **Engine:** 2.3L L4, 2.4L L4, 3.5L V6	**Engine Control Module (ECM) Powertrain Control Module (PCM) Power Source Circuit Unexpected Voltage:** The ECM/PCM operates for at least 5 seconds after the ignition switch is turned to LOCK (0). Battery voltage 10.1 volts. (IGP terminal of PCM) **NOTE: Before you troubleshoot, record all freeze data and any on-board snapshot.**
DTC: P0602 **1T PCM, MIL: Yes** **Year:** 2011, 2012 **Model:** MDX, RDX, RL, TL, TSX, ZDX **Engine:** 2.3L L4, 2.4L L4, 3.5L V6, 3.7L V6	**ECM/PCM Programming Error:** With ignition on the execution time is continuous and the duration time is 1 second or less. The ECM/PCM program update stops 1 second before it is finished. **NOTE: This DTC is indicated when a PCM update is not completed.** **WARNING:** Do not turn the ignition switch to LOCK (0) or ACC (I) while updating the PCM. If you turn the ignition switch to LOCK (0) before completion, the ECM/PCM can be damaged.
DTC: P0606 **1T PCM, MIL: Yes** **Year:** 2011, 2012 **Model:** RDX, TSX **Engine:** 2.3L L4, 2.4L L4, 3.5L V6	**ECM/PCM Processor Malfunction:** After 30 seconds have elapsed since start-up, or after the engine speed exceeds 1,000 rpm once. No signal from the DKS CPU is detected or is abnormal for at least 5 seconds. **NOTE: Before you troubleshoot, record all freeze data and any on-board snapshot, and review the general troubleshooting information.**
DTC: P060A **1T PCM, MIL: Yes** **Year:** 2011, 2012 **Model:** MDX, TL, TSX, ZDX **Engine:** 2.4L L4, 3.5L V6, 3.7L V6	**Powertrain Control Module (PCM) (A/T System) Internal Control Module Malfunction:** (Symptom 1) The execution time is continuous and the duration time 500 milliseconds or more The serial communication between the FI CPU and A/T CPU is abnormal or the serial communication is interrupted for at least 500 milliseconds. (Symptom 2) The execution time is continuous and the duration time , 30 milliseconds or more The watchdog timer that monitors the A/T CPU detects an abnormality, and the FI CPU receives the A/T CPU check signal for at least 30 milliseconds. **NOTE: Before you troubleshoot, record all freeze data and any on-board snapshot.**
DTC: P060A **1T PCM, MIL: Yes** **Year:** 2011, 2012 **Model:** RDX, ZDX **Engine:** 2.3L L4, 3.7L V6	**Powertrain Control Module (PCM) Internal Control Module Malfunction:** With the ingition ON and battery voltage at a minimum of 10. 0 volts. The execution time is continuous and the duration time is 200 milliseconds or more. With keyless access system 110 milliseconds or more. One of these 2 symptoms occurs: (1) Symptom 1: The internal communication between the FI CPU and the A/T CPU is abnormal or the internal communication is interrupted for at least 200 milliseconds. (2) Symptom 2: The watchdog timer that monitors the A/T CPU detects an abnormality, and the FI CPU receives the A/T CPU check signal for at least 110 milliseconds.
DTC: P060F **1T PCM, MIL: Yes** **Year:** 2011, 2012 **Model:** TSX **Engine:** 2.4L L4, 3.5L V6	**Powertrain Control Module (PCM) Internal Control Module Keep Alive Memory (KAM) Error :** With the ignition switch ON a malfunction is detected whenever the keep alive data retrieval and writing process is not completed normally. **NOTE: Before you troubleshoot, record all freeze data and any on-board snapshot with the HDS, and review General Troubleshooting Information.**
DTC: P0615 **1T PCM** **Year:** 2011, 2012 **Model:** MDX, ZDX **Engine:** 3.7L V6	**Starter Cut Relay STRLD Circuit Malfunction:** Withe the ignition switch ON (II). The diagnosis line (STRLD) input voltage is between 2.4 V to 2.6 V for at least 1 second. The execution time is continuous and the duration time is 1 seconds or more. **NOTE: Before you troubleshoot, record all freeze data and any on-board snapshot.**
DTC: P0627 **1T PCM, MIL: Yes** **Year:** 2011, 2012 **Model:** MDX, RDX, RL, ZDX **Engine:** 2.3L L4, 3.7L V6	**Fuel Pump Control Module System Malfunction:** When the diagnosis signal from the fuel pump control module system is low for a set time or more, the PCM detects a malfunction. The execution time is continuous, diagnosis signal is low for at least 2 seconds. **NOTE: Before you troubleshoot, record all freeze data and any on-board snapshot.**
DTC: P062F **1T PCM, TCIL: Yes** **Year:** 2011, 2012 **Model:** MDX, RDX, RL, TL, ZDX **Engine:** 2.3L L4, 3.5L V6, 3.7L V6	**ECM/PCM Internal Control Module Keep Alive Memory (KAM) Error :** A malfunction is detected whenever the keep alive data retrieval and writing process is not completed normally. **NOTE: Before you troubleshoot, record all freeze data and any on-board snapshot.**

DTC	Trouble Code Title and Conditions
DTC: P0630 **1T PCM, MIL: Yes** **Year:** 2011, 2012 **Model:** MDX, RDX, RL, TL, TSX, ZDX **Engine:** 2.3L L4, 2.4L L4, 3.5L V6, 3.7L V6	**VIN Not Programmed or Mismatch:** The VIN is not registered in the keep-alive memory in the PCM. **NOTE: Before you troubleshoot, record all freeze data and any on-board snapshot.**
DTC: P0641 **1T PCM, MIL: Yes** **Year:** 2011, 2012 **Model:** TL, ZDX **Engine:** 3.5L V6, 3.7L V6	**Sensor Reference Voltage A Malfunction:** With the ignition ON the execution time is continuous and the sensor power voltage is 5.2 V or more, or 4.5 V or less, for at least 2.0 seconds. **NOTE: Before you troubleshoot, record all freeze data and any on-board snapshot. It may be possible to locate the fault by disconnecting one component at a time from the 5-volt reference circuit while viewing the 5-Volt Reference circuit parameter on the scan tool. The scan tool parameter would change from Fault to OK when the source of the fault is disconnected. If all 5-volt reference components have been disconnected and a Fault is still indicated, the fault may exist in the wiring harness.**
DTC: P0651 **1T PCM, MIL: Yes** **Year:** 2012 **Model:** TL **Engine:** 3.5L V6	**Sensor Reference Voltage B Malfunction:** The sensor power voltage is 5.2 V or more, or 4.5 V or less, for at least 2.0 seconds.
DTC: P0651 **1T VCM, MIL: Yes** **Year:** 2011, 2012 **Model:** ZDX **Engine:** 3.7L V6	**Sensor Reference Voltage B Malfunction:** With the ignition ON the execution time is continuous and the sensor power voltage is 5.2 V or more, or 4.5 V or less, for at least 2.0 seconds. **NOTE: Before you troubleshoot, record all freeze data and any on-board snapshot. It may be possible to locate the fault by disconnecting one component at a time from the 5-volt reference circuit while viewing the 5-Volt Reference circuit parameter on the scan tool. The scan tool parameter would change from Fault to OK when the source of the fault is disconnected. If all 5-volt reference components have been disconnected and a Fault is still indicated, the fault may exist in the wiring harness.**
DTC: P0685 **2T PCM, MIL: Yes** **Year:** 2011, 2012 **Model:** TSX **Engine:** 2.4L L4, 3.5L V6	**ECM/PCM Power Control Circuit Malfunction:** When the voltage to the ECM/PCM is turned off and the ECM/PCM shuts down without the normal shut down procedure, a malfunction in the PGM-FI main relay 1 control circuit is detected. **NOTE: Before you troubleshoot, record all freeze data and any on-board snapshot.**
DTC: P0685 **2T PCM, MIL: Yes** **Year:** 2011, 2012 **Model:** MDX, RDX, RL, TL, ZDX **Engine:** 2.3L L4, 3.5L V6, 3.7L V6	**Powertrain Control Module (PCM) Power Control Circuit/Internal Circuit Malfunction:** When the voltage to the PCM is turned off and the PCM shuts down without the normal shut down procedure, a malfunction in the PGM-FI main relay 1 control circuit is detected. **NOTE: Before you troubleshoot, record all freeze data and any on-board snapshot.**
DTC: P0705 **1T PCM, MIL: Yes, TCIL: Yes** **Year:** 2011, 2012 **Model:** MDX, RL, TL, TSX, ZDX **Engine:** 2.4L L4, 3.5L V6, 3.7L V6	**Short in Transmission Range Switch Circuit (Multiple Shift-position Input):** One of 3 conditions occurs: (1) The PCM detects the selected range switch input and another range switch input simultaneously for at least 1 seconds. (2) The PCM detects the P, R, or N range switch input and the FWD switch input simultaneously for at least 1 seconds (3) The PCM detects the D or D3 range switch input and the RVS switch input simultaneously for at least 1 second. **NOTE: Before you troubleshoot, record all freeze data and any on-board snapshot. This code is caused by an electrical circuit problem and cannot be caused by a mechanical problem in the transmission.**
DTC: P0705 **1T PCM, MIL: Yes, TCIL: Yes** **Year:** 2011, 2012 **Model:** RDX **Engine:** 2.3L L4	**Transmission Range Switch Circuit (Multiple Shift-position Input):** Malfunction 1: The PCM detects the selected range switch input and another range switch (except L2 switch) input simultaneously for at least 1 second. Malfunction 2: The PCM detects the P,R or N range switch input and the L2 switch input simultaneously for at least 1 second. **NOTE: Record all freeze data and review General Troubleshooting Information before you troubleshoot. This code is caused by an electrical circuit problem and cannot be caused by a mechanical problem in the transmission.**
DTC: P0706 **2T PCM, MIL: Yes** **Year:** 2011, 2012 **Model:** MDX, RDX, RL, TL, ZDX **Engine:** 2.3L L4, 3.5L V6, 3.7L V6	**Open in Transmission Range Switch Circuit:** No FWD position signal is detected when the vehicle speed changes from 6 mph (10 km/h) 25 mph (40 km/h) 6 mph (10 km/h) in D or D3. DTCs P0705, P0721, P0722 are not active. **NOTE: This code is caused by an electrical circuit problem and cannot be caused by a mechanical problem in the transmission.**
DTC: P0711 **1T PCM, TCIL: Yes** **Year:** 2011, 2012 **Model:** MDX, RDX, RL, TL, TSX, ZDX **Engine:** 2.3L L4, 2.4L L4, 3.5L V6, 3.7L V6	**Problem in ATF Temperature Sensor Circuit:** The ATF temperature sensor signal does not change. Stuck at low temperature or stuck at high temperature is detected. **NOTE: This code is caused by an electrical circuit problem and cannot be caused by a mechanical problem in the transmission.**

DTC	Trouble Code Title and Conditions
DTC: P0712 **1T PCM, TCIL: Yes** **Year:** 2011, 2012 **Model:** MDX, RDX, RL, TL, TSX, ZDX **Engine:** 2.3L L4, 2.4L L4, 3.5L V6, 3.7L V6	**Short in ATF Temperature Sensor Circuit:** When the ATF temperature sensor signal voltage to the PCM is under the specification, indicating that the temperature is above the specification (a short to ground), a malfunction is detected. The ATF temperature sensor output voltage is less than 0.07 V for at least 10 seconds. **NOTE: This code is caused by an electrical circuit problem and cannot be caused by a mechanical problem in the transmission.**
DTC: P0713 **1T PCM, TCIL: Yes** **Year:** 2011, 2012 **Model:** MDX, RDX, RL, TL, TSX, ZDX **Engine:** 2.3L L4, 2.4L L4, 3.5L V6, 3.7L V6	**Open in ATF Temperature Sensor Circuit:** When the ATF temperature sensor signal voltage to the PCM is above the specification, indicating that the temperature is under the specification (open), a malfunction is detected. The ATF temperature sensor output voltage is 4.93 V or more for at least 10 seconds. **NOTE: This code is caused by an electrical circuit problem and cannot be caused by a mechanical problem in the transmission.**
DTC: P0714 **2T PCM, TCIL: Yes** **Year:** 2012 **Model:** TL **Engine:** 3.5L V6	**ATF Temperature Sensor Circuit Intermittent Failure:** The temperature in the ATF temperature sensor deviates from the temperature in ECT sensor 2, less than -43 ºF (-24 ºC) or greater than 57 ºF (32 ºC). **NOTE: Greater than 108 ºF (60 ºC) if engine block heater is equipped.**
DTC: P0716 **1T PCM, MIL: Yes, TCIL: Yes** **Year:** 2011, 2012 **Model:** MDX, RDX, RL, TL, TSX, ZDX **Engine:** 2.3L L4, 2.4L L4, 3.5L V6, 3.7L V6	**Problem in Input Shaft (Main shaft) Speed Sensor Circuit:** If no pulses occur with the input shaft (main shaft) rotating, the PCM detects a malfunction that may be caused by an open, a temporary open, or a short to ground. The vehicle speed measured by the input shaft (main shaft) speed sensor/(divided by) the vehicle speed measured by the output shaft (countershaft) speed sensor is less than 0.156 for at least 10 seconds. **NOTE: This code is caused by an electrical circuit problem and cannot be caused by a mechanical problem in the transmission.**
DTC: P0717 **1T PCM, MIL: Yes, TCIL: Yes** **Year:** 2011, 2012 **Model:** MDX, RDX, RL, TL, TSX, ZDX **Engine:** 2.3L L4, 2.4L L4, 3.5L V6, 3.7L V6	**Problem in Input Shaft (Main shaft) Speed Sensor Circuit (No Signal Input):** If no pulses occur with the input shaft (main shaft) rotating, the PCM detects a malfunction that may be caused by an open, a temporary open, or a short to ground. When the vehicle speed measured by the output shaft (countershaft) speed sensor is 13 mph (20 km/h) or more, the vehicle speed measured by the input shaft (main shaft) speed sensor is 1 mph (2 km/h) or less for at least 10 seconds. **NOTE: This code is caused by an electrical circuit problem and cannot be caused by a mechanical problem in the transmission.**
DTC: P0718 **2T PCM, MIL: Yes, TCIL: Yes** **Year:** 2011, 2012 **Model:** MDX, RDX, RL, TL, TSX, ZDX **Engine:** 2.3L L4, 2.4L L4, 3.5L V6, 3.7L V6	**Input Shaft (Main shaft) Speed Sensor Intermittent Failure:** If no pulses occur with the input shaft (main shaft) rotating, the PCM detects a malfunction that may be caused by an open, a temporary open, or a short to ground. The fluctuation of the vehicle speed measured by the input shaft (main shaft) speed sensor in 10 milliseconds is 4 mph (6km/h) or more, and it fluctuates at least six times within 500 milliseconds. **NOTE: This code is caused by an electrical circuit problem and cannot be caused by a mechanical problem in the transmission.**
DTC: P0720 **2T PCM, MIL: Yes** **Year:** 2011, 2012 **Model:** RDX, TSX **Engine:** 2.3L L4, 2.4L L4	**Output Shaft (Countershaft) Speed Sensor Circuit Malfunction :** With the engine speed at a minimum of 4,000 rpm and during fuel cut-off operation for deceleration. NO signal from the output shaft (Countershaft) speed sensor is detected for at least 5 seconds. **NOTE: Before you troubleshoot, record all freeze data and any on-board snapshot, and review the general troubleshooting information.**
DTC: P0721 **1T PCM, MIL: Yes, TCIL: Yes** **Year:** 2011, 2012 **Model:** MDX, RDX, RL, TL, ZDX **Engine:** 2.3L L4, 3.5L V6, 3.7L V6	**Problem in Output Shaft (Countershaft) Speed Sensor Circuit:** If pulse dropouts occur with the output shaft (countershaft) rotating, the PCM detects a malfunction that may be caused by an open, a temporary open, or a short to ground. The vehicle speed measured by the input shaft (main shaft) speed sensor/(divided by) the vehicle speed measured by the output shaft (countershaft) speed sensor is greater than 6.0 for at least 10 seconds. **NOTE: This code is caused by an electrical circuit problem and cannot be caused by a mechanical problem in the transmission.**
DTC: P0722 **1T PCM, MIL: Yes, TCIL: Yes** **Year:** 2011, 2012 **Model:** MDX, RDX, RL, TL, TSX, ZDX **Engine:** 2.3L L4, 2.4L L4, 3.5L V6, 3.7L V6	**Problem in Output Shaft (Countershaft) Speed Sensor Circuit (No Signal Input):** If pulse dropouts occur with the output shaft (countershaft) rotating, the PCM detects a malfunction that may be caused by an open, a temporary open, or a short to ground. When the vehicle speed measured by the input shaft (main shaft) speed sensor is 13 mph (20 km/h) or more, the vehicle speed measured by the output shaft (countershaft) speed sensor is 1 mph (2 km/h) or less for at least 10 seconds. **NOTE: This code is caused by an electrical circuit problem and cannot be caused by a mechanical problem in the transmission.**

DTC	Trouble Code Title and Conditions
DTC: P0723 **2T PCM, MIL: Yes, TCIL: Yes** **Year:** 2011, 2012 **Model:** MDX, RDX, RL, TL, TSX, ZDX **Engine:** 2.3L L4, 2.4L L4, 3.5L V6, 3.7L V6	**Output Shaft (Countershaft) Speed Sensor Intermittent Failure:** If pulse dropouts occur with the output shaft (countershaft) rotating, the PCM detects a malfunction that may be caused by an open, a temporary open, or a short to ground. Based on the fluctuation of the vehicle speed measured by the output shaft (countershaft) speed sensor, a malfunction is detected. The fluctuation of the vehicle speed measured by the output shaft (countershaft) speed sensor in 10 milliseconds is 4 mph (6km/h) or more, and it fluctuates at least six times within 500 milliseconds. **NOTE: This code is caused by an electrical circuit problem and cannot be caused by a mechanical problem in the transmission.**
DTC: P0729 **2T PCM, TCIL: Yes** **Year:** 2011, 2012 **Model:** TL, ZDX **Engine:** 3.5L V6, 3.7L V6	**Problem in 6th Clutch and 6th Clutch Hydraulic Circuit:** With the vehicle running in drive and engine speed at a minimum of 7 mph, battery voltage 11 volts. The actual gear ratio must match one of these conditions for at least 12 seconds with the 6th gear shift command: Actual gear ratio is greater than the 6th gear ratio by a factor of 1.25. Actual gear ratio is less than the 6th gear ratio by a factor of 0.8. **NOTE: Before you troubleshoot, record all freeze data and any on-board snapshots.**
DTC: P0731 **2T PCM, TCIL: Yes** **Year:** 2011, 2012 **Model:** MDX, RDX, RL, TL, TSX, ZDX **Engine:** 2.3L L4, 2.4L L4, 3.5L V6, 3.7L V6	**Problem in 1st Clutch and 1st Clutch Hydraulic Circuit (1st gear incorrect ratio):** The Powertrain Control Module (PCM) computes the ratio of the input shaft (main shaft) speed to the output shaft (countershaft) speed. When the ratio is not the 1st gear ratio, it is detected as a malfunction of the hydraulic circuit or the 1st clutch. (Symptom 1) The actual gear ratio must match one of these conditions for at least 12 seconds with the 1st gear command: * Actual gear ratio is greater than the 1st gear ratio by a factor of 1.2 * Actual gear ratio is less than the 1st gear ratio by a factor of 0.75 (Symptom 2) The actual gear position is neutral for at least 3 seconds and then the gear up-shifted from 2nd to 3rd, even though 1st gear shift is commanded. **NOTE: Before you troubleshoot, record all freeze data and any on-board. snapshot,**
DTC: P0732 **2T PCM, TCIL: Yes** **Year:** 2011, 2012 **Model:** MDX, RDX, RL, TL, TSX, ZDX **Engine:** 2.3L L4, 2.4L L4, 3.5L V6, 3.7L V6	**Problem in 2nd Clutch and 2nd Clutch Hydraulic Circuit (2nd gear incorrect ratio):** The Powertrain Control Module (PCM) computes the ratio of the input shaft (main shaft) speed to the output shaft (countershaft) speed. When the ratio is not the 2nd gear ratio, it is detected as a malfunction of the hydraulic circuit or the 2nd clutch. The actual gear ratio must match one of these conditions for at least 12 seconds with the 2nd gear command. * Actual gear ratio is greater than the 2nd gear ratio by a factor of 1.2. * Actual gear ratio is less than the 2nd gear ratio by a factor of 0.75. **NOTE: Before you troubleshoot, record all freeze data and any on-board snapshot.**
DTC: P0733 **2T PCM, TCIL: Yes** **Year:** 2011, 2012 **Model:** MDX, RDX, RL, TL, TSX, ZDX **Engine:** 2.3L L4, 2.4L L4, 3.5L V6, 3.7L V6	**Problem in 3rd Clutch and 3rd Clutch Hydraulic Circuit (3rd gear incorrect ratio):** The powertrain control module (PCM) computes the ratio of the input shaft (main shaft) speed to the output shaft (countershaft) speed. When the ratio is not the 3rd gear ratio, it is detected as a malfunction of the hydraulic circuit or the 3rd clutch. The actual gear ratio must match one of these conditions for at least 12 seconds with the 3rd gear command. * Actual gear ratio is greater than the 3rd gear ratio by a factor of 1.2. * Actual gear ratio is less than the 3rd gear ratio by a factor of 0.75. **NOTE: Before you troubleshoot, record all freeze data and any on-board snapshot.**
DTC: P0734 **2T PCM, TCIL: Yes** **Year:** 2011, 2012 **Model:** MDX, RDX, RL, TL, TSX, ZDX **Engine:** 2.3L L4, 2.4L L4, 3.5L V6, 3.7L V6	**Problem in 4th Clutch and 4th Clutch Hydraulic Circuit (4th gear incorrect ratio):** The Powertrain Control Module (PCM) computes the ratio of the input shaft (main shaft) speed to the output shaft (countershaft) speed. When the ratio is not the 4th gear ratio, it is detected as a malfunction of the hydraulic circuit or the 4th clutch. The actual gear ratio must match one of these conditions for at least 12 seconds with the 4th gear command. * Actual gear ratio is greater than the 4th gear ratio by a factor of 1.2. * Actual gear ratio is less than the 4th gear ratio by a factor of 0.75. **NOTE: Before you troubleshoot, record all freeze data and any on-board snapshot.**
DTC: P0735 **2T PCM, TCIL: Yes** **Year:** 2011, 2012 **Model:** MDX, RDX, RL, TL, TSX, ZDX **Engine:** 2.3L L4, 2.4L L4, 3.5L V6, 3.7L V6	**Problem in 5th Clutch and 5th Clutch Hydraulic Circuit (5th gear incorrect ratio):** The Powertrain Control Module (PCM) computes the ratio of the input shaft (main shaft) speed to the output shaft (countershaft) speed. When the ratio is not the 5th gear ratio, it is detected as a malfunction of the hydraulic circuit or the 5th clutch. The actual gear ratio must match one of these conditions for at least 12 seconds with the 5th gear command. * Actual gear ratio is greater than the 5th gear ratio by a factor of 1.2 * Actual gear ratio is less than the 5th gear ratio by a factor of 0.75 **NOTE: Before you troubleshoot, record all freeze data and any on-board snapshot**
DTC: P0741 **2T PCM, MIL: Yes** **Year:** 2011, 2012 **Model:** MDX, RDX, RL, TL, TSX, ZDX **Engine:** 2.3L L4, 2.4L L4, 3.5L V6, 3.7L V6	**Torque Converter Clutch Hydraulic Circuit Stuck OFF:** If the ratio of engine speed and input shaft (main shaft) speed is not about 1:1 while the PCM is issuing the command to turn shift solenoid valve D and A/T clutch pressure control solenoid valve C ON, the PCM detects a faulty lock-up control system. The ratio of the engine revolutions to the transmission input pulses does not reach about 100 % for at least 22 seconds. **NOTE: Before you troubleshoot, record all freeze data and any on-board snapshot.**

DTC	Trouble Code Title and Conditions
DTC: P0746 **2T PCM, MIL: Yes, TCIL: Yes** **Year:** 2011, 2012 **Model:** MDX, RDX, RL, TL, TSX, ZDX **Engine:** 2.3L L4, 3.5L V6, 3.7L V6	**A/T Clutch Pressure Control Solenoid Valve A Stuck OFF:** When an improper gear ratio is output compared to the predetermined gear ratio, an A/T clutch pressure control solenoid valve A OFF failure is detected. One of these symptoms occur: * Transmission is held in 1st gear. * The engine speed flares when upshifting to 2nd-3rd. * The engine speed flares when upshifting to 3rd-4th or 5th if applicable. **NOTE: Before you troubleshoot, record all freeze data and any on-board snapshot.**
DTC: P0747 **2T PCM, MIL: Yes, TCIL: Yes** **Year:** 2011, 2012 **Model:** MDX, RDX, RL, TL, TSX, ZDX **Engine:** 2.3L L4, 2.4L L4, 3.5L V6, 3.7L V6	**A/T Clutch Pressure Control Solenoid Valve A Stuck ON:** When an improper gear ratio is output compared to the predetermined gear ratio, an A/T clutch pressure control solenoid valve A ON failure is detected. The execution time is continuous and the duration time is 20 seconds. One of these conditions occur: * The transmission is held in 2nd gear against the 2nd-3rd gear upshift command as long as 20 seconds though there is no record of being neutral when the 1st gear shift is commanded. * The transmission is held in 4th gear against the 4th-5th gear upshift command as long as 20 seconds though there is no record of being neutral when the 1st gear shift is commanded. **NOTE: Before you troubleshoot, record all freeze data and any on-board snapshot.**
DTC: P0751 **2T PCM, MIL: Yes, TCIL: Yes** **Year:** 2011, 2012 **Model:** MDX, RDX, RL, TL, TSX, ZDX **Engine:** 2.3L L4, 2.4L L4, 3.5L V6, 3.7L V6	**Shift Solenoid Valve A Stuck OFF:** When an improper gear ratio is output compared to the predetermined gear change mode, a shift solenoid valve A OFF failure is detected and a DTC is stored. The execution time is continuous and the duration time is 2 seconds or more. The transmission is held in 5th gear against the 3rd gear command for at least 2 seconds. **NOTE: Before you troubleshoot, record all freeze data and any on-board snapshot.**
DTC: P0752 **2T PCM, MIL: Yes, TCIL: Yes** **Year:** 2011, 2012 **Model:** MDX, RDX, RL, TL, TSX, ZDX **Engine:** 2.3L L4, 2.4L L4, 3.5L V6, 3.7L V6	**Shift Solenoid Valve A Stuck ON:** When the wrong transmission fluid switch is turned on for a given speed change mode, a shift solenoid valve turn-on malfunction is detected. The execution time is continuous depending on the driving pattern. The 3rd clutch transmission fluid switch is ON against the 4th-5th gear upshift command for at least 11 seconds. **NOTE: Before you troubleshoot, record all freeze data and any on-board snapshot.**
DTC: P0756 **2T PCM, MIL: Yes, TCIL: Yes** **Year:** 2011, 2012 **Model:** MDX, RDX, RL, TL, TSX, ZDX **Engine:** 2.3L L4, 2.4L L4, 3.5L V6, 3.7L V6	**Shift Solenoid Valve B Stuck OFF:** When an improper gear ratio is output compared to the predetermined gear change mode, a shift solenoid valve B OFF failure is detected. The transmission is held in 4th gear against the 2nd gear command for at least 2 seconds. The execution time is continuous and the duration time is 2 seconds or more. **NOTE: Before you troubleshoot, record all freeze data and any on-board snapshot.**
DTC: P0757 **2T PCM, MIL: Yes, TCIL: Yes** **Year:** 2011, 2012 **Model:** MDX, RDX, RL, TL, TSX, ZDX **Engine:** 2.3L L4, 2.4L L4, 3.5L V6, 3.7L V6	**Shift Solenoid Valve B Stuck ON:** When the wrong gear ratio is output for a given speed change mode, or when the wrong transmission fluid pressure switch is turned-on, a shift solenoid valve turn-on malfunction is detected. The execution time is continuous depending on the driving pattern. One of these conditions occur: * The 2nd clutch transmission fluid switch is ON against the 3rd-4th gear upshift command for at least 11 seconds. * After the 3rd-4th gear upshift command is output, it is neutral for at least 2 seconds when the 5th gear shift command is output, though there is no history of being neutral. **NOTE: Before you troubleshoot, record all freeze data and any on-board snapshot.**
DTC: P0761 **2T PCM, MIL: Yes, TCIL: Yes** **Year:** 2011, 2012 **Model:** MDX, RDX, RL, TL, TSX, ZDX **Engine:** 2.3L L4, 2.4L L4, 3.5L V6, 3.7L V6	**Shift Solenoid Valve C Stuck OFF:** The execution time is continuous and the duration time is 20 seconds. When an improper gear ratio is output compared to the predetermined gear change mode, a shift solenoid valve C OFF failure is detected. One of these symptoms occurred when the actual gear position was neutral when the 1st gear shift is commanded: * The transmission is held in 2nd gear against the 2nd-3rd gear upshift command for at least 17 seconds. * The transmission is held in 4th gear against the 4th-5th gear upshift command for at least 17 seconds. **NOTE: Before you troubleshoot, record all freeze data and any on-board snapshot.**
DTC: P0762 **2T PCM, MIL: Yes, TCIL: Yes** **Year:** 2011, 2012 **Model:** MDX, RDX, RL, TL, ZDX **Engine:** 2.3L L4, 3.5L V6, 3.7L V6	**Shift Solenoid Valve C Stuck ON:** When an improper gear ratio is output compared to the predetermined gear change mode, a shift solenoid valve C ON failure is detected. The execution time is continuous and the duration time is 20 seconds. The transmission is held in 3rd gear against the 3rd-4th gear upshift command for as long as 20 seconds, without records that the gear change time was short when the 2nd-3rd gear upshift were commanded. **NOTE: Before you troubleshoot, record all freeze data and any on-board snapshot.**

DTC	Trouble Code Title and Conditions
DTC: P0766 **2T PCM, MIL: Yes, TCIL: Yes** **Year:** 2011, 2012 **Model:** MDX, RDX, RL, TL, TSX, ZDX **Engine:** 2.3L L4, 3.5L V6, 3.7L V6	**Shift Solenoid Valve D Stuck OFF:** When an improper gear ratio is output compared to the predetermined gear change mode, a shift solenoid valve D OFF failure is detected. **NOTE: Before you troubleshoot, record all freeze data and any on-board snapshot.**
DTC: P0767 **2T PCM, MIL: Yes, TCIL: Yes** **Year:** 2011, 2012 **Model:** MDX, RDX, RL, TSX, ZDX **Engine:** 2.3L L4, 3.5L V6, 3.7L V6	**Shift Solenoid Valve D Stuck ON:** When an improper gear ratio is output compared to the predetermined gear change mode, a shift solenoid valve D ON failure is detected. One of these conditions occur: * The actual gear position is neutral for at least 3 seconds when 1st gear in-gear is commanded, though there is no history of being neutral when reverse gear in-gear is commanded. * The actual gear position is neutral for at least 3 seconds, though 1st gear in-gear is commanded and reverse drive occurred during this driving cycle. **NOTE: Before you troubleshoot, record all freeze data and any on-board snapshot.**
DTC: P0771 **2T PCM, MIL: Yes, TCIL: Yes** **Year:** 2011, 2012 **Model:** TSX **Engine:** 2.4L L4	**Shift Solenoid Valve E Stuck OFF:** While driving the vehicle with the torque converter lock-up ON a maalfunction is detected when shifting from 3rd gear into 4th gear. **NOTE: Before you troubleshoot, record all freeze data and any on-board snapshot with the HDS, and review General Troubleshooting Information.**
DTC: P0776 **2T PCM, MIL: Yes, TCIL: Yes** **Year:** 2011, 2012 **Model:** MDX, RDX, RL, TL, TSX, ZDX **Engine:** 2.3L L4, 2.4L L4, 3.5L V6, 3.7L V6	**A/T Clutch Pressure Control Solenoid Valve B Stuck OFF:** When an improper gear ratio is output compared to the predetermined gear change mode, an A/T clutch pressure control solenoid valve B OFF failure is detected. The transmission is held in 3rd gear against the 3rd-4th gear upshift command for as long as 20 seconds, with records that the gear change time was short when the 2nd-3rd gear upshift was commanded. **NOTE: Before you troubleshoot, record all freeze data and any on-board snapshot.**
DTC: P0777 **2T PCM, MIL: Yes, TCIL: Yes** **Year:** 2011, 2012 **Model:** MDX, RDX, RL, TL, TSX, ZDX **Engine:** 2.3L L4, 2.4L L4, 3.5L V6, 3.7L V6	**A/T Clutch Pressure Control Solenoid Valve B Stuck ON:** When an improper gear ratio is output compared to the predetermined gear change mode, an A/T clutch pressure control solenoid valve B ON failure is detected. The engine speed flares during 2nd-3rd and 3rd-4th upshifts for at least 1 second. The execution time is continuous depending on the driving pattern. **NOTE: Before you troubleshoot, record all freeze data and any on-board snapshot.**
DTC: P0780 **2T PCM, MIL: Yes, TCIL: Yes** **Year:** 2011, 2012 **Model:** TSX **Engine:** 2.4L L4, 3.5L V6	**Shift Control System:** This code is stored whenever DTCs P1730, P1731, P1732, P1733, and P1734 are detected. Refer to specific DTC information. Before you troubleshoot, record all freeze data and any on-board snapshot with the HDS, and review General Troubleshooting Information.
DTC: P0796 **2T PCM, MIL: Yes, TCIL: Yes** **Year:** 2011, 2012 **Model:** MDX, RDX, RL, TL, TSX, ZDX **Engine:** 2.3L L4, 2.4L L4, 3.5L V6, 3.7L V6	**A/T Clutch Pressure Control Solenoid Valve C Stuck OFF:** When an improper gear ratio is output compared to the predetermined gear ratio, an A/T clutch pressure control solenoid valve C OFF failure is detected. The execution time is continuous depending on the driving pattern. **NOTE: Before you troubleshoot, record all freeze data and any on-board snapshot.**
DTC: P0797 **2T PCM, MIL: Yes, TCIL: Yes** **Year:** 2011, 2012 **Model:** MDX, RDX, RL, TL, TSX, ZDX **Engine:** 2.3L L4, 2.4L L4, 3.5L V6, 3.7L V6	**A/T Clutch Pressure Control Solenoid Valve C Stuck ON:** When the wrong transmission fluid pressure switch is turned on, an A/T clutch pressure control solenoid valve C turn-on malfunction is detected. The 2nd clutch transmission fluid switch is ON against the 3rd-4th gear upshift command for at least 11 seconds. The execution time is continuous and the duration time is 20 seconds. **NOTE: Before you troubleshoot, record all freeze data and any on-board snapshot,.**
DTC: P0812 **2T PCM, MIL: Yes, TCIL: Yes** **Year:** 2011, 2012 **Model:** MDX, RDX, RL, TL, TSX, ZDX **Engine:** 2.3L L4, 2.4L L4, 3.5L V6, 3.7L V6	**Open in Transmission Range Switch ATP RVS Switch Circuit:** If the R switch is OPEN with the shift lever in R, the PCM detects a switch OPEN failure. The RVS signal is detected but the R switch signal is not detected for at least 2 seconds. The execution time is continuous depending on the driving pattern. **NOTE: This code is caused by an electrical circuit problem and cannot be caused by a mechanical problem in the transmission.**

DTC	Trouble Code Title and Conditions
DTC: P0815 **1T PCM, MIL: Yes, TCIL: Yes** **Year:** 2011, 2012 **Model:** MDX, ZDX **Engine:** 3.7L V6	**Short in Transmission Gear Selection Switch Upshift Switch Circuit, or Transmission Gear Selection Switch Upshift Switch Stuck ON :** The S-UP switch signal is turned on in P, R, N, and D3 for at least 10 seconds. The execution time is continuous and the duration time is 10 seconds or more. **NOTE: This code is caused by an electrical circuit problem and cannot be caused by a mechanical problem in the transmission.**
DTC: P0816 **1T PCM, MIL: Yes, TCIL: Yes** **Year:** 2011, 2012 **Model:** MDX, ZDX **Engine:** 3.7L V6	**Short in Transmission Gear Selection Switch Downshift Switch Circuit, or Transmission Gear Selection Switch Downshift Switch Stuck ON :** The S-DN switch signal is turned on in P, R, N, and D3 for at least 10 seconds. DTCs P0705, P0706, P0815, P0957, P0958 are not active. The execution time is continuous and the duration time is 10 seconds or more. This code is caused by an electrical circuit problem and cannot be caused by a mechanical problem in the transmission.
DTC: P0842 **1T PCM, MIL: Yes, TCIL: Yes** **Year:** 2011, 2012 **Model:** MDX, RDX, RL, TL, TSX, ZDX **Engine:** 2.3L L4, 2.4L L4, 3.5L V6, 3.7L V6	**Short in 2nd Clutch Transmission Fluid Pressure Switch Circuit, or 2nd Clutch Transmission Fluid Pressure Switch Stuck ON:** The input signal from the 2nd clutch transmission fluid pressure switch to the PCM is low when driving in 1st gear, 3rd gear, or 5th gear. The execution time is continuous and the duration time is 2 seconds or more. **NOTE: This code is caused by an electrical circuit problem and cannot be caused by a mechanical problem in the transmission.**
DTC: P0843 **1T PCM, MIL: Yes, TCIL: Yes** **Year:** 2011, 2012 **Model:** MDX, RDX, RL, TL, TSX, ZDX **Engine:** 2.3L L4, 2.4L L4, 3.5L V6, 3.7L V6	**Open in 2nd Clutch Transmission Fluid Pressure Switch Circuit, or 2nd Clutch Transmission Fluid Pressure Switch Stuck OFF :** The input signal from the 2nd clutch transmission fluid pressure switch to the PCM is high when driving in 2nd gear. The execution time is continuous and the duration time is 2 seconds or more. **NOTE: Before you troubleshoot, record all freeze data and any on-board snapshot.**
DTC: P0847 **1T PCM, MIL: Yes, TCIL: Yes** **Year:** 2011, 2012 **Model:** MDX, RDX, RL, TL, TSX, ZDX **Engine:** 2.3L L4, 2.4L L4, 3.5L V6, 3.7L V6	**Short in 3rd Clutch Transmission Fluid Pressure Switch Circuit, or 3rd Clutch Transmission Fluid Pressure Switch Stuck ON:** The input signal from the 3rd clutch transmission fluid pressure switch to the PCM is low when driving in 1st gear, 2nd gear, 4th gear, or 5th gear. The execution time is continuous and the duration time is 2 seconds or more. **NOTE: This code is caused by an electrical circuit problem and cannot be caused by a mechanical problem in the transmission.**
DTC: P0848 **1T PCM, MIL: Yes, TCIL: Yes** **Year:** 2011, 2012 **Model:** MDX, RDX, RL, TL, TSX, ZDX **Engine:** 2.3L L4, 2.4L L4, 3.5L V6, 3.7L V6	**Open in 3rd Clutch Transmission Fluid Pressure Switch Circuit, or 3rd Clutch Transmission Fluid Pressure Switch Stuck OFF :** The input signal from the 3rd clutch transmission fluid pressure switch to the PCM is high when driving in 3rd gear. The execution time is continuous and the duration time is 2 seconds or more. **NOTE: Before you troubleshoot, record all freeze data and any on-board snapshot.**
DTC: P084C **2T PCM, MIL: Yes, TCIL: Yes** **Year:** 2011, 2012 **Model:** ZDX **Engine:** 3.7L V6	**Short in Line Pressure Switch Circuit, or Line Pressure Switch Stuck ON:** One of these 2 symptoms occurs: (1) Symptom 1: The input signal from the line pressure switch to the PCM is low during the self shut down mode after the signal is low when the normal line pressure mode was commanded. (2) Symptom 2: The input signal from the line pressure switch to the PCM is low. **NOTE: Before you troubleshoot, record all freeze data and any on-board snapshot**
DTC: P0872 **1T PCM, MIL: Yes, TCIL: Yes** **Year:** 2011, 2012 **Model:** MDX, RDX, RL, TL, TSX, ZDX **Engine:** 2.3L L4, 2.4L L4, 3.5L V6, 3.7L V6	**Short in 4th Clutch Transmission Fluid Pressure Switch Circuit, or 4th Clutch Transmission Fluid Pressure Switch Stuck ON :** The input signal from the 4th clutch transmission fluid pressure switch to the PCM is low when driving in 5th gear. The execution time is continuous and the duration time is 2 seconds or more. **NOTE: This code is caused by an electrical circuit problem and cannot be caused by a mechanical problem in the transmission.**
DTC: P0873 **1T PCM, MIL: Yes, TCIL: Yes** **Year:** 2011, 2012 **Model:** MDX, RDX, RL, TL, TSX, ZDX **Engine:** 2.3L L4, 2.4L L4, 3.5L V6, 3.7L V6	**Open in 4th Clutch Transmission Fluid Pressure Switch Circuit, or 4th Clutch Transmission Fluid Pressure Switch Stuck OFF:** The input signal from the 4th clutch transmission fluid pressure switch to the PCM is high when driving in 4th gear. The execution time is continuous and the duration time is 2 seconds or more. **NOTE: Before you troubleshoot, record all freeze data and any on-board snapshot.**

DTC	Trouble Code Title and Conditions
DTC: P0877 **2T PCM, MIL: Yes, TCIL: Yes** **Year:** 2011, 2012 **Model:** TL, ZDX **Engine:** 3.5L V6, 3.7L V6	**Short in Transmission Fluid Pressure Switch D (5th Clutch) Circuit, or Transmission Fluid Pressure Switch D (5th Clutch) Stuck ON:** One of these 3 conditions occurs: (1) The input signal from transmission fluid pressure switch D (5th clutch) to the PCM is low for at least 5 seconds when the shift lever is in P, R, or N. (2) The input signal from transmission fluid pressure switch D (5th clutch) to the PCM is low for at least 5 seconds when driving in 1st or 2nd gear. (3) The input signal from transmission fluid pressure switch D (5th clutch) to the PCM is low for at least 5 seconds when the 2nd gear shift is commanded or when the shift lever is in P, R, or N, after the signal is low for at least 3 seconds when driving in 3rd, 4th, or 6th gear. **NOTE: Before you troubleshoot, record all freeze data and any on-board snapshot. This code is caused by an electrical circuit problem and cannot be caused by a mechanical problem in the transmission.**
DTC: P0878 **2T PCM, MIL: Yes, TCIL: Yes** **Year:** 2011, 2012 **Model:** TL, ZDX **Engine:** 3.5L V6, 3.7L V6	**Open in Transmission Fluid Pressure Switch D (5th Clutch) Circuit, or Transmission Fluid Pressure Switch D (5th Clutch) Stuck OFF:** The input signal from transmission fluid pressure switch D (5th clutch) to the PCM is high when driving in 5th gear. The execution time is continuous and the duration time is 4 seconds or more. **NOTE: Before you troubleshoot, record all freeze data and any on-board snapshot. If DTC P0878 is stored in the PCM, the transmission does not shift to any gear other than 2nd through 5th gear because of the fail-safe function**
DTC: P0957 **1T PCM, MIL: Yes, TCIL: Yes** **Year:** 2011, 2012 **Model:** MDX, ZDX **Engine:** 3.7L V6	**Short in Transmission Gear Selection Switch Circuit, or Transmission Gear Selection Switch Stuck ON :** The S-MODE switch signal is turned on in P, R, N, and D3 for at least 10 seconds. The execution time is continuous and the duration time is 10 seconds or more. DTCs P0705, P0706, P0815, P0816, P0958 are not active. **NOTE: This code is caused by an electrical circuit problem and cannot be caused by a mechanical problem in the transmission.**
DTC: P0958 **1T PCM, MIL: Yes, TCIL: Yes** **Year:** 2011, 2012 **Model:** MDX, ZDX **Engine:** 3.7L V6	**Open in Transmission Gear Selection Switch Circuit, or Transmission Gear Selection Switch Stuck OFF:** The S-UP switch inputs or S-DN switch inputs are counted at least two times when the S-MODE switch signal is turned off in D. The execution time is continuous depending on the driving pattern. **NOTE: This code is caused by an electrical circuit problem and cannot be caused by a mechanical problem in the transmission.**
DTC: P0962 **1T PCM, MIL: Yes, TCIL: Yes** **Year:** 2011, 2012 **Model:** MDX, RDX, RL, TL, TSX, ZDX **Engine:** 2.3L L4, 2.4L L4, 3.5L V6, 3.7L V6	**Problem in A/T Clutch Pressure Control Solenoid Valve A Circuit:** If the measured current for the PCM output duty cycle is not within a specified range (open or short), a malfunctionis detected. The execution time is continuous and the duration time is 2 seconds or more. The measured current for the PCM command value is Duty 57-89 %, current less than 0.2, low input. **NOTE: This code is caused by an electrical circuit problem and cannot be caused by a mechanical problem in the transmission.**
DTC: P0963 **1T PCM, MIL: Yes, TCIL: Yes** **Year:** 2011, 2012 **Model:** MDX, RDX, RL, TL, TSX, ZDX **Engine:** 2.3L L4, 2.4L L4, 3.5L V6, 3.7L V6	**Problem in A/T Clutch Pressure Control Solenoid Valve A:** Duty cycle is less than 13-27 %, current (A) 0.6-0.9, high input failure. The execution time is continuous and the duration time is 2 seconds or more. DTCs P0962, P0966, P0967, P0970, P0971 are not active. **Note: This code is caused by an electrical circuit problem and cannot be caused by a mechanical problem in the transmission.**
DTC: P0966 **1T PCM, MIL: Yes, TCIL: Yes** **Year:** 2011, 2012 **Model:** MDX, RDX, RL, TL, TSX, ZDX **Engine:** 2.3L L4, 2.4L L4, 3.5L V6, 3.7L V6	**Problem in A/T Clutch Pressure Control Solenoid Valve B Circuit:** The measured current is less than 0.2, duty cycle 57-89 %, low input. The execution time is continuous and the duration time is 1 seconds or more. **Note: This code is caused by an electrical circuit problem and cannot be caused by a mechanical problem in the transmission.**
DTC: P0967 **1T PCM, MIL: Yes, TCIL: Yes** **Year:** 2011, 2012 **Model:** MDX, RDX, RL, TL, TSX, ZDX **Engine:** 2.3L L4, 2.4L L4, 3.5L V6, 3.7L V6	**Problem in A/T Clutch Pressure Control Solenoid Valve B:** The measured current for the PCMs command is 0.6-0.9, duty cycle 13-27%, high input failure. Engine running, DTCs P0962, P0963, P0966, P0970, P0971 are not active. The execution time is continuous and the duration time is 1 seconds or more. **NOTE: This code is caused by an electrical circuit problem and cannot be caused by a mechanical problem in the transmission.**

DTC	Trouble Code Title and Conditions
DTC: P0970 **1T PCM, MIL: Yes, TCIL: Yes** **Year:** 2011, 2012 **Model:** MDX, RDX, RL, TL, TSX, ZDX **Engine:** 2.3L L4, 2.4L L4, 3.5L V6, 3.7L V6	**Problem in A/T Clutch Pressure Control Solenoid Valve C Circuit:** The measured current for the PCMs command value is 0.2-0.4, duty cycle 57-89%, low imput failure. Engine is running with battery voltage at 11 volts, DTCs P0962, P0963, P0966, P0967, P0971 are not active. The execution time is continuous and the duration time is 1 seconds or more. **NOTE: This code is caused by an electrical circuit problem and cannot be caused by a mechanical problem in the transmission.**
DTC: P0971 **1T PCM, MIL: Yes, TCIL: Yes** **Year:** 2011, 2012 **Model:** MDX, RDX, RL, TL, TSX, ZDX **Engine:** 2.3L L4, 2.4L L4, 3.5L V6, 3.7L V6	**Problem in A/T Clutch Pressure Control Solenoid Valve C:** The measured current for the PCMs command value is 0.6-0.9, duty cycle 13-27, high imput failure. The execution time is continuous and the duration time is 1 seconds or more. **NOTE: This code is caused by an electrical circuit problem and cannot be caused by a mechanical problem in the transmission.**
DTC: P0973 **1T PCM, MIL: Yes, TCIL: Yes** **Year:** 2011, 2012 **Model:** MDX, RDX, RL, TL, TSX, ZDX **Engine:** 2.3L L4, 2.4L L4, 3.5L V6, 3.7L V6	**Short in Shift Solenoid Valve A Circuit:** The return signal does not match the command to turn ON shift solenoid valve A for at least 1 second. The execution time is continuous and the duration time is 2 seconds or more. DTCs P0974, P0982, P0983 are not active. **NOTE: This code is caused by an electrical circuit problem and cannot be caused by a mechanical problem in the transmission.**
DTC: P0974 **1T PCM, MIL: Yes, TCIL: Yes** **Year:** 2011, 2012 **Model:** MDX, RDX, RL, TL, TSX, ZDX **Engine:** 2.3L L4, 2.4L L4, 3.5L V6, 3.7L V6	**Open in Shift Solenoid Valve A Circuit:** The return signal does not match the command to turn OFF shift solenoid valve A for at least 1 second. The execution time is continuous and the duration time is 1 seconds or more. DTCs P0973, P0982, P0983 are not active. **NOTE: This code is caused by an electrical circuit problem and cannot be caused by a mechanical problem in the transmission.**
DTC: P0976 **1T PCM, MIL: Yes, TCIL: Yes** **Year:** 2011, 2012 **Model:** MDX, RDX, RL, TL, TSX, ZDX **Engine:** 2.3L L4, 2.4L L4, 3.5L V6, 3.7L V6	**Short in Shift Solenoid Valve B Circuit:** The return signal does not match the command to turn ON shift solenoid valve B for at least 1 second. The execution time is continuous and the duration time is 2 seconds or more. DTCs P0977, P0979, P0980 are not active. **NOTE: This code is caused by an electrical circuit problem and cannot be caused by a mechanical problem in the transmission.**
DTC: P0977 **1T PCM, MIL: Yes, TCIL: Yes** **Year:** 2011, 2012 **Model:** MDX, RDX, RL, TL, TSX, ZDX **Engine:** 2.3L L4, 2.4L L4, 3.5L V6, 3.7L V6	**Open in Shift Solenoid Valve B Circuit:** The return signal does not match the command to turn OFF shift solenoid valve B for at least 1 second. The execution time is continuous and the duration time is 1 seconds or more. Battery voltage minimum 11 volts, and DTCs P0976, P0979, P0980 are not active. **NOTE: This code is caused by an electrical circuit problem and cannot be caused by a mechanical problem in the transmission.**
DTC: P0979 **1T PCM, MIL: Yes, TCIL: Yes** **Year:** 2011, 2012 **Model:** MDX, RDX, RL, TL, TSX, ZDX **Engine:** 2.3L L4, 2.4L L4, 3.5L V6, 3.7L V6	**Short in Shift Solenoid Valve C Circuit:** The return signal does not match the command to turn ON shift solenoid valve C for at least 1 second. The execution time is continuous and the duration time is 1 seconds or more. Battery voltage is 11 volts and DTCs P0976, P0977, P0980 are not active. **NOTE: This code is caused by an electrical circuit problem and cannot be caused by a mechanical problem in the transmission.**
DTC: P0980 **1T PCM, MIL: Yes, TCIL: Yes** **Year:** 2011, 2012 **Model:** MDX, RDX, RL, TL, TSX, ZDX **Engine:** 2.3L L4, 2.4L L4, 3.5L V6, 3.7L V6	**Open in Shift Solenoid Valve C Circuit:** The return signal does not match the command to turn OFF shift solenoid valve C for at least 1 second. The execution time is continuous and the duration time is 1 seconds or more. Battery voltage is 11 V and DTCs P0976, P0977, P0979 are not active. **NOTE: This code is caused by an electrical circuit problem and cannot be caused by a mechanical problem in the transmission.**

DTC	Trouble Code Title and Conditions
DTC: P0982 **1T PCM, MIL: Yes, TCIL: Yes** **Year:** 2011, 2012 **Model:** MDX, RDX, RL, TL, TSX, ZDX **Engine:** 2.3L L4, 2.4L L4, 3.5L V6, 3.7L V6	**Short in Shift Solenoid Valve D Circuit:** The return signal does not match the command to turn ON shift solenoid valve D for at least 1 second. The execution time is continuous and the duration time is 1 seconds or more. Battery voltage is 11v, and DTCs P0973, P0974, P0983 are not active. **NOTE: This code is caused by an electrical circuit problem and cannot be caused by a mechanical problem in the transmission.**
DTC: P0983 **1T PCM, MIL: Yes, TCIL: Yes** **Year:** 2011, 2012 **Model:** MDX, RDX, RL, TL, TSX, ZDX **Engine:** 2.3L L4, 2.4L L4, 3.5L V6, 3.7L V6	**Open in Shift Solenoid Valve D Circuit:** The return signal does not match the command to turn OFF shift solenoid valve D for at least 1 second. The execution time is continuous and the duration time is seconds or more. Battery voltage is 11v and DTCs P0973, P0974, P0982 are not active. **NOTE: This code is caused by an electrical circuit problem and cannot be caused by a mechanical problem in the transmission.**
DTC: P0985 **1T PCM, MIL: Yes, TCIL: Yes** **Year:** 2011, 2012 **Model:** TSX **Engine:** 2.4L L4, 3.5L V6	**Short in Shift Solenoid Valve E Circuit:** With the engine started and in park the return signal does not match the command to turn ON shift solenoid valve E for at least 1 second. **NOTE: Before you troubleshoot, record all freeze data and any on-board snapshot with the HDS, and review General Troubleshooting Information. This code is caused by an electrical circuit problem and cannot be caused by a mechanical problem in the transmission.**
DTC: P0986 **1T PCM, MIL: Yes, TCIL: Yes** **Year:** 2011, 2012 **Model:** TSX **Engine:** 2.4L L4, 3.5L V6	**Open in Shift Solenoid Valve E Circuit:** With the engine running and driven in 2nd gear in the D position, the return signal does not match the command to turn OFF shift solenoid E for at least 1 second. **NOTE: Before you troubleshoot, record all freeze data and any on-board snapshot with the HDS, and review General Troubleshooting Information. This code is caused by an electrical circuit problem and cannot be caused by a mechanical problem in the transmission.**
DTC: P0989 **2T PCM, MIL: Yes, TCIL: Yes** **Year:** 2011, 2012 **Model:** TL, ZDX **Engine:** 3.5L V6, 3.7L V6	**Short in Transmission Fluid Pressure Switch E (6th Clutch) Circuit, or Transmission Fluid Pressure Switch E (6th Clutch) Stuck ON:** One of these 3 conditions occurs: (1) The input signal from transmission fluid pressure switch E (6th clutch) to the PCM is low for at least 4 seconds when the shift lever is in P, R, or N. (2) The input signal from transmission fluid pressure switch E (6th clutch) to the PCM is low for at least 4 seconds when driving in 1st, 2nd, or 3rd gear. (3) The input signal from transmission fluid pressure switch E (6th clutch) to the PCM is low for at least 4 seconds when the 3rd gear shift is commanded or when the shift lever is in P, R, or N, after the signal is low for at least 2 seconds when driving in 4th or 5th gear. **NOTE: Before you troubleshoot, record all freeze data and any on-board snapshot. This code is caused by an electrical circuit problem and cannot be caused by a mechanical problem in the transmission.**
DTC: P0990 **2T PCM, MIL: Yes, TCIL: Yes** **Year:** 2011, 2012 **Model:** TL, ZDX **Engine:** 3.5L V6, 3.7L V6	**Open in Transmission Fluid Pressure Switch E (6th Clutch) Circuit, or Transmission Fluid Pressure Switch E (6th Clutch) Stuck OFF:** The input signal from transmission fluid pressure switch E (6th clutch) to the PCM is high when driving in 6th gear. **NOTE: Before you troubleshoot, record all freeze data and any on-board snapshot. If DTC P0990 is stored in the PCM, the transmission does not shift to any gear other than 3rd or 6th gear because of the fail-safe function.**
DTC: P0A14 **1T PCM** **Year:** 2012 **Model:** TL **Engine:** 3.5L V6	**Front Engine Mount Actuator Circuit Malfunction:** Engine running, and DTCs P0A15, P0AB6. P0AB7, P16C4 and P16C6 are not active. The current valve during one cycle of the front engine mount actuator, which outputs in 1 TDC duration as one cycle, is 7.5 A or more as a maximum current or 5.3 A or more as a minimum current, or the engine mounts cycle during start up and continue to cycle for at least 20 seconds. **NOTE: When testing the engine mount actuator, be sure to use a VOM capable of reading 0.01 ohms change of resistance. If the vehicle has been driven, allow the engine mount actuator to cool down for about 2 hours before testing.**
DTC: P0A15 **1T PCM** **Year:** 2012 **Model:** TL **Engine:** 3.5L V6	**Front Engine Mount Actuator Control Circuit Low Current:** With the engine running and the DTCs P0A14, P0AB6, P0AB7, P15AB, P16C4 and P16C6 are not active. The duty cycle of the front engine mount actuator, which outputs in 1 TDC duration as one cycle, reads 100% all the time during engine the mount operation and continues at least 20 seconds. **NOTE: When testing the engine mount actuator, be sure to use a VOM capable of reading 0.01 ohms change of resistance. If the vehicle has been driven, allow the engine mount actuator to cool down for about 2 hours before testing.**
DTC: P0AB6 **1T PCM** **Year:** 2012 **Model:** TL **Engine:** 3.5L V6	**Rear Engine Mount Actuator Circuit Malfunction:** Engine running, and DTCs P0A15, P0AB6. P0AB7, P16C4 and P16C6 are not active. The current valve during one cycle of the front engine mount actuator, which outputs in 1 TDC duration as one cycle, is 7.5 A or more as a maximum current or 5.3 A or more as a minimum current, or the engine mounts cycle during start up and continue to cycle for at least 20 seconds. **NOTE: When testing the engine mount actuator, be sure to use a VOM capable of reading 0.01 ohms change of resistance. If the vehicle has been driven, allow the engine mount actuator to cool down for about 2 hours before testing.**

DTC	Trouble Code Title and Conditions
DTC: P0AB7 **1T PCM** **Year:** 2012 **Model:** TL **Engine:** 3.5L V6	**Rear Engine Mount Actuator Control Circuit Low Current:** With the engine running and the DTCs P0A14, P0AB6, P0AB7, P15AB, P16C4 and P16C6 are not active. The duty cycle of the rear engine mount actuator, which outputs in 1 TDC duration as one cycle, reads 100% all the time during engine the mount operation and continues at least 20 seconds. **NOTE: When testing the engine mount actuator, be sure to use a VOM capable of reading 0.01 ohms change of resistance. If the vehicle has been driven, allow the engine mount actuator to cool down for about 2 hours before testing.**
DTC: P0AB8 **1T PCM** **Year:** 2012 **Model:** TL **Engine:** 3.5L V6	**Rear Engine Mount Actuator Control Circuit High Current:** With the ignition switch ON and DTCs P15AF, P15B0, P15B1, P16C5, P16C9 not active. The minimum front ACM actuator current in one cycle is 6.6 A or more for at least 2.6 seconds, the front ACM actuator output cycle is one rotation of the crankshaft.

OBD II Trouble Code List (P1XXX Codes)

DTC	Trouble Code Title and Conditions
DTC: P1009 **1T PCM, MIL: Yes** **Year:** 2011, 2012 **Model:** RDX, TSX **Engine:** 2.3L L4, 2.4L L4, 3.5L V6	**Variable Valve Timing Control (VTC) Advance Malfunction :** With the engine idling at operating temperature, the camshaft phase value is not 24.0 degrees or less within the monitored area (camshaft phase control directed value plus failure judgment value) after 3 seconds have passed, or when the camshaft phase value is 5.0 degrees for M/T and 20.0 degrees for A/T or less and continues for 0.5 seconds or more, even when the VTC does not actuate. **NOTE: Before you troubleshoot, record all freeze data and any on-board snapshot, and review the general troubleshooting information. If DTC P0341 is stored at the same time as DTC P1009, troubleshoot DTC P1009 first, then recheck for DTC P0341.**
DTC: P1077 **2T PCM, MIL: Yes** **Year:** 2011, 2012 **Model:** MDX, ZDX **Engine:** 3.7L V6	**Intake Manifold Tuning (IMT) Valve Stuck in High RPM Position:** When the PCM sends a close (long runner) command, no long runner return signal is received for at least 7 seconds. Execution is once per drive cycle, engine speed, 3,600 rpm and intake air temperature a minimum of 5 ºF (-15 ºC). **NOTE: Before you troubleshoot, record all freeze data and any on-board snapshot.**
DTC: P1077 **2T PCM, MIL: Yes** **Year:** 2011, 2012 **Model:** RL, TL, TSX, ZDX **Engine:** 3.5L V6, 3.7L V6	**Intake Manifold Tuning (IMT) Valve Stuck in High RPM Position:** When the PCM sends a close (long runner) command, no long runner return signal is received for at least 7 seconds. The execution time is once per driving cycle. **NOTE: Before you troubleshoot, record all freeze data and any on-board snapshot.**
DTC: P1078 **2T PCM, MIL: Yes** **Year:** 2011, 2012 **Model:** MDX, TL, ZDX **Engine:** 3.5L V6, 3.7L V6	**Intake Manifold Tuning (IMT) Valve Stuck in Low RPM Position:** When the PCM sends an open (short runner) command, no short runner return signal is received for at least 3 seconds. Intake air temperature 5 ºF (-15 ºC), engine speed is 3,800 rpm and battery voltage minimum of 10.5v. DTCs P0107, P0108, P0112, P0113, P0117, P0118, P0563, P1128, P1129, P2227, P2228, P2229 are not active. **NOTE: Before you troubleshoot, record all freeze data and any on-board snapshot.**
DTC: P1078 **2T PCM, MIL: Yes** **Year:** 2011, 2012 **Model:** RL, TSX, ZDX **Engine:** 3.5L V6, 3.7L V6	**Intake Manifold Tuning (IMT) Valve Stuck in Low RPM Position:** When the PCM sends an open (short runner) command, no short runner return signal is received for at least 3 seconds. The execution time is once per driving cycle. **NOTE: Before you troubleshoot, record all freeze data and any on-board snapshot.**
DTC: P1109 **1T PCM, MIL: Yes** **Year:** 2011, 2012 **Model:** MDX, RDX, RL, TL, TSX, ZDX **Engine:** 2.3L L4, 2.4L L4, 3.5L V6, 3.7L V6	**Barometric Pressure (BARO) Sensor Circuit Out of Range High:** The BARO sensor output voltage is between 3.59 V to 4.49 V for at least 2 seconds. The execution time is continuous and DTCs P2228, P2229 are not active. **NOTE: Before you troubleshoot, record all freeze data and any on-board snapshot.**
DTC: P1116 **2T PCM, MIL: Yes** **Year:** 2011, 2012 **Model:** MDX, RDX, RL, TL, TSX, ZDX **Engine:** 2.3L L4, 2.4L L4, 3.5L V6, 3.7L V6	**Engine Coolant Temperature (ECT) Sensor 1 Circuit Range/Performance Problem:** A malfunction is detected if the following 3 conditions are present after the engine has stopped and the ignition switch has been turned to LOCK (0) for at least 6 hours before restarting the engine: (1) When the temperature difference between the IAT and ECT1 is 57 ºF (32 ºC) or more. (2) When the temperature difference between the IAT and ECT2 is 30 ºF (17 ºC) or more. (3) When the temperature difference between the ECT2 and ECT1 is 73 ºF (41 ºC) or more. **NOTE: If DTC P0111 is stored at the same time as DTC P1116, troubleshoot DTC P0111 first, then recheck for DTC P1116.**

DTC	Trouble Code Title and Conditions
DTC: P1128 **2T PCM, MIL: Yes** **Year:** 2011, 2012 **Model:** MDX, RDX, RL, TL, TSX, ZDX **Engine:** 2.3L L4, 2.4L L4, 3.5L V6, 3.7L V6	**Manifold Absolute Pressure (MAP) Sensor Signal Lower Than Expected:** The MAP sensor output is 33 kPa (9.8 in.Hg, 249 mmHg) or less for at least 2 seconds when atmospheric pressure is 52 kPa (15.4 in.Hg, 392 mmHg). The execution time is once per driving cycle. **NOTE: Before you troubleshoot, record all freeze data and any on-board snapshot.**
DTC: P1129 **2T PCM, MIL: Yes** **Year:** 2011, 2012 **Model:** MDX, RL, TL, TSX, ZDX **Engine:** 2.4L L4, 3.5L V6, 3.7L V6	**Manifold Absolute Pressure (MAP) Sensor Signal Higher Than Expected:** The MAP sensor output is 36 kPa (10.9 in.Hg, 277 mmHg) or more for at least 2 seconds. The execution time is once per driving cycle. **NOTE: Before you troubleshoot, record all freeze data and any on-board snapshot.**
DTC: P1129 **2T PCM, MIL: Yes** **Year:** 2011, 2012 **Model:** RDX **Engine:** 2.3L L4	**MAP Sensor Signal Higher Than Expected:** With the vehicle speed 1,750 or more and the throttle position is at 22º or more. The difference between the barometric pressure sensor measured during boost conditions and negative pressure (NO boost) is a specified output or less, an upward shift in the MAP sensor signal is detected. **NOTE: Before you troubleshoot, record all freeze data and any on-board snapshot, and review the general troubleshooting information.**
DTC: P1157 **1T PCM, MIL: Yes** **Year:** 2011, 2012 **Model:** RDX, TSX **Engine:** 2.3L L4, 2.4L L4, 3.5L V6	**Air Fuel Ratio (A/F) Sensor (Sensor 1) AFS Line High Voltage:** With the engine running and battery voltage between 10.5–16.0 V, the A/F sensor (SENSOR 1) heater element resistance is 250 ohms or more for at least 5 seconds. **NOTE: Before you troubleshoot, record all freeze data and any on-board snapshot, and review the general troubleshooting information.**
DTC: P1172 **1T PCM, MIL: Yes** **Year:** 2011, 2012 **Model:** RDX, TSX **Engine:** 2.3L L4, 2.4L L4	**Air/Fuel Ratio (A/F) Sensor (Sensor 1) Circuit Out of Range High:** A malfunction is detected when the rear A/F sensor (Sensor 1) output voltage is 4.9 V or more. The execution time is continuous and the duration time is 7 seconds or more. **NOTE: Before you troubleshoot, record all freeze data and any on-board snapshot.**
DTC: P1172 **2T PCM, MIL: Yes** **Year:** 2011, 2012 **Model:** MDX, RL, TL, TSX, ZDX **Engine:** 3.5L V6, 3.7L V6	**Rear Air/Fuel Ratio (A/F) Sensor (Bank 1, Sensor 1) Circuit Out of Range High:** A malfunction is detected when the rear A/F sensor (bank 1, sensor 1) output voltage is 4.7 V or more. The execution time is continuous and the duration time is 7 seconds or more. **NOTE: Before you troubleshoot, record all freeze data and any on-board snapshot.**
DTC: P1174 **2T PCM, MIL: Yes** **Year:** 2011, 2012 **Model:** MDX, RL, TL, ZDX **Engine:** 3.5L V6, 3.7L V6	**Front Air/Fuel Ratio (A/F) Sensor (Bank 2, Sensor 1) Circuit Out of Range High:** A malfunction is detected when the front A/F sensor (bank 2, sensor 1) output voltage is 4.7 V or more. The execution time is continuous and the duration time is 7 seconds or more. **NOTE: Before you troubleshoot, record all freeze data and any on-board snapshot.**
DTC: P1233 **2T PCM, MIL: Yes** **Year:** 2011, 2012 **Model:** RDX **Engine:** 2.3L L4	**Turbocharger Boost Sensor/BARO Sensor Incorrect Correlation:** Three seconds after start-up with the engine speed at a minimum of 550 rpm at idle. The difference between the BARO sensor output and the turbocharger boost sensor output is 26 kPa (7.8 in.Hg, 200 mmHG) or more for at least 5 seconds. **NOTE: Before you troubleshoot, record all freeze data and any on-board snapshot, and review the general troubleshooting information.**
DTC: P1234 **2T PCM, MIL: Yes** **Year:** 2011, 2012 **Model:** RDX **Engine:** 2.3L L4	**Turbocharger Boost Sensor/MAP Sensor Incorrect Correlation:** Engine speed at a minimum of 1,750 rpm, throttle position 8–22 º. The Map sensor output: LOW, or the turbocharger boost sensor: HIGH. The Map sensor output: HIGH, or the turbocharger boost sensor: LOW. **NOTE: Before you troubleshoot, record all freeze data and any on-board snapshot, and review the general troubleshooting information.**
DTC: P1243 **T BCM** **Year:** 2011, 2012 **Model:** RL, TL **Engine:** 3.5L V6, 3.7L V6	**Problem in the Passenger's Mode Control Linkage, Doors, or Motor Circuit:** Passenger's Mode Control Malfunction detected.

DTC	Trouble Code Title and Conditions
DTC: P1297 **2T PCM** **Year:** 2011, 2012 **Model:** MDX, RDX, RL, TL, TSX, ZDX **Engine:** 2.3L L4, 2.4L L4, 3.5L V6, 3.7L V6	**Electrical Load Detector (ELD) Circuit Low Voltage:** The ELD output voltage is 0.27 V or less for at least 5 seconds. The execution time is continuous and DTC P1298 is not active. **NOTE: Before you troubleshoot, record all freeze data and any on-board snapshot.**
DTC: P1298 **2T PCM** **Year:** 2011, 2012 **Model:** MDX, RDX, RL, TL, TSX, ZDX **Engine:** 2.3L L4, 2.4L L4, 3.5L V6, 3.7L V6	**Electrical Load Detector (ELD) Circuit High Voltage:** With the ignition switch ON (II) the ELD output voltage is 4.57 V or more for at least 5 seconds. The execution time is continuous and DTC P1297 is not active. **NOTE: Before you troubleshoot, record all freeze data and any on-board snapshot.**
DTC: P1454 **2T PCM, MIL: Yes** **Year:** 2011, 2012 **Model:** MDX, RDX, RL, TL, TSX, ZDX **Engine:** 2.3L L4, 2.4L L4, 3.5L V6, 3.7L V6	**Fuel Tank Pressure (FTP) Sensor Circuit Range/Performance Problem:** One of these 2 conditions is met. (1) The FTP sensor output fluctuates by 0.6 kPa (0.1 in.Hg, 5 mmHg) or more, or -0.6 kPa (-0.1 in.Hg, -5 mmHg) or less for at least 3 seconds. (2) The FTP sensor output value is -1.3 kPa (-0.3 in.Hg, -10 mmHg) or less for at least 3 seconds. DTCs P0452, P0453 are judged as OK. **NOTE: Before you troubleshoot, record all freeze data and any on-board snapshot.**
DTC: P1458 **2T PCM, MIL: Yes** **Year:** 2012 **Model:** TL **Engine:** 3.5L V6	**Fuel Tank Pressure (FTP) Sensor Circuit Range/Performance Problem:** Malfunction Threshold: The misalignment of zero point pressure inside the fuel tank is 0.6 kPa (0.1 inHg, 5 mmHg) or more. The output from the FTP sensor is flickering 3.1 seconds or more.
DTC: P145A **2T PCM, MIL: Yes** **Year:** 2011, 2012 **Model:** RDX **Engine:** 2.3L L4	**EVAP System Incorrect Purge Flow Detected:** Engine coolant temperature 140 ºF (60 ºC) or more. Map value 20 kPa (6.0 in.Hg, 150 mm). Charging pressure 14 kPa (4.0 in.Hg, 100 mmHg). Battery voltage a minimum of 10.5 V. Evap Canister purge valve duty 15–90%. Fuel trim 0.69–1.47. Vehicle in closed loop. When the P145D is determined as failure and when P0496 is determined as normal, abnormal boost area purge flow is judged. **NOTE: Perform the EVAP function test in the inspection menu with the HDS.**
DTC: P145B **2T PCM, MIL: Yes** **Year:** 2011, 2012 **Model:** RDX **Engine:** 2.3L L4	**EVAP System Purge Line Non-return Valve A Stuck Open:** Engine coolant temperature 140 ºF (60 ºC) or more. Map value 20 kPa (6.0 in.Hg, 150 mm). Charging pressure 14 kPa (4.0 in.Hg, 100 mmHg). Battery voltage a minimum of 10.5 V. Evap Canister purge valve duty 15–90%. Fuel trim 0.69–1.47. Vehicle in closed loop. The fuel tank pressure pulse phase misalignment duration is 20 milliseconds or more for at least 13 seconds. **NOTE: Perform the EVAP function test in the inspection menu with the HDS.**
DTC: P145C **2T PCM, MIL: Yes** **Year:** 2011, 2012 **Model:** MDX, RDX, RL, TL, TSX, ZDX **Engine:** 2.3L L4, 2.4L L4, 3.5L V6, 3.7L V6	**Evaporative Emission (EVAP) System Purge Flow Malfunction:** The pulses detected by the fuel tank pressure sensor are 1.0 % of the duty cycle or less for at least 31 seconds. The execution time is continuous. **NOTE: This DTC is representative of an EVAP system purge flow problem. If DTC P145C is indicated alone, troubleshoot P0496 and P0497 using the freeze data for P145C. If any of the DTCs listed below are indicated at the same time as DTC P145C, troubleshoot those DTC first, then recheck for P145C:**
DTC: P1549 **1T PCM** **Year:** 2011, 2012 **Model:** MDX, RDX, RL, TL, TSX, ZDX **Engine:** 2.3L L4, 2.4L L4, 3.5L V6, 3.7L V6	**Charging System High Voltage:** The IGP terminal voltage is 16.0 volts or more for at least 60 seconds. The execution time is continuous. **NOTE: If a high voltage battery (24 V, etc.) is connected to the vehicle, this DTC can be stored.**
DTC: P15AB **1T PCM** **Year:** 2012 **Model:** TL **Engine:** 3.5L V6	**Engine Mount Control Unit Power Source Circuit Low Voltage:** With the ignition on (II), the IG1 terminal voltage is 4.0 V or less for at least 2.6 seconds. **NOTE: Before you troubleshoot, record all freeze data and any on-board snapshots, and review the general troubleshooting information.**

DTC	Trouble Code Title and Conditions
DTC: P15AE **1T PCM** **Year:** 2012 **Model:** TL **Engine:** 3.5L V6	**Cylinder Pause Signal 1 Malfunction:** With the engine running, the SCPA signal input value differs from the PCM output value (Received by CAN communication) for at least 2.6 seconds. **NOTE: Select the VTEC TEST in the PGM-FI INSPECTION menu, and do the VPS TEST of the 3 CYLINDER ACTIVATION TEST with the HDS.**
DTC: P15B0 **1T PCM** **Year:** 2012 **Model:** TL **Engine:** 3.5L V6	**Crankshaft Position (CKP) Sensor Signal Malfunction:** With the engine running at idle, the CKP signal is not input to the engine mount control unit for at least 20 seconds. **NOTE: If PGM-FI system's DTC P0335 and/or P0339 is stored at the same time as DTC P15B0 troubleshoot P0335 and/or P0339 first, then recheck for P15B0.**
DTC: P15BD **1T PCM** **Year:** 2012 **Model:** TL **Engine:** 3.5L V6	**Cylinder Pause Signal 2 Malfunction:** With the engine running, the SCPA signal input value differs from the PCM output value (Received by CAN communication) for at least 2.6 seconds. **NOTE: Select the VTEC TEST in the PGM-FI INSPECTION menu, and do the VPS TEST.**
DTC: P15BE **1T PCM** **Year:** 2012 **Model:** TL **Engine:** 3.5L V6	**Camshaft Position (CMP) Sensor Signal Malfunction:** With the engine running at idle, the CMP signal is not input to the engine mount control unit for at least 20 seconds. **NOTE: If PGM-FI system's DTC P0365 and/or P0369 is stored at the same time as DTC P15BE, troubleshoot P0365 and/or P0369 first, then recheck for P15BE.**
DTC: P15BF **1T PCM** **Year:** 2012 **Model:** TL **Engine:** 3.5L V6	**Camshaft Position (CMP) Sensor Signal Intermittent Interruption:** With the engine running at idle, other than 10 CMP pulses are detected during two revolutions of the crankshaft. This condition has been detected for at least 127 times. **NOTE: If PGM-FI system's DTC P0365 and/or P0369 is stored at the same time as DTC P15BF, troubleshoot P0365 and/or P0369 first, then recheck for P15BF.**
DTC: P15C0 **1T PCM** **Year:** 2012 **Model:** TL **Engine:** 3.5L V6	**Crankshaft Position (CKP) Sensor Signal Intermittent Interruption:** With the engine running at idle, The number of CKP pulse is not 58 during intervals between reference pulses for each crank revolution. OR. No interval is detected while the CKP sensor pulse inputs 82.
DTC: P1648 **1T BCM** **Year:** 2011, 2012 **Model:** ZDX **Engine:** 3.7L V6	**Passenger's Door LF Antenna Circuit Open or Short:** Passenger's Door LF Antenna Circuit Open or Short condition present.
DTC: P1658 **1T PCM, MIL: Yes** **Year:** 2011, 2012 **Model:** MDX, RDX, TL, ZDX **Engine:** 2.3L L4, 3.5L V6, 3.7L V6	**Electronic Throttle Control System (ETCS) Control Relay ON Malfunction:** The communication signal is input from the throttle actuator control module for at least 2 seconds, after the throttle actuator control module relay is turned off. The execution time is once per driving cycle. DTCs P0122, P0123, P0222, P0223, P1659, P2101, P2118, P2122, P2123, P2127, P2128, P2135, P2138, P2176 are not active. **NOTE: Before you troubleshoot, record all freeze data and any on-board snapshot.**
DTC: P1659 **1T PCM, MIL: Yes** **Year:** 2011, 2012 **Model:** MDX, RDX, TL, ZDX **Engine:** 2.3L L4, 3.5L V6, 3.7L V6	**Electronic Throttle Control System (ETCS) Control Relay OFF Malfunction:** Ignition switch ON (II), battery voltage 10 volts minimum and DTCs P0122, P0123, P0222, P0223, P2101, P2118, P2122, P2123, P2127, P2128, P2135, P2138, P2176 are not active. The voltage is not applied from the throttle actuator controller relay for at least 200 milliseconds. The execution time is once per driving cycle. **NOTE: Before you troubleshoot, record all freeze data and any on-board snapshot.**
DTC: P1683 **1T PCM, MIL: Yes** **Year:** 2011, 2012 **Model:** MDX, RDX, RL, TL, TSX, ZDX **Engine:** 2.3L L4, 2.4L L4, 3.5L V6, 3.7L V6	**Throttle Valve Default Position Spring Performance Problem:** Ignition switch LOCK (0), battery voltage a minimum of 6 volts, engine coolant minimum 158 ºF (70 ºC) and DTCs P0117, P0118, P0122, P0123, P0222, P0223, P2101, P2118, P2122, P2123, P2127, P2128, P2135, P2138, P2176 not active. The throttle valve position is more than +5 º from fully closed, or less than +3 º from fully closed for at least 3 seconds. **NOTE: Before you troubleshoot, record all freeze data and any on-board snapshot.**

DTC	Trouble Code Title and Conditions
DTC: P1684 **1T PCM, MIL: Yes** **Year:** 2011, 2012 **Model:** MDX, RDX, RL, TL, TSX, ZDX **Engine:** 2.3L L4, 2.4L L4, 3.5L V6, 3.7L V6	**Throttle Valve Return Spring Performance Problem:** Ignition switch LOCK (0), coolant temperature minimum 158 ºF (70 ºC), battery voltage minimum 6 volts and DTCs P0117, P0118, P0122, P0123, P0222, P0223, P2101, P2118, P2122, P2123, P2127, P2128, P2135, P2138, P2176 not active. The throttle valve opening angle is 17 º or more, or 11 º or less, for at least 3 seconds. Before you troubleshoot, record all freeze data and any on-board snapshot.
DTC: P16BB **1T PCM** **Year:** 2011, 2012 **Model:** MDX, RDX, RL, TL, TSX, ZDX **Engine:** 2.3L L4, 2.4L L4, 3.5L V6, 3.7L V6	**Alternator B Terminal Circuit Low Voltage:** Engine speed 500–3,000 rpm, alternator control mode 14.5 volts. The IGP terminal voltage is 12.0 V or less, and the alternator power generation amount is within 1.0 % to 50.0 %, for at least 60 seconds. The execution time is continuous. **NOTE: Before you troubleshoot, record all freeze data and any on-board snapshot.**
DTC: P16BC **1T PCM** **Year:** 2011, 2012 **Model:** MDX, RDX, RL, TL, TSX, ZDX **Engine:** 2.3L L4, 2.4L L4, 3.5L V6, 3.7L V6	**Alternator FR Terminal Circuit/IGP Circuit Low Voltage:** Engine speed 500–3,000 rpm, alternator control mode 14.5 volts minimum. The IGP terminal voltage is 12.0 V or less, and the alternator power generation amount is 0.5 % or less, for at least 60 seconds. The execution time is continuous. **NOTE: Before you troubleshoot, record all freeze data and any on-board snapshot.**
DTC: P16BD **1T PCM** **Year:** 2011, 2012 **Model:** MDX, RL, TL, TSX, ZDX **Engine:** 3.5L V6, 3.7L V6	**Starter Cut Relay 2 Malfunction:** Engine running, starter switch OFF, DTC P16BE not active. The terminal voltage of the STRLD drops to 2.2 V for at least 5 seconds when the starter cut relay output is turned off. **NOTE: Before you troubleshoot, record all freeze data and any on-board snapshot.**
DTC: P16BE **1T PCM** **Year:** 2011, 2012 **Model:** MDX, RL, TL, TSX, ZDX **Engine:** 3.5L V6, 3.7L V6	**Starter Cut Relay 1 Malfunction:** Engine running, Starter switch OFF cut relay turn on switching failure, ON STRLD open circuit failure. One of these conditions occurs. (1) The terminal voltage of the STRLD exceeds 3.0 volts for at least 5 seconds when the starter cut relay output is turned off. (2) The terminal voltage of the STRLD is 2.4–2.6 volts for at least 0.8 second when the starter cut relay output is turned on. **NOTE: Before you troubleshoot, record all freeze data and any on-board snapshot.**
DTC: P16BF **1T PCM** **Year:** 2011, 2012 **Model:** MDX, RL, TL, TSX, ZDX **Engine:** 3.5L V6, 3.7L V6	**Starter Cut Relay STRLY Circuit Malfunction:** With ignition switch ON (II) the signal of the on command from the PCM to the starter and the return signal from the starter cut relay do not coincide for at least 5 seconds. The execution time is continuous. **NOTE: Before you troubleshoot, record all freeze data and any on-board snapshot.**
DTC: P16C0 **1T PCM, MIL: Yes** **Year:** 2011, 2012 **Model:** MDX, RDX, TL, TSX, ZDX **Engine:** 2.3L L4, 2.4L L4, 3.5L V6, 3.7L V6	**PCM A/T Control System Incomplete Update:** The program rewriting process does not finish normally due to an irregular process, such as turning off the PCM power during the A/T CPU rewriting process, and then turning the ignition switch to ON (II) again. The execution time is continuous and the duration time is 1 seconds or more. **NOTE: This code is indicated when PCM updating is incomplete.**
DTC: P16C4 **1T PCM** **Year:** 2012 **Model:** TL **Engine:** 3.5L V6	**Engine Mount Actuator Control Power Circuit (Stuck Off):** With the ignition switch ON (II), the IGSOL voltage is 4.0 V or less for at least 2.6 seconds during the active engine mount (ACM) control relay is on. **NOTE: Before you troubleshoot, record all freeze data and any on-board snapshots, and review the general troubleshooting information.**
DTC: P16C5 **1T PCM** **Year:** 2012 **Model:** TL **Engine:** 3.5L V6	**Engine Mount Actuator Control Power Circuit (Stuck ON):** With the ignition ON (II), the IGSOL voltage is 8.0 V or more for at least 2.6 seconds during the active control engine mount (ACM) control relay is off. **NOTE: Before you troubleshoot, record all freeze data and any on-board snapshots, and review the general troubleshooting information.**

DTC	Trouble Code Title and Conditions
DTC: P16C6 **1T PCM** **Year:** 2012 **Model:** TL **Engine:** 3.5L V6	**Engine Mount Actuator High Voltage During Function Test:** With the ignition switch ON (II), one of these 3 conditions occur: 1). The value during one cycle of the front engine mount actuator current is 7.5 A or more as a maximum current or 5.3 A or more as a minimum current. 2). The engine mount cycle occurs during current start up and continues. 3). The duty cycle of the front engine mount actuator eads 100% all the time during engine mount operation and continues at least 20 seconds. **NOTE: Before you troubleshoot, record all freeze data and any on-board snapshots, and review the general troubleshooting information.**
DTC: P1717 **2T PCM, MIL: Yes, TCIL: Yes** **Year:** 2012 **Model:** TL **Engine:** 3.5L V6	**Open in Transmission Range Switch ATPRVS Switch Circuit :** With the vehicle driven in reverse at a speed of 3 (5 km/h) mph or less for at least 2 seconds, no RVS signal is detected.
DTC: P1730 **2T PCM, MIL: Yes, TCIL: Yes** **Year:** 2011, 2012 **Model:** TSX **Engine:** 2.4L L4	**Problem in Shift Control System:** With the engine running and in Drive position allow trans mission to shift to 5th gear. Shift solenoid A or D stuck OFF, Shift solenoid B stuck ON, Shift Valves A, B or D stuck.
DTC: P1731 **2T PCM, MIL: Yes, TCIL: Yes** **Year:** 2011, 2012 **Model:** TSX **Engine:** 2.4L L4	**Problem in Shift Control System:** With the engine running and in Drive position allow transmission to shift to 5th gear. Shift solenoid E stuck ON, Shift solenoid E stuck, A/T Clutch pressure control solenoid A stuck OFF. **NOTE: Before you troubleshoot, record all freeze data and any on-board snapshot with the HDS, and review General Troubleshooting Information.**
DTC: P1732 **2T PCM, MIL: Yes, TCIL: Yes** **Year:** 2011, 2012 **Model:** TSX **Engine:** 2.4L L4	**Problem in Shift Control System: Shift Solenoid B or C Stuck ON:** With the engine running and in Drive position allow transmission to shift to 5th gear. Shift solenoid B or C stuck ON, Shift solenoid B or C Stuck. **NOTE: Before you troubleshoot, record all freeze data and any on-board snapshot with the HDS, and review General Troubleshooting Information.**
DTC: P1733 **2T PCM, MIL: Yes, TCIL: Yes** **Year:** 2011, 2012 **Model:** TSX **Engine:** 2.4L L4	**Problem in Shift Control System:** With the engine running and in Drive position allow transmission to shift to 5th gear. Shift solenoid D stuck ON, Shift solenoid D Stuck. A/T clutch pressure control switch valve C Stuck OFF. **NOTE: Before you troubleshoot, record all freeze data and any on-board snapshot with the HDS, and review General Troubleshooting Information.**
DTC: P1734 **2T PCM, MIL: Yes, TCIL: Yes** **Year:** 2011, 2012 **Model:** TSX **Engine:** 2.4L L4	**Problem in Shift Control System:** With the engine running and in Drive position allow transmission to shift to 5th gear. Shift solenoid B or C stuck OFF, Shift solenoid B or C Stuck. **NOTE: Before you troubleshoot, record all freeze data and any on-board snapshot with the HDS, and review General Troubleshooting Information.**
DTC: P1743 **2T PCM, TCIL: Yes** **Year:** 2011, 2012 **Model:** MDX, RDX, RL, TL, ZDX **Engine:** 2.3L L4, 3.5L V6, 3.7L V6	**Problem in Shift Control System; Shift Valve E Stuck OFF:** Vehicle speed is 5 mph (9 km/h), torque converter slip ratio 96–110 %. battery minimun of 11 volts. The transmission is neutral against the 5th gear shift command for at least 2 seconds, without records that the engine speed flares when upshifting to 3rd-4th. The execution time is continuous depending on the driving pattern. **NOTE: Before you troubleshoot, record all freeze data and any on-board snapshot,.**
DTC: P1744 **2T PCM, TCIL: Yes** **Year:** 2011, 2012 **Model:** MDX, RDX, RL, TL, ZDX **Engine:** 2.3L L4, 3.5L V6, 3.7L V6	**Problem in Shift Control System; Shift Valve E Stuck ON:** The transmission is held in 5th gear against the 1st gear shift command for at least 2 seconds. After driving the vehicle in 3rd gear in D for at least 2 seconds. The execution time is continuous depending on the driving pattern. **NOTE: Before you troubleshoot, record all freeze data and any on-board snapshot.**
DTC: P1745 **2T PCM, TCIL: Yes** **Year:** 2011, 2012 **Model:** MDX, RDX, RL, TL, TSX, ZDX **Engine:** 2.3L L4, 3.5L V6, 3.7L V6	**Problem in Shift Control System; Servo Control Valve Stuck OFF or Servo Valve Stuck OFF:** The engine speed flares when upshifting to 2nd-3rd, the engine speed flares when upshifting to 3rd-4th, driving in 5th gear neutral condition. The execution time is continuous and the duration time is 1–2 seconds or more. **NOTE: Before you troubleshoot, record all freeze data and any on-board snapshot.**

DTC	Trouble Code Title and Conditions
DTC: P177A **2T PCM, TCIL: Yes** **Year:** 2011, 2012 **Model:** TL, ZDX **Engine:** 3.5L V6, 3.7L V6	**Line Pressure Solenoid Valve A Stuck OFF, or Line Pressure Switch Stuck OFF:** The input signal from the line pressure switch to the PCM is high when the low line pressure mode is commanded. The execution time is continuous and the duration time is 4 seconds or more.
DTC: P1780 **2T PCM, MIL: Yes, TCIL: Yes** **Year:** 2011, 2012 **Model:** MDX, RDX, RL, TL, TSX, ZDX **Engine:** 2.3L L4, 3.5L V6, 3.7L V6	**Problem in Shift Control System (Transmission is in Default Mode):** The A/T control switches to the default mode due to a mechanical malfunction. The execution time is continuous depending on the driving pattern. **NOTE: DTC P1780 means there is one or more A/T DTCs about the shift control system. Before you troubleshoot, record all freeze data and any on-board snapshot.**

OBD II Trouble Code List (P2XXX Codes)

DTC	Trouble Code Title and Conditions
DTC: P2101 **1T PCM, MIL: Yes** **Year:** 2011, 2012 **Model:** MDX, RDX, RL, TL, TSX, ZDX **Engine:** 2.3L L4, 2.4L L4, 3.5L V6, 3.7L V6	**Electronic Throttle Control System (ETCS) Malfunction (Includes Hybrid Models):** Ignition switch ON (II), battery voltage minimum of 6 volts and DTCs P0122, P0123, P0222, P0223, P2118, P2122, P2123, P2127, P2128,P2135, P2138, P2176 are not active. Difference between throttle valve target position and actual throttle valve position, 4–6 º or more. The execution time is continuous and the duration time is 250–500 milliseconds or more. **NOTE: Before you troubleshoot, record all freeze data and any on-board snapshot.**
DTC: P2108 **1T PCM, MIL: Yes** **Year:** 2011, 2012 **Model:** RL **Engine:** 3.7L V6	**Throttle Actuator Control Module Problem:** Withe the ignition ON (II) one of these 4 conditions must be met for at least 200 milliseconds. (1) Data read from the ROM is abnormal. (2) Data read from the RAM is abnormal. (3) The A/D converter standard voltage is out of specified value. (4) The serial signals between the PCM and the throttle actuator control module do not agree. **NOTE: Before you troubleshoot, record all freeze data and any on-board snapshot.**
DTC: P2118 **1T PCM, MIL: Yes** **Year:** 2011, 2012 **Model:** MDX, RDX, RL, TL, TSX, ZDX **Engine:** 2.3L L4, 2.4L L4, 3.5L V6, 3.7L V6	**Throttle Actuator Current Range/Performance Problem (Includes Hybrid Models):** With ignition ON (II), battery voltage minimum of 6 volts and DTCs P0122, P0123, P0222, P0223, P2101, P2122, P2123, P2127, P2128, P2135, P2138, P2176 are not active. The current flow to the throttle actuator is 11 A or more for at least 200 milliseconds. The execution time is continuous. **NOTE: Before you troubleshoot, record all freeze data and any on-board snapshot.**
DTC: P2122 **1T PCM, MIL: Yes** **Year:** 2011, 2012 **Model:** MDX, RDX, RL, TL, TSX, ZDX **Engine:** 2.3L L4, 2.4L L4, 3.5L V6, 3.7L V6	**APP Sensor A or 1 (TP Sensor D) Circuit Low Voltage (Includes Hybrid Models):** With the ignition switch ON (II) the APP sensor A output voltage is 0.2 V or less for at least 200 milliseconds. The execution time is continuous. DTC P2123 is not active. **NOTE: Before you troubleshoot, record all freeze data and any on-board snapshot.**
DTC: P2123 **1T PCM, MIL: Yes** **Year:** 2011, 2012 **Model:** MDX, RDX, RL, TL, TSX, ZDX **Engine:** 2.3L L4, 2.4L L4, 3.5L V6, 3.7L V6	**APP Sensor A or 1 (TP Sensor D) Circuit High Voltage (Includes Hybrid Models):** With engine running the APP sensor A output voltage is 4.9 V or more for at least 200 milliseconds. The execution time is continuous. DTC P2122 is not active.
DTC: P2127 **1T PCM, MIL: Yes** **Year:** 2011, 2012 **Model:** MDX, RDX, RL, TL, TSX, ZDX **Engine:** 2.3L L4, 2.4L L4, 3.5L V6, 3.7L V6	**APP Sensor B or 2 (Throttle Position (TP) Sensor E) Circuit Low Voltage (Includes Hybrid Models):** With the ignition ON (II) the APP sensor B output voltage is 0.2 V or less for at least 200 milliseconds. The execution time is continuous. DTC P2128 is not active. **NOTE: Before you troubleshoot, record all freeze data and any on-board snapshot.**

DTC	Trouble Code Title and Conditions
DTC: P2128 **1T PCM, MIL: Yes** **Year:** 2011, 2012 **Model:** MDX, RDX, RL, TL, TSX, ZDX **Engine:** 2.3L L4, 2.4L L4, 3.5L V6, 3.7L V6	**APP Sensor B or 2 (Throttle Position (TP) Sensor E) Circuit High Voltage (Includes Hybrid Models):** With engine running the APP sensor B output voltage is 4 V or more for at least 200 milliseconds. The execution time is continuous. DTC P2127 is not active. **NOTE: Before you troubleshoot, record all freeze data and any on-board snapshot.**
DTC: P2135 **1T PCM, MIL: Yes** **Year:** 2011, 2012 **Model:** MDX, RDX, RL, TL, TSX, ZDX **Engine:** 2.3L L4, 2.4L L4, 3.5L V6, 3.7L V6	**Throttle Position (TP) Sensor A/B or 1/2 Incorrect Voltage Correlation (Includes Hybrid Models):** The difference between the throttle valve positions indicated by TP sensor A and TP sensor B exceeds the value as follows, for at least 200 milliseconds. Difference between TP sensor A and TP sensor B 1.8–14.7º or more, or 5º or less. **NOTE: Before you troubleshoot, record all freeze data and any on-board snapshot.**
DTC: P2138 **1T PCM, MIL: Yes** **Year:** 2011, 2012 **Model:** MDX, RDX, RL, TL, TSX, ZDX **Engine:** 2.3L L4, 2.4L L4, 3.5L V6, 3.7L V6	**APP Sensor A/B or 1/2 (Throttle Position (TP) Sensor D/E) Incorrect Voltage Correlation (Includes Hybrid Models):** With the ignition ON (II) one of these conditions must be met for at least 300 milliseconds: (1) APP sensor B voltage exceeds the range from 0 V or less to 0.37 V or more when the APP sensor A voltage is 0.37 V. (2) APP sensor B voltage exceeds the range from 2.31 V or less to 2.69 V or more when the APP sensor A voltage is 5 V. The execution time is continuous and DTCs P2122, P2123, P2127, P2128 are not active. **NOTE: Before you troubleshoot, record all freeze data and any on-board snapshot.**
DTC: P2176 **1T PCM, MIL: Yes** **Year:** 2011, 2012 **Model:** MDX, RDX, RL, TL, TSX, ZDX **Engine:** 2.3L L4, 2.4L L4, 3.5L V6, 3.7L V6	**Throttle Actuator Control System Idle Position Not Learned (Includes Hybrid Models):** One of these condition must be met for at least 0.7 seconds. (1) The registration of the throttle valve fully closed position is not completed within a predetermined time after the ignition switch is turned ON. (2) The registered value of the throttle valve fully closed position is 0.74 volts TP1, 1.61 volts TP2 or more, or 0.49 volts TP1, 1.37 volts TP2 or less. The execution time is once per driving cycle and DTCs P0122, P0123, P0222, P0223, P2101, P2118, P2135 are not active. **NOTE: If DTC P2135 is stored at the same time as DTC P2176, troubleshoot DTC P2135 first, then recheck for DTC P2176.**
DTC: P2183 **2T PCM, MIL: Yes** **Year:** 2011, 2012 **Model:** MDX, RDX, RL, TL, TSX, ZDX **Engine:** 2.3L L4, 2.4L L4, 3.5L V6, 3.7L V6	**Engine Coolant Temperature (ECT) Sensor 2 Circuit Range/Performance Problem (Includes Hybrid Models):** A malfunction is detected if the following three conditions are present after the engine has stopped and the ignition switch has been turned to LOCK (0) for at least 6 hours before restarting the engine: (1) When the temperature difference between the IAT and ECT1 is 57 ºF (32 ºC) or more. (2) When the temperature difference between the IAT and ECT2 is 30 ºF (17 ºC) or more. (3) When the temperature difference between the ECT2 and ECT1 is 73 ºF (41ºC) or more. The execution time is once per driving cycle and the duration time is 10 seconds or more. **NOTE: If DTC P0111 is stored at the same time as DTC P2183, troubleshoot DTC P0111 first, then recheck for DTC P2183.**
DTC: P2184 **1T PCM, MIL: Yes** **Year:** 2011, 2012 **Model:** MDX, RDX, RL, TL, TSX, ZDX **Engine:** 2.3L L4, 2.4L L4, 3.5L V6, 3.7L V6	**Engine Coolant Temperature (ECT) Sensor 2 Circuit Low Voltage (Includes Hybrid Models):** With ignition switch ON (II) the ECT sensor 2 output voltage is 0.08 V or less for at least 2 seconds. The execution time is continuous. DTC P2185 is not active. **NOTE: Before you troubleshoot, record all freeze data and any on-board snapshot.**
DTC: P2185 **1T PCM, MIL: Yes** **Year:** 2011, 2012 **Model:** MDX, RDX, RL, TL, TSX, ZDX **Engine:** 2.3L L4, 2.4L L4, 3.5L V6, 3.7L V6	**Engine Coolant Temperature (ECT) Sensor 2 Circuit High Voltage (Includes Hybrid Models):** With ignition ON (II) the ECT sensor 2 output voltage is 4.92 volts or more for at least 2 seconds. The execution time is continuous and DTC P2184 is not active. **NOTE: Before you troubleshoot, record all freeze data and any on-board snapshot.**
DTC: P2195 **1T PCM, MIL: Yes** **Year:** 2011, 2012 **Model:** RDX, TSX **Engine:** 2.3L L4, 2.4L L4, 3.5L V6	**A/F Sensor (Sensor 1) Signal Stuck Lean (Includes Hybrid Models):** The A/F sensor (Sensor 1) output voltage is 2.5 V or more for at least 7 seconds. The execution time is once per driving cycle and DTCs P0134, P0135, P0171, P0300, P0301, P0302, P0303, P0304, P0305, P0306, P0627, P1172, P2237, P2238, P2243, P2245, P2251, P2252 are not active. **NOTE: If DTC P2101, P2118, P2135, P2138, P2176, or a combination of P2122 and P2127, P2122 and P2138 or P2127 and P2138 is stored at the same time as DTC P2195, troubleshoot those DTCs first, then recheck for DTC P2195.**

DTC	Trouble Code Title and Conditions
DTC: P2195 **2T PCM, MIL: Yes** **Year:** 2011, 2012 **Model:** MDX, RL, TL, TSX, ZDX **Engine:** 3.5L V6, 3.7L V6	**Rear Air/Fuel Ratio (A/F) Sensor (Bank 1, Sensor 1) Signal Stuck Lean:** The rear A/F sensor (bank 1, sensor 1) output voltage is 3.48 V or more for at least 7 seconds. The execution time is once per driving cycle and DTCs P0134, P0135, P0171, P0300, P0301, P0302, P0303, P0304, P0305, P0306, P0627, P1172, P2237, P2238, P2243, P2245, P2251, P2252 are not active. **NOTE: If DTC P2101, P2118, P2135, P2138, P2176, or a combination of P2122 and P2127, P2122 and P2138 or P2127 and P2138 is stored at the same time as DTC P2195, troubleshoot those DTCs first, then recheck for DTC P2195.**
DTC: P2197 **2T PCM, MIL: Yes** **Year:** 2011, 2012 **Model:** MDX, RL, TL, ZDX **Engine:** 3.5L V6, 3.7L V6	**Front Air/Fuel Ratio (A/F) Sensor (Bank 2, Sensor 1) Signal Stuck Lean:** The front A/F sensor (bank 2, sensor 1) output voltage is 3.48 V or more for at least 7 seconds. The execution time is once per driving cycle and DTCs P0154, P0155, P0174, P0300, P0301, P0302, P0303, P0304, P0305, P0306, P0627, P1174, P2240, P2241, P2247, P2249, P2254, P2255 are not active. **NOTE: If DTC P2101, P2118, P2135, P2138, P2176, or a combination of P2122 and P2127, P2122 and P2138 or P2127 and P2138 is stored at the same time as DTC P2195, P2197, troubleshoot those DTCs first, then recheck for DTC P2197.**
DTC: P2199 **2T PCM, MIL: Yes** **Year:** 2011, 2012 **Model:** RDX **Engine:** 2.3L L4	**Intake Air Temperature (IAT) Sensor 1, 2 Correlation:** With the ignition ON for two seconds or OFF for 8 hours the voltage difference between IAT sensor 1 and IAT sensor 2 indicates a difference in intake air temperature of 77 ºF (25 ºC) or more. **NOTE: Before you troubleshoot, record all freeze data and any on-board snapshot, and review the general troubleshooting information.**
DTC: P2227 **2T PCM, MIL: Yes** **Year:** 2011, 2012 **Model:** MDX, RDX, RL, TL, TSX, ZDX **Engine:** 2.3L L4, 2.4L L4, 3.5L V6, 3.7L V6	**Barometric Pressure (BARO) Sensor Circuit Range/Performance Problem:** Throttle position 17.0–33.0 degrees at 1,100–3,000 rpm. The difference between the BARO sensor output and the MAP sensor output is 26 kPa (7.5 in.Hg, 190 mmHg) or more for at least 2.5 seconds. The execution time is once per driving cycle and the duration time is 2.5 seconds or more. **NOTE: If DTC P0107, P0108, P1128, and/or P1129 are stored at the same time as DTC P2227, troubleshoot those DTCs first, then recheck for DTC P2227.**
DTC: P2228 **1T PCM, MIL: Yes** **Year:** 2011, 2012 **Model:** MDX, RDX, RL, TL, TSX, ZDX **Engine:** 2.3L L4, 2.4L L4, 3.5L V6, 3.7L V6	**Barometric Pressure (BARO) Sensor Circuit Low Voltage:** With ignition ON (II) the BARO sensor output voltage is 1.31 V or less for at least 2 seconds. The execution time is continuous and DTCs P1109, P2229 are not active. **NOTE: Before you troubleshoot, record all freeze data and any on-board snapshot.**
DTC: P2229 **1T PCM, MIL: Yes** **Year:** 2011, 2012 **Model:** MDX, RDX, RL, TL, TSX, ZDX **Engine:** 2.3L L4, 2.4L L4, 3.5L V6, 3.7L V6	**Barometric Pressure (BARO) Sensor Circuit High Voltage:** With ignition switch ON (II) the BARO sensor output voltage is 4.49 V or more for at least 2 seconds. The execution time is continuous and DTCs P1109, P2228 are not active. **NOTE: Before you troubleshoot, record all freeze data and any on-board snapshot.**
DTC: P2237 **1T PCM, MIL: Yes** **Year:** 2011, 2012 **Model:** MDX, RL, TL, ZDX **Engine:** 3.5L V6, 3.7L V6	**Rear Air/Fuel Ratio (A/F) Sensor (Bank 1, Sensor 1) IP Circuit High Voltage:** With the engine running the IPB1 terminal voltage is 2 volts or less, or 5.6 volts or more, for at least 15 seconds. The execution time is continuous and DTCs P0135, P2195, P2238, P2243, P2245, P2251, P2252, P2627, P2628 are not active. **NOTE: Before you troubleshoot, record all freeze data and any on-board snapshot.**
DTC: P2238 **1T PCM, MIL: Yes** **Year:** 2011, 2012 **Model:** RDX **Engine:** 2.3L L4	**Air/Fuel Ratio (A/F) Sensor (Sensor 1) AFS+ Circuit Low Voltage:** With the engine running and battery voltage at a minimum of 10.5 V, the AFS+ voltage is 0.4 V or less for at least 4 seconds. **NOTE: Before you troubleshoot, record all freeze data and any on-board snapshot, and review the general troubleshooting information.**
DTC: P2238 **1T PCM, MIL: Yes** **Year:** 2011, 2012 **Model:** MDX, RL, TL, TSX, ZDX **Engine:** 3.5L V6, 3.7L V6	**Rear Air/Fuel Ratio (A/F) Sensor (Bank 1, Sensor 1) IP Circuit Low Voltage:** With engine running the IPB1 input terminal voltage is 1 volts or less for at least 5 seconds after the sensor becomes active and 85 seconds before the sensor becomes active. The execution time is continuous, DTCs P0135, P2195, P2237, P2243, P2245, P2251, P2252, P2627, P2628 are not active. **NOTE: Before you troubleshoot, record all freeze data and any on-board snapshot.**
DTC: P2240 **1T PCM, MIL: Yes** **Year:** 2011, 2012 **Model:** MDX, RL, TL, TSX, ZDX **Engine:** 3.5L V6, 3.7L V6	**Front Air/Fuel Ratio (A/F) Sensor (Bank 2, Sensor 1) IP Circuit High Voltage:** With the engine running the IPB2 terminal voltage is 5.6 volts or more, or 2 volts or less, for at least 15 seconds. The execution time is continuous and DTCs P0155, P2197, P2240, P2241, P2247, P2249, P2254, P2255, P2630, P2631 are not active. **NOTE: Before you troubleshoot, record all freeze data and any on-board snapshot.**

DTC	Trouble Code Title and Conditions
DTC: P2241 **1T PCM, MIL: Yes** **Year:** 2011, 2012 **Model:** MDX, RL, TL, ZDX **Engine:** 3.5L V6, 3.7L V6	**Front Air/Fuel Ratio (A/F) Sensor (Bank 2, Sensor 1) IP Circuit Low Voltage:** With engine running the IPB1 input terminal voltage is 1 volts or less for at least 5 seconds after the sensor becomes active and 85 seconds before the sensor becomes active. The execution time is continuous DTCs P0135, P2195, P2237, P2243, P2245, P2251, P2252, P2627, P2628 are not active. **NOTE: Before you troubleshoot, record all freeze data and any on-board snapshot.**
DTC: P2243 **1T PCM, MIL: Yes** **Year:** 2011, 2012 **Model:** MDX, RL, TL, TSX, ZDX **Engine:** 3.5L V6, 3.7L V6	**Rear Air/Fuel Ratio (A/F) Sensor (Bank 1, Sensor 1) VCENT Circuit High Voltage:** With engine running the VSB1 terminal voltage repeatedly fluctuates from a value above 4.8 volts to a value below 3.4 volts, at least 150 times. The execution time is continuous and DTCs P0135, P2195, P2237, P2238, P2245, P2251, P2252, P2627, P2628 are not active. **NOTE: Before you troubleshoot, record all freeze data and any on-board snapshot.**
DTC: P2245 **1T PCM, MIL: Yes** **Year:** 2011, 2012 **Model:** MDX, RL, TL, TSX, ZDX **Engine:** 3.5L V6, 3.7L V6	**Rear Air/Fuel Ratio (A/F) Sensor (Bank 1, Sensor 1) VCENT Circuit Low Voltage:** With engine running the IPB1 input terminal voltage is 1 volts or less for at least 5 seconds after the sensor becomes active, 85 seconds before the sensor becomes active. The execution time is continuous and DTCs P0135, P2195, P2237, P2238, P2243, P2251, P2252, P2627, P2628 are not active.
DTC: P2246 **1T PCM, MIL: Yes** **Year:** 2011, 2012 **Model:** MDX, ZDX **Engine:** 3.7L V6	**Rocker Arm Oil Pressure Switch Circuit Low Voltage:** When the engine rpm is 4,300 (High lift cam operation) the rocker arm oil control solenoid is ON, the rocker arm oil pressure switch remains ON. **NOTE: Before you troubleshoot, record all freeze data and any on-board snapshot.**
DTC: P2247 **1T PCM, MIL: Yes** **Year:** 2011, 2012 **Model:** MDX, RL, TL, TSX, ZDX **Engine:** 3.5L V6, 3.7L V6	**Front A/F Sensor (Bank 2, Sensor 1) VCENT Circuit High Voltage :** With engine running the VSB1 terminal voltage repeatedly fluctuates from a value above 4.8 volts to a value below 3.4 volts, at least 150 times. The execution time is continuous and DTCs P0135, P2195, P2237, P2238, P2245, P2251, P2252, P2627, P2628 are not active. **NOTE: Before you troubleshoot, record all freeze data and any on-board snapshot.**
DTC: P2249 **1T PCM, MIL: Yes** **Year:** 2011, 2012 **Model:** MDX, RL, TL, TSX, ZDX **Engine:** 3.5L V6, 3.7L V6	**Front A/F Sensor (Bank 2, Sensor 1) VCENT Circuit Low Voltage :** With the engine running the VSB2 terminal voltage between reads 0.3–1.5 volts.The execution time is continuous and DTCs P0155, P2197, P2240, P2241, P2247, P2254, P2255 are not active. The IPB2 input terminal voltage is 1.0 volts or less for at least 5 seconds, after the sensor becomes active, 85 seconds before the sensor becomes active. **NOTE: Before you troubleshoot, record all freeze data and any on-board snapshot.**
DTC: P2251 **1T PCM, MIL: Yes** **Year:** 2011, 2012 **Model:** MDX, RL, TL, TSX, ZDX **Engine:** 3.5L V6, 3.7L V6	**Rear Air/Fuel Ratio (A/F) Sensor (Bank 1, Sensor 1) VS Circuit High Voltage:** With engine running the VSB1 terminal voltage is 6 V or more, and the PCM internal signal voltage is 4.6 V or more, for at least 5 seconds. The execution time is continuous and DTCs P0135, P2195, P2237, P2238, P2243, P2245, P2252, P2627, P2628 are not active. **NOTE: If DTC P2251 is stored at the same time as DTC P0134, troubleshoot DTC P2251 first, then recheck for P0134.**
DTC: P2252 **1T PCM, MIL: Yes** **Year:** 2011, 2012 **Model:** RDX, TSX **Engine:** 2.3L L4, 2.4L L4	**Air/Fuel Ratio (A/F) Sensor (Sensor 1) AFS- Circuit Low Voltage :** With the engine running the AFS-terminal voltage is 0.4 V or less for at least 5 seconds. **NOTE: Before you troubleshoot, record all freeze data and any on-board snapshot, and review the general troubleshooting information.**
DTC: P2252 **1T PCM, MIL: Yes** **Year:** 2011, 2012 **Model:** MDX, RL, TL, TSX, ZDX **Engine:** 3.5L V6, 3.7L V6	**Rear Air/Fuel Ratio (A/F) Sensor (Bank 1, Sensor 1) VS Circuit Low Voltage:** With engine running the VSB1 terminal voltage is 0.3 V or less, and the IPB1 terminal voltage is 1 V or more, for at least 5 seconds after the sensor becomes active, 85 seconds before the sensor becomes active.
DTC: P2254 **1T PCM, MIL: Yes** **Year:** 2011, 2012 **Model:** MDX, RL, TL, TSX, ZDX **Engine:** 3.5L V6, 3.7L V6	**Front A/F Sensor (Bank 2, Sensor 1) VS Circuit High Voltage:** With engine running the VSB2 terminal voltage is 6 V or more, and the PCM internal signal voltage is 4.6 V or more, for at least 5 seconds. The execution time is continuous and DTCs P0155, P2197, P2240, P2241, P2247, P2249, P2255, P2630, P2631 are not active. **NOTE: If DTC P2254 is stored at the same time as DTC P0154, troubleshoot DTC P2254 first, then recheck for P0154.**
DTC: P2255 **1T PCM, MIL: Yes** **Year:** 2011, 2012 **Model:** MDX, RL, TL, TSX, ZDX **Engine:** 3.5L V6, 3.7L V6	**Front A/F Sensor (Bank 2, Sensor 1) VS Circuit Low Voltage :** With engine running the VSB2 terminal voltage is 0.3 V or less, and the IPB2 terminal voltage is 1 V or more, for at least 5 seconds after the sensor is active, 85 seconds before the sensor becomes active. **NOTE: Before you troubleshoot, record all freeze data and any on-board snapshot.**

DTC	Trouble Code Title and Conditions
DTC: P2261 **2T PCM, MIL: Yes** **Year:** 2011, 2012 **Model:** RDX **Engine:** 2.3L L4	**Turbocharger Bypass Valve Stuck Closed:** When the throttle from full open supercharging condition returns to fully closed the pulse method of releasing pressure accumulation from the turbocharger bypass control solenoid valve is 350 or more for at least 2.5 seconds. **NOTE: Before you troubleshoot, record all freeze data and any on-board snapshot, and review the general troubleshooting information.**
DTC: P2263 **2T PCM, MIL: Yes** **Year:** 2011, 2012 **Model:** RDX **Engine:** 2.3L L4	**Turbocharger Boost System Performance Problem:** With the engine at operating temperature and throttle not fully open, rpm 1,500 or more the degree of boost pressure rise delay is 0.1 or less for at least 2.0 seconds. **Note: If DTC P0229 is stored at the same time as DTC P2263, troubleshoot DTC P0229 first, then recheck for DTC P2263.**
DTC: P2270 **2T PCM, MIL: Yes** **Year:** 2011, 2012 **Model:** RDX **Engine:** 2.3L L4	**Secondary HO2S (Sensor 2) Circuit Signal Stuck Lean:** With the engine running, and in closed loop with a speed of 30 mph or more, the secondary HO2S output voltage is 0.650 V or less. **NOTE: Before you troubleshoot, record all freeze data and any on-board snapshot, and review the general troubleshooting information.**
DTC: P2270 **2T PCM, MIL: Yes** **Year:** 2011, 2012 **Model:** MDX, RL, TL, ZDX **Engine:** 3.5L V6, 3.7L V6	**Rear Secondary Heated Oxygen Sensor (Secondary HO2S (Bank 1, Sensor 2) Circuit Signal Stuck Lean:** The rear secondary HO2S (bank 1, sensor 2) output voltage is 0.650 V or less. The execution time is once per driving cycle and the duration time is 36.3 seconds or less. **NOTE: Before you troubleshoot, record all freeze data and any on-board snapshot.**
DTC: P2271 **2T PCM, MIL: Yes** **Year:** 2011, 2012 **Model:** RDX **Engine:** 2.3L L4	**Secondary HO2S (Sensor 2) Circuit Signal Stuck Rich :** The secondary HO2S (Sensor 2) output voltage is 0.293 V or more. The execution time is once per driving cycle and the duration time is 24.2 seconds or more. **NOTE: Before you troubleshoot, record all freeze data and any on-board snapshot.**
DTC: P2271 **2T PCM, MIL: Yes** **Year:** 2011, 2012 **Model:** MDX, RL, TL, ZDX **Engine:** 3.5L V6, 3.7L V6	**Rear Secondary HO2S (Bank 1, Sensor 2) Circuit Signal Stuck Rich :** If, after current is applied to the rear secondary HO2S heater, the rear secondary HO2S does not fluctuate and the output is stuck within the specified area, a malfunction is detected. The rear secondary HO2S (bank 1, sensor 2) output voltage is 0.293 V or more. The execution time is once per driving cycle and the duration time is 36.3 seconds or more. **NOTE: Before you troubleshoot, record all freeze data and any on-board snapshot.**
DTC: P2272 **2T PCM, MIL: Yes** **Year:** 2011, 2012 **Model:** MDX, RL, TL, ZDX **Engine:** 3.5L V6, 3.7L V6	**Front Secondary HO2S (Bank 2, Sensor 2) Circuit Signal Stuck Lean :** If, after current is applied to the front secondary HO2S heater, the front secondary HO2S does not fluctuate and the output is stuck within the specified area, a malfunction is detected. The front secondary HO2S (bank 2, sensor 2) output voltage is 0.650 V or less.
DTC: P2273 **2T PCM, MIL: Yes** **Year:** 2011, 2012 **Model:** MDX, RL, TL, ZDX **Engine:** 3.5L V6, 3.7L V6	**Front Secondary HO2S (Bank 2, Sensor 2) Circuit Signal Stuck Rich :** If, after current is applied to the front secondary HO2S heater, the front secondary HO2S does not fluctuate and the output is stuck within the specified area, a malfunction is detected. The front secondary HO2S (bank 2, sensor 2) output voltage is 0.293 V or more. **NOTE: Before you troubleshoot, record all freeze data and any on-board snapshot.**
DTC: P2279 **2T PCM, MIL: Yes** **Year:** 2011, 2012 **Model:** MDX, RL, TL, TSX, ZDX **Engine:** 2.4L L4, 3.5L V6, 3.7L V6	**Intake Air System Leak:** With the engine running in closed loop for a minimum of 15 seconds. Either of these 2 conditions is met: (1) The estimated volume of intake air is 310 l/min (327.6 US qt/min, 272.8 Imp qt/min) or more when the MAP value is 35 kPa (10.3 in. Hg, 260 mmHg). (2) The estimated volume of intake air is 302 l/min (319.2 US qt/min, 265.8 Imp qt/min) or more when the MAP value is 62 kPa (18.2 in. Hg, 460 mmHg). The execution time is once per driving cycle and the duration time is 22 seconds or more. **NOTE: If DTC P0443 is stored at the same time as DTC P2279, troubleshoot DTC P0443 first, then recheck for DTC P2279.**
DTC: P2413 **2T PCM, MIL: Yes** **Year:** 2011, 2012 **Model:** MDX, RL, TL, TSX, ZDX **Engine:** 3.5L V6, 3.7L V6	**Exhaust Gas Recirculation (EGR) System Malfunction:** With the engine running at 4,400 rpm, commanded EGR valve lift is 0.040 in. (1 mm). If the actual valve lift is 0.006 in. (0.15 mm) or less for at least 5 seconds, the valve is considered stuck closed. **NOTE: Before you troubleshoot, record all freeze data and any on-board snapshot.**

DTC	Trouble Code Title and Conditions
DTC: P2422 **2T PCM, MIL: Yes** **Year:** 2011, 2012 **Model:** MDX, RDX, RL, TL, TSX, ZDX **Engine:** 2.3L L4, 2.4L L4, 3.5L V6, 3.7L V6	**EVAP Canister Vent Shut Valve Close Malfunction :** The output from the fuel tank pressure sensor is -4 kPa (-1.0 in. Hg, -25 mmHg), -2 kPa (-0.4 in. Hg, -10 mmHg) or less for at least 1.04–8 seconds. Elapsed time after the FTP sensor output exceeds the malfunction threshold 1.04 seconds. Excessive negative pressure is detected 8 seconds.
DTC: P2446 **1T PCM, MIL: Yes** **Year:** 2012 **Model:** TL **Engine:** 3.5L V6	**Rocker Arm Oil Pressure Switch Circuit Low Voltage (J35A6 Engine):** With the engine running, engine speed a minimum of 4,400 rpm (Variable that is depending on the engine load). The rocker arm oil control solenoid is ON, the rocker arm oil pressure switch remains ON. The execution time is once per driving cycle and DTCs P0522, P0523, P2648, P2649 are not active. **NOTE: Before you troubleshoot, record all freeze data and any on-board snapshot.**
DTC: P2552 **1T PCM, MIL: Yes** **Year:** 2011, 2012 **Model:** RL **Engine:** 3.7L V6	**Throttle Actuator Control Module Relay Malfunction :** The serial signal is input from the throttle actuator control module for at least 2 seconds after the throttle actuator control module relay is turned OFF. **NOTE: Before you troubleshoot, record all freeze data and any on-board snapshot.**
DTC: P2610 **1T PCM, MIL: Yes** **Year:** 2011, 2012 **Model:** MDX, RDX, RL, TL, TSX, ZDX **Engine:** 2.3L L4, 2.4L L4, 3.5L V6, 3.7L V6	**ECM/PCM Ignition Off Internal Timer Malfunction:** The access process to the ignition off timer fails, or a malfunction is found in the read data for at least 10 seconds. Ignition switch ON (II) when a battery is disconnected and connected again is excluded. **NOTE: Before you troubleshoot, record all freeze data and any on-board snapshot.**
DTC: P2627 **1T PCM, MIL: Yes** **Year:** 2012 **Model:** TL **Engine:** 3.5L V6	**VTEC System Stuck ON (J35A7 Engine):** With the engine running, engine speed between 1,500–3,000 rpm, the rocker arm oil control solenoid is ON, the EOP sensor output remains HIGH (39 kPa 11.6 in. Hg, 293 mmHg) or more. The execution time is once per driving cycle and DTCs P0522, P0523, P2227, P2228, P2229 are not active. **NOTE: Before you troubleshoot, record all freeze data and any on-board snapshot.**
DTC: P2646 **1T PCM, MIL: Yes** **Year:** 2011, 2012 **Model:** MDX, RDX, RL, TL, TSX, ZDX **Engine:** 2.3L L4, 2.4L L4, 3.5L V6, 3.7L V6	**Rocker Arm Oil Pressure Switch (VTEC Oil Pressure Switch) Circuit Low Voltage :** With the engine running, engine speed a minimum of 4,800 rpm (Variable that is depending on the engine load). The rocker arm oil control solenoid is ON, the rocker arm oil pressure switch remains ON. The execution time is once per driving cycle and DTCs P2648, P2649 are not active. **NOTE: Before you troubleshoot, record all freeze data and any on-board snapshot.**
DTC: P2646 **1T PCM, MIL: Yes** **Year:** 2012 **Model:** TL **Engine:** 3.5L V6	**VTEC System Stuck OFF (J35A7 Engine):** With the engine running at idle, battery voltage a minimum of 10.5 V and DTCs P0522, P0523, P2227, P2228 and P2229 not active, the rocker arm oil control solenoid is OFF, the EOP sensor output remains LOW (39 kPa, 11.5 in. Hg, 292 mmHg) or less.
DTC: P2647 **1T PCM, MIL: Yes** **Year:** 2012 **Model:** TL **Engine:** 3.5L V6	**Rocker Arm Oil Pressure Switch Circuit High Voltage (J35A6 Engine):** With the engine running at idle, and battery voltage a minimum of 10.5 V. The rocker arm oil control solenoid is OFF, the rocker arm oil pressure switch remains OFF. The execution time is once per driving cycle and DTCs P0522, P0523, P2648, P2649 are not active. **NOTE: Before you troubleshoot, record all freeze data and any on-board snapshot.**
DTC: P2647 **1T PCM, MIL: Yes** **Year:** 2011, 2012 **Model:** MDX, RDX, RL, TL, TSX, ZDX **Engine:** 2.3L L4, 2.4L L4, 3.5L V6, 3.7L V6	**Rocker Arm Oil Pressure Switch (VTEC Oil Pressure Switch) Circuit High Voltage :** When the rocker arm oil control solenoid is OFF, the rocker arm oil pressure switch remains OFF. (Low lift cam operation) **NOTE: Before you troubleshoot, record all freeze data and any on-board snapshot.**

DTC	Trouble Code Title and Conditions
DTC: P2648 **1T PCM, MIL: Yes** **Year:** 2011, 2012 **Model:** MDX, RDX, RL, TL, TSX, ZDX **Engine:** 2.3L L4, 2.4L L4, 3.5L V6, 3.7L V6	**Rocker Arm Oil Control Solenoid (VTEC Solenoid Valve) Circuit Low Voltage :** The return signal is OFF (low) for at least 2 seconds when the PCM outputs the ON (high) signal to the rocker arm oil control solenoid. DTC P2649 is not active. **NOTE: Before you troubleshoot, record all freeze data and any on-board snapshot.**
DTC: P2648 **1T PCM, MIL: Yes** **Year:** 2012 **Model:** TL **Engine:** 3.5L V6	**Rocker Arm Oil Control Solenoid A (Bank 1) Circuit Low Voltage (J35A7 Engine):** With the vehicle driven in a lower gear and engine speed at 3,300 rpm, battery voltage a minimum of 10.5 V, the return signal is OFF (LOW) for at least 1 second when the PCM outputs the ON (HIGH) signal to the rocker arm oil control solenoid A (Bank 1).
DTC: P2648 **1T PCM, MIL: Yes** **Year:** 2012 **Model:** TL **Engine:** 3.5L V6	**Rocker Arm Oil Control Solenoid Circuit Low Voltage (J35A6 Engine):** With the vehicle driven in a lower gear at 4,400 rpm or more and battery voltage a minimum of 10.5 V, the return signal is OFF (LOW) for at least 2 seconds when the PCM outputs the ON (HIGH) signal to the rocker arm oil control solenoid.
DTC: P2649 **1T PCM, MIL: Yes** **Year:** 2011, 2012 **Model:** MDX, RDX, RL, TL, TSX, ZDX **Engine:** 2.3L L4, 2.4L L4, 3.5L V6, 3.7L V6	**Rocker Arm Oil Control Solenoid Circuit High Voltage :** The return signal is ON (High) for at least 2 seconds when the PCM outputs the OFF (Low) signal to the rocker arm oil control solenoid. With the engine running the execution time is continuous. DTC P2648 is not active. **NOTE: Before you troubleshoot, record all freeze data and any on-board snapshot.**
DTC: P2649 **1T PCM, MIL: Yes** **Year:** 2012 **Model:** TL **Engine:** 3.5L V6	**Rocker Arm Oil Control Solenoid A (Bank 1) Circuit High Voltage (J35A7 Engine):** The return signal is ON (High) for at least 2 seconds when the PCM outputs the OFF (Low) signal to the rocker arm oil control solenoid. With the engine running the execution time is continuous. DTC P2648 is not active. **NOTE: Before you troubleshoot, record all freeze data and any on-board snapshot.**
DTC: P2649 **1T PCM, MIL: Yes** **Year:** 2012 **Model:** TL **Engine:** 3.5L V6	**Rocker Arm Oil Control Solenoid Circuit High Voltage (J35A6 Engine):** With the engine running at idle for at least 7 seconds and battery voltage a minimum of 10.0 V. The return signal is ON (HIGH) for at least 2 seconds when the PCM outputs the OFF (LOW) signal to the rocker arm oil control solenoid. DTC P2648 is not active. **NOTE: Before you troubleshoot, record all freeze data and any on-board snapshot.**
DTC: P2653 **2T PCM, MIL: Yes** **Year:** 2012 **Model:** TL **Engine:** 3.5L V6	**Rocker Arm Oil Control Solenoid B (Bank 1) Circuit Low Voltage:** With the engine running at about 3,000 rpm for 10 seconds, the return signal is OFF (Low) for at least 1.0 second when the PCM outputs the ON (High) signal to the rocker arm oil control solenoid B (Bank 1). **NOTE: Before you troubleshoot, record all freeze data and any on-board snapshot, and review the general troubleshooting information.**
DTC: P2654 **2T PCM, MIL: Yes** **Year:** 2012 **Model:** TL **Engine:** 3.5L V6	**Rocker Arm Oil Control Solenoid B (Bank 1) Circuit High Voltage:** With the engine running at idle for at least 10 seconds, the return signal is ON (High) for at least 2.0 seconds when the PCM outputs the OFF (Low) signal to the rocker arm oil control solenoid B (Bank 1). **NOTE: Before you troubleshoot, record all freeze data and any on-board snapshot, and review the general troubleshooting information.**
DTC: P2658 **2T PCM, MIL: Yes** **Year:** 2012 **Model:** TL **Engine:** 3.5L V6	**Rocker Arm Oil Control Solenoid A (Bank 2) Circuit Low Voltage:** With the engine running at about 3,000 rpm for 10 seconds, the return signal is OFF (Low) for at least 1.0 second when the PCM outputs the ON (High) signal to the rocker arm oil control solenoid A (Bank 2). **NOTE: Before you troubleshoot, record all freeze data and any on-board snapshot, and review the general troubleshooting information.**
DTC: P2659 **2T PCM, MIL: Yes** **Year:** 2012 **Model:** TL **Engine:** 3.5L V6	**Rocker Arm Oil Control Solenoid A (Bank 2) Circuit High Voltage:** With the engine running at idle for at least 10 seconds, the return signal is ON (High) for at least 2.0 seconds when the PCM outputs the OFF (Low) signal to the rocker arm oil control solenoid A (Bank 2). **NOTE: Before you troubleshoot, record all freeze data and any on-board snapshot, and review the general troubleshooting information**

DTC	Trouble Code Title and Conditions
DTC: P2714 **2T PCM, MIL: Yes, TCIL: Yes** **Year:** 2011, 2012 **Model:** TL, ZDX **Engine:** 3.5L V6, 3.7L V6	**A/T Clutch Pressure Control Solenoid Valve D Stuck OFF:** Shift lever position D, S, R, with ATF the temperature between -4 ºF (-20 ºC), 248 ºF (120 ºC) improper gear ratio output has been detected for the shift schedule. The execution time is continuous and the duration time is between 3–10 seconds or more.
DTC: P2720 **2T PCM, MIL: Yes, TCIL: Yes** **Year:** 2011, 2012 **Model:** TL, ZDX **Engine:** 3.5L V6, 3.7L V6	**Problem in A/T Clutch Pressure Control Solenoid Valve D Circuit:** The measured current for the PCM's command value is less than 0.19 with duty cylce at 56.5–89%, with duty cycle at 89% or more it is Less than 0.27. **NOTE: Before you troubleshoot, record all freeze data and any on-board snapshot.This code is caused by an electrical circuit problem and cannot be caused by a mechanical problem in the transmission.**
DTC: P2721 **2T PCM, MIL: Yes, TCIL: Yes** **Year:** 2011, 2012 **Model:** TL, ZDX **Engine:** 3.5L V6, 3.7L V6	**Problem in A/T Clutch Pressure Control Solenoid Valve D:** The measured current for the PCM's command value is more than 0.6 duty cycle less than 13.7%. More than 0.9 with duty cycle at 13.7–27%. (High Imput) **NOTE: Before you troubleshoot, record all freeze data and any on-board snapshot. This code is caused by an electrical circuit problem and cannot be caused by a mechanical problem in the transmission.**
DTC: P2821 **2T PCM, MIL: Yes, TCIL: Yes** **Year:** 2011, 2012 **Model:** TL, ZDX **Engine:** 3.5L V6, 3.7L V6	**Line Pressure Solenoid Valve A Stuck ON:** The input signal from the line pressure switch to the PCM is high during the self shut down mode after the signal is low when the normal line pressure mode was commanded. **NOTE: Before you troubleshoot, record all freeze data and any on-board snapshot. When DTC P2821 is stored in the PCM, engine torque is restricted because of the fail-safe function.**
DTC: P2826 **2T PCM, TCIL: Yes** **Year:** 2011, 2012 **Model:** TL, ZDX **Engine:** 3.5L V6, 3.7L V6	**Short in Line Pressure Solenoid Valve A Circuit:** The return signal does not match the command to turn on line pressure solenoid valve A for at least 1 second. The circuit detects malfunctions such as a circuit short or open. **NOTE: Before you troubleshoot, record all freeze data and any on-board snapshot. This code is caused by an electrical circuit problem and cannot be caused by a mechanical problem in the transmission.**
DTC: P2827 **2T PCM, TCIL: Yes** **Year:** 2011, 2012 **Model:** TL, ZDX **Engine:** 3.5L V6, 3.7L V6	**Open in Line Pressure Solenoid Valve A Circuit:** The return signal does not match the command to turn off line pressure solenoid valve A for at least 1 second. **NOTE: Before you troubleshoot, record all freeze data and any on-board snapshot. This code is caused by an electrical circuit problem and cannot be caused by a mechanical problem in the transmission.**
DTC: P2A00 **2T PCM, MIL: Yes** **Year:** 2011, 2012 **Model:** MDX, RL, TL, TSX, ZDX **Engine:** 3.5L V6, 3.7L V6	**Rear Air/Fuel Ratio (A/F) Sensor (Bank 1, Sensor 1) Circuit Range/Performance Problem:** The rear A/F sensor (bank 1, sensor 1) output voltage is 2.55 V or less, or 4.50 V or more. The execution time is once per driving cycle and the duration time is 5.3 seconds or more. **NOTE: Before you troubleshoot, record all freeze data and any on-board snapshot.**
DTC: P2A00 **2T PCM, MIL: Yes** **Year:** 2011, 2012 **Model:** RDX, TSX **Engine:** 2.3L L4, 2.4L L4	**Air/Fuel Ratio (A/F) Sensor (Sensor 1) Circuit Range/Performance Problem:** The A/F sensor (Sensor 1) output voltage is 3.0 V or less, or 4.8 V or more. The execution time is once per driving cycle and the duration time is 3.5 seconds or more. **NOTE: Before you troubleshoot, record all freeze data and any on-board snapshot.**
DTC: P2A03 **2T PCM, MIL: Yes** **Year:** 2011, 2012 **Model:** MDX, RL, TL, TSX, ZDX **Engine:** 3.5L V6, 3.7L V6	**Front Air/Fuel Ratio (A/F) Sensor (Bank 2, Sensor 1) Circuit Range/Performance Problem:** The front A/F sensor (bank 2, sensor 1) output voltage is 2.55 V or less, or 4.50 V or more. The execution time is once per driving cycle and the duration time is 5.3 seconds or more.

OBD II Trouble Code List (P3XXX Codes)

DTC	Trouble Code Title and Conditions
DTC: P3400 **1T PCM, MIL: Yes** **Year:** 2012 **Model:** TL **Engine:** 3.5L V6	**Valve Pulse System (VPS) Stuck Off (Bank 1):** With the engine running in closed loop, and in 5th gear at a steady speed of 12 mph or more for 5 seconds. When the PCM sends ON (Open) or OFF (Close) command to the rocker arm oil control solenoid A (Bank 1) the rear rocker arm oil pressure switch detects a malfunction. **NOTE: Before you troubleshoot, record all freeze data and any on-board snapshot, and review the general troubleshooting information.** • Poor connections or loose terminals at rocker arm oil control solenoid A (Bank 1), rocker arm oil control solenoid B (Bank 1), the rear rocker arm oil pressure switch, and the PCM • Low or diluted engine oil • Clogged oil passage of the rear bank cylinder pause system • "Short" circuit between the PCM and the rear rocker arm oil pressure switch • Faulty rear rocker arm oil pressure switch
DTC: P3497 **1T PCM, MIL: Yes** **Year:** 2012 **Model:** TL **Engine:** 3.5L V6	**Valve Pause System (VPS) Stuck Off (Bank 2):** With the engine running in closed loop, and in 5th gear at a steady speed of 68 mph or more for 5 seconds. When the PCM sends ON (Open) or OFF (Close) command to the rocker arm oil control solenoid A (Bank 2) the rear rocker arm oil pressure switch detects a malfunction. • Poor connections or loose terminals at rocker arm oil control solenoid A (Bank 1), rocker arm oil control solenoid B (Bank 1), the rear rocker arm oil pressure switch, and the PCM • Low or diluted engine oil • Clogged oil passage of the rear bank cylinder pause system • "Short" circuit between the PCM and the rear rocker arm oil pressure switch • Faulty rear rocker arm oil pressure switch

HONDA

Accord • Crosstour

BRAKES 7-13

**ANTI-LOCK BRAKE
SYSTEM (ABS)** 7-13
General Information 7-13
  Precautions 7-13
Speed Sensors 7-13
  Removal & Installation 7-13
**BLEEDING THE BRAKE
SYSTEM** 7-15
Bleeding Procedure 7-15
  Manual 7-15
Brake Fluid 7-15
  Fluid Recommendations 7-15
  Level Check 7-15
FRONT DISC BRAKES 7-15
Brake Calipers 7-15
  Removal & Installation 7-15
Brake Pads 7-16
  Removal & Installation 7-16
PARKING BRAKE 7-20
Parking Brake Cable 7-20
  Adjustments 7-20
Parking Brake Shoes 7-21
  Adjustments 7-21
  Removal & Installation 7-21
REAR DISC BRAKES 7-19
Brake Calipers 7-19
  Removal & Installation 7-19
Brake Pads 7-19
  Removal & Installation 7-19

CHASSIS ELECTRICAL 7-23

**AIR BAG (SUPPLEMENTAL
RESTRAINT SYSTEM)** 7-23
Arming the System 7-24
Disarming the System 7-24
Precautions 7-23

DRIVE TRAIN 7-25

Automatic Transaxle Fluid 7-25
  Filter Replacement 7-25
Clutch 7-25
  Bleeding 7-25
Clutch Master Cylinder 7-25
  Removal & Installation 7-25

Clutch Slave (Release)
  Cylinder 7-27
  Removal & Installation 7-27
Front Halfshaft 7-28
  Removal & Installation 7-28
Intermediate Shaft 7-30
  Removal & Installation 7-30
Propeller Shaft 7-30
  Removal & Installation 7-30
Rear Differential 7-31
  Drain & Refill 7-31
  Level Check 7-31
  Removal & Installation 7-31
Rear Halfshaft 7-32
  Removal & Installation 7-32
Transfer Case Assembly 7-27
  Removal & Installation 7-27

ENGINE COOLING 7-33

Electric Engine Fan 7-34
  Removal & Installation 7-34
Engine Coolant 7-33
  Drain & Refill 7-33
  Fluid Recommendations 7-34
  Level Check 7-34
Radiator 7-35
  Removal & Installation 7-35
Thermostat 7-36
  Removal & Installation 7-36
Water Pump 7-37
  Removal & Installation 7-37

ENGINE ELECTRICAL 7-38

BATTERY SYSTEM 7-38
Battery 7-38
  Battery Reconnect/Relearn
   Procedure 7-38
  Removal & Installation 7-38
CHARGING SYSTEM 7-38
Alternator 7-38
  Removal & Installation 7-38
IGNITION SYSTEM 7-39
Firing Orders 7-39
Ignition Coil 7-39
  Removal & Installation 7-39
Ignition Timing 7-39

  Adjustment 7-40
  Inspection 7-39
Spark Plugs 7-40
  Removal & Installation 7-40
STARTING SYSTEM 7-41
Starter 7-41
  Removal & Installation 7-41

ENGINE MECHANICAL 7-42

Accessory Drive Belt System 7-42
  Accessory Belt Routing 7-42
  Adjustment 7-42
  Inspection 7-42
  Removal & Installation 7-42
Air Cleaner 7-42
  Air Filter Element
   Replacement 7-43
  Removal & Installation 7-42
Camshaft & Bearings 7-43
  Removal & Installation 7-43
Crankshaft Front Seal 7-44
  Removal & Installation 7-44
Crankshaft Pulley 7-44
  Removal & Installation 7-44
Crankshaft Rear Cover & Seal ... 7-45
  Removal & Installation 7-45
Cylinder Head 7-45
  Removal & Installation 7-45
Cylinder Head Cover 7-49
  Removal & Installation 7-49
Engine Oil & Filter 7-51
  Oil & Filter Change 7-51
  Oil Level Check 7-51
Exhaust Manifold 7-52
  Component Locations 7-52
Flexplate (Driveplate) 7-52
  Removal & Installation 7-52
Front Cover & Seal 7-52
  Removal & Installation 7-52
Intake Manifold 7-53
  Removal & Installation 7-53
Oil Filter Adapter 7-54
  Removal & Installation 7-54
Oil Pan 7-54
  Removal & Installation 7-54
Oil Pump 7-56
  Removal & Installation 7-56

Pistons & Rings 7-58
  Positioning 7-58
Rear Main Seal.....................7-58
  Removal & Installation.......7-58
Rocker Arms.........................7-59
  Removal & Installation.......7-59
Timing (Camshaft) Chain &
  Sprockets 7-65
  Removal & Installation.......7-65
Timing Belt & Sprockets 7-62
  Removal & Installation.......7-62
Timing Belt Front Cover 7-61
  Removal & Installation.......7-61
Timing Chain Cover 7-63
  Removal & Installation.......7-63
Valve Lash (Clearance)
  Adjustment.........................7-68
  Adjustment 7-68

**ENGINE PERFORMANCE &
EMISSION CONTROLS.......7-70**

Accelerator Pedal
  Position Sensor 7-70
  Location...............................7-70
  Removal & Installation.........7-70
Camshaft Position Sensor 7-70
  Removal & Installation.........7-70
Crankshaft Position
  Sensor 7-71
  CKP Pattern Clear/CKP
    Pattern Learn 7-71
  Removal & Installation.........7-71
EGR Valve 7-75
  Location...............................7-75
  Removal & Installation.........7-75
Engine Control Module
  (ECM)/Powertrain Control
  Module (PCM).......................7-71
  ECM/PCM Idle Learn
    Procedure 7-72
  ECM/PCM Update 7-72
  Location...............................7-71
  Removal & Installation.........7-71
Engine Coolant Temperature
  Sensor 7-73
  Removal & Installation.........7-73
EVAP Canister 7-74
  Location...............................7-74
  Removal & Installation.........7-74
EVAP Canister Purge
  Control Valve 7-75
  Removal & Installation.........7-75
Fuel Tank (EVAP) Pressure
  (FTP) Sensor 7-75
  Location...............................7-75
  Removal & Installation.........7-75

Heated Oxygen Sensor 7-76
  Location...............................7-76
  Removal & Installation.........7-76
Input Speed Sensor 7-77
  Removal & Installation.........7-77
Intake Air Temperature
  Sensor 7-78
  Location...............................7-78
  Removal & Installation.........7-78
Knock Sensor 7-78
  Removal & Installation.........7-78
Maintenance Reminder
  Lights..................................7-78
  Reset Procedure...................7-78
Manifold Absolute
  Pressure Sensor 7-79
  Removal & Installation.........7-79
Mass Airflow/Intake Air
  Temperature Sensor 7-79
  Location...............................7-79
  Removal & Installation.........7-79
PCV Valve 7-80
  Location...............................7-80
  Removal & Installation.........7-80
Throttle Control Actuator...........7-80
  Location...............................7-80
Throttle Position Sensor............7-80
  Location...............................7-80

FUEL...........................7-80

**GASOLINE FUEL INJECTION
SYSTEM........................7-80**
  Fuel Filter............................7-81
  Removal & Installation.........7-81
  Fuel Pressure Regulator 7-86
  Removal & Installation.........7-86
  Fuel Pump/Fuel Gauge
    Sending Unit........................7-86
  Removal & Installation.........7-86
  Fuel Rail & Injectors 7-86
  Removal & Installation.........7-86
  Fuel System Service
    Precautions.........................7-80
  Fuel Tank 7-81
    Draining..............................7-81
  Removal & Installation.........7-82
  Fuel Tank Unit 7-83
  Removal & Installation.........7-83
  Idle Speed 7-88
    Adjustment 7-88
  Relieving Fuel System
    Pressure.............................7-81
    With the HDS.....................7-81
    Without the HDS.................7-81
  Throttle Body.........................7-88
  Removal & Installation.........7-88

**HEATING & AIR
CONDITIONING SYSTEM 7-90**

Blower Motor 7-90
  Removal & Installation.........7-90
Heater Core 7-91
  Removal & Installation.........7-91

PRECAUTIONS 7-13

**SPECIFICATIONS AND
MAINTENANCE CHARTS 7-4**

Brake Specifications.................7-11
Camshaft and Bearing
  Specifications 7-6
Capacities 7-5
Crankshaft and Connecting
  Rod Specifications 7-7
Engine and Vehicle
  Identification 7-4
Engine Tune-Up Specifications ...7-4
Fluid Specifications...................7-5
General Engine Specifications.....7-4
Piston and Ring
  Specifications 7-7
Scheduled Maintenance
  Intervals...............................7-12
Tire Wheel and Ball Joint
  Specifications 7-9
Torque Specifications 7-8
Valve Specifications 7-6
Wheel Alignment......................7-9

STEERING 7-92

Power Rack & Pinion
  Steering Gear 7-92
  Removal & Installation.........7-92
Power Steering Fluid.................7-97
  Fluid Fill Procedure 7-97
  Fluid Level Check 7-97
  Fluid Recommendations 7-97
Power Steering Pump................7-96
  Removal & Installation.........7-96

SUSPENSION..................7-98

FRONT SUSPENSION..........7-98
Knuckle & Spindle 7-98
  Removal & Installation.........7-98
Lower Ball Joints.....................7-99
  Removal & Installation.........7-99
Lower Control Arms 7-100
  Removal & Installation.......7-100
Stabilizer Bar & Links 7-100
  Removal & Installation.......7-100

Strut Bar 7-105
   Removal & Installation........7-105
Struts.....................................7-105
   Overhaul 7-105
   Removal & Installation........7-105
Upper Ball Joints....................7-107
   Removal & Installation........7-107
Upper Control Arms 7-108
   Removal & Installation........7-108

Wheel Hubs & Bearings..........7-109
   Adjustment 7-109
   Removal & Installation........7-109
REAR SUSPENSION 7-111
Control
   Arms/Links 7-111
   Removal & Installation........7-111
Knuckles 7-113
   Removal & Installation........7-113

Stabilizer Bar 7-114
   Removal & Installation........7-114
Struts.....................................7-114
   Overhaul 7-115
   Removal & Installation........7-114
Wheel Hubs & Bearings..........7-116
   Adjustment 7-116
   Removal & Installation........7-116

SPECIFICATIONS AND MAINTENANCE CHARTS

ENGINE AND VEHICLE IDENTIFICATION

Engine						Model Year	
Code	Liters	Cu. In. (cc)	Cyl.	Fuel Sys.	Eng. Mfg.	Code	Year
J35Z2	3.5	212.0 (3471)	6	MPFI	Honda	B	2011
K24Z2	2.4	144.0 (2354)	4	MPFI	Honda	C	2012
K24Z3	2.4	144.0 (2354)	4	MPFI	Honda		

MPFI: Multiport Fuel Injection

71051_ACCD_C0001

GENERAL ENGINE SPECIFICATIONS

Year	Model	Engine Disp. Liters	Engine ID	Net Horsepower @ rpm	Net Torque @ rpm (ft. lbs.)	Bore x Stroke (in.)	Compression Ratio	Oil Pressure @ rpm
2011	Accord/Crosstour	2.4	K24Z2	177@6500	161@4300	3.43x3.90	10.5:1	44@3000
	Accord/Crosstour	2.4	K24Z3	190@7000	162@4400	3.43x3.90	10.5:1	44@3000
	Accord/Crosstour	3.5	J35Z2	271@6200	254@5000	3.50x3.70	10.5:1	44@3000
2012	Accord/Crosstour	2.4	K24Z2	177@6500	161@4300	3.43x3.90	10.5:1	44@3000
	Accord/Crosstour	2.4	K24Z3	190@7000	162@4400	3.43x3.90	10.5:1	44@3000
	Accord/Crosstour	3.5	J35Z2	271@6200	254@5000	3.50x3.70	10.5:1	44@3000

71051_ACCD_C0002

ENGINE TUNE-UP SPECIFICATIONS

Year	Engine Displacement Liters	Engine ID	Spark Plugs Gap (in.)	Ignition Timing (deg.) MT	Ignition Timing (deg.) AT	Fuel Pump (psi)	Idle Speed (rpm) MT	Idle Speed (rpm) AT	Valve Clearance In.	Valve Clearance Ex.
2011	2.4	K24Z2	0.039-0.043	6-10B	6-10B	48-55	720-830	750-850	0.008-0.010	0.010-0.011
	2.4	K24Z3	0.039-0.043	6-10B	6-10B	48-55	720-830	750-850	0.008-0.010	0.010-0.011
	3.5	J35Z2	0.039-0.043	3-7B	8-12B	57-64	600-700	600-700	0.008-0.009	0.011-0.013
2012	2.4	K24Z2	0.039-0.043	6-10B	6-10B	48-55	720-830	750-850	0.008-0.010	0.010-0.011
	2.4	K24Z3	0.039-0.043	6-10B	6-10B	48-55	720-830	750-850	0.008-0.010	0.010-0.011
	3.5	J35Z2	0.039-0.043	3-7B	8-12B	57-64	600-700	600-700	0.008-0.009	0.011-0.013

NOTE: The Vehicle Emission Control Information label often reflects specification changes made during production.

The label figures must be used if they differ from those in this chart

B: Before Top Dead Center

71051_ACCD_C0003

CAPACITIES

Year	Model	Engine Disp. Liters	Engine ID	Engine Oil with Filter	Transmission (pts.) 5-Spd	Transmission (pts.) Auto.	Rear Drive Axle (pts.)	Fuel Tank (gal.)	Cooling System (qts.)
2011	Accord/Crosstour	2.4	K24Z2	4.4	4.0	5.2	NA	18.5	①
	Accord/Crosstour	2.4	K24Z3	4.4	4.0	5.2	NA	18.5	①
	Accord/Crosstour	3.5	J35Z2	4.5	4.2	7.0	②	18.5	7.0
2012	Accord/Crosstour	2.4	K24Z2	4.4	4.0	5.2	NA	18.5	①
	Accord/Crosstour	2.4	K24Z3	4.4	4.0	5.2	NA	18.5	①
	Accord/Crosstour	3.5	J35Z2	4.5	4.2	7.0	②	18.5	7.0

NA: Not Applicable

NOTE: All capacities are approximate. Add fluid gradually and ensure a proper fluid level is obtained.

NOTE: Capacities given are service, not overhaul capacities

① Automatic Transaxle: 6.2

   Manual Transaxle: 6.4

② Crosstour 4WD: 2.6 pts.

71051_ACCD_C0004

FLUID SPECIFICATIONS

Year	Model	Engine Displ. Liters	Engine Oil	Man. Trans.	Auto. Trans.	Drive Axle Rear	Power Steering Fluid	Brake Master Cylinder	Cooling System
2011	Accord/ Crosstour	2.4	0W-20 Honda	Honda MTF	Honda ATF-DW1	NA	Honda PS Fluid	Honda DOT 3	①
	Accord/ Crosstour	3.5	0W-20 Honda	Honda MTF	Honda ATF-DW1	②	Honda PS Fluid	Honda DOT 3	①
2012	Accord/ Crosstour	2.4	0W-20 Honda	Honda MTF	Honda ATF-DW1	NA	Honda PS Fluid	Honda DOT 3	①
	Accord/ Crosstour	3.5	0W-20 Honda	Honda MTF	Honda ATF-DW1	②	Honda PS Fluid	Honda DOT 3	①

DOT: Department Of Transpotation

NA: Not Applicable

① Honda Long Life Antifreeze/Coolant-Type2

② Crosstour 4WD: Honda Dual Pump Fluid II

71051_ACCD_C0005

VALVE SPECIFICATIONS

Year	Engine Displacement Liters	Engine ID	Seat Angle (deg.)	Face Angle (deg.)	Spring Test Pressure (lbs. @ in.)	Spring Installed Height (in.)	Stem-to-Guide Clearance (in.)		Stem Diameter (in.)	
							Intake	Exhaust	Intake	Exhaust
2011	2.4	K24Z2	45	45	NA	NA	0.0012-0.0022	0.0022-0.0031	0.2156-0.2159	0.2146-0.2150
	2.4	K24Z3	45	45	NA	NA	0.0012-0.0022	0.0022-0.0031	0.2156-0.2159	0.2146-0.2150
	3.5	J35Z2	45	45	NA	NA	0.0008-0.0018	0.0022-0.0031	0.2159-0.2163	0.2146-0.2150
2012	2.4	K24Z2	45	45	NA	NA	0.0012-0.0022	0.0022-0.0031	0.2156-0.2159	0.2146-0.2150
	2.4	K24Z3	45	45	NA	NA	0.0012-0.0022	0.0022-0.0031	0.2156-0.2159	0.2146-0.2150
	3.5	J35Z2	45	45	NA	NA	0.0008-0.0018	0.0022-0.0031	0.2159-0.2163	0.2146-0.2150

NA: Information not available

71051_ACCD_C0007

CAMSHAFT AND BEARING SPECIFICATIONS
All measurements are given in inches.

Year	Engine Displacement Liters	Engine ID	Journal Diameter	Brg. Oil Clearance	Shaft End-play	Runout	Journal Bore	Lobe Height	
								Intake	Exhaust
2011	2.4	K24Z2	NA	①	0.0020-0.0080	0.0010	NA	②	1.3500
	2.4	K24Z3	NA	①	0.0020-0.0080	0.0010	NA	②	1.3500
	3.5	J35Z2	NA	0.0020-0.0035	0.0020-0.0080	0.0010	NA	③	④
2012	2.4	K24Z2	NA	①	0.0020-0.0080	0.0010	NA	②	1.3500
	2.4	K24Z3	NA	①	0.0020-0.0080	0.0010	NA	②	1.3500
	3.5	J35Z2	NA	0.0020-0.0035	0.0020-0.0080	0.0010	NA	③	④

NA: Information not available

① No. 1 journal: 0.0012-0.0027 inches

No. 2-5 journals: 0.0024-0.0039 inches

② Primary: 1.3285 inches

Mid: 1.3959 inches

Secondary: 1.3285 inches

Mid: 1.4348 in.

Secondary: 1.3891 in.

③ Nos. 1-4 cylinders: 1.3965 in.

Nos 5-6 cylinders: 1.3964 in.

④ Nos. 1-4 cylinders: 1.4481 in.

Nos 5-6 cylinders: 1.4472 in.

71051_ACCD_C0006

CRANKSHAFT AND CONNECTING ROD SPECIFICATIONS

All measurements are given in inches.

Year	Engine Displacement Liters	Engine ID	Crankshaft				Journal Diameter	Oil Clearance	Side Clearance
			Main Brg. Journal Dia.	Main Brg. Oil Clearance	Shaft End-play	Thrust on No.			
2011	2.4	K24Z2	①	②	0.0040-0.0140	4	1.8888-1.8898	0.0013-0.0026	0.0060-0.0140
	2.4	K24Z3	①	②	0.0040-0.0140	4	1.8888-1.8898	0.0013-0.0026	0.0060-0.0140
	3.5	J35Z2	2.8337-2.8346	0.0007-0.0018	0.0040-0.0140	3	2.1644-2.1654	0.0008-0.0017	0.0060-0.0140
2012	2.4	K24Z2	①	②	0.0040-0.0140	4	1.8888-1.8898	0.0013-0.0026	0.0060-0.0140
	2.4	K24Z3	①	②	0.0040-0.0140	4	1.8888-1.8898	0.0013-0.0026	0.0060-0.0140
	3.5	J35Z2	2.8337-2.8346	0.0007-0.0018	0.0040-0.0140	3	2.1644-2.1654	0.0008-0.0017	0.0060-0.0140

① Journals 1, 2 4 and 5: 2.1647-2.1657 inches
Journal 3: 2.1644-2.1654 inches

② Journals 1, 2 4 and 5: 0.0007-0.0016 inches
Journal 3: 0.0010-0.0019 inches

71051_ACCD_C0008

PISTON AND RING SPECIFICATIONS

All measurements are given in inches.

Year	Engine Disp. Liters	Engine ID	Piston Clearance	Ring Gap			Ring Side Clearance		
				Top Compression	Bottom Compression	Oil Control	Top Compression	Bottom Compression	Oil Control
2011	2.4	K24Z2	0.0008-0.0016	0.0080-0.0140	0.0200-0.0260	0.0080-0.0280	0.0024-0.0033	0.0016-0.0026	NA
	2.4	K24Z3	0.0008-0.0016	0.0080-0.0140	0.0200-0.0260	0.0080-0.0280	0.0024-0.0033	0.0016-0.0026	NA
	3.5	J35Z2	0.0006-0.0016	0.0080-0.0140	0.0160-0.0220	0.0080-0.0280	0.0022-0.0031	0.0012-0.0022	NA
2012	2.4	K24Z2	0.0008-0.0016	0.0080-0.0140	0.0200-0.0260	0.0080-0.0280	0.0024-0.0033	0.0016-0.0026	NA
	2.4	K24Z3	0.0008-0.0016	0.0080-0.0140	0.0200-0.0260	0.0080-0.0280	0.0024-0.0033	0.0016-0.0026	NA
	3.5	J35Z2	0.0006-0.0016	0.0080-0.0140	0.0160-0.0220	0.0080-0.0280	0.0022-0.0031	0.0012-0.0022	NA

NA: Information not available

71051_ACCD_C0009

TORQUE SPECIFICATIONS
All readings in ft. lbs.

Year	Engine Displacement Liters	Engine ID	Cylinder Head Bolts	Main Bearing Bolts	Rod Bearing Bolts	Crankshaft Damper Bolts	Flywheel Bolts	Manifold Intake	Spark Plugs	Oil Pan Drain Plug
2011	2.4	K24Z2	①	②	③	④	⑤	16	13	29
	2.4	K24Z3	①	②	③	④	⑤	16	13	29
	3.5	J35Z2	⑥	⑦	⑧	⑨	⑩	16	13	29
2012	2.4	K24Z2	①	②	③	④	⑤	16	13	29
	2.4	K24Z3	①	②	③	④	⑤	16	13	29
	3.5	J35Z2	⑥	⑦	⑧	⑨	⑩	16	13	29

① Step 1: 29 ft. lbs.

Step 2: Rotate 90 degrees

Step 3: Rotate 90 degrees

Step 4: If new bolt rotate

additional 90 degrees

② Step 1: 22 ft. lbs.

Step 2: Rotate 48 degrees

③ Step 1: 30 ft. lbs.

Step 2: Rotate 120 degrees

④ Old bolt:

Step 1: 36 ft. lbs

Step 2: Plus 90 degrees

New bolt:

Step 1: 130 ft. lbs

Step 2: Loosen fully

Step 3: 36 ft. lbs

Step 4: Plus 90 degrees

⑤ Automatic transaxle: 54 ft. lbs.

Manual transaxle: 90 ft. lbs.

⑥ Step 1: 22 ft. lbs.

Step 2: Rotate 90 degrees

Step 3: Rotate 90 degrees

Step 4: If new bolt rotate

additional 90 degrees

⑦ Step 1: Cap bolts, 56 ft. lbs.

Step 2: Side bolts, 36 ft. lbs.

⑧ Step 1: 14 ft. lbs.

Step 2: Rotate 90 degrees

⑨ Step 1: 47 ft. lbs.

Step 2: Rotate 60 degrees

⑩ Automatic transaxle: 54 ft. lbs.

Manual transaxle: 76 ft. lbs.

71051_ACCD_C0010

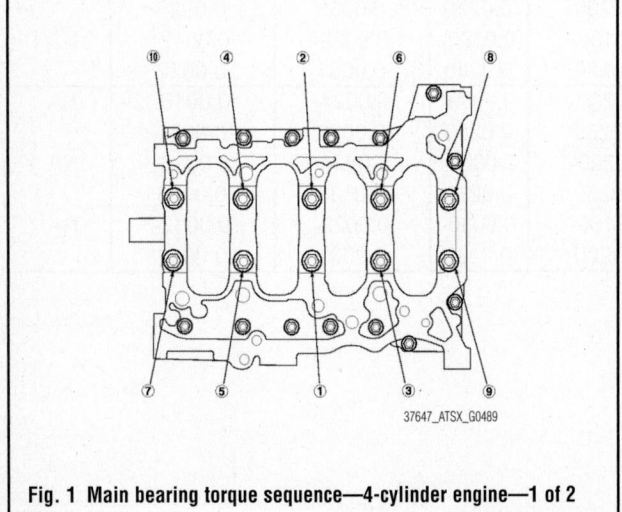

37647_ATSX_G0489

Fig. 1 Main bearing torque sequence—4-cylinder engine—1 of 2

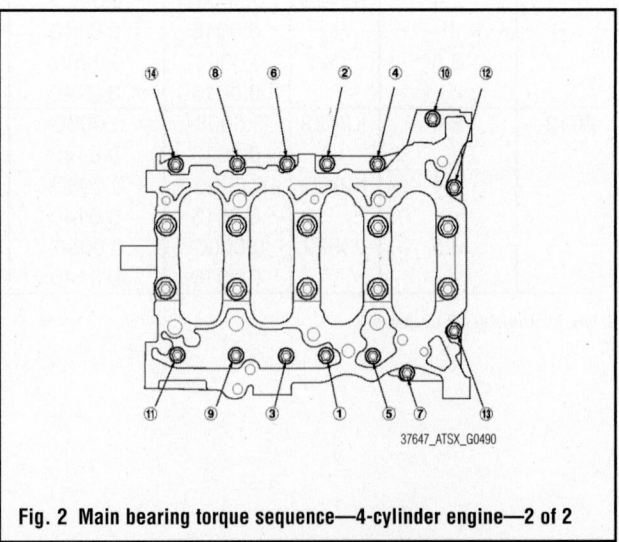

37647_ATSX_G0490

Fig. 2 Main bearing torque sequence—4-cylinder engine—2 of 2

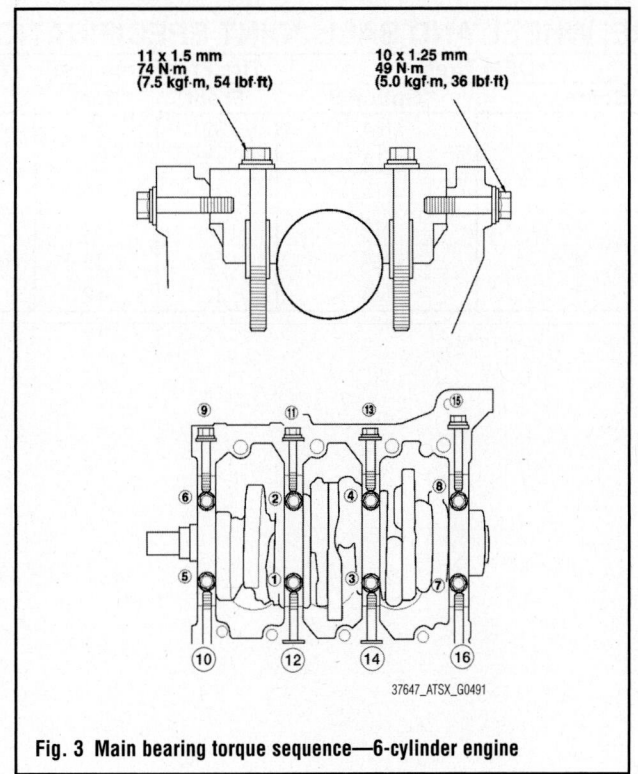

Fig. 3 Main bearing torque sequence—6-cylinder engine

WHEEL ALIGNMENT

Year	Model	Position	Caster Range (+/-Deg.)	Caster Preferred Setting (Deg.)	Camber Range (+/-Deg.)	Camber Preferred Setting (Deg.)	Toe-in (in.)
2011	Accord	F	+0.25/-1.05	①	②	0.00	0.00 +/-0.08
		R	—	—	③	-1 00'	0.08 +/-0.08
	Crosstour	F	+/-0.30	3.30	+/-0.30	0.00	0.00 +/-0.08
		R			+/-0.30	-1 00'	0.00 +/-0.08
2012	Accord	F	+0.25/-1.05	①	②	0.00	0.00 +/-0.08
		R	—	—	③	-1 00'	0.08 +/-0.08
	Crosstour	F	+/-0.30	3.30	+/-0.30	0.00	0.00 +/-0.08
		R			+/-0.30	-1 00'	0.00 +/-0.08

① 2-door: 3.47; 4-door: 3.48

② 17 inch wheels: +0.30 -0.45
   18 inch wheels: 0.05 +0.30 -0.45

③ 17 inch wheels: -+0.30 -0.45
   18 inch wheels: -1.19 +0.30 -0.45

TIRE, WHEEL AND BALL JOINT SPECIFICATIONS

Year	Model	OEM Tires		Tire Pressures (psi)		Wheel Size	Ball Joint Inspection	Lug Nut (ft. lbs.)
		Standard	Optional	Front	Rear			
2011	Accord Coupe	①	NA	32	32	②	NA	80
	Accord Sedan	③	NA	32	32	④	NA	80
	Crosstour	⑤	NA	32	32	⑥	NA	80
2012	Accord Coupe	①	NA	32	32	②	NA	80
	Accord Sedan	③	NA	32	32	④	NA	80
	Crosstour	⑤	NA	32	32	⑥	NA	80

OEM: Original Equipment Manufacturer

PSI: Pounds Per Square Inch

NA: Information not available

① LX-S/EX/EX-L: P225/50R17

   EX-L V6: P235/45R18

② LX-S/EX/EX-L: 17 inch

   EX-L V6: 18 inch

③ LX/LX-P/SE: P215/60R16

   EX/EX-L: P225/50R17

   EX V6/EX-L V6: P225/50R17

④ LX/LX-P/SE: 16 inch

   EX/EX-L: 17 inch

   EX V6/EX-L V6: 17 inch

⑤ EX/EX-L/EX V6: P225/65R17

   EX-L V6: P225/60R18

⑥ EX/EX-L/EX V6: 17 inch

   EX-L V6: 18 inch

71051_ACCD_C0012

BRAKE SPECIFICATIONS

All measurements in inches unless noted

| Year | Model | Position | Brake Disc | | | Minimum Lining Thickness | | Brake Caliper | |
			Original Thickness	Minimum Thickness	Maximum Runout	Front	Rear	Bracket Bolts (ft. lbs.)	Mounting Bolts (ft. lbs.)
2011	Accord	F	①	②	0.0016	③	—	80	25
	L4	R	0.350-0.358	0.315	0.0016	—	0.039	80	17
	Accord	F	1.098-1.106	1.024	0.0016	0.063	—	80	37
	V6	R	0.350-0.358	0.315	0.0016	—	0.039	80	17
	Crosstour	F	1.102	1.024	0.0016	0.063	—	101	37
	L4	R	0.354	0.315	0.0016	—	0.063	80	17
	Crosstour	F	1.102	1.024	0.0016	0.063	—	101	37
	V6	R	0.354	0.315	0.0016	—	0.063	80	17
2012	Accord	F	①	②	0.0016	③	—	80	25
	L4	R	0.350-0.358	0.315	0.0016	—	0.039	80	17
	Accord	F	1.098-1.106	1.024	0.0016	0.063	—	80	37
	V6	R	0.350-0.358	0.315	0.0016	—	0.039	80	17
	Crosstour	F	1.102	1.024	0.0016	0.063	—	101	37
	L4	R	0.354	0.315	0.0016	—	0.063	80	17
	Crosstour	F	1.102	1.024	0.0016	0.063	—	101	37
	V6	R	0.354	0.315	0.0016	—	0.063	80	17

F: Front

R: Rear

① NISSIN: 1.098-1.106 inches
   AKEBONO: 0.902-0.909 inches

② NISSIN: 1.024 inches
   AKEBONO: 0.827 inches

③ NISSIN: 0.063 inches
   AKEBONO: 0.063 inches

71051_ACCD_C0013

MAINTENANCE MINDER SCHEDULE
Honda Accord

All Honda's displays engine oil life and maintenance service items in the information display to indicate when to perform maintenance service. If the engine oil life is 15% or less, based on the onboard computer's caluculations, you will see SERVICE DUE SOON in the information display every time the ignition key is turned to ON. The maintenance minder indicator will also come on and the maintenance code(s) for other scheduled maintenance items needing service will be displayed below the message.

Symbol	Item	Service
A	Engine oil ①	Change
B	Engine oil and filter	Change
	Fluid levels	Inspect
	Brakes	Inspect
	Parking brake adjustment	Check
	Steering gear and linkage	Inspect
	Suspension components	Inspect
	Driveshaft boots	Inspect
	Brake hoses and lines	Inspect
	Exhaust system	Inspect
	Fuel lines and connections	Inspect
1	Tires	Rotate
2	Engine air filter ②	Replace
	Dust and pollen filter ③	Replace
	Accessory drive belt	Inspect
3	Transmission fluid ④	Replace
	Transfer case fluid ④	Replace
4	Spark plugs	Replace
	Timing belt ⑤	Replace
	Water pump	Inspect
	Valve clearance ⑥	Inspect
5	Engine coolant	Replace
6	VTM-4 rear differential fluid	Replace

① If the message SERVICE DUE NOW does not appear more than 12 months after the display is reset, change every year.

② If driven in dusty conditions, replace every 15,000 miles.

③ If driven in urban areas that have a high concentration of soot from industry and diesel, replace every 15,000 miles

④ If regularly driven in mountainous areas at very low speed or trailer towing, change the fluid every 30,000 miles.

⑤ If driven regularly in temperatures over 110 deg.F or below -20 deg.F, or towing a trailer, replace every 60,000 miles.

⑥ Adjust if necessary.

Additionally, replace the brake fluid every 3 years, and inspect the idle speed every 160,000 miles.
To reset the Engine Oil Life Display:
1. Turn the ignition switch to ON.
2. Press the SELECT button repeatedly until the engine oil life display or the service message is displayed.
3. Press the RESET button for about 10 seconds. You will see a MAINT RESET message.
4. Select the appropriate answer, MAINT RESET >N (NO) or MAINT RESET > y (YES) by pressing the SELECT button repeatedly.
   >N or >Y is displayed on the outside temperature >N or >Y is displayed on the outside temperature display.
5. Select the MAINT RESET > Y (YES), and press and hold the RESET button again to reset the engine oil life to 100%.

PRECAUTIONS

Before servicing any vehicle, please be sure to read all of the following precautions, which deal with personal safety, prevention of component damage, and important points to take into consideration when servicing a motor vehicle:

• Never open, service or drain the radiator or cooling system when the engine is hot; serious burns can occur from the steam and hot coolant.

• Observe all applicable safety precautions when working around fuel. Whenever servicing the fuel system, always work in a well-ventilated area. Do not allow fuel spray or vapors to come in contact with a spark, open flame, or excessive heat (a hot drop light, for example). Keep a dry chemical fire extinguisher near the work area. Always keep fuel in a container specifically designed for fuel storage; also, always properly seal fuel containers to avoid the possibility of fire or explosion. Refer to the additional fuel system precautions later in this section.

• Fuel injection systems often remain pressurized, even after the engine has been turned **OFF**. The fuel system pressure must be relieved before disconnecting any fuel lines. Failure to do so may result in fire and/or personal injury.

• Brake fluid often contains polyglycol ethers and polyglycols. Avoid contact with the eyes and wash your hands thoroughly after handling brake fluid. If you do get brake fluid in your eyes, flush your eyes with clean, running water for 15 minutes. If eye irritation persists, or if you have taken brake fluid internally, IMMEDIATELY seek medical assistance.

• The EPA warns that prolonged contact with used engine oil may cause a number of skin disorders, including cancer. You should make every effort to minimize your exposure to used engine oil. Protective gloves should be worn when changing oil. Wash your hands and any other exposed skin areas as soon as possible after exposure to used engine oil. Soap and water, or waterless hand cleaner should be used.

• All new vehicles are now equipped with an air bag system, often referred to as a Supplemental Restraint System (SRS) or Supplemental Inflatable Restraint (SIR) system. The system must be disabled before performing service on or around system components, steering column, instrument panel components, wiring and sensors. Failure to follow safety and disabling procedures could result in accidental air bag deployment, possible personal injury and unnecessary system repairs.

• Always wear safety goggles when working with, or around, the air bag system. When carrying a non-deployed air bag, be sure the bag and trim cover are pointed away from your body. When placing a non-deployed air bag on a work surface, always face the bag and trim cover upward, away from the surface. This will reduce the motion of the module if it is accidentally deployed. Refer to the additional air bag system precautions later in this section.

• Clean, high quality brake fluid from a sealed container is essential to the safe and proper operation of the brake system. You should always buy the correct type of brake fluid for your vehicle. If the brake fluid becomes contaminated, completely flush the system with new fluid. Never reuse any brake fluid. Any brake fluid that is removed from the system should be discarded. Also, do not allow any brake fluid to come in contact with a painted surface; it will damage the paint.

• Never operate the engine without the proper amount and type of engine oil; doing so WILL result in severe engine damage.

• Timing belt maintenance is extremely important. Many models utilize an interference-type, non-freewheeling engine. If the timing belt breaks, the valves in the cylinder head may strike the pistons, causing potentially serious (also time-consuming and expensive) engine damage. Refer to the maintenance interval charts for the recommended replacement interval for the timing belt, and to the timing belt section for belt replacement and inspection.

• Disconnecting the negative battery cable on some vehicles may interfere with the functions of the on-board computer system(s) and may require the computer to undergo a relearning process once the negative battery cable is reconnected.

• When servicing drum brakes, only disassemble and assemble one side at a time, leaving the remaining side intact for reference.

• Only an MVAC-trained, EPA-certified automotive technician should service the air conditioning system or its components.

BRAKES

GENERAL INFORMATION

PRECAUTIONS

• Certain components within the ABS system are not intended to be serviced or repaired individually.

• Do not use rubber hoses or other parts not specifically specified for and ABS system. When using repair kits, replace all parts included in the kit. Partial or incorrect repair may lead to functional problems and require the replacement of components.

• Lubricate rubber parts with clean, fresh brake fluid to ease assembly. Do not use shop air to clean parts; damage to rubber components may result.

• Use only DOT 3 brake fluid from an unopened container.

• If any hydraulic component or line is removed or replaced, it may be necessary to bleed the entire system.

• A clean repair area is essential. Always clean the reservoir and cap thoroughly before removing the cap. The slightest amount of dirt in the fluid may plug an orifice and impair the system function. Perform repairs after components have been thoroughly cleaned; use only denatured alcohol to clean components. Do not allow ABS components to come into contact with any substance containing mineral oil; this includes used shop rags.

• The Anti-Lock control unit is a microprocessor similar to other computer units in

ANTI-LOCK BRAKE SYSTEM (ABS)

the vehicle. Ensure that the ignition switch is **OFF** before removing or installing controller harnesses. Avoid static electricity discharge at or near the controller.

• If any arc welding is to be done on the vehicle, the control unit should be unplugged before welding operations begin.

SPEED SENSORS

REMOVAL & INSTALLATION

Front

See Figure 4.

1. Turn the ignition switch to LOCK (0).
2. Release the clamp, then disconnect the wheel speed sensor connector.

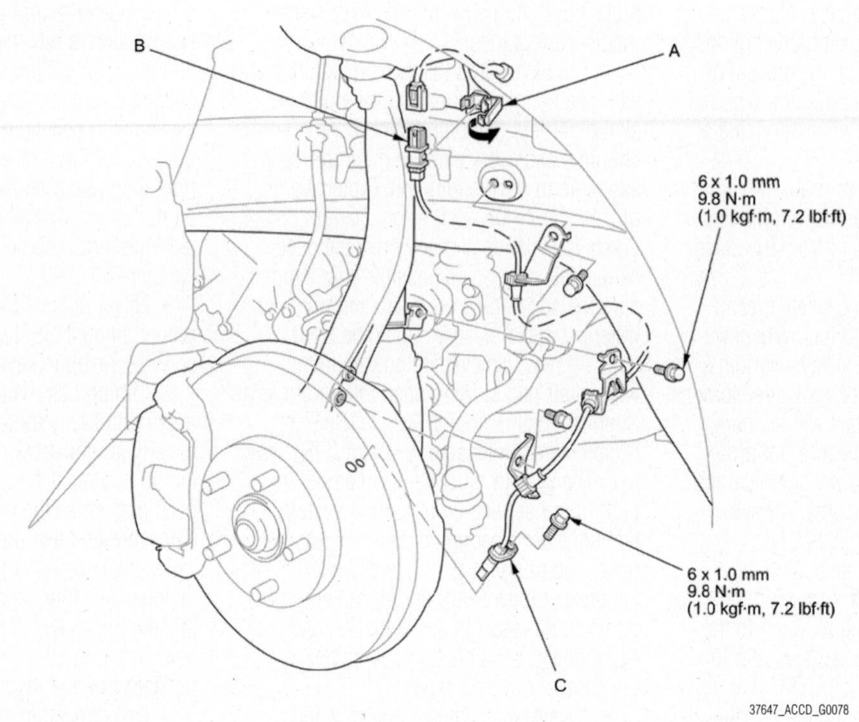

6 x 1.0 mm
9.8 N·m
(1.0 kgf·m, 7.2 lbf·ft)

6 x 1.0 mm
9.8 N·m
(1.0 kgf·m, 7.2 lbf·ft)

37647_ACCD_G0078

Fig. 4 Release the clamp (A), then disconnect the wheel speed sensor connector (B) and remove the wheel speed sensor (C)

3. Remove the bolts and the wheel speed sensor.

To install:

4. Install the wheel speed sensor in the reverse order of removal, and note these items:

- Do not twist the sensor wires.
- If the wheel speed sensor comes in contact with the wheel bearing, it is faulty.
- Make sure there is no debris in the sensor mounting hole.

5. Start the engine, and make sure the ABS and the VSA indicators go off.

6. Test-drive the vehicle, and make sure the ABS and the VSA indicators do not come on.

Rear

See Figure 5.

1. Turn the ignition switch to LOCK (0).

2. Release the clamp, then disconnect the wheel speed sensor connector.

3. Remove the clamps, the bolt, and the wheel speed sensor.

4. Install the wheel speed sensor in

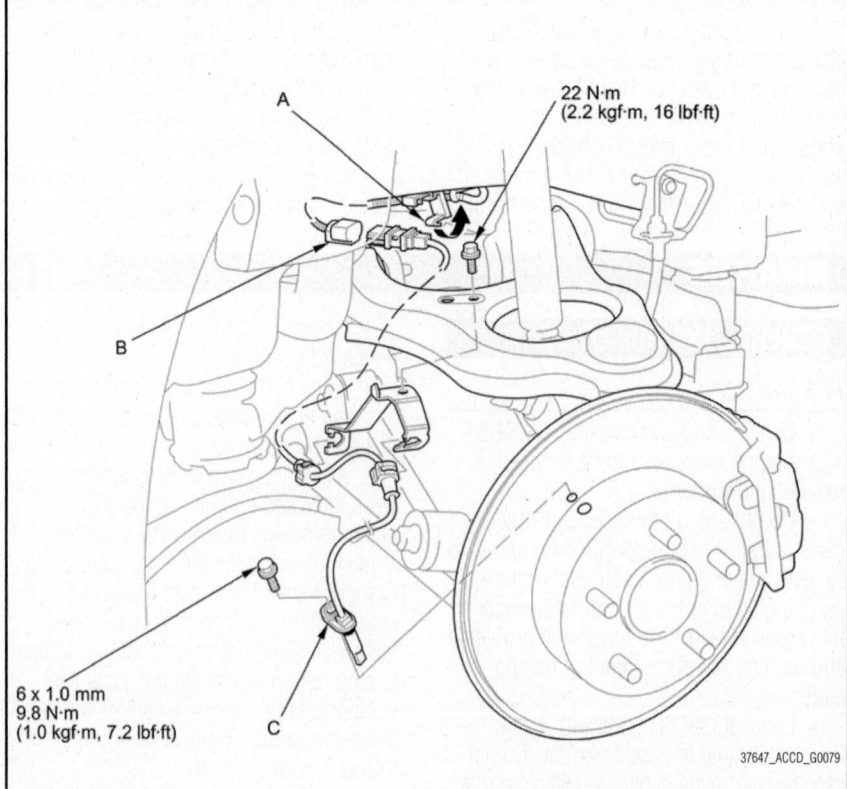

22 N·m
(2.2 kgf·m, 16 lbf·ft)

6 x 1.0 mm
9.8 N·m
(1.0 kgf·m, 7.2 lbf·ft)

37647_ACCD_G0079

Fig. 5 Release the clamp (A), then disconnect the wheel speed sensor connector (B) and remove the wheel speed sensor (C)

the reverse order of removal, and note these items:

- Do not twist the sensor wires.
- If the wheel speed sensor comes in contact with the hub bearing unit, it is faulty.
- Make sure there is no debris in the sensor mounting hole.

5. Start the engine, and make sure the ABS and the VSA indicators go off.

6. Test-drive the vehicle, and make sure the ABS and the VSA indicators do not come on.

BRAKES | BLEEDING THE BRAKE SYSTEM

BLEEDING PROCEDURE

MANUAL

See Figure 6.

➡Do not reuse the drained fluid. Use only clean Honda DOT 3 Brake Fluid from an unopened container. Using a non-Honda brake fluid can cause corrosion and shorten the life of the system.

➡Do not mix different brands of brake fluid; they may not be compatible.

➡Make sure no dirt or other foreign matter is allowed to contaminate the brake fluid.

※ WARNING

Do not spill brake fluid on the vehicle, it may damage the paint; if brake fluid does contact the paint, wash it off immediately with water.

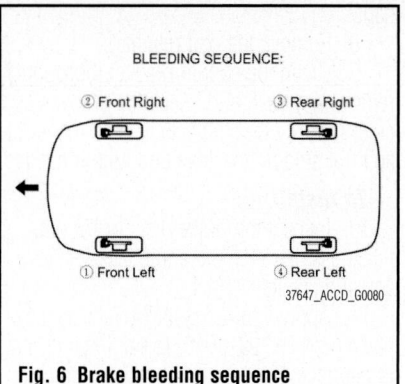

BLEEDING SEQUENCE:

② Front Right   ③ Rear Right

① Front Left   ④ Rear Left

37647_ACCD_G0080

Fig. 6 Brake bleeding sequence

➡The reservoir on the master cylinder must be at the MAX (upper) level mark at the start of the bleeding procedure and checked after bleeding each brake caliper. Add fluid as required.

1. Make sure the brake fluid level in the reservoir is at the MAX (upper) level line.
2. Attach a length of clear drain tube to the bleed screw.

3. Have someone slowly pump the brake pedal several times, and then apply steady pressure.

4. Starting at the left-front, loosen the brake bleed screw to allow air to escape from the system. Then tighten the bleed screw securely.

5. Repeat the procedure for each wheel in the sequence shown following until air bubbles no longer appear in the fluid.

6. Refill the master cylinder reservoir to the MAX (upper) level line.

BRAKE FLUID

FLUID RECOMMENDATIONS

Use only clean Honda DOT 3 Brake Fluid from an unopened container.

LEVEL CHECK

View the level in the reservoir to ensure the brake fluid is up to the "MAX line.

BRAKES | FRONT DISC BRAKES

※ CAUTION

Dust and dirt accumulating on brake parts during normal use may contain asbestos fibers from production or aftermarket brake linings. Breathing excessive concentrations of asbestos fibers can cause serious bodily harm. Exercise care when servicing brake parts. Do not sand or grind brake lining unless equipment used is designed to contain the dust residue. Do not clean brake parts with compressed air or by dry brushing. Cleaning should be done by dampening the brake components with a fine mist of water, then wiping the brake components clean with a dampened cloth. Dispose of cloth and all residue containing asbestos fibers in an impermeable container with the appropriate label. Follow practices prescribed by the Occupational Safety and Health Administration

(OSHA) and the Environmental Protection Agency (EPA) for the handling, processing, and disposing of dust or debris that may contain asbestos fibers.

BRAKE CALIPERS

REMOVAL & INSTALLATION
See Figure 7.

➡Keep grease away from the brake disc and the brake pads.

1. Raise and support the vehicle.
2. Remove the front wheel.
3. Remove the brake hose mounting bolt and disconnect the brake hose from the caliper.
4. Remove the brake caliper bracket mounting bolts, then remove the caliper assembly from the knuckle.
5. Installation is the reverse order of removal.

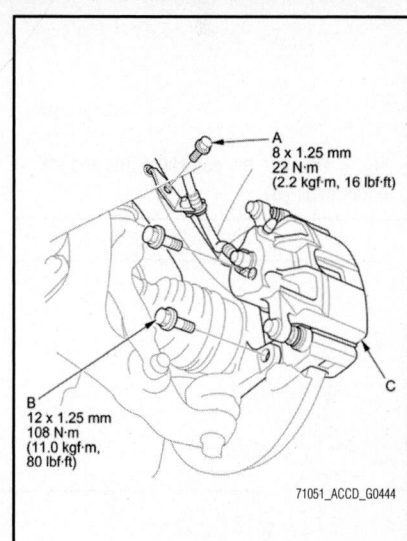

A
8 x 1.25 mm
22 N·m
(2.2 kgf·m, 16 lbf·ft)

B
12 x 1.25 mm
108 N·m
(11.0 kgf·m, 80 lbf·ft)

C

71051_ACCD_G0444

Fig. 7 Remove the brake hose mounting bolt (A), the brake caliper bracket mounting bolts (B), then remove the caliper assembly (C)

BRAKE PADS

REMOVAL & INSTALLATION

Accord

NISSIN Type

See Figures 8 through 11.

1. Remove some brake fluid from the master cylinder.
2. Raise and support the vehicle.
3. Remove the front wheels.
4. Remove the brake hose mounting bolt.
5. Remove the flange bolt while holding the caliper pin with a wrench. Be careful not to damage the pin boot, and pivot the caliper body up out of the way. Check the hose and the pin boots for damage and deterioration.

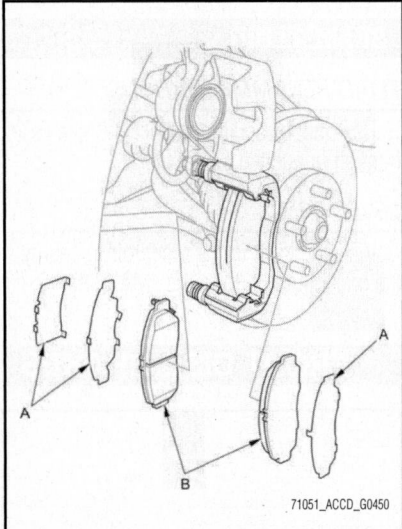

Fig. 8 Remove the pad shims (A) and the brake pads (B)

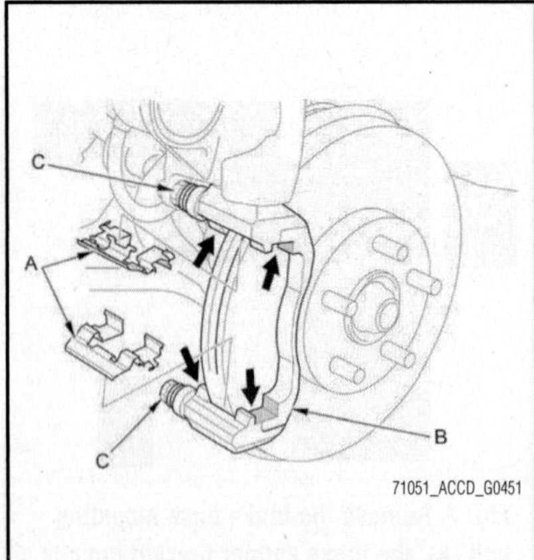

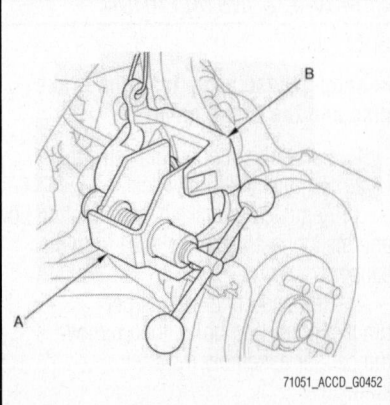

Fig. 9 Remove the pad retainers (A); caliper bracket (B), caliper pins (C)

6. For some models: Remove the pad return springs while holding the brake pads.

➡**The pad return springs are installed on the pads to prevent brake drag. Be careful when pivoting the caliper body up or the spring could pop out of position.**

7. Remove the pad shims and the brake pads.
8. Remove the pad retainers.
9. Clean the caliper bracket thoroughly; remove any rust, and check for grooves and cracks. Verify that the caliper pins move in and out smoothly. Clean and lube if needed.

To install:

10. Inspect the brake disc for runout, thickness, parallelism, and check for damage and cracks.
11. Apply a thin coat of M-77 assembly paste (P/N 08798-9010) to the retainer mating surface of the caliper bracket (indicated by the arrows).
12. Install the pad retainers. Wipe off the excess assembly paste from the retainers. Keep the assembly paste away from the brake disc and the brake pads.
13. Install the brake caliper piston compressor tool on the caliper body.
14. Press in the piston with the brake caliper piston compressor tool so the caliper body will fit over the brake pads. Make sure the piston boot is in position to prevent damaging it when pivoting the caliper body down.

➡**Be careful when pressing in the piston; brake fluid might overflow from the master cylinder's reservoir. If brake fluid gets on any painted surface, wash it off immediately with water.**

15. Remove the brake caliper piston compressor tool.

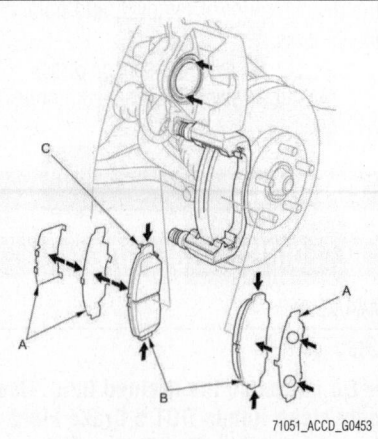

Fig. 11 Apply a thin coat of M-77 assembly paste to the pad side of the shims (A), the back of the brake pads (B) and the other areas indicated by the arrows; install wear indicator (C) on the upper inside position

16. Apply a thin coat of M-77 assembly paste (P/N 08798- 9010) to the pad side of the shims (A), the back of the brake pads (B) and the other areas indicated by the arrows. Wipe off the excess assembly paste from the pad shims and the brake pads friction material.

➡**Keep grease and assembly paste away from the brake disc and the brake pads. Contaminated brake disc or brake pads reduce stopping ability.**

17. Install the brake pads and the pad shims correctly. Install the brake pad with the wear indicator (C) on the upper inside position. If you are reusing the brake pads, always reinstall the brake pads in their original positions to prevent a temporary loss of braking efficiency.
18. For some models: Install the pad return springs while holding the brake pads.

➡**Insert the pad return spring ends into the pad installation holes securely.**

19. Pivot the caliper body down into position. Install the flange bolt, and tighten it to 37 ft. lbs. (50 Nm), while holding the caliper pin with a wrench being careful not to damage the pin boot.
20. Install the brake hose mounting bolt. Tighten to 16 ft. lbs. (22 Nm).
21. Clean the mating surfaces between the brake disc and the inside of the wheel, then install the front wheels.
22. Press the brake pedal several times to make sure the brakes work.

➡**Engagement may require a greater pedal stroke immediately after the**

Fig. 10 Install the brake caliper piston compressor tool (A) on the caliper body (B)

brake pads have been replaced as a set. Several applications of the brake pedal will restore the normal pedal stroke.

23. Add brake fluid as needed.

24. After installation, check for leaks at hose and line joints or connections, and retighten if necessary. Test-drive the vehicle, then recheck for leaks.

AKEBONO TYPE

See Figures 12 through 15.

1. Remove some brake fluid from the master cylinder.

2. Raise and support the vehicle.

3. Remove the front wheels.

4. Remove the brake hose mounting bolt.

5. Remove the flange bolt, and pivot the caliper body up out of the way. Check the hose and the pin boots for damage and deterioration.

6. For some models: Remove the pad return springs while holding the brake pads.

➡ **The pad return springs are installed on the pads to prevent brake drag. Be careful when pivoting the caliper body up or the spring could pop out of position.**

7. Remove the pad shims and the brake pads.

8. Remove the pad retainers.

9. Clean the caliper bracket thoroughly; remove any rust, and check for grooves and cracks. Verify that the caliper pins move in and out smoothly. Clean and lube if needed.

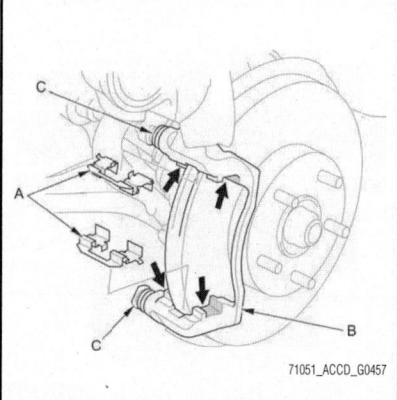

Fig. 13 Remove the pad retainers (A); Clean the caliper bracket (B); caliper pins (C)

To install:

10. Inspect the brake disc for runout, thickness, parallelism, and check for damage and cracks.

11. Apply a thin coat of M-77 assembly paste (P/N 08798-9010) to the retainer mating surface of the caliper bracket (indicated by the arrows).

12. Install the pad retainers. Wipe off the excess assembly paste from the retainers. Keep the assembly paste away from the brake disc and the brake pads.

13. Install the brake caliper piston compressor tool (A) on the caliper body (B).

14. Press in the piston with the brake caliper piston compressor tool so the caliper body will fit over the brake pads. Make sure the piston boot is in position to prevent damaging it when pivoting the caliper body down.

➡ **Be careful when pressing in the piston; brake fluid might overflow from the master cylinder's reservoir. If brake fluid gets on any painted surface, wash it off immediately with water.**

15. Remove the brake caliper piston compressor tool.

16. Apply a thin coat of M-77 assembly paste (P/N 08798-9010) to the pad side of the shims (A), the back of the brake pads (B) and the other areas indicated by the arrows. Wipe off the excess assembly paste from the pad shims and the brake pads friction material.

➡ **Keep grease and assembly paste away from the brake disc and the brake pads. Contaminated brake disc or brake pads reduce stopping ability.**

17. Install the brake pads and the pad shims correctly. Install the brake pad with the wear indicator (C) on the upper inside position. If you are reusing the brake pads, always reinstall the brake pads in their original positions to prevent a temporary loss of braking efficiency.

18. For some models: Install the pad return springs while holding the brake pads.

➡ **Insert the pad return spring ends into the pad installation holes securely.**

19. Pivot the caliper body down into position. Install the flange bolt, and tighten it to 25 ft. lbs. (34 Nm).

20. Install the brake hose mounting bolt. Tighten to 16 ft. lbs. (22 Nm).

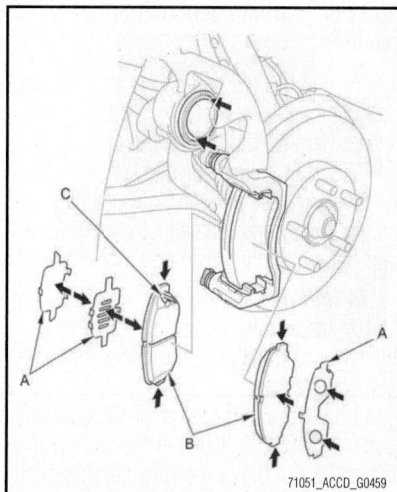

Fig. 15 Apply a thin coat of M-77 assembly paste to the pad side of the shims (A), the back of the brake pads (B) and the other areas indicated by the arrows; install the brake pad with the wear indicator (C) on the upper inside position

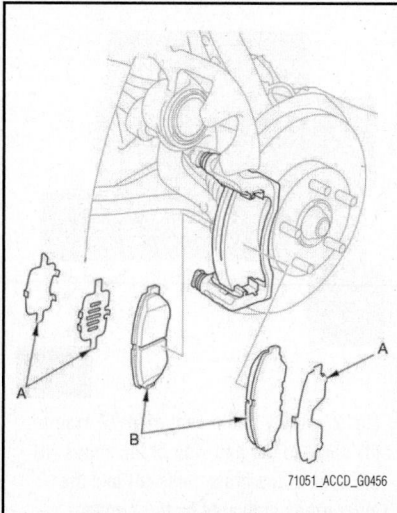

Fig. 12 Remove the pad shims (A) and the brake pads (B)

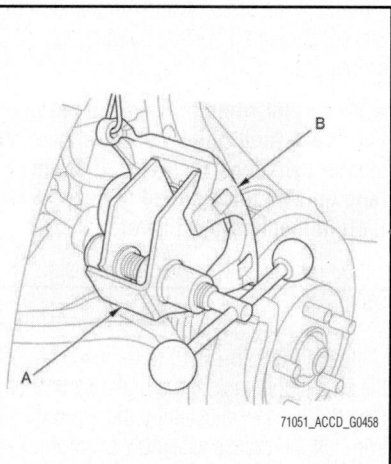

Fig. 14 Install the brake caliper piston compressor tool (A) on the caliper body (B)

21. Clean the mating surfaces between the brake disc and the inside of the wheel, then install the front wheels.

22. Press the brake pedal several times to make sure the brakes work.

➡**Engagement may require a greater pedal stroke immediately after the brake pads have been replaced as a set. Several applications of the brake pedal will restore the normal pedal stroke.**

23. Add brake fluid as needed.

24. After installation, check for leaks at hose and line joints or connections, and retighten if necessary. Test-drive the vehicle, then recheck for leaks.

Crosstour

See Figures 16 through 20.

1. Remove some brake fluid from the master cylinder.

2. Raise and support the vehicle.

3. Remove the front wheels.

4. Remove the brake hose mounting bolt.

5. Remove the flange bolt while holding the caliper pin with a wrench. Be careful not to damage the pin boot, and pivot the caliper body up out of the way. Check the hose and the pin boots for damage and deterioration.

6. Remove the pad return springs while holding the brake pads.

➡**The pad return springs are installed on the pads to prevent brake drag. Be careful when pivoting the caliper body up or the spring could pop out of position.**

7. Remove the pad shims and the brake pads.

8. Remove the pad retainers.

9. Clean the caliper bracket thoroughly; remove any rust, and check for grooves and cracks. Verify that the caliper pins move in and out smoothly. Clean and lube if needed.

To install:

10. Inspect the brake disc for runout, thickness, parallelism, and check for damage and cracks.

11. Apply a thin coat of M-77 assembly paste (P/N 08798-9010) to the retainer mating surface of the caliper bracket (indicated by the arrows).

12. Install the pad retainers. Wipe off the excess assembly paste from the retainers. Keep the assembly paste away from the brake disc and the brake pads.

13. Install the brake caliper piston compressor tool on the caliper body.

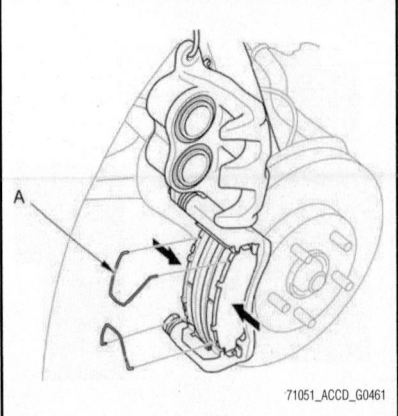

Fig. 16 Remove the pad return springs (A) while holding the brake pads

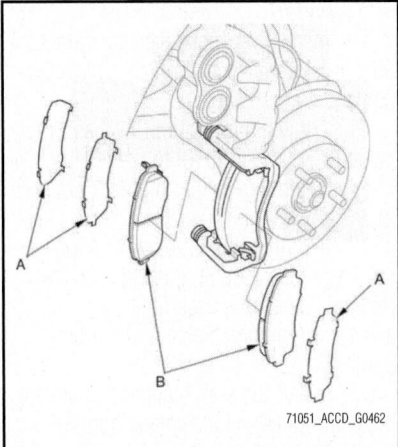

Fig. 17 Remove the pad shims (A) and the brake pads (B)

14. Press in the piston with the brake caliper piston compressor tool so the caliper body will fit over the brake pads. Make sure the piston boot is in position to prevent damaging it when pivoting the caliper body down.

➡**Be careful when pressing in the piston; brake fluid might overflow from the master cylinder's reservoir. If brake fluid gets on any painted surface, wash it off immediately with water.**

15. Remove the brake caliper piston compressor tool.

16. Apply a thin coat of M-77 assembly paste (P/N 08798-9010) to the pad side of the shims, the back of the brake pads and the other areas indicated by the arrows. Wipe off the excess assembly paste from the pad shims and the brake pads friction material.

➡**Keep grease and assembly paste away from the brake disc and the**

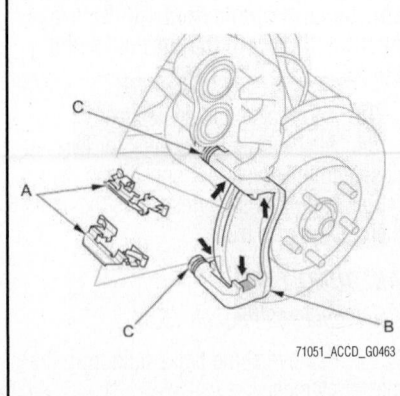

Fig. 18 Remove the pad retainers (A); Clean the caliper bracket (B); caliper pins (C)

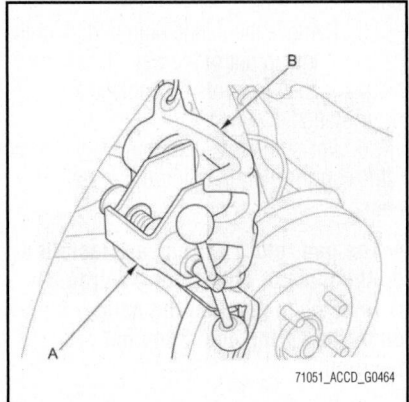

Fig. 19 Install the brake caliper piston compressor tool (A) on the caliper body (B)

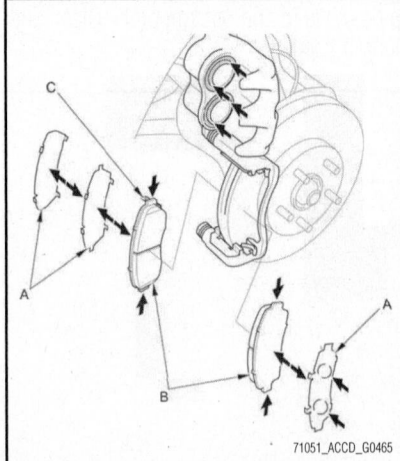

Fig. 20 Apply a thin coat of M-77 assembly paste to the pad side of the shims (A), the back of the brake pads (B) and the other areas indicated by the arrows; Install the brake pad with the wear indicator (C) on the upper inside position

brake pads. Contaminated brake disc or brake pads reduce stopping ability.

17. Install the brake pads and the pad shims correctly. Install the brake pad with the wear indicator on the upper inside position. If you are reusing the brake pads, always reinstall the brake pads in their original positions to prevent a temporary loss of braking efficiency.

18. Install the pad return springs while holding the brake pads.

➡**Insert the pad return spring ends into the pad installation holes securely.**

19. Pivot the caliper body down into position. Install the flange bolt, and tighten it to 37 ft. lbs. (50 Nm) while holding the caliper pin with a wrench being careful not to damage the pin boot.

20. Install the brake hose mounting bolt. Tighten to 16 ft. lbs. (22 Nm).

21. Clean the mating surfaces between the brake disc and the inside of the wheel, then install the front wheels.

22. Press the brake pedal several times to make sure the brakes work.

➡**Engagement may require a greater pedal stroke immediately after the brake pads have been replaced as a set. Several applications of the brake pedal will restore the normal pedal stroke.**

23. Add brake fluid as needed.

24. After installation, check for leaks at hose and line joints or connections, and retighten if necessary. Test-drive the vehicle, then recheck for leaks.

BRAKES

BRAKE CALIPERS

REMOVAL & INSTALLATION

See Figures 21 through 23.

1. Raise and support the vehicle.
2. Remove the rear wheel.

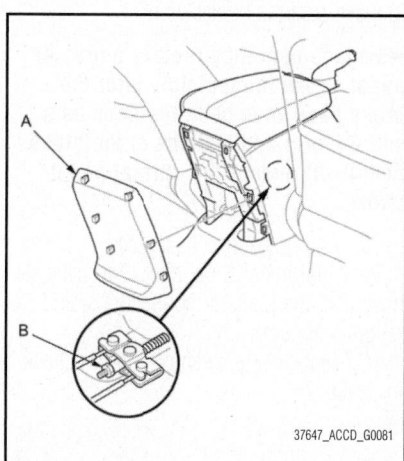

Fig. 21 Loosen the parking brake cable adjusting nut

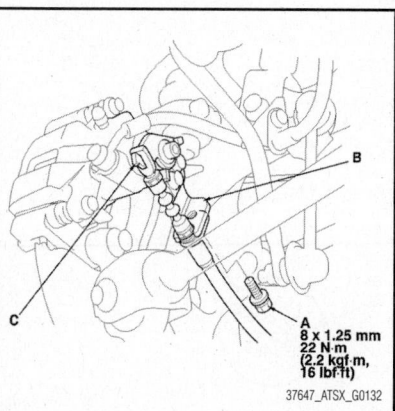

Fig. 22 Remove the flange bolt (A) from the arm (B) and disconnect the parking brake cable from the lever (C)

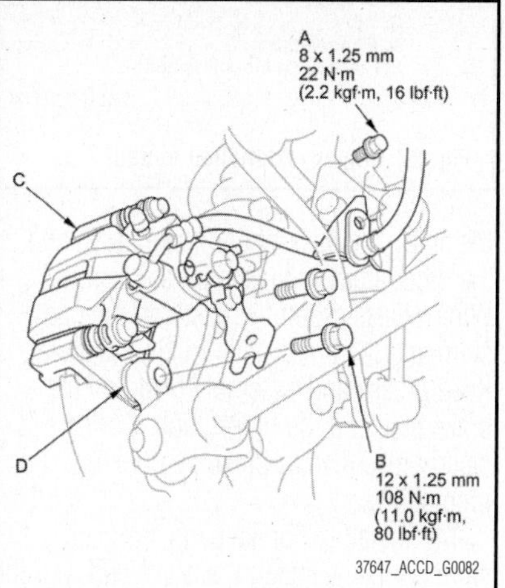

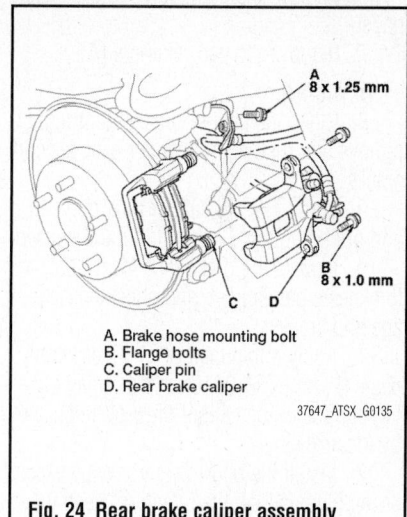

Fig. 23 Remove the brake caliper

3. Release the parking brake lever fully.
4. Loosen the parking brake cable adjusting nut.
5. Remove the flange bolt from the arm.
6. Disconnect the parking brake cable from the lever.
7. Remove the brake hose mounting bolt.
8. Remove the brake caliper bracket mounting bolts and remove the caliper assembly from the knuckle.
9. Installation is the reverse of removal.

BRAKE PADS

REMOVAL & INSTALLATION

See Figures 24 through 27.

1. Remove some brake fluid from the master cylinder.
2. Raise the vehicle on a lift.
3. Remove the rear wheels.
4. Remove the brake hose mounting bolt.
5. Remove the flange bolts while holding respective caliper pin with a wrench. Be

REAR DISC BRAKES

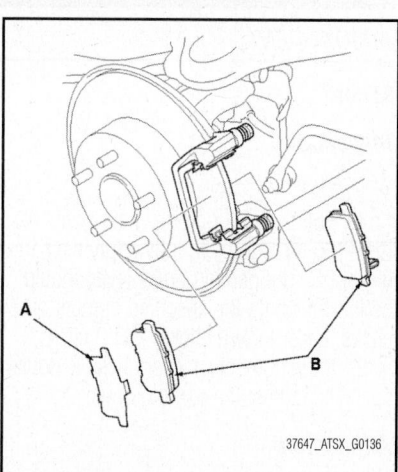

A. Brake hose mounting bolt
B. Flange bolts
C. Caliper pin
D. Rear brake caliper

Fig. 24 Rear brake caliper assembly

careful not to damage the pin boot, and remove the caliper. Check the hose, the pin boots, and the parking brake cable boots for damage and deterioration.

➡**Do not twist the brake hose and the parking brake cable to prevent damage.**

Fig. 25 Remove the pad shim (A) and the brake pads (B)

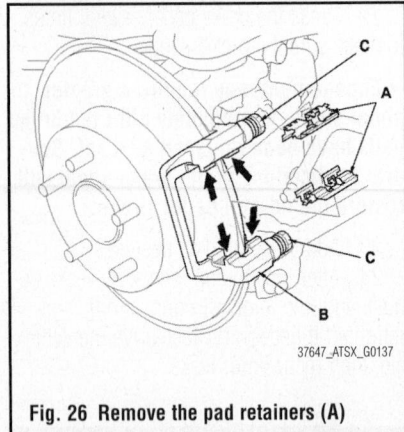

Fig. 26 Remove the pad retainers (A)

6. Remove the pad shim and the brake pads.

7. Remove the pad retainers (A).

To install:

8. Clean the caliper bracket thoroughly; remove any rust, and check for grooves and cracks.

9. Verify that the caliper pins move in and out smoothly. Clean and lube if needed.

10. Inspect the brake disc for runout, thickness, parallelism, and check for damage and cracks.

11. Apply a thin coat of M-77 assembly paste (P/N 08798-9010) to the retainer mating surface of the caliper bracket (indicated by the arrows).

12. Install the pad retainers. Wipe excess assembly paste off the retainers. Keep the assembly paste off the brake disc and the brake pads.

13. Apply a thin coat of M-77 assembly paste (P/N 08798-9010) to the pad side of

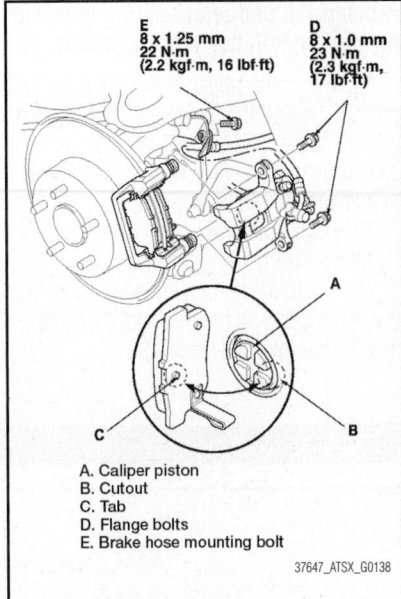

A. Caliper piston
B. Cutout
C. Tab
D. Flange bolts
E. Brake hose mounting bolt

Fig. 27 Proper installation of pads

the shim, the back of the brake pads, and the other areas indicated by the arrows. Wipe excess assembly paste off the pad shim and the brake pads friction material. Keep grease and assembly paste off the brake disc and the brake pads. Contaminated brake disc or brake pads reduce stopping ability.

14. Install the brake pads and pad shim correctly. Install the brake pad with the wear indicator on the bottom inside. If you are reusing the brake pads, always reinstall the brake pads in their original positions to prevent a temporary loss of braking efficiency.

15. Rotate the caliper piston clockwise into the cylinder, then align the cutout in the piston with the tab on the inner pad by turning the piston back. Lubricate the boot with rubber grease to avoid twisting the piston boot. If the piston boot is twisted, back it out so it is positioned properly.

➡ **Be careful when moving the piston back in the caliper; brake fluid might overflow from the master cylinder's reservoir. If brake fluid gets on any painted surface, wash it off immediately with water.**

16. Install the caliper. Install the flange bolts, and tighten it to the specified torque while holding the respective caliper pins with a wrench being careful not to damage the pin boots and parking brake cable boots.

17. Install the brake hose mounting bolt.

18. Clean the mating surfaces between the brake disc and the inside of the wheel, then install the rear wheels.

19. Press the brake pedal several times to make sure the brakes work.

➡ **Engagement may require a greater pedal stroke immediately after the brake pads have been replaced as a set. Several applications of the brake pedal will restore the normal pedal stroke.**

20. Add brake fluid as needed.

21. After installation, check for leaks at hose and line joints or connections, and retighten if necessary.

22. Test-drive the vehicle, then recheck for leaks.

BRAKES   PARKING BRAKE

PARKING BRAKE CABLE

ADJUSTMENTS

Accord

Inspection

See Figure 28.

1. Pull the parking brake lever with 44 lbs. (196 N) of force to fully apply the parking brake. The parking brake lever should be locked within the specified number of clicks. Lever locked clicks: 7 to 9 clicks.

2. If the number of lever clicks is not as specified, adjust the parking brake.

Adjustment

See Figures 29 and 30.

1. Release the parking brake lever fully.

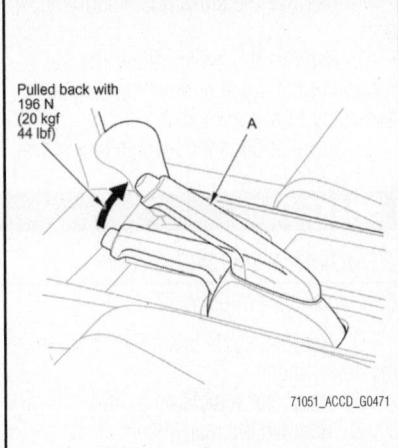

Fig. 28 Pull the parking brake lever (A) with 44 lbs. (196 N) of force to fully apply the parking brake

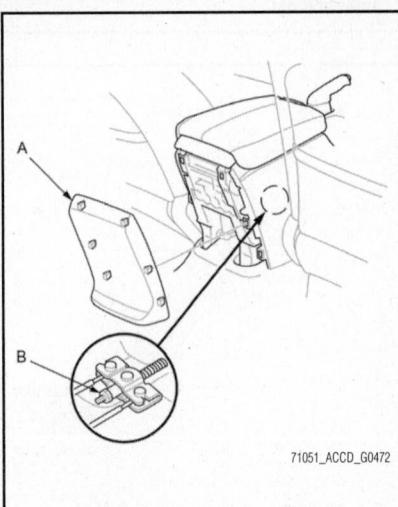

Fig. 29 Pull out the center console rear trim (A); loosen the adjusting nut (B)

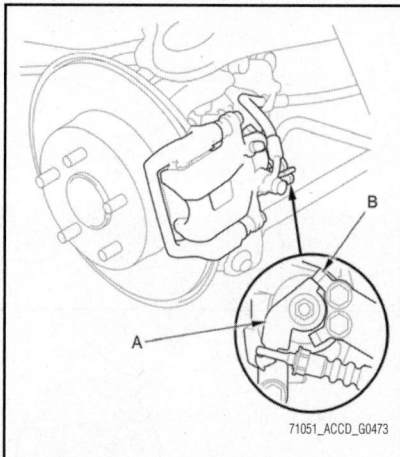

Fig. 30 Make sure the lever (A) on the rear brake caliper contacts the arm (B)

2. Pull out the center console rear trim.
3. Loosen the adjusting nut.
4. Raise and support the vehicle.
5. Remove the rear wheels.
6. Make sure the lever on the rear brake caliper contacts the arm.

➡**The lever will contact the arm when the adjusting nut is loosened.**

7. Clean the mating surfaces between the brake disc and the inside of the wheel, then install the rear wheels.
8. Pull the parking brake lever 1 click.
9. Tighten the adjusting nut until the parking brakes drag slightly when the rear wheels are turned.
10. Release the parking brake lever fully, and check that the parking brakes do not drag when the rear wheels are turned. Readjust if necessary.
11. Make sure the parking brake lever is within the specified number of clicks (7 to 9 clicks).
12. Install the center console rear trim.

Crosstour

Inspection

See Figure 28.

1. Pull the parking brake lever (A) with 44 lbs. (196 N) of force to fully apply the parking brake. The parking brake lever should be locked within the specified number of clicks. Lever locked clicks: 7 to 9 clicks.
2. If the number of lever clicks is not as specified, adjust the parking brake.

➡**Minor parking brake lever adjustments (1 to 2 clicks) can be made with the adjusting nut (see parking brake minor adjustment). If a larger adjustment is required, follow the major**

adjustment procedure using the adjuster nut at the parking brake drum (see parking brake major adjustment). After installing new parking brake shoes and/or new brake disc/drum, make sure you drive the vehicle for "break-in".

Minor Adjustment

See Figure 31.

1. Raise and support the vehicle.
2. Release the parking brake lever fully.
3. Pull out the center console rear trim.
4. Loosen the adjusting nut.
5. Pull the parking brake lever 1 click.
6. Tighten the adjusting nut until the parking brakes drag slightly when the rear wheels are turned.
7. Release the parking brake lever fully, and check that the parking brakes do not drag when the rear wheels are turned. Readjust if necessary.
8. Make sure the parking brake lever is within the specified number of clicks (7 to 9 clicks).
9. Install the center console rear trim.

Major Adjustment (To Be Done When Replacing Parking Brake Shoes)

See Figures 31 and 32.

1. Raise and support the vehicle.
2. Release the parking brake lever fully.
3. Pull out the center console rear trim.
4. Loosen the adjusting nut.
5. Remove the rear wheels.
6. Remove the access plug.
7. Turn the ratchet teeth on the adjuster assembly with a flat-tip screwdriver until the shoes lock against the parking brake drum.

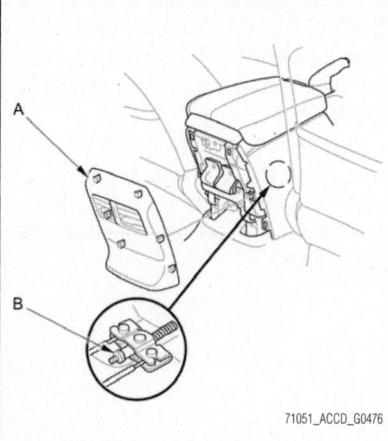

Fig. 31 Pull out the center console rear trim (A); loosen the adjusting nut (B)

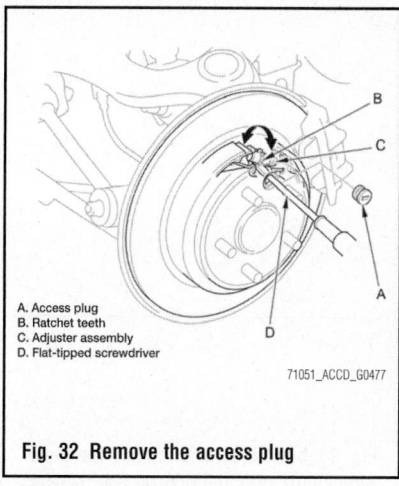

A. Access plug
B. Ratchet teeth
C. Adjuster assembly
D. Flat-tipped screwdriver

Fig. 32 Remove the access plug

Then back off the adjuster 8 clicks, and install the access plug.
8. Do the minor adjustment procedure.
9. Clean the mating surfaces between the brake disc/drum and the inside of the wheel, then install the rear wheels.

PARKING BRAKE SHOES

ADJUSTMENTS

See Figures 31 and 32.

1. Raise and support the vehicle.
2. Release the parking brake lever fully.
3. Pull out the center console rear trim.
4. Loosen the adjusting nut.
5. Remove the rear wheels.
6. Remove the access plug.
7. Turn the ratchet teeth on the adjuster assembly with a flat-tip screwdriver until the shoes lock against the parking brake drum. Then back off the adjuster 8 clicks, and install the access plug.
8. Do the minor adjustment procedure.
9. Clean the mating surfaces between the brake disc/drum and the inside of the wheel, then install the rear wheels.

REMOVAL & INSTALLATION

See Figures 33 through 38.

1. Raise and support the vehicle.
2. Remove the rear wheels.
3. Release the parking brake, and remove the brake disc/drum.
4. Disconnect and remove the upper return spring.
5. Remove the adjuster assembly.
6. Disconnect and remove the lower return spring.
7. Remove the parking brake lever assembly, and disconnect it from the parking brake cable end.
8. Remove the tension pins by pushing respective retainer springs, and turning the pin.

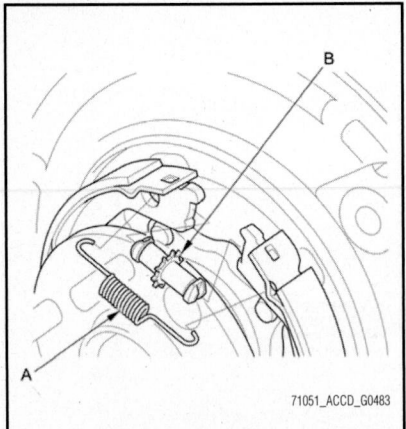

Fig. 33 Disconnect and remove the upper return spring (A) and remove the adjuster assembly (B)

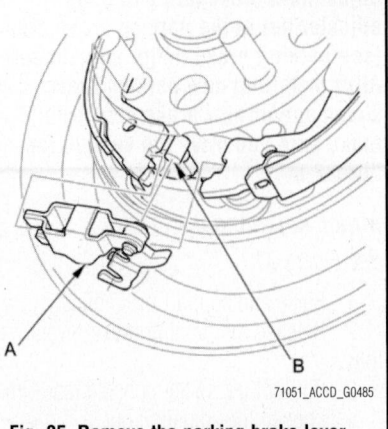

Fig. 35 Remove the parking brake lever assembly (A), and disconnect it from the parking brake cable end (B)

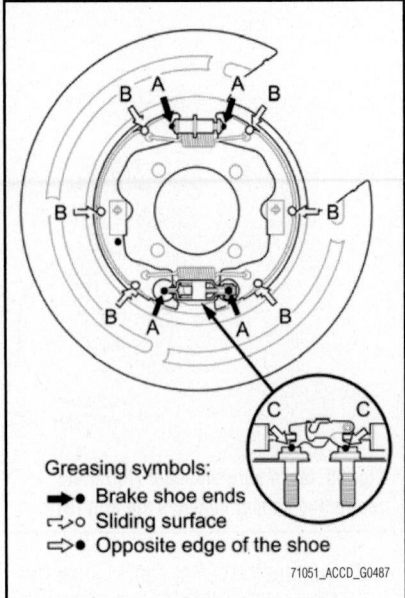

Greasing symbols:
➡● Brake shoe ends
↪○ Sliding surface
⇨● Opposite edge of the shoe

Fig. 37 Apply Molykote 44MA grease to the shoe ends (A), sliding surfaces (B), and opposite edges of the parking brake shoe (C) as shown

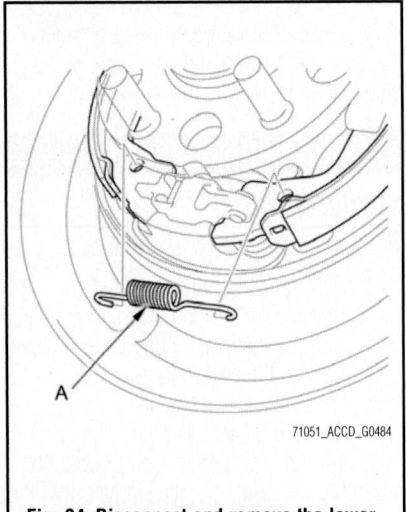

Fig. 34 Disconnect and remove the lower return spring (A)

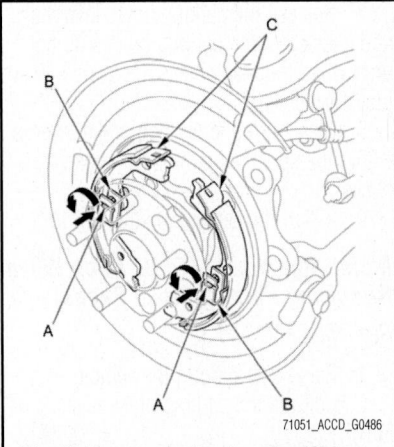

Fig. 36 Remove the tension pins (A) by pushing respective retainer springs (B), and turning the pin, then remove the parking brake shoes (C)

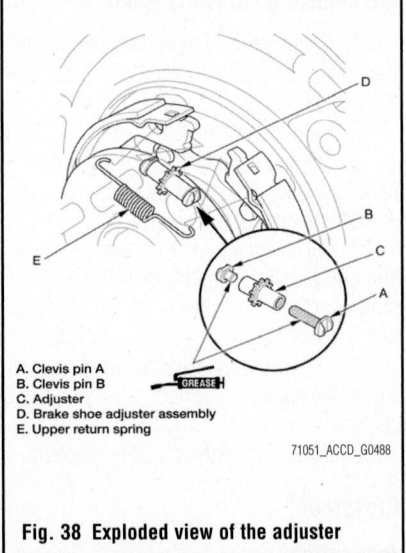

A. Clevis pin A
B. Clevis pin B
C. Adjuster
D. Brake shoe adjuster assembly
E. Upper return spring

Fig. 38 Exploded view of the adjuster

9. Remove the parking brake shoes.

To install:

10. Apply Molykote 44MA grease to the shoe ends, sliding surfaces, and opposite edges of the parking brake shoe as shown. Wipe off any excess. Keep grease away from the brake linings.

11. Reinstall the tension pins and retainer springs.

12. Connect the parking brake cable end to the parking brake lever assembly, and install the parking brake lever assembly.

13. Reinstall the lower return spring.

14. Clean the threaded portions of the clevis pin A, and coat the threads of the clevis pin A with grease. Clean the sliding surface of the clevis pin B, and coat the sliding surface of the clevis pin B with grease. Install the clevis pin A and B on the adjuster and shorten the clevis pin A by turning the adjuster.

15. Position the brake shoe adjuster assembly on the parking brake shoes.

16. Reinstall the upper return spring.

17. Install the brake disc/drum and the brake caliper.

18. Do the major parking brake adjustment.

19. Clean the mating surfaces between the brake disc/drum and the inside of the wheel, then install the rear wheels.

CHASSIS ELECTRICAL — AIR BAG (SUPPLEMENTAL RESTRAINT SYSTEM)

PRECAUTIONS

✷✷ CAUTION

These vehicles are equipped with an air bag system. The system must be disarmed before performing service on, or around, system components, the steering column, instrument panel components, wiring and sensors. Failure to follow the safety precautions and the disarming procedure could result in accidental air bag deployment, possible injury and unnecessary system repairs.

Disconnect and isolate the battery negative cable before beginning any airbag system component diagnosis, testing, removal, or installation procedures. Allow system capacitor to discharge for two minutes before beginning any component service. This will disable the airbag system. Failure to disable the airbag system may result in accidental airbag deployment, personal injury, or death.

Misalignment of the cable reel could cause an open in the wiring, making the SRS system, remote steering wheel controls, and the horn inoperative. Center the cable reel whenever you do the following.

• Installation of the steering wheel
• Installation of the cable reel
• Installation of the steering column
• Other steering-related adjustment or installation

Do not disassemble the cable reel.
Do not apply grease to the cable reel.
If the cable reel shows any signs of damage, replace it with a new one. For example, if the cable reel does not rotate smoothly, replace it.

Never do electrical tests on the airbags, such as measuring resistance.

Do not place an intact undeployed airbag face down on a solid surface. The airbag will propel into the air if accidentally deployed and may result in personal injury or death.

When carrying or handling an undeployed airbag, the trim side (face) of the airbag should be pointing towards the body to minimize possibility of injury if accidental deployment occurs. Failure to do this may result in personal injury or death.

Replace airbag system components with OEM replacement parts. Substitute parts may appear interchangeable, but internal differences may result in inferior occupant protection. Failure to do so may result in occupant personal injury or death.

Wear safety glasses, rubber gloves, and long sleeved clothing when cleaning powder residue from vehicle after an airbag deployment. Powder residue emitted from a deployed airbag can cause skin irritation. Flush affected area with cool water if irritation is experienced. If nasal or throat irritation is experienced, exit the vehicle for fresh air until the irritation ceases. If irritation continues, see a physician.

The side curtain airbag module assembly is a long, jointed part containing an inflator, a flexible bag, an adapter pipe, and a center bracket. When removing or installing the side curtain airbag assembly, never:

• Handle the flexible bag.
• Drop the airbag assembly.
• Cut, tear, or unwrap the tape strips.

Do not disassemble the SRS unit, the front impact sensors, the side impact sensors, the driver's seat position sensor, the rear safing sensor, and the front passenger's weight sensors.

Always install the SRS unit, front impact sensors, the rear safing sensor, and the side impact sensors with new TORX® bolts torqued to 7.2 ft. lbs. (9.8 Nm).

Do not spill water or oil on the SRS unit or any of the sensors.

Do not use a replacement airbag that is not in the original packaging. This may result in improper deployment, personal injury, or death.

Never attempt to modify, splice, or repair SRS wiring. If there is an open or damage in SRS wiring or terminals, replace the harness.

Make sure all SRS ground locations are clean and grounds are securely fastened for optimum metal-to-metal contact. Poor grounds can cause intermittent problems that are difficult to diagnose.

Do not use any silicone based cleaners or lubricants on any SRS connectors or terminals.

Make sure the battery is fully charged when doing electrical test. If the battery is not fully charged, the results of the tests may not be accurate.

When using electrical test equipment, insert the probe of the tester into the wire side of the connector (except water proof connector). Do not insert the probe of the tester into the terminal side of the connector, and do not tamper with the connector.

The factory installed fasteners, screws and bolts used to fasten airbag components have a special coating and are specifically designed for the airbag system. Do not use substitute fasteners. Use only original

equipment fasteners listed in the parts catalog when fastener replacement is required.

During, and following, any child restraint anchor service, due to impact event or vehicle repair, carefully inspect all mounting hardware, tether straps, and anchors for proper installation, operation, or damage. If a child restraint anchor is found damaged in any way, the anchor must be replaced. Failure to do this may result in personal injury or death.

Deployed and non-deployed airbags may or may not have live pyrotechnic material within the airbag inflator.

Do not dispose of driver/passenger/curtain airbags or seat belt tensioners unless you are sure of complete deployment. Refer to the Hazardous Substance Control System for proper disposal.

Dispose of deployed airbags and tensioners consistent with state, provincial, local, and federal regulations.

After any airbag component testing or service, do not connect the battery negative cable. Personal injury or death may result if the system test is not performed first.

If the vehicle is equipped with the Occupant Classification System (OCS), do not connect the battery negative cable before performing the OCS Verification Test using the scan tool and the appropriate diagnostic information. Personal injury or death may result if the system test is not performed properly.

Never replace both the Occupant Restraint Controller (ORC) and the Occupant Classification Module (OCM) at the same time. If both require replacement, replace one, then perform the Airbag System test before replacing the other.

Both the ORC and the OCM store Occupant Classification System (OCS) calibration data, which they transfer to one another when one of them is replaced. If both are replaced at the same time, an irreversible fault will be set in both modules and the OCS may malfunction and cause personal injury or death.

If equipped with OCS, the Seat Weight Sensor is a sensitive, calibrated unit and must be handled carefully. Do not drop or handle roughly. If dropped or damaged, replace with another sensor. Failure to do so may result in occupant injury or death.

If equipped with OCS, the front passenger seat must be handled carefully as well. When removing the seat, be careful when setting on floor not to drop. If dropped, the sensor may be inoperative, could result in occupant injury, or possibly death.

If equipped with OCS, when the passenger front seat is on the floor, no one should sit in the front passenger seat. This uneven force may damage the sensing ability of the seat weight sensors. If sat on and damaged, the sensor may be inoperative, could result in occupant injury, or possibly death.

➡**Some systems store data in memory that is lost when the battery is disconnected. Before disconnecting the battery, refer to Battery Terminal Disconnection and Reconnection.**

✳✳ CAUTION

Please read the following precautions carefully before servicing the airbag system. Observe the instructions described in this manual, or the airbags could accidentally deploy and cause damage or injuries.

- Except when doing electrical inspections that requires battery power, always turn the ignition switch to LOCK (0), disconnect the negative cable from the battery, then wait at least 3 minutes before starting work.

➡**The SRS memory is not erased even if the ignition switch is turned to LOCK (0) or the battery cables are disconnected from the battery.**

- Use replacement parts which are manufactured to the same standards and quality as the original parts. Do not install used SRS parts. Use only new parts when making SRS repairs.
- Carefully inspect any SRS part before you install it.

✳✳ CAUTION

Do not install any part that shows signs of being dropped or improperly handled, such as dents, cracks or deformation.

- Before disconnecting the SRS unit connectors, always disconnect the appropriate SRS parts connectors.

- Use only a digital multimeter to check the system. If it is not a Honda multimeter, make sure its output is 10 mA (0.01 A) or less when switched to the lowest value in the ohmmeter range. A tester with a higher output could cause accidental deployment and possible injury.
- Do not put objects on the front passenger's airbag.
- To prevent damage to the airbag, keep it away from any oil, grease, detergent, or water.
- Store the removed airbag on a secure, flat surface away from any high heat source exceeding 200°F (93°C).
- Never do electrical tests on the airbags, such as measuring resistance.
- Do not position yourself in front of the airbag during removal, inspection, or replacement.
- Never attempt to modify, splice, or repair SRS wiring. If there is an open or damage in SRS wiring, replace the harness.
- Be sure to install the harness wires so they do not get pinched or interfere with other parts.
- Make sure all SRS ground locations are clean, and grounds are securely fastened for optimum metal-to-metal contact. Poor grounds can cause intermittent problems that are difficult to diagnose.
- Do not use any silicone based cleaners or lubricants on any SRS connectors or terminals.

DISARMING THE SYSTEM

➡**Some systems store data in memory that is lost when the battery is disconnected. Do the following steps before disconnecting the battery.**

1. Make sure you have the anti-theft code(s) for the audio and/or the navigation system (if equipped).

2. For some models or if you're replacing the audio unit, it may be necessary to write down the audio presets (AM and FM), and the XM radio presets (if equipped), because the audio unit does not retain the presets after the battery is disconnected.

3. Make sure the ignition switch is in LOCK (0).

4. Disconnect and isolate the negative cable from the battery.

✳✳ CAUTION

Always disconnect the negative cable from the battery first.

5. Disconnect the positive cable from the battery.

6. Wait at least 3 minutes for the battery and SRS system to discharge before performing any repairs.

ARMING THE SYSTEM

➡**Some systems store data in memory that is lost when the battery is disconnected. Do the following steps to restore the systems back to normal operation.**

1. Clean the battery terminals.

2. Reconnect the positive cable to the battery first, then reconnect the negative cable to the battery.

✳✳ CAUTION

Always connect the positive cable to the battery first.

3. Apply multipurpose grease to the terminals to prevent corrosion.

4. Enter the anti-theft code(s) for the audio system and/or the navigation system (if equipped).

5. Enter the audio presets (if applicable), and enter the XM radio presets (if equipped).

6. Set the clock (for vehicles without navigation).

7. Do the steering column position memorization. See "Steering Column" in "STEERING" section.

DRIVE TRAIN

AUTOMATIC TRANSAXLE FLUID

FILTER REPLACEMENT

With ATF Warmer

See Figure 39.

➡The ATF filter is not a scheduled maintenance item. Replace the filter only if it is leaking, or contaminated, or when the transmission is being overhauled or replaced with a remanufactured unit.

1. Raise the vehicle on a lift, or apply the parking brake, block the rear wheels, and raise the front of the vehicle. Make sure it is securely supported.
2. Remove the front splash shield.
3. Remove the drain plug and sealing washer, and drain the ATF.
4. Reinstall the drain plug with a new sealing washer.
5. Remove the nut securing the underhood fuse/relay box, and swing it out of the way.
6. Remove the air cleaner housing and the intake air duct.
7. Disconnect the ATF hose from the ATF warmer. Turn the end of the ATF hose up to prevent ATF from flowing out, then plug the ATF hose and line.
8. Disconnect the ATF hose from the ATF filter. Turn the end of the ATF hose up to prevent ATF from flowing out, then plug the ATF hose.
9. Remove the bolt, then remove the ATF filter holder and the ATF filter.
10. Disconnect the ATF hose from the ATF filter.

11. Replace the ATF filter, then secure it with the ATF filter holder and the bolt.
12. Connect the ATF hoses, and secure the hoses with the clips.
13. Install the front splash shield.
14. Refill the transmission with ATF.

Without ATF Warmer

See Figure 40.

➡The ATF filter is not a scheduled maintenance item. Replace the filter only if it is leaking, or contaminated, or when the transmission is being overhauled or replaced with a remanufactured unit.

1. Raise the vehicle on a lift, or apply the parking brake, block the rear wheels, and raise the front of the vehicle. Make sure it is securely supported.
2. Remove the front splash shield.
3. Remove the drain plug and sealing washer, and drain the ATF.
4. Reinstall the drain plug with a new sealing washer.
5. Disconnect the ATF hoses from the ATF filter.
6. Remove the bolt, then remove the ATF filter holder and the ATF filter.
7. Replace the ATF filter, then secure it with the ATF filter holder and the bolt.
8. Connect the ATF hoses, and secure the hoses with the clips.
9. Install the front splash shield.
10. Refill the transmission with ATF.

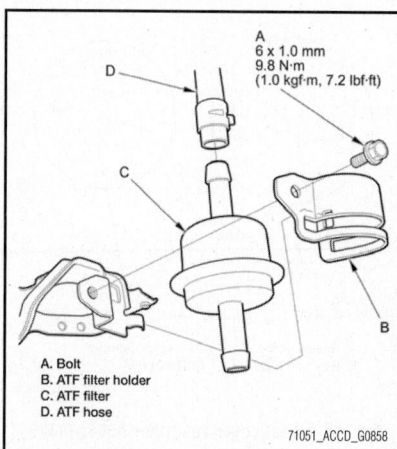

A. Bolt
B. ATF filter holder
C. ATF filter
D. ATF hose

71051_ACCD_G0858

Fig. 39 Remove the bolt, then remove the ATF filter holder and the ATF filter

C
6 x 1.0 mm
12 N·m
(1.2 kgf·m,
8.7 lbf·ft)

A. ATF hoses
B. ATF filter
C. Colt
D. ATF filter holder

71051_ACCD_G0859

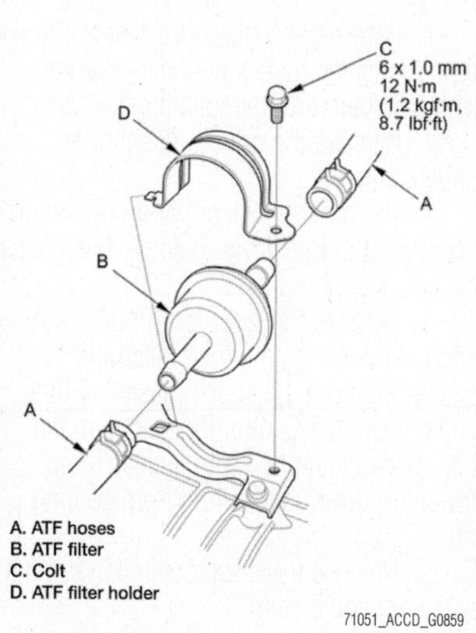

Fig. 40 Disconnect the ATF hoses from the ATF filter

CLUTCH

BLEEDING

Note the following:
• Do not reuse the drained fluid. Always use Honda DOT 3 Brake Fluid from an unopened container. Using a non-Honda brake fluid can cause corrosion and shorten the life of the system.
• Make sure no dirt or other foreign matter is allowed to contaminate the brake fluid.
• Do not spill brake fluid on the vehicle; it may damage the paint or plastic. If brake fluid does contact the paint or plastic, wash it off immediately with water.
• It may be necessary to limit the movement of the release fork with a block of wood to remove all the air from the system.

1. Do the battery removal procedure.
2. Make sure the brake fluid level in the clutch reservoir is at the MAX (upper) level line.
3. Attach one end of a clear tube to the bleeder screw, and put the other end into a container. Loosen the bleeder screw to allow air to escape from the system.
4. Make sure there is an adequate supply of fluid in the reservoir, then slowly push the clutch pedal all the way down. Before releasing the pedal, have an assistant temporarily tighten the bleeder screw. Loosen the bleeder screw, and push the clutch pedal down again. Repeat this step until no more bubbles appear at the clear tube.

➡Make sure the fluid level on the reservoir does not go below MIN.

5. Tighten the bleeder screw securely.
6. Refill the brake fluid in the reservoir to the MAX (upper) level line.
7. Do the battery installation procedure.

CLUTCH MASTER CYLINDER

REMOVAL & INSTALLATION

4-Cylinder Engine

See Figures 41 through 43.

1. Remove and discard the brake fluid from the clutch master cylinder reservoir with a syringe or other suitable device.
2. Pry out the lock pin, and pull the clevis pin out of the clevis. Remove the master cylinder mounting nuts.
3. Remove the reservoir mounting bolt and the retaining clip, then disconnect the clutch line from the clutch master cylinder,

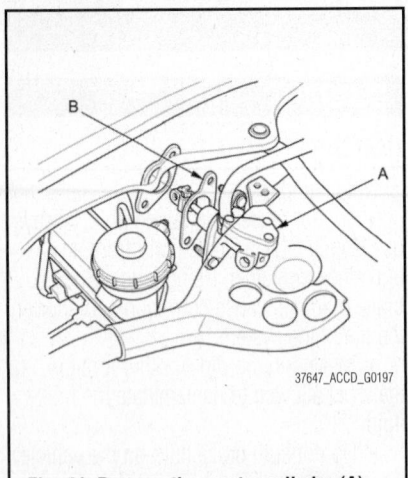

Fig. 41 Remove the master cylinder (A) and the clutch master cylinder seal (B)

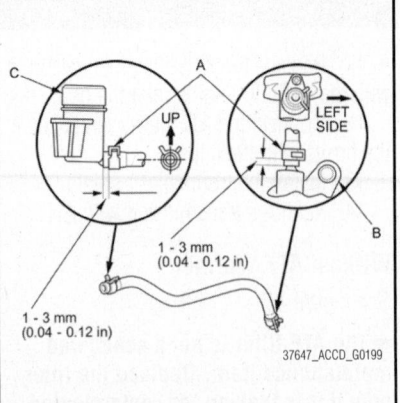

Fig. 43 Make sure the hose clamps (A) are positioned on the master cylinder (B) and reservoir (C) as shown

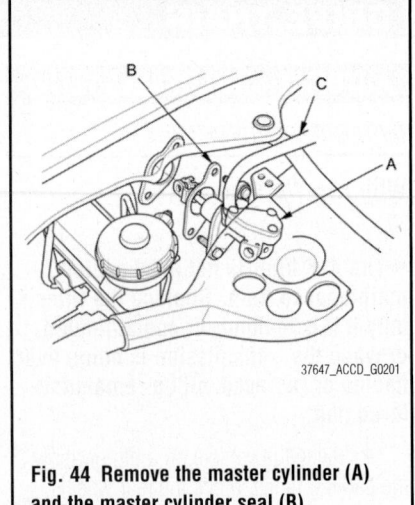

Fig. 44 Remove the master cylinder (A) and the master cylinder seal (B)

and remove the O-ring. Plug or wrap the end of the clutch line with a clean shop towel to prevent brake fluid from coming out.

4. Remove the master cylinder and the clutch master cylinder seal.

To install:

5. Install a new master cylinder seal, then install the master cylinder.

6. Install a new O-ring on the clutch line, then install the clutch line in the clutch master cylinder with a new retaining clip. Install the master cylinder reservoir mounting bolt.

➡**Apply silicone grease (P/N 08C30-B0234M) to the O-ring and the end of the clutch line. Make sure not to get any silicone grease on the terminal part of the connectors and switches, especially if you have silicone grease on your hands or gloves.**

7. To prevent the retaining clip from

coming off, pry apart the tip of the clip with a screwdriver.

8. Make sure the hose clamps are positioned on the master cylinder and reservoir as shown.

9. Install the master cylinder mounting nuts.

10. Apply multipurpose grease to the clevis pin and the mating surfaces of the clevis and the pedal. Slide the clevis pin into the clevis, then install the lock pin.

11. Bleed the clutch hydraulic system.

12. Adjust the clutch pedal, the clutch pedal position switch, and the clutch interlock switch.

13. Check the clutch operation, and check for leaks.

14. Test-drive the vehicle.

6-Cylinder Engine

See Figures 42, 44 and 45.

1. Remove and discard the brake fluid from the clutch master cylinder reservoir with a syringe or other suitable device.

2. Remove the driver's dashboard undercover.

3. Pry out the lock pin and pull the clevis pin out of the clevis. Remove the master cylinder mounting nuts.

4. Remove the reservoir mounting bolt and the retaining clip, then disconnect the clutch line from the clutch master cylinder, and remove the O-ring. Plug or wrap the end of the clutch line with a clean shop towel to prevent brake fluid from coming out.

5. Remove the master cylinder and the master cylinder seal.

6. Disconnect the reservoir hose.

➡**Inspect the hose. If the hose has damage, leaks, interference, or twisting, replace it.**

To install:

7. Connect the reservoir hose, make sure the hose clamps are positioned on the master cylinder and the reservoir as shown.

8. Install a new master cylinder seal, then install the master cylinder.

9. Install a new O-ring on the clutch line, then install the clutch line in the clutch master cylinder, with a new retaining clip. Install the master cylinder reservoir mounting bolt.

➡**Apply silicone grease (P/N 08C30-B0234M) to the O-ring and the end of the clutch line. Make sure not to get any silicone grease on the terminal part of the connectors and switches, especially if you have silicone grease on your hands or gloves.**

10. To prevent the retaining clip from coming off, pry apart the tip of the clip with a screwdriver.

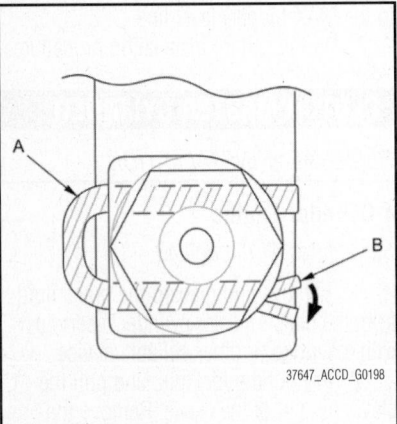

Fig. 42 To prevent the retaining clip (A) from coming off, pry apart the tip of the clip (B) with a screwdriver

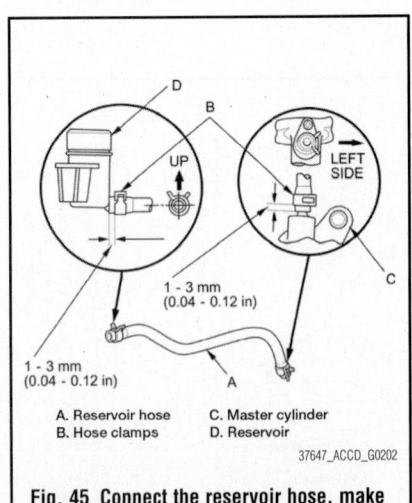

A. Reservoir hose   C. Master cylinder
B. Hose clamps   D. Reservoir

Fig. 45 Connect the reservoir hose, make sure the hose clamps are positioned on the master cylinder and the reservoir as shown

11. Install the master cylinder mounting nuts.

12. Apply multipurpose grease to the clevis pin and the mating surfaces of the clevis and the pedal. Slide the clevis pin into the clevis, then install the lock pin.

13. Bleed the clutch hydraulic system.

14. Adjust the clutch pedal, the clutch pedal position switch, and the clutch interlock switch.

15. Check the clutch operation, and check for leaks.

16. Install the driver's dashboard undercover.

17. Test-drive the vehicle.

CLUTCH SLAVE (RELEASE) CYLINDER

REMOVAL & INSTALLATION

4-Cylinder Engine

See Figures 46 and 47.

1. Do the battery removal procedure.

2. Remove the mounting bolts, the bracket mounting nut, and the slave cylinder.

3. Remove the roll pins. Disconnect the clutch line, and remove the O-ring. Plug or wrap the end of the clutch line with a clean shop towel to prevent brake fluid from coming out.

To install:

4. Install a new O-ring on the clutch line, install the clutch line in the slave cylinder, and install new roll pins.

5. Pull back the boot, and apply silicone grease (P/N 08C30-B0234M) to the boot and the slave cylinder. Reinstall the boot.

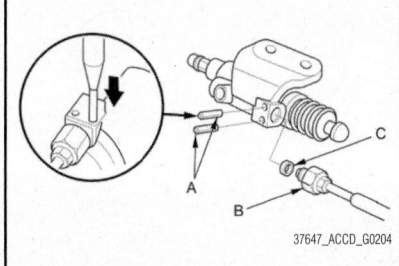

Fig. 47 Remove the roll pins (A), disconnect the clutch line (B), and remove the O-ring (C)

6. Apply a light coat of super high temp urea grease (P/N 08798-9002) to the end of the slave cylinder pushrod. Install the slave cylinder mounting bolts and the bracket mounting nut.

7. Bleed the clutch hydraulic system.

8. Check the clutch operation, and check for leaks.

9. Do the battery installation procedure.

10. Test-drive the vehicle.

6-Cylinder Engine

See Figures 48 and 49.

1. Remove the clip, then remove the splash separator.

2. Do the battery removal procedure.

3. Remove the mounting bolts, the bracket mounting nut, and the slave cylinder.

4. Remove the roll pins. Disconnect the clutch line, and remove the O-ring. Plug or wrap the end of the clutch line with a clean shop towel to prevent brake fluid from coming out.

To install:

5. Install a new O-ring on the clutch line, install the clutch line in the slave cylinder, and install new roll pins.

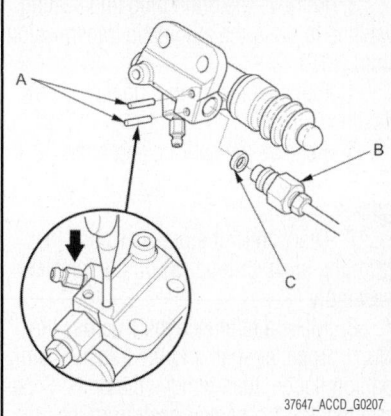

Fig. 49 Remove the roll pins (A), disconnect the clutch line (B), and remove the O-ring (C)

6. Pull back the boot, and apply silicone grease (P/N 08C30-B0234M) to the boot and the slave cylinder. Reinstall the boot.

7. Apply a light coat of super high temp urea grease (P/N 08798-9002) to the end of the slave pushrod. Install the slave cylinder mounting bolts and the bracket mounting nut.

8. Bleed the clutch hydraulic system.

9. Check the clutch operation, and check for leaks.

10. Do the battery installation procedure.

11. Install the splash separator with the clip.

12. Test-drive the vehicle.

TRANSFER CASE ASSEMBLY

REMOVAL & INSTALLATION

See Figures 50 through 52.

1. Raise the vehicle on a lift, and make sure it is supported securely.

2. Move the shift lever to N.

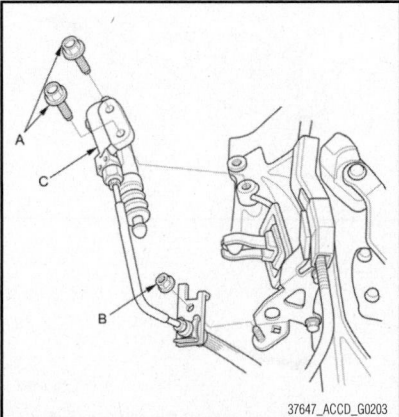

Fig. 46 Remove the mounting bolts (A), the bracket mounting nut (B), and the slave cylinder (C)

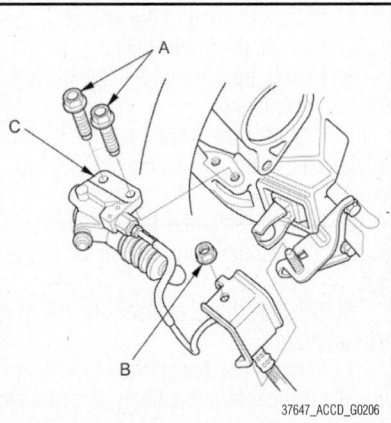

Fig. 48 Remove the mounting bolts (A), the bracket mounting nut (B), and the slave cylinder (C)

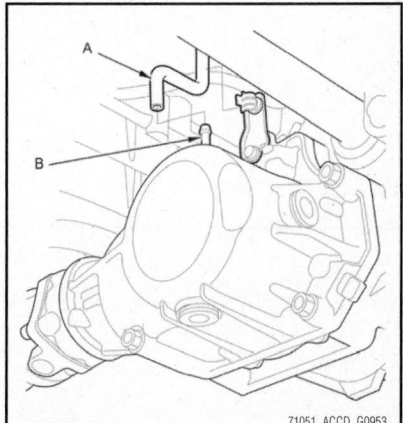

Fig. 50 Disconnect the breather hose (A) from the breather pipe (B) on the transfer assembly

3. Remove the drain plug and sealing washer, to drain the automatic transmission fluid (ATF).

4. Reinstall the drain plug with a new sealing washer.

5. Remove the self-locking nuts.

6. Remove exhaust pipe A and the gasket.

7. Disconnect the breather hose (A) from the breather pipe (B) on the transfer assembly.

8. Make a reference mark across the No. 1 propeller shaft and the transfer companion flange, then remove the bolts. Separate the No. 1 propeller shaft from the transfer companion flange.

9. Remove the transfer assembly and the dowel pin from the transmission.

10. Remove the transfer breather hose bracket from the transfer assembly.

To install:

11. Install the transfer breather hose bracket on the transfer assembly.

12. Clean the areas where the transfer assembly contacts the transmission with solvent, and dry with compressed air. Then apply ATF to the contact area.

13. Install the 14 x 20 mm dowel pin in the transmission, and install the transfer assembly on the transmission. Tighten the transfer mounting bolts to 38 ft. lbs. (51 Nm).

14. Install the No. 1 propeller shaft to the transfer companion flange by aligning the reference mark. Tighten the bolts to 24 ft. lbs. (32 Nm).

15. Connect the transfer breather hose over the breather pipe with the dot mark fac-

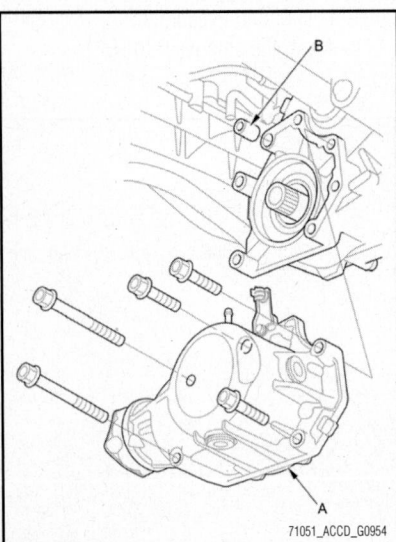

Fig. 51 Remove the transfer assembly (A) and the dowel pin (B) from the transmission

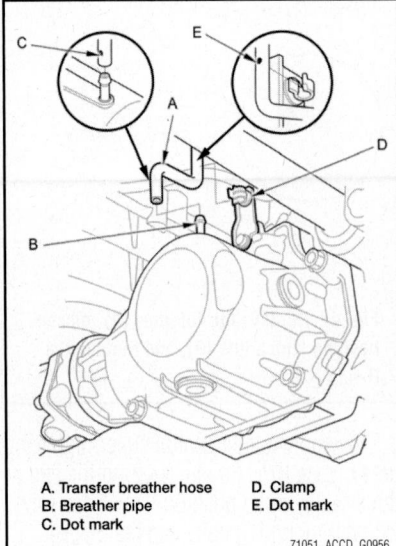

A. Transfer breather hose   D. Clamp
B. Breather pipe   E. Dot mark
C. Dot mark

71051_ACCD_G0956

Fig. 52 Connect the transfer breather hose over the breather pipe with the dot mark facing the rear of the vehicle, then install the hose on the clamp at the dot mark

ing the rear of the vehicle, then install the hose on the clamp at the dot mark.

16. Install exhaust pipe A with the new self-locking nuts and new gaskets.

17. Refill the transfer assembly with transfer fluid (hypoid gear oil), if necessary.

18. Move the shift lever to P.

19. Refill the transmission with ATF.

FRONT HALFSHAFT

REMOVAL & INSTALLATION

See Figures 53 through 58.

Special Tools Required:
- Ball Joint Thread Protector, 14 mm 07AAE-SJAA100
- Ball Joint Remover, 28 mm 07MAC-SL0A202

1. Raise and support the vehicle.

2. Remove the front wheels.

3. Pry up the stake on the spindle nut, then remove the nut.

4. Drain the transmission fluid, then reinstall the drain plug with a new sealing washer.

5. Hold the stabilizer link joint pin using a hex wrench, and remove the flange nut.

6. Separate the front stabilizer link from the lower arm.

7. Remove the damper fork mounting nut, the damper fork mounting bolt, and the damper pinch bolt, then remove the damper fork.

8. Remove the cotter pin from the knuckle ball joint, then remove the castle

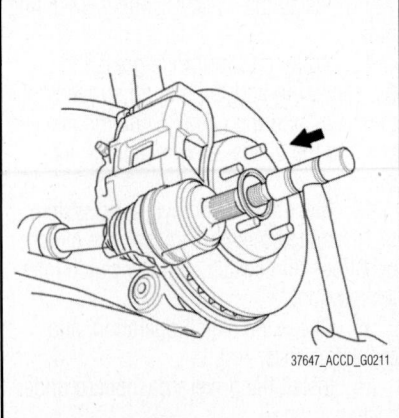

37647_ACCD_G0211

Fig. 53 Pull the knuckle outward, and separate the outboard joint from the front hub using a plastic hammer

nut. Separate the ball joint from the lower arm using the 28 mm ball joint remover and the 14 mm ball joint thread protector.

➡**Be careful not to damage the ball joint boot when installing the remover. Do not force or hammer on the lower arm, or pry between the lower arm and the knuckle. You could damage the ball joint.**

9. Pull the knuckle outward, and separate the outboard joint from the front hub using a plastic hammer.

10. Left driveshaft: Pry the inboard joint from the differential using a prybar. Remove the driveshaft as an assembly.

➡**Do not pull on the driveshaft, or the inboard joint may come apart. Pull the inboard joint straight out to avoid damaging the oil seal. Be careful not to**

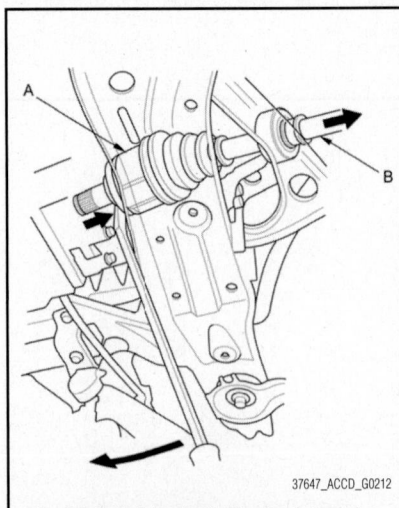

37647_ACCD_G0212

Fig. 54 Left driveshaft: Pry the inboard joint (A) from the differential using a prybar. Remove the driveshaft (B) as an assembly—4-Cylinder Engine

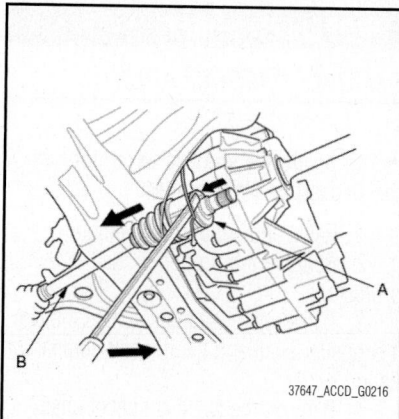

Fig. 55 Left driveshaft: Pry the inboard joint (A) from the differential using a prybar. Remove the driveshaft (B) as an assembly—6-cylinder engine

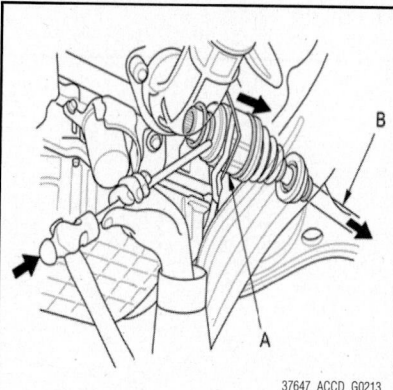

Fig. 56 Right driveshaft: Drive the inboard joint (A) off of the intermediate shaft using a drift punch and a hammer. Remove the driveshaft (B) as an assembly—4-Cylinder Engine

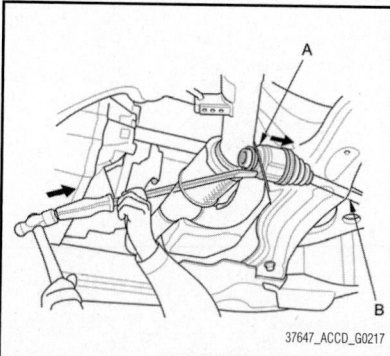

Fig. 57 Right driveshaft: Drive the inboard joint (A) off of the intermediate shaft using a drift punch and a hammer. Remove the driveshaft (B) as an assembly—6-cylinder engine

damage the oil seal or the end of the inboard joint using the prybar.

11. Right driveshaft: Drive the inboard joint off of the intermediate shaft using a drift punch and a hammer. Remove the driveshaft as an assembly.

➡**Do not pull on the driveshaft, or the inboard joint may come apart.**

12. Remove the set ring from the left driveshaft inboard joint.

13. Remove the set ring from the intermediate shaft.

To install:

➡**Before starting installation, make sure the mating surfaces of the joint and the splined section are clean.**

14. Apply Moly 60 paste (P/N 08734-0001) to the contact area of the outboard joint and the front wheel bearing.

➡**The paste helps prevent noise and vibration.**

15. Install a new set ring into the set ring groove of the left driveshaft inboard joint.

16. Install a new set ring into the set ring groove of the intermediate shaft.

17. Apply super high temp urea grease (P/N 08798-9002) to the whole splined surface of the right driveshaft. After applying grease, remove the grease from the splined grooves at intervals of 2–3 splines and from the set ring groove so that air can bleed from the intermediate shaft.

18. Clean the areas where the driveshaft contacts the differential thoroughly with solvent, and dry then with compressed air.

➡**Do not wash the rubber parts with solvent.**

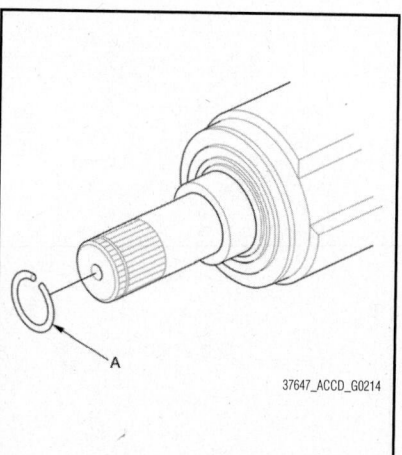

Fig. 58 Remove the set ring (A) from the left driveshaft inboard joint

19. Insert the inboard end of the driveshaft into the differential or intermediate shaft until the set ring locks in the groove.

➡**Insert the driveshaft horizontally to prevent damaging the oil seal.**

20. Install the outboard joint into the front hub on the knuckle.

21. Wipe off any grease contamination from the ball joint tapered section and threads, then install the knuckle onto the lower arm. Be careful not to damage the ball joint boot. Wipe off the grease before tightening the nut at the ball joint. Torque the castle nut to the lower torque specification, then tighten it only far enough to align the slot with the ball joint pin hole.

➡**Make sure the ball joint boot is not damaged or cracked. Do not align the nut by loosening it.**

22. Install a new cotter pin into the ball joint pin hole, and bend the cotter pin.

23. Install the damper fork over the driveshaft and onto the lower arm. Install the damper in the damper fork so the aligning tab is aligned with the slot in the damper fork. Loosely install the damper pinch bolt.

24. Loosely install a new damper fork mounting bolt and a new damper fork mounting nut.

25. Connect the front stabilizer link to the lower arm, and loosely install a new flange nut. Hold the stabilizer link joint pin using a hex wrench, and tighten the flange nut.

26. Place a floor jack under the lower arm, and raise the suspension to load it with the vehicle's weight.

➡**Do not put the floor jack under the ball joint.**

27. Tighten the damper pinch bolt to 36 ft. lbs. (49 Nm)and tighten the damper fork mounting nut while holding the damper fork mounting bolt to 47 ft. lbs.(64 Nm), then remove the floor jack.

28. Apply a small amount of engine oil to the seating surface of a new spindle nut.

29. Install the spindle nut and tighten it to 242 Ft. Lbs. (329 Nm). After tightening the spindle nut, use a drift to stake the spindle nut shoulder against the driveshaft.

30. Clean the mating surfaces of the brake disc and the wheel, then install the front wheels.

31. Turn the wheel by hand, and make sure there is no interference between the driveshaft and surrounding parts.

32. Refill the transmission with the recommended transmission fluid:

33. Lower the vehicle.

34. Check the wheel alignment, and adjust it if necessary.

35. Test-drive the vehicle.

INTERMEDIATE SHAFT

REMOVAL & INSTALLATION

4-Cylinder Engine

See Figure 59.

1. Drain the transmission fluid. Reinstall the drain plug using a new sealing washer.

2. Remove the right driveshaft.

3. Remove the flange bolt and the two dowel bolts.

4. Remove the intermediate shaft from the differential. Hold the intermediate shaft horizontally until it is clear of the differential to prevent damaging the oil seal.

To install:

5. Clean the areas where the intermediate shaft contacts the differential thoroughly with solvent, and dry then with compressed air.

➡**Do not wash the rubber parts with solvent.**

6. Install a new set ring onto the set ring groove of the intermediate shaft.

7. Insert the intermediate shaft into the differential correctly.

➡**Insert the intermediate shaft carefully to prevent damaging the oil seal.**

8. Install the flange bolt and the two dowel bolts. Tighten the bolts to 29 ft. lbs. (39 Nm).

9. Install the right driveshaft.

10. Refill the transmission with the recommended transmission fluid.

11. Check the wheel alignment, and adjust it if necessary.

12. Test-drive the vehicle.

6-Cylinder Engine

See Figure 60.

1. Drain the transmission fluid, then reinstall the drain plug with a new sealing washer.

2. Remove the right driveshaft.

3. Remove exhaust pipe A and the gaskets.

4. Remove the flange bolt and the two dowel bolts.

5. Remove the intermediate shaft from the differential. Hold the intermediate shaft horizontally until it is clear of the differential to prevent damaging the oil seal.

To install:

6. Clean the areas where the intermediate shaft contacts the differential thoroughly with solvent, and dry then with compressed air.

➡**Do not wash the rubber parts with solvent.**

7. Install a new set ring onto the set ring groove of the intermediate shaft.

8. Insert the intermediate shaft into the differential correctly.

➡**Insert the intermediate shaft carefully to prevent damaging the oil seal.**

9. Install the flange bolt and the two dowel bolts.

10. Install exhaust pipe A with new gaskets and new nuts.

11. Install the right driveshaft.

12. Refill the transmission with the recommended transmission fluid.

13. Check the wheel alignment, and adjust it if necessary.

14. Test-drive the vehicle.

PROPELLER SHAFT

REMOVAL & INSTALLATION

See Figures 61 and 62.

➡**The propeller shaft is used only on the Crosstour 4WD model.**

1. Raise and support the vehicle.

2. Remove the propeller shaft protector.

3. Make reference mark across the No. 1 propeller shaft and the transfer companion flange.

4. Remove the No. 1 propeller shaft flange bolts.

5. Remove the center support bearing mounting bolts.

6. Make reference mark across the No. 2 propeller shaft and the rear differential companion flange.

7. Remove the No. 2 propeller shaft flange bolts.

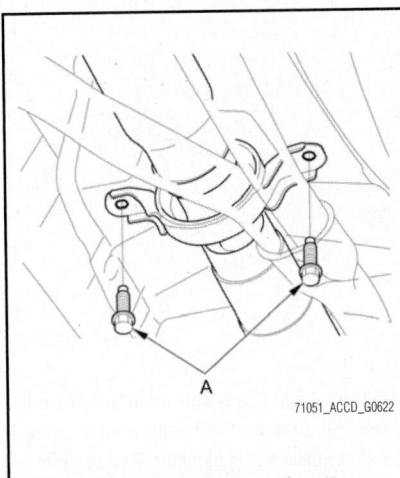

71051_ACCD_G0622

Fig. 61 Remove the center support bearing mounting bolts (A)

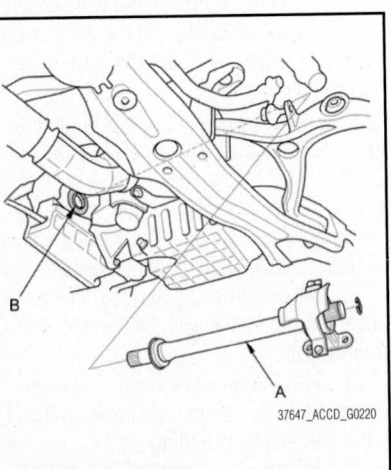

37647_ACCD_G0220

Fig. 59 Remove the intermediate shaft (A) from the differential

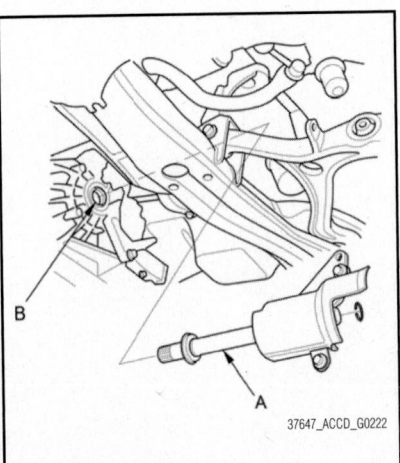

37647_ACCD_G0222

Fig. 60 Remove the intermediate shaft (A) from the differential

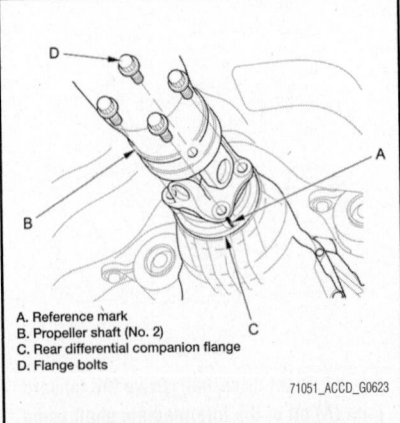

A. Reference mark
B. Propeller shaft (No. 2)
C. Rear differential companion flange
D. Flange bolts

71051_ACCD_G0623

Fig. 62 Make reference mark across the No. 2 propeller shaft and the rear differential companion flange

To install:

8. Set the No. 2 propeller shaft to the rear differential by aligning the reference mark you made during the removal procedure. Then install and tighten new flange bolts to 24 ft. lbs. (32 Nm).

9. Install new center support bearing mounting bolts. Tighten to 29 ft. lbs. (39 Nm).

10. Set the No. 1 propeller shaft to the transfer companion flange by aligning the reference mark you made during the removal procedure. Then install and tighten new flange bolts to 24 ft. lbs. (32 Nm).

11. Install the propeller shaft protector. Tighten the bolts to 16 ft. lbs. (22 Nm).

12. If you installed a new propeller shaft, test-drive the vehicle at 55 mph (88 km/h), and check for noise or vibration. If there is a noise or vibration, rotate the propeller shaft 180 degrees from its current alignment with the rear differential companion flange, then recheck.

REAR DIFFERENTIAL

LEVEL CHECK

See Figure 63.

➡**Always use Honda Dual Pump Fluid II.**

1. Raise and support the vehicle.
2. Use solvent and a brush to wash any oil and dirt off the oil filler plug, and dry them with compressed air.
3. Remove the oil filler plug and the sealing washer, then check the condition of

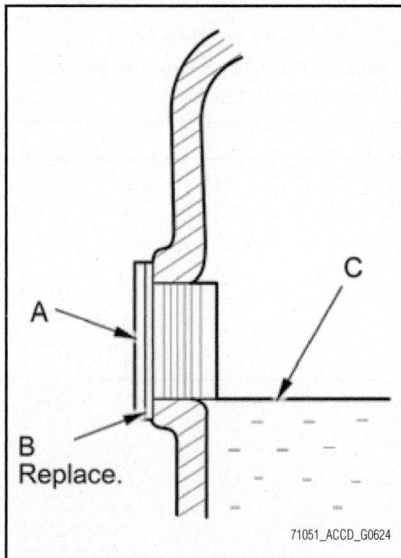

Fig. 63 Remove the oil filler plug (A) and the sealing washer (B), make sure the rear differential fluid is at the proper level (C)

the rear differential fluid, if the fluid is dirty or burnt, replace it.

4. Make sure the rear differential fluid is at the proper level. If the level is below the proper level, check for rear differential fluid leaks at the differential. If a problem is found, fix it before filling the differential with fluid.

5. The fluid level must reach up to the bottom of the oil filler plug hole. If necessary, add the rear differential fluid until it runs out, then reinstall the oil filler plug with a new sealing washer.

DRAIN & REFILL

See Figures 64 and 65.

➡**Always use Honda Dual Pump Fluid II.**

1. Raise and support the vehicle.
2. Remove the oil filler plug and the sealing washer.
3. Remove the drain plug and the sealing washer, then drain the rear differential fluid.

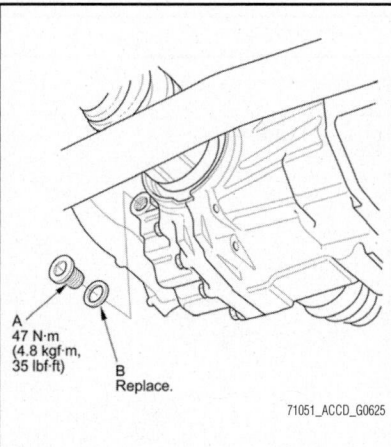

Fig. 64 Remove the oil filler plug (A) and the sealing washer (B)

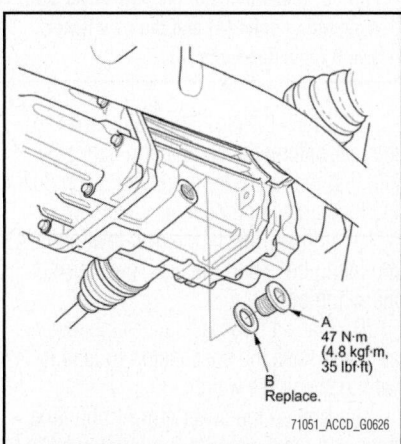

Fig. 65 Remove the drain plug (A) and the sealing washer (B), then drain the rear differential fluid

4. Clean the drain plug, then reinstall it with a new sealing washer.

5. Refill the differential with rear differential fluid to the proper level.

➡**If you disassemble the differential, check the rear differential fluid level again after the 4WD system check is finished. Add fluid if necessary.**

6. Reinstall the oil filler plug with a new sealing washer.

7. If the Maintenance Minder required to replace the rear differential fluid, reset the Maintenance Minder, and this procedure is complete. If the Maintenance Minder did not require to replace the rear differential fluid, go to step 9.

8. Lower the vehicle.

9. With the ignition switch in LOCK (0), then connect the HDS to the DLC located under the driver's side of the dashboard.

10. Turn the ignition switch to ON (II).

11. Make sure the HDS communicates with the vehicle and the PCM. If it does not communicate, troubleshoot the DLC circuit.

12. Select GAUGES in the BODY ELECTRICAL with the HDS.

13. Select ADJUSTMENT in the GAUGES with the HDS.

14. Select MAINTENANCE MINDER in the ADJUSTMENT with the HDS.

15. Select MAINTENANCE SUB ITEM 6 RESET with the HDS, and reset the rear differential fluid life.

REAR DIFFERENTIAL

REMOVAL & INSTALLATION

See Figures 66 through 69.

➡**The rear differential is used only on the Crosstour 4WD.**

1. Raise and support the vehicle.
2. Drain the rear differential fluid.
3. Remove the rear wheels.
4. Remove the muffler.
5. Remove the lower arm A mounting bolts and the lower arm B mounting bolts.
6. Make reference mark across the No. 2 propeller shaft and the rear differential companion flange.
7. Separate the No. 2 propeller shaft from the rear differential.

➡**Suspend the propeller shaft with an appropriate size nylon strap.**

8. Place a transmission jack under the rear differential assembly.
9. Remove the rear differential rear mount bolts and the front mount bolts.
10. Lower the rear differential assembly

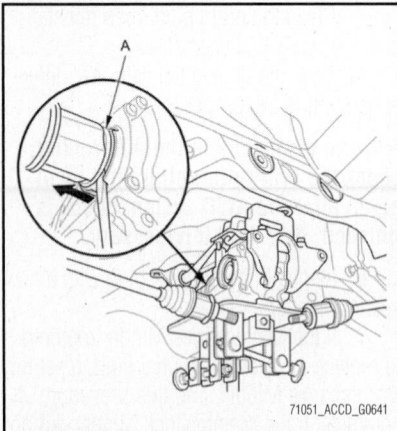

Fig. 66 Be careful not to damage the driveshaft ring (A) when prying out the rear driveshaft inboard joints

while pulling both rear driveshaft inboard joints out of the rear differential assembly.

➡**Be careful not to damage the driveshaft ring when prying out the rear driveshaft inboard joints.**

To install:

11. Raise and support the vehicle.

12. Install the rear differential front mounts to the rear subframe with new rear differential front mount bolts loosely.

13. Raise the rear differential with the transmission jack.

14. Install a new set ring into the groove of the rear driveshaft inboard joint on both sides.

15. Lift the rear differential up into position. Insert the driveshafts into the rear differential on both sides. Turn push on driveshaft to lock the set ring into place.

16. Tighten new rear differential front mount bolts to 61 ft. lbs. (83 Nm).

17. Tighten new rear differential rear mount bolts to 54 ft. lbs. (73 Nm).

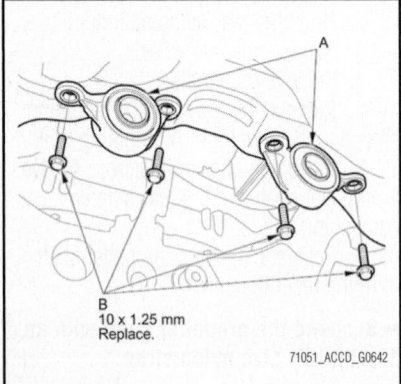

Fig. 67 Install the rear differential front mounts (A) to the rear subframe with new rear differential front mount bolts (B) loosely

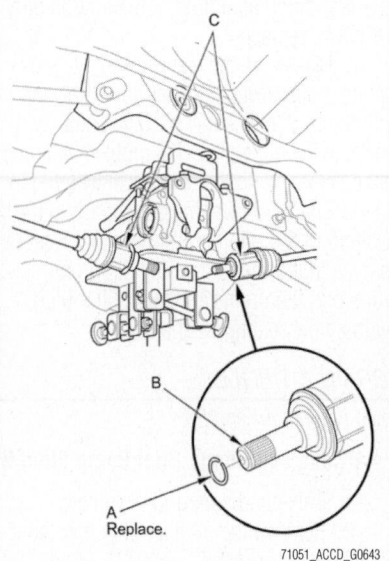

Fig. 68 Install a new set ring (A) into the groove (B) of the rear driveshaft inboard joint on both sides; Insert the driveshafts (C) into the rear differential on both sides

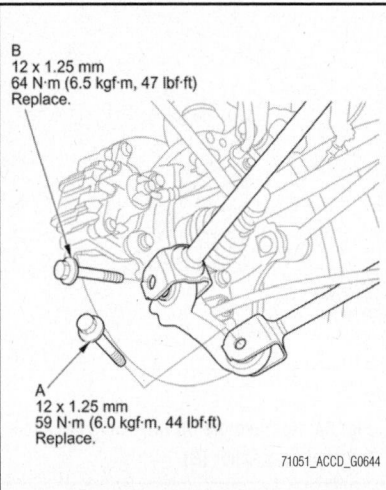

Fig. 69 Loosely install the new lower arm A mounting bolts (A) and the new lower arm B mounting bolts (B)

18. Attach the No. 2 propeller shaft to the rear differential companion flange by aligning the reference mark you made during the removal procedure.

19. Loosely install the new lower arm A mounting bolts and the new lower arm B mounting bolts.

20. Place a floor jack under the lower arm, and raise the suspension to load it with the vehicle's weight.

21. Tighten the lower arm A mounting bolts and the lower arm B mounting bolts to the specified torque.

22. Install the muffler.

23. Clean the mating surfaces between

the brake disc/drum and the inside of the wheel, then install the rear wheels.

24. Refill the rear differential with the recommended fluid.

25. Check the wheel alignment, and adjust it if necessary.

26. Test-drive the vehicle.

REAR HALFSHAFT

REMOVAL & INSTALLATION

See Figures 70 through 73.

➡**The rear halfshafts are used only on the Crosstour 4WD model.**

1. Raise and support the vehicle.

2. Remove the rear wheel.

3. Pry up the stake on the spindle nut, then remove the nut.

4. Drain the rear differential fluid.

5. Remove the rear driveshaft inboard joint from the rear differential assembly.

6. Pull the knuckle outward, and sepa-

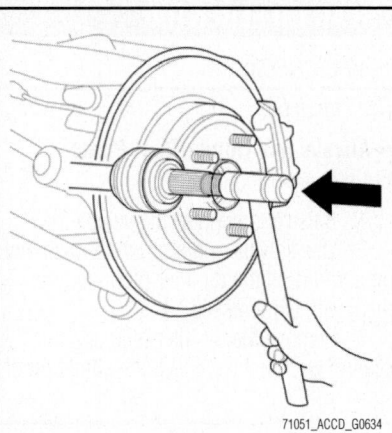

Fig. 70 Pull the knuckle outward, and separate the outboard joint from the rear hub using a soft face hammer

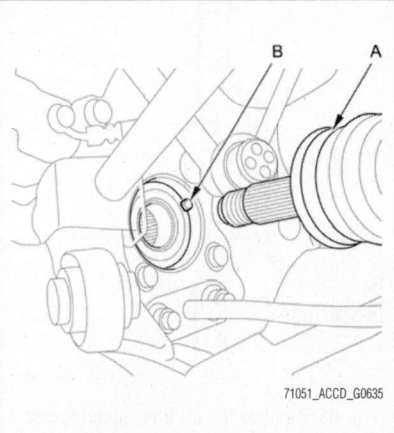

Fig. 71 Remove the outboard joint (A); wheel speed sensor (B)

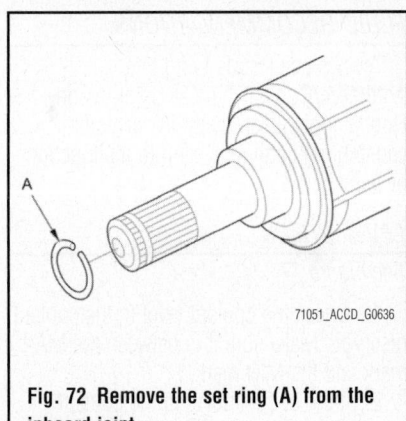

Fig. 72 Remove the set ring (A) from the inboard joint

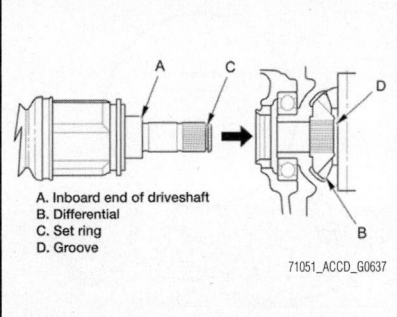

A. Inboard end of driveshaft
B. Differential
C. Set ring
D. Groove

Fig. 73 Insert the inboard end of the driveshaft into the differential until the set ring locks in the groove

rate the outboard joint from the rear hub using a soft face hammer.

7. Remove the outboard joint:
- Be careful not to damage the wheel speed sensor.
- Pull on the outboard joint. Do not pull on the driveshaft because the joint may come apart.

8. Remove the set ring from the inboard joint.

To install:

➡**Before starting installation, make sure the mating surfaces of the** joint and the splined section are clean.

9. Install a new set ring into the set ring groove of the rear driveshaft inboard joint.

10. Install the outboard joint into the rear hub.

➡**Be careful not to damage the wheel speed sensor.**

11. Clean the areas where the driveshaft contacts the differential thoroughly with solvent, and dry them with compressed air.

➡**Do not wash the rubber parts with solvent.**

12. Insert the inboard end of the driveshaft into the differential until the set ring locks in the groove.

➡**Insert the driveshaft horizontally to prevent damaging the oil seal.**

13. Install the rear differential.

14. Apply a small amount of engine oil to the seating surface of a new spindle nut (A).

15. Install the spindle nut, then tighten it. After tightening, use a drift to stake the spindle nut shoulder (B) against the driveshaft.

16. Clean the mating surfaces between the brake disc/drum and the inside of the wheel, then install the rear wheel.

17. Turn the wheel by hand, and make sure there is no interference between the driveshaft and surrounding parts.

18. Refill the rear differential with the recommended fluid.

19. Lower the vehicle.

20. Check the wheel alignment, and adjust it if necessary.

21. Test-drive the vehicle.

ENGINE COOLING

ENGINE COOLANT

DRAIN & REFILL

4-Cylinder Engine

See Figure 74.

1. Wait until the engine is cool, then carefully remove the radiator cap.

2. Loosen the drain plug, and drain the coolant.

3. After the coolant has drained, tighten the radiator drain plug securely.

4. Remove, drain, and reinstall the coolant reservoir.

5. Fill the coolant reservoir to the MAX mark with Honda Long Life Antifreeze/Coolant Type 2.

6. Pour Honda Long Life Antifreeze/Coolant Type 2 into the radiator up to the base of the filler neck.

➡**Important coolant information:**

- Always use Honda Long Life Antifreeze/Coolant Type 2. Using a non-Honda coolant can result in corrosion, causing the cooling system to malfunction or fail.
- Honda Long Life Antifreeze/Coolant

Type 2 is a mixture of 50 % antifreeze and 50 % water. Do not add water.

- If the vehicle is regularly driven in very low temperatures (under −31°F, −35°C) a 60 % concentration of coolant should be used. To accomplish this, pour 0.4 gal (1.6 L) of Honda Extreme Cold Weather

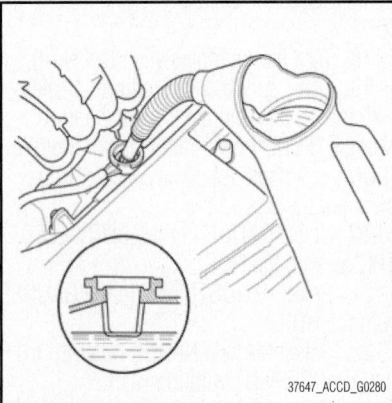

Fig. 74 Pour Honda Long Life Antifreeze/Coolant Type 2 into the radiator up to the base of the filler neck

Antifreeze/Coolant Type 2 into the radiator first, then add Honda Long Life antifreeze/Coolant Type 2 until the radiator is full.

7. Loosely install the radiator cap.

8. Start the engine, and let it run until it warms up (the radiator fan comes on at least twice).

9. Turn off the engine. Check the level in the radiator, and add Honda Long Life Antifreeze/Coolant Type 2, if needed.

10. Put the radiator cap securely, then start the engine again, and check for leaks.

11. Clean up any spilled engine coolant.

12. If the maintenance minder required engine coolant replacement, reset the maintenance minder, and this procedure is complete. If the maintenance minder did not require engine coolant replacement, go to the next step.

13. Turn the ignition switch to LOCK (0).

14. Connect the Honda Diagnostic System (HDS) to the Data Link Connector (DLC).

15. Turn the ignition switch to ON (II).

16. Make sure the HDS communicates with the vehicle and the Engine Control Module (ECM)/Powertrain Control Module

(PCM). If it does not communicate, troubleshoot the DLC circuit.

17. Select GAUGES in the BODY ELECTRICAL with the HDS.

18. Select ADJUSTMENT in the GAUGES with the HDS.

19. Select MAINTENANCE MINDER in the ADJUSTMENT with the HDS.

20. Select RESET in the MAINTENANCE MINDER with the HDS.

21. Select MAINTENANCE SUB ITEM 5 RESET with the HDS.

6-Cylinder Engine

See Figures 75 and 76.

1. Wait until the engine is cool, then carefully remove the radiator cap.

2. Loosen the drain plug and drain the coolant.

3. Install a rubber hose on the drain bolt located at the rear of the engine block, then loosen the drain bolt.

4. When the coolant stops draining, tighten the drain bolt. Remove the rubber hose.

5. Tighten the radiator drain plug securely.

6. Remove, drain, and reinstall the coolant reservoir.

7. Fill the coolant reservoir to the MAX mark with Honda Long Life Antifreeze/Coolant Type 2.

8. Pour Honda Long Life Antifreeze/Coolant Type 2 into the radiator up to the base of the filler neck.

➡Important coolant information:

- Always use Honda Long Life Antifreeze/Coolant Type 2. Using a non-Honda coolant can result in corrosion, causing the cooling system to malfunction or fail.

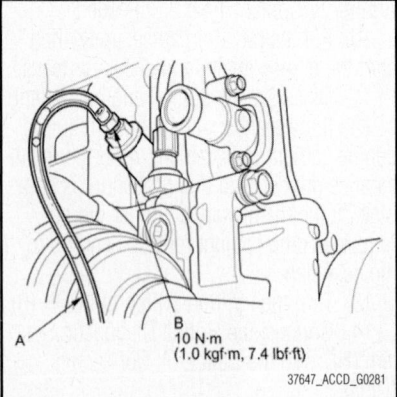

Fig. 75 Install a rubber hose (A) on the drain bolt (B) located at the rear of the engine block

10 N·m
(1.0 kgf·m, 7.4 lbf·ft)

37647_ACCD_G0281

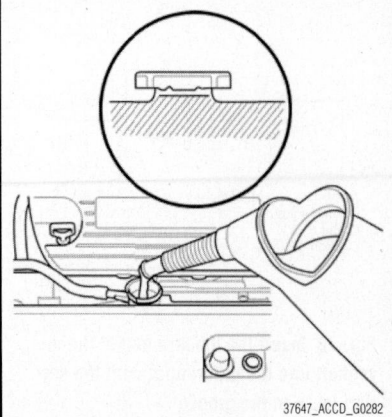

Fig. 76 Pour Honda Long Life Antifreeze/Coolant Type 2 into the radiator up to the base of the filler neck

37647_ACCD_G0282

- Honda Long Life Antifreeze/Coolant Type 2 is a mixture of 50 % antifreeze and 50 % water. Do not add water.

9. Loosely install the radiator cap.

10. Set the heater temperature control to maximum heat.

11. Start the engine, and let it run until it warms up (the radiator fan comes on at least twice).

12. Turn off the engine. Check the level in the radiator, and add Honda Long Life Antifreeze/Coolant Type 2, if needed.

13. Put the radiator cap on securely, then run the engine again, and check for leaks.

14. Clean up any spilled engine coolant.

15. If the maintenance minder required engine coolant replacement, reset the maintenance minder, and this procedure is complete. If the maintenance minder did not require engine coolant replacement, go to the next step.

16. Turn the ignition switch to LOCK (0).

17. Connect the Honda Diagnostic System (HDS) to the Data Link Connector (DLC).

18. Turn the ignition switch to ON (II).

19. Make sure the HDS communicates with the vehicle and the Engine Control Module (ECM)/Powertrain Control Module (PCM). If it does not communicate, troubleshoot the DLC circuit.

20. Select GAUGES in the BODY ELECTRICAL with the HDS.

21. Select ADJUSTMENT in the GAUGES with the HDS.

22. Select MAINTENANCE MINDER in the ADJUSTMENT with the HDS.

23. Select RESET in the MAINTENANCE MINDER with the HDS.

24. Select MAINTENANCE SUB ITEM 5 RESET with the HDS.

FLUID RECOMMENDATIONS

Always use Honda Long Life Antifreeze/Coolant Type 2. Using a non-Honda coolant can result in corrosion, causing the cooling system to malfunction or fail.

LEVEL CHECK

See Figure 77.

1. Check the coolant level in the coolant reservoir. Make sure it is between the MAX mark and the MIN mark.

2. If the coolant level in the coolant reservoir is at or below the MIN mark, add coolant to bring it between the MIN and MAX marks, then inspect the cooling system for leaks.

3. Check the coolant level in the radiator, and add Honda Long Life Antifreeze/Coolant Type 2 into the radiator up to the base of the filler neck, if needed.

- Always use Honda Long Life Antifreeze/Coolant Type 2. Using a non-Honda coolant can result in corrosion, causing the cooling system to malfunction or fail.

- Honda Long Life Antifreeze/Coolant Type 2 is a mixture of 50% antifreeze and 50% water. Do not add water.

ELECTRIC ENGINE FAN

REMOVAL & INSTALLATION

4-Cylinder Engine

See Figures 78 and 79.

1. Do the battery removal procedure.

2. Remove the harness clamps, then remove the battery base.

3. Remove the front grille cover:

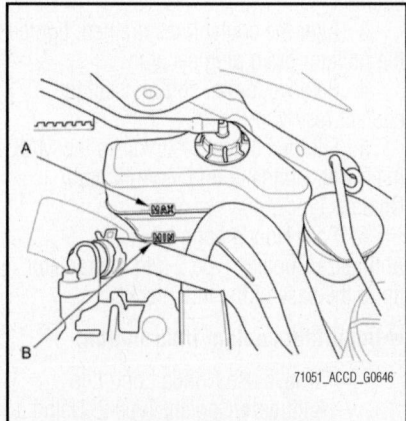

Fig. 77 Check the coolant level in the coolant reservoir. Make sure it is between the MAX mark (A) and the MIN mark (B)

71051_ACCD_G0646

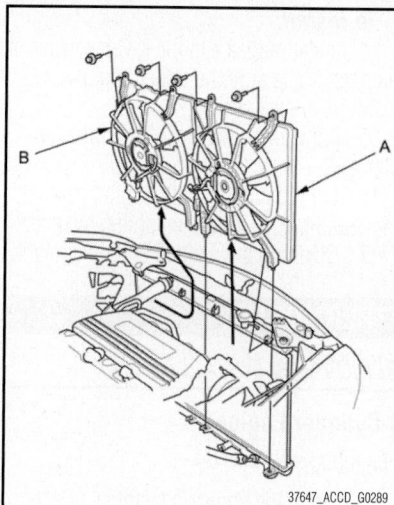

Fig. 78 Remove the A/C condenser fan shroud assembly (A), then remove the radiator fan shroud assembly (B)

4. Remove the water separator and the intake air duct.

5. Remove the coolant reservoir.

6. Disconnect the fan motor connectors, then remove the harness clamp.

7. Remove the A/C condenser fan shroud assembly, then remove the radiator fan shroud assembly.

➡ **Pull up the radiator, then move the radiator fan shroud assembly toward the right side of the vehicle to allow enough space to lift it up and away from the A/C condenser fan shroud assembly.**

8. Disassemble the fan shrouds.

To install:

9. Reassemble the fan shrouds.

10. Install the radiator fan shroud assembly, then install the A/C condenser fan shroud assembly.

11. Connect the fan motor connectors, then install the harness clamp.

12. Install the coolant reservoir.

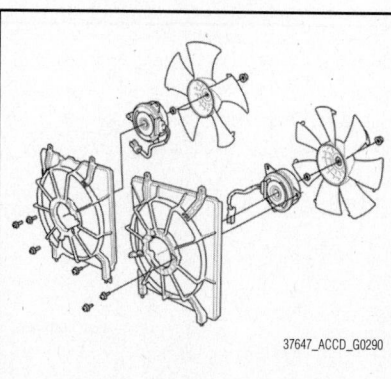

Fig. 79 Disassemble the fan shrouds

13. Install the water separator and the intake air duct.

14. Install the front grille cover.

15. Install the battery base, then install the harness clamps.

16. Do the battery installation procedure.

6-Cylinder Engine

See Figure 80.

1. Do the battery removal procedure.

2. Remove the battery base.

3. Remove the front grille cover.

4. Remove the splash separator and the intake air ducts.

5. Raise the vehicle on the lift.

6. Remove the splash shield.

7. Disconnect the A/C condenser fan motor connector, and remove the harness clamp.

8. Loosen the A/C condenser fan shroud mounting bolts.

9. Disconnect the radiator fan motor connector.

10. Lower the vehicle on the lift.

11. Remove the coolant reservoir.

12. Remove the upper brackets.

13. Remove the A/C condenser fan shroud assembly, then remove the radiator fan shroud assembly.

➡ **Pull-up the radiator, then move the radiator fan shroud assembly toward the right side of the vehicle to allow for enough space to lift it up and away from the A/C condenser fan shroud assembly.**

14. Disassemble the fan shrouds.

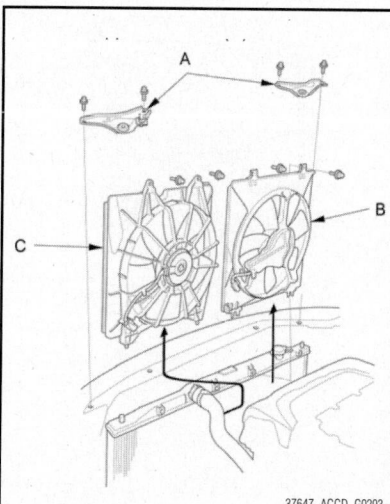

Fig. 80 Remove the upper brackets (A), the A/C condenser fan shroud assembly (B), then remove the radiator fan shroud assembly (C)

To install:

15. Assemble the fan shrouds.

16. Install the radiator fan shroud assembly, then install the A/C condenser fan shroud assembly.

17. Install the upper brackets.

18. Install the coolant reservoir.

19. Raise the vehicle on the lift.

20. Connector the radiator fan motor connector.

21. Tighten the A/C condenser fan shroud mounting bolts.

22. Connect the A/C condenser fan motor connector, and install the harness clamp.

23. Install the splash shield.

24. Lower the vehicle on the lift.

25. Install the splash separator and the intake air ducts.

26. Install the front grille cover.

27. Install the battery base.

28. Do the battery installation procedure.

RADIATOR

REMOVAL & INSTALLATION

4-Cylinder Engine

See Figure 81.

1. Do the battery removal procedure.

2. Remove the harness clamps, then remove the battery base.

3. Remove the front grille cover.

4. Remove the water separator and the intake air duct.

5. Drain the engine coolant.

6. Remove the splash shield.

7. A/T model: Disconnect the Automatic Transmission Fluid (ATF) cooler hoses, then plug the cooler hoses and the lines.

8. Disconnect the lower radiator hose from the radiator.

9. Disconnect the fan motor connectors and Engine Coolant Temperature (ECT) sensor 2 connector, then remove the harness clamps.

10. Disconnect the upper radiator hose and remove the upper brackets and the coolant reservoir.

11. Remove the radiator.

12. Remove the fan shroud assemblies from the radiator.

To install:

13. Install the radiator in the reverse order of removal. Make sure the upper and lower cushions (4) are set securely.

14. Do the battery installation procedure.

15. A/T model: Refill the transmission with ATF.

16. Fill the radiator with engine coolant, and bleed the air from the cooling system.

17. Clean up any spilled engine coolant.

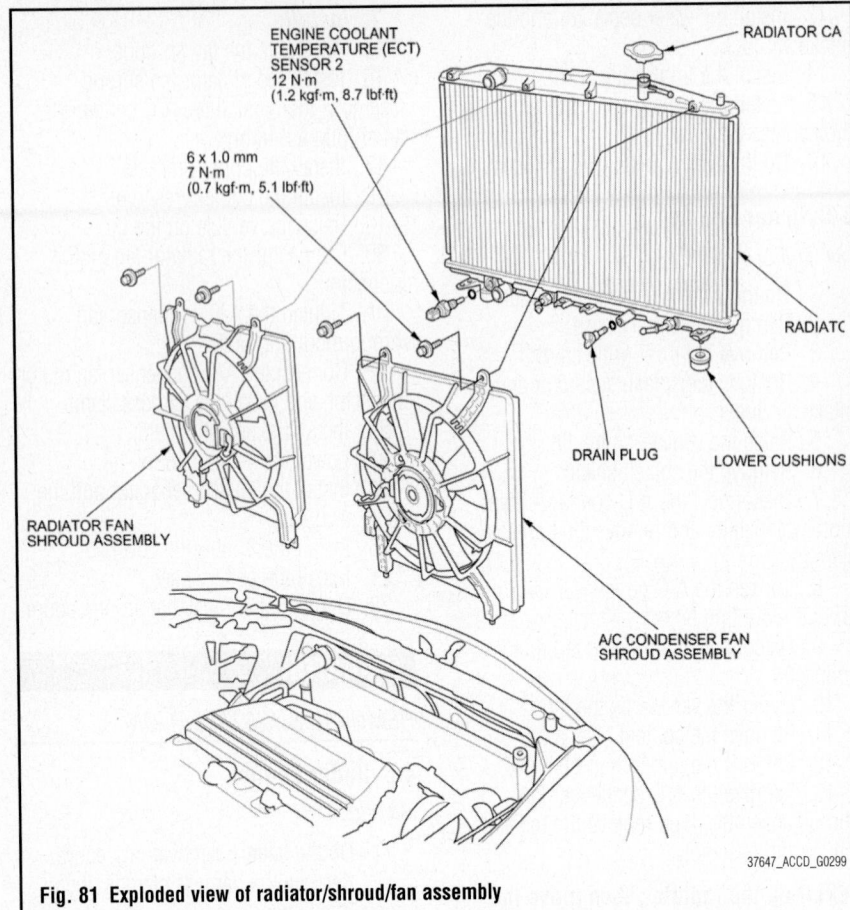

ENGINE COOLANT
TEMPERATURE (ECT)
SENSOR 2
12 N·m
(1.2 kgf·m, 8.7 lbf·ft)

RADIATOR CA

6 x 1.0 mm
7 N·m
(0.7 kgf·m, 5.1 lbf·ft)

RADIATC

DRAIN PLUG

LOWER CUSHIONS

A/C CONDENSER FAN
SHROUD ASSEMBLY

RADIATOR FAN
SHROUD ASSEMBLY

37647_ACCD_G0299

Fig. 81 Exploded view of radiator/shroud/fan assembly

6-Cylinder Engine

See Figure 82.

1. Remove the front grille cover:
2. Remove the splash separator.
3. Raise the vehicle on the lift.
4. Drain the engine coolant.
5. Remove the splash shield.
6. Disconnect the A/C condenser fan motor connector, and remove the harness clamp.
7. Disconnect the radiator fan motor connector and the Engine Coolant Temperature (ECT) sensor 2 connector, then remove the harness clamp.
8. Disconnect the lower radiator hose from the radiator.
9. A/T model: Disconnect the Automatic Transmission Fluid (ATF) cooler hoses from the radiator, then plug the cooler hoses and the lines.
10. Lower the vehicle on the lift.
11. Disconnect the upper radiator hose and remove the coolant reservoir.
12. Remove the upper brackets, then pull up the radiator with fan shrouds.
13. Remove the fan shrouds and other parts from the radiator.

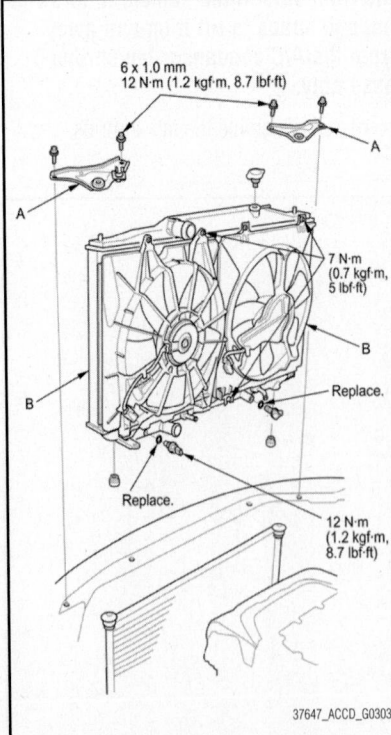

6 x 1.0 mm
12 N·m (1.2 kgf·m, 8.7 lbf·ft)

A

A

7 N·m
(0.7 kgf·m,
5 lbf·ft)

B

Replace.

B

Replace.

12 N·m
(1.2 kgf·m,
8.7 lbf·ft)

37647_ACCD_G0303

Fig. 82 Remove the upper brackets (A), then pull up the radiator with fan shrouds (B)

To install:

14. Install radiator in the reverse order or removal. Make sure the upper and lower cushions are set securely.
15. A/T model: Refill the transmission with ATF.
16. Fill the radiator with engine coolant, and bleed the air from the cooling system.
17. Clean up any spilled engine coolant.

THERMOSTAT

REMOVAL & INSTALLATION

4-Cylinder Engine

See Figures 83 and 84.

1. Drain the engine coolant.
2. Clean any dirt off the quick connector, thermostat cover, and lower radiator hose.
3. Pull out lock by hand, and then wiggle the quick connector loose, and remove it from the thermostat cover. Do not use any tools to remove the quick connector.
4. Remove the thermostat.

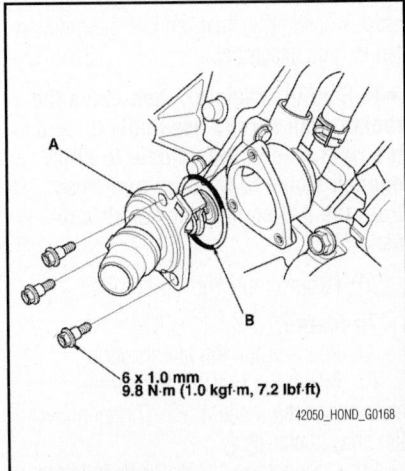

A

B

6 x 1.0 mm
9.8 N·m (1.0 kgf·m, 7.2 lbf·ft)

42050_HOND_G0168

Fig. 83 Always use a new O-ring (B) when installing the thermostat (A)

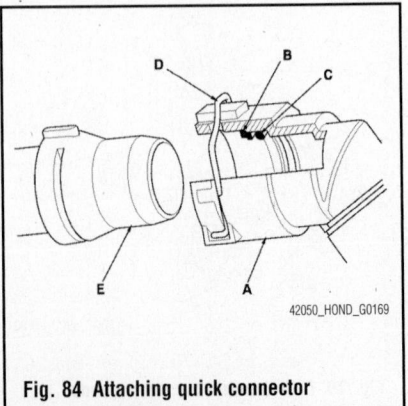

D   B
               C

E   A

42050_HOND_G0169

Fig. 84 Attaching quick connector

To install:

5. Install the thermostat with a new O-ring.

6. Check the quick connector and set ring for cracks or damage. If the connector and/or set ring are cracked or damaged, replace the connector.

7. Make sure the set ring is in place inside the quick connector. If the set ring is off the connector, replace the quick connector.

8. Replace the O-ring in the quick connector.

9. Check the lock. If the lock is damaged or deformed, replace it. When installing the new lock on the connector, press it straight down along the groove.

10. Clean the connecting surface of the thermostat cover, and then apply clean engine coolant around the connecting surface.

11. Push down the lock, and then push the quick connector onto the thermostat cover until you hear it click.

12. Refill the radiator with engine coolant, and bleed air from the cooling system with the heater valve open.

6-Cylinder Engine

See Figure 85.

1. Drain the engine coolant.
2. Do the battery removal procedure.
3. Remove the thermostat cover, then remove the thermostat.
4. Install the new thermostat with a new rubber seal, then install the thermostat cover.
5. Do the battery installation procedure.
6. Refill the radiator with engine coolant, and bleed the air from the cooling system.
7. Clean up any spilled engine coolant.

WATER PUMP

REMOVAL & INSTALLATION

4-Cylinder Engine

See Figure 86.

1. Remove the drive belt.
2. Drain the engine coolant.
3. Remove the tensioner pulley.
4. Remove the water pump pulley.
5. Remove the six bolts securing the water pump, then remove the water pump.

To install:

6. Inspect and clean the O-ring groove and the mating surface of the water passage.
7. Install the water pump with a new O-ring and the water pump pulley.

8. Clean up any spilled engine coolant.
9. Install the tensioner pulley.
10. Install the drive belt.
11. Refill the radiator with engine coolant, and bleed the air from the cooling system.

6-Cylinder Engine

See Figure 87.

1. Drain the engine coolant.
2. Remove the timing belt:
3. Remove the timing belt adjuster:
4. Remove the five bolts securing the water pump, then remove the water pump.
5. Inspect and clean the O-ring groove and the mating surface of the engine block.

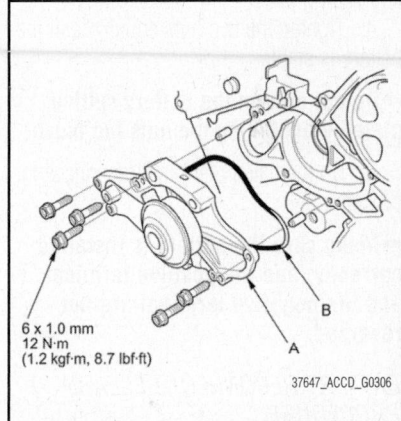

Fig. 87 Remove the five bolts securing the water pump (A), then remove the water pump and O-ring (B)

6. Install the water pump with a new O-ring.
7. Clean up any spilled engine coolant.
8. Install the timing belt adjuster:
9. Install the timing belt:
10. Refill the radiator with engine coolant, and bleed the air from the cooling system.

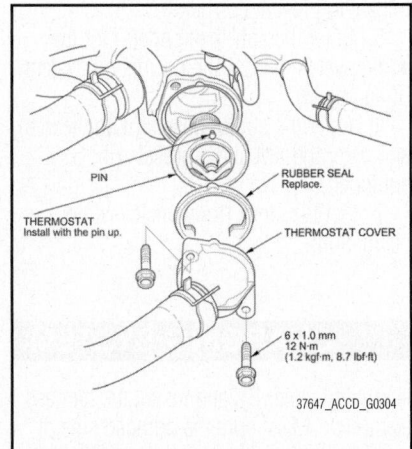

Fig. 85 Exploded view of thermostat housing assembly

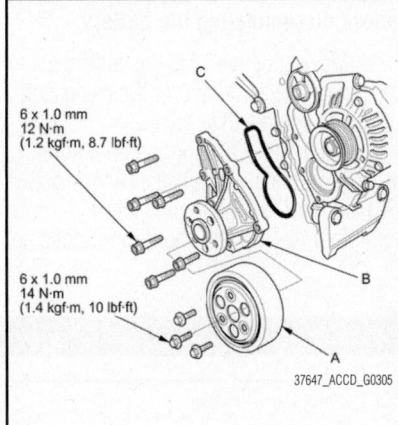

Fig. 86 Remove the water pump pulley (A), the water pump (B) and the O-ring (C)

ENGINE ELECTRICAL   BATTERY SYSTEM

BATTERY

REMOVAL & INSTALLATION

See Figure 88.

➡The battery terminal disconnection and reconnection procedure must be done before and after doing this procedure. Some systems store data in memory that is lost when the battery is disconnected.

1. Do the battery terminal disconnection procedure.

2. Remove the two nuts securing the battery setting plate, then remove the battery setting plate and the battery.

3. Install the battery, then install the battery setting plate.

4. Tighten the two nuts equally until the battery is stable.

➡Do not deform the battery setting plate by tightening the nuts too much.

5. Do the battery terminal reconnection procedure.

➡Make sure the battery is installed correctly, and the positive terminal and the negative terminal are not reversed.

BATTERY RECONNECT/RELEARN PROCEDURE

Disconnection

➡Some systems store data in memory that is lost when the battery is discon-

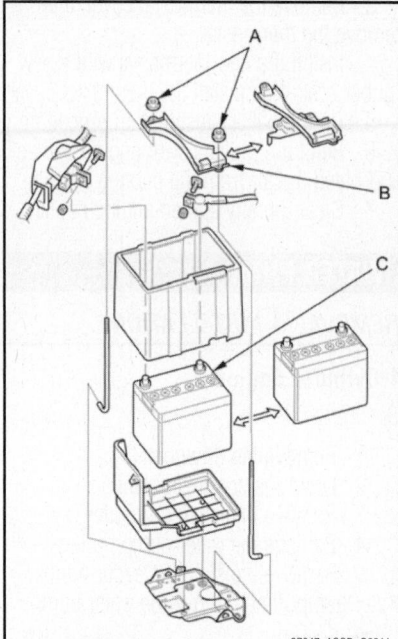

Fig. 88 Remove the two nuts (A), then remove the battery setting plate (B) and the battery (C)

nected. Do the following procedures before disconnecting the battery.

1. Make sure you have the anti-theft code(s) for the audio and/or the navigation system (if equipped).

2. If you are replacing the audio unit, write down the audio presets (AM and FM), and the XM audio presets (if equipped), because the audio unit does not

retain the presets after the battery is disconnected.

3. Make sure the ignition switch is in LOCK (0).

4. Disconnect and isolate the negative cable from the battery.

➡Always disconnect the negative cable from the battery first.

5. Disconnect the positive cable from the battery.

Reconnection

➡Some systems store data in memory that is lost when the battery is disconnected. Do the following procedures to restore the system back to normal operation.

1. Clean the battery terminals.
2. Test the battery.
3. Reconnect the positive cable to the battery first, then reconnect the negative cable to the battery.

➡Always connect the positive cable to the battery first.

4. Apply multipurpose grease to the terminals to prevent corrosion.

5. Enter the anti-theft code(s) for the audio system and/or the navigation system (if equipped).

6. Enter the audio presets (if applicable), and enter the XM audio presets (if equipped).

7. Set the clock (for vehicles without navigation).

ENGINE ELECTRICAL   CHARGING SYSTEM

ALTERNATOR

REMOVAL & INSTALLATION

4-Cylinder Engine

See Figure 89.

1. Do the battery terminal disconnection procedure.

2. Remove the drive belt.

3. Remove the two bolts securing the alternator.

4. Disconnect the alternator connector and the positive alternator cable, and remove the harness clamp, then remove the alternator.

To install:

5. Install the alternator, then connect the alternator connector (A) and the positive

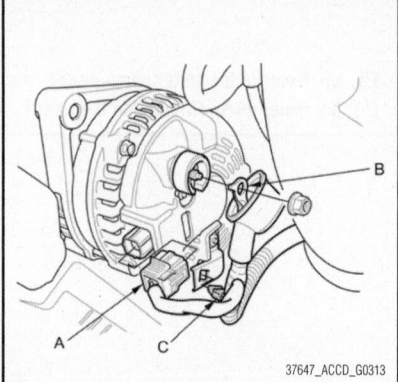

Fig. 89 Disconnect the alternator connector (A) and the positive alternator cable (B), and remove the harness clamp (C)

alternator cable (B), and install the harness clamp (C). Make sure the crimped side of the ring terminal faces away from the alternator when you connect it.

6. Tighten the two bolts securing the alternator. Tighten the 8 x 1.25 mm bolt to 16 ft. lbs. (22 Nm); tighten the 10 x 1.25 mm bolt to 33 ft. lbs. (44 Nm).

7. Install the drive belt.

8. Do the battery terminal reconnection procedure.

6-Cylinder Engine

See Figure 90.

1. Do the battery terminal disconnection procedure.

2. Raise the vehicle on the lift.

3. Remove the engine splash shield.

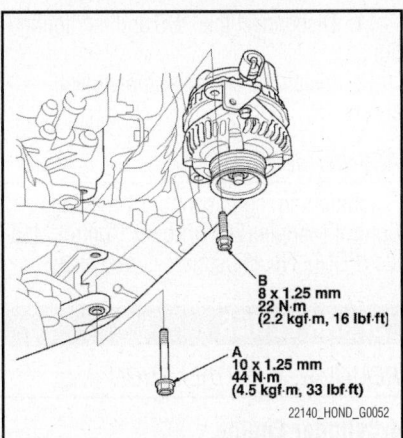

Fig. 90 Remove the mounting bolt (A) and the alternator bracket mounting bolt (B) to remove the alternator—3.5L Engines

4. Disconnect the A/C condenser fan motor connector and remove the harness clamp.

5. Loosen the A/C condenser fan shroud mounting bolts.

6. Lower the vehicle on the lift.

7. Remove the A/C condenser fan shroud assembly and the coolant reservoir.

8. Remove the accessory drive belt.

9. Disconnect the alternator connector and the positive alternator cable from the alternator.

10. Remove the harness clamp from the alternator and disconnect the A/C compressor clutch connector from the A/C compressor.

11. Remove the bolt securing the harness holder.

12. Remove the mounting bolt, the alternator bracket mounting bolt, and the alternator.

To install:

13. Install the alternator, then tighten the mounting bolt and the alternator bracket mounting bolt. Tighten the mounting bolt to 33 ft. lbs. (44 Nm). Tighten the bracket mounting bolt to 16 ft. lbs. (22 Nm).

14. Install the bolt securing the harness holder.

15. Install the harness clamp to the alternator and connect the A/C compressor clutch connector to the A/C compressor.

16. Connect the alternator connecter and the positive alternator cable to the alternator. Make sure the crimped side of the ring terminal faces away from the alternator when you connect it.

17. Install the drive belt.

18. Install the A/C condenser fan shroud assembly and the coolant reservoir.

19. Raise the vehicle on the lift.

20. Tighten the A/C compressor fan shroud mounting bolts.

21. Connect the A/C condenser fan motor connector and install the harness clamp.

22. Install the splash shield.

23. Lower the vehicle on the lift.

24. Do the battery terminal reconnection procedure.

ENGINE ELECTRICAL

IGNITION SYSTEM

FIRING ORDERS

4-Cylinder Engine

The firing order is 1–3–4–2.

6-Cylinder Engine

The firing order is 1–4–2–5–3–6.

IGNITION COIL

REMOVAL & INSTALLATION

4-Cylinder Engine

See Figure 91.

1. Remove the ignition coil cover.
2. Disconnect the ignition coil connectors, then remove the ignition coils.
3. Remove the spark plug and inspect them.
4. Apply a small amount of anti-seize compound to the plugs into the cylinder head, finger tight. Torque them to 13 ft. lbs. (18 Nm).
5. Install the ignition coils, then connect the ignition coil connectors.
6. Install the ignition coil cover.

6-Cylinder Engine

See Figure 92.

1. Remove the engine cover.

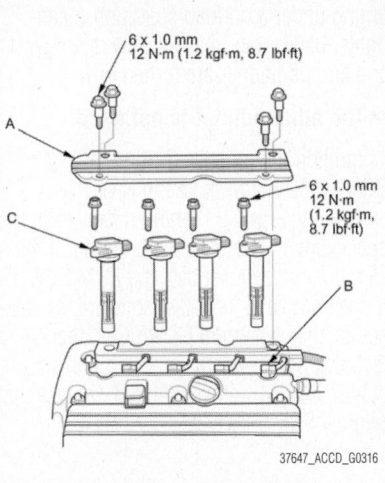

Fig. 91 Remove the ignition coil cover (A), disconnect the ignition coil connectors (B), then remove the ignition coils (C)

2. Disconnect the ignition coil connectors, then remove the ignition coils.

3. Remove the spark plugs and inspect them.

4. Apply a small amount of anti-seize compound to the plug threads, and screw the plugs into the cylinder head, finger-tight. Torque them to 13 ft. lbs. (18 Nm).

5. Install the ignition coils in the reverse order of removal.

IGNITION TIMING

INSPECTION

4-Cylinder Engine

1. Connect the Honda Diagnostic System (HDS) to the Data Link Connector (DLC).

2. Turn the ignition switch to ON (II).

3. Make sure the HDS communicates with the vehicle and the Engine Control Module (ECM)/Powertrain Control Module (PCM) If it does not communicate, troubleshoot the DLC circuit..

4. Check for DTCs. If a DTC is present, diagnose and repair the cause before continuing with this test.

5. Start the engine. Hold the engine speed at 3,000 rpm with no load (in N or P (A/T model) or neutral (M/T model)) until the radiator fan comes on, then let it idle.

6. Check the idle speed.

7. Jump the SCS line with the HDS.

➡**This step must be done to protect the ECM/PCM from damage.**

8. Connect the timing light to the service loop (white tape).

9. Aim the light toward the pointer on the cam chain case. Check the ignition timing under a no load condition (headlights, blower fan, rear window defogger, and air conditioner are turned off).

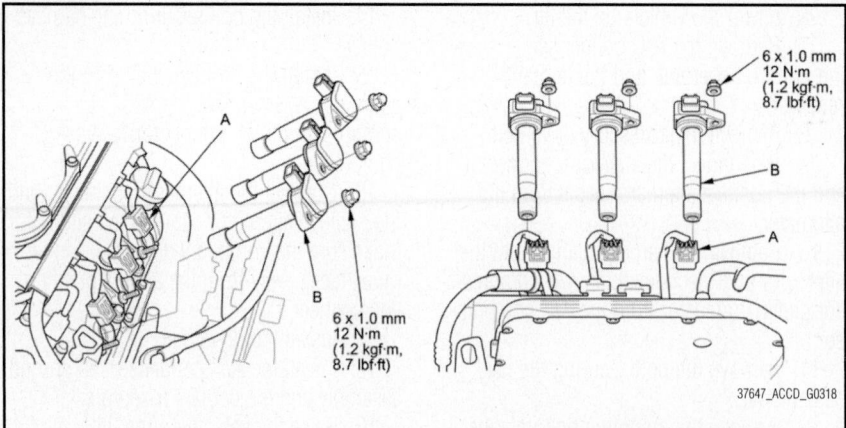

37647_ACCD_G0318

Fig. 92 Disconnect the ignition coil connectors (A), then remove the ignition coils (B)

➡ **The other pointer is not used.**

Ignition Timing specifications:
- M/T model: 8°±2° BTDC (RED mark) at idle in neutral
- A/T model: 8°±2° BTDC (RED mark) at idle in N or P

10. If the ignition timing differs from the specification, check the cam timing. If the cam timing is OK, update the ECM/PCM if it does not have the latest software, or substitute a known-good ECM/PCM, then recheck. If the system works properly, and the ECM/PCM was substituted, replace the original ECM/PCM.

11. Disconnect the HDS and the timing light.

6-Cylinder Engine

1. Connect the Honda Diagnostic System (HDS) to the Data Link Connector (DLC).

2. Turn the ignition switch to ON (II).

3. Make sure the HDS communicates with the vehicle and the Engine Control Module (ECM)/Powertrain Control Module (PCM). If it does not communicate, troubleshoot the DLC circuit.

4. Check for DTCs. If a DTC is present, diagnose and repair the cause before continuing with this test.

5. Start the engine. Hold the engine speed at 3,000 rpm with no load (in neutral (M/T model) or N or P (A/T model)) until the radiator fan comes on, then let it idle.

6. Check the idle speed.

7. Jump the SCS line with the HDS.

8. Connect the timing light to the No. 1 ignition coil harness.

9. Aim the light toward the pointer on the timing belt cover. Check the ignition timing under a no load condition (headlights, blower fan, rear window defogger, and air conditioner are turned off).

➡ **The other pointer is not used.**

Ignition timing specification: 10°±2° BTDC (RED mark) at idle in N or P

10. If the ignition timing differs from the specification, check the cam timing. If the cam timing is OK, update the ECM/PCM if it does not have the latest software, or substitute a known-good ECM/PCM, then recheck. If the system works properly, and the ECM/PCM was substituted, replace the original ECM/PCM.

11. Disconnect the HDS and the timing light.

12. Install the engine compartment covers.

ADJUSTMENT

Ignition timing is control by the Engine Control Module (ECM)/Power Control Module (PCM). No adjustment is necessary.

SPARK PLUGS

REMOVAL & INSTALLATION

4-Cylinder Engine

See Figure 91.

1. Remove the ignition coil cover.

2. Disconnect the ignition coil connectors, then remove the ignition coils.

3. Remove the spark plug and inspect them.

4. Apply a small amount of anti-seize compound to the plugs into the cylinder head, finger tight. Torque them to 13 ft. lbs. (18 Nm).

5. Install the ignition coils, then connect the ignition coil connectors.

6. Install the ignition coil cover.

6-Cylinder Engine

See Figure 92.

1. Remove the engine cover.

2. Disconnect the ignition coil connectors, then remove the ignition coils.

3. Remove the spark plugs and inspect them.

4. Apply a small amount of anti-seize compound to the plug threads, and screw the plugs into the cylinder head, finger-tight. Torque them to 13 ft. lbs. (18 Nm).

5. Install the ignition coils in the reverse order of removal.

STARTER

REMOVAL & INSTALLATION

4-Cylinder Engine

See Figures 93 and 94.

1. Do the battery removal procedure.
2. Remove the intake manifold.
3. Disconnect the positive starter cable and the S terminal connector.
4. Remove the harness clamp.
5. Remove the two bolts holding the starter, then remove the starter.

To install:

6. Install the starter, then tighten the two bolts.
7. Install the harness clamp.
8. Connect the positive starter cable and the S terminal connector. Make sure the crimped side of the ring terminal faces away from the starter when you connect it.
9. Install the intake manifold.
10. Do the battery installation procedure.
11. Start the engine to make sure the starter works properly.

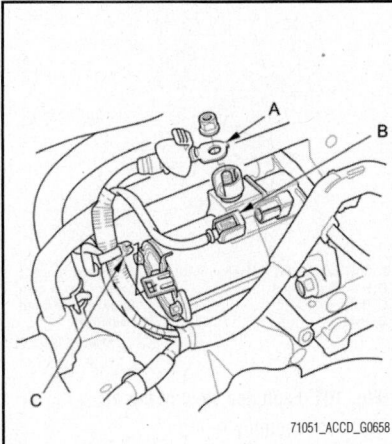

Fig. 93 Disconnect the positive starter cable (A) and the S terminal connector (B), then remove the harness clamp (C)

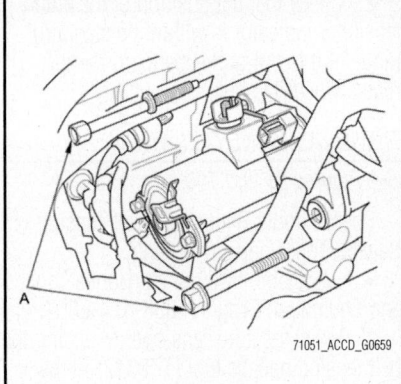

Fig. 94 Remove the two bolts (A) holding the starter, then remove the starter

6-Cylinder Engine

See Figures 95 and 96.

1. Do the battery removal procedure.
2. Remove the battery base.
3. Remove the splash separator and the intake air ducts:
4. Remove the harness clamp.
5. Disconnect the positive starter cable and the S terminal connector.
6. A/T: Remove the upper radiator hose bracket and the dipstick.
7. Remove the two bolts holding the starter, then remove the starter.

To install:

8. Install the starter, then tighten the mounting bolts.
   a. M/T: Tighten the starter mounting bolts to 54 ft. lbs. (74 Nm).
   b. A/T: Tighten the starter mounting bolts to 33 ft. lbs. (44 Nm).

➡**Always use a new gasket (A/T).**

9. A/T: Install the upper radiator hose bracket and the dipstick.
10. Connect the positive starter cable and the S terminal connector. Make sure the crimped side of the ring terminal faces away from the starter when you connect it.
11. Install the harness clamp.
12. Install the splash separator and the intake air ducts:

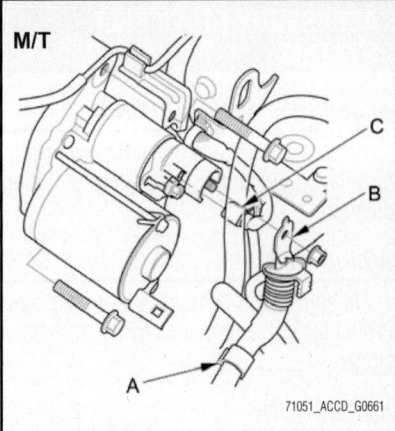

Fig. 95 Exploded view of the starter assembly showing the harness clamp (A), positive starter cable (B) and the S terminal connector (C) —M/T

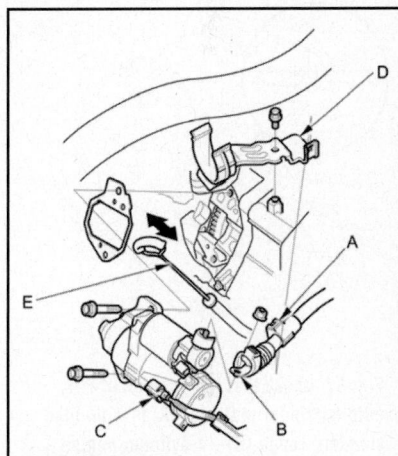

A. Harness clamp
B. Positive starter cable
C. S terminal connector
D. Upper radiator hose bracket
E. Dipstick

Fig. 96 Exploded view of the starter assembly—A/T

13. Install the battery base.
14. Do the battery installation procedure.
15. Start the engine to make sure the starter works properly.

ENGINE MECHANICAL

ACCESSORY DRIVE BELT SYSTEM

ACCESSORY BELT ROUTING

Refer to the graphic in the Removal and Installation procedure for the proper routing diagram.

ADJUSTMENT

No adjustment is necessary or possible. Proper tension is provided by the auto-tensioner.

INSPECTION

See Figures 97 and 98.

1. Inspect the belt for cracks or damage. If the belt is cracked or damaged, replace it.

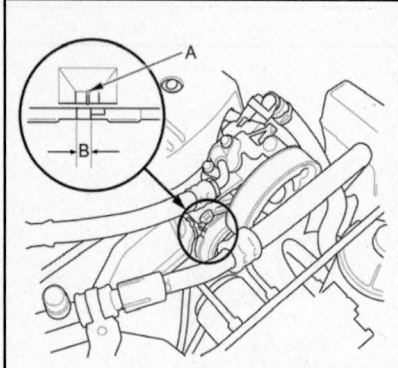

Fig. 97 Check that the position of the auto-tensioner indicator (A) is within the standard range (B)—4-cylinder engine

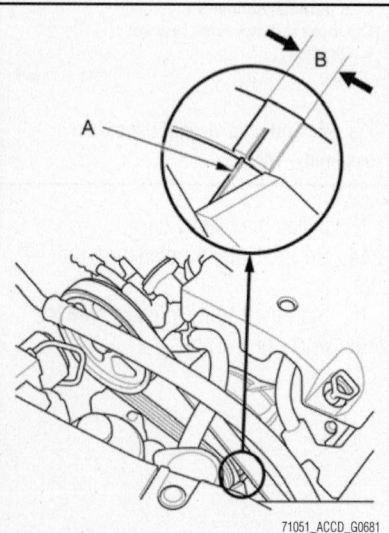

Fig. 98 Check that the position of the auto-tensioner indicator (A) is within the standard range (B)—6-cylinder engine

2. Check that the position of the auto-tensioner indicator is within the standard range. If it is out of the standard range, replace the drive belt.

REMOVAL & INSTALLATION

See Figures 99 and 100.

Special Tools Required*: Belt Tension Release Tool Snap-on TK9317
- *: Available through the Honda Tool and Equipment Program 888-424-6857.

1. Move the auto-tensioner (A) using the belt tension release tool (TK9317) in the direction of the arrow to relieve tension from the drive belt, then remove the drive belt.

2. Install the new drive belt in the reverse order of removal.

AIR CLEANER

REMOVAL & INSTALLATION

4-Cylinder Engine

See Figure 101.

1. Disconnect the MAF sensor/IAT sensor connector.

Fig. 101 Exploded view of the air cleaner assembly

2. Remove the harness clamps and the bolts.

3. Loosen the band, then remove the air cleaner housing.

4. Install the parts in the reverse order of removal.

➡ **When torquing the screw of the hose band, align the edge of the hose band**

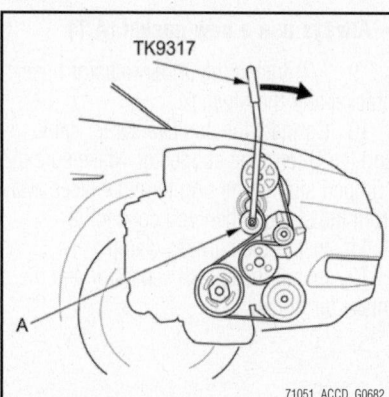

Fig. 99 Move the auto-tensioner (A) using the belt tension release tool (TK9317) in the direction of the arrow to relieve tension from the drive belt, then remove the drive belt—4-cylinder engine

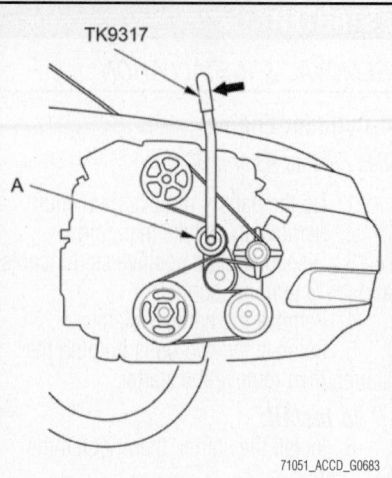

Fig. 100 Move the auto-tensioner (A) using the belt tension release tool (TK9317) in the direction of the arrow to relieve tension from the drive belt, then remove the drive belt—6-cylinder engine

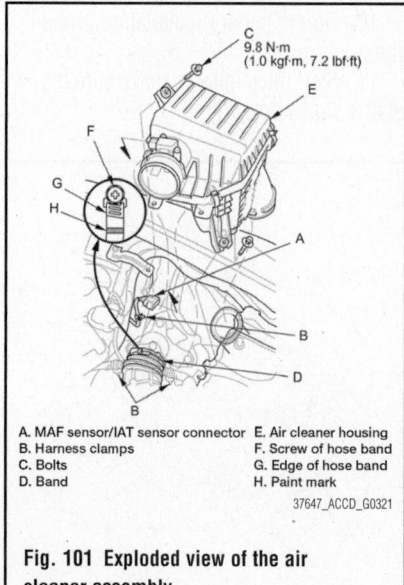

A. MAF sensor/IAT sensor connector
B. Harness clamps
C. Bolts
D. Band
E. Air cleaner housing
F. Screw of hose band
G. Edge of hose band
H. Paint mark

Fig. 101 Exploded view of the air cleaner assembly

with the mark painted on the hose band.

6-Cylinder Engine

See Figure 102.

1. Do the battery removal procedure.

2. Disconnect the MAF sensor/IAT sensor 5P connector.

3. Remove the harness clamps and the bolt.

4. Loosen the band, then remove the air cleaner housing.

5. Install the parts in the reverse order of removal.

6. Do the battery installation procedure.

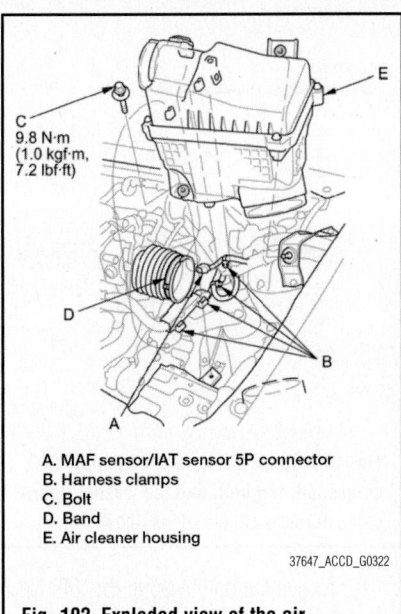

A. MAF sensor/IAT sensor 5P connector
B. Harness clamps
C. Bolt
D. Band
E. Air cleaner housing

37647_ACCD_G0322

Fig. 102 Exploded view of the air cleaner assembly

AIR FILTER ELEMENT REPLACEMENT

See Figures 103 and 104.

1. Open the air cleaner housing cover.

2. Remove the air cleaner element from the air cleaner housing.

3. Check the air cleaner element for damage or clogging. If it is damaged or clogged, replace it.

➡ **Do not use compressed air to clean the air cleaner element.**

4. Clean and remove any debris from inside the air cleaner.

To install:

5. Install the parts in the reverse order of removal.

   a. If you did not replace the air

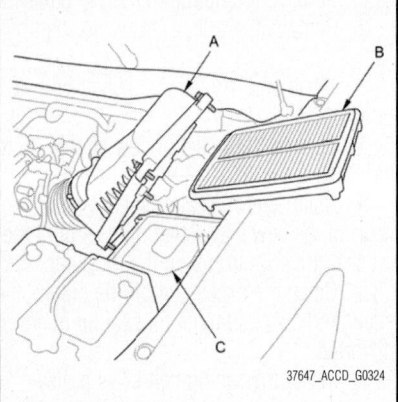

37647_ACCD_G0324

Fig. 104 Air cleaner housing cover (A); air cleaner element (B); air cleaner housing (C)—6-cylinder engine

cleaner element, this procedure is complete.

   b. If the maintenance minder required air cleaner replacement, reset the maintenance minder.

   c. If the idle speed fluctuates, do the idle speed inspection.

CAMSHAFT & BEARINGS

REMOVAL & INSTALLATION

6-Cylinder Engine

Front

See Figure 105.

1. Do the battery removal procedure.
2. Drain the engine coolant.
3. Disconnect the radiator hoses.
4. Remove the EGR valve.
5. Remove the EGR valve stud bolts.
6. Remove the timing belt.

7. Remove the front rocker arm assembly.

8. Remove the front camshaft pulley.

9. Remove the thrust cover, then remove the front camshaft.

To install:

10. Install the front camshaft in the reverse order of removal. Always use a new O-ring (C). Apply new engine oil to the journals and the cam lobes.

11. Apply new engine oil to the threads of the camshaft pulley mounting bolt, then install the front camshaft pulley.

12. Install the front rocker arm assembly, then tighten the mounting bolts.

13. Install the timing belt.

14. Adjust the valve clearance.

15. Install the EGR valve stud bolts, then install the EGR valve.

16. Connect the radiator hoses.

17. Do the battery installation procedure.

18. Fill the radiator with engine coolant, and bleed the air from the cooling system.

19. Do the CKP pattern clear/CKP pattern learn procedure.

Rear

See Figure 106.

1. Relieve the fuel pressure.
2. Do the battery removal procedure.
3. Remove the under-hood fuse/relay box.
4. Drain the engine coolant.
5. Remove the quick-connector fitting cover, then disconnect the fuel feed hose.
6. Disconnect the heater hoses and remove the EVAP canister purge joint with the bracket.
7. Remove the timing belt.
8. Remove the rear rocker arm assembly.
9. Remove the rear camshaft pulley.

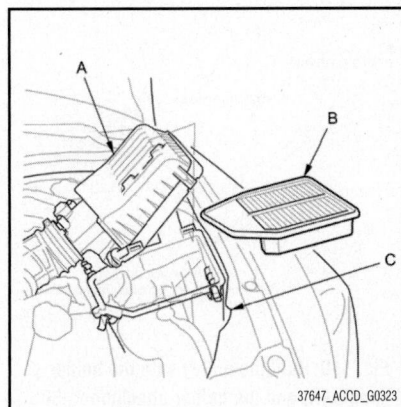

37647_ACCD_G0323

Fig. 103 Air cleaner housing cover (A); air cleaner element (B); air cleaner housing (C)—4-cylinder engine

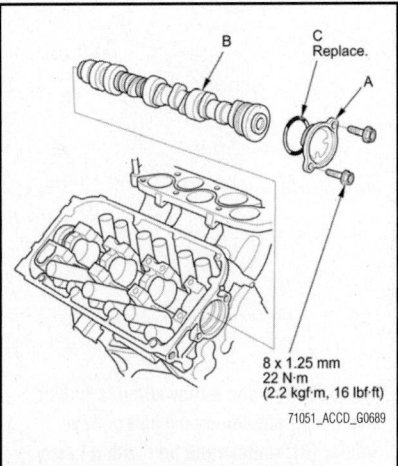

8 x 1.25 mm
22 N·m
(2.2 kgf·m, 16 lbf·ft)

71051_ACCD_G0689

Fig. 105 Remove the thrust cover (A), then remove the front camshaft (B)

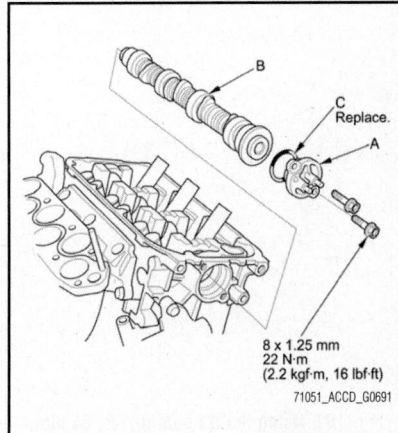

8 x 1.25 mm
22 N·m
(2.2 kgf·m, 16 lbf·ft)

71051_ACCD_G0691

Fig. 106 Remove the thrust cover (A), then remove the rear camshaft (B) and O-ring (C)

10. Remove the thrust cover, then remove the rear camshaft.

To install:

11. Install the rear camshaft in the reverse order of removal. Always use a new O-ring. Apply new engine oil to the journals and the cam lobes.

12. Apply new engine oil to the threads of the camshaft pulley mounting bolt, then install the rear camshaft pulley.

13. Install the rear rocker arm assembly, then tighten the mounting bolts.

14. Install the timing belt.

15. Adjust the valve clearance.

16. Connect the heater hoses and install the EVAP canister purge joint with the bracket.

17. Connect the fuel feed hose, then install the quick-connect fitting cover.

18. Install the under-hood fuse/relay box.

19. Do the battery installation procedure.

20. Inspect for fuel leaks. Turn the ignition switch to ON (II) (do not operate the starter) so the fuel pump runs for about 2 seconds and pressurizes the fuel line. Repeat this operation three times, then check for fuel leakage at any point in the fuel line.

21. Fill the radiator with engine coolant, and bleed the air from the cooling system.

22. Do the CKP pattern clear/CKP pattern learn procedure.

CRANKSHAFT FRONT SEAL

REMOVAL & INSTALLATION

6-Cylinder Engine

See Figure 107.

Special Tools Required: Oil Seal Driver, 64 mm 07OAD-RCAA100

1. Remove the timing belt drive pulley.

2. Remove the pulley end crankshaft oil seal.

3. Clean and dry the crankshaft oil seal housing.

4. Apply a light coat of new engine oil to the lip of the crankshaft oil seal.

5. Using the oil seal driver, 64 mm, drive in the new crankshaft oil seal until the oil seal driver bottoms on the oil pump.

6. Clean the excess oil off the crankshaft, and check that the oil seal lip is not distorted.

7. Install the timing belt drive pulley.

CRANKSHAFT PULLEY

REMOVAL & INSTALLATION

4-Cylinder Engine

See Figures 108 and 109.

Special Tools Required:
- Holder Handle 07JAB-001020B
- Crankshaft Pulley Holder 07AAB-RJAA100
- Socket, 19 mm 07JAA-001020A

1. Remove the right front wheel.

2. Remove the front splash shield.

3. Remove the drive belt.

4. Hold the pulley with the holder handle and the crankshaft pulley holder.

5. Remove the bolt with a heavy duty socket, 19 mm and a breaker bar, then remove the crankshaft pulley.

To install:

6. Clean the crankshaft pulley, the crankshaft, the bolt, and the washer. Lubricate with new engine oil as shown.

7. Install the crankshaft pulley, and hold the pulley with the holder handle and the crankshaft pulley holder.

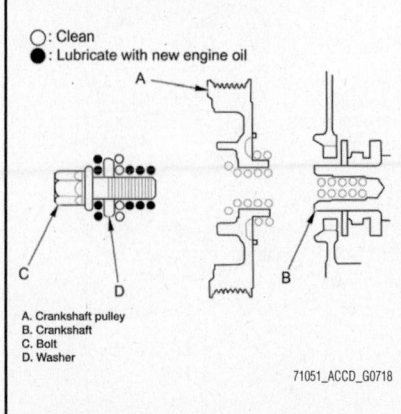

A. Crankshaft pulley
B. Crankshaft
C. Bolt
D. Washer

71051_ACCD_G0718

Fig. 109 Clean the crankshaft pulley, the crankshaft, the bolt, and the washer; lubricate with new engine oil as shown

8. Torque the bolt to 36 ft. lbs. (49 Nm) with a torque wrench and the heavy duty socket, 19 mm. Do not use an impact wrench. If the pulley bolt or crankshaft are new, torque the bolt to 130 ft. lbs. (177 Nm), then remove the bolt and torque it to 36 ft. lbs. (49 Nm).

9. Tighten the pulley bolt an additional 90°.

10. Install the drive belt.

11. Install the front splash shield.

12. Install the right front wheel.

6-Cylinder Engine

See Figures 110 through 112.

Special Tools Required:
- Holder Handle 07JAB-001020B
- Holder Attachment, 50 mm, Offset 07MAB-PY3010A
- Socket, 19 mm 07JAA-001020A

1. Raise the vehicle on the lift.

2. Remove the right front wheel.

3. Remove the front splash shield.

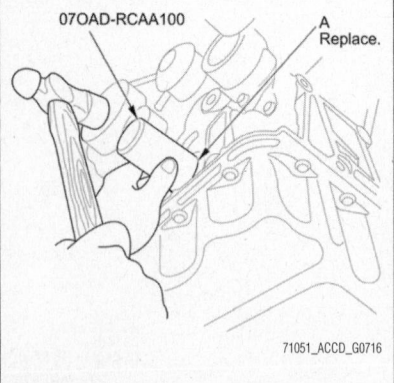

07OAD-RCAA100

A Replace.

71051_ACCD_G0716

Fig. 107 Using the oil seal driver, 64 mm, drive in the new crankshaft oil seal (A) until the oil seal driver bottoms on the oil pump

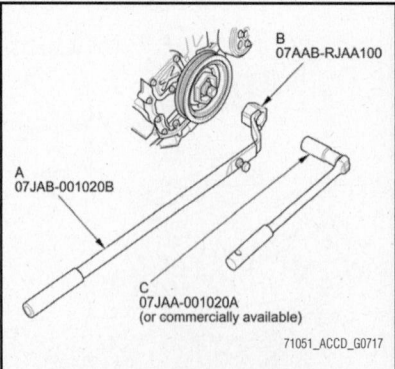

B 07AAB-RJAA100

A 07JAB-001020B

C 07JAA-001020A
(or commercially available)

71051_ACCD_G0717

Fig. 108 Hold the pulley with the holder handle (A) and the crankshaft pulley holder (B), remove the bolt with a heavy duty socket, 19 mm (C) and a breaker bar, then remove the crankshaft pulley

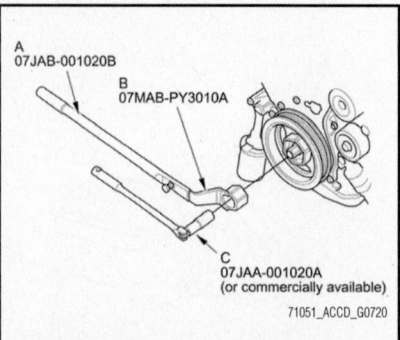

A 07JAB-001020B

B 07MAB-PY3010A

C 07JAA-001020A
(or commercially available)

71051_ACCD_G0720

Fig. 110 Hold the pulley with the holder handle (A) and the holder attachment, 50 mm, offset (B), remove the bolt with a heavy duty socket, 19 mm (C) and a breaker bar, then remove the crankshaft pulley

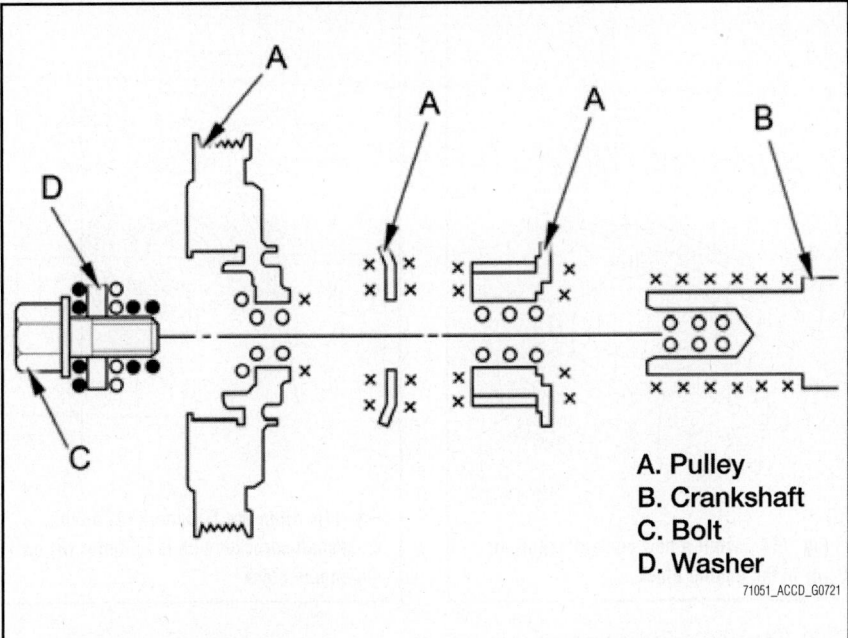

A. Pulley
B. Crankshaft
C. Bolt
D. Washer

71051_ACCD_G0721

Fig. 111 Remove any oil and clean the pulley, the crankshaft, the bolt, and the washer; lubricate with new engine oil as shown

4. Remove the drive belt:

5. Hold the pulley with the holder handle and the holder attachment, 50 mm, offset.

6. Remove the bolt with a heavy duty socket, 19 mm and a breaker bar, then remove the crankshaft pulley.

To install:

7. Remove any oil and clean the pulley, the crankshaft, the bolt, and the washer. Lubricate with new engine oil as shown.

8. Install the crankshaft pulley, then tighten the bolt. Do not use an impact wrench.

a. Hold the pulley with the holder handle and the holder attachment. Torque the bolt to 47 ft. lbs. (64 Nm) with a torque wrench and the heavy duty socket (19mm).

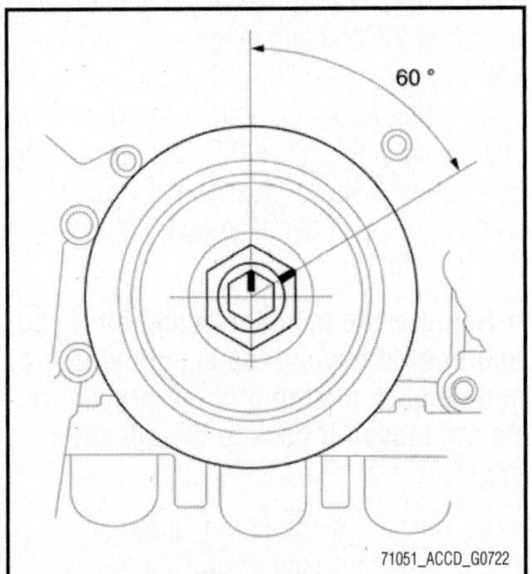

b. Tighten the bolt an additional 60°.

9. Install the drive belt:

10. Install the front splash shield.

11. Install the right front wheel.

CRANKSHAFT REAR COVER & SEAL

REMOVAL & INSTALLATION

4-Cylinder Engine

Special Tools Required:
• Driver handle, 15 x 135L 07749-0010000
• Oil seal driver attachment 96 mm 07ZAD-PNAA100

1. Remove the transmission:

2. M/T model: Remove the pressure plate, the clutch disk, and the flywheel.

3. A/T model: Remove the drive plate.

4. Clean and dry the crankshaft oil seal housing.

5. Use the driver handle, 15 x 135L and the oil seal driver attachment 96 mm to drive a new crankshaft oil seal squarely into the engine block to the specified installed height.

6. M/T model: Install the flywheel, the clutch disk, and the pressure plate.

7. A/T model: Install the drive plate.

8. Install the transmission:

6-Cylinder Engine

Special Tools Required:
• Driver handle, 15 x 135L 07749-0010000

• Oil seal driver attachment 106 mm 070AD-RCA0200

1. M/T model: Remove the transmission, and the flywheel.

2. A/T model: Remove the transmission, and the drive plate.

3. Remove the transmission end crankshaft oil seal.

4. Clean and dry the crankshaft oil seal housing.

5. Apply a light coat of new engine oil to the lip of the crankshaft oil seal.

6. Using the driver handle, 15 x 135 L and the oil seal driver attachment, 106 mm, drive in the new crankshaft oil seal until the oil seal driver attachment bottoms against the engine block end cover. Align the hole in the oil seal driver attachment with the pin on the crankshaft.

7. Clean any excess grease off the crankshaft, and check that the oil seal lip is not distorted.

8. M/T model: Install the flywheel, and the transmission.

9. A/T model: Install the drive plate, and the transmission.

CYLINDER HEAD

REMOVAL & INSTALLATION

4-Cylinder Engine

See Figures 113 through 119.

➡**Prior to starting the removal of the cylinder head, note the following:**

- Use fender covers to avoid damaging painted surfaces.
- To avoid damage, unplug the wiring connectors carefully while holding the connector portion.
- Connect the Honda Diagnostic System (HDS) to the Data Link Connector (DLC), and monitor the Engine Coolant Temperature (ECT)

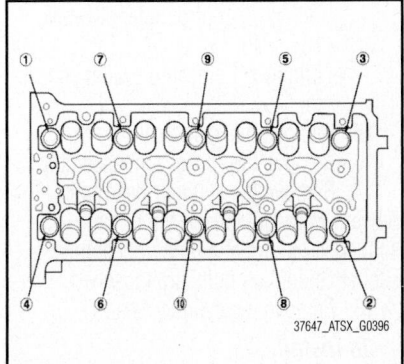

37647_ATSX_G0396

Fig. 113 Remove the cylinder head bolts in sequence

60°

71051_ACCD_G0722

Fig. 112 Tighten the bolt an additional 60°

sensor 1. To avoid damaging the cylinder head, wait until the ECT sensor 1 temperature drops below 100°F (38°C) before loosening the cylinder head bolts.

- Mark all wiring and hoses to avoid misconnection. Also, be sure that they do not contact other wiring or hoses, or interfere with other parts.

1. Remove the strut brace.
2. Relieve the fuel pressure.
3. Drain the engine coolant.
4. Remove the drive belt.
5. Remove the intake manifold.
6. Remove the catalytic converter.
7. Remove the Evaporative Emission (EVAP) canister hose.
8. Remove the quick-connect fitting cover, then disconnect the fuel feed hose.
9. Disconnect the four fuel injector connectors, the engine mount control solenoid connector, and remove the ground cables.
10. Remove the four bolts securing the EVAP canister purge valve bracket.
11. Disconnect the upper radiator hose, the heater hoses, and the water bypass hose.
12. Remove the two bolts securing the connecting pipe.
13. Disconnect the water bypass hose.
14. Disconnect the following engine wire harness connectors, and remove the wire harness clamps from the cylinder head:

- Engine Coolant Temperature (ECT) sensor 1 connector
- Camshaft Position (CMP) sensor A (Intake) connector
- Camshaft Position (CMP) sensor B (Exhaust) connector
- Rocker arm oil control solenoid connector
- Rocker arm oil pressure switch connector
- EVAP canister purge valve connector
- Variable Valve Timing Control (VTC) oil control solenoid valve connector
- Engine oil pressure switch connector

15. Remove the cam chain.
16. Remove the rocker arm assembly.
17. Remove the cylinder head bolts. To prevent warpage, loosen the bolts in sequence ⅓ turn at a time; repeat the sequence until all bolts are loosened.
18. Remove the cylinder head.

To install:

19. Install a new coolant separator in the engine block whenever the engine block is replaced.

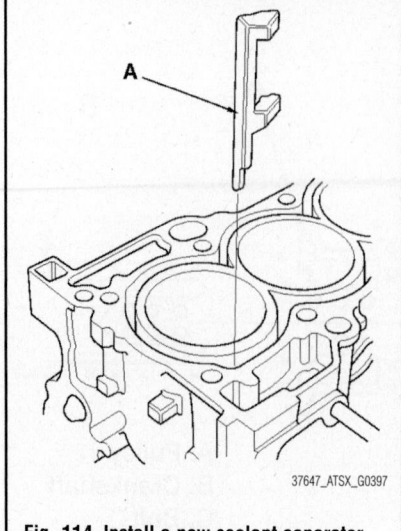

Fig. 114 Install a new coolant separator (A) in the engine block

20. Clean the cylinder head and the engine block surface.
21. Install the new cylinder head gasket and the dowel pins on the engine block. Always use a new cylinder head gasket.
22. Set the crankshaft to Top Dead Center (TDC). Align the TDC mark on the crankshaft sprocket with the pointer on the engine block.
23. Install the cylinder head on the engine block.
24. Measure the diameter of each cylinder head bolt at point A and point B.
25. If either diameter is less than 0.42 inches (10.6 mm), replace the cylinder head bolt.
26. Apply new engine oil to the threads and under the bolt heads of all cylinder head bolts.
27. Torque the cylinder head bolts in sequence to 29 ft. lbs. (39 Nm). Use a beam-type torque wrench. When using a

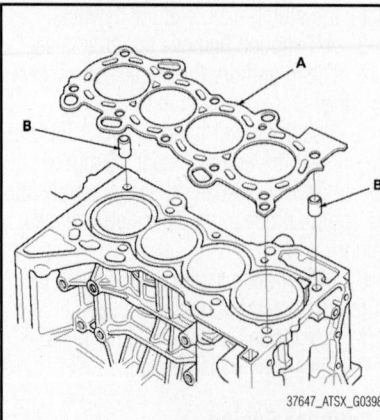

Fig. 115 Install the new cylinder head gasket (A) and the dowel pins (B) on the engine block

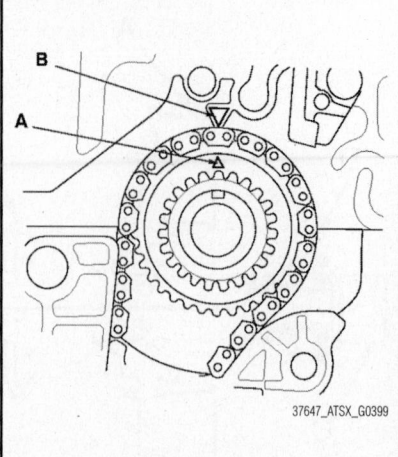

Fig. 116 Align the TDC mark (A) on the crankshaft sprocket with the pointer (B) on the engine block

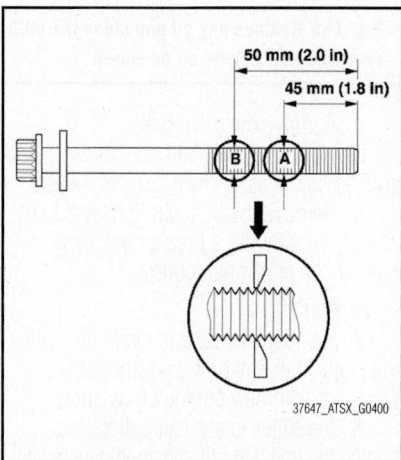

Fig. 117 Measure the diameter of each cylinder head bolt at point A and point B

preset click-type torque wrench, be sure to tighten slowly and do not overtighten. If a bolt makes any noise while you are torquing it, loosen the bolt and retighten it from the first step.

28. After torquing, tighten all cylinder head bolts in two steps (90° per step) using the sequence shown in step 9. If you are using a new cylinder head bolt, tighten the bolt an extra 90°.

➡ Remove the cylinder head bolt if you tightened it beyond the specified angle, and go back to step 6 of the procedure. Do not loosen it back to the specified angle.

29. Install the rocker arm assembly.
30. Install the cam chain.
31. Connect the following engine wire harness connectors, and install the wire harness clamps to the cylinder head:

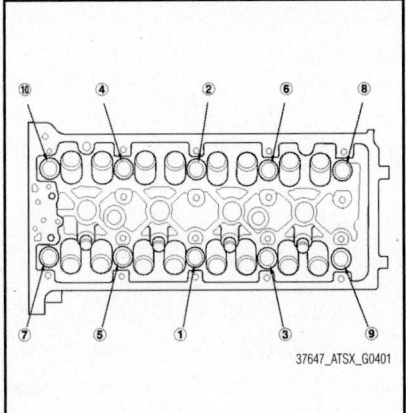

Fig. 118 Torque the cylinder head bolts in sequence

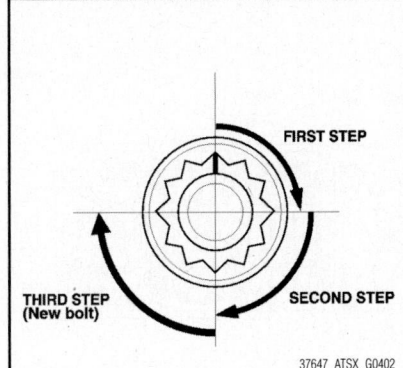

Fig. 119 After torquing, tighten all cylinder head bolts in two steps (90° per step); tighten new bolts an extra 90°

- Engine Coolant Temperature (ECT) sensor 1 connector
- Camshaft Position (CMP) sensor A (Intake) connector
- Camshaft Position (CMP) sensor B (Exhaust) connector
- Rocker arm oil control solenoid connector
- Rocker arm oil pressure switch connector
- Evaporative Emission (EVAP) canister purge valve connector
- Variable Valve Timing Control (VTC) oil control solenoid valve connector
- Engine oil pressure switch connector

32. Install the two bolts securing the connecting pipe.

33. Install the water bypass hose.

34. Connect the upper radiator hose, the heater hoses, and the water bypass hose.

35. Install the four bolts securing the EVAP canister purge valve bracket.

36. Connect the four fuel injector connectors, the engine mount control solenoid connector, and install the ground cables.

37. Connect the fuel feed hose, then install the quick-connect fitting cover.

38. Connect the EVAP canister hose.

39. Install the catalytic converter.

40. Install the intake manifold.

41. Install the drive belt.

42. Install the strut brace.

43. After installation, check that all tubes, hoses, and connectors are installed correctly.

44. Inspect for fuel leaks. Turn the ignition switch to ON (II) (do not operate the starter) so the fuel pump runs for about 2 seconds and pressurizes the fuel line. Repeat this operation three times, then check for fuel leakage at any point in the fuel line.

45. Refill the radiator with engine coolant, and bleed the air from the cooling system with the heater valve open.

46. Check for fluid leaks.

47. Do the Engine Control Module (ECM)/Powertrain Control Module (PCM) idle lean procedure.

48. Do the Crankshaft Position (CKP) pattern clear/CKP pattern lean procedure.

49. Inspect the idle speed.

50. Inspect the ignition timing.

6-Cylinder Engine

See Figures 120 through 128.

1. Relieve the fuel pressure.

2. Do the battery terminal disconnection procedure.

3. Drain the engine coolant.

4. Remove the alternator.

5. Remove the intake manifold.

6. Remove the six ignition coils.

7. Remove the timing belt.

8. Disconnect the following engine wire harness connectors, and remove the wire harness clamps from the cylinder head:

- Six injector connectors
- Knock sensor connector
- Engine Coolant Temperature (ECT) sensor 1 connector
- Engine mount control solenoid valve connector
- Camshaft Position (CMP) sensor connector
- Rocker arm oil control solenoid connector
- Rocker arm oil pressure switch connector
- Two Air Fuel Ratio (A/F) sensor connectors
- Two secondary Heated Oxygen Sensor (secondary HO2S) connectors

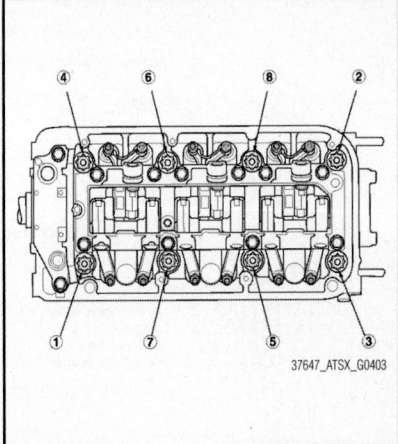

Fig. 120 Remove the front cylinder head bolts

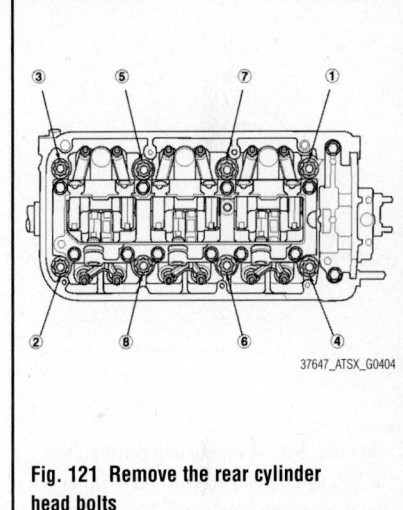

Fig. 121 Remove the rear cylinder head bolts

9. Remove the front Warm Up Three Way Catalytic Converter (front WU-TWC) and the rear Warm Up Three Way Catalytic Converter (rear WU-TWC).

10. Remove the quick-connect fitting cover, then disconnect the fuel feed hose.

11. Remove the connector bracket from the front cylinder head.

12. Remove the engine mount control solenoid valve bracket from the rear cylinder head.

13. Remove the Evaporative Emission (EVAP) canister joint with the bracket.

14. Remove the injector bases.

15. Remove the water passage.

16. Remove the camshaft pulleys and the back covers.

17. Remove the cylinder head covers.

18. Remove the cylinder head bolts. To prevent warpage, loosen the bolts in sequence ⅓ turn at a time; repeat the sequence until all bolts are loosened.

19. Remove the cylinder heads.

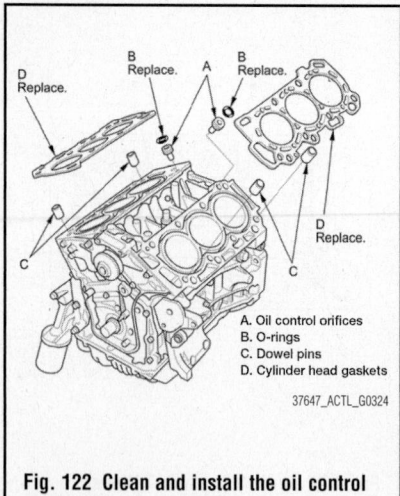

A. Oil control orifices
B. O-rings
C. Dowel pins
D. Cylinder head gaskets

37647_ACTL_G0324

Fig. 122 Clean and install the oil control orifices with new O-rings

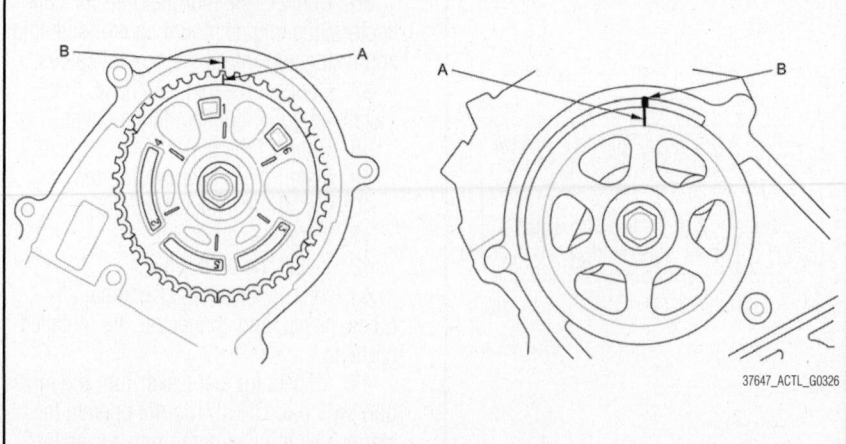

37647_ACTL_G0326

Fig. 124 Set the camshaft pulleys to TDC by aligning the TDC marks (A) on the camshaft pulleys with the pointers (B) on the back covers

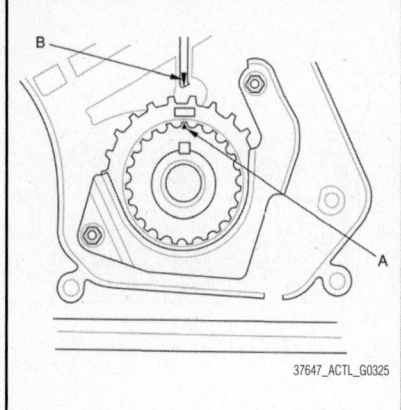

37647_ACTL_G0325

Fig. 123 Set the timing belt drive pulley to TDC by aligning the TDC mark (A) on the tooth of the timing belt drive pulley with the pointer (B) on the oil pump

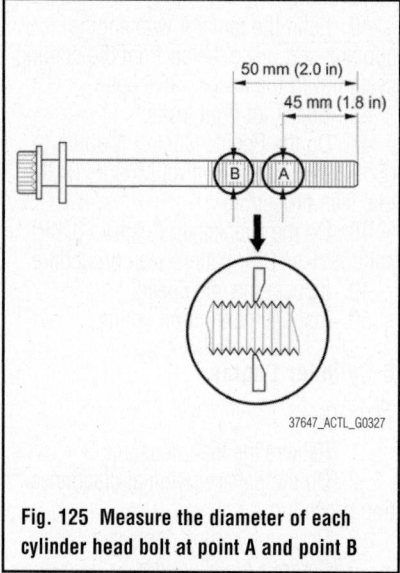
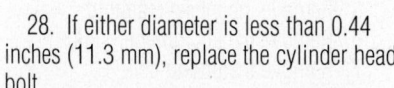

37647_ACTL_G0327

Fig. 125 Measure the diameter of each cylinder head bolt at point A and point B

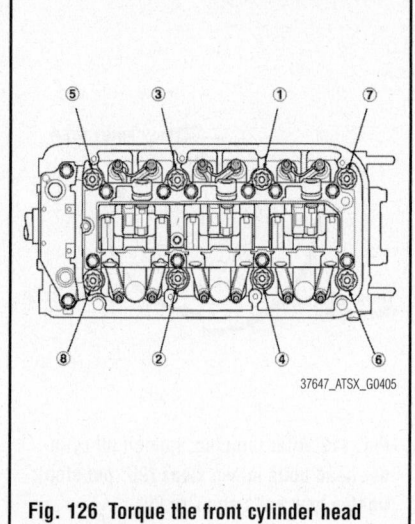

37647_ATSX_G0405

Fig. 126 Torque the front cylinder head bolts in sequence

To install:

20. Clean the cylinder head and the engine block surface.

21. Clean and install the oil control orifices with new O-rings.

22. Install the dowel pins and the new cylinder head gaskets.

23. Clean the timing belt pulleys, the timing belt guide plate, and the upper and lower covers.

24. Set the timing belt drive pulley to Top Dead Center (TDC) by aligning the TDC mark on the tooth of the timing belt drive pulley with the pointer on the oil pump.

25. Set the camshaft pulleys to TDC by aligning the TDC marks on the camshaft pulleys with the pointers on the back covers.

26. Install the cylinder heads on the engine block.

27. Measure the diameter of each cylinder head bolt at point A and point B.

28. If either diameter is less than 0.44 inches (11.3 mm), replace the cylinder head bolt.

29. Apply new engine oil to the threads and under the bolt heads of all cylinder head bolts.

30. Torque the cylinder head bolts in sequence to 22 ft. lbs. (29 Nm) using a beam-type torque wrench. When using a preset click-type torque wrench, be sure to tighten slowly and do not overtighten. If a bolt makes any noise while you are torquing it, loosen the bolt and retighten it from the first step.

31. After torquing, tighten all cylinder head bolts in two steps (90° per step) using the sequence shown in step 11. If you are using a new cylinder head bolt, tighten the bolt an extra 90°.

➡Remove the cylinder head bolt if you tightened it beyond the specified angle,

and go back to step 8 of the procedure. Do not loosen it back to the specified angle.

32. Install the timing belt.

33. Adjust the valve clearance.

34. Install the cylinder head covers.

35. Install the water passage.

36. Install the injector bases.

37. Install the connector bracket to the front cylinder head.

38. Install the Evaporative Emission (EVAP) canister joint with the bracket.

39. Install the engine mount control solenoid valve bracket to the rear cylinder head.

40. Connect the fuel feed hose, then install the quick-connect fitting cover.

41. Install the front Warm Up Three Way Catalytic Converter (front WU-TWC) and the rear Warm Up Three Way Catalytic Converter (rear WU-TWC).

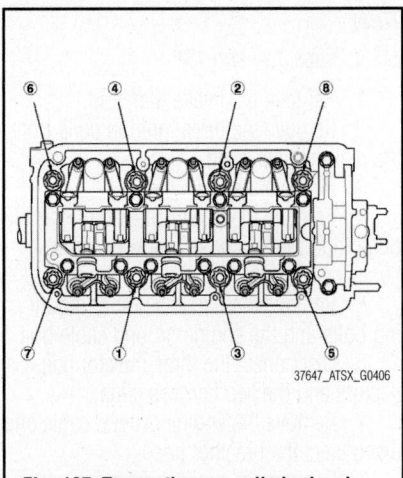

Fig. 127 Torque the rear cylinder head bolts in sequence

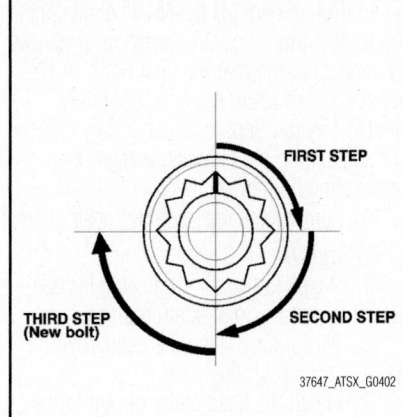

Fig. 128 After torquing, tighten all cylinder head bolts in two steps (90° per step); tighten new bolts an extra 90°

42. Connect the following engine wire harness connectors, and install the wire harness clamps to the cylinder head:
   - Six injector connectors
   - Knock sensor connector
   - Engine Coolant Temperature (ECT) sensor 1 connector
   - Engine mount control solenoid valve connector
   - Camshaft Position (CMP) sensor connector
   - Rocker arm oil control solenoid connector
   - Rocker arm oil pressure switch connector
   - Two Air Fuel Ratio (A/F) sensor connectors
   - Two secondary Heated Oxygen Sensor (secondary HO2S) connectors
43. Install the six ignition coils.
44. Install the intake manifold.

45. Install the alternator.
46. Do the battery terminal reconnection procedure.
47. After installation, check that all tubes, hoses, and connectors are installed correctly.
48. Inspect for fuel leaks. Turn the ignition switch to ON (II), (do not operate the starter) so the fuel pump runs for about 2 seconds and pressurizes the fuel line. Repeat this operation three times, then check for fuel leakage at any point in the fuel line.
49. Refill the radiator with engine coolant, and bleed the air from the cooling system.
50. Check for fluid leaks.
51. Do the Powertrain Control Module (PCM) idle learn procedure.
52. Do the Crankshaft Position (CKP) pattern clear/CKP pattern learn procedure.
53. Inspect the idle speed.
54. Inspect the ignition timing.

CYLINDER HEAD COVER

REMOVAL & INSTALLATION

4-Cylinder Engine

See Figures 129 and 130.

1. Remove the strut brace (if equipped).
2. Remove the dipstick and the Power Steering (P/S) hose bracket and disconnect the breather hose and the brake booster vacuum hose.
3. Remove the two bolts securing the Evaporative Emission (EVAP) canister purge valve bracket.
4. Remove the four ignition coils.
5. Remove the cylinder head cover.

To install:

6. Thoroughly clean the head cover gasket and the groove.
7. Install the head cover gasket in the groove of the cylinder head cover.
8. Check that the mating surfaces are clean and dry.
9. Apply liquid gasket, P/N 08717-0004, 08718-0003, or 08718-0009, on the

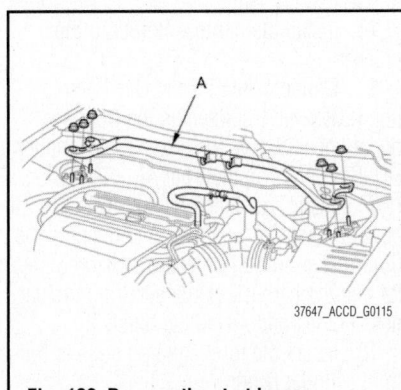

Fig. 129 Remove the strut brace

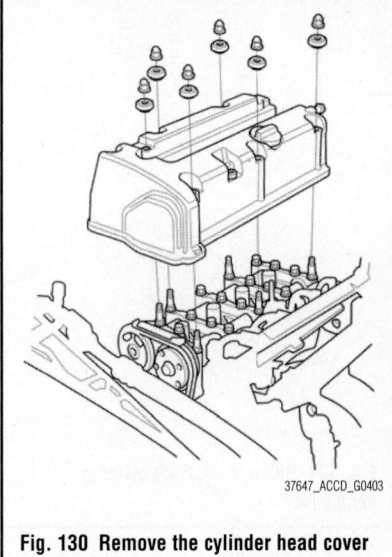

Fig. 130 Remove the cylinder head cover

chain case and the No. 5 rocker shaft holder mating areas. Install the component within 5 minutes of applying the liquid gasket.

➡ **If too much time has passed after applying the liquid gasket, remove the old liquid gasket and residue, then reapply new liquid gasket.**

10. Set the spark plug seals on the spark plug tubes. Place the cylinder head cover on the cylinder head, then slide the cover slightly back and forth to seat the head cover gasket.
11. Inspect the cover washers. Replace any washer that is damaged or deteriorated.
12. Torque the bolts in three steps. In the final step tighten all bolts, in sequence, to 9 ft. lbs. (12 Nm).

➡ **Wait at least 30 minutes before filling the engine with oil. Do not run the engine for at least 3 hours after installing the head cover.**

13. Install the two bolts securing the Evaporative Emission (EVAP) canister purge valve bracket.
14. Connect the breathe hose and the brake booster vacuum hose and install the Power Steering (P/S) hose bracket, and the dipstick.
15. Install the four ignition coils.
16. Install the engine cover.
17. Install the strut brace (if equipped).

6-Cylinder Engine

Front

See Figures 131 through 133.

1. Remove the intake manifold.
2. Remove the three ignition coils from the front cylinder head.

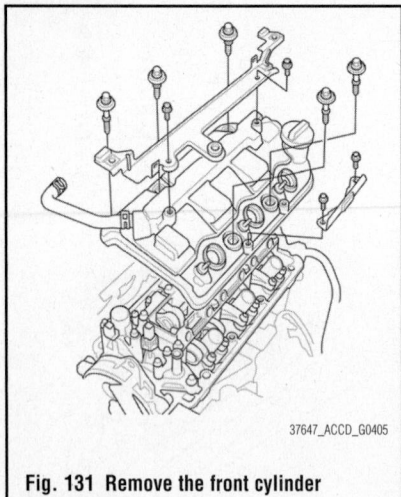

Fig. 131 Remove the front cylinder head cover

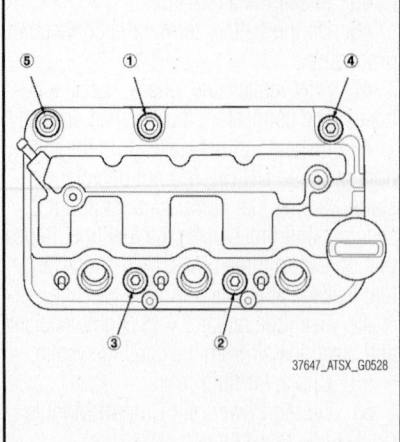

Fig. 133 Tighten the cylinder head cover bolts sequence

Rear

See Figures 134 and 135.

1. Remove the intake manifold.
2. Remove the three ignition coils from the rear cylinder head.
3. Remove the drive belt.
4. Remove the Power Steering (P/S) pump and the P/S hose bracket with its hoses connected.
5. Remove the harness holder mounting bolts and the engine ground cable bolt.
6. Disconnect the three injector connectors and the two harness clips.
7. Remove the engine ground cable and disconnect the breather hose.
8. Remove the harness from the upper cover.
9. Disconnect the rear rocker arm oil pressure switch connector, the rocker arm oil control solenoid B (BANK 1) connector, the rocker arm oil control solenoid A (BANK 1) connector, the rear Air Fuel Ratio (A/F) sensor 1 connector, the rear secondary Heated Oxygen Sensor 2 (secondary HO2S) connector, and the harness clamps, then remove the harness holder.
10. Remove the rear cylinder head cover.

To install:

11. Check the spark plug seals for damage. If any seal is damaged, replace it.
12. Thoroughly clean the head cover gasket and the groove.
13. Install the head cover gasket in the groove of the cylinder head cover. Make sure the head cover gasket is seated securely.
14. Remove all of the old liquid gasket from the rocker shaft holder and the cylinder head.
15. Clean the head cover contacting surfaces with a shop towel.

3. Disconnect the Exhaust Gas Recirculation (EGR) valve connector, the front secondary Heated Oxygen Sensor 2 (secondary HO2S) connector, the front Air Fuel Ratio (A/F) sensor 2 connector, the rocker arm oil control solenoid A (BANK 2) connector, the front rocker arm oil pressure switch connector and the harness clamp securing the harness holder, and remove the dipstick.
4. Remove the bolt securing the harness holder.
5. Remove the front cylinder head cover.

To install:

6. Check the spark plug seals for damage. If any seal is damaged, replace it.
7. Thoroughly clean the head cover gasket and the groove.
8. Install the head cover gasket in the groove of the cylinder head cover. Make sure the head cover gasket is seated securely.
9. Remove all of the old liquid gasket from the rocker shaft holder and the cylinder head.

10. Clean the head cover contacting surfaces with a shop towel.
11. Apply liquid gasket, P/N 08717-0004, 08718-0003, or 08718-0009, evenly to the rocker shaft holder mating areas. Install the component within 5 minutes of applying the liquid gasket.

➡**If you apply liquid gasket P/N 08718-0012, the component must be installed within 5 minutes. If too much time has passed after applying the liquid gasket, remove the old liquid gasket and residue, then reapply the new liquid gasket.**

12. Set the spark plug seals on the spark plug tubes, and install the front cylinder head cover.

➡**Wait at least 30 minutes before filling the engine with oil. Do not run the engine for at least 3 hours after installing the cylinder head cover.**

13. Inspect the spark plug seals for damage.
14. Inspect the cover washers. Replace any washer that is damaged or deteriorated.
15. Tighten the bolts in three steps. In the final step torque all bolts, in sequence to 9 ft. lbs. (12 Nm).
16. Install the harness holder to the bracket.
17. Connect the Exhaust Gas Recirculation (EGR) valve connector, the front secondary heated oxygen sensor 2 (secondary HO2S) connector, the front Air Fuel Ratio (A/F) sensor 2 connector, the rocker arm oil control solenoid A (BANK 2) connector, the front rocker arm oil pressure switch connector and the harness clamp securing the harness holder, and install the dipstick.
18. Install the three ignition coils to the front cylinder head.
19. Install the intake manifold.

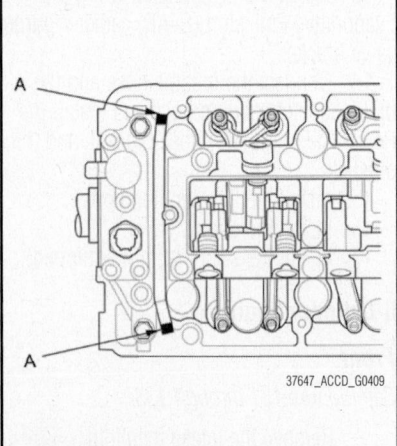

Fig. 132 Apply liquid gasket evenly to the rocker shaft holder mating areas (A)

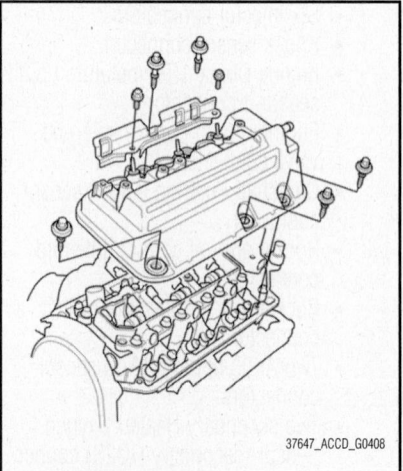

Fig. 134 Remove the rear cylinder head cover

16. Apply liquid gasket, P/N 08717-0004, 08718-0003, or 08718-0009, evenly to the rocker shaft holder mating areas. Install the component within 5 minutes of applying the liquid gasket.

➡**If you apply liquid gasket P/N 08718-0012, the component must be installed within 5 minutes. If too much time has passed after applying the liquid gasket, remove the old liquid gasket and residue, then reapply the new liquid gasket.**

17. Set the spark plug seals on the spark plug tubes, and install the rear cylinder head cover.

➡**Wait at least 30 minutes before filling the engine with oil. Do not run the engine for at least 3 hours after installing the cylinder head cover.**

18. Inspect the spark plug seals for damage.

19. Inspect the cover washers. Replace any washer that is damaged or deteriorated.

20. Tighten the bolts in three steps. In the final step torque all bolts, in sequence to 9 ft. lbs. (12 Nm).

21. Connect the rear rocker arm oil pressure switch connector, the rocker arm oil control solenoid B (BANK 1) connector, the rocker arm oil control solenoid A (BANK 1) connector, the rear Air Fuel Ratio (A/F) sensor 1 connector, the rear secondary Heated Oxygen Sensor 2 (secondary HO2S) connector, and the harness clamps, then install the harness holder.

22. Tighten the harness holder mounting bolts.

23. Install the engine ground cable and connect the breather hose, then install the harness to the upper cover.

24. Reconnect the three injector connectors and the two harness clips.

25. Install the Power Steering (P/S) pump and the P/S hose bracket.

26. Install the drive belt.

27. Install the three ignition coils to the rear cylinder head.

28. Install the intake manifold.

ENGINE OIL & FILTER

OIL LEVEL CHECK

4-Cylinder Engine

See Figure 136.

1. Park the vehicle on level ground, and start the engine. Hold the engine speed at 3,000 rpm with no load (M/T in neutral, A/T in P or N) until the radiator fan comes on, then turn off the engine, and wait a few minutes.

2. Remove the dipstick, and wipe off the dipstick, then reinstall the dipstick.

3. Remove the dipstick, and check the engine oil level. It should be between the upper mark and the lower mark.

4. If the engine oil level is near or below the lower mark, check for oil leakage, add engine oil to bring it to the upper mark.

6-Cylinder Engine

See Figure 137.

1. Park the vehicle on level ground, and start the engine. Hold the engine speed at 3,000 rpm with no load (M/T in neutral, A/T in P or N) until the radiator fan comes on, then turn off the engine, and wait a few minute.

2. Remove the dipstick, and wipe off the dipstick, then reinstall the dipstick.

3. Remove the dipstick, and check the

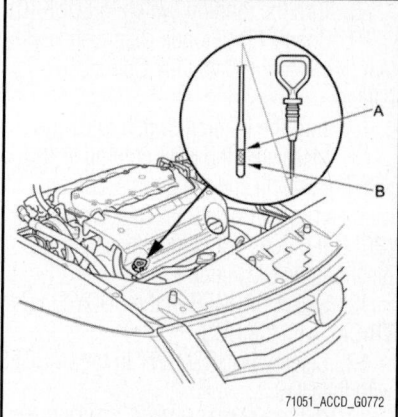

Fig. 137 Remove the dipstick, and check the engine oil level. It should be between the upper mark (A) and the lower mark (B)

engine oil level. It should be between the upper mark and the lower mark.

4. If the engine oil level is near or below the lower mark, check for oil leakage, and add engine oil to bring it to the upper mark.

OIL & FILTER CHANGE

4-Cylinder Engine

OIL

See Figure 138.

1. Warm up the engine.

2. Remove the engine undercover.

3. Remove the drain bolt, and drain the engine oil.

4. Reinstall the drain bolt with a new washer.

5. Refill the engine with the recommended oil.

6. Run the engine for more than 3 minutes, then check for oil leakage.

7. If the maintenance minder required engine oil replacement, reset the maintenance minder, and this procedure is complete. If the maintenance minder did not require engine oil replacement, go to step 8.

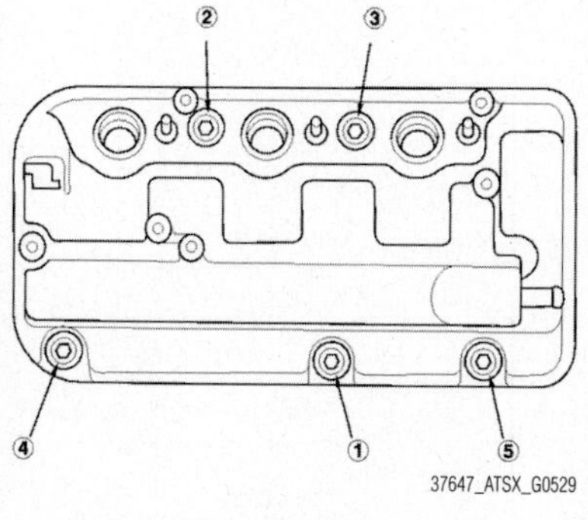

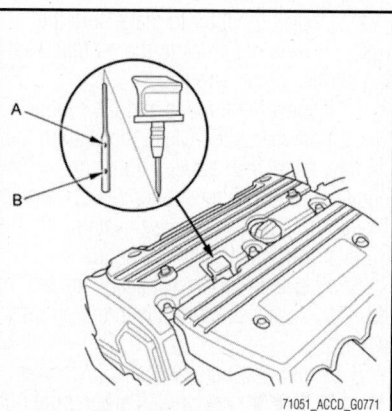

Fig. 135 Tighten the cylinder head cover bolts in sequence

Fig. 136 Remove the dipstick, and check the engine oil level. It should be between the upper mark (A) and the lower mark (B)

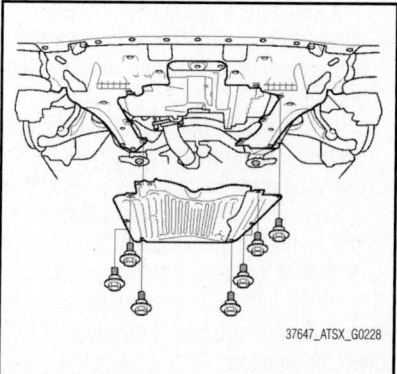

Fig. 138 Remove the engine undercover

8. Turn the ignition switch to LOCK (0).

9. Connect the Honda Diagnostic System (HDS) to the Data Link Connector (DLC).

10. Turn the ignition switch to ON (II).

11. Make sure the HDS communicates with the vehicle and the Engine Control Module (ECM)/Powertrain Control Module (PCM) If it does not communicate, troubleshoot the DLC circuit..

12. Select GAUGES in the BODY ELECTRICAL with the HDS.

13. Select ADJUSTMENT in the GAUGES with the HDS.

14. Select MAINTENANCE MINDER in the ADJUSTMENT with the HDS.

15. Select RESET in the MAINTENANCE MINDER with the HDS.

16. Select RESETTING THE ENGINE OIL LIFE with the HDS.

➡️**If you changed the Automatic Transmission Fluid (ATF) at the same time with the engine oil, select RESETTING THE ENGINE OIL LIFE AND ATF with the HDS instead.**

Filter

1. Remove the engine undercover.

2. Drain the engine oil.

3. Remove the oil filter with the oil filter wrench.

4. Make sure the rubber seal from the oil filter is not stuck to the oil filter seating surface of the engine.

5. Inspect the threads and the rubber seal on the new filter. Clean the seat on the oil filter base, then apply a light coat of new engine oil to the filter rubber seal. Use only filters with a built-in bypass system.

6. Install the oil filter by hand.

7. After the rubber seal seats, tighten the oil filter clockwise ¾ turn with the oil filter wrench.

8. If four numbers or marks (1 to 4 or ▼ to ▼▼▼▼) are printed around the outside of the filter, you can use the following procedure to tighten the filter.

   a. Spin the filter on until its seal lightly seats against the oil filter base, and note which number or mark is at the bottom.

   b. Tighten the filter by turning it clockwise three numbers or marks from the one you noted. For example, if mark ▼ is at the bottom when the seal is lightly seated, tighten the filter until the mark ▼▼▼▼ comes around to the bottom.

9. Refill with the recommended oil. Run the engine for more than 3 minutes, then check for oil leaks.

6-Cylinder Engine

Oil

1. Warm up the engine.

2. Remove the drain bolt, and drain the engine oil.

3. Reinstall the drain bolt with a new washer.

4. Refill with the recommended oil.

5. Run the engine for more than 3 minutes, then check for oil leakage.

6. If the maintenance minder required engine oil replacement, reset the maintenance minder, and this procedure is complete. If the maintenance minder did not require engine oil replacement, go to step 7.

7. Turn the ignition switch to LOCK (0).

8. Connect the Honda Diagnostic System (HDS) to the Data Link Connector (DLC).

9. Turn the ignition switch to ON (II).

10. Make sure the HDS communicates with the vehicle and the Powertrain Control Module (PCM). If it does not communicate, troubleshoot the DLC circuit.

11. Select GAUGES in the BODY ELECTRICAL with the HDS.

12. Select ADJUSTMENT in the GAUGES with the HDS.

13. Select MAINTENANCE MINDER in the ADJUSTMENT with the HDS.

14. Select RESET in the MAINTENANCE MINDER with the HDS.

15. Select RESETTING THE ENGINE OIL LIFE with the HDS.

➡️**If you changed the ATF at the same time with the engine oil, select RESETTING THE ENGINE OIL LIFE AND ATF with the HDS instead.**

Filter

1. Remove the oil filter with the oil filter wrench.

2. Inspect the filter to make sure the rubber seal is not stuck to the oil filter seating surface of the engine.

3. Inspect the threads and the rubber seal on the new filter. Clean the seat on the oil filter base, then apply a light coat of new engine oil to the filter rubber seal. Use only filters with a built-in bypass system.

4. Install the oil filter by hand.

5. After the rubber seal seats, tighten the oil filter clockwise ¾ turn with the oil filter wrench.

6. If four numbers or marks (1 to 4 or ▼ to ▼▼▼▼) are printed around the outside of the filter, you can use the following procedure to tighten the filter.

   a. Spin the filter on until its seal lightly seats against the oil filter base,

and note which number or mark is at the bottom.

   b. Tighten the filter by turning it clockwise three numbers or marks from the one you noted. For example, if mark ▼ is at the bottom when the seal is lightly seated, tighten the filter until the mark ▼▼▼▼ comes around to the bottom.

7. Refill with the recommended oil. Run the engine for more than 3 minutes, then check for oil leaks.

EXHAUST MANIFOLD

COMPONENT LOCATIONS

The exhaust manifold is cast as part of the cylinder head assembly. This is no longer a separate component.

FLEXPLATE (DRIVEPLATE)

REMOVAL & INSTALLATION

See Figure 139.

1. Remove the transaxle assembly.

2. Remove the drive plate and the washer from the engine crankshaft.

3. Install the drive plate and the washer on the engine crankshaft, and tighten the eight bolts in a crisscross pattern in at least two steps.

4. Install the transmission assembly.

FRONT COVER & SEAL

REMOVAL & INSTALLATION

4-Cylinder Engine

See Figures 140 and 141.

Special Tools Required:
• Driver Handle, 15 x 135L 07749-0010000

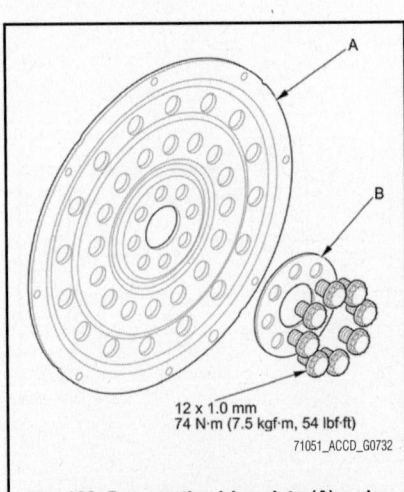

12 x 1.0 mm
74 N·m (7.5 kgf·m, 54 lbf·ft)

71051_ACCD_G0732

Fig. 139 Remove the drive plate (A) and the washer (B) from the engine crankshaft

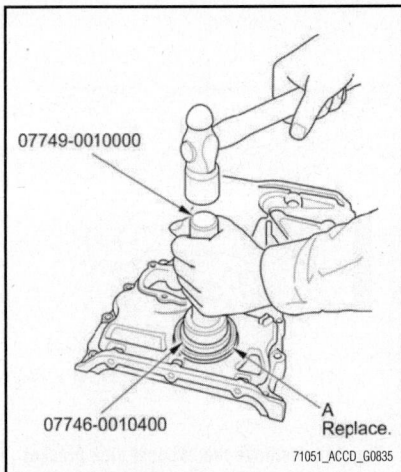

Fig. 140 Use the driver handle, and the bearing driver attachment, drive a new oil seal (A) squarely into the chain case to the specified installed height

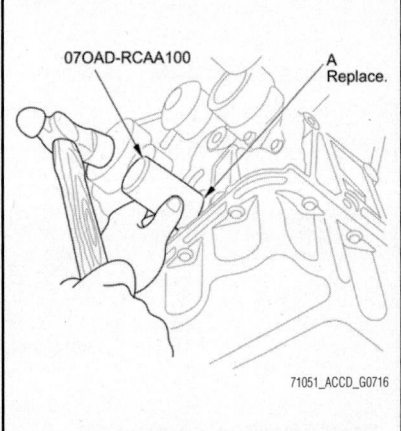

Fig. 142 Using the oil seal driver, 64 mm, drive in the new crankshaft oil seal (A) until the oil seal driver bottoms on the oil pump

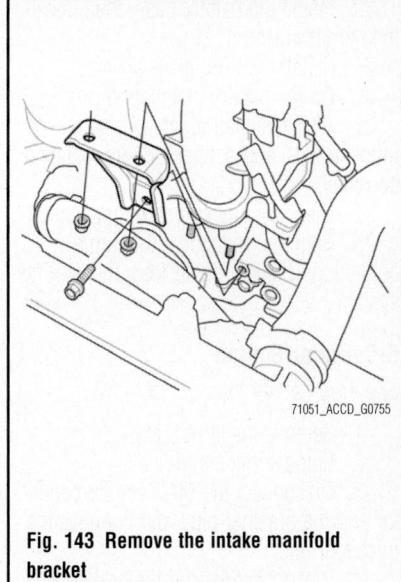

Fig. 143 Remove the intake manifold bracket

- Bearing Driver Attachment, 52 x 55 mm 07746-0010400
1. Clean and dry the crankshaft oil seal housing.
2. Apply a light coat of new engine oil to the lip of the chain case oil seal.
3. Use the driver handle, 15 x 135L and the bearing driver attachment, 52 x 55 mm to drive a new oil seal squarely into the chain case to the specified installed height.
4. Measure the distance between chain case surface and the oil seal.

➡ **Oil Seal Installed Height: 1.276–1.303 inches (32.4–33.1 mm)**

6-Cylinder Engine

See Figure 142.

Special Tools Required: Oil Seal Driver, 64 mm 07OAD-RCAA100
1. Remove the timing belt drive pulley.

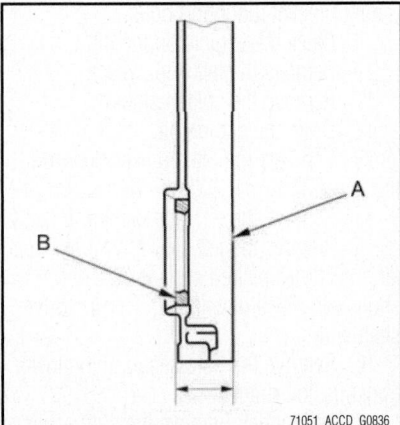

Fig. 141 Measure the distance between the chain case surface (A) and the oil seal (B)

2. Remove the pulley end crankshaft oil seal.
3. Clean and dry the crankshaft oil seal housing.
4. Apply a light coat of new engine oil to the lip of the crankshaft oil seal.
5. Using the oil seal driver, 64 mm, drive in the new crankshaft oil seal until the oil seal driver bottoms on the oil pump.
6. Clean the excess oil off the crankshaft, and check that the oil seal lip is not distorted.
7. Install the timing belt drive pulley.

INTAKE MANIFOLD

REMOVAL & INSTALLATION

4-Cylinder Engine

See Figures 143 and 144.

1. Do the battery removal procedure.
2. Remove the front grille cover:
3. Remove the harness clamps, then remove the battery base.
4. Remove the water separator and the intake air duct.
5. Disconnect the breather pipe, then remove the intake air duct.
6. Disconnect the EVAP canister hose and the brake booster vacuum hose.
7. Disconnect the water bypass hoses, then plug the water bypass hoses.
8. Disconnect the MAP sensor connector and the throttle body connector, then remove the wire harness clamps.
9. Remove the intake manifold bracket.
10. Disconnect the vacuum hose from the intake manifold.
11. Remove the intake manifold, then disconnect the PCV hose from the intake manifold.

To install:
12. Connect the PCV hose to the intake manifold, then install the intake manifold with new gaskets, and tighten the bolts and nuts in a crisscross pattern in three steps, beginning with the inner bolt to 16 ft. lbs. (22 Nm).
13. Connect the vacuum hose to the intake manifold.
14. Install the intake manifold bracket. Tighten the bolt and nuts to 16 ft. lbs. (22 Nm).
15. Connect the MAP sensor connector and the throttle body connector, then install the wire harness clamps.
16. Connect the water bypass hoses.
17. Connect the EVAP canister hose and the brake booster vacuum hose.
18. Install the intake air duct, then connect the breather pipe.
19. Install the water separator and the intake air duct.

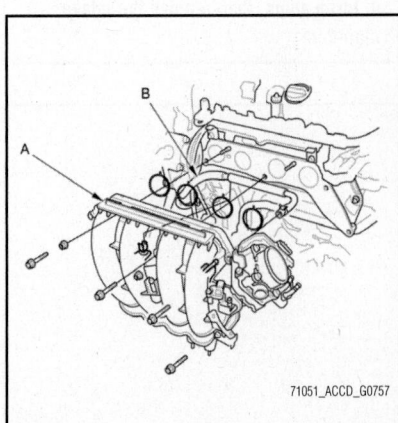

Fig. 144 Remove the intake manifold (A), then disconnect the PCV hose (B) from the intake manifold

20. Install the battery base, then install the harness clamps.

21. Install the front grille cover:

22. Do the battery installation procedure.

23. After installation, check that all tubes, hoses, and connectors are installed correctly.

24. Clean up any spilled engine coolant.

25. Refill the radiator with engine coolant, and bleed the air from the cooling system.

6-Cylinder Engine

See Figures 145 through 147.

1. Remove the strut brace.

2. Remove the engine cover.

3. Disconnect the MAP sensor connector and the breather pipe, then remove the intake air duct.

4. Disconnect the throttle body connector, the EVAP canister purge valve connector, the EVAP hose, and the brake booster vacuum hose.

5. Disconnect and plug the water bypass hoses.

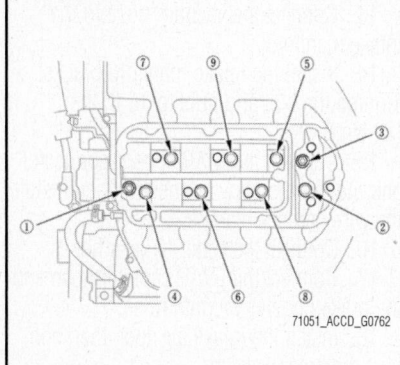

Fig. 145 Remove the intake manifold mounting bolts and nuts sequentially in three steps, then remove the intake manifold

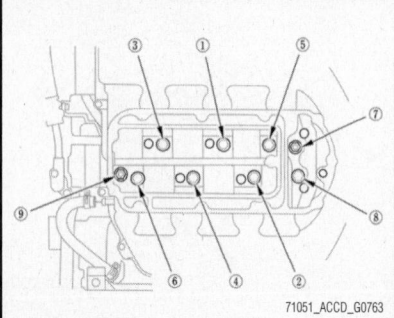

Fig. 146 Tighten the bolts and nuts sequentially in three steps

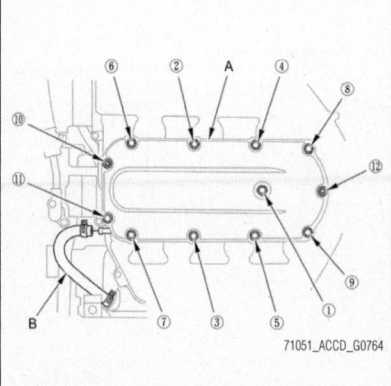

Fig. 147 Install the upper cover (A), tighten the bolts and nuts sequentially in three steps, then connect the PCV hose (B)

6. Disconnect the PCV hose, then remove the upper cover mounting bolts and nuts sequentially in three steps, then remove the upper cover.

7. Remove the intake manifold mounting bolts and nuts sequentially in three steps, then remove the intake manifold.

To install:

8. Install the intake manifold. Tighten the bolts and nuts sequentially in three steps. Always use a new intake manifold gasket. Specified Torque: 8 x 1.25 mm: 16 ft. lbs. (22 Nm)

9. Install the upper cover. Tighten the bolts and nuts sequentially in three steps. Always use a new gasket, then connect the PCV hose. Specified Torque: 6 x 1.0 mm: 9 ft. lbs. (12 Nm)

10. Connect the throttle body connector, the EVAP canister purge valve connector, the water bypass hoses, the EVAP hose, and the brake booster vacuum hose.

11. Install the intake air duct, then connect the breather pipe and the MAP sensor connector.

12. Clean up any spilled engine coolant.

13. After installation, check that all tubes, hoses, and connectors are installed correctly.

14. Install the engine cover.

15. Install the strut brace.

16. Refill the radiator with engine coolant, and bleed the air from the cooling system.

OIL FILTER ADAPTER

REMOVAL & INSTALLATION

4-Cylinder Engine

See Figure 148.

1. Remove the oil filter.

2. Remove the exhaust pipe bracket, then remove the oil filter base and O-rings.

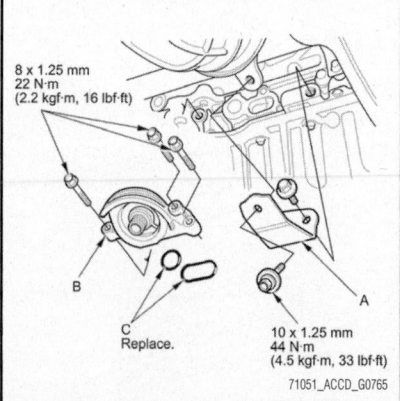

Fig. 148 Remove the exhaust pipe bracket (A), then remove the oil filter base (B) and O-rings (C)

3. Clean the O-ring grooves and the mating surface with the oil filter base.

4. Install the oil filter base with new O-rings.

5. Install the oil filter.

OIL PAN

REMOVAL & INSTALLATION

4-Cylinder Engine

See Figures 149 through 151.

1. If the engine is already out of the vehicle, go to step 19.

2. Remove the strut brace (if equipped).

3. Do the battery removal procedure.

4. Remove the air cleaner assembly.

5. Remove the harness clamps, then remove the battery base.

6. Remove the front engine mount stop, then remove the front engine mount bolt.

7. Loosen the rear engine mount mounting bolts.

8. Loosen the upper transmission mount bracket mounting bolts.

9. Raise the vehicle on the lift.

10. Remove the left front wheel.

11. Remove the splash shield.

12. Drain the engine oil.

13. Separate the left side knuckle from the lower arm.

14. Remove the left side damper fork.

15. Remove the left side driveshaft. Coat all precision-finished surface with new engine oil. Tie a plastic bag over the driveshaft end.

16. Remove the nuts securing the lower transmission mount.

17. A/T model: Remove the shift cable bracket.

18. Use a transmission jack to lift the transmission 1.2–1.6 inches (30–40 mm).

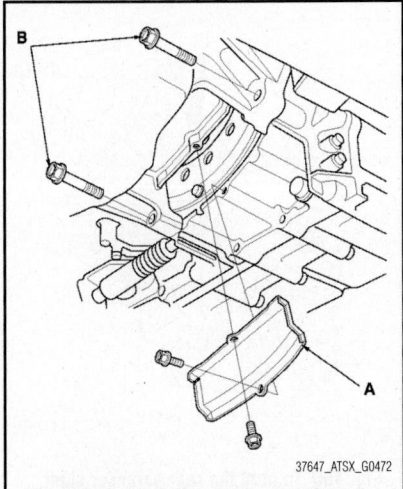

Fig. 149 Remove the clutch cover/torque converter cover (A) and the transmission mounting bolts (B)

19. Remove the clutch cover/torque converter cover and the transmission mounting bolts.

20. Remove the bolts securing the oil pan.

21. Using a flat blade screwdriver, separate the oil pan from the engine block in the places shown.

22. Remove the oil pan.

To install:

23. Remove all of the old liquid gasket from the oil pan mating surfaces, the bolts, and the bolt holes.

24. Clean and dry the oil pan mating surfaces.

25. Apply liquid gasket (P/N 08717-0004, 08718-0003, or 08718-0009) to the engine block mating surface of the oil pan and to the inside edge of the threaded bolt holes. Install the component within 5 minutes of applying the liquid gasket.

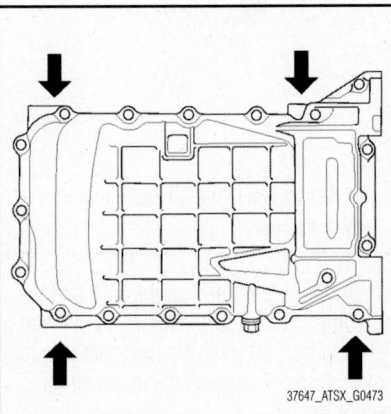

Fig. 150 Separate the oil pan from the engine block in the places shown

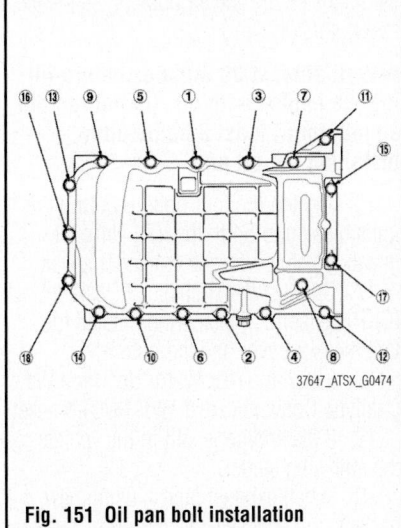

Fig. 151 Oil pan bolt installation sequence

➡**If too much time has passed after applying the liquid gasket, remove the old liquid gasket and residue, then reapply new liquid gasket.**

26. Install the oil pan.

27. Tighten the bolts in three steps. In the final step, torque all bolts, in sequence, to 9 ft. lbs. (12 Nm). Wipe off the excess liquid gasket on the each side of crankshaft pulley and the flywheel/drive plate.

➡**Wait at least 30 minutes before filling the engine with oil. Do not run the engine for at least 3 hours after installing the oil pan.**

28. Install the clutch cover/torque converter cover and the transmission mounting bolts. Tighten the transmission mounting bolts to 47 ft. lbs. (64 Nm).

29. If the engine is still in the vehicle, do steps 8 through 24.

30. Lower the transmission jack from the transmission.

31. A/T model: Install the shift cable bracket.

32. Tighten the nuts securing the lower transmission mount.

33. Install a new set ring on the end of driveshaft, then install the driveshaft. Make sure the ring "clicks" into place in the differential.

34. Connect the lower arm to the left side knuckle.

35. Install the left side damper fork.

36. Install the splash shield.

37. Install the left front wheel.

38. Lower the vehicle on the lift.

39. Tighten the upper transmission mount bracket mounting bolts to 43 ft. lbs. (59 Nm).

40. Tighten the rear engine mount mounting bolts. Tighten the 12 x 1.25 mm bolts to 58 ft. lbs. (78 Nm); tighten the 10 x 1.25 mm bolts to 40 ft. lbs. (54 Nm).

41. Tighten the front engine mount bolt to 40 ft. lbs. (54 Nm), then install the front engine mount stop. Tighten the nuts to 58 ft. lbs. (78 Nm).

42. Install the battery base, then install the harness clamps.

43. Install the air cleaner assembly.

44. Do the battery installation procedure.

45. Install the strut brace (if equipped).

46. Refill the engine with the recommended engine oil.

6-Cylinder Engine

See Figures 152 through 154.

➡**If the engine is already out of the vehicle, go to step 6.**

1. Raise the vehicle on the lift.
2. Drain the engine oil.

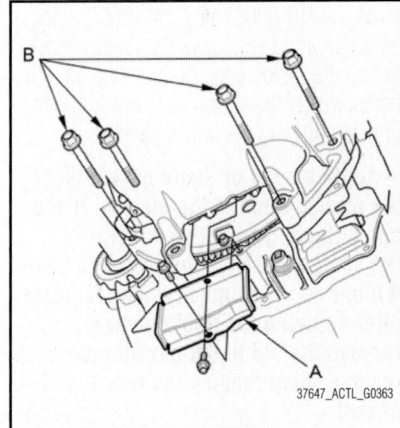

Fig. 152 Remove the clutch/torque converter cover (A) and the four bolts (B) securing the transmission

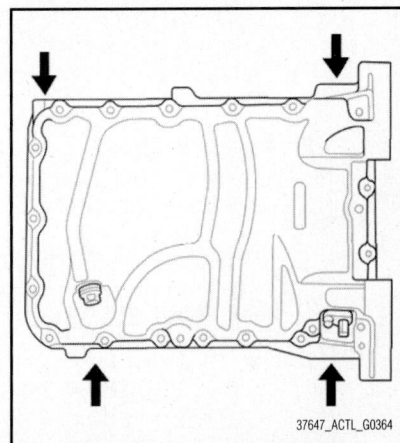

Fig. 153 Using a flat blade screwdriver, separate the oil pan from the engine block in the places shown

3. Remove the splash shield.

4. Remove exhaust pipe A.

5. Remove the rear Warm Up Three Way Catalytic Converter (rear WU-TWC) bracket.

6. Remove the Crankshaft Position (CKP) sensor cover and the bolt, then disconnect the CKP sensor connector.

7. Remove the clutch/torque converter cover and the four bolts securing the transmission.

8. Remove the bolts securing the oil pan.

9. Using a flat blade screwdriver, separate the oil pan from the engine block in the places shown.

10. Remove the oil pan.

To install:

11. Remove all of the old liquid gasket from the oil pan mating surfaces, the bolts, and the bolt holes.

12. Clean and dry the oil pan mating surfaces.

13. Apply liquid gasket, P/N 08717-0004, 08718-0003, or 08718-0009 to the oil pan mating surface of the engine block and to the inside edge of the threaded bolt holes. Install the component within 5 minutes of applying the liquid gasket.

➡**Apply a bead of liquid gasket along the mating edge of the oil pan. If you apply liquid gasket P/N 08718-0012, the component must be installed within 4 minutes. If too much time has passed after applying the liquid gasket, remove the old liquid gasket and residue, then reapply the new liquid gasket.**

14. Install the oil pan on the engine block.

15. Tighten the bolts in three steps. In the final step, torque all bolts, in sequence, to 9 ft. lbs. (12 Nm).

➡**Wait at least 30 minutes before filling the engine with oil. Do not run the engine for at least 3 hours after installing the oil pan.**

16. Tighten the four bolts securing the transmission to 54 ft. lbs. (74 Nm), then install the clutch/torque converter cover.

17. Connect the Crankshaft Position (CKP) sensor connector, then install the CKP sensor cover and the bolt.

18. Install the rear Warm Up Three Way Catalytic Converter (rear WU-TWC) bracket.

19. If the engine is still in the vehicle, do the following steps.

20. Install exhaust pipe A using new gaskets and new self-locking nuts.

21. Install the splash shield.

22. Refill the engine with the recommended engine oil.

OIL PUMP

REMOVAL & INSTALLATION

4-Cylinder Engine

See Figures 155 through 162.

1. Turn the crankshaft pulley so its Top Dead Center (TDC) mark lines up with the pointer.

➡**The other pointer is not used.**

2. Remove the oil pan.

3. To hold the rear balancer shaft, insert a 6 mm long pin punch (Snap-on® PPC108LA or equivalent) into the maintenance hole in the balancer shaft holder and through the rear balancer shaft.

4. Turn the crankshaft counterclockwise to compress the oil pump chain auto-tensioner.

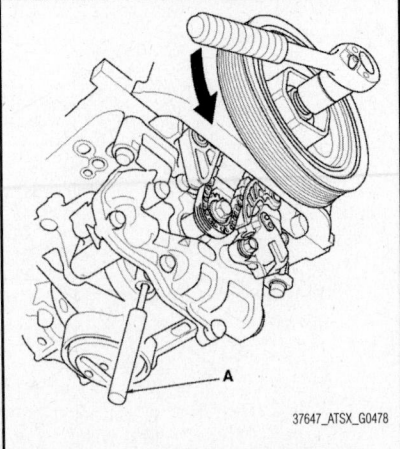

Fig. 156 To hold the rear balancer shaft, insert a 6 mm long pin punch (A)

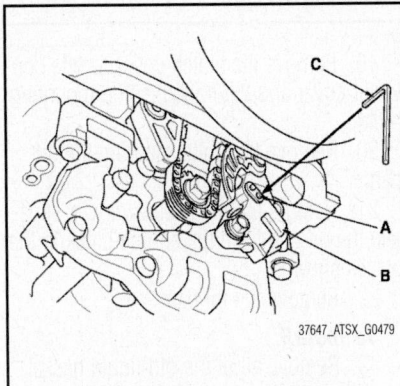

Fig. 157 Align the holes on the lock (A) and the oil pump chain auto-tensioner (B), then insert a 0.12 inches (3.0 mm) diameter pin (C) into the holes

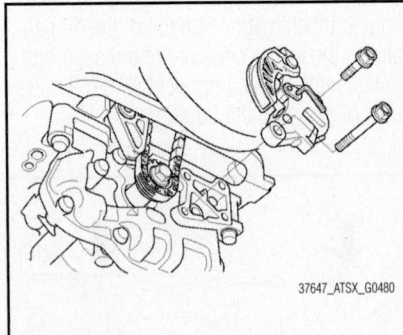

Fig. 158 Remove the oil pump chain auto-tensioner

5. Align the holes on the lock and the oil pump chain auto-tensioner, then insert a 0.12 inches (3.0 mm) diameter pin into the holes. Turn the crankshaft clockwise to secure the pin.

6. Remove the oil pump chain auto-tensioner.

7. Loosen the oil pump sprocket mounting bolt.

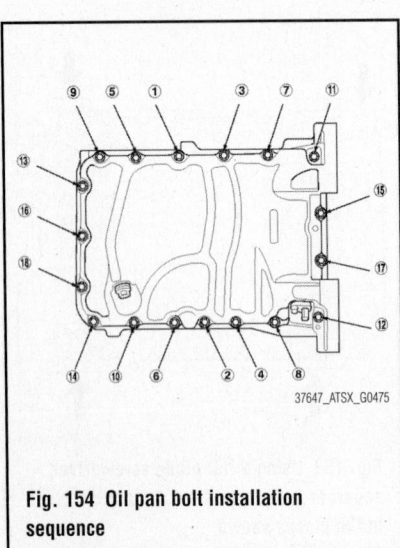

Fig. 154 Oil pan bolt installation sequence

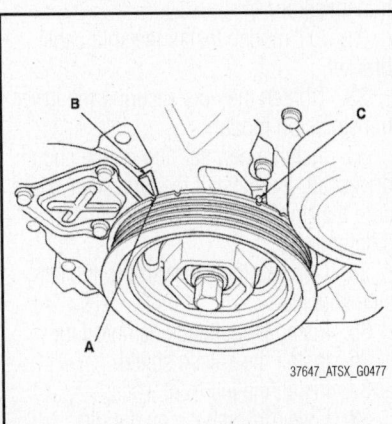

Fig. 155 Turn the crankshaft pulley so its Top Dead Center (TDC) mark (A) lines up with the pointer (B)

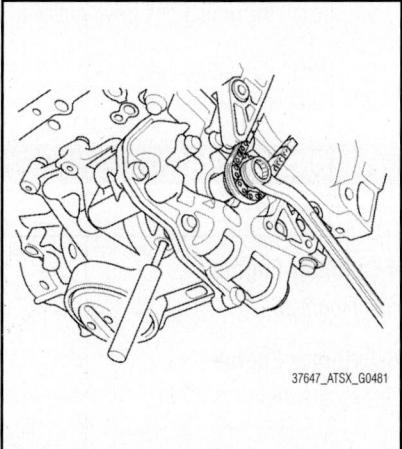

Fig. 159 Loosen the oil pump sprocket mounting bolt.

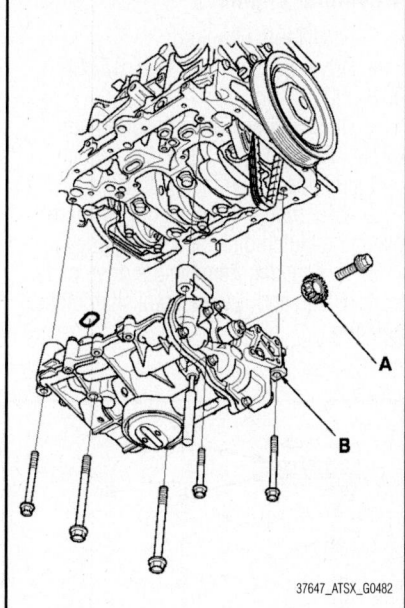

Fig. 160 Remove the oil pump sprocket (A), then remove the oil pump (B).

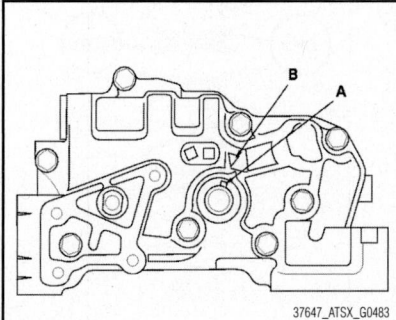

Fig. 161 Align the dowel pin (A) on the rear balancer shaft with the mark (B) on the oil pump.

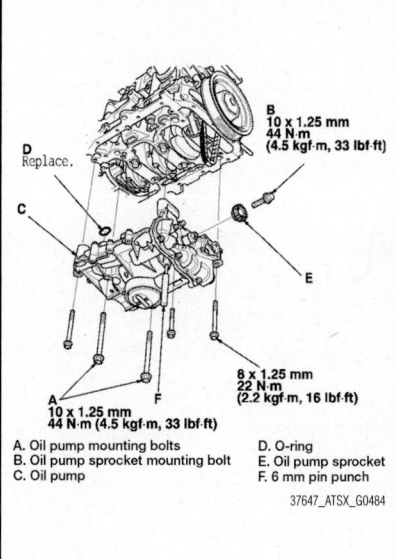

A. Oil pump mounting bolts
B. Oil pump sprocket mounting bolt
C. Oil pump
D. O-ring
E. Oil pump sprocket
F. 6 mm pin punch

Fig. 162 Install the oil pump with a new O-ring, then install the oil pump sprocket

8. Remove the oil pump sprocket, then remove the oil pump.

To install:

9. Make sure the No. 1 piston Top Dead Center (TDC) mark lines up with the pointer.

10. Align the dowel pin on the rear balancer shaft with the mark on the oil pump.

11. To hold the rear balancer shaft, insert a 6 mm long pin punch (Snap-on PPC108LA or equivalent) into the maintenance hole in the balancer shaft holder and through the rear balancer shaft.

12. Apply new engine oil to the threads of the oil pump mounting bolts and oil pump sprocket mounting bolt, then loosely install the oil pump with a new O-ring, then install the oil pump sprocket.

13. Tighten the oil pump mounting bolts and the oil pump sprocket mounting bolt.

14. Remove the 6 mm pin punch.

15. Install the oil pump chain auto-tensioner.

➡If the 1.2 inches (3.0 mm) diameter pin come off the auto-tensioner, compress the auto-tensioner.

16. Remove the pin from the oil pump chain auto-tensioner.

17. Install the oil pan.

6-Cylinder Engine

See Figures 163 through 166.

1. Drain the engine oil.
2. Remove the timing belt:
3. Remove the timing belt drive pulley from the crankshaft:

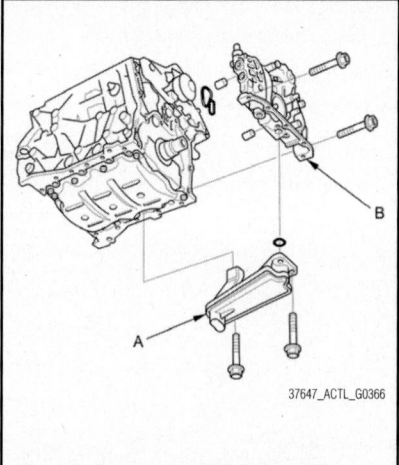

Fig. 163 Remove the oil strainer (A), then remove the oil pump assembly (B)

4. Attach the chain hoist to the engine hook on the engine hanger bracket.

5. Remove the rocker arm oil control solenoid/oil filter assembly.

6. Remove the oil pan.

7. Remove the oil strainer.

8. Remove the mounting bolts, then remove the oil pump assembly.

To install:

9. Remove the old oil seal from the oil pump.

10. Clean and dry the crankshaft oil seal housing.

11. Using the oil seal driver, 64 mm, drive in the new crankshaft oil seal until the oil seal driver bottoms on the pump.

12. Remove all of the old liquid gasket from the oil pump mating surfaces, the bolts, and the bolt holes.

13. Clean and dry the oil pump mating surfaces.

14. Apply liquid gasket, P/N 08717-0004, 08718-0003, or 08718-0009, to the engine block mating surface of the oil pump and to the inside edge of the threaded bolt holes. Install the component within 5 minutes of applying the liquid gasket.

➡Apply a bead of liquid gasket along the mating surface. If you apply liquid gasket P/N 08718-0012, the component must be installed within 5 minutes. If too much time has passed after applying the liquid gasket, remove all of the old liquid gasket and residue, then reapply the new liquid gasket.

15. Apply a light coat of new engine oil to the lip of the crankshaft oil seal, and apply new engine oil to the new O-ring.

16. Install the dowel pins, then align the inner rotor with the crankshaft, and install the oil pump.

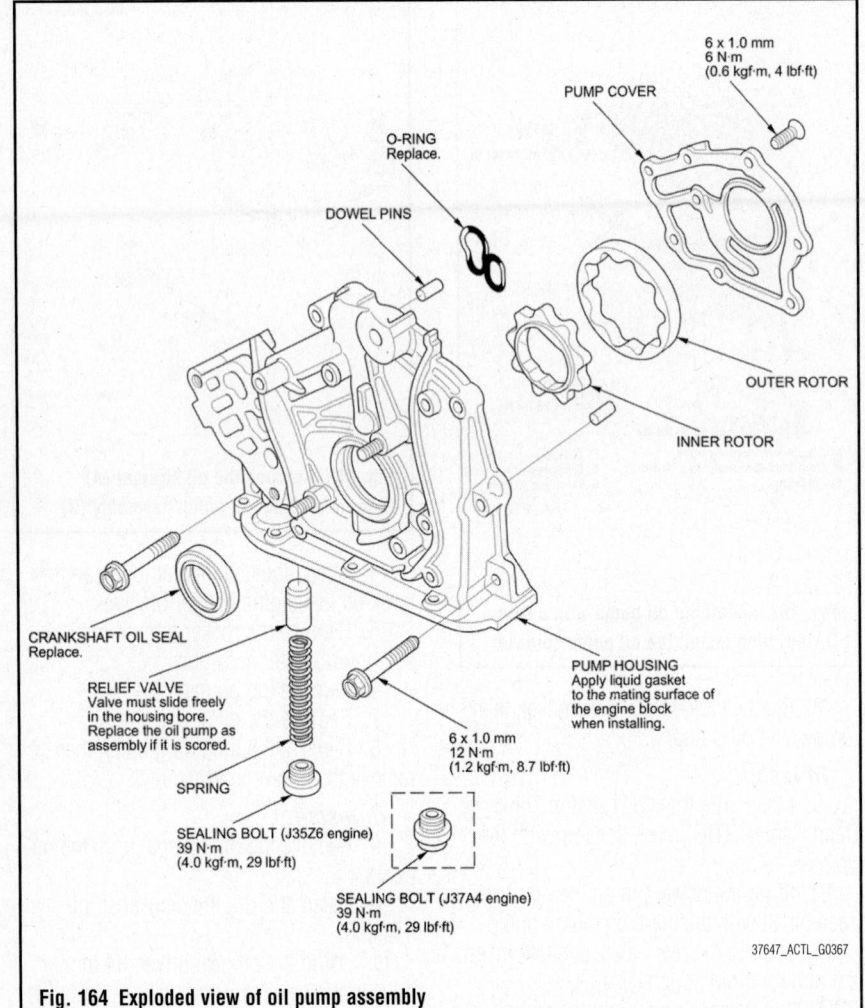

Fig. 164 Exploded view of oil pump assembly

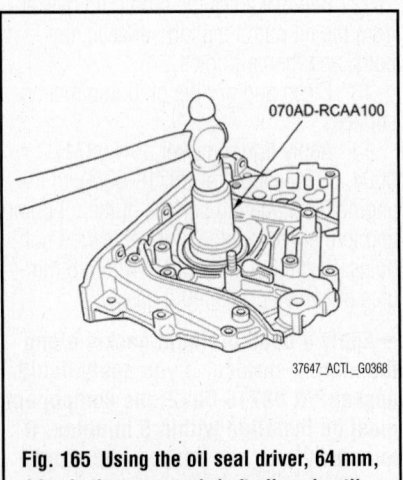

Fig. 165 Using the oil seal driver, 64 mm, drive in the new crankshaft oil seal until the oil seal driver bottoms on the pump

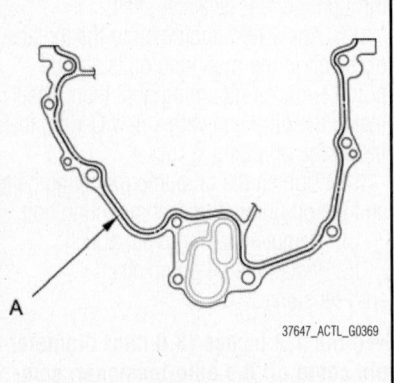

Fig. 166 Apply liquid gasket to the engine block mating surface of the oil pump and to the inside edge of the threaded bolt holes

➡**Wait at least 30 minutes before filling the engine with oil. Do not run the engine for at least 3 hours after installing the oil pump.**

17. Clean the excess grease off the crankshaft, and check the seal for distortion.

18. Install the oil strainer with a new O-ring.

19. Install the oil pan.

20. Install the rocker arm oil control solenoid/oil filter assembly with a new rocker arm oil control solenoid filter.

21. Install the timing belt drive pulley to the crankshaft:

22. Install the timing belt:

23. Remove the chain hoist.

PISTONS & RINGS

POSITIONING

4-Cylinder Engine

See Figure 167.

6-Cylinder Engine

See Figure 168.

REAR MAIN SEAL

REMOVAL & INSTALLATION

4-Cylinder Engine

Special Tools Required:
• Driver handle, 15 x 135L 07749-0010000
• Oil seal driver attachment 96 mm 07ZAD-PNAA100

1. Remove the transmission:

2. M/T model: Remove the pressure plate, the clutch disk, and the flywheel.

3. A/T model: Remove the drive plate.

4. Clean and dry the crankshaft oil seal housing.

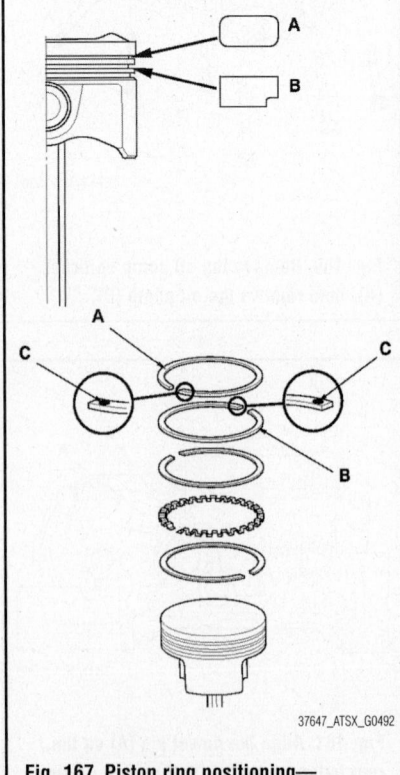

Fig. 167 Piston ring positioning— 4-cylinder engine

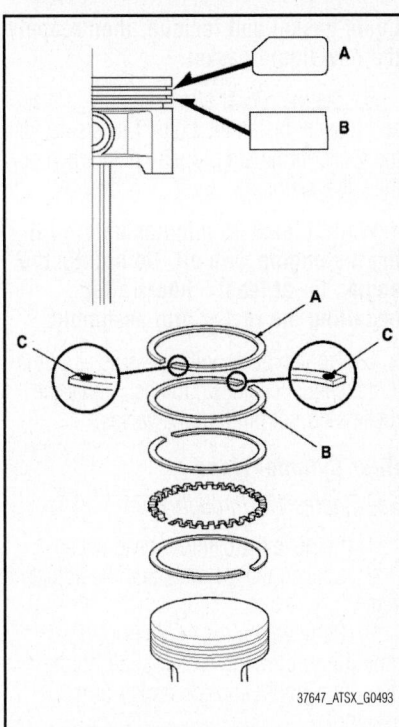

Fig. 168 Piston ring positioning—
6-cylinder engine

5. Use the driver handle, 15 x 135L and the oil seal driver attachment 96 mm to drive a new crankshaft oil seal squarely into the engine block to the specified installed height.

6. M/T model: Install the flywheel, the clutch disk, and the pressure plate.

7. A/T model: Install the drive plate.

8. Install the transmission:

6-Cylinder Engine

Special Tools Required:
• Driver handle, 15 x 135L 07749-0010000
• Oil seal driver attachment 106 mm 070AD-RCA0200

1. M/T model: Remove the transmission, and the flywheel.

2. A/T model: Remove the transmission, and the drive plate.

3. Remove the transmission end crankshaft oil seal.

4. Clean and dry the crankshaft oil seal housing.

5. Apply a light coat of new engine oil to the lip of the crankshaft oil seal.

6. Using the driver handle, 15 x 135 L and the oil seal driver attachment, 106 mm, drive in the new crankshaft oil seal until the oil seal driver attachment bottoms against the engine block end cover. Align the hole in the oil seal driver attachment with the pin on the crankshaft.

7. Clean any excess grease off the

crankshaft, and check that the oil seal lip is not distorted.

8. M/T model: Install the flywheel, and the transmission.

9. A/T model: Install the drive plate, and the transmission.

ROCKER ARMS

REMOVAL & INSTALLATION

4-Cylinder Engine

See Figures 169 through 173.

1. Remove the cam chain.

2. Loosen the rocker arm adjusting screws.

3. Remove the camshaft holder bolts. To prevent damaging the camshafts, loosen the bolts, in sequence, two turns at a time.

➡ **Bolt #1 is not on all engines.**

4. Remove cam chain guide B, the camshaft holders, and the camshafts.

5. Insert the bolts into the rocker shaft holder, then remove the rocker arm assembly.

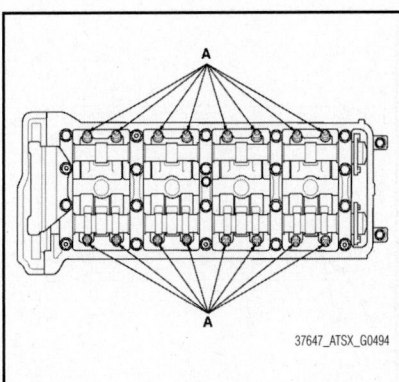

Fig. 169 Loosen the rocker arm adjusting screws (A)

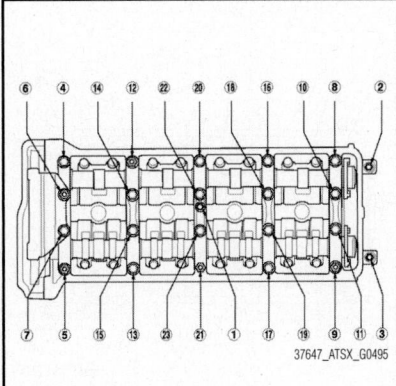

Fig. 170 Remove the camshaft holder bolts in sequence

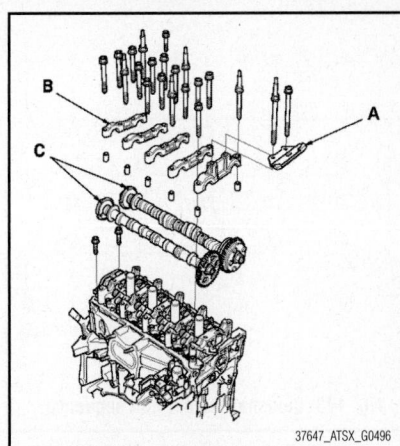

Fig. 171 Remove cam chain guide B (A), the camshaft holders (B), and the camshafts (C)

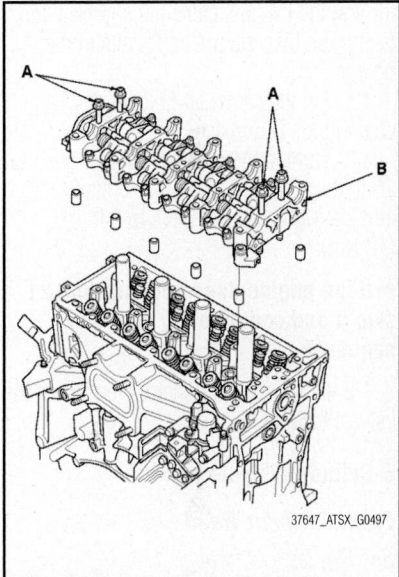

Fig. 172 Insert the bolts (A) into the rocker shaft holder, then remove the rocker arm assembly (B)

To install:

6. Reassemble the rocker arm assembly.

7. Clean and dry the No. 5 rocker shaft holder mating surface.

8. Apply liquid gasket (P/N 08717-0004, 08718-0003, or 08718-0009) to the cylinder head mating surface of the No. 5 rocker shaft holder and to the inside edge of the threaded bolt holes. Install the component within 5 minutes of applying the liquid gasket.

➡ **If too much time has passed after applying the liquid gasket, remove the old liquid gasket and residue, then reapply new liquid gasket.**

9. Insert the bolts into the rocker shaft holder, then install the rocker arm assembly on the cylinder head.

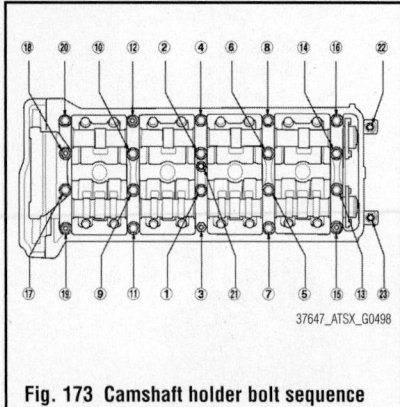

Fig. 173 Camshaft holder bolt sequence

10. Remove the bolts from the rocker shaft holder.

11. Make sure the punch marks on the Variable Valve Timing Control (VTC) actuator and the exhaust camshaft sprocket are facing up, then set the camshafts in the holder.

12. Set the camshaft holders and cam chain guide B in place

13. Tighten the camshaft holder bolts to 16 ft. lbs. (22 Nm) except for bolts 21, 22, and 23. Tighten these bolts to 9 ft. lbs. (12 Nm).

➡**If the engine does not have bolt 21, skip it and continue the torque sequence.**

14. Install the cam chain, then adjust the valve clearance.

6-Cylinder Engine

Front Cylinder Head

See Figures 174 through 176.

1. Remove the cylinder head cover.
2. Loosen the locknuts and the adjusting screws
3. Remove the rocker shaft bridge mounting bolts, the rocker shaft holder

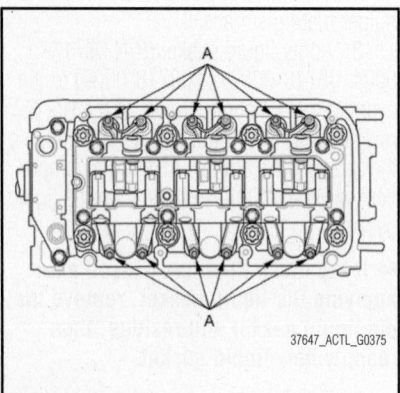

Fig. 174 Loosen the locknuts and the adjusting screws (A)

mounting bolts, and the rocker arm assembly.

   a. Loosen the rocker shaft bridge mounting bolts and the rocker shaft holder mounting bolts in sequence two turns at a time, to prevent damaging the valves or the rocker arm assembly.

   b. When removing the rocker arm assembly, do not remove the rocker shaft bridge mounting bolts and the rocker shaft holder mounting bolts. The bolts will keep the rocker arms on the shafts.

To install:

4. Apply liquid gasket, P/N 08717-0004, 08718-0003, or 08718-0009 to the rocker shaft holder mating surface of the cylinder head. Install the component within 5 minutes of applying the liquid gasket.

➡**Apply a bead of liquid gasket along the mating surfaces. If you apply liquid gasket P/N 08718-0012, the component must be installed within 4 minutes. If too much time has passed after applying the liquid gasket, remove the old**

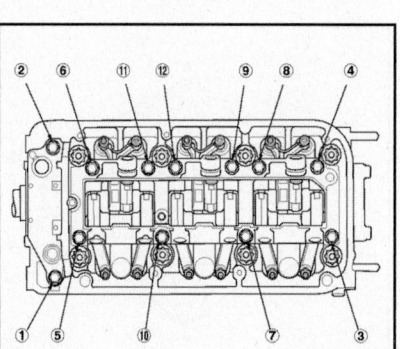

Fig. 175 Front rocker shaft bridge bolt loosening sequence

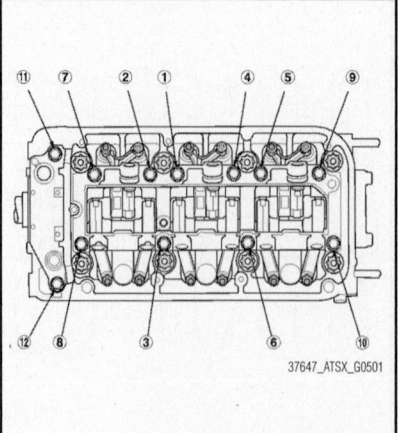

Fig. 176 Tighten each rocker shaft bridge bolt two turns at a time in sequence

liquid gasket and residue, then reapply the new liquid gasket.

5. Set the rocker arm assembly in place, and loosely install the bolts. Make sure that the rocker arms are properly positioned on the valve stems.

➡**Wait at least 30 minutes before filling the engine with oil. Do not run the engine for at least 3 hours after installing the rocker arm assembly.**

6. Tighten each bolt two turns at a time in the sequence shown to ensure that the rockers do not bind on the valves.

Rear Cylinder Head

See Figures 177 through 179.

1. Remove the cylinder head cover.
2. Loosen the locknuts and the adjusting screws.
3. Remove the rocker shaft bridge mounting bolts, the rocker shaft holder mounting bolts, and the rocker arm assembly.

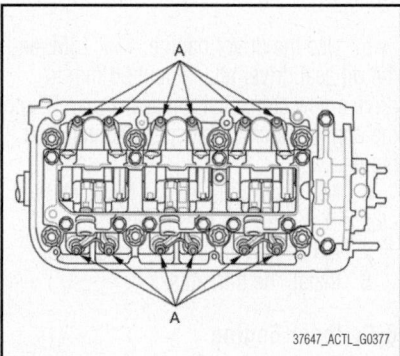

Fig. 177 Loosen the locknuts and the adjusting screws (A)

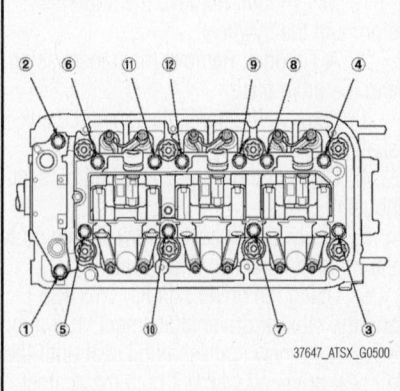

Fig. 178 Rear cylinder head bolt loosening sequence

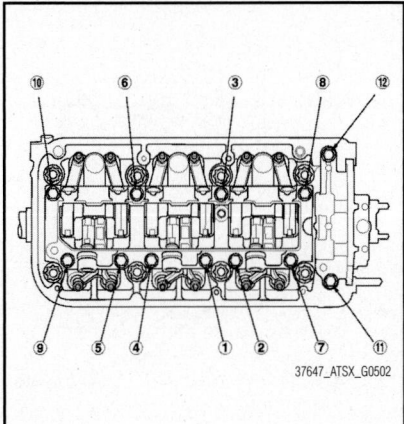

Fig. 179 Tighten each rocker shaft bridge bolt two turns at a time in sequence

a. Loosen the rocker shaft bridge mounting bolts and the rocker shaft holder mounting bolts in sequence two turns at a time, to prevent damaging the valves or the rocker arm assembly.

b. When removing the rocker arm assembly, do not remove the rocker shaft bridge mounting bolts and the rocker shaft holder mounting bolts. The bolts will keep the rocker arms on the shafts.

To install:

4. Apply liquid gasket, P/N 08717-0004, 08718-0003, or 08718-0009 to the rocker shaft holder mating surface of the cylinder head. Install the component within 5 minutes of applying the liquid gasket.

➡**Apply a bead of liquid gasket along the mating surfaces. If you apply liquid gasket P/N 08718-0012, the component must be installed within 4 minutes. If too much time has passed after applying the liquid gasket, remove the old liquid gasket and residue, then reapply the new liquid gasket.**

5. Set the rocker arm assembly in place, and loosely install the bolts. Make sure that the rocker arms are properly positioned on the valve stems.

➡**Wait at least 30 minutes before filling the engine with oil. Do not run the engine for at least 3 hours after installing the rocker arm assembly.**

6. Tighten each bolt two turns at a time in the sequence shown to ensure that the rockers do not bind on the valves.

TIMING BELT FRONT COVER

REMOVAL & INSTALLATION

6-Cylinder Engine

See Figures 180 through 184.

1. Turn the crankshaft so its white mark lines up with the pointer.

➡**The other pointer is not used.**

2. Check that the No. 1 piston Top Dead Center (TDC) mark on the front camshaft pulley and the pointer on the front upper cover are aligned.

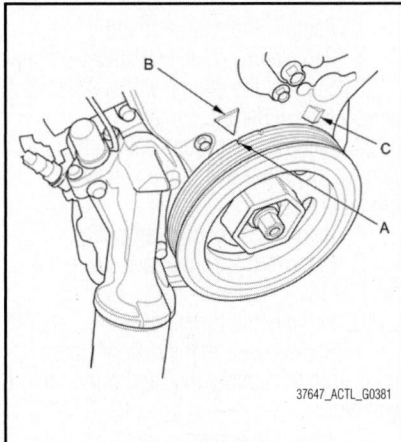

Fig. 180 Turn the crankshaft so its white mark (A) lines up with the pointer (B)

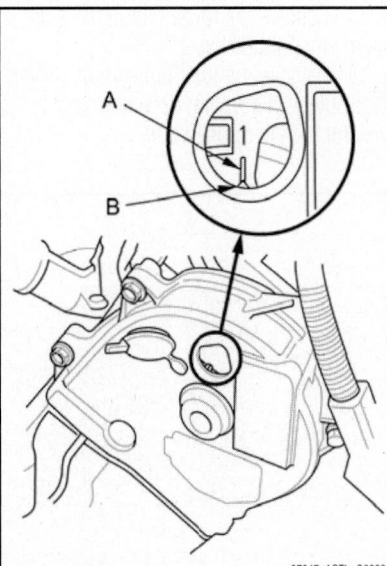

Fig. 181 Check that the No. 1 piston TDC mark (A) on the front camshaft pulley and the pointer (B) on the front upper cover are aligned

➡**If the marks are not aligned, rotate the crankshaft 360 degrees, and recheck the camshaft pulley mark.**

3. Raise the vehicle on the lift, then remove the right front wheel.

4. Remove the splash shield.

5. Remove the drive belt auto-tensioner.

6. Support the engine with a jack and a wood block under the oil pan.

7. Remove the ground cable, then remove the upper half of the side engine mount bracket.

8. Remove the crankshaft pulley.

9. Remove the front upper cover and the rear upper cover.

10. Remove the lower cover.

To install:

11. Install the lower cover.

12. Install the front upper cover and the rear upper cover.

13. Install the crankshaft pulley.

14. Rotate the crankshaft pulley about

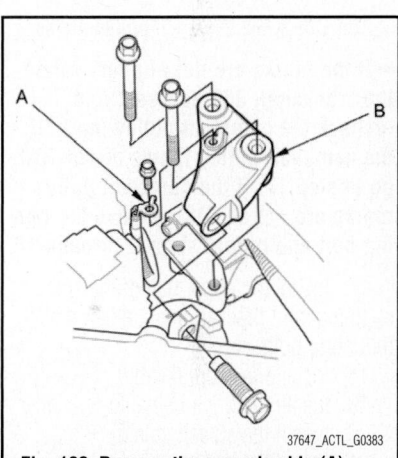

Fig. 182 Remove the ground cable (A), then remove the upper half of the side engine mount bracket (B)

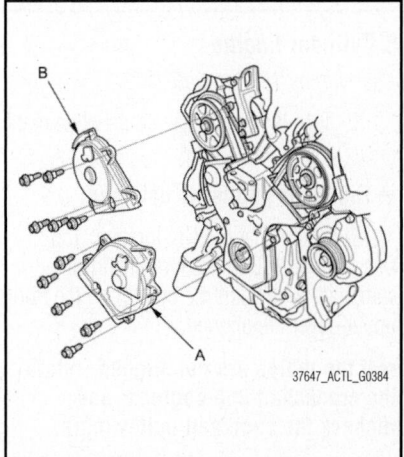

Fig. 183 Remove the front upper cover (A) and the rear upper cover (B)

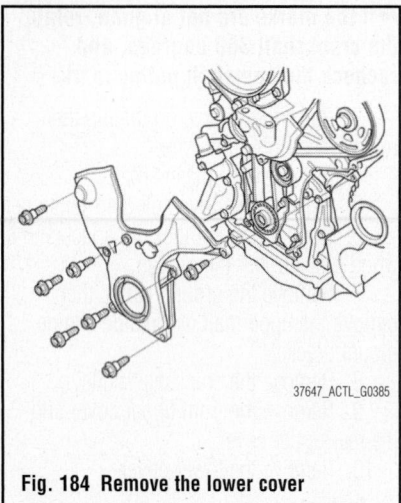

Fig. 184 Remove the lower cover

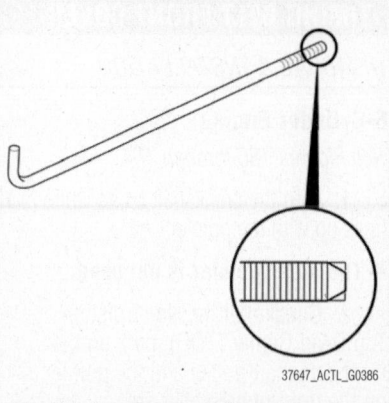

Fig. 185 Remove one of the battery clamp bolts from the battery tray, and grind the end of it

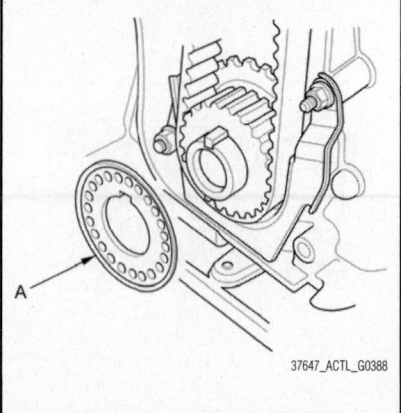

Fig. 187 Remove the timing belt guide plate (A)

six turns clockwise so the timing belt positions itself on the pulleys.

15. Turn the crankshaft pulley so its white mark lines up with the pointer.

➡The other pointer is not used.

16. Check the camshaft pulley marks.

➡If the marks are not aligned, rotate the crankshaft 360 degrees, and recheck the camshaft pulley mark. If the camshaft pulley marks are at TDC, go to step 17. If the camshaft pulley marks are not at TDC, remove the timing belt and repeat steps 2 through 16.

17. Install the upper half of the side engine mount bracket, then tighten the mounting bolts.

18. Install the ground cable.
19. Install the drive belt auto-tensioner.
20. Install the splash shield.
21. Install the right front wheel.

TIMING BELT & SPROCKETS

REMOVAL & INSTALLATION

6-Cylinder Engine

See Figures 180 through 192.

1. Turn the crankshaft so its white mark lines up with the pointer.

➡The other pointer is not used.

2. Check that the No. 1 piston Top Dead Center (TDC) mark on the front camshaft pulley and the pointer on the front upper cover are aligned.

➡If the marks are not aligned, rotate the crankshaft 360 degrees, and recheck the camshaft pulley mark.

3. Raise the vehicle on the lift, then remove the right front wheel.

4. Remove the splash shield.
5. Remove the drive belt auto-tensioner.
6. Support the engine with a jack and a wood block under the oil pan.
7. Remove the ground cable, then remove the upper half of the side engine mount bracket.
8. Remove the crankshaft pulley.
9. Remove the front upper cover and the rear upper cover.
10. Remove the lower cover.
11. Remove one of the battery clamp bolts from the battery tray, and grind the end of it.
12. Thread the battery clamp bolt into hold the timing belt adjuster in its current position. Tighten it by hand, do not use a wrench.
13. Remove the timing belt guide plate.
14. Remove the lower half of the side engine mount bracket.
15. Remove the idler pulley bolt and the idler pulley, then remove the timing belt. Discard the idler pulley bolt.

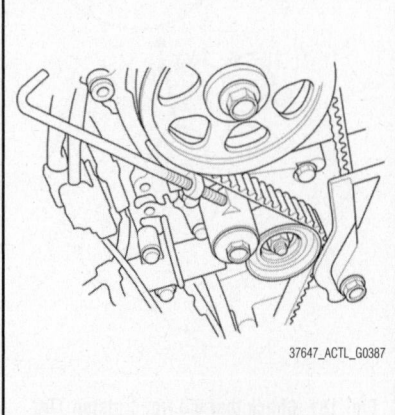

Fig. 186 Thread the battery clamp bolt into hold the timing belt adjuster in its current position

To install:

16. Clean the timing belt pulleys, the timing belt guide plate, and the upper and lower covers.

17. Set the timing belt drive pulley to Top Dead Center (TDC) by aligning the TDC mark on the tooth of the timing belt drive pulley with the pointer on the oil pump.

18. Set the camshaft pulleys to TDC by aligning the TDC marks on the camshaft pulleys with the pointers on the back covers.

19. Loosely install the idler pulley with a new idler pulley bolt so the pulley can move but does not come off.

20. If the auto-tensioner has extended and the timing belt cannot be installed, do the timing belt replacement procedure.

21. Install the timing belt in a counterclockwise sequence starting with the drive pulley. Take care not to damage the timing belt during installation.

22. Tighten the idler pulley bolt to 33 ft. lbs. (44 Nm).

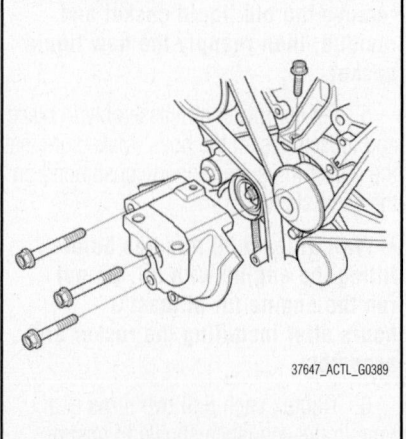

Fig. 188 Remove the lower half of the side engine mount bracket

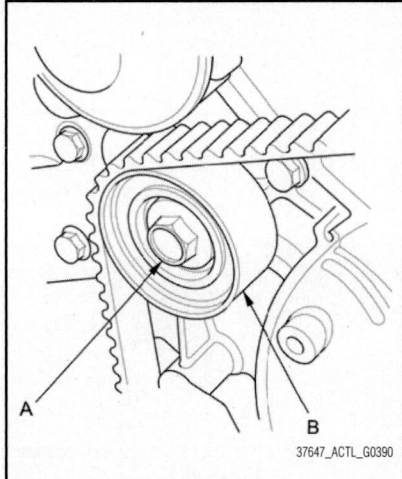

Fig. 189 Remove the idler pulley bolt (A) and the idler pulley (B), then remove the timing belt

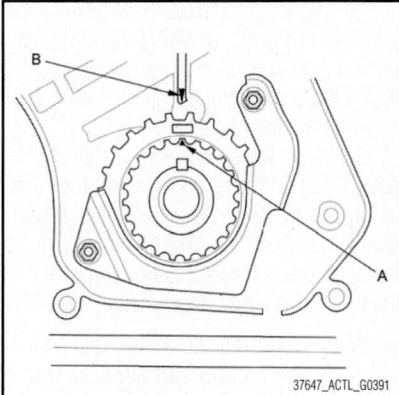

Fig. 190 Set the timing belt drive pulley to TDC by aligning the TDC mark (A) on the tooth of the timing belt drive pulley with the pointer (B) on the oil pump

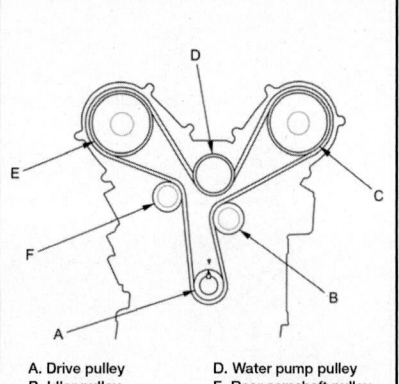

A. Drive pulley
B. Idler pulley
C. Front camshaft pulley
D. Water pump pulley
E. Rear camshaft pulley
F. Adjusting pulley

Fig. 192 Install the timing belt in a counterclockwise sequence starting with the drive pulley

23. Remove the battery clamp bolt from the back cover.

24. Install the lower half of the side engine mount bracket. Tighten the three long bolts to 33 ft. lbs. (44 Nm).

25. Install the timing belt guide plate.

26. Install the lower cover.

27. Install the front upper cover and the rear upper cover.

28. Install the crankshaft pulley.

29. Rotate the crankshaft pulley about six turns clockwise so the timing belt positions itself on the pulleys.

30. Turn the crankshaft pulley so its white mark lines up with the pointer.

➡**The other pointer is not used.**

31. Check the camshaft pulley marks.

➡If the marks are not aligned, rotate the crankshaft 360 degrees, and recheck the camshaft pulley mark. If the camshaft pulley marks are at TDC, go to step 17. If the camshaft pulley marks are not at TDC, remove the timing belt and repeat steps 2 through 16.

32. Install the upper half of the side engine mount bracket, then tighten the mounting bolts.

33. Install the ground cable.

34. Install the drive belt auto-tensioner.

35. Install the splash shield.

36. Install the right front wheel.

TIMING CHAIN COVER

REMOVAL & INSTALLATION

4-Cylinder Engine

See Figures 193 through 199.

➡**Keep the cam chain away from magnetic fields.**

1. Remove the front wheels.
2. Remove the splash shield.
3. Remove the drive belt.
4. Turn the crankshaft so its white mark lines up with the pointer.

➡**The other pointer is not used.**

5. Remove the cylinder head cover.
6. Set the No. 1 piston at Top Dead Center (TDC). The punch mark on the Variable Valve Timing Control (VTC) actuator and the punch mark on the exhaust camshaft sprocket should be at the top. Align the TDC marks on the VTC actuator and the exhaust camshaft sprocket.

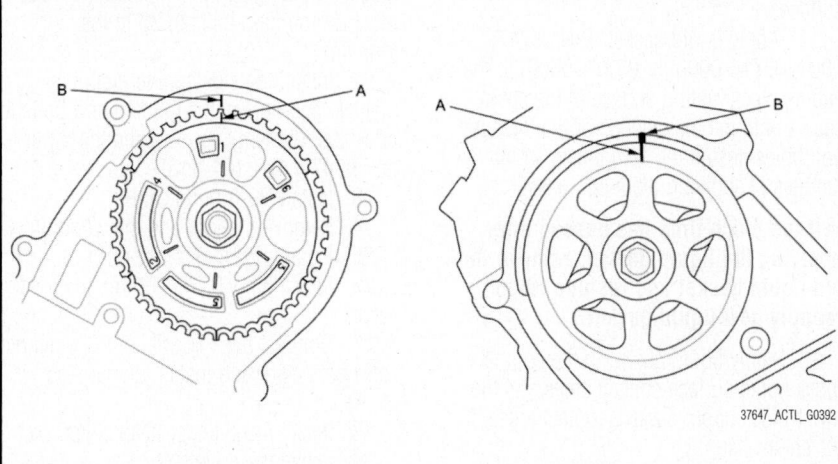

Fig. 191 Set the camshaft pulleys to TDC by aligning the TDC marks (A) on the camshaft pulleys with the pointers (B) on the back covers

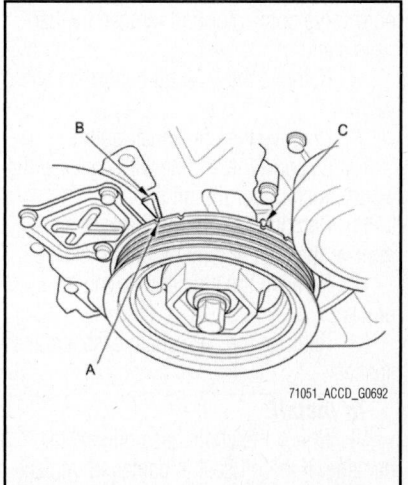

Fig. 193 Turn the crankshaft so its white mark (A) lines up with the pointer (B); the other pointer (C) is not used

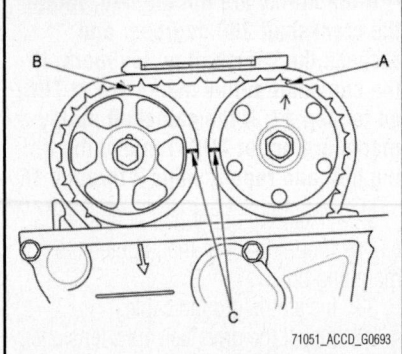

Fig. 194 The punch mark (A) on the VTC actuator and the punch mark (B) on the exhaust camshaft sprocket should be at the top, align the TDC marks (C) on the VTC actuator and the exhaust camshaft sprocket

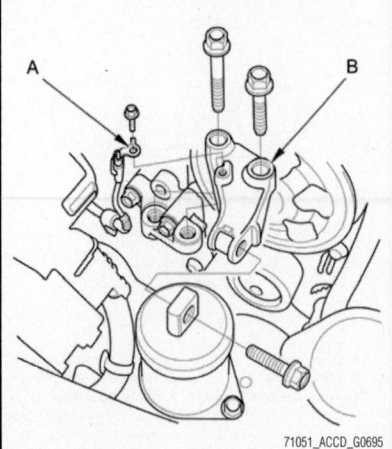

Fig. 196 Remove the ground cable (A), then remove the side engine mount bracket (B)

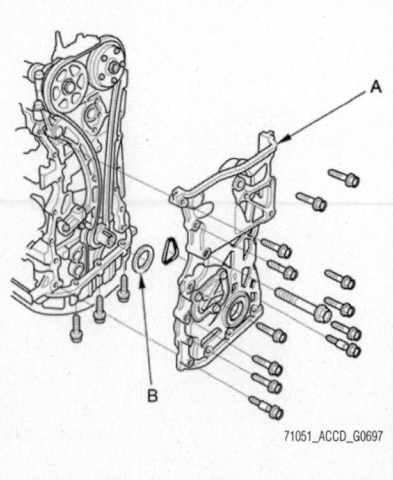

Fig. 198 Remove the cam chain case (A) and the spacer (B)

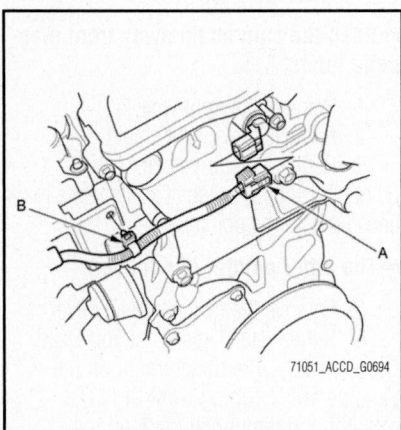

Fig. 195 Disconnect the VTC oil control solenoid valve connector (A) and remove the harness clamp (B)

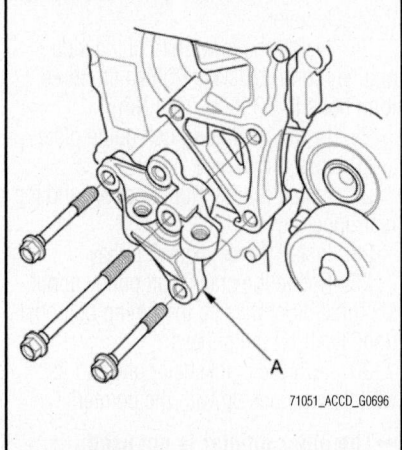

Fig. 197 Remove the side engine mount bracket (A)

7. Disconnect the VTC oil control solenoid valve connector and remove the harness clamp.

8. Remove the VTC oil control solenoid valve.

9. Remove the crankshaft pulley.

10. Support the engine with a jack and a wood block under the oil pan.

11. Remove the ground cable, then remove the side engine mount bracket.

12. Remove the side engine mount bracket.

13. Remove the cam chain case and the spacer.

To install:

14. Check the chain case oil seal for damage. If the oil seal is damaged, replace the chain case oil seal.

15. Remove the old liquid gasket from the chain case mating surfaces, the bolts, and the bolt holes.

16. Clean and dry the chain case mating surfaces.

17. Apply liquid gasket (P/N 08717-0004, 08718-0003, or 08718-0009) to the engine block mating surface of the chain case and to the inside edge of the threaded bolt holes. Install the component within 5 minutes of applying the liquid gasket.

➡️**If too much time has passed after applying the liquid gasket, remove the old liquid gasket and residue, then reapply new liquid gasket.**

18. Apply liquid gasket to the engine block upper surface contact areas and the lower block upper surface contact areas on the chain case.

➡️**Apply about 0.43 inches (11 mm) diameter and about 0.12 inches (3 mm) thickness of liquid gasket to the areas B and C.**

19. Apply liquid gasket (P/N 08717-0004, 08718-0003, or 08718-0009) to the oil pan mating surface of the oil pump. Install the component within 5 minutes of applying the liquid gasket.

20. Install the spacer, then install the new O-ring on the chain case. Set the edge of the chain case to the edge of the oil pan, then install the chain case on the engine block. Wipe off the excess liquid gasket on the oil pan and chain case mating area.

➡️**When installing the chain case, do not slide the bottom surface onto the oil pan mounting surface. Wait at least 30 minutes before filling the engine with oil. Do not run the engine for at least 3 hours after installing the chain case.**

21. Install the side engine mount bracket, then tighten the side engine mount bracket mounting bolts to 33 ft. lbs. (44 Nm).

22. Install the side engine mount bracket, and tighten the top two new bolts to 40 ft. lbs. (54 Nm), then tighten the lower bolt to 47 ft. lbs. (64 Nm).

23. Install the ground cable.

24. Remove the jack and the wood block.

25. Install the crankshaft pulley.

26. Install the VTC oil control solenoid valve.

27. Connect the VTC oil control solenoid valve connector and install the harness clamp.

28. Install the cylinder head cover.

29. Install the drive belt.

30. Install the splash shield.

31. Install the front wheels.

32. Do the Crankshaft Position (CKP) pattern clear/CKP pattern learn procedure.

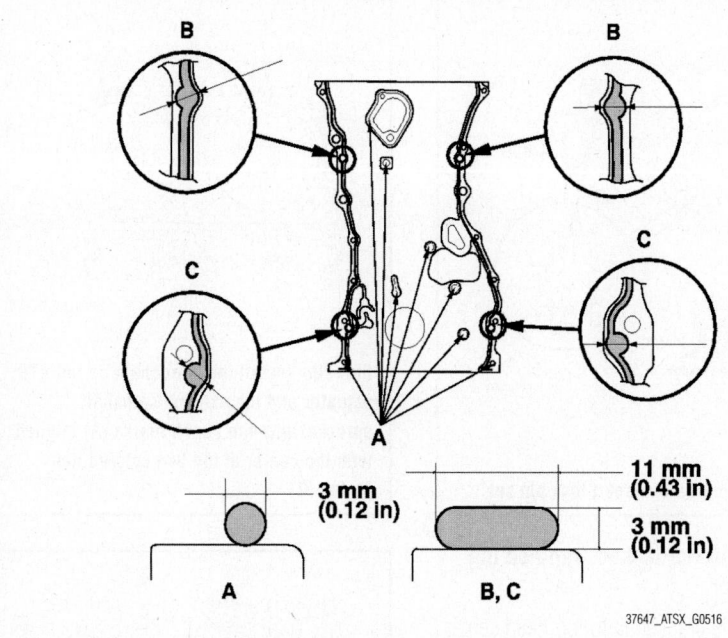

Fig. 199 Apply about 0.43 inches (11 mm) diameter and about 0.12 inches (3 mm) thickness of liquid gasket to the areas (B) and (C)

TIMING (CAMSHAFT) CHAIN & SPROCKETS

REMOVAL & INSTALLATION

4-Cylinder Engine

See Figures 193 through 198, 200 through 217.

➡**Keep the cam chain away from magnetic fields.**

1. Remove the right front wheel.
2. Remove the front splash shield.
3. Remove the drive belt.
4. Turn the crankshaft so its white mark lines up with the pointer.

➡**The other pointer is not used.**

5. Remove the cylinder head cover.
6. Check the No. 1 piston at top dead center (TDC). The punch mark on the VTC actuator and the punch mark on the exhaust camshaft sprocket should be at the top. Align the TDC marks on the VTC actuator and the exhaust camshaft sprocket.
7. Disconnect the VTC oil control solenoid valve connector and remove the harness clamp.
8. Remove the VTC oil control solenoid valve.
9. Remove the crankshaft pulley.
10. Lift and support the engine with a jack and a wood block under the oil pan.
11. Remove the ground cable, then remove the side engine mount bracket.

12. Remove the side engine mount bracket.
13. Remove the cam chain case and the spacer.
14. Loosely install the crankshaft pulley.
15. Turn the crankshaft counterclockwise to compress the auto-tensioner.
16. Rotate the crankshaft counterclockwise to align the holes on the lock and the

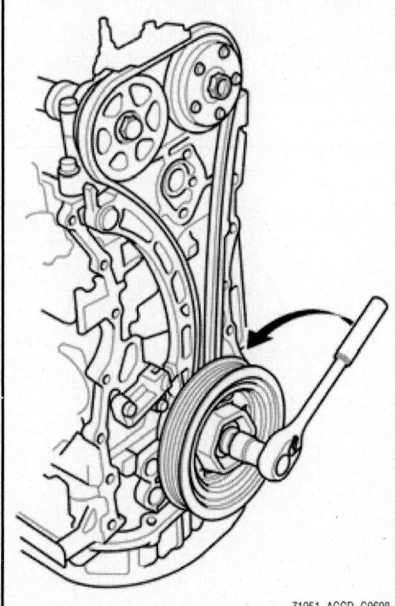

Fig. 200 Turn the crankshaft counterclockwise to compress the auto-tensioner

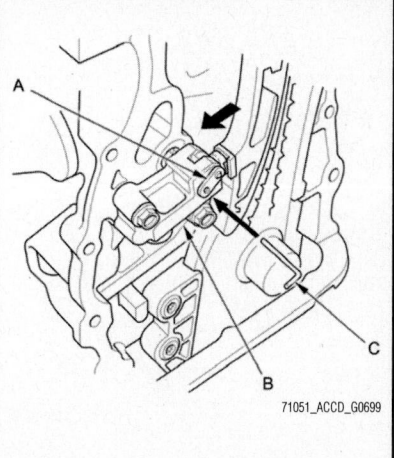

Fig. 201 Rotate the crankshaft counterclockwise to align the holes on the lock (A) and the auto-tensioner (B), then insert a pin (C) into the holes

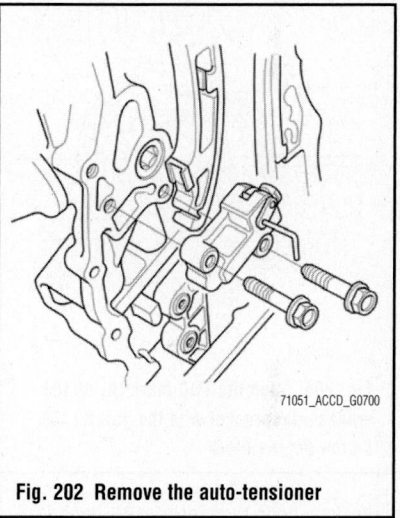

Fig. 202 Remove the auto-tensioner

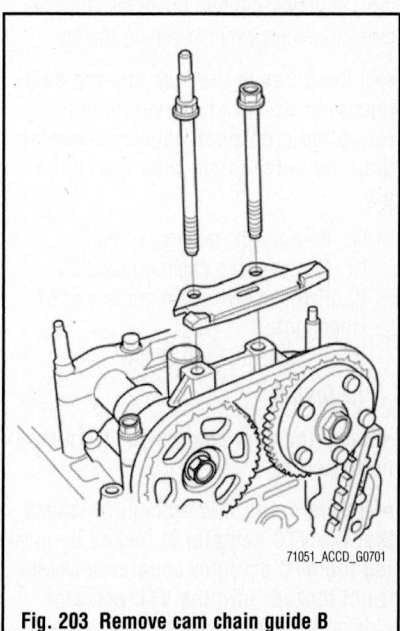

Fig. 203 Remove cam chain guide B

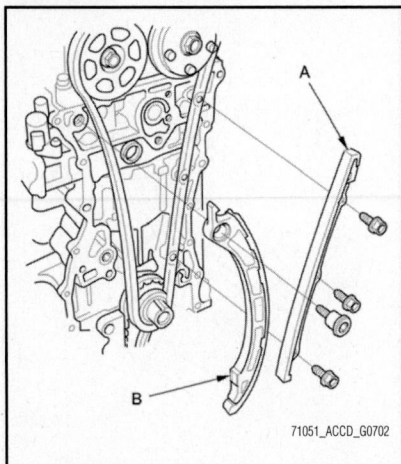

Fig. 204 Remove cam chain guide A and the tensioner arm (B)

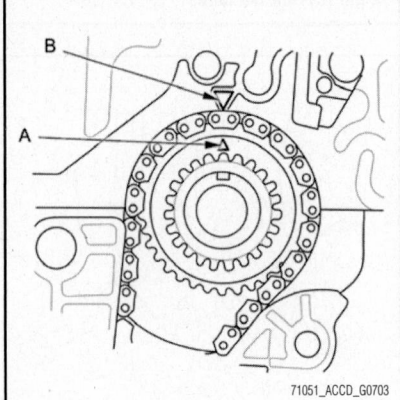

Fig. 205 Align the TDC mark (A) on the crankshaft sprocket with the pointer (B) on the engine block

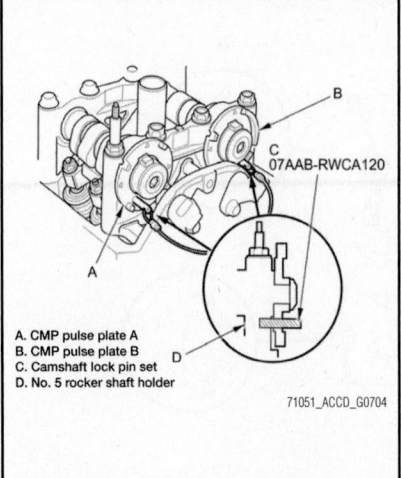

A. CMP pulse plate A
B. CMP pulse plate B
C. Camshaft lock pin set
D. No. 5 rocker shaft holder

Fig. 206 Insert a camshaft lock pin set

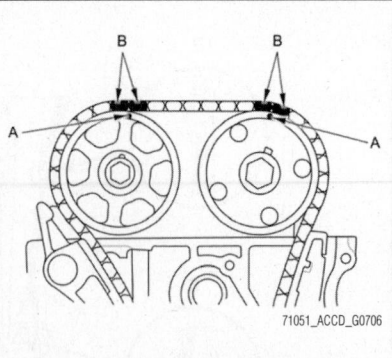

Fig. 208 Install the cam chain on the VTC actuator and the exhaust camshaft sprocket with the punch marks (A) aligned with the center of the two colored link plates (B)

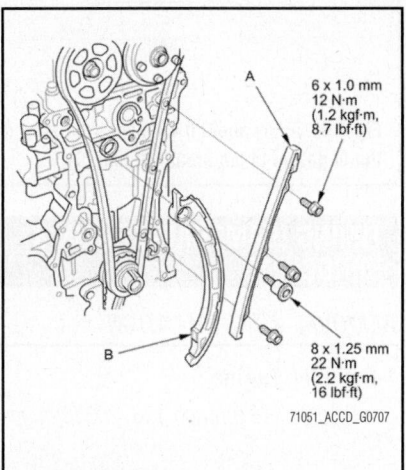

6 x 1.0 mm
12 N·m
(1.2 kgf·m, 8.7 lbf·ft)

8 x 1.25 mm
22 N·m
(2.2 kgf·m, 16 lbf·ft)

Fig. 209 Install cam chain guide A and the tensioner arm (B)

auto-tensioner, then insert a ³⁄₆₄ inch (1.2 mm) diameter pin into the holes. Turn the crankshaft clockwise to secure the pin.

➡️If the holes in the lock and the auto-tensioner do not align, continue to rotate the crankshaft counterclockwise until the holes align, then install the pin.

17. Remove the auto-tensioner.
18. Remove cam chain guide B.
19. Remove cam chain guide A and the tensioner arm.
20. Remove the cam chain.

To install:

➡️Keep the cam chain away from magnetic fields.

➡️Before doing this procedure, check that the VTC actuator is locked by turning the VTC actuator counterclockwise. If not locked, turn the VTC actuator clockwise until it stops, then recheck

it. If it is still not locked, replace the VTC actuator.

21. Set the crankshaft to top dead center (TDC). Align the TDC mark on the crankshaft sprocket with the pointer on the engine block.
22. Set the camshafts to TDC. The punch mark on the VTC actuator and the punch mark on the exhaust camshaft sprocket should be at the top. Align the TDC marks on the VTC actuator and the exhaust camshaft sprocket.
23. To hold the intake camshaft, insert a camshaft lock pin set into the maintenance hole in CMP pulse plate A and through the No. 5 rocker shaft holder.
24. To hold the exhaust camshaft, insert a camshaft lock pin set into the maintenance hole in CMP pulse plate B and through the No. 5 rocker shaft holder.
25. Install the cam chain on the crank-

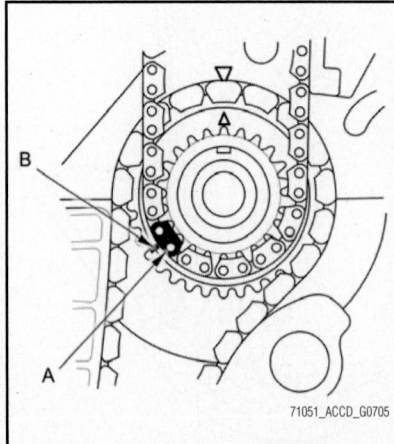

Fig. 207 Install the cam chain on the crankshaft sprocket with the colored link plate (A) aligned with the mark (B) on the crankshaft sprocket

shaft sprocket with the colored link plate aligned with the mark on the crankshaft sprocket.

26. Install the cam chain on the VTC actuator and the exhaust camshaft sprocket with the punch marks aligned with the center of the two colored link plates.
27. Install cam chain guide A and the tensioner arm.
28. Install cam chain guide B.
29. Compress the auto-tensioner when replacing the cam chain. Remove the diameter pin from the auto-tensioner that was installed during removal. Turn the plate counterclockwise, to release the lock, then press the rod, and set the first cam to the first edge of the rack. Insert the ³⁄₆₄ inch (1.2 mm) diameter pin into the holes.

➡️If the chain tensioner is not set up as described, the tensioner will become damaged.

30. Install the auto-tensioner.

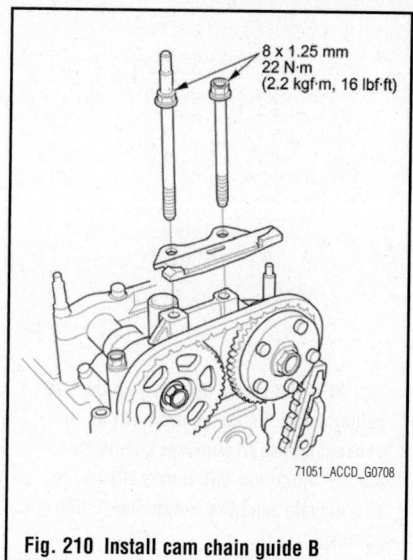

Fig. 210 **Install cam chain guide B**

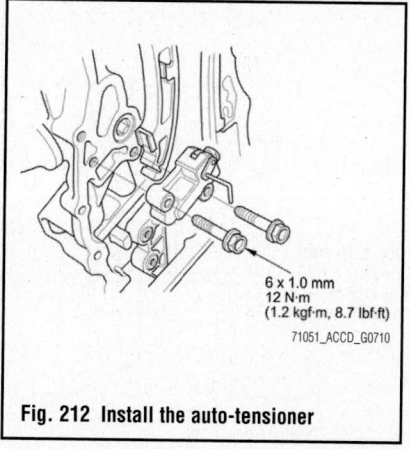

Fig. 212 **Install the auto-tensioner**

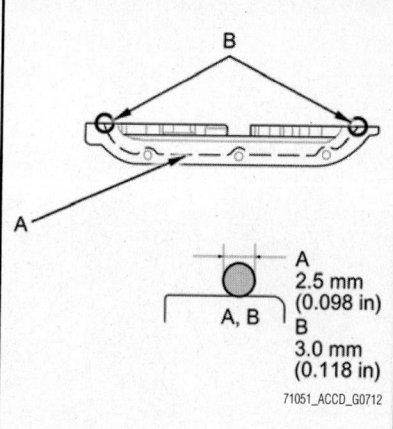

Fig. 214 **Apply a bead of liquid gasket along the broken lines**

31. Remove the diameter pin from the auto-tensioner.

32. Remove the camshaft lock pin set.

33. Check the chain case oil seal for damage. If the oil seal is damaged, replace the chain case oil seal.

34. Remove the old liquid gasket from the chain case mating surfaces, the bolts, and the bolt holes.

35. Clean and dry the chain case mating surfaces.

36. Apply liquid gasket, P/N 08717-0004, 08718-0003, or 08718-0009, to the engine block mating surface of the chain case, and to the inside edge of the threaded bolt holes. Install the component within 5 minutes of applying the liquid gasket.

- Apply a bead of liquid gasket along the broken line A and B.
- Apply a bead of liquid gasket along the broken line C.
- If too much time has passed after

applying the liquid gasket, remove the old liquid gasket and residue, then reapply new liquid gasket.

37. Apply liquid gasket, P/N 08717-0004, 08718-0003, or 08718-0009 to the oil pan mating surface of the chain case, and to the inside edge of the threaded bolt holes. Install the component within 5 minutes of applying the liquid gasket.

- Apply a bead of liquid gasket along the broken line A.
- Apply a bead of liquid gasket along the broken line B.
- If too much time has passed after applying the liquid gasket, remove the old liquid gasket and residue, then reapply new liquid gasket.

38. Install the spacer, then install a new O-ring on the chain case. Set the edge of the chain case to the edge of the oil pan,

then install the chain case on the engine block. Wipe off the excess liquid gasket on the oil pan and chain case mating surface.

- When installing the chain case, do not slide the bottom surface onto the oil pan mounting surface.
- Wait at least 30 minutes before filling the engine with oil.
- Do not run the engine for at least 3 hours after installing the chain case.

39. Install the side engine mount bracket, then tighten the side engine mount bracket mounting bolts.

40. Install the side engine mount bracket, then loosely install the new side

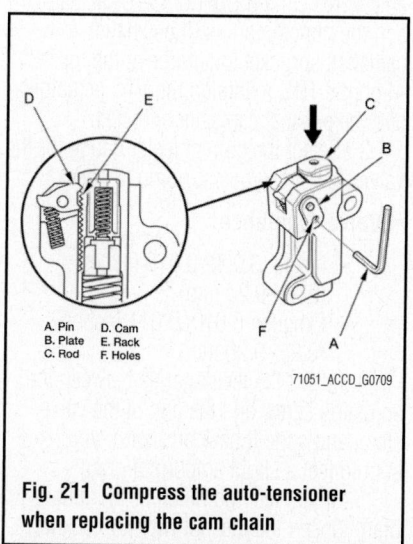

Fig. 211 **Compress the auto-tensioner when replacing the cam chain**

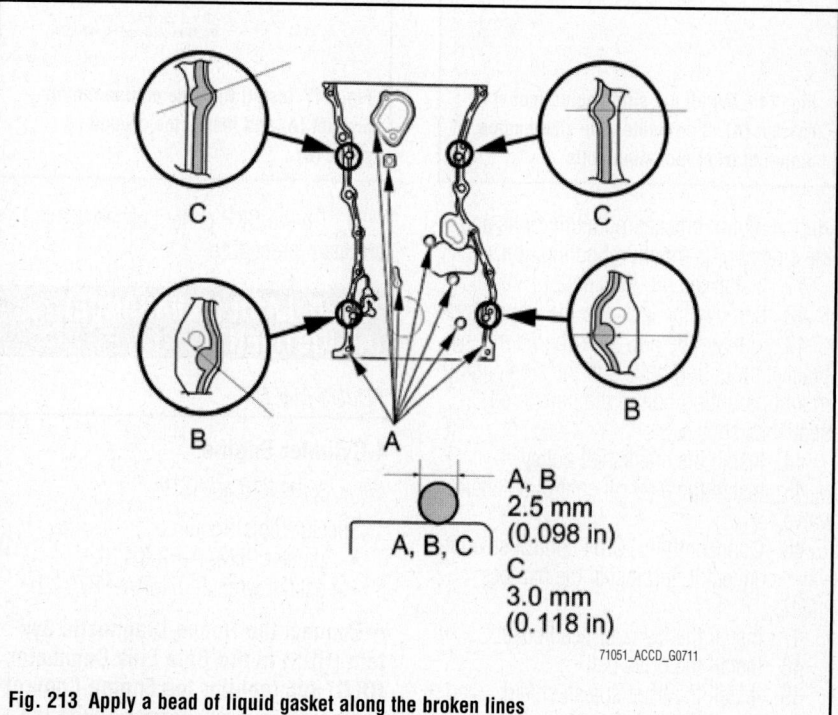

Fig. 213 **Apply a bead of liquid gasket along the broken lines**

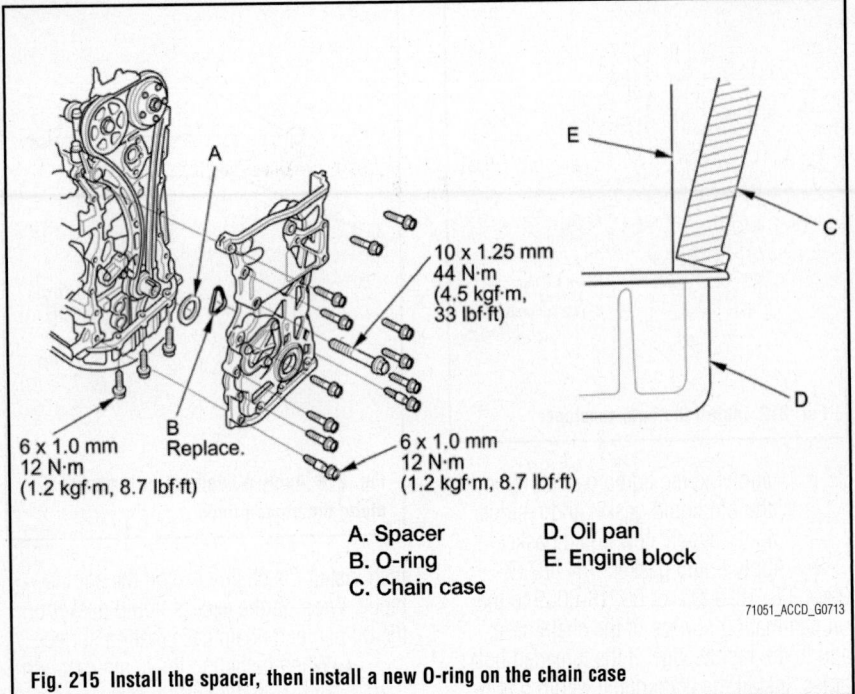

A. Spacer   D. Oil pan
B. O-ring   E. Engine block
C. Chain case

71051_ACCD_G0713

Fig. 215 Install the spacer, then install a new O-ring on the chain case

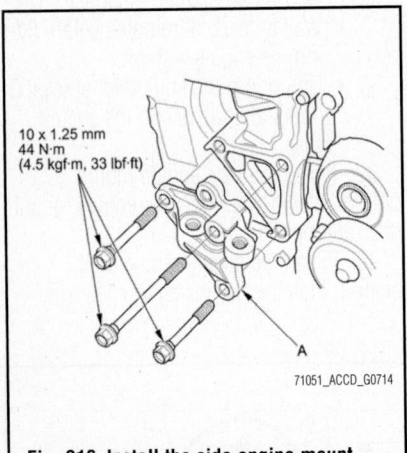

10 x 1.25 mm
44 N·m
(4.5 kgf·m, 33 lbf·ft)

71051_ACCD_G0714

Fig. 216 Install the side engine mount bracket (A), then tighten the side engine mount bracket mounting bolts

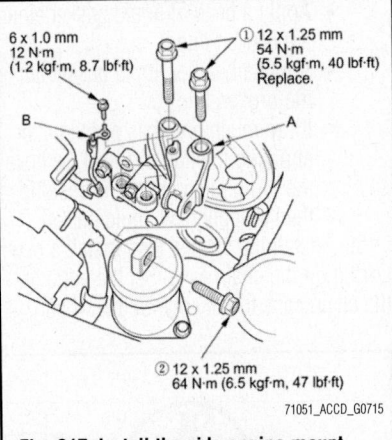

71051_ACCD_G0715

Fig. 217 Install the side engine mount bracket (A) and install the ground cable (B)

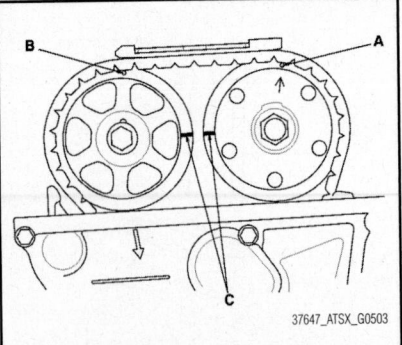

37647_ATSX_G0503

Fig. 218 The punch mark (A) on the VTC actuator and the punch mark (B) on the exhaust camshaft sprocket should be at the top. Align the TDC marks (C) on the VTC actuator and the exhaust camshaft sprocket

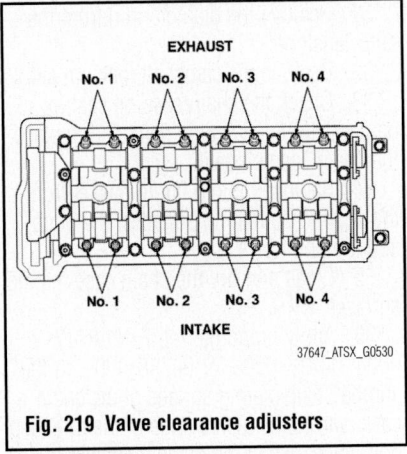

37647_ATSX_G0530

Fig. 219 Valve clearance adjusters

engine mount bracket mounting bolts and the side engine mount mounting bolt.

41. Install the ground cable.
42. Remove the jack and the wood block.
43. Tighten the new side engine mount bracket mounting bolts and the side engine mount mounting bolt in the numbered sequence shown.
44. Install the crankshaft pulley.
45. Install the VTC oil control solenoid valve.
46. Connect the VTC oil control solenoid valve connector and install the harness clamp.
47. Install the cylinder head cover.
48. Install the drive belt.
49. Install the front splash shield.
50. Install the right front wheel.

51. Do the CKP pattern clear/CKP pattern learn procedure.

VALVE LASH (CLEARANCE) ADJUSTMENT

ADJUSTMENT

4-Cylinder Engine

See Figures 218 and 219.

Special Tools Required:
• Adjuster 07MAA-PR70110
• Locknut wrench 07MAA-PR70120

➡Connect the Honda Diagnostic System (HDS) to the Data Link Connector (DLC) and monitor the Engine Coolant Temperature (ECT) sensor 1 with the

HDS. Adjust the valve clearance only when the ECT sensor 1 temperature is less than 100°F (38°C).

1. Remove the cylinder head cover.
2. Set the No. 1 piston at Top Dead Center (TDC). The punch mark on the Variable Valve Timing Control (VTC) actuator and the punch mark on the exhaust camshaft sprocket should be at the top. Align the TDC marks on the VTC actuator and the exhaust camshaft sprocket.
3. Select the correct feeler gauge for the valve clearance you are going to check.

➡Valve Clearance:

• Intake: 0.008–0.010 inches (0.21–0.25 mm)
• Exhaust: 0.010–0.011 inches (0.25–0.29 mm)

4. Insert the feeler gauge between the adjusting screw and the end of the valve stem, and slide it back and forth; you should feel a slight amount of drag.
5. If you feel too much or too little drag, loosen the locknut with the locknut

wrench and the adjuster, and turn the adjusting screw until the drag on the feeler gauge is correct.

6. Tighten the locknut to 10 ft. lbs. (14 Nm), and recheck the clearance. Repeat the adjustment if necessary.

7. Rotate the crankshaft 180° clockwise (camshaft pulley turns 90°).

8. Check and, if necessary, adjust the valve clearance on the No. 3 cylinder.

9. Rotate the crankshaft 180° clockwise (camshaft pulley turns 90°).

10. Check and, if necessary, adjust the valve clearance on the No. 4 cylinder.

11. Rotate the crankshaft 180° clockwise (camshaft pulley turns 90°).

12. Check and, if necessary, adjust the valve clearance on the No. 2 cylinder.

13. Install the cylinder head cover.

6-Cylinder Engine

See Figures 220 through 223.

➡Connect the Honda Diagnostic System (HDS) to the data link connector (DLC), and monitor the Engine Coolant Temperature (ECT) sensor 1. Adjust the valve clearance only when the ECT sensor 1 is less than 100°F (38°C).

1. Remove the cylinder head covers.

2. Set the No. 1 piston at Top Dead Center (TDC). Align the pointer on the front upper cover with the No. 1 piston TDC mark on the front camshaft pulley.

3. Select the correct feeler gauge for the valve clearance you are going to check.

➡Valve Clearance:

- Intake: 0.008–0.009 inches (0.20–0.24 mm)
- Exhaust: 0.011–0.013 inches (0.28–0.32 mm)

➡Apply new engine oil to the nut threads.

4. Insert the feeler gauge between the adjusting screw and the end of the valve stem on the No. 1 cylinder, and slide it back and forth; you should feel a slight amount of drag.

5. If you feel too much or too little drag, loosen the locknut, and turn the adjusting screw until the drag on the feeler gauge is correct.

6. While holding the adjusting screw with the screw driver, tighten the locknut, then recheck the clearance. Repeat the adjustment, if necessary.

➡Specified Torque:

- Intake: 14 ft. lbs. (20 Nm)
- Exhaust: 10 ft. lbs. (14 Nm)

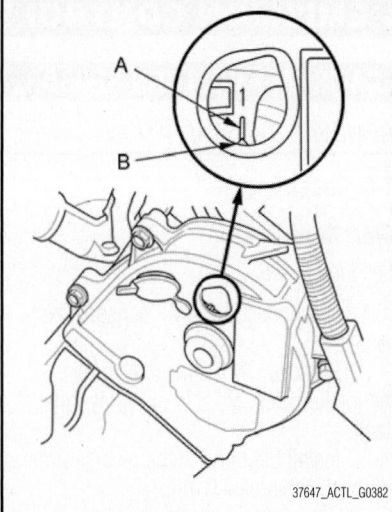

Fig. 220 Set the No. 1 piston at TDC. Align the pointer (A) on the front upper cover with the No. 1 piston TDC mark (B) on the front camshaft pulley

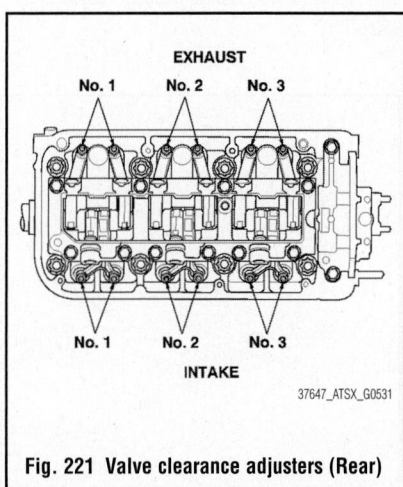

Fig. 221 Valve clearance adjusters (Rear)

➡Apply new engine oil to the nut threads.

7. Rotate the crankshaft clockwise. Align the pointer on the front upper cover with the No. 4 piston TDC mark on the front camshaft pulley.

8. Check and, if necessary, adjust the valve clearance on the No. 4 cylinder.

9. Rotate the crankshaft clockwise. Align the pointer on the front upper cover with the No. 2 piston TDC mark on the front camshaft pulley.

Check and, if necessary, adjust the valve clearance on the No. 2 cylinder.

10. Rotate the crankshaft clockwise. Align the pointer on the front upper cover with the No. 5 piston TDC mark on the front camshaft pulley.

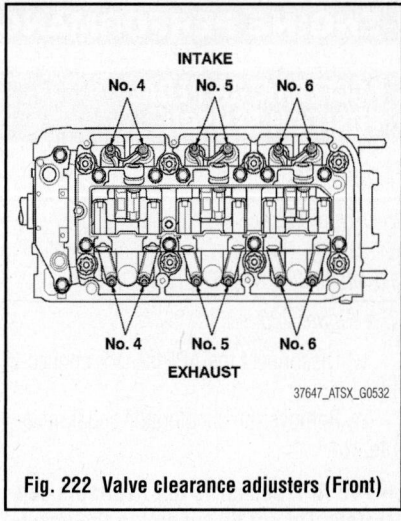

Fig. 222 Valve clearance adjusters (Front)

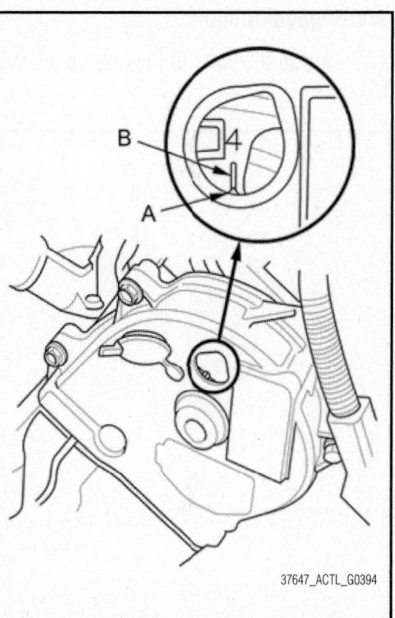

Fig. 223 Rotate the crankshaft clockwise. Align the pointer (A) on the front upper cover with the No. 4 piston TDC mark (B) on the front camshaft pulley

11. Check and, if necessary, adjust the valve clearance on the No. 5 cylinder.

12. Rotate the crankshaft clockwise. Align the pointer on the front upper cover with the No. 3 piston TDC mark on the front camshaft pulley.

13. Check and, if necessary, adjust the valve clearance on the No. 3 cylinder.

14. Rotate the crankshaft clockwise. Align the pointer on the front upper cover with the No. 6 piston TDC mark on the front camshaft pulley.

15. Check and, if necessary, adjust the valve clearance on the No. 6 cylinder.

16. Install the cylinder head covers.

ENGINE PERFORMANCE & EMISSION CONTROLS

ACCELERATOR PEDAL POSITION SENSOR

LOCATION

See Figure 224.

REMOVAL & INSTALLATION

See Figure 225.

1. Disconnect the APP sensor connector.

2. Remove the accelerator pedal module.

➡**The APP sensor is not available separately. Do not disassemble the accelerator pedal module.**

3. Install the parts in the reverse order of removal.

CAMSHAFT POSITION SENSOR

REMOVAL & INSTALLATION

4-Cylinder Engine

CMP Sensor A

See Figure 226.

1. Disconnect the CMP sensor A connector.

2. Remove CMP sensor A from the intake camshaft side of the cylinder head.

3. Install the parts in the reverse order of removal with a new O-ring.

CMP Sensor B

See Figure 227.

1. Disconnect the connector and hoses from the EVAP canister purge valve, then remove the EVAP canister purge valve assembly.

2. Disconnect the CMP sensor B connector.

3. Remove CMP sensor B.

4. Install the parts in the reverse order of removal with a new O-ring.

6-Cylinder Engine

See Figures 228 and 229.

1. Remove the timing belt.

2. Remove the front camshaft pulley (CMP sensor pulse plate).

3. Disconnect the CMP sensor connector, then remove the back cover.

4. Remove the CMP sensor from the back cover.

5. Install the parts in the reverse order of removal.

6. Install the timing belt.

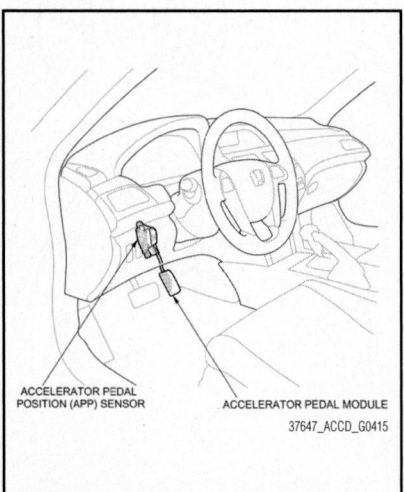

Fig. 224 Accelerator Pedal Position (APP) sensor location

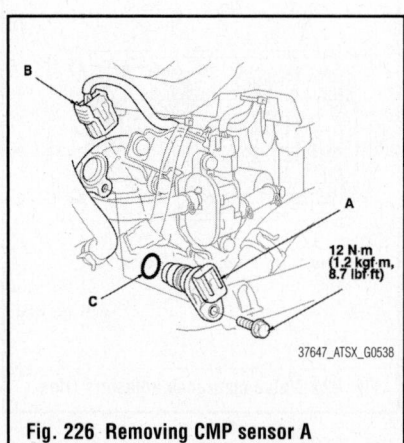

Fig. 226 Removing CMP sensor A

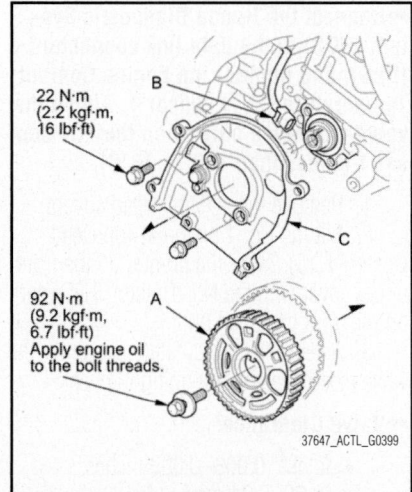

Fig. 228 Remove the front camshaft pulley (CMP sensor pulse plate) (A)

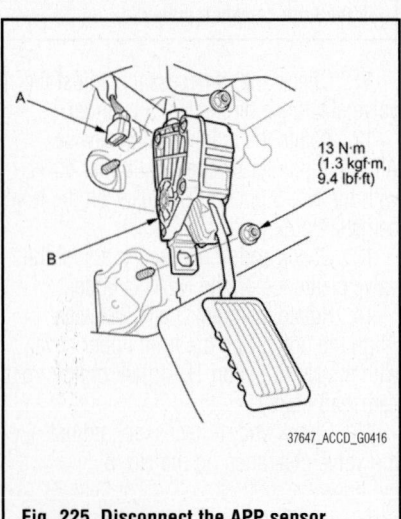

Fig. 225 Disconnect the APP sensor connector (A)

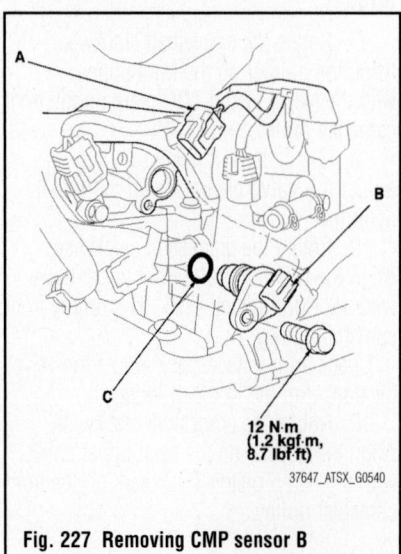

Fig. 227 Removing CMP sensor B

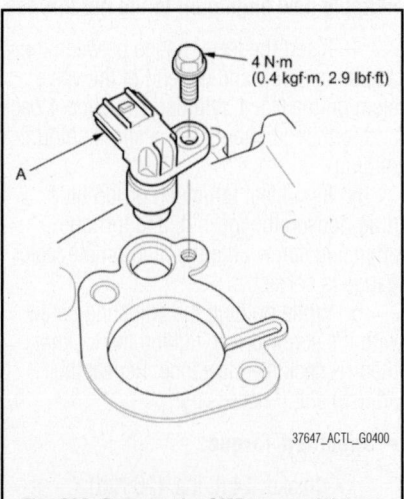

Fig. 229 Remove the CMP sensor (A) from the back cover

7. Do the CKP pattern clear/CKP pattern learn procedure.

CRANKSHAFT POSITION SENSOR

REMOVAL & INSTALLATION

4-Cylinder Engine

See Figure 230.

1. Raise the vehicle on a lift.

➡**Make sure the vehicle is level, because engine oil will drip out when you remove the sensor.**

2. Remove the CKP sensor cover.
3. Disconnect the CKP sensor connector.
4. Remove the CKP sensor.
5. Install the parts in the reverse order of removal with a new O-ring.
6. Do the CKP pattern clear/CKP pattern learn procedure.
7. Check the engine oil level, and add more oil if needed.

6-Cylinder Engine

See Figure 231.

1. Raise the vehicle on a lift.

➡**Make sure the vehicle is level, because engine oil will drip out when you remove the sensor.**

2. Remove the CKP sensor cover.
3. Disconnect the CKP sensor connector.
4. Remove the CKP sensor.
5. Install the parts in the reverse order of removal with a new O-ring.
6. Do the CKP pattern clear/CKP pattern learn procedure.

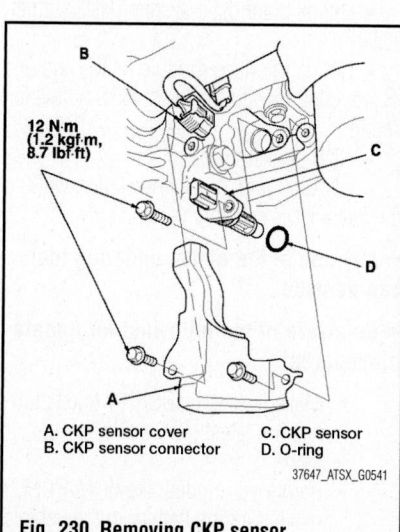

12 N·m
(1.2 kgf·m, 8.7 lbf·ft)

A. CKP sensor cover
B. CKP sensor connector
C. CKP sensor
D. O-ring

37647_ATSX_G0541

Fig. 230 Removing CKP sensor

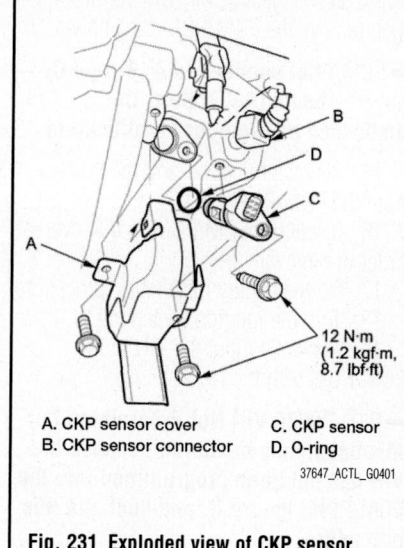

12 N·m
(1.2 kgf·m, 8.7 lbf·ft)

A. CKP sensor cover
B. CKP sensor connector
C. CKP sensor
D. O-ring

37647_ACTL_G0401

Fig. 231 Exploded view of CKP sensor assembly

7. Check for engine oil level, and add more oil if needed.

CKP PATTERN CLEAR/CKP PATTERN LEARN

Clear/Learn Procedure (with the HDS)

1. Connect the HDS to the Data Link Connector (DLC) located under the driver's side of the dashboard.
2. Turn the ignition switch to ON (II).
3. Make sure the HDS communicates with the ECM/PCM and other vehicle systems. If it doesn't, go to the DLC circuit troubleshooting.
4. Select CRANK PATTERN in the ADJUSTMENT MENU with the HDS.
5. Select CRANK PATTERN CLEAR, and clear the CKP pattern.
6. Select CRANK PATTERN LEARNING with the HDS, and follow the screen prompts.

Learn Procedure (without the HDS)

1. Start the engine. Hold the engine speed at 3,000 rpm without load (A/T in P or N, M/T in neutral) until the radiator fan comes on.
2. Test-drive the vehicle on a level road: Decelerate (with the throttle fully closed) from an engine speed of 2,500 rpm down to 1,000 rpm with the Transmission in 2nd.
3. Repeat step 2 several times.
4. Turn the ignition switch to LOCK (0).
5. Turn the ignition switch to ON (II), and wait 30 seconds.

ENGINE CONTROL MODULE (ECM)/POWERTRAIN CONTROL MODULE (PCM)

LOCATION

The ECM/PCM is located in the engine compartment in front of the air cleaner assembly along the left side.

REMOVAL & INSTALLATION

4-Cylinder Engine

See Figures 232 and 233.

Special Tools Required:
• Honda Diagnostic System (HDS) tablet tester
• Honda Interface Module (HIM) and an iN workstation with the latest HDS software version
• HDS pocket tester
• GNA600 and an iN workstation with the latest HDS software version

➡**Any one of above updating tools can be used.**

1. Connect the HDS to the Data Link Connector (DLC) located under the driver's side of the dashboard.
2. Turn the ignition switch to ON (II).
3. Make sure the HDS communicates with the ECM/PCM and other vehicle systems. If it doesn't, go to the DLC circuit troubleshooting. If you are returning from the DLC circuit troubleshooting, skip steps

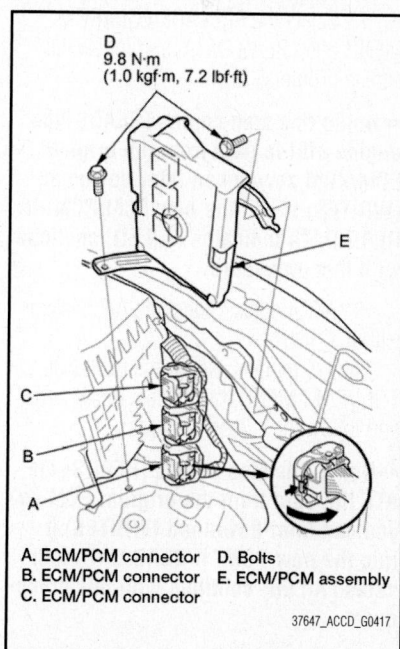

D
9.8 N·m
(1.0 kgf·m, 7.2 lbf·ft)

A. ECM/PCM connector
B. ECM/PCM connector
C. ECM/PCM connector
D. Bolts
E. ECM/PCM assembly

37647_ACCD_G0417

Fig. 232 Disconnect ECM/PCM connectors, then remove the ECM/PCM assembly

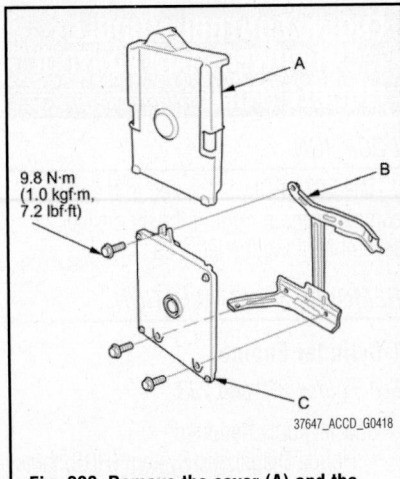

Fig. 233 Remove the cover (A) and the bracket (B) from the ECM/PCM (C)

9.8 N·m
(1.0 kgf·m,
7.2 lbf·ft)

37647_ACCD_G0418

4 through 9, 20 through 25, and 28 through 30, and do this after replacing the ECM/PCM:

- Replace the engine oil and the engine oil filter.
- Replace the ATF (A/T model).
- Clean the throttle body.

4. Select the PGM-FI system with the HDS.

5. Select the INSPECTION MENU with the HDS.

6. Select the ETCS TEST, then select the TP POSITION CHECK, and follow the screen prompts.

➡**If the TP POSITION CHECK indicates FAILED, continue with this procedure.**

7. Select the REPLACE ECM/PCM MENU, then READ DATA, and follow the screen prompts.

➡**Doing this step copies (READS) the engine oil life data from the original ECM/PCM so you can later download (WRITES) it into the new ECM/PCM. If READ DATA indicates FAILED, continue with this procedure.**

8. A/T models: Select the A/T system with the HDS.

9. A/T models: Select the REPLACE TCM/PCM MENU, then select READ DATA, and follow the screen prompts.

➡**Doing this step copies (READS) the ATF life data from the original PCM so you can later download (WRITES) it into the new PCM. If READ DATA indicates FAILED, continue with this procedure.**

10. Turn the ignition switch to LOCK (0).

11. Jump the SCS line with the HDS.

12. Do the battery removal procedure.

13. Remove the bolts.

14. Disconnect ECM/PCM connectors, then remove the ECM/PCM assembly.

➡**ECM/PCM connectors A, B, and C have symbols (A=□, B=△, C=○) embossed on them for identification.**

15. Remove the cover and the bracket from the ECM/PCM.

16. Install the ECM/PCM in the reverse order of removal.

17. Do the battery installation procedure.

18. Turn the ignition switch to ON (II).

19. Manually input the VIN to the ECM/PCM with the HDS.

➡**DTC P0630 VIN Not Programmed or Mismatch may be stored because the VIN has not been programmed into the ECM/PCM; ignore it, and continue this procedure.**

20. If the READ DATA (engine oil life) failed in step 7, go to step 23 (A/T model) or step 26 (M/T model). Otherwise, go to step 21.

21. Select the PGM-FI system with the HDS.

22. Select the REPLACE ECM/PCM MENU, then select WRITE DATA, and follow the screen prompts.

➡**If the WRITE DATA indicates FAILED, continue with this procedure.**

23. A/T models: If the READ DATA (ATF life) failed in step 9, go to step 26. Otherwise go to step 24.

24. A/T models: Select the A/T SYSTEM with the HDS.

25. A/T models: Select the REPLACE TCM/PCM MENU, then select WRITE DATA, and follow the screen prompts.

➡**If the WRITE DATA indicates FAILED, continue with this procedure.**

26. Select IMMOBI system with the HDS.

27. Enter the immobilizer ECM/PCM code that you got from the iN, and use the ECM/PCM replacement procedure in the IMMOBI MENU of the HDS; it allows you to start the engine.

28. If the TP POSITION CHECK failed in step 6, clean the throttle body, then go to step 29.

29. If the READ DATA failed in step 7 or the WRITE DATA failed in step 22, replace the engine oil and engine oil filter, then go to step 30 (A/T model) or step 31 (M/T model).

30. If the READ DATA failed in step 9 or the WRITE DATA failed in step 25, replace the ATF, then go to step 31.

31. Select PGM-FI system, and reset the ECM/PCM with the HDS.

32. Update the ECM/PCM if it does not have the latest software.

33. Do the ECM/PCM idle learn procedure.

34. Do the CKP pattern clear/CKP pattern learn procedure.

ECM/PCM IDLE LEARN PROCEDURE

The idle learn procedure must be done so the ECM/PCM can learn the engine idle characteristics.

Do the idle learn procedure whenever you do any of these actions:
- Replace the ECM/PCM.
- Reset the ECM/PCM.
- Update the ECM/PCM.
- Replace or clean the throttle body.
- Disassemble the engine or the transmission.

➡**Clearing the DTCs with the HDS does not require you to do the idle learn procedure.**

1. Make sure all electrical items (the A/C, the audio system, the lights, etc.) are off.

2. Reset the ECM/PCM with the HDS.

3. Turn the ignition switch to ON (II), and wait 2 seconds.

4. Start the engine. Hold the engine speed at 3,000 rpm without load (A/T in P or N, M/T in neutral) until the radiator fan comes on, or until the engine coolant temperature reaches 194°F (90°C).

5. Let the engine idle for about 5 minutes with the throttle fully closed.

➡**If the radiator fan comes on, do not include its running time in the 5 minutes.**

ECM/PCM UPDATE

Special Tools Required:
- Honda diagnostic system (HDS) tablet tester
- Honda interface module (HIM) and an iN workstation with the latest HDS software version
- HDS pocket tester
- GNA600 and an iN workstation with the latest HDS software version

➡**Any one of the above updating tools can be used.**

➡**Be aware of the following for update procedures:**

- Make sure the HDS/iN workstation has the latest HDS software version.
- Before you update the ECM/PCM, make sure the battery in the vehicle

is fully charged, and connect a jumper battery (not a battery charger) to maintain system voltage.

- Never turn the ignition switch to ACC (I) or LOCK (0) during the update. If there is a problem with the update, leave the ignition switch in ON (II).
- To prevent ECM/PCM damage, do not operate anything electrical (headlights, audio system, brakes, A/C, power windows, moonroof (if equipped), door locks, etc.) during the update.
- To ensure the latest program is installed, do an ECM/PCM update whenever the ECM/PCM is substituted or replaced.
- You cannot update an ECM/PCM with a program it already has. It will only accept a new program.
- High temperature in the engine compartment might cause the ECM/PCM to become too hot to run the update. If the engine has been running before this procedure, open the hood and cool the engine compartment.
- If you need to diagnose the Honda Interface Module (HIM) because the HIM's red (#3) light came on or was flashing during the update, leave the ignition switch in ON (II) when you disconnect the HIM from the Data Link Connector (DLC). This will prevent ECM/PCM damage.

1. Turn the ignition switch to ON (II), but do not start the engine.

2. Connect the HDS to the Data Link Connector (DLC) located under the driver's side of the dashboard.

3. Make sure the HDS communicates with the ECM/PCM and other vehicle systems. If it doesn't, go to the DLC circuit troubleshooting. If you are returning from the DLC circuit troubleshooting, skip steps 4 and 5, and clean the throttle body after updating the ECM/PCM.

4. Select the INSPECTION MENU with the HDS.

5. Select the ETCS TEST, then select the TP POSITION CHECK, and follow the HDS screen prompts.

➡**If the TP POSITION CHECK indicates FAILED, continue this procedure.**

6. Exit the HDS diagnostic system, then select the update mode, and follow the screen prompts to update the ECM/PCM.

7. If the software in the ECM/PCM is

the latest, disconnect the HDS/HIM/GNA600 from the DLC, and go back to the procedure that you were doing. If the software in the ECM/PCM is not the latest, follow the instructions on the screen. If prompted to choose the PGM-FI system or the A/T system (A/T), make sure you update both.

➡**If the ECM/PCM update system requires you to cool the ECM/PCM, follow the instructions on the screen. If you run into a problem during the update procedure (programming takes over 15 minutes, status bar goes over 100 %, D (A/T) or immobilizer indicator flashes, HDS tablet freezes, etc.), follow these steps to minimize the chance of damaging the ECM/PCM:**

- Leave the ignition switch in ON (II).
- Connect a jumper battery (do not connect a battery charger).
- Shut down the HDS.
- Disconnect the HDS from the DLC.
- Reboot the HDS.
- Reconnect the HDS to the DLC, and do the update again.

8. If the TP POSITION CHECK failed in step 5, clean the throttle body.

9. Do the ECM/PCM idle learn procedure.

10. Do the CKP pattern clear/CKP pattern learn procedure.

ENGINE COOLANT TEMPERATURE SENSOR

REMOVAL & INSTALLATION

4-Cylinder Engine

ECT Sensor 1

See Figure 234.

1. Drain the engine coolant.
2. Disconnect the ECT sensor 1 connector.
3. Remove ECT sensor 1.
4. Install the parts in the reverse order of removal with a new O-ring, then refill the radiator with engine coolant.

ECT Sensor 2

See Figure 235.

1. Remove the front splash shield.
2. Drain the engine coolant.
3. Disconnect the ECT sensor 2 connector, then remove ECT sensor 2.
4. Install ECT sensor 2 with a new O-ring.
5. Install the front splash shield.
6. Refill the radiator with engine coolant.

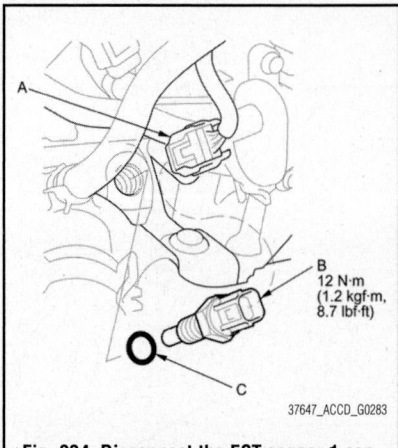

Fig. 234 Disconnect the ECT sensor 1 connector (A), then remove ECT sensor 1 (B) and O-ring (C)

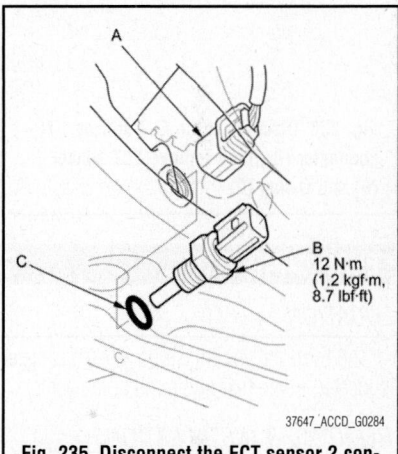

Fig. 235 Disconnect the ECT sensor 2 connector (A), then remove ECT sensor 2 (B) and O-ring (C)

6-Cylinder Engine

ECT Sensor 1

See Figure 236.

1. Drain the engine coolant.
2. Remove the engine cover.
3. Disconnect the ECT sensor 1 2P connector.
4. Remove ECT sensor 1.
5. Install the parts in the reverse order of removal with a new O-ring, then refill the radiator with engine coolant.

ECT Sensor 2

See Figure 235.

1. Remove the front splash shield.
2. Drain the engine coolant.
3. Disconnect the ECT sensor 2 connector, then remove ECT sensor 2.
4. Install ECT sensor 2 with a new O-ring.
5. Install the front splash shield.
6. Refill the radiator with engine coolant.

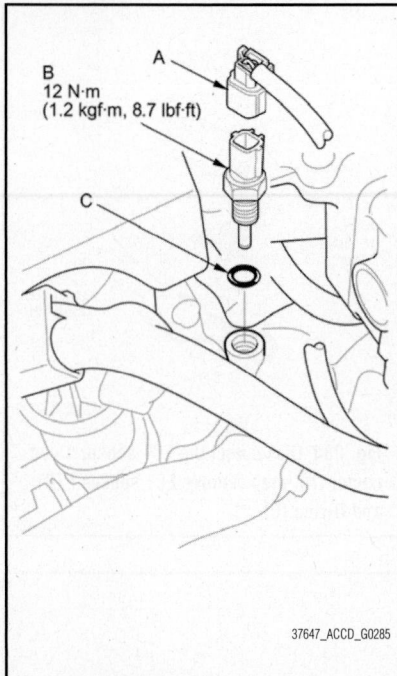

Fig. 236 Disconnect the ECT sensor 1 2P connector (A), then remove ECT sensor 1 (B) and O-ring (C)

EVAP CANISTER

LOCATION

The Evaporative Emissions (EVAP) canister is located in the rear subframe of the vehicle.

REMOVAL & INSTALLATION

4-Cylinder Engine

See Figures 237 and 238.

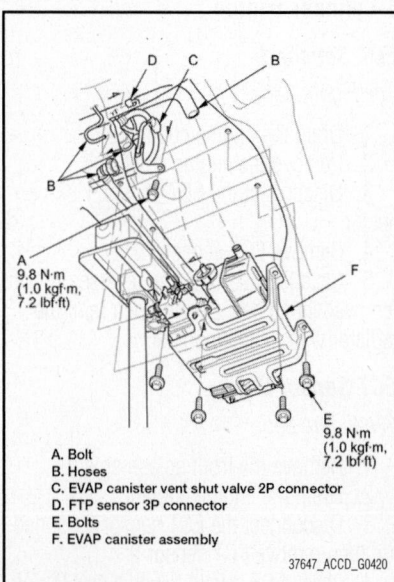

A. Bolt
B. Hoses
C. EVAP canister vent shut valve 2P connector
D. FTP sensor 3P connector
E. Bolts
F. EVAP canister assembly

Fig. 237 Remove the EVAP canister assembly

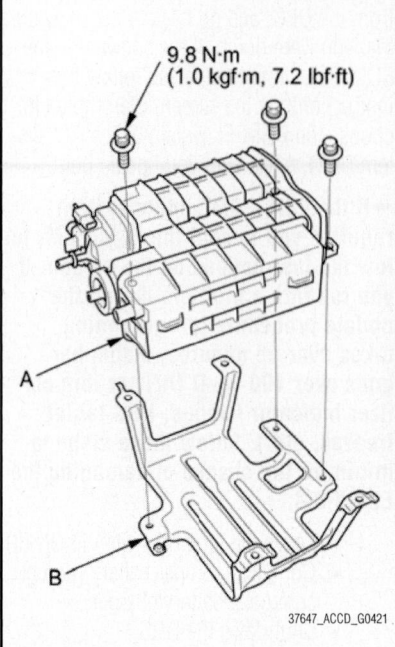

Fig. 238 Remove the EVAP canister (A) from the EVAP canister bracket (B)

1. Raise the vehicle on a lift.
2. Remove the wheel sensor harness clamps.
3. Support the rear subframe with a transmission jack and a wooden block as shown.
4. Remove the rear subframe mounting bolts.
5. Lower the transmission jack and the rear subframe about 50 mm.

➡**Be careful not to damage the connecting parts.**

6. Remove the bolt, and disconnect the hoses, the EVAP canister vent shut valve 2P connector, and the FTP sensor 3P connector.
7. Remove the bolts, then remove the EVAP canister assembly.
8. Remove the EVAP canister from the EVAP canister bracket.

To install:

9. Install the EVAP canister assembly to the body.

➡**Attach the bracket arm to the body.**

10. Install the parts in the reverse order of removal. Use new bolts when you install the rear subframe.
11. Check the wheel alignment.

6-Cylinder Engine

See Figures 238 and 239.

1. Raise the vehicle on a lift.
2. Remove the wheel sensor harness clamps.
3. Support the rear subframe with a transmission jack and a wooden block.
4. Remove the rear subframe mounting bolts.
5. Lower the transmission jack and the rear subframe about 50 mm.

➡**Be careful not to damage the connecting parts.**

6. Remove the bolt, and disconnect the hoses, the EVAP canister vent shut valve 2P connector, and the FTP sensor 3P connector.
7. Remove the bolts, then remove the EVAP canister assembly.
8. Remove the EVAP canister from the EVAP canister bracket.

To install:

9. Install the EVAP canister assembly to the body.
10. Install the parts in the reverse order of removal. Use new bolts when you install the rear subframe.
11. Check the wheel alignment.

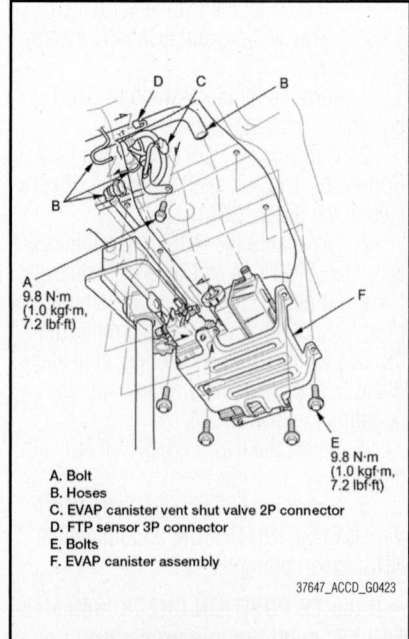

A. Bolt
B. Hoses
C. EVAP canister vent shut valve 2P connector
D. FTP sensor 3P connector
E. Bolts
F. EVAP canister assembly

Fig. 239 Remove the bolt, and disconnect the hoses, the EVAP canister vent shut valve 2P connector, and the FTP sensor 3P connector

EVAP CANISTER PURGE CONTROL VALVE

REMOVAL & INSTALLATION

4-Cylinder Engine

See Figure 240.

1. Disconnect the EVAP canister purge valve 2P connector.
2. Disconnect the hoses, then remove the EVAP canister purge valve.
3. Install the parts in the reverse order of removal.

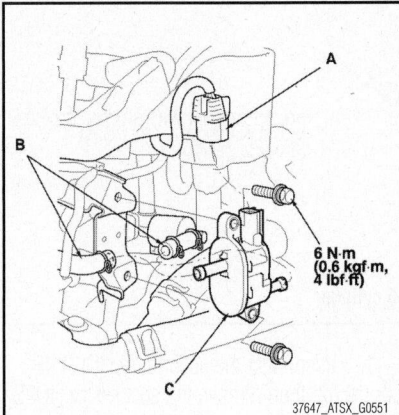

Fig. 240 Disconnect the EVAP canister purge valve 2P connector (A), the hoses (B), then remove the EVAP canister purge valve (C)

6-Cylinder Engine

See Figures 241 and 242.

1. Disconnect the hoses and the EVAP canister purge valve 2P connector.

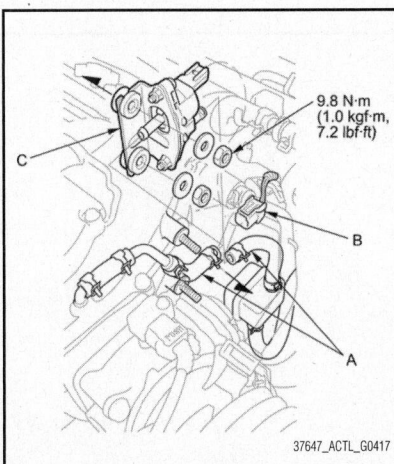

Fig. 241 Disconnect the hoses (A) and the EVAP canister purge valve 2P connector (B)

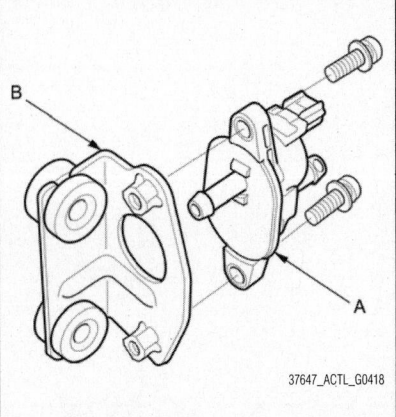

Fig. 242 Remove the EVAP canister purge valve (A) from the bracket (B)

2. Remove the EVAP canister purge valve assembly.
3. Remove the EVAP canister purge valve from the bracket.
4. Install the parts in the reverse order of removal.

FUEL TANK (EVAP) PRESSURE (FTP) SENSOR

LOCATION

The FTP sensor is attached to the EVAP canister.

REMOVAL & INSTALLATION

See Figures 243 and 244.

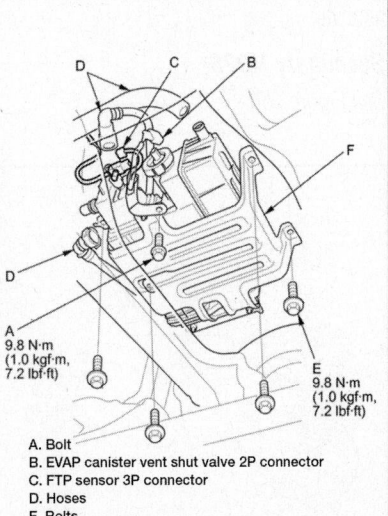

A. Bolt
B. EVAP canister vent shut valve 2P connector
C. FTP sensor 3P connector
D. Hoses
E. Bolts
F. EVAP canister assembly

Fig. 243 Remove the bolt, and disconnect the EVAP canister vent shut valve 2P connector, the FTP sensor 3P connector, and the hoses

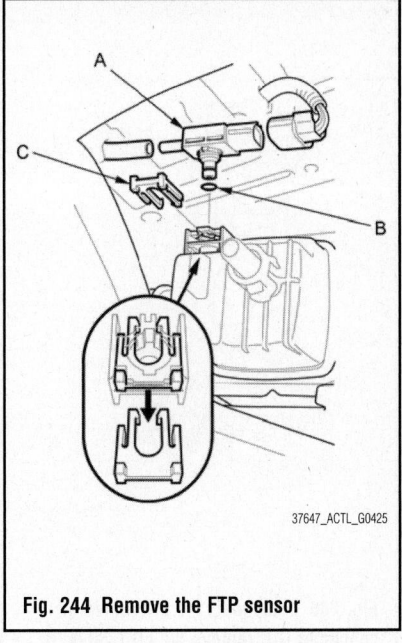

Fig. 244 Remove the FTP sensor

1. Remove the bolt, and disconnect the EVAP canister vent shut valve 2P connector, the FTP sensor 3P connector, and the hoses.
2. Remove the bolts, and move the EVAP canister assembly to the rear.
3. Remove the FTP sensor.
4. Install the parts in the reverse order of removal with a new O-ring and a new retainer.

EGR VALVE

LOCATION

6-Cylinder Engine

The EGR valve is located at the front left of the engine.

REMOVAL & INSTALLATION

6-Cylinder Engine

See Figures 245 and 246.

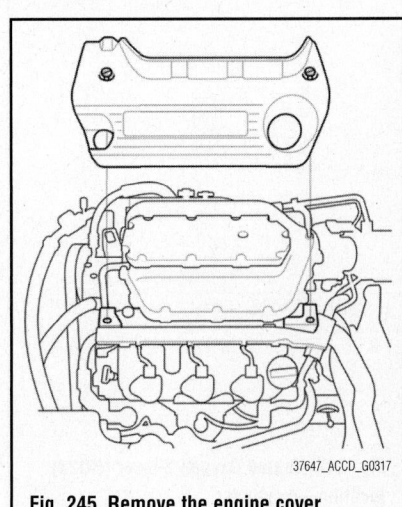

Fig. 245 Remove the engine cover

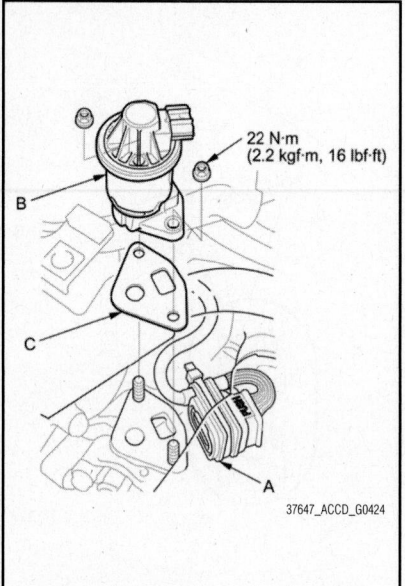

Fig. 246 Disconnect the EGR valve 5P connector (A), remove the EGR valve (B) and gasket (C)

1. Remove the engine cover.
2. Disconnect the EGR valve 5P connector.
3. Remove the EGR valve.
4. Install the parts in the reverse order of removal with a new gasket.

HEATED OXYGEN SENSOR

LOCATION

See Figures 247 and 248.

Refer to the accompanying illustrations.

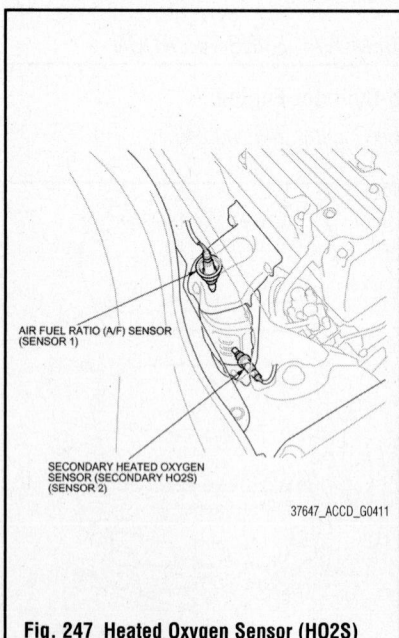

Fig. 247 Heated Oxygen Sensor (HO2S) location—4 cylinder

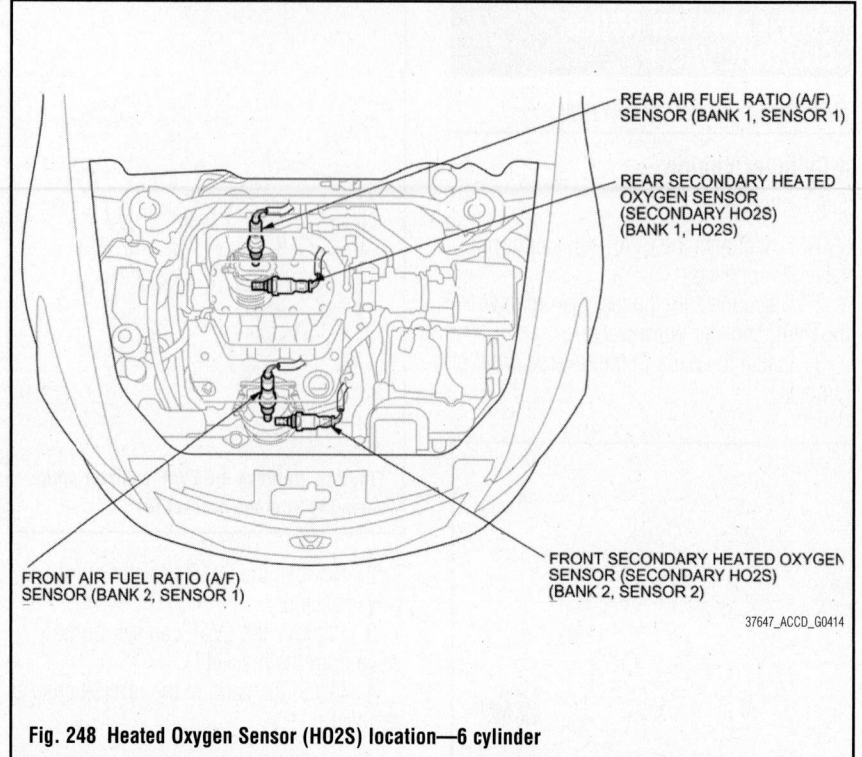

Fig. 248 Heated Oxygen Sensor (HO2S) location—6 cylinder

REMOVAL & INSTALLATION

4-Cylinder Engine

Air Fuel Ratio (AFR) Sensor

See Figure 249.

1. Disconnect the A/F sensor 4P connector, then remove the A/F sensor.
2. Install the parts in the reverse order of removal.

Secondary HO2S

See Figure 250.

1. Disconnect the secondary HO2S 4P connector, then remove the secondary HO2S.
2. Install the parts in the reverse order of removal.

6-Cylinder Engine

Air Fuel Ratio (AFR) Sensor—Front Bank (Bank2)

See Figure 251.

1. Disconnect the front A/F sensor connector, then remove the A/F sensor.
2. Install the parts in the reverse order of removal.

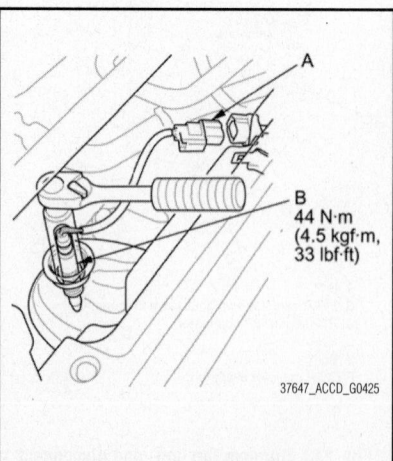

Fig. 249 Disconnect the A/F sensor 4P connector (A), then remove the A/F sensor (B)

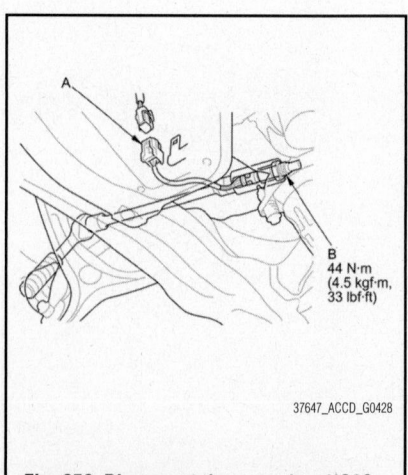

Fig. 250 Disconnect the secondary HO2S 4P connector (A), then remove the secondary HO2S (B)

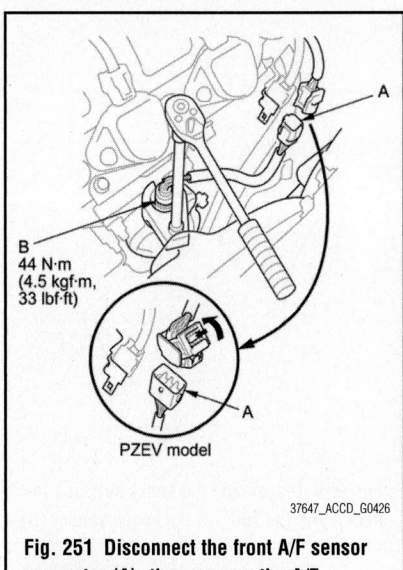

Fig. 251 Disconnect the front A/F sensor connector (A), then remove the A/F sensor (B)

Air Fuel Ratio (AFR) Sensor—Front Bank (Bank1)

See Figure 252.

1. Disconnect the rear A/F sensor connector, then remove the A/F sensor.

2. Install the parts in the reverse order of removal.

Secondary HO2S —Front Bank (Bank2)

See Figure 253 and 254.

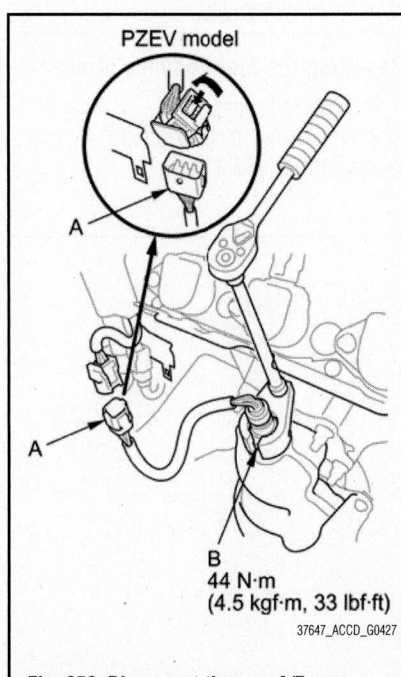

Fig. 252 Disconnect the rear A/F sensor connector (A), then remove the A/F sensor (B)

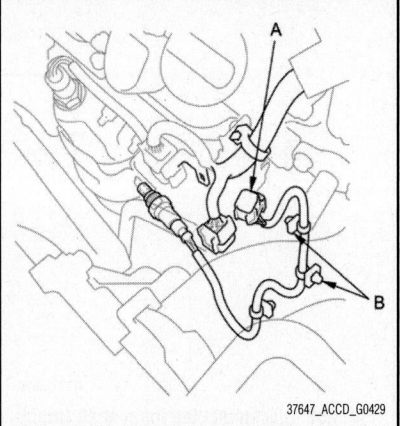

Fig. 253 Disconnect the front secondary HO2S 4P connector (A), and remove the harness clamps (B)

1. Disconnect the front secondary HO2S 4P connector, and remove the harness clamps.

2. Raise the vehicle on a lift.

3. Remove the harness clamp, then remove the front secondary HO2S.

4. Install the parts in the reverse order of removal.

Secondary HO2S —Rear Bank (Bank 1)

See Figure 255.

1. Raise the vehicle on a lift.

2. Disconnect the rear secondary HO2S 4P connector, and remove the harness clamps.

3. Remove the rear secondary HO2S.

4. Install the parts in the reverse order of removal.

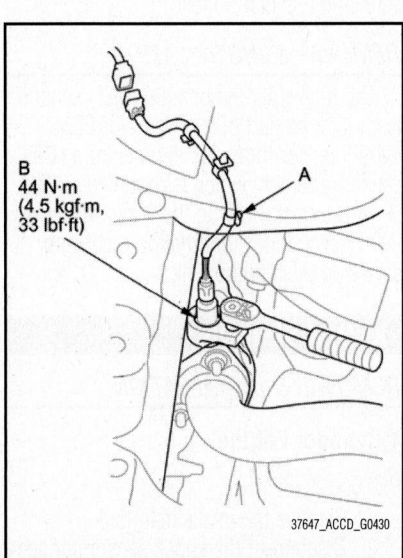

Fig. 254 Remove the harness clamp (A), then remove the front secondary HO2S (B)

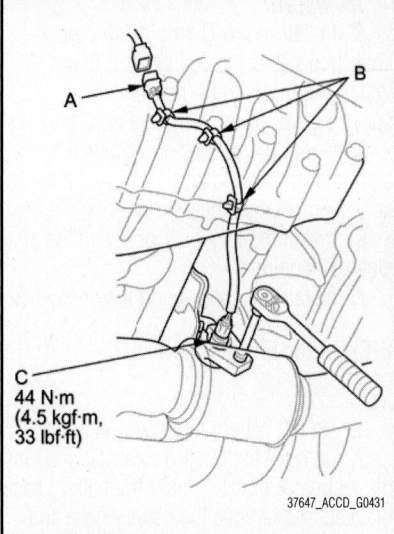

Fig. 255 Disconnect the rear secondary HO2S 4P connector (A), and remove the harness clamps (B) and the rear secondary HO2S (C)

INPUT SPEED SENSOR

REMOVAL & INSTALLATION

4-Cylinder Engine
See Figure 256.

1. Remove the nut securing the under-hood fuse/relay box, and swing it out of the way.

2. Remove the intake air duct and the air cleaner housing.

3. Disconnect the input shaft (mainshaft) speed sensor connector, and remove the input shaft (mainshaft) speed sensor and O-ring.

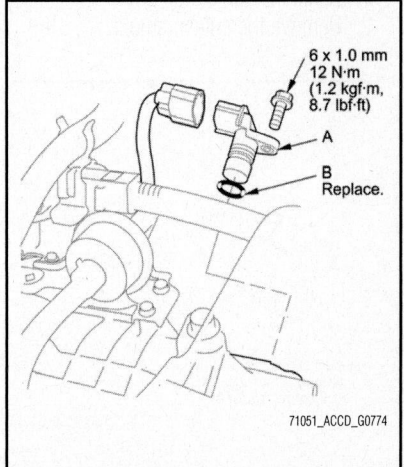

Fig. 256 Disconnect the input shaft (mainshaft) speed sensor connector, and remove the input shaft (mainshaft) speed sensor (A) and O-ring (B)

To install:

4. Install a new O-ring on the input shaft (mainshaft) speed sensor, then install the input shaft (mainshaft) speed sensor in the transmission housing.

5. Check the connector for corrosion, dirt, or oil, and clean or repair if necessary, then connect the connector securely.

6. Install the intake air duct and the air cleaner housing.

7. Install the under-hood fuse/relay box.

6-Cylinder Engine With A/T

See Figure 257.

1. Do the battery removal procedure.

2. Remove the 6 x 1.0 mm bolt securing the resonator bracket under the battery base bracket, and remove the battery base and the battery base bracket in the engine compartment.

3. Disconnect the input shaft (mainshaft) speed sensor connector, and remove the input shaft (mainshaft) speed sensor and O-ring.

To install:

4. Install a new O-ring on a new input shaft (mainshaft) speed sensor, then install the input shaft (mainshaft) speed sensor.

5. Check the connector for corrosion, dirt, or oil, and clean or repair if necessary, then connect the connector securely.

6. Install the battery base bracket and the battery base, and secure the resonator bracket under the battery base bracket with the bolt.

7. Do the battery installation procedure.

6-Cylinder Engine With M/T

See Figure 258.

1. Raise the vehicle on a lift.
2. Remove the splash shield.

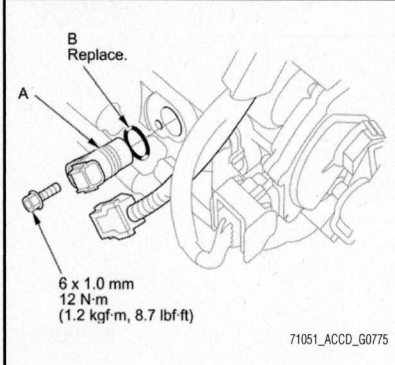

Fig. 257 Disconnect the input shaft (mainshaft) speed sensor connector, and remove the input shaft (mainshaft) speed sensor (A) and O-ring (B)

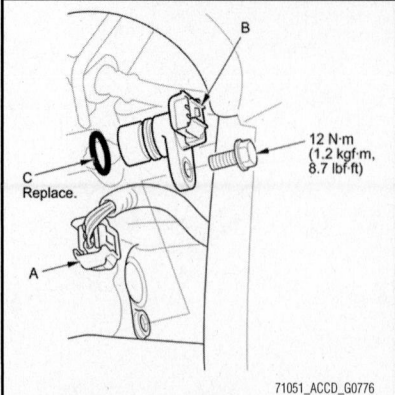

Fig. 258 Disconnect the input shaft (mainshaft) speed sensor connector (A), remove the input shaft (mainshaft) speed sensor (B) and O-ring (C)

3. Disconnect the input shaft (mainshaft) speed sensor connector.

4. Remove the input shaft (mainshaft) speed sensor and O-ring.

To install:

5. Install the parts in the reverse order of removal with a new O-ring.

6. If needed, refill the manual transmission fluid.

INTAKE AIR TEMPERATURE SENSOR

LOCATION

The Intake Air Temperature (IAT) sensor is an integral part of the Mass Air Flow (MAF) sensor/Intake Air Temperature (IAT) sensor assembly which is mounted on the air intake duct. Refer to the Mass Air Flow (MAF) section for information regarding servicing this component.

REMOVAL & INSTALLATION

The Intake Air Temperature (IAT) sensor is an integral part of the Mass Air Flow (MAF) sensor/Intake Air Temperature (IAT) sensor assembly which is mounted on the air intake duct. Refer to the Mass Air Flow (MAF) section for information regarding servicing this component.

KNOCK SENSOR

REMOVAL & INSTALLATION

4-Cylinder Engine

See Figure 259.

1. Remove the intake manifold.
2. Disconnect the knock sensor connector.
3. Remove the knock sensor.

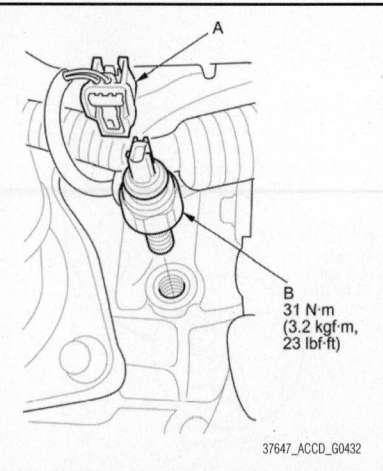

Fig. 259 Disconnect the knock sensor connector (A) and remove the knock sensor (B)

4. Install the parts in the reverse order of removal.

6-Cylinder Engine

See Figure 260.

1. Remove the intake manifold.
2. Remove the injector base.
3. Disconnect the knock sensor connector.
4. Remove the knock sensor.
5. Install the parts in the reverse order of removal.

MAINTENANCE REMINDER LIGHTS

RESET PROCEDURE

Resetting the Maintenance Minder

Note the following:
• The vehicle must be stopped to reset the Maintenance Minder.

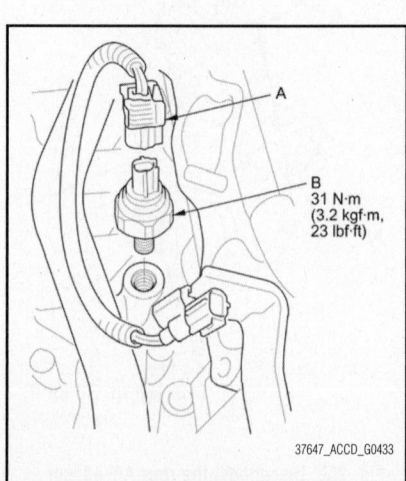

Fig. 260 Disconnect the knock sensor connector

• If a required service is done and the Maintenance Minder is not reset, or if the Maintenance Minder is reset without doing the service, the system will not show the proper maintenance timing. This can lead to serious mechanical problems because there will be no accurate record of when the required maintenance is needed.

• The engine oil life and the maintenance item(s) can be independently reset with the HDS.

1. Turn the ignition switch to ON (II).

2. Press the SELECT/RESET knob repeatedly until the engine oil life indicator is displayed.

3. Press and hold the SELECT/RESET knob for about 10 seconds. The engine oil life indicator and the maintenance item code(s) will blink.

➡**If you are resetting the Maintenance Minder when the engine oil life is more than 15 %, make sure any maintenance item(s) requiring service are done before resetting the display.**

4. Press and hold the SELECT/RESET knob for another 5 seconds. The maintenance item code(s) will disappear, and the engine oil life will reset to "100."

Resetting Individual Maintenance Items

1. Connect the Honda diagnostic System (HDS) to the data link connector (DLC).

2. Turn the ignition switch to ON (II).

3. Make sure the HDS communicates with the vehicle and the engine control module/powertrain control module (ECM/PCM). If it doesn't communicate, troubleshoot the DLC circuit.

4. Select GAUGES in the BODY ELECTRICAL with the HDS.

5. Select ADJUSTMENT in the GAUGES with the HDS.

6. Select MAINTENANCE MINDER in the ADJUSTMENT with the HDS.

7. Select RESET in the MAINTENANCE MINDER with the HDS.

8. Select the individual maintenance item you wish to reset with the HDS.

MANIFOLD ABSOLUTE PRESSURE SENSOR

REMOVAL & INSTALLATION

4-Cylinder Engine

See Figure 261.

1. Disconnect the MAP sensor connector.

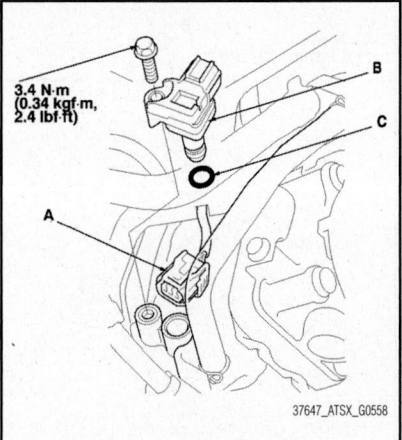

Fig. 261 Disconnect the MAP sensor connector (A) and remove the MAP sensor (B)

2. Remove the MAP sensor.

3. Install the parts in the reverse order of removal with a new O-ring.

6-Cylinder Engine

See Figure 262.

1. Disconnect the MAP sensor connector.

2. Remove the screw.

3. Remove the MAP sensor.

4. Install the parts in the reverse order of removal with a new O-ring.

MASS AIRFLOW/INTAKE AIR TEMPERATURE SENSOR

LOCATION

The MAF/IAT sensor is located on the engine air intake.

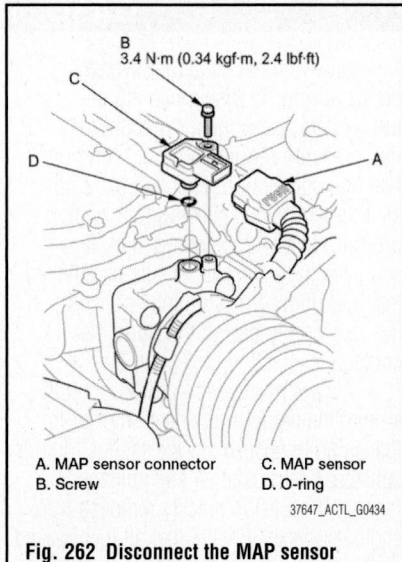

A. MAP sensor connector
B. Screw
C. MAP sensor
D. O-ring

37647_ACTL_G0434

Fig. 262 Disconnect the MAP sensor connector

REMOVAL & INSTALLATION

4-Cylinder Engine

See Figure 263.

1. Disconnect the MAF sensor/IAT sensor connector.

2. Remove the screws.

3. Remove the MAF sensor/IAT sensor.

4. Install the parts in the reverse order of removal with a new O-ring.

6-Cylinder Engine

See Figure 264.

1. Disconnect the MAF/IAT sensor 5P connector.

2. Remove the screws.

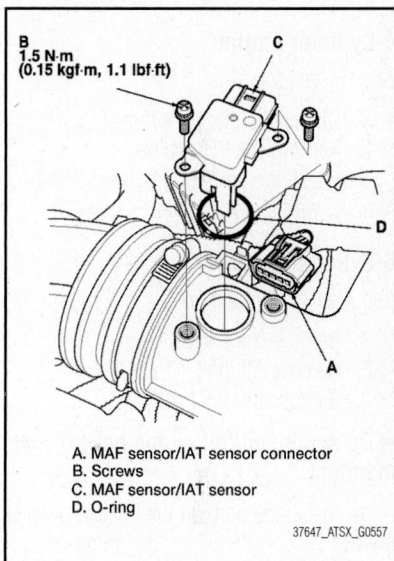

A. MAF sensor/IAT sensor connector
B. Screws
C. MAF sensor/IAT sensor
D. O-ring

37647_ATSX_G0557

Fig. 263 Disconnect the MAF sensor/IAT sensor connector

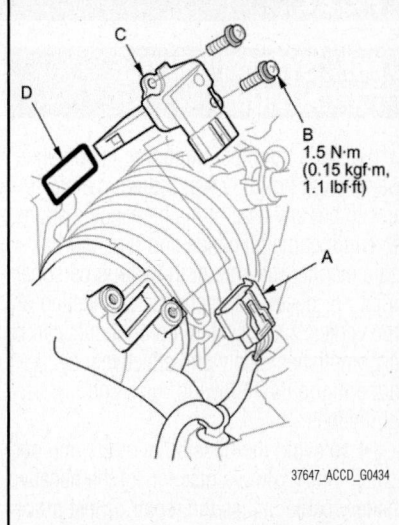

37647_ACCD_G0434

Fig. 264 Disconnect the MAF/IAT sensor 5P connector

3. Remove the MAF/IAT sensor.

4. Install the parts in the reverse order of removal with a new O-ring.

PCV VALVE

LOCATION

The PCV system consists of the breather tube and the PCV valve. The breather tube connects the crankcase to a contained fresh air source such as the air cleaner. Air passes into this tube and into the engine after being filtered through a spark arrestor screen in order to prevent a the possibility of an explosion within the engine in the case of a backfire.

REMOVAL & INSTALLATION

4-Cylinder Engine

See Figure 265.

1. Disconnect the PCV hose.
2. Remove the PCV valve.
3. Install the parts in the reverse order of removal with a new washer.

6-Cylinder Engine

See Figures 266 and 267.

1. Remove the engine cover.
2. Remove the bolt.
3. Remove the PCV valve.

➡**Do not to spill oil on the hot exhaust manifold.**

4. Install the parts in the reverse order of removal.

➡**When installing a new PCV valve, make sure the O-rings are in place.**

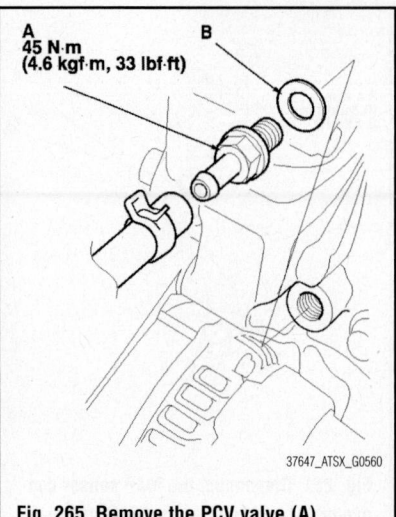

Fig. 265 Remove the PCV valve (A)

37647_ATSX_G0560

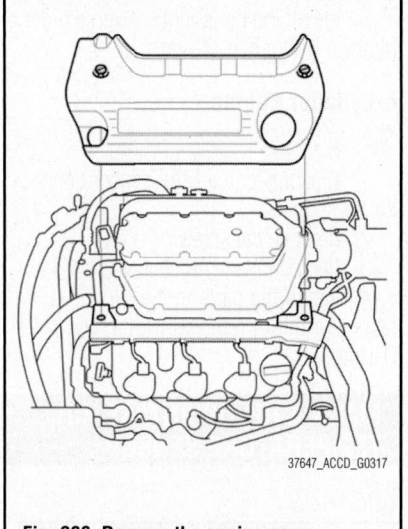

Fig. 266 Remove the engine cover

37647_ACCD_G0317

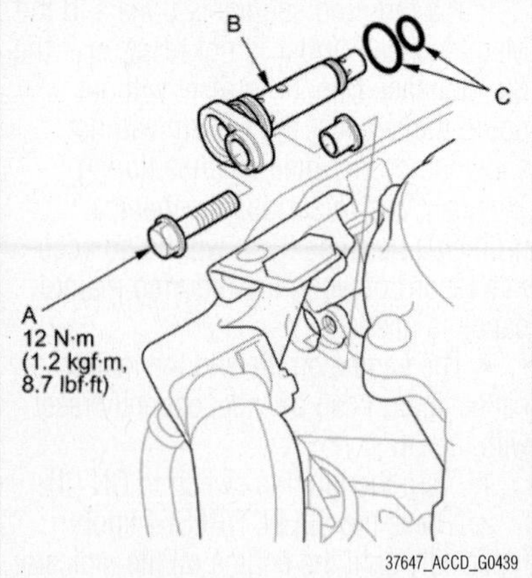

Fig. 267 Remove the bolt (A), the PCV valve (B) and the O-rings (C)

37647_ACCD_G0439

When installing a used PCV valve, use new O-rings.

THROTTLE CONTROL ACTUATOR

LOCATION

The Throttle Control Actuator is an integral part of the throttle body. Refer to Throttle Body in the Fuel Systems Section.

THROTTLE POSITION SENSOR

LOCATION

The Throttle Position Sensor is an integral part of the throttle body. Refer to Throttle Body in the Fuel Systems Section.

FUEL   GASOLINE FUEL INJECTION SYSTEM

FUEL SYSTEM SERVICE PRECAUTIONS

Safety is the most important factor when performing not only fuel system maintenance but any type of maintenance. Failure to conduct maintenance and repairs in a safe manner may result in serious personal injury or death. Maintenance and testing of the vehicle's fuel system components can be accomplished safely and effectively by adhering to the following rules and guidelines.

• To avoid the possibility of fire and personal injury, always disconnect the negative battery cable unless the repair or test procedure requires that battery voltage be applied.

• Always relieve the fuel system pressure prior to disconnecting any fuel system component (injector, fuel rail, pressure regulator, etc.), fitting or fuel line connection. Exercise extreme caution whenever relieving fuel system pressure to avoid exposing skin, face and eyes to fuel spray. Please be advised that fuel under pressure may penetrate the skin or any part of the body that it contacts.

• Always place a shop towel or cloth around the fitting or connection prior to loosening to absorb any excess fuel due to spillage. Ensure that all fuel spillage (should it occur) is quickly removed from engine surfaces. Ensure that all fuel soaked cloths or towels are deposited into a suitable waste container.

• Always keep a dry chemical (Class B) fire extinguisher near the work area.

• Do not allow fuel spray or fuel vapors to come into contact with a spark or open flame.

• Always use a back-up wrench when loosening and tightening fuel line connection fittings. This will prevent unnecessary stress and torsion to fuel line piping.

• Always replace worn fuel fitting O-rings with new Do not substitute fuel hose or equivalent where fuel pipe is installed.

Before servicing the vehicle, make sure to also refer to the precautions in the beginning of this section as well.

RELIEVING FUEL SYSTEM PRESSURE

WITH THE HDS

✱✱ CAUTION

Before disconnecting fuel lines or hoses, relieve pressure from the system by disabling the fuel pump and disconnecting the fuel line/quick connect fitting in the engine compartment.

1. Connect the HDS to the Data Link Connector (DLC) located under the driver's side of the dashboard.
2. Turn the ignition switch to ON (II).
3. Make sure the HDS communicates with the ECM/PCM. If it doesn't, go to the DLC circuit troubleshooting.
4. Turn the ignition switch to LOCK (0).
5. Remove the fuel fill cap to relieve the pressure in the fuel tank.
6. Turn the ignition switch to ON (II).
7. From the INSPECTION MENU of the HDS, select Fuel Pump OFF, then start the engine, and let it idle until it stalls.

➡**Do not allow the engine to idle above 1,000 rpm or the ECM/PCM will continue to operate the fuel pump. Pending or Confirmed DTC may be set during this procedure. Check for DTCs, and clear them as needed.**

8. Turn the ignition switch to LOCK (0).
9. Do the battery terminal disconnection procedure.
10. Remove the cover and the quick-connect fitting cover.
11. Check the quick-connect fitting for dirt, and clean it if needed.
12. Place a rag or a shop towel over the quick-connect fitting.
13. Disconnect the quick-connect fitting.
   a. Hold the connector with one hand, and squeeze the retainer tabs with the other hand to release them from the locking tabs.
   b. Pull the connector off.

➡**Be careful not to damage the line or other parts. Do not use tools. If the connector does not move, keep the retainer tabs pressed down, and alternately pull and push the connector until it comes off easily. Do not remove the retainer from the line; once removed, the retainer must be replaced with a new one.**

14. After disconnecting the quick-connect fitting, check it for dirt or damage.

15. Do the battery terminal reconnection procedure.

WITHOUT THE HDS

✱✱ CAUTION

Before disconnecting fuel lines or hoses, relieve pressure from the system by disabling the fuel pump and disconnecting the fuel line/quick connect fitting in the engine compartment.

1. Remove PGM-FI main relay 2 from the driver's under-dash fuse/relay box.
2. Start the engine, and let it idle until it stalls.

➡**If any DTCs are stored, clear and ignore them.**

3. Turn the ignition switch to LOCK (0).
4. Remove the fuel fill cap to relieve the pressure in the fuel tank.
5. Do the battery terminal disconnection procedure.
6. Remove the cover and the quick-connect fitting cover.
7. Check the quick-connect fitting for dirt, and clean it if needed.
8. Place a rag or a shop towel over the quick-connect fitting.
9. Disconnect the quick-connect fitting.
   a. Hold the connector with one hand, and squeeze the retainer tabs with the other hand to release them from the locking tabs.
   b. Pull the connector off.

➡**Be careful not to damage the line or other parts. Do not use tools. If the connector does not move, keep the retainer tabs pressed down, and alternately pull and push the connector until it comes off easily. Do not remove the retainer from the line; once removed, the retainer must be replaced with a new one.**

10. After disconnecting the quick-connect fitting, check it for dirt or damage.
11. Do the battery terminal reconnection procedure.

FUEL FILTER

REMOVAL & INSTALLATION

See Figure 268.

➡**The fuel filter should be replaced whenever the fuel pressure drops below the specified value, after making sure that the fuel pump and the fuel pressure regulator are OK.**

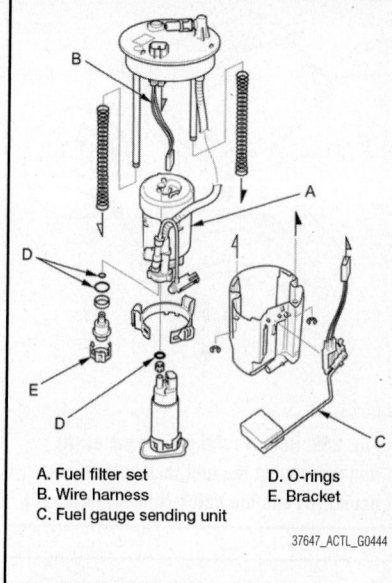

A. Fuel filter set
B. Wire harness
C. Fuel gauge sending unit
D. O-rings
E. Bracket

37647_ACTL_G0444

Fig. 268 Exploded view of fuel pump assembly

1. Remove the fuel tank unit.
2. Remove the fuel filter set.
3. Check these items before installing the fuel tank unit:
   • When connecting the wire harness, make sure the connection is secure and the connectors are firmly locked into place.
   • When installing the fuel gauge sending unit, make sure the connection is secure and the connector is firmly locked into place. Be careful not to bend or twist it excessively.
4. Install the parts in the reverse order of removal with new O-rings and a new bracket. When installing the fuel tank unit, align the marks on the unit and the fuel tank.

➡**Coat the O-rings with clean engine oil; do not use any other oil or fluid. Do not pinch the O-rings during installation. Use all the new parts supplied in the fuel filter replacement kit.**

FUEL TANK

DRAINING

Accord and Crosstour (FWD)

See Figures 269 through 271.

1. Remove the fuel tank unit.
Special Tools Required: Fuel sender wrench 07AAA-S0XA100
   a. Relieve the fuel pressure.
   b. Remove the rear seat cushion cover.
   c. Remove the access panel from the floor.

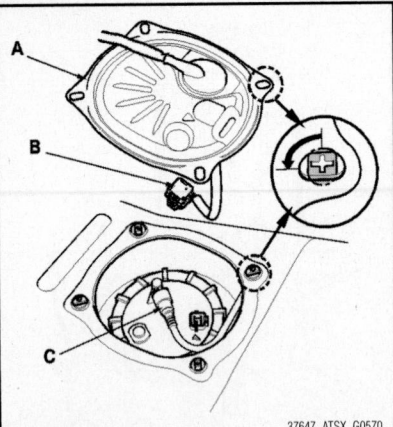

Fig. 269 Remove the access panel (A) and disconnect the fuel tank unit 4P connector (B) and the quick-connect fitting (C)

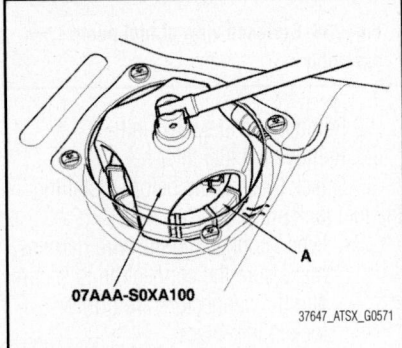

Fig. 270 Using the special tool, loosen the locknut (A)

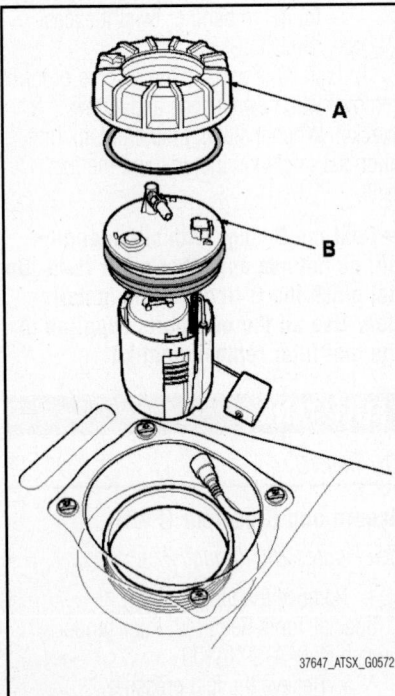

Fig. 271 Remove the locknut (A) and the fuel tank unit (B)

d. Disconnect the fuel tank unit 4P connector.

e. Disconnect the quick-connect fitting from the fuel tank unit.

f. Using the special tool, loosen the locknut.

g. Remove the locknut and the fuel tank unit.

2. Using a hand pump, a hose, and a container suitable for fuel, draw the fuel from the fuel tank.

3. Reinstall the fuel tank unit.

Crosstour (4WD)

1. Remove the fuel tank unit.

2. Remove the secondary fuel gauge sending unit.

3. Using a hand pump, a hose, and a container suitable for fuel, draw the fuel from the fuel tank.

4. Reinstall the fuel tank unit and the secondary fuel gauge sending unit.

REMOVAL & INSTALLATION

Accord and Crosstour (FWD)

See Figures 272 and 273.

1. Drain the fuel tank.

2. Reinstall the fuel tank unit without connecting the fuel tank unit 4P connector and the quick-connect fitting.

3. Remove the fuel fill pipe cover.

4. Disconnect the quick-connect fittings and the fuel fill tube from the fuel fill pipe. Slide back the clamps, then twist the hose as you pull to avoid damaging them.

5. Raise the vehicle on a lift.

6. Disconnect the hose from the EVAP canister.

7. Remove the hose from the clamp.

➡**Be careful not to damage the hose.**

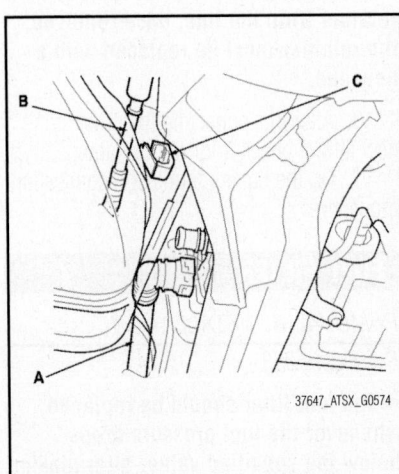

Fig. 272 Disconnect the hose (A) from the EVAP canister

8. Remove the exhaust pipe.

9. Remove the middle floor undercover.

10. Remove the fuel tank protector.

11. Place a jack or other support under the fuel tank.

12. Remove the strap bolts and the straps.

13. Remove the fuel tank.

To install:

14. Install the parts in the reverse order of removal.

➡**New fuel tanks have a ring pull at the fuel vapor hose connector. When you connect the hose and confirm that the connection is secure, remove the ring pull by pulling it down.**

Before connecting the fuel fill pipe and the quick-connect fitting, check for dirt, and clean them if needed, taking care not to damage the fuel fill pipe or other parts.

When installing the fuel tank protector, make sure to insert it into the clip in the direction shown.

Crosstour (4WD)

See Figures 274 through 276.

1. Drain the fuel tank.

2. Reinstall the fuel tank unit without connecting the fuel tank unit 4P connector and the quick-connect fitting. Then reinstall the secondary fuel gauge sending unit without connecting the secondary fuel tank unit 4P connector.

3. Remove the fuel fill pipe cover.

4. Disconnect the quick-connect fittings from the fuel fill pipe.

5. Raise the vehicle on a lift.

6. Remove the muffler.

7. Remove the propeller shaft.

8. Remove left and right the middle floor undercovers.

9. Support the rear differential with a transmission jack, then remove the rear differential mounting bolts, and lower the rear differential.

10. Place a jack or other support under the fuel tank.

11. Remove the strap bolts and the straps.

12. Slowly lower the fuel tank, and disconnect the hose and the quick-connect fitting.

13. Remove the fuel tank.

14. Remove the fuel tank protector from the fuel tank.

To install:

15. Install the parts in the reverse order of removal, noting the following:

- When installing the fuel tank protector, make sure to insert it into the clip (B) in the direction shown.

- Before connecting the fuel fill pipe and the quick-connect fittings, check for dirt, and clean them if needed, taking care not to damage the fuel fill pipe and other parts.

FUEL TANK UNIT

REMOVAL & INSTALLATION

Accord and Crosstour (FWD)

See Figures 277 through 282.

Special Tools Required: Fuel Sender Wrench 07AAA-S0XA100

1. Relieve the fuel pressure.
2. Remove the fuel fill cap.
3. Remove the rear seat cushion.
4. Remove the access panel from the floor.
5. Disconnect the fuel tank unit 4P connector.
6. Disconnect the quick-connect fitting from the fuel tank unit.
7. Using the special tool, loosen the locknut.
8. Remove the locknut and the fuel tank unit.

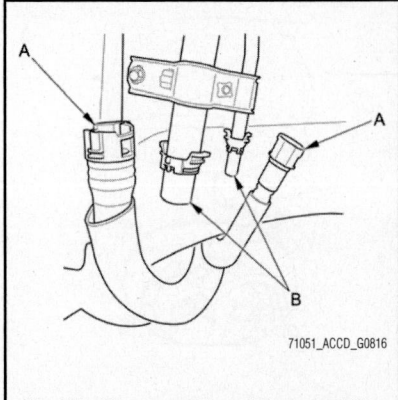

71051_ACCD_G0816

Fig. 274 Disconnect the quick-connect fittings (A) from the fuel fill pipe (B)

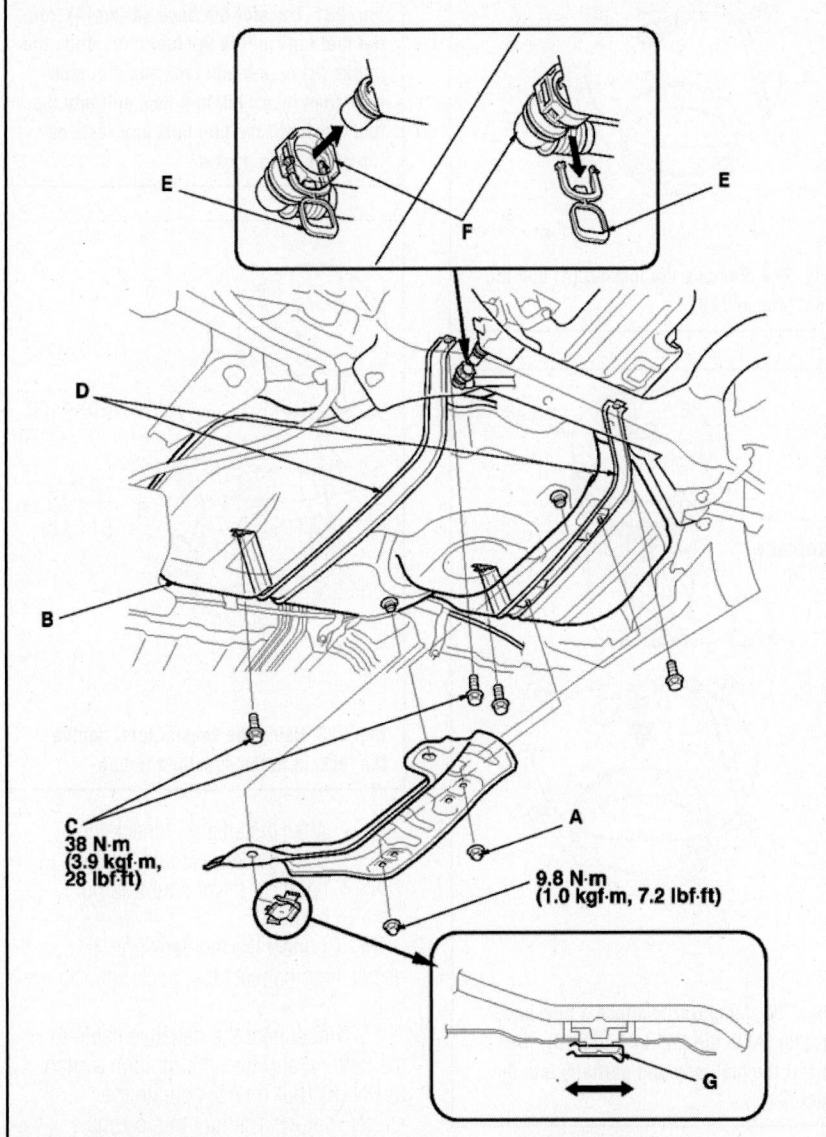

A. Fuel tank protector mounting nuts
B. Fuel tank
C. Strap bolts
D. Straps
E. Ring pull
F. Fuel vapor hose connector
G. Clip

37647_ATSX_G0577

Fig. 273 Removing the fuel tank

B
73 N·m
(7.5 kgf·m, 54 lbf·ft)
Replace.

A
54 N·m
(5.5 kgf·m, 40 lbf·ft)

71051_ACCD_G0817

Fig. 275 Support the rear differential with a transmission jack, then remove the rear differential mounting bolts (A and B), and lower the rear differential (C)

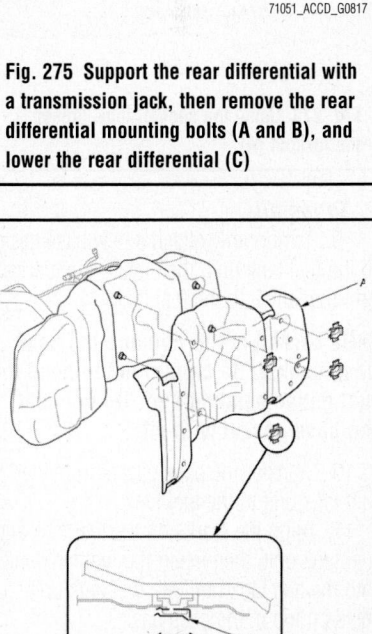

71051_ACCD_G0819

Fig. 276 Remove the fuel tank protector (A) from the fuel tank

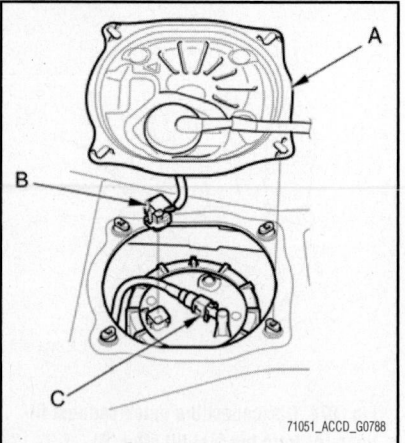

Fig. 277 Remove the access panel (A) from the floor, disconnect the fuel tank unit 4P connector (B), then disconnect the quick-connect fitting (C) from the fuel tank unit

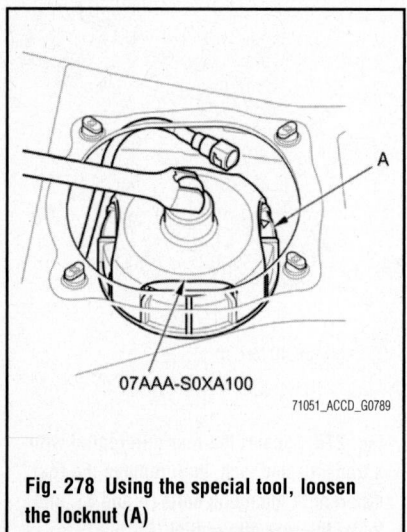

Fig. 278 Using the special tool, loosen the locknut (A)

To install:

9. Temporarily attach a new base gasket to the fuel tank unit, then insert the fuel tank unit partially into the fuel tank.

➡**Be careful not to damage the new base gasket. Be careful not to bend the fuel gauge sending unit. Do not coat the base gasket with oil.**

10. Transfer the base gasket from the fuel tank unit to the fuel tank.

11. Align the marks on the fuel tank and fuel tank unit, then insert the fuel tank unit into the fuel tank until the fuel tank unit rests on top of the base gasket.

➡**To avoid a fuel leak, check the base gasket, visually or by hand, to make sure it is not pinched.**

12. Tighten a new locknut by hand with a new locknut plate.

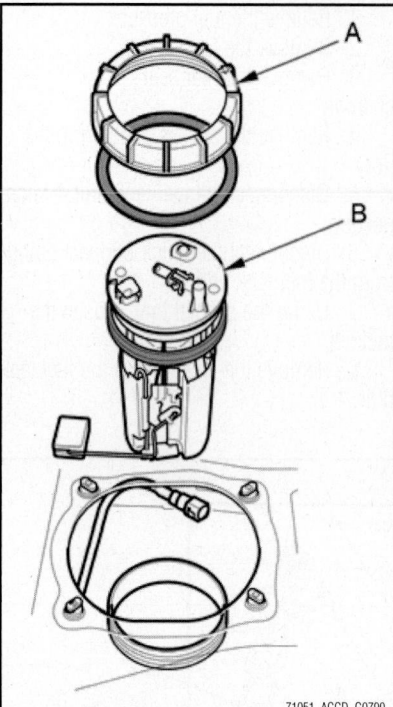

Fig. 279 Remove the locknut (A) and the fuel tank unit (B)

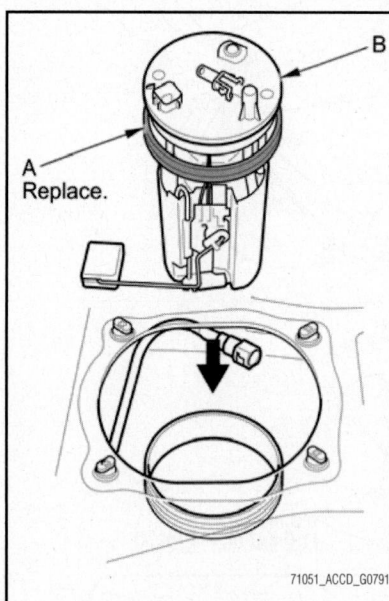

Fig. 280 Temporarily attach a new base gasket (A) to the fuel tank unit (B), then insert the fuel tank unit partially into the fuel tank

➡**Before tightening, align the mark on the locknut to the start of the thread.**

13. Using the special tool, tighten the locknut to the specified torque.
   • After tightening, make sure the marks (A) are still aligned.

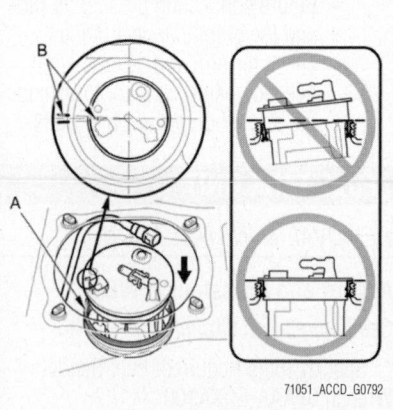

Fig. 281 Transfer the base gasket (A) from the fuel tank unit to the fuel tank, align the marks (B) on the fuel tank and fuel tank unit, then insert the fuel tank unit into the fuel tank until the fuel tank unit rests on top of the base gasket

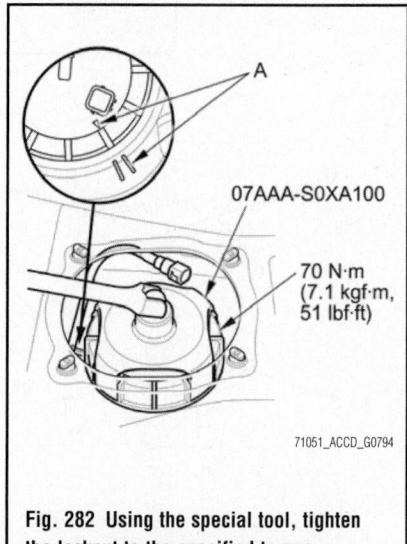

Fig. 282 Using the special tool, tighten the locknut to the specified torque

   • After installation, check the base gasket, visually or by hand, to make sure it is not pinched.

14. Connect the fuel tank unit 4P connector, then connect the quick-connect fitting.

15. Reconnect the negative cable to the battery, and turn the ignition switch to ON (II) (but do not operate the starter motor). The fuel pump runs for about 2 seconds, and fuel pressure rises. Repeat this two or three times, then make sure there are no fuel leaks.

16. Install the access panel.

17. Install the rear seat cushion.

18. Install the fuel fill cap.

Crosstour (4WD)

See Figures 283 through 287.

Special Tools Required: Fuel Sender Wrench 07AAA-S0XA100

1. Relieve the fuel pressure.
2. Remove the rear seat cushion.
3. Remove the access panel from the left side of the floor.
4. Disconnect the fuel tank unit 4P connector.
5. Disconnect the quick-connect fitting from the fuel tank unit.
6. Using the special tool, loosen the locknut.
7. Remove the locknut, and lift up the fuel tank unit.

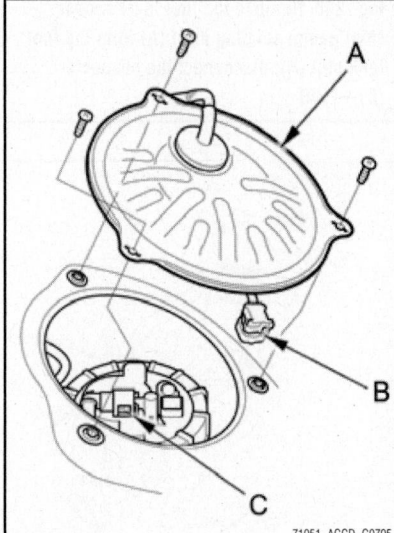

Fig. 283 Remove the access panel (A) from the left side of the floor, disconnect the fuel tank unit 4P connector (B), and disconnect the quick-connect fitting (C) from the fuel tank unit

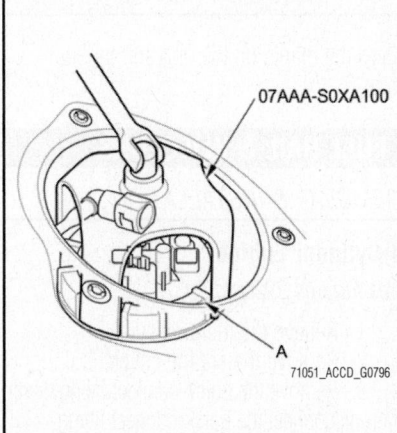

Fig. 284 Using the special tool, loosen the locknut (A)

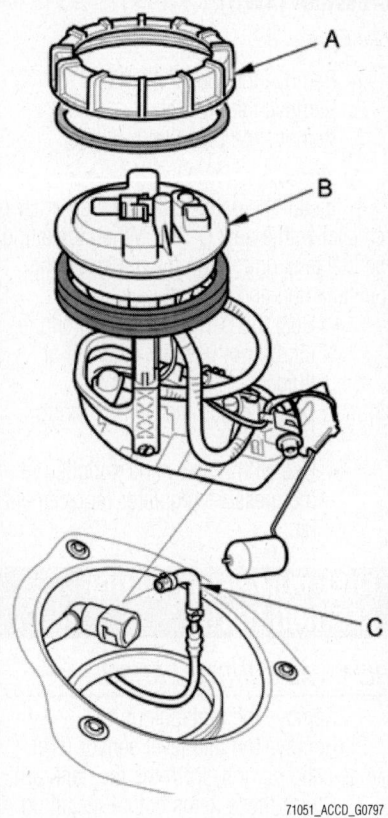

71051_ACCD_G0797

Fig. 285 Remove the locknut (A), and lift up the fuel tank unit (B), then disconnect the transfer tube (C), and remove the fuel tank unit

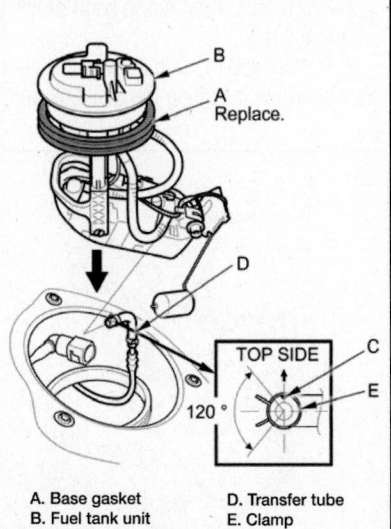

A. Base gasket
B. Fuel tank unit
C. Mark on transfer tube
D. Transfer tube
E. Clamp

71051_ACCD_G0798

Fig. 286 Temporarily attach a new base gasket to the fuel tank unit, position the mark of the transfer tube up, connect the transfer tube, then insert the fuel tank unit partially into the fuel tank

8. Disconnect the transfer tube, and remove the fuel tank unit.

To install:

9. Temporarily attach a new base gasket to the fuel tank unit, position the mark of the transfer tube up, connect the transfer tube, then insert the fuel tank unit partially into the fuel tank. Note the following:
- Be careful not to damage the new base gasket.
- Make sure the clamp is positioned as shown.
- Be careful not to bend the fuel gauge sending unit.
- Do not coat the base gasket with oil.

10. Transfer the base gasket from the fuel tank unit to the fuel tank.

11. Align the marks on the fuel tank and the fuel tank unit, then insert the fuel tank unit into the fuel tank until the fuel tank unit rests on top of the base gasket.

➡**To prevent a fuel leak, check the base gasket, visually or by hand, to make sure it is not pinched.**

12. Tighten a new locknut by hand with a new locknut plate.

➡**Before tightening, align the mark on the locknut with the start of the threads.**

13. Using the special tool, tighten a new locknut to the specified torque.
- After tightening, make sure the marks are still aligned.
- After installation, check the base gasket, visually or by hand, to be sure it is not pinched.

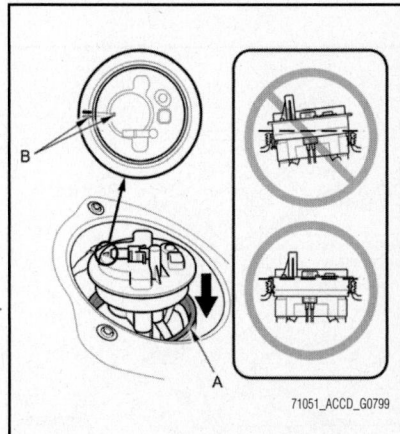

71051_ACCD_G0799

Fig. 287 Transfer the base gasket (A) from the fuel tank unit to the fuel tank, align the marks (B) on the fuel tank and the fuel tank unit, then insert the fuel tank unit into the fuel tank until the fuel tank unit rests on top of the base gasket

14. Connect the fuel tank unit 4P connector, then connect the quick-connect fitting.

15. Reconnect the negative cable to the battery, and turn the ignition switch to ON (II) (but do not operate the starter motor). The fuel pump will run for about 2 seconds, and fuel pressure will rise. Repeat two or three times, then make sure there are no fuel leaks.

16. Install the access panel.
17. Install the rear seat cushion.
18. Install the fuel fill cap.

FUEL PRESSURE REGULATOR

REMOVAL & INSTALLATION

Accord and Crosstour (FWD)

See Figure 288.

1. Remove the fuel tank unit.
2. Remove the reservoir.
3. Remove the bracket.
4. Remove the fuel pressure regulator.

To install:

5. Install the parts in the reverse order of removal with new O-rings and a new bracket. When installing the fuel tank unit, align the marks on the unit and the fuel tank.

- Coat the O-rings with clean engine oil; do not use any other oils or fluids.
- Do not pinch the O-rings during installation.
- Use all the new parts supplied in the pressure regulator replacement kit.

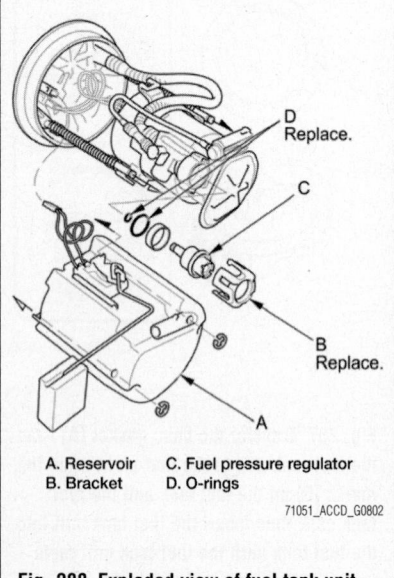

A. Reservoir C. Fuel pressure regulator
B. Bracket D. O-rings

71051_ACCD_G0802

Fig. 288 Exploded view of fuel tank unit

Crosstour (4WD)

See Figure 289.

1. Remove the fuel tank unit.
2. Remove the bracket.
3. Remove the pressure regulator.

To install:

4. Install the parts in the reverse order of removal with a new O-ring. When installing the fuel tank unit, align the marks on the fuel tank unit and the fuel tank.

- Coat the O-ring with clean engine oil; do not use any other oil or fluid.
- Do not pinch the O-ring during installation.
- Use all the new parts supplied in the pressure regulator replacement kit.

FUEL PUMP/FUEL GAUGE SENDING UNIT

REMOVAL & INSTALLATION

1. Remove the fuel tank unit.
2. Remove the fuel level sensor (fuel gauge sending unit) from the fuel tank unit.
3. Check these items before installing the fuel tank unit:

   a. When connecting the wire harness, make sure the connection is secure and the connector is firmly locked into place.

   b. When installing the fuel gauge sending unit, make sure the connection is secure. Be careful not to bend or twist it excessively.

4. Install the parts in the reverse order of removal. When installing the fuel tank unit,

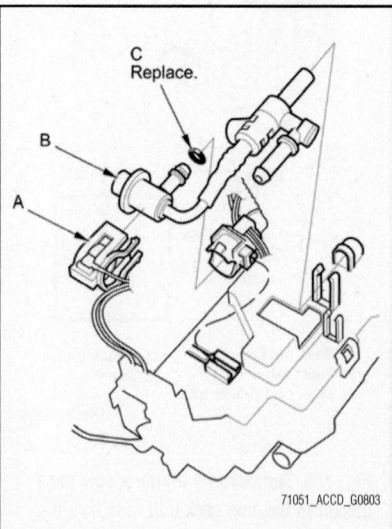

71051_ACCD_G0803

Fig. 289 Remove the bracket (A), the pressure regulator (B) and O-rings (C)

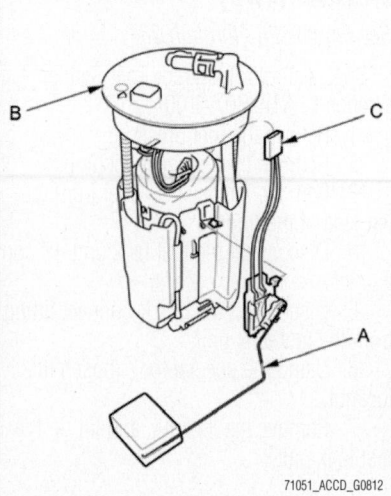

71051_ACCD_G0812

Fig. 290 Remove the fuel level sensor (fuel gauge sending unit) (A) from the fuel tank unit (B); disconnect the connector (C)—FWD

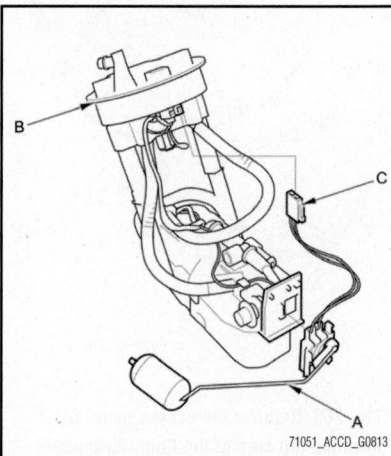

71051_ACCD_G0813

Fig. 291 Remove the fuel level sensor (fuel gauge sending unit) (A) from the fuel tank unit (B); disconnect the connector (C)—4WD

align the marks on the unit and the fuel tank.

FUEL RAIL & INJECTORS

REMOVAL & INSTALLATION

4-Cylinder Engine

See Figures 292 and 293.

1. Relieve the fuel pressure.
2. Remove the engine cover.
3. Remove the quick-connect fitting cover, then disconnect the quick-connect fitting.
4. Disconnect the injector connectors and the engine mount control solenoid valve connector.

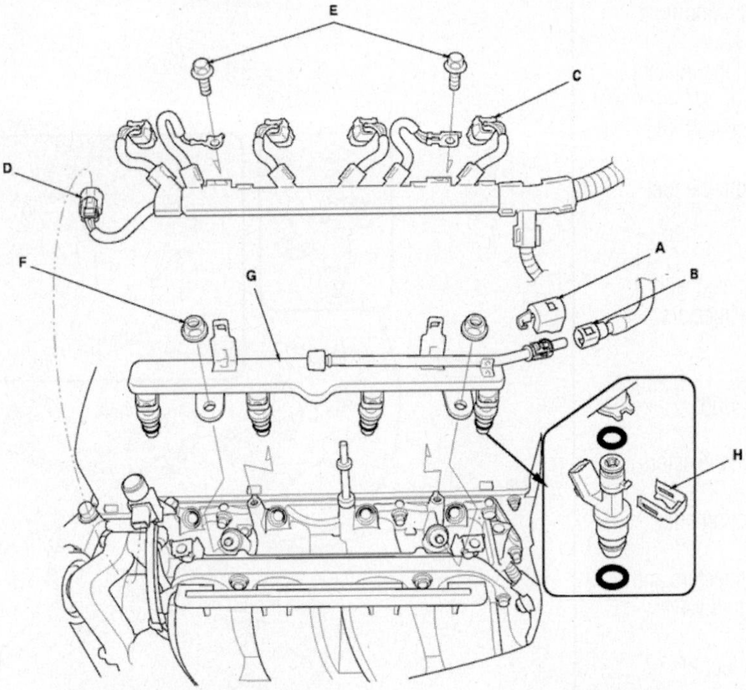

A. Quick-disconnect fitting cover
B. Quick-disconnect fitting
C. Injector connectors
D. Engine mount control solenoid valve connector

E. Ground cable bolts
F. Fuel rail mounting nuts
G. Fuel rail
H. Injector clips

37647_ATSX_G0568

Fig. 292 Fuel rail assembly

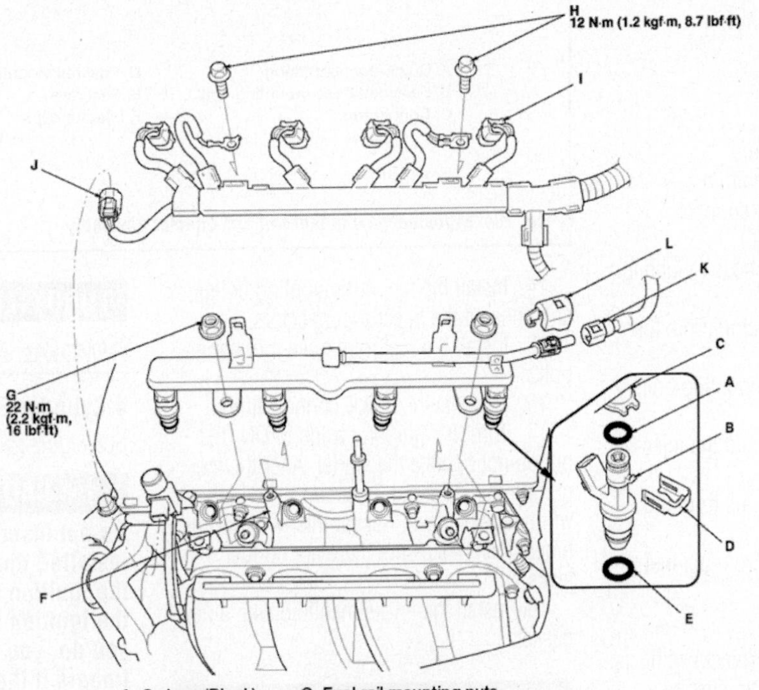

H 12 N·m (1.2 kgf·m, 8.7 lbf·ft)

G 22 N·m (2.2 kgf·m, 16 lbf·ft)

A. O-rings (Black)
B. Injectors
C. Fuel rail
D. Injector clips
E. O-rings (Brown)
F. Injector base

G. Fuel rail mounting nuts
H. Ground cable bolts
I. Injector connectors
J. Engine mount control solenoid valve connector
K. Quick-connect fitting
L. Quick-connect fitting cover

37647_ATSX_G0569

Fig. 293 Installation of injectors and fuel rail

5. Remove the ground cable bolts.

6. Remove the fuel rail mounting nuts from the fuel rail.

7. Remove the fuel rail and the injectors from the injector base.

8. Remove the injector clips from the fuel rail.

9. Remove the injectors from the fuel rail.

To install:

10. Coat the new O-rings (black) with clean engine oil, and insert the injectors into the fuel rail.

11. Install the injectors clips.

12. Coat the new injector O-rings (brown) with clean engine oil.

13. Install the fuel rail and the injectors in the injector base.

14. Install the fuel rail mounting nuts and the ground cable bolts.

15. Connect the injector connectors and the engine mount control solenoid valve connector.

16. Connect the quick-connect fitting, and quick-connect fitting 4cover.

17. Turn the ignition switch to ON (II), but do not operate the starter. After the fuel pump runs for about 2 seconds, the fuel rail will be pressurized. Repeat this two or three times, then make sure there are no fuel leaks.

18. Reinstall the engine cover.

6-Cylinder Engine

See Figures 294 and 295.

1. Relieve the fuel pressure.
2. Remove the intake manifold.
3. Disconnect the quick-connect fitting.
4. Remove the fuel joint hose mounting bolt.
5. Disconnect the connectors from the injectors.
6. Remove the fuel rail mounting bolts from the fuel rails.
7. Remove the fuel rails and the injectors from the injector base.
8. Remove the injector clips from the fuel rails.
9. Remove the injectors from the fuel rails.

To install:

10. Coat the new O-rings (black) with clean engine oil, and insert the injectors into the fuel rails.

11. Install the injector clips.

12. Coat the new injector O-rings (green) with clean engine oil.

13. Install the fuel rails and the injectors in the injector base.

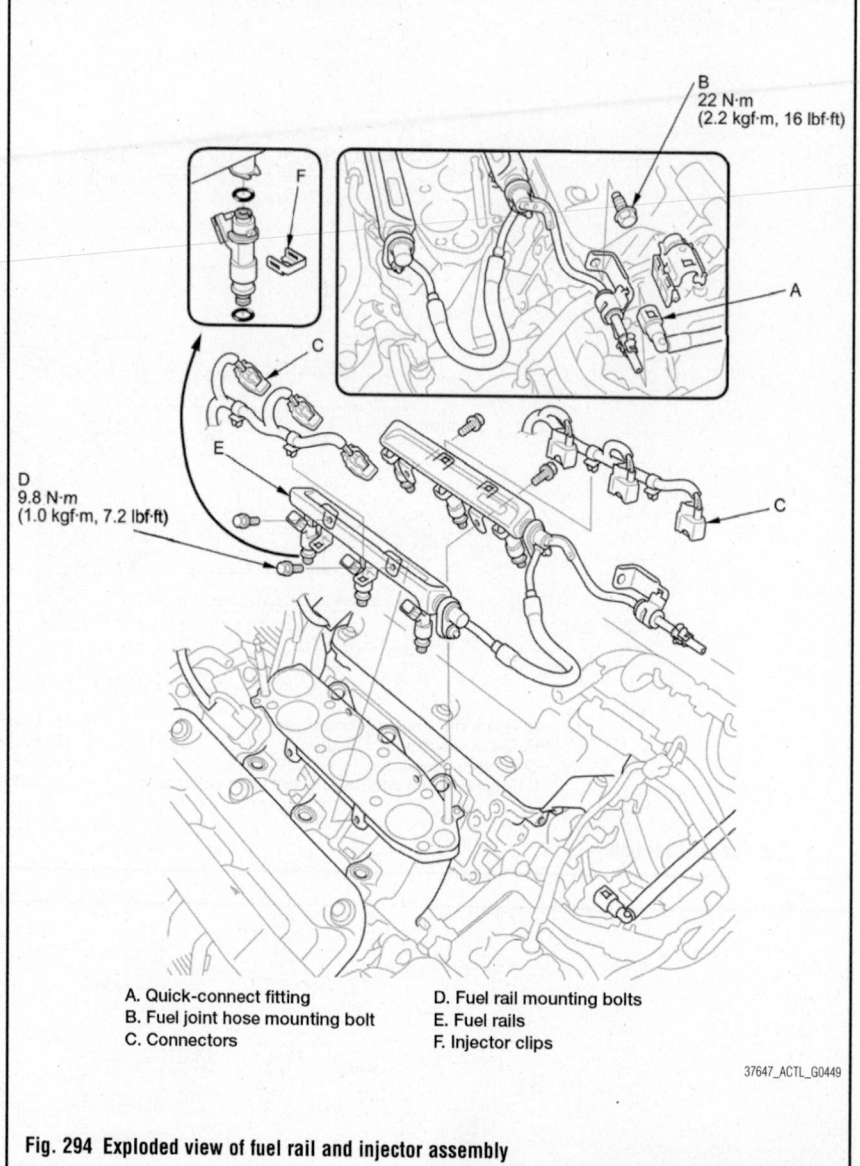

A. Quick-connect fitting
B. Fuel joint hose mounting bolt
C. Connectors
D. Fuel rail mounting bolts
E. Fuel rails
F. Injector clips

37647_ACTL_G0449

Fig. 294 Exploded view of fuel rail and injector assembly

14. Install the fuel rail mounting bolts, and connect the injector connectors.

15. Install the fuel joint hose mounting bolt.

16. Connect the quick-connect fitting.

17. Turn the ignition switch to ON (II), but do not operate the starter. After the fuel pump runs for about 2 seconds, the fuel rail will be pressurized. Repeat this two or three times, then make sure there are no fuel leaks.

18. Install the intake manifold with a new gasket.

IDLE SPEED

ADJUSTMENT

The idle speed is controlled by the ECM/PCM. No adjustment is possible or necessary.

THROTTLE BODY

REMOVAL & INSTALLATION

4-Cylinder Engine

See Figure 296.

✳✳ CAUTION

Do not insert your fingers into the installed throttle body when you turn the ignition switch to ON (II) or while the ignition switch is in ON (II). If you do, you will seriously injure your fingers if the throttle valve is activated.

1. Make sure the ignition switch turned LOCK (0).
2. Remove the intake air duct.
3. Disconnect the throttle body connector.

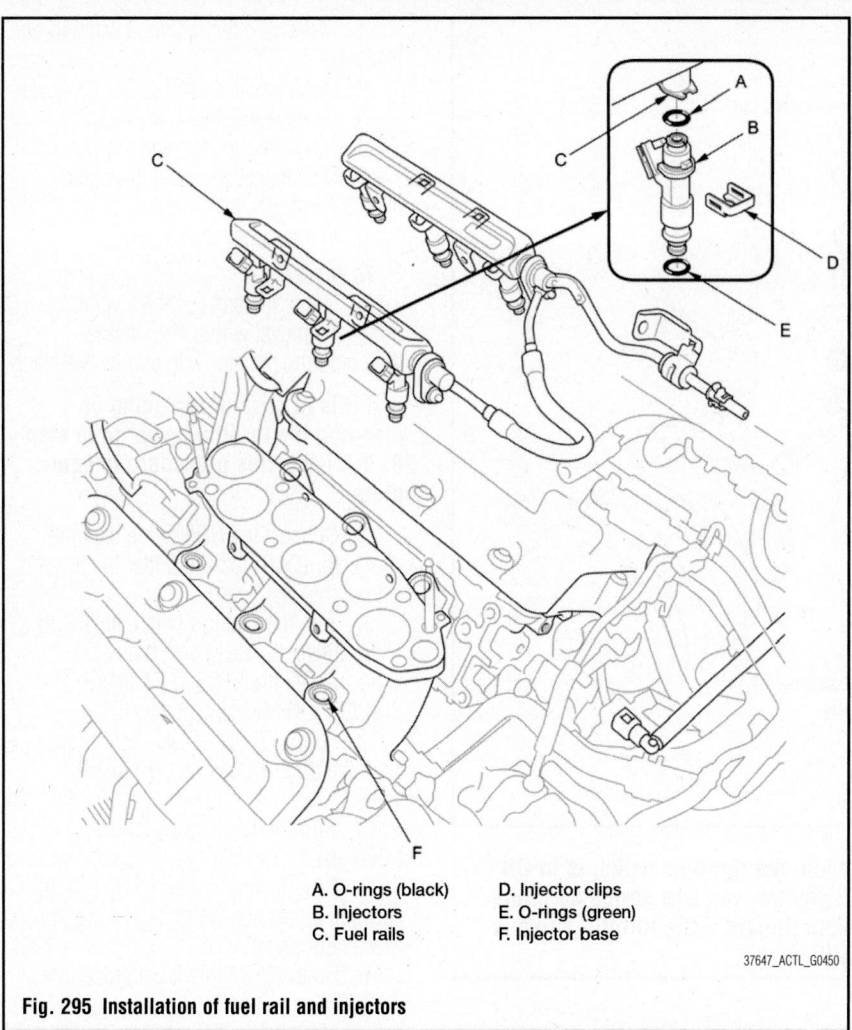

A. O-rings (black)
B. Injectors
C. Fuel rails
D. Injector clips
E. O-rings (green)
F. Injector base

37647_ACTL_G0450

Fig. 295 Installation of fuel rail and injectors

4. Disconnect and plug the water bypass hoses.
5. Remove the throttle body.

To install:

6. Install the parts in the reverse order of removal with a new gasket.

➡**After you torque the hose clamp screw, make sure the clearance is less than 0.04 inches (1.0 mm).**

7. Refill the radiator with engine coolant.

➡**If it is in case of replacing or cleaning the throttle body, go to step 8. If it isn't, this procedure is complete.**

8. Connect the HDS to the data link connector (DLC) located under the driver's side of the dashboard.
9. Turn the ignition switch to ON (II).
10. Reset the PCM with the HDS.
11. Select the ETCS TEST in the INSPECTION MENU with the HDS.
12. Select TP POSITION CHECK and clear the throttle position (TP) learned value.
13. Turn the ignition switch to LOCK (0).
14. Turn the ignition switch to ON (II), and wait 2 seconds without pressing the accelerator pedal.
15. Do the PCM idle learn procedure.

6-Cylinder Engine

See Figure 297.

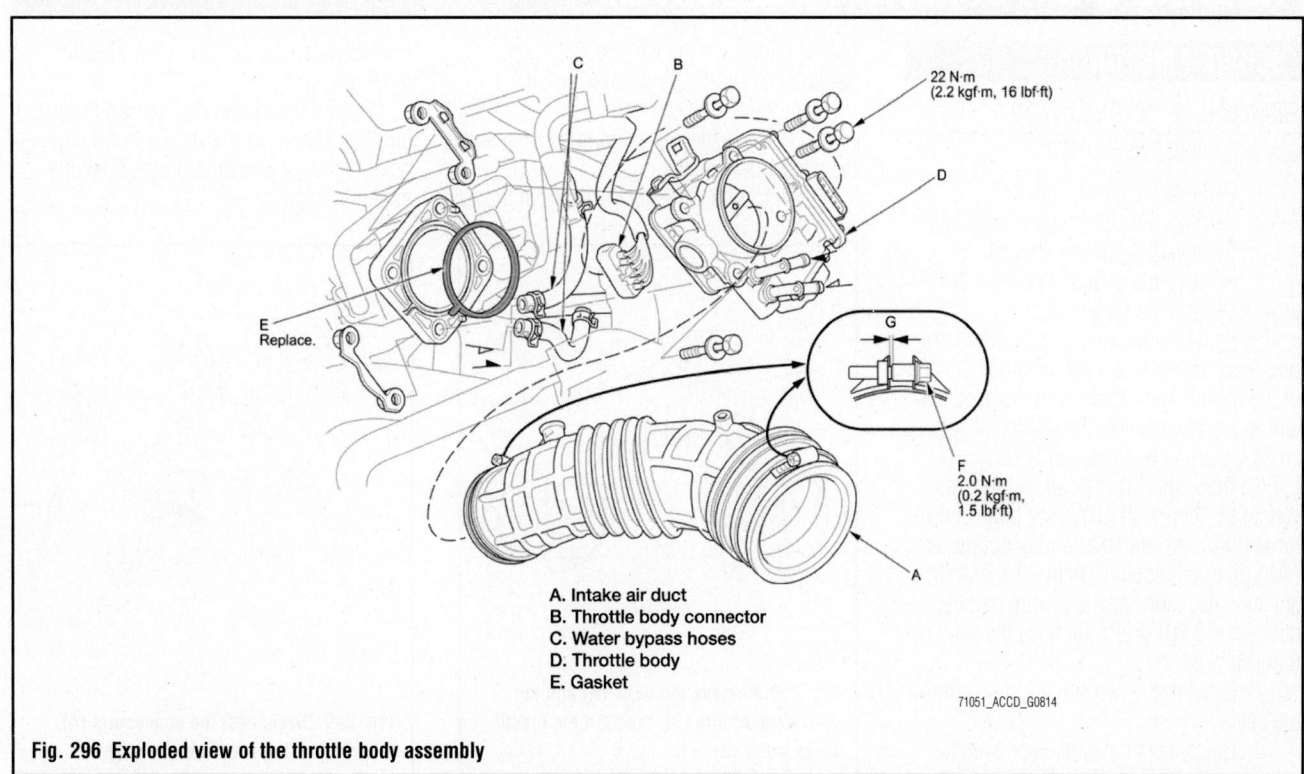

22 N·m (2.2 kgf·m, 16 lbf·ft)

2.0 N·m (0.2 kgf·m, 1.5 lbf·ft)

E Replace.

A. Intake air duct
B. Throttle body connector
C. Water bypass hoses
D. Throttle body
E. Gasket

71051_ACCD_G0814

Fig. 296 Exploded view of the throttle body assembly

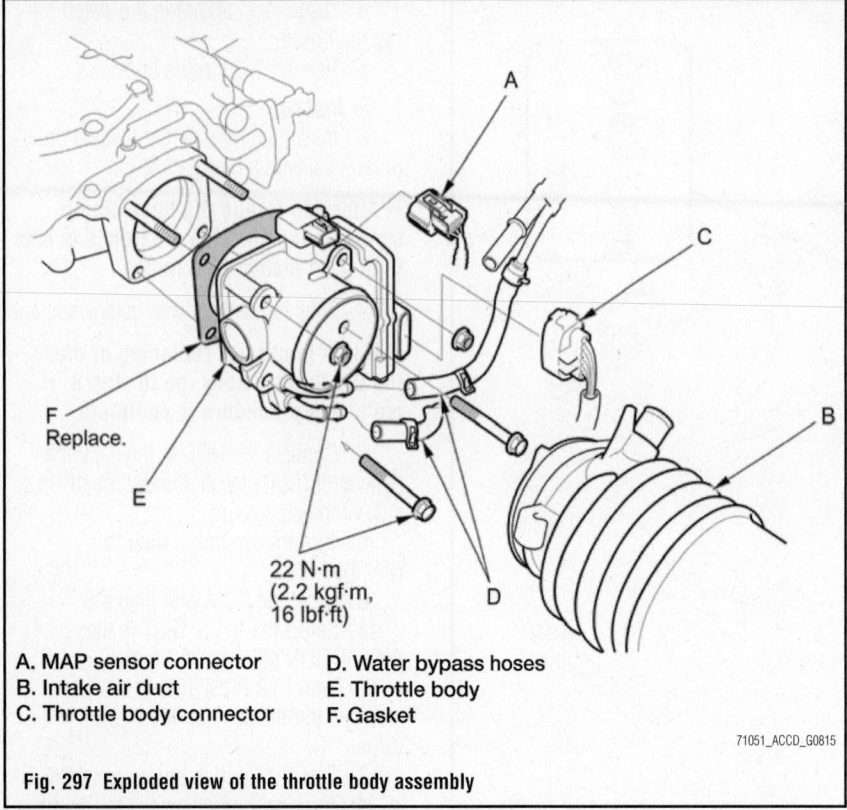

F Replace.

22 N·m
(2.2 kgf·m,
16 lbf·ft)

A. MAP sensor connector
B. Intake air duct
C. Throttle body connector
D. Water bypass hoses
E. Throttle body
F. Gasket

71051_ACCD_G0815

Fig. 297 Exploded view of the throttle body assembly

※ CAUTION

Do not insert your fingers into the installed throttle body when you turn the ignition switch to ON (II) or while the ignition switch is in ON (II). If you do, you will seriously injure your fingers if the throttle valve is activated.

1. Make sure the ignition switch turned LOCK (0).
2. Disconnect the MAP sensor connector.
3. Remove the intake air duct.
4. Disconnect the throttle body connector.
5. Disconnect and plug the water bypass hoses.
6. Remove the throttle body.

To install:

7. Install the parts in the reverse order of removal with a new gasket, then refill the radiator with engine $coolant.

➡If it is in case of replacing or cleaning the throttle body, go to step 8. If it isn't, this procedure is complete.

8. Connect the HDS to the data link connector (DLC) located under the driver's side of the dashboard.
9. Turn the ignition switch to ON (II).
10. Reset the PCM with the HDS.
11. Select the ETCS TEST in the INSPECTION MENU with the HDS.
12. Select TP POSITION CHECK and clear the throttle position (TP) learned value.
13. Turn the ignition switch to LOCK (0).
14. Turn the ignition switch to ON (II), and wait 2 seconds without pressing the accelerator pedal.
15. Do the PCM idle learn procedure.

HEATING & AIR CONDITIONING SYSTEM

BLOWER MOTOR

REMOVAL & INSTALLATION

See Figures 298 through 303.

1. Remove the glove box.
2. Remove the passenger's undercover.
3. Remove the right kick panel.
4. Remove the dust and pollen filter assembly from the blower unit.
5. Remove the bolts and the wire harness clip. Then cut the plastic cross brace in the glove box opening with diagonal cutters in the area shown. Retain the plastic cross brace to be reinstalled later.
6. Disconnect these connectors: Passenger's under-dash fuse/relay box connector D (28P), the stereo amplifier connectors (with premium audio system), the AM/FM antenna lead, and right side wire harness connector C410 (20P). Remove the wire harness clips.
7. Remove the two screws, then remove the cover.
8. Disconnect the connector from the blower motor and the wire harness clip. Remove the self-tapping screw and the mounting nut.
9. Remove the self-tapping screws, and the passenger's heater duct.
10. Disconnect the connector from the recirculation control motor. Remove the mounting nuts.
11. Pull the blower unit out while rotating it clockwise as shown, so that the glove box bracket passes through the dust and pollen filter area.

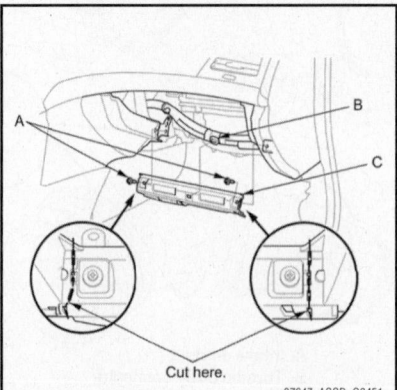

Cut here.

37647_ACCD_G0451

Fig. 298 Remove the bolts (A) and the wire harness clip (B), then cut the plastic cross brace (C)

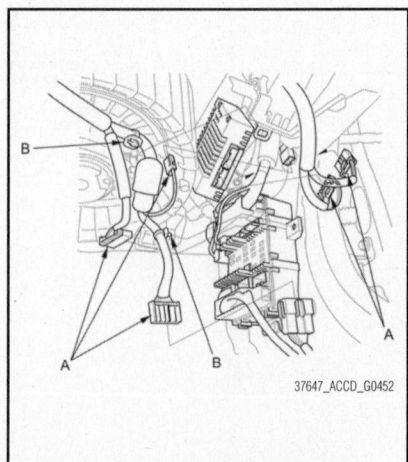

37647_ACCD_G0452

Fig. 299 Disconnect the connectors (A), and remove the wire harness clips (B)

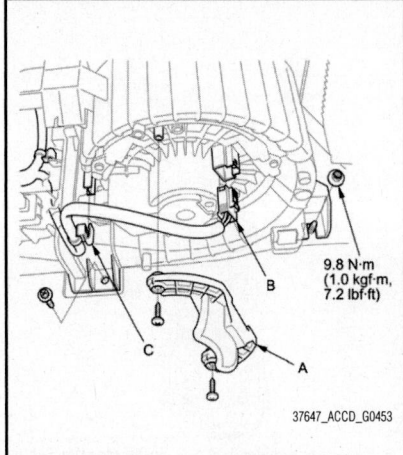

Fig. 300 Remove the cover (A), disconnect the connector (B) and wire harness (C)

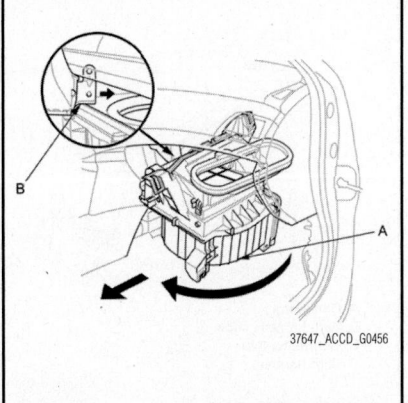

Fig. 303 Pull the blower unit (A) out while rotating it clockwise, so that the glove box bracket (B) passes through the dust and pollen filter area

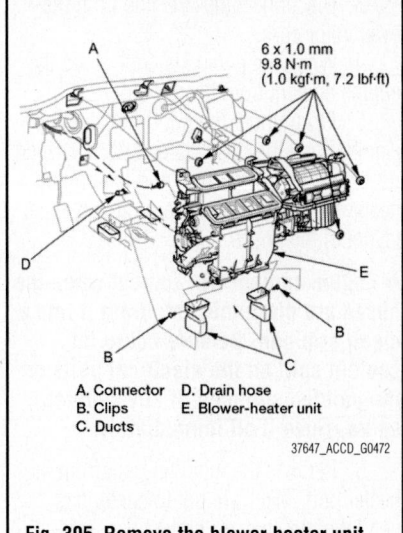

A. Connector   D. Drain hose
B. Clips   E. Blower-heater unit
C. Ducts

Fig. 305 Remove the blower-heater unit

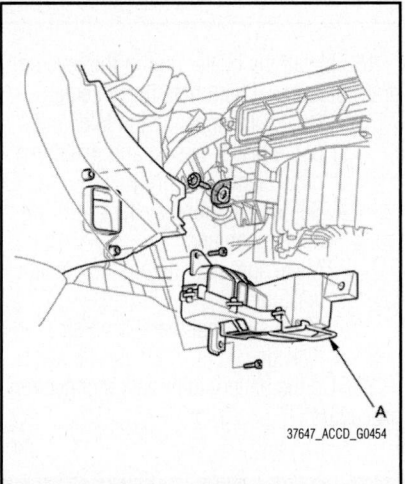

Fig. 301 Remove the self-tapping screws, and the passenger's heater duct (A)

12. Install the unit in the reverse order of removal. Make sure that there is no air leakage.

HEATER CORE

REMOVAL & INSTALLATION
See Figures 304 through 309.

❊❊ WARNING

SRS components are located in this area. Review the SRS component locations and the precautions and procedures before doing repairs or service.

1. Do the battery terminal disconnection procedure.
2. Recover the refrigerant with a recovery/recycling/ charging station.

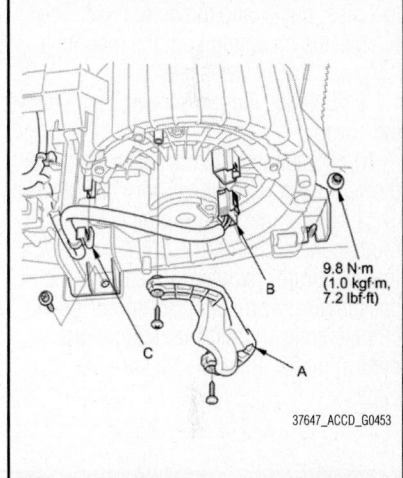

Fig. 306 Remove the cover (A), disconnect the connector (B) and wire harness (C)

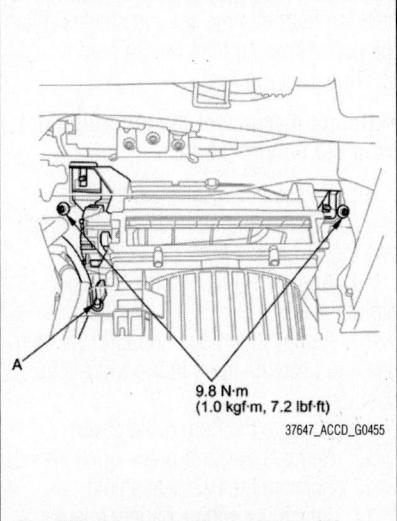

Fig. 302 Disconnect the connector (A) from the recirculation control motor

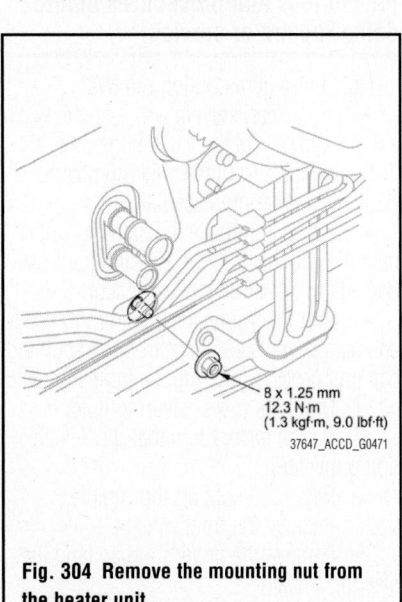

Fig. 304 Remove the mounting nut from the heater unit

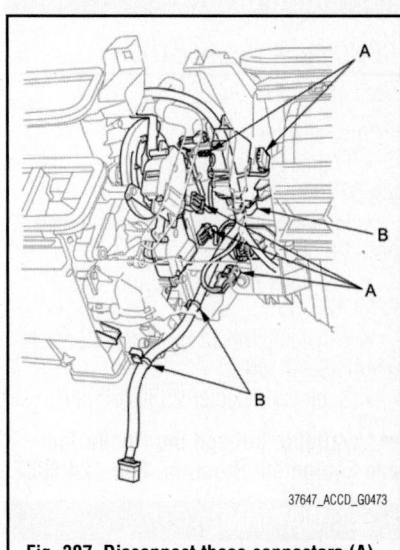

Fig. 307 Disconnect these connectors (A) and remove the wire harness clips (B)

3. Disconnect the A/C line from the evaporator core.

4. When the engine is cool, drain the engine coolant from the radiator.

5. From under the hood, remove the clamp. Slide the hose clamps back. Disconnect the inlet heater hose and the outlet heater hose from the heater unit. Note the layout of the hoses.

➥**Engine coolant will run out when the hoses are disconnected; drain it into a clean drip pan. Be sure not to let coolant spill on the electrical parts or the painted surfaces. If any coolant spills, rinse it off immediately.**

6. Remove the mounting nut from the heater unit. Take care not to damage or bend the fuel lines or brake lines, etc..

7. Remove the dashboard.

8. Disconnect the connector. Remove the clips, ducts, and the drain hose. Then remove the mounting bolt, the mounting nuts, and the blower-heater unit.

9. Remove the two screws, then remove the cover.

10. Disconnect the connector from the blower motor. Remove the wire harness clip.

11. Disconnect these connectors: The mode control motor, the power transistor, the evaporator temperature sensor, the passenger's air mix control motor (with climate control), and the recirculation control motor. Remove the wire harness clips.

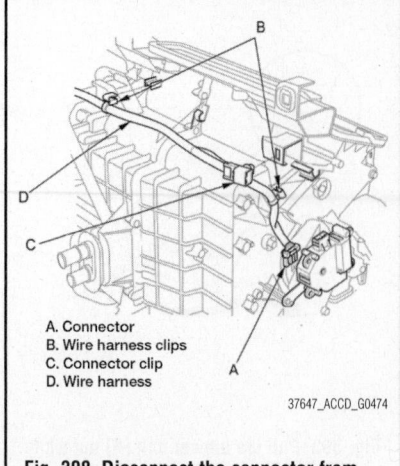

A. Connector
B. Wire harness clips
C. Connector clip
D. Wire harness

37647_ACCD_G0474

Fig. 308 Disconnect the connector from the air mix control motor

12. Disconnect the connector from the air mix control motor. Remove the wire harness clips, the connector clip, and the wire harness.

13. Remove the self-tapping screws and the passenger's heater duct. Remove the self-tapping screws and the expansion valve cover. Remove the self-tapping screw and the heater core cover. Remove the self-tapping screws, the heater pipe bracket, and the grommet, and carefully pull out the heater core.

To install:

14. Install the heater core and the evaporator core in the reverse order of removal.

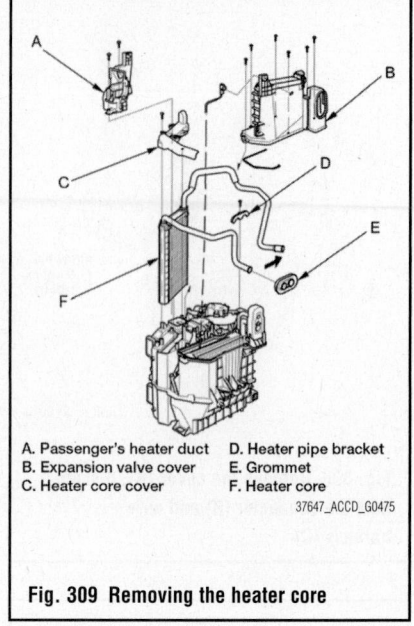

A. Passenger's heater duct   D. Heater pipe bracket
B. Expansion valve cover   E. Grommet
C. Heater core cover   F. Heater core

37647_ACCD_G0475

Fig. 309 Removing the heater core

15. Install the heater unit in the reverse order of removal, and note these items:

- Do not interchange the inlet and outlet heater hoses, and install the hose clamps securely.
- Refill the cooling system with engine coolant.
- Make sure that there is no coolant leakage.
- Make sure that there is no air leakage.

16. Do the battery terminal reconnection procedure.

STEERING

POWER RACK & PINION STEERING GEAR

REMOVAL & INSTALLATION

See Figures 310 through 329.

Special Tools Required*:
- Engine Hanger Adapter VSB02C000015
- Engine Support Hanger, A and Reds AAR-T1256
- Ball Joint Remover, 28 mm 07MAC-SL0A202
- Ball Joint Thread Protector, 12 mm 07AAF-SDAA100
- Subframe Adapter VSB02C000016

➥***Available through the Honda Tool and Equipment Program, 888-424-6857.**

✳✳ WARNING

SRS components are located in this area. Review the SRS component

locations: 4-Door, 2-Door and the precautions and procedures before doing repairs or service.

Note these items during removal:
- Using clean solvent and a brush, wash any oil and dirt off the valve body unit, it's lines, and the end of the steering gearbox. Blow dry with compressed air.
- Be sure to remove the steering wheel before disconnecting the steering joint, or damage to the cable reel can occur.
- Lower the front subframe from the body, and remove the steering gearbox through the gap produced by lowering the front subframe.

1. Drain the power steering fluid.

2. Do the battery terminal disconnection procedure.

3. Raise and support the vehicle.

4. Remove the front wheels.

5. Remove the driver's airbag, and steering wheel.

6. Remove the steering joint cover.

7. Loosen the upper steering joint bolt, and remove the lower steering joint bolt. Disconnect the steering joint by sliding the steering joint into the column shaft. Tighten the upper steering joint bolt to hold the steering joint in place.

➥**Do not disconnect the steering joint from the column shaft.**

8. Remove the center guide (if equipped) from the top of the pinion shaft, and discard it.

9. Apply vinyl tape to the splines on the pinion shaft.

10. Remove the hood support rod, then use it to prop the hood in the wide-open position.

11. Remove the front grille cover:

12. Remove the strut brace (if equipped).

13. Remove the P/S heat shield.

14. Attach the engine hanger adapter (VSB02C000015) to the threaded hole in the cylinder head.

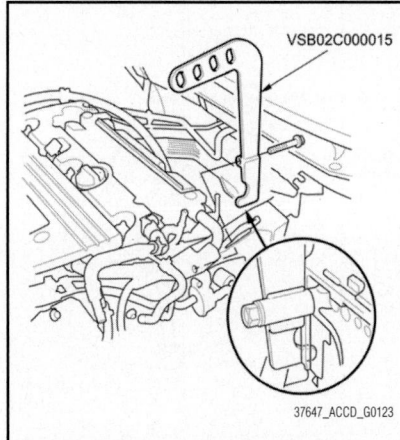

Fig. 310 Attach the engine hanger adapter to the threaded hole in the cylinder head

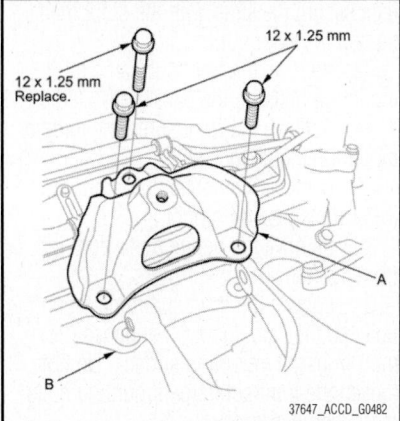

Fig. 312 A/T: Remove the rear engine mount upper bracket (A) from the base bracket (B)

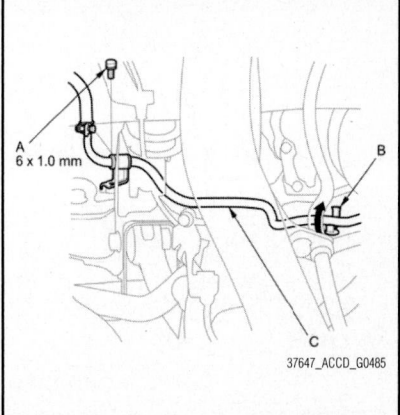

Fig. 314 Remove the return line clamp bolt (A), release the return line clamp (B), and remove the return line (C)

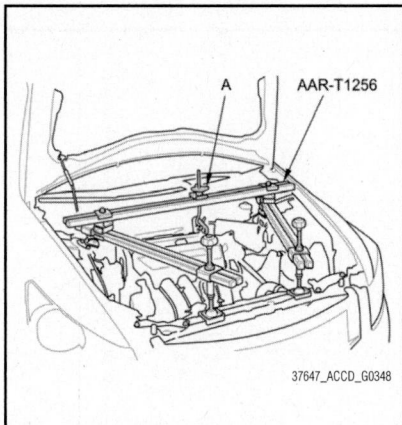

Fig. 311 Tighten the wing nut (A) by hand to lift and support the engine/transmission

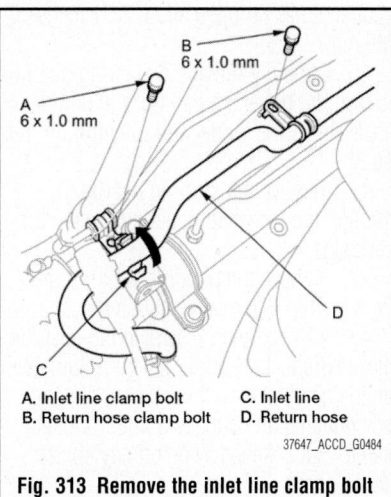

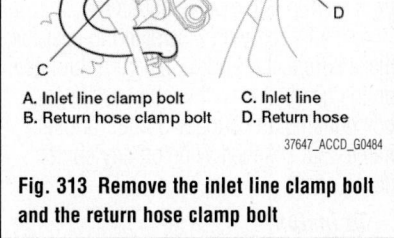

A. Inlet line clamp bolt   C. Inlet line
B. Return hose clamp bolt   D. Return hose

Fig. 313 Remove the inlet line clamp bolt and the return hose clamp bolt

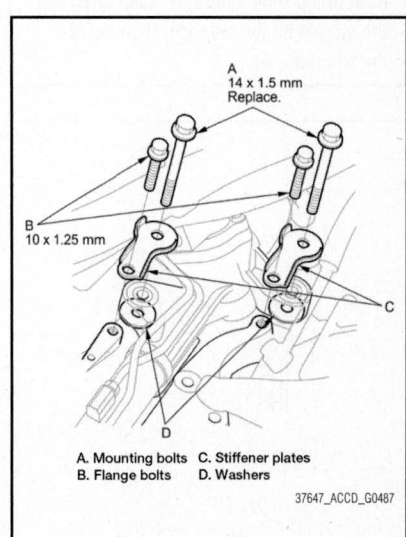

A. Mounting bolts   C. Stiffener plates
B. Flange bolts   D. Washers

Fig. 315 Remove the gearbox mounting bracket (A) and mounting cushion (B)

15. Install the engine support hanger (AAR-T1256), then attach the hook to the slotted hole in the hanger adapter. Tighten the wing nut by hand to lift and support the engine/transmission.

➡**Be careful when working around the windshield.**

16. Remove the rear engine mount.
17. A/T: Remove the rear engine mount upper bracket from the base bracket.
18. Remove the inlet line clamp bolt and the return line clamp bolt.
19. Loosen the flare nuts, and disconnect the inlet line and the return line.
20. Remove the inlet line clamp bolt and the return hose clamp bolt.
21. Release the return hose clamp, and remove the return hose.
22. Remove the return line clamp bolt.
23. Release the return line clamp, and remove the return line.
24. Remove the flange bolts from the passenger's side of the steering gearbox,

then remove the gearbox mounting bracket and mounting cushion.

25. Remove the mounting bolts and the flange bolts from the driver's side of the steering gearbox, and remove the stiffener plates and the washers.
26. Remove cotter pin from the tie-rod end ball joint, then remove the nut on both sides.
27. Disconnect the tie-rod end ball joint from the knuckle using the ball joint thread protector and the ball joint remover on both side.

➡**Be careful not to damage the ball joint boot when installing the remover.**

28. Raise the vehicle.
29. Remove the front splash shield.
30. Remove exhaust pipe A.
31. Remove the damper fork mounting nut and the mounting bolt.
32. Attach the front subframe adapter (VSB02C000016) to the subframe by looping the strap over the front of the subframe,

Fig. 316 Remove the mounting bolts and the flange bolts from the driver's side of the steering gearbox, and remove the stiffener plates and the washers

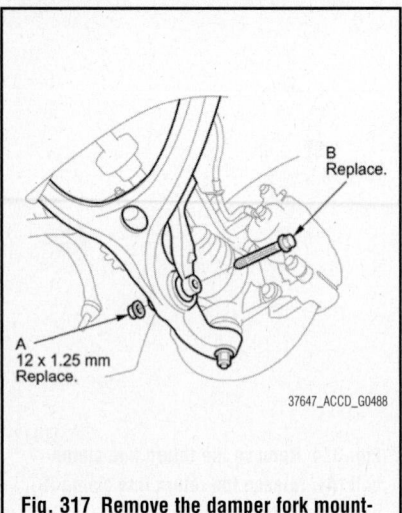

Fig. 317 Remove the damper fork mounting nut (A) and the mounting bolt (B)

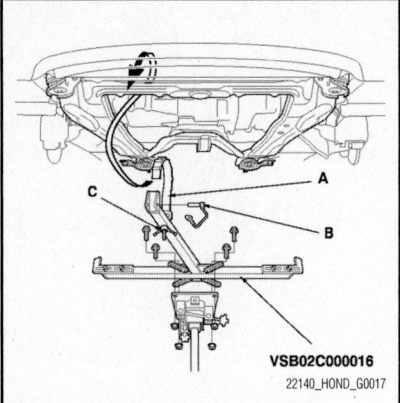

Fig. 318 Attach the front subframe adapter (VSB02C000016) to the front subframe by looping the strap (A) over the front of the front subframe, then secure the strap with the stop (B), then tighten the wing nut (C)

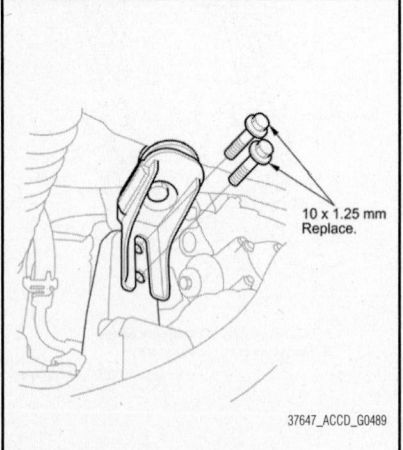

Fig. 319 Remove the front subframe middle mounting bolts on the passenger's side

then secure the strap with the stop, then tighten the wing nut.

33. Remove the front subframe middle mounting bolts on the passenger's side.

34. Remove the front subframe middle mount on the driver's side.

35. Remove the flange nuts from the lower transmission mount.

36. Remove the flange bolts from the front subframe front stiffeners.

37. Loosen the front subframe mounting bolts so they are about 0.79 inches (20 mm) from the mounting surface. Do not loosen the front subframe mounting bolts more than necessary.

38. Remove the flange bolts and front subframe mounting bolts from the front subframe rear stiffeners.

39. Lower the jack slowly until the front subframe has dropped about 2.71 inches (69 mm).

40. Carefully move the steering gearbox toward the driver's side until the pinion shaft clears the fenderwell opening on the body.

41. Remove the steering gearbox through the fenderwell opening on the driver's side.

42. Remove the pinion shaft grommet from the top of the valve housing.

43. After removing the steering gearbox, make sure that no power steering fluid gets on the gearbox mount cushions, the gearbox housing, the surface of the front subframe, and stiffener. Wipe off any spilled fluid at once.

To install:

44. Before installing the steering gearbox, make sure that no power steering fluid is on the mating surface of the steering gearbox and the front subframe. To prevent the gearbox mounting bolts from loosening after the installation, remove any power steering fluid from the mount cushions and the bolt holes.

45. Wrap vinyl tape over the splines on the pinion shaft.

46. Install the pinion shaft grommet. Align the slot in the pinion shaft grommet with the lug portion on the valve housing. Make sure there is no gap between the grommet and the valve housing.

47. Turn the lip of the pinion shaft grommet.

48. Slide the steering gearbox between the front subframe and the body from the driver's side.

49. Carefully move the steering gearbox toward the passenger's side until the pinion shaft clears the fenderwell opening on the body.

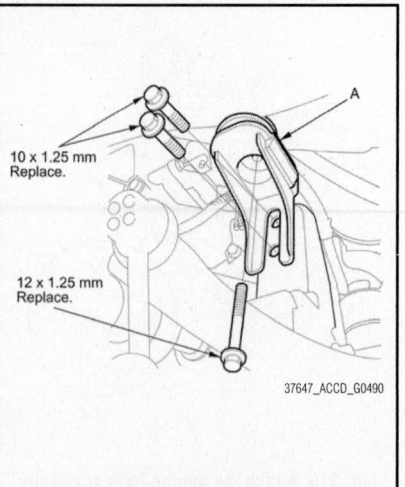

Fig. 320 Remove the front subframe middle mount (A) on the driver's side

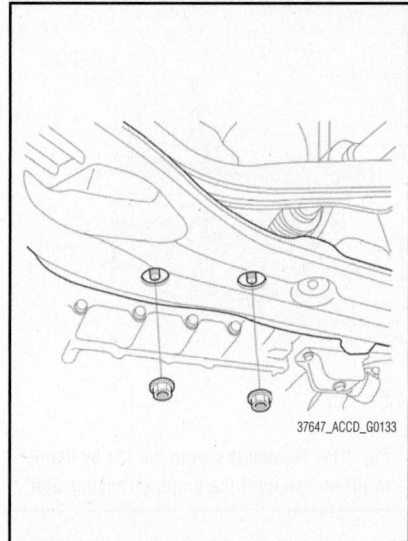

Fig. 321 Remove the flange nuts (A) from the lower transmission mount

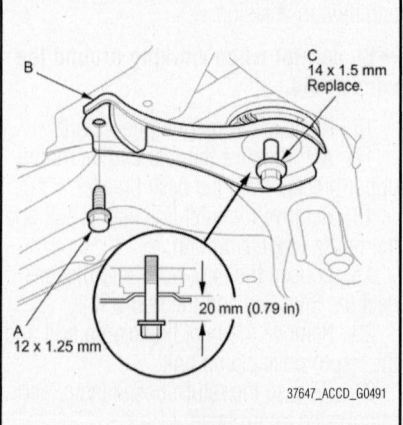

Fig. 322 Remove the flange bolts (A) from the front subframe front stiffeners (B), and loosen the front subframe mounting bolts (C)

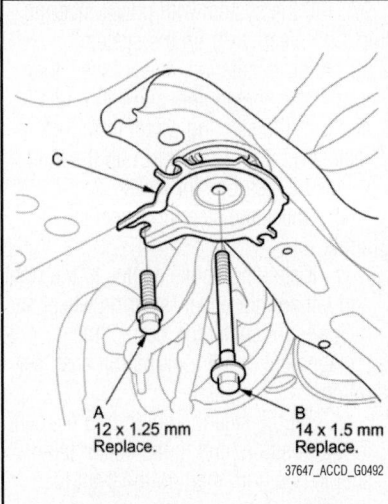

Fig. 323 Remove the flange bolts (A) and front subframe mounting bolts (B) from the front subframe rear stiffeners (C)

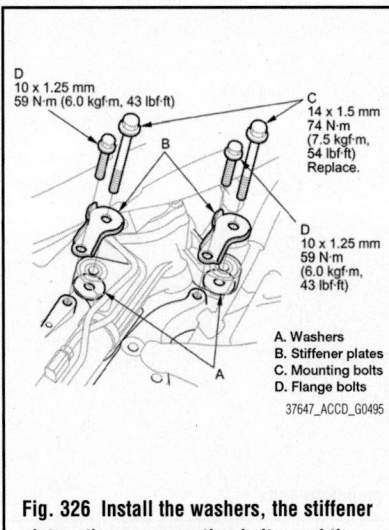

Fig. 326 Install the washers, the stiffener plates, the new mounting bolts, and the flange bolts on the driver's side of the gearbox

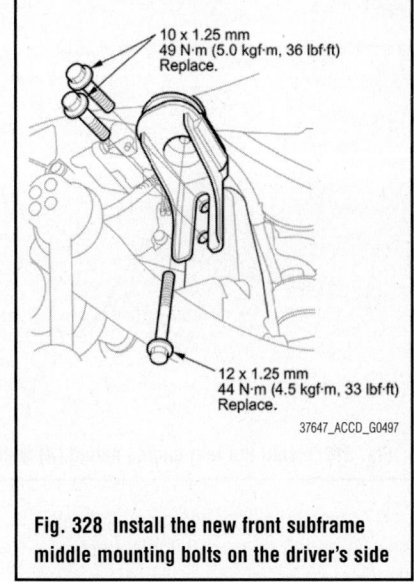

Fig. 328 Install the new front subframe middle mounting bolts on the driver's side

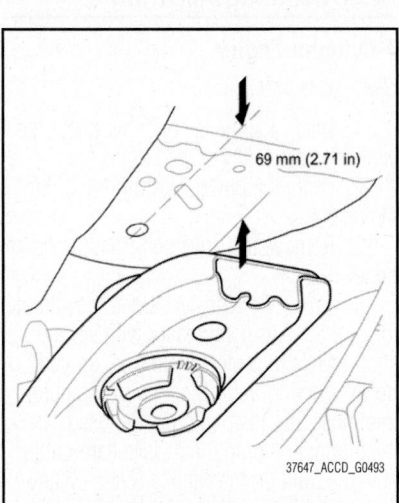

Fig. 324 Lower the jack slowly until the front subframe has dropped about 2.71 inches (69 mm)

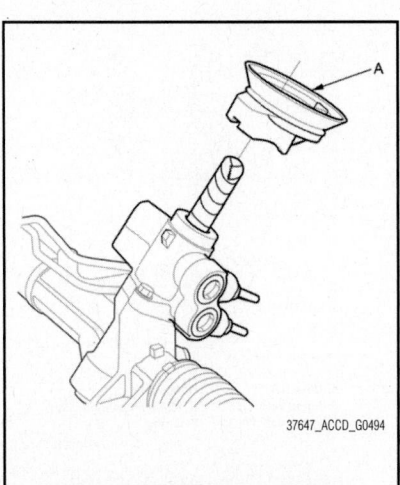

Fig. 325 Remove the pinion shaft grommet (A) from the top of the valve housing

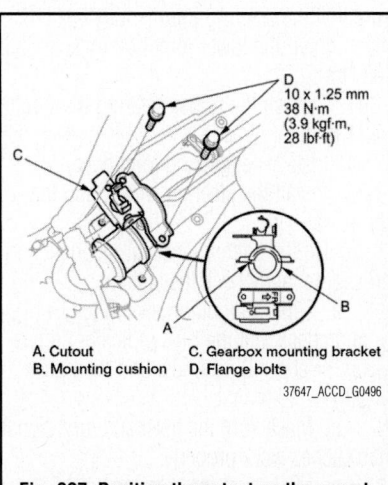

Fig. 327 Position the cutout on the mounting cushion

50. Continue moving the gearbox toward the passenger's side until the steering gearbox is in position.

51. Install the washers, the stiffener plates, the new mounting bolts, and the flange bolts on the driver's side of the gearbox. Then loosely install the mounting bolts and the flange bolts.

52. Position the cutout on the mounting cushion as shown, and install it on the passenger's side of the steering gearbox.

53. Install the gearbox mounting bracket over the mounting cushion, and tighten the flange bolts to 28 ft. lbs. (38 Nm).

54. Tighten the flange bolts on the driver's side of the steering gearbox to 28 ft. lbs. (38 Nm) alternately in two steps.

55. Install the front subframe front stiffeners and the front subframe rear stiffeners, then loosely install the new front subframe

mounting bolts, the new flange bolts, and the flange bolts.

56. Align the front subframe using the 15.7 mm end of the subframe alignment pin. Vertically install the subframe alignment pin, and align the right rear corner of the front subframe and the vehicle frame holes, then loosely tighten the subframe mounting bolt until the front subframe contacts the body frame.

57. Loosely tighten the left rear subframe mounting bolt with the same procedure as the right rear using the subframe alignment pin.

58. Tighten the right rear subframe mounting bolt to 76 ft. lbs. (103 Nm) with the subframe alignment pin installed.

59. Tighten the left rear subframe mounting bolt to 76 ft. lbs. (103 Nm) with the subframe alignment pin installed.

60. Tighten the front subframe mounting bolts to 76 ft. lbs. (103 Nm).

61. Tighten the flange bolts to 40 ft. lbs. (54 Nm).

62. Install the flange nuts to the lower transmission mount, and tighten them to 33 ft. lbs. (44 Nm).

63. Install the new front subframe middle mounting bolts on the driver's side, and tighten them to the specified torque.

64. Install the new front subframe middle mounting bolts on the passenger's side, and tighten them to the same torque as the driver's side.

65. Lower the transmission jack supporting the front subframe.

66. A/T: Install the rear engine mount upper bracket to the base bracket with a new mounting bolt and mounting bolts, and tighten them to 40 ft. lbs. (54 Nm).

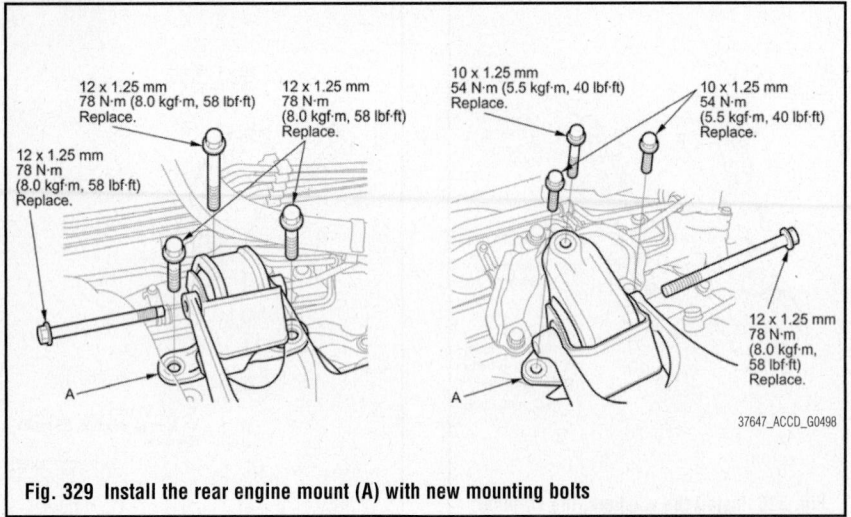

Fig. 329 Install the rear engine mount (A) with new mounting bolts

67. Install the rear engine mount with new mounting bolts, and lightly tighten them.

68. Remove the engine support hanger, the hanger balance bar, and the hanger adapter set.

69. Tighten the rear engine mount mounting bolts to the specified torque.

70. Install the new damper fork mounting bolt and the new mounting nut, and loosely tighten the nut.

71. Install exhaust pipe A.

72. Install the front splash shield.

73. Lower the vehicle.

74. On both sides, wipe off any grease contamination from the ball joint tapered section and threads. Reconnect the tie-rod end ball joint to the knuckle. Install the nut, and tighten it to 40 ft. lbs. (54 Nm).

75. Install a new cotter pin.

76. Loosely connect the return line and inlet line to the valve housing by hand.

77. Install the return line clamp bolt.

78. Install the return line to the return hose clamp, and clamp it.

79. Install the inlet line clamp bracket bolt and the return hose clamp bracket bolt.

80. Install the return hose to the return hose clamp, and clamp it.

81. Install the inlet line clamp bracket bolt and the return line clamp bolt.

82. Tighten the flare nuts.

83. Install the P/S heat shield with the flange bolts.

84. Install the strut brace.

85. Install the front grille cover.

86. Place a floor jack under the lower arm, and raise the suspension to load it with the vehicle's weight. Do not place the jack against the ball joint pin of the knuckle.

87. Tighten the damper fork mounting nut while holding the mounting bold to 47 ft. lbs. (64 Nm).

88. Install the front wheels, then set the wheels in the straight ahead position.

➡**Before installing the wheel, clean the mating surfaces of the brake disc and the inside of the wheel.**

89. Center the steering rack within its stroke.

90. Loosen the upper steering joint bolt, and slip the lower end of the steering joint onto the pinion shaft taking care to align the gap within the angle of 89 +/- 20 degrees.

➡**Pick up the tabs of the pinion shaft grommet, and turn up the lip of the pinion shaft grommet securely in place. Make sure that light does not enter from the space between the pinion shaft grommet and the body.**

91. Align the bolt hole on the steering joint with the groove around the pinion shaft, then loosely install the lower steering joint bolt. Be sure that the joint bolt is securely in the groove in the pinion shaft.

92. Pull on the steering joint to make sure that the steering joint is fully seated, then tighten the lower joint bolt to 21 ft. lbs. (28 Nm).

93. Tighten the upper steering joint bolt to 21 ft. lbs. (28 Nm).

94. Install the steering joint cover.

95. Install the steering wheel, and the driver's airbag.

96. Do the battery terminal reconnection procedure, and check these items:

a. Turn the ignition switch to ON (II) and check that the SRS indicator comes on for about 6 seconds, and then goes off.

b. Make sure the horn and turn signal switches work properly.

c. Make sure the steering wheel switches work properly.

97. Fill the system with power steering fluid, and bleed air from the system.

98. After installation, check these items.

a. Start the engine, allow it to idle, and turn the steering wheel from lock to lock several times to warm up the fluid. Check the gearbox for leaks.

b. Check the steering wheel spoke angle.

c. If steering spoke angles to the right and left are not equal (steering wheel and rack are not centered), correct the engagement of the joint/pinion shaft serrations.

d. Set the steering column to the center tilt position, and to the center telescopic position, then do the front toe inspection

POWER STEERING PUMP

REMOVAL & INSTALLATION

4-Cylinder Engine

See Figure 330.

1. Place a suitable container under the vehicle to catch any spilled fluid.

2. Drain the power steering fluid from the reservoir.

3. Remove the drive belt from the pump pulley.

4. Cover the auto-tensioner, the alternator, and the A/C compressor with several shop towels to protect them from spilled power steering fluid. Disconnect the pump inlet hose and the pump outlet hose from the pump, and plug them. Take care not to spill the fluid on the vehicle. Wipe off any spilled fluid at once. Do not turn the steering wheel with the pump removed.

5. Remove the pump mounting bolts, then remove the pump.

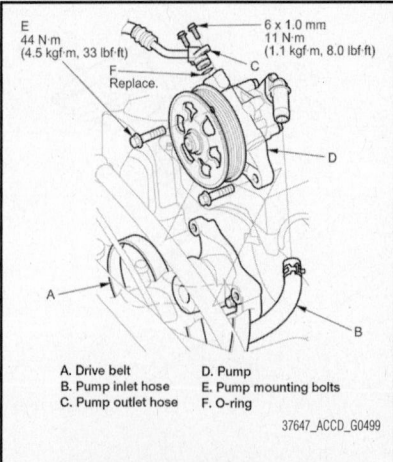

A. Drive belt
B. Pump inlet hose
C. Pump outlet hose
D. Pump
E. Pump mounting bolts
F. O-ring

Fig. 330 Disconnect the pump inlet hose and the pump outlet hose from the pump

6. Cover the opening of the pump with a piece of tape to prevent foreign material from entering the pump.

To install:

7. Transfer the pump inlet hose and the pump outlet hose from the original pump onto the new pump with a new O-ring.

8. Loosely install the pump in the pump bracket with the mounting bolts, then tighten the pump fittings.

9. Tighten the pump mounting bolts to 33 ft. lbs. (44 Nm).

10. Install the drive belt.

➡Note these items during drive belt installation:

- Inspect the belt for wear and cracks. Replace the belt if necessary.
- Make sure that the belt is properly positioned on the pulleys.
- Do not get power steering fluid or grease on the auto-tensioner, the alternator, the A/C compressor, and the drive belt, or the pulley faces. Clean off any fluid or grease before installation.

11. Fill the reservoir to the upper level line.

12. Start the engine, and check for leaks.

6-Cylinder Engine

See Figure 331.

1. Place a suitable container under the vehicle to catch any spilled fluid.

2. Drain the power steering fluid from the reservoir.

3. Remove the engine cover.

4. Remove the drive belt from the pump pulley.

5. Cover the auto-tensioner, the alternator, and the A/C compressor with several shop towels to protect them from spilled power steering fluid. Disconnect the pump inlet hose and the pump outlet hose from the pump, and the plug them. Take care not to spill the fluid on the vehicle. Wipe off any spilled fluid at once. Do not turn the steering wheel with the pump removed.

6. Remove the pump mounting bolts, then remove the pump.

7. Cover the opening of the pump with a piece of tape to prevent foreign material from entering the pump.

To install:

8. Transfer the pump inlet hose and the pump outlet hose from the original pump onto the new pump with a new O-ring.

9. Loosely install the pump in the pump bracket with the mounting bolts, then tighten the pump fittings.

10. Tighten the pump mounting bolts to 16 ft. lbs. (22 Nm).

11. Install the drive belt.

➡Note these items during drive belt installation:

- Inspect the belt for wear and cracks. Replace the belt if necessary.
- Make sure that the belt is properly positioned on the pulleys.
- Do not get power steering fluid or grease on the auto-tensioner, the alternator, the A/C compressor, and the drive belt or pulley faces. Clean off any fluid or grease before installation.

12. Install the engine cover.

13. Fill the reservoir to the upper level line.

14. Start the engine, and check for leaks.

POWER STEERING FLUID

FLUID RECOMMENDATIONS

Always use Honda Power Steering Fluid. Using any other type of power steering fluid or automatic transmission fluid can cause increased wear, fluid leaks, and poor steering in cold weather.

FLUID LEVEL CHECK

Check the reservoir at regular intervals, and add the recommended fluid as necessary. Always use Honda Power Steering Fluid. Using any other type of power steering fluid or automatic transmission fluid can cause increased wear, fluid leaks, and poor steering in cold weather.

➡If the fluid is contaminated, the screen in the reservoir may be partially blocked. Inspect the reservoir screen for any debris. If the reservoir screen is clogged, replace the reservoir, and check for the source of the contamination.

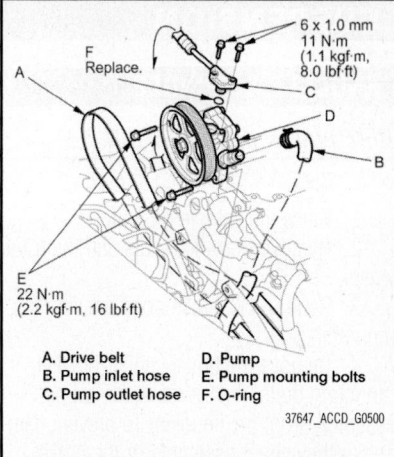

A. Drive belt
B. Pump inlet hose
C. Pump outlet hose
D. Pump
E. Pump mounting bolts
F. O-ring

37647_ACCD_G0500

Fig. 331 Disconnect the pump inlet hose and the pump outlet hose from the pump

FLUID FILL PROCEDURE

1. Remove the reservoir from its holder. Raise the reservoir, then disconnect the return hose to drain the reservoir. Take care not to spill the fluid on the vehicle. Wipe off any spilled fluid at once.

➡Inspect the reservoir screen for any debris. If the reservoir screen is clogged, replace the reservoir, and check for the source of the contamination.

2. Connect a hose of suitable diameter to the disconnected return hose, and put the hose end in a suitable container.

3. Start the engine, let it run at idle, and turn the steering wheel from lock to lock several times. When fluid stops running out of the hose, shut off the engine. Discard the fluid.

➡Stop the engine immediately once the fluid stops running out of the hose to prevent pump damage.

4. Reinstall the return hose on the reservoir.

5. Fill the reservoir to the upper level line.

6. Start the engine, and run it at idle. Turn the steering from lock to lock several times to bleed air from the system.

7. Recheck the fluid level and, add some if necessary. Do not fill the reservoir beyond the upper level line.

8. If the fluid is contaminated, dark, or discolored, repeat the procedure as necessary until the system is clean.

KNUCKLE & SPINDLE

REMOVAL & INSTALLATION

See Figures 332 through 335.

1. Raise the vehicle on a lift.
2. Remove the wheel nuts and the front wheel.
3. Remove the brake hose bracket mounting bolt.
4. Remove the brake caliper bracket mounting bolts, then remove the caliper assembly from the knuckle. To prevent damage to the caliper assembly or the brake hose, use a short piece of wire to hang the caliper assembly from the undercarriage. Do not twist the brake hose excessively.
5. Remove the wheel speed sensor harness bracket and the wheel speed sensor from the knuckle. Do not disconnect the wheel speed sensor connector.
6. Pry up the stake on the spindle nut, then remove the nut.
7. Remove the front brake disc.
8. Check the front hub for damage and cracks.
9. Remove the cotter pin from the tie-rod end ball joint, then remove the nut.

➥During installation, install the new cotter pin after tightening the nut, and bend its end.

10. Disconnect the tie-rod end ball joint from the knuckle using the ball joint thread protector and the ball joint remover.

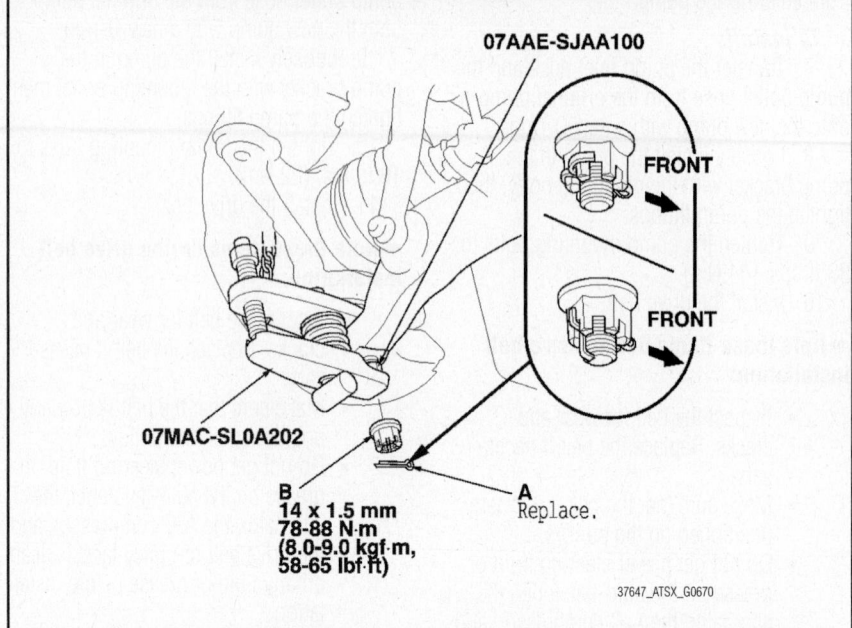

Fig. 333 Remove the cotter pin (A) from the knuckle ball joint, then remove the castle nut (B)

11. Remove the cotter pin from the knuckle ball joint, then remove the castle nut.

➥During installation, insert the new cotter pin into the ball joint pin hole from the front to the rear of the vehicle, and bend its end as shown. Check the ball joint pin hole direction before connecting the ball joint.

12. Disconnect the knuckle ball joint from the lower arm using the ball

joint thread protector and the ball joint remover.

➥Be careful not to damage the ball joint boot when installing the remover. Do not force or hammer on the lower arm, or pry between the lower arm and the knuckle. You could damage the ball joint.

13. Remove the cotter pin from the upper arm ball joint, then remove the castle nut.

➥During installation, insert the new cotter pin into the ball joint pin hole from the front to the rear of the vehicle, and bend its end as shown. Check the ball joint pin hole direction before connecting the ball joint.

14. Disconnect the upper arm ball joint from the knuckle using the ball joint thread protector and the ball joint remover.
15. Pull the knuckle outward, and separate the outboard joint from the front hub using a plastic hammer.

➥Do not pull the driveshaft end outward. The driveshaft inboard joint may come apart. During installation, apply grease to the mating surfaces of the wheel bearing and the driveshaft outboard joint.

To install:

16. Install the knuckle/hub in the reverse order of removal, and note these items:
   a. First install all of the components, and lightly tighten the bolts and the nuts,

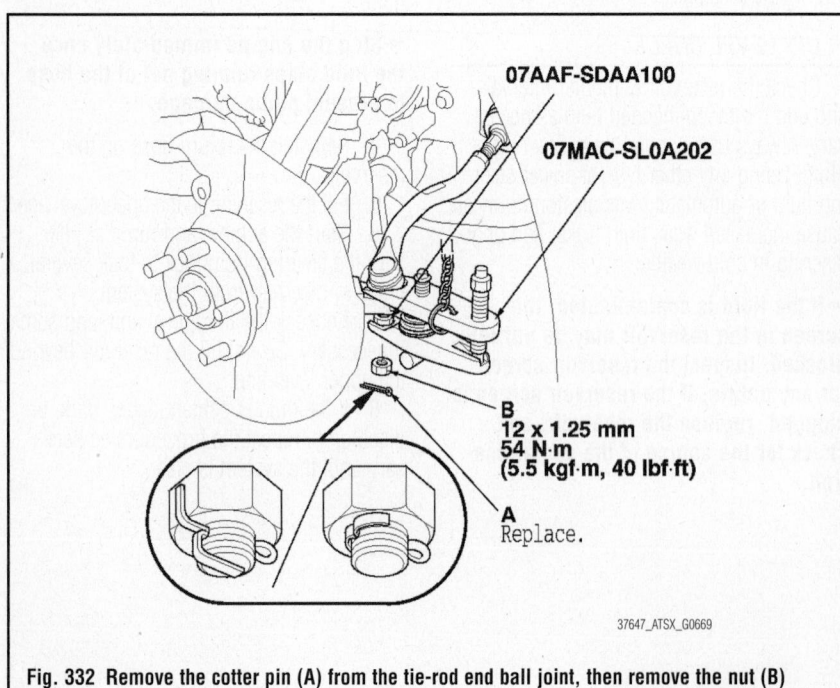

Fig. 332 Remove the cotter pin (A) from the tie-rod end ball joint, then remove the nut (B)

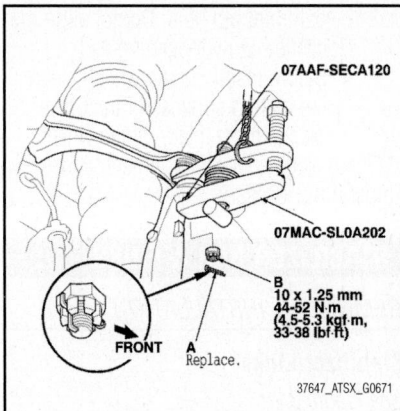

Fig. 334 Remove the cotter pin (A) from the upper arm ball joint, then remove the castle nut (B)

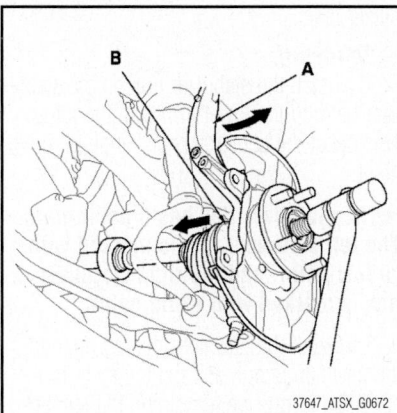

Fig. 335 Pull the knuckle (A) outward, and separate the outboard joint (B) from the front hub using a plastic hammer

then raise the suspension to load it with the vehicle's weight before fully tightening to the specified torque. Do not place the jack against the ball joint pin of the knuckle.

b. Be careful not to damage the ball joint boot when connecting the knuckle.

c. Before connecting the ball joint, degrease the threaded section and the tapered portion of the ball joint pin, the ball joint connecting hole, the threaded section, and the mating surfaces of the castle nut.

d. Torque the castle nut to the lower torque specification, then tighten it only far enough to align the slot with the ball joint pin hole. Do not align the castle nut by loosening it.

e. Use a new spindle nut on reassembly.

f. Before installing the spindle nut, apply a small amount of engine oil to the seating surface of the nut. After tighten-

ing, use a drift to stake the spindle nut shoulder against the driveshaft.

g. Before installing the brake disc, clean the mating surfaces of the front hub and the inside of the brake disc.

h. Before installing the wheel, clean the mating surfaces of the brake disc and the inside of the wheel.

17. Check the wheel alignment, and adjust it if necessary.

LOWER BALL JOINTS

REMOVAL & INSTALLATION

See Figures 336 through 338.

Special Tools Required:
• Ball joint thread protector, 14 mm 07AAE-SJAA100
• Ball joint thread protector, 12 mm 07AAF-SDAA100
• Ball joint thread protector, 10 mm 07AAF-SECA120
• Ball joint remover, 32 mm 07MAC-SL0A102
• Ball joint remover, 28 mm 07MAC-SL0A202

➡**Always use a ball joint remover to disconnect a ball joint. Do not strike the housing or any other part of the ball joint connection to disconnect it.**

1. Install a hex nut or the ball joint thread protector onto the threads of the ball joint.

➡**Using a hex nut, make sure the nut is flush with the ball joint pin end to prevent damage to the threaded end of the ball joint pin.**

2. Apply grease to the ball joint remover on the areas shown. This will ease the installation of the tool, and prevent damage to the pressure bolt threads.

3. Loosen the pressure bolt, and install

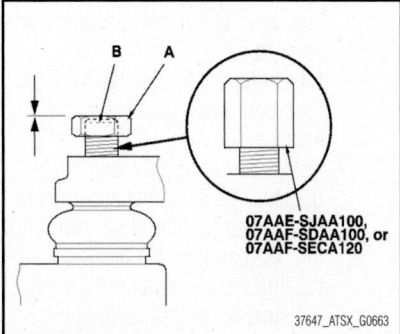

Fig. 336 Install a hex nut (A) or the ball joint thread protector onto the threads of the ball joint (B)

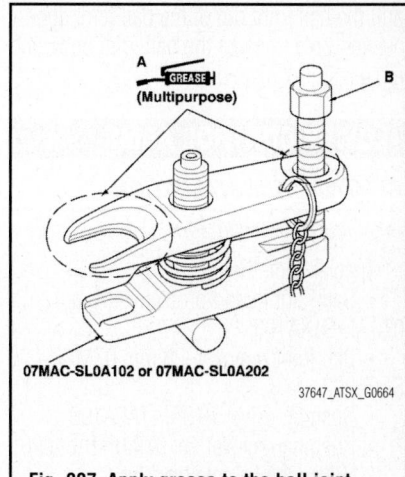

Fig. 337 Apply grease to the ball joint remover on the areas shown (A)

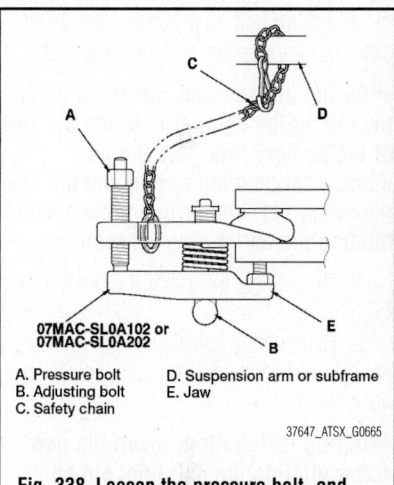

A. Pressure bolt
B. Adjusting bolt
C. Safety chain
D. Suspension arm or subframe
E. Jaw

Fig. 338 Loosen the pressure bolt, and install the ball joint remover as shown

the ball joint remover as shown. Insert the jaws carefully, making sure not to damage the ball joint boot. Adjust the jaw spacing by turning the adjusting bolt.

➡**Fasten the safety chain securely to a suspension arm or the subframe. Do not fasten it to a brake line or wire harness.**

4. After adjusting the adjusting bolt, make sure the head of the adjusting bolt is in the position shown to allow the jaw to pivot.

5. With a wrench, tighten the pressure bolt until the ball joint pin pops loose from the ball joint connecting hole. If necessary, apply penetrating type lubricant to loosen the ball joint pin.

➡**Do not use pneumatic or electric tools on the pressure bolt.**

6. Remove the ball joint remover, then remove the nut or the ball joint thread protector from the end of the ball joint pin, and

pull the ball joint out of the ball joint connecting hole. Inspect the ball joint boot, and replace it if damaged.

LOWER CONTROL ARMS

REMOVAL & INSTALLATION

See Figures 339 and 340.

Special Tools Required:
- Ball joint thread protector, 14 mm 07AAE-SJAA100
- Ball joint remover, 28 mm 07MAC-SL0A202
- Bushing driver 070AF-TA0A100
- Bushing receiver set 070AF-TA0A220

1. Raise the vehicle on a lift.
2. Remove the front wheel.
3. Remove the damper pinch bolt and the damper fork mounting nut while holding the mounting bolt, then remove the damper fork from the damper and the lower arm.

➡**During installation, insert the aligning tab on the damper unit into the slot of the damper fork. Use the new damper fork mounting bolt and the new mounting nut, and torque the nut while holding the bolt during reassembly.**

4. Disconnect the stabilizer link from the lower arm.
5. Remove the cotter pin from the knuckle ball joint, then remove the castle nut.

➡**During installation, insert the new cotter pin into the ball joint pin hole from the front to the rear of the vehicle, and bend it. Check the ball joint pin**

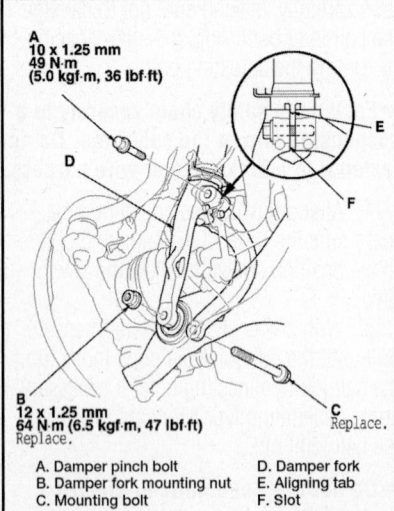

A.
10 x 1.25 mm
49 N·m
(5.0 kgf·m, 36 lbf·ft)

B.
12 x 1.25 mm
64 N·m (6.5 kgf·m, 47 lbf·ft)
Replace.

C. Replace.

A. Damper pinch bolt
B. Damper fork mounting nut
C. Mounting bolt
D. Damper fork
E. Aligning tab
F. Slot

37647_ATSX_G0666

Fig. 339 Remove the damper pinch bolt and the damper fork mounting nut while holding the mounting bolt

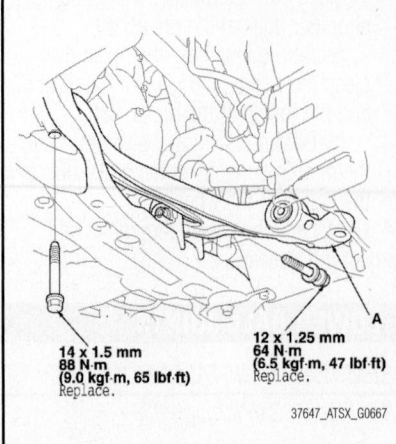

14 x 1.5 mm
88 N·m
(9.0 kgf·m, 65 lbf·ft)
Replace.

12 x 1.25 mm
64 N·m
(6.5 kgf·m, 47 lbf·ft)
Replace.

37647_ATSX_G0667

Fig. 340 Remove the lower arm mounting bolts, and remove the lower arm (A)

hole direction before connecting the ball joint.

6. Disconnect the knuckle ball joint from the lower arm using the ball joint thread protector and the ball joint remover.

➡**Be careful not to damage the ball joint boot when installing the remover. Do not force or hammer on the lower arm, or pry between the lower arm and the knuckle. You could damage the ball joint.**

7. Remove the lower arm mounting bolts, and remove the lower arm.

➡**Use new lower arm mounting bolts during reassembly.**

To install:

8. Install the lower arm in the reverse order of removal, and note these items:
- First install all of the components, and lightly tighten the bolts and the nuts, then raise the suspension to load it with the vehicle's weight before fully tightening it to the specified torque. Do not place the jack against the ball joint pin of the knuckle.
- Be careful not to damage the ball joint boot when connecting the knuckle.
- Before connecting the ball joint, degrease the threaded section and the tapered portion of the ball joint pin, the ball joint connecting hole, the threaded section, and the mating surfaces of the castle nut.
- Torque the castle nut to the lower torque specification, 58–65 ft. lbs. (78–88 Nm), then tighten it only far enough to align the slot with the

ball joint pin hole. Do not align the castle nut by loosening it.
- Before installing the wheel, clean the mating surfaces of the brake disc and the inside of the wheel.

9. Check the wheel alignment, and adjust it if necessary.

STABILIZER BAR & LINKS

REMOVAL & INSTALLATION

Stabilizer Links

See Figure 341.

1. Raise the vehicle on a lift.
2. Remove the front wheel.
3. Remove the self-locking nut and the flange nut while holding the respective joint pin with a hex wrench, then remove the stabilizer link.

To install:

4. Install the stabilizer link on the stabilizer bar and the lower arm with the joint pins set at the center of their range of movement.

➡**The stabilizer link has a paint mark. The left stabilizer link is marked with yellow paint, and the right stabilizer link is marked with white paint.**

5. Install the new self-locking nut and the new flange nut, and tighten them to the specified torque while holding the respective joint pin with a hex wrench.
6. Clean the mating surfaces of the brake disc and the inside of the wheel, then install the front wheel.
7. Test-drive the vehicle.
8. After 5 minutes of driving, tighten the self-locking nut again to the specified torque.

Stabilizer Bar

4-CYLINDER ENGINE

See Figures 310, 311, 342 through 350.

Special Tools Required*:
- Engine Hanger Adapter VSB02C000015
- Engine Support Hanger, A and Reds AAR-T1256
- Subframe Adapter VSB02C000016
- Subframe Alignment Pin 070AG-SJAA10S
- *Available through the Honda Tool and Equipment Program, 888-424-6857.

Note these items during replacement:
- Be sure to remove the steering wheel before disconnecting the steering joint. Damage to the cable reel can occur.

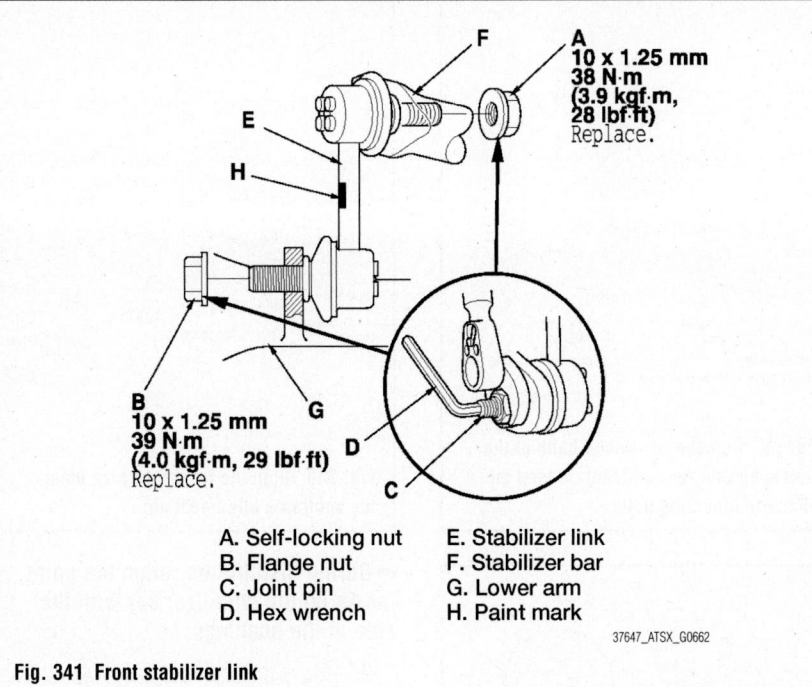

A
10 x 1.25 mm
38 N·m
(3.9 kgf·m,
28 lbf·ft)
Replace.

B
10 x 1.25 mm
39 N·m
(4.0 kgf·m, 29 lbf·ft)
Replace.

A. Self-locking nut E. Stabilizer link
B. Flange nut   F. Stabilizer bar
C. Joint pin   G. Lower arm
D. Hex wrench   H. Paint mark

37647_ATSX_G0662

Fig. 341 Front stabilizer link

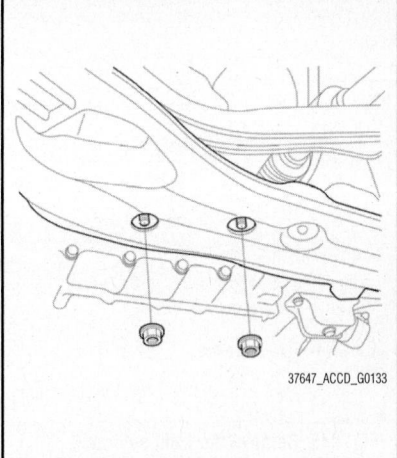

37647_ACCD_G0133

Fig. 343 Remove the nuts (A) securing of the lower transmission mount

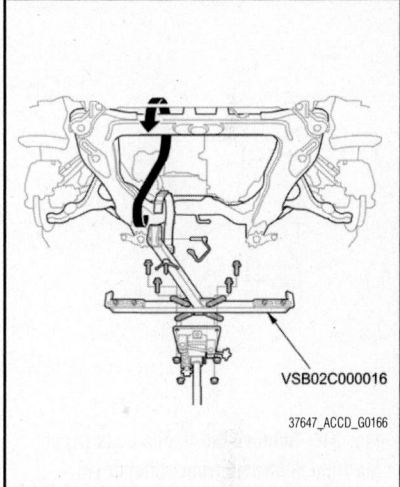

VSB02C000016

37647_ACCD_G0166

Fig. 344 Attach the subframe adapter to the subframe by looping the belt over the front of the subframe, then secure the belt with its stop

• Lower the front subframe from the body, and replace the front stabilizer bar through the gap created by lowering the front subframe.

1. Remove the hood support rod, then use it to prop the hood in the wide-open position.

2. Remove the front grille cover:

3. Do the battery terminal disconnection procedure.

4. Raise and support the vehicle.

5. Remove the front wheels.

6. Remove the driver's airbag and the steering wheel.

7. Remove the steering joint cover.

8. Loosen the steering joint upper bolt, and remove the steering joint lower bolt. Disconnect the steering joint by sliding the steering joint into the column shaft. Tighten the steering joint upper bolt to hold the column shaft.

➡**Do not disconnect the steering joint from the column shaft. If the center guide is in place and has not moved, leave it in place. If the center guide has moved or been removed, discard it.**

9. Disconnect the Power Steering Pressure (PSP) switch connector.

10. Remove the power steering pump outlet hose mounting bolt.

11. Remove the front strut brace (if equipped).

12. Attach the engine hanger adapter (VSB02C000015) to the threaded hole in the cylinder head.

13. Install the engine support hanger (AAR-T1256), then attach the hook to the slotted hole in the engine hanger adapter. Tighten the wing nut by hand to lift and support the engine/transmission.

➡**Be careful when working around the windshield.**

14. Remove the engine mount bolt from the rear engine mount and the rear engine mount bracket.

➡**Use a new engine mount bolt during reassembly.**

15. Raise the vehicle on the lift to full height.

16. Remove the front splash shield.

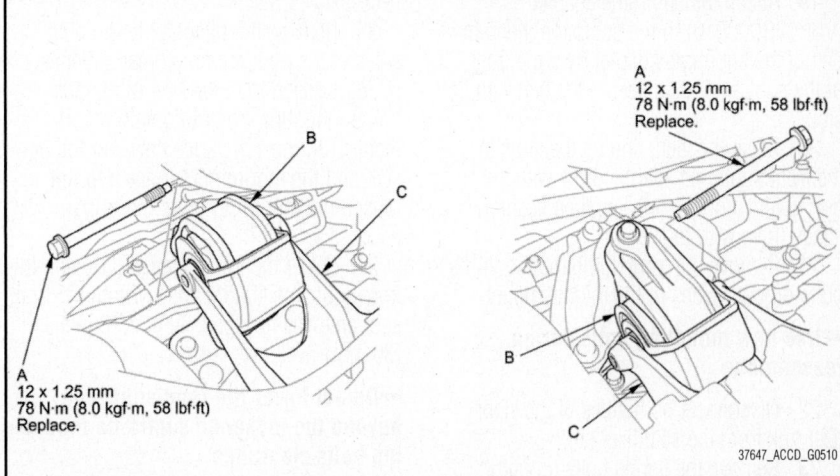

A
12 x 1.25 mm
78 N·m (8.0 kgf·m, 58 lbf·ft)
Replace.

A
12 x 1.25 mm
78 N·m (8.0 kgf·m, 58 lbf·ft)
Replace.

37647_ACCD_G0510

Fig. 342 Remove the engine mount bolt (A) from the rear engine mount (B) and the rear engine mount bracket (C)

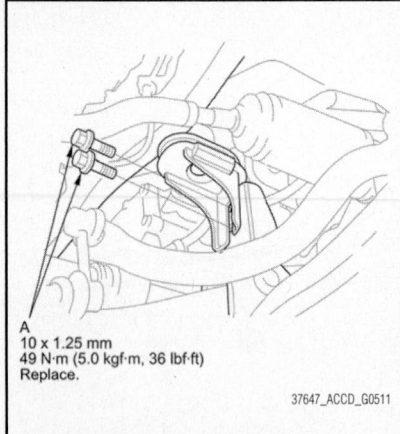

A
10 x 1.25 mm
49 N·m (5.0 kgf·m, 36 lbf·ft)
Replace.

37647_ACCD_G0511

Fig. 345 Remove the front subframe mounting bolts (A) on both sides of the middle mount

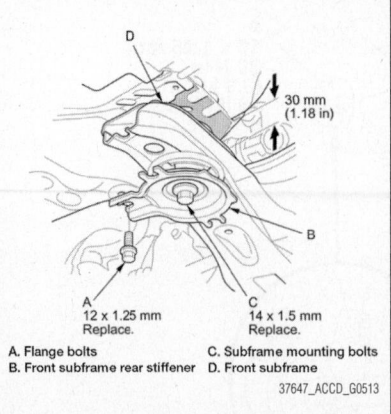

D

30 mm
(1.18 in)

B

A
12 x 1.25 mm
Replace.

C
14 x 1.5 mm
Replace.

A. Flange bolts
B. Front subframe rear stiffener
C. Subframe mounting bolts
D. Front subframe

37647_ACCD_G0513

Fig. 347 Remove the flange bolts of the front subframe rear stiffener, loosen the subframe mounting bolts

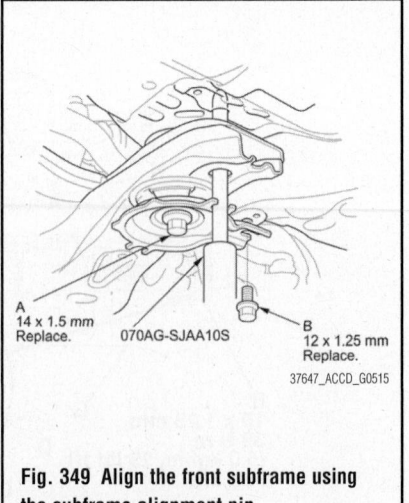

A
14 x 1.5 mm
Replace.

070AG-SJAA10S

B
12 x 1.25 mm
Replace.

37647_ACCD_G0515

Fig. 349 Align the front subframe using the subframe alignment pin

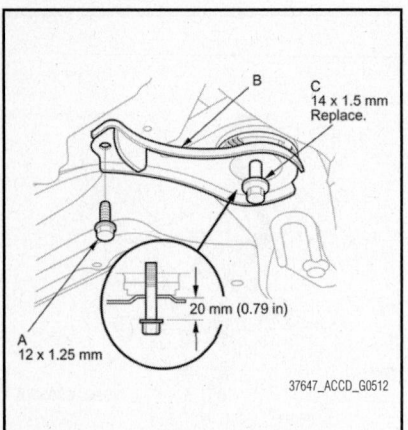

B

C
14 x 1.5 mm
Replace.

20 mm (0.79 in)

A
12 x 1.25 mm

37647_ACCD_G0512

Fig. 346 Remove the flange bolts (A) of the front subframe front stiffener (B), loosen subframe mounting bolts (C)

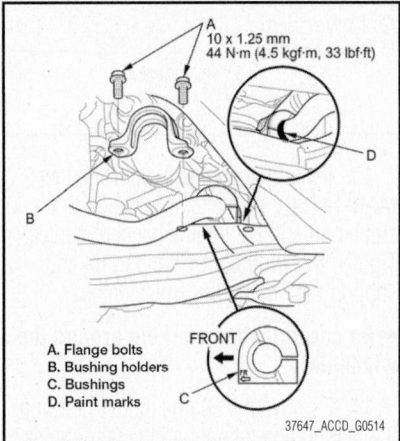

A
10 x 1.25 mm
44 N·m (4.5 kgf·m, 33 lbf·ft)

B

D

FRONT

A. Flange bolts
B. Bushing holders
C. Bushings
D. Paint marks

C

37647_ACCD_G0514

Fig. 348 Remove the flange bolts and the bushing holders, then remove the bushings

➠**During installation, align the paint marks on the stabilizer bar with the side of the bushings.**

29. Move the stabilizer bar toward the passenger's side, and remove the stabilizer bar.

To install:

30. Install the stabilizer bar.

➠**Note the right and left direction of the stabilizer bar. Note the direction of installation for the bushings.**

31. Align the front subframe using the subframe alignment pin. Vertically install the subframe alignment pin, and align the right-rear corner of the front subframe and vehicle frame holes, then loosely tighten the new subframe mounting bolt until the front subframe contacts the body frame.

17. Remove the nuts securing of the lower transmission mount.

18. Remove the exhaust pipe A hanger from the front subframe.

19. Attach the subframe adapter (VSB02C000016) to the subframe, hang the belt of the subframe adapter over the front of the subframe, then secure the belt with its stop.

20. Raise the jack, line up the slots in the front subframe adapter arms with the bolt holes on the jack base, then securely attach them with four bolts.

21. Remove the front subframe mounting bolts on both sides of the middle mount.

➠**Use new mounting bolts during reassembly.**

22. Disconnect both sides of the stabilizer link from the stabilizer bar.

23. Remove the flange bolts on both sides of the front subframe front stiffener.

24. Loosen the front side of the sub-

frame mounting bolts to obtain a 0.79 inches (20 mm) distance between the bolt seat and the mounting surface. Do not loosen the mounting bolts more than necessary.

25. Remove the flange bolts on both sides of the front subframe rear stiffener.

26. Loosen the rear side of the subframe mounting bolts to obtain a 1.18 inches (30 mm) distance between the bolt seat and the mounting surface. Do not loosen the mounting bolts more than necessary.

27. Lower the transmission jack with the front subframe adapter slowly until the front subframe has dropped about 1.18 inches (30 mm).

➠**Do not lower the front subframe beyond the loosened subframe mounting bolts clearance.**

28. Remove the flange bolts and the bushing holders, then remove the bushings.

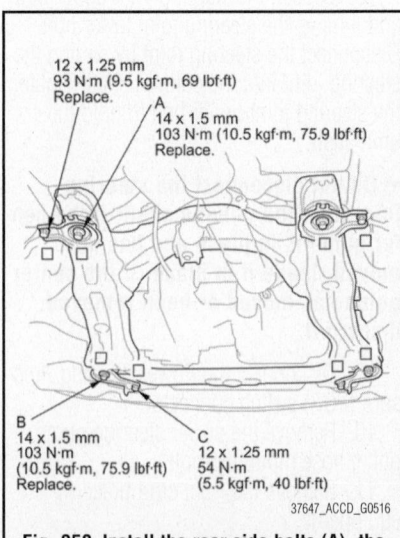

12 x 1.25 mm
93 N·m (9.5 kgf·m, 69 lbf·ft)
Replace.

14 x 1.5 mm
103 N·m (10.5 kgf·m, 75.9 lbf·ft)
Replace.

B
14 x 1.5 mm
103 N·m
(10.5 kgf·m, 75.9 lbf·ft)
Replace.

C
12 x 1.25 mm
54 N·m
(5.5 kgf·m, 40 lbf·ft)

37647_ACCD_G0516

Fig. 350 Install the rear side bolts (A), the front side subframe mounting bolts (B) and the flange bolts (C)

32. Loosely tighten the left-rear subframe mounting bolt using the same procedure as the right-rear with the subframe alignment pin.

33. Loosely install the new 12 mm flange bolts to the subframe rear stiffener.

34. Torque the subframe mounting bolts to 76 ft. lbs. (103 Nm) starting with the right-rear bolt. Use the subframe alignment pin when tightening the rear side bolts.

➡**Before tightening the new front side subframe mounting bolts, raise the jack and loosely install the 12 mm flange bolts to align the subframe front stiffener.**

35. Check all of the front subframe mounting bolts, and retighten if necessary.

36. Install all of the removed parts in the reverse order of removal, and note these items:

- Refer to stabilizer link removal/installation to connect the stabilizer bar to the links.
- If the center guide is in place, use it to determine the steering joint installation angle.
- If the center guide is gone, check the steering joint installation angle.
- Check the steering wheel installation.
- When connecting the rear engine mount to the rear engine mount bracket, first lightly tighten the mounting bolt, then remove the engine support hanger, and tighten it to the specified torque.
- Before installing the wheel, clean the mating surfaces of the brake disc and inside of the wheel.

37. Do the battery terminal reconnection procedure, then turn the ignition switch to ON (II) and check that the SRS indicator should come on for about 6 seconds and then go off.

38. Check the wheel alignment, and adjust it if necessary.

6-Cylinder Engine

See Figures 351 through 360

Special Tools Required*:
- 2008 V6 Attachment Arm SIL02C000033
- Engine Hanger Balance Bar VSB02C000019
- Engine Support Hanger, A and Reds AAR-T1256
- Subframe Adapter VSB02C000016
- Subframe Alignment Pin 070AG-SJAA10S
- *Available through the Honda Tool and Equipment Program, 888-424-6857.

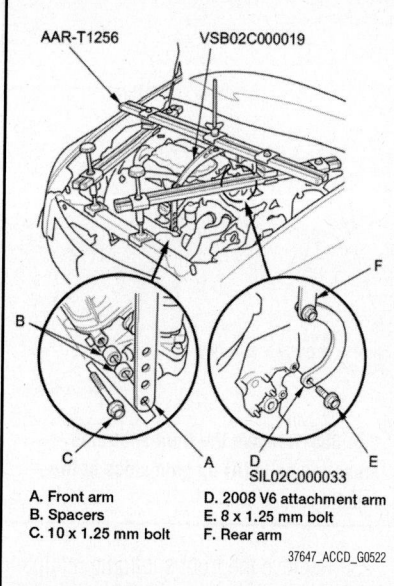

A. Front arm
B. Spacers
C. 10 x 1.25 mm bolt
D. 2008 V6 attachment arm
E. 8 x 1.25 mm bolt
F. Rear arm

37647_ACCD_G0522

Fig. 351 Lift and support the engine with engine support hanger and the engine hanger balance bar

Note these items during replacement:
- Be sure to remove the steering wheel before disconnecting the steering joint. Damage to the cable reel can occur.
- Lower the front subframe from the body, and replace the front stabilizer bar through the gap created by lowering the front subframe.

1. Remove the hood support rod, then use it as shown to prop the hood in the wide-open position.

2. Remove the front grille cover:

3. Do the battery terminal disconnection procedure.

4. Raise and support the vehicle.

5. Remove the front wheels.

6. Remove the driver's airbag and the steering wheel.

7. Remove the steering joint cover.

8. Loosen the steering joint upper bolt, and remove the steering joint lower bolt. Disconnect the steering joint by sliding the steering joint into the column shaft. Tighten the steering joint upper bolt to hold the column shaft.

➡**Do not disconnect the steering joint from the column shaft. If the center guide is in place and has not moved, leave it in place. If the center guide has moved or been removed, discard it.**

9. Disconnect the Power Steering Pressure (PSP) switch connector.

10. Remove the power steering pump outlet hose mounting bolt.

11. Remove the front strut brace.

12. Remove the connector bracket from the front cylinder head; use the bracket bolt hole to attach the engine hanger balance bar front arm.

13. Remove the harness clamp bracket from the rear cylinder head; use the bracket bolt hole to attach the 2008 V6 attachment arm (SIL02C000033).

14. Lift and support the engine with engine support hanger (AAR-T1256) and the engine hanger balance bar (VSB02C000019). Attach the front arm to the front cylinder head with several spacers and the bolt (10 x 1.25 mm). Attach the 2008 V6 attachment arm to the rear cylinder head and the bolt (8 x 1.25 mm), then attach the rear arm to the 2008 V6 attachment arm.

15. A/T: Remove the rear engine mount stop.

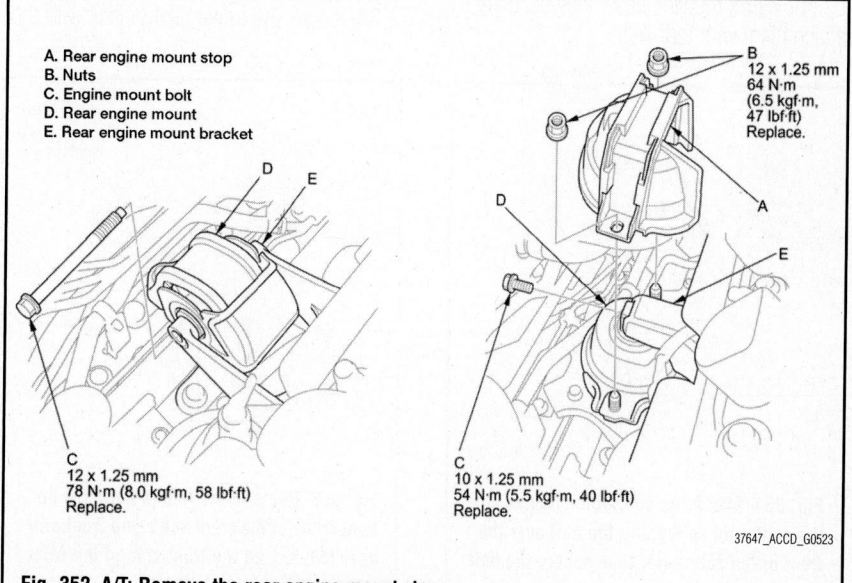

A. Rear engine mount stop
B. Nuts
C. Engine mount bolt
D. Rear engine mount
E. Rear engine mount bracket

B. 12 x 1.25 mm
64 N·m
(6.5 kgf·m,
47 lbf·ft)
Replace.

C. 12 x 1.25 mm
78 N·m (8.0 kgf·m, 58 lbf·ft)
Replace.

C. 10 x 1.25 mm
54 N·m (5.5 kgf·m, 40 lbf·ft)
Replace.

37647_ACCD_G0523

Fig. 352 A/T: Remove the rear engine mount stop

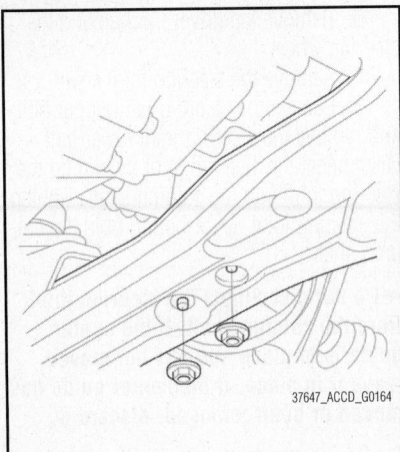

Fig. 353 Remove the nuts (A) securing the lower transmission mount

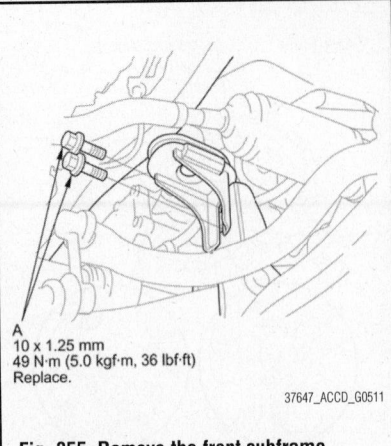

Fig. 355 Remove the front subframe mounting bolts (A) on both sides of the middle mount

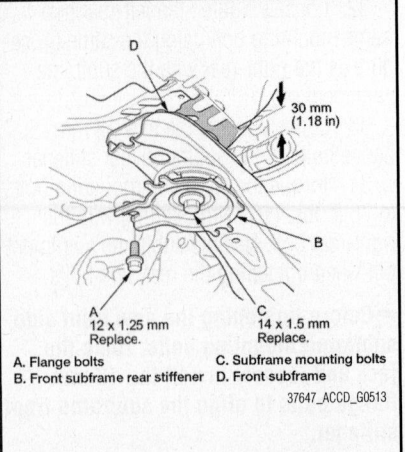

A. Flange bolts
B. Front subframe rear stiffener
C. Subframe mounting bolts
D. Front subframe

Fig. 357 Remove the flange bolts on both sides of the front subframe rear stiffener

➡ **Use new nuts during reassembly.**

16. Remove the engine mount bolt from the rear engine mount and the rear engine mount bracket.

➡ **Use a new engine mount bolt during reassembly.**

17. Raise the vehicle on the lift to full height.
18. Remove the front splash shield.
19. Remove the nuts securing the lower transmission mount.
20. Remove the exhaust pipe A hanger from the front subframe.
21. Attach the subframe adapter (VSB02C000016) to the subframe, hang the belt of the subframe adapter over the front of the subframe, then secure the belt with its stop.
22. Raise the jack, line up the slots in the front subframe adapter arms with the bolt holes on the jack base, then securely attach them with four bolts.

23. Remove the front subframe mounting bolts on both sides of the middle mount.

➡ **Use new mounting bolts during reassembly.**

24. Disconnect both sides of the stabilizer link from the stabilizer bar.
25. Remove the flange bolts on both sides of the front subframe front stiffener.
26. Loosen the front side of the front subframe mounting bolts to obtain a 0.79 inches (20 mm) distance between the bolt seat and the mounting surface. Do not loosen the mounting bolts more than necessary.
27. Remove the flange bolts on both sides of the front subframe rear stiffener.
28. Loose the rear side of the subframe mounting bolts to obtain a 1.18 inches (30 mm) distance between the bolt seat and the mounting surface. Do not loosen the mounting bolts more than necessary.
29. Lower the transmission jack with the

front subframe adapter slowly until the front subframe has dropped about 1.18 inches (30 mm).

➡ **Do not lower the front subframe beyond the loosened subframe mounting bolts clearance.**

30. Remove the flange bolts and the bushing holders, then remove the bushings.

➡ **During installation, align the paint marks on the stabilizer bar with the side of the bushings.**

31. Move the stabilizer bar toward the passenger's side, and remove the stabilizer bar.

To install:
32. Install the stabilizer bar.

➡ **Note the right and left direction of the stabilizer bar. Note the direction of installation for the bushings.**

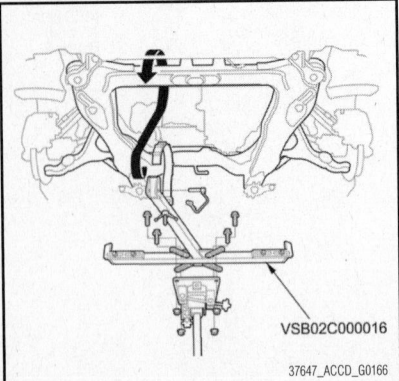

Fig. 354 Attach the subframe adapter to the subframe by looping the belt over the front of the subframe, then secure the belt with its stop

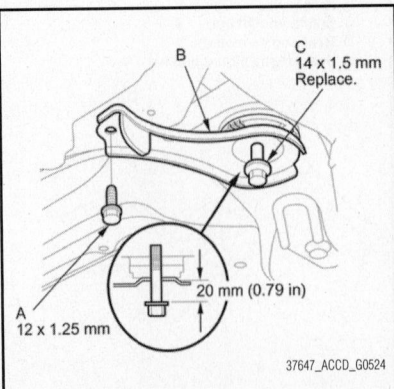

Fig. 356 Remove the flange bolts (A) on both sides of the front subframe front stiffener (B); loosen the front side of the front subframe mounting bolts (C)

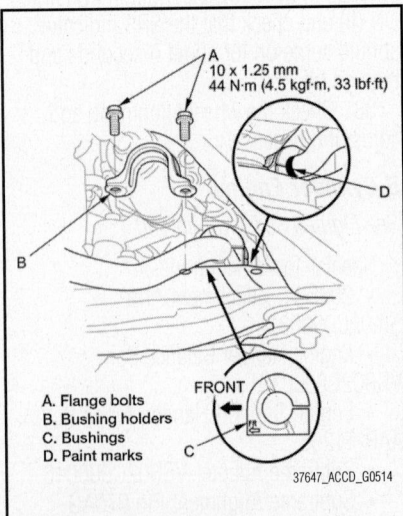

A. Flange bolts
B. Bushing holders
C. Bushings
D. Paint marks

Fig. 358 Remove the flange bolts and the bushing holders, then remove the bushings

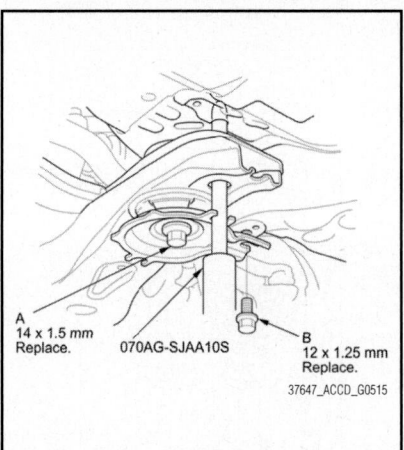

Fig. 359 Align the front subframe using the subframe alignment pin

33. Align the front subframe using the subframe alignment pin. Vertically install the subframe alignment pin, and align the right-rear corner of the front subframe and vehicle frame holes, then loosely tighten the new subframe mounting bolt until the front subframe contacts the body frame.

34. Loosely tighten the left-rear subframe mounting bolt using the same procedure as the right-rear with the subframe alignment pin.

35. Loosely install the new 12 mm flange bolts to the subframe rear stiffener.

36. Torque the subframe mounting bolts to the specified torque starting with the right-rear bolt. Use the subframe alignment pin when tightening the rear side bolts.

➡**Before tightening the new front side subframe mounting bolts, raise the jack and loosely install the 12 mm flange bolts to align the subframe front stiffener.**

37. Check all of the front subframe mounting bolts, and retighten if necessary.

38. Install all of the removed parts in the reverse order of removal, and note these items:

- Refer to stabilizer link removal/installation to connect the stabilizer bar to the links.
- If the center guide is in place, use it to determine the steering joint installation angle.
- If the center guide is gone, check the steering joint installation angle.
- Check the steering wheel installation.
- When connecting the rear engine mount to the rear engine mount bracket, first lightly tighten the mounting bolt, then remove the engine support hanger, and tighten it to the specified torque.
- Before installing the wheel, clean the mating surfaces of the brake disc and inside of the wheel.

39. Do the battery terminal reconnection procedure, then turn the ignition switch to ON (II) and check that the SRS indicator should come on for about 6 seconds and then goes off.

40. Check the wheel alignment, and adjust it if necessary.

STRUT BAR

REMOVAL & INSTALLATION
See Figure 361.

➡**The manufacturer does not provide a specific Removal and Installation procedure for this component. Refer to the graphic(s) when servicing this component.**

STRUTS

REMOVAL & INSTALLATION
See Figures 362 through 366.

1. Raise the vehicle on a lift.
2. Remove the front wheel.
3. Remove the wheel speed sensor harness bracket mounting bolt.
4. Remove the damper pinch bolt and the damper fork mounting nut while holding the mounting bolt, then remove the damper fork from the damper and the lower arm.
5. Remove the front strut brace mounting nuts.
6. Remove the damper mounting nuts from the top of the damper. Do not let the

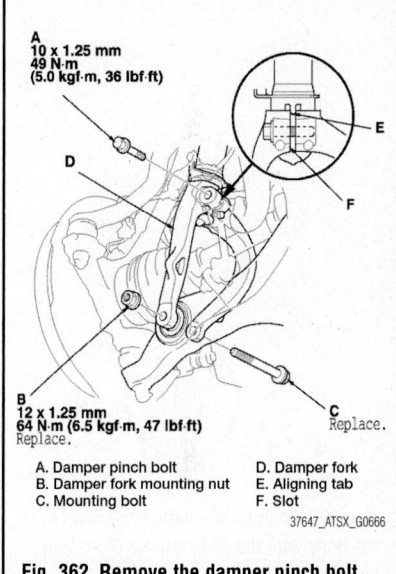

A. Damper pinch bolt
B. Damper fork mounting nut
C. Mounting bolt
D. Damper fork
E. Aligning tab
F. Slot

Fig. 362 Remove the damper pinch bolt and the damper fork mounting nut while holding the mounting bolt

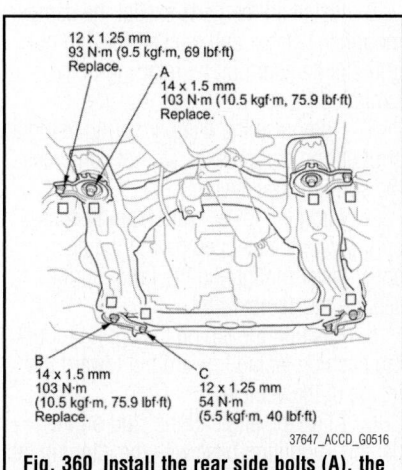

Fig. 360 Install the rear side bolts (A), the front side subframe mounting bolts (B) and the flange bolts (C)

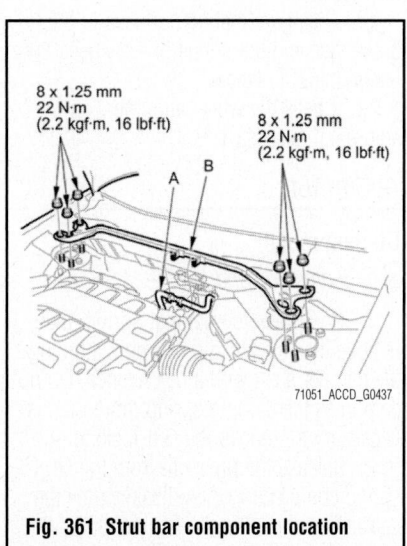

Fig. 361 Strut bar component location

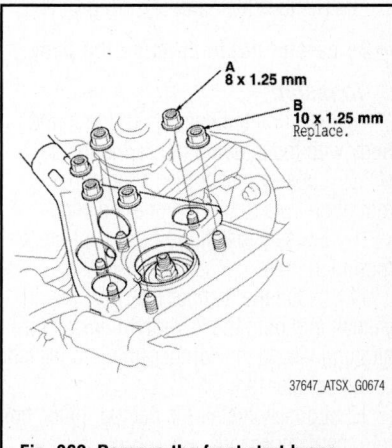

Fig. 363 Remove the front strut brace mounting nuts (A) and the damper mounting nuts (B)

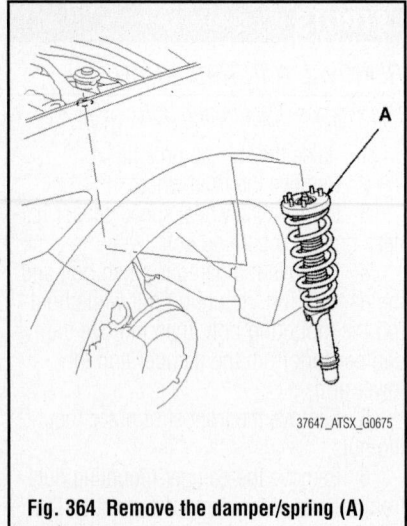

Fig. 364 Remove the damper/spring (A)

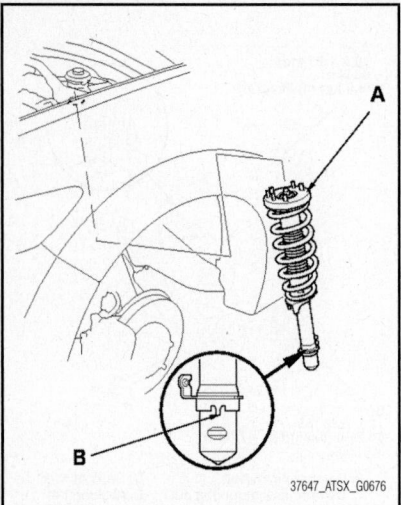

Fig. 365 Position the damper/spring (A) in the body with the aligning tab (B) facing inside

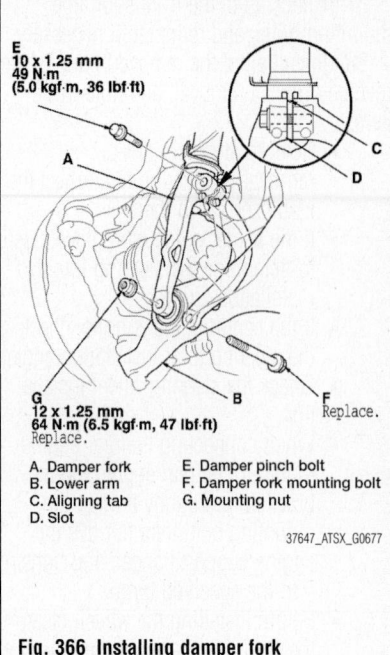

E
10 x 1.25 mm
49 N·m
(5.0 kgf·m, 36 lbf·ft)

G
12 x 1.25 mm
64 N·m (6.5 kgf·m, 47 lbf·ft)
Replace.

Replace.

A. Damper fork
B. Lower arm
C. Aligning tab
D. Slot
E. Damper pinch bolt
F. Damper fork mounting bolt
G. Mounting nut

Fig. 366 Installing damper fork

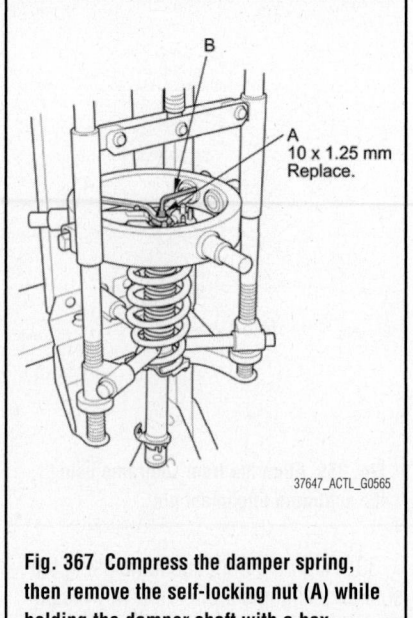

B

A
10 x 1.25 mm
Replace.

Fig. 367 Compress the damper spring, then remove the self-locking nut (A) while holding the damper shaft with a hex wrench (B)

damper/spring drop down under its own weight.

7. Remove the damper/spring.

➡**Be careful not to damage the body.**

To install:

8. Position the damper/spring in the body with the aligning tab facing inside.

9. Loosely install the new damper mounting nuts to the top of the damper.

10. Loosely install the front strut brace mounting nuts.

11. Install the damper fork over the driveshaft and onto the lower arm. Install the aligning tab on the damper unit into the slot of the damper fork.

12. Loosely install the damper pinch bolt into the damper fork.

13. Connect the damper fork and the lower arm with the new damper fork mount-ing bolt, then lightly tighten the new mount-ing nut.

14. Place a floor jack under the lower arm, and raise the suspension to load it with the vehicle's weight.

15. Tighten the damper pinch bolt and the damper fork mounting nut while holding the mounting bolt to the specified torque. Damper pinch bolt: tighten to 36 ft. lbs. (49 Nm); damper fork mounting nut: tighten to 47 ft. lbs. (64 Nm).

16. Tighten the damper mounting nuts and front strut brace mounting nuts on top of the damper to the specified torque. Damper mounting nuts: tighten to 41 ft. lbs. (55 Nm); strut brace mounting nuts: tighten to 16 ft. lbs. (22 Nm).

17. Install the wheel speed sensor harness bracket.

18. Clean the mating surfaces of the brake disc and the inside of the wheel, then install the front wheel.

19. Check the wheel alignment, and adjust it if necessary.

OVERHAUL

Disassembly

See Figures 367 and 368.

1. Compress the damper spring, then remove the self-locking nut while holding the damper shaft with a hex wrench. Do not compress the damper spring more than is necessary to remove the self-locking nut.

2. Release the pressure from the strut spring compressor, then disassemble the damper as shown in the Exploded View.

Inspection

1. Reassemble all the parts, except for the damper spring.

2. Compress the damper assembly by hand, and check for smooth operation through a full stroke, both compression and extension. The damper should extend smoothly and constantly when compression is released. If it does not, the gas is leaking and the damper should be replaced.

3. Check for oil leaks, abnormal noises, and binding during these tests.

Reassembly

See Figures 369 through 372.

1. Install the spring mounting cushion on the damper mounting base by aligning the tab and notch.

2. Install all the parts except the damper mounting washer and the self-locking nut onto the damper unit by referring to the Exploded View.

3. Compress the damper spring using a strut spring compressor. Do not compress the spring excessively.

4. Align the lower end of the damper spring with the stepped part of the dust cover lower mount and the lower spring seat on the damper unit.

5. Position the tab on the spring mounting cushion facing forward but toward the inside of the vehicle.

6. Align the angle of the stud on the damper mounting base with the aligning tab on the bottom of the damper unit.

7. Install the damper mounting washer and the new self-locking nut.

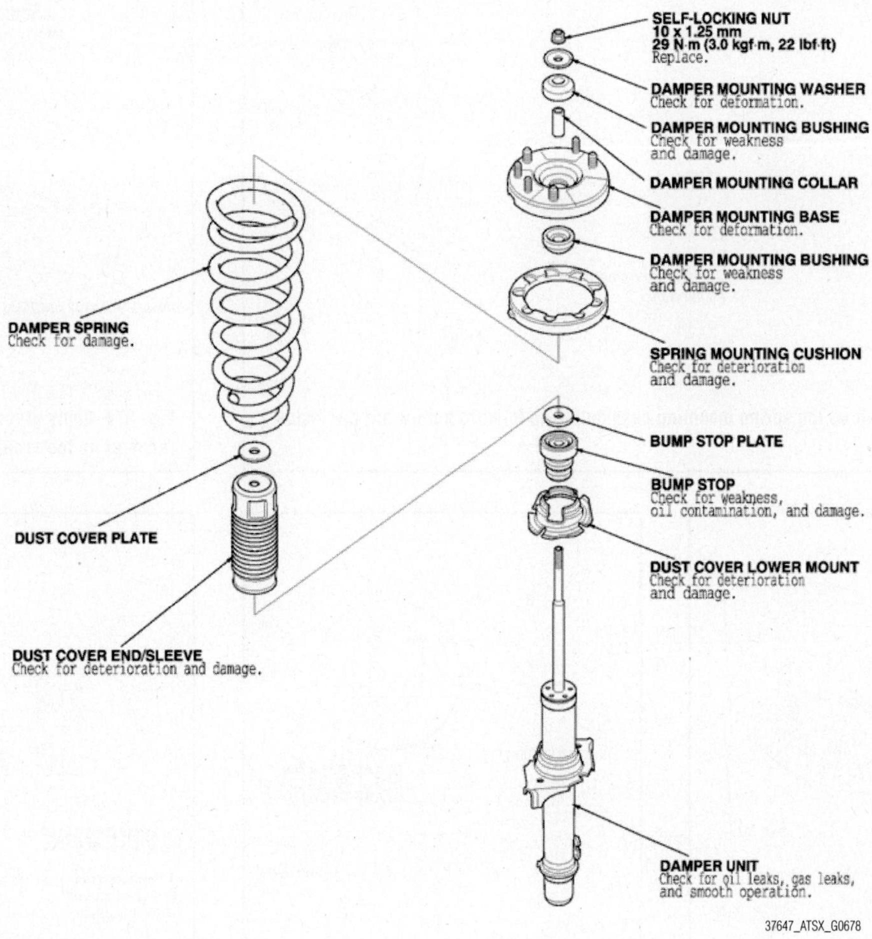

SELF-LOCKING NUT
10 x 1.25 mm
29 N·m (3.0 kgf·m, 22 lbf·ft)
Replace.

DAMPER MOUNTING WASHER
Check for deformation.

DAMPER MOUNTING BUSHING
Check for weakness
and damage.

DAMPER MOUNTING COLLAR

DAMPER MOUNTING BASE
Check for deformation.

DAMPER MOUNTING BUSHING
Check for weakness
and damage.

SPRING MOUNTING CUSHION
Check for deterioration
and damage.

DAMPER SPRING
Check for damage.

BUMP STOP PLATE

BUMP STOP
Check for weakness,
oil contamination, and damage.

DUST COVER PLATE

DUST COVER LOWER MOUNT
Check for deterioration
and damage.

DUST COVER END/SLEEVE
Check for deterioration and damage.

DAMPER UNIT
Check for oil leaks, gas leaks,
and smooth operation.

37647_ATSX_G0678

Fig. 368 Exploded view of damper/spring assembly

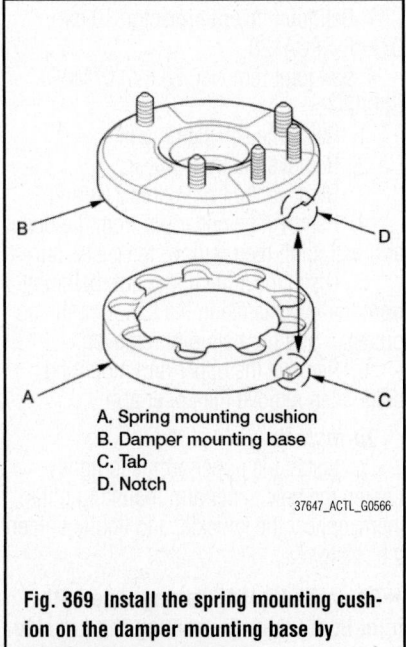

A. Spring mounting cushion
B. Damper mounting base
C. Tab
D. Notch

37647_ACTL_G0566

Fig. 369 Install the spring mounting cushion on the damper mounting base by aligning the tab and notch

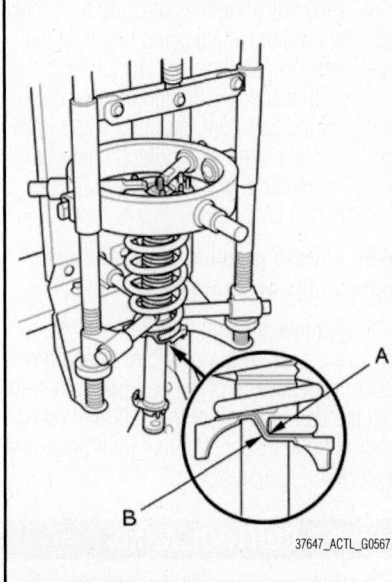

37647_ACTL_G0567

Fig. 370 Align the lower end (A) of the damper spring with the stepped part (B) of the dust cover lower mount

8. Hold the damper shaft with a hex wrench, and tighten the self-locking nut to 22 ft. lbs. (29 Nm).

9. Remove the damper/spring from the strut spring compressor.

UPPER BALL JOINTS

REMOVAL & INSTALLATION

See Figures 373 through 375.

Special Tools Required:
• Ball joint thread protector, 14 mm 07AAE-SJAA100
• Ball joint thread protector, 12 mm 07AAF-SDAA100
• Ball joint thread protector, 10 mm 07AAF-SECA120
• Ball joint remover, 32 mm 07MAC-SL0A102
• Ball joint remover, 28 mm 07MAC-SL0A202

➡**Always use a ball joint remover to disconnect a ball joint. Do not strike**

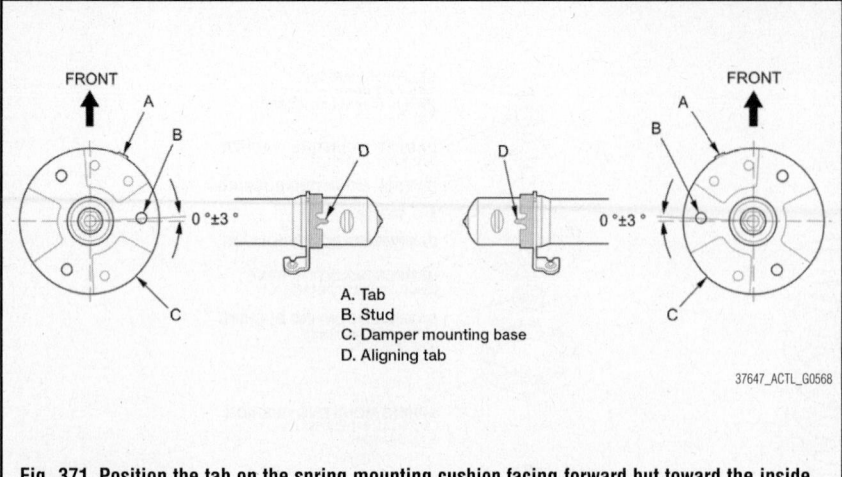

Fig. 371 Position the tab on the spring mounting cushion facing forward but toward the inside of the vehicle

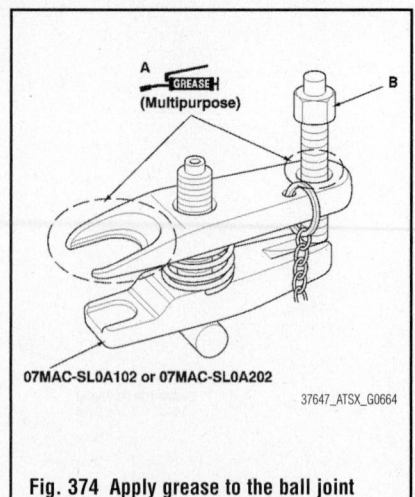

Fig. 374 Apply grease to the ball joint remover on the areas shown (A)

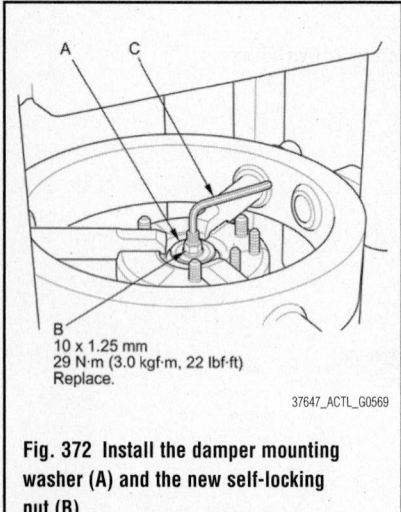

Fig. 372 Install the damper mounting washer (A) and the new self-locking nut (B)

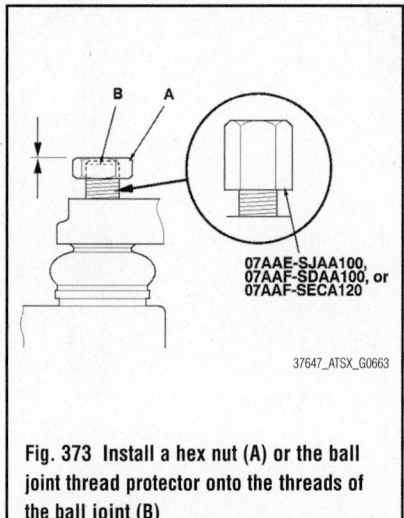

Fig. 373 Install a hex nut (A) or the ball joint thread protector onto the threads of the ball joint (B)

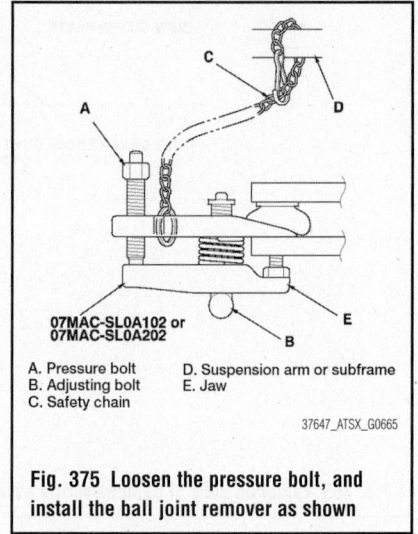

Fig. 375 Loosen the pressure bolt, and install the ball joint remover as shown

the housing or any other part of the ball joint connection to disconnect it.

1. Install a hex nut or the ball joint thread protector onto the threads of the ball joint.

➡ Using a hex nut, make sure the nut is flush with the ball joint pin end to prevent damage to the threaded end of the ball joint pin.

2. Apply grease to the ball joint remover on the areas shown. This will ease the installation of the tool, and prevent damage to the pressure bolt threads.

3. Loosen the pressure bolt, and install the ball joint remover as shown. Insert the jaws carefully, making sure not to damage the ball joint boot. Adjust the jaw spacing by turning the adjusting bolt.

➡ Fasten the safety chain securely to a suspension arm or the subframe. Do not fasten it to a brake line or wire harness.

4. After adjusting the adjusting bolt, make sure the head of the adjusting bolt is in the position shown to allow the jaw to pivot.

5. With a wrench, tighten the pressure bolt until the ball joint pin pops loose from the ball joint connecting hole. If necessary, apply penetrating type lubricant to loosen the ball joint pin.

➡ Do not use pneumatic or electric tools on the pressure bolt.

6. Remove the ball joint remover, then remove the nut or the ball joint thread protector from the end of the ball joint pin, and pull the ball joint out of the ball joint connecting hole. Inspect the ball joint boot, and replace it if damaged.

UPPER CONTROL ARMS

REMOVAL & INSTALLATION

See Figures 376 through 378.

Special Tools Required:

• Ball joint thread protector, 10 mm 07AAF-SECA120
• Ball joint remover, 28 mm 07MAC-SL0A202

1. Raise the vehicle on a lift.
2. Remove the front wheel.
3. Remove the front damper/spring.
4. Remove the cotter pin from the upper arm ball joint, then remove the castle nut.
5. Disconnect the upper arm ball joint from the knuckle using the ball joint thread protector and the ball joint remover.
6. Remove the upper arm mounting bolts, then remove the upper arm.

To install:

7. Install the upper arm, and lightly tighten the new upper arm mounting bolts, then connect the knuckle, and lightly tighten the castle nut.

➡ Be careful not to damage the ball joint boot when connecting the knuckle. Before connecting the ball joint,

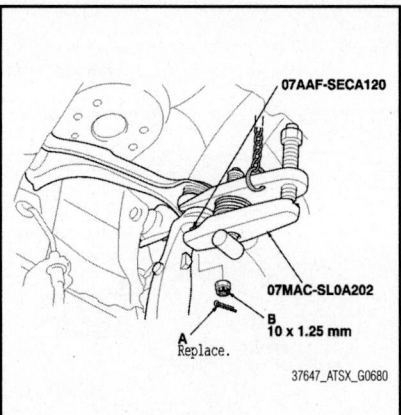

Fig. 376 Remove the cotter pin (A) from the upper arm ball joint, then remove the castle nut (B)

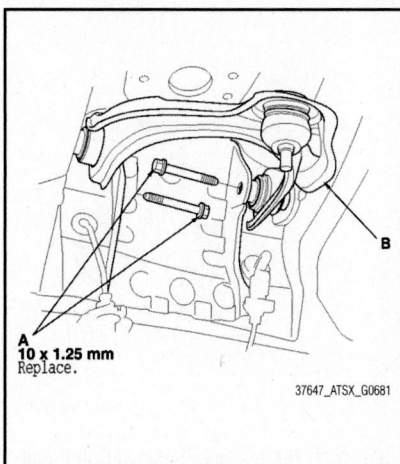

Fig. 377 Remove the upper arm mounting bolts (A), then remove the upper arm (B)

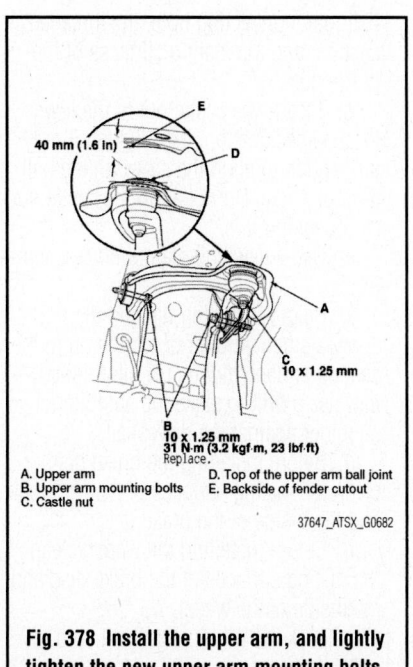

A. Upper arm
B. Upper arm mounting bolts
C. Castle nut
D. Top of the upper arm ball joint
E. Backside of fender cutout

Fig. 378 Install the upper arm, and lightly tighten the new upper arm mounting bolts

degrease the threaded section and the tapered portion of the ball joint pin, the ball joint connecting hole, the threaded section and the mating surfaces of the castle nut.

8. Place a floor jack under the lower arm, and raise the suspension until the clearance between the top of the upper arm ball joint and the backside of the fender cut out point is 1.6 inches (40 mm), then tighten the upper arm mounting bolts to 23 ft. lbs. (31 Nm).

➡**To measure the specified clearance, temporarily remove the front inner fender.**

9. Lower the floor jack
10. Install the front damper/spring.
11. Place the floor jack under the lower arm, and raise the suspension to load it with the vehicle's weight.
12. Tighten the castle nut on the upper arm ball joint to 33–38 ft. lbs. (44–52 Nm).

➡**Torque the castle nut to the lower torque specification, then tighten it only far enough to align the slot with the ball joint pin hole. Do not align the castle nut by loosening it. Insert the new cotter pin into the ball joint pin hole from the front to the rear of the vehicle, and bend its end. Check the ball joint pin hole direction before connecting the ball joint.**

13. Clean the mating surfaces of the brake disc and the inside of the wheel, then install the front wheel.

14. Check the wheel alignment, and adjust it if necessary.

WHEEL HUBS & BEARINGS

ADJUSTMENT

1. Raise and support the vehicle.
2. Remove the wheels.
3. Install suitable flat washers and the wheel nuts. Tighten the nuts to the specified torque to hold the brake disc securely against the hub.
4. Attach the dial gauge. Place the dial gauge against the hub flange.
5. Measure the bearing end play while moving the brake disc inward and outward.

➡**Wheel bearing end play: Front/Rear: 0–0.002 inches (0–0.05 mm)**

6. If the bearing end play measurement is more than the standard, replace the wheel bearing or the hub bearing unit.

REMOVAL & INSTALLATION

See Figures 379 through 382.

1. Raise the vehicle on a lift.
2. Remove the wheel nuts and the front wheel.
3. Remove the brake hose bracket mounting bolt.
4. Remove the brake caliper bracket mounting bolts, then remove the caliper assembly from the knuckle. To prevent damage to the caliper assembly or the brake hose, use a short piece of wire to hang the caliper assembly from the undercarriage. Do not twist the brake hose excessively.

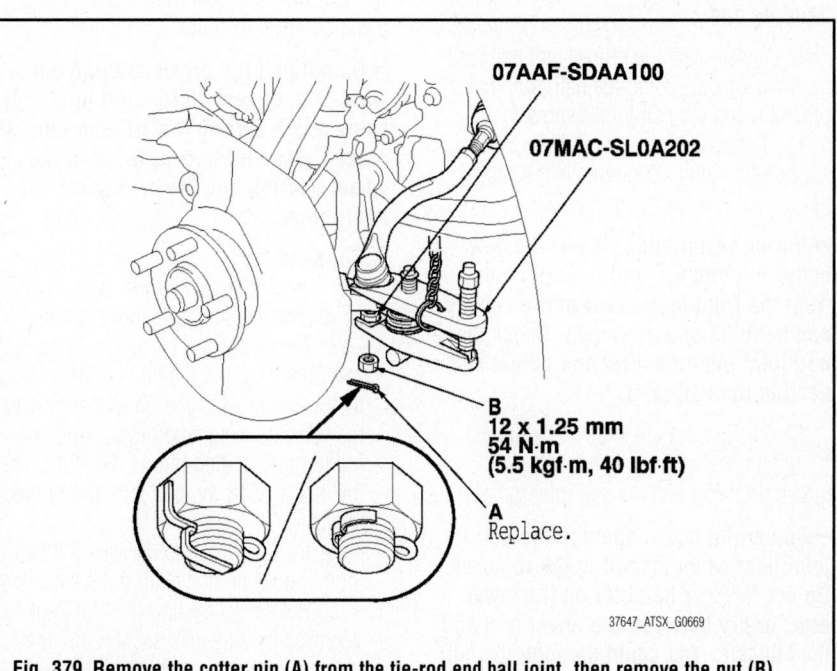

Fig. 379 Remove the cotter pin (A) from the tie-rod end ball joint, then remove the nut (B)

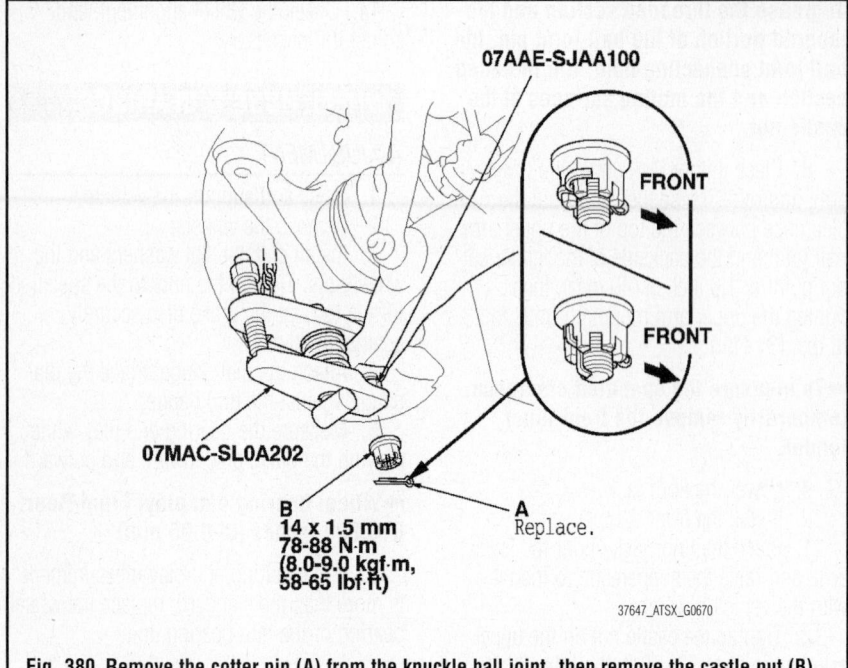

Fig. 380 Remove the cotter pin (A) from the knuckle ball joint, then remove the castle nut (B)

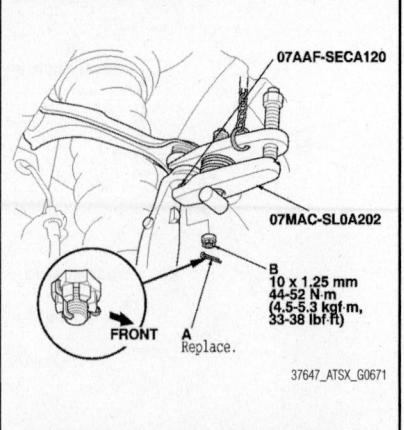

Fig. 381 Remove the cotter pin (A) from the upper arm ball joint, then remove the castle nut (B)

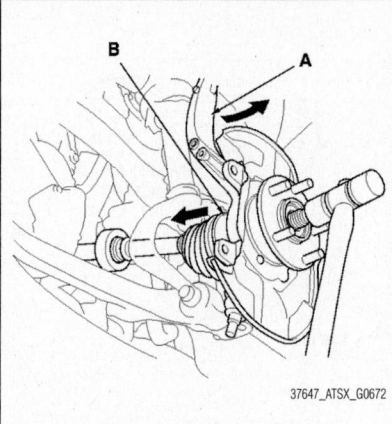

Fig. 382 Pull the knuckle (A) outward, and separate the outboard joint (B) from the front hub using a plastic hammer

5. Remove the wheel speed sensor harness bracket and the wheel speed sensor from the knuckle. Do not disconnect the wheel speed sensor connector.

6. Pry up the stake on the spindle nut, then remove the nut.

7. Remove the front brake disc.

8. Check the front hub for damage and cracks.

9. Remove the cotter pin from the tie-rod end ball joint, then remove the nut.

➡ During installation, install the new cotter pin after tightening the nut, and bend its end.

10. Disconnect the tie-rod end ball joint from the knuckle using the ball joint thread protector and the ball joint remover.

11. Remove the cotter pin from the knuckle ball joint, then remove the castle nut.

➡ During installation, insert the new cotter pin into the ball joint pin hole from the front to the rear of the vehicle, and bend its end as shown. Check the ball joint pin hole direction before connecting the ball joint.

12. Disconnect the knuckle ball joint from the lower arm using the ball joint thread protector and the ball joint remover.

➡ Be careful not to damage the ball joint boot when installing the remover. Do not force or hammer on the lower arm, or pry between the lower arm and the knuckle. You could damage the ball joint.

13. Remove the cotter pin from the upper arm ball joint, then remove the castle nut.

➡ During installation, insert the new cotter pin into the ball joint pin hole from the front to the rear of the vehicle, and bend its end as shown. Check the ball joint pin hole direction before connecting the ball joint.

14. Disconnect the upper arm ball joint from the knuckle using the ball joint thread protector and the ball joint remover.

15. Pull the knuckle outward, and separate the outboard joint from the front hub using a plastic hammer.

➡ Do not pull the driveshaft end outward. The driveshaft inboard joint may come apart. During installation, apply grease to the mating surfaces of the wheel bearing and the driveshaft outboard joint.

To install:

16. Install the knuckle/hub in the reverse order of removal, and note these items:

  a. First install all of the components, and lightly tighten the bolts and the nuts, then raise the suspension to load it with the vehicle's weight before fully tightening to the specified torque. Do not place the jack against the ball joint pin of the knuckle.

  b. Be careful not to damage the ball joint boot when connecting the knuckle.

  c. Before connecting the ball joint, degrease the threaded section and the tapered portion of the ball joint pin, the ball joint connecting hole, the threaded section, and the mating surfaces of the castle nut.

  d. Torque the castle nut to the lower torque specification, then tighten it only far enough to align the slot with the ball joint pin hole. Do not align the castle nut by loosening it.

  e. Use a new spindle nut on reassembly.

  f. Before installing the spindle nut, apply a small amount of engine oil to the seating surface of the nut. After tightening, use a drift to stake the spindle nut shoulder against the driveshaft.

  g. Before installing the brake disc, clean the mating surfaces of the front hub and the inside of the brake disc.

  h. Before installing the wheel, clean the mating surfaces of the brake disc and the inside of the wheel.

17. Check the wheel alignment, and adjust it if necessary.

CONTROL ARMS/LINKS

REMOVAL & INSTALLATION

Rear Control Arm

See Figure 383.

1. Raise the vehicle on a lift.
2. Remove the rear wheel.
3. Remove the control arm mounting self-locking nut and the washer from the knuckle side.

➡**Use a new self-locking nut during reassembly.**

4. Mark the cam positions of the adjusting bolt and the adjusting cam plate with the frame.
5. Remove the self-locking nut while holding the adjusting bolt, then remove the adjusting cam plate, the adjusting bolt, and the control arm.

To install:

➡**Use a new adjusting bolt and a new self-locking nut during reassembly.**

6. Install the control arm in the reverse order of removal, and note these items:
   • First install all of the components, and lightly tighten the bolts and the nuts, then raise the suspension to load it with the vehicle's weight before fully tightening to the specified torque.
   • Position the extended surfaces of the cam on the adjusting bolt and the adjusting cam plate facing down.
   • Align the cam positions of the adjusting bolt and the adjusting cam plate with the marked positions on the frame when tightening the self-locking nut.
   • Before installing the wheel, clean the mating surfaces on the brake disc and the inside of the wheel.

7. Check the wheel alignment, and adjust it if necessary.

Rear Lower Arm A

See Figure 384.

1. Raise the vehicle on a lift.
2. Remove the rear wheel.
3. Remove the parking brake cable mounting bolt.
4. Remove the lower arm A mounting bolts, then remove lower arm A.

To install:

➡**Use new mounting bolts during reassembly.**

5. Install lower arm A in the reverse order of removal, and note these items:
   • First install all of the components, and lightly tighten the bolts, then raise the suspension to load it with the vehicle's weight before fully tightening to the specified torque.
   • Before installing the wheel, clean the mating surfaces on the brake disc and the inside of the wheel.

6. Check the wheel alignment, and adjust it if necessary.

Rear Lower Arm B

See Figure 385.

1. Raise the vehicle on a lift.
2. Remove the rear wheel.
3. Remove the lower arm B mounting bolts, then remove lower arm B.

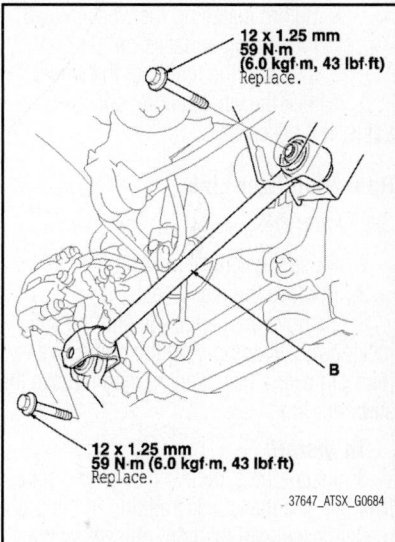

Fig. 385 Remove the lower arm B mounting bolts, then remove lower arm B

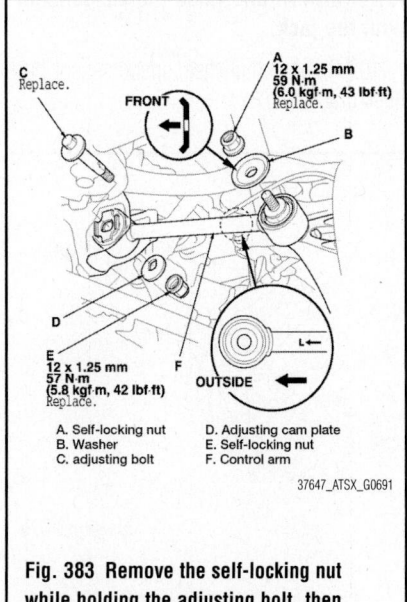

Fig. 383 Remove the self-locking nut while holding the adjusting bolt, then remove the adjusting cam plate, the adjusting bolt, and the control arm

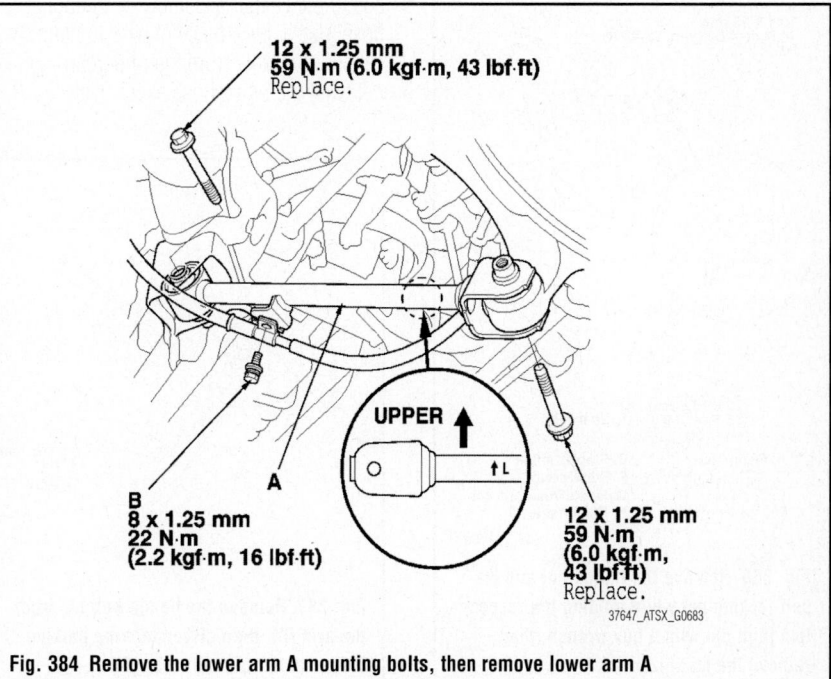

Fig. 384 Remove the lower arm A mounting bolts, then remove lower arm A

To install:

➡ **Use new mounting bolts during reassembly.**

4. Install lower arm B in the reverse order of removal, and note these items:
- First install all of the components, and lightly tighten the bolts, then raise the suspension to load it with the vehicle's weight before fully tightening to the specified torque.
- Make sure the clearance between lower arm B and the parking brake cable is more than 0.20 inches (5 mm).
- Before installing the wheel, clean the mating surfaces on the brake disc and the inside of the wheel.

5. Check the wheel alignment, and adjust it if necessary.

Rear Stabilizer Link

See Figure 386.

1. Raise the vehicle on a lift.
2. Remove the rear wheel.
3. Remove the flange nut and the self-locking nut while holding the respective joint pin with a hex wrench, then remove the stabilizer link.

To install:

4. Install the stabilizer link on the stabilizer bar and the knuckle adding in the brake hose bracket with the joint pins set at the center of their range of movement.

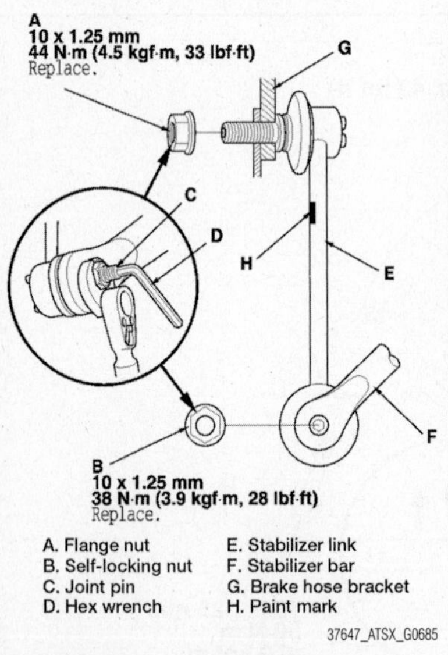

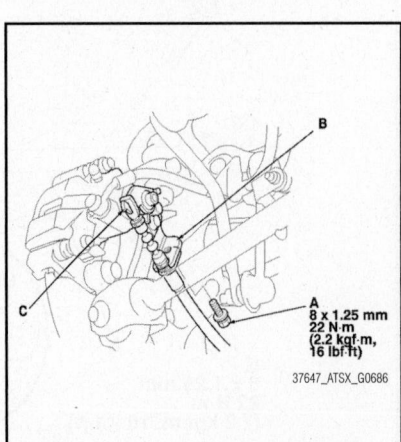

Fig. 386 Remove the flange nut and the self-locking nut while holding the respective joint pin with a hex wrench, then remove the stabilizer link

➡ **The stabilizer link has a paint mark. The paint mark indicates the difference between the left and right stabilizer links. Install the end of the stabilizer link with the paint mark in the upper position.**

5. Install the new flange nut and the new self-locking nut, and tighten them to the specified torque while holding the respective joint pin with a hex wrench.
6. Clean the mating surfaces of the brake disc and the inside of the wheel, then install the rear wheel.
7. Test-drive the vehicle.
8. After 5 minutes of driving, tighten the self-locking nuts again to the specified torque.

Rear Upper Arm

See Figures 387 through 390.

Special Tools Required:
- Ball joint thread protector, 14 mm 07AAE-SJAA100
- Ball joint remover, 32 mm 07MAC-SL0A102

1. Raise the vehicle on a lift.
2. Remove the rear wheel.
3. Remove the rear damper/spring.
4. Release the parking brake lever fully.
5. Loosen the parking brake cable adjusting nut.
6. Remove the flange bolt from the arm. Then disconnect the parking brake cable from the lever.
7. Remove the brake caliper bracket mounting bolts, then remove the caliper assembly from the knuckle. To prevent damage to the caliper assembly or the brake hose, use a short piece of wire to hang the caliper assembly from the undercarriage. Do not twist the brake hose excessively.

Fig. 387 Remove the flange bolt (A) from the arm (B), then disconnect the parking brake cable from the lever (C)

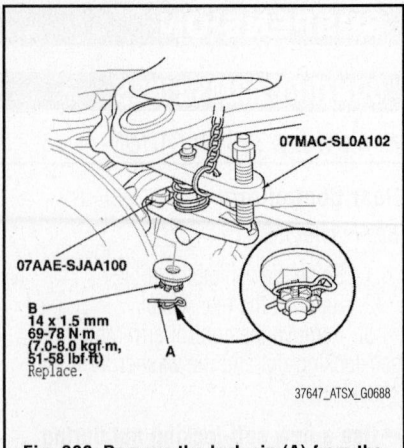

Fig. 388 Remove the lock pin (A) from the upper arm ball joint, then remove the castle nut (B)

➡ **Make sure the washers are reinstalled between the brake caliper bracket and the knuckle, if they are removed.**

8. Remove the lock pin from the upper arm ball joint, then remove the castle nut.

➡ **During installation, install the lock pin as shown after tightening the new castle nut.**

9. Disconnect the upper arm ball joint from the knuckle using the ball joint thread protector and the ball joint remover.

➡ **Be careful not to damage the ball joint boot when installing the remover. During installation, to connect the ball joint, position a floor jack under the connecting point of the knuckle and lower arm A, and raise the suspension with the jack.**

10. Remove the wheel speed sensor harness bracket.

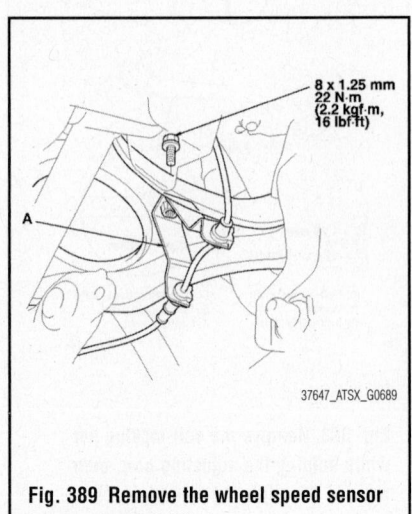

Fig. 389 Remove the wheel speed sensor harness bracket (A)

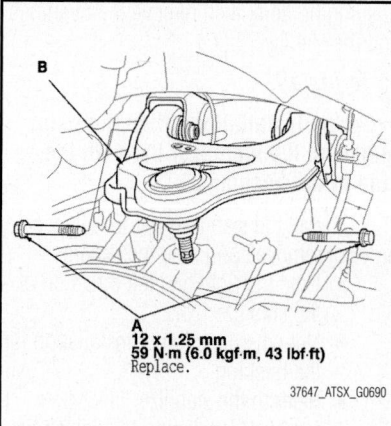

Fig. 390 Remove the upper arm mounting bolts (A), then remove the upper arm (B)

11. Remove the upper arm mounting bolts, then remove the upper arm.

To install:

➡ **Use new mounting bolts during reassembly.**

12. Install the upper arm in the reverse order of removal, and note these items:
- First install all of the components, and lightly tighten the bolts and the nuts, then raise the suspension to load it with the vehicle's weight before fully tightening to the specified torque.
- Be careful not to damage the ball joint boot when connecting the knuckle.
- Before connecting the ball joint, degrease the threaded section and the tapered portion of the ball joint pin, the ball joint connecting hole, the threaded section, and the mating surfaces of the castle nut.
- Torque the castle nut to the lower torque specification, then tighten it only far enough to align the slot with the ball joint pin hole. Do not align the castle nut by loosening it.
- After installing the brake caliper, make sure the clearance between lower arm B and the parking brake cable is more than 0.20 inches (5 mm).
- Do the parking brake adjustment.
- Before installing the wheel, clean the mating surfaces on the brake disc and the inside of the wheel.

13. Check the wheel alignment, and adjust it if necessary.

KNUCKLES

REMOVAL & INSTALLATION

See Figures 391 through 395.

1. Remove the hub bearing unit.
2. Remove the splash guard.
3. Remove the lock pin from the upper arm ball joint, then remove the castle nut.

➡ **During installation, install the lock pin after tightening the new castle nut.**

4. Disconnect the upper arm ball joint from the knuckle using the ball joint thread protector and the ball joint remover.

➡ **Be careful not to damage the ball joint boot when installing the remover.**

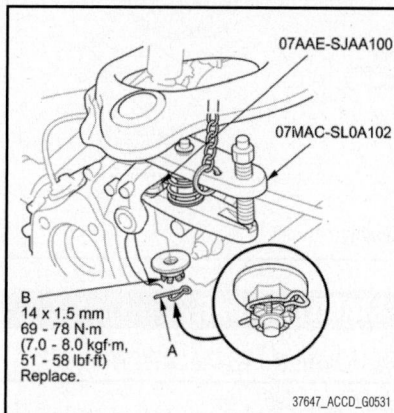

Fig. 391 Remove the lock pin (A) from the upper arm ball joint, then remove the castle nut (B)

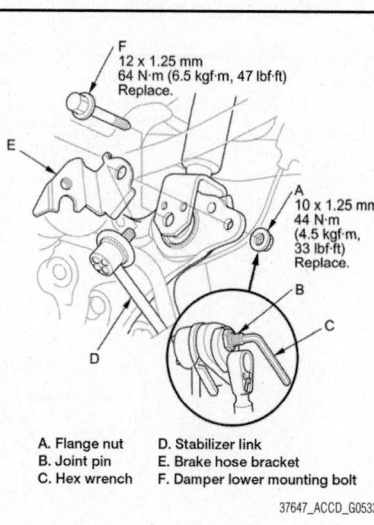

A. Flange nut
B. Joint pin
C. Hex wrench
D. Stabilizer link
E. Brake hose bracket
F. Damper lower mounting bolt

Fig. 392 Remove the flange nut while holding the joint pin with a hex wrench, then disconnect the stabilizer link from the knuckle, and remove the brake hose bracket

During installation, to connect the ball joint, raise the suspension with a jack.

5. Remove the wheel speed sensor from the knuckle. Do not disconnect the wheel speed sensor connector.
6. Remove the flange nut while holding the joint pin with a hex wrench, then disconnect the stabilizer link from the knuckle, and remove the brake hose bracket.

➡ **Use the new flange nut during reassembly.**

7. Remove the damper lower mounting bolt.

➡ **Use the new mounting bolt during reassembly.**

8. Remove the control arm mounting self-locking nut and the washer.

➡ **Use a new self-locking nut during reassembly.**

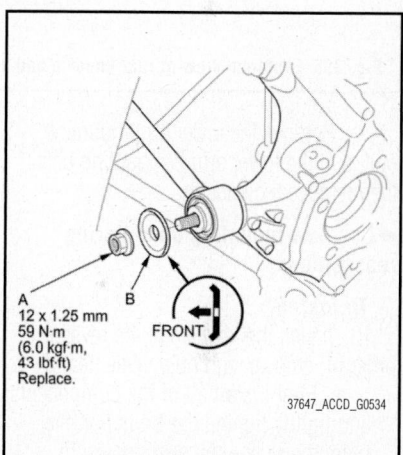

Fig. 393 Remove the control arm mounting self-locking nut (A) and the washer (B)

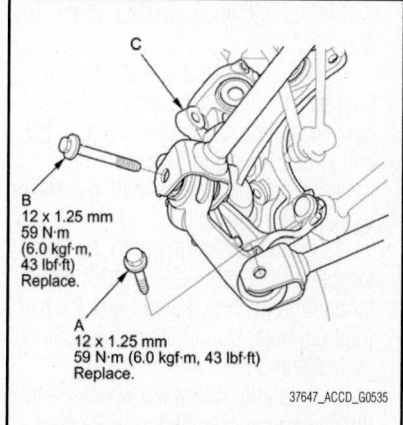

Fig. 394 Remove the lower arm mounting bolt (A), and the lower arm B mounting bolt (B), then remove the knuckle (C)

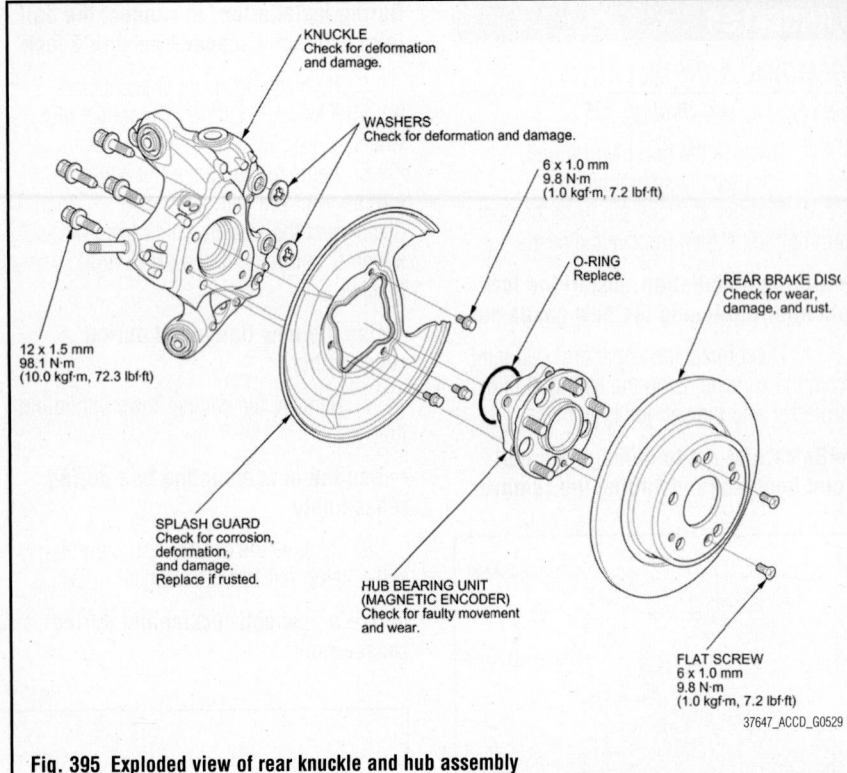

KNUCKLE
Check for deformation
and damage.

WASHERS
Check for deformation and damage.

6 x 1.0 mm
9.8 N·m
(1.0 kgf·m, 7.2 lbf·ft)

O-RING
Replace.

REAR BRAKE DISC
Check for wear,
damage, and rust.

12 x 1.5 mm
98.1 N·m
(10.0 kgf·m, 72.3 lbf·ft)

SPLASH GUARD
Check for corrosion,
deformation,
and damage.
Replace if rusted.

HUB BEARING UNIT
(MAGNETIC ENCODER)
Check for faulty movement
and wear.

FLAT SCREW
6 x 1.0 mm
9.8 N·m
(1.0 kgf·m, 7.2 lbf·ft)

37647_ACCD_G0529

Fig. 395 Exploded view of rear knuckle and hub assembly

9. Remove the lower arm mounting bolt, and the lower arm B mounting bolt, then remove the knuckle.

➡**Use new mounting bolts during reassembly.**

To install:

10. Install the knuckle in the reverse order of removal, and note these items:

a. First install all of the components, and lightly tighten the bolts and the nuts, then raise the suspension to load it with the vehicle's weight before fully tightening to the specified torque.

b. Be careful not to damage the ball joint boot when connecting the knuckle.

c. Before connecting the ball joint, degrease the threaded section and the tapered portion of the ball joint pin, the ball joint connecting hole, and the threaded section and the mating surfaces of the castle nut.

d. Torque the castle nut to the lower torque specification, then tighten it only far enough to align the slot with the ball joint pin hole. Do not align the castle nut by loosening it.

e. Before installing the wheel, clean the mating surfaces on the brake disc and the inside of the wheel.

11. Check the wheel alignment, and adjust it if necessary.

STABILIZER BAR

REMOVAL & INSTALLATION

See Figure 396.

1. Raise the vehicle on a lift.
2. Remove the rear wheels.
3. Disconnect both stabilizer links from the stabilizer bar.
4. Remove the flange bolts and the

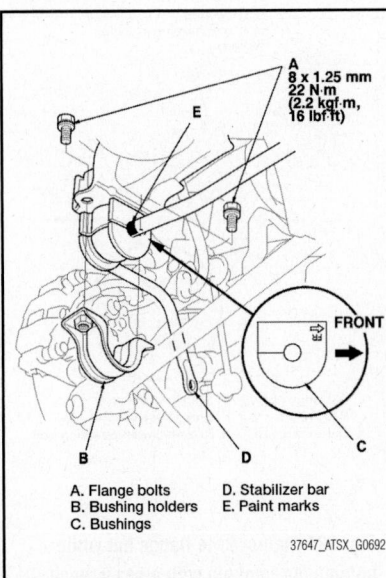

A
8 x 1.25 mm
22 N·m
(2.2 kgf·m,
16 lbf·ft)

E

FRONT

B   D   C

A. Flange bolts
B. Bushing holders
C. Bushings
D. Stabilizer bar
E. Paint marks

37647_ATSX_G0692

Fig. 396 Remove the flange bolts and the bushing holders, then remove the bushings and the stabilizer bar

bushing holders, then remove the bushings and the stabilizer bar.

To install:

➡**During installation, align the paint marks on the stabilizer bar with the side of the bushings.**

5. Install the stabilizer bar in the reverse order of removal, and note these items:

• Note the right and left direction of the stabilizer bar.
• Note the direction of installation for the bushing.
• Refer to the stabilizer link removal/installation to connect the stabilizer bar to the links.
• Before installing the wheel, clean the mating surfaces of the brake disc and the inside of the wheel.

STRUTS

REMOVAL & INSTALLATION

See Figures 397 through 400.

1. Raise the vehicle on a lift.
2. Remove the rear wheel.
3. Fold down the rear seat-back, then remove the lid.

➡**Lift up the tab inside underneath the lid first using a flat-tipped screwdriver, then release the hooks.**

4. Remove the damper mounting nuts from the top of the damper.

5. Remove the flange nut while holding the joint pin with a hex wrench, then disconnect the stabilizer link from the knuckle, and remove the brake hose bracket.

6. Remove the damper lower mounting bolt.

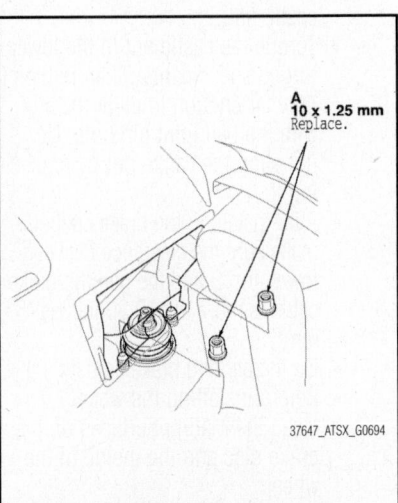

A
10 x 1.25 mm
Replace.

37647_ATSX_G0694

Fig. 397 Remove the damper mounting nuts (A) from the top of the damper

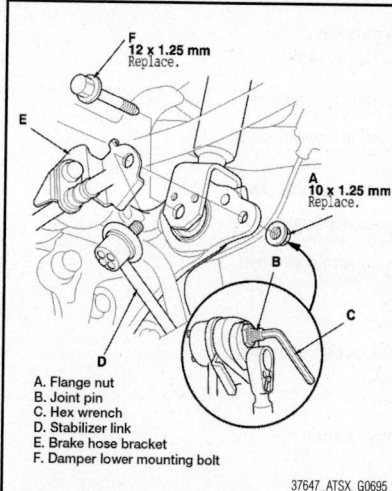

A. Flange nut
B. Joint pin
C. Hex wrench
D. Stabilizer link
E. Brake hose bracket
F. Damper lower mounting bolt

37647_ATSX_G0695

Fig. 398 Remove the flange nut while holding the joint pin with a hex wrench, then disconnect the stabilizer link from the knuckle, and remove the brake hose bracket

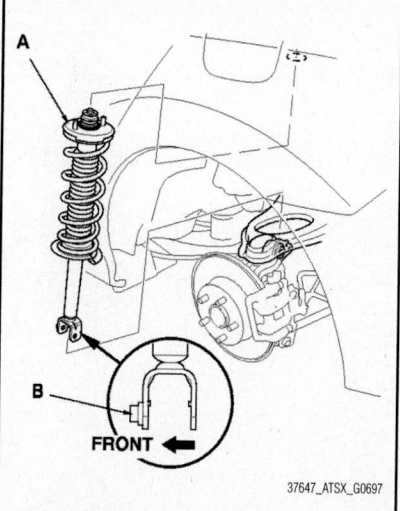

37647_ATSX_G0697

Fig. 400 Lower the rear suspension, and position the damper/spring (A) in the body with the welded nut (B) on the bottom of the damper facing forward

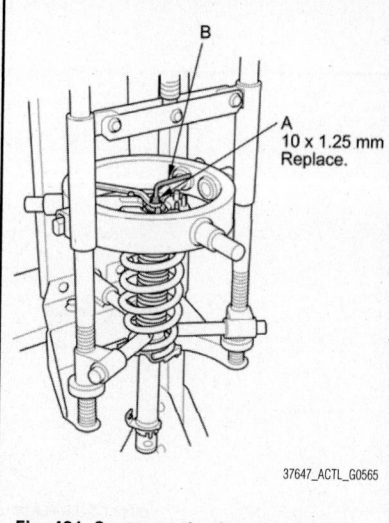

37647_ACTL_G0565

Fig. 401 Compress the damper spring, then remove the self-locking nut (A) while holding the damper shaft with a hex wrench (B)

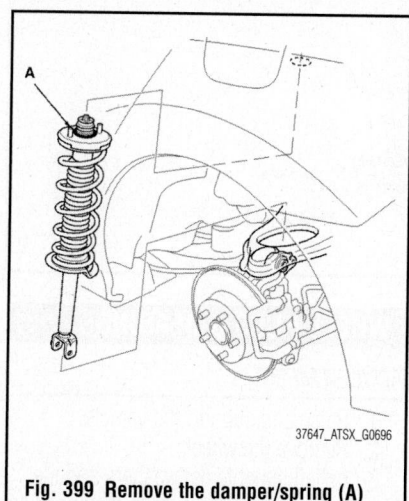

37647_ATSX_G0696

Fig. 399 Remove the damper/spring (A) by lowering the rear suspension

7. Remove the damper/spring by lowering the rear suspension.

➡**Be careful not to damage the body.**

To install:

8. Lower the rear suspension, and position the damper/spring in the body with the welded nut on the bottom of the damper facing forward.

➡**Be careful not to damage the body. Make sure the damper is installed in the correct direction.**

9. Loosely install the new damper mounting nuts to the top of the damper.

10. Loosely install the new damper lower mounting bolt on the bottom of the damper. Connect the stabilizer link to the brake hose bracket to the

knuckle, and loosely install the new flange nut.

11. Place a floor jack under the connecting point of the knuckle and lower arm A, and raise the suspension to load with the vehicle's weight.

12. Tighten the damper lower mounting bolt and the flange nut while holding the joint pin with the hex wrench to 33 ft. lbs. (44 Nm).

13. Tighten the damper mounting nuts on top of the damper to 41 ft. lbs. (55 Nm).

14. Install the lid, and set the rear seatback to the original position.

15. Clean the mating surfaces of the brake disc and the inside of the wheel, then install the rear wheel.

16. Check the wheel alignment, and adjust it if necessary.

OVERHAUL

Disassembly

See Figures 401 and 402.

1. Compress the damper spring, then remove the self-locking nut while holding the damper shaft with a hex wrench. Do not compress the damper spring more than is necessary to remove the self-locking nut.

2. Release the pressure from the strut spring compressor, then disassemble the damper as shown in the Exploded View.

Inspection

1. Reassemble all the parts, except for the damper spring.

2. Compress the damper assembly by hand, and check for smooth operation

through a full stroke, both compression and extension. The damper should extend smoothly and constantly when compression is released. If it does not, the gas is leaking and the damper should be replaced.

3. Check for oil leaks, abnormal noises, and binding during these tests.

Reassembly

See Figures 403 through 406.

1. Install the spring mounting cushion on the damper mounting base by aligning the tab and notch.

2. Install all the parts except the damper mounting washer and the self-locking nut onto the damper unit by referring to the Exploded View.

3. Compress the damper spring using a strut spring compressor. Do not compress the spring excessively.

4. Align the lower end of the damper spring with the stepped part of the dust cover lower mount and the lower spring seat on the damper unit.

5. Position the tab on the spring mounting cushion facing forward but toward the inside of the vehicle.

6. Align the angle of the stud on the damper mounting base with the aligning tab on the bottom of the damper unit.

7. Install the damper mounting washer and the new self-locking nut.

8. Hold the damper shaft with a hex wrench, and tighten the self-locking nut to 22 ft. lbs. (29 Nm).

9. Remove the damper/spring from the strut spring compressor.

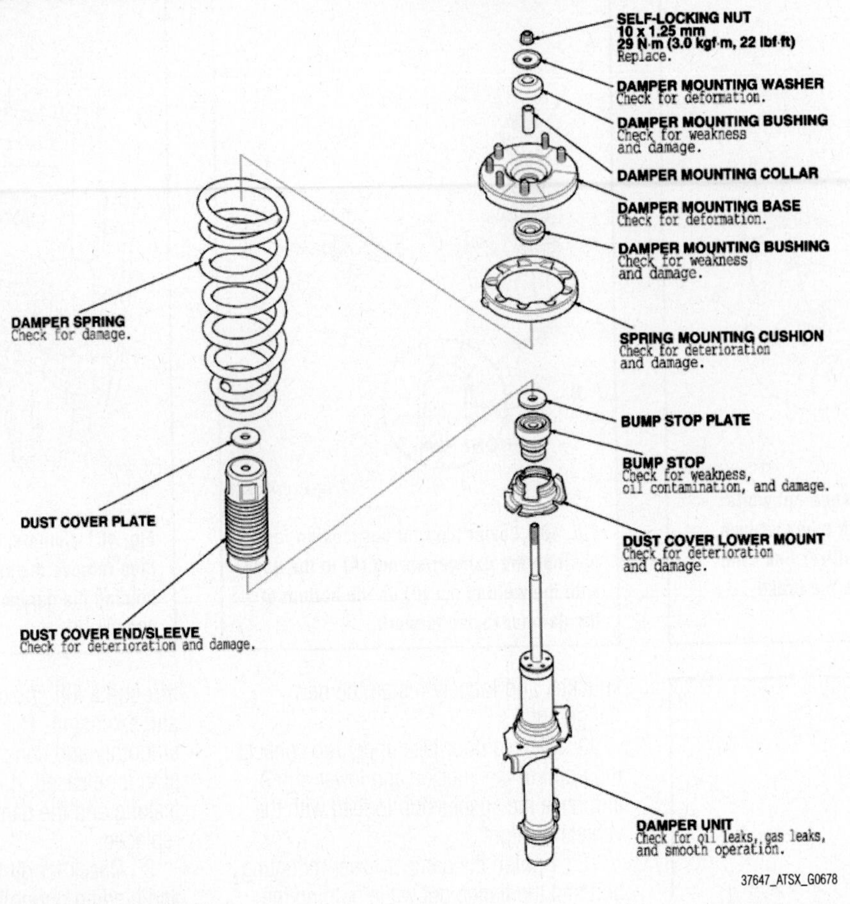

SELF-LOCKING NUT
10 x 1.25 mm
29 N·m (3.0 kgf·m, 22 lbf·ft)
Replace.

DAMPER MOUNTING WASHER
Check for deformation.

DAMPER MOUNTING BUSHING
Check for weakness
and damage.

DAMPER MOUNTING COLLAR

DAMPER MOUNTING BASE
Check for deformation.

DAMPER MOUNTING BUSHING
Check for weakness
and damage.

SPRING MOUNTING CUSHION
Check for deterioration
and damage.

DAMPER SPRING
Check for damage.

BUMP STOP PLATE

BUMP STOP
Check for weakness,
oil contamination, and damage.

DUST COVER PLATE

DUST COVER LOWER MOUNT
Check for deterioration
and damage.

DUST COVER END/SLEEVE
Check for deterioration and damage.

DAMPER UNIT
Check for oil leaks, gas leaks,
and smooth operation.

37647_ATSX_G0678

Fig. 402 Exploded view of damper/spring assembly

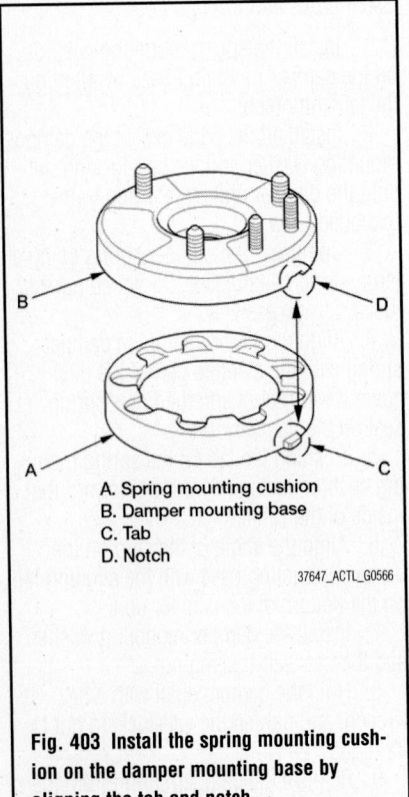

A. Spring mounting cushion
B. Damper mounting base
C. Tab
D. Notch

37647_ACTL_G0566

Fig. 403 Install the spring mounting cushion on the damper mounting base by aligning the tab and notch

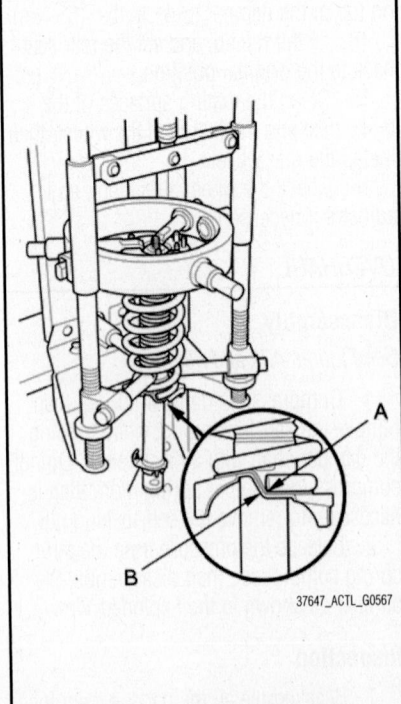

37647_ACTL_G0567

Fig. 404 Align the lower end (A) of the damper spring with the stepped part (B) of the dust cover lower mount

WHEEL HUBS & BEARINGS

ADJUSTMENT

1. Raise and support the vehicle.
2. Remove the wheels.
3. Install suitable flat washers and the wheel nuts. Tighten the nuts to the specified torque to hold the brake disc securely against the hub.
4. Attach the dial gauge. Place the dial gauge against the hub flange.
5. Measure the bearing end play while moving the brake disc inward and outward.

➡ **Wheel bearing end play: Front/Rear: 0–0.002 inches (0–0.05 mm)**

6. If the bearing end play measurement is more than the standard, replace the wheel bearing or the hub bearing unit.

REMOVAL & INSTALLATION

See Figures 407 through 411.

1. Raise and support the vehicle.
2. Remove the wheel nuts, and the rear wheel.
3. Release the parking brake lever fully.

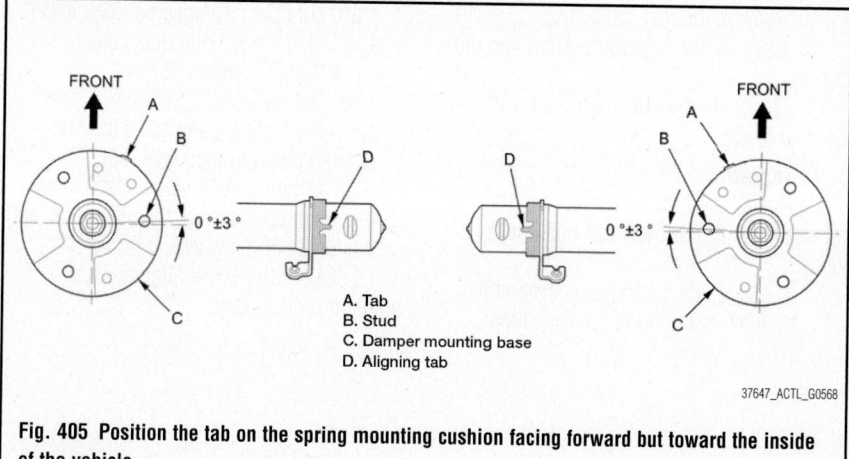

A. Tab
B. Stud
C. Damper mounting base
D. Aligning tab

37647_ACTL_G0568

Fig. 405 Position the tab on the spring mounting cushion facing forward but toward the inside of the vehicle

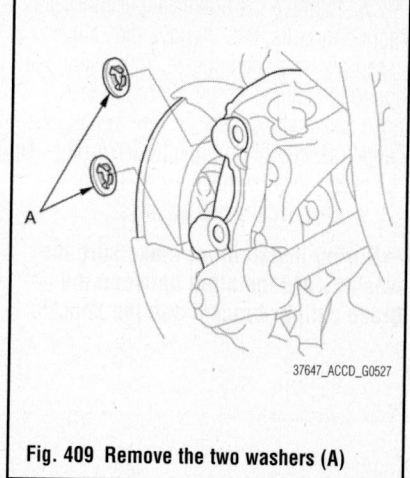

37647_ACCD_G0527

Fig. 409 Remove the two washers (A)

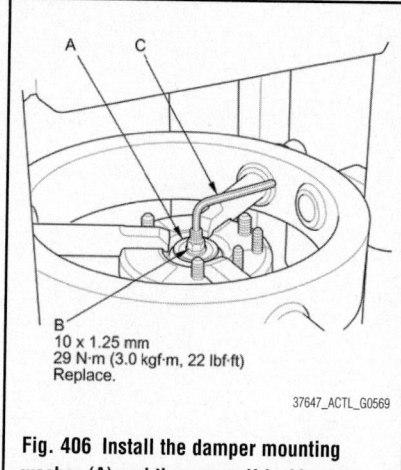

B
10 x 1.25 mm
29 N·m (3.0 kgf·m, 22 lbf·ft)
Replace.

37647_ACTL_G0569

Fig. 406 Install the damper mounting washer (A) and the new self-locking nut (B)

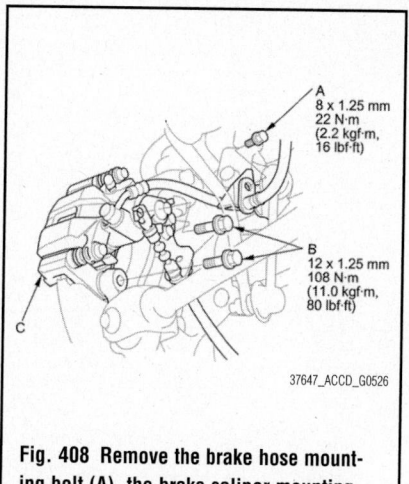

A
8 x 1.25 mm
22 N·m
(2.2 kgf·m,
16 lbf·ft)

B
12 x 1.25 mm
108 N·m
(11.0 kgf·m,
80 lbf·ft)

37647_ACCD_G0526

Fig. 408 Remove the brake hose mounting bolt (A), the brake caliper mounting bolts (B) and the caliper assembly (C)

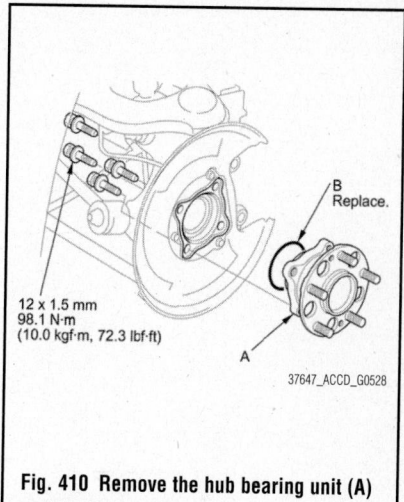

B
Replace.

12 x 1.5 mm
98.1 N·m
(10.0 kgf·m, 72.3 lbf·ft)

A

37647_ACCD_G0528

Fig. 410 Remove the hub bearing unit (A) and the O-ring (B)

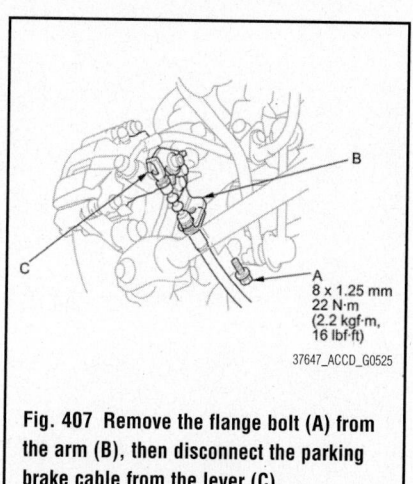

B

A
8 x 1.25 mm
22 N·m
(2.2 kgf·m,
16 lbf·ft)

37647_ACCD_G0525

Fig. 407 Remove the flange bolt (A) from the arm (B), then disconnect the parking brake cable from the lever (C)

4. Loosen the parking brake cable adjusting nut.

5. Remove the flange bolt from the arm. Then disconnect the parking brake cable from the lever.

6. Remove the brake hose mounting bolt.

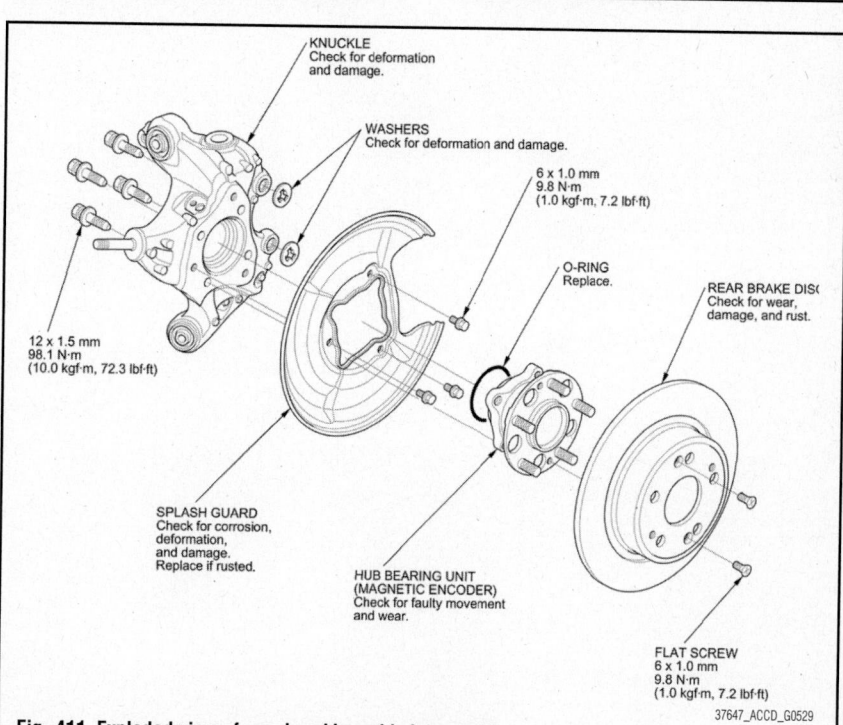

KNUCKLE
Check for deformation and damage.

WASHERS
Check for deformation and damage.

6 x 1.0 mm
9.8 N·m
(1.0 kgf·m, 7.2 lbf·ft)

O-RING
Replace.

REAR BRAKE DISC
Check for wear, damage, and rust.

12 x 1.5 mm
98.1 N·m
(10.0 kgf·m, 72.3 lbf·ft)

SPLASH GUARD
Check for corrosion, deformation, and damage.
Replace if rusted.

HUB BEARING UNIT
(MAGNETIC ENCODER)
Check for faulty movement and wear.

FLAT SCREW
6 x 1.0 mm
9.8 N·m
(1.0 kgf·m, 7.2 lbf·ft)

37647_ACCD_G0529

Fig. 411 Exploded view of rear knuckle and hub assembly

7. Remove the brake caliper bracket mounting bolts, then remove the caliper assembly from the knuckle. To prevent damage to the caliper assembly or the brake hose, use a short piece of wire to hang the caliper assembly from the undercarriage. Do not twist the brake hose excessively.

8. Remove the two washers.

➡**During installation, make sure the washers are installed between the brake caliper bracket and the knuckle.**

9. Remove the rear brake disc.

10. Remove the hub bearing unit and the O-ring.

11. Check the hub bearing unit for damage and cracks.

To install:

12. Install the hub bearing unit in the reverse order of removal, and note these items:

    a. Use a new O-ring on reassembly.

    b. After installing the brake caliper, make sure the clearance between lower arm B and the parking brake cable is more than 0.2 inches (5 mm).

    c. Before installing the brake disc, clean the mating surfaces of the hub bearing unit and the brake disc.

    d. Before installing the wheel, clean the mating surfaces of the brake disc and the inside of the wheel.

13. Check the wheel alignment, and adjust it if necessary.

HONDA

Civic • Civic Hybrid

8

BRAKES 8-13

**ANTI-LOCK BRAKE
SYSTEM (ABS)** 8-14
Wheel Speed Sensors 8-14
Removal & Installation..........8-14
**BLEEDING THE BRAKE
SYSTEM** 8-13
Bleeding Procedure..................8-13
Bleeding Procedure 8-13
FRONT DISC BRAKES 8-14
Brake Caliper.........................8-14
Removal & Installation.........8-14
Disc Brake Pads 8-19
Removal & Installation..........8-19
**INFORMATION AND
PRECAUTIONS** 8-13
Anti-lock Systems.................8-13
Disc and Drum Systems 8-13
PARKING BRAKE 8-25
Parking Brake Cables 8-25
Adjustment 8-25
REAR DISC BRAKES 8-21
Brake Caliper.........................8-21
Removal & Installation..........8-21
Brake Drum 8-22
Removal & Installation..........8-22
Brake Shoes 8-23
Removal & Installation..........8-23
Disc Brake Pads 8-22
Removal & Installation..........8-22

CHASSIS ELECTRICAL 8-25

**AIR BAG (SUPPLEMENTAL
RESTRAINT SYSTEM)** 8-25
General Information..................8-25
Arming the System 8-25
Disarming the System..........8-25
Precautions 8-25

DRIVE TRAIN 8-26

Clutch Hydraulic System
Bleeding...............................8-26
Procedure 8-26
Front Driveshaft......................8-26
Removal & Installation..........8-26

ENGINE COOLING 8-28

Battery.................................8-31
Removal & Installation..........8-31
Battery Cables 8-31
Disconnecting & Connecting...8-31
Radiator & Fan 8-28
Removal & Installation..........8-28
Thermostat 8-28
Removal & Installation..........8-28
Water Pump 8-30
Removal & Installation..........8-30

ENGINE ELECTRICAL 8-31

CHARGING SYSTEM 8-31
Alternator 8-31
Removal & Installation..........8-31
**HYBRID ENGINE
COMPONENTS** 8-32
IMA Battery Current Sensor 8-34
Removal & Installation..........8-34
IMA Motor Power Cable..........8-34
Removal & Intallation 8-34
IMA Motor Rotor Position
Sensor 8-36
Removal & Installation..........8-36
IMA Service Precautions...........8-32
Disconnecting the Motor
Power Cable Connector
From the Motor Stator........8-33
General Information 8-32
Handling of A Vehicle
Damaged in A Collision 8-33
Other Precautions 8-33
Precautions When
Working on High
Magnetic Area 8-33
Precautions When Working
on High Voltage Area..........8-32
Preparation Items.................8-32
Turning Off and on
Power To the High
Voltage Circuit...................8-33
IGNITION SYSTEM 8-36
Firing Order............................8-36
Idle Speed 8-37
Inspection 8-37

Ignition Coils & Spark Plugs 8-36
Removal & Installation..........8-36
Ignition Timing........................8-37
Inspection & Adjustment 8-37
STARTING SYSTEM 8-38
Starter 8-38
Removal & Installation..........8-38

ENGINE MECHANICAL 8-39

Accessory Drive Belts 8-39
Accessory Belt Routing........8-39
Adjustment 8-40
Inspection 8-39
Removal & Installation.........8-40
Camshaft 8-41
Removal & Installation.........8-41
Camshaft (Timing) Chain &
Sprockets 8-48
Removal & Installation.........8-48
Crankshaft Front Seal..............8-60
Removal & Installation.........8-60
Crankshaft Pulley 8-62
Removal & Installation..........8-62
Cylinder Head 8-64
Removal & Installation..........8-64
Cylinder Head (Valve) Cover 8-74
Removal & Installation.........8-74
Exhaust Manifold 8-75
Removal & Installation.........8-75
Intake Manifold 8-76
Removal & Installation.........8-76
Oil Pan 8-79
Removal & Installation.........8-79
Oil Pump...............................8-83
Removal & Installation.........8-83
Piston and Ring.......................8-88
Positioning 8-88
Rocker Arms..........................8-89
Removal & Installation.......8-89
Valve Lash (Clearance)............8-92
Adjustment 8-92

**ENGINE PERFORMANCE &
EMISSION CONTROLS** 8-95

Accelerator Pedal Position
(APP) Sensor 8-95

Location...................................8-95
Removal & Installation...........8-95
Camshaft Position (CMP)
Sensor..................................8-95
Removal & Installation..........8-95
Crankshaft Position (CKP)
Sensor..................................8-96
CKP Pattern Clear/CKP
Pattern LEAN Procedure.....8-97
Removal & Installation..........8-96
Engine Control
Module/Powertrain Control
Module (ECM/PCM)..............8-97
ECM/PCM Idle Learn
Procedure.........................8-99
ECM/PCM Update................8-98
Location...............................8-97
Removal & Installation..........8-97
Engine Coolant Temperature
(ECT) Sensor......................8-99
Location...............................8-99
Removal & Installation..........8-99
Evaporative Emissions
(EVAP) Canister..................8-101
Location...............................8-101
Removal & Installation.........8-101
Exhaust Gas Recirculation
(EGR) Valve........................8-102
Location...............................8-102
Removal & Installation.........8-102
Heated Oxygen Sensor
(HO2S)................................8-102
Removal & Installation.........8-102
Intake Air Temperature/Mass
Air Flow (IAT/MAF)
Sensor................................8-104
Location...............................8-104
Removal & Installation.........8-104
Knock Sensor (KS).................8-104
Location...............................8-104
Removal & Installation.........8-104
Manifold Absolute
Pressure (MAP) Sensor.......8-106
Removal & Installation.........8-106
Positive Crankcase
Ventilation (PCV) Valve.........8-107
Location...............................8-107
Removal & Installation........8-107

FUEL...........................8-108

GASOLINE FUEL
INJECTION SYSTEM........8-108
Fuel Filter..............................8-110
Removal & Installation.........8-110

Fuel Pump/ Fuel Gauge
Sending Unit........................8-111
Removal & Installation.........8-111
Fuel Rail and Injector..............8-111
Removal & Installation.........8-111
Fuel System Pressure.............8-109
Relieving.............................8-109
Fuel System Service
Precautions.........................8-108
Fuel Tank................................8-113
Draining..............................8-113
Removal & Installation.........8-114
Fuel Tank Unit........................8-114
Removal & Installation.........8-114
Idle Speed..............................8-117
Inspection & Adjustment....8-117
Throttle Body..........................8-118
Removal & Installation.........8-118

HEATING & AIR
CONDITIONING SYSTEM...8-121

Blower Motor..........................8-121
Removal & Installation.........8-121
Heater Unit & Core..................8-122
Removal & Installation.........8-122

SPECIFICATIONS AND
MAINTENANCE CHARTS......8-3

Brake Specifications................8-10
Camshaft, and Bearing
Specifications......................8-6
Capacities...............................8-4
Crankshaft and
Connecting Rod
Specifications......................8-7
Engine and Vehicle
Identification........................8-3
Engine Tune-Up Specifications...8-3
Fluid Specifications..................8-5
General Engine Specifications.....8-3
Piston and Ring Specifications...8-7
Maintenance Minder
Schedule.........................8-11, 12
Tire, Wheel and Ball Joint
Specifications......................8-10
Torque Specifications..............8-8
Valve Specifications.................8-5
Wheel Alignment.....................8-8

STEERING....................8-124

Electronic Power Steering
(EPS) Control Unit...............8-124
Removal & Installation........8-124

Electronic Power Steering
(EPS) Motor.........................8-124
Removal & Installation.........8-124
EPS System Resetting............8-126
EPS Torque Sensor
Neutral Position—
memorizing.....................8-126
Steering Angle Sensor
Neutral Position—clear....8-126
VSA Sensor Neutral
Position—
memorization....................8-126
Power Rack & Pinion
Steering Gear......................8-126
Removal & Installation.........8-126
Torque Sensor Neutral
Position Memorization.....8-133
Power Steering Pump.............8-134
Bleeding.............................8-135
Removal & Installation.........8-134

SUSPENSION................8-135

FRONT SUSPENSION........8-135
Lower Ball Joint.......................8-135
Removal & Installation.........8-135
Lower Control Arm..................8-136
Removal & Installation.........8-136
Stabilizer Bar & Links.............8-137
Removal & Installation.........8-137
Struts & Springs.....................8-138
Removal & Installation.........8-138
Wheel Hub/Knuckle
Assembly.............................8-138
Adjustment.........................8-140
Removal & Installation.........8-138
REAR SUSPENSION.........8-140
Coil Spring..............................8-140
Removal & Installation.........8-140
Knuckle/Hub Bearing Unit.......8-141
Removal & Installation.........8-141
Stabilizer Bar..........................8-145
Removal & Installation.........8-145
Struts (Dampers)....................8-146
Removal & Installation.........8-146
Upper Control Arm..................8-147
Removal & Installation.........8-147

SPECIFICATIONS AND MAINTENANCE CHARTS

ENGINE AND VEHICLE IDENTIFICATION

Engine						Model Year	
Code	Liters	Cu. In. (cc)	Cyl.	Fuel Sys.	Eng. Mfg.	Code	Year
LDA2	1.3	82.0 (1339)	4	PGM-FI	Honda	B	2011
R18A1	1.8	110.0 (1798)	4	PGM-FI	Honda	C	2012
R18A4	1.8	110.0 (1798)	4	CNG	Honda		
LEA2	1.8	110.0 (1798)	4	Hybrid	Honda		
R18Z1	1.8	110.0 (1798)	4	SMFI	Honda		
K20Z3	2.0	121.9 (1997)	4	PGM-FI	Honda		
K24Z7	2.4	143.6 (2354)	4	SMFI	Honda		

SMFI: Sequential multi-port fuel injection

CNG: Compressed Natural Gas

71051_CIVC_C0001

GENERAL ENGINE SPECIFICATIONS

Year	Model	Engine Displacement Liters	Engine ID	Net Horsepower @ rpm	Net Torque @ rpm (ft. lbs.)	Bore X Stroke (in.)	Compression Ratio	Oil Pressure @ rpm
2011	Civic Hybrid	1.3	LDA2	110@6000	123@2500	2.87x3.15	10.8:1	50@3000
	Civic	1.8	R18A1	140@4800	128@4800	3.19x3.44	10.5:1	50@3000
	Civic GX	1.8	R18A4	113@6300	109@4300	3.19x3.44	10.5:1	50@3000
	Civic	2.0	K20Z3	197@7800	139@6100	3.39x3.39	11.0:1	44@3000
2012	Civic	1.8	R18Z1	110@6500	106@4300	3.19x3.44	12.7:1	54.5@2000
	Civic Hybrid	1.8	LEA2	110@5500	127@1000-3500	2.87x3.52	10.8:1	54.5@2000
	Civic	2.4	K24Z7	201@7000	170@4400	3.43x3.90	11.0:1	43.5@3000

71051_CIVC_C0002

ENGINE TUNE-UP SPECIFICATIONS

Year	Engine Displacement Liters	Engine ID	Spark Plugs Gap (in.)	Ignition Timing (deg.) MT	Ignition Timing (deg.) AT	Fuel Pump (psi)	Idle Speed (rpm) MT	Idle Speed (rpm) AT	Valve Clearance In.	Valve Clearance Ex.
2011	1.3	LDA2	0.039-0.043	—	8-12B	38-46	—	770-870	0.006-0.007	0.009-0.011
	1.8	R18A1	0.039-0.043	6-10B	6-10B	48-55	620-720	620-720	0.007-0.009	0.009-0.011
	1.8	R18A4	0.039-0.043	6-10B	6-10B	—	①	①	0.007-0.009	0.009-0.011
	2.0	K20Z3	0.039-0.043	6-10B	—	48-55	700-800	—	0.008-0.010	0.010-0.011
2012	1.8	R18Z1	0.039-0.043	NA	6-10B	55-63	620-720	620-720	0.007-0.009	0.009-0.011
	1.8	LEA2	0.028-0.031	NA	6-10B	55-63	②	②	0.007-0.009	0.009-0.011
	2.4	K24Z7	0.039-0.043	6-10B	NA	48-55	700-800	NA	0.008-0.010	0.010-0.011

NOTE: The Vehicle Emission Control Information label often reflects specification changes made during production.

The label figures must be used if they differ from those in this chart

NA: Not Available

B: Before Top Dead Center

① Shown with no load; with full electrical load: 660-760 rpm

② No load: 620-720 rpm

   With full electrical load: 660-760 rpm

71051_CIVC_C0003

CAPACITIES

Year	Model	Engine Displacement Liters	Engine ID	Engine Oil with Filter	Transmission (pts.)			Fuel Tank (gal.)	Cooling System (qts.)
					5-Spd	6-Spd	Auto.		
2011	Civic Hybrid	1.3	LDA2	3.4	—	—	6.0	12.3	5.0
	Civic	1.8	R18A1	3.9	3.0	3.2	5.0	13.2	①
	Civic GX	1.8	R18A4	3.9	3.0	3.2	5.0	②	①
	Civic	2.0	K20Z3	4.6	3.2	3.2	—	13.2	③
2012	Civic	1.8	R18Z1	3.9	3.0	3.2	5.0	13.2	④
	Civic Hybrid	1.8	R18A9	3.9	3.0	3.2	6.0	⑤	⑥
	Civic	2.4	K24Z7	4.4	3.2	3.2	—	13.2	⑦

NOTE: All capacities are approximate. Add fluid gradually and ensure a proper fluid level is obtained.

NOTE: Capacities given are service, not overhaul capacities

① M/T 4 door with coolant change: 5.48

　M/T 4 door with engine overhaul: 6.88

　A/T 4 door with coolant change: 5.60

　A/T 4 door with engine overhaul: 7.10

　M/T 2 door with coolant change: 5.48

　M/T 2 door with engine overhaul: 6.88

　A/T 2 door with coolant change: 6.80

　A/T 2 door with engine overhaul: 7.52

② Fuel system is CNG

③ With coolant change: 4.76

　With engine overhaul: 7.20

④ M/T with coolant change: 5.91 (Denso); 5.81 (Trad)

　M/T with engine overhaul: 6.78 (Denso); 6.69 (Trad)

　A/T with coolant change: 5.91 (Denso); 5.81 (Trad)

　A/T with engine overhaul: 6.93 (Denso); 6.84 (Trad)

⑤ Info not shown; consult Owner's Manual for fuel tank capacity.

⑥ With coolant change: 5.02

　With engine overhaul: 6.36

⑦ With coolant change: 5.80

　With engine overhaul: 7.12

71051_CIVIC_C0004

FLUID SPECIFICATIONS

Year	Model	Engine Displacement Liters	Engine Oil	Man. Trans.	Auto. Trans.	Power Steering Fluid	Brake Master Cylinder	Cooling System
2011	Civic Hybrid	1.3	0W-20 Honda	NA	Honda CVTF	NA	Honda DOT 3	①
	Civic	1.8	5W-20 Honda	Honda MTF	Honda ATF-Z1	Honda PS Fluid	Honda DOT 3	①
	Civic	2.0	5W-20 Honda	Honda MTF	—	Honda PS Fluid	Honda DOT 3	①
2012	Civic	1.8	0W-20 Honda	—	Honda ATF DW-1	Honda PS Fluid	Honda DOT 3	①
	Civic Hybrid	1.8	0W-20 Honda	—	Honda CVTF	Honda PS Fluid	Honda DOT 3	①
	Civic	2.4	0W-20 Honda	Honda MTF	Honda ATF DW-1	Honda PS Fluid	Honda DOT 3	①

DOT: Department Of Transpotation

NA: Not Applicable

① Honda Long Life Antifreeze/Coolant-Type2

71051_CIVC_C0005

VALVE SPECIFICATIONS

Year	Engine Displacement Liters	Engine ID/VIN	Seat Angle (deg.)	Face Angle (deg.)	Spring Test Pressure (lbs. @ in.)	Spring Installed Height (in.)	Stem-to-Guide Clearance (in.) Intake	Stem-to-Guide Clearance (in.) Exhaust	Stem Diameter (in.) Intake	Stem Diameter (in.) Exhaust
2011	1.3	LDA2	45	45	NA	NA	0.0008-0.0020	0.0020-0.0031	0.2157-0.2161	0.2146-0.2150
	1.8	R18A1	45	45	NA	NA	0.0008-0.0020	0.0020-0.0031	0.2157-0.2161	0.2146-0.2150
	1.8	R18A4	45	45	NA	NA	0.0008-0.0020	0.0020-0.0031	0.2157-0.2161	0.2146-0.2150
	2.0	K20Z3	45	45	NA	NA	0.0012-0.0022	0.0022-0.0031	0.2156-0.2159	0.2146-0.2150
2012	1.8	R18Z1	45	45	NA	NA	0.0008-0.0020	0.0020-0.0031	0.2157-0.2161	0.2146-0.2150
	1.8	LEA2	45	45	NA	NA	0.0008-0.0020	0.0020-0.0031	0.2157-0.2161	0.2146-0.2150
	2.4	K24Z7	45	45	NA	NA	0.0012-0.0022	0.0022-0.0031	0.2156-0.2159	0.2146-0.2150

NA: Information not available

71051_CIVC_C0007

CAMSHAFT AND BEARING SPECIFICATIONS

All measurements are given in inches.

Year	Engine Displacement Liters	Engine VIN	Journal Diameter	Brg. Oil Clearance	Shaft End-play	Runout	Journal Bore	Lobe Height Intake	Exhaust
2011	1.3	LDA2	NA	0.0020-0.0035	0.0020-0.0060	0.0010	NA	①	②
	1.8	R18A1	NA	0.0018-0.0033	0.0020-0.0100	0.0010	NA	③	1.4100
	1.8	R18A4	NA	0.0018-0.0033	0.0020-0.0100	0.0010	NA	④	1.3538
	2.0	K20Z3	NA	⑤	0.0020-0.0080	0.0010	NA	⑥	⑦
2012	1.8	R18Z1	NA	0.0018-0.0033	0.0020-0.0100	0.0012	NA	⑧	1.4122
	1.8	LEA2	NA	0.0018-0.0033	0.0020-0.0100	NA	NA	④	1.3538
	2.4	K24Z7	NA	⑤	0.0020-0.0080	0.0012	NA	①	⑥

NA: Information not available

① 1st: 1.1692 in.
2nd: 1.4003 in.
3rd: 1.4196 in.

② 1st: 1.1771 in.
2nd: 1.4054 in.

③ Primary: 1.4076 in.
Secondary A: 1.3920 in.
Secondary B: 1.4184 in.

④ Primary: 1.3473 in.
Secondary A: 1.3338 in.
Secondary B: 1.3537 in.

⑤ No. 1 journal: 0.001-0.003 in.
No. 2-5 journals: 0.002-0.004 in

⑥ Primary: 1.2910 in.
Mid: 1.3990 in.
Secondary: 1.2865 in.

⑦ Primary: 1.2902 in.
Mid: 1.3688 in.
Secondary: 1.2859 in.

⑧ Primary: 1.1412 in.
Secondary A: 1.2133 in.
Secondary B: 1.4051 in.

71051_CIVC_C0006

CRANKSHAFT AND CONNECTING ROD SPECIFICATIONS

All measurements are given in inches.

Year	Engine Displacement Liters	Engine ID/VIN	Crankshaft				Connecting Rod		
			Main Brg. Journal Dia.	Main Brg. Oil Clearance	Shaft End-play	Thrust on No.	Journal Diameter	Oil Clearance	Side Clearance
2011	1.3	LDA2	1.9676-1.9685	0.0007-0.0014	0.0040-0.0140	4	1.5739-1.5748	0.0008-0.0015	0.0060-0.0120
	1.3	LDA2	1.9676-1.9685	0.0007-0.0014	0.0040-0.0140	4	1.5739-1.5748	0.0008-0.0015	0.0060-0.0120
	1.8	R18A1	2.1644-2.1654	0.0007-0.0013	0.0040-0.0140	4	1.7707-1.7716	0.0009-0.0017	0.0060-0.0140
	1.8	R18A4	2.1644-2.1654	0.0007-0.0013	0.0040-0.0140	4	1.7707-1.7716	0.0009-0.0017	0.0060-0.0140
	2.0	K20Z3	①	②	0.0040-0.0140	4	1.7707-1.7717	0.0013-0.0026	0.0060-0.0120
2012	1.8	R18Z1	NA	0.00071-0.00134	0.0039-0.0138	4	NA	0.0009-0.0017	0.0059-0.0138
	1.8	LEA2	NA	0.00071-0.00134	0.0039-0.0138	4	NA	0.0009-0.0017	0.0059-0.0138
	2.4	K24Z7	NA	③	0.0039-0.0138	4	1.7707-1.7717	0.00126-0.00260	0.0059-0.0138

① Journals 1, 2 4 and 5: 2.1648-2.1657

  Journal 3: 2.1644-2.1654

② Journals 1, 2 4 and 5: 0.0007-0.0016

  Journal 3: 0.0010-0.0019

③ Journals 1, 2 4 and 5: 0.0007-0.00161 in.

  Journal 3: 0.00098-0.00193 in.

71051_CIVC_C0008

PISTON AND RING SPECIFICATIONS

All measurements are given in inches.

Year	Engine Displacement Liters	Engine ID/VIN	Piston Clearance	Ring Gap			Ring Side Clearance		
				Top Compression	Bottom Compression	Oil Control	Top Compression	Bottom Compression	Oil Control
2011	1.3	LDA2	0.0004-0.0016	0.0060-0.0120	0.0140-0.0200	0.0080-0.0280	0.0025-0.0035	0.0012-0.0022	NA
	1.8	R18A1	0.0004-0.0014	0.0080-0.0140	0.0160-0.0020	①	0.0018-0.0028	0.0014-0.0024	NA
	1.8	R18A4	0.0004-0.0014	0.0080-0.0140	0.0160-0.0220	0.0080-0.0200	0.0018-0.0028	0.0014-0.0024	NA
	2.0	K20Z3	0.0008-0.0016	0.0080-0.0140	0.0200-0.0260	0.0080-0.0280	0.0018-0.0028	0.0016-0.0026	NA
2012	1.8	R18Z1	0.00039-0.00138	NA	NA	NA	NA	NA	NA
	1.8	LEA2	0.00039-0.00138	NA	NA	NA	NA	NA	NA
	2.4	K24Z7	0.0015-0.0016	0.0079-0.0118	0.0079-0.0126	0.0079-0.0197	NA	NA	NA

NA: Information not available

① Riken: 0.008-0.020 in.

  Except RIKEN: 0.008-0.028 in.

71051_CIVC_C0009

TORQUE SPECIFICATIONS
All readings in ft. lbs.

Year	Engine Disp. Liters	Engine ID/VIN	Cylinder Head Bolts	Main Bearing Bolts	Rod Bearing Bolts	Crankshaft Damper Bolts	Flywheel/ Drive Plate Bolts	Manifold Intake	Exhaust	Spark Plugs	Oil Pan Drain Plug
2011	1.3	LDA2	①	②	③	④	33	17	16	13	29
	1.8	R18A1	⑤	⑥	⑦	⑧	⑨	17	23	18	29
	1.8	R18A4	⑤	⑥	⑦	⑧	⑨	17	23	18	29
	2.0	K20Z3	⑤	⑥	⑦	⑧	⑨	16	23	18	29
2012	1.8	R18Z1	⑩	⑥	⑪	⑫	76 ⑬	18	—	13	29
	1.8	LEA2	⑩	⑥	⑪	⑫	—	18	—	13	29
	2.4	K24Z7	⑩	⑭	⑮	⑯	55 ⑬	16	—	13	29

① Step 1: 22 ft. lbs.
  Step 2: Rotate 130 degrees
② Step 1: 18 ft. lbs.
  Step 2: Rotate 40 degrees
③ Step 1: 7.2 ft. lbs.
  Step 2: Rotate 90 degrees
④ Old bolt:
  Step 1: 27 ft. lbs
  Step 2: Plus 90 degrees
  New bolt:
  Step 1: 130 ft. lbs
  Step 2: Loosen fully
  Step 3: 27 ft. lbs
  Step 4: Plus 90 degrees
⑤ Step 1: 29 ft. lbs.
  Step 2: Rotate 90 degrees
  Step 3: Rotate 90 degrees
  Step 4: If new bolt rotate
  additional 60 degrees
⑥ Step 1: 18 ft. lbs.
  Step 2: Rotate 57 degrees
⑦ Step 1: 14 ft. lbs.
  Step 2: Rotate 90 degrees
⑧ Old bolt:
  Step 1: 27 ft. lbs
  Step 2: Plus 90 degrees
  New bolt:
  Step 1: 130 ft. lbs
  Step 2: Loosen fully
  Step 3: 27 ft. lbs
  Step 4: Plus 90 degrees

⑨ Automatic transaxle: 54 ft. lbs.
  Manual transaxle: 76 ft. lbs.
⑩ Old bolt:
  Step 1: 36 ft. lbs
  Step 2: Plus 90 degrees
  New bolt:
  Step 1: 130 ft. lbs
  Step 2: Loosen fully
  Step 3: 36 ft. lbs
  Step 4: Plus 90 degrees
⑪ Step 1: 15 ft. lbs.
  Step 2: Rotate 90 degrees
⑫ Old bolt:
  Step 1: 133 ft. lbs
  Step 2: remove the bolt
  Step 3: 37 ft. lbs.
  Step 4: Plus 82 degrees
  New bolt:
  Step 1: 52 ft. lbs
  Step 2: Plus 90 degrees
⑬ Tighten in 2 or more steps
⑭ Step 1: 22 ft. lbs.
  Step 2: Rotate 48 degrees
⑮ Step 1: 30 ft. lbs.
  Step 2: Rotate 120 degrees
⑯ Old bolt:
  Step 1: 37 ft. lbs
  New bolt:
  Step 1: 131 ft. lbs
  Step 2: Loosen fully
  Step 3: 37 ft. lbs

71051_CIVIC_C0010

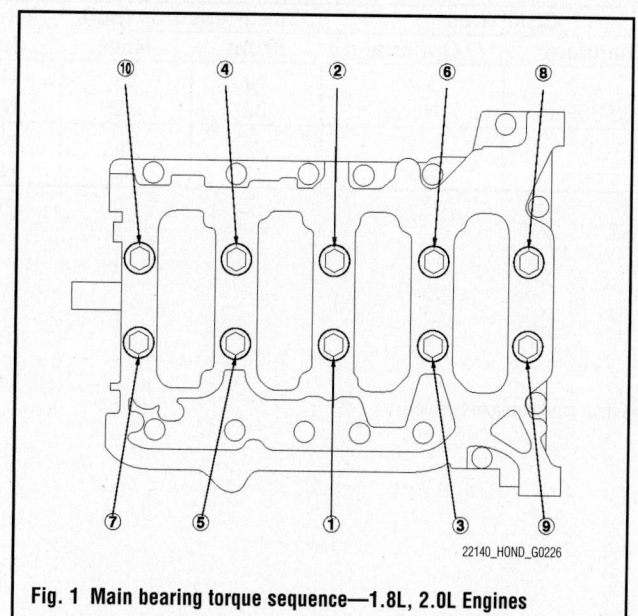

Fig. 1 Main bearing torque sequence—1.8L, 2.0L Engines

22140_HOND_G0226

WHEEL ALIGNMENT

Year	Model		Caster Range (+/-Deg.)	Caster Preferred Setting (Deg.)	Camber Range (+/-Deg.)	Camber Preferred Setting (Deg.)	Toe-in (in.)
2011	Civic	F	1.00	+7.00	0.30	0.00	0.00+/-0.08
		R	—	—	①	①	0.08 ②
	Civic Hybrid	F	1.00	+7.10	0.50	0.05	0.00+/-0.08
		R	—	—	0.183	-0.90	0.00+/-0.08
2012	Civic (Exc. Si)	F	0.5	+5.05	0.30	0.00	0.00+/-0.08
		R	0.5	-0.125	0.50	-0.75	0.08 ③
	Civic Si	F	0.5	+5.06	0.50	-0.1	0.00+/-0.08
		R	—	—	0.50	-0.14	0.08+0.08/-0.04
	Civic Hybrid	F	0.5	+5.06	0.50	-0.15	0.00+/-0.08
		R	—	—	0.183	-0.90	0.08+0.08/-0.04

① With "C" marks on upper arms: -0 degrees 45 minutes (preferred)

Range:

+1 degree 5 minutes

-0 degrees 45 minutes

Without "C" marks on upper arms: -1 degree 30 minutes (preferred)

Range:

+1 degree 5 minutes

-0 degrees 45 minutes

② Range: +0.08 to -0.45 in.

③ Range: +0.08 to -0.45 in.

TIRE, WHEEL AND BALL JOINT SPECIFICATIONS

Year	Model	OEM Tires		Tire Pressures (psi)		Wheel Size	Ball Joint Inspection	Lug Nut (ft. lbs.)
		Standard	Optional	Front	Rear			
2011	Civic	①	②	③	③	NA	NA	80
	Civic Hybrid	195/65R15	None	32	32	NA	NA	80
2012	Civic	①	②	③	③	NA	NA	80
	Civic Hybrid	195/65R15	None	32	32	NA	NA	80

OEM: Original Equipment Manufacturer

PSI: Pounds Per Square Inch

NA: Information not available

① DX, DX-VP, DX-G: P195/65R15

   LX, LX-S, EX, EX-L: P205/55R16

   Si: P215/45R17

② Multiple tire and wheel options available through aftermarket dealers.

③ 4-door:

   DX, DX-VP, DX-G: 30 psi

   LX, LX-S, EX, EX-L: 32 psi

   Si with ABS: Front 32, Rear 29 psi

   Si with VSA: Front 32, Rear 33 psi

   2-door:

   Except Si Model: 32 psi

   Si with ABS: Front 32, Rear 29 psi

   Si with VSA: 32 psi

71051_CIVC_C0012

BRAKE SPECIFICATIONS
All measurements in inches unless noted

Year	Model		Brake Disc			Brake Drum Diameter		Minimum Pad Thickness		Brake Caliper	
			Original Thickness	Minimum Thickness	Maximum Runout	Original Inside Diameter	Max. Wear Limit	Front	Rear	Bracket Bolts (ft. lbs.)	Mounting Bolts (ft. lbs.)
2011	Civic	F	0.82-0.83	0.750	0.0016	—	—	0.38-0.40	—	80	25
		R	0.35-0.36	0.310	0.0016	7.870-7.874	7.910	—	①	54	17
	Civic GX	F	0.90-0.91	0.830	0.0016	—	—	0.38-0.40	—	80	25
		R	—	—	—	8.657-8.661	8.700	—	0.180	54	17
	Civic Si	F	0.98-0.99	0.910	0.0016			0.35-0.38		80	25
		R	0.35-0.36	0.310	0.0016	—	—	—	0.33-0.37	54	17
	Civic Hybrid	F	0.82-0.83	0.750	0.002	NA	NA	0.37-0.41	—	80	17
		R	0.35-0.36	0.310	0.002	NA	NA	—	0.33-0.37	54	17
2012	Civic Hybrid	F	0.906	0.827	0.016	—	—	0.37-0.41	—	80	16
		R	0.354	0.315	0.016	8.660	8.700	—	0.79	—	17
	Civic GX	F	0.827	0.748	0.0016	—	—	0.38-0.40	—	80	16
		R	—	—	—	8.657-8.661	8.700	—	0.180	—	17
	Civic Si	F	0.984	0.906	0.0016			0.35-0.38		80	16
		R	0.35-0.36	0.310	0.0016	—	—	—	0.33-0.37	—	17

F: Front

R: Rear

NA: Information not available

① With rear disc: 0.33-0.37 in.

   With rear drum: 0.160

71051_CIVC_C0013

MAINTENANCE MINDER SCHEDULE
Honda—Civic (Except GX) & Civic Hybrid

All Honda's displays engine oil life and maintenance service items in the information display to indicate when to perform maintenance service. If the engine oil life is 15% or less, based on the onboard computer's caluculations, you will see SERVICE DUE SOON in the information display every time the ignition key is turned to ON. The maintenance minder indicator will also come on and the maintenance code(s) for other scheduled maintenance items needing service will be displayed below the message.

The maintenance minder is an important feature of the information display. Based on engine and transmission operating conditions, and accumulated engine revolutions, the Civic Hybrid's onboard computer (PCM) calculates the remaining engine oil and the CVT fluid life. The system also displays the remaining engine oil life along with the code(s) for other scheduled maintenance items needing service.

Symbol	Item	Service
A	Engine oil ①	Change
B	Engine oil and filter	Change
	Fluid levels	Inspect
	Brakes	Inspect
	Parking brake adjustment	Check
	Steering gear and linkage	Inspect
	Suspension components	Inspect
	Driveshaft boots	Inspect
	Brake hoses and lines	Inspect
	Exhaust system	Inspect
	Fuel lines and connections	Inspect
1	Tires	Rotate
2	Engine air filter ②	Replace
	Dust and pollen filter ③	Replace
	Accessory drive belt	Inspect
3	Transmission/CVT fluid ④	Replace
4	Spark plugs	Replace
	Valve clearance ⑥	Inspect
5	Engine coolant	Replace

① If the message SERVICE DUE NOW does not appear more than 12 months after the display is reset, change every year.

② If driven in dusty conditions, replace every 15,000 miles.

③ If driven in urban areas that have a high concentration of soot from industry and diesel, replace every 15,000 miles

④ If regularly driven in mountainous areas at very low speed or trailer towing, change the fluid every 30,000 miles.

⑤ If driven regularly in temperatures over 110 deg.F or below -20 deg.F, or towing a trailer, replace every 60,000 miles.

⑥ Adjust if necessary.

Additionally, replace the brake fluid every 3 years, and inspect the idle speed every 160,000 miles.
To reset the Engine Oil Life Display:
1. Turn the ignition switch to ON.
2. Press the SELECT button repeatedly until the engine oil life display or the service message is displayed.
3. Press the RESET button for about 10 seconds. You will see a MAINT RESET message.
4. Select the appropriate answer, MAINT RESET >N (NO) or MAINT RESET > y (YES) by pressing the SELECT button repeatedly.
   >N or >Y is displayed on the outside temperature >N or >Y is displayed on the outside temperature display.
5. Select the MAINT RESET > Y (YES), and press and hold the RESET button again to reset the engine oil life to 100%.

MAINTENANCE MINDER SCHEDULE
Honda—Civic GX

All Honda's displays engine oil life and maintenance service items in the information display to indicate when to perform maintenance service. If the engine oil life is 15% or less, based on the onboard computer's caluculations, you will see SERVICE DUE SOON in the information display every time the ignition key is turned to ON. The maintenance minder indicator will also come on and the maintenance code(s) for other scheduled maintenance items needing service will be displayed below the message.

Symbol	Item	Service
A	Engine oil ①	Change
B	Engine oil and filter	Change
	Fuel filter element B (low pressure)	Change
	Fuel filter element A (high pressure)	Drain
	Fluid levels	Inspect
	Brakes	Inspect
	Parking brake adjustment	Inspect
	Steering gear and linkage	Inspect
	Suspension components	Inspect
	Driveshaft boots	Inspect
	Brake hoses and lines	Inspect
	Exhaust system	Inspect
	Fuel lines and connections	Inspect
1	Tires	Rotate
2	Engine air filter ②	Replace
	Dust and pollen filter ③	Replace
	Fuel filter element A (high pressure)	Replace
	Fuel tank	Inspect
	Valve clearance ④	Inspect
	Accessory drive belt	Inspect
3	Transmission fluid	Replace
4	Spark plugs	Replace
5	Engine coolant	Replace

① If the message SERVICE DUE NOW does not appear more than 12 months after the display is reset, change every year.

② If driven in dusty conditions, replace every 15,000 miles.

③ If driven in urban areas that have a high concentration of soot from industry and diesel, replace every 15,000 miles

④ Adjust if necessary.

Additionally, replace the brake fluid every 3 years, and inspect the idle speed every 160,000 miles.

To reset the Engine Oil Life Display:

1. Turn the ignition switch to ON.

2. Press the SELECT button repeatedly until the engine oil life display or the service message is displayed.

3. Press the RESET button for about 10 seconds. You will see a MAINT RESET message.

4. Select the appropriate answer, MAINT RESET >N (NO) or MAINT RESET > y (YES) by pressing the SELECT button repeatedly.

   >N or >Y is displayed on the outside temperature >N or >Y is displayed on the outside temperature display.

5. Select the MAINT RESET > Y (YES), and press and hold the RESET button again to reset the engine oil life to 100%.

BRAKES | **INFORMATION AND PRECAUTIONS**

ANTI-LOCK SYSTEMS

• Certain components within the ABS system are not intended to be serviced or repaired individually.

• Do not use rubber hoses or other parts not specifically specified for and ABS system. When using repair kits, replace all parts included in the kit. Partial or incorrect repair may lead to functional problems and require the replacement of components.

• Lubricate rubber parts with clean, fresh brake fluid to ease assembly. Do not use shop air to clean parts; damage to rubber components may result.

• Use only DOT 3 brake fluid from an unopened container.

• If any hydraulic component or line is removed or replaced, it may be necessary to bleed the entire system.

• A clean repair area is essential. Always clean the reservoir and cap thoroughly before removing the cap. The slightest amount of dirt in the fluid may plug an ori- fice and impair the system function. Perform repairs after components have been thor- oughly cleaned; use only denatured alcohol to clean components. Do not allow ABS components to come into contact with any substance containing mineral oil; this includes used shop rags.

• The Anti-Lock control unit is a micro- processor similar to other computer units in the vehicle. Ensure that the ignition switch is **OFF** before removing or installing con- troller harnesses. Avoid static electricity dis- charge at or near the controller.

• If any arc welding is to be done on the vehicle, the control unit should be unplugged before welding operations begin.

DISC AND DRUM SYSTEMS

✳✳ CAUTION

Dust and dirt accumulating on brake parts during normal use may contain asbestos fibers from production or aftermarket brake linings. Breathing excessive concentrations of asbestos fibers can cause serious bodily harm. Exercise care when servicing brake parts. Do not sand or grind brake lin- ing unless equipment used is designed to contain the dust residue. Do not clean brake parts with com- pressed air or by dry brushing. Clean- ing should be done by dampening the brake components with a fine mist of water, then wiping the brake compo- nents clean with a dampened cloth. Dispose of cloth and all residue con- taining asbestos fibers in an imper- meable container with the appropriate label. Follow practices prescribed by the Occupational Safety and Health Administration (OSHA) and the Envi- ronmental Protection Agency (EPA) for the handling, processing, and dis- posing of dust or debris that may con- tain asbestos fibers.

BRAKES | **BLEEDING THE BRAKE SYSTEM**

BLEEDING PROCEDURE

BLEEDING PROCEDURE

See Figure 2.

➡ Do not reuse the drained fluid. Use only new Honda DOT 3 Brake Fluid from an unopened container. Using a non-Honda brake fluid can cause corrosion and shorten the life of the system.

➡ Make sure no dirt or other foreign matter gets in the brake fluid.

1. The reservoir connected to the master cylinder must be at the MAX (upper) level mark at the start of the bleeding procedure, and checked after bleeding each wheel loca- tion. Add fluid as required.

2. Make sure the brake fluid level in the reservoir is at the MAX (upper) level line.

3. Have someone slowly pump the brake pedal several times, then apply steady pres- sure.

4. Start the bleeding at the driver's side of the front brake system. Bleed the calipers or the wheel cylinders in the sequence shown.

5. Attach a length of clear drain tube to the bleed screw, then loosen the bleed screw to allow air to escape from the system. Then tighten the bleed screw securely.

6. Refill the master cylinder reservoir to the MAX (upper) level line.

7. Repeat the procedure for each brake circuit until there are no air bubbles in the fluid.

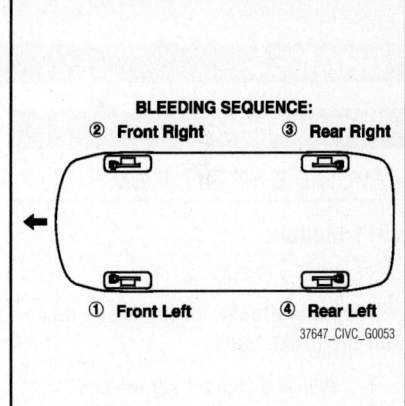

BLEEDING SEQUENCE:
② Front Right ③ Rear Right
① Front Left ④ Rear Left

37647_CIVIC_G0053

Fig. 2 Start the bleeding at the driver's side of the front brake system. Bleed the calipers or the wheel cylinders in the sequence shown

BRAKES   ANTI-LOCK BRAKE SYSTEM (ABS)

WHEEL SPEED SENSORS

REMOVAL & INSTALLATION

2011 Models

→**Required: guide pin tool (07AAG-SVBA100, or equivalent).**

1. Turn the ignition switch to LOCK (0).
2. Release the clamp, then disconnect the wheel speed sensor connector.
3. Remove the clips, the bolt, and the wheel speed sensor.

To install:

4. Install the wheel speed sensor in the reverse order of removal, and note these items:
5. Do not twist the sensor wires.
6. If the wheel speed sensor comes in contact with the wheel bearing, it is faulty.
7. Make sure there is no debris in the sensor mounting hole.
8. Start the engine, and make sure the ABS or VSA indicator goes off.
9. Test-drive the vehicle, and make sure the ABS or VSA indicator does not come on.

2012 Models

Front

1. Raise the vehicle on a lift, and make sure it is securely supported.
2. Remove the front wheel.
3. Disconnect the connector and remove the clips.
4. Remove the wheel speed sensor.

To install:

5. Install the wheel speed sensor.

→**Do not twist the sensor wires. If the wheel speed sensor comes in contact with the wheel bearing, it is faulty. Make sure there is no debris in the sensor mounting hole.**

6. Install the clips.
7. Connect the connector.
8. Install the front wheel.

→**Before installing the wheel, clean the mating surfaces between the brake disc and the inside of the wheel.**

Rear

1. Raise the vehicle on a lift, and make sure it is securely supported.

2. Remove the rear wheel.
3. Disconnect the connector.
4. On the left side, carefully remove the connector from the clamp.

→**The connector is securely attached to the clamp, and the clamp may break when you remove the connector.**

5. For all, remove the clips and remove the wheel speed sensor.

To install:

6. Install the wheel speed sensor.
   - Do not twist the sensor wires.
   - If the wheel speed sensor comes in contact with the hub bearing unit, it is faulty.
   - Make sure there is no debris in the sensor mounting hole.
7. Install the clips.
8. Left side: Install the connector to the clamp.
9. Connect the connector.
10. Install the rear wheel.

→**Before installing the wheel, clean the mating surfaces between the brake disc or the brake drum and the inside of the wheel.**

BRAKES   FRONT DISC BRAKES

BRAKE CALIPER

REMOVAL & INSTALLATION

2011 Models

See Figures 3 through 5.

→**Keep any grease off the brake disc and the brake pads.**

1. Raise and support the vehicle.
2. Remove the rear wheel.
3. Release the parking brake lever fully.
4. Remove the brake hose mounting bolt from the bracket.
5. Remove the brake caliper bracket mounting bolts, then remove the caliper assembly from the knuckle. To prevent damage to the caliper assembly or brake hose, use a short piece of wire to hang the caliper assembly from the undercarriage. Do not twist the brake hose and the parking brake cable excessively.

→**Make sure the washers are in position on reassembly, if they are removed.**

6. Installation is the reverse of the removal procedure.

2012 Models

See Figures 6 and 7.

1. Raise the vehicle on a lift, and make sure it is securely supported.
2. Remove the front wheel.

> ※ **CAUTION**
>
> **Do not spill brake fluid on the vehicle; it may damage the paint. If brake fluid does contact the paint, wash it off immediately with water.**

3. Remove the banjo bolt and disconnect the brake hose from the caliper body.
4. Remove the flange bolts and remove the caliper body.

→**Be careful when removing the caliper body or the spring could pop out of position.**

5. Remove the pad return springs.
6. Remove the pad shims and the brake pads.
7. Remove the caliper pins. Remove the pin boots.

To install:

8. Apply a thin coat of Honda silicone grease (P/N 08C30-B0234M) to the caliper pins and the new pin boots.
9. Install the pin boots.
10. Install the caliper pins.

→**Make sure that the caliper pins are installed correctly. The upper caliper pin and the lower caliper pin are different. If these caliper pins are installed in the wrong location, it will cause vibration, uneven or rapid brake pad wear, and possibly uneven tire wear.**

11. Make sure that the pin boots are properly positioned into the grooves of the caliper pin and the grooves of the caliper body. Remove air with the inside of the boots.
12. Apply a thin coat of M-77 assembly paste to the pad side of the shims, the back of the brake pads, and the other areas indicated by the arrows.
13. Wipe off the excess assembly paste from the pad shims and brake pads friction material. Install the brake pads and pad shims. Install the brake pad with the wear indicator on the upper inside position.

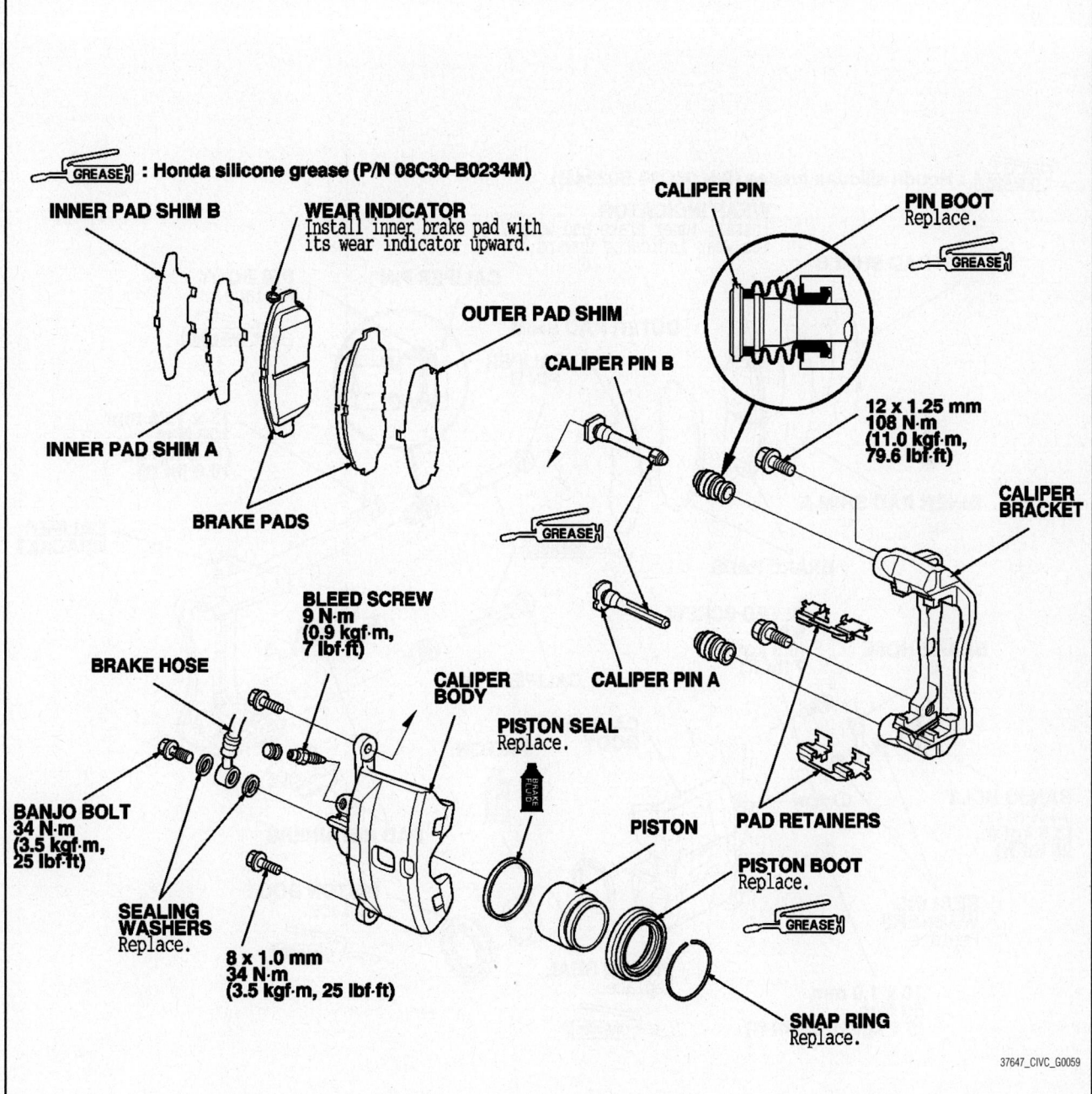

GREASE : Honda silicone grease (P/N 08C30-B0234M)

INNER PAD SHIM B

WEAR INDICATOR
Install inner brake pad with
its wear indicator upward.

OUTER PAD SHIM

INNER PAD SHIM A

BRAKE PADS

CALIPER PIN B

CALIPER PIN

PIN BOOT
Replace.

GREASE

**12 x 1.25 mm
108 N·m
(11.0 kgf·m,
79.6 lbf·ft)**

**CALIPER
BRACKET**

GREASE

CALIPER PIN A

**BLEED SCREW
9 N·m
(0.9 kgf·m,
7 lbf·ft)**

BRAKE HOSE

**CALIPER
BODY**

PISTON SEAL
Replace.

PAD RETAINERS

**BANJO BOLT
34 N·m
(3.5 kgf·m,
25 lbf·ft)**

PISTON

PISTON BOOT
Replace.

GREASE

**SEALING
WASHERS**
Replace.

**8 x 1.0 mm
34 N·m
(3.5 kgf·m, 25 lbf·ft)**

SNAP RING
Replace.

37647_CIVC_G0059

Fig. 3 Exploded view of the front brake caliper—with 1.8L except DX model

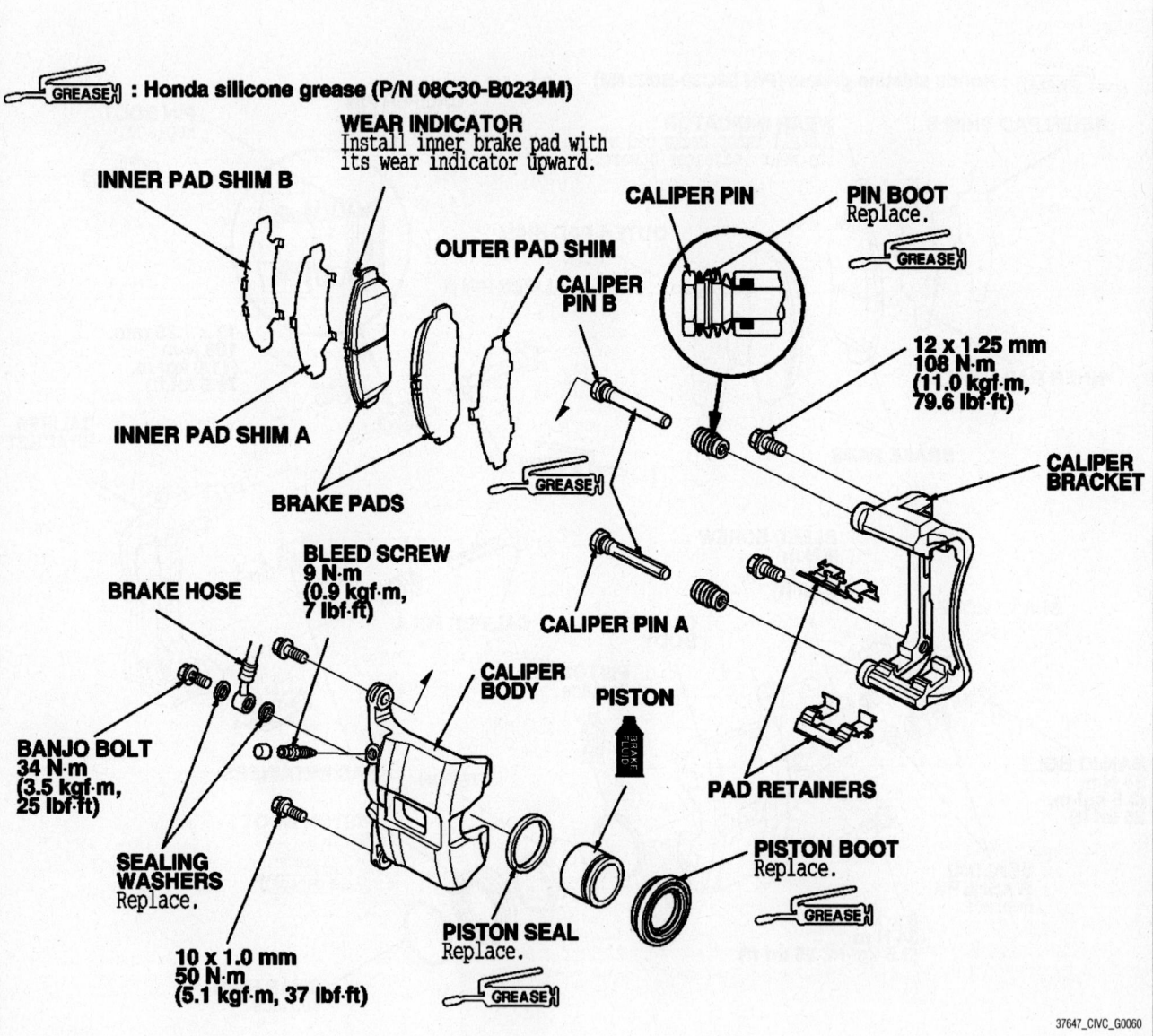

GREASE : Honda silicone grease (P/N 08C30-B0234M)

WEAR INDICATOR
Install inner brake pad with its wear indicator upward.

INNER PAD SHIM B

OUTER PAD SHIM

CALIPER PIN B

CALIPER PIN

PIN BOOT
Replace.

GREASE

INNER PAD SHIM A

12 x 1.25 mm
108 N·m
(11.0 kgf·m, 79.6 lbf·ft)

CALIPER BRACKET

BRAKE PADS

GREASE

CALIPER PIN A

BLEED SCREW
9 N·m
(0.9 kgf·m, 7 lbf·ft)

BRAKE HOSE

CALIPER BODY

PISTON

PAD RETAINERS

BANJO BOLT
34 N·m
(3.5 kgf·m, 25 lbf·ft)

PISTON BOOT
Replace.

GREASE

SEALING WASHERS
Replace.

PISTON SEAL
Replace.

GREASE

10 x 1.0 mm
50 N·m
(5.1 kgf·m, 37 lbf·ft)

GREASE

37647_CIVC_G0060

Fig. 4 Exploded view of the front brake caliper—with 1.8L DX model

GREASE : **Honda silicone grease (P/N 08C30-B0234M)**

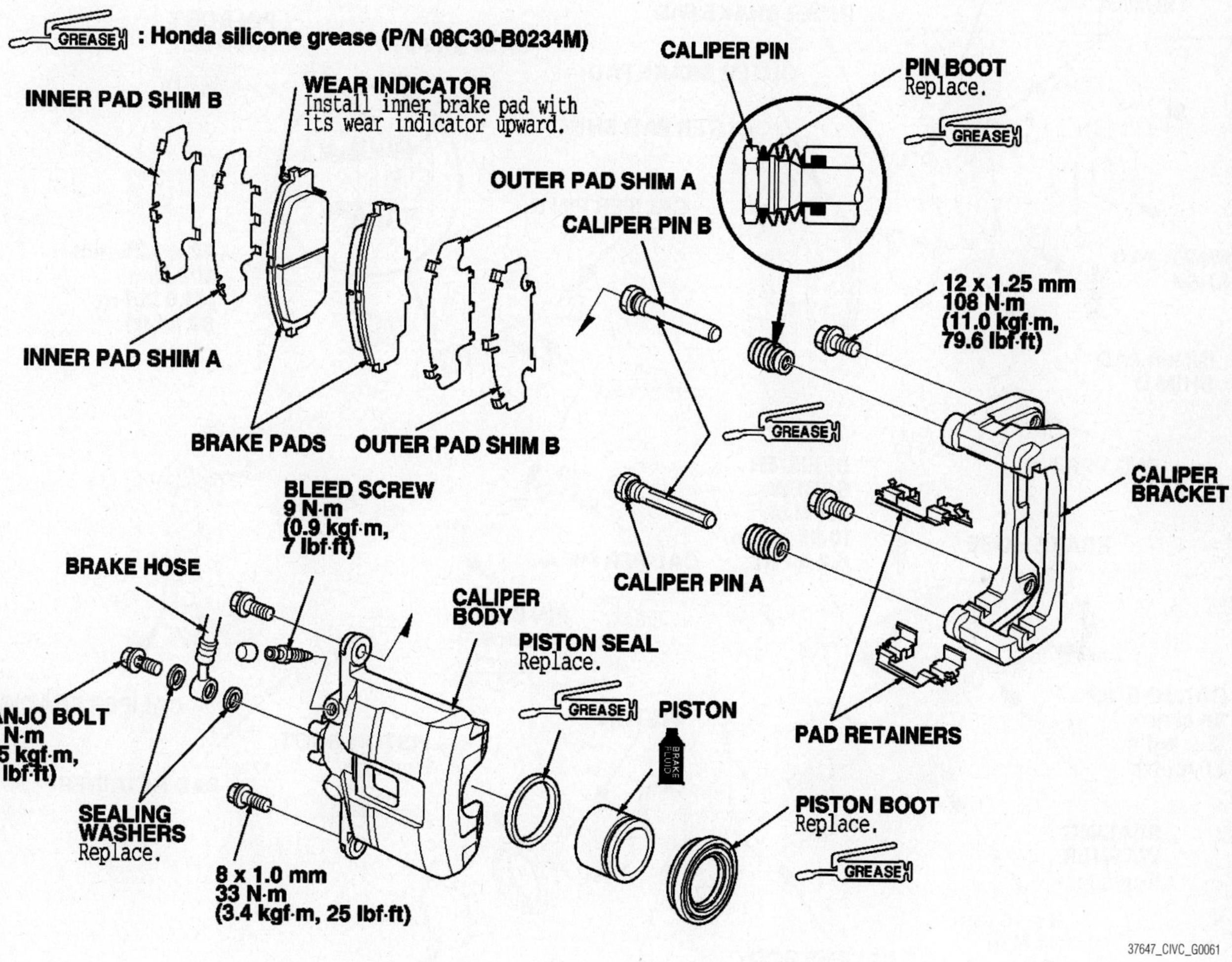

INNER PAD SHIM B

WEAR INDICATOR
Install inner brake pad with
its wear indicator upward.

CALIPER PIN

PIN BOOT
Replace.

GREASE

INNER PAD SHIM A

OUTER PAD SHIM A

CALIPER PIN B

12 x 1.25 mm
108 N·m
(11.0 kgf·m,
79.6 lbf·ft)

BRAKE PADS   **OUTER PAD SHIM B**

GREASE

**CALIPER
BRACKET**

BLEED SCREW
9 N·m
(0.9 kgf·m,
7 lbf·ft)

CALIPER PIN A

BRAKE HOSE

**CALIPER
BODY**

PISTON SEAL
Replace.

GREASE   **PISTON**

PAD RETAINERS

BANJO BOLT
34 N·m
(3.5 kgf·m,
25 lbf·ft)

**SEALING
WASHERS**
Replace.

8 x 1.0 mm
33 N·m
(3.4 kgf·m, 25 lbf·ft)

PISTON BOOT
Replace.

GREASE

37647_CIVIC_G0061

Fig. 5 Exploded view of the front brake caliper—with 2.0L Si model

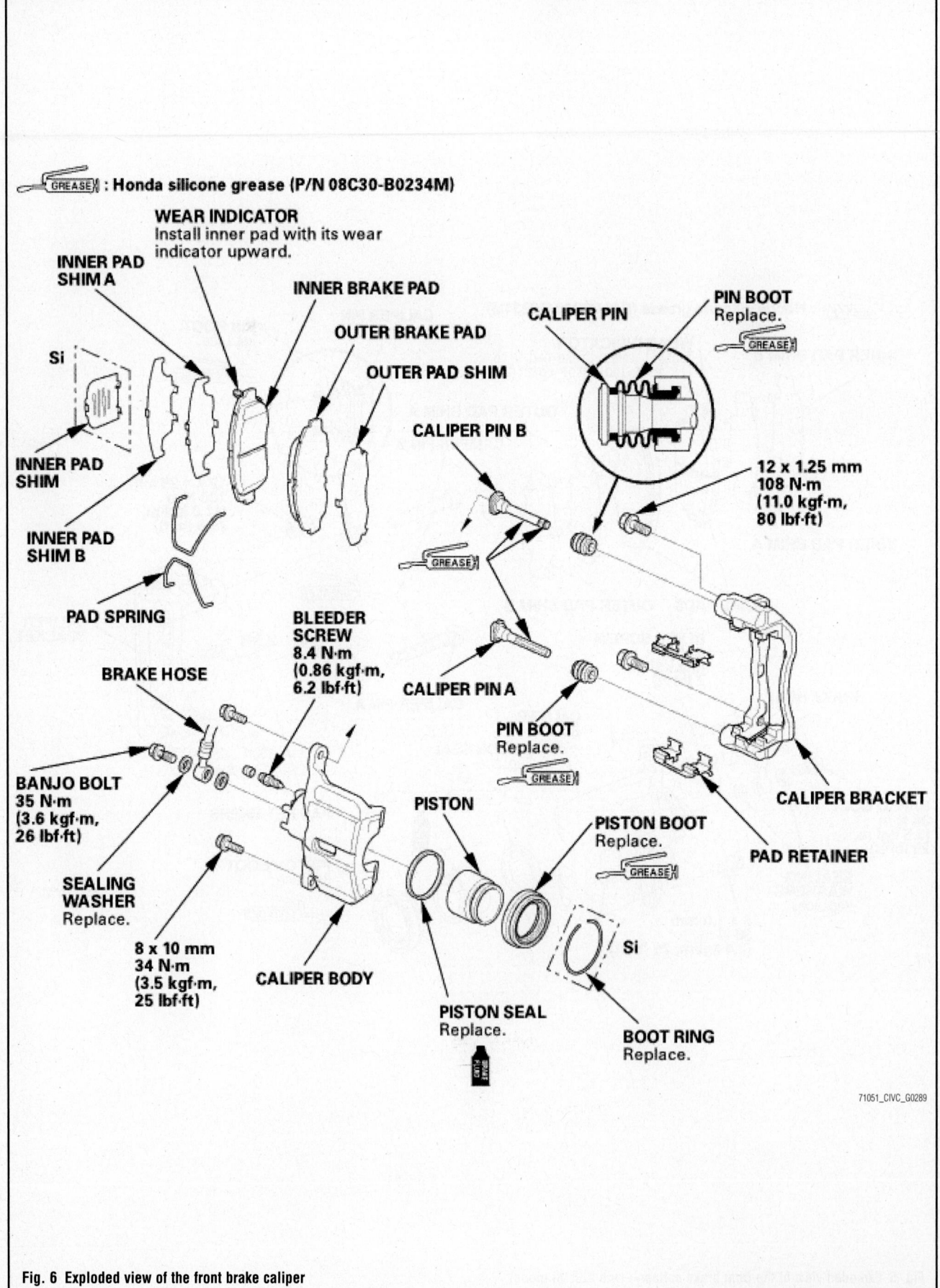

GREASE : Honda silicone grease (P/N 08C30-B0234M)

WEAR INDICATOR
Install inner pad with its wear
indicator upward.

INNER PAD
SHIM A

INNER BRAKE PAD

OUTER BRAKE PAD

OUTER PAD SHIM

Si

INNER PAD
SHIM

INNER PAD
SHIM B

PAD SPRING

BLEEDER
SCREW
8.4 N·m
(0.86 kgf·m,
6.2 lbf·ft)

CALIPER PIN B

CALIPER PIN A

CALIPER PIN

PIN BOOT
Replace.

GREASE

12 x 1.25 mm
108 N·m
(11.0 kgf·m,
80 lbf·ft)

PIN BOOT
Replace.

GREASE

CALIPER BRACKET

PAD RETAINER

BRAKE HOSE

BANJO BOLT
35 N·m
(3.6 kgf·m,
26 lbf·ft)

SEALING
WASHER
Replace.

8 x 10 mm
34 N·m
(3.5 kgf·m,
25 lbf·ft)

CALIPER BODY

PISTON

PISTON BOOT
Replace.

GREASE

PISTON SEAL
Replace.

Si

BOOT RING
Replace.

71051_CIVIC_G0289

Fig. 6 Exploded view of the front brake caliper

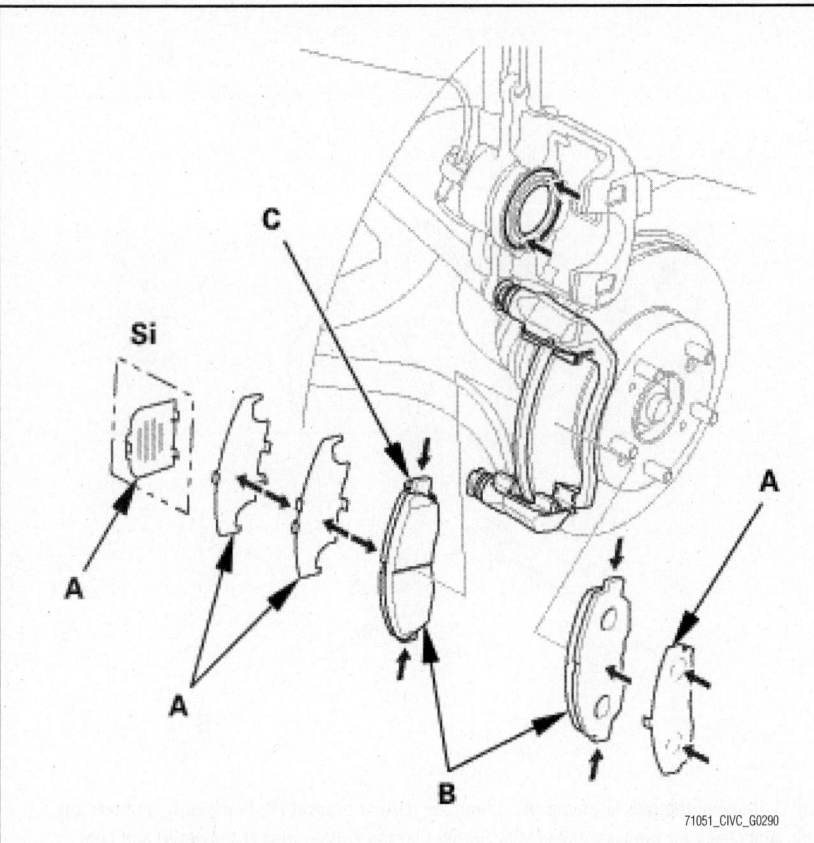

Fig. 7 Apply a thin coat of M-77 assembly paste to the pad side of the shims (A), the back of the brake pads (B), and the other areas indicated by the arrows. Wipe off the excess assembly paste from the pad shims and brake pads friction material. Install the brake pads and pad shims. Install the brake pad with the wear indicator (C) on the upper inside position

5. If any part of the brake pad thickness is less than the service limit, replace the front brake pads as a set.

6. Remove some brake fluid from the master cylinder.

7. Remove the flange bolt, and pivot the caliper up out of the way. Check the hose and pin boots for damage and deterioration.

8. Remove the pad shims and the brake pads.

9. Remove the pad retainers.

To install:

10. Clean the caliper bracket thoroughly; remove any rust, and check for grooves and cracks.

11. Verify that the caliper pins move in and out smoothly. Clean and lube if needed.

12. Inspect the brake disc for runout, thickness, parallelism, and check for damage and cracks. See "Brakes" specification chart in this section.

13. Apply a thin coat of M-77 assembly paste to the retainer mating surface of the caliper bracket (indicated by the arrows).

14. Install the pad retainers. Wipe excess assembly paste off the retainers. Keep the assembly paste off the brake disc and the brake pads.

15. Install the brake caliper piston compressor tool on the caliper body (B).

16. Press in the piston with the brake caliper piston compressor tool so the caliper will fit over the brake pads. Make

14. Install the pad return springs.
15. Install the caliper body.
16. Install the flange bolts.
17. Connect the brake hose to the caliper body with the banjo bolt and new sealing washers.
18. Bleed the brake system, as described in this section.
19. Install the front wheel.

➡ **Before installing the wheel, clean the mating surfaces between the brake disc and the inside of the wheel.**

DISC BRAKE PADS

REMOVAL & INSTALLATION

2011 Models

See Figure 8.

1. Raise and support the vehicle.
2. Remove the front wheels.
3. Check the thickness of the inner pad and the outer pad. Do not include the thickness of the backing plate.
4. Pad thickness should not be less than a service limit of 0.06 in. (1.6 mm).

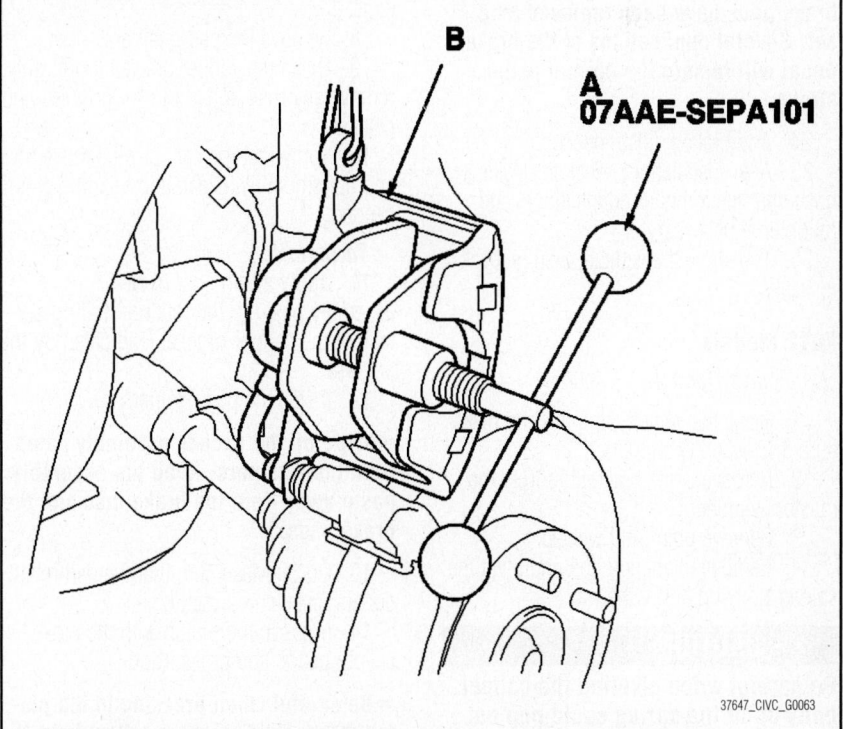

Fig. 8 Install the brake caliper piston compressor tool (A) on the caliper body (B)

sure the piston boot is in position to prevent damaging it when pivoting the caliper down.

➡ **Be careful when pressing in the piston; brake fluid might overflow from the master cylinder's reservoir. If brake fluid gets on any painted surface, wash it off immediately with water.**

18. Remove the brake caliper piston compressor tool.

19. Apply a thin coat of M-77 assembly paste to the pad side of the shims, the back of the brake pads and the other contact areas. Wipe excess assembly paste off the pad shims and the brake pads friction material. Keep grease and assembly paste off the brake disc and the brake pads. Contaminated brake disc or brake pads reduce stopping ability.

20. Install the brake pads and the pad shims correctly. Install the brake pad with the wear indicator on the upper inside. If you are reusing the brake pads, always reinstall the brake pads in their original positions to prevent a temporary loss of braking efficiency.

21. Pivot the caliper down into position. Install the flange bolt and tighten it to the specified torque of 25 ft. lbs. (34 Nm).

22. Clean the mating surfaces between the brake disc and the inside of the wheel, then install the front wheels.

23. Press the brake pedal several times to make sure the brakes work.

➡ **Engagement may require a greater pedal stroke immediately after the brake pads have been replaced as a set. Several applications of the brake pedal will restore the normal pedal stroke.**

24. Add brake fluid as needed.

25. After installation, check for leaks at hose and line joints or connections, and retighten if necessary.

26. Test-drive the vehicle, then recheck for leaks.

2012 Models

See Figures 7 and 9.

1. Raise the vehicle on a lift, and make sure it is securely supported.

2. Remove some brake fluid from the master cylinder.

3. Remove both front wheels.

4. Remove the flange bolt and pivot the caliper body up out of the way.

✳✳ CAUTION

Be careful when pivoting the caliper body up or the spring could pop out of position.

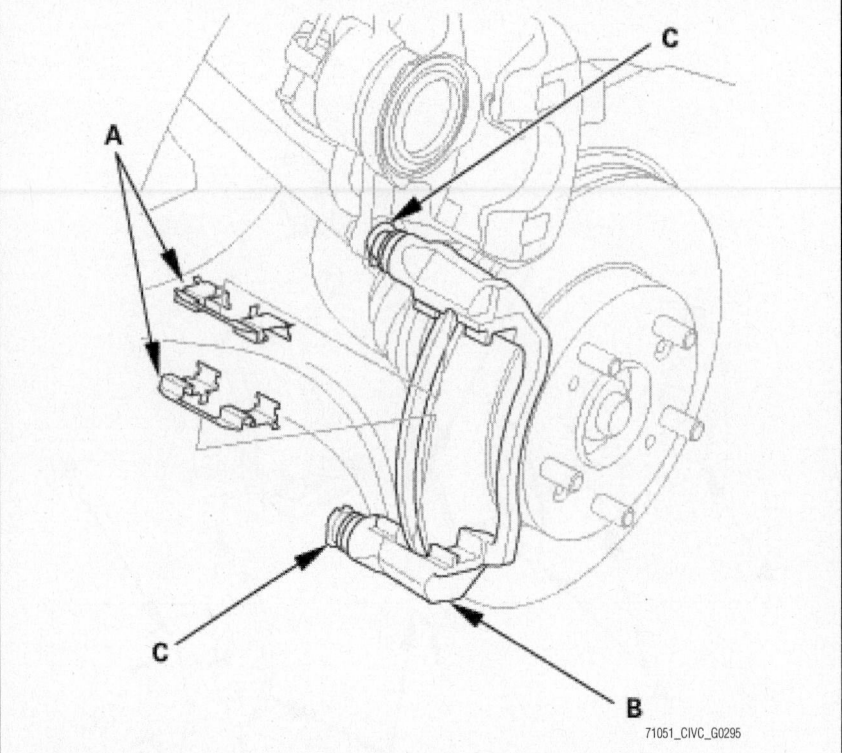

Fig. 9 Remove the pad retainers (A). Clean the caliper bracket (B) thoroughly; remove any rust, and check for grooves and cracks. Verify that the caliper pins (C) move in and out smoothly. Clean and lube the pins if needed

5. Check the hose and pin boots for damage and deterioration.

6. Remove the pad return springs.

7. Remove the pad shims and the brake pads.

8. Remove the pad retainers.

9. Clean the caliper bracket thoroughly; remove any rust, and check for grooves and cracks.

10. Verify that the caliper pins move in and out smoothly. Clean and lube the pins if needed.

To install:

11. Apply a thin coat of M-77 assembly paste to the retainer mating surface of the caliper bracket (indicated by the arrows).

12. Install the pad retainers.

➡ **Wipe off the excess assembly paste from the retainers. Keep the assembly paste away from the brake disc and the brake pads.**

13. On Si: Make sure that the pad retainers are installed correctly.

14. Press in the piston with the brake caliper piston compressor tool.

➡ **Be careful when pressing in the piston; brake fluid might overflow from the master cylinder's reservoir. If brake**

fluid gets on any painted surface, wash it off immediately with water.

15. Apply a thin coat of M-77 assembly paste to the pad side of the shims, the back of the brake pads, and the other areas indicated by the arrows.

➡ **Wipe off the excess assembly paste from the pad shims and brake pads friction material.**

16. Install the brake pads and the pad shims. Install the brake pad with the wear indicator on the upper inside position.

17. Install the pad return springs.

18. Pivot the caliper body down into position. And install and tighten the flange bolt to 25 ft. lbs. (34 Nm).

19. Press the brake pedal several times to make sure the brakes work.

➡ **Engagement may require a greater pedal stroke immediately after the brake pads have been replaced as a set. Several applications of the brake pedal will restore the normal pedal stroke.**

20. Install both front wheels.

➡ **Before installing the wheel, clean the mating surfaces between the brake disc and the inside of the wheel.**

21. Add brake fluid as needed.

BRAKE CALIPER

REMOVAL & INSTALLATION

2011 Models

See Figure 10.

➡**Keep any grease off the brake disc and the brake pads.**

1. Raise and support the vehicle.
2. Remove the rear wheel.
3. Release the parking brake lever fully.
4. Remove the brake hose mounting bolt from the bracket.
5. Remove the brake caliper bracket mounting bolts, then remove the caliper assembly from the knuckle. To prevent damage to the caliper assembly or brake hose, use a short piece of wire to hang the caliper assembly from the undercarriage. Do not twist the brake hose and the parking brake cable excessively.

➡**Make sure the washers are in position on reassembly, if they are removed.**

6. Installation is the reverse of the removal procedure.

2012 Models

See Figure 11.

1. Raise the vehicle on a lift, and make sure it is securely supported.

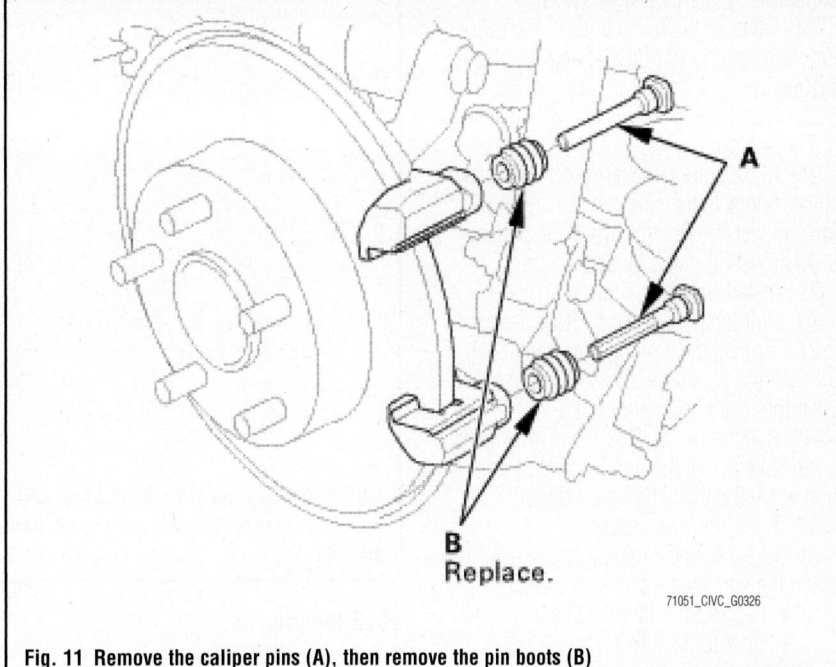

Fig. 11 Remove the caliper pins (A), then remove the pin boots (B)

2. Remove both rear wheels.
3. Remove the center console rear trim.
4. Release the parking brake lever fully.
5. Loosen the adjusting nut.
6. Remove the parking brake cable clip.
7. Disconnect the parking brake cable from the lever. Be careful not to bend or distort the cable and boot.

8. Remove the banjo bolt and disconnect the brake hose from the caliper body.
9. Remove the flange bolts while holding respective caliper pins with a wrench.
10. Remove the brake caliper body.
11. Remove the pad shims and the brake pads.
12. Remove the caliper pins, then remove the pin boots.

To install:

➡**The upper and lower caliper pins are different. During installation, make sure the caliper pins are in the proper positions.**

13. Apply a thin coat of Honda silicone grease to the caliper pins and the new pin boots.
14. Install the pin boots.
15. Install the caliper pins.
16. Make sure that the pin boots are properly positioned into the grooves of the caliper pin and the grooves of the caliper body.
17. Remove air with the inside of the boots.
18. Apply a thin coat of M-77 assembly paste to the pad side of the shims, the back of the brake pads, and the sliding other areas.

➡**Wipe off the excess assembly paste from the pad shims and brake pads friction material.**

19. Install the brake pads and the pad shims. Install the brake pad with the wear indicator on the bottom inside position.

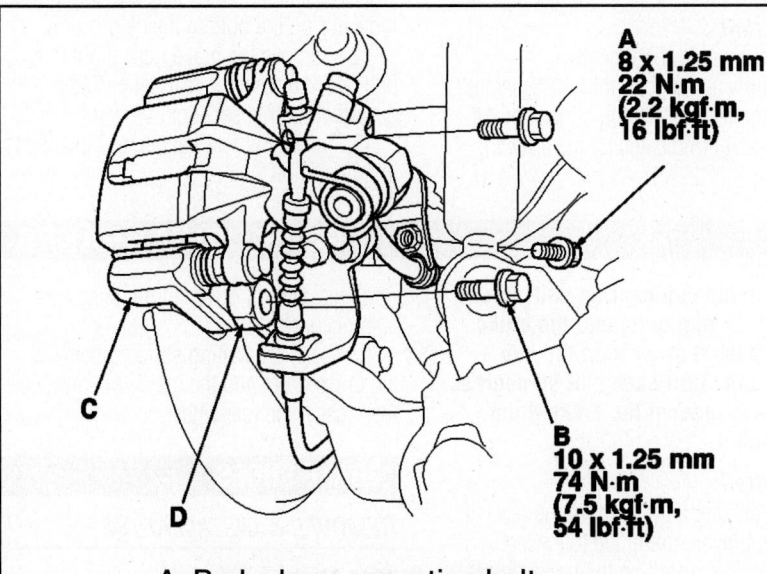

A
**8 x 1.25 mm
22 N·m
(2.2 kgf·m,
16 lbf·ft)**

B
**10 x 1.25 mm
74 N·m
(7.5 kgf·m,
54 lbf·ft)**

A. Brake hose mounting bolt
B. Caliper bracket mounting bolts
C. Caliper assembly
D. Washers

37647_CIVC_G0065

Fig. 10 Removing the rear brake caliper

20. Install the caliper body.

21. Install the flange bolts while holding respective caliper pins with a wrench.

22. Connect the brake hose to the caliper body with the banjo bolt and new sealing washers.

23. Bleed the brake system, as described in this section.

24. Make sure the lever on the rear brake caliper contacts the stop pin. The lever will only contact the stop pin when the parking brake adjusting nut is loosened.

25. Install both rear wheels.

26. Pull the parking brake lever 1 click.

27. Tighten the parking brake adjusting nut until the parking brakes drag slightly when the rear wheels are turned.

28. Release the parking brake lever fully.

29. Check that the parking brakes do not drag when the rear wheels are turned.

30. Readjust if necessary.

31. Make sure the parking brake lever is within the specified number of clicks:

- Except Si: 8 to 10 clicks
- Si: 8 to 9 clicks

32. Install the center console rear trim.

DISC BRAKE PADS

REMOVAL & INSTALLATION

2011 Models

1. Raise and support the vehicle.

2. Remove the rear wheels.

3. Check the thickness of the inner pad and the outer pad. Do not include the thickness of the backing plate.

4. Pad thickness should not be less than 0.06 in. (1.6 mm).

5. If any part of the brake pad thickness is less than the service limit, replace the rear brake pads as a set.

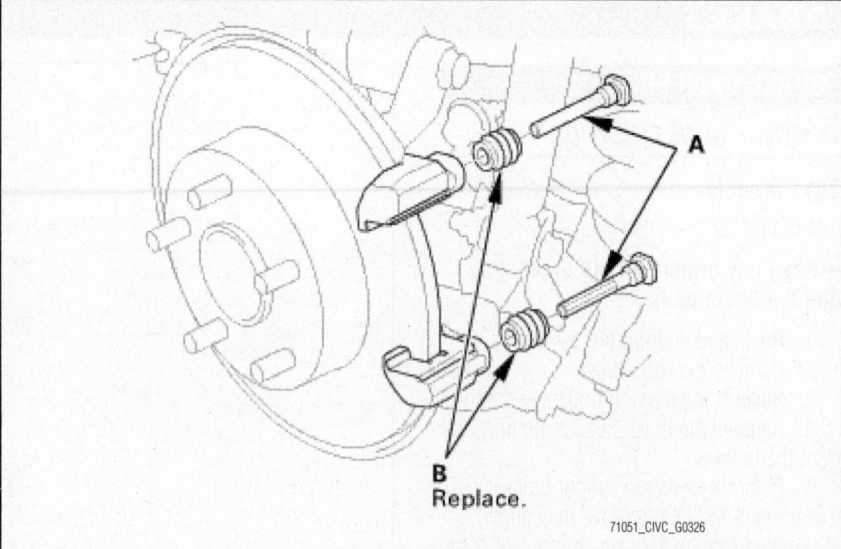

Fig. 12 Remove the pad retainers (A). Clean the caliper bracket (B) thoroughly; remove any rust, and check for grooves and cracks. Verify that the caliper pins (C) move in and out smoothly

2012 Models

See Figure 12.

1. Remove the rear caliper, as described in this section.

2. Remove the pad shims and the brake pads.

3. Remove the pad retainers.

4. Clean the caliper bracket thoroughly; remove any rust, and check for grooves and cracks.

5. Verify that the caliper pins move in and out smoothly.

To install:

6. Clean and lube if needed.

7. Apply a thin coat of M-77 assembly paste to the retainer mating surface of the caliper bracket (indicated by the arrows).

8. Install the pad retainers.

➡ Wipe off the excess assembly paste from the retainers. Keep the assembly paste away from the brake disc and the brake pads.

9. Apply a thin coat of M-77 assembly paste to the pad side of the shims, the back of the brake pads, and the mounting other areas.

10. Install the brake pads and the pad shims.

11. Install the brake pad with the wear indicator on the bottom inside position.

12. Remove the brake hose mounting bolt, then remove the rear caliper, as described in this section.

13. Install the rear caliper, as described in this section.

BRAKES

BRAKE DRUM

REMOVAL & INSTALLATION

See Figure 13.

➡ Keep any grease off the brake drum and the brake shoes.

1. Raise and support the vehicle.

2. Remove the rear wheel.

3. Release the parking brake, and remove the brake drum from the hub bearing unit.

➡ If necessary, turn the adjuster bolt with a flat-tip screwdriver until the shoes become loose. If the brake drum is stuck to the hub bearing unit, thread two 8 x 1.25 mm bolts into the brake drum to push it away from the hub bearing unit. Turn each bolt 90 degrees at a time to prevent the brake drum from binding.

To install:

4. Install the brake drum in the reverse order of removal, noting the following:

a. Before installing the brake drum, clean the mating surfaces between the hub bearing unit and the inside of the brake drum.

b. After installation, press the brake pedal several times to make sure the brakes work and self adjust the brake

REAR DRUM BRAKES

shoes. Do not drive before doing this procedure.

5. Clean the mating surfaces between the brake drum and the inside of the wheel, then install the rear wheel.

BRAKE SHOES

REMOVAL & INSTALLATION

2011 Models

See Figures 14 and 15.

1. Raise and support the vehicle.

2. Remove the rear wheels.

3. Release the parking brake, and remove the brake drum.

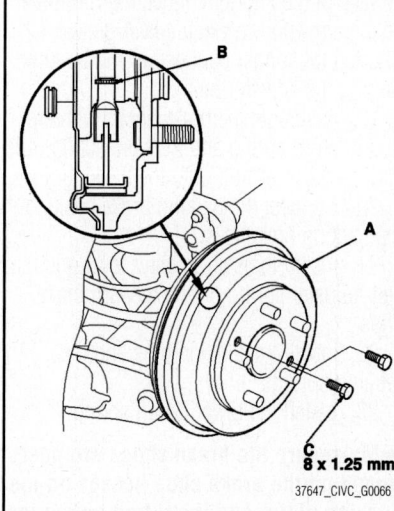

Fig. 13 Release the parking brake, and remove the brake drum (A) from the hub bearing unit. If necessary, turn the adjuster bolt (B) with a flat-tip screwdriver until the shoes become loose. If the brake drum is stuck to the hub bearing unit, thread two 8 x 1.25 mm bolts (C) into the brake drum to push it away from the hub bearing unit. Turn each bolt 90 degrees at a time to prevent the brake drum from binding

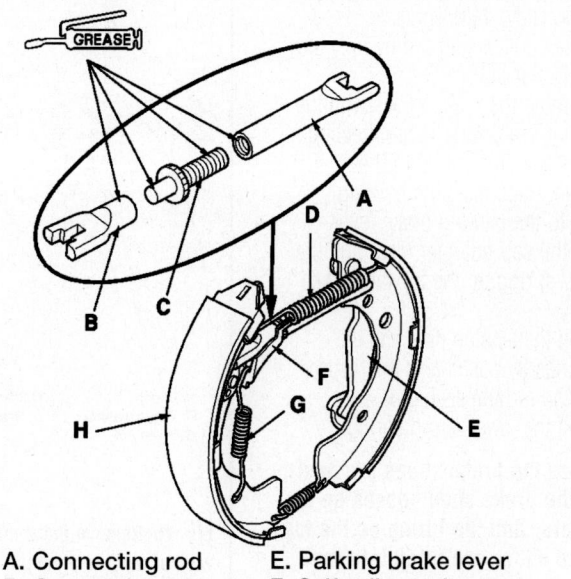

A. Connecting rod
B. Connecting rod
C. Adjuster bolt
D. Upper return spring

E. Parking brake lever
F. Self-adjuster lever
G. Self-adjuster spring

37647_CIVC_G0070

Fig. 15 Assemble the brake shoes with the upper return spring, with the connecting rods and the adjuster bolt onto the backing plate. Reconnect the parking brake cable to the parking brake lever, then install the self-adjuster lever and the self-adjuster spring on the forward brake shoe

4. Remove the tension pins by pushing the respective retainer springs, and turning the pin.

5. Remove the lower return spring, and remove the brake shoe assembly over the hub.

6. Remove the forward brake shoe by removing the upper return spring, and disassemble the brake shoe assembly.

7. Remove the rearward brake shoe by disconnecting the parking brake cable from the parking brake lever.

8. Remove the U-clip, the wave washer, and the pivot pin, and separate the parking brake lever from the brake shoe.

To install:

9. Apply Molykote®-44MA grease to the sliding surface of the pivot pin for the rearward brake shoe.

10. Install the parking brake lever and the wave washer on the pivot pin, and secure the pin with a new U-clip.

➡**Pinch the U-clip securely to prevent the parking brake lever from coming out of the brake shoe.**

11. Connect the parking brake cable to the parking brake lever.

12. Apply a thin coat of Molykote-44MA grease to the connecting rod ends, and the sliding surfaces. Wipe off any excess. Keep grease off the brake linings.

13. Apply a thin coat of Molykote-44MA grease to the shoe ends and to the edge of the shoe surfaces that make contact with the backing plate. Wipe off any excess. Keep grease off the brake linings.

14. Install the two connecting rods on the adjuster bolt.

15. Clean the threaded portions of longer connecting rod and the sliding

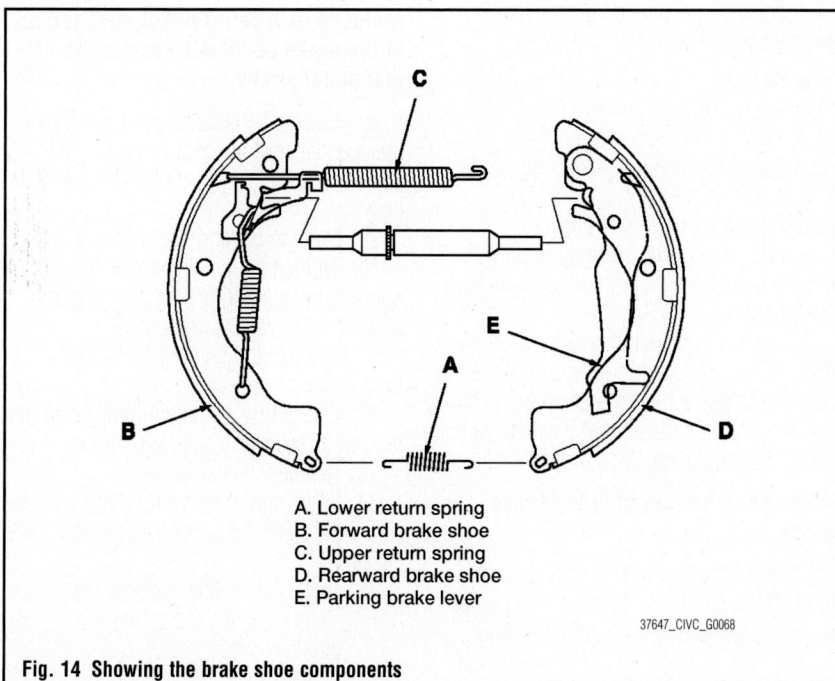

A. Lower return spring
B. Forward brake shoe
C. Upper return spring
D. Rearward brake shoe
E. Parking brake lever

37647_CIVC_G0068

Fig. 14 Showing the brake shoe components

surface of short connecting rod, then coat them with Molykote 44MA grease.

16. Shorten connecting rod by fully turning in the adjuster bolt.

17. Assemble the brake shoes with the upper return spring, with the connecting rods and the adjuster bolt onto the backing plate. Reconnect the parking brake cable to the parking brake lever, then install the self-adjuster lever and the self-adjuster spring on the forward brake shoe.

18. Install the tension pins and the retainer springs by pushing in the respective spring and turning each pin.

19. Install the lower return spring.

➡**Make sure the brake shoes are positioned on the brake shoe bosses on the backing plate, and the fitting on the top of the brake shoes are fitted into the wheel cylinder pistons.**

20. Install the brake drum.

➡**Before installing the brake drum, clean the mating surfaces between the hub bearing unit and the inside of the brake drum.**

21. Clean the mating surfaces between the brake drum and the inside of the wheel, then install the rear wheels.

22. Press the brake pedal several times to make sure the brakes work and to set the self-adjusting brake.

➡**Engagement of the brakes may require a greater pedal stroke immediately after the brake shoes have been replaced as a set. Several applications of the brake pedal will restore the normal pedal stroke.**

23. Adjust the parking brake.

2012 Models

See Figure 16.

1. Raise the vehicle on a lift, and make sure it is securely supported.

2. Remove the center console rear trim.

3. Release the parking brake lever fully and loosen the adjusting nut.

4. Remove both rear wheels.

5. Release the parking brake.

6. Remove the brake drum as described in this section.

   a. If necessary, turn the adjuster bolt with a flat-tip screwdriver until the shoes become loose.

   b. If the brake drum is stuck to the hub bearing unit, thread two 8 x 1.25 mm bolts into the brake drum to push it

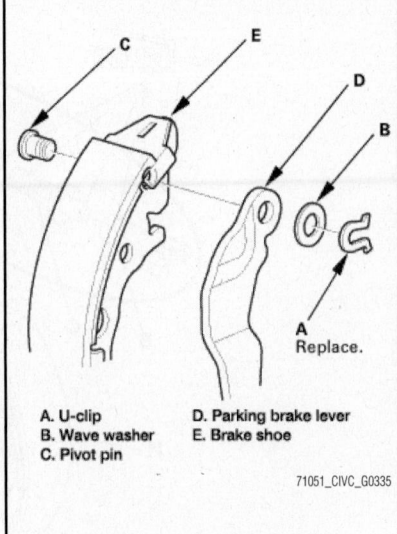

A. U-clip
B. Wave washer
C. Pivot pin

D. Parking brake lever
E. Brake shoe

71051_CIVIC_G0335

Fig. 16 Remove the U-clip, the wave washer and the pivot pin. Separate the parking brake lever from the brake shoe

away from the hub bearing unit. Turn each bolt 90 degrees at a time to prevent the brake drum from binding.

7. Remove the tension pins and the retainer springs.

8. Remove the lower return spring and the upper return spring.

9. Pull the brake shoe assembly forward.

10. Disassemble the brake shoe assembly.

11. Remove the rearward brake shoe by disconnecting the parking brake cable from the parking brake lever.

12. Remove the U-clip, the wave washer and the pivot pin.

13. Separate the parking brake lever from the brake shoe.

To install:

14. Apply Molykote 44MA grease to the sliding surface of the pivot pin.

15. Install the parking brake lever and the wave washer on the pivot pin, and secure the pin with a new U-clip.

16. Pinch the U-clip securely to prevent the parking brake lever from coming out of the brake shoe.

17. Apply a thin coat of Molykote 44MA grease to the connecting rod ends and the sliding surfaces, to the shoe ends and to the edge of the shoe surfaces that make contact with the backing plate. Wipe off any excess.

➡**Keep grease away from the brake linings.**

18. Install the connecting rods onto the adjuster bolt. Clean the threaded portions of the longer connecting rod and the sliding

surface of the shorter connecting rod, then coat them with Molykote 44MA grease. Shorten the longer connecting rod by fully turning the adjuster bolt.

19. Assemble the brake shoes with the upper return spring and the connecting rods and the adjuster bolt onto the backing plate.

20. Connect the parking brake cable to the parking brake lever.

21. Install the self-adjuster lever and the self-adjuster spring on the forward brake shoe.

22. Install the tension pins and the retainer springs.

23. Install the lower return spring.

➡**Make sure the brake shoes are positioned on the brake shoe bosses on the backing plate, and the fitting on the top of the brake shoes are fitted into the wheel cylinder pistons.**

24. Install the brake drum.

➡**Before installing the brake drum, clean the mating surfaces between the hub bearing unit and the inside of the brake drum.**

25. After installation, press the brake pedal several times to make sure the brakes work and self-adjust the brake shoes. Do not drive the vehicle before doing this procedure.

26. Install both rear wheels.

27. Press the brake pedal several times to make sure the brakes work and to set the self-adjusting brake.

➡**Engagement of the brakes may require a greater pedal stroke immediately after the brake shoes have been replaced as a set. Several applications of the brake pedal will restore the normal pedal stroke.**

28. Loosen the adjusting nut. Press the brake pedal several times to set the self-adjusting brake before adjusting the parking brake.

29. Pull the parking brake lever 1 click.

30. Tighten the parking brake adjusting nut until the parking brakes drag slightly when the rear wheels are turned.

31. Release the parking brake lever fully.

32. Check that the parking brakes do not drag when the rear wheels are turned.

33. Readjust if necessary.

34. Make sure the parking brake lever is within the specified number of clicks (7 to 9 clicks).

35. Install the center console rear trim.

BRAKES

PARKING BRAKE

PARKING BRAKE CABLES

ADJUSTMENT

With Rear Disc Brakes

See Figure 17.

1. Remove the center console panel.
2. Release the parking brake lever fully.
3. Loosen the parking brake adjusting nut at the lower end of the parking brake handle.
4. Raise and support the vehicle.
5. Remove the rear wheels.
6. Make sure the lever on the rear brake caliper contacts the stop pin.

➡**The lever will only contact the stop pin when the parking brake adjusting nut is loosened.**

7. Clean the mating surfaces between the brake disc and the inside of the wheel, then install the rear wheels.
8. Pull the parking brake lever 1 click.
9. Tighten the parking brake adjusting nut until the parking brakes drag slightly when the rear wheels are turned.

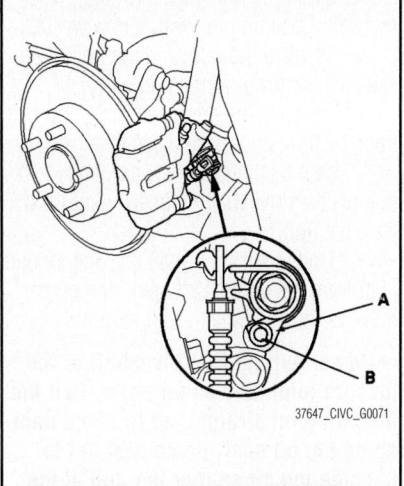

Fig. 17 Make sure the lever (A) on the rear brake caliper contacts the stop pin (B)

37647_CIVIC_G0071

10. Release the parking brake lever fully, and check that the parking brakes do not drag when the rear wheels are turned. Readjust if necessary.
11. Make sure the parking brake lever is within the specified number of clicks (8 to 10 clicks).

12. Install the center console panel.

With Rear Drum Brakes

2011 Models

1. Remove the center console panel.
2. Release the parking brake lever fully.
3. Loosen the parking brake adjusting nut at the lower end of the parking brake handle.
4. Press the brake pedal several times to set the self-adjusting brake before adjusting the parking brake.
5. Pull the parking brake lever 1 click. Raise and support the vehicle.
6. Tighten the parking brake adjusting nut until the parking brakes drag slightly when the rear wheels are turned.
7. Release the parking brake lever fully, and check that the parking brakes do not drag when the rear wheels are turned. Readjust if necessary.
8. Make sure the parking brakes are fully applied when the parking brake lever is pulled all the way (8 to 10 clicks for 2011 models, or 7–9 clicks for 2012 models).
9. Install the center console panel.

CHASSIS ELECTRICAL

AIR BAG (SUPPLEMENTAL RESTRAINT SYSTEM)

GENERAL INFORMATION

❊❊ CAUTION

These vehicles are equipped with an air bag system. The system must be disarmed before performing service on, or around, system components, the steering column, instrument panel components, wiring and sensors. Failure to follow the safety precautions and the disarming procedure could result in accidental air bag deployment, possible injury and unnecessary system repairs.

PRECAUTIONS

Disconnect and isolate the battery negative cable before beginning any airbag system component diagnosis, testing, removal, or installation procedures. Allow system capacitor to discharge for two minutes before beginning any component service. This will disable the airbag system. Failure to disable the airbag system may result in accidental airbag deployment, personal injury, or death.

DISARMING THE SYSTEM

➡**Some systems store data in memory (including seat position, mirror position, etc.) that is lost when the battery is disconnected. Do the following procedures before disconnecting the battery.**

1. Make sure you have the anti-theft code(s) for the audio system and/or navigation system (if equipped).

➡**For some models, it may be necessary to write down the audio the presets (XM, AM and FM), because the audio unit does not retain the presets after the battery is disconnected.**

2. Make sure the ignition switch is in LOCK (0).
3. Disconnect and isolate the negative cable from the battery.

➡**Always disconnect the negative battery cable from the battery first.**

4. Disconnect the positive cable from the battery.
5. Wait at least 3 minutes before starting any further work. This allows the SRS system to fully discharge.

ARMING THE SYSTEM

➡**Some systems store data in memory (including seat position, mirror position, etc) that is lost when the battery is disconnected. Do the following procedures to restore the system back to normal operation.**

1. Clean the battery terminals.
2. Test the battery, if necessary.
3. Reconnect the positive cable to the battery first, then reconnect the negative cable to the battery.

❊❊ CAUTION

Always connect the positive cable to the battery first.

4. Apply multipurpose grease to the terminals to prevent corrosion.
5. Enter the anti-theft code(s) for the audio system and/or navigation system (if equipped).
6. Enter the audio presets.
7. Set the clock (for vehicles without navigation).

DRIVE TRAIN

CLUTCH HYDRAULIC SYSTEM BLEEDING

PROCEDURE

1. Make sure the brake fluid level in the clutch reservoir is at the MAX (upper) level line.

2. Attach one end of a clear tube to the bleeder screw and put the other end into a container. Loosen the bleeder screw to allow air to escape from the system.

3. Make sure there is an adequate supply of fluid in the reservoir, then slowly push the clutch pedal all the way down. Before releasing the pedal, have an assistant temporarily tighten the bleeder screw. Loosen the bleeder screw, and push the pedal down again. Repeat this step until no more bubbles appear in the clear tube.

➡**Make sure the fluid level on the reservoir does not go below MIN (lower).**

4. Tighten the bleeder screw securely.

5. Refill the brake fluid in the reservoir to the MAX (upper) level line.

FRONT DRIVESHAFT

REMOVAL & INSTALLATION

2011 Models

See Figure 18.

1. Raise and support the vehicle.
2. Remove the front wheels.
3. Pry up the stake on the spindle nut, then remove the nut.

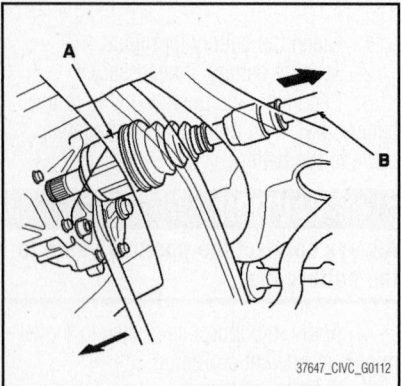

Fig. 18 On the left driveshaft, pry the inboard joint (A) from the differential using the prybar. Remove the driveshaft (B) as an assembly

37647_CIVIC_G0112

4. Drain the transmission fluid, then reinstall the drain plug with a new washer.

5. Remove the nuts and the bolt, then separate the lower arm using a prybar.

6. Separate the driveshaft outboard joint from the front hub using a soft face hammer.

7. Pull the knuckle outward, and remove the driveshaft outboard joint from the front hub.

8. On the left driveshaft, pry the inboard joint from the differential using the prybar. Remove the driveshaft as an assembly.

➡**Do not pull on the driveshaft or the inboard joint may come apart. Pull the inboard joint straight out to avoid damaging the oil seal. Be careful not to damage the oil seal or the end of the inboard joint using the prybar.**

9. On the right driveshaft, drive the inboard joint off of the intermediate shaft using a drift punch and a hammer (M/T model). Pry the inboard joint from the differential using the prybar (A/T model).

10. Remove the driveshaft as an assembly.

11. Do not pull on the driveshaft or the inboard joint may come apart. Pull the inboard joint straight out to avoid damaging the oil seal.

➡**Be careful not to damage the oil seal or the end of the inboard using the prybar.**

12. Remove the set ring from the inboard joint (except right driveshaft with M/T model).

13. Remove the set ring from the intermediate shaft (M/T model).

To install:

➡**Before starting installation, make sure the mating surfaces of the joint and the splined section are clean.**

14. Apply about 5 g (0.18 oz) of moly 60 paste to the contact area of the outboard joint and the front wheel bearing.

➡**The paste helps prevent noise and vibration.**

15. Install a new set ring into the set ring groove of the driveshaft.

16. Install a new set ring into the set ring groove of the intermediate shaft.

17. On M/T model, apply specified super high temp urea grease to the whole splined surface of the right driveshaft. After applying grease, remove the grease from the splined grooves at intervals of 2-3 splines and from the set ring groove so that air can bleed from the intermediate shaft.

18. Clean the areas where the driveshaft contacts the differential thoroughly with solvent, and dry them with compressed air.

➡**Do not wash the rubber parts with solvent.**

19. Insert the inboard end of the driveshaft into the differential or the intermediate shaft (M/T model) until the set ring locks in the groove.

➡**Insert the driveshaft horizontally to prevent damaging the oil seal.**

20. Install the outboard joint into the front hub.

21. Install the knuckle onto the lower arm. During installation, install a new flange bolt and new self-locking nuts. After lightly tightening all three fasteners, tighten them to the specified torque of 43 ft. lbs. (59 Nm) in the following order: the nut on the front, the nut on the rear, then the bolt.

22. Apply a small amount of engine oil to the seating surface of a new spindle nut.

23. Install the spindle nut, then tighten the nut. After tightening, use a drift to stake the spindle nut shoulder against the driveshaft.

24. Clean the mating surfaces of the brake discs and the wheels, then install the front wheels.

25. Turn the front wheels by hand, and make sure there is no interference between the driveshaft and surrounding parts.

26. Refill the transmission with the recommended transmission fluid:

27. Lower the vehicle.

28. Check the wheel alignment, and adjust it if necessary.

29. Test-drive the vehicle.

2012 Models

Left Side

See Figures 19 and 20.

1. Raise the vehicle on a lift, and make sure it is securely supported.

2. Remove the left front wheel.

3. Pry up the stake on the spindle nut and remove the nut.

4. Remove the engine undercover.

5. Remove the drain plug and the sealing washer and drain the MTF. Reinstall the drain plug with a new sealing washer.

6. Remove the drain plug and drain the ATF. Reinstall the drain plug using a new sealing washer.

7. Remove the flange bolt. Remove the

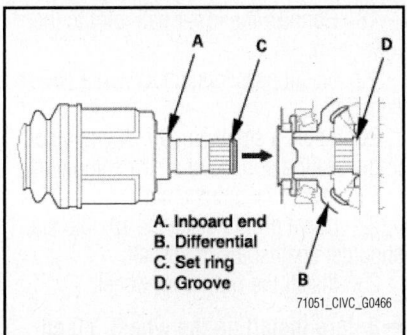

Fig. 19 Insert the inboard end of the driveshaft into the differential until the set ring locks in the groove

flange nuts. Disconnect the lower ball joint from the lower arm.

8. Separate the outboard joint from the front hub using a soft face hammer.

9. Pull the knuckle outward, and separate the outboard joint from the front hub.

10. Pry the inboard joint from the differential using a pry bar.

➡**Do not pull on the driveshaft, or the inboard joint may come apart. Pull the inboard joint straight out to avoid damaging the oil seal. Be careful not to damage the oil seal or the end of the inboard joint using the pry bar.**

11. Remove the set ring.

To install:

12. Install a new set ring.

13. Clean the areas where the driveshaft contacts the differential thoroughly with solvent.

➡**Do not wash the rubber parts with solvent.**

14. Dry the areas where the driveshaft contacts the differential thoroughly with compressed air.

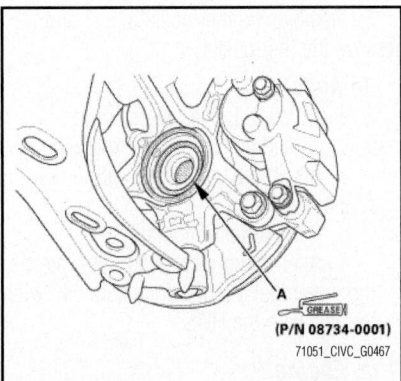

Fig. 20 Apply moly 60 paste to the contact area (A) of the outboard joint and the front wheel bearing

15. Insert the inboard end of the driveshaft into the differential until the set ring locks in the groove.

➡**Insert the driveshaft horizontally to prevent damaging the oil seal.**

16. Apply about 0.18 oz. (5 g) of moly 60 paste to the contact area of the outboard joint and the front wheel bearing.

➡**The paste helps to prevent noise and vibration.**

17. Install the outboard joint into the front hub on the knuckle.

18. Connect the lower ball joint to the lower arm.

19. Install the new flange nuts and tighten to 44 ft. lbs. (59 Nm), then install the new flange bolt and tighten to 44 ft. lbs. (59 Nm).

20. Apply a small amount of engine oil to the seating surface of a new spindle nut, then install the spindle nut. Use a drift to stake the spindle nut shoulder against the driveshaft.

21. Tighten the spindle nut as follows:
  • 1.8L engine: 133 ft. lbs. (181 Nm)
  • 2.4L engine: 181 ft. lbs. (245 Nm)

22. Install the left front wheel.

➡**Before installing the wheel, clean the mating surfaces between the brake disc and the inside of the wheel.**

23. Turn the wheel by hand, and make sure there is no interference between the driveshaft and surrounding parts.

24. Refill the ATF or MTF, as applicable.

➡**Always use Honda Manual Transaxle Fluid (MTF).**

25. Reinstall the filler plug with a new sealing washer.

26. Install the engine undercover.

27. Perform wheel alignment and turning angle check and adjustment. See "Wheel Alignment" in "Suspension" section.

28. Test-drive the vehicle.

29. After doing all procedures, select the individual maintenance item you wish to reset with the HDS.

Right Side

See Figures 21 through 23.

1. Raise the vehicle on a lift, and make sure it is securely supported.

2. Remove the right front wheel.

3. Pry up the stake on the spindle nut and remove the nut.

4. Remove the engine undercover.

5. Remove the drain plug and the sealing washer and drain the MTF or ATF, as applicable.

6. Remove the flange bolt. Remove the

flange nuts. Disconnect the lower ball joint from the lower arm.

7. Separate the outboard joint from the front hub using a soft face hammer. Pull the knuckle outward, and separate the outboard joint from the front hub.

8. On models with the intermediate front driveshaft, drive the inboard joint off of the intermediate shaft using a drift punch and a hammer.

➡**Do not pull on the driveshaft or the inboard joint may come apart.**

9. Remove the set ring from the end of the shaft.

❋❋ **CAUTION**

Do not pull on the driveshaft or the inboard joint may come apart. Pull the inboard joint straight out to avoid damaging the oil seal. Be careful not to damage the oil seal or the end of the inboard using the pry bar.

To install:

10. Install a new set ring.

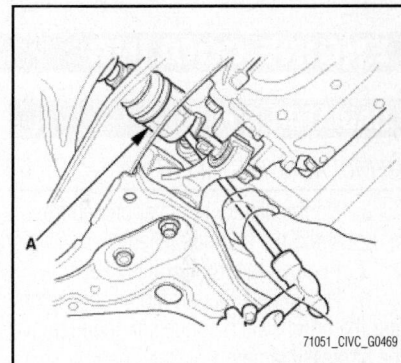

Fig. 21 Drive the inboard joint (A) off of the intermediate shaft using a drift punch and a hammer—with intermediate driveshaft

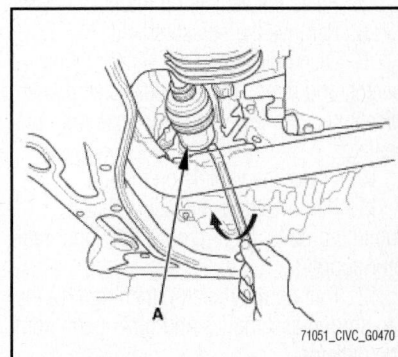

Fig. 22 Drive the inboard joint (A) off of the intermediate shaft using a drift punch and a hammer—without intermediate driveshaft

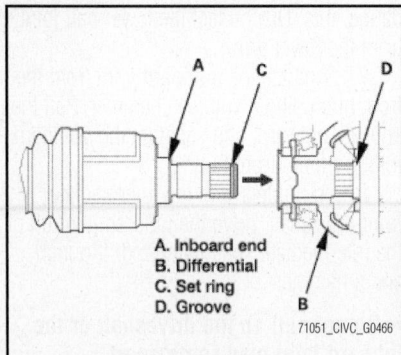

A. Inboard end
B. Differential
C. Set ring
D. Groove

71051_CIVIC_G0466

Fig. 23 Insert the inboard end of the driveshaft into the differential until the set ring locks in the groove

11. Clean the areas where the driveshaft contacts the differential thoroughly with solvent.

➡**Do not wash the rubber parts with solvent.**

12. Dry the areas where the driveshaft contacts the differential thoroughly with compressed air.

13. Insert the inboard end of the driveshaft into the differential until the set ring locks in the groove.

➡**Insert the driveshaft horizontally to prevent damaging the oil seal.**

14. Apply 0.5–1.0 g (0.018–0.035 oz.) of super high temp urea grease to the whole splined surface of the right driveshaft. After applying grease, remove the grease from the splined grooves at intervals of 2–3 splines and from the set ring groove so that air can bleed from the intermediate shaft.

15. Insert the inboard end of the driveshaft onto the intermediate shaft until the set ring locks in the groove.

16. Apply about 5 g (0.18 oz.) of moly 60 paste to the contact area of the outboard joint and the front wheel bearing.

➡**The paste helps to prevent noise and vibration.**

17. Install the outboard joint into the front hub on the knuckle.

18. Connect the lower ball joint to the lower arm.

19. Install new flange nuts and a new flange bolt.

20. Apply a small amount of engine oil to the seating surface of a new spindle nut.

21. Install the spindle nut.

22. Use a drift to stake the spindle nut shoulder against the driveshaft.

23. Install the right front wheel.

➡**Before installing the wheel, clean the mating surfaces between the brake disc and the inside of the wheel.**

24. Turn the wheel by hand, and make sure there is no interference between the driveshaft and surrounding parts.

25. Refill that ATF or MTF as applicable.

26. Install the engine undercover.

27. Perform wheel alignment and turning angle check.

28. Test-drive the vehicle.

29. After doing all procedures, select the individual maintenance item you wish to reset with the HDS.

ENGINE COOLING

RADIATOR & FAN

REMOVAL & INSTALLATION

1. Do the battery removal procedure.
2. Drain the engine coolant.
3. Remove the front grille cover.
4. Disconnect the fan motor connectors and the hood switch connector, then remove the harness clamps.
5. Disconnect the reservoir hose and remove the radiator cap base mounting bolts, then the radiator upper bracket bolts and brackets.
6. Disconnect the upper radiator hose.
7. Raise the vehicle on the lift.
8. Remove the splash shield.
9. Disconnect the ECT sensor 2 connector, and remove the harness clamp, then disconnect the lower radiator hose from the radiator.
10. Lower the vehicle on the lift.
11. Remove the condenser bracket mounting bolts, then remove the upper radiator support.
12. Pull up the radiator, then remove the fan shroud assemblies and other parts from the radiator.

To install:

13. Install the radiator in the reverse order of removal. Make sure the upper and lower cushions are set securely.

14. Install the bulkhead in the reverse order of removal. Apply body paint to the bulkhead mounting bolts.

15. Do the battery installation procedure.

16. Fill the radiator with engine coolant, and bleed the air from the cooling system.

THERMOSTAT

REMOVAL & INSTALLATION

2011 Models

1.8L Engine

1. Drain the engine coolant.
2. Remove the harness clamp bracket and the thermostat cover, then remove the thermostat.
3. Install the thermostat with a new rubber seal, then install the harness clamp bracket.
4. Refill the radiator with engine coolant, and bleed the air from the cooling system.

2.0L Engine

1. Drain the engine coolant.
2. Remove the splash shield.
3. Disconnect the lower radiator hose, then remove the thermostat.
4. Install the new thermostat assembly with a new O-ring, then install the lower radiator hose.

5. Install the splash shield.
6. Refill the radiator with engine coolant, and bleed the air from the cooling system.

2012 Models

1.8L Hybrid Engine

See Figure 24.

1. Wait until the engine is cool, then carefully remove the radiator cap.
2. Raise the vehicle on a lift, and make sure it is securely supported.
3. Loosen the drain plug and drain the coolant.
4. Remove the intake air pipe.
5. Remove the thermostat cover and remove the thermostat.

To install:

6. Installation is the reverse of the removal procedure, noting the following:
   a. Install the thermostat with a new rubber seal.
   b. Refill radiator coolant.
   c. After doing all procedures, select the individual maintenance item you wish to reset with the HDS.

1.8L Engine

See Figures 25 and 26.

1. Wait until the engine is cool, then carefully remove the radiator cap.

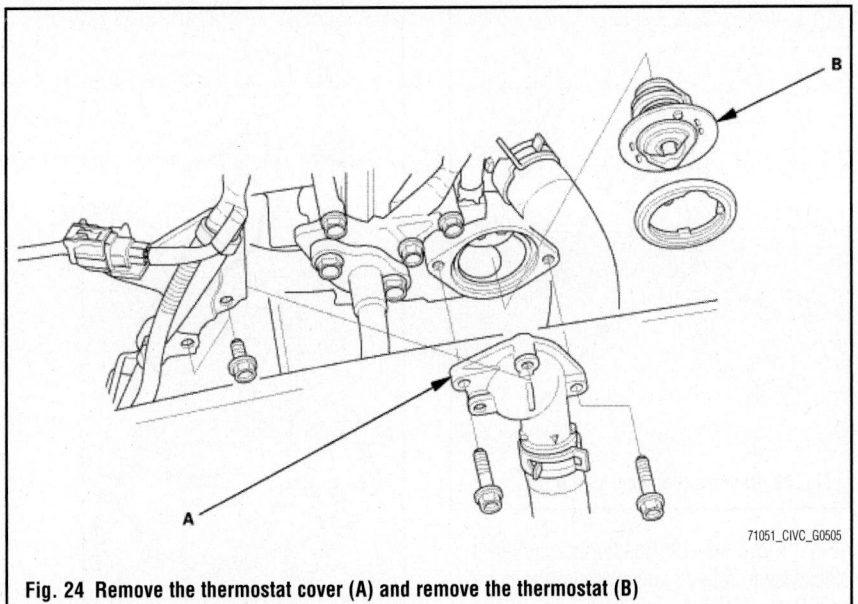

Fig. 24 Remove the thermostat cover (A) and remove the thermostat (B)

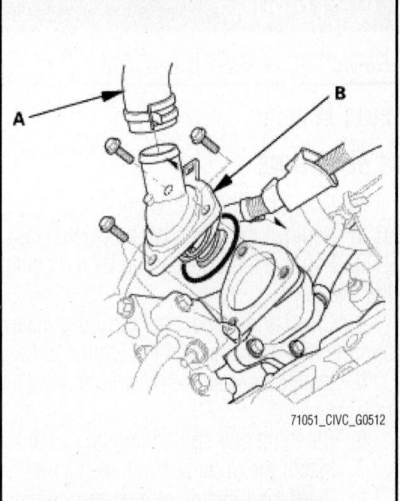

Fig. 27 Disconnect the lower radiator hose (A), then remove the thermostat assembly (B)

2. Raise the vehicle on a lift, and make sure it is securely supported.

3. Loosen the drain plug and drain the coolant.

4. Remove the intake air pipe.

5. Remove the EGR pipe.

6. Disconnect the connectors and remove the harness clamps.

7. Disconnect the connector and harness clamp, as shown.

8. Remove the harness bracket, then disconnect the lower radiator hose, the water bypass hose, and the heater hose.

9. Disconnect the connector and remove the EGR valve.

10. Remove the thermostat cover and remove the thermostat.

11. Remove the thermostat housing.

To install:

12. Installation is the reverse of the removal procedure, noting the following:

a. Use a new thermostat gasket.

b. Refill the radiator coolant.

c. After doing all procedures, select the individual maintenance item you wish to reset with the HDS

2.4L Engine

See Figure 27.

1. Wait until the engine is cool, then carefully remove the radiator cap.

2. Raise the vehicle on a lift, and make sure it is securely supported.

3. Loosen the drain plug and drain the coolant.

4. Remove the splash shield from under the vehicle.

5. Disconnect the lower radiator hose, then remove the thermostat assembly.

➡**Replace the thermostat if it is stuck in the open position at room temperature.**

To install:

6. Installation is the reverse of the removal procedure, noting the following:

a. Install the thermostat assembly with new O-ring.

b. Refill the engine coolant.

c. After doing all procedures, select the individual maintenance item you wish to reset with the HDS.

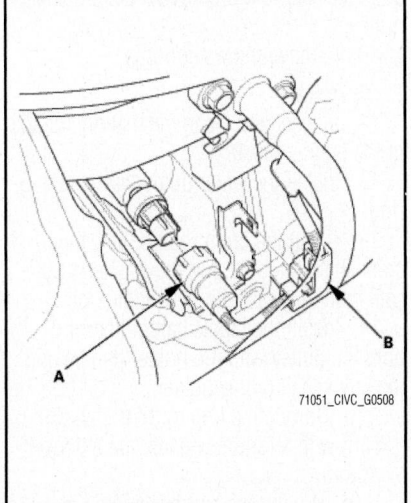

Fig. 25 Disconnect the connector (A) and harness clamp (B), as shown

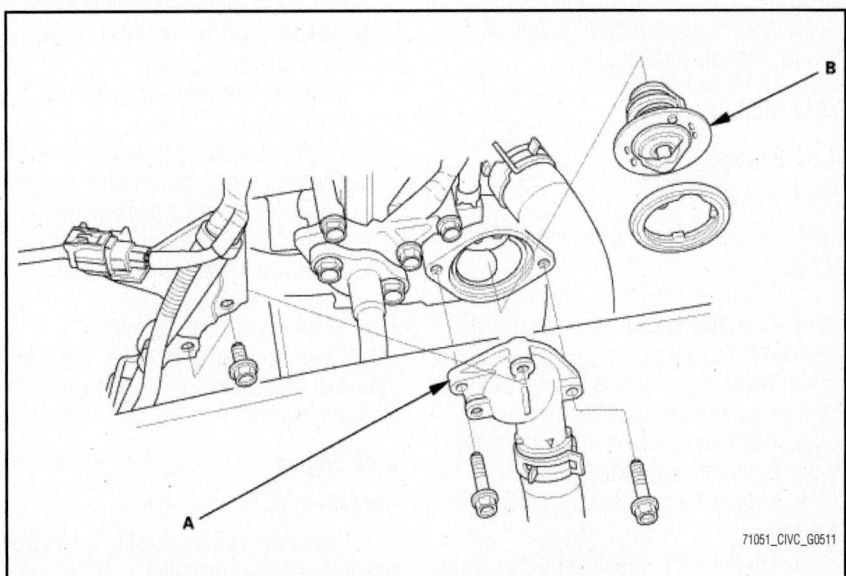

Fig. 26 Remove the thermostat cover (A) and remove the thermostat (B)

WATER PUMP

REMOVAL & INSTALLATION

2011 Models

1.8L ENGINE

1. Drain the engine coolant.
2. Remove the drive belt auto-tensioner.
3. Remove the water pump by removing the five bolts.
4. Clean and inspect the O-ring groove and the mating surface of the engine block.
5. Install the water pump with a new O-ring.
6. Clean up any spilled engine coolant.
7. Install the drive belt auto-tensioner.
8. Refill the radiator with engine coolant, and bleed the air from the cooling system.

2.0L Engine

1. Remove the drive belt.
2. Drain the engine coolant.
3. Remove the drive belt auto-tensioner pulley.
4. Remove the crankshaft pulley.
5. Remove the oil cooler joint pipe, then remove the seven bolts securing the water pump. Remove the water pump.

To install:

6. Inspect, and clean the O-ring groove and mating surface of the water passage.
7. Install the water pump with a new O-rings and the oil cooler joint pipe.
8. Clean up any spilled engine coolant.
9. Install the crankshaft pulley.
10. Install the drive belt auto-tensioner pulley.
11. Install the drive belt.
12. Refill the radiator with engine coolant, and bleed the air from the cooling system. See "Engine Coolant" in this section for bleeding procedure.

2012 Models

1.8L Engines

See Figure 28.

1. Make sure the ignition switch is in LOCK (0).
2. Disconnect and isolate the negative cable and battery sensor, then the positive cable from the battery.
3. Wait until the engine is cool, then carefully remove the radiator cap.
4. Raise the vehicle on a lift, and make sure it is securely supported.
5. Loosen the drain plug and drain the coolant.
6. Loosen the water pump pulley mounting bolts.

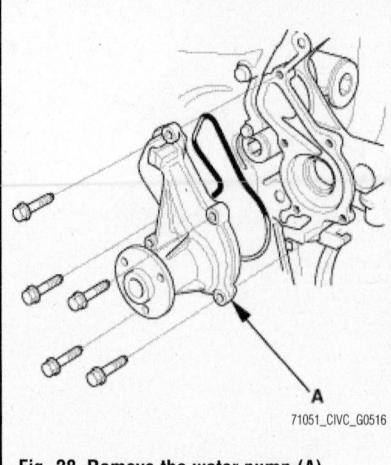

Fig. 28 Remove the water pump (A)

71051_CIVC_G0516

7. Move the auto-tensioner counter-clockwise to relieve tension from the drive belt. Remove the drive belt.
8. From the alternator, disconnect the connector and the cable. Remove the harness connector and the harness clamps. Remove the alternator.
9. Remove the water pump pulley.
10. Remove the drive belt auto-tensioner. Check for oil leaks from the drive belt auto-tensioner damper (B) and check for damage to the damper rubber.
11. Remove the water pump.

To install:

12. Installation is the reverse of the removal procedure, noting the following:
   a. Inspect and clean the O-ring groove and the mating surface of the engine block.
   b. Install the water pump with a new O-ring.
   c. Install the drive belt auto-tensioner. Tighten the tensioner end bolt to 17 ft. lbs. (24 Nm) and the tensioner center bolt to 41 ft. lbs. (55 Nm).
   d. Tighten the alternator mounting bolts to 17 ft. lbs. (24 Nm).
   e. Move the auto-tensioner counter-clockwise to install the drive belt.
   f. Tighten the water pump pulley mounting bolts to 9 ft. lbs. (12 Nm).
13. Clean the battery terminals and reconnect the battery.
14. Refill the coolant, as needed.
15. After doing all procedures, select the individual maintenance item you wish to reset with the HDS.

2.4L Engine

See Figure 29.

1. Raise the vehicle on a lift, and make sure it is securely supported.
2. Remove the right front wheel.

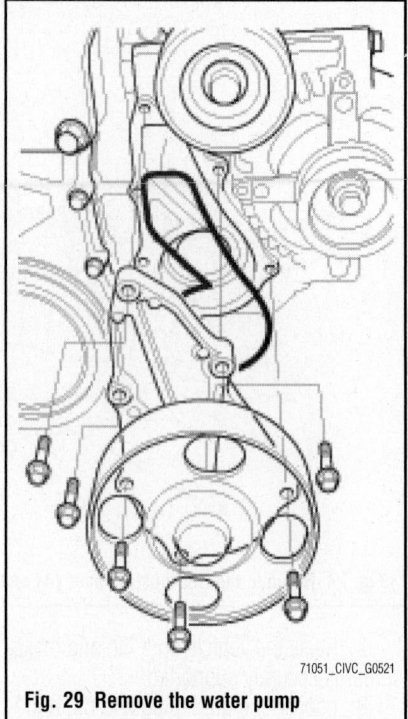

Fig. 29 Remove the water pump

71051_CIVC_G0521

3. Wait until the engine is cool, then carefully remove the radiator cap.
4. Loosen the drain plug and drain the coolant.
5. Move the auto-tensioner using the belt tension release tool in the direction of the arrow to relieve tension from the drive belt, then remove the drive belt.
6. Remove the idler pulley base.
7. Remove the auto-tensioner.
8. Remove the splash shield from under the vehicle.
9. Hold the pulley with the holder handle and the crankshaft pulley holder. Remove the bolt with a socket, 19 mm and a breaker bar, then remove the crankshaft pulley.
10. Remove the water pump.

To install:

11. Inspect and clean the mating surface of the engine block.
12. Install the water pump with a new O-ring.
13. Clean the crankshaft pulley, the crankshaft, the bolt, and the washer. Lubricate mating areas with new engine oil.
14. Install the crankshaft pulley, and hold the pulley with the holder handle and the crankshaft pulley holder.
   a. Torque the bolt to 37 ft. lbs. (50 Nm) with a torque wrench and a socket, 19mm.
   b. Do not use an impact wrench.
   c. If the pulley bolt or crankshaft are new, torque the bolt to 131ft. lbs. (177

Nm), then remove the bolt and torque it to 37 ft. lbs. (50 Nm).

    d. Tighten the pulley bolt an additional 90°.

15. Install the splash shield.

16. Install the auto-tensioner.

17. Install the idler pulley base.

18. Install the drive belt.

19. Replace the engine coolant mixture.

20. Install the right front wheel.

21. Before installing the wheel, clean the mating surfaces between the brake disc and the inside of the wheel.

22. After doing all procedures, select the individual maintenance item you wish to reset with the HDS.

BATTERY SYSTEM

BATTERY

REMOVAL & INSTALLATION

See Figure 30.

1. Make sure the ignition switch is in LOCK (0).

2. Disconnect and isolate the negative cable and battery sensor from the battery.

➡**Always disconnect the negative side first.**

3. Disconnect the positive cable from the battery.

4. Remove the battery setting plate and the battery.

To install:

5. Install the battery and the battery setting plate.

✳✳ CAUTION

Do not deform the battery setting plate by over-tightening the nuts.

6. Clean the battery terminals.

7. Connect the positive cable to the battery.

➡**Always connect the positive side first.**

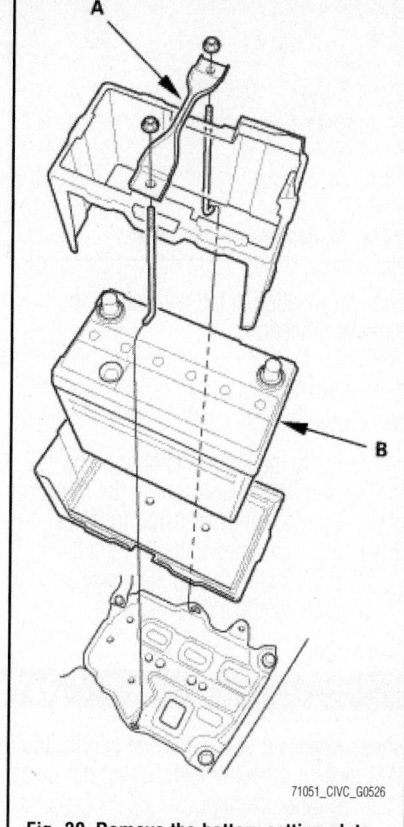

71051_CIVIC_G0526

Fig. 30 Remove the battery setting plate (A) and the battery (B)

8. Connect the negative cable and battery sensor to the battery.

9. Apply multipurpose grease to the terminals to prevent corrosion.

BATTERY CABLES

DISCONNECTING & CONNECTING

1. Make sure the ignition switch is in LOCK (0).

2. Disconnect and isolate the negative cable and battery sensor from the battery.

➡**Always disconnect the negative side first.**

3. Disconnect the positive cable from the battery.

To install:

4. Clean the battery terminals.

5. Connect the positive cable to the battery.

➡**Always connect the positive side first.**

6. Connect the negative cable and battery sensor to the battery.

7. Apply multipurpose grease to the terminals to prevent corrosion.

ENGINE ELECTRICAL

ALTERNATOR

REMOVAL & INSTALLATION

2011 Models

1.8L ENGINE

1. Do the battery terminal disconnect procedure, then wait at least 3 minutes before beginning work. See "Battery System" in "ENGINE ELECTRICAL" section.

2. Remove the drive belt.

3. Disconnect the alternator connector and the positive alternator cable from the alternator.

4. Remove the harness connector and the harness clamps.

5. Remove the retaining bolts and remove the alternator.

6. Installation is the reverse of the removal procedure.

7. Tighten the alternator mounting bolts to 17 ft. lbs. (24 Nm).

2.0L Engine

See Figure 32.

1. Do the battery terminal disconnect procedure, then wait at least 3 minutes before beginning work. See "Battery System" in "ENGINE ELECTRICAL" section.

2. Remove the drive belt.

3. Remove the front grille cover.

4. Disconnect the fan motor connectors

CHARGING SYSTEM

and the hood switch connector, then remove the harness clamps.

5. Remove the coolant reservoir hose, the radiator cap base mounting bolts, the clips and the radiator upper brackets.

6. Remove the condenser bracket mounting bolts, then remove the upper radiator support.

7. Remove the three bolts securing the alternator.

8. Disconnect the alternator connector and the positive alternator cable, and remove the harness clamp, then remove the alternator.

9. Installation is the reverse of the removal procedure.

10. Tighten the alternator mounting bolts to 16 ft. lbs. (22 Nm).

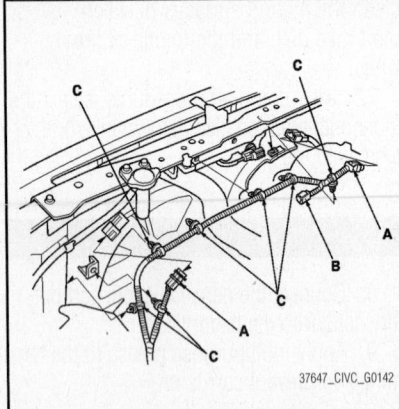

Fig. 32 Disconnect the fan motor connectors (A) and the hood switch connector (B), then remove the harness clamps (C)

2012 Models

1.8L Engine

See Figure 33.

1. Disconnect the battery cables, as described in this section.
2. Disconnect the alternator connector and the cable.
3. Remove the harness connector and the harness clamps.
4. Remove the alternator.

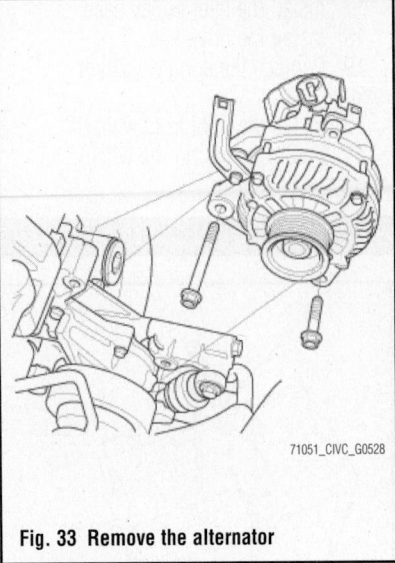

Fig. 33 Remove the alternator

5. Installation is the reverse of the removal procedure.

2.4L Engine

See Figure 34.

1. Move the auto-tensioner using the belt tension release tool in the direction of the arrow to relieve tension from the drive belt, then remove the drive belt.
2. Remove the idler pulley base.
3. Remove the auto-tensioner.

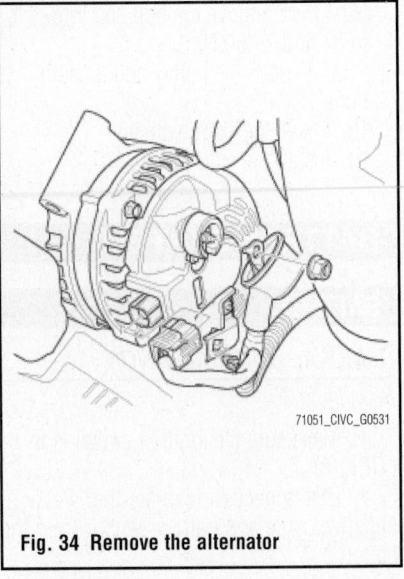

Fig. 34 Remove the alternator

4. Disconnect the connectors. Remove the harness clamps.
5. Remove the bulkhead.
6. Remove the A/C condenser fan shroud assembly.
7. Remove the three bolts securing the alternator.
8. Remove the alternator.
9. Installation is the reverse of the removal procedure.
10. Reconnect the battery cables, as described in this section.

ENGINE ELECTRICAL

IMA SERVICE PRECAUTIONS

GENERAL INFORMATION

The IMA system uses 144 V high voltage circuits and an organic solution (lithium-ion) contained IMA battery module. Improper handling may cause serious injury such as electrocution. Perform in the proper way following the steps below.

PREPARATION ITEMS

Note the following:
- Protective items (insulated gloves or rubber gloves, protective goggles, protective shoes)
- ABC fire extinguisher (available for both oil fires and electric fires)
- Shop towels (for wiping off the electrolyte)
- Insulating tape
- Voltmeter

PRECAUTIONS WHEN WORKING ON HIGH VOLTAGE AREA

1. Wear insulated gloves or rubber gloves, protective goggles, and protective shoes whenever you inspect or service the IMA system. Check the gloves for pin holes, tears, or other damage.
2. As a safety warning, attach a sign saying, WORKING ON HIGH VOLTAGE PARTS. DO NOT TOUCH! to the steering wheel.
3. The high voltage cables and their covers are identified by orange coloring. The caution labels are attached to high voltage and other related parts. When the system is energized, be careful not to touch these cables and parts without adequate protective gear.
4. Be sure to shut off the electrical circuits and isolate the IMA system and related parts before servicing the IMA system. (See "Turning Off and ON Power to the High Voltage Circuit" below).
5. Always use insulated tools when servicing, disassembling, or replacing items marked with in each procedure even if the IMA module switch is OFF.
6. Do not disassemble the IMA battery module when doing the repair or replacement procedure.

HYBRID ENGINE COMPONENTS

7. After disconnecting the high voltage terminals, busbars, etc., insulate the parts with insulating tape.
8. Do not touch bare wires that could carry a high voltage. If you must touch them, put on insulated gloves, and measure the voltage between the body ground. If the voltage is above the 12 V battery voltage, insulate the part with insulating tape before doing any operation since to prevent shorting the IMA battery (144 V).
9. When insulating a high voltage area with insulating tape, be sure to cover it thoroughly.
10. Check around the battery module for leaks of the organic solution. If you find a leak, avoid touching it, but if you must touch it, put on rubber gloves and protective goggles.
11. Keep sparks and flames away from the IMA battery, since the organic solution is flammable.
12. Seal-up the replaced IMA battery with a P.P. (Polypropylene) bag, and keep it in a well-ventilated area away from sparks and flames.

PRECAUTIONS WHEN WORKING ON HIGH MAGNETIC AREA

❊❊ WARNING

The rotor assembly contains very strong magnets and should be handled with special care. People with pacemakers or other magnetically sensitive medical devices should not handle the rotor assembly.

1. Use the special tool (rotor puller) to remove or install the rotor assembly.

2. Keep the rotor assembly away from magnetically sensitive devices.

HANDLING OF A VEHICLE DAMAGED IN A COLLISION

1. Wear insulated gloves or rubber gloves, protective goggles, and protective shoes whenever you inspect or service the IMA system.

2. Do not touch bare wires that could carry a high voltage. If you must touch them, put on insulated gloves, and measure the voltage between the body ground. If the voltage is above the 12 V battery voltage, insulate the part with insulating tape before doing any operation since to prevent shorting the IMA battery (144 V).

3. When insulating a high voltage area with insulating tape, be sure to cover it thoroughly.

4. If fire breaks out in the vehicle, extinguish it with an ABC fire extinguisher. Extinguishing fire with a small amount of water may be dangerous. Discharge a large amount of water from a hydrant or wait for firefighters.

5. If the vehicle is soaked in water, you may risk getting an electric shock. Do not touch the parts and the wires of the high voltage system. Extract the vehicle from the water completely before doing any operation.

6. Check around the battery module for leaks of the organic solution. If you find a leak, avoid touching it, but if you must touch it, put on rubber gloves and protective goggles.

7. Keep sparks and flames away from the IMA battery, since the organic solution is flammable.

8. Seal-up the replaced IMA battery with a P.P. (Polypropylene) bag, and keep it in a well-ventilated area away from sparks and flames.

OTHER PRECAUTIONS

1. If the vehicle is not used for 1 month or more, the 144 V battery life may be shortened. Drive the vehicle for at least 30 minutes a month to charge the IMA battery.

2. High temperature may damage the IMA battery. When drying paint in a heated paint booth, make sure the temperature does not exceed 150°F (65°C).

3. If the 12 V battery is discharged, its cables have been disconnected, or the MCM has been reset, the IMA battery level indicator does not display the SOC when the engine is first started. To display the level in the indicator, start the engine, and hold the engine speed between 3,500–4,000 rpm without load (in P or N) until the level gauge displays at least half charged level.

TURNING OFF AND ON POWER TO THE HIGH VOLTAGE CIRCUIT

See Figures 35 and 36.

The following procedure should be done before you work on or near any energized high voltage components. Follow the procedure exactly. Otherwise, you may be injured or may damage equipment.

1. Turn the ignition switch to LOCK (0), then remove the key from the ignition switch.

2. Remove the rear seat back.

3. Remove the battery module switch lid from the IPU cover.

4. Turn the battery module switch OFF, then check that the bolt is showing.

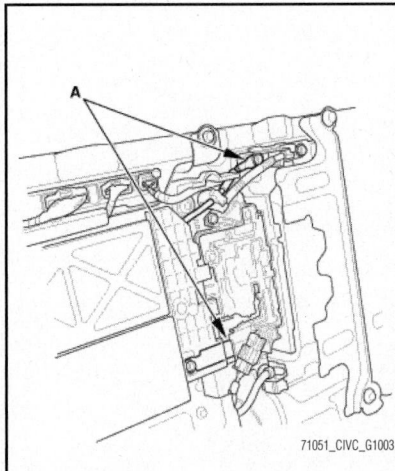

Fig. 35 Measure the voltage at the IMA battery module terminals (A). There should be less than 30 V

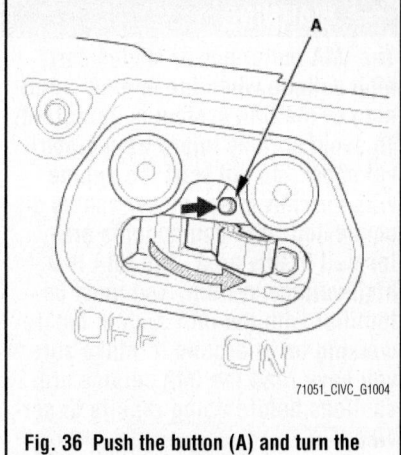

Fig. 36 Push the button (A) and turn the battery module switch ON

5. Wait at least 5 minutes to allow the PDU capacitors to discharge.

6. Remove the IPU cover.

7. Measure the voltage at the IMA battery module terminals. There should be less than 30 V.

8. After service or repairs are completed, make sure all high voltage circuits are connected properly. Install the IPU cover.

9. Push the button and turn the battery module switch ON.

10. Reinstall all remaining removed parts.

DISCONNECTING THE MOTOR POWER CABLE CONNECTOR FROM THE MOTOR STATOR

1. Turn the ignition switch to LOCK (0). Turn the battery module switch OFF.

2. Slide the protector in the direction of the arrow. Push the tab, then raise the lever. Remove the IMA motor power cable from the motor stator.

3. If the outside of the IMA motor power cable connector is dirty, clean it before you disconnect it.

4. Cover the disconnected connector with a plastic bag, and wrap the IMA motor power cable terminals with tape.

5. If the IMA motor power cable terminals are wet, dry them with a clean shop towel. Do not use compressed air.

IMA BATTERY CURRENT SENSOR

REMOVAL & INSTALLATION

See Figures 37 and 38.

➡This battery current sensor is used with Hybrid engines only.

❋❋ WARNING

The IMA motor power cables carry high voltage when the engine is running or the IMA system is energized. To avoid serious injury from electrical shock, do not start the engine with the IMA motor power cables disconnected. IMA components are located in this area. The IMA is a high-voltage system. You must be familiar with the IMA system before working on or around it. Make sure you have read the IMA service precautions before doing repairs or service.

1. Disconnect the 12V battery: negative cable (isolate on removal), battery sensor, then positive cable.

2. Remove the following (see applicable component section as needed):
- Rear Seat Cushion
- Both Rear Seat Side Bolsters
- Rear Seat-Back

3. Turn the IMA Battery Main Switch OFF as follows:
   a. Turn the ignition switch to LOCK (0).
   b. Remove the battery module switch lid from the IPU cover.
   c. Turn the battery module switch OFF, then check that the bolt is showing.
   d. Wait at least 5 minutes to allow the PDU capacitors to discharge.

4. Remove the IPU cover.

5. Remove the power cable terminal cover.

6. Check the position of the U phase, V phase, and W phase cables before disconnecting the IMA motor power cables.

7. Disconnect the connector. Remove the four cables from the DC-DC converter. Wrap the end of the cables with insulating tape.

❋❋ CAUTION

Use insulated tools when doing the following procedure.

8. Remove the cables. Disconnect the connectors.

9. Disconnect the connectors. Remove the ground cable.

10. Remove the PCU assembly.

11. Remove the junction board assembly.

12. Remove the high voltage contactor.

13. Remove the battery current sensor.

To install:

14. Install these components in the following order:
- Battery current sensor
- High voltage contactor
- Junction board assembly
- PCU assembly
- Ground cable
- Connectors
- Remaining cables

15. Install the four cables to the DC-DC converter. Connect the connector.

16. Install the power cable terminal cover.

17. Install the IPU cover.

18. Make sure the ignition switch is turned to LOCK (0).

19. Turn the battery module switch ON.

20. Install the battery module switch lid to the IPU cover.

21. Install the following:
- Rear Seat-Back
- Both Rear Seat Side Bolsters
- Rear Seat Cushion

22. Reconnect the 12V battery positive cable, the battery sensor, then the negative cable.

IMA MOTOR POWER CABLE

REMOVAL & INTALLATION

See Figures 39 through 42.

➡Take care not to scratch the surface of IMA motor power cable.

❋❋ CAUTION

IMA components are located in this area. The IMA is a high-voltage system. You must be familiar with the IMA system before working on or around it. Make sure you have read the IMA service precautions before doing repairs or service.

1. Raise the vehicle on a lift, and make sure it is securely supported.

2. Disconnect the 12 V battery:
   a. Make sure the ignition switch is in LOCK (0).
   b. Disconnect and isolate the negative cable and battery sensor from the 12 V battery.
   c. Disconnect the positive cable from the 12 V battery.

3. Remove the following components (refer to applicable components headings and sections as needed):
- Rear seat cushion
- Both rear seat side bolsters
- Rear seat-back

4. Ensure the IMA battery main switch is OFF:
   a. Turn the ignition switch to LOCK (0).
   b. Remove the battery module switch lid from the IPU cover.
   c. Turn the battery module switch OFF, then check that the bolt is showing.
   d. Wait at least 5 minutes to allow the PDU capacitors to discharge.

5. Remove the IPU cover.

6. Remove the power cable terminal cover.

7. Disconnect the connector. Remove the four cables from the DC-DC converter. Wrap the end of the cables with insulating tape.

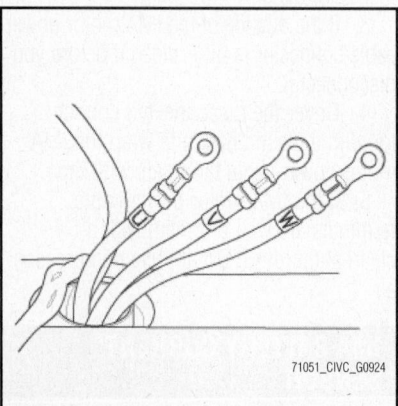

71051_CIVC_G0924

Fig. 37 Check the position of the U phase, V phase, and W phase cables before disconnecting the IMA motor power cables

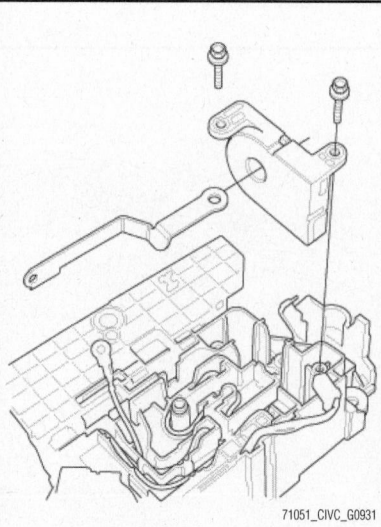

71051_CIVC_G0931

Fig. 38 Remove the battery current sensor

✳✳ **CAUTION**

Use insulated tools when doing the following procedure.

8. Remove the cables. Disconnect the connectors.

9. Disconnect the connectors. Remove the ground cable.

10. Remove the air cleaner and ducting.

11. Disconnect the IMA motor power cable connector:

    a. Slide the protector in the direction of the arrow.

    b. Push the tab, then raise the lever.

    c. Disconnect the IMA motor power cable from the motor stator.

➠**If the outside of the IMA motor power cable connector is dirty, clean it before you disconnect it.**

    d. Cover the disconnected connector with a plastic bag and wrap the IMA motor power cable terminals with tape.

➠**If the IMA motor power cable terminals are wet, dry them with a clean shop towel. Do not use compressed air.**

12. Remove the following (refer to component headings or sections as needed):
- Wiper Arm Assembly
- Both Side Cowl Covers
- Center Cowl Cover
- Under Cowl Panel

13. Remove the IMA motor power cable peripheral assembly:

    a. Remove the under-hood fuse/relay box terminal.

    b. Remove the ground terminal.

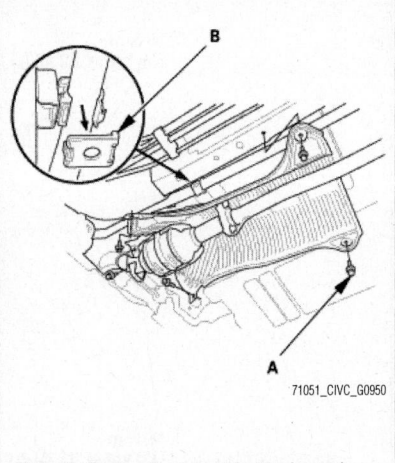

Fig. 40 Center floor heat baffle (move): Remove the bolts (A) and remove the clamp (B)

c. Remove the cover. Disconnect the connector.

d. Remove the harness clamps.

e. Remove the heat shield.

14. Remove the rear engine under cover as needed.

15. Remove the left floor undercover.

16. Remove the fuel tank undercover.

17. Center floor heat baffle (move):

    a. Remove the bolts.

    b. Remove the clamp.

18. Remove the motor power cable:

    a. Remove the bracket.

    b. Remove the brake wire bracket, then open the clamp holder cap.

    c. Remove the clamp holder cap.

    d. Remove the brake wire brackets.

    e. Pull out the IMA motor power cable from the body.

To install:

19. Install or connect the following (refer component headings or sections as needed):
- Motor power cable
- Center floor heat baffle move

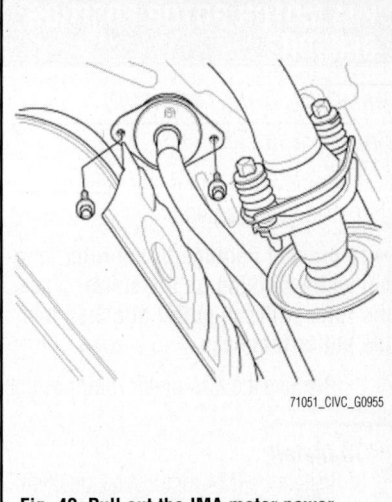

Fig. 42 Pull out the IMA motor power cable from the body

- IMA motor power cable peripheral assembly
- Under cowl panel
- Center cowl cover
- Both side cowl covers
- Wiper arm assembly
- Air cleaner

20. Install the four cables to the DC-DC converter and connect the connector.

21. Install the power cable terminal cover.

22. Install the IPU cover.

23. Make sure the ignition switch is turned to LOCK (0).

24. Turn the battery module switch ON.

25. Install the battery module switch lid to the IPU cover.

26. Install or connect the following (refer component headings or sections as needed):
- Rear seat-back
- Both rear seat side bolsters
- Rear seat cushion
- Battery terminal reconnection

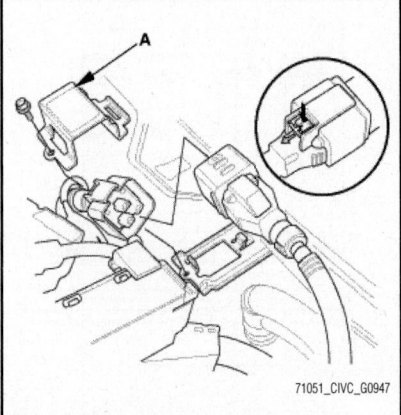

Fig. 39 Remove the cover (A). Disconnect the connector

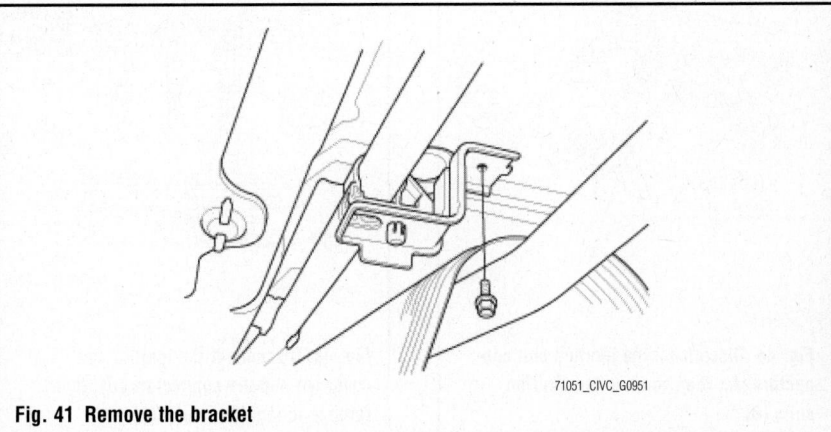

Fig. 41 Remove the bracket

IMA MOTOR ROTOR POSITION SENSOR

REMOVAL & INSTALLATION

See Figure 43.

1. Remove the IMA motor housing, as described in this section.

➡ **To prevent damage to the rotor magnet while working on the stator, place the rotor with the puller attached into the puller tool case.**

2. Remove the IMA motor rotor position sensor.

To install:

3. Install the IMA motor rotor position sensor in the numbered sequence shown.

4. Install the IMA motor, as described in this section.

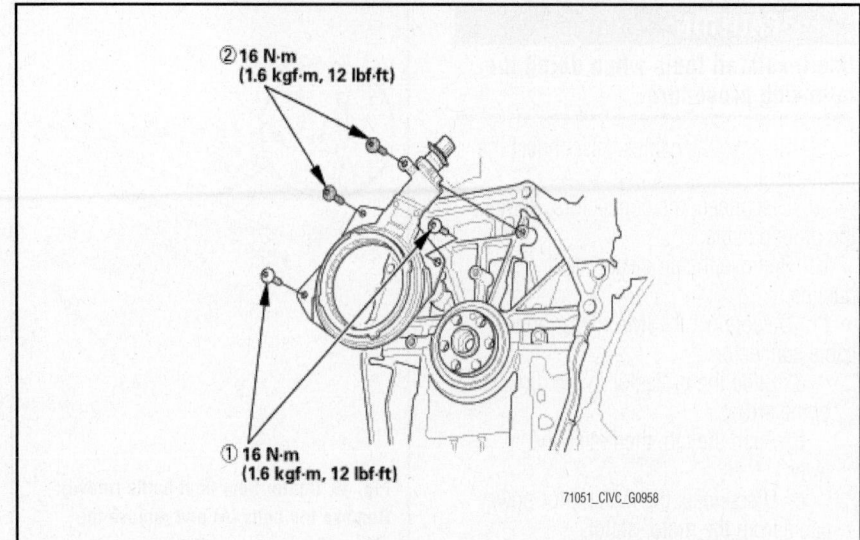

Fig. 43 Install the IMA motor rotor position sensor in the numbered sequence shown

ENGINE ELECTRICAL

FIRING ORDER

Firing order for all engines is 1-3-4-2.

IGNITION COILS & SPARK PLUGS

REMOVAL & INSTALLATION

1.8L Engine

See Figure 44.

1. Remove the engine cover, if applicable.
2. Disconnect the ignition coil connectors, then remove the ignition coils.
3. Remove the spark plugs and inspect them.
4. Apply a small amount of anti-seize compound to the plug threads, and screw the plugs into the cylinder head, finger tight. Torque them to 18 ft. lbs. (25 Nm) for 2011 models, or to 13 ft. lbs. (18 Nm) for 2012 models.
5. Install the ignition coils in the reverse order of removal.

2.0L Engine

See Figure 45.

1. Remove the ignition coil cover.
2. Disconnect the ignition coil connectors, then remove the ignition coils.

IGNITION SYSTEM

3. Remove the spark plugs and inspect them.
4. Apply a small amount of anti-seize compound to the plug threads, and screw the plugs into the cylinder head, finger tight. Torque them to 18 ft. lbs. (25 Nm).
5. Install the ignition coils in the reverse order of removal.

2.4L Engine

See Figure 46.

➡ **Set the wiper arms to the auto-stop position before removal.**

1. Remove the cowl top wiper covers. Remove the wiper arms.
2. Remove both side cowl covers.

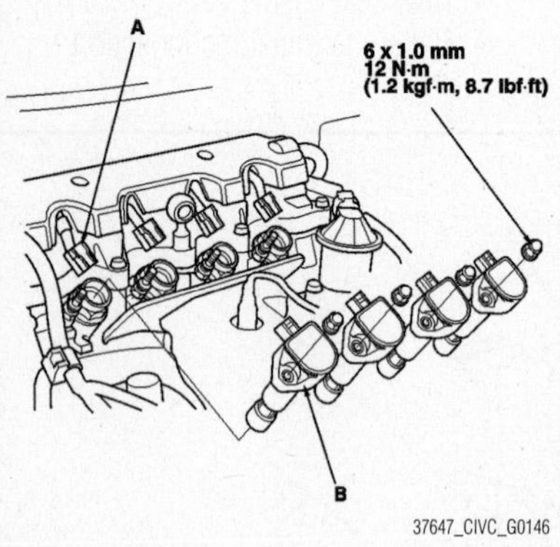

Fig. 44 Disconnect the ignition coil connectors (A), then remove the ignition coils (B)

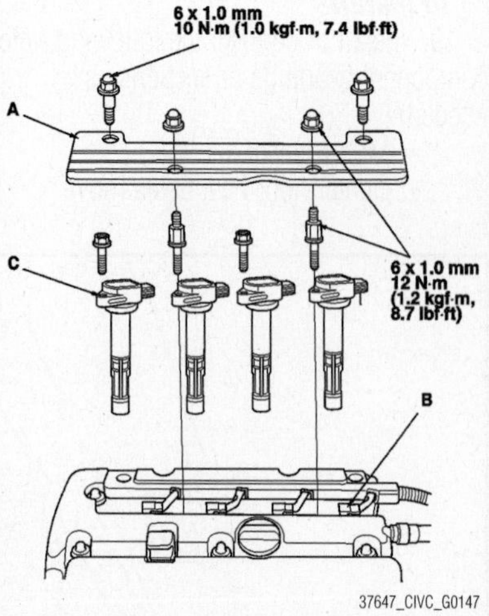

Fig. 45 Disconnect the ignition coil cover (A) and the connectors (B), then remove the ignition coils (C)

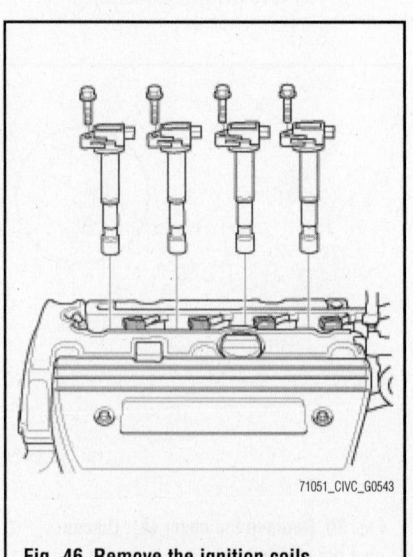

Fig. 46 Remove the ignition coils

3. Remove the center cowl cover. Disconnect the windshield washer tube. If necessary, remove the hood rear seal.

4. Remove the under cowl panel.

5. Remove the ignition coil cover. Remove the ignition coils.

6. Remove the spark plugs.

To install:

7. Apply a small amount of anti-seize compound to the plug threads, and screw the plugs into the cylinder head, finger tight, then tighten the plugs to 13 ft. lbs. (18 Nm).

8. Install the ignition coils.

9. Install the ignition coil cover.

10. Install the under cowl panel.

11. If necessary, install the hood rear seal.

12. Connect the windshield washer tube.

13. Install the center cowl cover.

14. Install the side cowl covers.

➡**Set the wiper arms to the auto-stop position before installation.**

15. Install the wiper arms.

16. Install the cowl top wiper covers.

IDLE SPEED

INSPECTION

1. Before checking the idle speed, check these items:
- The malfunction indicator lamp (MIL) has not been reported on, and there are no DTCs.
- Ignition timing
- Spark plugs
- Air cleaner
- PCV system

2. Apply the parking brake, and make sure the headlights are off.

3. Disconnect the evaporative emission (EVAP) canister purge valve connector.

4. Connect the scan tool to the data link connector (DLC) located under the driver's side of the dashboard.

5. Make sure the scan tool communicates with the ECM/PCM. If it doesn't, perform further DLC circuit troubleshooting.

6. Start the engine. Hold the engine speed at 3,000 RPM without load (in neutral) until the radiator fan comes on, then let it idle.

7. Check the idle speed without load conditions: headlights, blower fan, radiator fan, and air conditioner off. Idle speed should be 750±50 RPM.

8. Let the engine idle for 1 minute with high electric load (A/C on, temperature set to max cool, blower fan on high, headlights on high beam). Idle speed should be 750±50 RPM.

a. If the idle speed is not within specification, do the "ECM/PCM Idle Learn Procedure". See "Engine Control Module (ECM/PCM)" in "ENGINE PERFORMANCE & EMISSION CONTROLS" section.

b. If the idle speed is still not within specification, go to the symptom troubleshooting.

9. Reconnect the EVAP canister purge valve connector.

IGNITION TIMING

INSPECTION & ADJUSTMENT

2011 Models

1. Connect the Honda Diagnostic System (HDS), or equivalent scan tool, to the data link connector (DLC).

2. Turn the ignition switch to ON (II).

3. Make sure the scan tool communicates with the vehicle and the engine control module (ECM/PCM)/powertrain control module (PCM). If it does not communicate, troubleshoot the DLC circuit.

4. Check for DTCs. If a DTC is present, diagnose and repair the cause before continuing with this test. See "DIAGNOSTICS" section.

5. Start the engine. Hold the engine speed at 3,000 RPM with no load (neutral (M/T model or in N or P A/T model) until the radiator fan comes on, then let it idle.

6. Check the idle speed. See "Idle Speed" in this section.

7. Jump the SCS line with the scan tool.

8. Connect the timing light to the No. 1 ignition coil harness.

9. Aim the light toward the pointer on the cam chain case. Check the ignition timing under a no load condition (headlights, blower fan, rear window defogger, and air conditioner are turned off).

10. Ignition timing should be:
- M/T model: 8°±2° BTDC (RED mark) at idle in Neutral
- A/T model: 8°±2° BTDC (RED mark) at idle in N or P

➡**On 2.0L engine, the other pointer on the right side is not used.**

11. If the ignition timing differs from the specification, check the cam timing. If the cam timing is OK, update the ECM/PCM if it does not have the latest software, or substitute a known-good ECM/PCM, then recheck. If the system works properly, and the ECM/PCM was substituted, replace the original ECM/PCM.

12. Disconnect the scan tool and the timing light.

2012 Models

1. Connect the HDS to the data link connector (DLC) located under the driver's side of the dashboard.

2. Turn the ignition switch to ON (II).

3. Check for DTCs. If a DTC is present, diagnose and repair the cause before continuing with this inspection.

4. Start the engine. Hold the engine speed at 3,000 rpm without load (A/T in P or N, M/T in neutral) until the radiator fan comes on, then let it idle.

5. Check the idle speed without electrical load: headlights, blower fan, radiator fan, and air conditioner off.

6. Idle speed should be:
- 1.8L engines: 750±50 rpm (A/T in P or N, M/T in neutral).
- 2.4L engine: 720±50 rpm (A/T in P or N, M/T in neutral).

71051_CIVIC_G0544

Fig. 47 Connect the timing light to the service loop (white tape)

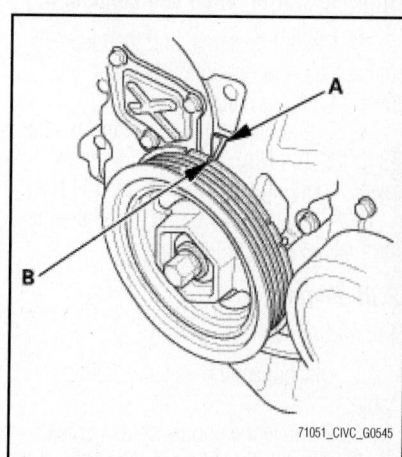

71051_CIVIC_G0545

Fig. 48 Aim the light toward the pointer (A) on the cam chain case. Ignition timing should be as indicated BTDC (RED mark (B)) at idle

7. Let the engine idle for 1 minute with high electric load (A/C switch on, temperature set to max cool, blower fan on High, and headlights on high beam).

8. Idle speed should be:
- 1.8L engines: 750±50 rpm (A/T in P or N, M/T in neutral).
- 2.4L engine: 720±50 rpm (A/T in P or N, M/T in neutral).

➡️ If the idle speed is not within specification, do the ECM/PCM idle learn procedure. If the idle speed is still not

within specification, go to symptom troubleshooting.

9. Jump the SCS line with the HDS.
10. Connect the timing light to the service loop (white tape).
11. Aim the light toward the pointer on the cam chain case.
12. Check the ignition timing under a no load condition (headlights, blower fan, rear window defogger, and air conditioner are turned off).
13. Ignition timing should be:

- 1.8L engine: 10±2° BTDC (RED mark) at idle.
- 2.4L engine: 8±2° BTDC (RED mark) at idle.

14. If the ignition timing differs from the specification, check the cam timing. If the cam timing is OK, update the ECM if it does not have the latest software, or substitute a known good ECM, then recheck. If the system works properly, and the ECM was substituted, replace the original ECM.

15. Disconnect the HDS and the timing light.

ENGINE ELECTRICAL

STARTER

REMOVAL & INSTALLATION

2011 Models

1.8L Engine

1. Disconnect the battery cables.
2. Remove the exhaust pipe.
3. Remove the intake manifold bracket.
4. Remove the wiring harness clamps and the harness connector from each clamp.
5. Remove the two bolts securing the starter, and then remove the starter from the engine.
6. Disconnect the starter wiring harnesses.

To install:

7. Connect the positive starter cable and the connector.

➡️ **Make sure the starter cable crimped side of the ring terminal faces away from the starter when you connect it.**

8. Install the starter, and then loosely install the upper mounting bolt and the lower mounting bolt.
9. Tighten the upper mounting bolt to 33 ft. lbs. (44 Nm), and then tighten the lower mounting bolt to 33 ft. lbs. (44 Nm).
10. The remainder of the installation is the reverse order of removal.

2.0L Engine

See Figure 49.

1. Disconnect the negative battery cable.
2. Remove the engine splash shield.
3. Remove the intake manifold bracket.
4. Remove the harness clamp, and the two bolts securing the starter, and then remove the starter from the engine.
5. Disconnect the positive starter cable and the S terminal connector.

6. Remove the harness clamp and remove the starter.

To install:

7. Install the wiring harness clamp.
8. Connect the positive starter cable and S terminal connector.

➡️ **Make sure the starter cable crimped side of the ring terminal faces away from the starter when you connect it.**

9. Install the starter and tighten the mounting bolts.
10. The remainder of the installation is the reverse order of removal.

2012 Models

1.8L Non-Hybrid Engine

See Figure 50.

1. Make sure the ignition switch is in LOCK (0).

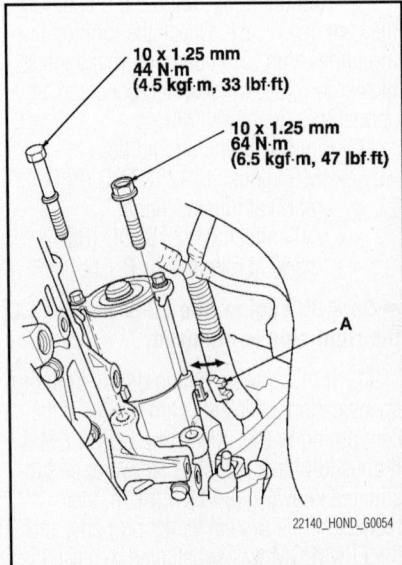

10 x 1.25 mm
44 N·m
(4.5 kgf·m, 33 lbf·ft)

10 x 1.25 mm
64 N·m
(6.5 kgf·m, 47 lbf·ft)

A

22140_HOND_G0054

Fig. 49 Starter mounting bolt torque specifications—2.0L Civic

STARTING SYSTEM

2. Disconnect and isolate the battery sensor and cables.

➡️ **Always disconnect the negative side first.**

3. Raise the vehicle on a lift, and make sure it is securely supported.
4. Remove the right front wheel.
5. Remove the engine undercover.
6. Remove the intake manifold bracket.
7. Disconnect the connector.
8. Remove the clamps and the connector.
9. Remove the starter.
10. Disconnect the cable and the connector.

To install:

11. Connect the cable and the connector. Make sure the starter cable crimped side of the ring terminal faces away from the starter when you connect it.
12. Install the starter, then loosely install the upper mounting bolt and the lower mounting bolt.
13. Tighten the upper mounting bolt, then tighten the lower mounting bolt to 33 ft. lbs. (44 Nm).
14. Connect the connector.

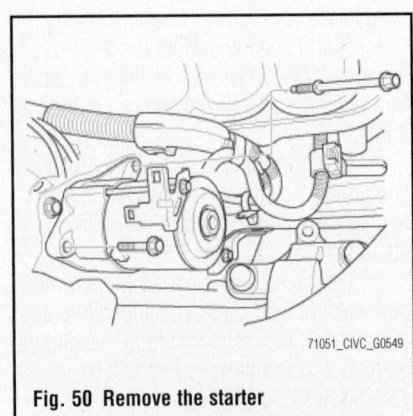

71051_CIVIC_G0549

Fig. 50 Remove the starter

15. Install the intake manifold bracket. Tighten the two upper bolts to 7 ft. lbs. (10 Nm) and the lower bolt to 17 ft. lbs. (24 Nm).

16. Install the engine undercover.

17. Install the right front wheel.

18. Clean the battery terminals.

19. Reconnect the cables and sensor to the battery.

➡**Always connect the positive side first.**

20. Start the engine to make sure the starter works properly.

1.8L Hybrid Engine

See Figure 51.

1. Remove the intake air duct, the intake air pipe, disconnect then remove the air cleaner.

2. Disconnect the cable and the connector.

3. Remove the upper radiator hose bracket.

4. Remove the starter.

To install:

5. Install the starter.

6. Install the upper radiator hose bracket.

7. Connect the connector and the cable. Make sure the crimped side of the ring terminal faces away from the starter when you connect it.

8. Install the air cleaner.

9. Connect the intake air duct to the air cleaner.

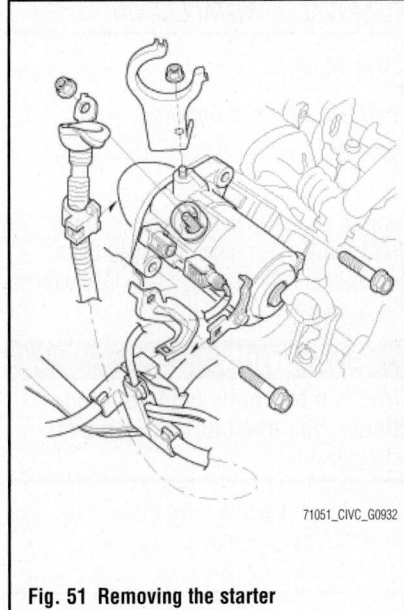

71051_CIVIC_G0932

Fig. 51 Removing the starter

10. Install the intake air pipe and the intake air duct.

2.4L Engine

See Figure 52.

1. Make sure the ignition switch is in LOCK (0).

2. Disconnect and isolate the battery sensor and cables.

➡**Always disconnect the negative side first.**

3. Raise the vehicle on a lift, and make sure it is securely supported.

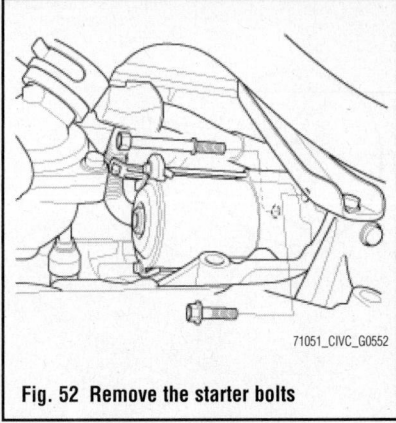

71051_CIVIC_G0552

Fig. 52 Remove the starter bolts

4. Remove the splash shield.

5. Remove the starter bolts.

6. Disconnect the cable and the connector.

7. Remove the starter.

To install:

8. Install the starter.

9. Connect the cable, and the connector. Make sure the crimped side of the ring terminal faces away from the starter when you connect it.

10. Tighten the lower starter bolt to 47 ft. lbs. (64 Nm) and the upper bolt to 33 ft. lbs. (44 Nm).

11. Install the splash shield.

12. Connect the cables and sensor to the battery.

➡**Always install the positive cable first.**

13. Start the engine to make sure the starter works properly.

ENGINE MECHANICAL

ACCESSORY DRIVE BELTS

ACCESSORY BELT ROUTING

➡**See "Removal & Installation" for belt routing figures.**

INSPECTION

Except Hybrid Engine

See Figures 53 and 54.

1. Inspect the belt for cracks or damage If the belt is cracked or damaged, replace it..

2. Check the position of the auto-tensioner indicator pointer is within the standard range. If it is out of the standard range, replace the drive belt.

1.8L Hybrid Engine

See Figure 56.

1. Check the position of the auto-tensioner indicator (A). Start the engine, then

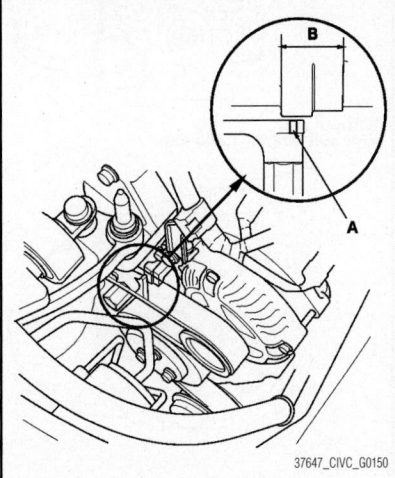

37647_CIVIC_G0150

Fig. 53 Check the position of the auto-tensioner indicator pointer (A) is within the standard range (B) as shown.—1.8L engine

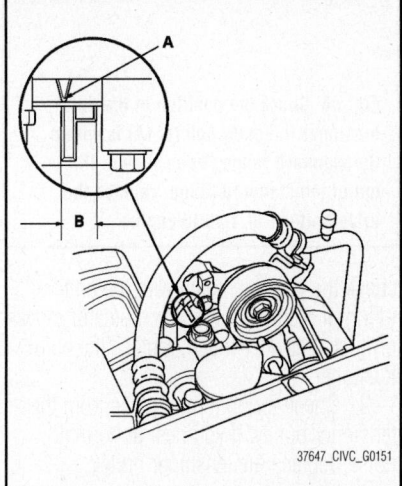

37647_CIVIC_G0151

Fig. 54 Check the position of the auto-tensioner indicator pointer (A) is within the standard range (B) as shown.—2.0L engine

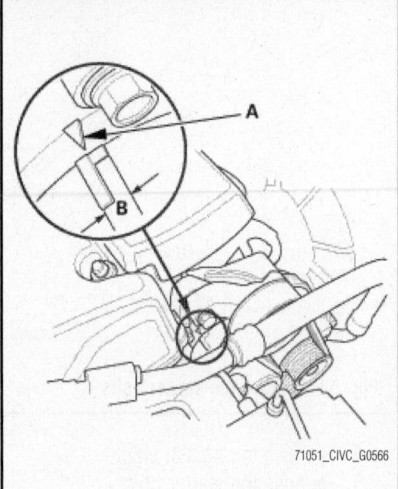

Fig. 55 Check the position of the auto-tensioner indicator pointer (A) is within the standard range (B) as shown. If it is out of the standard range, replace the drive belt—2.4L engine

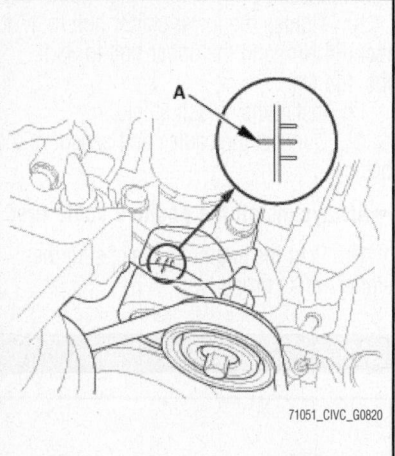

Fig. 56 Check the position of the auto-tensioner indicator pointer (A) is within the standard range (B) as shown. If it is out of the standard range, replace the drive belt—1.8L Hybrid Engine

check the position again with the engine idling. If the position of the indicator moves or fluctuates very much, replace the auto-tensioner.

2. Check for abnormal noise from the tensioner pulley. If you hear abnormal noise, replace the tensioner pulley.

ADJUSTMENT

➡These belt systems are equipped with automatic tensioners and no adjustment is required.

REMOVAL & INSTALLATION

2011 Models

1.8L Non-Hybrid Engine

See Figure 57.

1. Set a long-handled, boxed-end wrench on the drive belt auto-tensioner from above the engine. Slowly turn the wrench in the direction shown, then remove the drive belt.

✳✳ CAUTION

This is a hydraulic type auto-tensioner; you must turn the wrench slowly.

2. Install the new drive belt in the reverse order of removal.

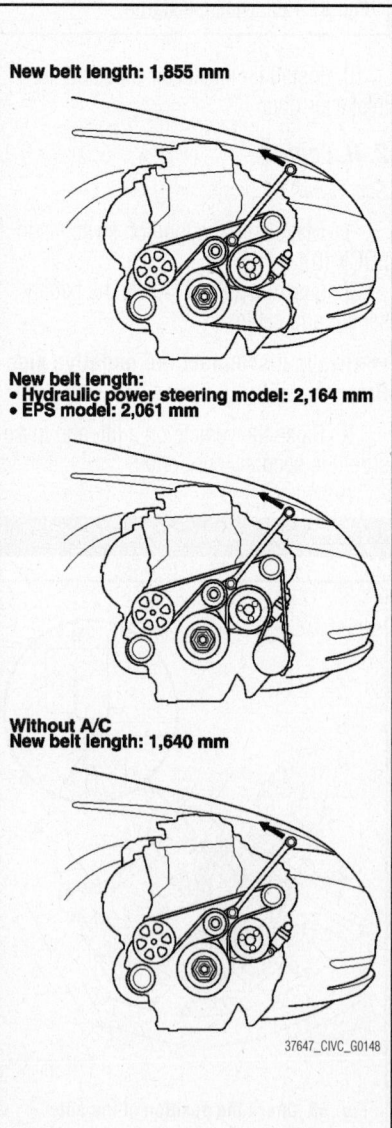

New belt length: 1,855 mm

New belt length:
• Hydraulic power steering model: 2,164 mm
• EPS model: 2,061 mm

Without A/C
New belt length: 1,640 mm

Fig. 57 Removing accessory drive belt—1.8L engine

1.8L Hybrid Engine

See Figure 58.

1. Move the auto-tensioner to relieve tension from the drive belt and remove the drive belt.

2. Install the new belt in the reverse order of removal.

2.0L Engine

See Figure 59.

1. Move the auto-tensioner using the belt tension release tool to relieve tension from the drive belt, then remove the drive belt.

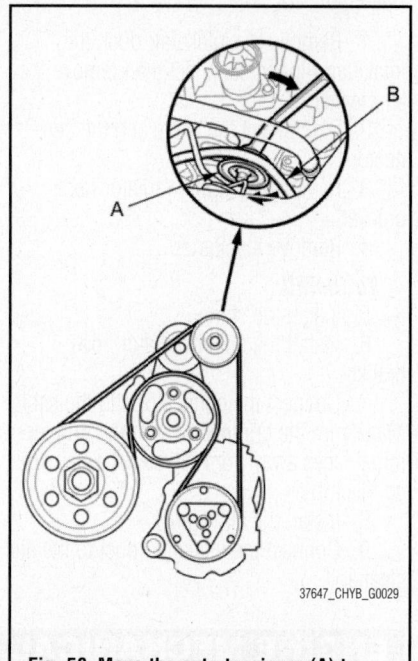

Fig. 58 Move the auto-tensioner (A) to relieve tension from the drive belt (B), and remove the drive belt

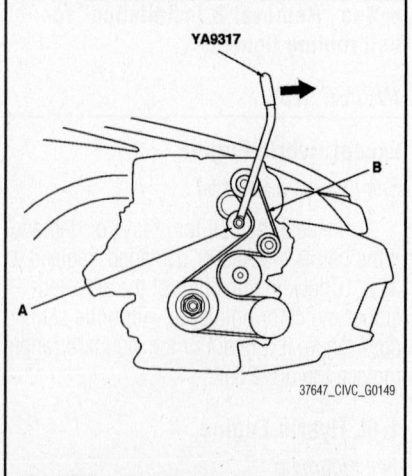

Fig. 59 Removing accessory drive belt—2.0L engine

2. Install the new drive belt in the reverse order of removal.

2012 Models

1.8L Non-Hybrid Engine

See Figure 60.

1. Move the auto-tensioner counter-clockwise to relieve tension from the drive belt.

2. Remove the drive belt.

3. Installation is the reverse of the removal procedure.

1.8L Hybrid Engine

See Figure 61.

1. Move the auto-tensioner (A) clockwise to relieve tension from the drive belt (B).

2. Remove the drive belt.

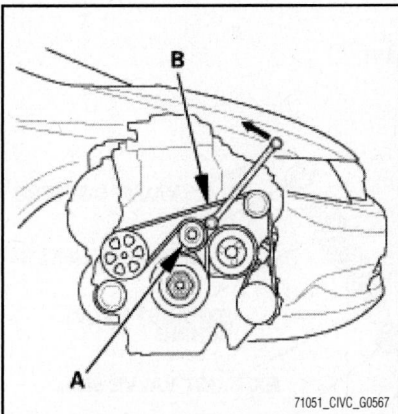

Fig. 60 Move the auto-tensioner (A) counterclockwise to relieve tension from the drive belt (B)

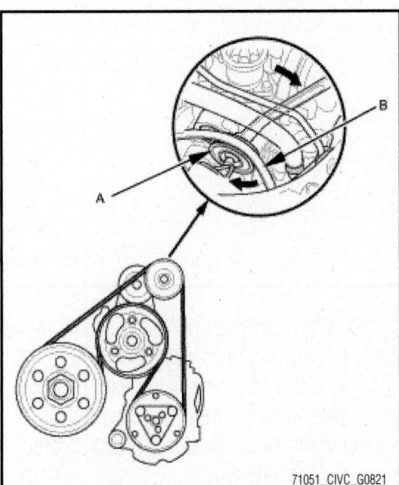

Fig. 61 Move the auto-tensioner (A) clockwise to relieve tension from the drive belt (B)

3. Installation is the reverse of the removal procedure.

2.4L Engines

See Figure 62.

Move the auto-tensioner using the belt tension release tool in the direction of the arrow to relieve tension from the drive belt, then remove the drive belt.

CAMSHAFT

REMOVAL & INSTALLATION

2011 Models

1.8L Engine

See Figure 63.

1. Remove the cylinder head. See "Cylinder Head" in this section.

2. Remove the rocker arm assembly. See "Rocker Arms" in this section.

3. Remove the camshaft sprocket:

   a. Remove the cam chain.

   b. Hold the camshaft with a 27 mm open-end wrench, then loosen the bolt.

   c. Remove the camshaft sprocket.

4. Remove the camshaft position (CMP) sensor.

5. Remove the camshaft thrust cover, then pull out the camshaft.

6. Installation is the reverse of the removal procedure.

2.0L Engine

See Figure 64.

➡**Manufacturer does not provide a specific removal and installation procedure for this component. Use the**

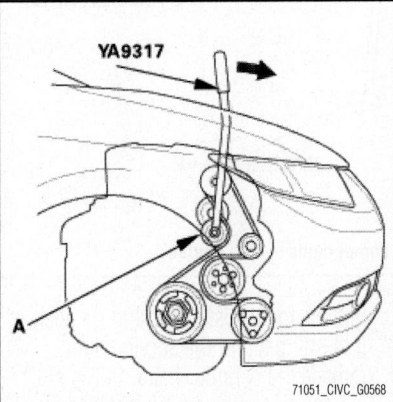

Fig. 62 Move the auto-tensioner (A) using the belt tension release tool in the direction of the arrow to relieve tension from the drive belt, then remove the drive belt

illustration as a guide when servicing this component.

2012 Models

1.8L Non-Hybrid Engine

See Figures 65 through 69.

1. Raise the vehicle on a lift, and make sure it is securely supported.

2. Remove the right front wheel.

3. Remove the splash shield.

4. Make sure the ignition switch is in LOCK (0).

5. Disconnect and isolate the negative cable, battery sensor and then the positive cable from the battery.

6. Remove the battery setting plate and the battery.

7. Loosen the water pump pulley mounting bolts.

8. Move the auto-tensioner counter-clockwise to relieve tension from the drive belt, then remove the drive belt.

9. Remove the alternator.

10. Remove the water pump pulley.

11. Remove the drive belt auto-tensioner, as described in this section.

12. Remove the engine cover.

13. Disconnect the connecters and remove the harness holders.

14. Remove the dipstick.

15. Disconnect the breather hose from the front of the cylinder head cover.

16. Remove the cylinder head cover.

17. Set the No. 1 piston at top dead center (TDC). The "UP" mark on the camshaft sprocket should be at the top, and the TDC grooves on the camshaft sprocket should line up with the top edge of the head.

18. Remove the crankshaft pulley, as described in this section.

19. Lift and support the engine with a jack and a wood block under the oil pan.

20. Remove the ground cable, the side engine mount bracket, and the side engine mount.

21. Disconnect the PCV hose.

22. Remove the oil pump.

23. Remove the cam (timing) chain, as described in this section.

24. Remove the rocker arm assembly as described in this section.

25. Hold the camshaft with a 27 mm open-end wrench and loosen the camshaft sprocket mounting bolt. Remove the camshaft sprocket.

26. Before removing the camshaft, perform a camshaft end play inspection:

   a. Put the rocker shaft holders and the lost motion holder on the cylinder head, then tighten the bolts, in sequence,

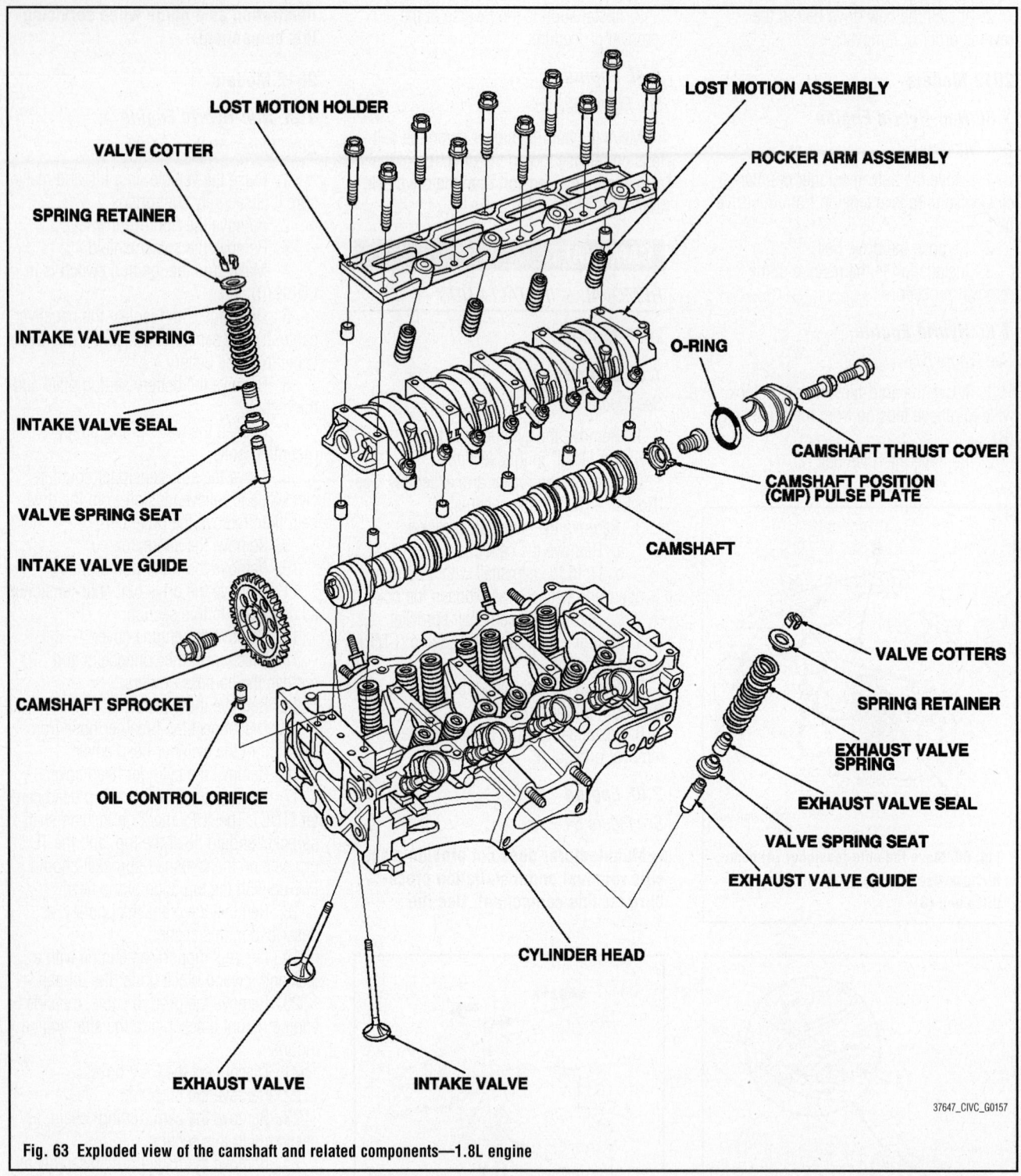

Fig. 63 Exploded view of the camshaft and related components—1.8L engine

to the specified torque of 11 ft. lbs. (15 Nm).

b. Seat the camshaft by pushing it away from the camshaft pulley end of the cylinder head.

c. Zero the dial indicator against the end of the camshaft, then push the camshaft back and forth, and read the end play. If the end play is beyond the service limit of 0.016 in. (0.4 mm),

replace the thrust cover and recheck. If it is still beyond the service limit, replace the cylinder head. If it is still beyond the service limit, replace the camshaft.

➡**Do not rotate the camshaft during inspection.**

27. Remove the camshaft thrust cover and pull out the camshaft.

28. With camshaft removed, measure the camshaft journal diameters as follows:

a. Wipe the camshaft clean, then inspect the lift ramps. Replace the camshaft if any lobes are pitted, scored, or excessively worn.

b. Measure the diameter of each camshaft journal.

c. Zero the gauge to the journal diameter.

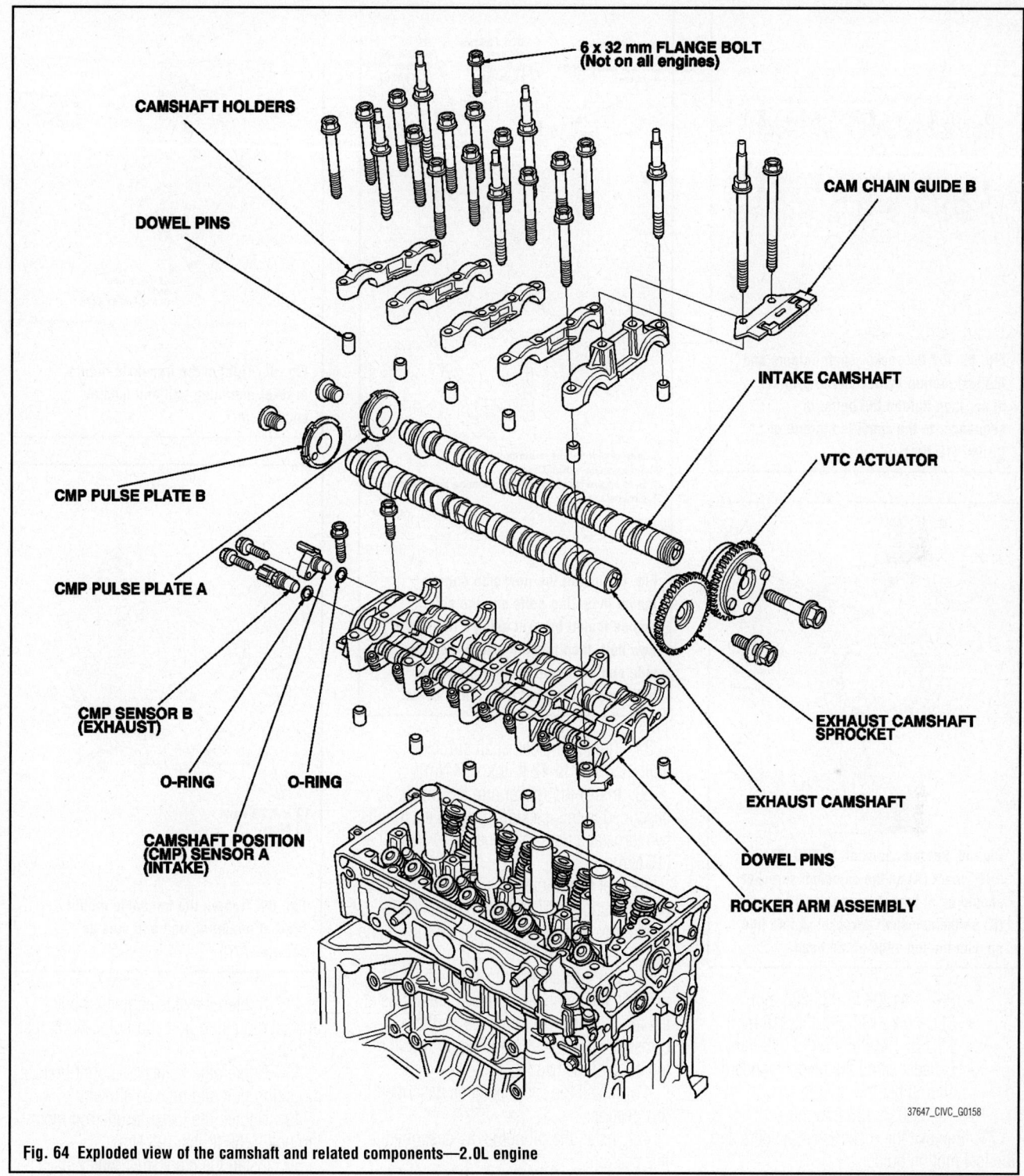

Fig. 64 Exploded view of the camshaft and related components—2.0L engine

d. Clean the camshaft bearing surfaces in the cylinder head. Measure the inside diameter of each camshaft bearing surface, and check for an out-of-round condition.

e. If the camshaft-to-holder clearance is within limits of 0.00177–0.00331 in. (0.45–0.084 mm), go to the next step.

f. If the camshaft-to-holder clearance is beyond the service limit of

0.0059 in. (0.15 mm), and the camshaft has been replaced, replace the cylinder head.

g. If the camshaft-to-holder clearance is beyond the service limit, and the camshaft has not been replaced, go to step 5.

29. Check the total runout with the camshaft supported on V-blocks as follows:

a. If the total runout of the camshaft is within the service limit of 0.0012 in. (0.03 mm), replace the cylinder head.

b. If the total runout is beyond the service limit, replace the camshaft, and recheck the oil clearance. If the oil clearance is still beyond service limit, replace the cylinder head.

30. Measure the cam lobe height. The intake cam lobe height standard should be:

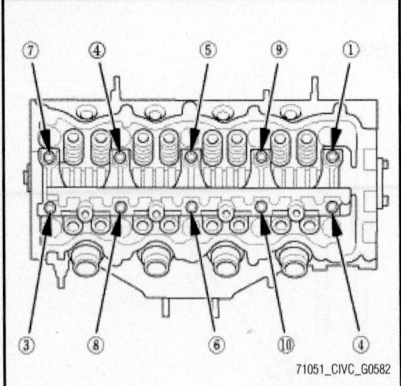

Fig. 65 Put the rocker shaft holders and the lost motion holder on the cylinder head, then tighten the bolts, in sequence, to the specified torque of 11 ft. lbs. (15 Nm)

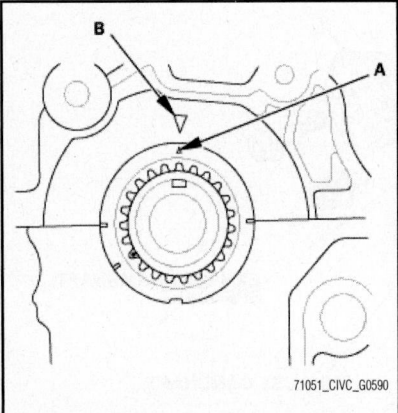

Fig. 66 Set the camshaft to TDC. The "UP" mark (A) on the camshaft sprocket should be at the top, and the TDC grooves (B) on the camshaft sprocket should line up with the top edge of the head

- PRI: 1.41204 in. (35.866 mm)
- SEC A: 1.21330 in. (30.818 mm)
- SEC B: 1.40508 in. (25.689 mm)
- Exhaust cam lobe height standard should be:
- 1.41220 in. (35.870 mm)

31. Remove the rocker shaft holders and the lost motion holder.

To install:

32. Install the camshaft and install the camshaft thrust cover with a new O-ring.

33. Install the camshaft sprocket.

34. Apply new engine oil to the threads of the camshaft sprocket mounting bolt.

35. Install the camshaft sprocket mounting bolt.

36. Hold the camshaft with a 27 mm open-end wrench.

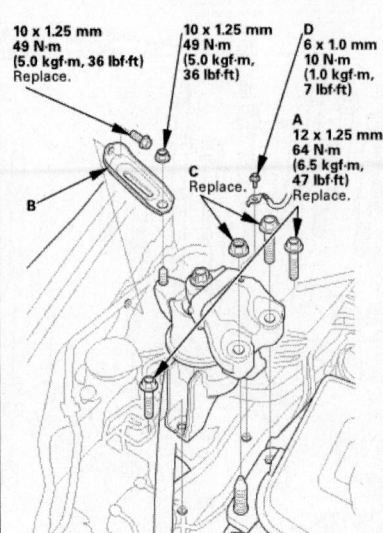

A. Side engine mount mounting bolts
B. Side engine mount bracket
C. Side engine mount bracket mounting bolts and nut
D. Ground cable

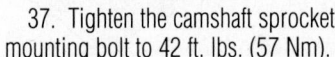

Fig. 67 Install the new side engine mount mounting bolts and the side engine mount bracket using a nut and a new bolt, then loosely install the new side engine mount bracket mounting bolts and nut. Install the ground cable

37. Tighten the camshaft sprocket mounting bolt to 42 ft. lbs. (57 Nm).

38. Install the rocker arm assembly, as described in this section. Tighten each bolts two turns at a time in sequence to 11 ft. lbs. (15 Nm).

39. Set the crankshaft to top dead center (TDC). Align the TDC mark on the crankshaft sprocket with the pointer on the engine block.

40. Set the camshaft to TDC. The "UP" mark on the camshaft sprocket should be at the top, and the TDC grooves on the camshaft sprocket should line up with the top edge of the head.

41. Install the cam chain, as described in this section.

42. Install the oil pump, as described in this section.

43. Install the new side engine mount mounting bolts and the side engine mount bracket using a nut and a new bolt, then loosely install the new side engine mount bracket mounting bolts and nut. Install the ground cable.

44. Remove the jack and the wood block.

45. Loosen the transaxle mount bracket mounting bolt and nuts.

46. Loosen the lower torque rod mounting bolt.

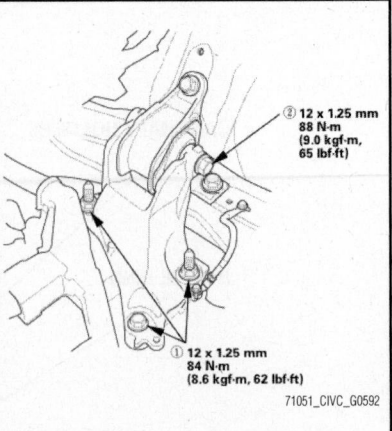

Fig. 68 Tighten the transaxle mount bracket mounting bolt and nuts as shown—M/T

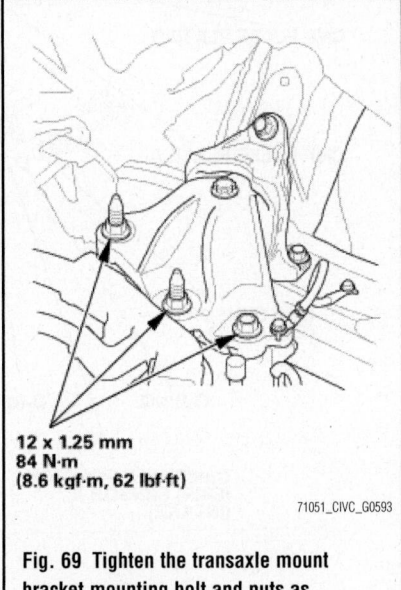

Fig. 69 Tighten the transaxle mount bracket mounting bolt and nuts as shown—A/T

47. Tighten the side engine mount bracket mounting bolt and nut to 69 ft. lbs. (93 Nm).

48. Tighten the transaxle mount bracket mounting bolt and nuts as shown.

49. Tighten the lower torque rod mounting bolt to 69 ft. lbs. (93 Nm).

50. Install the crankshaft pulley, as described in this section.

51. Perform valve clearance check and adjustment, as described in this section.

52. Install the cylinder head cover, as described in this section.

➡Do not run the engine for at least 3 hours after installing the head cover.

53. Connect the breather hose to the front of the cylinder head cover.

54. Install the dipstick.

55. Install the harness holders and connect the connectors on the injectors.

56. Install the engine cover.

57. Install the drive belt auto-tensioner.

58. Install the water pump pulley, as described in this section.

59. Install the alternator, as described in "Engine Electrical" section.

60. Move the auto-tensioner counter-clockwise and install the drive belt.

61. Tighten the water pump pulley mounting bolts to 10 ft. lbs. (14 Nm).

62. Install the splash shield.

63. Install the right front wheel. Tighten the nuts to 80 ft. lbs. (108 Nm).

64. Install the battery and cables, as described in "Engine Electrical" section.

65. Connect the HDS to the data link connector (DLC) located under the driver's side of the dashboard.

66. Turn the ignition switch to ON (II).

67. Perform the CKP Pattern Clear/CKP Learn procedure:

   a. Select CRANK PATTERN in the ADJUSTMENT MENU with the HDS.

   b. Select CRANK PATTERN CLEAR, and clear the CKP pattern.

   c. Select CRANK PATTERN LEARNING with the HDS, and follow the screen prompts.

1.8L Hybrid Engine

See Figures 70 through 73.

1. Raise the vehicle on a lift, and make sure it is securely supported.

2. Remove or disconnect the following:
   - Engine cover
   - Air cleaner
   - Intake air resonator, intake air duct and intake air pipe
   - Throttle body without disconnecting the hoses
   - Engine undercover
   - Intake manifold bracket mounting bolts
   - Intake manifold/chamber assembly
   - Ignition coils
   - Cylinder head cover

3. Make a reference mark across the camshaft sprocket and the cam chain.

4. Apply new engine oil to the slider surface of the cam chain tensioner slider through the oil return hole in the cylinder head.

5. Remove the cylinder head plug.

6. Hold the crankshaft pulley, and set the socket wrench on the camshaft sprocket bolt. Remove the maintenance bolt. Turn the camshaft clockwise to compress the cam chain tensioner. Install the 6 x 1.0 mm bolt in the bolt hole in the engine block through

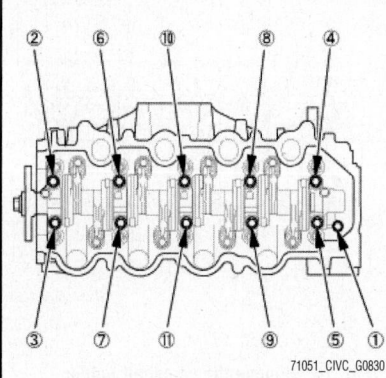

Fig. 70 Remove the camshaft holder bolts. To prevent damaging the camshaft, loosen the bolts in sequence two turns at a time

the maintenance hole and cam chain tensioner.

➡**The turning torque should not exceed 33 ft. lbs. (44 Nm) when turning the camshaft.**

❋❋ CAUTION

Do not turn the camshaft counter-clockwise.

7. Hold the camshaft with a 27 mm open-end wrench and remove the camshaft sprocket.

➡**Hang the cam chain with a wire.**

8. Loosen the rocker arm adjusting screws.

9. Remove the camshaft holder bolts. To prevent damaging the camshaft, loosen the bolts in sequence two turns at a time.

10. Remove the rocker arm assembly.

11. Remove the camshaft.

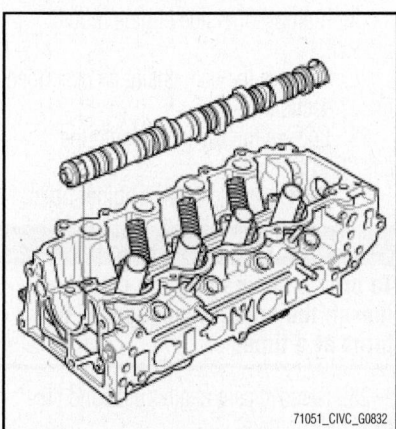

Fig. 71 Remove the camshaft

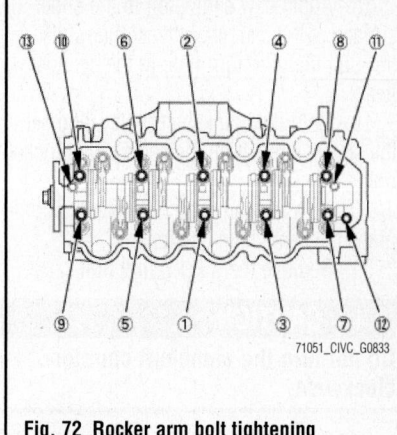

Fig. 72 Rocker arm bolt tightening sequence

To install:

12. Install the camshaft.

13. Install the rocker arm assembly.

14. Apply new engine oil to the bolt threads and flange.

15. Tighten each bolts two turns at a time in sequence to the following:
   - 8 mm Bolts: 16 ft. lbs. (22 Nm)
   - 8 mm Bolts (11, 13): 15 ft. lbs. (20 Nm)
   - 6 mm Bolt (12): 9 ft. lbs. (12 Nm)

16. Install the cam chain to the camshaft sprocket by aligning the reference mark, then install the camshaft sprocket on the camshaft.

17. Hold the camshaft with a 27 mm open-end wrench. Tighten the camshaft sprocket mounting bolt to 41 ft. lbs. (56 Nm).

NOTE: Apply new engine oil to the bolt threads and flange.

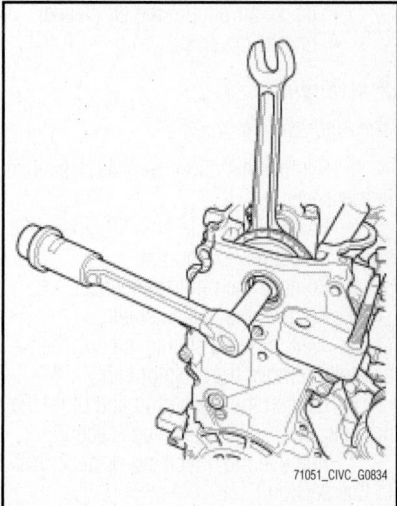

Fig. 73 Hold the camshaft with a 27 mm open-end wrench. Tighten the camshaft sprocket mounting bolt to 41 ft. lbs. (56 Nm)

18. Apply new engine oil to the slider surface of the cam chain tensioner slider through the oil return hole in the cylinder head.

19. Hold the crankshaft pulley, and set the socket wrench on the camshaft sprocket bolt.

20. Turn the camshaft clockwise to compress the cam chain tensioner.

21. Remove the 6 x 1.0 mm bolt .

✳✳ CAUTION

Do not turn the camshaft counter-clockwise.

22. Install the maintenance bolt with a new washer.

23. Install the new cylinder head plug.

24. Perform a valve clearance adjustment, as described in this section.

25. Install or connect the following (refer to applicable sections on these components as needed):

- Cylinder head cover
- Breather hose
- Harness holder
- Ignition coils
- Connectors
- PCV hose
- Intake manifold/chamber assembly
- Engine wire harness and connectors
- Intake manifold bracket mounting bolts.
- Engine undercover
- Throttle body with new gasket
- Intake air pipe, air duct and resonator
- Air cleaner
- Intake air duct to the air cleaner
- Engine cover.

2.4L Engine

See Figures 74 through 81.

1. Remove the air cleaner, as described in this section.

2. Raise the vehicle on a lift, and make sure it is securely supported.

3. Remove the right front wheel.

4. Remove the splash shield.

5. Wait until the engine is cool, then carefully remove the radiator cap.

6. Loosen the drain plug and drain the coolant. See "Engine Cooling" section.

7. Remove the drive belt, as described in this section.

8. Set the wiper arms to the auto-stop position. Remove the cowl top wiper covers. Remove the wiper arms.

9. Remove the side cowl cover, the center cowl cover, the windshield washer tube,

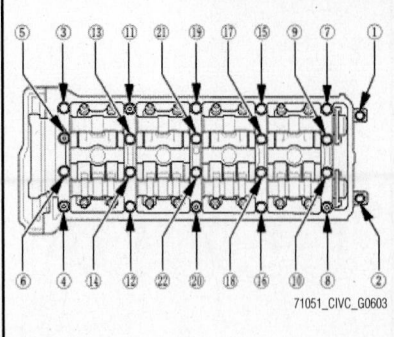

Fig. 74 Remove the camshaft holder bolts.

and, if necessary, remove the hood rear seal.

10. Remove the under cowl panel.

11. Remove the engine cover.

12. Remove the ignition coil cover. Remove the ignition coils.

13. Remove the cylinder head cover.

14. Turn the crankshaft so its white mark lines up with the pointer.

➡**The other pointer is not used.**

15. Set the No. 1 piston at top dead center (TDC). The punch mark on the VTC actuator and the punch mark on the exhaust camshaft sprocket should be at the top. Align the TDC marks on the VTC actuator and the exhaust camshaft sprocket.

16. Remove the harness clamp.

17. Remove the ground cable. Remove the VTC oil control solenoid valve.

18. Hold the pulley with the holder handle and the crankshaft pulley holder. Remove the bolt with a

19. Lift and support the engine with a jack and a wood block under the oil pan.

20. Remove the upper torque rod.

21. Remove the ground cable, the side engine mount bracket, and the side engine mount. Remove the side engine mount bracket.

22. Remove the cam chain, as described in this section.

23. Loosen the rocker arm and the adjusting screws.

24. Remove the camshaft holder bolts.

✳✳ CAUTION

To prevent damaging the camshafts, loosen the bolts, in sequence, two turns at a time.

25. Remove cam chain guide and the camshaft holders.

26. Remove the camshafts.

27. Insert the bolts into the rocker shaft

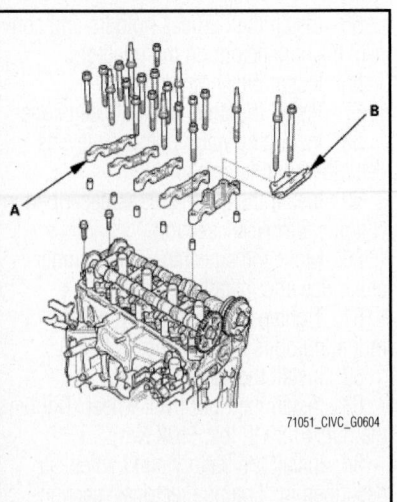

Fig. 75 Remove cam chain guide (B) and the camshaft holders (A)

holder, then remove the rocker arm assembly, as described in this section.

28. Check the camshaft end play as follows:

a. Put the rocker shaft holders, the camshaft, and the camshaft holders on the cylinder head, then tighten the bolts, in sequence, to the specified torque of:

- 8 x 1.25 mm: 16 ft. lbs. (22 Nm)
- 6 x 1.0 mm: 9 ft. lbs. (12 Nm)

➡**If the engine does not have bolt "21", skip it and continue the torque sequence.**

b. Seat the camshaft by pushing it away from the camshaft pulley end of the cylinder head.

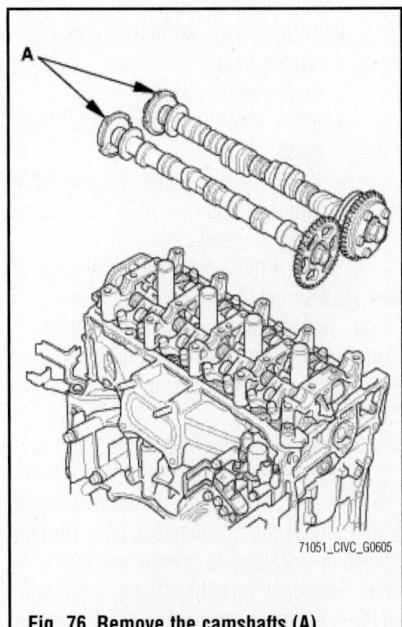

Fig. 76 Remove the camshafts (A)

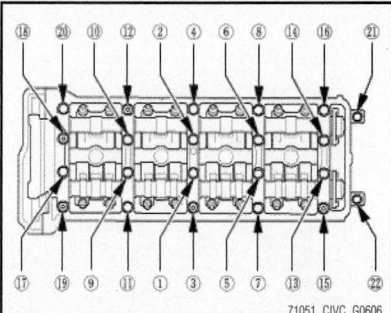

Fig. 77 Put the rocker shaft holders, the camshaft, and the camshaft holders on the cylinder head, then tighten the bots, in sequence

c. Zero the dial indicator against the end of the camshaft, then push the camshaft back and forth, and read the end play. If the end play is beyond the service limit of 0.0157 in. (0.4 mm), replace the cylinder head and recheck. If it is still beyond the service limit, replace the camshaft.

d. Loosen the camshaft holder bolts two turns at a time, in a crisscross pattern. Then remove the camshaft holders from the cylinder head.

e. Lift the camshafts out of the cylinder head, and wipe them clean, then inspect the lift ramps. Replace the camshaft if any lobes are pitted, scored, or excessively worn.

29. Check the camshaft-to-holder oil clearance as follows:

a. Clean the camshaft journal surfaces in the cylinder head, then set the camshafts back in place. Place a Plastigage strip across each journal.

b. Install the camshaft holders, then tighten the bolts to the specified torque as shown in step 1.

c. Remove the camshaft holders. Measure the widest portion of Plastigage on each journal.

d. If the camshaft-to-holder clearance is within the service limits, go to step 10.

e. If the camshaft-to-holder clearance is beyond the service limit, and the camshaft has been replaced, replace the cylinder head.

f. If the camshaft-to-holder clearance is beyond the service limit, and the camshaft has not been replaced, go to step 9.

g. Camshaft-to-Holder Oil Clearance should be:
- Standard (New): No. 1 Journal: 0.030–0.069 mm (0.00118–0.00272 in.)
- Standard (New): No. 2, 3, 4, 5

Journals: 0.060–0.099 mm (0.00236–0.00390 in.)
- Service Limit: 0.15 mm (0.00591 in.)

30. Check the camshaft total runout with the camshaft supported on V-blocks:

a. If the total runout of the camshaft is within the standard, replace the cylinder head.

b. If the total runout is beyond the standard of 0.0012 in. (0.03 mm), replace the camshaft and recheck the camshaft-to-holder oil clearance. If the oil clearance is still beyond the standard, replace the cylinder head.

31. Measure the intake cam lobe height (standard, new):
- PRI: 1.32850 (33.744 mm)
- MID: 1.39590 (35.456 mm)
- SEC: 1.32850 (33.744 mm)

32. Measure the exhaust cam lobe height (standard, new):
- 1.35004 (34.291 mm)

33. Clean and dry the No. 5 rocker shaft holder mating surface. Apply liquid gasket to the cylinder head mating surface of the No. 5 rocker shaft holder and to the inside edge of the threaded bolt holes. Install the component within 5 minutes of applying the liquid gasket.

34. Apply a 2.5 mm (0.098 in) diameter bead of liquid gasket along the broken line (A).

➡**If too much time has passed after applying the liquid gasket, remove the old liquid gasket and residue, then reapply new liquid gasket.**

35. Install the lost motion assemblies in the cylinder head. Apply new engine oil to the lost motion assembly. Insert the bolts into the rocker shaft holder, then install the rocker arm assembly on the cylinder head.

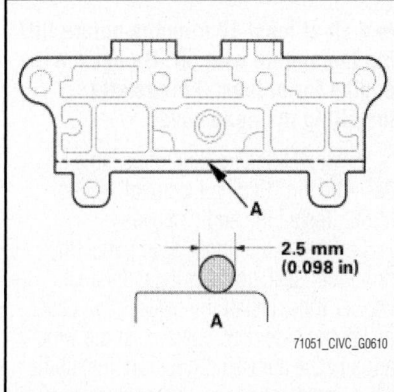

Fig. 78 Apply a 2.5 mm (0.098 in) diameter bead of liquid gasket along the broken line (A)

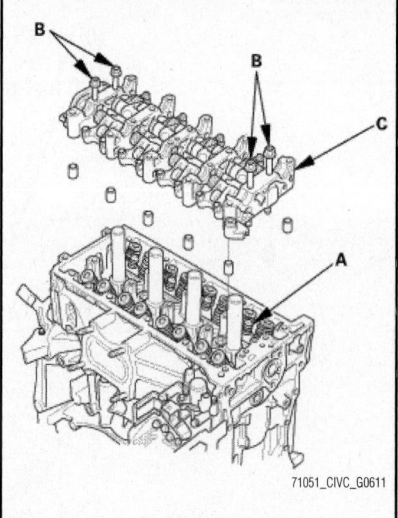

Fig. 79 Install the lost motion assemblies (A) in the cylinder head. Apply new engine oil to the lost motion assembly. Insert the bolts (B) into the rocker shaft holder, then install the rocker arm assembly (C) on the cylinder head

36. Remove the bolts from the rocker shaft holder.

To install:

37. Make sure the punch marks on the VTC actuator and the exhaust camshaft sprocket are facing up, then set the camshafts in the holder. Apply new engine oil to the camshaft journals and lobes.

38. Set the camshaft holders and cam chain guide in place. Tighten the bolts to the specified torque:
- 8 x 1.25 mm: 16 ft. lbs. (22 Nm)
- 6 x 1.0 mm: 9 ft. lbs. (12 Nm)

39. Install the cam chain and case, as described in this section.

40. Install the side engine mount bracket, then tighten the side engine mount bracket mounting bolts to 33 ft. lbs. (44 Nm).

41. Install the new side engine mount mounting bolts and the side engine mount bracket using a nut and a new bolt, then loosely install the new side engine mount bracket mounting bolts and nut. Install the ground cable.

42. Loosen the transaxle mount bracket bolt and nuts.

43. Loosen the lower torque rod mounting bolt.

44. Tighten the side engine mount bracket mounting bolt and nut to 69 ft. lbs. (93 Nm).

45. Tighten the transaxle mount bracket mounting bolt and nuts to 61 ft. lbs. (83 Nm).

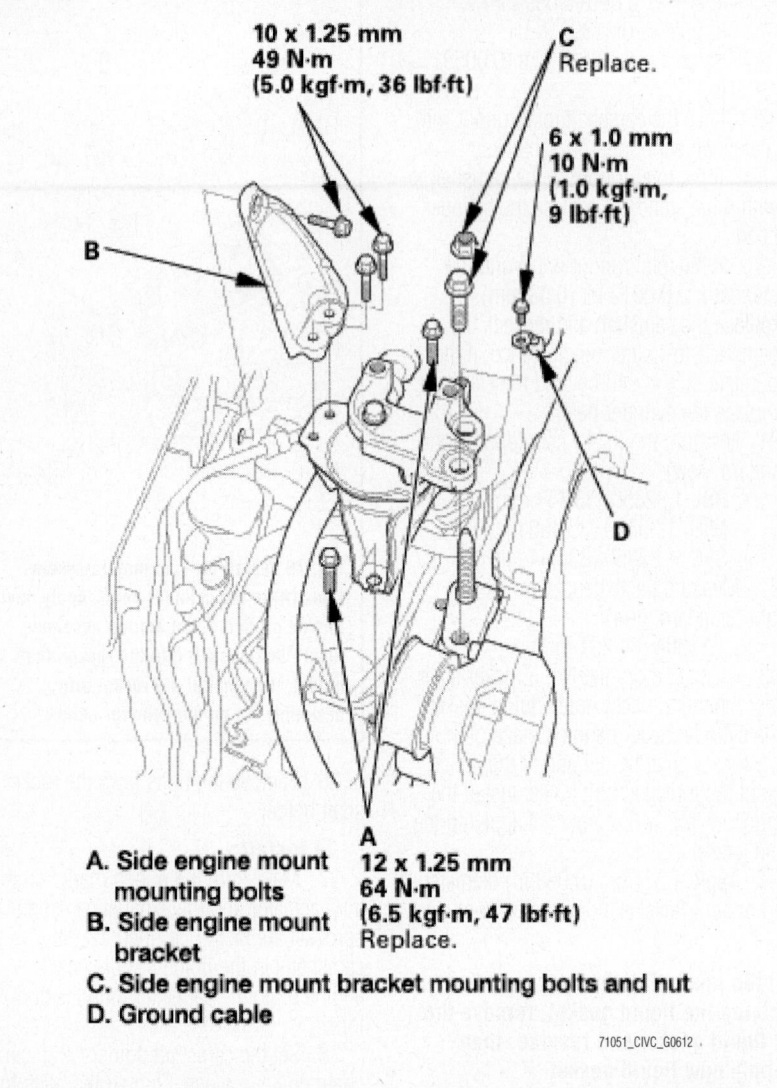

10 x 1.25 mm
49 N·m
(5.0 kgf·m, 36 lbf·ft)

6 x 1.0 mm
10 N·m
(1.0 kgf·m, 9 lbf·ft)

C
Replace.

A. Side engine mount mounting bolts
B. Side engine mount bracket
C. Side engine mount bracket mounting bolts and nut
D. Ground cable

12 x 1.25 mm
64 N·m
(6.5 kgf·m, 47 lbf·ft)
Replace.

71051_CIVC_G0612

Fig. 80 Install the new side engine mount mounting bolts and the side engine mount bracket using a nut and a new bolt, then loosely install the new side engine mount bracket mounting bolts and nut. Install the ground cable

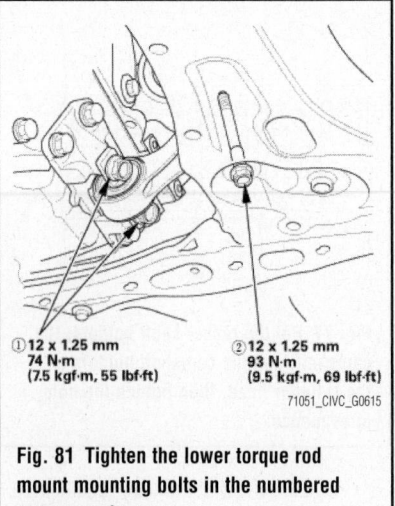

① **12 x 1.25 mm**
74 N·m
(7.5 kgf·m, 55 lbf·ft)

② **12 x 1.25 mm**
93 N·m
(9.5 kgf·m, 69 lbf·ft)

71051_CIVC_G0615

Fig. 81 Tighten the lower torque rod mount mounting bolts in the numbered sequence shown

46. Tighten the lower torque rod mount mounting bolts in the numbered sequence shown.

47. Install the air cleaner, as described in this section.

48. Install the upper torque rod using new bolts. Tighten the two shorter upper torque rod mounting bolts, then the longer single bolt to 47 ft. lbs. (64 Nm).

49. Install the crankshaft pulley, as described in this section.

50. Install the splash shield.

51. Install the VTC oil control solenoid valve with a new O-ring. Coat a new O-ring with engine oil. Clean and dry the mating surface of the valve. Install the ground cable.

52. Install the harness clamp.

53. Perform valve clearance adjustment, as described in this section.

54. Install the cylinder head cover, as described in this section.

➡ **Wait at least 30 minutes before filling the engine with oil. Do not run the engine for at least 3 hours after installing the head cover.**

55. Install the ignition coils, as described in "Engine Electrical" section.

56. Install the engine cover.

57. Install the under cowl panel, the hood rear seal, connect the windshield washer tube, install the center cowl cover, install the side cowl covers, set the wiper arms to the auto-stop position, install the wiper arms and the cowl top wiper covers.

58. Install the drive belt, as described in this section.

59. Refill the cooling system with proper coolant.

60. Install the right front wheel.

61. Connect the HDS to the data link connector (DLC) located under the driver's side of the dashboard.

62. Turn the ignition switch to ON (II).

63. Select CRANK PATTERN in the ADJUSTMENT MENU with the HDS.

64. Select CRANK PATTERN CLEAR, and clear the CKP pattern.

65. Select CRANK PATTERN LEARNING with the HDS, and follow the screen prompts.

66. After doing all procedures, select the individual maintenance item you wish to reset with the HDS.

CAMSHAFT (TIMING) CHAIN & SPROCKETS

REMOVAL & INSTALLATION

2011 Models

1.8L Engine

See Figures 82 through 86.

➡ **Keep the cam chain away from magnetic fields.**

1. Remove the front wheels.

2. Remove the splash shield.

3. Remove the drive belt auto-tensioner.

4. Remove the cylinder head cover.

5. Set the No. 1 piston at top dead center (TDC). The "UP" mark on the camshaft sprocket should be at the top, and the TDC grooves on the camshaft sprocket should line up with the top edge of the head.

6. Disconnect the positive crankcase ventilation (PCV) hose.

7. Remove the crankshaft pulley.

8. Support the engine with a jack and a wood block under the oil pan.

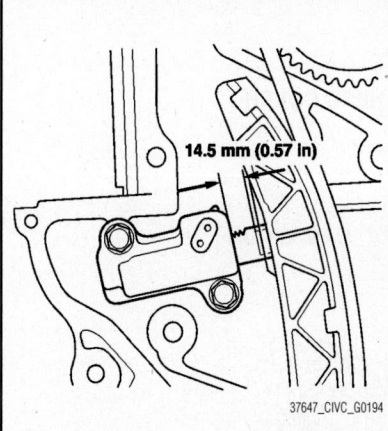

14.5 mm (0.57 in)

37647_CIVC_G0194

Fig. 82 Measure the tensioner rod length between the tensioner body and bottom of the flat surface section on the tensioner rod. If the length is more than the service limit, replace the cam chain

9. Remove the bolt securing the A/C line, then remove the upper torque rod.

10. Remove the ground cable, then remove the side engine mount bracket.

11. Remove the oil pump.

12. Measure the tensioner rod length between the tensioner body and bottom of the flat surface section on the tensioner rod. If the length is more than the service limit, replace the cam chain.

13. Tensioner rod length service limit is 0.57 in. (14.5 mm).

14. Loosely install the crankshaft pulley.

15. Turn the crankshaft counterclockwise to compress the auto-tensioner.

16. Align the holes on the lock and the

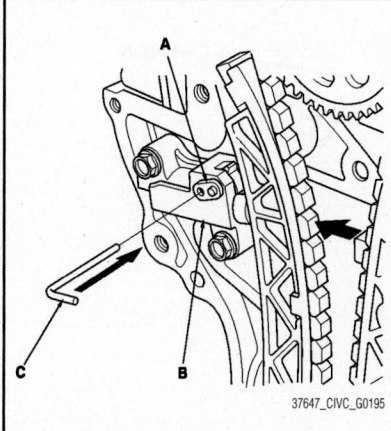

37647_CIVC_G0195

Fig. 83 Align the holes on the lock (A) and the auto-tensioner (B), then insert a 1.0 mm (0.04 in) diameter pin (C) into the holes. Turn the crankshaft clockwise to secure the pin

auto-tensioner, then insert a 0.04 in. (1.0 mm) diameter pin into the holes. Turn the crankshaft clockwise to secure the pin.

➡**Check the auto-tensioner cam position If the position is not aligned, set the first cam to the first edge of the rack.**

17. Remove the auto-tensioner.

18. Remove the crankshaft pulley.

19. Remove the cam chain guide and the cam chain tensioner arm.

20. Remove the cam chain.

To install:

➡**Keep the cam chain away from magnetic fields.**

21. Set the crankshaft to top dead center (TDC). Align the TDC mark on the crankshaft sprocket with the pointer on the engine block.

22. Set the camshaft to TDC. The "UP" mark on the camshaft sprocket should be at the top, and the TDC grooves on the camshaft sprocket should line up with the top edge of the head.

23. Install the cam chain on the crankshaft sprocket with the colored piece aligned with the mark on the crankshaft sprocket.

24. Install the cam chain on the camshaft sprocket with the colored link plate aligned with the mark on the camshaft sprocket.

25. Install the cam chain guide and the cam chain tensioner arm.

26. Compress the auto-tensioner when replacing the cam chain. Remove the pin from the auto-tensioner that was installed during removal. Turn the plate counterclockwise, to release the lock, then press the rod, and set the first cam to the first edge of the rack. Insert the 1.0 mm (0.04 in) diameter pin into the holes.

➡**If the chain tensioner is not set up as described, the tensioner will become damaged.**

27. Install the auto-tensioner.

28. Remove the pin or lock pin from the auto-tensioner.

29. Check the oil pump oil seal for damage If the oil seal is damaged, replace the oil seal..

30. Remove all of the old liquid gasket from the oil pump mating surfaces, the bolts, and the bolt holes.

31. Clean and dry the oil pump mating surfaces.

32. Apply liquid gasket to the engine block mating surface of the oil pump, and to the inside edge of the threaded bolt holes. Install the component within 5 minutes of applying the liquid gasket.

➡**If too much time has passed after applying the liquid gasket, remove the old liquid gasket and residue, then reapply new liquid gasket.**

33. Apply liquid gasket to the engine block upper surface contact areas on the oil pump and the lower block upper surface contact areas on the oil pump.

34. Apply liquid gasket to the oil pan mating surface of the oil pump, and to the inside edge of the threaded bolt holes. Install the component within 5 minutes of applying the liquid gasket. Apply a 3 mm (0.12 in) diameter bead of liquid gasket.

➡**If too much time has passed after applying the liquid gasket, remove the old liquid gasket and residue, then reapply new liquid gasket.**

35. Install new O-rings on the oil pump. Set the edge of the oil pump on the edge of the oil pan, then install the oil pump on the engine block. Loosely install the dowel bolts, then tighten the 8 mm bolts to 21 ft. lbs. (33 Nm), the 6 mm bolts and the dowel bolts. Wipe off the excess liquid gasket on the oil pan and oil pump mating surface.

➡**When installing the oil pump, do not slide the bottom surface onto the oil pan mounting surface.**

36. Wait at least 30 minutes to allow liquid gasket to cure before filling the engine with oil.

➡**Do not run the engine within 3 hours after installing the oil pump.**

37. Install the side engine mount bracket, then loosely tighten the new bolt and nut, then loosely tighten the bolt.

38. Install the ground cable.

39. Remove the air cleaner assembly.

40. Loosen the top transmission mounting bolt and nuts.

41. Raise the vehicle on the lift.

42. Loosen the lower torque rod mounting bolt.

43. Lower the vehicle on the lift.

44. Tighten the side engine mount bracket mounting bolts and nut to 52 ft. lbs. (72 Nm).

45. Tighten the transmission mounting bolt and nuts to 54 ft. lbs. (74 Nm).

46. Raise the vehicle on the lift.

47. Tighten the lower torque rod mounting bolt to 69 ft. lbs. (93 Nm).

48. Lower the vehicle on the lift.

49. Install the air cleaner assembly.

50. Install the upper torque rod, then tighten the new upper torque rod mounting bolts to 47 ft. lbs (64 Nm).

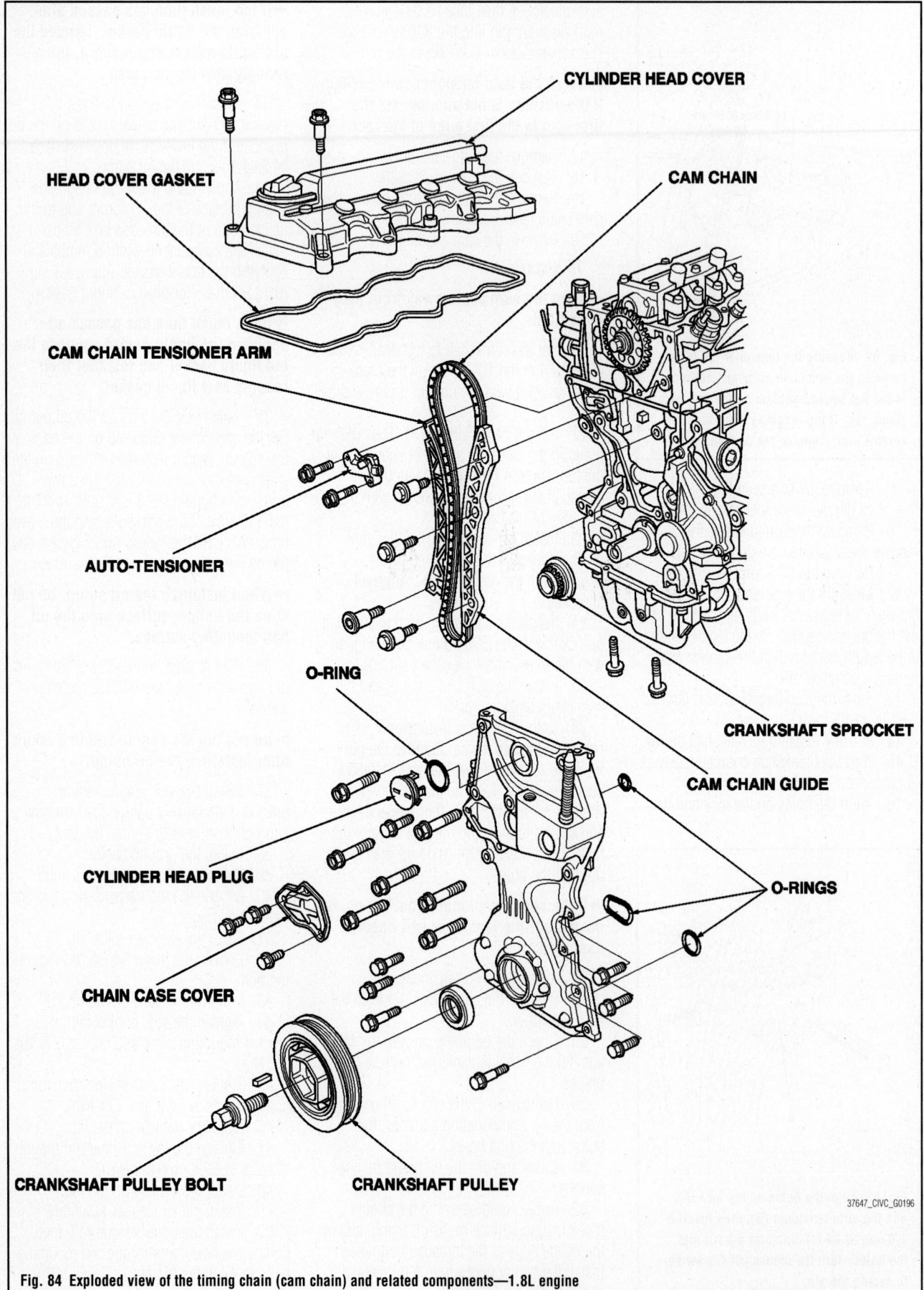

CYLINDER HEAD COVER

HEAD COVER GASKET

CAM CHAIN

CAM CHAIN TENSIONER ARM

AUTO-TENSIONER

O-RING

CRANKSHAFT SPROCKET

CAM CHAIN GUIDE

CYLINDER HEAD PLUG

O-RINGS

CHAIN CASE COVER

CRANKSHAFT PULLEY BOLT

CRANKSHAFT PULLEY

37647_CIVC_G0196

Fig. 84 Exploded view of the timing chain (cam chain) and related components—1.8L engine

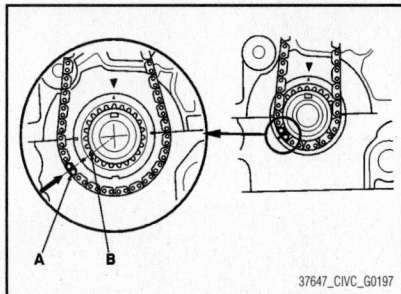

Fig. 85 Install the cam chain on the crankshaft sprocket with the colored piece (A) aligned with the mark (B) on the crankshaft sprocket

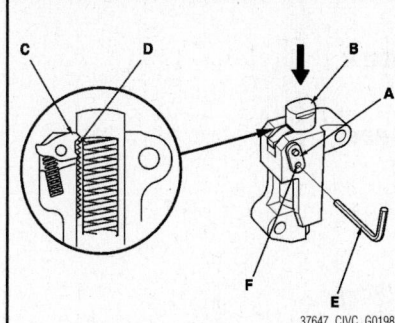

Fig. 86 Compress the auto-tensioner when replacing the cam chain. Remove the pin from the auto-tensioner that was installed during removal. Turn the plate counterclockwise, to release the lock, then press the rod, and set the first cam to the first edge of the rack. Insert the 1.0 mm (0.04 in) diameter pin into the holes

51. Install the bolt securing the A/C line.
52. Install the crankshaft pulley.
53. Connect the positive crankcase ventilation (PCV) hose.
54. Install the cylinder head cover.
55. Install the drive belt auto-tensioner.
56. Install the splash shield.
57. Install the front wheels.
58. Do the Crankshaft Position (CKP) Pattern Clear/CKP Pattern Learn Procedure. See "Crankshaft Position (CKP) Sensor" in "ENGINE PERFORMANCE & EMISSION CONTROLS" section.

2.0L Engine

See Figures 87 through 89.

➡**Keep the cam chain away from magnetic fields.**

1. Remove the front wheels.
2. Remove the splash shield.
3. Drain the engine coolant.

4. Remove the drive belt.
5. Remove the cylinder head cover.
6. Set the No. 1 piston at top dead center (TDC). The punch mark on the variable valve timing control (VTC) actuator and the punch mark on the exhaust camshaft sprocket should be at the top. Align the TDC marks on the VTC actuator and the exhaust camshaft sprocket.
7. Remove the oil cooler hose joint pipe from the water pump.
8. Disconnect the crankshaft position (CKP) sensor connector and the VTC oil control solenoid valve connector and remove the harness clamps from the front of the engine.
9. Remove the VTC oil control solenoid valve:
   a. Disconnect the VTC oil control solenoid valve 2P connector.
   b. Remove the ground wire.
   c. Remove the bolt and the VTC oil control solenoid valve.
10. Remove the crankshaft pulley.
11. Support the engine with a jack and a wood block under the oil pan.
12. Remove the upper torque rod.
13. Remove the ground cable, then remove the side engine mount bracket.
14. Remove the side engine mount bracket.
15. Remove the cam chain case, then remove the CKP pulse plate.
16. Loosely install the crankshaft pulley.
17. Turn the crankshaft counterclockwise to compress the auto-tensioner.
18. Align the holes on the lock and the auto-tensioner, then insert a 1.2 mm (0.05 in) diameter pin or lock pin into the holes. Turn the crankshaft clockwise to secure the pin.
19. Remove the auto-tensioner.
20. Remove the crankshaft pulley.
21. Remove cam chain guide from on top of the camshaft pulleys.
22. Remove cam chain guide and the tensioner arm.
23. Remove the cam chain.

To install:

➡**Keep the cam chain away from magnetic fields.**

24. Before doing this procedure, check that the variable valve timing control (VTC) actuator is locked by turning the VTC actuator counterclockwise. If not locked, turn the VTC actuator clockwise until it stops, then recheck it. If it is still not locked, replace the VTC actuator.
25. Set the crankshaft to top dead center (TDC). Align the TDC mark on the crank-

shaft sprocket with the pointer on the engine block.
26. Set the camshafts to TDC. The punch mark on the VTC actuator and the punch mark on the exhaust camshaft sprocket should be at the top. Align the TDC marks on the VTC actuator and the exhaust camshaft sprocket.
27. To hold the intake camshaft, insert a camshaft lock pin set (07AAB-RWCA120) into the maintenance hole in camshaft position (CMP) pulse plate A and through No. 5 rocker shaft holder.
28. To hold the exhaust camshaft, insert a camshaft lock pin set into the maintenance hole in CMP pulse plate B and through No. 5 rocker shaft holder.
29. Install the cam chain on the crankshaft sprocket with the colored link plate aligned with the mark on the crankshaft sprocket.
30. Install the cam chain on the VTC actuator and the exhaust camshaft sprocket with the punch marks aligned with the two colored link plates.
31. Install cam chain guide A and the tensioner arm.
32. Compress the auto-tensioner when replacing the cam chain. Remove the pin from the auto-tensioner that was installed during removal. Turn the plate counterclockwise, to release the lock, then press the rod, and set the first cam to the first edge of the rack. Insert the 1.2 mm (0.05 in) diameter pin into the holes.

➡**If the chain tensioner is not set up as described, the tensioner will be damaged.**

33. Install cam chain guide B on top of the cam pulleys.
34. Install the auto-tensioner.
35. Remove the pin or lock pin from the auto-tensioner.
36. Remove the camshaft lock pin set.
37. Install the crankshaft position (CKP) pulse plate.
38. Check the chain case oil seal for damage. If the oil seal is damaged, replace the chain case oil seal.
39. Remove all of the old liquid gasket from the chain case mating surfaces, the bolts, and the bolt holes.
40. Clean and dry the chain case mating surfaces.
41. Apply liquid gasket to the engine block mating surface of the chain case, and to the inside edge of the threaded bolt holes. Install the component within 5 minutes of applying the liquid gasket.
42. Apply a 3 mm (0.12 in) diameter bead of liquid gasket along these areas.

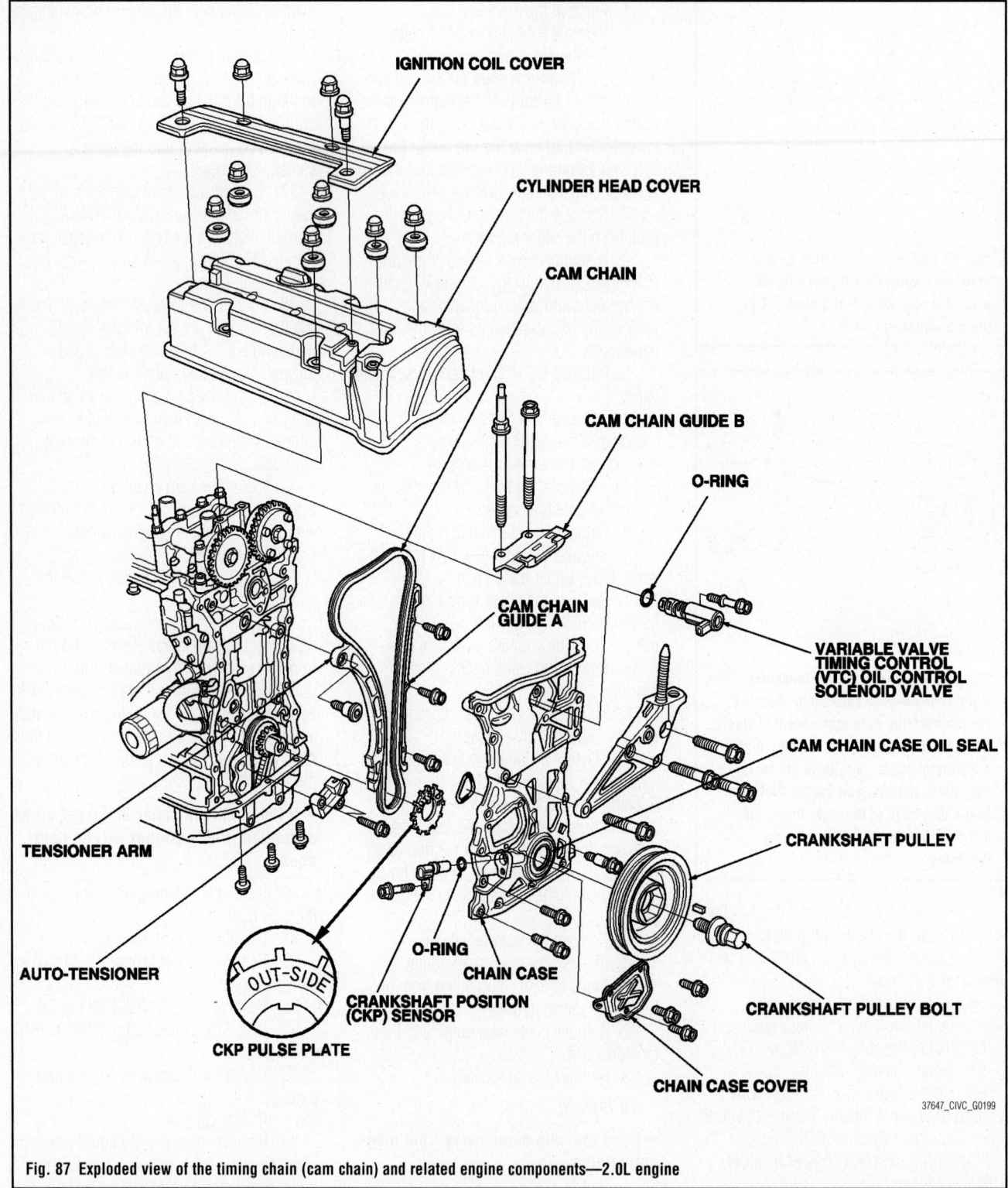

IGNITION COIL COVER

CYLINDER HEAD COVER

CAM CHAIN

CAM CHAIN GUIDE B

O-RING

CAM CHAIN GUIDE A

VARIABLE VALVE TIMING CONTROL (VTC) OIL CONTROL SOLENOID VALVE

CAM CHAIN CASE OIL SEAL

CRANKSHAFT PULLEY

TENSIONER ARM

CRANKSHAFT PULLEY BOLT

AUTO-TENSIONER

OUT-SIDE

O-RING

CHAIN CASE

CRANKSHAFT POSITION (CKP) SENSOR

CKP PULSE PLATE

CHAIN CASE COVER

37647_CIVC_G0199

Fig. 87 Exploded view of the timing chain (cam chain) and related engine components—2.0L engine

➡**If too much time has passed after applying the liquid gasket, remove the old liquid gasket and residue, then reapply new liquid gasket.**

43. Apply liquid gasket to the engine block upper surface contact areas and the lower block upper surface contact areas on the chain case.

44. Apply liquid gasket to the oil pan mating surface of the chain case, and to the inside edge of the threaded bolt holes. Install the component within 5 minutes of applying the liquid gasket.

45. Apply a 3 mm (0.12 in) diameter bead of liquid gasket along this area.

➡**If too much time has passed after applying the liquid gasket, remove the old liquid gasket and residue, then reapply new liquid gasket.**

46. Install a new O-ring on the chain case. Set the edge of the chain case on the edge of the oil pan, then install the chain case on the engine block. Wipe off the

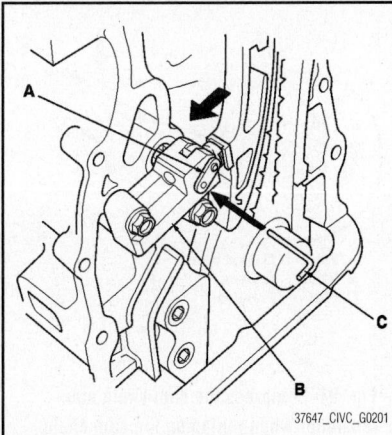

Fig. 88 Align the holes on the lock (A) and the auto-tensioner (B), then insert a 1.2 mm (0.05 in) diameter pin or lock pin (C) into the holes. Turn the crankshaft clockwise to secure the pin—2.0L engine

excess liquid gasket on the oil pan and chain case mating surface.

➡**When installing the chain case, do not slide the bottom surface onto the oil pan mounting surface. Wait at least 30 minutes before filling the engine with oil. Do not run the engine within 3 hours after installing the chain case.**

47. Install the side engine mount bracket. Tighten the bolts to 33 ft. lbs. (44 Nm).

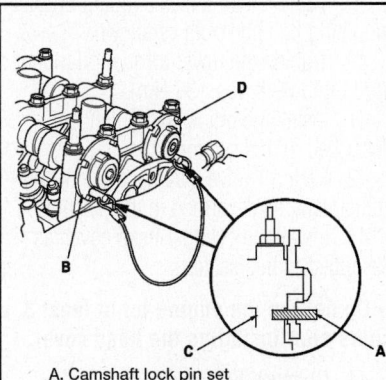

A. Camshaft lock pin set
B. Camshaft position pulse plate A
C. No. 5 rocker shaft holder
D. Camshaft position pulse plate B

Fig. 89 To hold the intake camshaft, insert a camshaft lock pin set (07AAB-RWCA120) into the maintenance hole in camshaft position (CMP) pulse plate A and through No. 5 rocker shaft holder. To hold the exhaust camshaft, insert a camshaft lock pin set into the maintenance hole in CMP pulse plate B and through No. 5 rocker shaft holder

48. Install the side engine mount bracket, then loosely tighten the new bolt and nut, and loosely tighten the bolt.

49. Install the ground cable.

50. Remove the air cleaner assembly.

51. Loosen the transmission mounting bolt and nuts.

52. Raise the vehicle on the lift.

53. Loosen the lower torque rod mounting bolt.

54. Loosen the front mount mounting bolt.

55. Lower the vehicle on the lift.

56. Tighten the side engine mount mounting bolts and nut to 52 ft. lbs. (72 Nm).

57. Tighten the transmission mounting bolt and nuts to 54 ft. lbs. (74 Nm).

58. Raise the vehicle on the lift.

59. Tighten the lower torque rod mounting bolt to 69 ft. lbs. (93 Nm).

60. Lower the vehicle on the lift.

61. Install the air cleaner assembly.

62. Install the upper torque rod, then tighten the new upper torque rod mounting bolts to 47 ft. lbs. (64 Nm).

63. Raise the vehicle on the lift.

64. Tighten the front mount mounting bolt to 47 ft. lbs. (64 Nm).

65. Install the crankshaft pulley.

66. Install the splash shield.

67. Lower the vehicle on the lift.

68. Install the variable valve timing control (VTC) oil control solenoid valve.

69. Install the harness clamps and connect the CKP sensor connector and the VTC oil control solenoid valve connector.

70. Install the oil cooler hose joint pipe with a new O-ring.

71. Install the cylinder head cover.

72. Install the drive belt.

73. Refill the radiator with engine coolant, and bleed the air from the cooling system.

74. Do the Crankshaft Position (CKP) Pattern Clear/CKP Pattern Learn Procedure. See "Crankshaft Position (CKP) Sensor" in "ENGINE PERFORMANCE & EMISSION CONTROLS" section.

2012 Models

1.8L Non-Hybrid Engine

See Figures 90 through 97.

1. Raise and support vehicle on the lift.

2. Remove the right front wheel.

3. Disconnect and isolate the negative battery cable, battery sensor, and then the positive cable.

4. Remove the battery. See "Engine Electrical" section.

5. Loosen the water pump pulley bolts.

6. Remove the drive belt.

7. Remove the alternator. See "Engine Electrical" section.

8. Remove the water pump pulley.

9. Remove the drive belt auto-tensioner assembly.

10. Remove the engine cover.

11. Remove the cylinder head cover peripheral assembly, then remove the cylinder head cover, as described in this section.

12. Check the No. 1 piston at top dead center (TDC). The "UP" mark on the camshaft sprocket should be at the top, and the TDC grooves (B) on the camshaft sprocket should line up with the top edge of the head.

➡**If the marks are not aligned, rotate the crankshaft 360 degrees, and recheck the camshaft pulley mark.**

13. Remove the crankshaft pulley, as described in this section.

14. Lift and support the engine with a jack and a wood block under the oil pan.

15. Remove the ground cable, the side engine mount bracket, and the side engine mount.

16. Disconnect the PCV hose.

17. Remove the oil pump as described in "Engine Lubrication" section.

18. Measure the tensioner rod length between the tensioner body and bottom of the flat surface section on the tensioner rod. If the length is more than the service limit of 0.571 in. (14.5 mm), replace the cam chain.

19. Loosely install the crankshaft pulley.

20. Turn the crankshaft counterclockwise to compress the auto-tensioner.

21. Rotate the crankshaft counterclockwise to align the holes on the lock and the auto-tensioner. Insert a 1/32 in. (1.0 mm) diameter pin into the holes.

22. Turn the crankshaft clockwise to secure the pin.

➡**If the holes in the lock and the auto-tensioner do not align, continue to rotating the crankshaft counterclockwise until the holes align, then install the pin.**

23. Remove the cam chain auto-tensioner.

24. Remove the crankshaft pulley.

25. Remove the cam chain guide and the cam chain tensioner arm. Remove the cam chain.

To install:

➡**Keep the cam chain away from magnetic fields.**

26. Set the crankshaft to top dead center (TDC). Align the TDC mark on the

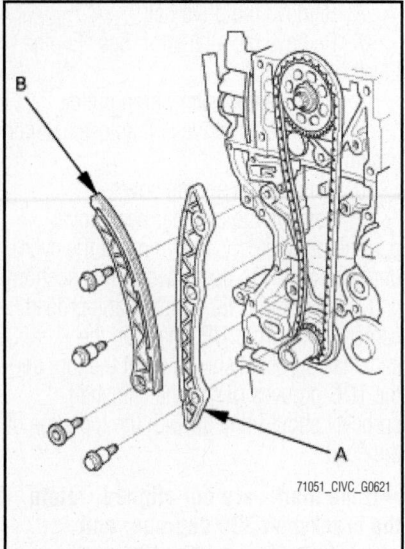

Fig. 90 Remove the cam chain guide (A) and the cam chain tensioner arm (B). Remove the cam chain

crankshaft sprocket with the pointer on the engine block.

27. Install the cam chain on the crankshaft sprocket with the colored piece aligned with the mark on the crankshaft sprocket.

28. Install the cam chain on the camshaft sprocket with the colored link plate aligned with the mark on the camshaft sprocket.

29. Install the cam chain guide and the cam chain tensioner arm.

30. Compress the cam chain auto-tensioner when replacing the cam chain. Remove the pin from the auto-tensioner that was installed during removal. Turn the plate counterclockwise, to release the lock, then press the rod, and set the first cam to the first edge of the rack. Insert the 1/32 in. (1.0 mm) diameter pin back into the holes.

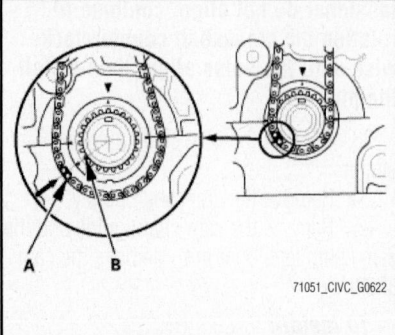

Fig. 91 Install the cam chain on the crankshaft sprocket with the colored piece (A) aligned with the mark (B) on the crankshaft sprocket

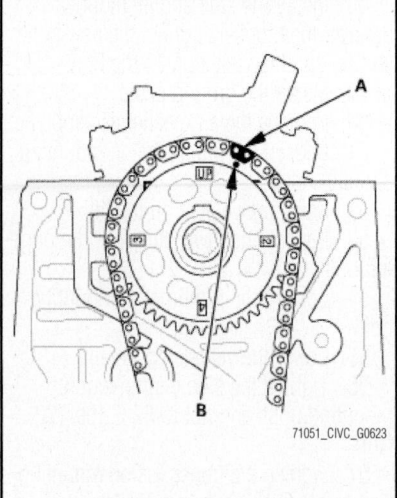

Fig. 92 Install the cam chain on the camshaft sprocket with the colored link plate (A) aligned with the mark (B) on the camshaft sprocket

✳✳ CAUTION

If the chain tensioner is not set up as described, the tensioner will be damaged.

31. Install the cam chain auto-tensioner.

32. Remove the pin from the cam chain auto-tensioner.

33. Install the oil pump. See "Engine Lubrication" section.

34. Install the new side engine mount mounting bolts and the side engine mount bracket using a nut and a new bolt, then

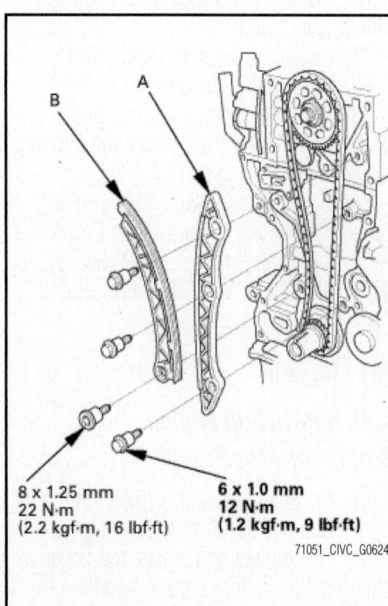

8 x 1.25 mm
22 N·m
(2.2 kgf·m, 16 lbf·ft)

6 x 1.0 mm
12 N·m
(1.2 kgf·m, 9 lbf·ft)

Fig. 93 Install the cam chain guide (A) and the cam chain tensioner arm (B)

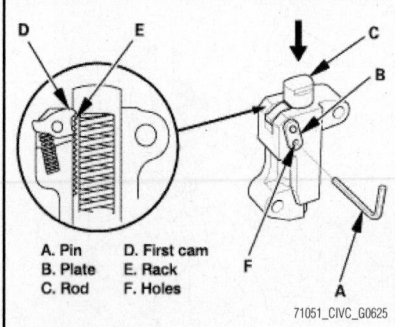

A. Pin   D. First cam
B. Plate   E. Rack
C. Rod   F. Holes

Fig. 94 Compress the cam chain auto-tensioner when replacing the cam chain. Remove the pin from the auto-tensioner that was installed during removal. Turn the plate counterclockwise, to release the lock, then press the rod, and set the first cam to the first edge of the rack. Insert the 1/32 in. (1.0 mm) diameter pin back into the holes

loosely install the new side engine mount bracket mounting bolts and nut. Install the ground cable.

35. Remove the jack and the wood block.

36. Loosen the transaxle mount bracket mounting bolt and nuts.

37. Loosen the lower torque rod mounting bolt.

38. Tighten the side engine mount bracket mounting bolt and nut to 69 ft. lbs. (93 Nm).

39. Tighten the transaxle mount bracket mounting bolt and nuts as shown.

40. Tighten the lower torque rod mounting bolt to 69 ft. lbs. (93 Nm).

41. Install the crankshaft pulley, as described in this section.

42. Perform valve clearance check and adjustment, as described in this section.

43. Install the cylinder head cover, as described in this section.

➡Do not run the engine for at least 3 hours after installing the head cover.

44. Connect the breather hose to the front of the cylinder head cover.

45. Install the dipstick.

46. Install the harness holders and connect the connectors on the injectors.

47. Install the engine cover.

48. Install the drive belt auto-tensioner.

49. Install the water pump pulley, as described in this section.

50. Install the alternator, as described in "Engine Electrical" section.

51. Move the auto-tensioner counterclockwise and install the drive belt.

52. Tighten the water pump pulley mounting bolts to 10 ft. lbs. (14 Nm).

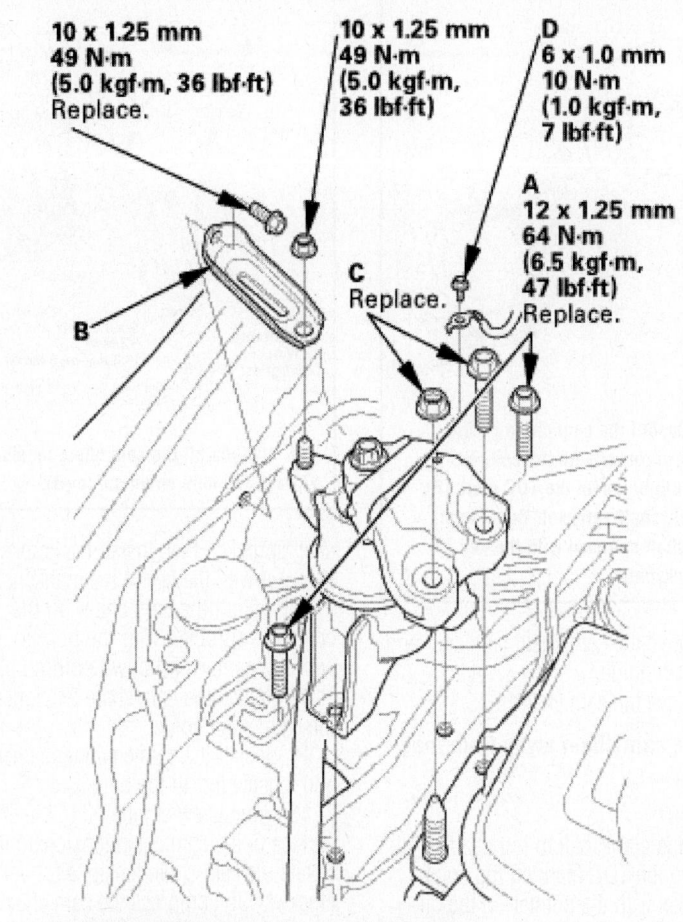

10 x 1.25 mm
49 N·m
(5.0 kgf·m, 36 lbf·ft)
Replace.

10 x 1.25 mm
49 N·m
(5.0 kgf·m,
36 lbf·ft)

D
6 x 1.0 mm
10 N·m
(1.0 kgf·m,
7 lbf·ft)

A
12 x 1.25 mm
64 N·m
(6.5 kgf·m,
47 lbf·ft)
Replace.

C
Replace.

B

A. Side engine mount mounting bolts
B. Side engine mount bracket
C. Side engine mount bracket mounting bolts and nut
D. Ground cable

71051_CIVIC_G0591

Fig. 95 Install the new side engine mount mounting bolts and the side engine mount bracket using a nut and a new bolt, then loosely install the new side engine mount bracket mounting bolts and nut. Install the ground cable

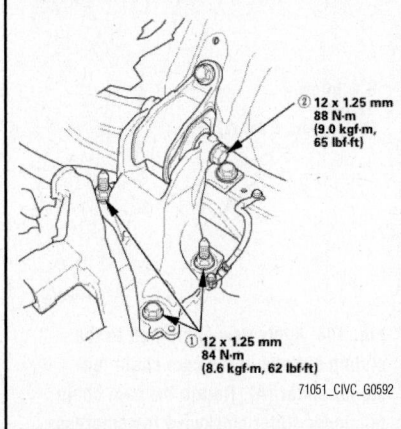

② 12 x 1.25 mm
88 N·m
(9.0 kgf·m,
65 lbf·ft)

① 12 x 1.25 mm
84 N·m
(8.6 kgf·m, 62 lbf·ft)

71051_CIVIC_G0592

Fig. 96 Tighten the transaxle mount bracket mounting bolt and nuts as shown—M/T

53. Install the splash shield.

54. Install the right front wheel. Tighten the nuts to 80 ft. lbs. (108 Nm).

55. Install the battery and cables, as described in "Engine Electrical" section.

56. Connect the HDS to the data link connector (DLC) located under the driver's side of the dashboard.

57. Turn the ignition switch to ON (II).

58. Perform the CKP Pattern Clear/CKP Learn procedure:

a. Select CRANK PATTERN in the ADJUSTMENT MENU with the HDS.

b. Select CRANK PATTERN CLEAR, and clear the CKP pattern.

c. Select CRANK PATTERN LEARN-ING with the HDS, and follow the screen prompts.

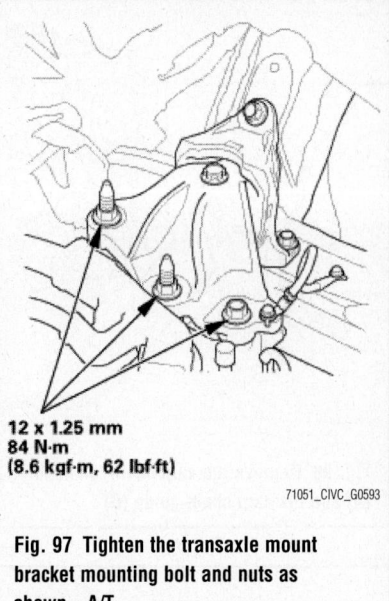

12 x 1.25 mm
84 N·m
(8.6 kgf·m, 62 lbf·ft)

71051_CIVIC_G0593

Fig. 97 Tighten the transaxle mount bracket mounting bolt and nuts as shown—A/T

1.8L Hybrid Engine

See Figures 98 through 106.

1. Raise the vehicle on a lift and make sure it is securely supported.

2. Remove or disconnect the following (refer to applicable component sections, as needed):

- Engine cover
- Ignition coils
- Cylinder head cover
- Right front wheel
- Splash shield
- Engine under cover
- Water pump pulley (loosen bolts only)
- Drive belt
- Water pump pulley
- Crankshaft pulley
- Support engine with jack and wood block under the oil pan
- Side engine mount

3. Set the No. 1 piston at top dead center (TDC). The "UP" mark on the camshaft sprocket should be at the top, and the TDC grooves (B) on the camshaft sprocket should line up with the top edge of the head.

4. Remove the cam chain case.

5. Measure the cam chain separation. If the distance is less than the service limit of 0.59 in. (15 mm), replace the cam chain and the cam chain tensioner.

6. Apply new engine oil to the sliding surface of the cam chain tensioner slider. Hold the cam chain tensioner slider with a screwdriver, then remove the upper bolt and loosen the lower bolt.

7. Remove the cam chain tensioner slider.

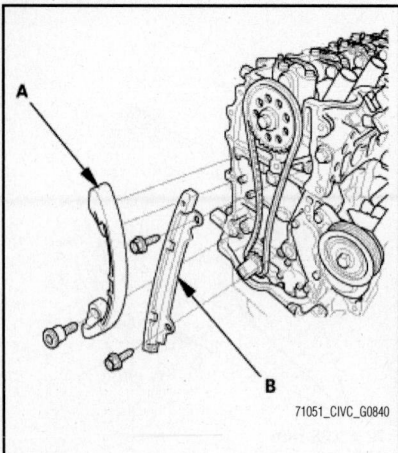

Fig. 98 Remove the cam chain tensioner (A) and the cam chain guide (B)

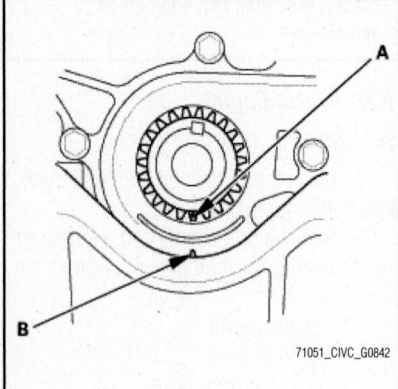

Fig. 99 Set the crankshaft to top dead center (TDC). Align the TDC mark (A) on the crankshaft sprocket with the pointer (B) on the oil pump.

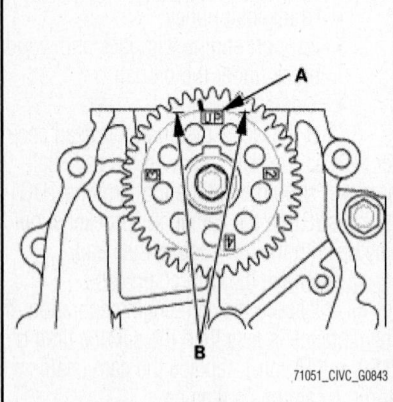

Fig. 100 Set the camshaft TDC. The "UP" mark (A) on the camshaft sprocket should be at the top, and the TDC grooves (B) on the camshaft sprocket should line up with the top edge of the head

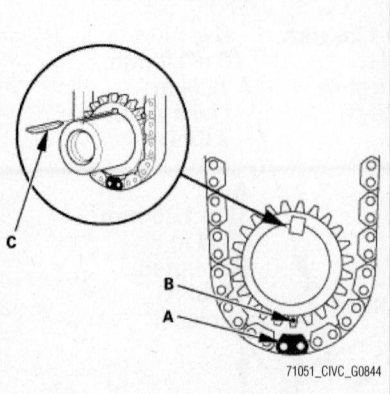

Fig. 101 Install the cam chain on the crankshaft sprocket with the colored piece (A) aligned with the TDC mark (B) on the crankshaft sprocket, then install the crankshaft sprocket with the key (C) to the crankshaft

8. Remove the cam chain tensioner and the cam chain guide.

9. Remove the cam chain.

➡ **Keep the cam chain away from magnetic fields.**

To install:

10. Set the crankshaft to top dead center (TDC). Align the TDC mark on the crankshaft sprocket with the pointer on the oil pump.

11. Set the camshaft TDC. The "UP" mark on the camshaft sprocket should be at the top, and the TDC grooves on the camshaft sprocket should line up with the top edge of the head.

12. Install the cam chain on the crankshaft sprocket with the colored piece aligned with the TDC mark on the crank-

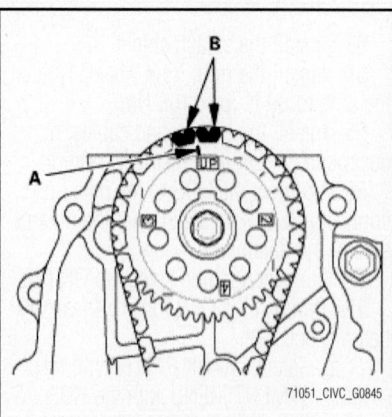

Fig. 102 Install the cam chain on the camshaft sprocket with the pointer (A) aligned with the center of the two colored pieces (B)

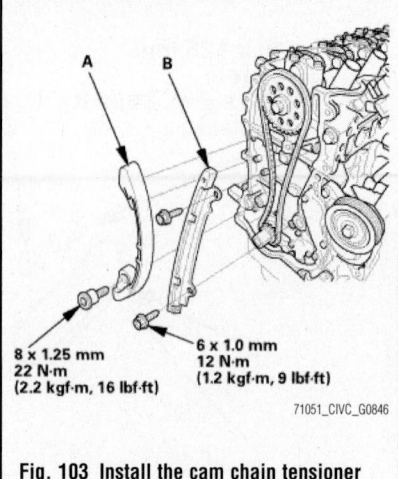

Fig. 103 Install the cam chain tensioner (A) and the cam chain guide (B)

shaft sprocket, then install the crankshaft sprocket with the key to the crankshaft.

13. Install the cam chain on the camshaft sprocket with the pointer aligned with the center of the two colored pieces.

14. Install the cam chain tensioner and the cam chain guide.

15. Install the cam chain tensioner slider, and loosely install the bolt.

16. Apply new engine oil to the sliding surface of the cam chain tensioner slider. Rotate the cam chain tensioner slider clockwise to compress the cam chain tensioner, and install the remaining bolt, then tighten the bolts to the specified torque.

17. Check the cam chain case oil seal for damage. If the oil seal is damaged, replace the cam chain case oil seal.

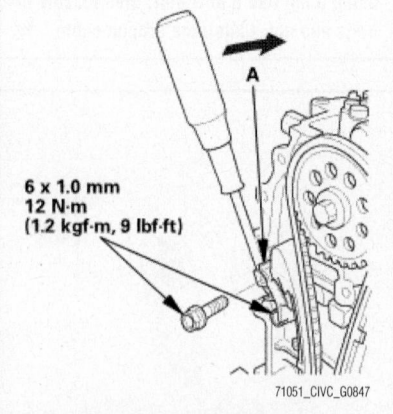

Fig. 104 Apply new engine oil to the sliding surface of the cam chain tensioner slider (A). Rotate the cam chain tensioner slider clockwise to compress the cam chain tensioner, and install the remaining bolt, then tighten the bolts to the specified torque

18. Remove all of the old liquid gasket from the cam chain case mating surfaces, the bolts, and the bolt holes.

19. Clean and dry the cam chain case mating surfaces.

20. Apply liquid gasket to the cylinder head and the engine block mating surface of the chain case and to the inside edge of the bolt holes. Install the component within 5 minutes of applying the liquid gasket.

21. Apply a 0.098 in. (2.5 mm) diameter bead of liquid gasket along the broken line. Apply a 0.118 in. (3.0 mm) diameter bead of liquid gasket to the upper surface contact areas of the engine block.

➡️**If too much time has passed after applying the liquid gasket, remove the old liquid gasket and residue, then reapply new liquid gasket.**

22. Apply liquid gasket to the oil pan mating surface of the chain case and to the inside edge of the bolt holes. Install the component within 5 minutes of applying the liquid gasket.

23. Apply a 0.098 in. (2.5 mm) diameter bead of liquid gasket along the broken line. Apply a 0.20 in. (5.0 mm) diameter bead of liquid gasket to the shaded area.

➡️**If too much time has passed after applying the liquid gasket, remove the old liquid gasket and residue, then reapply new liquid gasket.**

24. Remove or disconnect the following (refer to applicable component sections, as needed):

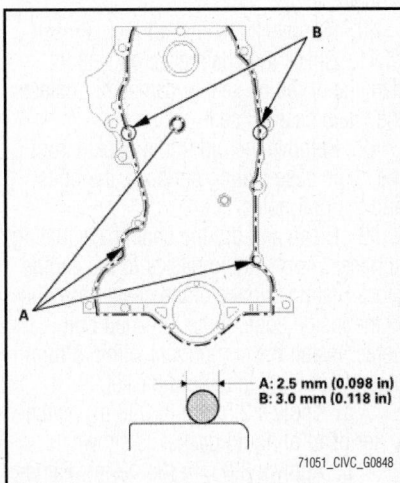

Fig. 105 Apply a 0.098 in. (2.5 mm) diameter bead of liquid gasket along the broken line (A). Apply a 0.118 in. (3.0 mm) diameter bead of liquid gasket to the upper surface contact areas of the engine block (B)

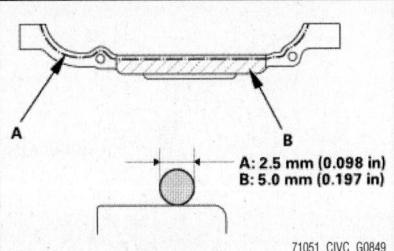

A: 2.5 mm (0.098 in)
B: 5.0 mm (0.197 in)

71051_CIVIC_G0849

Fig. 106 Apply a 0.098 in. (2.5 mm) diameter bead of liquid gasket along the broken line (A). Apply a 0.20 in. (5.0 mm) diameter bead of liquid gasket to the shaded area (B)

• Cylinder head cover
• Ignition coils
• Engine cover
• Transaxle mount bolts (loosen only)
• Lower torque rod
• Side engine mount
• Crankshaft pulley
• Water pump pulley (with loose bolts)
• Drive belt
• Water pump pulley (tighten bolts)
• Engine under cover
• Splash shield
• Right front wheel

25. Connect the HDS to the data link connector (DLC) located under the driver's side of the dashboard.

26. Turn the ignition switch to ON (II).

27. Make sure the HDS communicates with vehicle. If it does not communicate, go to the DLC circuit troubleshooting.

28. Select CRANK PATTERN in the ADJUSTMENT MENU with the HDS.

29. Select CRANK PATTERN CLEAR, and clear the CKP pattern.

30. Select CRANK PATTERN LEARNING with the HDS, and follow the screen prompts.

2.4L Engine

See Figures 107 through 116.

1. Remove the air cleaner, as described in this section.

2. Raise the vehicle on a lift, and make sure it is securely supported.

3. Remove the right front wheel.

4. Remove the splash shield.

5. Wait until the engine is cool, then carefully remove the radiator cap.

6. Loosen the drain plug and drain the coolant. See "Engine Cooling" section.

7. Remove the drive belt, as described in this section.

8. Set the wiper arms to the auto-stop position. Remove the cowl top wiper covers. Remove the wiper arms.

9. Remove the side cowl cover, the center cowl cover, the windshield washer tube, and, if necessary, remove the hood rear seal.

10. Remove the under cowl panel.

11. Remove the engine cover.

12. Remove the ignition coil cover. Remove the ignition coils.

13. Remove the cylinder head cover.

14. Turn the crankshaft so its white mark lines up with the pointer.

➡️**The other pointer is not used.**

15. Set the No. 1 piston at top dead center (TDC). The punch mark on the VTC actuator and the punch mark on the exhaust camshaft sprocket should be at the top. Align the TDC marks on the VTC actuator and the exhaust camshaft sprocket.

16. Remove the harness clamp.

17. Remove the ground cable. Remove the VTC oil control solenoid valve.

18. Hold the pulley with the holder handle and the crankshaft pulley holder. Remove the bolt with a socket, 19 mm and a breaker bar, then remove the crankshaft pulley.

19. Lift and support the engine with a jack and a wood block under the oil pan.

20. Remove the upper torque rod.

21. Remove the ground cable, the side engine mount bracket, and the side engine mount. Remove the side engine mount bracket.

22. Remove the cam chain case (A) and the spacer (B).

23. Loosely install the crankshaft pulley and turn the crankshaft counterclockwise to compress the auto-tensioner.

24. Rotate the crankshaft counterclockwise to align the holes on the lock and the auto-tensioner, then insert a 3/64 in. (1.2 mm) diameter pin into the holes. Turn the crankshaft clockwise to secure the pin.

➡️**If the holes in the lock and the auto-tensioner do not align, continue to rotating the crankshaft counterclockwise until the holes align, then install the pin.**

25. Remove the auto-tensioner.

26. Remove cam chain guide B.

27. Remove cam chain guide A and the tensioner arm.

28. Remove the cam chain.

To install:

➡️**Keep the cam chain away from magnetic fields.**

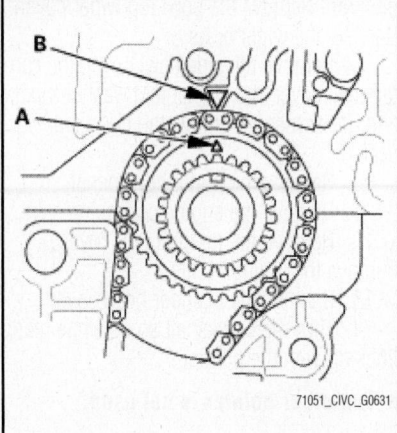

Fig. 107 Set the crankshaft to top dead center (TDC). Align the TDC mark (A) on the crankshaft sprocket with the pointer (B) on the engine block

29. Before doing this procedure, check that the VTC actuator is locked by turning the VTC actuator counterclockwise. If not locked, turn the VTC actuator clockwise until it stops, then recheck it. If it is still not locked, replace the VTC actuator.

30. Set the crankshaft to top dead center (TDC). Align the TDC mark on the crankshaft sprocket with the pointer on the engine block.

31. Set the camshafts to TDC. The punch mark on the VTC actuator and the punch mark on the exhaust camshaft sprocket should be at the top. Align the TDC marks on the VTC actuator and the exhaust camshaft sprocket.

32. To hold the intake camshaft, insert a camshaft lock pin set into the maintenance hole in CMP pulse plate A and through the

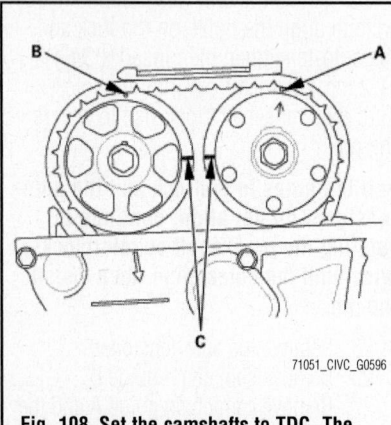

Fig. 108 Set the camshafts to TDC. The punch mark (A) on the VTC actuator and the punch mark (B) on the exhaust camshaft sprocket should be at the top. Align the TDC marks (C) on the VTC actuator and the exhaust camshaft sprocket

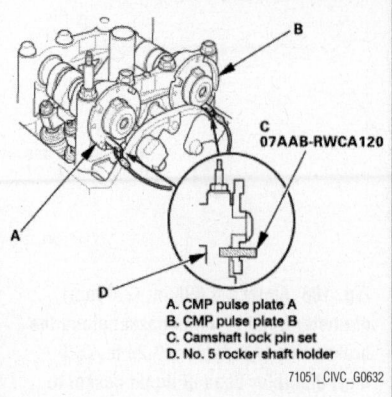

C
07AAB-RWCA120

A. CMP pulse plate A
B. CMP pulse plate B
C. Camshaft lock pin set
D. No. 5 rocker shaft holder

Fig. 109 To hold the intake camshaft, insert a camshaft lock pin set into the maintenance hole in CMP pulse plate A and through the No. 5 rocker shaft holder. Or, to hold the exhaust camshaft, insert a camshaft lock pin set into the maintenance hole in CMP pulse plate B and through the No. 5 rocker shaft holder

No. 5 rocker shaft holder. Or, to hold the exhaust camshaft, insert a camshaft lock pin set into the maintenance hole in CMP pulse plate B and through the No. 5 rocker shaft holder.

33. Install the cam chain on the crankshaft sprocket with the colored link plate aligned with the mark on the crankshaft sprocket.

34. Install the cam chain on the VTC actuator and the exhaust camshaft sprocket with the punch marks aligned with the center of the two colored link plates.

35. Install cam chain guide A and the tensioner arm.

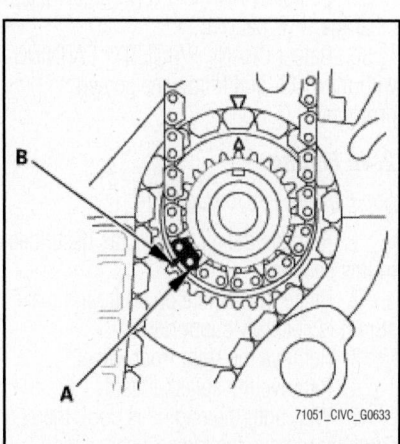

Fig. 110 Install the cam chain on the crankshaft sprocket with the colored link plate (A) aligned with the mark (B) on the crankshaft sprocket

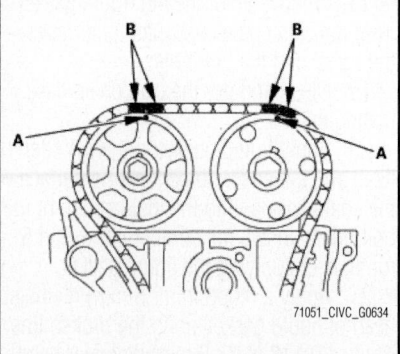

Fig. 111 Install the cam chain on the VTC actuator and the exhaust camshaft sprocket with the punch marks (A) aligned with the center of the two colored link plates (B)

36. Install cam chain guide B.

37. Compress the auto-tensioner when replacing the cam chain. Remove the pin from the auto-tensioner that was installed during removal. Turn the plate counterclockwise, to release the lock, then press the rod, and set the first cam to the first edge of the rack. Insert the 3/64 in. (1.2 mm) diameter pin back into the holes.

✷✷ CAUTION

If the chain tensioner is not set up as described, the tensioner will be damaged.

38. Install the auto-tensioner. Tighten the retaining bolts to 9 ft. lbs. (12 Nm).

39. Remove the pin from the auto-tensioner.

40. Remove the camshaft lock pin set.

41. Check the chain case oil seal for damage. If the oil seal is damaged, replace the chain case oil seal.

42. Remove the old liquid gasket from the chain case mating surfaces, the bolts, and the bolt holes.

43. Clean and dry the chain case mating surfaces. Apply liquid gasket to the engine block mating surface of the chain case and to the inside edge of the threaded bolt holes. Install the component within 5 minutes of applying the liquid gasket.

   a. Apply a 2.5 mm (0.098 in) diameter bead of liquid gasket as shown.

   b. Apply a 3.0 mm (0.118 in) diameter bead of liquid gasket as shown.

➡**If too much time has passed after applying the liquid gasket, remove the old liquid gasket and residue, then reapply new liquid gasket.**

44. Apply liquid gasket to the oil pan mating surface of the chain case,

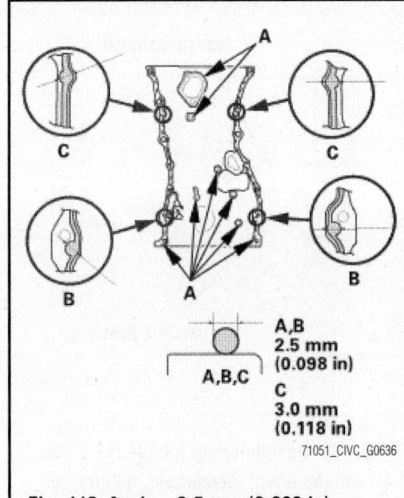

Fig. 112 Apply a 2.5 mm (0.098 in) diameter bead of liquid gasket along the broken line (A) and (B). Apply a 3.0 mm (0.118 in) diameter bead of liquid gasket along the broken line (C)

and to the inside edge of the threaded bolt holes. Install the component within 5 minutes of applying the liquid gasket.

   a. Apply a 2.5 mm (0.098 in) diameter bead of liquid gasket along the broken line indicated.

   b. Apply a 3.0 mm (0.118 in) diameter bead of liquid gasket along the broken line indicated.

➡ **If too much time has passed after applying the liquid gasket, remove the old liquid gasket and residue, then reapply new liquid gasket.**

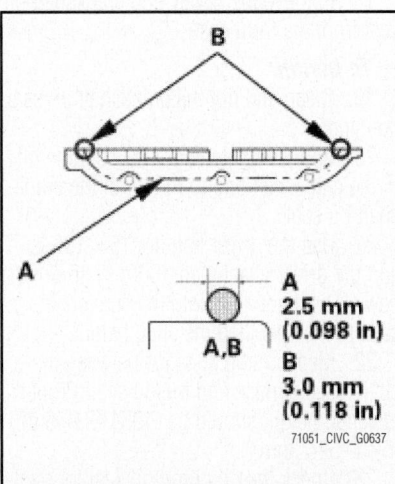

Fig. 113 Apply a 2.5 mm (0.098 in) diameter bead of liquid gasket along the broken line (A). Apply a 3.0 mm (0.118 in) diameter bead of liquid gasket along the broken line (B)

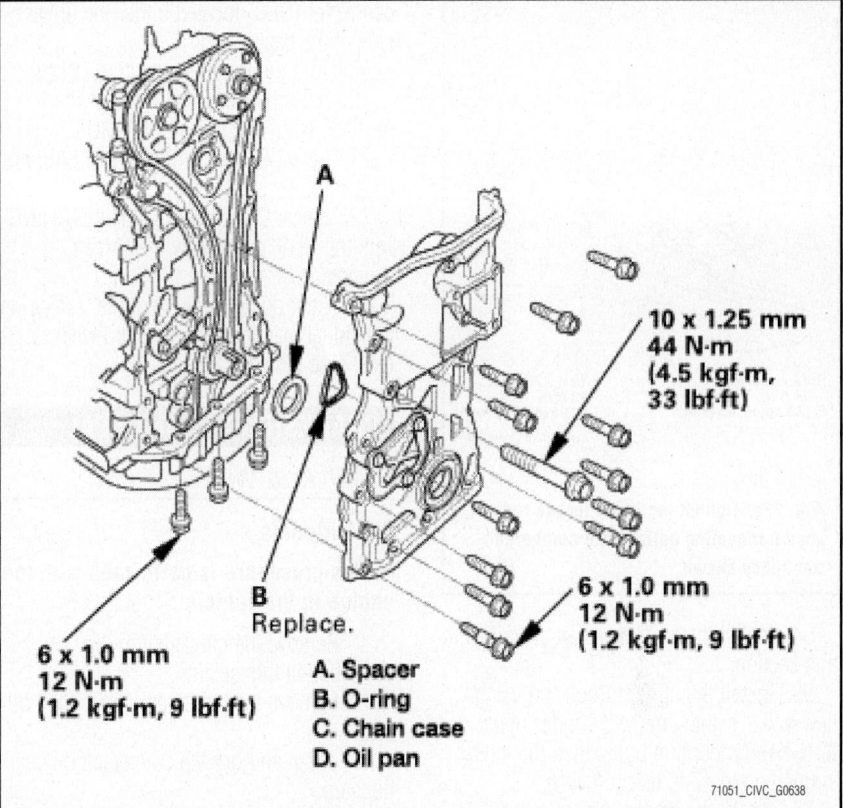

Fig. 114 Install the spacer, then install a new O-ring on the chain case. Set the edge of the chain case to the edge of the oil pan

   45. Install the spacer, then install a new O-ring on the chain case. Set the edge of the chain case to the edge of the oil pan.

   46. Then install the chain case on the engine block. Wipe off the excess liquid gasket on the oil pan and the chain case mating area.

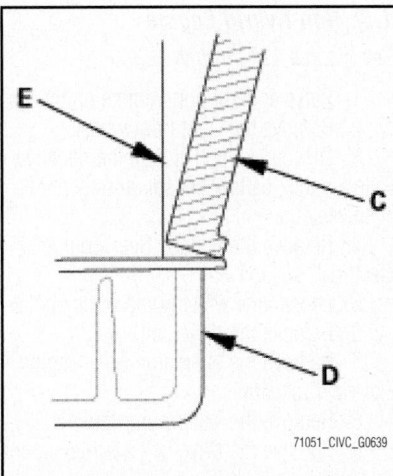

Fig. 115 Then install the chain case on the engine block (E), above the oil pan (D). Wipe off the excess liquid gasket on the oil pan and the chain case mating area

➡ **When installing the chain case, do not slide the bottom surface onto the oil pan mounting surface. Wait at least 30 minutes before filling the engine with oil. Do not run the engine for at least 3 hours after installing the chain case.**

   47. Install the side engine mount bracket, then tighten the side engine mount bracket mounting bolts to 33 ft. lbs. (44 Nm).

   48. Install the new side engine mount mounting bolts and the side engine mount bracket using a nut and a new bolt, then loosely install the new side engine mount bracket mounting bolts and nut. Install the ground cable.

   49. Loosen the transaxle mount bracket bolt and nuts.

   50. Loosen the lower torque rod mounting bolt.

   51. Tighten the side engine mount bracket mounting bolt and nut to 69 ft. lbs. (93 Nm).

   52. Tighten the transaxle mount bracket mounting bolt and nuts to 61 ft. lbs. (83 Nm).

   53. Tighten the lower torque rod mount mounting bolts in the numbered sequence shown.

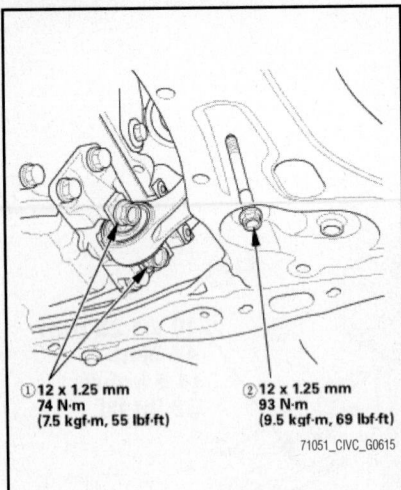

① 12 x 1.25 mm
74 N·m
(7.5 kgf·m, 55 lbf·ft)

② 12 x 1.25 mm
93 N·m
(9.5 kgf·m, 69 lbf·ft)

71051_CIVC_G0615

Fig. 116 Tighten the lower torque rod mount mounting bolts in the numbered sequence shown

54. Install the air cleaner, as described in this section.

55. Install the upper torque rod using new bolts. Tighten the two shorter upper torque rod mounting bolts, then the longer single bolt to 47 ft. lbs. (64 Nm).

56. Install the crankshaft pulley, as described in this section.

57. Install the splash shield.

58. Install the VTC oil control solenoid valve with a new O-ring. Coat a new O-ring with engine oil. Clean and dry the mating surface of the valve. Install the ground cable.

59. Install the harness clamp.

60. Perform valve clearance adjustment, as described in this section.

61. Install the cylinder head cover, as described in this section.

➡**Wait at least 30 minutes before filling the engine with oil. Do not run the engine for at least 3 hours after installing the head cover.**

62. Install the ignition coils, as described in "Engine Electrical" section.

63. Install the engine cover.

64. Install the under cowl panel, the hood rear seal, connect the windshield washer tube, install the center cowl cover, install the side cowl covers, set the wiper arms to the auto-stop position, install the wiper arms and the cowl top wiper covers.

65. Install the drive belt, as described in this section.

66. Refill the cooling system with proper coolant.

67. Install the right front wheel.

68. Connect the HDS to the data link

connector (DLC) located under the driver's side of the dashboard.

69. Turn the ignition switch to ON (II).

70. Select CRANK PATTERN in the ADJUSTMENT MENU with the HDS.

71. Select CRANK PATTERN CLEAR, and clear the CKP pattern.

72. Select CRANK PATTERN LEARNING with the HDS, and follow the screen prompts.

73. After doing all procedures, select the individual maintenance item you wish to reset with the HDS.

CRANKSHAFT FRONT SEAL

REMOVAL & INSTALLATION

2011 Models

➡**This procedure is performed with the engine in the vehicle.**

1. Remove the crankshaft pulley as described in this section.

2. Remove the pulley end crankshaft oil seal.

3. Clean and dry the crankshaft oil seal housing.

4. Apply a light coat of new engine oil to the crankshaft and to the lip of the crankshaft oil seal.

5. Using the oil seal driver, drive in the new crankshaft oil seal until the oil seal driver bottoms against the oil pump. When the seal is in place, clean any excess grease off the crankshaft, and check that the oil seal lip is not distorted.

6. Install the crankshaft pulley.

2012 Models

1.8L Non-Hybrid Engine

See Figures 117 through 123.

1. Raise and support vehicle on the lift.

2. Remove the right front wheel.

3. Disconnect and isolate the negative battery cable, battery sensor, and then the positive cable.

4. Remove the battery. See "Engine Electrical" section.

5. Loosen the water pump pulley bolts.

6. Remove the drive belt.

7. Remove the alternator. See "Engine Electrical" section.

8. Remove the water pump pulley.

9. Remove the drive belt auto-tensioner assembly.

10. Remove the engine cover.

11. Remove the cylinder head cover peripheral assembly, then remove the cylinder head cover, as described in this section.

12. Check the No. 1 piston at top dead

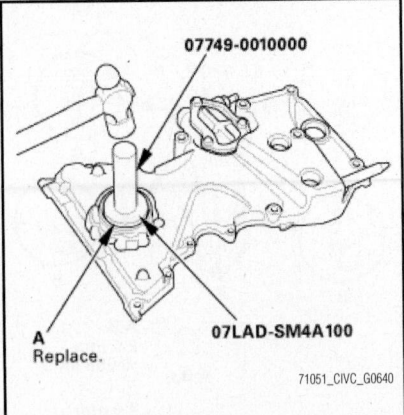

07749-0010000

A
Replace.

07LAD-SM4A100

71051_CIVC_G0640

Fig. 117 Use the driver handle, 15 x 135 L, and the driver attachment, 60 mm, to drive a new oil seal (A) squarely into the cam chain case to the specified installed height

center (TDC). The "UP" mark on the camshaft sprocket should be at the top, and the TDC grooves on the camshaft sprocket should line up with the top edge of the head.

➡**If the marks are not aligned, rotate the crankshaft 360 degrees, and recheck the camshaft pulley mark.**

13. Remove the crankshaft pulley, as described in this section.

14. Lift and support the engine with a jack and a wood block under the oil pan.

15. Remove the ground cable, the side engine mount bracket, and the side engine mount.

16. Disconnect the PCV hose.

17. Remove the oil pump as described in "Engine Lubrication" section.

18. Remove the pulley end crankshaft oil seal from the front case.

To install:

19. Clean and dry the crankshaft oil seal housing.

20. Apply a light coat of new engine oil to the crankshaft and to the lip of the crankshaft oil seal.

21. Use the driver handle, 15 x 135 L, and the driver attachment, 60 mm, to drive a new oil seal squarely into the cam chain case to the specified installed height.

22. Measure the distance between the chain case surface and the oil seal. Proper installed height should be 2.051–2.075 in. (52.1–52.7 mm).

23. Check the oil pump oil seal for damage. If the oil seal is damaged, replace the oil seal.

24. Remove all of the old liquid gasket from the oil pump mating surfaces, the bolts, and the bolt holes.

25. Clean and dry the oil pump mating surfaces.

26. Apply liquid gasket to the engine block mating surface of the oil pump, and to the inside edge of the threaded bolt holes. Install the component within 5 minutes of applying the liquid gasket.

27. Apply a 0.098 in. (2.5 mm) diameter bead of liquid gasket along the broken line.

28. If too much time has passed after applying the liquid gasket, remove the old liquid gasket and residue, then reapply new liquid gasket.

29. Apply liquid gasket to the oil pan mating surface of the oil pump, and to the inside edge of the threaded bolt holes. Install the component within 5 minutes of applying the liquid gasket.

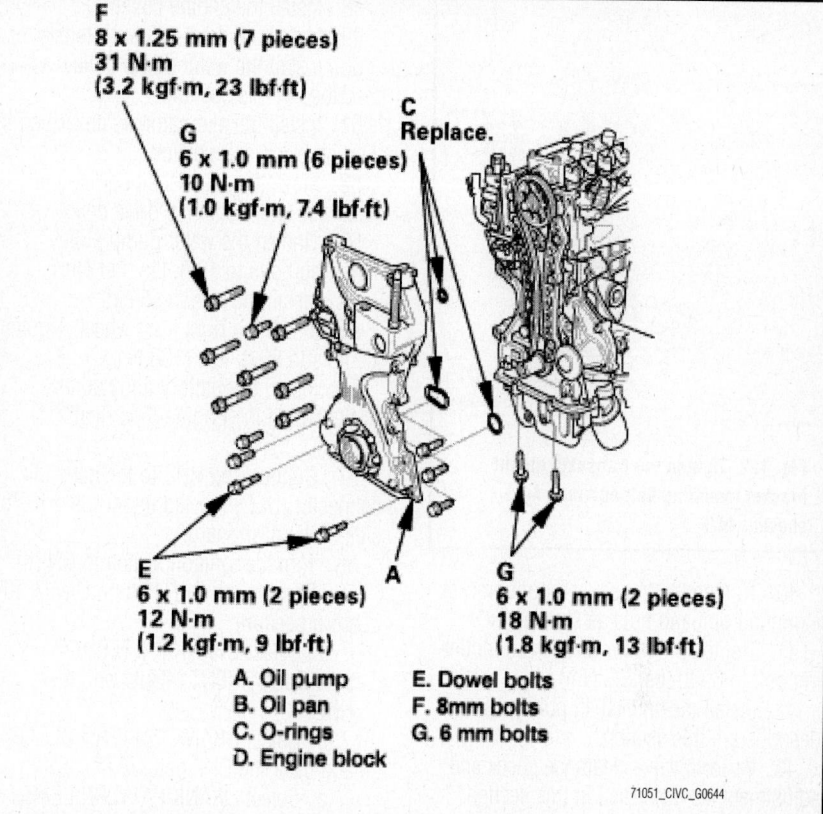

F
8 x 1.25 mm (7 pieces)
31 N·m
(3.2 kgf·m, 23 lbf·ft)

G
6 x 1.0 mm (6 pieces)
10 N·m
(1.0 kgf·m, 7.4 lbf·ft)

C
Replace.

E
6 x 1.0 mm (2 pieces)
12 N·m
(1.2 kgf·m, 9 lbf·ft)

G
6 x 1.0 mm (2 pieces)
18 N·m
(1.8 kgf·m, 13 lbf·ft)

A. Oil pump   E. Dowel bolts
B. Oil pan   F. 8mm bolts
C. O-rings   G. 6 mm bolts
D. Engine block

71051_CIVC_G0644

Fig. 120 Loosely install the dowel bolts, then tighten the 8mm bolts, the 6 mm bolts and the dowel bolts

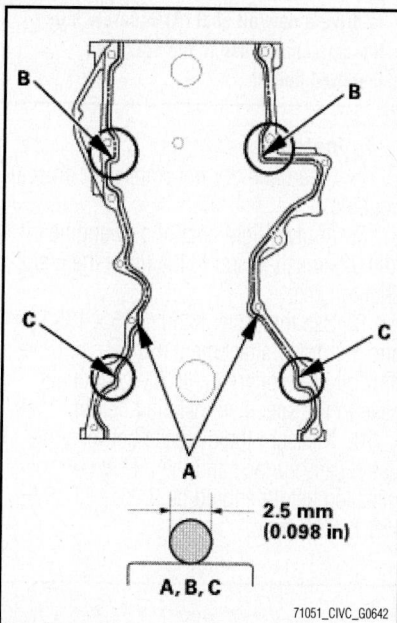

2.5 mm
(0.098 in)

A, B, C

71051_CIVC_G0642

Fig. 118 Apply a 0.098 in. (2.5 mm) diameter bead of liquid gasket along the broken line (A), (B), and (C)

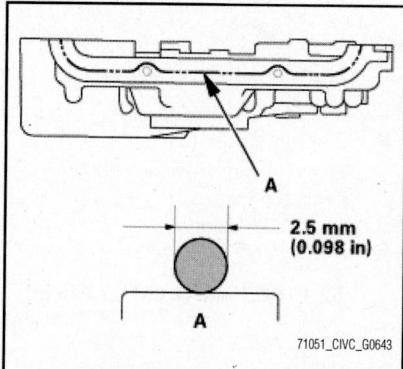

2.5 mm
(0.098 in)

A

71051_CIVC_G0643

Fig. 119 Apply a 0.098 in. (2.5 mm) diameter bead of liquid gasket along the broken line (A)

30. Apply a 0.098 in. (2.5 mm) diameter bead of liquid gasket along the broken line.

➡ **If too much time has passed after applying the liquid gasket, remove the old liquid gasket and residue, then reapply new liquid gasket.**

31. Set the edge of the oil pump on the edge of the oil pan with new O-rings.

32. Install the oil pump on the engine block.

33. Loosely install the dowel bolts, then tighten the 8mm bolts, the 6 mm bolts and the dowel bolts.

34. Wipe off the excess liquid gasket on the oil pan and oil pump mating surface.
Note the following:
• When installing the oil pump, do not slide the bottom surface onto the oil pan mounting surface.
• Wait at least 30 minutes before filling the engine with oil.
• Do not run the engine within 3 hours after installing the oil pump.

35. Install the new side engine mount mounting bolts and the side engine mount bracket using a nut and a new bolt, then loosely install the new side engine mount bracket mounting bolts and nut. Install the ground cable.

36. Remove the jack and the wood block.

37. Loosen the transaxle mount bracket mounting bolt and nuts.

38. Loosen the lower torque rod mounting bolt.

39. Tighten the side engine mount bracket mounting bolt and nut to 69 ft. lbs. (93 Nm).

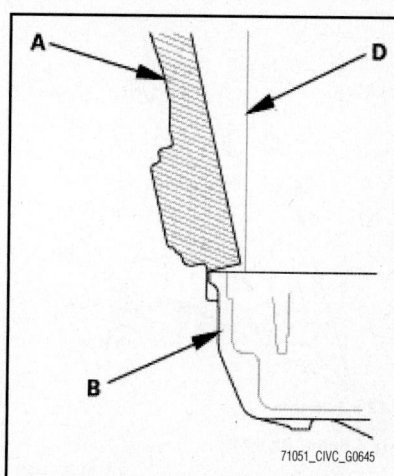

A   D

B

71051_CIVC_G0645

Fig. 121 Set the edge of the oil pump (A) on the edge of the oil pan (B) with new O-rings. Install the oil pump on the engine block (D)

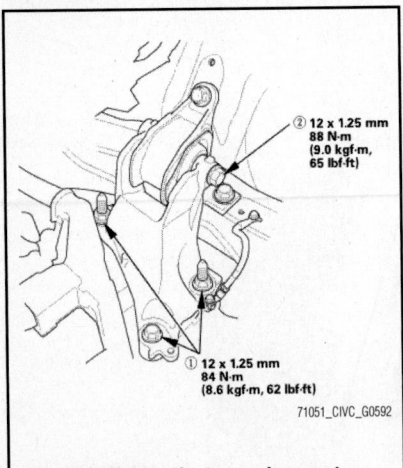

Fig. 122 Tighten the transaxle mount bracket mounting bolt and nuts as shown—M/T

40. Tighten the transaxle mount bracket mounting bolt and nuts as shown.

41. Tighten the lower torque rod mounting bolt to 69 ft. lbs. (93 Nm).

42. Install the crankshaft pulley, as described in this section.

43. Perform valve clearance check and adjustment, as described in this section.

44. Install the cylinder head cover, as described in this section.

➡️**Do not run the engine for at least 3 hours after installing the head cover.**

45. Connect the breather hose to the front of the cylinder head cover.

46. Install the dipstick.

47. Install the harness holders and connect the connectors on the injectors.

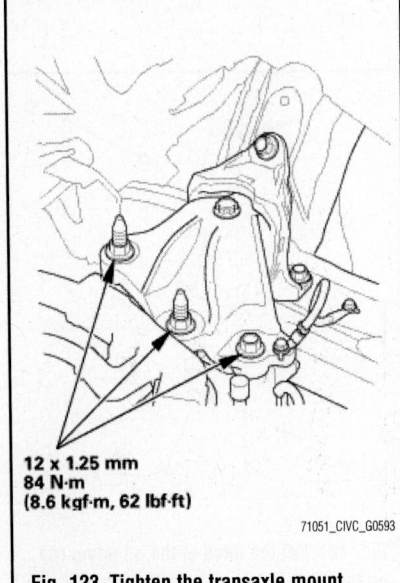

Fig. 123 Tighten the transaxle mount bracket mounting bolt and nuts as shown—A/T

48. Install the engine cover.

49. Install the drive belt auto-tensioner.

50. Install the water pump pulley, as described in this section.

51. Install the alternator, as described in "Engine Electrical" section.

52. Move the auto-tensioner counter-clockwise and install the drive belt.

53. Tighten the water pump pulley mounting bolts to 10 ft. lbs. (14 Nm).

54. Install the splash shield.

55. Install the right front wheel. Tighten the nuts to 80 ft. lbs. (108 Nm).

56. Install the battery and cables, as described in "Engine Electrical" section.

57. Connect the HDS to the data link connector (DLC) located under the driver's side of the dashboard.

58. Turn the ignition switch to ON (II).

59. Perform the CKP Pattern Clear/CKP Learn procedure:

   a. Select CRANK PATTERN in the ADJUSTMENT MENU with the HDS.

   b. Select CRANK PATTERN CLEAR, and clear the CKP pattern.

   c. Select CRANK PATTERN LEARNING with the HDS, and follow the screen prompts.

1.8L Hybrid Engine

See Figures 124 through 126.

1. Raise the vehicle on a lift, and make sure it is securely supported.

2. Remove the right front wheel.

3. Remove the splash shield.

4. Remove the drive belt, as described in this section.

5. Remove the crankshaft pulley, as described in this section.

6. Remove the chain case oil seal using a commercially available pry bar.

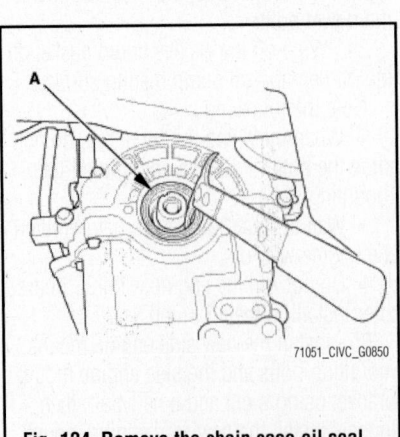

Fig. 124 Remove the chain case oil seal (A) using a commercially available pry bar

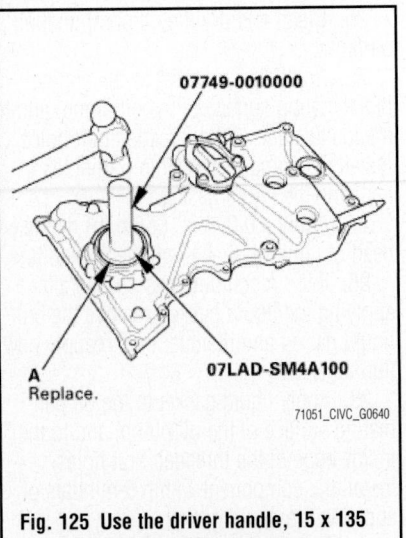

Fig. 125 Use the driver handle, 15 x 135 L, and the driver attachment, 60 mm, to drive a new oil seal (A) squarely into the cam chain case to the specified installed height

To install:

7. Clean and dry the crankshaft oil seal housing.

8. Apply a light coat of new engine oil to the crankshaft and to the lip of the crankshaft oil seal.

9. Use the driver handle, 15 x 135 L, and the driver attachment, 60 mm, to drive a new oil seal squarely into the cam chain case to the specified installed height.

10. Measure the distance between the chain case surface and the oil seal. Proper installed height should be 2.051–2.075 in. (52.1–52.7 mm).

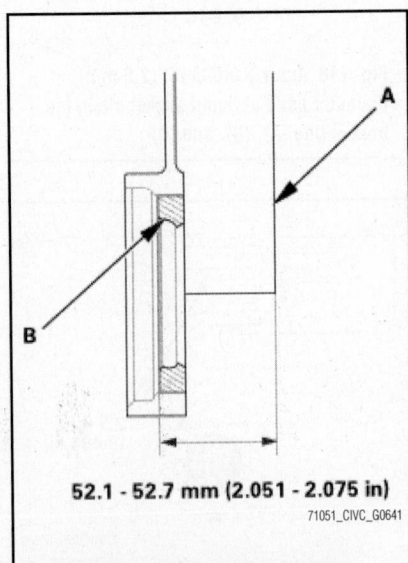

Fig. 126 Measure the distance between the chain case surface (A) and the oil seal (B). Proper installed height should be 2.051–2.075 in. (52.1–52.7 mm)

11. Install the remaining components as removed. Refer to applicable component sections as needed.

CRANKSHAFT PULLEY

REMOVAL & INSTALLATION

2011 Models

1. Remove the right front wheel.
2. Remove the drive belt.
3. Counterhold the pulley and remove the bolt, then remove the crankshaft pulley.
4. Clean the crankshaft pulley, the crankshaft mating surface, the bolt, and the washer.
5. Lubricate the bolt with engine oil prior to installation.
6. Install the crankshaft pulley onto the crankshaft by aligning the flat sides of the pulley with the flat sides of the inner oil pump gear.
7. On 1.8L engine, do the following:
   a. Counterhold the crankshaft pulley and install and torque the bolt to 51 ft. lbs. (69 Nm).

> ❊❊ **CAUTION**
>
> **Do not use an impact wrench.**

   b. If the pulley bolt or crankshaft are new, torque the bolt to 130 ft. lbs. (177 Nm), then remove the bolt and re-install it and torque it to 51 ft. lbs. (69 Nm).
   c. Mark the bolt head at 12 o'clock and then mark the crankshaft pulley with a reference line 90° clockwise from the bolt mark, then tighten the bolt an additional 90°. The mark on the bolt head lines up with the mark on the crankshaft pulley.
8. On 2.0L engine, do the following:
   a. Counterhold the crankshaft pulley and install and torque the bolt to 36 ft. lbs. (49 Nm).

> ❊❊ **CAUTION**
>
> **Do not use an impact wrench.**

   b. If the pulley bolt or crankshaft are new, torque the bolt to 130 ft. lbs. (177 Nm), then remove the bolt and re-install it and torque it to 36 ft. lbs. (49 Nm).
   c. Mark the bolt head at 12 o'clock and then mark the crankshaft pulley with a reference line 90° clockwise from the bolt mark, then tighten the bolt an additional 90°. The mark on the bolt head lines up with the mark on the crankshaft pulley.
9. Install the drive belt.
10. Install the right front wheel.

2012 Models

1.8L Non-Hybrid Engine

See Figures 127 and 128.

1. Raise the vehicle on a lift, and make sure it is securely supported.
2. Remove the right front wheel.
3. Remove the splash shield.
4. Move the auto-tensioner counterclockwise to relieve tension from the drive belt. Remove the drive belt.
5. Hold the pulley with the holder handle and the crankshaft pulley holder. Remove the bolt with a socket, 19 mm and a breaker bar.
6. Remove the crankshaft pulley.

To install:

7. Clean the crankshaft pulley, the crankshaft, the bolt, and the washer. Lubricate with new engine oil as shown.

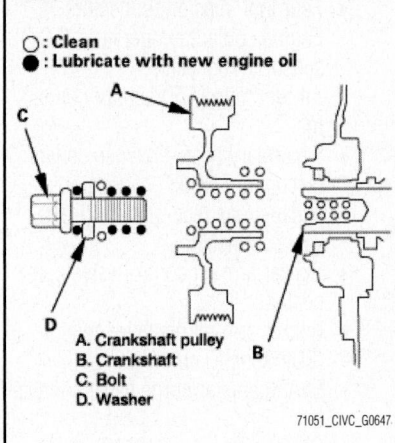

○ : Clean
● : Lubricate with new engine oil

A. Crankshaft pulley
B. Crankshaft
C. Bolt
D. Washer

71051_CIVC_G0647

Fig. 127 Clean the crankshaft pulley, the crankshaft, the bolt, and the washer. Lubricate with new engine oil as shown

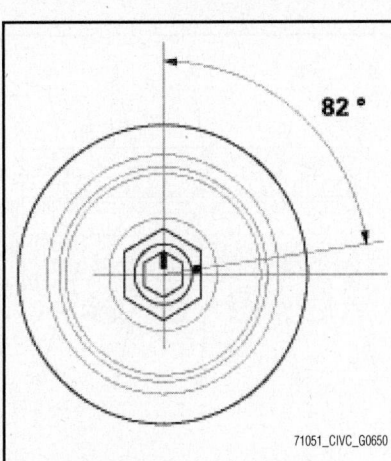

82°

71051_CIVC_G0650

Fig. 128 Tighten the bolt an additional 82 degrees

8. Install the crankshaft pulley onto the crankshaft by aligning the flat sides of the pulley with the flat sides of the inner oil pump gear.
9. When a new crankshaft or a new pulley bolt is installed:
   a. Hold the pulley with the holder handle and crankshaft pulley holder, torque the bolt to 133 ft. lbs. (180 Nm) with a torque wrench and a socket, then remove the bolt.
   b. Torque the bolt to 37 ft. lbs. (50 Nm) with a torque wrench and a socket.

➡ **Tighten the crankshaft pulley bolt. Do not use an impact wrench.**

   c. Tighten the bolt an additional 82 degrees.
10. When the crankshaft or the pulley bolt is reused:
   a. Hold the pulley with the holder handle and crankshaft pulley holder, then torque the bolt to 52 ft. lbs. (70 Nm) with a torque wrench and a socket.

➡ **Tighten the crankshaft pulley bolt. Do not use an impact wrench.**

11. Tighten the bolt an additional 90 degrees.
12. Move the auto-tensioner counterclockwise. Install the drive belt.
13. Install the splash shield.
14. Install the right front wheel.

1.8L Hybrid Engine

See Figure 127.

1. Raise the vehicle on a lift, and make sure it is securely supported.
2. Remove the right front wheel.
3. Remove the splash shield.
4. Hold the pulley with the holder handle and the crankshaft pulley holder. Remove the bolt with a socket, 19 mm and a breaker bar.
5. Remove the crankshaft pulley.

To install:

6. Clean the crankshaft pulley, the crankshaft, the bolt, and the washer. Lubricate with new engine oil as shown.
7. Install the pulley and pulley bolt. Tighten to 29 ft. lbs. (39 Nm), plus an additional 94 degrees.
8. Install the drive belt, splash shield and right front wheel.

2.4L Engine

See Figure 129.

1. Raise the vehicle on a lift, and make sure it is securely supported.
2. Remove the right front wheel.
3. Remove the splash shield.

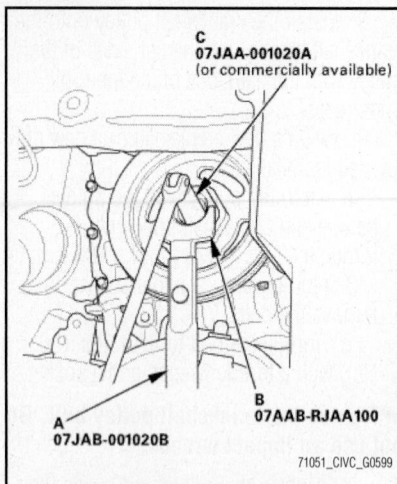

Fig. 129 Hold the pulley with the holder handle (A) and the crankshaft pulley holder (B). Remove the bolt with a socket, 19 mm (C) and a breaker bar, then remove the crankshaft pulley

4. Move the auto-tensioner using the belt tension release tool in the direction of the arrow to relieve tension from the drive belt, then remove the drive belt.

5. Hold the pulley with the holder handle and the crankshaft pulley holder. Remove the bolt with a socket, 19 mm and a breaker bar, then remove the crankshaft pulley.

6. Installation is the reverse of the removal procedure.

CYLINDER HEAD

REMOVAL & INSTALLATION

2011 Models

1.8L Engine (Except CNG Engine)
See Figures 130 and 131.

1. Before starting work, refer to the "Precautions" in this section.

2. Connect the Honda Diagnostic System (HDS), or equivalent scan tool, to the data link connecter (DLC), and monitor the engine coolant temperature (ECT) sensor 1. To avoid damaging the cylinder head, wait until the engine coolant temperature drops below 100°F (38°C) before loosening the cylinder head bolts.

3. Mark all wiring and hoses to avoid misconnection. Also, be sure that they do not contact other wiring or hoses, or interfere with other parts.

4. Relieve the fuel pressure. See "Relieving Fuel System Pressure" in the "FUEL SYSTEMS" section.

5. Do the battery terminal disconnect procedure, then wait at least 3 minutes

before beginning work. See "Battery System" in "ENGINE ELECTRICAL" section.

6. Drain the engine coolant.

7. Remove the air cleaner assembly.

8. Remove the drive belt.

9. Remove the intake manifold. See "Intake Manifold" in this section.

10. Remove the harness clamps and disconnect the positive crankcase ventilation (PCV) hose from the clamp.

11. Remove the air cleaner bracket, then remove the harness holder from the cylinder head.

12. On M/T model, disconnect the upper radiator hose and the heater hose from the cylinder head.

13. On A/T model, disconnect the upper radiator hose, the heater hose, and the water bypass hose.

14. Disconnect the following engine wire harness connectors, and remove the wire harness clamps from the cylinder head:
- Four fuel injector connectors
- Engine coolant temperature (ECT) sensor 1 connector
- Air fuel ratio (A/F) sensor connector
- Secondary heated oxygen sensor (secondary HO2S) connector
- Exhaust gas recirculation (EGR) valve connector
- Rocker arm oil control valve connector
- Rocker arm oil pressure sensor (EOP sensor) connector
- EVAP canister purge valve connector
- Camshaft position (CMP) sensor connector

15. Remove the four ignition coils.

16. Remove the three way catalytic converter (TWC).

17. Remove the thermostat housing.

18. Remove the cam chain. See "Timing Chain & Sprockets" in this section.

19. Remove the cylinder head bolts. To prevent warpage, loosen the bolts, in sequence, 1/3 turn at a time.

20. Repeat the sequence until all bolts are loosened.

21. Remove the cylinder head.

To install:

22. Clean the cylinder head and the engine block surface.

23. Install a new coolant separator in the engine block whenever the engine block is replaced.

24. Install the new cylinder head gasket and the dowel pins on the engine block. Always use a new cylinder head gasket.

25. Set the crankshaft to top dead center (TDC). Align the TDC mark on the crankshaft sprocket with the pointer on the engine block.

26. Set the camshaft to TDC. The "UP" mark on the camshaft sprocket should be at the top, and the TDC grooves on the camshaft sprocket should line up with the top edge of the head.

27. Install the cylinder head on the engine block.

28. Measure the diameter of each cylinder head bolt at multiple points.

29. If either diameter is less than 0.42 in. (10.6 mm), replace the cylinder head bolt.

30. Apply new engine oil to the threads and under the bolt heads of all cylinder head bolts.

31. Torque the cylinder head bolts in sequence to 29 ft. lbs. (39 Nm). Use a beam-type torque wrench. When using a preset click-type torque wrench, be sure to tighten slowly and do not overtighten. If a bolt makes any noise while you are torquing

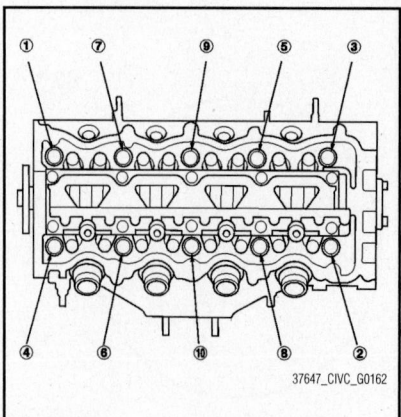

Fig. 130 Cylinder head bolt loosening sequence—1.8L engine

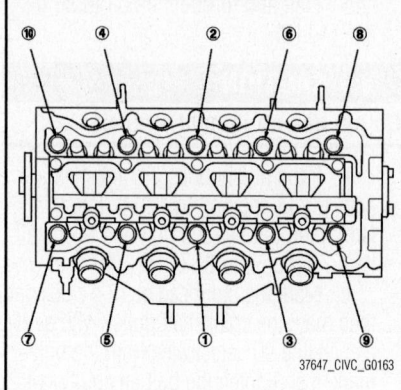

Fig. 131 Cylinder head bolt tightening sequence—1.8L engine

it, loosen the bolt and retighten it from the first step.

32. After torquing, tighten all cylinder head bolts in two steps, an additional 90° per step, using the sequence shown.

➡**If you are using a new cylinder head bolt, tighten the bolt an extra 60°.**

➡**Remove the cylinder head bolt if you tightened it beyond the specified angle, and go back to the procedure. Do not loosen it back to the specified angle.**

33. Install the cam chain. See "Timing Chain & Sprockets" in this section.

34. Adjust the valve clearance.

35. Install the cylinder head cover.

36. Install the thermostat housing.

37. Install the three way catalytic converter (TWC).

38. Install the four ignition coils.

39. Connect the following engine wire harness connectors, and install the wire harness clamps to the cylinder head:

- Four fuel injector connectors
- Engine coolant temperature (ECT) sensor 1 connector
- Air/Fuel ratio (A/F) sensor connector
- Secondary heated oxygen sensor (secondary HO2S) connector
- Exhaust gas recirculation (EGR) valve connector
- Rocker arm oil control valve connector
- Rocker arm oil pressure sensor (EOP sensor) connector
- EVAP canister purge valve connector
- Camshaft position (CMP) sensor connector

40. On M/T model, connect the upper radiator hose and the heater hose.

41. On A/T model, connect the upper radiator hose, the heater hose, and the water bypass hose.

42. Install the harness holder on the cylinder head, then install the air cleaner bracket.

43. Install the harness clamps, and connect the positive crankcase ventilation (PCV) hose to the clamp.

44. Install the intake manifold.

45. Install the drive belt.

46. Install the air cleaner assembly.

47. Do the battery reconnect procedure. See "Battery System" in "ENGINE ELECTRICAL" section.

48. After installation, check that all tubes, hoses, and connectors are installed correctly.

49. Inspect for fuel leaks:

a. Turn the ignition switch to ON (II) (do not operate the starter) so that the fuel pump runs for about 2 seconds and pressurizes the fuel line.

b. Repeat this operation three times, then check for fuel leakage at any point in the fuel line.

50. Refill the radiator with engine coolant, and bleed the air from the cooling system.

51. Check for fluid leaks.

52. Do the Engine Control Module (ECM/PCM) Idle Learn Procedure.

53. Do the Crankshaft Position (CKP) Pattern Clear/CKP Pattern Learn Procedure. See "Crankshaft Position (CKP) Sensor" in "ENGINE PERFORMANCE & EMISSION CONTROLS" section.

54. Inspect the idle speed.

55. Inspect the ignition timing.

1.8L CNG Engine
See Figures 132 through 135.

✳✳ CAUTION

To avoid damaging the wiring and terminals, unplug the wiring connectors carefully while holding the connector portion.

1. Connect the Honda Diagnostic System (HDS), or equivalent scan tool, to the data link connector (DLC), and monitor the engine coolant temperature (ECT) sensor 1.

✳✳ CAUTION

To avoid damaging the cylinder head, wait until the engine coolant temperature drops below 100°F (38°C) before loosening the cylinder head bolts.

2. Mark all wiring and hoses to avoid misconnection. Also, be sure that they do not contact other wiring or hoses, or interfere with other parts.

3. Turn off the manual shut-off valve.

4. To reduce pressure in the lines, start the engine, and run it until it stalls.

5. Do the battery terminal disconnect procedure, then wait at least 3 minutes before beginning work. See "Battery System" in "ENGINE ELECTRICAL" section.

6. Drain the engine coolant.

7. Remove the air cleaner assembly.

8. Remove the drive belt.

9. Remove the intake manifold. See "Intake Manifold" in this section.

10. Remove the harness clamps and remove the positive crankcase ventilation (PCV) hose from the clamp.

11. Remove the air cleaner bracket, then remove the harness holder from the cylinder head.

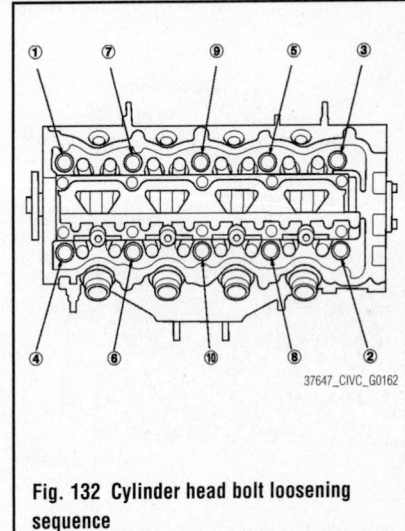

37647_CIVC_G0162

Fig. 132 Cylinder head bolt loosening sequence

12. Disconnect the upper radiator hose, the heater hose, and the water bypass hose.

13. Disconnect the following engine wire harness connectors, and remove the wire harness clamps from the cylinder head:

- Four fuel injector connectors
- Engine coolant temperature (ECT) sensor 1 connector
- Air fuel ratio (A/F) sensor connector
- Secondary heated oxygen sensor (secondary HO2S) connector
- Rocker arm oil control valve connector
- Rocker arm oil pressure sensor (EOP sensor) connector
- Camshaft position (CMP) sensor connector
- Remove the four ignition coils.
- Remove the three way catalytic converter (TWC).
- Remove the thermostat housing.
- Remove the cam chain. See "Timing Chain & Sprockets" in this section.

14. Remove the cylinder head bolts. To prevent warpage, unscrew the bolts, in sequence, 1/3 turn at a time. Repeat the sequence until all bolts are loosened.

15. Remove the cylinder head.

To install:

16. Clean the cylinder head and the block surface.

17. Install the new coolant separator in the engine block whenever the engine block is replaced.

18. Install the new cylinder head gasket and the dowel pins on the engine block. Always use a new cylinder head gasket.

19. Set the crankshaft to top dead center (TDC): Align the TDC mark on the crankshaft sprocket with the pointer on the engine block.

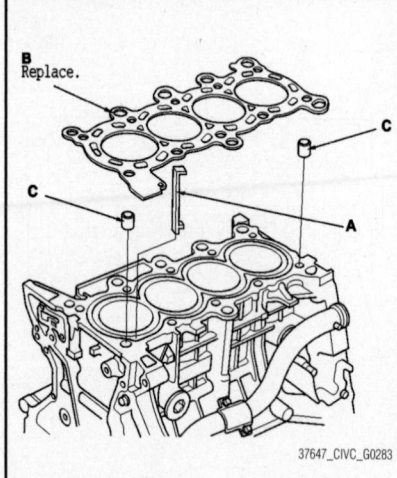

Fig. 133 Install the new coolant separator (A) in the engine block whenever the engine block is replaced. Install the new cylinder head gasket (B) and the dowel pins (C) on the engine block

20. Set the camshaft TDC. The "UP" mark on the camshaft sprocket should be at the top, and the TDC grooves on the camshaft sprocket should line up with the top edge of the head.

21. Install the cylinder head on the block.

22. Measure the diameter of each cylinder head bolt at point A and point B, as shown. If either diameter is less than 10.6 mm (0.42 in), replace the cylinder head bolt.

23. Apply new engine oil to the threads and under the bolt heads of all cylinder head bolts.

24. Torque the cylinder head bolts in

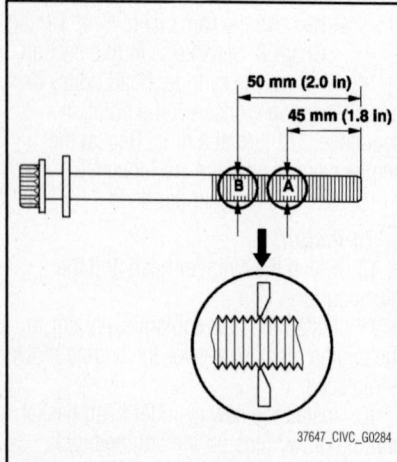

Fig. 134 Measure the diameter of each cylinder head bolt at point A and point B. If either diameter is less than 10.6 mm (0.42 in), replace the cylinder head bolt

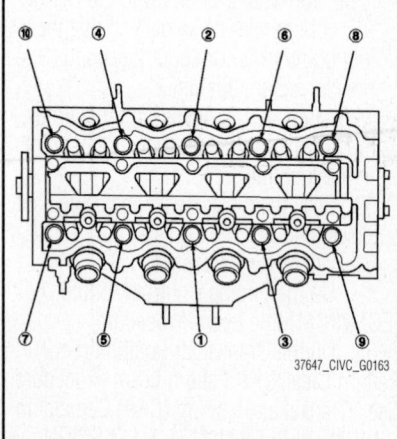

Fig. 135 Cylinder head bolt tightening sequence

sequence to 29 ft. lbs. (39 Nm). Use a beam-type torque wrench. When using a preset click-type torque wrench, be sure to tighten slowly and do not overtighten. If a bolt makes any noise while you are torquing it, loosen the bolt and retighten it from the first step.

25. After torquing, tighten all cylinder head bolts in two steps (90° per step) using the sequence shown. If you are using a new cylinder head bolt, tighten the bolt an extra 60°.

➡**Remove the cylinder head bolt if you tightened it beyond the specified angle, and go back and repeat the entire tightening procedure. Do not loosen it back to the specified angle.**

26. Install, adjust or connect the following:
- Cam chain
- Valve clearance
- Cylinder head cover
- Thermostat housing
- Three way catalytic converter (TWC)
- Four ignition coils

27. Connect the engine wire harness connectors, and install the wire harness clamps to the cylinder head:
- Four fuel injector connectors
- Engine coolant temperature (ECT) sensor 1 connector
- Air fuel ratio (A/F) sensor connector
- Secondary heated oxygen sensor (secondary HO2S) connector
- Rocker arm oil control valve connector
- Rocker arm oil pressure sensor (EOP sensor) connector
- Camshaft position (CMP) sensor connector

28. Connect the upper radiator hose, the heater hose, and the water bypass hose.

29. Install the harness holder on the cylinder head, then Install the air cleaner bracket.

30. Install the harness clamps and connect the positive crankcase ventilation (PCV) hose to the clamp.

31. Install the intake manifold.

32. Install the drive belt.

33. Install the air cleaner assembly.

34. Do the battery terminal reconnection procedure.

35. After installation, check that all tubes, hoses and connectors are installed correctly.

36. Do the leak inspection.

37. Refill the radiator with engine coolant, and bleed the air from the cooling system.

38. Do the Powertrain Control Module (PCM) Idle Learn Procedure. See "Engine Control Module/Powertrain Control Module (ECM/PCM)" in "ENGINE PERFORMANCE & EMISSION CONTROLS" section.

39. Do the Crankshaft Position (CKP) Pattern Clear/CKP Pattern Learn Procedure. See "Crankshaft Position (CKP) Sensor" in "ENGINE PERFORMANCE & EMISSION CONTROLS" section.

40. Inspect the idle speed.

41. Inspect the ignition timing.

2.0L Engine

See Figures 136 and 137.

1. Before starting work, refer to the "Precautions" in this section.

2. Connect the Honda Diagnostic System (HDS), or equivalent scan tool, to the data link connecter (DLC), and monitor the engine coolant temperature (ECT) sensor 1. To avoid damaging the cylinder head, wait until the engine coolant temperature drops below 100°F (38°C) before loosening the cylinder head bolts.

3. Mark all wiring and hoses to avoid misconnection. Also, be sure that they do not contact other wiring or hoses, or interfere with other parts.

4. Relieve the fuel pressure. See "Relieving Fuel System Pressure" in the "FUEL SYSTEMS" section.

5. Do the battery terminal disconnect procedure, then wait at least 3 minutes before beginning work. See "Battery System" in "ENGINE ELECTRICAL" section.

6. Drain the engine coolant.

7. Remove the air cleaner assembly.

8. Remove the drive belt.

9. Remove the intake manifold. See "Intake Manifold" in this section.

10. Remove the exhaust manifold. See "Exhaust Manifold" in this section.

11. Disconnect the evaporative emission (EVAP) canister hose and the brake booster vacuum hose.

12. Remove the quick-connect fitting cover, then disconnect the fuel feed hose.

13. Remove the harness holder from the bracket, then remove the harness holder bracket.

14. Disconnect the upper radiator hose and the heater hoses.

15. Remove the bolt securing the connecting pipe.

16. Disconnect the following engine wire harness connectors, and remove the wire harness clamps from the cylinder head:

- Four fuel injector connectors
- Engine coolant temperature (ECT) sensor 1 connector
- Camshaft position (CMP) sensor A (Intake) connector
- Camshaft position (CMP) sensor B (Exhaust) connector
- Rocker arm oil control valve connector
- Rocker arm oil pressure switch connector
- EVAP canister purge valve connector
- Engine oil pressure switch connector

17. Remove the cam chain. See "Timing Chain & Sprockets" in this section.

18. Remove the rocker arm assembly. See "Rocker Arms" in this section.

19. Remove the cylinder head bolts. To prevent warpage, loosen the bolts in sequence 1/3 turn at a time.

20. Repeat the sequence until all bolts are loosened.

21. Remove the cylinder head.

To install:

22. Install a new coolant separator in the engine block whenever the engine block is replaced.

23. Clean the cylinder head and the engine block surface.

24. Install the new cylinder head gasket and the dowel pins on the engine block. Always use a new cylinder head gasket.

25. Set the crankshaft to top dead center (TDC). Align the TDC mark on the crankshaft sprocket with the pointer on the engine block.

26. Set the camshaft pulley to TDC.

27. Install the cylinder head on the engine block.

28. Measure the diameter of each cylinder head bolt at multiple points.

29. If either diameter is less than 0.42 in. (10.6 mm), replace the cylinder head bolt.

30. Apply new engine oil to the threads and under the bolt heads of all cylinder head bolts.

31. Torque the cylinder head bolts in sequence to 29 ft. lbs. (39 Nm). Use a beam-type torque wrench. When using a preset click-type torque wrench, be sure to tighten slowly and do not overtighten. If a bolt makes any noise while you are torquing it, loosen the bolt and retighten it from the first step.

32. After torquing, tighten all cylinder head bolts in two steps, an additional 90° per step, using the sequence shown.

➡**If you are using a new cylinder head bolt, tighten the bolt an extra 60°.**

➡**Remove the cylinder head bolt if you tightened it beyond the specified angle, and go back to the procedure. Do not loosen it back to the specified angle.**

33. Install the rocker arm assembly.
34. Install the cam chain.
35. Connect the following engine wire harness connectors, and install the wire harness clamps to the cylinder head:

- Four fuel injector connectors
- Engine coolant temperature (ECT) sensor 1 connector

- Camshaft position (CMP) sensor A (Intake) connector
- Camshaft position (CMP) sensor B (Exhaust) connector
- Rocker arm oil control valve connector
- Rocker arm oil pressure switch connector
- EVAP canister purge valve connector
- Engine oil pressure switch connector

36. Install the bolt securing the connecting pipe.

37. Connect the upper radiator hose and the heater hoses.

38. Install the harness holder bracket, then install the harness holder.

39. Connect the fuel feed hose, then install the quick-connect fitting cover.

40. Connect the evaporative emission (EVAP) canister hose and the brake booster vacuum hose.

41. Install the exhaust manifold.
42. Install the intake manifold.
43. Install the drive belt.
44. Install the air cleaner assembly
45. Do the battery reconnect procedure. See "Battery System" in "ENGINE ELECTRICAL" section.

46. After installation, check that all tubes, hoses, and connectors are installed correctly.

47. Inspect for fuel leaks:

a. Turn the ignition switch to ON (II) (do not operate the starter) so that the fuel pump runs for about 2 seconds and pressurizes the fuel line.

b. Repeat this operation three times, then check for fuel leakage at any point in the fuel line.

48. Refill the radiator with engine coolant, and bleed the air from the cooling system.

49. Check for fluid leaks.

50. Do the Engine Control Module (ECM/PCM) Idle Learn Procedure.

51. Do the Crankshaft Position (CKP) Pattern Clear/CKP Pattern Learn Procedure. See "Crankshaft Position (CKP) Sensor" in "ENGINE PERFORMANCE & EMISSION CONTROLS" section.

52. Inspect the idle speed.
53. Inspect the ignition timing.

2012 Models

1.8L Non-Hybrid Engine

See Figures 138 through 143.

1. Raise the vehicle on a lift, and make sure it is securely supported.

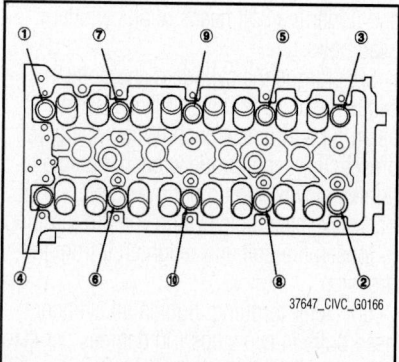

Fig. 136 Cylinder head bolt loosening sequence—2.0L engine

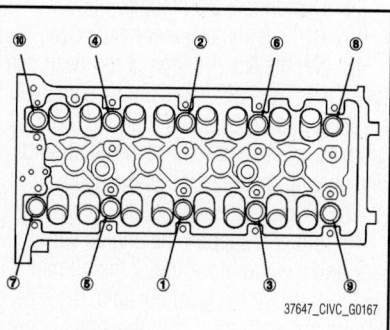

Fig. 137 Cylinder head bolt tightening sequence—2.0L engine

2. Remove the right front wheel.

3. Remove the fuel fill cap to relieve the pressure in the fuel tank.

4. Connect the HDS to the data link connector (DLC) located under the driver's side of the dashboard.

5. Turn the ignition switch to ON (II).

6. Make sure the HDS communicates with the vehicle. If it does not communicate, go to the DLC circuit troubleshooting.

7. Turn the ignition switch to ON (II).

8. From the INSPECTION MENU of the HDS, select Fuel Pump OFF, then start the engine, and let it idle until it stalls.

➡ **Do not allow the engine to idle above 1,000 rpm or the ECM/PCM will continue to operate the fuel pump.**

➡ **Pending or Confirmed DTC may be set during this procedure. Check for DTCs, and clear them as needed.**

9. Turn the ignition switch to LOCK (0).

10. Disconnect and isolate the negative cable, battery sensor and the positive cable.

11. Remove the battery. See "Engine Electrical" section.

12. Remove the quick-connect fitting cover (A).

   a. Check the fuel quick-connect fitting for dirt, and clean it if needed.

   b. Place a rag or shop towel over the quick-connect fitting.

   c. Disconnect the quick-connect fitting: Hold the connector with one hand, and squeeze the retainer tabs with the other hand to release them from the locking tabs. Pull the connector off. Be careful not to damage the line or other parts.
Note the following:
• Do not use tools.
• If the connector does not move, keep the retainer tabs pressed down, and alternately pull and push the connector until it comes off easily.
• Do not remove the retainer from the line; once removed, the retainer must be replaced with a new one.

13. Remove the engine cover.

14. Wait until the engine is cool, then carefully remove the radiator cap.

15. Drain the coolant. See "Engine Cooling" section.

16. Remove the air cleaner assembly, as described in this section.

17. Remove the wiper arms.

18. If necessary, remove the hood rear seal.

19. Remove the under cowl panel.

20. Remove the intake air duct.

21. Remove the throttle body without disconnecting the hoses.

22. Disconnect the EVAP canister hose, the brake booster vacuum hose, and the PCV hose.

23. Remove the heater hose clamp bracket.

24. Remove the intake manifold bracket.

25. Disconnect the connectors. Remove the harness clamps.

26. Remove the intake manifold assembly.

27. Disconnect the connectors and the harness holders.

28. Remove the harness clamp and disconnect the PCV hose from the clamp.

29. Remove the harness holder from the cylinder head.

30. Disconnect the upper radiator hose, the heater hose, and the water bypass hose.

31. Remove the ignition coils.

32. Disconnect the connector and remove the A/F sensor from the exhaust system attachment to the engine.

33. Remove the exhaust chamber cover.

34. Remove the EGR pipe.

35. Remove or disconnect the following:
• Splash shield.
• Engine undercover
• Secondary HO2S
• Exhaust pipe A
• Catalytic converter and cover

36. Disconnect the connectors and harness clamps.

37. Remove the thermostat housing.

38. Loosen the water pump pulley mounting bolts.

39. Move the drive belt auto-tensioner counterclockwise to relieve tension from the drive belt, then remove the drive belt.

40. Remove the alternator. See "Engine Electrical" section.

41. Remove the water pump and pulley. See "Engine Cooling" section.

42. Remove the drive belt auto-tensioner.

43. Remove the dipstick.

44. Disconnect the breather hose.

45. Remove the cylinder head cover.

46. Set the No. 1 piston at top dead center (TDC). The "UP" mark on the camshaft sprocket should be at the top, and the TDC grooves on the camshaft sprocket should line up with the top edge of the head.

47. Remove the crankshaft pulley.

48. Lift and support the engine with a jack and a wood block under the oil pan.

49. Remove the ground cable, the side engine mount bracket, and the side engine mount.

50. Disconnect the PCV hose.

51. Remove the oil pump. See "Engine Lubrication" section.

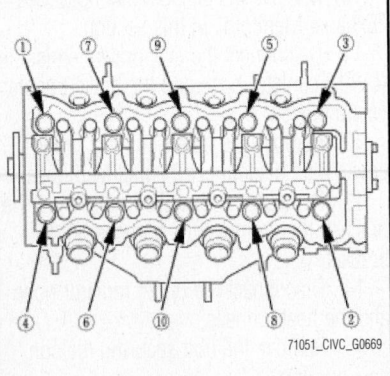

Fig. 138 Remove the cylinder head bolts. To prevent warpage, loosen the bolts in sequence, 1/3 turn at a time, repeat the sequence until all bolts are loosened. Remove the cylinder head

52. Remove the cam chain, as described in this section.

53. Remove the cylinder head bolts. To prevent warpage, loosen the bolts in sequence, 1/3 turn at a time, repeat the sequence until all bolts are loosened. Remove the cylinder head.

To install:

54. Set the crankshaft to top dead center (TDC). Align the TDC mark on the crankshaft sprocket with the pointer on the engine block.

55. Clean the cylinder head and the engine block surface.

56. Install a new coolant separator in the engine block whenever the engine block is replaced. Install the new cylinder head gasket and the dowel pins on the engine block. Always use a new cylinder head gasket. Install the cylinder head on the engine block.

57. Measure the diameter of each cylinder head bolt at point A and point B. If either diameter is less than 0.417 in. (10.6 mm), replace the cylinder head bolt.

58. Apply new engine oil to the threads and under the bolt heads of all cylinder head bolts.

59. Torque the cylinder head bolts in sequence to 30 ft. lbs. (40 Nm). Use a beam-type torque wrench. When using a preset click-type torque wrench, be sure to tighten slowly and do not overtighten. If a bolt makes any noise while you are torquing it, loosen the bolt and retighten it from the first step.

60. After torquing, tighten all cylinder head bolts in two steps (90 degrees per step using the sequence shown in previous step). If you are using a new cylinder head bolt, tighten the bolt an extra 60 degrees.

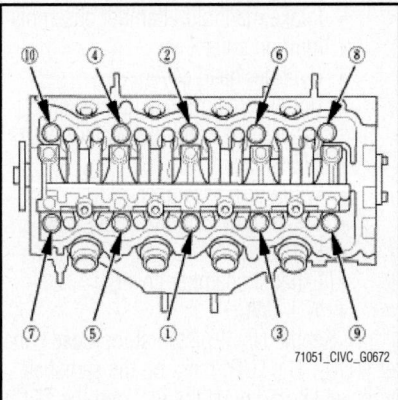

Fig. 139 Cylinder head bolt tightening sequence

➡ **Remove the cylinder head bolt if you tightened it beyond the specified angle, and go back to step 5 of the procedure. Do not loosen it back to the specified angle.**

61. Install the cam chain, as described in this section.

62. Install the oil pump on the engine block. See "Engine Lubrication" section.

63. Install the new side engine mount mounting bolts and the side engine mount bracket using a nut and a new bolt, then loosely install the new side engine mount bracket mounting bolts and nut. Install the ground cable.

64. Remove the jack and the wood block.

65. Loosen the transaxle mount bracket mounting bolt and nuts.

66. Loosen the lower torque rod mounting bolt.

67. Tighten the side engine mount bracket mounting bolt and nut to 69 ft. lbs. (93 Nm).

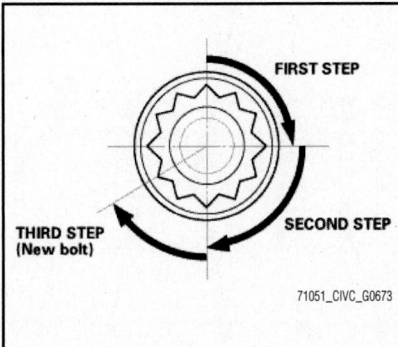

Fig. 140 After torquing, tighten all cylinder head bolts in two steps (90 degrees per step using the sequence shown in previous step). If you are using a new cylinder head bolt, tighten the bolt an extra 60 degrees

68. Tighten the transaxle mount bracket mounting bolt and nuts to specification shown.

69. Tighten the lower torque rod mounting bolt.

70. Install the crankshaft pulley, as described in this section.

71. Set the No. 1 piston at top dead center (TDC). The "UP" mark on the camshaft sprocket should be at the top, and the TDC grooves on the camshaft sprocket should line up with the top edge of the head.

72. Make the valve clearance adjustment, as described in this section.

73. Thoroughly clean the head cover gasket and the groove. If necessary, replace the head cover gasket.

74. Install the head cover gasket in the groove of the cylinder head cover and apply liquid gasket to the chain case contact areas.

75. Install the cylinder head cover, as described in this section.

76. Connect or install the following:
- Breather hose
- Dipstick
- Drive belt auto-tensioner
- Water pump pulley
- Alternator
- Harness connector and the harness clamps
- Drive belt

77. Tighten the water pump pulley mounting bolts to 10 ft. lbs. (14 Nm).

78. Install the thermostat housing. See "Engine Cooling" section.

79. Install the harness clamps and connectors, as removed.

80. Install the catalytic converter cover, and install the catalytic converter with

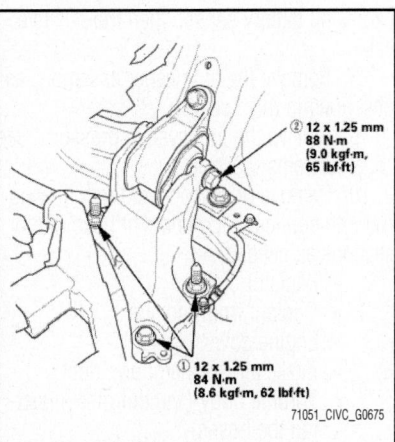

Fig. 141 Tighten the transaxle mount bracket mounting bolt and nuts to specification shown—M/T

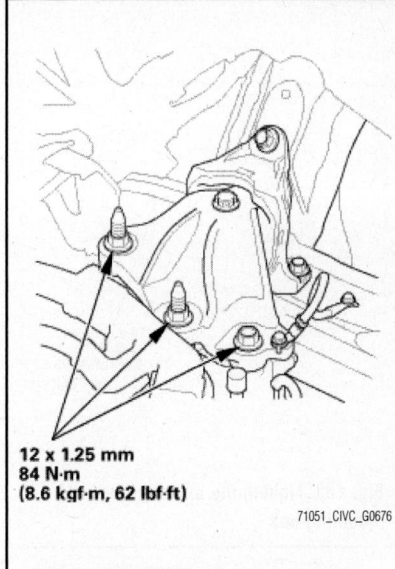

12 x 1.25 mm
84 N·m
(8.6 kgf·m, 62 lbf·ft)

Fig. 142 Tighten the transaxle mount bracket mounting bolt and nuts to specification shown—A/T

new gaskets. Tighten the nuts to 24 ft. lbs. (32 Nm).

81. Install exhaust pipe A with new gaskets and new self-locking nuts. Tighten the nuts to 10 ft. lbs. (14 Nm) and the bolts to 16 ft. lbs. (22 Nm).

82. Install or connect the following:
- Secondary HO2S to 33 ft. lbs. (44 Nm)
- Connector
- Engine undercover
- Splash shield
- EGR pipe with new gaskets; tighten fasteners to 16 ft. lbs. (22 Nm)
- Exhaust chamber cover.
- A/F sensor to 33 ft. lbs. (44 Nm)
- Connector
- Ignition coils
- Upper radiator hose
- Heater hose
- Water bypass hose
- Harness holder to the cylinder head
- Harness clamps and PCV hose to the clamp
- Harness holders and connectors

83. Install the intake manifold assembly, as described in this section.

84. Install the throttle body. See "Fuel Systems" section.

85. Install or connect the following:
- Intake air duct
- Under cowl panel
- Hood rear seal
- Windshield washer tube
- Center cowl cover
- Side cowl covers

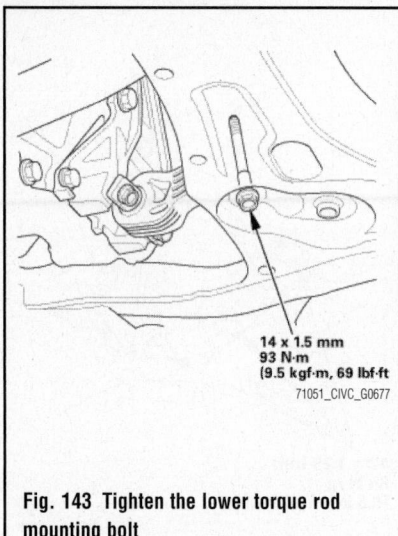

Fig. 143 Tighten the lower torque rod mounting bolt

14 x 1.5 mm
93 N·m
(9.5 kgf·m, 69 lbf·ft)
71051_CIVIC_G0677

➡ **Set the wiper arms to the auto-stop position before installation.**

- Wiper arms
- Cowl top wiper covers
- Air cleaner assembly
- Fuel feed hose
- Engine cover

86. Install the battery, then reconnect the battery cables. See "Engine Electrical" section.

87. Connect the HDS to the data link connector (DLC) located under the driver's side of the dashboard.

88. Turn the ignition switch to ON (II).

89. From the INSPECTION MENU of the HDS, select Fuel Pump ON.

➡ **Pending or Confirmed DTC may be set during this procedure. Check for DTCs, and clear them as needed.**

90. Turn the ignition switch to LOCK (0).

91. Turn the ignition switch to ON (II) (but do not operate the starter motor). The fuel pump runs for about 2 seconds, and fuel pressure rises. Repeat this two or three times, then make sure there are no fuel leaks.

92. Install the fuel fill cap.

93. Replace the radiator coolant. See "Engine Cooling" section.

94. Install the right front wheel.

95. Start the engine. Hold the engine speed at 3,000 rpm without load (A/T in P or N, M/T in neutral) until the radiator fan comes on, then let it idle.

96. Check the idle speed. See "Engine Electrical" section.

➡ **If the idle speed is not within specification, do the ECM/PCM idle learn procedure. If the idle speed is still not within specification, go to symptom troubleshooting.**

a. Select CRANK PATTERN in the ADJUSTMENT MENU with the HDS.

b. Select CRANK PATTERN CLEAR, and clear the CKP pattern.

c. Select CRANK PATTERN LEARN-ING with the HDS, and follow the screen prompts.

97. Jump the SCS line with the HDS.

98. Check the ignition timing. See "Engine Electrical" section.

99. After doing all procedures, select the individual maintenance item you wish to reset with the HDS.

1.8L Hybrid Engine

See Figures 144 and 145.

1. Raise the vehicle on a lift, and make sure it is securely supported.

➡ **For specific operations, refer to the user's manual that came with the Honda Diagnostic System (HDS). Make sure the HDS is loaded with the latest software.**

2. Connect the HDS to the data link connector (DLC) located under the driver's side of the dashboard.

3. Turn the ignition switch to ON (II).

4. Remove the fuel fill cap to relieve the pressure in the fuel tank.

5. . From the INSPECTION MENU of the HDS, select Fuel Pump OFF, then start the engine, and let it idle until it stalls.

➡ **Do not allow the engine to idle above 1,000 rpm or the PCM will continue to operate the fuel pump.**

➡ **Pending or Confirmed DTC may be set during this procedure. Check for DTCs, and clear them as needed.**

6. Turn the ignition switch to LOCK (0).

7. Disconnect and isolate the negative cable and battery sensor, then the positive cable.

8. Remove the air cleaner assembly, as described in this section.

9. Relieve the fuel system pressure. See "Fuel Systems" section.

10. Remove or disconnect the following (refer to applicable component headings or sections as needed):

- Radiator cap
- Coolant from radiator
- Engine cover
- Intake air resonator and duct
- Throttle body (without disconnecting the hoses)
- Splash shield and engine under-cover
- Intake manifold bracket and peripheral assembly

- Intake manifold/chamber assembly
- Ignition coils
- Cylinder head cover
- Upper radiator hose, the water bypass hose and the heater hose
- A/F sensor
- Wiper arm assembly
- Center cowl and the under cowl
- EGR pipe
- Exhaust chamber cover
- WU-TWC

11. Set the No. 1 piston at top dead center (TDC). The "UP" mark on the camshaft sprocket should be at the top, and the TDC grooves on the camshaft sprocket should line up with the top edge of the head.

12. Remove or disconnect the following (refer to applicable component headings or sections as needed):

- Right front wheel
- Water pump
- Crankshaft pulley
- Lift and support the engine with a jack and a wood block under the oil pan
- Side engine mount bracket and side engine mount assembly
- Cam chain case
- Cam chain

13. Remove the cylinder head bolts. To prevent warpage, loosen the bolts in sequence, 1/3 turn at a time, repeat the sequence until all bolts are loosened.

14. Remove the cylinder head.

To install:

15. Set the crankshaft to top dead center (TDC). Align the TDC mark on the crankshaft sprocket with the pointer on the oil pump.

16. Set the No. 1 piston at top dead center (TDC). The "UP" mark on the camshaft

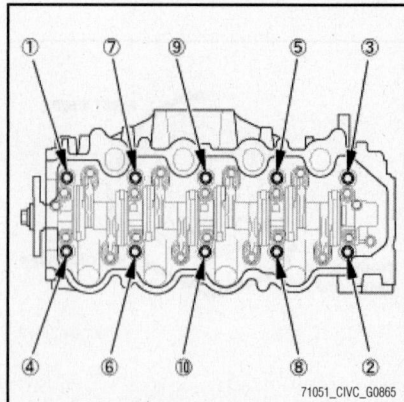

71051_CIVIC_G0865

Fig. 144 Remove the cylinder head bolts. To prevent warpage, loosen the bolts in sequence, 1/3 turn at a time, repeat the sequence until all bolts are loosened. Remove the cylinder head

sprocket should be at the top, and the TDC grooves on the camshaft sprocket should line up with the top edge of the head.

17. Clean the cylinder head and the engine block surface.

18. Install the cooling control spacer in the engine block whenever the engine block is replaced.

19. Check the play of the cooling control spacer. If there is play, replace the cooling control spacer.

20. Let the cooling control spacer contact a water jacket base as shown. The cooling control spacer upper end portions and the clearance of the cylinder block end face being more than 0.2 mm.

21. Install the new cylinder head gasket and the dowel pins on the engine block. Always use a new cylinder head gasket.

22. Install the cylinder head on the engine block.

23. Apply new engine oil to the threads and under the bolt heads of all cylinder head bolts.

24. Torque the cylinder head bolts in sequence to 21 ft. lbs. (29 Nm) with a beam-type torque wrench if possible.

➡**When using a preset-click-type torque wrench, be sure to tighten slowly and do not overtighten. If a bolt makes any noise while you are torquing it, loosen the bolt and retighten it from the first step.**

25. Tighten all cylinder head bolts an additional 130 degrees.

26. Install or connect the following (refer to applicable component headings or sections as needed):
- Cam chain and case
- Side engine mount and bracket (loose bolts and nuts)

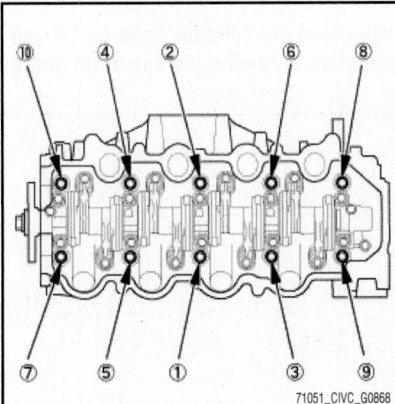

Fig. 145 Cylinder head bolt tightening sequence

- Ground cable
- A/C pipe bracket
- Transaxle mount bracket support bolt and nuts (loosen)
- Remove the jack and the wood block
- Lower torque rod mounting bolts (loosen)
- Side engine mount bracket mounting bolt and nut (tighten)
- Transaxle mount bracket support bolt and nuts (tighten)
- Lower torque rod mounting bolts (tighten)
- Crankshaft pulley
- Water pump
- Drive belt auto-tensioner
- Water pump pulley (loose bolts)
- Drive belt
- Water pump pulley mounting bolts (tighten)
- Right front wheel
- WU-TWC
- Exhaust chamber cover
- EGR pipe
- Under cowl panel and center cowl cover
- Side cowl covers
- Wiper arms assemblies
- A/F sensor
- Cylinder head peripheral components (as removed)
- Cylinder head cover

27. Ignition coils
- Intake manifold/chamber assembly
- Engine undercover and splash shield
- Throttle body with new gasket
- Intake air pipe, intake air duct and intake air resonator
- Engine cover
- Fuel fill cap

28. After installation, check that all tubes, hoses, and connectors are installed correctly.

29. Reconnect the battery positive cable, sensor and negative cable.

30. Connect the HDS to the data link connector (DLC) located under the driver's side of the dashboard.

31. Turn the ignition switch to ON (II).

32. From the INSPECTION MENU of the HDS, select Fuel Pump ON.

➡**Pending or Confirmed DTC may be set during this procedure. Check for DTCs, and clear them as needed.**

33. Turn the ignition switch to LOCK (0).

34. Turn the ignition switch to ON (II) (do not operate the starter motor).

➡**The fuel pump will run for about 2 seconds, and the fuel pressure will**

rise. **Repeat this two or three times, then make sure there are no fuel leaks.**

35. Refill the cooling system.

36. Select CRANK PATTERN in the ADJUSTMENT MENU with the HDS.

37. Select CRANK PATTERN CLEAR, and clear the CKP pattern.

38. Select CRANK PATTERN LEARNING with the HDS, and follow the screen prompts.

39. Check the idle speed without electrical load: headlights, blower fan, radiator fan, and air conditioner off. It should be 750±50 RPM in P or N.

➡**If the IMA battery level gauge displays no segments, hold the engine speed between 3,500–4,000 rpm without load (in P or N) until the level gauge will display at half charged level.**

40. Let the engine idle for 1 minute with high electric load (A/C switch on, temperature set to max cool, blower fan on High, and headlights on high beam). Idle speed should be 750±50 RPM in P or N.

➡**If the idle speed is not within specification, do the PCM idle learn procedure. If the idle speed is still not within specification, go to symptom troubleshooting.**

41. Check the ignition time. It should be 10±2° BTDC (RED mark) at idle or in P or N.

42. If the ignition timing differs from the specification, check the cam timing. If the cam timing is OK, update the PCM if it does not have the latest software, or substitute a known-good PCM, then recheck. If the system works properly, and the PCM was substituted, replace the original PCM.

43. After completing the entire procedure, select the individual maintenance item you wish to reset with the HDS.

2.4L Engine

See Figures 146 through 151.

1. Remove the splash shield.

2. Wait until the engine is cool, then carefully remove the radiator cap.

3. Raise the vehicle on a lift, and make sure it is securely supported.

4. Loosen the drain plug and drain the coolant.

5. Remove the fuel fill cap to relieve the pressure in the fuel tank.

➡**For specific operations, refer to the user's manual that came with the Honda Diagnostic System (HDS). Make sure the HDS is loaded with the latest software.**

6. Connect the HDS to the data link connector (DLC) located under the driver's side of the dashboard.

   a. Turn the ignition switch to ON (II).

   b. From the INSPECTION MENU of the HDS, select Fuel Pump OFF, then start the engine, and let it idle until it stalls.

➡**Do not allow the engine to idle above 1,000 rpm or the ECM/PCM will continue to operate the fuel pump.**

➡**Pending or Confirmed DTC may be set during this procedure. Check for DTCs, and clear them as needed.**

   c. Turn the ignition switch to LOCK (0).

7. Disconnect and isolate the negative cable, battery sensor and positive battery cable.

8. Relieve fuel system pressure. See "Fuel Systems" section.

9. Remove the air cleaner assembly. as described in this section.

10. Remove the drive belt.

11. Remove the engine cover.

12. Remove the intake air duct.

13. Remove the throttle body without disconnecting the hoses. See "Fuel Systems" section.

14. Remove the intake manifold bracket, then remove the intake manifold (A).

15. Disconnect the PCV hose.

16. Disconnect the ignition coil connectors. Move the engine harness holder.

17. Disconnect the connectors and remove the fuel injector and fuel rail assembly. See "Fuel Systems" section.

18. Remove the wiper arms.

19. Remove both side cowl covers, the center cowl cover, windshield washer tube, and the hood rear seal.

20. Remove the under cowl panel.

21. Disconnect the EVAP canister hose and the brake booster vacuum hose.

22. Remove the rocker arm oil control valve assembly and the rocker arm oil control valve filter. Remove the engine wire harness bracket.

23. Disconnect the connector and remove the A/F sensor from on top of the exhaust pipe.

24. Remove the warm-up three-way converter (WU-TWC) and cover.

25. Disconnect the upper radiator hose and the heater hoses.

26. Remove the EVAP canister purge valve bracket.

27. Remove the right front wheel.

28. Remove the ignition coil cover and remove the ignition coils.

29. Remove the cylinder head cover.

30. Turn the crankshaft so its white mark lines up with the pointer (the other pointer is not used).

31. Remove the cam chain case harness clamp.

32. Remove the ground cable and remove the VTC oil control solenoid valve.

33. Remove the crankshaft pulley, as described in this section.

34. Lift and support the engine with a jack and a wood block under the oil pan.

35. Remove the upper torque rod.

36. Remove the ground cable, the side engine mount bracket, and the side engine mount.

37. Remove the cam chain case and cam chain, as described in this section.

38. Remove the camshafts, as described in this section.

39. Insert the bolts into the rocker shaft holder, then remove the rocker arm assembly.

40. Remove the cylinder head bolts, loosening the bolts in sequence 1/3 turn at a time; repeat the sequence until all bolts are loosened.

41. Remove the cylinder head.

To install:

42. Set the crankshaft to top dead center (TDC). Align the TDC mark on the crankshaft sprocket with the pointer on the engine block.

43. Install a new coolant separator in the engine block whenever the engine block is replaced.

44. Clean the cylinder head and the engine block surface.

45. Install the new cylinder head gasket and the dowel pins on the engine block. Always use a new cylinder head gasket.

46. Install the cylinder head on the engine block.

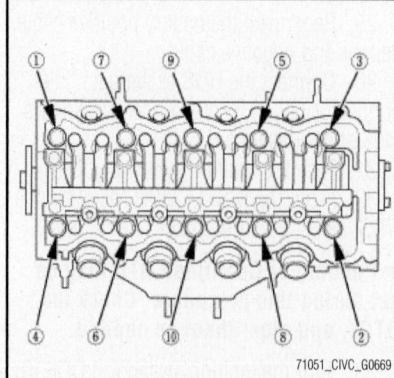

Fig. 146 Remove the cylinder head bolts, loosening the bolts in sequence 1/3 turn at a time; repeat the sequence until all bolts are loosened

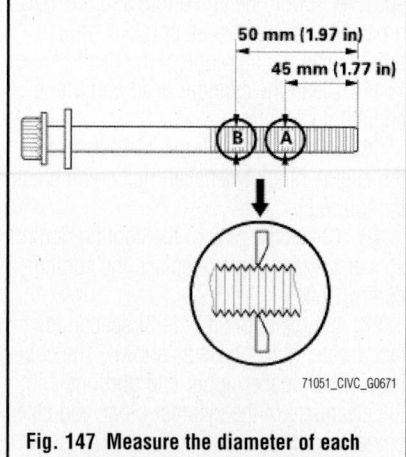

Fig. 147 Measure the diameter of each cylinder head bolt at point A and point B. If either diameter is less than 0.417 in. (10.6 mm), replace the cylinder head bolt

47. Measure the diameter of each cylinder head bolt at point A and point B. If either diameter is less than 0.417 in. (10.6 mm), replace the cylinder head bolt.

48. Apply new engine oil to the threads and under the bolt heads of all cylinder head bolts.

49. Torque the cylinder head bolts in sequence to 29 ft. lbs. (39 Nm).

➡**Use a beam-type torque wrench. When using a preset click-type torque wrench, be sure to tighten slowly and do not overtighten. If a bolt makes any noise while you are torquing it, loosen the bolt and retighten it from the first step.**

50. After torquing, tighten all cylinder head bolts in two steps (90 degrees per step) using the sequence shown. If you are using a new cylinder head bolt, tighten the bolt an extra 90 degrees.

➡**Remove the cylinder head bolt if you tightened it beyond the specified angle**

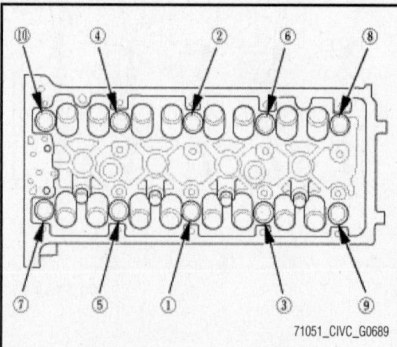

Fig. 148 Torque the cylinder head bolts in sequence to 29 ft. lbs. (39 Nm)

and repeat the previous step. Do not loosen it back to the specified angle.

51. Install the rocker arm assembly, as described in this section.

52. Install the camshafts, as described in this section.

53. Install the cam chain and case, as described in this section.

54. Install the side engine mount bracket, then tighten the side engine mount bracket mounting bolts.

55. Install the new side engine mount mounting bolts and the side engine mount bracket using a nut and a new bolt, then loosely install the new side engine mount bracket mounting bolts and nut. Install the ground cable .

56. Loosen the transaxle mount bracket bolt and nuts.

57. Loosen the lower torque rod mounting bolt.

58. Tighten the side engine mount bracket mounting bolt and nut to 69 ft. lbs. (93 Nm).

59. Tighten the transaxle mount bracket mounting bolt and nuts to 61 ft. lbs. (83 Nm).

60. Loosen the lower torque mount mounting bolt, then tighten the bolts in the numbered sequence shown.

61. Install the upper torque rod using new bolts. Tighten the upper torque rod mounting bolts in the numbered sequence shown.

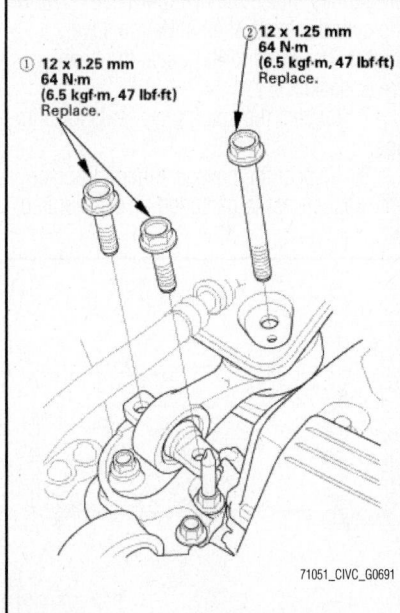

Fig. 149 Install the upper torque rod using new bolts. Tighten the upper torque rod mounting bolts in the numbered sequence shown

62. Install the crankshaft pulley, as described in this section.

63. Install the VTC oil control solenoid valve. Use a new O-ring coated with engine oil. Clean and dry the mating surface of the valve. Install the ground cable.

64. Install the cam chain case harness clamp.

65. Perform valve clearance adjustment, as described in this section.

66. Check the spark plug seals for damage. If any seals are damaged, replace them.

67. Thoroughly clean the head cover gasket and the groove. Replace the gasket as necessary.

68. Install the head cover gasket in the groove of the cylinder head cover. Make sure the head cover gasket is seated securely.

69. Remove all of the old liquid gasket from the chain case and the No.5 rocker shaft holder.

70. Clean the head cover contacting surfaces with a shop towel.

71. Apply liquid gasket, on the chain case and the No. 5 rocker shaft holder mating areas. Install the component within 5 minutes of applying the liquid gasket.

➡**If too much time has passed after applying the liquid gasket, remove the old liquid gasket and residue, then reapply new liquid gasket.**

72. Set the spark plug seals on the spark plug tubes. Place the cylinder head cover on the cylinder head, then slide the cover slightly back and forth to seat the head cover gasket.

73. Inspect the spark plug seals for damage.

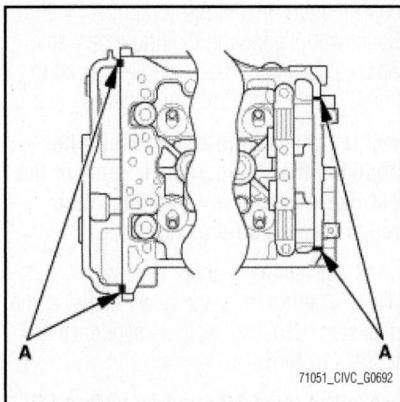

Fig. 150 Apply liquid gasket, on the chain case and the No. 5 rocker shaft holder mating areas (A). Install the component within 5 minutes of applying the liquid gasket

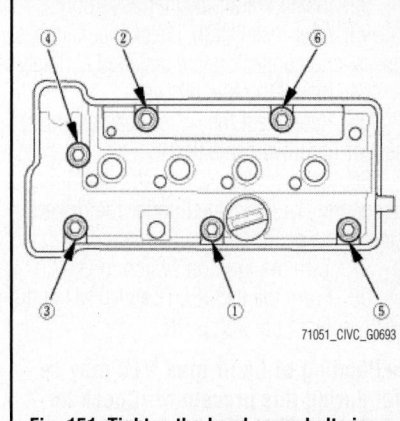

Fig. 151 Tighten the head cover bolts in two steps. In the final step torque bolts, in sequence, to 9 ft. lbs. (12 Nm)

74. Tighten the head cover bolts in two steps. In the final step torque bolts, in sequence, to 9 ft. lbs. (12 Nm).

➡**Wait at least 30 minutes before filling the engine with oil. Do not run the engine for at least 3 hours after installing the head cover.**

75. Install the ignition coils and cover.

76. Install the EVAP canister purge valve bracket.

77. Connect the upper radiator hose and the heater hoses.

78. Install the WU-TWC and cover with new gaskets.

79. Install the A/F sensor, tighten to 33 ft. lbs. (44 Nm) and attach the connector.

80. Install the engine wire harness bracket. Install the rocker arm oil control valve bracket and assembly, then install the rocker arm oil control valve filter.

81. Connect the EVAP canister hose and the brake booster vacuum hose.

82. Install the under cowl panel, hood rear seal, windshield washer tube, center cowl cover and the side cowl covers.

83. Install the wiper arms.

84. Install the cowl top wiper covers.

85. Install the injector base with a new gasket, then install the injectors and fuel rail together. Connect the quick-connect fitting.

86. Install the engine harness holder and attach the connectors.

87. Install the intake manifold, as described in this section.

88. Install the throttle body with new gasket. Install the harness clamps and connector.

89. Connect the EVAP canister hose and the brake booster vacuum hose.

90. Install the intake air duct.

91. Install the drive belt.

92. Install the air cleaner assembly.

93. After installation, check that all tubes, hoses, and connectors are installed correctly.

94. Install the fuel fill cap.

95. Reconnect the battery cables and sensor (positive cable first)

96. Connect the HDS to the data link connector (DLC) located under the driver's side of the dashboard.

97. Turn the ignition switch to ON (II).

98. From the INSPECTION MENU of the HDS, select Fuel Pump ON.

➡**Pending or Confirmed DTC may be set during this procedure. Check for DTCs, and clear them as needed.**

99. Turn the ignition switch to LOCK (0).

100. Turn the ignition switch to ON (II) (but do not operate the starter motor). The fuel pump runs for about 2 seconds, and fuel pressure rises. Repeat this two or three times, then make sure there are no fuel leaks.

101. Replace the radiator coolant. See "Engine Cooling" section.

102. Install the engine cover.

103. Install the right front wheel.

➡**Before installing the wheel, clean the mating surfaces between the brake disc and the inside of the wheel.**

104. Install the splash shield.

105. Check the idle speed. See "Engine Electrical" section.

106. Check the ignition timing. See "Engine Electrical" section.

➡**If the ignition timing differs from the specification, check the cam timing. If the cam timing is OK, update the ECM if it does not have the latest software, or substitute a known good ECM, then recheck. If the system works properly, and the ECM was substituted, replace the original ECM.**

107. Disconnect the HDS and the timing light.

108. After doing all procedures, select the individual maintenance item you wish to reset with the HDS.

CYLINDER HEAD (VALVE) COVER

REMOVAL & INSTALLATION

2012 Models

1.8L Non-Hybrid Engine

See Figure 152.

1. Remove the engine cover.

2. Disconnect the connecters, then remove the harness holders.

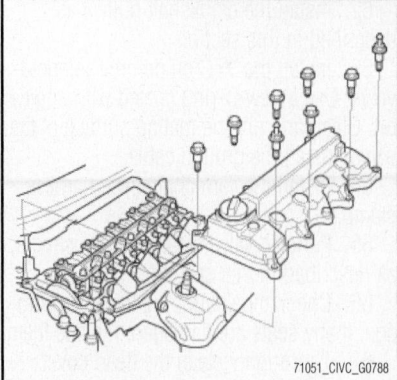

Fig. 152 Remove the cylinder head cover—1.8L non-hybrid shown; hybrid similar

3. Remove the alternator wiring harness clamp.

4. Remove the dipstick.

5. Disconnect the breather hose from the front of the cylinder head cover.

6. Remove the cylinder head cover.

To install:

7. Thoroughly clean the head cover gasket and the groove.

➡**Check and, if necessary, replace the head cover gasket.**

8. Install the head cover gasket in the groove of the cylinder head cover.

9. Make sure the head cover gasket is seated securely.

10. Clean the head cover contacting surfaces with a shop towel.

11. Remove the old liquid gasket from the top edge matting surfaces of the oil pump and cylinder head.

12. Apply liquid gasket to the chain case contact areas. Install the component within 5 minutes of applying the liquid gasket. Apply a 0.20 in. (5 mm) diameter bead of liquid gasket to the mating surfaces.

➡**If too much time has passed after applying the liquid gasket, remove the old liquid gasket and residue, then reapply new liquid gasket.**

13. Install the cylinder head cover.

14. Tighten the bolts in two steps. In the final step, torque all bolts in sequence to 7 ft. lbs. (10 Nm).

➡**Wait at least 30 minutes before filling the engine with oil. Do not run the engine for at least 3 hours after installing the head cover.**

15. Connect the breather hose to the front of the cylinder head cover.

16. Install the dipstick.

17. Install the alternator wiring harness clamp.

18. Install the harness holders and connecters as removed.

19. Install the engine cover.

2.4L Engine

See Figure 153.

1. Remove the following components (refer to individual component sections as needed):

- Wiper arm assemblies
- Side cowl covers
- Center cowl cover
- Lower cowl cover
- Engine cover
- Ignition coils

2. Remove the cylinder head cover.

To install:

3. Remove all of the old liquid gasket from the chain case and the No.5 rocker shaft holder.

4. Clean the head cover contacting surfaces with a shop towel.

5. Apply liquid gasket on the chain case and the No. 5 rocker shaft holder mating areas. Install the component within 5 minutes of applying the liquid gasket.

➡**If too much time has passed after applying the liquid gasket, remove the old liquid gasket and residue, then reapply new liquid gasket.**

6. Set the spark plug seals on the spark plug tubes. Place the cylinder head cover on the cylinder head, then slide the cover slightly back and forth to seat the head cover gasket.

7. Inspect the spark plug seals for damage.

8. Inspect the cover washers. Replace any washer that is damaged or deteriorated.

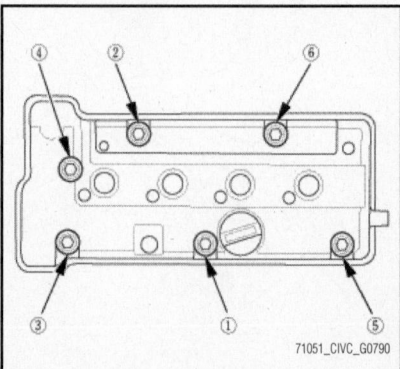

Fig. 153 Tighten the head cover bolts in two steps. In the final step torque bolts, in sequence, to 9 ft. lbs. (12 Nm)

9. Tighten the head cover bolts in two steps. In the final step torque bolts, in sequence, to 9 ft. lbs. (12 Nm).

➡**Wait at least 30 minutes before filling the engine with oil. Do not run the** engine for at least 3 hours after installing the head cover.

10. Continue installation of components in reverse order of removal.

EXHAUST MANIFOLD

REMOVAL & INSTALLATION

2.0L Engine

See Figure 154.

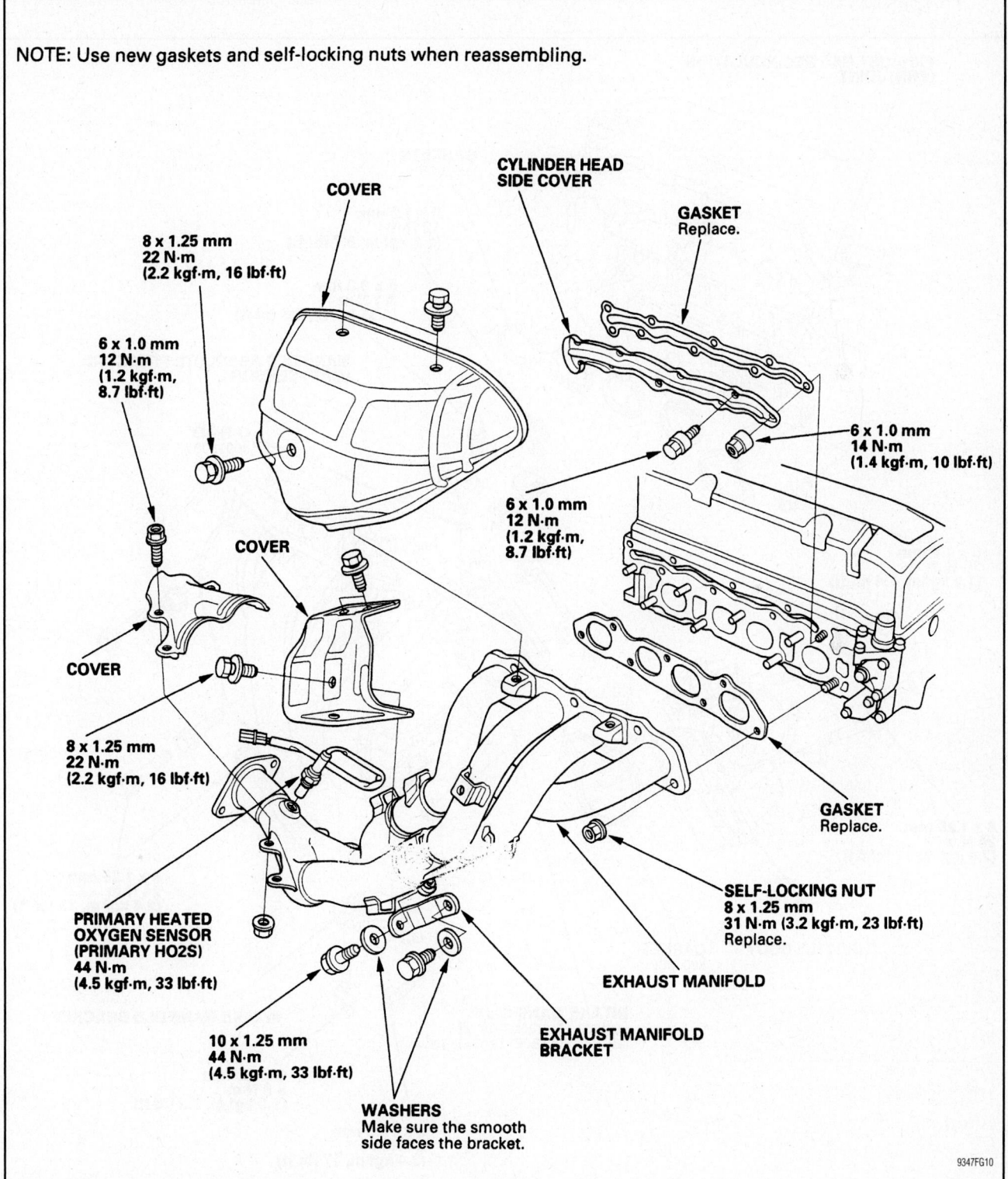

NOTE: Use new gaskets and self-locking nuts when reassembling.

COVER

8 x 1.25 mm
22 N·m
(2.2 kgf·m, 16 lbf·ft)

6 x 1.0 mm
12 N·m
(1.2 kgf·m, 8.7 lbf·ft)

COVER

COVER

8 x 1.25 mm
22 N·m
(2.2 kgf·m, 16 lbf·ft)

PRIMARY HEATED OXYGEN SENSOR (PRIMARY HO2S)
44 N·m
(4.5 kgf·m, 33 lbf·ft)

10 x 1.25 mm
44 N·m
(4.5 kgf·m, 33 lbf·ft)

WASHERS
Make sure the smooth side faces the bracket.

CYLINDER HEAD SIDE COVER

GASKET
Replace.

6 x 1.0 mm
14 N·m
(1.4 kgf·m, 10 lbf·ft)

6 x 1.0 mm
12 N·m
(1.2 kgf·m, 8.7 lbf·ft)

GASKET
Replace.

SELF-LOCKING NUT
8 x 1.25 mm
31 N·m (3.2 kgf·m, 23 lbf·ft)
Replace.

EXHAUST MANIFOLD

EXHAUST MANIFOLD BRACKET

9347FG10

Fig. 154 Exploded view of the exhaust manifold—2.0L engines

1. Before servicing the vehicle, refer to the precautions in the beginning of this section.

2. Drain the cooling system and relieve the fuel system pressure.

3. Remove or disconnect the following:
   - Negative battery cable
   - Catalytic converter
   - Exhaust manifold heat shields
   - Heated Oxygen (HO2S) sensor connector
   - Exhaust manifold bracket
   - Exhaust manifold

4. Installation is the reverse of the removal procedure.

INTAKE MANIFOLD

REMOVAL & INSTALLATION

2011 Models

1.8L Engine

See Figure 155.

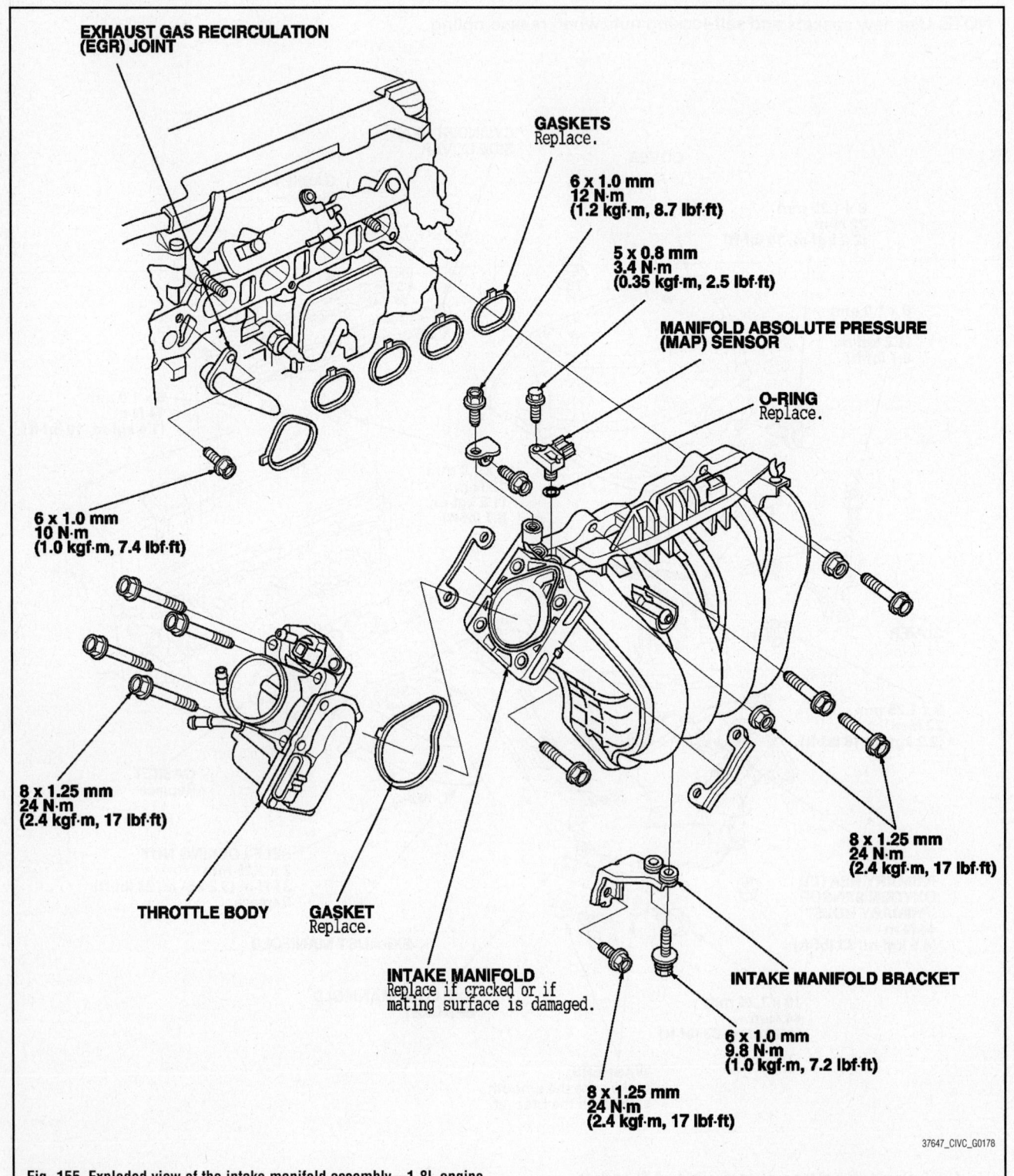

Fig. 155 Exploded view of the intake manifold assembly—1.8L engine

37647_CIVC_G0178

1. Remove the cowl cover and the under-cowl panel.

2. Relieve the fuel pressure. See "Fuel Pressure Relieving" in "FUEL SYSTEMS" section.

3. Remove the air cleaner assembly.

4. Remove the intake air duct.

5. Remove the injector cover.

6. Disconnect the evaporative emission (EVAP) canister hose, the brake booster vacuum hose, and the positive crankcase ventilation (PCV) hose, and remove the power steering (P/S) hose clamp.

7. Remove the quick-connect fitting cover, then disconnect the fuel feed hose.

8. Remove the following engine wire harness connectors, and remove the wire harness clamps from the intake manifold:
- Throttle actuator connector
- Manifold absolute pressure (MAP) sensor connector
- Evaporative emission (EVAP) canister purge valve connector
- Intake manifold tuning (IMT) valve actuator connector

9. Disconnect the water bypass hoses, then plug the water bypass hoses.

10. Remove the throttle body.

11. Remove the heater hose clamp bracket.

12. Raise the vehicle on the lift.

13. Remove the intake manifold bracket.

14. Lower the vehicle on the lift.

15. Remove the all intake manifold mounting bolts and nuts, then remove the intake manifold from the cylinder head.

16. Remove the harness clamps, then remove the intake manifold from the vehicle.

To install:

17. Install the harness clamps to the intake manifold.

18. Install the intake manifold with new gaskets and tighten the bolts and nuts in a crisscross pattern in three steps, beginning with the inner bolt. Final torque should be 17 ft. lbs. (24 Nm).

19. Raise the vehicle on the lift.

20. Install the intake manifold bracket. Tighten the lower bolt to 17 ft. lbs. (24 Nm) and the two upper bolts to 7 ft. lbs. (10 Nm).

21. Lower the vehicle on the lift.

22. Install the heater hose clamp bracket.

23. Install the throttle body with a new gasket. Torque the throttle body mounting bolts to 17 ft. lbs. (24 Nm).

24. Connect the water bypass hoses.

25. Connect the following engine wire harness connectors, and install the wire harness clamps to the intake manifold:
- Throttle actuator connector

- MAP sensor connector
- EVAP canister purge valve connector
- IMT valve actuator connector

26. Connect the fuel feed hose, then install the quick-connect fitting cover.

27. Connect the EVAP canister hose, the brake booster vacuum hose, and the PCV hose, and install the P/S hose clamp.

28. Install the injector cover.

29. Install the intake air duct.

30. Install the air cleaner assembly.

31. Install the cowl cover and the under-cowl panel.

32. Inspect for fuel leaks. Turn the ignition switch to ON (II) (do not operate the starter) so the fuel pump runs for about 2 seconds and pressurizes the fuel line. Repeat this operation three times, then check for fuel leakage at any point in the fuel line.

33. After installation, check that all tubes, hoses, and connectors are installed correctly.

34. Clean up any spilled engine coolant.

35. Refill the radiator with engine coolant, and bleed the air from the cooling system.

2.0L Engine

See Figure 156.

➡**Make sure to acquire the anti-theft code from the radio and write down the frequencies for the radio's preset buttons.**

1. Remove the engine cover.

2. Relieve the fuel system pressure.

3. Disconnect the negative battery cable.

4. Drain the cooling system.

5. Remove the vacuum hose, breather pipe and the intake air duct.

6. Remove the engine wire harness connectors and the wire harness clamps from the intake manifold.

7. Disconnect the four fuel injector connectors, the manifold absolute pressure connector and the throttle connector.

8. Remove the ground cable, harness clamp bracket and the harness holder from its mounting. Remove the PCV valve hose, evaporative emission canister hose and the power brake booster vacuum hose.

9. Remove the water bypass hoses. Remove the quick connect fitting cover. Disconnect the fuel feed hose.

10. Raise and support the vehicle safely. Remove the intake manifold connector cover. Remove the intake manifold bracket. Lower the vehicle.

11. Remove the intake manifold retaining bolts. Remove the intake manifold from the vehicle.

To install:

12. Installation is the reverse of the removal procedure, while using the following torque values:

13. Torque the manifold retaining bolts and nut to specification in a crisscross pattern in two or three steps, beginning with the inner bolt.

2012 Models

1.8L Non-Hybrid Engine

See Figure 157.

1. Remove the wiper arms.

2. Remove the fuel fill cap to relieve the pressure in the fuel tank.

3. Connect the HDS to the data link connector (DLC) located under the driver's side of the dashboard.

4. Turn the ignition switch to ON (II).

5. From the INSPECTION MENU of the HDS, select Fuel Pump OFF, then start the engine, and let it idle until it stalls.

➡**Do not allow the engine to idle above 1,000 rpm or the ECM/PCM will continue to operate the fuel pump.**

➡**Pending or Confirmed DTC may be set during this procedure. Check for DTCs, and clear them as needed.**

6. Turn the ignition switch to LOCK (0).

7. Disconnect and isolate the negative cable, battery sensor, and then the positive cable.

8. Relieve fuel system pressure. See "Fuel System" section.

9. Remove the air cleaner, as described in this section.

10. Remove the throttle body without disconnecting the hoses.

11. Disconnect the EVAP canister hose, the brake booster vacuum hose, and the PCV hose.

12. Remove the heater hose clamp bracket, the intake manifold bracket, disconnect the connectors, and remove the harness clamps.

13. Remove the intake manifold assembly.

To install:

14. Install the intake manifold assembly with new gaskets. Tighten the bolts/nuts in a crisscross pattern in three steps, beginning with the inner bolt.

15. Install the connectors, harness clamps, intake manifold bracket and heater hose clamp bracket. Tighten the two lower manifold bracket bolts to 7 ft. lbs.

(10 Nm) and the upper bolt to 18 ft. lbs. (24 Nm).

16. Connect the EVAP canister hose, the brake booster vacuum hose, and the PCV hose.

17. Install the throttle body with new O-ring. Tighten the bolts to 18 ft. lbs. (24 Nm).

18. Install the air cleaner assembly.

19. Connect the fuel feed hose.

20. Install the wiper arms.

21. Install the fuel fill cap.

22. Connect the positive cable, then the battery sensor and the negative cable to the battery.

23. Connect the HDS to the data link connector (DLC) located under the driver's side of the dashboard.

24. Turn the ignition switch to ON (II).

25. From the INSPECTION MENU of the HDS, select Fuel Pump ON.

➡Pending or Confirmed DTC may be set during this procedure. Check for DTCs, and clear them as needed.

26. Turn the ignition switch to LOCK (0).

27. Turn the ignition switch to ON (II) (but do not operate the starter motor). The fuel pump runs for about 2 seconds, and fuel pressure rises. Repeat this two or three times, then make sure there are no fuel leaks.

28. After installation, check that all tubes, hoses, and connectors are installed correctly.

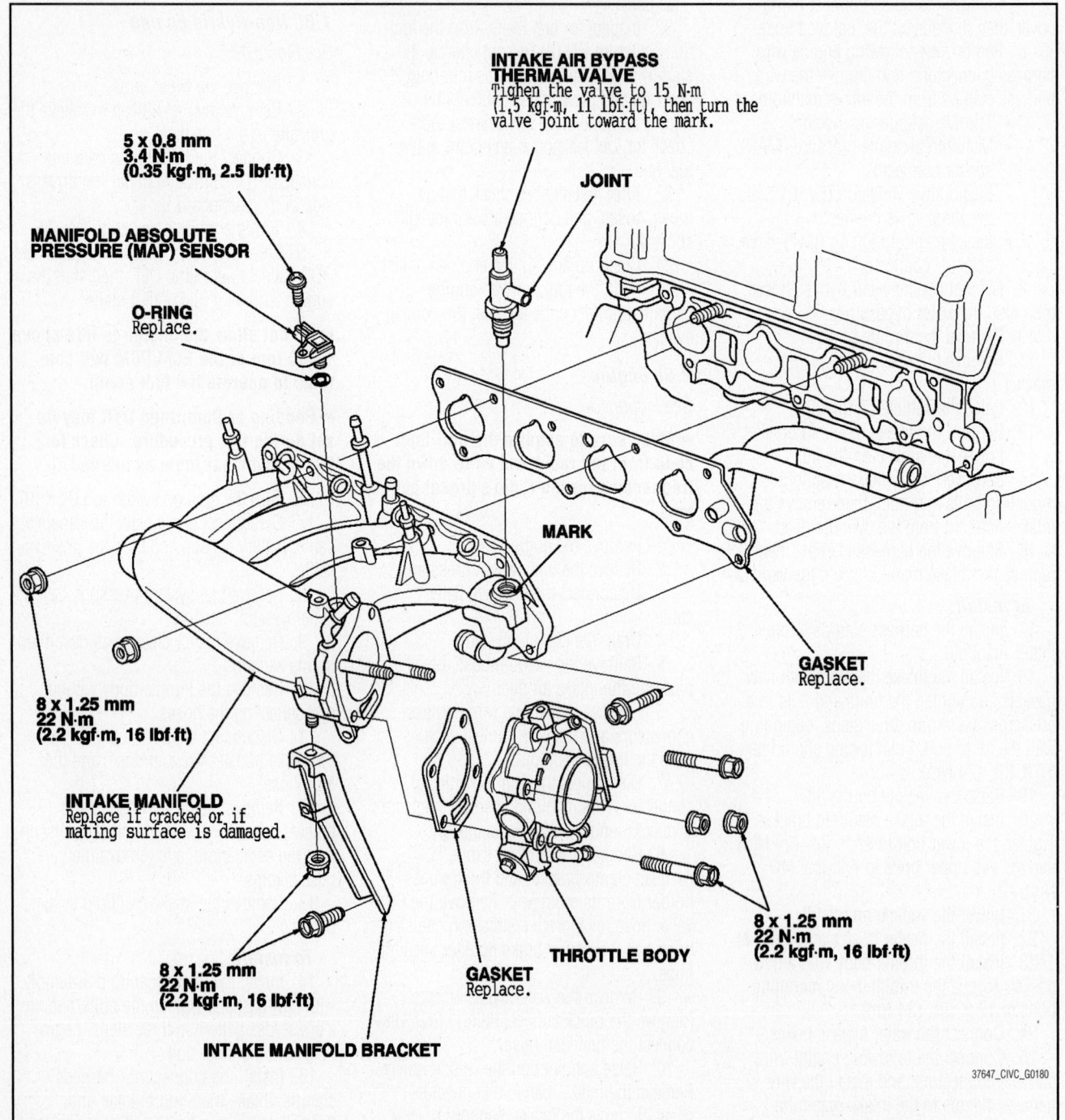

INTAKE AIR BYPASS THERMAL VALVE
Tighten the valve to 15 N·m (1.5 kgf·m, 11 lbf·ft), then turn the valve joint toward the mark.

JOINT

5 x 0.8 mm
3.4 N·m
(0.35 kgf·m, 2.5 lbf·ft)

MANIFOLD ABSOLUTE PRESSURE (MAP) SENSOR

O-RING
Replace.

MARK

GASKET
Replace.

8 x 1.25 mm
22 N·m
(2.2 kgf·m, 16 lbf·ft)

INTAKE MANIFOLD
Replace if cracked or if mating surface is damaged.

8 x 1.25 mm
22 N·m
(2.2 kgf·m, 16 lbf·ft)

THROTTLE BODY

GASKET
Replace.

8 x 1.25 mm
22 N·m
(2.2 kgf·m, 16 lbf·ft)

INTAKE MANIFOLD BRACKET

37647_CIVIC_G0180

Fig. 156 Exploded view of the intake manifold assembly—2.0L engine

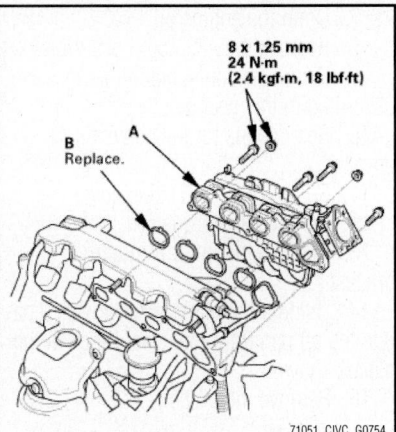

Fig. 157 Install the intake manifold assembly (A) with new gaskets (B). Tighten the bolts/nuts in a crisscross pattern in three steps, beginning with the inner bolt

1.8L Hybrid Engine

See Figure 158.

1. Raise the vehicle on a lift, and make sure it is securely supported.

2. Remove the following components (refer to applicable component headings or sections as needed):
   - Engine cover
   - Intake air resonator
   - Intake air duct
   - Intake air pipe
   - Throttle body (without disconnecting the fuel lines)
   - Engine under cover

3. Remove the intake manifold bracket mounting bolts.

4. Disconnect the connectors and the wire harness.

5. Disconnect the PCV hose and remove the intake manifold/chamber assembly.

To install:

6. Install the intake manifold/chamber assembly. Tighten the bolts to 18 ft. lbs. (24 Nm).

7. Connect the PCV hose.

8. Install the remaining component of reverse order of removal.

2.4L Engine

See Figure 159.

1. Remove the engine cover.

2. Remove the intake air duct.

3. Disconnect the EVAP canister hose, the brake booster vacuum hose, the connector and the harness clamps from near the throttle body.

4. Remove the throttle body without disconnecting the hoses.

5. Remove the intake manifold bracket.

6. Remove the intake manifold.

7. Disconnect the PCV hose.

To install:

8. Connect the PCV hose.

9. Install the intake manifold with new gaskets. Tighten the bolts to 16 ft. lbs. (22 Nm).

10. Install the intake manifold bracket. Tighten the bolts to 16 ft. lbs. (22 Nm).

11. Install the throttle body with new gasket. Tighten the bolts to 16 ft. lbs. (22 Nm).

12. Install the harness clamps, connect the connector, connect the EVAP canister hose and the brake booster vacuum hose near the throttle body.

13. Install the intake air duct.

14. Install the engine cover.

15. After installation, check that all tubes, hoses, and connectors are installed correctly.

OIL PAN

REMOVAL & INSTALLATION

2011 Models

1.8L Engine

See Figure 160.

1. Note the radio security code and the radio presets. Disconnect the negative battery cable.

2. Remove the drive belt. Remove the air conditioning condenser fan shroud.

3. Disconnect the compressor electrical connectors. Remove the compressor retaining bolts, and position it to the side without discharging the system.

4. Raise and support the vehicle safely. Remove the splash shield. Drain the engine oil.

5. Remove the exhaust pipe. Properly support the oil pan. Remove the lower torque rod. Remove the oil pan support tool.

6. Remove the lower torque rod bracket. Remove the air conditioning compressor bracket.

7. If equipped with automatic transaxle, remove the shift cable cover. Remove the torque converter cover.

8. Remove the clutch cover if equipped with manual transaxle.

9. Remove the oil pan retaining bolts. Using a flat bladed tool, carefully separate the oil pan from the engine block. Remove the oil pan from the vehicle.

To install:

10. Remove any old gasket from the mating surfaces. Be sure these surfaces are clean and dry.

11. Apply liquid gasket, part number 08717-004, 08718-0001, 08718-0003 or 08718-0009 evenly to the engine block mating surface of the oil pan.

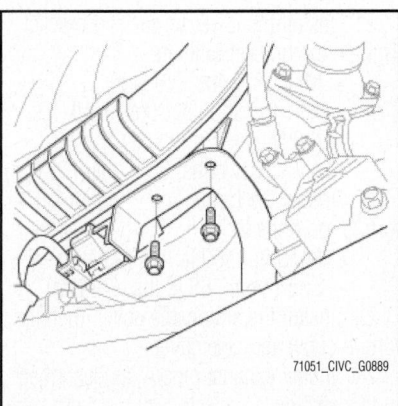

Fig. 158 Remove the intake manifold bracket mounting bolts

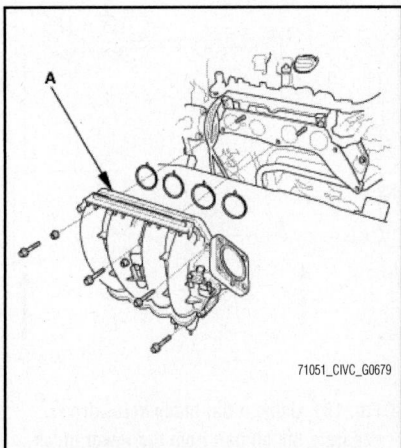

Fig. 159 Remove the intake manifold (A)

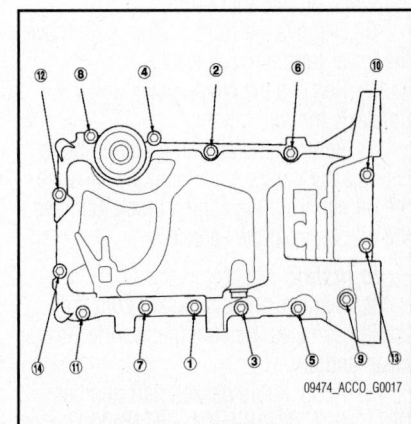

Fig. 160 Oil pan bolt tightening sequence—1.8L engine

➡**Do not install the parts if more than five minutes have elapsed since applying the liquid gasket. Instead, reapply after removing the previous coating material.**

12. Install the dowel pins, using new O-rings.

13. Position the oil pan in place. Tighten the oil pan retaining bolts in two or three steps to specification and in the proper sequence.

14. Continue the installation in the reverse order of the removal procedure.

15. After assembly, wait at least thirty minutes before filling the engine with clean engine oil.

16. Do not run the engine for at least three hours after installing the oil pan.

2.0L Engine

1. Note the radio security code and the radio presets. Disconnect the negative battery cable.

2. Raise and support the vehicle safely. Drain the engine oil. Remove the front tires.

3. Remove the splash shield. Separate the stabilizer links. Separate the knuckles from the lower arms.

4. Remove the steering gearbox bracket. Remove the steering gearbox mounting bolt, stiffener mounting bolt and stiffener.

5. Remove the gearbox mounting bolt, stiffener mounting bolt and stiffener. Remove the harness clamp from the subframe

6. Remove the lower torque rod. Remove the front mount mounting bolt.

7. Use a marker and make alignment marks on the reference lines that align with the centers of the rear subframe mounting bolts.

8. Loosen the mid-stiffener mounting bolts, on both sides. Support the subframe using the proper support tool.

9. Remove the front subframe. Remove the lower torque rod bracket.

10. Remove the clutch cover and the transaxle mounting bolts.

11. Remove the oil pan retaining bolts. Using a flat bladed tool, carefully separate the oil pan from the engine block. Remove the oil pan from the vehicle.

To install:

12. Remove any old gasket from the mating surfaces. Be sure these surfaces are clean and dry.

13. Apply liquid gasket, part number 08717-004, 08718-0001, 08718-0003 or 08718-0009 evenly to the engine block mating surface of the oil pan.

➡**Do not install the parts if more than five minutes have elapsed since applying the liquid gasket. Instead, reapply after removing the previous coating material.**

14. Position the oil pan in place. Tighten the oil pan retaining bolts in two or three steps to specification and in the proper sequence.

15. Continue the installation in the reverse order of the removal procedure.

16. After assembly, wait at least thirty minutes before filling the engine with clean engine oil.

17. Do not run the engine for at least three hours after installing the oil pan.

2012 Models

1.8L Non-Hybrid Engine

See Figures 161 and 162.

1. Remove the drive belt, as described in this section.

2. Remove the radiator mount upper brackets/cushions, the A/C condenser upper mount brackets, disconnect the connectors, remove the harness clamps, and remove the radiator bulkhead.

3. Remove the A/C condenser fan shroud assembly.

4. Remove the A/C compressor without disconnecting the A/C hoses. Do not bend the A/C hoses excessively. Hang the A/C compressor with a wire tie. See "Heating & Air Conditioning" section.

5. Raise the vehicle on a lift, and make sure it is securely supported.

6. Remove the splash shield and the engine undercover.

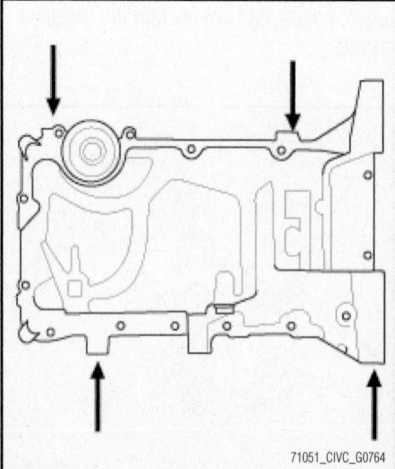

71051_CIVIC_G0764

Fig. 161 Using a flat blade screwdriver, separate the oil pan from the lower block in the places shown

7. Drain the engine oil.

8. Remove the A/C compressor bracket.

9. Remove exhaust pipe A, as described in this section.

10. Remove the torque converter case cover.

11. Remove the shift cable cover.

12. Remove the torque rod.

13. Remove the lower torque rod bracket.

14. Using a flat blade screwdriver, separate the oil pan from the lower block in the places shown.

15. Remove the oil pan.

To install:

16. Remove all of the old liquid gasket from the oil pan mating surfaces, the bolts, and the bolt holes.

17. Clean and dry the oil pan mating surfaces.

18. Apply liquid gasket to the engine block mating surface of the oil pan, and to the inside edge of the threaded bolt holes. Install the component within 5 minutes of applying the liquid gasket. Apply a 0.098 in. (2.5 mm) diameter bead of liquid gasket along the broken line.

➡**If too much time has passed after applying the liquid gasket, remove the old liquid gasket and residue, then reapply new liquid gasket.**

19. Install the dowel pins, then install the oil pan with new O-rings.

20. Tighten the bolts in three steps, in a crisscross pattern. In the final step, torque all bolts, in sequence, to 13 ft. lbs. (18 Nm). Wipe off the excess liquid gasket on the each side of crankshaft pulley and flywheel/drive plate.

➡**Wait at least 30 minutes before filling the engine with oil. Do not run the engine within 3 hours after installing the oil pan.**

21. Install the lower torque rod bracket. Tighten the bracket bolts to:
- M/T: 58 ft. lbs. (79 Nm)
- A/T: Top shoulder bolt: 29 ft. lbs. (39 Nm)
- A/T: Other bolts: 58 ft. lbs. (79 Nm)

22. Install the torque rod. Tighten the torque rod bolts to:
- Long bolt: 69 ft. lbs. (93 Nm)
- Short bolts: 65 ft. lbs. (74 Nm)

23. Install the shift cable cover, then the torque converter case cover.

24. Install exhaust pipe A, as described in this section.

25. Install the A/C compressor bracket. Tighten the bolts to 16 ft. lbs. (22 Nm).

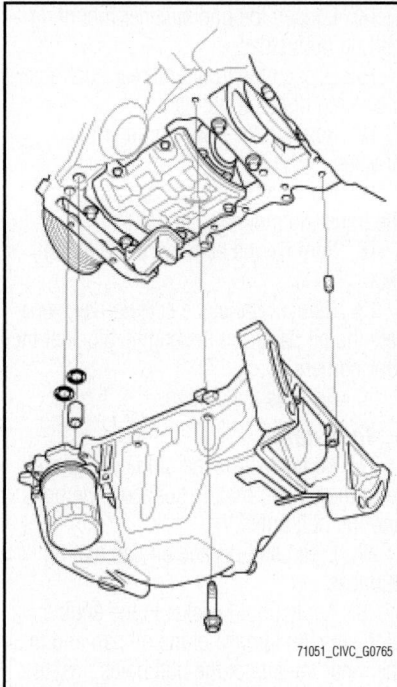

Fig. 162 Remove the oil pan

26. Install the engine undercover, then the splash shield.

27. Install the A/C compressor. Tighten the bolts to 16 ft. lbs. (22 Nm). Connect the connector and install the harness clamp.

28. Install the A/C condenser fan shroud assembly.

29. Install the radiator bulkhead, the harness clamps, and connect the connectors on the radiator.

30. Install the radiator mount upper bracket/cushion and the brackets.

31. Install the drive belt.

32. Refill the engine with the recommended engine oil.

33. After doing all procedures, select the individual maintenance item you wish to reset with the HDS.

1.8L Hybrid Engine

See Figures 163 through 167.

1. Raise the vehicle on a lift, and make sure it is securely supported.

2. Remove both front wheels.

3. Remove the splash shield.

4. Remove the engine under cover.

5. Drain the engine oil.

6. Remove the cooling guide.

7. Remove or disconnect the following (refer to applicable component headings or sections as needed):

- Condenser fan shroud
- A/C compressor (move without detaching hoses)
- Steering joint cover and pinch bolt

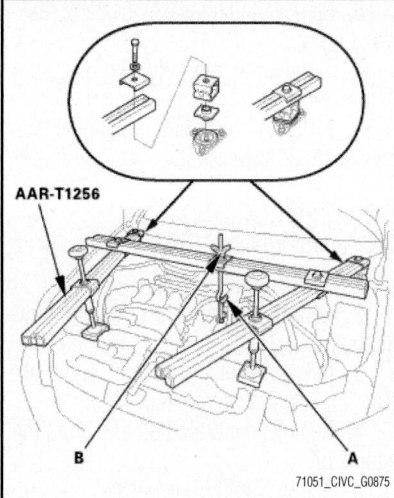

Fig. 163 Remove the interfering components and install the engine support hanger

- Steering joint (disconnect)
- Dipstick
- Intake manifold bracket
- Front stabilizer ball joints from lower arms
- Lower arms from steering knuckle
- Tie rod end ball joints (disconnect from knuckle)
- Radiator bulkhead (temporary installation)

8. Remove the interfering components and install the engine support hanger.

9. Remove the lower torque rod, as described in this section.

10. Remove the lower torque rod bracket.

11. Remove the driveshaft heat shield bracket.

12. Attach the subframe adapter to the front subframe by looping the belt over the

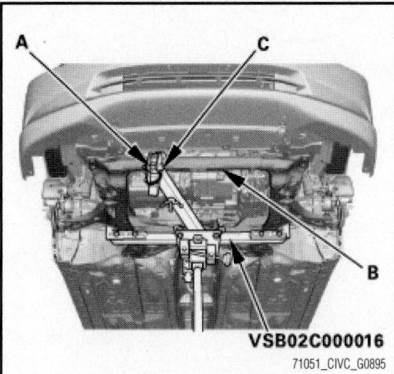

Fig. 164 Attach the subframe adapter to the front subframe by looping the belt (A) over the front of the front subframe (B), then secure the belt with its stop (C)

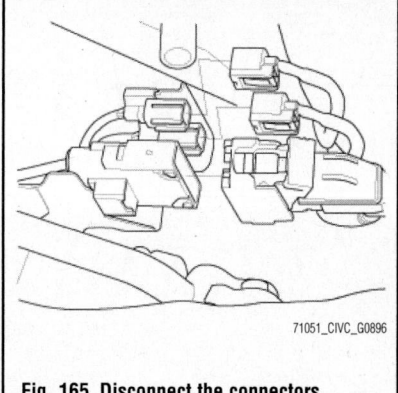

Fig. 165 Disconnect the connectors as shown

front of the front subframe, then secure the belt with its stop.

13. Raise the jack and line up the slots in the subframe adapter arms with the bolt holes on the jack base, then securely attach them with four bolts.

14. Disconnect the connectors as shown.

15. Loosen the mid-stiffener mounting bolts.

16. Lower the subframe.

17. Remove the dipstick tube.

18. Remove two bolts securing the transaxle.

19. Remove the oil pan bolts and carefully separate the oil pan at the corners.

To install:

20. Remove all of the old liquid gasket from the oil pan mating surfaces, the bolts, and the bolt holes.

21. Clean and dry the oil pan mating surfaces.

22. Apply liquid gasket to the engine block mating surface of the oil pan, and to the inside edge of the threaded bolt holes. Install the component within 5 minutes of

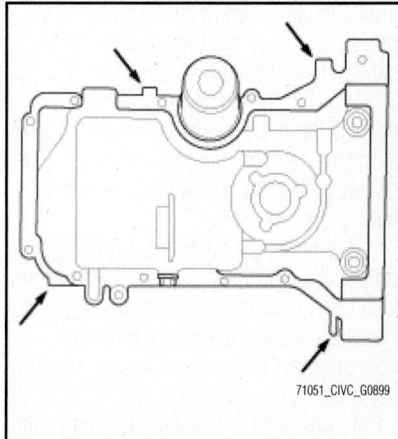

Fig. 166 Remove the oil pan bolts and carefully separate the oil pan at the corners

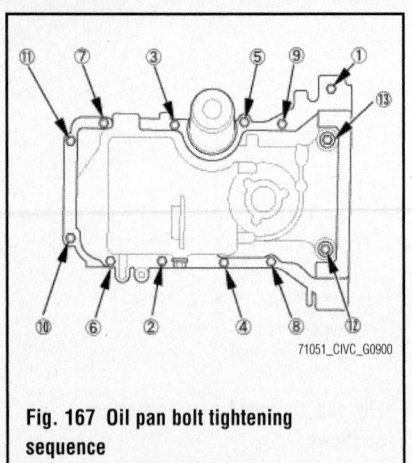

Fig. 167 Oil pan bolt tightening sequence

applying the liquid gasket. Apply a 0.098 in. (2.5 mm) diameter bead of liquid gasket along the broken line.

➡**If too much time has passed after applying the liquid gasket, remove the old liquid gasket and residue, then reapply new liquid gasket.**

23. Raise the oil pan carefully not to damage IMA motor (resolver stator).

24. Tighten the bolts in three steps. In the final step, torque all bolts, in sequence, to the specifications given. Wipe off the excess liquid gasket on the each side of crankshaft pulley and flywheel.

25. Torque specifications should be (final):
   - Bolt 1: 17 ft. lbs. (24 Nm)
   - Bolts 2–13: 9 ft. lbs. (12 Nm).

26. Install and tighten the transaxle bolts to 47 ft. lbs. (64 Nm)

27. Install the dipstick tube.

28. Raise the subframe into position.

29. Loosely install the subframe mounting bolts.

30. Insert the subframe alignment pin through the positioning hole on the right front subframe, and into the positioning hole on the body, then loosely tighten the subframe right front mounting bolt.

31. Insert the subframe alignment pin through the positioning slot on the left front subframe, and into the positioning hole on the body, then loosely tighten the subframe left front mounting bolt.

32. Tighten the subframe mounting bolts to the specified torque values starting with the right front subframe mounting bolt. Use the subframe alignment pin when tightening the front side subframe mounting bolts.

33. Check all of the subframe mounting bolts, and retighten if necessary.

34. Tighten the bolts to 77 ft. lbs. (104 Nm).

35. Remove the transaxle jack and the wood piece.

36. Tighten the mid-stiffeners bolts to 47 ft. lbs. (64 Nm).

37. Install the remaining components in reverse order of removal.

2.4L Engine

See Figures 168 and 169.

1. Warm up the engine.

2. Loosen the upper torque rod mounting bolt.

3. Raise the vehicle on a lift, and make sure it is securely supported.

4. Remove the drain bolt and drain the engine oil.

5. Remove both front wheels.

6. Remove the steering joint cover and pinch bolt.

7. Remove the splash shield.

8. Remove the flange nut while holding the respective joint in with a hex wrench. Remove the stabilizer link.

9. Remove the flange bolt and the self-locking nuts on both sides. Separate the lower arm using a pry bar on both sides.

10. Attach the universal lifting eyelet to the cylinder head.

11. Install the engine support hanger as shown.

12. Attach the hook to the slotted hole in the universal lifting eyelet. Tighten the wing nut by hand, and lift and support the engine/transaxle.

13. Remove the lower torque rod mounting bolts.

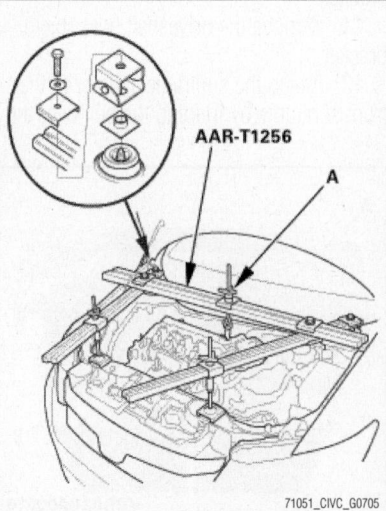

Fig. 168 Install the engine support hanger as shown. Attach the hook to the slotted hole in the universal lifting eyelet. Tighten the wing nut by hand, and lift and support the engine/transaxle

14. Loosen the mid stiffener mounting bolt on both sides.

15. Loosen the front and rear subframe mounting bolts on both sides.

16. Remove the lower torque rod bracket.

17. Remove the clutch case cover and the transaxle mounting bolts.

18. Remove the bolts securing the oil pan.

19. Using a flat blade screwdriver, separate the oil pan from the engine block at the four corners.

20. Remove the oil pan.

To install:

21. Remove all of the old liquid gasket from the oil pan mating surfaces, the bolts, and the bolt holes.

22. Clean and dry the oil pan mating surfaces.

23. Apply liquid gasket to the engine block mating surface of the oil pan and to the inner threads of the bolt holes. Install the component within 5 minutes of applying the liquid gasket.

➡**If too much time has passed after applying the liquid gasket, remove the old liquid gasket and residue, then reapply new liquid gasket.**

24. Install the oil pan on the engine block.

25. Tighten the bolts in three steps. In the final step, torque all bolts, in sequence, to 9 ft. lbs. (12 Nm). Wipe off the excess liquid gasket on the each side of crankshaft pulley and the flywheel/drive plate.

➡**Wait at least 30 minutes before filling the engine with oil. Do not run the engine for at least 3 hours after installing the oil pan.**

26. Install the clutch case cover and the transaxle mounting bolts. Tighten the case cover bolts to 9 ft. lbs. (12 Nm) and the

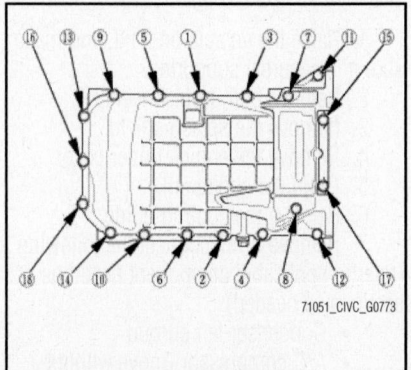

Fig. 169 Oil pan bolt tightening sequence

transaxle mounting bolts to 32 ft. lbs. (44 Nm).

27. Install the lower torque rod bracket using new bolts. Tighten bolt to 58 ft. lbs. (79 Nm).

28. Loosely install the new front subframe mounting bolts. Insert the subframe alignment pin through the positioning hole on the right front subframe, and into the positioning hole on the body, then loosely tighten the subframe right front mounting bolt.

29. Insert the subframe alignment pin through the positioning slot on the left front subframe, and into the positioning hole on the body, then loosely tighten the subframe left front mounting bolt.

30. Tighten the subframe mounting bolts to the specified torque values starting with the right front subframe mounting bolt. Use the subframe alignment pin when tightening the front side subframe mounting bolts.

31. Check all of the subframe mounting bolts, and retighten if necessary.

32. Loosely install the new lower torque rod mounting bolts.

33. Remove the engine support hanger.

34. Remove the universal lifting eyelet.

35. Loosen the lower torque mount mounting bolt.

36. Tighten the lower torque rod mount mounting bolts to 55 ft. lbs. (74 Nm) and the longer through bolt to 69 ft. lbs. (93 Nm).

37. Connect the lower arm to the knuckle and install the stabilizer link. See "Front Suspension" section.

38. Tighten the upper torque rod mounting bolt to 47 ft. lbs. (64 Nm).

39. Install the steering column components, as removed, and install the lower steering column pinch bolt and cover.

40. Install both front wheels.

41. Refill the engine with the recommended engine oil.

42. Install the splash shield.

43. Check the front wheel alignment. See "Wheel Alignment" section.

44. Connect the HDS to the data link connector (DLC) located under the driver's side of the dashboard.

45. Turn the ignition switch to ON (II).

46. Park the vehicle on a flat and level surface, with the steering wheel in the straight ahead position.

47. Select VSA ADJUSTMENT, then select ALL SENSORS with the HDS, and follow the screen prompts.

48. Select EPS ADJUSTMENT, then select EPS STEERING ANGLE SENSOR VALUE CLEAR and follow the screen prompts on the HDS.

49. After doing all procedures, select the individual maintenance item you wish to reset with the HDS.

OIL PUMP

REMOVAL & INSTALLATION

2011 Models

1.8L Engine

See Figures 170 and 171.

1. Note the radio security code and the radio presets.
2. Disconnect the negative battery cable.
3. Raise and support the vehicle safely.
4. Drain the engine oil.
5. Remove the front tires.
6. Remove the splash shield.
7. Lower the vehicle.
8. Remove the drive belt auto tensioner.
9. Remove the cylinder head cover.
10. Remove the PCV valve hose.
11. Remove the crankshaft pulley.
12. Properly support the engine, using a suitable jack and block of wood under the oil pan.
13. Remove the bolt securing the air conditioning line.
14. Remove the upper torque rod.
15. Remove the ground cable.

16. Remove the side engine mount bracket.

17. Remove the oil pump retaining bolts and remove the oil pump from the engine.

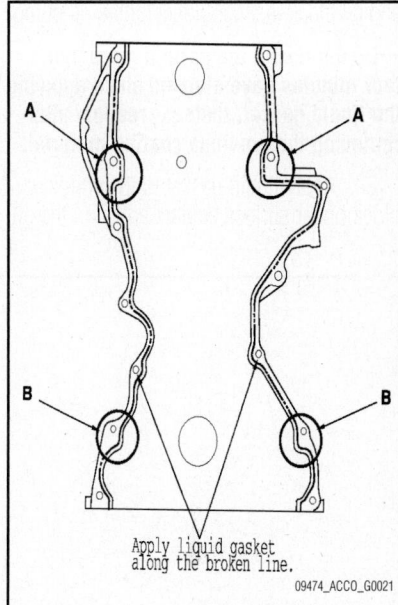

Apply liquid gasket along the broken line.

09474_ACCO_G0021

Fig. 171 Apply liquid gasket to the engine block upper surface contact areas (A) on the oil pump, lower block upper surface contact areas (B) of the oil pump—1.8L engine

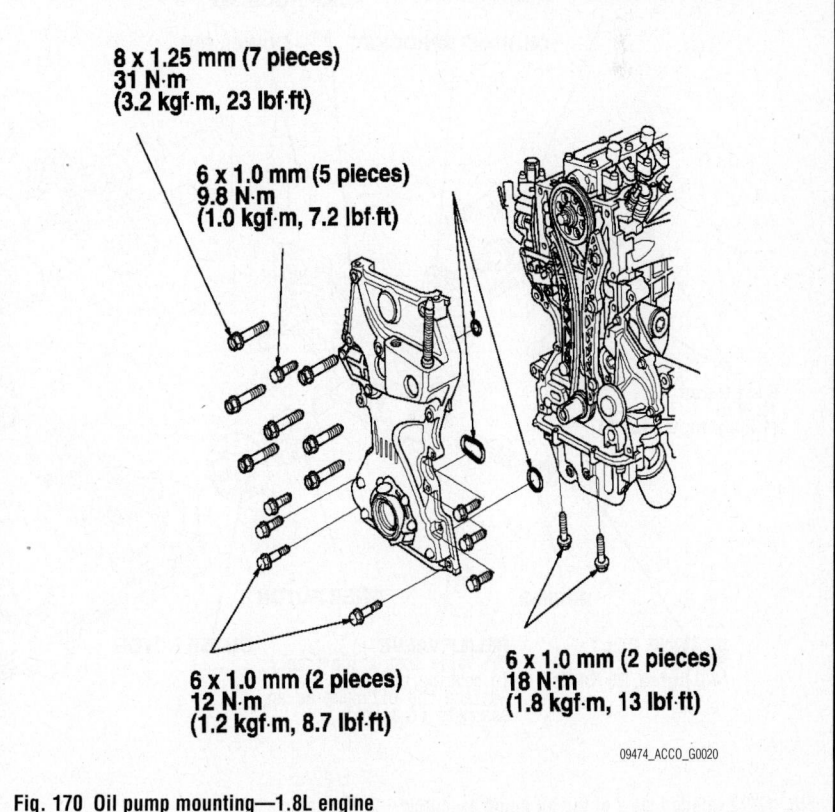

8 x 1.25 mm (7 pieces)
31 N·m
(3.2 kgf·m, 23 lbf·ft)

6 x 1.0 mm (5 pieces)
9.8 N·m
(1.0 kgf·m, 7.2 lbf·ft)

6 x 1.0 mm (2 pieces)
12 N·m
(1.2 kgf·m, 8.7 lbf·ft)

6 x 1.0 mm (2 pieces)
18 N·m
(1.8 kgf·m, 13 lbf·ft)

09474_ACCO_G0020

Fig. 170 Oil pump mounting—1.8L engine

To install:

18. Remove any old gasket from the mating surfaces. Be sure these surfaces are clean and dry.

19. Apply liquid gasket evenly to the engine block mating surface of the oil pump.

➡ **Do not install the parts if more than four minutes have elapsed since applying the liquid gasket. Instead, reapply after removing the previous coating material.**

20. Apply liquid gasket to the engine block upper surface contact areas on the oil pump, lower block upper surface contact areas of the oil pump.

21. Install a new O-ring on the oil pump. Set the edge of the oil pump on the edge of the oil pan. Install the oil pump on the engine block.

22. Loosely install the dowel bolts, and then tighten the 8mm bolts. Tighten the 6mm bolts and the dowel bolts.

23. The remainder of the installation is the reverse order of removal.

➡ **After assembly, wait at least thirty minutes before**

filling the engine with clean engine oil.

➡ **Do not run the engine for at least three hours after installing the oil pan.**

24. Refill the engine with oil to the correct level.

2.0L Engine

See Figures 172 through 174.

1. Note the radio security code and the radio presets.
2. Disconnect the negative battery cable.

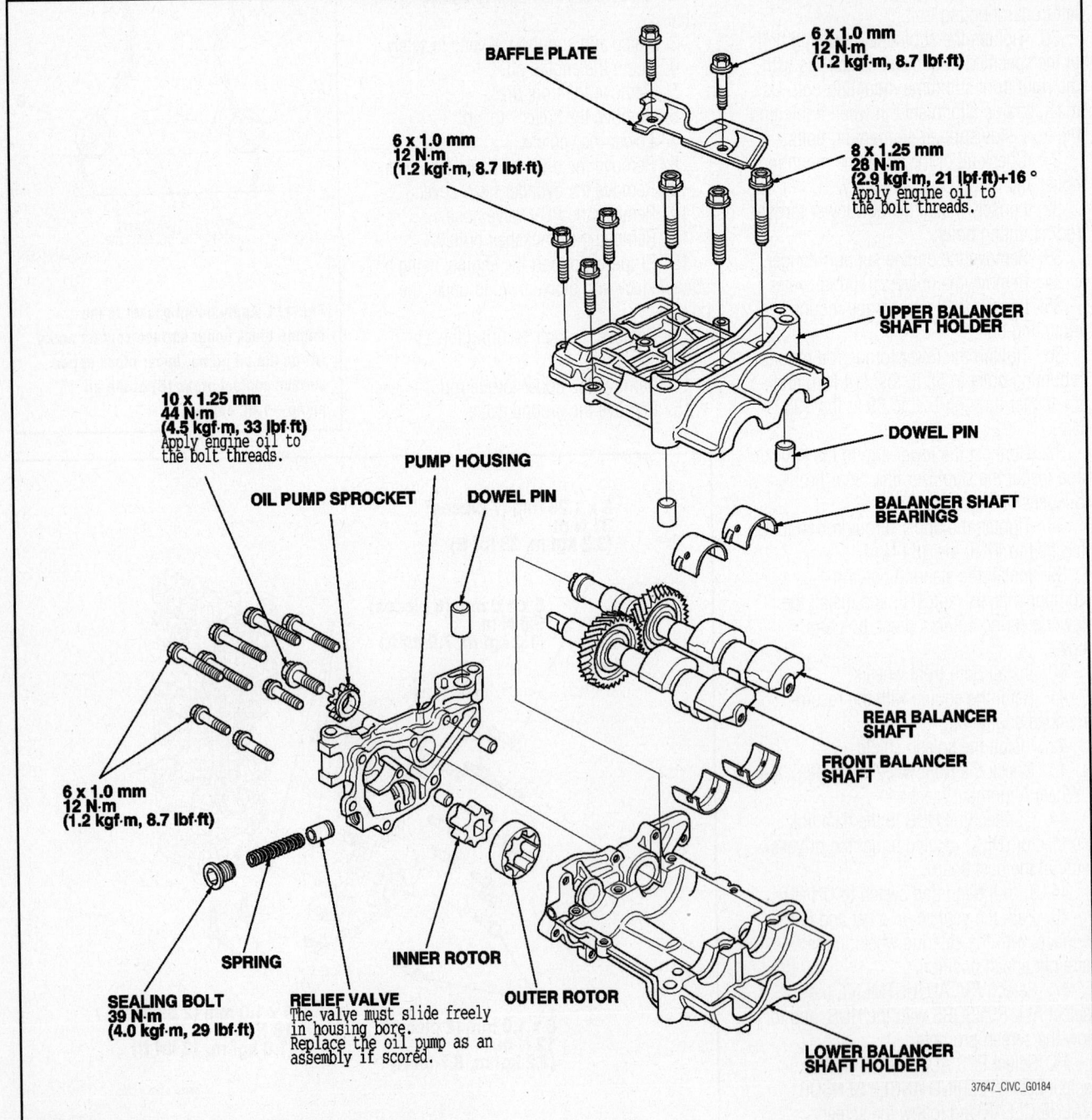

Fig. 172 Exploded view of the oil pump assembly—2.0L engine

37647_CIVC_G0184

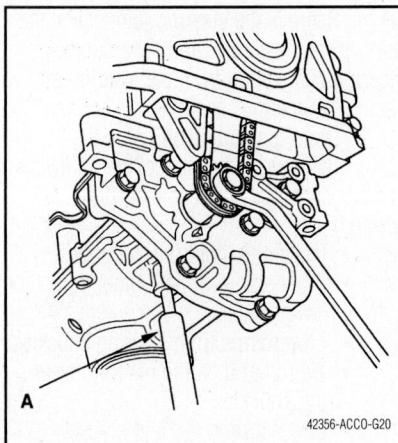

Fig. 173 Insert a 6mm pin into the maintenance hole in the lower balancer shaft holder and through the rear balancer shaft—2.0L engine

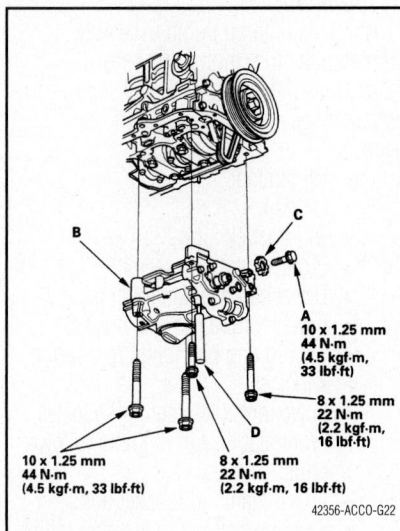

Fig. 174 Oil pump tightening specifications—2.0L engine

3. Position the number one piston at TDC.

4. Remove the oil pan.

5. Remove and discard the oil pump chain tensioner.

6. To hold the rear balancer shaft, insert a 6mm pin driver into the maintenance hole in the lower balancer shaft holder and through the rear balancer shaft.

7. Loosen the oil pump sprocket mounting bolt.

8. Remove the oil pump sprocket. Remove the oil pump.

To install:

9. Remove any old gasket from the mating surfaces. Be sure these surfaces are clean and dry.

10. Apply clean engine oil to the threads of the oil pump sprocket mounting bolt.

11. Loosely install the oil pump, and then install the oil pump sprocket. Remove the pin driver.

12. Tighten the pump retaining bolts to specification.

13. Squeeze the new oil pump chain tensioner and then install the set clip.

14. Install the oil pump chain tensioner. Remove the set clip from the pump chain tensioner.

15. Continue the installation in the reverse order of the removal procedure.

➡**After assembly, wait at least thirty minutes before filling the engine with clean engine oil.**

➡**Do not run the engine for at least three hours after installing the oil pan.**

16. Refill the engine with oil to the correct level.

2012 Models

1.8L Non-Hybrid Engine

See Figures 175 and 176.

1. Remove or disconnect the following components (see applicable sections, as needed):

- Battery cables
- Battery
- Raise and support the vehicle
- Right front wheel
- Splash shield
- Water pump (loosen mounting bolts only)
- Drive belt
- Alternator
- Water pump pulley
- Auto tensioner
- Idler pulley base
- Engine cover

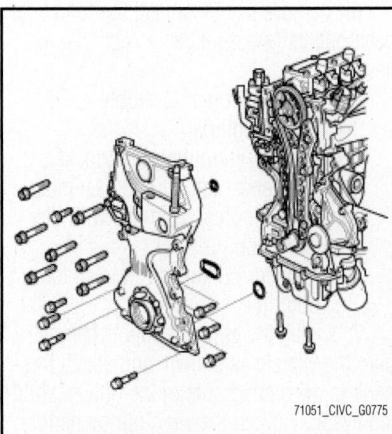

Fig. 175 Remove the oil pump assembly

- Cylinder head cover
- Crankshaft pulley

2. Lift and support the engine with a wood block under the oil pan.

3. Remove the side engine mount bracket and mount.

4. Disconnect the PCV hose.

5. Remove the oil pump assembly.

To install:

6. Check the oil pump oil seal for damage. If the oil seal is damaged, replace the oil seal.

7. Remove all of the old liquid gasket from the oil pump mating surfaces, the bolts, and the bolt holes.

8. Clean and dry the oil pump mating surfaces.

9. Apply liquid gasket to the engine block mating surface of the oil pump, and to the inside edge of the threaded bolt holes. Install the component within 5 minutes of applying the liquid gasket.

10. Apply a 0.098 in. (2.5 mm) diameter bead of liquid gasket along the broken line.

➡**If too much time has passed after applying the liquid gasket, remove the old liquid gasket and residue, then reapply new liquid gasket.**

11. Apply liquid gasket to the oil pan mating surface of the oil pump, and to the inside edge of the threaded bolt holes. Install the component within 5 minutes of applying the liquid gasket. Apply a 0.098 in. (2.5 mm) diameter bead of liquid gasket.

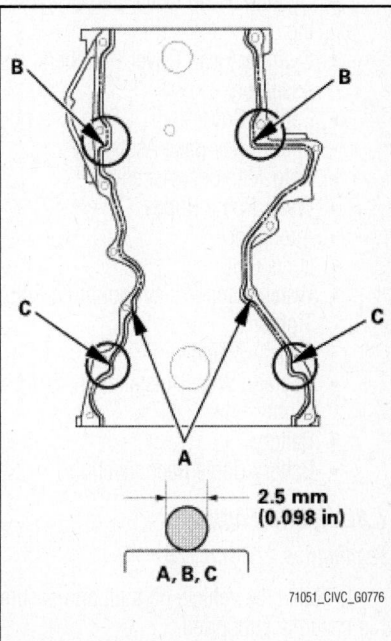

Fig. 176 Apply a 0.098 in. (2.5 mm) diameter bead of liquid gasket along the broken line (A), (B), and (C)

➡If too much time has passed after applying the liquid gasket, remove the old liquid gasket and residue, then reapply new liquid gasket.

12. Set the edge of the oil pump on the edge of the oil pan with new O-rings. Install the oil pump on the engine block. Loosely install the dowel bolts, then tighten the 8 mm bolts, the 6 mm bolts and the dowel bolts.

13. Wipe off the excess liquid gasket on the oil pan and oil pump mating surface.

➡When installing the oil pump, do not slide the bottom surface onto the oil pan mounting surface.

➡Wait at least 30 minutes before filling the engine with oil. Do not run the engine within 3 hours after installing the oil pump.

14. Install or reconnect the following (see applicable component sections, as needed):

- Mounting Bracket, Engine Side
- Remove the jack and the wood block.
- Transaxle Mount Bracket Mounting Bolt - Loosen
- Lower Torque Rod - Loosen
- Side Engine Mount - Tighten
- Transaxle Mount Bracket Mounting Bolt - Tighten
- Lower Torque Rod Mounting Bolt - Tighten
- Crankshaft Pulley
- Cylinder Head Cover and/or Packing
- Cylinder Head Cover Peripheral Assembly
- Engine Cover
- Idler Pulley Base Assembly
- Auto Tensioner Assembly
- Water Pump Pulley
- Alternator
- Drive Belt
- Water Pump Pulley Mounting Bolt - Tighten
- Splash Shield
- Tire and Wheel-Installation, Front Right
- Battery
- Battery Cable Reconnection

1.8L Hybrid Engine

See Figures 177 and 178.

1. Raise the vehicle on a lift and ensure it is securely supported.

2. Remove or disconnect the following components (see applicable sections, as needed):

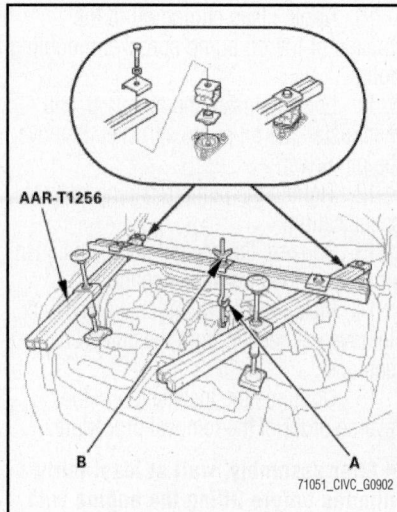

AAR-T1256

71051_CIVIC_G0902

Fig. 177 Install the engine support hanger onto the vehicle as shown, and attach the hook (A) to the slotted hole in the universal lifting eyelet. Tighten the wing nut (B) by hand, and lift and support the engine

- Engine cover
- Ignition coils
- Cylinder head cover

3. Set the No. 1 piston at top dead center (TDC). The "UP" mark on the camshaft sprocket should be at the top, and the TDC grooves on the camshaft sprocket should line up with the top edge of the head.

- Right front wheel
- Splash shield
- Engine under cover
- Water pump (loosen mounting bolts only)
- Drive belt
- Water pump pulley
- Crankshaft pulley

4. Lift and support the engine with a wood block under the oil pan.

5. Remove the cam chain case and cam chain, as described in this section.

6. Remove or disconnect the following components (see applicable sections, as needed):

- Auto tensioner assembly
- Cooling guide
- Radiator mounting top brackets
- Condenser fan shroud assembly
- A/C compressor (move with disconnecting hoses)
- Radiator bulkhead (reinstall temporarily)

7. Install the engine support hanger onto the vehicle as shown, and attach the hook to the slotted hole in the universal lifting eyelet. Tighten the wing nut by hand, and lift and support the engine.

8. Remove the steering joint cover, then remove the pinch bolt and separate the steering connection from the steering pinion. See "Steering" section.

9. Drain the engine oil.

10. Remove or disconnect the following components (see applicable sections, as needed):

- Intake manifold bracket bolts
- Front stabilizer ball joints from lower arms
- Lower arms from steering knuckles
- Tie rod end ball joints from steering knuckles
- Lower torque rod and bracket
- Driveshaft heat shield bracket
- Front subframe assembly
- Oil dipstick tube
- Oil pan
- Oil strainer

11. Remove the oil pump assembly.

To install:

12. Install the oil pump assembly. Tighten bolts to 9 ft. lbs. (12 Nm).

13. Install or connect the following components (see applicable sections, as needed):

- Oil strainer
- Oil pan
- Oil dipstick tube
- Front subframe assembly
- Driveshaft heat shield bracket
- Lower torque rod and bracket
- Tie rod end ball joints to steering knuckles
- Lower arms to steering knuckles
- Front stabilizer ball joints to lower arms
- Intake manifold bracket bolts
- Steering column to steering pinion and pinch bolt
- Engine support hanger
- Radiator bulkhead (remove)
- A/C compressor
- Condenser fan shroud

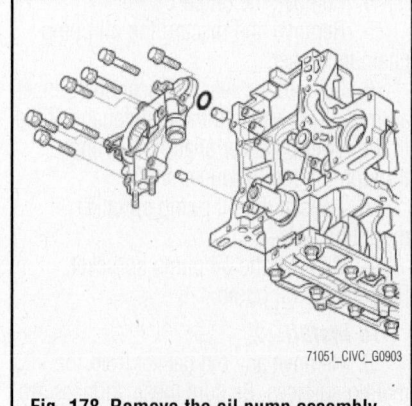

71051_CIVIC_G0903

Fig. 178 Remove the oil pump assembly

- Radiator bulkhead, mounting brackets and connections
- Cooling guide
- Auto tensioner assembly
- Cam chain and chain case
- Side engine mount
- Engine jack and wood block (remove)
- Crankshaft pulley
- Water pump pulley (loose bolts)
- Drive belt
- Water pump pulley (tighten bolts)
- Engine under cover
- Splash shield
- Cylinder head cover and related components
- Ignition coils
- Engine cover
- Engine oil (refill)
- Front wheels

14. Perform front wheel alignment. See "Wheel Alignment" section.

15. Connect the HDS to the data link connector (DLC) located under the driver's side of the dashboard.

16. Turn the ignition switch to ON (II).

17. Park the vehicle on a flat and level surface, with the steering wheel in the straight ahead position.

18. Select VSA ADJUSTMENT, then select ALL SENSORS with the HDS, and follow the screen prompts.

19. Select EPS ADJUSTMENT, then select EPS STEERING ANGLE SENSOR VALUE CLEAR and follow the screen prompts on the HDS.

20. After completing the entire procedure, select the individual maintenance item you wish to reset with the HDS.

2.4L Engine

See Figures 179 through 182.

1. Warm up the engine.
2. Raise the vehicle on a lift.
3. Drain the engine oil.
4. Remove or disconnect the following components (see applicable sections, as needed):
   - Steering joint and pinch bolt
   - Splash shield
   - Front lower ball joints
5. Install the engine lift eye on the cylinder head, then install the engine support hanger.
6. Remove or disconnect the following components (see applicable sections, as needed):
   - Lower torque rod
   - Front subframe assembly
7. Turn the crankshaft pulley so its top

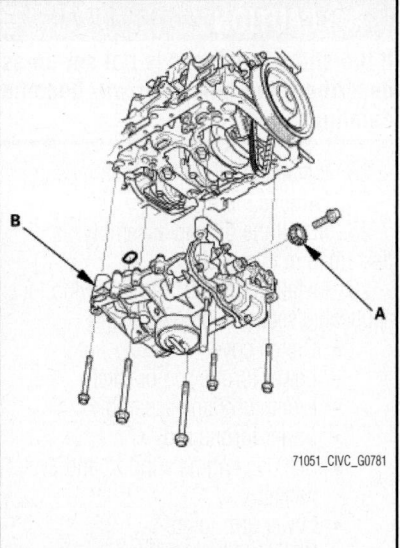

71051_CIVIC_G0781

Fig. 179 Remove the oil pump sprocket. Remove the oil pump

center (TDC) mark lines up with the pointer. (The other pointer is not used.)

8. Remove the lower torque rod bracket and the clutch cover (if equipped).

9. Remove the bolts securing the oil pan.

10. Using a flat blade screwdriver, separate the oil pan from the engine block at the four corners.

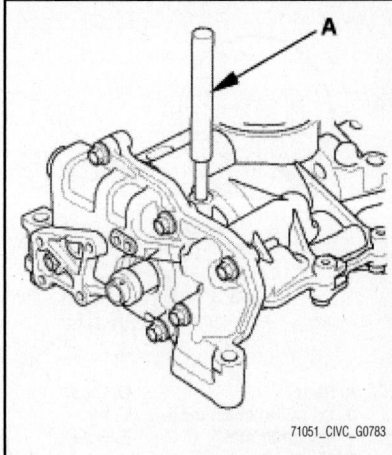

71051_CIVIC_G0783

Fig. 180 To hold the rear balancer shaft, insert a 6 mm long pin punch (A) (Snap-on PPC108LA or equivalent) into the maintenance hole in the balancer shaft holder and through the rear balancer shaft

11. Remove the oil pan.

12. To hold the rear balancer shaft, insert a 6 mm long pin punch into the maintenance hole in the balancer shaft holder and through the rear balancer shaft.

13. Turn the crankshaft counterclockwise to compress the oil pump chain auto-tensioner.

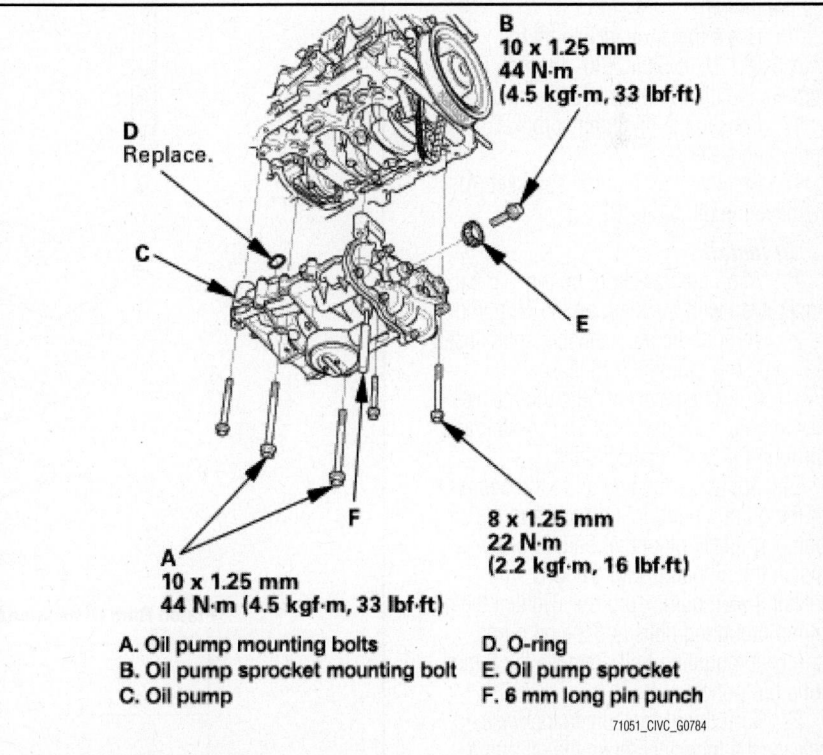

A. Oil pump mounting bolts
B. Oil pump sprocket mounting bolt
C. Oil pump
D. O-ring
E. Oil pump sprocket
F. 6 mm long pin punch

71051_CIVIC_G0784

Fig. 181 Apply new engine oil to the threads of the oil pump mounting bolts and the oil pump sprocket mounting bolt. Loosely install the oil pump with a new O-ring. Install the oil pump sprocket. Tighten the oil pump mounting bolts and the oil pump sprocket mounting bolt. Remove the 6 mm long pin punch

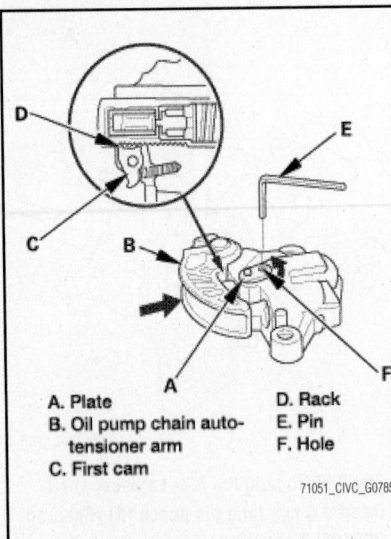

A. Plate
B. Oil pump chain auto-tensioner arm
C. First cam
D. Rack
E. Pin
F. Hole

71051_CIVC_G0785

Fig. 182 Turn the plate counterclockwise, to release the lock, then push the oil pump chain auto-tensioner arm, and set the first cam to the first edge of the rack. Insert a 7/64 in. (3.0 mm) diameter pin into the hole

14. Remove the crankshaft pulley.
15. Align the holes on the lock and the oil pump chain auto-tensioner, then insert a 3.0 mm (7/64 in.) diameter pin into the holes. Turn the crankshaft clockwise to secure the pin.
16. Use a transaxle jack to lift the transaxle 1.18–1.57 in. (30–40 mm), then remove the oil pump chain auto-tensioner.
17. Loosen the oil pump sprocket mounting bolt.
18. Remove the oil pump sprocket (A). Remove the oil pump (B).

To install:

19. Align the dowel pin on the rear balancer shaft with the mark on the oil pump.
20. To hold the rear balancer shaft, insert a 6 mm long pin punch (Snap-on PPC108LA or equivalent) into the maintenance hole in the balancer shaft holder and through the rear balancer shaft.
21. Apply new engine oil to the threads of the oil pump mounting bolts and the oil pump sprocket mounting bolt. Loosely install the oil pump with a new O-ring. Install the oil pump sprocket. Tighten the oil pump mounting bolts and the oil pump sprocket mounting bolt. Remove the 6 mm long pin punch.
22. Turn the plate counterclockwise, to release the lock, then push the oil pump chain auto-tensioner arm, and set the first cam to the first edge of the rack. Insert a 7/64 in. (3.0 mm) diameter pin into the hole.

✳✳ CAUTION

If the chain tensioner is not set up as described, the tensioner will become damaged.

23. Install the oil pump chain tensioner.
24. Remove the pin.
25. Install the oil pan assembly, as described in this section.
26. Install or connect the following (see applicable sections, as needed):
   - Clutch cover
   - Lower torque rod bracket
   - Front subframe assembly
   - Lower torque rod
   - Remove engine support and eyelet hanger.
   - Lower arm joints
   - Stabilizer ball joints
   - Steering column, wheel, pinch bolt and cover
   - Front wheels
27. Replace the engine oil.
28. Install the splash shield.

29. Perform front wheel alignment. See "Wheel Alignment" section.
30. Connect the HDS to the data link connector (DLC) located under the driver's side of the dashboard.
31. Turn the ignition switch to ON (II).
32. Park the vehicle on a flat and level surface, with the steering wheel in the straight ahead position.
33. Select VSA ADJUSTMENT, then select ALL SENSORS with the HDS, and follow the screen prompts.
34. Select EPS ADJUSTMENT, then select EPS STEERING ANGLE SENSOR VALUE CLEAR and follow the screen prompts on the HDS.
35. After doing all procedures, select the individual maintenance item you wish to reset with the HDS.

PISTON AND RING

POSITIONING

See Figures 183 through 185.

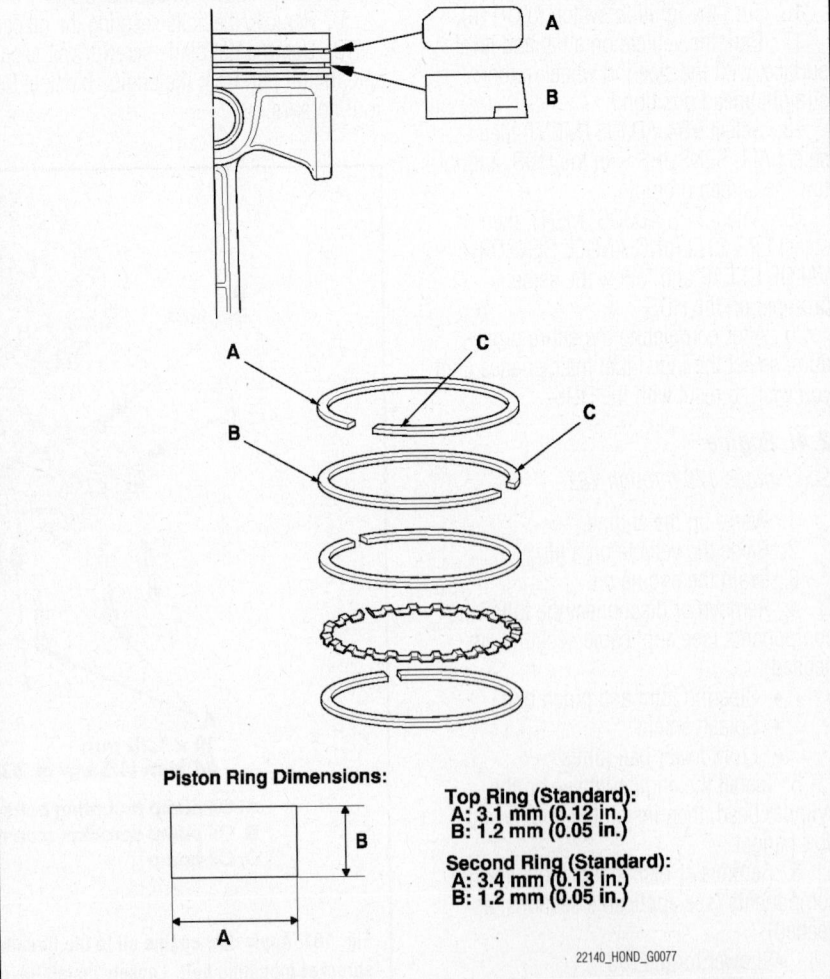

Piston Ring Dimensions:

Top Ring (Standard):
A: 3.1 mm (0.12 in.)
B: 1.2 mm (0.05 in.)

Second Ring (Standard):
A: 3.4 mm (0.13 in.)
B: 1.2 mm (0.05 in.)

22140_HOND_G0077

Fig. 183 Top ring (A), second ring (B) and the manufacturing marks (C) must face upward—2011 models

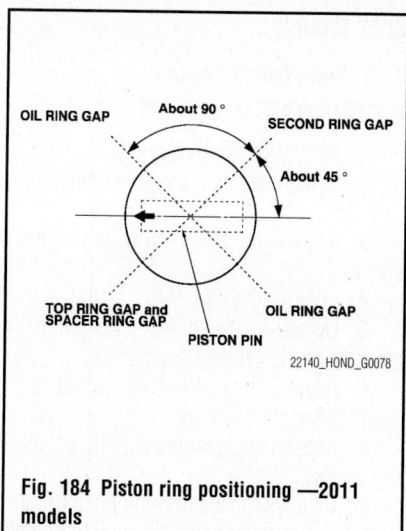

Fig. 184 Piston ring positioning —2011 models

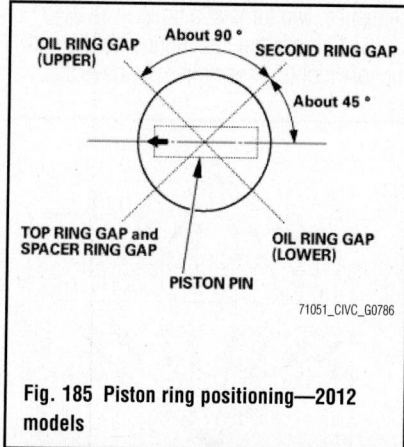

Fig. 185 Piston ring positioning—2012 models

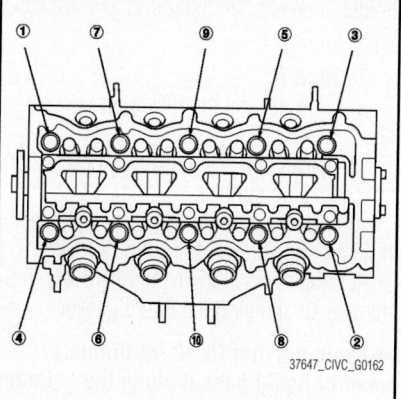

Fig. 186 Remove the lost motion holder bolts. To prevent damaging the lost motion holder and the rocker shaft, loosen the bolts, in sequence, two turns at a time, in the order shown

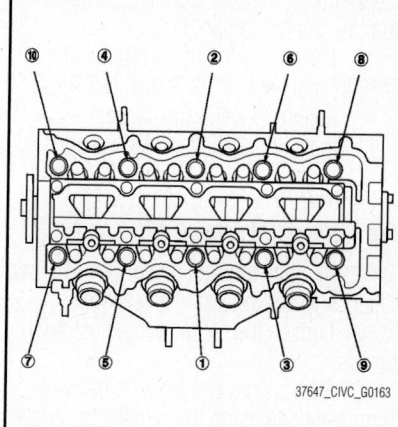

Fig. 188 Tighten each bolts two turns at a time, in sequence, to 11 ft. lbs. (15 Nm)

ROCKER ARMS

REMOVAL & INSTALLATION

2011 Models

1.8L Engine

See Figures 186 through 188.

1. Remove the cylinder head cover.
2. Loosen the rocker arm adjusting screws.
3. Remove the lost motion holder bolts. To prevent damaging the lost motion holder and the rocker shaft, loosen the bolts, in sequence, two turns at a time, in the order shown.
4. Remove the lost motion holder and the lost motion assemblies.
5. Remove the rocker arm assembly, then remove the oil control orifice.

To install:

6. If the rocker arm assembly is disassembled, reassemble the rocker arm assembly.
7. Install the oil control orifice with a new O-ring, then install the rocker arm assembly.

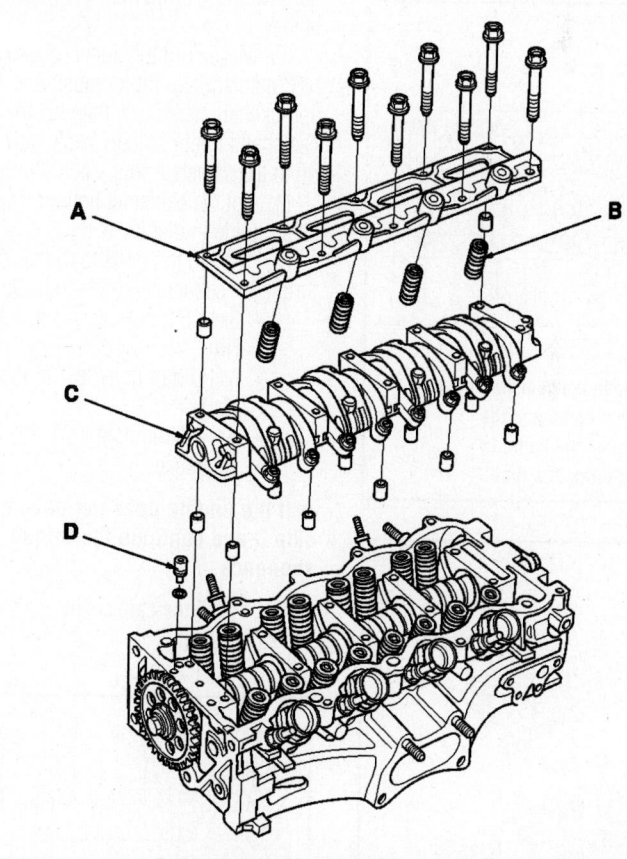

A. Lost motion holder
B. Lost motion assemblies
C. Rocker arm assembly
D. Oil control orifice

Fig. 187 Remove the lost motion holder and the lost motion assemblies, then remove the rocker arm assembly and remove the oil control orifice

8. Install the lost motion assembles and the lost motion holder.

9. Tighten each bolts two turns at a time, in sequence, to 11 ft. lbs. (15 Nm).

10. Adjust the valve clearance.

11. Install the cylinder head cover.

2.0L Engine

See Figures 189 through 191.

1. Remove the cam chain. See "Timing Chain & Sprockets" in this section.

2. Loosen the rocker arm adjusting screws.

3. Remove the camshaft holder bolts. To prevent damaging the camshafts, loosen the bolts, in sequence, two turns at a time.

➡**Bolt "1" is not on all engines.**

4. Remove cam chain guide, the camshaft holders, and the camshafts.

5. Insert the bolts into the rocker shaft

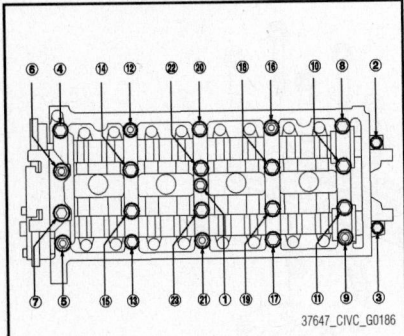

Fig. 189 Remove the camshaft holder bolts. To prevent damaging the camshafts, loosen the bolts, in sequence, two turns at a time

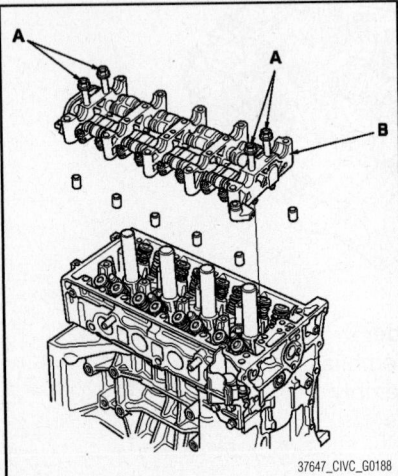

Fig. 190 Insert the bolts (A) into the rocker shaft holder, then remove the rocker arm assembly (B)

holder, then remove the rocker arm assembly.

To install:

6. Reassemble the rocker arm assembly.

7. Clean and dry the No. 5 rocker shaft holder mating surface.

8. Apply liquid gasket to the cylinder head mating surface of the No. 5 rocker shaft holder. Install the component within 5 minutes of applying the liquid gasket.

➡**Apply a 3 mm (0.12 in) diameter bead of liquid gasket along the cylinder head-to-rocker shaft holder mating surface. If too much time has passed after applying the liquid gasket, remove the old liquid gasket and residue, then reapply new liquid gasket.**

9. Insert the bolts used for removal into the rocker shaft holder, then install the rocker arm assembly on the cylinder head. Remove these bolts from the rocker shaft holder.

10. Make sure the punch marks on the VTC actuator and the exhaust camshaft sprocket are facing up, then set the camshafts in the holder. Apply new engine oil to the camshaft journals and lobes.

11. Set the camshaft holders and cam chain guide in place.

12. Tighten the bolts to the specified torque, in sequence, to the following:

- 8x1.25 mm bolts: 16 ft. lbs. (22 Nm)
- 6x1.0 mm bolts: 8.7 ft. lbs. (12 Nm)
- 6 x 1.0 mm bolts: 21, 22, 23 in sequence

➡**If the engine does not have bolt 21, skip it and continue the torque sequence.**

13. Install the cam chain, then adjust the valve clearance.

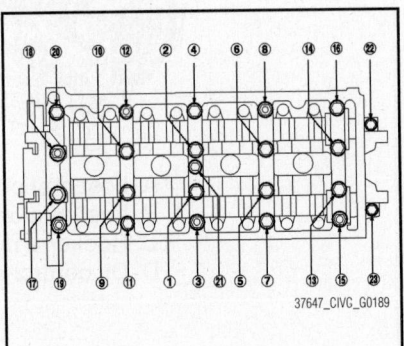

Fig. 191 Rocker arm bolt tightening sequence

2012 Models

1.8L Non-Hybrid Engine

See Figures 192 through 195.

1. Remove the engine cover.

2. Disconnect the connecters and the harness holders.

3. Remove the alternator wiring harness clamp.

4. Remove the dipstick.

5. Disconnect the breather hose from the front of the cylinder head cover.

6. Remove the cylinder head cover, as described in this section.

7. Loosen the locknuts and the adjusting screws.

8. Remove the lost motion holder bolts. To prevent damaging the lost motion holder and the rocker shaft, loosen the bolts, in sequence, two turns at a time.

9. Remove the lost motion holder and the lost motion assemblies. Remove the

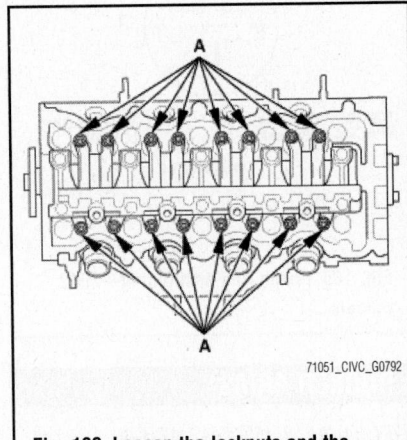

Fig. 192 Loosen the locknuts and the adjusting screws (A)

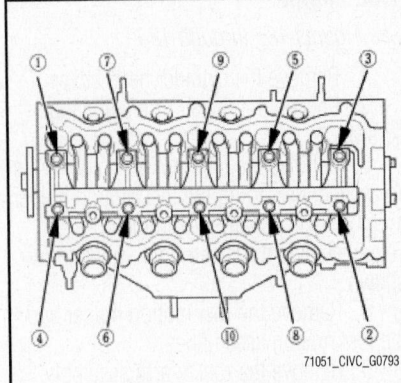

Fig. 193 Remove the lost motion holder bolts. To prevent damaging the lost motion holder and the rocker shaft, loosen the bolts, in sequence, two turns at a time

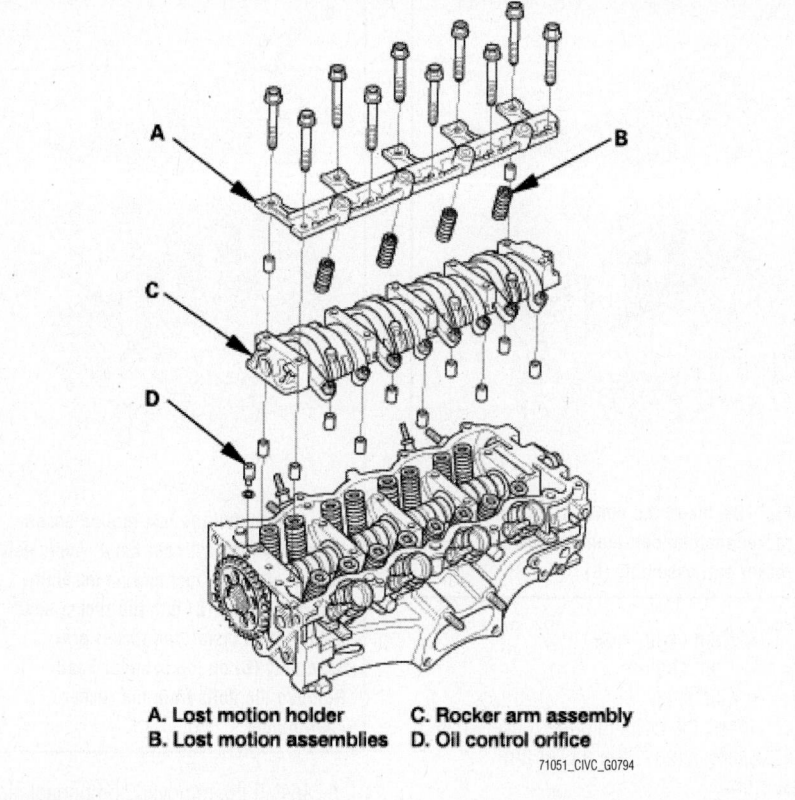

A. Lost motion holder   C. Rocker arm assembly
B. Lost motion assemblies   D. Oil control orifice

71051_CIVC_G0794

Fig. 194 Remove the lost motion holder and the lost motion assemblies. Remove the rocker arm assembly. Remove the oil control orifice

rocker arm assembly. Remove the oil control orifice.

10. Disassemble the rocker arm assembly if necessary. Keep all parts carefully ordered with their original components.

To install:

11. If the rocker arm assembly is disassembled, reassemble the rocker arm assembly.

12. Install the oil control orifice with a new O-ring. Install the rocker arm assembly. Install the lost motion assemblies and the lost motion holder.

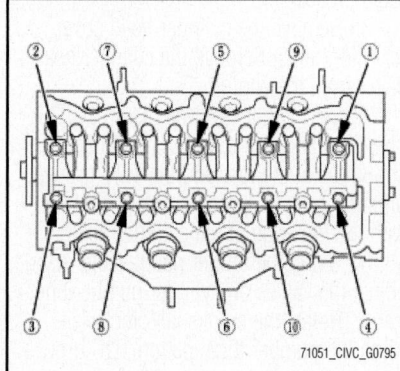

71051_CIVC_G0795

Fig. 195 Rocker arm bolt tightening sequence

➡**Apply new engine oil to the lost motion assembly.**

13. Tighten each bolts two turns at a time in sequence, to 11 ft. lbs. (15 Nm).

14. Perform a valve clearance adjustment, as described in this section.

15. Install the cylinder head cover, as described in this section.

16. Connect the breather hose, then install the dipstick.

17. Install the alternator wiring harness clamp

18. Install the cylinder head cover peripheral assembly.

19. Install the harness holders and the connectors.

20. Install the engine cover.

1.8L Hybrid Engine

See Figures 196 through 198.

1. Remove the engine cover, as described in this section.

2. Remove the ignition coils. See "Engine Electrical" section.

3. Remove the cylinder head cover, as described in this section.

4. Remove the cam chain sprocket. See "Cam Chain & Sprockets" in this section.

5. Loosen the rocker arm adjusting screws.

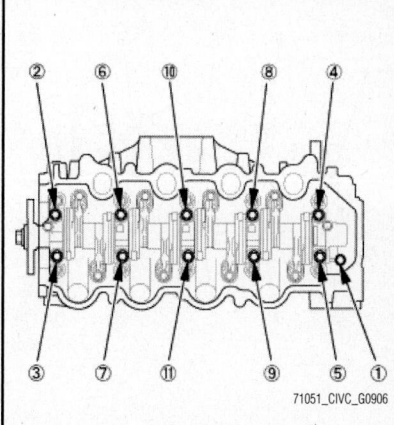

71051_CIVC_G0906

Fig. 196 Remove the camshaft holder bolts. To prevent damaging the camshaft, loosen the bolts in sequence two turns at a time

6. Remove the camshaft holder bolts. To prevent damaging the camshaft, loosen the bolts in sequence two turns at a time.

7. Remove the rocket arm assembly.

To install:

8. Position the rocker arm assembly into place.

9. Apply new engine oil to the bolt threads and flange

10. Tighten each bolts two turns at a time in sequence.

11. Tighten bolts as follows:
   • 8 mm bolts: 16 ft. lbs. (22 Nm)

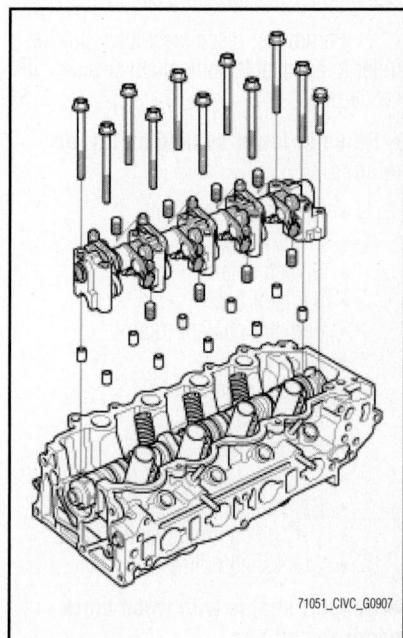

71051_CIVC_G0907

Fig. 197 Remove the rocker arm assembly

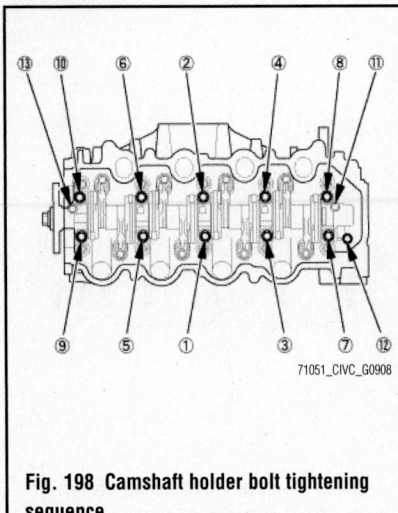

Fig. 198 Camshaft holder bolt tightening sequence

- 8 mm bolts 11 and 13: 15 ft. lbs. (20 Nm)
- 6 mm bolt 12: 9 ft. lbs. (12 Nm)

12. Install the cam chain to the camshaft sprocket by aligning the reference mark, then install the camshaft sprocket on the camshaft.

13. Continue with installation of the cam chain, sprockets and cam chain case, as described in this section.

14. Perform valve clearance adjustment, as described in this section.

15. Install the cylinder head cover, ignition coils and engine cover, as described in this section.

2.4L Engine

See Figures 199 through 201.

1. Remove or disconnect the following (refer to applicable component sections, if needed):

➡**Raise or lower vehicle on lift, as needed.**

- Air cleaner assembly
- Front wheels
- Splash shield
- Radiator cap
- Radiator coolant (drain)
- Drive belt
- Wiper arm assembly
- Engine cover
- Ignition coils
- Cylinder head cover
- Cam chain case peripherals
- VTC oil control solenoid valve
- Crankshaft pulley

➡**Support engine with wood block under the oil pan.**

- Upper torque rod
- Side engine mount and bracket

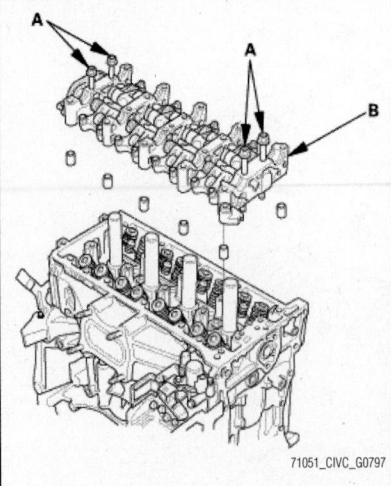

Fig. 199 Insert the bolts (A) into the rocker shaft holder, then remove the rocker arm assembly (B)

- Cam chain case
- Cam chain
- Camshaft

2. Insert the bolts into the rocker shaft holder, then remove the rocker arm assembly.

To install:

3. Apply a 2.5 mm (0.098 in) diameter bead of liquid gasket along the broken line.

➡**If too much time has passed after applying the liquid gasket, remove the old liquid gasket and residue, then reapply new liquid gasket.**

4. Install the lost motion assemblies in the cylinder head. Apply new engine oil to the lost motion assembly. Insert the bolts into the rocker shaft holder, then install the rocker arm assembly on the cylinder head. Remove the bolts from the rocker shaft holder.

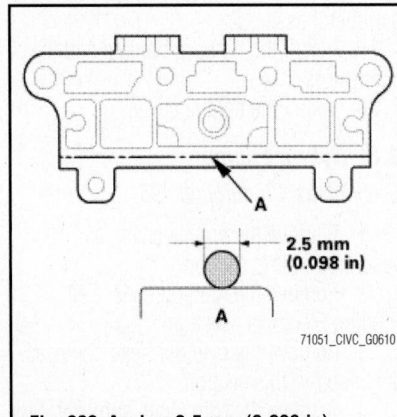

Fig. 200 Apply a 2.5 mm (0.098 in) diameter bead of liquid gasket along the broken line (A)

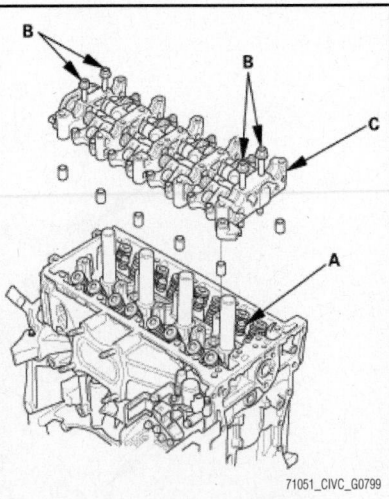

Fig. 201 Install the lost motion assemblies (A) in the cylinder head. Apply new engine oil to the lost motion assembly. Insert the bolts (B) into the rocker shaft holder, then install the rocker arm assembly (C) on the cylinder head. Remove the bolts from the rocker shaft holder

5. Install the removed components in reverse order of removal.

VALVE LASH (CLEARANCE)

ADJUSTMENT

2011 Models

1.8L Engine

➡**Adjust valves only when the cylinder head temperature is less than 100 degrees F.**

1. Before servicing the vehicle, refer to the precautions in the beginning of this section. Disconnect the negative battery cable.

2. Note the radio security code and the radio presets.

3. Remove the cylinder head cover retaining bolts. Remove the cylinder head cover from the engine.

4. Set the number one piston at TDC. The UP mark on the camshaft sprocket should be at the top, and the TDC grooves on the camshaft sprocket should line up with the top edge of the cylinder head.

5. Using the proper gauge feeler gauge, adjust the valves on cylinder number one.

6. Rotate the crankshaft clockwise. Align the number three piston TDC groove on the camshaft sprocket with the top edge of the cylinder head.

7. Using the proper gauge feeler gauge, adjust the valves on cylinder number three.

8. Rotate the crankshaft clockwise. Align the number four piston TDC groove on the camshaft sprocket with the top edge of the cylinder head.

9. Using the proper gauge feeler gauge, adjust the valves on cylinder number four.

10. Rotate the crankshaft clockwise. Align the number two piston TDC groove on the camshaft sprocket with the top edge of the cylinder head.

11. Using the proper gauge feeler gauge, adjust the valves on cylinder number two.

12. Install the cylinder head cover.

2.0L Engine

➡ **Adjust valves only when the cylinder head temperature is less than 100 degrees F.**

1. Before servicing the vehicle, refer to the precautions in the beginning of this section. Disconnect the negative battery cable.

2. Note the radio security code and the radio presets.

3. Remove the cylinder head cover retaining bolts. Remove the cylinder head cover from the engine.

4. Set the number one piston at TDC. The punch mark on the Variable Timing Control (VTC) actuator and the punch mark on the exhaust camshaft sprocket should be at the top. Align the TDC marks on the VTC actuator and exhaust camshaft sprocket.

5. Using the proper gauge feeler gauge, adjust the valves on cylinder number one.

6. Rotate the crankshaft 180 degrees. Using the proper gauge feeler gauge, adjust the valves on cylinder number three.

7. Rotate the crankshaft 180 degrees. Using the proper gauge feeler gauge, adjust the valves on cylinder number four.

8. Rotate the crankshaft 180 degrees. Using the proper gauge feeler gauge, adjust the valves on cylinder number two.

9. Install the cylinder head cover.

2012 Models

1.8L Engines

See Figures 202 through 205.

1. Remove the cylinder head cover, as described in this section.

➡ **Connect the HDS to the DLC and monitor ECT SENSOR 1 with the HDS. Adjust the valve clearance only when the engine coolant temperature is less than 100°F (38°C).**

2. Set the No. 1 piston at top dead center (TDC). The "UP" mark on the camshaft sprocket should be at the top, and the TDC

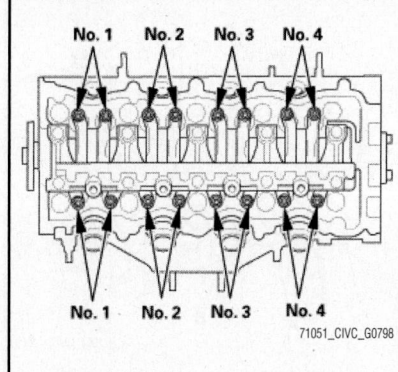

Fig. 202 Showing the valve numbering arrangement—non-hybrid engine

grooves on the camshaft sprocket should line up with the top edge of the head.

3. For non-hybrid engines, select the correct feeler gauge for the valve clearance you are going to check:
    *   Intake: 0.007–0.009 in. (0.18–0.22 mm)
    *   Exhaust: 0.009–0.010 in. (0.23–0.27 mm)

4. For hybrid engines, select the correct feeler gauge for the valve clearance you are going to check:
    *   Intake: 0.006–0.007 in. (0.15–0.19 mm)
    *   Exhaust: 0.010–0.011 in. (0.24–0.28 mm)

5. Insert the feeler gauge between the adjusting screw and the end of the valve stem on the No. 1 cylinder, and slide it back and forth; you should feel a slight amount of drag.

6. If you feel too much or too little drag, loosen the locknut, and turn the adjusting screw until the drag on the feeler gauge is correct.

7. Tighten the locknut to 10 ft. lbs. (14 Nm), and recheck the clearance. Repeat the

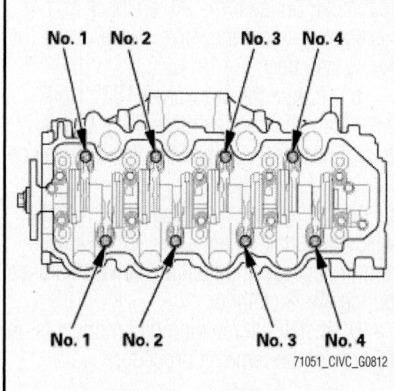

Fig. 203 Showing the valve numbering arrangement—hybrid engine

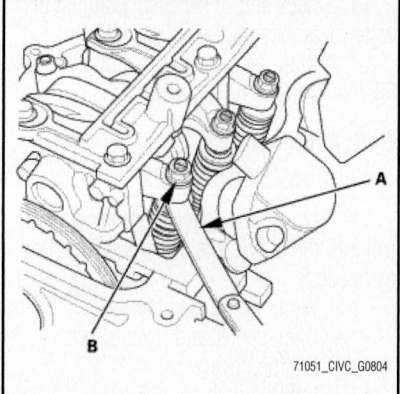

Fig. 204 Insert the feeler gauge (A) between the adjusting screw (B) and the end of the valve stem on the No. 1 cylinder, and slide it back and forth; you should feel a slight amount of drag— non-hybrid engine

adjustment if necessary. Apply new engine oil to the nut threads.

8. Rotate the crankshaft clockwise. Align the No. 3 piston TDC groove on the camshaft sprocket with the top edge of the head.

9. Check and, if necessary, adjust the valve clearance on the No. 3 cylinder.

10. Rotate the crankshaft clockwise. Align the No. 4 piston TDC groove on the camshaft sprocket with the top edge of the head.

11. Check and, if necessary, adjust the valve clearance on the No. 4 cylinder.

12. Rotate the crankshaft clockwise. Align the No. 2 piston TDC groove on the camshaft sprocket with the top edge of the head.

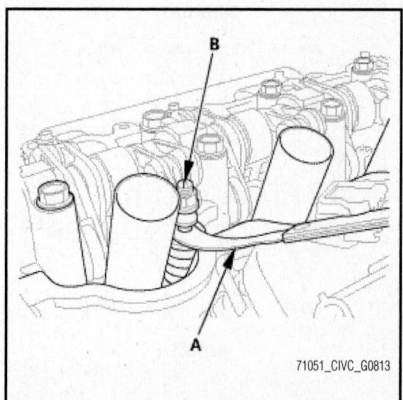

Fig. 205 Insert the feeler gauge (A) between the adjusting screw (B) and the end of the valve stem on the No. 1 cylinder, and slide it back and forth; you should feel a slight amount of drag— hybrid engine

13. Check and, if necessary, adjust the valve clearance on the No. 2 cylinder.

14. Install the remaining components, in reverse of the removal procedure.

2.4L Engine

See Figures 206 through 211.

1. Remove the following components (refer to the applicable component section as needed):
   - Wiper arm assembly
   - Side, center and lower cowls
   - Engine cover
   - Ignition coils
   - Cylinder head cover

2. Set the No. 1 piston at top dead center (TDC). The punch mark on the VTC actuator and the punch mark on the exhaust camshaft sprocket should be at the top. Align the TDC marks on the VTC actuator and the exhaust camshaft sprocket.

3. Select the correct feeler gauge for the valve clearance you are going to check:
   - Intake: 0.008–0.010 in. (0.21–0.25 mm)
   - Exhaust: 0.009–0.011 in. (0.23–0.27 mm)

4. Insert the feeler gauge between the adjusting screw and the end of the valve stem on the No. 1 cylinder, and slide it back and forth; you should feel a slight amount of drag.

5. If you feel too much or too little drag, loosen the locknut, and turn the adjusting screw until the drag on the feeler gauge is correct.

6. Tighten the locknut to 10 ft. lbs. (14 Nm), and recheck the clearance. Repeat the adjustment if necessary. Apply new engine oil to the nut threads.

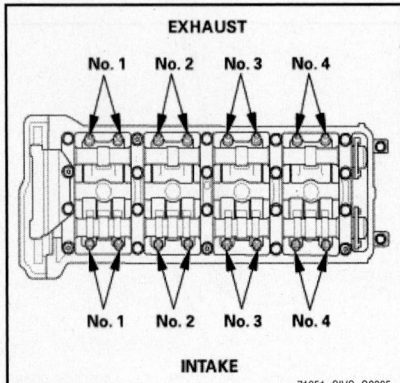

Fig. 206 Showing the valve numbering arrangement

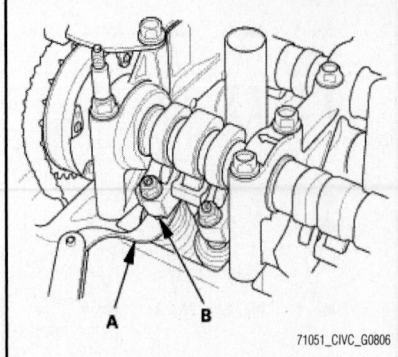

Fig. 207 Insert the feeler gauge (A) between the adjusting screw (B) and the end of the valve stem on the No. 1 cylinder, and slide it back and forth; you should feel a slight amount of drag

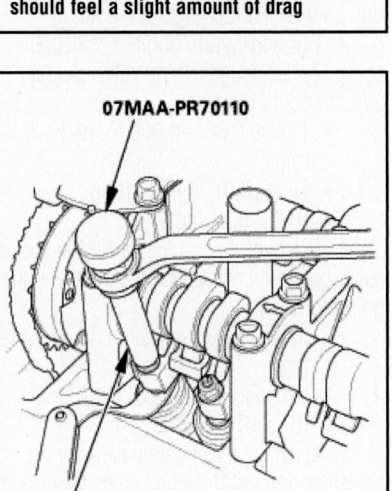

Fig. 208 If you feel too much or too little drag, loosen the locknut, and turn the adjusting screw until the drag on the feeler gauge is correct

7. Rotate the crankshaft 180° clockwise (camshaft pulley turns 90°). Check and, if necessary, adjust the valve clearance on the No. 3 cylinder.

8. Rotate the crankshaft 180° clockwise (camshaft pulley turns 90°). Check and, if necessary, adjust the valve clearance on the No. 4 cylinder.

9. Rotate the crankshaft 180° clockwise (camshaft pulley turns 90°). Check and, if necessary, adjust the valve clearance on the No. 2 cylinder.

10. Install the remaining components, in reverse of the removal procedure.

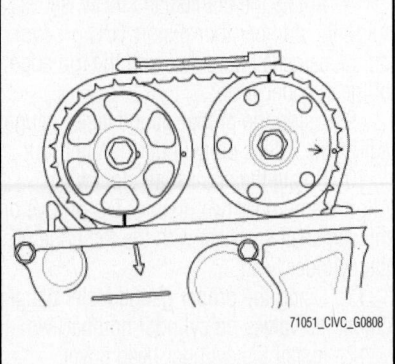

Fig. 209 Rotate the crankshaft 180° clockwise (camshaft pulley turns 90°). Check and, if necessary, adjust the valve clearance on the No. 3 cylinder

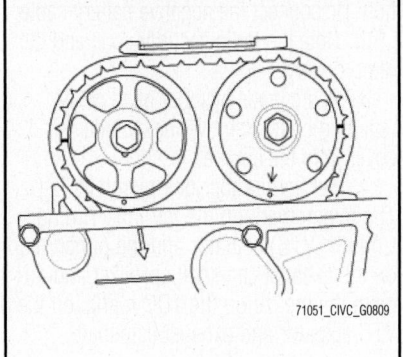

Fig. 210 Rotate the crankshaft 180° clockwise (camshaft pulley turns 90°). Check and, if necessary, adjust the valve clearance on the No. 4 cylinder

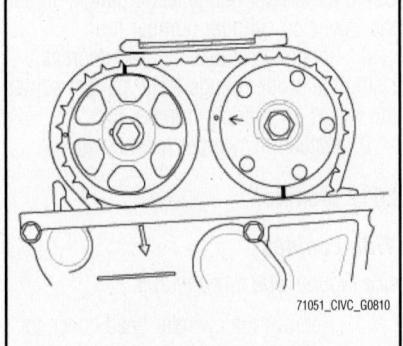

Fig. 211 Rotate the crankshaft 180° clockwise (camshaft pulley turns 90°). Check and, if necessary, adjust the valve clearance on the No. 2 cylinder

ENGINE PERFORMANCE & EMISSION CONTROLS

ACCELERATOR PEDAL POSITION (APP) SENSOR

LOCATION

The Accelerator Pedal Position (APP) sensor is located on the top of the accelerator pedal assembly.

REMOVAL & INSTALLATION

2011 Models

See Figure 212 and 213.

1. Disconnect the APP sensor 6P connector.
2. Remove the clip.

➡**Do not reuse the clip once it is removed.**

3. Push the tab, and remove the accelerator pedal pad from the pedal stop.
4. Remove the accelerator pedal module.

➡**The APP sensor is not available separately. Do not disassemble the accelerator pedal module.**

To install:

5. Set the accelerator pedal pad to the pedal stop.
6. Install the accelerator pedal module with a new clip.
7. Reconnect the APP sensor 6P connector.

2012 Models

See Figure 214.

1. Disconnect the connector and remove the accelerator pedal module.

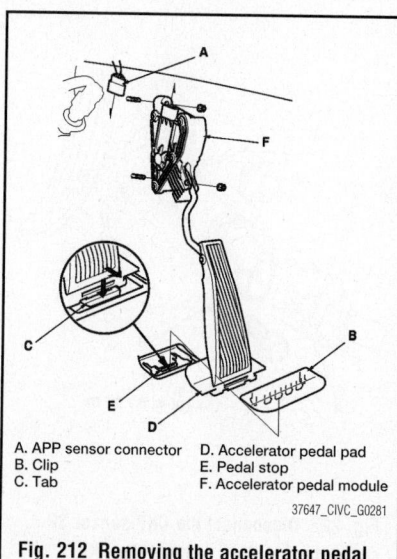

A. APP sensor connector   D. Accelerator pedal pad
B. Clip   E. Pedal stop
C. Tab   F. Accelerator pedal module

37647_CIVC_G0281

Fig. 212 Removing the accelerator pedal assembly—1.8L engine

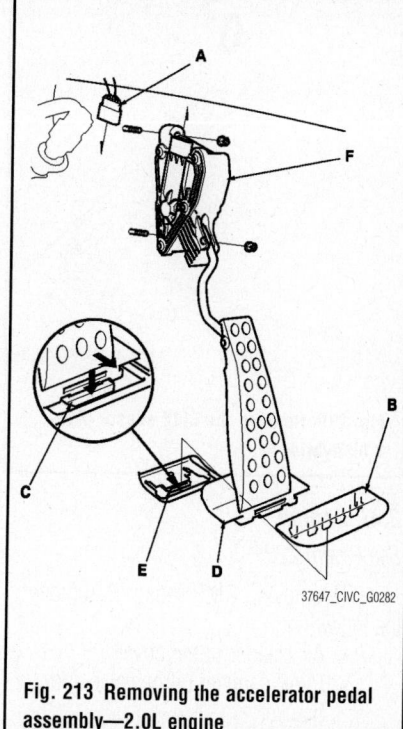

37647_CIVC_G0282

Fig. 213 Removing the accelerator pedal assembly—2.0L engine

CAMSHAFT POSITION (CMP) SENSOR

REMOVAL & INSTALLATION

2011 Models

1.8L Engines

See Figure 215.

1. Remove the cowl cover and the under-cowl panel.

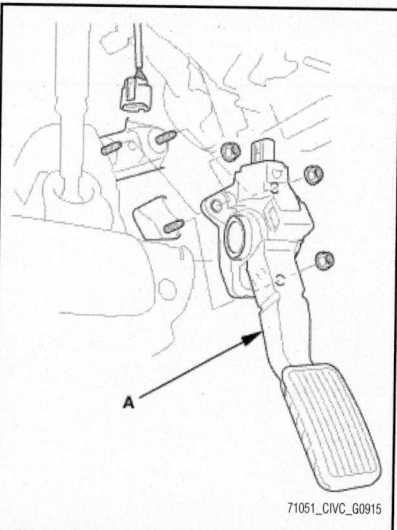

71051_CIVC_G0915

Fig. 214 Disconnect the connector and remove the accelerator pedal module

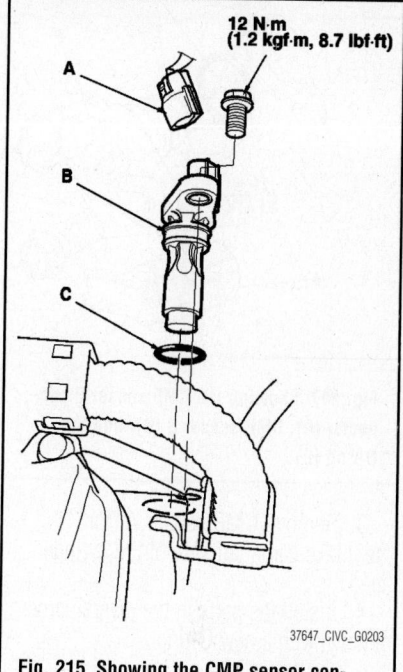

37647_CIVC_G0203

Fig. 215 Showing the CMP sensor connector (A), CMP sensor (B), and the O-ring (C)

2. Disconnect the CMP sensor 3P connector.
3. Remove the CMP sensor.
4. Install the parts in the reverse order of removal with a new O-ring.

2.0L Engine

See Figures 216 and 217.

1. Remove the air cleaner.
2. Disconnect the CMP sensor A 3P connector.

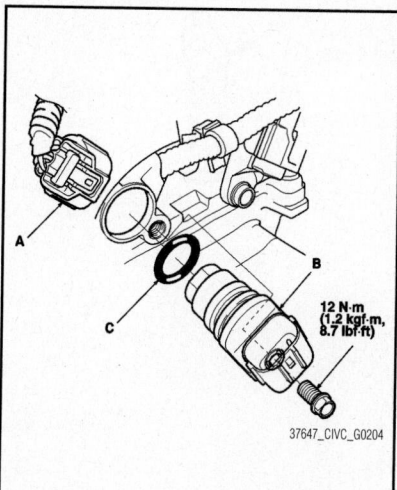

37647_CIVC_G0204

Fig. 216 Showing the CMP sensor A connector (A), CMP sensor A (B), and the O-ring (C)

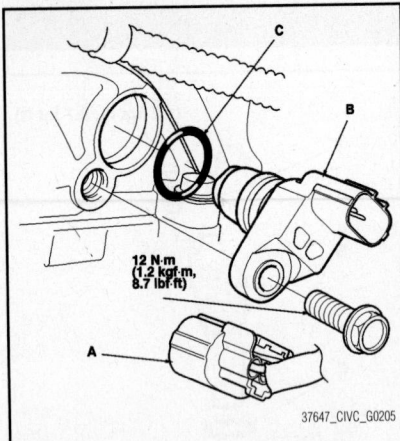

Fig. 217 Showing the CMP sensor B connector (A), CMP sensor B (B), and the O-ring (C)

3. Remove CMP sensor A from the intake camshaft side of the cylinder head.

4. Install the parts in the reverse order of removal with a new O-ring.

2012 Models

1.8L Engines

See Figures 218 and 219.

1. Remove the air cleaner, if needed for access.
2. Disconnect the connector.
3. Remove the CMP sensor.
4. Installation is the reverse of the removal procedure.

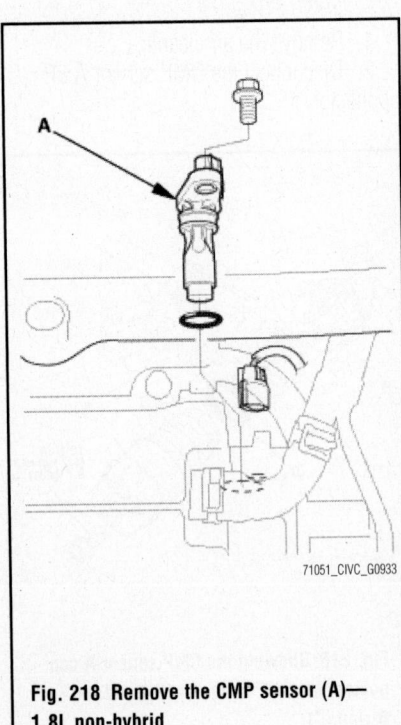

Fig. 218 Remove the CMP sensor (A)— 1.8L non-hybrid

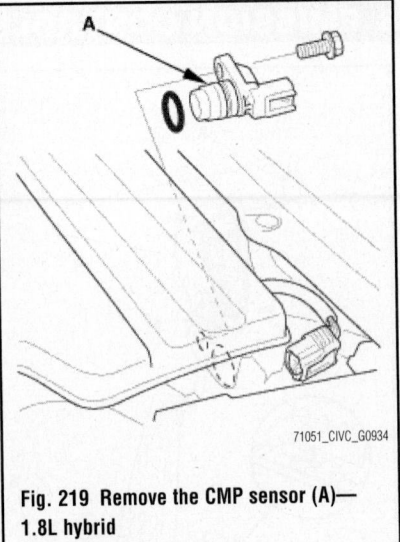

Fig. 219 Remove the CMP sensor (A)— 1.8L hybrid

2.4L Engine

See Figure 220.

1. If removing CMP sensor B, remove the following:
   - Air cleaner upper cover.
   - EVAP canister purge valve and bracket
2. Detach the connector and remove the CMP sensor A or B.
3. Installation is the reverse of the removal procedure.

CRANKSHAFT POSITION (CKP) SENSOR

REMOVAL & INSTALLATION

2011 Models

1.8L Engines

See Figure 221.

1. Remove the splash shield.

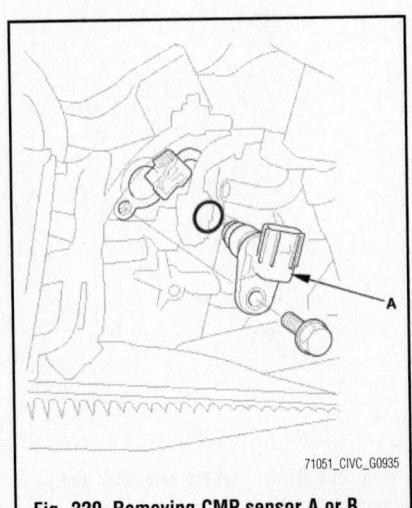

Fig. 220 Removing CMP sensor A or B

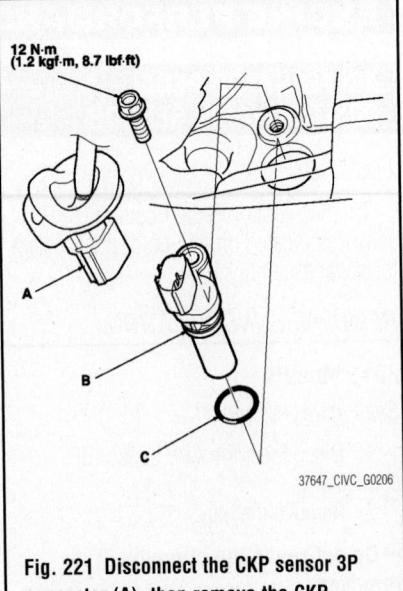

Fig. 221 Disconnect the CKP sensor 3P connector (A), then remove the CKP sensor (B) and O-ring (C)

2. Disconnect the CKP sensor 3P connector.
3. Remove the CKP sensor and O-ring.
4. Install the parts in the reverse order of removal with a new O-ring.
5. Do the "CKP Pattern Clear/CKP Pattern Learn Procedure".

2.0L Engine

See Figure 222.

1. Disconnect the CKP sensor 3P connector.
2. Remove the CKP sensor and O-ring.
3. Install the parts in the reverse order of removal with a new O-ring.

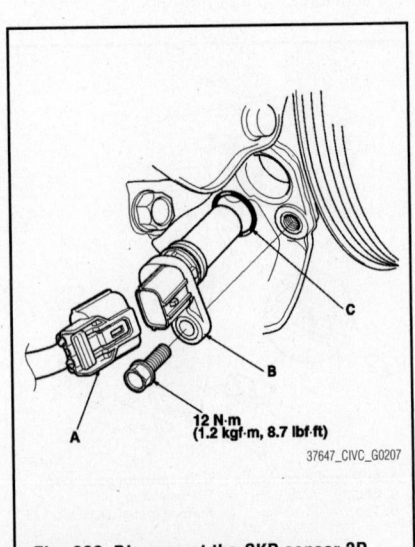

Fig. 222 Disconnect the CKP sensor 3P connector (A), then remove the CKP sensor (B) and O-ring (C)

4. Do the "CKP Pattern Clear/CKP Pattern Learn Procedure".

2012 Models

1.8L ENGINES

See Figures 223 and 224.

1. Raise vehicle on the lift and be sure it is securely supported.
2. Remove the engine undercover.
3. For 1.8L non-hybrid engine, loosen the CKP sensor cover top bolt,

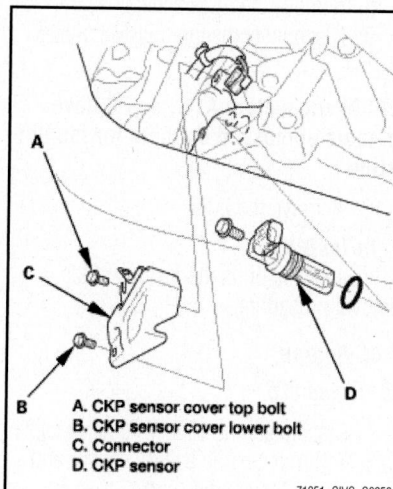

A. CKP sensor cover top bolt
B. CKP sensor cover lower bolt
C. Connector
D. CKP sensor

71051_CIVC_G0959

Fig. 223 For 1.8L non-hybrid engine, loosen the CKP sensor cover top bolt, remove the CKP sensor cover lower bolt, then slide cover down to remove it. Disconnect the connector. Remove the CKP sensor

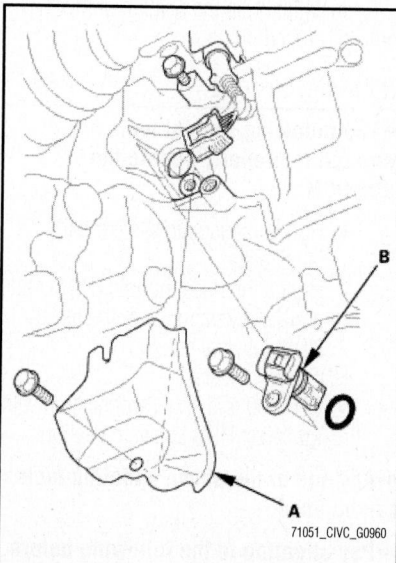

71051_CIVC_G0960

Fig. 224 For 1.8L hybrid engine, remove the CKP sensor cover (A), then remove the CKP sensor (B)

remove the CKP sensor cover lower bolt, then slide cover down to remove it. Disconnect the connector. Remove the CKP sensor.

4. For 1.8L hybrid engine, remove the CKP sensor cover, then remove the CKP sensor.
5. Installation is the reverse of the removal procedure.

2.4L Engine

See Figure 225.

1. Raise vehicle on the lift and be sure it is securely supported.
2. Remove the CKP sensor cover, detach the connector, then remove the CKP sensor.

CKP PATTERN CLEAR/CKP PATTERN LEARN PROCEDURE

1. Start the engine. Hold the engine speed at 3,000 RPM without load (in neutral) until the radiator fan comes on.
2. Test-drive the vehicle on a level road: Decelerate (with the throttle fully closed) from an engine speed of 2,500 RPM down to 1,000 RPM with the transmission in 1st.
3. Test-drive the vehicle on a level road: Decelerate (with the throttle fully closed) from an engine speed of 5,000 RPM down to 3,000 RPM with the transmission in 1st.
4. Repeat previous two steps several times.
5. Turn the ignition switch to LOCK (0).
6. Turn the ignition switch to ON (II), and wait 30 seconds.

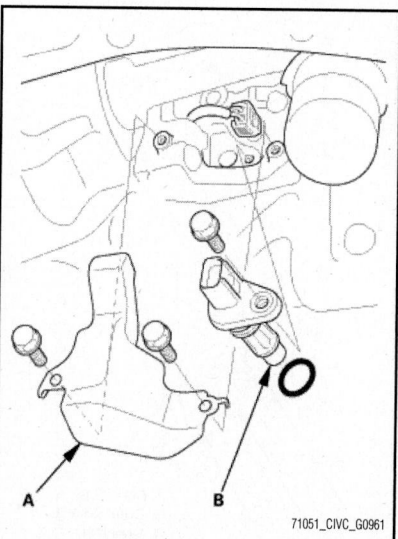

71051_CIVC_G0961

Fig. 225 Remove the CKP sensor cover (A), detach the connector, then remove the CKP sensor (B)

ENGINE CONTROL MODULE/POWERTRAIN CONTROL MODULE (ECM/PCM)

LOCATION

➡The ECM/PCM is located in the engine compartment. See "Removal & Installation" for figure references of component locations.

REMOVAL & INSTALLATION

2011 Models

➡The following special tools are needed to perform this operation:

- Honda diagnostic system (HDS) tablet tester
- Honda interface module (HIM) and an iN workstation with the latest HDS software version
- HDS pocket tester
- GNA600 and an iN workstation with the latest HDS software version

➡Any one of the above updating tools can be used.

1. Connect the HDS to the data link connector (DLC) (A) located under the driver's side of the dashboard.
2. Turn the ignition switch to ON (II).
3. Make sure the HDS communicates with the ECM/PCM and other vehicle systems. If it doesn't, go to the DLC circuit troubleshooting. If you are returning from DLC circuit troubleshooting, skip steps 4 through 9, 20 through 25, and 28 through 30, and do this after replacing the ECM/PCM:
    a. Replace the engine oil and the engine oil filter.
    b. Replace the ATF (A/T).
    c. Clean the throttle body.
4. Select the PGM-FI system with the HDS.
5. Select the INSPECTION MENU with the HDS.
6. Select the ETCS TEST, then select the TP POSITION CHECK, and follow the screen prompts.

➡If the TP POSITION CHECK indicates FAILED, continue with this procedure.

7. Select the REPLACE ECM/PCM MENU, then select READ DATA, and follow the screen prompts.

➡Doing this step copies (READS) the engine oil life data from the original ECM/PCM so you can later download (WRITES) it into the new ECM/PCM. If READ DATA indicates FAILED, continue with this procedure.

8. A/T: Select the A/T system with the HDS.

9. A/T: Select the REPLACE TCM/PCM MENU, then select READ DATA, and follow the screen prompts.

➡**Doing this step copies (READS) the ATF life data from the original PCM so you can later download (WRITES) it into the new PCM. If READ DATA indicates FAILED, continue with this procedure.**

10. Turn the ignition switch to LOCK (0).
11. Jump the SCS line with the HDS.
12. Do the battery removal procedure.
13. Remove the ECM/PCM cover.
14. Remove the bolts (D), then remove the ECM/PCM (E).
15. Disconnect ECM/PCM connectors A, B, and C.

➡**ECM/PCM connectors A, B, and C have symbols (A=□, B=△, C=○) embossed on them for identification.**

16. Install the ECM/PCM in the reverse order of removal.
17. Do the battery installation procedure.
18. Turn the ignition switch to ON (II).
19. Manually input the VIN to the ECM/PCM with the HDS.

➡**DTC P0630 VIN Not Programmed or Mismatch may be stored because the VIN has not been programmed into the ECM/PCM; ignore it, and continue this procedure.**

20. If the READ DATA (engine oil life) failed in step 7, go to step 23 (A/T) or step 26 (M/T). Otherwise, go to step 21.
21. Select the PGM-FI system with the HDS.
22. Select the REPLACE ECM/PCM MENU, then select WRITE DATA, and follow the screen prompts.

➡**If the WRITE DATA indicates FAILED, continue with this procedure.**

23. A/T: If the READ DATA (ATF life) failed in step 9, go to step 26. Otherwise go to step 24.
24. A/T: Select the A/T SYSTEM with the HDS.
25. A/T: Select the REPLACE TCM/PCM MENU, then select WRITE DATA, and follow the screen prompts.

➡**If the WRITE DATA indicates FAILED, continue with this procedure.**

26. Select IMMOBI system with the HDS.
27. Enter the immobilizer ECM/PCM code that you got from the iN, and use the

ECM/PCM replacement procedure in the IMMOBI MENU of the HDS; it allows you to start the engine.

28. If the TP POSITION CHECK failed in step 6, clean the throttle body, then go to step 29.
29. If the READ DATA failed in step 7 or the WRITE DATA failed in step 22, replace the engine oil and engine oil filter, then go to step 30 (A/T) or step 31 (M/T).
30. A/T: If the READ DATA failed in step 9 or the WRITE DATA failed in step 25, replace the ATF, then go to step 31.
31. Select PGM-FI system, and reset the ECM/PCM with the HDS.
32. Update the ECM/PCM if it does not have the latest software.
33. Do the "CKP Pattern Clear/CKP Pattern Learn Procedure".

2012 Models

1.8L Non-Hybrid Engine
See Figure 226.

➡**For specific operations, refer to the user's manual that came with the Honda Diagnostic System (HDS). Make sure the HDS is loaded with the latest software.**

1. Connect the HDS to the data link connector (DLC) located under the driver's side of the dashboard.
2. Turn the ignition switch to ON (II).
3. Select the REPLACE ECM/PCM MENU, then select READ DATA and follow the screen prompts.

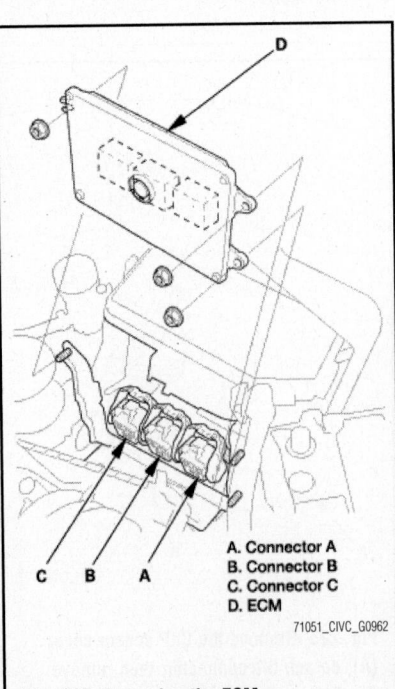

A. Connector A
B. Connector B
C. Connector C
D. ECM

71051_CIVIC_G0962

Fig. 226 Removing the ECM

a. Doing this step copies (READS) the engine oil life data from the original PCM so you can later download (WRITES) it into the new ECM/PCM.
b. If READ DATA indicates FAILED, continue with this procedure.
4. Turn the ignition switch to LOCK (0).
5. Jump the SCS line with the HDS.
6. On hybrid engines, move the under hood fuse/relay box.
7. Disconnect the battery negative cable (isolate it), remove the battery sensor and then remove the positive cable.
8. Remove the ECM cover.
9. Disconnect the three ECM connectors.

➡**ECM connectors A, B, and C have symbols embossed on them for identification.**

10. Remove the ECM.

To install:
11. Installation is the reverse of the removal procedure.

2.4L Engine
See Figure 226.

1. Disconnect the battery negative cable (isolate it), remove the battery sensor and then remove the positive cable.
2. Remove the ECM cover.
3. Disconnect the three ECM connectors.

➡**ECM connectors A, B, and C have symbols (A=□, B=△, C=○) embossed on them for identification.**

4. Remove the ECM.

To install:
5. Installation is the reverse of the removal procedure.

ECM/PCM UPDATE

➡**The following special tools are required to properly update the ECM/PCM:**

- Honda diagnostic system (HDS) tablet tester
- Honda interface module (HIM) and an iN workstation with the latest HDS software version
HDS pocket tester
- GNA600 and an iN workstation with the latest HDS software version

➡**Any one of the above updating tools can be used.**

➡**Pay attention to the following before starting the update:**

- Make sure the HDS/iN workstation has the latest HDS software version.

- Before you update the ECM/PCM, make sure the battery in the vehicle is fully charged, and connect a jumper battery (not a battery charger) to maintain system voltage.
- Never turn the ignition switch to ACC (I) or LOCK (0) during the update. If there is a problem with the update, leave the ignition switch in ON (II).
- To prevent ECM/PCM damage, do not operate anything electrical (headlights, audio system, brakes, A/C, power windows, moonroof (if equipped), door locks, etc.) during the update.
- To ensure the latest program is installed, do an ECM/PCM update whenever the ECM/PCM is substituted or replaced.
- You cannot update an ECM/PCM with a program it already has. It will only accept a new program.
- High temperature in the engine compartment might cause the ECM/PCM to become too hot to run the update. If the engine was running before this procedure, open the hood, and cool the engine compartment.
- If you need to diagnose the Honda interface module (HIM) because the HIM's red (#3) light came on or was flashing during the update, leave the ignition switch in ON (II) when you disconnect the HIM from the data link connector (DLC). This will prevent ECM/PCM damage.

1. Turn the ignition switch to ON (II), but do not start the engine.

2. Connect the HDS to the data link connector (DLC) (A) located under the driver's side of the dashboard.

3. Make sure the HDS communicates with the ECM/PCM and other vehicle systems. If it doesn't, go to the DLC circuit troubleshooting. If you are returning from the DLC circuit troubleshooting, skip steps 4 and 5 and clean the throttle body after updating the ECM/PCM.

4. Select the INSPECTION MENU with the HDS.

5. Select the ETCS TEST, then select the TP POSITION CHECK, and follow the HDS screen prompts.

➡**If the TP POSITION CHECK indicates FAILED, continue this procedure.**

6. Exit the HDS diagnostic system, then select the update mode, and follow the screen prompts to update the ECM/PCM.

7. If the software in the ECM/PCM is the latest, disconnect the HDS/HIM/GNA600 from the DLC, and go back to the procedure that you were doing. If the software in the ECM/PCM is not the latest, follow the instructions on the screen. If prompted to choose the PGM-FI system or the A/T system (A/T), make sure you update both.

➡**If the ECM/PCM update system requires you to cool the ECM/PCM, follow the instructions on the screen. If you run into a problem during the update procedure (programming takes over 15 minutes, status bar goes over 100 %, D (A/T) or immobilizer indicator flashes, HDS tablet freezes, etc.), follow these steps to minimize the chance of damaging the ECM/PCM:**

- Leave the ignition switch in ON (II).
- Connect a jumper battery (do not connect a battery charger).
- Shut down the HDS.
- Disconnect the HDS from the DLC.
- Reboot the HDS.
- Reconnect the HDS to the DLC, and do the update again.

8. If the TP POSITION CHECK failed in step 5, clean the throttle body.

9. Do the "ECM/PCM Idle Learn Procedure".

10. Do the "CKP Pattern Clear/CKP Pattern Learn Procedure".

ECM/PCM IDLE LEARN PROCEDURE

➡**The idle learn procedure must be done so the ECM/PCM can learn the engine idle characteristics.**

1. Do the idle learn procedure whenever you do any of these actions:
Replace the ECM/PCM.
- Reset the ECM/PCM.
- Update the ECM/PCM.
- Replace or clean the throttle body.
- Disassemble the engine or the transmission.

➡**Erasing DTCs with the HDS does not require you to do the idle learn procedure.**

2. Make sure all electrical items (the A/C, the audio system, the lights, etc.) are off.

3. Reset the ECM/PCM with the HDS.

4. Turn the ignition switch to ON (II), and wait 2 seconds.

5. Start the engine. Hold the engine speed at 3,000 RPM without load (A/T in P or N, M/T in neutral) until the radiator fan comes on, or until the engine coolant temperature reaches 194°F (90°C).

6. Let the engine idle for about 5 minutes with the throttle fully closed.

➡**If the radiator fan comes on, do not include its running time in the 5 minutes.**

ENGINE COOLANT TEMPERATURE (ECT) SENSOR

LOCATION

These engine use two Engine Coolant Temperature (ECT) sensors. See figures in "Removal & Installation" for identification of locations.

REMOVAL & INSTALLATION

ECT Sensor 1

2011 1.8L Engines
See Figure 227.

1. Drain the engine coolant.
2. Disconnect the ECT sensor 1 2P connector.
3. Remove ECT sensor 1 and O-ring.
4. Install the parts in the reverse order of removal with a new O-ring, then refill the radiator with engine coolant.

2011 2.0L Engine
See Figures 228 and 229.

1. Drain the engine coolant.
2. Remove the air cleaner.
3. Disconnect the CMP sensor A 3P connector, and the CMP sensor B 3P connector, then remove the harness cover.
4. Disconnect the ECT sensor 1 2P connector, then remove the ECT sensor 1 and O-ring.

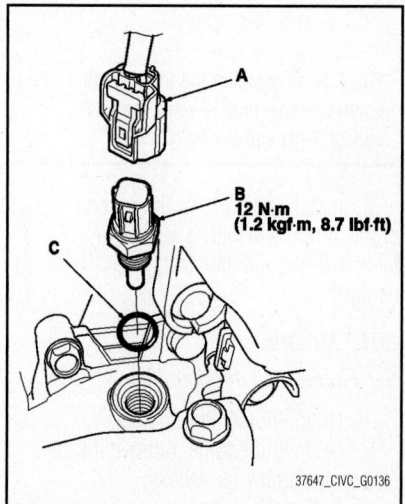

37647_CIVIC_G0136

Fig. 227 Disconnect the ECT sensor 1 2P connector (A), then remove the ECT sensor 1 (B) and O-ring (C)

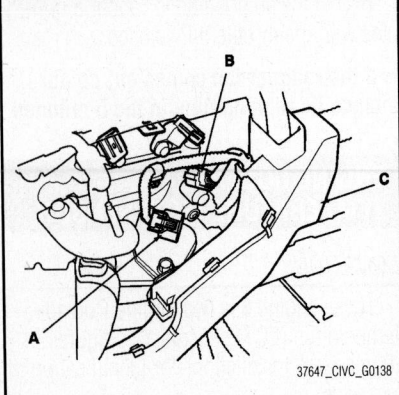

Fig. 228 Disconnect the CMP sensor A 3P connector (A), and the CMP sensor B 3P connector (B), then remove the harness cover (C)

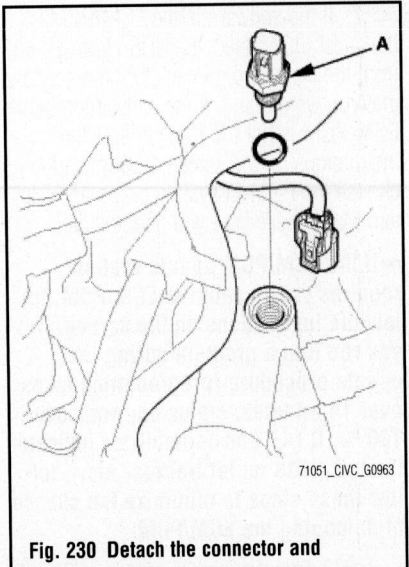

Fig. 230 Detach the connector and remove the ECT sensor (A)—1.8L engine

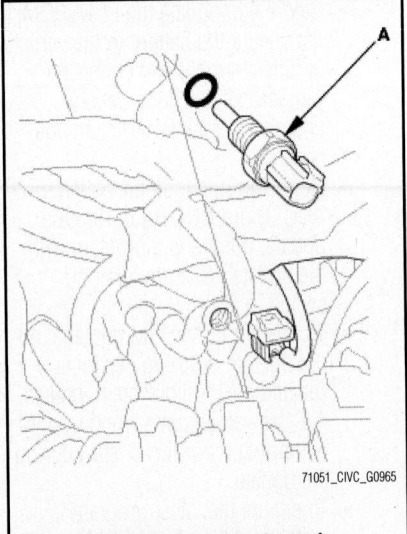

Fig. 232 Detach the connector and remove the ECT sensor (A)—2.4L engine

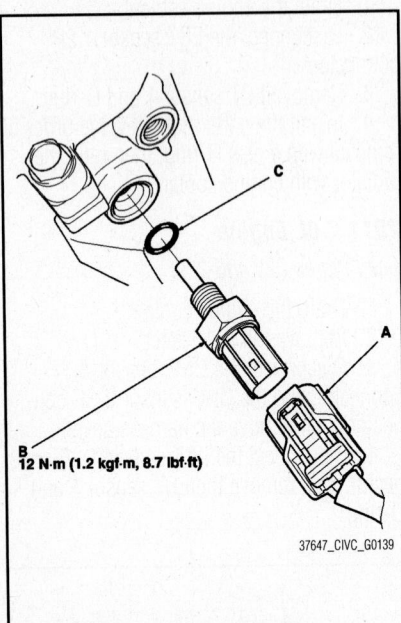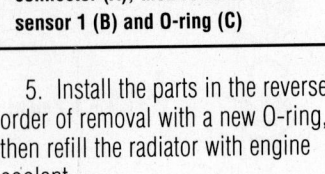

Fig. 229 Disconnect the ECT sensor 1 2P connector (A), then remove the ECT sensor 1 (B) and O-ring (C)

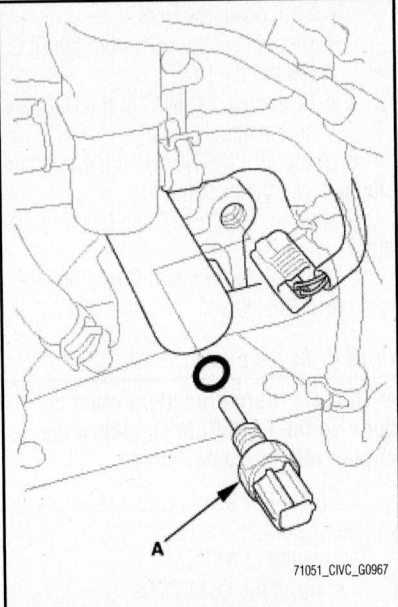

Fig. 231 Detach the connector and remove the ECT sensor (A)—1.8L hybrid engine

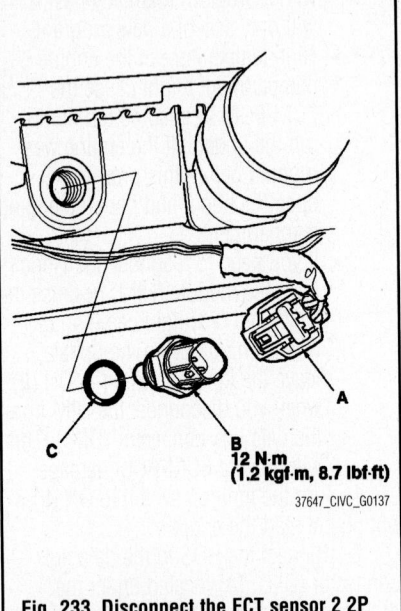

Fig. 233 Disconnect the ECT sensor 2 2P connector (A), then remove the ECT sensor 2 (B) and O-ring (C)

5. Install the parts in the reverse order of removal with a new O-ring, then refill the radiator with engine coolant.

2012 Models

See Figures 230 through 232.

1. Drain the radiator coolant.
2. On hybrid engine, remove the air cleaner assembly for access.
3. Detach the connector and remove the ECT sensor 1.
4. Installation is the reverse of the removal procedure.

ECT Sensor 2

2011 1.8L Engines

See Figure 233.

1. Drain the engine coolant.
2. Remove the splash shield.
3. Disconnect the ECT sensor 2 2P connector.
4. Remove ECT sensor 2 and O-ring.
5. Install the parts in the reverse order of removal with a new O-ring, then refill the radiator with engine coolant.
6. Install the splash shield.

2011 2.0L Engine

See Figure 234.

1. Drain the engine coolant.
2. Remove the splash shield.
3. Disconnect the ECT sensor 2 2P connector, then remove ECT sensor 2 and O-ring.
4. Install parts in the reverse order of removal with a new O-ring.
5. Install the splash shield.
6. Refill the radiator with engine coolant.

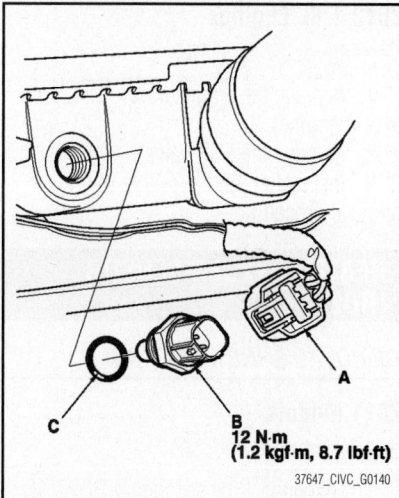

Fig. 234 Disconnect the ECT sensor 2 2P connector (A), then remove the ECT sensor 2 (B) and O-ring (C)

2012 Models

See Figures 235 and 236.

1. Drain the radiator coolant. See "Engine Cooling" section.
2. Remove the splash shield.
3. Detach the connector and remove the ECT sensor 2.
4. Installation is the reverse of the removal procedure.

EVAPORATIVE EMISSIONS (EVAP) CANISTER

LOCATION

The Evaporative Emissions (EVAP) canister is located under the vehicle, just forward of the fuel tank. See "Removal & Installation" for figure references of component locations.

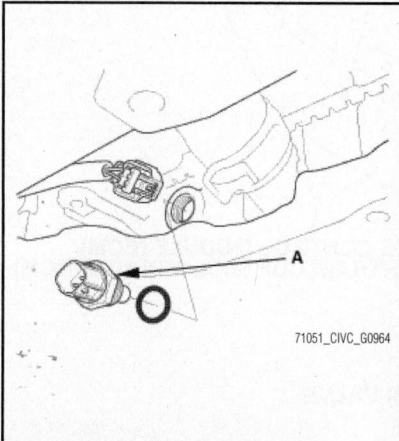

Fig. 235 Detach the connector and remove the ECT sensor 2 (A)—1.8L non-hybrid and hybrid engines

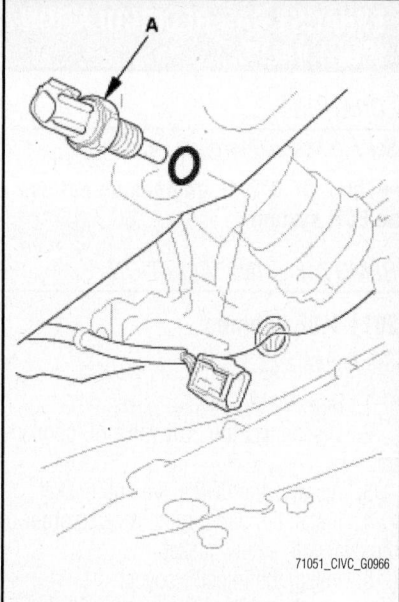

Fig. 236 Detach the connector and remove the ECT sensor 2 (A)—2.4L engine

REMOVAL & INSTALLATION

2011 Models

See Figures 237 and 238.

1. Raise the vehicle on a lift.
2. Remove the undercover at the fuel tank.
3. Remove the EVAP canister guard pipe.

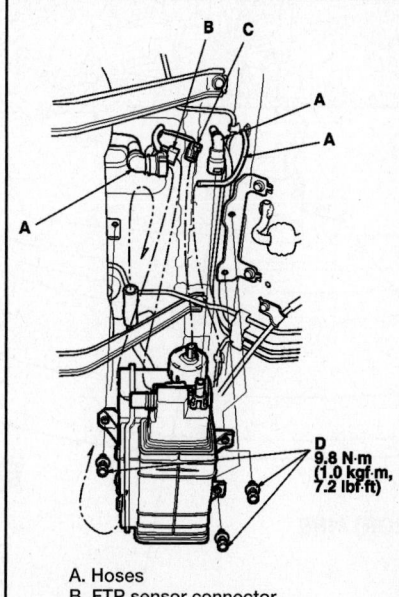

A. Hoses
B. FTP sensor connector
C. Vent shut valve connector
D. Bolts

Fig. 237 Removing the EVAP canister

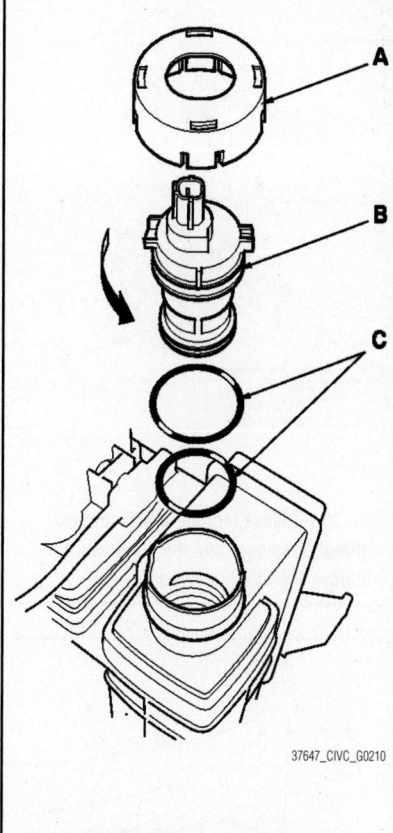

Fig. 238 Remove the cap (A), then remove the EVAP canister vent shut valve (B) and O-rings (C)

4. Remove the hoses, the FTP sensor connector, the EVAP canister vent shut valve connector, and the bolts.
5. Remove the EVAP canister.
6. Remove the cap.
7. Remove the EVAP canister vent shut valve and O-rings.
8. Install the EVAP canister vent shut valve in the new EVAP canister with new O-rings.

➡**Do not coat the O-rings with engine oil.**

9. Install the parts in the reverse order of removal.

2012 Models

See Figure 239.

1. Place the vehicle on a lift.
2. Remove the left underfloor cover.
3. Remove the EVAP canister:
    a. Disconnect the quick-connect fittings, the hose, and the connectors.
    b. Remove the EVAP canister.
4. Installation is the reverse of the removal procedure.

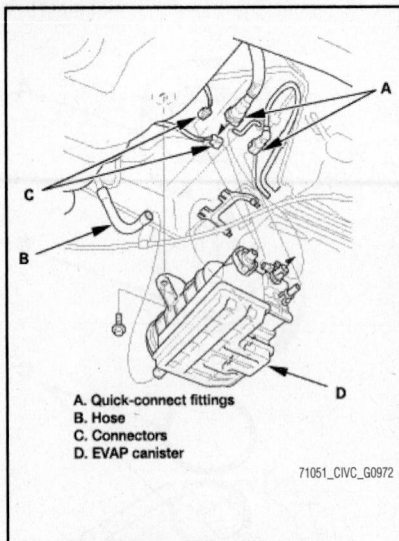

A. Quick-connect fittings
B. Hose
C. Connectors
D. EVAP canister

71051_CIVIC_G0972

Fig. 239 Disconnect the quick-connect fittings, the hose, and the connectors. Remove the EVAP canister

EXHAUST GAS RECIRCULATION (EGR) VALVE

LOCATION

See Figures 240 through 242.

➡The 2.0L & 2.4L engines do not use an EGR system.

REMOVAL & INSTALLATION

2011 1.8L Engines

See Figure 243.

1. Remove the injector cover.
2. Disconnect the EGR valve 6P connector.
3. Remove the EGR valve and gasket.
4. Install the parts in the reverse order of removal with a new gasket.
5. Install the injector cover.

2012 1.8L Engines

See Figures 244 and 245.

1. Remove the engine cover (1.8L non-hybrid engine).
2. Remove the EGR valve.
3. Installation is the reverse of the removal procedure.

HEATED OXYGEN SENSOR (HO2S)

REMOVAL & INSTALLATION

2011 Models

See Figures 246 and 247.

1. Disconnect the secondary HO2S 4P connector, then remove the secondary HO2S.
2. Install the parts in the reverse order of removal.

ENGINE CONTROL MODULE (ECM)/POWERTRAIN CONTROL MODULE (PCM)

EXHAUST GAS RECIRCULATION (EGR) PIPE

EXHAUST GAS RECIRCULATION (EGR) VALVE

37647_CIVIC_G0213

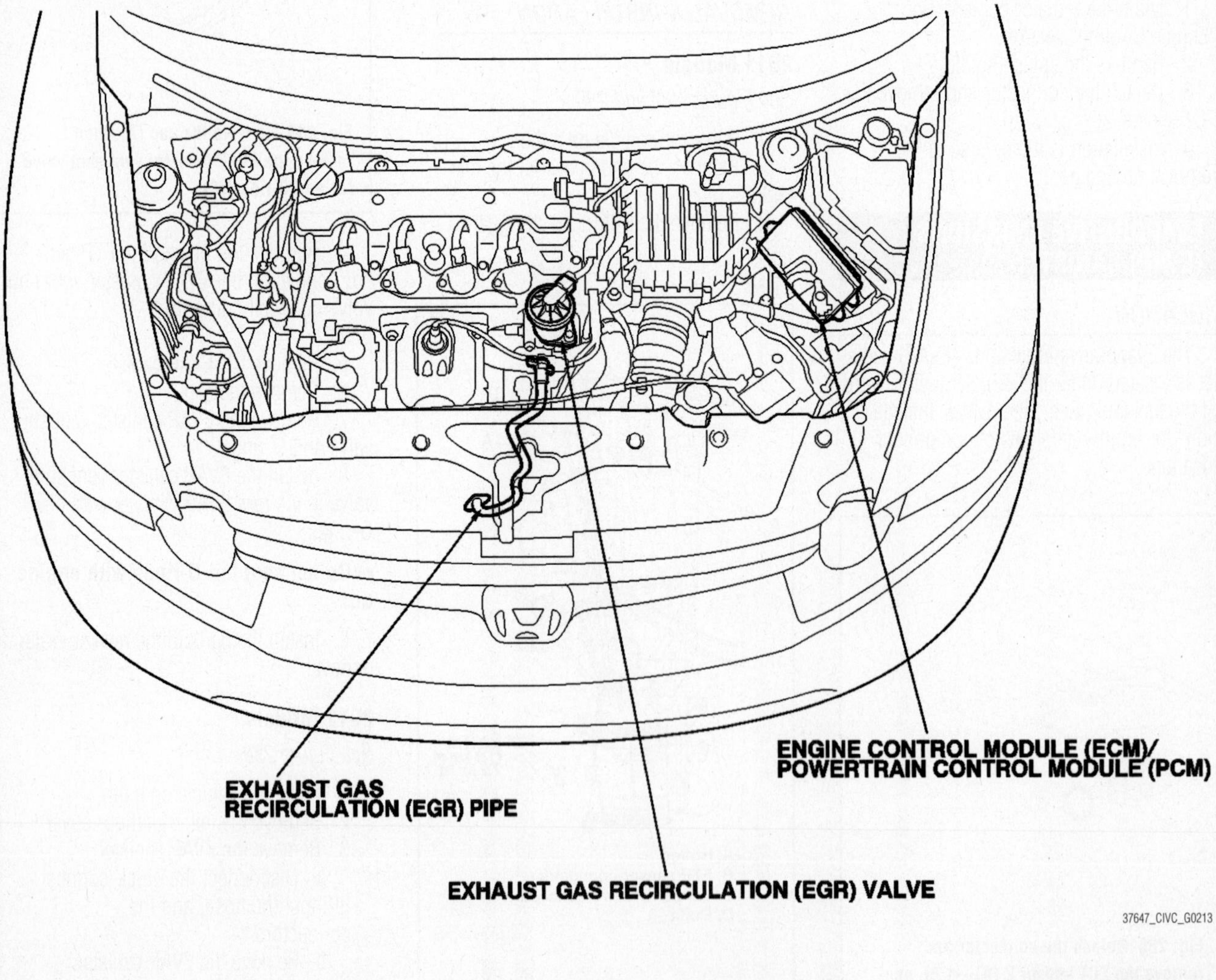

Fig. 240 Showing the location of the EGR valve and pipe—2011 1.8L engine

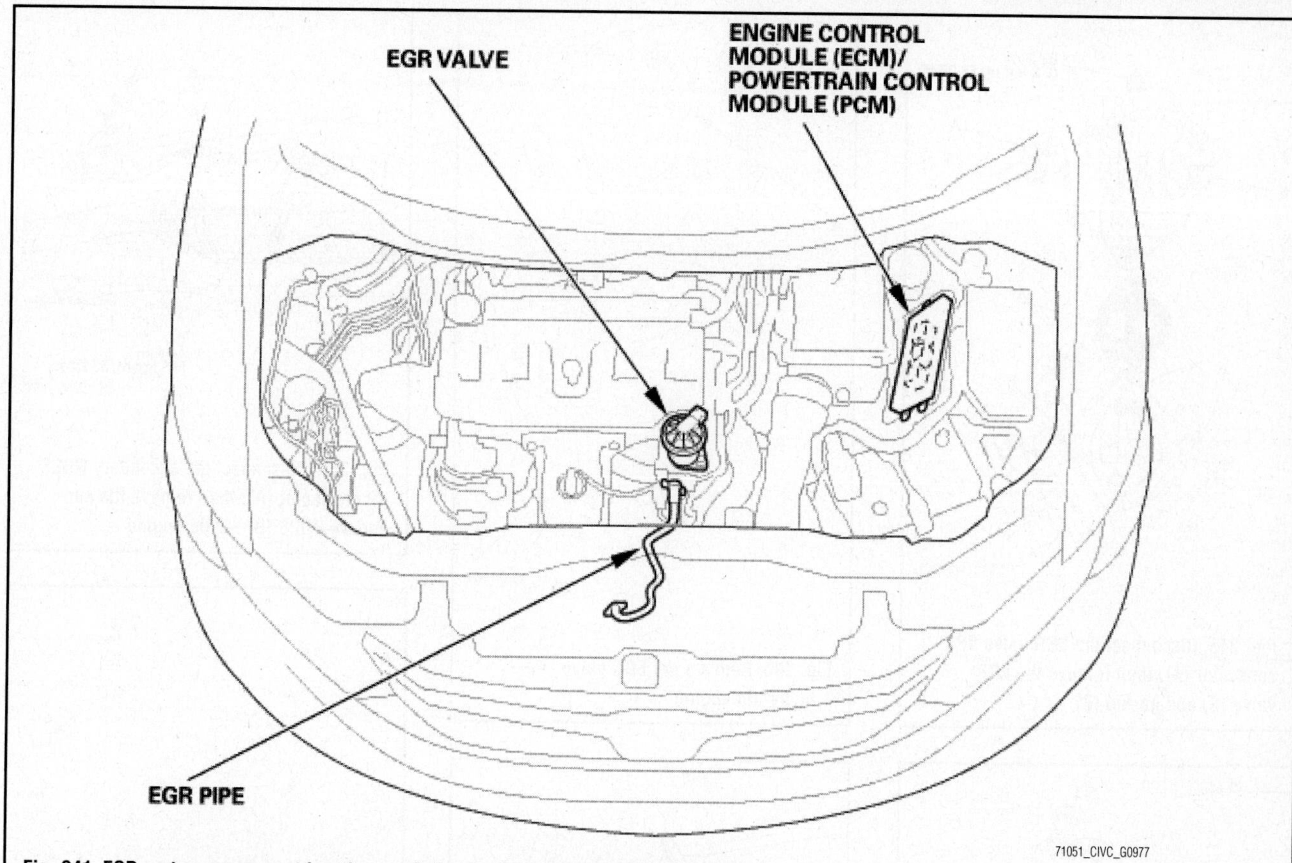

EGR VALVE

ENGINE CONTROL
MODULE (ECM)/
POWERTRAIN CONTROL
MODULE (PCM)

EGR PIPE

71051_CIVC_G0977

Fig. 241 EGR system component locations underhood—2012 1.8L non-hybrid engine

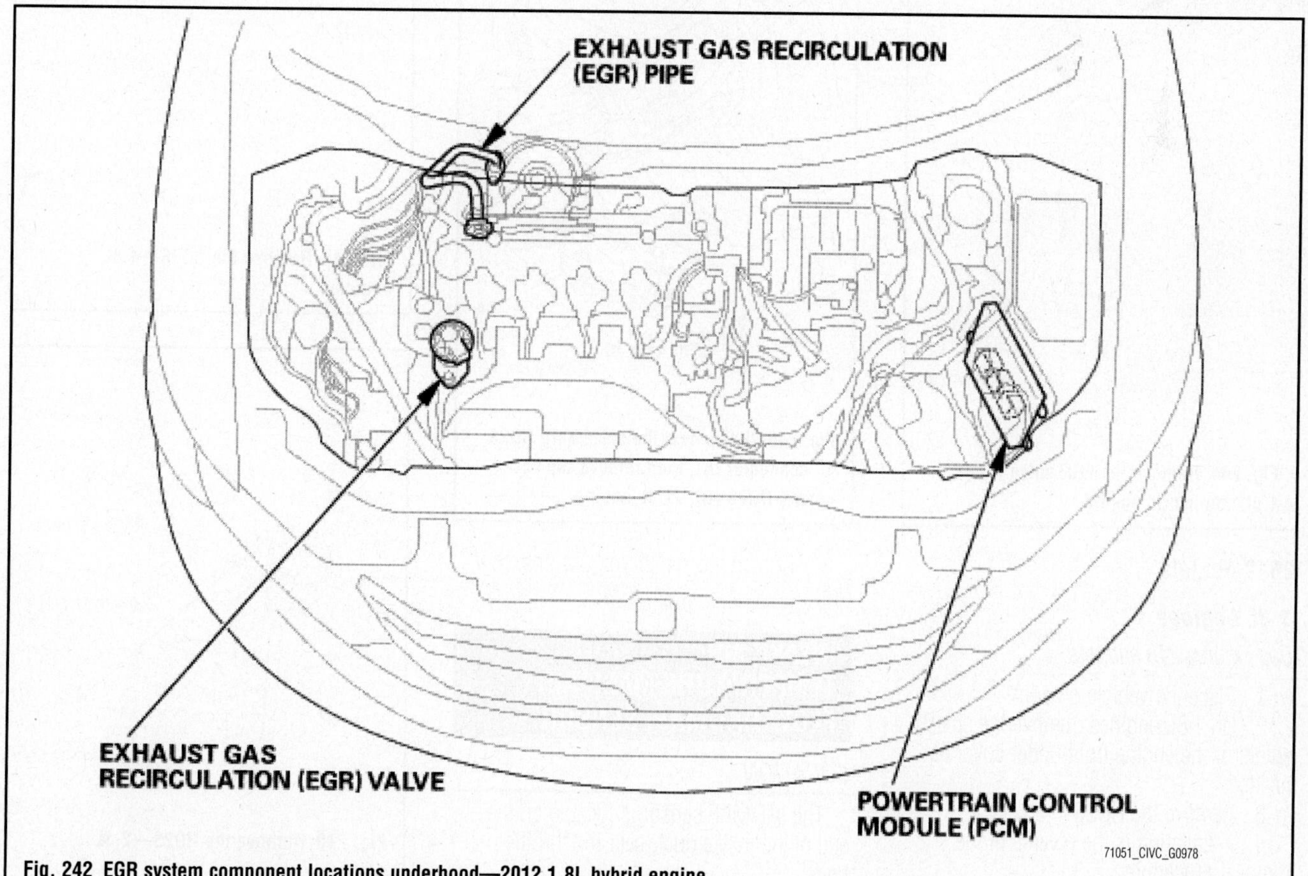

**EXHAUST GAS RECIRCULATION
(EGR) PIPE**

**EXHAUST GAS
RECIRCULATION (EGR) VALVE**

**POWERTRAIN CONTROL
MODULE (PCM)**

71051_CIVC_G0978

Fig. 242 EGR system component locations underhood—2012 1.8L hybrid engine

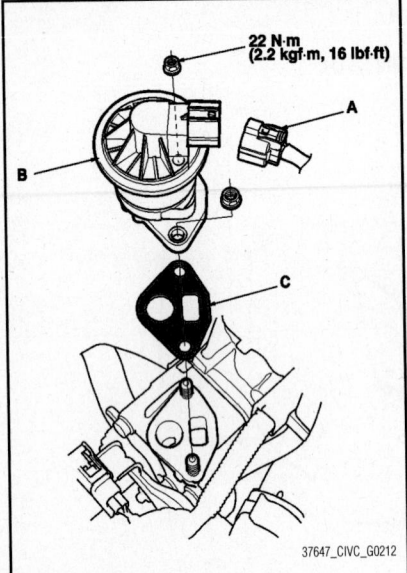

Fig. 243 Disconnect the EGR valve 6P connector (A), then remove the EGR valve (B) and gasket (C)

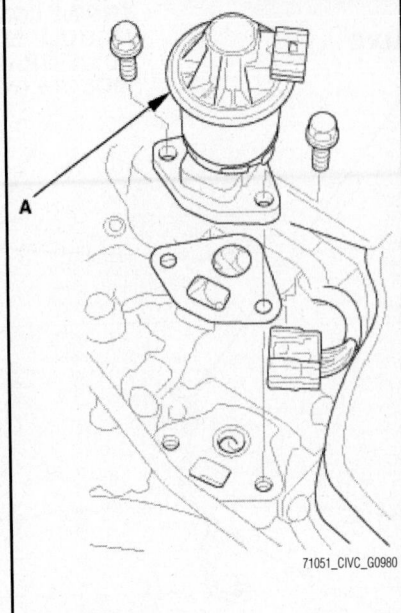

Fig. 245 Remove the EGR valve (A)— 1.8L hybrid engine

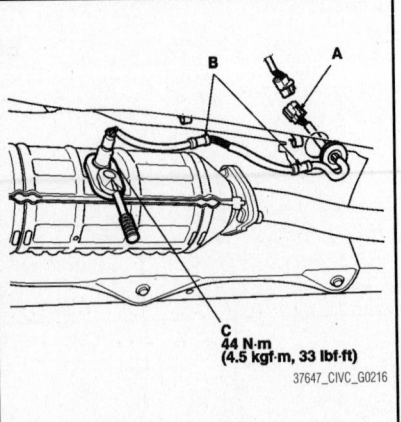

Fig. 247 Disconnect the secondary HO2S 4P connector (A), then remove the secondary HO2S (B)—2.0L engine

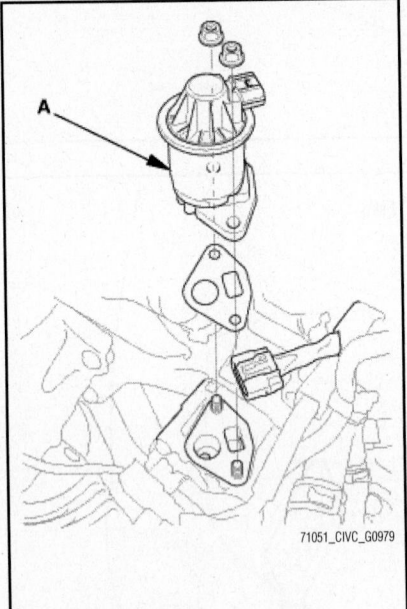

Fig. 244 Remove the EGR valve (A)— 1.8L non-hybrid engine

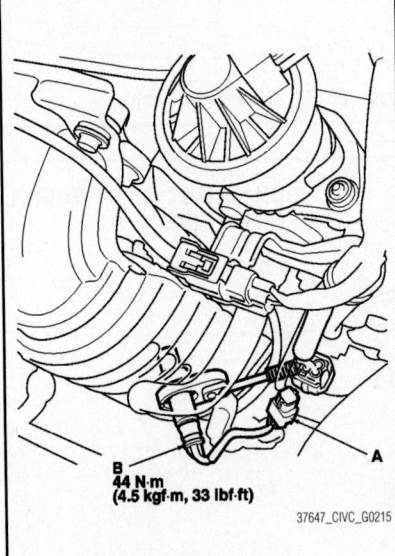

Fig. 246 Disconnect the secondary HO2S 4P connector (A), then remove the secondary HO2S (B)—1.8L engine

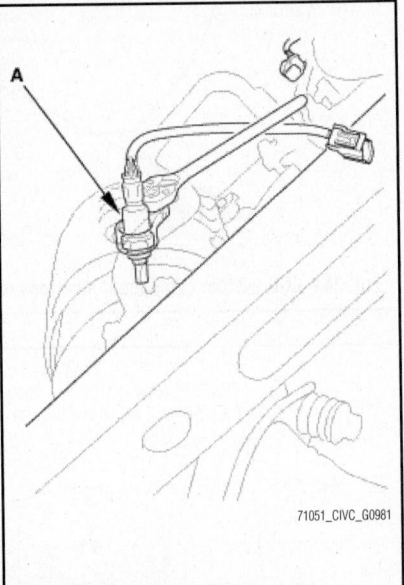

Fig. 248 Remove the HO2S—1.8L engines

2012 Models

1.8L Engines

See Figures 248 and 249.

1. Place the vehicle on a lift.
2. On 1.8L engines, remove the splash shield or the engine front under cover as needed.
3. Remove the HO2S.
4. Installation is the reverse of the removal procedure.

5. Tighten the sensor to 33 ft. lbs. (44 Nm).

INTAKE AIR TEMPERATURE/MASS AIR FLOW (IAT/MAF) SENSOR

LOCATION

The IAT/MAF sensor is located on the end of the intake duct, near the throttle body.

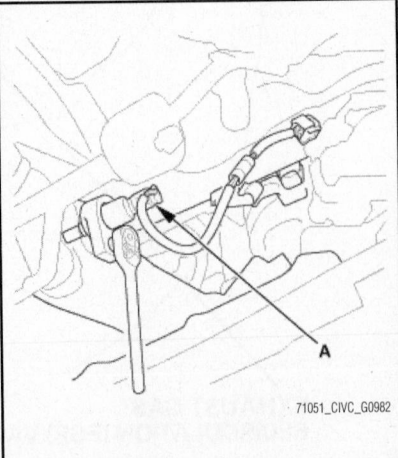

Fig. 249 Remove the HO2S—2.4L engine

REMOVAL & INSTALLATION

See Figures 250 through 252.

1. On the 1.8L non-hybrid engine, remove the air cleaner assembly if necessary.
2. Disconnect the MAF/IAT sensor connector.
3. Remove the bolts.
4. Remove the MAF/IAT sensor and O-ring.
5. Install the parts in the reverse order of removal with a new O-ring.

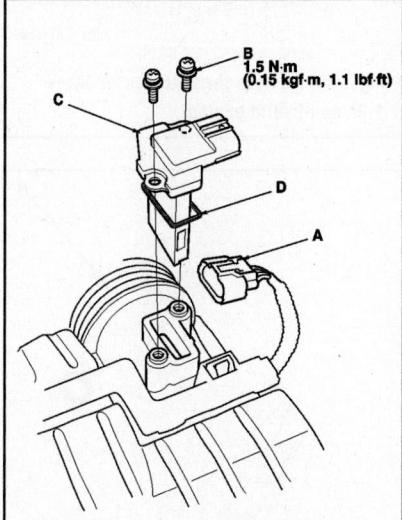

A. MAF/IAT connector C. MAF/IAT sensor
B. Bolts D. O-ring

37647_CIVC_G0217

Fig. 250 Removing the MAF/IAT sensor—2011 1.8L engine shown; 2.0L engine similar

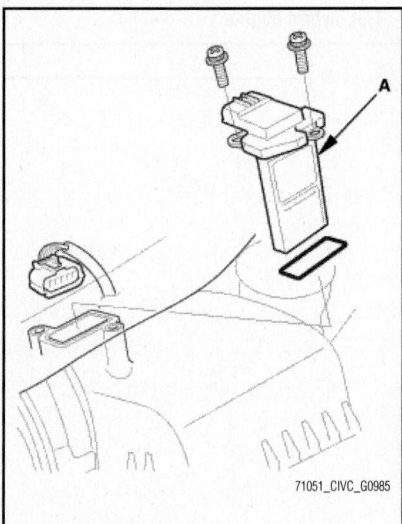

71051_CIVC_G0985

Fig. 251 Remove the MAF/IAT sensor (A)—2012 1.8L engines

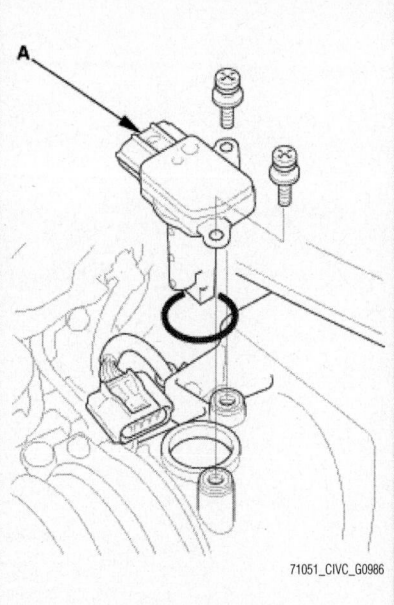

71051_CIVC_G0986

Fig. 252 Remove the MAF/IAT sensor (A)—2012 2.4L engine

KNOCK SENSOR (KS)

LOCATION

The Knock Sensor is threaded into the side of the engine block. For some engines, this is accessible when the intake manifold is removed.

REMOVAL & INSTALLATION

2011 Models

1.8L Engines

See Figure 253.

1. Remove the intake manifold.
2. Disconnect the knock sensor 1P connector.
3. Remove the knock sensor.
4. Install the parts in the reverse order of removal.

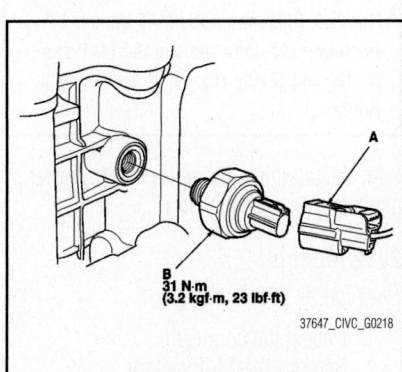

37647_CIVC_G0218

Fig. 253 Removing the connector (A) and the knock sensor (B)—1.8L engines

2.0L Engine

See Figure 254.

1. Disconnect the knock sensor 1P connector.
2. Remove the knock sensor.
3. Install the parts in the reverse order of removal.

2012 Models

1.8L Non-Hybrid Engine

See Figure 255.

1. Place the vehicle on a lift.
2. Remove the right front wheel.
3. Remove the engine undercover.
4. Remove the intake manifold bracket.
5. Detach the connector and remove the knock sensor (A).
6. Installation is the reverse of the removal procedure.

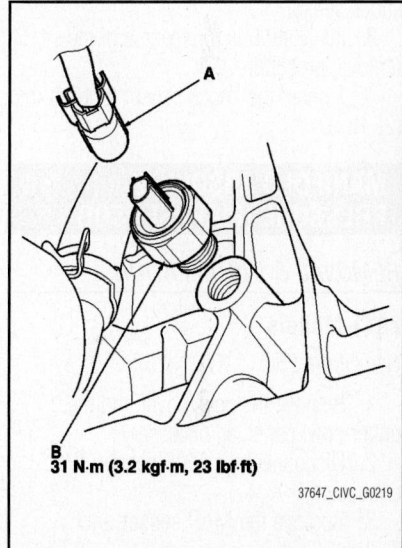

37647_CIVC_G0219

Fig. 254 Removing the connector (A) and the knock sensor (B)—2.0L engine

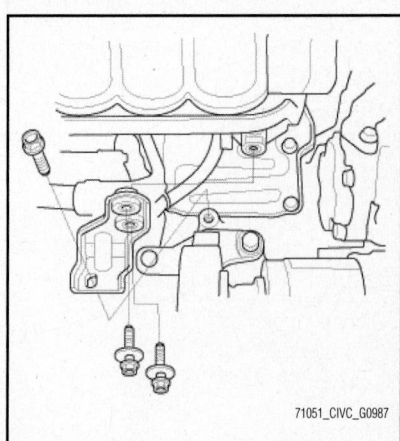

71051_CIVC_G0987

Fig. 255 Remove the intake manifold bracket

7. Tighten the knock sensor to 16 ft. lbs. (22 Nm).

1.8L Hybrid Engine

See Figure 256.

1. Remove the following (refer to component headings or sections as needed):
   - Engine cover
   - Intake air resonator, duct and pipe
   - Throttle body (move without disconnecting the hoses)
   - Engine undercover
   - Intake manifold/chamber assembly
2. Remove the knock sensor.
3. Installation is the reverse of the removal procedure.

2.4L Engine

See Figure 257.

1. Raise the vehicle on a lift.
2. Detach the connector and remove the knock sensor.
3. Installation is the reverse of the removal procedure.
4. Tighten the knock sensor to 16 ft. lbs. (22 Nm).

MANIFOLD ABSOLUTE PRESSURE (MAP) SENSOR

REMOVAL & INSTALLATION

2011 Models

See Figure 258.

1. Remove the cowl cover and the under-cowl panel, as necessary..
2. Disconnect the MAP sensor 3P connector.
3. Remove the MAP sensor and O-ring.

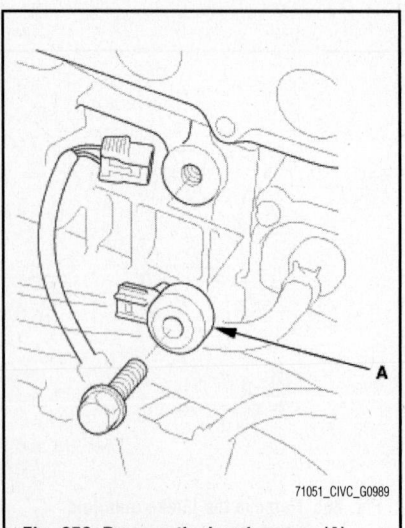

Fig. 256 Remove the knock sensor (A)

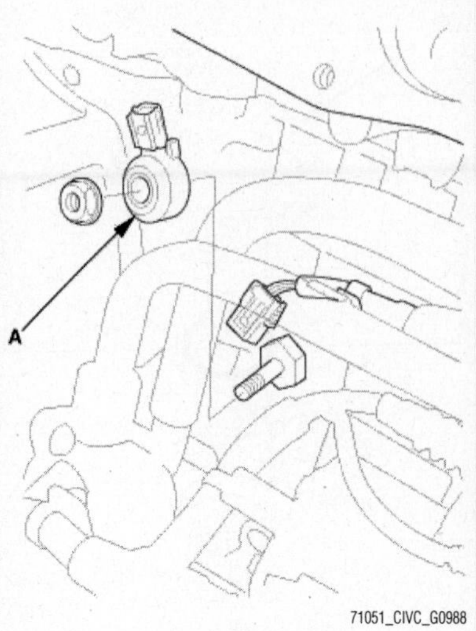

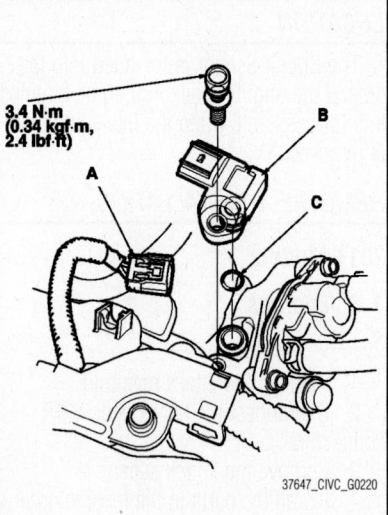

Fig. 257 Detach the connector and remove the knock sensor (A),

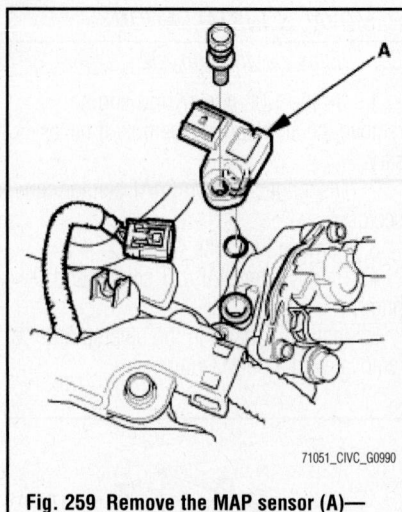

Fig. 258 Disconnect the MAP sensor 3P connector (A), then remove the MAP sensor (B) and O-ring (C)—1.8L shown; 2.0L similar

4. Install the parts in the reverse order of removal with a new O-ring

2012 Models

See Figures 259 through 261.

1. Detach the connector.
2. Remove the MAP sensor.
3. Installation is the reverse of the removal procedure.
4. Install a new O-ring.

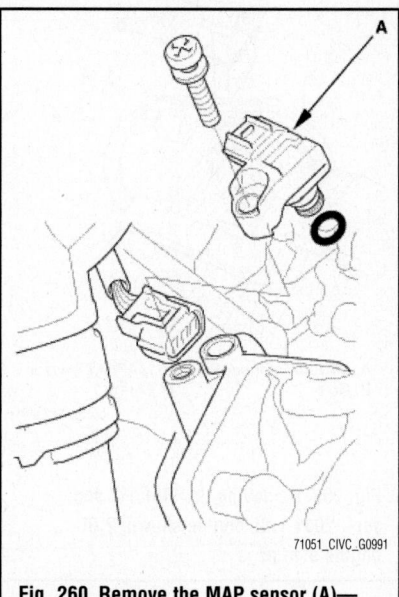

Fig. 259 Remove the MAP sensor (A)— 1.8L non-hybrid engine

Fig. 260 Remove the MAP sensor (A)— 1.8L hybrid engine

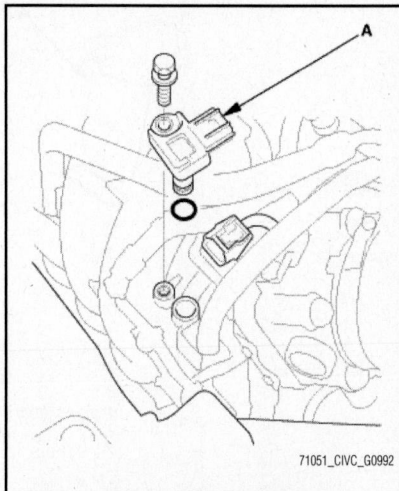

Fig. 261 Remove the MAP sensor (A)— 2.4L engine

POSITIVE CRANKCASE VENTILATION (PCV) VALVE

LOCATION

See Figure 262.

The Positive Crankcase Ventilation (PCV) valve is located on the passenger's side of the engine, as shown:

REMOVAL & INSTALLATION

1.8L Non-Hybrid & 2.4L Engines

See Figures 263 through 265.

1. Remove the harness holder.
2. Disconnect the PCV hose.
3. Remove the PCV valve and washer.
4. Install the parts in the reverse order of removal with a new washer.

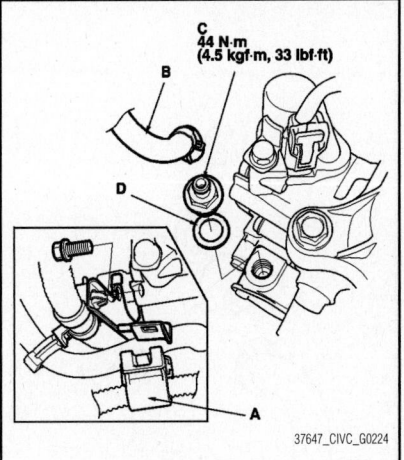

Fig. 263 Remove the harness holder (A), disconnect the PCV hose (B), and remove the PCV valve (C) and washer (D)—2011 1.8L engines

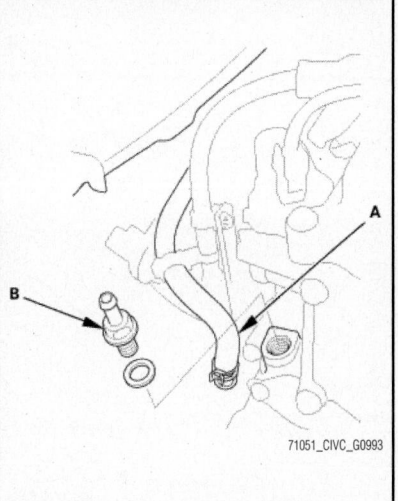

Fig. 264 Disconnect the PCV hose (A). Remove the PCV valve (B)—2012 1.8L non-hybrid engine

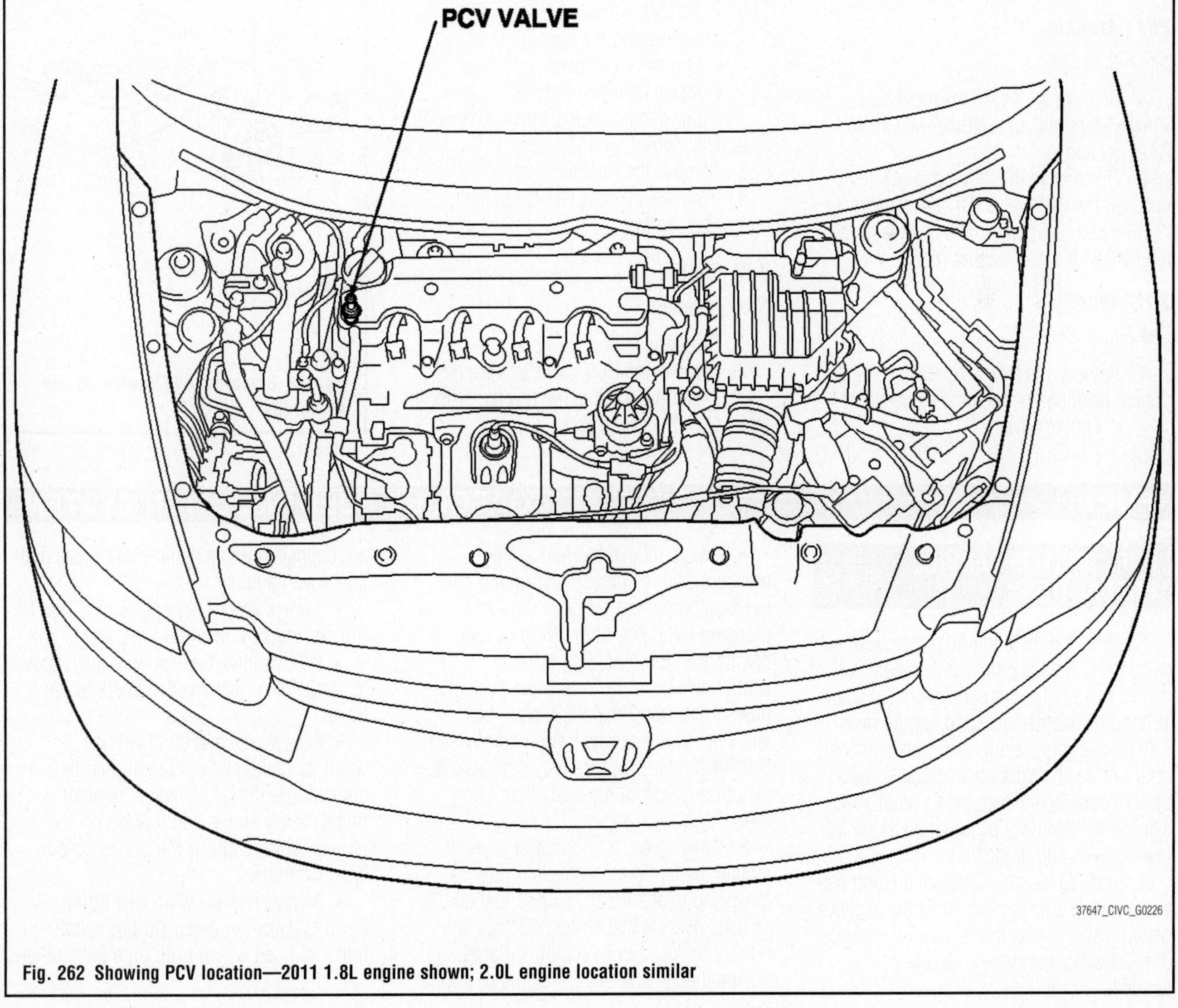

PCV VALVE

Fig. 262 Showing PCV location—2011 1.8L engine shown; 2.0L engine location similar

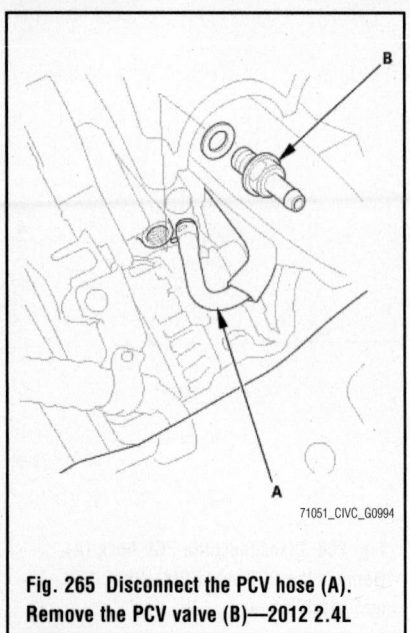

Fig. 265 Disconnect the PCV hose (A). Remove the PCV valve (B)—2012 2.4L engine

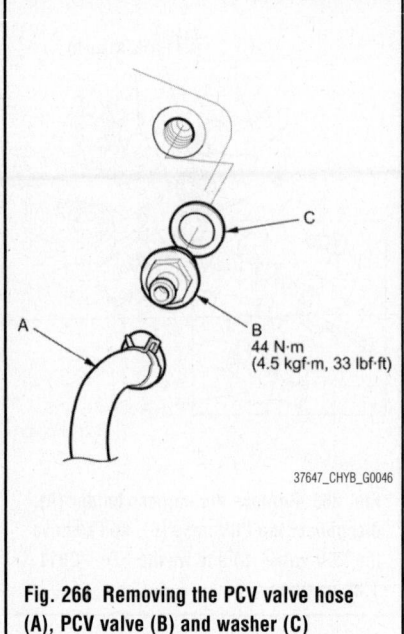

Fig. 266 Removing the PCV valve hose (A), PCV valve (B) and washer (C)

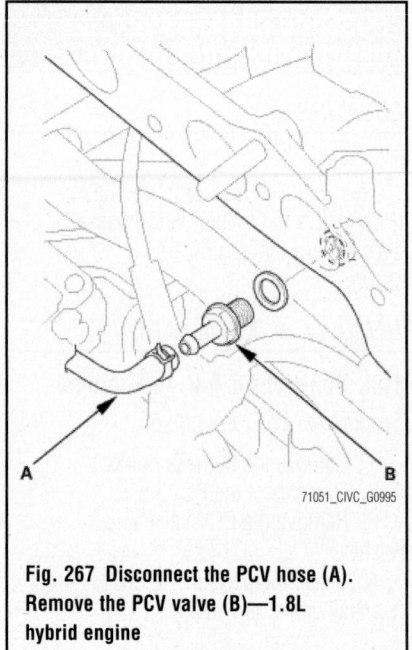

Fig. 267 Disconnect the PCV hose (A). Remove the PCV valve (B)—1.8L hybrid engine

1.8L Hybrid Engine

2011 Models

See Figure 266.

1. Remove the intake manifold. See "Intake Manifold" in "ENGINE MECHANICAL" section.

2. Disconnect the PCV hose (A), then unscrew the PCV valve (B), and remove it.

3. Install the parts in the reverse order of removal with a new washer (C).

2012 Models

See Figure 267.

1. Remove the following (refer to component headings or sections as needed):

- Engine cover

- Intake air resonator, duct and pipe
- Throttle body (move without disconnecting the hoses)
- Engine undercover
- Intake manifold bracket
- Intake manifold/chamber assembly

2. Disconnect the PCV hose.

3. Remove the PCV valve.

4. Installation is the reverse of the removal procedure.

2.0L Engine

See Figure 268.

1. Disconnect the PCV hose.

2. Remove the PCV valve and washer.

3. Install the parts in the reverse order of removal with a new washer.

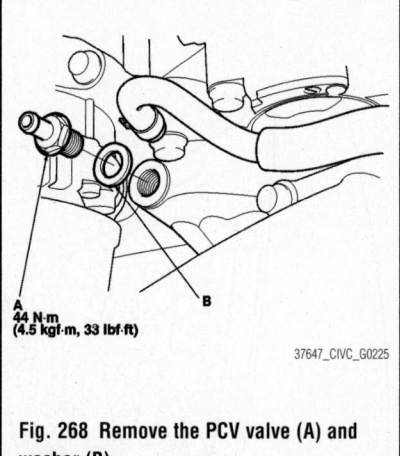

Fig. 268 Remove the PCV valve (A) and washer (B)

FUEL   GASOLINE FUEL INJECTION SYSTEM

FUEL SYSTEM SERVICE PRECAUTIONS

Safety is the most important factor when performing not only fuel system maintenance, but any type of maintenance. Failure to conduct maintenance and repairs in a safe manner may result in serious personal injury or death. Work on a vehicle's fuel system components can be accomplished safely and effectively by adhering to the following rules and guidelines.

- To avoid the possibility of fire and personal injury, always disconnect the negative battery cable unless the repair or test procedure requires that battery voltage be applied.

- Always relieve the fuel system pressure prior to disconnecting any fuel system component (injector, fuel rail, pressure regulator, etc.) fitting or fuel line connection. Exercise extreme caution whenever relieving fuel system pressure to avoid exposing skin, face and eyes to fuel spray. Please be advised that fuel under pressure may penetrate the skin or any part of the body that it contacts.

- Always place a shop towel or cloth around the fitting or connection prior to loosening to absorb any excess fuel due to spillage. Ensure that all fuel spillage is quickly removed from engine surfaces. Ensure that all fuel-soaked cloths or towels

are deposited into a flame-proof waste container with a lid.

- Always keep a dry chemical (Class B) fire extinguisher near the work area.

- Do not allow fuel spray or fuel vapors to come into contact with a spark or open flame.

- Always use a second wrench when loosening or tightening fuel line connection fittings. This will prevent unnecessary stress and torsion on fuel piping. Always follow the proper torque specifications.

- Always replace worn fuel fitting O-rings with new ones. Do not substitute fuel hose where rigid pipe is installed.

FUEL SYSTEM PRESSURE

RELIEVING

2011 Models

Except 1.8L CNG Engine

1. Remove the under-dash fuse/relay box, then remove PGM-FI main relay 2 (FUEL PUMP) from the under-dash fuse/relay box.
2. Reinstall the under-dash fuse/relay box.
3. Start the engine, and let it idle until it stalls.

➡ **If any DTCs are stored, clear and ignore them.**

4. Turn the ignition switch to LOCK (0).
5. Remove the fuel fill cap to relieve the pressure in the fuel tank.
6. Do the battery terminal disconnect procedure, then wait at least 3 minutes before beginning work. See "Battery System" in "ENGINE ELECTRICAL" section.
7. Remove the quick-connect fitting cover.
8. Check the fuel quick-connect fitting for dirt, and clean it if needed.
9. Place a rag or shop towel over the quick-connect fitting.
10. Disconnect the quick-connect fitting: Hold the connector with one hand, and squeeze the retainer tabs with the other hand to release them from the locking tabs. Pull the connector off.

✳✳ CAUTION

Be careful not to damage the line or other parts. Do not use tools.

11. If the connector does not move, keep the retainer tabs pressed down, and alternately pull and push the connector until it comes off easily.
12. Do not remove the retainer from the line; once removed, the retainer must be replaced with a new one.
13. After disconnecting the quick-connect fitting, check it for dirt or damage.
14. Do the battery reconnect procedure. See "Battery System" in "ENGINE ELECTRICAL" section.

1.8L CNG Engine—At The Fuel Tank

See Figures 269 and 270.

This procedure degrades the integrity of the fuel tank. Do it only if you are replacing the fuel tank.

✳✳ WARNING

Compressed natural gas is flammable and highly explosive. You could

be killed or seriously injured if leaking natural gas is ignited. Stop the engine, and keep heat, sparks, and flames away.

✳✳ CAUTION

This procedure should be done outside in a well-ventilated area or in a properly equipped CNG shop.

1. Do the battery disconnection procedure.
2. Raise the vehicle on a lift.
3. Close the manual shut-off valve at the filler door.
4. Slowly loosen the fuel supply line nut (B) going into fuel filter A. Continue to slowly loosen the nut until it is disconnected from fuel filter A.

➡ **The nut may be difficult to turn at first, because of fuel pressure in the line. The fuel eventually escapes with a hiss. Always use two wrenches when removing or installing fuel line nuts.**

5. Loosen the fuel supply line (C) at the manual shut-off valve, move the end of the line away from fuel filter A, and connect a fuel vent tube (D) (available from the AH special tools department) to the end of the line.
6. Vent the fuel in the line between the fuel tank and the manual shut-off valve by fully opening the manual shut-off valve. The small amount of fuel in the line should vent quickly with a hissing sound. The fuel flow should stop after a few seconds.
   a. If the fuel flow stops within a few seconds, close the manual shut-off valve, and go to the next step.
   b. If the fuel flow does not stop within

a few seconds, close the manual shut-off valve, then do steps 7 thru 17 of At the fuel tank (If the fuel tank internal valve is stuck open).

7. Attach one end of the ground wire (included with the new fuel tank) to the fuel supply line. Route the other end of the ground wire into the vehicle, where it will not interfere with vehicle movement.
8. Lower the vehicle to the ground.
9. Slowly remove the manual lock-down valve (A) from the fuel line side of the fuel tank, and install a manual override vent tool (B) (available from the AH special tools department) in its place. Do not remove the valve (C) from the receptacle side of the fuel tank.
10. With the help of an assistant, push the vehicle outside, near a metal water pipe or a chain-link fence, where the fuel can be safely vented.
11. Attach the other end of the ground wire to the water pipe or to a chain-link fence post buried in the ground. This disperses any static electricity into the ground.
12. Using an Allen wrench, slowly turn the bolt in the manual override vent tool clockwise until the bolt stops.
13. Vent the fuel from the tank by opening the manual shut-off valve. The fuel tank is empty when the hissing is gone, or when fuel cannot be felt coming from the vent pipe. Once the tank is empty, close the manual shut-off valve.

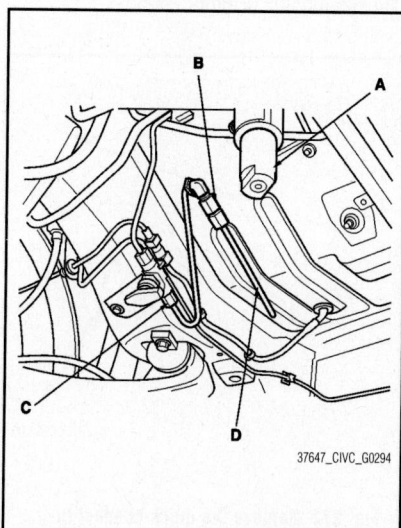

37647_CIVIC_G0294

Fig. 269 Loosen the fuel supply line

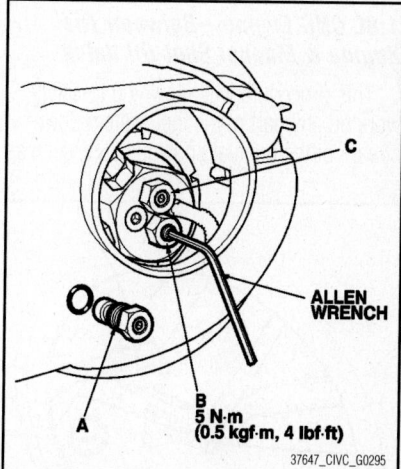

ALLEN WRENCH

B
5 N·m
(0.5 kgf·m, 4 lbf·ft)

37647_CIVIC_G0295

Fig. 270 Slowly remove the manual lock-down valve (A) from the fuel line side of the fuel tank, and install a manual override vent tool (B) (available from the AH special tools department) in its place. Do not remove the valve (C) from the receptacle side of the fuel tank

➡ **Depending on the amount of fuel in the tank, fuel venting can take several minutes, up to a few hours.**

14. Remove the manual override vent tool, and reinstall the manual lock-down valve.

15. Reinstall the fuel line to fuel filter A with a new O-ring, then retighten the fuel line at the manual shut-off valve.

16. Do a leak inspection after replacing the fuel tank.

17. Do the battery reconnection procedure.

1.8L CNG Engine—Between Fuel Tank & Engine

See Figure 271.

This procedure will allow you to safely work on any part of the fuel system downstream of the fuel tank such as the fuel joint block or the manual shut-off valve.

> ❈❈ **CAUTION**

This procedure should be done outside in a well-ventilated area or in a properly equipped CNG shop.

➡ **Make sure the manual shut-off valve is open.**

1. Remove the rear seat.

2. Disconnect the fuel subharness 6P connector (this prevents the tank from supplying fuel to the system).

3. Start the engine, and let it idle. After a few minutes, the engine will stall.

4. Turn the ignition switch to LOCK (0).

1.8L CNG Engine—Between The Engine & Manual Shut-Off Valve

This procedure will allow you to safely work on any part of the fuel system downstream of the manual shut-off valve, such as

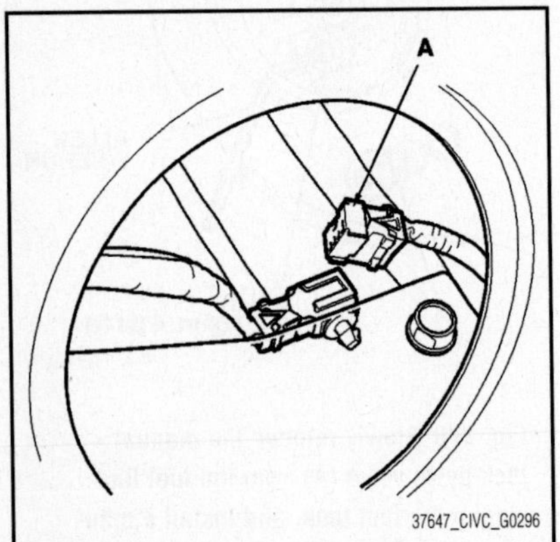

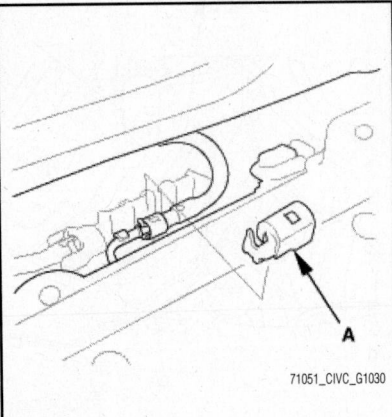

Fig. 271 Disconnect the fuel subharness 6P connector (A) (this prevents the tank from supplying fuel to the system)

fuel pressure regulator P1 or the fuel injectors.

1. Raise the vehicle on a lift.

2. Close the manual shut-off valve.

3. Start the engine, and let it idle. After a few minutes, the engine will stall.

4. Turn the ignition switch to LOCK (0).

2012 Models

See Figures 272 and 273.

1. Remove the fuel fill cap to relieve tank pressure.

2. Connect the HDS to the data link connector (DLC) located under the driver's side of the dashboard.

3. Turn the ignition switch to ON (II).

4. Turn the ignition switch to ON (II).

5. From the INSPECTION MENU of the HDS, select Fuel Pump OFF, then start the engine, and let it idle until it stalls.

➡ **Do not allow the engine to idle above 1,000 rpm or the ECM/PCM will continue to operate the fuel pump.**

➡ **Pending or Confirmed DTC may be set during this procedure. Check for DTCs, and clear them as needed.**

6. Turn the ignition switch to LOCK (0).

7. Disconnect the negative battery cable, the battery sensor, then the positive cable.

8. Remove the quick-connect fitting cover.

9. Check the fuel quick-connect fitting for dirt, and clean it if needed.

10. Place a rag or shop towel over the quick-connect fitting.

11. Disconnect the quick-connect fitting: Hold the connector with one hand, and squeeze the retainer tabs with the other hand to release them from the locking tabs. Pull the connector off. Be careful not to damage the line or other parts.

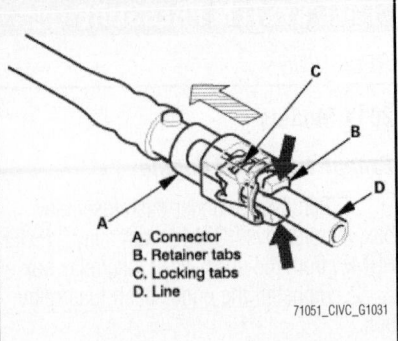

Fig. 273 Disconnect the quick-connect fitting: Hold the connector with one hand, and squeeze the retainer tabs with the other hand to release them from the locking tabs. Pull the connector off. Be careful not to damage the line or other parts

➡ **Do not use tools.**

12. If the connector does not move, keep the retainer tabs pressed down, and alternately pull and push the connector until it comes off easily.

➡ **Do not remove the retainer from the line; once removed, the retainer must be replaced with a new one.**

> **FUEL FILTER**

REMOVAL & INSTALLATION

See Figure 274.

The fuel filter should be replaced whenever the fuel pressure drops below the specified value, after making sure that the fuel pump and the fuel pressure regulator are OK.

1. Remove the fuel tank unit.

2. Remove the fuel filter.

3. Check these items before installing the fuel tank unit:

   a. When connecting the wire harness, make sure the connection is secure and the connectors are firmly locked into place.

   b. When installing the fuel gauge sending unit, make sure the connection is secure and the connector is firmly locked into place. Be careful not to bend or twist it excessively.

To install:

4. Install the parts in the reverse order of removal with new O-rings.

5. When installing the fuel tank unit, align the marks on the unit and the fuel tank.

➡ **Coat the O-rings with clean engine oil; do not use any other type oils or**

Fig. 272 Remove the quick-connect fitting cover (A)

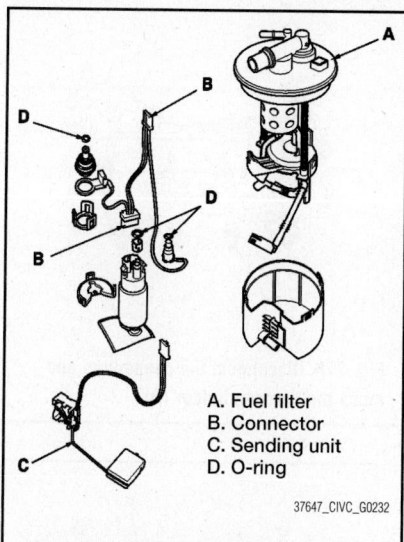

A. Fuel filter
B. Connector
C. Sending unit
D. O-ring

37647_CIVIC_G0232

Fig. 274 Exploded view of the fuel pump assembly, showing the fuel filter

fluids. **Do not pinch the O-rings during installation.**

6. Use all new parts supplied in the fuel filter replacement kit.

FUEL PUMP/ FUEL GAUGE SENDING UNIT

REMOVAL & INSTALLATION

See Figure 275.

1. Remove the fuel tank unit.
2. Remove the fuel level sensor (fuel sending unit) from the fuel tank unit.

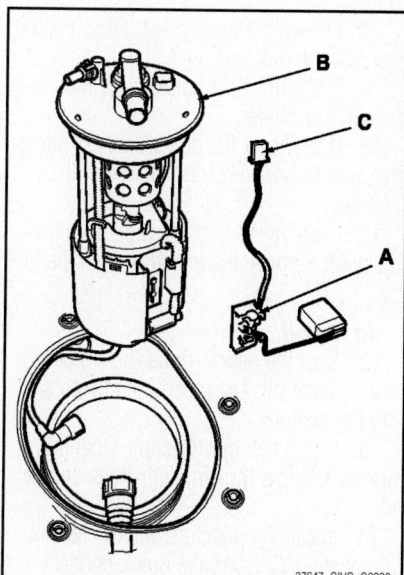

37647_CIVIC_G0233

Fig. 275 Remove the connector (C), fuel level sensor (fuel sending unit) (A) from the fuel tank unit (B)

3. Check these items before installing the fuel tank unit:

   a. When connecting the wire harness, make sure the connection is secure and the connector is firmly locked into place.

   b. When installing the fuel gauge sending unit, make sure the connection is secure. Be careful not to bend or twist it excessively.

4. Install the parts in the reverse order of removal. When installing the fuel tank unit, align the marks on the unit and the fuel tank.

FUEL RAIL AND INJECTOR

REMOVAL & INSTALLATION

2011 Models

1.8L ENGINE
See Figure 276.

1. Relieve the fuel pressure.
2. Remove the cowl cover and the under-cowl panel.
3. Remove the fuel line cover.
4. Disconnect the injector connectors from the injectors and the rocker arm oil control valve connector.

5. Disconnect the quick-connect fitting.
6. Remove the fuel rail mounting nuts from the fuel rail.
7. Remove the injector clips from the injectors.
8. Remove the injectors from the fuel rail.

To install:
9. Coat the new O-rings with clean engine oil, and insert the injectors into the fuel rail.
10. Install the injector clips.
11. Coat the new injector O-rings with clean engine oil.
12. Install the fuel rail and the injectors into the cylinder head.
13. Install the fuel rail mounting nuts.
14. Connect the connectors on the injectors.
15. Connect the quick-connect fitting.
16. Turn the ignition switch to ON (II), but do not operate the starter. After the fuel pump runs for about 2 seconds, the fuel rail is pressurized. Repeat this two or three times, then make sure there are no fuel leaks.
17. Install the fuel line cover.
18. Install the cowl cover and the under-cowl panel.

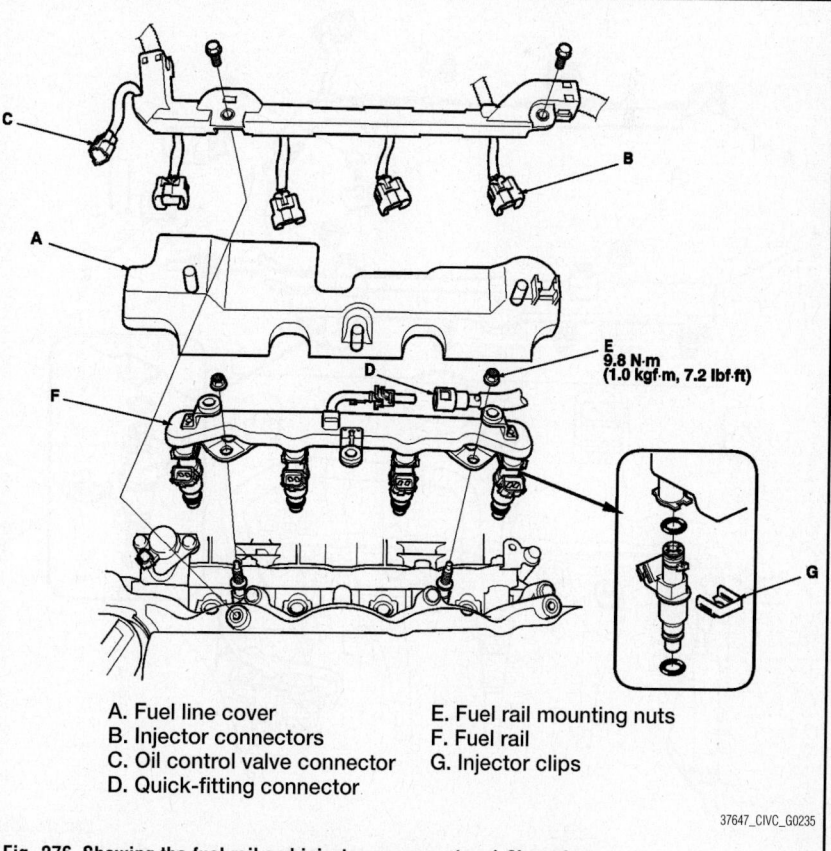

A. Fuel line cover
B. Injector connectors
C. Oil control valve connector
D. Quick-fitting connector
E. Fuel rail mounting nuts
F. Fuel rail
G. Injector clips

37647_CIVIC_G0235

Fig. 276 Showing the fuel rail and injector components—1.8L engine

2.0L Engine

See Figure 277.

1. Relieve fuel pressure.
2. Remove the engine cover.
3. Disconnect the injector connectors from the injectors.
4. Remove the ground cable bolt.
5. Disconnect the quick-connect fitting.
6. Remove the fuel rail mounting nuts from the fuel rail.
7. Remove the fuel rail and the injectors from the injector base.
8. Remove the injector clips from the injectors.
9. Remove the injectors from the fuel rail.
10. Coat the new O-rings with clean engine oil, and insert the injectors into the fuel rail.
11. Install the injector clips.
12. Coat the injector O-rings with clean engine oil.
13. Install the fuel rail and the injectors in the injector base.
14. Install the fuel rail mounting nuts and tighten to 16 ft. lbs. (22 Nm).
15. Connect the connectors on the injectors, and reinstall the ground cable bolt.

16. Connect the quick-connect fitting.
17. Turn the ignition switch to ON (II), but do not operate the starter. After the fuel pump runs for about 2 seconds, the fuel rail is pressurized. Repeat this two or three times, then make sure there are no fuel leaks.
18. Install the engine cover.

2012 Models

See Figures 278 through 282.

1. Remove the fuel fill cap to relieve tank pressure.
2. Connect the HDS to the data link connector (DLC) located under the driver's side of the dashboard.
3. Turn the ignition switch to ON (II).
4. Turn the ignition switch to ON (II).
5. From the INSPECTION MENU of the HDS, select Fuel Pump OFF, then start the engine, and let it idle until it stalls.

➡**Do not allow the engine to idle above 1,000 rpm or the ECM/PCM will continue to operate the fuel pump.**

➡**Pending or Confirmed DTC may be set during this procedure. Check for DTCs, and clear them as needed.**

6. Turn the ignition switch to LOCK (0).

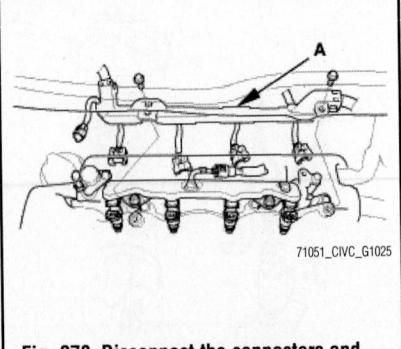

Fig. 278 Disconnect the connectors and move the harness holder (A)

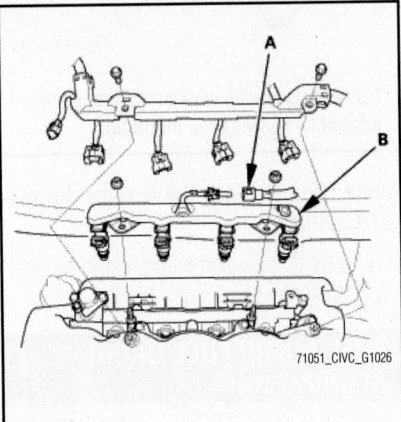

Fig. 279 Disconnect the quick-connect fitting (A), then remove the injectors and fuel rail (B) together

7. Disconnect the negative battery cable, the battery sensor, then the positive cable.
8. Relieve the fuel system pressure, as described in this section.
9. Disconnect the connectors and move the harness holder.
10. Disconnect the quick-connect fitting, then remove the injectors and fuel rail together.
11. Remove the injector clips from the injectors. Remove the injectors from the fuel rail.

To install:

12. Coat the new O-rings (black) with clean engine oil, and insert the injectors into the fuel rail.
13. Install the injector clips. Coat the injector O-rings (brown) with clean engine oil.
14. Install the injectors and the fuel rail together. Connect the quick-connect fitting.
15. Install the harness holder. Connect the connectors.
16. Reconnect any fuel connections.
17. Restore battery connection.

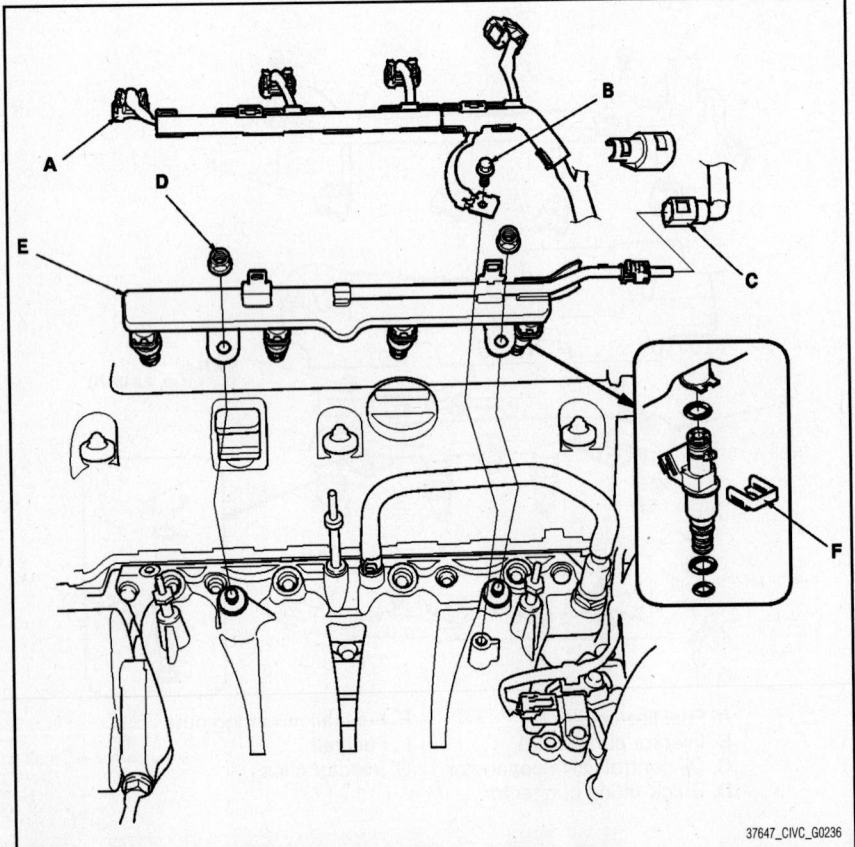

Fig. 277 Showing the fuel rail and injector components—2.0L engine

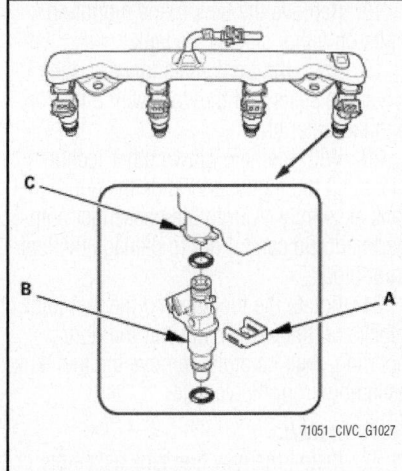

Fig. 280 Remove the injector clips (A) from the injectors (B). Remove the injectors from the fuel rail (C)

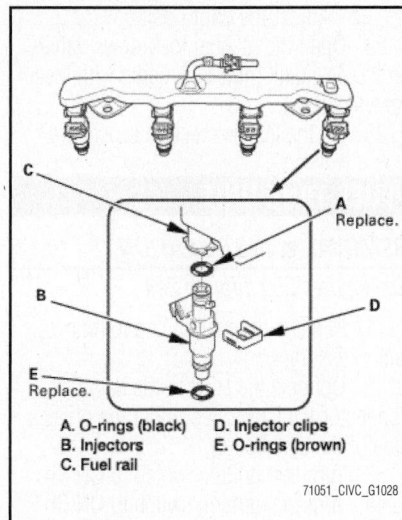

A. O-rings (black)
B. Injectors
C. Fuel rail
D. Injector clips
E. O-rings (brown)

71051_CIVC_G1028

Fig. 281 Coat the new O-rings (black) with clean engine oil, and insert the injectors into the fuel rail. Install the injector clips. Coat the injector O-rings (brown) with clean engine oil

18. Connect the HDS to the data link connector (DLC) located under the driver's side of the dashboard.
19. Turn the ignition switch to ON (II).
20. From the INSPECTION MENU of the HDS, select Fuel Pump ON.

➡**Pending or Confirmed DTC may be set during this procedure. Check for DTCs, and clear them as needed.**

21. Turn the ignition switch to LOCK (0).
22. Turn the ignition switch to ON (II) (but do not operate the starter motor). The fuel pump runs for about 2 seconds, and fuel pressure rises. Repeat this two or three

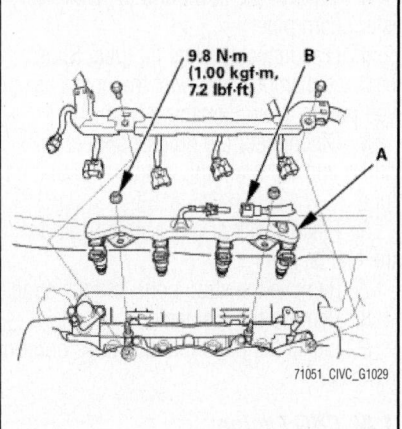

Fig. 282 Install the injectors and the fuel rail (A) together. Connect the quick-connect fitting (B)

times, then make sure there are no fuel leaks.

FUEL TANK

DRAINING

2011 Models

1. Remove the fuel tank unit. See "Fuel Pump/Fuel Gauge Sending Unit" in this section.
2. Using a hand pump, a hose, and a container suitable for fuel, draw the fuel from the fuel tank.
3. Reinstall the fuel tank unit.

2012 Models

1.8L Non-Hybrid & 2.4L Engines
See Figures 283 and 284.

1. Remove the fuel fill cap to relieve tank pressure.
2. Connect the HDS to the data link connector (DLC) located under the driver's side of the dashboard.
3. Turn the ignition switch to ON (II).
4. Turn the ignition switch to ON (II).
5. From the INSPECTION MENU of the HDS, select Fuel Pump OFF, then start the engine, and let it idle until it stalls.

➡**Do not allow the engine to idle above 1,000 rpm or the ECM/PCM will continue to operate the fuel pump.**

➡**Pending or Confirmed DTC may be set during this procedure. Check for DTCs, and clear them as needed.**

6. Turn the ignition switch to LOCK (0).
7. Disconnect the negative battery cable, the battery sensor, then the positive cable.

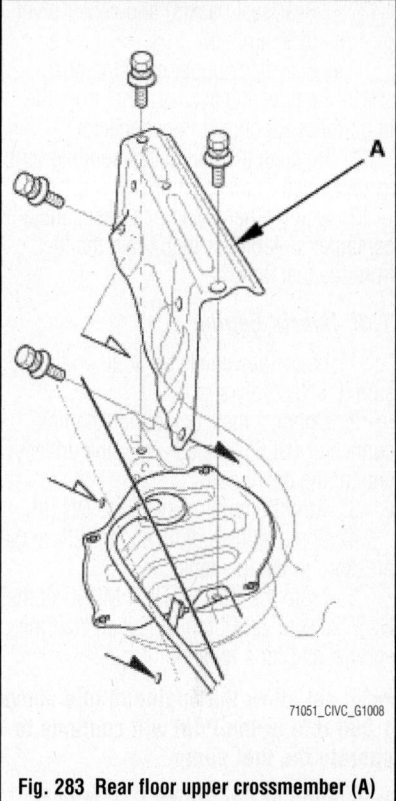

Fig. 283 Rear floor upper crossmember (A)

8. Relieve the fuel system pressure, as described in this section.
9. Remove the following (refer to component headings or sections as needed):
   • Rear seat cushion
   • Trunk floor cover

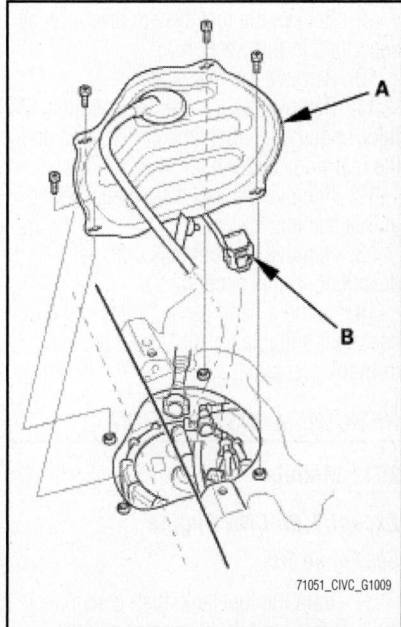

Fig. 284 Remove the access panel (A) from the floor, then disconnect the connector (B)

- Rear seat back(s) and center pivot (if equipped)
- Rear floor upper crossmember.

10. Remove the access panel from the floor, then disconnect the connector.

11. Remove the fuel gauge sending unit, as described in this section.

12. Using a hand pump, a hose, and a container suitable for fuel, drain the fuel from the fuel tank.

1.8L Hybrid Engine

1. Raise the vehicle on a lift, and make sure it is securely supported.

2. Connect the HDS to the data link connector (DLC) located under the driver's side of the dashboard.

3. Turn the ignition switch to ON (II).

4. Remove the fuel fill cap to relieve the pressure in the fuel tank.

5. From the INSPECTION MENU of the HDS, select Fuel Pump OFF, then start the engine, and let it idle until it stalls.

➡Do not allow the engine to idle above 1,000 rpm or the PCM will continue to operate the fuel pump.

➡Pending or Confirmed DTC may be set during this procedure. Check for DTCs, and clear them as needed.

6. Turn the ignition switch to LOCK (0).

7. Disconnect and isolate the negative cable and battery sensor, then the positive cable.

8. Disconnect the intake air duct from the air cleaner, then remove the air cleaner.

9. Relieve the fuel system pressure, as described in this section.

10. Remove the rear wheel.

11. Remove the left rear inner fender, left floor undercover, fuel tank undercover, and the rear floor under bar.

12. Remove the fuel tank, as described in this section.

13. Remove the fuel tank unit, as described in this section.

14. Using a hand pump, a hose, and a container suitable for fuel, drain the fuel from the fuel tank.

REMOVAL & INSTALLATION

2011 Models

Except 1.8L CNG Engine
See Figure 285.

1. Drain the fuel tank, then disconnect the fuel tank unit 4P connector and the quick-connect fitting.

2. Raise the vehicle on a lift.

3. Remove the cover and the EVAP canister guard pipe.

4. Disconnect the fuel fill tube. Slide back the clamps, then twist the hoses as you pull to avoid damaging them.

5. Disconnect the quick-connect fitting and the fuel tank vapor recirculation tube.

6. Place a jack or other support under the fuel tank.

7. Remove the strap bolts and the strap.

8. Remove the fuel tank.

9. Install the parts in the reverse order of removal.

1.8L CNG Engine
See Figures 286 through 289.

✺✺ WARNING

Compressed natural gas is flammable and highly explosive. You could be killed or seriously injured if leaking natural gas is ignited. Stop the engine, and keep heat, sparks, and flames away.

➡You must be an SCI-certified technician to do fuel tank inspection or replacement work.

✺✺ CAUTION

This procedure degrades the integrity of the fuel tank. If you do it, do not reinstall the original fuel tank. Install a new one.

1. Relieve fuel pressure between the fuel tank and the engine.

2. Raise the vehicle on a lift.

3. Remove the under-floor cover.

4. Remove the three fuel pipes at the fuel joint block and the two fuel joint block mounting bolts.

5. Lower the vehicle.

6. Remove the left-rear door, then disconnect the connector.

7. Remove the rear seat.

8. Remove the rear seat belt.

9. Remove the C-pillar trim.

10. Remove the rear shelf and the trunk floor lid.

11. Remove the trunk partition.

12. Remove the gussets.

13. Remove the harness clip and the rear parcel cover.

14. Remove the harness clip and the rear shelf gussets.

15. Remove the rear floor upper crossmember gusset.

16. Remove the bolts indicated (A).

17. Disconnect the vent hose from the fuel tank.

18. Remove the tank frame mounting bolts on either side of the tank.

19. Push the front seats forward, then cover the seats and the floor with a tarp or similar material.

20. With the help of two other technicians, lift the frame, and slide the entire fuel tank assembly (A) into the passenger compartment. Be careful not to damage the fuel joint block.

21. Rotate the fuel tank so the fuel joint block and the frame can clear the door opening, then carefully remove the fuel tank assembly from the vehicle.

To install:

22. Install the new fuel tank assembly in the reverse order of removal.

23. Do this when you install the left rear door:

  a. Tighten the door checker mounting bolt.

  b. Adjust the door position.

24. Open the manual lock-down valves on the new tank (new tank come with these valves closed).

25. Do the leak inspection procedure.

FUEL TANK UNIT

REMOVAL & INSTALLATION
See Figures 290 through 294.

1. Remove the fuel fill cap to relieve tank pressure.

2. Connect the HDS to the data link connector (DLC) located under the driver's side of the dashboard.

3. Turn the ignition switch to ON (II).

4. Turn the ignition switch to ON (II).

5. From the INSPECTION MENU of the HDS, select Fuel Pump OFF, then start the engine, and let it idle until it stalls.

➡Do not allow the engine to idle above 1,000 rpm or the ECM/PCM will continue to operate the fuel pump.

➡Pending or Confirmed DTC may be set during this procedure. Check for DTCs, and clear them as needed.

6. Turn the ignition switch to LOCK (0).

7. Disconnect the negative battery cable, the battery sensor, then the positive cable.

8. Relieve the fuel system pressure, as described in this section.

9. Remove the following (refer to component headings or sections as needed):

- Rear seat cushion
- Trunk floor cover
- Rear seat back(s) and center pivot (if equipped)

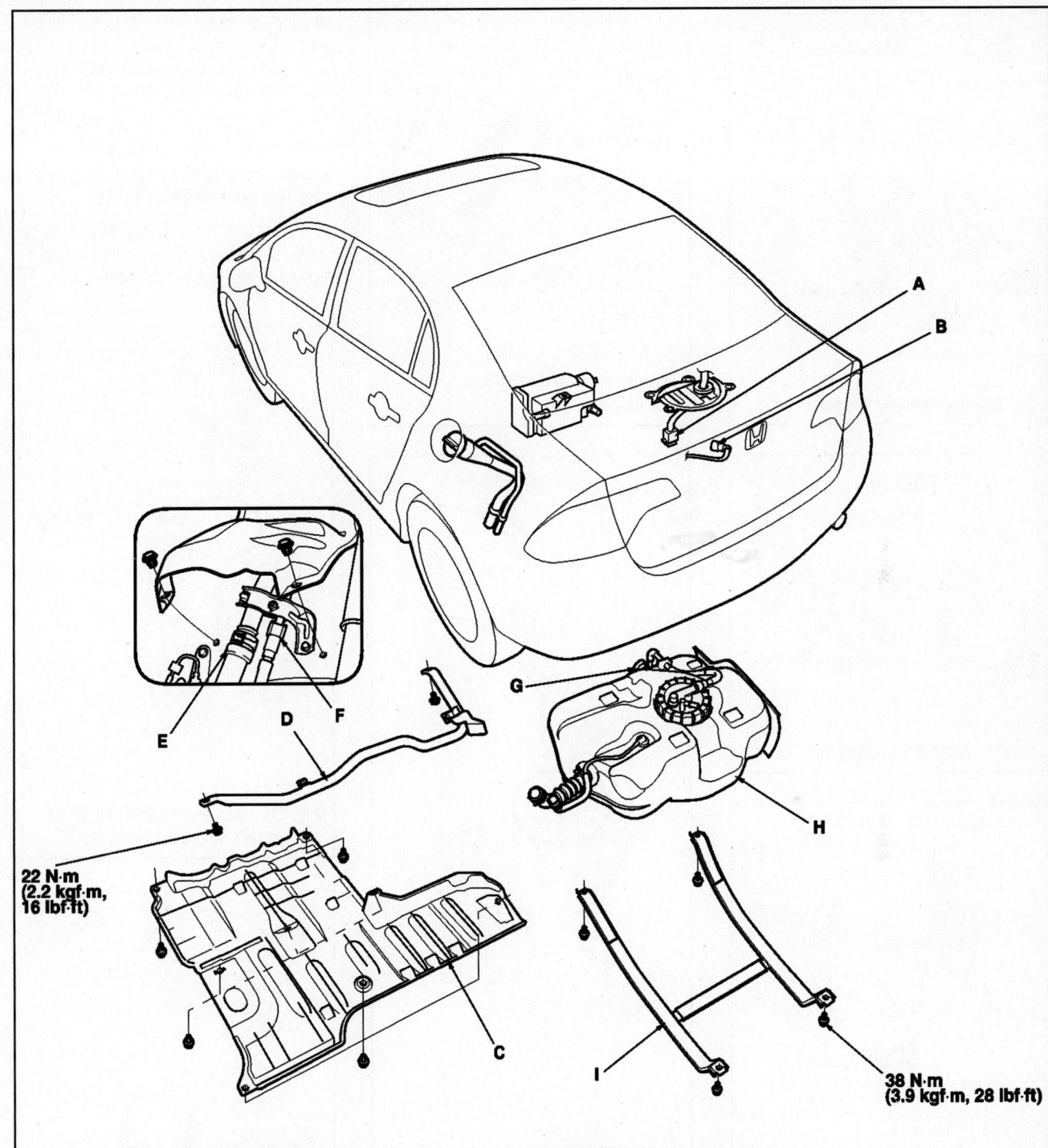

22 N·m
(2.2 kgf·m,
16 lbf·ft)

38 N·m
(3.9 kgf·m, 28 lbf·ft)

A. Connector
B. Quick-connect fitting
C. Cover
D. EVAP canister guard pipe
E. Fuel fill tube

F. Quick-connect fitting
G. Vapor recirculation tube
H. Fuel tank
I. Tank strap

37647_CIVC_G0237

Fig. 285 Showing the fuel tank and related components

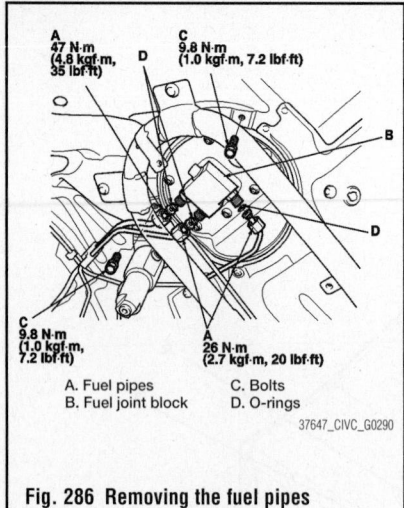

A
47 N·m
(4.8 kgf·m,
35 lbf·ft)

C
9.8 N·m
(1.0 kgf·m, 7.2 lbf·ft)

D

B

D

C
9.8 N·m
(1.0 kgf·m,
7.2 lbf·ft)

A
26 N·m
(2.7 kgf·m, 20 lbf·ft)

A. Fuel pipes   C. Bolts
B. Fuel joint block D. O-rings

37647_CIVIC_G0290

Fig. 286 Removing the fuel pipes

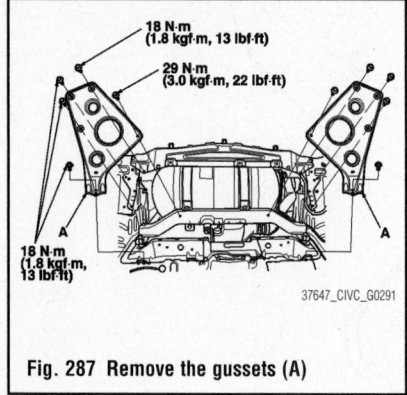

18 N·m
(1.8 kgf·m, 13 lbf·ft)

29 N·m
(3.0 kgf·m, 22 lbf·ft)

A

A

18 N·m
(1.8 kgf·m,
13 lbf·ft)

37647_CIVIC_G0291

Fig. 287 Remove the gussets (A)

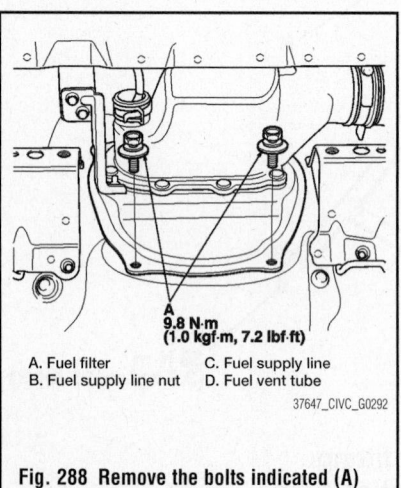

9.8 N·m
(1.0 kgf·m, 7.2 lbf·ft)

A. Fuel filter   C. Fuel supply line
B. Fuel supply line nut D. Fuel vent tube

37647_CIVIC_G0292

Fig. 288 Remove the bolts indicated (A)

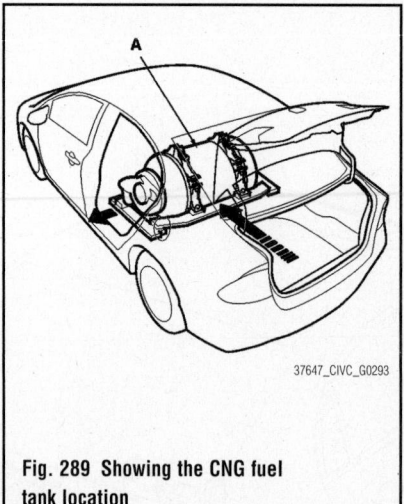

A

37647_CIVIC_G0293

Fig. 289 Showing the CNG fuel tank location

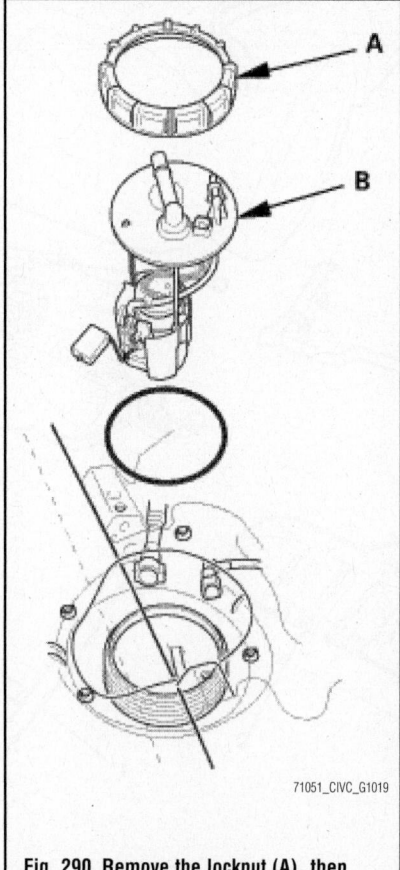

A

B

71051_CIVIC_G1019

Fig. 290 Remove the locknut (A), then remove the fuel tank unit (B)

ening, align the mark on the locknut to the start of the thread.

16. Using the special tool, tighten the locknut to the specified torque. After tightening, make sure the marks are still aligned.

➡**After installation, check the base gasket, visually or by hand, to make sure it is not pinched.**

17. Reconnect the quick-connect fittings.

18. Install the remaining components in reverse of the removal procedure.

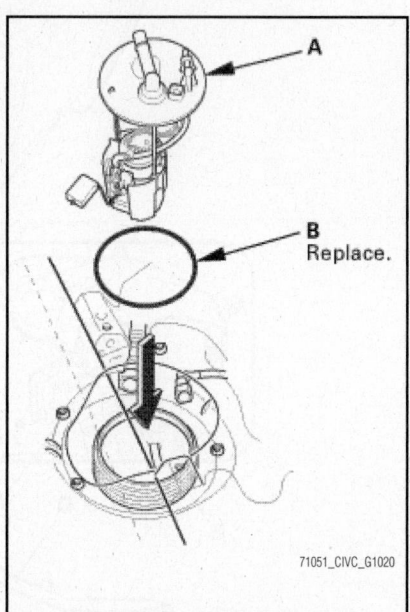

A

B
Replace.

71051_CIVIC_G1020

Fig. 291 Insert the fuel tank unit (A) into the fuel tank with a new O-ring (B)

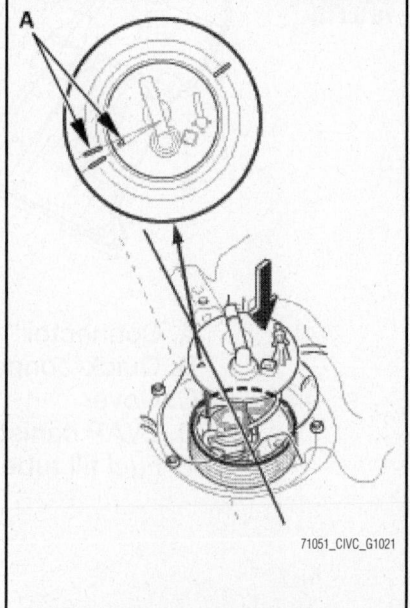

A

71051_CIVIC_G1021

Fig. 292 Align the marks (A) on the fuel tank and the fuel tank unit, then insert the fuel tank unit into the fuel tank

• Rear floor upper crossmember.

10. Remove the access panel from the floor, then disconnect the connector.

11. Using the special tool, loosen the locknut.

12. Remove the locknut, then remove the fuel tank unit.

To install:

13. Insert the fuel tank unit into the fuel tank with a new O-ring.

14. Align the marks on the fuel tank and the fuel tank unit, then insert the fuel tank unit into the fuel tank.

➡**To prevent a fuel leak, check the base gasket, visually or by hand, to make sure it is not pinched.**

15. Tighten a new locknut by hand holding the fuel tank unit vertically. Before tight-

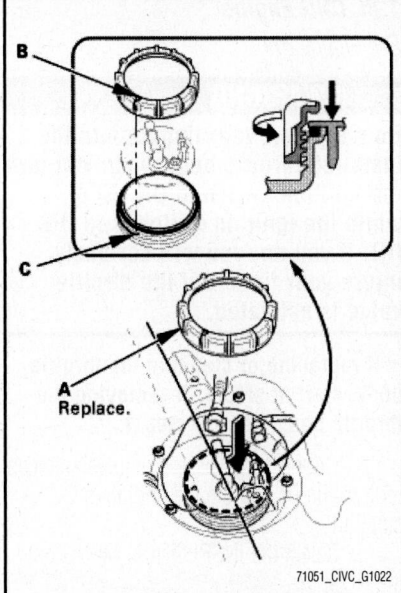

Fig. 293 Tighten a new locknut (A) by hand holding the fuel tank unit vertically. Before tightening, align the mark (B) on the locknut to the start of the thread (C)

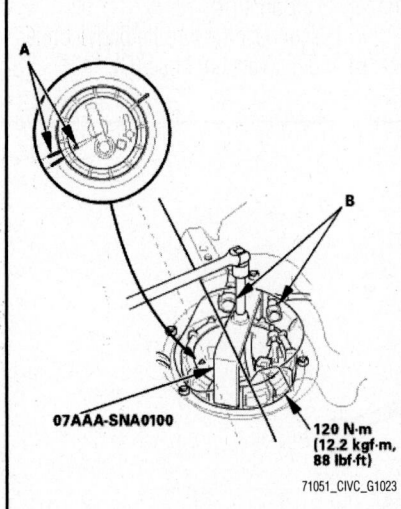

Fig. 294 Using the special tool, tighten the locknut to the specified torque. After tightening, make sure the marks (A) are still aligned. Reconnect the quick-connect fittings (B)

IDLE SPEED

INSPECTION & ADJUSTMENT

2011 Models

➡The idle speed is not adjustable except by performing the "ECM/PCM Idle Learn Procedure". The following information will help determine if the idle speed is not to specification.

1. Before checking the idle speed, check these items:
   - The malfunction indicator lamp (MIL) has not been reported on, and there are no DTCs.
   - Ignition timing
   - Spark plugs
   - Air cleaner
   - PCV system

2. Apply the parking brake, and make sure the headlights are off.

3. Disconnect the evaporative emission (EVAP) canister purge valve connector.

4. Connect the scan tool to the data link connector (DLC) located under the driver's side of the dashboard.

5. Make sure the scan tool communicates with the ECM/PCM. If it doesn't, go to the DLC circuit troubleshooting.

6. Start the engine. Hold the engine speed at 3,000 RPM without load (A/T in P or N, M/T in neutral) until the radiator fan comes on, then let it idle.

7. Check the idle speed without load conditions: headlights, blower fan, radiator fan, and air conditioner off. Idle speed should be:
   - 1.8L engine
   - M/T: 670±50 RPM
   - A/T: 670±50 RPM (in P or N)
   - 2.0L engine
   - M/T: 750±50 RPM
   - A/T: 750±50 RPM (in P or N)

8. Let the engine idle for 1 minute with high electric load (A/C on, temperature set to max cool, blower fan on high, headlights on high beam). Idle speed should be:
   - 1.8L engine
   - M/T: 710±50 RPM
   - A/T: 710±50 RPM (in P or N)
   - 2.0L engine
   - M/T: 750±50 RPM
   - A/T: 750±50 RPM (in P or N)

➡If the idle speed is not within specification, do the "ECM Idle Learn Procedure". If the idle speed is still not within specification, go to the "DIAGNOSTICS" section.

9. Reconnect the EVAP canister purge valve connector.

2012 Models

See Figures 295 and 296.

1. Connect the HDS to the data link connector (DLC) located under the driver's side of the dashboard.

2. Turn the ignition switch to ON (II).

3. Check for DTCs. If a DTC is present, diagnose and repair the cause before continuing with this inspection.

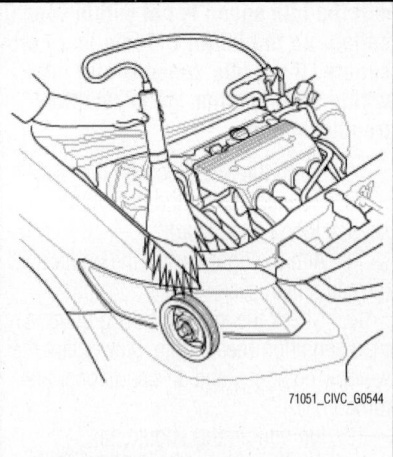

Fig. 295 Connect the timing light to the service loop (white tape)

4. Start the engine. Hold the engine speed at 3,000 rpm without load (A/T in P or N, M/T in neutral) until the radiator fan comes on, then let it idle.

5. Check the idle speed without electrical load: headlights, blower fan, radiator fan, and air conditioner off.

6. Idle speed should be:
   - 1.8L engines: 750±50 rpm (A/T in P or N, M/T in neutral).
   - 2.4L engine: 720±50 rpm (A/T in P or N, M/T in neutral).

7. Let the engine idle for 1 minute with high electric load (A/C switch on, temperature set to max cool, blower fan on High, and headlights on high beam).

8. Idle speed should be:
   - 1.8L engines: 750±50 rpm (A/T in P or N, M/T in neutral).
   - 2.4L engine: 720±50 rpm (A/T in P or N, M/T in neutral).

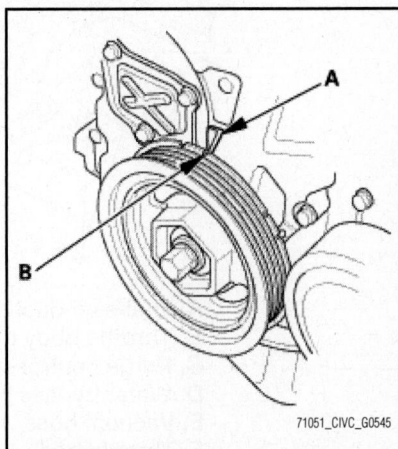

Fig. 296 Aim the light toward the pointer (A) on the cam chain case. Ignition timing should be as indicated BTDC (RED mark (B)) at idle

➡ **If the idle speed is not within specification, do the ECM/PCM idle learn procedure. If the idle speed is still not within specification, go to symptom troubleshooting.**

9. Jump the SCS line with the HDS.

10. Connect the timing light to the service loop (white tape).

11. Aim the light toward the pointer on the cam chain case.

12. Check the ignition timing under a no load condition (headlights, blower fan, rear window defogger, and air conditioner are turned off).

13. Ignition timing should be:
- 1.8L engine: 10±2° BTDC (RED mark) at idle.
- 2.4L engine: 8±2° BTDC (RED mark) at idle.

14. If the ignition timing differs from the specification, check the cam timing. If the cam timing is OK, update the ECM if it does not have the latest software, or substitute a known good ECM, then recheck. If the system works properly, and the ECM was substituted, replace the original ECM.

15. Disconnect the HDS and the timing light.

THROTTLE BODY

REMOVAL & INSTALLATION

2011 Models

1.8L (Except CNG) & 2.0L Engine

See Figures 297 and 298.

1. Turn the ignition switch to LOCK (0).

2. Remove the cowl cover and the under-cowl panel.

3. Remove the air cleaner and the intake air duct.

4. Disconnect the throttle body connector and the EVAP purge control solenoid valve connector.

5. Disconnect and plug the water bypass hoses.

6. Disconnect the vacuum hose.

7. Remove the throttle body.

8. Install the parts in the reverse order of removal with a new gasket, then do this.

9. Refill the radiator with engine coolant.

10. Do the "ECM Idle Learn Procedure". See "Engine Control Module/Powertrain Control Module (ECM/PCM)" in this section.

1.8L CNG Engine

See Figure 299.

✱✱ WARNING

Do not insert your fingers into the installed throttle body when you turn the ignition switch to ON (II) or while the ignition switch is in ON (II). If you do, you will seriously injure your fingers if the throttle valve is activated.

➡ **If replacing or cleaning the throttle body, start at step 1. If removing the throttle body, start at step 4.**

1. Connect the HDS, or equivalent scan tool, to the DLC while the engine is stopped.

2. Select the INSPECTION MENU on the HDS.

3. Do the TP POSITION CHECK in the ETCS TEST.

4. Turn the ignition switch to LOCK (0).

5. Remove the cowl cover and the under-cowl panel.

6. Remove the air cleaner, then remove the intake air duct (A).

7. Disconnect the throttle body connector (B) and the vacuum hose (C).

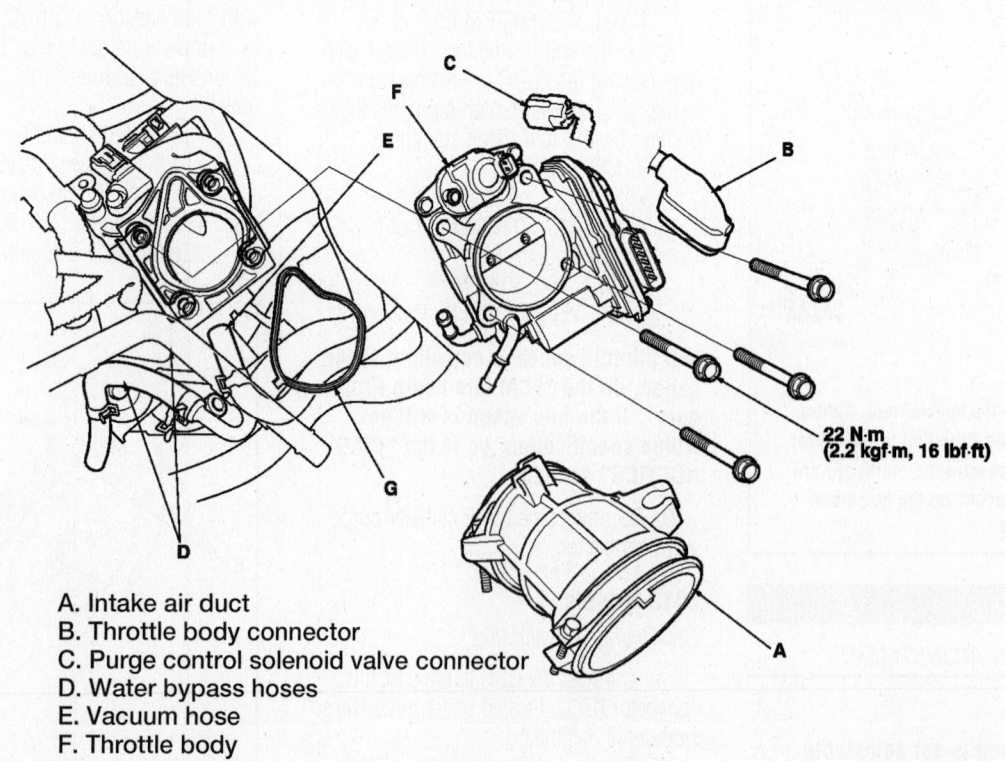

A. Intake air duct
B. Throttle body connector
C. Purge control solenoid valve connector
D. Water bypass hoses
E. Vacuum hose
F. Throttle body

22 N·m
(2.2 kgf·m, 16 lbf·ft)

37647_CIVIC_G0238

Fig. 297 Showing the throttle body assembly components—1.8L engine (except CNG)

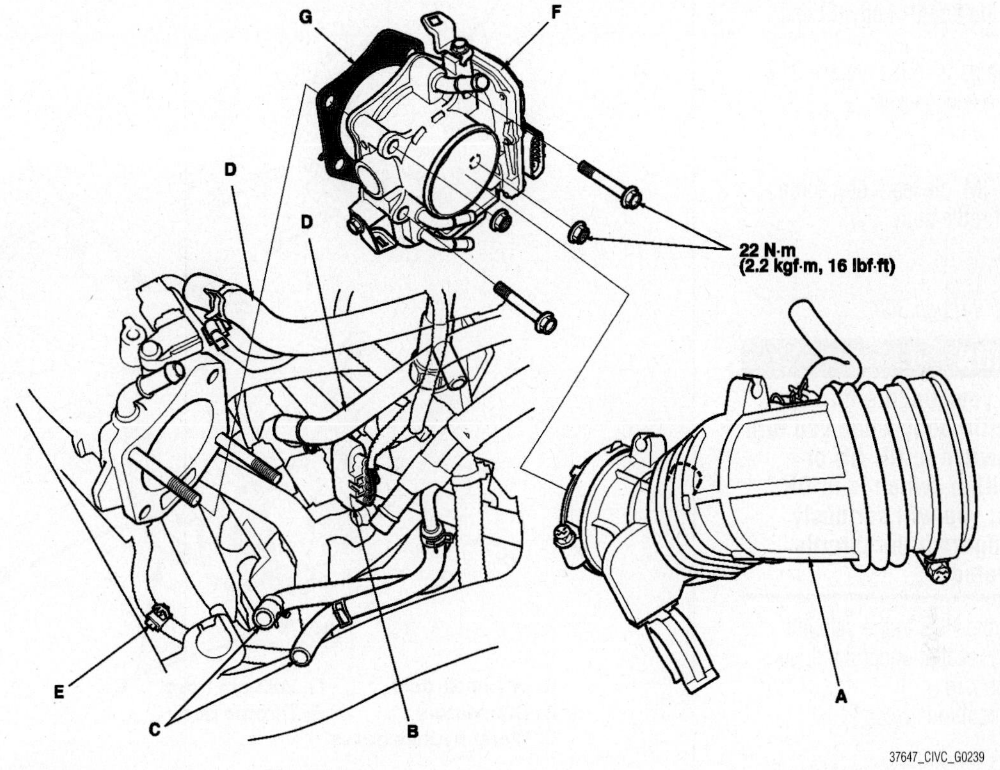
G
F
D
D
22 N·m
(2.2 kgf·m, 16 lbf·ft)
E
C
B
A
37647_CIVC_G0239
Fig. 298 Showing the throttle body assembly components—2.0L engine

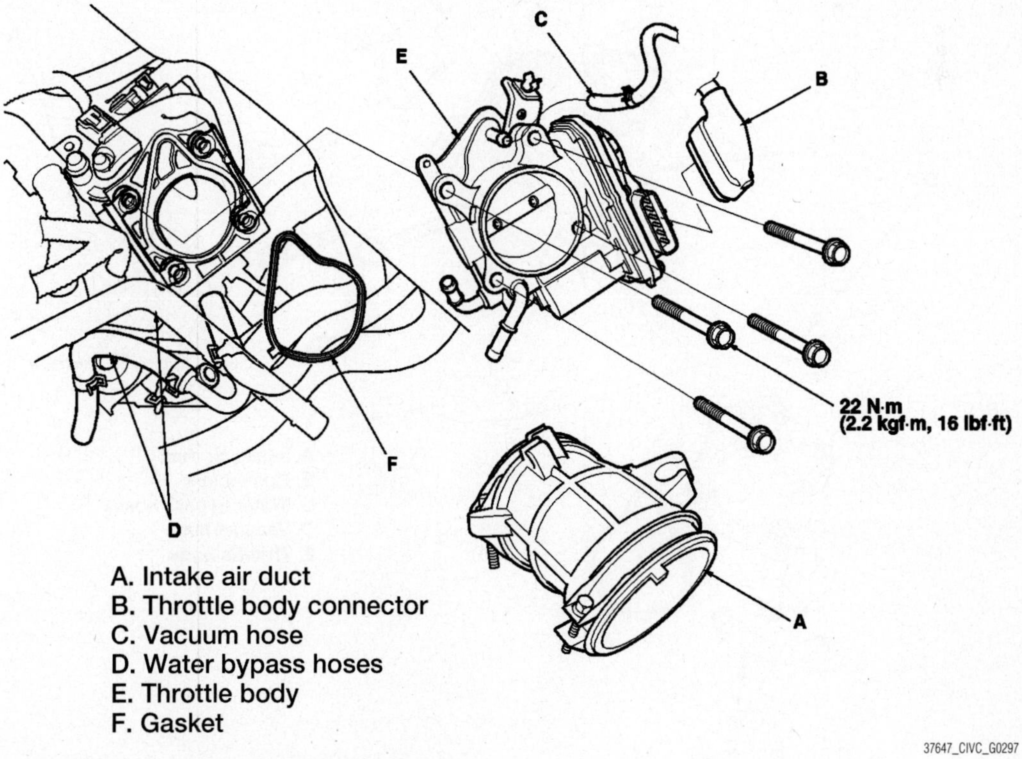
C
E
B
22 N·m
(2.2 kgf·m, 16 lbf·ft)
F
D
A
A. Intake air duct
B. Throttle body connector
C. Vacuum hose
D. Water bypass hoses
E. Throttle body
F. Gasket
37647_CIVC_G0297
Fig. 299 Removing/installing the throttle body assembly—1.8L CNG engine

8. Disconnect and plug the water bypass hoses (D).

9. Remove the throttle body (E) and gasket (F).

10. Install the parts in the reverse order of removal with a new gasket.

11. Refill the radiator with engine coolant.

12. Do the PCM idle learn procedure after replacing throttle body.

2012 Models

See Figures 300 through 302.

> **※ CAUTION**
>
> **Do not insert your fingers into the installed throttle body when you turn the ignition switch to ON (II) or while the ignition switch is in ON (II). If you do, you will seriously injure your fingers if the throttle valve is activated.**

1. Connect the HDS to the data link connector (DLC) located under the driver's side of the dashboard.

2. Turn the ignition switch to ON (II).

3. Select the INSPECTION MENU on the HDS.

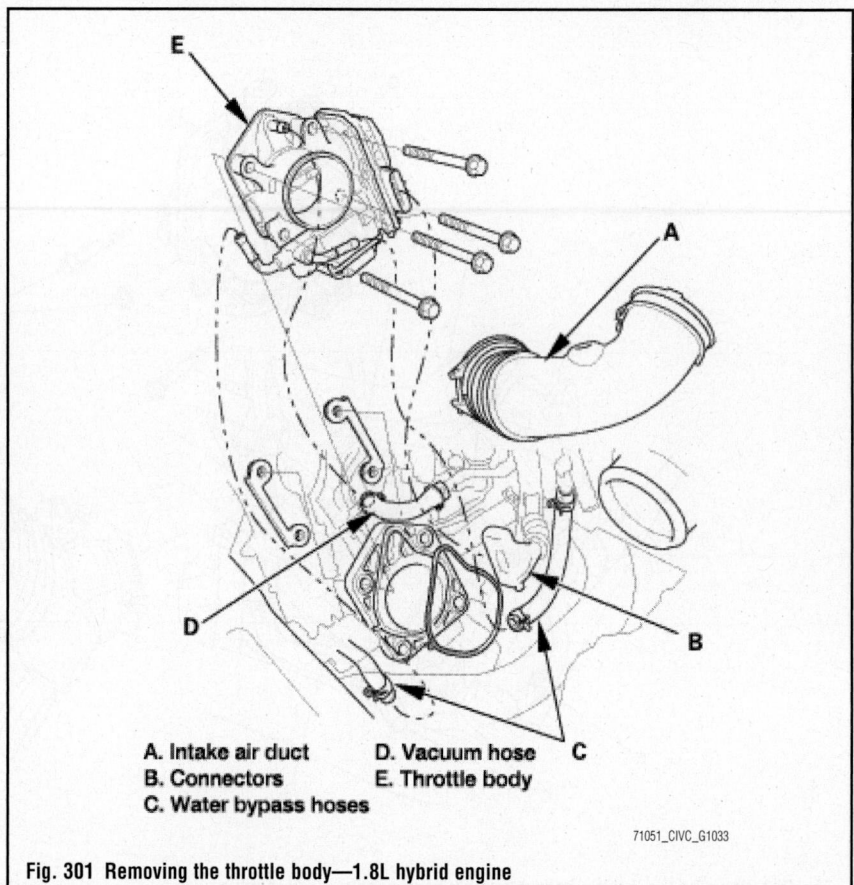

A. Intake air duct   D. Vacuum hose
B. Connectors   E. Throttle body
C. Water bypass hoses

71051_CIVC_G1033

Fig. 301 Removing the throttle body—1.8L hybrid engine

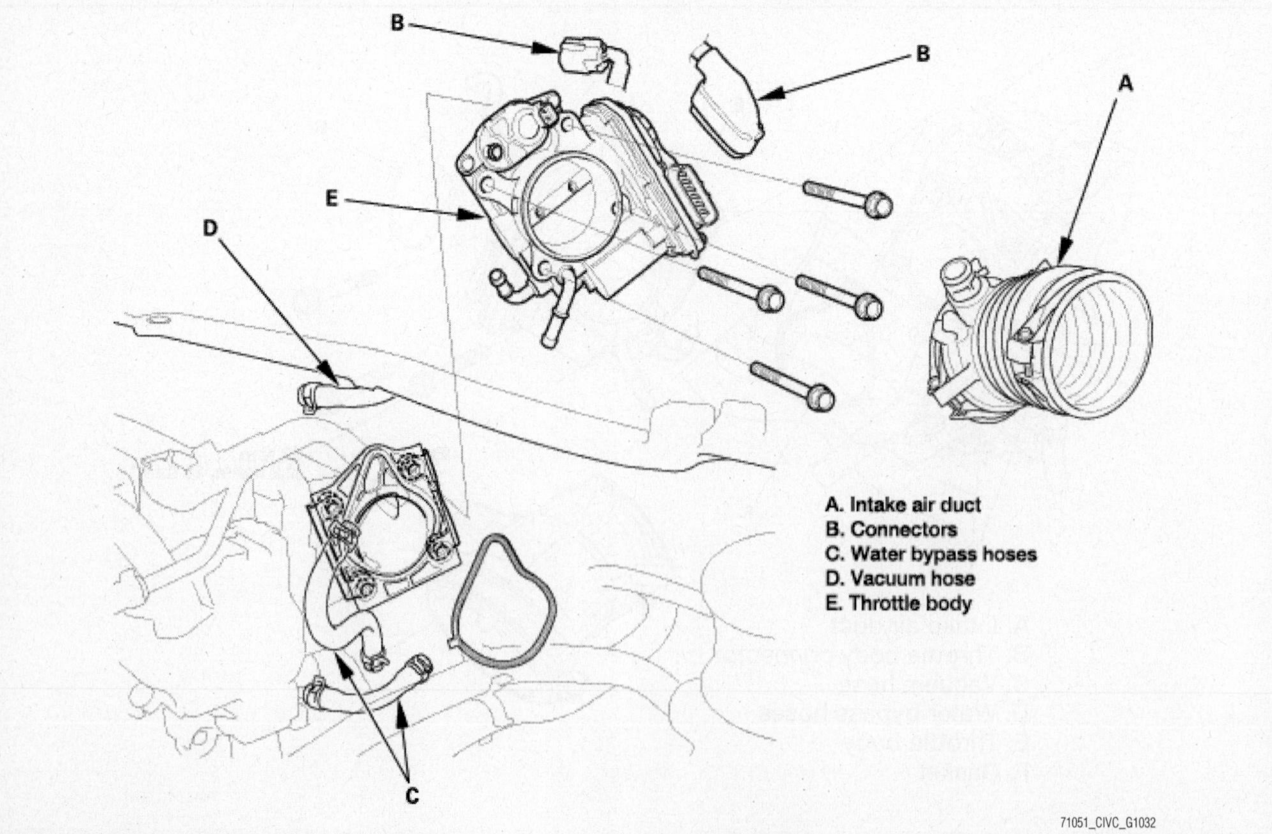

A. Intake air duct
B. Connectors
C. Water bypass hoses
D. Vacuum hose
E. Throttle body

71051_CIVC_G1032

Fig. 300 Removing the throttle body—1.8L engine

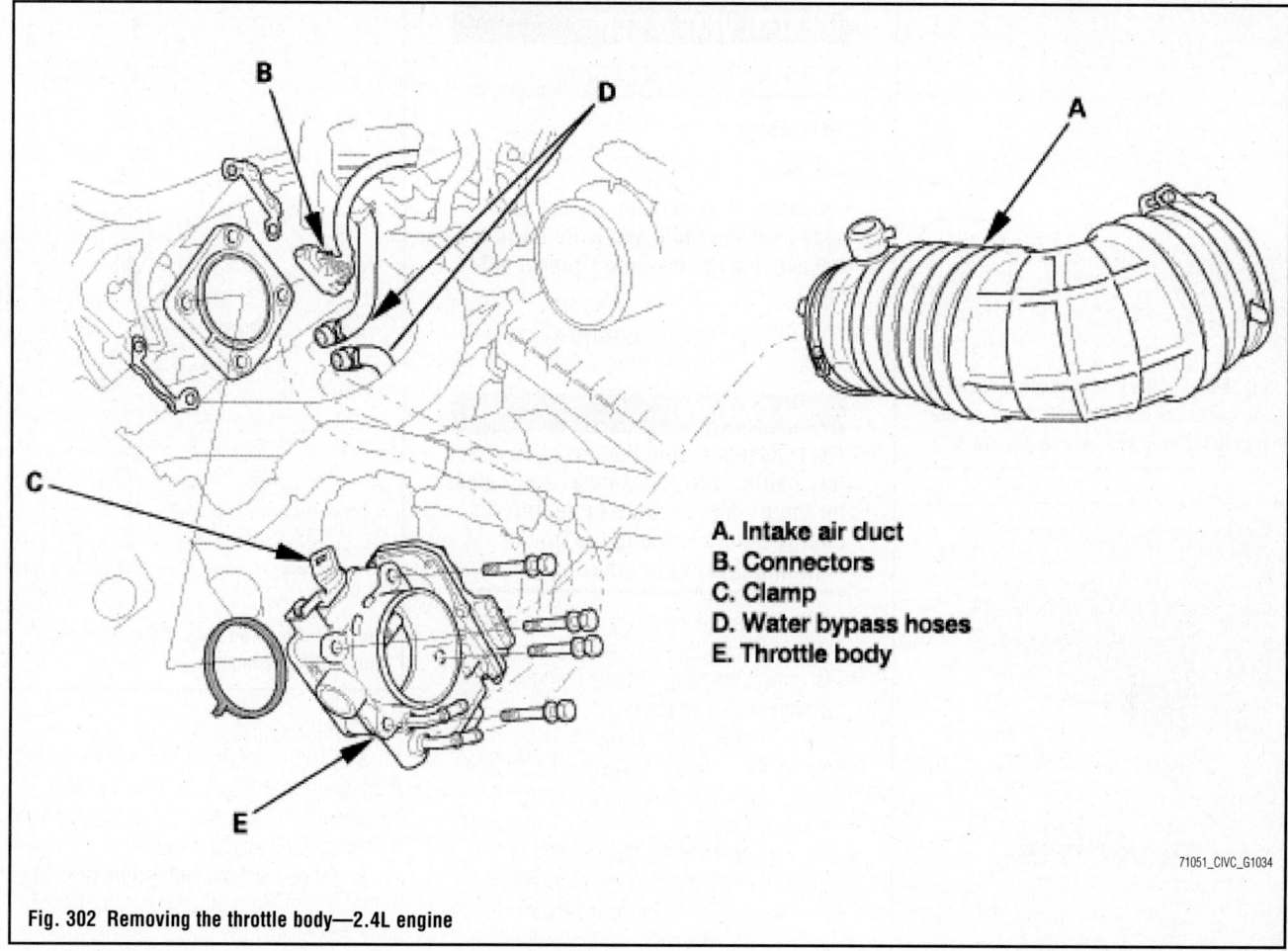

A. Intake air duct
B. Connectors
C. Clamp
D. Water bypass hoses
E. Throttle body

71051_CIVIC_G1034

Fig. 302 Removing the throttle body—2.4L engine

4. Do the TP POSITION CHECK in the ETCS TEST.

5. Turn the ignition switch to LOCK (0).

6. Remove the intake air pipe (1.8L only) and duct.

7. For 1.8L, remove the air cleaner and remaining duct.

8. Disconnect and plug the water bypass hoses and vacuum hose from the throttle body.

9. Remove the throttle body.

To install:

10. Installation is the reverse of the removal procedure.

11. Fill the coolant reservoir to the MAX mark with Honda Long Life Antifreeze/Coolant Type 2.

12. Reset the ECM/PCM with the HDS while the engine is stopped:

   a. Turn the ignition switch to LOCK (0).

   b. Turn the ignition switch to ON (II) and wait 30 seconds.

   c. Start the engine. Hold the engine speed at 3,000 rpm without load (A/T in P or N, M/T in neutral) until the radiator fan comes on, then let it idle.

13. For ECM/PCM relearn, let the engine idle for about 5 minutes with the throttle fully closed.

➡**If the radiator fan comes on, do not include its running time in the 5 minutes.**

HEATING & AIR CONDITIONING SYSTEM

BLOWER MOTOR

REMOVAL & INSTALLATION

2011 Models

See Figure 303.

1. Remove the glove box.

2. Cut the plastic cross brace in the glove box opening with diagonal cutters in the area shown, and discard it.

3. Remove the bolts and the glove box frame.

4. Remove the wire harness clip, the self-tapping screws, and the passenger's heater duct.

5. Disconnect the connector from the blower motor. Remove the wire harness clip.

6. Disconnect the connector from the recirculation control motor. Remove the self-tapping screws, the bolt, the mounting nuts, and the blower unit.

7. Installation is the reverse order of

removal. Make sure that there is no air leakage.

2012 Models

See Figures 304 and 305.

1. Remove the blower motor cover, connector and clip.

2. Remove the blower motor.

3. Installation is the reverse order of removal. Make sure that there is no air leakage.

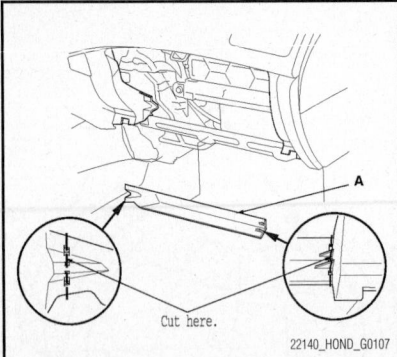

Fig. 303 Cut the plastic cross brace (A) in the glove box opening with diagonal cutters in the area shown, and discard it

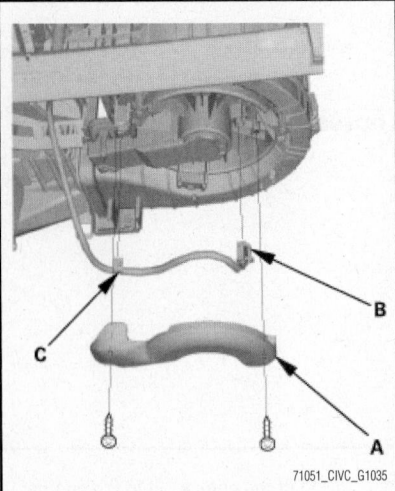

Fig. 304 Remove the blower motor cover (A), connector (B) and clip (C)

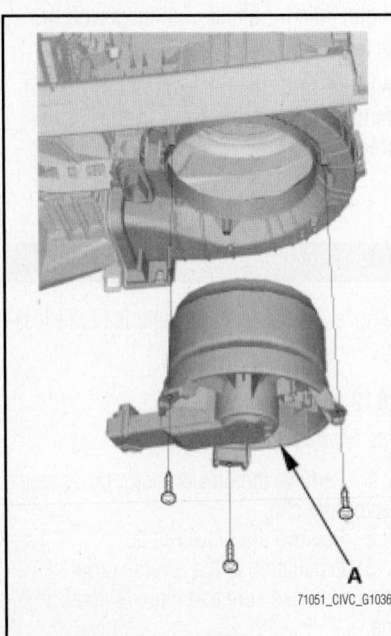

Fig. 305 Remove the blower motor

HEATER UNIT & CORE

REMOVAL & INSTALLATION

2011 Models

See Figures 306 and 307.

➡ **Make sure to acquire the anti-theft code from the radio and write down the frequencies for the radio's preset buttons.**

1. Disconnect the negative battery cable.

※ CAUTION

After disconnecting the negative battery cable, wait for at least 3 minutes for the air bag module to deplete its energy before working the on the instrument panel or steering wheel.

2. Discharge the air conditioning system.
3. Remove the air cleaner assembly. Disconnect the heater hoses. Remove the mounting nut from the heater unit.
4. Remove the sub-display visor. Remove the navigation system, if equipped. On vehicles without navigation system remove the radio.
5. Remove the dashboard retaining screws. Detach the retaining clips along the lower edge of the instrument panel.
6. Detach the retaining clips along the upper edge of the instrument panel. Gently pull forward to release the hooks from the holder of the gauge module.

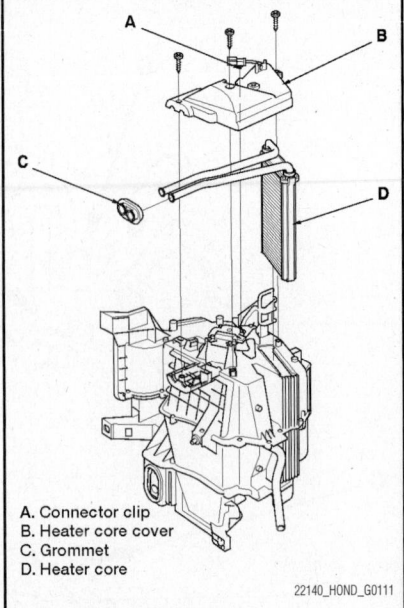

A. Connector clip
B. Heater core cover
C. Grommet
D. Heater core

Fig. 307 Exploded view of the heater unit—All Civic models

7. Remove the instrument panel.
8. Disconnect the electrical connector from the blower motor. Remove the wire harness clip. Disconnect the connector from the recirculation control motor.
9. Disconnect the connectors from the mode control motor, the evaporator temperature sensor and the power transistor. Remove the wire clip harness.
10. Disconnect the connectors from the air mix control motor and the air condition-

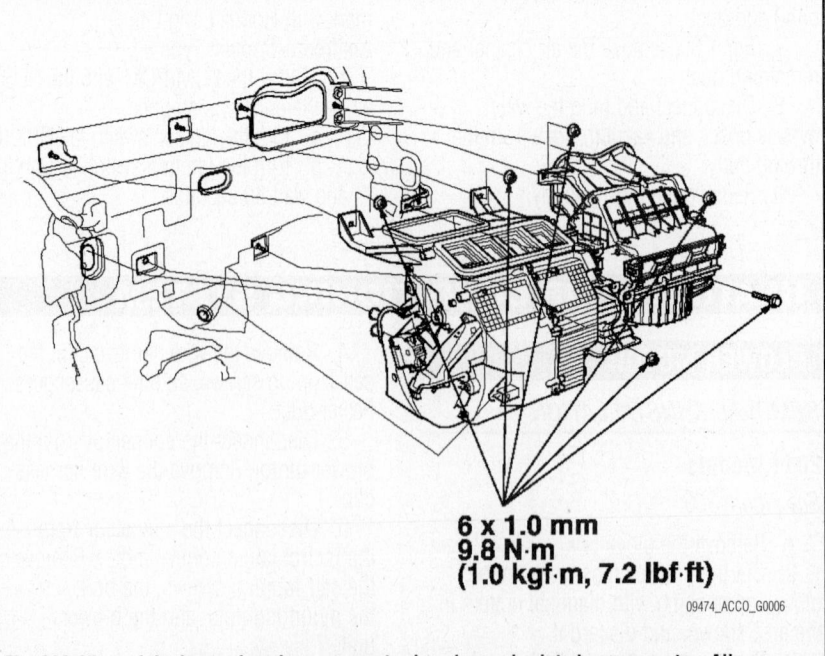

**6 x 1.0 mm
9.8 N·m
(1.0 kgf·m, 7.2 lbf·ft)**

Fig. 306 View of the heater housing, evaporator housing and related components—All Civic Models

ing wire harness. Remove the connector clip and the wire harness clips.

11. Remove the mounting bolt, mounting nuts and the heater/blower unit from the vehicle.

12. Remove the self-tapping screws, remove the grommet and carefully pry out the heater core.

To install:

13. Installation is the reverse of the removal procedure.

14. Evacuate, charge and leak test the air conditioning system refrigerant, as required.

15. Operate the engine to normal operating temperatures; then, check the climate control operation and check for leaks.

16. Enter the antitheft codes for the radio and the navigation system.

2012 Models

See Figures 308 through 313.

1. Remove or perform the following (refer to component headings or sections as needed):

- Evacuate the A/C system
- Wiper arms (2.4L)

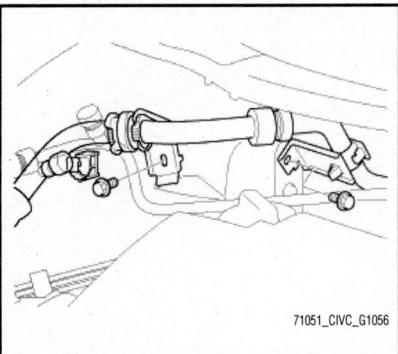

Fig. 308 On 1.8L engine, disconnect the bolts as shown

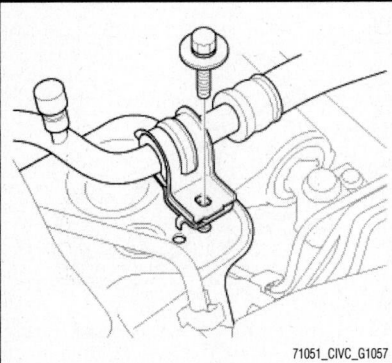

Fig. 309 On 2.4L engine, remove the bolt

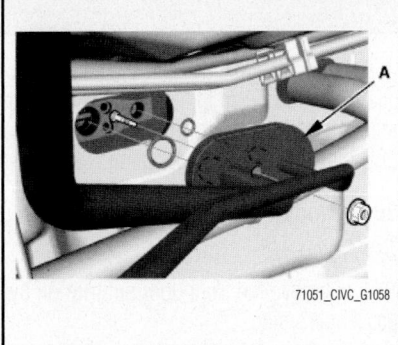

Fig. 310 On all, remove the A/C lines from the firewall connection

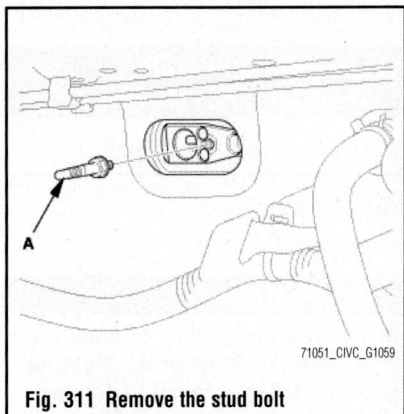

Fig. 311 Remove the stud bolt

- Side cowls (2.4L)
- Center and under cowls (2.4L)

2. On 1.8L engine, disconnect the bolts as shown.

3. On 2.4L engine, remove the bolt.

4. On all, remove the A/C lines from the firewall connection.

5. Remove the stud bolt.

6. Remove or perform the following (refer to component headings or sections as needed):

- Radiator cap
- Radiator coolant
- Front door opening seal
- A-pillar trims
- Driver's dashboard lower cover
- Front door sill trims
- Kick panels
- Glove box
- Shift lever knob
- Center console
- Rear heater joint duct
- Driver's airbag
- Steering wheel
- Combination switch
- Steering joint cover
- Steering joint (disconnect)
- Steering column

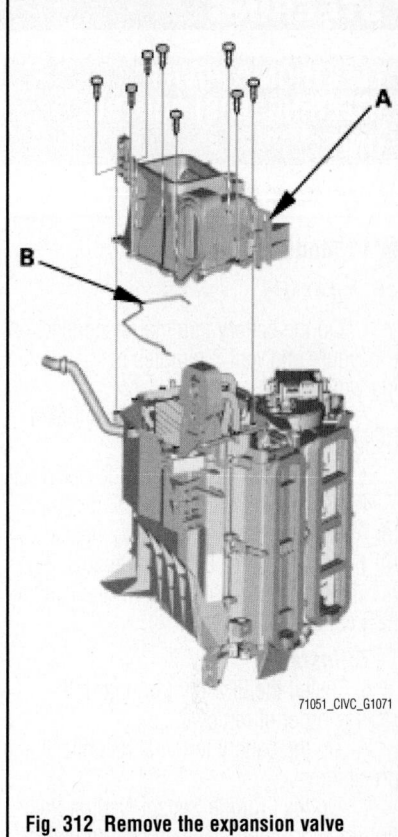

Fig. 312 Remove the expansion valve cover (A) and the seal (B)

- A-pillar corner trims
- Dashboard/steering hanger beam
- Blower-heater unit
- A/C wire harness
- Blower unit
- Passenger's heater duct
- Heater core cover

7. Remove the expansion valve cover and the seal.

8. Remove the heater pipe bracket, remove the grommet, carefully pull out the heater core.

9. Installation is the reverse of the removal procedure.

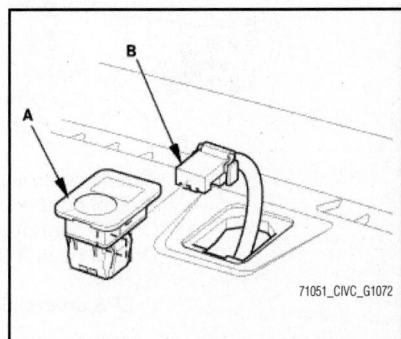

Fig. 313 Remove the heater pipe bracket (A), remove the grommet (B), carefully pull out the heater core (C)

STEERING

ELECTRONIC POWER STEERING (EPS) CONTROL UNIT

REMOVAL & INSTALLATION

2011 Models

See Figure 314.

1. Do the battery terminal disconnection procedure and wait 3 minutes before starting work.

2. Remove the passenger's dashboard under cover.

3. Remove the passenger's side kick panel.

4. Disconnect EPS control unit connector A (2P), connector B (2P), connector C (2P), and connector D (28P). Remove the nuts from the EPS control unit. Remove the EPS control unit.

To install:

6. Install the EPS control unit in the reverse order of removal.

7. Do the battery terminal reconnection procedure.

8. Do the "Torque Sensor Neutral Position Memorization". See procedure under "Power Rack & Pinion Steering Gear" for EPS application.

9. After installation, start the engine, allow it to idle. Turn the steering wheel from lock to lock several times. Check that the EPS indicator does not come on.

2012 Models

See Figure 315.

1. Remove the front door sill trim on the passenger's side.

2. Remove the passenger's kick panel.

3. Detach the connectors and remove the EPS control unit.

4. Installation is the reverse of the removal procedure.

ELECTRONIC POWER STEERING (EPS) MOTOR

REMOVAL & INSTALLATION

2011 Models

See Figures 316 and 317.

✳✳ CAUTION

Do not allow dust, dirt, or other foreign materials to enter the steering gearbox. Make sure not to get any silicone grease on the terminal part

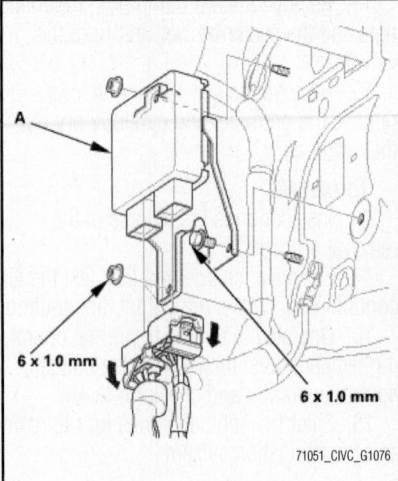

Fig. 315 Detach the connectors and remove the EPS control unit (A)

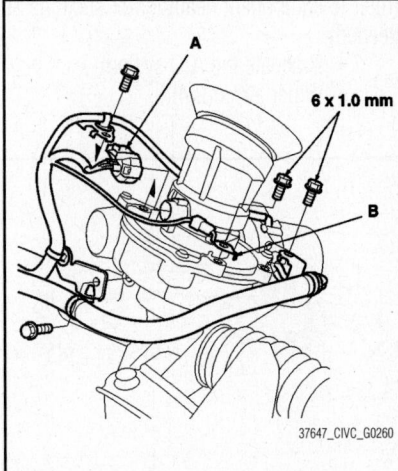

Fig. 316 Disconnect the torque sensor 3P connector (A) from the steering gearbox, then remove the wire harness clamp bolts and the ground terminal (B)

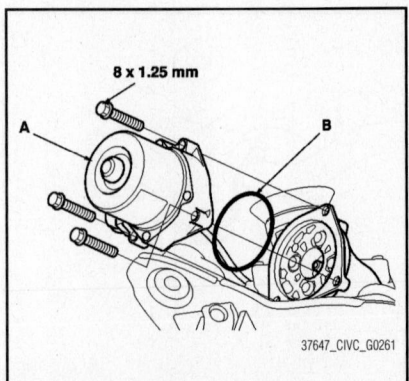

Fig. 317 Remove the EPS motor (A) from the steering gearbox, then remove the O-ring (B)

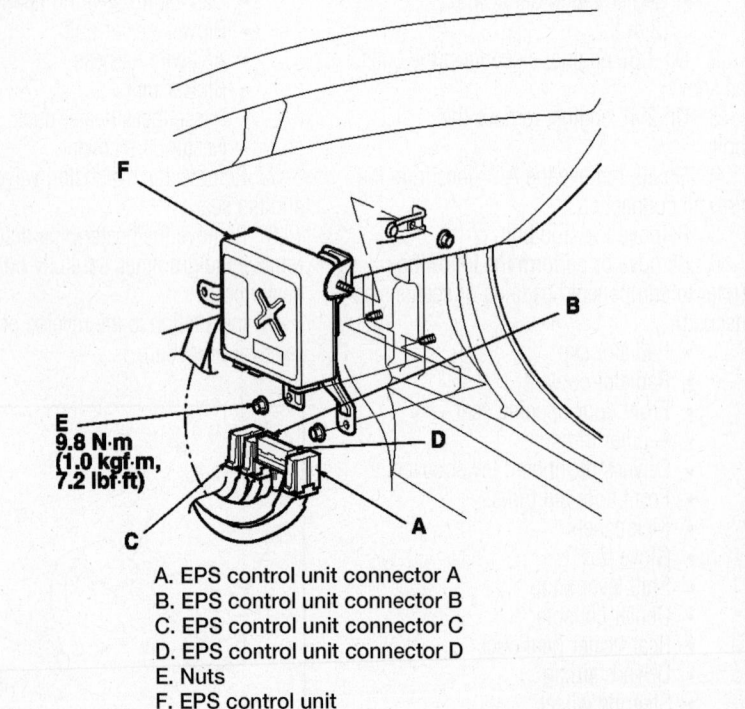

A. EPS control unit connector A
B. EPS control unit connector B
C. EPS control unit connector C
D. EPS control unit connector D
E. Nuts
F. EPS control unit

Fig. 314 Disconnect EPS control unit connector A (2P), connector B (2P), connector C (2P), and connector D (28P). Remove the nuts from the EPS control unit. Remove the EPS control unit

of the connectors and switches, especially if you have silicone grease on your hands or gloves.

1. Remove the steering gearbox. See "Power Rack & Pinion Steering Gear" in this section.

2. Disconnect the torque sensor 3P connector from the steering gearbox, then remove the wire harness clamp bolts and the ground terminal.

3. Remove the EPS motor from the steering gearbox, then remove the O-ring.

To install:

4. Clean the mating surface of the EPS motor and the steering gearbox.

5. Apply a thin coat of silicone grease to the O-ring and carefully fit it on the EPS motor.

6. Apply steering grease into the EPS motor shaft.

7. Set the motor on the steering gearbox by engaging the EPS motor shaft and the worm shaft.

8. Turn the EPS motor two or three times to the right and left about 45 degrees. Make sure the EPS motor is evenly seated on the steering gearbox, and that the O-ring is not pinched between the mating surfaces.

9. Loosely install the EPS motor mount bolts, then turn the steering wheel two or three times to the right and left about 45 degrees.

10. Tighten the EPS motor mount bolts to 14 ft. lbs. (20 Nm).

2012 Models

See Figures 318 through 323.

➡Do not allow dust, dirt, or other foreign materials to enter the steering gearbox.

1. Raise the vehicle on a lift, and make sure it is securely supported.

2. Remove both front wheels.

3. Remove the steering joint cover.

4. Set the steering column to the center tilt position, and to the center telescopic position.

5. Hold the lower slide shaft on the column with a piece of wire between the joint yoke of the lower slide shaft and the joint yoke of the upper shaft to prevent the lower slider shaft from pulling out.

6. Center the steering wheel spoke angle.

7. Install a commercially available steering wheel holder tool.

8. Loosen the steering joint bolt.

9. Remove the steering joint bolt. Disconnect the steering joint from the pinion

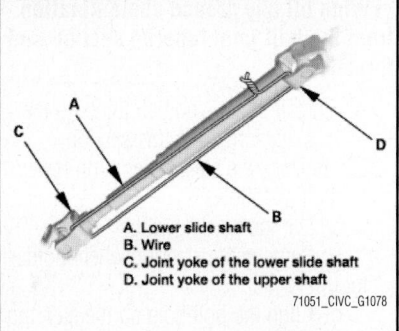

Fig. 318 Hold the lower slide shaft on the column with a piece of wire between the joint yoke of the lower slide shaft and the joint yoke of the upper shaft to prevent the lower slider shaft from pulling out

shaft. If the center guide is in place and has not moved, leave it in place. If the center guide has come off, discard it.

10. Remove the ball joint from the lower arm.

11. Disconnect the tie rod end ball joints from both sides as follows:

a. Remove the cotter pin from the tie-rod end ball joint. Remove the nut.

b. Disconnect the tie-rod end ball joint.

➡Be careful not to damage the ball joint boot when installing the remover.

12. Remove the engine undercover and the splash shield.

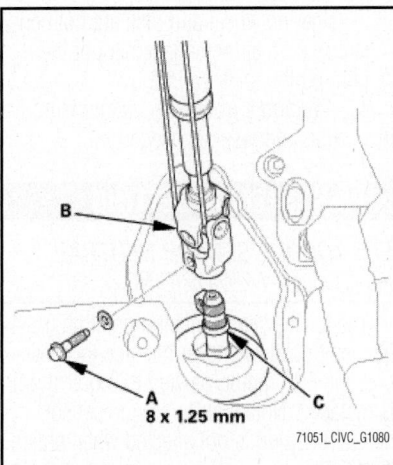

Fig. 319 Remove the steering joint bolt (A). Disconnect the steering joint (B) from the pinion shaft. If the center guide (C) is in place and has not moved, leave it in place. If the center guide has come off, discard it

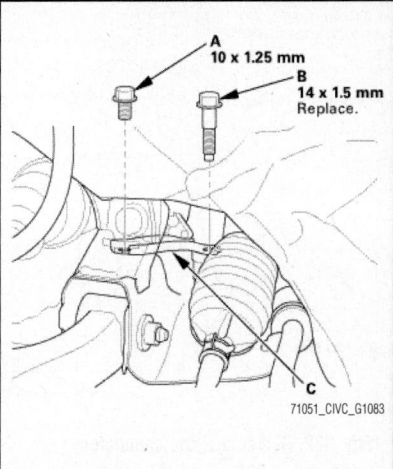

Fig. 320 Remove the steering gearbox mounting bolts (A), remove the stiffener mounting bolts (B), and remove the steering gearbox stiffeners (C)—driver's side

13. Remove the power steering gearbox and the peripheral components as follows:

a. Disconnect the connector and remove the EPS motor ground terminal.

b. Remove the steering gearbox mounting bolts.

c. Remove the stiffener mounting bolts.

d. Remove the steering gearbox stiffeners.

14. Move the steering gearbox aside.

15. Disconnect the connectors. Remove the EPS motor and the O-ring.

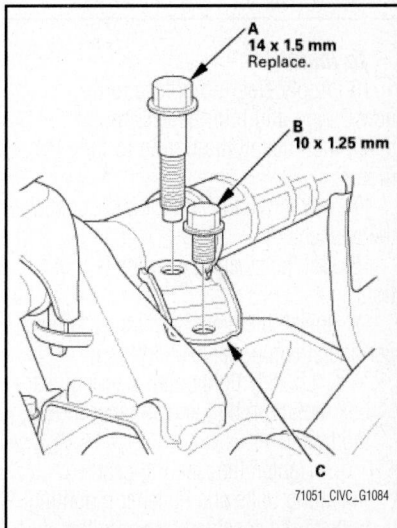

Fig. 321 Remove the steering gearbox mounting bolts (A), remove the stiffener mounting bolts (B), and remove the steering gearbox stiffeners (C)—passenger's side

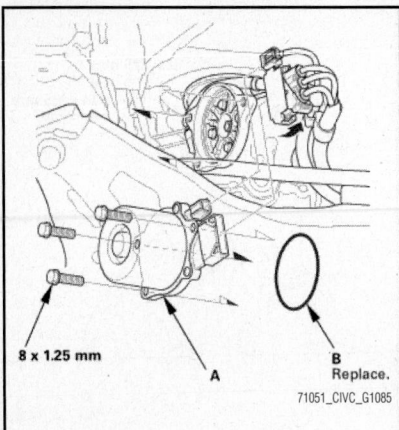

Fig. 322 Disconnect the connectors. Remove the EPS motor (A) and the O-ring (B)

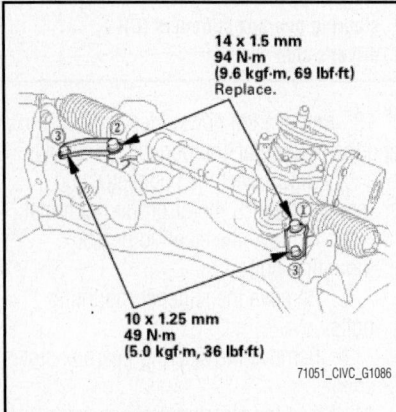

Fig. 323 Tighten the steering gearbox mounting bolts and stiffener mounting bolts to the specified torque in the sequence shown

To install:

16. Apply steering grease to the new O-ring and into the EPS motor shaft, then install the O-ring to the EPS motor.
17. Install the EPS motor and connect the connectors.
18. Set the steering gearbox is in position.
19. Install the power steering gearbox peripheral components as follows:
    a. Loosely tighten the steering gearbox mounting bolt and stiffener mounting bolt.
    b. Tighten the steering gearbox mounting bolts and stiffener mounting bolts to the specified torque in the sequence shown.
20. Connect the EPS motor connector and ground terminal.
21. Install the splash shield and engine undercover.

➡Wipe off any grease contamination from the ball joint tapered section and threads.

22. Install the tie-rod ball joints to the knuckle, as described in this section.
23. Release the steering column lower slide shaft as follows:
    a. Cut the holding wire.
    b. Remove the steering wheel holder tool.
    c. Align the bolt hole on the steering joint with the groove around the pinion shaft and loosely install the steering joint bolt.

➡Be sure that the joint bolt is securely in the groove in the pinion shaft.

    d. Pull on the steering joint to make sure that the steering joint is fully seated.
    e. Tighten the steering joint bolt to 21 ft. lbs. (29 Nm).
    f. Install the steering joint cover, tightening the lower screws first, then the upper screw.
24. Install both front wheels.
25. With the tires raised off the ground, check for the following symptoms by turning the steering wheel fully to the right and left several times.
26. After installation, check these items:
    a. Start the engine, allow it to idle, and turn the steering wheel from lock to lock several times.
    b. Check that the EPS indicator does not come on.
    c. Check the steering wheel spoke angle. If steering spoke angles to the right and left are not equal (steering wheel and rack are not centered), correct the engagement of the joint/pinion shaft splines.
27. Perform a front wheel alignment inspection and reset as needed.

EPS SYSTEM RESETTING

EPS TORQUE SENSOR NEUTRAL POSITION—MEMORIZING

The torque sensor neutral position must be memorized whenever the steering gearbox, the EPS motor, or the EPS control unit is replaced. Note that the torque sensor neutral position is not affected when erasing the DTC.

➡The torque sensor is temperature sensitive. When memorizing the torque sensor neutral position, the ambient temperature must be above 68°F (20°C).

1. Connect the HDS to the data link connector (DLC) located under the driver's side of the dashboard.
2. Turn the ignition switch to ON (II).
3. Select EPS ADJUSTMENT, then select TORQUE SENSOR LEARN with HDS, and follow the screen prompts.

VSA SENSOR NEUTRAL POSITION—MEMORIZATION

1. Park the vehicle on a flat and level surface, with the steering wheel in the straight ahead position.
2. Select VSA ADJUSTMENT, then select ALL SENSORS with the HDS, and follow the screen prompts.

STEERING ANGLE SENSOR NEUTRAL POSITION—CLEAR

Select EPS ADJUSTMENT, then select EPS STEERING ANGLE SENSOR VALUE CLEAR and follow the screen prompts on the HDS.

POWER RACK & PINION STEERING GEAR

REMOVAL & INSTALLATION

2011 Models

Without EPS

✳✳ WARNING

SRS components are located in this area. Review the SRS component locations and the precautions and procedures before doing repairs or service. See "AIR BAG (SUPPLEMENTAL RESTRAINT SYSTEM)" section.

➡The radio may contain a coded theft protection circuit. Always obtain the code number before disconnecting the battery.

1. Before servicing the vehicle, refer to the precautions in the beginning of this section.
2. Disconnect the negative and positive battery cables. Wait 3 minutes before working around the air bags.
3. Raise and support the vehicle safely. Remove the front tires. Drain the power steering fluid.
4. Remove the driver's side airbag. See "AIR BAG (SUPPLEMENTAL RESTRAINT SYSTEM)" SECTION.
5. Remove the steering wheel.

➡Be sure to remove the steering wheel before disconnecting the steering joint as damage to the cable reel can occur.

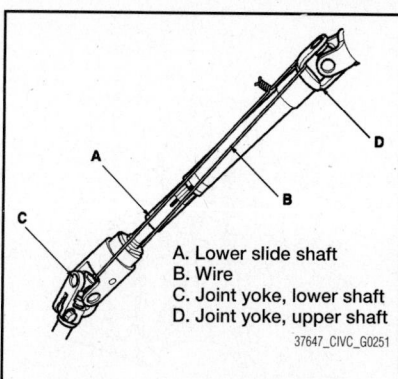

A. Lower slide shaft
B. Wire
C. Joint yoke, lower shaft
D. Joint yoke, upper shaft

37647_CIVIC_G0251

Fig. 324 Hold the lower slide shaft on the column with a piece of wire between the joint yoke on the lower slide shaft to the joint yoke on the upper shaft

6. Remove the driver's dashboard lower cover. Remove the driver's dashboard under cover.

7. Remove the steering joint bolts. Disconnect the steering joint by moving the joint toward the column.

8. Hold the lower slide shaft on the column with a piece of wire between the joint yoke on the lower slide shaft to the joint yoke on the upper shaft.

9. Remove the center guide, if equipped and discard it.

➡**The center guide is for factory assembly only.**

10. Remove the harness holder and the air cleaner housing bracket.

11. Remove the breather pipe and the harness holder bracket.

12. Attach special tool 07AAK-SNAA120, or equivalent to the cylinder head. Install the engine support hanger, AAR-12566 or equivalent. Attach the hook to the tool and tighten the wing nut. Lift and support the engine.

13. Remove the cotter pin from the tie rod ball joint and loosen the nut. Separate the tie rod ball joint and the knuckle, using the proper tool.

14. Remove the pump outlet hose clamp from the intake manifold. Remove the inlet line clamp bolt. Open the return line holder and remove the return line clamp bolt.

15. Loosen the adjustable hose clamp and disconnect the return hose. Loosen the flare nut and disconnect the inlet line.

16. Remove the splash shield. Remove the lower ball joint mounting bolt and flange nuts from the lower arm. Disconnect the lower arm from the lower ball housing.

17. Note the reference marks on both sides of the subframe that line up with the body (these are for installation reference).

18. Attach tool VSB02C000016 to the front subframe and a transaxle jack or powertrain lift tool. Make sure that the subframe is securely supported.

19. Remove the front subframe middle mount bolt from the left side. Remove the front subframe middle mount bolt from the right side. Remove the two 12mm flange bolts from the lower torque rod and bracket.

20. Remove the front subframe front mounting bolts from the right and left sides of the vehicle. Discard them.

21. Remove the front subframe rear mounting bolts from the right and left sides of the vehicle. Discard them.

22. Lower the front subframe and steering gearbox as an assembly. Remove the pinion shaft grommet from the top of the valve body unit.

23. Remove the two 10mm bolts from the right side of the steering gearbox. Remove the mounting bracket and mounting cushion.

24. Remove the four 10mm flange bolts from the left side of the steering gearbox. Remove the stiffener plates.

25. Remove the steering gearbox from the front subframe

To install:

26. Position the assembly to its mounting on the subframe.

27. Loosely install the stiffener plates and gearbox mounting bolts on the left side of the steering gearbox.

28. Position the cutout on the mounting cushion and install it on the right side of the steering gearbox.

29. Install the gearbox mounting bracket over the mounting cushion and loosely install the two 10mm bolts.

30. Tighten the 10mm bolts on both sides of the steering gearbox to 40 ft. lbs.

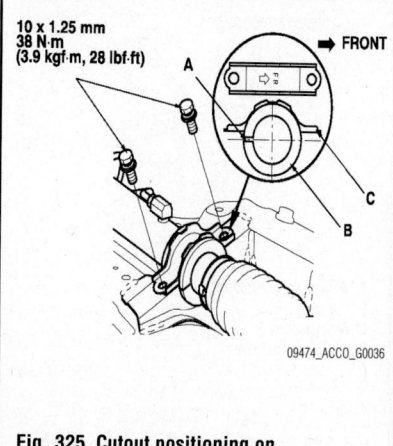

10 x 1.25 mm
38 N·m
(3.9 kgf·m, 28 lbf·ft)

➡ FRONT

A

C

B

09474_ACCO_G0036

Fig. 325 Cutout positioning on mounting cushion

right side and 28 ft. lbs. left side, alternately in two or more steps.

31. Install the pinion shaft grommet. Align the slot in the pinion shaft grommet with the lug portion on the valve housing. The grommet must not have a gap at the mating surface of the grommet and valve housing.

32. Carefully raise the front subframe in position.

33. Continue the installation in the reverse order of the removal procedure.

34. Be sure to center the cable reel by first rotating it clockwise until it stops. Then rotate it counterclockwise (approximately two and a half turns) until the arrow mark on the label points straight up. Install the steering wheel.

35. Fill the system with the proper grade and type power steering fluid.

36. Start the engine, allow it to idle, and turn the steering wheel from lock-to-lock several times to warm up the fluid. Check and adjust the fluid level.

37. Check the gearbox for fluid leaks, correct as necessary.

38. Check front end alignment.

39. Check the steering wheel spoke angle. Adjust by turning the right and left tie rods equally, if necessary.

With EPS

See Figures 326 through 334.

➡**Note these items during removal:**

• Use solvent and a brush, wash any oil and dirt off the end of the steering gearbox. Avoid any electrical parts. Blow dry with compressed air. Wear eye protection.

• Make sure to remove the steering wheel before disconnecting the steering joint, or damage to the cable reel occur.

1. Do the battery terminal disconnect procedure, then wait at least 3 minutes before beginning work. See "Battery System" in "ENGINE ELECTRICAL" section.

2. Raise and support the vehicle.

3. Remove the front wheels.

4. Remove the driver's airbag, and the steering wheel. See "AIR BAG (SUPPLEMENTAL RESTRAINT SYSTEM)" section.

5. Remove the driver's dashboard undercover.

6. Remove the steering joint cover at the floorboard.

7. Release the lock lever, and adjust the steering column to the full tilt up position, and to the full telescopic in position. Tighten the lock lever.

8. Hold the lower slide shaft on the col-

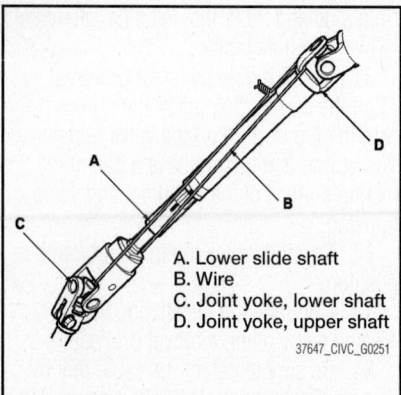

A. Lower slide shaft
B. Wire
C. Joint yoke, lower shaft
D. Joint yoke, upper shaft

37647_CIVC_G0251

Fig. 326 Hold the lower slide shaft on the column with a piece of wire between the joint yoke on the lower slide shaft to the joint yoke on the upper shaft

umn with a piece of wire between the joint yoke on the lower slide shaft to the joint yoke on the upper shaft.

9. Remove the steering joint bolt and disconnect the steering joint by moving the steering joint toward the column.

10. Remove the center guide (if equipped) from the top of the pinion shaft and discard it.

➡**The center guide is for factory assembly use only.**

11. Remove the cowl cover and the under-cowl panel.

12. Remove the air cleaner housing.

13. On non-Si models, remove the air cleaner mounting bracket and install the 1.8 support eyelet (07AAK-SNAA400) behind the breather pipe and down to the threaded hole on the cylinder head. Install a retaining bolt.

14. On Si models, attach the engine hanger adapter (VSB02C000015) to the threaded hole in the cylinder head.

15. Install the front leg assembly, hook, and the wing nut to an "A and Reds" engine support hanger (AAR-T1256) onto the engine hanger. Carefully position the engine hanger on the vehicle, and attach the hook to the forward hole in the engine hanger adapter (Si model) or the slotted hole in the 1.8 support eyelet (Except Si model). Tighten the wing nut by hand to lift and support the engine/transmission.

16. Remove the cotter pin from the tie-rod end ball joint, then remove the nut on both sides.

17. Disconnect the tie-rod end ball joint from the knuckle using the ball joint thread protector and the ball joint remover.

➡**Be careful not to damage the ball joint boot when installing the remover.**

18. Remove the front splash shield.

19. Disconnect the EPS motor 2P connector, the EPS motor 1P connector, torque sensor 4P connector, the EPS motor angle sensor 6P connector from passenger's side of the steering gearbox. Wrap the connectors with vinyl tape to avoid contamination from grease or water.

20. Remove the lower ball joint mounting bolt and the self-locking nuts from the lower arm on both sides.

21. Disconnect the lower arm from the lower ball joint housing.

22. On Si models, remove the exhaust hanger from the three way catalytic converter (TWC).

23. Note the reference marks on both sides of the subframe that line up with the body (these are for installation reference).

24. Attach a proper front subframe adaptor to the front subframe and the transmission jack or powertrain lift, then tighten the front subframe adaptor screw. Make sure the

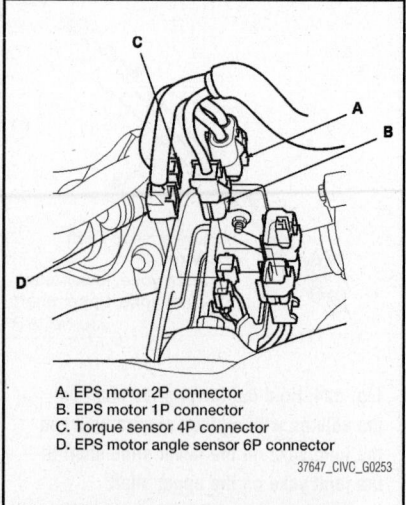

A. EPS motor 2P connector
B. EPS motor 1P connector
C. Torque sensor 4P connector
D. EPS motor angle sensor 6P connector

37647_CIVC_G0253

Fig. 329 Disconnect the EPS motor 2P connector, the EPS motor 1P connector, torque sensor 4P connector, and the EPS motor angle sensor 6P connector from passenger's side of the steering gearbox. Wrap the connectors with vinyl tape to avoid contamination from grease or water

front subframe is securely supported by the jack with the front subframe adaptor.

25. Remove the front subframe middle mount bolt from the left and right sides.

26. Remove the lower torque rod mounting bolts.

27. On Si models, remove the lower radiator hose from the radiator hose stay, then remove the front mounting bolt and nut.

28. Remove the front subframe front mounting bolts on both sides.

29. Remove the front subframe rear mounting bolts on both sides.

30. Lower the front subframe and steering gearbox as an assembly by lowering the jack slowly.

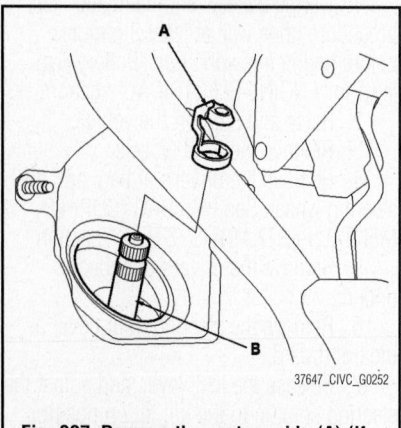

37647_CIVC_G0252

Fig. 327 Remove the center guide (A) (if equipped) from the top of the pinion shaft (B), and discard it

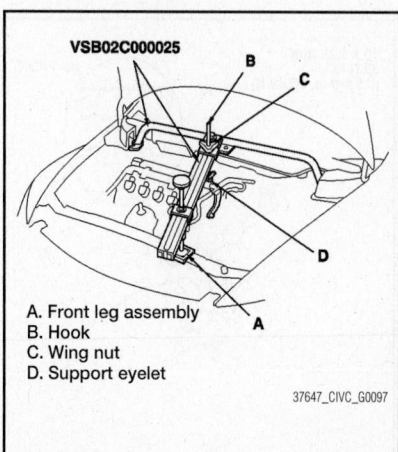

VSB02C000025

A. Front leg assembly
B. Hook
C. Wing nut
D. Support eyelet

37647_CIVC_G0097

Fig. 328 Installed position of the engine support hanger

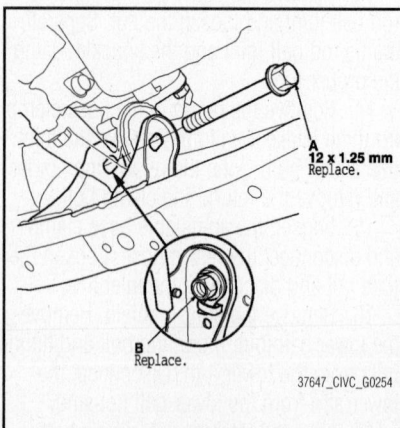

12 x 1.25 mm
Replace.

Replace.

37647_CIVC_G0254

Fig. 330 Remove the front mounting bolt (A) and nut (B)—Si models

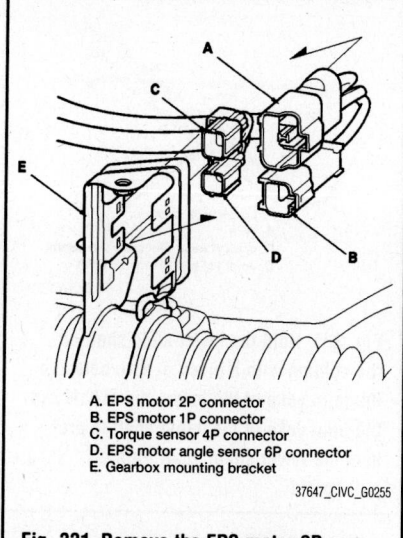

A. EPS motor 2P connector
B. EPS motor 1P connector
C. Torque sensor 4P connector
D. EPS motor angle sensor 6P connector
E. Gearbox mounting bracket

37647_CIVIC_G0255

Fig. 331 Remove the EPS motor 2P connector, the EPS motor 1P connector, torque sensor 4P connector, the EPS motor angle sensor 6P connector from the passenger's side of the gearbox mounting bracket

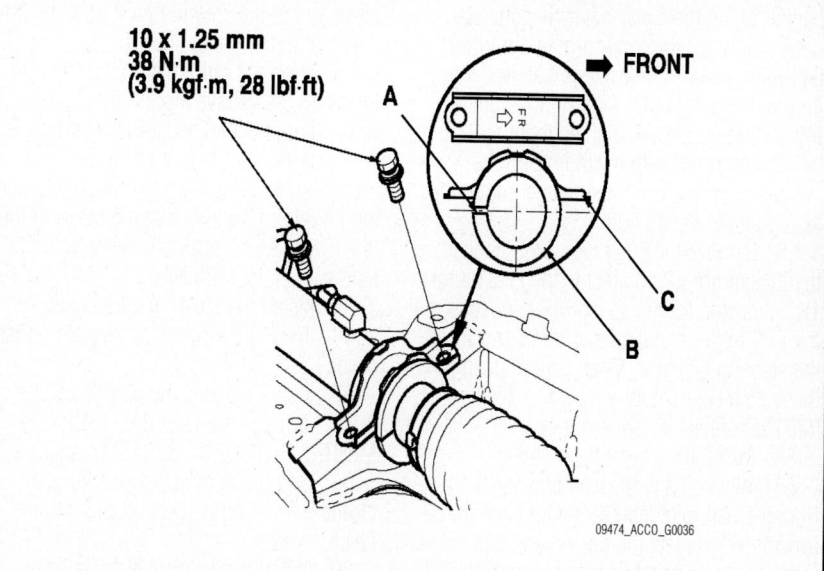

10 x 1.25 mm
38 N·m
(3.9 kgf·m, 28 lbf·ft)

→ FRONT

09474_ACCO_G0036

Fig. 332 Position the cutout (A) on the mounting cushion (B), as shown, and install it on the right side of the steering gearbox securely. Install the gearbox mounting bracket (C) over the mounting cushion, and loosely install the two 10 mm bolts

31. On non-Si models, remove the harness clips from the harness clamp brackets on the steering gear, then remove the harness clamp brackets.

32. On Si models, remove the harness clamp bracket from the front subframe, then remove the harness clips.

33. Remove the EPS motor 2P connector, the EPS motor 1P connector, torque sensor 4P connector, the EPS motor angle sensor 6P connector from the passenger's side of the gearbox mounting bracket.

34. Remove the pinion shaft grommet from the top of torque sensor.

35. Remove the two 10 mm bolts from the right side of the steering gearbox, then remove the gearbox mounting bracket and the mounting cushion.

36. Remove the four 10 mm flange bolts from the left side of the steering gearbox, then remove the stiffener plates.

37. Remove the steering gearbox from the front subframe.

To install:

38. Place the steering gearbox in position on the front subframe.

39. Loosely install the stiffener plates and the gearbox mounting bolts on the left side of the steering gearbox.

40. Position the cutout on the mounting cushion, as shown, and install it on the right side of the steering gearbox securely. Install the gearbox mounting bracket over the mounting cushion, and loosely install the two 10 mm bolts.

41. Tighten the 10 mm bolts on both

sides of the steering gearbox to 28 ft. lbs. (38 Nm), alternately in two or more steps.

42. Install the pinion shaft grommet. Align the slot in the pinion shaft grommet with the lug portion on the torque sensor. The grommet must not have a gap at the mating surface of the grommet and the torque sensor.

43. Install the EPS motor 2P connector, the EPS motor 1P connector, torque sensor 4P connector, the EPS motor angle sensor 6P connector on the right side of the gearbox mounting bracket.

44. On non-Si models, install the harness clamp bracket to the front subframe, then install the harness clips to the harness clamp bracket.

45. On Si models, install the harness clips to the harness clamp bracket, then install the harness clamp bracket to the front subframe.

46. Carefully raise the front subframe with the front subframe adapter and the transmission jack or the powertrain lift until the front subframe is in position, then loosely install new front subframe mounting bolts.

➡**Be sure that the pinion shaft grommet is in place securely. Make sure the pinion shaft grommet is not turned up. Incorrect installation can cause leakage of water, mud, and noise.**

47. Position the subframe, then align the front subframe reference marks to the body, as noted during removal.

48. Tighten the new front and rear subframe mounting bolts to the 76 ft. lbs. (103 Nm) on both sides.

49. On Si models, install the new nut and the new front mounting bolt and tighten the mounting bolt to 47 ft. lbs. (64 Nm). Install the lower radiator hose to the radiator hose stay.

50. Install the new lower torque rod mounting bolts and tighten them to 47 ft. lbs. (64 Nm).

51. Install the new front subframe middle mount bolt on the left and right sides and tighten it to 47 ft. lbs. (64 Nm).

52. Lower the transmission jack supporting the front subframe.

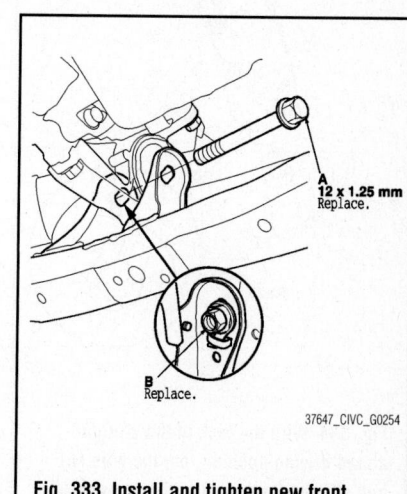

A
12 x 1.25 mm
Replace.

B
Replace.

37647_CIVIC_G0254

Fig. 333 Install and tighten new front mounting bolt (A) and nut (B)—Si models

53. On Si models, install the exhaust hanger to the three way catalytic converter.

54. Connect the lower arm to the lower ball joint. Install a new flange bolt and the new self-locking nuts. After lightly tightening all three fasteners, tighten them to 47 ft. lbs. (64 Nm) in the following order; the flange nut on the front, the flange nut on the rear, then the flange bolt.

55. Remove the vinyl tape, then connect the EPS motor 2P connector, the EPS motor 1P connector, torque sensor 4P connector, the EPS motor angle sensor 6P connector to the steering gearbox. Make sure to push these connectors until you hear a click so that the connectors are secured.

56. Install the front splash shield.

57. Wipe off any grease contamination from the ball joint tapered section and the threads. Reconnect the tie-rod end ball joints to the steering knuckles. Install the 12 mm nut and tighten it to 40 ft. lbs. (54 Nm).

58. Install the new cotter pin and bend it over.

59. Install the front wheel, then set the wheels in the straight ahead position.

➡**Before installing the wheel, clean the mating surfaces between the brake disc and inside of the wheel.**

60. Lower the vehicle.

61. Remove the engine hanger, support hanger, and 1.8 support eyelet (except Si model) or engine hanger adapter (Si model).

62. Center the steering rack within its stroke in the steering joint connection.

63. With the rack in the straight ahead

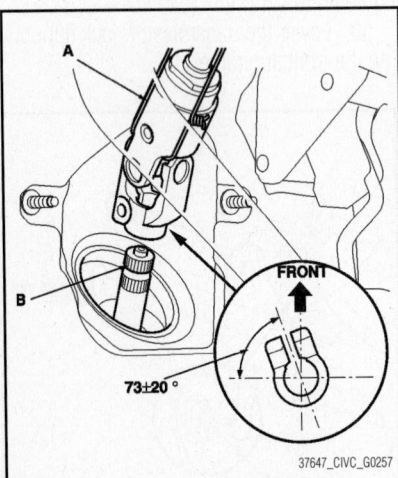

Fig. 334 With the rack in the straight ahead driving position, cut the wire (A) and slip the lower end of the steering joint onto the pinion shaft (B) in the range shown

driving position, cut the wire and slip the lower end of the steering joint onto the pinion shaft in the range shown.

64. Align the bolt hole on the steering joint with the groove around the pinion shaft. Then loosely install the joint bolt. Be sure that the joint bolt is securely in the groove in the pinion shaft. Pull on the steering joint to make sure that the steering joint is fully seated. Tighten the steering joint bolt to 21 ft. lbs. (28 Nm).

65. Install the steering joint cover.

66. Install the driver's dashboard under cover.

67. Install the steering wheel, and the driver's airbag. See "AIR BAG (SUPPLEMENTAL RESTRAINT SYSTEM)" section.

68. On non-Si models, install the air cleaner mounting bracket on the cylinder head.

69. Install the air cleaner housing.

70. Install the cowl cover and the under-cowl panel.

71. Do the battery terminal reconnection procedure, and check these items:

    a. Turn the ignition switch to ON (II) and check that the SRS indicator comes on for about 6 seconds, and then goes off.

    b. Make sure the horn and turn signal switches work properly.

    c. Make sure the steering wheel switches work properly.

72. After installation, check these items:

    a. Check the steering wheel spoke angle. If steering spoke angles to the right and left are not equal (steering wheel and rack are not centered), correct the engagement of the joint/pinion shaft splines.

    b. Set the steering column to the center tilt position, and to the center telescopic position, then check the wheel alignment and adjust.

    c. Make sure the steering wheel spokes are centered.

73. Do the "Torque Sensor Neutral Position Memorization".

74. Start the engine, and let it idle. Turn the steering wheel from lock to lock several times. Check that the EPS indicator does not come on.

2012 Models

See Figures 335 through 349 and 351.

1. Raise the vehicle on a lift, and make sure it is securely supported.

2. Open the hood, and secure it with hood support rod in the wide-open position.

3. Remove both front wheels.

4. Remove the steering joint cover.

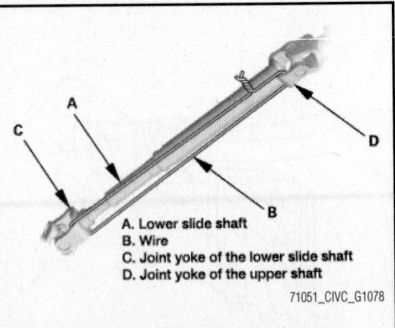

A. Lower slide shaft
B. Wire
C. Joint yoke of the lower slide shaft
D. Joint yoke of the upper shaft

71051_CIVIC_G1078

Fig. 335 Hold the lower slide shaft on the column with a piece of wire between the joint yoke of the lower slide shaft and the joint yoke of the upper shaft to prevent the lower slider shaft from pulling out

5. Set the steering column to the center tilt position, and to the center telescopic position.

6. Hold the lower slide shaft on the column with a piece of wire between the joint yoke of the lower slide shaft and the joint yoke of the upper shaft to prevent the lower slider shaft from pulling out.

7. Center the steering wheel spoke angle.

8. Install a commercially available steering wheel holder tool.

9. Loosen the steering joint bolt.

10. Remove the steering joint bolt. Disconnect the steering joint from the pinion shaft. If the center guide is in place and has not moved, leave it in place. If the center guide has come off, discard it.

11. Remove the ball joint from the lower arm.

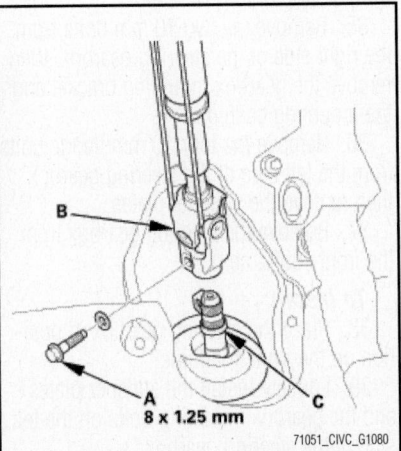

8 x 1.25 mm

71051_CIVIC_G1080

Fig. 336 Remove the steering joint bolt (A). Disconnect the steering joint (B) from the pinion shaft. If the center guide (C) is in place and has not moved, leave it in place. If the center guide has come off, discard it

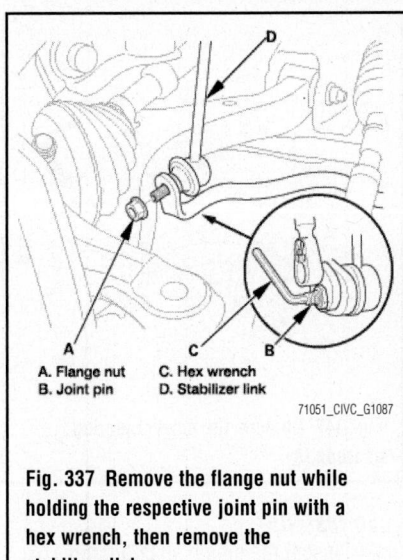

Fig. 337 Remove the flange nut while holding the respective joint pin with a hex wrench, then remove the stabilizer link

A. Flange nut
B. Joint pin
C. Hex wrench
D. Stabilizer link

71051_CIVC_G1087

12. Disconnect the tie rod end ball joints from both sides as follows:

   a. Remove the cotter pin from the tie-rod end ball joint. Remove the nut.

   b. Disconnect the tie-rod end ball joint.

➥**Be careful not to damage the ball joint boot when installing the remover.**

13. Disconnect both front stabilizer ball joints from the lower arm:

   a. Remove the flange nut while holding the respective joint pin with a hex wrench.

   b. Remove the stabilizer link.

14. Remove the engine undercover and the splash shield.

15. Remove the power steering gearbox peripheral assembly:

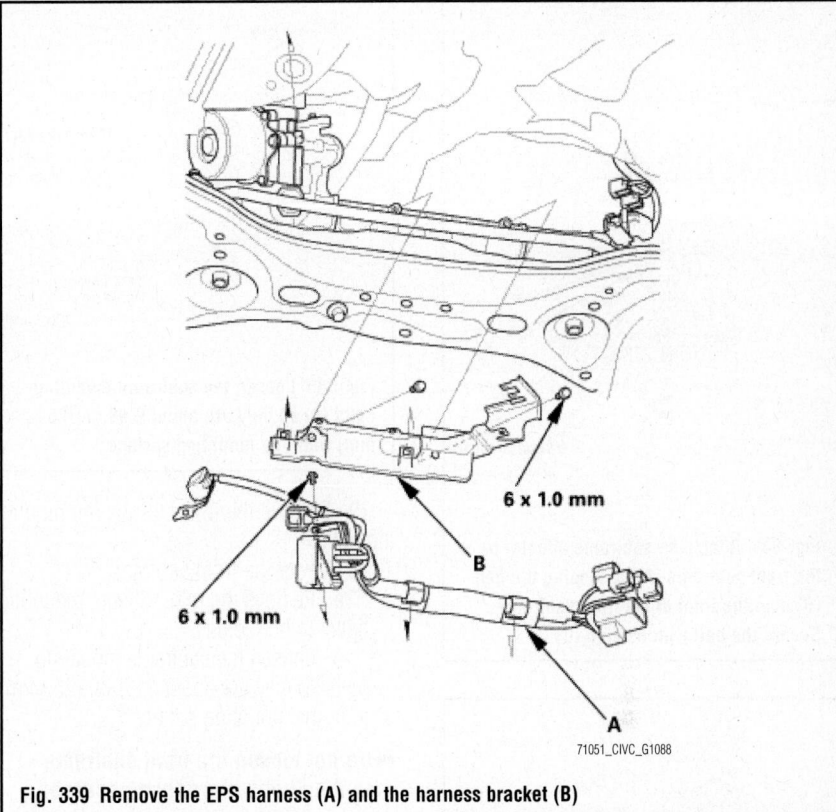

Fig. 339 Remove the EPS harness (A) and the harness bracket (B)

6 x 1.0 mm

6 x 1.0 mm

71051_CIVC_G1088

   a. Disconnect the connector and remove the EPS motor ground terminal.

   b. Remove the EPS harness and the harness bracket.

   c. Remove the steering gearbox mounting bolts.

   d. Remove the stiffener mounting bolts.

   e. Remove the steering gearbox stiffeners.

16. Attach the subframe adapter to the front subframe by looping the belt over the front of the front subframe. Secure the belt with its stop.

17. Raise the transaxle jack.

18. Line up the slots in the front subframe adapter arms with the bolt holes on the transaxle jack base.

19. Securely attach the front subframe adapter and the transaxle jack with four bolts.

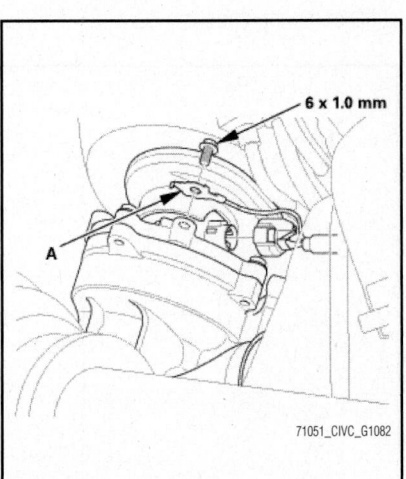

6 x 1.0 mm

A

71051_CIVC_G1082

Fig. 338 Disconnect the connector and remove the EPS motor ground terminal (A)

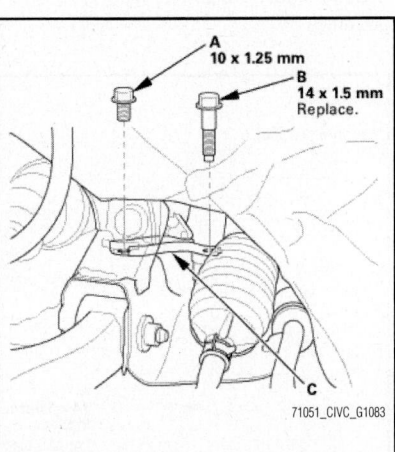

A
10 x 1.25 mm

B
14 x 1.5 mm
Replace.

C

71051_CIVC_G1083

Fig. 340 Remove the steering gearbox mounting bolts (A), remove the stiffener mounting bolts (B), and remove the steering gearbox stiffeners (C)—driver's side

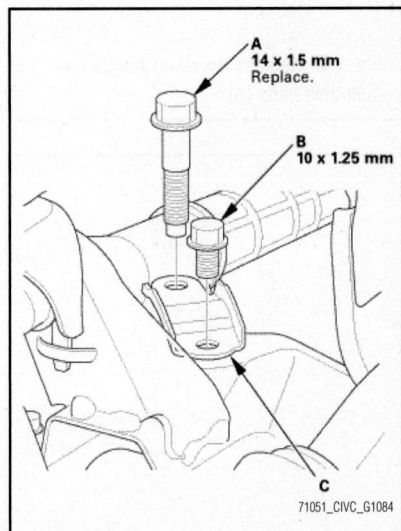

A
14 x 1.5 mm
Replace.

B
10 x 1.25 mm

C

71051_CIVC_G1084

Fig. 341 Remove the steering gearbox mounting bolts (A), remove the stiffener mounting bolts (B), and remove the steering gearbox stiffeners (C)—passenger's side

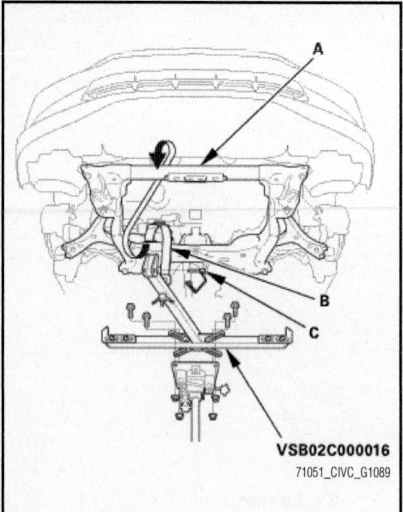

VSB02C000016
71051_CIVC_G1089

Fig. 342 Attach the subframe adapter to the front subframe (A) by looping the belt (B) over the front of the front subframe. Secure the belt with its stop (C)

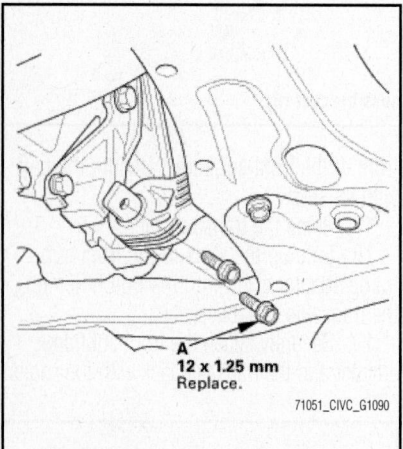

A
12 x 1.25 mm
Replace.

71051_CIVC_G1090

Fig. 343 Remove the lower torque rod mounting bolts (A)

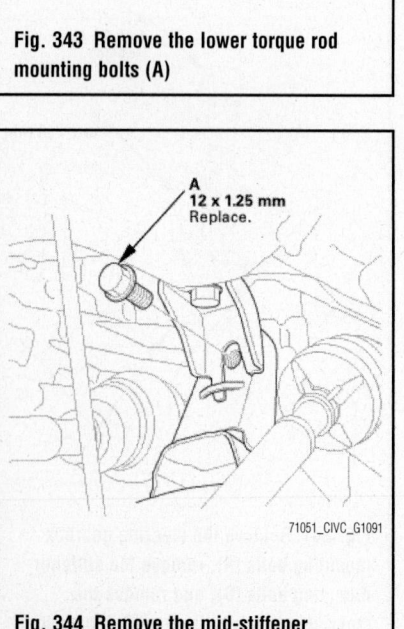

A
12 x 1.25 mm
Replace.

71051_CIVC_G1091

Fig. 344 Remove the mid-stiffener mounting bolts (A) on both sides

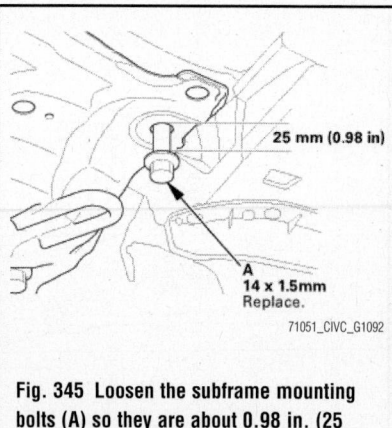

25 mm (0.98 in)

A
14 x 1.5mm
Replace.

71051_CIVC_G1092

Fig. 345 Loosen the subframe mounting bolts (A) so they are about 0.98 in. (25 mm) from the mounting surface

20. Remove the lower torque rod mounting bolts.

21. Lower the front subframe:

   a. Remove the mid-stiffener mounting bolts on both sides.

   b. Loosen the subframe mounting bolts so they are about 0.98 in. (25 mm) from the mounting surface.

➡**Do not loosen the front subframe mounting bolts more than necessary.**

   c. Remove the subframe mounting bolts on both sides.

   d. Lower the jack slowly until the front subframe has dropped about 3.15 in. (80 mm).

22. Remove the power steering gearbox.

➡**Be careful not to damage the EPS motor connector when placing it on the floor. The EPS motor connector must face up.**

23. Remove the pinion shaft grommet.

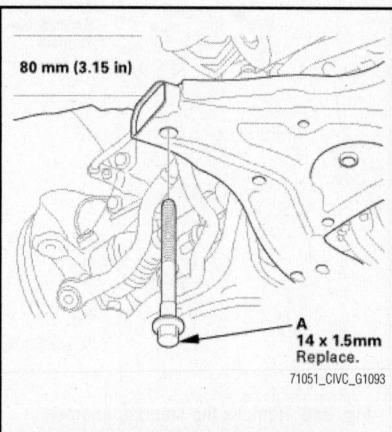

80 mm (3.15 in)

A
14 x 1.5mm
Replace.

71051_CIVC_G1093

Fig. 346 Remove the subframe mounting bolts (A) on both sides. Lower the jack slowly until the front subframe has dropped about 3.15 in. (80 mm)

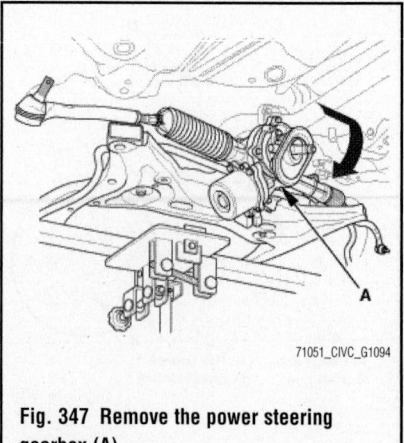

A

71051_CIVC_G1094

Fig. 347 Remove the power steering gearbox (A)

To install:

24. Install the pinion shaft grommet. Align the cutout portions with the lug portions. The grommet must not have a gap at the mating surface of the grommet and torque sensor.

25. Install the steering gearbox and raise the front subframe up to the body.

26. Install the front subframe, as described in this section.

27. Tighten the new mid-stiffener mounting bolts on both sides to 47 ft. lbs. (64 Nm).

28. Loosely install the new lower torque rod mounting bolts, loosen the lower torque rod bolt, then tighten the two inner lower torque rod mounting bolts to 54 ft. lbs. (73 Nm), and the outer lower torque rod (vertical) bolt to 69 ft. lbs. (93 Nm).

29. Lower the transaxle jack supporting the front subframe.

30. Install the steering gearbox stiffeners, loosely tighten the steering gearbox mounting bolts, then loosely tighten the stiffener mounting bolts.

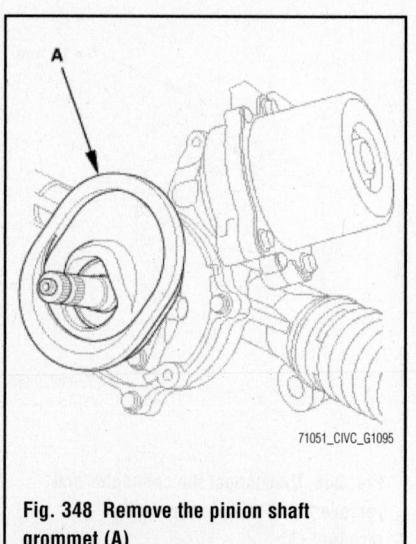

A

71051_CIVC_G1095

Fig. 348 Remove the pinion shaft grommet (A)

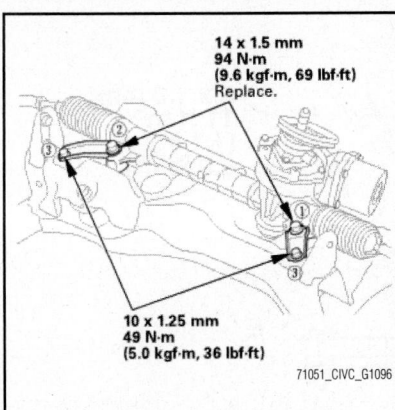

Fig. 339 Tighten the steering gearbox mounting bolts and the stiffener mounting bolts to the specified torque in the

31. Tighten the steering gearbox mounting bolts and the stiffener mounting bolts to the specified torque in the sequence shown.

32. Install the harness bracket, then install the EPS harness in the sequence shown.

33. Connect the EPS motor connector and the ground terminal.

34. Install the splash shield and the engine undercover.

35. Reconnect the front stabilizer ball joints (both lower arm side).

36. Install the stabilizer link on the stabilizer bar with the joint pins set at the center of their range of movement on both sides. Tighten the retaining nuts to 27 ft. lbs. (37 Nm).

37. Install the tie rod end ball joints. Tighten the nuts to 40ft. lbs. (54 Nm).

38. Release the steering column by cutting the wire.

39. Remove the steering wheel holder tool.

40. Reconnect the steering joint:

   a: Set the rack in the straight ahead driving position.

   b. Slip the lower end of the steering joint onto the pinion shaft.

   c. Pinion shaft with center guide: Install the steering joint by aligning the center guide.

   d. Pinion shaft without center guide: Position the steering column by aligning the gap within the angle.

   e. Align the bolt hole on the steering joint with the groove around the pinion shaft.

   f. Loosely install the steering joint bolt.

   g. Be sure that the joint bolt is securely in the groove in the pinion shaft.

   h. Pull on the steering joint to make sure that the steering joint is fully seated.

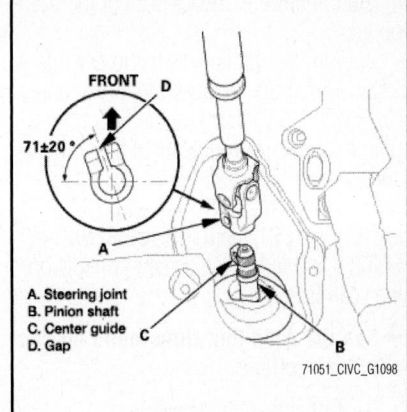

Fig. 351 Slip the lower end of the steering joint onto the pinion shaft. Pinion shaft with center guide: Install the steering joint by aligning the center guide. Pinion shaft without center guide: Position the steering column by aligning the gap within the angle

   i. Tighten the steering joint bolt to 21 ft. lbs. (29 Nm).

41. Install the steering joint cover.

42. Install both front wheels.

43. After installation, check these items:

   a. Start the engine, allow it to idle, and turn the steering wheel from lock to lock several times.

   b. Check that the EPS indicator does not come on.

   c. Check the steering wheel spoke angle. If steering spoke angles to the right and left are not equal (steering wheel and rack are not centered), correct the engagement of the joint/pinion shaft splines.

44. Perform a wheel alignment check and adjustment, as needed.

45. Perform "EPS Resetting" in this section.

TORQUE SENSOR NEUTRAL POSITION MEMORIZATION

The torque sensor neutral position must be memorized whenever the steering gearbox, the torque sensor, the EPS motor, or the EPS control unit is replaced. Note that the torque sensor neutral position is not affected when erasing the DTC.

➡**The torque sensor is temperature sensitive. This procedure should be performed within the range of 68°F±18°F (20°C±10°C).**

1. With the ignition switch in LOCK (0), connect the Honda Diagnostic System (HDS), or equivalent scan tool and software, to the data link connector (DLC),

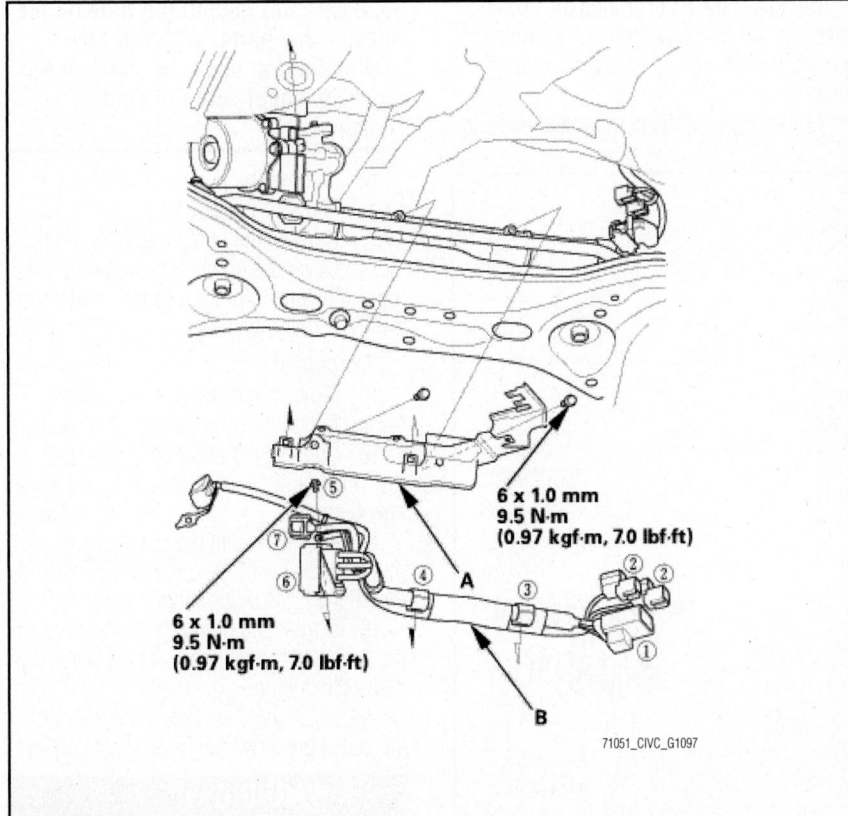

Fig. 350 Install the harness bracket (A), then install the EPS harness (B) in the sequence shown

located under the driver's side of the dash-board.

2. Turn the ignition switch to ON (II).

3. Make sure the scan tool communicates with the vehicle and the EPS control unit. If it doesn't, troubleshoot the DLC circuit.

4. From the EPS MENU, select MISCELLANEOUS TEST, then TORQUE SENSOR LEARN, and follow the screen prompts on the scan tool.

➡ **See the scan tool Help menu for specific instructions.**

5. Turn the ignition switch to LOCK (0).

POWER STEERING PUMP

REMOVAL & INSTALLATION

2011 Models Without EPS

See Figures 352 and 353.

1. Place a suitable container under the vehicle.

2. Drain the power steering fluid from the reservoir.

3. Remove the cowl cover and under-cowl panel.

4. Remove the air cleaner.

5. Remove the front splash shield.

6. If equipped with a manual transaxle, remove the shift cable bracket.

7. If equipped with an automatic transaxle, disconnect the shift cable from the control lever.

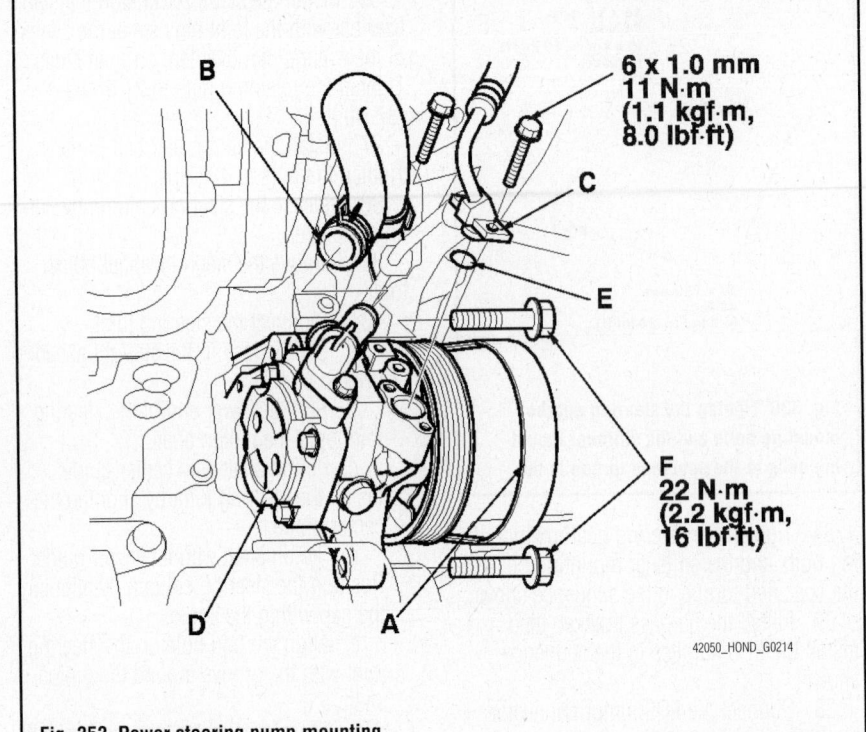

Fig. 353 Power steering pump mounting

8. Remove the upper torque rod from the body.

9. Remove the drive belt from the pump pulley.

10. Cover the parts around the power steering pump with several shop towels to protect them from spilled power steering fluid.

11. Disconnect the pump inlet hose and pump outlet hose from the pump, and plug them.

✳✳ WARNING

Take care not to spill the fluid on the body or any parts. Wipe off any spilled fluid at once. Do not turn the steering wheel with the pump removed.

12. Remove the pump outlet hose O-ring, and discard it.

13. Remove the pump mounting bolts.

14. Cover the opening of the pump with a piece of tape to prevent foreign material from entering the pump.

To install:

15. Move the power steering pump toward the driver's side, and then raise it.

16. Connect the pump inlet hose and pump outlet hose onto the new pump with new O-ring.

17. Loosely install the pump in the pump bracket with the mounting bolts, and then tighten the pump fittings securely.

18. Tighten the pump mounting bolts to the specified torque shown in the accompanying illustration.

19. Install the drive belt. Make sure that the belt is properly positioned on the pulleys.

✳✳ WARNING

Do not get power steering fluid or grease on any parts around the

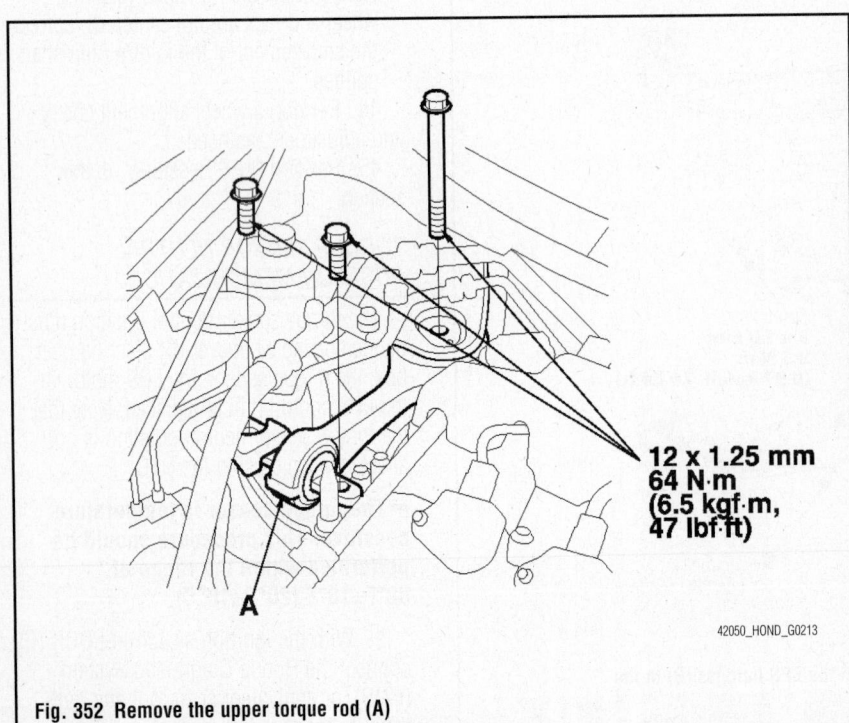

Fig. 352 Remove the upper torque rod (A)

power steering pump, drive belt, or pulley faces. Clean off any fluid or grease before installation.

20. Fill the reservoir to the upper level line and bleed the system.

BLEEDING

Check the reservoir at regular intervals, and add the recommended fluid as necessary. Always use Honda Power Steering Fluid. Using any other type of power steering fluid or automatic transmission fluid can cause increased wear and poor steering in cold weather.

➡ **If the fluid is contaminated, the screen in the reservoir may be partially** blocked. **Replace the reservoir if necessary.**

1. Remove the reservoir from its holder. Raise the reservoir, and then disconnect the return hose to drain the reservoir. Take care not to spill the fluid on the body and parts. Wipe off any spilled fluid at once.

➡ **Inspect the reservoir screen for any debris. If the reservoir screen is clogged, replace the reservoir.**

2. Connect a hose of suitable diameter to the disconnected return hose, and put the hose end in a suitable container.

3. Start the engine, let it run at idle, and turn the steering wheel from lock-to-lock several times. When fluid stops running out of the hose, shut off the engine. Discard the fluid.

4. Reinstall the return hose on the reservoir.

5. Fill the reservoir to the upper level line.

6. Start the engine and run it at fast idle, and then turn the steering from lock-to-lock several times to bleed air from the system.

7. Recheck the fluid level and add some if necessary. Do not fill the reservoir beyond the upper level line.

8. If the fluid is contaminated, dark, or discolored, repeat the procedure as necessary.

SUSPENSION

FRONT SUSPENSION

LOWER BALL JOINT

REMOVAL & INSTALLATION

2011 Models

See Figure 354.

1. Install a hex nut onto the threads of the ball joint. Make sure the nut is flush with the ball joint pin end to prevent damage to the threaded end of the ball joint pin.

2. Apply grease to the ball joint remover on the areas shown. This will ease installation of the tool and prevent damage to the pressure bolt threads.

3. Loosen the pressure bolt, and install the ball joint remover as shown. Insert the jaws carefully, making sure not to damage the ball joint boot. Adjust the jaw spacing by turning the adjusting bolt.

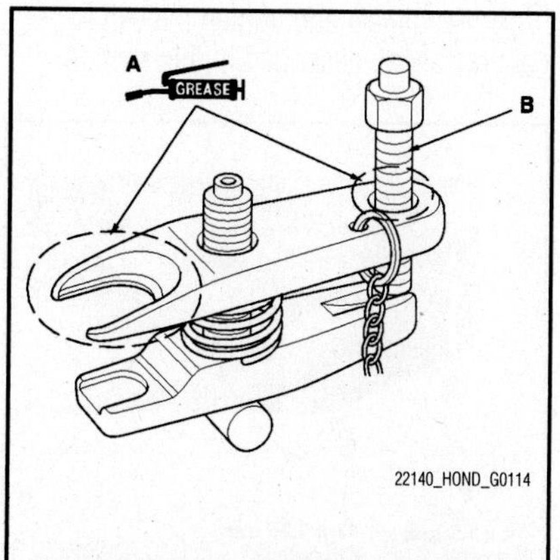

4. After adjusting the adjusting bolt, make sure the head of the adjusting bolt is in the position shown to allow the jaw to pivot.

5. With a wrench, tighten the pressure bolt until the ball joint pin pops loose from the ball joint pin hole. If necessary, apply penetrating type lubricant to loosen the ball joint pin.

6. Remove the ball joint remover, then remove the nut from the end of the ball joint pin, and pull the ball joint out of the ball joint pin hole. Inspect the ball joint boot, and replace it if damaged.

2012 Models

See Figures 355 through 359.

1. Raise the vehicle on a lift, and make sure it is securely supported.

2. Remove the front wheel.

3. Remove the flange bolt and nuts and disconnect the lower ball joint from the lower arm.

➡ **Always use a ball joint remover to disconnect a ball joint. Do not strike the housing or any other part of the ball joint connection to disconnect it.**

4. Pry up the stake on the spindle nut and remove the spindle nut.

5. Separate the outboard joint from the front hub using a soft face hammer on the end of the spindle.

6. Pull the knuckle outward, and separate the outboard joint from the front hub.

➡ **Always use a ball joint remover to disconnect a ball joint. Do not strike the housing or any other part of the ball joint connection to disconnect it.**

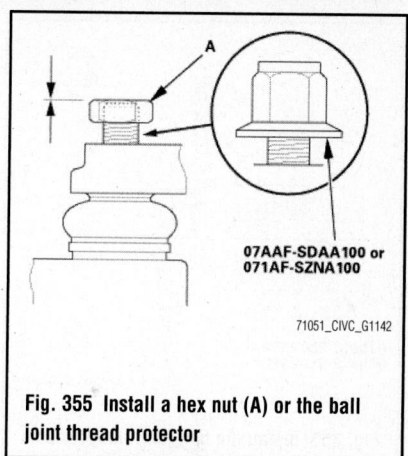

Fig. 355 Install a hex nut (A) or the ball joint thread protector

7. Install a hex nut or the ball joint thread protector.

➡ **When using a hex nut, make sure the nut is flush with the ball joint pin end to prevent damage to the threaded end of the ball joint pin.**

8. Apply grease to the ball joint remover on the areas shown.

9. Install the ball joint remover as shown. Fasten the safety chain securely to a suspension arm or the subframe.

10. Tighten the pressure bolt until the ball joint pin pops loose from the ball joint connecting hole.

➡ **Do not use pneumatic or electric tools on the pressure bolt.**

11. Remove the ball joint remover.

12. Remove the hex nut or the ball joint thread protector.

13. Pull the ball joint out of the ball joint connecting hole.

14. Remove the lock pin and the castle nut, then remove the lower ball joint.

Fig. 354 Apply grease to the ball joint remover on the areas shown (A). This will ease installation of the tool and prevent damage to the pressure bolt (B) threads

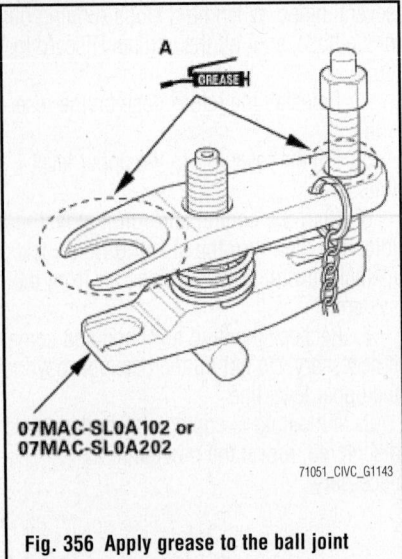

Fig. 356 Apply grease to the ball joint remover on the areas shown (A)

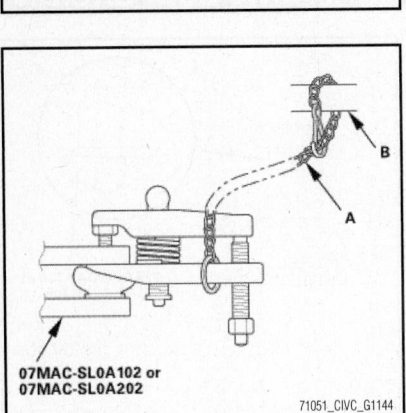

Fig. 357 Install the ball joint remover as shown. Fasten the safety chain (A) securely to a suspension arm or the sub-frame (B)

To install:

15. Install the lower ball joint with a new castle nut. Torque range for the nut is 41–51 ft. lbs. (59–69 Nm).

16. Torque the castle nut to the lower torque specification, then tighten it only far enough to align the slot with the ball joint pin hole. Do not align the castle nut by loosening it.

17. Install the lock pin in the proper position.

18. Apply about 5 g (0.18 oz.) of moly 60 paste to the contact area of the outboard joint and the front wheel bearing.

➡**The paste helps to prevent noise and vibration.**

19. Install the outboard joint into the front hub on the knuckle.

20. Apply a small amount of engine

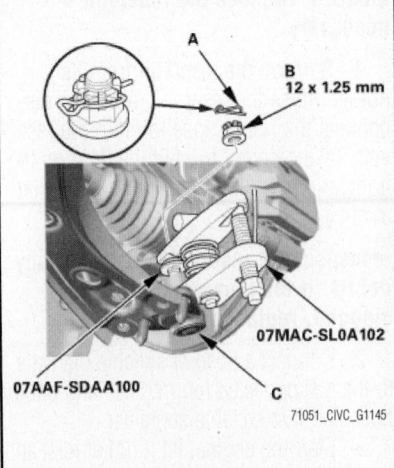

Fig. 358 Remove the lock pin (A) and the castle nut (B), then remove the lower ball joint (C)

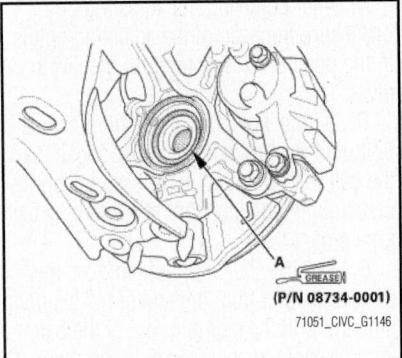

Fig. 359 Apply about 5 g (0.18 oz.) of moly 60 paste to the contact area (A) of the outboard joint and the front wheel bearing

oil to the seating surface of a new spindle nut.

21. Install the spindle nut and torque as follows:
- 1.8L engines: 133 ft. lbs. (181 Nm)
- 2.4L engines: 181 ft. lbs. (245 Nm)

22. Use a drift to stake the spindle nut shoulder against the driveshaft

23. Connect the lower ball joint to the lower arm.

24. Install the new flange nuts and the new flange bolt and tighten to 44 ft. lbs. (59 Nm).

25. Install the front wheel. Tighten the wheel nuts to 80 ft. lbs. (108 Nm).

26. Install the center cap.

27. Perform a front wheel alignment check.

LOWER CONTROL ARM

REMOVAL & INSTALLATION

2011 Models

1. Raise and support the vehicle safely. Remove the front tires.

2. Remove the flange nut while holding the ball joint pin. Disconnect the stabilizer links from the lower arm.

3. Turn the stabilizer bar backward to gain access to the front side of the lower arm mounting bolt.

4. Remove the flange bolt and nuts from the lower arm.

5. Disconnect the lower ball joint from the lower arm.

6. Remove the lower arm mounting bolts. Remove the lower arm from the front suspension subframe.

To install:

7. Installation is in the reverse order of the removal procedure.

8. Be sure to use new flange bolts, castle nut and lock pin. Tighten the flange nut to 25 ft. lbs. (34 Nm).

9. Check and adjust the wheel alignment, as necessary.

2012 Models

See Figure 360.

1. Raise and support the vehicle.

2. Remove the front wheel.

3. Disconnect the lower ball joint from the control arm.

4. Remove the lower control arm.

To install:

5. Install the lower control arm, using new bolts. Tighten the front (horizontal) bolt to 47 ft. lbs. (64 Nm). Tighten the rear (vertical) bolt to 61 ft. lbs. (83 Nm).

6. Reconnect and tighten the ball joint connection, as described in this section.

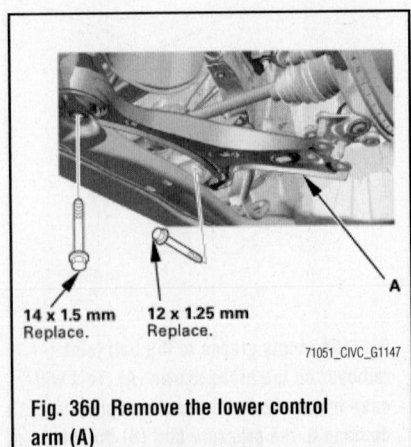

Fig. 360 Remove the lower control arm (A)

7. Install the front wheel. Tighten the wheel nuts to 80 ft. lbs. (108 Nm).

STABILIZER BAR & LINKS

REMOVAL & INSTALLATION

2011 Models

1. Raise and support the vehicle safely. Remove the front tires.
2. Disconnect both stabilizer links from the stabilizer bar.
3. Remove the flange bolts and the bushing holders. Remove the bushings and the stabilizer bar from the front suspension subframe.

To install:

4. Position the assembly to its mounting on the vehicle.
5. Install the mounting bolts.
6. Continue the installation in the reverse order of the removal procedure.

2012 Models

Stabilizer Bar

See Figures 361 through 364.

1. Raise the vehicle on a lift, and make sure it is securely supported.
2. Remove both front wheels.
3. Remove the steering joint cover and lower column joint from pinion shaft.
4. Remove the lower ball joint.
5. Disconnect the tie rod end ball joints.
6. Disconnect the stabilizer link at the stabilizer bar side:

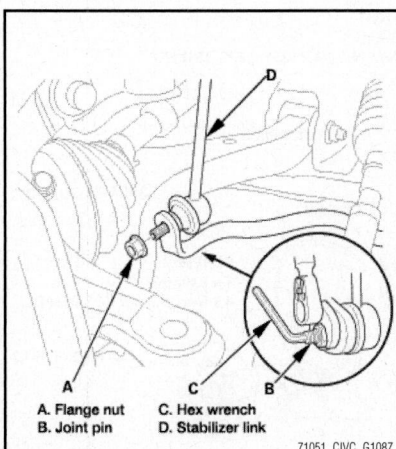

A. Flange nut   C. Hex wrench
B. Joint pin   D. Stabilizer link

71051_CIVC_G1087

Fig. 361 Remove the self-locking nut while holding the respective joint pin with a hex wrench. Disconnect both sides of the stabilizer link from the stabilizer bar

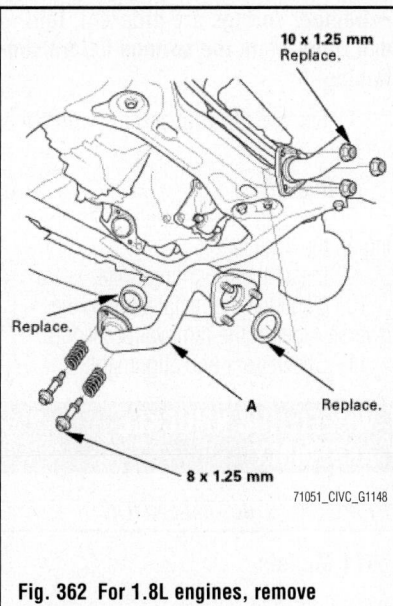

10 x 1.25 mm
Replace.

Replace.

Replace.

A

8 x 1.25 mm

71051_CIVC_G1148

Fig. 362 For 1.8L engines, remove exhaust pipe A

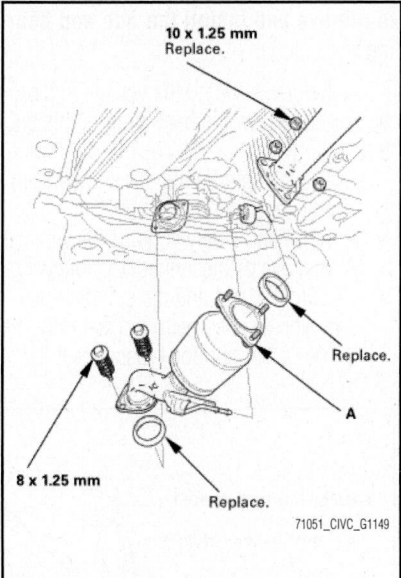

10 x 1.25 mm
Replace.

Replace.

A

8 x 1.25 mm

Replace.

71051_CIVC_G1149

Fig. 363 For 2.4L engines, remove the under-floor TWC (A)

a. Remove the self-locking nut while holding the respective joint pin with a hex wrench.
b. Disconnect both sides of the stabilizer link from the stabilizer bar.

7. For 1.8L engines, remove exhaust pipe A.
8. For 2.4L engines, remove the under-floor TWC.
9. Remove the steering gearbox mounting bolts and stiffener mounting bolts on both sides.
10. Remove the gearbox stiffeners.
11. Remove the bushing holder and the bushing.

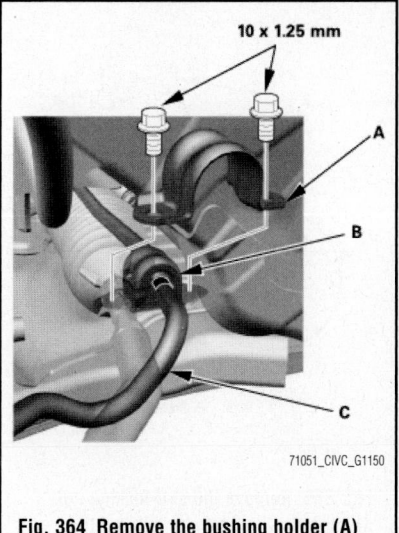

10 x 1.25 mm

A

B

C

71051_CIVC_G1150

Fig. 364 Remove the bushing holder (A) and the bushing (B). Move the steering gearbox upward and remove the stabilizer bar (C) from the driver's side

12. Move the steering gearbox upward and remove the stabilizer bar from the driver's side.

To install:

13. Move the steering gearbox upward, and install the stabilizer bar from the driver's side. Install the bushing and tighten the retaining bolts to 29 ft. lbs. (39 Nm).

➡**Note the direction of installation for the bushing (split facing forward). Align the stabilizer band or the paint mark on the stabilizer bar with the side of the bushing.**

14. Install the gearbox stiffeners. Loosely install the new steering gearbox mounting bolts.
15. Loosely install the new gearbox stiffener mounting bolts.
16. Tighten the steering gearbox mounting bolts (rear bolts on stiffeners) to 69 ft. lbs. (94 Nm) and the gearbox stiffener mounting bolts (gear side first) to 36 ft. lbs. (49 Nm).
17. Install the under-floor TWC or exhaust pipe A with new gasket(s). Tighten as follows:
   - Exhaust nuts: 24 ft. lbs. (33 Nm)
   - Exhaust bolts: 16 ft. lbs. (22 Nm)
18. Connect both sides of the stabilizer link to the stabilizer bar.
2. Install the new self-locking nut and tighten them to 27 ft. lbs. (37 Nm) while holding the respective joint pin with a hex wrench.
19. Reconnect the tie rod ends, as described in "Steering" section.

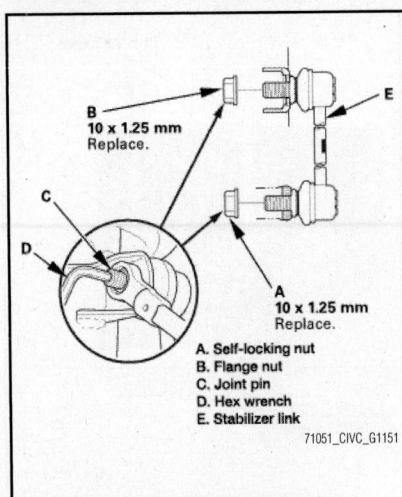

B
10 x 1.25 mm
Replace.

E

C

D

A
10 x 1.25 mm
Replace.

A. Self-locking nut
B. Flange nut
C. Joint pin
D. Hex wrench
E. Stabilizer link

71051_CIVC_G1151

Fig. 365 Remove the self-locking nut and the flange nut while holding the respective joint pin with a hex wrench. Remove the stabilizer link

20. Reconnect pinion shaft and lower column joint.

21. Install both front wheels. Tighten the wheel nuts to 80 ft. lbs. (108 Nm).

22. Perform a wheel alignment check.

Stabilizer Link

1. Raise and support the vehicle.

2. Remove the front wheels.

3. Remove the self-locking nut and the flange nut while holding the respective joint pin with a hex wrench.

4. Remove the stabilizer link.

5. Installation is the reverse of the removal procedure.

6. Using new nuts, tighten the lower link nut to 27 ft. lbs. (37 Nm) and the upper nut to 25 ft. lbs. (34 Nm).

STRUTS & SPRINGS

REMOVAL & INSTALLATION

1. Raise and support the vehicle safely.

2. Remove the front tires.

3. Remove the wheel sensor harness bracket and brake hose bracket from the damper. Do not disconnect the wheel sensor connector.

4. Remove the damper pinch bolts from the bottom of the damper. Do not allow the knuckle to rotate too far outward as this may cause the inner CV joint bearing to unseat.

5. Turn the ignition switch to the ON position. Turn the windshield wipers on; turn the ignition switch off leaving the wipers near the A-pillars.

6. Remove the service cap and lid. Remove the three flange nuts at the top of the damper.

➡**Damper springs are different, left and right. Mark the springs before continuing.**

7. Remove the strut assembly from the vehicle.

To install:

8. Position the assembly to its mounting on the vehicle.

9. Install the mounting bolts.

10. Continue the installation in the reverse order of the removal procedure.

11. Check front end alignment.

WHEEL·HUB/KNUCKLE ASSEMBLY

REMOVAL & INSTALLATION

2011 Models

See Figure 366.

➡**A hydraulic press and several bearing drivers and attachments are needed to remove and install the hub and bearing.**

1. Before servicing the vehicle, refer to the precautions in the beginning of this section.

2. Pry the spindle nut stake away from the spindle, and then loosen the nut.

3. Raise and safely support the vehicle.

4. Remove or disconnect the following:
   • Front wheel and the spindle nut
   • Wheel sensor wire bracket from the knuckle, but don't disconnect it.

• Caliper mounting bolts and the caliper. Support the caliper out of the way with a length of wire. Do not let the caliper hang from the brake hose.
• 6mm brake disc retaining screws. Screw two 12mm bolts into the disc to push it away from the hub.
• Tie rod castle nut
• Tie rod ball joint using a suitable ball joint remover.
• Cotter pin and loosen the lower arm ball joint nut half the length of the joint threads.
• Ball joint and lower arm using a suitable puller with the pawls applied to the lower arm.

➡**Avoid damaging the ball joint boot. If necessary, apply penetrating type lubricant to loosen the ball joint.**

• Ball joint nut cover.
• Cotter pin and the upper ball joint nut.
• Upper ball joint and knuckle using a ball joint remover.

5. Use a plastic mallet to free the halfshaft from the knuckle. Pull the knuckle out to remove it.

➡**A new wheel bearing must be used when the hub is removed.**

6. Place the knuckle in a press and use a base and pilot to press the hub assembly out of the wheel bearing.

7. Remove the knuckle ring seal and

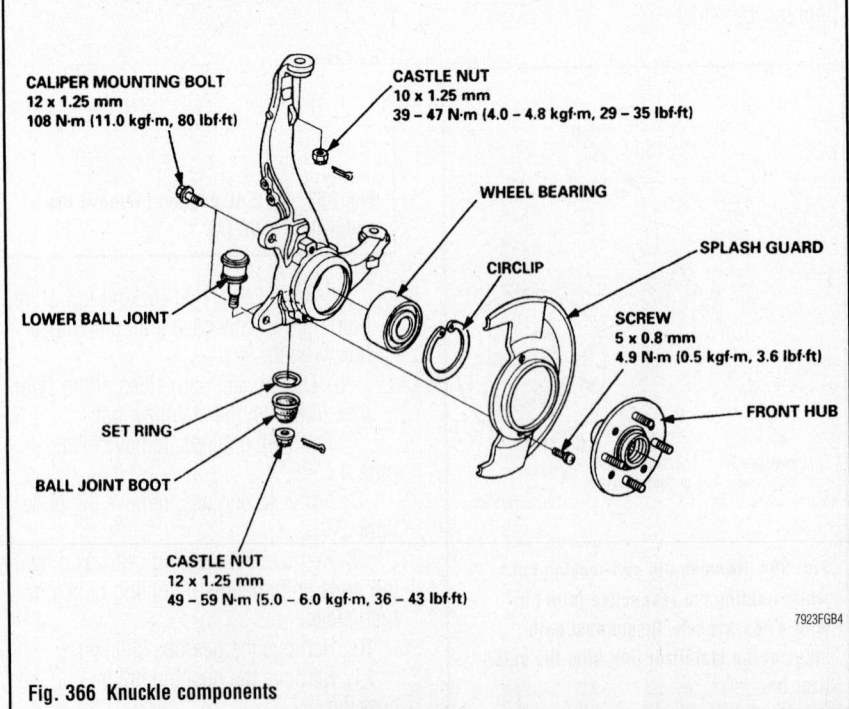

CALIPER MOUNTING BOLT
12 x 1.25 mm
108 N·m (11.0 kgf·m, 80 lbf·ft)

CASTLE NUT
10 x 1.25 mm
39 – 47 N·m (4.0 – 4.8 kgf·m, 29 – 35 lbf·ft)

WHEEL BEARING

CIRCLIP

SPLASH GUARD

LOWER BALL JOINT

SCREW
5 x 0.8 mm
4.9 N·m (0.5 kgf·m, 3.6 lbf·ft)

FRONT HUB

SET RING

BALL JOINT BOOT

CASTLE NUT
12 x 1.25 mm
49 – 59 N·m (5.0 – 6.0 kgf·m, 36 – 43 lbf·ft)

7923FGB4

Fig. 366 Knuckle components

circlip. Remove the splash guard from the knuckle.

8. Press the wheel bearing out of the knuckle using a driving attachment.

To install:

9. Clean the knuckle and hub assembly and inspect them for damage.

10. Install or connect the following:
- New wheel bearing into the hub using a driving tool
- Circlip in the outer groove of the knuckle
- Splash guard
- Hub assembly into the steering knuckle using a base and a driving and guide tool
- Knuckle ring seal
- Knuckle onto the spindle
- Knuckle onto the upper and lower ball joints and tighten the castle nuts
- Tie rod ball joint onto the steering knuckle

11. Tighten the upper ball joint nut and tie rod nut to 29–35 ft. lbs. (40–48 Nm) and the lower ball joint castle nut to 36–43 ft. lbs. (50–60 Nm).

12. Install or connect the following:
- Anti-lock Brake System (ABS) wheel sensor wire brackets onto the knuckle. Tighten the mounting bolts to 84 inch lbs. (10 Nm).
- Brake disc; use 2 lug nuts to evenly draw the disc onto the hub
- Retainer screws: 84 inch lbs. (10 Nm)
- Spindle washer and nut. Don't tighten the nut until the vehicle is on the ground.
- Brake caliper and tighten the bolts to 80 ft. lbs. (110 Nm)
- Front wheels and lower the vehicle

13. Tighten the spindle nut to 134 ft. lbs. (185 Nm), stake the nut, and install the grease cap.

14. Check and adjust the vehicle's front wheel alignment.

➡**Avoid damaging the ball joint boot. If necessary, apply penetrating-type lubricant to loosen the ball joint.**

2012 Models

See Figures 367 and 368.

1. Raise the vehicle on a lift, and make sure it is securely supported.

2. Remove the front wheel(s).

3. Remove the brake hose mounting bolt and remove the caliper assembly.

➡**To prevent damage to the caliper assembly or brake hose, use a**

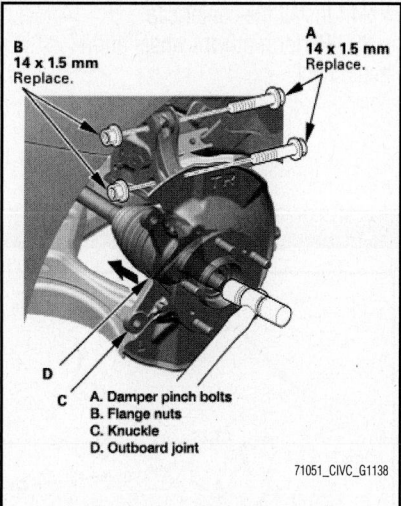

Fig. 367 Remove the damper pinch bolts and the flange nuts. Pull the knuckle outward, and separate the outboard joint from the front hub, then remove the knuckle/hub

A. Damper pinch bolts
B. Flange nuts
C. Knuckle
D. Outboard joint

71051_CIVIC_G1138

short piece of wire to hang the caliper assembly from the undercarriage. Do not twist the brake hose excessively.

4. Remove the wheel speed sensor, but do not disconnect the wheel speed sensor connector.

5. Pry up the stake on the spindle nut and remove the spindle nut.

6. Remove the brake disc.

7. Disconnect the ball joint from the knuckle, as described in this section.

8. Remove the cotter pin, the nut, and disconnect the tie-rod end ball joint.

9. Remove the flange bolt and the flange nuts and disconnect the lower ball joint from the lower arm.

10. Remove the front knuckle/hub assembly:
   a. Remove the damper pinch bolts and the flange nuts.
   b. Pull the knuckle outward, and separate the outboard joint from the front hub, then remove the knuckle/hub.

➡**Do not pull the driveshaft end outward. The inner driveshaft inboard joint may come apart.**

11. Remove the front lower ball joint:
   a. Remove the lock pin.
   b. Remove the castle nut.
   c. Remove the lower ball joint.

To install:

12. Install the front lower ball joint:
   a. Install the lower ball joint.
   b. Install the new castle nut.

c. Torque the castle nut to the lower torque specification, then tighten it only far enough to align the slot with the ball joint pin hole. Do not align the castle nut by loosening it.
   d. Install the lock pin as shown.

13. Install the knuckle/hub (A). Tighten the retaining bolts to 85 ft. lbs. (115 Nm).

➡**Apply grease to the mating surfaces of the wheel bearing and the driveshaft outboard joint**

14. Connect the lower ball joint to the lower arm.

15. Install the new flange nuts and the new flange bolt, tightening them to 44 ft. lbs. (59 Nm).

16. Connect the tie-rod end ball joint, install the nut and the cotter pin. Tighten the nut to 40 ft. lbs. (54 Nm).

17. Install the brake disc. Tighten the retaining screws to 7 ft. lbs. (9.5 Nm).

➡**Before installing the brake disc, clean the mating surfaces between the front hub and the inside of the brake disc.**

18. Apply a small amount of engine oil to the seating surface of a new spindle nut (A) and install the nut. Tighten the nut to 133 ft. lbs. (181 Nm).

19. Use a drift to stake the spindle nut shoulder against the driveshaft.

20. Install the wheel speed sensor.

21. Install the caliper assembly. Tighten the caliper bolts to 80 ft. lbs. (108 Nm).

22. Install the brake hose mounting bolt and tighten to 16 ft. lbs. (22 Nm).

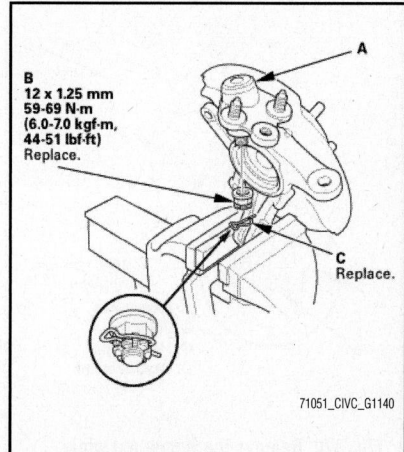

Fig. 368 Install the lower ball joint (A). Install the new castle nut (B). Install the lock pin (C) as shown

71051_CIVIC_G1140

23. Install the front wheel. Tighten the wheel nuts to 80 ft. lbs. (108 Nm).

➡**Before installing the wheel, clean the mating surfaces between the brake disc and the inside of the wheel.**

24. Install the center cap.
25. Perform a front wheel alignment check.

ADJUSTMENT

➡**The wheel bearings are not adjustable. If they are not within specification, the wheel bearings must be replaced.**

SUSPENSION

REAR SUSPENSION

COIL SPRING

REMOVAL & INSTALLATION

2012 Models

See Figures 369 through 374.

1. Raise and support the vehicle on a lift.
2. Remove the rear wheel.
3. Remove the rear wheel speed sensor and brake hose mounting bracket.
4. Position a floor jack at the connecting point of the trailing arm and the knuckle.

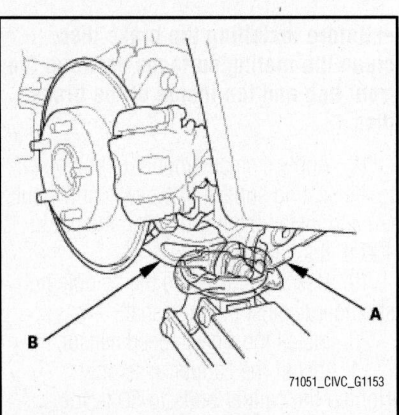

71051_CIVIC_G1153

Fig. 369 Position a floor jack at the connecting point of the trailing arm (A) and the knuckle (B)

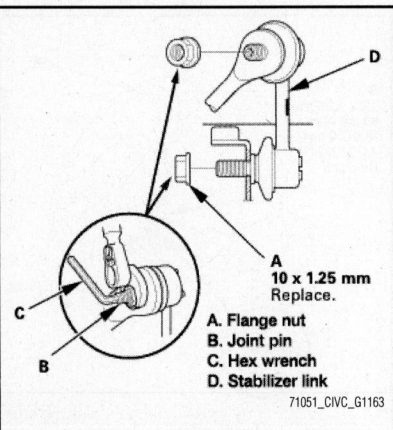

71051_CIVIC_G1163

A
10 x 1.25 mm
Replace.
A. Flange nut
B. Joint pin
C. Hex wrench
D. Stabilizer link

Fig. 370 Remove the flange nut while holding the respective joint pin with a hex wrench, then disconnect the stabilizer link from the trailing arm

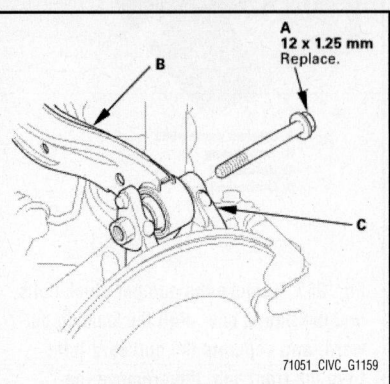

A
12 x 1.25 mm
Replace.

71051_CIVIC_G1159

Fig. 371 Remove the flange bolt (A), then disconnect the upper arm (B) from the knuckle (C)

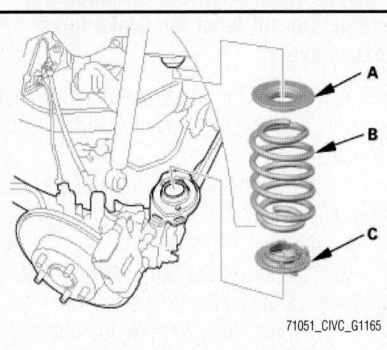

71051_CIVIC_G1165

Fig. 372 Remove the spring mounting cushion (A), remove the spring (B), then remove the lower spring seat (C)

5. Raise the floor jack until the suspension begins to compress.
6. Remove the damper (strut) lower mounting bolt.
7. Remove the flange nut while holding the respective joint pin with a hex wrench, then disconnect the stabilizer link from the trailing arm.
8. Remove the flange bolt, then disconnect the upper arm from the knuckle.
9. Remove the rear trailing arm front mounting bolts, then lower the jack slowly.
10. Remove the spring mounting cushion, remove the spring, then remove the lower spring seat.

To install:

11. Install the spring on the spring mounting cushion by aligning the upper

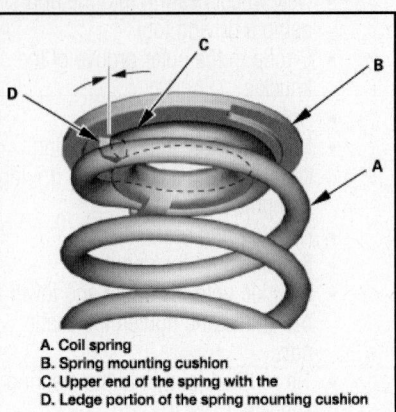

A. Coil spring
B. Spring mounting cushion
C. Upper end of the spring with the
D. Ledge portion of the spring mounting cushion

71051_CIVIC_G1166

Fig. 373 Install the spring on the spring mounting cushion by aligning the upper end of the spring with the ledge portion of the spring mounting cushion

end of the spring with the ledge portion of the spring mounting cushion.

12. Install the lower spring seat, then install the spring mounting cushion and the spring.

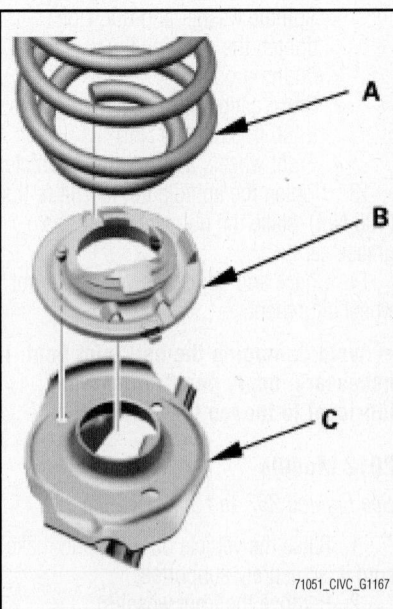

71051_CIVIC_G1167

Fig. 374 Align the bottom of the spring (A) and the lower spring seat (B) with the trailing arm (C) as shown

13. Align the bottom of the spring and the lower spring seat with the trailing arm as shown.

14. Loosely install the new trailing arm front mounting bolts.

15. Connect the upper arm to the knuckle. Loosely install the new flange bolt.

16. Position a floor jack at the connecting point of the trailing arm and the knuckle.

17. Raise the floor jack until the suspension begins to compress.

18. Loosely install the new damper lower mounting bolt.

19. Connect the stabilizer link to the trailing arm. Install the new flange nut, and tighten them to 29 ft. lbs. (39 Nm) while

holding the respective joint pin with a hex wrench.

20. Raise the rear suspension with a floor jack to load the vehicle weight.

21. Tighten all mounting hardware to the specified torque:
- Knuckle to trailing arm self-locking nut on adjusting bolt and cam plate: 51 ft. lbs. (65 Nm)
- Upper arm to knuckle flange bolt: 80 ft. lbs. (108 Nm)
- Damper lower mounting bolt: 44 ft. lbs. (59 Nm)

22. Install the rear wheel. Tighten the nuts to 80 ft. lbs. (108 Nm).

23. Perform a wheel alignment check.

KNUCKLE/HUB BEARING UNIT

REMOVAL & INSTALLATION

With Rear Disc Brakes

2011 Models

See Figures 375 and 376.

1. Raise and support the vehicle.
2. Remove the wheel nuts and the rear wheel.
3. Release the parking brake lever fully.
4. Remove the brake hose mounting bolt from the bracket.
5. Remove the brake caliper bracket

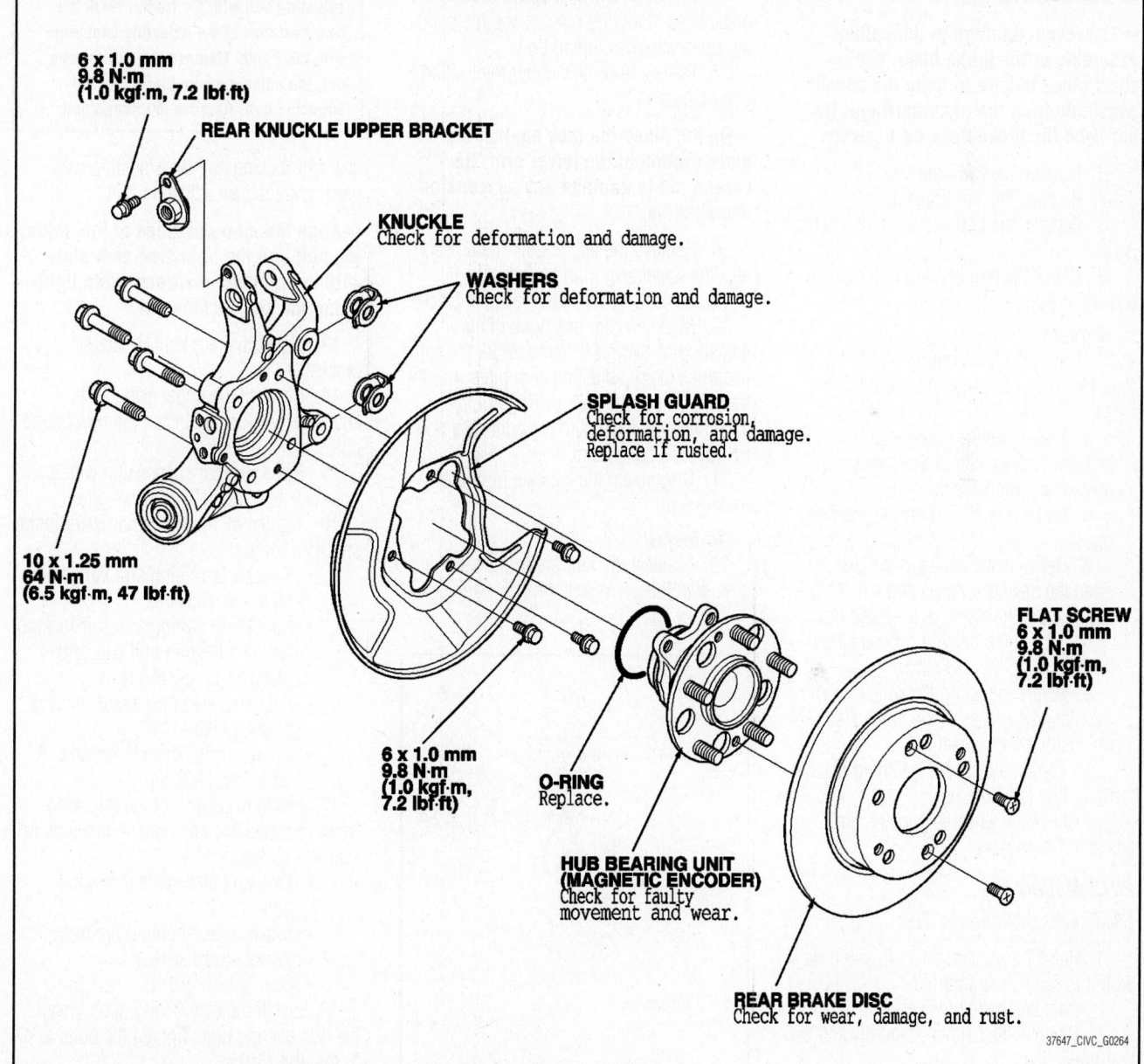

6 x 1.0 mm
9.8 N·m
(1.0 kgf·m, 7.2 lbf·ft)

REAR KNUCKLE UPPER BRACKET

KNUCKLE
Check for deformation and damage.

WASHERS
Check for deformation and damage.

SPLASH GUARD
Check for corrosion, deformation, and damage.
Replace if rusted.

10 x 1.25 mm
64 N·m
(6.5 kgf·m, 47 lbf·ft)

6 x 1.0 mm
9.8 N·m
(1.0 kgf·m, 7.2 lbf·ft)

FLAT SCREW
6 x 1.0 mm
9.8 N·m
(1.0 kgf·m, 7.2 lbf·ft)

O-RING
Replace.

HUB BEARING UNIT (MAGNETIC ENCODER)
Check for faulty movement and wear.

REAR BRAKE DISC
Check for wear, damage, and rust.

37647_CIVIC_G0264

Fig. 375 Exploded view of the knuckle and hub bearing assembly—with rear disc brakes

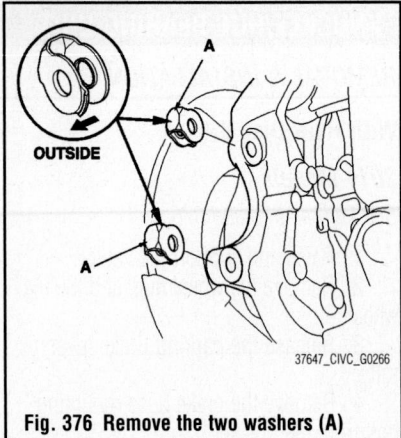

Fig. 376 Remove the two washers (A)

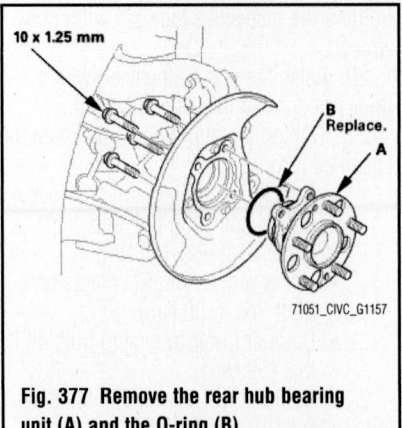

Fig. 377 Remove the rear hub bearing
unit (A) and the O-ring (B)

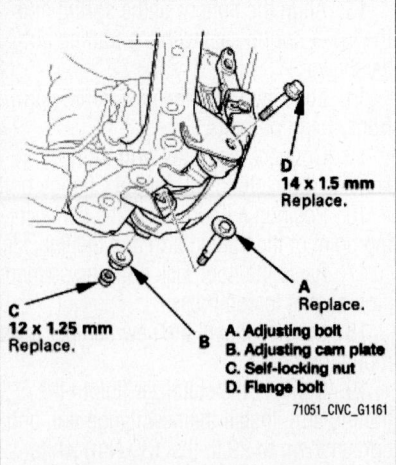

Fig. 379 Mark the cam positions of the
adjusting bolt with the frame. Mark the
cam positions of the adjusting cam plate
with the frame. Remove the self-locking
nut, the adjusting cam plate, and the
adjusting bolt. Remove the flange bolt

mounting bolts, then remove the caliper
assembly from the knuckle.

➡ **To prevent damage to the caliper
assembly or the brake hose, use a
short piece of wire to hang the caliper
assembly from the undercarriage. Do
not twist the brake hose excessively.**

6. Remove the two washers.
7. Remove the rear brake disc.
8. Remove the hub bearing unit and the
O-ring.
9. Check the hub bearing unit for damage and cracks.

To install:

10. Install the hub bearing unit in the
reverse order of removal, and note these
items:
 a. Make sure the washers are
 installed between the brake caliper
 bracket and the knuckle.
 b. Use a new O-ring during reassembly.
 c. Before installing the brake disc,
 clean the mating surfaces of the hub bearing unit and the inside of the brake disc.
 d. Torque the bearing hub retaining
 bolts to 47 ft. lbs. (64 Nm).
 e. Before installing the wheel, clean
 the mating surfaces of the brake disc and
 the inside of the wheel.
 f. Torque the caliper retaining bolts to
 54 ft. lbs. (74 Nm).
11. Check the wheel alignment, and
adjust it if necessary.

2012 Models

See Figures 377 through 379.

1. Raise the vehicle on a lift, and make
sure it is securely supported.
2. Remove the rear wheel.
3. Remove the caliper assembly and the
rear brake disc.
4. Remove the rear hub bearing unit
and the O-ring.

5. Remove the rear splash guard.
6. Remove the wheel speed sensor,
brake hose mounting bracket and the parking brake cable mounting bolt.
7. Place a floor jack under the trailing
arm to support it.

➡ **Do not place the jack against the
plate section of the lower arm. Be
careful not to damage any suspension
components.**

8. Remove the flange bolt, then disconnect the upper arm from the knuckle.
9. Remove the rear knuckle upper bracket.
10. Mark the cam positions of the
adjusting bolt with the frame. Mark the cam
positions of the adjusting cam plate with the
frame. Remove the self-locking nut, the
adjusting cam plate, and the adjusting bolt.
Remove the flange bolt.
11. Disconnect the knuckle from the
trailing arm.

To install:

12. Connect the knuckle to the trailing
arm, then loosely install the new flange bolt,

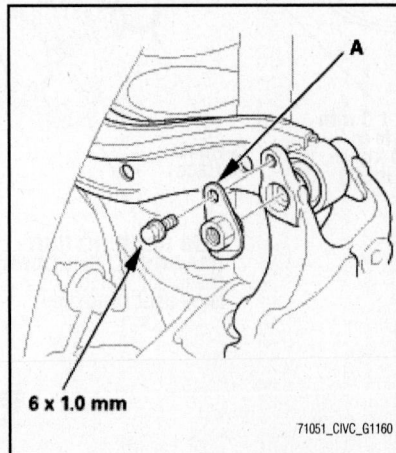

Fig. 378 Remove the rear knuckle upper
bracket (A)

new self-locking nut, the adjusting cam
plate, and the new adjusting bolt.

➡ **Align the cam positions of the adjusting bolt and the adjusting cam plate
with the marked positions when tightening the self-locking nut.**

13. Install the rear knuckle upper
bracket.
14. Connect the upper arm to the
knuckle, then loosely install the new flange
bolt.
15. Raise the rear suspension with a
floor jack to load the vehicle weight.
16. Tighten all mounting hardware to the
specified torque:
 • Knuckle to trailing arm flange bolt:
 65 ft. lbs. (88 Nm)
 • Knuckle to trailing arm self-locking
 nut on adjusting bolt and cam
 plate: 51 ft. lbs. (65 Nm)
 • Rear knuckle upper bracket nut: 7
 ft. lbs. (9 Nm).
 • Upper arm to knuckle flange bolt:
 80 ft. lbs. (108 Nm)
17. Install or reconnect the following
(refer to applicable component headings or
sections as needed):
 • Parking brake cable mounting
 bolt
 • Brake hose mounting bracket
 • Wheel speed sensor
 • Rear splash guard
18. Install the new O-ring, then install
the hub bearing unit. Tighten the bolts to 47
ft. lbs. (64 Nm).
19. Install the brake disc and caliper
assembly.

20. Install the rear wheel. Tighten the nuts to 80 ft. lbs. (108 Nm).

21. Perform rear wheel alignment.

With Rear Drum Brakes

2011 Models

See Figure 380.

1. Raise and support the vehicle.

2. Remove the wheel nuts and the rear wheel.

3. Release the parking brake lever fully.

4. Remove the hub bearing unit and the O-ring.

5. Check the hub bearing unit for damage and cracks.

To install:

6. Install the hub bearing unit in the reverse order of removal, and note these items:

   a. Use a new O-ring during reassembly.

   b. Before installing the brake drum, clean the mating surfaces of the hub bearing unit and the inside of the brake drum.

   c. Torque the bearing hub retaining bolts to 47 ft. lbs. (64 Nm).

   d. Before installing the wheel, clean the mating surfaces of the brake drum and the inside of the wheel.

7. Check the wheel alignment, and adjust it if necessary.

2012 Models

See Figure 381.

1. Raise the vehicle on a lift, and make sure it is securely supported.

2. Remove the rear wheel.

3. Remove the following components (refer to applicable component headings or sections as needed):

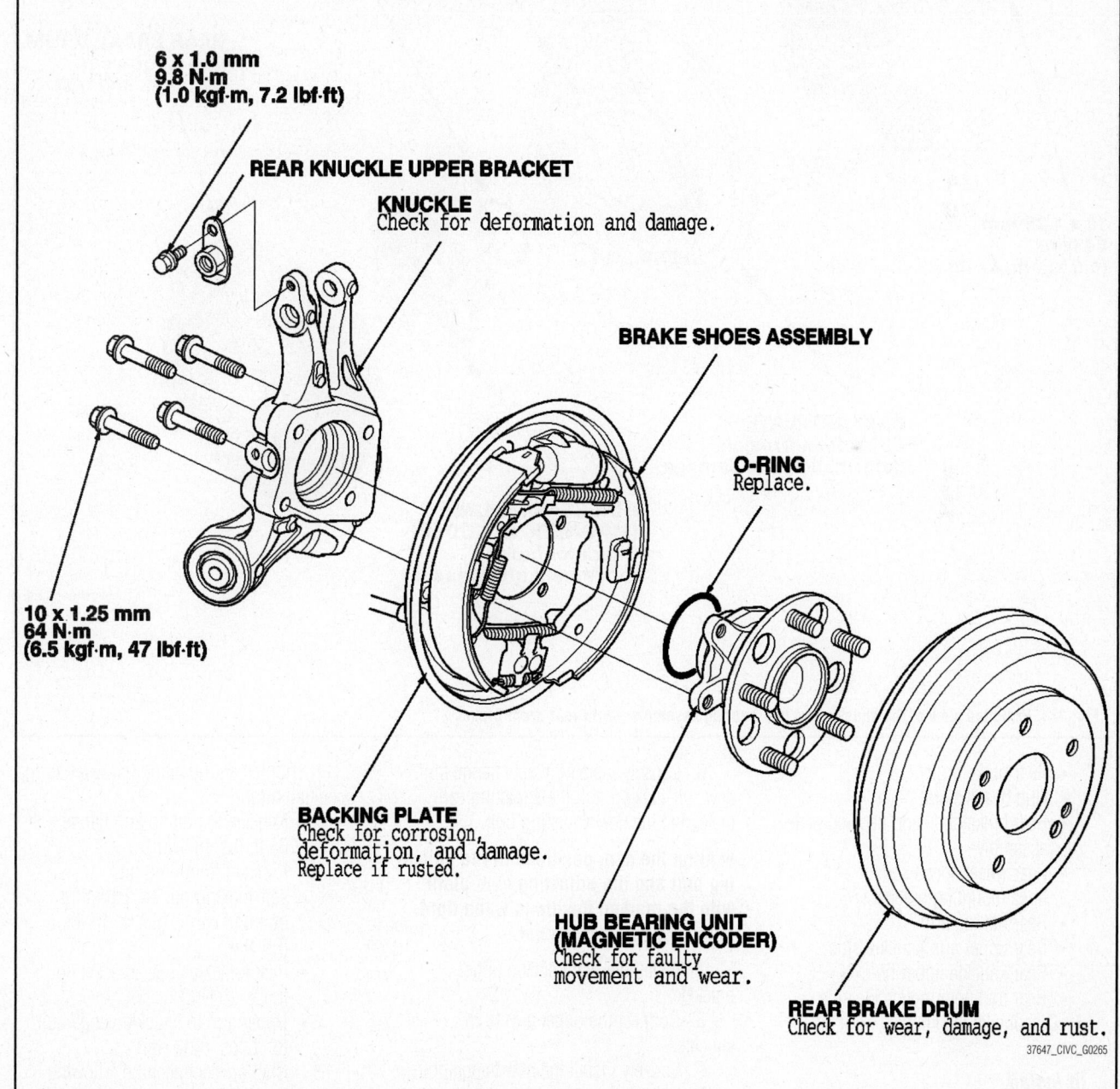

6 x 1.0 mm
9.8 N·m
(1.0 kgf·m, 7.2 lbf·ft)

REAR KNUCKLE UPPER BRACKET

KNUCKLE
Check for deformation and damage.

BRAKE SHOES ASSEMBLY

O-RING
Replace.

10 x 1.25 mm
64 N·m
(6.5 kgf·m, 47 lbf·ft)

BACKING PLATE
Check for corrosion, deformation, and damage. Replace if rusted.

HUB BEARING UNIT (MAGNETIC ENCODER)
Check for faulty movement and wear.

REAR BRAKE DRUM
Check for wear, damage, and rust.

37647_CIVC_G0265

Fig. 380 Exploded view of the knuckle and hub bearing assembly—with rear drum brakes

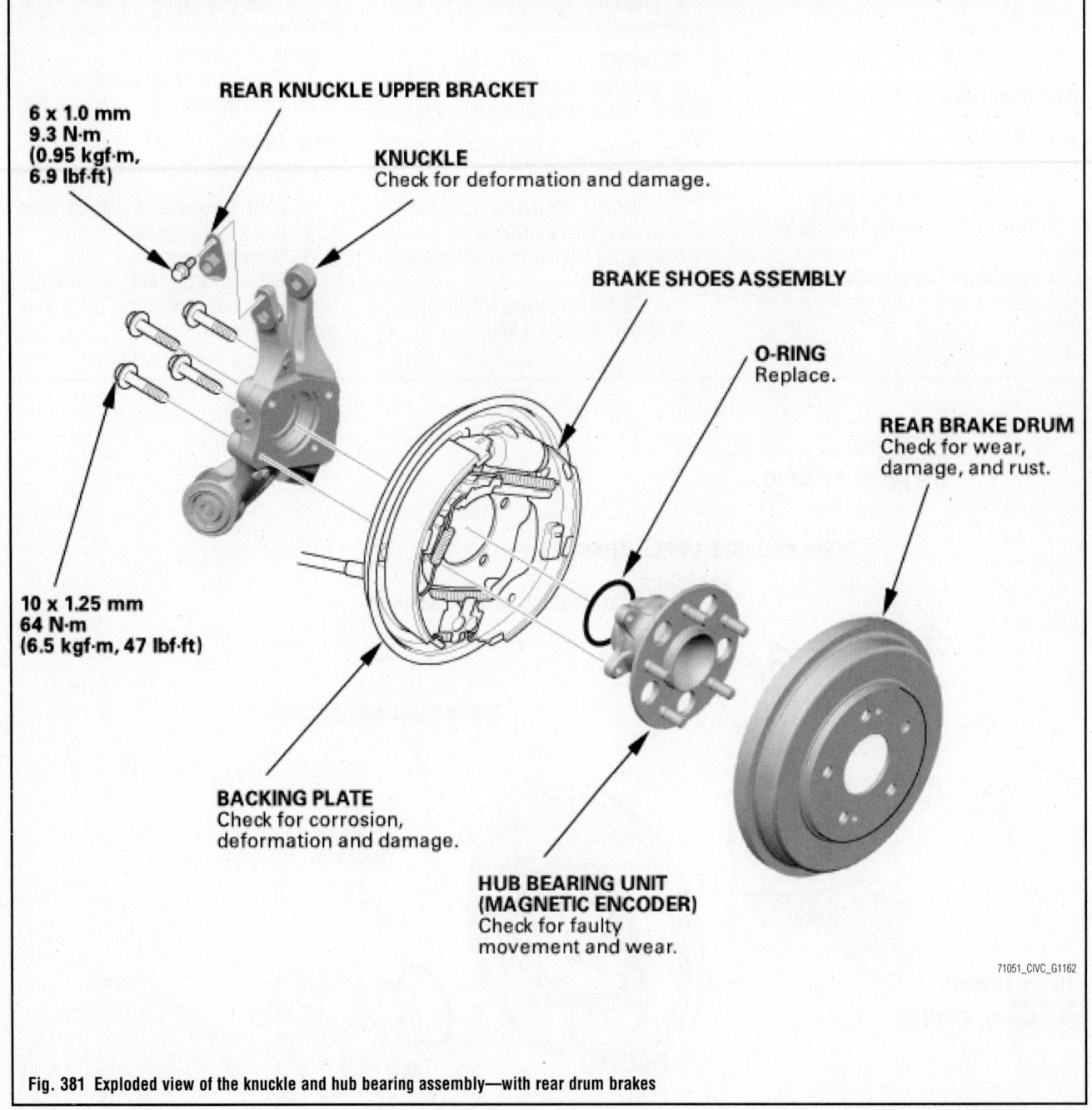

6 x 1.0 mm
9.3 N·m
(0.95 kgf·m,
6.9 lbf·ft)

REAR KNUCKLE UPPER BRACKET

KNUCKLE
Check for deformation and damage.

BRAKE SHOES ASSEMBLY

O-RING
Replace.

REAR BRAKE DRUM
Check for wear,
damage, and rust.

10 x 1.25 mm
64 N·m
(6.5 kgf·m, 47 lbf·ft)

BACKING PLATE
Check for corrosion,
deformation and damage.

HUB BEARING UNIT
(MAGNETIC ENCODER)
Check for faulty
movement and wear.

71051_CIVC_G1162

Fig. 381 Exploded view of the knuckle and hub bearing assembly—with rear drum brakes

- Rear brake drum
- Hub bearing unit
- Rear brake shoe and backing plate assembly
- Rear wheel speed sensor and brake hose mounting bracket
- Rear trailing arm
- Rear upper arm knuckle side
- Rear knuckle upper bracket
- Rear trailing arm knuckle side

4. Disconnect the knuckle from the trailing arm.

To install:

5. Connect the knuckle to the trailing arm.

6. Loosely install the new flange bolt, new self-locking nut, the adjusting cam plate, and the new adjusting bolt.

➡**Align the cam positions of the adjusting bolt and the adjusting cam plate with the marked positions when tightening the self-locking nut.**

7. Install the rear knuckle upper bracket.

8. Connect the upper arm to the knuckle.

9. Loosely install the new flange bolt.

10. Raise the rear suspension with a floor jack to load the vehicle weight.

11. Tighten all mounting hardware to the specified torque:
- Knuckle to trailing arm flange bolt: 65 ft. lbs. (88 Nm)
- Knuckle to trailing arm self-locking nut on adjusting bolt and cam plate: 51 ft. lbs. (65 Nm)
- Rear knuckle upper bracket nut: 7 ft. lbs. (9 Nm).
- Upper arm to knuckle flange bolt: 80 ft. lbs. (108 Nm)

12. Install or reconnect the following (refer to applicable component headings or sections as needed):

- Rear wheel speed sensor and brake hose mounting bracket
- Rear brake shoe and backing plate assembly

13. Connect the brake line to the wheel cylinder.

14. Install the hub bearing unit, as described in this section.

15. Install the brake drum.

16. After installation, press the brake pedal several times to make sure the brakes work and self-adjust the brake shoes. Do not drive the vehicle before doing this procedure.

17. Bleed the brake system.

18. Install the rear wheel. Tighten the nuts to 80 ft. lbs. (108 Nm).

19. Check the wheel alignment, and adjust it if necessary.

STABILIZER BAR

REMOVAL & INSTALLATION

2011 Models

See Figure 382.

1. Raise and support the vehicle.
2. Remove the rear wheels.
3. Disconnect both stabilizer links from the stabilizer bar.
4. Remove the flange bolts and the bushing holders, then remove the bushings and the stabilizer bar.

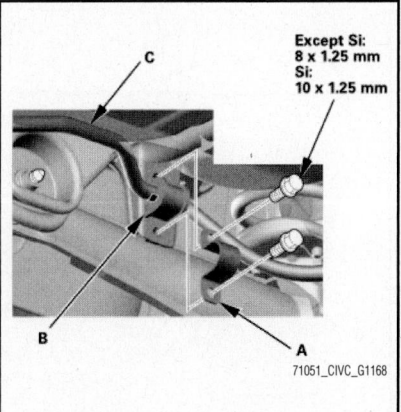

Fig. 383 Remove the bushing holder (A), the bushing (B), and the stabilizer bar (C)

To install:

➡**During installation, align the paint marks on the stabilizer bar with the sides of the bushings.**

6. Replace the stabilizer bar bracket, if necessary. Tighten the bracket bolts to 29 ft. lbs. (39 Nm).

7. Install the stabilizer bar in the reverse order of removal.

8. Tighten the stabilizer bar bolts as follows:

- Except Si: 16 ft. lbs. (22 Nm)
- Si: 36 ft. lbs. (49 Nm)

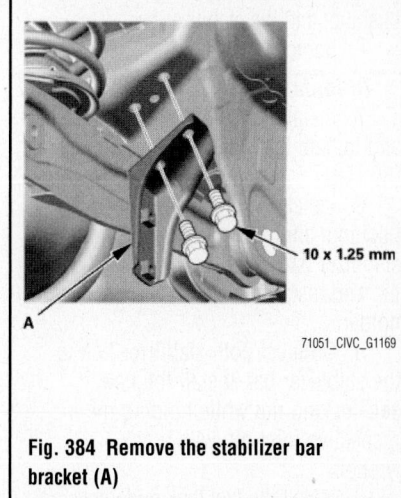

Fig. 384 Remove the stabilizer bar bracket (A)

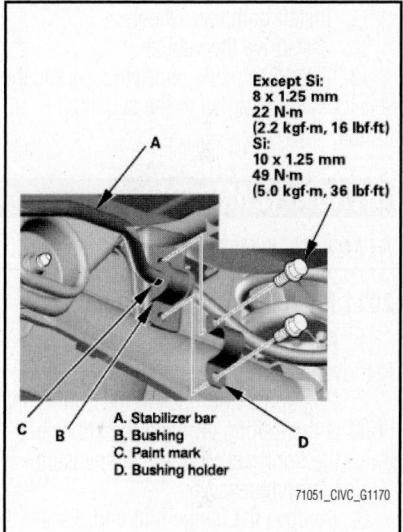

Fig. 385 Install the stabilizer bar, the bushing, align the paint mark on the stabilizer bar with the side of the bushing and install the bushing holder

➡**Note the right and left direction of the stabilizer bar.**

9. Before installing the wheel, clean the mating surfaces on the brake disc or the brake drum and inside of the wheel.

2012 Models

1. Raise the vehicle on a lift, and make sure it is securely supported.

2. Remove both rear wheels.

3. Remove the fuel tank undercover (if necessary).

4. Remove the self-locking nut while holding the respective joint pin with a hex wrench. Disconnect stabilizer link from the stabilizer bar.

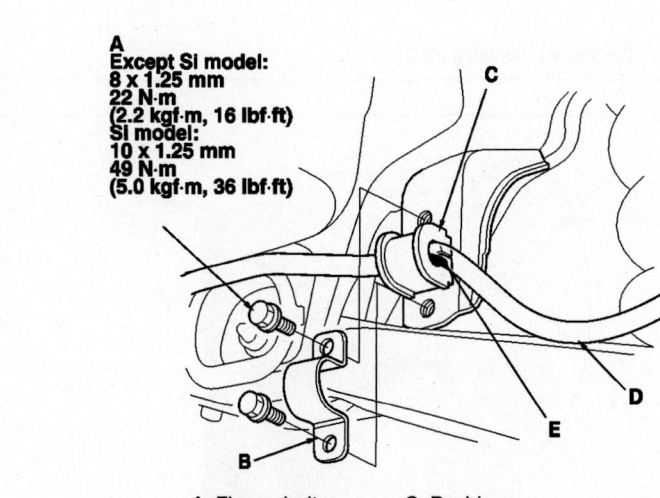

Fig. 382 Remove the flange bolts (A) and the bushing holders (B), then remove the bushings (C) and the stabilizer bar (D)

5. Remove the bushing holder, the bushing, and the stabilizer bar.

6. Remove the stabilizer bar bracket.

To install:

7. Install the stabilizer bar bracket and tighten the bolts to 29 ft. lbs. (39 Nm).

8. Install the stabilizer bar, the bushing, align the paint mark on the stabilizer bar with the side of the bushing and install the bushing holder.

9. Connect both stabilizer link to the stabilizer bar. Install the new self-locking nut while holding the respective joint pin with a hex wrench.

10. Install the fuel tank undercover (if removed).

11. Install both rear wheels.

12. Test-drive the vehicle.

13. After 5 minutes of driving, torque the self-locking nut again to the specified torque.

STRUTS (DAMPERS)

REMOVAL & INSTALLATION

2011 Models

1. Raise and support the vehicle safely. Remove the rear tires.

2. Position a floor jack at the connecting point of the trailing arm and the knuckle. Raise the floor jack until the suspension begins to compress.

3. Remove the flange bolt and discard it.

4. Remove the trunk side trim panel.

5. Remove the self-locking nut while holding the damper shaft. Compress the damper unit, by hand, and remove it from the vehicle.

To install:

6. Position the damper assembly in the vehicle.

7. Position a floor jack under the trailing to support the suspension. Install a new damper mounting bolt.

8. Loosely tighten the damper mounting bolt. Raise the floor jack until the suspension begins to compress. Tighten the damper mounting bolt.

9. Continue the installation in the reverse order of the removal procedure.

2012 Models

See Figures 386 through 388.

1. Raise the vehicle on a lift, and make sure it is securely supported.

2. Remove the rear wheel.

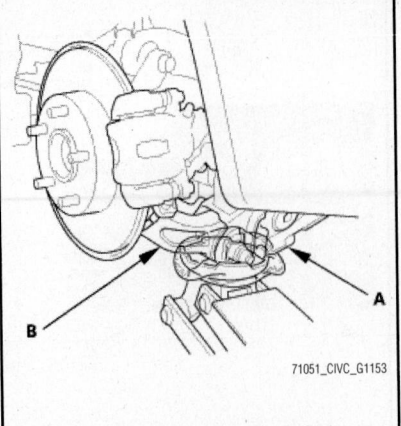

71051_CIVIC_G1153

Fig. 386 Position a floor jack at the connecting point of the trailing arm (A) and the knuckle (B)

3. Position a floor jack at the connecting point of the trailing arm and the knuckle.

4. Raise the floor jack until the suspension begins to compress.

5. Remove the damper mounting bolt.

6. Remove the following components from the rear trunk area (refer to applicable component headings or sections as needed):
- Trunk floor cover
- Trunk lid weatherstrip
- Trunk rear trim panel
- Trunk side trim panel

7. Remove the self-locking nut while holding the damper shaft with a hex wrench. Remove the damper mounting washer. Remove the damper mounting bushing.

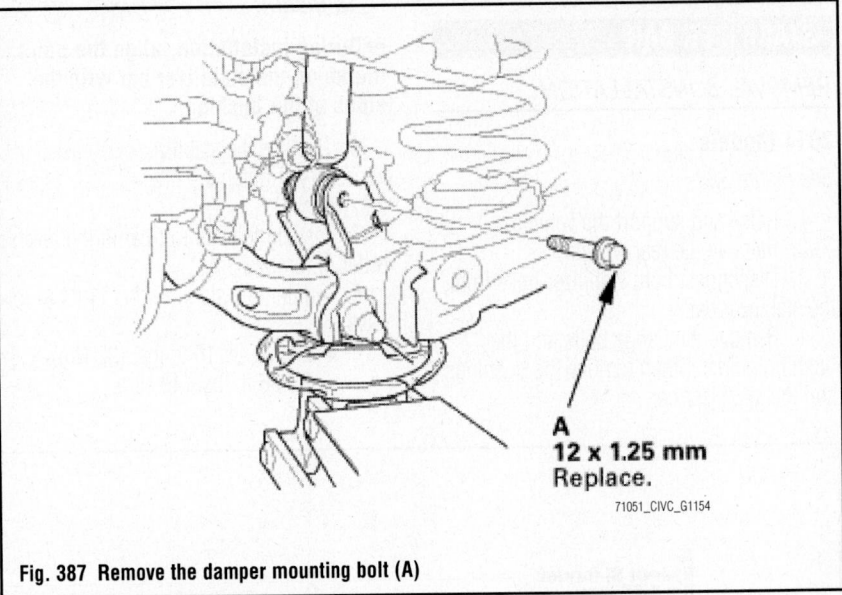

A
12 x 1.25 mm
Replace.

71051_CIVIC_G1154

Fig. 387 Remove the damper mounting bolt (A)

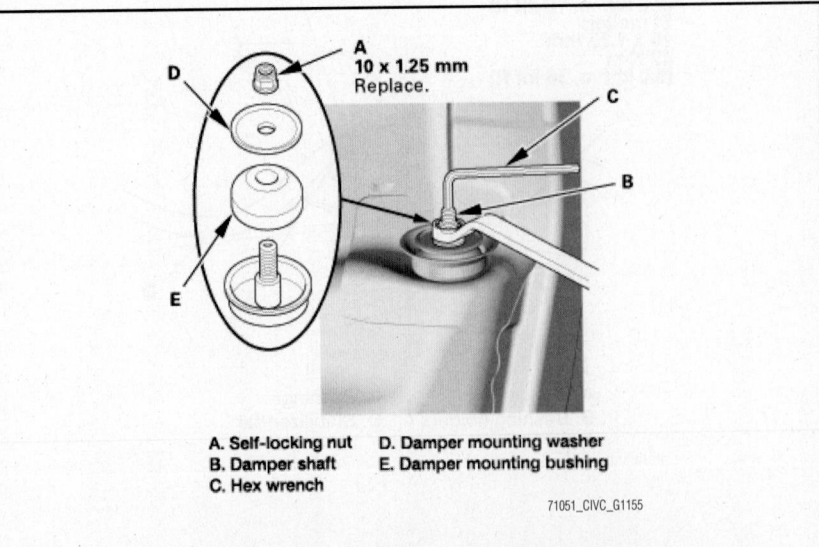

A
10 x 1.25 mm
Replace.

A. Self-locking nut   D. Damper mounting washer
B. Damper shaft   E. Damper mounting bushing
C. Hex wrench

71051_CIVIC_G1155

Fig. 388 Remove the self-locking nut while holding the damper shaft with a hex wrench. Remove the damper mounting washer. Remove the damper mounting bushing

8. Remove the damper assembly, then remove the damper mounting bushing.

To install:

9. Install the damper mounting bushing (new, if needed) and position the damper assembly.

10. Position a floor jack at the connecting point of the trailing arm and the knuckle.

11. Raise the floor jack until the suspension begins to compress.

12. Loosely install the new damper lower mounting bolt.

13. Raise the rear suspension with the jack until the vehicle just lifts off of the safety stands, then tighten the damper mounting bolt to 44 ft. lbs. (59 Nm).

14. Install the damper mounting bushing, mounting washer and loosely install the new self-locking nut.

15. Tighten the self-locking nut to 22 ft. lbs. (30 Nm) while holding the damper shaft with a hex wrench.

16. Install the trunk trim components, as removed.

17. Install the rear wheel. Tighten the nuts to 80 ft. lbs. (108 Nm).

18. Perform rear wheel alignment.

UPPER CONTROL ARM

REMOVAL & INSTALLATION

2011 Models

1. Raise and support the vehicle safely.
2. Remove the rear tires.
3. Place a jack under the trailing arm and the knuckle.
4. Remove the upper arm mounting bolts and the flange bolt. Remove the upper arm from the vehicle.

To install:

5. Position the upper arm on the vehicle. Be sure to use new retaining bolts and nuts.

6. Install all the suspension components and lightly tighten the bolts and nuts. Position a jack under the trailing arm and raise the suspension to load it with the vehicles weight, before fully tightening the bolts and nuts.

7. Continue the installation in the reverse order of the removal procedure.

8. Check and adjust the wheel alignment, as required.

2012 Models

See Figures 389 and 390.

1. Raise the vehicle on a lift, and make sure it is securely supported.

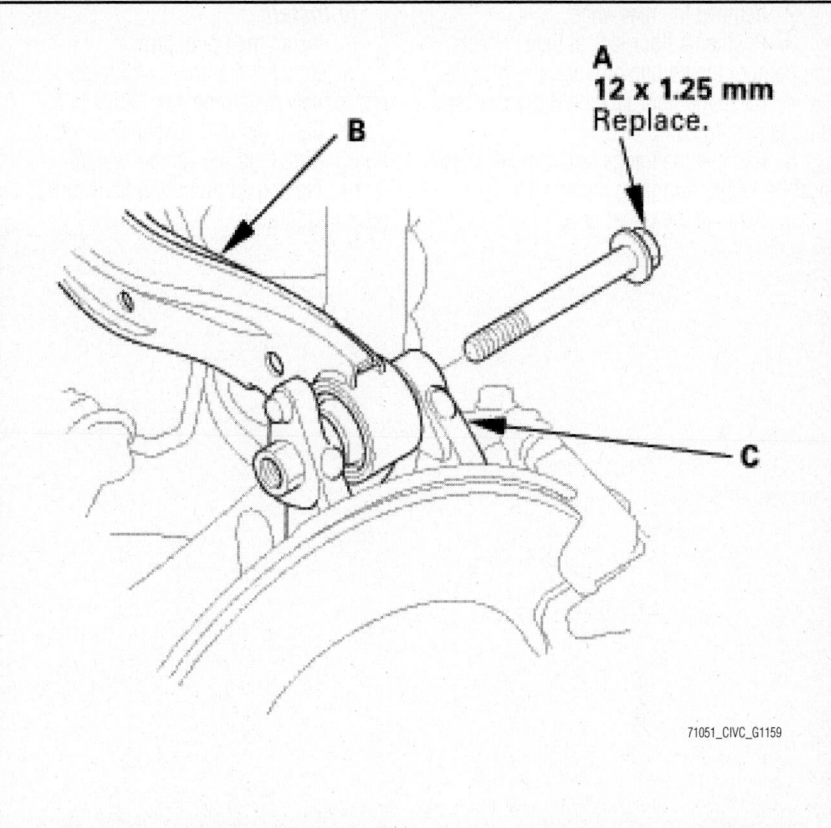

Fig. 389 Remove the flange bolt (A), then disconnect the upper arm (B) from the knuckle (C)

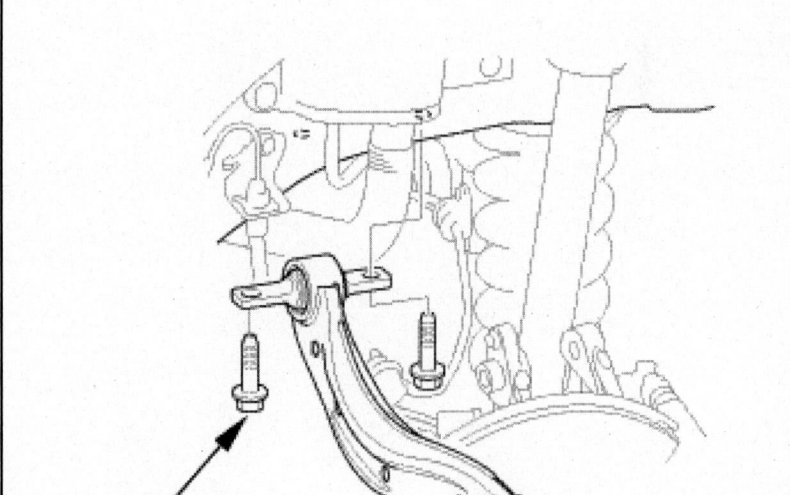

Fig. 390 Remove the upper arm (A)

2. Remove the rear wheel.

3. Position a floor jack at the connecting point of the trailing arm and the knuckle.

4. Raise the floor jack until the suspension begins to compress.

5. Remove the flange bolt, then disconnect the upper arm from the knuckle.

6. Remove the upper arm.

To install:

7. Install the upper arm.

8. Connect the upper arm to the knuckle and loosely install the new flange bolt.

9. Raise the rear suspension with a floor jack to load the vehicle weight.

10. Tighten all mounting hardware to the specified torque:

- Front upper arm mounting bolts: 44 ft. lbs. (59 Nm)
- Upper arm to knuckle mounting bolt: 80 ft. lbs. (108 Nm)

11. Install the rear wheel. Tighten the wheel nuts to 80 ft. lbs. (108 Nm).

12. Perform a wheel alignment check.

HONDA

CR-V

BRAKES 9-11

ANTI-LOCK BRAKE SYSTEM (ABS) 9-12
Wheel Speed Sensors 9-12
Removal & Installation 9-12
BLEEDING THE BRAKE SYSTEM 9-11
Bleeding Procedure 9-11
Bleeding Procedure 9-11
Bleeding the ABS System 9-11
FRONT DISC BRAKES 9-13
Brake Caliper 9-13
Removal & Installation 9-13
Disc Brake Pads 9-15
Removal & Installation 9-15
INFORMATION AND PRECAUTIONS 9-11
Anti-lock Systems 9-11
Disc and Drum Systems 9-11
PARKING BRAKE 9-20
Parking Brake Cables 9-20
Adjustment 9-20
Parking Brake Shoes 9-21
Removal & Installation 9-21
REAR DISC BRAKES 9-17
Brake Caliper 9-17
Removal & Installation 9-17
Disc Brake Pads 9-18
Removal & Installation 9-18

CHASSIS ELECTRICAL 9-24

AIR BAG (SUPPLEMENTAL RESTRAINT SYSTEM) 9-24
General Information 9-24
Arming the System 9-25
Disarming the System 9-25
Precautions 9-24

DRIVE TRAIN 9-25

Automatic Transaxle 9-25
Drain & Refill 9-25
Fluid Level Check 9-25
Fluid Recommendations 9-25

Front Driveshaft 9-25
Removal & Installation 9-25
Intermediate Shaft 9-28
Removal & Installation 9-28
Propeller Shaft 9-28
Removal & Installation 9-28
Rear Differential Carrier 9-30
Removal & Installation 9-30
Rear Driveshaft 9-30
Removal & Installation 9-30
Transfer Case Assembly 9-34
Removal & Installation 9-34

ENGINE COOLING 9-35

Engine Coolant 9-35
Drain & Refill Procedure 9-35
Fluid Recommendations 9-36
Level Check 9-36
Engine Fan 9-36
Removal & Installation 9-36
Radiator 9-36
Removal & Installation 9-36
Thermostat 9-38
Removal & Installation 9-38
Water Pump 9-38
Removal & Installation 9-38

ENGINE ELECTRICAL 9-39

BATTERY SYSTEM 9-39
Battery 9-39
Battery Reconnect/reconnection Procedure 9-39
Removal & Installation 9-39
CHARGING SYSTEM 9-40
Alternator 9-40
Removal & Installation 9-40
IGNITION SYSTEM 9-40
Firing Order 9-40
Ignition Coil 9-40
Removal & Installation 9-40
Ignition Timing 9-40
Inspection & Adjustment 9-40
Spark Plugs 9-40
Removal & Installation 9-40
STARTING SYSTEM 9-41
Starter 9-41
Removal & Installation 9-41

ENGINE MECHANICAL 9-41

Accessory Drive Belts 9-41
Accessory Belt Routing 9-41
Adjustment 9-41
Inspection 9-41
Removal & Installation 9-41
Air Cleaner 9-42
Removal & Installation 9-42
Cam Chain & Sprockets 9-45
Removal & Installation 9-45
Cam Chain Case Seal 9-50
Removal & Installation 9-50
Cam Chain Front Cover 9-44
Removal & Installation 9-44
Camshaft and Valve Lifters 9-42
Removal & Installation 9-42
Catalytic Converter 9-48
Removal & Installation 9-48
Crankshaft Pulley 9-50
Removal & Installation 9-50
Cylinder Head 9-51
Removal & Installation 9-51
Cylinder Head (Valve) Cover 9-52
Removal & Installation 9-52
Engine Oil & Filter 9-54
OIL Filter 9-54
OIL Level Check 9-54
Intake Manifold 9-54
Removal & Installation 9-54
Oil Pan 9-57
Removal & Installation 9-57
Oil Pump 9-58
Removal & Installation 9-58
Piston and Ring 9-60
Positioning 9-60
Rocker Arms 9-60
Removal & Installation 9-60
Valve Lash 9-61
Adjustment 9-61

ENGINE PERFORMANCE & EMISSION CONTROLS 9-61

Accelerator Pedal Position (APP) Sensor 9-61
Location 9-61
Removal & Installation 9-61

Camshaft Position (CMP)
  Sensor 9-61
    Location 9-61
    Removal & Installation..........9-61
Crankshaft Position (CKP)
  Sensor 9-62
    CKP Pattern Clear/CKP
    Pattern Learn 9-62
    Location 9-62
    Removal & Installation..........9-62
Engine Coolant Temperature
  (ECT) Sensor 9-62
    Location 9-62
    Removal & Installation..........9-62
Evaporative Emissions (EVAP)
  Canister 9-63
    Location 9-63
    Removal & Installation..........9-63
Exhaust Gas Recirculation
  (EGR) Valve 9-64
    Location 9-64
    Removal & Installation..........9-64
Heated Oxygen Sensor
  (HO2S) 9-64
    Location 9-64
    Removal & Installation..........9-64
Intake Air Temperature (IAT)
  Sensor 9-65
    Location 9-65
    Removal & Installation..........9-65
Knock Sensor (KS)...................9-65
    Location 9-65
    Removal & Installation..........9-65
Maintenance Minder 9-65
    Operation 9-65
    Reset Procedure.................9-65
Malfunction Indicator Light
  (MIL)...............................9-65
    Reset Procedure.................9-65
Manifold Absolute Pressure
  (MAP) Sensor 9-65
    Location 9-65
    Removal & Installation..........9-66
Mass Air Flow (MAF) Sensor....9-66
    Location 9-66
    Removal & Installation..........9-66
Positive Crankcase Ventilation
  (PCV) Valve 9-66
    Location 9-66
    Removal & Installation..........9-66
Powertrain Control Module
  (PCM) 9-66
    Location 9-66
    Removal & Installation..........9-66

Throttle Position Sensor
  (TPS) 9-67
    Removal & Installation..........9-67

FUEL.............................9-68

**GASOLINE FUEL INJECTION
SYSTEM........................9-68**
    Fuel Filter.........................9-69
      Removal & Installation..........9-69
    Fuel Level Sending Unit............9-69
      Location 9-69
      Removal & Installation..........9-69
    Fuel Rail and Injector 9-69
      Removal & Installation..........9-69
    Fuel System Pressure..............9-68
      Relieving 9-68
    Fuel System Service
      Precautions 9-68
    Fuel Tank.........................9-70
      Draining 9-70
      Removal & Installation..........9-70
    Fuel Tank Unit 9-71
      Removal & Installation..........9-71
    Idle Speed 9-71
      Adjustment 9-71
    Throttle Body 9-71
      Removal & Installation..........9-71

**HEATING & AIR CONDITIONING
SYSTEM........................9-72**

    Blower Motor 9-72
      Removal & Installation..........9-72
    Heater Core 9-73
      Removal & Installation..........9-73

**SPECIFICATIONS AND
MAINTENANCE CHARTS 9-3**

    Brake Specifications...............9-8
    Camshaft Specifications............9-5
    Capacities 9-4
    Crankshaft and Connecting
      Rod Specifications 9-5
    Engine and Vehicle
      Identification 9-3
    Engine Tune-Up Specifications.....9-3
    Fluid Specifications...............9-4
    General Engine Specifications.....9-3
    Scheduled Maintenance Minder
      Intervals 9-9
    Piston and Ring Specifications.....9-6
    Tire Wheel and Ball Joint
      Specifications 9-8

Torque Specifications.................9-6
Valve Specifications 9-5
Wheel Alignment......................9-7

STEERING 9-74

    EPS Control Unit...................9-74
      Removal & Installation..........9-74
    EPS Motor.........................9-74
      Removal & Installation..........9-74
    Power Rack & Pinion Steering
      Gear 9-74
      Removal & Installation..........9-74
    Power Steering Pump..............9-77
      Bleeding 9-77
      Removal & Installation..........9-77

SUSPENSION...................9-78

FRONT SUSPENSION 9-78
    Damper & Spring 9-78
      Removal & Installation..........9-78
    Lower Ball Joint 9-79
      Removal & Installation..........9-79
    Lower Control Arm 9-80
      Removal & Installation..........9-80
    Stabilizer Bar......................9-82
      Removal & Installation..........9-82
    Stabilizer Link 9-84
      Removal & Installation..........9-84
    Steering Knuckle & Wheel Hub
      Assembly 9-84
      Removal & Installation..........9-84
    Wheel Bearing 9-85
      Adjustment 9-87
      Removal & Installation..........9-85
REAR SUSPENSION 9-87
    Damper & Spring 9-87
      Removal & Installation..........9-87
    Hub Bearing Unit..................9-89
      Adjustment 9-89
      Removal & Installation..........9-89
    Knuckle 9-89
      Removal & Installation..........9-89
    Stabilizer Bar......................9-91
      Removal & Installation..........9-91
    Stabilizer Link 9-92
      Removal & Installation..........9-92
    Trailing Arm 9-92
      Removal & Installation..........9-92
    Upper Arm........................9-94
      Removal & Installation..........9-94
    Wheel Bearings 9-94
      Removal & Installation..........9-94

SPECIFICATIONS AND MAINTENANCE CHARTS

ENGINE AND VEHICLE IDENTIFICATION CHART

		Engine Code					Model Year	
Code	Liters (cc)	Cu. In.	Cyl.	Fuel Sys.	Engine Type	Eng. Mfg.	Code ①	Year
K24Z6	2.4 (2354)	144	4	MPFI	DOHC	Honda	B	2011
K24Z7	2.4 (2354)	143.6	4	SMFI	DOHC	Honda	C	2012

DOHC: Dual Overhead Cam

SMFI: Sequential Multi-port Fuel Injection

① 10th position of VIN

71051_HCRV_C0001

GENERAL ENGINE SPECIFICATIONS

Year	Model	Engine Displacement Liters (ID)	Net Horsepower @ rpm	Net Torque @ rpm (ft. lbs.)	Bore x Stroke (in.)	Com-pression Ratio	Oil Pressure @ rpm
2011	CR-V	2.4 (K24Z6)	166@5800	161@4200	3.43 x 3.90	9.7:1	44@3000
2012	CR-V	2.4 (K24Z7)	185@7000	163@4100	3.43x3.90	11.0:1	43.5@3000

71051_HCRV_C0002

ENGINE TUNE-UP SPECIFICATIONS

Year	Engine Displacement Liters (ID)	Spark Plug Gap (in.)	Ignition Timing (deg.) MT	Ignition Timing (deg.) AT	Fuel Pump (psi)	Idle Speed (rpm) MT	Idle Speed (rpm) AT	Valve Clearance (in.) In.	Valve Clearance (in.) Ex.
2011	2.4 (K24Z6)	0.039-0.043	—	6-10B	47-54	—	600-700	0.008-0.010	0.010-0.011
2012	2.4 (K24Z7)	0.039-0.043	—	6-10B	47-54	—	710-810	0.008-0.010	0.010-0.011

NOTE: The Vehicle Emission Control Information label often reflects changes made during production and must be used if they differ from this chart.

NOTE: The fuel pressure readings are given with the vacuum hose disconnected

B: Before top dead center

71051_HCRV_C0003

CAPACITIES

Year	Model	Engine Displacement Liters (ID)	Engine Oil with Filter (qts.)	Transmission (qts.) 5-Spd	Transmission (qts.) Auto.	Transfer Case (pts.)	Drive Axle Front (pts.)	Drive Axle Rear (pts.)	Fuel Tank (gal.)	Cooling System (qts.)
2011	CR-V	2.4 (K24Z6)	4.4	—	2.6 ①	②	③	2.6	15.3	6.2
2012	CR-V	2.4 (K24Z7)	④	—	⑤	②	③	⑥	⑦	⑧

NOTE: All capacities are approximate. Add fluid gradually and check to be sure a proper fluid level is obtained.

① Drain and refill

② Transfer assembly draws from same fluid source as transmission.

③ Front driveshaft assembly draws from same fluid source as transmission.

④ At oil change w/filter: 4.2 qts.

   At overhaul: 5.3 qts.

⑤ 2WD Model:

   Fluid change: 2.3 qts.

   Overhaul: 6.18 qts.

   AWD Model:

   Fluid change: 2.7 qts.

   Overhaul: 6.8 qts.

⑥ AWD Model (differential capacity):

   Fluid change: 2.6 pts.

   Overhaul: 3.2 pts.

⑦ Consult Owner's Manual for specified fuel tank capacity.

⑧ Fluid change: 6.58 qts.

   Overhaul: 7.86 qts.

71051_HCRV_C0004

FLUID SPECIFICATIONS

Year	Model	Engine Displacement Liters (ID)	Engine Oil	Auto. Trans.	Drive Axle	Power Steering Fluid	Brake Master Cylinder	Cooling System
2011	CR-V	2.4 (K24Z6)	①	②	③	④	⑤	⑥
2012	CR-V	2.4 (K24Z7)	API 0W-20	ATF DW-1	⑦	N/A	DOT 3	⑥

Note: If specification disagrees with specification in owners manual, use specification in owners manaual

N/A: Not applicable

DOT: Department Of Transportation

① Honda Motor Oil: American Honda P/N 08798-9023 (5W-20), Honda Canada P/N CA66806 (5W-20)

② Honda Automatic Transmission Fluid (ATF-Z1): American Honda P/N 08200-9001, Honda Canada P/N CA66689

③ Rear differential (4WD) Honda Dual Pump Fluid II: P/N 08200-9007

④ Honda Power Steering Fluid: P/N 08206-9002

⑤ Honda DOT 3 Brake Fluid: P/N 08798-9008

⑥ Honda Long Life Antifreeze/Coolant Type 2: P/N OL 999-9001; Honda Coolant Concentrate: P/N OL 999-9020

⑦ Front driveshaft uses same fluid source as transmission.

71051_HCRV_C0005

VALVE SPECIFICATIONS

Year	Engine Displacement Liters (ID)	Seat Angle (deg.)	Face Angle (deg.)	Spring Free Length (in.)	Stem-to-Guide Clearance (in.)		Stem Diameter (in.)	
					Intake	Exhaust	Intake	Exhaust
2011	2.4 (K24Z6)	NA	NA	①	0.0012-0.0022	0.0022-0.0031	0.2156-0.2159	0.2146-0.2150
2012	2.4 (K24Z7)	45	45	NA	0.0012-0.0022	0.0022-0.0032	0.2156-0.2159	0.2146-0.2150

NA: Not Available

① Valve spring free length:

  Intake: 1.873 in.

  Exhaust: 1.954 in.

71051_HCRV_C0006

CAMSHAFT SPECIFICATIONS
All measurements in inches unless noted

Year	Engine Displacement Liters (ID)	Journal Diameter	Brg. Oil Clearance	Shaft End-play	Circle Runout	Lobe Height	
						Intake	Exhaust
2011	2.4 (K24Z6)	NA	①	0.0020-0.0079	0.001 max.	②	1.35004
2012	2.4 (K24Z7)	NA	①	0.0020-0.0079	0.0012	②	1.35004

NA: Not Available

① Journal No. 1: 0.0012-0.0027 inches

  Journals No. 2, 3, 4, 5: 0.0024-0.0039 inches

② Intake primary: 1.3285 inches

  Intake mid: 1.3959 inches

  Intake secondary: 1.3285 inches

71051_HCRV_C0007

CRANKSHAFT AND CONNECTING ROD SPECIFICATIONS
All measurements are given in inches

Year	Engine Displacement Liters (ID)	Crankshaft				Connecting Rod		
		Main Brg. Journal Dia.	Main Brg. Oil Clearance	Shaft End-play	Thrust on No.	Journal Diameter	Oil Clearance	Side Clearance
2011	2.4 (K24Z6)	①	②	0.0040-0.0140	NA	2.0100	0.0002-0.0006	0.0060-0.0140
2012	2.4 (K24Z7)	NA	②	0.0040-0.0140	NA	NA	0.0013-0.0025	0.0060-0.0130

NA: Not Available

① Except No. 3: 2.1648-2.1657

  No. 3: 2.1644-2.1654

② Except No. 3: 0.0007-0.0016 inches

  No. 3: 0.0010-0.0019 inches

71051_HCRV_C0008

PISTON AND RING SPECIFICATIONS

All measurements are given in inches

Year	Engine Displacement Liters (ID)	Piston Clearance	Ring Gap			Ring Side Clearance		
			Top Compression	Bottom Compression	Oil Control	Top Compression	Bottom Compression	Oil Control
2011	2.4 (K24Z6)	0.0008-0.0016	0.0080-0.0140	0.0200-0.0260	0.0080-0.0280	0.0024-0.0033	0.0016-0.0026	NA
2012	2.4 (K24Z7)	0.0012	NA	NA	NA	0.0024-0.0033	0.0008-0.0018	NA

NA: not available.

71051_HCRV_C0009

TORQUE SPECIFICATIONS

All readings in ft. lbs.

Year	Engine Displacement Liters (ID)	Cylinder Head Bolts	Main Bearing Bolts	Rod Bearing Bolts	Crankshaft Pulley Bolt	Drive Plate Bolts	Manifold		Spark Plugs	Oil Pan Drain Plug
							Intake	Exhaust		
2011	2.4 (K24Z6)	①	②	③	④	⑤	⑥	⑦	13	29
2012	2.4 (K24Z7)	①	②	③	⑧	55	16	—	13	29

NOTE: Dip cylinder head bolts, main bearing bolts, and crankshaft damper bolt in clean engine oil prior to tightening.

① Step 1: 29 ft. lbs.

  Step 2: plus 90 degrees

  Step 3: plus 90 degrees

  Step 4: NEW BOLT ONLY plus 90 degrees

② Tighten bearing cap bolts in sequence:

  Step 1: 22 ft. lbs.

  Step 2: plus 48 degrees

  Tighten the 8mm bolts to 16 ft. lbs., in sequence

③ Step 1: Tighten to 30 ft. lbs.

  Step 2: plus 120 degrees

④ Step 1: Tighten to 36 ft. lbs.

  Step 2: plus 90 degrees

⑤ Tighten in sequence to 54 ft. lbs. in at least 2 steps

⑥ Tighten in sequence to 16 ft. lbs. in 3 steps

⑦ Tighten in sequence to 33 ft. lbs. in 3 steps

⑧ Used bolt: 37 ft. lbs.

  New bolt: 131 ft. lbs., loosen and retighten to 37 ft. lbs.

71051_HCRV_C0010

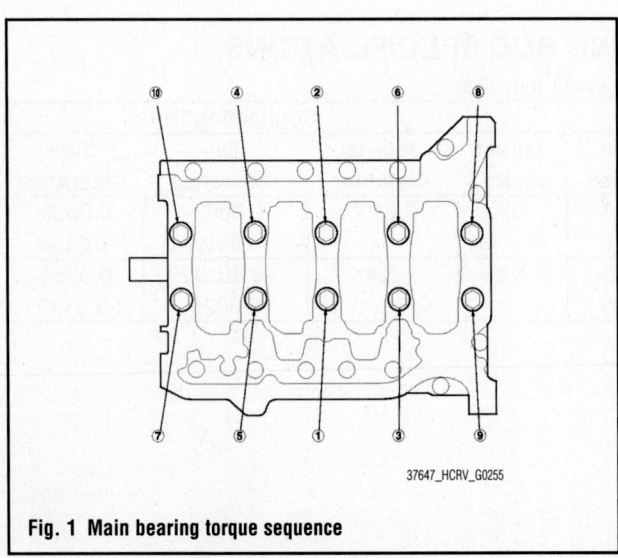

37647_HCRV_G0255

Fig. 1 Main bearing torque sequence

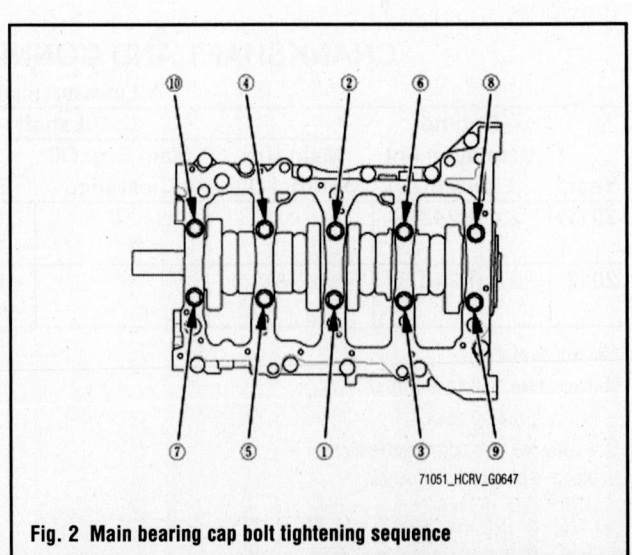

71051_HCRV_G0647

Fig. 2 Main bearing cap bolt tightening sequence

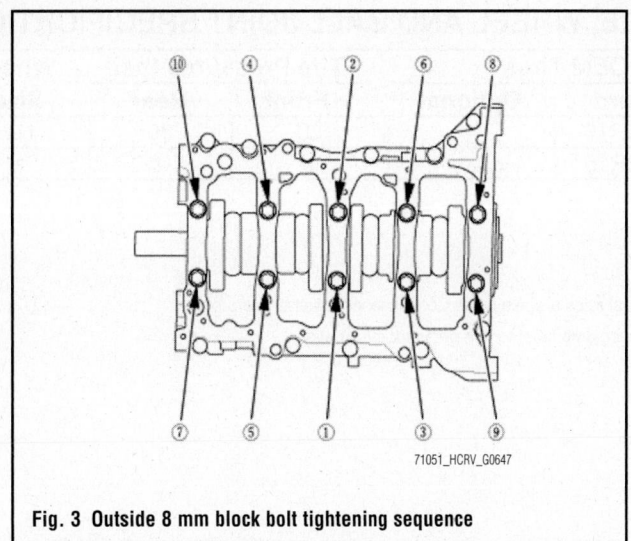

Fig. 3 Outside 8 mm block bolt tightening sequence

71051_HCRV_G0647

WHEEL ALIGNMENT

Year	Model		Caster Range (+/-Deg.)	Caster Preferred Setting (Deg.)	Camber Range (+/-Deg.)	Camber Preferred Setting (Deg.)	Toe-in (in.)
2011	CR-V	Front	1.00	+2.44	0.45	0	0+/-0.08
		Rear	—	—	0.45	-1.00	0.08+0.08/-0.04
2012	CR-V	Front	30	①	1	②	0+/-0.08
		Rear	—	—	0.45	③	0.10+/-0.08

① 2WD Model:
  With 16" wheel: 2° 44'
  Without 16" wheel: 2° 50'
  AWD Model:
  With 16" wheel: 2° 38'
  Without 16" wheel: 2° 44'

② 2WD Model:
  With 16" wheel: 0° 01'
  Without 16" wheel: 0° 10'
  AWD Model:
  With 16" wheel: 0° 08'
  Without 16" wheel: 0° 0'

③ 2WD Model:
  With 16" wheel: 1° 06'
  Without 16" wheel: 1° 12'
  AWD Model:
  With 16" wheel: 0° 50'
  Without 16" wheel: 1° 0'

71051_HCRV_C0011

TIRE, WHEEL AND BALL JOINT SPECIFICATIONS

Year	Model	OEM Tires		Tire Pressures (psi)		Wheel Size	Ball Joint Inspection	Lug Nut (ft. lbs.)
		Standard	Optional	Front	Rear			
2011	CR-V	P225/65R17	—	30	30	NA	NA	80
2012	CR-V	P215/70R16	①	②	②	①	NA	80

OEM: Original Equipment Manufacturer

PSI: Pounds Per Square Inch

NA: Not Available

① A wide variety of optional tire and wheel sizes are available. Consult local dealer or tire provider.

② Consult the Owner's Manual or the door jamb label for tire pressure information..

71051_HCRV_C0012

BRAKE SPECIFICATIONS
All measurements in inches unless noted

Year	Model		Brake Disc			Minimum Lining Thickness	Brake Caliper	
			Original Thickness	Minimum Thickness	Maximum Runout		Bracket Bolts (ft. lbs.)	Mounting Bolts (ft. lbs.)
2011	CR-V	Front	1.09-1.11	1.020	0.0016	0.060	101	37
		Rear	0.35-0.36	0.300	0.0016	0.060	80	17
2012	CR-V	Front	0.984	0.906	0.0016	0.063	NA	25
	2WD	Rear	0.354	0.295	0.0016	0.063	NA	17
2012	CR-V	Front	1.102	1.024	0.0016	0.063	NA	37
	AWD	Rear	0.354	0.295	0.0016	0.063	NA	17

NA: Not Available

71051_HCRV_C0013

MAINTENANCE MINDER SCHEDULE
Honda CR-V

The Maintenance Minder is an important feature of the information display. Based on engine and transmission operating conditions, and accumulated engine revolutions, the CR-V's onboard computer (ECM/PCM) calculates the remaining engine oil and the transmission fluid life. The system also displays the remaining engine oil life along with the code(s) for other scheduled maintenance items needing service.

Symbol	Item	Service
A	Engine oil ①	Change
B	Engine oil and filter	Change
	Fluid levels	Inspect
	Brakes	Inspect
	Parking brake adjustment	Check
	Steering gear and linkage	Inspect
	Suspension components	Inspect
	Driveshaft boots	Inspect
	Brake hoses and lines	Inspect
	Exhaust system	Inspect
	Fuel lines and connections	Inspect
1	Tires	Rotate
2	Engine air filter ②	Replace
	Dust and pollen filter ③	Replace
	Accessory drive belt	Inspect
3	Transmission fluid ④	Replace
	Transfer case fluid ④	Replace
4	Spark plugs	Replace
	Timing belt ⑤	Replace
	Water pump	Inspect
	Valve clearance ⑥	Inspect
5	Engine coolant	Replace
6	Rear differential fluid	Replace

① If the message SERVICE DUE NOW does not appear more than 12 months after the display is reset, change every year.

② If driven in dusty conditions, replace every 15,000 miles.

③ If driven in urban areas that have a high concentration of soot from industry and diesel, replace every 15,000 miles

④ If regularly driven in mountainous areas at very low speed or trailer towing, change the fluid every 30,000 miles.

⑤ If driven regularly in temperatures over 110 deg. F or below -20 deg. F, or towing a trailer, replace every 60,000 miles.

⑥ Adjust if necessary during services A, B, 1, 2, or 3 if they are noisy.

Additionally, replace the brake fluid every 3 years, and inspect the idle speed every 160,000 miles.

To reset the Maintenance Minder:

1. Turn the ignition switch to ON (II).

2. Press the SELECT/RESET button repeatedly until the engine oil life indicator is displayed.

3. Press and hold the SELECT/RESET button for about 10 seconds. The engine oil life indicator and maintenance item code(s) will blink.

4. Press and hold the SELECT/RESET knob for another 5 seconds. The maintenance item code(s) will disappear, and the engine oil life will reset to "100".

MAINTENANCE MINDER SCHEDULE
Honda CR-V (cont.)

To reset the Individual Maintenance Items:

1. Connect the Honda diagnostic System (HDS) to the data link connector (DLC).
2. Turn the ignition switch to ON (II).
3. Make sure the HDS communicates with the vehicle and the engine control module/powertrain control module (ECM/PCM) If it doesn't communicate, troubleshoot the DLC circuit.
4. Select GAUGES in the BODY ELECTRICAL with the HDS.
5. Select ADJUSTMENT in the GAUGES with the HDS.
6. Select MAINTENANCE MINDER in the ADJUSTMENT with the HDS.
7. Select RESET in the MAINTENANCE MINDER with the HDS.
8. Select the individual maintenance item you wish to reset with the HDS.

NOTES:

* If you are resetting the Maintenance Minder when the engine oil life is more than 15 %, make sure any maintenance items(s) requiring service are done before resetting the display.
* If a required service is done and the Maintenance Minder is not reset, or if the Maintenance Minder is reset without doing the service, the system will not show the proper maintenance timing.
* This can lead to serious mechanical problems because there will be no accurate record of when the required maintenance is needed.
* The engine oil life and the maintenance item(s) can be independently reset with the HDS.

71051_HCRV_C0015

BRAKES   **INFORMATION AND PRECAUTIONS**

ANTI-LOCK SYSTEMS

- Certain components within the ABS system are not intended to be serviced or repaired individually.
- Do not use rubber hoses or other parts not specifically specified for and ABS system. When using repair kits, replace all parts included in the kit. Partial or incorrect repair may lead to functional problems and require the replacement of components.
- Lubricate rubber parts with clean, fresh brake fluid to ease assembly. Do not use shop air to clean parts; damage to rubber components may result.
- Use only DOT 3 brake fluid from an unopened container.
- If any hydraulic component or line is removed or replaced, it may be necessary to bleed the entire system.
- A clean repair area is essential. Always clean the reservoir and cap thoroughly before removing the cap. The slightest amount of dirt in the fluid may plug an ori-

fice and impair the system function. Perform repairs after components have been thoroughly cleaned; use only denatured alcohol to clean components. Do not allow ABS components to come into contact with any substance containing mineral oil; this includes used shop rags.

- The Anti-Lock control unit is a microprocessor similar to other computer units in the vehicle. Ensure that the ignition switch is **OFF** before removing or installing controller harnesses. Avoid static electricity discharge at or near the controller.
- If any arc welding is to be done on the vehicle, the control unit should be unplugged before welding operations begin.

DISC AND DRUM SYSTEMS

> **✳✳ CAUTION**
>
> **Dust and dirt accumulating on brake parts during normal use may contain asbestos fibers from production or**

aftermarket brake linings. Breathing excessive concentrations of asbestos fibers can cause serious bodily harm. Exercise care when servicing brake parts. Do not sand or grind brake lining unless equipment used is designed to contain the dust residue. Do not clean brake parts with compressed air or by dry brushing. Cleaning should be done by dampening the brake components with a fine mist of water, then wiping the brake components clean with a dampened cloth. Dispose of cloth and all residue containing asbestos fibers in an impermeable container with the appropriate label. Follow practices prescribed by the Occupational Safety and Health Administration (OSHA) and the Environmental Protection Agency (EPA) for the handling, processing, and disposing of dust or debris that may contain asbestos fibers.

BRAKES   **BLEEDING THE BRAKE SYSTEM**

BLEEDING PROCEDURE

BLEEDING PROCEDURE

Note the following:
- Do not reuse the drained fluid. Use only clean Honda DOT 3 Brake Fluid from an unopened container. Using a non-Honda brake fluid can cause corrosion and shorten the life of the system.
- Make sure no dirt or other foreign matter is allowed to contaminate the brake fluid.
- Do not spill brake fluid on the vehicle, it may damage the paint. If brake fluid does contact the paint, wash it off immediately with water.
- The reservoir connected to the master cylinder must be at the

MAX (upper) level mark at the start of the bleeding procedure and checked after bleeding each brake system. Add fluid as required.

1. Make sure the brake fluid level in the reservoir is at the MAX (upper) level line.
2. Have someone slowly pump the brake pedal several times, then apply steady pressure.
3. Start the bleeding at the driver's side of the front brake system.
4. Bleed the calipers or the wheel cylinders in the following sequence:
   - Front left
   - Front right
   - Rear right
   - Rear left
5. Attach a length of clear drain tube to

the bleed screw, then loosen the bleed screw to allow air to escape from the system. Then tighten the bleed screw securely.

➡**Do not loosen the special bolt on the rear caliper.**

6. Refill the master cylinder reservoir to the MAX (upper) level line.
7. Repeat the procedure for each brake circuit until no air bubbles are in the fluid.

BLEEDING THE ABS SYSTEM

The bleeding procedure for the ABS System is the same as the conventional bleeding procedure. Refer to Brakes, Bleeding the Brake System, Bleeding Procedure.

WHEEL SPEED SENSORS

REMOVAL & INSTALLATION

2011 Models

Front Wheels

See Figure 4.

1. Before servicing the vehicle, refer to the Precautions Section.
2. Turn the ignition switch to LOCK (0).
3. Release the connector holding clamps, then disconnect the wheel speed sensor connector.
4. Remove the clips, the bolt, and the wheel speed sensor.

To install:

5. Install the wheel speed sensor in the reverse order of removal.
6. Install the sensor carefully to avoid twisting the wires.
7. If the wheel speed sensor comes in contact with the wheel bearing, it is faulty. Investigate the cause before replacing the sensor.
8. Start the engine and check that the ABS and the VSA indicators go off.
9. Test-drive the vehicle, and check that the ABS and the VSA indicators do not come on.

Rear Wheels

See Figure 5.

1. Before servicing the vehicle, refer to the Precautions Section.
2. Turn the ignition switch to LOCK (0).
3. Release the connector holding clamps, then disconnect the wheel speed sensor connector.
4. Remove the clips, the bolt, and the wheel speed sensor.

To install:

5. Install the wheel speed sensor in the reverse order of removal.
6. Apply multipurpose grease to the O-ring.
7. Install the sensor carefully to avoid twisting the wires.
8. If the wheel speed sensor comes in contact with the hub bearing unit, it is faulty. Investigate the cause before replacing the sensor.
9. Start the engine and check that the ABS and the VSA indicators go off.

A
B
6 x 1.0 mm
9.8 N·m
(1.0 kgf·m,
7.2 lbf·ft)
C
GREASE
22140_HCRV_G0190

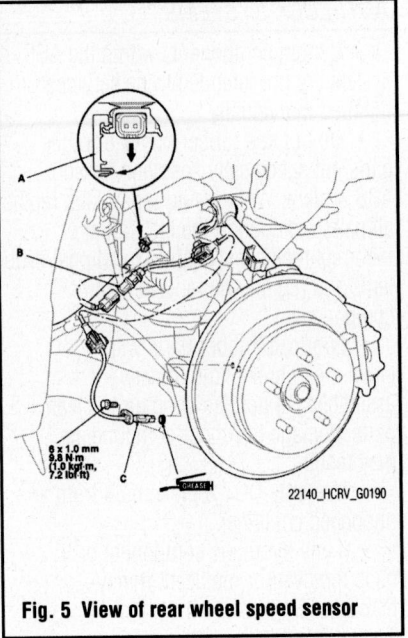

Fig. 5 View of rear wheel speed sensor

B
A
6 x 1.0 mm
9.8 N·m
(1.0 kgf·m, 7.2 lbf·ft)
C
22140_HCRV_G0189

Fig. 4 View of front wheel speed sensor

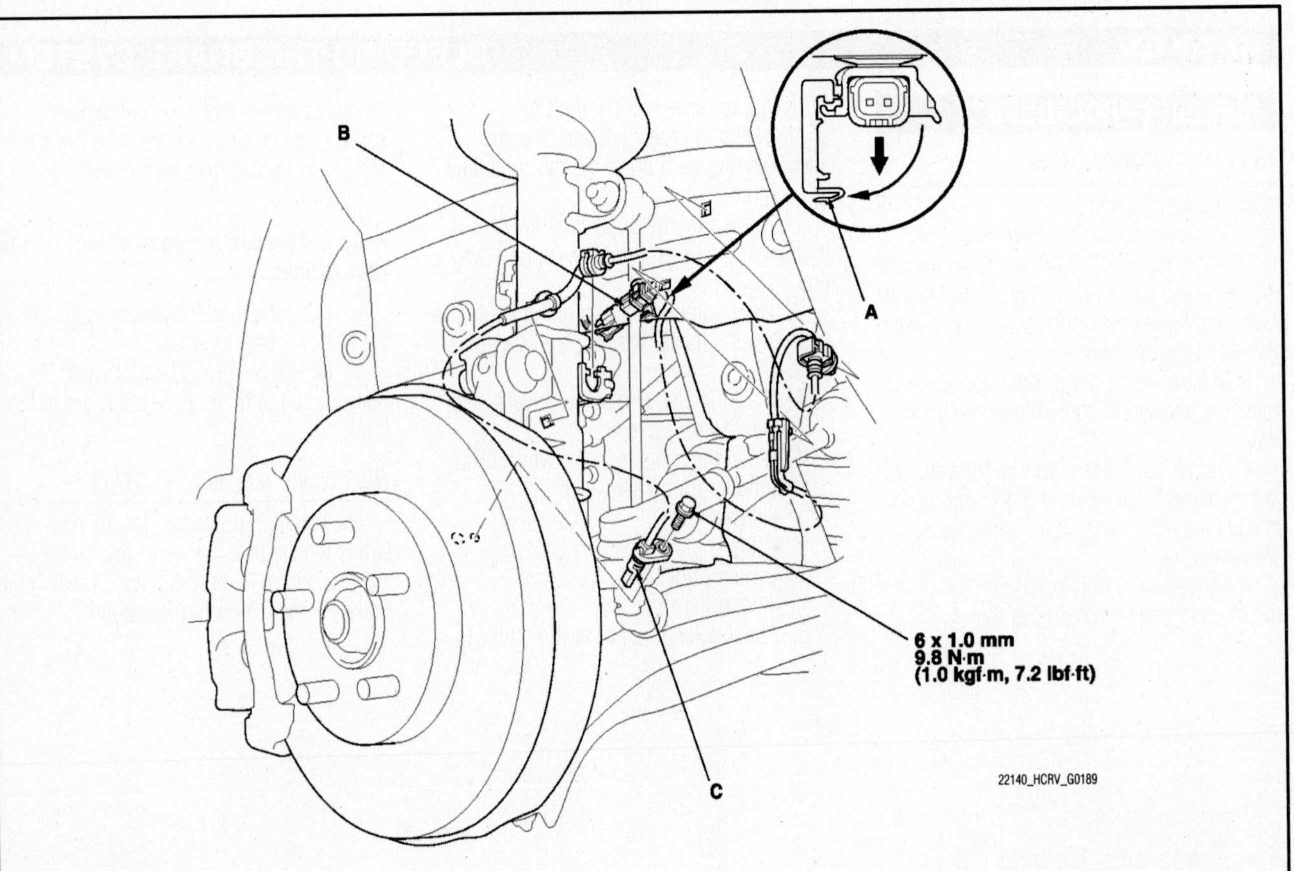

10. Test-drive the vehicle and check that the ABS and the VSA indicators do not come on.

2012 Models

See Figures 6 and 7.

1. Raise the vehicle on a lift, and make sure it is securely supported.
2. Remove the front wheel.
3. Turn the ignition switch to LOCK (0).
4. Release the connector holding clamps. Disconnect the connector. Remove the clips.
5. Remove the wheel speed sensor.

To install:

6. Install in reverse of the removal procedure.

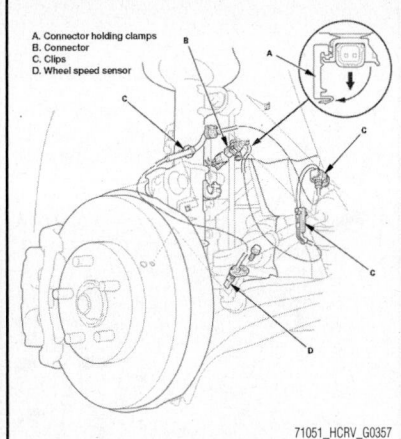

Fig. 6 Front wheel speed sensor: Release the connector holding clamps. Disconnect the connector. Remove the clips. Remove the wheel speed sensor

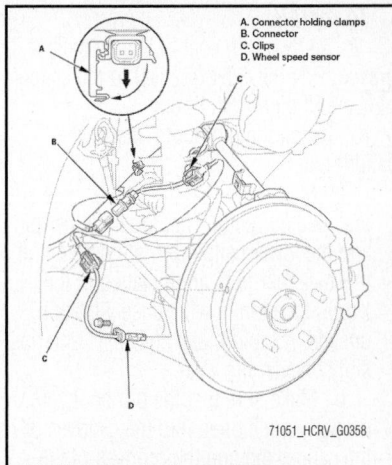

Fig. 7 Rear wheel speed sensor: Release the connector holding clamps. Disconnect the connector. Remove the clips. Remove the wheel speed sensor

BRAKES

FRONT DISC BRAKES

BRAKE CALIPER

REMOVAL & INSTALLATION

2011 Models

See Figure 8.

➡**Keep any grease off the brake disc and brake pads.**

1. Raise and support the vehicle.
2. Remove the front wheels.
3. Remove the brake hose mounting bolt.
4. Remove the brake caliper bracket mounting bolts, then remove the caliper assembly from the knuckle.

To install:

5. Install the brake caliper in the reverse order of removal.

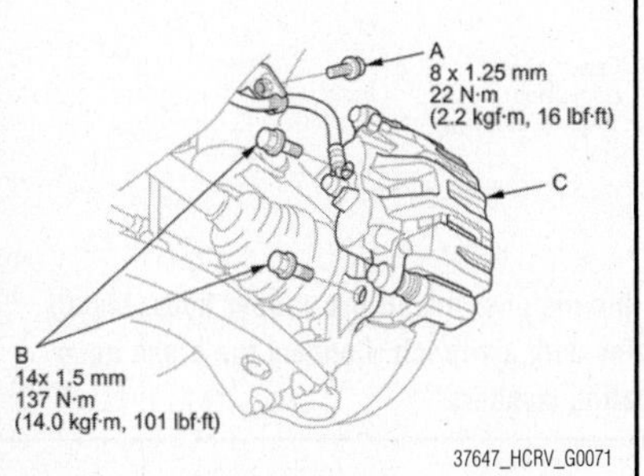

6. Clean the mating surfaces of the brake disc and the inside of the wheel, then install the front wheel.

2012 Models

See Figures 9 through 15.

1. Raise the vehicle on a lift, and make sure it is securely supported.
2. Remove the front wheel.
3. Pull out the hood rear seal, then remove the master cylinder maintenance lid. Remove the reservoir cap.
4. Remove the brake fluid from the master cylinder reservoir with a syringe.

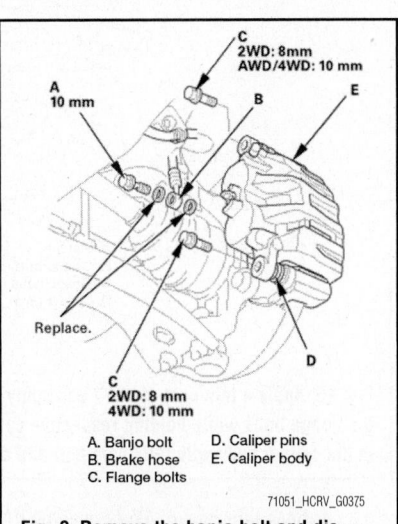

A. Banjo bolt
B. Brake hose
C. Flange bolts
D. Caliper pins
E. Caliper body

Fig. 9 Remove the banjo bolt and disconnect the brake hose from the caliper body. Remove the flange bolts while holding respective caliper pins with a wrench. Remove the caliper body

5. Remove the banjo bolt and disconnect the brake hose from the caliper body. Remove the flange bolts while holding respective caliper pins with a wrench. Remove the caliper body.

➡**Be careful when removing the caliper body or the spring could pop out of position.**

6. Remove the pad return springs while holding the brake pads.
7. Remove the pad shims and the brake pads.
8. Remove the caliper pins. Remove the pin boots.

➡**The upper and lower caliper pins are different. During installation, make sure the caliper pins are in the proper positions.**

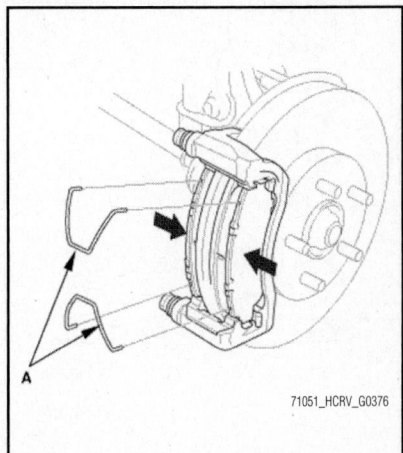

Fig. 10 Remove the pad return springs (A) while holding the brake pads

Fig. 8 Remove the brake hose mounting bolt (A), the brake caliper bracket mounting bolts (B), then remove the caliper assembly (C)

To install:

9. Apply a thin coat of Honda caliper grease (P/N 08C30-B0224M) to the caliper pins and the new pin boots.

10. Install the pin boots. Install the caliper pins.

a. Make sure that the caliper pins are installed correctly. The upper caliper pin and the lower caliper pin are different. If these caliper pins are installed in the wrong location, it will cause vibration, uneven or rapid brake pad wear, and possibly uneven tire wear.

b. Make sure that the pin boots are properly positioned into the grooves of the caliper pin and the grooves of the caliper body.

c. Remove air from the inside of the pin boots by gently squeezing them.

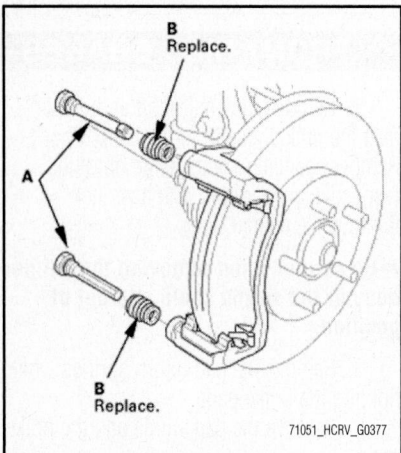

Fig. 11 Remove the caliper pins (A). Remove the pin boots (B)

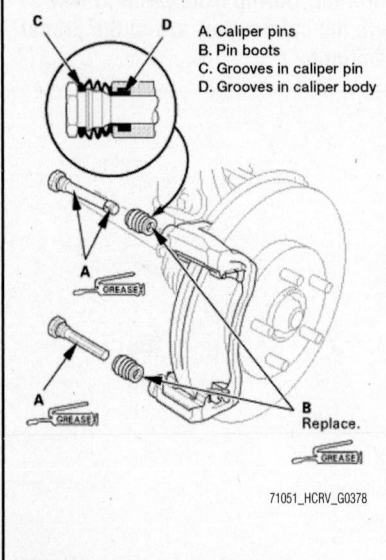

Fig. 12 Installing the caliper pins and boots

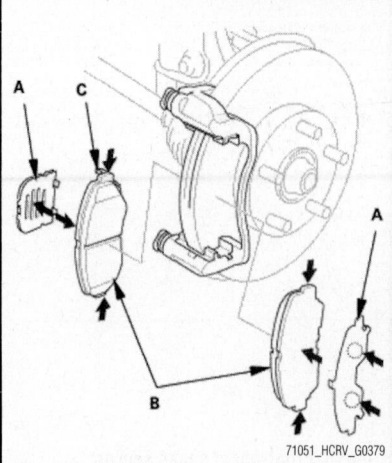

Fig. 13 2WD Models: Apply a thin coat of M-77 assembly paste to the pad side of the shims (A), the back of the brake pads (B), and the other areas indicated by the arrows. Wipe off the excess assembly paste from the pad shims and brake pads friction material. Install the brake pads and pad the shims. Install the brake pad with the wear indicator (C) on the upper inside position

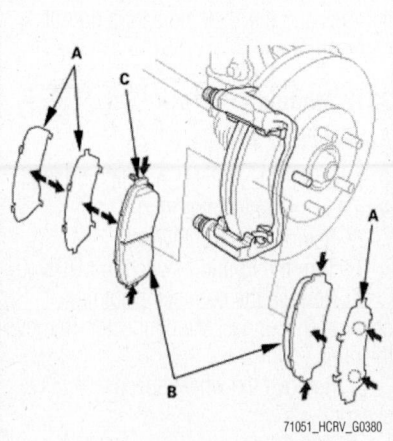

Fig. 14 AWD Models: Apply a thin coat of M-77 assembly paste to the pad side of the shims (A), the back of the brake pads (B), and the other areas indicated by the arrows. Wipe off the excess assembly paste from the pad shims and brake pads friction material. Install the brake pads and pad the shims. Install the brake pad with the wear indicator (C) on the upper inside position

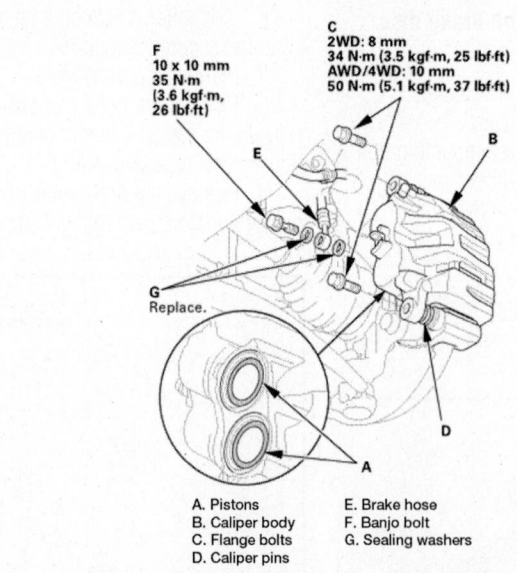

A. Pistons
B. Caliper body
C. Flange bolts
D. Caliper pins
E. Brake hose
F. Banjo bolt
G. Sealing washers

Fig. 15 Apply a thin coat of M-77 assembly paste the pistons. Install the caliper body. Install the flange bolts while holding respective caliper pins with a wrench. Connect the brake hose to the caliper body with the banjo bolt and new sealing washers

11. Apply a thin coat of M-77 assembly paste (P/N 08798-9010) to the pad side of the shims, the back of the brake pads, and the other areas indicated by the arrows. Wipe off the excess assembly paste from the pad shims and brake pads friction material.

12. Install the brake pads and pad the shims. Install the brake pad with the wear indicator on the upper inside position.

13. Install the pad return springs while holding the brake pads inward.

14. Apply a thin coat of M-77 assembly paste to the pistons.

15. Install the caliper body.

16. Install the flange bolts while holding respective caliper pins with a wrench.

17. Connect the brake hose to the caliper body with the banjo bolt and new sealing washers.

18. Bleed the brake system, as described in this section.

19. Install the front wheel.

DISC BRAKE PADS

REMOVAL & INSTALLATION

2011 Models

See Figures 16 through 19.

1. Remove some of the brake fluid from the master cylinder.

2. Raise and support the vehicle.

3. Remove the front wheels.

4. Remove the flange bolt while holding the caliper pin with a wrench. Be careful not to damage the pin boot, and pivot the caliper up out of the way. Check the hose and the pin boots for damage and deterioration.

5. Remove the brake pads and the pad shims.

6. Remove the pad retainers.

To install:

7. Clean the caliper bracket thoroughly; remove any rust, and check for grooves and cracks.

➡**Verify that the caliper pins move in and out smoothly. Clean and lube as needed.**

8. Apply a thin coat of M-77 assembly paste (P/N 08798-9010) to the retainer mating surface of the caliper bracket (indicated by the arrows and shaded area).

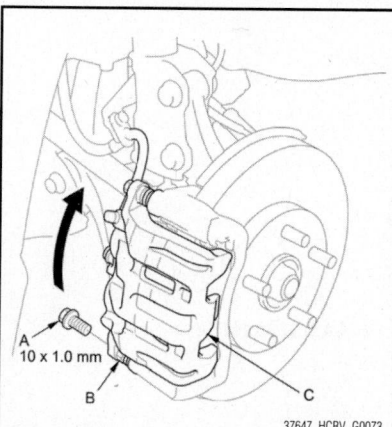

Fig. 16 Remove the flange bolt (A) while holding the caliper pin (B) with a wrench, and pivot the caliper (C)

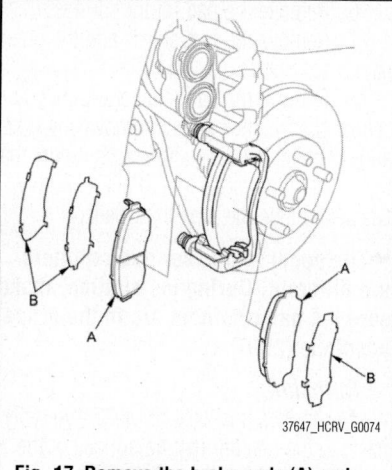

Fig. 17 Remove the brake pads (A) and the pad shims (B)

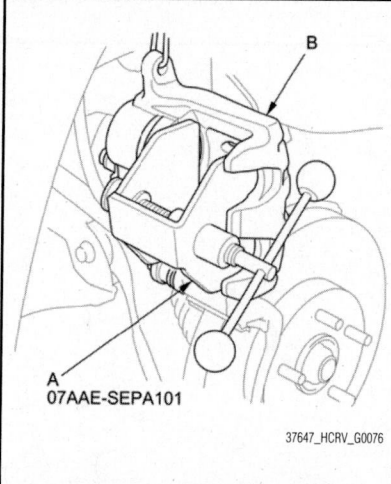

Fig. 18 Mount the brake caliper piston compressor (A) on the caliper body (B)

9. Install the pad retainers. Wipe excess assembly paste off the retainers. Keep any assembly paste off the brake discs and the brake pads.

10. Mount the brake caliper piston compressor on the caliper body.

11. Press in the piston with the brake caliper piston compressor tool so the caliper will fit over the brake pads. Make sure the piston boot is in position to prevent damaging it when pivoting the caliper down.

➡**Be careful when pressing in the piston; brake fluid might overflow from the master cylinder's reservoir. If brake fluid gets on any painted surface, wash it off immediately with water.**

12. Remove the brake caliper piston compressor tool.

13. Apply Molykote M-77 assembly paste (P/N 08798-9010) to the pad side of the shims, the back of the brake pads, and

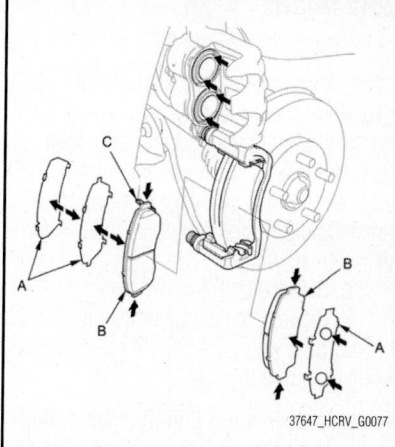

Fig. 19 Apply Molykote M-77 assembly paste (P/N 08798-9010) to the pad side of the shims (A), the back of the brake pads (B), and other areas indicated by the arrows

other areas indicated by the arrows. Wipe excess assembly paste off the pad shims and the brake pads friction material. Keep grease and assembly paste off the brake discs and the brake pads.

❋❋ CAUTION

Contaminated brake discs or brake pads reduce stopping ability.

14. Install the brake pads and pad shims correctly. Install the brake pad with the wear indicator on the upper inside. If you are reusing the brake pads, always reinstall the brake pads in their original positions to prevent a temporary loss of braking efficiency.

15. Pivot the caliper down into position. Install the flange bolt, and tighten it to 37 ft. lbs. (50 Nm) while holding the caliper pin with a wrench. Be careful not to damage the pin boot.

16. Clean the mating surface of the brake disc and the inside of the wheel, then install the front wheels.

17. Press the brake pedal several times to make sure the brakes work.

➡**Engagement may require a greater pedal stroke immediately after the brake pads have been replaced as a set. Several applications of the brake pedal will restore the normal pedal stroke.**

18. Add brake fluid as needed.

19. After installation, check for leaks at hose and line joints or connections, and retighten if necessary.

20. Test-drive the vehicle, then check for leaks.

2012 Models

See Figures 20 through 25.

1. Raise the vehicle on a lift, and make sure it is securely supported.

2. Pull out the hood rear seal, then remove the master cylinder maintenance lid. Remove the reservoir cap. Remove some brake fluid from the master cylinder reservoir with a syringe.

3. Remove both front wheels.

4. Remove the flange bolt while holding caliper pins with a wrench. Pivot the caliper body up out of the way.

➡️**Be careful when pivoting the caliper body up or the spring could pop out of position.**

5. Check the hose and pin boots for damage and deterioration.

6. Remove the pad return springs.

7. Remove the pad shims and the brake pads.

8. Remove the pad retainers. Clean the caliper bracket thoroughly; remove any rust, and check for grooves and cracks. Verify that the caliper pins move in and out smoothly. Clean and lube the pins if needed.

➡️**The upper and lower pad retainers are different. During installation, make sure the pad retainers are in the proper positions (2WD).**

To install:

9. Apply a thin coat of M-77 assembly paste to the retainer mating surface of the caliper bracket (indicated by the arrows). Install the pad retainers.

- Wipe off the excess assembly paste from the retainers. Keep the assem-

bly paste away from the brake disc and the brake pads.

- 2WD: Make sure that the pad retainers are installed correctly.

10. Press in the piston with the brake caliper piston compressor tool.

➡️**Be careful when pressing in the piston; brake fluid might overflow from the master cylinder's reservoir. If brake fluid gets on any painted surface, wash it off immediately with water.**

11. Apply a thin coat of M-77 assembly paste to the pad side of the shims, the back of the brake pads and the other areas. Wipe off the excess assembly paste from the pad shims and brake pads friction material.

12. Install the brake pads and pad shims. Install the brake pad with the

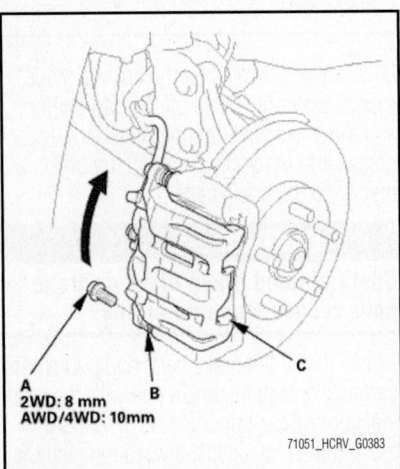
Fig. 20 Remove the flange bolt (A) while holding caliper pins (B) with a wrench. Pivot the caliper body (C) up out of the way

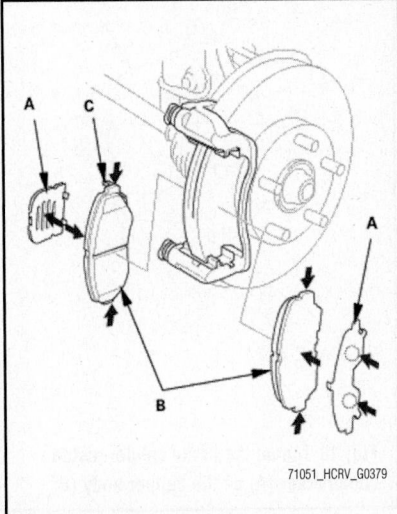
Fig. 22 2WD Models: Remove the pad shims (A) and the brake pads (B)

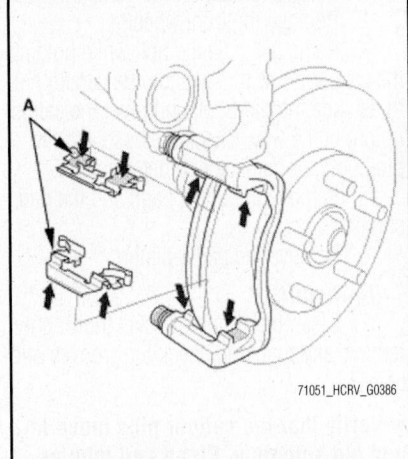
Fig. 24 Apply a thin coat of M-77 assembly paste to the retainer mating surface of the caliper bracket (indicated by the arrows). Install the pad retainers (A)

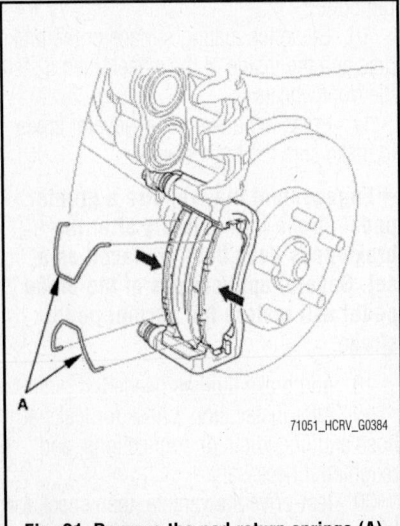
Fig. 21 Remove the pad return springs (A)

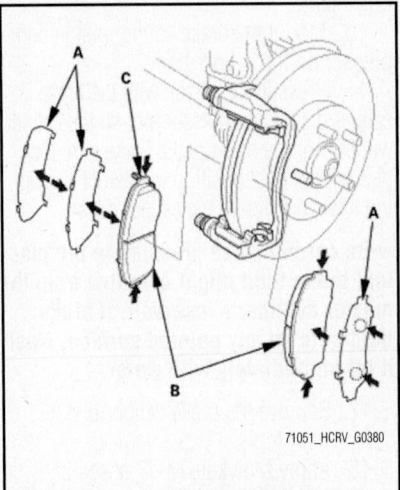
Fig. 23 AWD Models: Remove the pad shims (A) and the brake pads (B)

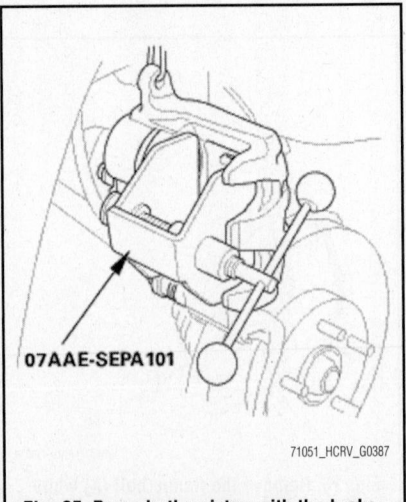
Fig. 25 Press in the piston with the brake caliper piston compressor tool

wear indicator on the upper inside position.

13. Install the pad return springs.

14. Pivot the caliper body down into position. Install the flange bolt. Tighten the caliper bolt to:
- 2WD: 25 ft. lbs. (34 Nm)

- AWD: 37 ft. lbs. (50 Nm)

15. Press the brake pedal several times to make sure the brakes work.

➡**Engagement may require a greater pedal stroke immediately after the brake pads have been replaced as a set. Several**

applications of the brake pedal will restore the normal pedal stroke.

16. Install both front wheels.
17. Add brake fluid as needed. Install the reservoir cap and the master cylinder maintenance lid.

BRAKES

REAR DISC BRAKES

BRAKE CALIPER

REMOVAL & INSTALLATION

2011 Models
See Figure 26.

➡**Keep any grease off the brake disc and brake pads.**

1. Raise and support the vehicle.
2. Remove the rear wheel.
3. Remove the brake hose bracket mounting bolt from the knuckle.
4. Remove the brake caliper bracket mounting bolts, and remove the caliper assembly from the knuckle.

To install:

5. Install the brake caliper in the reverse order of removal, and note these items:
- Before installing the brake disc/drum, clean the mating surface of the hub bearing unit and the inside of the brake disc/drum.
- Adjust the parking brake.

6. Clean the mating surfaces of the brake disc/drum and the inside of the wheel, then install the rear wheel.

2012 Models
See Figures 27 through 30.

1. Raise the vehicle on a lift, and make sure it is securely supported.
2. Remove both rear wheels.
3. Pull out the hood rear seal, then remove the master cylinder maintenance lid. Remove the reservoir cap. Remove the brake fluid from the master cylinder reservoir with a syringe.
4. Remove the banjo bolt and disconnect the brake hose from the caliper body. Remove the flange bolts while holding respective caliper pins with a wrench.
5. Remove the brake caliper body.
6. Remove the outer pad shim and the outer brake pad.
7. Remove the inner pad from the caliper body.
8. Remove the caliper pins. Remove the pin boots.

To install:

➡**The upper and lower caliper pins are different. During installation, make sure the caliper pins are in the proper positions.**

9. Apply a thin coat of Honda caliper grease (P/N 08C30-B0224M) to the caliper pins and the new pin boots.
10. Install the pin boots. Install the caliper pins.
   a. Make sure that the caliper pins are installed correctly. The upper caliper pin and the lower caliper pin are different. If

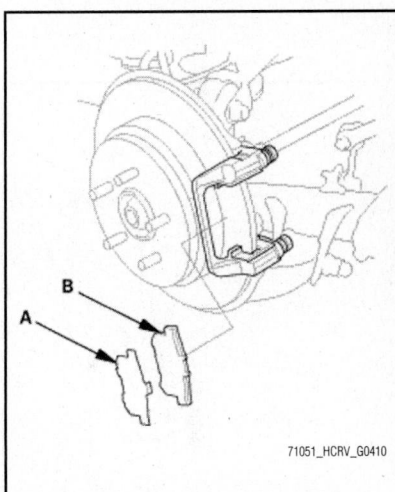

71051_HCRV_G0410

Fig. 28 Remove the outer pad shim (A) and the outer brake pad (B)

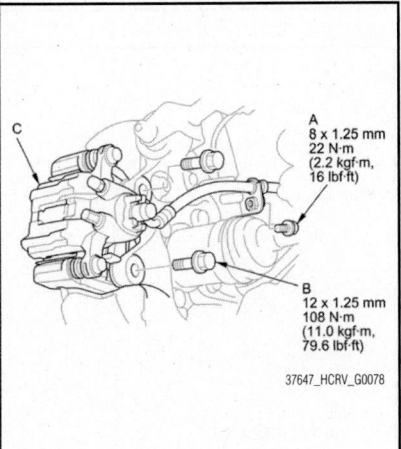

37647_HCRV_G0078

Fig. 26 Remove the brake hose bracket mounting bolt (A), the brake caliper bracket mounting bolts (B), and remove the caliper assembly (C) from the knuckle

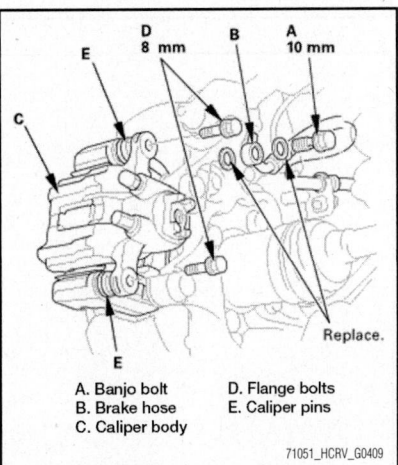

A. Banjo bolt
B. Brake hose
C. Caliper body
D. Flange bolts
E. Caliper pins

71051_HCRV_G0409

Fig. 27 Remove the banjo bolt and disconnect the brake hose from the caliper body. Remove the flange bolts while holding respective caliper pins with a wrench

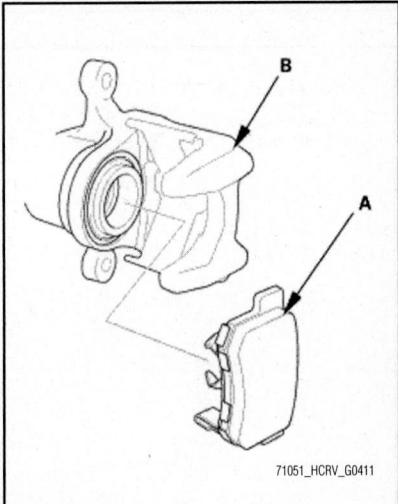

71051_HCRV_G0411

Fig. 29 Remove the inner pad (A) from the caliper body (B)

these caliper pins are installed in the wrong location, it will cause vibration, uneven or rapid brake pad wear, and possibly uneven tire wear.

b. Make sure that the pin boots are properly positioned into the grooves of the caliper pin and the grooves of the caliper body.

c. Remove air from within the pin boots.

11. Apply a thin coat of M-77 assembly paste to the pad side of the shims, the back of the brake pads, and the other contact (non-friction material) areas.

➡**Wipe off the excess assembly paste from the pad shims and brake pads friction material.**

12. Install the brake pads and the pad shims, installing the brake pad with the wear indicator on the bottom inside position.

13. Install the caliper body. Install the flange bolts while holding respective caliper pins with a wrench. Tighten the bolts to 17 ft. lbs. (23 Nm).

14. Connect the brake hose to the caliper body with the banjo bolt and new sealing washers. Tighten the bolt to 26 ft. lbs. (35 Nm).

15. Bleed the brake system, as described in this section.

16. Install remaining components.

DISC BRAKE PADS

REMOVAL & INSTALLATION

2011 Models

See Figures 32 through 38.

1. Remove some brake fluid from the master cylinder.
2. Raise and support the vehicle.
3. Remove the rear wheels.

4. Remove the brake hose from the brake hose bracket.

5. Remove the flange bolts while holding the caliper pin with a wrench. Be careful not to damage the pin boot, and remove the caliper. Check the hose and pin boots for damage and deterioration.

6. Remove the outer pad and the pad shim.

7. Remove the inner pad from the caliper.

8. Remove the pad retainers.

To install:

9. Clean the caliper bracket thoroughly; remove any rust, and check for grooves and cracks. Verify that the caliper pins move in and out smoothly. Clean and lube as needed.

10. Inspect the brake disc/drum for damage and cracks. Inspect for runout, thickness, and parallelism.

11. Apply Molykote M-77 assembly paste (P/N 08798-9010) to the retainers on

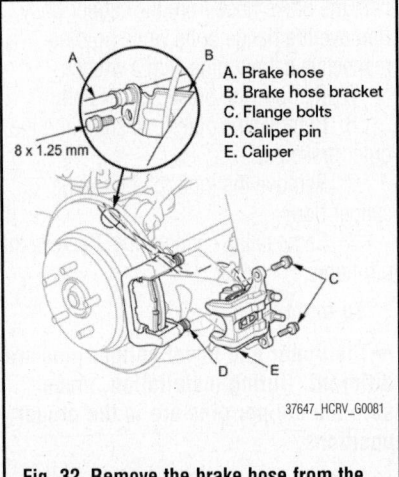

A. Brake hose
B. Brake hose bracket
C. Flange bolts
D. Caliper pin
E. Caliper

8 x 1.25 mm

37647_HCRV_G0081

Fig. 32 Remove the brake hose from the brake hose bracket

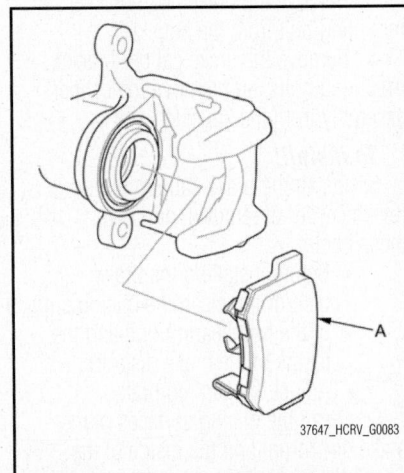

37647_HCRV_G0083

Fig. 34 Remove the inner pad (A) from the caliper

71051_HCRV_G0412

Fig. 30 Remove the caliper pins (A). Remove the pin boots (B).

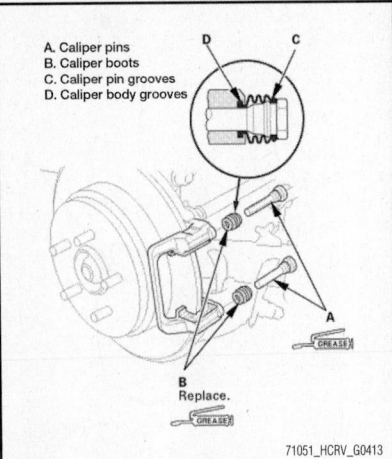

A. Caliper pins
B. Caliper boots
C. Caliper pin grooves
D. Caliper body grooves

B Replace.

71051_HCRV_G0413

Fig. 31 Installing the caliper boots and pins

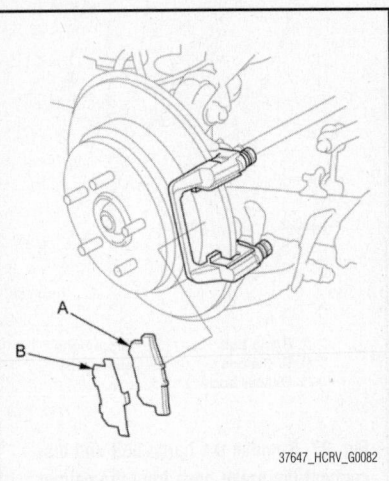

37647_HCRV_G0082

Fig. 33 Remove the outer pad (A) and the pad shim (B)

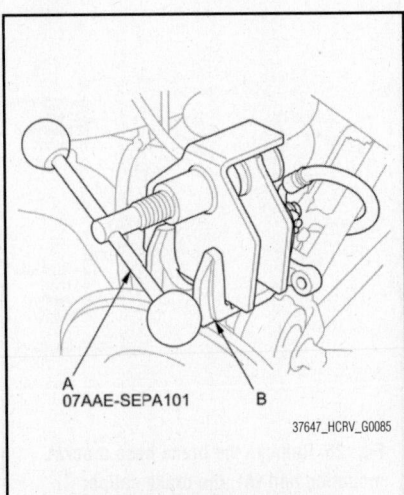

07AAE-SEPA101

37647_HCRV_G0085

Fig. 35 Mount the brake caliper piston compressor (A) on the caliper body (B)

their mating surfaces (indicated by the arrows) against the caliper bracket.

12. Install the pad retainers. Wipe excess assembly paste off the retainers. Keep any assembly paste off the brake discs and the brake pads.

13. Mount the brake caliper piston compressor on the caliper body.

14. Press in the piston with the brake caliper piston compressor tool so the caliper will fit over the brake pads. Make sure the piston boot is in position to prevent damaging it when pivoting the caliper down.

➡**Be careful when pressing in the piston; brake fluid might overflow from the master cylinder's reservoir. If brake fluid gets on any painted surface, wash it off immediately with water.**

15. Remove the brake caliper piston compressor tool.

16. Apply Molykote M-77 assembly paste (P/N 08798-9010) to the pad side of the shims, the back of the brake pads, and other areas indicated by the arrows. Wipe excess assembly paste off the pad shims and the brake pads friction material. Keep grease and assembly paste off the brake discs and the brake pads.

✳✳ CAUTION

Contaminated brake discs or brake pads reduce stopping ability.

17. Install the brake pads and pad shims correctly. Install the brake pad with the wear indicator on the bottom inside. If you are reusing the brake pads, always reinstall the

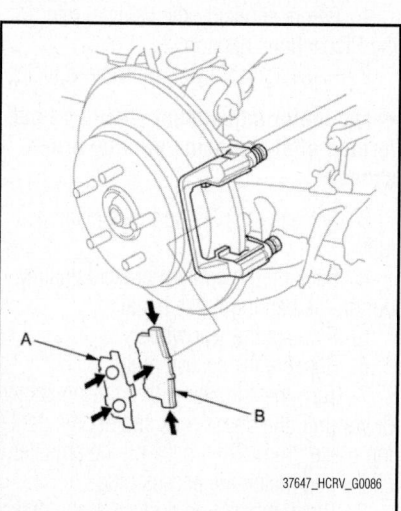

Fig. 36 Apply Molykote M-77 assembly paste to the pad side of the shims (A), the back of the brake pads (B), and other areas indicated by the arrows

brake pads in their original positions to prevent a temporary loss of braking efficiency.

18. Install the caliper into position. Install the flange bolts, and tighten it to 17 ft. lbs. (23 Nm) while holding the caliper pin with a wrench. Be careful not to damage the pin boots.

19. Install the brake hose to the brake hose bracket.

20. Clean the mating surface of the brake disc/drum and the inside of the wheel, then install the rear wheels.

21. Press the brake pedal several times to make sure the brakes work.

➡**Engagement may require a greater pedal stroke immediately after the brake pads have been replaced as a set. Several applications of the brake**

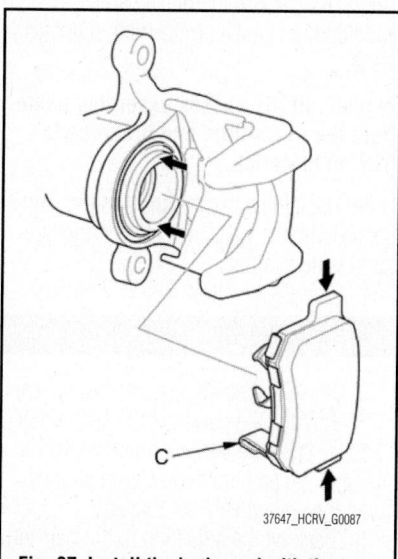

Fig. 37 Install the brake pad with the wear indicator (C) on the bottom inside

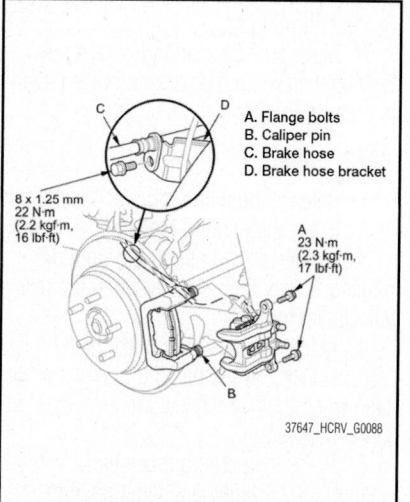

Fig. 38 Install the caliper into position

pedal will restore the normal pedal stroke.

22. Add brake fluid as needed.

23. After installation, check for leaks at hose and line joints or connections, and retighten if necessary.

24. Test-drive the vehicle, then check for leaks.

2012 Models

See Figures 39 through 42.

1. Raise the vehicle on a lift, and make sure it is securely supported.

2. Pull out the hood rear seal, then remove the master cylinder maintenance lid. Remove the reservoir cap. Remove some brake fluid from the master cylinder reservoir with a syringe.

3. Remove both rear wheels.

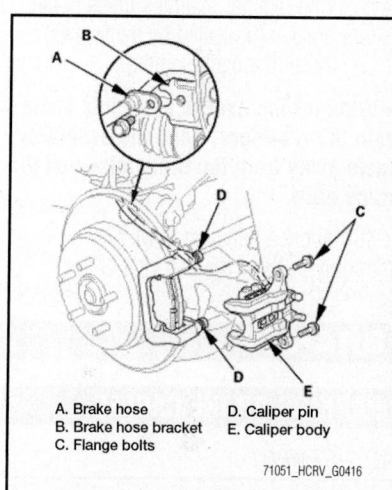

A. Brake hose   D. Caliper pin
B. Brake hose bracket   E. Caliper body
C. Flange bolts

Fig. 39 Remove the brake hose from the brake hose bracket. Remove the flange bolts while holding the caliper pin with a wrench. Remove the caliper body

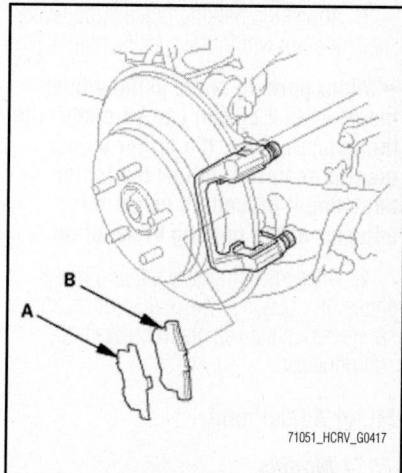

Fig. 40 Remove the outer pad shim (A) and the outer brake pad (B)

4. Remove the brake hose from the brake hose bracket. Remove the flange bolts while holding the caliper pin with a wrench. Remove the caliper body.

✳✳ WARNING

Do not twist the brake hose to prevent damage.

5. Remove the outer pad shim and the outer brake pad.
6. Remove the inner pad from the caliper body.
7. Remove the pad retainers. Clean the caliper bracket thoroughly; remove any rust, and check for grooves and cracks. Verify that the caliper pins move in and out smoothly. Clean and lube if needed.

To install:

8. Apply a thin coat of M-77 assembly paste to the retainer mating surface of the caliper bracket (indicated by the arrows).
9. Install the pad retainers.

➡**Wipe off the excess assembly paste from the retainers. Keep the assembly paste away from the brake disc and the brake pads.**

10. Apply a thin coat of M-77 assembly paste to the pad side of the

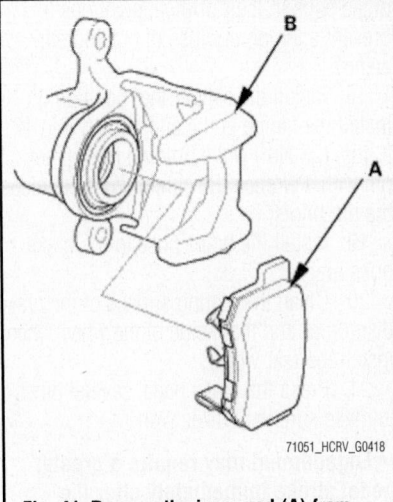

Fig. 41 Remove the inner pad (A) from the caliper body (B)

71051_HCRV_G0418

shims, the back of the brake pads, and the other contact (non-friction lining) areas.

➡**Wipe off the excess assembly paste from the pad shims and brake pads friction material.**

11. Install the brake pads and the pad shims, with the wear indicator on the bottom inside position.

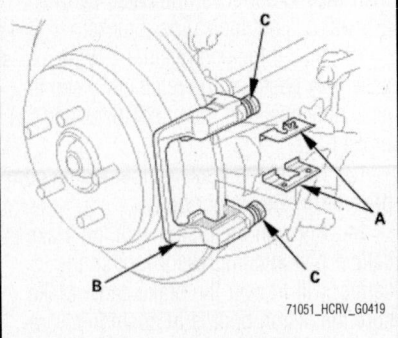

Fig. 42 Remove the pad retainers (A). Clean the caliper bracket (B) thoroughly; remove any rust, and check for grooves and cracks. Verify that the caliper pins (C) move in and out smoothly. Clean and lube if needed

71051_HCRV_G0419

12. Position the caliper body. Install the flange bolts while holding respective caliper pins with a wrench. Tighten the flange bolts to 17 ft. lbs. (23 Nm).
13. Install the brake hose and tighten the retaining bolt to 16 ft. lbs. (22 Nm).
14. Install both rear wheels.
15. Add brake fluid as needed.
16. Install the reservoir cap and the master cylinder maintenance lid.

BRAKES

PARKING BRAKE CABLES

ADJUSTMENT

1. Before servicing the vehicle, refer to the Precautions Section.
2. Press the parking brake pedal with 66 lbs. (294 N) of force to fully apply the parking brake. The parking brake lever should be locked within 6–7 clicks.
3. Adjust the parking brake if the lever clicks are not within the specification.

➡**Minor parking brake pedal adjustments (1 to 2 clicks) can be made with the adjusting nut. If a larger adjustment is required, follow the major adjustment procedure using the adjuster at the parking brake drum.**

4. After installing new parking brake shoes or a new rear brake disc/drum, make sure to drive the vehicle to break-in the components.

Minor Adjustment

2011 Models

1. Before servicing the vehicle, refer to the Precautions Section.

2. Raise and safely support the vehicle.
3. Release the parking brake pedal fully.
4. Tighten the parking brake adjusting nut until the parking brakes drag slightly when the rear wheels are rotated.
5. Back off the adjusting nut in half-turn increments.
6. Release the parking brake pedal fully, and check that the parking brakes do not drag when the rear wheels are rotated. Readjust if necessary.
7. Make sure the parking brakes are fully applied when the parking brake pedal is pressed in all the way.

2012 Models

1. Release the parking brake pedal fully.
2. Press the parking brake pedal 1 click.
3. Tighten the adjusting nut until the parking brakes drag slightly when the rear wheels are turned.
4. Release the parking brake pedal fully.
5. Check that the parking brakes do not drag when the rear wheels are turned. Readjust if necessary.
6. Press the parking brake pedal with enough force to fully apply the parking brake.

PARKING BRAKE

7. Make sure the parking brake pedal is fully engaged within 5 to 6 clicks.

Major Adjustment

2011 Models

1. Before servicing the vehicle, refer to the Precautions Section.
2. Raise and safely support the vehicle.

➡**The major adjustment should be performed after replacing parking brake shoes.**

3. Release the parking brake pedal fully.
4. Back off the parking brake adjusting nut on the parking brake pedal.
5. Remove the rear wheels.
6. Remove the access plug.
7. Turn the adjuster with a flat-tip screwdriver until the shoes lock against the parking brake drum. Then back off the adjuster 8 clicks and install the access plug.
8. Clean the mating surface of the brake disc/drum and the inside of the wheel, then install the rear wheels.
9. Do the minor adjustment procedure, as necessary.

2012 Models

1. Remove both rear wheels.
2. Remove the access plug from the rear wheel.
3. Turn the ratchet teeth on the adjuster nut with a brake adjuster tool or a flat-tip screwdriver until the shoes lock against the parking brake drum. Then back off the adjuster 7 clicks, and install the access plug.
4. Install the rear wheel on both sides.
5. Release the parking brake pedal fully.
6. Press the parking brake pedal 1 click.
7. Tighten the adjusting nut until the parking brakes drag slightly when the rear wheels are turned.
8. Release the parking brake pedal fully.
9. Check that the parking brakes do not drag when the rear wheels are turned. Readjust if necessary.
10. Press the parking brake pedal with enough force to fully apply the parking brake.

PARKING BRAKE SHOES

REMOVAL & INSTALLATION

2011 Models

See Figures 43 through 49.

1. Before servicing the vehicle, refer to the Precautions Section.
2. Raise and safely support the vehicle.
3. Remove the rear wheels.
4. Release the parking brake and remove the rear brake caliper and brake disc.

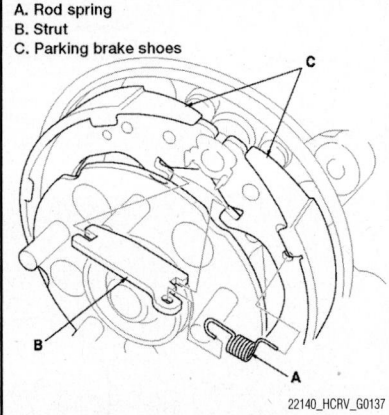

A. Rod spring
B. Strut
C. Parking brake shoes

22140_HCRV_G0137

Fig. 44 Disconnect the rod spring (A) and remove the strut (B). Remove the parking brake shoes (C)

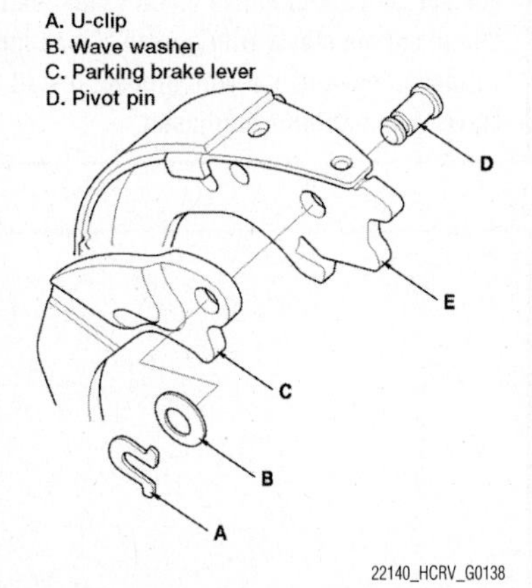

A. U-clip
B. Wave washer
C. Parking brake lever
D. Pivot pin

22140_HCRV_G0138

Fig. 45 Remove the U-clip, wave washer, parking brake lever, and pivot pin from the brake shoe

5. Disconnect and remove the upper brake spring.
6. Disconnect and remove the lower brake spring.
7. Remove the tension pins by pushing the retainer springs and turning the pins.
8. Remove the adjuster assembly by moving the forward brake shoe.
9. Remove the rearward brake shoe by disconnecting the parking brake cable from the parking brake lever.
10. Disconnect the rod spring, and remove the strut.
11. Remove the parking brake shoes.
12. Remove the U-clip, wave washer, parking brake lever, and pivot pin from the brake shoe.

To install:

13. Apply Molykote 44 MA grease to the sliding surface of the pivot pin and insert the pin into the rearward brake shoe.
14. Install the parking brake lever and wave washer on the pivot pin and secure with a new U-clip.
15. Install the wave washer with its convex side facing out.
16. Pinch the U-clip securely to prevent the pivot pin from coming out of the brake shoe.
17. Position the parking brake shoes, then hook the rod spring on the strut first with the spring end pointing downward. Then, hook the rod spring to the parking brake shoe and install the strut on the parking brake shoes.
18. Apply Molykote 44 MA grease to the sliding surfaces, the opposite edges of the parking brake shoe, and the pivot of the

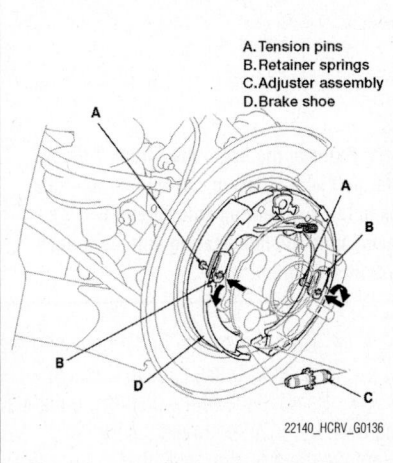

A. Tension pins
B. Retainer springs
C. Adjuster assembly
D. Brake shoe

22140_HCRV_G0136

Fig. 43 Remove the tension pins by pushing the retainer springs and turning the pins. Remove the adjuster assembly by moving the forward brake shoe

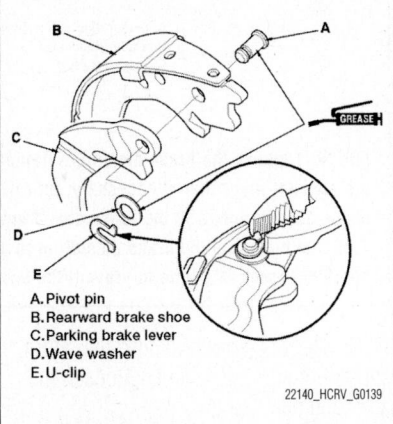

A. Pivot pin
B. Rearward brake shoe
C. Parking brake lever
D. Wave washer
E. U-clip

22140_HCRV_G0139

Fig. 46 Apply Molykote 44 MA grease to the sliding surface of the pivot pin and insert the pin into the rearward brake shoe. Install the parking brake lever and wave washer on the pivot pin and secure with a new U-clip

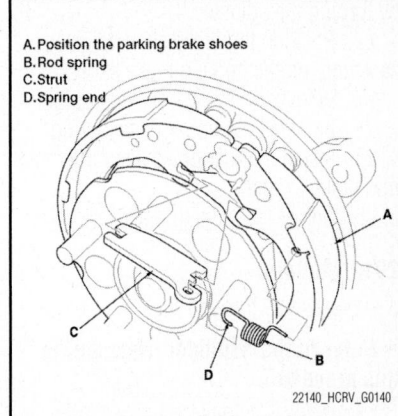

A. Position the parking brake shoes
B. Rod spring
C. Strut
D. Spring end

22140_HCRV_G0140

Fig. 47 Position the parking brake shoes, then hook the rod spring on the strut first with the spring end pointing downward. Then, hook the rod spring to the parking brake shoe and install the strut on the parking brake shoes

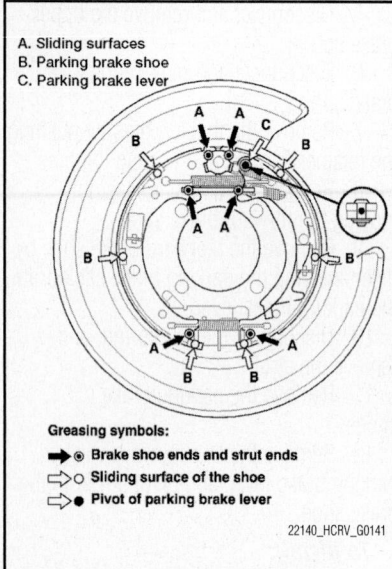

A. Sliding surfaces
B. Parking brake shoe
C. Parking brake lever

Greasing symbols:
➡◉ Brake shoe ends and strut ends
▷○ Sliding surface of the shoe
▷● Pivot of parking brake lever

22140_HCRV_G0141

Fig. 48 Apply Molykote 44 MA grease to the sliding surfaces, the opposite edges of the parking brake shoe, and the pivot of the parking brake lever

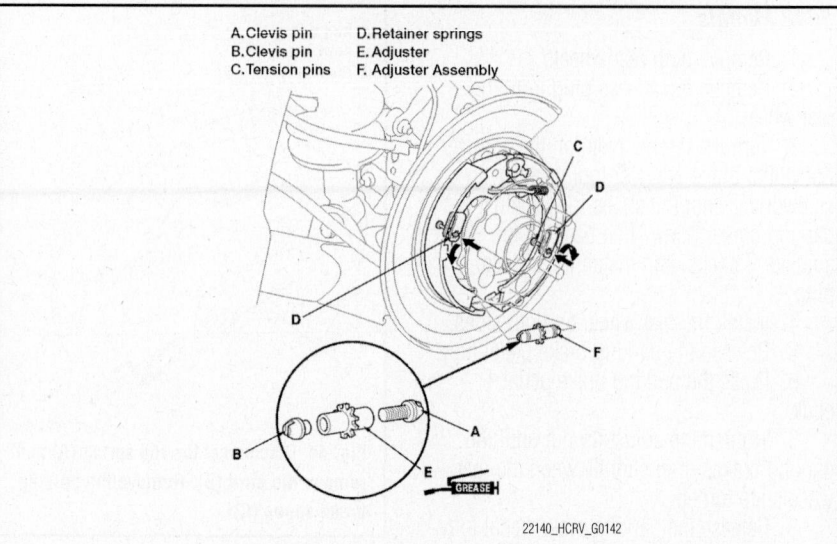

A. Clevis pin   D. Retainer springs
B. Clevis pin   E. Adjuster
C. Tension pins   F. Adjuster Assembly

22140_HCRV_G0142

Fig. 49 Reinstall the tension pins and retainer springs. Make sure the tension pin does not contact the parking brake lever. Clean the threaded portions of the clevis A, and coat the threads of the clevis with grease. Clean the sliding surface of the clevis B, and coat the sliding surface of the clevis B with grease. Install the clevis A and B on the adjuster and shorten the clevis A by turning the adjuster

parking brake lever as shown. Wipe off any excess. Keep grease off the brake linings.

19. Connect the parking brake cable to the parking brake lever. Apply silicone grease to the cable contact surface on the backing plate.

20. Reinstall the tension pins and retainer springs. Make sure the tension pin does not contact the parking brake lever.

21. Clean the threaded portions of the clevis A, and coat the threads of the clevis with grease. Clean the sliding surface of the clevis B, and coat the sliding surface of clevis B with grease. Install the clevis A and B on the adjuster and shorten the clevis A by turning the adjuster.

22. Position the brake shoe adjuster assembly on the parking brake shoes.

23. Reinstall the lower brake spring.

24. Reinstall the upper brake spring.

25. Install the rear brake disc/drum and rear brake caliper.

26. Adjust the parking brake.

2012 Models

See Figures 50 through 55.

➡**Refer to the exploded view during this procedure.**

1. Raise the vehicle on a lift, and make sure it is securely supported.

2. Remove both rear wheels.

3. Pull back the left rear carpet/cutout/insulation, under the driver's seat. Remove the equalizer cover.

4. Release the parking brake pedal fully.

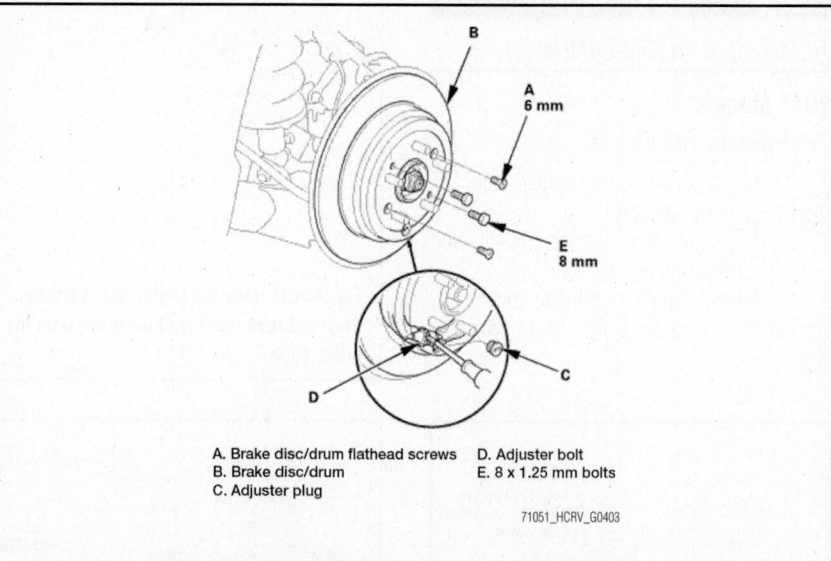

A. Brake disc/drum flathead screws   D. Adjuster bolt
B. Brake disc/drum   E. 8 x 1.25 mm bolts
C. Adjuster plug

71051_HCRV_G0403

Fig. 50 Remove the brake disc/drum flathead screws. Remove the brake disc/drum. Remove the adjuster plug, then if necessary, turn the adjuster bolt with a flat-tip screwdriver until the shoes become loose. If the brake disc/drum is stuck to the hub bearing unit, thread two 8 x 1.25 mm bolts into the brake disc/drum to push it away from the hub bearing unit. Turn each bolt 90 degrees at a time to prevent the brake disc/drum from binding

5. Disconnect and remove the return spring from the equalizer and loosen the adjusting nut.

6. Remove the brake hose mounting bolt. Remove the caliper assembly.

• To prevent damage to the caliper assembly or brake hose, use a short piece of wire to hang the caliper assembly from the undercarriage.

• Do not twist the brake hose and the parking brake cable excessively.

7. Remove the brake disc/drum flathead screws. Remove the brake disc/drum. Remove the adjuster plug, then if necessary, turn the adjuster bolt with a flat-tip screwdriver until the shoes become loose. If the brake disc/drum is stuck to the hub bearing unit, thread two 8 x 1.25 mm bolts into the brake disc/drum to push it away from the

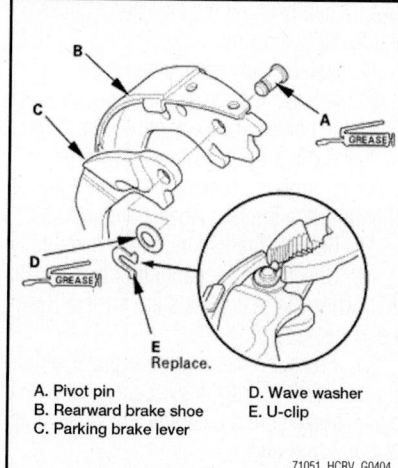

A. Pivot pin
B. Rearward brake shoe
C. Parking brake lever
D. Wave washer
E. U-clip

71051_HCRV_G0404

Fig. 51 Apply Molykote 44MA grease to the sliding surface of the pivot pin and insert the pin into the rearward brake shoe from the outside. Install the parking brake lever and wave washer on the pivot pin, and secure with a new U-clip

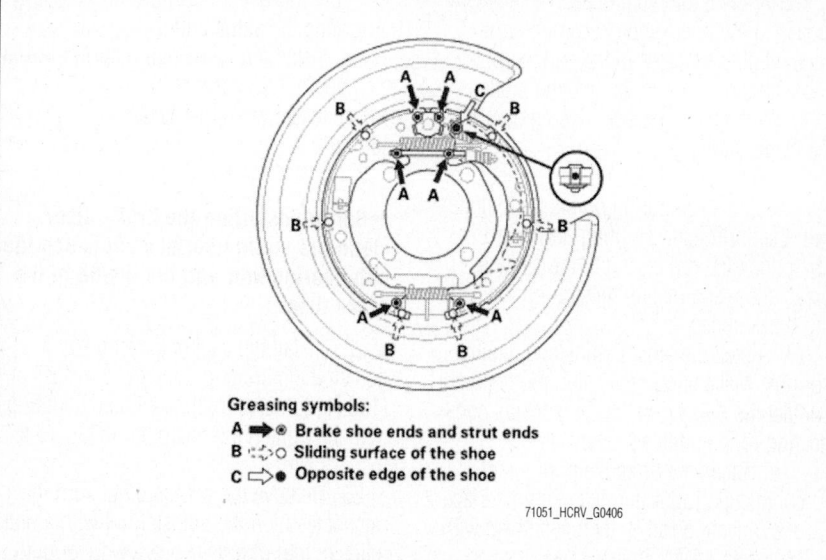

Greasing symbols:
A ➡ⓧ Brake shoe ends and strut ends
B ⤡○ Sliding surface of the shoe
C ⇨● Opposite edge of the shoe

71051_HCRV_G0406

Fig. 53 Apply a thin coat of Molykote 44MA grease to the brake shoe ends and strut ends (A), sliding surfaces of the parking brake shoe (B), and the pivot of the parking brake lever (C) as shown. Wipe off any excess. Keep grease off the brake linings

hub bearing unit. Turn each bolt 90 degrees at a time to prevent the brake disc/drum from binding.

8. Disconnect and remove upper brake spring and remove the upper return spring.

9. Disconnect and remove the lower return spring.

10. Remove the tension pins by pushing the respective retainer springs and turning the pin. Remove the adjuster assembly by moving forward brake shoe.

11. Disconnect the rod spring and remove the strut. Remove the forward brake shoe.

12. Remove the rearward brake shoe by disconnecting the rear parking brake cable from the parking brake lever.

13. Remove the U-clip, the wave washer, the parking brake lever and the pivot pin from the brake shoe.

To install:

14. Apply Molykote 44MA grease to the sliding surface of the pivot pin and insert the pin into the rearward brake shoe from the outside. Install the parking brake lever

and wave washer on the pivot pin, and secure with a new U-clip.

a. Install the wave washer with its convex side facing out.

b. Pinch the U-clip securely to prevent the parking brake lever from coming out of the brake shoe.

15. Connect the rear parking brake cable to the parking brake lever. Apply Molykote 44MA grease to the cable contact surface on the backing plate.

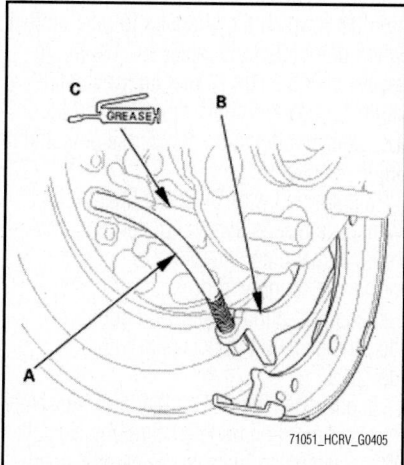

71051_HCRV_G0405

Fig. 52 Connect the rear parking brake cable (A) to the parking brake lever (B). Apply Molykote 44MA grease to the cable contact surface (C) on the backing plate

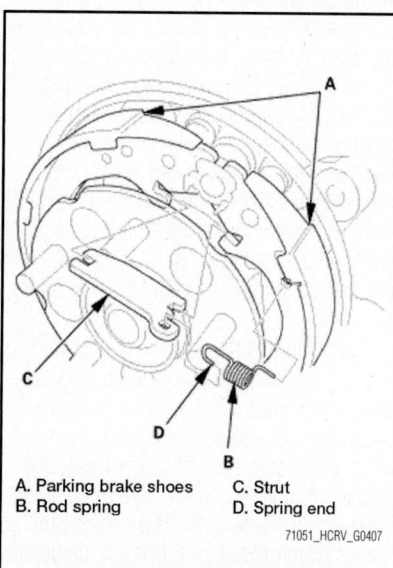

A. Parking brake shoes
B. Rod spring
C. Strut
D. Spring end

71051_HCRV_G0407

Fig. 54 Position the parking brake shoes, then hook the rod spring on the strut, first with the spring end pointing downward. Then hook the rod spring to the parking brake shoe, and install the strut on the parking brake shoes.

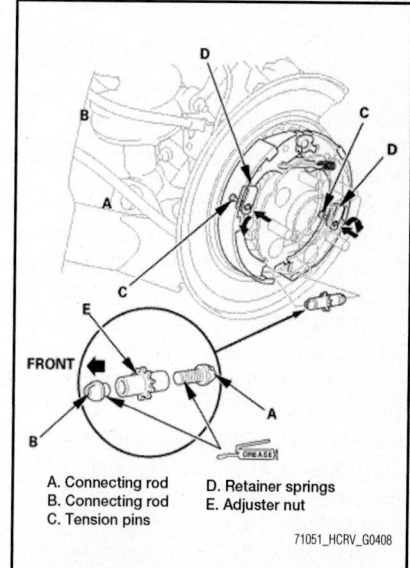

A. Connecting rod
B. Connecting rod
C. Tension pins
D. Retainer springs
E. Adjuster nut

71051_HCRV_G0408

Fig. 55 Install the tension pins and retainer springs. Make sure the tension pin does not contact the parking brake lever. Install connecting rods A and B on the adjuster nut

16. Apply a thin coat of Molykote 44MA grease to the brake shoe ends and strut ends, sliding surfaces of the parking brake shoe, and the pivot of the parking brake lever. Wipe off any excess. Keep grease off the brake linings.

17. Position the parking brake shoes, then hook the rod spring on the strut, first with the spring end pointing downward. Then hook the rod spring to the parking brake shoe, and install the strut on the parking brake shoes.

18. Install the tension pins and retainer springs. Make sure the tension pin does not contact the parking brake lever. Install connecting rods on the adjuster nut.

    a. Clean the threaded portions of connecting rod A and the sliding surface of connecting rod B, then coat them with Molykote 44MA grease.

    b. Shorten connecting rod A by fully turning the adjuster nut.

19. Install the lower return spring. Install the upper return spring.

20. Install the upper brake spring.

21. Install the brake disc.

➡ **Before installing the brake disc, clean the mating surfaces between the hub bearing unit and the inside of the brake disc.**

22. Install the caliper assembly and tighten the mounting bolts to 80 ft. lbs. (108 Nm). Install the brake hose mounting bolt and tighten the retaining bolt to 16 ft. lbs. (22 Nm).

23. Remove the access plug. Turn the ratchet teeth on the adjuster nut with a brake adjuster tool or a flat-tip screwdriver until the shoes lock against the parking brake drum. Then back off the adjuster 7 clicks, and install the access plug.

24. Install both rear wheels.

25. Release the parking brake pedal fully. Press the parking brake pedal 1 click.

26. Tighten the adjusting nut on the equalizer until the parking brakes drag slightly when the rear wheels are turned.

27. Release the parking brake pedal fully.

28. Check that the parking brakes do not drag when the rear wheels are turned. Readjust if necessary.

29. Press the parking brake pedal with enough force to fully apply the parking brake. Make sure the parking brake pedal is within 5 to 6 clicks.

30. Install the equalizer return spring. Install the equalizer cover.

31. Fold back into place the left rear carpet/cutout/insulation.

CHASSIS ELECTRICAL

AIR BAG (SUPPLEMENTAL RESTRAINT SYSTEM)

GENERAL INFORMATION

PRECAUTIONS

Disconnect and isolate the battery negative cable before beginning any airbag system component diagnosis, testing, removal, or installation procedures. Allow system capacitor to discharge for two minutes before beginning any component service. This will disable the airbag system. Failure to disable the airbag system may result in accidental airbag deployment, personal injury, or death.

Do not place an intact undeployed airbag face down on a solid surface. The airbag will propel into the air if accidentally deployed and may result in personal injury or death.

When carrying or handling an undeployed airbag, the trim side (face) of the airbag should be pointing towards the body to minimize possibility of injury if accidental deployment occurs. Failure to do this may result in personal injury or death.

Replace airbag system components with OEM replacement parts. Substitute parts may appear interchangeable, but internal differences may result in inferior occupant protection. Failure to do so may result in occupant personal injury or death.

Wear safety glasses, rubber gloves, and long sleeved clothing when cleaning powder residue from vehicle after an airbag deployment. Powder residue emitted from a deployed airbag can cause skin irritation. Flush affected area with cool water if irritation is experienced. If nasal or throat irritation is experienced, exit the vehicle for fresh air until the irritation ceases. If irritation continues, see a physician.

Do not use a replacement airbag that is not in the original packaging. This may result in improper deployment, personal injury, or death.

The factory installed fasteners, screws and bolts used to fasten airbag components have a special coating and are specifically designed for the airbag system. Do not use substitute fasteners. Use only original equipment fasteners listed in the parts catalog when fastener replacement is required.

During, and following, any child restraint anchor service, due to impact event or vehicle repair, carefully inspect all mounting hardware, tether straps, and anchors for proper installation, operation, or damage. If a child restraint anchor is found damaged in any way, the anchor must be replaced. Failure to do this may result in personal injury or death.

Deployed and non-deployed airbags may or may not have live pyrotechnic material within the airbag inflator.

Do not dispose of driver/passenger/curtain airbags or seat belt tensioners unless you are sure of complete deployment. Refer to the Hazardous Substance Control System for proper disposal.

Dispose of deployed airbags and tensioners consistent with state, provincial, local, and federal regulations.

After any airbag component testing or service, do not connect the battery negative cable. Personal injury or death may result if the system test is not performed first.

If the vehicle is equipped with the Occupant Classification System (OCS), do not connect the battery negative cable before performing the OCS Verification Test using the scan tool and the appropriate diagnostic information. Personal injury or death may result if the system test is not performed properly.

Never replace both the Occupant Restraint Controller (ORC) and the Occupant Classification Module (OCM) at the same time. If both require replacement, replace one, then perform the Airbag System test before replacing the other.

Both the ORC and the OCM store Occupant Classification System (OCS) calibration data, which they transfer to one another when one of them is replaced. If both are replaced at the same time, an irreversible fault will be set in both modules and the OCS may malfunction and cause personal injury or death.

If equipped with OCS, the Seat Weight Sensor is a sensitive, calibrated unit and must be handled carefully. Do not drop or handle roughly. If dropped or damaged, replace with another sensor. Failure to do so may result in occupant injury or death.

If equipped with OCS, the front passenger seat must be handled carefully as well. When removing the seat, be careful when setting on floor not to drop. If dropped, the sensor may be inoperative, could result in occupant injury, or possibly death.

If equipped with OCS, when the passenger front seat is on the floor, no one should sit in the front passenger seat.

This uneven force may damage the sensing ability of the seat weight sensors. If sat on and damaged, the sensor may be inoperative, could result in occupant injury, or possibly death.

DISARMING THE SYSTEM

1. Before servicing the vehicle, refer to the Precautions Section.
2. Turn the ignition switch to **OFF** or LOCK (0).
3. Disconnect the negative battery cable and isolate it from accidental reconnection.

Insulate the cable end with high-quality electrical tape or a similar non-conductive wrapping.

➡**Always disconnect the negative side first.**

4. Disconnect the positive cable from the battery.
5. Wait at least 3 minutes for the system capacitor to discharge before performing any service. The air bag system is designed to retain enough voltage to deploy the air bag for a short

period of time after the battery has been disconnected.

ARMING THE SYSTEM

1. Before servicing the vehicle, refer to the Precautions Section.
2. Reconnect the positive battery cable.
3. Reconnect the negative battery cable.
4. To confirm proper system operation, turn the ignition switch to the **ON** position. The SRS indicator light should light for at least 7 seconds and then go off.

DRIVE TRAIN

AUTOMATIC TRANSAXLE

DRAIN & REFILL

➡**Keep all foreign particles out of the transaxle.**

1. Warm up the engine to normal operating temperature (the radiator fan comes on), and turn the engine off.
2. Remove the drain plug and drain the Automatic Transmission Fluid (ATF).
3. Reinstall the drain plug with a new sealing washer.
4. Refill the transaxle with the recommended fluid through the dipstick guide tube opening. Always use Honda ATF-Z1 Automatic Transmission Fluid (ATF). Using a non-Honda ATF can affect shift quality.
5. Insert the dipstick into the guide tube aligning the notch with the guide tab.
6. Remove the dipstick.
7. Check that the fluid level is between the upper mark and the lower mark on the dipstick.
8. If the maintenance minder indicated to replace the ATF, reset the maintenance information display. If the maintenance minder did not require to replace the ATF, go to the next step and reset the ATF life with the HDS.
9. Connect the HDS to the DLC.
10. Turn the ignition switch to ON (II). Make sure the HDS communicates with the PCM. If it does not, go to the DLC circuit troubleshooting.
11. Select BODY ELECTRICAL with the HDS.
12. Select ADJUSTMENT in the GAUGES MENU with the HDS.
13. Select RESET in the MAINTENANCE MINDER with the HDS.
14. Select MAINTENANCE SUB ITEM 3 RESET to reset the ATF life with the HDS.

FLUID LEVEL CHECK

✳✳ WARNING

If the vehicle is driven when the ATF level is below the lower mark, one or more of these symptoms may occur:

- Transmission damage.
- Vehicle does not move in any gear.
- Vehicle accelerates poorly, and engine revs up abnormally high when starting off in forward and reverse positions.
- Engine vibrates at idle.

✳✳ CAUTION

If the vehicle is driven when the ATF level is above the upper mark, the vehicle may creep forward while in N, or have shifting problems.

1. Start the engine. Hold the engine speed at 3,000 rpm without load in P or N until the radiator fan comes on.
2. Park the vehicle on level ground.
3. Turn the ignition switch to LOCK (0).
4. Remove the ATF dipstick (yellow loop) and wipe it with a clean cloth.
5. Insert the dipstick into the guide tube aligning the notch with the guide tab.
6. Remove the dipstick, and check the ATF level. It should be between the upper mark and the lower mark.
7. If the ATF level is below the lower mark, check for fluid leaks at the transaxle, the hoses, and the line joints. If a problem is found, fix it before filling the transaxle with ATF.
8. If the ATF level is above the upper mark, drain the ATF to proper level.
9. If necessary, fill the transaxle with ATF through the dipstick guide tube open-

ing to bring the fluid level between the upper mark and the lower mark of the dipstick. Do not fill the fluid above the upper mark. Always use ATF DW-1 (genuine Honda automatic transaxle fluid).

➡**Using a non-Honda ATF can affect shift quality.**

10. Insert the dipstick into the guide tube aligning the notch with the guide tab.

FLUID RECOMMENDATIONS

ATF DW-1 (genuine Honda automatic transaxle fluid).

FRONT DRIVESHAFT

REMOVAL & INSTALLATION

2011 Models

See Figures 56 through 58.

1. Raise and support the vehicle.
2. Remove the front wheels.

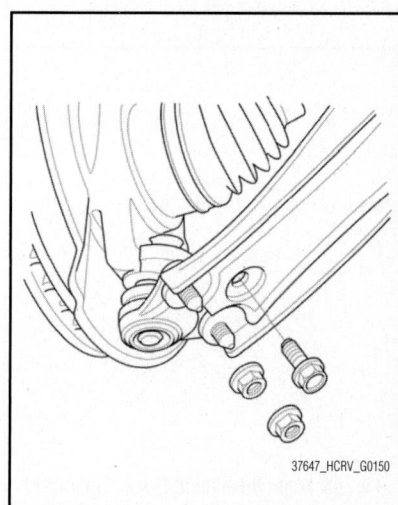

37647_HCRV_G0150

Fig. 56 Remove the nuts and bolt, then separate the lower arm using a prybar

3. Pry up the stake on the spindle nut, then remove the nut.

4. Drain the transmission fluid, then reinstall the drain plug with a new sealing washer.

5. Remove the nuts and bolt, then separate the lower arm using a prybar.

6. Pull the knuckle outward, and separate the outboard joint from the front hub using a soft face hammer.

7. Left driveshaft: Pry the inboard joint from the differential with a prybar. Remove the driveshaft as an assembly.

➡ **Do not pull on the driveshaft, or the inboard joint may come apart. Pull the inboard joint straight out to avoid damaging the oil seal. Be careful not to damage the oil seal or the end of the inboard joint using a prybar.**

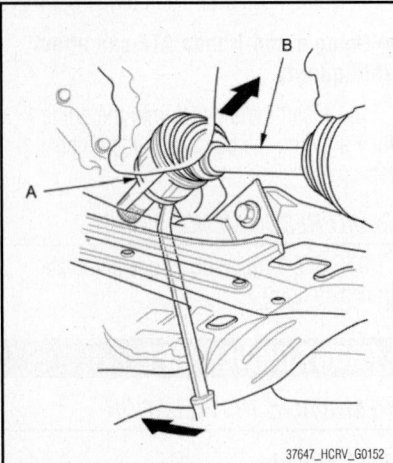

Fig. 57 Left driveshaft: Pry the inboard joint (A) from the differential with a prybar; remove the driveshaft (B) as an assembly

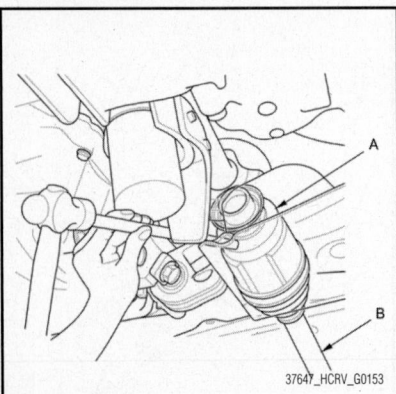

Fig. 58 Right driveshaft: Drive the inboard joint (A) off of the intermediate shaft using a drift punch and a hammer. Remove the driveshaft (B) as an assembly

8. Right driveshaft: Drive the inboard joint off of the intermediate shaft using a drift punch and a hammer. Remove the driveshaft as an assembly.

➡ **Do not pull on the driveshaft or the inboard joint may come apart. Be careful not to damage the end of the inboard joint with the drift.**

9. Remove the set ring from the left driveshaft inboard joint.

10. Remove the set ring from the intermediate shaft.

To install:

➡ **Before starting installation, make sure the mating surfaces of the joint and the splined section are clean.**

11. Apply grease 0.18 oz. (5 g) of Moly 60 paste (P/N 08734-0001) to the contact area of the outboard joint and the front wheel bearing.

➡ **The paste helps to prevent noise and vibration.**

12. Install a new set ring into the set ring groove of the left driveshaft inboard joint.

13. Install a new set ring into the set ring groove of the intermediate shaft.

14. Apply 0.02–0.04 oz. (0.5–1.0 g) of super high temp urea grease (P/N 08798-9002) to the whole splined surface of the right driveshaft. After applying grease, remove the grease from the splined grooves at intervals of 2–3 splines and from the set ring groove so that air can bleed from the intermediate shaft.

15. Clean the areas where the driveshaft contacts the differential thoroughly with solvent, and dry with compressed air.

➡ **Do not wash the rubber parts with solvent.**

16. Insert the inboard end of the driveshaft into the differential or the intermediate shaft until the set ring locks in the groove.

➡ **Insert the driveshaft horizontally to prevent damaging the oil seal.**

17. Install the outboard joint into the front hub on the knuckle.

18. Install the knuckle onto the lower arm. Then tighten the nuts and bolt to 38 ft. lbs. (52 Nm).

➡ **During installation, loosely install a new flange bolt and new self-locking nuts. After lightly tightening all three fasteners, tighten them to the specified torque in the following order; the nut on the front, the nut on the rear, then the bolt.**

19. Apply a small amount of engine oil to the seating surface of a new spindle nut.

20. Install the spindle nut, then tighten it to 242 ft. lbs. (328 Nm). After tightening, use a drift to stake the spindle nut shoulder against the driveshaft.

21. Clean the mating surfaces of the brake discs and the front wheels, then install the front wheels.

22. Turn the front wheels by hand, and make sure there is no interference between the driveshaft and the surrounding parts.

23. Lower the vehicle.

24. Refill the transmission with the recommended automatic transmission fluid.

25. Check the wheel alignment, and adjust it if necessary.

26. Test-drive the vehicle.

2012 Models

See Figures 59 through 64.

1. Raise the vehicle on a lift, and make sure it is securely supported.

2. Remove the front wheel.

3. Pry up the stake on the spindle nut. Remove the spindle nut.

4. Remove the front splash shield.

5. Remove the drain plug and drain the ATF. Reinstall the drain plug using a new sealing washer. Tighten the plug to 36 ft. lbs. (49 Nm).

6. Remove the flange bolt. Remove the flange nuts.

7. Disconnect the lower arm from the front lower ball joint.

8. Separate the outboard joint from the front hub using a soft face hammer.

9. Pull the knuckle outward, and separate the outboard joint from the front hub.

10. Pry the inboard joint from the differential using a pry bar.

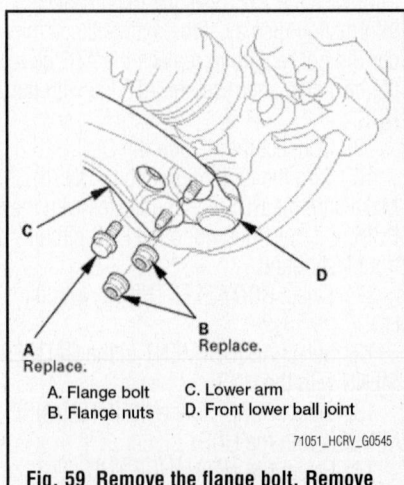

A. Flange bolt   C. Lower arm
B. Flange nuts   D. Front lower ball joint

Fig. 59 Remove the flange bolt. Remove the flange nuts. Disconnect the lower arm from the front lower ball joint

※※ **CAUTION**

Do not pull on the driveshaft, or the inboard joint may come apart. Pull the inboard joint straight out to avoid damaging the oil seal.

➡Be careful not to damage the oil seal or the end of the inboard joint using the pry bar.

11. Remove the set ring.

12. Drive the right front inboard joint off of the intermediate shaft using a drift punch and a hammer.

※※ **CAUTION**

Do not pull on the driveshaft or the inboard joint may come apart.

13. Remove the set ring.

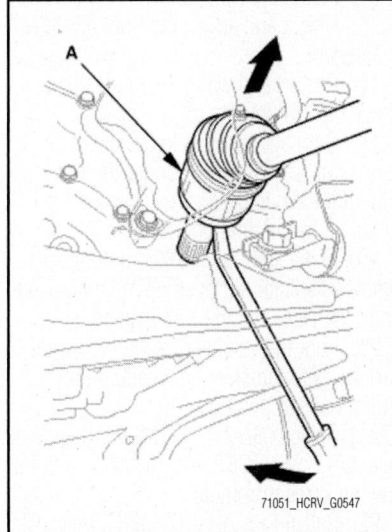

Fig. 60 Pry the inboard joint (A) from the differential using a pry bar

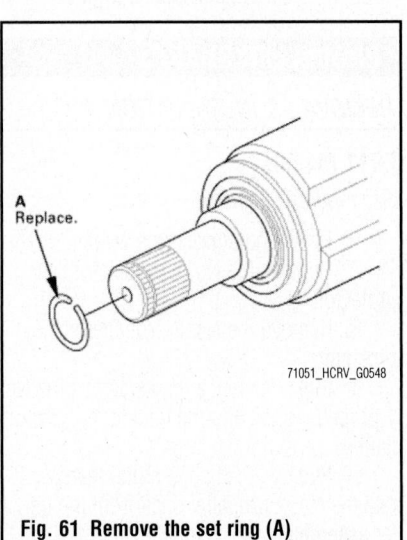

Fig. 61 Remove the set ring (A)

To install:

14. On the right front driveshaft inboard end, install a new set ring.

15. Apply a small amount of super high temp urea grease to the whole splined surface of the right driveshaft.

16. After applying grease, remove the grease from the splined grooves at intervals of every second or third spline and from the set ring groove so that air can bleed from the intermediate shaft during reassembly.

17. Insert the inboard end of the driveshaft onto the intermediate shaft until the set ring locks in the groove.

18. Install a new set ring.

19. Clean the areas where the driveshaft contacts the differential thoroughly with solvent.

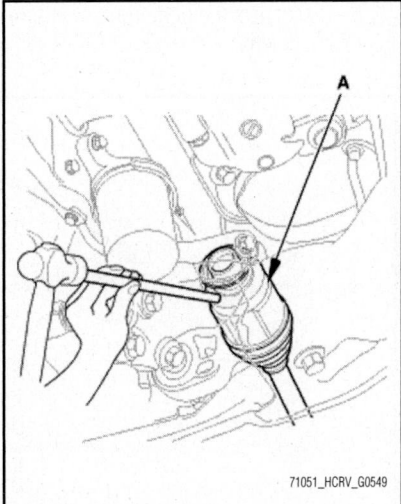

Fig. 62 Drive the right front inboard joint (A) off of the intermediate shaft using a drift punch and a hammer

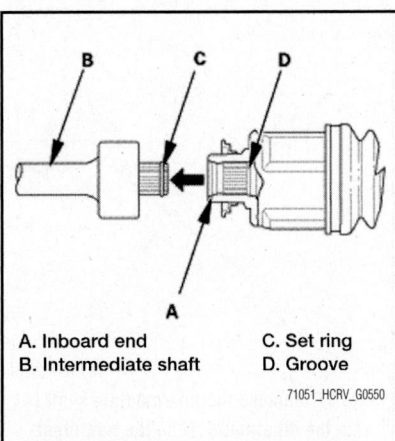

A. Inboard end   C. Set ring
B. Intermediate shaft   D. Groove

Fig. 63 Insert the inboard end of the driveshaft onto the intermediate shaft until the set ring locks in the groove

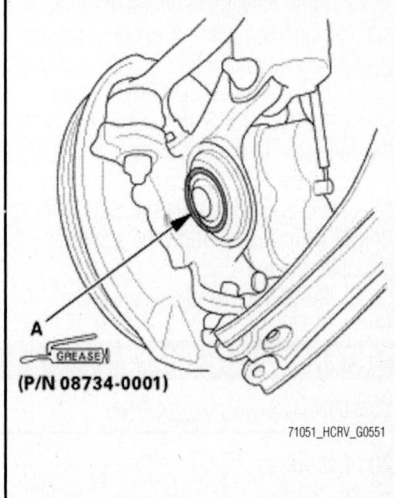

(P/N 08734-0001)

Fig. 64 Apply a small amount of moly 60 paste to the contact area of the outboard joint and the front wheel bearing

➡Do not wash the rubber parts with solvent.

20. Dry the areas where the driveshaft contacts the differential thoroughly with compressed air.

21. Insert the inboard end of the driveshaft into the differential until the set ring locks in the groove.

➡Insert the driveshaft horizontally to prevent damaging the oil seal.

22. Apply a small amount of moly 60 paste to the contact area of the outboard joint and the front wheel bearing.

➡The paste helps to prevent noise and vibration.

23. Install the outboard joint into the front hub on the knuckle.

24. Install the lower arm to the ball joint. Tighten the nuts to 38 ft. lbs. (52 Nm).

25. Install a new spindle nut and tighten to 242 ft. lbs. (328 Nm). Stake the nut.

26. Install the front wheels.

27. Turn the wheel by hand to ensure there are no interferences with the driveshaft.

28. Check and refill the ATF.

29. Install the front splash shield.

30. Check the front wheel alignment. Adjust as needed.

31. Restart and warm the engine and check the ATF level.

32. Connect the HDS to the data link connector (DLC) located under the driver's side of the dashboard.

33. Turn the ignition switch to ON (II).

34. Make sure the HDS communicates with the vehicle. If it does not communicate, go to the DLC circuit troubleshooting.

35. Park the vehicle on a flat and level surface, with the steering wheel in the straight ahead position.

36. Select VSA ADJUSTMENT, then select ALL SENSORS with the HDS, and follow the screen prompts.

37. Select EPS ADJUSTMENT, then select EPS STEERING ANGLE SENSOR VALUE CLEAR and follow the screen prompts on the HDS.

38. Reset the Maintenance Minder.

INTERMEDIATE SHAFT

REMOVAL & INSTALLATION

2011 Models

See Figure 65.

1. Drain the automatic transmission fluid. Reinstall the drain plug with a new sealing washer.

2. Remove the front right driveshaft.

3. Remove the flange bolt and the two dowel bolts.

➡**Remove the set ring from the intermediate shaft if it is installed.**

4. Remove the intermediate shaft from the differential. Hold the intermediate shaft horizontal until it is clear of the differential to prevent damaging the oil seal.

To install:

5. Clean the areas where the intermediate shaft contact the differential thoroughly with solvent, and dry with compressed air.

➡**Do not wash the rubber parts with solvent.**

6. Insert the intermediate shaft into differential correctly.

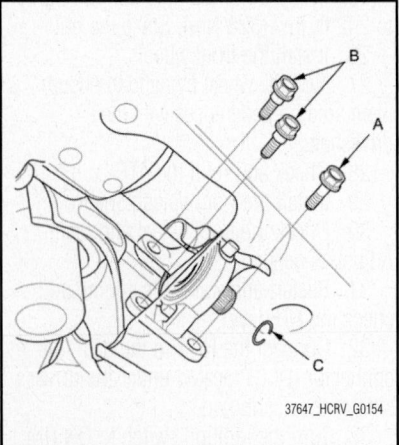

Fig. 65 Remove the flange bolt (A) and the two dowel bolts (B)

37647_HCRV_G0154

➡**Hold the intermediate shaft horizontally to prevent damaging the oil seal.**

7. Install the flange bolt and two dowel bolts, then install a new set ring. Tighten the bolts to 29 ft. lbs. (39 Nm).

8. Install the front right driveshaft.

9. Refill the transmission with the recommended automatic transmission fluid.

10. Check the wheel alignment, and adjust it if necessary.

11. Test-drive the vehicle.

2012 Models

See Figures 66 and 67.

1. Raise and support the vehicle on a lift.

2. Remove the front splash shield.

3. Drain the ATF. Install the plug with a new washer. Tighten to 36 ft. lbs. (49 Nm).

4. Remove the right front wheel.

5. Disconnect the right front lower arm from the ball joint.

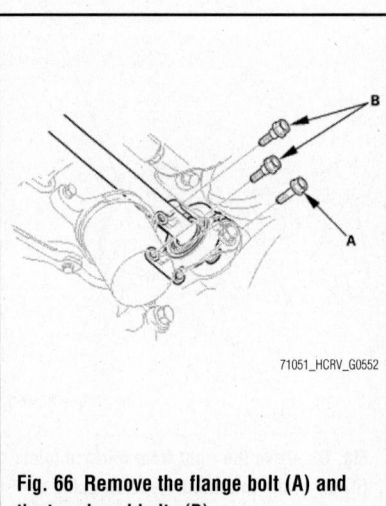

71051_HCRV_G0552

Fig. 66 Remove the flange bolt (A) and the two dowel bolts (B)

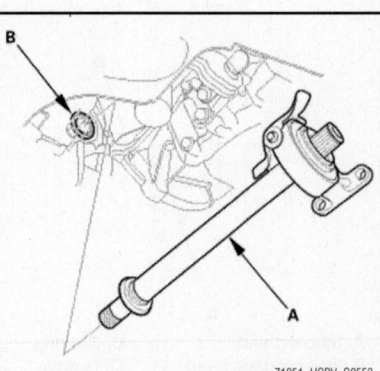

71051_HCRV_G0553

Fig. 67 Remove the intermediate shaft (A) from the differential. Hold the intermediate shaft horizontal until it is clear of the differential to prevent damaging the oil seal (B)

6. Drive the inboard joint off of the intermediate shaft using a drift punch and a hammer. See "Front Driveshaft" in this section.

7. Remove the set ring.

8. Remove the flange bolt and the two dowel bolts.

9. Remove the intermediate shaft from the differential. Hold the intermediate shaft horizontal until it is clear of the differential to prevent damaging the oil seal.

To install:

10. Install the intermediate driveshaft. Tighten the flange bolt and dowel bolts to 29 ft. lbs. (39 Nm).

11. Install the right front driveshaft inboard side, as described in this section.

12. Install the right front wheel.

13. Refill the ATF.

14. Install the front splash shield.

15. Check and adjust the front alignment as needed.

16. Restart and warm the engine and check the ATF level.

17. Connect the HDS to the data link connector (DLC) located under the driver's side of the dashboard.

18. Turn the ignition switch to ON (II).

19. Make sure the HDS communicates with the vehicle. If it does not communicate, go to the DLC circuit troubleshooting.

20. Park the vehicle on a flat and level surface, with the steering wheel in the straight ahead position.

21. Select VSA ADJUSTMENT, then select ALL SENSORS with the HDS, and follow the screen prompts.

22. Select EPS ADJUSTMENT, then select EPS STEERING ANGLE SENSOR VALUE CLEAR and follow the screen prompts on the HDS.

23. Reset the Maintenance Minder.

PROPELLER SHAFT

REMOVAL & INSTALLATION

2011 Models

See Figures 68 and 69.

1. Raise and support the vehicle.

2. Remove the No. 1 propeller shaft protector.

3. Remove the No. 2 propeller shaft protector.

4. Make reference marks across the No. 1 propeller shaft and the transfer companion flange.

5. Remove the flange bolts, then separate the No. 1 propeller shaft from the transfer assembly.

6. Remove the center support bearing mounting bolts.

7. Make reference marks across the No. 2 propeller shaft and the rear differential companion flange.

8. Remove the flange bolts, then separate the No. 2 propeller shaft from the rear differential, then remove the propeller shaft assembly.

To install:

9. Set the No. 2 propeller shaft to the rear differential by aligning the reference mark you made during the removal procedure. Then install new flange bolts and tighten to 24 ft. lbs. (32 Nm).

10. Install the center support bearing with new bolts. Tighten the bolts to 29 ft. lbs. (39 Nm).

11. Set the No. 1 propeller shaft to the transfer companion flange by aligning the reference mark you made during the

removal procedure. Then install new flange bolts and tighten to 24 ft. lbs. (32 Nm).

12. Install the No. 2 propeller shaft protector.

13. Install the No. 1 propeller shaft protector.

14. If you installed a new propeller shaft assembly, test-drive the vehicle at 55 mph (88 km/h), and check for noise or vibration. If there is a noise or vibration,

rotate the propeller shaft 180 degrees from its current alignment with the rear differential companion flange, then recheck.

2012 Models

See Figures 70 through 72.

1. Raise the vehicle on a lift, and make sure it is securely supported.

2. Remove the bolts and the clips

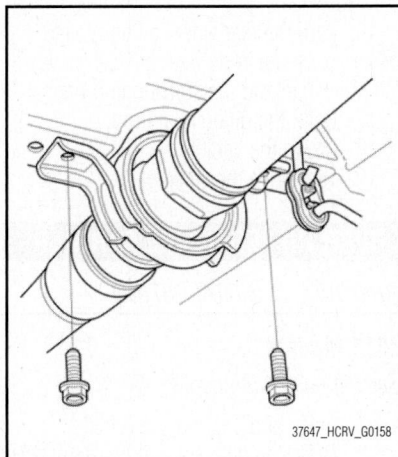

Fig. 68 Remove the center support bearing mounting bolts

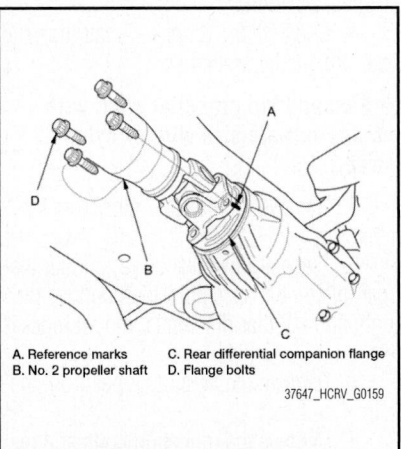

A. Reference marks C. Rear differential companion flange
B. No. 2 propeller shaft D. Flange bolts

37647_HCRV_G0159

Fig. 69 Make reference marks across the No. 2 propeller shaft and the rear differential companion flange

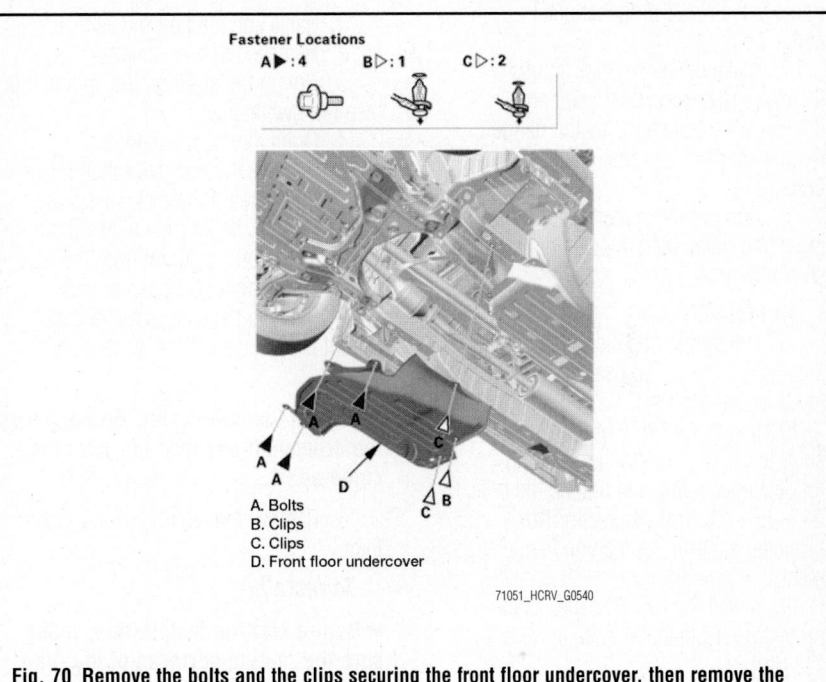

A. Bolts
B. Clips
C. Clips
D. Front floor undercover

71051_HCRV_G0540

Fig. 70 Remove the bolts and the clips securing the front floor undercover, then remove the front floor undercover

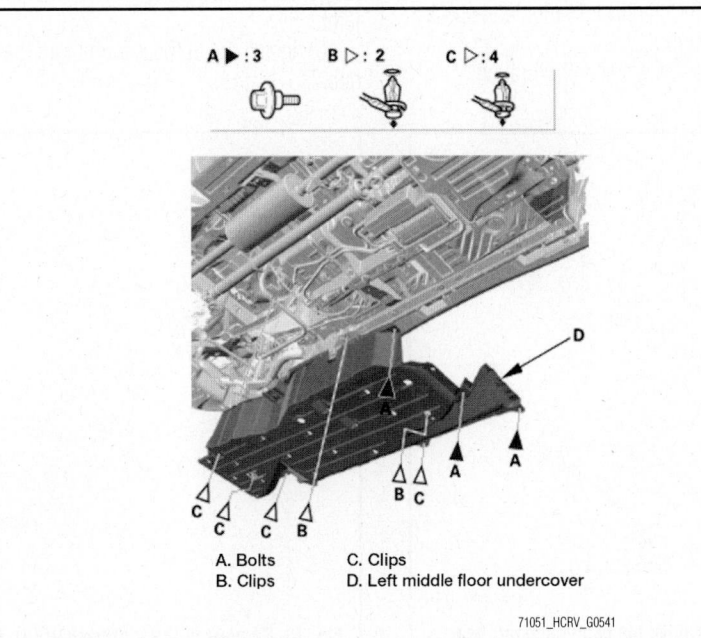

A. Bolts C. Clips
B. Clips D. Left middle floor undercover

71051_HCRV_G0541

Fig. 71 Remove the bolts and clips, then remove the left middle floor undercover

securing the front floor undercover, then remove the front floor undercover.

3. Remove the bolts and clips, then remove the left middle floor undercover.

4. Remove the No. 1 propeller shaft protector.

5. Remove the No. 2 propeller shaft protector.

6. Make reference mark across the No. 1 (front) propeller shaft and the transfer companion flange. Remove the No. 1 propeller shaft flange bolts.

7. Make reference mark across the No. 2 (rear) propeller shaft and the rear differential companion flange. Remove the No. 2 propeller shaft flange bolts.

8. Remove the center support bearing mounting bolts, then remove the propeller shaft assembly.

To install:

9. Install the propeller shaft and install and tighten the center bearing support bolts to 29 ft. lbs. (39 Nm).

10. Install the No. 2 propeller shaft-to-companion flange, aligning the reference mark, and tighten the bolts to 24 ft. lbs. (32 Nm). Repeat for No. 1 propeller shaft-to-companion flange attachment.

11. Install the two propeller shaft protectors. Tighten the bolts to 16 ft. lbs. (22 Nm).

12. Reinstall the middle and front floor undercovers.

13. If you installed a new propeller shaft, test-drive the vehicle at 55 mph and check for noise or vibration.

14. If there is a noise or vibration,

rotate the propeller shaft 180 degrees from its current alignment with the rear differential companion flange, then recheck.

REAR DRIVESHAFT

REMOVAL & INSTALLATION

2011 Models

See Figure 73.

1. Raise and support the vehicle.
2. Remove the rear wheels.
3. Pry up the stake on the spindle nut, then remove the nut.
4. Drain the differential fluid.
5. Remove the rear driveshaft inboard joint from the rear differential assembly.
6. Pull up the knuckle outward, and separate the outboard joint from the rear wheel hub using a soft face hammer.
7. Remove the rear driveshaft. Be careful not to damage the wheel speed sensor.

➡**Pull on the outer joint. Do not pull on the driveshaft because the joint may come apart.**

8. Remove the set ring from inboard joint.

To install:

➡**Before starting installation, make sure the mating surfaces of the joint and the splined section are clean.**

9. Install a new set ring onto the set ring groove of the rear driveshaft inboard joint.

10. Install the outboard joint into the rear hub.

➡**Be careful not to damage the wheel speed sensor.**

11. Clean the areas where the driveshaft contacts the differential thoroughly with solvent, and dry with compressed air.

➡**Do not wash the rubber parts with solvent.**

12. Insert the inboard end of the driveshaft into the differential until the set ring locks in the groove.

➡**Insert the driveshaft horizontally to prevent damaging the oil seal.**

13. Apply a small amount of engine oil to the seating surface of a new spindle nut.

14. Install the new spindle nut, then tighten it to 181 ft. lbs. (245 Nm). After tightening, use a drift to stake the spindle nut shoulder against the driveshaft.

15. Clean the mating surfaces of the brake discs and the rear wheels, then install the rear wheels.

16. Turn the rear wheel by hand, and make sure there is no interference between the driveshaft and the surrounding parts.

17. Refill the differential fluid.

18. Lower the vehicle.

19. Test-drive the vehicle.

REAR DIFFERENTIAL CARRIER

REMOVAL & INSTALLATION

2011 Models

See Figures 74 through 76.

1. Raise and support the vehicle.

2. Drain the differential fluid. Reinstall the drain plug with a new sealing washer.

3. Mark the propeller shaft and the companion flange of the rear differential assembly so they can be reinstalled in their original positions.

4. Separate the propeller shaft from the rear differential assembly.

➡**Suspend the propeller shaft with an appropriate size wire or nylon strap.**

5. Remove the breather tube from the clip.

6. Place a transmission jack under the rear differential assembly, then remove the right and left rear differential mount brackets.

7. Remove the breather tubes from the clip.

8. Remove the mounting bolts and the plates.

9. Disconnect the breather tube from the breather tube fitting.

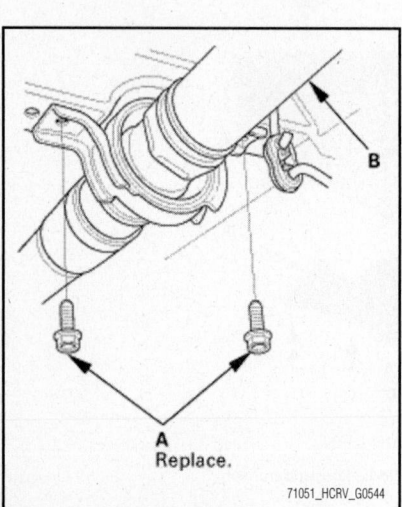

Fig. 72 Remove the center support bearing mounting bolts (A), then remove the propeller shaft assembly (B)

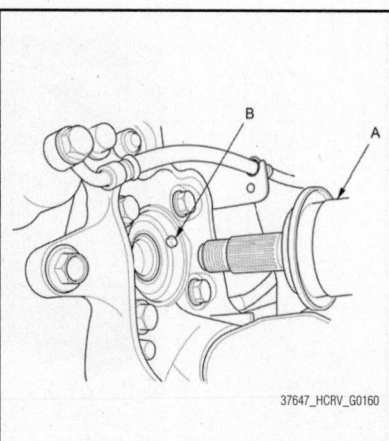

Fig. 73 Remove the rear driveshaft (A); be careful not to damage the wheel speed sensor (B)

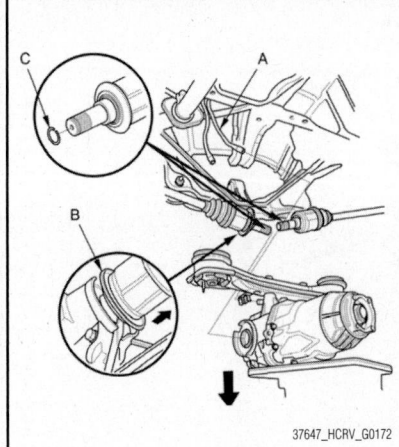

Fig. 74 Lowering the differential assembly showing the breather tube (A), driveshaft ring (B), and set rings (C)

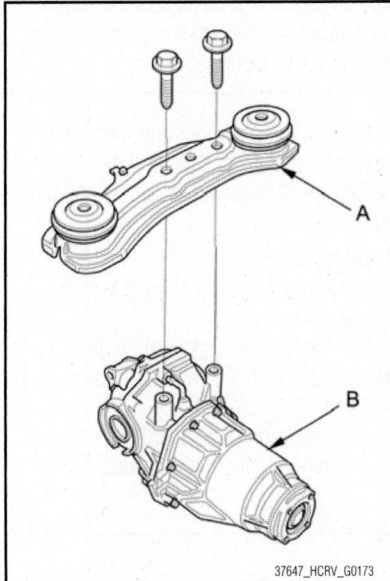

Fig. 75 Remove the rear differential mount assembly (A) from the rear differential assembly (B)

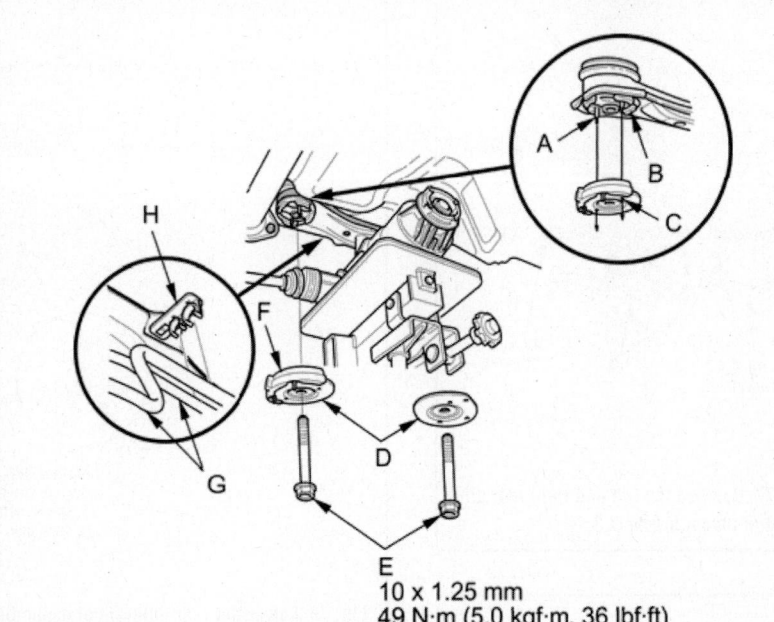

E.
10 x 1.25 mm
49 N·m (5.0 kgf·m, 36 lbf·ft)

A. Tab   E. Rear differential mount assembly mounting bolts
B. Rubber mount   F. Rubber heat insulator
C. Hole   G. Breather tubes
D. Plates   H. Clips

Fig. 76 Aligning the rear differential mount assembly

10. Lower the rear differential assembly while pulling both driveshaft inboard joints out of the rear differential assembly.

➡**Be careful not to damage the driveshaft ring when prying out the driveshaft inboard joints.**

11. Remove the set rings.
12. Remove the rear differential mount assembly from the rear differential assembly.

To install:
13. Install the rear differential mount assembly to the rear differential assembly

with new bolts. Tighten the bolts to 51 ft. lbs. (69 Nm).
14. Raise the rear differential with the transmission jack.
15. Install new set rings into the groove of the rear driveshaft inboard joints, then insert the driveshafts into the rear differential.
16. Lift the rear differential up into position, then push on both driveshafts to lock the set rings into place. Connect the breather tube.
17. Align the tab of the rubber mount with the hole of the plates, then install the plates and torque the rear differential mount assembly mounting bolts.

➡**The rubber heat insulator is installed only in the right side.**

18. Install the breather tubes to the clips.
19. Install the right and left rear differential mount brackets, then torque the new bolts. Tighten the two 12 x 1.25 mm bolts to 44 ft. lbs. (59 Nm) and the 10 x 1.25 mm bolt to 47 ft. lbs. (64 Nm).
20. Install the breather tube to the clip.
21. Install the No. 2 propeller shaft onto the rear differential by aligning the reference marks made during removal.

Make sure you use new mounting bolts. Tighten the bolts to 24 ft. lbs. (32 Nm).
22. Refill the rear differential with the recommended fluid.
23. Test-drive the vehicle.

2012 Models

FWD Models

➡**Removal of the differential on 2WD models requires complete disassembly of the transaxle.**

AWD Models

See Figures 77 through 86.

1. Raise the vehicle on a lift, and make sure it is securely supported.
2. Remove the oil filler plug and the sealing washer.
3. Remove the drain plug and the sealing washer, then drain the rear differential fluid.
4. Make a reference mark across the No. 2 propeller shaft and the rear differential companion flange. Remove the No. 2 propeller shaft flange bolts. Separate the No. 2 propeller shaft from the rear differential.

➡**Suspend the propeller shaft with an appropriate size nylon strap.**

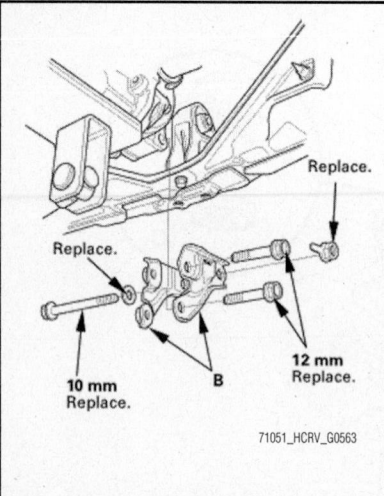

Fig. 77 Remove the left and right rear differential mount brackets B

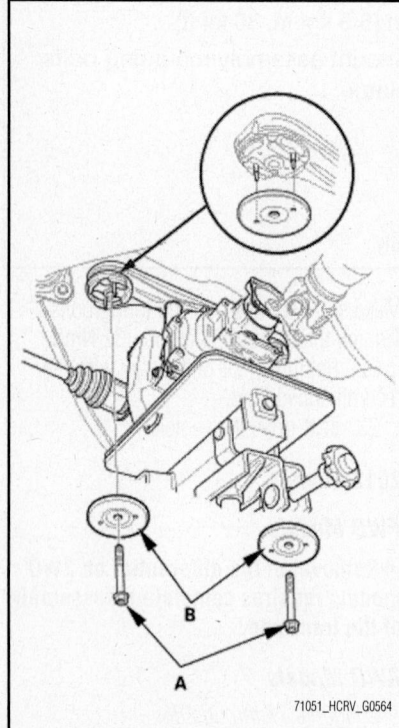

Fig. 78 Remove the mounting bolts (A) and the plates (B)

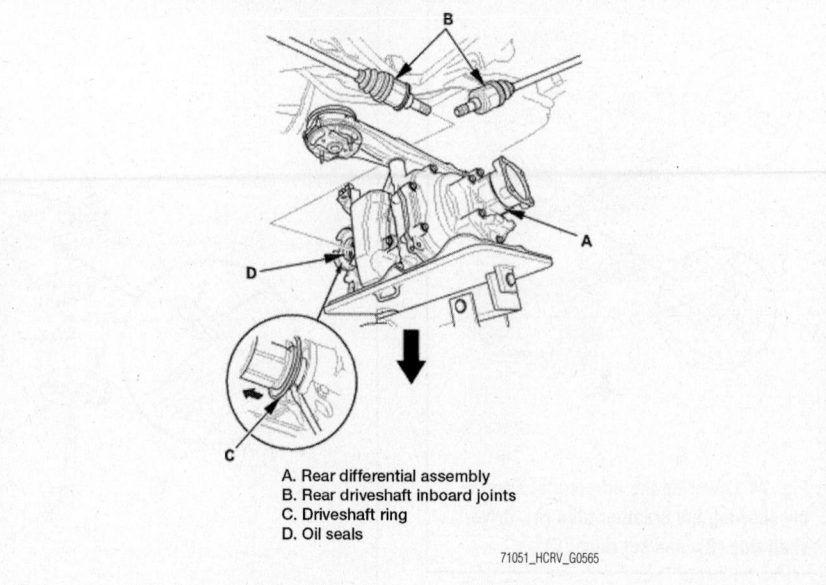

A. Rear differential assembly
B. Rear driveshaft inboard joints
C. Driveshaft ring
D. Oil seals

Fig. 79 Lower the rear differential assembly while pulling both rear driveshaft inboard joints out of the rear differential assembly. Be careful not to damage the driveshaft ring when prying out the driveshaft inboard joints. Be careful not to damage the oil seals

5. Remove the connector cover. Disconnect the connectors.

6. Release the harness clips.

7. Disconnect the breather hoses from the breather pipes and clamps.

8. Disconnect the breather hose from the clamp.

9. Place a transaxle jack under the rear differential assembly.

10. Remove the left and right rear differential mount brackets B.

11. Remove the mounting bolts and the plates.

12. Lower the rear differential assembly while pulling both rear driveshaft inboard joints out of the rear differential assembly. Be careful not to damage the driveshaft ring when prying out the driveshaft inboard joints. Be careful not to damage the oil seals.

13. Remove the set ring from inboard joint.

14. Remove the rear differential mount assembly A from the rear differential assembly.

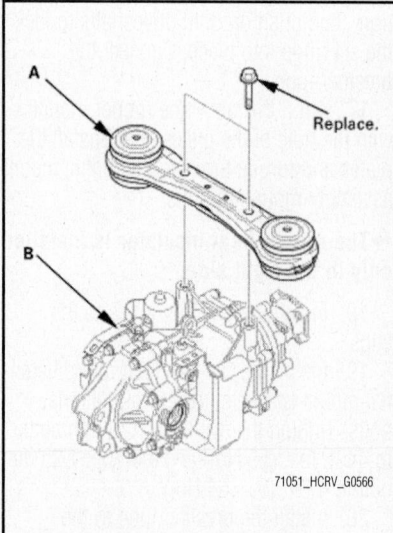

Fig. 80 Remove the rear differential mount assembly A from the rear differential assembly (B)

15. Remove the mounting bolts in a crisscross pattern in several steps. Remove the differential housing assembly and the oil seals.

To install:

16. Remove the dirt and oil from the sealing surfaces.

17. Apply liquid gasket (P/N 08718-0004 or 08718-0009) evenly to the sealing surface.

18. Install the component within 5 minutes of applying the liquid gasket. Make

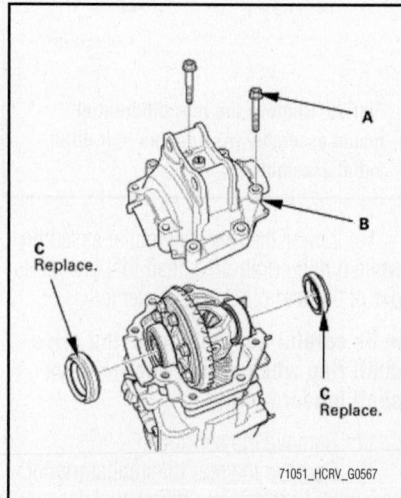

Fig. 81 Remove the mounting bolts (A) in a crisscross pattern in several steps. Remove the differential housing assembly (B) and the oil seals (C)

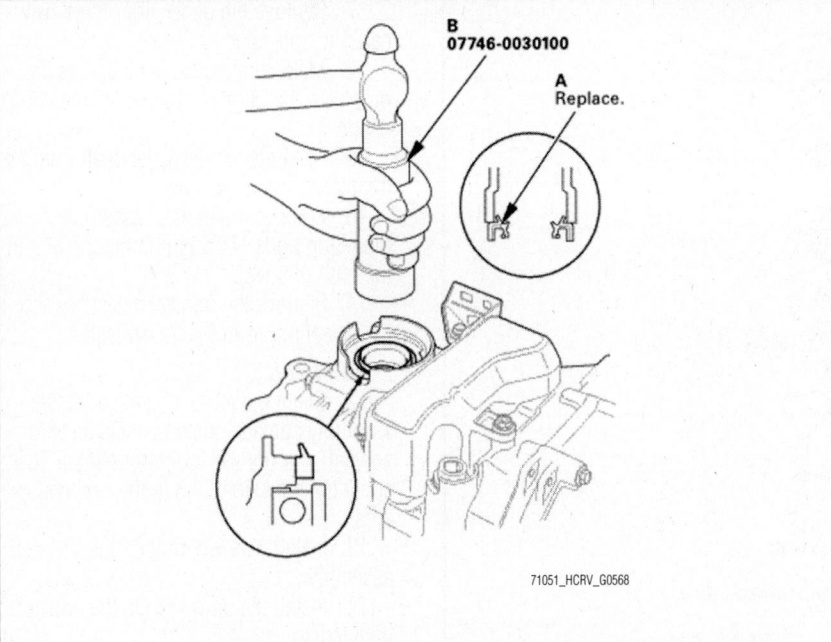

Fig. 82 Install a new rear differential right side oil seal (A) even and flush with the machined edge of the differential carrier assembly using the 40 mm driver handle (B)

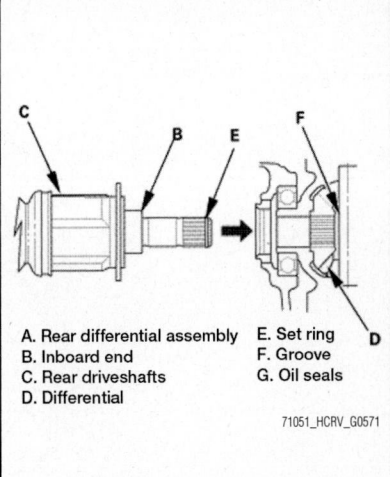

A. Rear differential assembly   E. Set ring
B. Inboard end   F. Groove
C. Rear driveshafts   G. Oil seals
D. Differential

71051_HCRV_G0571

Fig. 84 View 2: Lift the rear differential assembly up into position. Insert the inboard end of the rear driveshafts into the differential until the set ring locks in the groove. Insert the driveshaft horizontally to prevent damaging the oil seals

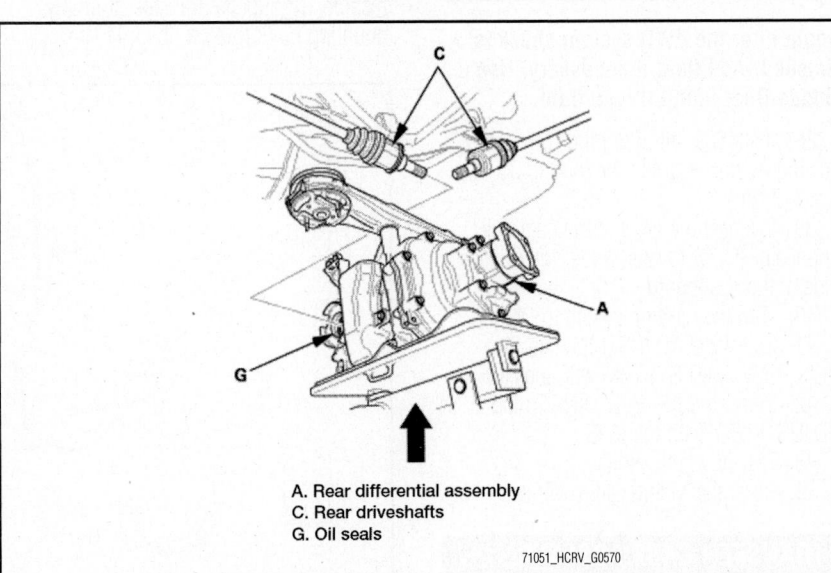

A. Rear differential assembly
C. Rear driveshafts
G. Oil seals

71051_HCRV_G0570

Fig. 83 View 1: Lift the rear differential assembly up into position. Insert the inboard end of the rear driveshafts into the differential until the set ring locks in the groove. Insert the driveshaft horizontally to prevent damaging the oil seals

sure you seal the entire circumference of the bolt holes to prevent fluid leakage.

- If you apply liquid gasket P/N 08718-0012, the component must be installed within 4 minutes.
- If too much time has passed after applying the liquid gasket, remove the old liquid gasket and residue, then reapply the new liquid gasket.
- Allow it to cure at least 30 minutes after assembly before filling the differential with the recommended fluid.

19. Install the differential housing assembly. Torque the mounting bolts, in a crisscross pattern in several steps, to 20 ft. lbs. (27 Nm).

20. Install a new rear differential right side oil seal even and flush with the machined edge of the differential carrier assembly using the 40 mm driver handle.

21. Install the rear differential mount assembly A to the rear differential assembly with new bolts tightened to 44 ft. lbs. (59 Nm).

22. Install new set ring into the groove of the rear driveshaft inboard joint on both sides.

23. Lift the rear differential assembly up into position.

24. Insert the inboard end of the rear driveshafts into the differential until the set ring locks in the groove. Insert the driveshaft horizontally to prevent damaging the oil seals.

25. Align the tabs of the rubber mount with the holes of the plates. Install the rear differential mounting bolts. Tighten the bolts to 32 ft. lbs. (44 Nm).

26. Install the left and right rear differential mount brackets.

27. Tighten new differential mount brackets mounting bolts.

28. Loosely install new differential mount brackets mounting bolt.

29. Lower the transaxle jack.

30. Tighten differential mount brackets B mounting bolt.

31. Connect the breather hoses to the breather pipes and clamps.

32. Connect the breather hose to the clamp.

33. Install the harness clips.

34. Connect the connectors. Install the connector cover.

35. Attach the No. 2 propeller shaft to the rear differential companion flange by aligning the reference mark you made during the removal procedure. Install new flange bolts to the specified torque of 24 ft. lbs. (32 Nm).

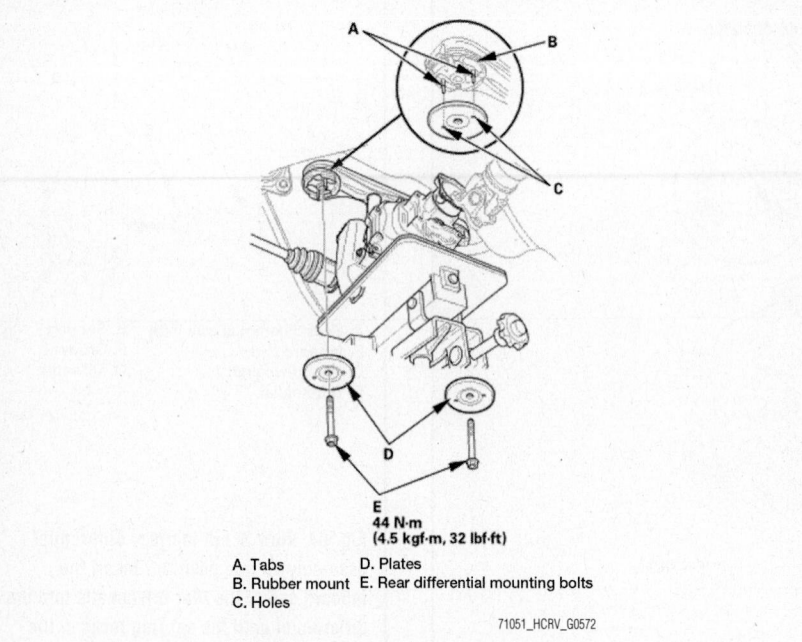

A. Tabs   D. Plates
B. Rubber mount   E. Rear differential mounting bolts
C. Holes

E
44 N·m
(4.5 kgf·m, 32 lbf·ft)

71051_HCRV_G0572

Fig. 85 Align the tabs of the rubber mount with the holes of the plates. Install the rear differential mounting bolts. Tighten the bolts to 32 ft. lbs. (44 Nm)

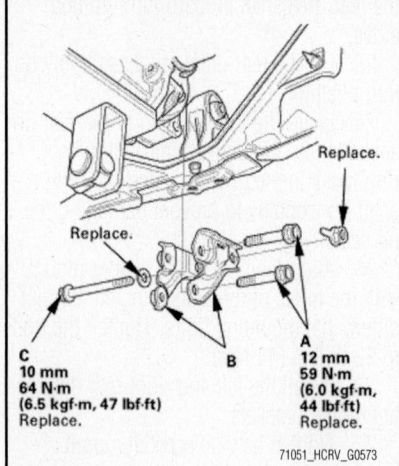

Replace.

Replace.

C
10 mm
64 N·m
(6.5 kgf·m, 47 lbf·ft)
Replace.

B

A
12 mm
59 N·m
(6.0 kgf·m,
44 lbf·ft)
Replace.

71051_HCRV_G0573

Fig. 86 Install the left and right rear differential mount brackets. Tighten new differential mount brackets B mounting bolts (A) to the specified torque. Loosely install new differential mount brackets (B) mounting bolt (C). Lower the transaxle jack. Tighten differential mount brackets B mounting bolt to the specified torque

36. Clean the drain plug, then install it with a new sealing washer. Tighten the plug to 35 ft. lbs. (47 Nm).

37. Refill the differential with rear differential fluid to the proper level.

➡ **If you disassemble the differential, check the rear differential fluid level**

again after the AWD system check is finished. Add fluid if necessary. Use Honda Dual Pump II type fluid.

38. Install the oil filler plug with a new sealing washer. Tighten the plug to 35 ft. lbs. (47 Nm).

39. Connect the HDS to the data link connector (DLC) located under the driver's side of the dashboard.

40. Turn the ignition switch to ON (II).

41. Select the ADJUSTMENT in the REAL TIME AWD SYSTEM with the HDS.

42. Select the AIR BLEEDING in the ADJUSTMENT with the HDS.

43. Test-drive the vehicle.

44. Reset the Maintenance Minder.

TRANSFER CASE ASSEMBLY

REMOVAL & INSTALLATION

2011 Models

See Figures 87 and 88.

1. Shift the transmission into N.

2. Raise the vehicle on a lift, and make sure it is supported securely.

3. Remove the drain plug, and drain the Automatic Transmission Fluid (ATF).

4. Reinstall the drain plug with a new sealing washer.

5. Disconnect the A/F sensor connector (if necessary) and the secondary HO2S connector.

6. Remove the secondary HO2S harness from the harness clamp.

7. Remove the under-floor three-way catalytic converter.

8. Make a reference mark across the propeller shaft and the transfer companion flange.

9. Separate the propeller shaft from the transfer companion flange.

10. Remove the transfer assembly mounting bolts (4), and pull out the bolt to the limit of travel.

11. Remove the transfer assembly and the dowel pin from the transmission.

To install:

12. Clean the areas where the transfer assembly contacts the transmission with solvent, and dry with compressed air. Then apply transmission fluid to the seal contact area.

13. Install a new O-ring on the transfer assembly.

14. Install the dowel pin in the transfer housing.

15. Install the one bolt part-way in the rear lower of the transfer housing, and install the transfer assembly on the transmission. Tighten the transfer assembly mounting bolts to 33 ft. lbs. (44 Nm).

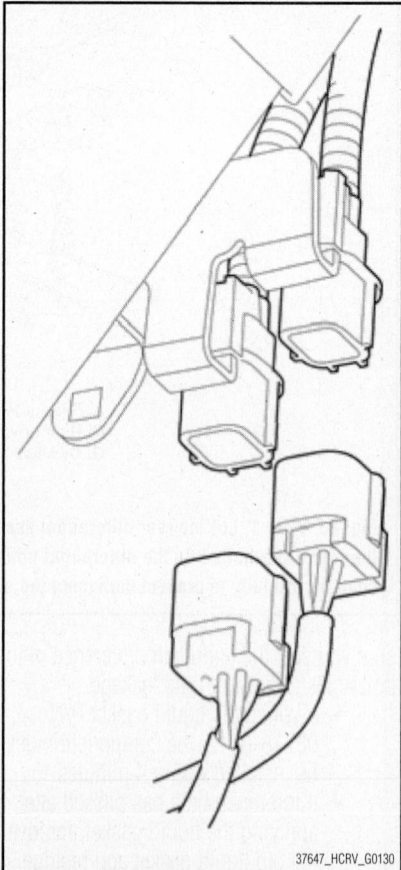

37647_HCRV_G0130

Fig. 87 Disconnect the A/F sensor connector (K24Z1 engine) and the secondary HO2S connector

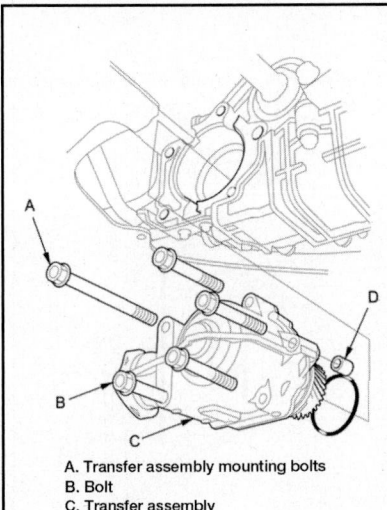

A. Transfer assembly mounting bolts
B. Bolt
C. Transfer assembly
D. Dowel pin

37647_HCRV_G0177

Fig. 88 Remove the transfer assembly mounting bolts (4), and pull out the bolt to the limit of travel

16. Install the propeller shaft to the transfer companion flange by aligning the reference mark.

17. Install the under-floor three-way catalytic converter with the bolts, new self-locking nuts, and new gaskets (A).

18. Connect the A/F sensor connector (if necessary) and the secondary HO2S connector.

19. Install the secondary HO2S harness in the harness clamp.

20. Refill the transmission with ATF.

2012 Models

See Figure 89.

1. Raise and support the vehicle.
2. Remove the splash shield.
3. Remove the front floor undercover.
4. Drain the ATF, as described in this section.
5. Remove the under floor TWC. See "Catalytic Converter" in "ENGINE MECHANICAL" section.
6. Disconnect the propeller shaft from the transfer assembly:
   a. Make reference mark across the No. 1 propeller shaft and the transfer companion flange.
   b. Separate the No. 1 propeller shaft from the transfer companion flange.
   c. Suspend the propeller shaft with an appropriate size nylon strap.
7. Remove the transfer assembly and the dowel pin.
8. Remove the O-ring from the transfer assembly.

To install:

9. Install a new O-ring on the transfer assembly. Install the transfer assembly and the dowel pin. Tighten the mounting bolts to 32 ft. lbs. (44 Nm).

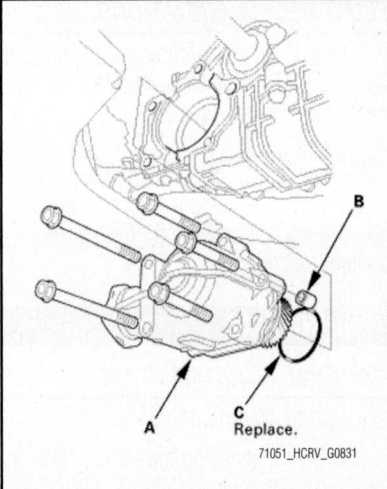

71051_HCRV_G0831

Fig. 89 Remove the transfer assembly (A) and the dowel pin (B). Remove the O-ring (C) from the transfer assembly

10. Set the No. 1 propeller shaft to the transfer companion flange by aligning the reference mark you made during the removal procedure. Then install new flange bolts to 24 ft. lbs. (32 Nm).

11. Install the TWC with new gaskets and flange nuts. Tighten the nuts to 25 ft. lbs. (33 Nm) and the bolts to 16 ft. lbs. (22 Nm).

12. Install the front floor undercover and the splash shield.

13. Refill and confirm the ATF to proper level.

ENGINE COOLING

ENGINE COOLANT

DRAIN & REFILL PROCEDURE

1. Start the engine. Set the heater temperature control dial to maximum heat, then turn off the ignition switch. Make sure the engine and radiator are cool to the touch.
2. Remove the radiator cap.
3. Loosen the drain plug and drain the coolant.
4. After the coolant has drained, tighten the radiator drain plug.
5. Remove the coolant reservoir and drain the coolant, then reinstall the coolant reservoir.
6. Fill the coolant reservoir to the MAX mark with Honda Long Life Antifreeze/Coolant Type 2 (P/N OL999-9001).
7. Pour Honda Long Life Antifreeze/Coolant Type 2 into the radiator up to the base of the filler neck.
   • Always use Honda Long Life Antifreeze/Coolant Type 2. Using a non-Honda coolant can result in corrosion, causing the cooling system to malfunction or fail.
   • Honda Long Life Antifreeze/Coolant Type 2 is a mixture of 50 % antifreeze and 50 % water. Do not add water.
   • If the vehicle is regularly driven in very low temperatures (under –31°F, –35°C) a 60 % concentration of coolant should be used. To accomplish this, pour 0.4 gal (1.6 L) of Honda Extreme Cold Weather Antifreeze/Coolant Type 2 into the radiator first, then add Honda Long Life antifreeze/Coolant Type 2 until the radiator is full.
8. Loosely install the radiator cap.
9. Start the engine, and let it run until it warms up (the radiator fan comes on at least twice).
10. Turn off the engine. Check the level in the radiator and add Honda Long Life Antifreeze/Coolant Type 2 if needed.

➡ Removing the radiator cap while the engine is hot can cause the coolant to spray out. Always let the engine and radiator cool before removing the cap.

11. Put the radiator cap on tightly, then run the engine again, and check for leaks.

12. Clean up any spilled engine coolant.

13. Connect the Honda Diagnostic System (HDS) to the Data Link Connector (DLC).

14. Turn the ignition switch to ON (II).

15. Make sure the HDS communicates with the vehicle and the Powertrain Control Module (PCM). If it doesn't communicate, troubleshoot the DLC circuit.

16. Select BODY ELECTRICAL with the HDS.

17. Select ADJUSTMENT in the GAUGES MENU with the HDS.

18. Select RESET in the MAINTENANCE MINDER with the HDS.

19. Select MAINTENANCE SUB ITEM 5 RESET with the HDS.

FLUID RECOMMENDATIONS

Honda Long Life Antifreeze/Coolant Type 2 or Honda Extreme Cold Weather Antifreeze/Coolant Type 2.

LEVEL CHECK

Always check coolant level after engine has cooled, and accessing only via the coolant reservoir.

ENGINE FAN

REMOVAL & INSTALLATION

See Figures 90 through 93.

1. Before servicing the vehicle, refer to the Precautions Section.

2. Remove the hood support rod, then use it to prop the hood in the wide-open position.

3. Remove the bulkhead cover:

   a. Put on gloves to protect hands. Take care not to scratch the front bumper and the body.

   b. Remove the clips by carefully pulling the front grille cover up, then remove the cover by releasing the front edge of the cover from the grille.

➡**To remove the clips, pry the inner clip up at the edge near the line on its head.**

4. Disconnect the fan motor connectors and the hood switch connector, then remove the harness clips.

5. Remove the radiator upper brackets, then remove the front bulkhead.

6. Remove the condenser fan shroud

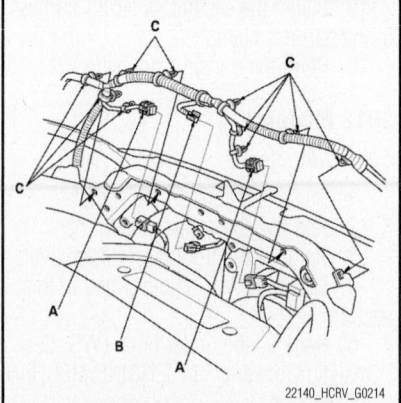

Fig. 91 Disconnect the fan motor connectors (A) and the hood switch connector (B), then remove the harness clips (C)

22140_HCRV_G0214

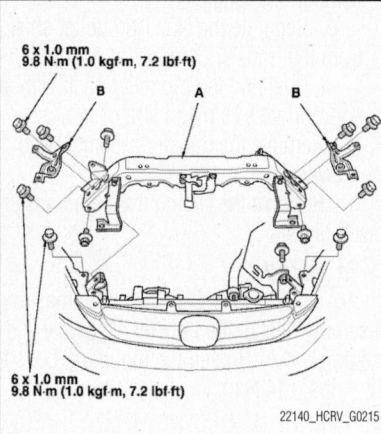

6 x 1.0 mm
9.8 N·m (1.0 kgf·m, 7.2 lbf·ft)

6 x 1.0 mm
9.8 N·m (1.0 kgf·m, 7.2 lbf·ft)

22140_HCRV_G0215

Fig. 92 Remove the radiator upper brackets (A), then remove the front bulkhead (B)

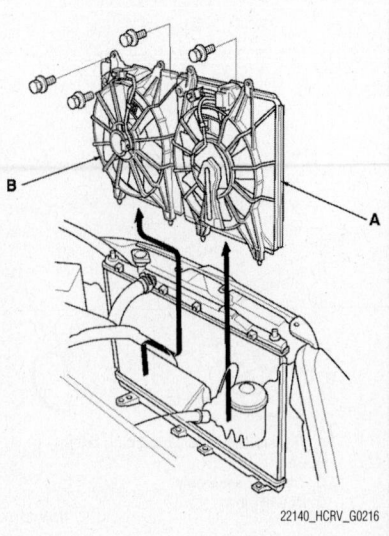

22140_HCRV_G0216

Fig. 93 Remove the condenser fan shroud assembly (A), then remove the radiator fan shroud assembly (B) from the condenser fan shroud side

assembly, then remove the radiator fan shroud assembly from the condenser fan shroud side.

7. Disassemble the fan shrouds, as necessary.

To install:

8. Assemble the fan shrouds.

9. Install the radiator fan shroud assembly, then install the condenser fan shroud assembly.

10. Install the front bulkhead, then install the radiator upper brackets.

11. Apply body paint to the bulkhead mounting bolts.

12. Connect the fan motor connectors and the hood switch connector, then install the harness clips.

13. Install the bulkhead cover in the reverse order of removal, and note these items:

- If the clips are damaged or stress-whitened, replace them with new ones
- Push the clips and the hooks into place securely

RADIATOR

REMOVAL & INSTALLATION

See Figures 90 through 92, 94 through 96.

❊❊ CAUTION

Never open a radiator cap or cooling system when the coolant temperature is above 100°F (37°C). Always drain

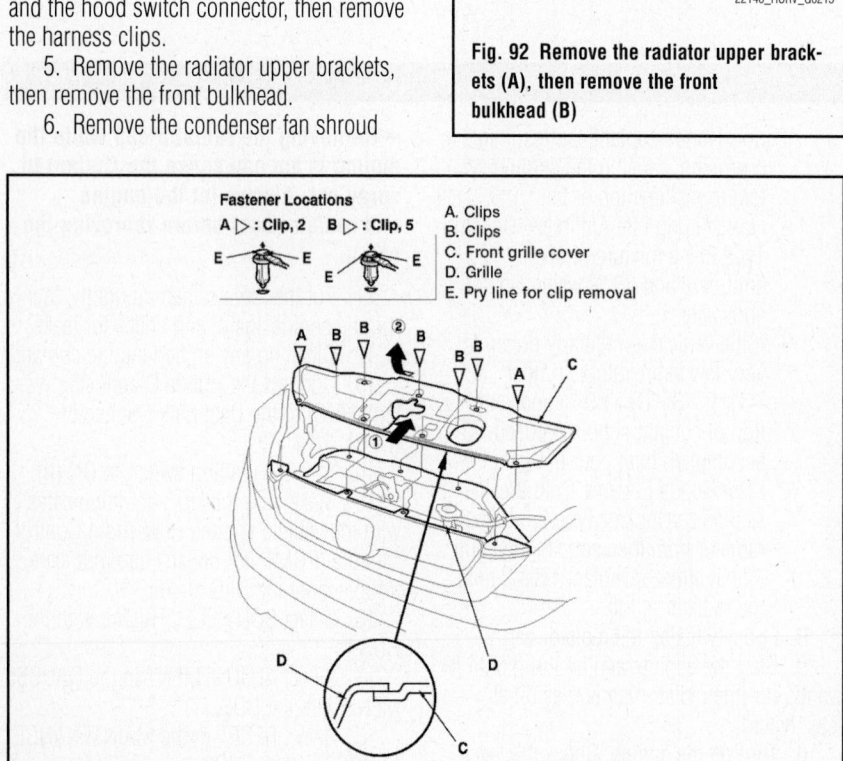

Fastener Locations
A ▷ : Clip, 2   B ▷ : Clip, 5

A. Clips
B. Clips
C. Front grille cover
D. Grille
E. Pry line for clip removal

22140_HCRV_G0213

Fig. 90 Remove the bulkhead cover

coolant into a sealable container. If spillage occurs, take care to clean the spill as quickly as possible.

1. Before servicing the vehicle, refer to the Precautions Section.

2. Disconnect the negative battery cable.

3. Drain the engine coolant.

4. Remove the hood support rod, then use it to prop the hood in the wide-open position.

5. Remove the bulkhead cover:

   a. Put on gloves to protect hands. Take care not to scratch the front bumper and the body.

   b. Remove the clips by carefully pulling the front grille cover up, then remove the cover by releasing the front edge of the cover from the grille.

➡**To remove the clips, pry the inner clip up at the edge near the line on its head.**

6. Disconnect the coolant reservoir hose and the upper radiator hose.

7. Raise and safely support the vehicle.

8. Remove the splash shield.

9. Disconnect the Engine Coolant Temperature (ECT) sensor 2 connector, and remove the harness clip.

10. Clean any dirt off the quick connector, the radiator, and the lower radiator hose.

11. Pull out the lock by hand, then wiggle the quick connector loose, and remove it

from the radiator. Do not use any tools to remove the quick connector.

12. Disconnect the Automatic Transmission Fluid (ATF) cooler hoses, then plug the hoses and the lines.

13. Lower the vehicle.

14. Disconnect the fan motor connectors and the hood switch connector, then remove the harness clips.

15. Remove the radiator upper brackets, then remove the front bulkhead.

16. Pull up the radiator, then remove the radiator fan shroud assembly, the A/C condenser fan shroud assembly, the radiator cap, the ETC sensor 2, and the drain plug.

To install:

17. Reassemble the radiator with new O-rings.

18. Install the radiator. Make sure the lower cushions are set securely.

19. Install the front bulkhead, then install the radiator upper brackets.

20. Apply body paint to the bulkhead mounting bolts.

21. Connect the fan motor connectors and the hood switch connector, then install the harness clips.

22. Raise and safely support the vehicle.

23. Check the quick connector and the set ring for cracks or damage. If the connector and/or set ring are cracked or damaged, replace the connector.

24. Make sure the set ring is in place inside the quick connector. If the set ring is displaced or not properly seated

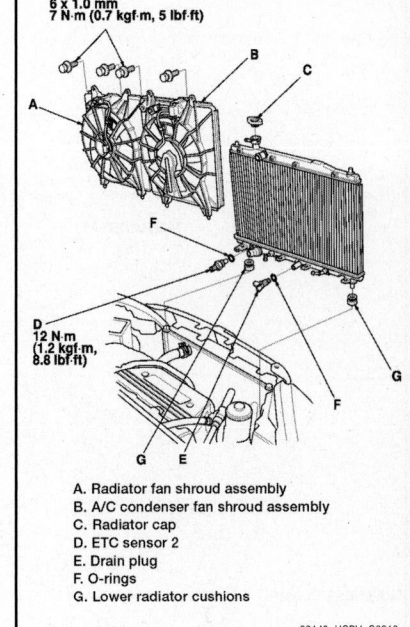

A. Radiator fan shroud assembly
B. A/C condenser fan shroud assembly
C. Radiator cap
D. ETC sensor 2
E. Drain plug
F. O-rings
G. Lower radiator cushions

22140_HCRV_G0218

Fig. 95 Expanded view of radiator removal

in the connector, replace the quick connector.

25. Replace the O-ring in the quick connector.

26. Check the lock. If the lock is damaged or deformed, replace it. When installing the new lock to the connector, slide it straight down along the groove.

27. Install a new lower radiator hose on the quick connector and install the clamp.

28. Clean the connecting surface of the radiator, then apply clean engine coolant around the connecting surface.

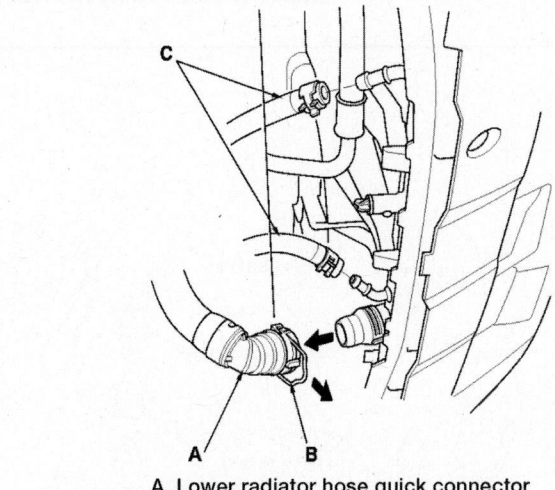

A. Lower radiator hose quick connector
B. Quick connector lock
C. Automatic Transmission Fluid (ATF) cooler hoses

22140_HCRV_G0217

Fig. 94 View of the lower radiator hose quick connector

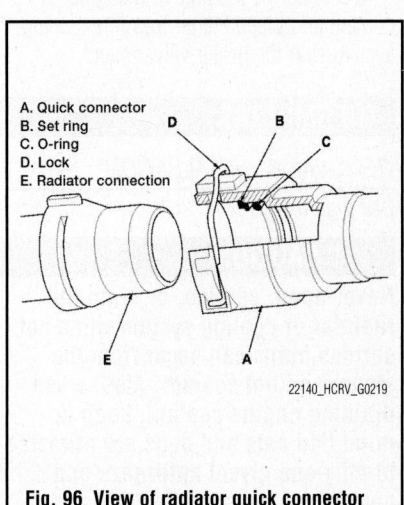

A. Quick connector
B. Set ring
C. O-ring
D. Lock
E. Radiator connection

22140_HCRV_G0219

Fig. 96 View of radiator quick connector fitting

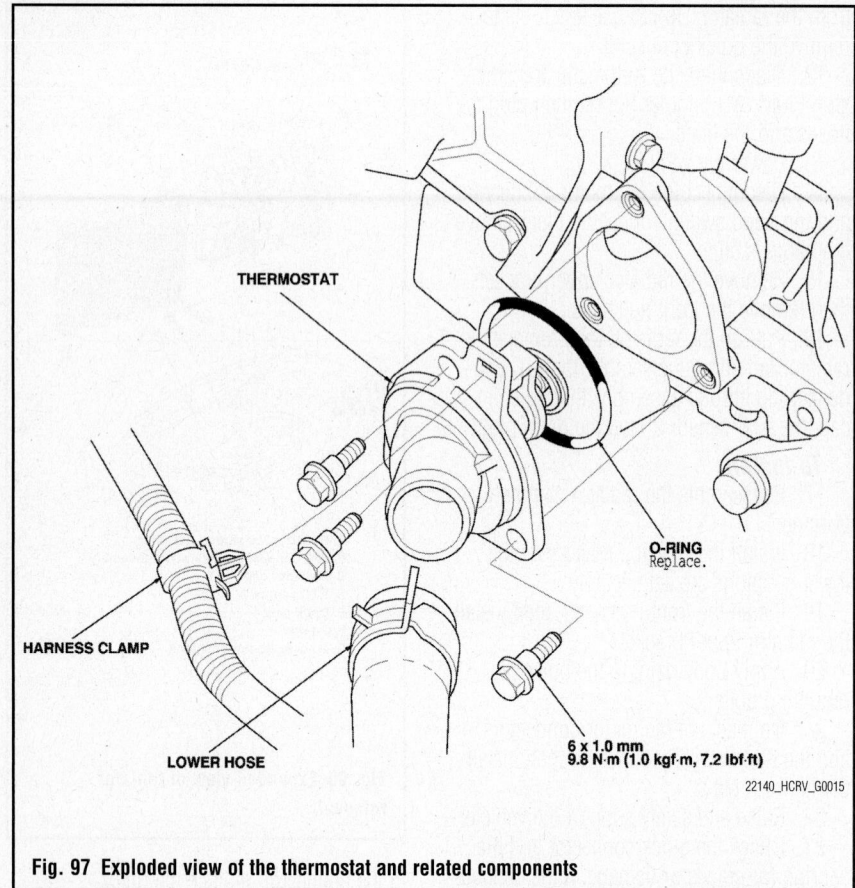

Fig. 97 Exploded view of the thermostat and related components

29. Push down the lock, then push the quick connector onto the radiator until you hear it click.

30. Install the ATF cooler hoses.

31. Connect the ECT sensor 2 connector and install the harness clip.

32. Install the splash shield.

33. Lower the vehicle.

34. Install the coolant reservoir hose and the upper radiator hose.

35. Install the bulkhead cover.

36. Refill the radiator with engine coolant and bleed the air from the cooling system with the heater valve open.

THERMOSTAT

REMOVAL & INSTALLATION
See Figure 97.

✳✳ CAUTION

Never open, service, or drain the radiator or cooling system when hot; serious burns can occur from the steam and hot coolant. Also, when draining engine coolant, keep in mind that cats and dogs are attracted to ethylene glycol antifreeze and could drink any that is left in an uncovered container or in puddles on the ground. This will prove fatal in sufficient quantities. Always drain coolant into a sealable container. Coolant should be reused unless it is contaminated or is several years old.

1. Before servicing the vehicle, refer to the Precautions Section.

2. Drain the engine coolant.

3. Remove the splash shield.

4. Remove the lower coolant hose.

5. Remove the thermostat.

To install:

6. Install the thermostat with a new O-ring. Tighten the bolts to 84 inch lbs. (10 Nm).

7. Install the lower coolant hose.

8. Install the splash shield.

9. Refill the radiator with engine coolant and bleed the air from the cooling system while running the engine with the heater valve open.

WATER PUMP

REMOVAL & INSTALLATION
See Figure 98.

✳✳ CAUTION

Never open, service, or drain the radiator or cooling system when hot; serious burns can occur from the steam and hot coolant. Also, when draining engine coolant, keep in mind that cats and dogs are attracted to ethylene glycol antifreeze and could drink any that is left in an uncovered container or in puddles on the ground. This will prove fatal in sufficient quantities. Always drain coolant into a sealable container. Coolant should be reused unless it is contaminated or is several years old.

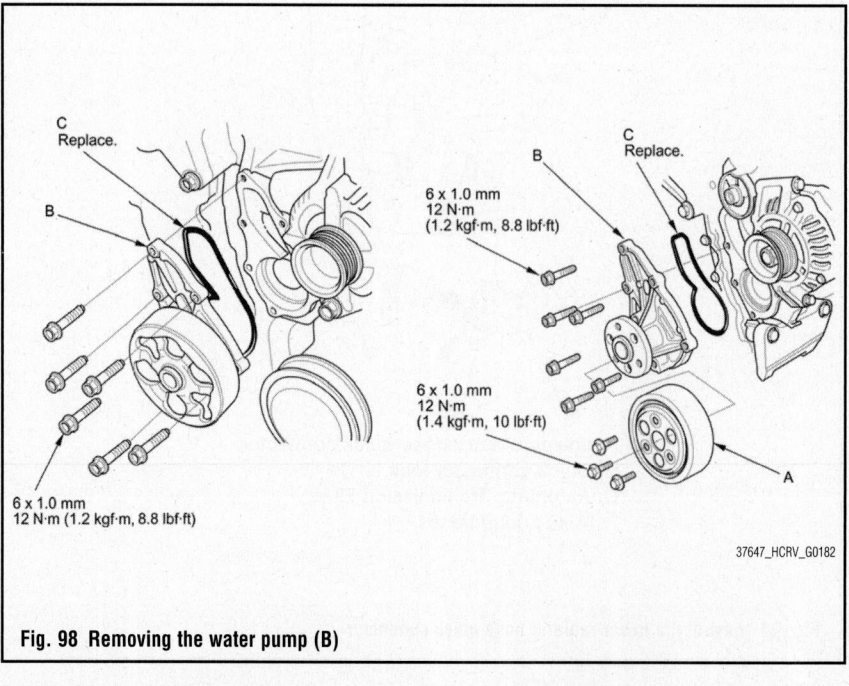

Fig. 98 Removing the water pump (B)

1. Before servicing the vehicle, refer to the Precautions Section.
2. Remove the drive belt.
3. Drain the engine coolant.
4. Remove the drive belt auto-tensioner pulley.
5. K24Z6 engine: Remove the water pump pulley.
6. Remove the 6 bolts securing the water pump, then remove the water pump.

To install:

7. Inspect and clean the O-ring groove and mating surface of the water passage.
8. Install the water pump with new O-ring and tighten the 6 bolts to 106 inch lbs. (12 Nm).

9. Clean up any spilled engine coolant.
10. Install the drive belt auto-tensioner pulley.
11. Refill the radiator with engine coolant and bleed the air from the cooling system while running the engine with the heater valve open.

ENGINE ELECTRICAL

BATTERY

REMOVAL & INSTALLATION

See Figure 99.

➡**This procedure is for 2012 models, 2011 is similar.**

➡**The battery terminal disconnection/reconnection procedure must be done before and after doing this procedure. Some systems store data in memory that is lost when the battery is disconnected.**

1. Do the battery terminal disconnection procedure.
2. Remove the two nuts securing the battery setting plate, then remove the battery setting plate and the battery.

To install:

3. Install the battery, then install the battery setting plate.
4. Tighten the two nuts equally until the battery is stable.

➡**Do not deform the battery setting plate by over-tightening the nuts.**

5. Do the battery terminal reconnection procedure.

➡**Make sure the battery is installed correctly, and the positive terminal and the negative terminal are not reversed.**

BATTERY DISCONNECT/RECONNECT PROCEDURE

Disconnection Procedure

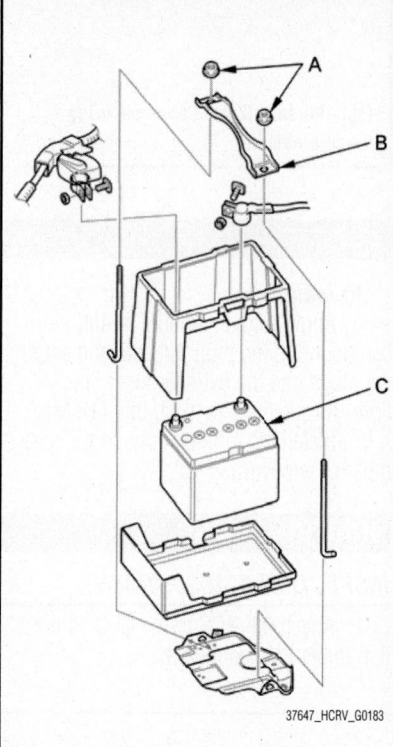

37647_HCRV_G0183

Fig. 99 Remove the two nuts (A) securing the battery setting plate, then remove the battery setting plate (B) and the battery (C)

➡**Some systems store data in memory that is lost when the battery is disconnected. Do the following steps before disconnecting the battery.**

1. Make sure you have the anti-theft code(s) for the audio and/or the navigation system (if equipped).

BATTERY SYSTEM

2. If you are replacing the audio unit, write down the audio presets (AM and FM), and the XM radio presets (if equipped), because the audio unit does not retain the presets after the battery is disconnected.
3. Make sure the ignition switch is in LOCK (0).
4. Disconnect and isolate the negative cable from the battery.

➡**Always disconnect the negative cable from the battery first.**

5. Disconnect the positive cable from the battery.

Reconnection Procedure

➡**Some systems store data in memory that is lost when the battery is disconnected. Do the following steps to restore the systems back to normal operation.**

1. Clean the battery terminals.
2. Test the battery.
3. Reconnect the positive cable to the battery first, then reconnect the negative cable to the battery.

➡**Always connect the positive cable to the battery first.**

4. Apply multipurpose grease to the terminals to prevent corrosion.
5. Enter the anti-theft code(s) for the audio system and/or the navigation system (if equipped).
6. Enter the audio presets (if applicable), and enter the XM radio presets (if equipped).
7. Set the clock (for vehicles without navigation).

ENGINE ELECTRICAL | CHARGING SYSTEM

ALTERNATOR

REMOVAL & INSTALLATION

See Figure 100.

1. Before servicing the vehicle, refer to the Precautions Section.
2. Disconnect the negative battery terminal.
3. Remove the accessory drive belt. Refer to Accessory Drive Belts, Removal and Installation.
4. Remove the drive belt auto-tensioner.
5. Remove the 3 bolts securing the alternator.
6. Disconnect the alternator electrical connectors and the harness clamp from the alternator.

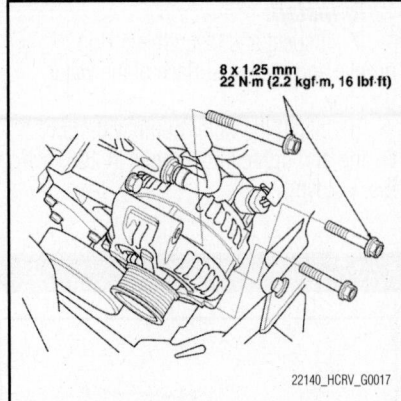

8 x 1.25 mm
22 N·m (2.2 kgf·m, 16 lbf·ft)

22140_HCRV_G0017

Fig. 100 Install the 3 bolts securing the alternator

7. Remove the alternator.

To install:

8. Place the alternator into position.
9. Connector the alternator electrical connectors and harness clamp to the alternator.
10. Install the 3 bolts securing the alternator and tighten to 16 ft. lbs. (22 Nm).
11. Install the drive belt auto-tensioner.
12. Install the drive belt. Refer to Accessory Drive Belts, Removal and Installation.
13. Connect the negative battery terminal.

ENGINE ELECTRICAL | IGNITION SYSTEM

FIRING ORDER

2.4L Engine, Firing order: 1–3–4–2

IGNITION COIL

REMOVAL & INSTALLATION

See Figure 101.

1. Remove the ignition coil cover.
2. Disconnect the ignition coil connectors, then remove the ignition coils.
3. Remove the spark plugs and inspect them.

To install:

4. Apply a small amount of anti-seize compound to the plug threads, and screw the plugs into the cylinder head, finger-tight. Torque them to 13 ft. lbs. (18 Nm).
5. Install the ignition coils in the reverse order of removal.

IGNITION TIMING

INSPECTION & ADJUSTMENT

1. Attach the HDS to the DLC connection under the dashboard.

2. Turn the ignition switch to ON.
3. Jump the SCS line with the HDS.
4. Connect the timing light to the service loop (white tape).
5. Aim the light toward the pointer (A) on the cam chain case. Check the ignition timing under a no load condition (headlights, blower fan, rear window defogger, and air conditioner are turned off).
6. Ignition timing should be 8±2° BTDC (RED mark) at idle.
7. If the ignition timing differs from the specification, check the cam timing. If the cam timing is OK, update the PCM if it does not have the latest software, or substitute a known-good PCM, then recheck. If the system works properly, and the PCM was substituted, replace the original PCM.
8. Disconnect the HDS and the timing light.

SPARK PLUGS

REMOVAL & INSTALLATION

See Figure 101.

1. Remove the ignition coil cover.
2. Disconnect the ignition coil connectors, then remove the ignition coils.
3. Remove the spark plugs and inspect them.

To install:

4. Apply a small amount of anti-seize compound to the plug threads, and screw the plugs into the cylinder head, finger-tight. Torque them to 13 ft. lbs. (18 Nm).

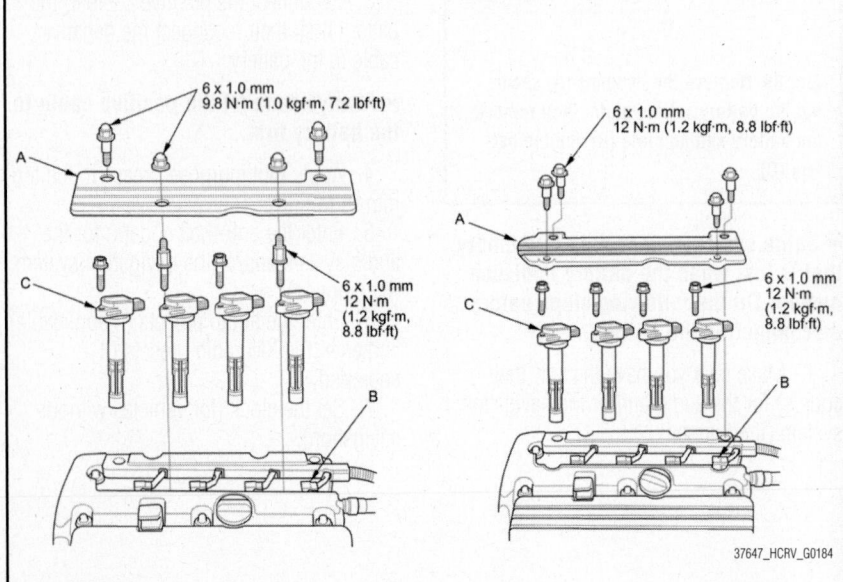

6 x 1.0 mm
9.8 N·m (1.0 kgf·m, 7.2 lbf·ft)

6 x 1.0 mm
12 N·m (1.2 kgf·m, 8.8 lbf·ft)

6 x 1.0 mm
12 N·m (1.2 kgf·m, 8.8 lbf·ft)

6 x 1.0 mm
12 N·m (1.2 kgf·m, 8.8 lbf·ft)

37647_HCRV_G0184

Fig. 101 Remove the ignition coil cover (A), disconnect the ignition coil connectors (B), then remove the ignition coils (C)

ENGINE ELECTRICAL

STARTER

REMOVAL & INSTALLATION

See Figures 102, 103 and 104.

1. Do the battery terminal disconnection procedure.
2. Remove the splash shield.
3. Remove the intake manifold bracket.
4. Disconnect the knock sensor connector.
5. Remove the harness clamp and remove the two bolts securing the starter, then remove the starter from the engine.
6. Disconnect the starter cable from the B terminal, then disconnect the BLK/WHT wire from the S terminal.
7. Remove the harness clamp, then remove the starter.

To install:

8. Install the starter cable and BLK/WHT wire. Make sure the starter cable

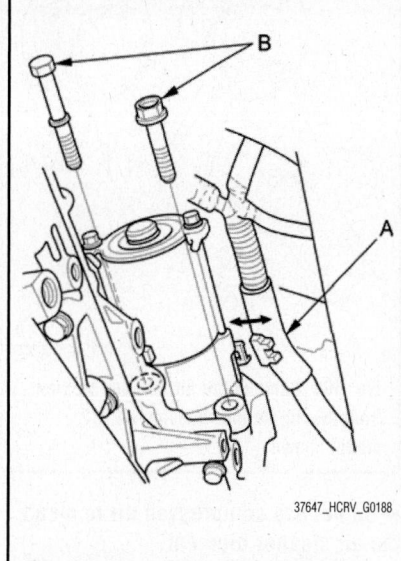

Fig. 103 Remove the harness clamp (A), and remove the two bolts (B) securing the starter

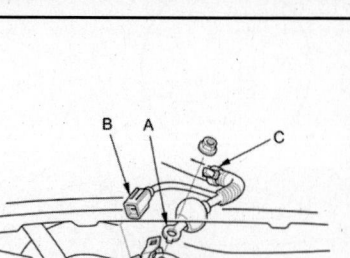

Fig. 104 Disconnect the starter cable (A) from the B terminal, then disconnect the BLK/WHT wire (B) from the S terminal; remove the harness clamp (C)

crimped side of the ring terminal faces away from the starter when you connect it.

9. Install the harness clamp.
10. Install the starter, tighten the bolts, then install the harness clamp.
11. Reconnect the knock sensor connector.
12. Install the intake manifold bracket.
13. Install the splash shield.
14. Do the battery terminal reconnection procedure.
15. Start the engine to make sure the starter works properly.

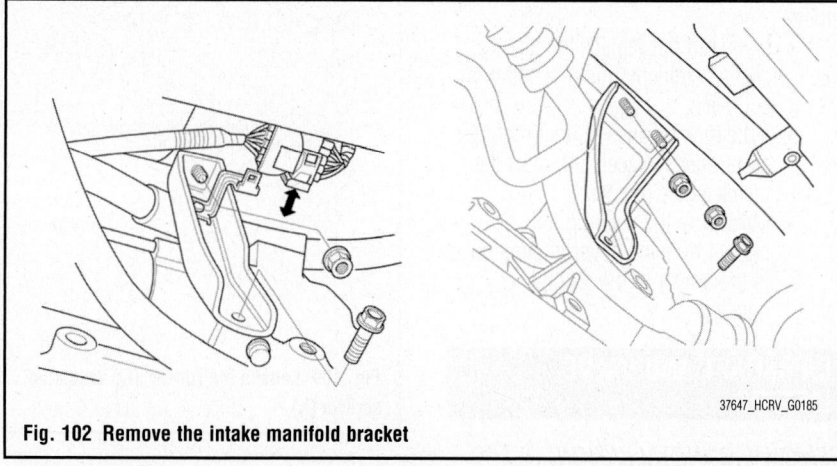

Fig. 102 Remove the intake manifold bracket

ENGINE MECHANICAL

ACCESSORY DRIVE BELTS

ACCESSORY BELT ROUTING

➡ **See illustration in "Removal & Installation".**

INSPECTION

See Figure 105.

1. Inspect the belt for cracks or damage. If the belt is cracked or damaged, replace it.
2. Check that the auto-tensioner indicator is within the standard range as shown. If it is out of the standard range, replace the drive belt.

ADJUSTMENT

Belt tension is maintained via the automatic belt tensioner. No routine adjustment is required.

REMOVAL & INSTALLATION

See Figures 106 and 107.

Special Tools Required: Belt Tension Release Tool Snap-on YA9317 or equivalent, commercially available

1. Move the auto-tensioner using the belt tension release tool in the direction of the arrow to relieve tension from the drive belt, then remove the drive belt.

2. Install the new belt in the reverse order of removal.

AIR CLEANER

REMOVAL & INSTALLATION

Air Cleaner Element

See Figure 108.

1. Open the air cleaner housing cover.
2. Remove the air cleaner element from the air cleaner housing.
3. Check the air cleaner element for damage or clogging. If it is damaged or clogged, replace it.

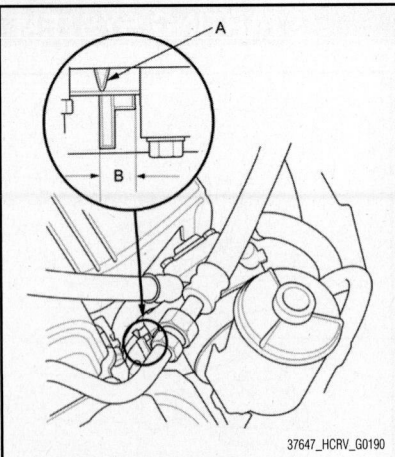

Fig. 105 Check that the auto-tensioner indicator (A) is within the standard range (B)

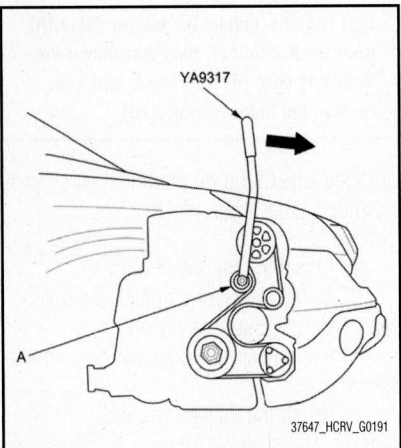

Fig. 106 Move the auto-tensioner (A) using the belt tension release tool to relieve tension from the drive belt, then remove the drive belt—2011 model

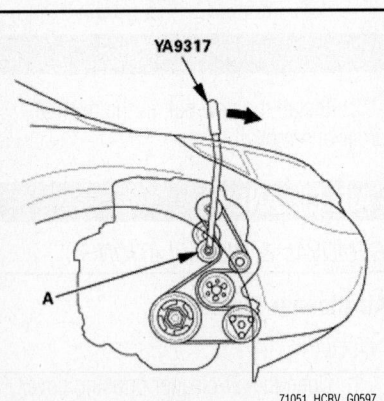

Fig. 107 Move the auto-tensioner (A) using the belt tension release tool in the direction of the arrow to relieve tension from the drive belt, then remove the drive belt—2012 model

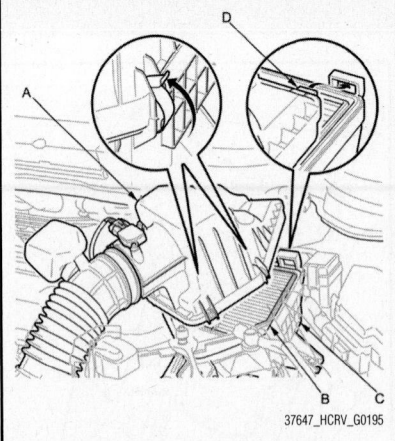

Fig. 108 Remove the air cleaner element from the air cleaner housing—2012 model shown

➡**Do not use compressed air to clean the air cleaner element.**

4. Clean and remove any debris from inside the air cleaner.

To install:

5. Install the parts in the reverse order of removal:

- If you did not replace the air cleaner element, this procedure is complete.
- If the maintenance minder required air cleaner replacement, reset the maintenance minder.
- Make sure the detent of the air cleaner housing cover is positioned in the holder of the air cleaner housing.

CAMSHAFT AND VALVE LIFTERS

REMOVAL & INSTALLATION

2011 Models

See Figures 109 through 114.

1. Remove the cam chain, as outlined in this section.

2. Loosen the rocker arm adjusting screws (A).

3. Remove the camshaft holder bolts. To prevent damaging the camshafts, loosen the bolts, in sequence, two turns at a time.

➡**Bolt 21 is not on all engines.**

4. Remove cam chain guide B, the camshaft holders, and the camshafts.

5. Insert the bolts into the rocker shaft holder, then remove the rocker arm assembly.

To install:

6. Reassemble the rocker arm assembly.

Clean and dry the No. 5 rocker shaft holder mating surface.

7. Apply liquid gasket, P/N 08717-0004, 08718-0001, 08718-0003, or 08718-0009, evenly to the cylinder head mating surface of the No. 5 rocker shaft holder, and to the inside edge of the threaded bolt holes. Install the component within 5 minutes of applying the liquid gasket.

➡**Apply a 0.12 inches (3 mm) diameter bead of liquid gasket along the broken line. If too much time has passed after applying the liquid gasket, remove the old liquid gasket and residue, then reapply new liquid gasket.**

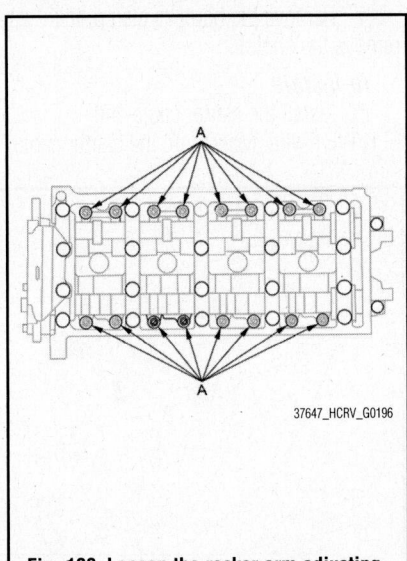

Fig. 109 Loosen the rocker arm adjusting screws (A)

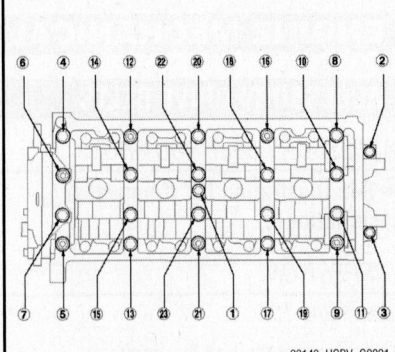

Fig. 110 Remove the camshaft holder bolts; to prevent damage to the camshafts, loosen the bolts , in sequence, 2 turns at a time

8. Insert the bolts into the rocker shaft holder, then install the rocker arm assembly on the cylinder head.

9. Remove the bolts from the rocker shaft holder.

10. Make sure the punch marks on the Variable Valve Timing Control (VTC) actuator and the exhaust camshaft sprocket are facing up, then set the camshafts in the holder. Apply new engine oil to the camshaft journals and lobes.

11. Set the camshaft holders and the cam chain guide B in place.

12. Tighten the bolts to the specified torque.

    a. Bolts 8 x 1.25mm: 16 ft. lbs. (22 Nm).

    b. Bolts 6 x 1.0mm: 9 ft. lbs. (12 Nm)—bolts 21, 22, and 23.

➡**If the engine does not have bolt 21, skip it and continue the torque sequence.**

13. Install the cam chain, then adjust the valve clearance.

2012 Models

See Figures 115 through 120.

1. Remove the cylinder head cover, as described in this section.

2. Turn the crankshaft so its white mark lines up with the pointer. (The other pointer is not used.)

3. Set the No. 1 piston at top dead center (TDC). The punch mark on the VTC actuator and the punch mark on the exhaust camshaft sprocket should be

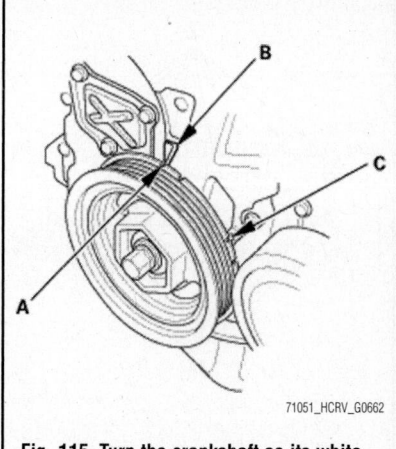

Fig. 115 Turn the crankshaft so its white mark (A) lines up with the pointer (B). (The other pointer (C) is not used.)

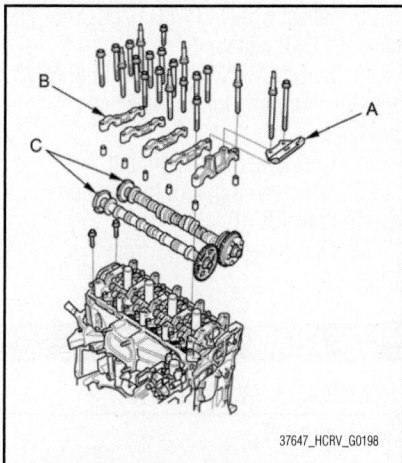

Fig. 111 Remove cam chain guide B (A), the camshaft holders (B), and the camshafts (C)

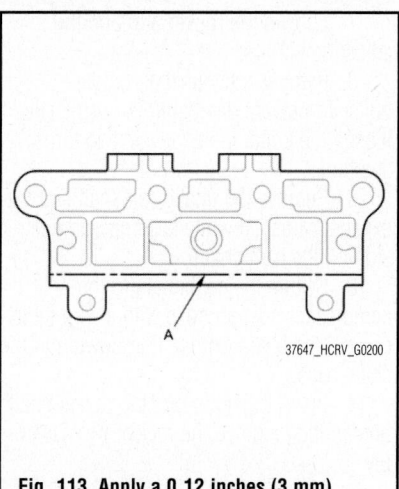

Fig. 113 Apply a 0.12 inches (3 mm) diameter bead of liquid gasket along the broken line (A)

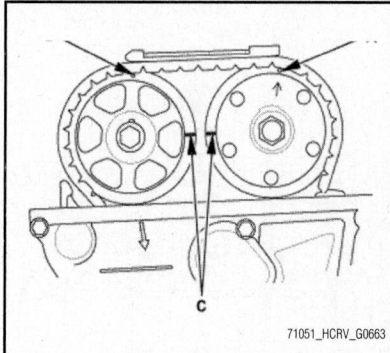

Fig. 116 Set the No. 1 piston at top dead center (TDC). The punch mark (A) on the VTC actuator and the punch mark (B) on the exhaust camshaft sprocket should be at the top. Align the TDC marks (C) on the VTC actuator and the exhaust camshaft sprocket

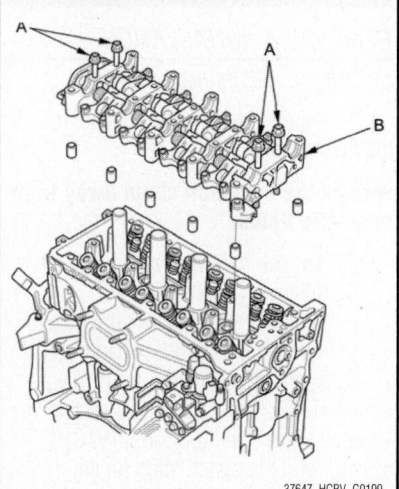

Fig. 112 Insert the bolts (A) into the rocker shaft holder, then remove the rocker arm assembly (B)

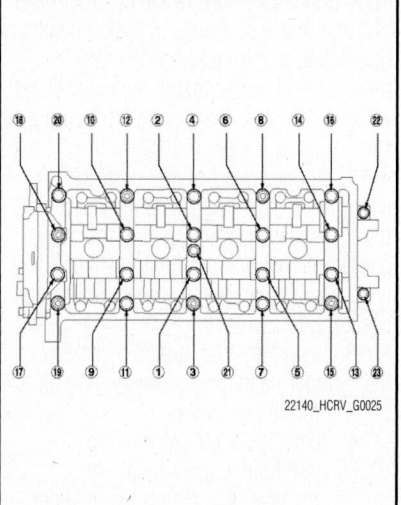

Fig. 114 Tightening sequence for camshaft holder bolts

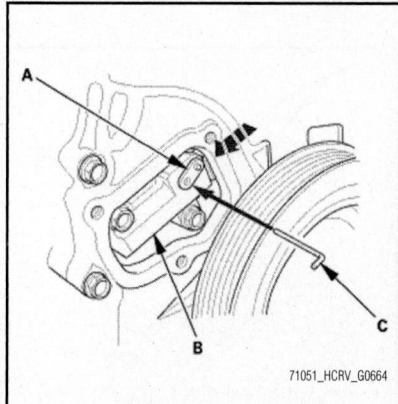

Fig. 117 Rotate the crankshaft counter-clockwise to align the holes on the lock (A) and the auto-tensioner (B), then insert a 1.2 mm (3/64 in) diameter pin (C) into the holes. Turn the crankshaft clockwise to secure the pin

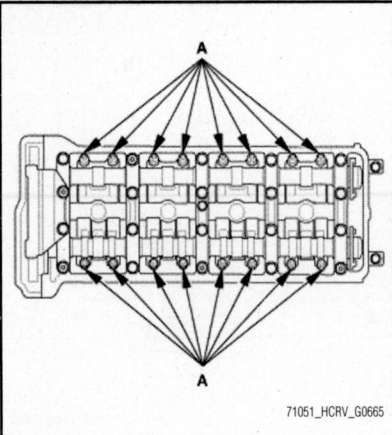

Fig. 118 Loosen the rocker arm and the adjusting screws (A)

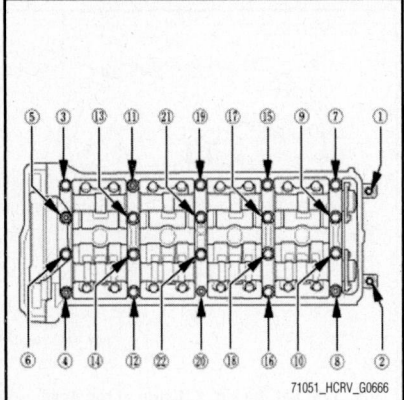

Fig. 119 Remove the camshaft holder bolts. To prevent damaging the camshafts, loosen the bolts, in sequence, two turns at a time

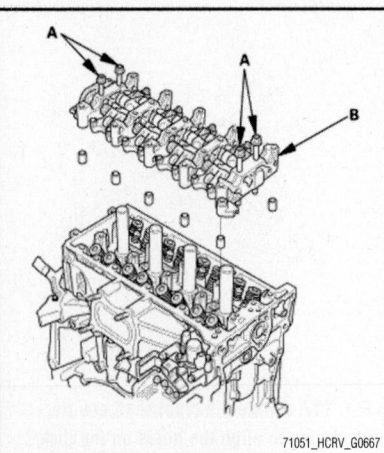

Fig. 120 Insert the bolts (A) into the rocker shaft holder, then remove the rocker arm assembly (B)

at the top. Align the TDC marks on the VTC actuator and the exhaust camshaft sprocket.

4. Remove the chain case cover.

5. Rotate the crankshaft counterclockwise to align the holes on the lock and the auto-tensioner, then insert a 1.2 mm (3/64 in) diameter pin into the holes. Turn the crankshaft clockwise to secure the pin.

➡**If the holes in the lock and the auto-tensioner do not align, continue to rotating the crankshaft counterclockwise until the holes align, then install the pin.**

6. Turn the crankshaft clockwise to secure the pin and the TDC on the No. 1 cylinder.

7. Loosen the rocker arm and the adjusting screws.

8. Remove the camshaft holder bolts. To prevent damaging the camshafts, loosen the bolts, in sequence, two turns at a time.

9. Remove the intake camshaft and the exhaust camshaft while keeping some tension on the cam chain.

10. Secure the cam chain to the A/C compressor suction hose with a wire tie to prevent the chain from falling down into the chain case.

11. Insert the bolts into the rocker shaft holder, then remove the rocker arm assembly.

To install:

12. Clean and dry the No. 5 rocker shaft holder mating surface.

13. Apply liquid gasket (P/N 08718-0004 or 08718-0009) to the cylinder head mating surface of the No. 5 rocker shaft holder and to the inside edge of the threaded bolt holes. Install the component within 5 minutes of applying the liquid gasket.

14. Apply a bead of liquid gasket along the broken line (A). If too much time has passed after applying the liquid gasket, remove the old liquid gasket and residue, then reapply new liquid gasket.

15. Install the lost motion assembly in the cylinder head. Apply new engine oil to the lost motion assembly.

16. Insert the bolts (A) into the rocker shaft holder, then install the rocker arm assembly (B) on the cylinder head.

17. Remove the bolts from the rocker shaft holder.

18. Install the exhaust camshaft. Slide the exhaust camshaft in at an angle to allow the cam chain to slip over the VTC actuator's teeth.

19. Install the cam chain, as described in this section.

20. Once the cam chain cover is installed, perform a valve clearance check, as described in this section.

21. Install the cylinder head cover and the ignition coils and cover.

22. Connect the HDS to the data link connector (DLC) located under the driver's side of the dashboard.

23. Turn the ignition switch to ON (II).

24. Select CRANK PATTERN in the ADJUSTMENT MENU with the HDS.

25. Select CRANK PATTERN CLEAR, and clear the CKP pattern.

26. Select CRANK PATTERN LEARNING with the HDS, and follow the screen prompts.

27. Check the ignition timing.

28. After learning, turn the ignition switch to LOCK (0).

29. Disconnect the HDS from the DLC.

CAM CHAIN FRONT COVER

REMOVAL & INSTALLATION
See Figure 121.

For service information, refer to Cam Chain and Sprockets, Removal and Installation.

CAM CHAIN & SPROCKETS

REMOVAL & INSTALLATION
See Figures 122 through 133.

1. Before servicing the vehicle, refer to the Precautions Section.

➡**Keep the camshaft chain away from magnetic fields.**

2. Remove the front wheels.

3. Remove the splash shield.

4. Remove the accessory drive belt.

5. Remove the cylinder head cover.

6. Set the No. 1 piston at Top Dead Center (TDC). The punch mark on the Variable Valve Timing Control (VTC) actuator and the punch mark on the exhaust camshaft sprocket should be at the top. Align the TDC marks on the VTC actuator and exhaust camshaft sprocket.

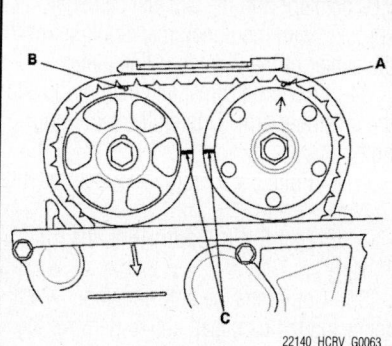

22140_HCRV_G0063

Fig. 121 Set the No. 1 piston at TDC. The punch mark (A) on the VTC actuator and the punch mark (B) on the exhaust camshaft sprocket should be at the top. Align the TDC marks (C) on the VTC actuator and exhaust camshaft sprocket.

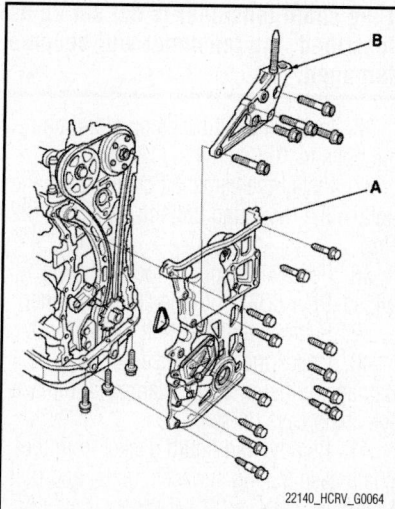

22140_HCRV_G0064

Fig. 122 Remove the camshaft chain case (A) and the side engine mount bracket (B)

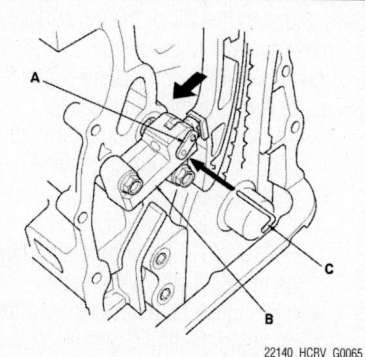

22140_HCRV_G0065

Fig. 123 Align the holes on the lock (A) and the auto-tensioner (B), then insert a 0.05 inches (1.2mm) diameter pin or lock pin (P/N 14511-PNA-003) (C) into the holes. Turn the crankshaft clockwise to secure the pin

7. Disconnect the Crankshaft Position (CKP) sensor connector and the VTC oil control solenoid valve connector.

8. Remove the VTC oil control solenoid valve.

9. Remove the crankshaft damper pulley.

10. Support the engine with a jack and a wood block under the oil pan.

11. Remove the upper torque rod.

12. Remove the ground cable, then remove the side engine mount bracket.

13. Remove the side engine mount bracket mounting bolts.

14. Remove the camshaft chain case and the side engine mount bracket.

15. Loosely install the crankshaft damper pulley.

16. Turn the crankshaft counterclockwise to compress the auto-tensioner.

17. Align the holes on the lock and the auto-tensioner, then insert a 0.05 inches (1.2mm) diameter pin or lock pin (P/N 14511-PNA-003) into the holes. Turn the crankshaft clockwise to secure the pin.

18. Remove the auto-tensioner.

19. Remove camshaft chain guides and the tensioner arm.

20. Remove the camshaft chain.

21. Hold the camshaft with an open-end wrench, then loosen the VTC actuator mounting bolt and the exhaust camshaft sprocket mounting bolt.

22. If the VTC actuator will be reused, perform these steps:

    a. Remove the intake camshaft, and seal the advance holes and the retard holes in the No 1 camshaft journal with tape.

22140_HCRV_G0066

Fig. 124 Removing the camshaft sprockets holding the camshaft with an open-end wrench

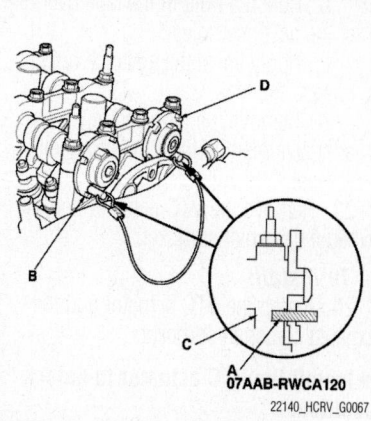

07AAB-RWCA120

22140_HCRV_G0067

Fig. 125 Insert a camshaft lock pin (07AAB-RWCA120) (A) into the maintenance hole in the CMP pulse plate (B) and through the No. 5 rocker shaft holder (C)

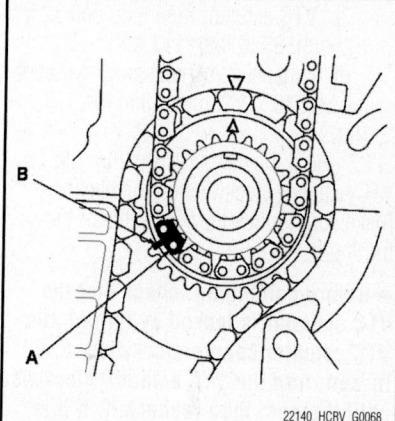

22140_HCRV_G0068

Fig. 126 Install the camshaft chain on the crankshaft sprocket with the colored link plate (A) aligned with the mark (B) on the crankshaft sprocket

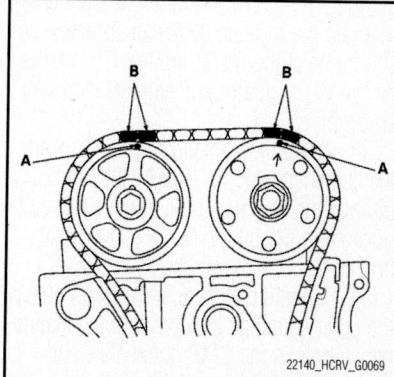

22140_HCRV_G0069

Fig. 127 Install the camshaft chain on the VTC actuator and the exhaust camshaft sprocket with the punch marks (A) aligned with the center of the 2 colored link plates (B)

b. Punch a hole in the tape over one of the advance holes.

c. Apply air to the advance hole to release the lock.

d. Remove the tape and any adhesive residue from the camshaft journal.

23. Remove the VTC actuator and the exhaust camshaft sprocket.

To install:

24. Install the VTC actuator and the exhaust camshaft sprocket.

➡Install the VTC actuator to unlock position.

25. Apply engine oil to the threads of the VTC actuator mounting bolt and exhaust camshaft mounting bolt, then install the bolts.

26. Hold the camshaft with an open-end wrench, then tighten the bolts:

a. VTC actuator mounting bolt 12 x 1.25mm: 83 ft. lbs. (113 Nm).

b. Exhaust camshaft sprocket mounting bolt 10 x 1.25mm: 53 ft. lbs. (72 Nm).

27. Hold the camshaft and turn the VTC actuator clockwise until it clicks. Make sure to lock the VTC actuator by turning it.

➡Before continuing, check that the VTC actuator is locked by turning the VTC actuator counterclockwise. If not locked, turn the VTC actuator clockwise until it stops, then recheck it. If it is still not locked, replace the VTC actuator.

28. Set the crankshaft to TDC. Align the TDC mark on the crankshaft sprocket with the pointer on the engine block.

29. Set the camshafts to TDC. The punch mark on the VTC actuator and the punch mark on the exhaust camshaft sprocket should be at the top. Align the TDC marks on the VTC actuator and exhaust camshaft sprocket.

30. To hold the intake camshaft, insert the camshaft lock pin (07AAB-RWCA120) into the maintenance hole in the CMP pulse plate and through the No. 5 rocker shaft holder.

31. To hold the exhaust camshaft, insert the other camshaft lock pin into the maintenance hole in the CMP pulse plate and through No. 5 rocker shaft holder.

32. Install the camshaft chain on the crankshaft sprocket with the colored link plate aligned with the mark on the crankshaft sprocket.

33. Install the camshaft chain on the

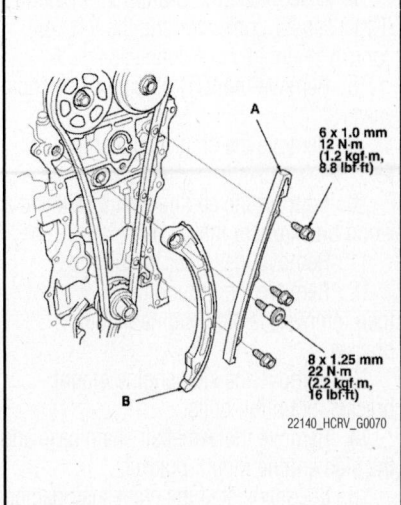

Fig. 128 Install the camshaft chain guide (A) and the tensioner arm (B)

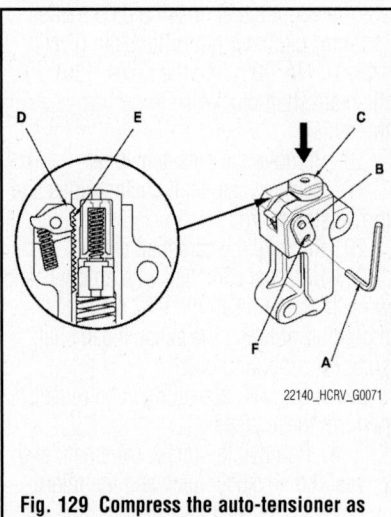

Fig. 129 Compress the auto-tensioner as illustrated using a lock pin

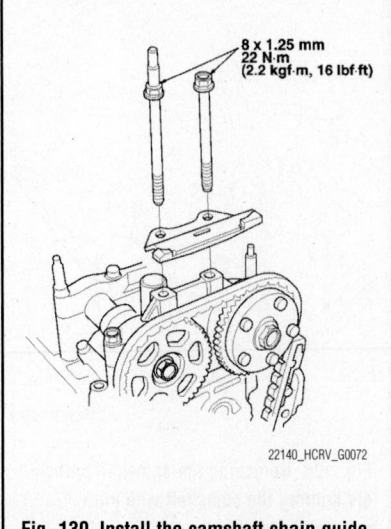

Fig. 130 Install the camshaft chain guide

VTC actuator and the exhaust camshaft sprocket with the punch marks aligned with the center of the 2 colored link plates.

34. Install the camshaft chain guide and the tensioner arm. Tighten the mounting bolts:

a. Bolts 6 x 1.0mm: 106 inch lbs. (12 Nm).

b. Bolt 8 x 1.25mm: 16 ft. lbs. (22 Nm).

35. Compress the auto-tensioner when replacing the camshaft chain. Remove the pin (P/N 14511-PNA-003) from the auto-tensioner that was installed during removal. Turn the plate counterclockwise, to release the lock, then press the rod, and set the first cam to the edge of the rack. Insert the 0.05 inches (1.2mm) diameter pin or lock pin into the holes.

✳✳ WARNING

If the chain tensioner is not set up as described, the tensioner will become damaged.

36. Install the auto-tensioner and tighten the bolts to 106 inch lbs. (12 Nm).

37. Install the camshaft chain guide and tighten the mounting bolts to 16 ft. lbs. (22 Nm).

38. Remove the pin or lock pin (P/N 14511-PNA-003) from the auto-tensioner.

39. Remove the camshaft lock pin set.

40. Check the chain case oil seal for damage. If the oil seal is damaged, replace the chain case oil seal.

41. Remove old liquid gasket from the chain case mating surfaces, bolts, and bolt holes.

42. Clean and dry the chain case mating surfaces.

43. Apply liquid gasket (P/N 08717-0004, 08718-0001, 08718-0003, or 08718-0009) to the engine block mating surface of the chain case and to the inside edge of the bolt holes. Install the component within 5 minutes of applying the liquid gasket.

- Apply a bead of liquid gasket about 0.12 inches (3mm) in diameter along the broken line.
- If you apply liquid gasket P/N 08718-0012, the component must be installed within 4 minutes.
- If too much time has passed after applying the liquid gasket, remove the old liquid gasket and residue, then reapply new liquid gasket.

44. Apply liquid gasket to the engine block upper surface contact areas and the lower block upper surface contact areas on the chain case.

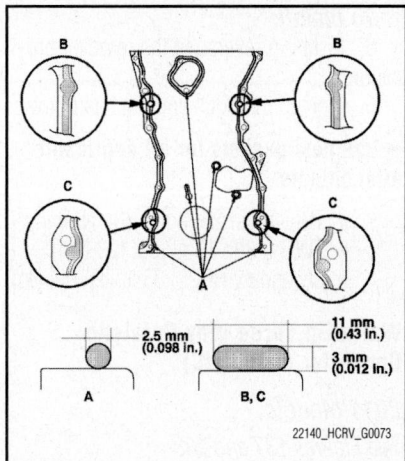

Fig. 131 Apply liquid gasket to the engine block mating surface of the chain case and to the inside edge of the bolt holes as illustrated

➡️**Apply about 0.43 inches (11mm) diameter and about 0.12 inches (3mm) thickness of liquid gasket to the areas.**

45. Apply liquid gasket (P/N 08717-0004, 08718-0001, 08718-0003, or 08718-0009) to the oil pan mating surface of the oil pump. Install the component within 5 minutes of applying the liquid gasket.

46. Install a new O-ring, the side engine mount bracket, and the mounting bolts on the chain case. Set the edge of the chain case to the edge of the oil pan, then install the chain case on the engine block. Tighten the mounting bolts to 106 inch lbs. (12 Nm).

47. Wipe off the excess liquid gasket on the oil pan and chain case mating area.

- When installing the chain case, do not slide the bottom surface onto the oil pan mounting surface.
- Wait at least 30 minutes to allow the liquid gasket to cure before filling the engine with oil.
- Do not run the engine for at least 3 hours after installing the chain case.

48. Tighten the side engine mount bracket mounting bolts to 33 ft. lbs. (44 Nm).

49. Install the side engine mount bracket, then loosely tighten the new bolt and nut.

50. Install the ground cable.

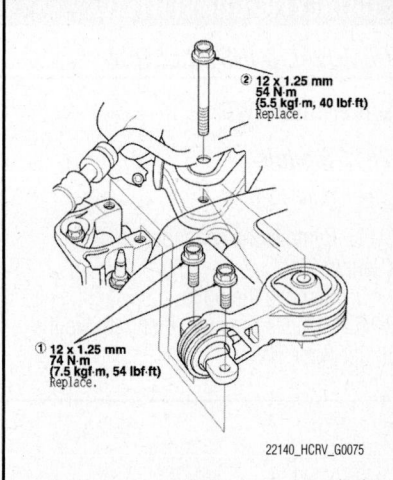

Fig. 133 Install the upper torque rod, then tighten the mounting bolts in the numbered sequence shown

51. Remove the air cleaner housing assembly.

52. Loosen the transaxle mounting bolt and nuts.

53. Raise the vehicle.

54. Loosen the lower torque rod mounting bolt.

55. Lower the vehicle.

56. Tighten the side engine mount mounting bolts and nut.

    a. Bolt 12 x 1.25mm: 52 ft. lbs. (72 Nm).

    b. Bolt 14 x 1.5mm and nut: 54 ft. lbs. (74 Nm).

57. Tighten the transaxle mounting bolt and nuts to 54 ft. lbs. (74 Nm).

58. Raise the vehicle.

59. Tighten the lower torque rod mounting bolt to 69 ft. lbs. (93 Nm).

60. Lower the vehicle.

61. Install the air cleaner housing assembly.

62. Install the upper torque rod, then tighten the new upper torque rod mounting bolts in the numbered sequence shown to the torque illustrated.

63. Install the crankshaft damper pulley.

64. Install the VTC oil control solenoid valve.

65. Connect the CKP sensor connector and VTC oil control solenoid valve connector.

66. Install the cylinder head cover.

67. Install the accessory drive belt.

68. Perform the CKP pattern clear/CKP learn procedure.

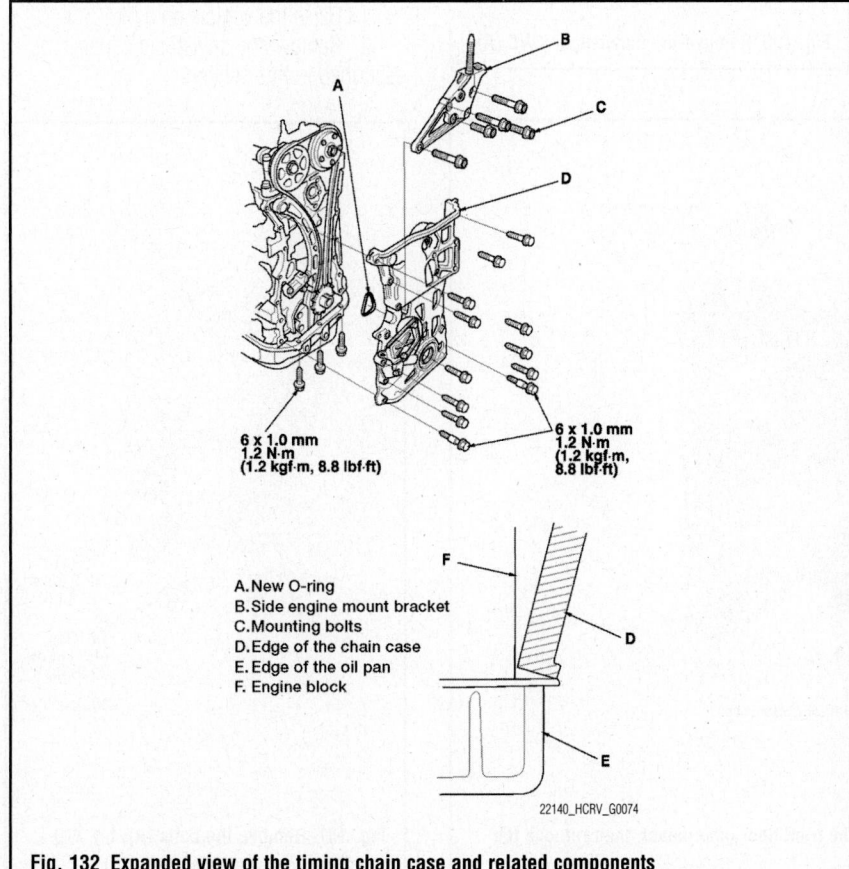

A. New O-ring
B. Side engine mount bracket
C. Mounting bolts
D. Edge of the chain case
E. Edge of the oil pan
F. Engine block

Fig. 132 Expanded view of the timing chain case and related components

CATALYTIC CONVERTER

REMOVAL & INSTALLATION

Under-floor TWC

2011 Models

See Figure 134.

1. Remove the secondary HO2S (Sensor 2).
2. Remove the bolts.
3. Remove the nuts, then remove the under-floor TWC.

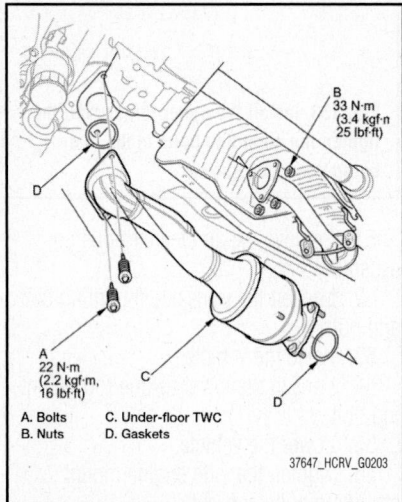

A. Bolts   C. Under-floor TWC
B. Nuts   D. Gaskets

37647_HCRV_G0203

Fig. 134 Exploded view of the under-floor TWC assembly

4. Install the parts in the reverse order of removal with new gaskets and new self-locking nuts.

2012 Models

See Figures 135 and 136.

1. Raise the vehicle on a lift, and make sure it is securely supported.
2. Remove the bolts and the clips securing the front floor undercover, then remove the front floor undercover.
3. Remove the secondary HO2S.
4. Remove the under-floor TWC.

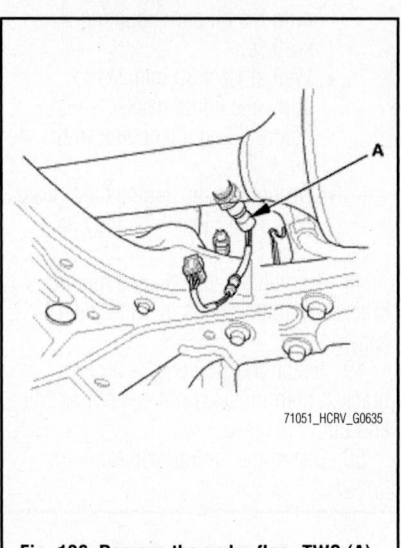

71051_HCRV_G0635

Fig. 136 Remove the under-floor TWC (A)

To install:

5. Install in reverse of the removal procedure.
6. Tighten the bolts and nuts as follows:

➡**Use new gaskets for all applicable attachments.**

- TWC nuts: 25 ft. lbs. (34 Nm)
- TWC bolts: 16 ft. lbs. (22 Nm)
- Secondary HO2S: 33 ft. lbs. (44 Nm)

Warm Up Three Way Catalytic Converter (WU-TWC)

2011 Models

See Figures 137 and 138.

1. Remove the A/F sensor (Sensor 1).
2. Remove the under-cowl panel.
3. Raise the vehicle on a lift.
4. Remove the bolts and the WU-TWC bracket.
5. Remove the bolts.
6. Lower the vehicle.
7. Remove the upper converter cover.
8. Remove the WU-TWC and the gaskets.
9. Install the parts in the reverse order of removal with new gaskets.

2012 Models

See Figures 139 through 143.

1. Raise the vehicle on a lift.
2. Remove the splash shield, as described in this section.

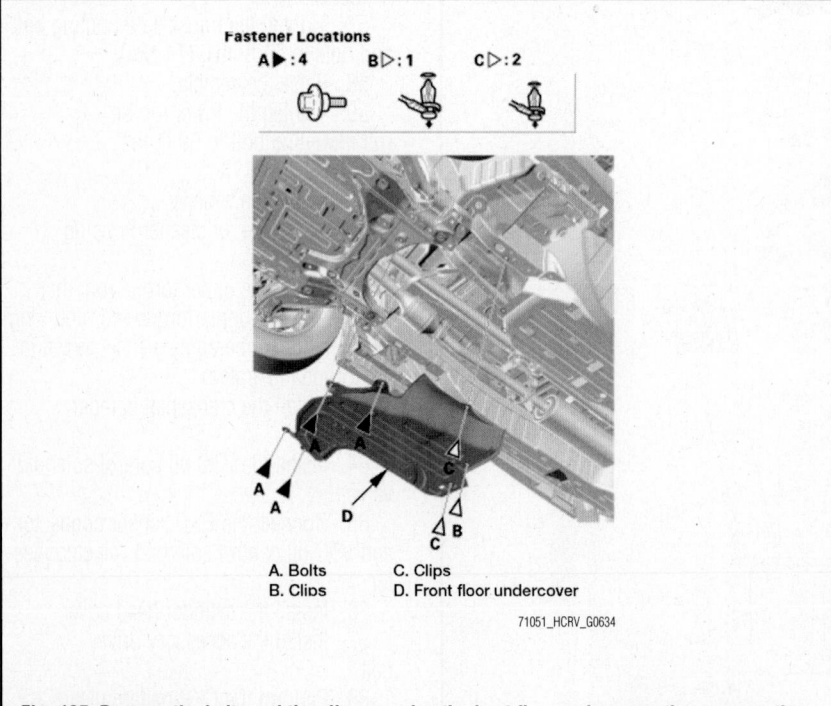

A. Bolts   C. Clips
B. Clips   D. Front floor undercover

71051_HCRV_G0634

Fig. 135 Remove the bolts and the clips securing the front floor undercover, then remove the front floor undercover

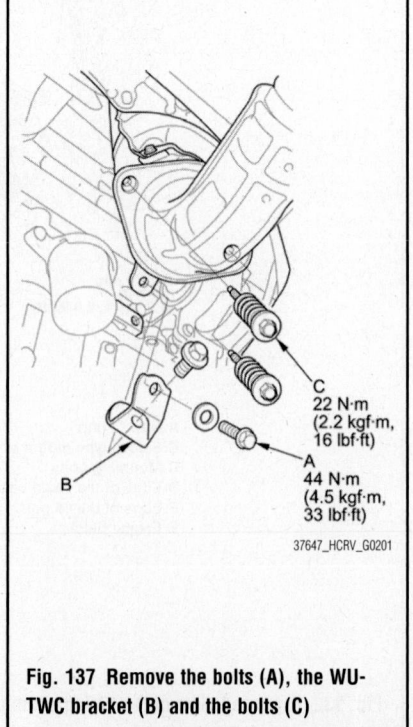

37647_HCRV_G0201

Fig. 137 Remove the bolts (A), the WU-TWC bracket (B) and the bolts (C)

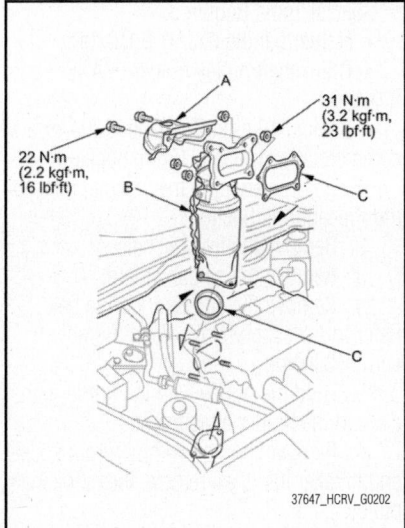

Fig. 138 Remove the upper converter cover (A), the WU-TWC (B) and the gaskets (C)

3. Remove both windshield wiper arm assemblies.

➡**Where applicable, repeat steps on both sides of the vehicle.**

4. Remove the clips. Pull up the front fender trim by hand to detach the clips, then remove the front fender trim.

5. Remove the clip. Pull out the hood rear seal. Detach the hooks, and remove the cowl top cover.

6. Detach the hook at the inside of the cowl cover. Release the hooks from the cowl

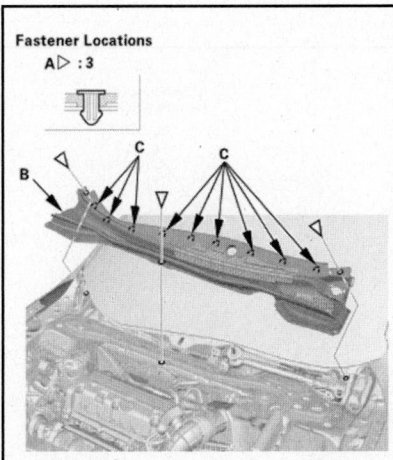

Fig. 139 Remove the clips (A). Carefully pull the center cowl cover (B) forward to release the hooks (C) from the windshield and the body

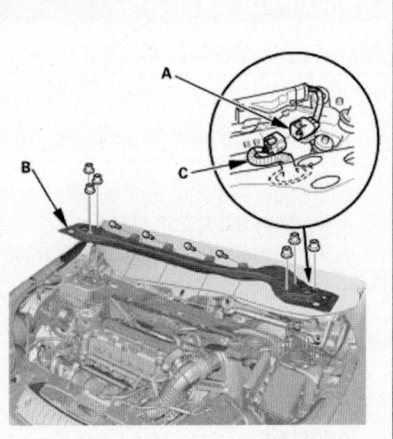

Fig. 140 Disconnect the connector (A). Remove the bolts and the nuts, then remove the under-cowl panel (B). Remove the wire harness (C) from the under-cowl panel

cover and the front fender, then remove the side cowl cover.

7. Right side; Detach the windshield washer tube.

8. Remove the clips. Carefully pull the center cowl cover forward to release the hooks from the windshield and the body.

9. Disconnect the connector. Remove the bolts and the nuts, then remove the under-cowl panel. Remove the wire harness from the under-cowl panel.

10. Remove the A/F sensor.

11. Remove the exhaust chamber cover.

12. Separate the lower side of the WU-TWC from the under-floor TWC.

13. Remove the WU-TWC. Remove the WU-TWC cover.

To install:

14. Install in reverse of the removal procedure.

➡**Use new gaskets for all applicable attachments.**

15. Tighten the bolts and nuts as follows:
- Upper side WU-TWC: 23 ft. lbs. (31 Nm)
- Lower side WU-TWC: 16 ft. lbs. (22 Nm)
- Lower side WU-TWC bracket: 33 ft. lbs. (45 Nm)
- Exhaust chamber cover: 16 ft. lbs. (22 Nm)
- A/F sensor: 33 ft. lbs. (44 Nm)

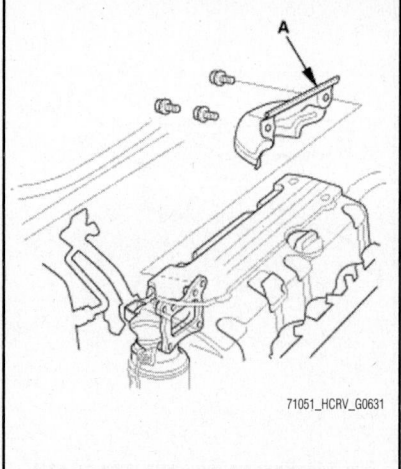

Fig. 141 Remove the exhaust chamber cover (A)

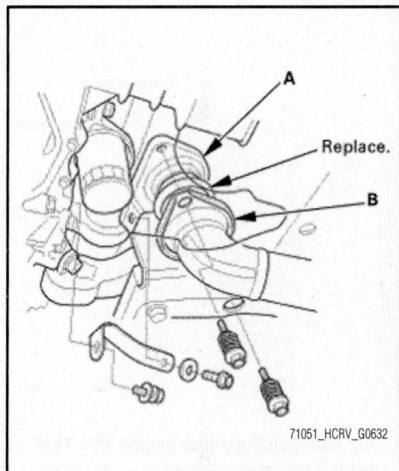

Fig. 142 Separate the lower side of the WU-TWC (A) from the under-floor TWC (B)

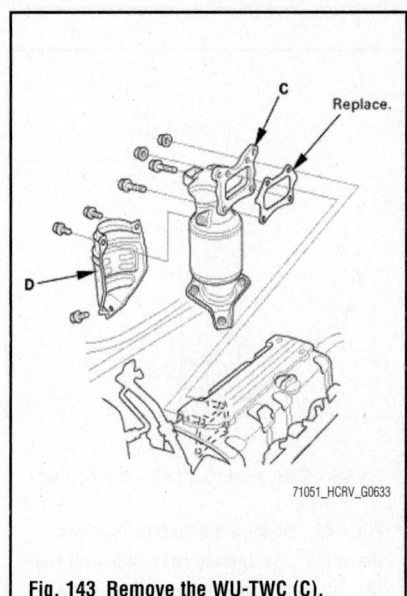

Fig. 143 Remove the WU-TWC (C). Remove the WU-TWC cover (D)

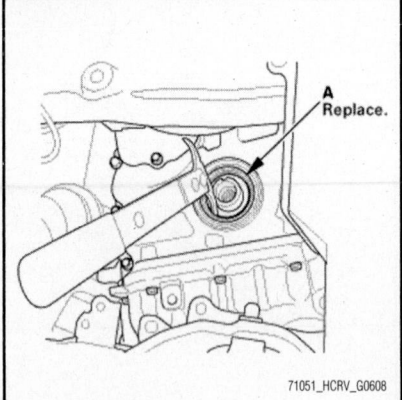

Fig. 144 Remove the chain case oil seal (A) using a commercially available pry bar

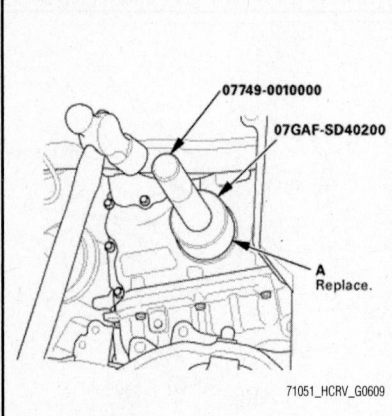

Fig. 145 Use the driver handle 15 x 135L and the hub dis/assembly tool, 42 mm to drive a new oil seal (A) squarely into the chain case to the specified installed height

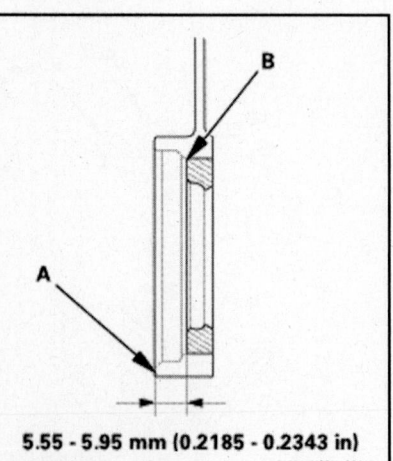

5.55 - 5.95 mm (0.2185 - 0.2343 in)

71051_HCRV_G0610

Fig. 146 Measure the distance between the chain case surface (A) and the oil seal (B). Oil seal installed height should be 0.2185–0.2343 in. (5.55–5.95 mm)

CAM CHAIN CASE SEAL

REMOVAL & INSTALLATION

See Figures 144 through 146.

1. Raise the vehicle on a lift, and make sure it is securely supported.
2. Remove the right front wheel.
3. Remove the splash shield.
4. Remove the drive belt, as described in this section.
5. Remove the crankshaft pulley, as described in this section.
6. Remove the chain case oil seal using a commercially available pry bar.

To install:

7. Clean and dry the crankshaft oil seal housing.
8. Apply light coat of new engine oil to the lip of the chain case oil seal.
9. Use the driver handle 15 x 135L and the hub dis/assembly tool, 42 mm to drive a new oil seal squarely into the chain case to the specified installed height.
10. Measure the distance between the chain case surface and the oil seal. Oil seal installed height should be 0.2185–0.2343 in. (5.55–5.95 mm).
11. Install the crankshaft pulley and drive belt, as described in this section.
12. Install the splash shield.
13. Install the front wheel.

CRANKSHAFT PULLEY

REMOVAL & INSTALLATION

2011 Models

See Figure 147.

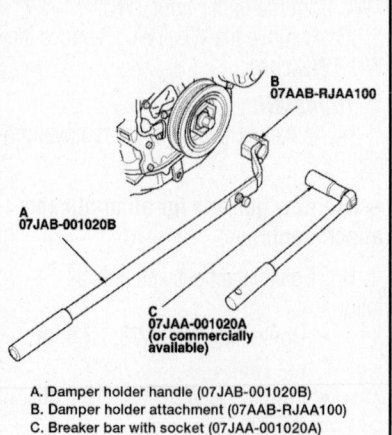

Fig. 147 View of special tools used to remove the crankshaft damper

A. Damper holder handle (07JAB-001020B)
B. Damper holder attachment (07AAB-RJAA100)
C. Breaker bar with socket (07JAA-001020A)

22140_HCRV_G0226

Special Tools Required:
• Holder handle 07JAB-001020B
• Crankshaft pulley holder 07AAB-RJAA100
• Socket, 19mm 07JAA-001020A or a commercially available 19mm socket

1. Before servicing the vehicle, refer to the Precautions Section.
2. Remove the front wheels.
3. Remove the splash shield.
4. Remove the accessory drive belt. Refer to Accessory Drive Belts, Removal and Installation.
5. Hold the pulley with the holder handle and the holder attachment.
6. Remove the bolt with a 19mm socket and breaker bar, then remove the crankshaft pulley.

To install:

7. Clean the crankshaft damper pulley, crankshaft, bolt, and washer. Lubricate crankshaft damper bolt with new engine oil.
8. Install the crankshaft damper pulley and hold the pulley with holder handle and holder attachment.
9. Tighten the crankshaft damper bolt to 36 ft. lbs. (49 Nm).
10. Tighten the crankshaft damper bolt an additional 90°.
11. Install the accessory drive belt. Refer to Accessory Drive Belts, Removal and Installation.
12. Install the splash shield.
13. Install the front wheels.

2012 Models

See Figures 148 and 149.

1. Raise the vehicle on a lift, and make sure it is securely supported.

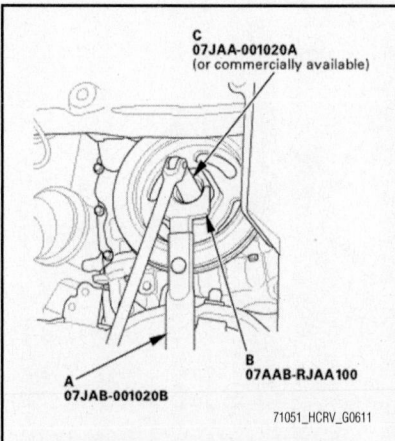

71051_HCRV_G0611

Fig. 148 Hold the pulley with the holder handle (A) and the crankshaft pulley holder (B). Remove the bolt with a socket, 19 mm (C) and a breaker bar, then remove the crankshaft pulley

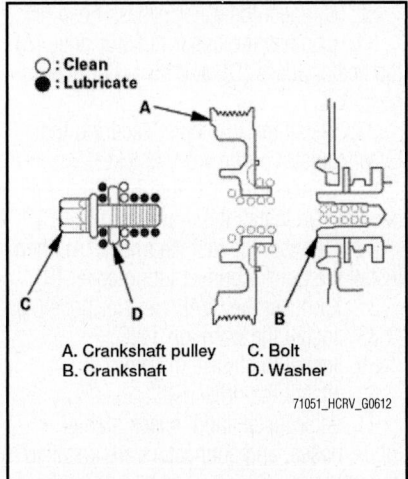

Fig. 149 Clean the crankshaft pulley, the crankshaft, the bolt, and the washer. Lubricate with new engine oil as shown

○ : Clean
● : Lubricate

A. Crankshaft pulley
B. Crankshaft
C. Bolt
D. Washer

71051_HCRV_G0612

2. Remove the right front wheel.
3. Remove the splash shield.
4. Remove the drive belt, as described in this section.
5. Hold the pulley with the holder handle and the crankshaft pulley holder. Remove the bolt with a socket, 19 mm and a breaker bar, then remove the crankshaft pulley.

To install:

6. Clean the crankshaft pulley, the crankshaft, the bolt, and the washer. Lubricate with new engine oil.
7. Install the crankshaft pulley, and hold the pulley with the holder handle and the crankshaft pulley holder. Torque the bolt to 37 ft. lbs. (50 Nm) with a torque wrench and a socket, 19 mm. Do not use an impact wrench. If the pulley bolt or crankshaft are new, torque the bolt to 131 ft. lbs. (177 Nm), then remove the bolt and torque it to 37 ft. lbs. (50 Nm). Tighten the pulley bolt an additional 90 degrees.
8. Install the drive belt, as described in this section.
9. Install the front splash shield and the right front wheel.

CYLINDER HEAD

REMOVAL & INSTALLATION

See Figures 150 through 156.

1. Relieve the fuel pressure.
2. Drain the engine coolant.
3. Remove the drive belt.
4. Remove the intake manifold.
5. Remove the warm up TWC.
6. Disconnect the Evaporative Emission (EVAP) canister hose.
7. Remove the quick-connect fitting cover, then disconnect the fuel feed hose.

8. Disconnect the four fuel injector connectors and remove the ground cables.
9. Remove the four bolts securing the EVAP canister purge valve bracket.
10. Disconnect the upper radiator hose, the heater hoses, and the water bypass hose.
11. Remove the two bolts securing the connecting pipe.
12. Disconnect the water bypass hose.
13. Disconnect the following engine wire harness connectors and remove the wire harness clamps from the cylinder head:
- Engine Coolant Temperature (ECT) sensor 1 connector
- Camshaft Position (CMP) sensor A (Intake) connector
- Camshaft Position (CMP) sensor B (Exhaust) connector

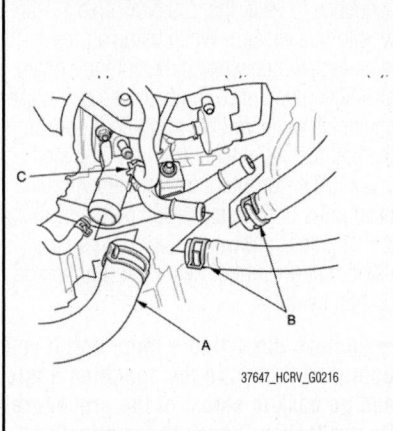

Fig. 150 Disconnect the upper radiator hose (A), the heater hoses (B), and the water bypass hose (C)

37647_HCRV_G0216

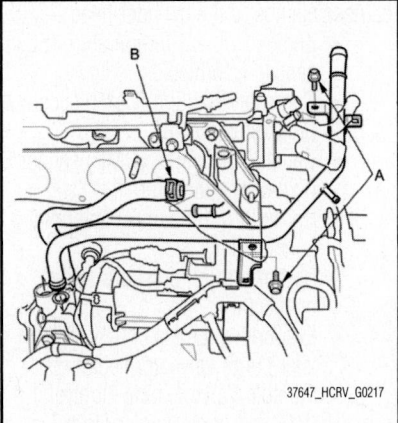

Fig. 151 Remove the two bolts (A) securing the connecting pipe, disconnect the water bypass hose (B)

37647_HCRV_G0217

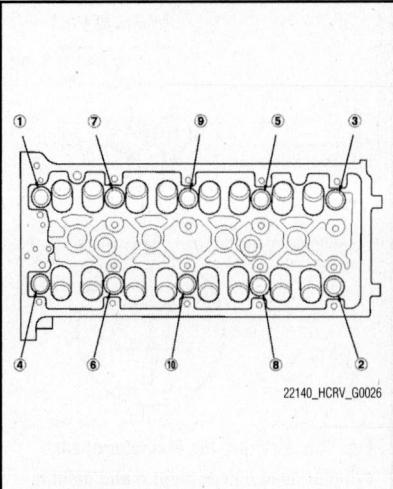

Fig. 152 Remove the cylinder head bolts, loosen the bolts in sequence

22140_HCRV_G0026

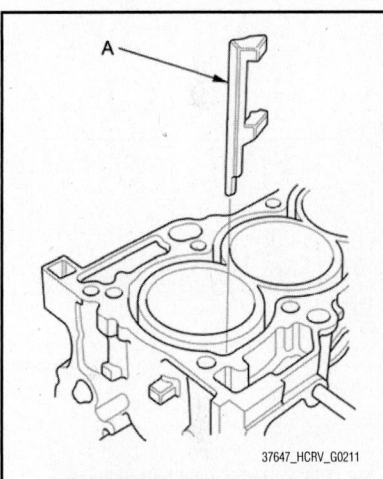

Fig. 153 Install a new coolant separator (A) in the engine block whenever the engine block is replaced

37647_HCRV_G0211

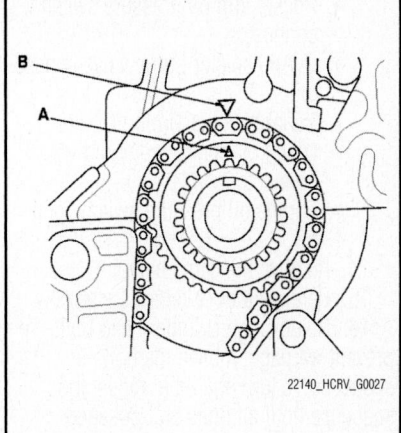

Fig. 154 Align the TDC mark (A) on the crankshaft sprocket with the pointer (B) on the engine block

22140_HCRV_G0027

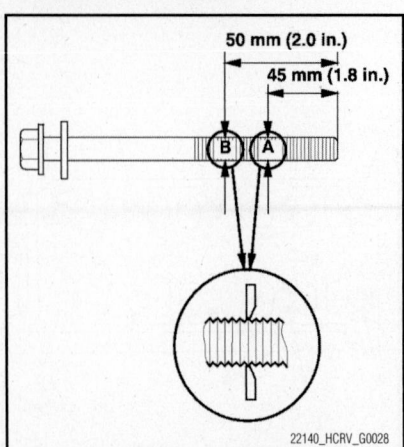

Fig. 155 Measure the diameter of each cylinder head bolt at point A and point B. Replace with new bolts if required

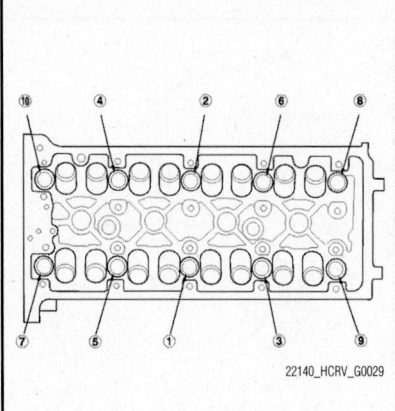

Fig. 156 Cylinder head bolt tightening sequence

- Rocker arm oil control valve connector
- Rocker arm oil pressure switch connector
- EVAP canister purge valve connector
- Variable Valve Timing Control (VTC) oil control solenoid valve connector
- Engine oil pressure switch connector

14. Remove the cam chain.
15. Remove the rocker arm assembly.
16. Remove the cylinder head bolts. To prevent warpage, loosen the bolts in sequence ⅓ turn at a time; repeat the sequence until all bolts are loosened.
17. Remove the cylinder head.

To install:

18. Clean the cylinder head and the engine block surface.

19. Install a new coolant separator in the engine block whenever the engine block is replaced.
20. Install the new cylinder head gasket and the dowel pins on the engine block. Always use a new cylinder head gasket.
21. Set the crankshaft to Top Dead Center (TDC). Align the TDC mark on the crankshaft sprocket with the pointer on the engine block.
22. Install the cylinder head on the engine block.
23. Measure the diameter of each cylinder head bolt at point A and point B.
24. If either diameter is less than 0.42 inches (10.6 mm), replace the cylinder head bolt.
25. Apply new engine oil to the threads and under the bolt heads of all cylinder head bolts.
26. Tighten the cylinder head bolts in sequence to 29 ft. lbs. (39 Nm). Use a beam-type torque wrench. When using a preset click-type torque wrench, be sure to tighten slowly and do not overtighten. If a bolt makes any noise while you are torquing it, loosen the bolt and retighten it from the first step.
27. After torquing, tighten all cylinder head bolts in two steps (90° per step) using the sequence shown in step 9. If you are using a new cylinder head bolt, tighten the bolt an extra 90°.

➡**Remove the cylinder head bolt if you tightened it beyond the specified angle, and go back to step 6 of the procedure. Do not loosen it back to the specified angle.**

28. Install the rocker arm assembly.
29. Install the cam chain.
30. Connect the following engine wire harness connectors, and install the wire harness clamps to the cylinder head:

- Engine Coolant Temperature (ECT) sensor 1 connector
- Camshaft Position (CMP) sensor A (Intake) connector
- Camshaft Position (CMP) sensor B (Exhaust) connector
- Rocker arm oil control solenoid connector
- Rocker arm oil pressure switch connector
- Evaporative Emission (EVAP) canister purge valve connector
- Variable Valve Timing Control (VTC) oil control solenoid valve connector
- Engine oil pressure switch connector

31. Install the two bolts (A) securing the connecting pipe.

32. Connect the water bypass hose (B).
33. Connect the upper radiator hose (A), the heater hoses (B), and the water bypass hose (C).
34. Install the four bolts securing the EVAP canister purge valve bracket.
35. Connect the four fuel injector connectors (A), install the ground cables (B).
36. Connect the fuel feed hose (A), then install the quick-connect fitting cover (B).
37. Connect the EVAP canister hose (A).
38. Install the warm up TWC.
39. Install the intake manifold.
40. Install the drive belt.
41. After installation, check that all tubes, hoses, and connectors are installed correctly.
42. Inspect for fuel leaks. Turn the ignition switch to ON (II) (do not operate the starter) so the fuel pump runs for about 2 seconds and pressurizes the fuel line. Repeat this operation three times, then check for fuel leakage at any point in the fuel line.
43. Refill the radiator with engine coolant, and bleed the air from the cooling system.
44. Check for fluid leaks.
45. Do the Powertrain Control Module (PCM) idle learn procedure.
46. Do the Crankshaft Position (CKP) pattern clear/CKP pattern learn procedure.
47. Inspect the idle speed.
48. Inspect the ignition timing.

CYLINDER HEAD (VALVE) COVER

REMOVAL & INSTALLATION

2011 Models

See Figures 157 through 159.

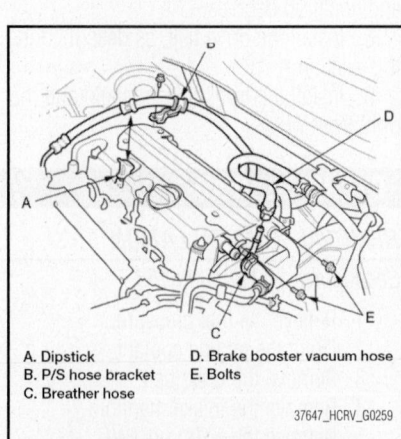

A. Dipstick
B. P/S hose bracket
C. Breather hose
D. Brake booster vacuum hose
E. Bolts

Fig. 157 Remove the dipstick and the Power Steering (P/S) hose bracket and disconnect the breather hose and the brake booster vacuum hose

1. Remove the engine cover.
2. Remove the four ignition coils.
3. Remove the dipstick and the Power Steering (P/S) hose bracket and disconnect the breather hose and the brake booster vacuum hose.
4. Remove the two bolts securing the install pipe component.
5. Remove the cylinder head cover.

To install:

6. Thoroughly clean the head cover gasket and the groove.
7. Install the cylinder head cover gasket in the groove of the cylinder head cover.
8. Check that the mating surfaces are clean and dry.
9. Apply liquid gasket, P/N 08717-0004, 08718-0003, or 08718-0009, on the chain case and the No. 5 rocker shaft holder mating areas. Install the component within 5 minutes of applying the liquid gasket.

➥**If too much time has passed after applying the liquid gasket, remove the old liquid gasket and residue, then reapply new liquid gasket.**

10. Set the spark plug seals on the spark plug tubes. Place the cylinder head cover on the cylinder head, then slide the cover slightly back and forth to seat the head cover gasket.
11. Inspect the cover washers. Replace any washer that is damaged or deteriorated.
12. Tighten the bolts in three steps. In the final step tighten all bolts, in a criss-cross pattern starting in the middle, to 9 ft. lbs. (12 Nm).

➥**Wait at least 30 minutes before filling the engine with oil. Do not run the engine for at least 3 hours after installing the head cover.**

13. Install the two bolts securing the install pipe component.
14. Connect the breathe hose and the brake booster vacuum hose and install the power steering (P/S) hose bracket, and the dipstick.
15. Install the four ignition coils.
16. Install the engine cover.

2012 Models

See Figures 160 through 163.

1. Remove the ignition coil cover.
2. Remove the ignition coils.
3. Remove the dipstick. Disconnect the breather hose. Remove the two bolts.
4. Remove the cylinder head cover.

To install:

5. Check the spark plug seals for damage. If any seals are damaged, replace it.

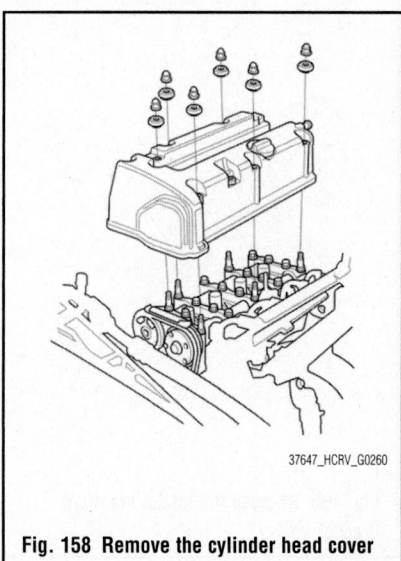

37647_HCRV_G0260

Fig. 158 Remove the cylinder head cover

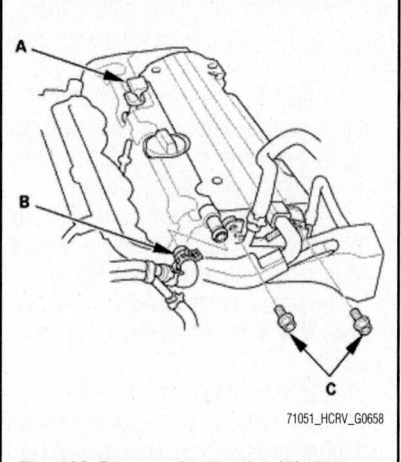

71051_HCRV_G0658

Fig. 160 Remove the dipstick (A). Disconnect the breather hose (B). Remove the two bolts (C)

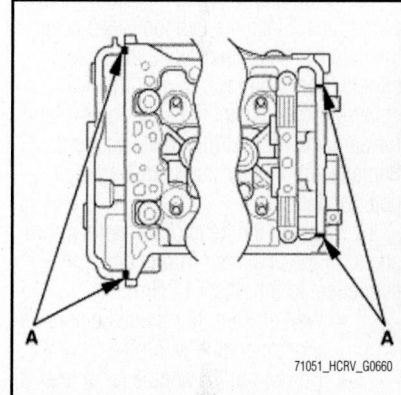

71051_HCRV_G0660

Fig. 162 Apply liquid gasket, (P/N 08718-0004 or 08718-0009) on the chain case and the No. 5 rocker shaft holder mating areas (A). Install the component within 5 minutes of applying the liquid gasket

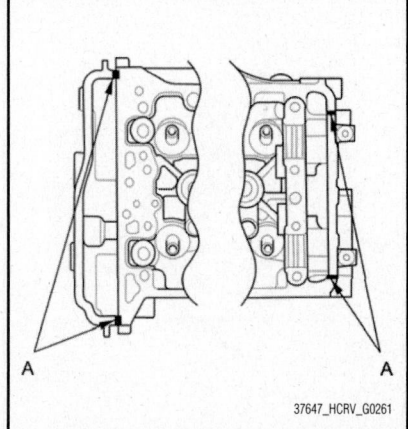

37647_HCRV_G0261

Fig. 159 Apply liquid gasket on the chain case and the No. 5 rocker shaft holder mating areas (A)

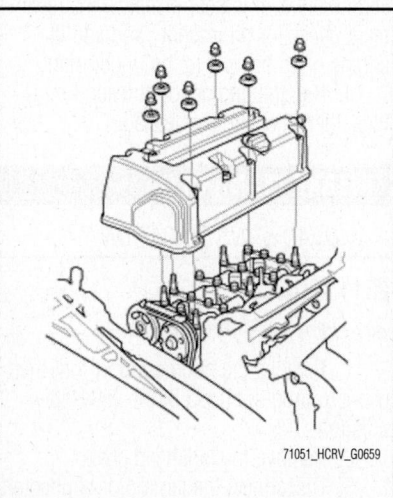

71051_HCRV_G0659

Fig. 161 Remove the cylinder head cover

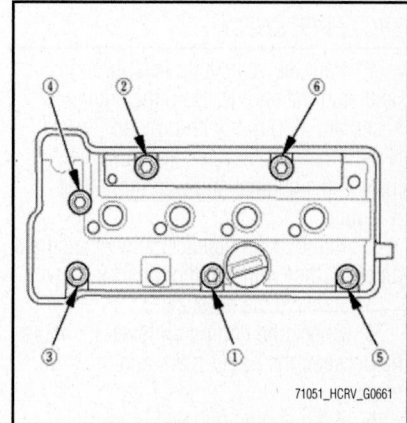

71051_HCRV_G0661

Fig. 163 Tighten the head cover bolts in two steps. In the final step torque bolts, in sequence, to 9 ft. lbs. (12 Nm)

6. Thoroughly clean the head cover gasket and the groove.

7. Check and, if necessary, replace the head cover gasket.

8. Install the head cover gasket in the groove of the cylinder head cover. Make sure the head cover gasket is seated securely.

9. Remove all of the old liquid gasket from the chain case and the No.5 rocker shaft holder.

10. Clean the head cover contacting surfaces with a shop towel.

11. Apply liquid gasket, (P/N 08718-0004 or 08718-0009) on the chain case and the No. 5 rocker shaft holder mating areas. Install the component within 5 minutes of applying the liquid gasket.

➡️If too much time has passed after applying the liquid gasket, remove the old liquid gasket and residue, then reapply new liquid gasket.

12. Set the spark plug seals on the spark plug tubes. Place the cylinder head cover on the cylinder head, then slide the cover slightly back and forth to seat the head cover gasket. Inspect the spark plug seals for damage. Inspect the cover washers. Replace any washer that is damaged or deteriorated.

13. Tighten the head cover bolts in two steps. In the final step torque bolts, in sequence, to 9 ft. lbs. (12 Nm).

- Wait at least 30 minutes before filling the engine with oil.
- Do not run the engine for at least 3 hours after installing the head cover.

14. Install the two bolts. Connect the breather hose. Install the dipstick.

15. Install the ignition coils. Install the ignition coil cover.

ENGINE OIL & FILTER

OIL LEVEL CHECK

1. Park the vehicle on level ground, and start the engine. Hold the engine speed at 3,000 rpm with no load P or N until the radiator fan comes on, then turn off the engine, and wait more than 3 minute.

2. Remove the dipstick, and wipe off the dipstick, then reinstall the dipstick.

3. Remove the dipstick.

4. Check the engine oil level. It should be between the upper mark and the lower mark.

5. If the engine oil level is near or below the lower mark check for oil leakage, and add the engine oil to bring it to the upper mark.

OIL FILTER

1. Raise the vehicle on a lift, and make sure it is securely supported.

2. Remove the engine oil drain lid.

3. Remove the oil filter with the oil filter wrench.

To install:

4. Inspect the filter to make sure the rubber seal is not stuck to the oil filter seating surface of the engine.

5. Inspect the threads (A) and the rubber seal (B) on the new filter.

6. Clean the seat on the oil filter base.

7. Apply a light coat of new engine oil to the oil filter rubber seal.

➡️Use only filters with a built-in bypass system.

8. Install the oil filter by hand.

9. After the rubber seal seats, tighten the oil filter clockwise with the oil filter wrench.

10. If four numbers or marks (1 to 4 or ▼ to ▼▼▼▼) are printed around the outside of the filter, use the following procedure to tighten the filter.

   a. Spin the filter on until its seal lightly seats against the oil pan, and note which number or mark is at the bottom.

   b. Tighten the filter by turning it clockwise three numbers or marks from the one you noted. For example, if mark ▼ is at the bottom when the seal is lightly seated, tighten the filter until the mark ▼▼▼▼ comes around to the bottom.

11. Install the engine oil drain lid.

12. Park the vehicle on level ground, and start the engine. Hold the engine speed at 3,000 rpm with no load P or N until the radiator fan comes on, then turn off the engine, and wait more than 3 minute.

13. Check the engine oil level. If the engine oil level is near or below the lower mark check for oil leakage, and add the engine oil to bring it to the upper mark.

14. Run the engine for at least 3 minutes, then check for oil leakage.

INTAKE MANIFOLD

REMOVAL & INSTALLATION

2011 Models

See Figures 164 through 168.

1. Remove the hood support rod, then use it to prop the hood in the wide-open position.

2. Remove the bulkhead cover.

3. Disconnect the fan motor connectors and the hood switch connector, then remove the harness clips.

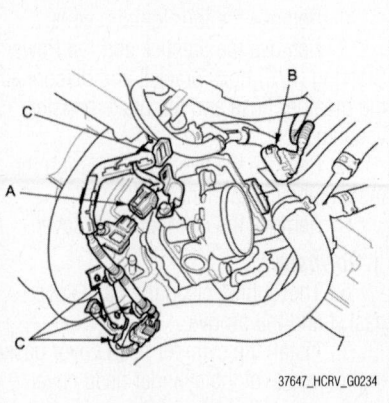

Fig. 164 Disconnect the Manifold Absolute Pressure (MAP) sensor connector (A) and the throttle actuator connector (B), then remove the wire harness clamps (C)

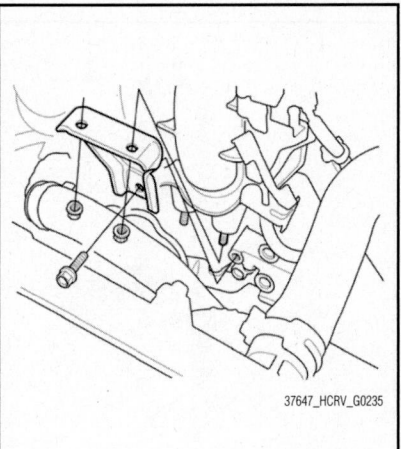

Fig. 165 Remove the intake manifold bracket

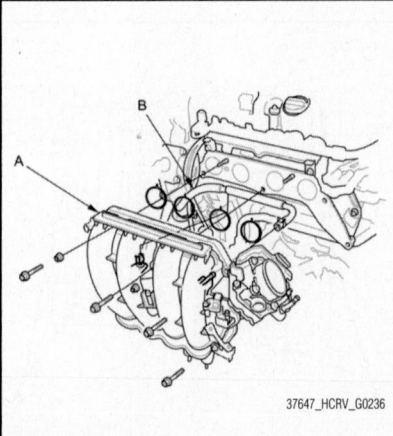

Fig. 166 Remove the intake manifold (A), then disconnect the PCV hose (B) from the intake manifold

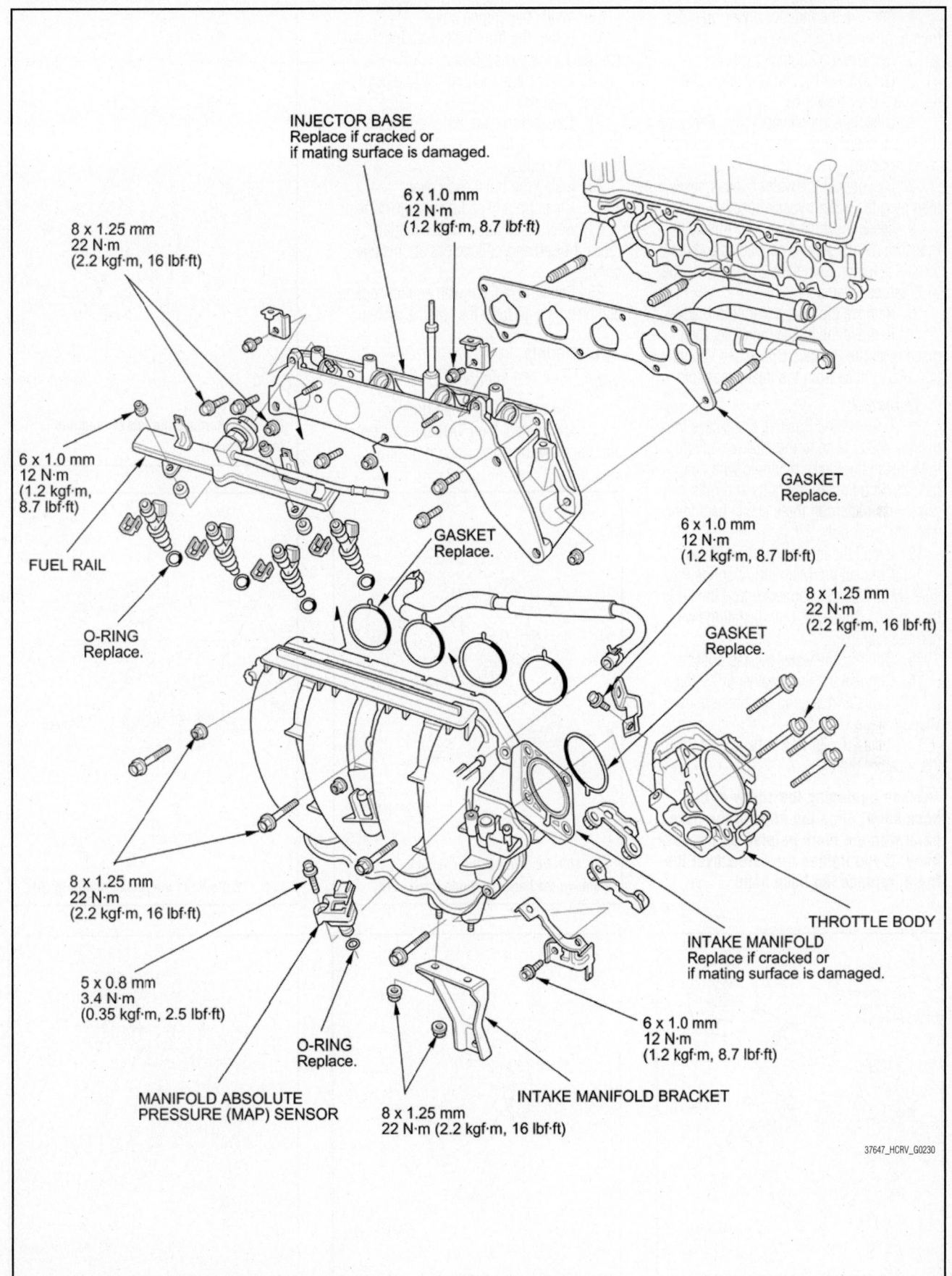

INJECTOR BASE
Replace if cracked or
if mating surface is damaged.

6 x 1.0 mm
12 N·m
(1.2 kgf·m, 8.7 lbf·ft)

8 x 1.25 mm
22 N·m
(2.2 kgf·m, 16 lbf·ft)

6 x 1.0 mm
12 N·m
(1.2 kgf·m,
8.7 lbf·ft)

FUEL RAIL

O-RING
Replace.

GASKET
Replace.

GASKET
Replace.

6 x 1.0 mm
12 N·m
(1.2 kgf·m, 8.7 lbf·ft)

GASKET
Replace.

8 x 1.25 mm
22 N·m
(2.2 kgf·m, 16 lbf·ft)

THROTTLE BODY

8 x 1.25 mm
22 N·m
(2.2 kgf·m, 16 lbf·ft)

INTAKE MANIFOLD
Replace if cracked or
if mating surface is damaged.

5 x 0.8 mm
3.4 N·m
(0.35 kgf·m, 2.5 lbf·ft)

O-RING
Replace.

MANIFOLD ABSOLUTE
PRESSURE (MAP) SENSOR

6 x 1.0 mm
12 N·m
(1.2 kgf·m, 8.7 lbf·ft)

INTAKE MANIFOLD BRACKET

8 x 1.25 mm
22 N·m (2.2 kgf·m, 16 lbf·ft)

37647_HCRV_G0230

Fig. 167 Exploded view of the intake manifold assembly

4. Remove the radiator upper brackets, then remove the front bulkhead.

5. Remove the engine cover.

6. Disconnect the breather pipe, then remove the air flow tube.

7. Disconnect the Evaporative Emission (EVAP) canister hose and the brake booster vacuum hose.

8. Disconnect the water bypass hoses, then plug the water bypass hoses.

9. Disconnect the Manifold Absolute Pressure (MAP) sensor connector and the throttle actuator connector, then remove the wire harness clamps.

10. Remove the intake manifold bracket.

11. Remove the intake manifold, then disconnect the Positive Crankcase Ventilation (PCV) hose from the intake manifold.

To install:

12. Connect the Positive Crankcase Ventilation (PCV) hose to the intake manifold, then install the intake manifold with new gaskets, and tighten the bolts and nuts in a crisscross pattern in three steps, beginning with the inner bolt.

13. Install the intake manifold bracket.

14. Connect the Manifold Absolute Pressure (MAP) sensor connector and the throttle actuator connector, then install the wire harness clamps.

15. Connect the water bypass hoses.

16. Connect the Evaporative Emission (EVAP) canister hose and the brake booster vacuum hose.

17. Install the air flow tube, then connect the breather pipe.

➡**When tightening the screw of the hose band, align the edge of the hose band with the mark painted on the hose band. If you tighten the screw over the mark, replace the hose band.**

18. Install the engine cover.

19. Install the front bulkhead, then install the radiator upper brackets.

20. Apply body paint to the bulkhead mounting bolts.

21. Connect the fan motor connectors and the hood switch connector, then install the harness clips.

22. Install the bulkhead cover.

23. Clean up any spilled engine coolant.

24. After installation, check that all tubes, hoses, and connectors are installed correctly.

25. Refill the radiator with engine coolant, and bleed the air from the cooling system.

2012 Models

See Figures 169 through 173.

1. Remove the intake air duct.

2. Disconnect the EVAP canister hose and the brake booster vacuum hose.

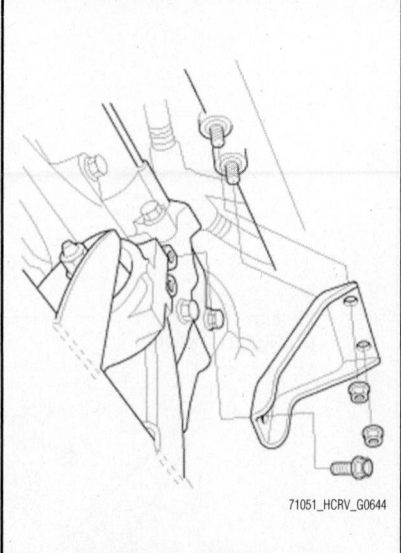

Fig. 171 Remove the intake manifold lower bracket

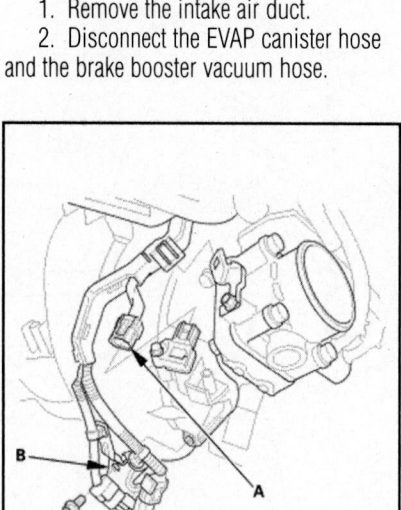

Fig. 169 Disconnect the connector (A). Remove the harness clamp bracket (B)

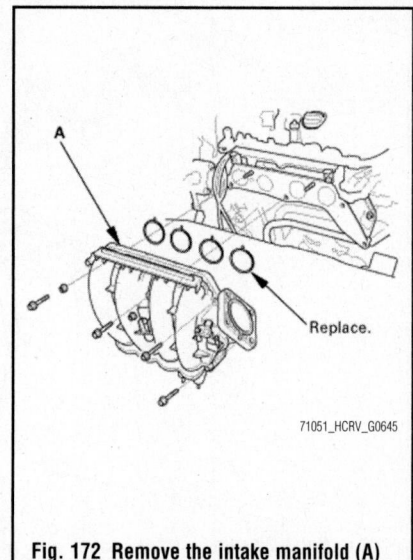

Fig. 172 Remove the intake manifold (A)

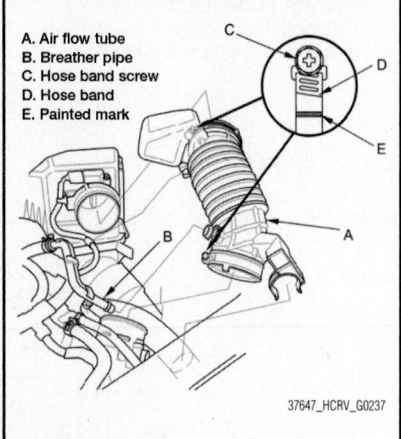

A. Air flow tube
B. Breather pipe
C. Hose band screw
D. Hose band
E. Painted mark

Fig. 168 Install the air flow tube, then connect the breather pipe

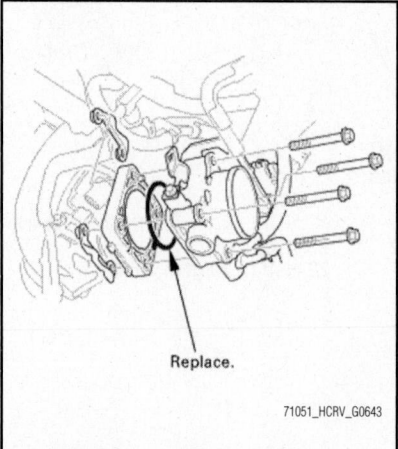

Fig. 170 Move the throttle body without disconnecting the hoses

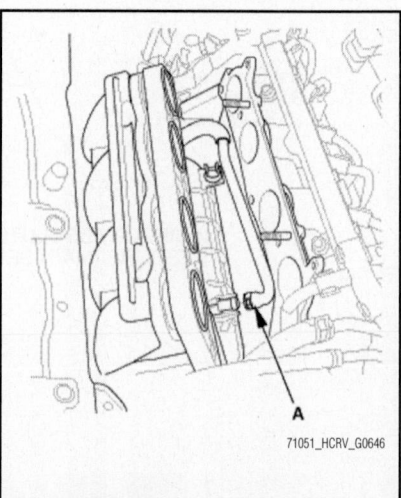

Fig. 173 Disconnect the PCV hose (A)

3. Disconnect the connector. Remove the harness clamp bracket.

4. Move the throttle body without disconnecting the hoses.

5. Remove the intake manifold lower bracket.

6. Remove the intake manifold.

7. Disconnect the PCV hose.

To install:

8. Install the intake manifold with new gaskets.

9. Tighten the intake manifold and bracket bolts to 16 ft. lbs. (22 Nm).

10. Reposition the throttle body and tighten the bolts to 16 ft. lbs. (22 Nm).

11. Install the throttle body connector and harness clamp bracket.

12. Install the remaining hoses.

13. Install the intake air duct.

OIL PAN

REMOVAL & INSTALLATION

See Figures 174 through 178.

1. If the engine is already out of the vehicle, go to step 21.

2. Raise the vehicle on the lift.

3. Drain the engine oil.

4. Remove the front wheels.

5. Remove the splash shield.

6. Disconnect the Air Fuel Ratio (A/F) sensor connector and secondary Heated Oxygen Sensor (secondary HO2S) connector, then remove the Three Way Catalytic Converter (TWC).

7. Remove the shift cable.

8. Separate the stabilizer links from the stabilizer bar.

9. Separate the knuckles from the lower arms.

10. 4WD model: Remove the propeller shaft.

11. Remove a bolt securing the P/S fluid line bracket, and unclamp the P/S fluid line clamps on the front subframe.

12. Remove the bolts securing the left steering gearbox mounting bracket.

13. Remove the bolts securing the right steering gearbox mounting brackets.

14. Remove the bolt securing the Automatic Transmission Fluid (ATF) filter.

15. Remove the lower torque rod.

16. Make the appropriate reference lines at both ends of the subframe that line up with the body.

17. Remove the subframe mounting bolts on both side.

18. Attach the subframe adapter to the subframe and hang the belt of the subframe adapter over the front of the subframe, then secure the belt with its stop.

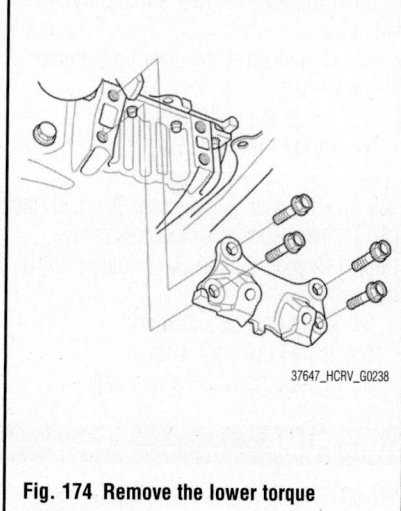

Fig. 174 Remove the lower torque rod bracket

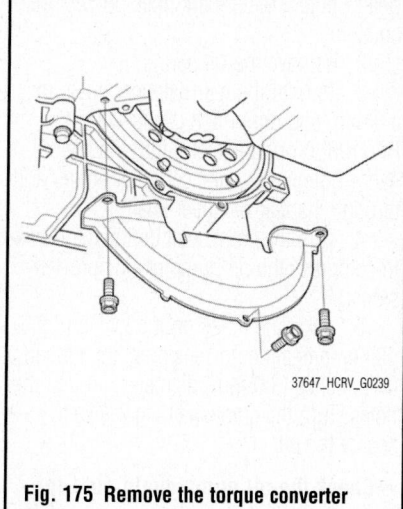

Fig. 175 Remove the torque converter cover

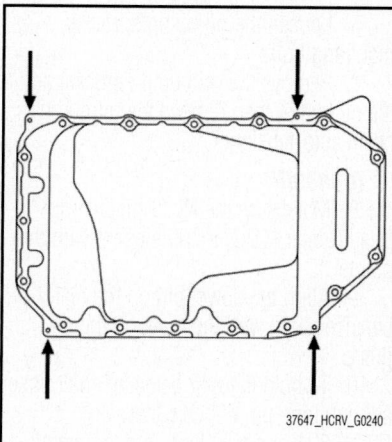

Fig. 176 Using a flat blade screwdriver, separate the oil pan from the block in the places shown

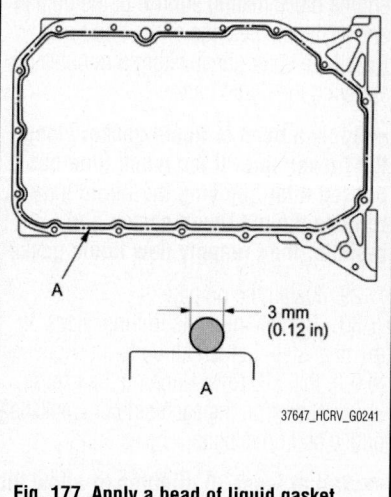

Fig. 177 Apply a bead of liquid gasket along the broken line (A)

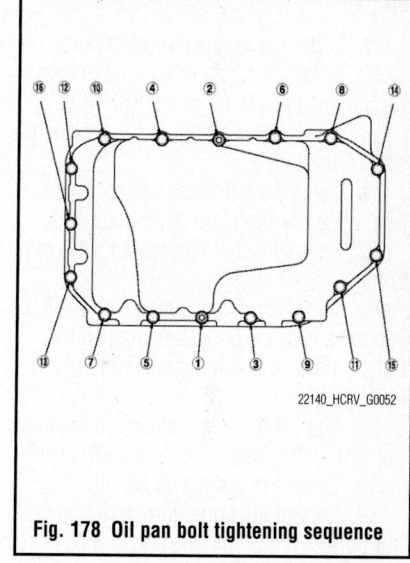

Fig. 178 Oil pan bolt tightening sequence

19. Raise the jack and line up the slots in the arms with the bolt holes on the corner of the jack base, then attach them with bolts securely.

20. Remove the subframe.

21. Remove the lower torque rod bracket.

22. Remove the torque converter cover.

23. Remove the bolts securing the oil pan.

24. Using a flat blade screwdriver, separate the oil pan from the block in the places shown.

25. Remove the oil pan.

To install:

26. Remove all of the old liquid gasket from the oil pan mating surfaces, bolts, and bolt holes.

27. Clean and dry the oil pan mating surfaces.

28. Apply liquid gasket (P/N 08717-0004, 08718-0003, or 08718-0009) to the

engine block mating surface of the oil pan and to the inside edge of the bolt holes. Install the component within 5 minutes of applying the liquid gasket.

➡**Apply a bead of liquid gasket along the broken line. If too much time has passed after applying the liquid gasket, remove the old liquid gasket and residue, then reapply new liquid gasket.**

29. Install the oil pan.

30. Tighten the bolts in three steps. In the final step, tighten all bolts, in sequence, to 9 ft. lbs. (12 Nm). Wipe off the excess liquid gasket on the each side of crankshaft pulley and drive plate.

➡**Wait at least 30 minutes to allow the liquid gasket to cure before filling the engine with oil. Do not run the engine for at least 3 hours after installing the oil pan.**

31. Install the torque converter cover.

32. Install the lower torque rod bracket. Tighten the bolts to 60 ft. lbs. (81 Nm).

33. If the engine is still in the vehicle, do steps 9 through 31.

34. Using the subframe adapter and a jack, raise the subframe up to body.

35. Loosely install the new 14 x 1.5 mm bolts.

36. Align all reference marks on the front subframe with the body, then tighten the bolts on the front subframe to 76 ft. lbs. (103 Nm).

37. Tighten the new subframe mounting bolts on both sides to 76 ft. lbs. (103 Nm).

38. Lower the vehicle on the lift.

39. Loosen the upper torque rod mounting bolt.

40. Raise the vehicle on the lift.

41. Install the lower torque rod, then tighten the new lower torque rod mounting bolts in the numbered sequence shown.

42. Lower the vehicle on the lift.

43. Tighten the upper torque rod mounting bolt.

44. Install the Automatic Transmission Fluid (ATF) filter.

45. Install the bolts securing the left steering gearbox mounting bracket.

46. Install the bolts securing the right steering gearbox mounting bracket.

47. Install the Power Steering (P/S) fluid line bracket and secure the hose with the hose clamps.

48. 4WD model: Install the propeller shaft.

49. Install a new set ring on the end of each driveshaft, then install the driveshafts. Make sure each ring "clicks" into place in the differential and intermediate shaft.

50. Connect the lower arms to the knuckles.

51. Connect the stabilizer links to the stabilizer bar.

52. Install the shift cable.

53. Install the Three Way Catalytic Converter (TWC). Use new gaskets and new self-locking nuts. Connect the Air Fuel Ratio (A/F) sensor connector and secondary Heated Oxygen Sensor (secondary HO2S) connector.

54. Install the splash shield.

55. Install the front wheels.

56. Check the wheel alignment.

OIL PUMP

REMOVAL & INSTALLATION

See Figures 179 through 185.

1. Turn the crankshaft pulley so its Top Dead Center (TDC) mark lines up with the pointer.

2. Remove the oil pan.

3. To hold the rear balancer shaft, insert a 6 mm long pin punch (Snap-On® PPC108LA or equivalent) into the maintenance hole in the balancer shaft holder and through the rear balancer shaft.

4. Turn the crankshaft counterclockwise to compress the oil pump chain auto-tensioner.

5. Align the holes on the lock and the oil pump chain auto-tensioner, then insert a 0.12 inches (3.0 mm) diameter pin into the holes. Turn the crankshaft clockwise to secure the pin.

➡**Check the oil pump chain auto-tensioner cam position. If the position is not aligned, set the first cam to the first edge of the rack.**

6. Loosen the oil pump sprocket mounting bolt.

7. Remove the oil pump sprocket and the oil pump, then remove the oil pump chain auto-tensioner.

To install:

8. Make sure the No. 1 piston Top Dead Center (TDC) mark lines up with the pointer.

9. Align the dowel pin on the rear balancer shaft with the mark on the oil pump.

10. To hold the rear balancer shaft, insert a 6 mm long pin punch (Snap-On® PPC108LA or equivalent) into the maintenance hole in the balancer shaft holder and through the rear balancer shaft.

11. Turn the plate counterclockwise, to release the lock, then push the oil pump chain auto-tensioner arm, and set the first

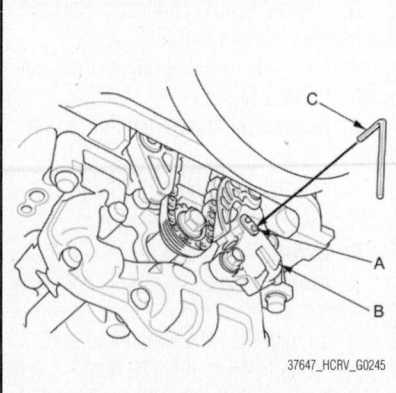

Fig. 179 Align the holes on the lock (A) and the oil pump chain auto-tensioner (B), then insert a 0.12 inches (3.0 mm) diameter pin (C) into the holes

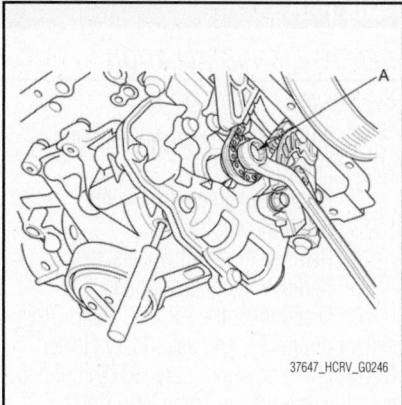

Fig. 180 Loosen the oil pump sprocket mounting bolt (A)

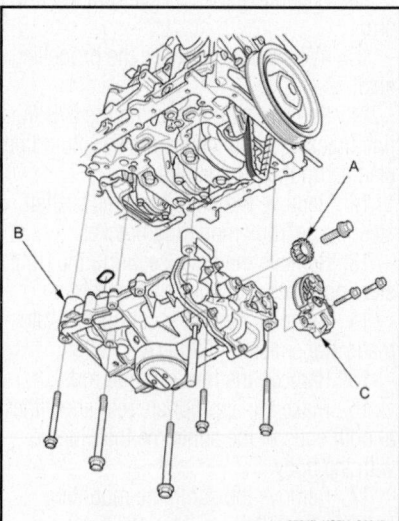

Fig. 181 Remove the oil pump sprocket (A) and the oil pump (B), then remove the oil pump chain auto-tensioner (C)

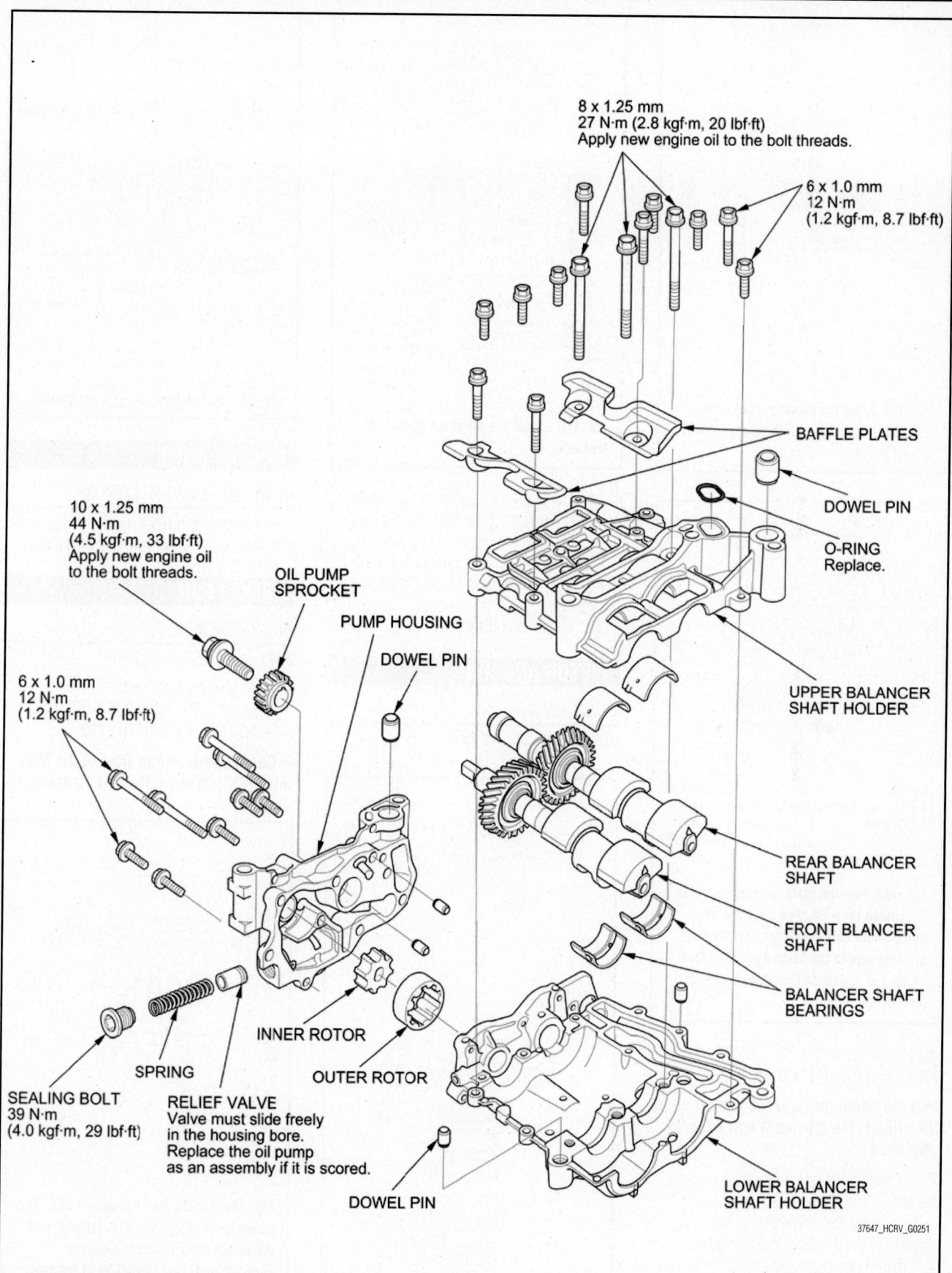

8 x 1.25 mm
27 N·m (2.8 kgf·m, 20 lbf·ft)
Apply new engine oil to the bolt threads.

6 x 1.0 mm
12 N·m
(1.2 kgf·m, 8.7 lbf·ft)

BAFFLE PLATES

DOWEL PIN

O-RING
Replace.

10 x 1.25 mm
44 N·m
(4.5 kgf·m, 33 lbf·ft)
Apply new engine oil
to the bolt threads.

OIL PUMP
SPROCKET

PUMP HOUSING

DOWEL PIN

UPPER BALANCER
SHAFT HOLDER

6 x 1.0 mm
12 N·m
(1.2 kgf·m, 8.7 lbf·ft)

REAR BALANCER
SHAFT

FRONT BLANCER
SHAFT

BALANCER SHAFT
BEARINGS

INNER ROTOR

OUTER ROTOR

SEALING BOLT
39 N·m
(4.0 kgf·m, 29 lbf·ft)

SPRING

RELIEF VALVE
Valve must slide freely
in the housing bore.
Replace the oil pump
as an assembly if it is scored.

DOWEL PIN

LOWER BALANCER
SHAFT HOLDER

37647_HCRV_G0251

Fig. 182 Exploded view of oil pump assembly

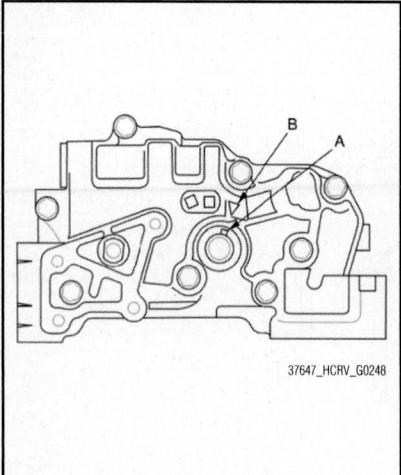

Fig. 183 Align the dowel pin (A) on the rear balancer shaft with the mark (B) on the oil pump

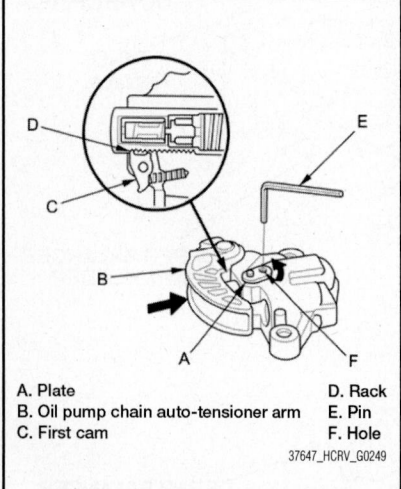

A. Plate
B. Oil pump chain auto-tensioner arm
C. First cam
D. Rack
E. Pin
F. Hole
37647_HCRV_G0249

Fig. 184 Turn the plate counterclockwise, to release the lock, then push the oil pump chain auto-tensioner arm, and set the first cam to the first edge of the rack; insert a 0.12 inches (3.0 mm) diameter pin into the hole

cam to the first edge of the rack. Insert a 0.12 inches (3.0 mm) diameter pin into the hole.

➡️**If the chain tensioner is not set up as described, the tensioner will become damaged.**

12. Install the oil pump chain auto-tensioner.

13. Apply new engine oil to the threads of the oil pump mounting bolts and the oil pump sprocket mounting bolt, then Loosely install the oil pump with a new O-ring, then install the oil pump sprocket.

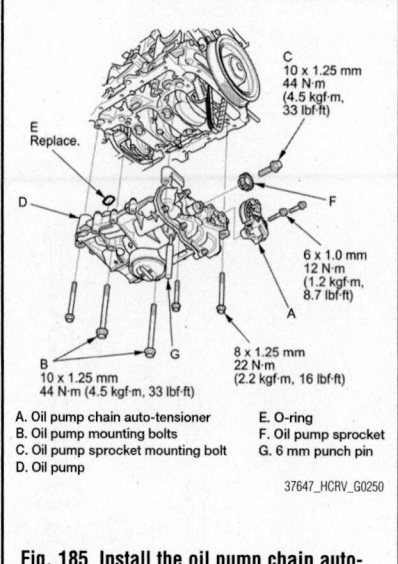

A. Oil pump chain auto-tensioner
B. Oil pump mounting bolts
C. Oil pump sprocket mounting bolt
D. Oil pump
E. O-ring
F. Oil pump sprocket
G. 6 mm punch pin
37647_HCRV_G0250

Fig. 185 Install the oil pump chain auto-tensioner

14. Tighten the oil pump mounting bolts and the oil pump sprocket mounting bolt.

15. Remove the 6 mm pin punch.

16. Remove the 0.12 inches (3.0 mm) diameter pin from the oil pump chain auto-tensioner.

17. Install the oil pan.

PISTON AND RING

POSITIONING

See Figures 186 and 187.

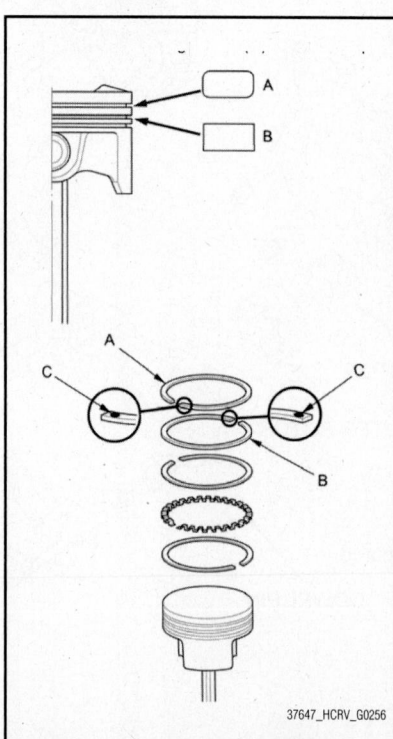

Fig. 186 Piston and ring positioning

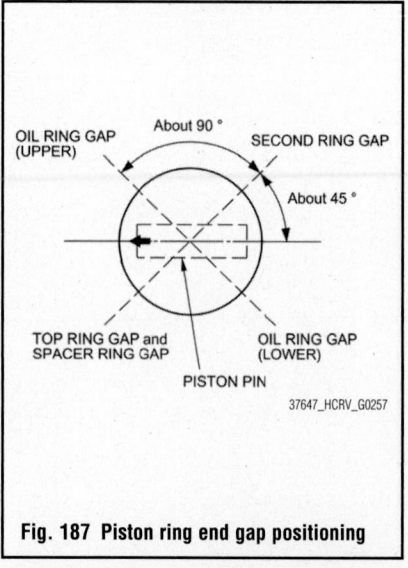

Fig. 187 Piston ring end gap positioning

ROCKER ARMS

REMOVAL & INSTALLATION

For service information, refer to Camshaft and Valve Lifters, Removal and Installation.

VALVE LASH

ADJUSTMENT

See Figures 188 through 190.

Special Tools Required:
- Locknut Wrench 07MAA-PR70120
- Adjuster 07MAA-PR70110

➡️**Connect the Honda Diagnostic System (HDS) to the Data Link Connector**

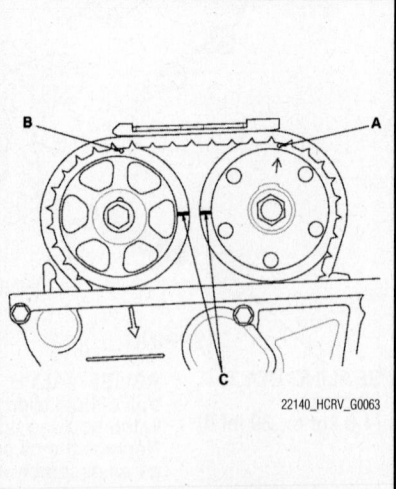

Fig. 188 Set the No. 1 piston at TDC. The punch mark (A) on the VTC actuator and the punch mark (B) on the exhaust camshaft sprocket should be at the top. Align the TDC marks (C) on the VTC actuator and exhaust camshaft sprocket.

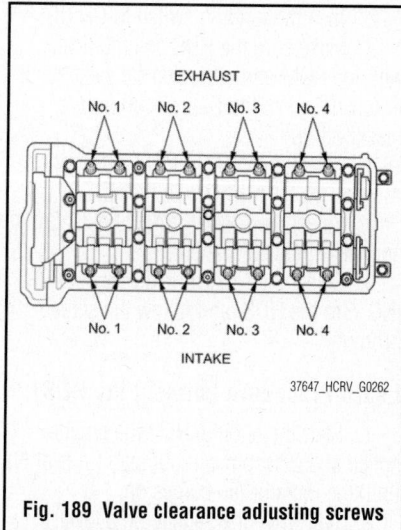

Fig. 189 Valve clearance adjusting screws

(DLC) and monitor the Engine Coolant Temperature (ECT) sensor 1 with the HDS. Adjust the valve clearance only when the ECT sensor 1 temperature is less than 100°F(38°C).

1. Remove the cylinder head cover.

2. Set the No. 1 piston at Top Dead Center (TDC). The punch mark on the Variable Valve Timing Control (VTC) actuator and the punch mark on t he exhaust camshaft sprocket should be at the top. Align the TDC marks on the

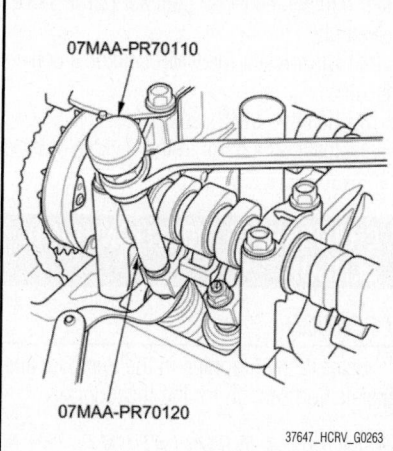

Fig. 190 Loosen the locknut with the locknut wrench and adjuster, and turn the adjusting screw

VTC actuator and the exhaust camshaft sprocket.

3. Select the correct feeler gauge for the valve clearance you are going to check.

- Intake: 0.008–0.010 inches (0.21–0.25 mm)
- Exhaust: 0.010–0.011 inches (0.25–0.29 mm)

4. Insert the feeler gauge between the adjusting screw and the end of the valve

stem, and slide it back and forth; you should feel a slight amount of drag.

5. If you feel too much or too little drag, loosen the locknut with the locknut wrench and adjuster, and turn the adjusting screw until the drag on the feeler gauge is correct.

➡**Apply new engine oil to the nut threads.**

6. Tighten the locknut to the specified torque, and recheck the clearance. Repeat the adjustment if necessary.

- Intake: 7 x 0.75 mm: 10 ft. lbs. (14 Nm)
- Exhaust: 7 x 0.75 mm: 10 ft. lbs. (14 Nm)

7. Rotate the crankshaft 180 degrees clockwise (camshaft pulley turns 90 degrees).

8. Check and, if necessary, adjust the valve clearance on the No. 3 cylinder.

9. Rotate the crankshaft 180 degrees clockwise (camshaft pulley turns 90 degrees).

10. Check and, if necessary, adjust the valve clearance on the No. 4 cylinder.

11. Rotate the crankshaft 180 degrees clockwise (camshaft pulley turns 90 degrees).

12. Check and, if necessary, adjust the valve clearance on the No. 2 cylinder.

13. Install the cylinder head cover.

ENGINE PERFORMANCE & EMISSION CONTROLS

ACCELERATOR PEDAL POSITION (APP) SENSOR

LOCATION

The Accelerator Pedal Position (APP) sensor is an integral part of the accelerator pedal module. The APP cannot be service separately.

REMOVAL & INSTALLATION

See Figure 191.

1. Disconnect the accelerator pedal module connector.

2. Remove the accelerator pedal module.

➡**The APP sensor is not available separately. Do not disassemble the accelerator pedal module.**

3. Install the parts in the reverse order of removal.

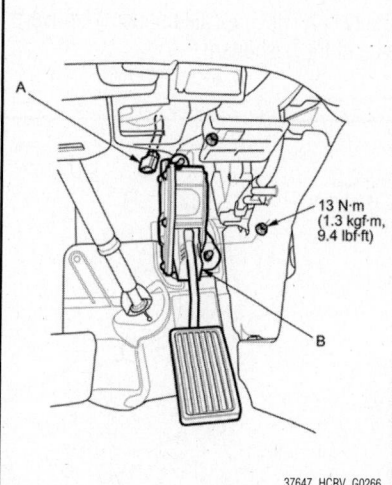

Fig. 191 Disconnect the accelerator pedal module connector (A); remove the accelerator pedal module (B)

CAMSHAFT POSITION (CMP) SENSOR

LOCATION

Refer to the graphics in the Removal and Installation section for the location(s).

REMOVAL & INSTALLATION

CMP Sensor A

See Figure 192.

1. Disconnect the CMP sensor A 3P connector.

2. Remove CMP sensor A from the intake camshaft side of the cylinder head.

3. Install the parts in the reverse order of removal with a new O-ring.

CMP Sensor B

See Figures 193 and 194.

1. Disconnect the connector and the hoses from the EVAP canister purge valve,

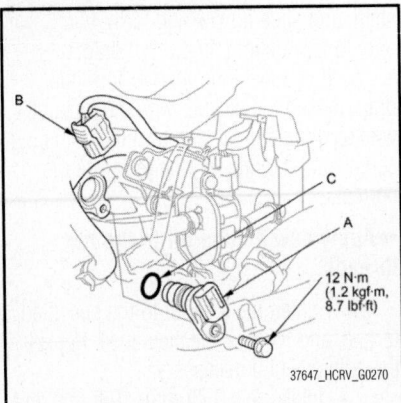

Fig. 192 Disconnect the CMP sensor A 3P connector (B); remove the CMP sensor A (A) and the O-ring (C)

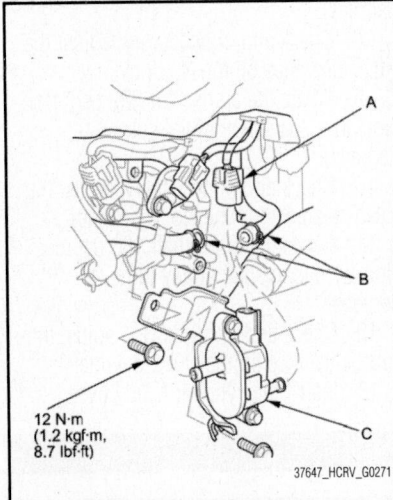

Fig. 193 Disconnect the connector (A) and the hoses (B) from the EVAP canister purge valve (C)

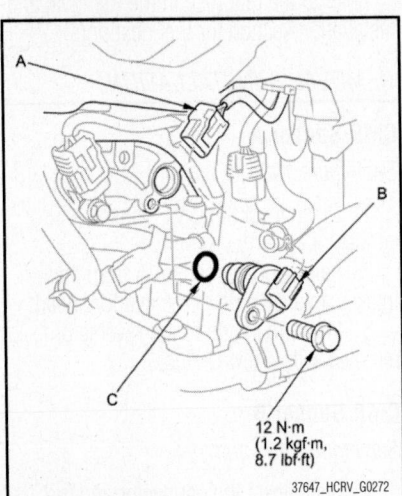

Fig. 194 Disconnect the CMP sensor B connector (A); remove the CMP sensor B (B) and the O-ring (C)

then remove the EVAP canister purge valve assembly.

2. Disconnect the CMP sensor B connector.

3. Remove CMP sensor B.

4. Install the parts in the reverse order of removal with a new O-ring.

CRANKSHAFT POSITION (CKP) SENSOR

LOCATION

Refer to the graphics in the Removal and Installation section for the location(s).

REMOVAL & INSTALLATION

See Figure 195.

1. Raise the vehicle on a lift.

➡**Make sure the vehicle is level, because engine oil will drip out when you remove the sensor.**

2. Remove the CKP sensor cover.

3. Disconnect the CKP sensor connector.

4. Remove the CKP sensor.

5. Install the parts in the reverse order of removal with a new O-ring.

6. Do the CKP pattern clear/CKP pattern learn procedure.

7. Check the engine oil level, and add more oil if needed.

CKP PATTERN CLEAR/CKP PATTERN LEARN

Clear/Learn Procedure (with the HDS)

1. Connect the HDS to the Data Link Connector (DLC) located under the driver's side of the dashboard.

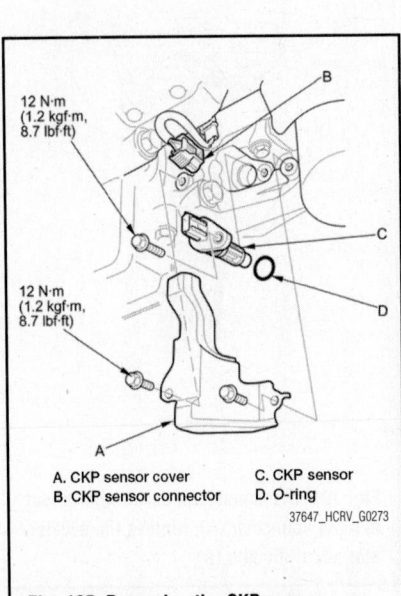

A. CKP sensor cover
B. CKP sensor connector
C. CKP sensor
D. O-ring

37647_HCRV_G0273

Fig. 195 Removing the CKP sensor

2. Turn the ignition switch to ON (II).

3. Make sure the HDS communicates with the PCM and other vehicle systems. If it doesn't, go to the DLC circuit troubleshooting.

4. Select the CRANK PATTERN in the ADJUSTMENT MENU with the HDS.

5. Select the CRANK PATTERN CLEAR, and clear the CKP pattern.

6. Select the CRANK PATTERN LEARNING with the HDS, and follow the screen prompts.

Learn Procedure (without the HDS)

1. Start the engine. Hold the engine speed at 3,000 rpm without load (in P or N) until the radiator fan comes on.

2. Test-drive the vehicle on a level road: Decelerate (with the throttle fully closed) from an engine speed of 2,500 rpm down to 1,000 rpm with the transmission in 2 position.

3. Repeat the previous step several times.

4. Turn the ignition switch to LOCK (0).

5. Turn the ignition switch to ON (II), and wait 30 seconds.

ENGINE COOLANT TEMPERATURE (ECT) SENSOR

LOCATION

Refer to the graphics in the Removal and Installation section for the location(s).

REMOVAL & INSTALLATION

ECT Sensor 1
See Figure 196.

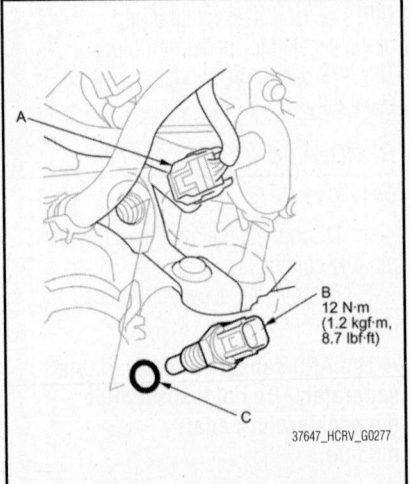

Fig. 196 Disconnect the ECT sensor 1 connector (A); remove ECT sensor 1 (B) and the O-ring (C)

1. Drain the engine coolant.
2. Disconnect the ECT sensor 1 connector (A).
3. Remove ECT sensor 1.
4. Install the parts in the reverse order of removal with a new O-ring, then refill the radiator with engine coolant.

ECT Sensor 2

See Figure 197.

1. Remove the front splash shield.
2. Drain the engine coolant.
3. Disconnect the ECT sensor 2 connector, then remove ECT sensor 2.
4. Install ECT sensor 2 with a new O-ring.
5. Install the front splash shield.
6. Refill the radiator with engine coolant.

EVAPORATIVE EMISSIONS (EVAP) CANISTER

LOCATION

The EVAP canister is mounted under the vehicle between the muffler and the rear wheel.

REMOVAL & INSTALLATION

See Figures 198 through 201.

1. Raise the vehicle on a lift.
2. Remove the cover.
3. Remove the EVAP canister baffle cover.
4. Disconnect the hoses and the fuel subharness 6P connector.
5. Remove the bolts and the EVAP canister bracket.
6. Remove the EVAP canister from the EVAP canister bracket.
7. Remove the cap (A).

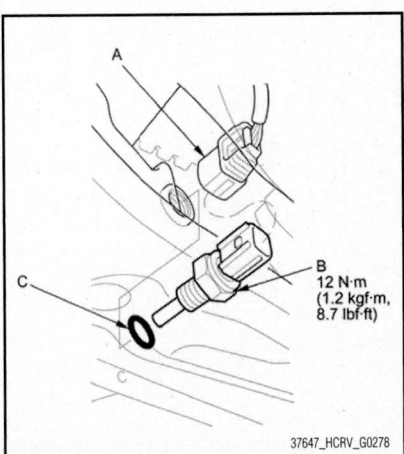

Fig. 197 Disconnect the ECT sensor 2 connector (A), then remove ECT sensor 2 (B) and the O-ring (C)

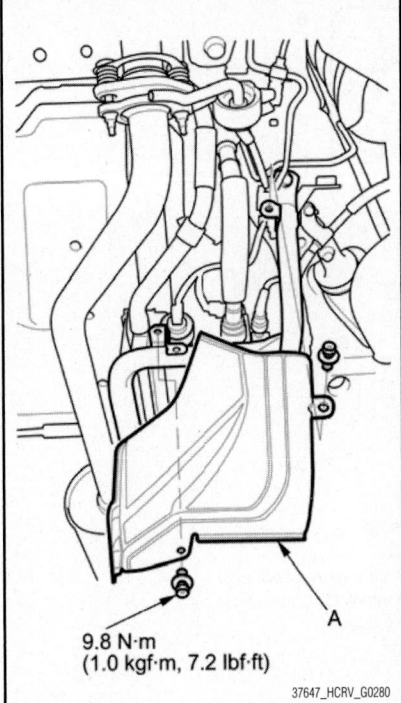

9.8 N·m
(1.0 kgf·m, 7.2 lbf·ft)

37647_HCRV_G0280

Fig. 198 Remove the EVAP canister baffle cover (A)

8. Remove the EVAP canister vent shut valve.
9. Install the EVAP canister vent shut valve into the new EVAP canister with new O-rings.

➡ **Do not coat the O-rings with oil.**

10. Install the parts in the reverse order of removal.

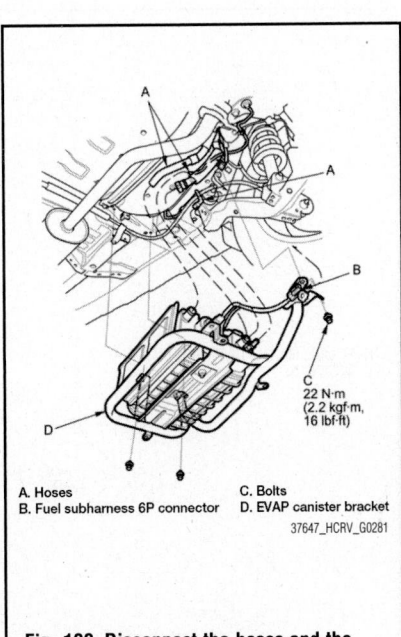

A. Hoses
B. Fuel subharness 6P connector
C. Bolts
D. EVAP canister bracket

22 N·m
(2.2 kgf·m, 16 lbf·ft)

37647_HCRV_G0281

Fig. 199 Disconnect the hoses and the fuel subharness 6P connector

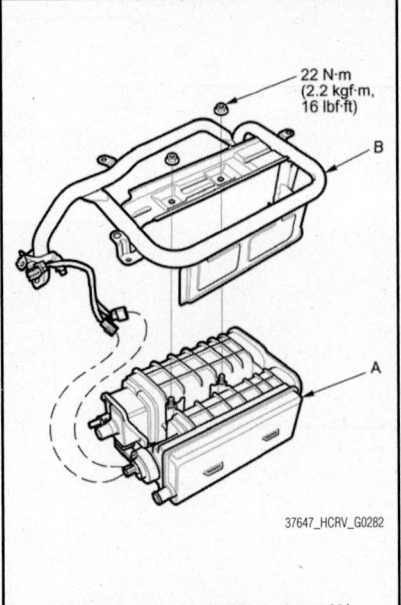

22 N·m
(2.2 kgf·m, 16 lbf·ft)

37647_HCRV_G0282

Fig. 200 Remove the EVAP canister (A) from the EVAP canister bracket (B)

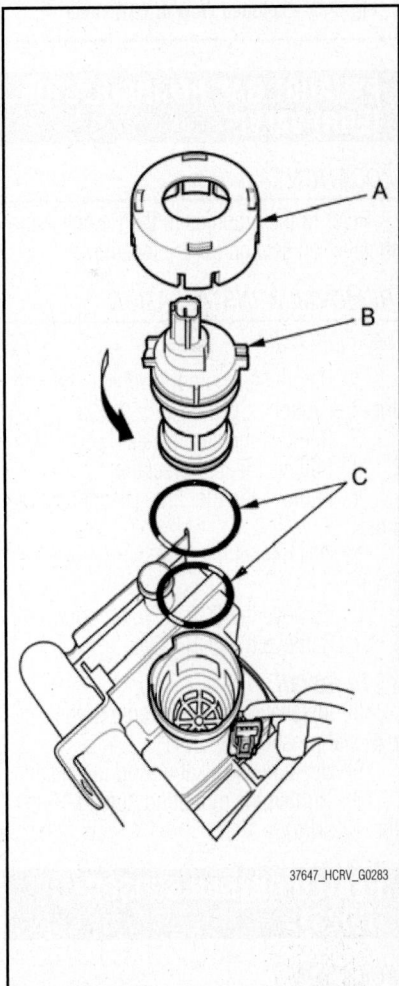

37647_HCRV_G0283

Fig. 201 Remove the cap (A), the EVAP canister vent shut valve (B), and the O-rings (C)

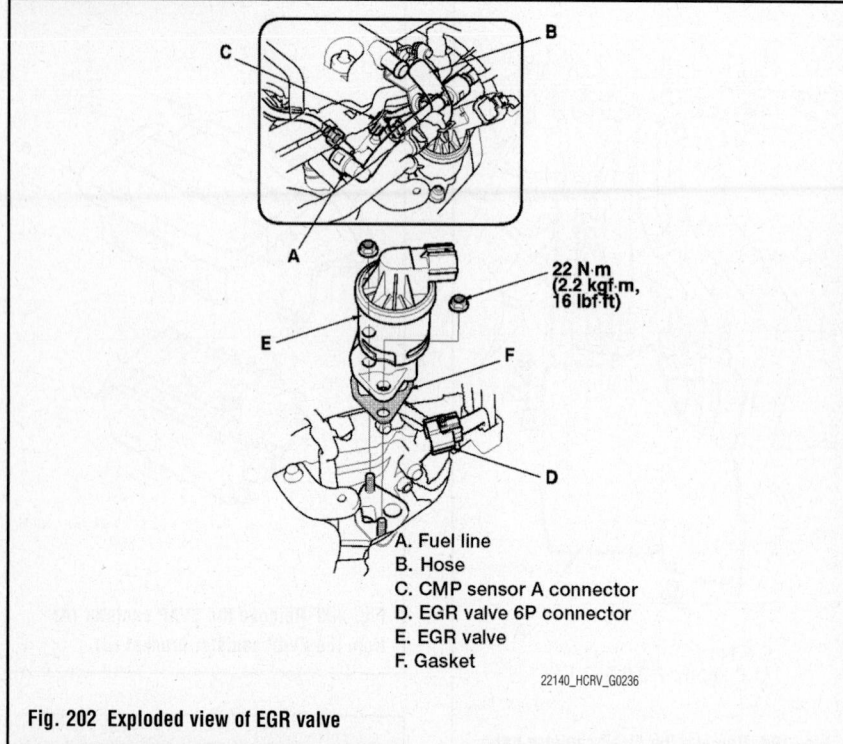

22 N·m
(2.2 kgf·m,
16 lbf·ft)

A. Fuel line
B. Hose
C. CMP sensor A connector
D. EGR valve 6P connector
E. EGR valve
F. Gasket

22140_HCRV_G0236

Fig. 202 Exploded view of EGR valve

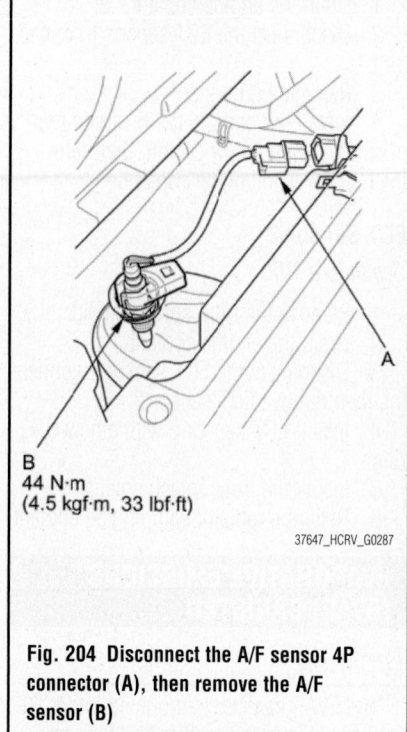

B
44 N·m
(4.5 kgf·m, 33 lbf·ft)

37647_HCRV_G0287

Fig. 204 Disconnect the A/F sensor 4P connector (A), then remove the A/F sensor (B)

EXHAUST GAS RECIRCULATION (EGR) VALVE

LOCATION

Refer to the graphics in the Removal and Installation section for the location(s).

REMOVAL & INSTALLATION

See Figure 202.

1. Before servicing the vehicle, refer to the Precautions Section.
2. Remove the air cleaner.
3. Relieve the fuel pressure.
4. Remove the fuel line. Disconnect the hose.
5. Disconnect the CMP sensor A connector.
6. Remove the EGR valve 6P connector.
7. Remove the EGR valve.

To install:

8. Installation is the reverse of the removal procedure.
9. Use a new gasket during installation.
10. Tighten the mounting nuts to 16 ft. lbs. (22 Nm).

HEATED OXYGEN SENSOR (HO2S)

LOCATION

See Figure 203.

Refer to the accompanying illustration.

REMOVAL & INSTALLATION

Special Tools Required: O2 Sensor Wrench Snap-On® S3176 or equivalent, commercially available

Air Fuel Ratio (A/F) Sensor

See Figure 204.

1. Disconnect the A/F sensor 4P connector, then remove the A/F sensor.

2. Install the parts in the reverse order of removal.

Secondary HO2S

See Figure 205.

1. Raise the vehicle on a lift.
2. Disconnect the secondary HO2S 4P connector.
3. Remove the secondary HO2S (Sensor 2).

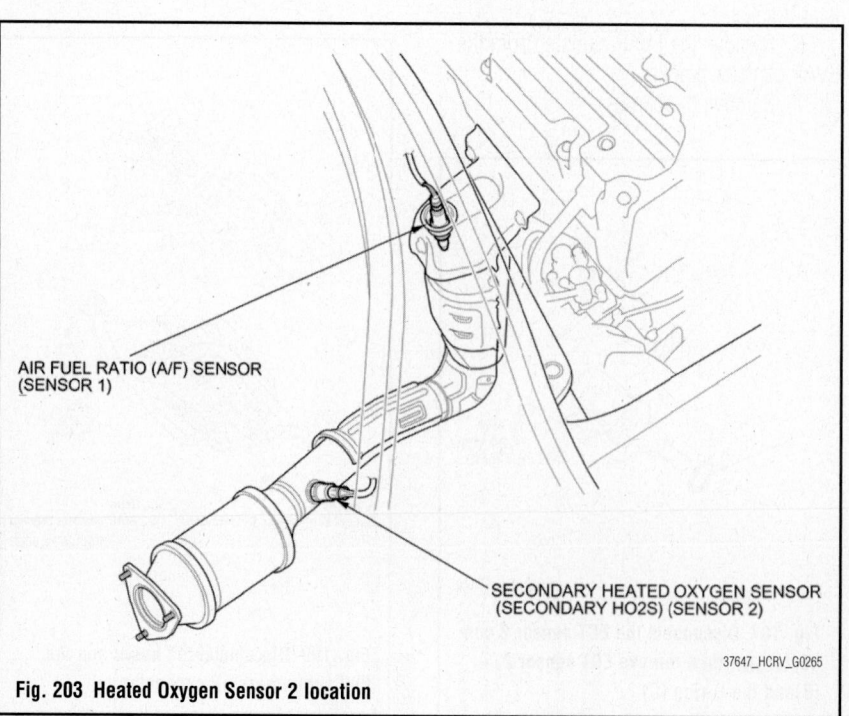

AIR FUEL RATIO (A/F) SENSOR
(SENSOR 1)

SECONDARY HEATED OXYGEN SENSOR
(SECONDARY HO2S) (SENSOR 2)

37647_HCRV_G0265

Fig. 203 Heated Oxygen Sensor 2 location

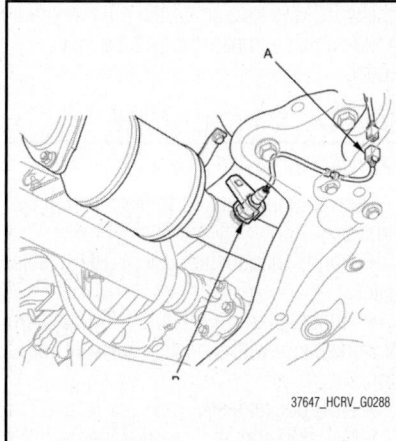

Fig. 205 Disconnect the secondary HO2S 4P connector (A) and remove the secondary HO2S (Sensor 2) (B)

4. Install the parts in the reverse order of removal.

INTAKE AIR TEMPERATURE (IAT) SENSOR

LOCATION

The Intake Air Temperature (IAT) sensor is an integral part of the Mass Air Flow (MAF) sensor/Intake Air Temperature (IAT) sensor assembly which is mounted on the air intake duct. Refer to the Mass Air Flow section for information regarding servicing this component.

REMOVAL & INSTALLATION

The Intake Air Temperature (IAT) sensor is an integral part of the Mass Air Flow (MAF) sensor/Intake Air Temperature (IAT) sensor assembly which is mounted on the air intake duct. Refer to the Mass Air Flow section for information regarding servicing this component.

KNOCK SENSOR (KS)

LOCATION

Refer to the graphics in the Removal and Installation section for the location(s).

REMOVAL & INSTALLATION

See Figure 206.

1. Remove the intake manifold.
2. Disconnect the knock sensor connector.
3. Remove the knock sensor.
4. Install the parts in the reverse order of removal.

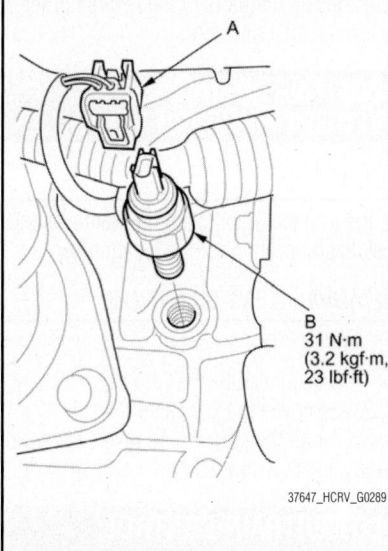

Fig. 206 Disconnect the knock sensor connector (A); remove the knock sensor (B)

MAINTENANCE MINDER

OPERATION

The Maintenance Minder is an important feature of the information display. Based on engine and transaxle operating conditions, and accumulated engine revolutions, the CR-V's onboard computer (PCM) calculates the remaining engine oil and the transaxle fluid life. The system also displays the remaining engine oil life along with the code(s) for other scheduled maintenance items needing service.

RESET PROCEDURE

1. Connect the Honda Diagnostic System (HDS) to the data link connector (DLC).
2. Turn the ignition switch to ON (II).
3. Make sure the HDS communicates with the vehicle and the powertrain control module (PCM). If it does not communicate, troubleshoot the DLC circuit.
4. Select GAUGES in the BODY ELECTRICAL with the HDS.
5. Select ADJUSTMENT in the GAUGES with the HDS.
6. Select MAINTENANCE MINDER in the ADJUSTMENT with the HDS.
7. Select RESET in the MAINTENANCE MINDER with the HDS.
8. Select the individual maintenance item you wish to reset with the HDS.

MALFUNCTION INDICATOR LIGHT (MIL)

RESET PROCEDURE

1. Proper operation of the Malfunction Indicator Light (MIL):
   - The MIL will illuminate with the ignition switch ON and the engine OFF
   - The MIL will turn OFF when the engine is started
   - The MIL will remain ON if the self-diagnostic system has detected a malfunction
   - The MIL may turn OFF if the malfunction is no longer present
   - If the MIL is illuminated and then the engine stalls, the MIL will remain illuminated as long as the ignition switch is ON
   - If the MIL is not illuminated and the engine stalls, the MIL will not illuminate until the ignition switch is cycled OFF, then ON
2. Resetting the MIL:
   - The control module turns OFF the MIL after 3 consecutive ignition cycles that the diagnostic system runs and does not fail
   - The control module turns OFF the MIL after a current Diagnostic Trouble Code (DTC) clears when the diagnostic cycle runs and passes
   - There may still be a history of DTC's stored in the system. These will clear after 40 consecutive warm-up cycles, if no failures are reported by any other related diagnostic system
   - Manual resetting of the MIL and any DTC stored in the system, requires the use of an OBD2 scan tool connected to the Data Link Connector (DLC) for communication with the vehicle. Follow the instructions of the scan tool for both retrieval and resetting of DTC's. The Honda Diagnostic System (HDS) can be used to command the MIL off.

➡**If the error symptoms causing the MIL to illuminate have been corrected, the MIL will return to normal operation.**

MANIFOLD ABSOLUTE PRESSURE (MAP) SENSOR

LOCATION

Refer to the graphics in the Removal and Installation section for the location(s).

REMOVAL & INSTALLATION

See Figure 207.

1. Disconnect the MAP sensor connector.
2. Remove the MAP sensor.
3. Install the parts in the reverse order of removal with a new O-ring.

MASS AIR FLOW (MAF) SENSOR

LOCATION

The Mass Air Flow (MAF) sensor/Intake Air Temperature (IAT) sensor assembly is mounted on the air intake duct.

REMOVAL & INSTALLATION

See Figure 208.

1. Disconnect the MAF sensor/IAT sensor connector.
2. Remove the screws.
3. Remove the MAF sensor/IAT sensor.

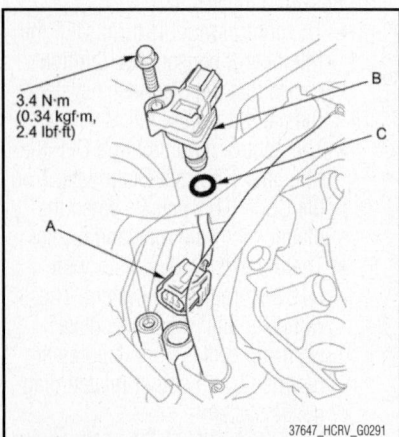

3.4 N·m
(0.34 kgf·m,
2.4 lbf·ft)

37647_HCRV_G0291

Fig. 207 Disconnect the MAP sensor connector (A); remove the MAP sensor (B) and the O-ring (C)

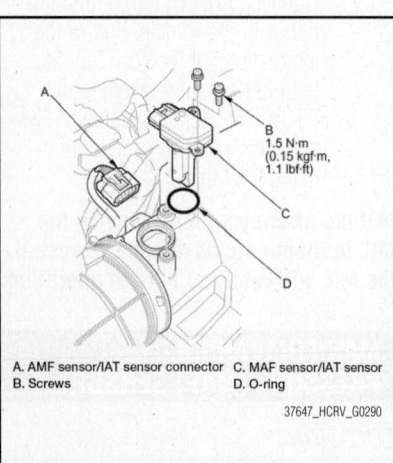

B
1.5 N·m
(0.15 kgf·m,
1.1 lbf·ft)

A. AMF sensor/IAT sensor connector
B. Screws
C. MAF sensor/IAT sensor
D. O-ring

37647_HCRV_G0290

Fig. 208 Disconnect the MAF sensor/IAT sensor connector

4. Install the parts in the reverse order of removal with a new O-ring.

POSITIVE CRANKCASE VENTILATION (PCV) VALVE

LOCATION

Refer to the graphics in the Removal and Installation section for the location(s).

REMOVAL & INSTALLATION

See Figure 209.

1. Disconnect the PCV hose.
2. Remove the PCV valve.
3. Install the parts in the reverse order of removal with a new washer.

POWERTRAIN CONTROL MODULE (PCM)

LOCATION

See Figure 210.

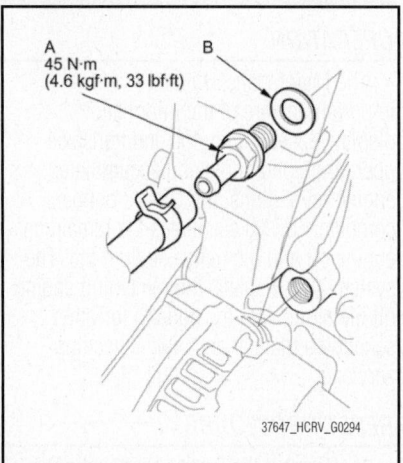

A
45 N·m
(4.6 kgf·m, 33 lbf·ft)

B

37647_HCRV_G0294

Fig. 209 Remove the PCV valve (A) and the washer (B)

The PCM is located on the driver's side in the engine compartment on the inner fender.

REMOVAL & INSTALLATION

See Figures 210 and 211.

Special tools required (one of the following):
- Honda Diagnostic System (HDS) tablet tester
- Honda Interface Module (HIM) and an iN workstation with the latest software version
- HDS pocket tester
- GNA600 and an iN workstation with the latest software version

➡**Make sure the HDS is loaded with the latest software version. If you are replacing the PCM after substituting a known-good PCM, reinstall the original PCM, then do this procedure. During the procedure, if any READ DATA, WRITE DATA, or other data checks fail, note the failure, then continue.**

1. Before servicing the vehicle, refer to the Precautions Section.
2. Connect the HDS to the Data Link Connector (DLC) located under the driver's side of the dashboard.
3. Turn the ignition switch to ON (II).
4. Make sure the HDS communicates with the PCM and other vehicles systems.
5. Select the PGM-FI system with the HDS.
6. Select the INSPECTION MENU with the HDS.
7. Select the ETCS TEST, then select the TP POSITION CHECK, and follow the screen prompts.

➡**If the TP POSITION CHECK indicates FAILED, continue with this procedure.**

8. Select the REPLACE PCM MENU, then select READ DATA and follow the screen prompts.

➡**Doing this step copies (READS) the engine oil life data from the original PCM so you can later download (WRITES) it into the new PCM. If READ DATA indicates FAILED, continue with this procedure.**

9. Select the A/T system with the HDS.
10. Select the REPLACE TCM/PCM MENU, then select READ DATA and follow the screen prompts.

➡**Doing this step copies (READS) the ATF life data from the original PCM so you can later download (WRITES) it into the new PCM. If READ DATA indi-**

Fig. 210 caption:

A. Under-hood fuse/relay box
B. Harness bracket
C. PCM cover
D. Battery hold down bolt

22140_HCRV_G0120

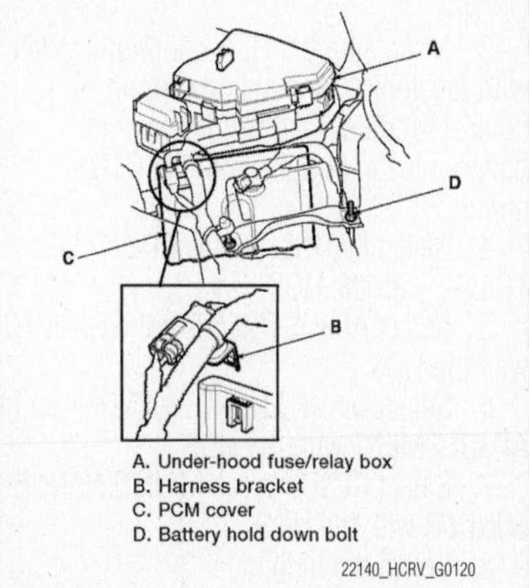

Fig. 210 Powertrain Control Module (PCM) location

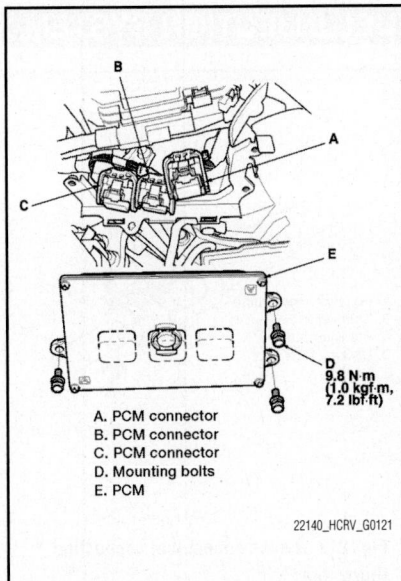

A. PCM connector
B. PCM connector
C. PCM connector
D. Mounting bolts
E. PCM

D
9.8 N·m
(1.0 kgf·m,
7.2 lbf·ft)

22140_HCRV_G0121

Fig. 211 Disconnect the PCM connectors and remove the PCM

cates FAILED, continue with this procedure.

11. Turn the ignition switch to LOCK (0).

12. Remove the under-hood fuse/relay box.

13. Remove the harness bracket.

14. Loosen the battery hold down bolt, and reposition the battery away from the PCM.

15. Remove the PCM cover.

16. Remove the bolts, then remove the PCM.

17. Disconnect the PCM connectors.

➡ **The PCM connectors A, B, and C have symbols embossed on them for identification. See illustration.**

To install:

18. Installation is the reverse of the removal procedure.

19. Turn the ignition switch to ON (II).

20. Manually input the VIN to the PCM with the HDS.

➡ **DTC P0630 VIN Not Programmed or Mismatch may be stored because the VIN has not been programmed into the PCM; ignore it, and continue this procedure.**

21. If the READ DATA (engine oil life) failed in step 9 of removal, go to step 7 below. Otherwise, go to the next step.

22. Select the PGM-FI system with the HDS.

23. Select the REPLACE PCM MENU, then select WRITE DATA and follow the screen prompts.

➡ **If the WRITE DATA indicates FAILED, continue with this procedure.**

24. If the READ DATA (ATF life) failed in step 11 of removal, go to step 9 below. Otherwise go to the next step.

25. Select the A/T SYSTEM with the HDS.

26. Select the REPLACE TCM/PCM MENU, then select WRITE DATA and follow the screen prompts.

➡ **If the WRITE DATA indicates FAILED, continue with this procedure.**

27. Select IMMOBI system with the HDS.

28. Enter the immobilizer code with the PCM replacement procedure in the HDS; this allows you to start the engine.

29. If the TP POSITION CHECK failed in step 7 of removal, clean the throttle body, then go to the next step.

30. If the READ DATA failed in step 8 of removal or the WRITE DATA failed in step 6 in installation, replace the engine oil and engine oil filter, then go to the next step.

31. If the READ DATA failed in step 11 of removal or the WRITE DATA failed in step 9 of installation, replace the ATF, then go to the next step.

32. Select PGM-FI system and reset the PCM with the HDS.

33. Update the PCM if it does not have the latest software.

34. Perform the PCM idle learn procedure.

35. Perform the CKP pattern learn procedure.

PCM Idle Learn Procedure with HDS

1. Connect the Honda Diagnostic System (HDS) to the Data Link Connector (DLC) located under the driver's side of the dashboard.

2. Turn the ignition switch to ON.

3. Make sure the HDS communicates with the PCM and other vehicle systems. If it does not, go to the DLC circuit troubleshooting.

4. Select CRANK PATTERN in the ADJUSTMENT MENU with the HDS.

5. Select CRANK PATTERN LEARNING with the HDS and follow the screen prompts.

PCM Idle Learn Procedure without HDS

1. Start the engine.

2. Hold the engine speed at 3,000 RPM without load (in P or N) until the radiator fan comes on.

3. Test-drive the vehicle on a level road: decelerate (with the throttle fully closed) from an engine speed of 2,500 RPM down to 1,000 RPM with the transaxle in 2 position.

4. Repeat step 3 several times.

5. Turn the ignition switch to LOCK.

6. Turn the ignition switch to ON and wait 30 seconds.

THROTTLE POSITION SENSOR (TPS)

REMOVAL & INSTALLATION

The TPS is a component of the electronic throttle control system. This sensor is located with the throttle actuator and is integral to the throttle body. Refer to Throttle Body, removal & installation.

FUEL SYSTEM SERVICE PRECAUTIONS

Safety is the most important factor when performing not only fuel system maintenance, but any type of maintenance. Failure to conduct maintenance and repairs in a safe manner may result in serious personal injury or death. Work on a vehicle's fuel system components can be accomplished safely and effectively by adhering to the following rules and guidelines.

• To avoid the possibility of fire and personal injury, always disconnect the negative battery cable unless the repair or test procedure requires that battery voltage be applied.

• Always relieve the fuel system pressure prior to disconnecting any fuel system component (injector, fuel rail, pressure regulator, etc.) fitting or fuel line connection. Exercise extreme caution whenever relieving fuel system pressure to avoid exposing skin, face and eyes to fuel spray. Please be advised that fuel under pressure may penetrate the skin or any part of the body that it contacts.

• Always place a shop towel or cloth around the fitting or connection prior to loosening to absorb any excess fuel due to spillage. Ensure that all fuel spillage is quickly removed from engine surfaces. Ensure that all fuel-soaked cloths or towels are deposited into a flame-proof waste container with a lid.

• Always keep a dry chemical (Class B) fire extinguisher near the work area.

• Do not allow fuel spray or fuel vapors to come into contact with a spark or open flame.

• Always use a second wrench when loosening or tightening fuel line connection fittings. This will prevent unnecessary stress and torsion on fuel piping. Always follow the proper torque specifications.

• Always replace worn fuel fitting O-rings with new ones. Do not substitute fuel hose where rigid pipe is installed.

FUEL SYSTEM PRESSURE

RELIEVING

With Honda Diagnostics System (HDS)

See Figure 212.

1. Before servicing the vehicle, refer to the Precautions Section.

2. Remove the fuel fill cap, to relieve the pressure in the fuel tank.

3. Turn the ignition switch to ON.

4. From the INSPECTION MENU of the Honda Diagnostics System (HDS), select Fuel Pump OFF, then start the engine, and let it idle until it stalls.

➡**Do not allow the engine to idle above 1,000 RPM or the PCM will continue to operate the fuel pump. A DTC or a Temporary DTC may be set during this procedure. Check for DTC's, and clear them as needed.**

5. Turn the ignition switch to LOCK.

6. Perform the battery terminal disconnection procedure. Some systems store data in memory that is lost when the battery is disconnected. Do the following steps before disconnecting the battery:

   a. Make sure to record the anti-theft code(s) for the audio and/or the navigation system (if equipped).

   b. If replacing the audio unit, write down the audio presets (AM and FM), and the XM audio presets (if equipped), because the audio unit does not retain the presets after the battery is disconnected.

   c. Make sure the ignition switch is in LOCK position.

   d. Disconnect and isolate the negative cable from the battery.

➡**Always disconnect the negative cable from the battery first.**

   e. Disconnect the positive cable from the battery.

7. Remove the quick-connect fitting cover.

8. Check the fuel quick-connect fitting for dirt, and clean it if needed.

9. Place a rag or shop towel over the quick-connect fitting.

10. Disconnect the quick-connect fitting: hold the connector with one hand and squeeze the retainer tabs with the other hand to release them from the locking tabs. Pull the connector off.

   • Be careful not to damage the line or other parts.
   • Do not use tools.
   • If the connector does not move, keep the retainer tabs pressed down, and alternately pull and push the connector until it comes off easily.
   • Do not remove the retainer from the line; once removed, the retainer

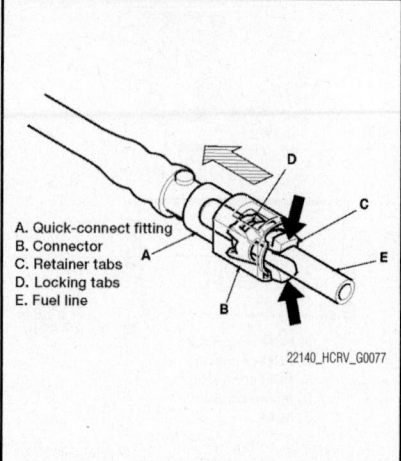

A. Quick-connect fitting
B. Connector
C. Retainer tabs
D. Locking tabs
E. Fuel line

22140_HCRV_G0077

Fig. 212 Quick-connect fuel connection illustrated

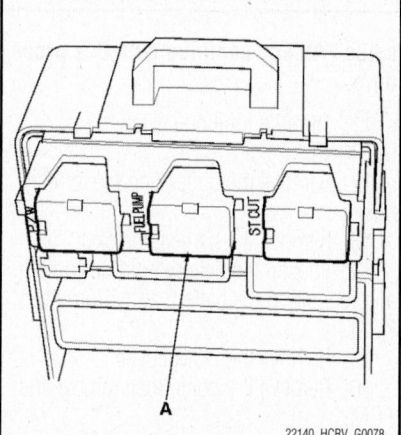

22140_HCRV_G0078

Fig. 213 Remove the PGM-FI main relay 2 (FUEL PUMP) (A) from the under-dash fuse/relay box

must be replaced with a new one.

11. After disconnecting the quick-connect fitting, check it for dirt or damage.

12. Reconnect the battery as needed.

Without Honda Diagnostics System (HDS)

See Figures 213 and 214.

1. Before servicing the vehicle, refer to the Precautions Section.

2. Remove the under-dash fuse/relay box, then remove the PGM-FI main relay 2 (FUEL PUMP) (A) from the under-dash fuse/relay box.

3. Reinstall the under-dash fuse/relay box.

4. Start the engine, and let it idle until it stalls.

➡ **If any DTC's are stored, clear and ignore them.**

5. Turn the ignition switch to LOCK.

6. Remove the fuel fill cap.

7. Perform the battery terminal disconnection procedure. Some systems store data in memory that is lost when the battery is disconnected. Do the following steps before disconnecting the battery:

   a. Make sure to record the anti-theft code(s) for the audio and/or the navigation system (if equipped).

   b. If replacing the audio unit, write down the audio presets (AM and FM), and the XM audio presets (if equipped), because the audio unit does not retain the presets after the battery is disconnected.

   c. Make sure the ignition switch is in LOCK position.

   d. Disconnect and isolate the negative cable from the battery.

➡ **Always disconnect the negative cable from the battery first.**

   e. Disconnect the positive cable from the battery.

   f. Disconnect the positive cable from the battery.

8. Remove the quick-connect fitting cover.

9. Check the fuel quick-connect fitting for dirt, and clean it if needed.

10. Place a rag or shop towel over the quick-connect fitting.

11. Disconnect the quick-connect fitting: hold the connector with one hand and squeeze the retainer tabs with the other hand to release them from the locking tabs. Pull the connector off.

   • Be careful not to damage the line or other parts
   • Do not use tools
   • If the connector does not move,

keep the retainer tabs pressed down, and alternately pull and push the connector until it comes off easily

   • Do not remove the retainer from the line; once removed, the retainer must be replaced with a new one

12. After disconnecting the quick-connect fitting, check it for dirt or damage.

13. Reconnect the battery as needed.

FUEL FILTER

REMOVAL & INSTALLATION

See Figure 215.

The fuel filter should be replaced whenever the fuel pressure drops below the specified value (47–54 psi), after making sure that the fuel pump and the fuel pressure regulator are functioning properly.

1. Before servicing the vehicle, refer to the Precautions Section.

2. Relieve the fuel pressure. Refer to Relieving Fuel System Pressure.

3. Remove the fuel pump unit. Refer to Fuel Pump, removal & installation.

4. Remove the fuel filter set.

To install:

5. Check these items before installing the fuel pump tank unit:

   a. When connecting the wire harness, make sure the connection is secure and the connectors are firmly locked into place.

   b. When installing the fuel gauge sending unit, make sure the connection is secure and the connector is firmly locked into place. Be careful not to bend or twist it excessively.

6. Install the parts in the reverse order of removal with new O-rings.

➡ **Coat the O-rings with clean engine oil.**

7. Install the fuel pump tank unit.

FUEL LEVEL SENDING UNIT

LOCATION

The fuel level sensor is part of the fuel tank unit.

REMOVAL & INSTALLATION

See Figure 216.

1. Before servicing the vehicle, refer to the Precautions Section.

2. Remove the fuel tank unit. Refer to Fuel Pump, removal & installation.

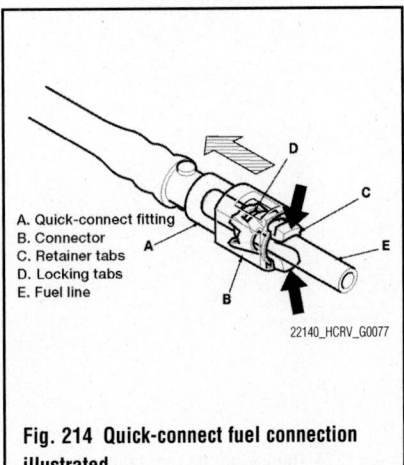

A. Quick-connect fitting
B. Connector
C. Retainer tabs
D. Locking tabs
E. Fuel line

22140_HCRV_G0077

Fig. 214 Quick-connect fuel connection illustrated

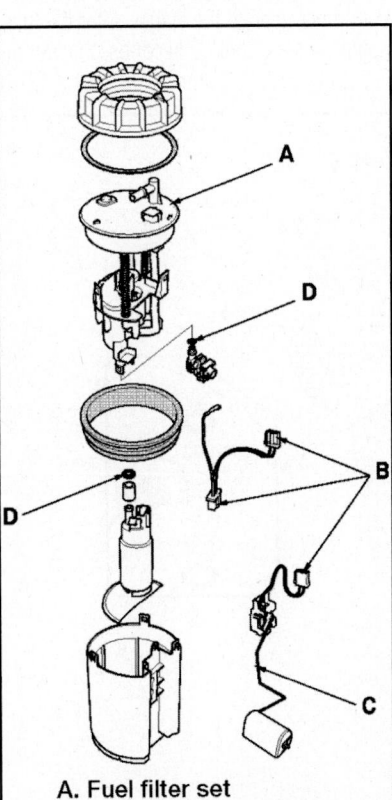

A. Fuel filter set
B. Electrical connectors
C. Fuel gauge sending unit
D. O-rings

22140_HCRV_G0079

Fig. 215 Exploded view of the fuel pump tank unit

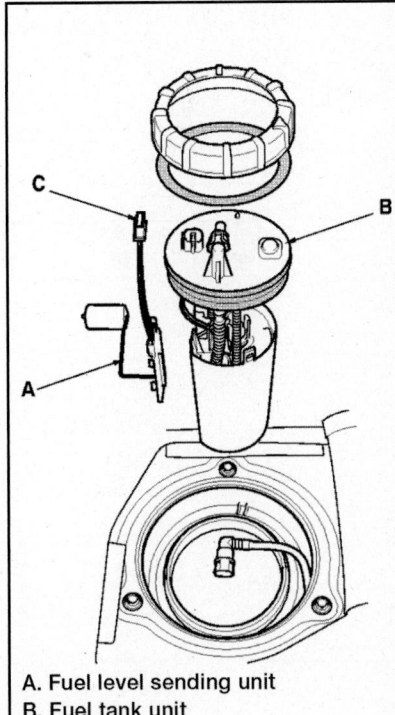

A. Fuel level sending unit
B. Fuel tank unit
C. 4P connector

22140_HCRV_G0245

Fig. 216 Exploded view of fuel pump and fuel level sending unit

3. Remove the fuel level sensor (fuel sending unit) from the fuel tank unit.

To install:

4. When connecting the wire harness, make sure the connection is secure and the connector is firmly locked into place.

5. When installing the fuel gauge sending unit, make sure the connection is secure and the connector is firmly locked into place. Be careful not to bend or twist it excessively.

6. Install the parts in the reverse order of removal.

7. Install the fuel tank unit. Refer to Fuel Pump, removal & installation.

FUEL RAIL AND INJECTOR

REMOVAL & INSTALLATION

See Figure 217.

1. Relieve the fuel pressure.
2. Remove the engine cover.
3. Disconnect the quick-connect fitting.
4. Disconnect the injector connectors and the engine mount control solenoid valve connector.

5. Remove the ground cable bolts.
6. Remove the fuel rail mounting nuts from the fuel rail.
7. Remove the fuel rail and the injectors from the injector base.
8. Remove the injector clips from the fuel rail.
9. Remove the injectors from the fuel rail.

To install:

10. Coat the new O-rings (black) with clean engine oil, and insert the injectors into the fuel rail.
11. Install the injector clips.
12. Coat the new injector O-rings (brown) with clean engine oil.
13. Install the fuel rail and the injectors in the injector base.
14. Install the fuel rail mounting nuts and the ground cable bolts.
15. Connect the injector connectors and the engine mount control solenoid valve connector.
16. Connect the quick-connect fitting.
17. Turn the ignition switch to ON (II), but do not operate the starter. After the fuel pump runs for about 2 seconds, the fuel rail

is pressurized. Repeat this two or three times, then make sure there are no fuel leaks.
18. Reinstall the engine cover.

FUEL TANK

DRAINING

1. Remove the fuel tank unit.
2. Using a hand pump, a hose, and a container suitable for fuel, draw the fuel from the fuel tank.
3. Reinstall the fuel tank unit.

REMOVAL & INSTALLATION

See Figure 218.

1. Remove the fuel tank unit, then drain the fuel tank.
2. Raise the vehicle on a lift.
3. Remove the EVAP canister.
4. Remove the cover and the fuel tank guard.
5. 4WD model: Remove the propeller shaft.
6. Remove the exhaust pipe.
7. Disconnect the hoses. Slide back the clamps, then twist the hoses as you pull to avoid damaging them.
8. Place a jack or other support under the fuel tank.
9. Remove the strap bolts and the straps.
10. Remove the fuel tank.
11. Install the parts in the reverse order of removal.

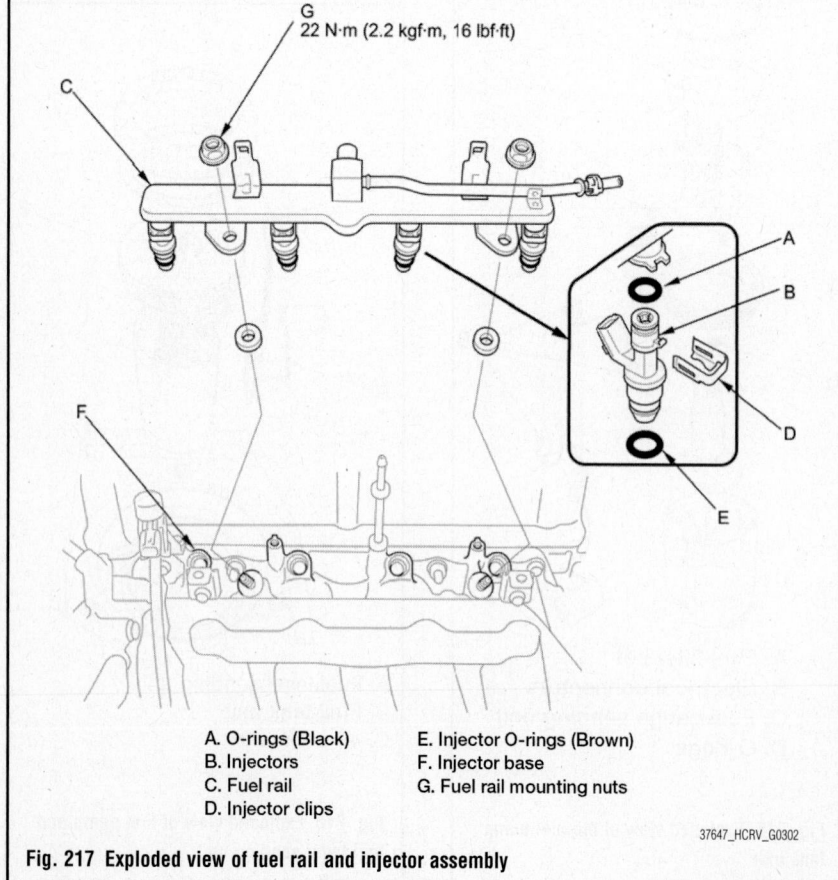

A. O-rings (Black)
B. Injectors
C. Fuel rail
D. Injector clips
E. Injector O-rings (Brown)
F. Injector base
G. Fuel rail mounting nuts

37647_HCRV_G0302

Fig. 217 Exploded view of fuel rail and injector assembly

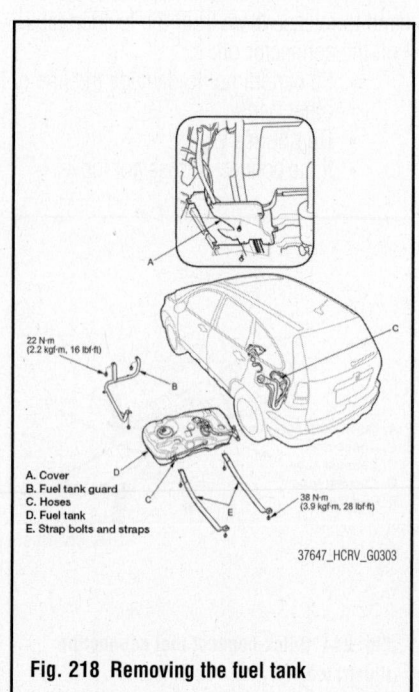

A. Cover
B. Fuel tank guard
C. Hoses
D. Fuel tank
E. Strap bolts and straps

37647_HCRV_G0303

Fig. 218 Removing the fuel tank

FUEL TANK UNIT

REMOVAL & INSTALLATION

See Figures 219 and 220.

Special tools required: Fuel sender wrench 07AAA-S0XA100

1. Before servicing the vehicle, refer to the Precautions Section.
2. Relieve the fuel pressure. Refer to Relieving Fuel System Pressure.
3. Disconnect the negative battery cable.
4. Remove the fuel fill cap.
5. Fold the left side rear seat forward and pull back the carpet to expose the access panel.

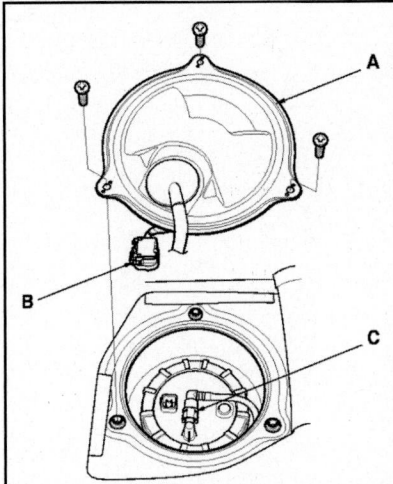

22140_HCRV_G0081

Fig. 219 Remove the access panel (A) from the floor, disconnect the 4P connector (B) and the quick-connect fitting (C) from the fuel pump tank unit

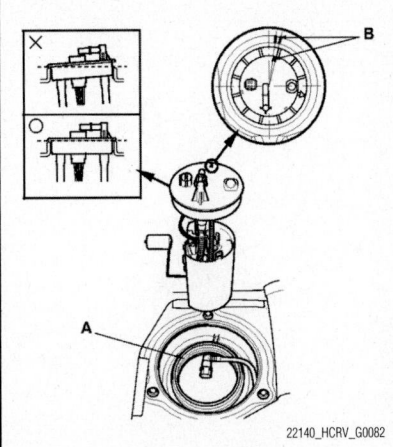

22140_HCRV_G0082

Fig. 220 Install the base gasket (A). Align the marks (B) on the fuel tank and the fuel tank unit, then insert the fuel tank unit until it sits on the base gasket

6. Remove the access panel from the floor.
7. Disconnect the fuel tank unit 4P connector.
8. Disconnect the quick-connect fitting from the fuel tank unit.
9. Using the special tool, loosen the locknut.
10. Remove the locknut, the base gasket, and the fuel pump tank unit.

To install:

11. Temporarily attach a new base gasket to the fuel pump tank unit, then insert the fuel tank unit partially into the fuel tank.
- Be careful not to damage the new base gasket
- Be careful not to bend the fuel gauge sending unit
- Do not coat the base gasket with oil
12. Transfer the base gasket from the fuel tank unit to the fuel tank.
13. Align the marks on the fuel tank and the fuel tank unit, then insert the fuel tank unit until it sits on the base gasket.

✳✳ CAUTION

To prevent a fuel leak, check the base gasket, visually or by hand, to make sure it is not pinched.

14. Using the special tool, tighten the new locknut to 52 ft. lbs. (70 Nm).
- After tightening, make sure the marks are still aligned
- After installation, check the base gasket visually or by hand to be sure it is not pinched
15. Connect the fuel tank unit 4P connector.
16. Reconnect the negative cable to the battery and turn the ignition switch to **ON** (but do not operate the starter motor). The fuel pump will run for about 2 seconds, and fuel pressure will rise. Repeat this 2–3 times, and check that there is no leakage in the fuel supply system.
17. Install the parts in the reverse order of removal.

IDLE SPEED

ADJUSTMENT

Idle speed is maintained by the Powertrain Control Module (PCM). No adjustment is necessary or possible.

THROTTLE BODY

REMOVAL & INSTALLATION

See Figure 221.

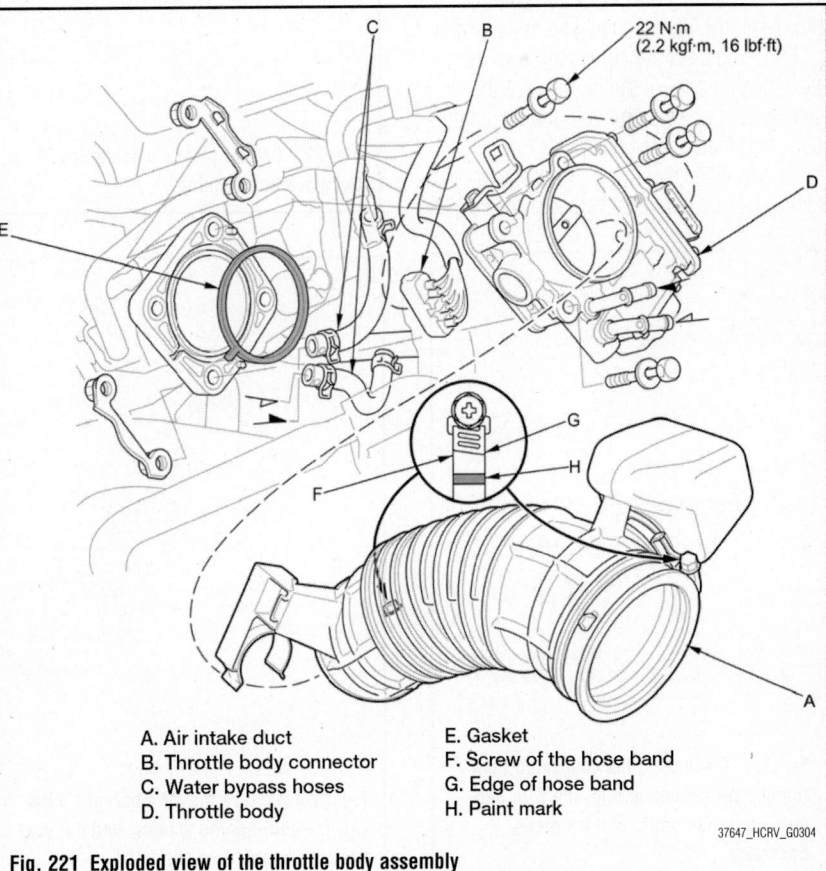

A. Air intake duct
B. Throttle body connector
C. Water bypass hoses
D. Throttle body
E. Gasket
F. Screw of the hose band
G. Edge of hose band
H. Paint mark

37647_HCRV_G0304

Fig. 221 Exploded view of the throttle body assembly

✱✱ CAUTION

Do not insert your fingers into the installed throttle body when you turn the ignition switch to ON (II) or while the ignition switch is in ON (II). If you do, you will seriously injure your fingers if the throttle valve is activated.

➡ **If you are replacing the throttle body, start at step 1. If you are removing the throttle body, start at step 4.**

1. Connect the HDS to the DLC while the engine is stopped.
2. Select the INSPECTION MENU on the HDS.
3. Do the TP POSITION CHECK in the ETCS TEST.
4. Turn the ignition switch to LOCK (0).
5. Remove the intake air duct.
6. Disconnect the throttle body connector.
7. Disconnect and plug the water bypass hoses.

8. Remove the throttle body.
9. Install the parts in the reverse order of removal with a new gasket.

➡ **When torquing the screw of the hose band, align the edge of the hose band with the mark painted on the hose band.**

10. After installing the throttle body, do these items:
   • Do the PCM idle learn procedure.
   • Refill the radiator with engine coolant.

HEATING & AIR CONDITIONING SYSTEM

BLOWER MOTOR

REMOVAL & INSTALLATION

See Figures 222 through 226.

1. Remove the glove box.
2. Disconnect the connector, then remove the connector clip, the wire harness clips, the bolts, and the glove box frame.
3. Cut the plastic cross brace in the glove box opening with diagonal cutters in the area shown. Retain the plastic cross brace it will be reinstalled.
4. Remove the wire harness clips, the self-tapping screws, and the passenger's heater duct.
5. Disconnect the connector from the blower motor. Remove the wire harness clip.
6. Disconnect the connector from the recirculation control motor. Remove the self-tapping screws, the bolt, the mounting nuts, and the blower unit.
7. Install the unit in the reverse order of removal. Make sure that there are no air leaks.

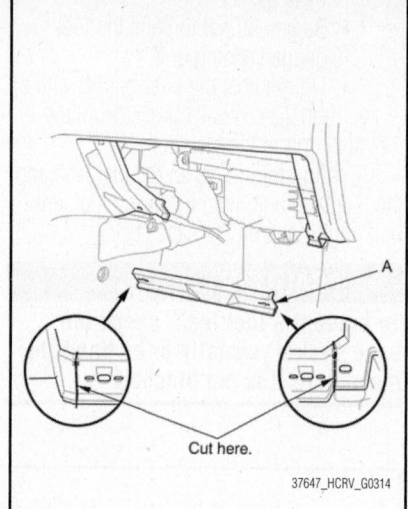

Fig. 223 Cut the plastic cross brace (A) in the glove box opening

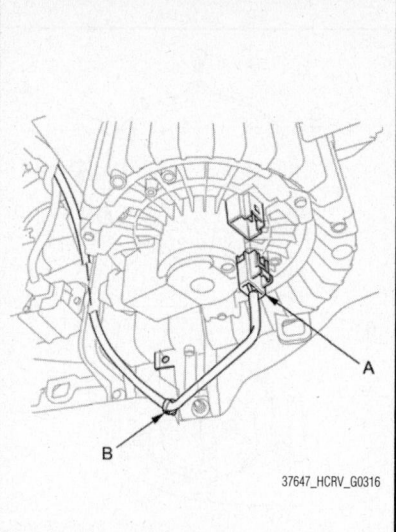

Fig. 225 Disconnect the connector (A) from the blower motor; remove the wire harness clip (B)

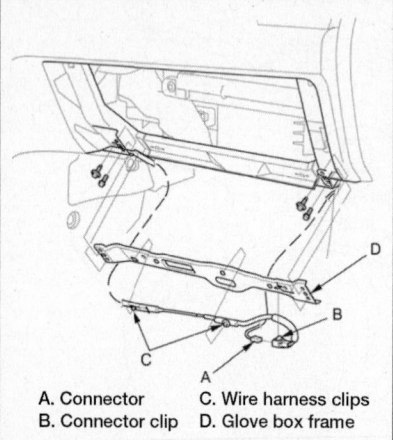

A. Connector   C. Wire harness clips
B. Connector clip D. Glove box frame

Fig. 222 Disconnect the connector, then remove the connector clip, the wire harness clips, the bolts, and the glove box frame

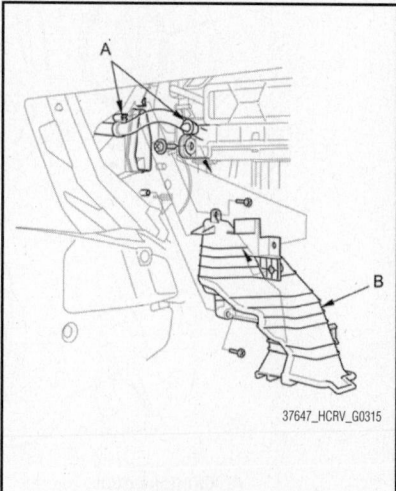

Fig. 224 Remove the wire harness clips (A), the self-tapping screws, and the passenger's heater duct (B)

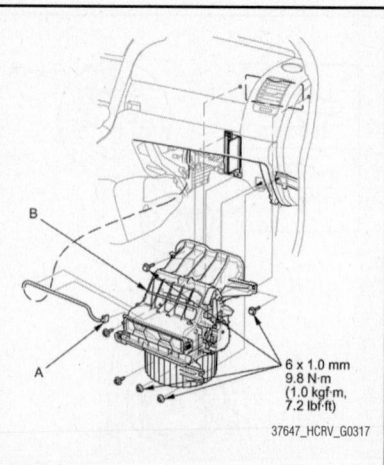

Fig. 226 Disconnect the connector (A) from the recirculation control motor; remove the self-tapping screws, the bolt, the mounting nuts, and the blower unit (B)

HEATER CORE

REMOVAL & INSTALLATION

See Figures 227 through 234.

✳ WARNING

SRS components are located in this area. Review the SRS component locations and the precautions and procedures before doing repairs or service.

1. Do the battery terminal disconnection procedure.
2. Recover the refrigerant with a recovery/recycling/charging station
3. Disconnect the A/C line from the evaporator core.

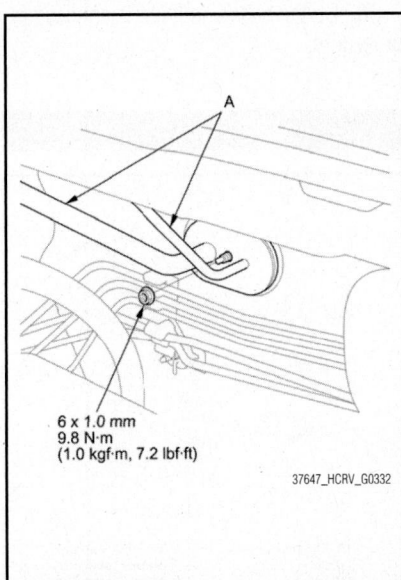

6 x 1.0 mm
9.8 N·m
(1.0 kgf·m, 7.2 lbf·ft)

37647_HCRV_G0332

Fig. 229 Remove the nut, then disconnect the A/C lines (A) from the evaporator core

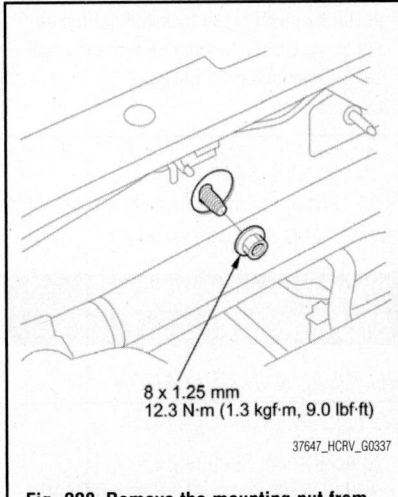

8 x 1.25 mm
12.3 N·m (1.3 kgf·m, 9.0 lbf·ft)

37647_HCRV_G0337

Fig. 228 Remove the mounting nut from the heater unit

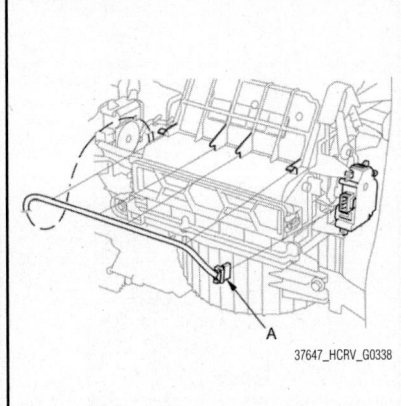

37647_HCRV_G0338

Fig. 229 Disconnect the connector (A) from the recirculation control motor

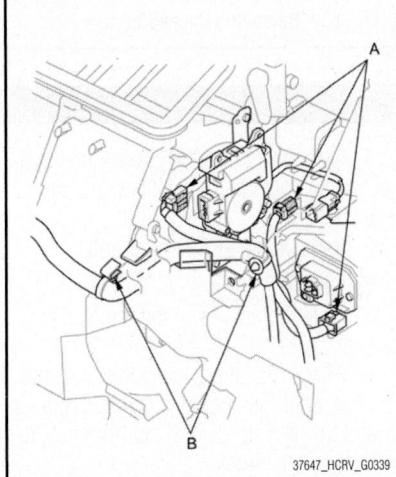

37647_HCRV_G0339

Fig. 230 Disconnect these connectors (A) from the mode control motor, the evaporator temperature sensor, and the power transistor; remove the wire harness clips (B)

4. When the engine is cool, drain the engine coolant from the radiator.
5. From under the hood, slide the hose clamps back. Disconnect the inlet heater hose and the outlet heater hose from the heater unit. Note the layout of the hoses.

➡**Engine coolant will run out when the hoses are disconnected; drain it into a clean drip pan. Be sure not to let coolant spill on the electrical parts or the painted surfaces. If any coolant spills, rinse it off immediately.**

6. Remove the mounting nut from the heater unit. Take care not to damage or bend the fuel lines or the brake lines.
7. Remove the dashboard.
8. Disconnect the connector from the blower motor. Remove the wire harness clip.
9. Disconnect the connector from the recirculation control motor.
10. Disconnect these connectors from the mode control motor, the evaporator temperature sensor, and the power transistor. Remove the wire harness clips.
11. Disconnect the connectors from the air mix control motor and A/C wire harness.
12. Remove the connector clip, the wire harness clips, and the wire harness.
13. Remove the clip and the duct.
14. Slide the clamp, then remove the drain hose.
15. Remove the mounting bolt, mounting nuts, and blower-heater unit.
16. Remove the self-tapping screws and the driver's duct.
17. Remove the self-tapping screws,

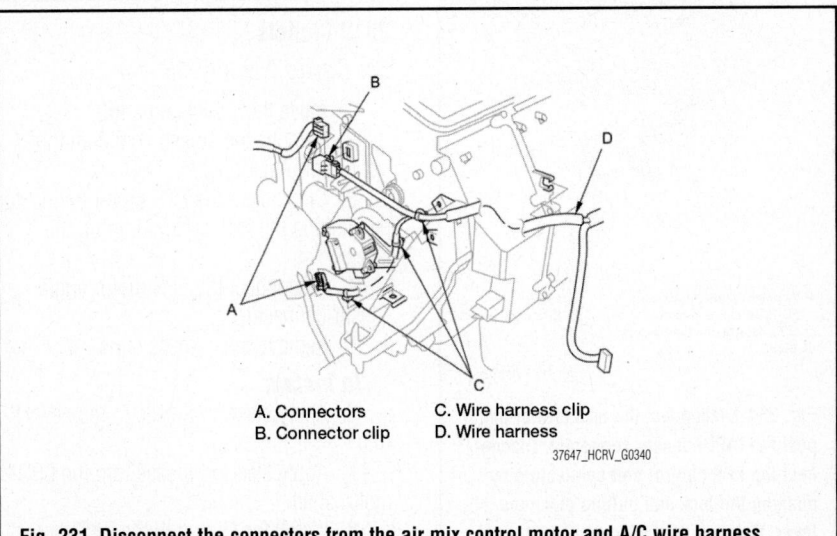

A. Connectors   C. Wire harness clip
B. Connector clip   D. Wire harness

37647_HCRV_G0340

Fig. 231 Disconnect the connectors from the air mix control motor and A/C wire harness

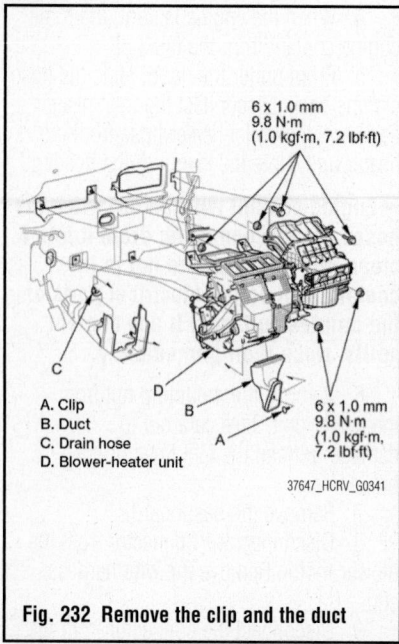

A. Clip
B. Duct
C. Drain hose
D. Blower-heater unit

6 x 1.0 mm
9.8 N·m
(1.0 kgf·m, 7.2 lbf·ft)

6 x 1.0 mm
9.8 N·m
(1.0 kgf·m, 7.2 lbf·ft)

37647_HCRV_G0341

Fig. 232 Remove the clip and the duct

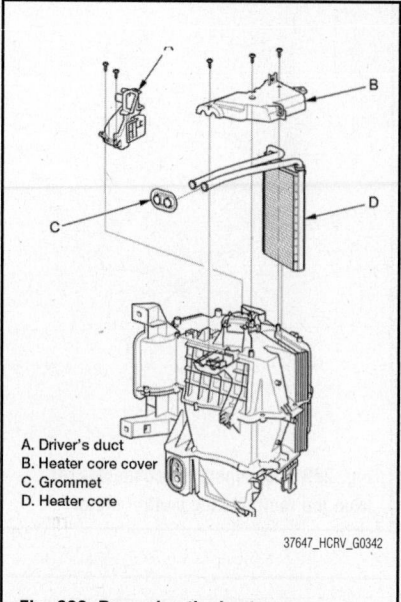

A. Driver's duct
B. Heater core cover
C. Grommet
D. Heater core

37647_HCRV_G0342

Fig. 233 Removing the heater core

the heater core cover, the grommet, and carefully pull out the heater core.

To install:

18. Install the heater core and the evaporator core in the reverse order of removal.

19. Install the heater unit in the reverse order of removal, and note these items:

- Do not interchange the inlet and outlet heater hoses, and install the hose clamps securely.
- Refill the cooling system with engine coolant.
- Make sure that there is no coolant leakage.
- Make sure that there is no air leakage.
- Charge the system.

20. Do the battery terminal reconnection procedure.

STEERING

EPS CONTROL UNIT

REMOVAL & INSTALLATION

2012 Models

See Figure 234.

1. Disconnect the battery cables.

2. Remove the driver's side dashboard panel, the under cover, and the lower cover.

3. Disconnect the accelerator pedal position (APP) sensor connector.

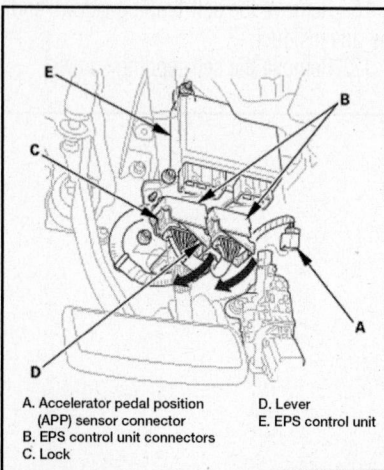

A. Accelerator pedal position (APP) sensor connector
B. EPS control unit connectors
C. Lock
D. Lever
E. EPS control unit

71051_HCRV_G0711

Fig. 234 Disconnect the accelerator pedal position (APP) sensor connector. Disconnect the EPS control unit connectors by pushing the lock and pulling down the lever. Remove the EPS control unit

4. Disconnect the EPS control unit connectors by pushing the lock and pulling down the lever.

5. Remove the EPS control unit.

To install:

6. Install in reverse of the removal procedure.

7. After installation, check these items:

a. Start the engine, allow it to idle, and turn the steering wheel from lock to lock several times.

b. Check that the EPS indicator does not come on.

EPS MOTOR

REMOVAL & INSTALLATION

2012 Models

See Figures 235 and 236.

1. Raise the vehicle on a lift.

2. Remove the splash shield and the front floor under cover.

3. Disconnect the EPS motor connector by pushing the lock and pulling up the lever.

4. Disconnect the EPS motor angle sensor connector.

5. Remove the harness clips.

To install:

6. Apply steering grease to the new O-ring.

7. Apply steering grease into the EPS motor shaft.

8. Install the O-ring to the EPS motor.

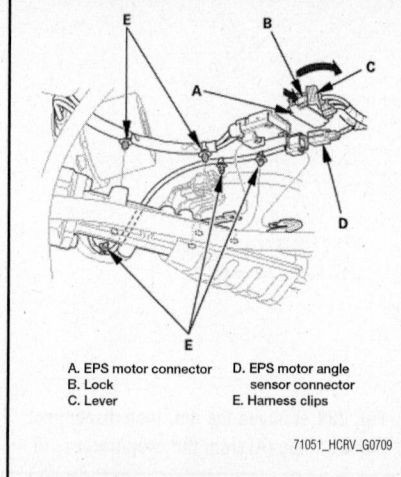

A. EPS motor connector
B. Lock
C. Lever
D. EPS motor angle sensor connector
E. Harness clips

71051_HCRV_G0709

Fig. 235 Disconnect the EPS motor connector by pushing the lock and pulling up the lever. Disconnect the EPS motor angle sensor connector. Remove the harness clips

9. Install the EPS motor.

10. Install the remaining components in reverse of the removal procedure.

POWER RACK & PINION STEERING GEAR

REMOVAL & INSTALLATION

See Figures 237 through 247.

Special Tools Required:
- Ball Joint Remover, 28 mm 07MAC-SL0A202

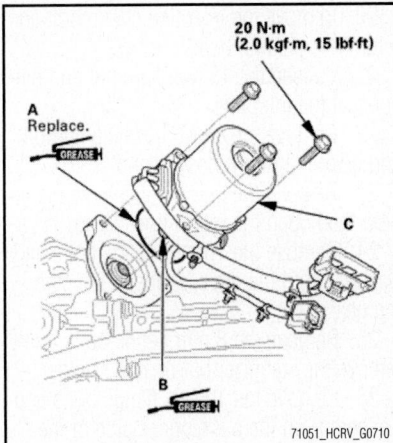

Fig. 236 Apply steering grease to the new O-ring (A). Apply steering grease into the EPS motor shaft (B). Install the O-ring to the EPS motor (C)

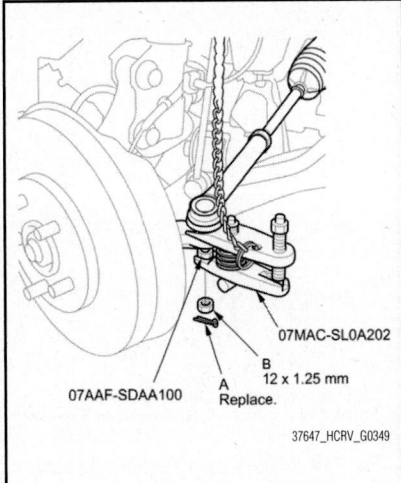

Fig. 239 Remove the cotter pin (A) from the 12 mm nut (B), and loosen the nut

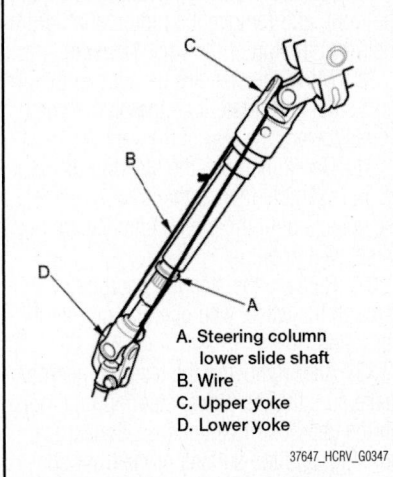

A. Steering column lower slide shaft
B. Wire
C. Upper yoke
D. Lower yoke

Fig. 237 Hold the steering column lower slide shaft together with a piece wire between the upper yoke and the lower yoke

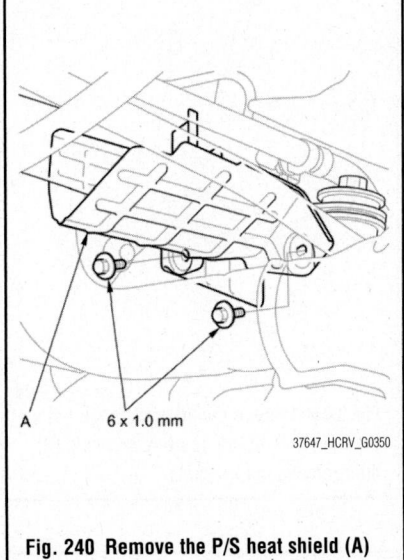

Fig. 240 Remove the P/S heat shield (A)

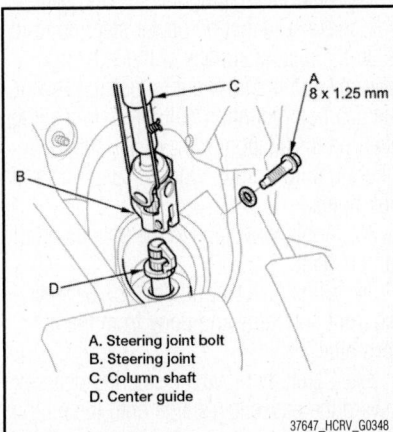

A. Steering joint bolt
B. Steering joint
C. Column shaft
D. Center guide

Fig. 238 Remove the steering joint bolt; disconnect the steering joint by sliding the steering joint into the column

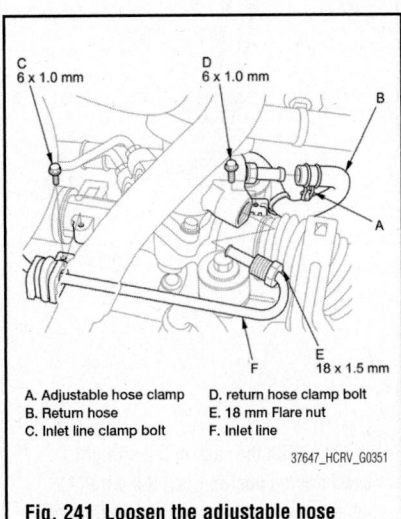

A. Adjustable hose clamp
B. Return hose
C. Inlet line clamp bolt
D. return hose clamp bolt
E. 18 mm Flare nut
F. Inlet line

Fig. 241 Loosen the adjustable hose clamp and disconnect the return hose

• Ball Joint Thread Protector, 12 mm 07AAF-SDAA100

➡**Note these items during removal:**

• Using solvent and a brush, wash any oil and dirt off the valve body unit, its lines, and the end of the steering gearbox. Blow dry with compressed air.
• Be sure to remove the steering wheel before disconnecting the steering joint. Damage to the cable reel can occur.
• The illustrations show a 2WD model.

1. Do the battery terminal disconnection procedure.
2. Drain the power steering fluid.
3. Raise and support the vehicle.
4. Remove the front wheels.
5. Remove the steering wheel.
6. Remove the driver's dashboard undercover.
7. Remove the steering joint cover.
8. Release the tilt/telescopic lock lever, and adjust the steering column to the full tilt up position, and to the full telescopic in position.
9. Tighten the tilt/telescopic lock lever.
10. Hold the steering column lower slide shaft together with a piece wire between the upper yoke and the lower yoke.
11. Release the tilt/telescopic lock lever, and adjust the steering column to the full telescopic out position, then tighten the tilt/telescopic lock lever.

➡**Do not release the tilt/telescopic lock lever when removing the steering column from the frame.**

12. Remove the steering joint bolt. Disconnect the steering joint by sliding the steering joint into the column. Do not disconnect the steering joint from the column shaft.

➡**If the center guide is in place and has not moved, leave it in place. If the center guide has come off, discard it.**

13. Apply vinyl tape to the splines on the pinion shaft.
14. Remove the cotter pin from the 12 mm nut, and loosen the nut.
15. Disconnect the tie-rod end ball joint from the knuckle using the ball joint thread protector and the ball joint remover.
16. Remove the air cleaner housing.
17. Remove the under-floor TWC.
18. Remove the P/S heat shield.
19. Loosen the adjustable hose clamp and disconnect the return hose.

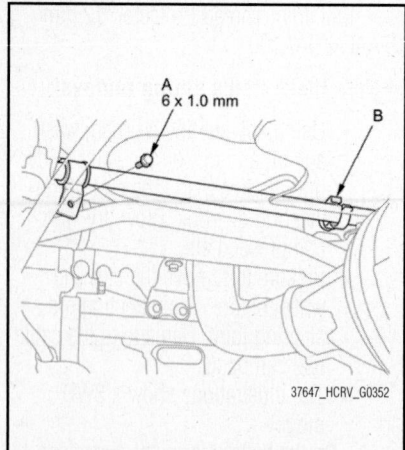

Fig. 242 Remove the return hose clamp bolt (A) and open the return hose clamp (B)

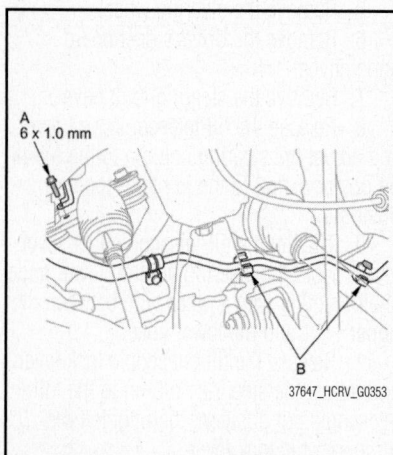

Fig. 243 Remove the pump outlet hose clamp bolt (A), then open the return line clamps (B)

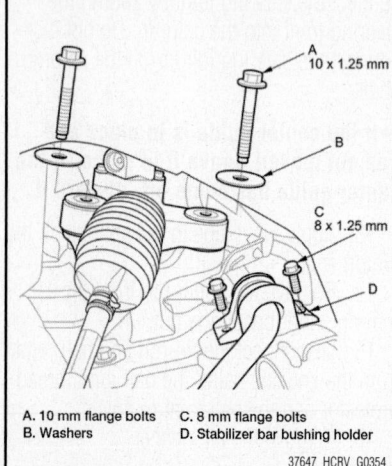

A. 10 mm flange bolts   C. 8 mm flange bolts
B. Washers   D. Stabilizer bar bushing holder

37647_HCRV_G0354

Fig. 244 Remove the 10 mm flange bolts and washers from the driver's side of steering gearbox

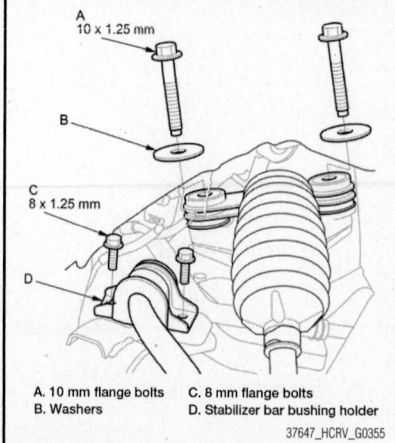

A. 10 mm flange bolts   C. 8 mm flange bolts
B. Washers   D. Stabilizer bar bushing holder

37647_HCRV_G0355

Fig. 245 Remove the 10 mm flange bolts and washers from the passenger's side of the steering gearbox

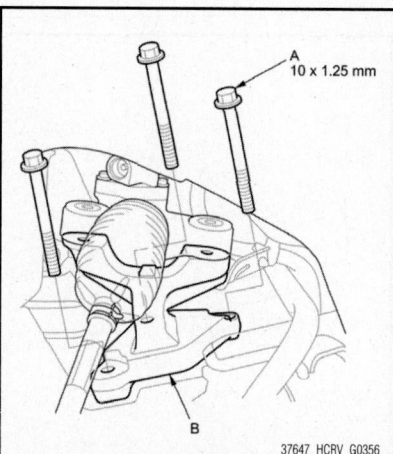

Fig. 246 Remove the 10 mm flange bolts (A) from both of the gearbox brackets (B), then remove the brackets

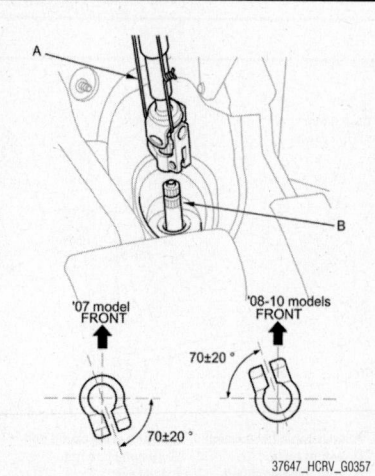

Fig. 247 With the rack in the straight ahead driving position, cut the wire (A), and slip the lower end of the steering joint onto the pinion shaft (B) in the range shown

20. Remove the inlet line clamp bolt and return hose clamp bolt.

21. Loosen the 18 mm flare nut and disconnect the inlet line.

22. Remove the return hose clamp bolt and open the return hose clamp.

23. Remove the pump outlet hose clamp bolt, then open the return line clamps.

24. Remove the 10 mm flange bolts and washers from the driver's side of the steering gearbox.

25. Remove the 8 mm flange bolts, then remove the stabilizer bar bushing holder.

26. Remove the 10 mm flange bolts and washers from the passenger's side of the steering gearbox.

27. Remove the 8 mm flange bolts, then remove the stabilizer bar bushing holder.

28. Remove the 10 mm flange bolts from both of the gearbox brackets, then remove the brackets.

29. Move the steering gearbox toward the front, and remove the pinion shaft grommet from the top of the valve housing.

30. Move the steering gearbox to the driver's side, and rotate it so the pinion shaft points toward the rear of the vehicle.

31. Carefully move the steering gearbox as an assembly toward the driver's side of the vehicle until the pinion shaft clears the wheel well opening.

32. Remove the steering gearbox through the wheel well opening on the driver's side.

33. After removing the steering gearbox, make sure that no power steering fluid gets on the gearbox mount cushions, gearbox housing, and the surface of the front subframe. Wipe off any spilled fluid at once.

To install:

➡ The illustrations show a 2WD model.

34. Before installing the steering gearbox, make sure that no power steering fluid is on the mating surface of the steering gearbox and the front subframe. To prevent the gearbox mounting bolts from loosening after the installation, remove any power steering fluid from the mount cushions and bolt holes.

35. Apply vinyl tape to the splines on the pinion shaft.

36. Slide the steering gearbox between the front subframe and body from the driver's side.

37. Carefully move the steering gearbox toward the passenger's side until the pinion shaft clears the wheel well opening on the body.

38. Rotate the steering gearbox so the pinion shaft points upward.

39. Continue moving the steering gearbox toward the passenger's side until the steering gearbox is in position.

40. Remove the vinyl tape from pinion shaft, and install the pinion shaft grommet. Align the slot in the pinion shaft grommet with the lug portion on the valve housing. The grommet must not have a gap at the mating surface of the grommet and valve housing.

41. Install the gearbox brackets, and loosely install the steering gearbox bracket mounting bolts.

42. Install the gearbox brackets with the 10 mm flange bolts. Tighten the bolts to 48 ft. lbs. (66 Nm).

43. Loosely install the 10 mm flange bolts and washers on the driver's side of the steering gearbox.

44. Install the stabilizer bar bushing holder.

45. Loosely install the 10 mm flange bolts and washers on the passenger's side of the steering gearbox.

46. Install the stabilizer bar bushing holder.

47. Tighten the 10 mm flange bolts to 52 ft. lbs. (71 Nm).

48. Install the pump outlet hose clamp bolt, then clamp the return line with the clamps.

49. Install the return hose clamp bolt, then clamp the return hose with the clamp.

50. Connect the inlet line, and tighten the 18 mm flare nut to 35 ft. lbs. (47 Nm).

51. Install the inlet line clamp bolt and return hose clamp bolt.

52. Connect the return hose securely, and tighten the adjustable hose clamp.

53. Install the P/S heat shield.

54. Install the under-floor TWC.

55. Install the air cleaner housing.

56. Wipe off any grease contamination from the ball joint tapered section and threads. Reconnect the tie-rod ends to the steering knuckles. Install the 12 mm nut, and tighten it to 40 ft. lbs. (54 Nm).

57. Install a new cotter pin, and bend it.

➡**Check the boot for damage and deterioration. If there is damage or deterioration, replace the boot.**

58. Center the steering rack within its stroke, then remove the vinyl tape from the pinion shaft.

59. With the rack in the straight ahead driving position, cut the wire, and slip the lower end of the steering joint onto the pinion shaft in the range shown.

60. Align the bolt hole on the steering joint with the groove around the pinion shaft, and loosely install the joint bolt. Be sure that the joint bolt is securely in the groove in the pinion shaft. Pull on the steering joint to make sure that the steering joint is fully seated. Tighten the steering joint bolt to 21 ft. lbs. (28 Nm).

61. Install the steering joint cover.

62. Install the driver's dashboard undercover.

63. Install the front wheels, then set the wheels in the straight ahead position.

64. Install the steering wheel and the driver's airbag.

65. Do the battery terminal reconnection procedure, and do these tasks:

   a. Turn the ignition switch to ON (II) and check that the SRS indicator comes on for about 6 seconds and goes off.

   b. Make sure the horn and turn signal switches work properly.

   c. Make sure the steering wheel switches work properly.

66. Fill the system with power steering fluid, and bleed air from the system.

67. After installation, and do these tasks:

   a. Start the engine, allow it to idle, and turn the steering wheel from lock to lock several times to warm up the fluid. Check the steering gearbox for leaks.

   b. Check the steering wheel spoke angle. If steering spoke angles to the right and left are not equal (steering wheel and rack are not centered), correct the engagement of the joint/pinion shaft serrations.

   c. Set the steering column to the center tilt position, and to the center telescopic position, then inspect the front toe.

POWER STEERING PUMP

REMOVAL & INSTALLATION

2011 Models

See Figure 248.

1. Place a suitable container under the vehicle to catch any spilled fluid.

2. Drain the power steering fluid from the reservoir.

3. Remove the drive belt from the pump pulley.

4. Cover the auto-tensioner, the alternator, and the A/C compressor with several shop towels to protect them from spilled power steering fluid. Disconnect the pump inlet hose and pump outlet hose from the pump, and plug them. Take care not to spill

the fluid on the vehicle. Wipe off any spilled fluid at once. Do not turn the steering wheel with the pump removed.

5. Remove the pump mounting bolts.

6. Cover the opening of the pump to prevent spillage.

7. Transfer the pump inlet hose and the pump outlet hose from the original pump onto the new pump with a new O-ring.

8. Loosely install the pump in the pump bracket with the mounting bolts, then tighten the pump fittings to the specified torque.

9. Tighten the pump mounting bolts to 16 ft. lbs. (22 Nm).

10. Install the drive belt, noting the following:

   • Inspect the drive belt for wear and cracks. Replace the belt if necessary.

   • Make sure that the drive belt is properly positioned on the pulleys.

   • Do not get power steering fluid or grease on the auto-tensioner, the alternator, the A/C compressor, and drive belt, or pulley faces. Clean off any fluid or grease before installation.

11. Fill the reservoir to the upper level line.

12. Start the engine, and check for fluid leaks.

BLEEDING

2011 Models

1. Before servicing the vehicle, refer to the Precautions Section.

2. Stop the engine.

3. Turn the steering wheel fully to the right and left several times.

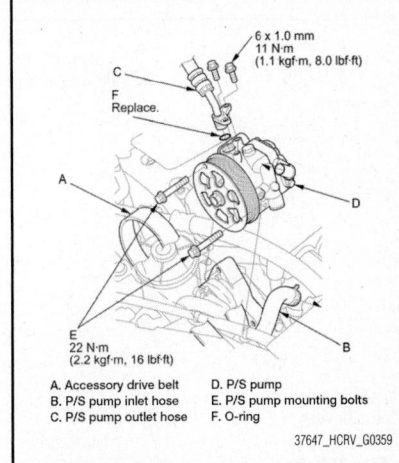

A. Accessory drive belt
B. P/S pump inlet hose
C. P/S pump outlet hose
D. P/S pump
E. P/S pump mounting bolts
F. O-ring

37647_HCRV_G0359

Fig. 248 Removing the hydraulic power steering pump

➡Do not allow the fluid level in the reservoir tank to go below the MIN level line. Check and add fluid as needed.

4. Run the engine at idle speed. Turn the steering wheel fully to the right and then fully to the left. Hold for about 3 seconds. Check for fluid leakage.

5. Repeat the above step several times at 3 second intervals.

✳✳ WARNING

Do not hold the steering wheel in the locked position for more than 10 seconds. Damage to the pump may occur.

6. Check for air bubbles or cloudy fluid. If found, repeat the bleeding procedure.

7. Stop the engine and check the fluid level. Correct as required.

SUSPENSION

DAMPER & SPRING

REMOVAL & INSTALLATION

2011 Models

See Figures 249 through 251

1. Before servicing the vehicle, refer to the Precautions Section.

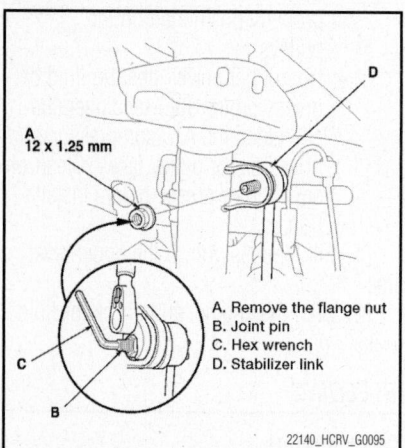

A
12 x 1.25 mm

A. Remove the flange nut
B. Joint pin
C. Hex wrench
D. Stabilizer link

22140_HCRV_G0095

Fig. 249 Remove the flange nut, while holding the joint pin with a hex wrench, and disconnect the stabilizer link from the strut

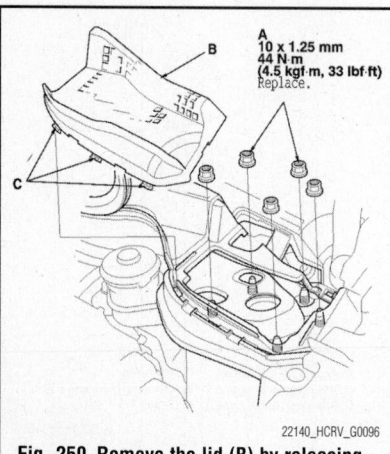

A
10 x 1.25 mm
44 N·m
(4.5 kgf·m, 33 lbf·ft)
Replace.

22140_HCRV_G0096

Fig. 250 Remove the lid (B) by releasing the hooks (C) and remove the flange nuts (A) from the top of the strut

2. Turn the ignition switch to ON, then turn on the windshield wipers. Turn the ignition switch to LOCK when the wipers are near the A-pillars.

3. Raise and safely support the vehicle.

4. Remove the front wheel.

5. Remove the wheel speed sensor harness guide, the harness clip, and the brake hose from the strut. Do not disconnect the wheel speed sensor connector.

6. Make an alignment marking on the camber adjusting bolt and strut for approximate installation alignment later.

7. Remove the flange nut, while holding the joint pin with a hex wrench, and disconnect the stabilizer link from the strut.

8. Remove the strut mounting bolts and self-locking nuts from the strut.

9. Remove the lid by releasing the hooks.

10. Remove the flange nuts from the top of the strut.

11. Remove the strut assembly.

➡The strut springs are different. Mark the springs L and R before removal.

To install:

12. Install the strut assembly on to the frame. Note the direction of the strut mounting base so that the stamp on it is toward the outside of the vehicle.

B

OUTSIDE

A

22140_HCRV_G0097

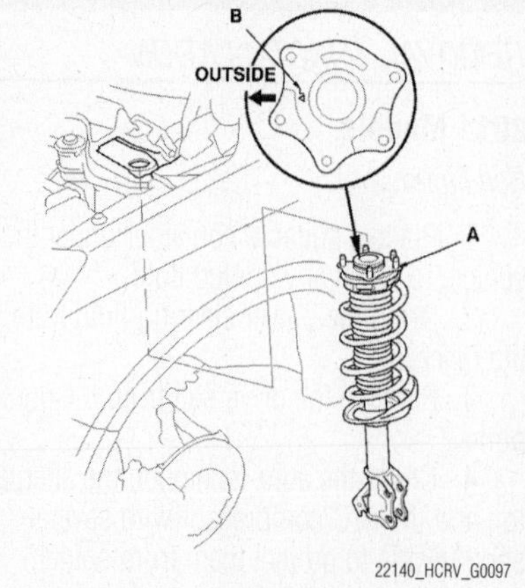

Fig. 251 View of strut assembly (A) and directional stamp for mounting (B)

FRONT SUSPENSION

13. Loosely install new flange nuts to the top of the strut.

14. Loosely install new strut mounting bolts and new self-locking nuts to the strut.

15. Connect the stabilizer link to the strut, and loosely install the flange nut, while holding the joint pin with a hex wrench.

16. Install the wheel speed sensor harness guide, the harness clip, and the brake hose to the strut.

17. Raise the front suspension with a floor jack to load the suspension with the vehicle's weight.

18. Tighten the strut mounting bolts and the self-locking nuts to 116 ft. lbs. (157 Nm).

19. Tighten the flange nuts on top of the strut to 33 ft. lbs. (44 Nm).

20. Install the lid by pushing the hooks into place securely.

21. Clean the mating surface of the brake disc and the inside of the wheel, then install the front wheel.

22. Check the wheel alignment and adjust if necessary.

2012 Models

See Figures 252 through 254.

1. Raise the vehicle on a lift, and make sure it is securely supported.

2. Remove the front wheel.

3. Remove the wheel speed sensor harness clips and the brake hose bracket from the damper.

4. Remove the flange nut while holding the joint pin with a hex wrench. Disconnect the stabilizer link from the damper.

5. Remove the damper pinch bolts and flange nuts from the lower end of the damper.

➡Do not allow the knuckle to rotate too far outward. This may allow the driveshaft inboard joint to come apart.

6. Remove the clip. Pull out the hood rear seal. Detach the hooks, and remove the cowl top lid. Remove the flange nuts.

7. Remove the damper/spring.

To install:

8. Install the damper/spring assembly, with the spring oriented properly.

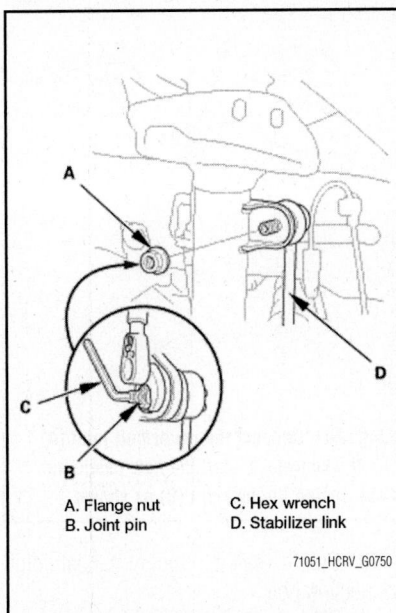

Fig. 252 Remove the flange nut while holding the joint pin with a hex wrench. Disconnect the stabilizer link from the damper

A. Flange nut
B. Joint pin
C. Hex wrench
D. Stabilizer link

71051_HCRV_G0750

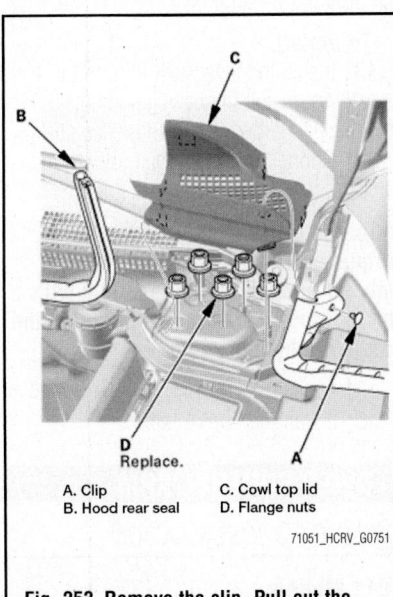

A. Clip
B. Hood rear seal
C. Cowl top lid
D. Flange nuts

Replace.

71051_HCRV_G0751

Fig. 253 Remove the clip. Pull out the hood rear seal. Detach the hooks, and remove the cowl top lid. Remove the flange nuts

9. Loosely install new flange nuts on the upper end of the damper/spring.

10. Loosely install new pinch bolts and nuts on the lower end of the damper/spring.

11. Use a floor jack to raise the suspension and load to vehicle weight. Then, tighten the retainers as follows:
- Upper flange nuts: 33 ft. lbs. (44 Nm)
- Pinch bolts and nuts: 116 ft. lbs. (157 Nm)

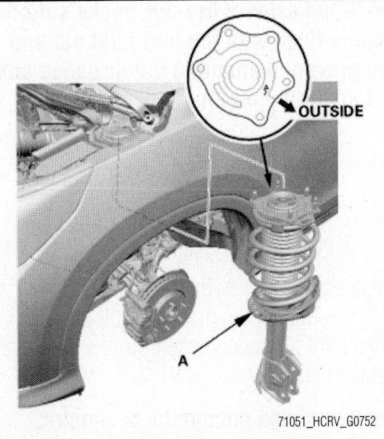

OUTSIDE

71051_HCRV_G0752

Fig. 254 Install the damper/spring assembly, with the spring oriented as shown

12. Connect the stabilizer link to the damper and tighten the nut to 58 ft. lbs. (78 Nm).

13. Install the wheel speed sensor line to the clips and the brake hose to its bracket.

14. Install the cowl top lid, hood rear seal and clip screw.

15. Install the front wheel.

16. Check and adjust wheel alignment.

17. Perform the following with the HDS:
- VSA sensor neutral position memorization
- Steering angle sensor neutral position clear

LOWER BALL JOINT

REMOVAL & INSTALLATION

2011 Models

See Figures 255 and 256.

Special Tools Required:

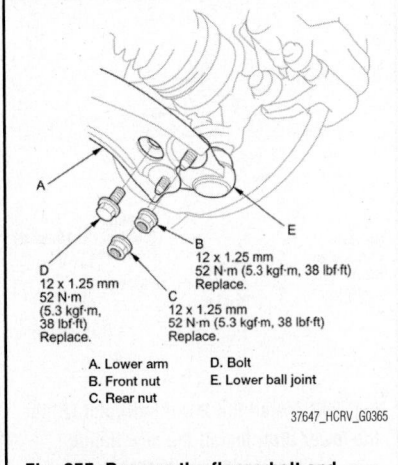

D
12 x 1.25 mm
52 N·m
(5.3 kgf·m,
38 lbf·ft)
Replace.

B
12 x 1.25 mm
52 N·m (5.3 kgf·m, 38 lbf·ft)
Replace.

C
12 x 1.25 mm
52 N·m (5.3 kgf·m, 38 lbf·ft)
Replace.

A. Lower arm
B. Front nut
C. Rear nut
D. Bolt
E. Lower ball joint

37647_HCRV_G0365

Fig. 255 Remove the flange bolt and flange nuts from the lower arm

- Ball Joint Remover, 32 mm 07MAC-SL0A102
- Ball Joint Thread Protector, 14 mm 07AAE-SJAA100

1. Raise and support the vehicle.

2. Remove the front wheel.

3. Remove the flange bolt and flange nuts from the lower arm.

➡ **During installation, install a new flange bolt and new flange nuts. After lightly tightening all three fasteners, tighten them to 38 ft. lbs. (52 Nm) in the following order; the nut on the front, the nut on the rear, then the bolt.**

4. Disconnect the lower ball joint from the lower arm.

5. Remove the lock pin from the lower ball joint pin, then remove the castle nut.

➡ **During installation, install the lock pin after tightening a new castle nut.**

6. Disconnect the lower ball joint from the knuckle using the ball joint thread protector and the ball joint remover.

To install:

7. Install the lower ball joint in the reverse order of removal, and note these items:
- First install all the components, and lightly tighten the bolts and nuts, then tighten the lower ball joint to the lower arm to the specified torque. Raise the suspension to load it with the vehicle's weight before fully tightening the lower ball joint to the knuckle to the specified torque.
- Torque the castle nut to the lower torque specification, then tighten it only far enough to align the slot with the ball joint pin hole. Do not align the castle nut by loosening it.

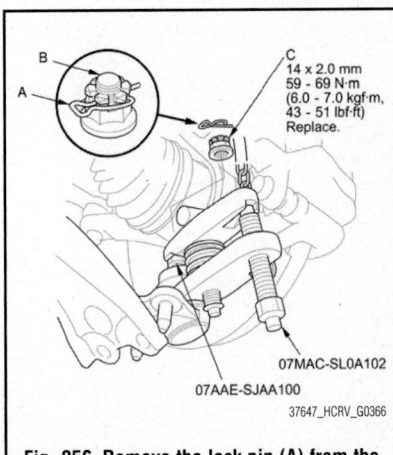

C
14 x 2.0 mm
59 - 69 N·m
(6.0 - 7.0 kgf·m,
43 - 51 lbf·ft)
Replace.

07MAC-SL0A102
07AAE-SJAA100

37647_HCRV_G0366

Fig. 256 Remove the lock pin (A) from the lower ball joint pin (B), then remove the castle nut (C)

- Before installing the wheel, clean the mating surface of the brake disc and the inside of the wheel.
- Check the wheel alignment, and adjust it if necessary.

2012 Models

See Figures 257 through 261.

1. Raise the vehicle on a lift, and make sure it is securely supported.
2. Remove the front wheel.

➡**Always use a ball joint remover to disconnect a ball joint. Do not strike the housing or any other part of the ball joint connection to disconnect it.**

3. Install a hex nut or the ball joint thread protector.

➡**When using a hex nut, make sure the nut is flush with the ball joint pin end to prevent damage to the threaded end of the ball joint pin.**

4. Apply grease to the ball joint remover on the bolt and contact areas.
5. Install the ball joint remover.
6. Fasten the safety tie securely to a suspension arm or the subframe.
7. Tighten the pressure bolt until the ball joint pin pops loose from the ball joint connecting hole.

➡**Do not use pneumatic or electric tools on the pressure bolt.**

8. Remove the ball joint remover.
9. Remove the hex nut or the ball joint thread protector.

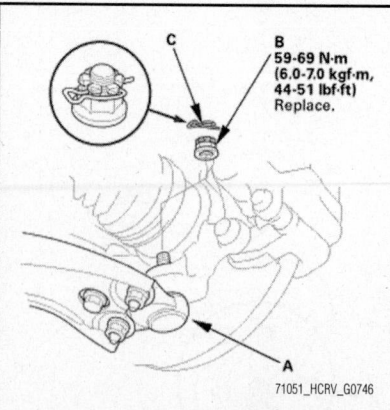

Fig. 261 Connect the lower ball joint (A) to the knuckle. Install the new castle nut (B). Install the lock pin (C) as shown

10. Pull the ball joint out of the ball joint connecting hole.
11. Remove the lock pin. Remove the castle nut. Disconnect the lower ball joint from the knuckle.
12. From under the control arm, remove the flange bolt and the flange nuts. Remove the front lower ball joint.

To install:

13. Install the lower ball joint to the lower arm. Install the new flange nuts and the new flange bolt in the sequence shown.
14. Connect the lower ball joint to the knuckle. Install the new castle nut.

➡**Torque the castle nut to the lower torque specification, then tighten it only far enough to align the slot with the ball joint pin hole. Do not align the castle nut by loosening it.**

15. Install the lock pin.
16. Install the front wheel.

LOWER CONTROL ARM

REMOVAL & INSTALLATION

2011 Models

See Figures 262 through 264.

1. Raise and support the vehicle.
2. Remove the front wheel.
3. Remove the flange bolts and the bushing holder.
4. Remove the flange bolt and self-locking nuts from the lower arm.

➡**During installation, loosely install a new flange bolt and new self-locking nuts. Then tighten them in the following order; the nut on the front, the nut on the rear, then the bolt.**

5. Disconnect the lower ball joint from the lower arm.

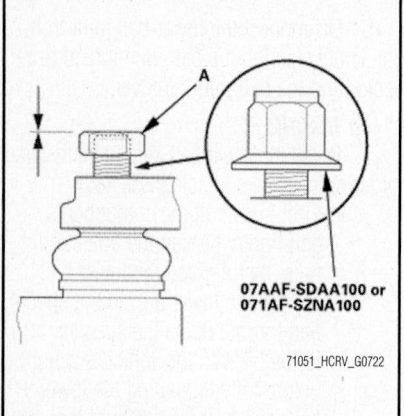

Fig. 257 Install a hex nut (A) or the ball joint thread protector

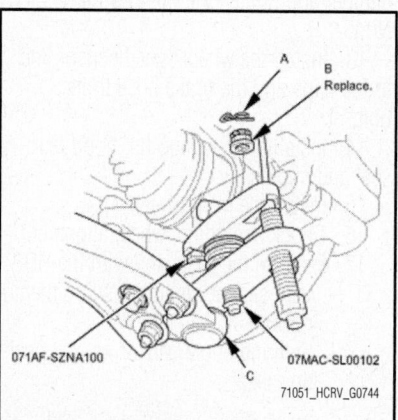

Fig. 259 Remove the lock pin (A). Remove the castle nut (B). Disconnect the lower ball joint (C) from the knuckle

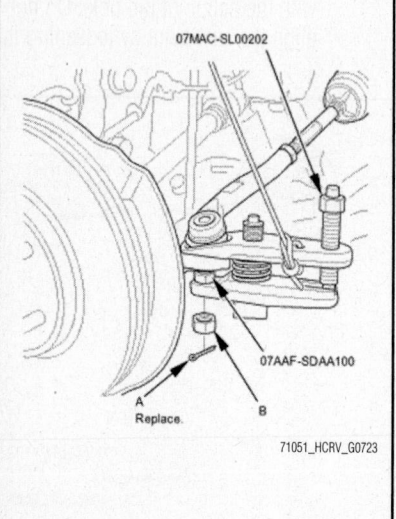

Fig. 258 Fasten the safety tie (A) securely to a suspension arm or the subframe (B)

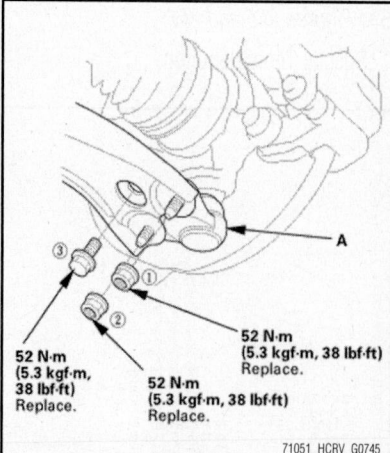

Fig. 260 Install the lower ball joint (A) to the lower arm. Install the new flange nuts and the new flange bolt in the sequence shown

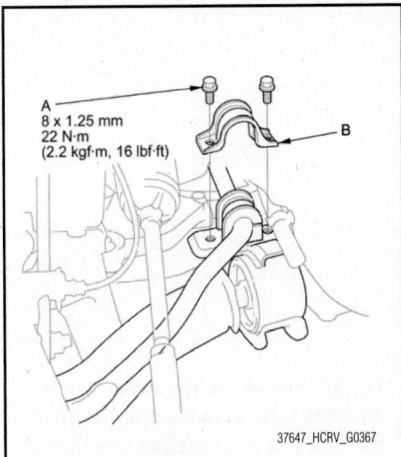

Fig. 262 Remove the flange bolts (A) and the bushing holder (B)

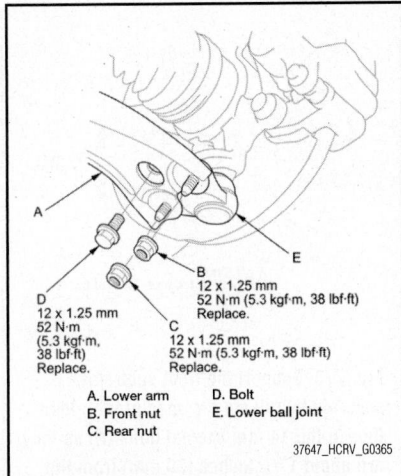

Fig. 263 Remove the flange bolt and flange nuts from the lower arm

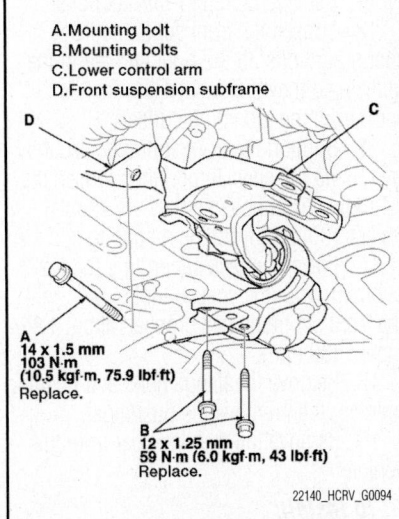

Fig. 264 Remove the lower arm mounting bolts, then remove the lower arm from the front suspension subframe

6. Remove the lower arm mounting bolt.

➡**During installation, install a new mounting bolt.**

7. Remove the lower arm mounting bolts, then remove the lower arm from the front suspension subframe.

To install:

➡**During installation, install new mounting bolts.**

8. Install the lower arm in the reverse order of removal, and note these items:

- First install all the components, and lightly tighten the bolts and nuts, then raise the suspension to load it with the vehicle's weight before fully tightening to the specified torque.
- Before installing the wheel, clean the mating surface of the brake disc and the inside of the wheel.
- Check the wheel alignment, and adjust it if necessary.

2012 Models

See Figures 265 through 268.

1. Raise the vehicle on a lift, and make sure it is securely supported.
2. Remove the front wheel.
3. Remove the front stabilizer bar bushing holder.
4. Remove the flange bolt. Remove the flange nuts. Disconnect the lower arm from the front lower ball joint.
5. Remove the lower arm. Remove the bushing stopper bracket from the lower arm.

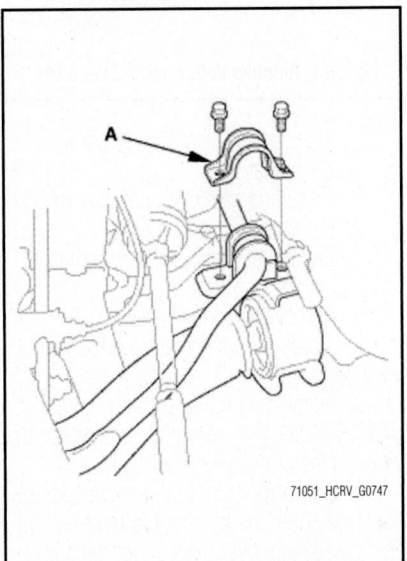

Fig. 265 Remove the front stabilizer bar bushing holder (A)

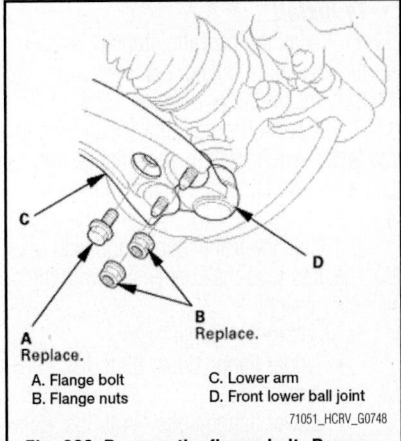

Fig. 266 Remove the flange bolt. Remove the flange nuts. Disconnect the lower arm from the front lower ball joint

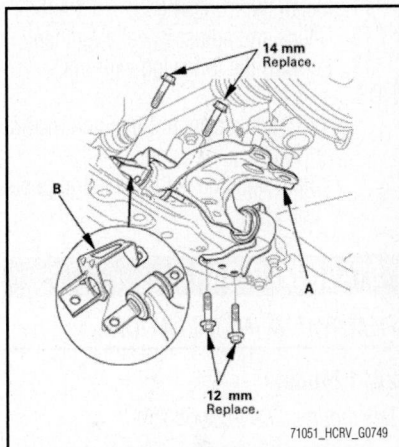

Fig. 267 Remove the lower arm (A). Remove the bushing stopper bracket (B) from the lower arm

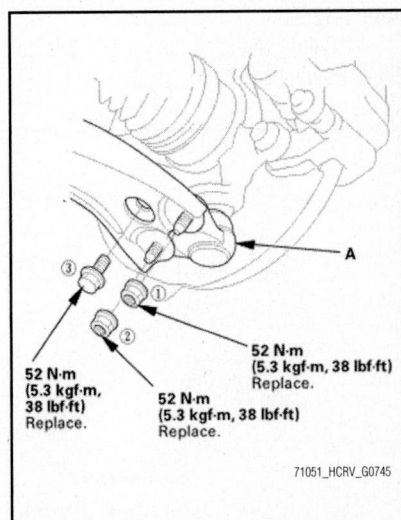

Fig. 268 Connect the front lower ball joint to the lower arm. Install the new flange nuts and the new flange bolt in the sequence shown

To install:

6. Install the bushing stopper bracket on the lower arm. Install the lower arm. Loosely install the new flange bolts.

7. Connect the front lower ball joint to the lower arm. Install the new flange nuts and the new flange bolt in the sequence shown.

8. Raise the front suspension with a floor jack to load the suspension with the vehicle's weight.

9. Tighten the flange bolts:
- Upper flange bolts: 83 ft. lbs. (113 Nm)
- Lower flange bolts: 44 ft. lbs. (59 Nm)

10. Install the front stabilizer bar bushing holder. Tighten the bolts to 16 ft. lbs. (22 Nm).

11. Install the front wheel.

12. Check and adjust wheel alignment.

13. Perform the following with the HDS:
- VSA sensor neutral position memorization
- Steering angle sensor neutral position clear

STABILIZER BAR

REMOVAL & INSTALLATION

2011 Models

See Figures 269 through 274.

Special Tools Required:
- Universal eyelet 07AAK-SNAA120
- Front subframe adapter VSB02C000016
- Engine support hanger, A and Reds AAR-T-12566

1. Before servicing the vehicle, refer to the Precautions Section.

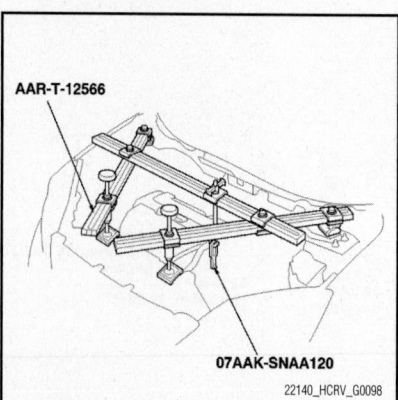

Fig. 269 Install the engine support hanger (AAR-T-12566) to the vehicle and attach the hook to the universal eyelet

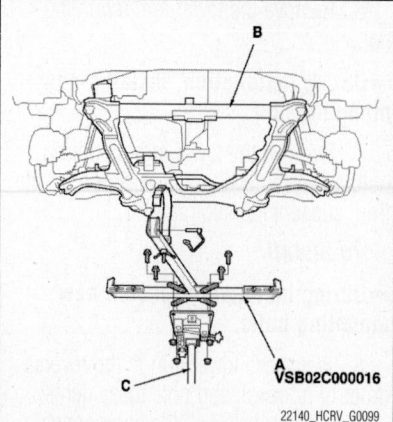

Fig. 270 Attach the front subframe adapter (A) to the front subframe (B) by hanging the hook of the special tool over the front of the subframe, then tighten the special tool screw. Raise the jack (C) and line up the slots in the arms with the bolt holes on the corner of the jack base

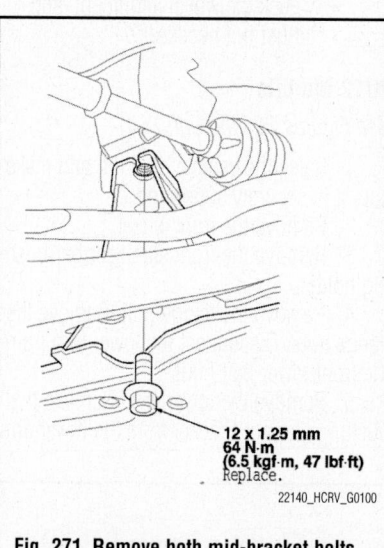

Fig. 271 Remove both mid-bracket bolts

2. Matchmark the stabilizer bar for proper reinstallation.

3. Raise and safely support the vehicle.

4. Remove the front wheels.

5. Disconnect both stabilizer control links from the stabilizer bar.

6. Remove the cowl cover.

7. Attach the universal eyelet to the cylinder head.

8. Install the engine support hanger (AAR-T-12566) to the vehicle and attach the hook to the universal eyelet.

9. Attach the front subframe adapter to the front subframe by hanging the hook of the special tool over the front of the subframe, then tighten the special tool screw.

10. Raise the jack and line up the slots in the arms with the bolt holes on the corner

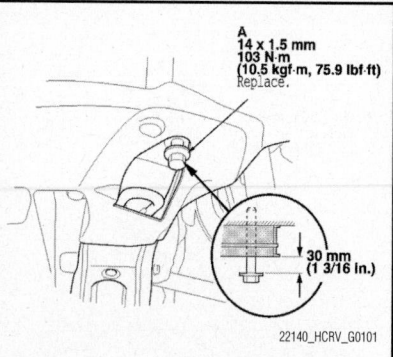

Fig. 272 Loosen the front subframe front mounting bolts (A) on the right and left of the vehicle so they are about 1 3/16 inches (30 mm) from the mounting surface

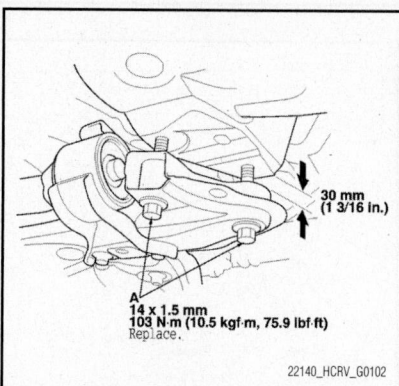

Fig. 273 Support the front subframe securely by raising the special tool, then loosen the 14 mm special bolts (A) so they are about 1 3/16 inches (30 mm) from the mounting surface

of the jack base, then securely attach them with bolts.

11. Remove both mid-bracket bolts.

12. Loosen the front subframe front mounting bolts on the right and left of the vehicle so they are about 1 3/16 inches (30 mm) from the mounting surface.

13. Support the front subframe securely by raising the special tool, then loosen the 14 mm special bolts so they are about 1 3/16 inches (30 mm) from the mounting surface.

14. Lower the jack supporting the front subframe with the special tool slowly until the front subframe has dropped about 1 3/16 inches (30 mm).

15. Remove the flange bolts and bushing holders, then remove the bushings.

16. Remove the stabilizer bar from the vehicle.

To install:

17. Install the stabilizer bar into position aligning the paint marks on the stabilizer bar with the sides of the bushings.

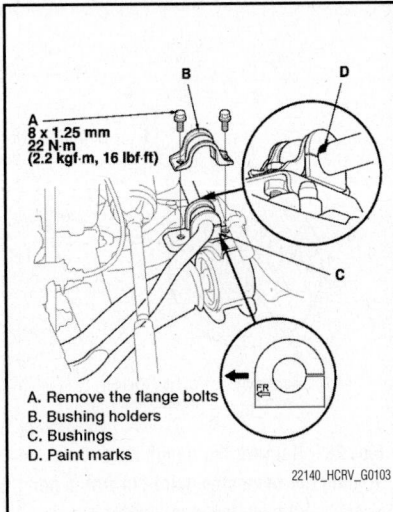

Fig. 274 Remove the flange bolts and bushing holders, then remove the bushings. (Align the paint marks during installation)

➡**Ensure the right and left direction of the stabilizer bar is correct before installation. Make sure the fore and aft direction of the bushings is correct.**

18. Install the bushings, bushing holders, and the flange bolts. Tighten the bolts to 16 ft. lbs. (22 Nm).

19. Raise the jack supporting the front subframe with the special tool slowly until the front subframe makes contact with the body frame.

20. Installation continues in the reverse of the removal procedure.

21. Tighten the bolts to the specified torque:

a. Bolts 14 x 1.5mm: 76 ft. lbs. (103 Nm).

b. Bolts 12 x 1.25mm: 47 ft. lbs. (64 Nm).

22. Check the front wheel alignment and make adjustments as necessary.

2012 Models

See Figures 275 through 279.

1. Raise and support the vehicle.
2. Remove the front wheels.
3. Remove or disconnect the following (refer to applicable component sections as needed):

- Steering column—lower slide shaft and lower joint connection
- Tie rod end ball joints
- Stabilizer bar links
- Front splash shield
- Front floor undercover

- TWC assembly
- Propeller shaft
- Subframe support
- Lower torque rod bolts

4. Move the front subframe as follows:

a. Remove the mid mounting bolts on both sides.

b. Loosen the front subframe mounting bolts so they are about 1.18 in. (30 mm) from the mounting surface on both sides.

➡**Do not loosen the front subframe mounting bolts more than necessary.**

c. Remove the front subframe mounting bolts on both side. Lower the jack slowly until the front subframe has dropped about 5.12 in. (130 mm).

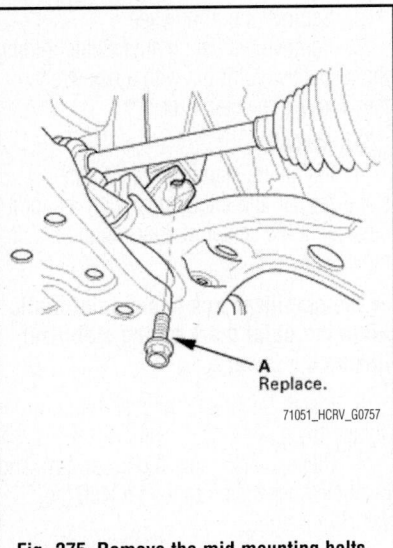

Fig. 275 Remove the mid mounting bolts (A) on both sides

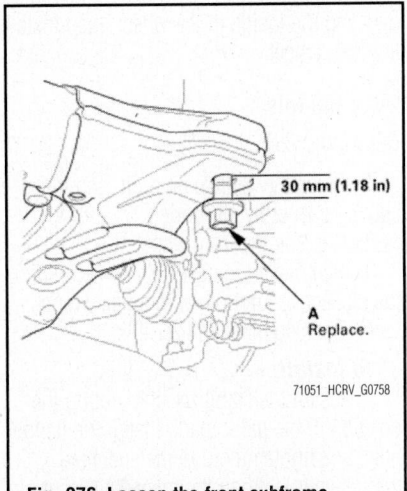

Fig. 276 Loosen the front subframe mounting bolts (A) so they are about 1.18 in. (30 mm) from the mounting surface on both sides

5. Remove the front stabilizer bar bushing holders. Remove the bushings. Remove the front stabilizer bar.

To install:

6. Install the stabilizer bar, with the bushing slit toward the rear. Tighten the bracket bolts to 16 ft. lbs. (22 Nm).

7. Turn the lip of the pinion shaft grommet. Lift the front subframe up to the body, then restore the pinion shaft grommet back to proper position.

➡**Be sure that the pinion shaft grommet is securely in place. Make sure the lip of the pinion shaft grommet is not turned up. Incorrect installation can cause leakage of water and mud, or noise.**

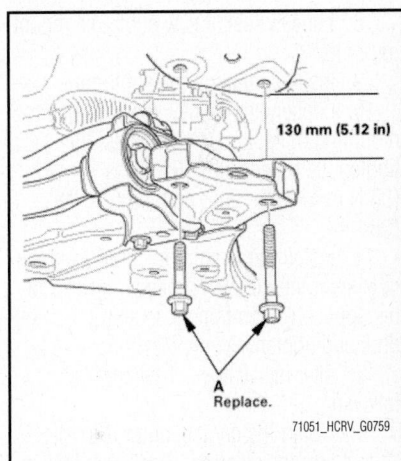

Fig. 277 Remove the front subframe mounting bolts (A) on both side. Lower the jack slowly until the front subframe has dropped about 5.12 in. (130 mm)

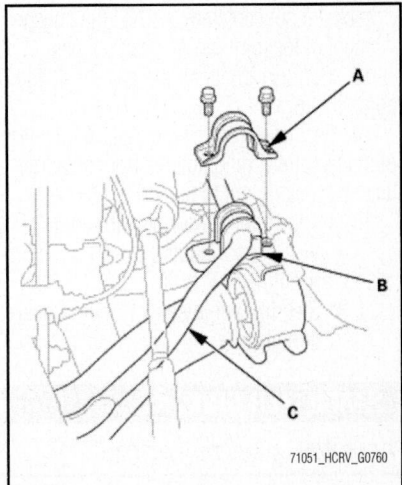

Fig. 278 Remove the front stabilizer bar bushing holders (A). Remove the bushings (B). Remove the front stabilizer bar (C)

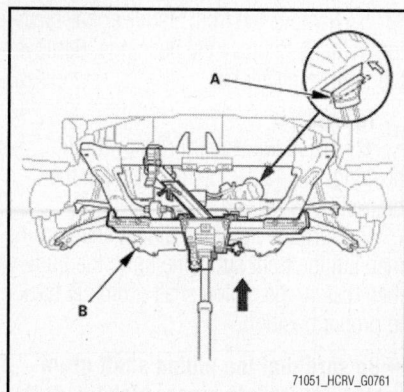

Fig. 279 Turn the lip (A) of the pinion shaft grommet. Lift the front subframe (B) up to the body, then restore the pinion shaft grommet back to proper position

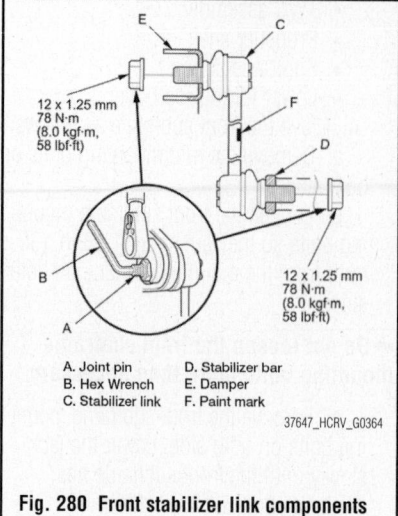

A. Joint pin
B. Hex Wrench
C. Stabilizer link
D. Stabilizer bar
E. Damper
F. Paint mark

37647_HCRV_G0364

Fig. 280 Front stabilizer link components

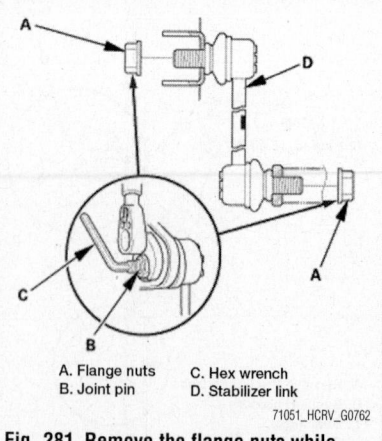

A. Flange nuts
B. Joint pin
C. Hex wrench
D. Stabilizer link

71051_HCRV_G0762

Fig. 281 Remove the flange nuts while holding the respective joint pin with a hex wrench. Remove the stabilizer link

8. Loosely install new subframe mounting bolts.

9. Install and align the subframe.

10. Loosely install new lower torque rod mounting bolts. Loosen the bolts, then tighten the outer bolt to 69 ft. lbs. (93 Nm) and the inner bolt to 65 ft. lbs. (88 Nm).

11. Install the remaining components in reverse of the removal procedure. Refer to the applicable component section for procedure and tightening specifications.

12. After installation, check the following:

a. Start the engine, allow it to idle, and turn the steering wheel from lock to lock several times.

b. Check that the EPS indicator does not come on.

c. Check the steering wheel spoke angle. If steering spoke angles to the right and left are not equal (steering wheel and rack are not centered), correct the engagement of the joint/pinion shaft splines.

13. Perform wheel alignment check and adjustment.

14. Test drive the vehicle. After 5 minutes of the test drive, retighten the front stabilizer link self-locking nut to 58 ft. lbs. (78 Nm).

15. Perform the following with the HDS:
- VSA sensor neutral position memorization
- Steering angle sensor neutral position clear

STABILIZER LINK

REMOVAL & INSTALLATION

2011 Models

See Figure 280.

1. Raise and support the vehicle.

2. Remove the front wheel.

3. Remove the flange nuts while holding the respective joint pin with a hex wrench, then remove the stabilizer link.

To install:

4. Install the stabilizer link on the stabilizer bar and the damper with the joint pins set at the center of their range of movement.

➡The stabilizer link has a paint mark. Align the paint mark on the stabilizer link facing inward.

5. Install the flange nuts, and lightly tighten them.

6. Place a jack under the lower arm, and raise the suspension to load it with the vehicle's weight.

7. Tighten the flange nuts to 58 ft. lbs. (78 Nm) while holding the respective joint pin with a hex wrench.

8. Clean the mating surface of the brake disc and the inside of the wheel, then install the front wheel.

2012 Models

See Figure 281.

1. Raise the vehicle on a lift, and make sure it is securely supported.

2. Remove the front wheel.

3. Remove the flange nuts while holding the respective joint pin with a hex wrench.

4. Remove the stabilizer link.

To install:

5. Install the stabilizer link, noting the position of the paint mark. Install the flange nuts, and tighten them to the specified torque while holding the respective joint pin with a hex wrench.

6. Install the front wheel.

7. Test-drive the vehicle.

8. After 5 minutes of driving, tighten the

self-locking nut again to the specified torque value.

STEERING KNUCKLE & WHEEL HUB ASSEMBLY

REMOVAL & INSTALLATION

See Figure 282.

Special Tools Required:
- Ball joint remover, 32mm 07MAC-SL0A102
- Ball joint remover, 28mm 07MAC-SL0A202
- Ball joint thread protector, 14mm 071AF-S3VA000

1. Before servicing the vehicle, refer to the Precautions Section.

2. Raise and safely support the vehicle.

3. Remove the front wheels.

4. Remove the brake hose mounting bolt.

5. Remove the brake caliper bracket mounting bolts and the caliper assembly from the knuckle.

➡To prevent damage to the caliper assembly or the brake hose, use a short piece of wire to hang the caliper assembly from the undercarriage. Do not twist the brake hose excessively.

6. Remove the wheel speed sensor from the knuckle. Do not disconnect the wheel speed sensor connector.

7. Raise the stake, then remove the spindle nut.

8. Remove the front brake disc rotor.

9. Check the front hub for damage and cracks.

10. Remove the cotter pin from the tie-rod end ball joint, then remove the nut.

11. Disconnect the tie-rod end ball joint

from the knuckle using a ball joint remover.

12. Remove the lock pin from the lower ball joint pin, then remove the castle nut.

13. Disconnect the lower ball joint from the lower arm.

14. Remove the strut mounting bolts and the self-locking nuts from the strut.

15. Remove the driveshaft outboard joint from the knuckle by tapping the driveshaft end with a plastic hammer while drawing the hub outward.

16. Remove the knuckle/hub.

➡**Do not pull the driveshaft end outward as the driveshaft inboard joint may come apart.**

To install:

Install all the components and lightly tighten the bolts and nuts, then raise the suspension to load it with the vehicle weight before fully tightening to the specified torque values.

• Be careful not to damage the ball joint boot when installing the knuckle

• Before connecting the lower ball joint to the knuckle, degrease the threaded section and tapered portion of the ball joint pin, the knuckle connecting hole, the threaded section, and the mating surface of the castle nut

• Torque the castle nut to the lower torque specification, then tighten it only far enough to align the slot with the ball joint pin hole. Do not align the castle nut by loosening it

• Use a new spindle nut during reassembly

• Before installing the spindle nut, apply

a small amount of engine oil to the seating surface of the nut. After tightening, use a drift to stake the spindle nut shoulder against the driveshaft.

• Before installing the brake disc, clean the mating surface of the front hub and the inside of the brake disc rotor.

17. Apply grease to the mating surface of the wheel bearing and the driveshaft outboard joint.

18. Install the knuckle/hub into position.

19. Install the driveshaft outboard joint to the knuckle.

20. Loosely install the strut mounting bolts and the self-locking nuts to the strut.

21. With the weight of the vehicle on the suspension, tighten the strut mounting bolts and nuts to 116 ft. lbs. (157 Nm).

22. Connect the lower ball joint to the lower arm. Tighten the castle nut to 43–51 ft. lbs. (59–69 Nm).

23. Install the lock pin into the lower ball joint.

24. Connect the tie-rod end ball joint to the knuckle and tighten the nut to 40 ft. lbs. (54 Nm). Install the cotter pin.

25. Install the front brake disc rotor.

26. Install the spindle nut and tighten to 242 ft. lbs. (328 Nm).

27. Stake the spindle nut with a drift tool.

28. Install the wheel speed sensor to the knuckle and tighten to 86 inch lbs. (10 Nm).

29. Install the caliper assembly to the knuckle and tighten the brake caliper bracket mounting bolts to 101 ft. lbs. (137 Nm).

30. Install the brake hose mounting bolt and tighten to 16 ft. lbs. (22 Nm).

31. Install the wheel and tighten to 80 ft. lbs. (108 Nm).

32. Check the wheel alignment and adjust as necessary.

WHEEL BEARING

REMOVAL & INSTALLATION

2011 Models

See Figures 283 through 286.

➡**For removal of wheel hub assembly, see "Steering Knuckle & Wheel Hub Assembly" in this section.**

Special Tools Required:
• Hub disassembly/assembly tool 07GAF-SD40100
• Attachment, 72 x 75mm 07746-0010600
• Driver 07749-0010000
• Attachment, 96mm 07948-SB00101
• Support base 07965-SD90100

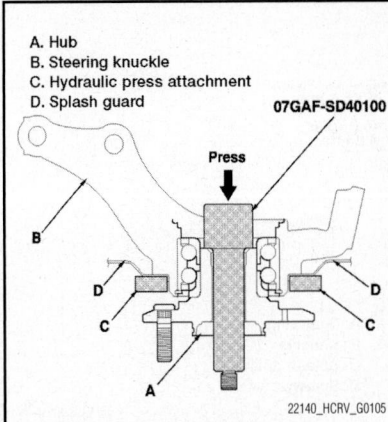

A. Hub
B. Steering knuckle
C. Hydraulic press attachment
D. Splash guard

22140_HCRV_G0105

Fig. 283 Separate the hub from the knuckle using the hub disassembly/assembly tool and a hydraulic press. Hold the knuckle with the attachment of the hydraulic press or equivalent tool. Be careful not to deform the splash guard

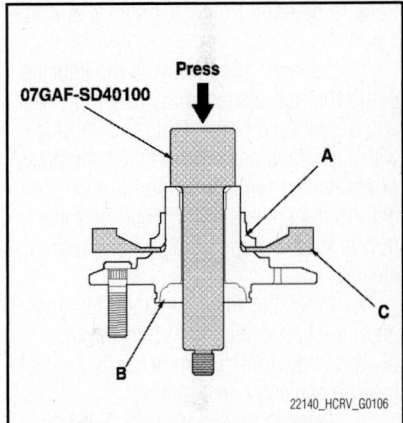

22140_HCRV_G0106

Fig. 284 Press the wheel bearing inner race (A) off of the hub (B) using the hub disassembly/assembly tool, a commercially available bearing separator (C), and a press

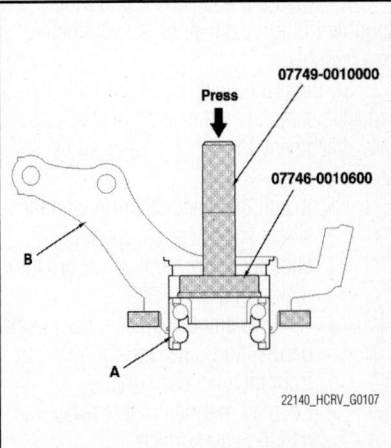

22140_HCRV_G0107

Fig. 285 Press the wheel bearing (A) out of the knuckle (B) using the attachment, the driver, and a press

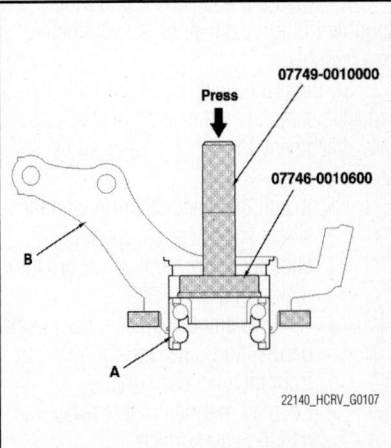

A. Strut mounting bolts
B. Self-locking nuts
C. Driveshaft outboard joint
D. Knuckle
E. Driveshaft end

22140_HCRV_G0104

Fig. 282 Remove the strut mounting bolts and the self-locking nuts from the strut. Remove the driveshaft outboard joint from the knuckle by tapping the driveshaft end with a plastic hammer while drawing the hub outward

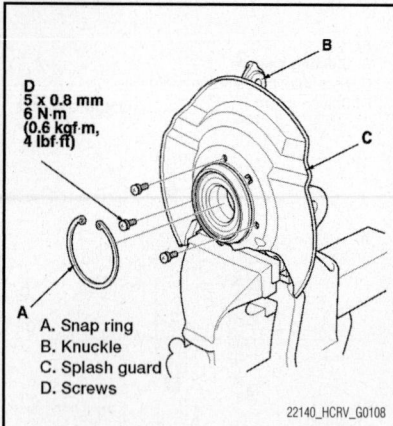

D
5 x 0.8 mm
6 N·m
(0.6 kgf·m,
4 lbf·ft)

A. Snap ring
B. Knuckle
C. Splash guard
D. Screws

22140_HCRV_G0108

Fig. 286 Install the snap ring securely in the knuckle. Install the splash guard and tighten the screws

1. Before servicing the vehicle, refer to the Precautions Section.

2. Remove the knuckle/hub assembly. Refer to Steering Knuckle, removal & installation.

3. Separate the hub from the knuckle using the hub disassembly/assembly tool and a hydraulic press. Hold the knuckle with the attachment of the hydraulic press or equivalent tool. Be careful not to deform the splash guard. Hold onto the hub to keep it from falling when pressed clear.

4. Press the wheel bearing inner race off of the hub using the hub disassembly/assembly tool, a commercially available bearing separator, and a press.

5. Remove the splash guard and the snap ring from the knuckle.

6. Press the wheel bearing out of the knuckle using the attachment, the driver, and a press.

To install:

7. Wash the knuckle and hub thoroughly in high flash-point solvent before reassembly.

8. Press a new wheel bearing into the knuckle using the old bearing, a steel plate, the attachment, the support base, and a press.

- Install the wheel bearing with the wheel speed sensor magnetic encoder (brown color) toward the inside of the knuckle
- Remove any oil, grease, dust, metal debris, and other foreign material from the encoder surface
- Keep all magnetic tools away from the encoder surface
- Be careful not to damage the encoder surface when inserting the wheel bearing

9. Install the snap ring securely in the knuckle.

10. Install the splash guard and tighten the screws to 48 inch lbs. (6 Nm).

11. Install the hub onto the knuckle using the attachment, the driver, the support base, and a hydraulic press. Be careful not to distort the splash guard.

12. Install the knuckle/hub assembly. Refer to Steering Knuckle, removal & installation.

2012 Models

See Figures 287 through 291.

➡**For removal of wheel hub assembly, see "Steering Knuckle & Wheel Hub Assembly" in this section.**

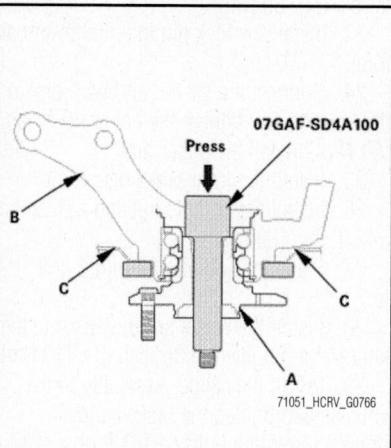

07GAF-SD4A100

Press

B

C

C

A

71051_HCRV_G0766

Fig. 287 Separate the hub (A) from the knuckle (B), using care not to damage or deform the splash guard (C)

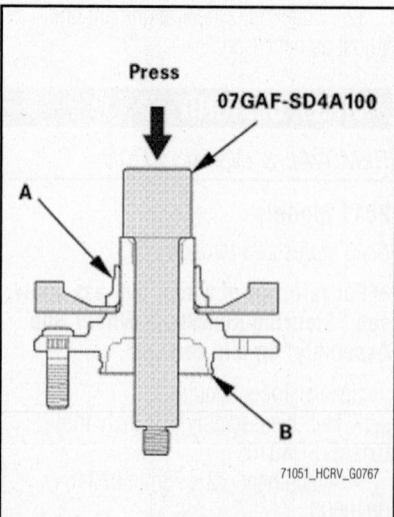

Press

07GAF-SD4A100

A

B

71051_HCRV_G0767

Fig. 288 Press out the wheel bearing inner race (A) of the hub (B)

1. Separate the hub from the knuckle, using care not to damage or deform the splash guard.

2. Press out the wheel bearing inner race of the hub.

3. Remove the splash guard. Remove the snap ring.

4. Press the wheel bearing out of the knuckle.

To install:

5. Wash the knuckle and the hub thoroughly in high flash point solvent before reassembly.

6. Press a new wheel bearing into the knuckle.

a. Install the wheel bearing with the wheel speed sensor magnetic encoder (brown color), toward the inside of the knuckle.

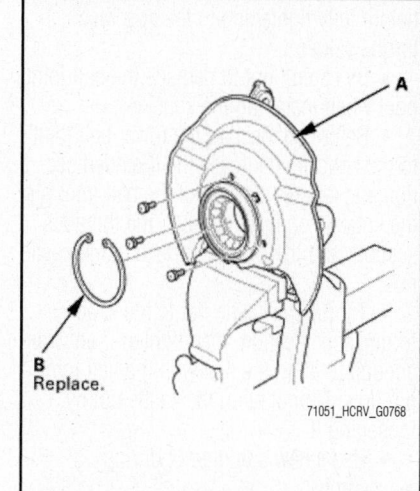

A

B
Replace.

71051_HCRV_G0768

Fig. 289 Remove the splash guard (A). Remove the snap ring (B)

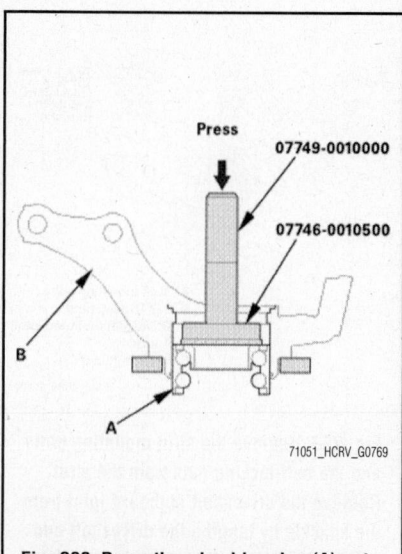

Press

07749-0010000

07746-0010500

B

A

71051_HCRV_G0769

Fig. 290 Press the wheel bearing (A) out of the knuckle (B)

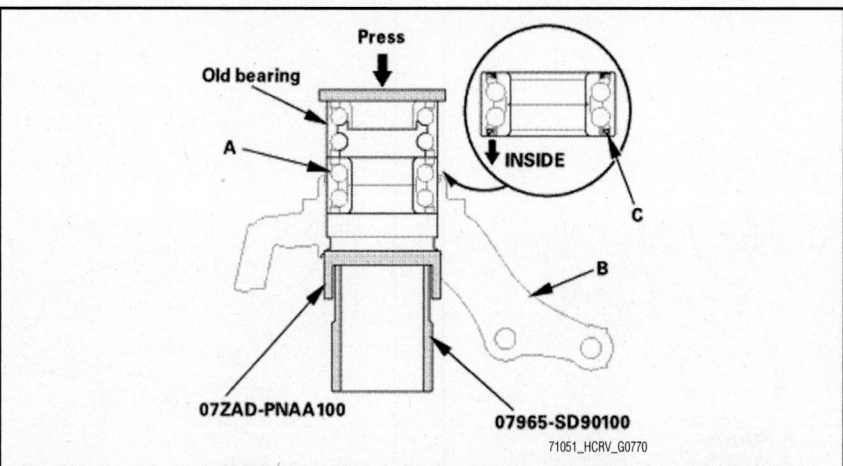

Fig. 291 Press a new wheel bearing (A) into the knuckle (B). Install the wheel bearing with the wheel speed sensor magnetic encoder (C) (brown color), toward the inside of the knuckle

b. Keep the wheel bearing and magnetic encoder away from any magnetic field. The encoder can be damaged or erased.

7. Install the new snap ring, then install the splash guard.

8. Install the hub onto the knuckle.

9. Install the wheel hub and knuckle assembly to the vehicle.

ADJUSTMENT

The front wheel bearings are not adjustable. If the bearings are noisy or become loose, they must be replaced.

SUSPENSION

REAR SUSPENSION

DAMPER & SPRING

REMOVAL & INSTALLATION

2011 Models

See Figures 292 through 294.

1. Before servicing the vehicle, refer to the Precautions Section.

2. Raise and safely support the vehicle.

3. Remove the rear wheel.

4. Remove the lid on the cargo area side trim panel by releasing the hooks.

5. Remove the flange nuts from the top of the strut.

6. Remove the flange bolts, then remove the brake hose bracket.

7. Position a floor jack under the trailing arm. Raise the floor jack until the suspension begins to compress.

8. Disconnect the stabilizer link from the trailing arm.

9. Remove the flange bolt from the bottom of the strut.

10. Remove the flange bolt, then remove the parking brake cable.

11. Remove the trailing arm mounting bolts.

12. Lower the rear suspension, then remove the strut assembly from the vehicle.

To install:

13. Position the strut assembly between the body and the trailing arm. Note the direction of the strut mounting base so that

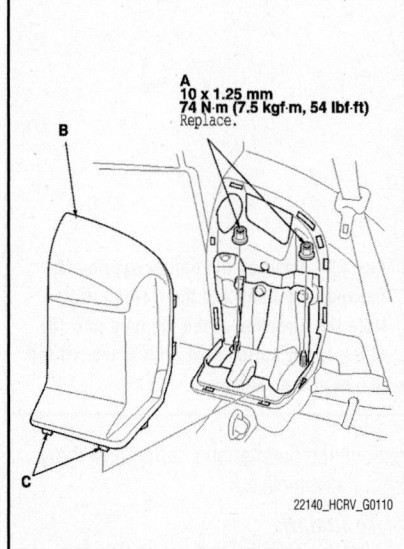

Fig. 292 Remove the lid (B) on the cargo area side trim panel by releasing the hooks (C) and remove the flange nuts (A) from the top of the strut

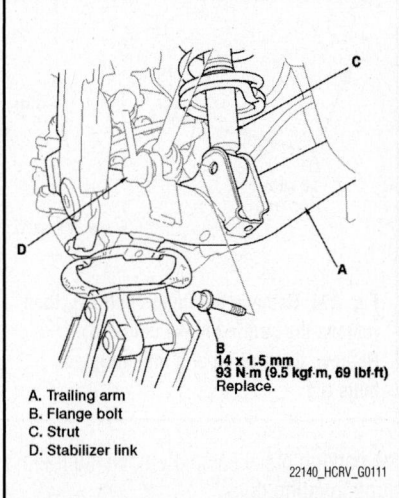

A. Trailing arm
B. Flange bolt
C. Strut
D. Stabilizer link

Fig. 293 Position a floor jack under the trailing arm and raise it until the suspension begins to compress. Disconnect the stabilizer link from the trailing arm. Remove the flange bolt from the bottom of the strut

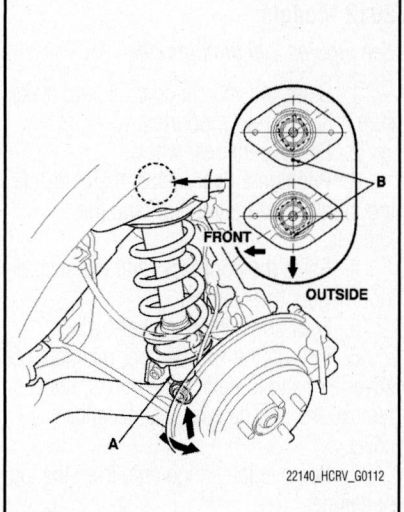

Fig. 294 Position the strut assembly (A) between the body and the trailing arm. Note the direction of the strut mounting base and the hook or the paint mark (B)

the hook or the paint mark on it is toward the outside of the vehicle.

14. Loosely install new flange nuts on the top of the strut.

15. Loosely install new trailing arm mounting bolts.

16. Install the parking brake cable, then install the flange bolt.

17. Position a floor jack under the trailing arm. Raise the floor jack until the hole in the trailing arm aligns with the hole in the strut.

18. Loosely install a new flange bolt on the bottom of the strut.

19. Connect the stabilizer link to the trailing arm.

20. Raise the rear suspension with a floor jack to load the suspension with the vehicle's weight.

21. Tighten new flange nuts, new flange bolts, and other fasteners to the specified torque values.

    a. Flange nuts and bolts: 54 ft. lbs. (74 Nm).

    b. Parking brake cable bolt: 16 ft. lbs. (22 Nm).

    c. New flange bolt at bottom of strut: 69 ft. lbs. (93 Nm).

    d. Brake hose bracket bolts: 16 ft. lbs. (22 Nm).

22. Install the strut cover lid.

23. Clean the mating surface of the brake disc and the inside of the wheel, then install the rear wheel.

24. Check the wheel alignment and adjust it if necessary.

2012 Models

See Figures 295 through 298.

1. Raise the vehicle on a lift, and make sure it is securely supported.

2. Remove the rear wheel.

3. Position a floor jack at the connecting point of the trailing arm and the knuckle.

4. Raise the floor jack until the suspension begins to compress.

5. Fold down the rear seat-back.

6. Pull up the rear damper maintenance cover by hand to detach the hooks, then remove the rear damper maintenance cover.

7. Remove the flange nuts from the top of damper.

8. Remove the flange nut while holding the respective joint pin with a hex wrench. Disconnect the stabilizer link from the trailing arm.

9. Remove the brake hose from the brake hose mounting bracket.

10. Remove the flange bolt, then remove

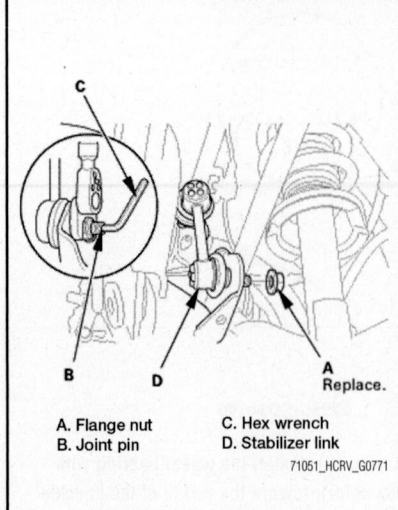

A. Flange nut   C. Hex wrench
B. Joint pin   D. Stabilizer link

71051_HCRV_G0771

Fig. 295 Remove the flange nut while holding the respective joint pin with a hex wrench. Disconnect the stabilizer link from the trailing arm

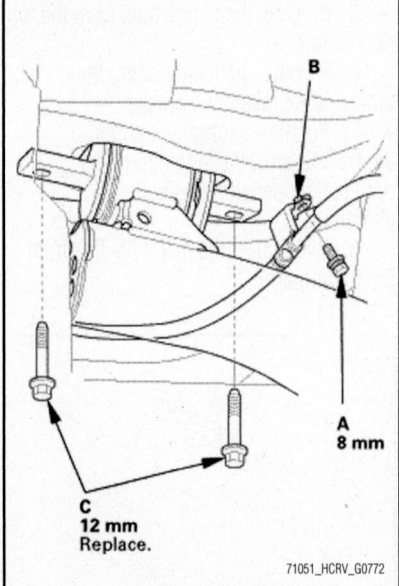

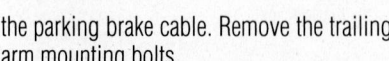

71051_HCRV_G0772

Fig. 296 Remove the flange bolt (A), then remove the parking brake cable (B). Remove the trailing arm mounting bolts (C)

the parking brake cable. Remove the trailing arm mounting bolts.

11. Remove the flange bolt, then remove the parking brake cable. Remove the trailing arm mounting bolts.

➡**Note the damper mounting bolt direction for installation in the same direction.**

12. Lower the rear suspension, then

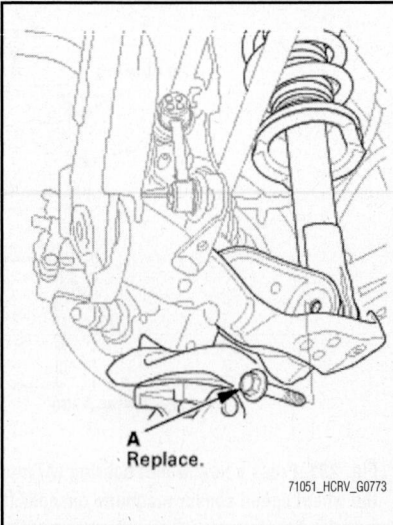

71051_HCRV_G0773

Fig. 297 Remove the flange bolt (A), then remove the parking brake cable (B). Remove the trailing arm mounting bolts (C)

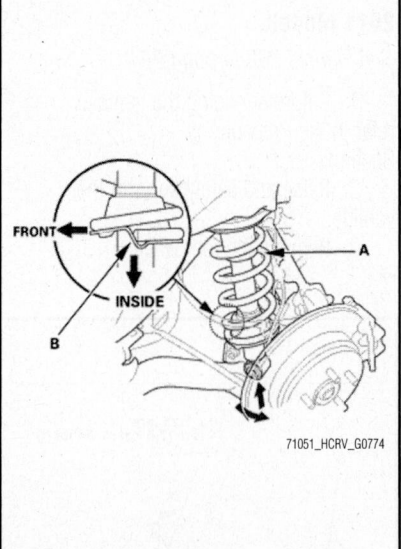

71051_HCRV_G0774

Fig. 298 Position the damper/spring (A) between the body and the trailing arm. Note the direction of the stepped part (B) of the lower spring seat on it is toward the inside of the vehicle

remove the damper and spring assembly from the vehicle.

To install:

13. Position the damper/spring between the body and the trailing arm. Note the direction of the stepped part of the lower spring seat on it is toward the inside of the vehicle.

14. Position a floor jack at the connect-

ing point of the trailing arm and the knuckle. Raise the floor jack until the suspension begins to compress.

15. Loosely install the new damper mounting bolt in the same direction as removal.

16. Loosely install new trailing arm mounting bolts. Install the parking brake cable. Install the parking brake cable mounting bolt.

17. Connect the stabilizer link to the trailing arm. Install the new flange nut and tighten them to 28 ft. lbs. (38 Nm) while holding the respective joint pin with a hex wrench.

18. Loosely install new flange nuts to the top of the damper.

19. Raise the rear suspension with the jack until the vehicle just lifts off of the safety stands.

20. Tighten the damper mounting bolt to 69 ft. lbs. (93 Nm).

21. Tighten the flange nuts to 54 ft. lbs. (74 Nm).

22. Fold down the rear seat-back. Install the rear damper maintenance cover. Make sure the hooks snap into place securely.

23. Install the rear wheel.

24. Install the remaining components in reverse of the removal procedure.

25. Check and adjust rear wheel alignment.

26. Perform the following with the HDS:
  • VSA sensor neutral position memorization
  • Steering angle sensor neutral position clear

HUB BEARING UNIT

REMOVAL & INSTALLATION

See Figures 299 through 302.

1. Raise and support the vehicle.
2. Remove the wheel nuts and the rear wheel.
3. Remove the brake hose bracket mounting bolt from the knuckle.
4. Remove the brake caliper bracket mounting bolts, and remove the caliper assembly from the knuckle. To prevent damage to the caliper assembly or brake hose, use a short piece of wire to hang the caliper assembly from the undercarriage. Do not twist the brake hose excessively.
5. Remove the two washers.
6. 4WD model: Pry up the stake on the spindle nut, then remove the nut.
7. Release the parking brake, and remove the brake disc/drum.
8. 4WD model: Remove the flange

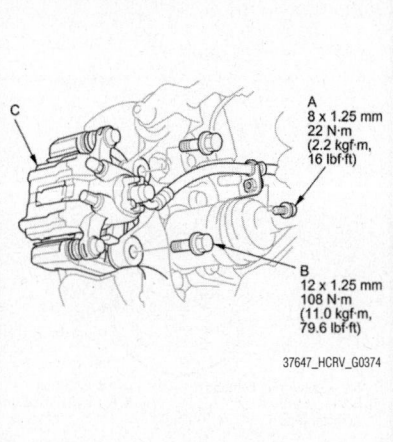

Fig. 299 Remove the brake hose bracket mounting bolt (A), the brake caliper bracket mounting bolts (B), and the caliper assembly (C)

bolts, and remove the hub bearing unit by tapping the driveshaft end with a soft face hammer while drawing the hub bearing unit outward.

9. 2WD model: Remove the hub bearing unit and the O-ring.

10. Check the hub bearing unit for damage and cracks.

11. Install the hub bearing unit in the reverse order of removal, and note these items:
  • 2WD model: Use a new O-ring on reassembly.
  • 4WD model: Use a new spindle nut on reassembly.
  • 4WD model: Before installing the spindle nut, apply a small amount of engine oil to the seating surface of the nut. After tightening, use a drift to stake the spindle nut shoulder against the driveshaft.

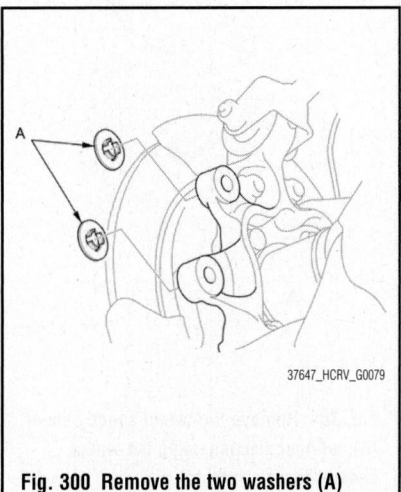

Fig. 300 Remove the two washers (A)

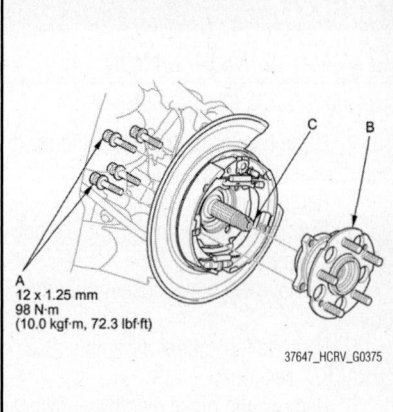

Fig. 301 4WD model: Remove the flange bolts (A), and remove the hub bearing unit (B) by tapping the driveshaft end (C) with a soft face hammer while drawing the hub bearing unit outward

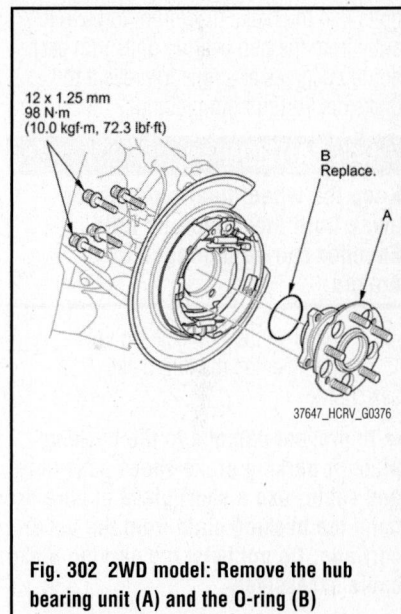

Fig. 302 2WD model: Remove the hub bearing unit (A) and the O-ring (B)

  • Before installing the brake disc/drum, clean the mating surface of the hub bearing unit and the inside of the brake disc/drum.
  • Before installing the wheel, clean the mating surface of the brake disc/drum and the inside of the wheel.
  • After installation, press the brake pedal several times to make sure the brakes work.

12. Check the wheel alignment, and adjust it if necessary.

ADJUSTMENT

The front wheel bearings are not adjustable. If the bearings are noisy or become loose, they must be replaced.

KNUCKLE

REMOVAL & INSTALLATION

See Figures 303 through 306.

1. Raise the vehicle on a lift, and make sure it is securely supported.

2. Remove the rear wheel.

3. Pry up the stake on the spindle nut, then remove the nut.

4. Remove the brake hose mounting bolt.

5. Remove the caliper assembly. See "BRAKES" section.

6. Remove the brake disc/drum flathead screws. Remove the brake disc/drum. Remove the adjuster plug, then if necessary, turn the adjuster bolt with a flat-tip screwdriver until the shoes become loose.

7. If the brake disc/drum is stuck to the hub bearing unit, thread two 8 x 1.25 mm bolts into the brake disc/drum to push it away from the hub bearing unit. Turn each bolt 90 degrees at a time to prevent the brake disc/drum from binding.

❊❊ CAUTION

Keep the wheel bearing/encoder away from any magnetic field. Encoder can be damaged or erased.

8. Remove the hub bearing unit.

9. Remove the parking brake assembly.

➡**To prevent damage to the backing plate or parking brake shoes assembly and cable, use a short piece of wire to hang the backing plate from the undercarriage. Do not twist the parking brake cable excessively.**

10. Remove the wheel speed sensor without disconnecting the wheel speed sensor connector.

11. Place a floor jack under the trailing arm to support it.

12. Remove the flange bolt. Disconnect the upper arm from the knuckle.

13. Mark the cam positions of the adjusting bolt with the frame. Mark the cam positions of the adjusting cam plate with the frame. Remove the self-locking nut, the adjusting cam plate, and the adjusting bolt. Remove the flange bolt. Disconnect the knuckle from the trailing arm.

14. Remove the hub and the brake hose mounting bracket, if needed.

To install:

15. If removed, install the brake hose

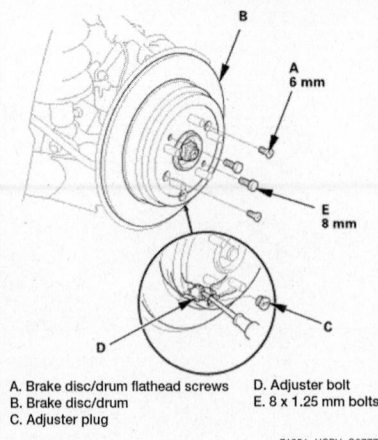

A. Brake disc/drum flathead screws
B. Brake disc/drum
C. Adjuster plug
D. Adjuster bolt
E. 8 x 1.25 mm bolts

71051_HCRV_G0777

Fig. 303 Remove the brake disc/drum flathead screws. Remove the brake disc/drum. Remove the adjuster plug, then if necessary, turn the adjuster bolt with a flat-tip screwdriver until the shoes become loose. If the brake disc/drum is stuck to the hub bearing unit, thread two 8 x 1.25 mm bolts into the brake disc/drum to push it away from the hub bearing unit. Turn each bolt 90 degrees at a time to prevent the brake disc/drum from binding

mounting bracket. Tighten the bolts to 16 ft. lbs. (22 Nm).

16. Connect the knuckle to the trailing arm.

17. Loosely install the new flange bolt. Loosely install the new self-locking nut, the adjusting cam plate, and the new adjusting bolt.

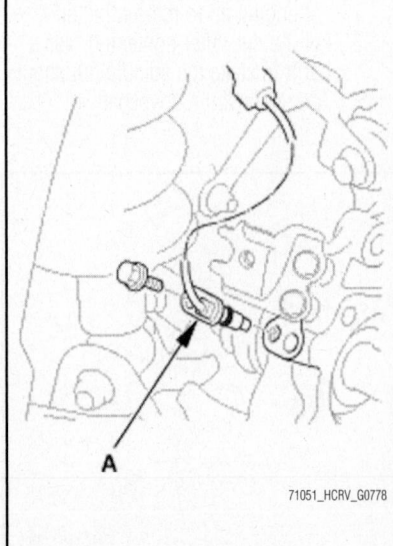

71051_HCRV_G0778

Fig. 304 Remove the wheel speed sensor (A), without disconnecting the wheel speed sensor connector

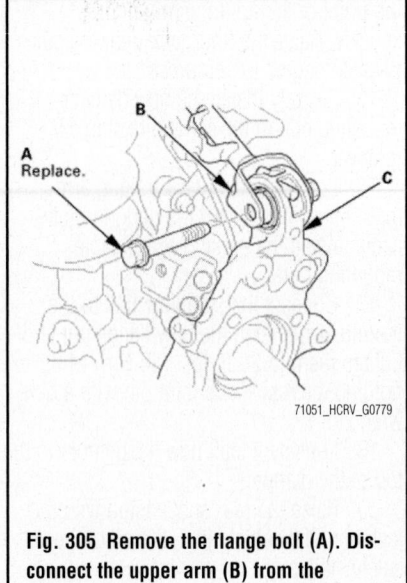

71051_HCRV_G0779

Fig. 305 Remove the flange bolt (A). Disconnect the upper arm (B) from the knuckle (C)

18. Align the cam positions of the adjusting bolt and the adjusting cam plate with the marked positions when tightening the self-locking nut.

19. Connect the upper arm to the knuckle. Loosely install the new flange bolt.

20. Raise the rear suspension with a floor jack to load the vehicle weight.

21. Tighten all mounting hardware to the specified torque:

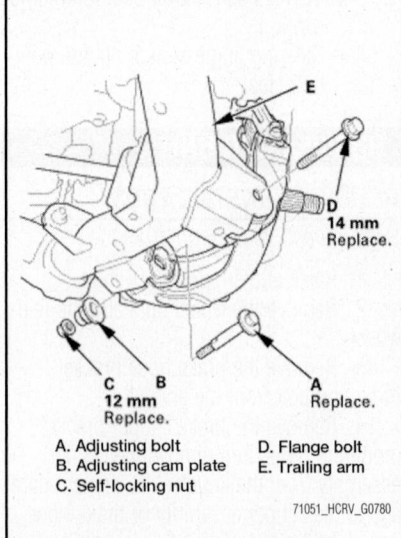

A. Adjusting bolt
B. Adjusting cam plate
C. Self-locking nut
D. Flange bolt
E. Trailing arm

71051_HCRV_G0780

Fig. 306 Mark the cam positions of the adjusting bolt with the frame. Mark the cam positions of the adjusting cam plate with the frame. Remove the self-locking nut, the adjusting cam plate, and the adjusting bolt. Remove the flange bolt. Disconnect the knuckle from the trailing arm

- Flange bolt: 51 ft. lbs. (69 Nm)
- Self-locking nut on cam plate: 42 ft. lbs. (57 Nm)
- Upper arm to knuckle flange bolt: 69 ft. lbs. (93 Nm)

22. Install the wheel speed sensor. Tighten the screw to 7.2 ft. lbs. (9.8 Nm).

23. Install the parking brake assembly. Tighten the nuts to 101 ft. lbs. (137 Nm).

✳✳ CAUTION

Keep the wheel bearing/encoder away from any magnetic field. Encoder can be damaged or erased.

24. Install the hub bearing unit. Tighten the retaining bolts to 72 ft. lbs. (98 Nm).

25. Install the brake disc. Tighten the flathead screws to 7 ft. lbs. (9.5 Nm).

➡**Before installing the brake disc, clean the mating surfaces between the hub bearing unit and the inside of the brake disc.**

26. Install the caliper assembly, then install the brake hose mounting bolt. See "BRAKES" section.

27. Apply a small amount of engine oil to the seating surface of a new spindle nut. Install the spindle nut, then tighten it to the following:
- With dual-pump system: 181 ft. lbs. (245 Nm)
- With AWD: 133 ft. lbs. (181 Nm)

28. After tightening, use a drift to stake the spindle nut shoulder against the driveshaft.

29. Install the rear wheel.

30. Install the remaining components in reverse of the removal procedure.

31. Check and adjust rear wheel alignment.

32. Perform the following with the HDS:
- VSA sensor neutral position memorization
- Steering angle sensor neutral position clear

STABILIZER BAR

REMOVAL & INSTALLATION

2011 Models

See Figure 307.

1. Raise and support the vehicle.
2. Remove the rear wheels.
3. Disconnect both stabilizer links from the stabilizer bar.

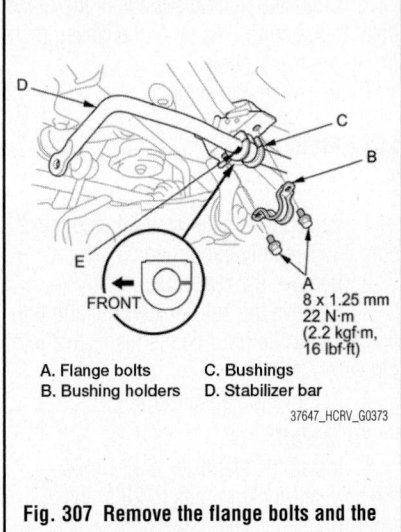

A. Flange bolts   C. Bushings
B. Bushing holders   D. Stabilizer bar

37647_HCRV_G0373

Fig. 307 Remove the flange bolts and the bushing holders, then remove the bushings and the stabilizer bar

4. Remove the flange bolts and the bushing holders, then remove the bushings and the stabilizer bar.

To install:

5. Install the stabilizer bar in the reverse order of removal, and note these items:
- Note the right and left direction of the stabilizer bar.
- Align the paint marks (E) on the stabilizer bar with the sides of the bushings.
- Note the fore/aft direction of the bushings.
- Refer to the stabilizer link removal/installation to connect the stabilizer bar to the links.
- Clean the mating surface of the brake disc/drum and the inside of the wheel, then install the rear wheel.

6. Check the wheel alignment, and adjust it if necessary.

2012 Models

See Figures 308 and 309.

1. Raise the vehicle on a lift, and make sure it is securely supported.
2. Remove both rear wheels.
3. Remove the self-locking nut while holding the respective joint pin with a hex wrench. Disconnect stabilizer link from the stabilizer bar. See "Stabilizer Link" in this section.
4. With dual pump system, do the following:
    a. Remove the bushing holders.
    b. Remove the bushings.
    c. Remove the stabilizer bar.

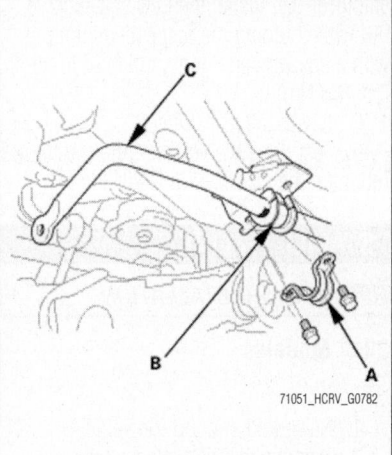

71051_HCRV_G0782

Fig. 308 With dual pump system, do the following: Remove the bushing holders (A), remove the bushings (B), then remove the stabilizer bar (C)

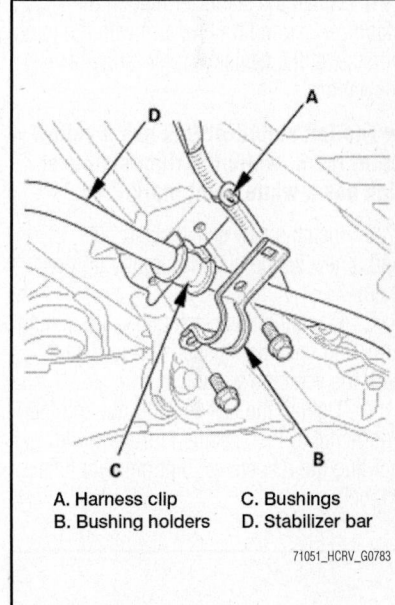

A. Harness clip   C. Bushings
B. Bushing holders   D. Stabilizer bar

71051_HCRV_G0783

Fig. 309 With Real Time AWD system, do the following: Remove the harness clip from the bushing holders, remove the bushings, then remove the stabilizer bar

5. With Real Time AWD system, do the following:
    a. Remove the harness clip from the bushing holders.
    b. Remove the bushings.
    c. Remove the stabilizer bar.

To install:

6. Install the stabilizer bar. Ensure the bushing slit is toward the rear of the vehicle. Tighten the bushing bracket bolts to 16 ft. lbs. (22 Nm).

7. Connect both stabilizer link to the

stabilizer bar. Install the new self-locking nut while holding the respective joint pin with a hex wrench. Tighten the nut to 28 ft. lbs. (38 Nm).

8. Install the front wheel.

9. Test drive the vehicle. After 5 minutes, retighten the stabilizer link self-locking.

STABILIZER LINK

REMOVAL & INSTALLATION

2011 Models

See Figure 310.

1. Raise and support the vehicle.
2. Remove the rear wheel.
3. Remove the self-locking nut and the flange nut while holding the respective joint pin with a hex wrench, then remove the stabilizer link.

To install:

4. Install the stabilizer link on the stabilizer bar and trailing arm with the joint pins set at the center of their range of movement.

➡ **The left stabilizer link has a yellow paint mark, while the right stabilizer link has a white paint mark.**

5. Install a new self-locking nut and a new flange nut, and lightly tighten them.

6. Place the floor jack under the trailing arm, and raise the suspension to load it with the vehicle's weight.

7. Tighten the self-locking nut and the flange nut to the specified torque while holding the respective joint pin with a hex wrench.

8. Clean the mating surface of the brake disc/drum and the inside of the wheel, then install the rear wheel.

9. Test-drive the vehicle.

2012 Models

See Figure 311.

1. Raise the vehicle on a lift, and make sure it is securely supported.
2. Remove the rear wheel.
3. Remove the self-locking nut and the flange nut while holding the respective joint pin with a hex wrench.
4. Remove the stabilizer link.

To install:

5. Install the stabilizer link. Tighten the new self-locking nuts to 28 ft. lbs. (38 Nm).

6. Install the rear wheels.

7. Test drive the vehicle. After 5 minutes, retighten the stabilizer link self-locking.

TRAILING ARM

REMOVAL & INSTALLATION

2011 Models

See Figures 312 through 314.

1. Raise and support the vehicle.
2. Remove the rear wheel.
3. Remove the knuckle.
4. Place the floor jack under the trailing arm to support it.
5. Remove the flange nut, while holding the joint pin with a hex wrench, and disconnect the stabilizer link from the trailing arm.

➡ **During installation, install a new flange nut.**

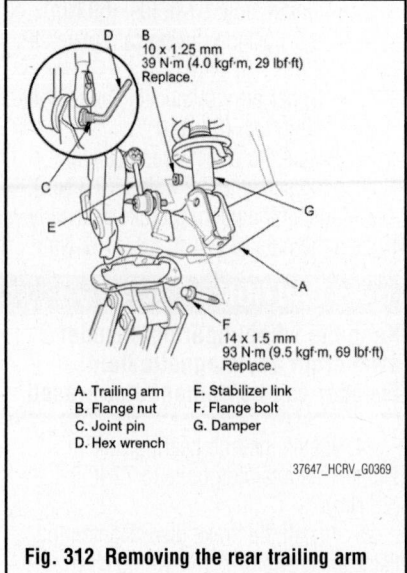

A. Trailing arm E. Stabilizer link
B. Flange nut   F. Flange bolt
C. Joint pin   G. Damper
D. Hex wrench

B 10 x 1.25 mm 39 N·m (4.0 kgf·m, 29 lbf·ft) Replace.

F 14 x 1.5 mm 93 N·m (9.5 kgf·m, 69 lbf·ft) Replace.

37647_HCRV_G0369

Fig. 312 Removing the rear trailing arm

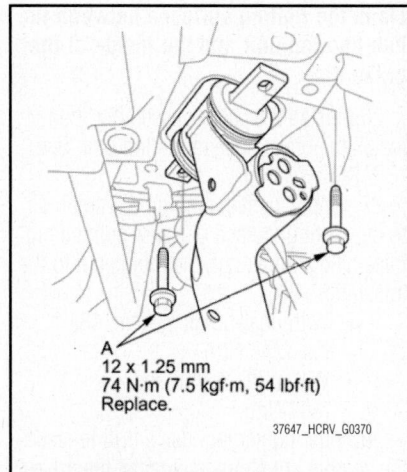

A 12 x 1.25 mm 74 N·m (7.5 kgf·m, 54 lbf·ft) Replace.

37647_HCRV_G0370

Fig. 313 Remove the trailing arm front mounting bolts (A)

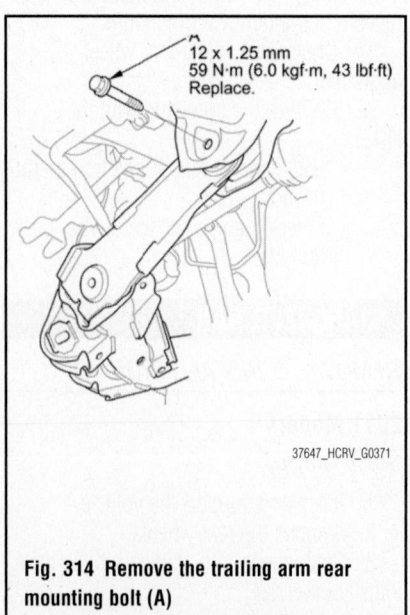

A 12 x 1.25 mm 59 N·m (6.0 kgf·m, 43 lbf·ft) Replace.

37647_HCRV_G0371

Fig. 314 Remove the trailing arm rear mounting bolt (A)

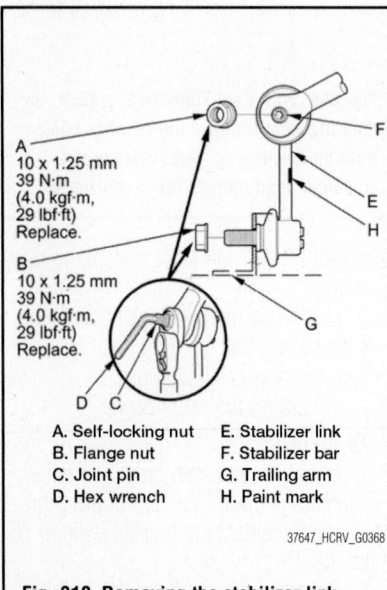

A 10 x 1.25 mm 39 N·m (4.0 kgf·m, 29 lbf·ft) Replace.

B 10 x 1.25 mm 39 N·m (4.0 kgf·m, 29 lbf·ft) Replace.

A. Self-locking nut E. Stabilizer link
B. Flange nut   F. Stabilizer bar
C. Joint pin   G. Trailing arm
D. Hex wrench   H. Paint mark

37647_HCRV_G0368

Fig. 310 Removing the stabilizer link

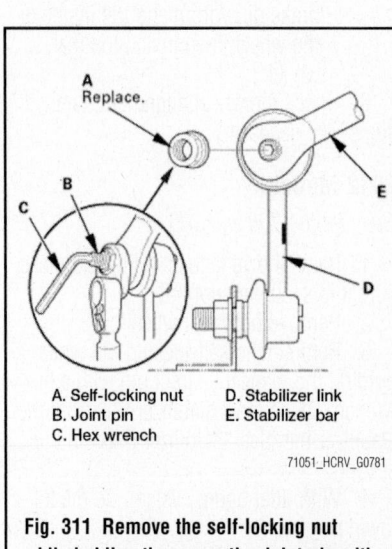

A Replace.

A. Self-locking nut D. Stabilizer link
B. Joint pin   E. Stabilizer bar
C. Hex wrench

71051_HCRV_G0781

Fig. 311 Remove the self-locking nut while holding the respective joint pin with a hex wrench. Disconnect stabilizer link from the stabilizer bar

6. Remove the flange bolt, and disconnect the damper from the trailing arm.

7. Remove the trailing arm front mounting bolts.

8. Remove the trailing arm rear mounting bolt.

9. Lower the jack, and remove the trailing arm.

To install:

10. Install the trailing arm in the reverse order of removal, and note these items:

- First install all the suspension components, and lightly tighten the bolts and nuts, then place a jack under the trailing arm, and raise the suspension to load it with the vehicle's weight before fully tightening the bolts and nuts to the specified torque.
- Use new flange bolts during reassembly.
- Before installing the wheel, clean the mating surface of the brake disc/drum and the inside of the wheel.

11. Check the wheel alignment, and adjust it if necessary.

2012 Models

See Figures 315 through 318.

1. Raise the vehicle on a lift, and make sure it is securely supported.

2. Remove the rear wheel.

3. Position a floor jack at the connecting point of the trailing arm and the knuckle.

4. Raise the floor jack until the suspension begins to compress.

5. Remove the damper lower mounting bolt.

6. Mark the cam positions of the adjusting bolt with the frame. Mark the cam positions of the adjusting cam plate with the frame. Remove the self-locking nut, the adjusting cam plate, and the adjusting bolt. Remove the flange bolt. Disconnect the knuckle from the trailing arm.

7. Remove the flange nut while holding the respective joint pin with a hex wrench. Disconnect the stabilizer link from the trailing arm.

8. Remove the trailing arm front mounting bolts.

9. Remove the trailing arm rear mounting bolt. Lower the jack, and remove the trailing arm.

To install:

10. Install the trailing arm and loosely install new mounting bolts.

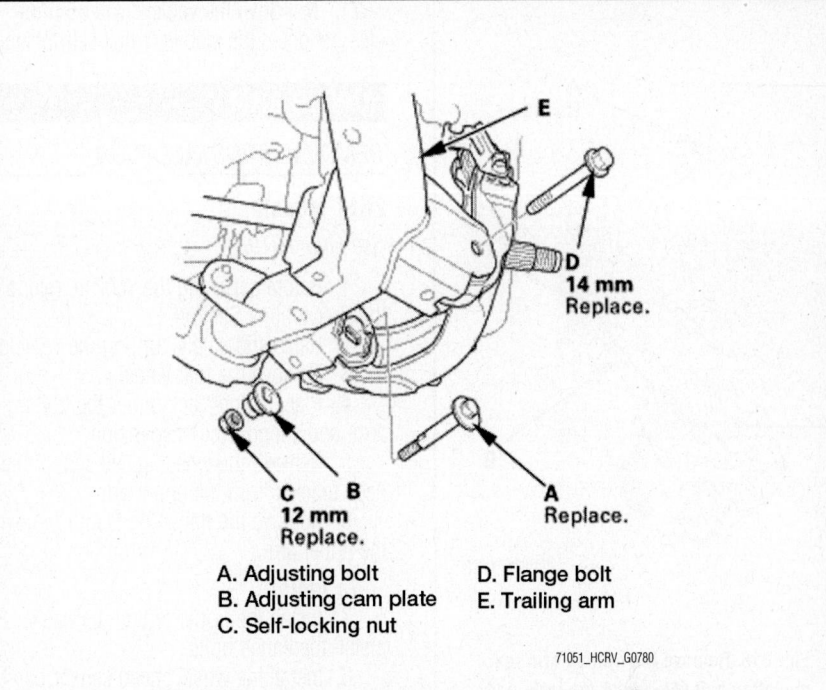

A. Adjusting bolt   D. Flange bolt
B. Adjusting cam plate E. Trailing arm
C. Self-locking nut

71051_HCRV_G0780

Fig. 315 Mark the cam positions of the adjusting bolt with the frame. Mark the cam positions of the adjusting cam plate with the frame. Remove the self-locking nut, the adjusting cam plate, and the adjusting bolt. Remove the flange bolt. Disconnect the knuckle from the trailing arm

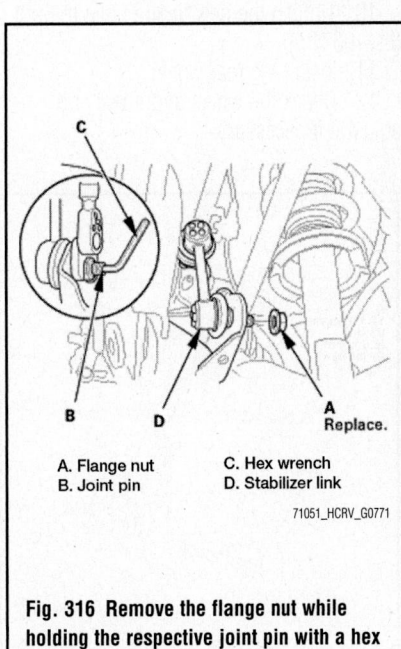

A. Flange nut   C. Hex wrench
B. Joint pin   D. Stabilizer link

71051_HCRV_G0771

Fig. 316 Remove the flange nut while holding the respective joint pin with a hex wrench. Disconnect the stabilizer link from the trailing arm

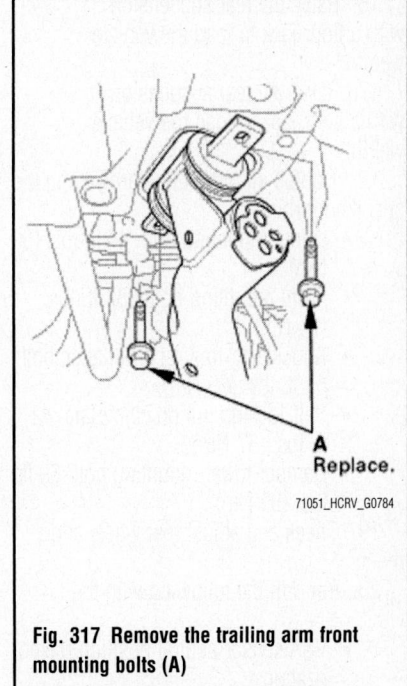

71051_HCRV_G0784

Fig. 317 Remove the trailing arm front mounting bolts (A)

11. Reconnect the stabilizer link to the trailing arm. Tighten the nut to 28 ft. lbs. (38 Nm).

12. Connect the knuckle to the trailing arm.

13. Loosely install the new flange bolt. Loosely install the new self-locking nut, the

adjusting cam plate, and the new adjusting bolt.

14. Align the cam positions of the adjusting bolt and the adjusting cam plate with the marked positions when tightening the self-locking nut.

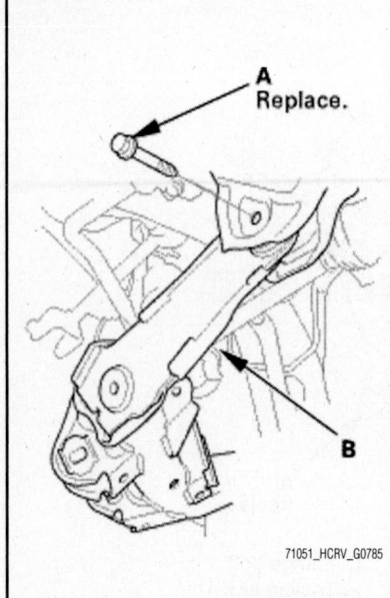

Fig. 318 Remove the trailing arm rear mounting bolt (A). Lower the jack, and remove the trailing arm (B)

15. Connect the upper arm to the knuckle. Loosely install the new flange bolt.

16. Raise the rear suspension with a floor jack to load the vehicle weight.

17. Raise the rear suspension with a floor jack to load the vehicle weight.

18. Tighten all mounting hardware to the specified torque:
- Upper mounting bolt: 44 ft. lbs. (59 Nm)
- Front mounting bolts: 54 ft. lbs. (74 Nm)
- Knuckle-to-trailing arm flange bolt: 51 ft. lbs. (69 Nm)
- Self-locking nut on cam plate: 42 ft. lbs. (57 Nm)
- Damper lower mounting bolt: 69 ft. lbs. (93 Nm)

19. Check and adjust rear wheel alignment.

20. Perform the following with the HDS:
- VSA sensor neutral position memorization
- Steering angle sensor neutral position clear

21. Test drive the vehicle. After 5 minutes, retighten the stabilizer link self-locking.

UPPER ARM

REMOVAL & INSTALLATION

2011 Models

See Figure 319.

1. Before servicing the vehicle, refer to the Precautions Section.

2. Raise and safely support the vehicle.

3. Remove the rear wheel.

4. Place a floor jack under the trailing arm, and support the suspension.

5. Remove the wheel speed sensor harness bracket from the upper arm.

6. Remove the flange bolts and remove the upper arm.

To install:

7. Install the upper arm and loosely install the flange bolts.

8. Install the wheel speed sensor harness bracket to the upper arm.

9. Place a floor jack under the trailing arm and raise the suspension to load it with the vehicle weight before tightening the flange bolts.

10. Tighten the new flange bolts to 69 ft. lbs. (93 Nm).

11. Install the rear wheel.

12. Check the wheel alignment and adjust it if necessary.

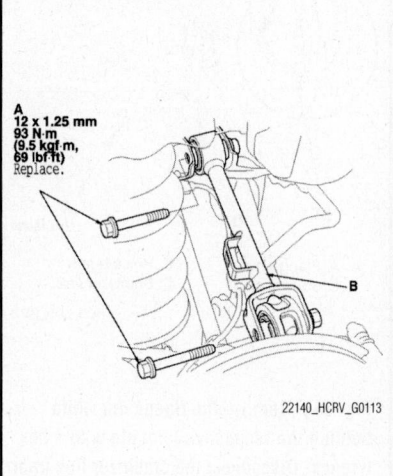

Fig. 319 Remove the flange bolts (A) and remove the upper arm (B)

2012 Models

See Figure 320.

1. Raise the vehicle on a lift, and make sure it is securely supported.

2. Remove the rear wheel.

3. Position a floor jack at the connecting point of the trailing arm and the knuckle.

4. Raise the floor jack until the suspension begins to compress.

5. Remove the wheel speed sensor harness bracket from the upper arm.

6. Remove the flange bolts and remove the upper arm.

To install:

7. Position the upper arm loosely install the retaining bolts.

8. Install the wheel speed sensor harness bracket.

9. Raise the rear suspension with the floor jack to load vehicle weight.

10. Tighten the upper arm bolts to 68 ft. lbs. (93 Nm).

11. Install the rear wheel.

WHEEL BEARINGS

REMOVAL & INSTALLATION

➡ **The rear wheel hub and bearing unit are an integral assembly. If wheel bearing is defective, the unit must be replaced as an assembly.**

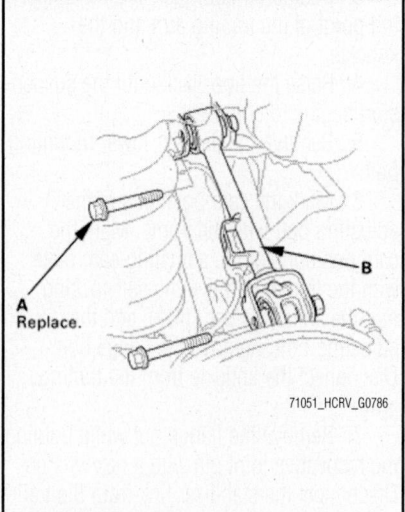

Fig. 320 Remove the flange bolts (A), and remove the upper arm (B)

BRAKES 10-8

**ANTI-LOCK BRAKE SYSTEM
(ABS)** **10-8**
General Information 10-8
  Precautions 10-8
Wheel Speed Sensors 10-8
  Removal & Installation 10-8
**BLEEDING THE BRAKE
SYSTEM** **10-9**
Bleeding Procedure 10-9
FRONT DISC BRAKES **10-10**
Brake Calipers 10-10
  Removal & Installation 10-10
Brake Pads 10-10
  Removal & Installation 10-10
PARKING BRAKE **10-13**
Adjustments 10-13
Cables 10-13
  Removal & Installation 10-13
REAR DISC BRAKES **10-11**
Brake Calipers 10-11
  Removal & Installation 10-11
Brake Pads 10-11
  Removal & Installation 10-11

CHASSIS ELECTRICAL **10-14**

**AIR BAG (SUPPLEMENTAL
RESTRAINT SYSTEM)** **10-14**
Arming the System 10-14
Disarming the System 10-14
General Information 10-14
  Service Precautions 10-14

DRIVELINE **10-15**

Clutch Hydraulic System
  Bleeding 10-16
  Bleeding Procedure 10-16
Clutch Master Cylinder 10-16
  Removal & Installation 10-16
Clutch Slave (Release)
  Cylinder 10-17
  Removal & Installation 10-17
CVT Transaxle Fluid 10-15
  Drain & Refill 10-15
  Filter Replacement 10-15

Fluid Level Check 10-15
Fluid Recommendations 10-16
Front Halfshaft 10-18
  Removal & Installation 10-18
Rear Axle Beam 10-19
  Removal & Installation 10-19
Rear Differential Carrier Final
  Driven Gear 10-19
  Removal & Installation 10-19

ENGINE COOLING **10-20**

Condenser & Radiator
  Fans 10-20
  Removal & Installation 10-20
Engine Coolant 10-21
  Drain & Refill 10-21
  Fluid Recommendations 10-21
Radiator 10-21
  Removal & Installation 10-21
Thermostat 10-22
  Removal & Installation 10-22
Water Pump 10-22
  Removal & Installation 10-22

ENGINE ELECTRICAL **10-23**

BATTERY SYSTEM **10-23**
Battery 10-23
  Battery Cable Reset 10-23
  Charging 10-23
  Removal & Installation 10-23
Battery Cables 10-23
  Disconnection &
  Reconnection 10-23
IGNITION SYSTEM **10-24**
Firing Orders 10-24
Ignition Coils & Spark Plugs .. 10-24
  Removal & Installation 10-24
Ignition Timing 10-24
  Inspection & Adjustment 10-24
Spark Plugs 10-24
  Removal & Installation 10-24
**INTEGRATED MOTOR
ASSIST (IMA)** **10-25**
Direct Current/Direct Current
  Converter 10-26
  Removal & Installation 10-26

High Voltage Circuit 10-25
  Disabling & Enabling
  Power to the High Voltage
  Circuit 10-25
IMA Motor 10-27
  Removal & Installation 10-27
IMA Motor Power Cable 10-26
  Disconnecting the
  IMA Motor Power Cable
  Connector FROM
  the Motor Stator 10-26
Intelligent Power Unit 10-29
  Removal & Installation 10-29
Precautions 10-25
STARTING SYSTEM **10-32**
Starter 10-32
  Removal & Installation 10-32

ENGINE MECHANICAL **10-32**

Accessory Drive Belt
  System 10-32
  Adjustment 10-32
  Inspection 10-32
  Removal & Installation 10-32
Air Cleaner 10-33
  Removal & Installation 10-33
Camshaft & Bearings 10-33
  Removal & Installation 10-33
Crankshaft Pulley 10-34
  Removal & Installation 10-34
Crankshaft Rear Cover &
  Seal 10-35
  Removal & Installation 10-35
Cylinder Head 10-35
  Removal & Installation 10-35
Cylinder Head (Valve) Cover ... 10-40
  Removal & Installation 10-40
Engine Oil & Filter 10-36
  OIL & Filter Change 10-36
  OIL Level Check 10-36
Exhaust Manifold 10-37
  Removal & Installation 10-37
Front Cover Seal (Cam Chain
  Case Seal) 10-37
  Removal & Installation 10-37
Intake Manifold 10-37
  Removal & Installation 10-37

Oil Pan 10-38
 Removal & Installation........10-38
Oil Pump 10-39
 Removal,
  Overhaul & Installation.....10-39
Rear Main Seal 10-40
 Removal & Installation........10-40
Rocker Arm Assembly 10-40
 Removal & Installation........10-40
Timing (Camshaft) Chain
 Cover, Chain, &
 Sprockets 10-41
 Removal & Installation........10-41
Valve Clearance 10-45
 Adjustment 10-45

**ENGINE PERFORMANCE &
EMISSION CONTROLS 10-46**

Air-Fuel Ratio Sensor 10-46
 Location 10-46
 Removal & Installation........10-46
Camshaft Position Sensor 10-46
 Location 10-46
 Removal & Installation........10-46
Crankshaft Position
 Sensor 10-46
 Removal & Installation........10-46
Electronic Control
 Module/Powertrain Control
 Module (ECM/PCM).............10-46
 ECM/PCM IDLE Learn
  Procedure 10-48
 ECM/PCM Updating 10-47
 Removal & Installation........10-46
Engine Coolant Temperature
 Sensor 10-48
 Location 10-48
 Removal & Installation........10-48
EVAP Canister 10-49
 Location 10-49
 Removal & Installation........10-48
Exhaust Gas Recirculation
 (EGR) Valve 10-49
 Location 10-49
 Removal & Installation........10-49
Knock Sensor (KS) 10-50
 Location 10-50
 Removal & Installation........10-50
Maintenance Reminder Lights
 (Maintenance Minder)...........10-50
 Description 10-50
 Reset Procedure..................10-50

Mass Air Flow/Intake Air
 Temperature(MAF/IAT) 10-49
 Location 10-49
 Removal & Installation........10-49
Positive Crankcase
 Ventilation (PCV) Valve.........10-51
 Location 10-51
 Removal & Installation........10-51
Vehicle Speed Sensor 10-51
 Location 10-51
 Removal & Installation........10-51

FUEL 10-52

**GASOLINE FUEL INJECTION
SYSTEM 10-52**
Fuel Filter 10-53
 Removal & Installation........10-53
Fuel Injectors & Fuel Rail.......10-55
 Removal & Installation........10-55
Fuel Pump/Fuel Gauge
 Sending Unit 10-54
 Removal & Installation........10-54
Fuel System Service
 Precautions 10-52
Fuel Tank...............................10-55
 Draining 10-55
 Removal & Installation........10-55
Fuel Tank Unit 10-53
 Removal & Installation........10-53
Idle Speed 10-56
 Inspection 10-56
Relieving Fuel System
 Pressure 10-52
Throttle Body 10-56
 Removal & Installation........10-56

**HEATING & AIR CONDITIONING
SYSTEM 10-57**

Blower Unit 10-57
 Removal & Installation........10-57
Heater Unit & Heater Core.......10-58
 Removal & Installation........10-58

PRECAUTIONS 10-8

**SPECIFICATIONS AND
MAINTENANCE CHARTS.....10-3**

Brake Specifications.................10-6
Camshaft Specifications...........10-4
Capacities 10-3

Crankshaft and Connecting
 Rod Specifications 10-4
Engine & Model Year
 Identification 10-3
Engine Tune-Up Specifications...10-3
Fluid Specifications..................10-4
General Engine Specifications...10-3
Piston and Ring Specifications...10-5
Scheduled Maintenance
 Intervals 10-7
Tire Wheel and Ball Joint
 Specifications 10-6
Torque Specifications...............10-5
Valve Specifications 10-4
Wheel Alignment.....................10-6

STEERING 10-59

Control Unit 10-59
 Removal & Installation........10-59
EPS Motor.............................10-59
 Removal & Installation........10-59
Steering Gear 10-59
 Removal & Installation........10-59

SUSPENSION 10-62

FRONT SUSPENSION 10-62
Damper (Strut) & Spring.........10-66
 Removal & Installation........10-66
Knuckle/Hub & Wheel
 Bearing 10-62
 Removal & Installation........10-62
Lower Ball Joints....................10-64
 Removal & Installation........10-64
Lower Control Arms 10-65
 Removal & Installation........10-65
Stabilizer Bar 10-66
 Removal & Installation........10-66
Stabilizer Links......................10-66
 Removal & Installation........10-66
REAR SUSPENSION 10-68
Axle Beam 10-68
 Removal & Installation........10-68
Coil Springs...........................10-68
 Removal & Installation........10-68
Dynamic Damper....................10-69
 Removal & Installation........10-69
Hub & Bearing Unit................10-70
 Removal & Installation........10-70
Shock Absorbers (Dampers) ...10-70
 Removal & Installation........10-70
Stabilizer Bushing 10-71
 Removal & Installation........10-71

SPECIFICATIONS AND MAINTENANCE CHARTS

ENGINE AND VEHICLE IDENTIFICATION

Engine							Model Year	
Code	Liters (cc)	Cu. In.	Cyl.	Fuel Sys.	Engine Type	Eng. Mfg.	Code ①	Year
LEA1	1.5 (1496)	91.00	I4	MPFI	SOHC	Honda	B	2011
							C	2012

MPFI: Multi-Point Fuel Injection

DOHC: Double Overhead Camshafts

① 10th digit of VIN

71051_HCRZ_C0001

GENERAL ENGINE SPECIFICATIONS

All measurements are given in inches.

Year	Model	Engine Displacement Liters	Engine Series ID	Net Horsepower @ rpm	Net Torque @ rpm (ft. lbs.)	Bore x Stroke (in.)	Com- pression Ratio	Oil Pressure @ rpm
2011	CR-Z	1.5	LEA1	122@6000	128@1000	2.87 x 3.50	10.4:1	①
2012	CR-Z	1.5	LEA1	122@6000	128@1000	2.87 x 3.50	10.4:1	①

NA: Not Available

① 10 psi (69 kPa) @ idle

49.8 psi (343 kPa) @ 3000 RPM

71051_HCRZ_C0002

GASOLINE ENGINE TUNE-UP SPECIFICATIONS

Year	Engine Displacement Liters	Engine ID	Spark Plug Gap (in.)	Ignition Timing (deg.) MT	AT	Fuel Pump (psi)	Idle Speed (rpm) MT	AT	Valve Clearance In.	Ex.
2011	1.5	LEA1	0.047-0.051	①	①	47-54	②	②	HYD	HYD
2012	1.5	LEA1	0.047-0.051	①	①	47-54	②	②	HYD	HYD

NOTE: The Vehicle Emission Control Information label reflects specification changes made during production.

Follow the figures on the label if they differ from those in this chart.

HYD: Hydraulic

① Ignition timing should be 10 ±2 ° BTDC; refer to Inspection procedure in service information.

② Idle speed should be 750 ± 50 rpm, but is maintained by the Electronic Control Module (ECM).

71051_HCRZ_C0003

CAPACITIES

Year	Model	Engine Displacement Liters	Engine VIN	Engine Oil with Filter (qts.)	Transmission (pts.) Manual	CVT ①	Fuel Tank (gal.)	Cooling System (qts.)
2011	CR-Z	1.5	LEA1	3.8	3.0	5.6	10.6	②
2012	CR-Z	1.5	LEA1	3.8	3.0	5.6	10.6	②

NOTE: All capacities are approximate. Add fluid gradually and check to be sure a proper fluid level is obtained.

① Drain and refill

② Drain and refill with M/T: 4.68 qts.

Drain and refill with CVT: 4.6 qts.

71051_HCRZ_C0004

FLUID SPECIFICATIONS

Year	Model	Engine Displacement Liters	Engine ID	Engine Oil	Auto. Trans.	Manual Trans.	Brake Master Cylinder
2011	CR-Z	1.5	LEA1	0W-20	CVTF	MTF	②
2012	CR-Z	1.5	LEA1	0W-20	CVTF	MTF	②

DOT: Department Of Transportation

② DOT 3 Brake Fluid

71051_HCRZ_C0005

VALVE SPECIFICATIONS

Year	Engine Displacement Liters	Engine ID	Seat Angle (deg.)	Face Angle (deg.)	Stem-to-Guide Clearance (in.) Intake	Stem-to-Guide Clearance (in.) Exhaust	Stem Diameter (in.) Intake	Stem Diameter (in.) Exhaust
2011	1.5	LEA1	45	45	0.0008-0.0020	0.0020-0.0031	0.2157-0.2161	0.2146-0.2150
2012	1.5	LEA1	45	45	0.0008-0.0020	0.0020-0.0031	0.2157-0.2161	0.2146-0.2150

71051_HCRZ_C0006

CAMSHAFT AND BEARING SPECIFICATIONS CHART

All measurements are given in inches.

Year	Engine Displ. Liters	Engine ID	Journal Dia.	Brg. Oil Clearance	Shaft End-play	Runout	Lobe Height Intake	Lobe Height Exhaust
2011	1.5	LEA1	NA	0.0018-0.0033	0.0020-0.0980	0.0012	①	②
2012	1.5	LEA1	NA	0.0018-0.0033	0.0020-0.0980	0.0012	①	②

NA: Not Available

① Intake Primary: 1.37578 inch

   Intake Secondary: 1.20193 inch

② Exhaust: 1.40212 inch

71051_HCRZ_C0007

CRANKSHAFT AND CONNECTING ROD SPECIFICATIONS

All measurements are given in inches.

Year	Engine Displacement Liters	Engine ID	Crankshaft Main Brg. Journal Dia.	Crankshaft Main Brg. Oil Clearance	Crankshaft Shaft End-play	Crankshaft Thrust on No.	Connecting Rod Journal Diameter	Connecting Rod Oil Clearance	Connecting Rod Side Clearance
2011	1.5	LEA1	1.9676-1.9685	0.00063-0.0013	0.0039-0.0138	—	1.5739-1.5748	0.0008-0.0015	0.006-0.0130
2012	1.5	LEA1	1.9676-1.9685	0.00063-0.0013	0.0039-0.0138	—	1.5739-1.5748	0.0008-0.0015	0.006-0.0130

71051_HCRZ_C0008

PISTON AND RING SPECIFICATIONS
All measurements are given in inches.

Year	Engine Displ. Liters	Engine ID	Piston Clearance	Ring Gap			Ring Side Clearance		
				Top Compression	Bottom Compression	Oil Control	Top Compression	Bottom Compression	Oil Control
2011	1.5	LEA1	0.0004-0.0014	0.006-0.011	①	0.008-0.027	②	0.0118-0.0217	NA
2012	1.5	LEA1	0.0004-0.0014	0.006-0.011	①	0.008-0.027	②	0.0118-0.0217	NA

NA: Not Available

① Riken: 0.012-0.016 in.

Nippon: 0.014-0.019 in.

② Riken: 0.0256-0.0354 in.

Nippon: 0.024-0.0354 in.

71051_HCRZ_C0009

TORQUE SPECIFICATIONS
All readings in ft. lbs.

Year	Engine Displacement Liters (ID)	Cylinder Head Bolts	Main Bearing Bolts	Rod Bearing Bolts	Crankshaft Damper Bolts	Driveplate Bolts (A/T)	Flywheel Bolts (M/T)	Intake Manifold	Spark Plugs	Oil Pan Drain Plug
2011	1.5 (LEA1)	①	②	③	④	32	NA	⑤	13	30
2012	1.5 (LEA1)	①	②	③	④	32	NA	⑤	13	30

NA: Not Available

① Step 1: 22 ft. lbs.

Step 2: Plus 130 degrees

② Step 1: 18 ft. lbs.

Step 2: Plus 40 degrees

③ Step 1: 7 ft. lbs.

Step 2: Plus 90 degrees

④ NEW BOLT:

Step 1: 130 ft. lbs.

Step 2: Loosen bolt

Step 3: 29 ft. lbs.

Step 4: Plus 94 degrees

USED BOLT:

Step 1: 27 ft. lbs.

Step 2: Plus 90 degrees

⑤ Bolt/nut torques vary by location; refer to "Intake Manifold" in "Engine Mechanical" service information.

71051_HCRZ_C0010

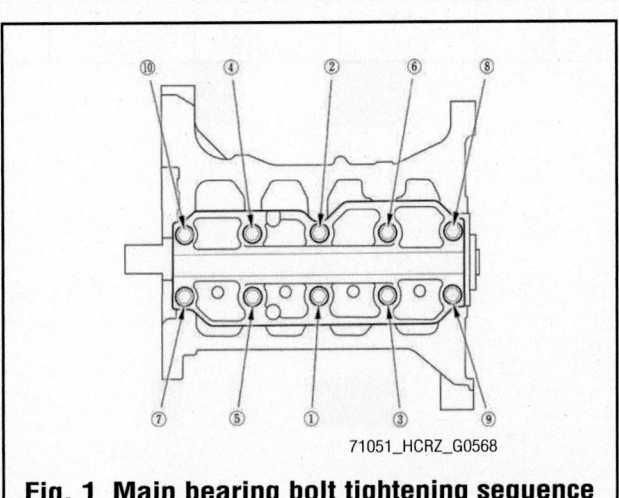

71051_HCRZ_G0568

Fig. 1 Main bearing bolt tightening sequence

WHEEL ALIGNMENT

Year	Model		Caster Range (+/-Deg.)	Caster Preferred Setting (Deg.)	Camber Range (+/-Deg.)	Camber Preferred Setting (Deg.)	Toe-in (mm)
2011	CR-Z	Front	1°	3° 22'	1°	0.0	0 ± 3
		Rear	—	—	1°	-1.5	3 +2/-3
2012	CR-Z	Front	1°	3° 22'	1°	0.0	0 ± 3
		Rear	—	—	1°	-1.5	3 +2/-3

71051_HCRZ_C0011

TIRE, WHEEL AND BALL JOINT SPECIFICATIONS

Year	Model	OEM Tires Standard	OEM Tires Optional	Tire Pressures (psi) Front	Tire Pressures (psi) Rear	Wheel Size	Ball Joint Inspection	Lug Nut Torque (ft. lbs.)
2011	CR-Z	P195/55R16	P205/45R17	①	①	②	③	80
2012	CR-Z	P195/55R16	P205/45R17	①	①	②	③	80

① Refer to placard on vehicle for proper inflation pressure.

② Wheel size varies according to tires selected.

③ Replace if any measurable movement is found.

71051_HCRZ_C0012

BRAKE SPECIFICATIONS

All measurements in inches unless noted

Year	Model		Brake Disc Original Thickness	Brake Disc Minimum Thickness	Brake Disc Maximum Runout	Brake Pads Original Thickness	Brake Pads Minimum Thickness	Brake Caliper Bracket Bolts (ft. lbs.)	Brake Caliper Mounting Bolts (ft. lbs.)
2011	CR-Z	Front	0.827	0.748	0.0016	0.394	0.063	80	—
		Rear	0.354	0.315	0.0016	0.354	0.063	41	—
2012	CR-Z	Front	0.827	0.748	0.0016	0.394	0.063	80	—
		Rear	0.354	0.315	0.0016	0.354	0.063	41	—

71051_HCRZ_C0013

MAINTENANCE MINDER SCHEDULE
Honda CR-Z

All Honda's displays engine oil life and maintenance service items in the information display to indicate when to perform maintenance service. If the engine oil life is 15% or less, based on the onboard computer's caluculations, you will see SERVICE DUE SOON in the information display every time the ignition key is turned to ON. The maintenance minder indicator will also come on and the maintenance code(s) for other scheduled maintenance items needing service will be displayed below the message.

Symbol	Item	Service
A	Engine oil ①	Change
B	Engine oil and filter	Change
	Fluid levels	Inspect
	Brakes	Inspect
	Parking brake adjustment	Check
	Steering gear and linkage	Inspect
	Suspension components	Inspect
	Driveshaft boots	Inspect
	Brake hoses and lines	Inspect
	Exhaust system	Inspect
	Fuel lines and connections	Inspect
1	Tires	Rotate
2	Engine air filter ②	Replace
	Dust and pollen filter ③	Replace
	Accessory drive belt	Inspect
3	CVT or MT fluid	Replace
4	Spark plugs	Replace
	Valve clearance ④	Inspect
5	Engine coolant	Replace

① If the message SERVICE DUE NOW does not appear more than 12 months after the display is reset, change every year.

② If driven in dusty conditions, replace every 15,000 miles.

③ If driven in urban areas that have a high concentration of soot from industry and diesel, replace every 15,000 miles

⑥ Adjust if necessary.

Additionally, replace the brake fluid every 3 years, and inspect the idle speed every 160,000 miles.
To reset the Engine Oil Life Display:
1. Turn the ignition switch to ON.
2. Press the SELECT button repeatedly until the engine oil life display or the service message is displayed.
3. Press the RESET button for about 10 seconds. You will see a MAINT RESET message.
4. Select the appropriate answer, MAINT RESET >N (NO) or MAINT RESET > y (YES) by pressing the SELECT button repeatedly.
    >N or >Y is displayed on the outside temperature >N or >Y is displayed on the outside temperature display.
5. Select the MAINT RESET > Y (YES), and press and hold the RESET button again to reset the engine oil life to 100%.

PRECAUTIONS

Before servicing any vehicle, please be sure to read all of the following precautions, which deal with personal safety, prevention of component damage, and important points to take into consideration when servicing a motor vehicle:

• Never open, service or drain the radiator or cooling system when the engine is hot; serious burns can occur from the steam and hot coolant.

• Observe all applicable safety precautions when working around fuel. Whenever servicing the fuel system, always work in a well-ventilated area. Do not allow fuel spray or vapors to come in contact with a spark, open flame, or excessive heat (a hot drop light, for example). Keep a dry chemical fire extinguisher near the work area. Always keep fuel in a container specifically designed for fuel storage; also, always properly seal fuel containers to avoid the possibility of fire or explosion. Refer to the additional fuel system precautions later in this section.

• Fuel injection systems often remain pressurized, even after the engine has been turned **OFF**. The fuel system pressure must be relieved before disconnecting any fuel lines. Failure to do so may result in fire and/or personal injury.

• Brake fluid often contains polyglycol ethers and polyglycols. Avoid contact with the eyes and wash your hands thoroughly after handling brake fluid. If you do get brake fluid in your eyes, flush your eyes with clean, running water for 15 minutes. If eye irritation persists, or if you have taken brake fluid internally, IMMEDIATELY seek medical assistance.

• The EPA warns that prolonged contact with used engine oil may cause a number of skin disorders, including cancer. You should make every effort to minimize your exposure to used engine oil. Protective gloves should be worn when changing oil. Wash your hands and any other exposed skin areas as soon as possible after exposure to used engine oil. Soap and water, or waterless hand cleaner should be used.

• All new vehicles are now equipped with an air bag system, often referred to as a Supplemental Restraint System (SRS) or Supplemental Inflatable Restraint (SIR) system. The system must be disabled before performing service on or around system components, steering column, instrument panel components, wiring and sensors. Failure to follow safety and disabling procedures could result in accidental air bag deployment, possible personal injury and unnecessary system repairs.

• Always wear safety goggles when working with, or around, the air bag system. When carrying a non-deployed air bag, be sure the bag and trim cover are pointed away from your body. When placing a non-deployed air bag on a work surface, always face the bag and trim cover upward, away from the surface. This will reduce the motion of the module if it is accidentally deployed. Refer to the additional air bag system precautions later in this section.

• Clean, high quality brake fluid from a sealed container is essential to the safe and proper operation of the brake system. You should always buy the correct type of brake fluid for your vehicle. If the brake fluid becomes contaminated, completely flush the system with new fluid. Never reuse any brake fluid. Any brake fluid that is removed from the system should be discarded. Also, do not allow any brake fluid to come in contact with a painted surface; it will damage the paint.

• Never operate the engine without the proper amount and type of engine oil; doing so WILL result in severe engine damage.

• Timing belt maintenance is extremely important. Many models utilize an interference-type, non-freewheeling engine. If the timing belt breaks, the valves in the cylinder head may strike the pistons, causing potentially serious (also time-consuming and expensive) engine damage. Refer to the maintenance interval charts for the recommended replacement interval for the timing belt, and to the timing belt section for belt replacement and inspection.

• Disconnecting the negative battery cable on some vehicles may interfere with the functions of the on-board computer system(s) and may require the computer to undergo a relearning process once the negative battery cable is reconnected.

• When servicing drum brakes, only disassemble and assemble one side at a time, leaving the remaining side intact for reference.

• Only an MVAC-trained, EPA-certified automotive technician should service the air conditioning system or its components.

BRAKES

GENERAL INFORMATION

PRECAUTIONS

• Certain components within the ABS system are not intended to be serviced or repaired individually.

• Do not use rubber hoses or other parts not specifically specified for and ABS system. When using repair kits, replace all parts included in the kit. Partial or incorrect repair may lead to functional problems and require the replacement of components.

• Lubricate rubber parts with clean, fresh brake fluid to ease assembly. Do not use shop air to clean parts; damage to rubber components may result.

• Use only DOT 3 brake fluid from an unopened container.

• If any hydraulic component or line is removed or replaced, it may be necessary to bleed the entire system.

• A clean repair area is essential. Always clean the reservoir and cap thoroughly before removing the cap. The slightest amount of dirt in the fluid may plug an orifice and impair the system function. Perform repairs after components have been thoroughly cleaned; use only denatured alcohol to clean components. Do not allow ABS components to come into contact with any substance containing mineral oil; this includes used shop rags.

• The Anti-Lock control unit is a microprocessor similar to other computer units in the vehicle. Ensure that the ignition switch is **OFF** before removing or installing controller harnesses. Avoid static electricity discharge at or near the controller.

ANTI-LOCK BRAKE SYSTEM (ABS)

• If any arc welding is to be done on the vehicle, the control unit should be unplugged before welding operations begin.

WHEEL SPEED SENSORS

REMOVAL & INSTALLATION

Front

See Figure 3.

1. Turn the ignition switch to LOCK (0).
2. Remove the grommet, then disconnect the wheel speed sensor 2P connector.
3. Remove the bolt and the bracket, the clip and the wire guide grommet.
4. Remove the bolt and the wheel speed sensor.

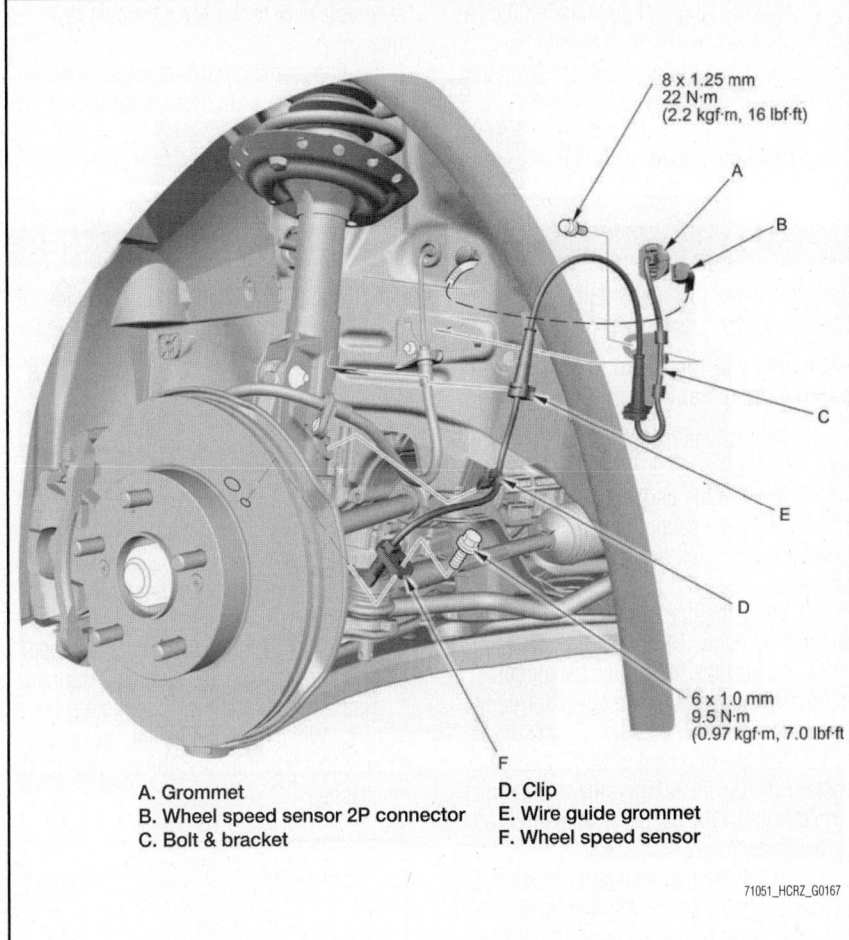

8 x 1.25 mm
22 N·m
(2.2 kgf·m, 16 lbf·ft)

6 x 1.0 mm
9.5 N·m
(0.97 kgf·m, 7.0 lbf·ft

A. Grommet
B. Wheel speed sensor 2P connector
C. Bolt & bracket

D. Clip
E. Wire guide grommet
F. Wheel speed sensor

71051_HCRZ_G0167

Fig. 3 Remove the grommet, then disconnect the wheel speed sensor 2P connector. Remove the bolt and the bracket, the clip and the wire guide grommet. Remove the bolt and the wheel speed sensor.

To install:

5. Install the wheel speed sensor in the reverse order of removal, and note these items:

   a. Do not twist the sensor wires.

   b. If the wheel speed sensor comes in contact with the hub bearing, it is faulty.

   c. Make sure there is no debris in the sensor mounting hole.

6. Start the engine, and make sure the VSA indicators go off.

7. Test-drive the vehicle, and make sure the VSA indicators do not come on.

Rear

1. Turn the ignition switch to LOCK (0).

2. Release the connector holding clamps, then disconnect the wheel speed sensor 2P connector.

3. Remove the wire guide grommets and the clip.

4. Remove the bolt and the wheel speed sensor.

➡ **No illustration is available for this component.**

To install:

5. Install the wheel speed sensor in the reverse order of removal, and note these items:

- Do not twist the sensor wires.
- If the wheel speed sensor comes in contact with the hub bearing unit, it is faulty.
- Make sure there is no debris in the sensor mounting hole.

6. Start the engine, and make sure the VSA indicators go off.

7. Test-drive the vehicle, and make sure the VSA indicators do not come on.

BRAKES

BLEEDING THE BRAKE SYSTEM

BLEEDING PROCEDURE

See Figures 4 and 5.

Note the following:

- Do not reuse the drained fluid. Use only new Honda DOT 3 Brake Fluid from an unopened container. Using a non-Honda brake fluid can cause corrosion and shorten the life of the system.
- Make sure no dirt or other foreign matter gets in the brake fluid.
- The reservoir tank connected to the master cylinder must be at the MAX (upper) level mark at the start of the bleeding procedure and checked after bleeding each wheel location. Add fluid as required.

1. Make sure the brake fluid level in the reservoir tank is at the MAX (upper) level line.

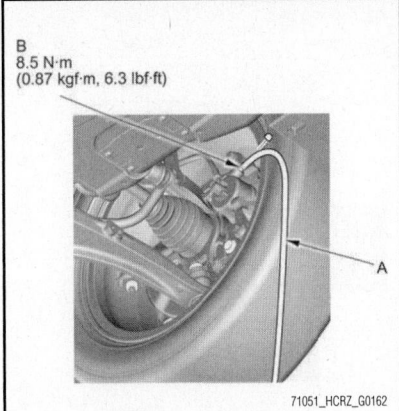

B
8.5 N·m
(0.87 kgf·m, 6.3 lbf·ft)

71051_HCRZ_G0162

Fig. 4 Attach a length of clear drain tube (A) to the bleed screw (B), then loosen the bleed screw to allow air to escape from the system, then tighten the bleed screw securely—front wheels

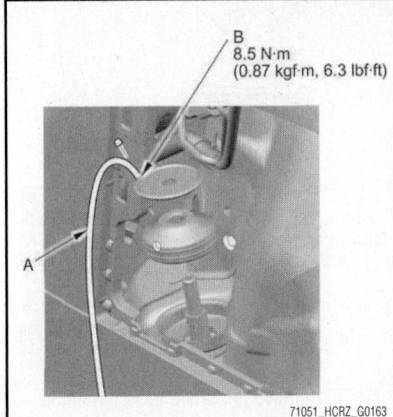

B
8.5 N·m
(0.87 kgf·m, 6.3 lbf·ft)

71051_HCRZ_G0163

Fig. 5 Attach a length of clear drain tube (A) to the bleed screw (B), then loosen the bleed screw to allow air to escape from the system, then tighten the bleed screw securely—rear wheels

2. Have someone slowly pump the brake pedal several times, then apply steady pressure.

3. Start the bleeding at the driver's side of the front brake system.

4. Attach a length of clear drain tube (A) to the bleed screw (B), then loosen the bleed screw to allow air to escape from the system. Then tighten the bleed screw securely.

5. Refill the master cylinder

reservoir tank to the MAX (upper) level line.

6. Repeat the procedure for each brake circuit until there are no air bubbles in the fluid.

BRAKES

BRAKE CALIPERS

REMOVAL & INSTALLATION

➡ **Refer to "Brake Pads" in this section.**

BRAKE PADS

REMOVAL & INSTALLATION

See Figures 6 through 9.

1. Remove some brake fluid from the master cylinder.

2. Raise and support the vehicle.

3. Remove the rear wheels.

4. Remove the parking brake cable mounting nut.

5. Remove the brake hose mounting bolt.

6. Remove the flange bolts while holding the respective caliper pins with a wrench. Be careful not to damage the pin boot, and remove the caliper. Check the

hose, the pin boots, and the parking brake cable boots for damage and deterioration.

➡ **Do not twist the brake hose and the parking brake cable to prevent damage.**

7. Remove the pad shims and the brake pads.

8. Remove the pad retainers.

9. Clean the caliper bracket thoroughly; remove any rust, and check for grooves and cracks.

10. Verify that the caliper pins move in and out smoothly. Clean and lube if needed.

11. Inspect the brake disc for runout, thickness, parallelism, and check for damage and cracks.

12. Apply a thin coat of M-77 assembly paste (P/N 08798-9010) to the retainer mating surfaces of the caliper bracket (indicated by the arrows).

13. Install the pad retainers. Wipe off the excess assembly paste from the retainers. Keep the assembly paste away from the brake disc and the brake pads.

14. Apply a thin coat of M-77 assembly

FRONT DISC BRAKES

paste (P/N 08798- 9010) to the pad side of the shims, the back of the brake pads and the other areas indicated by the arrows. Wipe off the excess assembly paste from the pad shims and the brake pads friction material. Keep grease and assembly paste away from the brake disc and the brake pads.

a. The grease between the mating surfaces of the shim and the pad must be applied lightly on one place on the inner pad and two places on the outer pad as shown in the illustration.

b. After the shim is attached, remove the shim from the pad and make sure that grease is not spread on the whole surface.

c. When the shim is attached, make sure that the shim and the pad are tightly attached.

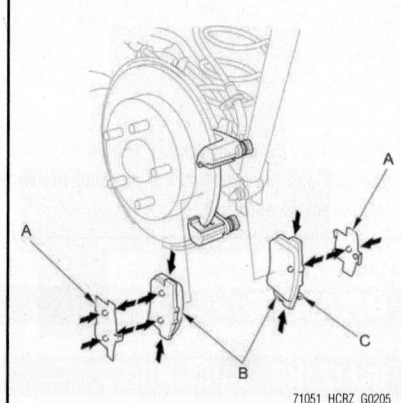

71051_HCRZ_G0205

Fig. 8 Apply a thin coat of M-77 assembly paste (P/N 08798- 9010) to the pad side of the shims (A), the back of the brake pads (B), and the other areas indicated by the arrows. Wipe off the excess assembly paste from the pad shims and the brake pads friction material. The grease between the mating surfaces of the shim and the pad must be applied lightly on one place on the inner pad and two places on the outer pad as shown in the illustration. After the shim is attached, remove the shim from the pad and make sure that grease is not spread on the whole surface. Install the brake pads and pad shims correctly. Install the brake pad with the wear indicator (C) on the bottom inside.

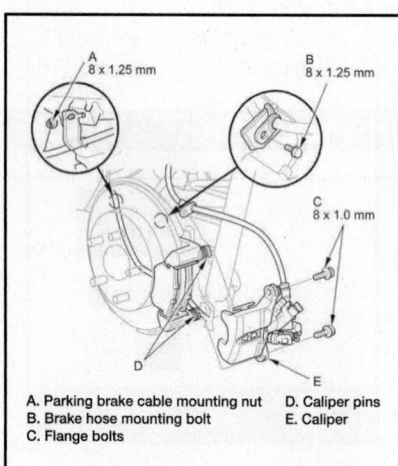

A. Parking brake cable mounting nut
B. Brake hose mounting bolt
C. Flange bolts
D. Caliper pins
E. Caliper

71051_HCRZ_G0202

Fig. 6 Remove the brake hose mounting bolt. Remove the flange bolts while holding the respective caliper pins with a wrench. Be careful not to damage the pin boot, and remove the caliper. Check the hose, the pin boots, and the parking brake cable boots for damage and deterioration.

71051_HCRZ_G0204

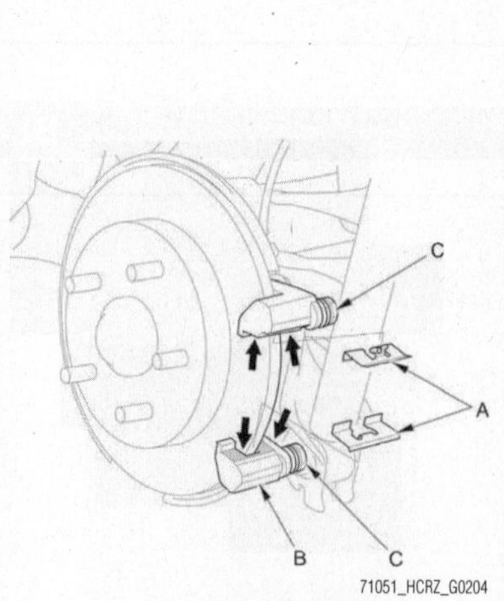

Fig. 7 Remove the pad retainers (A). Clean the caliper bracket (B) thoroughly; remove any rust, and check for grooves and cracks. Verify that the caliper pins (C) move in and out smoothly. Clean and lube if needed.

15. Install the brake pads and pad shims correctly. Install the brake pad with the wear indicator on the bottom inside. If you are reusing the brake pads, always reinstall the brake pads in their original positions to prevent a temporary loss of braking efficiency.

16. Rotate the caliper piston clockwise into the cylinder, then align the cutout in the piston with the tab on the inner pad by turning the piston back. Lubricate the boot with silicone grease to avoid twisting the piston boot. If the piston boot is twisted, back it out so it is positioned properly.

➡**Be careful when moving the piston back in the caliper; brake fluid might overflow from the master cylinder reservoir tank. If brake fluid gets on any painted surface, wash it off immediately with water.**

17. Install the caliper. Install the flange bolts and tighten it to the specified torque while holding the respective caliper pins

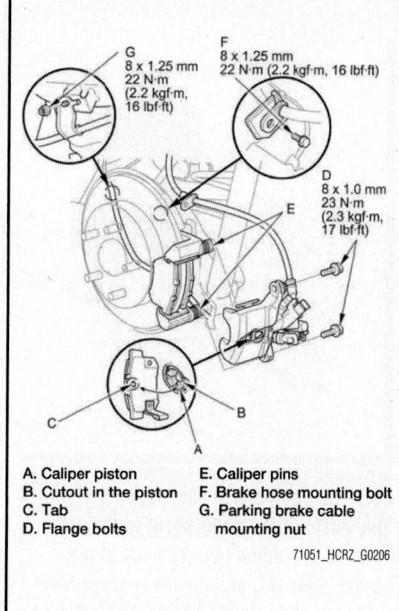

A. Caliper piston
B. Cutout in the piston
C. Tab
D. Flange bolts
E. Caliper pins
F. Brake hose mounting bolt
G. Parking brake cable mounting nut

71051_HCRZ_G0206

Fig. 9 Installing the caliper

with a wrench being careful not to damage the pin boots and parking brake cable boots.

18. Install the brake hose mounting ttbolt.

19. Install the parking brake cable mounting nut.

20. Clean the mating surfaces between the brake disc and the inside of the wheel, then install the rear wheels.

21. Press the brake pedal several times to make sure the brakes work. Engagement may require a greater pedal stroke immediately after the brake pads have been replaced as a set. Several applications of the brake pedal will restore the normal pedal stroke.

22. Add brake fluid as needed.

23. After installation, check for leaks at hose and line joints or connections, and retighten if necessary. Test-drive the vehicle, then recheck for leaks.

24. Adjust the parking brake.

BRAKES

REAR DISC BRAKES

BRAKE CALIPERS

REMOVAL & INSTALLATION

➡**Refer to "Brake Pads" in this section.**

BRAKE PADS

REMOVAL & INSTALLATION

See Figures 10 through 13.

1. Remove some brake fluid from the master cylinder.
2. Raise and support the vehicle.
3. Remove the rear wheels.
4. Remove the parking brake cable mounting nut.
5. Remove the brake hose mounting bolt.
6. Remove the flange bolts while holding the respective caliper pins with a wrench. Be careful not to damage the pin boot, and remove the caliper.
7. Check the hose, the pin boots, and the parking brake cable boots for damage and deterioration.

➡**Do not twist the brake hose and the parking brake cable to prevent damage.**

8. Remove the pad shims and the brake pads.
9. Remove the pad retainers.
10. Clean the caliper bracket thoroughly; remove any rust, and check for grooves and

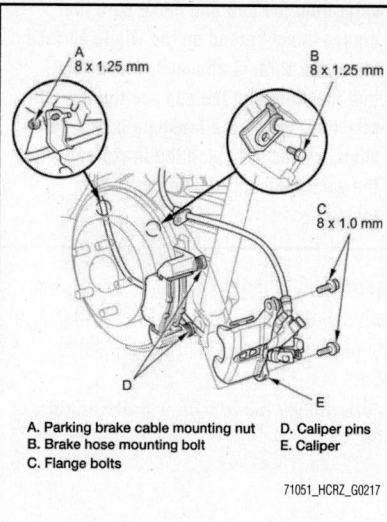

A. Parking brake cable mounting nut
B. Brake hose mounting bolt
C. Flange bolts
D. Caliper pins
E. Caliper

71051_HCRZ_G0217

Fig. 10 Remove the parking brake cable mounting nut. Remove the brake hose mounting bolt. Remove the flange bolts while holding the respective caliper pins with a wrench. Be careful not to damage the pin boot, and remove the caliper.

cracks. Verify that the caliper pins move in and out smoothly. Clean and lube if needed.

11. Inspect the brake disc for runout, thickness, parallelism, and check for damage and cracks.

12. Apply a thin coat of M-77 assembly paste (P/N 08798-9010) to the retainer mating surfaces of the caliper bracket (indicated by the arrows).

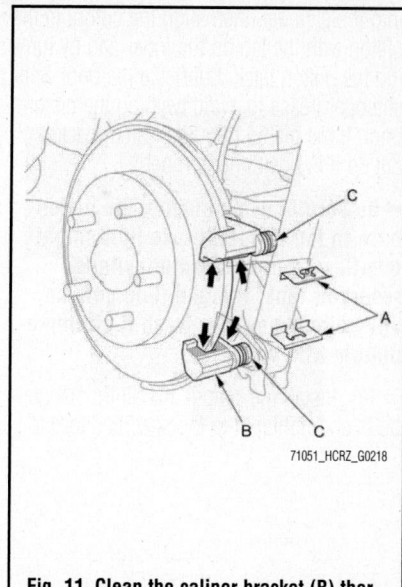

71051_HCRZ_G0218

Fig. 11 Clean the caliper bracket (B) thoroughly; remove any rust, and check for grooves and cracks. Verify that the caliper pins (C) move in and out smoothly. Clean and lube if needed.

13. Install the pad retainers. Wipe off the excess assembly paste from the retainers. Keep the assembly paste away from the brake disc and the brake pads.

14. Apply a thin coat of M-77 assembly paste (P/N 08798-9010) to the pad side of the shims, the back of the brake pads and the other areas indicated by the arrows. Wipe off the excess assembly paste from

the pad shims and the brake pads friction material. Keep grease and assembly paste away from the brake disc and the brake pads.

- Contaminated brake disc or brake pads reduce stopping ability.
- The grease between the mating surfaces of the shim and the pad must be applied lightly on one place on the inner pad and two places on the outer pad as shown in the illustration.
- After the shim is attached, remove the shim from the pad and make sure that grease is not spread on the whole surface.
- When the shim is attached, make sure that the shim and the pad are tightly attached.

15. Install the brake pads and pad shims correctly. Install the brake pad with the wear indicator on the bottom inside. If you are reusing the brake pads, always reinstall the brake pads in their original positions to prevent a temporary loss of braking efficiency.

16. Rotate the caliper piston clockwise into the cylinder, then align the cutout in the piston with the tab on the inner pad by turning the piston back. Lubricate the boot with silicone grease to avoid twisting the piston boot. If the piston boot is twisted, back it out so it is positioned properly.

➡**Be careful when moving the piston back in the caliper; brake fluid might overflow from the master cylinder reservoir tank. If brake fluid gets on any painted surface, wash it off immediately with water.**

17. Install the caliper. Install the flange bolts and tighten it to the specified torque

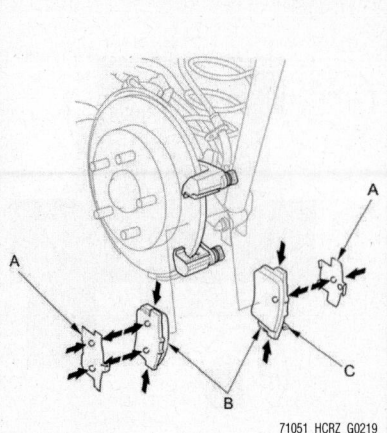

71051_HCRZ_G0219

Fig. 12 Apply a thin coat of M-77 assembly paste (P/N 08798- 9010) to the pad side of the shims (A), the back of the brake pads (B), and the other areas indicated by the arrows. The grease between the mating surfaces of the shim and the pad must be applied lightly on one place on the inner pad and two places on the outer pad as shown in the illustration. After the shim is attached, remove the shim from the pad and make sure that grease is not spread on the whole surface. When the shim is attached, make sure that the shim and the pad are tightly attached. Install the brake pads and pad shims correctly. Install the brake pad with the wear indicator (C) on the bottom inside.

while holding the respective caliper pins with a wrench being careful not to damage the pin boots and parking brake cable boots.

18. Install the brake hose mounting bolt.

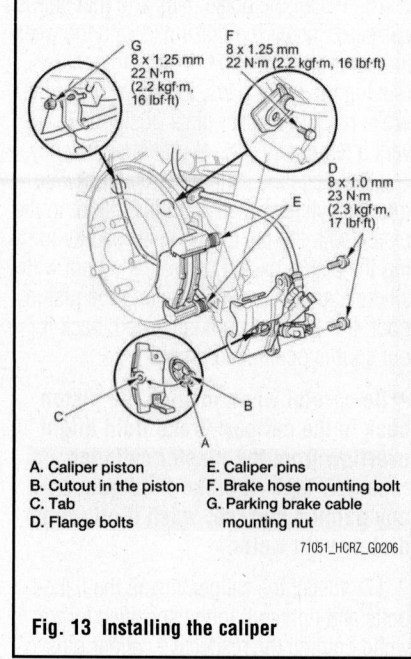

A. Caliper piston
B. Cutout in the piston
C. Tab
D. Flange bolts
E. Caliper pins
F. Brake hose mounting bolt
G. Parking brake cable mounting nut

71051_HCRZ_G0206

Fig. 13 Installing the caliper

19. Install the parking brake cable mounting nut.

20. Clean the mating surfaces between the brake disc and the inside of the wheel, then install the rear wheels.

21. Press the brake pedal several times to make sure the brakes work. Engagement may require a greater pedal stroke immediately after the brake pads have been replaced as a set. Several applications of the brake pedal will restore the normal pedal stroke.

22. Add brake fluid as needed.

23. After installation, check for leaks at hose and line joints or connections, and retighten if necessary. Test-drive the vehicle, then recheck for leaks.

24. Adjust the parking brake.

BRAKES PARKING BRAKE

ADJUSTMENTS

See Figures 14 and 15.

1. Pull the parking brake lever with 44 lbs. (196 N) of force to fully apply the parking brake. The parking brake lever should be locked within 5–7 clicks.

2. If the number of lever clicks is not as specified, adjust the parking brake.

3. After servicing the rear brake caliper, loosen the parking brake adjusting nut, start the engine, and press the brake pedal several times to set the self-adjusting brake before adjusting the parking brake.

4. Raise and support the vehicle.

5. Release the parking brake lever fully.

6. Pull out the rear console.

7. Loosen the parking brake adjusting nut.

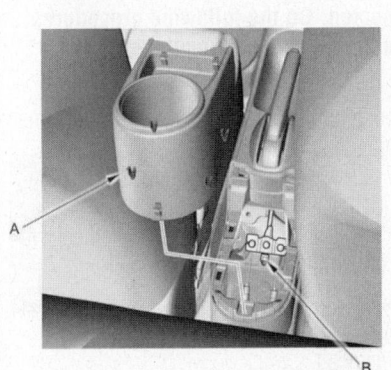

Fig. 14 Pull out the rear console (A). Loosen the parking brake adjusting nut (B).

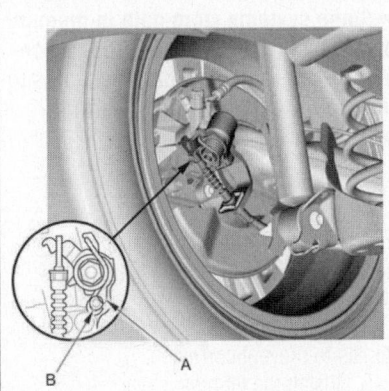

Fig. 15 Make sure the lever (A) on the rear brake caliper contacts the stop pin (B).

8. Make sure the lever on the rear brake caliper contacts the stop pin.

➡**When the parking brake lever does not contact the stop pin, loosen the adjusting nut in the center console until the parking brake lever contacts the stop pin, then start the engine, and press the brake pedal several times.**

9. Pull the parking brake lever 1 click.

10. Tighten the parking brake adjusting nut until the parking brakes drag slightly when the rear wheels are turned.

11. Release the parking brake lever fully, and check that the parking brakes do not drag when the rear wheels are turned. Readjust if necessary.

12. Make sure the parking brake lever within the specified number of clicks (5 to 7 clicks).

13. Install the trim in the reverse order of removal, and note these items:

    a. If the clips are damaged or stress-whitened, replace them with new ones.

    b. Push the clips into place securely.

CABLES

REMOVAL & INSTALLATION

See Figures 16 and 17.

✳✳ WARNING

The parking brake cables must not be bent or distorted. This will lead to stiff operation and premature cable failure.

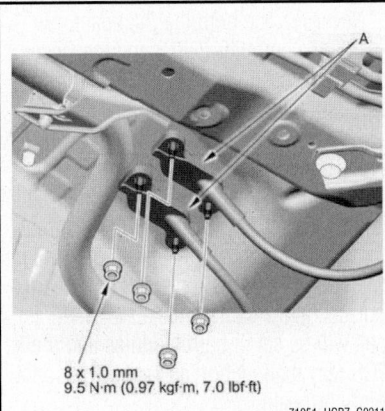

Fig. 16 Remove the parking brake cable grommets (A), and pull out the parking brake cable from the body floor.

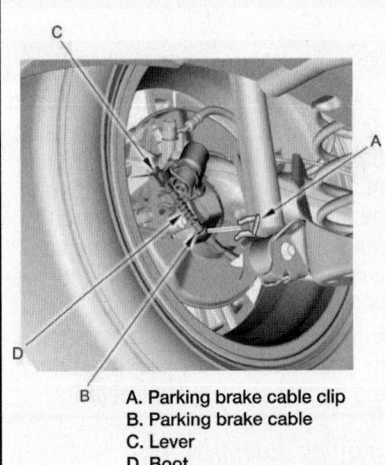

A. Parking brake cable clip
B. Parking brake cable
C. Lever
D. Boot

Fig. 17 Remove the parking brake cable clip from the parking brake cable. Disconnect the parking brake cable from the lever. Remove the parking brake cable mounting hardware, then remove the cable. Be careful not to bend or distort the cable, and boot.

1. Release the parking brake lever fully.

2. Remove the center console.

3. Loosen the parking brake adjusting nut.

4. Disconnect the parking brake cables from the equalizer.

5. Remove the middle floor undercover.

6. Remove the parking brake cable grommets and pull out the parking brake cable from the body floor.

7. Remove the parking brake cable clip from the parking brake cable.

8. Disconnect the parking brake cable from the lever.

9. Remove the parking brake cable mounting hardware, then remove the cable. Be careful not to bend or distort the cable, and boot.

To install:

10. Install the parking brake cable in the reverse order of removal, and note these items:

    a. Make sure the parking brake cable clip is fully seated on the cable housing.

    b. Adjust the parking brake.

CHASSIS ELECTRICAL
AIR BAG (SUPPLEMENTAL RESTRAINT SYSTEM)

GENERAL INFORMATION

✳✳ CAUTION

These vehicles are equipped with an air bag system. The system must be disarmed before performing service on, or around, system components, the steering column, instrument panel components, wiring and sensors. Failure to follow the safety precautions and the disarming procedure could result in accidental air bag deployment, possible injury and unnecessary system repairs.

SERVICE PRECAUTIONS

Disconnect and isolate the battery negative cable before beginning any airbag system component diagnosis, testing, removal, or installation procedures. Allow system capacitor to discharge for two minutes before beginning any component service. This will disable the airbag system. Failure to disable the airbag system may result in accidental airbag deployment, personal injury, or death.

Do not place an intact undeployed airbag face down on a solid surface. The airbag will propel into the air if accidentally deployed and may result in personal injury or death.

When carrying or handling an undeployed airbag, the trim side (face) of the airbag should be pointing AWAY FROM the body to minimize possibility of injury if accidental deployment occurs. Failure to do this may result in personal injury or death.

Replace airbag system components with OEM replacement parts. Substitute parts may appear interchangeable, but internal differences may result in inferior occupant protection. Failure to do so may result in occupant personal injury or death.

Wear safety glasses, rubber gloves, and long sleeved clothing when cleaning powder residue from vehicle after an airbag deployment. Powder residue emitted from a deployed airbag can cause skin irritation. Flush affected area with cool water if irritation is experienced. If nasal or throat irritation is experienced, exit the vehicle for fresh air until the irritation ceases. If irritation continues, see a physician.

Do not use a replacement airbag that is not in the original packaging. This may result in improper deployment, personal injury, or death.

The factory installed fasteners, screws and bolts used to fasten airbag components have a special coating and are specifically designed for the airbag system. Do not use substitute fasteners. Use only original equipment fasteners listed in the parts catalog when fastener replacement is required.

During, and following, any child restraint anchor service, due to impact event or vehicle repair, carefully inspect all mounting hardware, tether straps, and anchors for proper installation, operation, or damage. If a child restraint anchor is found damaged in any way, the anchor must be replaced. Failure to do this may result in personal injury or death.

Deployed and non-deployed airbags may or may not have live pyrotechnic material within the airbag inflator.

Do not dispose of driver/passenger/curtain airbags or seat belt tensioners unless you are sure of complete deployment. Refer to the Hazardous Substance Control System for proper disposal.

Dispose of deployed airbags and tensioners consistent with state, provincial, local, and federal regulations.

After any airbag component testing or service, do not connect the battery negative cable. Personal injury or death may result if the system test is not performed first.

If the vehicle is equipped with the Occupant Classification System (OCS), do not connect the battery negative cable before performing the OCS Verification Test using the scan tool and the appropriate diagnostic information. Personal injury or death may result if the system test is not performed properly.

Never replace both the Occupant Restraint Controller (ORC) and the Occupant Classification Module (OCM) at the same time. If both require replacement, replace one, then perform the Airbag System test before replacing the other.

Both the ORC and the OCM store Occupant Classification System (OCS) calibration data, which they transfer to one another when one of them is replaced. If both are replaced at the same time, an irreversible fault will be set in both modules and the OCS may malfunction and cause personal injury or death.

If equipped with OCS, the Seat Weight Sensor is a sensitive, calibrated unit and must be handled carefully. Do not drop or

handle roughly. If dropped or damaged, replace with another sensor. Failure to do so may result in occupant injury or death.

If equipped with OCS, the front passenger seat must be handled carefully as well. When removing the seat, be careful when setting on floor not to drop. If dropped, the sensor may be inoperative, could result in occupant injury, or possibly death.

If equipped with OCS, when the passenger front seat is on the floor, no one should sit in the front passenger seat. This uneven force may damage the sensing ability of the seat weight sensors. If sat on and damaged, the sensor may be inoperative, could result in occupant injury, or possibly death.

DISARMING THE SYSTEM

➡**Some systems store data in memory that is lost when the battery is disconnected. Do the following procedures before disconnecting the battery.**

1. Make sure you have the anti-theft code for the audio system or the audio-navigation unit.
2. Make sure the ignition switch is in LOCK (0).
3. Disconnect and isolate the negative cable from the battery.

➡**Always disconnect the negative battery cable from the battery first.**

4. Disconnect the positive cable from the battery.
5. Wait at least 3 minutes before proceeding with any SRS work.

ARMING THE SYSTEM

➡**Some systems store data in memory that is lost when the battery is disconnected. Do the following procedures to restore the system back to normal operation.**

1. Clean the battery terminals.
2. Test the battery.
3. Reconnect the positive cable to the battery first, then reconnect the negative cable to the battery.
4. Apply multipurpose grease to the terminals to prevent corrosion.
5. Install the terminal cover.
6. Enter the anti-theft code for the audio system or audio-navigation unit.
7. Set the clock (for vehicles without navigation).

DRIVELINE

CVT TRANSAXLE FLUID

DRAIN & REFILL

See Figure 18.

☀ WARNING

Keep all foreign particles out of the transaxle.

1. Start the engine, and warm it up to normal operating temperature (the radiator fan comes on twice).

2. Park the vehicle on level ground, and turn the engine off.

3. Raise the vehicle on a lift, and make sure it is securely supported.

4. Remove the lid for maintenance.

5. Remove the drain plug and drain the CVTF.

Note the following:

• Remove metal particles from the magnetic surface of the drain plug.

• If a cooler cleaning is necessary, refer to the CVTF cooler cleaning.

6. Reinstall the drain plug and a new sealing washer.

7. Install the lid for maintenance.

8. Remove the dipstick (yellow loop) and wipe it with a clean cloth.

9. Refill the transaxle with the recommended fluid through the dipstick tube opening to bring it to the upper mark on the dipstick. Always use Honda CVTF. Using a non-Honda CVTF can affect shift quality.

10. CVT fluid capacity is:

• 2.8 L (3.0 US qts.) at change
• 5.2 L (5.5 US qts.) at overhaul

11. Check that the CVT fluid level.

12. Insert the dipstick.

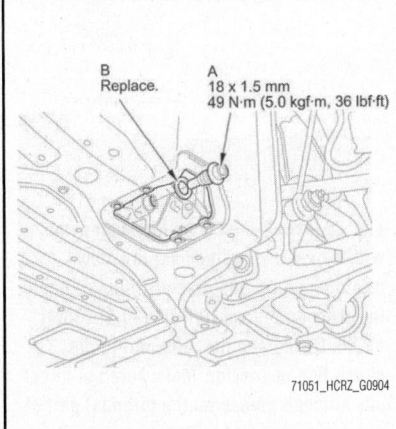

Fig. 18 Remove the drain plug (A) and washer (B) and drain the CVTF.

13. If the Maintenance Minder required replacing the CVTF, reset the Maintenance Minder. Even if the Maintenance Minder did not require you to replace the CVTF, go to the next step.

14. Connect the HDS to the DLC located under the driver's side of the dashboard.

15. Go to the Individual Maintenance Items Reset, and do the MAINTENANCE SUB ITEM 3 RESET with the HDS.

FILTER REPLACEMENT

See Figure 19.

1. Raise the vehicle on a lift, or apply the parking brake, block the rear wheels, and raise the front of the vehicle. Make sure it is securely supported.

2. Remove the splash shield.

3. Remove the lid for maintenance.

4. Remove the drain plug and washer and drain the CVTF.

    a. Remove metal particles from the magnetic surface of the drain plug.

5. Reinstall the drain plug and a new sealing washer.

6. Disconnect the CVTF cooler hoses from the CVTF filter.

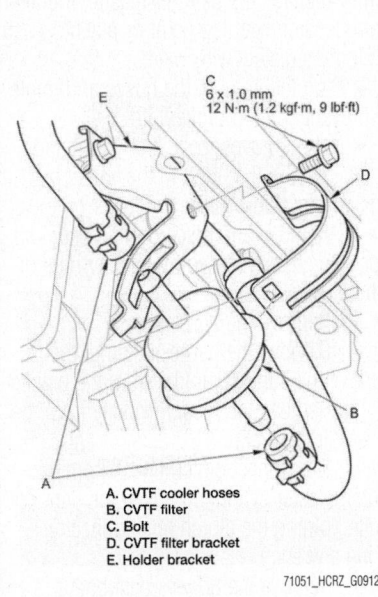

Fig. 19 Disconnect the CVTF cooler hoses from the CVTF filter. Remove the bolt and the CVTF filter bracket, then remove the CVTF filter. Secure the new CVTF filter on holder bracket with the CVTF filter holder and the bolt.

7. Remove the bolt and the CVTF filter bracket, then remove the CVTF filter.

8. Secure the new CVTF filter on holder bracket with the CVTF filter holder and the bolt.

9. Slide the CVTF cooler hoses with the hose clips on the CVTF filter.

10. Refill the transaxle with CVTF to the proper level.

11. Install the splash shield.

FLUID LEVEL CHECK

➥**Keep all foreign particles out of the transaxle.**

1. Park the vehicle on the level ground.

2. Start the engine, and warm it up to normal operating temperature (the radiator fan comes on twice), and turn the engine off.

➥**Check the CVT fluid level within 60–90 seconds after turning the engine off. Higher fluid level may be indicated if the radiator fan comes on twice or more.**

3. Remove the dipstick (yellow loop) and wipe it with a clean cloth.

4. Insert the dipstick back into the tube.

5. Remove the dipstick, and check the CVT fluid level. It should be between the upper mark and the lower mark on the HOT level in tip of the dipstick.

6. If the CVT fluid level is below the lower mark, check for fluid leaks at the transaxle, the CVT fluid filter, the CVT fluid cooler hoses, the CVT fluid cooler lines, and the CVT fluid cooler line joints. If a problem is found, fix it before filling the transaxle with CVT fluid.

➥**If the vehicle is driven when the CVT fluid level is below the lower mark, one or more of these symptoms may occur:**

• Noise from transaxle in D and R.
• Engine runs, but vehicle does not move in any position.
• No shift to a higher ratio or lower ratio.
• Flares while driving.
• Excessive shock when accelerating and decelerating.
• Vehicle does not creep on a flat road in D.
• Late shift after shifting from N to D or N to R.
• Unstable rpm while driving.
• Stall speed high.

7. If the level is above the upper mark, drain the fluid to proper level.

8. If necessary, fill the transaxle with CVTF through the dipstick tube opening to bring the fluid level between the upper mark and the lower mark of the dipstick. Do not fill the fluid above the upper mark.

9. Insert the dipstick.

FLUID RECOMMENDATIONS

Manufacturer recommends always use Honda CVTF. Using a non-Honda CVTF can affect shift quality.

CLUTCH HYDRAULIC SYSTEM BLEEDING

BLEEDING PROCEDURE

See Figure 20.

Note the following:

• Do not reuse the drained fluid. Use Honda DOT 3 Brake Fluid from an unopened container. Using a non-Honda brake fluid can cause corrosion and shorten the life of the system.

• Make sure no dirt or other foreign matter is allowed to contaminate the brake fluid.

• Do not spill brake fluid on the vehicle; it may damage the paint or plastic. If brake fluid does contact the paint or plastic, wash it off immediately with water.

• It may be necessary to limit the movement of the release fork with a block of wood to remove all the air from the system.

• Use fender covers to avoid damaging painted surfaces.

1. Make sure the brake fluid level in the clutch reservoir is at the MAX (upper) level line.

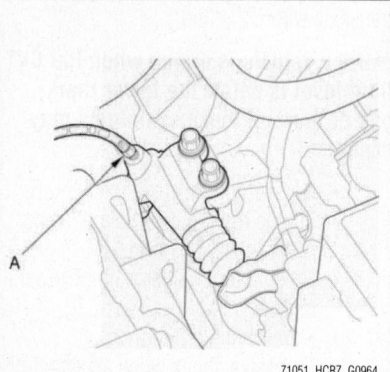

Fig. 20 Attach one end of a clear tube to the bleeder screw (A) and put the other end into a container. Loosen the bleeder screw to allow air to escape from the system.

2. Remove the coolant reservoir.

3. Attach one end of a clear tube to the bleeder screw and put the other end into a container. Loosen the bleeder screw to allow air to escape from the system.

4. Make sure there is an adequate supply of fluid in the reservoir, then slowly push the clutch pedal all the way down. Before releasing the pedal, have an assistant temporarily tighten the bleeder screw. Loosen the bleeder screw, and push the clutch pedal down again. Repeat this step until no more bubbles appear at the clear tube.

➡**Make sure the fluid level on the reservoir does not go below MIN.**

5. Tighten the bleeder screw securely.

6. Refill the brake fluid in the reservoir to the MAX (upper) level line.

7. Install the coolant reservoir.

8. Check the clutch operation, and check for leaks.

CLUTCH MASTER CYLINDER

REMOVAL & INSTALLATION

See Figures 21 through 25.

Note the following:

• Use fender covers to avoid damaging painted surfaces.

• Do not spill brake fluid on the vehicle; it may damage the paint or plastic. If brake fluid does contact the paint or plastic, wash it off immediately with water.

• Plug the ends of the hoses and joints to prevent spilling brake fluid.

• Put on gloves to protect your hands.

1. Secure the hood in the wide open position (support rod in the lower hole).

2. Remove and discard the brake fluid from the clutch reservoir with a syringe or other suitable device.

3. Remove the air cleaner.

4. Remove the harness clamp, then remove the harness holder from the bracket.

5. Remove the reservoir hose from the reservoir hose clip.

6. Disconnect clutch line A from the clutch master cylinder. Loosen the flare nut while holding the clutch line connector using a wrench.

7. Remove the driver's dashboard undercover.

8. Pry out the lock pin and pull the clevis pin out of the clevis. Remove the master cylinder mounting nuts.

9. Remove the clutch master cylinder and the clutch master cylinder seal from the body.

10. Disconnect the reservoir hose.

➡**Inspect the hose. If the hose has damage, leaks, interference, or twisting, replace it.**

11. Remove the retaining clip, then remove the clutch line connector and the O-ring from the master cylinder.

12. Inspect the length of the clutch master cylinder push rod to the figure as shown.

To install:

13. Install the clutch line connector with a new O-ring, then set in a new retaining clip to the master cylinder:

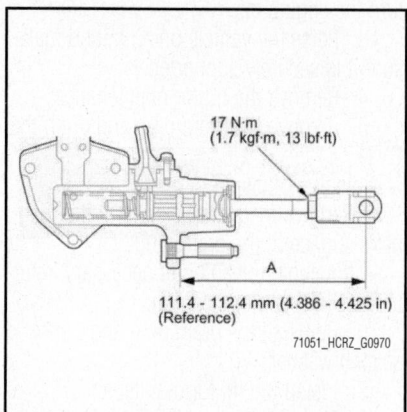

Fig. 21 Inspect the length (A) of the clutch master cylinder push rod to the figure as shown.

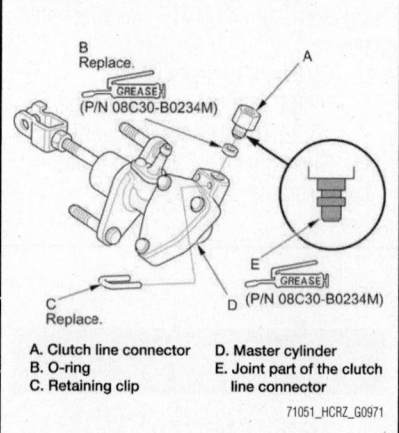

A. Clutch line connector
B. O-ring
C. Retaining clip
D. Master cylinder
E. Joint part of the clutch line connector

Fig. 22 Install the clutch line connector with a new O-ring, then set in a new retaining clip to the master cylinder. Apply the silicone grease (P/N 08C30-B0234M) on the O-ring and the joint part of the clutch line connector. Make sure not to get any silicone grease on the terminal part of the connectors and switches, especially if you have silicone grease on your hands or gloves.

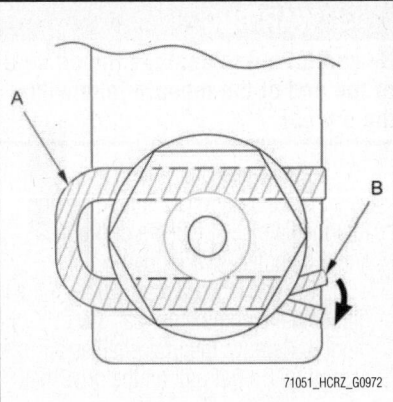

Fig. 23 To prevent the retaining clip (A) from coming off, pry apart the tip (B) of the clip with a flat-tipped screwdriver.

a. Apply the silicone grease (P/N 08C30-B0234M) on the O-ring and the joint part of the clutch line connector.

b. Make sure not to get any silicone grease on the terminal part of the connectors and switches, especially if you have silicone grease on your hands or gloves.

14. To prevent the retaining clip from coming off, pry apart the tip of the clip with a flat-tipped screwdriver.

15. Install the reservoir hose and adjust the hose clamp positions and its direction. When attaching the reservoir

hose, align the blue mark on the hose to the rib.

16. Install a new master cylinder seal and the clutch master cylinder to the body.

17. Install the master cylinder mounting nuts. Connect the clevis to the clutch pedal, then apply multipurpose grease between the clutch pedal and the clevis. Wipe off any excess grease.

18. Apply multipurpose grease to the clevis pin. Slide the clevis pin into the clevis, then install a lock pin.

19. Connect clutch line A to the clutch master cylinder. Tighten the flare nut to 11 ft. lbs. (15 Nm) while holding the clutch line connector, using a wrench.

20. Install the harness holder to the bracket, then install the harness clamp.

21. Install the reservoir hose to the reservoir hose clip.

22. Install the air cleaner.

23. Bleed the clutch hydraulic system.

24. Adjust the clutch pedal, clutch pedal position switch A, clutch pedal position switch B, and clutch pedal position switch C.

25. Check the clutch operation, and check for leaks.

26. Install the driver's dashboard undercover.

27. Test-drive the vehicle.

CLUTCH SLAVE (RELEASE) CYLINDER

REMOVAL & INSTALLATION

See Figures 26 and 27.

Note the following:
• Use fender covers to avoid damaging painted surfaces.
• Do not spill brake fluid on the vehicle; it may damage the paint or plastic. If brake fluid does contact the paint or plastic, wash it off immediately with water.
• Make sure not to get any silicone grease on the terminal part of the connectors and switches, especially if you have silicone grease on your hands or gloves.

1. Disconnect clutch line B. Plug the end of clutch line B with a shop towel to prevent brake fluid from coming out. Loosen the flare nut while holding the clutch line connector with a wrench.

2. Remove the mounting bolts, then remove the slave cylinder.

3. Pull back the boot and apply silicone grease (P/N 08C30-B0234M) to the boot and the slave cylinder. Reinstall the boot.

4. Remove the roll pins, then remove the clutch line connector and the O-ring from the slave cylinder.

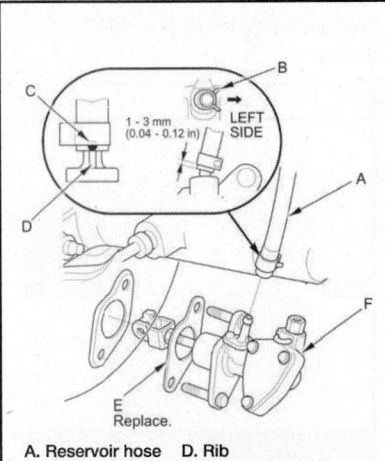

A. Reservoir hose   D. Rib
B. Hose clamp   E. Master cylinder seal
C. Blue mark   F. Clutch master cylinder

Fig. 24 Install the reservoir hose and adjust the hose clamp positions and its direction. When attaching the reservoir hose, align the blue mark on the hose to the rib. Install a new master cylinder seal and the clutch master cylinder to the body.

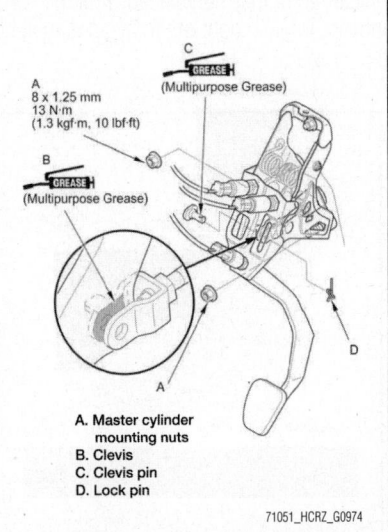

A. Master cylinder mounting nuts
B. Clevis
C. Clevis pin
D. Lock pin

Fig. 25 Install the master cylinder mounting nuts. Connect the clevis to the clutch pedal, then apply multipurpose grease between the clutch pedal and the clevis. Wipe off any excess grease. Apply multipurpose grease to the clevis pin. Slide the clevis pin into the clevis, then install a lock pin.

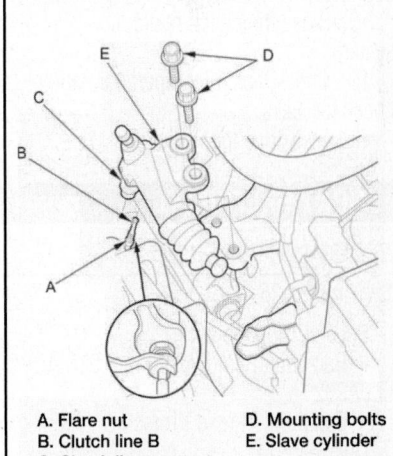

A. Flare nut   D. Mounting bolts
B. Clutch line B   E. Slave cylinder
C. Clutch line connector

Fig. 26 Disconnect clutch line B. Plug the end of clutch line B with a shop towel to prevent brake fluid from coming out. Loosen the flare nut while holding the clutch line connector with a wrench. Remove the mounting bolts, then remove the slave cylinder.

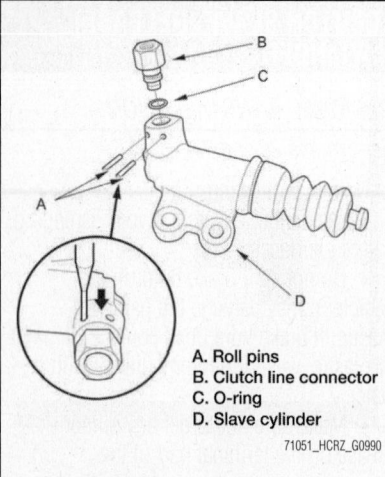

A. Roll pins
B. Clutch line connector
C. O-ring
D. Slave cylinder

71051_HCRZ_G0990

Fig. 27 Remove the roll pins, then remove the clutch line connector and the O-ring from the slave cylinder.

To install:

5. Install the clutch line connector with a new O-ring, then set in new roll pins to the slave cylinder.

6. Apply a light coat of super high temp urea grease (P/N 08798-9002) to the push rod of the slave cylinder.

7. Connect clutch line B to the slave cylinder, and loosely tighten the flare nut.

8. Install the slave cylinder to the transaxle housing, then tighten the clutch line flare nut to the specified torque. Tighten the flare nut while holding the clutch line connector with a wrench.

9. Bleed the clutch hydraulic system.

10. Check the clutch operation, and check for leaks.

11. Test-drive the vehicle.

FRONT HALFSHAFT

REMOVAL & INSTALLATION

See Figures 28 through 33

Special Tools Required:
- Ball Joint Remover, 28 mm 07MAC-SL0A202
- Ball Joint Thread Protector, 14 mm 071AF-S3VA000

1. Raise and support the vehicle.

2. Remove the front wheel.

3. Pry up the stake on the spindle nut, then remove the nut.

4. Remove the splash shield.

5. Drain the transaxle fluid, then reinstall the drain plug with a new sealing washer:
- Manual transaxle: 29 ft. lbs. (39 Nm)

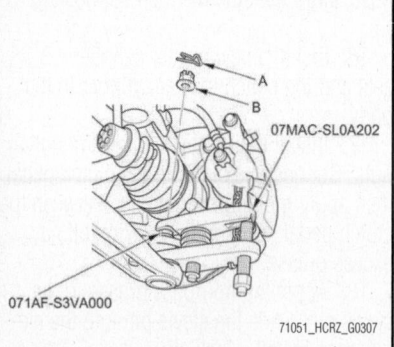

07MAC-SL0A202

071AF-S3VA000

71051_HCRZ_G0307

Fig. 28 Remove the lock pin (A) from the lower arm ball joint, then remove the castle nut (B). Separate the ball joint from the knuckle using the 14 mm ball joint thread protector and the 28 mm ball joint remover.

- CVT: 36 ft. lbs. (49 Nm)

6. Remove the lock pin from the lower arm ball joint, then remove the castle nut. Separate the ball joint from the knuckle using the 14 mm ball joint thread protector and the 28 mm ball joint remover.

7. Pull the knuckle outward, and separate the outboard joint from the front hub using a soft face hammer.

8. Left driveshaft: Pry the inboard joint from the differential using a pry bar. Remove the driveshaft as an assembly. Do not pull on the driveshaft, or the inboard joint may come apart. Pull the inboard joint straight out to avoid damaging the oil seal.

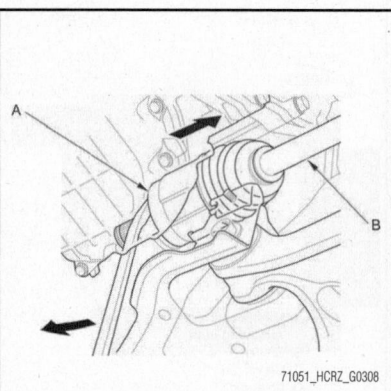

71051_HCRZ_G0308

Fig. 29 Left driveshaft: Pry the inboard joint (A) from the differential using a pry bar. Remove the driveshaft as an assembly. Do not pull on the driveshaft (B) or the inboard joint may come apart. Pull the inboard joint straight out to avoid damaging the oil seal.

✳✳ WARNING

Be careful not to damage the oil seal or the end of the inboard joint with the pry bar.

9. Right driveshaft:

a. Drive the inboard joint off of the intermediate shaft using a drift punch and a hammer (M/T model).

b. Pry the inboard joint (B) from the differential using the pry bar (CVT model). Remove the driveshaft as an assembly. Do not pull on the driveshaft

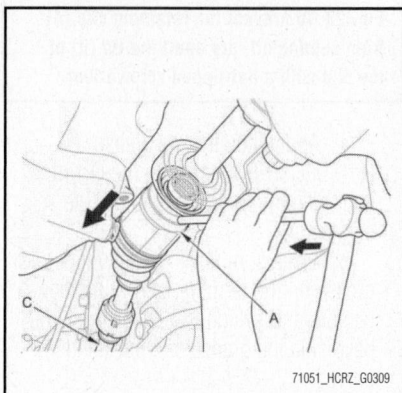

71051_HCRZ_G0309

Fig. 30 Drive the inboard joint (A) off of the intermediate shaft using a drift punch and a hammer. Do not pull on the driveshaft (C) or the inboard joint may come apart. Pull the inboard joint straight out to avoid damaging the oil seal—(M/T model).

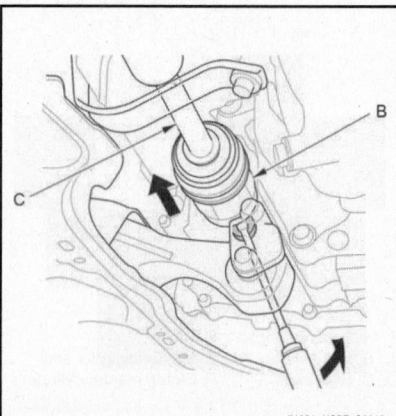

71051_HCRZ_G0310

Fig. 31 Pry the inboard joint (B) from the differential using the pry bar. Remove the driveshaft as an assembly. Do not pull on the driveshaft (C) or the inboard joint may come apart. Pull the inboard joint straight out to avoid damaging the oil seal—(CVT model).

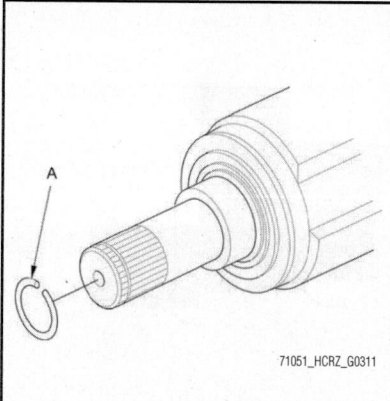

Fig. 32 Remove the set ring (A) from the inboard joint (except M/T model right driveshaft).

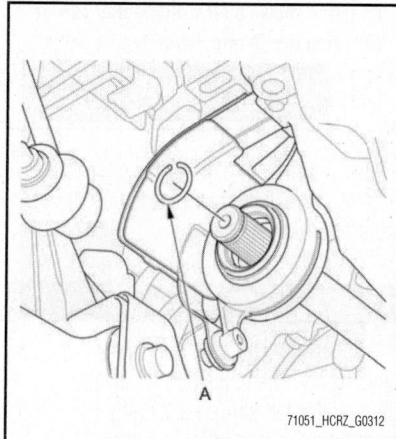

71051_HCRZ_G0312

Fig. 33 Remove the set ring (A) from the intermediate shaft (M/T model).

(C) or the inboard joint may come apart. Pull the inboard joint straight out to avoid damaging the oil seal.

10. Remove the set ring from the inboard joint (except M/T model right driveshaft).

12. Remove the set ring (A) from the intermediate shaft (M/T model).

REAR AXLE BEAM

REMOVAL & INSTALLATION

See Figure 34.

1. Raise and support the vehicle.
2. Remove the rear wheels.
3. Remove the rear hub bearing unit.
4. Remove the rear spring seat lower cover.
5. Remove the splash guard from the axle beam.
6. Remove the wheel speed sensor, the wheel speed sensor clip, and the wire guide grommet from the axle beam. Do not dis-

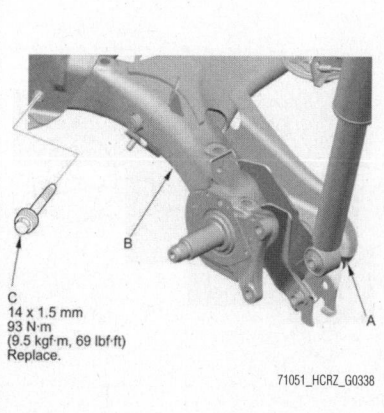

C
14 x 1.5 mm
93 N·m
(9.5 kgf·m, 69 lbf·ft)
Replace.

71051_HCRZ_G0338

Fig. 34 Place a floor jack under the lower spring seat (A) on both sides of the axle beam (B) and support it by raising the floor jack. Do not place the floor jack under the center of the axle beam. Remove the axle beam mounting bolt (C) on both sides.

connect the wheel speed sensor 2P connector.

7. Remove the rear spring. Refer to "Rear Suspension" section.

8. Place a floor jack under the lower spring seat on both sides of the axle beam and support it by raising the floor jack. Do not place the floor jack under the center of the axle beam.

9. Remove the axle beam mounting bolt on both sides.

➡**Use new axle beam mounting bolts during reassembly.**

10. Lower the jack slowly, then remove the axle beam.

To install:

11. Install the axle beam in the reverse order of removal, and note these items:

   a. First install all the components, and lightly tighten the bolts, and place a jack under the lower spring seat of the axle beam on both sides, then raise the suspension to load it with the vehicle's weight before fully tightening to the specified torque.

   b. After installing the brake hose and the parking brake cable, check for interference and twisting of other parts.

   c. Before installing the brake disc, clean the mating surfaces between the hub bearing unit and the inside of the brake disc.

   d. Check the brake hose joint for leaks, and tighten if necessary.

   e. Before installing the wheel, clean the mating surfaces between the brake disc and the inside of the wheel.

12. Check the wheel alignment, and adjust it if necessary.

REAR DIFFERENTIAL CARRIER FINAL DRIVEN GEAR

REMOVAL & INSTALLATION

Manual Transaxle

See Figure 35.

1. Loosen the bolts in a crisscross pattern in several steps, then remove the final driven gear from the differential carrier.

2. Install the final driven gear with the chamfer on the inside diameter facing the carrier. Tighten the bolts in a crisscross pattern in several steps.

CVT

See Figures 36 through 38.

1. Remove the carrier bearings as follows:

   a. Remove the carrier bearings (transaxle housing side and flywheel housing side) using a commercially available puller and a spacer and replace the carrier bearings.

2. Remove the final driven gear from the differential carrier and replace the differential carrier or the final driven gear. The final driven gear bolts have left-hand threads.

3. Install the final driven gear in the direction shown on the differential carrier.

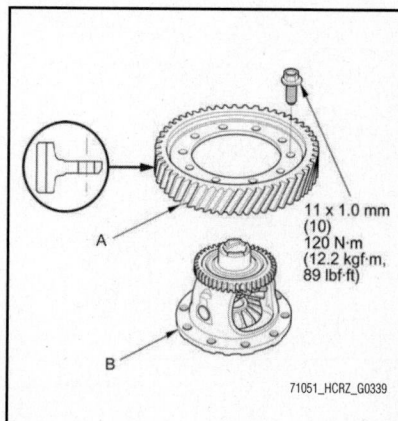

11 x 1.0 mm
(10)
120 N·m
(12.2 kgf·m,
89 lbf·ft)

A

B

71051_HCRZ_G0339

Fig. 35 Loosen the bolts in a crisscross pattern in several steps, then remove the final driven gear (A) from the differential carrier (B).

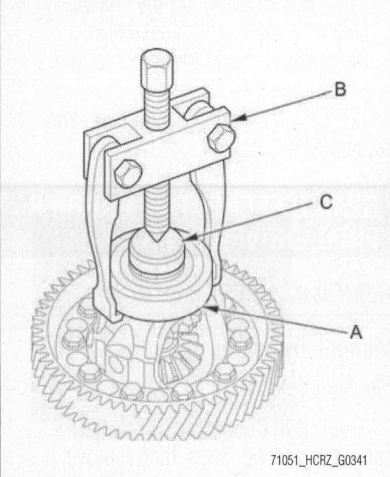

71051_HCRZ_G0341

Fig. 36 Remove the carrier bearings (transaxle housing side and flywheel housing side) (A) using a commercially available puller (B) and a spacer (C) and replace the carrier bearings.

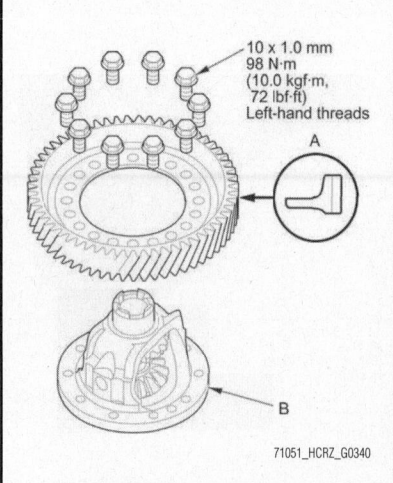

10 x 1.0 mm
98 N·m
(10.0 kgf·m,
72 lbf·ft)
Left-hand threads

71051_HCRZ_G0340

Fig. 37 Remove the final driven gear (A) from the differential carrier (B) and replace the differential carrier or the final driven gear. The final driven gear bolts have left-hand threads.

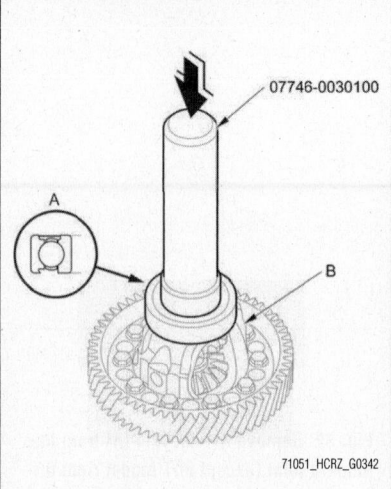

07746-0030100

71051_HCRZ_G0342

Fig. 38 Install new bearings (A) in the direction shown on the differential carrier (B) using the 40 mm driver handle and a press. Press the bearing on until it bottoms.

4. Tighten the bolts to the specified torque in a crisscross pattern in at least two or three steps.

5. Install the carrier bearings:
   a. Install new bearings in the direction shown on the differential carrier

using the 40 mm driver handle and a press. Press the bearing on until it bottoms.

ENGINE COOLING

CONDENSER & RADIATOR FANS

REMOVAL & INSTALLATION

See Figures 39 and 40.

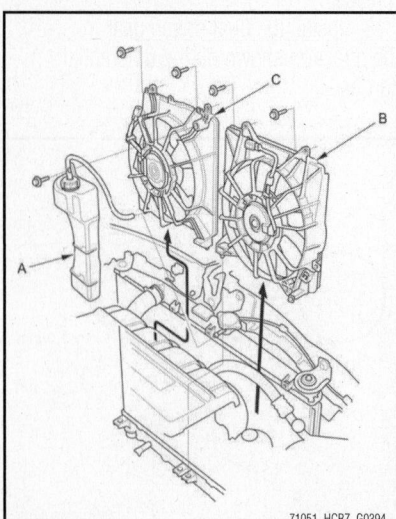

71051_HCRZ_G0394

Fig. 39 Remove the coolant reservoir (A). Remove the A/C condenser fan shroud assembly (B), then remove the radiator fan shroud assembly (C) from the A/C condenser fan shroud side.

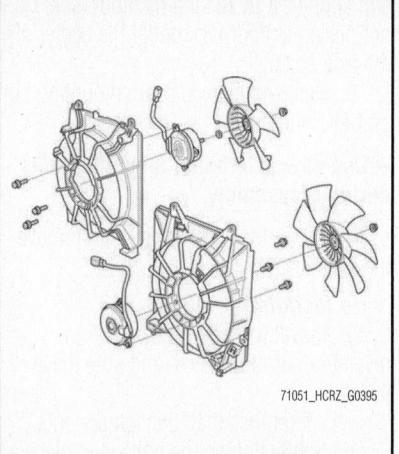

71051_HCRZ_G0395

Fig. 40 Disassemble the fan shrouds.

1. Raise the vehicle on the lift.
2. Remove the splash shield.
3. Remove the A/C compressor clutch connector from the clamp, then remove the harness clamps.
4. Lower the vehicle on the lift.
5. Remove the intake air tube and the resonator chamber joint.
6. Disconnect the fan motor connectors and remove the harness clamps.
7. Remove the coolant reservoir.

8. Remove the A/C condenser fan shroud assembly, then remove the radiator fan shroud assembly from the A/C condenser fan shroud side.
9. Disassemble the fan shrouds.

To install:
10. Assemble the fan shrouds.
11. Install the radiator fan shroud assembly onto the A/C condenser fan shroud side, then install the A/C condenser fan shroud assembly.
12. Install the coolant reservoir.
13. Connect the fan motor connectors and install the harness clamps.
14. Install the resonator chamber joint and the intake air tube.
   a. When tightening the screw of the hose band, align the edge of the hose band with the mark painted on the hose band.
   b. If you tighten the screw over the mark, replace the hose band.
15. Raise the vehicle on the lift.
16. Install the A/C compressor clutch connector to the clamp, then install the harness clamps.
17. Install the splash shield.
18. Lower the vehicle on the lift.

ENGINE COOLANT

DRAIN & REFILL

See Figure 41.

1. Wait until the engine is cool, then carefully remove the radiator cap.

2. Remove the splash shield.

3. Loosen the drain plug and drain the coolant.

4. Remove the drain bolt from the rear side of the engine block.

5. After the coolant has drained, reinstall the drain bolt with a new washer.

6. Tighten the radiator drain plug securely.

7. Remove, drain, and reinstall the coolant reservoir.

8. Fill the coolant reservoir to the MAX mark with Honda Long Life Antifreeze/Coolant Type 2.

9. Pour Honda Long Life Antifreeze/Coolant Type 2 into the radiator up to the base of the filler neck.

Note the following:

• Always use Honda Long Life Antifreeze/Coolant Type 2. Using a non-Honda coolant can result in corrosion, causing the cooling system to malfunction or fail.

• Honda Long Life Antifreeze/Coolant Type 2 is a mixture of 50 % antifreeze and 50 % water. Do not add water.

10. Engine coolant refill capacities are:
  • Reservoir capacity of 0.116 US gal. (0.44 L):
  • M/T model at coolant change: 1.168 US gal. (4.42 L)
  • M/T model after engine overhaul: 1.300 US gal. (4.92 L)
  • CVT model at coolant change: 1.149 US gal. (4.35 L)
  • CVT model after engine overhaul: 1.279 US gal. (4.84 L)

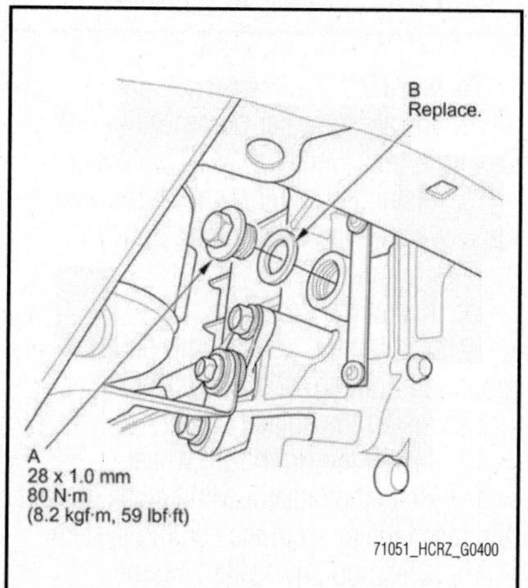

11. Loosely install the radiator cap.

12. Start the engine, and let it run until it warms up (the radiator fan comes on at least twice).

13. Turn off the engine. Check the level in the radiator, and add the recommended coolant, if needed.

14. Put the radiator cap on securely, then start the engine again, and check for leaks.

15. Clean up any spilled engine coolant.

16. If the Maintenance Minder required replacing the engine coolant, reset the Maintenance Minder, and this procedure is complete. If the Maintenance Minder did not require engine coolant replacement, go to step 16.

17. Turn the ignition switch to LOCK (0).

18. Connect the HDS to the DLC.

19. Turn the ignition switch to ON (II).

20. Make sure the HDS communicates with the vehicle and ECM/PCM. If it does not communicate, troubleshoot the DLC circuit.

21. Select GAUGES in the BODY ELECTRICAL with the HDS.

22. Select ADJUSTMENT in the GAUGE with the HDS.

23. Select MAINTENANCE MINDER in the ADJUSTMENT with the HDS.

24. Select RESET in the MAINTENANCE MINDER with the HDS.

25. Select MAINTENANCE SUB ITEM 5 RESET with the HDS.

26. Install the splash shield.

FLUID RECOMMENDATIONS

Honda Long Life Antifreeze/Coolant Type 2.

RADIATOR

REMOVAL & INSTALLATION

See Figures 42 and 43.

1. Raise the vehicle on the lift.

2. Drain the engine coolant.

3. Remove the splash shield.

4. Disconnect the ECT sensor 2 connector and remove the A/C compressor clutch connector from the clamp.

5. Remove the harness clamps.

6. Disconnect the lower radiator hose.

7. CVT model: Disconnect the CVT fluid cooler hoses, then plug the CVTF cooler hoses and lines.

8. Lower the vehicle on the lift.

9. Remove the intake air tube and the resonator chamber joint.

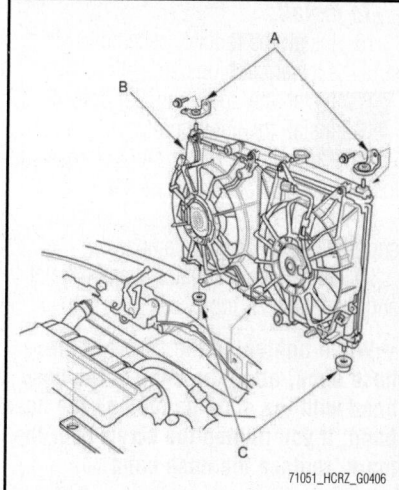

Fig. 42 Remove the upper radiator brackets (A), then pull up the radiator assembly (B) with the lower cushions (C).

10. Disconnect the fan motor connectors and remove the harness clamps.

11. Disconnect the upper radiator hose and remove the air intake tube bracket.

12. Remove the coolant reservoir.

13. Remove the upper radiator brackets, then pull up the radiator assembly with the lower cushions.

14. Remove the radiator fan shroud assembly, the A/C condenser fan shroud assembly, the radiator cap, the drain plug and ECT sensor 2 from the radiator.

15. Reassemble the radiator with new O-rings.

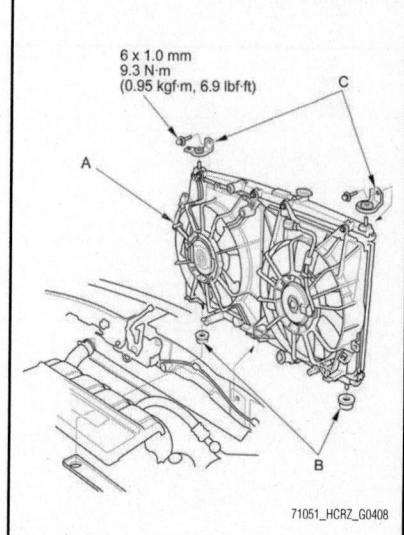

Fig. 43 Install the radiator assembly (A). Make sure the lower cushions (B) are set securely. Install the upper radiator brackets (C).

Fig. 41 Remove the drain bolt (A) from the rear side of the engine block. After the coolant has drained, reinstall the drain bolt with a new washer (B).

To install:

16. Install the radiator assembly. Make sure the lower cushions are set securely.

17. Install the upper radiator brackets.

18. Install the coolant reservoir.

19. Connect the upper radiator hose and install the air intake tube bracket.

20. Connect the fan motor connectors and install the harness clamps.

21. Install the resonator chamber joint and the intake air tube.

➡ **When tightening the screw of the hose band, align the edge of the hose band with the mark painted on the hose band. If you tighten the screw over the mark, replace the hose band.**

22. Raise the vehicle on the lift.

23. Connect the lower radiator hose.

24. CVT model: Connect the CVT fluid cooler hoses.

25. Connect the ECT sensor 2 connector and install the A/C compressor clutch connector to the clamp.

26. Install the harness clamps.

27. Install the splash shield.

28. Lower the vehicle on the lift.

29. CVT model: Refill the transaxle with CVT fluid.

30. Refill the radiator with engine coolant, and bleed the air from the cooling system.

31. Clean up any spilled engine coolant.

THERMOSTAT

REMOVAL & INSTALLATION

See Figures 44 and 45.

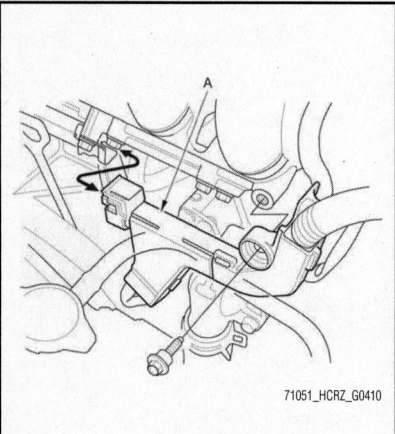

Fig. 44 Remove the harness holder (A).

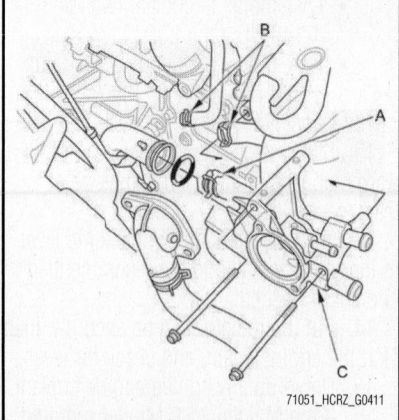

Fig. 45 Disconnect the heater hose (A) and the water bypass hoses (B). Remove the thermostat housing (C).

1. Drain the engine coolant.

2. Remove the coolant reservoir.

3. Remove the harness holder.

4. Remove the thermostat.

5. Disconnect the heater hose and the water bypass hoses.

6. Remove the thermostat housing.

7. Remove the thermostat.

To install:

8. Install the thermostat housing with a new connecting pipe O-ring.

9. Connect the heater hose and the water bypass hoses.

10. Install the thermostat.

11. Install the harness holder.

12. Install the coolant reservoir.

13. Refill the radiator with engine coolant, and bleed the air from the cooling system.

14. Clean up any spilled engine coolant.

WATER PUMP

REMOVAL & INSTALLATION

See Figures 46 and 47.

1. Remove the right front wheel.

2. Remove the splash shield.

3. Drain the engine coolant.

4. Loosen the water pump pulley mounting bolts.

5. Remove the drive belt.

6. Remove the water pump pulley.

7. Remove the five bolts securing the water pump, then remove the water pump.

8. Inspect and clean the mating surface with the engine block.

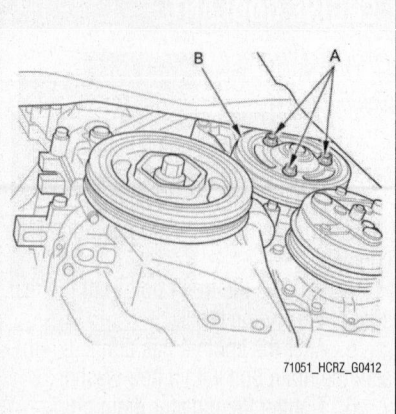

Fig. 46 Loosen the water pump pulley mounting bolts (A). Remove the drive belt. Remove the water pump pulley (B).

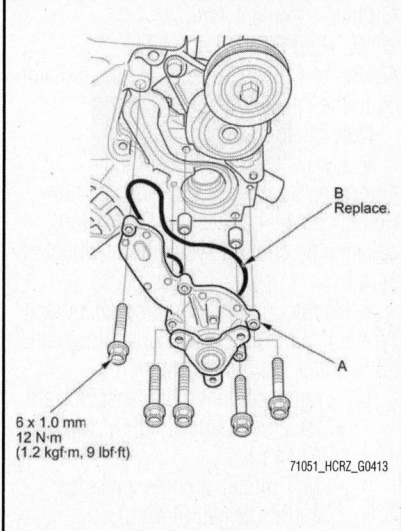

Fig. 47 Remove the five bolts securing the water pump (A), then remove the water pump. Replace the O-ring (B).

To install:

9. Install the water pump with a new O-ring in the reverse order of removal.

10. Install the water pump pulley and loosely install the water pump pulley mounting bolts.

11. Install the drive belt.

12. Tighten the water pump pulley mounting bolts to 12 ft. lbs. (16 Nm).

13. Install the splash shield.

14. Install the right front wheel.

15. Refill the radiator with engine coolant, and bleed the air from the cooling system.

16. Clean up any spilled engine coolant.

ENGINE ELECTRICAL

BATTERY SYSTEM

BATTERY

CHARGING

> ❄❄ **CAUTION**
>
> **The chemical reaction which takes place in all batteries generates explosive hydrogen gas. A spark can cause the battery to explode and splash acid. To avoid personal injury, be sure there is proper ventilation and take appropriate fire safety precautions when working with or near a battery.**

A battery should be charged at a slow rate to keep the plates inside from getting too hot. However, if some maintenance-free batteries are allowed to discharge until they are almost "dead," they may have to be charged at a high rate to bring them back to "life." Always follow the charger manufacturer's instructions on charging the battery.

REMOVAL & INSTALLATION

See Figure 48.

➡**The 12 volt battery terminal disconnection/reconnection procedure must be done before and after doing this procedure. Some systems store data in memory that is lost when the battery disconnected.**

1. Do the 12 volt battery terminal disconnection procedure.
2. Remove the two nuts securing the battery setting plate, then remove the battery setting plate and the battery.

To install:

3. Install the battery, then install the battery setting plate.
4. Tighten the two nuts equally until the battery is stable.

➡**Do not deform the battery hold down plate by over-tightening the nuts.**

5. Do the 12 volt battery terminal reconnection procedure.

➡**Make sure the battery is installed correctly, and the positive terminal and the negative terminal are not connected in reverse.**

BATTERY CABLE RESET

➡**Also Refer to "Battery Cables" in this section.**

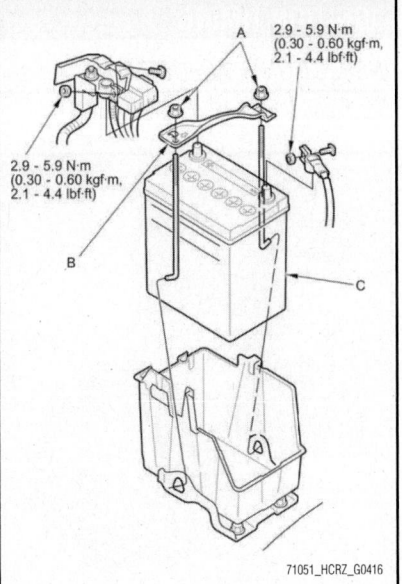

71051_HCRZ_G0416

Fig. 48 Remove the two nuts (A) securing the battery setting plate (B), then remove the battery setting plate and the battery (C).

Many automotive systems today rely on software. Should that software develop errors, which can happen from something like a low battery or interrupting a learn procedure, it can cause all sorts of odd electrical problems that can be hard to troubleshoot. If you've got such a problem, try doing a battery cable reset; it just might do the trick.

A battery cable reset forces all capacitors to discharge faster, and clears and resets most control unit volatile memory. It's like removing and reinstalling the battery in a cell phone that's acting weird or Before you start, make sure the battery is fully charged. A low battery can cause problems with electronics.

1. Get the anti-theft code and write down your customer's audio presets.
2. Turn the ignition switch to LOCK (0), and remove the key. This lessens the chance of voltage spikes.
3. Disconnect the battery cables: negative cable first, then positive.
4. Short the battery cables together with a jumper wire.
5. Turn the ignition switch to ON (II), and wait 10 minutes.

6. Turn the ignition switch to LOCK (0), and remove the key.
7. Remove the jumper wire, and reconnect the battery cables—positive cable first, then negative.
8. Check ISIS, and do the required reset/learn procedures for the vehicle you're working on.
9. Enter the anti-theft code, and restore your customer's settings.

BATTERY CABLES

DISCONNECTION & RECONNECTION

Disconnection

➡**Some systems store data in memory that is lost when the battery is disconnected. Do the following procedures before disconnecting the battery.**

1. Make sure you have the anti-theft code for the audio system or the audio-navigation unit.
2. Make sure the ignition switch is in LOCK (0).
3. Disconnect and isolate the negative cable from the battery.

> ❄❄ **CAUTION**
>
> **Always disconnect the negative battery cable from the battery first.**

4. Disconnect the positive cable from the battery.

Reconnection

1. Clean the battery terminals.
2. Test the battery.
3. Apply multipurpose grease to the terminals to prevent corrosion.
4. Reconnect the positive cable to the battery first, then reconnect the negative cable to the battery.

> ❄❄ **CAUTION**
>
> **Always connect the positive cable to the battery first.**

5. Install the terminal cover.
6. Enter the anti-theft code for the audio system or audio-navigation unit.
7. Set the clock (for vehicles without navigation).

ENGINE ELECTRICAL

IGNITION SYSTEM

FIRING ORDERS

Firing order is: 1–3–4–2.

IGNITION COILS & SPARK PLUGS

REMOVAL & INSTALLATION

See Figure 49.

1. Remove the cowl cover and the under-cowl panel.
2. Disconnect the ignition coil connectors, then remove the ignition coils.
3. Remove the spark plugs and inspect them.
4. Apply a small amount of anti-seize compound to the plug threads, and screw the plugs into the cylinder head, finger tight, then tighten the plugs to the specified torque of 13 ft. lbs. (18 Nm).
5. Install all remaining parts in the reverse order of removal.

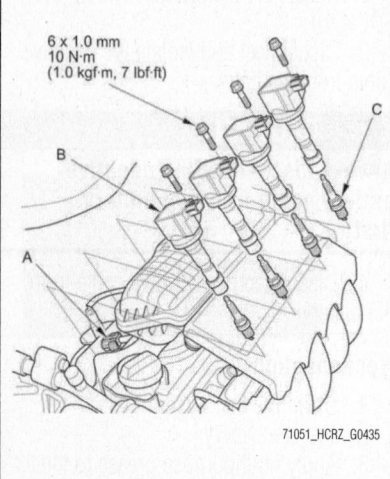

6 x 1.0 mm
10 N·m
(1.0 kgf·m, 7 lbf·ft)

Fig. 49 Disconnect the ignition coil connectors (A), then remove the ignition coils (B). Remove the spark plugs (C) and inspect them.

IGNITION TIMING

INSPECTION & ADJUSTMENT

See Figures 50 and 51.

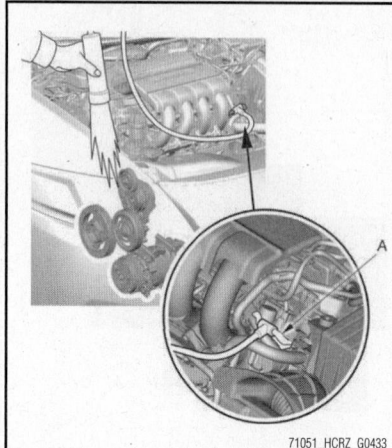

71051_HCRZ_G0433

Fig. 50 Connect the timing light to the service loop (A).

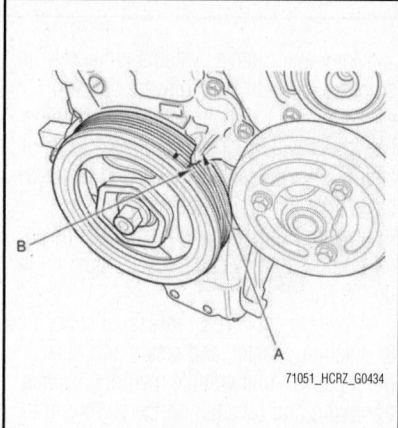

71051_HCRZ_G0434

Fig. 51 Aim the light toward the pointer (A) on the chain case. Check the ignition timing under no load condition (headlights, blower fan, rear window defogger, and air conditioner are turned off). Red mark (B) shows timing setting.

1. Connect the HDS (or equivalent scan tool) to the DLC.
2. Turn the ignition switch to ON (II).
3. Make sure the HDS communicates with the vehicle and the ECM/PCM. If it does not communicate, troubleshoot the DLC circuit.
4. Check for DTCs. If a DTC is present, diagnose and repair the cause before continuing with this test.
5. Start the engine. Hold the engine speed at 3,000 rpm with no load (M/T in neutral, CVT in P or N) until the radiator fan comes on, then let it idle.
6. Check the idle speed.
7. Jump the SCS line with the HDS.
8. Connect the timing light to the service loop.
9. Aim the light toward the pointer on the chain case. Check the ignition timing under no load condition (headlights, blower fan, rear window defogger, and air conditioner are turned off).
10. Ignition timing specifications should be:
    - M/T model: 10 ±2° BTDC (RED mark) at idle in neutral
    - CVT model: 10 ±2° BTDC (RED mark) at idle in P or N
11. If the ignition timing differs from the specification, check the camshaft timing.
    a. If the camshaft timing is OK, update the ECM/PCM if it does not have the latest software, or substitute a known-good ECM/PCM, then recheck.
    b. If the system works properly, and the ECM/PCM was substituted, replace the original ECM/PCM.
12. Disconnect the HDS and the timing light.

SPARK PLUGS

REMOVAL & INSTALLATION

➡Refer to "Ignition Coils & Spark Plugs" in this section.

PRECAUTIONS

Before working on any part of the hybrid high voltage system, observe the following precautions:

• The IMA (integrated motor assist) system uses high voltage (100 V) circuits. Be sure to shut off the electrical circuits and isolate the IMA system and related parts before servicing the IMA system.

• The high voltage cables and their covers are identified by orange coloring. The caution labels are attached to high voltage and other related parts. When the system is energized, be careful not to touch these cables and parts without adequate protective gear.

• The front floor under-cover protecting the high voltage cables is marked .

• If the 12 V battery is discharged, its cables have been disconnected, or the MCM (Motor Control Module) has been reset, the IMA battery level indicator does not display the state-of-charge (SOC) when the engine is first started. To display the level in the indicator, start the engine, and hold the engine speed between 3,500 and 4,000 rpm without load (in P or N (CVT model)) or (in neutral (M/T model)) until the level in the indicator is half full.

• When the IMA system is energized, servicing, disassembling, or replacing items marked with in each procedure requires insulated tools.

• When the IMA system indicator is on, do the IMA system troubleshooting first.

• Wear insulated gloves whenever you inspect or service the IMA system. Be sure to check the gloves for pin holes, tears, or other damage.

• To make sure the system is not energized, turn the battery module switch OFF, and secure the switch in the OFF position with the locking cover before servicing the IMA system (see Turning off and on power to the high voltage circuit).

• Wait at least 5 minutes after turning off the battery module switch, then disconnect the negative cable from the 12 V battery (it takes about 5 minutes for the PDU capacitor to discharge).

• Before disconnecting the high voltage cable terminals, use a voltmeter to make sure that the voltage between the terminals is below 30 V.

• When the system is energized and you are servicing parts without an insulated sheath, be sure to use insulated tools to prevent short circuiting.

• The IMA rotor assembly contains very strong magnets and should be handled with special care. People with pacemakers or other magnetically sensitive medical devices should not handle the rotor assembly.

Use the special tool (rotor puller) to remove or install the rotor assembly.

• The IMA motor power cables carry high voltage when the engine is running or the IMA system is energized. To avoid serious injury from electrical shock, do not start the engine with the IMA motor power cables disconnected.

• The IMA motor rotor contains very strong magnets and should be handled with special care. People with pacemakers or other sensitive medical devices should not handle the IMA motor rotor.

• If the IMA motor rotor is installed by hand, it may suddenly be pulled toward the stator with great force, causing serious hand or finger injury. Always use the special tool (rotor puller) to remove or install the rotor assembly.

• Keep the motor rotor away from magnetically sensitive devices.

• Do not blow air near the rotor, as metal particles may get on the magnets.

• Store the rotor in the designated storage box, and keep it away from sensitive devices during storage.

HIGH VOLTAGE CIRCUIT

DISABLING & ENABLING POWER TO THE HIGH VOLTAGE CIRCUIT

See Figures 52 through 55.

❋ CAUTION

The following procedure should be done before you work on or near any energized high voltage components. Follow the procedure exactly. Otherwise, you may be injured or may damage equipment.

1. Turn the ignition switch to LOCK (0), then remove the key from the ignition switch.

2. Remove the cargo floor lid, the cargo floor box, and the spare tire.

3. Loosen the bolt and remove the bolt.

4. Remove the battery module switch lid from the IPU cover.

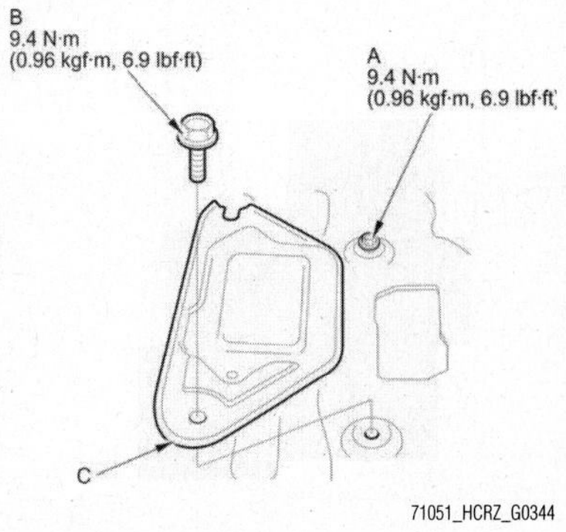

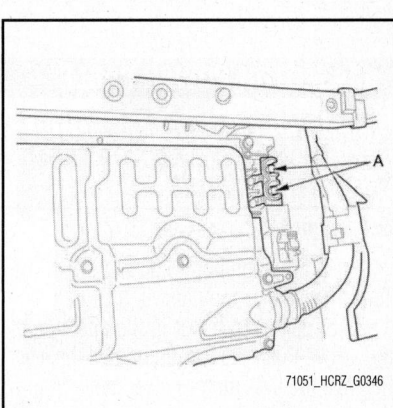

B
9.4 N·m
(0.96 kgf·m, 6.9 lbf·ft)

A
9.4 N·m
(0.96 kgf·m, 6.9 lbf·ft)

C

71051_HCRZ_G0344

Fig. 52 Loosen the bolt (A) and remove the bolt (B). Remove the battery module switch lid (C) from the IPU cover.

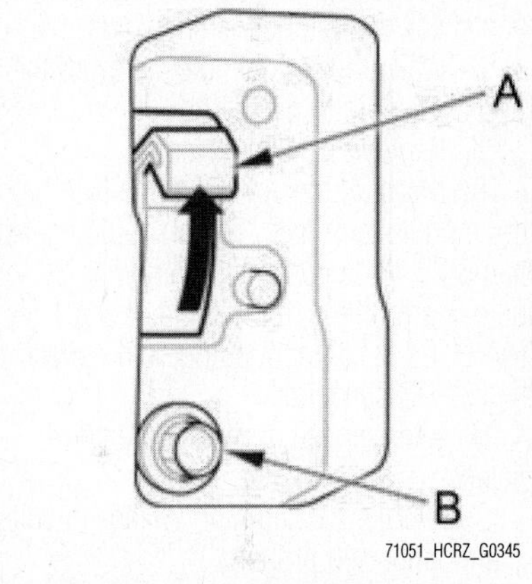

A

B

71051_HCRZ_G0345

Fig. 53 Turn the battery module switch (A) OFF, then check that the bolt (B) is showing.

A

71051_HCRZ_G0346

Fig. 54 Measure the voltage at the IMA battery module terminals (A). There should be less than 30 V. If the voltage is more than 30 V, there is a problem in the system; check for IMA DTCs before continuing, and replace the DC-DC converter.

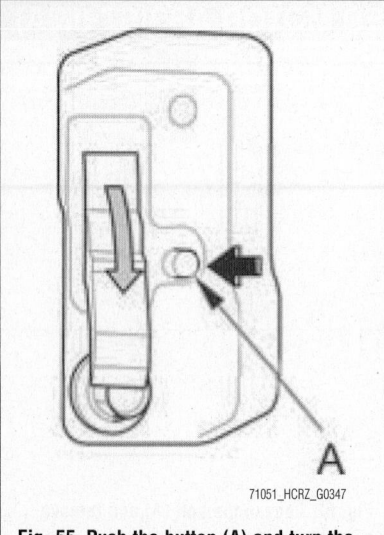

Fig. 55 Push the button (A) and turn the battery module switch ON.

5. Turn the battery module switch OFF, then check that the bolt is showing.

6. Wait at least 5 minutes to allow the PDU capacitors to discharge.

7. Remove the IPU cover.

8. Measure the voltage at the IMA battery module terminals. There should be less than 30 V. If the voltage is more than 30 V, there is a problem in the system; check for IMA DTCs before continuing, and replace the DC-DC converter.

9. After service or repairs are completed:

  a. Make sure all high voltage circuits are connected properly.

  b. Install the IPU cover.

10. Push the button and turn the battery module switch ON.

11. Reinstall all remaining removed parts.

IMA MOTOR POWER CABLE

DISCONNECTING THE IMA MOTOR POWER CABLE CONNECTOR FROM THE MOTOR STATOR

See Figure 56.

1. Turn the ignition switch to LOCK (0). Turn the battery module switch OFF.

2. Slide the protector in the direction of the arrow. Push the tab, then raise the lever. Remove the IMA motor power cable from the motor stator.

3. Turn the battery module switch OFF before disconnecting the motor power cable.

➡If the outside of the IMA motor power cable connector is dirty, clean it before you disconnect it.

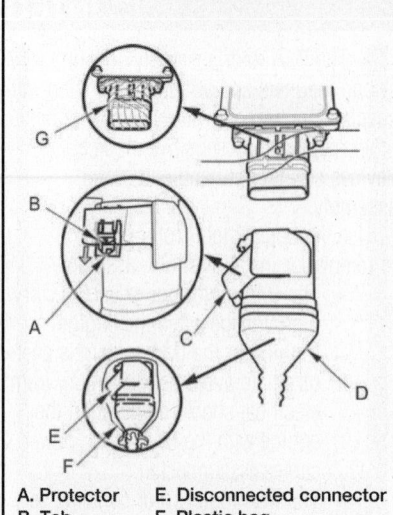

A. Protector
B. Tab
C. Lever
D. IMA motor power cable
E. Disconnected connector
F. Plastic bag
G. Insulating tape

Fig. 56 Disconnecting the IMA Motor Power Cable

4. Cover the disconnected connector with a plastic bag and wrap the IMA motor power cable terminals with insulating tape.

➡If the IMA motor power cable is wet, dry it with a clean shop towel. Do not use compressed air.

DIRECT CURRENT/DIRECT CURRENT CONVERTER

REMOVAL & INSTALLATION

See Figures 57 and 58.

✳ CAUTION

IMA components are located in this area. The IMA is a high-voltage system. You must be familiar with the IMA system before working on or around it. Make sure you have read the IMA service precautions before doing repairs or service.

✳ CAUTION

The IMA motor power cables carry high voltage when the engine is running or the IMA system is energized. To avoid serious injury from electrical shock, do not start the engine with the IMA motor power cables disconnected.

1. Make sure the ignition switch is in LOCK (0).

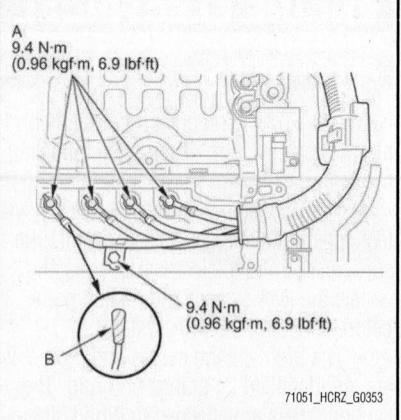

Fig. 57 Disconnect the four IMA motor power cables (A) from the phase motor current sensor, then wrap the end of the 12 volt power cable (B) with insulating tape.

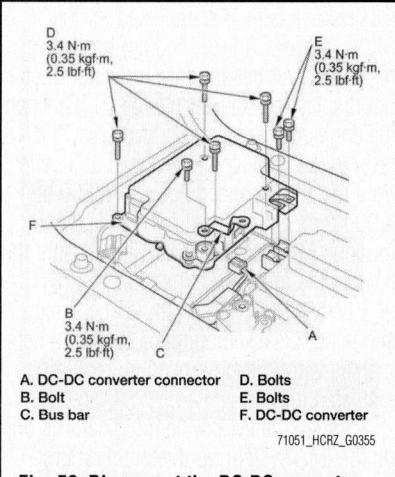

A. DC-DC converter connector
B. Bolt
C. Bus bar
D. Bolts
E. Bolts
F. DC-DC converter

Fig. 58 Disconnect the DC-DC converter connector. Remove the bolt and the bus bar. Remove the remaining bolts and the DC-DC converter.

2. Do the 12 volt battery terminal disconnection procedure.

3. Remove the MCM.

4. Disconnect the four IMA motor power cables from the phase motor current sensor, then wrap the end of the 12 volt power cable with insulating tape.

➡Check the position of the U phase, the V phase, and the W phase cables before disconnecting them.

5. Remove the bolts and the terminal cover.

6. Disconnect the DC-DC converter connector.

7. Remove the bolt and the bus bar.

8. Remove the remaining bolts and the DC-DC converter.

9. Install the parts in the reverse order of removal.

➡**Make sure the IMA motor power cables are correctly positioned before you reconnect them.**

10. Do the 12 volt battery terminal reconnection procedure.

IMA MOTOR

REMOVAL & INSTALLATION

Drain Cover

See Figure 59.

1. Remove the splash shield.
2. Remove the drain cover.
3. Install the parts in the reverse order of removal.

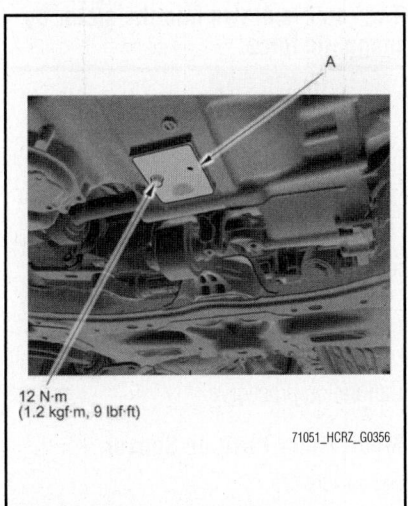

12 N·m
(1.2 kgf·m, 9 lbf·ft)

71051_HCRZ_G0356

Fig. 59 Remove the drain cover (A).

Motor Housing

See Figures 60 and 61.

1. Remove the IMA motor rotor, as described in this section.
2. Remove the connector and the harness clamps from the bracket.
3. Remove the bolt and the IMA motor housing.
4. Set the dowel pins in the IMA motor housing before installing the motor stator on the engine.
5. Remove the bracket from the IMA motor housing.

To install:
6. Install the parts in the reverse order of removal.
7. Install the IMA motor rotor, as described in this section.

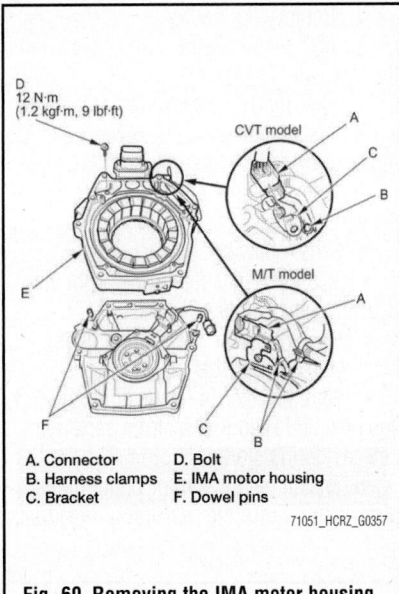

D
12 N·m
(1.2 kgf·m, 9 lbf·ft)

CVT model

M/T model

A. Connector
B. Harness clamps
C. Bracket
D. Bolt
E. IMA motor housing
F. Dowel pins

71051_HCRZ_G0357

Fig. 60 Removing the IMA motor housing

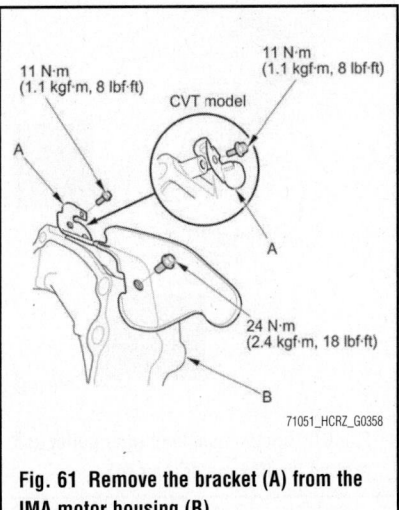

11 N·m
(1.1 kgf·m, 8 lbf·ft)

11 N·m
(1.1 kgf·m, 8 lbf·ft)

CVT model

24 N·m
(2.4 kgf·m, 18 lbf·ft)

71051_HCRZ_G0358

Fig. 61 Remove the bracket (A) from the IMA motor housing (B).

8. Do the IMA motor rotor position calibration procedure. Refer to "Motor Rotor".

Motor Power Cable

See Figure 62.

1. Remove the IMA motor housing.
2. Remove the terminal cover and the IMA motor power cable terminal.
3. Install the parts in the reverse order of removal.
4. Do the IMA Motor Rotor Position Calibration procedure.

Motor Rotor

See Figures 63 through 68.

Special Tools Required: Rotor Puller 07YAC-PHM010C.

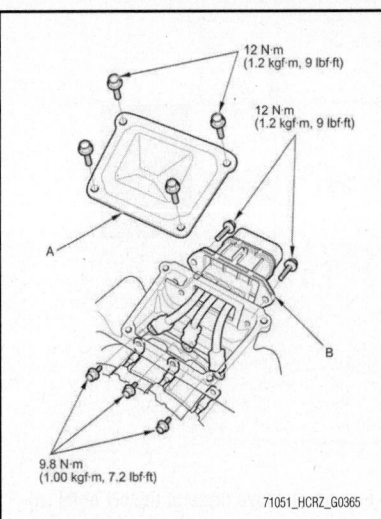

12 N·m
(1.2 kgf·m, 9 lbf·ft)

12 N·m
(1.2 kgf·m, 9 lbf·ft)

9.8 N·m
(1.00 kgf·m, 7.2 lbf·ft)

71051_HCRZ_G0365

Fig. 62 Remove the terminal cover (A) and the IMA motor power cable terminal (B).

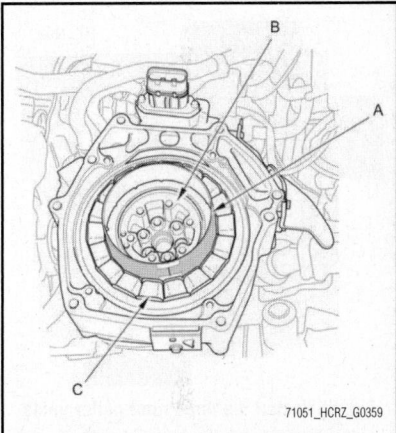

71051_HCRZ_G0359

Fig. 63 Install a plastic film (A) between the IMA motor rotor (B) and the motor stator (C).

❋❋ CAUTION

The motor rotor contains very strong magnets and should be handled with special care. People with pacemakers or other sensitive medical devices should not handle the IMA motor rotor. Read "Precautions" at the start of this section.

❋❋ CAUTION

If the rotor is installed by hand, it may suddenly be pulled toward the stator with great force, causing serious hand or finger injury. Always use the special tool (rotor puller) to remove or install the rotor assembly.

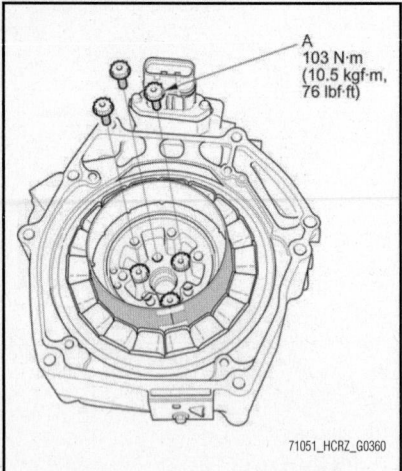

Fig. 64 Remove three of the six bolts (A) as shown.

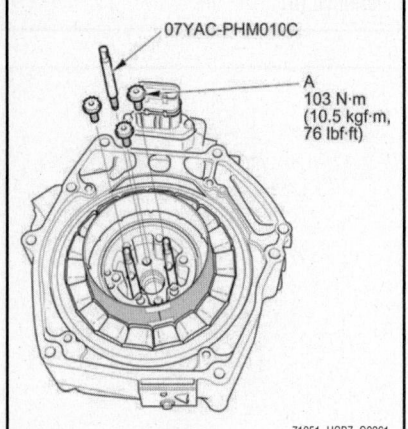

Fig. 65 Install the three rotor puller guide pins, then remove the remaining three bolts (A).

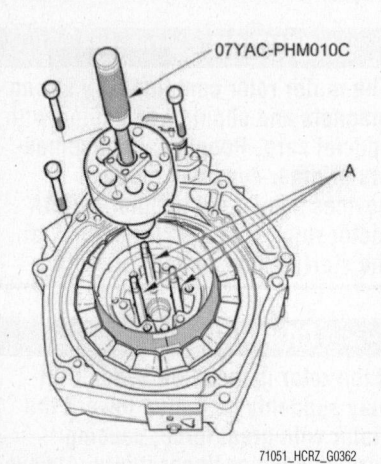

Fig. 66 Attach the rotor puller with its supplied bolts. When installing the rotor puller, position the puller to fit over the guide pins (A).

1. Remove the transaxle.
2. M/T model: Remove the clutch and the flywheel.
3. CVT model: Remove the drive plate.
4. Install a plastic film between the IMA motor rotor and the motor stator.
5. Remove three of the six bolts as shown.

Note the following:

• Keep the motor rotor away from magnetically sensitive devices.

• Do not blow air near the rotor, as metal particles may get on the magnets.

• Store the rotor in the designated storage box, and keep it away from sensitive devices during storage.

6. Install the three rotor puller guide pins, then remove the remaining three bolts.

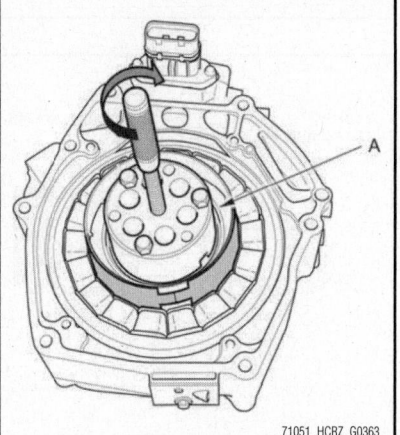

Fig. 67 Turn the handle of rotor puller and remove the IMA motor rotor (A).

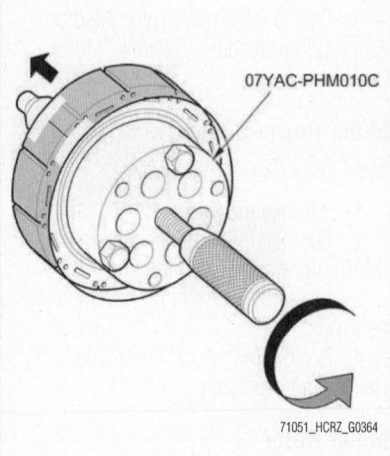

Fig. 68 Turn the handle of the special tool slowly when inserting the rotor into the stator.

7. Attach the rotor puller with its supplied bolts. When installing the rotor puller, position the puller to fit over the guide pins.
8. Turn the handle of rotor puller and remove the IMA motor rotor.

➡**To prevent damage to the rotor magnets while working on the stator, place the rotor, with puller attached, into the puller case.**

To install:

9. Install the parts in the reverse order of removal.

Note the following:

• When installing the rotor, make sure the special tool is attached with its handle turned out several threads.

• Turn the handle of the special tool slowly when inserting the rotor into the stator.

✱✱ CAUTION

The rotor is drawn into the stator by magnetic force.

• When installing the rotor puller, position the puller to fit over the guide pins.

10. Remove the plastic film.
11. CVT model: Install the drive plate.
12. M/T model: Install the flywheel and the clutch.
13. Install the transaxle. See applicable "Transmission" section.
14. Do the IMA Motor Rotor Position Calibration procedure.

Motor Rotor Position Sensor

See Figure 69.

1. Remove the IMA motor housing.
2. Remove the bolts and the IMA motor rotor position sensor.
3. Install the parts in the reverse order of removal.
4. Tighten the side bolts first, then tighten the upper bolts.
5. Do the IMA Motor Rotor Position Calibration procedure.

IMA Motor Rotor Position Calibration

➡**Do the IMA motor rotor position calibration whenever any of these actions are done:**

• The MCM is replaced.
• The IMA motor rotor position sensor is replaced or removed during service.
• The IMA motor is replaced or removed during service.
• The engine assembly is replaced or removed during service.

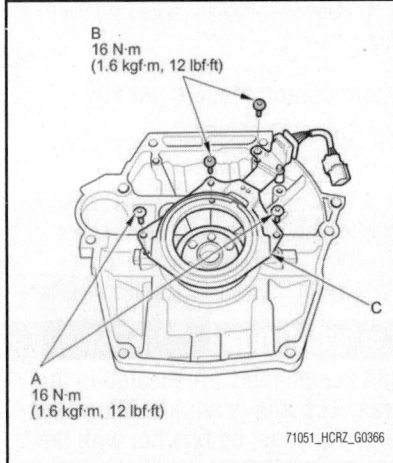

Fig. 69 Remove the bolts (A, B) and the IMA motor rotor position sensor (C).

- The transaxle is replaced or removed during service.

1. Connect the HDS to the data link connector (DLC) located under the driver's side of the dashboard.

2. Turn the ignition switch to ON (II).

3. Make sure the HDS communicates with the vehicle and the MCM (IMA system). If it doesn't, troubleshoot the DLC circuit.

4. Select IMA SYSTEM on the HDS.

5. Select the MOTOR ROTOR POSITION CALIBRATION in the ADJUSTMENT MENU of the HDS.

6. Turn the ignition switch to LOCK (0), and disconnect the HDS from the DLC.

INTELLIGENT POWER UNIT

REMOVAL & INSTALLATION

IPU Case

See Figures 70 and 71.

✳✳ CAUTION

IMA components are located in this area. The IMA is a high-voltage system. You must be familiar with the IMA system before working on or around it. Make sure you have read the IMA service precautions before doing repairs or service.

1. Make sure the ignition switch to LOCK (0).

2. Do the 12 volt battery terminal disconnection procedure.

3. Remove the IPU module air duct.

4. Remove the IPU cover.

5. Remove the PCU lid.

6. Disconnect the four IMA motor power cables from the phase motor current sensor, then wrap the end of the 12 volt power cable with insulating tape.

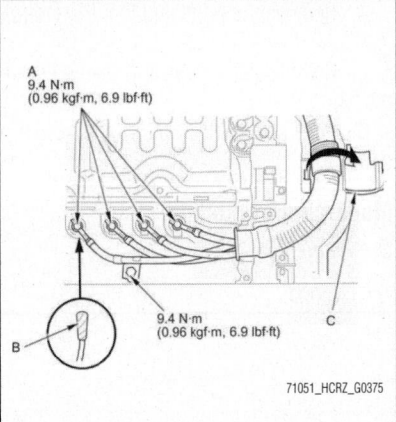

Fig. 70 Disconnect the four IMA motor power cables (A) from the phase motor current sensor, then wrap the end of the 12 volt power cable (B) with insulating tape. Open the clamp (C).

➡**Check the position of the U phase, the V phase, and the W phase cables before disconnecting them.**

7. Open the clamp.

8. Disconnect the connector.

9. Remove the bolts, then remove the IPU assembly.

10. Remove the harness clamp, the clips, and the IPU case. The IPU frame seals must be replaced with new ones when the IPU case is removed.

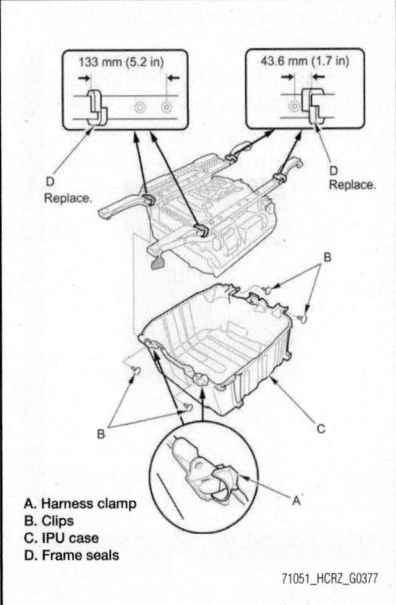

Fig. 71 Remove the harness clamp, the clips, and the IPU case. The IPU frame seals must be replaced with new ones when the IPU case is removed.

To install:

11. Install the parts in the reverse order of removal.

➡**Make sure the IMA motor power cables are correctly positioned before you reconnect them.**

12. Do the 12 volt battery terminal reconnection procedure.

IPU Cover

See Figure 72.

✳✳ CAUTION

IMA components are located in this area. The IMA is a high-voltage system. You must be familiar with the IMA system before working on or around it. Make sure you have read the IMA service precautions before doing repairs or service.

1. Remove the cargo floor lid, the IPU duct cover, and the spare tire beam.

2. Turn the battery module switch OFF.

3. Remove the bolts, the clips, and the IPU cover.

To install:

4. Install the IPU cover, the clips, and the bolts.

5. Turn the battery module switch ON.

6. Install the spare tire beam, the IPU duct cover, and the cargo floor lid.

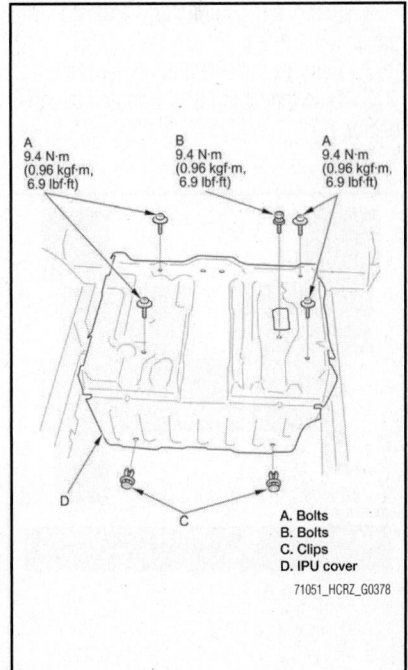

Fig. 72 Remove the bolts, the clips, and the IPU cover.

IPU Module Air Duct

See Figure 73.

1. Remove the cargo floor lid, the IPU duct cover, and the cargo area side trim panel.

2. Remove the IPU module air ducts.

3. Install the parts in the reverse order of removal.

4. Remove the bolts and the IPU module fan.

5. Remove the screws and the IPU module fan.

6. Install the parts in the reverse order of removal.

Mode Control Module

See Figure 75.

4. After installation, make sure the motor runs smoothly.

Motor Control Module (MCM)

See Figures 76 through 78.

Note the following:

• Wear insulated gloves and use insulated tools to protect you from electrical shock.

• When removing or installing items marked with , never fail to use insulated tools.

✲✲ CAUTION

IMA components are located in this area. The IMA is a high-voltage system. You must be familiar with the IMA system before working on or around it. Make sure you have read the IMA service precautions before doing repairs or service.

1. Connect the HDS to the data link connector (DLC) located under the driver's side of the dashboard.

2. Turn the ignition switch to ON (II).

3. Make sure the HDS communicates with the vehicle and the MCM. If it doesn't, troubleshoot the DLC circuit.

4. Select the IMA SYSTEM with the HDS.

5. Select the REPLACEMENT MCM MENU, then select READ DATA, and follow the HDS prompts.

➡ **If the READ DATA indicates FAILED, continue this procedure.**

6. Make sure the ignition switch to LOCK (0).

7. Remove the IPU cover.

8. Remove the bolts and the PCU lid.

9. Remove the bolts and the PCU bus plate.

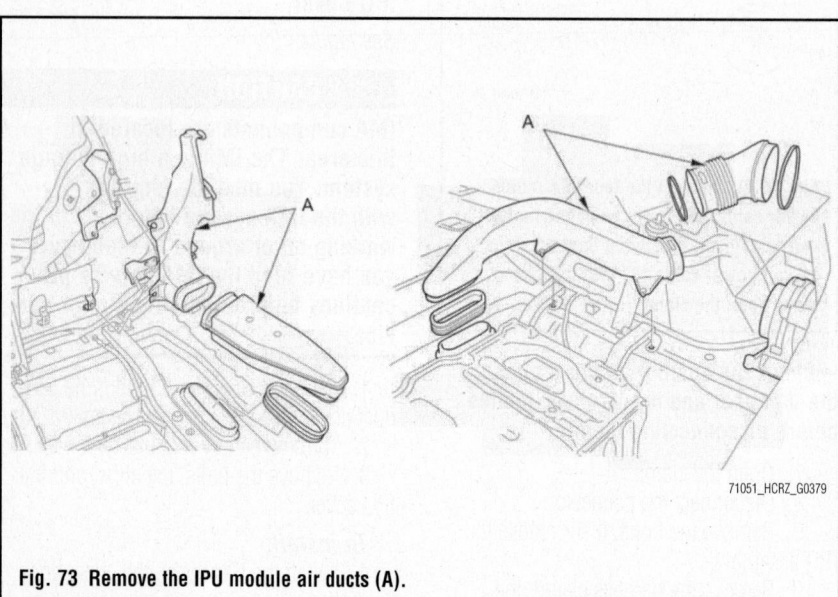

Fig. 73 Remove the IPU module air ducts (A).

IPU Module Fan

See Figure 74.

1. Remove the right cargo area side trim panel.

2. Remove the IPU module air duct.

3. Disconnect the IPU module fan connector.

1. Remove the blower unit.

2. Disconnect the 7P connector from the mode control motor. Remove the self-tapping screws and the mode control motor from the heater unit.

3. Install the motor in the reverse order of removal. Make sure the pin on the motor is properly engaged with the linkage.

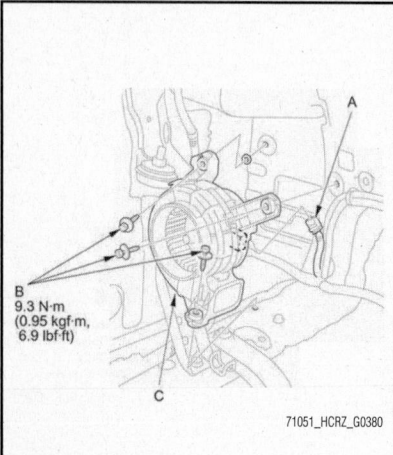

Fig. 74 Disconnect the IPU module fan connector (A). Remove the bolts (B) and the IPU module fan (C).

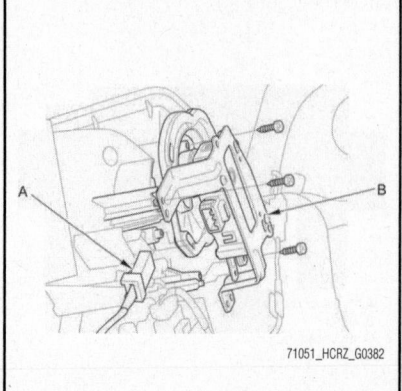

Fig. 75 Disconnect the 7P connector (A) from the mode control motor (B). Remove the self-tapping screws and the mode control motor from the heater unit.

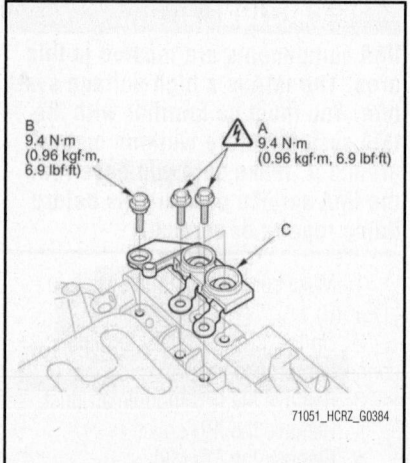

Fig. 76 Remove the bolts (A, B) and the PCU bus plate (C).

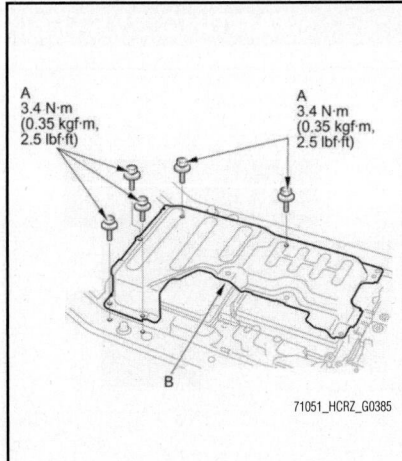

Fig. 77 Remove the bolts (A) and the PCU cover (B).

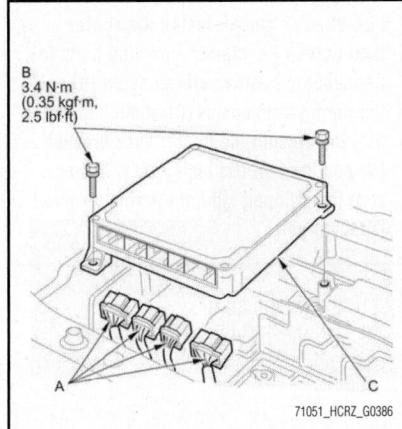

Fig. 78 Disconnect the MCM connectors (A). Remove the bolts (B) and the MCM (C).

10. Remove the bolts and the PCU cover.

11. Disconnect the MCM connectors.

12. Remove the bolts and the MCM.

To install:

13. Install the parts in the reverse order of removal.

14. Do the IMA motor rotor position calibration procedure.

15. Turn the ignition switch to ON (II).

16. Select the IMA SYSTEM with the HDS.

17. Select the REPLACEMENT MCM MENU, then select WRITE DATA, and follow the HDS prompts.

➡**If the WRITE DATA indicates FAILED, continue this procedure.**

Motor Power Inverter (MPI) Module

See Figures 79 and 80.

❋❋ **CAUTION**

IMA components are located in this area. The IMA is a high-voltage system. You must be familiar with the IMA system before working on or around it. Make sure you have read the IMA service precautions before doing repairs or service.

1. Remove the DC-DC converter.

2. Disconnect the phase motor current sensor connector.

3. Remove the bolts and the phase motor current sensor.

4. Remove the bolts.

❋❋ **CAUTION**

The IMA motor power cables carry high voltage when the engine is running or the IMA system is energized. To avoid serious injury from electrical shock, do not start the engine with the IMA motor power cables disconnected.

5. Disconnect the connectors, then remove the PCU wire harness.

6. Remove the bolts and the MPI module.

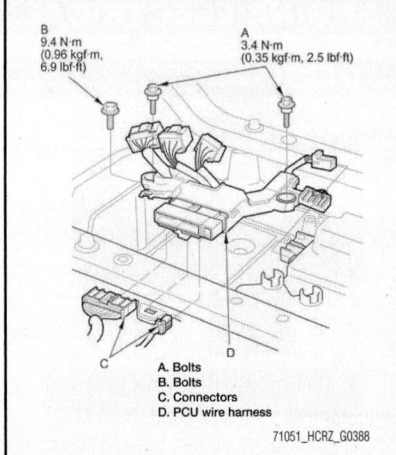

A. Bolts
B. Bolts
C. Connectors
D. PCU wire harness

Fig. 79 Remove the bolts. Disconnect the connectors, then remove the PCU wire harness.

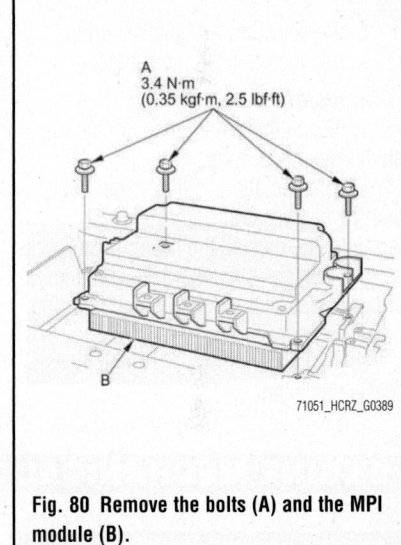

Fig. 80 Remove the bolts (A) and the MPI module (B).

7. Install the parts in the reverse order of removal.

8. Do the IMA motor rotor position calibration procedure.

ENGINE ELECTRICAL STARTING SYSTEM

STARTER

REMOVAL & INSTALLATION

See Figures 81 and 82.

1. Do the 12 volt battery terminal disconnection procedure. Refer to "Battery Cables" in "Battery System" section.

2. Remove the air cleaner.

3. M/T model: Remove the clutch hose bracket bolts and nut.

4. CVT model: Disconnect the solenoid wire harness connector and the CVT input shaft (drive pulley) speed sensor connector.

5. Disconnect the positive starter cable and the harness connector from the S terminal, then remove the heater hose bracket and the harness clamps.

6. Remove the two bolts securing the starter, then remove the starter.

To install:

7. Install the starter, then tighten the starter mounting bolts.

8. Connect the positive starter cable and the harness connector to the S terminal, then install the heater hose bracket and the harness clamps. Make sure the crimped side of the ring terminal is facing out.

9. M/T model: Install the clutch hose bracket bolts and nut.

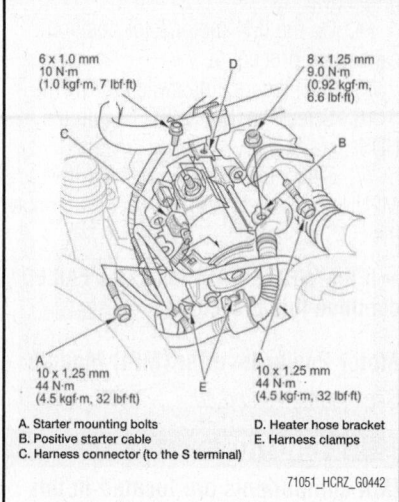

A. Starter mounting bolts
B. Positive starter cable
C. Harness connector (to the S terminal)
D. Heater hose bracket
E. Harness clamps

71051_HCRZ_G0442

Fig. 81 M/T model: Install the starter, then tighten the starter mounting bolts (A). Connect the positive starter cable (B) and the harness connector (C) to the S terminal, then install the heater hose bracket (D) and the harness clamps (E). Make sure the crimped side of the ring terminal is facing out.

10. CVT model: Install the solenoid wire harness connector and the CVT input shaft (drive pulley) speed sensor connector.

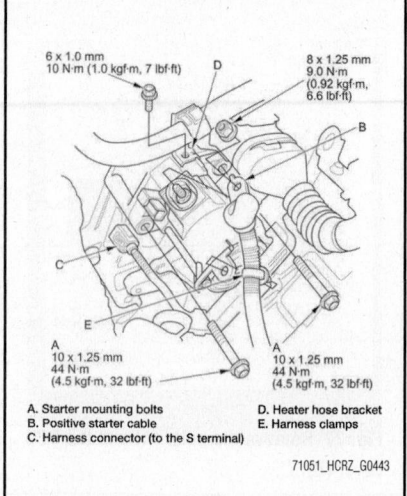

A. Starter mounting bolts
B. Positive starter cable
C. Harness connector (to the S terminal)
D. Heater hose bracket
E. Harness clamps

71051_HCRZ_G0443

Fig. 82 CVT model: Install the starter, then tighten the starter mounting bolts (A). Connect the positive starter cable (B) and the harness connector (C) to the S terminal, then install the heater hose bracket (D) and the harness clamps (E). Make sure the crimped side of the ring terminal is facing out.

11. Install the air cleaner.

12. Do the 12 volt battery terminal reconnection procedure.

ENGINE MECHANICAL

ACCESSORY DRIVE BELT SYSTEM

ADJUSTMENT

➡**Drive belt is automatically adjusted by the belt tensioner. No manual adjustment is required.**

INSPECTION

See Figure 83.

1. Inspect the belt for cracks or damage.

➡**If the belt is cracked or damaged, replace it.**

2. Check the position of the drive belt auto-tensioner indicator is within the standard range as shown. If it is out of the standard range, replace the drive belt, then check the drive belt auto-tensioner indicator position.

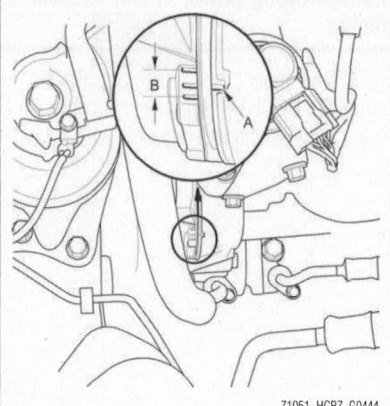

71051_HCRZ_G0444

Fig. 83 Check the position of the drive belt auto-tensioner indicator (A) is within the standard range (B) as shown. If it is out of the standard range, replace the drive belt, then check the drive belt auto-tensioner indicator position.

REMOVAL & INSTALLATION

Drive Belt

See Figure 84.

Special Tools Required: Belt Tension Release Tool Snap-on YA9317, commercially available.

1. Move the drive belt auto-tensioner using the belt tension release tool in the direction shown to relieve tension from the drive belt, then remove the drive belt.

2. Install a new belt in the reverse order of removal.

3. Check the position of the drive belt auto-tensioner indicator is within the standard range. If it is out of the standard range, check the drive belt auto-tensioner.

Fig. 84 Move the drive belt auto-tensioner (A) using the belt tension release tool (YA9317) in the direction shown to relieve tension from the drive belt, then remove the drive belt.

AIR CLEANER

REMOVAL & INSTALLATION

Air Cleaner Assembly

See Figure 85.

1. Disconnect the MAF sensor/IAT sensor connector and remove the harness clips.

➡**Be careful not to damage the MAF sensor/IAT sensor on the air cleaner cover.**

2. Disconnect the breather hose.
3. Loosen the hose bands and remove the bolts, then remove the air cleaner from the throttle body and the intake air tube.

To install:

4. Install the parts in the reverse order of removal.
5. When tightening the screw of the hose band, align the edge of the hose band with the mark painted on the hose band.

➡**If you tighten the screw over the mark, replace the hose band.**

Air Cleaner Element

See Figure 86.

1. Open the air cleaner housing cover.
2. Remove the air cleaner element from the air cleaner housing.
3. Check the air cleaner element for damaged, dirt, or clogging. If it is damaged or clogged, replace it.

➡**Do not use compressed air to clean the air cleaner element.**

4. Clean and remove any debris from inside the air cleaner housing.
5. Install the parts in the reverse order of removal.
6. If the idle speed fluctuates, do the idle speed inspection.

CAMSHAFT & BEARINGS

REMOVAL & INSTALLATION

Camshaft

1. Remove the camshaft sprocket.
2. Remove the rocker arm assembly.
3. Remove the CMP sensor.
4. Remove the CMP pulse plate.

5. Remove the camshaft.
6. Installation is the reverse of the removal procedure.

Camshaft Sprocket

See Figures 87 through 90.

➡**Keep the cam chain away from magnetic fields.**

1. Remove the cylinder head cover.
2. Make a reference mark across the camshaft sprocket and the cam chain.

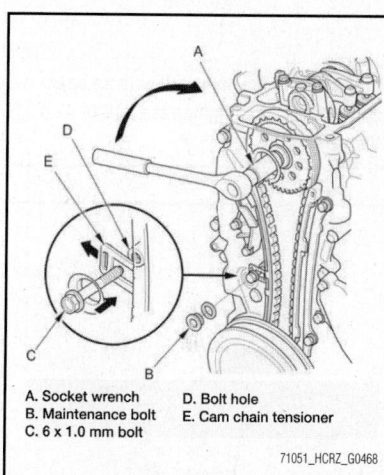

A. Socket wrench
B. Maintenance bolt
C. 6 x 1.0 mm bolt
D. Bolt hole
E. Cam chain tensioner

Fig. 87 Hold the crankshaft pulley, and set the socket wrench on the camshaft sprocket bolt. Remove the maintenance bolt and turn the camshaft clockwise to compress the cam chain tensioner, then install the 6 x 1.0 mm bolt in the bolt hole in the engine block through the maintenance hole and cam chain tensioner.

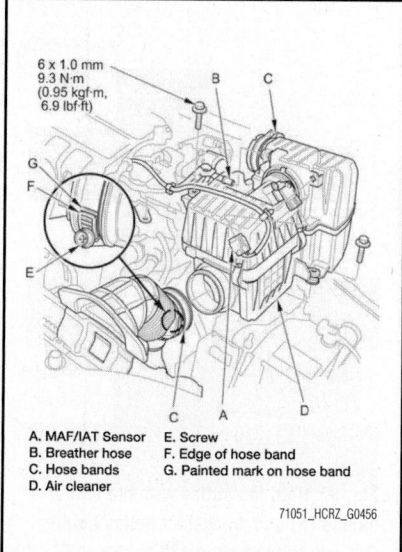

A. MAF/IAT Sensor
B. Breather hose
C. Hose bands
D. Air cleaner
E. Screw
F. Edge of hose band
G. Painted mark on hose band

Fig. 85 Air cleaner assembly components identified

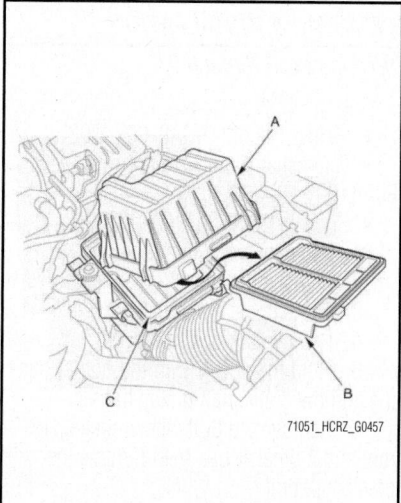

Fig. 86 Open the air cleaner housing cover (A). Remove the air cleaner element (B) from the air cleaner housing (C).

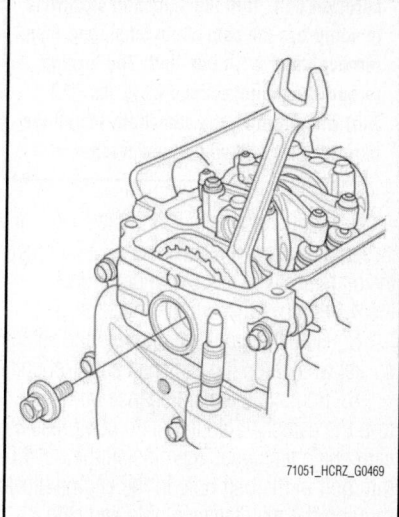

Fig. 88 Hold the camshaft with an open-end wrench, then remove the camshaft sprocket.

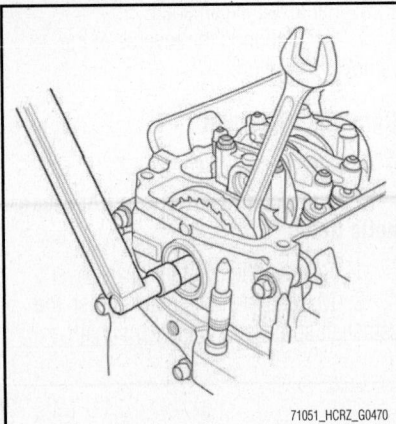

Fig. 89 Hold the camshaft with an open-end wrench, then tighten the bolt to 42 ft. lbs. (57 Nm).

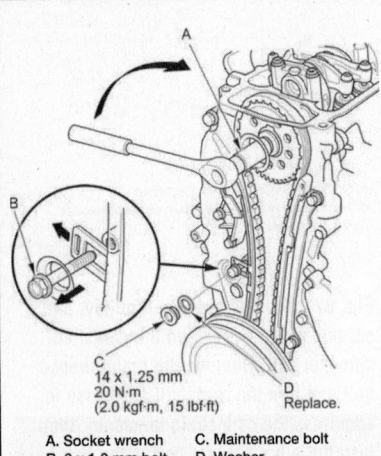

A. Socket wrench
B. 6 x 1.0 mm bolt
C. Maintenance bolt
D. Washer

14 x 1.25 mm
20 N·m
(2.0 kgf·m, 15 lbf·ft)
D. Replace.

Fig. 90 Hold the crankshaft pulley, and set the socket wrench on the camshaft sprocket bolt. Turn the camshaft clockwise to compress the cam chain tensioner, then remove the 6 x 1.0 mm bolt. The turning torque should not exceed 42 ft. lbs. (57 Nm) when turning the camshaft. Install the maintenance bolt with a new washer.

3. Apply new engine oil to the slider surface of the cam chain tensioner slider through the oil return hole in the cylinder head.

4. Remove the cylinder head plug.

5. Hold the crankshaft pulley, and set the socket wrench on the camshaft sprocket bolt.

6. Remove the maintenance bolt and turn the camshaft clockwise to compress the cam chain tensioner, then install the 6 x 1.0 mm bolt in the bolt hole in the engine block through the maintenance hole and cam chain tensioner.

7. The turning torque should not exceed 42 ft. lbs. (57 Nm) when turning the camshaft.

➥Do not turn the camshaft counter-clockwise.

8. Hold the camshaft with an open-end wrench, then remove the camshaft sprocket.

➥Hang the cam chain with a wire.

To install:

➥Keep the cam chain away from magnetic fields.

9. Install the cam chain to the camshaft sprocket by aligning the reference mark, then install the camshaft sprocket on the camshaft.

10. Apply new engine oil to the threads and flange of camshaft sprocket bolt.

11. Hold the camshaft with an open-end wrench, then tighten the bolt to 42 ft. lbs. (57 Nm).

12. Apply new engine oil to the slider surface of the cam chain tensioner slider through the oil return hole in the cylinder head.

13. Hold the crankshaft pulley, and set the socket wrench on the camshaft sprocket bolt.

14. Turn the camshaft clockwise to compress the cam chain tensioner, then remove the 6 x 1.0 mm bolt. The turning torque should not exceed 42 ft. lbs. (57 Nm) when turning the camshaft.

➥Do not turn the camshaft counter-clockwise.

15. Install the maintenance bolt with a new washer.

16. Install a new cylinder head plug.

17. Install the cylinder head cover.

CRANKSHAFT PULLEY

REMOVAL & INSTALLATION

See Figures 91 through 93.

Special Tools Required:
- Holder Handle 07JAB-001020B
- Crankshaft Pulley Holder 07AAB-RJAA100
- Socket, 19 mm 07JAA-001020A

1. Raise the vehicle on the lift.
2. Remove the right front wheel.
3. Remove the splash shield.
4. Remove the drive belt.
5. Hold the pulley with the holder handle and the crankshaft pulley holder.
6. Remove the bolt with a socket, 19 mm and a breaker bar, then remove the crankshaft pulley.

To install:

7. Remove any oil and clean the crankshaft pulley, the crankshaft, the bolt, and the washer. Lubricate with new engine oil as shown.

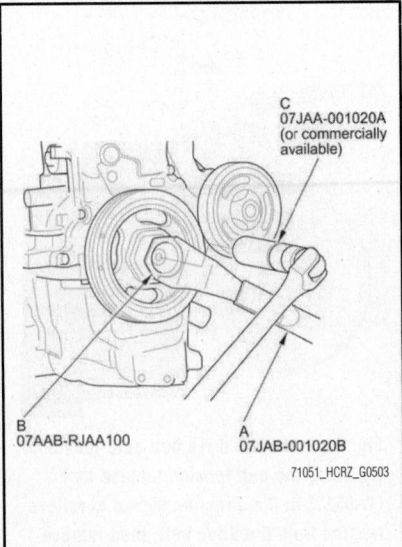

C
07JAA-001020A
(or commercially available)

B
07AAB-RJAA100

A
07JAB-001020B

Fig. 91 Hold the pulley with the holder handle (A) and the crankshaft pulley holder (B). Remove the bolt with a socket, 19 mm (C) and a breaker bar, then remove the crankshaft pulley.

8. Install the crankshaft pulley.

9. When a new crankshaft or a new pulley bolt is installed: Tighten the crankshaft pulley bolt. Do not use an impact wrench.

a. Hold the pulley with the holder handle and crankshaft pulley holder, torque the bolt to 130 ft. lbs. (177 Nm) with a torque wrench and a socket, then remove the bolt.

b. Torque the bolt to 29 ft. lbs. (39 Nm) with a torque wrench and a socket.

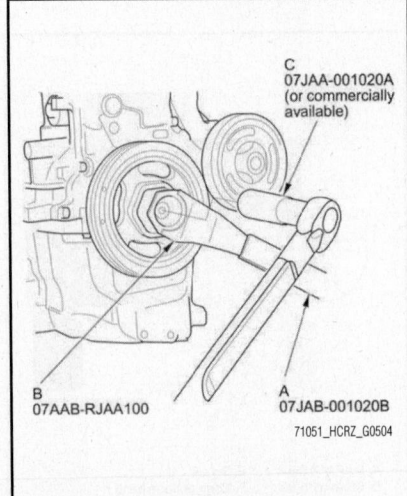

C
07JAA-001020A
(or commercially available)

B
07AAB-RJAA100

A
07JAB-001020B

Fig. 92 Hold the pulley with the holder handle (A) and crankshaft pulley holder (B), torque the bolt to 130 ft. lbs. (177 Nm) with a torque wrench and a socket (C), then remove the bolt.

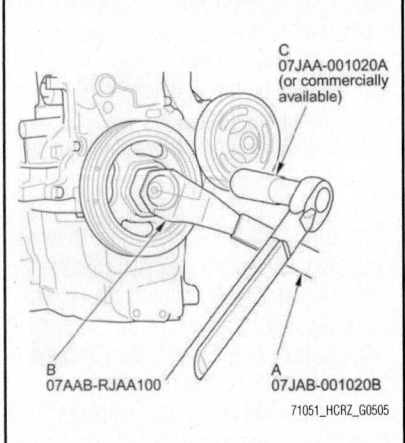

Fig. 93 Hold the pulley with the holder handle (A) and crankshaft pulley holder (B), then torque the bolt to 27 ft. lbs. (37 Nm) with a torque wrench and a socket (C).

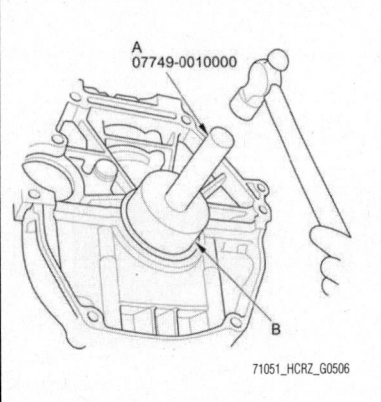

Fig. 94 Use the driver handle, 15 ? 135 L and the oil seal driver attachment, 96 mm to drive a new crankshaft oil seal squarely into the block to the specified installed height.

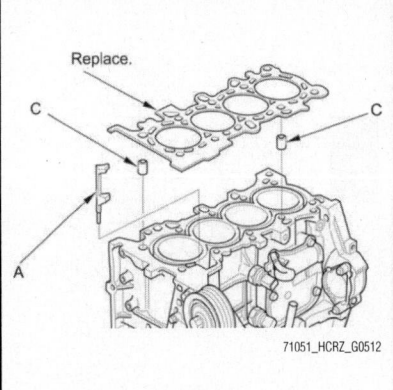

Fig. 96 Install a new coolant separator (A) in the engine block whenever the engine block is replaced. Install a new cylinder head gasket (B) and the dowel pins (C) on the engine block. Always use a new cylinder head gasket.

c. Tighten the bolt an additional 94 degrees.

10. When the crankshaft or the pulley bolt is reused: Tighten the crankshaft pulley bolt. Do not use an impact wrench.

a. Hold the pulley with the holder handle and crankshaft pulley holder, then torque the bolt to 27 ft. lbs. (37 Nm) with a torque wrench and a socket.

b. Tighten the bolt an additional 90 degrees.

11. Install the drive belt.

12. Install the splash shield.

13. Install the right front wheel.

CRANKSHAFT REAR COVER & SEAL

REMOVAL & INSTALLATION

See Figures 94 and 95.

Special Tools Required:
• Driver Handle, 15 x 135L 07749-0010000
• Oil Seal Driver Attachment, 96 mm 07ZAD-PNAA100

1. Remove the transaxle. Refer to "Transmission" section.

2. Remove the IMA motor rotor, the IMA motor housing, and the IMA motor rotor position sensor.

3. Clean and dry the crankshaft oil seal housing.

4. Apply a light coat of new engine oil to the lip of the crankshaft oil seal.

5. Use the driver handle, 15 ? 135 L and the oil seal driver attachment, 96 mm to drive a new crankshaft oil seal squarely into the block to the specified installed height.

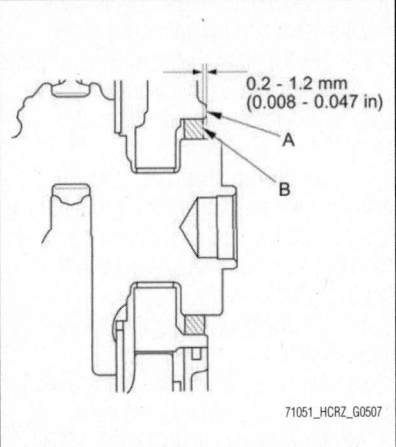

Fig. 95 Measure the distance between the cylinder block (A) and the oil seal (B). Oil Seal Installed Height should be 0.008–0.047 in. (0.2–1.2 mm).

6. Measure the distance between the cylinder block and the oil seal. Oil Seal Installed Height should be 0.008–0.047 in. (0.2–1.2 mm).

7. Install the IMA motor rotor position sensor, the IMA motor housing, and the IMA motor rotor.

8. Install the transaxle:

CYLINDER HEAD

REMOVAL & INSTALLATION

See Figures 96 through 99.

Note the following:
• Use fender covers to avoid damaging painted surfaces.

• To avoid damage, unplug the wiring connectors carefully while holding the connector portion.

• Connect the HDS to the DLC, and monitor ECT SENSOR 1. To avoid damaging the cylinder head, wait until the engine coolant temperature drops below 100°F (38°C) before loosening the cylinder head bolts.

• Mark all wiring and hoses to avoid misconnection. Also, be sure that they do not contact other wiring or hoses, or interfere with other parts.

• Keep the cam chain away from magnetic fields.

1. Remove the cowl cover and the under-cowl panel.

Relieve the fuel pressure.

2. Drain the engine coolant.

3. Do the 12 volt battery removal procedure.

4. Remove the throttle body. Refer to "Gasoline Fuel Injection" section.

5. Remove the intake manifold/chamber assembly. Refer to "Intake Manifold" in this section.

6. Remove the injectors. Refer to "Gasoline Fuel Injection" section.

7. Remove the four ignition coils. Refer to "Engine Electrical" section.

8. Remove the warm-up TWC.

9. Disconnect the CMP sensor connector and the secondary HO2S connector.

10. Disconnect the oil pressure switch connector and the knock sensor connector.

11. Remove the heater hose bracket and the ground cable.

12. Remove the harness holder bolt, then remove the harness holder from the bracket.

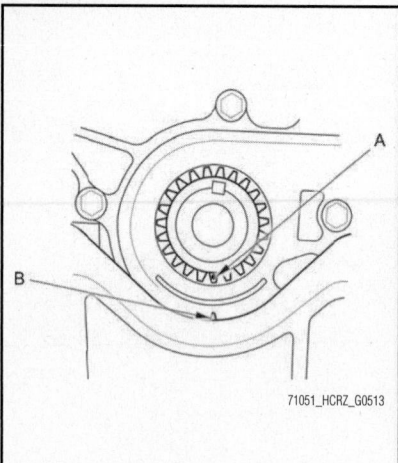

Fig. 97 Set the crankshaft to top dead center (TDC). Align the TDC mark (A) on the crankshaft sprocket with the pointer (B) on the oil pump.

71051_HCRZ_G0513

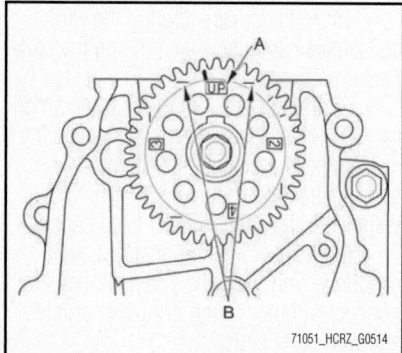

Fig. 98 Set the camshaft TDC. The "UP" mark (A) on the camshaft sprocket should be at the top, and the TDC grooves (B) on the camshaft sprocket should line up with the top edge of the head.

71051_HCRZ_G0514

13. Disconnect the upper radiator hose, the water bypass hose and the heater hose.

14. Disconnect the ECT sensor 1 connector.

15. Remove the cylinder head cover.

16. Remove the cam chain. Refer to "Timing (Camshaft) Chain & Sprocket" in this section.

17. Remove the cylinder head bolts. To prevent warping, loosen the bolts in an alternating sequence 1/3 turn at a time; repeat the sequence until all bolts are loosened.

18. Remove the cylinder head.

To install:

19. Clean the cylinder head and the engine block surface.

20. Install a new coolant separator in the engine block whenever the engine block is replaced.

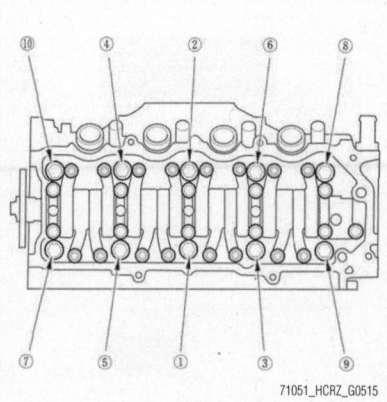

71051_HCRZ_G0515

Fig. 99 Torque the cylinder head bolts in sequence to 22 ft. lbs. (29 Nm) with a beam-type torque wrench if possible. Tighten all cylinder head bolts an additional 130 degrees.

21. Install a new cylinder head gasket and the dowel pins on the engine block. Always use a new cylinder head gasket.

22. Set the crankshaft to top dead center (TDC). Align the TDC mark on the crankshaft sprocket with the pointer on the oil pump.

23. Set the camshaft TDC. The "UP" mark on the camshaft sprocket should be at the top, and the TDC grooves on the camshaft sprocket should line up with the top edge of the head.

24. Install the cylinder head on the engine block.

25. Apply new engine oil to the threads and flange of all cylinder head bolts.

26. Torque the cylinder head bolts in sequence to 22 ft. lbs. (29 Nm) with a beam-type torque wrench if possible.

➡When using a preset-click-type torque wrench, be sure to tighten slowly and do not overtighten. If a bolt makes any noise while you are torquing it, loosen the bolt and retighten it from the first step.

27. Tighten all cylinder head bolts an additional 130 degrees.

28. Install the cam chain. Refer to "Timing (Camshaft) Chain & Sprocket" in this section.

29. Install the cylinder head cover.

30. Connect the upper radiator hose, the water bypass hose, and the heater hose.

31. Connect the ECT sensor 1 connector.

32. Install the harness holder.

33. Install the heater hose bracket and the ground cable.

34. Connect the oil pressure switch connector and the knock sensor connector.

35. Connect the CMP sensor connector and the secondary HO2S connector.

36. Install the warm-up TWC.

37. Install the four ignition coils.

38. Install the injectors.

39. Install the intake manifold/chamber assembly.

40. Install the throttle body.

41. After installation, check that all tubes, hoses, and connectors are installed correctly.

42. Do the 12 volt battery installation procedure.

43. Inspect for fuel leaks. Turn the ignition switch to ON (II) (do not operate the starter) so the fuel pump runs for about 2 seconds and pressurizes the fuel line. Repeat this operation three times, then check for fuel leakage at any point in the fuel line.

44. Refill the radiator with engine coolant, and bleed the air from the cooling system.

45. Do the CKP pattern clear/CKP pattern learn procedure.

46. Inspect the idle speed. Refer to "Gasoline Fuel Injection System" section.

47. Inspect the ignition timing. Refer to "Engine Electrical" section.

ENGINE OIL & FILTER

OIL LEVEL CHECK

1. Park the vehicle on level ground, and start the engine. Hold the engine at 3,000 rpm with no load (M/T in neutral, A/T in P or N) until the radiator fan comes on, then turn off the engine, and wait a few minutes.

2. Remove the dipstick, and wipe off the dipstick, then reinstall the dipstick.

3. Remove the dipstick, and check the engine oil level. It should be between the upper mark and lower mark.

4. If the engine oil level is near or below the lower mark, check for oil leakage, and add engine oil to bring it to the upper mark.

OIL & FILTER CHANGE

Engine Oil

1. Allow the engine to reach operating temperature (fan comes on at least twice).

2. Remove the splash shield.

3. Remove the drain bolt and drain the engine oil.

4. Reinstall the drain bolt with a new washer and torque to 30 ft. lbs. (40 Nm).

5. Install the splash shield.

6. Refill the engine with the recommended engine oil:

- At Oil Change: 3.6 US qt. (3.4 L)
- At Oil Change Including Filter: 3.8 US qt. (3.6 L)
- After Engine Overhaul: 4.4 US qt. (4.2 L)

7. Run the engine for more than 3 minutes, then check the oil level and for any oil leakage.

8. If the Maintenance Minder required replacing the engine oil, reset the Maintenance Minder, when this procedure is complete. If the Maintenance Minder did not require engine oil replacement, go to step 9.

9. Turn the ignition switch to LOCK.

10. Connect the HDS to the DLC.

11. Turn the ignition switch to ON (II).

12. Make sure the HDS communicates with the vehicle and ECM/PCM. If it does not communicate, troubleshoot the D LC circuit.

13. Select GAUGE in the BODY ELECTRICAL with the HDS.

14. Select ADJUSTMENT in GAUGES with the HDS.

15. Select MAINTENANCE MINDER in the ADJUSTMENT with the HDS.

16. Select RESET in the MAINTENANCE MINDER with the HDS.

17. Select RESETTING THE ENGINE OIL LIFE with the HDS.

Engine Oil Filter

See Figure 100.

1. Drain the engine oil.
2. Remove the oil filter with the oil filter wrench.

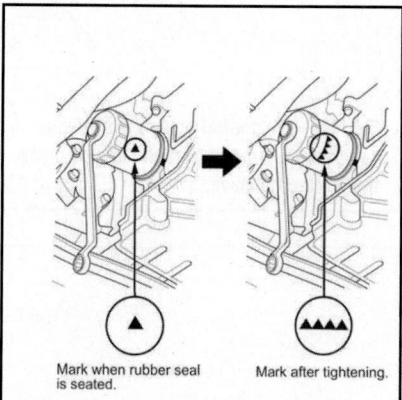

Mark when rubber seal is seated.   Mark after tightening.

Number or Mark when rubber seal is seated	1 or ▼	2 or ▼▼	3 or ▼▼▼	4 or ▼▼▼▼
Number or Mark after tightening	4 or ▼▼▼▼	1 or ▼	2 or ▼▼	3 or ▼▼▼

71051_HCRZ_G0623

Fig. 100 Showing oil filter markings and tightening settings

3. Inspect the filter to make sure the rubber seal is not stuck to the oil filter seating surface of the engine.

4. Inspect the threads and the rubber seal on the new filter. Clean the seat on the oil pan, then apply a light coat of new engine oil to the filter rubber seal. Use only filters with a built-in bypass system.

5. Install the oil filter by hand.

6. If oil filter has none of the markings indicated below, after the rubber seal seats, tighten the oil filter clockwise with the oil filter wrench about 3/4 turn.

7. If four numbers or marks (1 to 4 or ▼ to ▼▼▼▼) are printed around the outside of the filter, you can use the following procedure to tighten the filter.

   a. Spin the filter on until its seal lightly seats against the oil pan, and note which number or mark is at the bottom.

   b. Tighten the filter by turning it clockwise three numbers or marks from the one you noted. For example, if mark ▼ is at the bottom when the seal is lightly seated, tighten the filter until the mark ▼▼▼▼ comes around the bottom.

8. After installation, fill the engine with the engine oil up to the specified level, run the engine for more than 3 minutes, then check for oil leakage.

EXHAUST MANIFOLD

REMOVAL & INSTALLATION

➡This vehicle does not use a separate exhaust manifold. The front exhaust pipe flange bolts directly to the engine block.

FRONT COVER SEAL (CAM CHAIN CASE SEAL)

REMOVAL & INSTALLATION

See Figures 101 and 102.

Special Tools Required:
• Bearing Driver Attachment, 52 x 55 mm 07746-0010400
• Driver Handle, 15 x 135L 07749-0010000

1. Clean and dry the cam chain case oil seal housing.

2. Apply a light coat of new engine oil to the lip of the cam chain case oil seal.

3. Use the driver handle, 15 x 135 L, and the bearing driver attachment, 52 x 55 mm, to drive a new oil seal squarely into the chain case to the specified installed height.

4. Measure the distance between the chain case surface and the oil seal. Oil seal installed height should be: 1.193–1.220 in. (30.3–31.0 mm)

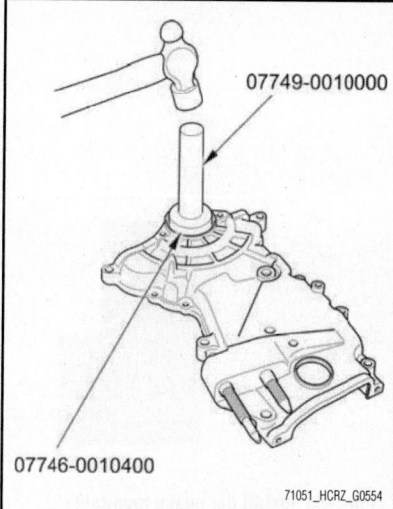

07749-0010000

07746-0010400

71051_HCRZ_G0554

Fig. 101 Use the driver handle, 15 x 135 L, and the bearing driver attachment, 52 x 55 mm, to drive a new oil seal squarely into the chain case to the specified installed height.

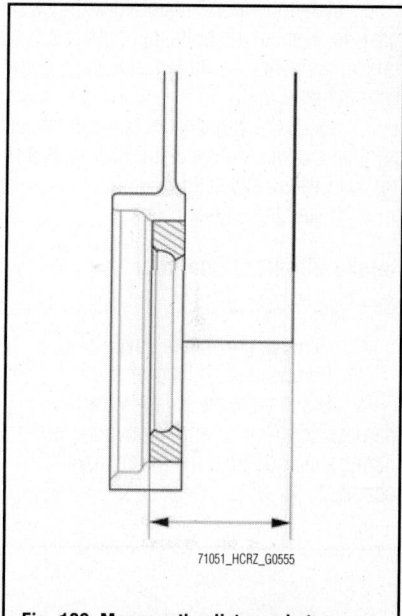

71051_HCRZ_G0555

Fig. 102 Measure the distance between the chain case surface and the oil seal.

INTAKE MANIFOLD

REMOVAL & INSTALLATION

Intake Manifold

See Figure 103.

1. Remove the engine cover.
2. Disconnect the EGR valve connector and the PCV hose. Remove the engine wire harness holder and the dipstick.
3. Remove the intake manifold.
4. Disassemble the intake manifold.

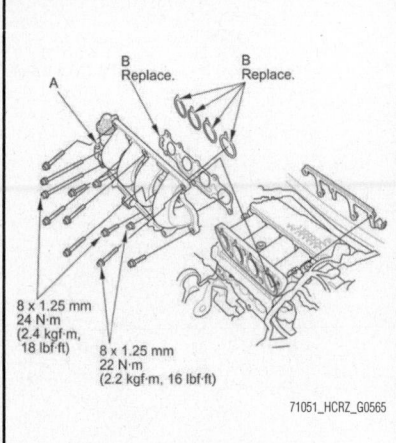

Fig. 103 Install the intake manifold (A) with new gaskets (B). Tighten the bolts and nuts in a crisscross pattern in three steps, beginning with the inner bolt.

To install:

5. Reassemble the intake manifold.

6. Install the intake manifold with new gaskets. Tighten the bolts and nuts in a crisscross pattern in three steps, beginning with the inner bolt.

7. Install the engine wire harness holder and the dipstick. Connect the PCV hose and the EGR valve connector.

8. Install the engine cover.

Intake Manifold Chamber

See Figures 104 and 105.

1. Remove the engine cover.
2. Remove the air cleaner.
3. Disconnect the engine wire harness connectors, and remove the wire harness clamps from the intake manifold chamber.

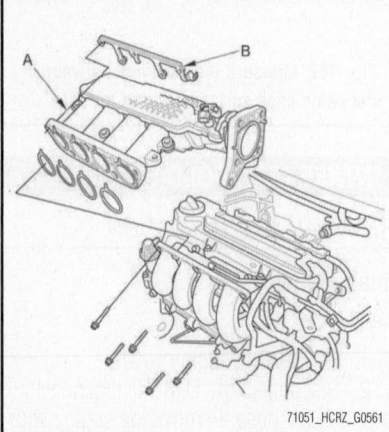

Fig. 104 Remove the intake manifold chamber (A) and the intake manifold flange plate (B).

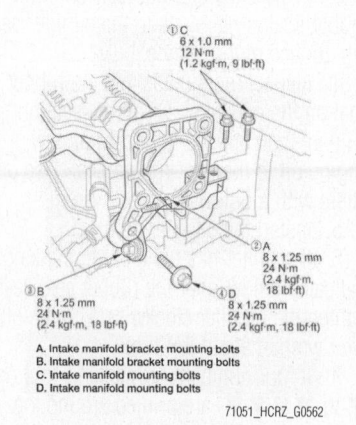

Fig. 105 Loosen the intake manifold bracket mounting bolts, then loosely install the intake manifold mounting bolts. Tighten the bolts in the numbered sequence shown.

4. Disconnect the brake booster vacuum hose. Remove the water bypass hose from the clamp.

5. Remove the throttle body without disconnecting the water bypass hoses.

6. Remove the intake manifold mounting bolts.

7. Remove the intake manifold chamber and the intake manifold flange plate.

8. Disassemble the intake manifold chamber.

To install:

9. Reassemble the intake manifold chamber.

10. Install the intake manifold chamber with new gaskets and the intake manifold flange plate. Tighten the bolts in a crisscross pattern in three steps, beginning with the inner bolt. Tighten to 16 ft. lbs. (22 Nm).

11. Loosen the intake manifold bracket mounting bolts, then loosely install the intake manifold mounting bolts.

12. Tighten the bolts in the numbered sequence shown.

13. Install the throttle body with a new gasket.

14. Connect the brake booster vacuum hose and install the water bypass hose to the clamp.

15. Connect the engine wire harness connectors, and install the wire harness clamps to the intake manifold chamber:
- Throttle actuator connector
- MAP sensor connector

16. Install the air cleaner.

17. Install the engine cover.

OIL PAN

REMOVAL & INSTALLATION

See Figures 106 through 109.

1. If the engine is already out of the vehicle, go to step 7.

2. Remove the splash shield.

3. Drain the engine oil.

4. Remove the drive belt.

5. Remove the A/C compressor without disconnecting the A/C hoses.

6. Remove the torque rod.

7. Remove the dipstick, then remove the dipstick tube.

8. Remove the CKP sensor cover, then disconnect the CKP sensor connector.

9. CVT model: Remove the driveshaft heat shield.

10. Remove the torque rod bracket.

11. Remove the transaxle mounting bolts.

12. Remove the oil pan bolts.

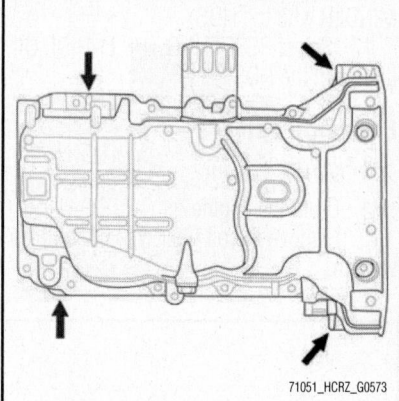

Fig. 106 Using a flat blade screwdriver, separate the oil pan from the engine block in the places shown.

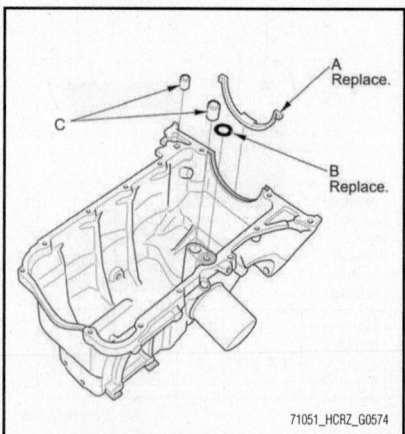

Fig. 107 Install a new oil pan gasket (A), a new O-ring (B), and the dowel pins (C) on the oil pan.

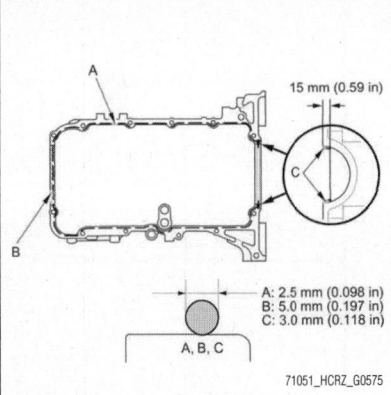

Fig. 108 Apply a 2.5 mm (0.098 in) diameter bead of liquid gasket along the broken line (A). Apply a 5.0 mm (0.197 in) diameter bead of liquid gasket to the shaded area (B). Apply a 3.0 mm (0.118 in) diameter bead of liquid gasket along the broken line (C).

13. Using a flat blade screwdriver, separate the oil pan from the engine block in the places shown.

14. Remove the oil pan.

➥**Lower the oil pan carefully so as not to damage the IMA motor (resolver stator).**

To install:

15. Remove all of the old liquid gasket from the oil pan mating surfaces, the bolts, and the bolt holes.

16. Clean and dry the oil pan mating surfaces and the O-ring groove.

17. Install a new oil pan gasket, a new O-ring, and the dowel pins on the oil pan.

18. Apply liquid gasket (P/N 08718-0004) to the engine block mating surface of

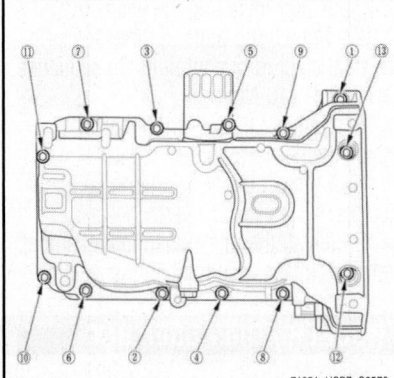

Fig. 109 Tighten the bolts in three steps. Wipe off the excess liquid gasket from crankshaft pulley end and the drive plate end.

the oil pan and to the inside edge of the bolt holes. Install the component within 5 minutes of applying the liquid gasket.

    a. Apply a 2.5 mm (0.098 in) diameter bead of liquid gasket along the broken line.

    b. Apply a 5.0 mm (0.197 in) diameter bead of liquid gasket to the shaded area.

    c. Apply a 3.0 mm (0.118 in) diameter bead of liquid gasket along the broken line.

    d. If too much time has passed after applying the liquid gasket, remove the old liquid gasket and residue, then reapply new liquid gasket.

19. Install the oil pan. Note the following:

• Raise the oil pan carefully not to damage IMA motor rotor position sensor.

• Wait at least 30 minutes before filling the engine with oil.

• Do not run the engine for at least 3 hours after installing the oil pan.

• Make sure to install the bolts in the correct locations according to size.

20. Tighten the bolts in three steps. Wipe off the excess liquid gasket from crankshaft pulley end and the drive plate end.

    • Bolt 1: 18 ft. lbs. (24 Nm)
    • Bolts 2–13: 9 ft. lbs. (12 Nm)

21. Install the transaxle mounting bolts and tighten to 47 ft. lbs. (64 Nm).

22. Install the torque rod bracket.

23. CVT model: Install the driveshaft heat shield. Tighten the bolts to 16 ft. lbs. (22 Nm).

24. Connect the CKP sensor connector, then install the CKP sensor cover.

25. Install the dipstick tube with a new O-ring, then install the dipstick.

26. If the engine is still in the vehicle, do steps 14 through 17.

27. Install the torque rod.

28. Install the A/C compressor.

29. Install the drive belt.

30. Install the splash shield.

31. Refill the engine with engine oil.

OIL PUMP

REMOVAL, OVERHAUL & INSTALLATION

See Figures 110 and 111.

Special Tools Required:
• Support Eyelet 07AAK-SNAA600
• Engine Support Hanger, A and Reds AAR-T1256
• Engine Hanger Adapter VSB02C000026

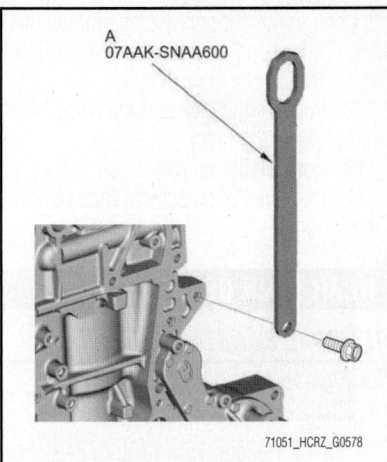

Fig. 110 Attach the support eyelet (A) to the cylinder block.

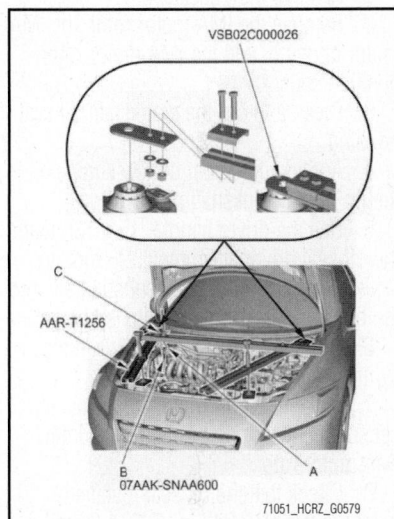

Fig. 111 Install the engine hanger adapter (VSB02C000026) to the engine support hanger (AAR-T1256), then install the engine support hanger onto the vehicle, and attach the hook (A) to the support eyelet (B). Tighten the wing nut (C) by hand, and lift and support the engine/IMA motor/transaxle.

1. Open the hood, and secure it with the hood support rod in the wide-open position.

2. Drain the engine oil.

3. Remove the cowl cover and the under-cowl panel.

4. Support the engine with a jack and a wood block under the oil pan.

5. Remove the cam chain.

6. Remove the drive belt auto-tensioner.

7. Attach the support eyelet to the cylinder block.

8. Install the engine hanger adapter (VSB02C000026) to the engine support hanger (AAR-T1256), then install the engine support hanger onto the vehicle, and attach

the hook to the support eyelet. Tighten the wing nut by hand, and lift and support the engine/IMA motor/transaxle.

9. Remove the jack and the wood block from under the oil pan.

10. Remove the oil pan.

11. Remove the oil screen, then remove the oil pump.

REAR MAIN SEAL

REMOVAL & INSTALLATION

See Figures 112 and 113.

Special Tools Required:
- Driver Handle, 15 x 135L 07749-0010000
- Oil Seal Driver Attachment, 96 mm 07ZAD-PNAA100

1. Remove the transaxle:

2. Remove the IMA motor rotor, the IMA motor housing, and the IMA motor rotor position sensor.

3. Clean and dry the crankshaft oil seal housing.

4. Apply a light coat of new engine oil to the lip of the crankshaft oil seal.

5. Use the driver handle, 15 x 135 L and the oil seal driver attachment, 96 mm, to drive a new crankshaft oil seal squarely into the block to the specified installed height.

6. Measure the distance between the cylinder block and the oil seal.

7. Install the IMA motor rotor position sensor, the IMA motor housing, and the IMA motor rotor.

8. Check that the oil seal installed height is 0.008–0.047 in. (0.2–1.2 mm)

9. Install the transaxle:

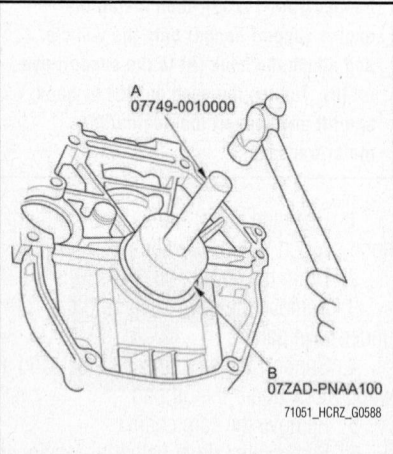

Fig. 112 Use the driver handle, 15 x 135 L (A), and the oil seal driver attachment, 96 mm (B), to drive a new crankshaft oil seal squarely into the block to the specified installed height.

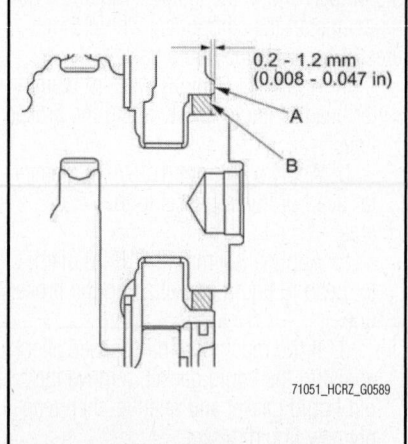

Fig. 113 Measure the distance between the cylinder block (A) and the oil seal (B).

CYLINDER HEAD (VALVE) COVER

REMOVAL & INSTALLATION

See Figures 114 and 115.

1. Remove the intake manifold chamber. Refer to "Intake Manifold" in this section.

2. Disconnect the four ignition coil connectors.

3. Remove the harness holder and disconnect the breather hose.

4. Remove the cylinder head cover.

To install:

5. Thoroughly clean the head cover gasket and the groove of the cylinder head cover.

6. Check and if necessary, replace the head cover gasket.

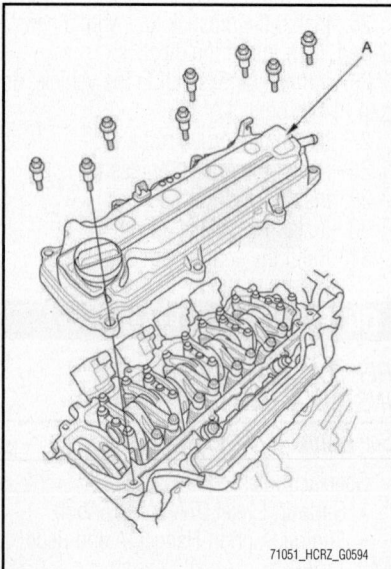

Fig. 114 Remove the cylinder head cover (A).

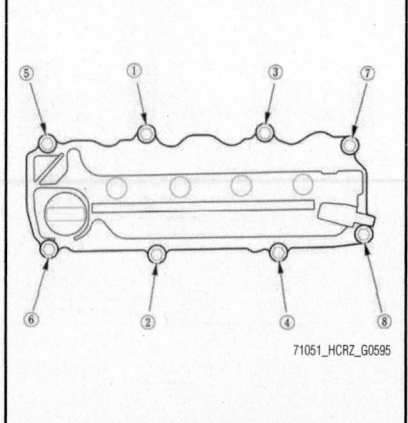

Fig. 115 Tighten the bolts in three steps. In the final step, torque all bolts, in sequence, to 7 ft. lbs. (10 Nm).

7. Install the head cover gasket in the groove of the cylinder head cover. Make sure the head cover gasket is seated securely.

8. Remove all of the old liquid gasket from the cylinder head.

9. Clean the head cover contacting surfaces with a shop towel.

10. Apply liquid gasket (P/N 08718-0004) to the chain case contact areas. Install the component within 5 minutes of applying the liquid gasket.

  a. Apply a 3.0 mm (0.118 in) diameter bead of liquid gasket to the chain case contact areas.

  b. If too much time has passed after applying the liquid gasket, remove the old liquid gasket and residue, then reapply new liquid gasket.

11. Place the cylinder head cover on the cylinder head, then slide the cover slightly back and forth to seat the head cover gasket.

  a. Wait at least 30 minutes before filling the engine with oil.

  b. Do not run the engine for at least 3 hours after installing the head cover.

12. Tighten the bolts in three steps. In the final step, torque all bolts, in sequence, to 7 ft. lbs. (10 Nm).

13. Connect the breather hose and install the harness holder.

14. Connect the four ignition coil connectors.

15. Install the intake manifold chamber. Refer to "Intake Manifold" in this section.

ROCKER ARM ASSEMBLY

REMOVAL & INSTALLATION

See Figures 116 through 118.

1. Remove the cylinder head cover.

2. Loosen the rocker arm adjusting screws.

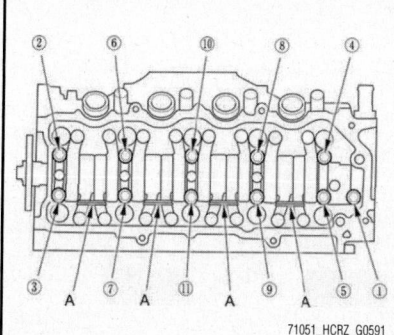

Fig. 116 Unscrew the rocker shaft mounting bolts two turns at a time, in the sequence shown. Bundle the intake rocker arms with rubber bands (A) to keep them together as a set.

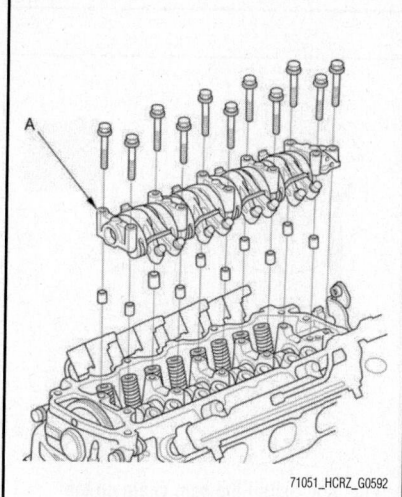

Fig. 117 Remove the rocker shaft mounting bolts, then remove the rocker arm assembly (A).

3. Unscrew the rocker shaft mounting bolts two turns at a time, in the sequence shown.

4. Bundle the intake rocker arms with rubber bands to keep them together as a set.

5. Remove the rocker shaft mounting bolts, then remove the rocker arm assembly.

To install:

6. If the rocker arm assembly is disassembled, reassemble the rocker arm assembly.

7. Apply engine oil to the end of the valve stem.

8. Install the rocker arm assembly, then remove the rubber bands.

9. Tighten each bolt two turns at a time in sequence to the following:

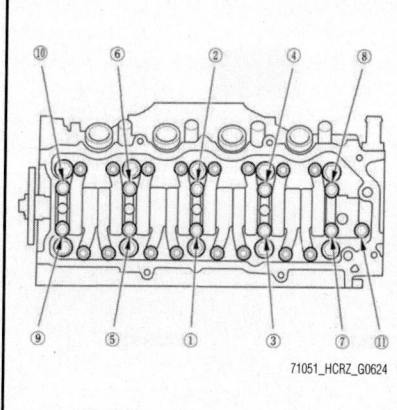

Fig. 118 Tighten each bolt two turns at a time in sequence.

- Bolts 1–10: 11 ft. lbs. (15 Nm)
- Bolt 11: 7 ft. lbs. (10 Nm)

10. Adjust the valve clearance. Refer to "Valve Clearance" in this section.

11. Install the cylinder head cover. Tighten bolts to 7 ft. lbs. (10 Nm).

TIMING (CAMSHAFT) CHAIN COVER, CHAIN, & SPROCKETS

REMOVAL & INSTALLATION

See Figures 119 through 145.

➡ **Keep the cam chain away from magnetic fields.**

1. Turn the crankshaft pulley so its top dead center (TDC) mark lines up with the pointer.

2. Remove the cylinder head cover.

3. Check the No. 1 piston at TDC. The "UP" mark on the camshaft sprocket should

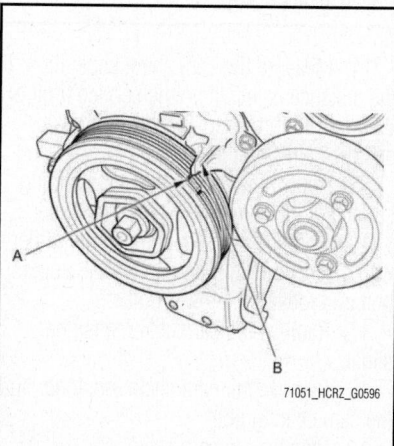

Fig. 119 Turn the crankshaft pulley so its top dead center (TDC) mark (A) lines up with the pointer (B).

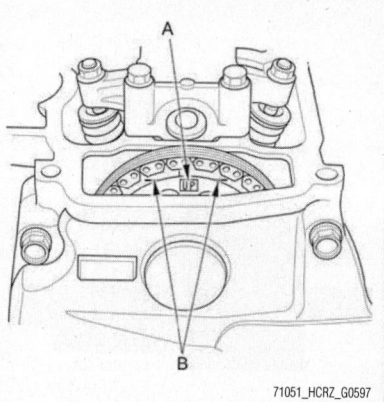

Fig. 120 Check the No. 1 piston at TDC. The "UP" mark (A) on the camshaft sprocket should be at the top, and the TDC grooves (B) on the camshaft sprocket should line up with the top edge of the head.

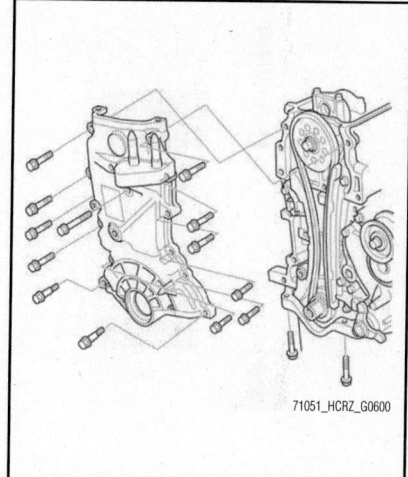

Fig. 121 Remove the chain case.

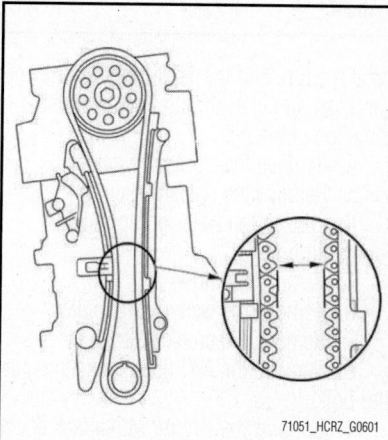

Fig. 122 Measure the cam chain separation. If the distance is less than the service limit of 0.59 in. (15 mm), replace the cam chain and the cam chain tensioner.

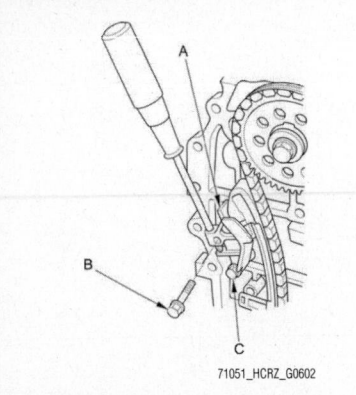

71051_HCRZ_G0602

Fig. 123 Apply new engine oil to the sliding surface of the cam chain tensioner slider. Hold the cam chain tensioner slider with a screwdriver, then remove the upper bolt (B), and loosen the lower bolt (C).

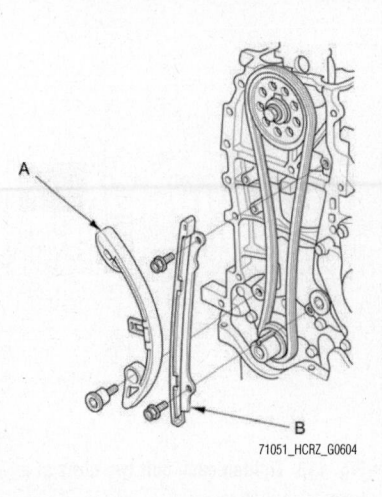

71051_HCRZ_G0604

Fig. 125 Remove the cam chain tensioner (A) and the cam chain guide (B).

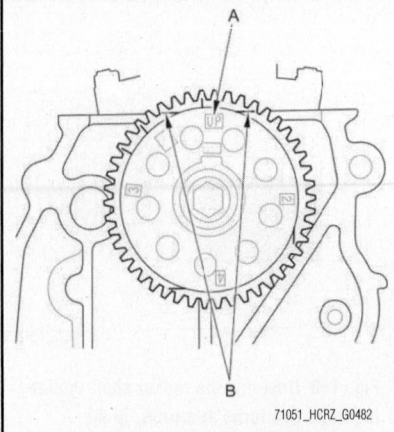

71051_HCRZ_G0482

Fig. 127 Set the camshaft to TDC. The "UP" mark (A) on the camshaft sprocket should be at the top, and the TDC grooves (B) on the camshaft sprocket should line up with the top edge of the head.

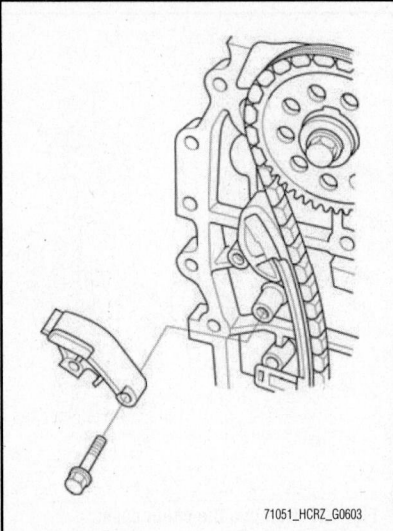

71051_HCRZ_G0603

Fig. 124 Remove the cam chain tensioner slider.

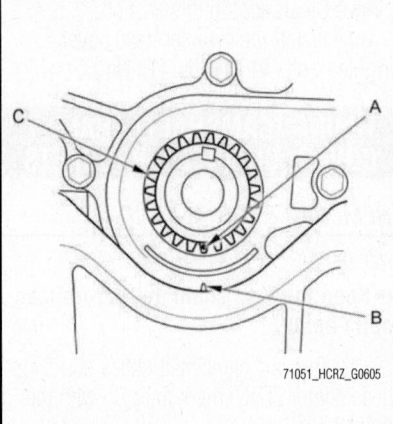

71051_HCRZ_G0605

Fig. 126 Set the crankshaft to top dead center (TDC). Align the TDC mark (A) on the crankshaft sprocket with the pointer (B) on the oil pump. Remove the crankshaft sprocket (C).

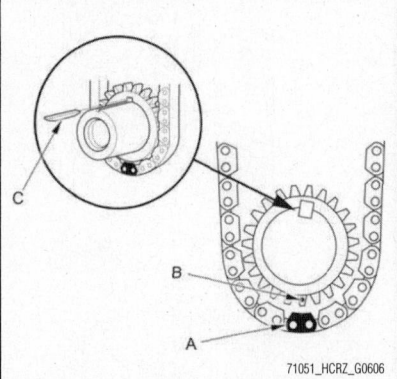

71051_HCRZ_G0606

Fig. 128 Install the cam chain on the crankshaft sprocket with the colored piece (A) aligned with the TDC mark (B) on the crankshaft sprocket, then install the crankshaft sprocket with the special key (C) to the crankshaft.

be at the top, and the TDC grooves on the camshaft sprocket should line up with the top edge of the head.

4. Remove the right front wheel.

5. Remove the splash shield.

6. Loosen the water pump pulley mounting bolts.

7. Remove the drive belt.

8. Remove the water pump pulley.

9. Remove the crankshaft pulley.

10. Remove the A/C line bracket mounting bolt.

11. Support the engine with a jack and a wood block under the oil pan.

12. Remove the ground cable, then remove the side engine mount/bracket assembly.

13. Remove the chain case.

14. Measure the cam chain separation. If the distance is less than the service limit of 0.59 in. (15 mm), replace the cam chain and the cam chain tensioner.

15. Apply new engine oil to the sliding surface of the cam chain tensioner slider.

16. Hold the cam chain tensioner slider with a screwdriver, then remove the upper bolt and loosen the lower bolt.

17. Remove the cam chain tensioner slider.

18. Remove the cam chain tensioner and the cam chain guide.

19. Remove the cam chain.

To install:

➡**Keep the cam chain away from magnetic fields.**

20. Set the crankshaft to top dead center (TDC). Align the TDC mark on the crankshaft sprocket with the pointer on the oil pump. Remove the crankshaft sprocket.

21. Set the camshaft to TDC. The "UP" mark on the camshaft sprocket should be at the top, and the TDC grooves on the camshaft sprocket should line up with the top edge of the head.

22. Install the cam chain on the crankshaft sprocket with the colored piece aligned with the TDC mark on the crankshaft sprocket, then install the crankshaft sprocket with the special key to the crankshaft.

23. Install the cam chain on the camshaft sprocket with the pointers aligned with the three colored pieces as shown.

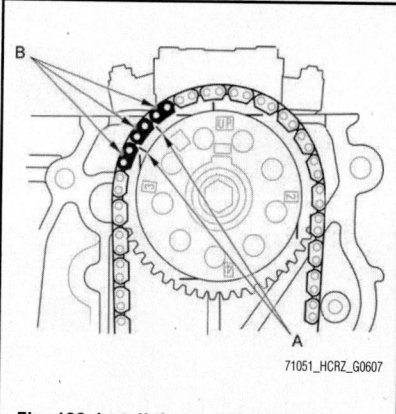

Fig. 129 Install the cam chain on the camshaft sprocket with the pointers (A) aligned with the three colored pieces (B) as shown.

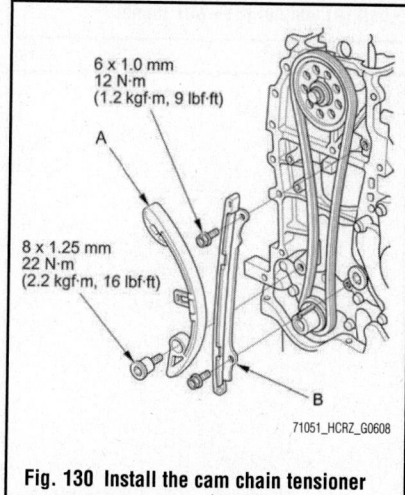

Fig. 130 Install the cam chain tensioner (A) and the cam chain guide (B).

24. Install the cam chain tensioner and the cam chain guide.

25. Install the cam chain tensioner slider and loosely install the bolt.

26. Apply new engine oil to the sliding surface of the cam chain tensioner slider.

27. Rotate the cam chain tensioner slider clockwise to compress the cam chain tensioner, and install the remaining bolt, then tighten the bolts to 9 ft. lbs. (12 Nm).

28. Check the cam chain case oil seal for damage. If the oil seal is damaged, replace the cam chain case oil seal.

29. Remove all of the old liquid gasket from the chain case mating surfaces, the bolts, and the bolt holes.

30. Clean and dry the chain case mating surfaces.

31. Apply liquid gasket (P/N 08718-0004) to the cylinder head and the engine block mating surfaces of the chain case and to the inside edge of the bolt holes. Install

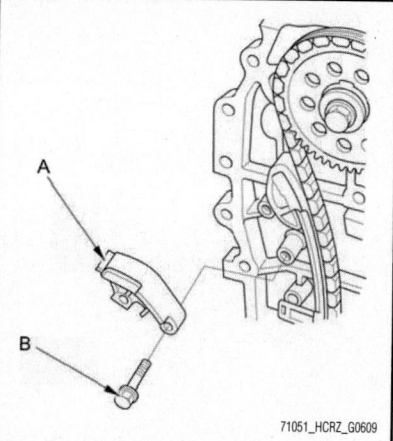

Fig. 131 Install the cam chain tensioner slider (A) and loosely install the bolt (B).

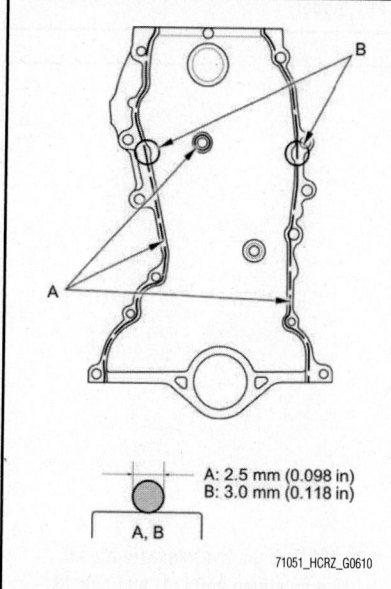

Fig. 132 Apply a 2.5 mm (0.098 in) diameter bead of liquid gasket along the broken line. Apply a 3.0 mm (0.118 in) diameter bead of liquid gasket to the upper surface contact areas of the engine block.

the component within 5 minutes of applying the liquid gasket.

a. Apply a 2.5 mm (0.098 in) diameter bead of liquid gasket along the broken line.

b. Apply a 3.0 mm (0.118 in) diameter bead of liquid gasket to the upper surface contact areas of the engine block.

c. If too much time has passed after applying the liquid gasket, remove the old liquid gasket and residue, then reapply new liquid gasket.

32. Apply liquid gasket (P/N 08718-0004) to the oil pan mating surface of the

chain case and to the inside edge of the bolt holes. Install the component within 5 minutes of applying the liquid gasket.

a. Apply a 2.5 mm (0.098 in) diameter bead of liquid gasket along the broken line.

b. Apply a 5.0 mm (0.197 in) diameter bead of liquid gasket to the shaded area.

c. If too much time has passed after applying the liquid gasket, remove the old liquid gasket and residue, then reapply new liquid gasket.

33. Set the edge of the chain case on the edge of the oil pan, then install the chain case on the engine block.

a. When installing the chain case, do not slide the bottom surface onto the oil pan mounting surface.

b. Wait at least 30 minutes before filling the engine with oil.

c. Do not run the engine for at least 3 hours after installing the chain case.

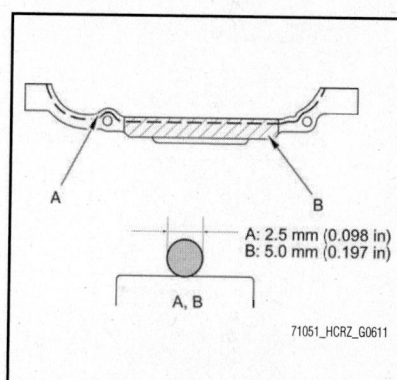

Fig. 133 Apply a 2.5 mm (0.098 in) diameter bead of liquid gasket along the broken line (A). Apply a 5.0 mm (0.197 in) diameter bead of liquid gasket to the shaded area (B).

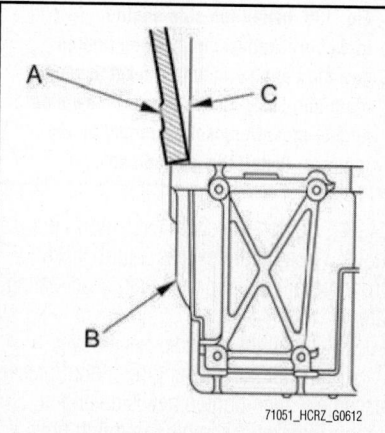

Fig. 134 Set the edge of the chain case (A) on the edge of the oil pan (B), then install the chain case on the engine block (C).

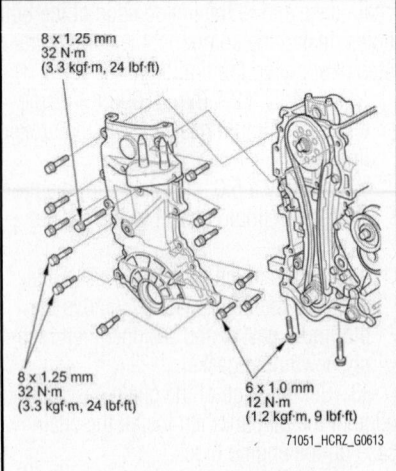

8 x 1.25 mm
32 N·m
(3.3 kgf·m, 24 lbf·ft)

8 x 1.25 mm
32 N·m
(3.3 kgf·m, 24 lbf·ft)

6 x 1.0 mm
12 N·m
(1.2 kgf·m, 9 lbf·ft)

71051_HCRZ_G0613

Fig. 135 Tighten the chain case mounting bolts. Wipe off the excess liquid gasket from the oil pan and the chain case mating area.

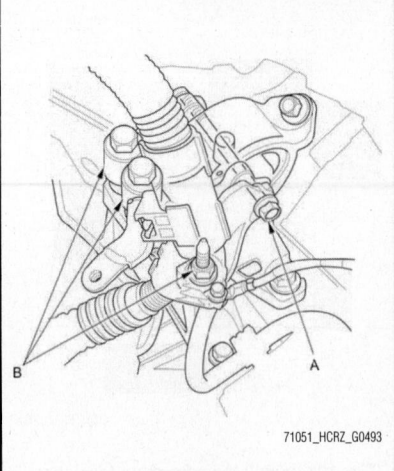

71051_HCRZ_G0493

Fig. 137 Loosen the transaxle mount bracket mounting bolts (A) and nuts (B)—M/T model

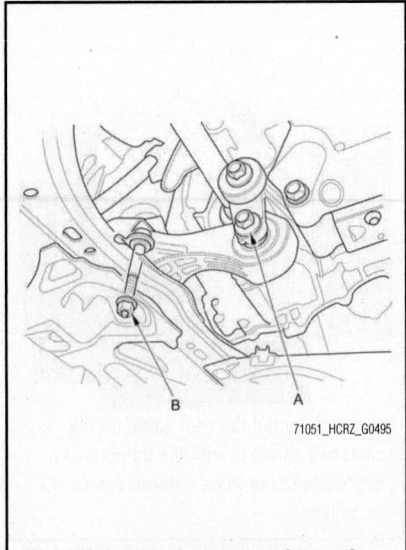

71051_HCRZ_G0495

Fig. 139 Loosen the torque rod mounting bolt (A) and nut (B)—M/T model

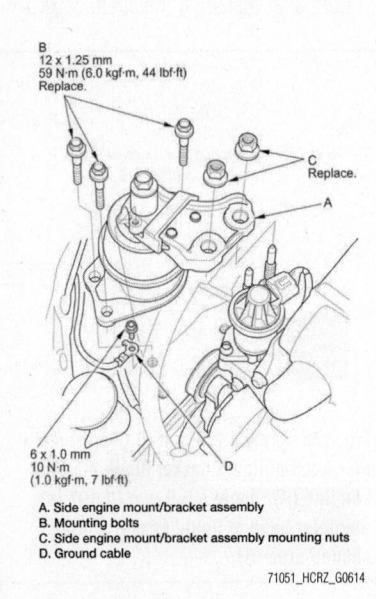

B
12 x 1.25 mm
59 N·m (6.0 kgf·m, 44 lbf·ft)
Replace.

C
Replace.

A

6 x 1.0 mm
10 N·m
(1.0 kgf·m, 7 lbf·ft)

D

A. Side engine mount/bracket assembly
B. Mounting bolts
C. Side engine mount/bracket assembly mounting nuts
D. Ground cable

71051_HCRZ_G0614

Fig. 136 Install the side engine mount/bracket assembly, then tighten new side engine mount/bracket assembly mounting bolts. Loosely install new side engine mount/bracket assembly mounting nuts. Install the ground cable.

34. Tighten the chain case mounting bolts. Wipe off the excess liquid gasket from the oil pan and the chain case mating area.

35. Install the cylinder head cover.

36. Install the side engine mount/bracket assembly, then tighten new side engine mount/bracket assembly mounting bolts.

37. Loosely install new side engine mount/bracket assembly mounting nuts.

38. Install the ground cable.

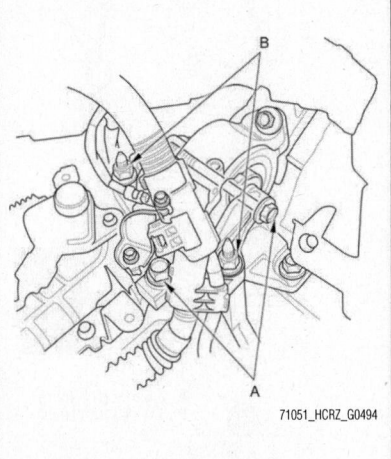

B

A

71051_HCRZ_G0494

Fig. 138 Loosen the transaxle mount bracket mounting bolts (A) and nuts (B)—CVT model

39. Remove the jack and the wood block from under the oil pan.

40. Remove the air cleaner.

41. Loosen the transaxle mount bracket mounting bolts and nuts.

42. Raise the vehicle on the lift.

43. Loosen the torque rod mounting bolts and nuts.

44. Lower the vehicle on the lift.

45. Tighten the side engine mount/bracket assembly mounting nuts to 69 ft. lbs. (93 Nm).

46. Tighten the transaxle mount mounting bolts and nuts in the numbered sequence shown.

47. Raise the vehicle on the lift.

48. Tighten the torque rod mounting bolt and nut in the numbered sequence shown.

49. Lower the vehicle on the lift.

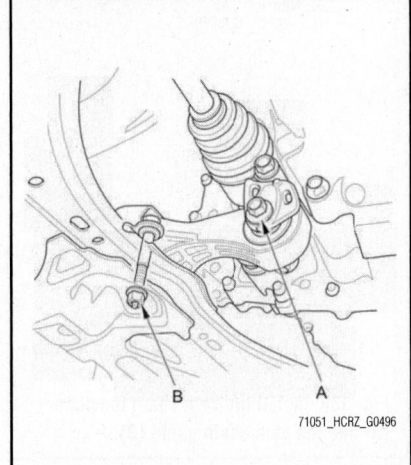

B

A

71051_HCRZ_G0496

Fig. 140 Loosen the torque rod mounting bolt (A) and nut (B)—CVT model

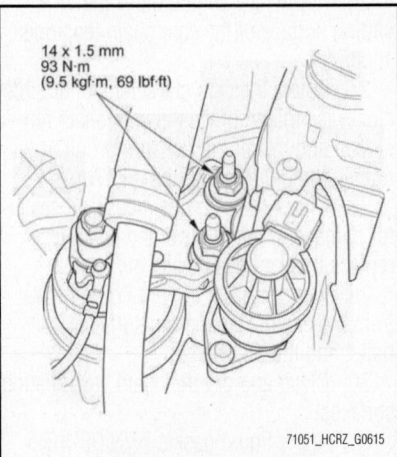

14 x 1.5 mm
93 N·m
(9.5 kgf·m, 69 lbf·ft)

71051_HCRZ_G0615

Fig. 141 Tighten the side engine mount/bracket assembly mounting nuts.

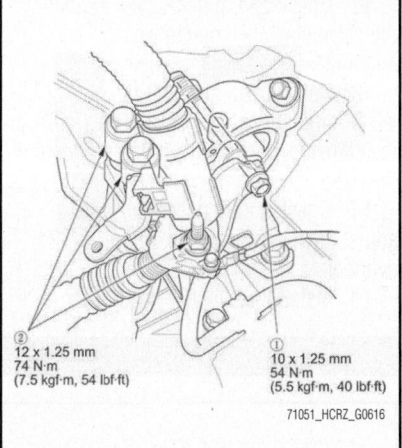

Fig. 142 Tighten the transaxle mount mounting bolts and nuts in the numbered sequence shown and to the indicated torque settings—M/T model

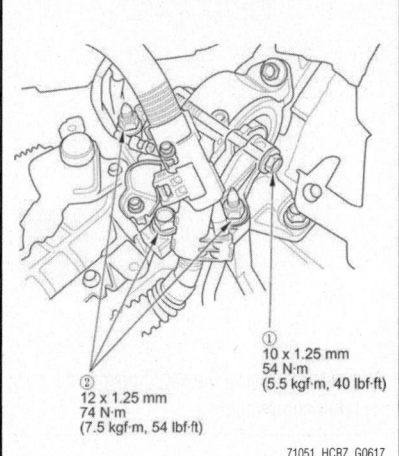

Fig. 143 Tighten the transaxle mount mounting bolts and nuts in the numbered sequence shown and to the indicated torque settings—CVT model

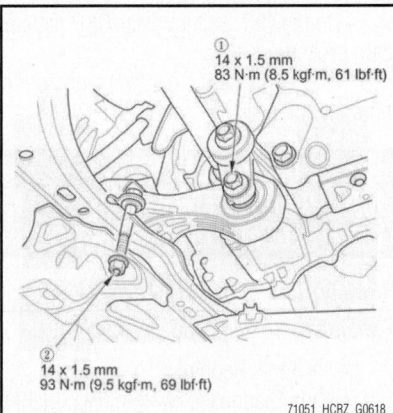

Fig. 144 Tighten the torque rod mounting bolt and nut in the numbered sequence shown—M/T model

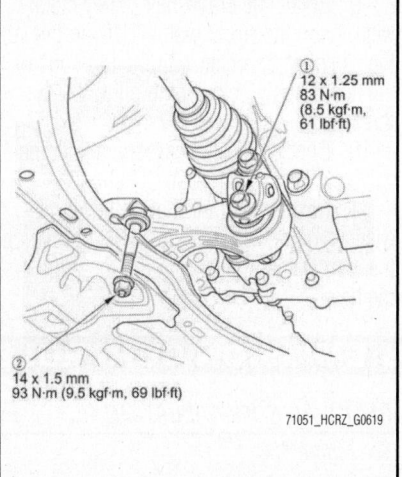

Fig. 145 Tighten the torque rod mounting bolt and nut in the numbered sequence shown—CVT model

50. Install the air cleaner.
51. Install the A/C line bracket mounting bolt.
52. Install the crankshaft pulley.
53. Install the water pump pulley and loosely install the water pump pulley mounting bolts.
54. Install the drive belt.
55. Tighten the water pump pulley mounting bolts to 12 ft. lbs. (16 Nm).
56. Install the splash shield.
57. Install the right front wheel.
58. Do the CKP pattern clear/CKP pattern learn procedure.

VALVE CLEARANCE

ADJUSTMENT

See Figures 146 through 148.

1. Connect the HDS to the DLC, and monitor ECT SENSOR 1. Adjust the valve clearance only when the engine coolant temperature is less than 100°F (38°C).
2. Remove the cylinder head cover.
3. Set the No. 1 piston at top dead center (TDC). The "UP" mark on the camshaft sprocket should be at the top and the TDC grooves on the camshaft sprocket should line up with the top edge of the head.
4. Select the correct feeler gauge for the valve clearance you are going to check.
5. Insert the appropriate size feeler gauge between the adjusting screw and the end of the valve stem on No. 1 cylinder, and slide it back and forth; you should feel a slight amount of drag.
6. If you feel too much or too little drag, loosen the locknut, and turn the

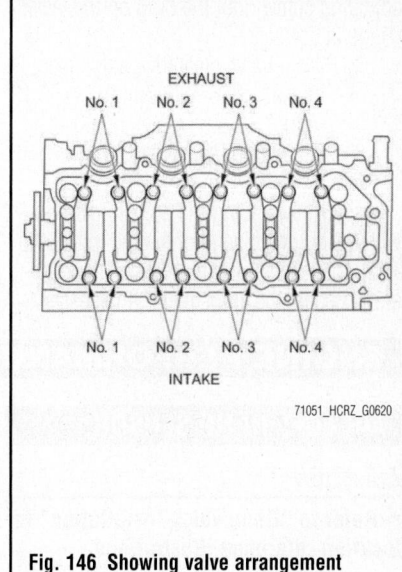

Fig. 146 Showing valve arrangement

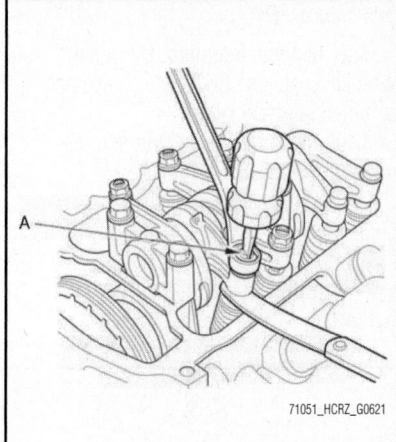

Fig. 147 If you feel too much or too little drag, loosen the locknut, and turn the adjusting screw (A) until the drag on the feeler gauge is correct.

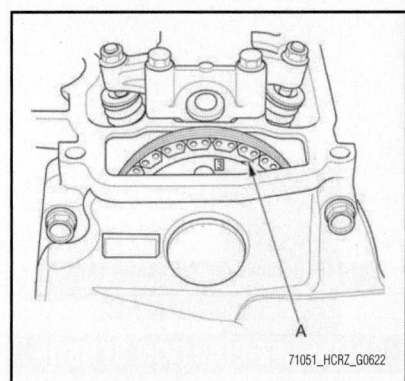

Fig. 148 Rotate the crankshaft pulley clockwise. Align the No. 3 piston TDC groove (A) on the camshaft sprocket with the top edge of the head. Apply new engine oil to the nut threads.

adjusting screw until the drag on the feeler gauge is correct.

7. Proper valve clearance adjustment is:
- Intake: 0.006–0.007 in. (0.15–0.19 mm)
- Exhaust: 0.010–0.012 in. (0.26–0.30 mm)

8. While holding the adjusting screw with the screw driver, tighten the locknut to 10 ft. lbs. (14 Nm), then recheck the clearance. Repeat the adjustment, if necessary.

9. Rotate the crankshaft pulley clockwise. Align the No. 3 piston TDC groove on the camshaft sprocket with the top edge of the head. Apply new engine oil to the nut threads.

10. Check and, if necessary, adjust the valve clearance on the No. 3 cylinder.

11. Rotate the crankshaft pulley clockwise. Align the No. 4 piston TDC groove on the camshaft sprocket with the top edge of the head.

12. Check and, if necessary, adjust the valve clearance on the No. 4 cylinder.

13. Rotate the crankshaft pulley clockwise. Align the No. 2 piston TDC groove on the camshaft sprocket with the top edge of the head.

14. Check and, if necessary, adjust the valve clearance on the No. 2 cylinder.

15. Install the cylinder head cover.

ENGINE PERFORMANCE & EMISSION CONTROLS

AIR-FUEL RATIO SENSOR

LOCATION

➡ Refer to "Removal & Installation" for location reference illustrations.

REMOVAL & INSTALLATION

See Figure 149.

Special Tools Required: O2 Sensor Wrench Snap-on® S6176 or equivalent, commercially available.

1. Remove the cowl cover and the under-cowl panel.
2. Disconnect the A/F sensor connector.
3. Remove the A/F sensor.
4. Install the parts in the reverse order of removal.

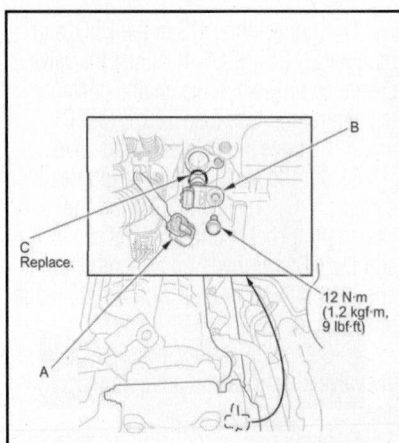

A
44 N·m (4.5 kgf·m, 32 lbf·ft)

71051_HCRZ_G0627

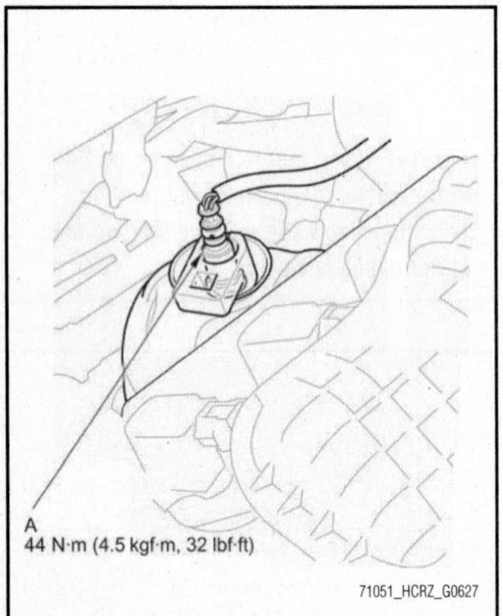

Fig. 149 Remove the A/F sensor (A).

CAMSHAFT POSITION SENSOR

LOCATION

The Camshaft Position (CMP) sensor is located on the left end of the engine, near the camshaft.

REMOVAL & INSTALLATION

See Figure 150.

1. Remove the air cleaner.
2. Remove the cowl cover and the under-cowl panel.
3. Disconnect the CMP sensor connector.
4. Remove the CMP sensor.
5. Install the sensor in the reverse order of removal with a new O-ring.

CRANKSHAFT POSITION SENSOR

REMOVAL & INSTALLATION

See Figure 151.

1. Raise the vehicle on a lift.

➡ Make sure the vehicle is level, because engine oil will drip out when you remove the sensor.

2. Remove the splash shield.
3. Loosen the bolt. Remove the bolt and the CKP sensor cover.

B

C
Replace.

12 N·m
(1.2 kgf·m,
9 lbf·ft)

A

71051_HCRZ_G0628

Fig. 150 Disconnect the CMP sensor connector (A). Remove the CMP sensor (B). Install the sensor in the reverse order of removal with a new O-ring (C).

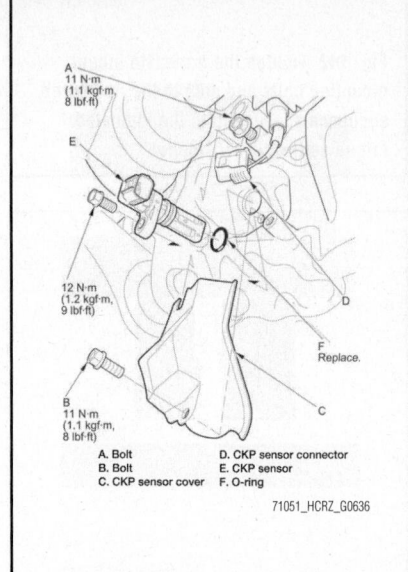

A
11 N·m
(1.1 kgf·m,
8 lbf·ft)

E

12 N·m
(1.2 kgf·m,
9 lbf·ft)

D

F
Replace.

C

B
11 N·m
(1.1 kgf·m,
8 lbf·ft)

A. Bolt
B. Bolt
C. CKP sensor cover
D. CKP sensor connector
E. CKP sensor
F. O-ring

71051_HCRZ_G0636

Fig. 151 Showing the CKP sensor and related components

4. Disconnect the CKP sensor connector.
5. Remove the CKP sensor.

To install:

6. Install the parts in the reverse order of removal with a new O-ring.
7. Do the CKP pattern clear/CKP pattern learn procedure.
8. Check the engine oil level, and add more oil if needed.

ELECTRONIC CONTROL MODULE/POWERTRAIN CONTROL MODULE (ECM/PCM)

REMOVAL & INSTALLATION

See Figures 152 and 153.

Special Tools Required:
- Honda Diagnostic System (HDS) tablet tester
- Honda Interface Module (HIM) and an iN workstation with the latest HDS software version

- HDS pocket tester
- GNA600 and an iN workstation with the latest HDS software version
- MVCI unit with the latest control module (CM) update software installed
- Any one of the above updating tools can be used.
- Make sure the HDS/iN workstation or the MVCI has the latest HDS software version.
- The lifetime points of the Eco guide cannot be carried over to the replacement ECM/PCM.

1. Connect the HDS to the data link connector (DLC) located under the driver's side of the dashboard.

2. Turn the ignition switch to ON (II).

3. Make sure the HDS communicates with the ECM/PCM and all other vehicle systems. If it doesn't, go to the DLC circuit troubleshooting. If you are returning from the DLC circuit troubleshooting, skip steps 4 through 7, 18 through 20, and 23 through 24, and do these procedures after replacing the ECM/PCM:

   a. Replace the engine oil and the engine oil filter.

   b. Clean the throttle body.

4. Select the PGM-FI system with the HDS.

5. Select the INSPECTION MENU with the HDS.

6. Select the ETCS TEST, then select the TP POSITION CHECK, and follow the screen prompts.

➡If the TP POSITION CHECK indicates FAILED, continue with this procedure.

7. Select the REPLACE ECM/PCM MENU, then select READ DATA, and follow the screen prompts.

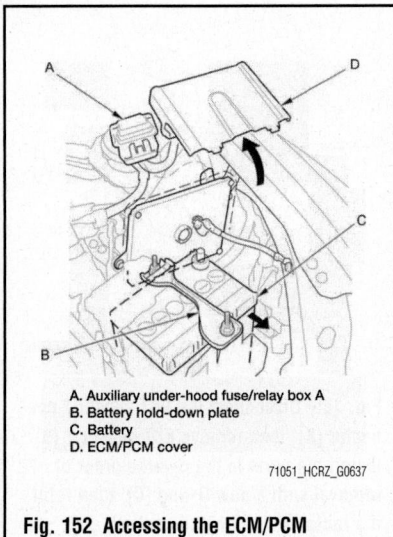

A. Auxiliary under-hood fuse/relay box A
B. Battery hold-down plate
C. Battery
D. ECM/PCM cover

71051_HCRZ_G0637

Fig. 152 Accessing the ECM/PCM

   a. Doing this step copies (READS) the engine oil life data from the original ECM/PCM so you can later download (WRITES) it into the new ECM/PCM.

   b. If the READ DATA indicates FAILED, continue with this procedure.

8. Jump the SCS line with the HDS.

9. Turn the ignition switch to LOCK (0).

10. Remove auxiliary under-hood fuse/relay box A.

11. Remove the battery hold-down plate, then move the battery forward.

12. Remove the ECM/PCM cover.

13. Remove the bolts.

14. Disconnect ECM/PCM connectors A, B, and C, then remove the ECM/PCM.

➡ECM/PCM connectors A, B, and C have symbols embossed on them for identification.

To install:

15. Install all parts in the reverse order of removal.

16. Turn the ignition switch to ON (II).

17. Manually input the VIN to the ECM/PCM with the HDS.

➡DTC P0630 VIN Not Programmed or Mismatch may be stored because the VIN has not been programmed into the ECM/PCM; ignore it, and continue this procedure.

18. If the READ DATA (engine oil life) failed in step 7, go to step 21. Otherwise, go to step 19.

19. Select the PGM-FI system with the HDS.

20. Select the REPLACE ECM/PCM MENU, then select WRITE DATA, and follow the screen prompts.

➡If the WRITE DATA indicates FAILED, continue with this procedure.

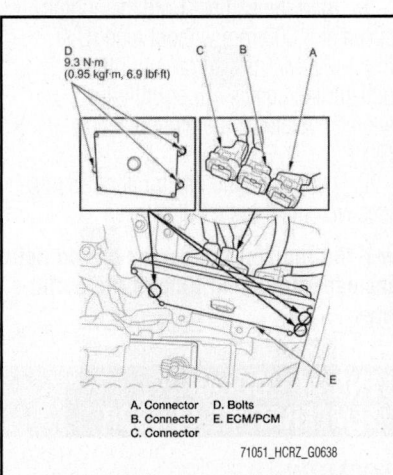

D
9.3 N·m
(0.95 kgf·m, 6.9 lbf·ft)

A. Connector
B. Connector
C. Connector
D. Bolts
E. ECM/PCM

71051_HCRZ_G0638

Fig. 153 Disconnect ECM/PCM connectors A, B, and C, then remove the ECM/PCM.

21. Select the IMMOBI system with the HDS.

22. Enter the immobilizer ECM/PCM code that you got from the iN, and use the ECM/PCM replacement procedure in the IMMOBI MENU of the HDS; it allows you to start the engine.

23. If the TP POSITION CHECK failed in step 6, clean the throttle body, then go to step 24.

24. If the READ DATA failed in step 7 or the WRITE DATA failed in step 20, replace the engine oil, and the engine oil filter, then go to step 25.

25. Select the PGM-FI system, and reset the ECM/PCM with the HDS.

26. Update the ECM/PCM if it does not have the latest software.

27. Do the ECM/PCM idle learn procedure.

➡If the IMA battery level gauge displays no segments, start the engine, and hold it between 3,500–4,000 rpm without load (CVT in P or N, M/T in neutral) until the level gauge displays at least half of its segments.

28. Do the CKP pattern clear/CKP pattern learn procedure.

29. CVT: Do the start clutch control calibration procedure.

ECM/PCM UPDATING

Special Tools Required:
- Honda Diagnostic System (HDS) tablet tester
- Honda Interface Module (HIM) and an iN workstation with the latest HDS software version
- HDS pocket tester
- GNA600 and an iN workstation with the latest HDS software version
- MVCI unit with the latest control module (CM) update software installed
- Any one of the above updating tools can be used.
- Make sure the HDS/iN workstation or the MVCI has the latest HDS software version.
- Before you update the ECM/PCM, make sure the 12 volt battery in the vehicle is fully charged, and connect a jumper battery (not a battery charger) to maintain system voltage.
- Never turn the ignition switch to ACCESSORY (I) or to LOCK (0) during the update. If there is a problem with the update, leave the ignition switch turned to ON (II).
- To prevent ECM/PCM damage, do not operate anything electrical (headlights, navigation, brakes, A/C, power windows, door locks, etc.) during the update.

- To ensure the latest program is installed, do a ECM/PCM update whenever the ECM/PCM is substituted or replaced.

- You cannot update an ECM/PCM with a program it already has. It will only accept a new program.

- High temperature in the engine compartment might cause the ECM/PCM to become too hot to run the update. If the engine was running before the update, open the hood and cool the engine compartment.

- If you need to diagnose the Honda interface module (HIM) because the HIM's red (#3) light came on or was flashing during the update, leave the ignition switch in ON (II) when you disconnect the HIM from the data link connector (DLC). This will prevent damage to the ECM/PCM.

1. Turn the ignition switch to ON (II), but do not start the engine.

2. Connect the HDS to the data link connector (DLC) located under the driver's side of the dashboard.

3. Make sure the HDS communicates with the ECM/PCM and all other vehicle systems. If it doesn't, go to the DLC circuit troubleshooting. If you are returning from the DLC circuit troubleshooting, skip step 4 and 5, then clean the throttle body after updating the ECM/PCM.

4. Select the INSPECTION MENU with the HDS.

5. Select the ETCS TEST, then select the TP POSITION CHECK, and follow the HDS screen prompts.

➡**If the TP POSITION CHECK indicates FAILED, continue this procedure.**

6. Exit the HDS diagnostic system, then select the update mode, and follow the screen prompts to update the ECM/PCM.

7. If the software in the ECM/PCM is the latest, disconnect the updating tool from the DLC, then go back to the procedure that you were doing. If the software in the ECM/PCM is not the latest, follow the instructions on the screen. If prompted to choose the PGM-FI system or the CVT system, make sure you update both.

➡**If the ECM/PCM update system requires you to cool the ECM/PCM, follow the instructions on screen. If you have a problem during the update procedure (programming takes over 15 minutes, status bar goes over 100 %, D or immobilizer indicator flashes, HDS tablet freezes, etc.), follow these steps to minimize the chance of damaging the ECM/PCM.**

8. Leave the ignition switch in ON (II).

9. Connect a jumper battery (do not connect a battery charger).

10. Shut down the updating tool.

11. Disconnect the updating tool from the DLC.

12. Reboot the updating tool.

13. Reconnect the updating tool to the DLC, and try the update procedure again.

14. If the TP POSITION CHECK failed in step 5, clean the throttle body.

15. Do the ECM/PCM idle learn procedure.

16. Do the CKP pattern clear/CKP pattern learn procedure.

17. Select the CVT SYSTEM, then reset the PCM with the HDS.

ECM/PCM IDLE LEARN PROCEDURE

➡**The idle learn procedure must be done so the ECM/PCM can learn the engine idle characteristics.**

1. Do the idle learn procedure whenever you do any of these actions:

   a. Replace the ECM/PCM.
   b. Reset the ECM/PCM.
   c. Update the ECM/PCM.
   d. Replace or clean the throttle body.
   e. Disassemble the engine or the transaxle.

➡**Clearing DTCs with the HDS does not require you to do the idle learn procedure.**

2. Make sure all electrical items (A/C, navigation, lights, etc.) are off.

3. Reset the ECM/PCM with the HDS.

4. Turn the ignition switch to ON (II), and wait 2 seconds.

5. Start the engine. Hold the engine speed at 3,000 rpm without load (CVT in P or N, M/T in neutral) until the radiator fan comes on, or until the engine coolant temperature reaches 194°F (90°C).

6. Let the engine idle for about 5 minutes with the throttle fully closed.

➡**If the radiator fan comes on, do not include its running time in the 5 minutes.**

ENGINE COOLANT TEMPERATURE SENSOR

LOCATION

See "Removal & Installation" illustrations.

REMOVAL & INSTALLATION

ECT Sensor 1

See Figure 154.

1. Remove the air cleaner.
2. Remove the bolt.
3. Disconnect the ECT sensor 1 connector.
4. Remove ECT sensor 1.
5. Install the parts in the reverse order of removal with a new O-ring, then refill the radiator with engine coolant.

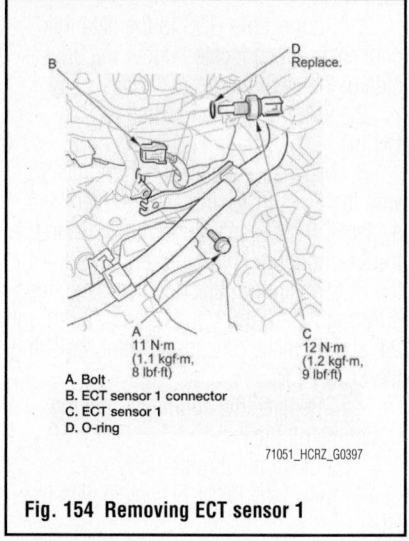

A. Bolt
B. ECT sensor 1 connector
C. ECT sensor 1
D. O-ring

71051_HCRZ_G0397

Fig. 154 Removing ECT sensor 1

ECT Sensor 2

See Figure 155.

1. Drain the engine coolant.
2. Raise the vehicle on a lift.
3. Remove the splash shield.
4. Disconnect the ECT sensor 2 connector, then remove ECT sensor 2.

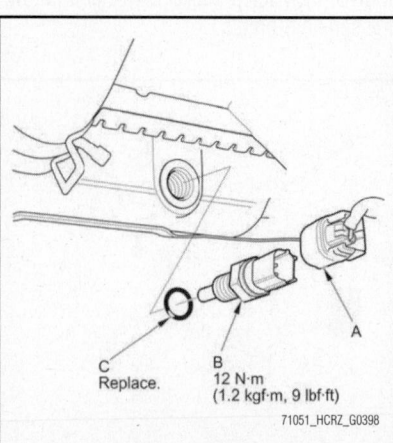

71051_HCRZ_G0398

Fig. 155 Disconnect the ECT sensor 2 connector (A), then remove ECT sensor 2 (B). Install the parts in the reverse order of removal with a new O-ring (C), then refill the radiator with engine coolant.

5. Install the parts in the reverse order of removal with a new O-ring, then refill the radiator with engine coolant.

EVAP CANISTER

LOCATION

See Figure 156.

1. Raise the vehicle on a lift.
2. Remove the bolts, then remove the EVAP canister guard.
3. Remove the EVAP canister cover.
4. Disconnect the hoses.
5. Disconnect the quick-connect fittings from the EVAP canister.

6. Disconnect the FTP sensor 3P connector and the EVAP canister vent shut valve 2P connector, then remove the harness clip.
7. Remove the bolts, then remove the EVAP canister assembly.
8. Disconnect the drain tube, then remove the EVAP canister from the EVAP canister bracket.
9. Remove the FTP sensor.
10. Remove the EVAP canister vent shut valve.
11. Install the parts in the reverse order of removal.

EXHAUST GAS RECIRCULATION (EGR) VALVE

LOCATION

See Figure 158.

REMOVAL & INSTALLATION

See Figure 159.

1. Disconnect the EGR valve 5P connector.
2. Remove the EGR valve.
3. Install the parts in the reverse order of removal with a new gasket.

MASS AIR FLOW/INTAKE AIR TEMPERATURE(MAF/IAT) SENSOR

LOCATION

The MAF/IAT sensor is located on the air intake duct, prior to the throttle body.

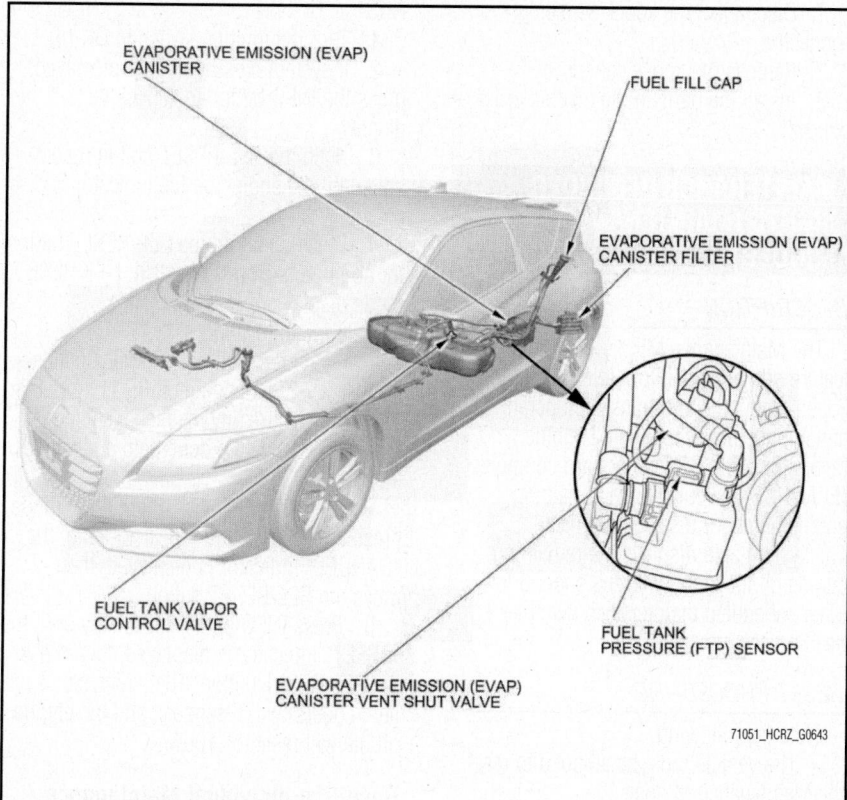

Fig. 156 EVAP canister location

REMOVAL & INSTALLATION

See Figure 157.

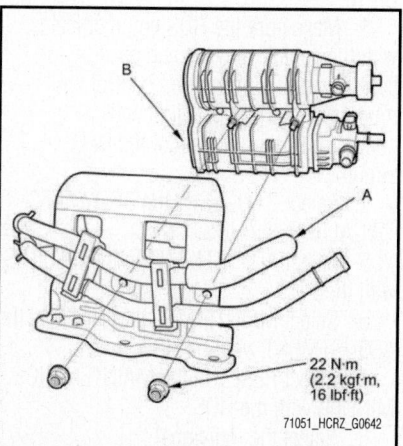

22 N·m
(2.2 kgf·m,
16 lbf·ft)

71051_HCRZ_G0642

Fig. 157 Disconnect the drain tube (A), then remove the EVAP canister (B) from the EVAP canister bracket.

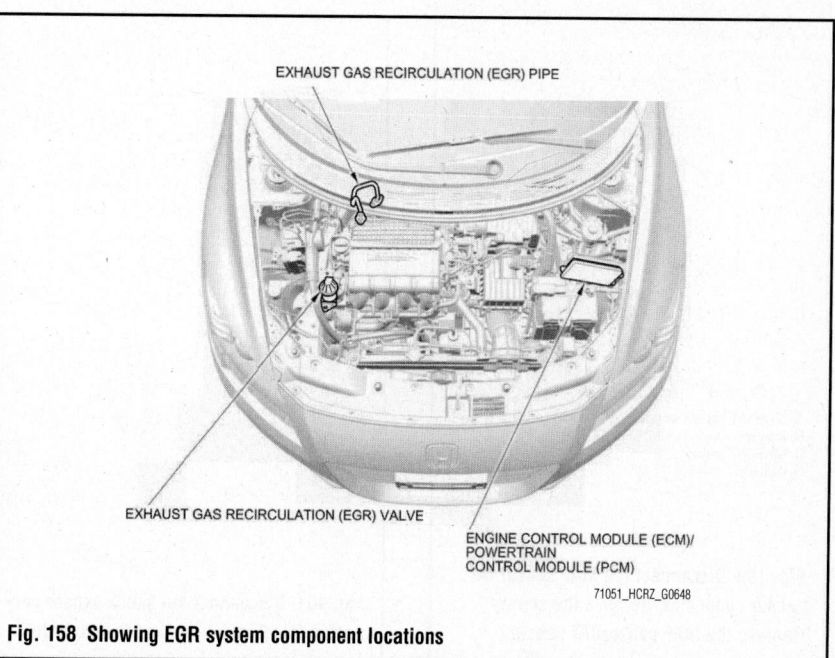

Fig. 158 Showing EGR system component locations

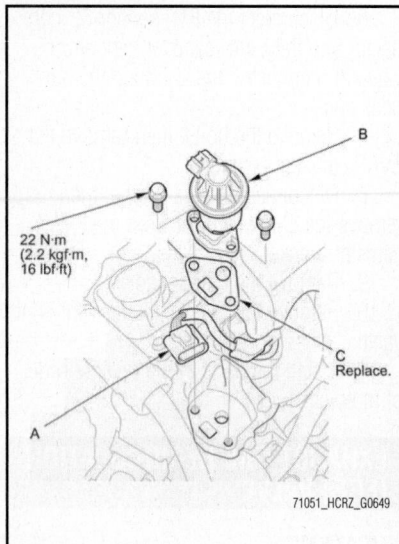

Fig. 159 Disconnect the EGR valve 5P connector (A). Remove the EGR valve (B). Install the parts in the reverse order of removal with a new gasket (C).

REMOVAL & INSTALLATION

See Figure 160.

1. Disconnect the MAF sensor/IAT sensor connector.
2. Remove the screws. Remove the MAF sensor/IAT sensor.
3. Install the parts in the reverse order of removal with a new gasket.

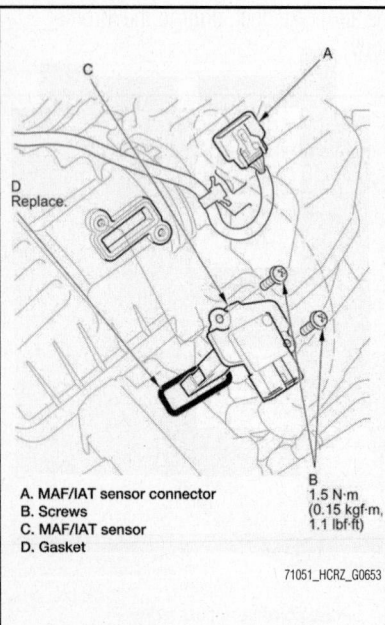

A. MAF/IAT sensor connector
B. Screws
C. MAF/IAT sensor
D. Gasket

B 1.5 N·m
(0.15 kgf·m,
1.1 lbf·ft)

71051_HCRZ_G0653

Fig. 160 Disconnect the MAF sensor/IAT sensor connector. Remove the screws. Remove the MAF sensor/IAT sensor.

KNOCK SENSOR (KS)

LOCATION

See "Removal & Installation" illustration.

REMOVAL & INSTALLATION

See Figure 161.

1. Disconnect the knock sensor connector.
2. Remove the knock sensor.
3. Install the parts in the reverse order of removal.

MAINTENANCE REMINDER LIGHTS (MAINTENANCE MINDER)

DESCRIPTION

The Maintenance Minder is an important feature of the multi-information display. Based on engine and transaxle operating conditions, and accumulated engine revolutions, the CR-Z's onboard computer (ECM/PCM) calculates the remaining engine oil and transaxle fluid life. The system also displays the remaining engine oil life along with the code(s) for other scheduled maintenance items needing service.

RESET PROCEDURE

Note the following:
- The vehicle must be stopped to reset the Maintenance Minder.
- If a required service is done and the Maintenance Minder is not reset, or if

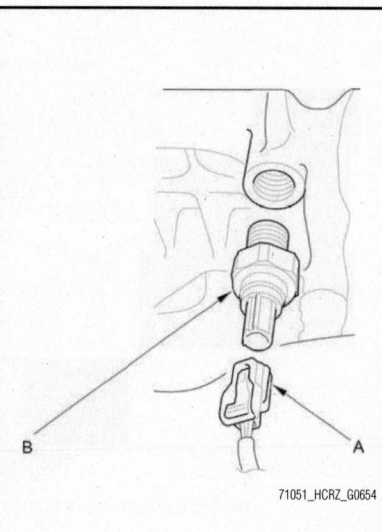

71051_HCRZ_G0654

Fig. 161 Disconnect the knock sensor connector (A). Remove the knock sensor (B).

the Maintenance Minder is reset without doing the service, the system will not show the proper maintenance timing. This can lead to serious mechanical problems because there will be no accurate record of when the required maintenance is needed.

- The engine oil life and maintenance item(s) can be independently reset with the HDS.

1. Turn the ignition switch to ON (II).
2. If system message(s) are displayed, press the INFO button to cancel the display.
3. Push the SEL/RESET button repeatedly until the engine oil life indicator is displayed.
4. Press and hold the SEL/RESET button for about 10 seconds, the "OIL LIFE RESET" mode display appears.

Note the following:
- If you are resetting the Maintenance Minder when the engine oil life is more than 15 %, make sure any maintenance item(s) requiring service are done before resetting the display.
- To cancel the "OIL LIFE RESET" mode, press the INFO button repeatedly until the "CANCEL" indicator is displayed, then press the SEL/RESET button.

5. Press INFO button repeatedly until the "RESET" indicator is displayed, then press the SEL/RESET button. The maintenance item code(s) will disappear, and the engine oil life will reset to "100%."

Resetting Individual Maintenance Items

1. Connect the Honda Diagnostic System (HDS) to the data link connector (DLC).
2. Turn the ignition switch to ON (II).
3. Make sure the HDS communicates with the vehicle and the engine control module/powertrain control module (ECM/PCM). If it doesn't communicate, troubleshoot the DLC circuit.
4. Select GAUGES in the BODY ELECTRICAL with the HDS.
5. Select ADJUSTMENT in the GAUGES with the HDS.
6. Select MAINTENANCE MINDER in the ADJUSTMENT with the HDS.
7. Select RESET in the MAINTENANCE MINDER with the HDS.
8. Select the individual maintenance item you wish to reset with the HDS.

POSITIVE CRANKCASE VENTILATION (PCV) VALVE

LOCATION
See Figure 162.

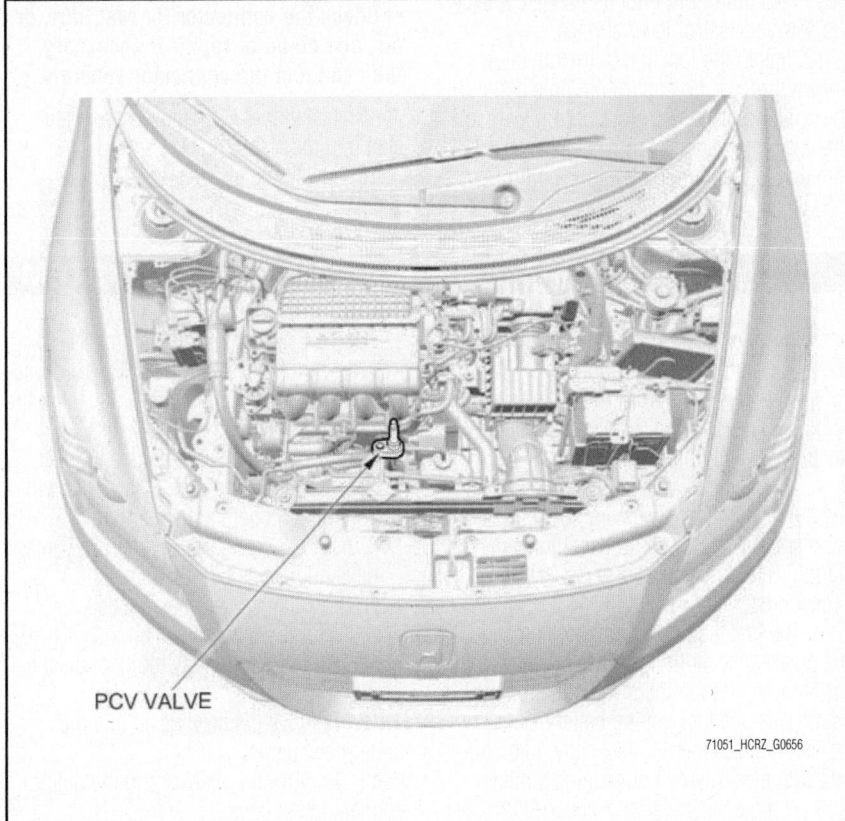

PCV VALVE

71051_HCRZ_G0656

Fig. 162 Showing the location of the PCV valve

REMOVAL & INSTALLATION
See Figure 163.

1. Remove the splash shield.

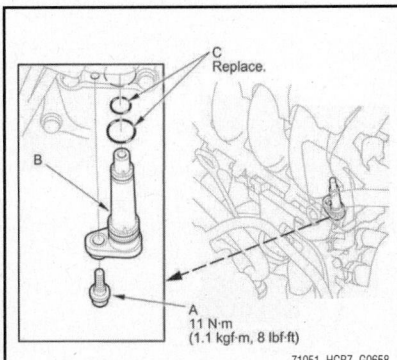

C
Replace.

B

A

11 N·m
(1.1 kgf·m, 8 lbf·ft)

71051_HCRZ_G0658

Fig. 163 Remove the bolt (A). Remove the PCV valve (B). When installing a new or reused PCV valve, make sure new O-rings (C) are in place.

2. Remove the bolt.
3. Remove the PCV valve.
4. Install the parts in the reverse order of removal.
5. When installing a new or reused PCV valve, make sure new O-rings are in place.

VEHICLE SPEED SENSOR

LOCATION

➡See "Removal & Installation" illustrations.

REMOVAL & INSTALLATION
See Figures 164 through 166.

1. Remove the air cleaner.
2. Remove the IMA motor power cable clamp from the clamp bracket.
3. Disconnect the transaxle range switch 8P connector, the CVT output shaft (driven pulley) speed sensor 3P connector, and the vehicle speed sensor 3P connector.
4. Remove the harness holder from the clamp bracket and remove the harness cover from the clamp bracket.
5. Remove the clamp bracket.
6. Remove the lock pin and the control pin from the selector control lever.

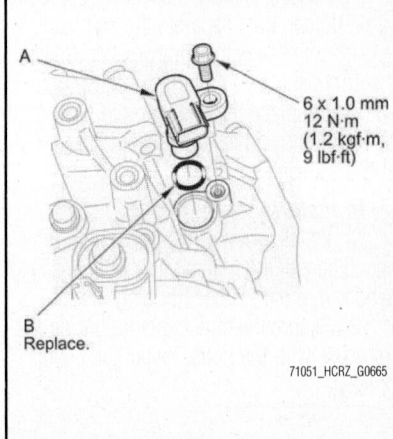

A

6 x 1.0 mm
12 N·m
(1.2 kgf·m,
9 lbf·ft)

B
Replace.

71051_HCRZ_G0665

Fig. 164 Remove the vehicle speed sensor (A) and the O-ring (B).

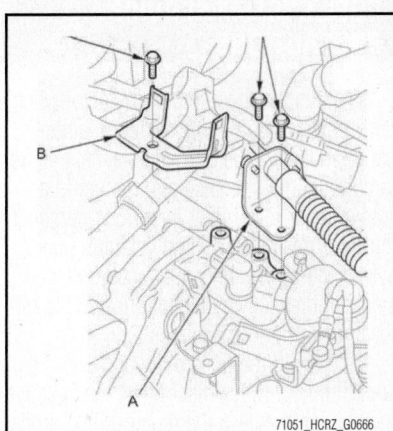

B

A

71051_HCRZ_G0666

Fig. 165 Secure the shift cable holder (A) on the transaxle with the bolts. Install the clamp bracket (B).

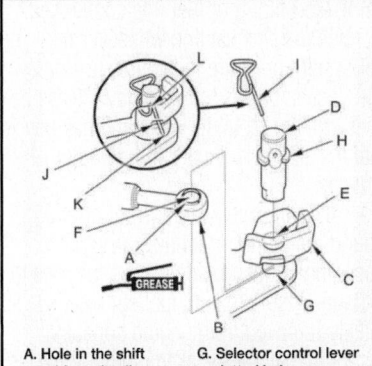

A. Hole in the shift cable end collar
B. Shift cable end
C. Selector control lever
D. Control pin
E. Selector control lever hole
F. Shift cable end collar hole
G. Selector control lever slotted hole
H. Control pin flange
I. Lock pin
J. Selector control lever opening
K. Lock pin hooked end
L. Control pin countersunk hole

71051_HCRZ_G0667

Fig. 166 Lubricating and reassembling the selector control lever shift cable end

7. Remove the bolts securing the shift cable holder, then separate the shift cable end from the selector control lever. Check the shift cable end collar for wear or damage, and replace the shift cable if necessary.

8. Remove the vehicle speed sensor and the O-ring.

To install:

9. Install a new O-ring on a new vehicle speed sensor, then install the vehicle speed sensor in the transaxle housing.

10. Secure the shift cable holder on the transaxle with the bolts. Install the clamp bracket.

11. Apply molybdenum grease to the hole in the shift cable end collar. Attach the shift cable end to the selector control lever. Insert the control pin through the selector control lever hole, through the shift cable end collar hole, and into the selector control lever slotted hole in the direction shown. Push the control pin until its flange contacts the selector control lever surface.

12. Insert the lock pin in the direction shown through the control pin hole and out the opening of the selector control lever so that the hooked end of the lock pin locks into the countersunk hole of the control pin.

13. Install the harness cover to the clamp bracket and install the harness holder to the clamp bracket.

14. Connect the transaxle range switch 8P connector, the CVT output shaft (driven pulley) speed sensor 3P connector, and the vehicle speed sensor 3P connector.

➡**Check the connector for rust, dirt, or oil, and clean or repair if necessary, then connect the connector securely.**

15. Install the IMA motor power cable clamp to the clamp bracket.

16. Install the air cleaner.

FUEL   GASOLINE FUEL INJECTION SYSTEM

FUEL SYSTEM SERVICE PRECAUTIONS

Safety is the most important factor when performing not only fuel system maintenance, but any type of maintenance. Failure to conduct maintenance and repairs in a safe manner may result in serious personal injury or death. Work on a vehicle's fuel system components can be accomplished safely and effectively by adhering to the following rules and guidelines.

• To avoid the possibility of fire and personal injury, always disconnect the negative battery cable unless the repair or test procedure requires that battery voltage be applied.

• Always relieve the fuel system pressure prior to disconnecting any fuel system component (injector, fuel rail, pressure regulator, etc.) fitting or fuel line connection. Exercise extreme caution whenever relieving fuel system pressure to avoid exposing skin, face and eyes to fuel spray. Please be advised that fuel under pressure may penetrate the skin or any part of the body that it contacts.

• Always place a shop towel or cloth around the fitting or connection prior to loosening to absorb any excess fuel due to spillage. Ensure that all fuel spillage is quickly removed from engine surfaces. Ensure that all fuel-soaked cloths or towels are deposited into a flame-proof waste container with a lid.

• Always keep a dry chemical (Class B) fire extinguisher near the work area.

• Do not allow fuel spray or fuel vapors to come into contact with a spark or open flame.

• Always use a second wrench when loosening or tightening fuel line connection fittings. This will prevent unnecessary stress and torsion on fuel piping. Always follow the proper torque specifications.

• Always replace worn fuel fitting O-rings with new ones. Do not substitute fuel hose where rigid pipe is installed.

• The fuel feed hose, the fuel line and the quick-connect fittings are not heat-resistant; be careful not to damage them during welding or other heat-generating procedures.

• The fuel feed hose, the fuel line and the quick-connect fittings are not acid-proof; do not touch them with a shop towel which was used for wiping battery electrolyte. Replace them if they came into contact with electrolyte or something similar.

• When connecting or disconnecting the fuel feed hose, the fuel line and the quick-connect fittings, be careful not to bend or twist them excessively. Replace them if they are damaged.

RELIEVING FUEL SYSTEM PRESSURE

With the HDS

See Figures 167 and 168.

Before disconnecting fuel lines or hoses, relieve pressure from the system by disabling the fuel pump and disconnecting the fuel line/quick connect fitting in the engine compartment.

1. Connect the HDS to the data link connector (DLC) located under the driver's side of the dashboard.

2. Turn the ignition switch to ON (II).

3. Make sure the HDS communicates with the ECM/PCM. If it doesn't, go to the DLC circuit troubleshooting.

4. Turn the ignition switch to LOCK (0).

5. Remove the fuel fill cap to relieve the pressure in the fuel tank.

6. Turn the ignition switch to ON (II).

7. From the INSPECTION MENU of the HDS, select "Fuel Pump OFF", then start the engine, and let it idle until it stalls.

a. Do not allow the engine to idle above 1,000 rpm or the ECM/PCM will continue to operate the fuel pump.

b. A Confirmed or Pending DTC may be set during this procedure. Check for DTCs, and clear them as needed.

8. Turn the ignition switch to LOCK (0).

9. Do the 12V battery terminal disconnection procedure.

10. Remove the cowl cover and the under-cowl panel.

11. Remove the bracket and the quick-connect fitting cover.

12. Check the fuel quick-connect fitting for dirt, and clean it if needed.

13. Place a rag or shop towel over the quick-connect fitting.

14. Disconnect the quick-connect fitting:

a. Hold the connector with one hand, and squeeze the retainer tabs with the

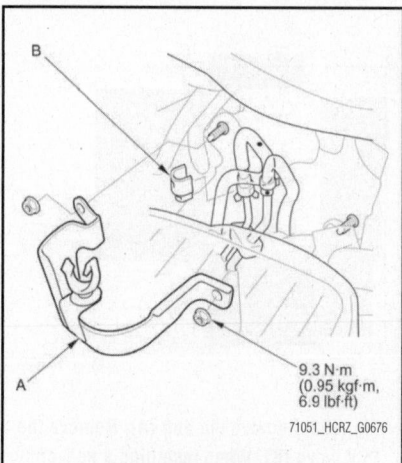

9.3 N·m
(0.95 kgf·m,
6.9 lbf·ft)

71051_HCRZ_G0676

Fig. 167 Remove the bracket (A) and the quick-connect fitting cover (B).

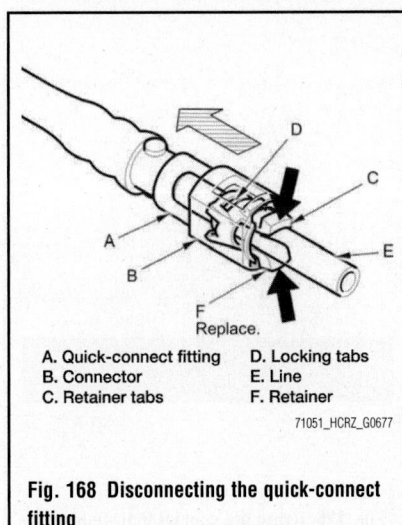

Fig. 168 Disconnecting the quick-connect fitting

A. Quick-connect fitting   D. Locking tabs
B. Connector   E. Line
C. Retainer tabs   F. Retainer

71051_HCRZ_G0677

other hand to release them from the locking tabs. Pull the connector off.

b. Be careful not to damage the line or other parts. Do not use tools.

c. If the connector does not move, keep the retainer tabs pressed down, and alternately pull and push the connector until it comes off easily.

d. Do not remove the retainer from the line; once removed, the retainer must be replaced with a new one.

15. After disconnecting the quick-connect fitting, check it for dirt or damage.

16. Do the 12V battery terminal reconnection procedure.

Without HDS

See Figure 169.

Before disconnecting fuel lines or hoses, relieve pressure from the system by disabling the fuel pump and disconnecting the fuel line/quick connect fitting in the engine compartment.

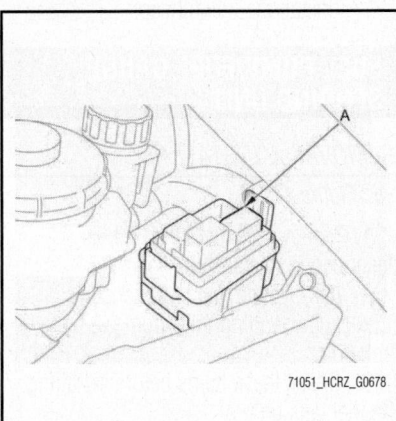

Fig. 169 Remove PGM-FI main relay 2 (A) from auxiliary under-hood relay box A.

71051_HCRZ_G0678

➡️**During this procedure, use the "With HDS" illustrations where applicable.**

1. Remove PGM-FI main relay 2 from auxiliary under-hood relay box A.
2. Start the engine, and let it idle until it stalls.

➡️**If any DTCs are stored, clear and ignore them.**

3. Turn the ignition switch to LOCK (0).
4. Remove the fuel fill cap to relieve the pressure in the fuel tank.
5. Do the 12 volt battery terminal disconnection procedure.
6. Remove the cowl cover and the under-cowl panel.
7. Remove the bracket and the quick-connect fitting cover.
8. Check the fuel quick-connect fitting for dirt, and clean it if needed.
9. Place a rag or shop towel over the quick-connect fitting.
10. Disconnect the quick-connect fitting:

a. Hold the connector with one hand, and squeeze the retainer tabs with the other hand to release them from the locking tabs. Pull the connector off.

b. Be careful not to damage the line or other parts. Do not use tools.

c. If the connector does not move, keep the retainer tabs pressed down, and alternately pull and push the connector until it comes off easily.

d. Do not remove the retainer from the line; once removed, the retainer must be replaced with a new one.

11. After disconnecting the quick-connect fitting, check it for dirt or damage.

12. Do the 12 volt battery terminal reconnection procedure.

FUEL FILTER

REMOVAL & INSTALLATION
See Figure 170.

➡️**The fuel filter should be replaced whenever the fuel pressure drops below the specified value, after making sure that the fuel pump and the fuel pressure regulator are OK.**

1. Remove the fuel tank unit.
2. Remove the fuel filter set.
3. Check these items before installing the fuel tank unit:

a. When connecting the wire harness, make sure the connection is secure and the connectors are firmly locked into place.

b. When installing the fuel gauge sending unit, make sure the connection is secure and the connector is firmly

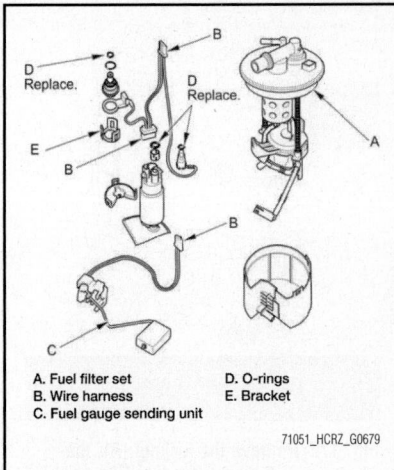

Fig. 170 Showing the fuel tank unit, fuel filter and related components

A. Fuel filter set   D. O-rings
B. Wire harness   E. Bracket
C. Fuel gauge sending unit

71051_HCRZ_G0679

locked into place. Be careful not to bend or twist it excessively.

4. Install the parts in the reverse order of removal with new O-rings and a new bracket. When installing the fuel tank unit, align the marks on the unit and the fuel tank.

a. Coat the O-rings with clean engine oil; do not use any other oils or fluids.

b. Do not pinch the O-rings during installation.

c. Use all the new parts supplied in the fuel filter replacement kit.

FUEL TANK UNIT

REMOVAL & INSTALLATION
See Figures 171 through 176.

Special Tools Required: Fuel Pump Module Locknut Wrench 07AAA-SNAA100.

1. Relieve the fuel pressure. See "Relieving Fuel System Pressure" in this section.

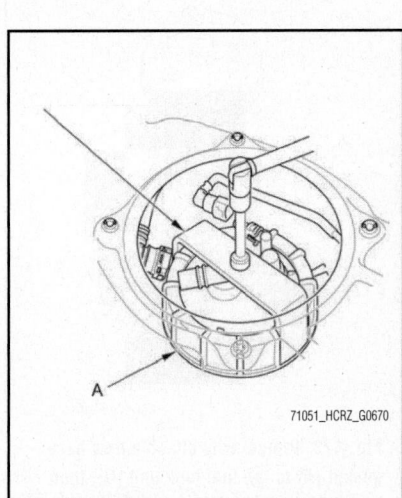

Fig. 171 Using the special tool, loosen the locknut (A).

71051_HCRZ_G0670

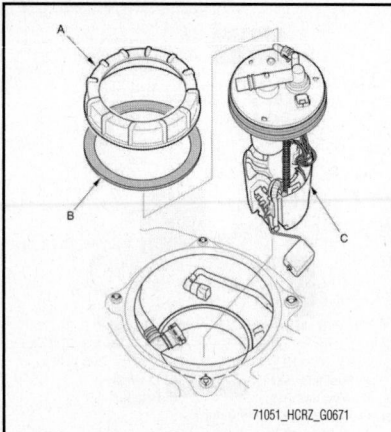

Fig. 172 Remove the locknut (A), the locknut plate (B), and the fuel tank unit (C).

2. Remove the rear tray.

3. Remove the access panel from the floor.

4. Disconnect the fuel tank unit 4P connector.

5. Disconnect the quick-connect fittings from the fuel tank unit.

6. Using the special tool, loosen the locknut.

7. Remove the locknut, the locknut plate, and the fuel tank unit.

To install:

8. Temporarily attach a new base gasket to the fuel tank unit, then insert the fuel tank unit partially into the fuel tank.

Note the following:

• Be careful not to damage the new base gasket.

• Be careful not to bend the fuel gauge sending unit.

• Do not coat the base gasket with oil.

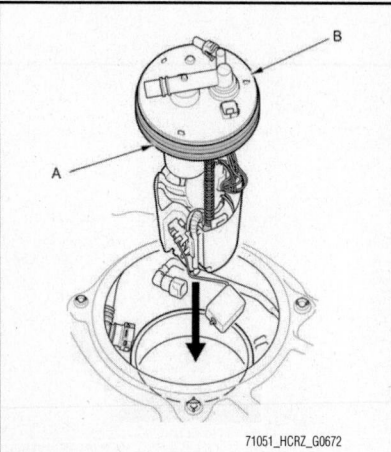

Fig. 173 Temporarily attach a new base gasket (A) to the fuel tank unit (B), then insert the fuel tank unit partially into the fuel tank.

9. Transfer the base gasket from the fuel tank unit to the fuel tank.

10. Align the marks on the fuel tank and fuel tank unit, then insert the fuel tank unit into the fuel tank until the fuel tank unit rests on top of the base gasket.

> ※ **CAUTION**
>
> **To avoid a fuel leak, check the base gasket, visually or by hand, to make sure it is not pinched.**

11. Tighten a new locknut by hand with a new locknut plate. Before tightening, align the mark on the locknut plate to the start of the thread.

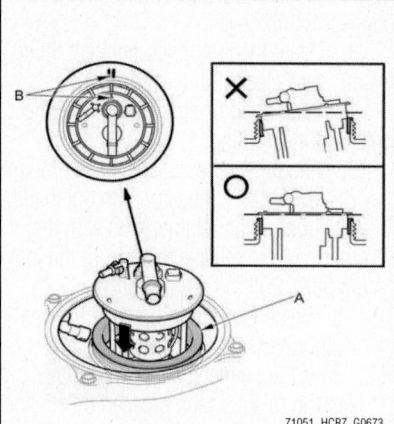

Fig. 174 Transfer the base gasket (A) from the fuel tank unit to the fuel tank. Align the marks (B) on the fuel tank and fuel tank unit, then insert the fuel tank unit into the fuel tank until the fuel tank unit rests on top of the base gasket.

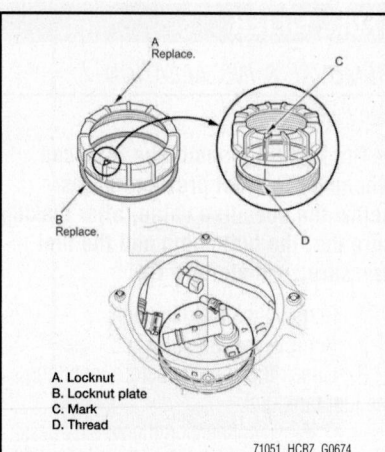

A. Locknut
B. Locknut plate
C. Mark
D. Thread

Fig. 175 Tighten a new locknut (A) by hand with a new locknut plate (B). Before tightening, align the mark (C) on the locknut plate to the start of the thread (D).

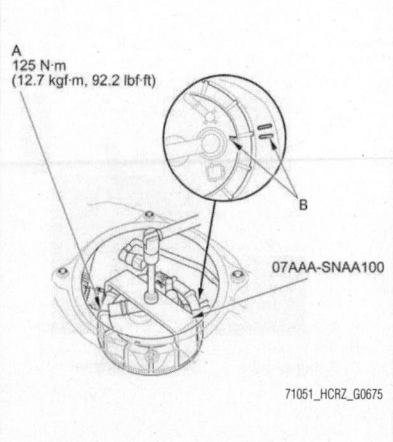

Fig. 176 Using the special tool, tighten the locknut (A) to 92 ft. lbs. (125 Nm). After tightening, make sure the marks (B) are still aligned.

12. Using the special tool, tighten the locknut to 92 ft. lbs. (125 Nm).

Note the following:

• After tightening, make sure the marks are still aligned.

  • After installation, check the base gasket, visually or by hand, to make sure it is not pinched.

13. Connect the fuel tank unit 4P connector, then connect the quick-connect fitting.

14. Reconnect the negative cable to the 12 volt battery, and turn the ignition switch to ON (II). The fuel pump will run for about 2 seconds, and fuel pressure will rise. Repeat this two or three times, then make sure there are no fuel leaks.

15. Install the access panel.

16. Install the rear tray.

17. Install the fuel fill cap.

FUEL PUMP/FUEL GAUGE SENDING UNIT

REMOVAL & INSTALLATION

See Figure 177.

1. Remove the fuel tank unit, as described in this section.

2. Remove the fuel level sensor (fuel gauge sending unit) from the fuel tank unit.

3. Check these items before installing the fuel tank unit:

   a. When connecting the wire harness, make sure the connection is secure and the connector is firmly locked into place.

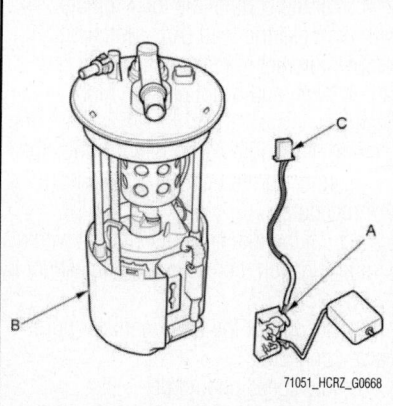

Fig. 177 Remove the fuel level sensor (fuel gauge sending unit) (A) from the fuel tank unit (B). When connecting the wire harness, make sure the connection is secure and the connector (C) is firmly locked into place.

b. When installing the fuel gauge sending unit, make sure the connection is secure. Be careful not to bend or twist it excessively.

4. Install the parts in the reverse order of removal. When installing the fuel tank unit, align the marks on the unit and the fuel tank.

FUEL INJECTORS & FUEL RAIL

REMOVAL & INSTALLATION

See Figure 178.

1. Relieve the fuel pressure as described in this section.
2. Remove the intake manifold/chamber assembly. Refer to "Engine Mechanical" section.
3. Remove the nut.
4. Disconnect the injector connectors, the rocker arm oil control solenoid connector, and the EGR valve connector.
5. Remove the fuel rail mounting nuts and the bolt from the fuel rail, then remove the injectors and the fuel rail together.
6. Remove the injector clips from the injectors.
7. Remove the injectors from the fuel rail.
8. Coat the new O-rings (black) with clean engine oil, then insert the injectors into the fuel rail.
9. Install the injector clips.
10. Coat the injector O-rings (brown) with clean engine oil.
11. Install the fuel rail and the injectors in the cylinder head.
12. Install the fuel rail mounting nuts and the bolt.

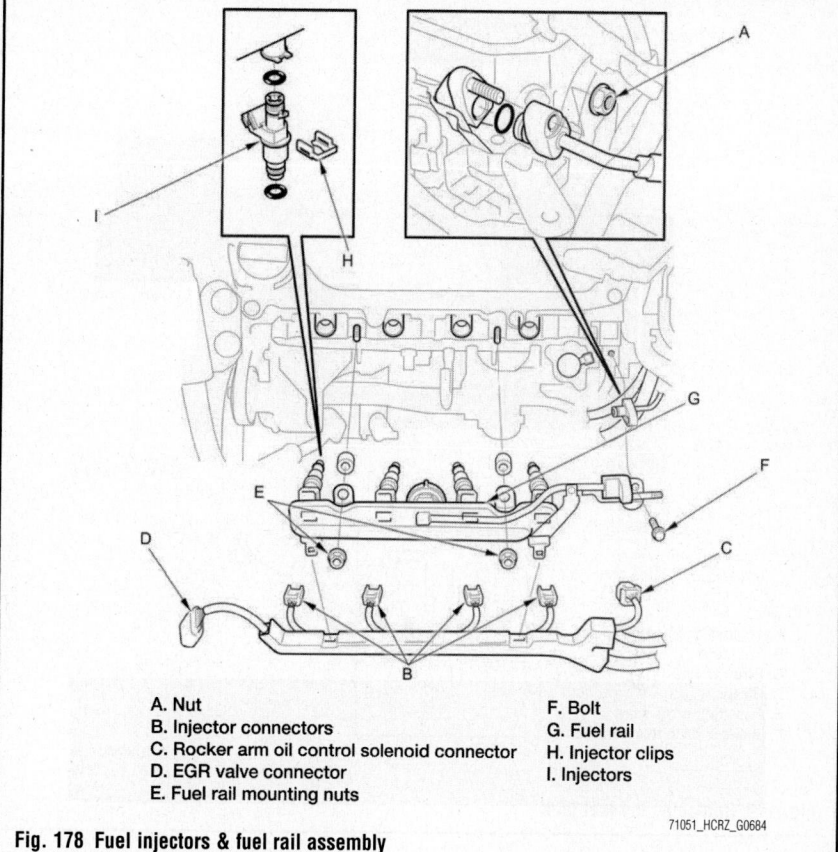

A. Nut
B. Injector connectors
C. Rocker arm oil control solenoid connector
D. EGR valve connector
E. Fuel rail mounting nuts
F. Bolt
G. Fuel rail
H. Injector clips
I. Injectors

Fig. 178 Fuel injectors & fuel rail assembly

13. Install the nut with a new O-ring.
14. Connect the injector connectors, the rocker arm oil control solenoid connector, and the EGR valve connector.
15. Install the intake manifold/chamber assembly.
16. Turn the ignition switch to ON (II), but do not operate the starter. After the fuel pump runs for about 2 seconds, the fuel rail is pressurized. Repeat this process two or three times, then check for fuel leaks.

FUEL TANK

DRAINING

1. Remove the fuel tank unit.
2. Using a hand pump, a hose, and a container suitable for fuel, draw the fuel from the fuel tank.
3. Reinstall the fuel tank unit.

REMOVAL & INSTALLATION

See Figures 179 and 180.

1. Drain the fuel tank until it is less than half full.
2. Reinstall the fuel tank unit without connecting the fuel tank unit 4P connector, and the quick-connect fittings.

3. Raise the vehicle on a lift.
4. Disconnect the quick-connect fittings and open the clamp.
5. Remove the middle floor undercover.
6. Remove the trailing arm braces.
7. Place a jack or other support under the fuel tank.
8. Remove the strap bolts and the straps.
9. Remove the fuel tank.
10. Remove the fuel tank protectors. When installing the fuel tank protectors, make sure to insert the clips in the direction shown.

To install:

11. Install the parts in the reverse order of removal.

a. New fuel tank have a ring pull at the fuel tank vapor recirculation tube and the fuel fill neck tube. When you connect the hose and confirm that the connection is secure, remove the ring pull by pulling it down.

b. Before connecting the fuel fill pipe and the quick-connect fittings, check for dirt, and clean them if needed, taking care not to damage the fuel fill pipe and other parts.

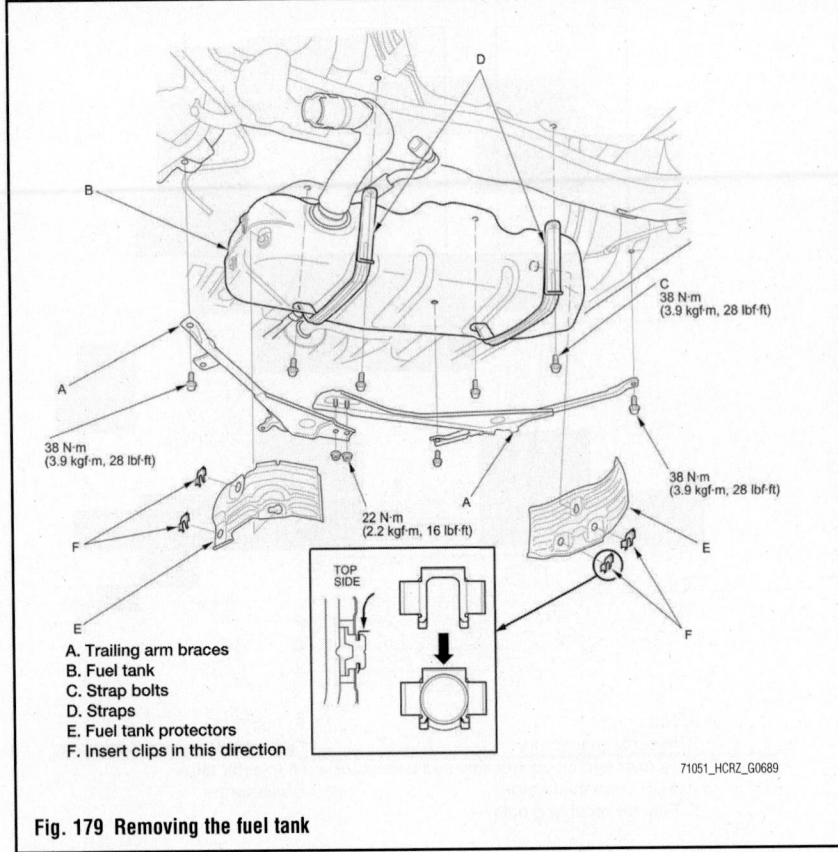

A. Trailing arm braces
B. Fuel tank
C. Strap bolts
D. Straps
E. Fuel tank protectors
F. Insert clips in this direction

71051_HCRZ_G0689

Fig. 179 Removing the fuel tank

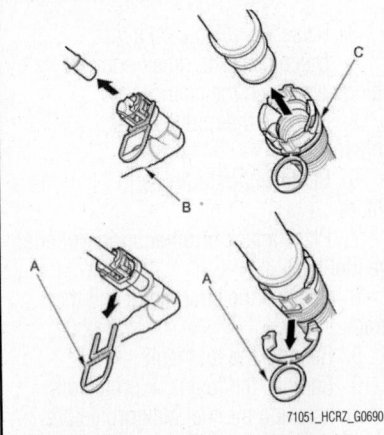

71051_HCRZ_G0690

Fig. 180 New fuel tank have a ring pull (A) at the fuel tank vapor recirculation tube (B) and the fuel fill neck tube (C). When you connect the hose and confirm that the connection is secure, remove the ring pull by pulling it down.

IDLE SPEED

INSPECTION

Note the following:
• Before checking the idle speed, check these items:
• The malfunction indicator lamp (MIL) has not been reported on, and there are no DTCs.

• Ignition timing
• Spark plugs
• Air cleaner
• PCV system
• Apply the parking brake, and make sure the headlights are off.

1. Disconnect the evaporative emission (EVAP) canister purge valve connector.

2. Connect the HDS to the data link connector (DLC) located under the driver's side of the dashboard.

3. Make sure the HDS communicates with the ECM/PCM. If it doesn't, go to the DLC circuit troubleshooting.

4. Start the engine. Hold the engine speed at 3,000 rpm without load (CVT in P or N, M/T in neutral) until the radiator fan comes on, then let it idle.

5. Check the idle speed under no load conditions: headlights, blower fan, radiator fan, audio system, and A/C off.

➡ **Check the state-of-charge (SOC) in the IMA system DATA LIST with the HDS. If the IMA battery charge level is less than 40%, start the engine, and hold it between 3,500–4,000 rpm without load (CVT in P or N, M/T in neutral) until the SOC indicator displays at least half of its segments.**

6. Let the engine idle for 1 minute with high electric load (A/C switch on, temperature set to max cool, blower fan on high, and headlights on high beam).

a. If the idle speed is not within specification, do the ECM/PCM idle learn procedure.

b. If the idle speed is still not within specification, continue to troubleshoot by symptoms.

7. Reconnect the EVAP canister purge valve connector.

8. Idle speed should be:
• CVT: 700–800 rpm (in P or N)
• M/T: 700–800 ± 50 rpm

THROTTLE BODY

REMOVAL & INSTALLATION
See Figure 181.

✳✳ CAUTION

Do not insert your fingers into the installed throttle body when you turn the ignition switch to ON (II) or while the ignition switch is in ON (II). If you do, you will seriously injure your fingers if the throttle valve is activated.

➡ **If you are replacing or cleaning the throttle body, start at step 1. If you are removing the throttle body temporarily, begin at step 4.**

1. Connect the HDS to the DLC while the engine is stopped.

2. Select the INSPECTION MENU on the HDS.

3. Do the TP POSITION CHECK in the ETCS TEST.

4. Turn the ignition switch to LOCK (0).

5. Remove the air cleaner.

6. Disconnect the EVAP canister purge valve connector.

7. Disconnect the throttle body connector, then disconnect the EVAP canister purge valve hose.

8. Disconnect and plug the water bypass hoses.

9. Disconnect the purge hoses.

10. Remove the throttle body.

11. Remove the purge pipe.

12. Install the parts in the reverse order of removal with a new gasket, then refill the radiator with engine coolant.

a. If you replace or clean the throttle body, go to step 13.

b. If you did not replace or clean the throttle body, this procedure is complete.

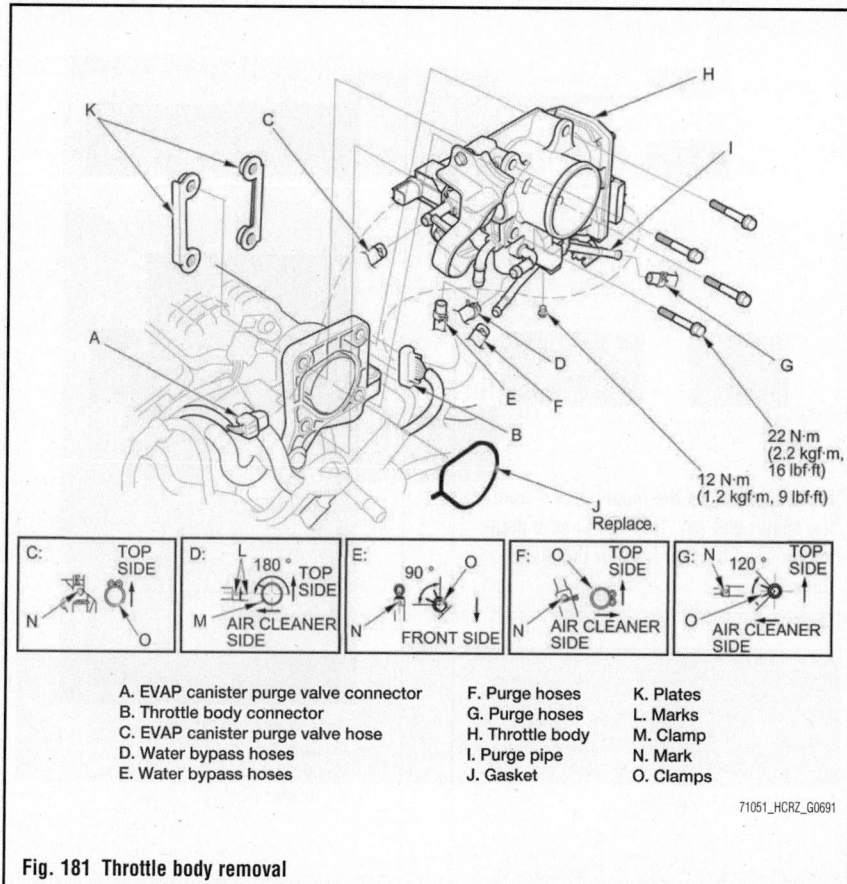

Fig. 181 Throttle body removal

A. EVAP canister purge valve connector
B. Throttle body connector
C. EVAP canister purge valve hose
D. Water bypass hoses
E. Water bypass hoses
F. Purge hoses
G. Purge hoses
H. Throttle body
I. Purge pipe
J. Gasket
K. Plates
L. Marks
M. Clamp
N. Mark
O. Clamps

71051_HCRZ_G0691

c. Be careful not to drop or damage the plates.

d. Align the marks on the hoses and the throttle body, then insert the hoses. Make sure the clamp is positioned as shown.

e. Align the mark on the hoses as shown, then insert the hoses. Make sure the clamps are positioned as shown.

13. Connect the HDS to the data link connector (DLC) located under the driver's side of the dashboard.

14. Turn the ignition switch to ON (II).

15. Reset the ECM/PCM with the HDS.

16. Select the ETCS TEST in the INSPECTION MENU with the HDS.

17. Select the TP POSITION CHECK, then clear the throttle position (TP) learned value.

18. Turn the ignition switch to LOCK (0).

19. Turn the ignition switch to ON (II), and wait 2 seconds without pressing the accelerator pedal.

20. Do the ECM/PCM idle learn procedure.

HEATING & AIR CONDITIONING SYSTEM

BLOWER UNIT

REMOVAL & INSTALLATION

See Figures 182 through 184.

1. Remove the passenger's dashboard undercover and the glove box.

2. Cut the plastic cross brace in the glove box opening with diagonal cutters in the area shown, and discard it.

3. Disconnect the connectors from the blower motor and the recirculation control motor. Remove the wire harness clips.

4. Remove the self-tapping screws and the duct.

5. Remove the self-tapping screws, the mounting nuts, and the blower unit.

6. Note these items when overhauling the blower unit:

a. The recirculation control motor, the blower motor, and the dust and pollen filter can be replaced without removing the blower unit.

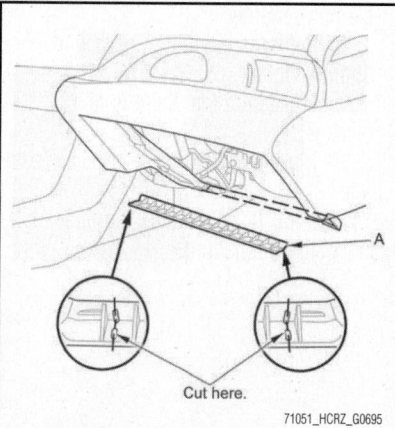

Fig. 182 Cut the plastic cross brace (A) in the glove box opening with diagonal cutters in the area shown, and discard it.

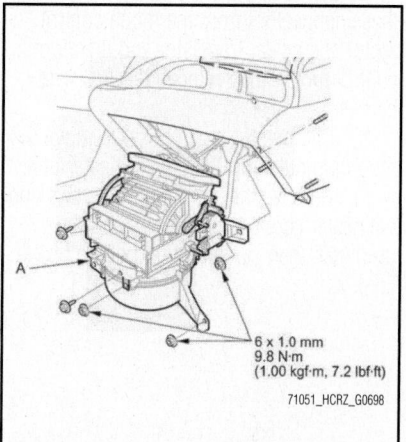

Fig. 183 Remove the self-tapping screws, the mounting nuts, and the blower unit (A).

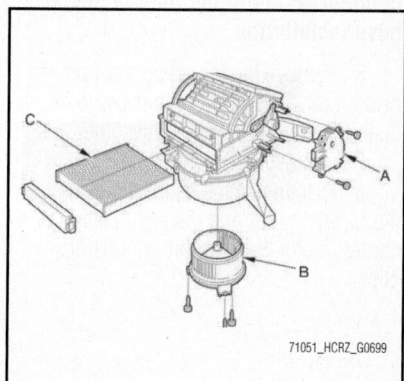

Fig. 184 The recirculation control motor (A), the blower motor (B), and the dust and pollen filter (C) can be replaced without removing the blower unit.

b. Before reassembly, make sure that the recirculation control linkage and door move smoothly without binding.

c. After reassembly, make sure the recirculation control motor runs smoothly.

d. Install the unit in the reverse order of removal. Make sure that there is no air leakage.

HEATER UNIT & HEATER CORE

REMOVAL & INSTALLATION

See Figures 185 through 187.

☀ CAUTION

SRS components are located in this area. Review the SRS component locations, and the precautions and procedures before doing repairs or service. See "Air Bags (Supplemental Restraint System)" section.

1. Do the 12 volt battery terminal disconnection procedure.

2. Recover the refrigerant with a recovery/recycling/charging station.

3. Disconnect the A/C line from the evaporator core.

4. Drain the engine coolant from the radiator.

5. From the inlet heater hose, remove the clip. Slide the hose clamps back, then disconnect the inlet heater hose and the outlet heater hose from the heater unit. Note the layout of the hoses. Engine coolant will run out when the hoses are disconnected; drain it into a clean drip pan. Be sure not to let coolant spill on the electrical parts or the painted surfaces. If any coolant spills, rinse it off immediately.

➡**Discard the removed clip because it is unnecessary at the time of heater hose installation.**

6. Remove the mounting nut from the heater unit. Take care not to damage or bend the fuel lines or the brake lines, etc.

7. Remove the dashboard.

8. Disconnect the connectors from the blower motor and the recirculation control motor. Remove the wire harness clips.

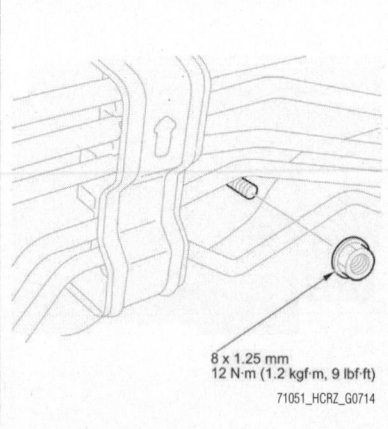

8 x 1.25 mm
12 N·m (1.2 kgf·m, 9 lbf·ft)
71051_HCRZ_G0714

Fig. 185 Remove the mounting nut from the heater unit (A). Take care not to damage or bend the fuel lines or the brake lines, etc.

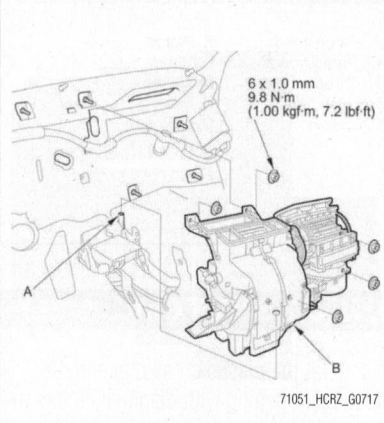

6 x 1.0 mm
9.8 N·m
(1.00 kgf·m, 7.2 lbf·ft)
71051_HCRZ_G0717

Fig. 186 Remove the drain hose (A). Remove the mounting nuts and the blower-heater unit (B).

9. Disconnect the connector from the evaporator temperature sensor. Disconnect these connectors from the mode control motor, the power transistor and the air mix control motor, then remove the A/C wire harness.

10. Remove the drain hose. Remove the mounting nuts and the blower-heater unit.

11. Remove the self-tapping screws and the heater core cover, the cap, and the grommet, then carefully pull out the heater core.

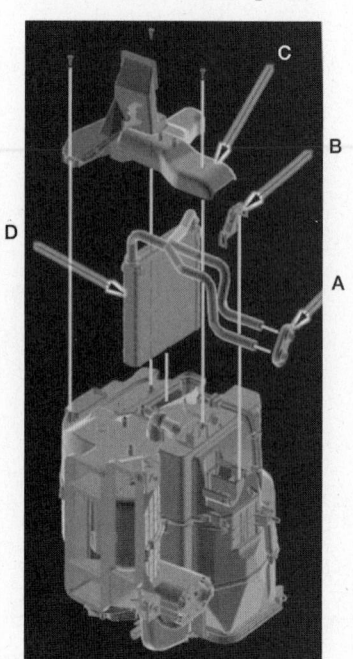

A. Heater core cover
B. Cap
C. Grommet
D. Heater core

71051_HCRZ_G0718

Fig. 187 Remove the self-tapping screws and the heater core cover, the cap, and the grommet, then carefully pull out the heater core.

To install:

12. Install the heater core and the evaporator core in the reverse order of removal.

13. Install the heater unit in the reverse order of removal, and note these items:

a. Do not interchange the inlet and outlet heater hoses, and install the hose clamps securely.

b. Refill the cooling system with engine coolant.

c. Make sure that there is no coolant leakage.

d. Make sure that there is no air leakage.

14. Do the 12 volt battery terminal reconnection procedure.

STEERING

CONTROL UNIT

REMOVAL & INSTALLATION

See Figure 188.

1. Do the 12 volt battery terminal disconnection procedure.
2. Remove the driver's kick panel.
3. Disconnect EPS control unit 11P connector and the 16P connector by pushing the lock and pulling down the lever from the EPS control unit.
4. Remove the EPS control unit.

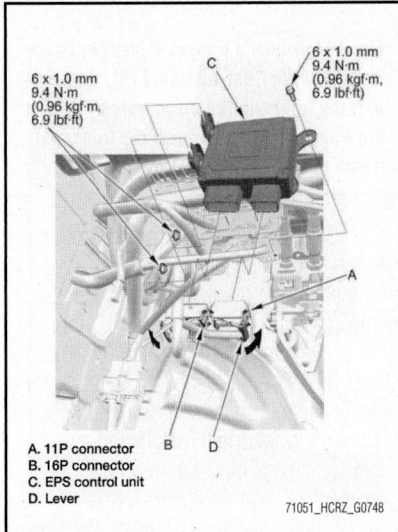

A. 11P connector
B. 16P connector
C. EPS control unit
D. Lever

71051_HCRZ_G0748

Fig. 188 Removing/installing the EPS control unit

To install:

5. Install the EPS control unit.
6. Pull up the lever of the EPS control unit 11P connector and the 16P connector, then confirm the connector is fully seated to the EPS control unit.
7. Install the driver's kick panel.
8. Do the 12 volt battery terminal reconnection procedure.
9. After installation, start the engine, allow it to idle, and turn the steering wheel from lock to lock several times.
10. Check that the EPS indicator does not come on.

EPS MOTOR

REMOVAL & INSTALLATION

See Figures 189 and 190.

➡Do not allow dust, dirt, or other foreign materials to enter the steering gearbox.

1. Remove the steering gearbox, as described in this section.
2. Remove the EPS motor 3P connector and the EPS motor angle sensor 8P connector from the connector bracket. Remove the wire harness clips.
3. Remove the EPS motor and the O-ring from the steering gearbox.

To install:

4. Clean the mating surfaces of the EPS motor and the steering gearbox.
5. Apply the steering grease (08C35-B0534L) to a new O-ring and carefully fit it on the EPS motor.
6. Apply the steering grease (08C35-B0534L) to the EPS motor shaft.
7. Install the EPS motor on the steering gearbox by engaging the EPS motor shaft and the worm shaft.
8. Before tightening the bolts, turn the

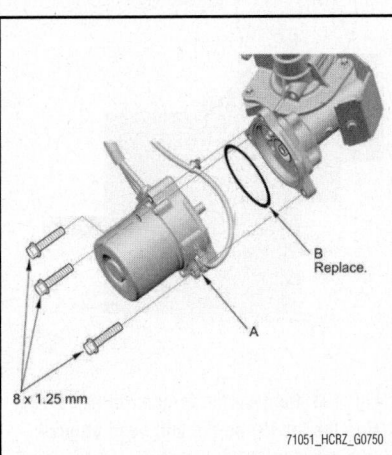

B
Replace.

A

8 x 1.25 mm

71051_HCRZ_G0750

Fig. 189 Remove the EPS motor (A) and the O-ring (B) from the steering gearbox.

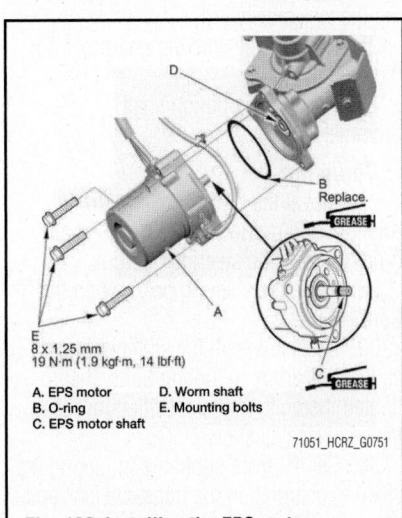

D

B
Replace.
GREASE

A

E
8 x 1.25 mm
19 N·m (1.9 kgf·m, 14 lbf·ft)

C
GREASE

A. EPS motor   D. Worm shaft
B. O-ring   E. Mounting bolts
C. EPS motor shaft

71051_HCRZ_G0751

Fig. 190 Installing the EPS motor

motor two or three times to the right and left about 45 degrees. Make sure the EPS motor is evenly seated on the steering gearbox, and that the O-ring is not pinched between the mating surfaces.

9. Tighten the EPS motor mounting bolts to 14 ft. lbs. (19 Nm).
10. Install the wire harness clips.
11. Install the EPS motor 3P connector and the EPS motor angle sensor 8P connector to the connector bracket.
12. Install the steering gearbox as described in this section.

STEERING GEAR

REMOVAL & INSTALLATION

See Figures 191 through 201.

Special Tools Required:
• Ball Joint Thread Protector, 12 mm 07AAF-SDAA100
• Ball Joint Remover, 28 mm 07MAC-SL0A202
• Ball Joint Thread Protector, 14 mm 071AF-S3VA000

1. Note these items during removal:
   a. Using solvent and a brush, wash any oil and dirt off the end of the steering gearbox. Avoid any electrical parts. Blow dry with compressed air.
   b. Be sure to hold the steering wheel before disconnecting the steering joint or damage to the cable reel can occur.
2. Do the 12 volt battery terminal disconnection procedure.
3. Raise and support the vehicle.
4. Remove the front wheels.
5. Tilt the steering column all the way down and move it all the way in.

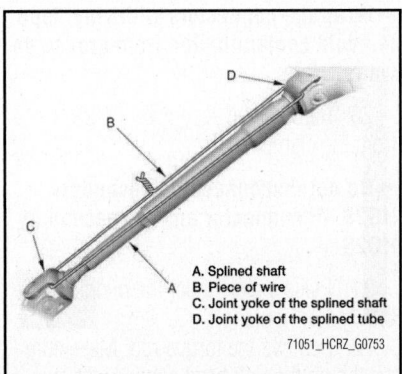

D
B

C

A

A. Splined shaft
B. Piece of wire
C. Joint yoke of the splined shaft
D. Joint yoke of the splined tube

71051_HCRZ_G0753

Fig. 191 Hold the splined shaft on the column with a piece of wire between the joint yoke of the splined shaft and joint yoke of the splined tube to prevent the splined shaft from pulling out.

6. Release the hooks, then remove the steering joint cover.

7. Hold the splined shaft on the column with a piece of wire between the joint yoke of the splined shaft and joint yoke of the splined tube to prevent the splined shaft from pulling out.

8. Remove the steering joint bolt.

9. Center the steering wheel spokes, and install a commercially available steering wheel holder tool.

10. Disconnect the steering joint by moving the steering joint upward toward the column.

11. Remove the center guide (if equipped) from the top of the pinion shaft and discard it.

➡**The center guide is for factory assembly use only.**

12. Remove the cotter pin from the tie-rod end ball joint, then remove the nut on both sides.

13. Disconnect the tie-rod end ball joint from the knuckle using the ball joint remover and the ball joint thread protector on both sides.

14. Remove the lock pin from the lower arm ball joint, and remove the castle nut from both sides.

15. Disconnect the lower arm ball joint from the knuckle using the ball joint remover and the ball joint thread protector on both sides.

16. Remove the stabilizer links from the stabilizer bar on both sides.

17. Remove the splash shield.

18. Disconnect the EPS motor angle sensor 8P connector and the torque sensor 6P connector.

19. Disconnect the EPS motor 3P connector by pushing the lock and pulling up the lever.

➡**Wrap the connectors with vinyl tape to avoid contamination from grease or water.**

20. Remove the secondary HO2S wire from the clamp.

➡**Do not disconnect the secondary HO2S 4P connector and the secondary HO2S.**

21. Place a jack under the middle of the IMA motor housing.

22. Remove the torque rod. Make sure the tab on the bolt head aligned with the guide on the front subframe.

23. Attach a transaxle jack to the middle of the front subframe and support the front subframe securely by raising the transaxle jack. Remove the front subframe mounting bolts.

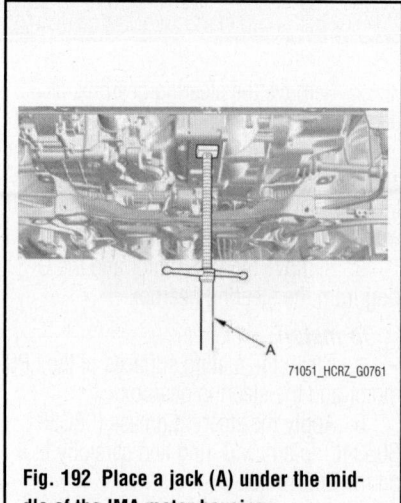

Fig. 192 Place a jack (A) under the middle of the IMA motor housing.

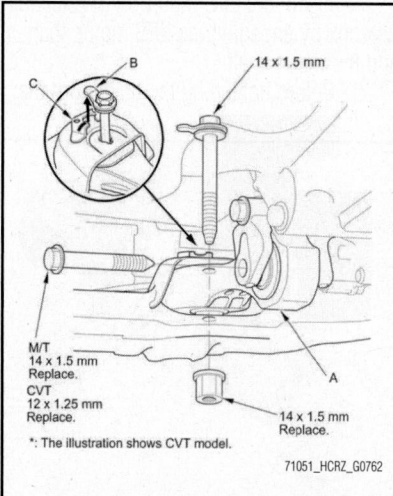

Fig. 193 Remove the torque rod (A). Make sure the tab (B) on the bolt head aligned with the guide (C) on the front subframe.

24. Lower the front subframe and the steering gearbox as an assembly by lowering the jack slowly.

25. Remove the steering gearbox from the front subframe.

26. Remove the pinion shaft grommet.

To install:

27. Install the pinion shaft grommet. Align the lug portion in the pinion shaft grommet with the steering gearbox.

28. Place the steering gearbox on the front subframe.

29. Loosely install the stiffener plates and new gearbox mounting bolts, then tighten the bolts to the specified torque in the sequence as shown.

30. Set the front subframe mounting the steering gearbox on the transaxle jack and support it.

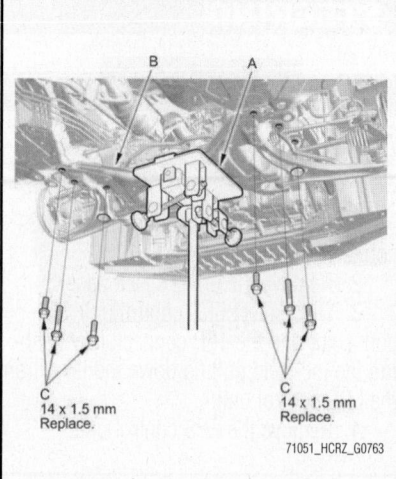

Fig. 194 Attach a transaxle jack (A) to the middle of the front subframe (B), and support the front subframe securely by raising the transaxle jack. Remove the front subframe mounting bolts (C).

31. Carefully raise the front subframe with the transaxle jack, and pass the pinion shaft into the passenger's compartment.

   a. Be sure that the pinion shaft grommet is securely in place. Make sure the lip of the pinion shaft grommet is not turned up. Incorrect installation can cause leakage of water and mud, or noise.

   b. Take care not to damage the lower arm ball joint boot with the edge of the knuckle, etc.

32. Install the front subframe, then loosely tighten the new front subframe mounting bolts.

33. Align the front subframe.

34. Install the torque rod.

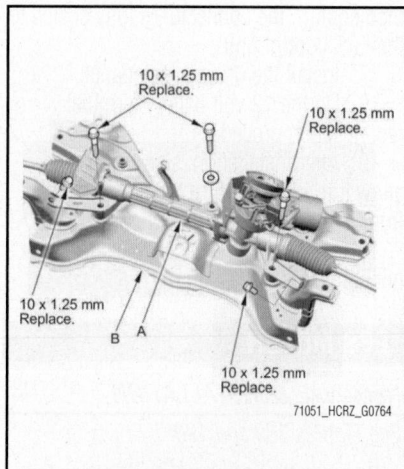

Fig. 195 Remove the steering gearbox (A) from the front subframe (B).

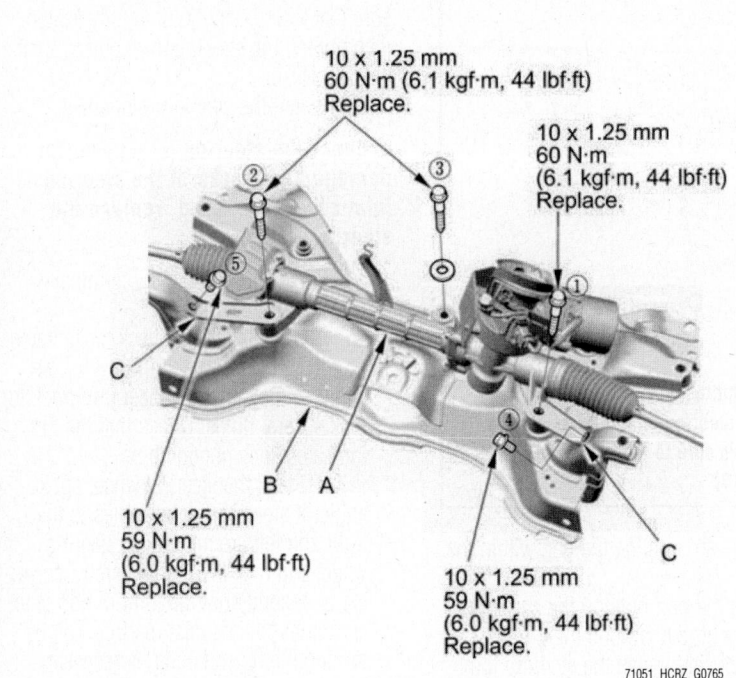

10 x 1.25 mm
60 N·m (6.1 kgf·m, 44 lbf·ft)
Replace.

10 x 1.25 mm
60 N·m
(6.1 kgf·m, 44 lbf·ft)
Replace.

10 x 1.25 mm
59 N·m
(6.0 kgf·m, 44 lbf·ft)
Replace.

10 x 1.25 mm
59 N·m
(6.0 kgf·m, 44 lbf·ft)
Replace.

71051_HCRZ_G0765

Fig. 196 Place the steering gearbox (A) on the front subframe (B). Loosely install the stiffener plates (C), and new gearbox mounting bolts, then tighten the bolts to the specified torque in the sequence as shown.

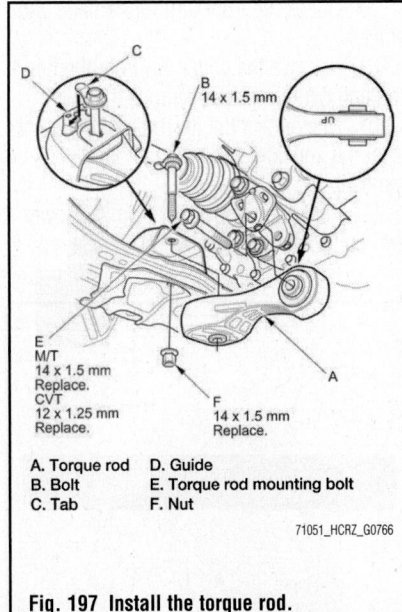

E
M/T
14 x 1.5 mm
Replace.
CVT
12 x 1.25 mm
Replace.

F
14 x 1.5 mm
Replace.

A. Torque rod   D. Guide
B. Bolt   E. Torque rod mounting bolt
C. Tab   F. Nut

71051_HCRZ_G0766

Fig. 197 Install the torque rod.

➡ **Be sure to install the torque rod with the "UP" mark facing up.**

35. Install the bolt with the tab on the bolt head aligned with the guide on the front subframe, then loosely install a new torque rod mounting bolt and nut.

36. Remove the jack from under the IMA motor.

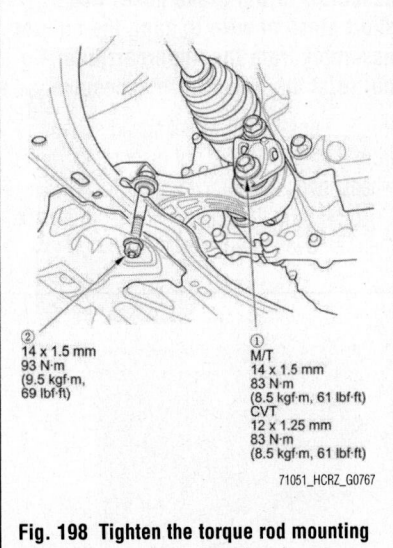

②
14 x 1.5 mm
93 N·m
(9.5 kgf·m,
69 lbf·ft)

①
M/T
14 x 1.5 mm
83 N·m
(8.5 kgf·m, 61 lbf·ft)
CVT
12 x 1.25 mm
83 N·m
(8.5 kgf·m, 61 lbf·ft)

71051_HCRZ_G0767

Fig. 198 Tighten the torque rod mounting bolt and nut in the numbered sequence shown.

37. Tighten the torque rod mounting bolt and nut in the numbered sequence shown.

38. Install the secondary HO2S wire in the clamp.

39. Remove the vinyl tape from the connectors.

40. Connect the torque sensor 6P connector and the EPS motor angle sensor 8P connector.

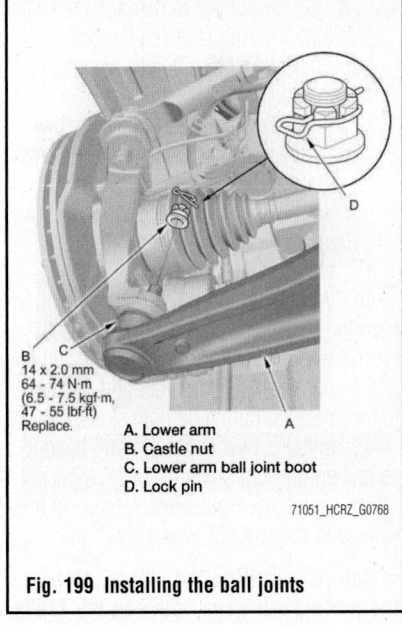

B
C
14 x 2.0 mm
64 - 74 N·m
(6.5 - 7.5 kgf·m,
47 - 55 lbf·ft)
Replace.

A. Lower arm
B. Castle nut
C. Lower arm ball joint boot
D. Lock pin

71051_HCRZ_G0768

Fig. 199 Installing the ball joints

41. Pull down the lever of the EPS motor 3P connector, then confirm the connector is fully seated.

42. Install the stabilizer links to the stabilizer bar on both sides.

43. Install the splash shield.

44. Wipe off any grease contamination from the tapered section and threads of the ball joint. Connect the lower arm to the knuckle. Install a new castle nut and tighten it to the specified torque on both sides.

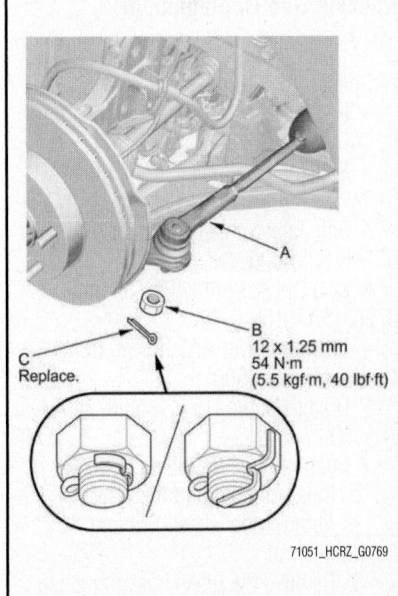

C
Replace.

B
12 x 1.25 mm
54 N·m
(5.5 kgf·m, 40 lbf·ft)

71051_HCRZ_G0769

Fig. 200 Connect the tie-rod end (A) to the knuckles. Install the nut (B), and tighten it to the specified torque on both sides. Install a new cotter pin (C), and bend it as shown on both sides.

a. Be careful not to damage the lower arm ball joint boot. Check the ball joint boot for deformation before connecting the knuckle.

b. Torque the castle nut to the lower torque specification, then tighten it only far enough to align the slot with the joint pin clip hole. Do not align the castle nut by loosening it.

45. Install the lock pin on both sides.

46. Wipe off any grease contamination from the tapered section and threads of the tie-rod end ball joint. Connect the tie-rod end to the knuckles. Install the nut and tighten it to the specified torque on both sides.

47. Install a new cotter pin and bend it as shown on both sides.

48. Install the front wheels, then set the wheels in the straight ahead position.

➡**Before installing the wheel, clean the mating surfaces between the brake disc and the inside of the wheel.**

49. Remove the steering wheel holder tool.

50. Cut the wire and slip the lower end of the steering joint onto the pinion

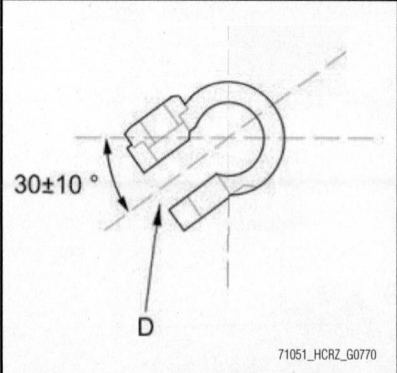

Fig. 201 Cut the wire and slip the lower end of the steering joint onto the pinion shaft, taking care to align the gap within the angle (D).

shaft, taking care to align the gap within the angle.

51. Align the bolt hole on the steering joint with the groove around the pinion shaft, then loosely install the steering joint bolt. Be sure that the joint bolt is securely in the groove in the pinion shaft.

52. Pull on the steering joint to make sure that the steering joint is fully seated, then tighten the steering joint bolt to 21 ft. lbs. (28 Nm).

53. Install the steering joint cover.

➡**Check the steering joint cover for damage and cracks. If the steering joint cover is cracked, replace the steering joint cover.**

54. Do the 12 volt battery terminal reconnection procedure.

55. After installation, check these items:

a. Start the engine, allow it to idle, and turn the steering wheel from lock to lock several times. Check that the EPS indicator does not come on.

b. Check the steering wheel spoke angle. If steering spoke angles to the right and left are not equal (steering wheel and rack are not centered), correct the engagement of the joint/pinion shaft serrations, then adjust the front toe by turning the tie-rod ends, if necessary.

56. Check the wheel alignment, and adjust it if necessary.

SUSPENSION

FRONT SUSPENSION

KNUCKLE/HUB & WHEEL BEARING

REMOVAL & INSTALLATION

Knuckle/Hub Replacement
See Figures 202 through 204.

Special Tools Required:
- Ball Joint Thread Protector, 12 mm 07AAF-SDAA100
- Ball Joint Remover, 28 mm 07MAC-SL0A202
- Ball Joint Thread Protector, 14 mm 071AF-S3VA000
- Hub Dis/Assembly Pin, 34 mm 07965-SA70100
- Bearing Driver Attachment, 62 x 68 mm 07746-0010500
- Driver Handle, 15 x 135L 07749-0010000
- Support Base 07965-SD90100

1. Raise and support the vehicle.

2. Remove the wheel nuts and the front wheel.

3. Remove the brake hose mounting bolt from the damper bracket.

4. Remove the brake caliper bracket mounting bolts, then remove the caliper assembly from the knuckle.

➡**To prevent damage to the caliper assembly or the brake hose, use a short piece of wire to hang the caliper assembly from the undercarriage. Do not twist the brake hose excessively.**

5. Remove the wheel speed sensor from the knuckle. Do not disconnect the wheel speed sensor 2P connector.

6. Pry up the stake on the spindle nut, then remove the nut.

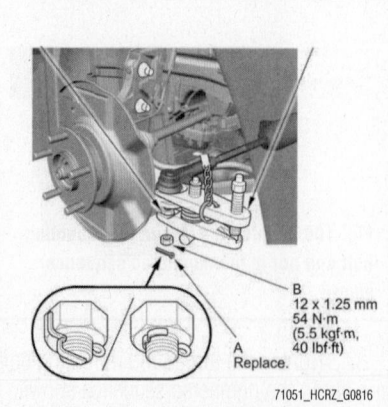

Fig. 202 Remove the cotter pin (A) from the tie-rod end ball joint, then remove the nut (B). Disconnect the tie-rod end ball joint from the knuckle using the ball joint remover.

7. Remove the front brake disc.

8. Check the front hub for damage and cracks.

9. Remove the cotter pin from the tie-rod end ball joint, then remove the nut.

10. Disconnect the tie-rod end ball joint from the knuckle using the ball joint remover.

11. Remove the lock pin from the lower arm ball joint, then remove the castle nut.

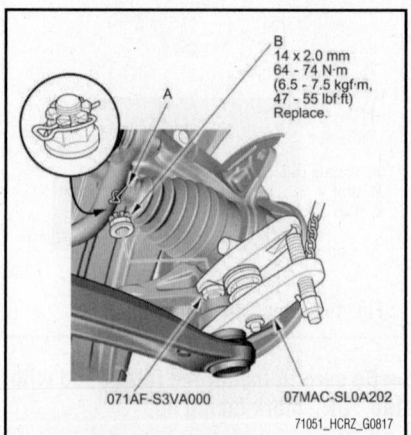

Fig. 203 Remove the lock pin (A) from the lower arm ball joint, then remove the castle nut (B). Disconnect the lower ball joint from the knuckle using the ball joint remover.

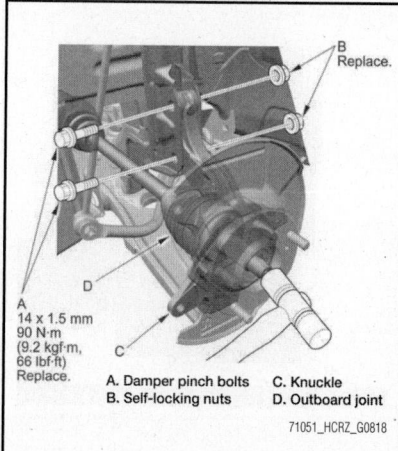

Fig. 204 Remove the damper pinch bolts and the self-locking nuts from the damper. Pull the knuckle outward, and separate the outboard joint from the front hub using a soft face hammer.

12. Disconnect the lower ball joint from the knuckle using the ball joint remover.

13. Remove the damper pinch bolts and the self-locking nuts from the damper.

14. Pull the knuckle outward, and separate the outboard joint from the front hub using a soft face hammer.

➡ **Do not pull the driveshaft end outward. The driveshaft inboard joint may come apart.**

To install:

15. Install the knuckle/hub in the reverse order of removal, and note these items:

a. Be careful not to damage the ball joint boot when connecting the knuckle.

➡ **During installation, apply grease to the mating surfaces of the wheel bearing and the driveshaft outboard joint.**

b. Before connecting the ball joint, degrease the threaded section and the tapered portion of the ball joint pin, the ball joint connecting hole, and the threaded section and the mating surfaces of the castle nut.

c. During installation, install a new cotter pin after tightening the nut, and bend its end as shown.

d. Use new damper pinch bolts and new self-locking nuts during reassembly.

e. Torque the castle nut to the lower torque specification, then tighten it only far enough to align the slot with the ball joint pin hole. Do not align the castle nut by loosening it.

f. Use a new spindle nut on reassembly.

g. Before installing the spindle nut, apply a small amount of engine oil to the seating surface of the nut. After tightening, use a drift to stake the spindle nut shoulder against the driveshaft.

h. During installation, install the lock pin as shown after tightening a new castle nut.

i. Before installing the brake disc, clean the mating surfaces between the front hub and the inside of the brake disc.

j. Before installing the wheel, clean the mating surfaces between the brake disc and the inside of the wheel.

16. Check the wheel alignment, and adjust it if necessary.

Wheel Bearing Replacement

See Figures 205 through 210.

Special Tools Required:
- Ball Joint Thread Protector, 12 mm 07AAF-SDAA100
- Ball Joint Remover, 28 mm 07MAC-SL0A202
- Ball Joint Thread Protector, 14 mm 071AF-S3VA000
- Hub Dis/Assembly Pin, 34 mm 07965-SA70100
- Bearing Driver Attachment, 62 x 68 mm 07746-0010500
- Driver Handle, 15 x 135L 07749-0010000
- Support Base 07965-SD90100

1. Separate the hub from the knuckle using the hub dis/assembly pin and a hydraulic press. Hold the knuckle with the attachment of the hydraulic press or equivalent tool. Be careful not to damage or deform the splash guard.

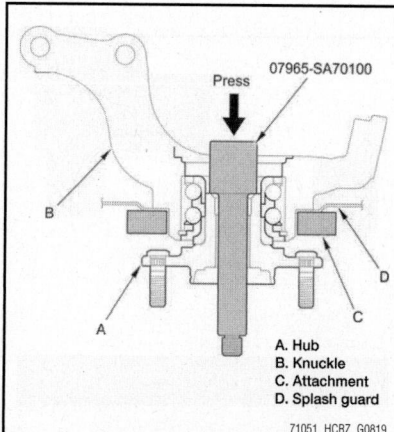

Fig. 205 Separate the hub from the knuckle using the hub dis/assembly pin and a hydraulic press. Hold the knuckle with the attachment of the hydraulic press or equivalent tool. Be careful not to damage or deform the splash guard.

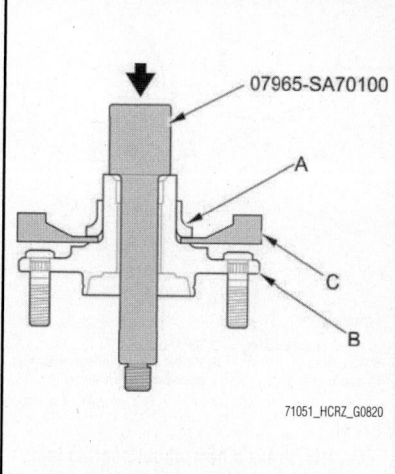

Fig. 206 Press the wheel bearing inner race (A) off of the hub (B) using the hub dis/assembly pin, a commercially available bearing separator (C), and a press.

➡ **Hold onto the hub to keep it from falling when pressed clear.**

2. Press the wheel bearing inner race off of the hub using the hub dis/assembly pin, a commercially available bearing separator, and a press.

3. Remove the splash guard and the snap ring from the knuckle.

4. Press the wheel bearing out of the knuckle using the bearing driver attachment, the driver handle, and a press.

5. Wash the knuckle and the hub thoroughly in high flash point solvent before reassembly.

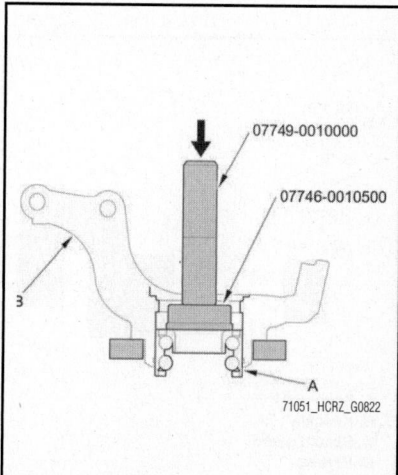

Fig. 207 Press the wheel bearing (A) out of the knuckle (B) using the bearing driver attachment, the driver handle, and a press.

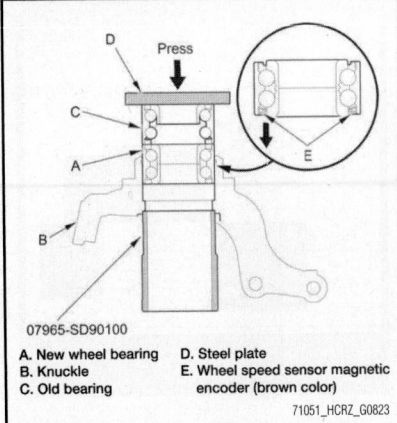

Fig. 208 Press a new wheel bearing into the knuckle using the old bearing, a steel plate, the support base, and a press. Install the wheel bearing with the wheel speed sensor magnetic encoder (brown color) toward the inside of the knuckle.

6. Press a new wheel bearing into the knuckle using the old bearing, a steel plate, the support base, and a press.

  a. Install the wheel bearing with the wheel speed sensor magnetic encoder (brown color) toward the inside of the knuckle.

  b. Remove any oil, grease, dust, metal debris, and other foreign material from the magnetic encoder surface.

  c. Keep any magnetic tools away from the encoder surface.

  d. Be careful not to damage the encoder surface when you insert the wheel bearing.

7. Install the snap ring securely in the knuckle.

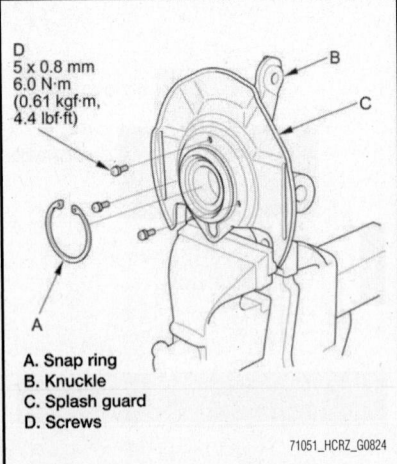

Fig. 209 Install the snap ring securely in the knuckle. Install the splash guard and tighten the screws to the specified torque.

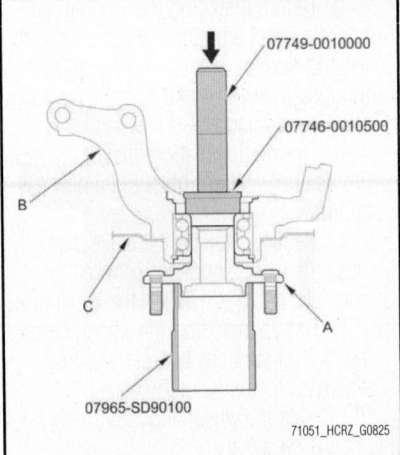

Fig. 210 Install the hub (A) onto the knuckle (B) using the bearing driver attachment, the driver handle, the support base, and a hydraulic press. Be careful not to damage the splash guard (C).

8. Install the splash guard and tighten the screws to the specified torque.

9. Install the hub onto the knuckle using the bearing driver attachment, the driver handle, the support base, and a hydraulic press. Be careful not to damage the splash guard.

LOWER BALL JOINTS

REMOVAL & INSTALLATION

Removal

See Figures 211 through 213.

  Special Tools Required:
  • Ball Joint Thread Protector, 12 mm 07AAF-SDAA100

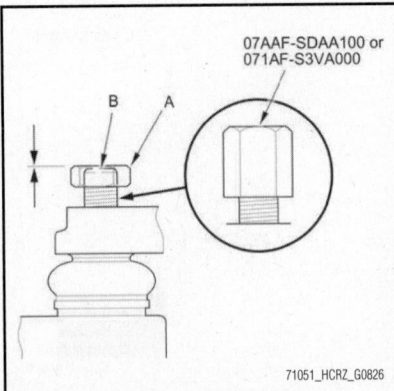

Fig. 211 Install a hex nut (A) or the ball joint thread protector onto the threads of the ball joint (B). When using a hex nut, make sure the nut is flush with the ball joint pin end to prevent damage to the threaded end of the ball joint pin.

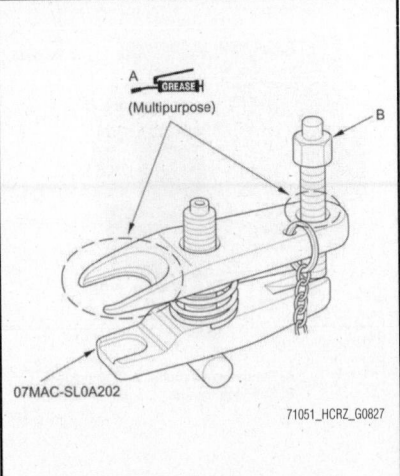

Fig. 212 Apply grease to the ball joint remover on the areas shown (A). This will ease installation of the tool and prevent damage to the pressure bolt (B) threads.

  • Ball Joint Thread Protector, 14 mm 071AF-S3VA000
  • Ball Joint Remover, 28 mm 07MAC-SL0A202

1. Install a hex nut or the ball joint thread protector onto the threads of the ball joint. When using a hex nut, make sure the nut is flush with the ball joint pin end to prevent damage to the threaded end of the ball joint pin.

2. Apply grease to the ball joint remover on the areas shown. This will ease installation of the tool and prevent damage to the pressure bolt threads.

3. Loosen the pressure bolt and install the ball joint remover as shown. Insert the jaws carefully, making sure not to damage the ball joint boot. Always use a ball joint remover to disconnect a ball joint. Do not strike the housing or any other part of the ball joint connection to disconnect it. Adjust the jaw spacing by turning the adjusting bolt.

4. Fasten the safety chain securely to a suspension arm or the subframe. Do not fasten it to a brake line or wire harness.

5. After adjusting the adjusting bolt, make sure the head of the adjusting bolt is in the position shown to allow the jaw to pivot.

6. With a wrench, tighten the pressure bolt until the ball joint pin pops loose from the ball joint connecting hole. If necessary, apply penetrating type lubricant to loosen the ball joint pin.

➡Do not use pneumatic or electric tools on the pressure bolt.

7. Remove the ball joint remover, then remove the nut from the end of the ball joint

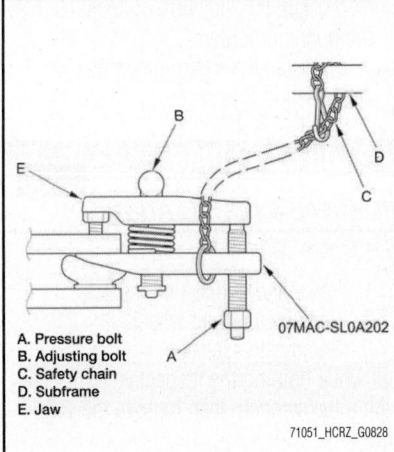

A. Pressure bolt
B. Adjusting bolt
C. Safety chain
D. Subframe
E. Jaw

07MAC-SL0A202

71051_HCRZ_G0828

Fig. 213 Loosen the pressure bolt and install the ball joint remover as shown. Insert the jaws carefully, making sure not to damage the ball joint boot. Always use a ball joint remover to disconnect a ball joint. Do not strike the housing or any other part of the ball joint connection to disconnect it. Adjust the jaw spacing by turning the adjusting bolt. Fasten the safety chain securely to a suspension arm or the subframe. Do not fasten it to a brake line or wire harness. After adjusting the adjusting bolt, make sure the head of the adjusting bolt is in the position shown to allow the jaw to pivot.

pin, and pull the ball joint out of the ball joint connecting hole. Inspect the ball joint boot, and replace it if damaged.

Inspection & Installation

See Figures 214 and 215.

Special Tools Required: Tie-Rod Boots Driver, 37.8 mm 07974-6790001.

1. Check the ball joint boot for weakness, damage, cracks, and grease leaks. Note the following:
   • If the ball joint boot is damaged and leaks grease, replace the lower arm.
   • If the ball joint boot is weak and cracked but does not leak grease, go to step 2. Replace the ball joint boot.
2. Remove the lower arm.
3. Remove the ball joint boot. Be careful not to damage the mating surface with ball joint boot of ball joint housing.
4. Pack the interior and lip of a new ball joint boot with grease. Keep the grease off of the ball joint boot-to-lower arm mating surfaces.

➡**Use the grease that comes with the ball joint kit.**

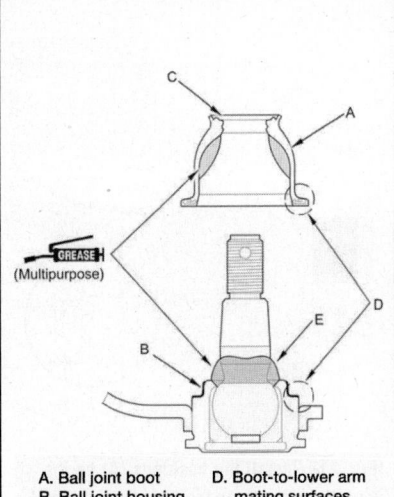

A. Ball joint boot
B. Ball joint housing
C. Interior and lip
D. Boot-to-lower arm mating surfaces
E. Base

71051_HCRZ_G0829

Fig. 214 Remove the ball joint boot. Be careful not to damage the mating surface with ball joint boot of ball joint housing. Pack the interior and lip of a new ball joint boot with grease. Keep the grease off of the ball joint boot-to-lower arm mating surfaces. Pack fresh grease into the base. Do not let dirt or other foreign materials get into the ball joint boot.

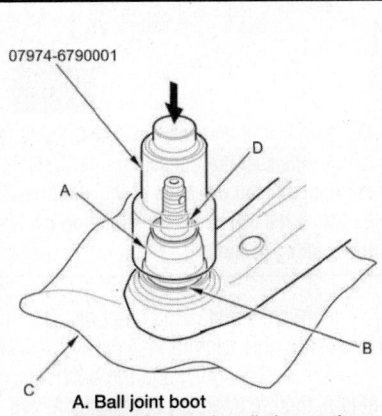

07974-6790001

A. Ball joint boot
B. Ball joint boot installation sections
C. Clean shop towel
D. Ball joint pin

71051_HCRZ_G0830

Fig. 215 Install the ball joint boot using the tie-rod boots driver by hand. The ball joint boot must not have a gap at the ball joint boot installation sections. Press the ball joint boot on a clean shop towel. After installing the ball joint boot on the ball joint, squeeze it gently to force out any air, then wipe the grease off the tapered portion of the ball joint pin.

5. Pack fresh grease into the base. Do not let dirt or other foreign materials get into the ball joint boot.
6. Install the ball joint boot using the tie-rod boots driver by hand. The ball joint boot must not have a gap at the ball joint boot installation sections.
7. Press the ball joint boot on a clean shop towel. After installing the ball joint boot on the ball joint, squeeze it gently to force out any air, then wipe the grease off the tapered portion of the ball joint pin.
8. After installing the ball joint boot, check the ball pin tapered section for grease contamination, and wipe it if necessary.
9. Install the lower arm.

LOWER CONTROL ARMS

REMOVAL & INSTALLATION

See Figure 216.

Special Tools Required:
• Ball Joint Remover, 28 mm 07MAC-SL0A202
• Ball Joint Thread Protector, 14 mm 071AF-S3VA000
1. Raise and support the vehicle.
2. Remove the front wheel.
3. Remove the splash shield.
4. Remove the lock pin from the lower arm ball joint, then remove the castle nut.
5. Disconnect the lower ball joint from the knuckle using the ball joint remover.
6. Remove the lower arm mounting bolts, then remove the lower arm from the front subframe.

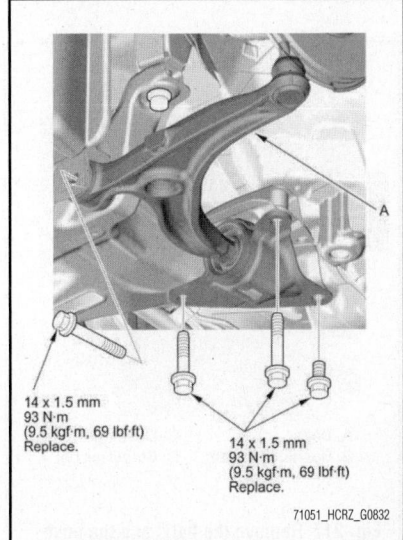

14 x 1.5 mm
93 N·m
(9.5 kgf·m, 69 lbf·ft)
Replace.

14 x 1.5 mm
93 N·m
(9.5 kgf·m, 69 lbf·ft)
Replace.

71051_HCRZ_G0832

Fig. 216 Remove the lower arm mounting bolts, then remove the lower arm (A) from the front subframe.

➡**Do not remove the lower arm from both sides at the same time. The lower arm mounting bolts also secure the subframe to the vehicle.**

To install:

7. Install the lower arm in the reverse order of removal, and note these items:

a. First install all the components, and lightly tighten the bolts and the nuts, then raise the suspension to load it with the vehicle's weight before fully tightening to the specified torque. Do not place the jack against the ball joint of the lower arm.

b. During installation, install the lock pin as shown after tightening a new castle nut.

c. Be careful not to damage the ball joint boot when connecting to the knuckle.

d. Before connecting the ball joint, degrease the threaded section and the tapered portion of the ball joint pin, the ball joint connecting hole, and the threaded section and the mating surfaces of the castle nut.

e. Torque the castle nut to the lower torque specification, then tighten it only far enough to align the slot with the ball joint pin hole. Do not align the castle nut by loosening it.

f. Before installing the wheel, clean the mating surfaces between the brake disc and the inside of the wheel.

8. Check the wheel alignment, and adjust it if necessary.

STABILIZER BAR

REMOVAL & INSTALLATION

See Figures 217 through 219.

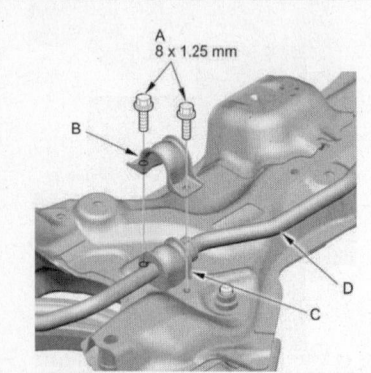

A. Bolts   C. Bushings
B. Bushing holders   D. Stabilizer bar

71051_HCRZ_G0833

Fig. 217 Remove the bolts and the bushing holders, then remove the bushings. Remove the stabilizer bar from the front subframe.

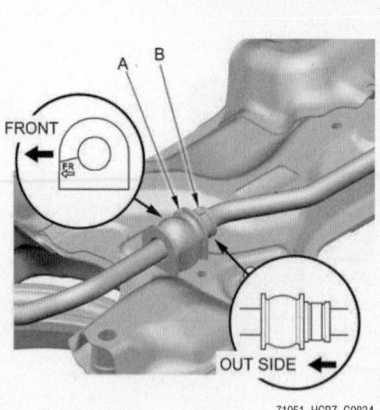

Fig. 218 Install the bushings (A) on the stabilizer bar. Note the direction of installation for the bushings. Align the stabilizer band clamp (B) with the side of the bushings.

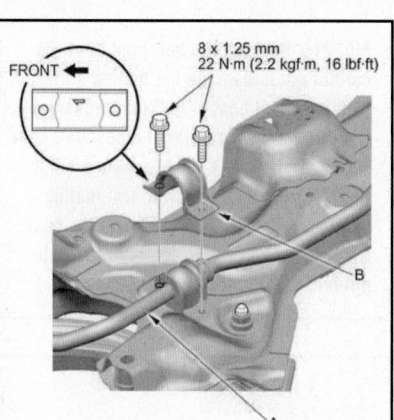

8 x 1.25 mm
22 N·m (2.2 kgf·m, 16 lbf·ft)

71051_HCRZ_G0835

Fig. 219 Install the stabilizer bar (A) with the bushings on the front subframe. Note the right and left direction of the stabilizer bar. Note the direction of installation for the bushing holders.

1. Remove the steering gearbox, as described in this section.
2. Remove the bolts and the bushing holders, then remove the bushings.
3. Remove the stabilizer bar from the front subframe.

To install:

4. Install the bushings on the stabilizer bar.

a. Note the direction of installation for the bushings.

b. Align the stabilizer band clamp with the side of the bushings.

5. Install the stabilizer bar with the bushings on the front subframe.

a. Note the right and left direction of the stabilizer bar.

b. Note the direction of installation for the bushing holders.

c. Install the steering gearbox, as described in this section.

STABILIZER LINKS

REMOVAL & INSTALLATION

See Figure 220.

1. Raise and support the vehicle.
2. Remove the front wheel.
3. Remove the self-locking nut and the nut while holding the respective joint pin with a hex wrench, then remove the stabilizer link.

To install:

4. Install the stabilizer link on the stabilizer bar and the damper with the joint pins set at the center of their range of movement.

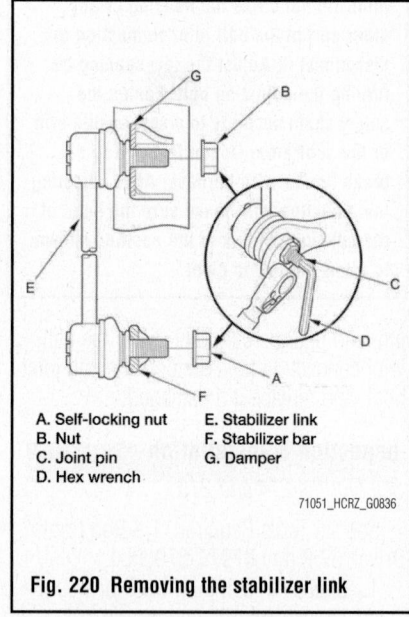

A. Self-locking nut   E. Stabilizer link
B. Nut   F. Stabilizer bar
C. Joint pin   G. Damper
D. Hex wrench

71051_HCRZ_G0836

Fig. 220 Removing the stabilizer link

5. Install a new self-locking nut and a new nut, and tighten them to the specified torque while holding the respective joint pin with a hex wrench.

6. Clean the mating surfaces between the brake disc and the inside of the wheel, then install the front wheel.

7. Test-drive the vehicle.

8. After 5 minutes of driving, tighten the self-locking nut again to the specified torque.

DAMPER (STRUT) & SPRING

REMOVAL & INSTALLATION

See Figures 221 through 224.

71051_HCRZ_G0839

Fig. 221 Remove the damper pinch bolts (A) and the self-locking nuts (B) from the damper.

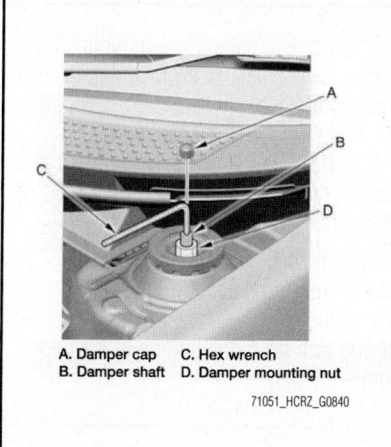

A. Damper cap   C. Hex wrench
B. Damper shaft   D. Damper mounting nut

71051_HCRZ_G0840

Fig. 222 Remove the damper cap from the top of the damper. Hold the damper shaft using a hex wrench and loosen the damper mounting nut.

1. Raise and support the vehicle.
2. Remove the front wheel.
3. Remove the wheel speed sensor from the knuckle. Do not disconnect the wheel speed sensor 2P connector.
4. Disconnect the stabilizer link from the damper. Refer to "Stabilizer Links" in this section, if needed.
5. Remove the wheel speed sensor clip,

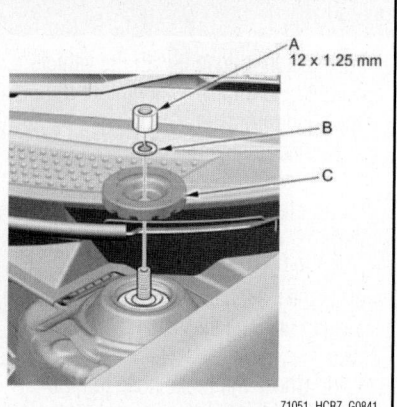

71051_HCRZ_G0841

Fig. 223 Remove the damper mounting nut (A) and the wave washer (B), then remove the damper mounting base (C) from the top of the damper.

the wire guide grommet, and the brake hose bracket from the damper. Do not disconnect the wheel speed sensor 2P connector.

6. Remove the damper pinch bolts and the self-locking nuts from the damper.

➡Do not allow the knuckle to rotate too far outward. This may allow the driveshaft inboard joint come apart.

7. Remove the damper cap from the top of the damper.
8. Hold the damper shaft using a hex wrench and loosen the damper mounting nut.
9. Remove the damper mounting nut and the wave washer, then remove the damper mounting base from the top of the damper.
10. Remove the damper/spring and the damper rubber mount.

➡Be careful not to damage the body.

To install:
11. Install all of the removed parts in the reverse order of removal.
12. During installation, install new damper pinch bolts and new self-locking nuts, then tighten the damper pinch bolts to 66 ft. lbs. (90 Nm).
13. Before installing the wheel, clean the mating surfaces between the brake disc and the inside of the wheel.
14. Check the wheel alignment, and adjust it if necessary.

71051_HCRZ_G0842

Fig. 224 Remove the damper/spring (A) and the damper rubber mount (B).

AXLE BEAM

REMOVAL & INSTALLATION

See Figure 225.

1. Raise and support the vehicle.
2. Remove the rear wheels.
3. Remove the rear hub bearing unit.
4. Remove the rear spring seat lower cover.
5. Remove the splash guard from the axle beam.
6. Remove the wheel speed sensor, the wheel speed sensor clip, and the wire guide grommet from the axle beam. Do not disconnect the wheel speed sensor 2P connector.
7. Remove the rear spring.
8. Place a floor jack under the lower spring seat on both sides of the axle beam and support it by raising the floor jack. Do not place the floor jack under the center of the axle beam.
9. Remove the axle beam mounting bolt on both sides.
10. Lower the jack slowly, then remove the axle beam.

To install:

11. Install the axle beam in the reverse order of removal, and note these items:

    a. First install all the components, and lightly tighten the bolts, and place a jack under the lower spring seat of the axle beam on both sides, then raise the suspension to load it with the vehicle's weight before fully tightening to the specified torque.

    b. After installing the brake hose and the parking brake cable, check for interference and twisting of other parts.

    c. Before installing the brake disc, clean the mating surfaces between the hub bearing unit and the inside of the brake disc.

    d. Check the brake hose joint for leaks, and tighten if necessary.

    e. Before installing the wheel, clean the mating surfaces between the brake disc and the inside of the wheel.

12. Check the wheel alignment, and adjust it if necessary.

COIL SPRINGS

REMOVAL & INSTALLATION

See Figures 226 through 231.

1. Raise and support the vehicle.
2. Remove the rear wheels.
3. Remove the wire guide grommet and the brake hose mounting bolt from both sides of the axle beam.
4. Position the floor jack under spring seat on both sides of the axle beam. Raise the floor jack until the suspension begins to compress.

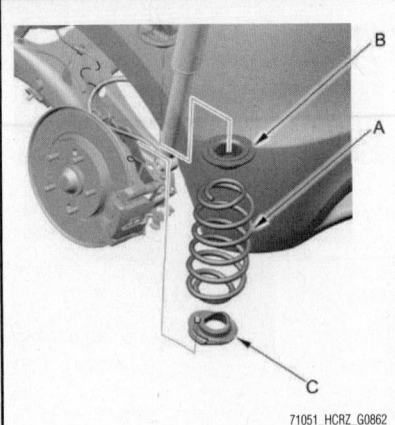

71051_HCRZ_G0862

Fig. 227 Remove the spring (A), the upper rubber mount (B), and the lower rubber mount (C). Do not lower the jack more than necessary.

5. Remove the damper mounting bolt that connects the axle beam and the damper from both sides.
6. Lower the floor jack gradually.
7. Remove the spring, the upper rubber mount, and the lower rubber mount. Do not lower the jack more than necessary.

To install:

8. Install the upper rubber mount on the spring by aligning the upper end of the spring with the ledge portion of the upper rubber mount.

71051_HCRZ_G0853

Fig. 225 Place a floor jack under the lower spring seat (A) on both sides of the axle beam (B), and support it by raising the floor jack. Do not place the floor jack under the center of the axle beam. Remove the axle beam mounting bolt (C) on both sides.

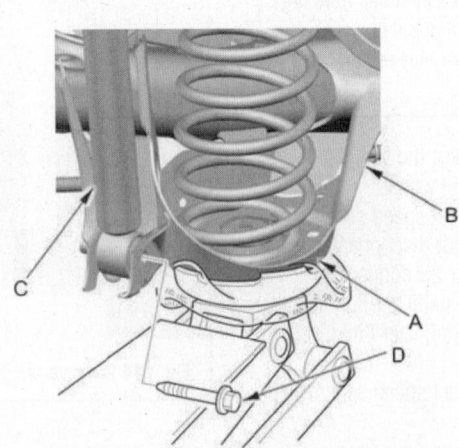

A. Spring seat   C. Damper mounting bolt
B. Axle beam   D. Damper

71051_HCRZ_G0861

Fig. 226 Position the floor jack under spring seat on both sides of the axle beam. Raise the floor jack until the suspension begins to compress. Remove the damper mounting bolt that connects the axle beam and the damper from both sides.

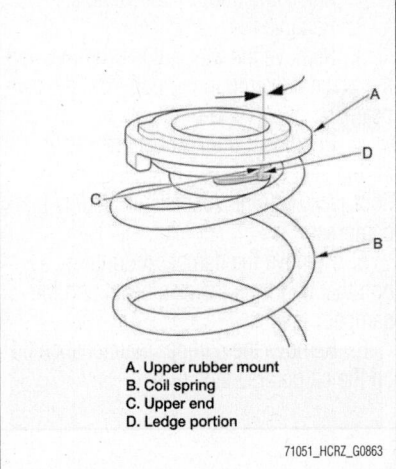

A. Upper rubber mount
B. Coil spring
C. Upper end
D. Ledge portion

71051_HCRZ_G0863

Fig. 228 Install the upper rubber mount on the spring by aligning the upper end of the spring with the ledge portion of the upper rubber mount.

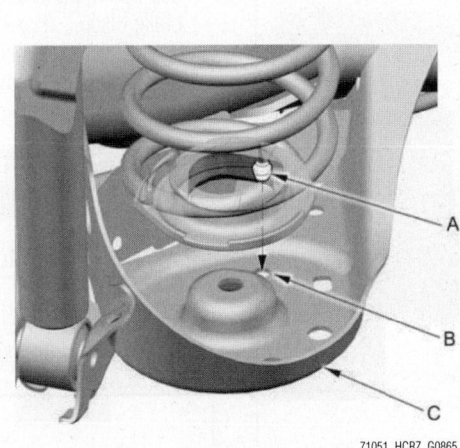

71051_HCRZ_G0865

Fig. 230 Install the tab (A) of the lower rubber mount into the groove (B) of the lower spring seat. Make sure that the tab of the lower rubber mount is properly installed into the axle beam (C).

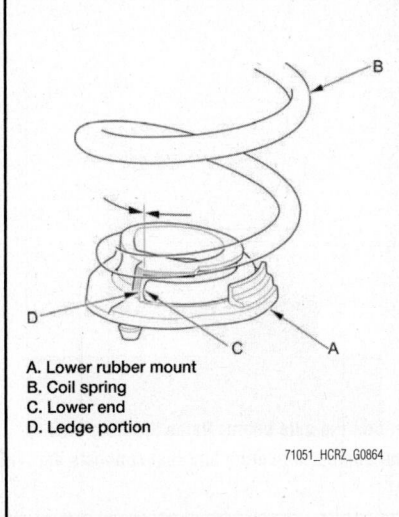

A. Lower rubber mount
B. Coil spring
C. Lower end
D. Ledge portion

71051_HCRZ_G0864

Fig. 229 Install the lower rubber mount on the spring by aligning the lower end of the spring with the ledge portion of the lower rubber mount.

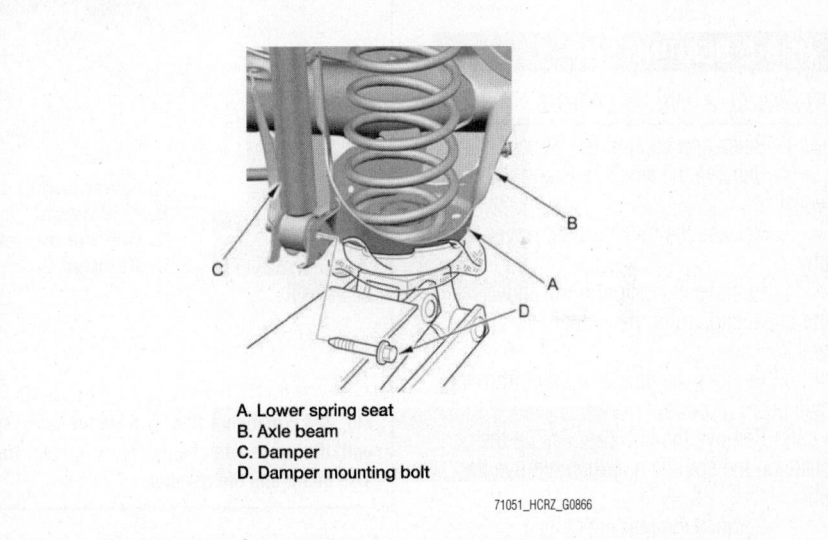

A. Lower spring seat
B. Axle beam
C. Damper
D. Damper mounting bolt

71051_HCRZ_G0866

Fig. 231 Position a floor jack under the lower spring seat on both sides of the axle beam. Slowly raise the jacks until you can align the bolt hole with the holes in the axle beam and the damper, then loosely tighten a new damper mounting bolt on both sides.

9. Install the lower rubber mount on the spring by aligning the lower end of the spring with the ledge portion of the lower rubber mount.

10. Install the tab of the lower rubber mount into the groove of the lower spring seat.

    a. Make sure that the tab of the lower rubber mount is properly installed into the axle beam.

    b. Make sure that the spring is installed correctly.

11. Position a floor jack under the lower spring seat on both sides of the axle beam.

12. Slowly raise the jacks until you can align the bolt hole with the holes in the axle beam and the damper, then loosely tighten a new damper mounting bolt on both sides.

13. Raise the rear suspension with the jacks until the vehicle just lifts off of the lift, then tighten the damper mounting bolts to the specified torque.

14. Install the wire guide grommet and the brake hose bracket mounting bolt on both sides of the axle beam.

15. Clean the mating surfaces between the brake disc and the inside of the wheel, then install the rear wheel.

DYNAMIC DAMPER

REMOVAL & INSTALLATION

See Figure 232.

1. Raise and support the vehicle.

2. Remove the dynamic damper from the axle beam.

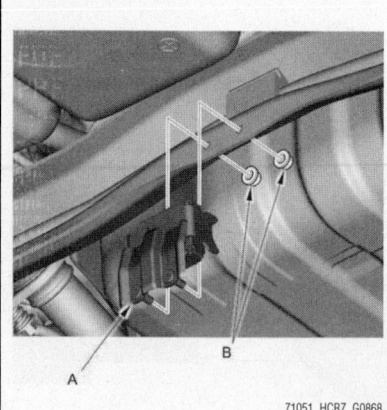

Fig. 232 Remove the dynamic damper (A) from the axle beam. Use new self-locking nuts (B) during reassembly.

3. Use new self-locking nuts during reassembly.

4. Install the new dynamic damper in the reverse order of removal.

HUB & BEARING UNIT

REMOVAL & INSTALLATION

1. Raise and support the vehicle.

2. Remove the wheel nuts and the rear wheel.

3. Release the parking brake lever fully.

4. Remove the caliper body and the brake pads from the caliper bracket.

5. Remove the caliper bracket from the axle beam (2 bolts).

6. Remove the hub cap. Pry up the stake on the spindle nut, then remove the nut.

7. Remove the rear brake disc.

8. Remove the hub bearing unit from the spindle.

9. Check the hub bearing unit for damage and cracks.

To install:

10. Install the hub bearing unit in the reverse order of removal, and note these items:

a. Tighten all mounting hardware to the specified torque.

b. Use a new spindle nut and hub cap on reassembly.

c. Before installing the spindle nut, apply a small amount of engine oil to the seating surface of the nut. After tightening, use a drift to stake the spindle nut shoulder against the spindle.

d. Before installing the brake disc,

clean the mating surfaces between the hub bearing unit and the inside of the brake disc.

e. Before installing the wheel, clean the mating surfaces between the brake disc and the inside of the wheel.

SHOCK ABSORBERS (DAMPERS)

REMOVAL & INSTALLATION

Removal

See Figures 233 through 236.

1. Raise and support the vehicle.

2. Remove the rear wheels.

3. Remove the wire guide grommet and the brake hose mounting bolt from the axle beam.

4. Position a floor jack under lower spring seat on the axle beam. Raise the floor jack until the suspension begins to compress.

5. Remove the damper mounting bolt that connects the axle beam and the damper.

6. Remove the damper maintenance lid on the cargo area side trim.

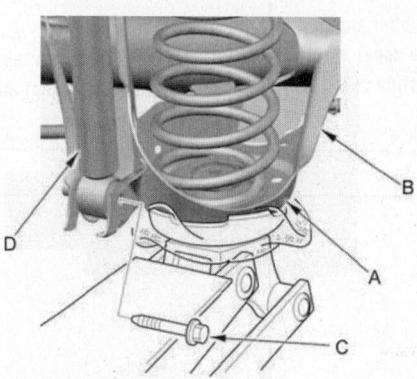

A. Lower spring seat
B. Axle beam
C. Damper mounting bolt
D. Damper

Fig. 233 Position a floor jack under lower spring seat on the axle beam. Raise the floor jack until the suspension begins to compress. Remove the damper mounting bolt that connects the axle beam and the damper.

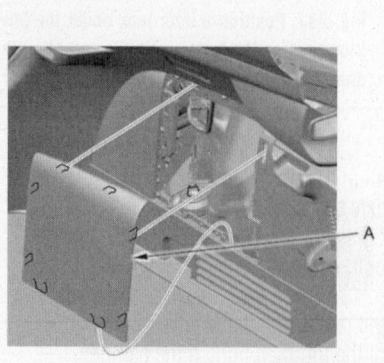

Fig. 234 Remove the damper maintenance lid (A) on the cargo area side trim—driver's side

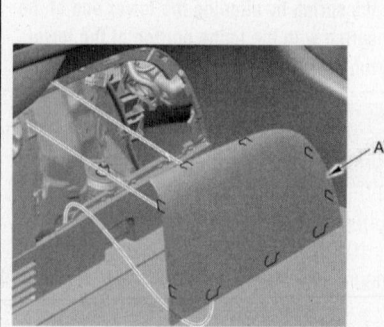

Fig. 235 Remove the damper maintenance lid (A) on the cargo area side trim—passenger's side

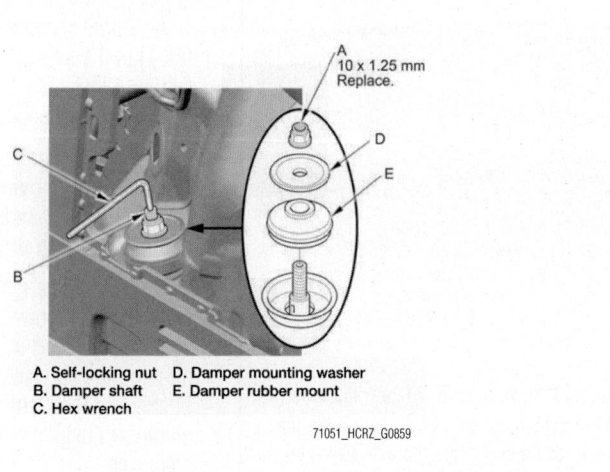

A. Self-locking nut   D. Damper mounting washer
B. Damper shaft   E. Damper rubber mount
C. Hex wrench

71051_HCRZ_G0859

Fig. 236 Remove the self-locking nut while holding the damper shaft with a hex wrench. Remove the damper mounting washer and the damper rubber mount from the top of the damper.

7. Remove the self-locking nut while holding the damper shaft with a hex wrench.

8. Remove the damper mounting washer and the damper rubber mount from the top of the damper.

9. Compress the damper assembly by hand, and remove it from the vehicle. Remove the damper rubber mount.

Inspection & Installation

See Figures 237 and 238.

1. Install the flange nut on the damper shaft end, and set the socket wrench and T-handle on the nut.

2. Compress the damper unit by hand, and check for smooth operation through a full stroke, both compression and extension.

   a. The damper should extend smoothly and constantly when compression is released.

   b. If it does not, the gas is leaking and the damper should be replaced.

3. Check for oil leaks, abnormal noises, and binding during these tests.

To install:

4. Install the damper rubber mount onto the damper unit. Position the damper assembly (B) between the body and the axle beam.

5. Position the floor jack under lower spring seat on the axle beam.

6. Slowly raise the jack until you can align the bolt hole with the holes in the axle beam and the damper, then loosely tighten a new damper mounting bolt.

7. Raise the rear suspension with the jack until the vehicle just lifts off of the lift, then tighten the damper mounting bolts to the specified torque.

8. Install the damper rubber mount, the damper mounting washer, and the self-locking nut on the top end of damper shaft.

➡**During installation, note the direction of the damper rubber mount and the damper mounting washer.**

9. Tighten the self-locking nut to 22 ft. lbs. (30 Nm) while holding the damper shaft with a hex wrench.

10. Install the damper maintenance lid on the cargo area side trim.

11. Install the wire guide grommet and the brake hose mounting bolt on the axle beam.

12. Clean the mating surfaces between the brake disc and the inside of the wheel, then install the rear wheel.

STABILIZER BUSHING

REMOVAL & INSTALLATION

See Figures 239 and 240.

1. Raise and support the vehicle.

2. Remove the dynamic damper from the axle beam, as described in this section.

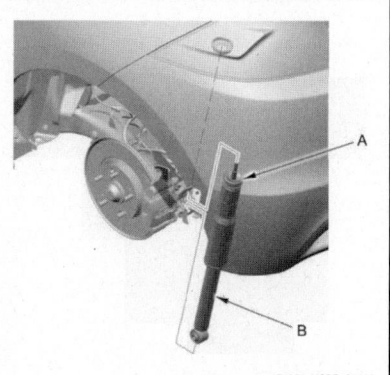

71051_HCRZ_G0860

Fig. 237 Install the damper rubber mount (A) onto the damper unit. Position the damper assembly (B) between the body and the axle beam.

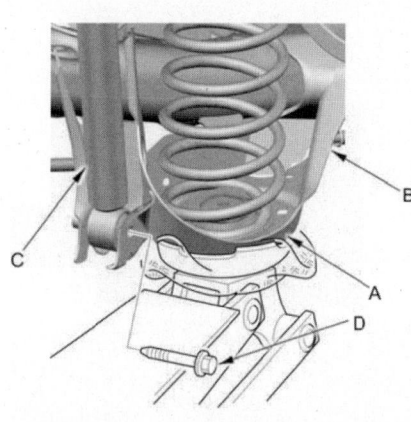

A. Spring seat   C. Damper mounting bolt
B. Axle beam   D. Damper

71051_HCRZ_G0861

Fig. 238 Position the floor jack under lower spring seat on the axle beam. Slowly raise the jack until you can align the bolt hole with the holes in the axle beam and the damper, then loosely tighten a new damper mounting bolt.

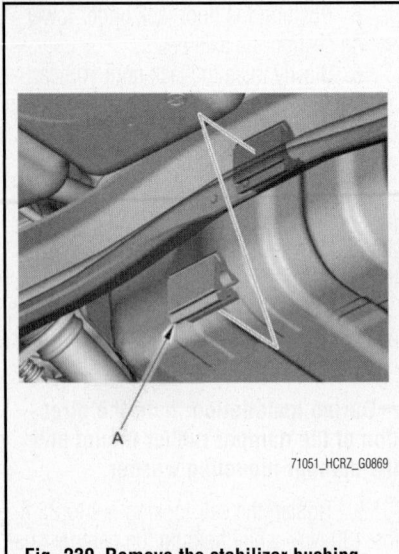

71051_HCRZ_G0869

Fig. 239 Remove the stabilizer bushing (A) from the axle beam.

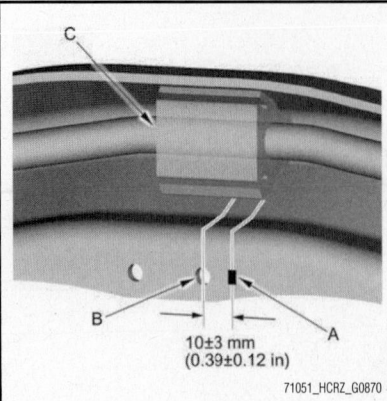

10±3 mm
(0.39±0.12 in)

71051_HCRZ_G0870

Fig. 240 Make a mark (A) near the hole (B) of the axle beam as shown. Install the stabilizer bushing (C) into the axle beam by aligning the end face of the stabilizer bushing to the plant mark. Make sure the stabilizer bushing is installed into the stabilizer

3. Remove the stabilizer bushing from the axle beam.

4. Apply silicone spray (P/N 08209-0001) to the new stabilizer bushing. This will ease installation of the bushing into the stabilizer and the axle beam.

➥**When spraying any agents that contain silicone, cover all the connectors, terminals, and switches in area with a protective cloth or plastic sheet.**

5. Make a mark near the hole of the axle beam as shown. Install the stabilizer bushing into the axle beam by aligning the end face of the stabilizer bushing to the plant mark. Make sure the stabilizer bushing is installed into the stabilizer.

6. Install the dynamic damper on the axle beam, as described in this section.

HONDA

Element

11

BRAKES **11-9**

BLEEDING THE BRAKE
SYSTEM......................**11-9**
Bleeding Procedure...................11-9
Bleeding Procedure 11-9
FRONT DISC BRAKES........**11-10**
Brake Caliper...............11-10
Removal & Installation........11-10
Disc Brake Pads 11-10
Removal & Installation........11-10
INFORMATION AND
PRECAUTIONS **11-9**
Anti-Lock Systems.................11-9
Disc and Drum Systems 11-9
PARKING BRAKE..............**11-10**
Parking Brake Cables 11-10
Adjustment 11-10
Parking Brake Shoes 11-11
Adjustment 11-11
REAR DISC BRAKES **11-10**
Brake Caliper....................11-10
Removal & Installation........11-10
Disc Brake Pads 11-10
Removal & Installation........11-10

CHASSIS ELECTRICAL **11-11**

AIR BAG (SUPPLEMENTAL
RESTRAINT SYSTEM) **11-11**
General Information.................11-11
Arming the System 11-12
Disarming the System.........11-11
Service Precautions............11-11

DRIVETRAIN **11-12**

Clutch Driven Disc and
Pressure Plate.....................11-12
Adjustments...................11-12
Removal & Installation........11-12
Clutch Master Cylinder 11-12
Removal & Installation........11-12
Clutch Slave Cylinder.............11-13
Removal & Installation........11-13
Front Halfshaft....................11-15
CV-joint Overhaul 11-15
Removal & Installation........11-15

Hydraulic Clutch System.........11-14
Bleeding 11-14
Rear Halfshaft......................11-15
CV-joint Overhaul 11-15
Removal & Installation........11-15
Rear Pinion Seal.....................11-17
Removal & Installation........11-17
Transfer Assembly 11-15
Removal & Installation........11-15

ENGINE COOLING **11-17**

Thermostat 11-17
Removal & Installation........11-17
Water Pump 11-17
Removal & Installation........11-17

ENGINE ELECTRICAL **11-18**

CHARGING SYSTEM **11-18**
Alternator 11-18
Removal & Installation........11-18
IGNITION SYSTEM **11-18**
Ignition Coil 11-18
Removal & Installation........11-18
Ignition Timing.......................11-18
Adjustment 11-19
Inspection 11-18
Spark Plugs............................11-19
Removal & Installation........11-19
STARTING SYSTEM **11-19**
Starter 11-19
Removal & Installation........11-19

ENGINE MECHANICAL......**11-20**

Accessory Drive Belts 11-20
Accessory Belt Routing.......11-20
Adjustment 11-20
Inspection 11-20
Removal & Installation........11-20
Camshaft and Valve Lifters......11-20
Removal & Installation........11-20
Crankshaft Damper..................11-20
Removal & Installation........11-20
Crankshaft Front Seal.............11-21
Removal & Installation........11-21
Cylinder Head 11-21
Removal & Installation........11-21

Driveplate 11-22
Removal & Installation........11-22
Exhaust Manifold 11-22
Removal & Installation........11-22
Flywheel 11-22
Removal & Installation........11-22
Intake Manifold 11-23
Removal & Installation........11-23
Oil Pan 11-24
Removal & Installation........11-24
Oil Pump 11-24
Removal & Installation........11-24
Piston and Ring......................11-24
Positioning 11-24
Rear Main Seal.......................11-24
Removal & Installation........11-24
Rocker Arms/Shafts................11-25
Removal & Installation........11-25
Timing Chain & Front Seal 11-26
Removal & Installation........11-26
Valve (Rocker Arm) Covers 11-27
Removal & Installation........11-27
Valve Lash..............................11-28
Adjustment 11-28

ENGINE PERFORMANCE &
EMISSION CONTROL **11-28**

Air Fuel (A/F) Ratio Sensor 11-28
Location........................11-28
Removal & Installation........11-28
Camshaft Position (CMP)
Sensor 11-29
Location........................11-29
Removal & Installation........11-29
Crankshaft Position (CKP)
Sensor 11-29
Location........................11-29
Removal & Installation........11-29
Electronic Control Module
(ECM) Powertrain
Control Module.....................11-29
Location........................11-29
Removal & Installation........11-29
Engine Coolant
Temperature (ECT) Sensor 11-31
Location........................11-31
Removal & Installation........11-31

Heated Oxygen (HO2S)
  Sensor 11-31
    Location 11-31
    Removal & Installation.......11-31
Intake Air Temperature (IAT)
  Sensor 11-31
    Location 11-31
    Removal & Installation.......11-31
Knock Sensor (KS).................11-32
    Location 11-32
    Removal & Installation.......11-32
Manifold Absolute
  Pressure (MAP) Sensor 11-32
    Removal & Installation.......11-32
Mass Air Flow (MAF) Sensor ...11-32
    Removal & Installation.......11-32
Output Shaft Speed (OSS)
  Sensor 11-32
    Location 11-32
    Removal & Installation.......11-32

FUEL 11-33

GASOLINE FUEL
  INJECTION SYSTEM 11-33
  Fuel Filter................................11-33
    Removal & Installation.......11-33
  Fuel Pump................................11-34
    Removal & Installation.......11-34
  Fuel Rail & Injectors 11-34
    Removal & Installation.......11-34
  Fuel System Service
    Precautions...........................11-33
  Idle Speed 11-35
    Inspection.............................11-35
  Relieving Fuel System
    Pressure................................11-33
  Throttle Body...........................11-34
    Removal & Installation.......11-34

HEATING & AIR CONDITIONING SYSTEM 11-35

Blower Motor 11-35
    Removal & Installation.......11-35
Heater Core 11-36
    Removal & Installation.......11-36

SPECIFICATIONS AND MAINTENANCE CHARTS.....11-3

Brake Specifications...................11-7
Camshaft Specifications............11-5
Capacities 11-4
Crankshaft Specifications..........11-5
Engine and Vehicle
  Identification 11-3
Engine Tune-Up
  Specifications 11-3
Fluid Specifications...................11-4
General Engine
  Specifications 11-3
Piston Specifications.................11-5
Scheduled Maintenance
  Intervals 11-8
Tire Wheel and Ball Joint
  Specifications 11-7
Torque Specifications................11-6
Valve Specifications 11-4
Wheel Alignment......................11-6

STEERING 11-38

Power Rack & Pinion
  Steering Gear 11-38
    Removal & Installation.......11-38
Power Steering Pump..............11-39
    Removal & Installation.......11-39

SUSPENSION 11-40

FRONT SUSPENSION 11-40
Coil Spring..............................11-40
    Removal & Installation.......11-40
Lower Ball Joint 11-40
    Removal & Installation.......11-40
Lower Control Arm...................11-40
  Control Arm Bushing
    Replacement 11-41
    Removal & Installation.......11-40
Stabilizer Bar 11-41
    Removal & Installation.......11-41
Stabilizer Link 11-41
    Removal & Installation.......11-41
Steering Knuckle 11-41
    Removal & Installation.......11-41
Strut/Damper...........................11-42
    Removal & Installation.......11-42
Wheel Bearings 11-42
    Adjustment 11-42
    Removal & Installation.......11-42
REAR SUSPENSION 11-43
Coil Spring..............................11-43
    Removal & Installation.......11-43
Strut/Damper...........................11-43
    Removal & Installation.......11-43
Upper Ball Joint 11-43
    Removal & Installation.......11-43
Upper Control Arm...................11-43
  Control Arm Bushing
    Replacement 11-44
    Removal & Installation.......11-43
Wheel Bearings 11-44
    Adjustment 11-44
    Removal & Installation.......11-44

SPECIFICATIONS AND MAINTENANCE CHARTS

ENGINE AND VEHICLE IDENTIFICATION CHART

		Engine Code					Model Year	
Code	Liters (cc)	Cu. In.	Cyl.	Fuel Sys.	Engine Type	Eng. Mfg.	Code ①	Year
K24A8	2.4 (2354)	144	4	SMFI	DOHC	Honda	B	2011

DOHC: Double Overhead Cam

SMFI: Sequential Multi-port Fuel Injection

① 10th position of VIN

71051_ELEM_C0001

GENERAL ENGINE SPECIFICATIONS

Year	Model	Engine Displacement Liters (ID)	Net Horsepower @ rpm	Net Torque @ rpm (ft. lbs.)	Bore x Stroke (in.)	Compression Ratio	Oil Pressure @ rpm
2011	Element	2.4 (K24A8)	166@5800	161@4500	3.42x3.90	9.7:1	44@3000

SMFI: Sequential Multi-port Fuel Injection

71051_ELEM_C0002

ENGINE TUNE-UP SPECIFICATIONS

Year	Engine Displacement Liters (ID)	Spark Plug Gap (in.)	Ignition Timing (deg.) MT	Ignition Timing (deg.) AT	Fuel Pump (psi)	Idle Speed (rpm) MT	Idle Speed (rpm) AT	Valve Clearance (in.) In.	Valve Clearance (in.) Ex.
2011	2.4 (K24A8)	0.039-0.043	6-10B	6-10B	48-55	650-750	650-750	0.008-0.010	0.011-0.013

NOTE: The Vehicle Emission Control Information label often reflects changes made during production and must be used if they differ from this chart.

NOTE: The fuel pressure readings are given with the vacuum hose connected to the regulator and the engine running

B: Before top dead center

HYD: Hydraulic

71051_ELEM_C0003

CAPACITIES

Year	Model	Engine Displacement Liters (ID)	Engine Oil with Filter (qts.)	Transaxle (pts.)		Transfer Case (pts.)	Drive Axle		Fuel Tank (gal.)	Cooling System (qts.)
				5-Spd	Auto.		Front (pts.)	Rear (pts.)		
2011	Element	2.4 (K24A8)	4.4	4.0	①	②	②	2.2	15.9	③

NOTE: All capacities are approximate. Add fluid gradually and check to be sure a proper fluid level is obtained.

① 2WD: 6.6 pts. for fluid change, 15.2 for overhaul

   4WD: 6.3 pts. for fluid change, 13.8 for overhaul

② Included in transaxle refill figure

③ Manual trans: 7.6 qts

   Auto trans: 7.5 qts.

71051_ELEM_C0004

FLUID SPECIFICATIONS

Year	Model	Engine Displacement Liters (ID)	Engine Oil	Man. Trans.	Auto. Trans.	Power Steering Fluid	Brake Master Cylinder	Cooling System
2011	Element	2.4 (K24A8)	5W-20 Honda	Honda MTF	Honda ATF-Z1	Honda PS Fluid	Honda DOT 3	①

DOT: Department Of Transpotation

① Honda Long Life Antifreeze/Coolant-Type2

71051_ELEM_C0014

VALVE SPECIFICATIONS

Year	Engine Displacement Liters (ID)	Seat Angle (deg.)	Face Angle (deg.)	Spring Test Pressure (lbs. @ in.)	Spring Installed Height (in.)	Stem-to-Guide Clearance (in.)		Stem Diameter (in.)	
						Intake	Exhaust	Intake	Exhaust
2011	2.4 (K24A8)	NA	NA	NA	①	0.0012-0.0022	0.0022-0.0031	0.2156-0.2159	0.2146-0.2150

NA: Not Available

① Valve spring free length:

   Intake: 1.668 in.

   Exhaust: 1.745 in.

71051_ELEM_C0005

CAMSHAFT SPECIFICATIONS
All measurements are given in inches.

Year	Engine Displacement Liters	Engine VIN	Journal Diameter	Brg. Oil Clearance	Shaft End-play	Runout	Journal Bore	Lobe Height	
								Intake	Exhaust
2011	2.4	K24A8	NA	①	0.0020-0.0080	0.0010	NA	②	1.3422

NA: Information not available

① No. 1 journal: 0.001-0.003 in.

   No. 2-5 journals: 0.002-0.004 in

② Primary: 1.3356 in.

   Secondary: 1.1668 in.

71051_ELEM_C0013

CRANKSHAFT SPECIFICATIONS
All measurements are given in inches

Year	Engine Displacement Liters (ID)	Crankshaft				Connecting Rod		
		Main Brg. Journal Dia.	Main Brg. Oil Clearance	Shaft End-play	Thrust on No.	Journal Diameter	Oil Clearance	Side Clearance
2011	2.4 (K24A8)	①	②	0.0040-0.0140	3	1.8888-1.8898	0.0008-0.0019	0.016

① Except No. 3: 2.1648-2.1657

   No. 3: 2.1644-2.1654

② Except No. 3: 0.0007-0.0016

   No. 3: 0.0010-0.0019

71051_ELEM_C0006

PISTON SPECIFICATIONS
All measurements are given in inches

Year	Engine Displacement Liters (ID)	Piston Clearance	Ring Gap			Ring Side Clearance		
			Top Compression	Bottom Compression	Oil Control	Top Compression	Bottom Compression	Oil Control
2011	2.4 (K24A8)	0.0008-0.0016	0.0080-0.0140	0.0160-0.0220	0.0080-0.0280	0.0018-0.0028	0.0020-0.0030	NA

NA: Not Applicable

71051_ELEM_C0007

TORQUE SPECIFICATIONS
All readings in ft. lbs.

Year	Engine Displacement Liters (ID)	Cylinder Head Bolts	Main Bearing Bolts	Rod Bearing Bolts	Crankshaft Damper Bolts	Flywheel Bolts	Manifold		Spark Plugs	Oil Pan Drain Plug
							Intake	Exhaust		
2011	2.4 (K24A8)	①	②	③	④	76	16	33	13	33

NOTE: Dip main bearing bolts and crankshaft damper bolt in clean engine oil prior to tightening.

① Step 1: 29 ft. lbs.

   Step 2: +90 degrees

   Step 3: +90 degrees

   Step 4: NEW BOLT ONLY +90 degrees

② 22 ft. lbs. +56 degrees

③ 14 ft. lbs. +90 degrees

④ 36 ft. lbs. +90 degrees

71051_ELEM_C0008

WHEEL ALIGNMENT

Year	Model		Caster		Camber		Toe-in (in.)
			Range (+/-Deg.)	Preferred Setting (Deg.)	Range (+/-Deg.)	Preferred Setting (Deg.)	
2011	Element	F	1.00	+1.75	0.75	0	0+/-0.08
		R	—	—	0.75	-1.00	0.08+/-0.08

71051_ELEM_C0009

TIRE, WHEEL AND BALLJOINT SPECIFICATIONS

| Year | Model | OEM Tires | | Tire Pressures (psi) | | Wheel Size | Ball Joint Inspection | Lug Nut (ft. lbs.) |
		Standard	Optional	Front	Rear			
2011	Element	P215/70R16	None	26	26	6JJ	NS	80

OEM: Original Equipment Manufacturer

PSI: Pounds Per Square Inch

NS: Not specified by manufacturer

71051_ELEM_C0010

BRAKE SPECIFICATIONS
All measurements in inches unless noted

| Year | Model | | Brake Disc | | | Minimum Lining Thickness | | Brake Caliper | |
			Original Thickness	Minimum Thickness	Maximum Runout	Front	Rear	Bracket Bolts (ft. lbs.)	Mounting Bolts (ft. lbs.)
2011	Element	Front	0.910	0.830	0.004	0.060	—	80	25
		Rear	0.350	0.300	0.004	—	0.060	41	16

71051_ELEM_C0011

MAINTENANCE MINDER SCHEDULE
Honda Element

The Element displays engine oil life and maintenance service items in the information display to indicate when to perform maintenance service. If the engine oil life is 15% or less, based on the onboard computer's caluculations, you will see SERVICE DUE SOON in the information display every time the ignition key is turned to ON. The maintenance minder indicator will also come on and the maintenance code(s) for other scheduled maintenance items needing service will be displayed below the message.

Symbol	Item	Service
A	Engine oil ①	Change
B	Engine oil and filter	Change
	Brakes	Inspect
	Parking brake adjustment	Check
	Steering gear, boots and linkage	Inspect
	Suspension components	Inspect
	Driveshaft boots	Inspect
	Brake hoses and linings	Inspect
	Exhaust system	Inspect
	All fluid levels and fluid condition	Inspect
	Exhaust system components	Inspect
	Fuel lines and connections	Inspect
1	Tires: condition and pressures	Rotate
2	Engine air filter ②	Replace
	Dust and pollen filter ③	Replace
	Accessory drive belt	Inspect
3	Transmission fluid	Replace
	Transfer case fluid	Replace
4	Spark plugs	Replace
	Timing belt ④	Replace
	Water pump	Inspect
	Valve clearance ⑤	Inspect
5	Engine coolant	Replace
6	Rear differential fluid ⑥	Replace

① If the message SERVICE DUE NOW does not appear more than 12 months after the display is reset, change every year.

② If driven in dusty conditions, replace every 15,000 miles.

③ If driven in urban areas that have a high concentration of soot from industry and diesel, replace every 15,000 miles

④ If driven regularly in temperatures over 110 deg.F or below -20 deg.F, or towing a trailer, replace every 60,000 miles.

⑤ Adjust if necessary.

⑥ Driving in mountainous areas at very low vehicle speeds or trailer towing results in higher level of mechanical (shear) stress to fluid. This requires differential fluid changes more frequently than recommended by the maintenance minder. If the vehicle i regularly driven under these conditions, changed have the differential fluid at 7,500 miles, then every 15,000 miles.

Additionally, replace the brake fluid every 3 years, and inspect the idle speed every 160,000 miles.

To reset the Engine Oil Life Display:

1. Turn the ignition switch to ON.

2. Press the SELECT button repeatedly until the engine oil life display or the service message is displayed.

3. Press the TRIP/RESET button for about 10 seconds. The engine oil life indicator and maintenance item code(s) will blink.

NOTE: If you are resetting the display when the engine oil life is more than 15 %, make sure any maintenance item(s) requiring service are done before resetting the display.

4. Press TRIP/RESET button for another 5 seconds. The maintenance item code(s) will disappear, and the engine oil life will reset to "100".

71051_ELEM_C0012

BRAKES   INFORMATION AND PRECAUTIONS

ANTI-LOCK SYSTEMS

- Certain components within the ABS system are not intended to be serviced or repaired individually.
- Do not use rubber hoses or other parts not specifically specified for and ABS system. When using repair kits, replace all parts included in the kit. Partial or incorrect repair may lead to functional problems and require the replacement of components.
- Lubricate rubber parts with clean, fresh brake fluid to ease assembly. Do not use shop air to clean parts; damage to rubber components may result.
- Use only DOT 3 brake fluid from an unopened container.
- If any hydraulic component or line is removed or replaced, it may be necessary to bleed the entire system.
- A clean repair area is essential. Always clean the reservoir and cap thoroughly before removing the cap. The slightest amount of dirt in the fluid may plug an ori-

fice and impair the system function. Perform repairs after components have been thoroughly cleaned; use only denatured alcohol to clean components. Do not allow ABS components to come into contact with any substance containing mineral oil; this includes used shop rags.

- The Anti-Lock control unit is a microprocessor similar to other computer units in the vehicle. Ensure that the ignition switch is **OFF** before removing or installing controller harnesses. Avoid static electricity discharge at or near the controller.
- If any arc welding is to be done on the vehicle, the control unit should be unplugged before welding operations begin.

DISC AND DRUM SYSTEMS

✳✳ CAUTION

Dust and dirt accumulating on brake parts during normal use may contain asbestos fibers from production or

aftermarket brake linings. Breathing excessive concentrations of asbestos fibers can cause serious bodily harm. Exercise care when servicing brake parts. Do not sand or grind brake lining unless equipment used is designed to contain the dust residue. Do not clean brake parts with compressed air or by dry brushing. Cleaning should be done by dampening the brake components with a fine mist of water, then wiping the brake components clean with a dampened cloth. Dispose of cloth and all residue containing asbestos fibers in an impermeable container with the appropriate label. Follow practices prescribed by the Occupational Safety and Health Administration (OSHA) and the Environmental Protection Agency (EPA) for the handling, processing, and disposing of dust or debris that may contain asbestos fibers.

BRAKES   BLEEDING THE BRAKE SYSTEM

BLEEDING PROCEDURE

BLEEDING PROCEDURE

See Figure 1.

When bleeding the brake system, observe the following:

- Do not reuse the drained fluid. Use only clean Honda DOT 3 Brake Fluid from an unopened container. Using a non-Honda brake fluid can cause corrosion and shorten the life of the system. Do not mix different brands of brake fluid; they may not be compatible.
- Make sure no dirt or other foreign matter is allowed to contaminate the brake fluid.
- Do not spill brake fluid on the vehicle, it may damage the paint; if brake fluid does

contact the paint, wash it off immediately with water.

1. The reservoir on the master cylinder must be at the MAX (upper) level mark at the start of the bleeding procedure and checked after bleeding each brake caliper. Add fluid as required.

2. Make sure the brake fluid level in the reservoir is at the MAX (upper) level line.

3. Slide a piece of clear plastic hose over the bleed screw, and submerge the other end in a container of new brake fluid.

4. Have someone slowly pump the brake pedal several times, then apply steady pressure.

5. Loosen the left-front brake bleed screw to allow air to escape from the system. Then tighten the bleed screw securely.

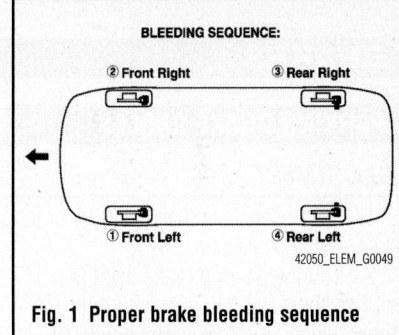

BLEEDING SEQUENCE:

② Front Right   ③ Rear Right

① Front Left   ④ Rear Left

42050_ELEM_G0049

Fig. 1 Proper brake bleeding sequence

6. Repeat the procedure for each caliper until no air bubbles are in the fluid. Bleed the calipers in the sequence shown.

7. Refill the master cylinder reservoir to the MAX (upper) level line.

BRAKES

BRAKE CALIPER

REMOVAL & INSTALLATION

1. Remove the wheel
2. Remove the brake hose banjo bolt and washers. Discard the washers.
3. Remove the upper and lower bolts.
4. Lift off the caliper.
5. Remove the pad springs.
6. Remove the pads and shims.
7. Remove the pad retainers.

BRAKES

BRAKE CALIPER

REMOVAL & INSTALLATION

1. Remove the wheel
2. Remove the brake hose banjo bolt and washers. Discard the washers.
3. Remove the caliper pin bolts.
4. Lift off the caliper and suspend it safely.
5. Remove the pads and shims.
6. Remove the pad retainers.

BRAKES

PARKING BRAKE CABLES

ADJUSTMENT

1. Before servicing the vehicle, refer to the precautions section.
2. Pull the parking brake lever (A) with 44 lbs. (196 N) of force to fully apply the parking brake. The parking brake lever should be locked within 4–7 clicks.
3. Adjust the parking brake if the lever clicks are not within the specification.

Minor Adjustment

1. Raise the rear of the vehicle, and support it with safety stands in the proper locations.
2. Release the parking brake lever fully.
3. Remove the center console, by pulling the front edge of the center console to release the clips, and remove the console by releasing the hooks.
4. Pull the parking brake lever 1 click.
5. Tighten the adjusting nut (A) until the parking brakes drag slightly when the rear wheels are turned.
6. Release the parking brake lever fully, and check that the parking brakes do not

8. Installation is the reverse of removal. Coat both sides of the shims and the backs of the pads with brake grease. Torque the bolts to 25 ft. lbs. (34 Nm). Install new washers and torque the banjo bolt to 25 ft. lbs. (34 Nm).

DISC BRAKE PADS

REMOVAL & INSTALLATION

1. Remove the lower bolt.

7. Installation is the reverse of removal. Coat both sides of the shims and the backs of the pads with brake grease. Torque the bolts to 16 ft. lbs. (22 Nm). Install new washers and torque the banjo bolt to 25 ft. lbs. (34 Nm).

DISC BRAKE PADS

REMOVAL & INSTALLATION

1. Remove the caliper pin bolts.

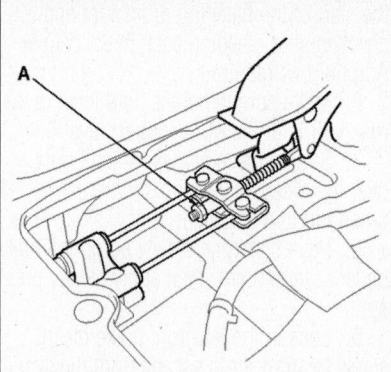

42050_ELEM_G0053

Fig. 2 Tighten the adjusting nut (A) until the parking brakes drag slightly when the rear wheels are turned

drag when the rear wheels are turned. Readjust if necessary.
7. Make sure the parking brakes are fully applied when the parking brake lever is pulled all the way.
8. Install the console in the reverse order of removal. During installation, check for damaged or stress-whitened clips, and replace them with new ones. Push the clip portions into place securely.

FRONT DISC BRAKES

2. Pivot the caliper up and hold the pads.
3. Remove the pad springs.
4. Remove the pads and shims.
5. Remove the pad retainers.
6. Installation is the reverse of removal. Coat both sides of the shims and the backs of the pads with brake grease. Torque the lower bolt to 25 ft. lbs. (34 Nm).

REAR DISC BRAKES

2. Lift off the caliper and suspend it safely.
3. Remove the pads and shims.
4. Remove the pad retainers.
5. Installation is the reverse of removal. Coat both sides of the shims and the backs of the pads with brake grease. Torque the bolts to 16 ft. lbs. (22 Nm).

PARKING BRAKE

Major Adjustment

See Figures 3 and 4.

➡**This adjustment should be done when replacing parking brake shoes and after lining surface break-in.**

1. Raise the rear of the vehicle, and support it with safety stands in the proper locations.
2. Release the parking brake lever fully.

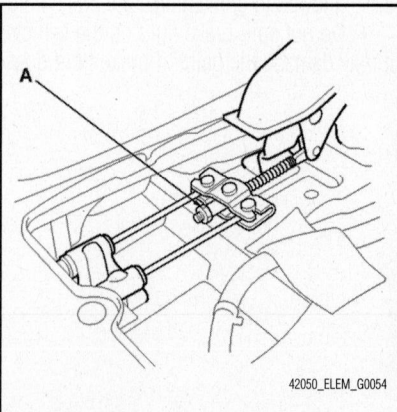

42050_ELEM_G0054

Fig. 3 Back off the adjusting nut (A) in the equalizer

3. Remove the center console, by pulling the front edge of the center console to release the clips, and remove the console by releasing the hooks.

4. Back off the adjusting nut (A) in the equalizer.

5. Remove the rear wheels.

6. Remove the access plug (A).

7. Turn the adjuster (B) with a flat-tip screwdriver (C) until the shoes lock against the drum. Then back off 8 clicks, and install the access plug.

8. Perform the minor adjustment procedure.

9. Install the rear wheels.

10. Install the console in the reverse order of removal. During installation, check for damaged or stress-whitened clips, and replace them with new ones. Push the clip portions into place securely.

PARKING BRAKE SHOES

ADJUSTMENT

1. Before servicing the vehicle, refer to the precautions section.

2. Pull the parking brake lever (A) with 44 lbs. (196 N) of force to fully apply the parking brake. The parking brake lever should be locked within 4–7 clicks.

3. Adjust the parking brake if the lever clicks are not within the specification.

Minor Adjustment

See Figure 3.

1. Raise the rear of the vehicle, and support it with safety stands in the proper locations.

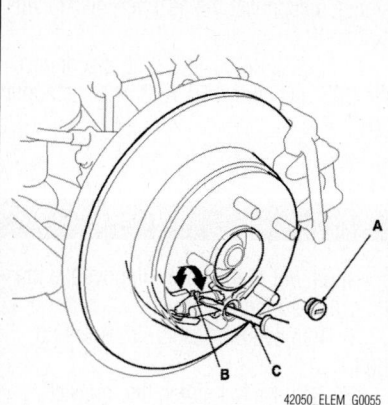

Fig. 4 Remove the access plug (A). Turn the adjuster (B) with a flat-tip screwdriver (C) until the shoes lock against the drum. Then back off 8 clicks, and install the access plug.

2. Release the parking brake lever fully.

3. Remove the center console, by pulling the front edge of the center console to release the clips, and remove the console by releasing the hooks.

4. Pull the parking brake lever 1 click.

5. Tighten the adjusting nut (A) until the parking brakes drag slightly when the rear wheels are turned.

6. Release the parking brake lever fully, and check that the parking brakes do not drag when the rear wheels are turned. Readjust if necessary.

7. Make sure the parking brakes are fully applied when the parking brake lever is pulled all the way.

8. Install the console in the reverse order of removal. During installation, check for damaged or stress-whitened clips, and replace them with new ones. Push the clip portions into place securely.

Major Adjustment

See Figures 3 and 4.

➡ **This adjustment should be done when replacing parking brake shoes and after lining surface break-in.**

1. Raise the rear of the vehicle, and support it with safety stands in the proper locations.

2. Release the parking brake lever fully.

3. Remove the center console, by pulling the front edge of the center console to release the clips, and remove the console by releasing the hooks.

4. Back off the adjusting nut (A) in the equalizer.

5. Remove the rear wheels.

6. Remove the access plug (A).

7. Turn the adjuster (B) with a flat-tip screwdriver (C) until the shoes lock against the drum. Then back off 8 clicks, and install the access plug.

8. Perform the minor adjustment procedure.

9. Install the rear wheels.

10. Install the console in the reverse order of removal. During installation, check for damaged or stress-whitened clips, and replace them with new ones. Push the clip portions into place securely.

CHASSIS ELECTRICAL

AIR BAG (SUPPLEMENTAL RESTRAINT SYSTEM)

GENERAL INFORMATION

✳✳ CAUTION

These vehicles are equipped with an air bag system. The system must be disarmed before performing service on, or around, system components, the steering column, instrument panel components, wiring and sensors. Failure to follow the safety precautions and the disarming procedure could result in accidental air bag deployment, possible injury and unnecessary system repairs.

SERVICE PRECAUTIONS

✳✳ CAUTION

Disconnect and isolate the battery negative cable before beginning any airbag system component diagnosis, testing, removal, or installation procedures. Allow system capacitor to discharge for two minutes before beginning any component service. This will disable the airbag system. Failure to disable the airbag system may result in accidental airbag deployment, personal injury, or death.

DISARMING THE SYSTEM

1. Disconnect and isolate the negative battery cable. Wait 3 minutes for the system capacitor to discharge before performing any service.

2. To disarm the driver's airbag, remove the access panel from the steering wheel, then disconnect the driver's airbag 4P connector from the cable reel.

3. To disarm the front passenger's airbag, remove the glove box, then disconnect the passenger's airbag 4P connector from dashboard wire harness B.

4. To disarm the side airbag, disconnect the side airbag 2P connector from the floor wire harness.

5. To disarm the seat belt tensioner, disconnect the seat belt tensioner 2P-connector from the rear door wire harness.

6. To disarm the seat belt buckle ten-sioner, disconnect the seat belt buckle tensioner 4P connector.

7. To disarm the SRS unit, disconnect the SRS unit connector A, B or C, as applicable.

ARMING THE SYSTEM

1. To rearm, connect the electrical connector(s) as necessary, then connect the negative battery cable.

DRIVETRAIN

CLUTCH DRIVEN DISC AND PRESSURE PLATE

ADJUSTMENTS

The Element is equipped with a hydraulic clutch system. No adjustment is necessary.

REMOVAL & INSTALLATION

See Figure 5.

1. Before servicing the vehicle, refer to the precautions section.

2. Remove or disconnect the following:
- Negative battery cable
- Transaxle
- Pressure plate. Loosen the bolts evenly in a crossing pattern.
- Clutch disc

To install:

3. Install the clutch disc and pressure plate. Tighten the pressure plate bolts in a crisscross pattern, in several steps to 19 ft. lbs. (26 Nm).

4. Install or connect the following:
- Transaxle
- Negative battery cable

CLUTCH MASTER CYLINDER

REMOVAL & INSTALLATION

See Figures 6 through 10.

✳✳ WARNING

Always use fender covers to avoid damaging painted surfaces. Do not spill brake fluid on the vehicle; it may damage the paint; if brake fluid does contact the paint, wash it off immediately with water.

1. Remove the brake fluid from the clutch master cylinder reservoir with a syringe.

2. Make sure you have the anti-theft code for the radio, then write down the audio presets. Disconnect the negative cable from the battery first, then disconnect the positive cable. Remove the battery.

3. Remove the air cleaner housing.

4. Remove the battery base.

5. Pry out the lock pin, and pull the pedal pin out of the yoke. Remove the master cylinder mounting nuts.

6. Remove the reservoir mounting bolt.

7. Remove the clutch line bracket.

8. Remove the clutch master cylinder from the vehicle.

9. Disconnect the reservoir hose, then remove the retaining clip and clutch line from the clutch master cylinder. Plug the end of the reservoir hose and clutch line with a shop towel to prevent brake fluid from coming out.

10. Remove the O-ring and clutch master cylinder seal from the clutch master cylinder.

To install:

11. Install the clutch master cylinder in the reverse order of removal, and note these items:

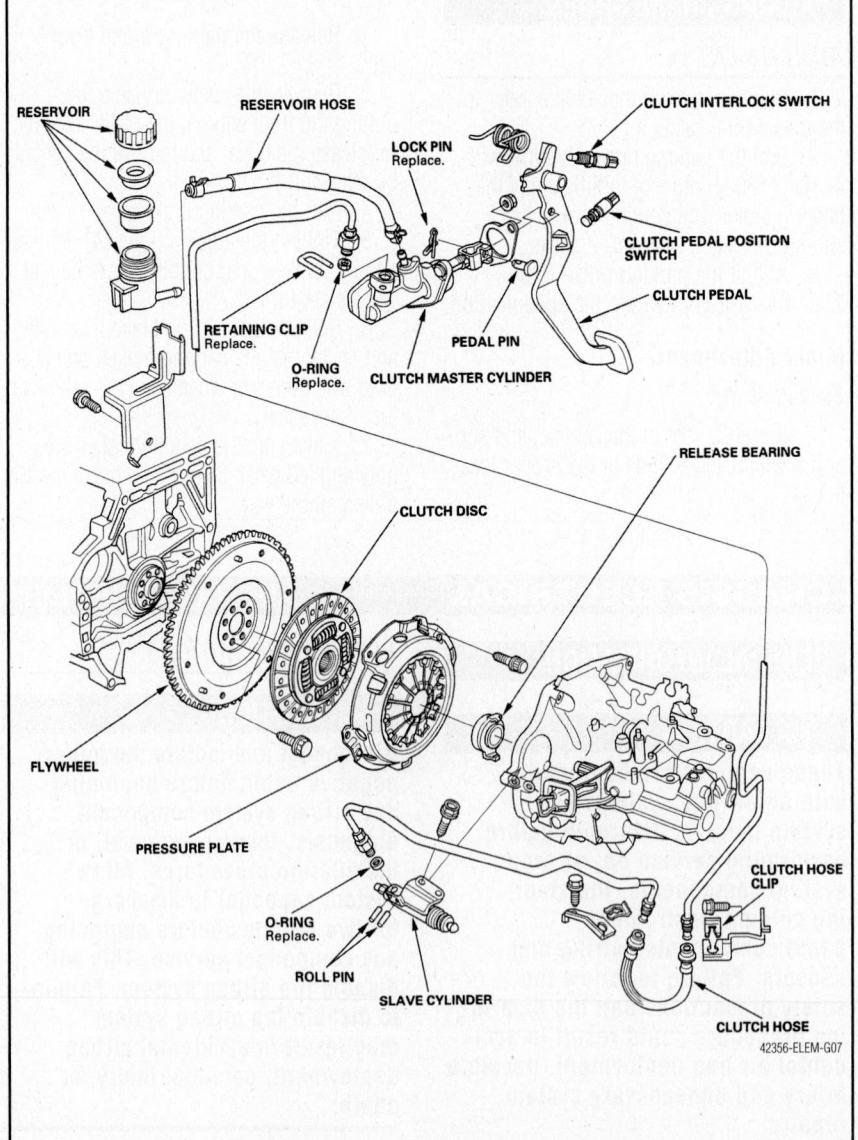

Fig. 5 Exploded view of the clutch system components

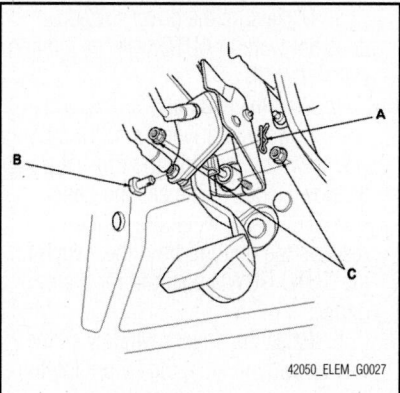

Fig. 6 Pry out the lock pin (A), and pull the pedal pin (B) out of the yoke. Remove the master cylinder mounting nuts (C)

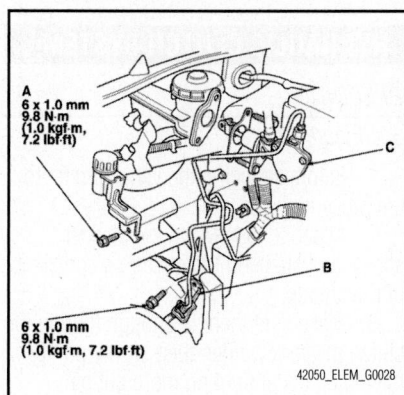

Fig. 7 Remove the reservoir mounting bolt (A), the clutch line bracket (B), then remove the clutch master cylinder (C)

a. Apply brake assembly lube to the clutch line, and install a new O-ring.

b. Tighten the master cylinder mounting nuts to 9 ft. lbs. (13 Nm).

12. Install the battery base.

13. Install the air cleaner housing.

14. Install the battery. Clean the battery posts and cable terminals with sandpaper. Connect the positive cable to the battery first, then connect the negative cable, and apply grease to prevent corrosion.

15. Enter the anti-theft code for the radio, then enter the audio presets, and set the clock.

16. Perform the power window control unit reset procedure, as follows.

➡ **Resetting the power window control unit is required after performing the following procedures:**

Loss of battery power, Loss of power from the No. 23 (20 A) fuse in the under-dash fuse/relay box,

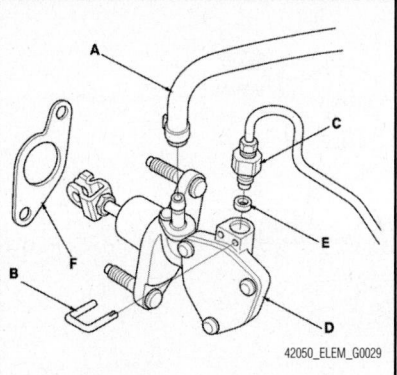

Fig. 8 View of the reservoir hose (A), retaining clips (B), clutch line (C), clutch master cylinder (D), O-ring (E) and clutch master cylinder seal (F)

Open circuit caused by disconnecting the 14P connector from the power window master switch

a. Make sure the driver's window does not work in AUTO with the ignition switch ON (II).

b. Start the engine.

c. Lower the driver's window all the way down by pushing the driver's power window switch to the second detent (AUTO DOWN); when the window reaches the bottom, hold the switch in the AUTO DOWN position for 2 seconds.

d. Raise the driver's window all the way up without stopping by pulling the driver's power window switch to the UP position; when the window reaches the top, hold the switch in the UP position for 2 seconds.

e. If the window does not work in AUTO, repeat steps b through e.

17. Make sure the hose clamps are positioned on the master cylinder and reservoir as shown.

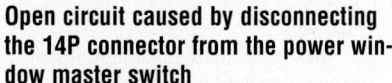

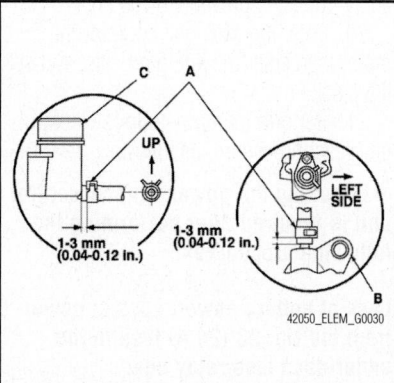

Fig. 9 The hose clamps (A) must be properly positioned on the master cylinder (B) and reservoir (C)

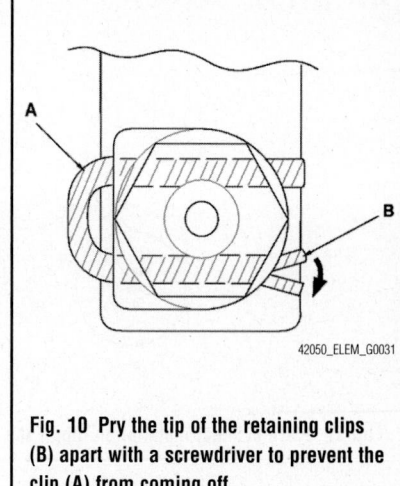

Fig. 10 Pry the tip of the retaining clips (B) apart with a screwdriver to prevent the clip (A) from coming off

18. To prevent the retaining clip from coming off, pry apart the tip of the retaining clip with a screwdriver.

➡ **Reservoir filling is covered in the bleeding procedure.**

19. Bleed the clutch hydraulic system.

CLUTCH SLAVE CYLINDER

REMOVAL & INSTALLATION
See Figures 11 through 13.

✳✳ WARNING

Always use fender covers to avoid damaging painted surfaces. Do not spill brake fluid on the vehicle; it may damage the paint; if brake fluid does contact the paint, wash it off immediately with water.

1. Remove the brake fluid from the clutch master cylinder reservoir with a syringe.

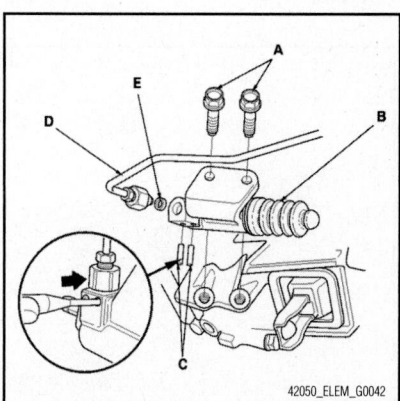

Fig. 11 View of the slave cylinder mounting bolts (A), slave cylinder (B), roll pins (C), clutch line (D) and O-ring (E)

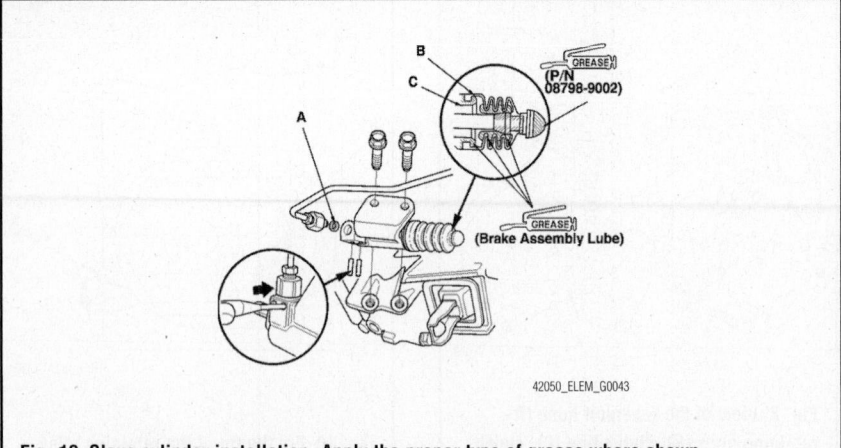

Fig. 12 Slave cylinder installation. Apply the proper type of grease where shown

2. Make sure you have the anti-theft code for the radio, then write down the audio presets. Disconnect the negative cable from the battery first, then disconnect the positive cable. Remove the battery.

3. Remove the air cleaner housing.

4. Remove the battery base.

5. Remove the clutch line bracket.

6. Remove the mounting bolts and the slave cylinder.

7. Remove the roll pins. Disconnect the clutch line, and remove the O-ring. Plug the end of the clutch line with a shop towel to prevent brake fluid from coming out.

To install:

8. Install the slave cylinder in the reverse order of removal. Install a new O-ring (A).

9. Pull back the boot (B), and apply brake assembly lube to the boot and slave cylinder rod (C). Reinstall the boot.

10. Apply super high temp urea grease (P/N 08798-9002) to the slave cylinder push rod. Tighten the slave cylinder mounting bolts to 16 ft. lbs. (22 Nm).

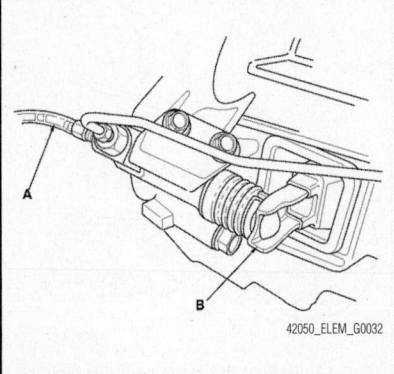

Fig. 13 View of the hose (A) attached to the bleeder screw and the release fork (B)

11. Attach a hose (A) to the bleeder screw, and suspend the hose in a container of brake fluid.

12. Make sure there is enough fluid in the clutch master cylinder, then slowly pump the clutch pedal until no more bubbles appear at the bleeder hose.

13. It may be necessary to limit the movement of the release fork (B) with a block of wood to remove all the air from the system.

14. Tighten the bleeder screw to 70 inch lbs. (8 Nm); do not overtighten the screw.

15. Refill the clutch master cylinder with fluid when done. Use only Honda DOT 3 Brake Fluid from an unopened container.

16. Make sure the fluid level in the reservoir is at the MAX (upper) level line.

17. Install the air cleaner housing.

18. Install the battery base.

19. Install the battery. Clean the battery posts and cable terminals with sandpaper. Connect the positive cable to the battery first, then connect the negative cable, and apply grease to prevent corrosion.

20. Enter the anti-theft code for the radio, then enter the audio presets, and set the clock.

21. Perform the power window control unit reset procedure, as follows.

➡**Resetting the power window control unit is required after performing the following procedures:**

Loss of battery power, Loss of power from the No. 23 (20 A) fuse in the under-dash fuse/relay box,

Open circuit caused by disconnecting the 14P connector from the power window master switch

a. Make sure the driver's window does not work in AUTO with the ignition switch ON (II).

b. Start the engine.

c. Lower the driver's window all the way down by pushing the driver's power window switch to the second detent (AUTO DOWN); when the window reaches the bottom, hold the switch in the AUTO DOWN position for 2 seconds.

d. Raise the driver's window all the way up without stopping by pulling the driver's power window switch to the UP position; when the window reaches the top, hold the switch in the UP position for 2 seconds.

e. If the window does not work in AUTO, repeat steps b through e.

HYDRAULIC CLUTCH SYSTEM

BLEEDING

See Figure 13.

1. Before servicing the vehicle, refer to the precautions section.

2. Attach a hose (A) to the bleeder screw, and suspend the hose in a container of brake fluid.

3. Make sure there is enough fluid in the clutch master cylinder, then slowly pump the clutch pedal until no more bubbles appear at the bleeder hose.

4. It may be necessary to limit the movement of the release fork (B) with a block of wood to remove all the air from the system.

5. Tighten the bleeder screw to 70 inch lbs. (8 Nm); do not overtighten the screw.

6. Refill the clutch master cylinder with fluid when done. Use only Honda DOT 3 Brake Fluid from an unopened container.

7. Make sure the fluid level in the reservoir is at the MAX (upper) level line.

TRANSFER ASSEMBLY

REMOVAL & INSTALLATION

See Figure 14.

1. Before servicing the vehicle, refer to the precautions section.

2. Drain the transaxle fluid. Install the drain plug with a new gasket and tighten to 36 ft. lbs. (49 Nm).

3. Disconnect the negative battery cable.

4. Matchmark the installed position of the propeller shaft and transfer companion flange.

5. Remove or disconnect the following:
   • Propeller shaft from the transfer assembly

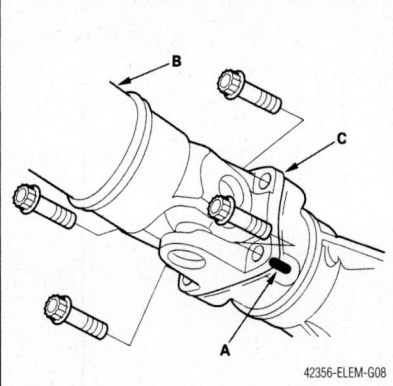

Fig. 14 Matchmark (A) the installed position of the propeller shaft (B) and transfer companion flange (C)

- Mounting bolts and transfer assembly

To install:

6. Clean the transfer assembly mating surfaces, then apply clean transmission fluid to the mating surfaces.

7. Install or connect the following:
- New O-ring seal on the transfer assembly
- 4 bolts in the transfer housing, then the transfer assembly with the dowel pin. Tighten the 10mm bolts to 33 ft. lbs. (44 Nm).
- Propeller shaft to the transfer companion flange, aligning the mark made during removal. Tighten the 8mm bolts to 24 ft. lbs. (33 Nm).
- Negative battery cable

8. Fill the transaxle to the correct level and check for leaks.

FRONT HALFSHAFT

REMOVAL & INSTALLATION

1. Before servicing the vehicle, refer to the precautions section.
2. Drain the transaxle.
3. Remove or disconnect the following:
- Negative battery cable
- Front wheels
- Spindle nut
- Stabilizer bar
- Lower ball joint from the control arm

4. On the left side, pry the inboard joint from the case with a prybar.
5. On the right side, drive the inboard shaft off the intermediate shaft with a drift and hammer.
6. Installation is the reverse of removal. Observe the following torques:
- Ball stud nuts: 40 ft. lbs. (54 Nm)

- Stabilizer link nuts: 29 ft. lbs. (39 Nm)
- Spindle nut: 181 ft. lbs. (245 Nm)

CV-JOINT OVERHAUL

Outboard Joint

1. Before servicing the vehicle, refer to the precautions section.
2. Remove or disconnect the following:
- Axle halfshaft from the vehicle and place it in a vise
- Outboard joint boot clamps and push the boot back
- Outboard joint by driving it off the axle shaft with a brass drift and hammer
- Outboard joint boot

To install:

→**Use new circlips and boot clamps for assembly.**

3. Install the outboard joint boot and clamps to the axle shaft.
4. Fill the outboard joint with grease. Install the outboard joint to the axle shaft. Tap the stub shaft with a brass hammer to seat the circlip.
5. Fill the outboard joint boot with grease and install the boot clamps.
6. Install the axle halfshaft to the vehicle.

Inboard Joint

1. Before servicing the vehicle, refer to the precautions section.
2. Remove or disconnect the following:
- Axle halfshaft from the vehicle.
- Inboard joint boot clamps and push the boot back
- Inboard joint housing from the axle
- Rollers from the spider
- Snapring and the spider from the axle shaft
- Inboard joint boot

To install:

→**Use new circlips and boot clamps for assembly.**

3. Install or connect the following:
- Inboard joint boot and clamps to the axle shaft
- Spider with a new snapring
- Rollers to the spider

4. Fill the joint housing with grease and install it.
5. Fill the inboard joint boot with grease and install the boot clamps.
6. Install the axle halfshaft to the vehicle.

REAR HALFSHAFT

REMOVAL & INSTALLATION

1. Before servicing the vehicle, refer to the precautions section.
2. Drain the differential.
3. Remove or disconnect the following:
- Negative battery cable
- Rear wheels
- Spindle nut

4. Pry the inboard joint from the differential.
5. Remove the outer CV-joint stub shaft from the hub by tapping the stub shaft with a plastic hammer.

To install:

→**Use new circlips and self-locking nuts for assembly.**

6. Install the outer CV-joint stub shaft into the hub.
7. Install the inner CV-joint to the differential until the circlip locks in the retaining groove.
8. Install or connect the following:
- Spindle nut. Tighten the nut to 134 ft. lbs. (181 Nm).
- Rear wheels
- Negative battery cable

9. Fill the differential to the correct level and check for leaks.

CV-JOINT OVERHAUL

See Figure 15.

1. Before servicing the vehicle, refer to the precautions section.
2. Remove or disconnect the following:
- Axle halfshaft from the vehicle
- Joint boot clamps and push the boot back
- Joint housing from the axle
- Rollers from the spider
- Snapring and the spider from the axle shaft
- Joint boot

To install:

→**Use new circlips and boot clamps for assembly.**

3. Install or connect the following:
- Joint boot and clamps to the axle shaft
- Spider with a new snapring
- Rollers to the spider

4. Fill the joint housing with grease and install it.
5. Fill the joint boot with grease and install the boot clamps.
6. Install the axle halfshaft to the vehicle.

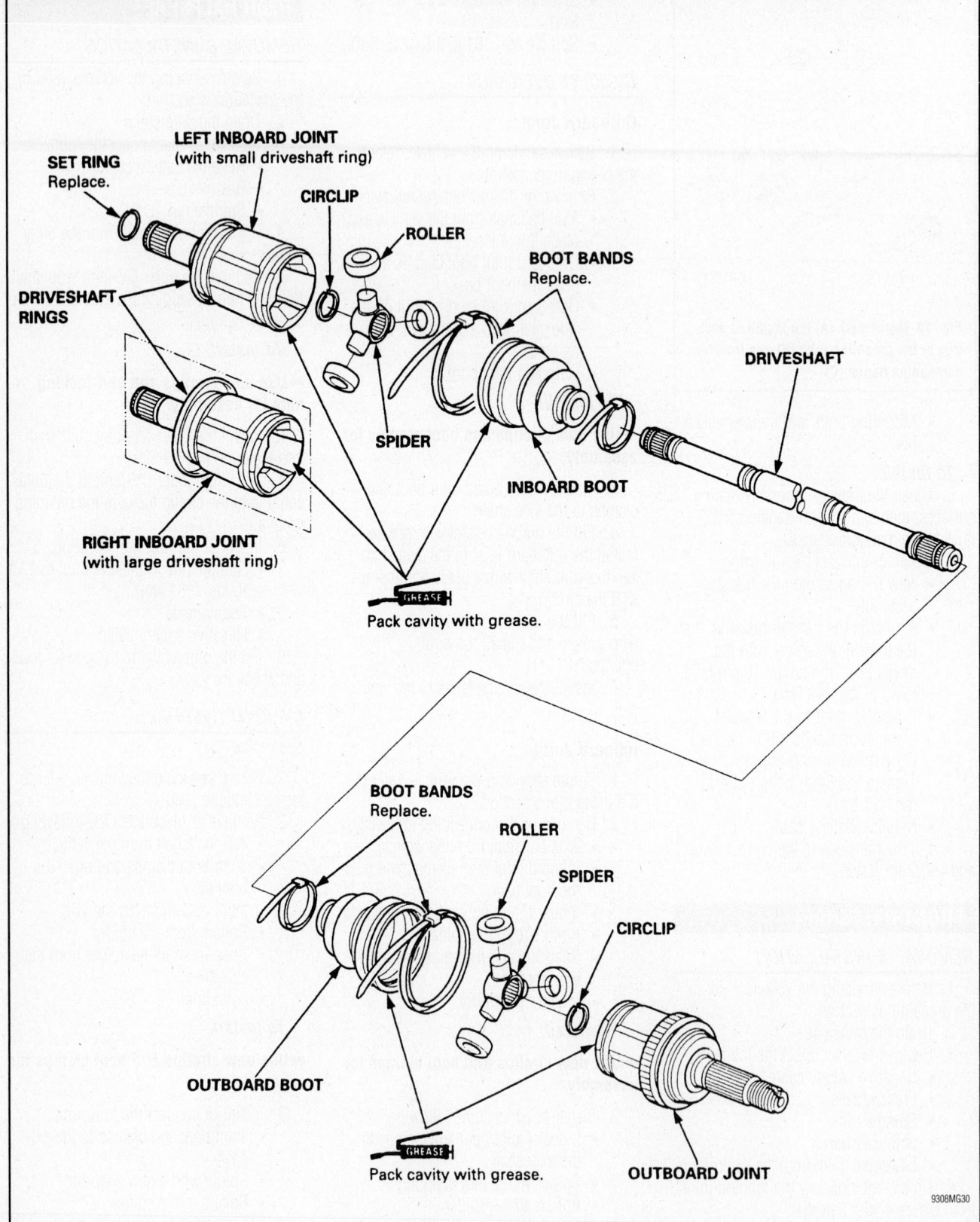

SET RING
Replace.

LEFT INBOARD JOINT
(with small driveshaft ring)

CIRCLIP

ROLLER

BOOT BANDS
Replace.

DRIVESHAFT

DRIVESHAFT RINGS

SPIDER

INBOARD BOOT

RIGHT INBOARD JOINT
(with large driveshaft ring)

GREASE
Pack cavity with grease.

BOOT BANDS
Replace.

ROLLER

SPIDER

CIRCLIP

OUTBOARD BOOT

GREASE
Pack cavity with grease.

OUTBOARD JOINT

9308MG30

Fig. 15 Exploded view of the rear axle

REAR PINION SEAL

REMOVAL & INSTALLATION

See Figure 16.

1. Before servicing the vehicle, refer to the precautions section.
2. Remove or disconnect the following:
   - Driveshaft
   - Companion flange
   - Pinion seal

To install:

➡**Use a new locknut and O-ring for assembly.**

3. Install or connect the following:
   - Pinion seal. Drive the seal square into the bore.
   - Companion flange. Tighten the locknut to 108 ft. lbs. (147 Nm).
   - Driveshaft. Tighten the flange bolts to 24 ft. lbs. (32 Nm).

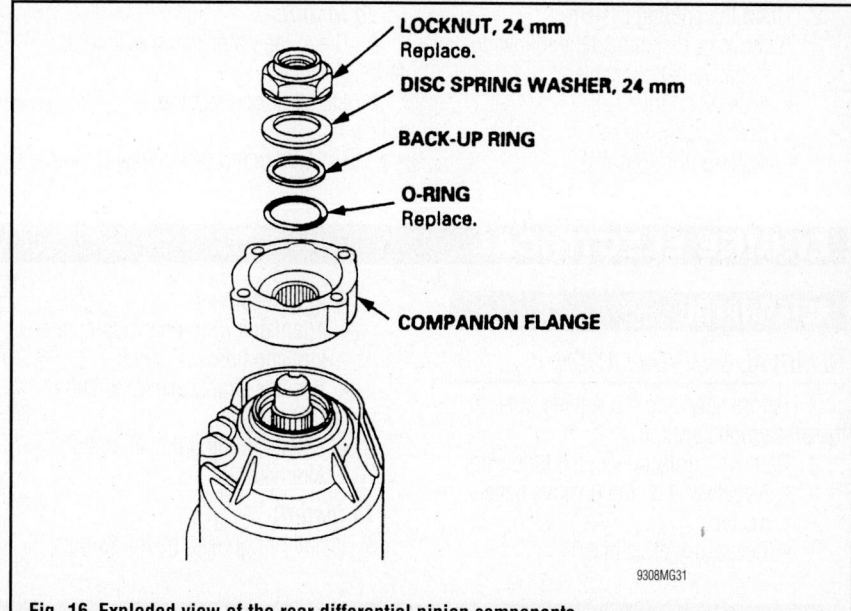

Fig. 16 Exploded view of the rear differential pinion components

ENGINE COOLING

THERMOSTAT

REMOVAL & INSTALLATION

See Figures 17 and 18.

1. Before servicing the vehicle, refer to the precautions section.
2. Drain the engine coolant.
3. Disconnect the negative battery cable.
4. Clean any dirt from quick connector, thermostat cover, and lower radiator hose.
5. Pull the lock out by hand, then wiggle the quick connector loose, and remove it from the thermostat cover. Do not use any tools to remove the quick connector.
6. Remove the thermostat.

To install:

7. Install the thermostat with a new O-ring.
8. Check the quick connector (A) and set ring (B) for cracks or damage. If the connector and/or set ring are cracked or damaged, replace the connector.

➡**Make sure the set ring is in place inside the quick connector. If the set ring is off the connector, replace the quick connector.**

9. Replace the O-ring (C) in the quick connector.
10. Check the lock (D). If the lock is damaged or deformed, replace it. When installing the new lock to the connector, push it straight down along the groove.

11. Clean the connecting surface of the thermostat cover (E), then apply clean engine coolant around the connecting surface.
12. Push the lock down, then push the quick connector onto the thermostat cover until you hear an audible click.
13. Refill the radiator with engine coolant, and bleed air from the cooling system with the heater valve open.

WATER PUMP

REMOVAL & INSTALLATION

See Figure 19.

1. Before servicing the vehicle, refer to the precautions section.

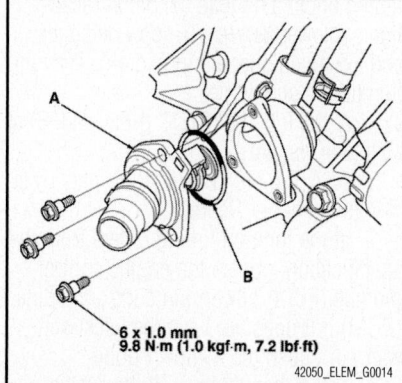

Fig. 17 View of the thermostat (A) and O-ring (B). Always use a new O-ring during thermostat installation

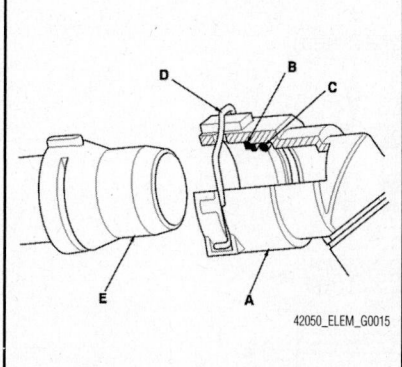

Fig. 18 View of the quick connector (A), set ring (B), O-ring (C), lock (D) and thermostat cover (E)

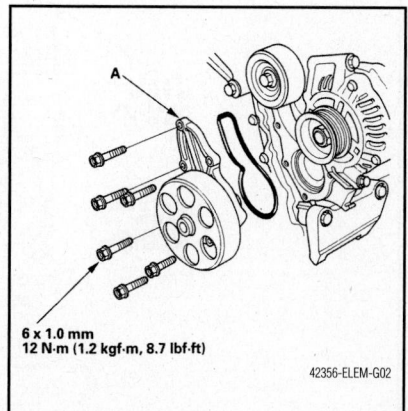

Fig. 19 Exploded view of the water pump mounting

2. Drain the cooling system.
3. Remove or disconnect the following:
- Negative battery cable
- Accessory drive belt
- Crankshaft pulley
- Water pump (6 bolts)

To install:

4. Clean the water pump mating surfaces.
5. Install or connect the following:
- Water pump with a new O-ring.

Torque the bolts to 8.7 ft. lbs. (12 Nm).
- Crankshaft pulley
- Accessory drive belt
- Negative battery cable
6. Refill the engine cooling system.

ENGINE ELECTRICAL

ALTERNATOR

REMOVAL & INSTALLATION

1. Before servicing the vehicle, refer to the precautions section.
2. Remove or disconnect the following:
- Negative, then the positive battery cables
- Accessory drive belt

- Auto-tensioner
- Alternator wiring harness connectors and harness clamp
- Positive Crankcase Ventilation (PCV) valve
- 3 bolts holding the alternator
- Alternator

To install:

3. Install or connect the following:

CHARGING SYSTEM

- Alternator. Tighten the bolts to 16 ft. lbs. (22 Nm).
- PCV valve
- Alternator wiring harness connectors and harness clamp
- Auto tensioner
- Accessory drive belt
- Negative battery cable

ENGINE ELECTRICAL

IGNITION SYSTEM

IGNITION COIL

REMOVAL & INSTALLATION

See Figure 20.

1. Before servicing the vehicle, refer to the precautions section.
2. Disconnect the negative battery cable.
3. Remove the ignition coil cover, disconnect the ignition coil connectors, then remove the ignition coils.

To install:

4. Install the ignition coils and tighten the retainers to 8.7 ft. lbs. (12 Nm).
5. Attach the ignition coil electrical connectors.
6. Install the ignition coil cover and tighten the retainers to 7.2 ft. lbs. (9.8 Nm).
7. Connect the negative battery cable.

IGNITION TIMING

INSPECTION

See Figure 21.

1. Connect the Honda Diagnostic System (HDS) to the data link connector (DLC), and check for DTCs. If a DTC is present, diagnose and repair the cause before inspecting the ignition timing.
2. Start the engine. Hold the engine speed at 3,000 rpm without load (in Park or Neutral) until the radiator fan comes on, then let it idle.
3. Check the idle speed, as outlined in the Fuel System Section.
4. Jump the SCS line with the HDS.
5. Free the service loop from the wire harness, then connect the timing light to the service loop.

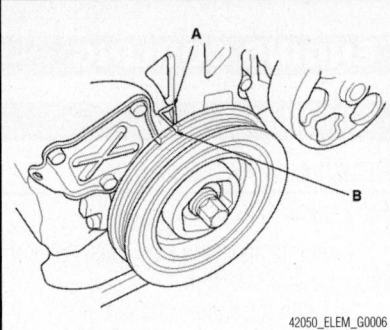

42050_ELEM_G0006

Fig. 21 Aim the light toward the pointer (A) on the cam chain case. Check the ignition timing under a no load condition (headlights, blower fan, rear window defogger, and air conditioner are turned off)

6. Aim the light toward the pointer (A) on the cam chain case. Check the ignition timing under a no load condition (headlights, blower fan, rear window defogger, and air conditioner are turned off). The ignition timing should be:

a. M/T: 6–10° BTDC (RED mark B) at idle in Neutral

b. A/T: 6–10° BTDC (RED mark B) at idle in Park or Neutral

7. If the ignition timing differs from the specification, update the engine control module (ECM)/powertrain control module (PCM) if it does not have the latest software, or substitute a known-good ECM/PCM, then recheck. If the system works properly, and the ECM/PCM was substituted, replace the original ECM/PCM.

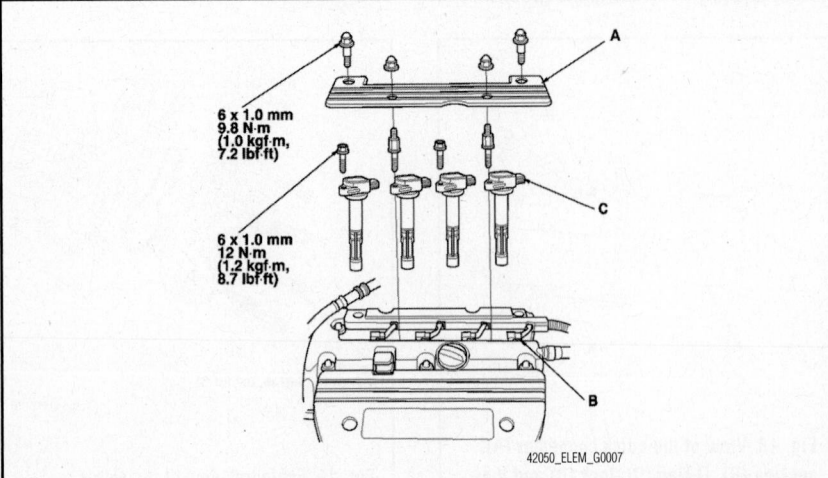

6 x 1.0 mm
9.8 N·m
(1.0 kgf·m,
7.2 lbf·ft)

6 x 1.0 mm
12 N·m
(1.2 kgf·m,
8.7 lbf·ft)

42050_ELEM_G0007

Fig. 20 Exploded view of the ignition coil cover (A), coil connectors (B) and ignition coils (C)

8. Disconnect the HDS and the timing light.

9. Secure the service loop to the wire harness with wire ties.

ADJUSTMENT

The ignition timing is controlled by the Powertrain Control Module (PCM). No adjustment is necessary or possible.

SPARK PLUGS

REMOVAL & INSTALLATION

See Figure 22.

1. Disconnect the negative battery cable.
2. Remove the ignition coil, as outlined later in this section.
3. Remove the spark plug.
4. Inspect the spark plug.

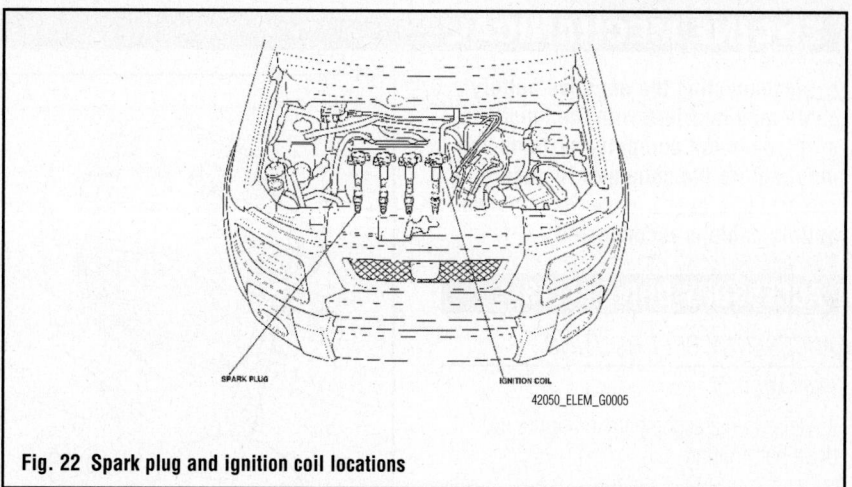

Fig. 22 Spark plug and ignition coil locations

To install:

5. Install the spark plug and tighten to 13 ft. lbs. (18 Nm).

6. Install the ignition coil.
7. Connect the negative battery cable.

ENGINE ELECTRICAL   STARTING SYSTEM

STARTER

REMOVAL & INSTALLATION

See Figure 23.

➡The factory sound system has a coded theft protection system. It is recommended that you know your reset code before you begin.

1. Before servicing the vehicle, refer to the precautions section.
2. Remove or disconnect the following:
   - Negative then the positive battery cables
   - Intake manifold
   - Starter cable from the B terminal
   - Black/white wire from the S (solenoid) terminal
   - Harness clamp and holder
   - Two bolts that mount the starter to the transaxle assembly
   - Starter

To install:

3. Install in the reverse order of removal. Refer to the illustration for torque specifications.

➡When installing the heavy gauge starter cable, make sure the crimped side of the terminal end is facing out.

4. Enter the anti-theft code and radio presets.

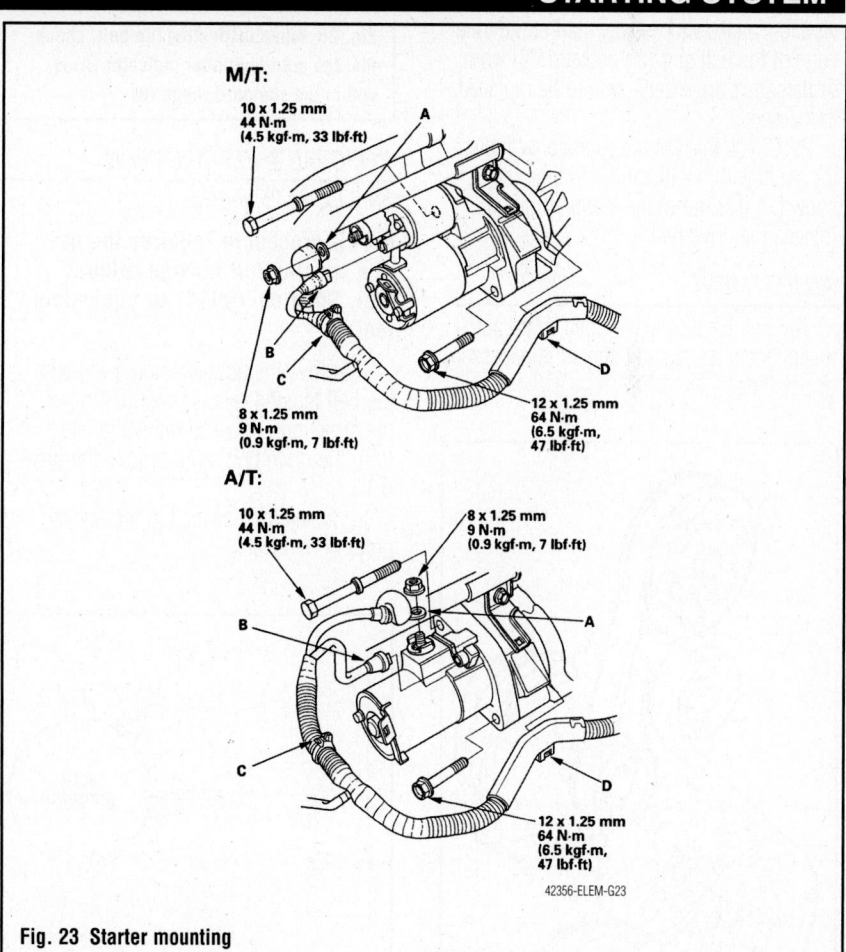

Fig. 23 Starter mounting

ENGINE MECHANICAL

→Disconnecting the negative battery cable may interfere with the functions of the on board computer systems and may require the computer to undergo a relearning process, once the negative battery cable is reconnected.

ACCESSORY DRIVE BELTS

ACCESSORY BELT ROUTING

See Figure 24.

Refer to the accompanying figure for drive belt routing.

INSPECTION

See Figure 25.

1. Inspect the drive belt for signs of glazing or cracking. A glazed belt will be perfectly smooth from slippage, while a good belt will have a slight texture of fabric visible. Cracks will usually start at the inner edge of the belt and run outward. All worn or damaged drive belts should be replaced immediately.

2. Check that the auto-tensioner indicator (A) is within the standard range (B) as shown. If it is out of the standard range, replace the drive belt.

ADJUSTMENT

The belt tension maintained by an automatic tensioner. No adjustment is necessary or possible.

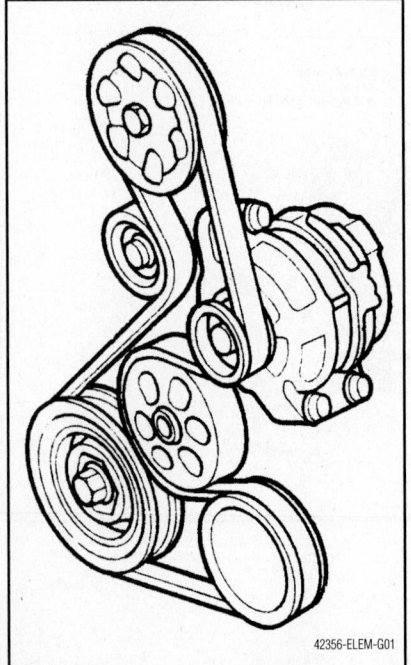

Fig. 24 Accessory drive belt routing

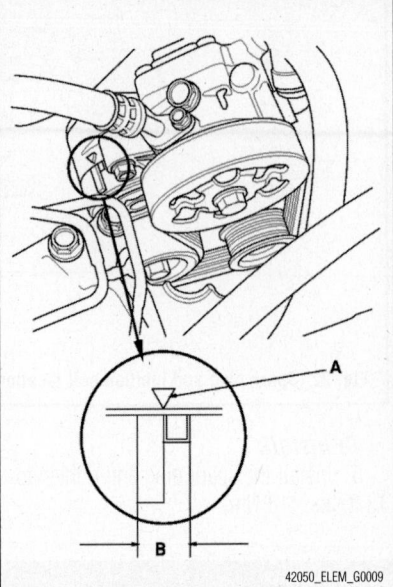

Fig. 25 When inspecting the belt, check that the auto-tensioner indicator (A) is within the standard range (B)

REMOVAL & INSTALLATION

See Figures 25 and 26.

→This procedure requires the use of a special Belt tension release tool, Snap-on YA9317 or equivalent tool.

1. Move the auto-tensioner (A) with the belt tension release tool (B) in the direction shown to relieve tension from the drive belt, and remove the drive belt.

2. Install the new belt in the reverse order of removal.

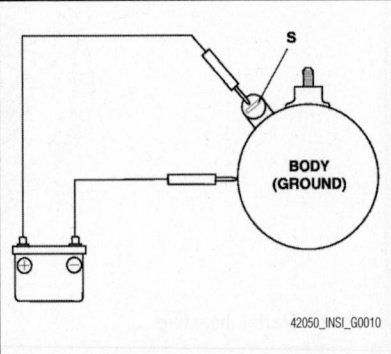

Fig. 26 To remove the belt, move the auto-tensioner (A) with the belt tension release tool (B) in the direction shown to relieve tension from the drive belt, and remove the drive belt

CAMSHAFT AND VALVE LIFTERS

REMOVAL & INSTALLATION

See the Rocker Arm Shaft Removal & Installation procedure.

CRANKSHAFT DAMPER

REMOVAL & INSTALLATION

See Figures 27 through 29.

→This procedure requires the use of the following special tools, or their equivalents: Holder handle 07JAB-001020B, 50mm Holder attachment 07NAB-001040A, and 19mm Socket 07JAA-001020A.

1. Before servicing the vehicle, refer to the precautions section.
2. Remove the right front wheel.
3. Remove the splash shield.
4. Remove the drive belt.
5. Hold the pulley with holder handle (A) and holder attachment (B).
6. Remove the bolt with a 19mm socket (C) and breaker bar, then remove the crankshaft pulley.

To install:

7. Clean the crankshaft pulley (A), crankshaft (B), bolt (C), and washer (D). Lubricate with the new engine oil as shown.

8. Install the crankshaft pulley, and hold the pulley with holder handle (A) and holder attachment (B).

9. Tighten the bolt to 36 ft. lbs. (49 Nm) with a torque wrench and 19mm socket (C). Never use an impact wrench to tighten the crankshaft pulley bolt.

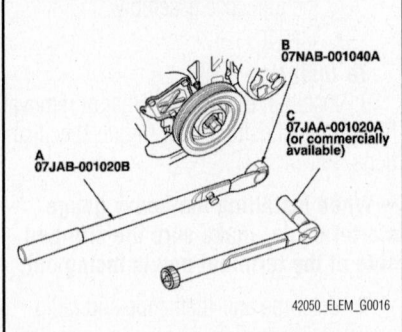

Fig. 27 To remove the crankshaft pulley, hold the pulley with holder handle (A) and holder attachment (B), then remove the bolt with a 19mm socket (C) and breaker bar

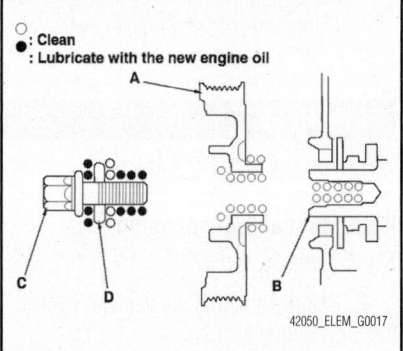

Fig. 28 Clean the crankshaft pulley (A), crankshaft (B), bolt (C), and washer (D). Lubricate with the new engine oil as shown

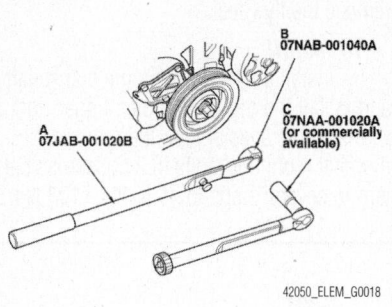

Fig. 29 Install the crankshaft pulley, and hold the pulley with holder handle (A) and holder attachment (B). Tighten the bolt to specifications with a torque wrench and 19mm socket

10. Tighten the pulley bolt an additional 90°.
11. Install the drive belt.
12. Install the splash shield.
13. Install the right front wheel.

CRANKSHAFT FRONT SEAL

REMOVAL & INSTALLATION

For the 2.4L engine, see the Timing Chain Removal & Installation procedure.

CYLINDER HEAD

REMOVAL & INSTALLATION

See Figures 30 through 33.

1. Before servicing the vehicle, refer to the precautions section.
2. Drain the cooling system.
3. Relieve the fuel system pressure.
4. Remove or disconnect the following:
   • Negative battery cable

• Accessory drive belt
• Intake Air Temperature (IAT) sensor connector
• Vacuum hoses and breather pipe and air intake duct
• Fuel feed hose
• Bolt securing the connecting pipe support bracket to the engine block
• Evaporative emission (EVAP) canister hose and brake booster vacuum hose
• Intake manifold
• Exhaust manifold
• Cam chain
• Positive Crankcase Ventilation (PCV) hose and ground cable
• Upper radiator hose, heater hoses and water bypass hose

5. Remove the following engine wire harness connectors and wire harness clamps from the cylinder head:
   • Four injector connector
   • Engine Coolant Temperature (ECT) sensor connector
   • Camshaft Position (CMP) sensor A & B (intake & exhaust) connectors
   • VTEC solenoid valve connector
   • Engine Oil Pressure (EOP) sensor connector

6. Remove or disconnect the following:
   • 3 bolts holding the EVAP canister purge valve bracket and remove the two bolts (B) securing the harness bracket
   • Timing (cam) chain
   • Rocker arm assembly

7. Loosen the cylinder head bolts in sequence and ⅓ turns until all bolts are loose.
8. Remove the cylinder head.

To install:

9. Be sure all cylinder head and block gasket surfaces are clean. Check the

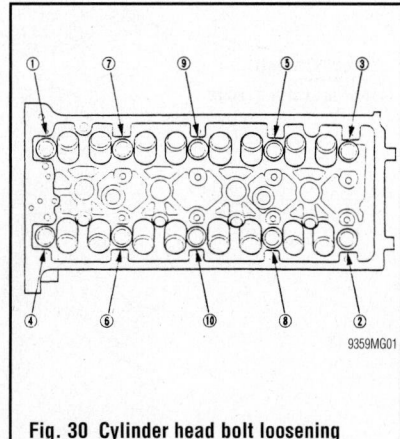

Fig. 30 Cylinder head bolt loosening sequence

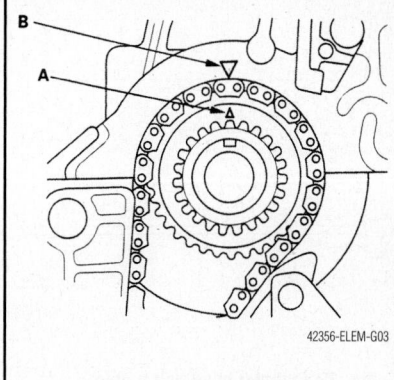

Fig. 31 Set the crankshaft to TDC by aligning the mark (A) on the crankshaft sprocket with the pointer (B) on the cylinder block

cylinder head for warpage. If warpage is less than 0.002 in. (0.05mm), cylinder head resurfacing is not required. Maximum resurface limit is 0.008 in. (0.2mm) based on a cylinder head height of 3.94 in. (100mm).

10. Install or connect the following:
   • New gasket and dowel pins on the cylinder block

11. Set the crankshaft to Top Dead Center (TDC). Align the TDC mark (A) on the crankshaft sprocket with the pointer (B) on the cylinder block.

12. Measure the diameter of each cylinder head bolt at points A & B, as shown in the illustration. If either diameter is less than 0.42 in. (10.6mm), replace the head bolt

13. Apply engine oil to the threads and under the bolt heads of all of the bolts.

14. Install the cylinder head. Tighten the bolts in sequence as follows:
   a. Step 1: 29 ft. lbs. (39 Nm).
   b. Step 2: Plus 90 degrees.
   c. Step 3: Plus 90 degrees.

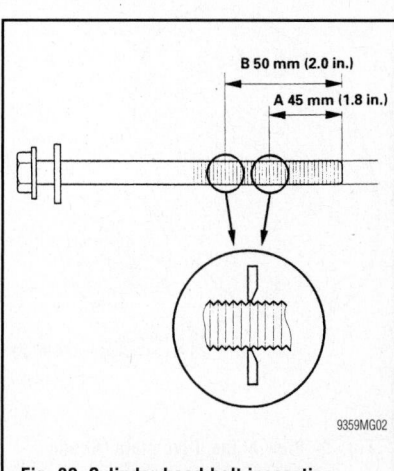

Fig. 32 Cylinder head bolt inspection

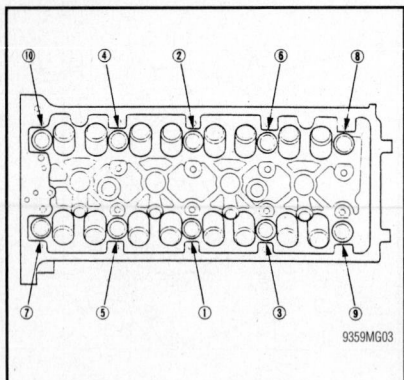

Fig. 33 Cylinder head bolt torque sequence

d. Step 4: If using new cylinder head bolts, add an additional 90 degrees.

15. The remainder of installation is the reverse of removal.

16. Fill the cooling system.

17. Connect the negative battery cable and enter the radio security code.

18. Start the engine and check carefully for any leaks.

DRIVEPLATE

REMOVAL & INSTALLATION

With Automatic Transmission

See Figure 34.

1. Before servicing the vehicle, refer to the precautions section.

2. Remove the transmission assembly, as outlined in the Drive Train Section.

3. Remove the drive plate (A) and washer (B) from the engine crankshaft.

To install:

4. Install the drive plate and washer on the engine crankshaft, and tighten the eight

bolts in a crisscross pattern in two or more steps to a final torque of 54 ft. lbs. (74 Nm).

5. Install the transmission assembly, as outlined in the Drive Train Section.

EXHAUST MANIFOLD

REMOVAL & INSTALLATION

See Figure 35.

1. Before servicing the vehicle, refer to the precautions section.

2. Raise and safely support the vehicle.

3. Remove or disconnect the following:
 • VTEC solenoid valve
 • Intermediate shaft heat cover
 • Cover and exhaust manifold bracket
 • Exhaust manifold

To install:

4. Clean the mounting surfaces.

5. Install or connect the following:
 • New gasket on the cylinder head
 • Exhaust manifold. Tighten the nuts, in a criss-cross pattern starting with the inner nut, to 33 ft. lbs. (45 Nm).

 • Exhaust manifold bracket and cover
 • Intermediate shaft heat cover
 • VTEC solenoid valve

FLYWHEEL

REMOVAL & INSTALLATION

With Manual Transmission

See Figures 36 and 37.

1. Before servicing the vehicle, refer to the precautions section.

2. Remove the pressure plate and clutch disc, as outlined in the Drive Train Section.

3. Install the special tool on the flywheel, as shown in the accompanying illustration.

4. Remove the flywheel mounting bolts in a crisscross pattern in several steps, then remove the flywheel.

To install:

5. Install the flywheel on the crankshaft, and install the mounting bolts, finger-tight.

Install the special tool, then torque the flywheel mounting bolts in a crisscross pattern in several steps to 76 ft. lbs. (103 Nm).

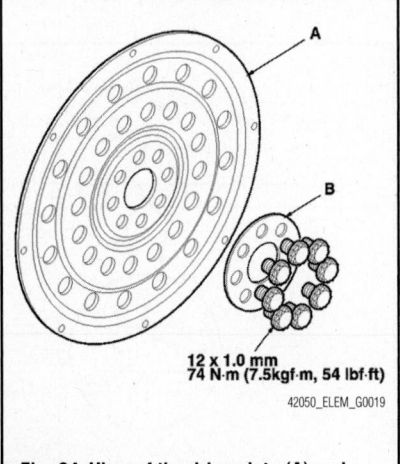

Fig. 34 View of the drive plate (A) and washer

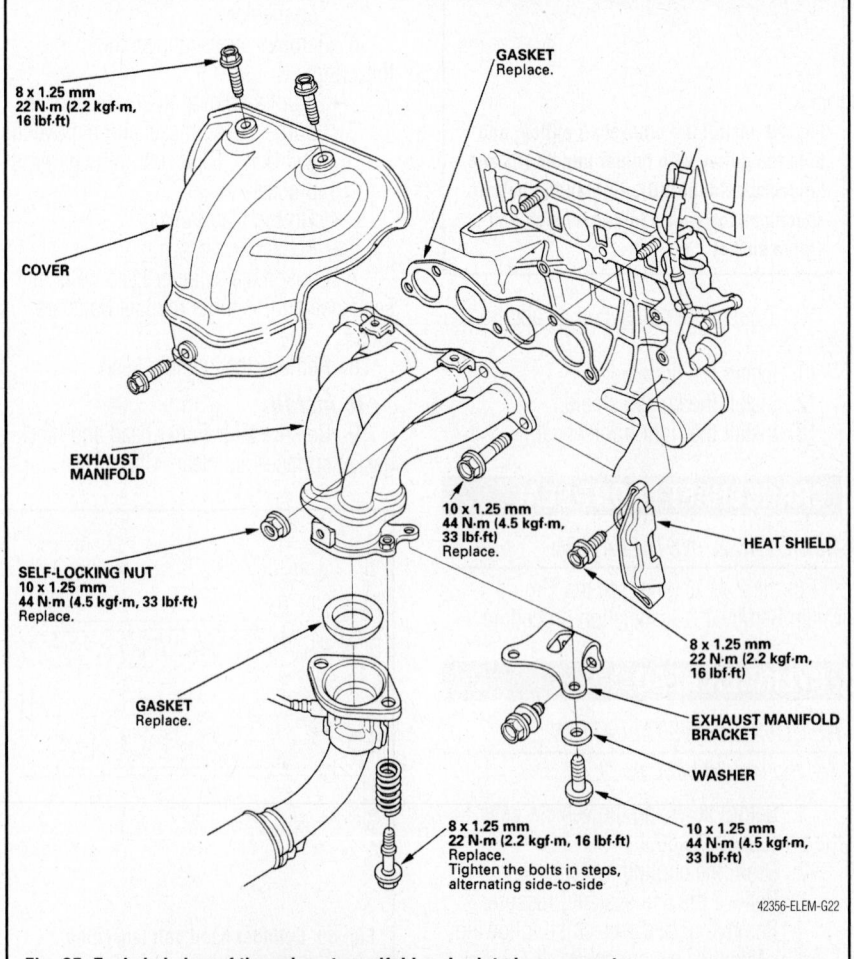

Fig. 35 Exploded view of the exhaust manifold and related components

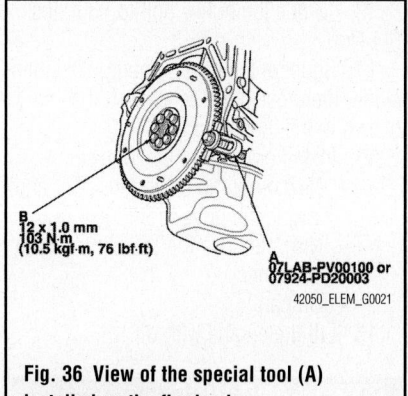

B
12 x 1.0 mm
103 N·m
(10.5 kgf·m, 76 lbf·ft)

07LAB-PV00100 or
07924-PD20003

42050_ELEM_G0021

Fig. 36 View of the special tool (A) installed on the flywheel

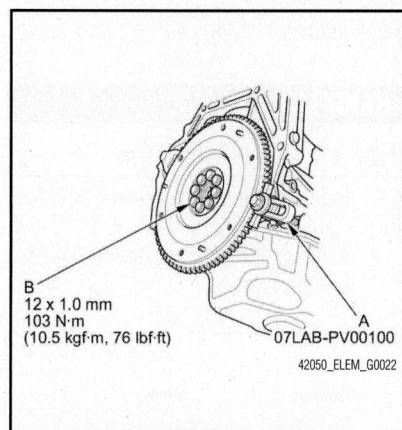

B
12 x 1.0 mm
103 N·m
(10.5 kgf·m, 76 lbf·ft)

A
07LAB-PV00100

42050_ELEM_G0022

Fig. 37 After installing the special tool (A), tighten the flywheel mounting bolts (B) to specifications in several steps, using a criss-cross pattern

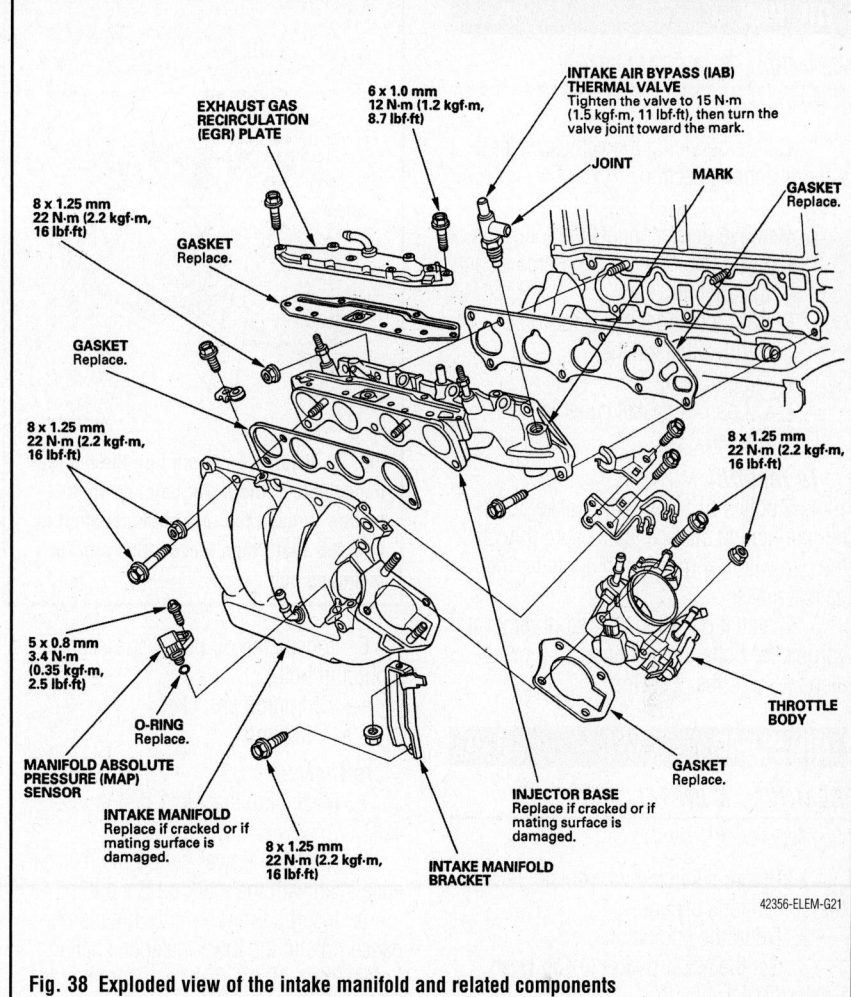

Fig. 38 Exploded view of the intake manifold and related components

6. Install the clutch disc and pressure plate, as outlined in the Drive Train Section.

INTAKE MANIFOLD

REMOVAL & INSTALLATION

See Figure 38.

1. Before servicing the vehicle, refer to the precautions section.
2. Disconnect the negative battery cable.
3. Drain the engine coolant into a sealable container.
4. Remove or disconnect the following:
   • Intake Air Temperature (IAT) sensor electrical connector
   • Vacuum hose and breather pipe and the air intake duct
   • Intake manifold cover
   • Throttle and cruise control cables by loosening the locknuts, then slipping the cable ends out of the accelerator linkage.

➡**Do not bend the cables during removal. Always replace any throttle or**

cruise control cables that get kinked during removal.

   • Evaporative emission (EVAP) canister hose and brake booster vacuum hose
   • Idle Air Control (IAC) valve connectors
   • Throttle Position (TP) sensor connector
   • Manifold Absolute Pressure (MAP) sensor connector
   • Necessary engine wire harness connectors and wire harness clamps from the intake manifold
   • Bolt securing the harness holder and remove the harness clamps
   • Water bypass hoses, then plug them
   • Harness clamp and harness connector from the intake manifold bracket
   • Intake manifold bracket
   • A/T vacuum hose
   • Retainer and intake manifold

To install:
5. Clean the mounting surfaces.
6. Install or connect the following:
   • New gasket
   • Intake manifold. Tighten the bolts, in a criss-cross pattern beginning with the inner bolt, to 16 ft. lbs. (22 Nm).
   • A/T vacuum hose
   • Intake manifold bracket
   • Harness clamp and connector to the intake manifold bracket
   • Water bypass hoses
   • Bolt securing the harness holder and tighten to 8.7 ft. lbs. (12 Nm)
   • Harness clamps
   • EVAP canister hose and brake booster vacuum hose
   • Throttle and cruise control cables
   • Intake manifold cover
   • Intake air duct
   • IAT sensor connector, vacuum hose and breather pipe
7. Refill the cooling system.
8. Connect the negative battery cable, start the engine, and check for leaks.

OIL PAN

REMOVAL & INSTALLATION

See Figure 39.

1. Before servicing the vehicle, refer to the precautions section.
2. Drain the engine oil.
3. Remove or disconnect the following:
   - Subframe. See Engine Removal and Installation.
   - With a manual transmission, remove the stiffener
   - Oil pan bolts
   - Oil pan. A gasket cutter will be needed.

To install:

4. Apply a bead of liquid gasket to the oil pan mating surface. make sure to install the pan within 4 minutes of applying the gasket maker.
5. Installation is the reverse of removal. Torque the bolts, in sequence, in 2 or 3 steps, to 9 ft. lbs. (12 Nm).

OIL PUMP

REMOVAL & INSTALLATION

See Figures 40 and 41.

1. Before servicing the vehicle, refer to the precautions section.
2. Drain the engine oil.
3. Set the No. 1 piston to Top Dead Center (TDC).
4. Remove or disconnect the following:
   - Negative battery cable
   - Oil pan
   - Oil pump chain tensioner and discard
5. Insert a 6mm pin driver into the maintenance hole in the lower balance shaft holder and through the rear balancer shaft to hold the rear balancer shaft.

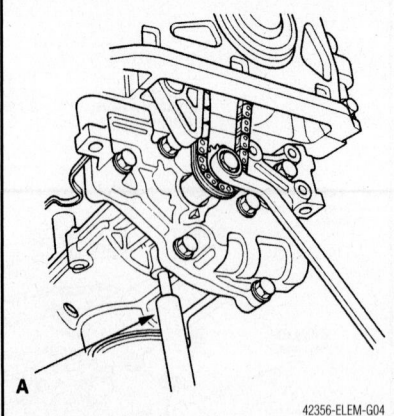

Fig. 40 Insert a 6mm pin into the maintenance hole in the lower balance shaft holder, through the rear balancer shaft to hold the shaft, then loosen the sprocket mounting bolt

6. Loosen the oil pump sprocket mounting bolt.
   - Oil pump sprocket
   - Oil pump

To install:

7. Make sure that No.1 piston is at TDC.
8. Align the dowel pin on the rear balance shaft with the mark on the pump.
9. Insert a 6mm pin into the maintenance hole in the lower balance shaft holder, through the rear balancer shaft to hold the shaft.
10. Install or connect the following:
    - Engine oil to the threads of the oil pump sprocket mounting bolt
    - Oil pump and sprocket loosely
11. Remove the balance shaft holding pin.
12. Torque the 10mm mounting bolts to 33 ft. lbs. (44 Nm); the 8mm bolts to 16 ft. lbs. (22 Nm).

13. Torque the pulley bolt to 33 ft. lbs. (44 Nm).
14. Squeeze the new oil pump chain tensioner then install the set clip on it as shown in the illustration.
15. Install or connect the following:
    - New oil pump chain tensioner and torque the bolts to 9 ft. lbs. (12 Nm). Remove the set clip from the tensioner.
    - Oil pan
16. Fill the engine with oil.

PISTON AND RING

POSITIONING

See Figures 42 through 44.

REAR MAIN SEAL

REMOVAL & INSTALLATION

1. Before servicing the vehicle, refer to the precautions section.
2. Remove or disconnect the following:
   - Transaxle

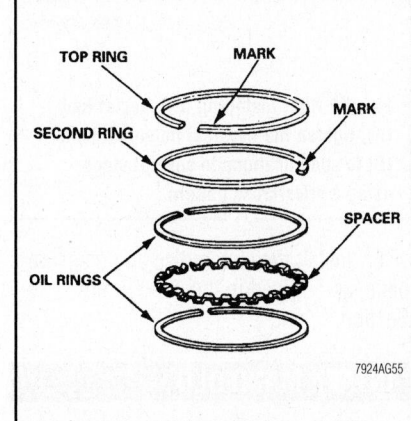

Fig. 42 Piston ring positioning and top mark location

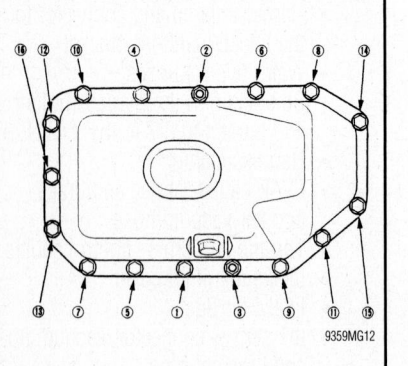

Fig. 39 Oil pan fastener tightening sequence

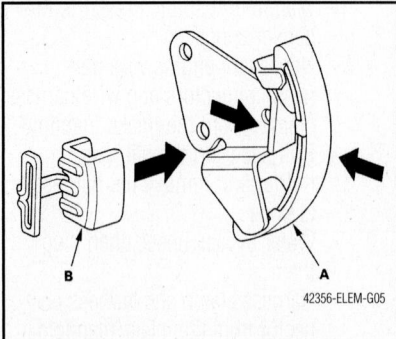

Fig. 41 Squeeze the new oil pump chain tensioner (A) then install the set clip (A) on it as shown. The clip is supplied with the new tensioner

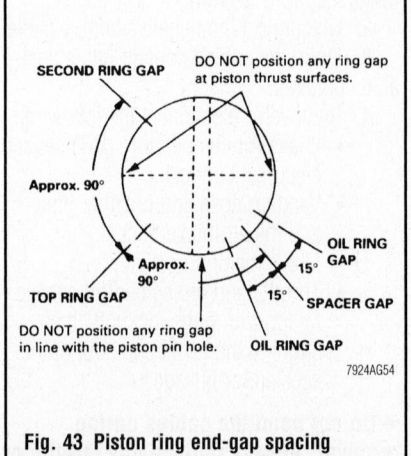

Fig. 43 Piston ring end-gap spacing

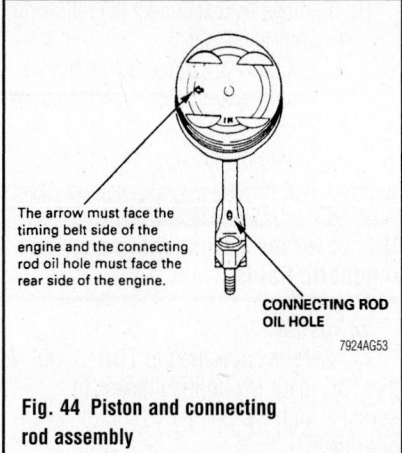

Fig. 44 Piston and connecting rod assembly

- Clutch pressure plate and disc, if equipped
- Flywheel
- Oil seal

To install:

3. Install or connect the following:
- Oil seal. Drive the seal square into the seal case.
- Flywheel. Tighten the bolts in a crossing pattern to 76 ft. lbs. (103 Nm).
- Clutch pressure plate and disc, if equipped
- Transaxle
4. Check the fluid levels.
5. Start the engine and check for leaks.

ROCKER ARMS/SHAFTS

REMOVAL & INSTALLATION

See Figures 45 through 49.

1. Before servicing the vehicle, refer to the precautions section.
2. Remove or disconnect the following:
- Timing (cam) chain
- Loosen the rocker arm adjusting screws

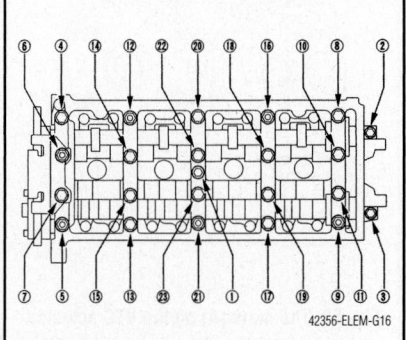

Fig. 45 Camshaft holder bolt loosening sequence. Note that bolt 1 in the illustration is not on all engines.

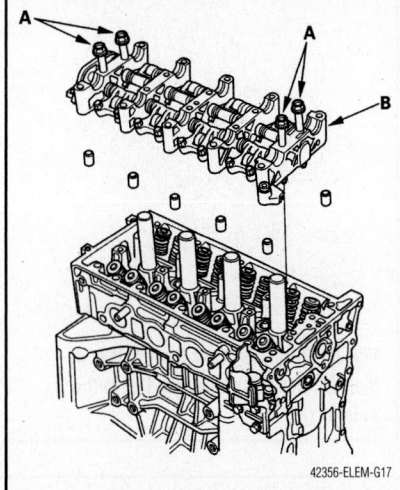

Fig. 46 Insert the bolts (A) into the rocker shaft holder, then remove the rocker arm assembly (B)

- Camshaft holder bolts, two turns at a time in sequence
- Timing chain guide (B), camshaft holders and camshafts
3. Insert the bolts (A) into the rocker shaft holder, then remove the rocker arm assembly (B)

To install:

4. Clean and dry the No. 5 rocker shaft holding mating surface.
5. Apply a suitable liquid gasket P/N 08718-0009, or equivalent, evenly to the cylinder head mating surface of the No. 5 rocker shaft holder.

➡**The parts must be installed within 5 minutes of applying the liquid gasket.**

6. Reassemble the rocker arm assembly, as necessary.
7. Install or connect the following
- Bolts (A) into the rocker shaft holder, then the rocker arm assembly on the cylinder head. Remove the bolts from the rocker shaft holder.
8. Make sure the punch marks on the variable valve timing control (VTC) actuator and exhaust camshaft sprocket are facing up, then set the camshafts (A) in the holder
9. Set the camshaft holders (B) and timing chain guide B (C) in place.
10. Tighten the bolts, in sequence, to the following specification:
   a. 8mm bolts: 16 ft. lbs. (22 Nm).

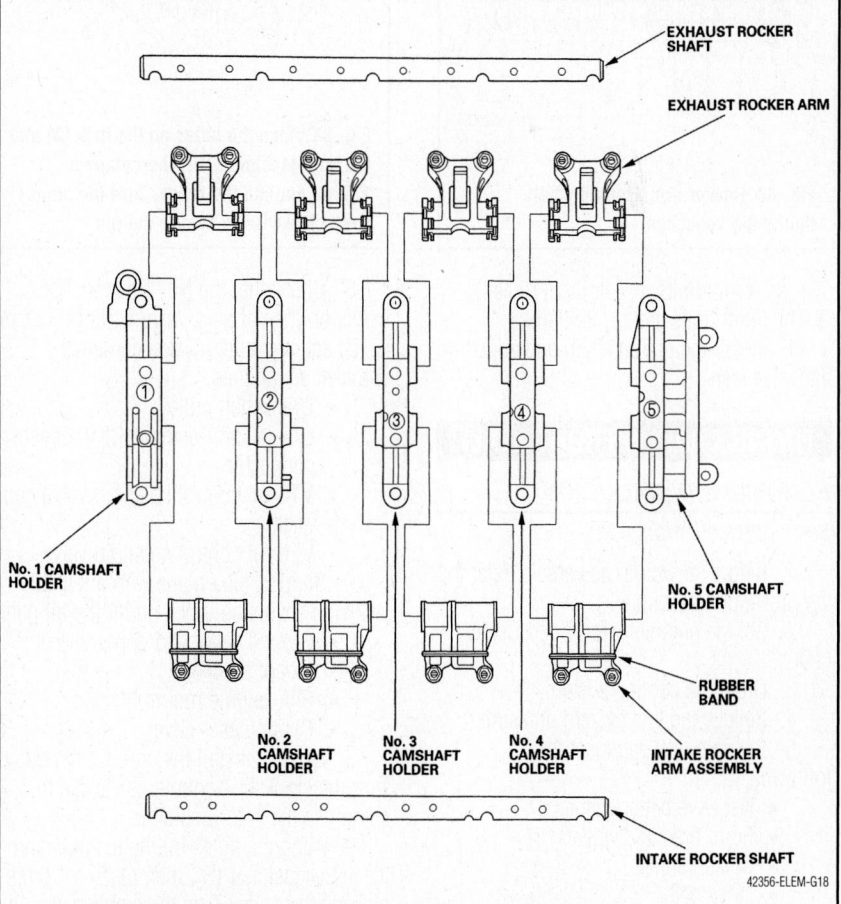

Fig. 47 Exploded view of the rocker arms and related components

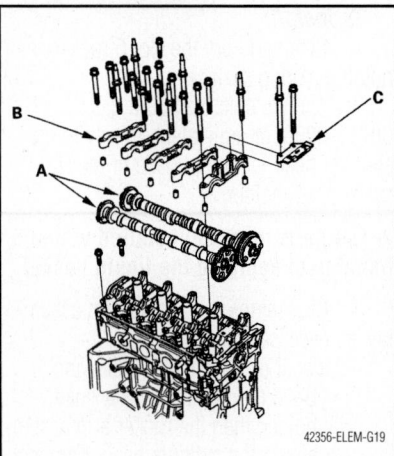

Fig. 48 When installing the camshafts (A) make sure the punch marks on the VTC actuator and exhaust cam sprockets are facing up

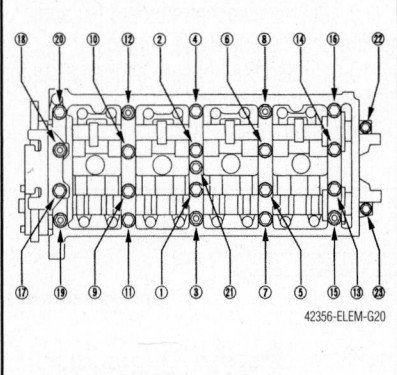

Fig. 49 Rocker arm assembly bolt tightening sequence

b. 6mm bolts: 8.7 ft. lbs. (12 Nm). The 6mm bolts are 21, 22 and 23.

11. Install the timing chain and adjust the valve lash.

TIMING CHAIN & FRONT SEAL

REMOVAL & INSTALLATION

See Figures 50 through 58.

1. Before servicing the vehicle, refer to the precautions section.
2. Set the engine to Top Dead Center (TDC).
3. Drain the cooling system.
4. Relieve the fuel system pressure.
5. Remove or disconnect the following:
   - Negative battery cable
   - Front tires and wheels
   - Splash shield
   - Drive belt
   - Cylinder head cover.

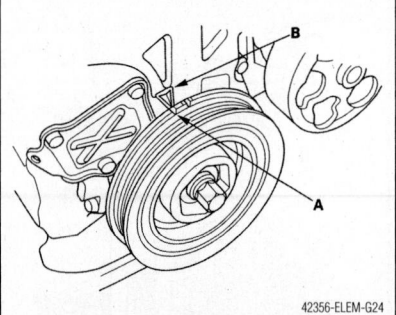

Fig. 50 Turn the crankshaft pulley so the TDC mark (A) is aligned with the pointer (B)

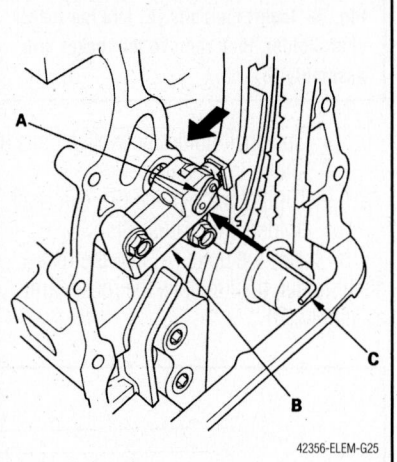

Fig. 51 Align the holes on the lock (A) and the auto-tensioner (B), then place a 1.5mm pin into the holes. Turn the crankshaft clockwise to secure the pin

6. Check that the No. 1 piston TDC marks on the Variable Valve Timing Control (VTC) actuator and exhaust camshaft sprocket are aligned.
   - Crankshaft pulley
   - Crankshaft Position (CKP) sensor connector
   - VTC oil control solenoid valve connector
   - VTC oil control solenoid valve
7. Support the engine with a suitable jack with a wooden block under the oil pan.
   - Ground cable and upper engine mount bracket
   - Side engine mount bracket
   - Chain (case) cover
8. Loosely install the crankshaft pulley. Turn the crankshaft counterclockwise to compress the auto-tensioner.
9. Align the holes on the lock (A) and the auto-tensioner (B), then place a 1.5mm pin into the holes. Turn the crankshaft clockwise to secure the pin.

10. Remove or disconnect the following:
   - Auto-tensioner
   - Timing chain guide B (top guide)
   - Timing chain guide A and tensioner arm
   - Timing chain

※※ WARNING

Do not let the timing chain near any magnetic fields.

To install:

11. Set the crankshaft to TDC. Align the TDC mark (A) on the crankshaft sprocket with the pointer (B) on the cylinder block.

12. Set the camshafts to TDC. The punch mark (A) on the VTC actuator and the punch mark (B) on the exhaust camshaft (C) should be at the top. Align the TDC marks (C) on the VTC actuator and exhaust camshaft sprockets.

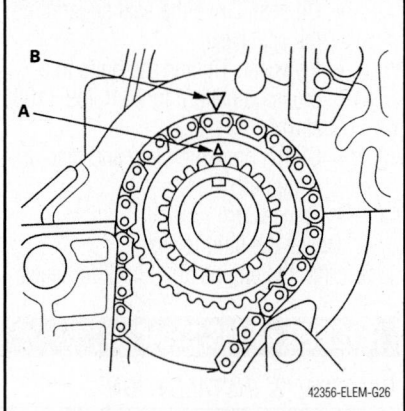

Fig. 52 Set the crankshaft to TDC. Align the TDC mark (A) on the crankshaft sprocket with the pointer (B) on the cylinder block.

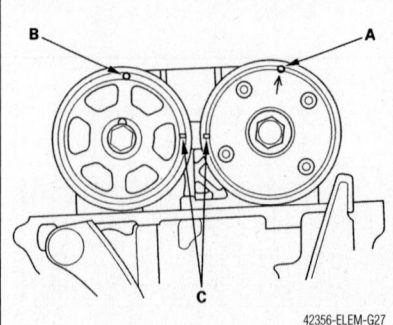

Fig. 53 The mark (A) on the VTC actuator and the mark (B) on the exhaust cam (C) should be at the top. Align the TDC marks (C) on the VTC actuator and exhaust cam sprockets.

13. Install or connect the following:
- Timing chain on the crankshaft sprocket with the colored link of the chain aligned with the mark on the crank sprocket
- Timing chain on the VTC actuator and exhaust camshaft sprocket with the punch marks aligned with the center of the 2 colored links
- Timing chain guide A and tensioner arm. Tighten the guide bolts to 8.7 ft. lbs. (12 Nm) and the tensioner arm retainer to 16 ft. lbs. (22 Nm).
- Auto-tensioner and tighten the bolts to 8.7 ft. lbs. (12 Nm)
- Timing chain guide B and tighten the retainers to 16 ft. lbs. (22 Nm)

14. Remove the pin from the auto-tensioner.

15. Inspect the chain cover seal for damage and replace if necessary. Clean and dry the chain cover mating surfaces.

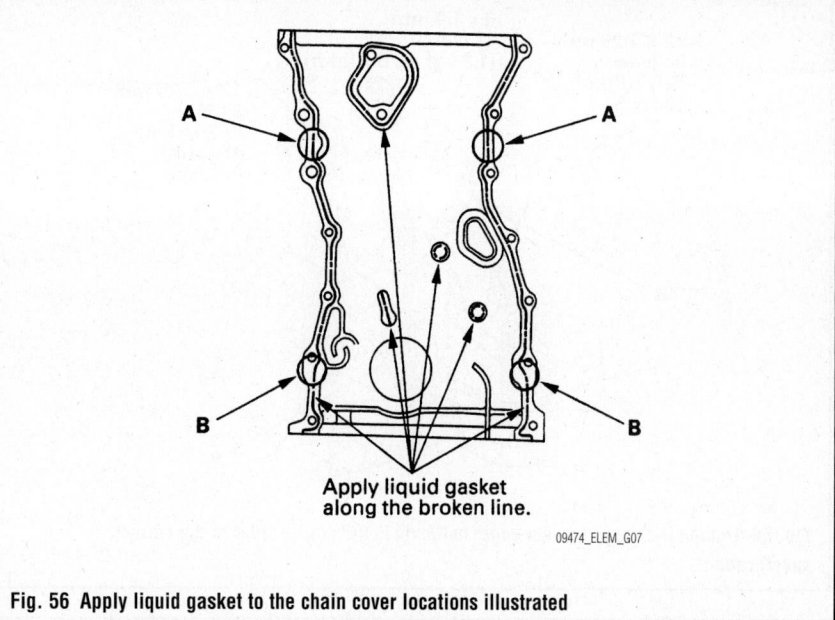

Fig. 56 Apply liquid gasket to the chain cover locations illustrated

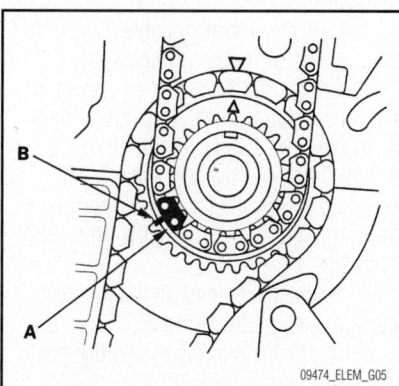

Fig. 54 Install the timing chain on the crankshaft sprocket with the colored link of the chain aligned with the mark on the crank sprocket

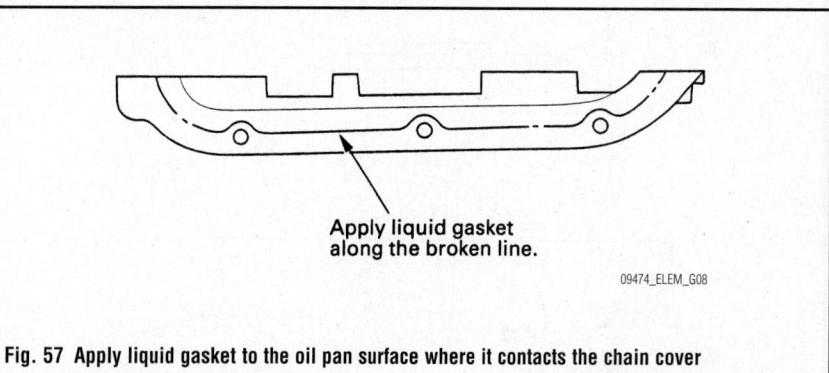

Fig. 57 Apply liquid gasket to the oil pan surface where it contacts the chain cover

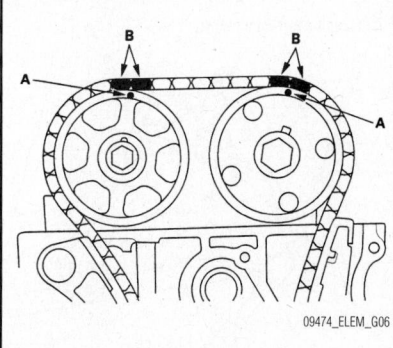

Fig. 55 Install the timing chain on the VTC actuator and exhaust camshaft sprocket with the punch marks aligned with the center of the 2 colored links

16. Install or connect the following:
- Liquid gasket, P/N 08718-0009 evenly to the cylinder block mating surface of the timing chain cover and the inner threads of the holes
- Liquid gasket to the cylinder block upper surface contact areas on the chain cover and the oil pan mating surface of the chain cover in the inner threads of the holes
- Liquid gasket to the oil pan surface where it contacts the chain cover

➡ **Make sure to install the components within 4 minutes of applying the sealer.**

- New O-ring the timing chain cover. Set the edge of the cover to the edge of the oil pan, then install the cover on the engine block. Tighten the retainers to 8.7 ft. lbs. (12 Nm).

➡ **When installing the chain case, do not slide the bottom surface on the oil pan mounting surface.**

- Side engine mounting bracket and tighten the retainers to 33 ft. lbs. (44 Nm)
- Upper mount, then tighten the bolts/nuts as shown in the illustration
- Ground cable
- VTC oil control solenoid valve
- CKP sensor and VTC oil control solenoid valve connectors
- Crankshaft pulley. Tighten the bolt to 36 ft. lbs. (49 Nm), then tighten an additional 90 degrees.
- Cylinder head cover
- Drive belt
- Splash shield

17. Fill the engine cooling system and connect the negative battery cable.

VALVE (ROCKER ARM) COVERS

REMOVAL & INSTALLATION

1. Before servicing the vehicle, refer to the precautions section.
2. Disconnect the negative battery cable.
3. Remove the intake manifold cover.

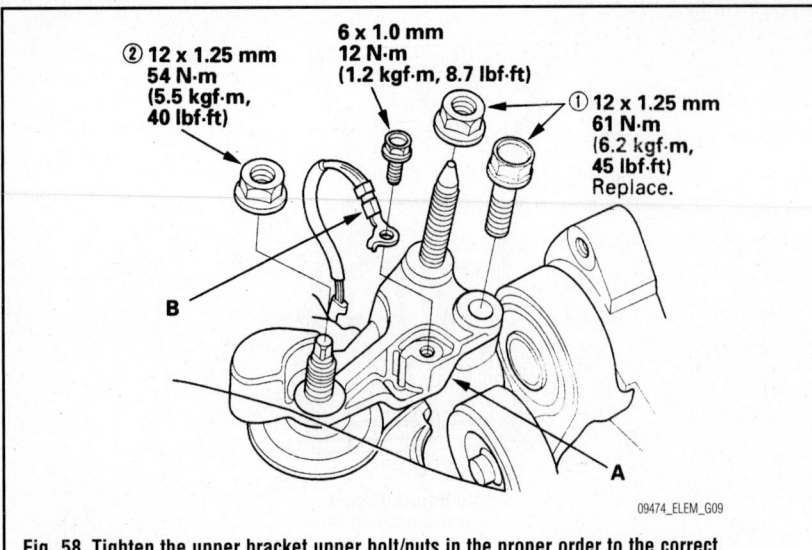

Fig. 58 Tighten the upper bracket upper bolt/nuts in the proper order to the correct specification

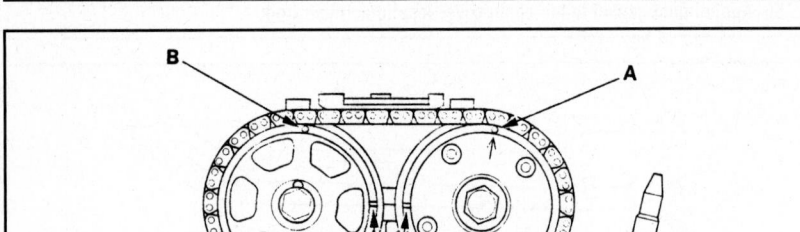

Fig. 59 Align the timing marks

4. Remove the four ignition coils.

5. Remove the two bolts securing the vacuum line.

6. Remove the bolt securing the power steering hose bracket.

7. Remove the dipstick and breather hose.

8. Remove the retainers and the cylinder head cover.

9. Installation is the reverse of the removal procedure.

VALVE LASH

ADJUSTMENT

See Figure 59.

Adjust the valves only when the cylinder head temperature is less than 100°F (38°C).

1. Before servicing the vehicle, refer to the precautions section.

2. Remove or disconnect the following:

- Negative battery cable
- Valve (rocker arm) cover

3. Set the timing marks as shown in the illustration with NO.1 at TDC. Check all clearances. Intake should be 0.008–0.010 in.; exhaust should be 0.011–0.013 in. Intake locknut torque is 14 ft. lbs. (19 Nm); exhaust is 10 ft. lbs. (14 Nm).

4. Rotate the crankshaft 180 degrees clockwise and recheck No.3.

5. Rotate the crankshaft 180 degrees clockwise and recheck No.4.

6. Rotate the crankshaft 180 degrees clockwise and recheck No.2.

ENGINE PERFORMANCE & EMISSION CONTROL

AIR FUEL (A/F) RATIO SENSOR

LOCATION

See Figure 60.

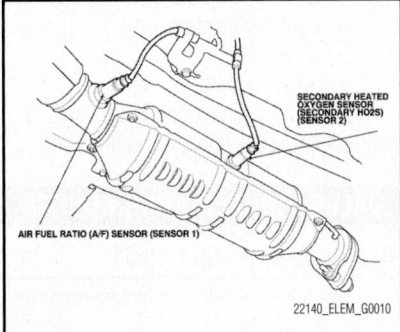

Fig. 60 A/F ratio sensor and Secondary HO2S locations

Refer to the accompanying illustration for sensor location.

REMOVAL & INSTALLATION

See Figure 61.

➡ **This procedure requires the use of O2 sensor socket wrench, Snap-on YA8875, SP Tools 93750, or equivalent, commercially available O2 sensor removal tool.**

1. Disconnect the A/F sensor 4P connector (A), then remove the A/F sensor (B), using the proper tool.

2. Install the parts in the reverse order of removal. Tighten the sensor to 33 ft. lbs. (44 Nm).

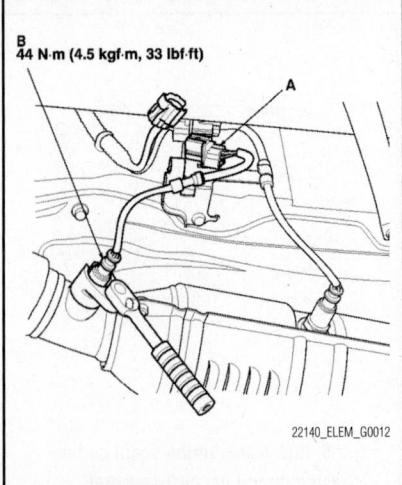

Fig. 61 View of the A/F sensor connector (A) and sensor (B)

CAMSHAFT POSITION (CMP) SENSOR

REMOVAL & INSTALLATION

See Figure 62.

1. Remove the air cleaner.
2. Remove the EVAP canister purge valve.
3. Disconnect the CMP sensor B connector (A).
4. Remove CMP sensor B (B).
5. Install the parts in the reverse order of removal with a new O-ring (C) coated with clean engine oil. Tighten the sensor bolt to 8.7 ft. lbs. (12 Nm).

CRANKSHAFT POSITION (CKP) SENSOR

REMOVAL & INSTALLATION

See Figures 63 and 64.

1. Disconnect the CKP sensor connector (A).
2. Remove the CKP sensor (B).

To install:

3. Install the parts in the reverse order of removal with a new O-ring (C) coated with clean engine oil. Tighten the sensor bolt to 8.7 ft. lbs. (12 Nm).
4. Perform the CKP pattern clear/pattern learn procedure:

    a. Connect the HDS to the data link connector (DLC) (A) located under the driver's side of the dashboard.
    b. Turn the ignition switch ON (II).
    c. Make sure the HDS communicates

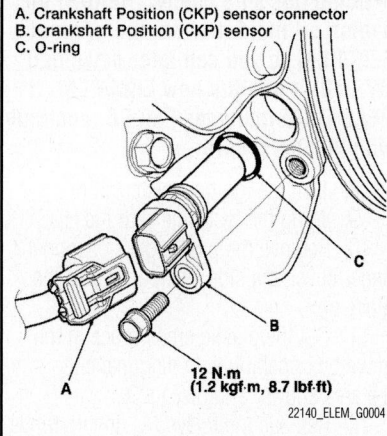

A. Crankshaft Position (CKP) sensor connector
B. Crankshaft Position (CKP) sensor
C. O-ring

12 N·m
(1.2 kgf·m, 8.7 lbf·ft)

22140_ELEM_G0004

Fig. 63 View of the CKP sensor connector (A) and sensor (B)

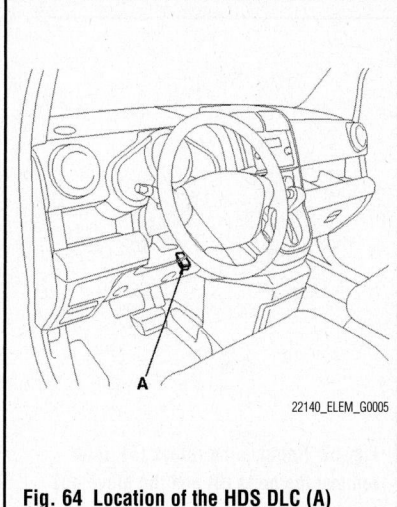

22140_ELEM_G0005

Fig. 64 Location of the HDS DLC (A)

with the ECM/PCM and other vehicle system.

d. Select CRANK PATTERN in the ADJUSTMENT MENU with the HDS.

e. Select CRANK PATTERN LEARNING with the HDS, and follow the screen prompts.

ELECTRONIC CONTROL MODULE (ECM) POWERTRAIN CONTROL MODULE

LOCATION

See Figure 65.

➡**On manual transaxle equipped vehicles, it is referred to as the Engine Control Module (ECM). On automatic transaxle equipped vehicles, it is referred to as the Powertrain Control Module (PCM).**

Refer to the accompanying illustration for ECM/PCM location.

REMOVAL & INSTALLATION

See Figures 64, 66 through 68.

➡**On manual transaxle equipped vehicles, it is referred to as the Engine Control Module (ECM). On automatic transaxle equipped vehicles, it is referred to as the Powertrain Control Module (PCM).**

This procedure requires the following special tools (or their equivalents):
• Honda diagnostics system (HDS) tablet tester

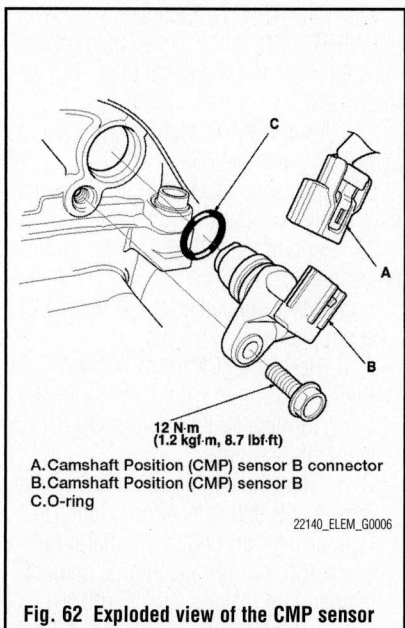

12 N·m
(1.2 kgf·m, 8.7 lbf·ft)

A. Camshaft Position (CMP) sensor B connector
B. Camshaft Position (CMP) sensor B
C. O-ring

22140_ELEM_G0006

Fig. 62 Exploded view of the CMP sensor B connector (A) and sensor (B)

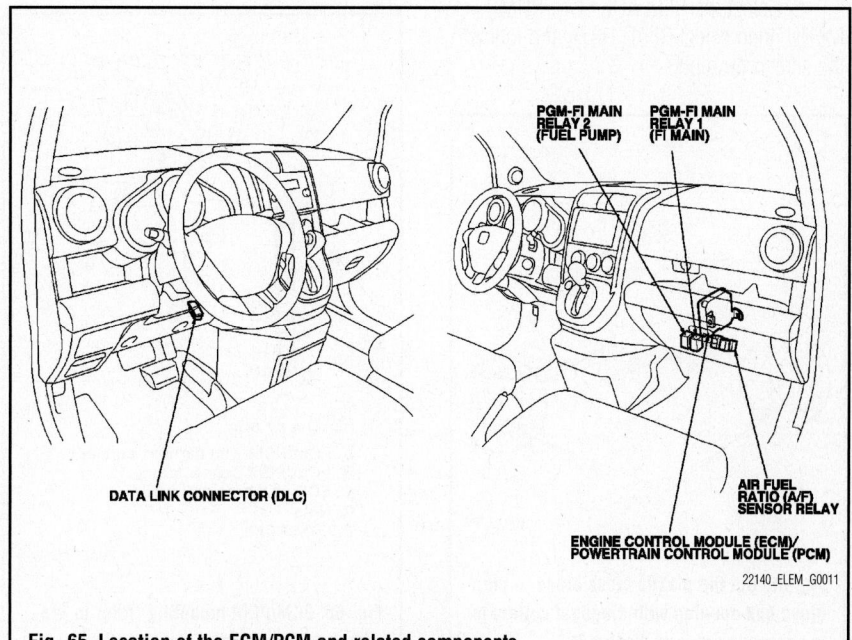

DATA LINK CONNECTOR (DLC)

PGM-FI MAIN RELAY 2 (FUEL PUMP)
PGM-FI MAIN RELAY 1 (FI MAIN)

AIR FUEL RATIO (A/F) SENSOR RELAY

ENGINE CONTROL MODULE (ECM)/ POWERTRAIN CONTROL MODULE (PCM)

22140_ELEM_G0011

Fig. 65 Location of the ECM/PCM and related components

- Honda interface module (HIM) and an iN workstation with HDS and CM update software
- HDS pocket tester
- GNA 600 and an iN workstation with HDS and CM update software

➡**Make sure the HDS is loaded with the latest software version.**

- If you are replacing the ECM/PCM after substituting a known-good ECM/PCM, reinstall the original ECM/PCM, then do this procedure.
- During the procedure, if any READ DATA, WRITE DATA, or other data checks fail, note the failure, then continue.

1. Connect the HDS to the data link connector (DLC) (A) located under the driver's side of the dashboard.
2. Turn the ignition switch ON (II).
3. Make sure the HDS communicates with the ECM/PCM and other vehicle system. If you are returning from DLC circuit troubleshooting, skip steps 4 through 7, 21 through 23, and 26 through 27, and do this after replacing the ECM/PCM:
  a. Replace the engine oil and the engine oil filter.
  b. Clean the throttle body.
4. Select the PGM-FI system with the HDS.
5. Select the INSPECTION MENU with the HDS.
6. Select the ETCS TEST, then select the TP POSITION CHECK, and follow the screen prompts.

➡**If the TP POSITION CHECK indicates FAILED, continue with this procedure.**

7. Select the REPLACE ECM/PCM MENU, then select READ DATA and follow the screen prompts.

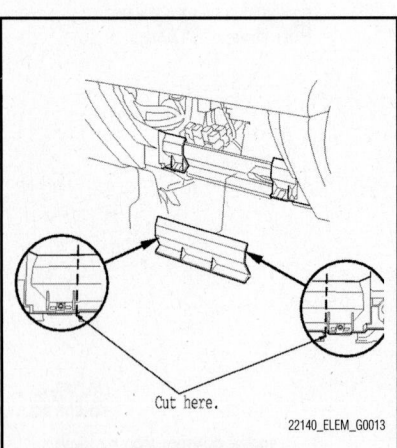

Cut here.

22140_ELEM_G0013

Fig. 66 Cut the plastic cross brace in the glove box opening with diagonal cutters in the area shown, and discard it

➡**Doing this step copies (READS) the engine oil life data from the original ECM/PCM so you can later download (WRITE) it into the new ECM/PCM. If READ DATA indicates FAILED, continue with this procedure.**

8. Turn the ignition switch OFF.
9. Jump the SCS line with the HDS.
10. Remove the passenger's dashboard under cover, the side kick panel, and the glove box.
11. Cut the plastic cross brace in the glove box opening with diagonal cutters in the area shown, and discard it.
12. Remove the relays (A), then remove the bolts (B) and the glove box frame (C).
13. Remove the gray 20P ECM/PCM wire harness connector (A) from the ECM/PCM mounting bracket.

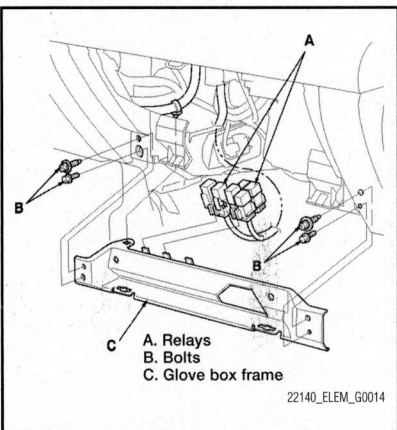

A. Relays
B. Bolts
C. Glove box frame

22140_ELEM_G0014

Fig. 67 Remove the relays (A), then remove the bolts (B) and the glove box frame (C)

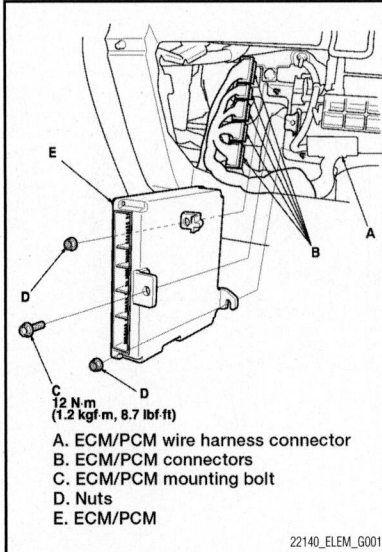

C 12 N·m
(1.2 kgf·m, 8.7 lbf·ft)
A. ECM/PCM wire harness connector
B. ECM/PCM connectors
C. ECM/PCM mounting bolt
D. Nuts
E. ECM/PCM

22140_ELEM_G0015

Fig. 68 ECM/PCM mounting (refer to procedure for component identification)

14. Disconnect the ECM/PCM connectors (B).
15. Remove the ECM/PCM mounting bolt (C) and the bracket.
16. Remove the nuts (D), then remove the ECM/PCM (E).

To install:

17. Install the parts in the reverse order of removal.
18. Open the SCS line with the HDS.
19. Turn the ignition switch ON (II).
20. Manually input the VIN to the ECM/PCM with the HDS.

➡**DTC P0630 "VIN Not Programmed or Mismatch" may be stored because the VIN has not been programmed into the ECM/PCM; ignore it, and continue this procedure.**

If the READ DATA (engine oil life) failed in step 7, go to step 24. Otherwise, go to step 22.

21. Select the PGM-FI system with the HDS.
22. Select the REPLACE ECM/PCM MENU, then select WRITE DATA and follow the screen prompts.

➡ **If the WRITE DATA indicates FAILED, continue with this procedure.**

23. Select IMMOBI system with the HDS.
24. Enter the immobilizer code with the ECM/PCM replacement procedure in the HDS; it allows you to start the engine.
25. If the TP POSITION CHECK failed in step 6 clean the throttle body, then go to step 27.
26. If the READ DATA failed in step 7 or the WRITE DATA failed in step 23, replace the engine oil and engine oil filter, then go to step 28.
27. Select PGM-FI system, and reset the ECM/PCM with the HDS.
28. Update the ECM/PCM if it does not have the latest software.
29. Perform the ECM/PCM idle learn procedure, as follows:
  a. Make sure all electrical items (A/C, audio, lights, etc.) are off.
  b. Reset the ECM/PCM with the HDS.
  c. Turn the ignition switch ON (II), and wait 2 seconds.
  d. Start the engine. Hold the engine speed at 3,000 rpm without load (in Park or neutral) until the radiator fan comes on, or until the engine coolant temperature reaches 194 °F (90 °C).
  e. Let the engine idle for about 5 minutes with the throttle fully closed.

➡If the radiator fan comes on, do not include its running time in the 5 minutes

30. Perform the CKP pattern learn procedure, as follows:

a. Connect the HDS to the data link connector (DLC) (A) located under the driver's side of the dashboard.

b. Turn the ignition switch ON (II).

c. Make sure the HDS communicates with the ECM/PCM and other vehicle system.

d. Select CRANK PATTERN in the ADJUSTMENT MENU with the HDS.

e. Select CRANK PATTERN LEARNING with the HDS, and follow the screen prompts.

ENGINE COOLANT TEMPERATURE (ECT) SENSOR

REMOVAL & INSTALLATION

Sensor 1

See Figure 69.

1. Remove the air cleaner.
2. Remove the EVAP canister purge valve.
3. Unbolt the under-hood fuse/relay box bolt, and move the assembly aside.
4. Drain the engine coolant.
5. Disconnect the ECT sensor 1 connector (A).
6. Remove the ECT sensor 1 (B).

7. Install the parts in the reverse order of removal with a new O-ring (C) coated with clean engine oil, then refill the radiator with engine coolant.

To install:

8. Install the parts in the reverse order of removal with a new O-ring (C) coated with clean engine oil. Tighten to 8.7 ft. lbs. (12 Nm).

9. Refill the radiator with engine coolant.

Sensor 2

See Figure 70.

1. Drain the engine coolant.
2. Remove the splash shield.
3. Disconnect the ECT sensor 2 connector (A).
4. Remove ECT sensor 2 (B).

To install:

5. Install the parts in the reverse order of removal with a new O-ring (C) coated with clean engine oil. Tighten to 8.7 ft. lbs. (12 Nm).

6. Refill the radiator with engine coolant.

HEATED OXYGEN (HO2S) SENSOR

LOCATION

See Figure 71.

Refer to the accompanying illustration for sensor location.

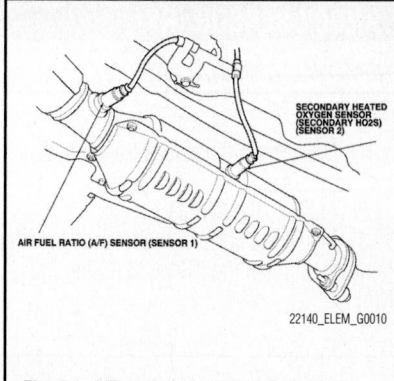

Fig. 71 A/F ratio sensor and Secondary HO2S locations

REMOVAL & INSTALLATION

See Figure 72.

➡This procedure requires the use of O2 sensor socket wrench, Snap-on YA8875, SP Tools 93750, or equivalent, commercially available O2 sensor removal tool.

1. Disconnect the secondary HO2S 4P connector (A), then remove the secondary HO2S (B).

2. Install the parts in the reverse order of removal. Tighten the sensor to 33 ft. lbs. (44 Nm).

INTAKE AIR TEMPERATURE (IAT) SENSOR

LOCATION

Refer to Mass Air Flow (MAF) sensor.

REMOVAL & INSTALLATION

Refer to Mass Air Flow (MAF) sensor.

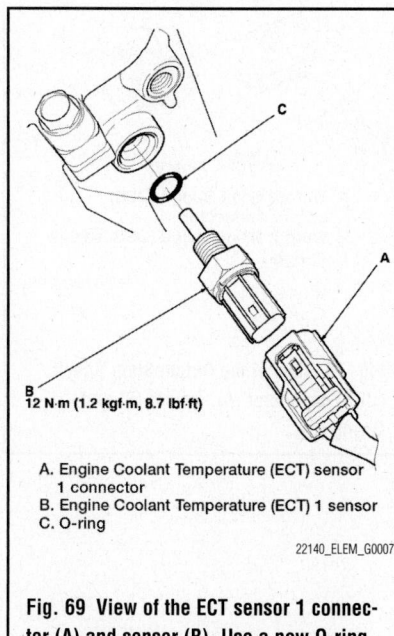

A. Engine Coolant Temperature (ECT) sensor 1 connector
B. Engine Coolant Temperature (ECT) 1 sensor
C. O-ring

22140_ELEM_G0007

Fig. 69 View of the ECT sensor 1 connector (A) and sensor (B). Use a new O-ring (C) coated with clean engine oil during installation

A. Engine Coolant Temperature (ECT) sensor 2 connector
B. Engine Coolant Temperature (ECT) 2 sensor
C. O-ring

22140_ELEM_G0008

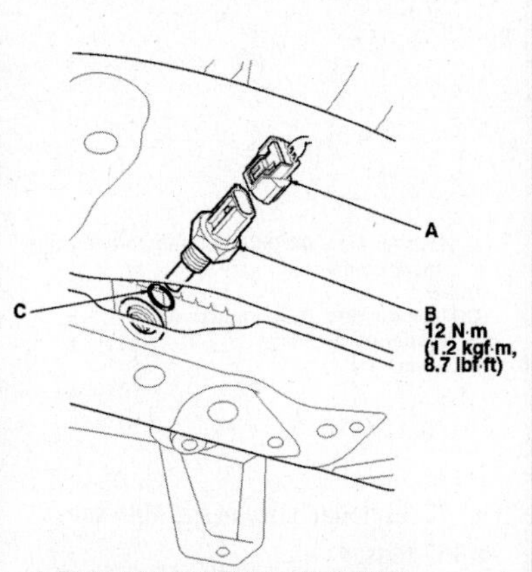

Fig. 70 View of the ECT sensor 2 connector (A) and sensor (B). Use a new O-ring (C) coated with clean engine oil during installation

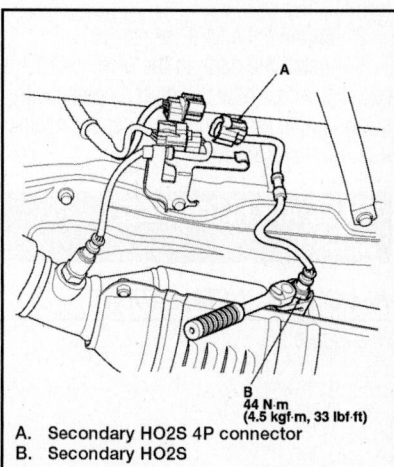

A. Secondary HO2S 4P connector
B. Secondary HO2S

22140_ELEM_G0016

Fig. 72 Disconnect the secondary HO2S 4P connector (A), then remove the secondary HO2S (B)

KNOCK SENSOR (KS)

REMOVAL & INSTALLATION

See Figure 73.

1. Disconnect the Knock Sensor (KS) connector (A).
2. Remove the KS (B).
3. Install the parts in the reverse order of removal. Tighten the KS to 23 ft. lbs. (32 Nm).

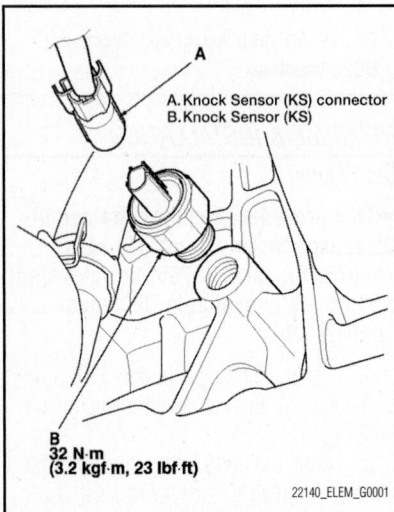

Fig. 73 View of the Knock Sensor (KS) connector (A) and KS (B)

MANIFOLD ABSOLUTE PRESSURE (MAP) SENSOR

REMOVAL & INSTALLATION

See Figure 74.

1. Disconnect the MAP sensor connector (A).
2. Remove the MAP sensor (B).
3. Install the parts in the reverse order of removal with a new O-ring (C) coated with clean engine oil. Tighten the sensor retainer to 2.4 ft. lbs. (3.4 Nm).

MASS AIR FLOW (MAF) SENSOR

REMOVAL & INSTALLATION

See Figure 75.

➥The Intake Air Temperature Sensor is integrated with the Mass Air Flow (MAF) sensor.

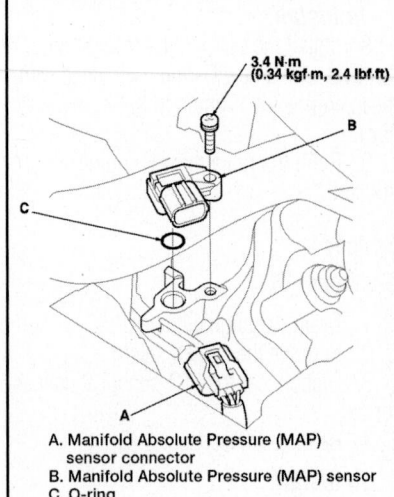

A. Manifold Absolute Pressure (MAP) sensor connector
B. Manifold Absolute Pressure (MAP) sensor
C. O-ring

Fig. 74 View of the MAP sensor connector (A) and sensor (B)

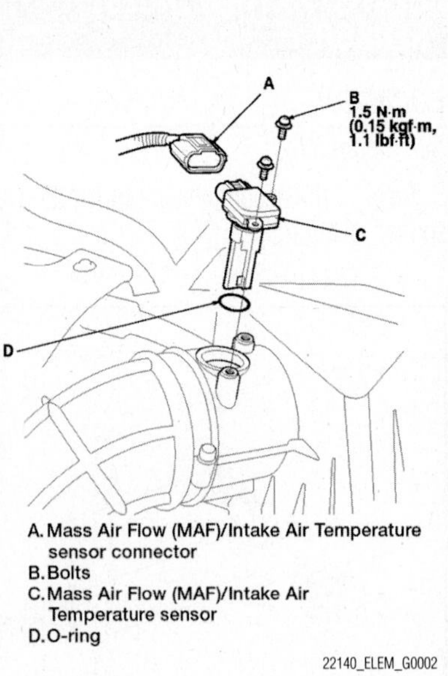

A. Mass Air Flow (MAF)/Intake Air Temperature sensor connector
B. Bolts
C. Mass Air Flow (MAF)/Intake Air Temperature sensor
D. O-ring

Fig. 75 Exploded view of the MAF sensor/IAT sensor

1. Disconnect the MAF sensor/IAT sensor connector (A).
2. Remove the bolts (B).
3. Remove the MAF sensor/IAT sensor (C).
4. Install the parts in the reverse order of removal with a new O-ring (D) coated with clean engine oil.

OUTPUT SHAFT SPEED (OSS) SENSOR

REMOVAL & INSTALLATION

With Manual Transaxle

See Figure 76.

1. Remove the air cleaner.
2. Disconnect the output shaft (countershaft) speed sensor connector (A).
3. Remove the output shaft (countershaft) speed sensor (B).
4. Install the parts in the reverse order of removal with a new O-ring (C) coated with clean engine oil. Tighten the sensor retainer to 8.7 ft. lbs. (12 Nm).

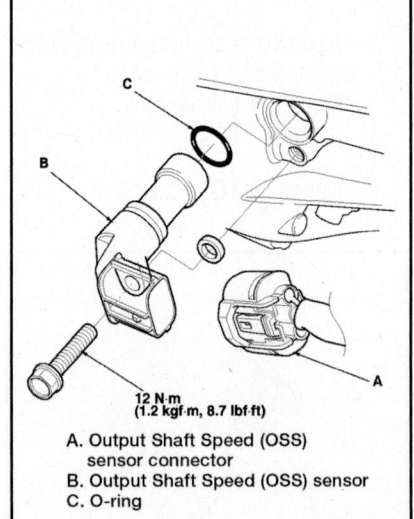

A. Output Shaft Speed (OSS) sensor connector
B. Output Shaft Speed (OSS) sensor
C. O-ring

Fig. 76 View of the Output Shaft Speed (OSS) connector (A), sensor (B) and O-ring (C)

FUEL   **GASOLINE FUEL INJECTION SYSTEM**

FUEL SYSTEM SERVICE PRECAUTIONS

Safety is the most important factor when performing not only fuel system maintenance but any type of maintenance. Failure to conduct maintenance and repairs in a safe manner may result in serious personal injury or death. Maintenance and testing of the vehicle's fuel system components can be accomplished safely and effectively by adhering to the following rules and guidelines.

• To avoid the possibility of fire and personal injury, always disconnect the negative battery cable unless the repair or test procedure requires that battery voltage be applied.

• Always relieve the fuel system pressure prior to disconnecting any fuel system component (injector, fuel rail, pressure regulator, etc.), fitting or fuel line connection. Exercise extreme caution whenever relieving fuel system pressure to avoid exposing skin, face and eyes to fuel spray. Please be advised that fuel under pressure may penetrate the skin or any part of the body that it contacts.

• Always place a shop towel or cloth around the fitting or connection prior to loosening to absorb any excess fuel due to spillage. Ensure that all fuel spillage (should it occur) is quickly removed from engine surfaces. Ensure that all fuel soaked cloths or towels are deposited into a suitable waste container.

• Always keep a dry chemical (Class B) fire extinguisher near the work area.

• Do not allow fuel spray or fuel vapors to come into contact with a spark or open flame.

• Always use a back-up wrench when loosening and tightening fuel line connection fittings. This will prevent unnecessary stress and torsion to fuel line piping.

• Always replace worn fuel fitting O-rings with new Do not substitute fuel hose or equivalent where fuel pipe is installed.

Before servicing the vehicle, make sure to also refer to the precautions in the beginning of this section as well.

RELIEVING FUEL SYSTEM PRESSURE

See Figure 77.

❋❋ CAUTION

The fuel injection system remains under pressure after the engine has been turned OFF. Properly relieve

fuel pressure before disconnecting any fuel lines. Failure to do so may result in fire or personal injury.

➡**The radio may contain a coded theft protection circuit. Always obtain the code number before disconnecting the battery.**

1. Before servicing the vehicle, refer to the precautions section.

2. Remove the glove box, then remove the PGM-FI main relay (FUEL PUMP) from the fuse/relay box. Start the engine and let it run until it stalls.

3. Turn the engine OFF.

4. Disconnect the negative battery cable.

5. Remove the fuel filler cap.

6. Remove the quick-connect fitting cover.

7. Clean any dirt from the quick-connect fitting.

8. Place a rag or shop towel over quick-connect fitting.

9. Detach the quick-connect fitting by holding the connector with one hand, then squeeze the retainer tabs with the other hand to release them from the locking pawls. Pull the connector off.

❋❋ CAUTION

Do not allow fuel spray or fuel vapors to come in contact with a spark or open flame. Keep a dry chemical fire extinguisher nearby. Never store fuel in an open container due to risk of fire or explosion.

➡**A fuel pressure gauge may be attached at the quick-connect location.**

10. Connect the quick-connect fitting, making sure the locking pawls are properly engaged,

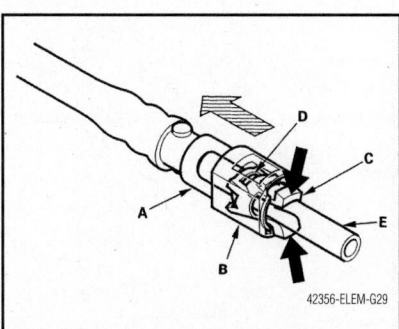

Fig. 77 Hold the quick-connect (A) connector (B) with one hand, then squeeze the retainer tabs (C) with the other hand to release them from the locking pawls (D)

11. Clean up any fuel spilled on the engine and intake manifold.

12. Install the fuel pump relay to the under dash fuel/relay box and install the glove box.

13. Install the fuel filler cap.

14. Reconnect the negative battery cable.

15. Turn the ignition **ON**, but don't start the engine. Repeat this 2 or 3 times to pressurize the fuel system. Check for fuel leaks.

16. Enter the radio security code.

FUEL FILTER

REMOVAL & INSTALLATION

See Figure 78.

➡**The fuel filter should be replaced whenever the fuel pressure drops below 48 psi, after making sure that the fuel pump and fuel pressure regulator are okay.**

1. Before servicing the vehicle, refer to the precautions section.

2. Relieve the fuel system pressure.

3. Remove or disconnect the following:
• Negative battery cable
• Fuel pump
• Fuel filter carrier (A)
• Fuel filter

To install:

4. Install or connect the following:
• Fuel filter
• Fuel lines
• New gasket (B)

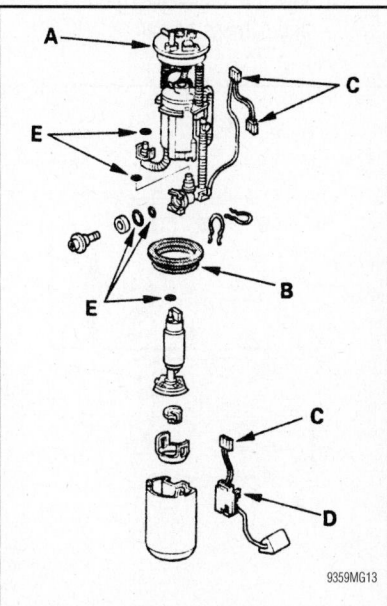

Fig. 78 Exploded view of the fuel filter mounting

- New o-rings (E)
- Connectors (C)
- Sending unit (D)

5. Start the engine and check for leaks.

FUEL PUMP

REMOVAL & INSTALLATION

1. Before servicing the vehicle, refer to the precautions section.

2. Relieve the fuel system pressure.

3. Remove or disconnect the following:
- Negative battery cable
- Fuel filler cap
- Center console, then both track floor covers and sill trims.

4. Fold back the floor covering until you can get to the access panel
- Access panel from the floor
- Fuel pump connector
- Fuel supply and return line quick-connect fittings
- Fuel pump locknut, using special tool No. 07XAA-001010A
- Fuel pump/sending unit assembly

5. Installation is the reverse of removal.

FUEL RAIL & INJECTORS

REMOVAL & INSTALLATION

See Figure 79.

1. Before servicing the vehicle, refer to the precautions section.

2. Relieve the fuel system pressure.

3. Remove or disconnect the following:
- Negative battery cable
- Engine cover
- Injector connectors, ground cable and harness holder

- Fuel line quick-connect fittings
- Fuel rail mounting nuts
- Injector clip(s) from the injector(s)
- Fuel injectors from the fuel rail

To install:

4. Install or connect the following:
- Injectors to the fuel rail with new O-rings coated with clean engine oil.
- Injector clips
- Injectors in the injector base
- Fuel rail and injector assembly. Tighten the nuts to 16 ft. lbs. (22 Nm).
- Ground cable bolt
- Injector connectors
- Fuel lines
- Negative battery cable

5. Start the engine and check for leaks.

THROTTLE BODY

REMOVAL & INSTALLATION

See Figures 80, and 82.

1. Before servicing the vehicle, refer to the precautions section.

2. Relieve the fuel system pressure.

3. Disconnect the negative battery cable.

4. Remove the engine cover.

5. Detach the electrical connectors and cables from the throttle body.

6. Remove the retainers, then remove the throttle body.

7. Remove and discard the throttle body gasket.

To install:

8. Position a new throttle body gasket, then install the throttle body. Tighten the retainers to 16 ft. lbs. (22 Nm).

9. Attach the cables and electrical connectors to the throttle body.

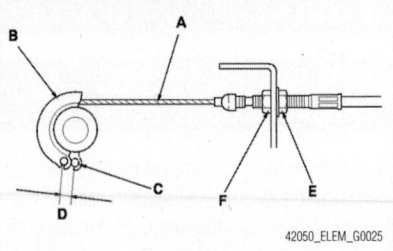

Fig. 81 View of the actuator cable (A). Measure the movement of the output linkage (B) which should first be located at the fully closed position (C). If the freeplay (D) is not within specs, loosen the locknut (E) and turn the adjusting nut (F) until the free play is correct

10. Adjust the actuator cable, as follows:
a. Make sure the actuator cable moves smoothly with no binding or sticking.
b. Measure the amount of movement of the output linkage until the engine speed starts to increase. At first, the output linkage should be located at the fully closed position. The free play should be 0.13–0.17 in. (3.25–4.25mm).
c. If the free play is not within specs, loosen the locknut, and turn the adjusting nut until the free play is as specified, then retighten the locknut.

11. Adjust the throttle cable, as follows:
a. Check cable free play at the throttle linkage. Cable free play should be ⅜–½ in. (10–12mm). If the free play is not within specifications, loosen the locknut, turn the adjusting nut until the deflection is as specified, then retighten the locknut.
b. With the cable properly adjusted, check the throttle valve to be sure it

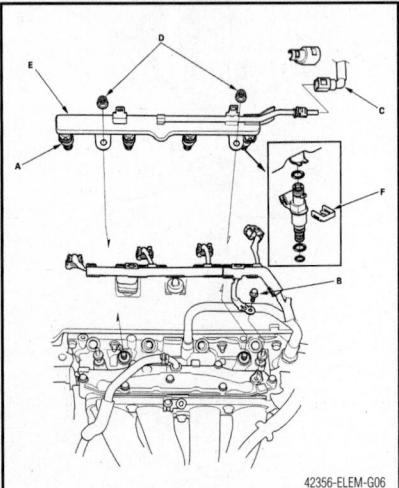

Fig. 79 Exploded view of the fuel rail (E), injectors (A) and related components

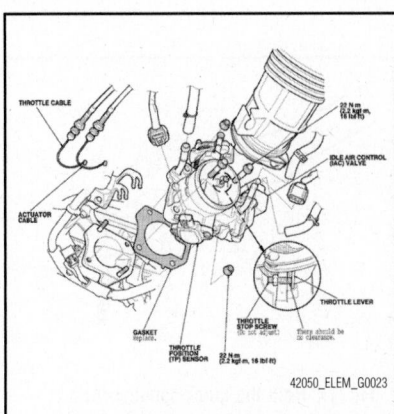

Fig. 80 Exploded view of the throttle body and related components

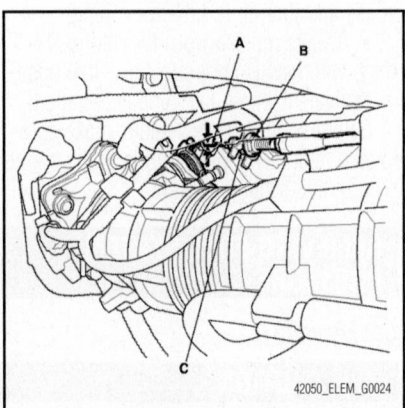

Fig. 82 Throttle cable free play (A) should be 3/8–1/2 in. (10–12mm).

opens fully when you push the accelerator pedal to the floor. Also check the throttle valve to be sure it returns to the idle position whenever you release the accelerator pedal.

12. Connect the negative battery cable.

IDLE SPEED

INSPECTION

See Figure 83.

➡**Leave the idle air control (IAC) valve connected.**

Before checking the idle speed, check these items:

• The malfunction indicator lamp (MIL) has not been reported on.
• Ignition timing
• Spark plugs
• Air cleaner
• PCV system

1. Pull the parking brake lever up.
2. Disconnect the evaporative emission (EVAP) canister purge valve connector.
3. Connect a suitable OBD II compliant scan tool to the Data Link Connector (DLC) located under the driver's side of the dashboard.
4. Start the engine. Hold the engine speed at 3,000 rpm without load (in Park or neutral) until the radiator fan comes on, then let it idle.
5. Check the idle speed without load

conditions: headlights, blower fan, radiator fan, and air conditioner off.

6. Idle speed should be:
   a. M/T: 650–750 rpm
   b. A/T: 650–750 rpm (in Park or neutral)
7. Let the engine idle for 1 minute with a high electrical load (A/C switch on, temperature set to MAX cool, blower fan on high, rear window defogger on, and headlights on high beam).
8. Idle speed should be:
   a. M/T: 670–770 rpm
   b. A/T: 670–770 rpm (in Park or neutral)
9. If the idle speed is not within specification, do the ECM/PCM idle learn procedure. The idle learn procedure must be done so the ECM/PCM can learn the engine idle characteristics.

The idle learn procedure must be performed whenever you do any of the following:
• Replace ECM/PCM.
• Reset ECM/PCM.
• Update ECM/PCM.

➡**Erasing DTCs with the HDS does not require you to do the idle learn procedure.**

• Clean or replace the throttle body.
   a. Make sure all electrical items (A/C, audio, rear window defogger, lights, etc.) are off.
   b. Reset the ECM/PCM with the HDS.

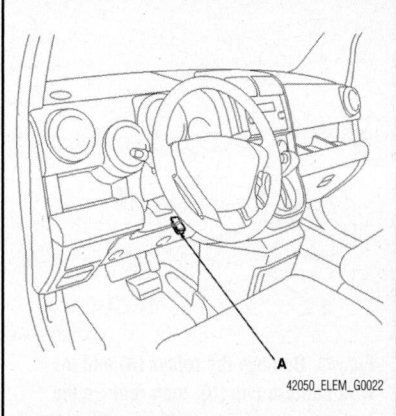

Fig. 83 The Data Link Connector (DLC) is located under the driver's side of the dashboard

   c. Turn the ignition switch ON (II), and wait 2 seconds.
   d. Start the engine. Hold the engine speed at 3,000 rpm without load (in Park or neutral) until the radiator fan comes on, or until the engine coolant temperature reaches 194 °F (90 °C).
   e. Let the engine idle for about 5 minutes with the throttle fully closed.

➡**If the radiator fan comes on, do not include its running time in the 5 minutes.**

10. Reconnect the EVAP canister purge valve connector.

HEATING & AIR CONDITIONING SYSTEM

BLOWER MOTOR

REMOVAL & INSTALLATION

See Figures 84 through 88.

1. Remove the passenger's dashboard under cover, as follows:
   a. Gently pull down the rear edge to release the clips.
   b. Pull the cover away to release the pins (B) from the holders (C).
2. Use a suitable trim panel removal tool to remove the passenger's kick panel.

✳ WARNING

Be careful not to scratch or damage the dash when removing the glove box.

3. Remove the glove box, as follows:
   a. While holding the glove box, remove the glove box stop on each side.
   b. Remove the bolts, then remove the glove box.

4. Cut the plastic cross brace in the glove box opening with diagonal cutters in the area shown. Remove and discard the plastic cross brace.

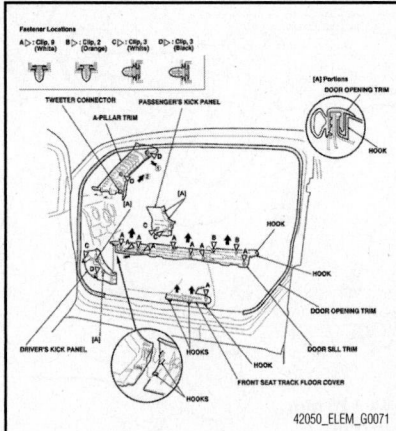

Fig. 84 Passenger side kick panel and related interior trim panels

5. Remove the relays (A) and the wire harness clip (B), then remove the bolts and the glove box frame (C).
6. Remove the ECM/PCM.

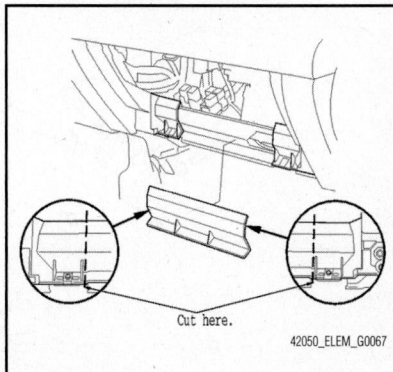

Fig. 85 Cut the plastic cross brace in the glove box opening with diagonal cutters in the area shown. Remove and discard the plastic cross brace

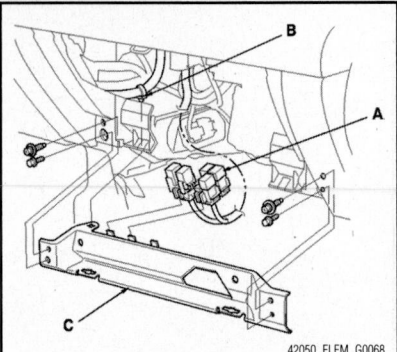

Fig. 86 Remove the relays (A) and the wire harness clip (B), then remove the bolts and the glove box frame (C)

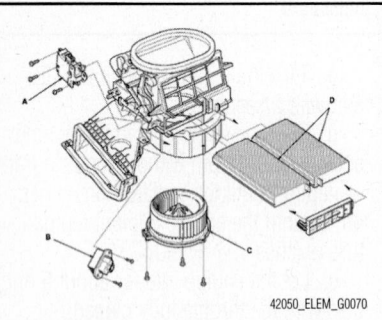

Fig. 87 The recirculation control motor (A), the power transistor (B), the blower motor (C), and the dust and pollen filters (with A/C) (D) can be replaced without removing the blower unit

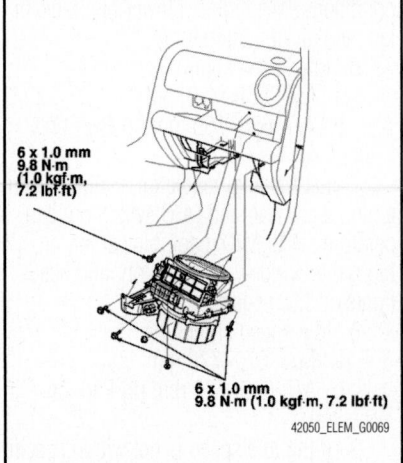

6 x 1.0 mm
9.8 N·m
(1.0 kgf·m, 7.2 lbf·ft)

6 x 1.0 mm
9.8 N·m (1.0 kgf·m, 7.2 lbf·ft)

Fig. 88 Blower unit mounting and fastener tightening specifications.

7. Disconnect the connectors from the blower motor, the power transistor, and the recirculation control motor, then remove the wire harness clips.

8. If blower motor replacement is necessary, it can be removed at this time. You do not have to remove the entire blower unit to replace the blower motor.

9. Fold the floor covering and pad back toward you. Remove the mounting bolts, the mounting nut, and the blower unit.

10. Install the unit in the reverse order of

removal. Make sure that there is no air leakage.

HEATER CORE

REMOVAL & INSTALLATION

See Figures 89 through 92.

1. Before servicing the vehicle, refer to the precautions section.

2. Disconnect the negative battery cable.

3. Drain the cooling system into a clean container for reuse.

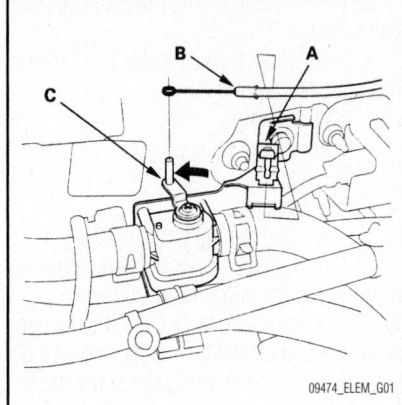

Fig. 89 Open the cable clamp (A) and disconnect the heater valve cable (B) from the heater valve arm (C). Turn the heater valve arm to the full open position

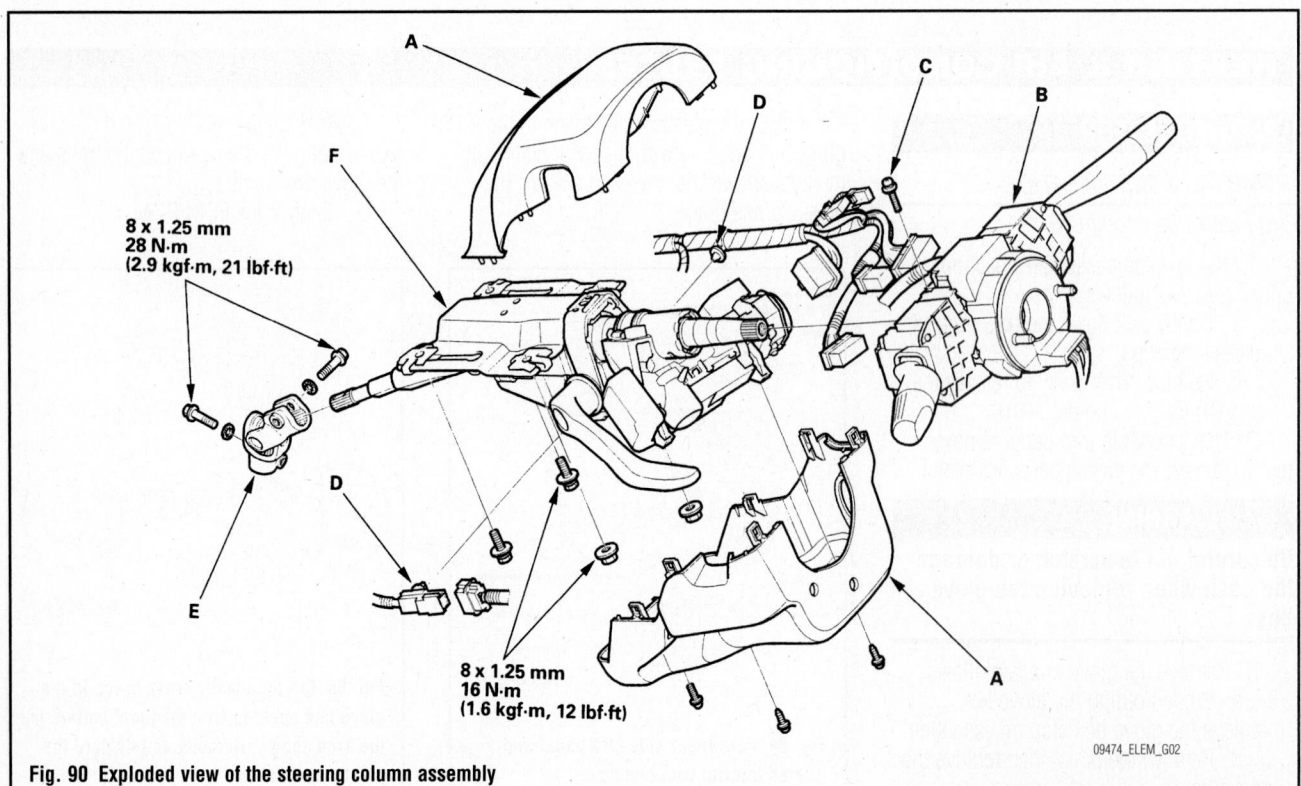

8 x 1.25 mm
28 N·m
(2.9 kgf·m, 21 lbf·ft)

8 x 1.25 mm
16 N·m
(1.6 kgf·m, 12 lbf·ft)

Fig. 90 Exploded view of the steering column assembly

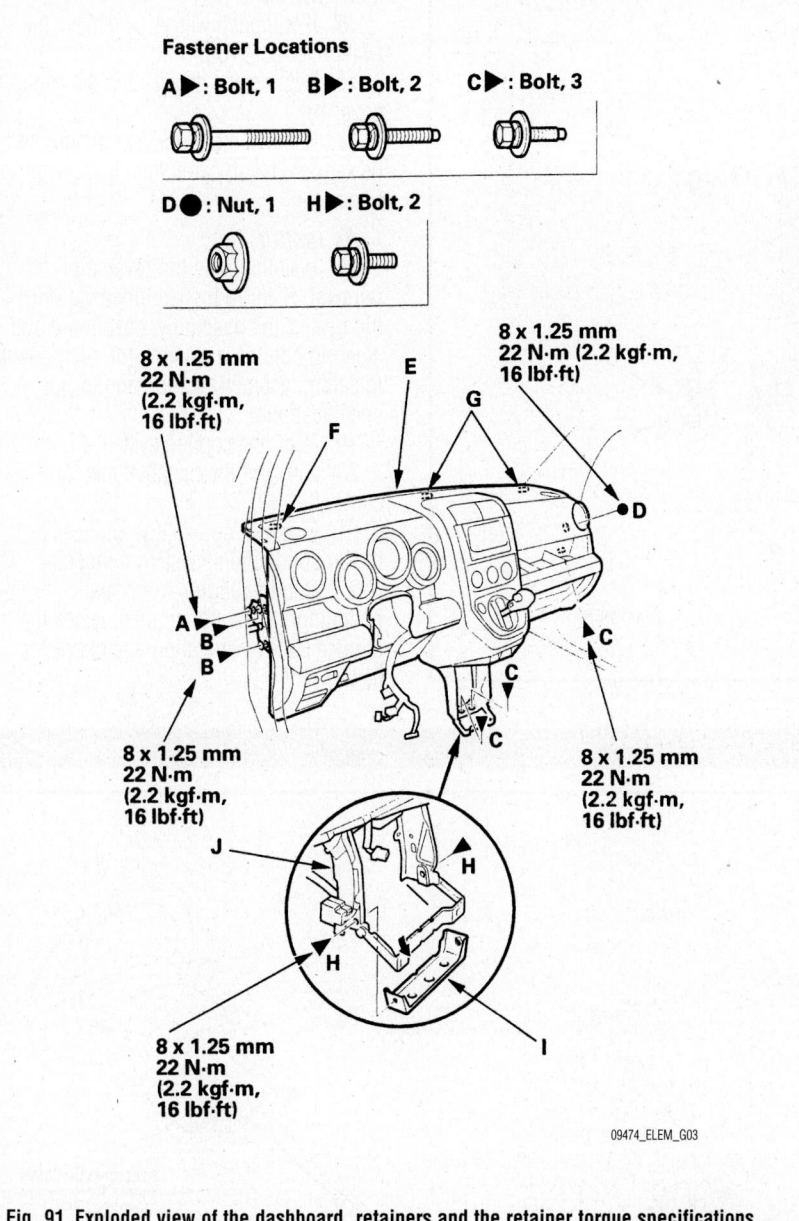

Fastener Locations

A ▶ : Bolt, 1   B ▶ : Bolt, 2   C ▶ : Bolt, 3

D ● : Nut, 1   H ▶ : Bolt, 2

8 x 1.25 mm
22 N·m
(2.2 kgf·m,
16 lbf·ft)

8 x 1.25 mm
22 N·m (2.2 kgf·m,
16 lbf·ft)

8 x 1.25 mm
22 N·m
(2.2 kgf·m,
16 lbf·ft)

8 x 1.25 mm
22 N·m
(2.2 kgf·m,
16 lbf·ft)

8 x 1.25 mm
22 N·m
(2.2 kgf·m,
16 lbf·ft)

09474_ELEM_G03

Fig. 91 Exploded view of the dashboard, retainers and the retainer torque specifications

4. Disconnect the A/C lines from the evaporator core if equipped with A/C.

5. Open the cable clamp (A) and disconnect the heater valve cable (B) from the heater valve arm (C). Turn the heater valve arm to the full open position as illustrated.

6. Remove the heater hoses from the heater core

7. Remove the mounting bolt and heater valve. Remove the nut from the heater unit being careful of any lines, hoses and wiring in the vicinity.

8. Remove the dashboard as follows:

a. Remove the driver's side lower instrument panel cover clips and remove the lower cover.

b. Remove the glove box stops from each side of the glove box.

c. Remove the glove box-to-instrument panel bolts and the glove box.

d. Remove the passenger's side lower instrument panel cover clips and remove the lower cover.

e. Remove the center lower cover clips and remove the lower cover.

f. Remove the passenger vent.

g. Remove the A–trim on both sides.

h. Remove the passenger side kick panel.

i. Remove the steering wheel.

j. Remove the steering column covers.

k. Disconnect the wiring from the combination switch and remove the assembly by removing the screw on top of the switch.

l. Disconnect the ignition switch connectors and release the wire harness clips from the column.

m. Disconnect the steering joint bolt and disconnect it from the column shaft.

n. Remove the steering column retainers and the column.

o. Control cable if equipped with an automatic transmission or shift cable if equipped with a manual transaxle.

p. Woofer, if equipped.

q. On the driver's side disconnect the following:

- Tweeter connector
- Drivers door wiring connectors
- Brake switch connector
- Clutch switch connector, if equipped
- Engine compartment harness connectors from the fuse/relay box

r. In the middle of the dashboard, disconnect the SRS control unit connector, floor harness connector and engine compartment harness connectors.

s. On the passenger side, disconnect the following:

- Passenger door wiring connectors
- Antenna lead
- PCM connectors
- Engine wire harness connectors
- Heater sub-harness connectors
- Passenger airbag connectors.
- Amplifier connectors
- Wire harness protector from the amplifier, if equipped

t. Remove any remaining harness and connector clips.

u. Dashboard bolts. Refer to the exploded view for bolt location and torque values.

v. Remove the dashboard.

9. Remove the PCM.

10. Disconnect the following connectors:

- Dashboard wiring harness
- Air mixture control motor
- Evaporator temperature sensor
- Power transistor
- Mode control motor
- Blower motor

11. Disconnect the following clips:

- Wire harness clips
- Connector clips

12. Remove the wire harness, heater duct and clip.

13. Remove the drain hose and the mounting nuts and the heater unit.

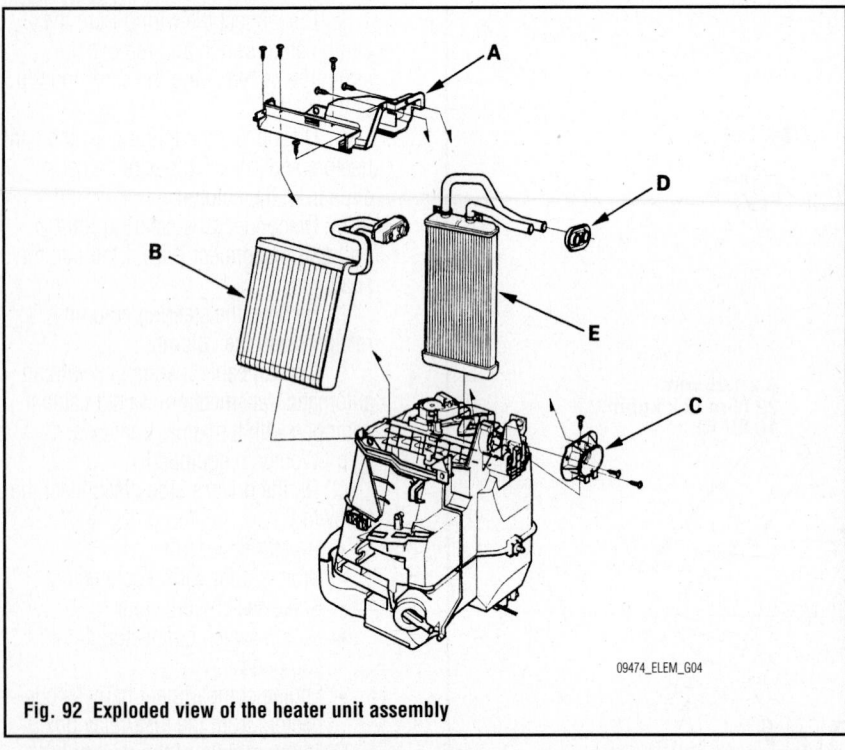

Fig. 92 Exploded view of the heater unit assembly

14. Remove the screws and the expansion valve cover (A).

15. If equipped with A/C, remove the evaporator core (B).

16. Remove the screws and the flange cover (C).

17. Remove the grommet (D) and the heater core being careful not to damage any lines.

To install:

18. Installation is the reverse of removal. Refer to the exploded views of the heater unit assembly, dashboard and steering column assembly for component location, fastener location and torque specifications.

19. Refill the cooling system.

20. Connect the negative battery cable.

21. Evacuate and charge and leak test the air conditioning system refrigerant.

22. Run the engine to normal operating temperatures; then, check the climate control operation and check for leaks.

STEERING

POWER RACK & PINION STEERING GEAR

REMOVAL & INSTALLATION

See Figure 93.

❋❋ WARNING

Do not permit the steering wheel to turn whenever the steering gear is disconnected from the steering column. Damage to the air bag wiring can result.

1. Before servicing the vehicle, refer to the precautions section.

2. Center the steering wheel and lock it in position.

3. Remove or disconnect the following:
   - Negative battery cable and wait at least 3 minutes before continuing
   - Front wheels

4. Remove the air bag and steering wheel as follows:

   a. Align the front wheels in the straight ahead position

   b. Remove the access panel from the steering wheel and disconnect the drivers airbag 4P connector.

   c. Remove the two Torx® bolts using a T30 bit.

   d. Remove the airbag.

   e. Disconnect the cruise control connector and horn switch connector.

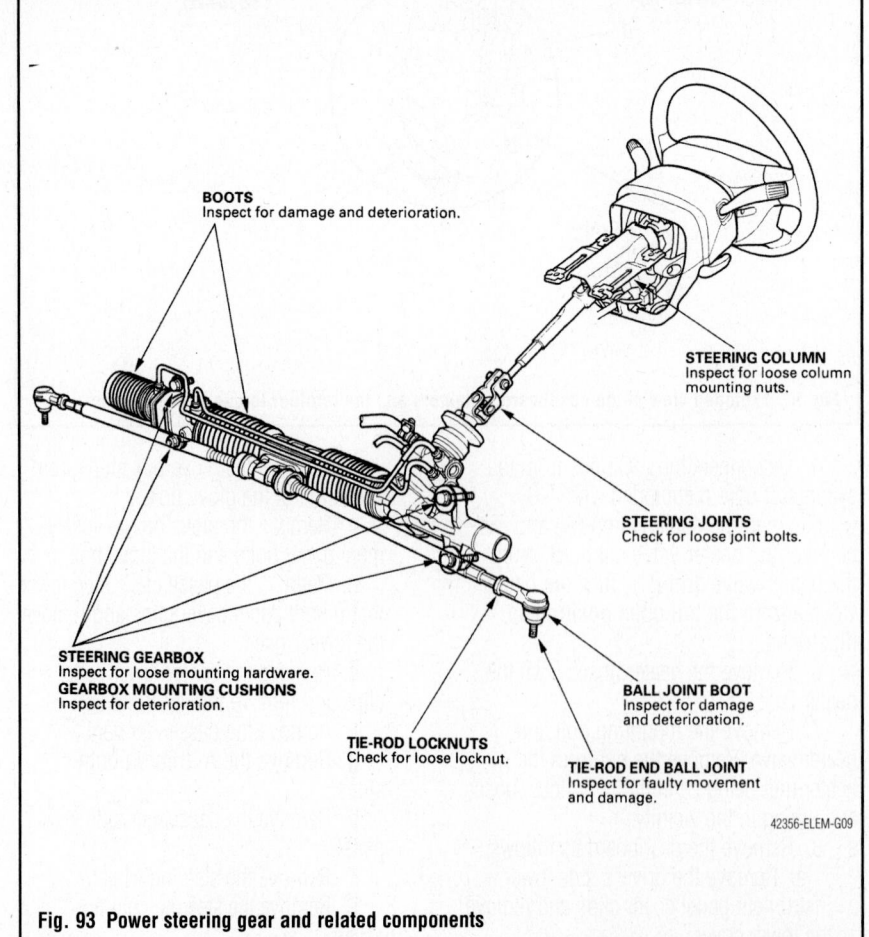

Fig. 93 Power steering gear and related components

BOOTS
Inspect for damage and deterioration.

STEERING COLUMN
Inspect for loose column mounting nuts.

STEERING JOINTS
Check for loose joint bolts.

STEERING GEARBOX
Inspect for loose mounting hardware.
GEARBOX MOUNTING CUSHIONS
Inspect for deterioration.

TIE-ROD LOCKNUTS
Check for loose locknut.

BALL JOINT BOOT
Inspect for damage and deterioration.

TIE-ROD END BALL JOINT
Inspect for faulty movement and damage.

f. Loosen the steering wheel bolt and using a suitable puller, free the steering wheel.

➡**Do not tap on the steering wheel or column shaft during removal. If you thread the puller bolts more than 5 threads into the wheel hub you will hit the cable reel and damage it. To prevent damage, insert a pair of jam nuts 5 threads up on each puller bolt.**

g. Remove the puller, steering wheel bolt and wheel.

5. Remove or disconnect the following:

- Driver's side dashboard lower cover and undercover
- Steering joint bolts and disconnect the joint by moving the joint towards the column
- Center pin from the top of the pinion shaft, if equipped and discard the pin
- Tie rod ends
- Power steering heat baffle plate
- Engine wiring harness clamp and clip from their brackets
- Loosen the adjustable hose clamp and disconnect the return hose
- Loosen the 14mm flare nut and disconnect the feed line
- Open the hose holders on the return hose and remove the clamp
- Power steering pressure switch connector
- Feed line on the power steering line mounting bracket and set it aside
- Body stiffener
- Left, then right side flange bolts and washers
- Mounting brackets

6. Lower the unit so the pinion shaft points upward. Remove the pinion shaft grommet. The steering gear is removed through the driver's side.

To install:

7. Installation of the steering gear is the reverse of removal. Observe the following torques:

- Mounting bracket and side flange bolts: 46 ft. lbs. (62 Nm)
- Supply line flare nut: 27 ft. lbs. (37 Nm)
- Tie rod ball stud nuts: 40 ft. lbs. (54 Nm)
- Steering joint bolts: 21 ft. lbs. (28 Nm)

8. Install the steering wheel and air bag as follows:

a. First make sure the front wheels are aligned straight ahead. Center the cable reel by rotating the cable reel clockwise until it stops, then rotate it counterclockwise about 2 ½ turns. The arrow mark on the cable reel should point straight up.

b. Position the tabs on the turn signal canceling sleeve, install the steering wheel and make sure the wheel hub engages the pins of the cable reel and tabs of the canceling sleeve. Do not tap on the wheel or column.

c. Install the steering wheel bolt and tighten to 29 ft. lbs. (39 Nm). Connect the horn switch, cruise control switch and ensure the wiring is routed correctly and properly secured.

d. Install the driver's side air bag and tighten the Torx ® bolts to 7 ft. lbs. (9 Nm).

e. Connect the cable reel to the airbag 4P connector and install the access panel.

f. Connect the negative battery cable.

g. Turn the ignition switch on and ensure the airbag light illuminates for about 6 seconds and then goes out.

h. Ensure proper operation of the horn and cruise control.

POWER STEERING PUMP

REMOVAL & INSTALLATION

See Figures 94 through 96.

1. Place a suitable container under the vehicle.

2. Drain the power steering fluid from the reservoir.

3. Remove the drive belt (A) from the pump pulley, as follows:

➡**This procedure requires the use of a special Belt tension release tool, Snap-on YA9317 or equivalent tool.**

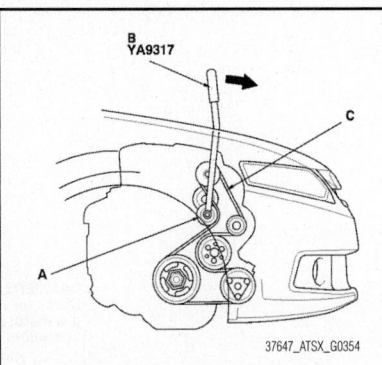

Fig. 94 To remove the belt, move the auto-tensioner (A) with the belt tension release tool (B) in the direction shown to relieve tension from the drive belt, and remove the drive belt

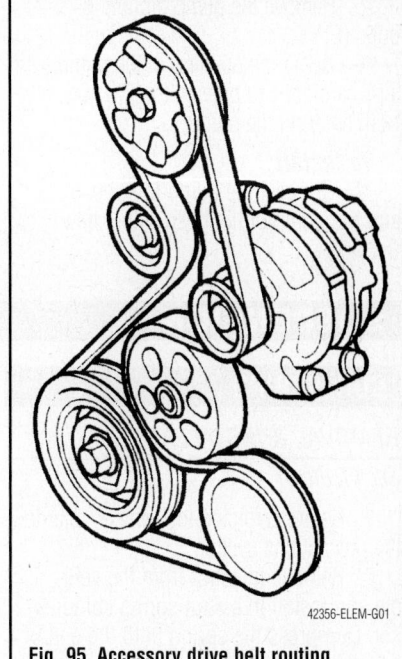

Fig. 95 Accessory drive belt routing

a. Move the auto-tensioner (A) with the belt tension release tool (B) in the direction shown to relieve tension from the drive belt, and remove the drive belt.

b. Install the new belt in the reverse order of removal.

4. Cover the auto-tensioner, alternator, and A/C compressor with several shop towels to protect them from spilled power steering fluid. Disconnect the pump inlet hose (B) and the pump outlet hose (C) from the pump (D), and plug them. Take care not to spill the fluid on the body or parts. Wipe off any spilled fluid at once. Do not turn the steering wheel with the pump removed.

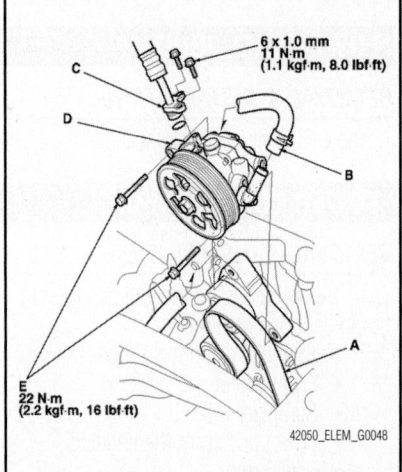

Fig. 96 View of the driver belt (A), power steering pump inlet hose (B), outlet hose (C), power steering pump (D) and mounting bolts (E)

5. Remove the pump mounting bolts (E).

6. Cover the opening of the pump with a piece of tape to prevent foreign material from entering the pump.

To install:

7. Connect the pump inlet hose and the pump outlet hose with a new O-ring.

8. Loosely install the pump in the pump bracket with the mounting bolts, then tighten the pump fittings securely.

✳✳ WARNING

Do not get power steering fluid or grease on the auto-tensioner, alternator, A/C compressor, and drive belt or pulley faces. Clean off any fluid or grease before installation.

9. Install the drive belt. Make sure that the belt is properly positioned on the pulleys.

10. Tighten the pump mounting bolts to 16 ft. lbs. (22 Nm).

11. Fill the power steering fluid reservoir to the upper level line.

SUSPENSION

FRONT SUSPENSION

COIL SPRING

REMOVAL & INSTALLATION

See Figure 97.

1. Before servicing the vehicle, refer to the precautions section.

2. Remove the strut from the vehicle and install in a strut spring compressor. Compress the spring until the end of the spring comes away from the spring seat.

3. Remove the upper strut mount, spring seat and related components.

4. Remove the coil spring from the strut spring compressor.

To install:

➡ **Use a new self-locking nut.**

5. Compress the spring and position the strut so that the end of the spring aligns with the notch in the spring seat.

6. Install the upper strut mounting components and tighten the nut to 33 ft. lbs. (44 Nm).

7. Install the strut to the vehicle.

8. Check the wheel alignment and adjust as necessary.

LOWER BALL JOINT

REMOVAL & INSTALLATION

The ball joint is not replaceable.

LOWER CONTROL ARM

REMOVAL & INSTALLATION

1. Before servicing the vehicle, refer to the precautions section.

2. Remove or disconnect the following:
   - Front wheel
   - Stabilizer link
   - Lower arm from the knuckle
   - Lower arm

To install:

3. Install all suspension components and fasteners and hand tighten them.. Place a jack under the suspension, raise the sus-

pension with the jack and load the jack with the vehicle weight.

4. Installation is the reverse of removal. Observe the following torques:
   - Lower arm bolts: 61 ft. lbs. (83 Nm)

- Ball stud nut: 51 ft. lbs. (69 Nm). Install the cotter pin into the ball joint from the inside to the outside of the vehicle.
- Stabilizer link: 29 ft. lbs. (39 Nm)

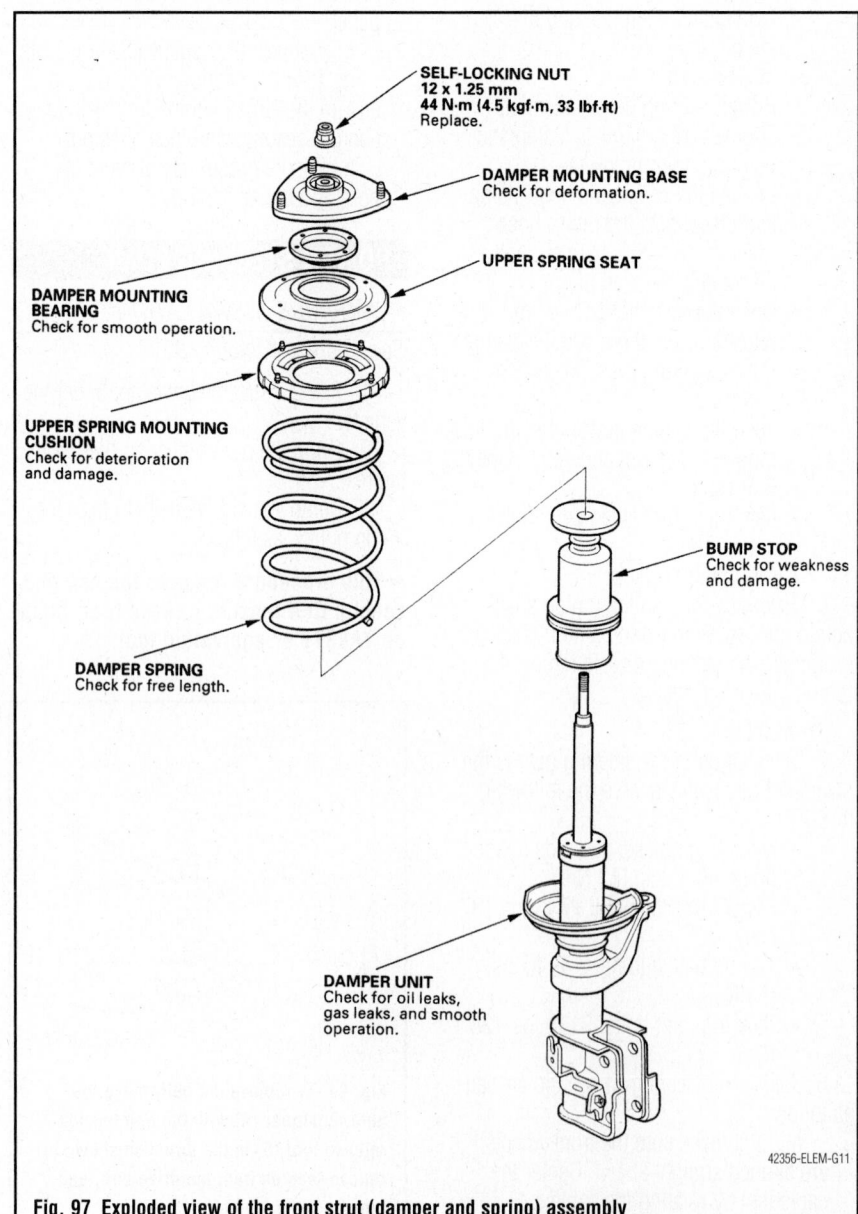

Fig. 97 Exploded view of the front strut (damper and spring) assembly

SELF-LOCKING NUT
12 x 1.25 mm
44 N·m (4.5 kgf·m, 33 lbf·ft)
Replace.

DAMPER MOUNTING BASE
Check for deformation.

UPPER SPRING SEAT

DAMPER MOUNTING BEARING
Check for smooth operation.

UPPER SPRING MOUNTING CUSHION
Check for deterioration and damage.

DAMPER SPRING
Check for free length.

BUMP STOP
Check for weakness and damage.

DAMPER UNIT
Check for oil leaks, gas leaks, and smooth operation.

42356-ELEM-G11

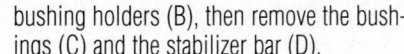

CONTROL ARM BUSHING REPLACEMENT

The lower control arm front inner bushing and the damper fork bushing are serviced with the control arm as an assembly.

STABILIZER BAR

REMOVAL & INSTALLATION

See Figures 98 and 99.

1. Raise the front of the vehicle, and support it with safety stands in the proper locations.
2. Remove the front wheels.
3. Disconnect the stabilizer links from the stabilizer bar on the right and left sides. Refer to the Stabilizer Link procedure in this section.
4. Remove the flange bolts (A) and

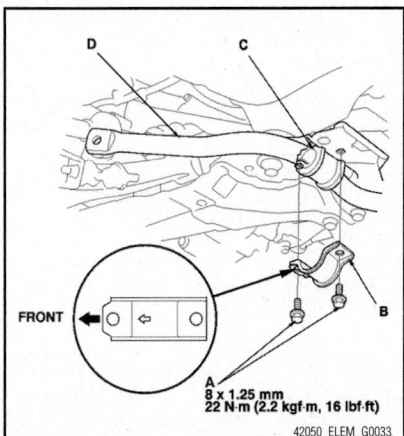

Fig. 98 Remove the flange bolts (A) and bushing holders (B), then remove the bushings (C) and the stabilizer bar (D)

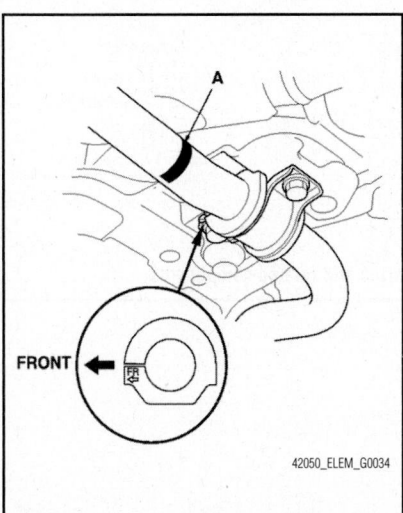

Fig. 99 The paint mark (A) indicates the right side of the stabilizer bar

bushing holders (B), then remove the bushings (C) and the stabilizer bar (D).

To install:

5. Install the stabilizer bar in the reverse order of removal, and note these items:

   a. Note the right and left direction of the stabilizer bar. The paint mark (A) on the stabilizer bar shows the right side.

   b. Do not set the bushings on the bent or curved part of the stabilizer bar.

   c. Note the fore/aft direction of the bushing holders.

   d. Refer to stabilizer link removal/installation to connect the stabilizer bar to the links.

STABILIZER LINK

REMOVAL & INSTALLATION

See Figures 100 through 102.

1. Raise the front of the vehicle, and support it with safety stands in the proper locations.
2. Remove the front wheel.
3. Remove the self-locking nut (A) and flange nut (B) while holding the respective joint pin (C) with a hex wrench (D), and remove the stabilizer link (E).

To install:

4. Install the stabilizer link (A) on the stabilizer bar (B) and lower arm (C) with the joint pins (D) set at the center of their range of movement.

➥**The left stabilizer has a yellow paint mark (E), while the right stabilizer link has a white paint mark.**

5. Install a new self-locking nut and flange nut, and lightly tighten them.

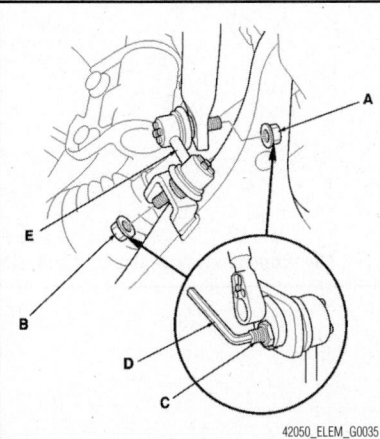

Fig. 100 Remove the self-locking nut (A) and flange nut (B) while holding the respective joint pin (C) with a hex wrench (D), and remove the stabilizer link (E)

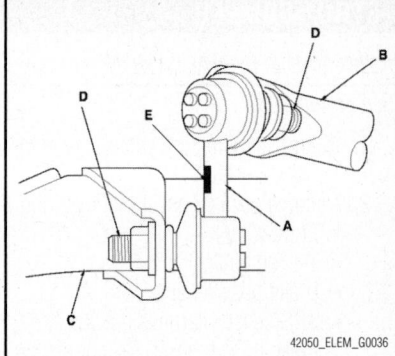

Fig. 101 View of the stabilizer link (A), stabilizer bar (B), lower arm (C), joint pins (D) and left stabilizer yellow paint mark (E)

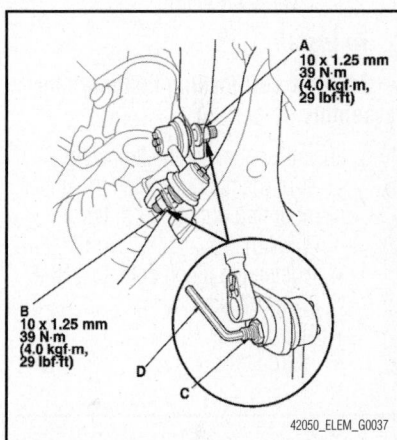

Fig. 102 View of the self-locking nut (A) and flange nut (B), joint pin (C) and hex wrench (D)

➥**Use a new self-locking nut during installation.**

✳✳ WARNING

Do not place the jack against the ball joint pin.

6. Tighten the self-locking nut (A) and flange nut (B) to 29 ft. lbs. (39 Nm) while holding the respective joint pin (C) with a hex wrench (D).
7. Reinstall the front wheel and test-drive the vehicle.
8. After 5 minutes of driving, retighten the self-locking nut to 29 ft. lbs. (39 Nm).

STEERING KNUCKLE

REMOVAL & INSTALLATION

Refer to the Wheel Bearing Removal & Installation procedure.

STRUT/DAMPER

REMOVAL & INSTALLATION

See Figure 103.

1. Before servicing the vehicle, refer to the precautions section.
2. Remove or disconnect the following:
   - Front wheel
   - Tie rod end
   - Brake hose retainer
   - ABS sensor harness bracket and brake hose bracket. Do not disconnect the wheel sensor connector.
   - Pinch bolts from the damper, while holding the nuts
   - Flange nuts from the top of the damper
   - Strut (damper), after lowering the lower control arm

To install:

➡**Use new self-locking fasteners for assembly.**

3. Install or connect the following:
   - Strut (damper). Tighten the upper mounting nuts to 33 ft. lbs. (44 Nm).
   - Tighten the pinch bolts to 116 ft. lbs. (157 Nm)
   - ABS sensor
   - Tie rod end

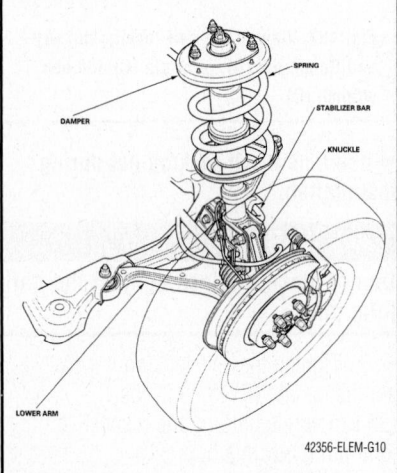

Fig. 103 Front suspension components

- Brake hose retainer
- Front wheel

WHEEL BEARINGS

ADJUSTMENT

The wheel bearings are sealed units and are not adjustable.

REMOVAL & INSTALLATION

See Figure 104.

1. Before servicing the vehicle, refer to the precautions section.
2. Remove or disconnect the following:
   - Front wheel
   - Brake hose bracket
   - Brake caliper and rotor. Forcing screws are needed to remove the rotor.
   - Spindle nut
   - ABS sensor
   - Stabilizer link
   - Lower arm from the knuckle
   - Strut-to-knuckle bolts
   - Steering hub/knuckle assembly
3. Press the hub from the knuckle. The bearings and races can now be pressed out and replaced.

➡**With ABS, install the bearing with the magnetic encoder (brown color) toward the inside of the knuckle.**

4. Observe the following torques:
   - Strut bolts: 116 ft. lbs. (157 Nm)
   - Ball stud nuts: 51 ft. lbs. (69 Nm)
   - Stabilizer bar link: 29 ft. lbs. (39 Nm)
   - Spindle nut: 181 ft. lbs. (245 Nm)

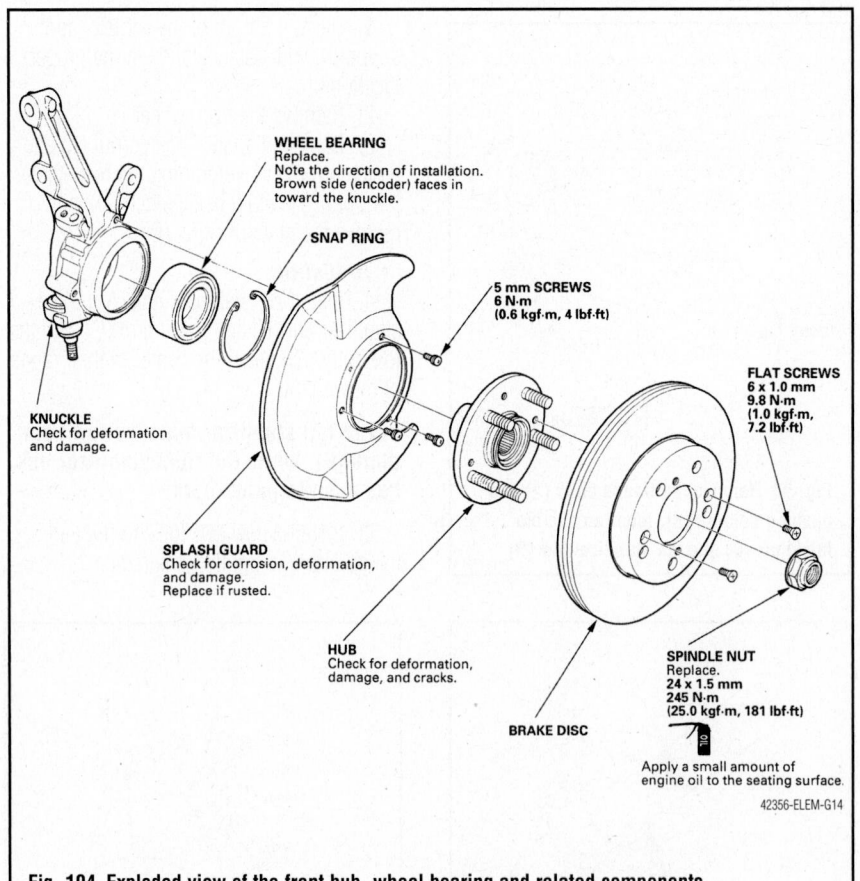

WHEEL BEARING
Replace.
Note the direction of installation.
Brown side (encoder) faces in toward the knuckle.

SNAP RING

5 mm SCREWS
6 N·m
(0.6 kgf·m, 4 lbf·ft)

FLAT SCREWS
6 x 1.0 mm
9.8 N·m
(1.0 kgf·m, 7.2 lbf·ft)

KNUCKLE
Check for deformation and damage.

SPLASH GUARD
Check for corrosion, deformation, and damage.
Replace if rusted.

HUB
Check for deformation, damage, and cracks.

BRAKE DISC

SPINDLE NUT
Replace.
24 x 1.5 mm
245 N·m
(25.0 kgf·m, 181 lbf·ft)
Apply a small amount of engine oil to the seating surface.

42356-ELEM-G14

Fig. 104 Exploded view of the front hub, wheel bearing and related components

SUSPENSION

COIL SPRING

REMOVAL & INSTALLATION

See Figure 104.

1. Before servicing the vehicle, refer to the precautions section.
2. Remove the strut from the vehicle and install in a strut spring compressor. Compress the spring until the end of the spring comes away from the spring seat.
3. Remove or disconnect the following:
   - Upper strut mount, spring seat and related components
   - Coil spring from the strut spring compressor

To install:

→**Use a new self-locking nut.**

4. Compress the spring and position the strut so that the end of the spring aligns with the notch in the spring seat.

5. Install or connect the following:
   - Upper strut mounting components and tighten the nut to 22 ft. lbs. (29 Nm).
   - Strut to the vehicle
6. Check the wheel alignment and adjust as necessary.

STRUT/DAMPER

REMOVAL & INSTALLATION

See Figure 106.

1. Before servicing the vehicle, refer to the precautions section.
2. Support the vehicle under the lower control arm.
3. Remove or disconnect the following:
   - Rear wheel
   - Flange bolt from the bottom of the damper (strut)

REAR SUSPENSION

   - Evaporative emission (EVAP) canister bolts, and loosen the EVAP canister mounting (left side only)
   - Interior access panel, if necessary
   - Flange nuts from the top of the damper in the cargo area
   - Strut

To install:

4. Install or connect the following:
   - Strut. Position the damper mounting base so the indent mark is toward the inside of the vehicle,
   - Upper flange nuts, hand-tight only
   - Bottom flange bolt, hand-tight only
5. With the suspension raised with a jack to load it with the vehicles weight, tighten the bottom bolt to 69 ft. lbs. and the top nuts to 54 ft. lbs. (74 Nm).
   - Interior access panel, if necessary
   - EVAP canister mounting bolts
   - Rear wheel

UPPER BALL JOINT

REMOVAL & INSTALLATION

The upper ball joints are replaced with the upper control arms as an assembly.

UPPER CONTROL ARM

REMOVAL & INSTALLATION

1. Before servicing the vehicle, refer to the precautions section.
2. Support the lower control arm assembly with a floor jack.

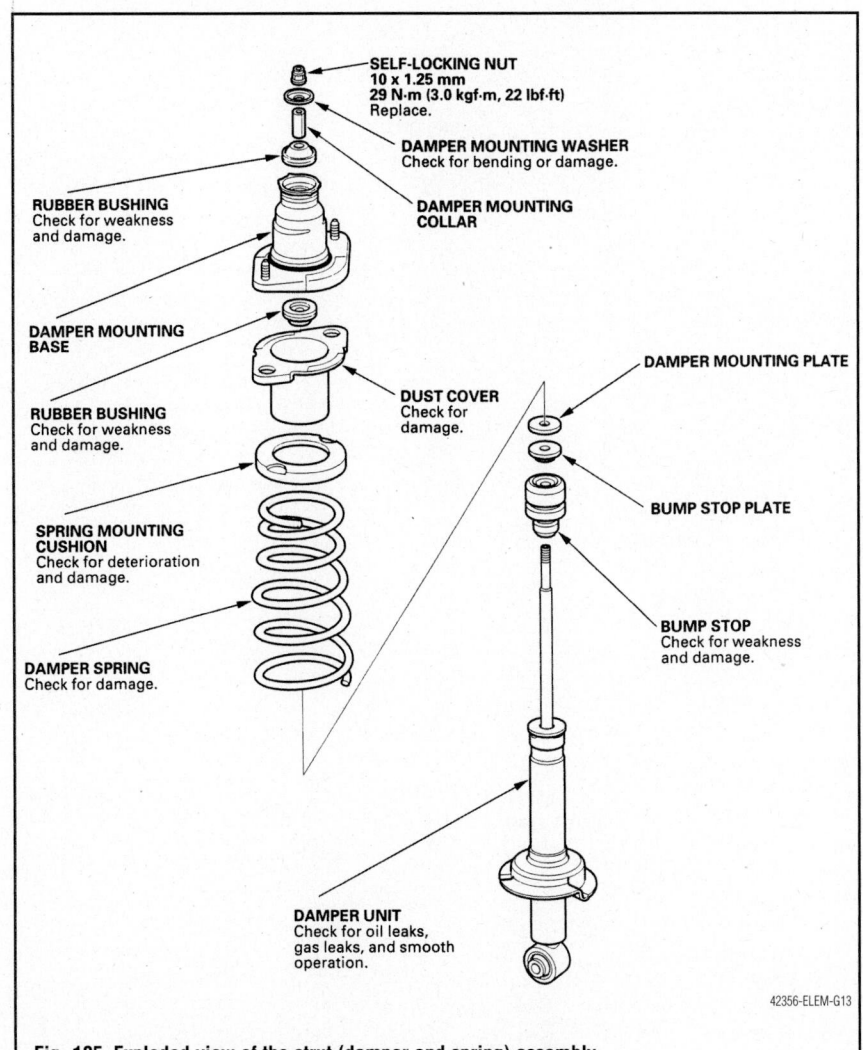

SELF-LOCKING NUT
10 x 1.25 mm
29 N·m (3.0 kgf·m, 22 lbf·ft)
Replace.

DAMPER MOUNTING WASHER
Check for bending or damage.

DAMPER MOUNTING COLLAR

RUBBER BUSHING
Check for weakness and damage.

DAMPER MOUNTING BASE

RUBBER BUSHING
Check for weakness and damage.

DAMPER MOUNTING PLATE

DUST COVER
Check for damage.

BUMP STOP PLATE

SPRING MOUNTING CUSHION
Check for deterioration and damage.

BUMP STOP
Check for weakness and damage.

DAMPER SPRING
Check for damage.

DAMPER UNIT
Check for oil leaks, gas leaks, and smooth operation.

42356-ELEM-G13

Fig. 105 Exploded view of the strut (damper and spring) assembly

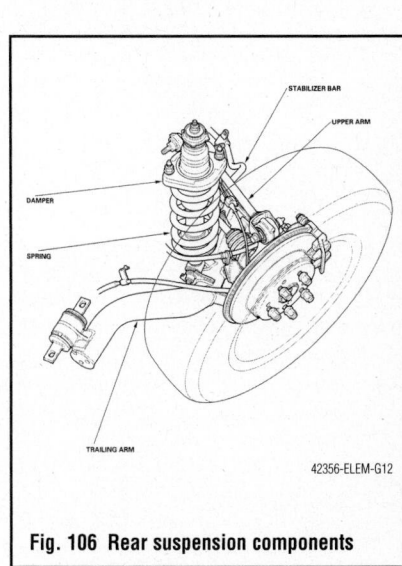

STABILIZER BAR

UPPER ARM

DAMPER

SPRING

TRAILING ARM

42356-ELEM-G12

Fig. 106 Rear suspension components

3. Remove or disconnect the following:
- Wheel speed sensor harness bracket, if equipped
- Flange bolts and the control arm

To install:

4. Install all suspension components and fasteners and hand tighten them.. Place a jack under the trailing arm, raise the suspension with the jack and load the jack with the vehicle weight.

5. Tighten the upper control arm flange bolts to 69 ft. lbs. (93 Nm).

6. Clean the wheel, mating surface of the brake disc or drum and inside of the wheel.

7. Check and adjust the wheel alignment as needed.

CONTROL ARM BUSHING REPLACEMENT

The upper control arm bushings are serviced with the upper control arm as an assembly.

WHEEL BEARINGS

ADJUSTMENT

The wheel bearings are sealed units and are not adjustable.

REMOVAL & INSTALLATION

1. Before servicing the vehicle, refer to the precautions section.

2. Remove or disconnect the following:
- Rear wheel
- Brake caliper
- Rotor
- Spindle nut
- Axle shaft (4wd)
- Parking brake shoes
- Parking brake cable
- Wheel sensor, if equipped

3. Support the trailing arm.

4. Remove or disconnect the following:
- Upper arm from the knuckle

5. Matchmark the trailing arm cam adjusting bolt and cam. Remove the bolt. Discard the nut.

6. Remove the flange bolt.

7. Remove the knuckle assembly.

8. Press the hub from the knuckle. The bearings and races can now be pressed out and replaced.

To install:

9. Installation is the reverse of the removal procedure.

➡️ **With ABS, install the bearing with the magnetic encoder (brown color) toward the inside of the knuckle.**

10. Observe the following torques:
- Flange bolt: 69 ft. lbs. (93 Nm)
- Cam bolts: 43 ft. lbs. (59 Nm)
- Spindle nut: 134 ft. lbs. (181 Nm)
- Caliper mounting bolts: 41 ft. lbs. (55 Nm)

HONDA

Fit

12

BRAKES 12-9

**ANTI-LOCK BRAKE
SYSTEM (ABS)** 12-9
General Information...................12-9
Precautions 12-9
Wheel Speed Sensors 12-9
Removal & Installation..........12-9
**BLEEDING THE BRAKE
SYSTEM** 12-11
Bleeding Procedure 12-11
FRONT DISC BRAKES 12-11
Brake Calipers 12-11
Removal & Installation...12-11
Brake Pads 12-11
Removal & Installation........12-11
PARKING BRAKE.............12-14
Adjustments 12-14
Cables 12-14
Cables 12-14
Removal & Installation........12-14
REAR DRUM BRAKES........12-13
Brake Drums 12-13
Removal & Installation........12-13
Brake Shoes 12-13
Removal & Installation........12-13

CHASSIS ELECTRICAL 12-14

**AIR BAG (SUPPLEMENTAL
RESTRAINT SYSTEM)** 12-14
Arming the System..................12-16
Disarming the System 12-16
General Information.................12-14
Service Precautions 12-14

DRIVE TRAIN 12-16

Automatic Transaxle 12-16
Filter Replacement 12-16
Clutch Hydraulic System
Bleeding..............................12-16
Bleeding Procedure 12-16
Clutch Master Cylinder 12-16
Removal & Installation........12-16
Front Driveshaft....................12-18
CV-boots Inspection 12-19
Removal & Installation........12-18

Slave (Release) Cylinder 12-18
Removal & Installation........12-18

ENGINE COOLING 12-19

Engine Coolant......................12-20
Drain & Refill....................12-20
Engine Cooling Fan................12-19
Removal & Installation........12-19
Radiator............................12-21
Removal & Installation........12-21
Radiator & A/C Condenser
Fan, Motor & Shroud...........12-19
Removal & Installation........12-19
Thermostat 12-22
Removal & Installation........12-22
Water Pump 12-22
Removal & Installation........12-22

ENGINE ELECTRICAL 12-23

BATTERY SYSTEM...........12-23
Battery 12-23
Removal & Installation........12-23
Battery Disconnection/
Reconnection 12-23
Disconnection Procedure....12-23
Reconnection Procedure.....12-23
CHARGING SYSTEM 12-24
Alternator 12-24
Removal & Installation........12-24
IGNITION SYSTEM 12-24
Firing Orders.........................12-24
Ignition Coils & Spark
Plugs.................................12-24
Removal & Installation........12-24
Ignition Timing......................12-24
Inspection & Adjustment 12-24
STARTING SYSTEM 12-25
Starter 12-25
Removal & Installation........12-25

ENGINE MECHANICAL......12-25

Accessory Drive Belt System...12-25
Adjustment 12-25
Belt Routings 12-25
Inspection 12-25
Removal & Installation........12-26

Air Cleaner 12-26
Removal & Installation........12-26
Camshaft 12-26
Removal & Installation........12-26
Camshaft Front Case, Chain &
Sprockets 12-28
Removal & Installation........12-28
Camshaft Front Cover Seal 12-30
Removal & Installation........12-30
Camshaft Sprocket 12-27
Removal & Installation........12-27
Crankshaft Pulley 12-30
Removal & Installation........12-30
Crankshaft Rear Cover &
Seal.................................12-31
Removal & Installation........12-31
Cylinder Head 12-32
Removal & Installation........12-32
Cylinder Head Covers 12-33
Removal & Installation........12-33
Engine Oil & Filter 12-33
Oil & Filter Change.............12-33
Oil Level Check.................12-33
Front Cover & Seal 12-34
Removal & Installation........12-34
Intake Manifold/Chamber
Assembly 12-34
Removal & Installation........12-34
Oil Pan 12-34
Removal & Installation........12-34
Oil Pump 12-35
Removal & Installation........12-35
Pistons & Rings 12-36
Positioning 12-36
Rear Main Seal 12-36
Removal & Installation........12-36
Rocker Arms........................12-36
Removal & Installation........12-36
Valve Lash (Valve Clearance) ..12-37
Adjustment 12-37

**ENGINE PERFORMANCE &
EMISSION CONTROLS** 12-38

Camshaft Position (CMP)
Sensor 12-38
Location...........................12-38
Removal & Installation........12-38

Crankshaft Position (CKP)
Sensor 12-38
CKP Pattern Clear/CKP
Pattern Learn 12-38
Location 12-38
Removal & Installation........12-38
Engine Control Module/
Powertrain Control Module
(ECM/PCM) 12-38
ECM/PCM Idle Learn
Procedure 12-40
ECM/PCM Update 12-39
Removal & Installation........12-38
Engine Coolant Temperature
(ECT) Sensors....................12-40
Removal & Installation........12-40
Heated Oxygen Sensor
(HO2S)..............................12-40
Location 12-40
Removal & Installation........12-40
Input Shaft (Mainshaft)
Speed Sensor......................12-41
Location 12-41
Removal & Installation........12-41
Knock Sensor (KS).................12-42
Location 12-42
Removal & Installation........12-42
Maintenance Minder
Indicator...........................12-42
Resetting Individual
Maintenance Item 12-42
Resetting the Maintenance
Minder..........................12-42
Manifold Absolute
Pressure (MAP) Sensor 12-42
Location 12-42
Removal & Installation........12-42
Mass Airflow Sensor/Intake
Air Temperature (MAF/IAT) ...12-42
Location 12-42
Removal & Installation........12-42
Output Shaft Speed (OSS)
Sensor 12-42
Location 12-42
Removal & Installation........12-43
Positive Crankcase
Ventilation (PCV) Valve.........12-43
Location 12-43
Removal & Installation........12-43

FUEL 12-43

**GASOLINE FUEL INJECTION
SYSTEM 12-43**
Fuel Filter..........................12-44
Removal & Installation........12-44
Fuel Level Sending Unit..........12-44
Removal & Installation........12-44
Fuel Rail & Injector 12-45
Removal & Installation........12-45
Fuel Tank 12-46
Draining..........................12-46
Removal & Installation........12-46
Fuel Tank Unit 12-44
Removal & Installation........12-44
Idle Speed 12-43
Inspection........................12-43
Relieving Fuel System
Pressure............................12-47
Relieving 12-47
Throttle Body.......................12-47
Removal & Installation........12-47

**HEATING & AIR CONDITIONING
SYSTEM 12-49**

Blower Motor 12-49
Removal & Installation........12-49
Heater Core 12-50
Removal & Installation........12-50

PRECAUTIONS 12-9

**SPECIFICATIONS AND
MAINTENANCE CHARTS.....12-3**

Brake Specifications 12-7
Camshaft Specifications............12-5
Capacities 12-4
Crankshaft and Connecting
Rod Specifications 12-5
Engine and Model Year
Identification 12-3
Engine Tune-Up
Specifications 12-3
Fluid Specifications...................12-4
General Engine Specifications...12-3
Piston and Ring
Specifications 12-5

Scheduled Maintenance
Intervals 12-8
Tire, Wheel, & Ball Joint
Specifications 12-7
Torque Specifications 12-6
Valve Specifications 12-4
Wheel Alignment.......................12-6

STEERING 12-50

Control Unit 12-50
Removal & Installation........12-50
Electrical Power Steering
(EPS) Motor 12-50
Removal & Installation........12-50
Power Rack & Pinion
Steering Gear 12-51
Removal & Installation........12-51

SUSPENSION 12-54

FRONT SUSPENSION 12-54
Lower Ball Joints.....................12-54
Removal & Installation........12-54
Lower Control Arms 12-54
Removal & Installation........12-54
Stabilizer Bar & Links 12-55
Removal & Installation........12-55
Steering Knuckle 12-56
Removal & Installation........12-56
Struts (Damper & Spring) 12-57
Removal & Installation........12-57
Wheel Hubs & Bearings..........12-58
End Play Inspection 12-59
Removal & Installation........12-58
REAR SUSPENSION 12-60
Ball Joints 12-60
Removal & Installation........12-60
Coil Springs.........................12-60
Removal & Installation........12-60
Damper (Shock Absorber)........12-61
Removal & Installation........12-61
Stabilizer Bushing 12-62
Replacement 12-62
Wheel Hubs & Bearings..........12-62
End Play Inspection 12-62
Removal & Installation........12-62

SPECIFICATIONS AND MAINTENANCE CHARTS

ENGINE AND VEHICLE IDENTIFICATION

		Engine					Model Year	
Code ①	Liters (cc)	Cu. In.	Cyl.	Fuel Sys.	Engine Type	Eng. Mfg.	Code ①	Year
CD3	1.5 (1496)	91.3	I4	MPFI	SOHC	Honda	B	2011
							C	2012

MPFI: Multi-Point Fuel Injection

DOHC: Double Overhead Camshafts

① 10th digit of VIN

71051_HFIT_C0001

GENERAL ENGINE SPECIFICATIONS
All measurements are given in inches.

Year	Model	Engine Displacement Liters	Engine Series ID	Net Horsepower @ rpm	Net Torque @ rpm (ft. lbs.)	Bore x Stroke (in.)	Com-pression Ratio	Oil Pressure @ rpm
2011	Fit	1.5	CD3	117@6600	106@4800	3.11 x 3.52	10.4:1	10 psi@Idle
2012	Fit	1.5	CD3	117@6600	106@4800	3.11 x 3.52	10.4:1	10 psi@Idle

NA: Not Available

A71051_HFIT_C0002

GASOLINE ENGINE TUNE-UP SPECIFICATIONS

Year	Engine Displacement Liters	Engine ID	Spark Plug Gap (in.)	Ignition Timing (deg.) MT	AT	Fuel Pump (psi)	Idle Speed (rpm) MT	AT	Valve Clearance In.	Ex.
2011	1.5	CD3	0.047-0.051	①	①	47-54	②	②	HYD	HYD
2012	1.5	CD3	0.047-0.051	①	①	47-54	②	②	HYD	HYD

NOTE: The Vehicle Emission Control Information label reflects specification changes made during production.

Follow the figures on the label if they differ from those in this chart.

HYD: Hydraulic

① Ignition timing is preset and cannot be adjusted

② Idle speed is maintained by the Electronic Control Module (ECM)

71051_HFIT_C0003

CAPACITIES

Year	Model	Engine Displacement Liters	Engine ID	Engine Oil with Filter (qts.)	Transaxle (pts.)		Fuel Tank (gal.)	Cooling System (qts.)
					Manual	Auto. ①		
2011	Fit	1.5	CD3	3.8	3.2	5.2	10.6	②
2012	Fit	1.5	CD3	3.8	3.2	5.2	10.6	②

NOTE: All capacities are approximate. Add fluid gradually and check to be sure a proper fluid level is obtained.

① Drain and refill

② At radiator drain and refill:

  M/T: 4.2 qts.

  A/T: 4.56 qts.

  At engine overhaul:

  M/T: 4.9 qts.

  A/T: 5.2 qts.

71051_HFIT_C0004

FLUID SPECIFICATIONS

Year	Model	Engine Displacement Liters	Engine ID	Engine Oil	Auto. Trans.	Manual Trans.	Power Steering Fluid	Brake Master Cylinder
2011	Fit	1.5	CD3	5W-20	DW-1	MTF	—	DOT 3
2012	Fit	1.5	CD3	5W-20	DW-1	MTF	—	DOT 3

DOT: Department Of Transportation

71051_HFIT_C0005

VALVE SPECIFICATIONS

Year	Engine Displacement Liters	Engine ID	Seat Angle (deg.)	Face Angle (deg.)	Spring Installed Height (in.)	Stem-to-Guide Clearance (in.)		Stem Diameter (in.)	
						Intake	Exhaust	Intake	Exhaust
2011	1.5	CD3	45	45	NA	0.0008-0.0020	0.0020-0.0031	0.2157-0.2161	0.2146-0.2150
2012	1.5	CD3	45	45	NA	0.0008-0.0020	0.0020-0.0031	0.2157-0.2161	0.2146-0.2150

NA: Information not available.

71051_HFIT_C0006

CAMSHAFT AND BEARING SPECIFICATIONS CHART

All measurements are given in inches.

Year	Engine Displ. Liters	Engine ID	Journal Dia.	Brg. Oil Clearance	Shaft End-play	Runout	Journal Bore	Lobe Height Intake	Lobe Height Exhaust
2011	1.5	CD3	NA	0.0018-0.0033	0.002-0.0098	0.0012	NA	①	①
2012	1.5	CD3	NA	0.0018-0.0033	0.002-0.0098	0.0012	NA	①	①

NA: Not Available

① Intake Primary: 1.38744 inch

Intake Secondary: 1.42413 inch

Exhaust: 1.39649 inch

71051_HFIT_C0007

CRANKSHAFT AND CONNECTING ROD SPECIFICATIONS

All measurements are given in inches.

Year	Engine Displacement Liters	Engine ID	Crankshaft Main Brg. Journal Dia.	Crankshaft Main Brg. Oil Clearance	Crankshaft Shaft End-play	Crankshaft Thrust on No.	Connecting Rod Journal Diameter	Connecting Rod Oil Clearance	Connecting Rod Side Clearance
2011	1.5	CD3	1.9676-1.9685	0.0020	0.018	—	1.5739-1.5748	0.0008-0.0015	0.006-0.0120
2012	1.5	CD3	1.9676-1.9685	0.0020	0.018	—	1.5739-1.5748	0.0008-0.0015	0.006-0.0120

71051_HFIT_C0008

PISTON AND RING SPECIFICATIONS

All measurements are given in inches.

Year	Engine Displ. Liters	Engine ID	Piston Clearance	Ring Gap Top Compression	Ring Gap Bottom Compression	Ring Gap Oil Control	Ring Side Clearance Top Compression	Ring Side Clearance Bottom Compression	Ring Side Clearance Oil Control
2011	1.5	CD3	0.0004-0.0014	0.006-0.0110	0.012-0.016 ①	0.008-0.0270	0.0030 ②	0.0030 ②	0.002 ③
2012	1.5	CD3	0.0004-0.0014	0.006-0.0110	0.012-0.016 ①	0.008-0.0270	0.0030 ②	0.0030 ②	0.002 ③

NA: Not Available

① With Riken ring shown

With Nippon ring: 0.014-0.019 in.

Service limit for both: 0.025 in.

② Service limit: 0.005 in.

③ Service limit: 0.004 in.

71051_HFIT_C0009

TORQUE SPECIFICATIONS
All readings in ft. lbs.

Year	Engine Disp. Liters	Engine ID	Cylinder Head Bolts	Main Bearing Bolts	Rod Bearing Bolts	Crankshaft Damper Bolts	Flywheel Bolts	Manifold Intake	Manifold Exhaust	Spark Plugs	Oil Pan Drain Plug
2011	1.5	CD3	①	②	③	④	87	17	33	13	29
2012	1.5	CD3	①	②	③	④	87	17	33	13	29

NA: Not Available

① Step 1: 22 ft. lbs.
  Step 2: Plus 130 degrees

② Step 1: 18 ft. lbs.
  Step 2: Plus 40 degrees

③ Step 1: 7.2 ft. lbs.
  Step 2: Plus 90 degrees

④ Step 1: New bolt: 130 ft. lbs.
  Step 2: 27 ft. lbs.

71051_HFIT_C0010

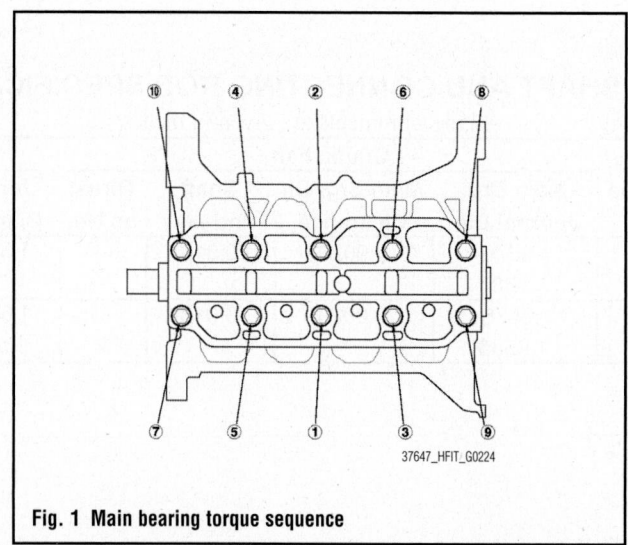

Fig. 1 Main bearing torque sequence

WHEEL ALIGNMENT

Year	Model		Caster Range (+/-Deg.)	Caster Preferred Setting (Deg.)	Camber Range (+/-Deg.)	Camber Preferred Setting (Deg.)	Toe-in (mm)
2011	Fit	Front	1.0	3.33	1.0	0.0	0.0 +/- 3.0
		Rear	—	—	1.0	-1.0	2.5 +/- 2.5
2012	Fit	Front	1.0	3.33	1.0	0.0	0.0 +/- 3.0
		Rear	—	—	1.0	-1.0	2.5 +/- 2.5

71051_HFIT_C0011

TIRE, WHEEL AND BALL JOINT SPECIFICATIONS

| Year | Model | OEM Tires | | Tire Pressures (psi) | | Wheel Size | Ball Joint Inspection | Lug Nut Torque (ft. lbs.) |
		Standard	Optional	Front	Rear			
2011	Fit	P175/65R14	P185/55R16	①	①	N/A	②	80
2012	Fit	P175/65R14	P185/55R16	①	①	N/A	②	80

① Refer to placard on vehicle for proper inflation pressure.

② Replace if any measurable movement is found.

71051_HFIT_C0012

BRAKE SPECIFICATIONS
All measurements in inches unless noted

| Year | Model | | Brake Disc | | | Brake Drum Diameter | | | Minimum Pad/Lining Thickness | Brake Caliper | |
			Original Thickness	Minimum Thickness	Maximum Runout	Original Inside Diameter	Max. Wear Limit	Maximum Machine Diameter		Bracket Bolts (ft. lbs.)	Mounting Bolts (ft. lbs.)
2011	Fit	F	0.827	0.748	0.0016	—	—	—	0.063	62-69	17
		R	—	—	—	7.874	7.91	7.91	0.039	—	—
2012	Fit	F	0.827	0.748	0.0016	—	—	—	0.063	62-69	17
		R	—	—	—	7.874	7.91	7.91	0.039	—	—

F: Front

R: Rear

71051_HFIT_C0013

MAINTENANCE MINDER SCHEDULE
Honda FIT

The Maintenance Minder is an important feature of the information display. Based on engine and transmission operating conditions, and accumulated engine revolutions, the onboard computer calculates remaining engine oil and transmission fluid life. The system also displays the remaining engine oil life along with the code(s) for other scheduled maintenance items needing service.

Symbol	Item	Service
A	Engine oil ①	Change
B	Engine oil and filter	Change
	Fluid levels	Inspect
	Brakes	Inspect
	Parking brake adjustment	Check
	Steering gear and linkage	Inspect
	Suspension components	Inspect
	Driveshaft boots	Inspect
	Brake hoses and lines	Inspect
	Exhaust system	Inspect
	Fuel lines and connections	Inspect
1	Tires	Rotate
2	Engine air filter ②	Replace
	Dust and pollen filter ③	Replace
	Accessory drive belt	Inspect
3	Transmission fluid	Replace
4	Spark plugs	Replace
	Valve clearance ④	Inspect
5	Engine coolant	Replace

① If the message SERVICE DUE NOW does not appear more than 12 months after the display is reset, change every year.

② If driven in dusty conditions, replace every 15,000 miles.

③ If driven in urban areas with a high concentration of soot from industry and diesel, replace every 30,000 miles.

③ If regularly driven in moutainous areas at very low speed or trailer towing, replace every 30,000 miles.

⑤ Adjust if necessary. Adjust the valves during services A, B, 1, 2, or 3 only if they are noisy.

Additionally, replace the brake fluid every 3 years, and inspect the idle speed every 160,000 miles.

** Resetting the Maintenance Minder:

1. Turn the ignition switch to the ON (II) position.

2. Press the select/reset knob repeatedly until the engine oil life is displayed.

3. Press the select/reset knob for about 10 seconds. The information display shows the reset mode initial display as shown.

4. Select the "OIL LIFE" indicator by turning the select/reset knob. The display begins to blink. Push the same knob to enter this setting.

5. The engine oil life and the maintenance item code(s) will begin to blink. Press the select/reset knob for another 5 seconds.

6. The maintenance item code(s) will disappear, and the engine oil life will reset to "100."

** Resetting individual Maintenance Item:

1. Connect the Honda Diagnostic System (HDS) to the data link connector (DLC.

2. Turn the ignition switch to ON (II.

3. Make sure the HDS communicates with the vehicle and the engine control module/powertrain control module (ECM/PCM). If it does not communicate, troubleshoot the DLC circuit.

4. Select GAUGES in the BODY ELECTRICAL with the HDS.

5. Select ADJUSTMENT in the GAUGES with the HDS.

6. Select MAINTENANCE MINDER in the ADJUSTMENT with the HDS.

7. Select RESET in the MAINTENANCE MINDER with the HDS.

8. Select the individual maintenance item you wish to reset with the HDS.

If a required service is done and the Maintenance Minder is not reset, or if the Maintenance Minder is reset without doing the service, the system will not show the proper maintenance timing. This can lead to serious mechanical problems because there will be no accurate record of when the required maintenance is needed.

PRECAUTIONS

Before servicing any vehicle, please be sure to read all of the following precautions, which deal with personal safety, prevention of component damage, and important points to take into consideration when servicing a motor vehicle:

• Never open, service or drain the radiator or cooling system when the engine is hot; serious burns can occur from the steam and hot coolant.

• Observe all applicable safety precautions when working around fuel. Whenever servicing the fuel system, always work in a well-ventilated area. Do not allow fuel spray or vapors to come in contact with a spark, open flame, or excessive heat (a hot drop light, for example). Keep a dry chemical fire extinguisher near the work area. Always keep fuel in a container specifically designed for fuel storage; also, always properly seal fuel containers to avoid the possibility of fire or explosion. Refer to the additional fuel system precautions later in this section.

• Fuel injection systems often remain pressurized, even after the engine has been turned **OFF**. The fuel system pressure must be relieved before disconnecting any fuel lines. Failure to do so may result in fire and/or personal injury.

• Brake fluid often contains polyglycol ethers and polyglycols. Avoid contact with the eyes and wash your hands thoroughly after handling brake fluid. If you do get brake fluid in your eyes, flush your eyes with clean, running water for 15 minutes. If eye irritation persists, or if you have taken

brake fluid internally, IMMEDIATELY seek medical assistance.

• The EPA warns that prolonged contact with used engine oil may cause a number of skin disorders, including cancer. You should make every effort to minimize your exposure to used engine oil. Protective gloves should be worn when changing oil. Wash your hands and any other exposed skin areas as soon as possible after exposure to used engine oil. Soap and water, or waterless hand cleaner should be used.

• All new vehicles are now equipped with an air bag system, often referred to as a Supplemental Restraint System (SRS) or Supplemental Inflatable Restraint (SIR) system. The system must be disabled before performing service on or around system components, steering column, instrument panel components, wiring and sensors. Failure to follow safety and disabling procedures could result in accidental air bag deployment, possible personal injury and unnecessary system repairs.

• Always wear safety goggles when working with, or around, the air bag system. When carrying a non-deployed air bag, be sure the bag and trim cover are pointed away from your body. When placing a non-deployed air bag on a work surface, always face the bag and trim cover upward, away from the surface. This will reduce the motion of the module if it is accidentally deployed. Refer to the additional air bag system precautions later in this section.

• Clean, high quality brake fluid from a sealed container is essential to the safe and

proper operation of the brake system. You should always buy the correct type of brake fluid for your vehicle. If the brake fluid becomes contaminated, completely flush the system with new fluid. Never reuse any brake fluid. Any brake fluid that is removed from the system should be discarded. Also, do not allow any brake fluid to come in contact with a painted surface; it will damage the paint.

• Never operate the engine without the proper amount and type of engine oil; doing so WILL result in severe engine damage.

• Timing belt maintenance is extremely important. Many models utilize an interference-type, non-freewheeling engine. If the timing belt breaks, the valves in the cylinder head may strike the pistons, causing potentially serious (also time-consuming and expensive) engine damage. Refer to the maintenance interval charts for the recommended replacement interval for the timing belt, and to the timing belt section for belt replacement and inspection.

• Disconnecting the negative battery cable on some vehicles may interfere with the functions of the on-board computer system(s) and may require the computer to undergo a relearning process once the negative battery cable is reconnected.

• When servicing drum brakes, only disassemble and assemble one side at a time, leaving the remaining side intact for reference.

• Only an MVAC-trained, EPA-certified automotive technician should service the air conditioning system or its components.

BRAKES

ANTI-LOCK BRAKE SYSTEM (ABS)

GENERAL INFORMATION

PRECAUTIONS

• Certain components within the ABS system are not intended to be serviced or repaired individually.

• Do not use rubber hoses or other parts not specifically specified for and ABS system. When using repair kits, replace all parts included in the kit. Partial or incorrect repair may lead to functional problems and require the replacement of components.

• Lubricate rubber parts with clean, fresh brake fluid to ease assembly. Do not use shop air to clean parts; damage to rubber components may result.

• Use only DOT 3 brake fluid from an unopened container.

• If any hydraulic component or line is removed or replaced, it may be necessary to bleed the entire system.

• A clean repair area is essential. Always clean the reservoir and cap thoroughly before removing the cap. The slightest amount of dirt in the fluid may plug an orifice and impair the system function. Perform repairs after components have been thoroughly cleaned; use only denatured alcohol to clean components. Do not allow ABS components to come into contact with any substance containing mineral oil; this includes used shop rags.

• The Anti-Lock control unit is a microprocessor similar to other computer units in the vehicle. Ensure that the ignition switch is **OFF** before removing or installing con-

troller harnesses. Avoid static electricity discharge at or near the controller.

• If any arc welding is to be done on the vehicle, the control unit should be unplugged before welding operations begin.

WHEEL SPEED SENSORS

REMOVAL & INSTALLATION

Front

See Figure 3.

1. Turn the ignition switch to LOCK (0).
2. Remove the grommet, then disconnect the wheel speed sensor connector.
3. Remove the bracket, the clip, and the wire guide grommet.

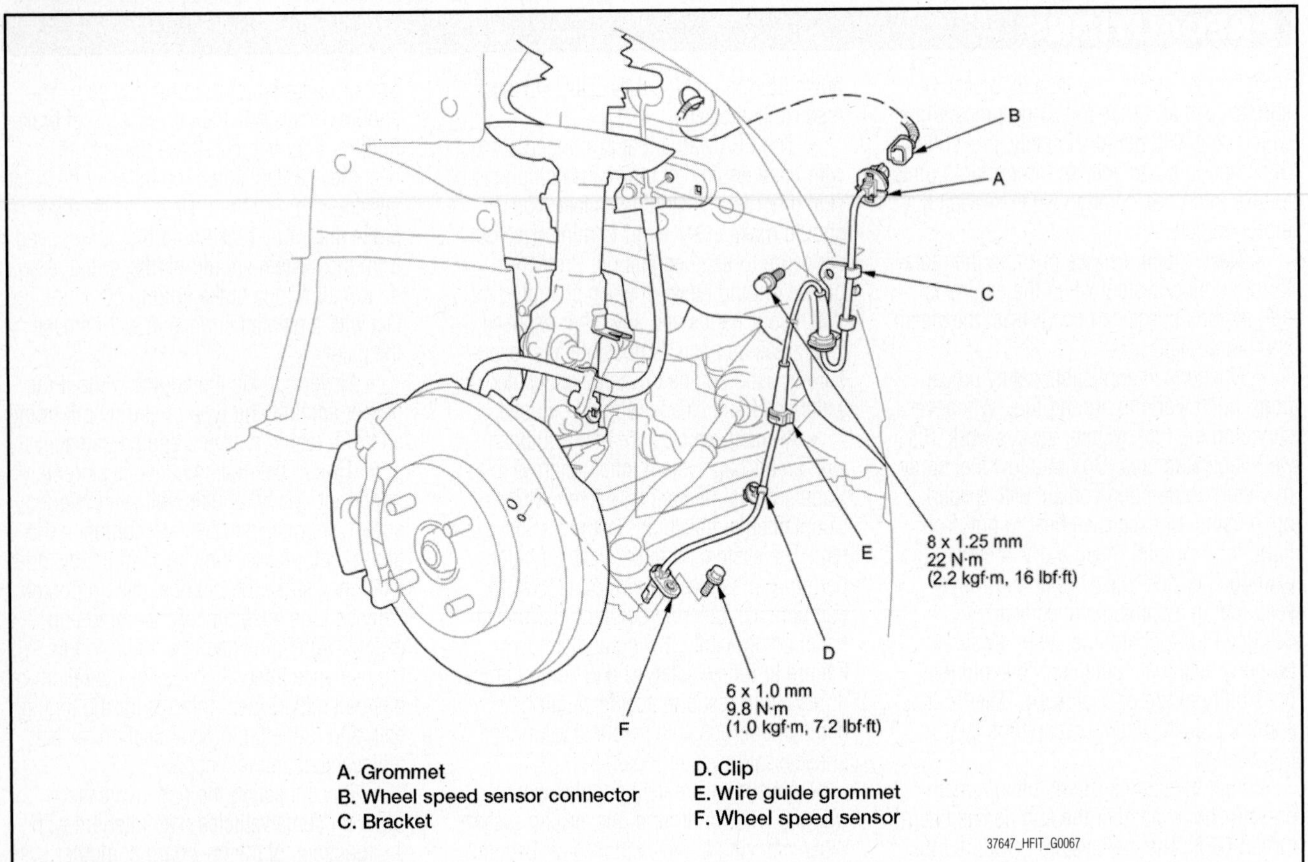

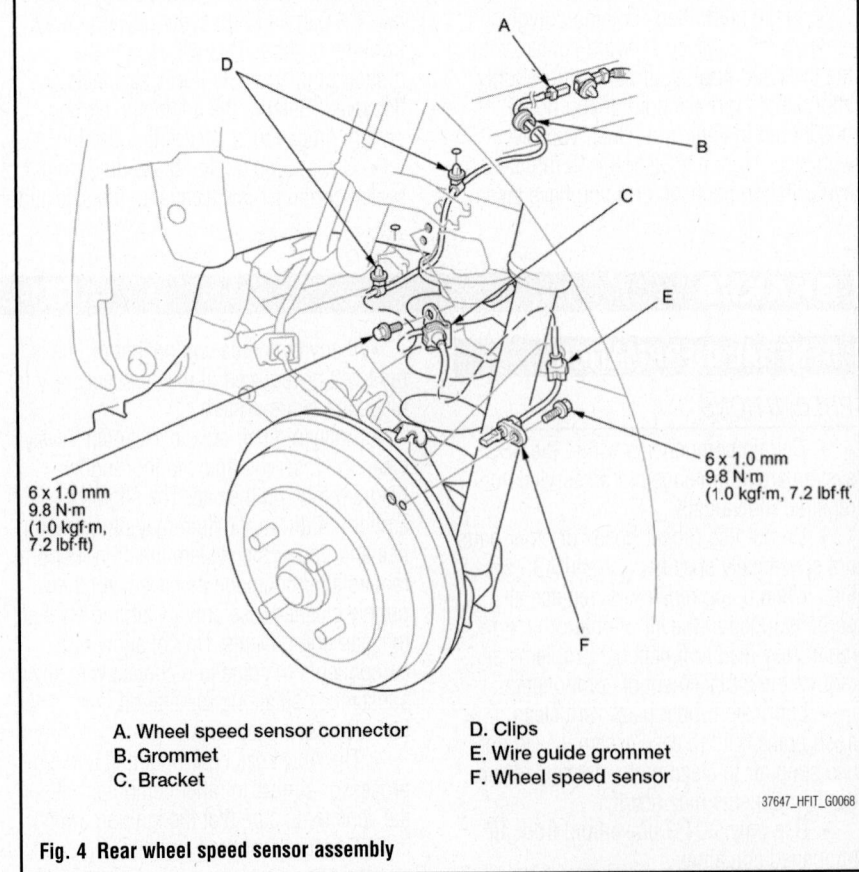

8 x 1.25 mm
22 N·m
(2.2 kgf·m, 16 lbf·ft)

6 x 1.0 mm
9.8 N·m
(1.0 kgf·m, 7.2 lbf·ft)

A. Grommet	D. Clip
B. Wheel speed sensor connector	E. Wire guide grommet
C. Bracket	F. Wheel speed sensor

37647_HFIT_G0067

Fig. 3 Front wheel speed sensor assembly

4. Remove the bolts and the wheel speed sensor.

To install:

5. Install the wheel speed sensor in the reverse order of removal, and note these items:

- Do not twist the sensor wires.
- If the wheel speed sensor comes in contact with the hub bearing unit, it is faulty.
- Make sure the grommet is installed properly.
- Make sure there is no debris in the sensor mounting hole.

6. Start the engine, and make sure the ABS and the VSA indicators go off.

7. Test-drive the vehicle, and make sure the ABS and the VSA indicators do not come on.

Rear

See Figure 4.

1. Turn the ignition switch to LOCK (0).

2. Pull back the carpet under the rear seat, then disconnect the wheel speed sensor connector.

3. Remove the grommet, the bracket, the clips, and the wire guide grommet.

4. Remove the bolt and the wheel speed sensor.

6 x 1.0 mm
9.8 N·m
(1.0 kgf·m,
7.2 lbf·ft)

6 x 1.0 mm
9.8 N·m
(1.0 kgf·m, 7.2 lbf·ft)

A. Wheel speed sensor connector	D. Clips
B. Grommet	E. Wire guide grommet
C. Bracket	F. Wheel speed sensor

37647_HFIT_G0068

Fig. 4 Rear wheel speed sensor assembly

To install:

5. Install the wheel speed sensor in the reverse order of removal, and note these items:

- Do not twist the sensor wires.
- If the wheel speed sensor comes in

contact with the hub bearing unit, it is faulty.

- Make sure the grommet is installed properly.
- Make sure there is no debris in the sensor mounting hole.

6. Start the engine, and make sure the ABS and the VSA indicators go off.

7. Test-drive the vehicle, and make sure the ABS and the VSA indicators do not come on.

BRAKES   BLEEDING THE BRAKE SYSTEM

BLEEDING PROCEDURE

➡Brake fluid replacement and air bleeding procedures are the same for vehicles with or without ABS.

✷✷ WARNING

Do not reuse the drained fluid. Use only clean Honda DOT 3 Brake Fluid from an unopened container. Using a non-Honda brake fluid can cause corrosion and shorten the life of the system. Do not mix different brands of brake fluid; they may not be compatible. Make sure no dirt or other foreign matter is allowed to

contaminate the brake fluid. Do not spill brake fluid on the vehicle, it may damage the paint; if brake fluid does contact the paint, wash it off immediately with water.

1. Ensure the reservoir connected to the master cylinder is at the MAX (upper) level mark at the start of the bleeding procedure and checked after bleeding each brake system. Add fluid as required.

2. Have someone slowly pump the brake pedal several times, then apply steady pressure.

3. Start the bleeding at the driver's side of the front brake system.

4. Bleed the calipers or the wheel cylinders in the following sequence:

- Left front
- Right front
- Right rear
- Left rear

5. Attach a length of clear drain tube to the bleed screw, then, loosen the bleed screw to allow air to escape from the system. Then tighten the bleed screw securely.

6. Refill the master cylinder reservoir to the MAX (upper) level line.

7. Repeat the procedure for each brake circuit until no air bubbles are in the fluid.

BRAKES   FRONT DISC BRAKES

BRAKE CALIPERS

REMOVAL & INSTALLATION

See Figure 5.

1. Remove the wheel nuts and front wheel.

2. Remove the brake hose mounting bolt.

3. Remove the brake caliper bracket mounting bolts, and remove the caliper assembly from the knuckle.

4. Detach the brake caliper from the brake hose, if the caliper is being replaced.

5. If the caliper is not being replaced, it can be moved out of the way and suspended with wire, so the hose is not under strain.

✷✷ WARNING

To prevent damage to the caliper assembly or brake hose, use a short piece of wire to hang the caliper assembly from the undercarriage. Do not twist the brake hose excessively.

To install:

6. Position the caliper in place.

7. If removed, reattach the brake line to the caliper.

8. Install and tighten the caliper mounting bolts to 80 ft. lbs. (108 Nm).

9. Install the brake hose clip retaining bolt.

10. If the brake hose was removed from the caliper, the system, perform "Bleeding Procedure" in this section.

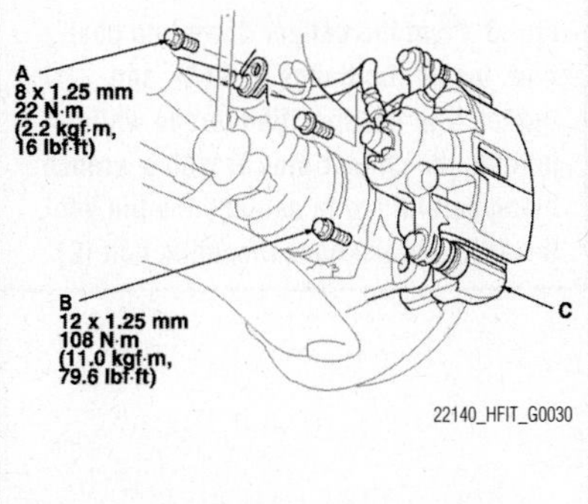

BRAKE PADS

REMOVAL & INSTALLATION

See Figures 6 through 9.

➡**Special Tools Required: Brake Caliper Piston Compressor 07AAE-SEPA101.**

1. Remove some of the brake fluid from the master cylinder.

2. Raise and support the vehicle.

3. Remove the front wheels.

4. Remove the brake hose mounting bolt. Remove the flange bolt while holding the caliper pin with a wrench. Be careful not to damage the pin boot, and pivot the caliper up out of the way. Check the hose and the pin boots for damage and deterioration.

5. If equipped, remove the pad return spring.

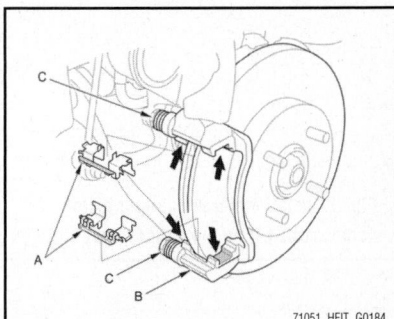

71051_HFIT_G0184

Fig. 6 Remove the pad retainers (A). Clean the caliper bracket (B) thoroughly; remove any rust, and check for grooves and cracks. Verify that the caliper pins (C) move in and out smoothly. Clean and lube if needed.

A
8 x 1.25 mm
22 N·m
(2.2 kgf·m,
16 lbf·ft)

B
12 x 1.25 mm
108 N·m
(11.0 kgf·m,
79.6 lbf·ft)

C

22140_HFIT_G0030

Fig. 5 Showing the front brake caliper brake hose mounting bolt (A), caliper mounting bolts (B), and caliper assembly (C)

➡The pad springs are installed on the pads to prevent brake drag. Be careful when fully pivoting up the caliper body or the spring could be flipped out of position.

6. Remove the pad shims and the brake pads.

7. Remove the pad retainers. Clean the caliper bracket thoroughly; remove any rust, and check for grooves and cracks. Verify that the caliper pins move in and out smoothly. Clean and lube if needed.

8. Inspect the brake disc for runout, thickness, parallelism, and check for damage and cracks.

9. Apply a thin coat of M-77 assembly paste (P/N 08798-9010) to the retainer mating surface of the caliper bracket (indicated by the arrows).

10. Install the pad retainers. Wipe excess assembly paste from the retainers. Keep the assembly paste away from the brake disc and the brake pads.

11. Install the brake caliper piston compressor on the caliper body.

12. Press in the piston with the brake caliper piston compressor tool so the caliper body will fit over the brake pads.

13. Make sure the piston boot is in position to prevent damaging it when pivoting the caliper body down.

➡Be careful when pressing in the piston; brake fluid might overflow from the

master cylinder's reservoir. If brake fluid gets on any painted surface, wash it off immediately with water.

14. Remove the brake caliper piston compressor tool.

To install:

15. Apply a thin coat of M-77 assembly paste (P/N 08798-9010) to the pad side of the shims, the back of the brake pads, and the other areas indicated by the arrows. Wipe off the excess assembly paste from the pad shims and brake pads friction material. Keep grease and assembly paste away from the brake disc and brake pads.

➡Contaminated brake disc or brake pads reduce stopping ability.

16. Install the brake pads and the pad shims correctly. Install the brake pad with the wear indicator (C) on the upper inside.

➡If you are reusing the brake pads, always reinstall the brake pads in their original positions to prevent a temporary loss of braking efficiency.

17. If equipped, install the pad return spring while holding the brake pads. Insert the pad return spring ends into the pad installation holes securely.

18. Pivot the caliper down into position. Install the flange bolt and tighten it to the specified torque while holding the caliper pin with a wrench being careful not to damage the pin boot. Install the brake hose mounting bolt.

19. Clean the mating surfaces between the brake disc and the inside of the wheel, then install the front wheels.

20. Press the brake pedal several times to make sure the brakes work.

➡Engagement may require a greater pedal stroke immediately after the brake pads have been replaced as a set. Several applications of the brake pedal will restore the normal pedal stroke.

21. Add brake fluid as needed.
22. After installation, check for leaks at hose and line joints or connections, and retighten if necessary.
23. Test-drive the vehicle, then recheck for leaks.

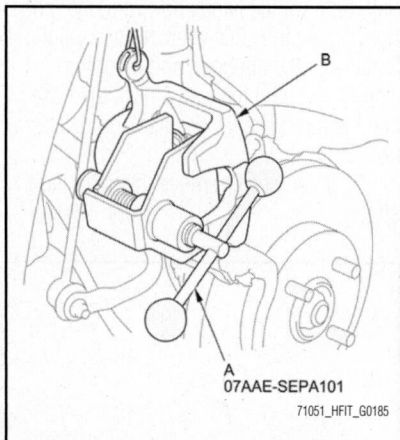

07AAE-SEPA101
71051_HFIT_G0185

Fig. 7 Install the brake caliper piston compressor (A) on the caliper body (B).

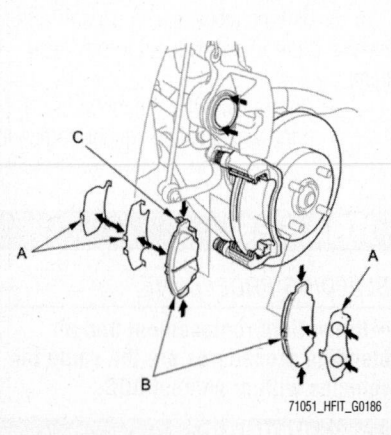

71051_HFIT_G0186

Fig. 8 Apply a thin coat of M-77 assembly paste (P/N 08798- 9010) to the pad side of the shims (A), the back of the brake pads (B), and the other areas indicated by the arrows. Wipe off the excess assembly paste from the pad shims and brake pads friction material. Keep grease and assembly paste away from the brake disc and brake pads.

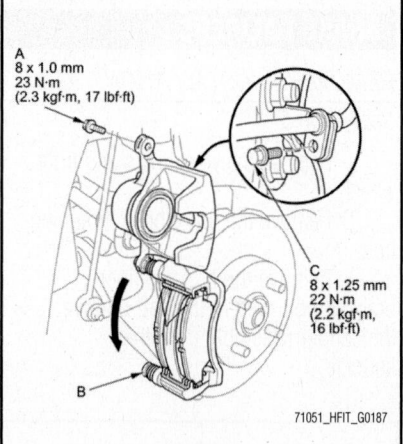

A
8 x 1.0 mm
23 N·m
(2.3 kgf·m, 17 lbf·ft)

C
8 x 1.25 mm
22 N·m
(2.2 kgf·m, 16 lbf·ft)

71051_HFIT_G0187

Fig. 9 Pivot the caliper down into position. Install the flange bolt (A), and tighten it to the specified torque while holding the caliper pin (B) with a wrench being careful not to damage the pin boot. Install the brake hose mounting bolt (C).

BRAKES

BRAKE DRUMS

REMOVAL & INSTALLATION

See Figure 10.

❊❊ CAUTION

Keep any grease off the brake drum and brake shoes.

1. Raise the rear of the vehicle, and support it with safety stands in the proper locations.
2. Remove the rear wheel.
3. Remove the parking brake, and remove the brake drum from the hub bearing unit. If necessary, turn the adjuster bolt with a flat-tip screwdriver until the shoes become loose.
4. If the brake drum has clung to the hub bearing unit, thread two 8 x 1.25 mm bolts into the brake drum to push it away from the hub bearing unit. Turn each bolt 90 degrees at a time to prevent cocking the brake drum.

To install:

5. Install the brake drum in the reverse order of removal.

   a. After installation, press the brake pedal several times to make sure the brakes work and self-adjust the brake shoes.

   b. Before installing the brake drum, clean the mating surfaces of the rear hub and the inside of the brake drum.

   c. Clean the mating surfaces of the brake drum and the inside of the wheel, then install the rear wheel.

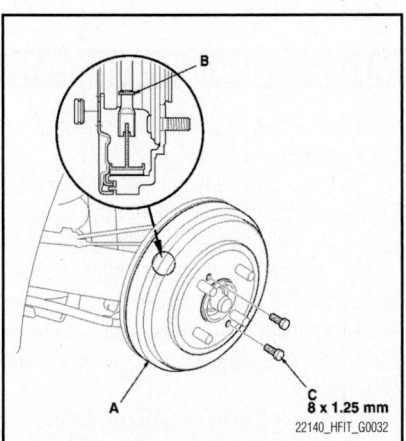

Fig. 10 Remove the parking brake, then the brake drum (A), turn the adjuster bolt (B) until the shoes become loose

BRAKE SHOES

REMOVAL & INSTALLATION

See Figures 11 through 13.

1. Raise the rear of the vehicle, and support it with safety stands in the proper locations.
2. Remove the rear wheels.
3. Release the parking brake, and remove the brake drum.
4. Remove the tension pins by pushing respective retainer spring and turning the pin.
5. Remove the lower return spring, and remove the brake shoe assembly over the hub.
6. Remove the forward brake shoe by removing the upper return spring, and disassemble the brake shoe assembly.

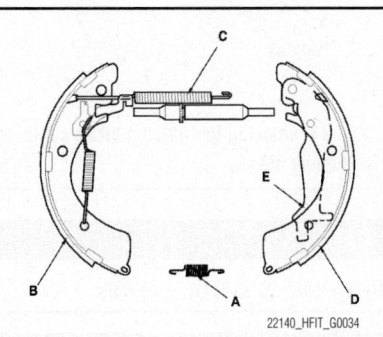

Fig. 11 Showing the lower return spring, forward brake shoe, upper return spring, and rear brake shoe, and parking brake lever locations

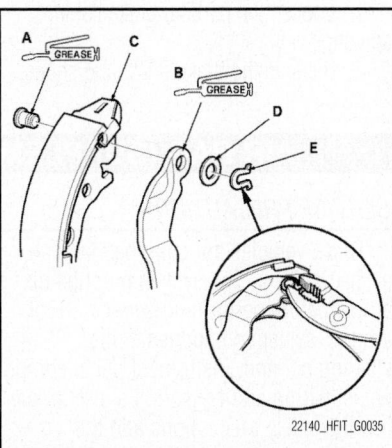

Fig. 12 Apply rubber grease to the sliding surface of the pivot pin and parking brake lever for the rearward brake shoe, then install the parking brake lever and the wave washer on the pivot pin, and secure with a new U-clip

7. Remove the rearward brake shoe by disconnecting the parking brake cable from the parking brake lever.
8. Remove the U-clip, wave washer, and pivot pin, and separate the parking brake lever from the brake shoe.

To install:

9. Apply rubber grease to the sliding surface of the pivot pin and parking brake lever for the rearward brake shoe.
10. Install the parking brake lever and the wave washer on the pivot pin, and secure with a new U-clip.

➡**Pinch the U-clip securely to prevent the parking brake lever from coming out of the brake shoe.**

11. Connect the parking brake cable to the parking brake lever.
12. Apply a thin coat of rubber grease to the connecting rod ends and the sliding surfaces. Wipe off any excess. Keep grease off the brake linings.
13. Apply a thin coat of Molykote 44 MA grease to the shoe ends and the edge of the shoe surfaces that contact the backing plate. Wipe off any excess. Keep grease off the brake linings.
14. Install connecting rods A and B on the adjuster bolt.

➡**Clean the threaded portions of connecting rod A and the sliding surface of connecting rod B, then coat them with rubber grease.**

15. Shorten connecting rod A by fully turning the adjuster bolt.
16. Assemble the brake shoes, the upper return spring, and with the connecting rods the adjuster bolt on the backing

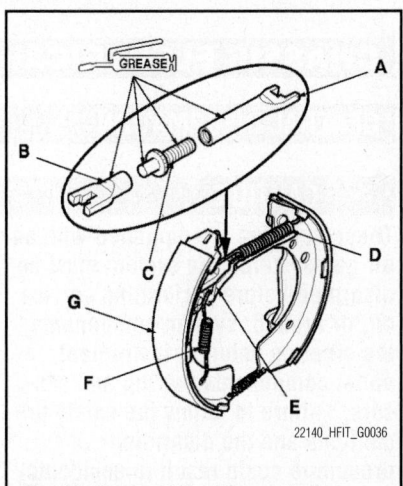

Fig. 13 Reassembling the rear brake shoes

plate, then install the self-adjuster lever and the self-adjuster spring on the forward brake shoe.

17. Install the tension pins and the retainer springs by pushing in respective spring and turning each pin.

Install the lower return spring.

→**Make sure the brake shoe positioning on the brake shoe bosses of the backing plate, and fitting the top of the brake shoes onto the wheel cylinder pistons.**

18. Before installing the brake drum, clean the mating surface of the rear hub and the inside of the brake drum.

19. Install the brake drum.
20. Install the rear wheels.
21. Press the brake pedal several times to make sure the brakes work and to set the self-adjusting brake.

→**Engagement of the brakes may require a greater pedal stroke immediately after the brake shoes have been replaced as a set. Several applications of the brake pedal will restore the normal pedal stroke.**

22. Adjust the parking brake.

BRAKES

ADJUSTMENTS

CABLES

See Figure 14.

1. Pull the parking brake lever with about 44 lbs. of force to fully apply the parking brake. The parking brake lever should be locked within 6–8 clicks. If the number of lever clicks is excessive, adjust the parking brake.
2. Remove the center console.
3. Release the parking brake lever fully.
4. Loosen the parking brake cable adjusting nut.
5. Press the brake pedal several times to set the self-adjusting brake before adjusting the parking brake.
6. Pull the parking brake lever 1 click.
7. Tighten the parking brake cable adjusting nut until the parking brakes drag slightly when the rear wheels are turned.
8. Release the parking brake lever fully, and check that the parking brakes do not drag when the rear wheels are turned. Readjust if necessary.
9. Make sure the parking brakes are fully applied when the parking brake lever is pulled all the way.
10. Install the center console.

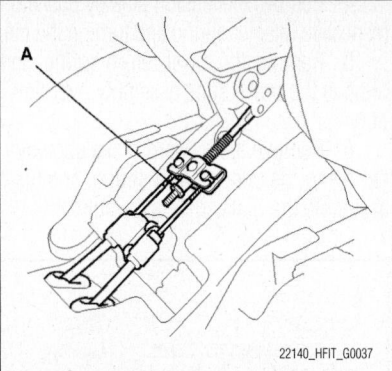

22140_HFIT_G0037

Fig. 14 Showing the parking brake cable adjusting nut (A)

CABLES

REMOVAL & INSTALLATION

✳✳ CAUTION

The parking brake cables must not be bent or distorted. This will lead to stiff operation and premature failure.

1. Loosen the parking brake cable adjusting nut.
2. Remove the brake drum and shoes,

PARKING BRAKE

and disconnect the parking brake cable from the parking brake lever.

3. Remove the cable clip from the cable insertion part at the reverse side of the backing plate.
4. Pull the parking brake cable, and remove it from the backing plate.
5. Reinstall the parking brake cable in the reverse order of removal, and note these items:

a. Be careful not to bend or distort the cable.

b. Align the projection of the parking cable holder with the cutout at the backing plate insertion part.

c. Insert until the groove in the cable holder aligns with the cutout at the cable insertion part.

d. Install the cable clip by inserting the straight end of the clip into the cutout at the cable insertion part, and secure the parking cable holder securely.

e. Connect the parking brake cable to the brake lever, and install the brake shoes and drum.

f. Do the parking brake adjustment. Apply the parking brake firmly 10 times then adjust it again.

CHASSIS ELECTRICAL

GENERAL INFORMATION

✳✳ CAUTION

These vehicles are equipped with an air bag system. The system must be disarmed before performing service on, or around, system components, the steering column, instrument panel components, wiring and sensors. Failure to follow the safety precautions and the disarming procedure could result in accidental air bag deployment, possible injury and unnecessary system repairs.

AIR BAG (SUPPLEMENTAL RESTRAINT SYSTEM)

SERVICE PRECAUTIONS

These vehicles are equipped with an air bag system. The system must be disarmed before performing service on, or around, system components, the steering column, instrument panel components, wiring and sensors. Failure to follow the safety precautions and the disarming procedure could result in accidental air bag deployment, possible injury and unnecessary system repairs.

Except when performing electrical inspections, always turn the ignition switch OFF, then disconnect the negative cable from the battery, then wait for 3 minutes before starting work.

The SRS memory is not erased even if the ignition switch is turned OFF or the battery cables are disconnected from the battery.

Use replacement parts which are manufactured to the same standards and quality as the original parts. Do not install used SRS parts. Use only new parts when making SRS repairs.

Carefully inspect any SRS part before you install it.

Do not install any part that shows signs of being dropped or improperly handled, such as dents, cracks, or deformation.

Before disconnecting the SRS unit connectors, always disconnect the appropriate SRS parts connectors.

Use only a digital multimeter to check the system. If it is not a Honda multimeter, make sure its output is 10 mA (0.01 A) or less when switched to the lowest value in the ohmmeter range. A tester with a higher output could cause accidental deployment and possible injury.

Do not put objects on the front passenger's air bag.

The original audio system has a coded theft protection circuit. Make sure you have the anti-theft codes for the audio system.

Before returning the vehicle to the customer, enter the anti-theft codes for the audio system.

Set the clock

Misalignment of the cable reel could cause an open in the wiring, making the SRS system, remote steering wheel controls, and the horn inoperative. Center the cable reel whenever the following is performed:

- Installation of the steering wheel
- Installation of the cable reel
- Installation of the steering column
- Other steering-related adjustment or installation

Do not disassemble the cable reel.

Do not apply grease to the cable reel.

If the cable reel shows any signs of damage, replace it with a new one. For example, if it does not rotate smoothly, replace the cable reel.

Do not disassemble an air bag. It has no serviceable parts. Once an air bag has been deployed, it cannot be repaired or reused.

For temporary storage of an air bag during service, store the removed air bag with the pad surface up. Never put anything on the air bag.

To prevent damage to the air bag, keep it away from any oil, grease, detergent, or water.

Store the removed air bag on a secure, flat surface away from any high heat source (exceeding 200°F).

Never perform electrical inspections to the air bags, such as measuring resistance.

Do not position yourself in front of the air bag during removal, inspection, or replacement.

Wear safety glasses, rubber gloves, and long sleeved clothing when cleaning powder residue from vehicle after an air bag deployment. Powder residue emitted from a deployed air bag can cause skin irritation. Flush affected area with cool water if irritation is experienced. If nasal or throat irritation is experienced, exit the vehicle for fresh air until the irritation ceases. If irritation continues, see a physician.

Do not use a replacement air bag that is not in the original packaging. This may result in improper deployment, personal injury, or death.

The factory installed fasteners, screws and bolts used to fasten air bag components have a special coating and are specifically designed for the air bag system. Do not use substitute fasteners. Use only original equipment fasteners listed in the parts catalog when fastener replacement is required.

During, and following, any child restraint anchor service, due to impact event or vehicle repair, carefully inspect all mounting hardware, tether straps, and anchors for proper installation, operation, or damage. If a child restraint anchor is found damaged in any way, the anchor must be replaced. Failure to do this may result in personal injury or death.

Deployed and non-deployed air bags may or may not have live pyrotechnic material within the air bag inflator.

The side curtain air bag inflator assembly is a long, jointed part containing an inflator, a flexible bag, and brackets.

When removing or installing the side curtain air bag inflator assembly, never handle the flexible bag.

Do not dispose of driver/passenger/curtain air bags or seat belt tensioners unless you are sure of complete deployment. Refer to the Hazardous Substance Control System for proper disposal.

Dispose of deployed air bags and tensioners consistent with state, provincial, local, and federal regulations.

After any air bag component testing or service, do not connect the battery negative cable. Personal injury or death may result if the system test is not performed first.

Be careful not to bump or impact the SRS unit, front impact sensors, or side impact sensors whenever the ignition switch is ON (II), or for at least 3 minutes after the ignition switch is turned OFF.

During installation or replacement, be careful not to bump (by impact wrench, hammer, etc.) the area around the SRS unit, front impact sensors, or the side impact sensors. The air bags could accidentally deploy and cause damage or injury.

After a collision where the air bags the side air bags, the side curtain air bags, or seat belt tensioners were deployed, many components must be inspected. After a collision where the air bags, the side air bags, or the side curtain air bags did not deploy, inspect for any damage or any deformation on the SRS unit, front impact sensors, and the side impact sensors. If there is any damage, replace the SRS unit and/or the sensors.

Do not disassemble the SRS unit, front impact sensors, side impact sensors, driver's seat position sensor, or front passenger's weight sensors.

Turn the ignition switch OFF, disconnect the negative cable from the battery, then wait for 3 minutes before starting installation or replacement of the SRS unit, or disconnecting the connectors from the SRS unit.

Be sure the SRS unit, front impact sensors, and side impact sensors are installed securely with the mounting bolts torqued to 7.2 ft. lbs. (9.8 Nm).

Do not spill water or oil on the SRS unit or the side impact sensors, and keep them away from dust.

Store the SRS unit, front impact sensors, and side impact sensors in a cool (less than 104°F) and dry (less than 80 percent relative humidity, no moisture) area.

Some of the SRS wiring can be identified by special yellow outer covering, and the SRS connectors can be identified by their yellow color. Observe the applicable procedures.

Never attempt to modify, splice, or repair SRS wiring. If there is an open or damage in SRS wiring, replace the harness.

Be sure to install the harness wires so they do not get pinched or interfere with other parts.

Make sure all SRS ground locations are clean, and grounds are securely fastened for optimum metal-to-metal contact. Poor grounds can cause intermittent problems that are difficult to diagnose.

When using electrical test equipment, insert the probe of the tester into the wire side of the connector. Do not insert the probe of the tester into the terminal side of the connector, and do not tamper with the connector.

Use a U-shaped probe. Do not insert the probe forcibly.

Using improper tools could cause an error in inspection due to poor metal-to-metal contact.

When cleaning the seats with side air

bag, use a damp cloth to clean the seat. Do not soak the seat with liquid, and do not spray steam on the seat.

Do not repair a torn or frayed seat-back cover. Replace the seat-back cover.

After a collision where the side air bag was deployed, replace the side air bag and seat frame with new parts. If the seat-back cushion is split, it must be replaced.

Never put aftermarket accessories on the seat (covers, pads, seat heaters, lights, etc.).

DISARMING THE SYSTEM

1. Before servicing the vehicle, refer to the Precautions Section.
2. Turn the ignition switch to the **LOCK** position.
3. Disconnect the negative battery cable.
4. Wait three minutes for the battery power to fully discharge from the system.

ARMING THE SYSTEM

1. Before servicing the vehicle, refer to the Precautions Section.
2. Connect the negative battery cable.
3. Turn the ignition switch **ON**.
4. Verify that the air bag indicator illuminates for 4–8 seconds, then goes off.

DRIVE TRAIN

AUTOMATIC TRANSAXLE

FILTER REPLACEMENT

See Figure 15.

➡**Manufacturer recommends use of ATF DW-1.**

1. Remove the air cleaner assembly.
2. Disconnect the ATF cooler hoses from the ATF filter.
3. Remove the ATF filter holder.
4. Remove the ATF filter, and replace it.

To install:

5. Install the new ATF filter on the filter mounting bracket (D), and secure it with its holder and bolt.
6. Slide the ATF cooler hose on the ATF filter until the hose end contacts the filter housing, and secure the hose with the clip (E) at 0.24–0.31 in. (6–8 mm) (F) from the filter housing. Install the cooler hose on the

other side of the ATF filter in the same manner.
7. Install the air cleaner assembly.

CLUTCH HYDRAULIC SYSTEM BLEEDING

BLEEDING PROCEDURE

See Figure 16.

Note the following:
• Do not reuse the drained fluid. Always use Honda DOT 3 Brake Fluid from an unopened container. Using a non-Honda brake fluid can cause corrosion and shorten the life of the system.
• Do not mix different brands of brake fluid; they may not be compatible.
• Make sure no dirt or other foreign matter is allowed to contaminate the brake fluid.
• Do not spill brake fluid on the vehicle; it may damage the paint or plastic. If brake fluid does contact the paint or plastic, wash it off immediately with water.
• If may be necessary to limit the movement of the release fork with a block of wood to remove all the air from the system.
• Use fender covers to avoid damaging painted surfaces.

1. Make sure the brake fluid level in the clutch reservoir is at the MAX (upper) level line.
2. Attach one end of a clear tube to the bleeder screw and put the other end into a container. Loosen the bleeder screw to allow air to escape from the system.
3. Make sure there is an adequate supply of fluid in the reservoir, then slowly push the clutch pedal all the way down.
4. Before releasing the pedal, have an assistant temporarily tighten the bleeder screw. Loosen the bleeder screw, and push the clutch pedal down again. Repeat this step until no more bubbles appear at the clear tube.

➡**Make sure the fluid level on the reservoir does not go below MIN (lower).**

5. Tighten the bleeder screw securely.
6. Refill the brake fluid in the reservoir to the MAX (upper) level line.
7. Check the clutch operation, and check for leaks.

CLUTCH MASTER CYLINDER

REMOVAL & INSTALLATION

See Figures 17 through 20.

➡**Use fender covers to avoid damaging painted surfaces.**

✳✳ WARNING

Do not spill brake fluid on the vehicle; it may damage the paint or plastic. If brake fluid does contact the paint or plastic, wash it off immediately with water.

➡**Without cruise control: Clutch pedal adjusting bolt is substituted as clutch pedal position switch.**

➡**Put on gloves to protect your hands.**

1. Secure the hood in the wide open position (support rod in the lower hole).

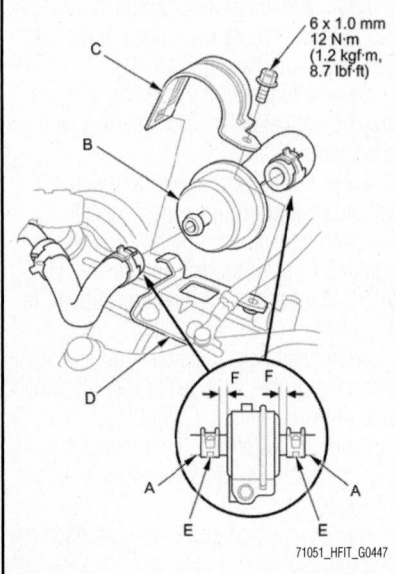

6 x 1.0 mm
12 N·m
(1.2 kgf·m,
8.7 lbf·ft)

71051_HFIT_G0447

Fig. 15 Disconnect the ATF cooler hoses (A) from the ATF filter (B). Remove the ATF filter holder (C).

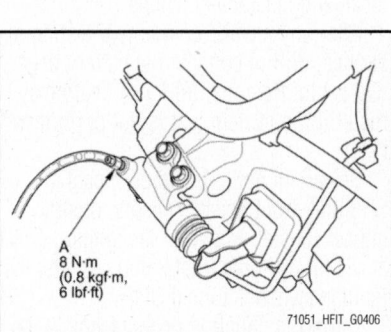

A
8 N·m
(0.8 kgf·m,
6 lbf·ft)

71051_HFIT_G0406

Fig. 16 Attach one end of a clear tube to the bleeder screw (A), and put the other end into a container. Loosen the bleeder screw to allow air to escape from the system.

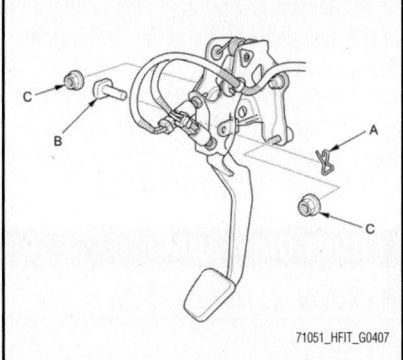

Fig. 17 Pry out the lock pin (A) and pull the clevis pin (B) out of the clevis. Remove the master cylinder mounting nuts (C).

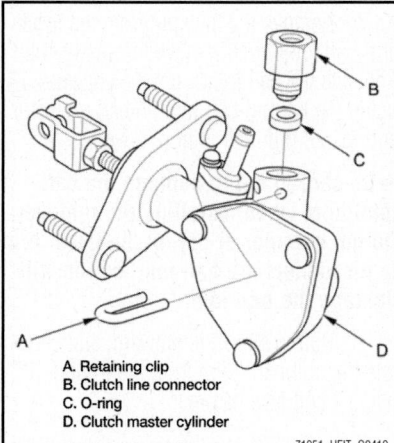

A. Retaining clip
B. Clutch line connector
C. O-ring
D. Clutch master cylinder

Fig. 18 Remove the retaining clip, then remove the clutch line connector and the O-ring from the master cylinder.

2. Remove the wiper arms.
3. Remove the cowl cover and hood hinge cover.
4. Remove the wiper motor.
5. Remove the under-cowl panel.
6. Remove the air cleaner housing.
7. Remove and discard the brake fluid from the clutch master cylinder reservoir with a syringe or other suitable device.
8. Remove the driver's dashboard lower cover.
9. Pry out the lock pin and pull the clevis pin out of the clevis. Remove the master cylinder mounting nuts.
10. Remove the clutch line bracket bolts.
11. Remove the reservoir hose from two clips.
12. Disconnect the reservoir hose, then remove the clutch line from the clutch master cylinder. Plug or wrap the end of the reservoir hose and clutch line with a shop towel to prevent brake fluid from coming out.

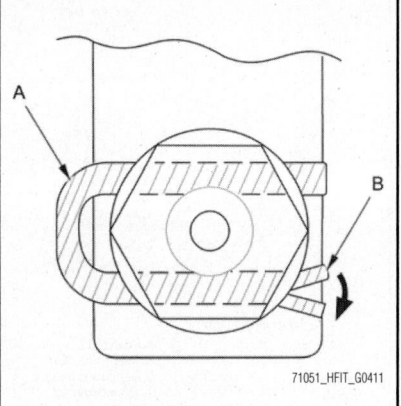

Fig. 19 To prevent the retaining clip (A) from coming off, pry apart the tip (B) of the clip with a screwdriver.

a. Loosen the flare nut while holding the clutch line connector with a wrench.
b. Inspect the hose. If the hose has damage, leaks interference, or twisting, replace it.
c. Remove the clutch master cylinder seal from the clutch master cylinder.
13. Remove the retaining clip, then remove the clutch line connector and the O-ring from the master cylinder.

To install:
14. Install the clutch line connector with a new O-ring, then set in a new retaining clip to the master cylinder.
a. Apply the silicone grease (P/N 08C30-B0234M) on the O-ring and the joint part of the clutch line connector.
b. Make sure not to get any silicone grease on the terminal part of the con-

nectors and switches, especially if you have silicone grease on your hands or gloves.
15. To prevent the retaining clip from coming off, pry apart the tip of the clip with a screwdriver.
16. Install a new clutch master cylinder seal to the clutch master cylinder.
17. Connect the clutch line and reservoir hose. Tighten to 11 ft. lbs. (15 Nm).
18. Tighten the flare nut while holding the clutch line connector with a wrench.
19. Make sure the hose clamps are positioned on the master cylinder and reservoir as shown.
20. Install the two reservoir hose clips.
21. Install the clutch line bracket bolts.
22. Install and tighten the master cylinder mounting nuts.
23. Apply multipurpose grease to the clevis pin and slide it into the clevis, then install a lock pin.
24. Adjust the clutch pedal, the clutch pedal position switch, and the clutch interlock switch.
25. Bleed the clutch hydraulic system.
26. Check the clutch operation, and check for leaks.
27. Install the driver's dashboard lower cover.
28. Install the air cleaner housing.
29. Install the under-cowl panel.
30. Install the wiper motor.
31. Install the cowl cover and hood hinge cover.
32. Install the wiper arms.
33. Test-drive the vehicle.

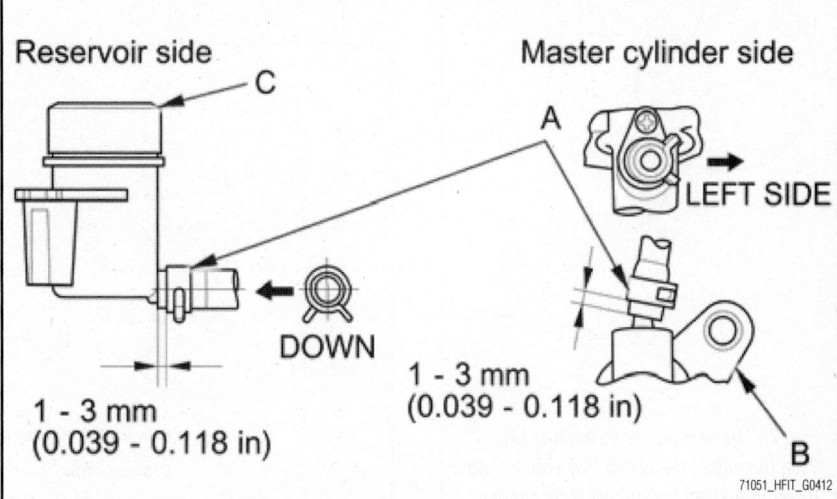

Fig. 20 Make sure the hose clamps (A) are positioned on the master cylinder (B) and reservoir (C) as shown.

SLAVE (RELEASE) CYLINDER

REMOVAL & INSTALLATION

See Figures 21 and 22.

➡ **Use fender covers to avoid damaging painted surfaces.**

✳✳ WARNING

Do not spill brake fluid on the vehicle; it may damage the paint or plastic. If brake fluid does contact the paint or plastic, wash it off immediately with water.

➡ **Without cruise control: Clutch pedal adjusting bolt is substituted as clutch pedal position switch.**

➡ **Make sure not to get any silicone grease on the terminal part of the connectors and switches, especially if you have silicone grease on your hands or gloves.**

1. Do the battery removal procedure.
2. Remove the air cleaner housing.
3. Remove the mounting bolts, clutch line bracket, and clutch line clip, then remove the slave cylinder.
4. Disconnect the clutch line. Plug the end of the clutch line with a shop towel to prevent brake fluid from coming out. Loosen the flare nut while holding the clutch line connector with a wrench.

To install:

5. Connect the clutch line. Tighten the flare nut to 11 ft. lbs. (15 Nm), while holding the clutch line connector with a wrench.

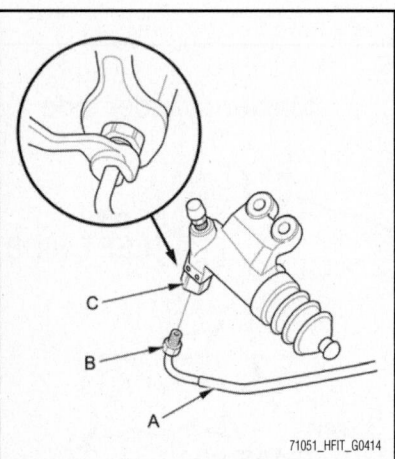

Fig. 21 Disconnect the clutch line (A). Plug the end of the clutch line with a shop towel to prevent brake fluid from coming out. Loosen the flare nut (B) while holding the clutch line connector (C) with a wrench.

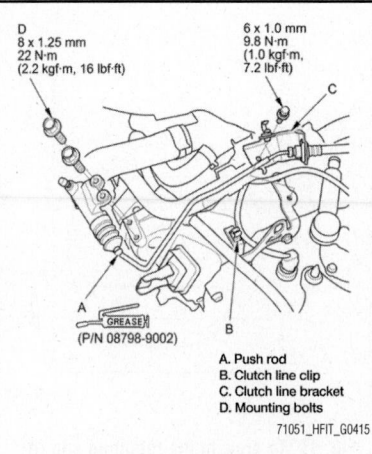

D
8 x 1.25 mm
22 N·m
(2.2 kgf·m, 16 lbf·ft)

6 x 1.0 mm
9.8 N·m
(1.0 kgf·m,
7.2 lbf·ft)

GREASE
(P/N 08798-9002)

A. Push rod
B. Clutch line clip
C. Clutch line bracket
D. Mounting bolts

71051_HFIT_G0415

Fig. 22 Apply a light coat of super high temp urea grease (P/N 08798-9002) to the push rod of the slave cylinder. Install the clutch line clip, the clutch line bracket and the mounting bolts.

6. Apply a light coat of super high temp urea grease (P/N 08798-9002) to the push rod of the slave cylinder.
7. Install the clutch line clip, the clutch line bracket and the mounting bolts.
8. Bleed the clutch hydraulic system, as described in this section.
9. Check the clutch operation, and check for leaks.
10. Install the air cleaner housing.
11. Do the battery installation procedure.
12. Test-drive the vehicle.

O-Ring Replacement

See Figure 23.

1. Remove the slave cylinder, as described above.

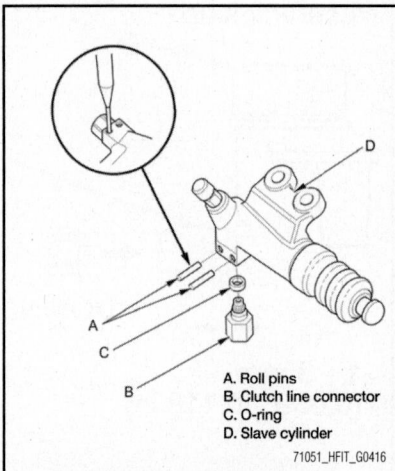

A. Roll pins
B. Clutch line connector
C. O-ring
D. Slave cylinder

71051_HFIT_G0416

Fig. 23 Remove the roll pins, then remove the clutch line connector and the O-ring from the slave cylinder.

2. Remove the roll pins, then remove the clutch line connector and the O-ring from the slave cylinder.
3. Install the clutch line connector with a new O-ring, then set in new roll pins to the slave cylinder.
4. Install the slave cylinder.

FRONT DRIVESHAFT

REMOVAL & INSTALLATION

See Figures 24 and 25.

1. Raise and support the vehicle.
2. Remove the front wheels.
3. Pry up the stake on the spindle nut, then remove the nut.
4. Drain the transmission fluid. Reinstall the drain plug using a new sealing washer:
5. Remove the lock pin from the lower arm ball joint, then remove the castle nut. Separate the ball joint from the knuckle using the 14 mm ball joint thread protector and the 28 mm ball joint remover .

➡ **Be careful not to damage the ball joint boot when installing the remover. Do not hammer or pry on the lower arm to disconnect the ball joint or you will damage the ball joint.**

6. Pull the knuckle outward, and separate the outboard joint from the front hub using a soft face hammer.

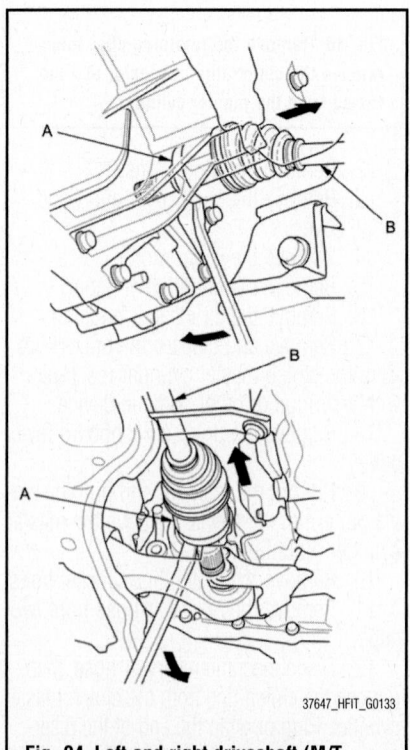

37647_HFIT_G0133

Fig. 24 Left and right driveshaft (M/T model)/left driveshaft (A/T model)

7. Left and right driveshaft (M/T model)/left driveshaft (A/T model): Pry the inboard joint from the differential with a prybar. Remove the driveshaft as an assembly.

➡**Do not pull the assembly by the driveshaft, or the inboard joint may come apart. Pull the inboard joint straight out to avoid damaging the oil seal. Be careful not to damage the oil seal or the end of the inboard joint with the prybar.**

8. Right driveshaft (A/T model): Drive the inboard joint off of the intermediate shaft using a drift punch and a hammer. Remove the driveshaft as an assembly.

➡**Do not pull the assembly by the driveshaft, or the inboard joint may come apart.**

9. Remove the set ring from the inboard joint (Except A/T model right driveshaft).

10. Remove the set ring from the intermediate shaft (A/T model right driveshaft).

To install:

➡**Before starting installation, make sure the mating surfaces of the joint and the splined section are clean.**

11. Apply Moly 60 paste (P/N 08734-0001) to the contact area of the outboard joint and the front wheel bearing.

➡**The paste helps to prevent noise and vibration.**

12. Install a new set ring onto the set ring groove of the driveshaft inboard joint (except A/T model right driveshaft).

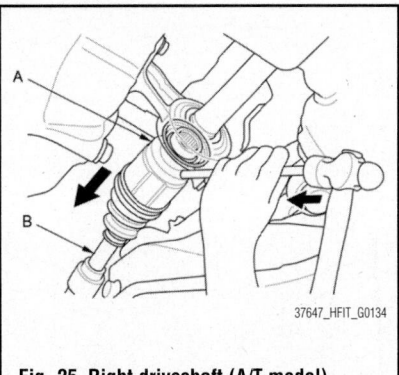

37647_HFIT_G0134

Fig. 25 Right driveshaft (A/T model)

13. Install a new set ring onto the set ring groove of the intermediate shaft (A/T model).

14. Apply super high temp urea grease (P/N 08798-9002) to the whole splined surface of the right driveshaft. After applying grease, remove the grease from the splined grooves at intervals of 2–3 splines and from the set ring groove so that air can bleed from the intermediate shaft.

15. Clean the areas where the driveshaft contacts the differential thoroughly with solvent or brake cleaner, and dry them with compressed air.

➡**Do not wash the rubber parts with solvent.**

16. Insert the inboard end of the driveshaft into the differential or intermediate shaft until the set ring locks in the groove.

➡**Insert the driveshaft horizontally to prevent damaging the oil seal.**

17. Install the outboard joint into the front hub on the knuckle.

18. Wipe off any grease contamination from the ball joint tapered section and threads, then install the knuckle onto the lower arm. Be careful not to damage the ball joint boot. Wipe off the grease before tightening the nut at the ball joint. Torque the new castle nut to the lower torque specification, then tighten it only far enough to align the slot with the ball joint pin hole.

➡**Make sure the ball joint boot is not damaged or cracked. Do not align the nut by loosening it.**

19. Install the lock pin into the ball joint pin hole.

20. Apply a small amount of engine oil to the seating surface of a new spindle nut.

21. Install the spindle nut, then tighten it. After tightening, use a drift to stake the spindle nut shoulder against the driveshaft.

22. Clean the mating surfaces of the brake disc and the wheel, then install the front wheels.

23. Turn the front wheel by hand, and make sure there is no interference between the driveshaft and surrounding parts.

24. Refill the transmission with the recommended transmission fluid:

25. Lower the vehicle.

26. Check the wheel alignment, and adjust it if necessary.

27. Test-drive the vehicle.

CV-BOOTS INSPECTION

Inspect the CV boots for cracks, wear, leakage and for damaged mounting bands.

ENGINE COOLING

ENGINE COOLING FAN

REMOVAL & INSTALLATION

See Figures 26 and 27.

1. Remove the coolant reservoir.
2. Disconnect the radiator fan motor connector, then remove the harness clamp.
3. With A/C: Disconnect the condenser fan motor connector and the A/C compressor clutch connector, then remove the harness clamps.
4. Remove the radiator upper brackets.
5. With A/C: Remove the condenser fan shroud assembly and the radiator fan shroud assembly from the radiator, then remove the condenser fan shroud assembly from the vehicle.

6. Remove the radiator fan shroud assembly from right side of the engine compartment.
7. Disassemble the fan shrouds.

To install:

8. Assemble the fan shrouds.
9. Install the radiator fan shroud assembly.
10. With A/C: Install the condenser fan shroud assembly.
11. Install the radiator upper brackets.
12. Connect the radiator fan motor connector, then install the harness clamp.
13. With A/C: Connect the condenser fan motor connector and the A/C compressor clutch connector, then install the harness clamps.
14. Install the coolant reservoir.

RADIATOR & A/C CONDENSER FAN, MOTOR & SHROUD

REMOVAL & INSTALLATION

See Figures 28 through 30.

1. Remove the coolant reservoir.
2. Disconnect the radiator fan motor connector, then remove the harness clamp.
3. With A/C: Disconnect the A/C condenser fan motor connector and the A/C compressor clutch connector, then remove the harness clamps.
4. Remove the radiator upper brackets.
5. With A/C: Remove the A/C condenser fan shroud assembly and the radiator fan shroud assembly from

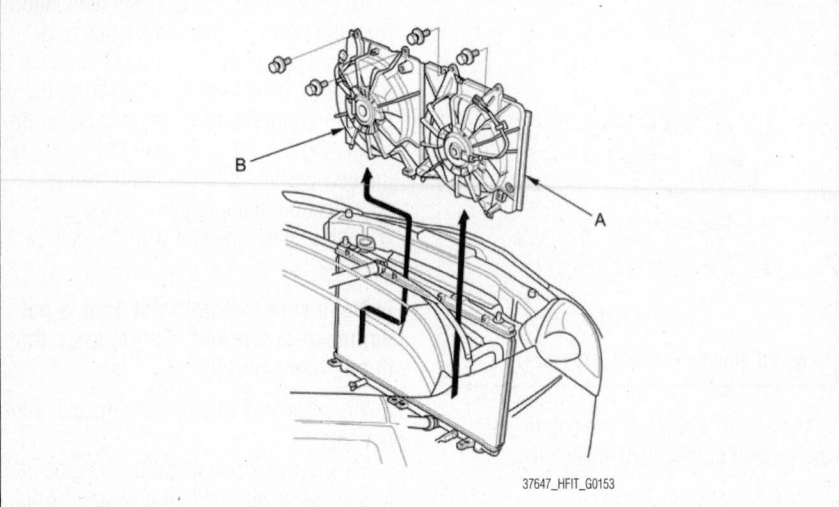

Fig. 26 Remove the condenser fan shroud assembly (A) and the radiator fan shroud assembly (B) from the radiator, then remove the condenser fan shroud assembly from the vehicle

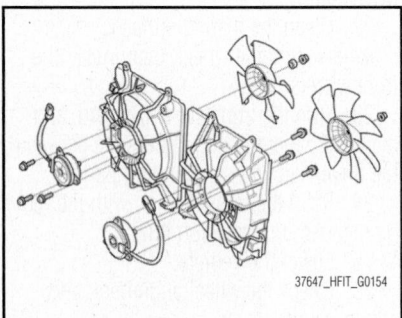

Fig. 27 Disassemble the fan shrouds

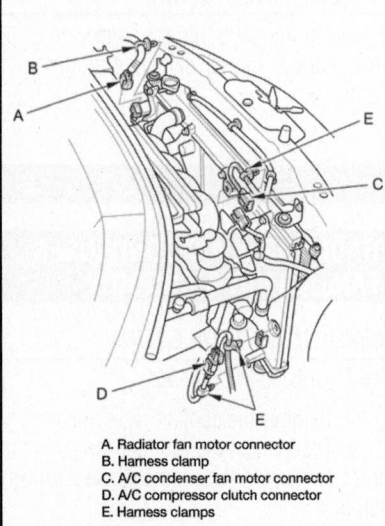

A. Radiator fan motor connector
B. Harness clamp
C. A/C condenser fan motor connector
D. A/C compressor clutch connector
E. Harness clamps

71051_HFIT_G0285

Fig. 28 Disconnect the radiator fan motor connector, then remove the harness clamp. With A/C: Disconnect the A/C condenser fan motor connector and the A/C compressor clutch connector, then remove the harness clamps.

the radiator, then remove the A/C condenser fan shroud assembly from the vehicle.

6. Remove the radiator fan shroud assembly from right side of the engine compartment.

7. Disassemble the fan shrouds.

To install:

8. Assemble the fan shrouds:

a. Install the radiator fan shroud assembly.

b. With A/C: Install the A/C condenser fan shroud assembly.

9. Install the radiator upper brackets.

10. Connect the radiator fan motor connector, then install the harness clamp.

11. With A/C: Connect the A/C condenser fan motor connector and the A/C

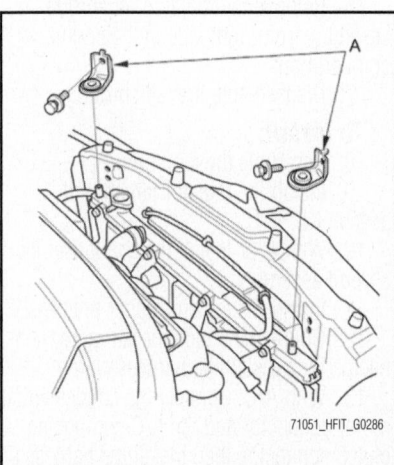

71051_HFIT_G0286

Fig. 29 Remove the radiator upper brackets (A).

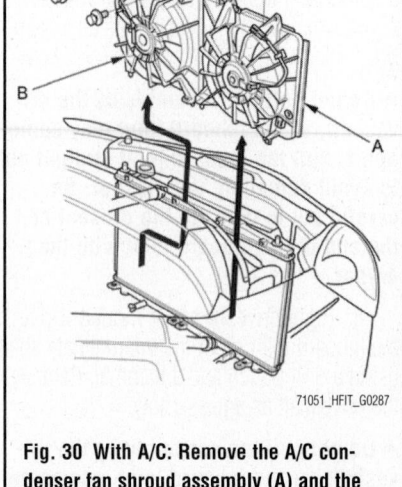

Fig. 30 With A/C: Remove the A/C condenser fan shroud assembly (A) and the radiator fan shroud assembly (B) from the radiator, then remove the A/C condenser fan shroud assembly from the vehicle.

compressor clutch connector, then install the harness clamps.

12. Install the coolant reservoir.

ENGINE COOLANT

DRAIN & REFILL

See Figures 31 and 32.

➡Always use Honda Long Life Antifreeze/Coolant Type 2. Using a non-Honda coolant can result in corrosion, causing the cooling system to malfunction or fail. Honda Long Life Antifreeze/Coolant Type 2 is a mixture of 50% antifreeze and 50% water. Do not add water.

1. Wait until the engine is cool, then carefully remove the radiator cap.

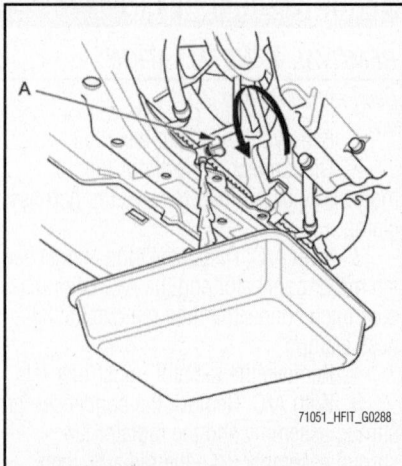

71051_HFIT_G0288

Fig. 31 Loosen the drain plug (A) and drain the coolant.

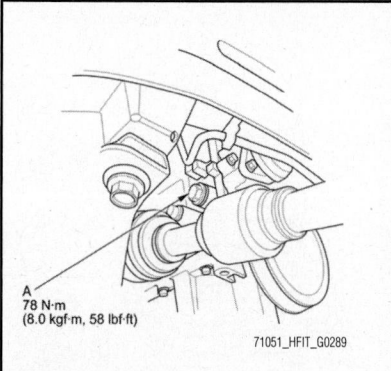

Fig. 32 Loosen the drain plug (A) and drain the coolant.

2. Remove the splash shield.

3. Loosen the drain plug and drain the coolant.

4. Remove the drain bolt located at the rear of the engine block.

5. After the coolant has drained, reinstall the drain bolt with a new washer.

6. Tighten the radiator drain plug securely.

7. Remove, drain, and reinstall the coolant reservoir.

8. Fill the coolant reservoir to the MAX mark with Honda Long Life Antifreeze/Coolant Type 2.

9. Pour Honda Long Life Antifreeze/Coolant Type 2 into the radiator up to the base of the filler neck.

10. Loosely install the radiator cap.

11. Start the engine, and let it run until it warms up (the radiator fan comes on at least twice).

12. Turn off the engine. Check the level in the radiator, and add the recommended coolant, if needed.

13. Put the radiator cap on tightly, then start the engine again, and check for leaks.

14. Clean up any spilled engine coolant.

15. If the Maintenance Minder required engine coolant replacement, reset the Maintenance Minder, and this procedure is complete. If the Maintenance Minder did not require engine coolant replacement, go to the next step.

   a. Turn the ignition switch to LOCK (0).

   b. Connect the HDS to the DLC.

   c. Turn the ignition switch to ON (II).

   d. Make sure the HDS communicates with the vehicle and the ECM/PCM. If it does not communicate, troubleshoot the DLC circuit.

   e. Select GAUGES in the BODY ELECTRICAL with the HDS.

   f. Select ADJUSTMENT in the GAUGE with the HDS.

   g. Select MAINTENANCE MINDER in the ADJUSTMENT with the HDS.

   h. Select RESET in the MAINTENANCE MINDER with the HDS.

   i. Select MAINTENANCE SUB ITEM 5 RESET with the HDS.

RADIATOR

REMOVAL & INSTALLATION

See Figure 33.

1. Drain the engine coolant.

2. Raise the vehicle on the lift to full height.

3. Disconnect the Engine Coolant Temperature (ECT) sensor 2 connector, then remove the harness clamp.

4. With A/C: Remove the A/C compressor clutch connector from the clamp, then remove the harness clamps.

5. Disconnect the lower radiator hose.

6. A/T model: Remove the Automatic Transmission Fluid (ATF) cooler hoses, then plug the hose and line.

7. Lower the vehicle on the lift.

8. Remove the coolant reservoir.

9. Disconnect the radiator fan motor connector, then remove the harness clamp.

10. With A/C: Disconnect the condenser fan motor connector, then remove the harness clamp.

11. Remove the upper radiator hose.

12. Remove the radiator upper brackets.

13. Pull up the radiator.

14. With A/C: Remove the A/C condenser fan shroud assembly.

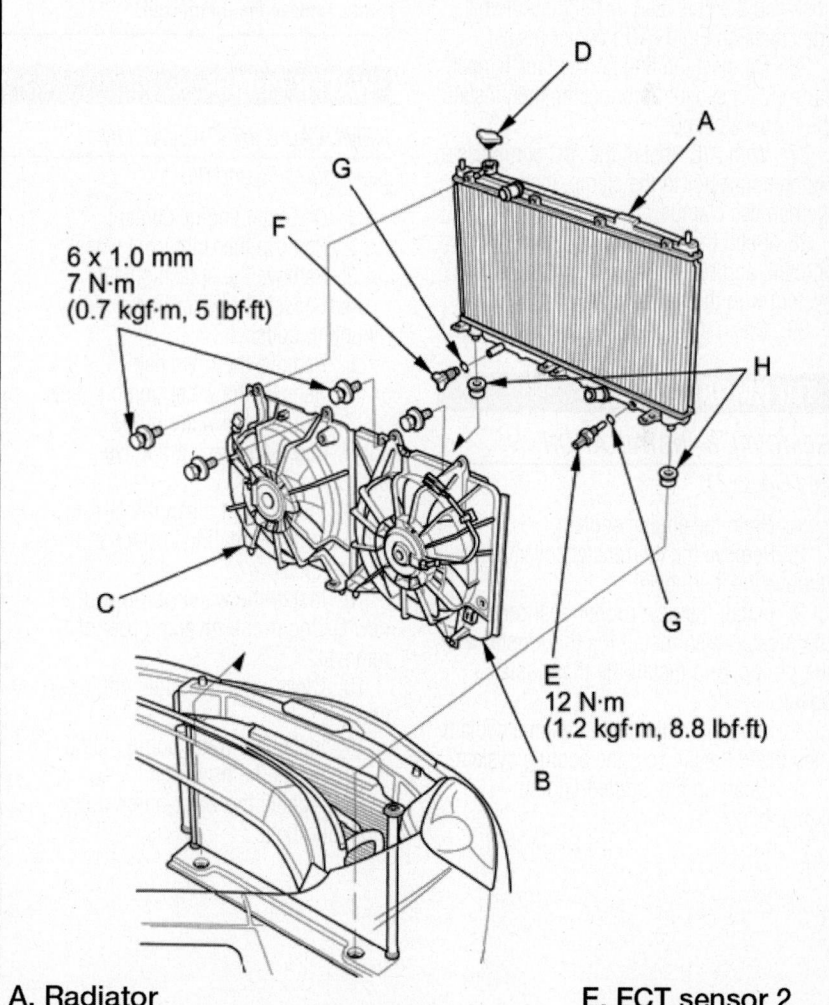

A. Radiator
B. A/C condenser fan shroud assembly
C. Radiator fan shroud assembly
D. Radiator cap

E. ECT sensor 2
F. Drain plug
G. O-rings
H. Lower cushions

Fig. 33 Pull up the radiator

15. Remove the radiator fan shroud assembly, radiator cap, the ECT sensor 2, and the drain plug.

To install:

16. Reassemble the radiator with new O-rings.

17. Install the radiator. Make sure the lower cushions are set securely.

18. Install the radiator upper brackets.

19. Install the upper radiator hose.

20. Connect the radiator fan motor connector, then install the harness clamp.

21. With A/C: Connect the condenser fan motor connector, then install the harness clamp.

22. Install the coolant reservoir.

23. Raise the vehicle on the lift to full height.

24. Install the lower radiator hose.

25. A/T model: Remove the plug from the hose and the line, then install the Automatic Transmission Fluid (ATF) cooler hoses.

26. Connect the Engine Coolant Temperature (ECT) sensor 2 connector, then install the harness clamp.

27. With A/C: Install the A/C compressor clutch connector to the clamp, then install the harness clamps.

28. Refill the radiator with engine coolant, and bleed the air from the cooling system with the heater valve open.

29. Clean up any spilled engine coolant.

THERMOSTAT

REMOVAL & INSTALLATION

See Figure 34.

1. Drain the engine coolant.

2. Remove the thermostat cover, then remove the thermostat.

3. Install the new rubber seal onto the thermostat, then install the thermostat with the pin up, and install the thermostat cover.

4. Refill the radiator with engine coolant, then bleed the air from the cooling system.

5. Clean up any spilled engine coolant.

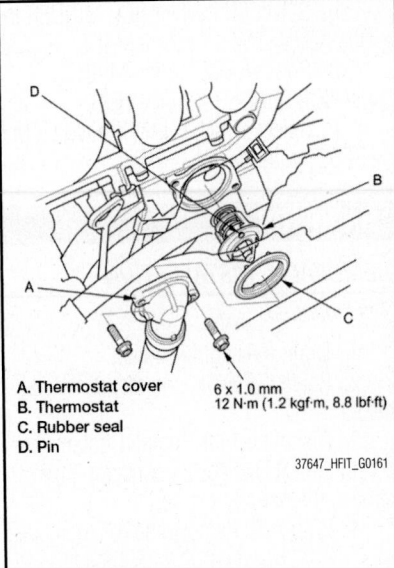

A. Thermostat cover
B. Thermostat
C. Rubber seal
D. Pin

6 x 1.0 mm
12 N·m (1.2 kgf·m, 8.8 lbf·ft)

37647_HFIT_G0161

Fig. 34 Remove the thermostat cover, then remove the thermostat

WATER PUMP

REMOVAL & INSTALLATION

See Figures 35 and 36.

1. Drain the engine coolant.

2. Remove the right front wheel.

3. Remove the splash shield.

4. Loosen the water pump pulley mounting bolts.

5. Remove the drive belt.

6. Remove the water pump pulley.

7. Remove the water pump and O-ring by removing the five bolts.

8. Inspect and clean the O-ring groove and the mating surface of the engine block.

9. Install the water pump with a new O-ring in the reverse order of removal.

10. Clean up any spilled engine coolant.

11. Install the water pump pulley.

12. Install the drive belt.

13. Tighten the water pump pulley mounting bolts.

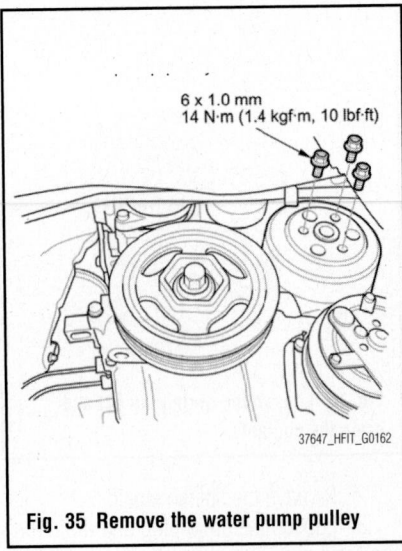

6 x 1.0 mm
14 N·m (1.4 kgf·m, 10 lbf·ft)

37647_HFIT_G0162

Fig. 35 Remove the water pump pulley

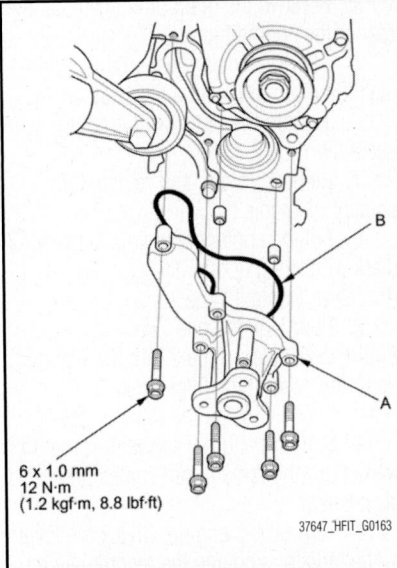

6 x 1.0 mm
12 N·m
(1.2 kgf·m, 8.8 lbf·ft)

37647_HFIT_G0163

Fig. 36 Remove the water pump (A) and O-ring (B) by removing the five bolts

14. Install the splash shield.

15. Install the right front wheel.

16. Refill the radiator with engine coolant, and bleed the air from the cooling system with the heater valve open.

BATTERY

REMOVAL & INSTALLATION

See Figure 37.

➡**The battery terminal disconnection/reconnection procedure must be done before and after doing this procedure. Some systems store data in memory that is lost when the battery is disconnected.**

1. Do the battery terminal disconnection procedure.
2. Remove the two nuts securing the battery setting plate, then remove the battery setting plate and the battery.

To install:

3. Install the battery, then install the battery setting plate.
4. Tighten the two nuts equally until the battery is stable.

➡**Do not deform the battery setting plate by over-tightening the nuts.**

5. Do the battery terminal reconnection procedure.

➡**Make sure the battery is installed correctly, and the positive terminal and the negative terminal are not connected in reverse.**

BATTERY DISCONNECTION/ RECONNECTION

DISCONNECTION PROCEDURE

➡**Some systems store data in memory that is lost when the battery is disconnected. Do the following procedures before disconnecting the battery.**

1. Make sure you have the anti-theft code for the audio system or the audio-navigation unit.
2. Make sure the ignition switch is in LOCK (0).

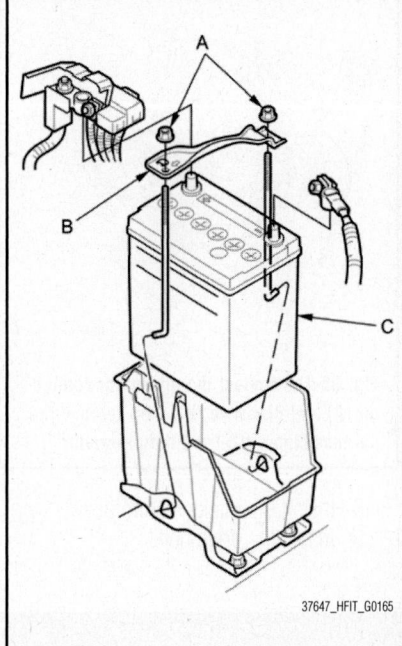

Fig. 37 Remove the two nuts (A) securing the battery setting plate (B), then remove the battery setting plate and the battery (C)

3. Disconnect and isolate the negative cable from the battery.

➡**Always disconnect the negative cable from the battery first.**

4. Disconnect the positive cable from the battery.

RECONNECTION PROCEDURE

See Figure 38.

➡**Some systems store data in memory that is lost when the battery is disconnected. Do the following procedures to restore the system back to normal operation.**

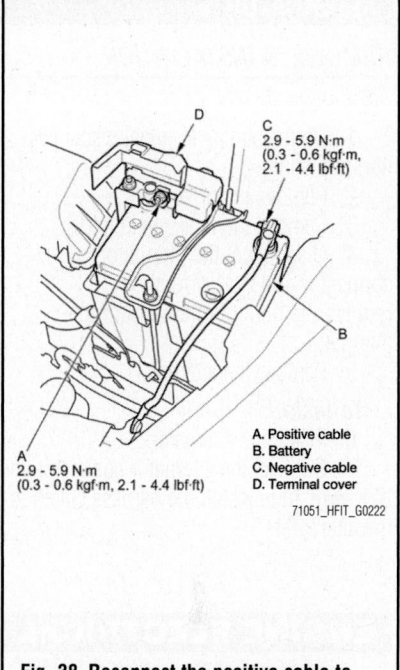

C
2.9 - 5.9 N·m
(0.3 - 0.6 kgf·m,
2.1 - 4.4 lbf·ft)

A
2.9 - 5.9 N·m
(0.3 - 0.6 kgf·m, 2.1 - 4.4 lbf·ft)

A. Positive cable
B. Battery
C. Negative cable
D. Terminal cover

71051_HFIT_G0222

Fig. 38 Reconnect the positive cable to the battery first, then reconnect the negative cable to the battery. Apply multipurpose grease to the terminals to prevent corrosion. Install the terminal cover.

1. Clean the battery terminals.
2. Test the battery.
3. Reconnect the positive cable to the battery first, then reconnect the negative cable to the battery.

➡**Always connect the positive cable to the battery first.**

4. Apply multipurpose grease to the terminals to prevent corrosion.
5. Install the terminal cover.
6. Enter the anti-theft code for the audio system or the audio-navigation unit.
7. Set the clock (for vehicles without navigation).

ENGINE ELECTRICAL   CHARGING SYSTEM

ALTERNATOR

REMOVAL & INSTALLATION

See Figures 39 and 40.

1. Do the battery terminal disconnection procedure.
2. Remove the drive belt.
3. Remove the intake manifold.
4. Disconnect the alternator connector and BLK wire, then remove the harness clamp from the alternator.
5. Remove the alternator.

To install:

6. Install the alternator.
7. Connect the alternator connector and BLK wire, then install the harness clamp to the alternator.

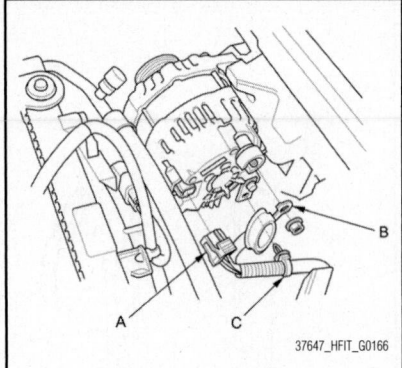

Fig. 39 Disconnect the alternator connector (A) and BLK wire (B), then remove the harness clamp (C) from the alternator

8. Install the intake manifold.
9. Install the drive belt.

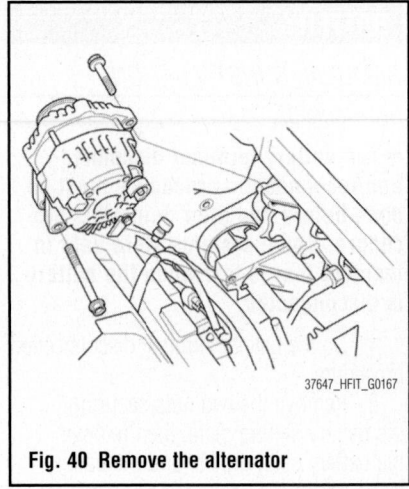

Fig. 40 Remove the alternator

10. Do the battery terminal reconnection procedure.

ENGINE ELECTRICAL   IGNITION SYSTEM

FIRING ORDERS

The firing order is: 1–3–4–2.

IGNITION COILS & SPARK PLUGS

REMOVAL & INSTALLATION

1. Remove the under-cowl panel.
2. Disconnect the ignition coil connectors, then remove the ignition coils.
3. Remove the spark plugs, and inspect them.
4. Apply a small amount of anti-seize compound to the plug threads, and screw the plugs into the cylinder head, finger tight, then tighten the plugs to the specified torque of 13 ft. lbs. (18 Nm).

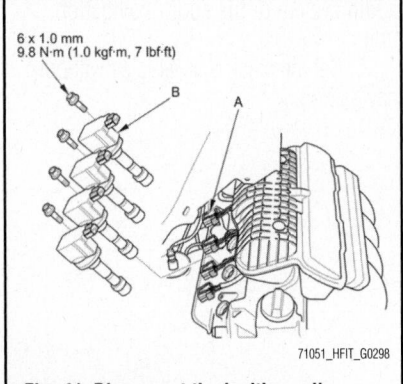

Fig. 41 Disconnect the ignition coil connectors (A), then remove the ignition coils (B).

5. Install all remaining parts in the reverse order of removal.

IGNITION TIMING

INSPECTION & ADJUSTMENT

Adjustment

Timing is controlled by the Engine Control Module (ECM)/Powertrain Control Module (PCM). No timing adjustment is possible or necessary.

Inspection

See Figures 42 and 43.

1. Connect the HDS to the DLC.
2. Turn the ignition switch to ON (II).

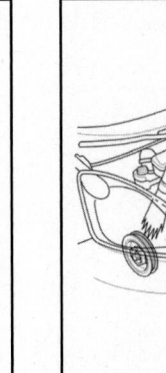

Fig. 42 Connect the timing light to the service loop (white tape).

3. Check for DTCs. If a DTC is present, diagnose and repair the cause before continuing with this test.
4. Start the engine. Hold the engine speed at 3,000 rpm with no load (M/T in neutral, A/T in P or N) until the radiator fan comes on, then let it idle.
5. Check the idle speed.
6. Jump the SCS line with the HDS.
7. Connect the timing light to the service loop (white tape).
8. Aim the light toward the pointer on

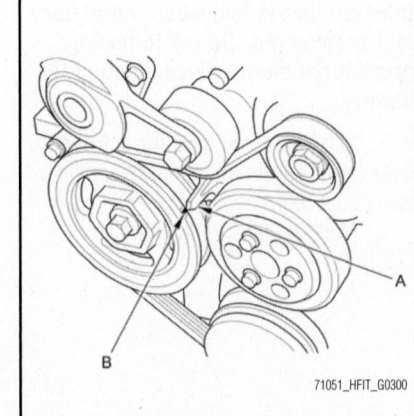

Fig. 43 Aim the light toward the pointer (A) on the chain case. Check the ignition timing under no load condition (headlights, blower fan, rear window defogger, and air conditioner are turned off). Indicating White mark (B) for timing.

the chain case. Check the ignition timing under no load condition (headlights, blower fan, rear window defogger, and air conditioner are turned off).

9. Timing specifications should be:
- M/T model: 0 ±2 ° BTDC (WHITE mark (B)) at idle in neutral

- A/T model: 0 ±2 ° BTDC (WHITE mark (B)) at idle in P or N

10. If the ignition timing differs from the specification, check the camshaft timing. If the camshaft timing is OK, update the ECM/PCM, if it does not have the latest software, or substitute a known-good ECM/PCM, then recheck. If the system works properly, and the ECM/PCM was substituted, replace the original ECM/PCM.

11. Disconnect the HDS and the timing light.

ENGINE ELECTRICAL

STARTER

REMOVAL & INSTALLATION

See Figure 44.

1. Do the battery terminal disconnection procedure.
2. Remove the coolant reservoir.
3. Remove the intake manifold.
4. Remove the dipstick, then remove the dipstick tube.
5. Disconnect the oil pressure switch connector.
6. Disconnect the starter cable from the B terminal and disconnect the connector from the S terminal.
7. Remove the two bolts securing the starter, then remove the starter from under the vehicle.

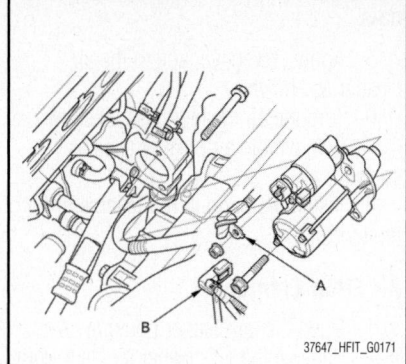

Fig. 44 Disconnect the starter cable (A) from the B terminal and disconnect the connector (B) from the S terminal

STARTING SYSTEM

To install:

8. Install the starter to the engine. Connect the starter cable and connector. Make sure the crimped side of the starter cable terminal faces away from the starter when you connect it.
9. Install the dipstick tube with a new O-ring.
10. Connect the oil pressure switch connector.
11. Install the coolant reservoir.
12. Install the intake manifold.
13. Do the battery terminal reconnection procedure.
14. Start the engine to make sure the starter works properly.

ENGINE MECHANICAL

ACCESSORY DRIVE BELT SYSTEM

ADJUSTMENT

➡**The drive belt tension is automatically maintained by the tensioner. No manual adjustment is needed.**

BELT ROUTINGS

See Figures 45 and 46.

INSPECTION

See Figure 47.

1. Inspect the belt for cracks or damage. If the belt is cracked or damaged, replace it.
2. Check that the position of the auto-tensioner indicator's pointer is within the standard range as shown.

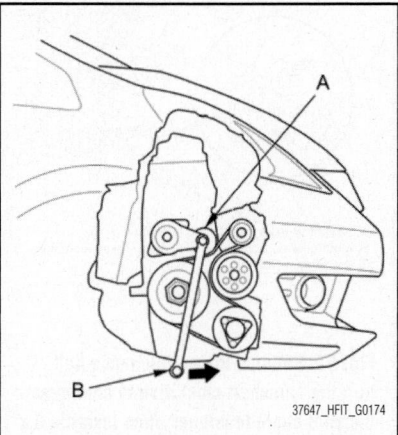

Fig. 45 Accessory drive belt routing with A/C

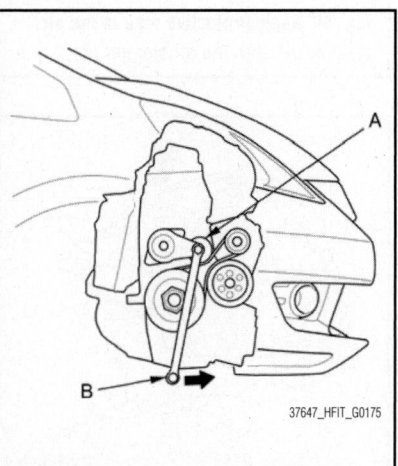

Fig. 46 Accessory drive belt routing without A/C

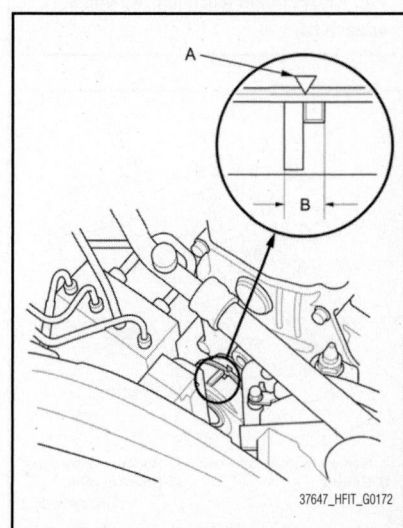

Fig. 47 Auto-tensioner indicator's pointer (A) is within the standard range (B)

If it is out of the standard range, replace the drive belt.

ADJUSTMENT

Accessory drive belt tension is provided by the auto-tensioner. No adjustment is necessary.

REMOVAL & INSTALLATION

See Figure 48.

1. Remove the splash shield.
2. Move the tensioner with a wrench in the direction shown to relieve tension from the drive belt, then remove the drive belt.
3. Install the new belt in the reverse order of removal.

AIR CLEANER

REMOVAL & INSTALLATION

Air Cleaner Assembly

See Figures 49 through 51.

1. Disconnect the MAF sensor/ IAT sensor connector, then remove the air cleaner cover and the air cleaner element.

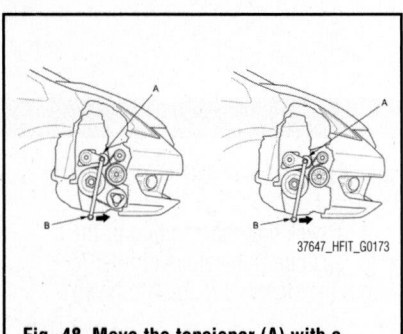

Fig. 48 Move the tensioner (A) with a wrench (B)

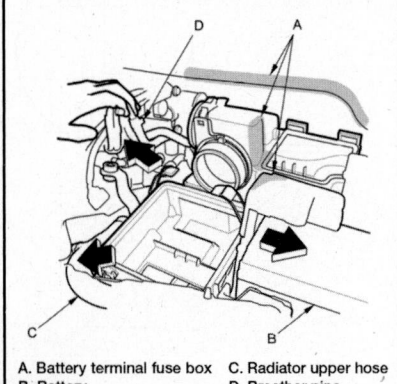

A. Battery terminal fuse box   C. Radiator upper hose
B. Battery   D. Breather pipe

37647_HFIT_G0177

Fig. 49 Apply protective tape to the air cleaner, the cowl cover, and the battery terminal fuse box

➡**Be careful not to damage the MAF sensor/IAT sensor on the air cleaner cover.**

2. Remove the air cleaner chamber.
3. Apply protective tape to the air cleaner, the cowl cover, and the battery terminal fuse box as shown.
4. Remove the battery hold down, and reposition the battery. Reposition the radiator upper hose, and the breather pipe as shown.

➡**Do not disconnect the battery terminals.**

5. Apply protective tape to the air cleaner as shown.
6. Turn the air cleaner case as shown.
7. Remove the air cleaner case as shown.
8. Install the parts in the reverse order of removal.

Air Filter Element

1. Open the air cleaner housing cover.
2. Remove the air cleaner element from the air cleaner housing.
3. Check the air cleaner element for damage, dirt, or clogging. If it is damaged or clogged, replace it.

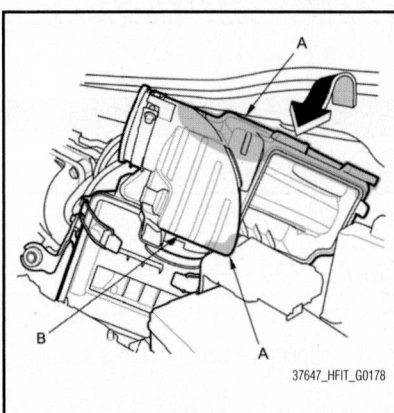

Fig. 50 Apply protective tape to the air cleaner (A); turn the air cleaner (B)

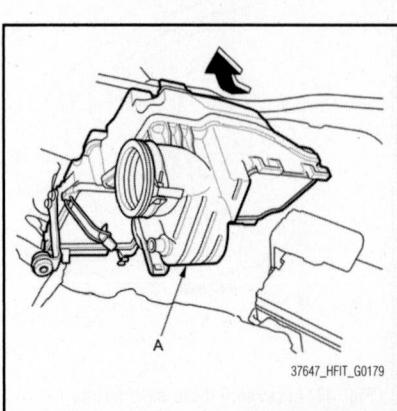

Fig. 51 Remove the air cleaner case (A)

➡**Do not use compressed air to clean the air cleaner element.**

4. Clean and remove any debris from inside the air cleaner housing.
5. Install the parts in the reverse order of removal.

   a. If you did not replace the air cleaner element, this procedure is complete.
   b. If the maintenance minder required air cleaner element replacement, reset the maintenance minder.
   c. If the idle speed fluctuates, do the idle speed inspection.

CAMSHAFT

REMOVAL & INSTALLATION

See Figures 51 and 52.

1. Remove the air cleaner assembly.
2. Remove the camshaft sprocket as follows:

   a. Remove the cylinder head cover.
   b. Make a reference mark in one position across the camshaft sprocket and cam chain.
   c. Apply new engine oil to the slider surface of the cam chain tensioner slider through the oil return hole in the cylinder head.
   d. Remove the cylinder head plug.
   e. Hold the crankshaft pulley and set the socket wrench on the camshaft sprocket bolt.
   f. Remove the maintenance bolt, and turn the camshaft clockwise to compress

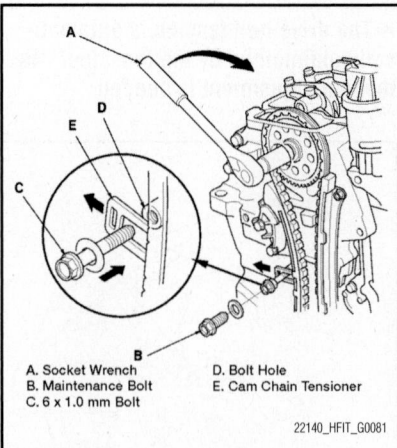

A. Socket Wrench   D. Bolt Hole
B. Maintenance Bolt   E. Cam Chain Tensioner
C. 6 x 1.0 mm Bolt

22140_HFIT_G0081

Fig. 52 Remove the maintenance bolt, turn the camshaft clockwise to compress the cam chain tensioner, then install a 6 x 1.0 mm bolt in the bolt hole on the engine block through the maintenance hole and cam chain tensioner

the cam chain tensioner, then install the 6 x 1.0 mm bolt in the bolt hole on the engine block through the maintenance hole and cam chain tensioner.

❉❉ WARNING

Turning torque should not exceed 41 ft. lbs. (56 Nm), when turning the camshaft. Do not turn the camshaft counterclockwise.

g. Hold the camshaft with a 27 mm open-end wrench, then remove the camshaft sprocket.

3. Remove the rocker arm assembly. See "Rocker Arm" in this section.

4. Remove the air cleaner housing bracket, ground cable and harness clamps, then remove the harness holder from the bracket.

5. Disconnect the Camshaft Position (CMP) sensor connector, then remove the CMP sensor.

6. Remove the camshaft thrust cover, then pull out the camshaft.

To install:

7. Install the camshaft into the cylinder head, then install the camshaft thrust cover with new O-ring. Tighten the bolts to 7.2 ft. lbs. (9.8 Nm).

8. Install the CMP sensor with new O-ring, then connect the CMP sensor connector.

9. Install the harness holder, then install harness clamps, ground cable and air cleaner housing bracket.

10. Install the camshaft sprocket as follows:

❉❉ WARNING

Keep the cam chain away from magnetic fields.

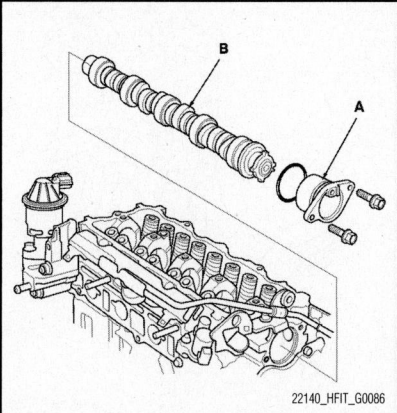

Fig. 53 Remove the camshaft thrust cover (A), then pull out the camshaft (B)

a. Install the cam chain to the camshaft sprocket by alignment the reference mark made during removal, then install the camshaft sprocket on the camshaft.

b. Hold the camshaft with a 27 mm open-end wrench, then tighten the bolt to 41 ft. lbs. (56 Nm).

c. Apply new engine oil to the slider surface of the cam chain tensioner slider through the oil return hole in the cylinder head.

d. Hold the crankshaft pulley and set the socket wrench on the camshaft sprocket bolt.

e. Turn the camshaft clockwise to compress the cam chain tensioner, then remove the 6 x 1.0 mm bolt.

❉❉ WARNING

Turning torque should not exceed 41 ft. lbs. (56 Nm), when turning the camshaft. Do not turn the camshaft counterclockwise.

f. Install the maintenance bolt with a new washer to 14 ft. lbs. (20 Nm).

g. Install the new cylinder head plug.

h. Install the cylinder head cover.

11. Install the rocker arm assembly. See "Rocker Arm" in this section.

12. Install the air cleaner assembly.

CAMSHAFT SPROCKET

REMOVAL & INSTALLATION
See Figures 54 and 55.

➡**This procedure is simply removal and installation of the camshaft sprockets. For complete chain assembly removal and installation, see "Camshaft Front Case, Camshaft Chain & Sprockets" in this section.**

❉❉ WARNING

Keep the cam chain away from magnetic fields.

1. Remove the drive belt.

2. Remove the drive belt auto-tensioner.

3. Remove the cylinder head cover.

4. Make a reference mark across the camshaft sprocket and the cam chain.

5. Apply new engine oil to the slider surface of the cam chain tensioner slider through the oil return hole in the cylinder head.

6. Remove the cylinder head plug.

7. Hold the crankshaft pulley and set a socket wrench on the camshaft sprocket bolt.

8. Remove the maintenance bolt and

turn the camshaft clockwise to compress the cam chain tensioner, then install a 6 x 1.0 mm bolt in the bolt hole on the engine block through the maintenance hole and cam chain tensioner.

❉❉ WARNING

Turning torque should not exceed 41 ft. lbs. (56 Nm) when turning the camshaft.

❉❉ WARNING

Do not turn the camshaft counterclockwise.

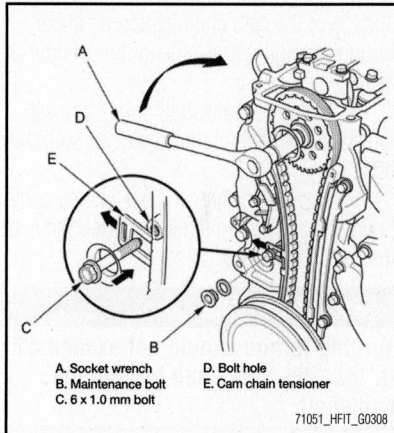

A. Socket wrench   D. Bolt hole
B. Maintenance bolt   E. Cam chain tensioner
C. 6 x 1.0 mm bolt

71051_HFIT_G0308

Fig. 54 Hold the crankshaft pulley and set a socket wrench on the camshaft sprocket bolt. Remove the maintenance bolt and turn the camshaft clockwise to compress the cam chain tensioner, then install a 6 x 1.0 mm bolt in the bolt hole on the engine block through the maintenance hole and cam chain tensioner.

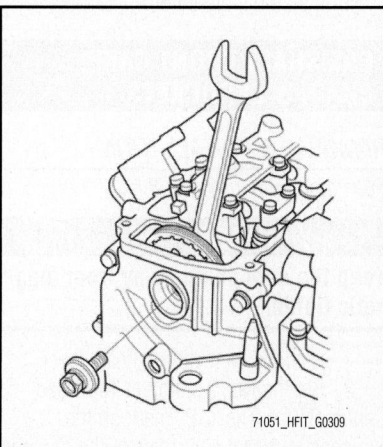

71051_HFIT_G0309

Fig. 55 Hold the camshaft with an open-end wrench, then remove the camshaft sprocket. Hang the cam chain with a wire.

9. Hold the camshaft with an open-end wrench, then remove the camshaft sprocket.

➡ Hang the cam chain with a wire.

To install:

➡ Keep the cam chain away from magnetic fields.

10. Install the cam chain to the camshaft sprocket by alignment the reference mark, then install the camshaft sprocket on the camshaft.

11. Apply new engine oil to the threads and flange of camshaft sprocket bolt.

12. Hold the camshaft with an open-end wrench, then tighten the bolt.

13. Apply new engine oil to the slider surface of the cam chain tensioner slider through the oil return hole in the cylinder head.

14. Hold the crankshaft pulley and set the socket wrench on the camshaft sprocket bolt.

15. Turn the camshaft clockwise to compress the cam chain tensioner, then remove the 6 x 1.0 mm bolt (B).

❋❋ WARNING

Turning torque should not exceed 41 ft. lbs. (56 Nm) when turning the camshaft.

❋❋ WARNING

Do not turn the camshaft counterclockwise.

16. Install the maintenance bolt with a new washer.

17. Install the new cylinder head plug.

18. Install the cylinder head cover.

19. Install the drive belt auto-tensioner.

20. Install the drive belt.

CAMSHAFT FRONT CASE, CHAIN & SPROCKETS

REMOVAL & INSTALLATION

See Figures 56 through 68.

❋❋ WARNING

Keep the cam chain away from magnetic fields.

1. Remove the cylinder head cover.

2. Set the No. 1 piston at Top Dead Center (TDC). The "UP" mark on the camshaft sprocket should be at the top, and the TDC grooves on the camshaft sprocket should line up with the top edge of the head.

3. Remove the right front wheel.

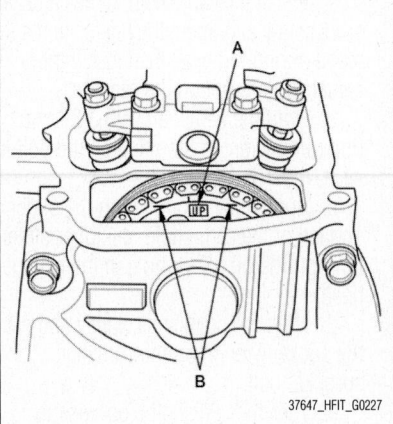

Fig. 56 Set the No. 1 piston at Top Dead Center (TDC). The "UP" mark (A) on the camshaft sprocket should be at the top, and the TDC grooves (B) on the camshaft sprocket should line up with the top edge of the head

4. Remove the splash shield.

5. Loosen the water pump pulley mounting bolts.

6. Remove the drive belt.

7. Remove the water pump pulley.

8. Remove the crankshaft pulley.

9. Remove the drive belt auto-tensioner.

10. Support the engine with a jack and a wood block under the oil pan.

11. Remove the ground cable, then remove the side engine mount/bracket assembly.

12. Remove the cam chain case.

13. Measure the cam chain separation. If the distance is less than the service limit of 0.59 inches (15 mm), replace the cam chain and cam chain tensioner.

14. Apply new engine oil to the sliding surface of the cam chain tensioner slider.

15. Hold the cam chain tensioner slider

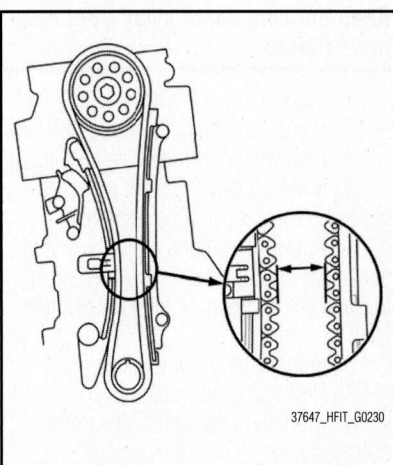

Fig. 57 Measure the cam chain separation

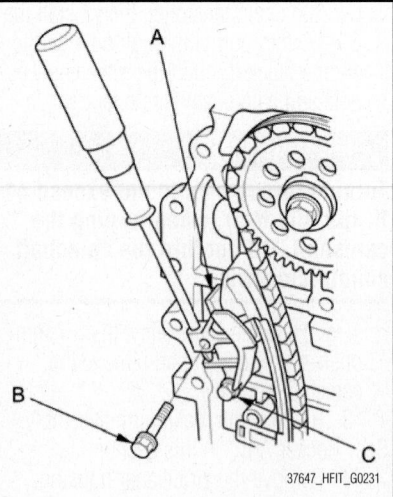

Fig. 58 Apply new engine oil to the sliding surface of the cam chain tensioner slider (A); remove the bolt (B), and loosen the bolt (C)

with the screwdriver, then remove the bolt, and loosen the bolt.

16. Remove the cam chain tensioner slider.

17. Remove the cam chain tensioner and the cam chain guide.

18. Remove the cam chain.

To install:

❋❋ WARNING

Keep the cam chain away from magnetic fields.

19. Set the crankshaft to top dead center (TDC). Align the TDC mark on the crankshaft sprocket with the pointer on the oil pump.

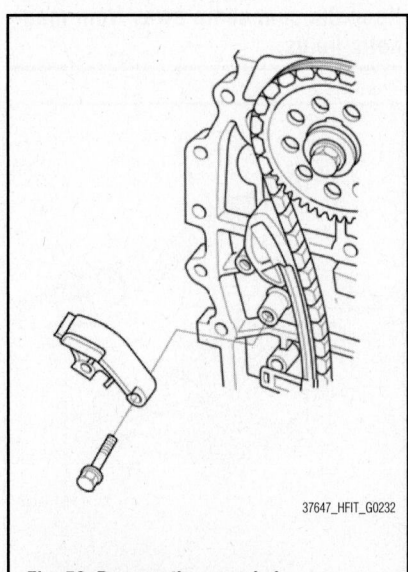

Fig. 59 Remove the cam chain tensioner slider

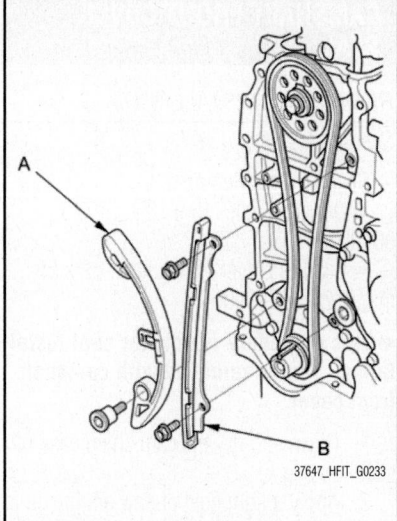

Fig. 60 Remove the cam chain tensioner (A) and the cam chain guide (B)

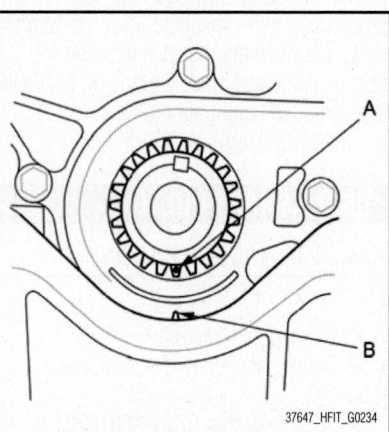

Fig. 61 Align the TDC mark (A) on the crankshaft sprocket with the pointer (B) on the oil pump

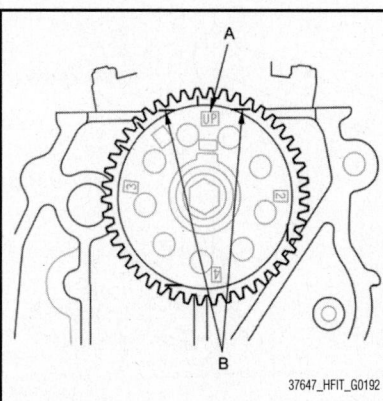

Fig. 62 The "UP" mark (A) on the camshaft sprocket should be at the top, and the TDC grooves (B) on the camshaft sprocket should line up with the top edge of the head

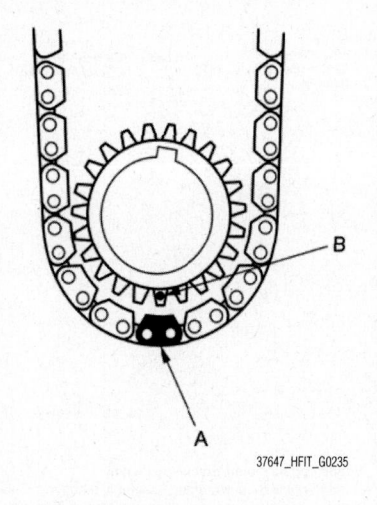

Fig. 63 Install the cam chain on the crankshaft sprocket with the colored piece (A) aligned with the TDC mark (B) on the crankshaft sprocket, then install the crankshaft sprocket to the crankshaft

20. Remove the crankshaft sprocket.
21. Set the camshaft to TDC. The "UP" mark on the camshaft sprocket should be at the top, and the TDC grooves on the camshaft sprocket should line up with the top edge of the head.
22. Install the cam chain on the crankshaft sprocket with the colored piece aligned with the TDC mark on the crankshaft sprocket, then install the crankshaft sprocket to the crankshaft.
23. Install the cam chain on the camshaft sprocket with the pointers aligned with the three colored pieces as shown.

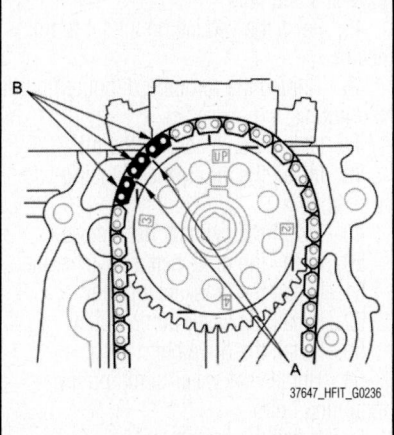

Fig. 64 Install the cam chain on the camshaft sprocket with the pointers (A) aligned with the three colored pieces (B)

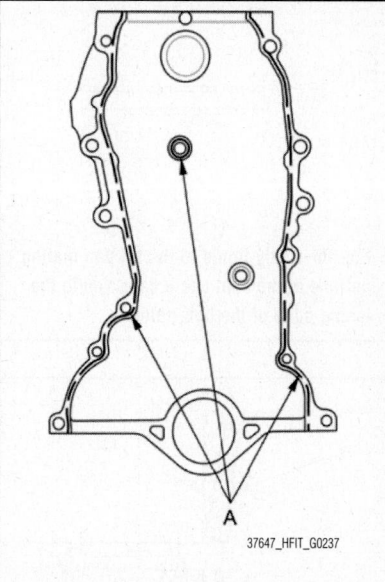

Fig. 65 Apply liquid gasket to the cylinder head and the engine block mating surfaces of the cam chain case and to the inside edge of the bolt holes

24. Install the cam chain tensioner and the cam chain guide.
25. Install the cam chain tensioner slider, and loosely tighten the bolt.
26. Apply new engine oil to the sliding surface of the cam chain tensioner slider.
27. Rotate the cam chain tensioner slider clockwise to compress the cam chain tensioner, and install the remaining bolt, then tighten the bolts.
28. Check the chain case oil seal for damage. If the oil seal is damaged, replace the chain case oil seal.
29. Remove the all of the old liquid gasket from the chain case mating surfaces, the bolts, and the bolt holes.
30. Clean and dry the chain case mating surfaces.
31. Apply liquid gasket (P/N 08717-0004, 08718-0003, or 08718-0009) to the cylinder head and the engine block mating surfaces of the cam chain case and to the inside edge of the bolt holes. Install the component within 5 minutes of applying the liquid gasket.

➡If you apply liquid gasket P/N 08718-0012, the component must be installed within 4 minutes. If too much time has passed after applying the liquid gasket, remove the old liquid gasket and residue, then reapply new liquid gasket.

32. Apply liquid gasket (P/N 08717-0004, 08718-0003, or 08718-0009) to the oil pan mating surface of the cam chain

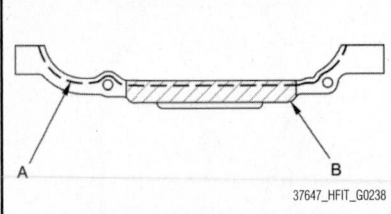

Fig. 66 Apply liquid to the oil pan mating surface of the cam chain case and to the inside edge of the bolt holes

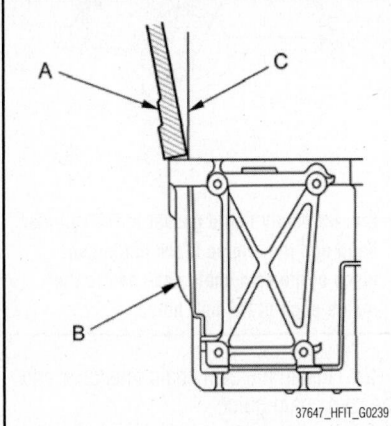

Fig. 67 Set the edge of the chain case (A) to the edge of the oil pan (B), then install the chain case on the engine block (C)

case and to the inside edge of the bolt holes. Install the component within 5 minutes of applying the liquid gasket.

➤If you apply liquid gasket P/N 08718-0012, the component must be installed within 4 minutes. If too much time has passed after applying the liquid gasket, remove the old liquid gasket and residue, then reapply new liquid gasket.

33. Set the edge of the chain case to the edge of the oil pan, then install the chain case on the engine block.

➤When installing the chain case, do not slide the bottom surface onto the oil pan mounting surface. Wait at least 30 minutes before filling the engine with oil. Do not run the engine for at least 3 hours after installing the chain case.

34. Tighten the chain case mounting bolts. Wipe off the excess liquid gasket on the oil pan and the chain case mating area.

35. Install the side engine mount/bracket assembly, then tighten the new side engine mount/bracket assembly mounting bolts.

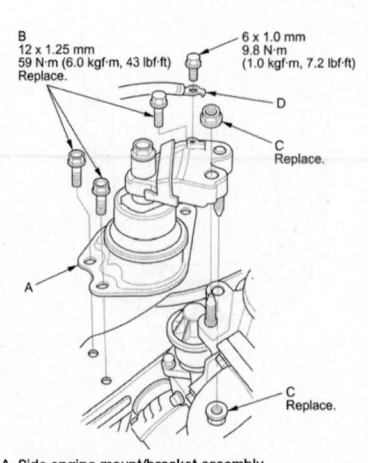

A. Side engine mount/bracket assembly
B. Side engine mount/bracket assembly mounting bolts
C. Side engine mount/bracket assembly mounting nuts
D. Ground cable

Fig. 68 Install the side engine mount/bracket assembly

36. Loosely tighten the new side engine mount/bracket assembly mounting nuts.

37. Install the ground cable.

38. Remove the air cleaner housing assembly.

39. Loosen the transaxle mount bracket mounting bolts and nuts.

40. Raise the vehicle on the lift to full height.

41. Loosen the torque rod mounting bolt and nut.

42. Lower the vehicle on the lift.

43. Tighten the side engine mount/bracket assembly mounting nuts.

44. Tighten the transaxle mount mounting bolts and nuts.

45. Raise the vehicle on the lift to full height.

46. Tighten the torque rod mounting bolt and nut.

47. Lower the vehicle on the lift.

48. Install the air cleaner housing assembly.

49. Install the cylinder head cover.

50. Install the drive belt auto-tensioner.

51. Install the crankshaft pulley.

52. Install the water pump pulley.

53. Install the drive belt.

54. Tighten the water pump pulley mounting bolts.

55. Install the splash shield.

56. Install the right front wheel.

57. Do the Crankshaft Position (CKP) pattern clear/CKP pattern learn procedure.

CAMSHAFT FRONT COVER SEAL

REMOVAL & INSTALLATION

See Figures 69 and 70.

Special Tools Required:
• Driver Handle, 15 x 135L 07749-0010000
• Bearing Driver Attachment, 52 x 55 mm 07746-0010400

➤This procedure is only for seal installation. Seal is removed with camshaft front cover.

1. Clean and dry the cam chain case oil seal housing.

2. Apply a light coat of new engine oil to the lip of the cam chain case oil seal.

3. Use the driver handle, 15 x 135L and the bearing driver attachment, 52 x 55 mm to drive a new oil seal squarely into the cam chain case to the specified installed height.

4. Measure the distance between the cam chain case surface and the oil seal. Oil seal installed height should be 1.193–1.220 in. (30.3–31.0 mm).

CRANKSHAFT PULLEY

REMOVAL & INSTALLATION

See Figures 71 and 72.

Special Tools Required:
• Crankshaft Pulley Holder 07AAB-RJAA100
• Socket, 19 mm 07JAA-001020A or equivalent
• Holder Handle 07JAB-001020B

1. Remove the drive belt.

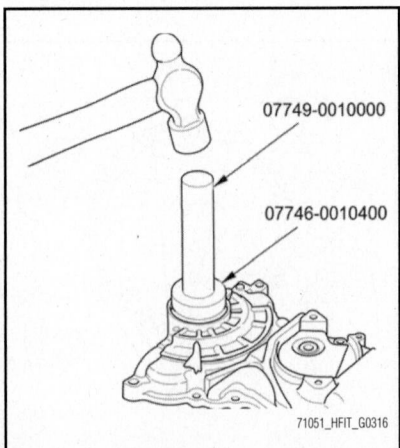

Fig. 69 Use the driver handle, 15 x 135L and the bearing driver attachment, 52 x 55 mm to drive a new oil seal squarely into the cam chain case to the specified installed height.

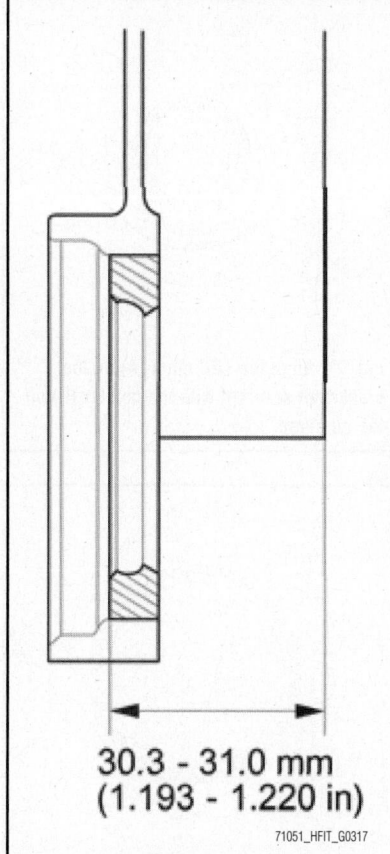

30.3 - 31.0 mm
(1.193 - 1.220 in)

71051_HFIT_G0317

Fig. 70 Measure the distance between the cam chain case surface and the oil seal. Oil seal installed height should be 1.193–1.220 in. (30.3–31.0 mm).

2. Hold the pulley with the handle and the crankshaft pulley holder.

3: Remove the bolt with a 19 mm socket and a breaker bar, then remove the crankshaft pulley.

To install:

4. Clean the crankshaft pulley, the crankshaft, the bolt, and the washer. Lubricate with new engine oil.

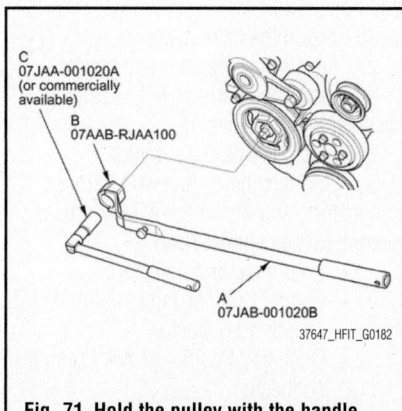

Fig. 71 Hold the pulley with the handle (A) and the crankshaft pulley holder (B)

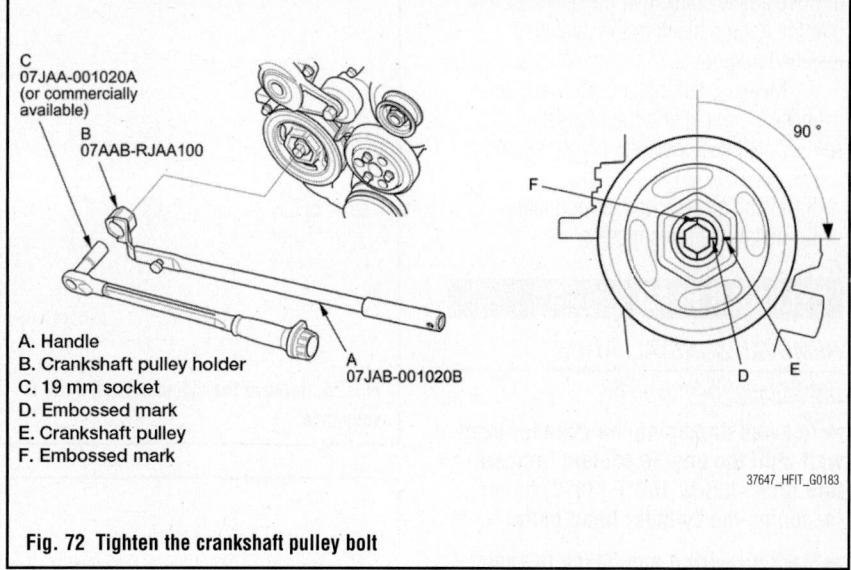

A. Handle
B. Crankshaft pulley holder
C. 19 mm socket
D. Embossed mark
E. Crankshaft pulley
F. Embossed mark

37647_HFIT_G0183

Fig. 72 Tighten the crankshaft pulley bolt

5. Install the crankshaft pulley.

6. Tighten the crankshaft pulley bolt. Do not use an impact wrench.

a. Hold the pulley with the handle and crankshaft pulley holder, then tighten the bolt to 27 ft. lbs. (37 Nm) with a torque wrench and a 19 mm socket. If the pulley bolt or crankshaft are new, tighten the bolt to 130 ft. lbs. (177 Nm), then remove the bolt and tighten it to 27 ft. lbs. (37 Nm).

b. Mark the embossed mark on the bolt flange and the crankshaft pulley as shown, then tighten the bolt an additional 90 degrees (The mark on the bolt head should line up with the next embossed mark on the bolt flange).

7. Install the drive belt.

CRANKSHAFT REAR COVER & SEAL

REMOVAL & INSTALLATION
See Figures 73 and 74.

➡ **Special Tools Required:**

- Driver Handle, 15 x 135L 07749-0010000
- Oil Seal Driver Attachment, 96 mm 07ZAD-PNAA100

1. Remove the transmission. See "Transmission" section.

2. On M/T model, remove the pressure plate, the clutch disc, and the flywheel.

3. On A/T model, remove the drive plate.

4. Clean and dry the crankshaft oil seal housing.

5. Apply a light coat of new engine oil to the lip of the crankshaft oil seal.

6. Using the driver handle, 15 x 135 L and the oil seal driver attachment, 96 mm,

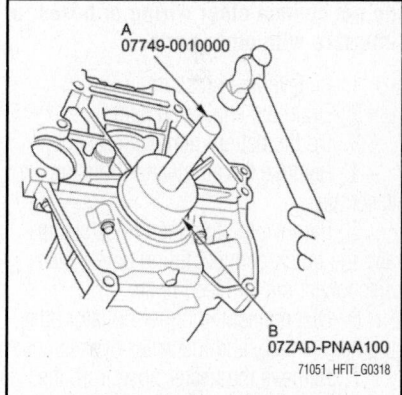

71051_HFIT_G0318

Fig. 73 Using the driver handle, 15 x 135 L (A) and the oil seal driver attachment, 96 mm (B), to drive a new crankshaft oil seal squarely into the engine block to the specified installed height.

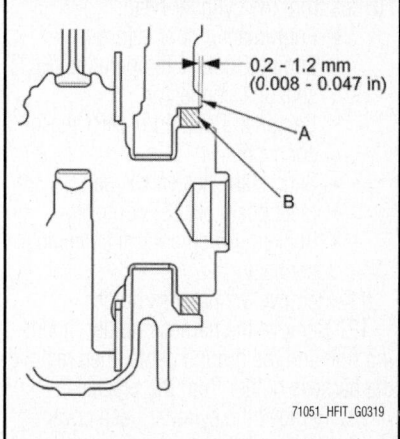

71051_HFIT_G0319

Fig. 74 Measure the distance between the cylinder block (A) and the crankshaft oil seal (B). Oil seal installed height should be 0.008–0.047 in. (0.2–1.2 mm).

to drive a new crankshaft oil seal squarely into the engine block to the specified installed height.

7. Measure the distance between the cylinder block (A) and the crankshaft oil seal (B). Oil seal installed height should be 0.008–0.047 in. (0.2–1.2 mm).

8. Install all removed components.

9. Install the transmission.

CYLINDER HEAD

REMOVAL & INSTALLATION

See Figures 75 through 80.

➡**To avoid damaging the cylinder head, wait until the engine coolant temperature drops below 100°F (38°C) before loosening the cylinder head bolts.**

➡**Mark all wiring and hoses to avoid misconnection. Also, be sure that they do not contact other wiring or hoses, or interfere with other parts.**

1. Relieve fuel pressure.
2. Drain the engine coolant.
3. Do the battery removal procedure.
4. Remove the air cleaner housing assembly.
5. Remove the fuel feed hose clamp and the quick-connect fitting cover, then disconnect the fuel feed hose.
6. Disconnect the upper radiator hose, the heater hose, and the water bypass hose.
7. Remove the heater hose from the clamps.
8. Remove the intake manifold/chamber assembly.
9. Remove the Warm Up Three Way Catalytic Converter (WU-TWC).
10. Remove the following engine wire harness connectors and wire harness clamps from the cylinder head:
    - Four injector connectors
    - Engine Coolant Temperature (ECT) sensor 1 connector
    - Camshaft Position (CMP) sensor connector
    - Secondary Heated Oxygen Sensor (secondary HO2S) connector
    - Rocker arm oil control solenoid connector
11. Remove the harness holder.
12. Remove the harness holder mounting bolt and the ground cable, then remove the harness holder from the bracket.
13. Remove the cylinder head cover.
14. Remove the cam chain.
15. Remove the cylinder head bolts. To prevent warpage, loosen the bolts in sequence 1/3 turn at a time; repeat the sequence until all bolts are loosened.

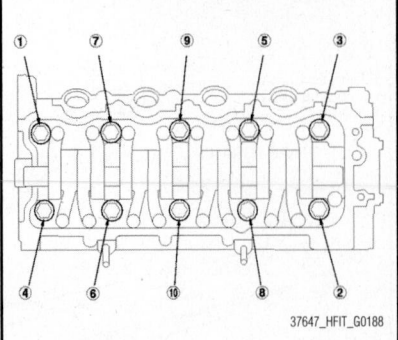

Fig. 75 Remove the cylinder head bolts in sequence

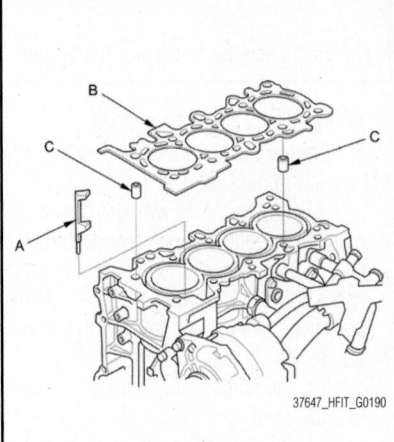

Fig. 76 Install a new coolant separator (A), new cylinder head gasket (B) and the dowel pins (C)

16. Remove the cylinder head.

To install:

17. Clean the cylinder head and the block surface.

18. Install a new coolant separator in the engine block whenever the engine block is replaced.

19. Install the new cylinder head gasket and the dowel pins on the engine block. Always use a new cylinder head gasket.

20. Set the crankshaft to Top Dead Center (TDC). Align the TDC mark on the crankshaft sprocket with the pointer on the oil pump.

21. Set the camshaft TDC. The "UP" mark on the camshaft sprocket should be at the top, and the TDC grooves on the camshaft sprocket should line up with the top edge of the head.

22. Install the cylinder head on the engine block.

23. Apply new engine oil to the threads and under the bolt heads of all cylinder head bolts.

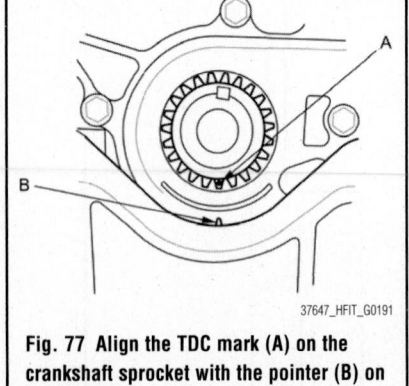

Fig. 77 Align the TDC mark (A) on the crankshaft sprocket with the pointer (B) on the oil pump

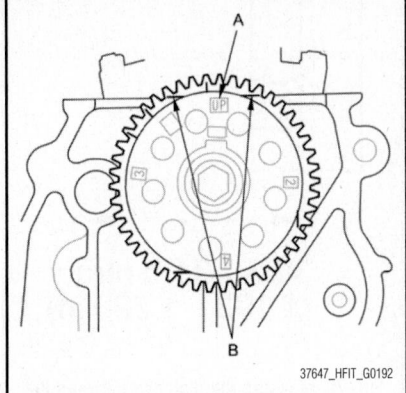

Fig. 78 The "UP" mark (A) on the camshaft sprocket should be at the top, and the TDC grooves (B) on the camshaft sprocket should line up with the top edge of the head

24. Tighten the cylinder head bolts in sequence to 22 ft. lbs. (29 Nm). Using a beam-type torque wrench. When using a pre-set-type torque wrench, be sure to tighten slowly and do not overtighten. If a bolt makes any noise while you are torquing it, loosen the bolt and retighten it from the first step.

25. Tighten all cylinder head bolts an additional 130°.

26. Install the cam chain.

27. Install the cylinder head cover.

28. Install the harness holder, then install the ground cable.

29. Install the harness holder.

30. Connect the engine wire harness connectors, and install the wire harness clamps to the cylinder head:
    - Four injector connectors
    - Engine Coolant Temperature (ECT) sensor 1 connector
    - Camshaft Position (CMP) sensor connector
    - Secondary Heated Oxygen Sensor (secondary HO2S) connector

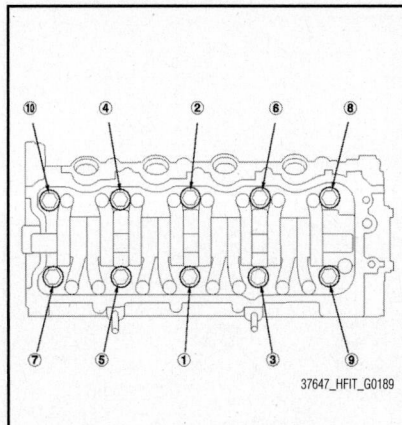

Fig. 79 Tighten the cylinder head bolts in sequence

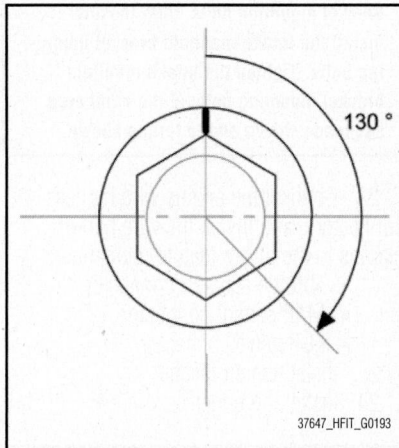

Fig. 80 Tighten all cylinder head bolts an additional 130°

- Rocker arm oil control solenoid connector

31. Install the Warm Up Three Way Catalytic Converter (WU-TWC).

32. Install the intake manifold/chamber assembly.

33. Install the upper radiator hose, the heater hose, and the water bypass hose.

34. Install the heater hose to the clamps.

35. Connect the fuel feed hose, then install the quick-connect fitting cover and the fuel feed hose clamp.

36. Install the air cleaner housing assembly.

37. Do the battery installation procedure.

38. Inspect for fuel leaks. Turn the ignition switch to ON (II) (do not operate the starter) so the fuel pump runs for about 2 seconds and pressurizes the fuel line. Repeat this operation three times, then check for fuel leakage at any point in the fuel line.

39. Refill the radiator with engine coolant, and bleed air from the cooling system with the heater valve open.

40. Do the Crankshaft Position (CKP) pattern clear/CKP pattern learn procedure.

41. Inspect the idle speed.

42. Inspect the ignition timing.

CYLINDER HEAD COVERS

REMOVAL & INSTALLATION

See Figure 81.

1. Remove the intake manifold chamber.

2. Disconnect the four ignition coil connectors.

3. Remove the harness holder and the breather hose.

4. Remove the cylinder head cover.

To install:

5. Thoroughly clean the head cover gasket and the groove.

6. Install the head cover gasket in the groove of the cylinder head cover.

7. Check that the mating surfaces are clean and dry.

8. Apply liquid gasket (P/N 08717-0004, 08718-0003, or 08718-0009) to the chain case contact areas (A). Install the component within 5 minutes of applying the liquid gasket.

➡**If you apply liquid gasket P/N 08718-0012, the component must be installed within 4 minutes. If too much time has passed after applying the liquid gasket, remove the old liquid gasket and residue, then reapply new liquid gasket.**

9. Place the cylinder head cover on the cylinder head, then slide the cover slightly back and forth to seat the head cover gasket.

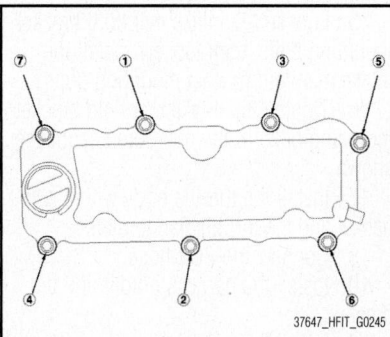

Fig. 81 Tighten the bolts in three steps in sequence

10. Tighten the bolts in three steps. In the final step tighten all bolts, in sequence, to 7 ft. lbs. (9.8 Nm).

➡**Wait at least 30 minutes before filling the engine with oil. Do not run the engine for at least 3 hours after installing the head cover.**

11. Install the harness holder and the breather hose.

12. Connect the four ignition coil connectors.

13. Install the intake manifold chamber.

ENGINE OIL & FILTER

OIL LEVEL CHECK

Park the vehicle on level ground, and start the engine. Hold the engine at 3,000 rpm with no load (M/T in neutral, A/T in P or N) until the radiator fan comes on, then turn off the engine, and wait a few minutes.

Remove the dipstick, and wipe off the dipstick, then reinstall the dipstick.

Remove the dipstick, and check the engine oil level. It should be between the upper mark and the lower mark.

If the engine oil level is near or below the lower mark, check for oil leakage, and add engine oil to bring it to the upper mark.

OIL & FILTER CHANGE

Engine Oil

1. Warm up the engine.

2. Remove the drain bolt and drain the engine oil.

3. Reinstall the drain bolt with a new washer.

4. Refill the engine with the recommended oil.

5. Connect the Honda Diagnostic System (HDS) to the Data Link Connector (DLC).

6. Turn the ignition switch to ON (II).

7. Make sure the HDS communicates with the vehicle and the engine control module (ECM)/powertrain control module (PCM). If it does not communicate, troubleshoot the DLC circuit.

8. Select BODY ELECTRICAL with the HDS.

9. Select ADJUSTMENT in GAUGE MENU with the HDS.

10. Select RESET in the MAINTENANCE MINDER with the HDS.

11. Select RESETTING THE ENGINE OIL LIFE with the HDS.

→If you changed the Automatic Transmission Fluid (ATF) at the same time with the engine oil, select RESETTING THE ENGINE OIL LIFE AND ATF with the HDS instead.

12. Run the engine for more than 3 minutes, then check the oil level and check for oil leaks.

Engine Oil Filter

1. Drain the engine oil.
2. Remove the oil filter with the oil filter wrench.
3. Inspect the filter to make sure the rubber seal is not stuck to the oil filter seating surface of the engine.
4. Inspect the threads (A) and the rubber seal (B) on the new filter. Clean the seat on the oil pan, then apply a light coat of new engine oil to the filter rubber seal. Use only filters with a built-in bypass system.
5. Install the oil filter by hand.
6. After the rubber seal seats, tighten the oil filter clockwise with the oil filter wrench.

→**Tighten: ¾ Turn Clockwise**

If four numbers or marks (1 to 4 or ▼ to ▼▼▼▼) are printed around the outside of the filter, you can use the following procedure to tighten the filter.

  a. Spin the filter on until its seal lightly seats against the oil pan, and note which number or mark is at the bottom.
  b. Tighten the filter by turning it clockwise three numbers or marks from the one you noted. For example, if mark ▼ is at the bottom when the seal is lightly seated, tighten the filter until the mark ▼▼▼▼ comes around to the bottom.
7. Refill the new engine oil, run the engine for more than 3 minutes, then check for oil leakage.

FRONT COVER & SEAL

REMOVAL & INSTALLATION

→**See "Camshaft Front Case, Chain & Sprockets" in this section.**

INTAKE MANIFOLD/CHAMBER ASSEMBLY

REMOVAL & INSTALLATION

See Figures 82 and 83.

1. Remove the under-cowl panel.
2. Remove the air cleaner.
3. Disconnect the engine wire harness connectors, and remove the wire harness clamps from the intake manifold chamber:

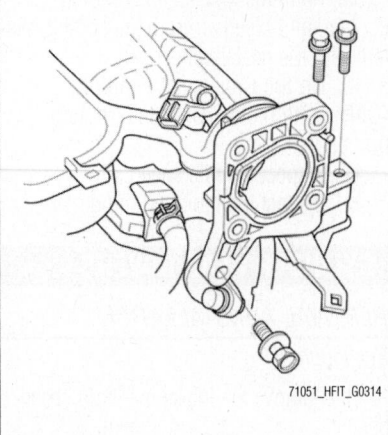

Fig. 82 Remove the intake manifold bracket mounting bolts.

  • Throttle actuator connector
  • MAP sensor connector
  • EGR valve connector
4. Disconnect the brake booster vacuum hose and remove the harness clamp.
5. Remove the water bypass hose from the clamp.
6. Remove the throttle body without disconnecting the water bypass hoses.
7. Remove the throttle flange plates.
8. Remove the harness holder and the dipstick.
9. Disconnect the PCV hose.
10. Remove the intake manifold bracket mounting bolts.
11. Remove the intake manifold/chamber assembly from the cylinder head.
12. Disassemble the intake manifold/chamber assembly.

To install:
13. Reassemble the intake manifold/chamber assembly.
14. Install the intake manifold/chamber assembly with a new gasket. Tighten the bolts and nuts in a crisscross pattern in three steps, beginning with the inner bolt. Tighten to 17 ft. lbs. (24 Nm).
15. Loosen the intake manifold bracket mounting bolts, then loosely install the intake manifold bracket mounting bolts.
16. Tighten the intake manifold bracket mounting bolts in the numbered sequence shown.
17. Install the throttle body with a new gasket and the throttle flange plates.
18. Connect the PCV hose.
19. Install the harness holder and the dipstick.
20. Install the water bypass hose to the clamp.
21. Connect the brake booster vacuum hose and install the harness clamp.

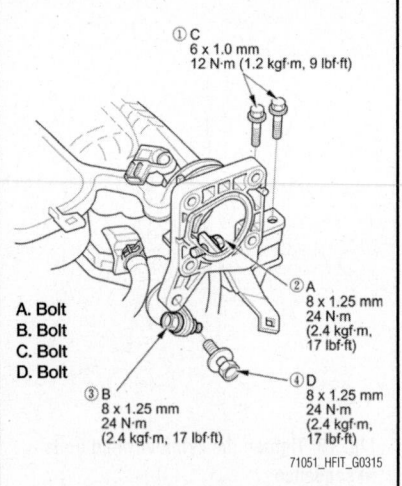

A. Bolt
B. Bolt
C. Bolt
D. Bolt

① C
6 x 1.0 mm
12 N·m (1.2 kgf·m, 9 lbf·ft)

② A
8 x 1.25 mm
24 N·m
(2.4 kgf·m,
17 lbf·ft)

③ B
8 x 1.25 mm
24 N·m
(2.4 kgf·m, 17 lbf·ft)

④ D
8 x 1.25 mm
24 N·m
(2.4 kgf·m,
17 lbf·ft)

Fig. 83 Loosen the intake manifold bracket mounting bolts, then loosely install the intake manifold bracket mounting bolts. Tighten the intake manifold bracket mounting bolts in the numbered sequence shown and to torque shown.

22. Connect the engine wire harness connectors, and install the wire harness clamps to the intake manifold chamber:
  • Throttle actuator connector
  • MAP sensor connector
  • EGR valve connector
23. Install the air cleaner.
24. Install the under-cowl panel.

OIL PAN

REMOVAL & INSTALLATION

See Figures 84 through 87.

1. If the engine is already out of the vehicle, go to step 7.
2. Drain the engine oil.
3. Remove the drive belt.
4. Remove the driveshaft heat shield.
5. Remove the A/C compressor without disconnecting the A/C hoses.
6. A/T model: Remove the shift cable cover.
7. M/T model: Remove the torque rod bracket.
8. Remove the dipstick, then remove the dipstick tube.
9. Remove the crankshaft position (CKP) sensor cover, then disconnect the CKP sensor connector.
10. Remove the clutch cover/torque converter cover, and transaxle mounting bolts.
11. Remove the oil pan bolts. Note the bolt locations by their size.
12. Using a flat blade screwdriver, separate the oil pan from the block in the places shown.
13. Remove the oil pan.

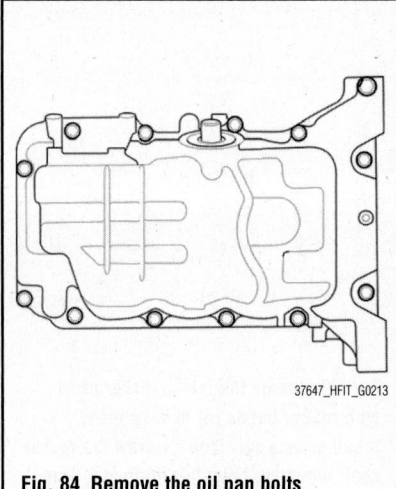

Fig. 84 Remove the oil pan bolts

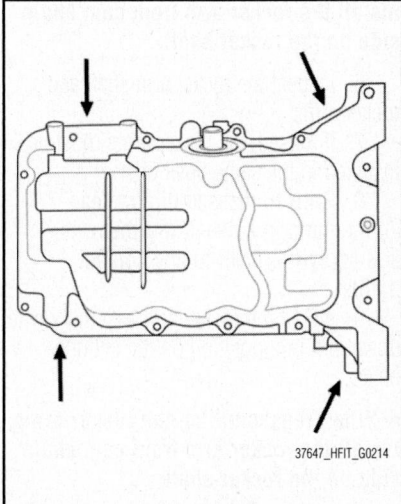

Fig. 85 Using a flat blade screwdriver, separate the oil pan from the block

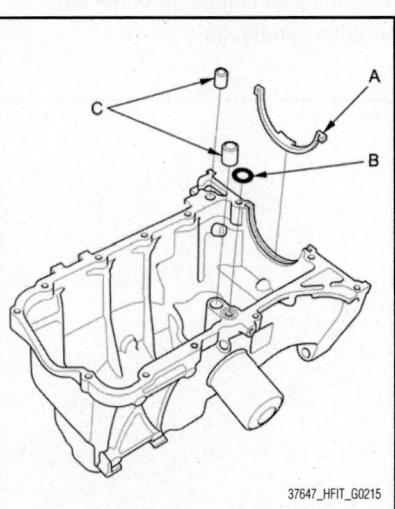

Fig. 86 Install the new oil pan gasket (A), new O-ring (B), and dowel pins (C) on the oil pan

To install:

14. Remove any old liquid gasket from the oil pan mating surfaces bolts, and bolt holes.

15. Clean and dry the oil pan mating surfaces.

16. Install the new oil pan gasket, new O-ring, and dowel pins on the oil pan.

17. Apply liquid gasket (P/N 08717-0004, 08718-0003, or 08718-0009) to the engine block mating surface of the oil pan and to the inside edge of the bolt holes. Install the component within 5 minutes of applying the liquid gasket.

18. Install the oil pan.

➡**Note the following:**

- Wait at least 30 minutes before filling the engine with oil.
- Do not run the engine for at least 3 hours after installing the oil pan.
- Make sure to install the bolts in the correct locations according to size.

19. Tighten the bolts in three steps. Wipe off the excess liquid gasket on the each side of crankshaft pulley and the flywheel/drive plate.

20. Install the clutch cover/torque converter cover, and install the transaxle mounting bolts.

21. Connect the CKP sensor connector, then install the CKP sensor cover.

22. Install the dipstick tube with a new O-ring, then install the dipstick.

23. M/T model: Install the torque rod bracket.

24. If the engine is still in the vehicle, do the remaining steps.

25. A/T model: Install the shift cable cover.

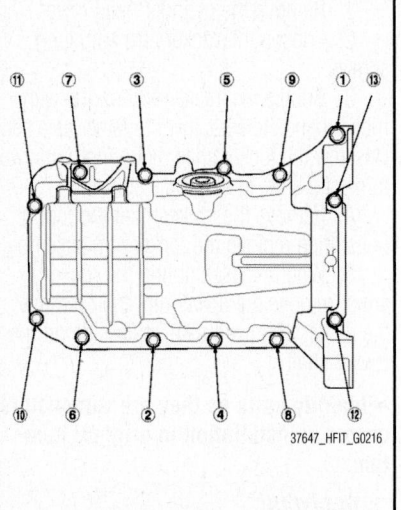

Fig. 87 Tighten the bolts in three steps in sequence

26. Install the A/C compressor.

27. Install the driveshaft heat shield.

28. Install the drive belt.

29. Refill the engine with engine oil.

OIL PUMP

REMOVAL & INSTALLATION

See Figures 88 and 89.

1. Remove the under-cowl panel.

2. Remove the cam chain.

3. Remove the alternator mounting bolt, then install the support eyelet.

4. Install the A and Reds engine support hanger (AAR-T1256), then attach the hook to the slotted hole in the support eyelet. Tighten the wing nut by hand to lift and support the engine.

5. Remove the oil pan.

6. Remove the oil screen, then remove the oil pump.

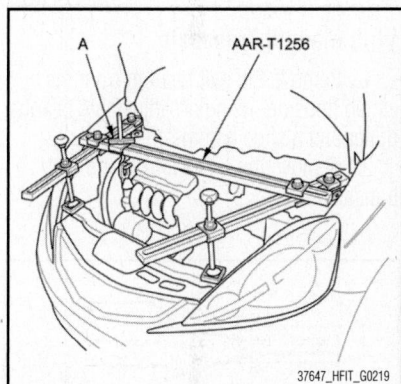

Fig. 88 Install the A and Reds engine support hanger and tighten the wing nut (A) by hand

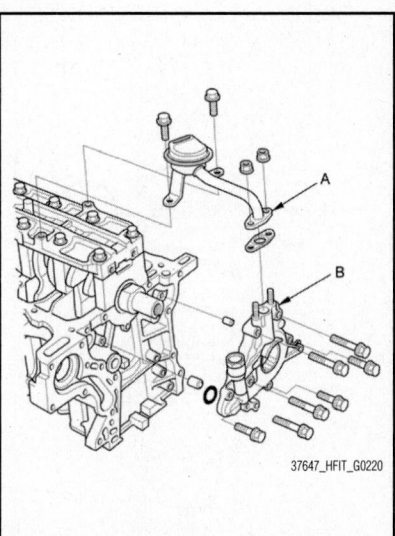

Fig. 89 Remove the oil screen (A), then remove the oil pump (B)

To install:

7. Clean the O-ring groove and the mating surface of the engine block.

8. Install the oil pump with a new O-ring.

9. Install the oil screen with a new gasket.

10. Install the oil pan.

11. Support the engine with a jack and a wood block under the oil pan.

12. Remove the engine hanger and support eyelet, then tighten the alternator mounting bolt.

13. Install the cam chain.

PISTONS & RINGS

POSITIONING

See Figures 90 and 91.

REAR MAIN SEAL

REMOVAL & INSTALLATION

With Manual Transaxle

1. Remove the ball bearing from the clutch housing, using an adjustable bearing puller and a slide hammer.

2. Remove the oil seal from the transaxle side.

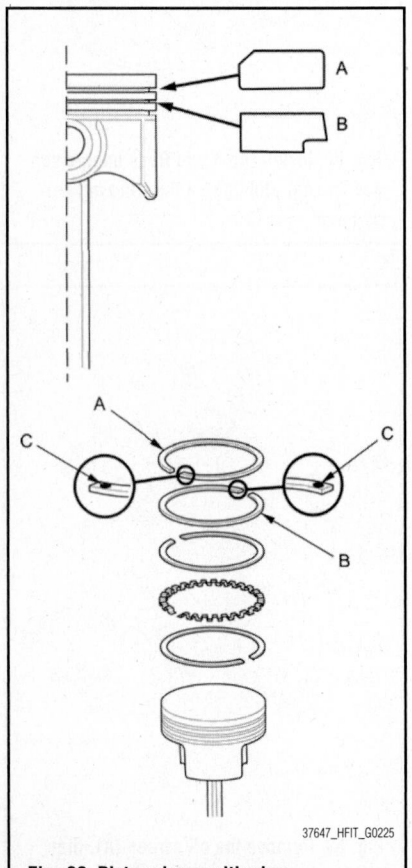

Fig. 90 Piston ring positioning

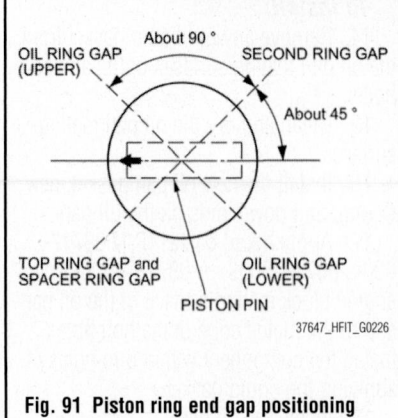

Fig. 91 Piston ring end gap positioning

To install:

3. Drive in the new oil seal from the transaxle side using the driver and 37 x 40 mm attachment.

4. Drive in the new ball bearing from the transaxle side using the driver and 52 x 55 mm attachment.

With Automatic Transaxle

1. Remove the mainshaft bearing and the oil seal using the adjustable bearing puller and a slide hammer.

To install:

2. Install the new mainshaft bearing until it bottoms in the torque converter housing using the driver and the attachment (62 x 68 mm).

3. Install the new oil seal flush with the housing using the driver and the attachment (72 x 75 mm).

ROCKER ARMS

REMOVAL & INSTALLATION

See Figures 92 through 95.

1. Remove the cylinder head cover.

2. Loosen the rocker arm adjusting screws.

3. Bundle the intake rocker arms with rubber bands to keep them together as a set. Unscrew the rocker shaft mounting bolts two turns at a time, in sequence shown.

4. Remove the rocker shaft mounting bolts, then remove the rocker arm assembly.

5. When disassembling the rocker arms, remove the dowel pin, then remove the rocker arm from cam chain side on the rocker shaft.

➡**Identify parts as they are removed to ensure reinstallation in original location.**

To install:

➡**When disassembling or reassembling the rocker arms, remove or**

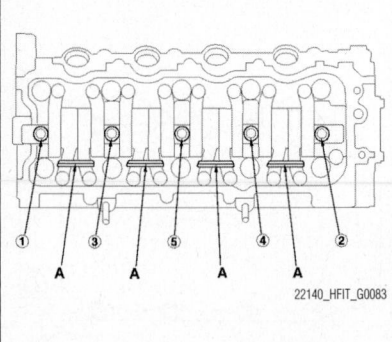

Fig. 92 Bundle the intake rocker arms with rubber bands (A) to keep them together as a set, then unscrew the rocker shaft mounting bolts two turns at a time, in sequence shown

install the rocker arm from cam chain side on the rocker shaft.

6. Inspect the rocker arm shaft and rocker arms.

7. If reused, the rocker arms must be installed in the same positions.

8. Prior to reassembling, clean all the parts in solvent, dry them, and apply lubricant to any contact points.

9. Install the rocker arm assembly with dowel pin into position on the cylinder head.

➡**When reassembling the rocker arms, Install the rocker arm from cam chain side on the rocker shaft**

10. Tighten each bolt, two turns at a time, in the sequence shown. Torque bolts to 22 ft. lbs. (29 Nm).

➡**Apply new engine oil to the bolt threads and flange.**

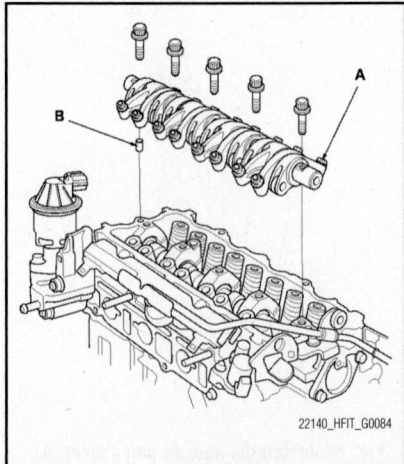

Fig. 93 Removing the dowel pin (B), then the rocker arm assembly (A)

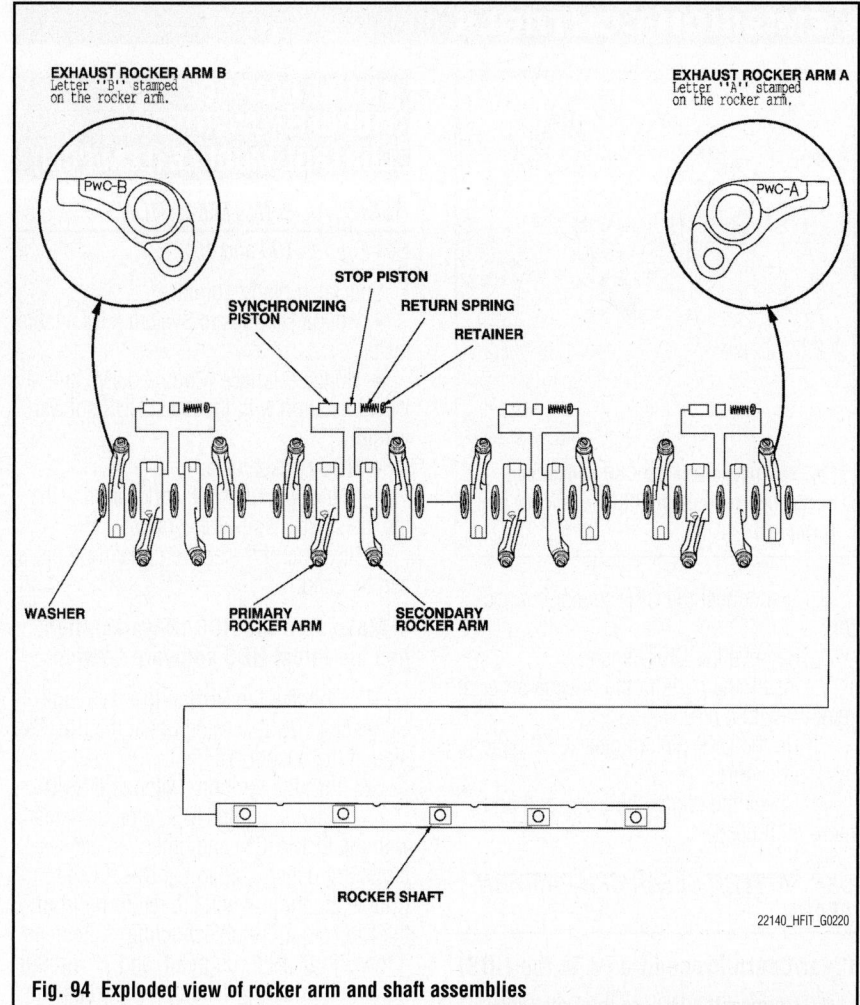

Fig. 94 Exploded view of rocker arm and shaft assemblies

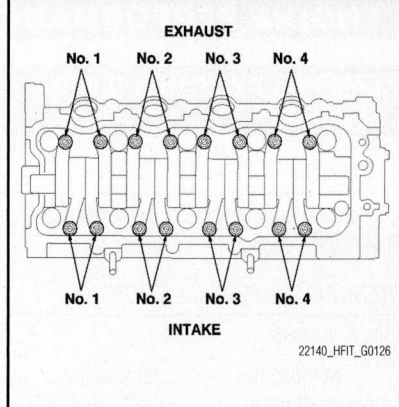

Fig. 96 Showing valve positions for reference during adjustment

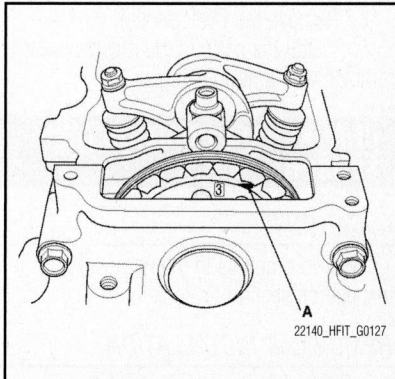

Fig. 97 Rotate the crankshaft clockwise. Align the No. 3 piston TDC groove (A) on the camshaft sprocket with the top edge of the head

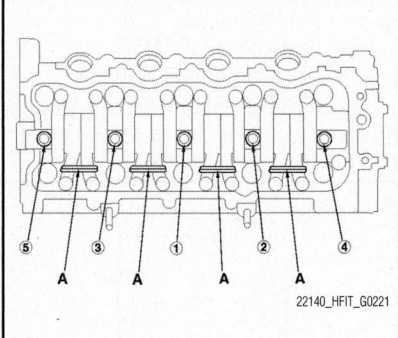

Fig. 95 Showing rocker arm bolt tightening sequence

11. Remove the rubber bands from the intake rocker arms.
12. Adjust the valve clearance.
13. Install the cylinder head cover.

VALVE LASH (VALVE CLEARANCE)

ADJUSTMENT

➡**Valves should be adjusted only when the cylinder head temperature is less than 100°F (38°C).**

1. Remove the cylinder head cover.
2. Set the No. 1 piston at top dead center (TDC). The "UP" mark on the camshaft sprocket should be at the top, and the TDC grooves on the camshaft sprocket should line up with the top edge of the head.
3. Select the correct thickness feeler gauge for the valves to check.
4. Proper valve clearance should be:
   • Intake: 0.006–0.007 inches (0.15–0.19 mm)

• Exhaust: 0.010–0.012 inches (0.26–0.30 mm)
5. Insert the feeler gauge between the adjusting screw and the end of the valve stem and slide it back and forth; a slight amount of drag should be felt.
6. If too much or too little drag is present, loosen the locknut, and turn the adjusting screw until the drag on the feeler gauge is correct.
7. Tighten the locknut to 10 ft. lbs. (14 Nm) and recheck the clearance. Repeat the adjustment if necessary.
8. Rotate the crankshaft clockwise. Align the No. 3 piston TDC groove on the camshaft sprocket with the top edge of the head.
9. Check and if necessary, adjust the valve clearance on No. 3 cylinder.
10. Repeat this procedure for cylinder No. 4 and then for cylinder No. 2.

ENGINE PERFORMANCE & EMISSION CONTROLS

CAMSHAFT POSITION (CMP) SENSOR

LOCATION

Refer to graphics in the Removal and Installation section for location.

REMOVAL & INSTALLATION

See Figure 98.

1. Remove the cowl cover and the under-cowl panel.
2. Remove the air cleaner.
3. Disconnect the CMP sensor connector.
4. Remove the CMP sensor.
5. Install the parts in the reverse order of removal with a new O-ring.

CRANKSHAFT POSITION (CKP) SENSOR

LOCATION

Refer to graphics in the Removal and Installation section for location.

REMOVAL & INSTALLATION

See Figure 99.

1. Raise the vehicle on a lift.

➡**Make sure the vehicle is level, because engine oil will drip out when you remove the sensor.**

2. Loosen the bolt. Remove the bolt and the CKP sensor cover.

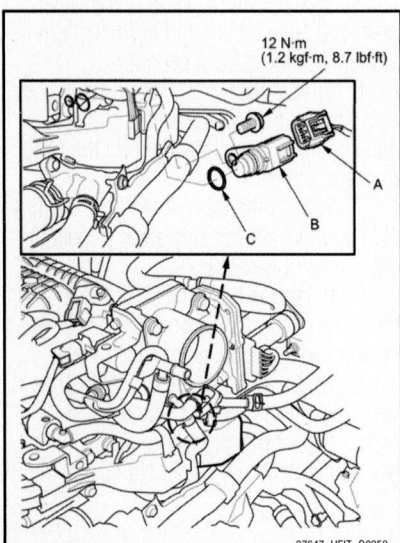

Fig. 98 Disconnect the CMP sensor connector (A) and remove the CMP sensor (B) and O-ring (C)

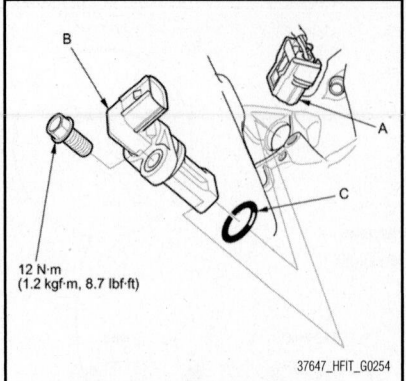

12 N·m
(1.2 kgf·m, 8.7 lbf·ft)

37647_HFIT_G0254

Fig. 99 Disconnect the CKP sensor connector (A), remove the CKP sensor (B) and O-ring (C)

3. Disconnect the CKP sensor connector.
4. Remove the CKP sensor.
5. Install the parts in the reverse order of removal with a new O-ring.
6. Do the CKP pattern clear/CKP pattern learn procedure.
7. Check the engine oil level, and add more oil if needed.

CKP PATTERN CLEAR/CKP PATTERN LEARN

Clear/Learn Procedure (with the HDS)

1. Connect the HDS to the Data Link Connector (DLC) located under the driver's side of the dashboard.
2. Turn the ignition switch to ON (II).
3. Make sure the HDS communicates with the ECM/PCM and other vehicle systems. If it doesn't, go to the DLC circuit troubleshooting.
4. Select CRANK PATTERN in the ADJUSTMENT MENU with the HDS.
5. Select CRANK PATTERN LEARNING with the HDS, and follow the screen prompts.

Learn Procedure (without the HDS)

1. Start the engine. Hold the engine speed at 3,000 rpm without load (A/T in P or N, M/T in neutral) until the radiator fan comes on.
2. Test-drive the vehicle on a level road: Decelerate (with the throttle fully closed) from an engine speed of 2,500 rpm down to 1,000 rpm with the A/T in 2, or the M/T in 2nd.
3. Repeat step 2 several times.
4. Turn the ignition switch to LOCK (0).
5. Turn the ignition switch to ON (II), and wait 30 seconds.

ENGINE CONTROL MODULE/POWERTRAIN CONTROL MODULE (ECM/PCM)

REMOVAL & INSTALLATION

See Figures 100 and 101.

Special Tools Required:
- Honda Diagnostic System (HDS) tablet tester
- Honda Interface Module (HIM) and an iN workstation with the latest HDS software version
- HDS pocket tester
- GNA600 and an iN workstation with the latest HDS software version
- Any one of the above updating tools can be used.

➡**Make sure the HDS/iN workstation has the latest HDS software version.**

1. Connect the HDS to the Data Link Connector (DLC) located under the driver's side of the dashboard.
2. Turn the ignition switch to ON (II).
3. Make sure the HDS communicates with the ECM/PCM and other vehicle systems. If it doesn't, go to the DLC circuit troubleshooting. If you are returning from the DLC circuit troubleshooting, skip steps 4 through 9, 19 through 24, and 27 through 29, and do these procedures after replacing the ECM/PCM:
   - Replace the engine oil and the engine oil filter.
   - Replace the ATF (A/T).
   - Clean the throttle body.
4. Select the PGM-FI system with the HDS.

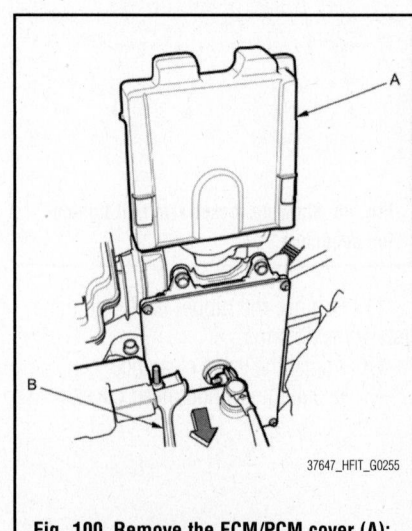

37647_HFIT_G0255

Fig. 100 Remove the ECM/PCM cover (A); remove the battery holder (B)

5. Select the INSPECTION MENU with the HDS.

6. Select the ETCS TEST, then select the TP POSITION CHECK, and follow the screen prompts.

➡**If the TP POSITION CHECK indicates FAILED, continue with this procedure.**

7. Select the REPLACE ECM/PCM MENU, then select READ DATA, and follow the screen prompts.

➡**Doing this step copies (READS) the engine oil life data from the original ECM/PCM so you can later download (WRITES) it into the new ECM/PCM. If READ DATA indicates FAILED, continue with this procedure.**

8. A/T: Select the A/T system with the HDS.

9. A/T: Select the REPLACE TCM/PCM MENU, then READ DATA, and follow the screen prompts.

➡**A/T: Doing this step copies (READS) the ATF life data from the original PCM so you can later download (WRITES) it into the new PCM. A/T: If READ DATA indicates FAILED, continue with this procedure.**

10. Jump the SCS line with the HDS.
11. Turn the ignition switch to LOCK (0).
12. Remove the ECM/PCM cover.
13. Remove the battery holder, and reposition the battery away from the ECM/PCM.

➡**Do not disconnect the battery terminals.**

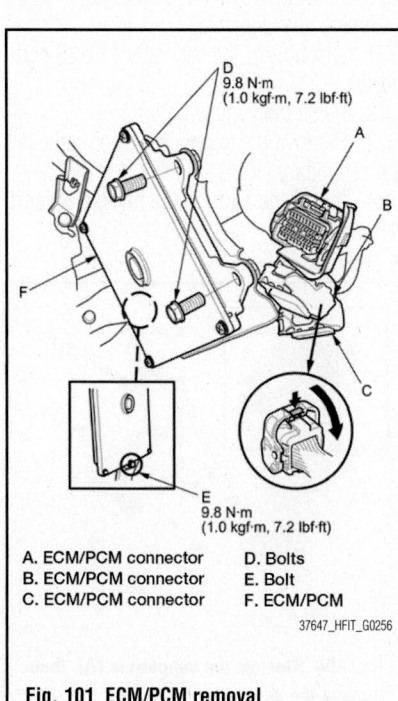

A. ECM/PCM connector	D. Bolts
B. ECM/PCM connector	E. Bolt
C. ECM/PCM connector	F. ECM/PCM

37647_HFIT_G0256

Fig. 101 ECM/PCM removal

14. Remove the bolts, and loosen the bolt.
15. Disconnect ECM/PCM connectors A, B, and C, then remove the ECM/PCM.

➡**ECM/PCM connectors A, B, and C have symbols (A=□, B=△, C=○) embossed on them for identification.**

16. Install a known-good ECM/PCM in the reverse order of removal.

17. Turn the ignition switch to ON (II).

18. Manually input the VIN to the ECM/PCM with the HDS.

➡**DTC P0630 (VIN Not Programmed or Mismatch) may be stored because the VIN has not been programmed into the ECM/PCM; ignore it, and continue this procedure.**

19. If the READ DATA (engine oil life) failed in step 7, go to step 22 (A/T) or step 25 (M/T). Otherwise, go to step 20.

20. Select the PGM-FI system with the HDS.

21. Select the REPLACE ECM/PCM MENU, then WRITE DATA, and follow the screen prompts.

➡**If the WRITE DATA indicates FAILED, continue with this procedure.**

22. A/T: If the READ DATA (ATF life) failed in step 8, go to step 25. Otherwise go to step 23.

23. A/T: Select the A/T SYSTEM with the HDS.

24. A/T: Select the REPLACE ECM/PCM MENU, then WRITE DATA, and follow the screen prompts.

➡**A/T: If the WRITE DATA indicates FAILED, continue with this procedure.**

25. Select IMMOBI system with the HDS.
26. Enter the immobilizer ECM/PCM code that you got from the iN, and use the ECM/PCM replacement procedure in the HDS; it allows you to start the engine.

27. If the TP POSITION CHECK failed in step 6 clean the throttle body, then go to step 28.

28. If the READ DATA failed in step 7 or the WRITE DATA failed in step 21, replace the engine oil and the engine oil filter, then go to step 29 (A/T) or step 30 (M/T).

29. If the READ DATA failed in step 9 or the WRITE DATA failed in step 24, replace the ATF, then go to step 30.

30. Select the PGM-FI system, and reset the ECM/PCM with the HDS.

31. Update the ECM/PCM if it does not have the latest software.

32. Do the ECM/PCM idle learn procedure.

33. Do the CKP pattern learn procedure.

ECM/PCM UPDATE

Special Tools Required:
• Honda Diagnostic System (HDS) tablet tester
• Honda Interface Module (HIM) and an iN workstation with the latest HDS software version
• HDS pocket tester
• GNA600 and an iN workstation with the latest HDS software version
• Any one of the above updating tools can be used.

➡**Note the following:**

• Use this procedure when you need to update the ECM/PCM at any time.
• Make sure the HDS/iN workstation has the latest HDS software version.
• Before you update the ECM/PCM, make sure the battery in the vehicle is fully charged, and connect a jumper battery (not a battery charger) to maintain system voltage.
• Never turn the ignition switch to ACC (I) or LOCK (0) during the update. If there is a problem with the update, leave the ignition switch ON (II).
• To prevent ECM/PCM damage, do not operate anything electrical (headlights, audio system, brakes, A/C, power windows, door locks, etc.) during the update.
• To ensure the latest program is installed, do an ECM/PCM update whenever the ECM/PCM is substituted or replaced.
• You cannot update an ECM/PCM with a program it already has. It will only accept a new program.
• High temperature in the engine compartment might cause the ECM/PCM to become too hot to run the update. If the engine has been running before this procedure, open the hood and cool the engine compartment.
• If you need to diagnose the Honda interface module (HIM) because the HIM's red (#3) light came on or was flashing during the update, leave the ignition switch in ON (II) when you disconnect the HIM from the Data Link Connector (DLC). This prevents damage to the ECM/PCM.

1. Turn the ignition switch to ON (II), but do not start the engine.

2. Connect the HDS to the Data Link Connector (DLC) located under the driver's side of the dashboard.

3. Make sure the HDS communicates with the ECM/PCM and other vehicle systems. If it doesn't, go to the DLC circuit troubleshooting. If you are returning from the DLC circuit troubleshooting, skip steps 4 and 5, and clean the throttle body after updating the ECM/PCM.

4. Select the INSPECTION MENU with the HDS.

5. Select the ETCS TEST, then select the TP POSITION CHECK, and follow the HDS screen prompts.

➡**If the TP POSITION CHECK indicates FAILED, continue this procedure.**

6. Exit the HDS diagnostic system, then select the update mode, and follow the screen prompts to update the ECM/PCM.

7. If the software in the ECM/PCM is the latest, disconnect the HDS/HIM from the DLC, and go back to the procedure that you were doing. If the software in the ECM/PCM is not the latest, follow the instructions on the screen. If prompted to choose the PGM-FI system or the A/T system, make sure you update both.

➡**If the ECM/PCM update system requires you to cool the ECM/PCM, follow the instructions on screen. If you have a problem during the update procedure (programming takes over 15 minutes, status bar goes over 100 %, D (A/T) or immobilizer indicator flashes, HDS tablet freezes, etc.), follow these steps to minimize the chance of damaging the ECM/PCM:**

- Leave the ignition switch in the ON (II) position.
- Connect a jumper battery (do not connect a battery charger).
- Shut down the HDS.
- Disconnect the HDS from the DLC.
- Reboot the HDS.
- Reconnect the HDS to the DLC, and try the update procedure again.

8. If the TP POSITION CHECK failed in step 5, clean the throttle body.

9. Do the ECM/PCM idle learn procedure.

10. Do the CKP pattern learn procedure.

ECM/PCM IDLE LEARN PROCEDURE

The idle learn procedure must be done so the ECM/PCM can learn the engine idle characteristics.

Do the idle learn procedure whenever you do any of these actions:
- Replace ECM/PCM.

- Reset ECM/PCM.
- Update ECM/PCM.
- Replace or clean the throttle body.
- Disassemble the engine or the transaxle.

➡**Clearing DTCs with the HDS does not require that you to do the idle learn procedure.**

1. Make sure all electrical items (A/C, audio, lights, etc.) are off.

2. Reset the ECM/PCM with the HDS.

3. Turn the ignition switch to ON (II), and wait 2 seconds.

4. Start the engine. Hold the engine speed at 3,000 rpm without load (A/T in P or N, M/T in neutral) until the radiator fan comes on, or until the engine coolant temperature reaches 194°F (90°C).

5. Let the engine idle for about 5 minutes with the throttle fully closed.

➡**If the radiator fan comes on, do not include its running time in the 5 minutes.**

ENGINE COOLANT TEMPERATURE (ECT) SENSORS

REMOVAL & INSTALLATION

ECT Sensor 1
See Figure 102.

1. Drain the engine coolant.
2. Remove the air cleaner.
3. Remove the hose clamp stay bolt.
4. Disconnect the ECT sensor 1 connector.
5. Remove ECT sensor 1.
6. Install the parts in the reverse order of removal with a new O-ring, then refill the radiator with engine coolant.

ECT Sensor 2
See Figure 103.

1. Drain the engine coolant.
2. Raise the vehicle on a lift.

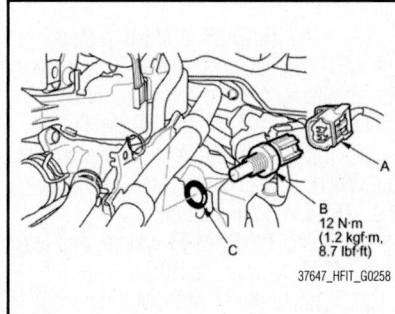

Fig. 102 Disconnect the ECT sensor 1 connector (A); remove the ECT sensor 1 (B) and the O-ring (C)

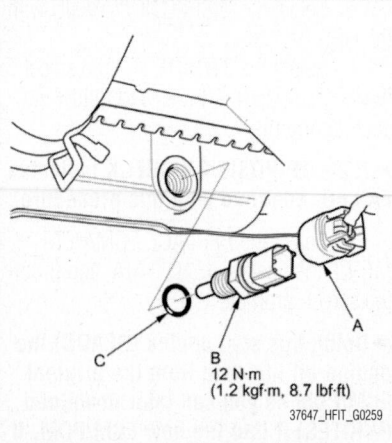

37647_HFIT_G0259

Fig. 103 Disconnect the ECT sensor 2 connector (A), then remove ECT sensor 2 (B) and the O-ring (C)

3. Disconnect the ECT sensor 2 connector, then remove ECT sensor 2.

4. Install the parts in the reverse order of removal with a new O-ring, then refill the radiator with engine coolant.

HEATED OXYGEN SENSOR (HO2S)

LOCATION

The secondary Heated Oxygen Sensor (HO2S) is located on the under-floor TWC.

REMOVAL & INSTALLATION
See Figure 104.

Special Tools Required: O2 Sensor Wrench, Snap-on S6176 or equivalent, commercially available

1. Disconnect the secondary HO2S connector.

2. Raise the vehicle on a lift.

3. Remove the supporters, then remove the secondary HO2S.

4. Install the parts in the reverse order of removal.

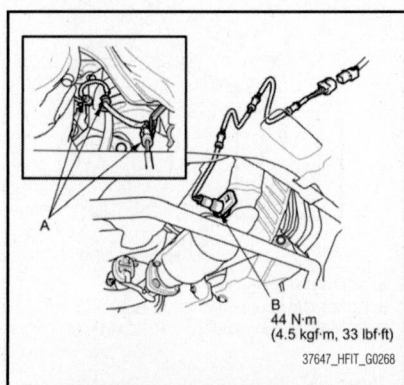

37647_HFIT_G0268

Fig. 104 Remove the supporters (A), then remove the secondary HO2S (B)

INPUT SHAFT (MAINSHAFT) SPEED SENSOR

LOCATION

See Figure 105.

REMOVAL & INSTALLATION

1. Remove the air cleaner assembly.
2. Disconnect the input shaft (mainshaft) speed sensor connector, and remove the input shaft (mainshaft) speed sensor.
3. Install a new O-ring on the new input shaft (mainshaft) speed sensor, then install the input shaft (mainshaft) speed sensor.
4. Check the connector for rust, dirt, or oil, then connect the connector securely.
5. Install the air cleaner assembly.

B
Replace.
A
6 x 1.0 mm
12 N·m
(1.2 kgf·m, 8.7 lbf·ft)
71051_HFIT_G0335

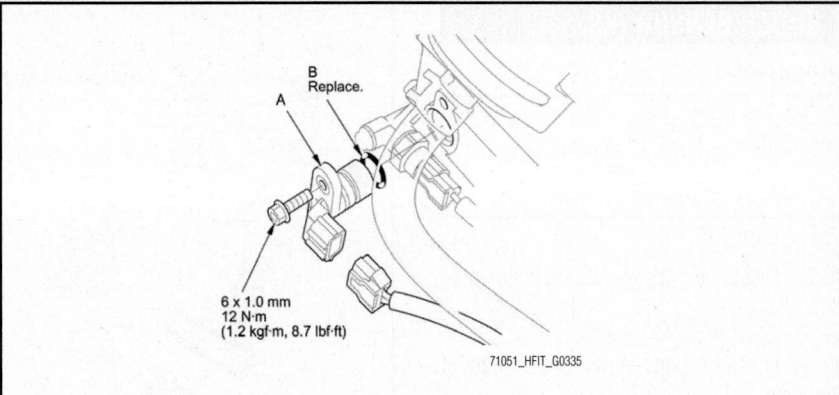

Fig. 106 Disconnect the input shaft (mainshaft) speed sensor connector, and remove the input shaft (mainshaft) speed sensor (A). Install a new O-ring (B) on the new input shaft (mainshaft) speed sensor, then install the input shaft (mainshaft) speed sensor.

A/T CLUTCH PRESSURE CONTROL SOLENOID VALVE A
POWERTRAIN CONTROL MODULE (PCM)
INPUT SHAFT (MAINSHAFT) SPEED SENSOR
TRANSMISSION FLUID PRESSURE SWITCH A (2ND CLUTCH)
TRANSMISSION RANGE SWITCH
TRANSMISSION FLUID PRESSURE SWITCH B (3RD CLUTCH)
OUTPUT SHAFT (COUNTERSHAFT) SPEED SENSOR
SHIFT SOLENOID VALVE D (Behind shift solenoid valve cover)
A/T CLUTCH PRESSURE CONTROL SOLENOID VALVE C
SHIFT SOLENOID HARNESS CONNECTOR
SHIFT SOLENOID VALVE C (Behind shift solenoid valve cover)
A/T CLUTCH PRESSURE CONTROL SOLENOID VALVE B
ATF TEMPERATURE SENSOR (Behind shift solenoid valve cover)
SHIFT SOLENOID VALVE B (Behind shift solenoid valve cover)
SHIFT SOLENOID VALVE A (Behind shift solenoid valve cover)
71051_HFIT_G0336

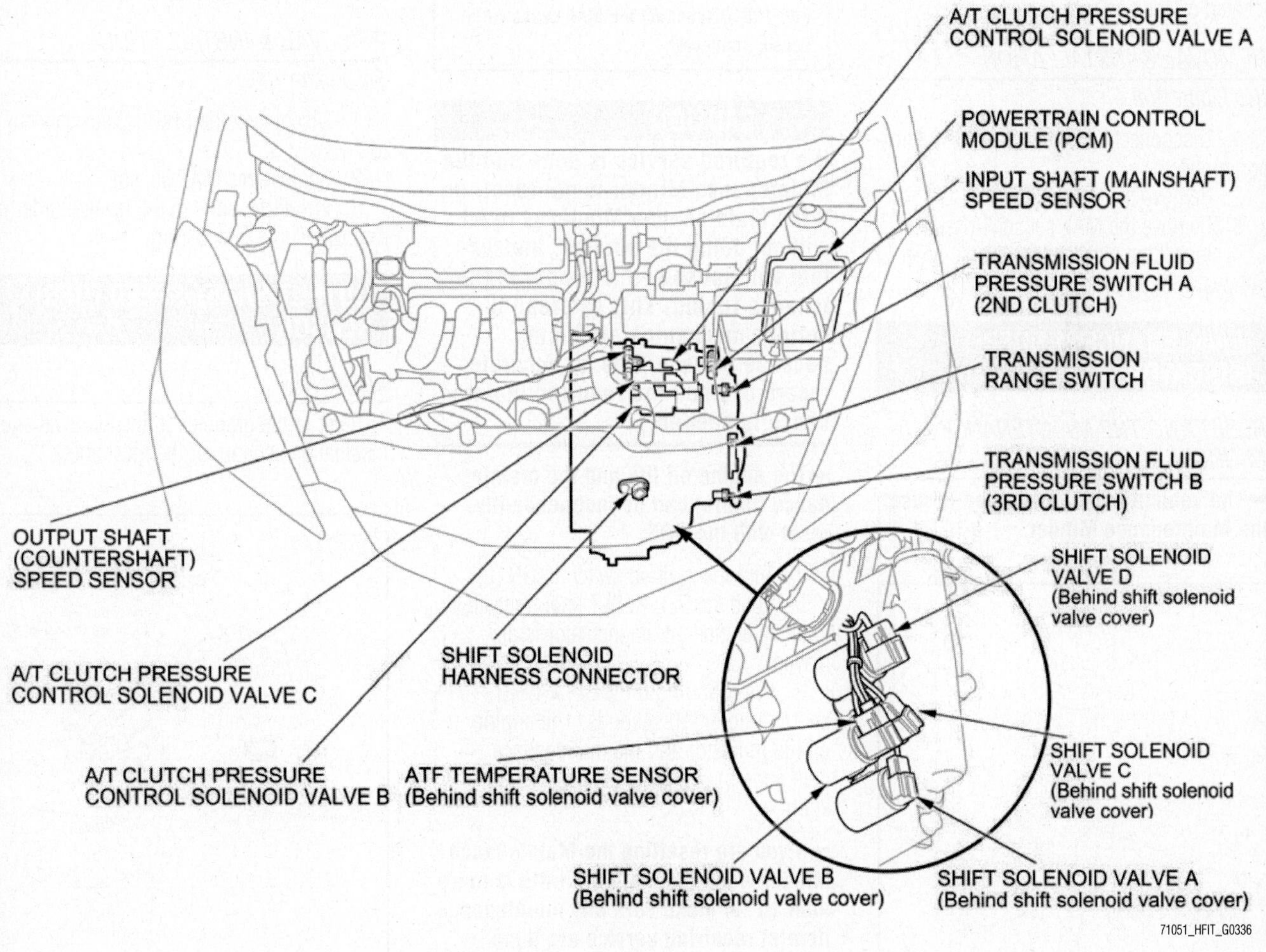

Fig. 105 Sensor and valve locations on the 5-speed automatic transmission

KNOCK SENSOR (KS)

LOCATION

Refer to the graphics in the Removal and Installation section for the location(s).

REMOVAL & INSTALLATION

See Figure 107.

1. Disconnect the knock sensor connector.
2. Remove the knock sensor.
3. Install the parts in the reverse order of removal.

MASS AIRFLOW SENSOR/INTAKE AIR TEMPERATURE (MAF/IAT)

LOCATION

The Intake Air Temperature (IAT) sensor is an integral part of the Mass Air Flow/Intake Air Temperature (MAF/IAT) located on the engine air intake.

REMOVAL & INSTALLATION

See Figure 108.

1. Disconnect the MAF sensor/IAT sensor connector.
2. Remove the screws.
3. Remove the MAF sensor/IAT sensor.
4. Install the parts in the reverse order of removal with a new gasket.

MAINTENANCE MINDER INDICATOR

RESETTING THE MAINTENANCE MINDER

➡The vehicle must be stopped to reset the Maintenance Minder.

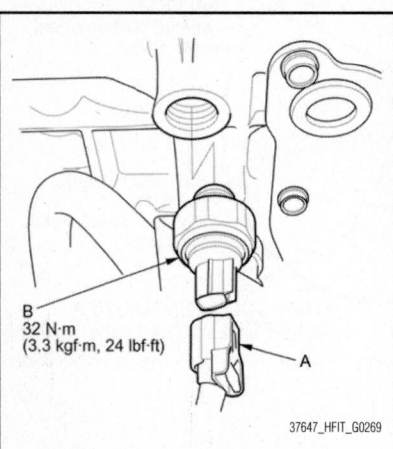

B
32 N·m
(3.3 kgf·m, 24 lbf·ft)

A

37647_HFIT_G0269

Fig. 107 Disconnect the knock sensor connector (A) and remove the knock sensor (B)

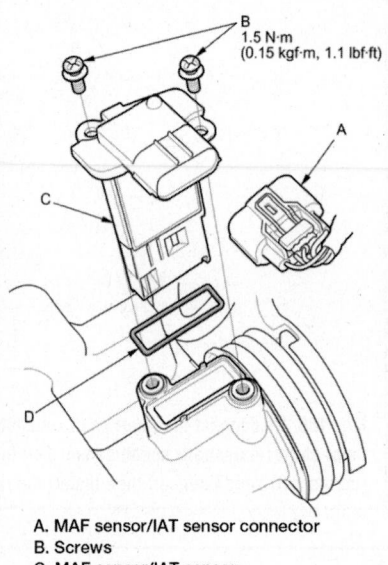

B
1.5 N·m
(0.15 kgf·m, 1.1 lbf·ft)

A

C

D

A. MAF sensor/IAT sensor connector
B. Screws
C. MAF sensor/IAT sensor
D. Gasket

37647_HFIT_G0270

Fig. 108 Disconnect the MAF sensor/IAT sensor connector

❋❋ CAUTION

If a required service is done and the Maintenance Minder is not reset, or if the Maintenance Minder is reset without doing the service, the system will not show the proper maintenance timing. This can lead to serious mechanical problems because there will be no accurate record of when the required maintenance is needed.

➡The engine oil life and the maintenance item(s) can be independently reset with the HDS.

1. Turn the ignition switch to ON (II).
2. Push the SEL/RESET knob repeatedly until the engine oil life indicator is displayed.
3. Press and hold the SEL/RESET knob for about 10 seconds. The engine oil life indicator and the maintenance item code(s) will blink then release the knob.

➡If you are resetting the Maintenance Minder when the engine oil life is more than 15 %, make sure any maintenance item(s) requiring service are done before resetting the display.

4. Press and hold the SEL/RESET knob for another 5 seconds. The maintenance item code(s) will disappear, and the engine oil life will reset to "100 %."

RESETTING INDIVIDUAL MAINTENANCE ITEM

1. Connect the Honda Diagnostic System (HDS) to the data link connector (DLC).
2. Turn the ignition switch to ON (II).
3. Select GAUGES in the BODY ELECTRICAL with the HDS.
4. Select ADJUSTMENT in the GAUGES with the HDS.
5. Select MAINTENANCE MINDER in the ADJUSTMENT with the HDS.
6. Select RESET in the MAINTENANCE MINDER with the HDS.
7. Select the individual maintenance item you wish to reset with the HDS.

MANIFOLD ABSOLUTE PRESSURE (MAP) SENSOR

LOCATION

Refer to the graphics in the "Removal & Installation" for the location(s).

REMOVAL & INSTALLATION

See Figure 109.

1. Disconnect the MAP sensor connector.
2. Remove the MAP sensor.
3. Install the parts in the reverse order of removal with a new O-ring.

OUTPUT SHAFT SPEED (OSS) SENSOR

LOCATION

Refer to the graphics in the Removal and Installation section for the location(s).

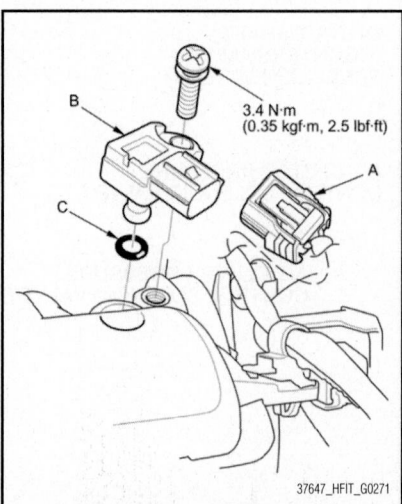

B

3.4 N·m
(0.35 kgf·m, 2.5 lbf·ft)

A

C

37647_HFIT_G0271

Fig. 109 Disconnect the MAP sensor connector (A); remove the MAP sensor (B) and the O-ring (C)

REMOVAL & INSTALLATION

A/T

See Figure 110.

1. Remove the air cleaner assembly.
2. Disconnect the output shaft (countershaft) speed sensor connector, and remove the output shaft (countershaft) speed sensor (A).
3. Install a new O-ring on the new output shaft (countershaft) speed sensor, then install the output shaft (countershaft) speed sensor.
4. Check the connector for rust, dirt, or oil, then connect the connector securely.
5. Install the air cleaner assembly.

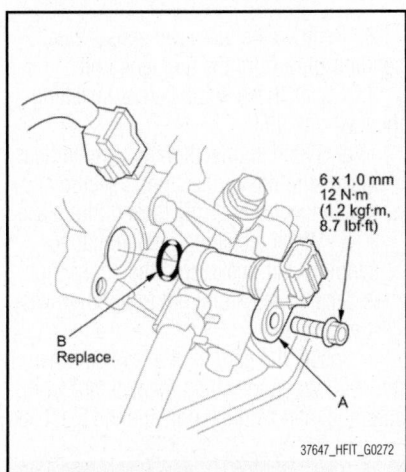

Fig. 110 Disconnect the output shaft (countershaft) speed sensor connector, and remove the output shaft (countershaft) speed sensor (A) and O-ring (B)

M/T

See Figure 111.

1. Raise the vehicle on a lift.
2. Disconnect the output shaft (countershaft) speed sensor connector.
3. Remove the output shaft (countershaft) speed sensor.
4. Install the parts in the reverse order of removal with a new O-ring.

POSITIVE CRANKCASE VENTILATION (PCV) VALVE

LOCATION

Refer to the graphics in the Removal and Installation section for the location(s).

REMOVAL & INSTALLATION

See Figures 112 and 113.

1. Remove the harness holder.
2. Disconnect the hose.

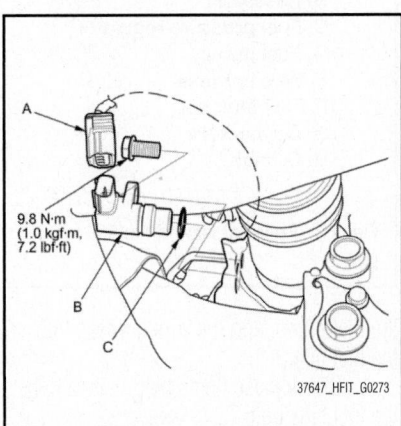

Fig. 111 Disconnect the output shaft speed sensor connector (A); remove the sensor (B) and the O-ring (C)

3. Remove the PCV valve.
4. Install the parts in the reverse order of removal with a new 14 mm washer.

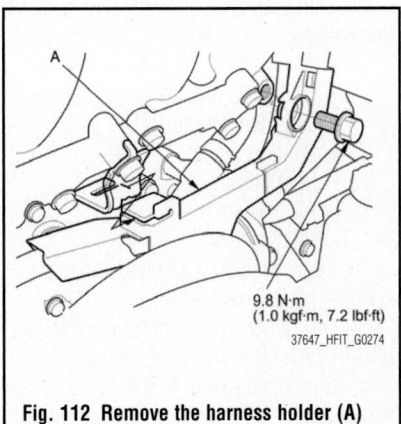

Fig. 112 Remove the harness holder (A)

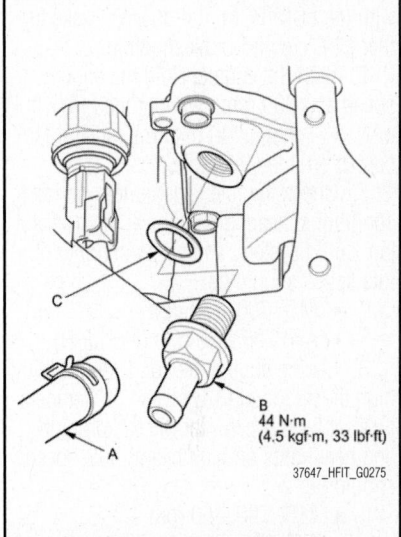

Fig. 113 Disconnect the hose (A); remove the PCV valve (B) and washer (C)

FUEL   GASOLINE FUEL INJECTION SYSTEM

FUEL SYSTEM SERVICE PRECAUTIONS

Safety is the most important factor when performing not only fuel system maintenance, but any type of maintenance. Failure to conduct maintenance and repairs in a safe manner may result in serious personal injury or death. Work on a vehicle's fuel system components can be accomplished safely and effectively by adhering to the following rules and guidelines.

• To avoid the possibility of fire and personal injury, always disconnect the negative battery cable unless the repair or test procedure requires that battery voltage be applied.

• Always relieve the fuel system pressure prior to disconnecting any fuel system component (injector, fuel rail, pressure regulator, etc.) fitting or fuel line connection. Exercise extreme caution whenever relieving fuel system pressure to avoid exposing skin, face and eyes to fuel spray. Please be advised that fuel under pressure may penetrate the skin or any part of the body that it contacts.

• Always place a shop towel or cloth around the fitting or connection prior to loosening to absorb any excess fuel due to spillage. Ensure that all fuel spillage is quickly removed from engine surfaces. Ensure that all fuel-soaked cloths or towels are deposited into a flame-proof waste container with a lid.

• Always keep a dry chemical (Class B) fire extinguisher near the work area.

• Do not allow fuel spray or fuel vapors to come into contact with a spark or open flame.

• Always use a second wrench when loosening or tightening fuel line connection fittings. This will prevent unnecessary stress and torsion on fuel piping. Always follow the proper torque specifications.

• Always replace worn fuel fitting O-rings with new ones. Do not substitute fuel hose where rigid pipe is installed.

IDLE SPEED

INSPECTION

1. Before checking the idle speed, check these items:

- The malfunction indicator lamp (MIL) has not been reported on, and there are no DTCs.
- Ignition timing
- Spark plugs
- Air cleaner
- PCV system

2. Apply the parking brake, and make sure the headlights are off.

3. Disconnect the evaporative emission (EVAP) canister purge valve connector.

4. Connect the HDS to the data link connector (DLC) (A) located under the driver's side of the dashboard.

5. Make sure the HDS communicates with the ECM/PCM. If it doesn't, make further DLC circuit troubleshooting.

6. Start the engine. Hold the engine speed at 3,000 rpm without load (A/T in P or N, M/T in neutral) until the radiator fan comes on, then let it idle.

7. Check the idle speed under no load conditions: headlights, blower fan, radiator fan, audio system, and air conditioner off. Idle speed should be:

- M/T: 700 ± 50 rpm
- A/T: 700 ± 50 rpm (in P or N)

8. Let the engine idle for 1 minute with high electric load (A/C switch on, temperature set to max cool, blower fan on high, and headlights on high beam). Idle speed should be:

- M/T: 790 ± 50 rpm
- A/T: 790 ± 50 rpm (in P or N)

a. If the idle speed is not within specification, do the ECM/PCM idle learn procedure.

b. If the idle speed is still not within specification, go to symptom troubleshooting.

9. Reconnect the EVAP canister purge valve connector.

FUEL FILTER

REMOVAL & INSTALLATION

See Figure 114.

The fuel filter should be replaced whenever the fuel pressure drops below 47–54 psi (320–370 kPa), after making sure that the fuel pump and the fuel pressure regulator are OK.

1. Remove the fuel tank unit.

2. Remove the fuel gauge sending unit, and the reservoir.

3. Remove the fuel pressure regulator,

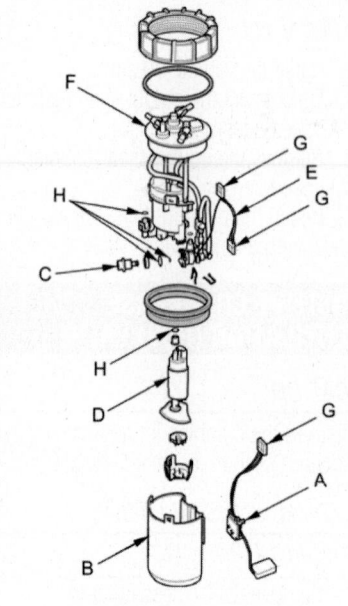

A. Fuel gauge sending unit
B. Reservoir
C. Fuel pressure regulator
D. Fuel pump
E. Wire harness
F. Fuel filter set
G. Connectors
H. O-rings

37647_HFIT_G0281

Fig. 114 Exploded view of the fuel tank unit

the fuel pump, and the wire harness from the fuel filter set.

4. Check these items before installing the fuel tank unit:

a. When connecting the wire harness, make sure the connection is secure and the connectors are firmly locked into place.

b. When installing the fuel gauge sending unit, make sure the connection is secure and the connector is firmly locked into place. Be careful not to bend or twist it excessively.

5. Install the parts in the reverse order of removal with new O-rings. When installing the fuel tank unit, align the marks on the unit and the fuel tank.

➡️**Coat the O-rings with clean engine oil; do not use any other oil or fluid. Do not pinch the O-rings during installation. Use all the new parts supplied in the fuel filter replacement kit.**

FUEL LEVEL SENDING UNIT

REMOVAL & INSTALLATION

See Figure 115.

1. Remove the fuel tank unit.

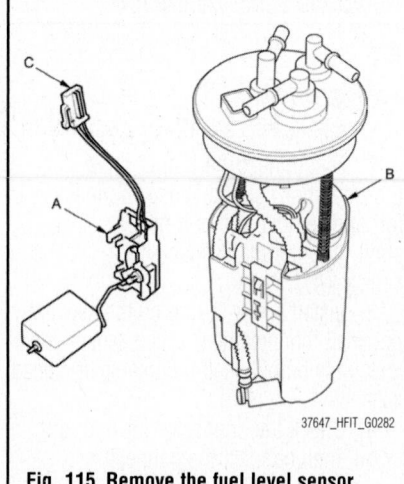

37647_HFIT_G0282

Fig. 115 Remove the fuel level sensor (fuel sending unit) (A) from the fuel tank unit (B)

2. Remove the fuel level sensor (fuel sending unit) from the fuel tank unit.

3. Check these items before installing the fuel tank unit:

a. When connecting the wire harness, make sure the connection is secure and the connector is firmly locked into place.

b. When installing the fuel gauge sending unit, make sure the connection is secure. Be careful not to bend or twist it excessively.

4. Install the parts in the reverse order of removal. When installing the fuel tank unit, align the marks on the unit and the fuel tank.

FUEL TANK UNIT

REMOVAL & INSTALLATION

See Figures 116 through 119.

Special Tools Required: Fuel Sender Wrench 07AAA-S0XA100

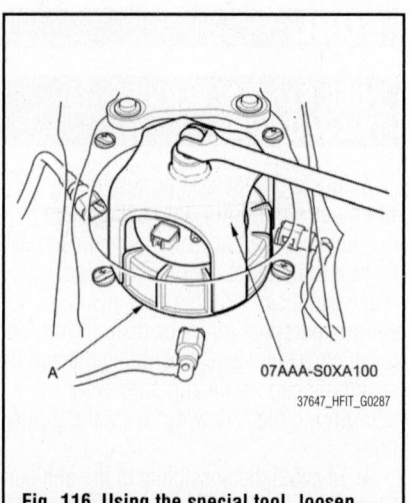

07AAA-S0XA100

37647_HFIT_G0287

Fig. 116 Using the special tool, loosen the locknut (A)

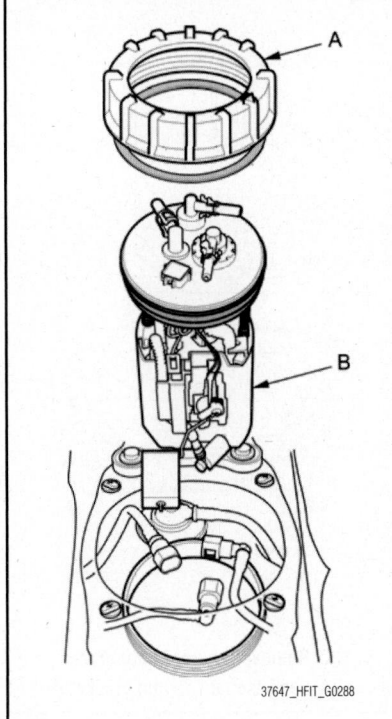

Fig. 117 Remove the locknut (A) and the fuel tank unit (B)

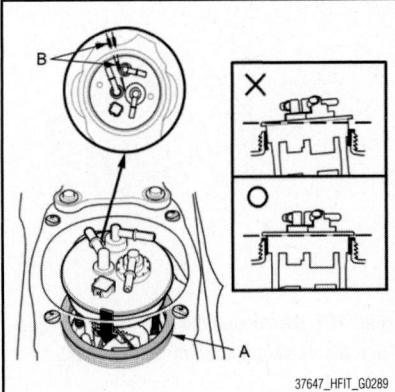

Fig. 118 Transfer the base gasket (A) and align the marks (B)

1. Relieve the fuel pressure.
2. Remove the center console.
3. Remove the bolts and the wire harness, then remove the parking brake lever.
4. Remove the access panel from the floor.
5. Disconnect the fuel tank unit 4P connector.
6. Disconnect the quick-connect fittings from the fuel tank unit.
7. Using the special tool, loosen the locknut.
8. Remove the locknut and the fuel tank unit.

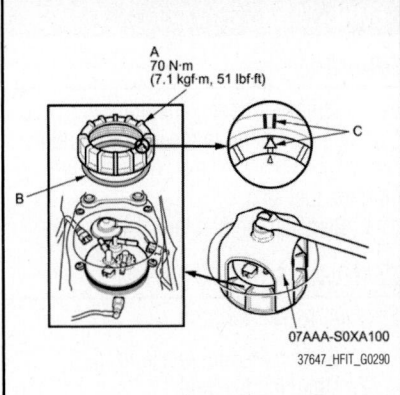

Fig. 119 Using the special tool, tighten a new locknut (A) with a new locknut plate (B); ensure the marks (C) are aligned

To install:

9. Temporarily attach a new base gasket to the fuel tank unit, then insert the fuel tank unit partially into the fuel tank.

➡️**Be careful not to damage the new base gasket. Be careful not to bend the fuel gauge sending unit. Do not coat the base gasket with oil.**

10. Transfer the base gasket from the fuel tank unit to the fuel tank.
11. Align the marks on the fuel tank and fuel tank unit, then insert the fuel tank unit into the fuel tank until the fuel tank unit rests on top of the base gasket.

➡️To avoid a fuel leak, check the base gasket, visually or by hand, to make sure it is not pinched.

12. Using the special tool, tighten a new locknut with a new locknut plate to 51 ft. lbs. (70 Nm).

➡️**Before tightening, align the marks on the fuel tank and the locknut. After tightening, make sure the marks are still aligned. After installation, check the base gasket, visually or by hand, to make sure it is not pinched.**

13. Connect the fuel tank unit 4P connector, then connect the quick-connect fitting, the suction hose, and the fuel vapor hose.
14. Reconnect the negative cable to the battery, and turn the ignition switch to ON (II) (but do not operate the starter motor). The fuel pump will run for about 2 seconds, and fuel pressure rises. Repeat this two or three times, then check for fuel leaks.
15. Reinstall the access panel, the parking brake lever, and the center console.

FUEL RAIL & INJECTOR

REMOVAL & INSTALLATION
See Figures 120 and 121.

1. Relieve the fuel pressure.
2. Remove the intake manifold and the intake manifold chamber.

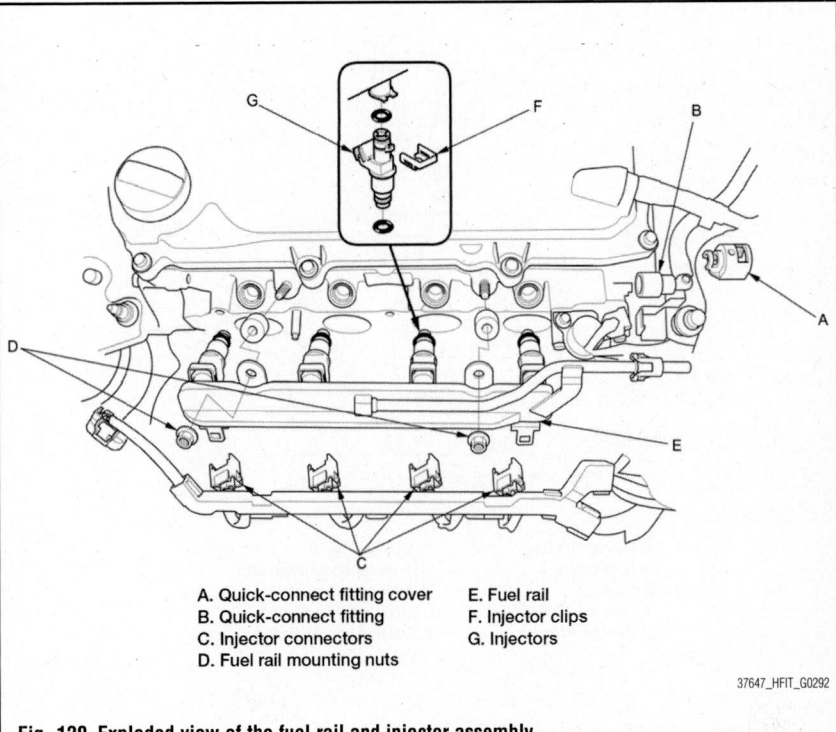

A. Quick-connect fitting cover
B. Quick-connect fitting
C. Injector connectors
D. Fuel rail mounting nuts
E. Fuel rail
F. Injector clips
G. Injectors

Fig. 120 Exploded view of the fuel rail and injector assembly

3. Remove the quick-connect fitting cover, then disconnect the quick-connect fitting.

4. Disconnect the injector connectors.

5. Remove the fuel rail mounting nuts from the fuel rail.

6. Remove the fuel rail and the injectors from the cylinder head.

7. Remove the injector clips from the fuel rail.

8. Remove the injectors from the fuel rail.

To install:

9. Coat the new O-rings (black) with clean engine oil, and insert the injectors into the fuel rail.

10. Install the injector clips.

11. Coat the new injector O-rings (brown) with clean engine oil.

12. Install the fuel rail and the injectors in the cylinder head.

13. Install the fuel rail mounting nuts.

14. Connect the injector connectors.

15. Connect the quick-connect fitting and the quick-connect fitting cover.

16. Turn the ignition switch to ON (II), but do not operate the starter. After the fuel pump runs for about 2 seconds, the fuel rail is pressurized. Repeat this two or three times, then check for fuel leaks.

17. Reinstall the intake manifold and the intake manifold chamber.

FUEL TANK

DRAINING

1. Remove the fuel tank unit.

2. Using a hand pump, a hose, and a container suitable for fuel, draw the fuel from the fuel tank.

3. Reinstall the fuel tank unit.

REMOVAL & INSTALLATION

See Figures 122 through 125.

1. Relieve the fuel pressure.

2. Drain the fuel tank.

3. Reinstall the fuel tank unit without connecting the fuel tank unit 4P connector and the quick-connect fitting (feed line).

4. Raise the vehicle on a lift.

5. Remove the fuel tank guard, and the fuel tank protector.

6. M/T model: Remove the fuel tank cover.

7. A/T model: Remove the floor under cover assembly.

8. Remove the front floor cross beam and the tank mount bracket.

9. Disconnect the fuel fill hose and the quick-connect fitting (fuel suction tube).

10. Disconnect the fuel tank vapor control valve hose.

11. Place a jack or other support under the fuel tank, then remove the strap bolts.

12. Remove the fuel tank.

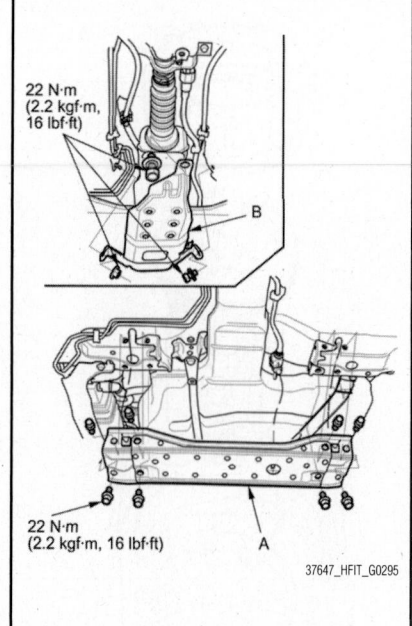

Fig. 122 Remove the front floor cross beam (A), and the tank mount bracket (B)

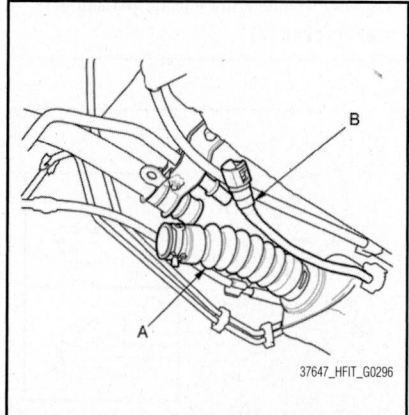

Fig. 123 Disconnect the fuel fill hose (A) and the quick-connect fitting (fuel suction tube) (B)

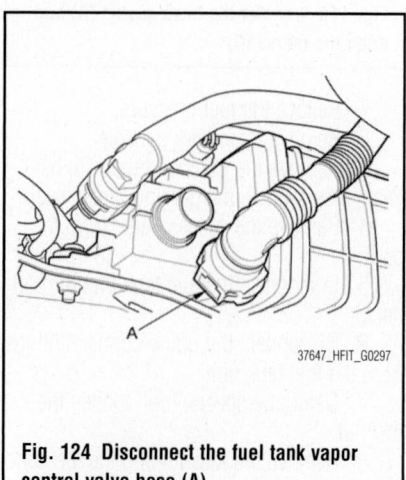

Fig. 124 Disconnect the fuel tank vapor control valve hose (A)

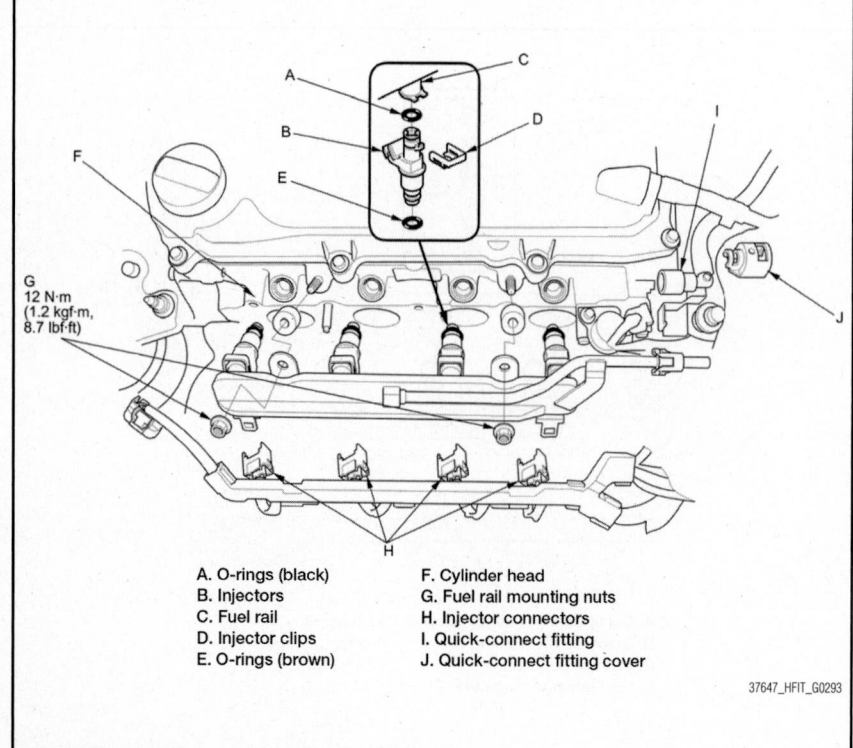

A. O-rings (black)
B. Injectors
C. Fuel rail
D. Injector clips
E. O-rings (brown)
F. Cylinder head
G. Fuel rail mounting nuts
H. Injector connectors
I. Quick-connect fitting
J. Quick-connect fitting cover

Fig. 121 Installation of injectors and fuel rail

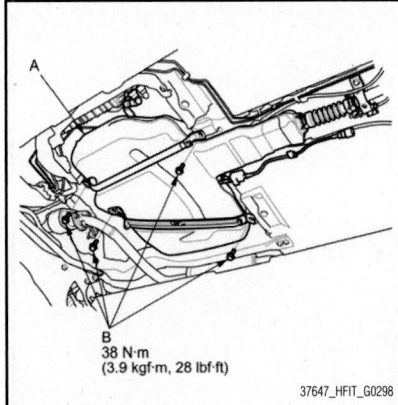

Fig. 125 Place a jack or other support under the fuel tank (A), then remove the strap bolts (B)

13. Install the parts in the reverse order of removal.

RELIEVING FUEL SYSTEM PRESSURE

RELIEVING

Before disconnecting fuel lines or hoses, relieve pressure from the system by disabling the fuel pump and then disconnecting the fuel tube/quick connect fitting in the engine compartment.

With the HDS

1. Connect the HDS to the Data Link Connector (DLC) located under the driver's side of the dashboard.
2. Turn the ignition switch to ON (II).
3. Make sure the HDS communicates with the ECM/PCM. If it doesn't, go to the DLC circuit troubleshooting.
4. Turn the ignition switch to LOCK (0).
5. Remove the fuel fill cap to relieve the pressure in the fuel tank.
6. Turn the ignition switch to ON (II).
7. From the INSPECTION MENU of the HDS, select Fuel Pump OFF, then start the engine, and let it idle until it stalls.

➡**Do not allow the engine to idle above 1,000 rpm or the ECM/PCM will continue to operate the fuel pump. A Confirmed or Pending DTC may be set during this procedure. Check for DTCs, and clear them as needed.**

8. Turn the ignition switch to LOCK (0).
9. Do the battery terminal disconnection procedure.
10. Remove the quick-connect fitting cover.
11. Check the fuel quick-connect fitting for dirt, and clean it if needed.

12. Place a rag or shop towel over the quick-connect fitting.
13. Disconnect the quick-connect fitting: Hold the connector with one hand, and squeeze the retainer tabs with the other hand to release them from the locking tabs. Pull the connector off.

➡**Be careful not to damage the line or other parts. Do not use tools. If the connector does not move, keep the retainer tabs pressed down, and alternately pull and push the connector until it comes off easily. Do not remove the retainer from the line; once removed, the retainer must be replaced with a new one.**

14. After disconnecting the quick-connect fitting, check it for dirt or damage.
15. Do the battery terminal reconnection procedure.

Without the HDS

See Figure 126.

1. Open the fuse access panel.
2. Remove PGM-FI main relay 2 from the under-dash fuse/relay box.
3. Start the engine, and let it idle until it stalls.

➡**If any DTCs are stored, clear and ignore them.**

4. Turn the ignition switch to LOCK (0).
5. Remove the fuel fill cap to relieve the pressure in the fuel tank.
6. Do the battery terminal disconnection procedure.
7. Remove the quick-connect fitting cover.
8. Check the fuel quick-connect fitting for dirt, and clean it if needed.
9. Place a rag or shop towel over the quick-connect fitting.

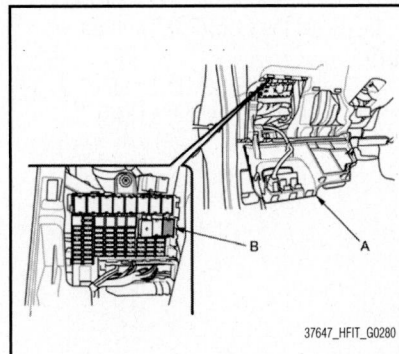

Fig. 126 Open the fuse access panel (A) and remove PGM-FI main relay 2 (B) from the under-dash fuse/relay box

10. Disconnect the quick-connect fitting: Hold the connector with one hand, and squeeze the retainer tabs with the other hand to release them from the locking tabs. Pull the connector off.

➡**Be careful not to damage the line or other parts. Do not use tools. If the connector does not move, keep the retainer tabs pressed down, and alternately pull and push the connector until it comes off easily. Do not remove the retainer from the line; once removed, the retainer must be replaced with a new one.**

11. After disconnecting the quick-connect fitting, check it for dirt or damage.
12. Do the battery terminal reconnection procedure.

THROTTLE BODY

REMOVAL & INSTALLATION

See Figure 127.

✳✳ CAUTION

Do not insert your fingers into the installed throttle body when you turn the ignition switch to ON (II) or while the ignition switch is in ON (II). If you do, you will seriously injure your fingers if the throttle valve is activated.

➡**If you are replacing or cleaning the throttle body, start at step 1. If you are removing the throttle body, start at step 4.**

1. Connect the HDS to the DLC while the engine is stopped.
2. Select the INSPECTION MENU on the HDS.
3. Do the TP POSITION CHECK in the ETCS TEST.
4. Turn the ignition switch to LOCK (0).
5. Remove the air cleaner.
6. Remove the harness clamp.
7. Disconnect the throttle body connector and the EVAP canister purge valve connector.
8. Disconnect and plug the water bypass hoses and the purge hose.
9. Remove the throttle body.
10. Remove the purge pipe, the water bypass hoses, and the purge hose from the throttle body.

To install:

11. Install the parts in the reverse order of removal with a new gasket, then refill the radiator with engine coolant. Note the following during installation:
   • If you replace or clean the throttle body, go to the next step.

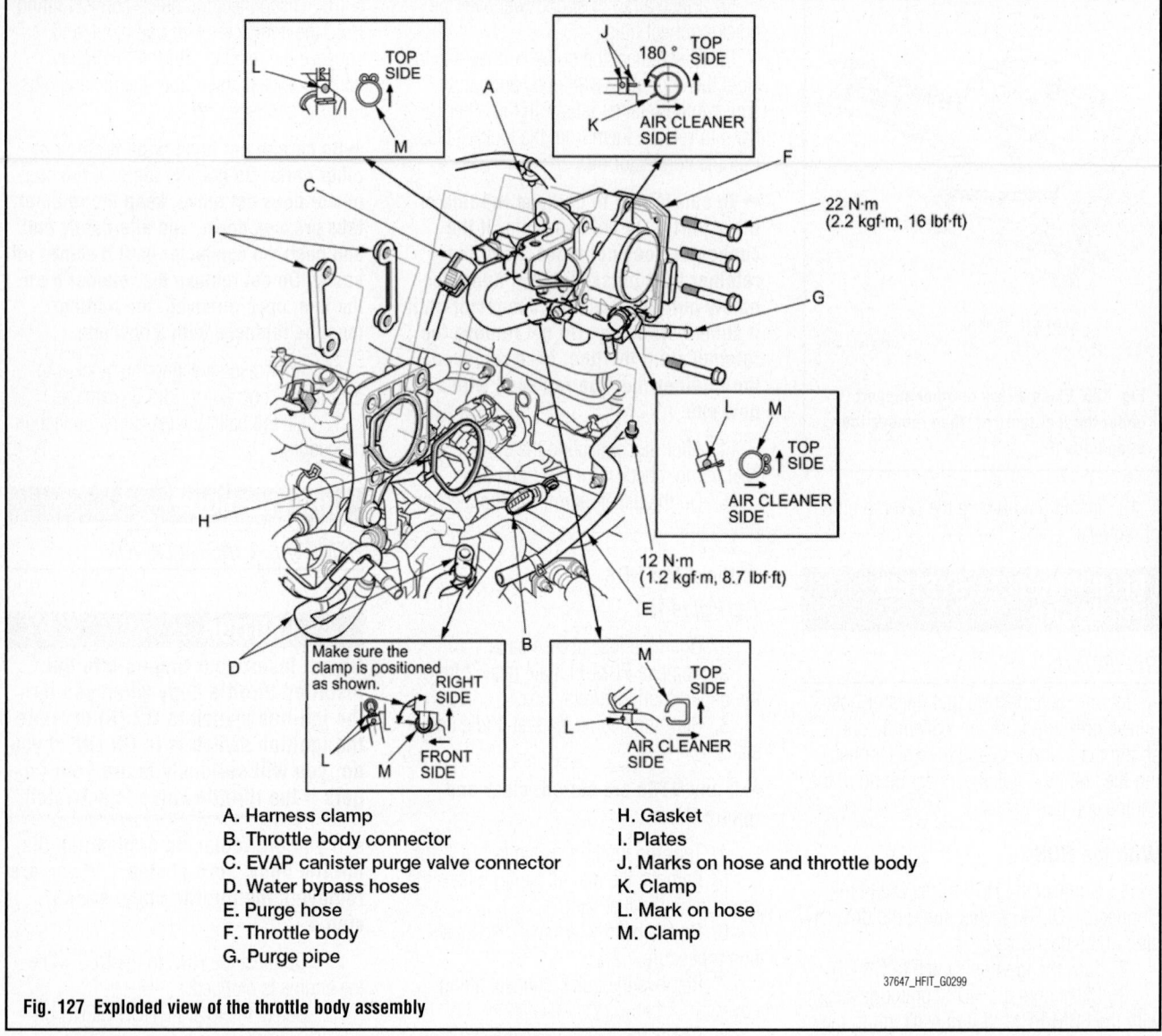

A. Harness clamp
B. Throttle body connector
C. EVAP canister purge valve connector
D. Water bypass hoses
E. Purge hose
F. Throttle body
G. Purge pipe
H. Gasket
I. Plates
J. Marks on hose and throttle body
K. Clamp
L. Mark on hose
M. Clamp

37647_HFIT_G0299

Fig. 127 Exploded view of the throttle body assembly

- If you did not replace or clean the throttle body, this procedure is complete.
- Be careful not to drop or damage the plates.
- Align the marks on the hose and throttle body, then insert the hose. Make sure the clamp is positioned as shown.

- Align the mark on the hose as shown, then insert the hose. Make sure the clamp is positioned as shown.

12. Turn the ignition switch to ON (II).
13. Reset the ECM/PCM with the HDS.
14. Select the ETCS TEST in the INSPECTION MENU with the HDS.

15. Select the TP POSITION CHECK, then clear the Throttle Position (TP) learned value.
16. Turn the ignition switch to LOCK (0).
17. Turn the ignition switch to ON (II), and wait 2 seconds without pressing the accelerator pedal.
18. Do the ECM/PCM idle learn procedure.

HEATING & AIR CONDITIONING SYSTEM

BLOWER MOTOR

REMOVAL & INSTALLATION

See Figures 128 through 133.

1. Remove the passenger's dashboard undercover.

2. Remove the glove box.

3. Remove the recirculation control cable from the blower unit.

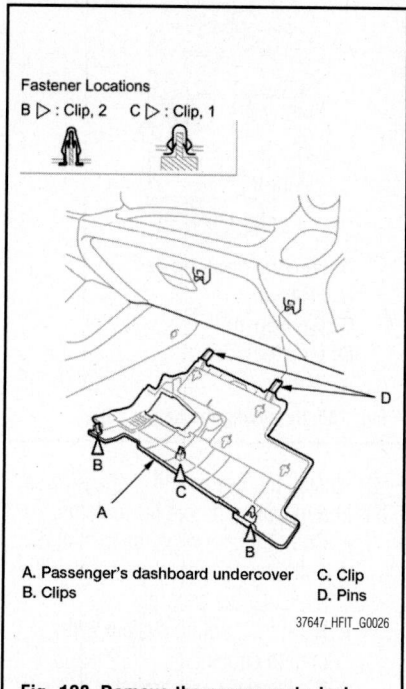

A. Passenger's dashboard undercover C. Clip
B. Clips D. Pins

37647_HFIT_G0026

Fig. 128 Remove the passenger's dashboard undercover

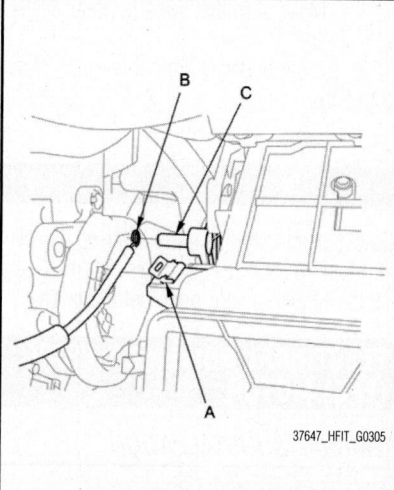

37647_HFIT_G0305

Fig. 129 Detach the recirculation control cable housing from the clamp (A), and disconnect the inner cable (B) from the recirculation control linkage (C)

a. Set the recirculation control lever to FRESH.

b. Detach the recirculation control cable housing from the clamp, and disconnect the inner cable from the recirculation control linkage.

4. Remove the bolts and the glove box frame.

5. Cut the plastic cross brace in the glove box opening with diagonal cutters in the area. Remove and discard the plastic cross brace.

6. Disconnect the connector from the blower motor.

7. Remove the self-tapping screws and the harness clip, then remove the passenger's heater duct.

8. Remove the connector clip, the self-tapping screws and the mounting nuts, then pull the blower unit out.

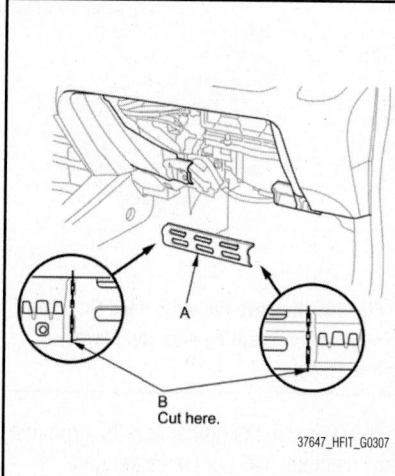

B
Cut here.

37647_HFIT_G0307

Fig. 130 Cut the plastic cross brace (A) in the glove box opening with diagonal cutters in the area (B)

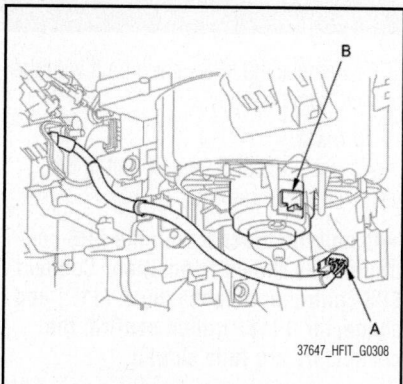

37647_HFIT_G0308

Fig. 131 Disconnect the connector (A) from the blower motor (B)

9. Install the unit in the reverse order of removal. Make sure that there is no air leakage.

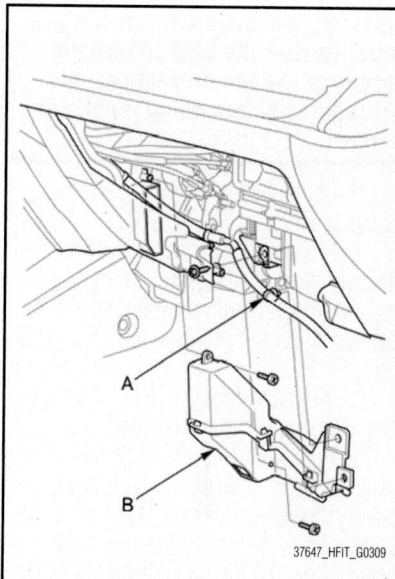

37647_HFIT_G0309

Fig. 132 Remove the self-tapping screws and the harness clip (A), then remove the passenger's heater duct (B)

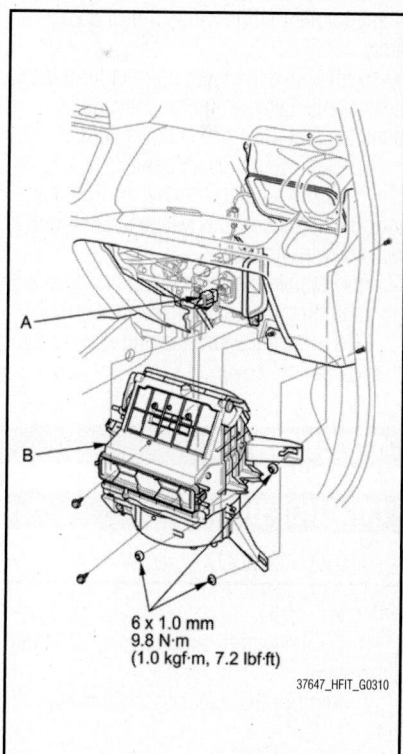

6 x 1.0 mm
9.8 N·m
(1.0 kgf·m, 7.2 lbf·ft)

37647_HFIT_G0310

Fig. 133 Remove the connector clip (A), the self-tapping screws and the mounting nuts, then pull the blower unit (B) out

HEATER CORE

REMOVAL & INSTALLATION

See Figures 134 through 136.

✳✳ WARNING

SRS components are located in this area. Review the SRS component locations and the precautions and procedures before doing repairs or service.

1. Do the battery terminal disconnection procedure.

2. Models with air conditioning: Recover the A/C refrigerant with a recovery/recycling/charging station, then disconnect the A/C lines from the evaporator core.

3. When the engine is cool, drain the engine coolant from the radiator.

4. Remove the clip from the inlet heater hose, and discard it. The clip is for factory assembly use only. Slide the hose clamps back, then disconnect the inlet heater hose and the outlet heater hose from the heater unit. Note the layout of the hoses. Engine coolant will run out when the hoses are disconnected; drain it into a clean drip pan. Be sure not to let coolant spill on electrical parts or painted surfaces. If any coolant spills, rinse it off immediately.

5. Remove the mounting nut from the heater unit. Take care not to damage or bend the fuel lines or the brake lines.

6. Remove the dashboard.

7. Disconnect the connector from the evaporator sensor, and remove the blower resistor wire harness.

8. Remove the drain hose, then remove the mounting nuts and blower-heater unit.

9. Remove the self-tapping screws and the heater core cover.

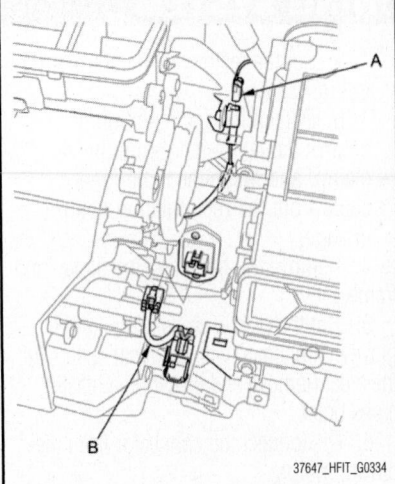

Fig. 134 Disconnect the connector (A) from the evaporator sensor, and remove the blower resistor wire harness (B)

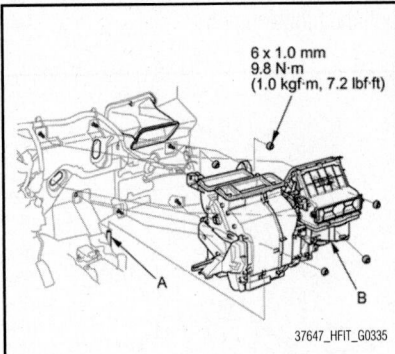

Fig. 135 Remove the drain hose (A), then remove the mounting nuts and blower-heater unit (B)

10. Remove the holder and the grommet, then carefully pull out the heater core.

To install:

11. Install the heater core in the reverse order of removal.

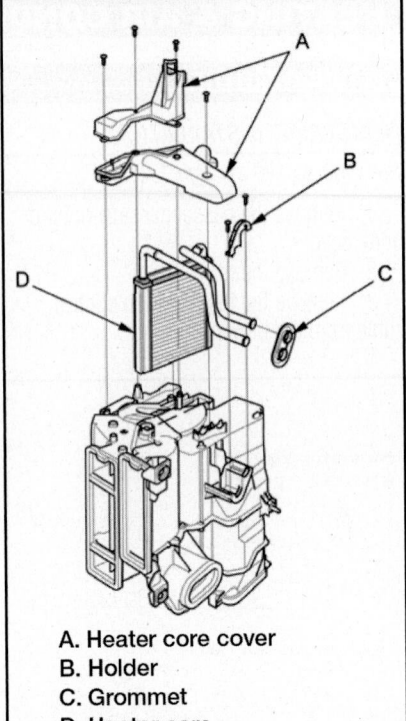

A. Heater core cover
B. Holder
C. Grommet
D. Heater core

Fig. 136 Removing the heater core

12. Install the heater unit in the reverse order of removal, and note these items:
  • Do not interchange the inlet and outlet heater hoses, and install the hose clamps securely.
  • Refill the cooling system with engine coolant.
  • If necessary, recharge the A/C system.
  • Make sure that there is no coolant leakage.
  • Make sure that there is no air leakage.

13. Do the battery terminal reconnection procedure.

STEERING

CONTROL UNIT

REMOVAL & INSTALLATION

See Figure 137.

1. Do the battery terminal disconnection procedure.

2. Remove the under-dash fuse/relay box.

3. Disconnect EPS control unit connector A (11P) and connector B (16P).

4. Loosen the bolt.

5. Remove the bolt and nut from the EPS control unit bracket.

6. Remove the EPS control unit with the bracket from the body.

To install:

7. Install the EPS control unit in the reverse order of removal.

➡Install the bracket with the EPS control unit as shown in position. Connect EPS control unit connector A (11P) and connector B (16P), then confirm the connectors are fully seated.

8. Do the battery terminal reconnection procedure.

9. After installation, start the engine, allow it to idle, and turn the steering wheel from lock to lock several times. Make sure that the EPS indicator does not come on.

ELECTRICAL POWER STEERING (EPS) MOTOR

REMOVAL & INSTALLATION

See Figures 138 and 139.

➡Do not allow dust, dirt, or other foreign materials to enter the steering gearbox.

1. Remove the steering gearbox.

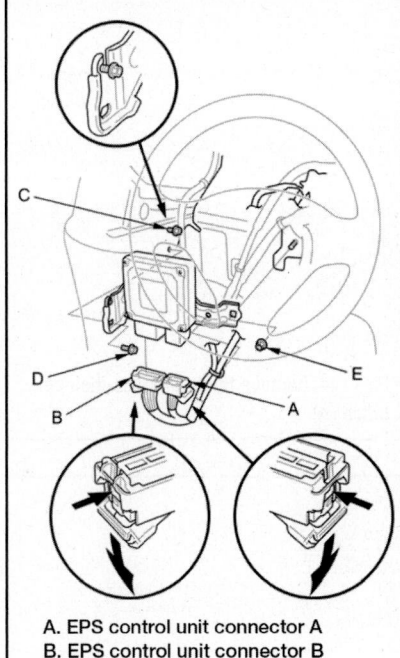

A. EPS control unit connector A
B. EPS control unit connector B
C. Bolt
D. Bolt
E. Nut

37647_HFIT_G0338

Fig. 137 Disconnect EPS control unit connector A (11P) and connector B (16P)

2. Remove the EPS motor 3P connector and EPS motor angle sensor 8P connector from the connector bracket, then remove clips.

3. Remove the EPS motor from the steering gearbox, then remove the O-rings.

➡**Do not discard the O-ring C.**

To install:

4. Clean the mating surface of the EPS motor and the steering gearbox.

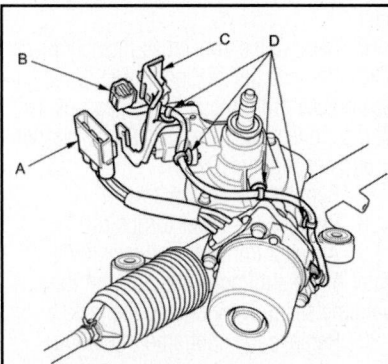

A. EPS motor 3P connector
B. EPS motor angle sensor 8P connector
C. Connector bracket
D. Clips

37647_HFIT_G0339

Fig. 138 Remove the EPS motor 3P connector and EPS motor angle sensor 8P connector from the connector bracket

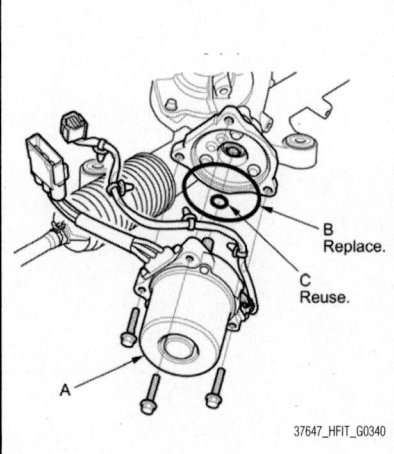

B. Replace.
C. Reuse.

37647_HFIT_G0340

Fig. 139 Remove the EPS motor (A) from the steering gearbox, then remove the O-rings (B, C)

5. Apply grease included in the motor set to the new O-ring, and carefully fit it on the EPS motor.

6. Install the O-ring to the EPS motor shaft.

7. Apply steering grease to the EPS motor shaft.

8. Install the EPS motor on the steering gearbox by engaging the EPS motor shaft and the worm shaft.

9. Before tightening the bolts, turn the motor two or three times to the right and left about 45 degrees. Make sure the EPS motor is evenly seated on the steering gearbox, and that the O-ring is not pinched between the mating surfaces.

10. Loosely install the EPS motor mounting bolts, then turn the pinion shaft two or three times to the right and left about 45 degrees.

11. Tighten the EPS motor mounting bolts to 14 ft. lbs. (20 Nm).

12. Install the EPS motor 3P connector and EPS motor angle sensor 8P connector to the connector bracket.

13. Install the wire harness clips.

14. Finish the installation, and note these items:
- Make sure the EPS motor 3P connector and EPS motor angle sensor 8P connector are properly connected.
- Make sure the EPS motor and the EPS wires are not caught or pinched by any parts.

15. Install the steering gearbox.

POWER RACK & PINION STEERING GEAR

REMOVAL & INSTALLATION

See Figures 140 through 150.

Special Tools Required*:
- Ball Joint Remover, 28 mm 07MAC-SL0A202
- Universal Lifting Eyelet 07AAK-SNAA120
- 1.8 Support Bolt 07AAK-SNAA500
- Engine Support Hanger, A and Reds AAR-T1256

➡*** Available through the Honda Tool and Equipment Program 888-424-6857.**

Note these items during removal:
- Use solvent and a brush, wash any oil and dirt off the end of the steering gearbox, but avoid any electrical parts. Blow dry with compressed air.
- Be sure to remove the steering wheel before disconnecting the steering joint or damage to the cable reel can occur.
- Lower the front subframe from the body then remove the steering gearbox.

1. Do the battery terminal disconnection procedure.

2. Raise the vehicle on a lift.

3. Remove the front wheels.

4. Release the lock lever, tilt the steering column all the way up slide it all the way in, then tighten the lock lever.

5. Remove the driver's airbag and the steering wheel.

6. Remove the steering joint cover.

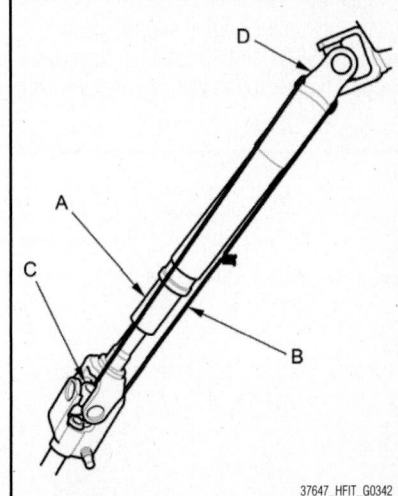

37647_HFIT_G0342

Fig. 140 Hold the lower slide shaft on the column with a piece of wire between the joint yoke of the lower slide shaft and joint yoke of the upper shaft to prevent the lower slide from pulling out

7. Hold the lower slide shaft on the column with a piece of wire between the joint yoke of the lower slide shaft and joint yoke of the upper shaft to prevent the lower slide from pulling out.

8. Release the lock lever, slide it all the way out, then tighten the lock lever.

9. Remove the steering joint bolt.

10. Disconnect the steering joint by sliding the steering joint into the column.

11. Remove the center guide (if equipped) from the top of the pinion shaft, and discard it. The center guide is for factory assembly use only.

12. Remove the cotter pin from the tie-rod end ball joint, then remove the nut on both sides.

13. Separate the tie-rod end ball joint and knuckle using the ball joint remover on both sides.

14. Remove the clip from the lower arm ball joint castle nut, and remove the nut from both sides.

15. Separate the lower arm ball joint and the knuckle using the ball joint remover on both sides.

16. Remove the stabilizer link from the stabilizer bar on both sides.

17. Remove the air cleaner housing.

18. Remove the cowl cover and the under-cowl panel.

19. Install the universal lifting eyelet (07AAK-SNAA120) to the bolt hole (A) at the air cleaner housing mounting bracket with the 1.8 support bolt (07AAK-SNAA500).

20. Remove the harness clamp from its clamp bracket located in front of the left damper top.

21. Set up the engine support hanger (AAR-T1256). Carefully position the engine support hanger to the vehicle; position both

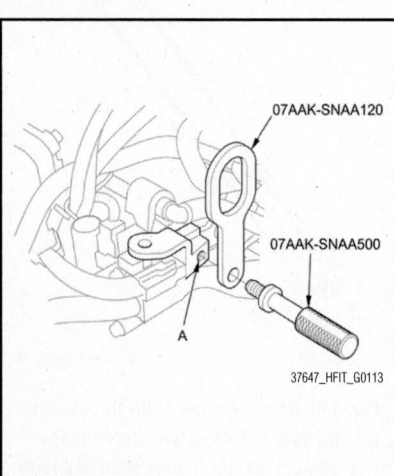

Fig. 141 Install the universal lifting eyelet to the bolt hole (A)

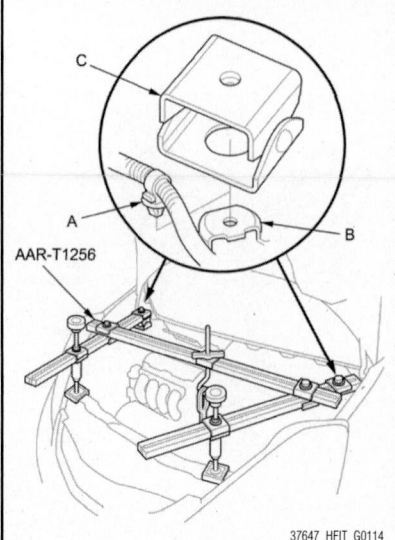

Fig. 142 Remove the harness clamp (A) from its clamp bracket (B) located in front of the left damper top; position both cross-arm foot bases (C) over the harness clamp brackets on both sides

cross-arm foot bases over the harness clamp brackets on both sides, and position both front stands on the front bulkhead. Attach the hook to the universal eyelet, tighten the wing nut by hand, and lift and support the engine.

22. Disconnect the EPS motor angle sensor 8P connector and torque sensor 6P connector from the steering gearbox.

23. Disconnect the EPS motor 3P connector by pushing the lock and pulling up the lever.

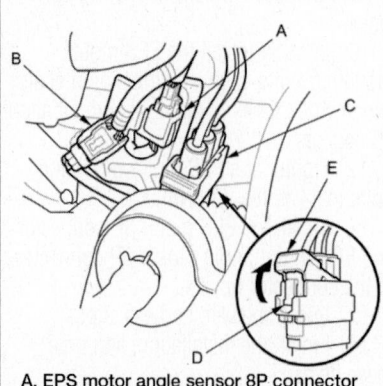

A. EPS motor angle sensor 8P connector
B. Torque sensor 6P connector
C. EPS motor 3P connector
D. Lock
E. Lever

Fig. 143 Disconnect the EPS motor angle sensor 8P connector and torque sensor 6P connector from the steering gearbox

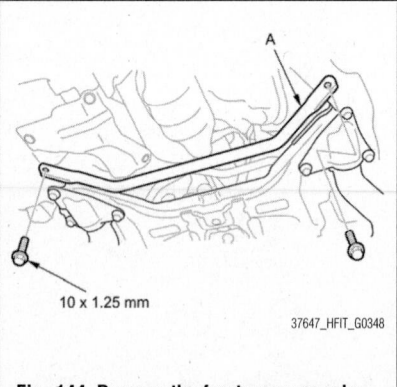

10 x 1.25 mm

Fig. 144 Remove the front cross-member brace (A)

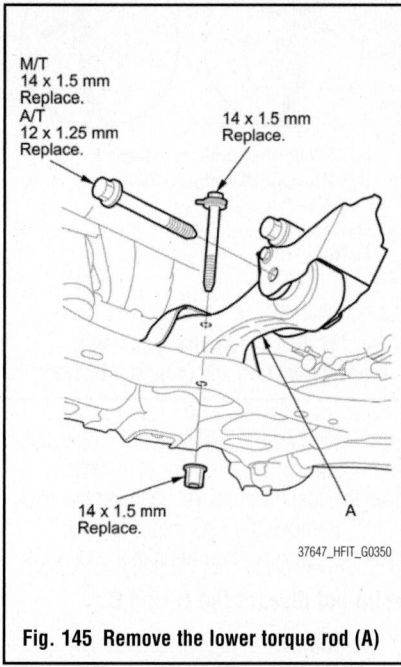

M/T
14 x 1.5 mm
Replace.
A/T
12 x 1.25 mm
Replace.

14 x 1.5 mm
Replace.

14 x 1.5 mm
Replace.

Fig. 145 Remove the lower torque rod (A)

24. Wrap the connectors with the vinyl tape to avoid contamination from grease or water.

25. Remove the front cross-member brace.

26. Remove the secondary HO2S harness bracket from the steering gearbox. Do not disconnect secondary HO2S 4P connector and secondary HO2S.

27. Remove the splash shield.

28. Remove the lower torque rod.

29. Attach a transmission jack to the center of the subframe, and support the subframe securely by raising the jack.

30. Remove the subframe mounting bolts.

31. Lower the subframe and steering gearbox as an assembly by lowering the jack slowly.

32. Remove the steering gearbox from the subframe.

33. Remove the pinion shaft grommet from the top of the torque sensor.

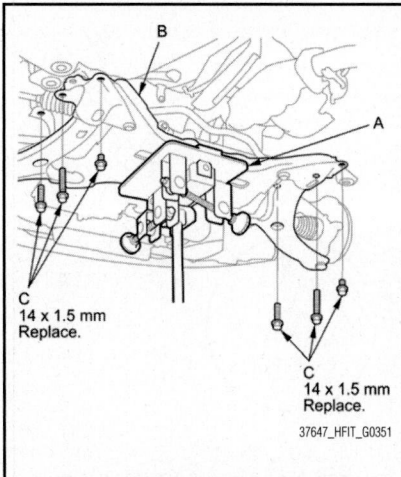

Fig. 146 Attach a transmission jack (A) to the center of the subframe (B), and remove the subframe mounting bolts (C)

To install:

34. Before installing the steering gearbox, make sure that no grease is on the mating surface of the steering gearbox and the front subframe. To prevent the gearbox mounting bolts from loosening after the installation, remove any grease from the bolt holes.

35. Install the pinion shaft grommet. Align the ledge portion in the pinion shaft grommet with the lug portion. The grommet must not have a gap at the mating surface of the grommet and the torque sensor.

36. Place the steering gearbox on the subframe.

37. Loosely install the stiffener plates, and new gearbox mounting bolts, then tighten the gearbox mounting bolts to 44 ft. lbs. (60 Nm). Tighten the three long gearbox mounting bolts first, starting in the middle then tightening the outside bolts. After the gearbox mounting bolts are tightened, then tighten the stiffener mounting bolts to 43 ft. lbs. (59 Nm).

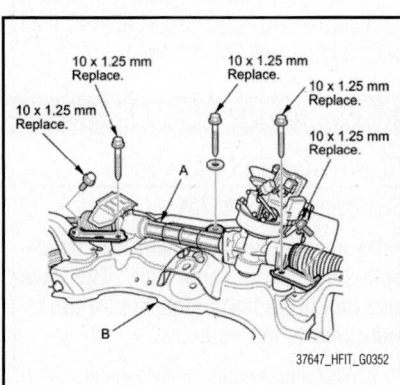

Fig. 147 Remove the steering gearbox (A) from the subframe (B)

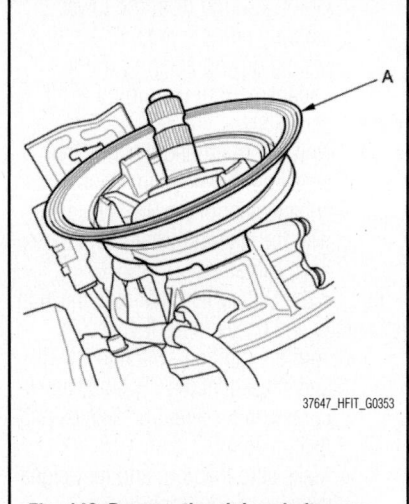

Fig. 148 Remove the pinion shaft grommet (A) from the top of the torque sensor

38. Turn the lip of the pinion shaft grommet to ease installation.

39. Set the subframe with the steering gearbox on the transmission jack and support it.

40. Carefully raise the subframe with the transmission jack, and pass the pinion shaft into the passenger's compartment. Return the lip of the pinion shaft grommet.

➡**Be sure that the pinion shaft grommet is securely in place. Make sure the lip of the pinion shaft grommet is not turned up. Incorrect installation can cause leakage of water or mud, and noise. Take care not to damage the lower arm ball joint boot with the edge of the knuckle, etc.**

41. Loosely install new subframe mounting bolts, then tighten the bolts to 69 ft. lbs. (93 Nm).

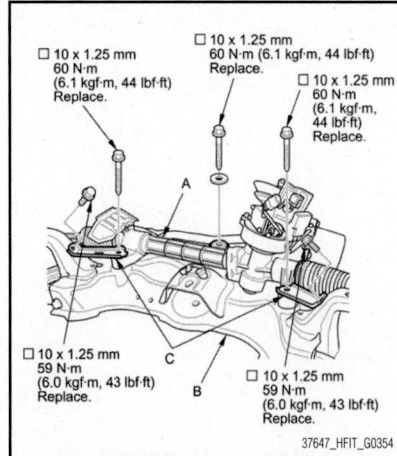

Fig. 149 Place the steering gearbox (A) on the subframe (B) and install the stiffener plates (C)

42. Install the lower torque rod with new mounting bolts and a new mounting nut, and tighten the bolt to 61 ft. lbs. (83 Nm) and tighten the nut to 69 ft. lbs. (93 Nm).

➡**Make sure the stopper of the flange bolt is securely installed.**

43. Install the splash shield.

44. Lower the vehicle.

45. Remove the engine support hanger (AAR-T1256) from the vehicle.

46. Remove the cross arm foot, then install the harness clip to the harness clamp bracket on the left side.

47. Remove the universal lifting eyelet (07AAK-SNAA120) and 1.8 support bolt (07AAK-SNAA500).

48. Install the secondary HO2S harness bracket to the steering gearbox.

49. Install the front cross-member brace. Tighten the mounting bolts to 40 ft. lbs. (54 Nm).

50. Remove the vinyl tape from the connectors.

51. Connect the EPS motor angle sensor 8P connector, torque sensor 6P connector to the steering gearbox.

52. Pull down the lever of the EPS motor 3P connector, then confirm the connector is fully seated.

53. Install the air cleaner housing.

54. Install the under cowl panel and cowl cover.

55. Connect the stabilizer links.

56. Wipe off any grease contamination from the ball joint tapered section and threads. Reconnect the tie-rod end ball joints to the knuckles. Install the nuts, and tighten to 32 ft. lbs. (43 Nm) on both sides.

57. Install a new cotter pin and bend it.

58. Wipe off any grease contamination from the lower arm ball joint tapered section and thread. Then reconnect the lower arm to the knuckle. Install the new castle nut and tighten it to 47–54 ft. lbs. (64–74 Nm) on both sides.

➡**Be careful not to damage the lower ball joint boot. Check the ball joint boot for deformation before connecting the knuckle. Torque the castle nut to the lower torque specification, then tighten it only far enough to align the slot with the joint pin clip hole. Do not align the castle nut by loosening it.**

59. Install the clip.

60. Install the front wheels, then set the wheels in the straight ahead position.

➡**Before installing the wheel, clean the mating surfaces of the brake disc and the inside of the wheel.**

61. Cut the wire, while holding the lower slide shaft on the steering column.

62. Slip the lower end of the steering joint onto the pinion shaft taking care to align the gap within the angle shown.

63. Release the lock lever, tilt the steering column all the way down, side it all the way out, then tighten the lock lever.

64. Align the bolt hole on the steering joint with the groove around the pinion shaft, then loosely install the lower steering joint bolt. Be sure that the joint bolt is securely in the groove in the pinion shaft.

65. Pull on the steering joint to make sure that the steering joint is fully seated, then tighten the lower joint bolt to the specified torque.

66. Install the steering joint cover.

67. Install the steering wheel and the driver's airbag .

68. With the tires raised off the ground, check for the following symptoms and probable causes by turning the steering wheel fully to the right and left several times.

- Rubbing sound coming from the lower steering column area. Steering column joint is contacting the cover.

- Grating sound from the lower steering column area, or a rough feeling during steering. Poor engagement of the pinion shaft serrations.

- Noise from around the steering wheel during steering. Poor engagement of the SRS cable reel with the steering wheel, or a damaged cable reel.

69. Do the battery terminal reconnection procedure, and check these items:

- Turn the ignition switch to to ON (II), and check that the SRS indicator comes on for about 6 seconds and then goes off.
- Make sure the horn and turn signal switches work properly.
- Make sure the steering wheel switches work properly.

70. After installation, check these items:

- Start the engine, allow it to idle, and turn the steering wheel from lock to lock several times.
- Check that the EPS indicator does not come on.
- Check the steering wheel spoke angle. If steering spoke angles to

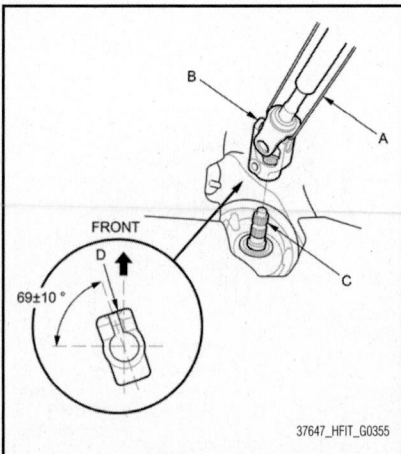

Fig. 150 Cut the wire, while holding the lower slide shaft on the steering column

the right and left are not equal (steering wheel and rack are not centered), correct the engagement of the joint/pinion shaft serrations, then adjust the front toe by turning the tie-rod ends, if necessary.

71. Check the wheel alignment, and adjust it if necessary.

SUSPENSION

FRONT SUSPENSION

LOWER BALL JOINTS

REMOVAL & INSTALLATION

See Figures 151 and 152.

Special Tools Required:
- Ball Joint Thread Protector, 10 mm 07AAF-SECA120
- Ball Joint Thread Protector, 14 mm 071AF-S3VA000
- Ball Joint Remover, 28 mm 07MAC-SL0A202

➡**Always use a ball joint remover to disconnect a ball joint. Do not strike the housing or any other part of the ball joint connection to disconnect it.**

1. Install a hex nut or the ball joint thread protector onto the threads of the ball joint.

➡**Using a hex nut, make sure the nut is flush with the ball joint pin end to prevent damage to the threaded end of the ball joint pin.**

2. Apply grease to the ball joint remover on the areas shown. This will ease installation of the tool and prevent damage to the pressure bolt threads.

3. Loosen the pressure bolt, and install

the ball joint remover as shown. Insert the jaws carefully, making sure not to damage the ball joint boot. Adjust the jaw spacing by turning the adjusting bolt.

➡**Fasten the safety chain securely to a suspension arm or the subframe. Do not fasten it to a brake line or wire harness.**

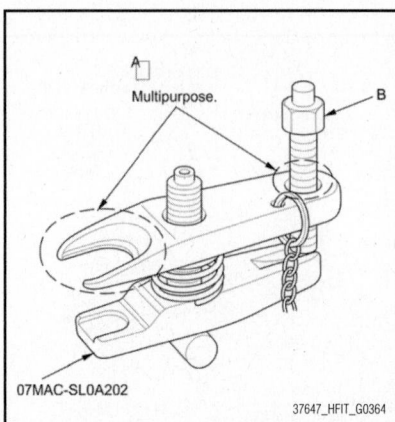

Fig. 151 Apply grease to the ball joint remover on the areas shown (A) to ease installation of the tool and prevent damage to the pressure bolt (B) threads

4. After adjusting the adjusting bolt, make sure the head of the adjusting bolt is in the position shown to allow the jaw to pivot.

5. With a wrench, tighten the pressure bolt until the ball joint pin pops loose from the ball joint connecting hole. If necessary, apply penetrating type lubricant to loosen the ball joint pin.

➡**Do not use pneumatic or electric tools on the pressure bolt.**

6. Remove the ball joint remover, then remove the nut from the end of the ball joint pin, and pull the ball joint out of the ball joint connecting hole. Inspect the ball joint boot, and replace it if damaged.

LOWER CONTROL ARMS

REMOVAL & INSTALLATION

See Figures 153 and 154.

➡**Do not remove the lower arm from both sides at the same time. The lower arm mounting bolts also secure the subframe to the vehicle.**

1. Raise and support the vehicle.
2. Remove the front wheel.
3. Remove the lock pin from the lower arm ball joint, then remove the castle nut.

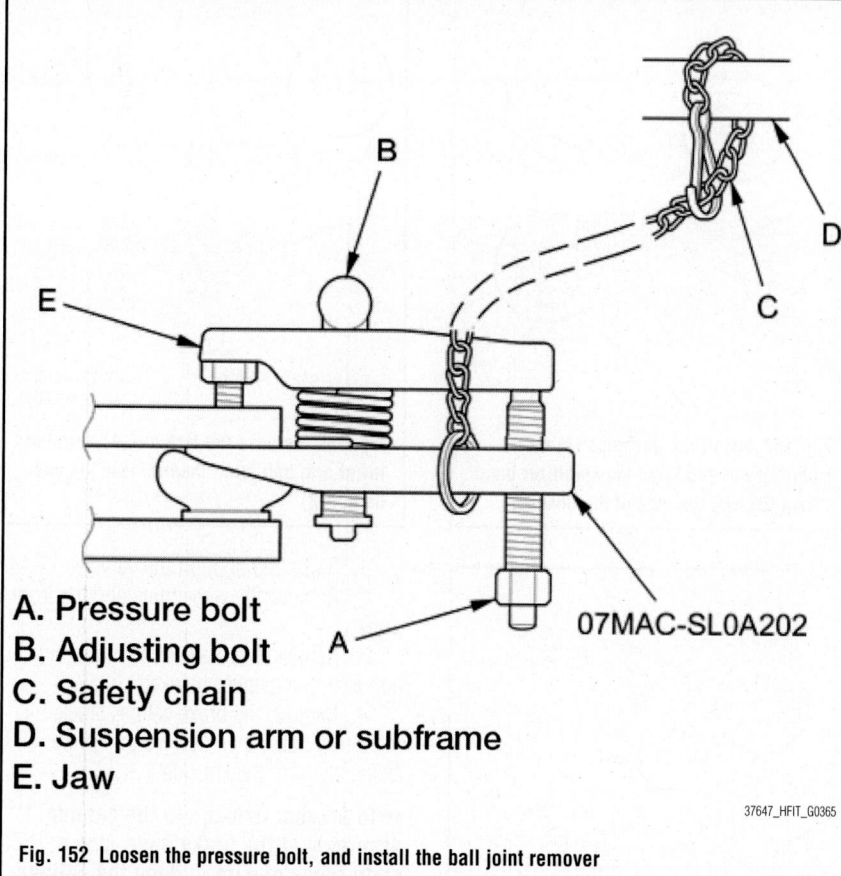

A. Pressure bolt
B. Adjusting bolt
C. Safety chain
D. Suspension arm or subframe
E. Jaw

37647_HFIT_G0365

Fig. 152 Loosen the pressure bolt, and install the ball joint remover

➡**During installation, install the lock pin as shown after tightening the new castle nut.**

4. Disconnect the lower ball joint from the knuckle using the ball joint remover.

➡**Be careful not to damage the ball joint boot when installing the remover. Do not force or hammer on the lower arm, or pry between the lower arm and the knuckle. You could damage the ball joint.**

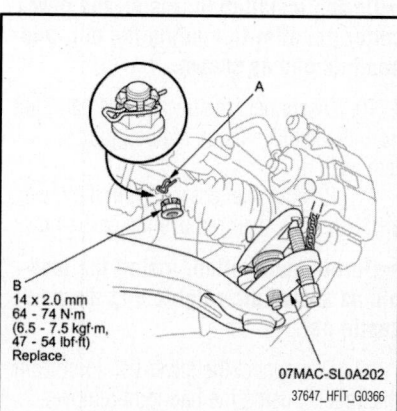

B
14 x 2.0 mm
64 - 74 N·m
(6.5 - 7.5 kgf·m,
47 - 54 lbf·ft)
Replace.

07MAC-SL0A202
37647_HFIT_G0366

Fig. 153 Remove the lock pin (A) from the lower arm ball joint, then remove the castle nut (B)

5. Remove the lower arm mounting bolts, then remove the lower arm from the front subframe.

To install:

➡**Use new lower arm mounting bolts during reassembly.**

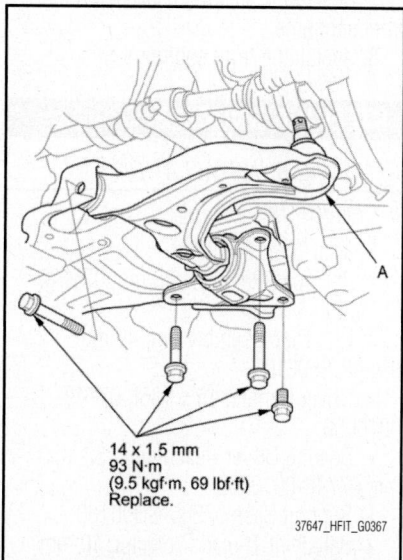

14 x 1.5 mm
93 N·m
(9.5 kgf·m, 69 lbf·ft)
Replace.

37647_HFIT_G0367

Fig. 154 Remove the lower arm mounting bolts, then remove the lower arm (A) from the front subframe

6. Install the lower arm in the reverse order of removal, and note these items:
- First install all the components, and lightly tighten the bolts and the nuts, then raise the suspension to load it with the vehicle's weight before fully tightening it to the specified torque values. Do not place the jack against the ball joint of the lower arm.
- Be careful not to damage the ball joint boot when connecting the knuckle.
- Before connecting the ball joint, degrease the threaded section and the tapered portion of the ball joint pin, the ball joint connecting hole, and the threaded section and the mating surfaces of the castle nut.
- Torque the castle nut to the lower torque specification, then tighten it only far enough to align the slot with the ball joint pin hole. Do not align the castle nut by loosening it.
- Before installing the wheel, clean the mating surfaces of the brake disc and the inside of the wheel.

7. Check the wheel alignment, and adjust it if necessary.

STABILIZER BAR & LINKS

REMOVAL & INSTALLATION

Front Stabilizer Link

See Figure 155.

1. Raise the front of the vehicle, and support it with safety stands in the proper locations.

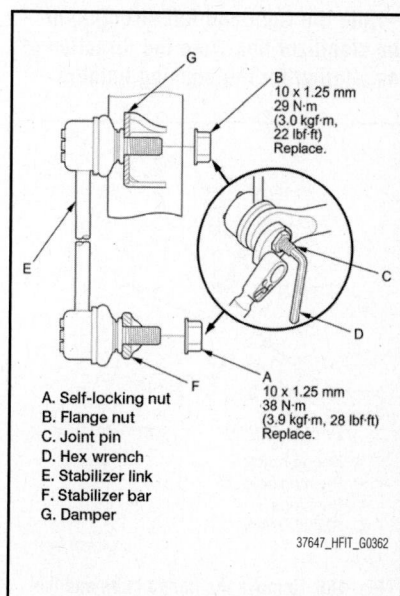

B
10 x 1.25 mm
29 N·m
(3.0 kgf·m,
22 lbf·ft)
Replace.

A
10 x 1.25 mm
38 N·m
(3.9 kgf·m, 28 lbf·ft)
Replace.

A. Self-locking nut
B. Flange nut
C. Joint pin
D. Hex wrench
E. Stabilizer link
F. Stabilizer bar
G. Damper

37647_HFIT_G0362

Fig. 155 Removing/installing the stabilizer link

2. Remove the front wheel.

3. Remove the self-locking nut and the flange nut while holding the respective joint pin with a hex wrench, then remove the stabilizer link.

To install:

4. Install the stabilizer link on the stabilizer bar and damper with the joint pins set at the center of their range of movement.

5. Install a new self-locking nut and a new flange nut, and lightly tighten them.

6. Tighten the lower self-locking nut to 28 ft. lbs. (38 Nm) and the upper flange nut to 22 ft. lbs. (29 Nm), while holding the respective joint pin with a hex wrench.

7. Reinstall all removed parts, and test-drive the vehicle.

Stabilizer Bar

See Figures 156 through 158.

1. Remove the front subframe.

2. Remove the steering gearbox from the front subframe.

3. Remove the flange bolts and the bushing holders, then remove the bushings.

4. Remove the stabilizer bar from the front subframe.

➡**Be careful not to damage the steering gearbox.**

To install:

5. Install the bushings on the stabilizer bar.

➡**Note the direction of installation for the bushings. Align the stabilizer band clamp with the side of the bushings.**

6. Install the stabilizer bar with the bushings on the front subframe.

➡**Note the right and left direction of the stabilizer bar. Note the direction of installation for the bushing holders.**

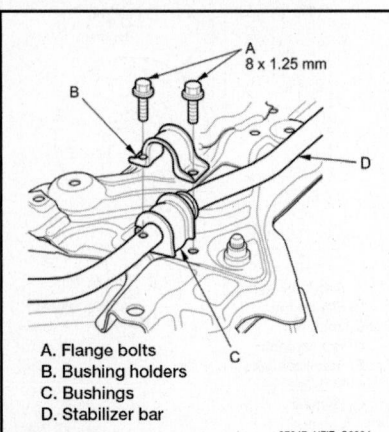

A. Flange bolts
B. Bushing holders
C. Bushings
D. Stabilizer bar

37647_HFIT_G0384

Fig. 156 Remove the flange bolts and the bushing holders, then remove the bushings

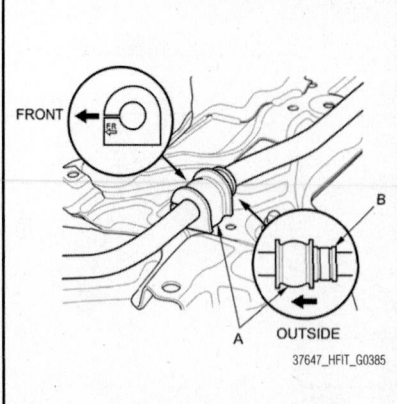

37647_HFIT_G0385

Fig. 157 Install the bushings (A) on the stabilizer bar and align the stabilizer band clamp (B) with the side of the bushings

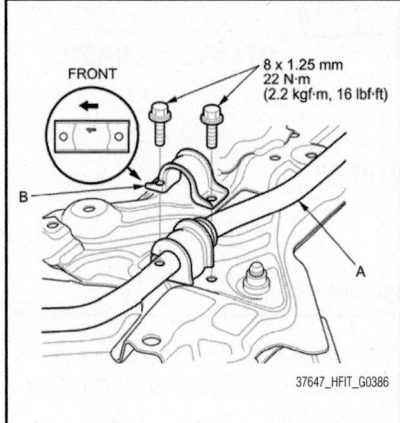

37647_HFIT_G0386

Fig. 158 Install the stabilizer bar (A) with the bushings on the front subframe

7. Install the steering gearbox on the front subframe.

8. Install the front subframe.

STEERING KNUCKLE

REMOVAL & INSTALLATION

See Figures 159 and 160.

Special Tools Required:
- Ball Joint Remover, 28 mm 07MAC-SL0A202
- Hub Dis/Assembly Pin, 40 mm 07GAF-SE00100
- Driver Handle, 15 x 135L 07749-0010000
- Bearing Driver Attachment, 52 x 55 mm 07746-0010400
- Support Base 07965-SD90100
- Ball Joint Thread Protector, 10 mm 07AAF-SECA120
- Ball Joint Thread Protector, 14 mm 071AF-S3VA000

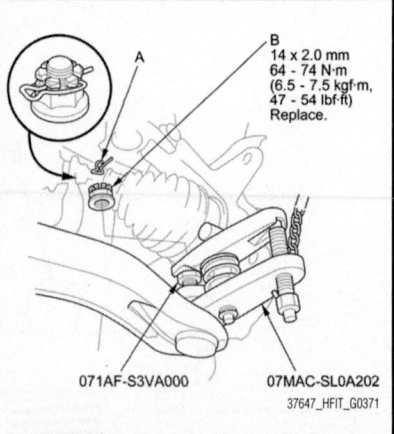

071AF-S3VA000   07MAC-SL0A202

37647_HFIT_G0371

Fig. 159 Remove the lock pin (A) from the lower arm ball joint, then remove the castle nut (B)

1. Raise and support the vehicle.

2. Remove the wheel nuts and the front wheel.

3. Remove the brake hose mounting bolt from the damper bracket.

4. Remove the brake caliper bracket mounting bolts, then remove the caliper assembly from the knuckle.

➡**To prevent damage to the caliper assembly or the brake hose, use a short piece of wire to hang the caliper assembly from the undercarriage. Do not twist the brake hose excessively.**

5. Pry up the stake on the spindle nut, then remove the nut.

6. Remove the front brake disc.

7. Check the front hub for damage and cracks.

8. Remove the wheel speed sensor from the knuckle. Do not disconnect the wheel speed sensor connector.

9. Remove the cotter pin from the tie-rod end ball joint, then remove the nut.

➡**During installation, install the new cotter pin after tightening the nut, and bend its end as shown.**

10. Disconnect the tie-rod end ball joint from the knuckle using the ball joint remover.

11. Remove the lock pin from the lower arm ball joint, then remove the castle nut.

➡**During installation, install the lock pin as shown after tightening the new castle nut.**

12. Disconnect the lower ball joint from the knuckle using the ball joint remover.

➡**Be careful not to damage the ball joint boot when installing the remover. Do not force or hammer on the lower**

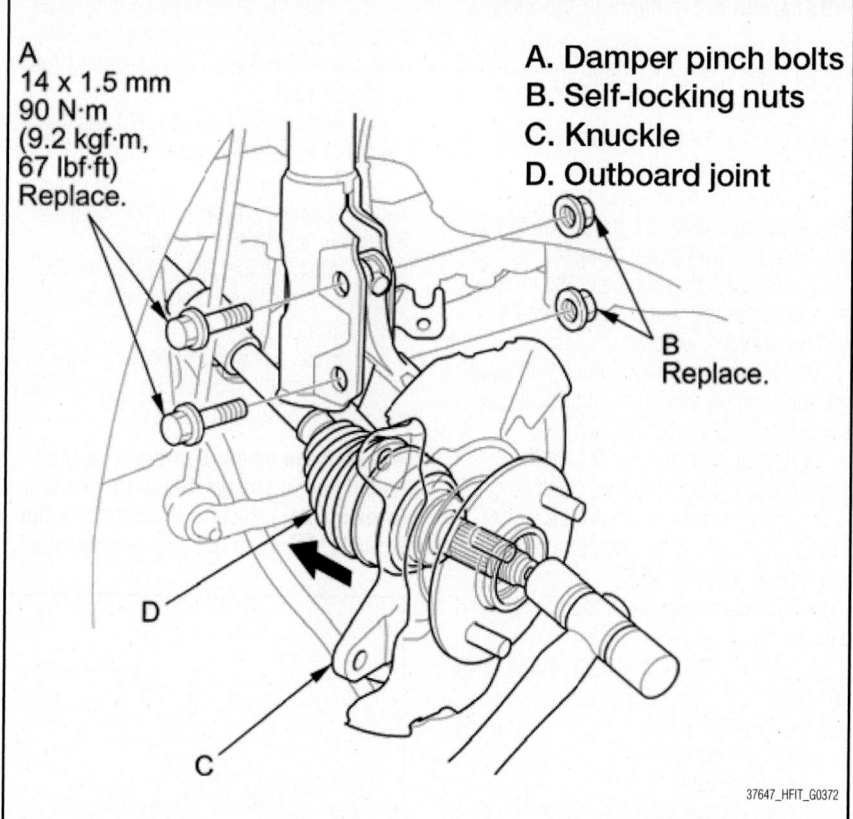

A
14 x 1.5 mm
90 N·m
(9.2 kgf·m,
67 lbf·ft)
Replace.

A. Damper pinch bolts
B. Self-locking nuts
C. Knuckle
D. Outboard joint

B
Replace.

37647_HFIT_G0372

Fig. 160 Remove the damper pinch bolts and the self-locking nuts from the damper

arm, or pry between the lower arm and the knuckle. You could damage the ball joint.

13. Remove the damper pinch bolts and the self-locking nuts from the damper.

➡**Use new damper pinch bolts and new self-locking nuts during reassembly.**

14. Pull the knuckle outward, and separate the outboard joint from the front hub using a soft face hammer.

➡**Do not pull the driveshaft end outward. The driveshaft inboard joint may come apart. During installation, apply grease to the mating surfaces of the wheel bearing and the driveshaft outboard joint.**

To install:

15. Install the knuckle/hub in the reverse order of removal, and note these items:

- First install all the components, and lightly tighten the bolts and the nuts, then raise the suspension to load it with the vehicle's weight before fully tightening to the specified torque values. Do not place the jack against the ball joint of the lower arm.

- Be careful not to damage the ball joint boot when connecting the knuckle.
- Before connecting the ball joint, degrease the threaded section and the tapered portion of the ball joint pin, the ball joint connecting hole, and the threaded section and the mating surfaces of the castle nut.
- Torque the castle nut to the lower torque specification, then tighten it only far enough to align the slot with the ball joint pin hole. Do not align the castle nut by loosening it.
- Use a new spindle nut on reassembly.
- Before installing the spindle nut, apply a small amount of engine oil to the seating surface of the nut. After tightening, use a drift to stake the spindle nut shoulder against the driveshaft.
- Before installing the brake disc, clean the mating surfaces of the front hub and the inside of the brake disc.
- Before installing the wheel, clean the mating surfaces of the brake disc and the inside of the wheel.

16. Check the wheel alignment, and adjust it if necessary.

STRUTS (DAMPER & SPRING)

REMOVAL & INSTALLATION

See Figures 161 through 164.

1. Raise and support the vehicle.
2. Remove the front wheel.
3. Remove the wheel speed sensor from the knuckle. Do not disconnect the wheel speed sensor connector.
4. Disconnect the stabilizer link from the damper.
5. Remove the wheel speed sensor clip, the wire guide grommet, and the brake hose bracket from the damper. Do not disconnect the wheel speed sensor connector.
6. Remove the damper pinch bolts and the self-locking nuts from the damper.

➡**Do not allow the knuckle to rotate too far outward. This may allow the driveshaft inboard joint to come apart.**

7. Passenger's side removal: Remove the lid.

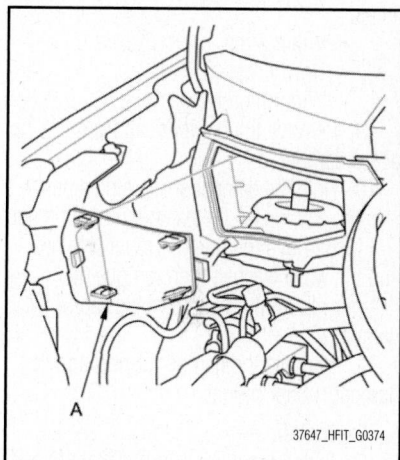

A

37647_HFIT_G0374

Fig. 161 Passenger's side removal: Remove the lid (A)

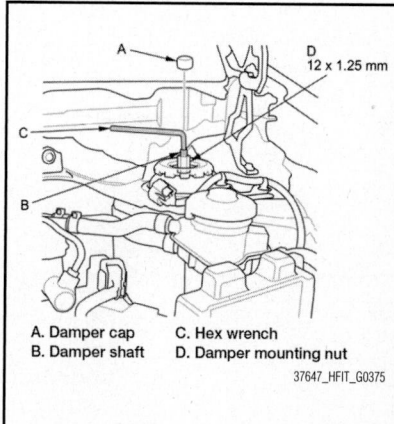

A
D
12 x 1.25 mm
C
B

A. Damper cap   C. Hex wrench
B. Damper shaft   D. Damper mounting nut

37647_HFIT_G0375

Fig. 162 Remove the damper cap from the top of the damper

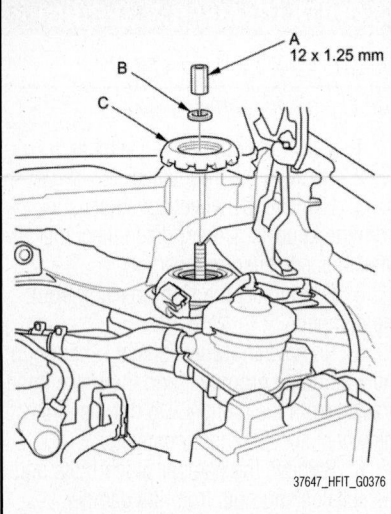

Fig. 163 Remove the damper mounting nut (A) and the wave washer (B), then remove the damper mounting base (C) from the top of the damper

8. Driver's side removal: Remove these items:

- Wiper arms
- Cowl cover
- Wiper motor

9. Remove the damper cap from the top of the damper.

10. Hold the damper shaft using a hex wrench, and loosen the damper mounting nut.

11. Remove the damper mounting nut and the wave washer, then remove the damper mounting base from the top of the damper.

12. Remove the damper/spring and the damper rubber mount.

To install:

13. Install the damper rubber mount and the damper/spring onto the upper mount.

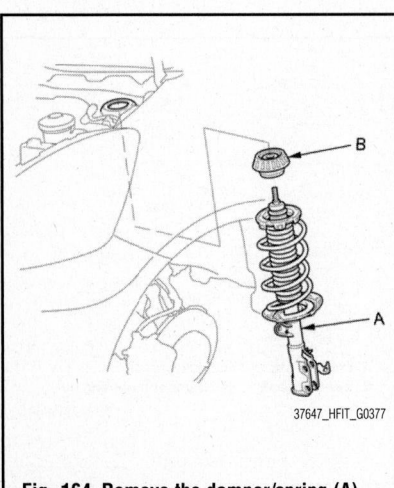

Fig. 164 Remove the damper/spring (A) and the damper rubber mount (B)

➡**Be careful not to damage the body.**

14. Install the damper mounting base and the wave washer, then loosely install the damper mounting nut.

15. Loosely install new damper pinch bolts and new self-locking nuts.

16. Install the wheel speed sensor harness clip, the wire guide grommet and the brake hose bracket to the damper.

17. Place a floor jack under the lower arm, and raise the suspension to load it with the vehicle's weight.

18. Tighten the damper pinch bolts and self-locking nuts while holding the damper pinch bolt to 67 ft. lbs. (90 Nm).

19. Install the stabilizer link to the damper and tighten to the specified torque.

20. Hold the damper shaft with a hex wrench, and tighten the damper mounting nut to 33 ft. lbs. (44 Nm).

21. Install the cap to the top of the damper.

22. Driver's side installation: Install these items:

- Wiper motor
- Cowl cover
- Wiper arms

23. Passenger's side installation: Install the lid.

24. Clean the mating surface of the brake disc and the inside of the wheel, then install the front wheels.

25. Check the wheel alignment, and adjust it if necessary.

WHEEL HUBS & BEARINGS

REMOVAL & INSTALLATION

See Figures 165 through 170.

Special Tools Required:
- Ball Joint Remover, 28 mm 07MAC-SL0A202
- Hub Dis/Assembly Pin, 40 mm 07GAF-SE00100
- Driver Handle, 15 x 135L 07749-0010000
- Bearing Driver Attachment, 52 x 55 mm 07746-0010400
- Support Base 07965-SD90100
- Ball Joint Thread Protector, 10 mm 07AAF-SECA120
- Ball Joint Thread Protector, 14 mm 071AF-S3VA000

1. Remove the steering knuckle.

2. Separate the hub from the knuckle using the hub dis/assembly tool and a hydraulic press. Hold the knuckle with the attachment of the hydraulic press or equivalent tool. Be careful not to damage or deform the splash guard. Hold onto the hub to keep it from falling when pressed clear.

3. Press the wheel bearing inner race off of the hub using the hub dis/assembly tool, a commercially available bearing separator, and a press.

4. Remove the splash guard and the snap ring from the knuckle.

5. Press the wheel bearing out of the knuckle using the bearing driver attachment, the driver handle, and a press.

6. Wash the knuckle and the hub thoroughly in high flash point solvent before reassembly.

7. Press a new wheel bearing into the knuckle using the old bearing, a steel plate, the support base, and a press.

➡**Install the wheel bearing with the wheel speed sensor magnetic encoder (brown color), toward the inside of the knuckle. Remove any oil, grease, dust,**

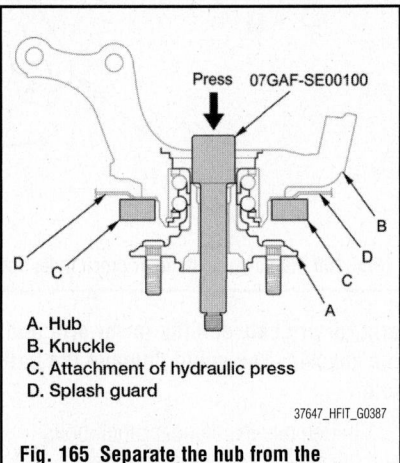

A. Hub
B. Knuckle
C. Attachment of hydraulic press
D. Splash guard

Fig. 165 Separate the hub from the knuckle using the hub dis/assembly tool and a hydraulic press

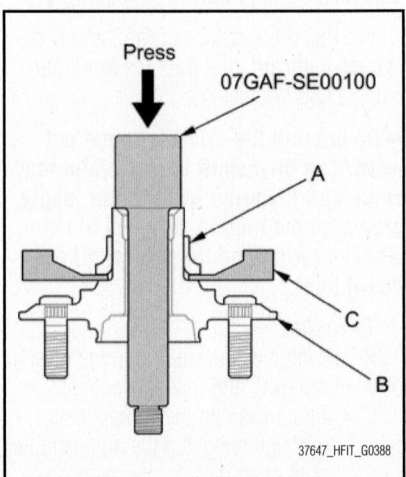

Fig. 166 Press the wheel bearing inner race (A) off of the hub (B) using the hub dis/assembly tool, a commercially available bearing separator (C), and a press

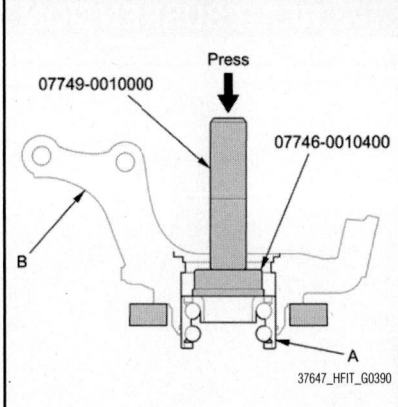

Fig. 167 Press the wheel bearing (A) out of the knuckle (B) using the bearing driver attachment, the driver handle, and a press

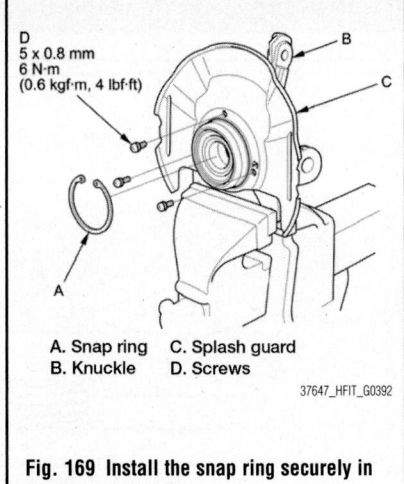

A. Snap ring   C. Splash guard
B. Knuckle   D. Screws

Fig. 169 Install the snap ring securely in the knuckle

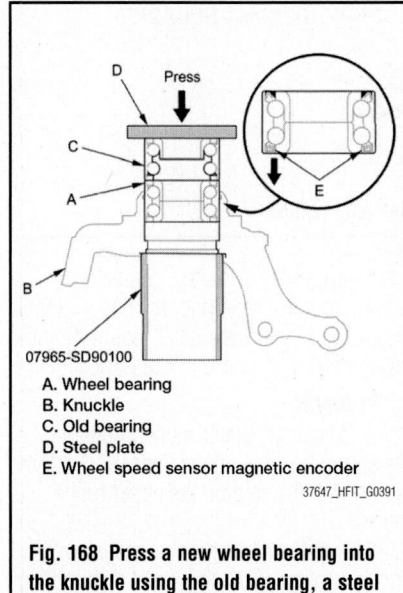

A. Wheel bearing
B. Knuckle
C. Old bearing
D. Steel plate
E. Wheel speed sensor magnetic encoder

Fig. 168 Press a new wheel bearing into the knuckle using the old bearing, a steel plate, the support base, and a press

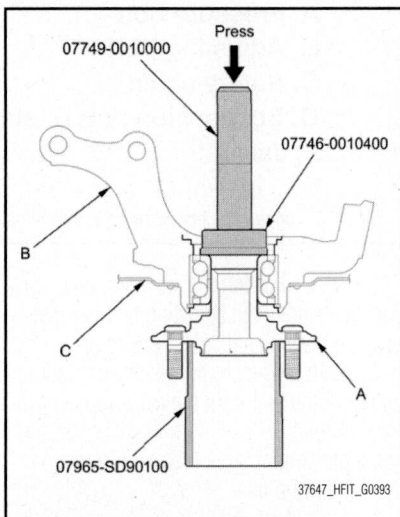

Fig. 170 Install the hub (A) onto the knuckle (B) using the bearing driver attachment, the driver handle, the support base, and a hydraulic press

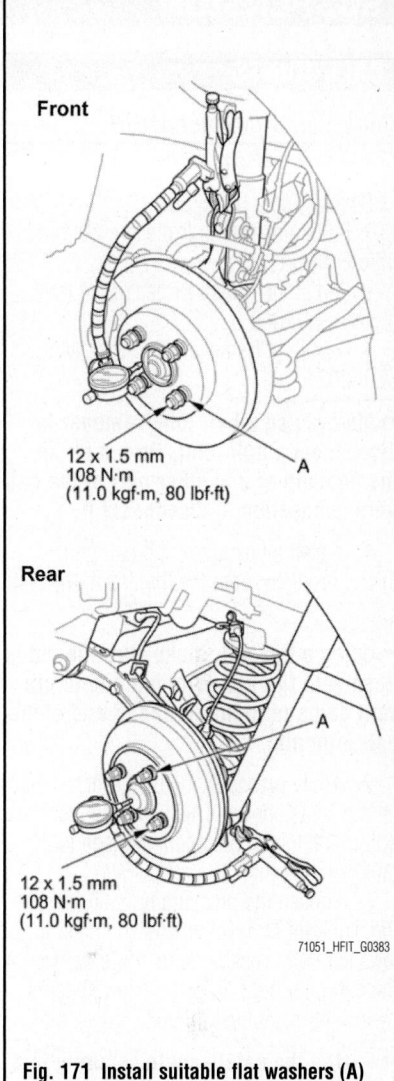

Front

12 x 1.5 mm
108 N·m
(11.0 kgf·m, 80 lbf·ft)

Rear

12 x 1.5 mm
108 N·m
(11.0 kgf·m, 80 lbf·ft)

Fig. 171 Install suitable flat washers (A) and the wheel nuts. Tighten the nuts to the specified torque to hold the brake disc securely against the hub.

metal debris, and other foreign material from the magnetic encoder surface. Keep any magnetic tools away from the encoder surface. Be careful not to damage the encoder surface when you insert the wheel bearing.

8. Install the snap ring securely in the knuckle.

9. Install the splash guard, and tighten the screws to the specified torque value.

10. Install the hub onto the knuckle using the bearing driver attachment, the driver handle, the support base, and a hydraulic press. Be careful not to damage the splash guard.

11. Install the steering knuckle.

END PLAY INSPECTION
See Figure 171.

1. Raise and support the vehicle.
2. Remove the wheels.
3. Install suitable flat washers and the wheel nuts. Tighten the nuts to the specified torque to hold the brake disc securely against the hub.

4. Attach the dial gauge. Place the dial gauge against the hub flange.

5. Measure the bearing end play while moving the brake disc or drum inward and outward.

6. If the bearing end play measurement is more than the standard, replace the wheel bearing or the hub bearing unit. End play maximum to 0–0.0020 in. (0–0.05 mm).

BALL JOINTS

REMOVAL & INSTALLATION

See Figures 172 and 173.

Special Tools Required:
- Ball Joint Thread Protector, 10 mm 07AAF-SECA120
- Ball Joint Thread Protector, 14 mm 071AF-S3VA000
- Ball Joint Remover, 28 mm 07MAC-SL0A202

➡**Always use a ball joint remover to disconnect a ball joint. Do not strike the housing or any other part of the ball joint connection to disconnect it.**

1. Install a hex nut or the ball joint thread protector onto the threads of the ball joint.

➡**Using a hex nut, make sure the nut is flush with the ball joint pin end to prevent damage to the threaded end of the ball joint pin.**

2. Apply grease to the ball joint remover on the areas shown. This will ease installation of the tool and prevent damage to the pressure bolt threads.

3. Loosen the pressure bolt, and install the ball joint remover as shown. Insert the jaws carefully, making sure not to damage the ball joint boot. Adjust the jaw spacing by turning the adjusting bolt.

➡**Fasten the safety chain securely to a suspension arm or the subframe. Do not fasten it to a brake line or wire harness.**

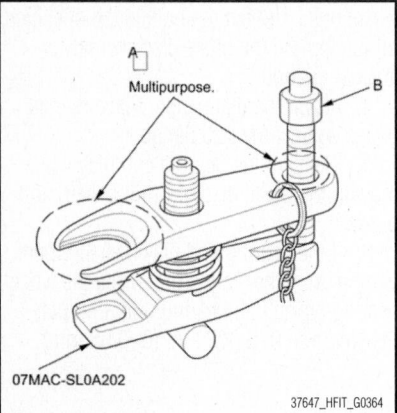

Fig. 172 Apply grease to the ball joint remover on the areas shown (A) to ease installation of the tool and prevent damage to the pressure bolt (B) threads

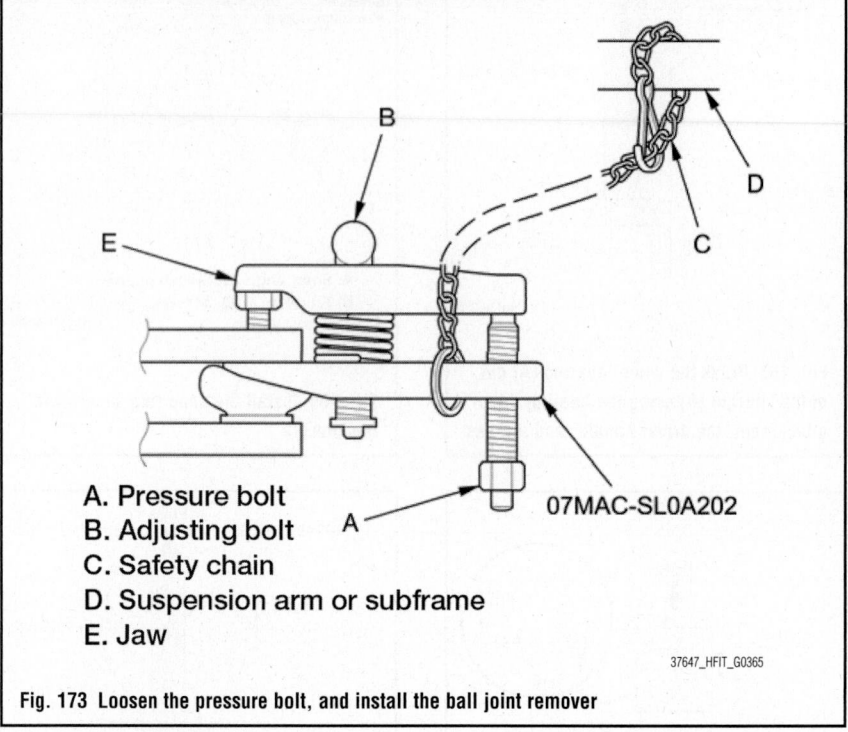

A. Pressure bolt
B. Adjusting bolt
C. Safety chain
D. Suspension arm or subframe
E. Jaw

37647_HFIT_G0365

Fig. 173 Loosen the pressure bolt, and install the ball joint remover

4. After adjusting the adjusting bolt, make sure the head of the adjusting bolt is in the position shown to allow the jaw to pivot.

5. With a wrench, tighten the pressure bolt until the ball joint pin pops loose from the ball joint connecting hole. If necessary, apply penetrating type lubricant to loosen the ball joint pin.

➡**Do not use pneumatic or electric tools on the pressure bolt.**

6. Remove the ball joint remover, then remove the nut from the end of the ball joint pin, and pull the ball joint out of the ball joint connecting hole. Inspect the ball joint boot, and replace it if damaged.

COIL SPRINGS

REMOVAL & INSTALLATION

See Figures 174 through 178.

1. Raise and support the vehicle.
2. Remove the rear wheel.
3. Remove the wheel speed sensor and the wire guide grommet from both sides of the axle beam. Do not disconnect the wheel speed sensor connector.
4. Position the floor jack under spring seat on both sides of the axle beam. Raise the floor jack until the suspension begins to compress.
5. Remove the damper mounting bolt from both sides.

6. Lower the floor jack gradually.
7. Remove the spring, the upper rubber mount, and the lower rubber mount. Do not lower the jack more than necessary.

To install:

8. Install the upper rubber mount on the spring by aligning the upper end of the spring with the stop on the upper rubber mount.

9. Install the lower rubber mount on the spring by aligning the lower end of the spring with the stop on the lower rubber mount.

10. Install the tab of the lower rubber mount into the groove of the lower spring seat.

➡**Make sure that the tab of the lower rubber mount is properly installed into the axle beam. Make sure that the spring is installed correctly.**

11. Position a floor jack under the lower spring seat on both sides of the axle beam.

12. Slowly raise the floor jacks until you can align the bolt hole with the holes in the axle beam and the damper, then loosely tighten the new damper mounting bolt on both sides.

13. Raise the rear suspension with the floor jacks until the vehicle just lifts off of the lift, then tighten the damper mounting bolts to the specified torque.

14. Install the wheel speed sensor and

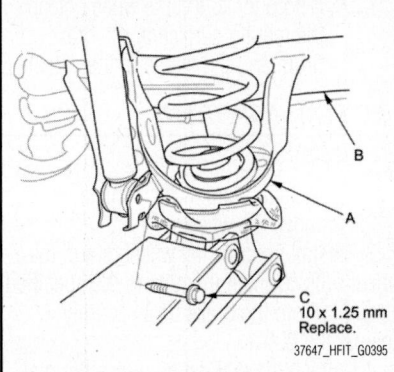

Fig. 174 Position the floor jack under spring seat (A) on both sides of the axle beam (B); remove the damper mounting bolt (C) from both sides

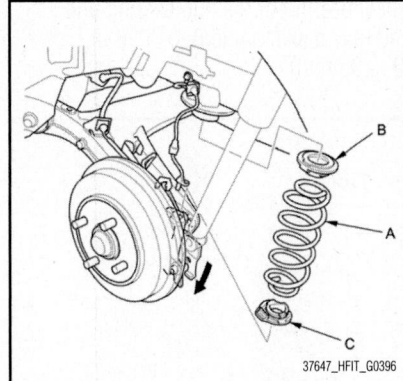

Fig. 175 Remove the spring (A), the upper rubber mount (B), and the lower rubber mount (C)

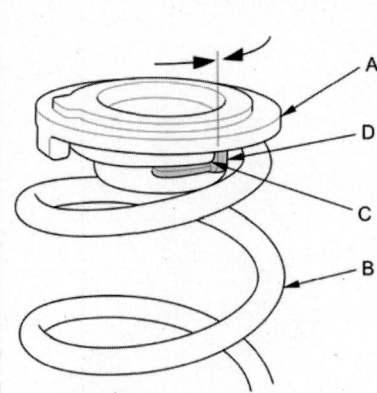

A. Upper rubber mount
B. Spring
C. Upper end
D. Stop

37647_HFIT_G0397

Fig. 176 Install the upper rubber mount on the spring by aligning the upper end of the spring with the stop on the upper rubber mount

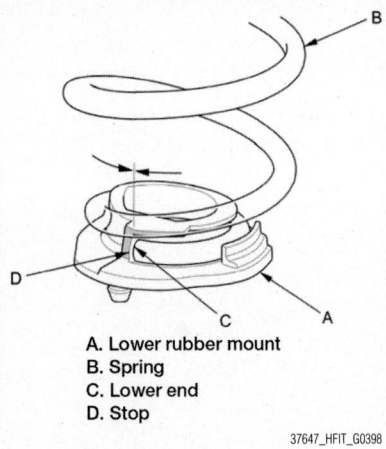

A. Lower rubber mount
B. Spring
C. Lower end
D. Stop

37647_HFIT_G0398

Fig. 177 Install the lower rubber mount on the spring by aligning the lower end of the spring with the stop on the lower rubber mount

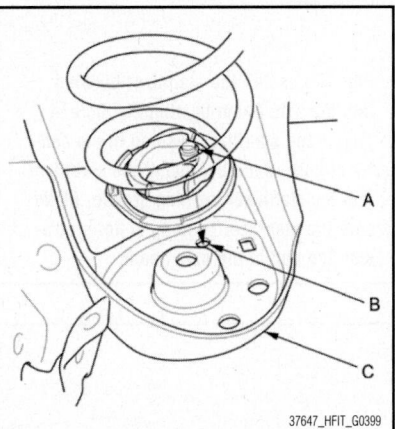

37647_HFIT_G0399

Fig. 178 Install the tab (A) of the lower rubber mount into the groove (B) of the lower spring seat

the wire guide grommet on both sides of the axle beam.

15. Clean the mating surfaces of the brake drum and the inside of the wheel, then install the rear wheel.

16. Check the wheel alignment, and adjust it if necessary.

DAMPER (SHOCK ABSORBER)

REMOVAL & INSTALLATION

See Figures 174 and 179.

1. Raise and support the vehicle.
2. Remove the rear wheel.
3. Remove the wire guide grommet (A) from both sides of the axle beam.
4. Position the floor jack under lower spring seat on both sides of the axle beam. Raise the floor jack until the suspension begins to compress.

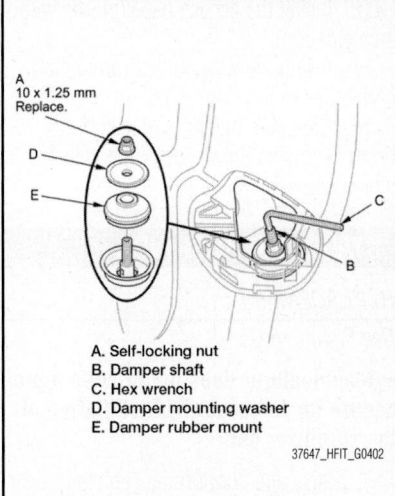

A. Self-locking nut
B. Damper shaft
C. Hex wrench
D. Damper mounting washer
E. Damper rubber mount

37647_HFIT_G0402

Fig. 179 Remove the self-locking nut while holding the damper shaft with a hex wrench

5. Remove the damper mounting bolt from both sides.

6. Remove the access panel from the cargo area side trim.

7. Remove the self-locking nut while holding the damper shaft with a hex wrench.

8. Remove the damper mount washer and the damper rubber mount from the top of the damper.

9. Compress the damper unit (A) by hand, and remove it from the vehicle.

To install:

10. Install the damper rubber mount onto the damper unit. Position the damper assembly between the body and the axle beam.

➡**Be careful not to damage the body.**

11. Position the floor jack under lower spring seat on both sides of the axle beam.

12. Slowly raise the jack until you can align the bolt hole with the holes in the axle beam and the damper, then loosely tighten the new damper mounting bolt on both sides.

13. Raise the rear suspension with the jack until the vehicle just lifts off of the lift, then tighten the damper mounting bolts to the specified torque.

14. Install the damper rubber mount, the damper mounting washer, and the new self-locking nut on the damper shaft.

➡**During installation, note the direction of the damper rubber mount and the damper mounting washer.**

15. Tighten the self-locking nut to 22 ft. lbs. (29 Nm) while holding the damper shaft with a hex wrench.

16. Install the access panel on the cargo area side trim.

17. Install the wire guide grommet on both sides of the axle beam.

18. Clean the mating surfaces of the brake drum and the inside of the wheel, then install the rear wheel.

STABILIZER BUSHING

REPLACEMENT

See Figure 180.

➡ **Manufacturer does not provide a procedure for removal and installation of the stabilizer bar.**

1. Raise and support the vehicle.
2. Remove the stabilizer bushing from the axle beam.
3. Apply silicone spray (P/N08209-0001) to the new stabilizer bushing. This will ease installation of the bushing into the stabilizer and the axle beam.

➡ **When spraying any agents that contain silicone, cover all the connectors, terminals, and switches in area with a protective cloth or plastic sheet.**

4. Install the stabilizer bushing into the axle beam by aligning the end face of the stabilizer bushing to the center of hole.
5. Note the following:
   a. Securely install the stabilizer into the stabilizer bushing groove.
   b. Make sure the stabilizer bushing is not protrusion the end of the axle beam.

WHEEL HUBS & BEARINGS

REMOVAL & INSTALLATION

See Figure 181.

1. Raise and support the vehicle.
2. Remove the wheel nuts and the rear wheel.
3. Remove the brake drum.
4. Remove the hub cap. Raise the stake, then remove the spindle nut.
5. Remove the hub bearing unit from the spindle.
6. Check the hub bearing unit for damage and cracks.
7. Install the hub bearing unit in the reverse order of removal, and note these items:
   • Tighten all mounting hardware to the specified torque values.
   • Use a new spindle nut and hub cap on reassembly.

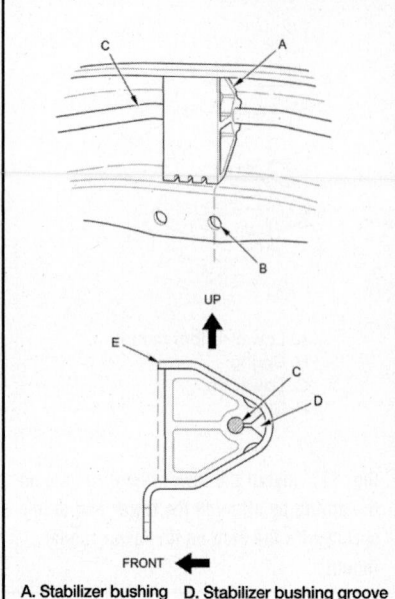

A. Stabilizer bushing
B. Center of hole
C. Stabilizer
D. Stabilizer bushing groove
E. Axle beam

71051_HFIT_G0382

Fig. 180 Install the stabilizer bushing into the axle beam by aligning the end face of the stabilizer bushing to the center of hole. Securely install the stabilizer into the stabilizer bushing groove. Make sure the stabilizer bushing is not protrusion the end of the axle beam.

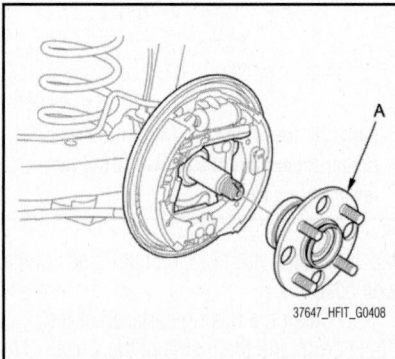

37647_HFIT_G0408

Fig. 181 Remove the hub bearing unit (A) from the spindle

• Before installing the spindle nut, apply a small amount of engine oil to the seating surface of the nut. After tightening, use a drift to stake the spindle nut shoulder against the spindle.
• Before installing the brake drum, clean the mating surface of the hub bearing unit and the inside of the brake drum.

• Before installing the wheel, clean the mating surface of the brake drum and the inside of the wheel.

END PLAY INSPECTION

See Figure 182.

1. Raise and support the vehicle.
2. Remove the wheels.
3. Install suitable flat washers and the wheel nuts. Tighten the nuts to the specified torque to hold the brake disc securely against the hub.
4. Attach the dial gauge. Place the dial gauge against the hub flange.
5. Measure the bearing end play while moving the brake disc or drum inward and outward.
6. If the bearing end play measurement is more than the standard, replace the wheel bearing or the hub bearing unit. End play maximum to 0–0.0020 in. (0–0.05 mm).

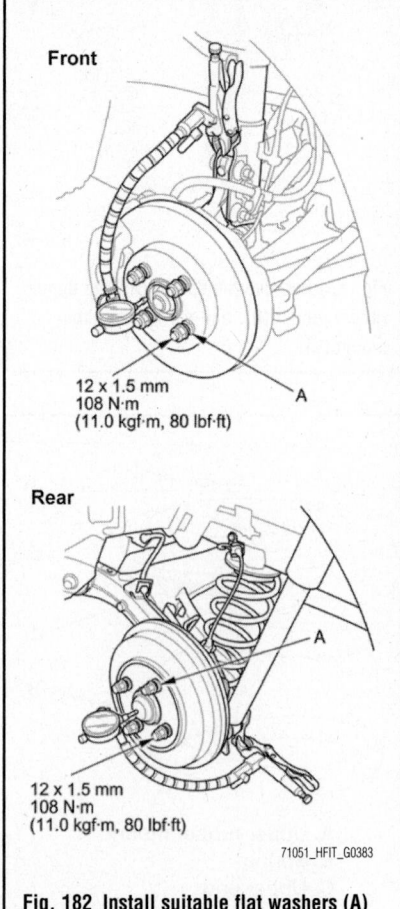

Front

12 x 1.5 mm
108 N·m
(11.0 kgf·m, 80 lbf·ft)

Rear

12 x 1.5 mm
108 N·m
(11.0 kgf·m, 80 lbf·ft)

71051_HFIT_G0383

Fig. 182 Install suitable flat washers (A) and the wheel nuts. Tighten the nuts to the specified torque to hold the brake disc securely against the hub.

HONDA

Insight Hybrid

13

BRAKES 13-9

ANTI-LOCK BRAKE SYSTEM (ABS) **13-9**
General Information...................13-9
  Precautions............................13-9
Wheel Speed Sensors 13-9
  Removal & Installation...........13-9
BLEEDING THE BRAKE SYSTEM **13-10**
Bleeding Procedure.................13-10
Brake Fluid...........................13-10
  Fluid Recommendations 13-10
  Level Check 13-10
FRONT DISC BRAKES **13-11**
Brake Calipers.......................13-11
  Removal & Installation........13-11
Brake Pads 13-11
  Removal & Installation........13-11
PARKING BRAKE..............**13-14**
Adjustments 13-14
REAR DRUM BRAKES........**13-12**
Brake Drums 13-12
  Removal & Installation........13-12
Brake Shoes 13-10
  Removal & Installation........13-12

CHASSIS ELECTRICAL 13-14

AIR BAGS (SUPPLEMENTAL RESTRAINT SYSTEM) **13-14**
Arming the System..................13-15
Disarming the System.............13-15
Precautions 13-14

DRIVE SHAFT.................13-15

Continually Variable
  Transmission (CVT) 13-15
  Checking Fluid Level..........13-15
  Drain & Refill.....................13-16
  Filter Replacement 13-16
Front Halfshaft.......................13-16
  CV-boots Inspection 13-18
  Removal & Installation.......13-16

ENGINE COOLING 13-18

Engine Coolant.......................13-18
  Bleeding 13-18

Drain & Refill Procedure......13-18
  Level Check 13-19
Engine Fan 13-19
  Removal & Installation........13-19
Radiator.................................13-20
  Removal & Installation........13-20
Thermostat 13-21
  Removal & Installation........13-21
Water Pump 13-21
  Removal & Installation........13-21

ENGINE ELECTRICAL **13-22**

BATTERY SYSTEM............**13-22**
Battery..................................13-22
  Battery Cable Reset............13-22
  Battery Disconnect/
    Reconnect Procedure........13-22
  Charging.............................13-22
  Removal & Installation........13-22
CHARGING SYSTEM **13-23**
Alternator 13-23
  Removal & Installation........13-23
DC-DC Converter...................13-25
  Removal & Installation........13-25
IMA Motor Drain Cover..........13-25
  Removal & Installation........13-25
IMA Motor Housing 13-25
  Removal & Installation........13-25
IMA Motor Power Cable.........13-26
  Removal & Installation........13-26
IMA Motor Rotor Position
  Sensor 13-27
  IMA Rotor Position
    Calibration.....................13-27
  Removal & Installation........13-27
IMA Service Precautions 13-23
  Disconnecting the Motor
    Power Cable Connector
    From the Motor Stator......13-24
  Turning Off and on Power
    To the High Voltage
    Circuit 13-24
IMA System Description..........13-23
  Operating Conditions..........13-23
IPU Case 13-28
  Removal & Installation........13-28
IPU Cover..............................13-29
  Removal & Installation........13-29

IPU Module Air Duct..............13-29
  Removal & Installation........13-29
IPU Module Fan 13-29
  Removal & Installation........13-29
Motor Control Module (MCM)..13-29
  IMA Rotor Position
    Calibration.....................13-29
  Motor Control Module
    (MCM) Update 13-29
  Removal & Installation........13-29
Motor Power Inverter (MPI)
  Module................................13-30
  IMA Rotor Position
    Calibration.....................13-31
  Removal & Installation........13-30
IGNITION SYSTEM **13-31**
Firing Orders.........................13-31
Ignition Coil(s) 13-31
  Removal & Installation........13-31
Ignition Timing......................13-32
  Adjustment 13-32
  Inspection 13-32
STARTING SYSTEM **13-32**
Starter 13-32
  Removal & Installation........13-32

ENGINE MECHANICAL **13-32**

Accessory Drive Belt System...13-32
  Accessory Belt Routing.......13-32
  Adjustment 13-32
  Inspection 13-32
  Removal & Installation........13-32
Air Cleaner 13-32
  Removal & Installation........13-32
Camshaft & Valve Lifters.........13-33
  Removal & Installation........13-33
Crankshaft Pulley 13-34
  Removal & Installation........13-34
Crankshaft Rear Cover
  & Seal 13-35
  Removal & Installation........13-35
Cylinder Head 13-35
  Removal & Installation........13-35
Cylinder Head Covers 13-37
  Removal & Installation........13-37
Engine Oil & Filter 13-38
  OIL & Filter Change............13-38
  OIL Level Check.................13-38

Intake Manifold 13-38
  Removal & Installation......13-38
Oil Pan 13-39
  Removal & Installation......13-39
Oil Pump.................................13-41
  Removal & Installation......13-41
Pistons & Rings 13-42
  Positioning.........................13-42
Rocker Arms...........................13-43
  Removal & Installation......13-43
Timing (Cam) Chain &
  Sprockets 13-44
  Removal & Installation......13-44
Valve Lash (Valve Clearance)..13-47
  Adjustment 13-47

ENGINE PERFORMANCE & EMISSION CONTROLS 13-48

Camshaft Position (CMP)
  Sensor 13-48
  Removal & Installation......13-48
Crankshaft Position (CKP)
  Sensor 13-48
  CKP Pattern Clear/CKP
    Pattern Learn Procedure ...13-48
  Removal & Installation......13-48
Engine Coolant Temperature
  (ECT) Sensor 13-49
  Location..............................13-49
  Removal & Installation......13-49
Evaporative Emissions (EVAP)
  Canister 13-49
  Removal & Installation......13-49
Exhaust Gas Recirculation
  (EGR) Valve.........................13-49
  Removal & Installation......13-49
Heated Oxygen Sensor (HO2S)..13-49
  Removal & Installation......13-49
Intake Air Temperature (IAT)
  Sensor 13-50
  Location..............................13-50
  Removal & Installation......13-50
Knock Sensor (KS)..................13-50
  Removal & Installation......13-50
Maintenance Minder 13-50
  Resetting Individual
    Maintenance Items 13-51
  Resetting the Maintenance
    Minder..............................13-50
Manifold Absolute Pressure
  (MAP) Sensor 13-51
  Removal & Installation......13-51
Mass Airflow/Intake Air
  Temperature Sensor 13-51
  Location..............................13-51
  Removal & Installation......13-51

Positive Crankcase Ventilation
  (PCV) Valve 13-51
  Removal & Installation........13-51
Powertrain Control Module
  (PCM) 13-51
  PCM Idle Learn Procedure....13-52
  PCM Update 13-52
  Removal & Installation........13-51
Throttle Position (TP)
  Sensor 13-53
  Location..............................13-53
  Removal & Installation........13-53
Vehicle Speed Sensor 13-53
  Removal & Installation........13-53

FUEL 13-55

GASOLINE FUEL INJECTION SYSTEM 13-55
Adjustments 13-55
  IDLE Speed 13-55
Fuel Filter...............................13-55
  Removal & Installation........13-55
Fuel Level Sensor/Fuel
  Gauge Sending Unit.............13-56
  Removal & Installation........13-56
Fuel Pressure Regulator 13-56
  Removal & Installation........13-56
Fuel Rail & Injectors 13-56
  Removal & Installation........13-56
Fuel System Service
  Precautions 13-55
Fuel Tank...............................13-57
  Draining 13-57
  Removal & Installation........13-57
Fuel Tank Unit 13-58
  Removal & Installation........13-58
Relieving Fuel System
  Pressure 13-55
Throttle Body..........................13-59
  Removal & Installation........13-59

HEATING & AIR CONDITIONING SYSTEM...13-61

Blower Motor 13-61
  Removal & Installation........13-61
Heater Core 13-61
  Removal & Installation........13-61

SPECIFICATIONS 13-3

Brake Specifications.................13-7
Camshaft & Bearing
  Specifications Chart...............13-5
Capacities 13-4
Crankshaft & Connecting Rod
  Specifications 13-5

Engine and Vehicle
  Identification 13-3
Fluid Specifications..................13-4
Gasoline Engine Tune-Up
  Specifications 13-3
General Engine Specifications...13-3
Minder Schedule Maintenance ...13-8
Piston & Ring Specifications 13-5
Tire, Wheel, & Ball Joint
  Specifications 13-7
Torque Specifications 13-6
Valve Specifications 13-4
Wheel Alignment
  Specifications 13-7

STEERING 13-63

EPS Control Unit.....................13-63
  Removal & Installation........13-63
EPS Motor..............................13-63
  Removal & Installation........13-63
Power Rack & Pinion Steering
  Gear 13-63
  Removal & Installation........13-63
Torque Sensor Neutral
  Position Memorization 13-66

SUSPENSION 13-66

FRONT SUSPENSION 13-66
Control Links 13-66
  Removal & Installation........13-66
Lower Ball Joint 13-66
  Removal & Installation........13-66
Lower Control Arm...................13-67
  Removal & Installation........13-67
Stabilizer Bar 13-68
  Removal & Installation........13-68
Stabilizer Link 13-69
  Removal & Installation........13-69
Steering Knuckle, Wheel Hub
  & Bearing.............................13-70
  Removal & Installation........13-70
Strut (Damper/Spring).............13-69
  Removal & Installation........13-69

REAR SUSPENSION 13-72
Axle Beam 13-72
  Removal & Installation........13-72
Coil Spring..............................13-73
  Removal & Installation........13-73
Dynamic Damper.....................13-74
  Removal & Installation........13-74
Shock Absorber (Damper).......13-74
  Removal & Installation........13-74
  Testing 13-75
Wheel Hub & Bearing 13-75
  Adjustment 13-76
  Removal & Installation........13-75

SPECIFICATIONS

ENGINE AND VEHICLE IDENTIFICATION

		Engine						Model Year	
Code ①	Liters (cc)	Cu. In.	Cyl.	Fuel Sys.	Engine Type	Eng. Mfg.	Code ②		Year
LDA3	1.3 (1.339)	82	4	SMFI	SOHC	Honda	B		2011
							C		2012

SMFI: Sequential Multi-Port Fuel Injection

DOHC: Double Overhead Camshaft

① Stamped into the front of the engine block and can be seen through the window next to the "H" logo on the front grill

② 10th digit of the Vehicle Identification Number (VIN)

71051_INHY_C0001

GENERAL ENGINE SPECIFICATIONS
All measurements are given in inches.

Year	Model	Engine Displacement Liters	Engine Series ID	Net Horsepower @ rpm	Net Torque @ rpm (ft. lbs.)	Bore x Stroke (in.)	Compression Ratio	Oil Pressure @ rpm
2011	Insight Hybrid	1.3	LDA3	①	②	NA	NA	NA
2012	Insight Hybrid	1.3	LDA3	①	②	NA	NA	NA

NA: Not Available

① Engine+Electric: 98@5800

   Motor: 13@1500

② Engine+Electric: 123@1000

   Motor: 58@1000

71051_INHY_C0002

GASOLINE ENGINE TUNE-UP SPECIFICATIONS

Year	Engine Displacement Liters	Engine Series ID	Spark Plug Gap (in.)	Ignition Timing (deg.)	Fuel Pump (psi)	Idle Speed (rpm)	Valve Clearance (in.) Intake	Valve Clearance (in.) Exhaust
2011	1.3	LDA3	0.039-0.043	8-12B	38-46	700-800	0.006-0.007	0.009-0.011
2012	1.3	LDA3	0.039-0.043	8-12B	38-46	700-800	0.006-0.007	0.009-0.011

NOTE: The Vehicle Emission Control Information label reflects specification changes made during production.

Follow the figures on the label if they differ from those in this chart.

B: Before top dead center

71051_INHY_C0003

CAPACITIES

Year	Model	Engine Displacement Liters	Engine Series ID	Engine Oil with Filter (qts.)	Transmission (pts.) Auto. ①	Drive Axle Front (pts.)	Drive Axle Rear (pts.)	Fuel Tank (gal.)	Cooling System (qts.)
2011	Insight Hybrid	1.3	LDA3	3.2	3.0	NA	NA	10.6	4.8
2012	Insight Hybrid	1.3	LDA3	3.2	3.0	NA	NA	10.6	4.8

NOTE: All capacities are approximate. Add fluid gradually and check to be sure a proper fluid level is obtained.

NA: Not Available

① Drain and refill

71051_INHY_C0004

FLUID SPECIFICATIONS

Year	Model	Engine Displacement Liters	Engine Series ID	Engine Oil	Auto. Trans.	Drive Axle	Transfer Case	Brake Master Cylinder	Cooling System
2011	Insight	1.3	LDA3	①	②	NA	NA	③	④
2012	Insight	1.3	LDA3	①	②	NA	NA	③	④

DOT: Department Of Transportation

NA: Not Available

① Honda Motor Oil: American Honda P/N 08798-9022 (0W-20), Honda Canada P/N 08798-8023C (0W-20)

② Honda CVTF: Honda P/N 08200-9006

③ Honda DOT 3 Brake Fluid: P/N 08798-9008

④ Honda Long Life Antifreeze/Coolant Type 2: P/N OL 999-9001

71051_INHY_C0005

VALVE SPECIFICATIONS

Year	Engine Displacement Liters	Engine Series ID	Seat Angle (deg.)	Face Angle (deg.)	Spring Test Pressure (lbs. @ in.)	Spring Installed Height (in.)	Stem-to-Guide Clearance (in.) Intake	Stem-to-Guide Clearance (in.) Exhaust	Stem Diameter (in.) Intake	Stem Diameter (in.) Exhaust
2011	1.3	LDA3	NA	NA	NA	①	0.0008-0.0020	0.0020-0.0031	0.2157-0.2161	0.2146-0.2150
2012	1.3	LDA3	NA	NA	NA	①	0.0008-0.0020	0.0020-0.0031	0.2157-0.2161	0.2146-0.2150

NA: Information not available

① Valve spring free length:

   Intake: 2.096-2.097 inches

   Exhaust: 2.256-2.257 inches

71051_INHY_C0006

CAMSHAFT AND BEARING SPECIFICATIONS CHART

All measurements are given in inches.

Year	Engine Displ. Liters	Engine Series ID	Journal Dia.	Brg. Oil Clearance	Shaft End-play	Runout	Journal Bore	Lobe Height Intake	Lobe Height Exhaust
2011	1.3	LDA3	NA	0.0020-0.0035	0.002-0.006	0.001 max.	NA	①	②
2012	1.3	LDA3	NA	0.0020-0.0035	0.002-0.006	0.001 max.	NA	①	②

NA: Not Available

① Intake PRI: 1.1693 inches

Intake SEC: 1.4116 inches

② Exhaust PRI: 1.1772 inches

Exhaust SEC: 1.3965 inches

71051_INHY_C0007

CRANKSHAFT AND CONNECTING ROD SPECIFICATIONS

All measurements are given in inches.

Year	Engine Disp. Liters	Engine Series ID/VIN	Crankshaft Main Brg. Journal Dia.	Crankshaft Main Brg. Oil Clearance	Crankshaft Shaft End-play	Crankshaft Thrust on No.	Connecting Rod Journal Diameter	Connecting Rod Oil Clearance
2011	1.3	LDA3	1.9676-1.9685	0.0007-0.0014	0.0040-0.0140	4	1.5739-1.5748	0.0002-0.0006
2012	1.3	LDA3	1.9676-1.9685	0.0007-0.0014	0.0040-0.0140	4	1.5739-1.5748	0.0002-0.0006

71051_INHY_C0008

PISTON AND RING SPECIFICATIONS

All measurements are given in inches.

Year	Engine Displacement Liters (ID)	Piston Clearance	Ring Gap Top Compression	Ring Gap Bottom Compression	Ring Gap Oil Control	Ring Side Clearance Top Compression	Ring Side Clearance Bottom Compression	Ring Side Clearance Oil Control
2011	1.3 (LDA3)	0.0008-0.0018	0.0060-0.0120	0.0120-0.0170	0.0080-0.0280	0.0026-0.0035	0.0012-0.0022	NA
2012	1.3 (LDA3)	0.0008-0.0018	0.0060-0.0120	0.0120-0.0170	0.0080-0.0280	0.0026-0.0035	0.0012-0.0022	NA

71051_INHY_C0009

TORQUE SPECIFICATIONS
All readings in ft. lbs.

Year	Engine Displacement Liters (ID)	Cylinder Head Bolts	Main Bearing Bolts	Rod Bearing Bolts	Crankshaft Damper Bolts	Flywheel Bolts	Manifold		Spark Plugs	Oil Pan Drain Plug
							Intake	Exhaust		
2011	1.3 (LDA3)	①	②	③	④	33	17	NA	13	⑤
2012	1.3 (LDA3)	①	②	③	④	33	17	NA	13	⑤

NOTE: Dip cylinder head bolts, main bearing bolts, and crankshaft damper bolt in clean engine oil prior to tightening.

NA: Not Available

① Step 1: 22 ft. lbs.

　Step 2: plus 130 degrees

② Tighten bearing cap bolts in sequence:

　Step 1: 18 ft. lbs.

　Step 2: plus 40 degrees

③ Step 1: Tighten to 7.2 ft. lbs.

　Step 2: plus 90 degrees

④ Step 1: Tighten to 27 ft. lbs.

　Step 2: plus 90 degrees

⑤ Tighten in sequence to 17 ft. lbs. in 3 steps

71051_INHY_C0010

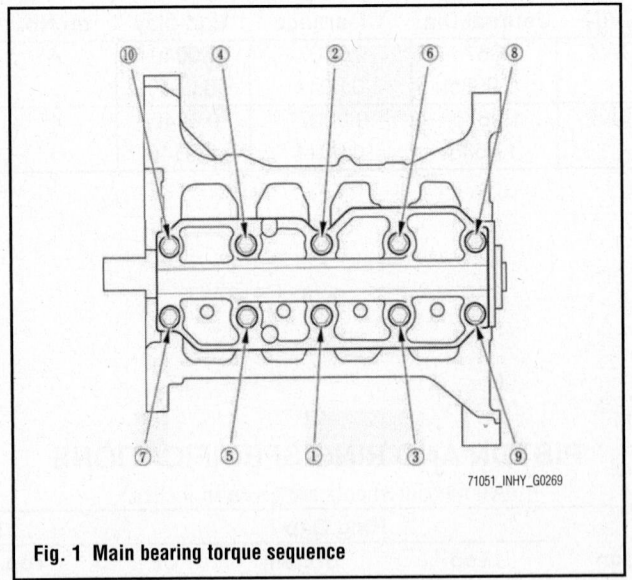

71051_INHY_G0269

Fig. 1 Main bearing torque sequence

WHEEL ALIGNMENT

Year	Model		Caster Range (+/-Deg.)	Caster Preferred Setting (Deg.)	Camber Range (+/-Deg.)	Camber Preferred Setting (Deg.)	Toe-in (Deg.)
2011	Insight Hybrid	Front	1.00	+3.20	1.0	0	0+/-0.12
		Rear	—	—	1.0	-1.00	0.10+/-0.10
2012	Insight Hybrid	Front	1.00	+3.20	1.0	0	0+/-0.12
		Rear	—	—	1.0	-1.00	0.10+/-0.10

NOTE: Measurements are given for unladen vehicle: fuel, engine coolant, and fluid levels are full.

Spare tire, jack, hand tools, and mats are in designated positions.

71051_INHY_C0011

TIRE, WHEEL AND BALL JOINT SPECIFICATIONS

Year	Model	OEM Tires Standard	OEM Tires Optional	Tire Pressures (psi) Front	Tire Pressures (psi) Rear	Wheel Size	Lug Nut Torque (ft. lbs.)
2011	Insight Hybrid	SBRP175/65R15	—	33	33	15	80
2012	Insight Hybrid	SBRP175/65R15	—	33	33	15	80

OEM: Original Equipment Manufacturer

PSI: Pounds Per Square Inch

NA: Information not available

71051_INHY_C0012

BRAKE SPECIFICATIONS

All measurements in inches unless noted

Year	Model		Brake Disc Original Thickness	Brake Disc Minimum Thickness	Brake Disc Maximum Runout	Brake Drum Diameter Original Inside Diameter	Brake Drum Diameter Max. Wear Limit	Brake Drum Diameter Max. Machine Diameter	Minimum Lining Thickness	Brake Caliper Bracket Bolts (ft. lbs.)	Brake Caliper Mounting Bolts (ft. lbs.)
2011	Insight	F	0.830	0.750	0.0016	—	—	—	0.060	80	25
		R	—	—	—	7.874	7.91	NA	0.040	—	—
2012	Insight	F	0.830	0.750	0.0016	—	—	—	0.060	80	25
		R	—	—	—	7.874	7.91	NA	0.040	—	—

F: Front

R: Rear

NA: Not Available

71051_INHY_C0013

MAINTENANCE MINDER SCHEDULE
Honda Insight Hybrid

All Honda's displays engine oil life and maintenance service items in the information display to indicate when to perform maintenance service. If the engine oil life is 15% or less, based on the onboard computer's caluculations, you will see SERVICE DUE SOON in the information display every time the ignition key is turned to ON. The maintenance minder indicator will also come on and the maintenance code(s) for other scheduled maintenance items needing service will be displayed below the message.

Symbol	Item	Service
A	Engine oil ①	Change
B	Engine oil and filter	Change
	Fluid levels	Inspect
	Brakes	Inspect
	Parking brake adjustment	Check
	Steering gear and linkage	Inspect
	Suspension components	Inspect
	Driveshaft boots	Inspect
	Brake hoses and lines	Inspect
	Exhaust system	Inspect
	Fuel lines and connections	Inspect
1	Tires	Rotate
2	Engine air filter ②	Replace
	Dust and pollen filter ③	Replace
	Accessory drive belt	Inspect
3	CVT fluid	Replace
4	Spark plugs	Replace
	Valve clearance ④	Inspect
5	Engine coolant	Replace

① If the message SERVICE DUE NOW does not appear more than 12 months after the display is reset, change every year.

② If driven in dusty conditions, replace every 15,000 miles.

③ If driven in urban areas that have a high concentration of soot from industry and diesel, replace every 15,000 miles

⑥ Adjust if necessary.

Additionally, replace the brake fluid every 3 years, and inspect the idle speed every 160,000 miles.

To reset the Engine Oil Life Display:

1. Turn the ignition switch to ON.

2. Press the SELECT button repeatedly until the engine oil life display or the service message is displayed.

3. Press the RESET button for about 10 seconds. You will see a MAINT RESET message.

4. Select the appropriate answer, MAINT RESET >N (NO) or MAINT RESET > y (YES) by pressing the SELECT button repeatedly.

>N or >Y is displayed on the outside temperature >N or >Y is displayed on the outside temperature display.

5. Select the MAINT RESET > Y (YES), and press and hold the RESET button again to reset the engine oil life to 100%.

BRAKES   **ANTI-LOCK BRAKE SYSTEM (ABS)**

GENERAL INFORMATION

PRECAUTIONS

- Certain components within the ABS system are not intended to be serviced or repaired individually.
- Do not use rubber hoses or other parts not specifically specified for and ABS system. When using repair kits, replace all parts included in the kit. Partial or incorrect repair may lead to functional problems and require the replacement of components.
- Lubricate rubber parts with clean, fresh brake fluid to ease assembly. Do not use shop air to clean parts; damage to rubber components may result.
- Use only DOT 3 brake fluid from an unopened container.
- If any hydraulic component or line is removed or replaced, it may be necessary to bleed the entire system.
- A clean repair area is essential. Always clean the reservoir and cap thoroughly before removing the cap. The slightest amount of dirt in the fluid may plug an orifice and impair the system function. Perform repairs after components have been thoroughly cleaned; use only denatured alcohol to clean components. Do not allow ABS components to come into contact with any substance containing mineral oil; this includes used shop rags.
- The Anti-Lock control unit is a microprocessor similar to other computer units in the vehicle. Ensure that the ignition switch is **OFF** before removing or installing controller harnesses. Avoid static electricity discharge at or near the controller.
- If any arc welding is to be done on the vehicle, the control unit should be unplugged before welding operations begin.

WHEEL SPEED SENSORS

REMOVAL & INSTALLATION

Front

See Figure 2.

1. Turn the ignition switch to LOCK (0).
2. Remove the front wheels.
3. Remove the grommet, then disconnect the wheel speed sensor connector.
4. Remove the bolt and the bracket, the clip, and the grommet.

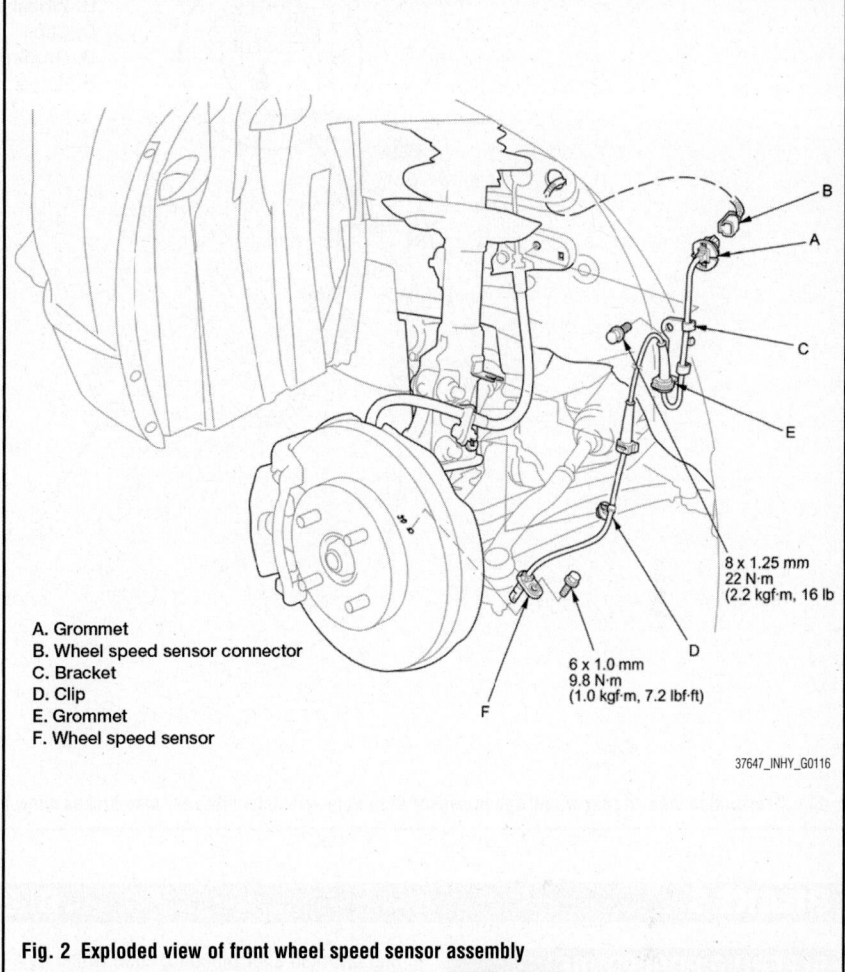

A. Grommet
B. Wheel speed sensor connector
C. Bracket
D. Clip
E. Grommet
F. Wheel speed sensor

8 x 1.25 mm
22 N·m
(2.2 kgf·m, 16 lb

6 x 1.0 mm
9.8 N·m
(1.0 kgf·m, 7.2 lbf·ft)

37647_INHY_G0116

Fig. 2 Exploded view of front wheel speed sensor assembly

5. Remove the bolt and the wheel speed sensor.

To install:

6. Install the wheel speed sensor in the reverse order of removal, and note these items:
- Do not twist the sensor wires.
- If the wheel speed sensor comes in contact with the wheel bearing unit, it is faulty.
- Make sure the grommet is installed properly.
- Make sure there is no debris in the sensor mounting hole.

7. Start the engine, and make sure the ABS, indicator go off.
8. Test-drive the vehicle, and make sure the ABS, indicator do not come on.

Rear

See Figure 3.

1. Turn the ignition switch to LOCK (0).
2. Remove the rear wheels.

3. Release the connector holding clamps, then disconnect the wheel sensor connector.
4. Remove the clip and the grommets.
5. Remove the bolt and the wheel speed sensor.

To install:

6. Install the wheel speed sensor in the reverse order of removal, and note these items:
- Do not twist the sensor wires.
- If the wheel speed sensor comes in contact with the hub bearing unit, it is faulty.
- Make sure the grommet is installed properly.
- Make sure there is no debris in the sensor mounting hole.

7. Start the engine, and make sure the ABS indicator go off.
8. Test-drive the vehicle, and make sure the ABS indicator do not come on.

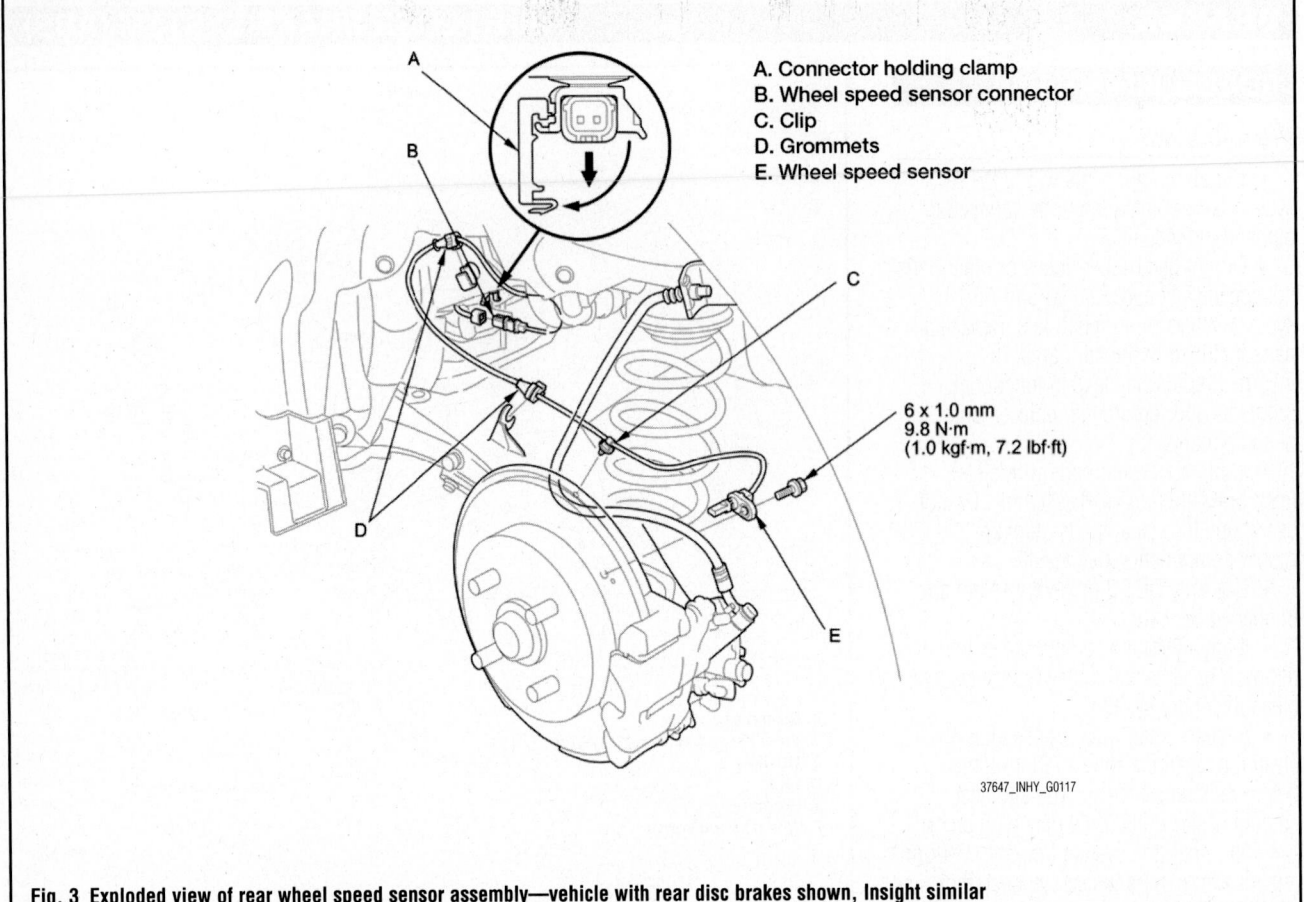

A. Connector holding clamp
B. Wheel speed sensor connector
C. Clip
D. Grommets
E. Wheel speed sensor

6 x 1.0 mm
9.8 N·m
(1.0 kgf·m, 7.2 lbf·ft)

37647_INHY_G0117

Fig. 3 Exploded view of rear wheel speed sensor assembly—vehicle with rear disc brakes shown, Insight similar

BRAKES   BLEEDING THE BRAKE SYSTEM

BLEEDING PROCEDURE

Note the following while bleeding the brake system:

• Do not reuse the drained fluid. Use only new Honda DOT 3 Brake Fluid from an unopened container. Using a non-Honda brake fluid can cause corrosion and shorten the life of the system.

• Make sure no dirt or other foreign matter is allowed to contaminate the brake fluid.

• Do not spill brake fluid on the vehicle; it may damage the paint. If brake fluid does contact the paint, wash it off immediately with water.

• The reservoir connected to the master cylinder must be at the MAX (upper) level mark at the start of the bleeding procedure and checked after bleeding each wheel location. Add fluid as required.

1. Make sure the brake fluid level in the reservoir is at the MAX (upper) level line.

2. Have someone slowly pump the brake pedal several times, then apply steady pressure.

3. Start the bleeding at the driver's side of the front brake system.

4. Bleed the calipers or the wheel cylinders in the following sequence:

• Front left
• Front right
• Rear right
• Rear left

5. Attach a length of clear drain tube to the bleed screw, then loosen the bleed screw to allow air to escape from the system. Then tighten the bleed screw securely.

6. Refill the master cylinder reservoir to the MAX (upper) level line.

7. Repeat the procedure for each brake circuit until there are no air bubbles are in the fluid.

BRAKE FLUID

FLUID RECOMMENDATIONS

Manufacturer recommends only use of DOT 3 brake fluid.

LEVEL CHECK

Always maintain brake fluid level between the "MAX" and "MIN" markings on the brake fluid reservoir.

BRAKES   **FRONT DISC BRAKES**

BRAKE CALIPERS

REMOVAL & INSTALLATION

See Figure 4.

➡**Keep any grease off the brake disc and brake pads.**

1. Raise the vehicle on a lift.
2. Remove the front wheel.
3. Remove the brake hose mounting bolt.
4. Remove the brake caliper bracket mounting bolts, then remove the caliper assembly from the knuckle.
5. Disconnect the brake hose from the caliper body.
6. Installation is the reverse of removal.
7. Tighten the caliper bracket mounting bolts to 80 ft. lbs. (108 Nm).

BRAKE PADS

REMOVAL & INSTALLATION

See Figures 5 through 9.

Special Tools Required: Brake Caliper Piston Compressor 07AAE-SEPA101

❈❈ CAUTION

Frequent inhalation of brake pad dust, regardless of material composition, could be hazardous to your health. Avoid breathing dust particles. Never use an air hose or brush to clean brake assemblies. Use an OSHA-approved vacuum cleaner.

1. Remove some of the brake fluid from the master cylinder.

2. Raise the vehicle on a lift.
3. Remove the front wheels.
4. Remove the brake hose mounting bolt.
5. Remove the flange bolt, be careful not to damage the pin boot, and pivot the caliper up out of the way.

➡**Check the hose and the pin boots for damage and deterioration.**

6. Remove the pad shims and the brake pads.
7. Remove the pad retainers.

To install:

8. Clean the caliper bracket thoroughly; remove any rust, and check for grooves and cracks. Verify that the caliper pins move in and out smoothly. Clean and lube the pins if needed.

9. Inspect the brake disc for runout, thickness, parallelism, and check for damage and cracks.
10. Apply a thin coat of M-77 assembly paste (P/N 08798-9010) to the retainer mating surface of the caliper bracket (indicated by the arrows and shaded area).
11. Install the pad retainers. Wipe excess assembly paste off the retainers. Keep the assembly paste off the brake disc and the brake pads.
12. Mount a brake caliper piston compressor on the caliper body.
13. Press in the piston with the brake caliper piston compressor tool so the caliper will fit over the brake pads. Make sure the piston boot is in position to prevent damaging it when pivoting the caliper down.

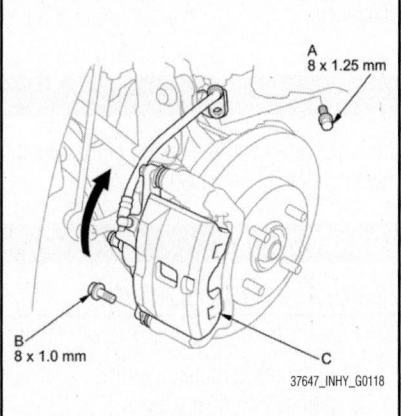

Fig. 5 Remove the brake hose mounting bolt (A), the flange bolt (B), and pivot the caliper (C) up out of the way

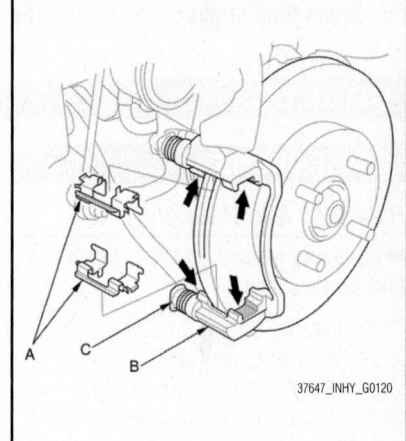

Fig. 7 Pad retainers (A), caliper bracket (B), and caliper pins (C)

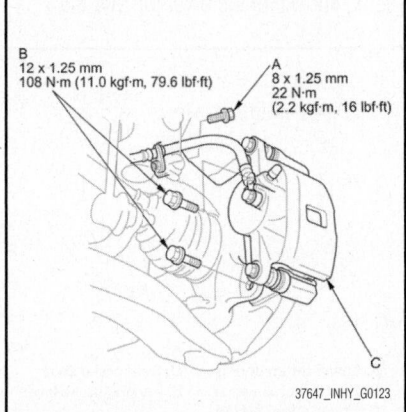

Fig. 4 Remove the brake hose mounting bolt (A), the brake caliper bracket mounting bolts (B), then remove the caliper assembly (C) from the knuckle

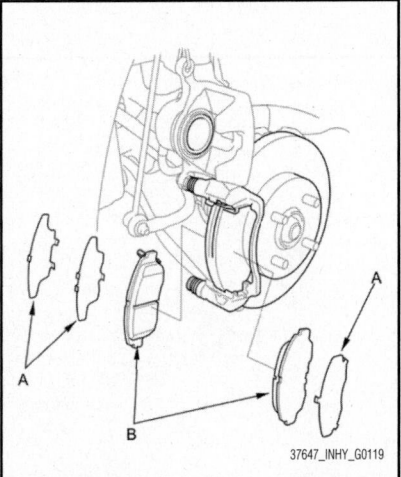

Fig. 6 Remove the pad shims (A) and the brake pads (B)

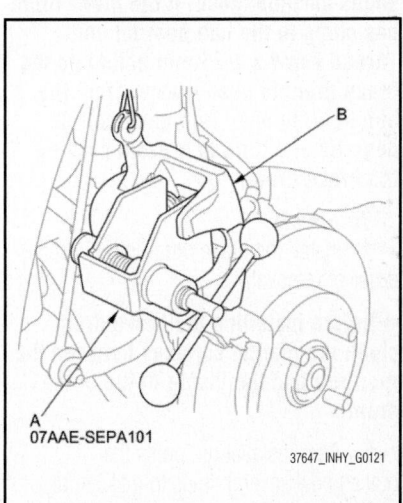

Fig. 8 Mount a brake caliper piston compressor (A) on the caliper body (B)

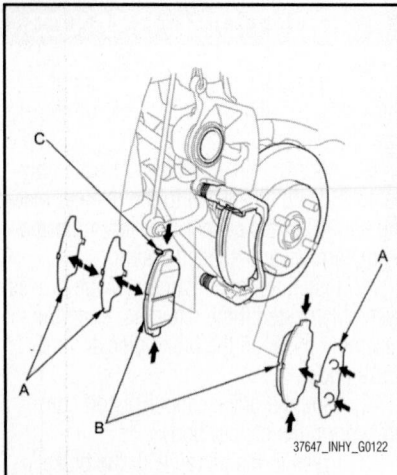

Fig. 9 Apply Molykote M-77 assembly paste to the pad side of the shims (A), the back of the brake pads (B), and other areas indicated by the arrows

➡Be careful when pressing in the piston; brake fluid might overflow from the master cylinder's reservoir. If brake fluid gets on any painted surface, wash it off immediately with water.

14. Remove the brake caliper piston compressor tool.

15. Apply Molykote M-77 assembly paste (P/N 08798-9010) to the pad side of the shims, the back of the brake pads, and other areas indicated by the arrows. Wipe excess assembly paste off the pad shims and the brake pads friction material. Keep grease and assembly paste off the brake discs and the brake pads. Contaminated brake discs or brake pads reduce stopping ability.

16. Install the brake pads and the pad shims correctly. Install the brake pad with the wear indicator on the upper inside. If you are reusing the brake pads, always reinstall the brake pads in their original positions to prevent a temporary loss of braking efficiency.

17. Pivot the caliper down into position. Install the flange bolt and tighten it to 25 ft. lbs. (34 Nm).

18. Install the brake hose mounting bolt.

19. Clean the mating surfaces of the brake disc and the inside of the wheel, then install the front wheels.

20. Press the brake pedal several times to make sure the brakes work.

➡**Engagement may require a greater pedal stroke immediately after the brake pads have been replaced as a set. Several applications of the brake pedal will restore the normal pedal stroke.**

21. Add brake fluid as needed.

22. After installation, check for leaks at hose and line joints or connections, and retighten if necessary.

23. Test-drive the vehicle, then check for leaks.

BRAKES

BRAKE DRUMS

REMOVAL & INSTALLATION

➡**Keep any grease off the brake drum and brake shoes.**

1. Raise the vehicle on a lift.
2. Remove the rear wheel.
3. Remove the parking brake, and remove the brake drum from the hub bearing unit.

➡**If necessary, turn the adjuster bolt with a flat-tip screwdriver until the shoes become loose. If the brake drum has clung to the hub bearing unit. Thread two 8 x 1.25 mm bolts into the brake drum to push it away from the hub bearing unit. Turn each bolt 90 degrees at a time to prevent cocking the brake drum.**

To install:

4. Install the brake drum in the reverse order of removal.

➡**Before installing the brake drum, clean the mating surfaces between the rear hub and the inside of the brake drum.**

5. After installation, press the brake pedal several times to make sure the brakes work and self-adjust the brake shoes.

6. Clean the mating surfaces between the brake drum and the inside of the wheel, then install the rear wheel.

BRAKE SHOES

REMOVAL & INSTALLATION

See Figures 10 through 16.

1. Raise the vehicle on a lift.
2. Remove the rear wheels.
3. Release the parking brake, and remove the brake drum.
4. Remove the tension pins by pushing the respective retainer spring and turning the pin.

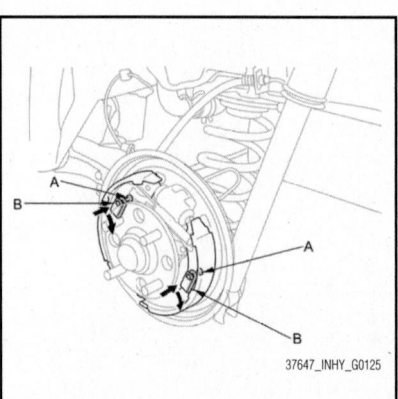

Fig. 10 Remove the tension pins (A) by pushing the respective retainer spring (B) and turning the pin

REAR DRUM BRAKES

5. Remove the lower return spring, and remove the brake shoe assembly over the hub.

6. Remove the forward brake shoe by removing the upper return spring, and disassemble the brake shoe assembly.

7. Remove the rearward brake shoe by disconnecting the parking brake cable from the parking brake lever.

8. Remove the U-clip, the wave washer, and the pivot pin, and separate the parking brake lever from the brake shoe.

To install:

9. Apply Molykote 44MA grease to the sliding surface of the pivot pin and the

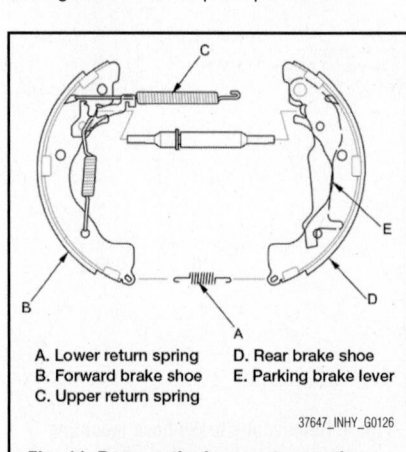

A. Lower return spring
B. Forward brake shoe
C. Upper return spring
D. Rear brake shoe
E. Parking brake lever

Fig. 11 Remove the lower return spring, and remove the brake shoe assembly over the hub

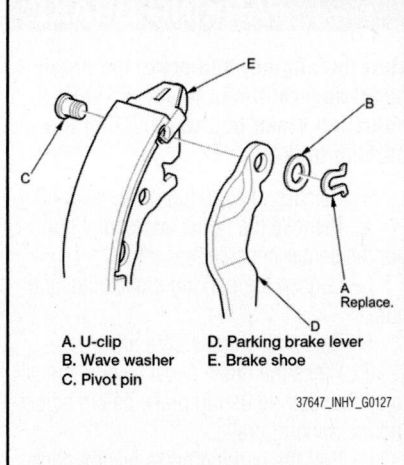

A. U-clip
B. Wave washer
C. Pivot pin
D. Parking brake lever
E. Brake shoe

37647_INHY_G0127

Fig. 12 Remove the U-clip, the wave washer, and the pivot pin, and separate the parking brake lever from the brake shoe

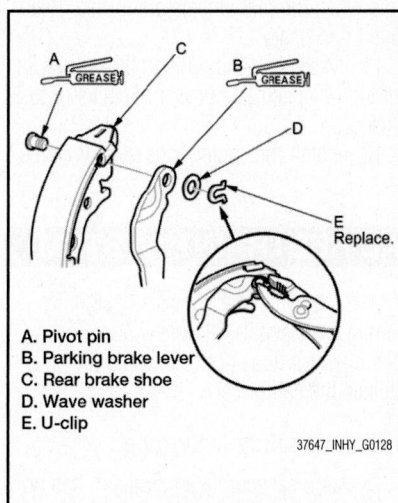

A. Pivot pin
B. Parking brake lever
C. Rear brake shoe
D. Wave washer
E. U-clip

37647_INHY_G0128

Fig. 13 Apply Molykote 44MA grease to the sliding surface of the pivot pin and the parking brake lever for the rearward brake shoe

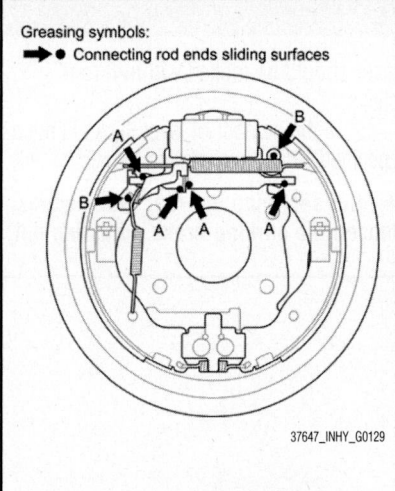

Greasing symbols:
➡● Connecting rod ends sliding surfaces

37647_INHY_G0129

Fig. 14 Apply a thin coat of Molykote 44MA grease to the connecting rod ends (A) and the sliding surfaces (B)

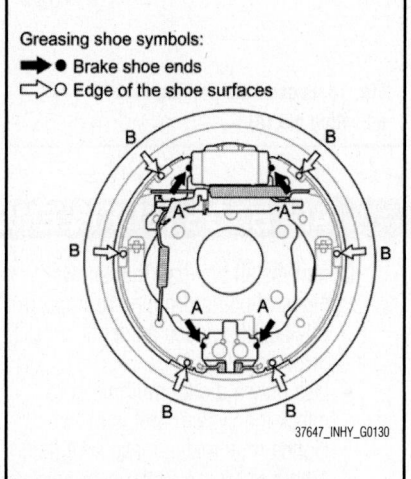

Greasing shoe symbols:
➡● Brake shoe ends
▷○ Edge of the shoe surfaces

37647_INHY_G0130

Fig. 15 Apply a thin coat of Molykote 44MA grease to the shoe ends (A) and the edge of the shoe surfaces (B) that contact the backing plate

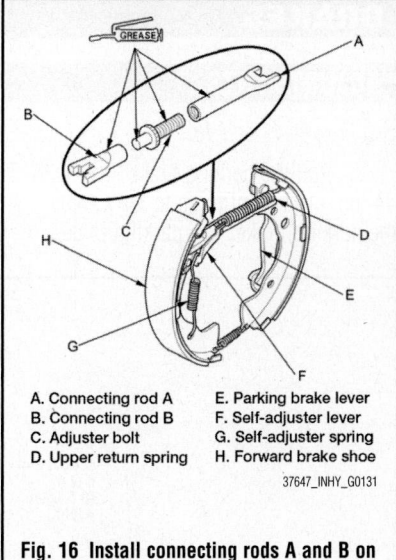

A. Connecting rod A
B. Connecting rod B
C. Adjuster bolt
D. Upper return spring
E. Parking brake lever
F. Self-adjuster lever
G. Self-adjuster spring
H. Forward brake shoe

37647_INHY_G0131

Fig. 16 Install connecting rods A and B on the adjuster bolt

parking brake lever for the rearward brake shoe.

10. Install the parking brake lever and the wave washer on the pivot pin, and secure with a new U-clip.

➡**Pinch the U-clip securely to prevent the parking brake lever from coming out of the brake shoe.**

11. Connect the parking brake cable to the parking brake lever.

12. Apply a thin coat of Molykote 44MA grease to the connecting rod ends and the sliding surfaces as shown. Wipe off any excess. Keep grease off the brake linings.

13. Apply a thin coat of Molykote 44MA grease to the shoe ends and the edge of the shoe surfaces that contact the backing plate as shown. Wipe off any excess. Keep grease off the brake linings.

14. Install connecting rods A and B on the adjuster bolt.

➡**Clean the threaded portions of connecting rod A and the sliding surface of connecting rod B, then coat them with Molykote 44MA grease. Shorten connecting rod A by fully turning in the adjuster bolt.**

15. Assemble the brake shoes with the upper return spring, and with the connecting rods and the adjuster bolt onto the backing plate. Reconnect the parking brake cable to the parking brake lever, then install the self-adjuster lever and the self-adjuster spring on the forward brake shoe.

16. Install the tension pins and the retainer springs by pushing in the respective spring and turning each pin.

17. Install the lower return spring.

➡**Make sure the brake shoes are positioned on the brake shoe bosses on the backing plate, and the fittings on the top of the brake shoes are fitted into the wheel cylinder pistons.**

18. Install the brake drum.

➡**Before installing the brake drum, clean the mating surface between the rear hub and the inside of the brake drum.**

19. Clean the mating surfaces between the brake drum and the inside of the wheel, then install the rear wheels.

20. Press the brake pedal several times to make sure the brakes work and to set the self-adjusting brake.

➡**Engagement of the brakes may require a greater pedal stroke immediately after the brake shoes have been replaced as a set. Several applications of the brake pedal will restore the normal pedal stroke.**

21. Do the parking brake adjustment.

BRAKES

ADJUSTMENTS

See Figures 17 and 18.

1. Pull the parking brake lever with 44 lbs. (196 N) of force to fully apply the parking brake. The parking brake

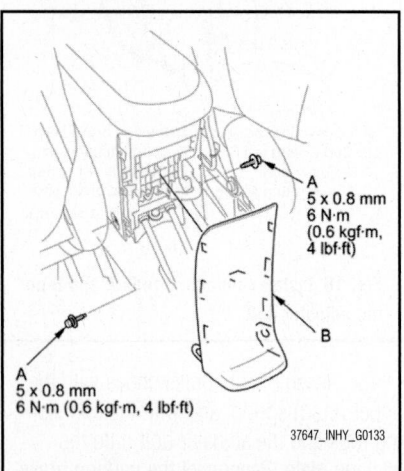

Fig. 17 Remove the bolts (A), and gently pull out the center console rear trim (B)

lever should be locked within 6 to 8 clicks.

2. If the number of lever clicks is not as specified, adjust the parking brake.

➡**After servicing the rear brake shoes, loosen the parking brake adjusting nut,**

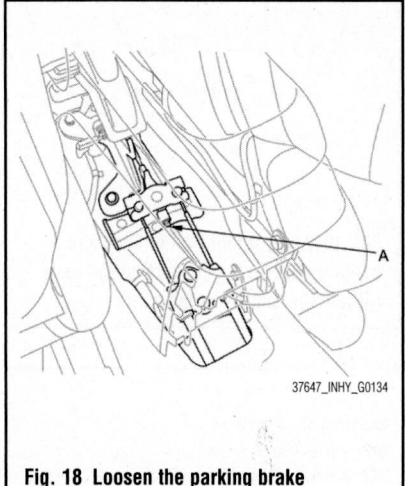

Fig. 18 Loosen the parking brake adjusting nut (A)

PARKING BRAKE

start the engine, and press the brake pedal several times to set the self-adjusting brake before adjusting the parking brake.

3. Release the parking brake lever fully.

4. Remove the bolts, and gently pull out the center console rear trim.

5. Loosen the parking brake adjusting nut.

6. Raise the vehicle on a lift.

7. Press the brake pedal several times to set the self-adjusting brake before adjusting the parking brake.

8. Pull the parking brake lever 1 click.

9. Tighten the parking brake adjusting nut until the parking brakes drag slightly when the rear wheels are turned.

10. Release the parking brake lever fully, and check that the parking brakes do not drag when the rear wheels are turned. Readjust if necessary.

11. Make sure the parking brake lever is within the specified number of clicks (6 to 8 clicks).

12. Install the center console rear cover.

CHASSIS ELECTRICAL

PRECAUTIONS

➡**Some systems store data in memory that is lost when the 12 volt battery is disconnected. Before disconnecting the 12 volt battery, refer to 12 Volt Battery Terminal Disconnection and Reconnection.**

Please read the following precautions carefully before servicing the airbag system. Observe the instructions described in this manual, or the airbags could accidentally deploy and cause damage or injuries.

• Except when doing electrical inspections, always turn the ignition switch to LOCK (0), disconnect the negative cable from the 12 volt battery, then wait at least 3 minutes before starting work.

NOTE: The SRS memory is not erased even if the ignition switch is turned to LOCK (0) or the battery cables are disconnected from the 12 volt battery.

• Use replacement parts which are manufactured to the same standards and quality as the original parts. Do not install used SRS parts. Use only new parts when making SRS repairs.

• Carefully inspect any SRS part before you install it.

AIR BAGS (SUPPLEMENTAL RESTRAINT SYSTEM)

Do not install any part that shows signs of being dropped or improperly handled, such as dents, cracks or deformation.

• Use only a digital multimeter to check the system. If it is not a Honda multimeter, make sure its output is 10 mA (0.01 A) or less when switched to the lowest value in the ohmmeter range. A tester with a higher output could cause accidental deployment and possible injury.

• Do not put objects on the front passenger's airbag.

Steering-Related Precautions

Cable Reel Alignment:

• Misalignment of the cable reel could cause an open in the wiring, making the SRS system, remote steering wheel controls, and the horn inoperative. Center the cable reel whenever you do the following.

• Installation of the steering wheel
• Installation of the cable reel
• Installation of the steering column
• Other steering-related adjustment or installation
• Do not disassemble the cable reel.
• Do not apply grease to the cable reel.

• If the cable reel shows any signs of damage, replace it with a new one. For example, if it does not rotate smoothly, replace the cable reel.

Airbag Handling & Storage

Do not disassemble an airbag. It has no serviceable parts. Once an airbag has been deployed, it cannot be repaired or reused.

For temporary storage of an airbag during service, observe the following precautions.

• Store the removed airbag with the pad surface up. Never put anything on the airbag.

• To prevent damage to the airbag, keep it away from any oil, grease, detergent, or water.

• Store the removed airbag on a secure, flat surface away from any high heat source exceeding 200°F (93°C).

• Never do electrical tests on the airbags, such as measuring resistance.

• Do not position yourself in front of the airbag during removal, inspection, or replacement.

• For proper disposal of a damaged airbag, refer to airbag disposal.

The side curtain airbag module assembly is a long, jointed part containing an inflator, a flexible bag, and brackets. When removing or installing the side curtain airbag inflator assembly, never do these things:

- Drop the curtain airbag.
- Cut, tear, or unwrap the tape strips.
- Handle the flexible bag.

SRS Unit, Front & Side Impact Sensors, Rear Safing Sensor, Driver's Seat Position Sensor & Front Passenger's Weight Sensors

Some systems store data in memory that is lost when the 12 volt battery is disconnected. Before disconnecting the 12 volt battery, refer to 12 Volt Battery Terminal Disconnection and Reconnection.

- Turn the ignition switch to LOCK (0), disconnect the negative cable from the 12 volt battery, then wait at least 3 minutes before beginning installation or replacement of the SRS unit or disconnecting the connectors from the SRS unit.
- Be careful not to bump or impact the SRS unit, front impact sensors or side impact sensors when the ignition switch is at ON (II), or for at least 3 minutes after the ignition switch is turned to LOCK (0).
- During installation or replacement, be careful not to bump (by impact wrench, hammer, etc.) the area around the SRS unit, front impact sensors or side impact sensors. The airbags could accidentally deploy and cause damage or injury.
- After a collision where a front airbag, side airbags, side curtain airbags, or a seat belt tensioner deployed, go to Component

Replacement/Inspection after Deployment. After a collision where the airbags or the side airbags did not deploy, inspect for any damage or any deformation on the SRS unit, front impact sensors and side impact sensors. Replace all damaged parts.

- Do not disassemble the SRS unit, front impact sensors and side impact sensors.
- Be sure the SRS unit, front impact sensors and side impact sensors are installed securely with the mounting bolts torqued to 7 ft. lbs. (9.8 Nm) whenever you remove or replace the SRS unit, all impact sensors, always install the components with new Torx® bolts.
- Do not spill water or oil on the SRS unit or the side impact sensors.

Wiring Precautions

Some of the SRS wiring can be identified by special yellow outer covering, and the SRS connectors can be identified by their yellow color. Observe the instructions.

- Never attempt to modify, splice, or repair SRS wiring.

If there is an open or damage in SRS wiring, replace the harness.

- Be sure to install the harness wires so they do not get pinched or interfere with other parts.
- Make sure all SRS ground locations are clean, and grounds are securely fastened for optimum metal-to-metal contact. Poor grounds can cause intermittent

problems that are difficult to diagnose.

- Do not use any silicone based cleaners or lubricants on any SRS connectors or terminals.

Precautions for Electrical Inspections

Special Tools Required: Back Probe Adapter, 17 mm 07TAZ-001020A

Make sure the 12 volt battery is fully charged when doing electrical tests. If the 12 volt battery is not fully charged, the results of the tests may not be accurate.

When using electrical test equipment, insert the probe of the tester into the wire side of the connector (except waterproof connector). Do not insert the probe of the tester into the terminal side of the connector, and do not tamper with the connector.

Use back probe adapter 07TAZ-001020A. Do not insert the probe forcibly.

Use specified service connectors in troubleshooting.

Using improper tools could cause an error in inspection due to poor metal-to-metal contact.

DISARMING THE SYSTEM

Disconnect the negative battery cable and isolate it. Wait at least 3 minutes before starting any work.

ARMING THE SYSTEM

Reconnect the negative battery cable.

DRIVE SHAFT

CONTINUALLY VARIABLE TRANSMISSION (CVT)

CHECKING FLUID LEVEL

❊❊ CAUTION

Keep all foreign particles out of the transmission.

1. Park the vehicle on the level ground.
2. Start the engine, and warm it up to normal operating temperature (the radiator fan comes on twice), and turn the engine off.

➡**Check the CVT fluid level within 60–90 seconds after turning the engine off. Higher fluid level may be indicated if the radiator fan comes on twice or more.**

3. Remove the dipstick (yellow loop) and wipe it with a clean cloth.

4. Insert the dipstick back into the tube.
5. Remove the dipstick, and check the CVT fluid level. It should be between the upper mark and the lower mark on the HOT level in tip of the dip stick.
6. If the CVT fluid level is below the lower mark, check for fluid leaks at the transmission, the CVTF filter, the CVTF cooler hoses, the CVTF cooler lines, and the CVTF cooler line joints. If a problem is found, fix it before filling the transmission with CVTF.
7. If the vehicle is driven when the CVT fluid level is below the lower mark, one or more of these symptoms may occur:

- Noise from transmission in D, S, L, and R.
- Engine runs, but vehicle does not move in any position.
- No shift to a higher ratio or lower ratio.

- Engine revs up abnormally high while driving.
- Excessive shock when accelerating and decelerating.
- Vehicle does not creep on a flat road in D, S, and L.
- Late shift after shifting from N to D or N to R.
- Unstable rpm while driving.
- Stall speed high.

8. If the level is above the upper mark, drain the CVTF to proper level.
9. If necessary, fill the transmission with CVTF through the dipstick tube opening to bring the fluid level between the upper mark and the lower mark of the dipstick. Do not fill the fluid above the upper mark. Always use Honda CVTF.

➡**Using a non-Honda CVTF can affect shift quality.**

10. Insert the dipstick.

DRAIN & REFILL

See Figure 19.

✳✳ CAUTION

Keep all foreign particles out of the transmission.

1. Start the engine, and warm it up to normal operating temperature (the radiator fan comes on twice).

2. Park the vehicle on level ground, and turn the engine off.

3. Raise the vehicle on a lift, and make sure it is securely supported.

4. Remove the lid for maintenance.

5. Remove the drain plug and drain the CVTF. Reinstall the drain plug with a new sealing washer.

   a. Remove metal particles from the magnetic surface of the drain plug.

   b. If a cooler cleaning is necessary, refer to the CVTF cooler cleaning.

6. Install the lid for maintenance.

7. Remove the dipstick (yellow loop) and wipe it with a clean cloth.

8. Refill the transmission with the recommended fluid through the dipstick tube opening to bring it to the upper mark on the dipstick. Always use Honda CVTF. Using a non-Honda CVTF can affect shift quality.

9. CVT fluid capacity should be:
   • 3.0 US qts. (2.8 L) at change
   • 5.5 US qts. (5.2 L) at overhaul

10. Check that the CVT fluid level is between the upper mark and the lower mark of the dipstick.

11. Insert the dipstick.

12. If the Maintenance Minder required to replace the CVTF, reset the Maintenance Minder, and this procedure is complete. If the Maintenance Minder did not require you to replace the CVTF, go to the next step.

13. Connect the HDS to the DLC located under the driver's side of the dashboard.

14. Turn the ignition switch to ON (II). Make sure the HDS communicates with the PCM. If it does not, go to the DLC circuit troubleshooting.

15. Select GAUGES in the BODY ELECTRICAL with the HDS.

16. Select ADJUSTMENT in the GAUGES with the HDS.

17. Select MAINTENANCE MINDER in the ADJUSTMENT with the HDS.

18. Select RESET in the MAINTENANCE MINDER with the HDS.

19. Select MAINTENANCE SUB ITEM 3 RESET, and reset the CVTF life with the HDS.

FILTER REPLACEMENT

See Figures 19 and 20.

1. Remove the splash shield.

2. Remove the drain plug and drain the CVT fluid (CVTF). Remove metal particles from the magnetic surface of the drain plug.

3. Reinstall the drain plug and a new sealing washer.

4. Disconnect the CVTF cooler hoses from the CVTF filter.

5. Remove the bolt securing the CVTF filter holder and remove the CVTF filter.

6. Slide the CVTF cooler hose on the CVTF filter until the hose end contacts the filter housing, and secure the hose with the clip at 0.24–0.31 in. (6–8 mm) from the filter housing. Install the cooler hose on the other side of the CVTF filter in the same manner.

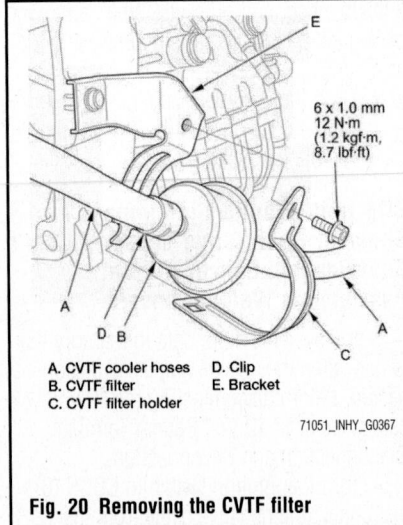

A. CVTF cooler hoses   D. Clip
B. CVTF filter   E. Bracket
C. CVTF filter holder

6 x 1.0 mm
12 N·m
(1.2 kgf·m,
8.7 lbf·ft)

71051_INHY_G0367

Fig. 20 Removing the CVTF filter

7. Secure the new CVTF filter on its bracket with the CVTF filter holder and the bolt.

8. Refill the transmission with CVTF to the proper level.

9. Install the splash shield.

➡ If a cooler cleaning is necessary, refer to the CVTF cooler cleaning.

FRONT HALFSHAFT

REMOVAL & INSTALLATION

See Figures 21 through 27.

Special Tools Required:
• Ball Joint Remover, 28 mm 07MAC-SL0A202
• Ball Joint Thread Protector, 14 mm 071AF-S3VA000

1. Raise the vehicle on a lift.

2. Remove the front wheels.

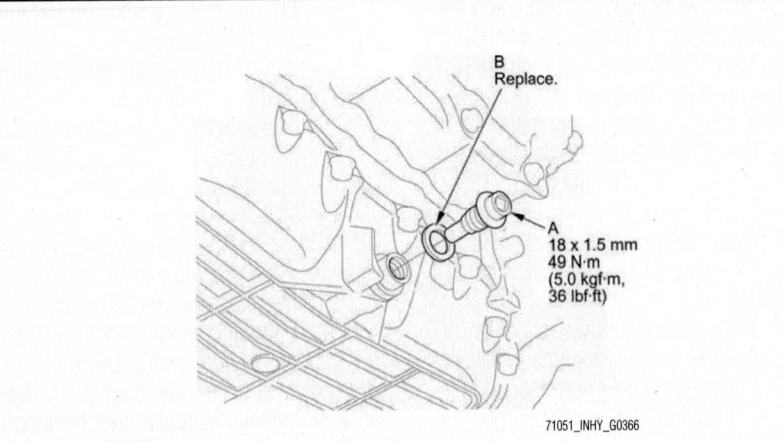

B
Replace.

A
18 x 1.5 mm
49 N·m
(5.0 kgf·m,
36 lbf·ft)

71051_INHY_G0366

Fig. 19 Remove the drain plug (A) and drain the CVT fluid (CVTF). Remove metal particles from the magnetic surface of the drain plug. Reinstall the drain plug and a new sealing washer (B).

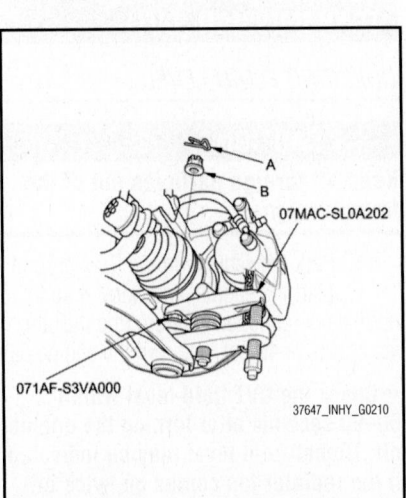

A
B
07MAC-SL0A202

071AF-S3VA000

37647_INHY_G0210

Fig. 21 Remove the lock pin (A) from the lower arm ball joint, then remove the castle nut (B)

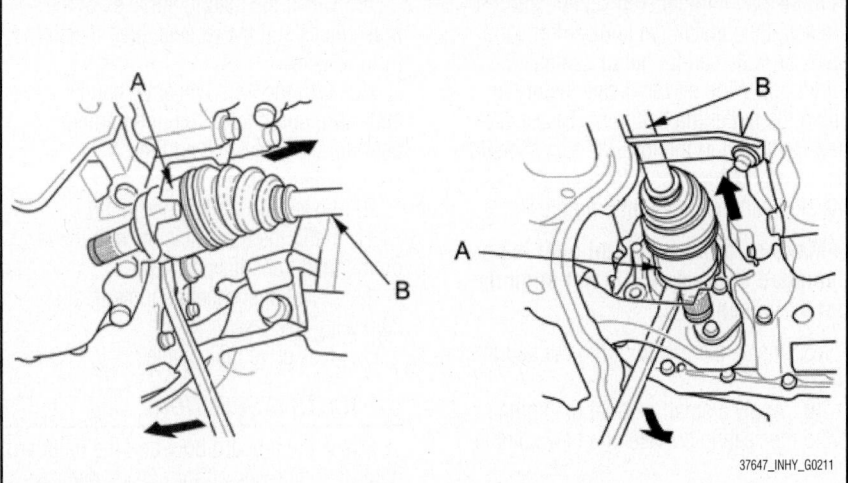

Fig. 22 Pry the inboard joint (A) from the differential using a prybar: do not pull on the driveshaft (B)

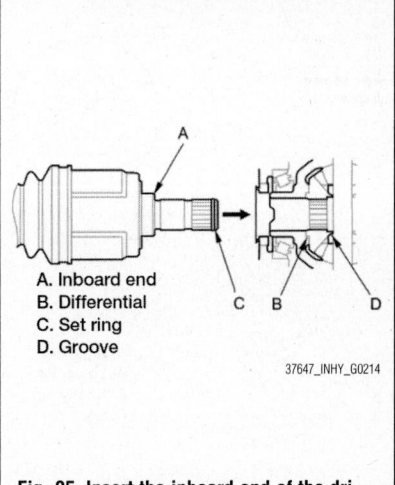

A. Inboard end
B. Differential
C. Set ring
D. Groove

Fig. 25 Insert the inboard end of the driveshaft into the differential until the set ring locks in the groove

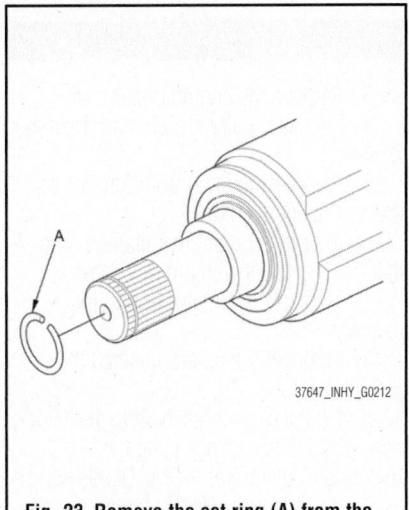

Fig. 23 Remove the set ring (A) from the inboard joint

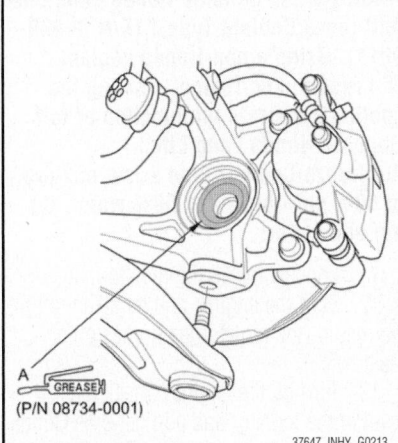

(P/N 08734-0001)

Fig. 24 Apply Moly 60 paste (P/N 08734-0001) to the contact area (A) of the outboard joint and the front wheel bearing

3. Pry up the stake on the spindle nut, then remove the nut.

4. Drain the transmission fluid, then reinstall the drain plug with a new sealing washer.

5. Remove the lock pin from the lower arm ball joint, then remove the castle nut. Separate the ball joint from the knuckle using the 14 mm ball joint thread protector and the 28 mm ball joint remover.

➡**Be careful not to damage the ball joint boot when installing the remover. Do not force or hammer on the lower arm, or pry between the lower arm and the knuckle. You could damage the ball joint.**

6. Pull the knuckle outward, and separate the outboard joint from the front hub using a plastic hammer.

7. Pry the inboard joint from the differential using a prybar. Remove the driveshaft as an assembly.

➡**Do not pull on the driveshaft, or the inboard joint may come apart. Pull the inboard joint straight out to avoid damaging the oil seal. Be careful not to damage the oil seal or the end of the inboard joint with the prybar.**

8. Remove the set ring from the inboard joint.

To install:

➡**Before starting installation, make sure the mating surfaces of the joint and the splined section are clean.**

9. Apply Moly 60 paste (P/N 08734-0001) to the contact area of the outboard joint and the front wheel bearing.

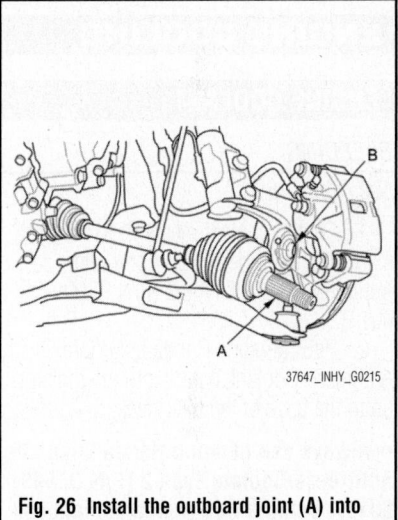

Fig. 26 Install the outboard joint (A) into the front hub (B) on the knuckle

➡**The paste helps to prevent noise and vibration.**

10. Install a new set ring into the set ring groove of the driveshaft inboard joint.

11. Clean the areas where the driveshaft contacts the differential thoroughly with solvent or brake cleaner, and dry with compressed air.

➡**Do not wash the rubber parts with solvent.**

12. Insert the inboard end of the driveshaft into the differential until the set ring locks in the groove.

➡**Insert the driveshaft horizontally to prevent damaging the oil seal.**

13. Install the outboard joint into the front hub on the knuckle.

14. Wipe off any grease contamination

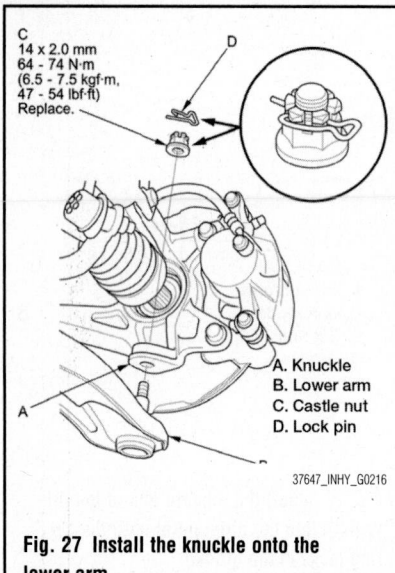

C
14 x 2.0 mm
64 - 74 N·m
(6.5 - 7.5 kgf·m,
47 - 54 lbf·ft)
Replace.

A. Knuckle
B. Lower arm
C. Castle nut
D. Lock pin

37647_INHY_G0216

Fig. 27 Install the knuckle onto the lower arm

from the ball joint tapered section and threads, then install the knuckle onto the lower arm. Be careful not to damage the ball joint boot. Wipe off the grease before tightening the nut at the ball joint. Torque the new castle nut to the lower torque specification, then tighten it only far enough to align the slot with the ball joint pin hole.

➡**Make sure the ball joint boot is not damaged or cracked. Do not align the nut by loosening it.**

15. Install the lock pin into the ball joint pin hole.

16. Apply a small amount of engine oil to the seating surface of a new spindle nut.

17. Install the spindle nut, then tighten it. After tightening, use a drift to stake the spindle nut shoulder against the driveshaft.

18. Clean the mating surfaces of the brake discs and the wheels, then install the front wheels.

19. Turn the front wheel by hand, and make sure there is no interference between the driveshaft and the surrounding parts.

20. Lower the vehicle on the lift.

21. Refill the transmission with the recommended transmission fluid.

22. Check the wheel alignment, and adjust it if necessary.

23. Test-drive the vehicle.

CV-BOOTS INSPECTION

Check the inboard boot and the outboard boot on the driveshaft for cracks, damage, leaking grease, and loose boot bands. If any damage is found, replace the boot and the boot bands.

ENGINE COOLING

ENGINE COOLANT

BLEEDING

See Figure 28.

1. Fill the coolant reservoir to the MAX mark with genuine Honda Long Life Antifreeze/Coolant Type 2 (P/N OL999-9011).

2. Pour genuine Honda Long Life Antifreeze/Coolant Type 2 into the radiator up to the base of the filler neck.

➡**Always use genuine Honda Long Life Antifreeze/Coolant Type 2 (P/N OL999-9011). Using a non-Honda coolant can result in corrosion, causing the cooling system to malfunction or fail. Genuine Honda Long Life Antifreeze/Coolant**

Type 2 is a mixture of 50% antifreeze and 50% water. Do not add water.

3. Loosely install the radiator cap.

4. Start the engine, and let it run until it warms up (the radiator fan comes on at least twice).

5. Turn off the engine. Check the level in the radiator and add genuine Honda Long Life Antifreeze/ Coolant Type 2, if needed.

6. Put the radiator cap on securely, then start the engine again, and check for leaks.

DRAIN & REFILL PROCEDURE

See Figures 29 through 30.

1. Make sure the engine and radiator are cool to the touch.

2. Remove the radiator cap.

3. Remove the engine undercover.

4. Loosen the drain plug, and drain the coolant.

5. Remove the drain bolt from the rear side of the engine block.

6. After the coolant has drained, reinstall the drain bolt with a new washer.

7. Tighten the radiator drain plug securely.

8. Remove, drain, and reinstall the coolant reservoir.

9. Fill the coolant reservoir to the MAX mark with genuine Honda Long Life Antifreeze/Coolant Type 2 (P/N OL999-9011).

10. Pour genuine Honda Long Life Antifreeze/Coolant Type 2 into the radiator up to the base of the filler neck.

➡**Always use genuine Honda Long Life Antifreeze/Coolant Type 2 (P/N OL999-9011). Using a non-Honda coolant can result in corrosion, causing the cooling system to malfunction or fail. Genuine Honda Long Life Antifreeze/Coolant Type 2 is a mixture of 50% antifreeze and 50% water. Do not add water.**

11. Loosely install the radiator cap.

12. Start the engine, and let it run until it warms up (the radiator fan comes on at least twice).

13. Turn off the engine. Check the level in the radiator and add genuine Honda Long Life Antifreeze/ Coolant Type 2, if needed.

14. Put the radiator cap on securely, then start the engine again, and check for leaks.

15. Clean up any spilled engine coolant.

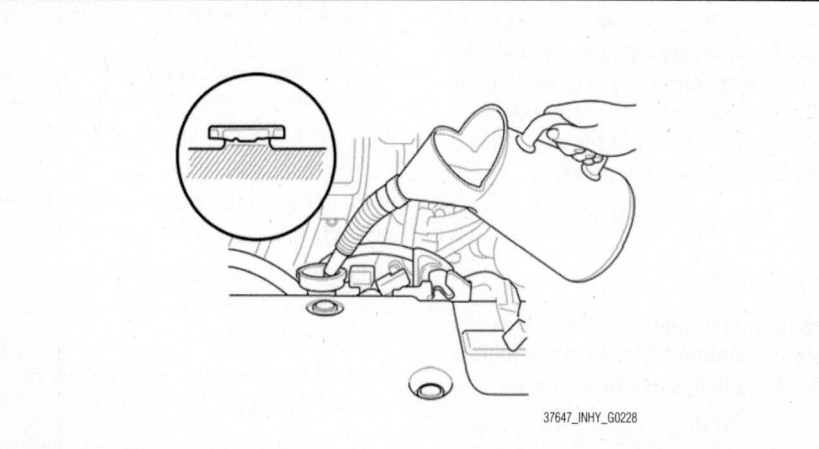

37647_INHY_G0228

Fig. 28 Pour genuine Honda Long Life Antifreeze/Coolant Type 2 into the radiator up to the base of the filler neck

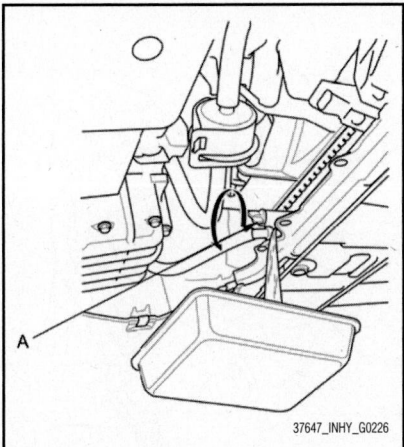

Fig. 29 Loosen the drain plug (A), and drain the coolant

16. Connect the Honda Diagnostic System (HDS) to the Data Link Connector (DLC).

17. Make sure the HDS communicates with the vehicle and Powertrain Control Module (PCM). If it does not communicate, troubleshoot the DLC circuit.

18. Turn the ignition switch to ON (II).

19. Select BODY ELECTRICAL with the HDS.

20. Select ADJUSTMENT in the GAUGE MENU with the HDS.

21. Select RESET in the MAINTENANCE MINDER with the HDS.

22. Select MAINTENANCE SUB ITEM 5 RESET with the HDS.

23. Install the engine undercover.

LEVEL CHECK

1. Check the coolant level in the coolant reservoir. Make sure it is between the MAX mark and the MIN mark.

2. If the coolant level in the coolant

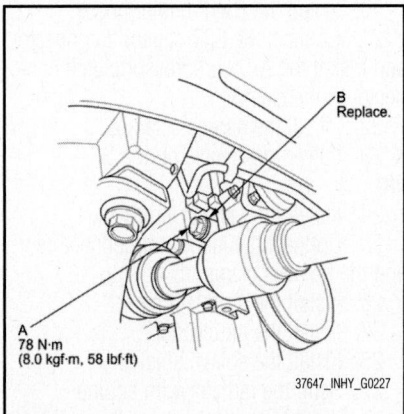

Fig. 30 Remove the drain bolt (A) and washer (B) from the rear side of the engine block

reservoir is at or below the MIN mark, add coolant to bring it between the MIN and MAX marks, then inspect the cooling system for leaks.

3. Check the coolant level in the radiator, and add Honda Long Life Antifreeze/ Coolant Type 2 into the radiator up to the base of the filler neck, if needed.

　a. Always use Honda Long Life Antifreeze/Coolant Type 2. Using a non-Honda coolant can result in corrosion, causing the cooling system to malfunction or fail.

　b. Honda Long Life Antifreeze/Coolant Type 2 is a mixture of 50% antifreeze and 50% water. Do not add water.

ENGINE FAN

REMOVAL & INSTALLATION

See Figures 31 through 34.

1. Remove the splash shield.

2. Raise the vehicle on the lift to full height.

3. Remove the A/C compressor clutch connector from the clamp, then remove the harness clamps.

4. Lower the vehicle on the lift.

5. Remove the air cleaner.

6. Remove the air duct bracket.

7. Disconnect the fan motor connectors and the harness clamp.

8. Remove the coolant reservoir.

9. Remove the A/C condenser fan shroud assembly and the radiator fan shroud assembly from the radiator, then remove the A/C condenser fan shroud assembly from the vehicle.

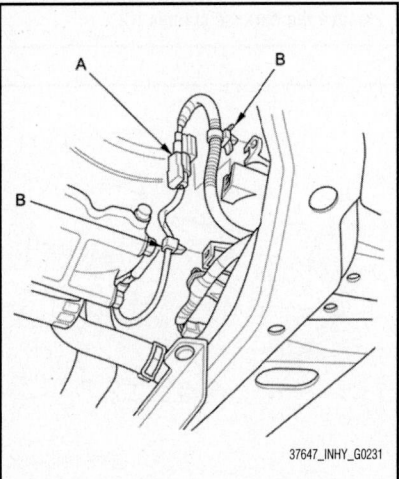

Fig. 31 Remove the A/C compressor clutch connector (A) from the clamp, then remove the harness clamps (B)

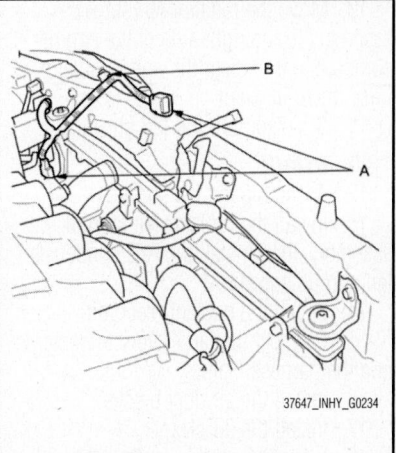

Fig. 32 Disconnect the fan motor connectors (A) and the harness clamp (B)

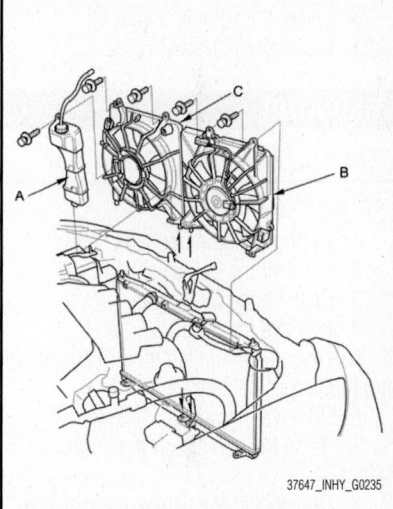

Fig. 33 Remove the coolant reservoir (A), the A/C condenser fan shroud assembly (B) and the radiator fan shroud assembly (C)

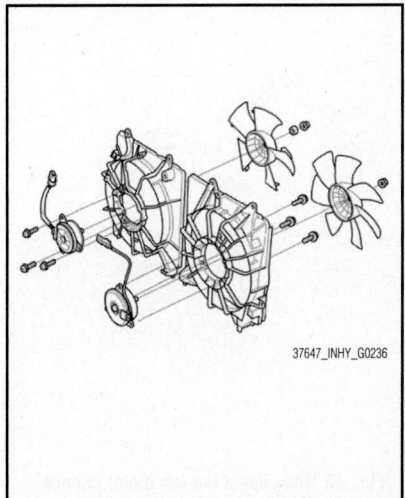

Fig. 34 Disassemble the fan shrouds

10. Move the radiator fan shroud assembly to the right side of the engine compartment to clear the upper radiator hose, then lift it out.

11. Disassemble the fan shrouds.

To install:

12. Assemble the fan shrouds.

13. Install the radiator fan shroud assembly, then install the A/C condenser fan shroud assembly.

14. Install the coolant reservoir.

15. Connect the fan motor connectors and the harness clamp.

16. Install the air duct bracket.

17. Install the air cleaner.

18. Raise the vehicle on the lift to full height.

19. Install the A/C compressor clutch connector to the clamp, then install the harness clamps.

20. Lower the vehicle on the lift.

21. Install the splash shields.

RADIATOR

REMOVAL & INSTALLATION

See Figures 35 through 40.

1. Drain the engine coolant.

2. Remove the splash shield.

3. Remove the air cleaner.

4. Remove the air duct bracket.

5. Disconnect the fan motor connectors and the harness clamp.

6. Remove the radiator brackets.

7. Raise the vehicle on the lift to full height.

8. Disconnect the engine coolant temperature (ECT) sensor 2 connector, and remove the A/C compressor clutch connector from the clamp.

9. Remove the harness clamps.

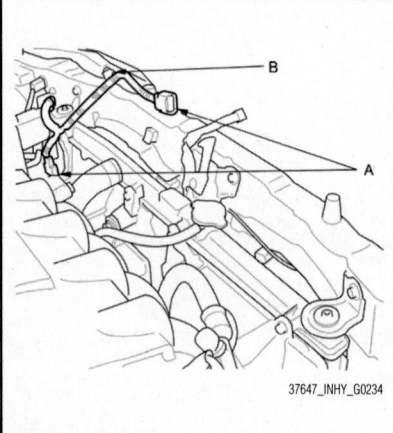

Fig. 35 Disconnect the fan motor connectors (A) and the harness clamp (B)

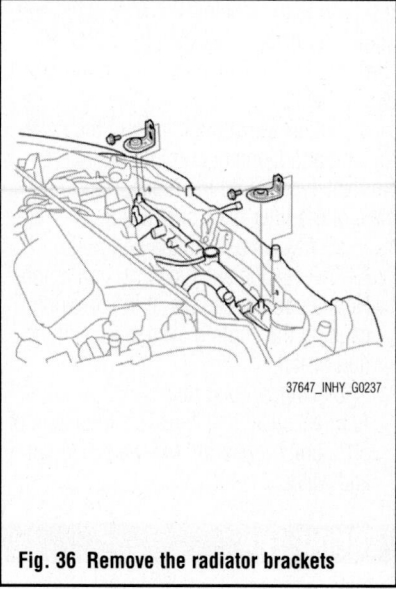

Fig. 36 Remove the radiator brackets

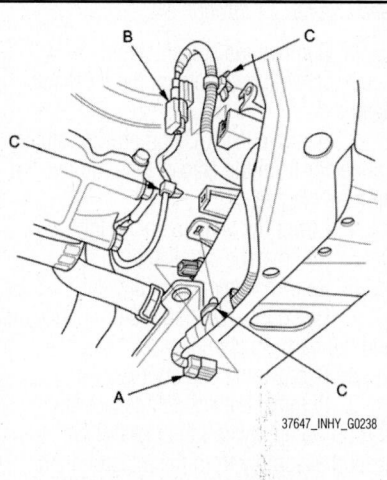

Fig. 37 Disconnect the ECT sensor 2 connector (A), and remove the A/C compressor clutch connector (B) from the clamp; remove the harness clamps (C)

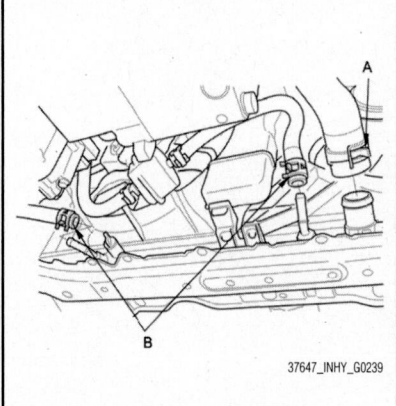

Fig. 38 Disconnect the lower radiator hose (A); remove the CVT fluid (CVTF) cooler hoses (B)

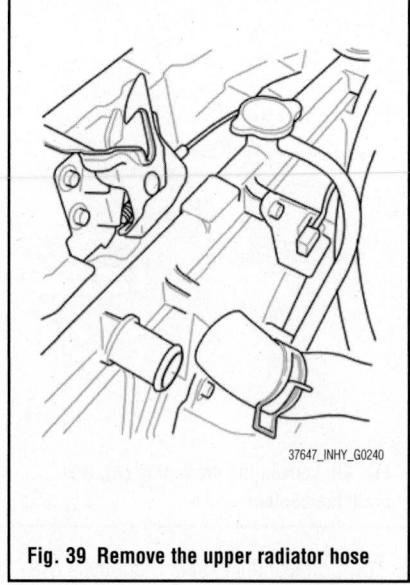

Fig. 39 Remove the upper radiator hose

10. Disconnect the lower radiator hose.

11. Remove the CVT fluid (CVTF) cooler hoses, then plug the hose and line.

12. Lower the vehicle on the lift.

13. Remove the upper radiator hose.

14. Pull up the radiator, then remove the coolant reservoir, the radiator fan shroud assembly, the A/C condenser fan shroud assembly, the radiator cap, the ECT sensor 2, and the drain plug.

To install:

15. Reassemble the radiator with new O-rings.

16. Install the radiator. Make sure the lower cushions are set securely.

17. Install the upper radiator hose.

18. Raise the vehicle on the lift to full height.

19. Install the lower radiator hose.

20. Install the CVTF cooler hoses.

21. Connect the ECT sensor 2 connector, and install the A/C compressor clutch connector to the clamp.

22. Install the harness clamps.

23. Lower the vehicle on the lift.

24. Install the radiator brackets.

25. Connect the fan motor connectors and the harness clamp.

26. Install the air duct bracket.

27. Install the air cleaner.

28. Install the splash shields.

29. Refill the radiator with engine coolant, and bleed the air from the cooling system.

30. Clean up any spilled engine coolant.

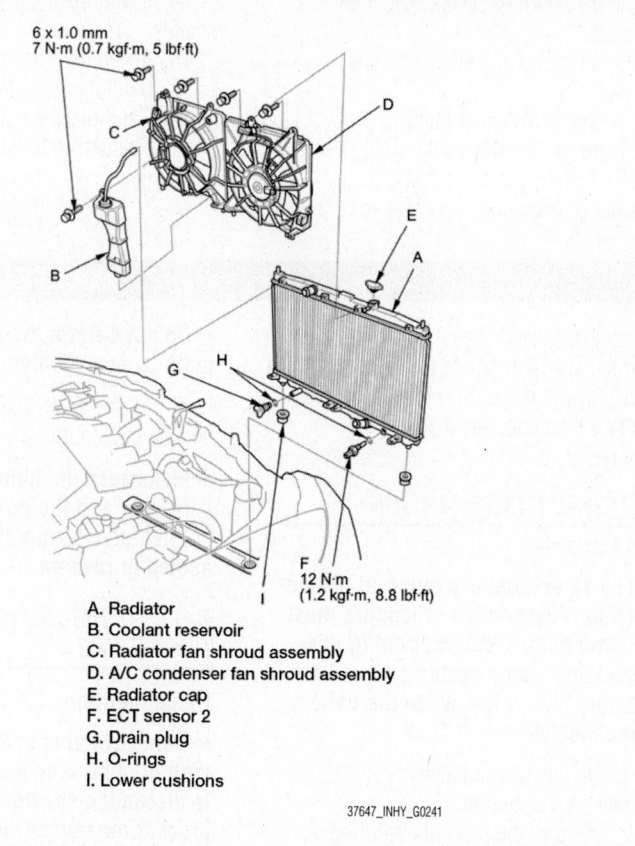

A. Radiator
B. Coolant reservoir
C. Radiator fan shroud assembly
D. A/C condenser fan shroud assembly
E. Radiator cap
F. ECT sensor 2
G. Drain plug
H. O-rings
I. Lower cushions

6 x 1.0 mm
7 N·m (0.7 kgf·m, 5 lbf·ft)

F
12 N·m
(1.2 kgf·m, 8.8 lbf·ft)

37647_INHY_G0241

Fig. 40 Pull up the radiator, then remove the coolant reservoir, the radiator fan shroud assembly, the A/C condenser fan shroud assembly, the radiator cap, the ECT sensor 2, and the drain plug

THERMOSTAT

REMOVAL & INSTALLATION

See Figure 41.

1. Drain the engine coolant.
2. Remove the thermostat cover, then remove the thermostat.

3. Install the new rubber seal on the thermostat, then install the thermostat with the pin up, and install the thermostat cover.
4. Refill the radiator with engine coolant, then bleed the air from the cooling system.

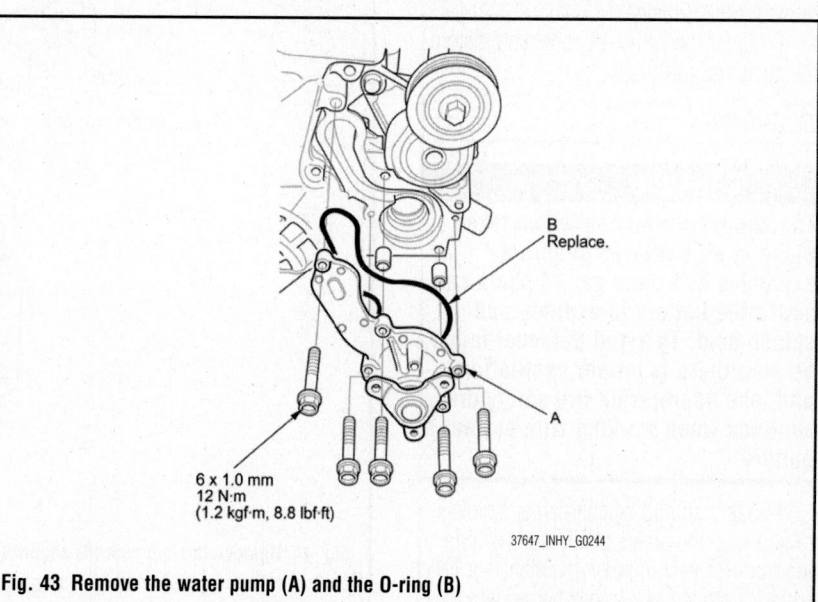

A. Thermostat cover
B. Thermostat
C. Rubber seal
D. Pin

6 x 1.0 mm
12 N·m(1.2 kgf·m, 8.8 lbf·ft)

37647_INHY_G0242

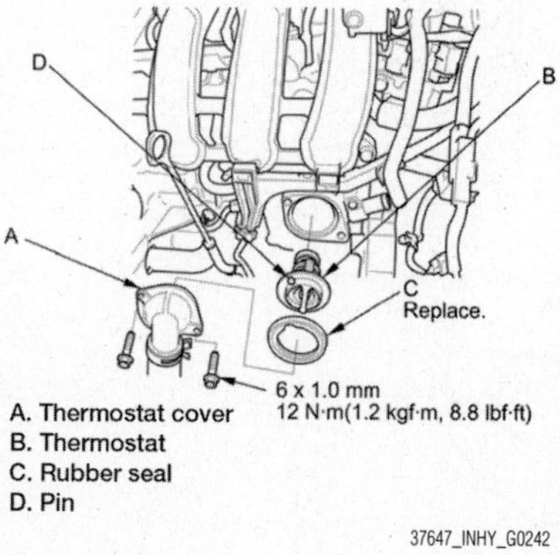

Fig. 41 Remove the thermostat cover, then remove the thermostat

5. Clean up any spilled engine coolant.

WATER PUMP

REMOVAL & INSTALLATION

See Figures 42 and 43.

1. Remove the front wheel.
2. Remove the splash shields.
3. Drain the engine coolant.
4. Loosen the water pump pulley mounting bolts.
5. Remove the drive belt.
6. Remove the water pump pulley.

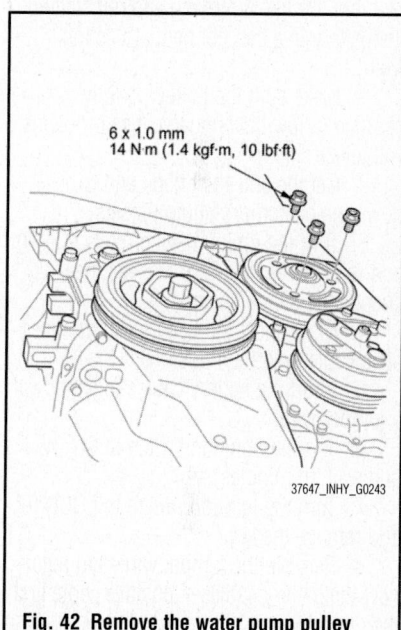

6 x 1.0 mm
14 N·m (1.4 kgf·m, 10 lbf·ft)

37647_INHY_G0243

Fig. 42 Remove the water pump pulley

B
Replace.

A

6 x 1.0 mm
12 N·m
(1.2 kgf·m, 8.8 lbf·ft)

37647_INHY_G0244

Fig. 43 Remove the water pump (A) and the O-ring (B)

7. Remove the water pump by removing the five bolts.

8. Inspect and clean the O-ring groove and the mating surface with the engine block.

To install:

9. Install the water pump with a new O-ring in the reverse order of removal.

10. Clean up any spilled engine coolant.

11. Install the water pump pulley.

12. Install the drive belt.

13. Tighten the water pump pulley mounting bolts.

14. Install the splash shields.

15. Install the front wheel.

16. Refill the radiator with engine coolant, and bleed the air from the cooling system with the heater valve open.

ENGINE ELECTRICAL

BATTERY

BATTERY CABLE RESET

A battery cable reset forces all capacitors to discharge faster, and clears and resets most control unit volatile memory. It's like removing and reinstalling the battery in a cell phone that's acting weird.

1. Make sure the battery is fully charged (a low battery can cause problems with electronics).

2. Get the anti-theft code and write down the customer's audio presets.

3. Turn the ignition switch to LOCK (0) and remove the key. This lessens the chance of voltage spikes.

4. Disconnect the battery cables—negative cable first, then positive.

5. Short the battery cables together with a jumper wire.

6. Turn the ignition switch to ON (II) and wait 10 minutes.

7. Turn the ignition switch to LOCK (0) and remove the key.

8. Remove the jumper wire, and reconnect the battery cables—positive cable first, then negative.

9. Do any required reset/learn procedures for the vehicle.

10. Enter the anti-theft code, and restore the customer's settings.

CHARGING

> ✳✳ **CAUTION**
>
> **The chemical reaction which takes place in all batteries generates explosive hydrogen gas. A spark can cause the battery to explode and splash acid. To avoid personal injury, be sure there is proper ventilation and take appropriate fire safety precautions when working with or near a battery.**

A battery should be charged at a slow rate to keep the plates inside from getting too hot. However, if some maintenance-free batteries are allowed to discharge until they are almost "dead," they may have to be charged at a high rate to bring them back to "life." Always follow the charger manufacturer's instructions on charging the battery.

REMOVAL & INSTALLATION

See Figure 44.

➡**The 12 volt battery terminal disconnection/reconnection procedure must be done before and after doing this procedure. Some systems store data in memory that is lost when the battery disconnected.**

1. Do the 12 volt battery terminal disconnection procedure.

2. Remove the two nuts securing the battery hold down plate, then remove the battery hold down plate and the battery.

To install:

3. Install the battery, then install the battery hold down plate.

4. Tighten the two nuts equally until the battery is stable.

BATTERY SYSTEM

➡**Do not deform the battery hold down plate by over-tightening the nuts.**

5. Do the 12 volt battery terminal reconnection procedure.

➡**Make sure the battery is installed correctly, and the positive terminal and the negative terminal are not connected in reverse.**

BATTERY DISCONNECT/RECONNECT PROCEDURE

Disconnection

➡**Some systems store data in memory that is lost when the battery is disconnected. Do the following procedures before disconnecting the battery.**

1. Make sure you have the anti-theft code(s) for the audio system and/or navigation system (if equipped).

2. If you are replacing the audio unit or the audio-navigation unit (if equipped), write down the audio presets (AM and FM).

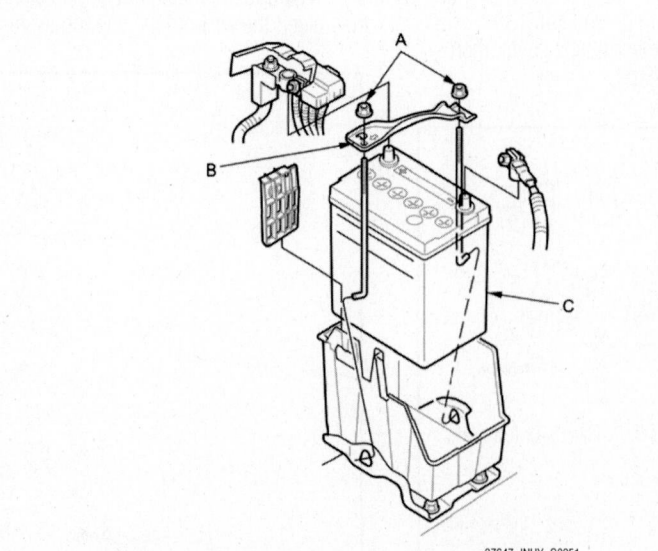

37647_INHY_G0251

Fig. 44 Remove the two nuts (A) securing the battery hold down plate (B), then remove the battery hold down plate and the battery (C)

3. Make sure the ignition switch is in LOCK (0).

4. Disconnect and isolate the negative cable from the battery.

➡**Always disconnect the negative battery cable from the battery first.**

5. Disconnect the positive cable from the battery.

Reconnection

➡**Some systems store data in memory that is lost when the battery is disconnected. Do the following procedures to restore the system back to normal operation.**

1. Clean the battery terminals.
2. Test the battery.
3. Reconnect the positive cable to the battery first, then reconnect the negative cable to the battery.

➡**Always connect the positive cable to the battery first.**

4. Install the terminal cover.
5. Apply multipurpose grease to the terminals to prevent corrosion.
6. Enter the anti-theft code(s) for the audio system and/or navigation system (if equipped).
7. Set the clock (for vehicles without navigation).

ENGINE ELECTRICAL

CHARGING SYSTEM

ALTERNATOR

REMOVAL & INSTALLATION

Refer to Integrated Motor Assisted (IMA) components in the Hybrid section regarding the vehicle Charging System.

ENGINE ELECTRICAL   HYBRID INTEGRATED MOTOR ASSISTED (IMA) SYTEM

IMA SYSTEM DESCRIPTION

The Integrated Motor Assisted (IMA) system is a highly-efficient parallel hybrid drive system with a main power unit (gasoline engine) and an assist unit (electric motor).

The engine is an in-line, 4-cylinder, 8-valve power plant with a displacement of 82 cu. in (1.339 liters). To reduce fuel use, the engine is equipped with i-DSI and a valve pause system that reduces engine pumping loss and increases the generation of electric energy during deceleration.

The IMA motor, directly connected to the engine crankshaft, functions as a generator during deceleration, an engine starter, and a motor to assist the engine that drives the wheels.

The IMA system contains a 100 V DC battery, a control system, and related parts. For safety, the Intelligent Power Unit (IPU) is located under the cargo compartment.

The IMA system improves fuel economy by capturing and storing energy during deceleration.

OPERATING CONDITIONS

1. Engine start: The IMA system drives the IMA motor and starts the engine during normal starts and when re-starting from an idle stop. The IMA motor is linked directly to the engine crankshaft.
2. Start running: The IMA motor assists the engine.
3. Slow acceleration: Runs only with the engine.
4. Low speed cruise: At a constant speed of about 25 mph (40 km/h) with light load, all engine cylinders are deactivated, and the vehicle runs only with the IMA motor. When the State-Of-Charge (SOC) drops, the engine starts and the IMA motor begins to charge the battery module.
5. Acceleration: The IMA motor assists the engine.
6. High acceleration: The IMA motor assists the engine.
7. High speed cruise: Runs only with the engine. When the state-of-charge drops, the IMA motor begins to charge the battery module.
8. Deceleration: All engine cylinders are deactivated, and the IMA motor captures the deceleration energy and charges the battery module.
9. Stop: When conditions are satisfied, the PCM stops the engine automatically (auto stop).

IMA SERVICE PRECAUTIONS

The IMA (integrated motor assisted) system uses high voltage (100 V) circuits. Be sure to shut off the electrical circuits and isolate the IMA system and related parts before servicing the IMA system.

The high voltage cables and their covers are identified by orange coloring. The caution labels are attached to high voltage and other related parts. When the system is energized, be careful not to touch these cables and parts without adequate protective gear.

The front floor under-cover protecting the high voltage cables is marked .

If the 12 V battery is discharged, its cables have been disconnected, or the MCM (Motor Control Module) has been reset, the IMA battery level indicator does not display the State-Of-Charge (SOC) when the engine is first started. To display the level in the indicator, start the engine, and hold it between 3,500 and 4,000 rpm without load (in P or N) until the level in the indicator is half full.

Observe the following instructions when inspecting or servicing the IMA system:

When the system is energized, servicing, disassembling, or replacing items marked in each procedure requires insulated tools.

When the IMA system indicator is on, do the IMA system troubleshooting first.

Wear insulated gloves whenever you inspect or service the IMA system. Be sure to check the gloves for pin holes, tears, or other damage.

To make sure the system is not energized, turn the battery module switch OFF, and secure the switch in the OFF position with the locking cover before servicing the IMA system (see Turning off and on power to the high voltage circuit).

Wait at least 5 minutes after turning off the battery module switch, then disconnect the negative cable from the 12 V battery (it takes about 5 minutes for the PDU capacitor to discharge).

Before disconnecting the high voltage cable terminals, use a voltmeter to make sure that the voltage between the terminals is below 30 V.

When the system is energized and you are servicing parts without an insulated sheath, be sure to use insulated tools to prevent short circuiting.

The rotor assembly contains very strong magnets and should be handled with special care. People with pacemakers or other magnetically sensitive medical devices should not handle the rotor assembly.

Use the special tool (rotor puller) to remove or install the rotor assembly.

❊❊ WARNING

If the rotor is installed by hand, it may suddenly be pulled toward the stator with great force, causing serious hand or finger injury. Always use the special tool (rotor puller) to remove or install the rotor assembly.

Keep the rotor assembly away from magnetically sensitive devices.

After disconnecting the high voltage terminals, busbars, etc., insulate the parts with insulated tape.

As a safety warning, attach a sign saying, WORKING ON HIGH VOLTAGE PARTS. DO NOT TOUCH! to the steering wheel.

DISCONNECTING THE MOTOR POWER CABLE CONNECTOR FROM THE MOTOR STATOR

See Figure 45.

❊❊ CAUTION

IMA components are located in this area. The IMA is a high-voltage sys-tem. You must be familiar with the IMA system before working on or around it. Make sure you have read the IMA service precautions before doing repairs or service.

1. Turn the ignition switch to LOCK (0). Turn the battery module switch OFF.

2. Slide the protector in the direction of the arrow. Push the tab, then raise the lever. Remove the IMA motor power cable from the motor stator.

➡Note the following:

- If the outside of the IMA motor power cable connector is dirty, clean it before you disconnect it.
- Cover the disconnected connector with a plastic bag, and wrap the IMA motor power cable terminals with insulating tape.
- If the IMA motor power cable is wet, dry them with a clean shop towel. Do not use compressed air.

TURNING OFF AND ON POWER TO THE HIGH VOLTAGE CIRCUIT

See Figures 46 through 49.

❊❊ CAUTION

IMA components are located in this area. The IMA is a high-voltage sys-tem. You must be familiar with the IMA system before working on or around it. Make sure you have read

the IMA service precautions before doing repairs or service.

The following procedure should be done before you work on or near any energized high voltage components. Follow the procedure exactly. Otherwise, you may be injured or may damage equipment.

1. Turn the ignition switch to LOCK (0), then remove the key from the ignition switch.

2. Remove the cargo floor lid, the cargo floor box, and the spare tire.

3. Loosen the bolt, and remove the bolt.

4. Remove the battery module switch lid from the IPU cover.

5. Turn the battery module switch OFF, then check that the bolt is showing.

6. Wait at least 5 minutes to allow the PDU capacitors to discharge.

7. Remove the IPU cover.

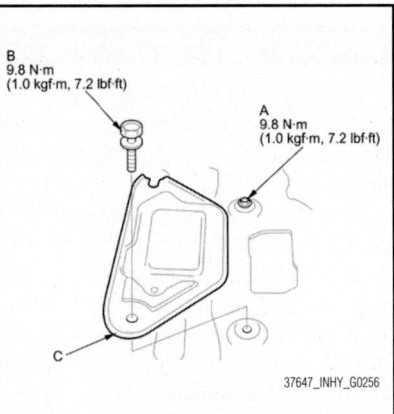

Fig. 46 Loosen the bolt (A), and remove the bolt (B) and the battery module switch lid (C)

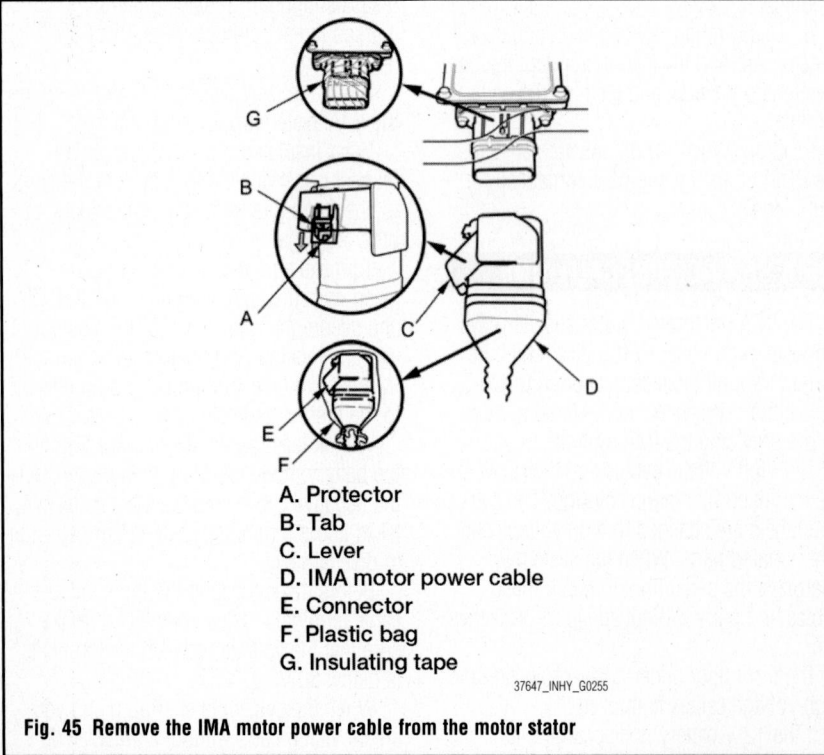

A. Protector
B. Tab
C. Lever
D. IMA motor power cable
E. Connector
F. Plastic bag
G. Insulating tape

Fig. 45 Remove the IMA motor power cable from the motor stator

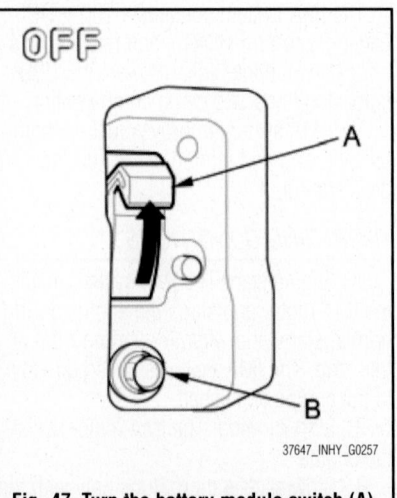

Fig. 47 Turn the battery module switch (A) OFF, then check that the bolt (B) is showing

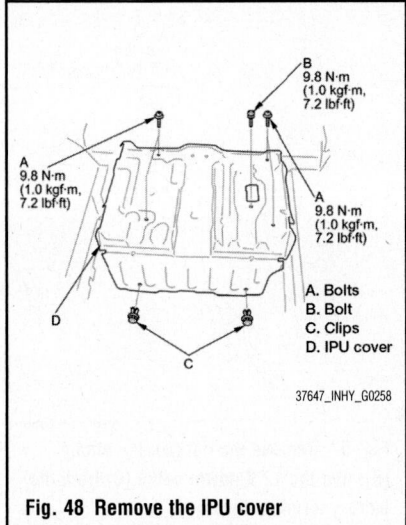

Fig. 48 Remove the IPU cover

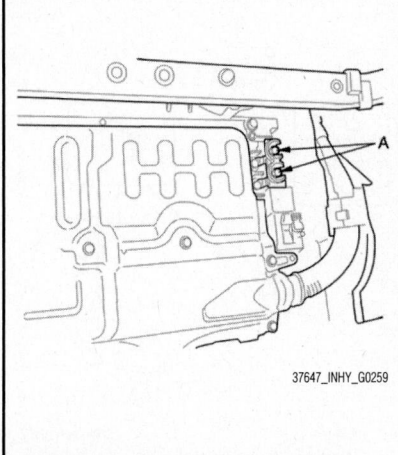

Fig. 49 Measure the voltage at the IMA battery module terminals (A)

8. Measure the voltage at the IMA battery module terminals. There should be less than 30 V. If the voltage is more than 30 V, there is a problem in the system; check for IMA DTCs before continuing.

9. After service or repairs are completed:

- Make sure all high voltage circuits are connected properly.
- Install the IPU cover.

10. Push the button, and turn the battery module switch ON.

11. Reinstall all remaining removed parts.

DC-DC CONVERTER

REMOVAL & INSTALLATION
See Figures 50 and 51.

❈❈ CAUTION

IMA components are located in this area. The IMA is a high-voltage

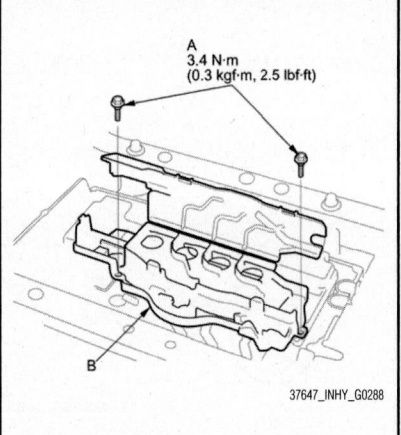

Fig. 50 Remove the bolts (A) and the terminal cover (B)

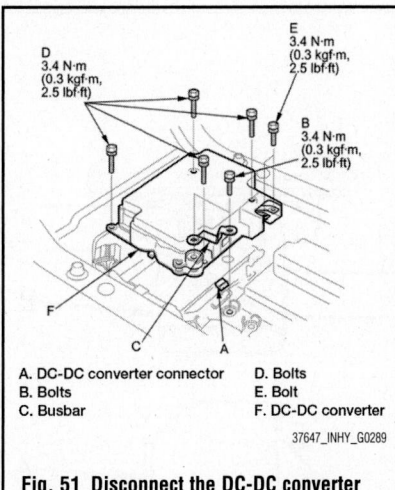

A. DC-DC converter connector
B. Bolts
C. Busbar
D. Bolts
E. Bolt
F. DC-DC converter

Fig. 51 Disconnect the DC-DC converter connector

system. You must be familiar with the IMA system before working on or around it. Make sure you have read the IMA service precautions before doing repairs or service.

1. Make sure the ignition switch is in LOCK (0).

2. Do the 12 volt battery terminal disconnection procedure.

3. Remove the MCM.

4. Remove the bolts and the terminal cover.

5. Disconnect the DC-DC converter connector.

6. Remove the bolts and the busbar.

7. Remove the bolts and the DC-DC converter.

8. Install the parts in the reverse order of removal.

9. Do the 12 volt battery terminal reconnection procedure.

IMA MOTOR DRAIN COVER

REMOVAL & INSTALLATION
See Figure 52.

❈❈ CAUTION

IMA components are located in this area. The IMA is a high-voltage system. You must be familiar with the IMA system before working on or around it. Make sure you have read the IMA service precautions before doing repairs or service.

1. Remove the splash shields.

2. Remove the drain cover.

3. Install the parts in the reverse order of removal.

IMA MOTOR HOUSING

REMOVAL & INSTALLATION
See Figure 53.

❈❈ CAUTION

IMA components are located in this area. The IMA is a high-voltage system. You must be familiar with the IMA system before working on or around it. Make sure you have read the IMA service precautions before doing repairs or service.

1. Remove the IMA motor rotor.

2. Remove the connector and the bracket.

3. Remove the bolt and the IMA motor housing.

➡Set the dowel pins in the IMA motor housing before installing the motor stator on the engine.

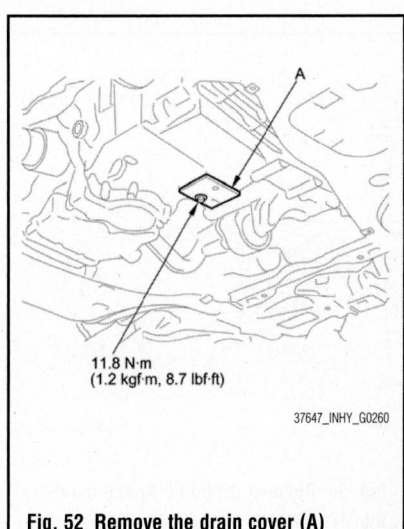

Fig. 52 Remove the drain cover (A)

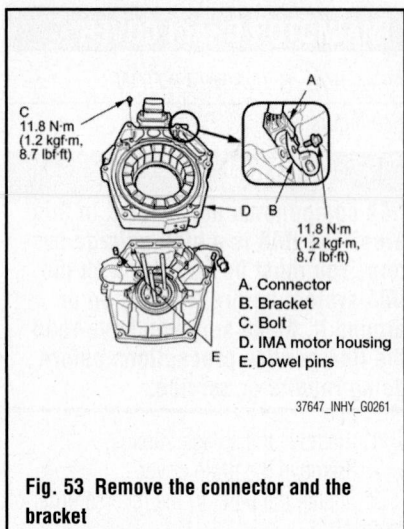

Fig. 53 Remove the connector and the bracket

4. Install the parts in the reverse order of removal.

5. Install the IMA motor rotor.

6. Do the motor rotor position calibration procedure.

IMA MOTOR POWER CABLE

REMOVAL & INSTALLATION

See Figures 54 through 62.

❋❋ CAUTION

IMA components are located in this area. The IMA is a high-voltage system. You must be familiar with the IMA system before working on or around it. Make sure you have read the IMA service precautions before doing repairs or service.

1. Make sure the ignition switch is in LOCK (0).

2. Do the 12 volt battery terminal disconnection procedure.

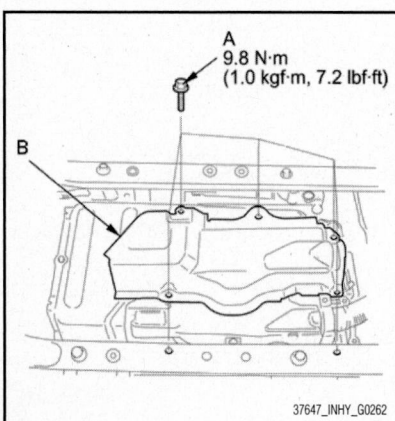

Fig. 54 Remove the bolts (A) and the PCU lid (B)

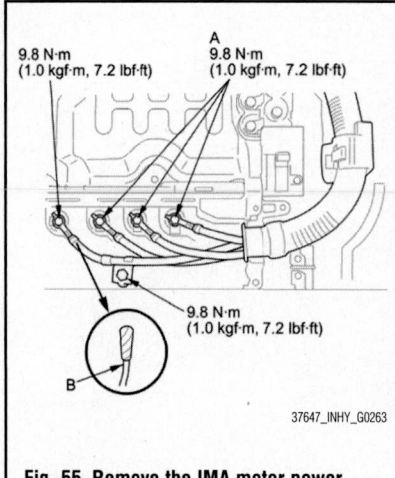

Fig. 55 Remove the IMA motor power cables (A)

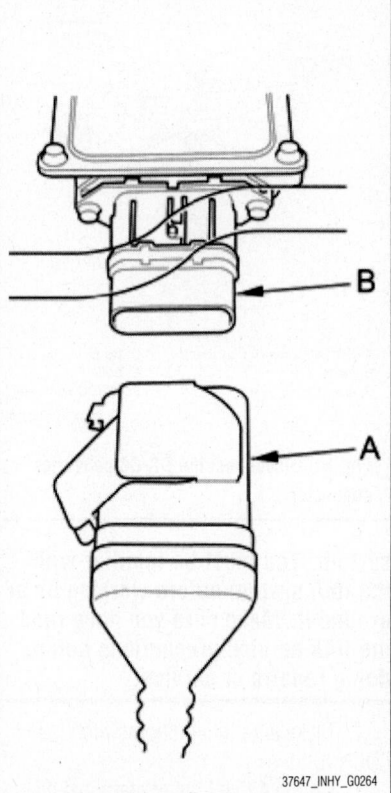

Fig. 56 Disconnect the IMA motor power cable connector (A) from the motor stator (B)

3. Remove the IPU cover.

4. Remove the bolts and the PCU lid.

5. Remove the IMA motor power cables.

➡**Check the position of the U phase, V phase, and W phase cables before disconnecting the IMA motor power cables.**

6. Remove the 12 V power cable, and wrap it with insulating tape.

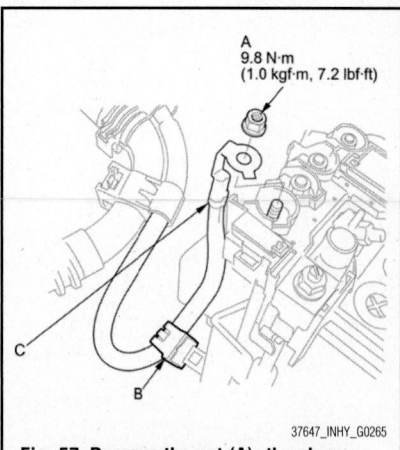

Fig. 57 Remove the nut (A), the clamp (B), and the 12 V power cable (C) from the battery terminal fuse box

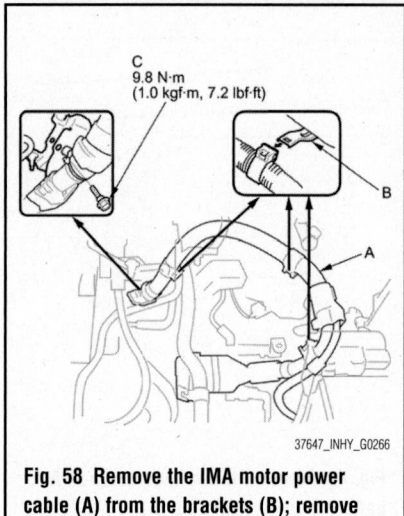

Fig. 58 Remove the IMA motor power cable (A) from the brackets (B); remove the bolt (C)

7. Remove the air cleaner.

8. Disconnect the IMA motor power cable connector from the motor stator.

➡**Refer to disconnecting the IMA motor power cable connector from the motor stator. If the IMA motor power cable terminals are wet, dry them with a clean shop towel. Do not use compressed air.**

9. Remove the nut, the clamp, and the 12 V power cable from the battery terminal fuse box.

10. Remove the cowl cover and the under-cowl panel.

11. Remove the IMA motor power cable from the brackets.

12. Remove the bolt.

13. Remove the fuel tank.

14. Remove the exhaust pipe and the muffler.

15. Remove the under-floor TWC.

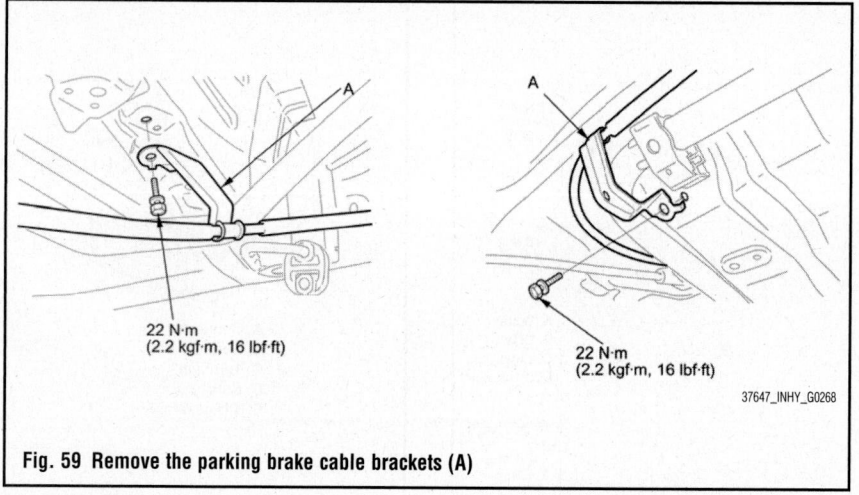

Fig. 59 Remove the parking brake cable brackets (A)

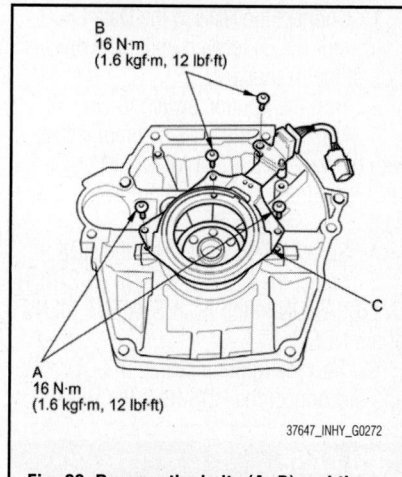

Fig. 63 Remove the bolts (A, B) and the
IMA motor rotor position sensor (C)

16. Remove the heat shield.
17. Remove the parking brake cable brackets.
18. Remove the bolts and the cover.
19. Remove the bolts and the nut then remove the clamps.

➡**Do not reuse the clamps. Replace them with new ones.**

To install:

20. Install the parts in the reverse order of removal.
21. Do the 12 volt battery terminal reconnection procedure.

IMA MOTOR ROTOR POSITION SENSOR

REMOVAL & INSTALLATION

See Figure 63.

1. Remove the IMA motor housing.
2. Remove the bolts and the IMA motor rotor position sensor.

To install:

3. Install the parts in the reverse order of removal.

➡**Tighten the A bolts first, then tighten the B bolts.**

4. Do the motor rotor position calibration procedure.

IMA ROTOR POSITION CALIBRATION

Do the IMA motor rotor position calibra-

tion whenever any of these actions are done:

- The MCM is replaced.
- The IMA motor rotor position sensor is replaced or removed during service.
- The IMA motor is replaced.
- The engine assembly is replaced.
- The transmission is replaced.

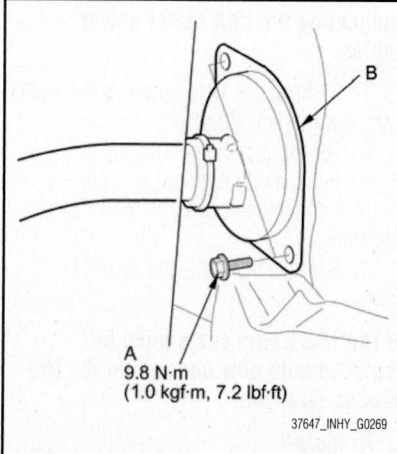

Fig. 60 Remove the bolts (A) and the cover (B)

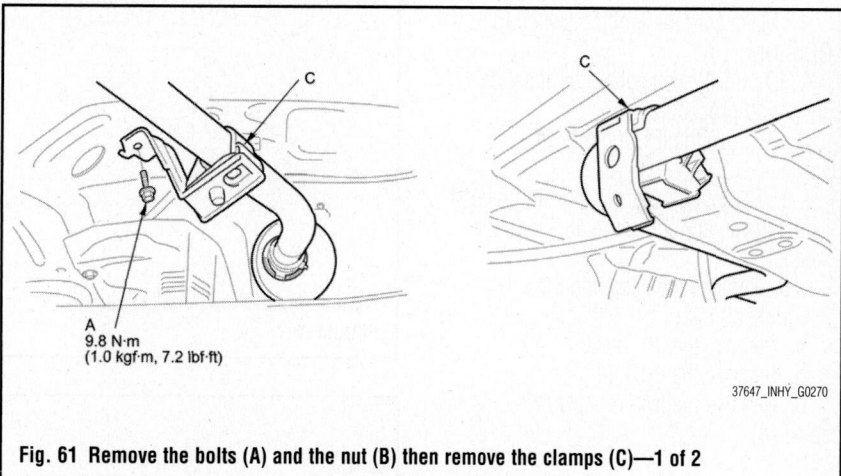

Fig. 61 Remove the bolts (A) and the nut (B) then remove the clamps (C)—1 of 2

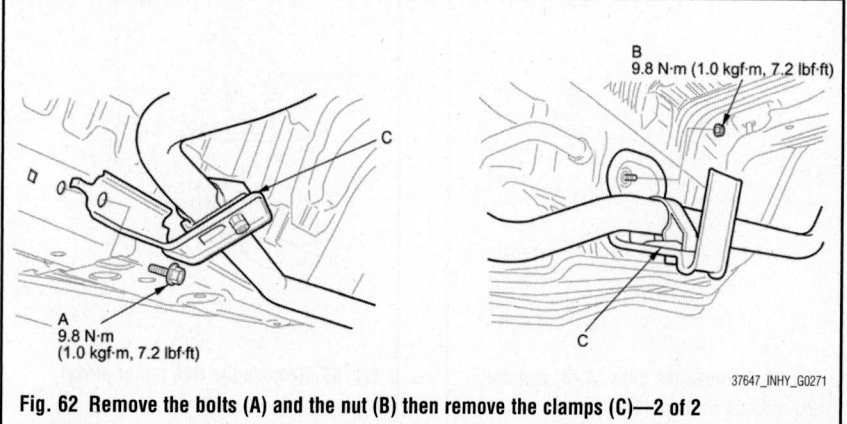

Fig. 62 Remove the bolts (A) and the nut (B) then remove the clamps (C)—2 of 2

1. Connect the HDS to the Data Link Connector (DLC) located under the driver's side of the dashboard.

2. Turn the ignition switch to ON (II).

3. Make sure the HDS communicates with the vehicle and the MCM (IMA system). If it doesn't, troubleshoot the DLC circuit.

4. Select IMA SYSTEM on the HDS.

5. Select the MOTOR ROTOR POSITION CALIBRATION in the ADJUSTMENT MENU of the HDS.

6. Turn the ignition switch to LOCK (0), and disconnect the HDS from the DLC.

IPU CASE

REMOVAL & INSTALLATION

See Figures 64 through 69.

❊❊ CAUTION

IMA components are located in this area. The IMA is a high-voltage system. You must be familiar with the IMA system before working on or around it. Make sure you have read the IMA service precautions before doing repairs or service.

1. Make sure the ignition switch to LOCK (0).

2. Do the 12 volt battery terminal disconnection procedure.

3. Remove the IPU module air duct.

  a. Remove the cargo floor lid, the IPU duct cover, and the cargo area side trim panel.

  b. Remove the clips and the IPU module air duct.

4. Remove the IPU cover.

5. Remove the PCU lid.

6. Remove the IMA motor power cables (A).

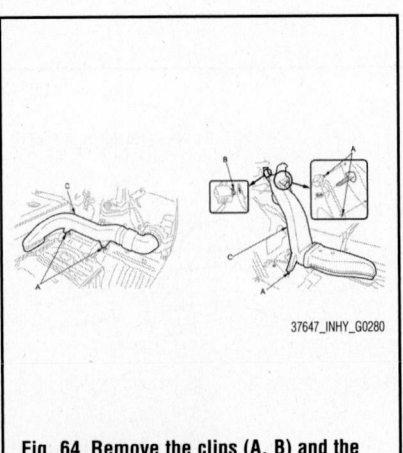

Fig. 64 Remove the clips (A, B) and the IPU module air duct (C)

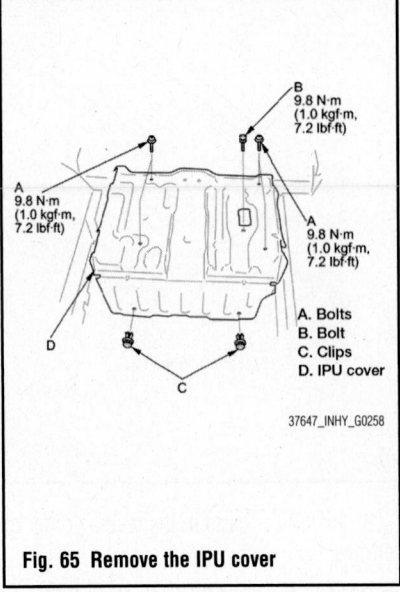

Fig. 65 Remove the IPU cover

A. Bolts
B. Bolt
C. Clips
D. IPU cover

37647_INHY_G0258

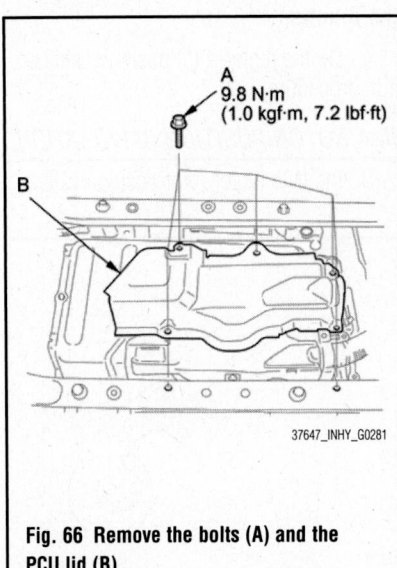

Fig. 66 Remove the bolts (A) and the PCU lid (B)

37647_INHY_G0281

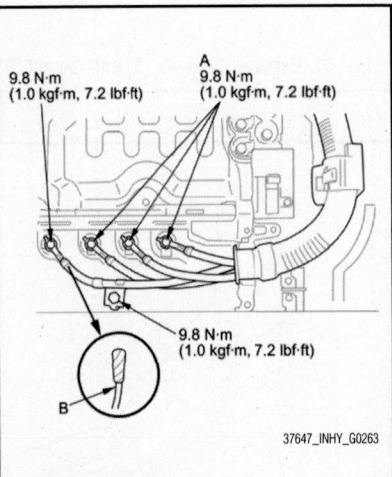

Fig. 67 Remove the IMA motor power cables

37647_INHY_G0263

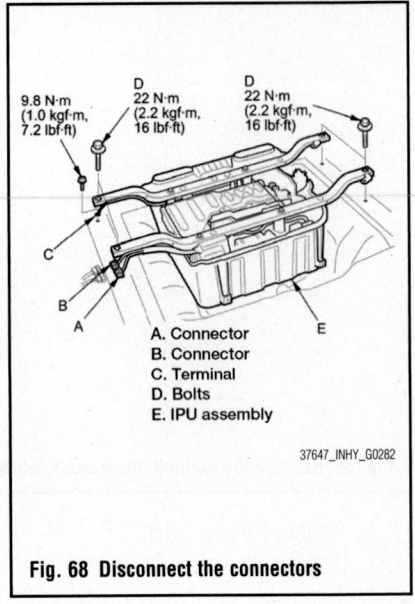

Fig. 68 Disconnect the connectors

A. Connector
B. Connector
C. Terminal
D. Bolts
E. IPU assembly

37647_INHY_G0282

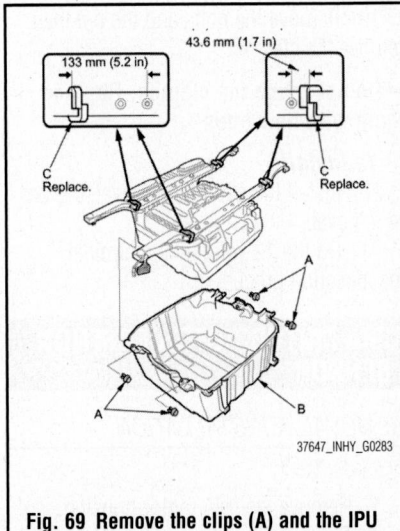

Fig. 69 Remove the clips (A) and the IPU case (B); replace the IPU frame seals (C)

37647_INHY_G0283

➡**Check the position of the U phase, V phase, and W phase cables before disconnecting the IMA motor power cables.**

7. Remove the 12 V power cable, and wrap it with insulating tape.

8. Disconnect the connectors.

9. Remove the terminal.

10. Remove the bolts and the IPU assembly.

11. Remove the clips and the IPU case.

➡**The IPU frame seals must be replaced with new ones when the IPU case is removed.**

To install:

12. Install the parts in the reverse order of removal.

13. Do the 12 volt battery terminal reconnection procedure.

IPU COVER

REMOVAL & INSTALLATION

See Figure 65.

❊❊ CAUTION

IMA components are located in this area. The IMA is a high-voltage system. You must be familiar with the IMA system before working on or around it. Make sure you have read the IMA service precautions before doing repairs or service.

1. Remove the cargo floor lid, the IPU duct cover, and the spare tire beam.
2. Turn the battery module switch OFF.
3. Remove the bolts and the clips.
4. Remove the IPU cover.
5. Install the parts in the reverse order of removal.

➡**Before the battery module switch is turned ON, make sure all the high voltage circuits are connected properly. Then push the button, and turn the battery module switch ON.**

IPU MODULE AIR DUCT

REMOVAL & INSTALLATION

See Figure 64.

1. Remove the cargo floor lid, the IPU duct cover, and the cargo area side trim panel.
2. Remove the clips and the IPU module air duct.
3. Installation is the reverse of removal.

IPU MODULE FAN

REMOVAL & INSTALLATION

See Figure 74.

1. Remove the right cargo area side trim panel.
2. Disconnect the IPU module fan connector.
3. Remove the bolts, the nut, and the IPU module fan.
4. Install the parts in the reverse order of removal.

MOTOR CONTROL MODULE (MCM)

REMOVAL & INSTALLATION

See Figures 66 through 71, 73.

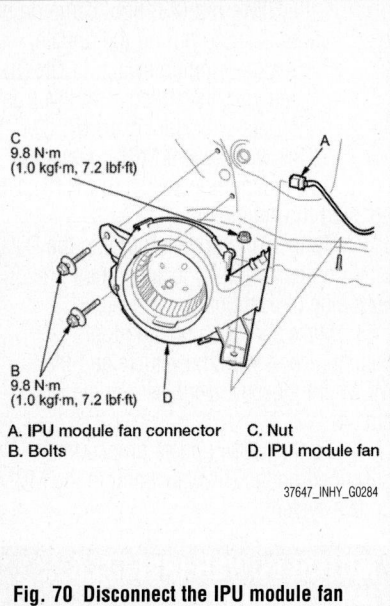

A. IPU module fan connector   C. Nut
B. Bolts   D. IPU module fan

37647_INHY_G0284

Fig. 70 Disconnect the IPU module fan connector

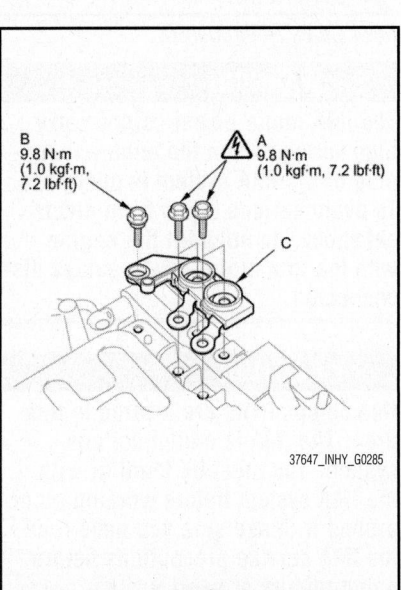

37647_INHY_G0285

Fig. 71 Remove the bolts (A, B) and the PCU busplate (C)

❊❊ CAUTION

IMA components are located in this area. The IMA is a high-voltage system. You must be familiar with the IMA system before working on or around it. Make sure you have read the IMA service precautions before doing repairs or service.

1. Remove the IPU cover.
2. Remove the bolts and the PCU lid.
3. Remove the bolts and the PCU busplate.
4. Remove the bolts and the PCU cover.
5. Disconnect the MCM connectors.

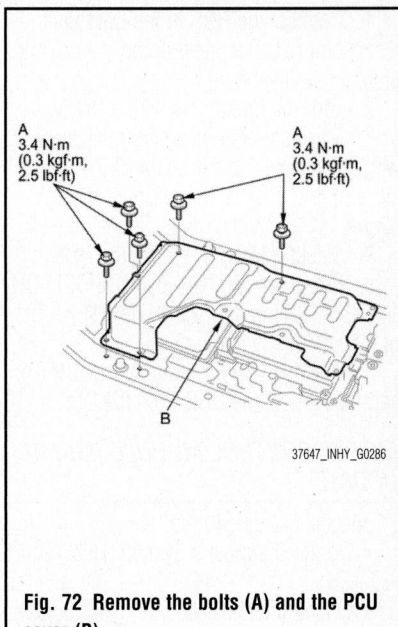

37647_INHY_G0286

Fig. 72 Remove the bolts (A) and the PCU cover (B)

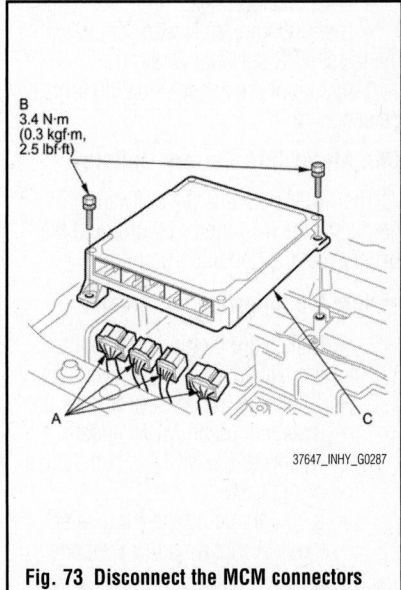

37647_INHY_G0287

Fig. 73 Disconnect the MCM connectors (A); remove the bolts (B) and the MCM (C)

6. Remove the bolts and the MCM.
7. Install the parts in the reverse order of removal.
8. Do the motor rotor position calibration procedure.

IMA ROTOR POSITION CALIBRATION

Do the IMA motor rotor position calibration whenever any of these actions are done:
• The MCM is replaced.
• The IMA motor rotor position sensor is replaced or removed during service.
• The IMA motor is replaced.
• The engine assembly is replaced.
• The transmission is replaced.

1. Connect the HDS to the Data Link Connector (DLC) located under the driver's side of the dashboard.

2. Turn the ignition switch to ON (II).

3. Make sure the HDS communicates with the vehicle and the MCM (IMA system). If it doesn't, troubleshoot the DLC circuit.

4. Select IMA SYSTEM on the HDS.

5. Select the MOTOR ROTOR POSITION CALIBRATION in the ADJUSTMENT MENU of the HDS.

6. Turn the ignition switch to LOCK (0), and disconnect the HDS from the DLC.

MOTOR CONTROL MODULE (MCM) UPDATE

Special Tools Required*:
- Honda Diagnostic System (HDS) tablet tester
- Honda Interface Module (HIM) and an iN workstation with the latest HDS software version
- HDS pocket tester
- GNA600 and an iN workstation with the latest HDS software version

*Any one of the above updating tools can be used.

IMA Motor/IMA Battery Update

The MCM contains the software programs for the IMA motor control and the battery module condition monitor.

➡**Note the following for update:**

- Make sure the updating tool has the latest HDS software version.
- To ensure the latest programs are installed, do an MCM update whenever the MCM is substituted or replaced.
- If you are using the HIM, select the IMA motor and/or the battery module in the HIM MCM update menu.
- You cannot update an MCM with the program it already has. It will only accept a new program.
- Before you update the MCM, make sure the vehicle's 12 V battery is fully charged.
- Do not turn the ignition switch to ACC (I) or to LOCK (0) while updating the MCM. If you do, the MCM can be damaged.
- To prevent MCM damage, do not operate anything electrical (audio system, brakes, A/C, power windows, door locks, etc.) during the update.
- If you need to diagnose the Honda interface module (HIM) because

the HIM's red (#3) light came on or was flashing during the update, leave the ignition switch in ON (II) when you disconnect the HIM from the Data Link Connector (DLC). This will prevent MCM damage.

1. Turn the ignition switch to ON (II). Do not start the engine.

2. Connect the updating tool to the Data Link Connector (DLC) located under the driver's side of dashboard.

3. Make sure the updating tool communicates with the vehicle and the MCM. If it doesn't, troubleshoot the DLC circuit.

4. Do the MCM update procedure as described on the HIM label and in the MCM update system.

MOTOR POWER INVERTER (MPI) MODULE

REMOVAL & INSTALLATION
See Figures 74 through 78.

✳✳ WARNING

The IMA motor power cables carry high voltage when the engine is running or the IMA system is energized. To avoid serious injury from electrical shock, do not start the engine with the IMA motor power cables disconnected.

✳✳ CAUTION

IMA components are located in this area. The IMA is a high-voltage system. You must be familiar with the IMA system before working on or around it. Make sure you have read the IMA service precautions before doing repairs or service.

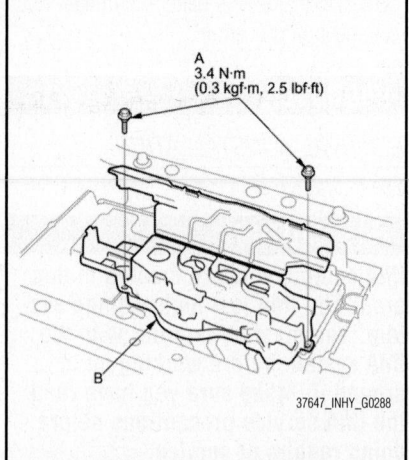

Fig. 75 Remove the bolts (A) and the terminal cover (B)

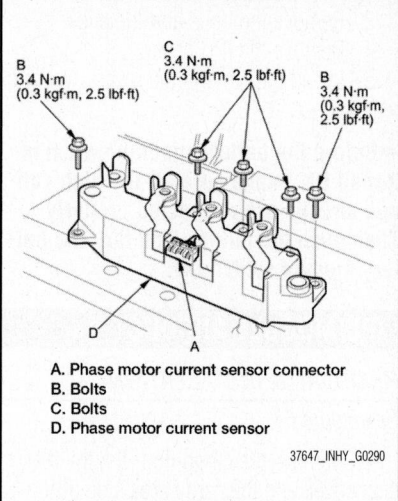

A. Phase motor current sensor connector
B. Bolts
C. Bolts
D. Phase motor current sensor

Fig. 76 Disconnect the phase motor current sensor connector

1. Remove the MCM.
2. Remove the DC-DC converter.
3. Remove the IMA motor power cables.

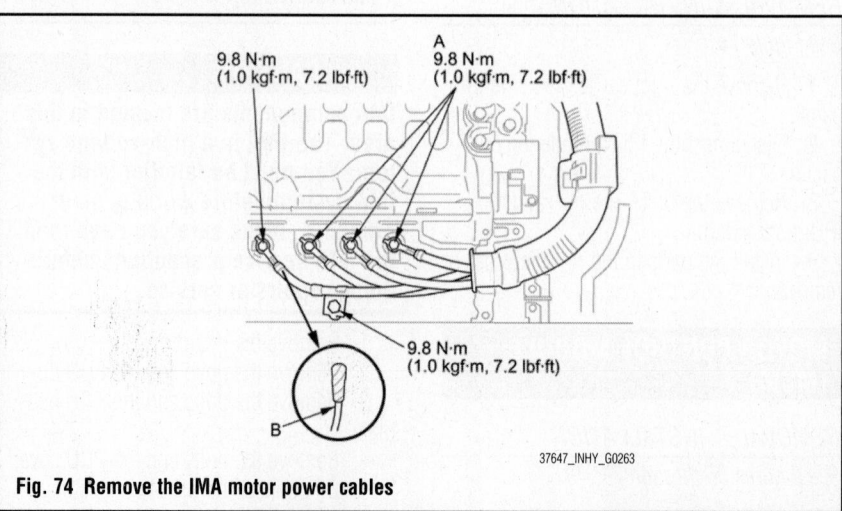

Fig. 74 Remove the IMA motor power cables

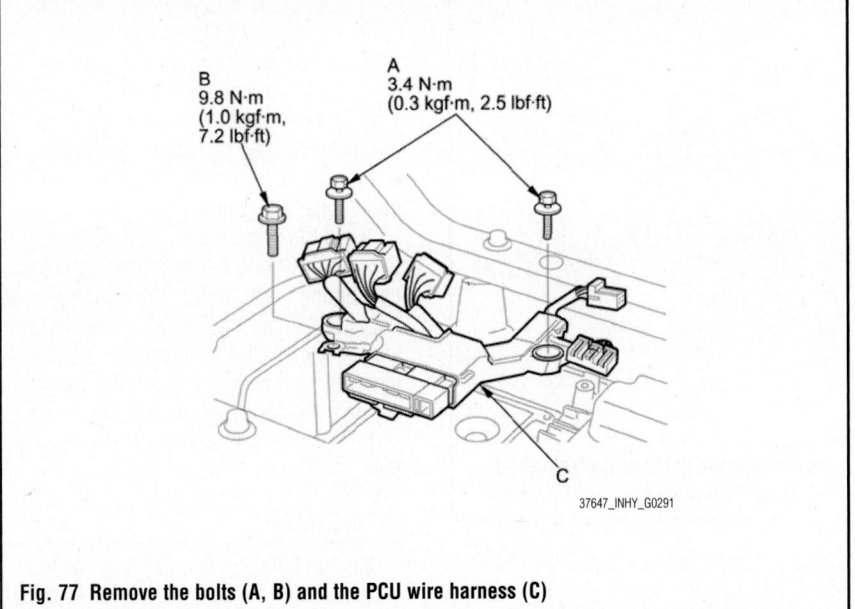

Fig. 77 Remove the bolts (A, B) and the PCU wire harness (C)

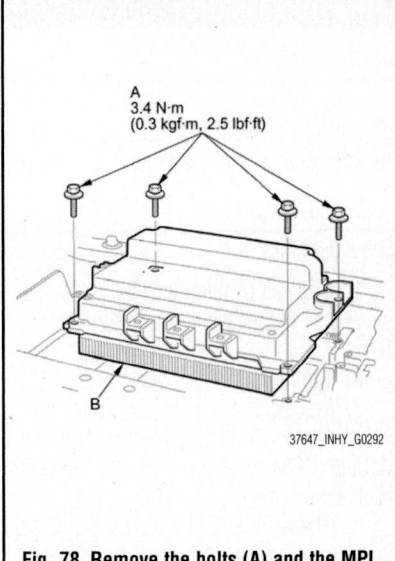

Fig. 78 Remove the bolts (A) and the MPI module (B)

➡**Check the position of the U phase, V phase, and W phase cables before disconnecting the IMA motor power cables.**

4. Remove the 12 V power cable, and wrap it with insulating tape.

5. Remove the bolts and the terminal cover.

6. Disconnect the phase motor current sensor connector.

7. Remove the bolts and the phase motor current sensor.

8. Remove the bolts and the PCU wire harness.

9. Remove the bolts and the MPI module.

10. Install the parts in the reverse order of removal.

11. Do the motor rotor position calibration procedure.

IMA ROTOR POSITION CALIBRATION

Do the IMA motor rotor position calibration whenever any of these actions are done:

- The MCM is replaced.
- The IMA motor rotor position sensor is replaced or removed during service.
- The IMA motor is replaced.
- The engine assembly is replaced.
- The transmission is replaced.

1. Connect the HDS to the Data Link Connector (DLC) located under the driver's side of the dashboard.

2. Turn the ignition switch to ON (II).

3. Make sure the HDS communicates with the vehicle and the MCM (IMA system). If it doesn't, troubleshoot the DLC circuit.

4. Select IMA SYSTEM on the HDS.

5. Select the MOTOR ROTOR POSITION CALIBRATION in the ADJUSTMENT MENU of the HDS.

6. Turn the ignition switch to LOCK (0), and disconnect the HDS from the DLC.

ENGINE ELECTRICAL

IGNITION SYSTEM

FIRING ORDERS

Firing order for the 1.3L LDA3 engine is 1–3–4–2.

IGNITION COIL(S)

REMOVAL & INSTALLATION

See Figures 79 and 80.

1. Remove the under-cowl panel.

2. Remove the engine cover.

3. Disconnect the ignition coil connectors, then remove the intake side ignition coils and the exhaust side ignition coils.

4. Install the ignition coils in the reverse order of removal.

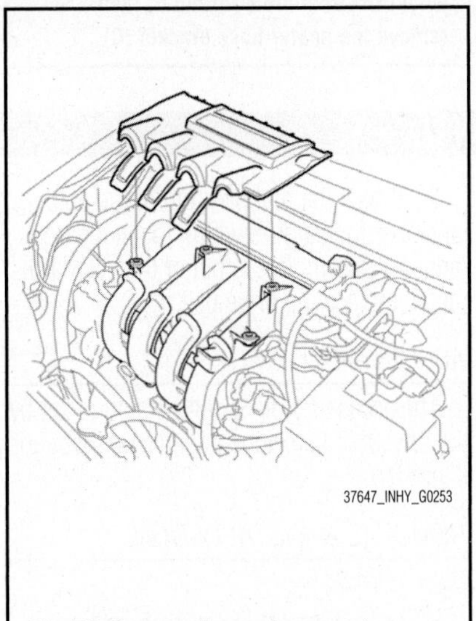

Fig. 79 Remove the engine cover

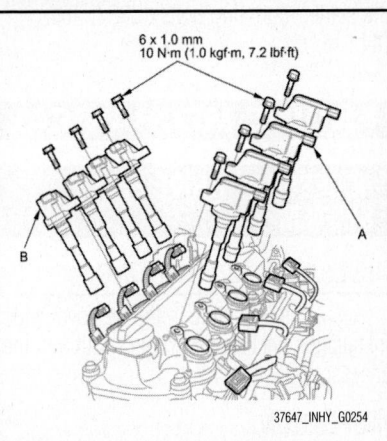

Fig. 80 Disconnect the ignition coil connectors, then remove the intake side ignition coils (A) and the exhaust side ignition coils (B)

IGNITION TIMING

ADJUSTMENT

Ignition timing is controlled by the Powertrain Control Module (PCM). No adjustment is necessary or possible.

INSPECTION

See Figure 81.

1. Connect the Honda Diagnostic System (HDS) to the Data Link Connector (DLC).
2. Turn the ignition switch to ON (II).
3. Make sure the HDS communicates with the vehicle and the Powertrain Control Module (PCM). If it does not communicate, troubleshoot the DLC circuit.
4. Check for DTCs. If a DTC is present, diagnose and repair the cause before inspecting the ignition timing.
5. Start the engine. Hold the engine speed at 3,000 rpm with no load (in P or N)

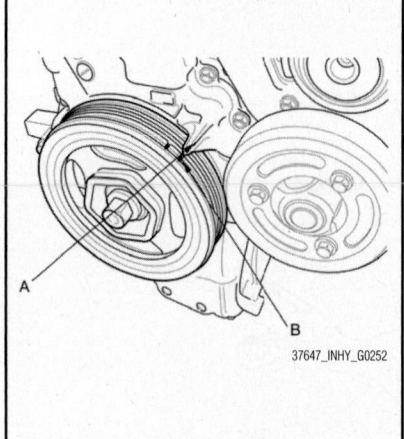

37647_INHY_G0252

Fig. 81 Aim the light toward the pointer (A) on the cam chain case; RED mark (B)

until the radiator fan comes on, then let it idle.

6. Check the idle speed.

7. Jump the SCS line with the HDS.
8. Connect the timing light to the exhaust side No. 1 ignition coil harness.
9. Aim the light toward the pointer on the cam chain case. Check the ignition timing under no load condition (headlights, blower fan, rear window defogger, and air conditioner are turned off).
10. If the ignition timing differs from 8–12° BTDC, check the cam timing. If the cam timing is OK, update the PCM if it does not have the latest software, or substitute a known-good PCM, then recheck. If the system works properly, and the PCM was substituted, replace the original PCM.
11. Disconnect the HDS and the timing light.

ADJUSTMENT

Ignition timing is controlled by the Powertrain Control Module (PCM). No adjustment is necessary or possible.

ENGINE ELECTRICAL

STARTER

REMOVAL & INSTALLATION

See Figure 82.

1. Do the 12 volt battery terminal disconnection procedure.
2. Remove the air cleaner.
3. Disconnect the positive starter cable and the BLK/WHT harness connector from the S terminal, then remove the heater hose bracket.
4. Remove the two bolts securing the starter, then remove the starter.

To install:

5. Install the starter, then tighten the starter mounting bolts to 33 ft. lbs, (44 Nm).

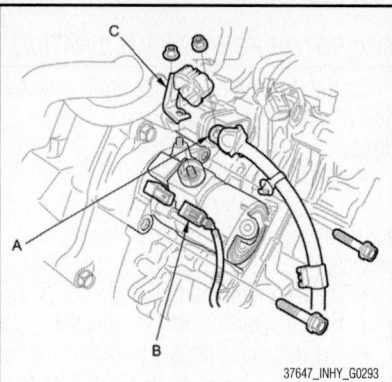

37647_INHY_G0293

Fig. 82 Disconnect the positive starter cable (A) and the BLK/WHT harness connector (B) from the S terminal, then remove the heater hose bracket (C)

STARTING SYSTEM

6. Connect the positive starter cable and the BLK/WHT harness connector to the S terminal, then install the heater hose bracket. Make sure the crimped side of the ring terminal is facing out.
7. Install the air cleaner.
8. Do the 12 volt battery terminal reconnection procedure.
9. If the IMA battery level indicator displays no level, start the engine, and hold it between 3,500 and 4,000 rpm without load (in P or N) until the level in the indicator is half full.

ENGINE MECHANICAL

ACCESSORY DRIVE BELT SYSTEM

ACCESSORY BELT ROUTING

Refer to the graphic in the Removal and Installation section for proper routing of the accessory drive belt.

INSPECTION

See Figure 88.

1. Inspect the belt for cracks or damage. If the belt is cracked or damaged, replace it.

2. Check the that position of the auto-tensioner indicator is within the standard range as shown. If it is out of the standard range, replace the drive belt.

ADJUSTMENT

The accessory drive belt tension is set by the drive belt auto-tensioner. No adjustment is needed.

REMOVAL & INSTALLATION

See Figure 84.

1. Move the auto-tensioner with a

wrench in the direction shown to relieve tension from the drive belt, then remove the drive belt.

2. Install the new belt in the reverse order of removal.

AIR CLEANER

REMOVAL & INSTALLATION

Air Cleaner Assembly

See Figures 85 and 86.

1. Remove the clips.

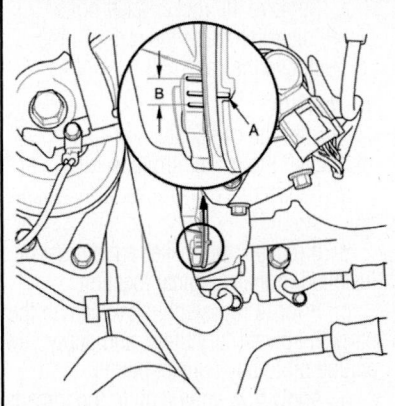

Fig. 83 Check the that position of the auto-tensioner indicator (A) is within the standard range (B)

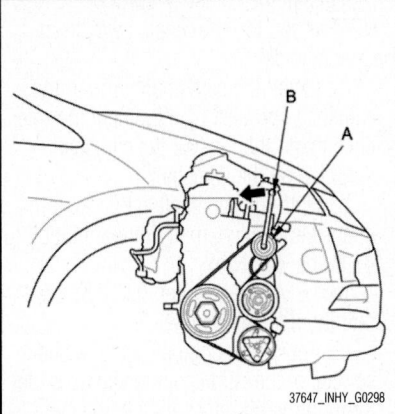

Fig. 84 Move the auto-tensioner (A) with a wrench (B) in the direction shown to relieve tension from the drive belt, then remove the drive belt

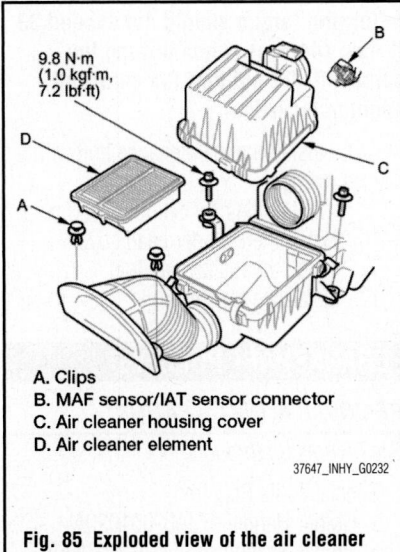

A. Clips
B. MAF sensor/IAT sensor connector
C. Air cleaner housing cover
D. Air cleaner element

37647_INHY_G0232

Fig. 85 Exploded view of the air cleaner assembly

37647_INHY_G0299

Fig. 86 Remove the air cleaner housing (A)

2. Disconnect the MAF sensor/IAT sensor connector, then remove the air cleaner housing cover, and the air cleaner element.

➡**Be careful not to damage the MAF sensor/IAT sensor on the air cleaner housing cover.**

3. Remove the air cleaner housing.
4. Install the parts in the reverse order of removal.

Air Filter Element

See Figure 87.

1. Open the air cleaner housing cover.
2. Remove the air cleaner element from the air cleaner housing.
3. Check the air cleaner element for damaged, dirt, or clogging. If it is damaged or clogged, replace it.

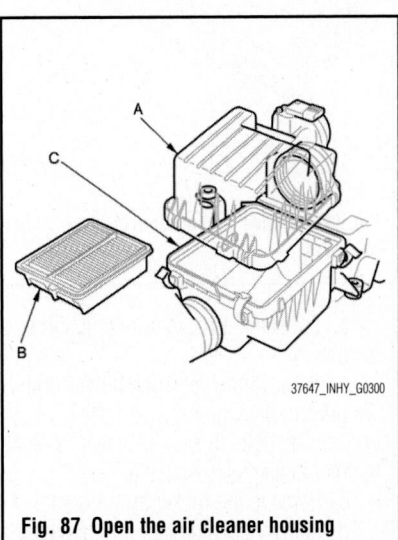

37647_INHY_G0300

Fig. 87 Open the air cleaner housing cover (A); remove the air cleaner element (B) from the air cleaner housing (C)

➡**Do not use compressed air to clean the air cleaner element.**

4. Clean and remove any debris from inside the air cleaner housing.
5. Install the parts in the reverse order of removal.
- If you did not replace the air cleaner element, this procedure is complete.
- If the maintenance minder required air cleaner element replacement, reset the maintenance minder.
- If the idle speed fluctuates, do the idle speed inspection.

CAMSHAFT & VALVE LIFTERS

REMOVAL & INSTALLATION

See Figures 88 through 92.

1. Remove the air cleaner.
2. Remove the intake manifold.
3. Remove the cylinder head cover.
4. Remove the camshaft sprocket.
   a. Make a reference mark across the camshaft sprocket and cam chain.
   b. Apply new engine oil to the slider surface of the cam chain tensioner slider through the oil return hole in the cylinder head.
   c. Remove the cylinder head plug.
   d. Hold the crankshaft pulley, and set the socket wrench on the camshaft sprocket bolt.
   e. Remove the maintenance bolt, and turn the camshaft clockwise to compress the cam chain tensioner, then install the 6 x 1.0 mm bolt in the bolt hole in the engine block through the maintenance hole and cam chain tensioner.

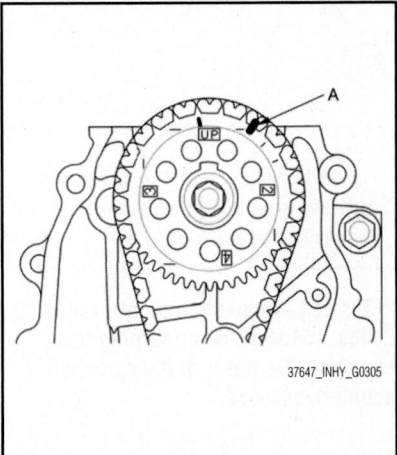

37647_INHY_G0305

Fig. 88 Make a reference mark (A) across the camshaft sprocket and cam chain

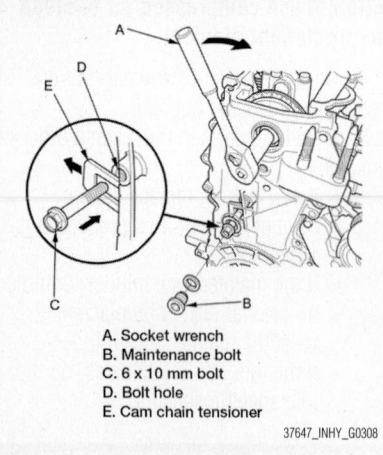

A. Socket wrench
B. Maintenance bolt
C. 6 x 10 mm bolt
D. Bolt hole
E. Cam chain tensioner

37647_INHY_G0308

Fig. 89 Hold the crankshaft pulley, and set the socket wrench on the camshaft sprocket bolt

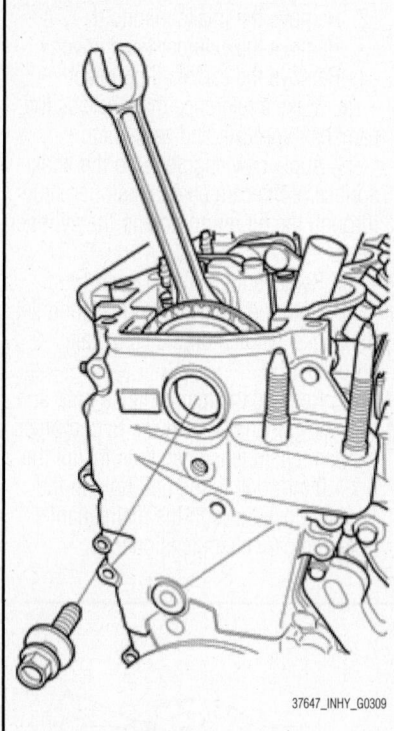

37647_INHY_G0309

Fig. 90 Hold the camshaft with a 27 mm open-end wrench, then remove the camshaft sprocket

➡️**Turning torque should not exceed 33 ft. lbs. (44 Nm), when turning the camshaft. Do not turn the camshaft counterclockwise.**

  f. Hold the camshaft with a 27 mm open-end wrench, then remove the camshaft sprocket.

➡️**Hang the cam chain with a wire.**

  5. Remove the rocker arm assembly.

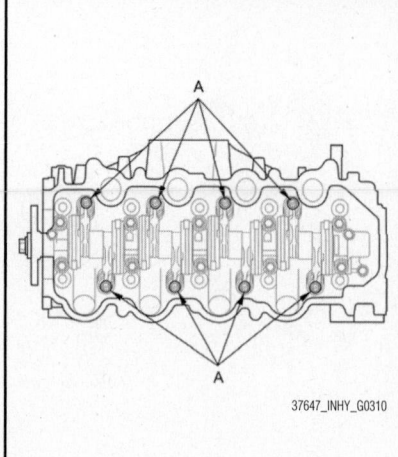

37647_INHY_G0310

Fig. 91 Loosen the rocker arm adjusting screws (A)

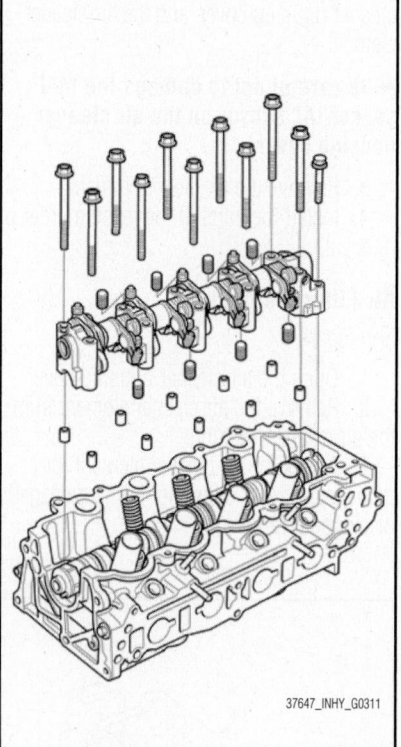

37647_INHY_G0311

Fig. 92 Remove the rocker arm assembly

  a. Loosen the rocker arm adjusting screws.
  b. Remove the camshaft holder bolts. To prevent damaging the camshaft, loosen the bolts in sequence two turns at a time, in a crisscross pattern.
  c. Remove the rocker arm assembly.
  d. Identify each part as it is removed so that each item can be reinstalled in their original location.

  e. Remove the rocker shaft bolts before disassembling the rocker arms.
  6. Remove the camshaft.

To install:
  7. Install the camshaft.
  8. Install the rocker arm assembly.
  a. Reassemble the rocker arm assembly.
  b. If reused, the rocker arms must be installed in their original location.
  c. Prior to reassembling, clean all the parts in solvent, dry them, and apply new engine oil to any contact points.
  d. Apply new engine oil to the threads of the rocker shaft bolts when installing them.
  e. When replacing the rocker arm assembly, remove the fastening hardware from the new rocker arm assembly.
  9. Install the camshaft sprocket.
NOTE: Keep the cam chain away from magnetic fields.
  a. Install the cam chain around the camshaft sprocket by alignment the reference mark, then install the camshaft sprocket on the camshaft.
  b. Hold the camshaft with a 27 mm open-end wrench, then tighten the bolt to 41 ft. lbs. (56 Nm).
  NOTE: Apply new engine oil to the bolt threads and flange.
  c. Apply new engine oil to the slider surface of the cam chain tensioner slider through the oil return hole in the cylinder head.
  d. Hold the crankshaft pulley, and set the socket wrench on the camshaft sprocket bolt.
  e. Turn the camshaft clockwise to compress the cam chain tensioner, then remove the 6 x 1.0 mm bolt.

➡️**Turning torque should not exceed 33 ft. lbs. (44 Nm), when turning the camshaft. Do not turn the camshaft counterclockwise.**

  f. Install the maintenance bolt with a new washer.
  g. Install the new cylinder head plug.
  10. Install the cylinder head cover.
  11. Install the intake manifold.
  12. Install the air cleaner.

CRANKSHAFT PULLEY

REMOVAL & INSTALLATION
See Figures 93 through 96.

Special Tools Required:
• Holder Handle 07JAB-001020A
• Holder Attachment, 50 mm 07NAB-001040A

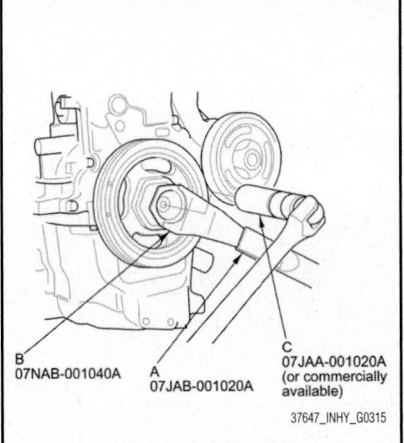

Fig. 93 For removal, hold the pulley with the holder handle (A) and the pulley holder attachment (B); remove the bolt with a 19 mm socket wrench (C) and a breaker bar

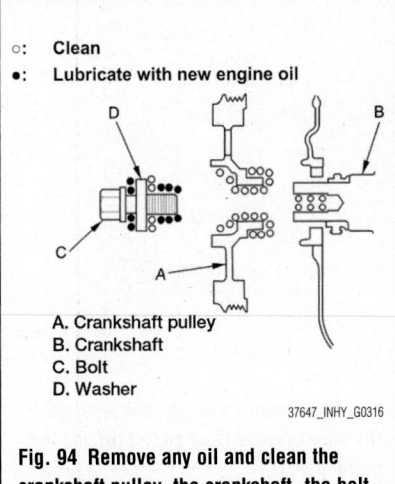

A. Crankshaft pulley
B. Crankshaft
C. Bolt
D. Washer

Fig. 94 Remove any oil and clean the crankshaft pulley, the crankshaft, the bolt, and the washer; lubricate with new engine oil as shown

• Socket, 19 mm 07JAA-001020A or equivalent
1. Remove the front wheels.
2. Remove the splash shields.
3. Remove the accessory drive belt.
4. Hold the pulley with the holder handle and the pulley holder attachment.
5. Remove the bolt with a 19 mm socket wrench and a breaker bar, then remove the crankshaft pulley.

To install:
6. Remove any oil and clean the crankshaft pulley, the crankshaft, the bolt, and the washer. Lubricate with new engine oil as shown.
7. Install the crankshaft pulley.
8. Tighten the crankshaft pulley bolt. Do not use an impact wrench.

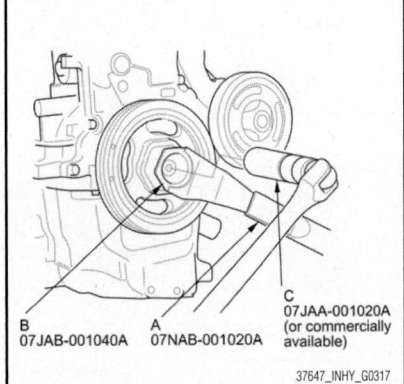

Fig. 95 To install, hold the pulley with the holder handle (A) and the pulley holder attachment (B), then tighten the bolt with a torque wrench and heavy duty 19 mm socket wrench (C)

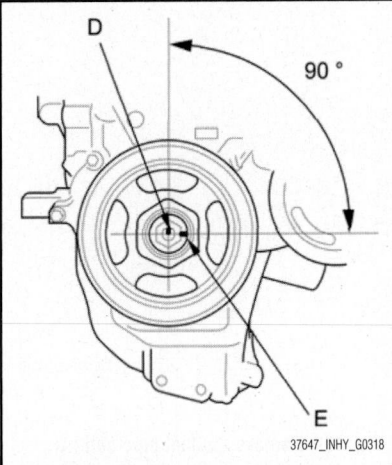

Fig. 96 Mark the bolt head (D) and the crankshaft pulley (E), then tighten the bolt an additional 90 degrees

a. Hold the pulley with the holder handle and the pulley holder attachment, then tighten the bolt to 27 ft. lbs. (37 Nm) with a torque wrench and heavy duty 19 mm socket wrench. If the pulley bolt or crankshaft is new, tighten the bolt to 130 ft. lbs. (177 Nm), then remove the bolt and tighten it to 27 ft. lbs. (37 Nm).

b. Mark the bolt head and the crankshaft pulley, then tighten the bolt an additional 90 ° (The mark on the bolt head lines up with the mark on the crankshaft pulley).
9. Install the drive belt.
10. Install the splash shields.
11. Install the front wheels.

CRANKSHAFT REAR COVER & SEAL

REMOVAL & INSTALLATION
See Figures 97 and 98.

➤**Special Tools Required:**

- Driver Handle, 15 x 135L 07749-0010000
- Oil Seal Driver Attachment, 96 mm 07ZAD-PNAA100

1. Remove the transmission. See "Constant Velocity Transmission" section.
2. Remove the IMA motor rotor, the IMA motor housing, and the IMA motor rotor position sensor.
3. Clean and dry the crankshaft oil seal housing.
4. Apply a light coat of new engine oil of the crankshaft oil seal.
5. Use the driver handle, 15 x 135L, and the oil seal driver attachment, 96 mm, to drive a new crankshaft oil seal squarely into the block to the specified installed height of 0.008–0.047 in. (0.2–1.2 mm).
6. Measure the distance between the cylinder block and the oil seal.
7. Install the IMA motor rotor position sensor, the IMA motor housing, and the IMA motor rotor.
8. Install the transmission.

CYLINDER HEAD

REMOVAL & INSTALLATION
See Figures 99 through 105.

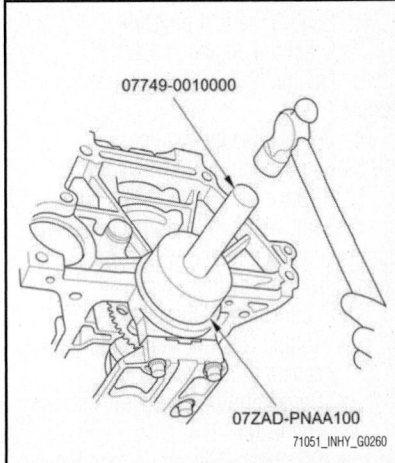

Fig. 97 Use the driver handle, 15 x 135L, and the oil seal driver attachment, 96 mm, to drive a new crankshaft oil seal squarely into the block to the specified installed height of 0.008–0.047 in. (0.2–1.2 mm).

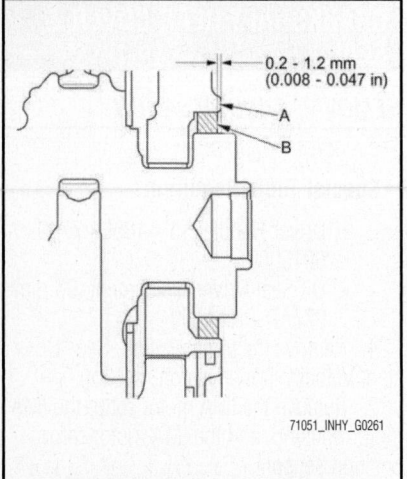

Fig. 98 Measure the distance between the cylinder block (A) and the oil seal (B).

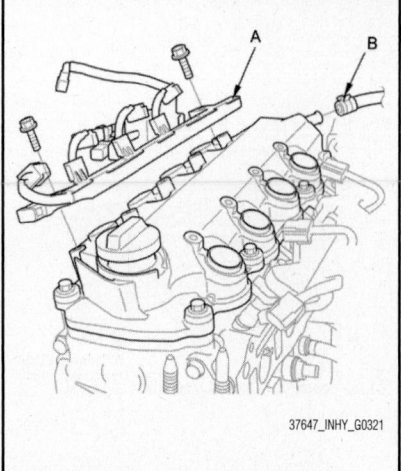

Fig. 99 Remove the harness holder (A), and disconnect the breather hose (B)

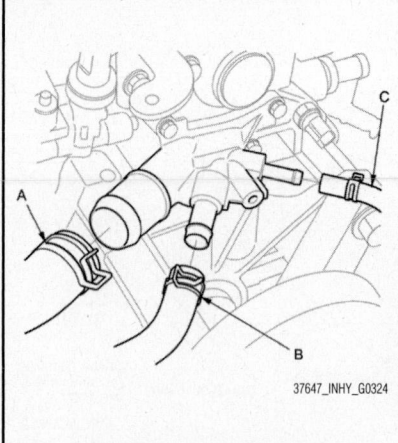

Fig. 102 Remove the upper radiator hose (A), the water bypass hose (B), and the heater hose (C)

➡**Note the following when servicing this component:**

- Use fender covers to avoid damaging painted surfaces.
- To avoid damage, unplug the wiring connectors carefully while holding the connector portion.
- To avoid damaging the cylinder head, wait until the engine coolant temperature drops below 100 °F (38 °C) before loosening the cylinder head bolts.
- Mark all wiring and hoses to avoid misconnection. Also, be sure that they do not contact other wiring or hoses, or interfere with other parts.
- Keep the cam chain away from magnetic fields.

1. Relieve the fuel pressure.
2. Drain the engine coolant.
3. Do the 12 volt battery removal procedure.
4. Remove the air cleaner.
5. Remove the intake manifold.
6. Remove the eight ignition coils.
7. Remove the following engine wire harness connectors and wire harness clamps from the cylinder head:
- Four injector connectors
- Engine Coolant Temperature (ECT) sensor 1 connector
- Camshaft Position (CMP) sensor connector
- Secondary Heated Oxygen Sensor (secondary HOS2) connector
- Rocker arm oil control solenoid connector

8. Remove the harness holder, and disconnect the breather hose.
9. Remove the fuel pipe bolt and the fuel pipe clamp.
10. Remove the harness holder mount-

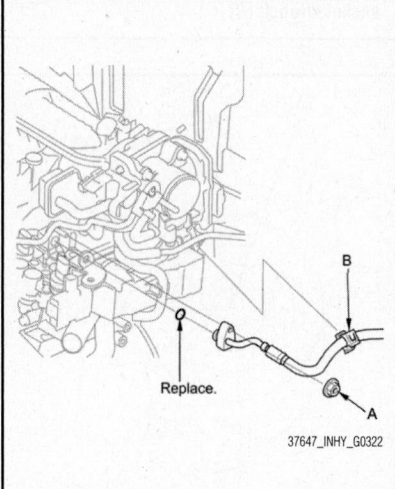

Fig. 100 Remove the fuel pipe bolt (A) and the fuel pipe clamp (B)

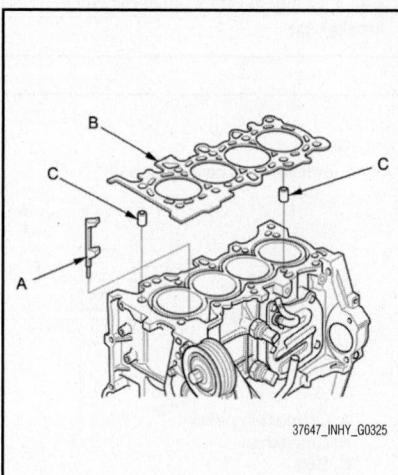

Fig. 103 Install a new coolant separator (A), new cylinder head gasket (B) and the dowel pins (C)

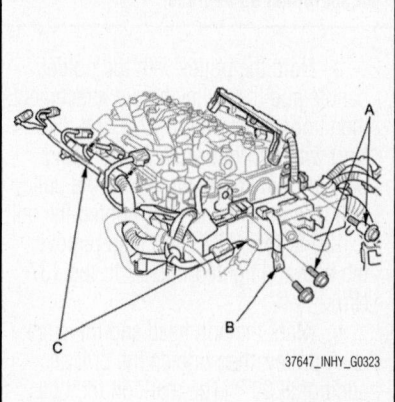

Fig. 101 Remove the harness holder mounting bolt (A) and the ground cable (B), then remove the harness holder (C)

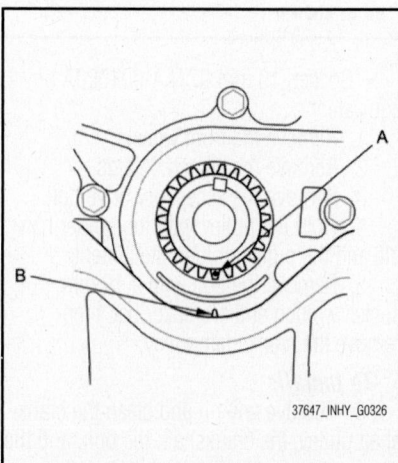

Fig. 104 Align the TDC mark (A) on the crankshaft sprocket with the pointer (B) on the oil pump

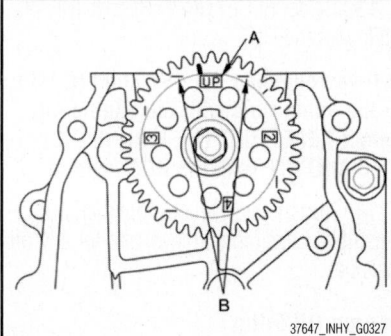

Fig. 105 The UP mark (A) on the camshaft sprocket should be at the top, and the TDC grooves (B) on the camshaft sprocket should line up with the top edge of the head

ing bolt and the ground cable, then remove the harness holder.

11. Remove the upper radiator hose, the water bypass hose, and the heater hose.

12. Remove the drive belt.

13. Remove the water pump.

14. Remove the cylinder head cover.

15. Remove the warm-up three way catalytic converter (WU-TWC).

16. Remove the cam chain.

17. Remove the cylinder head bolts. To prevent warpage, loosen the bolts in a crisscross sequence ⅓ turn at a time starting from each end; repeat the sequence until all bolts are loosened.

18. Remove the cylinder head.

To install:

19. Clean the cylinder head and the engine block surface.

20. Install a new coolant separator in the engine block whenever the engine block is replaced.

21. Install the new cylinder head gasket and the dowel pins on the engine block. Always use a new cylinder head gasket.

22. Set the crankshaft to Top Dead Center (TDC). Align the TDC mark on the crankshaft sprocket with the pointer on the oil pump.

23. Set the camshaft TDC. The UP mark on the camshaft sprocket should be at the top, and the TDC grooves on the camshaft sprocket should line up with the top edge of the head.

24. Install the cylinder head on the engine block.

25. Apply new engine oil to the threads and flange of all cylinder head bolts.

26. Tighten the cylinder head bolts in a criss cross sequence staring in the middle and working to each end to 22 ft. lbs. (29 Nm), use a beam-type torque wrench. When

using a preset-click-type torque wrench, be sure to tighten slowly and do not over-tighten. If a bolt makes any noise while you are torquing it, loosen the bolt and retighten it from the first step.

27. Tighten all cylinder head bolts an additional 130°.

28. Install the cam chain.

29. Install the warm-up three way catalytic converter (WU-TWC).

30. Install the cylinder head cover.

31. Install the water pump.

32. Install the drive belt.

33. Install the upper radiator hose, the water bypass hose, and the heater hose.

34. Install the harness holder and the ground cable.

35. Install the fuel pipe bolt and the fuel pipe clamp.

36. Install the breather hose and the harness holder.

37. Connect the engine wire harness connectors, and install the wire harness clamps to cylinder head.

- Four injector connectors
- Engine Coolant Temperature (ECT) sensor 1 connector
- Camshaft Position (CMP) sensor connector
- Secondary Heated Oxygen Sensor (secondary HO2S) connector
- Rocker arm oil control solenoid connector

38. Install the eight ignition coils.

39. Install the intake manifold.

40. Install the air cleaner.

41. After installation, check that all tubes, hoses, and connectors are installed correctly.

42. Do the 12 volt battery installation procedure.

43. Inspect for fuel leaks. Turn the ignition switch to ON (II) (do not operate the starter) so the fuel pump runs for about 2 seconds and pressurizes the fuel line. Repeat this operation three times, then check for fuel leakage at any point in the fuel line.

44. Refill the radiator with engine coolant, and bleed the air from the cooling system with the heater valve open.

45. Do the Crankshaft Position (CKP) pattern clear/CKP pattern learn procedure.

46. Inspect the idle speed.

47. Inspect the ignition timing.

CYLINDER HEAD COVERS

REMOVAL & INSTALLATION

See Figures 106 and 107.

1. Remove the intake manifold.
2. Remove the eight ignition coils.

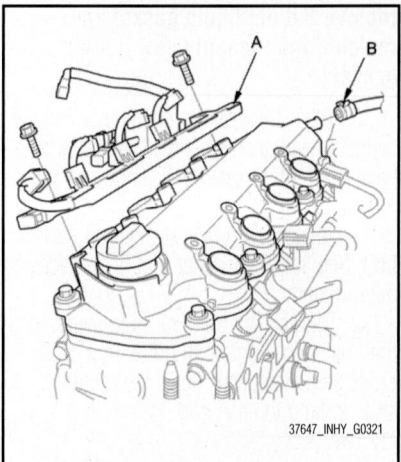

Fig. 106 Remove the harness holder (A), and disconnect the breather hose (B)

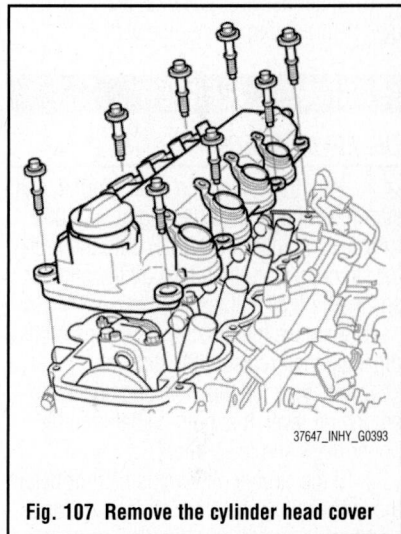

Fig. 107 Remove the cylinder head cover

3. Remove the harness holder, and disconnect the breather hose.

4. Remove the cylinder head cover.

To install:

5. Check the spark plug seals for damage. If the seal is damaged, replace the spark plug seals.

6. Thoroughly clean the head cover gasket and the groove.

7. Install the head cover gasket in the groove of the cylinder head cover.

8. Check that the mating surfaces are clean and dry.

9. Apply liquid gasket (P/N 08717-0004, 08718-0003, or 08718-0009) to the cam chain case contact areas. Install the component within 5 minutes of applying the liquid gasket.

➡ If you apply liquid gasket P/N 08718-0012, the component must be installed within 4 minutes. If too much time has passed after applying the liquid gasket, remove the old liquid gasket and residue, then reapply new liquid gasket.

10. Place the cylinder head cover) on the cylinder head, then slide the cover slightly back and forth to seat the head cover gasket.

11. Inspect the cover washer on each bolt. Replace any washer that is damaged or deteriorated.

12. Tighten the bolts in three steps. In the final step, tighten all bolts, in a criss-cross sequence, starting in the middle and working to each end, to 9 ft. lbs. (12 Nm).

13. Connect the breather hose, and install the harness holder.

14. Install the eight ignition coils.

15. Install the intake manifold.

16. After assembly, wait at least 30 minutes before filling the engine with oil.

ENGINE OIL & FILTER

OIL LEVEL CHECK

1. Park the vehicle on level ground, and start the engine. Hold the engine at 3,000 rpm with no load (in P or N) until the radiator fan comes on, then turn off the engine, and wait a few minutes.

2. Remove the dipstick, and wipe off the dipstick, then reinstall the dipstick.

3. Remove the dipstick, and check the engine oil level. It should be between the upper mark and lower mark.

4. If the engine oil level is near or below the lower mark, check for oil leakage, and add engine oil to bring it to the upper mark.

OIL & FILTER CHANGE

Engine Oil

See Figures 108 through 110.

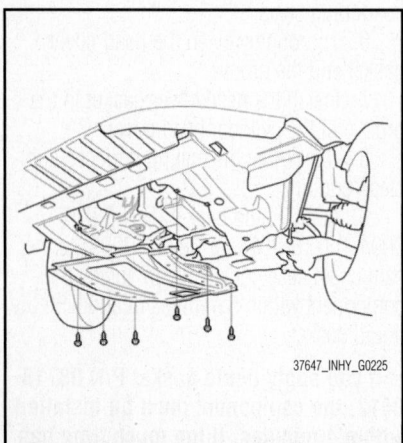

Fig. 108 Remove the engine undercover

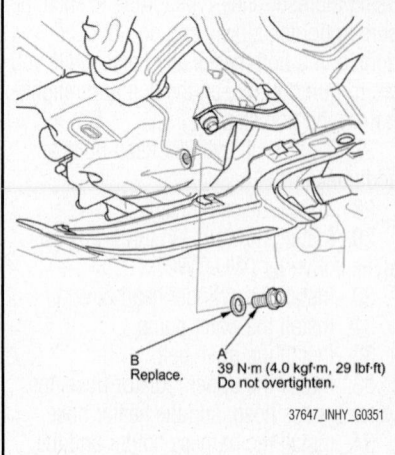

37647_INHY_G0351

Fig. 109 Remove the drain bolt (A) and washer (B)

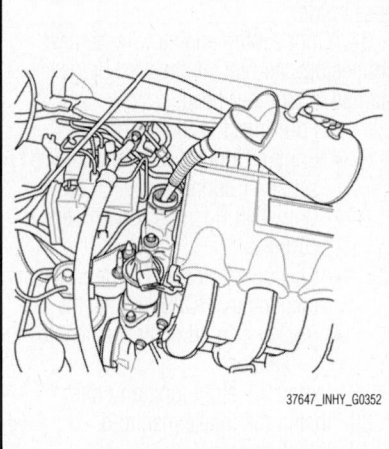

37647_INHY_G0352

Fig. 110 Refill the engine with the recommended engine oil

1. Warm up the engine.
2. Remove the engine undercover.
3. Remove the drain bolt and drain the engine oil.
4. Reinstall the drain bolt with a new washer.
5. Install the engine undercover.
6. Refill the engine with the recommended engine oil.
7. Connect the Honda Diagnostic System (HDS) to the Data Link Connector (DLC).
8. Turn the ignition switch to ON (II).
9. Make sure the HDS communicates with the vehicle and Powertrain Control Module (PCM). If it does not communicate, troubleshoot the DLC circuit.
10. Select BODY ELECTRICAL with the HDS.
11. Select ADJUSTMENT in GAUGE MENU with the HDS.
12. Select RESET in the MAINTENANCE MINDER with the HDS.

13. Select RESETTING THE ENGINE OIL LIFE with the HDS.

➡**If you changed the CVT Fluid (CVTF) at the same time with the engine oil, select RESETTING THE ENGINE OIL LIFE AND ATF with the HDS instead.**

14. Run the engine for more than 3 minutes, then check the oil level and for any oil leakage.

Engine Oil Filter

Special Tools Required: Oil Filter Wrench 07HAA-PJ70101

1. Drain the engine oil.
2. Remove the oil filter with the oil filter wrench.
3. Inspect the filter to make sure the rubber seal is not stuck to the oil filter seating surface of the engine.
4. Inspect the threads and the rubber seal on the new filter. Clean the seat on the oil pan, then apply a light coat of new engine oil to the filter rubber seal. Use only filters with a built-in bypass system.
5. Install the oil filter by hand.
6. After the rubber seal seats, tighten the oil filter clockwise with the oil filter wrench. Tighten: ¾ Turn Clockwise; Tightening Torque: 9 ft. lbs. (12 Nm)
7. If four numbers or marks (1 to 4 or ▼ to ▼▼▼▼) are printed around the outside of the filter, you can use the following procedure to tighten the filter.

   a. Spin the filter on until its seal lightly seats against the oil pan, and note which number or mark is at the bottom.

   b. Tighten the filter by turning it clockwise three numbers or marks from the one you noted. For example, if mark ▼ is at the bottom when the seal is lightly seated, tighten the filter until the mark ▼▼▼▼ comes around the bottom.

8. After installation, fill the engine with oil up to the specified level, run the engine for more than 3 minutes, then check for oil leakage.

INTAKE MANIFOLD

REMOVAL & INSTALLATION

See Figures 111 through 118.

1. Remove the engine cover.
2. Remove the air cleaner.
3. Remove the engine wire harness connectors and wire harness clamps from the intake manifold:

   • Throttle actuator connector
   • Manifold Absolute Pressure (MAP) sensor connector

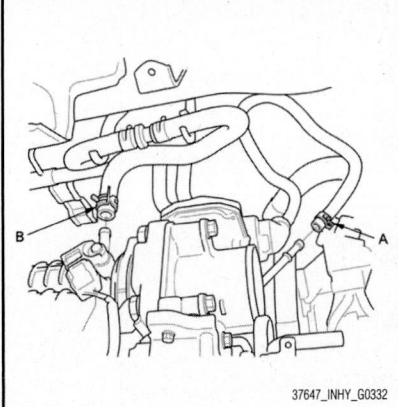

Fig. 111 Disconnect the Evaporative Emission (EVAP) canister hose (A) and the brake booster vacuum hose (B)

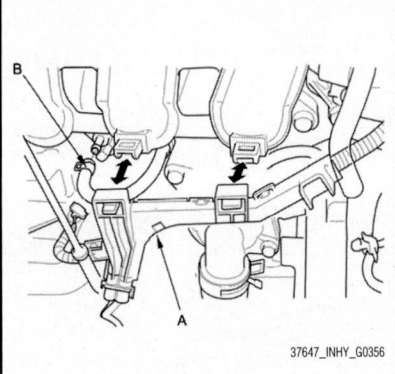

Fig. 114 Remove the harness holder (A) and the Positive Crankcase Ventilation (PCV) hose (B)

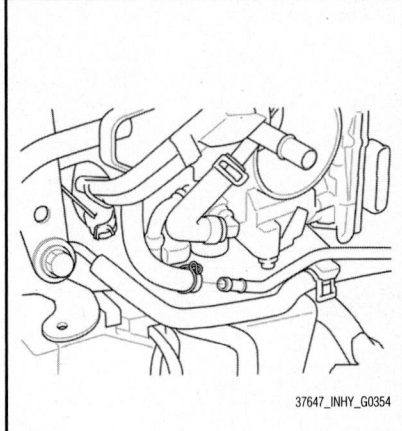

Fig. 112 Remove the EVAP canister purge hose

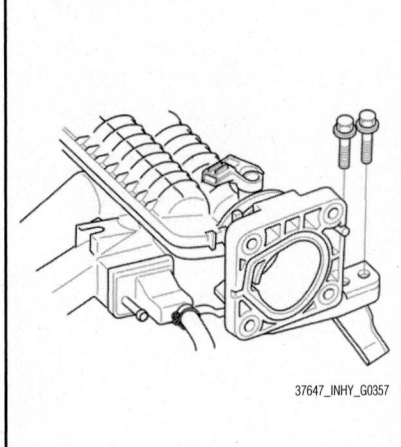

Fig. 115 Remove the intake manifold bracket mounting bolts

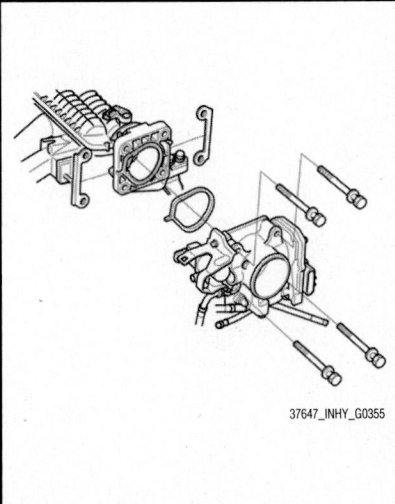

Fig. 113 Remove the throttle body without disconnecting the water bypass hoses

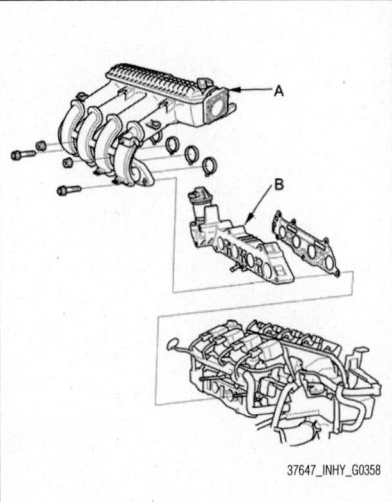

Fig. 116 Remove the intake manifold (A) and the EGR plate (B)

- Exhaust Gas Recirculation (EGR) valve connector
- Evaporative Emission (EVAP) canister purge valve connector

4. Disconnect the Evaporative Emission (EVAP) canister hose and the brake booster vacuum hose.

5. Remove the EVAP canister purge hose.

6. Remove the throttle body without disconnecting the water bypass hoses.

7. Remove the harness holder from the intake manifold.

8. Remove the Positive Crankcase Ventilation (PCV) hose.

9. Remove the dipstick.

10. Remove the intake manifold bracket mounting bolts.

11. Remove the intake manifold.

12. Remove the EGR plate.

To install:

13. Install the EGR plate with a new gasket.

14. Install the intake manifold with new gaskets, and tighten the bolts/nuts in a crisscross pattern in three steps, beginning with the inner bolt.

15. Loosen the intake manifold bracket mounting bolt. Tighten the mounting bolts, then tighten the mounting bolt.

16. Install the throttle body with a new gasket. Tighten the throttle body mounting bolts to 17 ft. lbs. (24 Nm).

17. Install the dipstick.

18. Install the PCV hose and the harness holder.

19. Install the EVAP canister purge hose.

20. Install the brake booster vacuum hose and the harness clamp.

21. Install the engine wire harness connectors and the wire harness clamps to the intake manifold:

- Throttle actuator connector
- MAP sensor connector
- EGR valve connector
- EVAP canister purge valve connector

22. Install the air cleaner.

23. Install the engine cover.

OIL PAN

REMOVAL & INSTALLATION

See Figures 119 through 123.

1. If the engine is already out of the vehicle, go to step 7.

2. Remove the splash shields.

3. Drain the engine oil.

4. Remove the drive belt.

5. Remove the driveshaft heat shield.

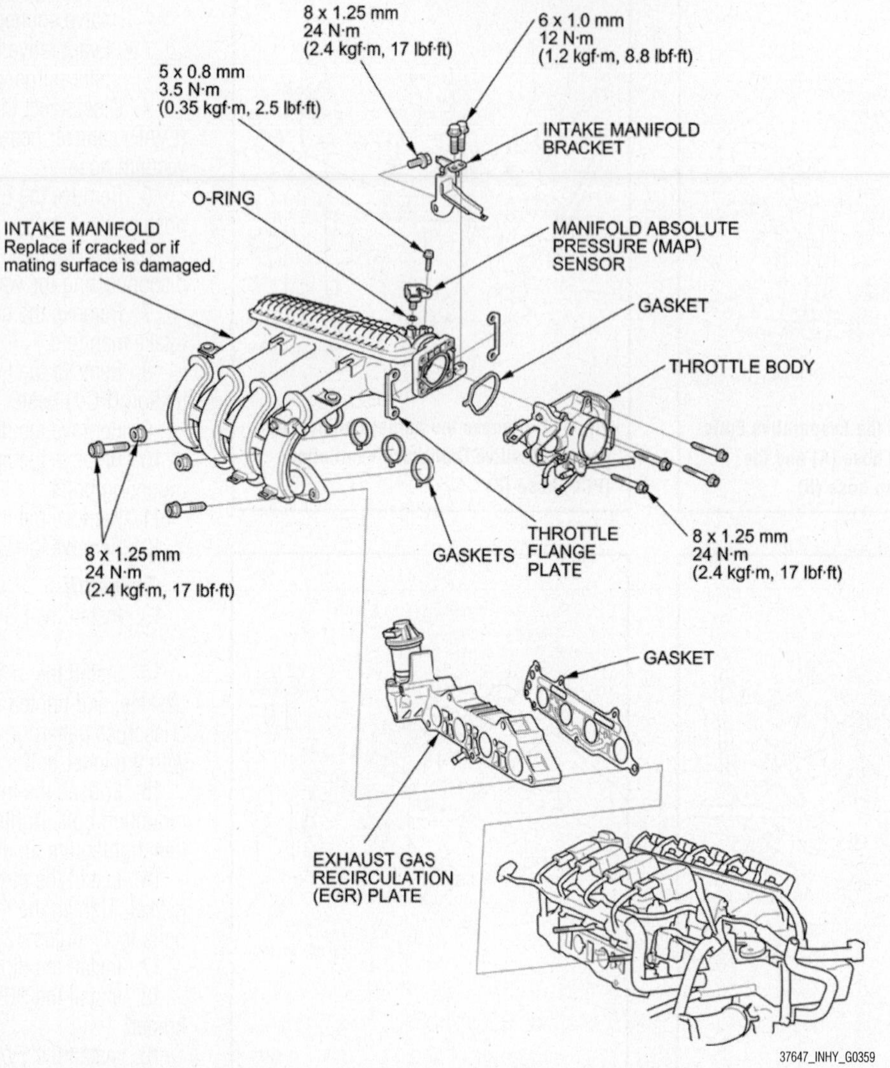

8 x 1.25 mm
24 N·m
(2.4 kgf·m, 17 lbf·ft)

6 x 1.0 mm
12 N·m
(1.2 kgf·m, 8.8 lbf·ft)

5 x 0.8 mm
3.5 N·m
(0.35 kgf·m, 2.5 lbf·ft)

INTAKE MANIFOLD
BRACKET

O-RING

INTAKE MANIFOLD
Replace if cracked or if
mating surface is damaged.

MANIFOLD ABSOLUTE
PRESSURE (MAP)
SENSOR

GASKET

THROTTLE BODY

8 x 1.25 mm
24 N·m
(2.4 kgf·m, 17 lbf·ft)

THROTTLE
FLANGE
PLATE

GASKETS

8 x 1.25 mm
24 N·m
(2.4 kgf·m, 17 lbf·ft)

GASKET

EXHAUST GAS
RECIRCULATION
(EGR) PLATE

37647_INHY_G0359

Fig. 117 Exploded view of the intake manifold assembly

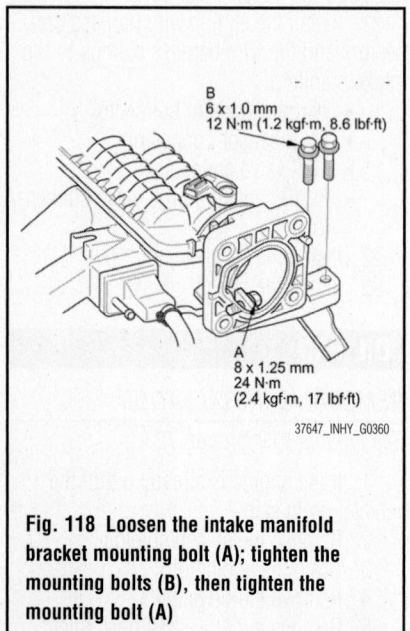

B
6 x 1.0 mm
12 N·m (1.2 kgf·m, 8.6 lbf·ft)

A
8 x 1.25 mm
24 N·m
(2.4 kgf·m, 17 lbf·ft)

37647_INHY_G0360

Fig. 118 Loosen the intake manifold bracket mounting bolt (A); tighten the mounting bolts (B), then tighten the mounting bolt (A)

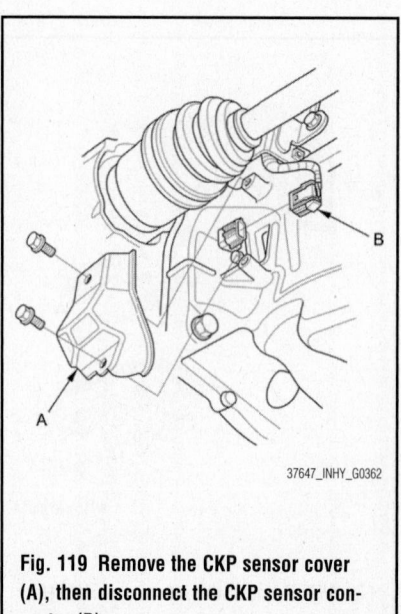

37647_INHY_G0362

Fig. 119 Remove the CKP sensor cover (A), then disconnect the CKP sensor connector (B)

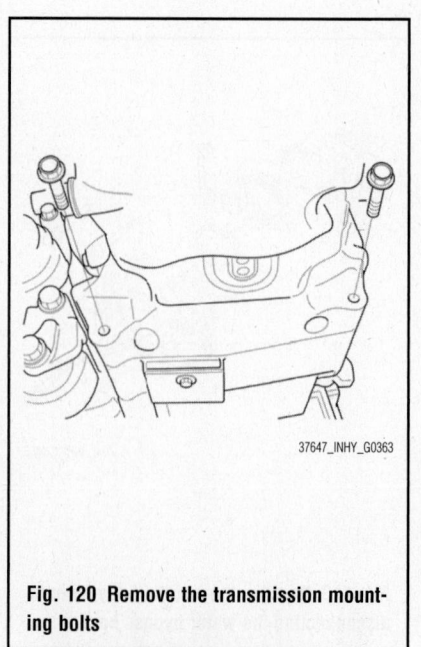

37647_INHY_G0363

Fig. 120 Remove the transmission mounting bolts

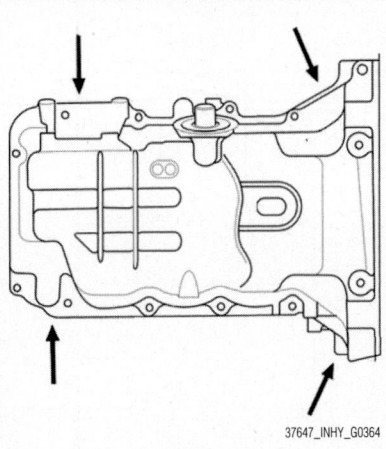

Fig. 121 Insert a flat blade screwdriver where shown, and separate the oil pan from the engine block

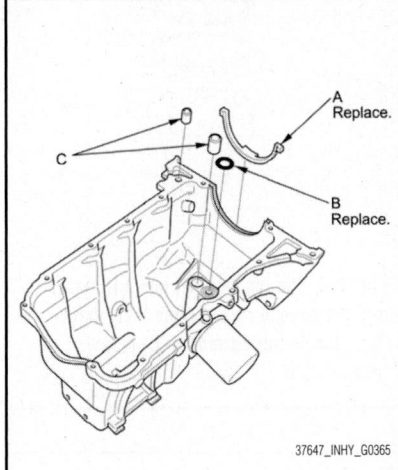

Fig. 122 Install the new oil pan gasket (A), the new O-ring (B), and the dowel pins (C) on the oil pan

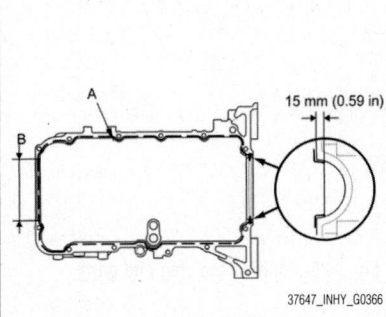

Fig. 123 Apply a bead of liquid gasket along the broken line (A); and an extra bead of liquid gasket to the shaded area (B)

6. Remove the A/C compressor without disconnecting the A/C hoses.

7. Remove the dipstick, then remove the dipstick tube.

8. Remove the Crankshaft Position (CKP) sensor cover (A), then disconnect the CKP sensor connector (B).

9. Remove the transmission mounting bolts.

10. Remove the oil pan bolts. Note the bolt locations by their size.

11. Insert a flat blade screwdriver where shown, and separate the oil pan from the engine block.

12. Remove the oil pan.

➡**Lower the oil pan carefully not to damage the IMA motor rotor position sensor.**

To install:

13. Remove all of the old liquid gasket from the oil pan mating surfaces, the bolts, and the bolt holes.

14. Clean and dry the oil pan mating surfaces and the O-ring groove.

15. Install the new oil pan gasket, the new O-ring, and the dowel pins on the oil pan.

16. Apply liquid gasket (P/N 08717-0004, 08718-0003, or 08718-0009) to the engine block mating surface of the oil pan and to the inside edge of the bolt holes. Install the component within 5 minutes of applying the liquid gasket.

    a. Apply a bead of liquid gasket along the broken line.

    b. Apply an extra bead of liquid gasket to the shaded area.

    c. If you apply liquid gasket P/N 08718-0012, the component must be installed within 4 minutes.

    d. If too much time has passed after applying the liquid gasket, remove the old liquid gasket and residue, then reapply new liquid gasket.

17. Install the oil pan.

    a. Raise the oil pan carefully not to damage IMA motor rotor position sensor.

    b. Wait at least 30 minutes before filling the engine with oil.

    c. Do not run the engine for at least 3 hours after installing the oil pan.

    d. Make sure to install the bolts in the correct locations according to size.

18. Tighten the bolts in three steps in a crisscross pattern starting in the middle of the oil pan and working toward each end. Wipe off the excess liquid gasket on the each side of crankshaft pulley and the drive plate.

19. Install the transmission mounting bolts and tighten them to 47 ft. lbs. (64 Nm).

20. Connect the CKP sensor connector, then install the CKP sensor cover.

21. Install the dipstick tube with new O-ring, then install the dipstick.

22. If the engine is still in the vehicle, do steps 11 through 15.

23. Install the A/C compressor.

24. Install the driveshaft heat shield.

25. Install the drive belt.

26. Install the splash shields.

27. Refill the engine with engine oil.

OIL PUMP

REMOVAL & INSTALLATION

See Figures 124 through 127.

Special Tools Required:
- Support Eyelet 07AAK-SNAA600
- Engine Support Hanger, A and Reds AAR-T1256

1. Remove the cam chain.

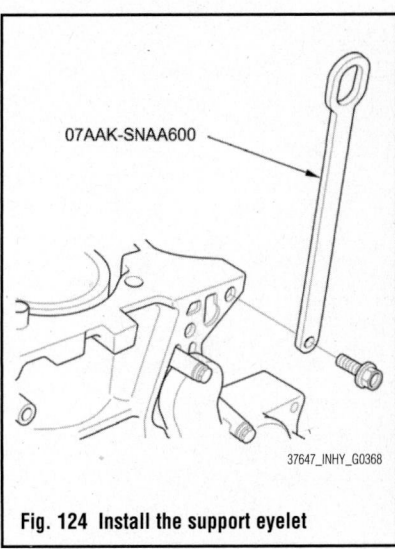

Fig. 124 Install the support eyelet

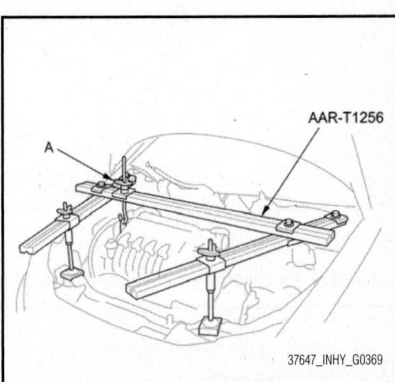

Fig. 125 Install the engine support hanger, then attach the hook to the slotted hole in the support eyelet; tighten the wing nut (A) by hand to lift and support the engine

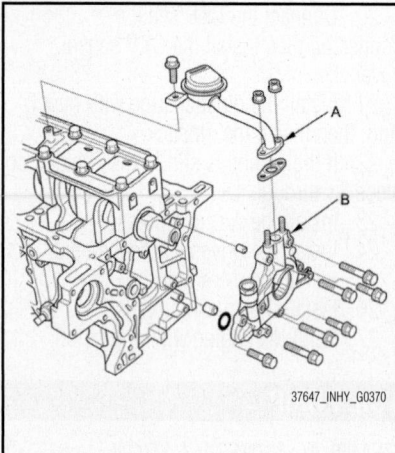

Fig. 126 Remove the oil screen (A), then remove the oil pump (B)

5. Remove the oil screen, then remove the oil pump.

6. Inspect both rotors and the pump housing for scoring or other damage. Replace parts, if necessary.

7. Check that the oil pump turns freely.

To install:

8. Clean the O-ring groove and the mating surface of the engine block.

9. Install the oil pump with a new O-ring.

10. Install the oil screen with a new gasket.

11. Install the oil pan.

12. Support the engine with a jack and a wood block under the oil pan.

13. Remove the engine hanger and support eyelet, then install the auto-tensioner.

14. Install the cam chain.

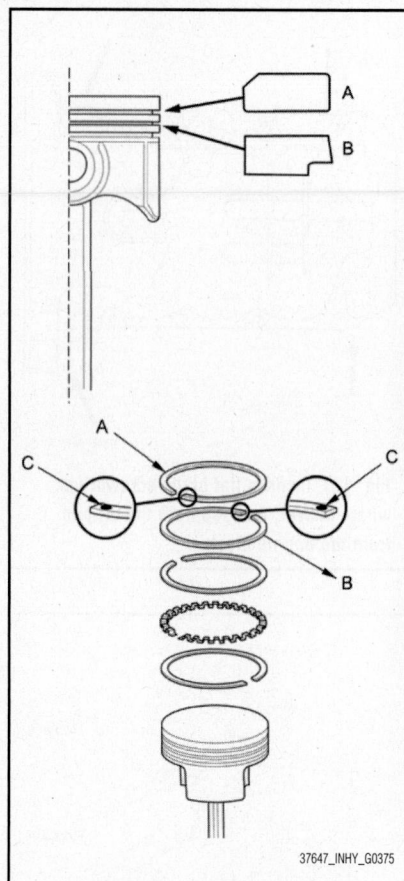

Fig. 128 The top ring (A) has a R mark, and the second ring (B) has a 2R mark. The manufacturing marks (C) must face upward

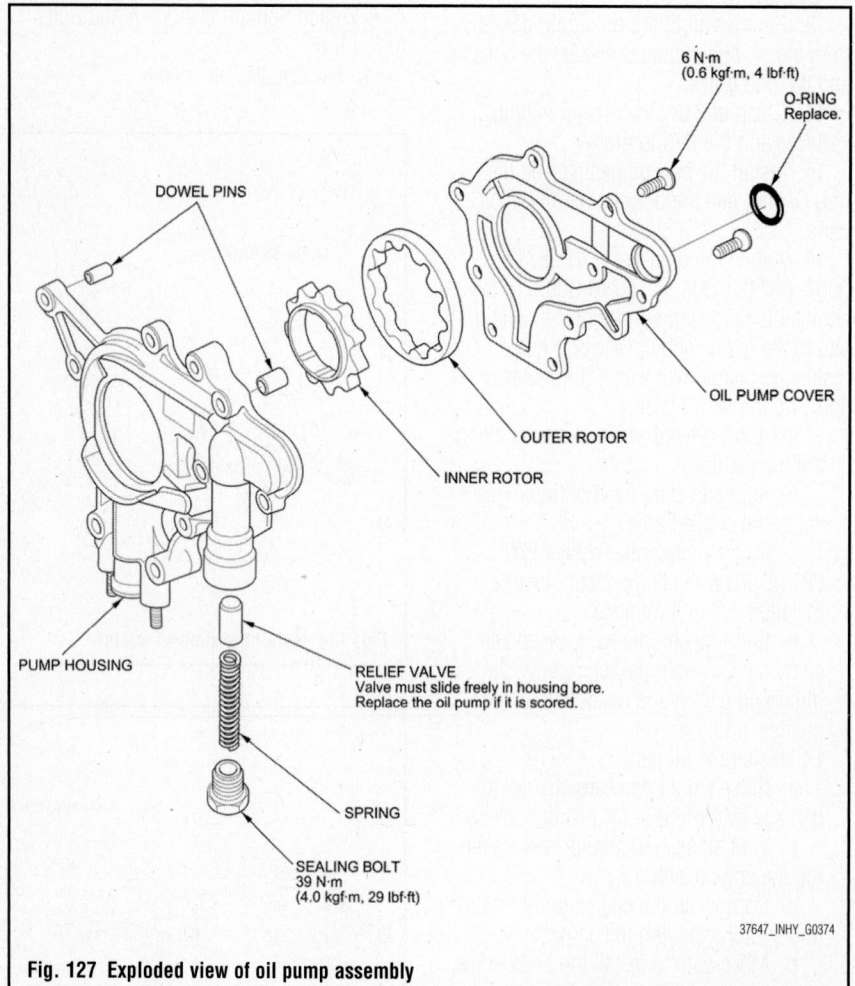

Fig. 127 Exploded view of oil pump assembly

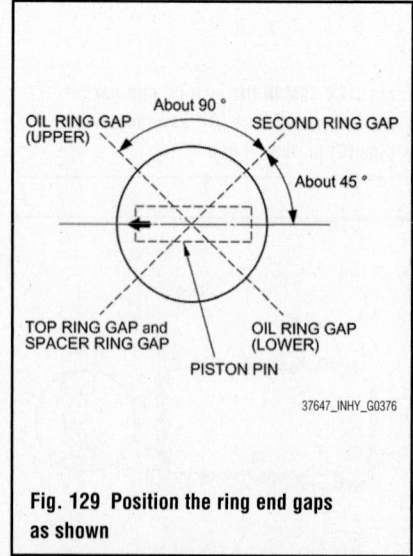

Fig. 129 Position the ring end gaps as shown

2. Remove the auto-tensioner, then install the support eyelet.

3. Install the engine support hanger (AAR-T1256), then attach the hook to the slotted hole in the support eyelet. Tighten the wing nut by hand to lift and support the engine.

4. Remove the oil pan.

PISTONS & RINGS

POSITIONING

See Figures 128 and 129.

1. Install the rings as shown. The top ring has a R mark, and the second ring has a 2R mark. The manufacturing marks must face upward.

2. Rotate the rings in their grooves to make sure they do not bind.

3. Position the ring end gaps as shown.

ROCKER ARMS

REMOVAL & INSTALLATION

See Figures 130 through 134.

1. Remove the camshaft sprocket.

a. Make a reference mark across the camshaft sprocket and cam chain.

b. Apply new engine oil to the slider surface of the cam chain tensioner slider through the oil return hole in the cylinder head.

c. Remove the cylinder head plug.

d. Hold the crankshaft pulley, and set the socket wrench on the camshaft sprocket bolt.

e. Remove the maintenance bolt, and turn the camshaft clockwise to compress the cam chain tensioner, then install the 6 x 1.0 mm bolt in the bolt hole in the engine block through the maintenance hole and cam chain tensioner.

➡ **Turning torque should not exceed 33 ft. lbs. (44 Nm), when turning the camshaft. Do not turn the camshaft counterclockwise.**

f. Hold the camshaft with a 27 mm open-end wrench, then remove the camshaft sprocket.

➡ **Hang the cam chain with a wire.**

2. Remove the rocker arm assembly.

a. Loosen the rocker arm adjusting screws.

b. Remove the camshaft holder bolts. To prevent damaging the camshaft, loosen the bolts in sequence two turns at a time, in a crisscross pattern.

c. Remove the rocker arm assembly.

d. Identify each part as it is removed so that each item can be reinstalled in their original location.

e. Remove the rocker shaft bolts before disassembling the rocker arms.

To install:

3. Install the rocker arm assembly.

a. Reassemble the rocker arm assembly.

b. If reused, the rocker arms must be installed in their original location.

c. Prior to reassembling, clean all the parts in solvent, dry them, and apply new engine oil to any contact points.

d. Apply new engine oil to the threads of the rocker shaft bolts when installing them.

e. When replacing the rocker arm assembly, remove the fastening hardware from the new rocker arm assembly.

4. Install the camshaft sprocket.

➡ **Keep the cam chain away from magnetic fields.**

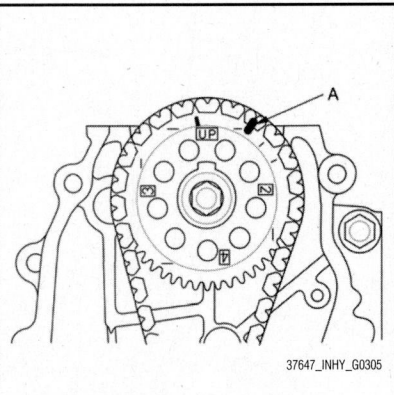

37647_INHY_G0305

Fig. 130 Make a reference mark (A) across the camshaft sprocket and cam chain

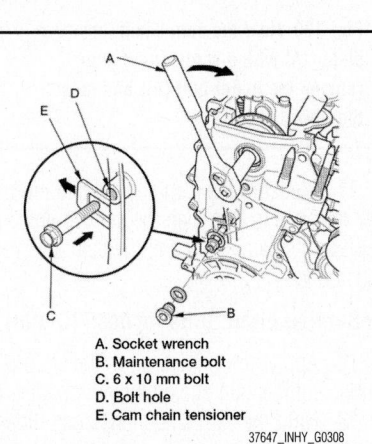

A. Socket wrench
B. Maintenance bolt
C. 6 x 10 mm bolt
D. Bolt hole
E. Cam chain tensioner

37647_INHY_G0308

Fig. 131 Hold the crankshaft pulley, and set the socket wrench on the camshaft sprocket bolt

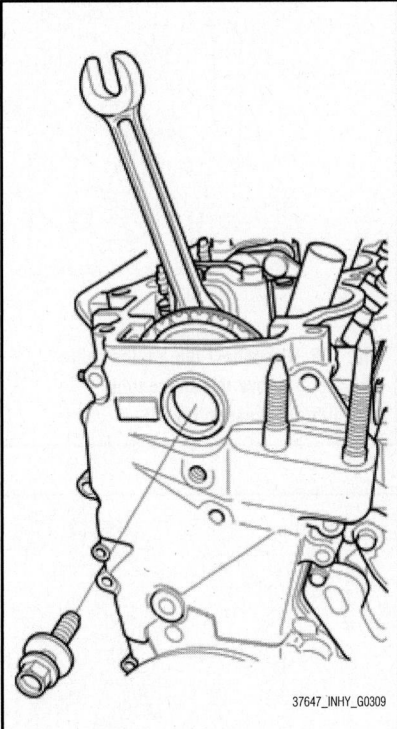

37647_INHY_G0309

Fig. 132 Hold the camshaft with a 27 mm open-end wrench, then remove the camshaft sprocket

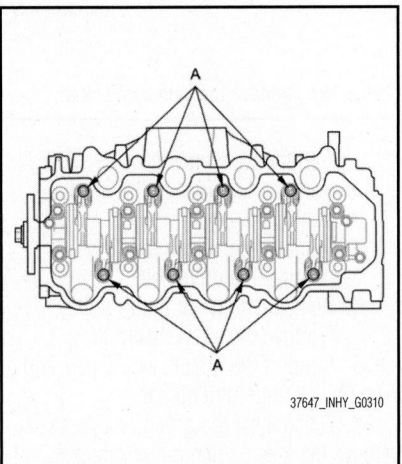

37647_INHY_G0310

Fig. 133 Loosen the rocker arm adjusting screws (A)

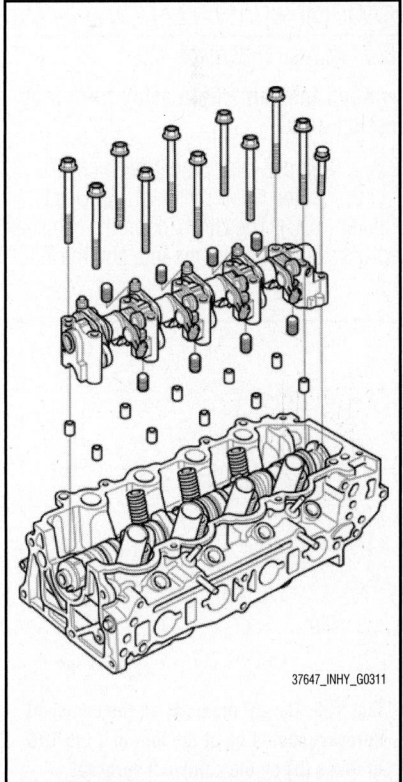

37647_INHY_G0311

Fig. 134 Remove the rocker arm assembly

a. Install the cam chain around the camshaft sprocket by alignment the reference mark, then install the camshaft sprocket on the camshaft.

b. Hold the camshaft with a 27 mm open-end wrench, then tighten the bolt to 41 ft. lbs. (56 Nm).

➡**Apply new engine oil to the bolt threads and flange.**

c. Apply new engine oil to the slider surface of the cam chain tensioner slider through the oil return hole in the cylinder head.

d. Hold the crankshaft pulley, and set the socket wrench on the camshaft sprocket bolt.

e. Turn the camshaft clockwise to compress the cam chain tensioner, then remove the 6 x 1.0 mm bolt.

➡**Turning torque should not exceed 33 ft. lbs. (44 Nm), when turning the camshaft. Do not turn the camshaft counterclockwise.**

f. Install the maintenance bolt with a new washer.

g. Install the new cylinder head plug.

TIMING (CAM) CHAIN & SPROCKETS

REMOVAL & INSTALLATION

See Figures 135 through 152.

➡**Keep the cam chain away from magnetic fields.**

1. Remove the cylinder head cover.
2. Set the No. 1 piston at Top Dead Center (TDC). The UP mark on the camshaft sprocket should be at the top, and the TDC

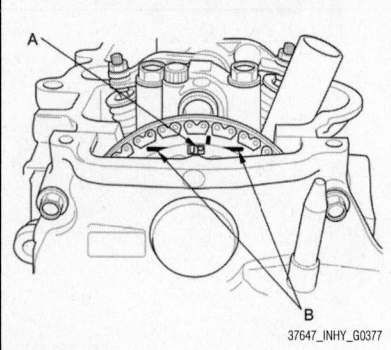

Fig. 135 The UP mark (A) on the camshaft sprocket should be at the top, and the TDC grooves (B) on the camshaft sprocket should line up with the top edge of the head

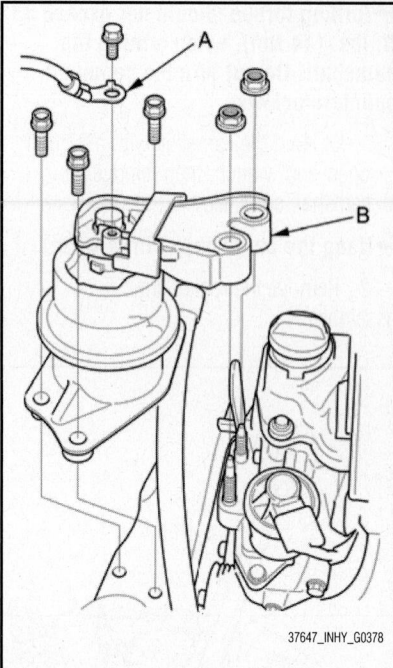

Fig. 136 Disconnect the ground cable (A), then remove the side engine mount/bracket assembly (B)

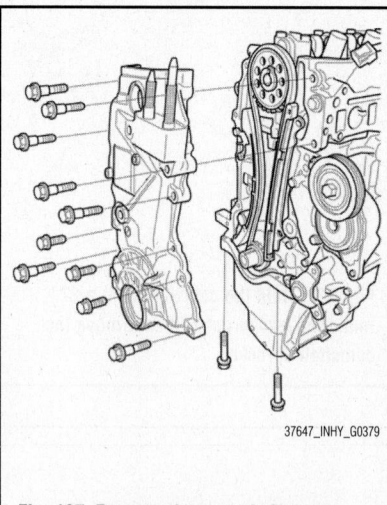

Fig. 137 Remove the cam chain case

grooves on the camshaft sprocket should line up with the top edge of the head.

3. Remove the front wheels.
4. Remove the splash shields.
5. Remove the drive belt.
6. Remove the water pump pulley.
7. Remove the crankshaft pulley.
8. Support the engine with a jack and a wood block under the oil pan.
9. Disconnect the ground cable, then remove the side engine mount/bracket assembly.
10. Remove the cam chain case.

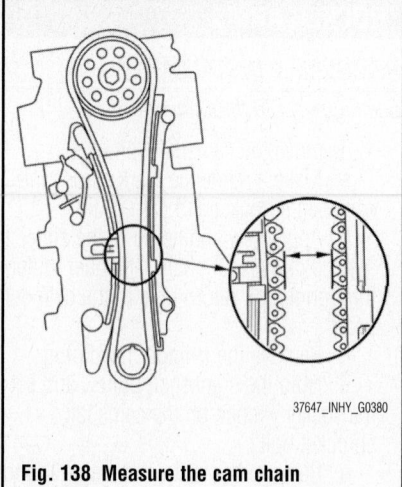

Fig. 138 Measure the cam chain separation

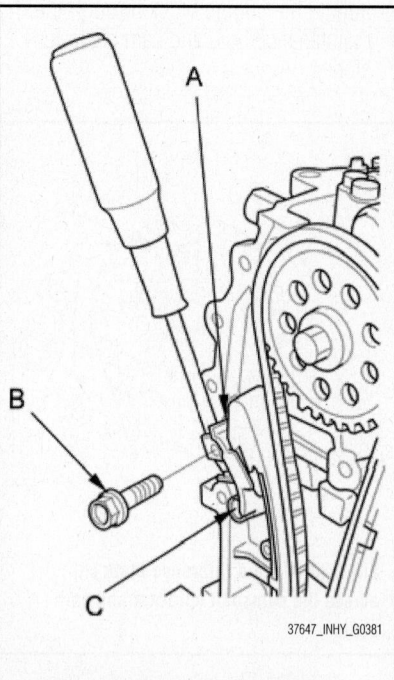

Fig. 139 Hold the cam chain tensioner slider (A) with a screwdriver, then remove the upper bolt (B), and loosen the lower bolt (C)

11. Measure the cam chain separation. If the distance is less than the service limit, replace the cam chain and the cam chain tensioner.

➡**Service Limit: 0.59 inches (15 mm)**

12. Apply new engine oil to the sliding surface of the cam chain tensioner slider.
13. Hold the cam chain tensioner slider with a screwdriver, then remove the upper bolt, and loosen the lower bolt.
14. Remove the cam chain tensioner slider.

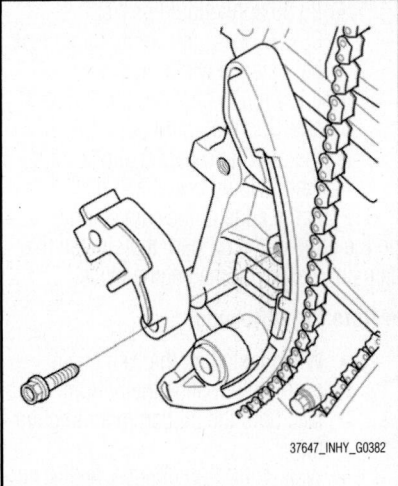

Fig. 140 Remove the cam chain tensioner slider

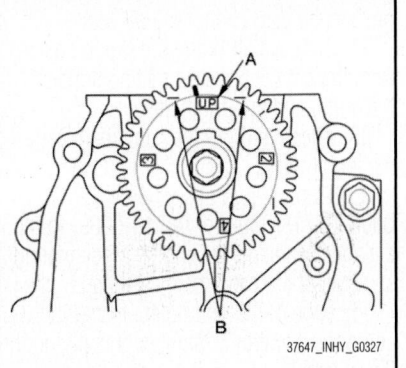

Fig. 143 The UP mark (A) on the camshaft sprocket should be at the top, and the TDC grooves (B) on the camshaft sprocket should line up with the top edge of the cylinder head

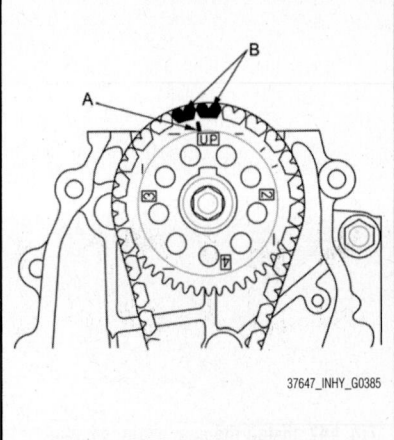

Fig. 145 Install the cam chain on the camshaft sprocket with the pointer (A) centered between the two colored pieces (B)

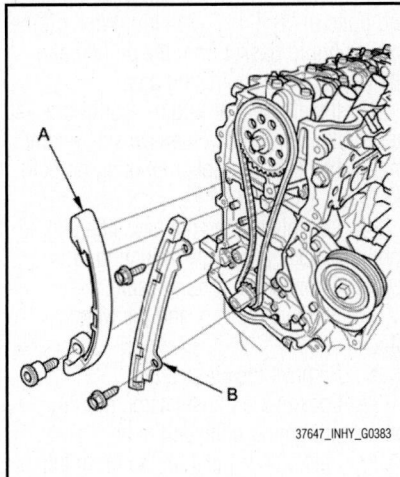

Fig. 141 Remove the cam chain tensioner (A) and the cam chain guide (B)

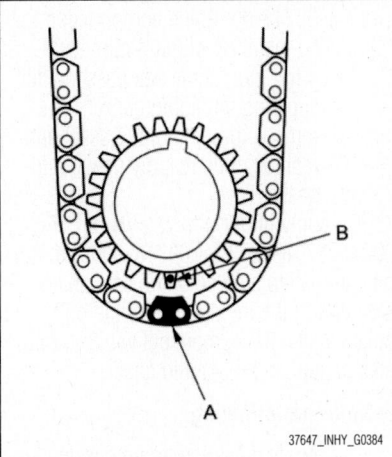

Fig. 144 Install the cam chain on the crankshaft sprocket with the colored piece (A) aligned with the TDC mark (B) on the crankshaft sprocket

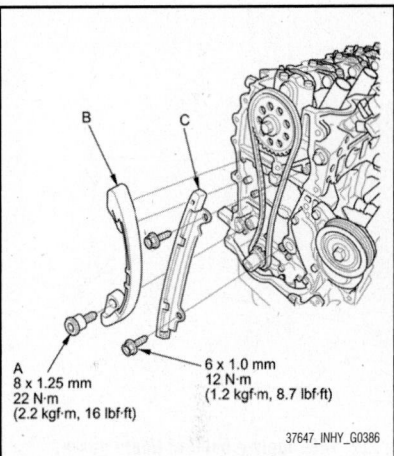

Fig. 146 Apply new engine oil to the threads of the cam chain tensioner mounting bolt (A); install the cam chain tensioner (B) and the cam chain guide (C)

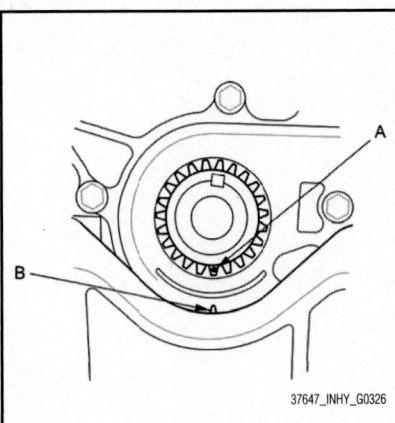

Fig. 142 Align the TDC mark (A) on the crankshaft sprocket with the pointer (B) on the oil pump

15. Remove the cam chain tensioner and the cam chain guide. Inspect the tensioner and the guide, and replace them if needed.

16. Remove the cam chain.

To install:

➡**Keep the cam chain away from magnetic fields.**

17. Set the crankshaft to Top Dead Center (TDC). Align the TDC mark on the crankshaft sprocket with the pointer on the oil pump.

18. Set the No. 1 piston at TDC. The UP mark on the camshaft sprocket should be at the top, and the TDC grooves on the camshaft sprocket should line up with the top edge of the cylinder head.

19. Install the cam chain on the crank-

shaft sprocket with the colored piece aligned with the TDC mark on the crankshaft sprocket.

20. Install the cam chain on the camshaft sprocket with the pointer centered between the two colored pieces.

21. Apply new engine oil to the threads of the cam chain tensioner mounting bolt.

22. Install the cam chain tensioner and the cam chain guide.

23. Install the cam chain tensioner slider, and loosely tighten the lower side bolt.

24. Apply new engine oil to the sliding surface of the cam chain tensioner slider.

25. Rotate the cam chain tensioner slider clockwise to compress the cam chain tensioner, and install the remaining bolt, then tighten the two bolts.

26. Check the cam chain case oil seal

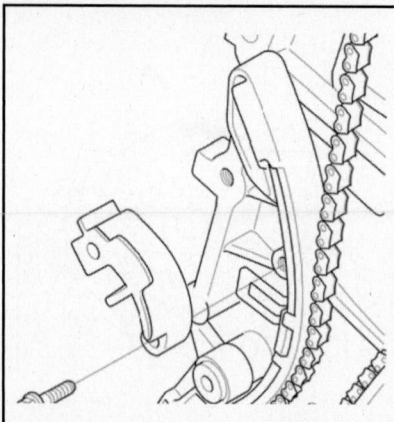

Fig. 147 Install the cam chain tensioner slider, and loosely tighten the lower side bolt

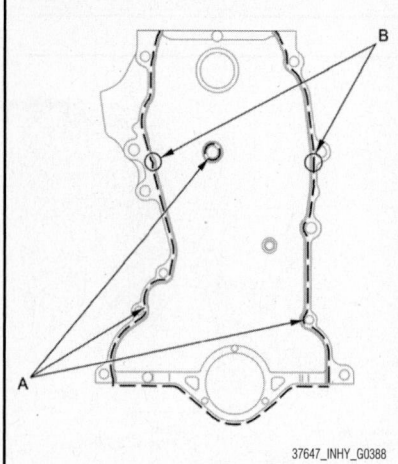

Fig. 148 Apply a bead of liquid gasket along the broken line (A) and to the upper surface contact areas of the engine block (B)

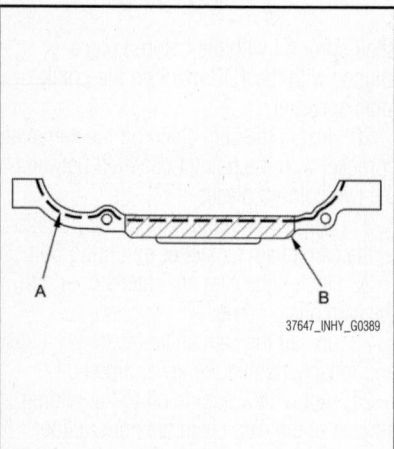

Fig. 149 Apply a bead of liquid gasket along the broken line (A); and an additional bead to the shaded area (B)

for damage. If the oil seal is damaged, replace the cam chain case oil seal.

27. Remove all of the old liquid gasket from the cam chain case mating surfaces, the bolts, and the bolt holes.

28. Clean and dry the cam chain case mating surfaces.

29. Apply liquid gasket (P/N 08717-0004, 08718-0003, or 08718-0009) to the cylinder head and the engine block mating surfaces of the cam chain case and to the inside edge of the bolt holes. Install the component within 5 minutes of applying the liquid gasket.

➡Note the following:

- Apply a bead of liquid gasket along the broken line.
- Apply a bead of liquid gasket to the upper surface contact areas of the engine block.
- If you apply liquid gasket P/N 08718-0012, the component must be installed within 4 minutes.
- If too much time has passed after applying the liquid gasket, remove the old liquid gasket and residue, then reapply new liquid gasket.

30. Apply liquid gasket (P/N 08717-0004, 08718-0003, or 08718-0009) to the oil pan mating surface of the cam chain case and to the inside edge of the bolt holes. Install the component within 5 minutes of applying the liquid gasket.

➡Note the following:

- Apply a bead of liquid gasket along the broken line.
- Apply an additional bead of liquid gasket to the shaded area.

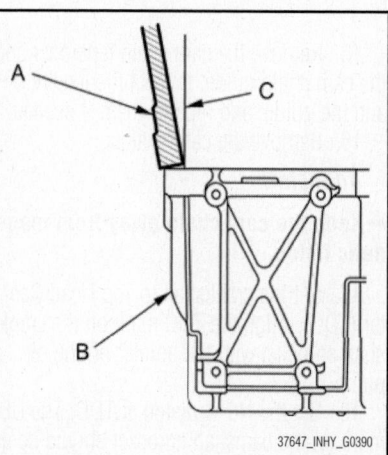

Fig. 150 Set the edge of the cam chain case (A) on the edge of the oil pan (B), then install the cam chain case on the engine block (C)

- If you apply liquid gasket P/N 08718-0012, the component must be installed within 4 minutes.
- If too much time has passed after applying the liquid gasket, remove the old liquid gasket and residue, then reapply new liquid gasket.

31. Set the edge of the cam chain case on the edge of the oil pan, then install the cam chain case on the engine block.

➡Note the following:

- When installing the cam chain case, do not slide the bottom surface onto the oil pan mounting surface.
- Wait at least 30 minutes before filling the engine with oil.
- Do not run the engine for at least 3 hours after installing the cam chain case.

32. Tighten the cam chain case mounting bolts to 23 ft. lbs. (31 Nm). Wipe off the excess liquid gasket from the oil pan and the cam chain case mating area.

33. Install the side engine mount/bracket assembly, then tighten the new side engine mount/bracket assembly mounting bolts to 43 ft. lbs. (59 Nm).

34. Loosely tighten the new side engine mount/bracket assembly mounting nuts.

35. Connect the ground cable.

36. Remove the jack and the wood block.

37. Remove the air cleaner.

38. Loosen the transmission mount bracket mounting bolts and nuts.

39. Raise the vehicle on the lift to full height.

40. Loosen the torque rod mounting bolt and nut.

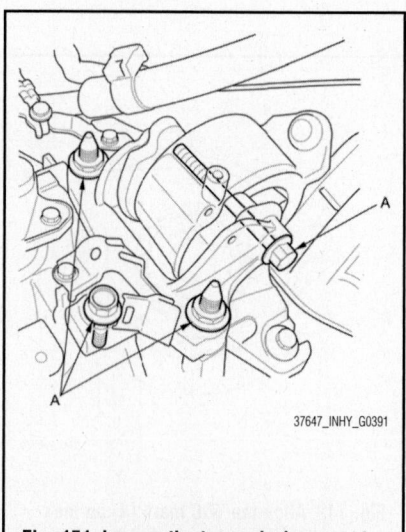

Fig. 151 Loosen the transmission mount bracket mounting bolts and nuts (A)

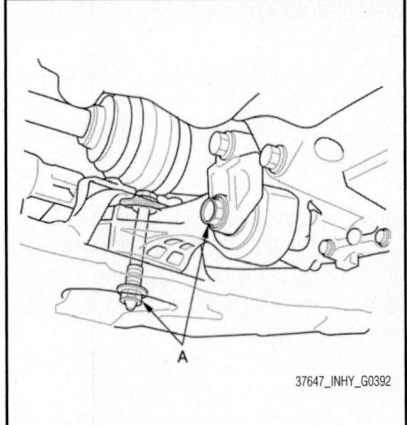

Fig. 152 Loosen the torque rod mounting bolt and nut (A)

41. Lower the vehicle on the lift.
42. Tighten the side engine mount/bracket assembly mounting nuts to 36 ft. lbs. (49 Nm).
43. Tighten the transmission mount mounting bolts and nuts. Tighten the 10 x 1.25 mm through bolt to 40 ft. lbs. (54 Nm), then tighten the 12 x 1.25 mm bolt and nuts to 54 ft. lbs. (74 Nm).
44. Raise the vehicle on the lift to full height.
45. Tighten the torque rod mounting bolt to 61 ft. lbs. (83 Nm),and then tighten the nut to 69 ft. lbs. (93 Nm).
46. Lower the vehicle on the lift.
47. Install the air cleaner.
48. Install the cylinder head cover.
49. Install the crankshaft pulley.
50. Install the water pump pulley.
51. Install the drive belt.
52. Install the splash shields.
53. Install the front wheel.
54. Do the Crankshaft Position (CKP) pattern clear/CKP pattern learn procedure.

VALVE LASH (VALVE CLEARANCE)

ADJUSTMENT

See Figures 153 through 158.

➡**Adjust the valves only when the cylinder head temperature is less than 100 °F (38 °C). Check the engine coolant temperature with the HDS if you are not sure.**

1. Remove the cylinder head cover.
2. Set the No. 1 piston at top dead center (TDC). The UP mark on the camshaft sprocket should be at the top, and the TDC grooves on the camshaft sprocket should

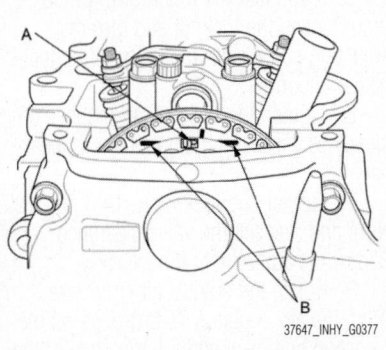

Fig. 153 The UP mark (A) on the camshaft sprocket should be at the top, and the TDC grooves (B) on the camshaft sprocket should line up with the top edge of the cam chain case

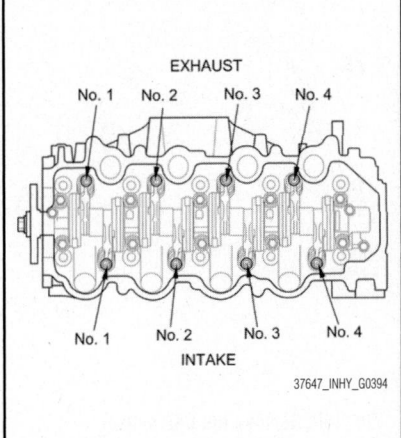

Fig. 154 Select the correct feeler gauge for the valves you are going to check

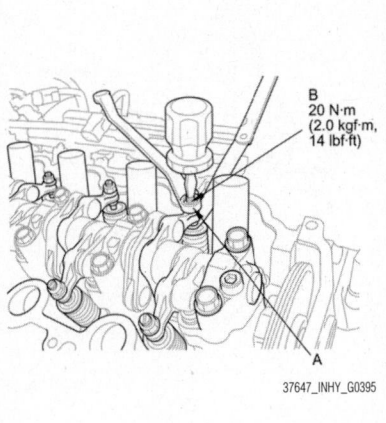

Fig. 155 Loosen the locknut (A), and turn the adjusting screw (B) until the drag on the feeler gauge is correct

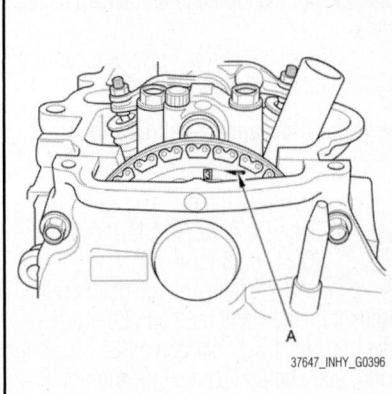

Fig. 156 Align the No. 3 piston TDC groove (A) on the camshaft sprocket with the top edge of the cam chain case

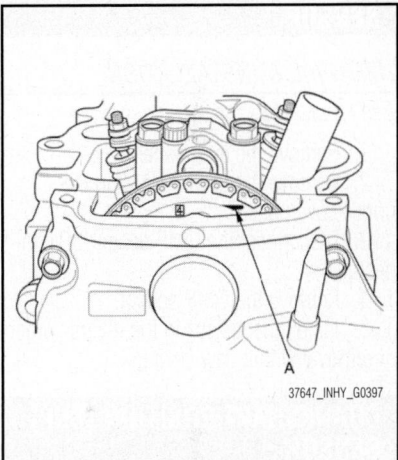

Fig. 157 Align the No. 4 piston TDC groove (A) on the camshaft sprocket with the top edge of the cam chain case

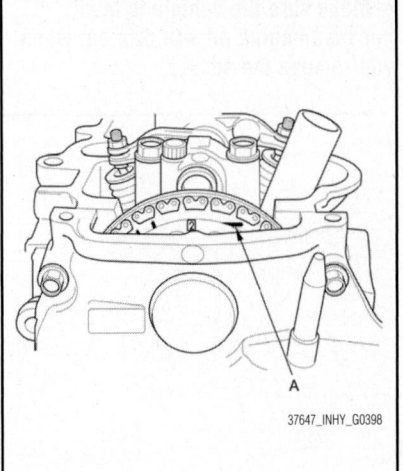

Fig. 158 Align the No. 2 piston TDC groove (A) on the camshaft sprocket with the top edge of the cam chain case

line up with the top edge of the cam chain case.

3. Select the correct feeler gauge for the valves you are going to check.

➥Valve Clearance Specifications:

- Intake: 0.006–0.007 inches (0.15–0.19 mm)
- Exhaust: 0.009–0.011 inches (0.24–0.28 mm)

4. Insert the feeler gauge between the adjusting screw and the end of the valve stem on No. 1 cylinder and slide it back and forth; you should feel a slight amount of drag.

5. If you feel too much or too little drag, loosen the locknut and turn the adjusting screw until the drag on the feeler gauge is correct.

6. Tighten the locknut, and recheck the clearance. Repeat the adjustment if necessary.

7. Tighten the locknut to 14 ft. lbs. (20 Nm), and recheck the valve clearance. Repeat the adjustment if necessary.

8. Rotate the crankshaft clockwise. Align the No. 3 piston TDC groove on the camshaft sprocket with the top edge of the cam chain case.

9. Check and, if necessary, adjust the valve clearance on the No. 3 cylinder.

10. Rotate the crankshaft clockwise. Align the No. 4 piston TDC groove on the camshaft sprocket with the top edge of the cam chain case.

11. Check and, if necessary, adjust the valve clearance on the No. 4 cylinder.

12. Rotate the crankshaft clockwise. Align the No. 2 piston TDC groove on the camshaft sprocket with the top edge of the cam chain case.

13. Check and, if necessary, adjust the valve clearance on the No. 2 cylinder.

14. Install the cylinder head cover.

ENGINE PERFORMANCE & EMISSION CONTROLS

CAMSHAFT POSITION (CMP) SENSOR

REMOVAL & INSTALLATION

See Figure 159.

1. Remove the air cleaner.
2. Remove the cowl cover and the under-cowl panel.
3. Disconnect the CMP sensor 3P connector.
4. Remove the CMP sensor.
5. Install the sensor in the reverse order of removal with a new O-ring.

CRANKSHAFT POSITION (CKP) SENSOR

REMOVAL & INSTALLATION

See Figures 160 and 161.

1. Raise the vehicle on a lift.

➥Make sure the vehicle is level, because engine oil will drip out when you remove the sensor.

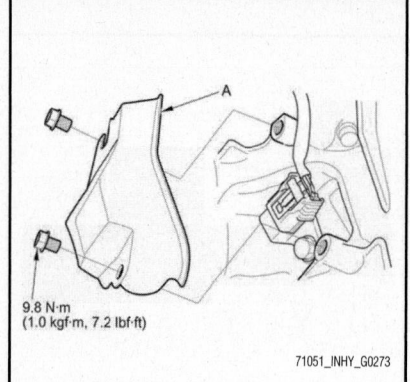

9.8 N·m
(1.0 kgf·m, 7.2 lbf·ft)

71051_INHY_G0273

Fig. 160 Remove the CKP sensor cover (A).

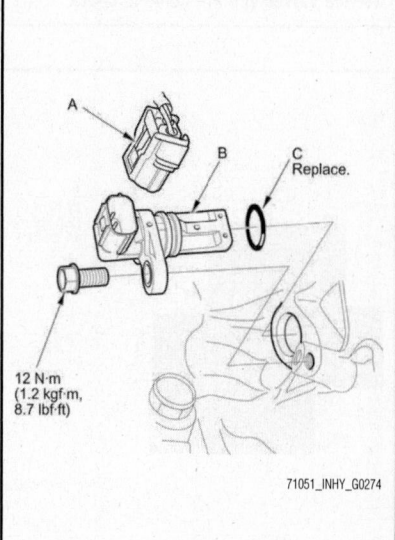

12 N·m
(1.2 kgf·m, 8.7 lbf·ft)

71051_INHY_G0274

Fig. 161 Disconnect the CKP sensor connector (A). Remove the CKP sensor (B) and a new O-ring (C).

2. Remove the engine undercover.
3. Remove the CKP sensor cover.
4. Disconnect the CKP sensor connector.
5. Remove the CKP sensor.
6. Install the parts in the reverse order of removal with a new O-ring.
7. Do the CKP pattern clear/CKP pattern learn procedure.
8. Check the engine oil level, and add more oil if needed.

CKP PATTERN CLEAR/CKP PATTERN LEARN PROCEDURE

Clear/Learn Procedure (with the HDS)

1. Connect the HDS to the Data Link Connector (DLC) located under the driver's side of the dashboard.
2. Turn the ignition switch to ON (II).
3. Make sure the HDS communicates with the PCM and all other vehicle systems. If it doesn't, go to the DLC circuit troubleshooting.
4. Select CRANK PATTERN in the ADJUSTMENT MENU with the HDS.
5. Select CRANK (CKP) PATTERN CLEAR, and clear the CKP pattern.
6. Select CRANK PATTERN LEARNING with the HDS, and follow the screen prompts.

Learn Procedure (without the HDS)

1. Start the engine. Hold the engine speed at 3,000 rpm without load (in P or N) until the radiator fan comes on.
2. Test-drive the vehicle on a level road: Decelerate (with the throttle fully closed) from an engine speed of 2,500 rpm down to 1,000 rpm with the transmission in L.
3. Repeat step 2 several times.
4. Turn the ignition switch to LOCK (0).
5. Turn the ignition switch to ON (II), and wait 30 seconds.

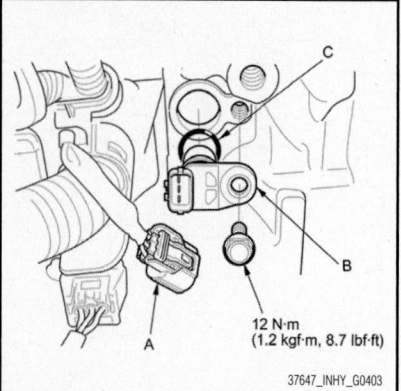

12 N·m
(1.2 kgf·m, 8.7 lbf·ft)

37647_INHY_G0403

Fig. 159 Disconnect the CMP sensor 3P connector (A); remove the CMP sensor (B) and the O-ring (C)

ENGINE COOLANT TEMPERATURE (ECT) SENSOR

LOCATION

ECT sensor 1 is located near the thermostat housing. ECT sensor 2 is located on the bottom tank of the radiator.

REMOVAL & INSTALLATION

ECT Sensor 1

See Figure 162.

1. Drain the engine coolant.
2. Remove the air cleaner.
3. Disconnect the ECT sensor 1 connector (A).
4. Remove ECT sensor 1.
5. Install the parts in the reverse order of removal with a new O-ring, then refill the radiator with engine coolant.

ECT Sensor 2

See Figure 163.

1. Drain the engine coolant.
2. Raise the vehicle on a lift.
3. Disconnect the ECT sensor 2 connector, then remove ECT sensor 2.
4. Install the parts in the reverse order of removal with a new O-ring, then refill the radiator with engine coolant.

EVAPORATIVE EMISSIONS (EVAP) CANISTER

REMOVAL & INSTALLATION

See Figures 164 through 166.

1. Raise the vehicle on a lift.
2. Disconnect the hoses from the EVAP canister filter.
3. Remove the bolts, then remove the canister guard.

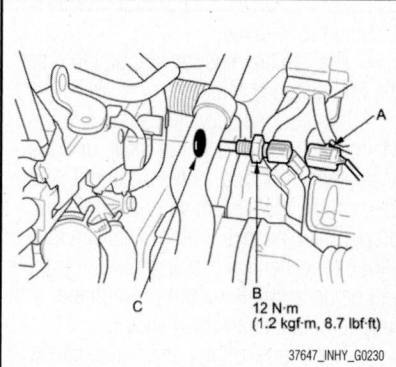

Fig. 162 Disconnect the ECT sensor 1 connector (A); remove ECT sensor 1 (B) and O-ring (C)

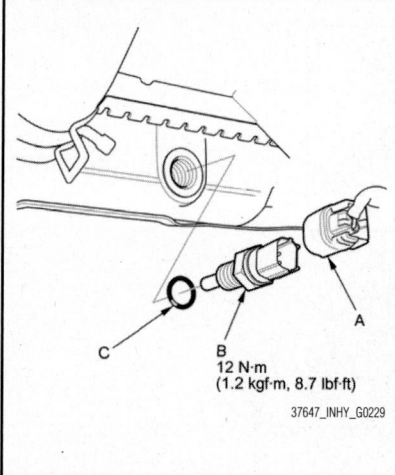

Fig. 163 Disconnect the ECT sensor 2 connector (A), then remove ECT sensor 2 (B) and O-ring (C)

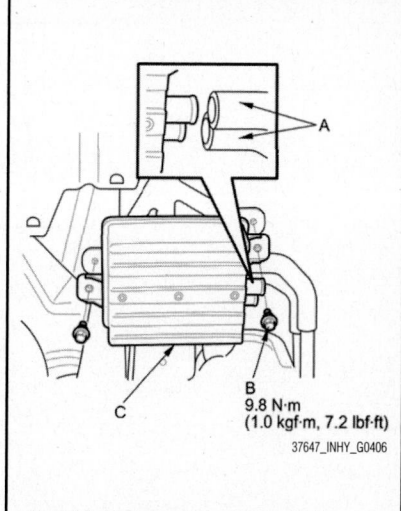

Fig. 164 Disconnect the hoses (A) from the EVAP canister filter (C)

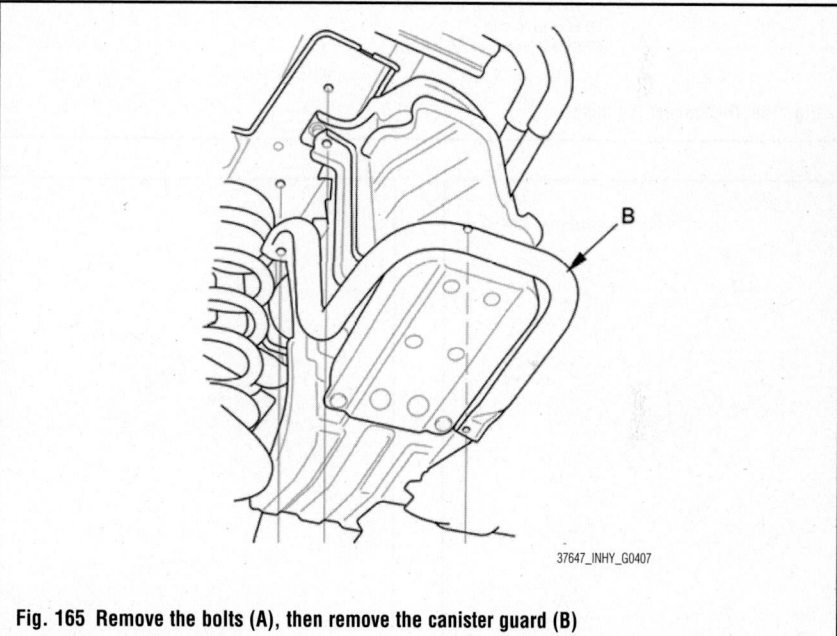

Fig. 165 Remove the bolts (A), then remove the canister guard (B)

4. Disconnect the hoses.
5. Disconnect the FTP sensor 3P connector and the EVAP canister vent shut valve 2P connector, then remove the harness clip.
6. Remove the bolts, then remove the EVAP canister assembly.
7. Install the parts in the reverse order of removal.

EXHAUST GAS RECIRCULATION (EGR) VALVE

REMOVAL & INSTALLATION

See Figure 167.

1. Disconnect the EGR valve 5P connector.

2. Remove the EGR valve.
3. Install the parts in the reverse order of removal with a new gasket.

HEATED OXYGEN SENSOR (HO2S)

REMOVAL & INSTALLATION

See Figures 168 and 169.

Special Tools Required: O2 Sensor Wrench, Snap-on® S6176 or equivalent, commercially available

1. Remove the cowl cover and the under-cowl panel.
2. Disconnect the secondary HO2S connector.
3. Raise the vehicle on a lift.

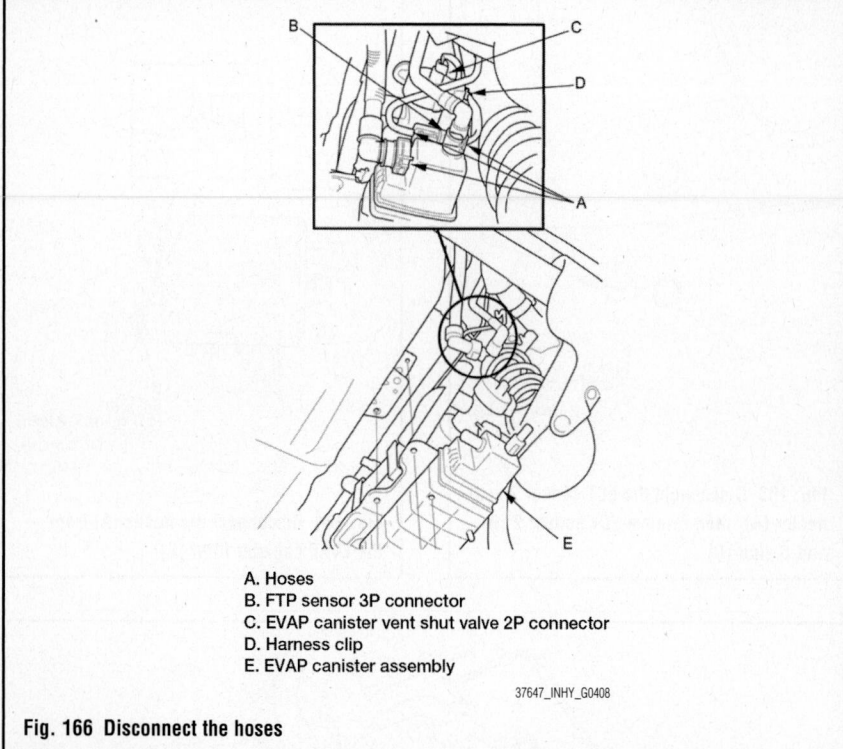

A. Hoses
B. FTP sensor 3P connector
C. EVAP canister vent shut valve 2P connector
D. Harness clip
E. EVAP canister assembly

37647_INHY_G0408

Fig. 166 Disconnect the hoses

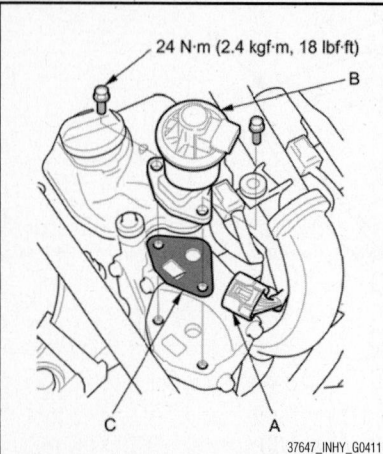

37647_INHY_G0411

Fig. 167 Disconnect the EGR valve 5P connector (A); Remove the EGR valve (B) and the gasket (C)

4. Remove the wire clips, then remove the secondary HO2S.

5. Install the parts in the reverse order of removal.

INTAKE AIR TEMPERATURE (IAT) SENSOR

LOCATION

The Intake Air Temperature (IAT) sensor is an integral part of the Mass Air Flow (MAF)/Intake Air Temperature (IAT) sensor located on the engine air intake duct.

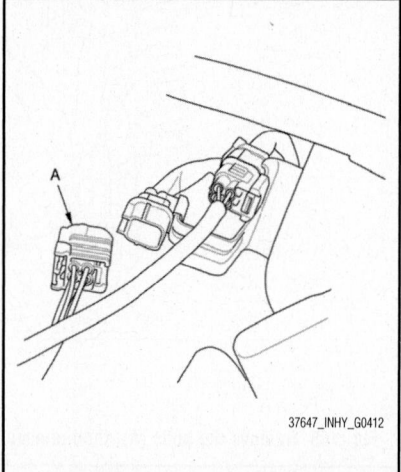

37647_INHY_G0412

Fig. 168 Disconnect the secondary HO2S connector (A)

REMOVAL & INSTALLATION

Refer to the Mass Air Flow (MAF)/Intake Air Temperature (IAT) sensor section when servicing this component.

KNOCK SENSOR (KS)

REMOVAL & INSTALLATION

See Figure 170.

1. Remove the engine oil dipstick.
2. Disconnect the knock sensor 1P connector.
3. Remove the knock sensor.

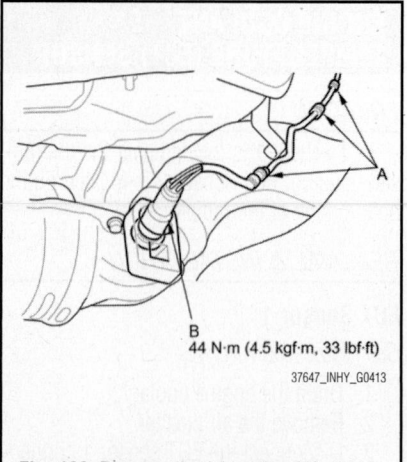

44 N·m (4.5 kgf·m, 33 lbf·ft)

37647_INHY_G0413

Fig. 169 Remove the wire clips (A), then remove the secondary HO2S (B)

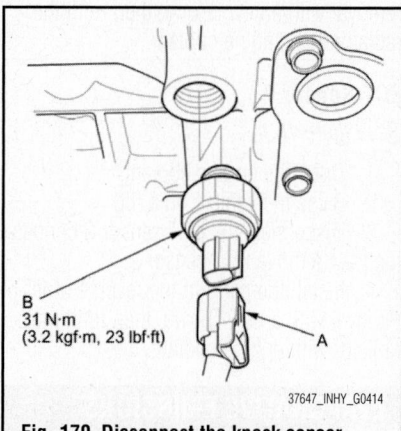

B
31 N·m
(3.2 kgf·m, 23 lbf·ft)

37647_INHY_G0414

Fig. 170 Disconnect the knock sensor 1P connector (A); remove the knock sensor (B)

4. Install the parts in the reverse order of removal.

MAINTENANCE MINDER

RESETTING THE MAINTENANCE MINDER

Note the following:
• The vehicle must be stopped to reset the Maintenance Minder.

• If a required service is done and the Maintenance Minder is not reset, or if the Maintenance Minder is reset without doing the service, the system will not show the proper maintenance timing. This can lead to serious mechanical problems because there will be no accurate record of when the required maintenance is needed.

• The engine oil life and maintenance item(s) can be independently reset with the HDS.

1. Turn the ignition switch to ON (II).
2. If system message(s) are displayed, press the INFO button to cancel the display.

3. Push the SEL/RESET button repeatedly until the engine oil life indicator is displayed.

4. Press and hold the SEL/RESET button for about 10 seconds, the "OIL LIFE RESET" mode display appears.

Note the following:

• If you are resetting the Maintenance Minder when the engine oil life is more than 15%, make sure any maintenance item(s) requiring service are done before resetting the display.

• To cancel the "OIL LIFE RESET" mode, press the INFO button repeatedly until the "CANCEL" indicator is
displayed, then press the SEL/RESET button.

5. Press the INFO button repeatedly until the "RESET" indicator is displayed, then press the SEL/RESET button. The maintenance item code(s) will disappear, and the engine oil life will reset to "100%."

RESETTING INDIVIDUAL MAINTENANCE ITEMS

1. Connect the Honda Diagnostic System (HDS) to the data link connector (DLC).

2. Turn the ignition switch to ON (II).

3. Make sure the HDS communicates with the vehicle and the powertrain control module (PCM). If it doesn't communicate, troubleshoot the DLC circuit.

4. Select BODY ELECTRICAL with the HDS.

5. Select ADJUSTMENT in the GAUGES with the HDS.

6. Select MAINTENANCE MINDER in the ADJUSTMENT with the HDS.

7. Select RESET in the MAINTENANCE MINDER with the HDS.

8. Select the individual maintenance item you wish to reset with the HDS.

MANIFOLD ABSOLUTE PRESSURE (MAP) SENSOR

REMOVAL & INSTALLATION

See Figure 171.

1. Remove the cowl cover and the under-cowl panel.

2. Disconnect the MAP sensor 3P connector.

3. Remove the MAP sensor.

4. Install the parts in the reverse order of removal with a new O-ring.

MASS AIRFLOW/INTAKE AIR TEMPERATURE SENSOR

LOCATION

The MAF/IAT sensor is located on the intake tube of the air cleaner assembly.

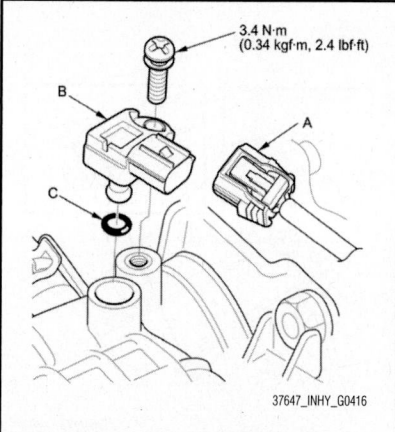

Fig. 171 Disconnect the MAP sensor 3P connector (A); remove the MAP sensor (B) and the O-ring (C)

REMOVAL & INSTALLATION

See Figure 172.

1. Disconnect the MAF sensor/IAT sensor 5P connector.

2. Remove the screws.

3. Remove the MAF sensor/IAT sensor.

4. Install the parts in the reverse order of removal with a new gasket.

POSITIVE CRANKCASE VENTILATION (PCV) VALVE

REMOVAL & INSTALLATION

See Figure 173.

1. Disconnect the hose.

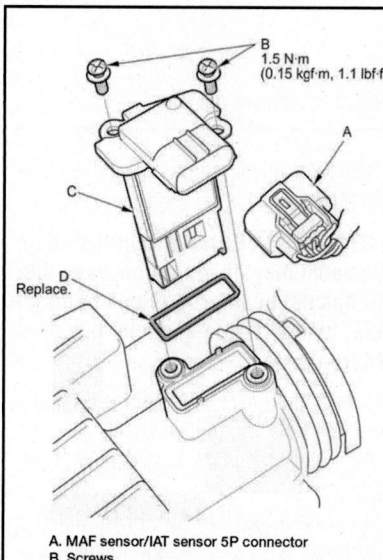

A. MAF sensor/IAT sensor 5P connector
B. Screws
C. MAF sensor/IAT sensor
D. Gasket

Fig. 172 Disconnect the MAF sensor/IAT sensor 5P connector. Remove the screws. Remove the MAF sensor/IAT sensor.

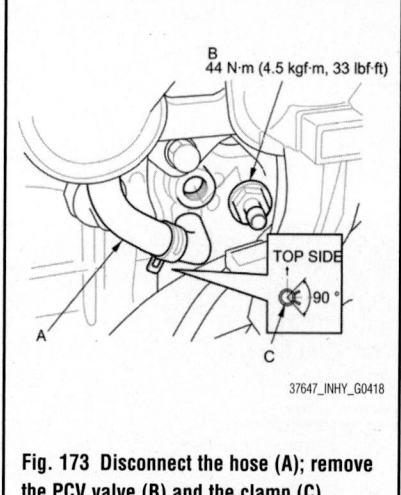

Fig. 173 Disconnect the hose (A); remove the PCV valve (B) and the clamp (C)

2. Remove the PCV valve.

3. Install the parts in the reverse order of removal.

➡**Make sure the hose clamp is positioned as shown.**

POWERTRAIN CONTROL MODULE (PCM)

REMOVAL & INSTALLATION

See Figures 174 through 176.

Special Tools Required:

• Honda Diagnostic System (HDS) tablet tester

• Honda Interface Module (HIM) and an iN workstation with the latest HDS software version

• HDS pocket tester

• GNA600 and an iN workstation with the latest HDS software version

➡**Any one of the above updating tools can be used. Make sure the HDS/iN workstation has the latest HDS software version. The lifetime points of the Eco guide cannot be carried over to the replacement PCM.**

1. Connect the HDS to the Data Link Connector (DLC) located under the driver's side of the dashboard.

2. Turn the ignition switch to ON (II).

3. Make sure the HDS communicates with the PCM and all other vehicle systems. If it doesn't, go to the DLC circuit troubleshooting. If you are returning from the DLC circuit troubleshooting, skip steps 4 through 7, 17 through 19, and 20 through 23, and do these procedures after replacing the PCM:

• Replace the engine oil and the engine oil filter.

• Clean the throttle body.

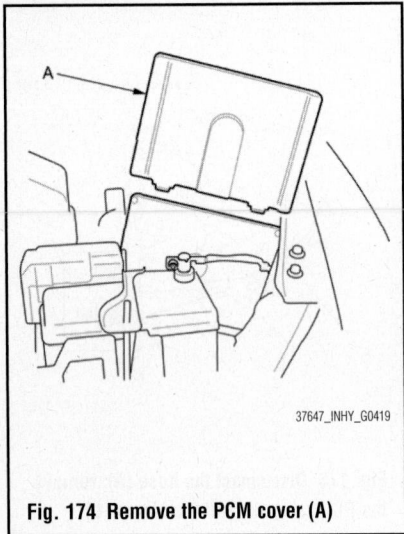

Fig. 174 Remove the PCM cover (A)

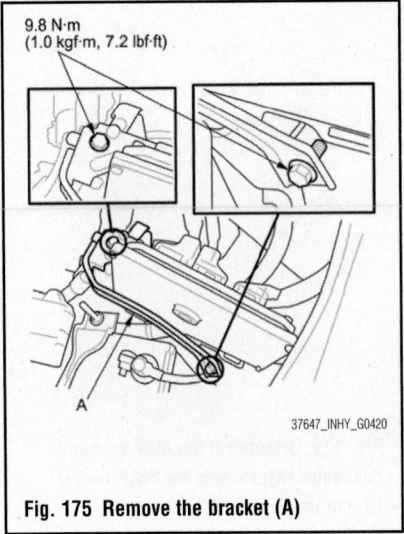

Fig. 175 Remove the bracket (A)

4. Select the PGM-FI system with the HDS.

5. Select the INSPECTION MENU with the HDS.

6. Select the ETCS TEST, then select the TP POSITION CHECK, and follow the screen prompts.

➡ **If the TP POSITION CHECK indicates FAILED, continue with this procedure.**

7. Select the REPLACE PCM MENU, then select READ DATA, and follow the screen prompts.

➡ **Doing this step copies (READS) the engine oil life data from the original PCM so you can later download (WRITES) it into the new PCM. If the READ DATA indicates FAILED, continue with this procedure.**

8. Jump the SCS line with the HDS.

9. Turn the ignition switch to LOCK (0).

10. Remove the PCM cover, then go to step 11 (KC model) or to step 12 (except KC model).

11. Remove the bracket.

12. Remove the bolts.

13. Disconnect PCM connectors A, B, and C, then remove the PCM.

➡ **PCM connectors A, B, and C have symbols embossed on them for identification.**

To install:

14. Install all parts in the reverse order of removal.

➡ **If the IMA battery level indicator displays no level, start the engine, and hold it between 3,500 and 4,000 rpm without load (in P or N) until the level in the indicator is half full.**

15. Turn the ignition switch to ON (II).

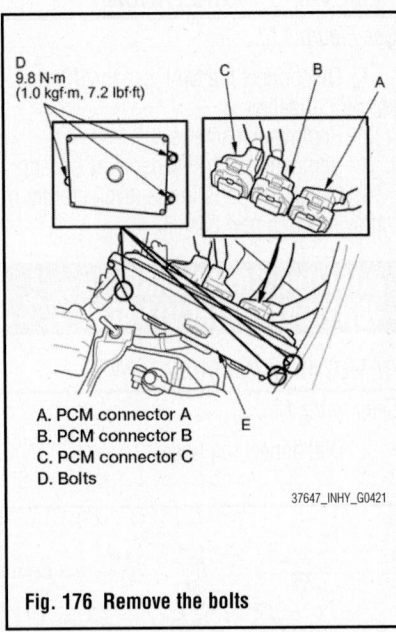

A. PCM connector A
B. PCM connector B
C. PCM connector C
D. Bolts

Fig. 176 Remove the bolts

16. Manually input the VIN to the PCM with the HDS.

➡ **DTC P0630 VIN Not Programmed or Mismatch may be stored because the VIN has not been programmed into the PCM; ignore it, and continue this procedure.**

17. If the READ DATA (engine oil life) failed in step 7, go to step 18.

18. Select the PGM-FI system with the HDS.

19. Select the REPLACE PCM MENU, then select WRITE DATA, and follow the screen prompts.

➡ **If the WRITE DATA indicates FAILED, continue with this procedure.**

20. Select the IMMOBI system with the HDS.

21. Enter the immobilizer PCM code that you got from the iN, and use the PCM replacement procedure in the IMMOBI MENU of the HDS; it allows you to start the engine.

22. If the TP POSITION CHECK failed in step 6, clean the throttle body, then go to step 23.

23. If the READ DATA failed in step 7 or the WRITE DATA failed in step 19, replace the engine oil, and the engine oil filter, then go to step 24.

24. Select the PGM-FI system, and reset the PCM with the HDS.

25. Update the PCM if it does not have the latest software.

26. Do the PCM idle learn procedure.

27. Do the CKP pattern clear/CKP pattern learn procedure.

28. Do the start clutch control calibration procedure.

PCM IDLE LEARN PROCEDURE

The idle learn procedure must be done so the PCM can learn the engine idle characteristics.

Do the idle learn procedure whenever you do any of these actions:
- Replace the PCM.
- Reset the PCM.
- Update the PCM.
- Replace or clean the throttle body.
- Disassemble the engine or the transmission.

➡ **Clearing DTCs with the HDS does not require you to do the idle learn procedure.**

1. Make sure all electrical items (A/C, navigation, lights, etc.) are off.

2. Reset the PCM with the HDS.

3. Turn the ignition switch to ON (II), and wait 2 seconds.

4. Start the engine. Hold the engine speed at 3,000 rpm without load (in P or N) until the radiator fan comes on, or until the engine coolant temperature reaches 194 °F (90 °C).

5. Let the engine idle for about 5 minutes with the throttle fully closed.

➡ **If the radiator fan comes on, do not include its running time in the 5 minutes.**

PCM UPDATE

Special Tools Required:
- Honda Diagnostic System (HDS) tablet tester
- Honda Interface Module (HIM) and an iN workstation with the latest HDS software version

- HDS pocket tester
- GNA600 and an iN workstation with the latest HDS software version

➡️**Any one of the above updating tools can be used.**

➡️**Note the following:**

- Make sure the HDS/iN workstation has the latest HDS software version.
- Before you update the PCM, make sure the 12 volt battery in the vehicle is fully charged, and connect a jumper battery (not a battery charger) to maintain system voltage.
- Never turn the ignition switch to ACC (I) or to LOCK (0) during the update. If there is a problem with the update, leave the ignition switch to ON (II).
- To prevent PCM damage, do not operate anything electrical (headlights, navigation system, brakes, A/C, power windows, door locks, etc.) during the update.
- To ensure the latest program is installed, do a PCM update whenever the PCM is substituted or replaced.
- You cannot update a PCM with a program it already has. It will only accept a new program.
- High temperature in the engine compartment might cause the PCM to become too hot to run the update. If the engine was running before the update, open the hood and cool the engine compartment.
- If you need to diagnose the Honda interface module (HIM) because the HIM's red (#3) light came on or was flashing during the update, leave the ignition switch in ON (II) when you disconnect the HIM from the data link connector (DLC). This will prevent damage to the PCM.

1. Turn the ignition switch to ON (II), but do not start the engine.

2. Connect the HDS to the Data Link Connector (DLC) located under the driver's side of the dashboard.

3. Make sure the HDS communicates with the PCM and all other vehicle systems. If it doesn't, go to the DLC circuit troubleshooting. If you are returning from the DLC circuit troubleshooting, skip step 4 and 5, then clean the throttle body after updating the PCM.

4. Select the INSPECTION MENU with the HDS.

5. Select the ETCS TEST, then select the TP POSITION CHECK, and follow the HDS screen prompts.

➡️**If the TP POSITION CHECK indicates FAILED, continue this procedure.**

6. Exit the HDS diagnostic system, then select the update mode, and follow the screen prompts to update the PCM.

7. If the software in the PCM is the latest, disconnect the HDS/HIM from the DLC, and go back to the procedure that you were doing. If the software in the PCM is not the latest, follow the instructions on the screen. If prompted to choose the PGM-FI system or the CVT system, make sure you update both.

➡️**If the PCM update system requires you to cool the PCM, follow the instructions on screen. If you have a problem during the update procedure (programming takes over 15 minutes, status bar goes over 100%, D or immobilizer indicator flashes, HDS tablet freezes, etc.), follow these steps to minimize the chance of damaging the PCM.**

a. Leave the ignition switch in the ON (II) position.

b. Connect a jumper battery (do not connect a battery charger).

c. Shut down the HDS.

d. Disconnect the HDS from the DLC.

e. Reboot the HDS.

f. Reconnect the HDS to the DLC, and try the update procedure again.

8. If the TP POSITION CHECK failed in step 5, clean the throttle body.

9. Do the PCM idle learn procedure.

10. Do the CKP pattern clear/CKP pattern learn procedure.

11. Select the AT SYSTEM, then reset the TCM with the HDS.

THROTTLE POSITION (TP) SENSOR

LOCATION

See Figure 177.

Refer to the accompanying illustration.

REMOVAL & INSTALLATION

➡️**The manufacturer does not provide a specific Removal and Installation procedure for this component. Refer to the location graphic(s) when servicing this component.**

VEHICLE SPEED SENSOR

REMOVAL & INSTALLATION

See Figures 178 through 183.

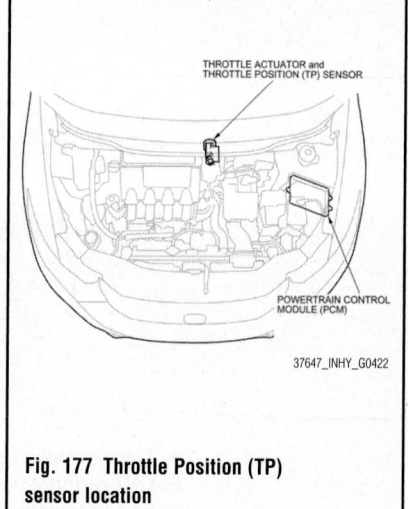

37647_INHY_G0422

Fig. 177 Throttle Position (TP) sensor location

1. Remove the air cleaner.

2. Remove the heater hose clamps and the IMA motor power cable clamps from the bracket.

3. Disconnect the transmission range switch connector and the CVT output shaft (driven pulley) speed sensor connector.

4. Remove the harness clamp and the harness holder clamp from the harness clamp bracket.

5. Remove the ground terminal bracket from the transmission mount.

6. Remove the harness clamp bracket.

7. Remove the snap pin and the control pin from the control lever.

8. Remove the bolts securing the shift cable holder, then separate the shift cable from the control lever.

9. Disconnect the vehicle speed sensor connector, and remove the vehicle speed sensor.

To install:

10. Install a new O-ring on a new vehicle

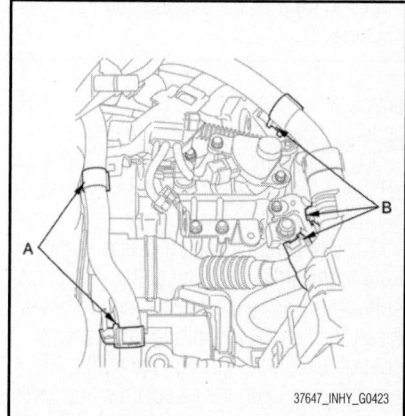

37647_INHY_G0423

Fig. 178 Remove the heater hose clamps (A) and the IMA motor power cable clamps (B) from the bracket

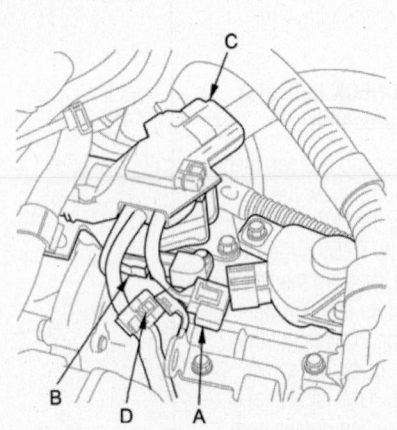

A. Transmission range switch connector
B. CVT output shaft (driven pulley)
   speed sensor connector
C. Harness clamp
D. Harness holder clamp

37647_INHY_G0424

Fig. 179 Disconnect the transmission range switch connector and the CVT output shaft (driven pulley) speed sensor connector

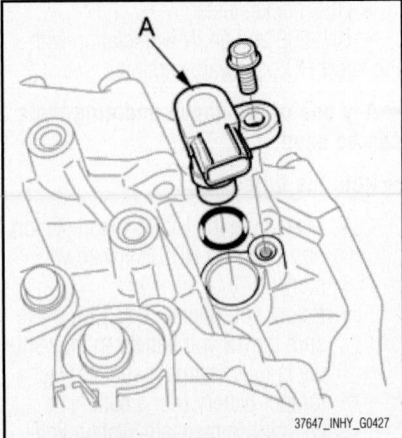

37647_INHY_G0427

Fig. 182 Disconnect the vehicle speed sensor connector, and remove the vehicle speed sensor (A) and the O-ring (B)

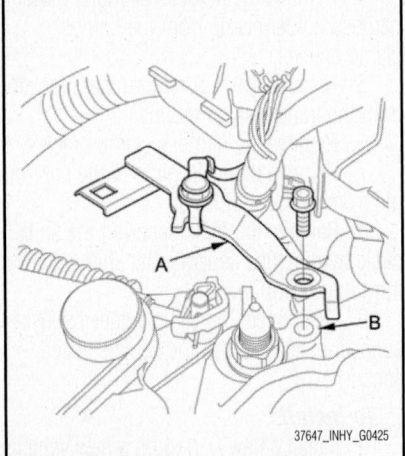

37647_INHY_G0425

Fig. 180 Remove the ground terminal bracket (A) from the transmission mount (B)

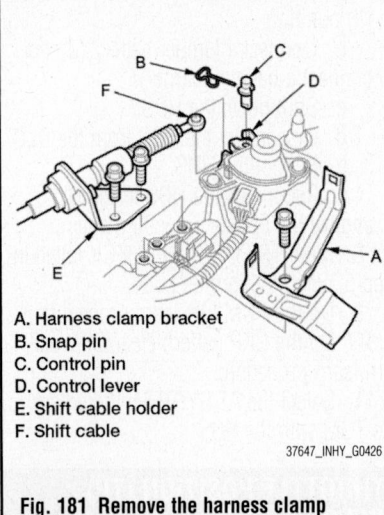

A. Harness clamp bracket
B. Snap pin
C. Control pin
D. Control lever
E. Shift cable holder
F. Shift cable

37647_INHY_G0426

Fig. 181 Remove the harness clamp bracket

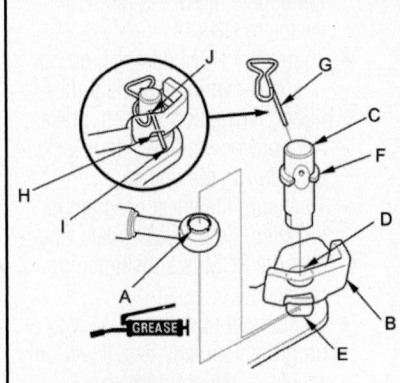

A. Shift cable end collar
B. Selector control lever
C. Control pin
D. Selector control lever hole
E. Selector control lever slotted hole
F. Flange
G. Snap pin
H. Opening of the selector control lever
I. Hooked end
J. Countersunk hole

37647_INHY_G0428

Fig. 183 Apply molybdenum grease to the hole in the shift cable end collar

speed sensor, then install the vehicle speed sensor in the transmission housing.

11. Check the connector for rust, dirt, or oil, and clean or repair if necessary, then connect the connector securely.

12. Apply molybdenum grease to the hole in the shift cable end collar. Attach the shift cable end to the selector control lever. Insert the control pin through the selector control lever hole, through the shift cable end hole, and into the selector control lever

slotted hole in the direction shown. Push the control pin until its flange contacts the selector control lever surface.

13. Insert the snap pin in the direction shown through the control pin hole and out the opening of the selector control lever so that the hooked end of the snap pin snaps into the countersunk hole of the control pin.

14. Secure the shift cable holder on the transmission with the bolts.

15. Install the harness clamp bracket.

16. Install the ground terminal bracket.

17. Install the harness clamps and the harness hold clamp.

18. Connect the transmission range switch connector, and the CVT output shaft (driven pulley) speed sensor connector.

19. Install the heater hose clamps and IMA motor power cable clamp.

20. Install the air cleaner.

FUEL   GASOLINE FUEL INJECTION SYSTEM

FUEL SYSTEM SERVICE PRECAUTIONS

Safety is the most important factor when performing not only fuel system maintenance but any type of maintenance. Failure to conduct maintenance and repairs in a safe manner may result in serious personal injury or death. Maintenance and testing of the vehicle's fuel system components can be accomplished safely and effectively by adhering to the following rules and guidelines.

• To avoid the possibility of fire and personal injury, always disconnect the negative battery cable unless the repair or test procedure requires that battery voltage be applied.

• Always relieve the fuel system pressure prior to disconnecting any fuel system component (injector, fuel rail, pressure regulator, etc.), fitting or fuel line connection. Exercise extreme caution whenever relieving fuel system pressure to avoid exposing skin, face and eyes to fuel spray. Please be advised that fuel under pressure may penetrate the skin or any part of the body that it contacts.

• Always place a shop towel or cloth around the fitting or connection prior to loosening to absorb any excess fuel due to spillage. Ensure that all fuel spillage (should it occur) is quickly removed from engine surfaces. Ensure that all fuel soaked cloths or towels are deposited into a suitable waste container.

• Always keep a dry chemical (Class B) fire extinguisher near the work area.

• Do not allow fuel spray or fuel vapors to come into contact with a spark or open flame.

• Always use a back-up wrench when loosening and tightening fuel line connection fittings. This will prevent unnecessary stress and torsion to fuel line piping.

• Always replace worn fuel fitting O-rings with new Do not substitute fuel hose or equivalent where fuel pipe is installed.

Before servicing the vehicle, make sure to also refer to the precautions in the beginning of this section as well.

RELIEVING FUEL SYSTEM PRESSURE

See Figures 184 through 186.

Before disconnecting fuel lines or hoses, relieve pressure from the system by dis-

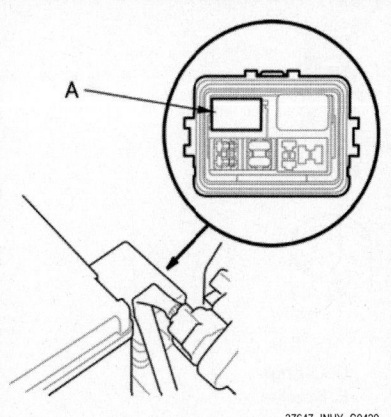

Fig. 184 Remove PGM-FI main relay 2 (A) from the auxiliary under-hood fuse/relay box

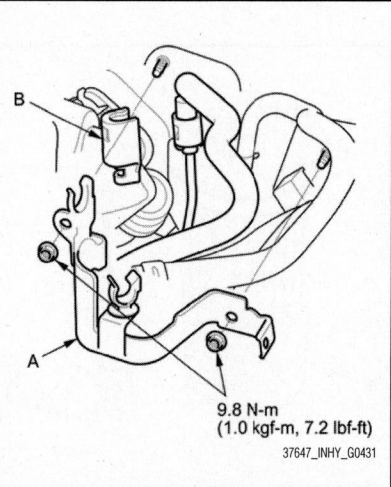

9.8 N-m
(1.0 kgf-m, 7.2 lbf-ft)
37647_INHY_G0431

Fig. 185 Remove the bracket (A) and the quick-connect fitting cover (B)

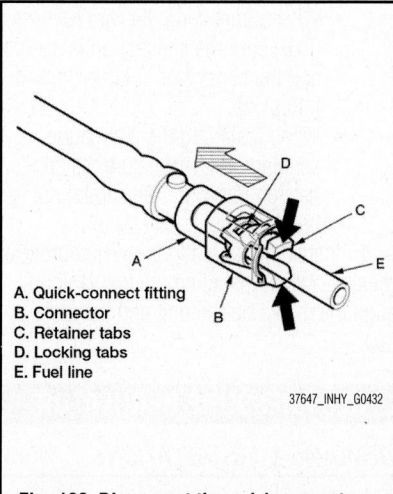

A. Quick-connect fitting
B. Connector
C. Retainer tabs
D. Locking tabs
E. Fuel line

37647_INHY_G0432

Fig. 186 Disconnect the quick-connect fitting

abling the fuel pump and then disconnecting the fuel tube/quick connect fitting in the engine compartment.

1. Remove PGM-FI main relay 2 from the auxiliary under-hood fuse/relay box (to the left of the 12 volt battery).

2. Start the engine, and let it idle until it stalls.

➡**If any DTCs are stored, clear and ignore them.**

3. Turn the ignition switch to LOCK (0).

4. Remove the fuel fill cap to relieve the pressure in the fuel tank.

5. Do the 12 volt battery terminal disconnection procedure.

6. Remove the bracket and the quick-connect fitting cover.

7. Check the fuel quick-connect fitting for dirt, and clean it if needed.

8. Place a rag or shop towel over the quick-connect fitting.

9. Disconnect the quick-connect fitting: Hold the connector with one hand, and squeeze the retainer tabs with the other hand to release them from the locking tabs. Pull the connector off.

a. Be careful not to damage the fuel line or other parts.

b. Do not use tools.

c. If the connector does not move, keep the retainer tabs pressed down, and alternately pull and push the connector until it comes off easily.

d. Do not remove the retainer from the line; once removed, the retainer must be replaced with a new one.

10. After disconnecting the quick-connect fitting, check it for dirt or damage.

11. Do the 12 volt battery terminal reconnection procedure.

ADJUSTMENTS

IDLE SPEED

The idle speed is controlled by the Powertrain Control Module (PCM). No adjustment is necessary or possible.

FUEL FILTER

REMOVAL & INSTALLATION

See Figure 187.

The fuel filter should be replaced whenever the fuel pressure drops below the specified value, after making sure that the fuel pump and the fuel pressure regulator are OK.

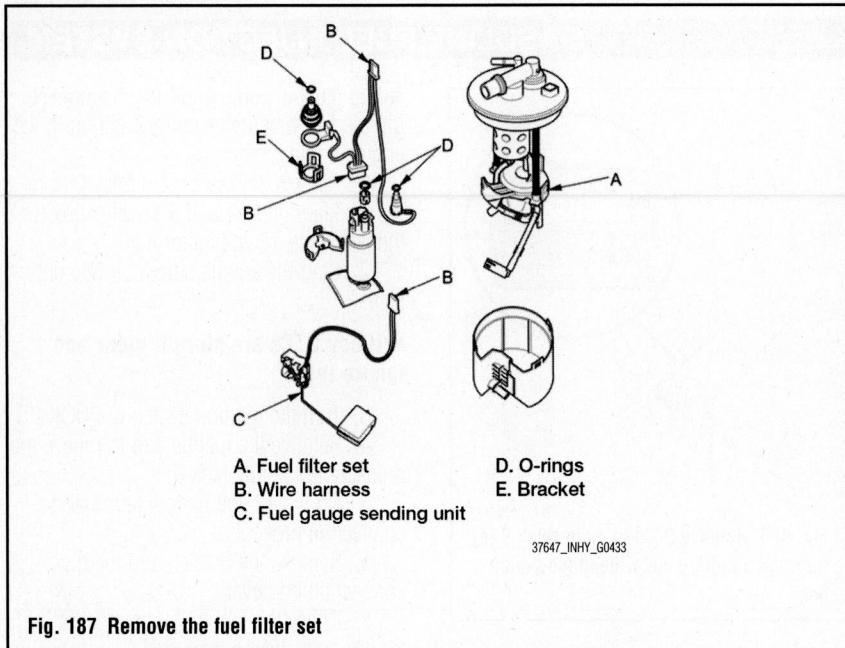

A. Fuel filter set   D. O-rings
B. Wire harness   E. Bracket
C. Fuel gauge sending unit

37647_INHY_G0433

Fig. 187 Remove the fuel filter set

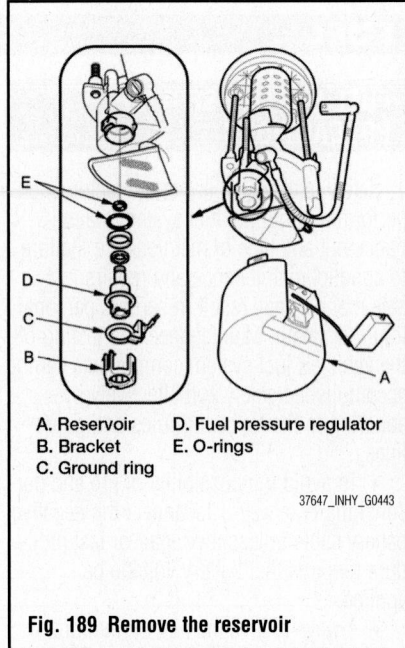

A. Reservoir   D. Fuel pressure regulator
B. Bracket   E. O-rings
C. Ground ring

37647_INHY_G0443

Fig. 189 Remove the reservoir

1. Remove the fuel tank unit.
2. Remove the fuel filter set.

To install:

3. Check these items before installing the fuel tank unit:

- When connecting the wire harness, make sure the connection is secure and the connectors are firmly locked into place.
- When installing the fuel gauge sending unit, make sure the connection is secure and the connector is firmly locked into place. Be careful not to bend or twist it excessively.

4. Install the parts in the reverse order of removal with new O-rings and a new bracket. When installing the fuel tank unit, align the marks on the unit and the fuel tank.

   a. Coat the O-rings with clean engine oil; do not use any other oils or fluids.

   b. Do not pinch the O-rings during installation.

   c. Use all the new parts supplied in the fuel filter replacement kit.

FUEL LEVEL SENSOR/FUEL GAUGE SENDING UNIT

REMOVAL & INSTALLATION

See Figure 188.

1. Remove the fuel tank unit.
2. Remove the fuel level sensor (fuel sending unit) from the fuel tank unit.

To install:

3. Check these items before installing the fuel tank unit:

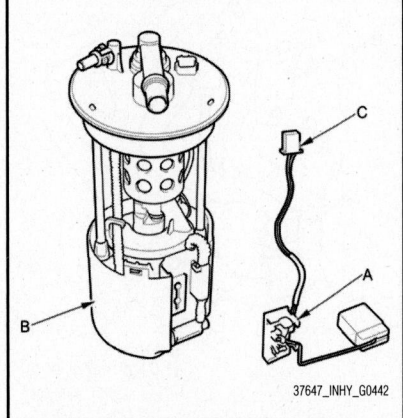

37647_INHY_G0442

Fig. 188 Remove the fuel level sensor (fuel sending unit) (A) from the fuel tank unit (B) and the connector (C)

- When connecting the wire harness, make sure the connection is secure and the connector is firmly locked into place.
- When installing the fuel gauge sending unit, make sure the connection is secure. Be careful not to bend or twist it excessively.

4. Install the parts in the reverse order of removal. When installing the fuel tank unit, align the marks on the unit and the fuel tank.

FUEL PRESSURE REGULATOR

REMOVAL & INSTALLATION

See Figure 189.

1. Remove the fuel tank unit.

2. Remove the reservoir.
3. Remove the bracket.
4. Remove the ground ring.
5. Remove the fuel pressure regulator.
6. Install the parts in the reverse order of removal with new O-rings and a new bracket. When installing the fuel tank unit, align the marks on the unit and the fuel tank.

   a. Coat the O-rings with clean engine oil; do not use any other oils or fluids.

   b. Do not pinch the O-rings during installation.

   c. Use all the new parts supplied in the pressure regulator replacement kit.

FUEL RAIL & INJECTORS

REMOVAL & INSTALLATION

See Figures 190 and 191.

1. Relieve the fuel pressure.
2. Remove the air cleaner.
3. Remove the intake manifold.
4. Remove the nut.
5. Disconnect the connectors from the injectors, the intake side ignition coils, and the EGR valve.
6. Remove the fuel rail mounting nuts from the fuel rail, then remove the injectors and the fuel rail together.
7. Remove the injector clips from the injector.
8. Remove the injectors from the fuel rail.

To install:

9. Coat the new O-rings (black) with

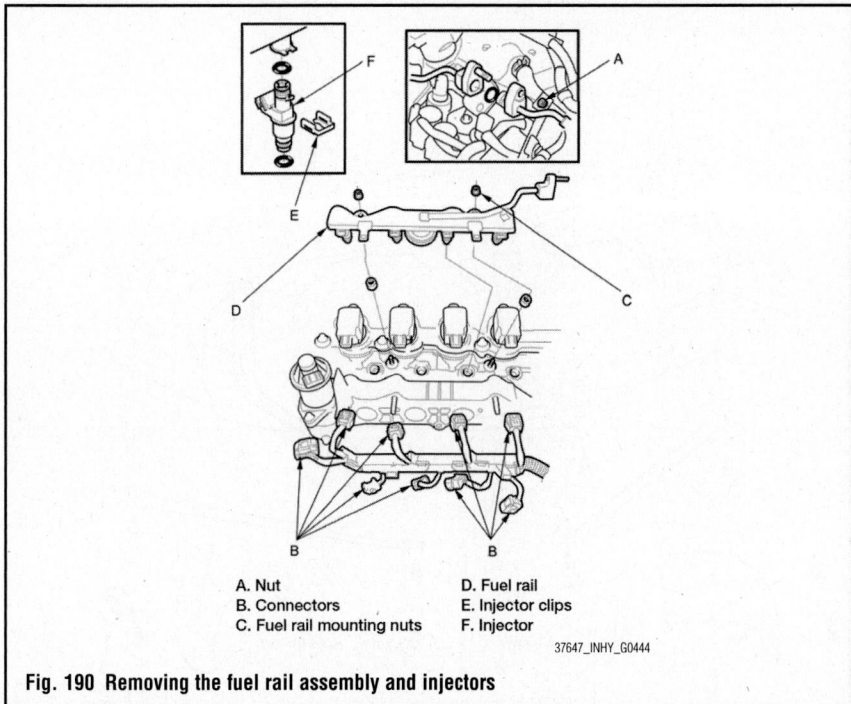

A. Nut
B. Connectors
C. Fuel rail mounting nuts
D. Fuel rail
E. Injector clips
F. Injector

37647_INHY_G0444

Fig. 190 Removing the fuel rail assembly and injectors

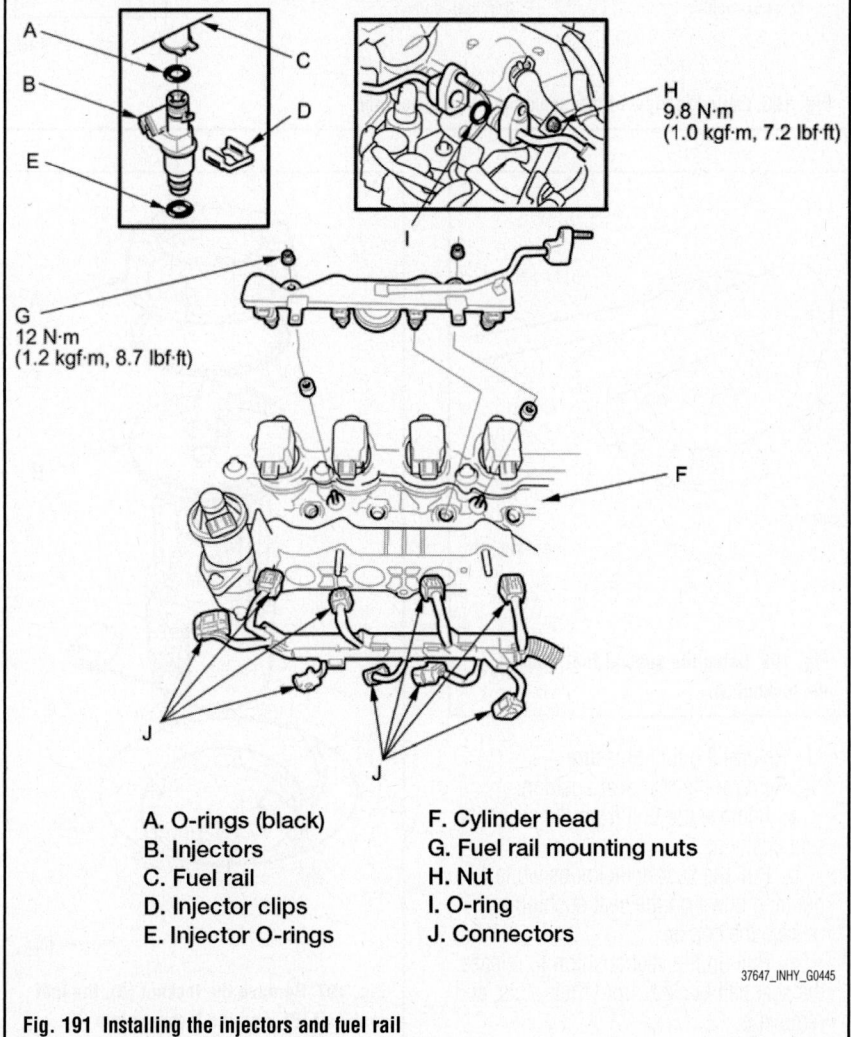

A. O-rings (black)
B. Injectors
C. Fuel rail
D. Injector clips
E. Injector O-rings
F. Cylinder head
G. Fuel rail mounting nuts
H. Nut
I. O-ring
J. Connectors

G
12 N·m
(1.2 kgf·m, 8.7 lbf·ft)

H
9.8 N·m
(1.0 kgf·m, 7.2 lbf·ft)

37647_INHY_G0445

Fig. 191 Installing the injectors and fuel rail

clean engine oil, and insert the injectors into the fuel rail.

10. Install the injector clips.

11. Coat the injector O-rings with clean engine oil.

12. Install the fuel rail and the injectors in the cylinder head.

13. Install the fuel rail mounting nuts.

14. Install the nut with a new O-ring.

15. Connect the connectors on the injectors, the intake side ignition coils, and the EGR valve.

16. Install the intake manifold.

17. Install the air cleaner.

18. Turn the ignition switch to ON (II), but do not operate the starter. After the fuel pump runs for about 2 seconds, the fuel rail will be pressurized. Repeat this two or three times, then check for fuel leakage.

FUEL TANK

DRAINING

1. Remove the fuel tank unit.

2. Using a hand pump, a hose, and a container suitable for fuel, draw the fuel from the fuel tank.

3. Reinstall the fuel tank unit.

REMOVAL & INSTALLATION

See Figures 192 through 195.

1. Drain the fuel tank until it is less than half full.

2. Reinstall the fuel tank unit without connecting the fuel tank unit 4P connectors, the quick-connect fitting or the vent tube.

3. Raise the vehicle on a lift.

4. Disconnect the quick-connect fittings.

5. Remove the hose from the clamp.

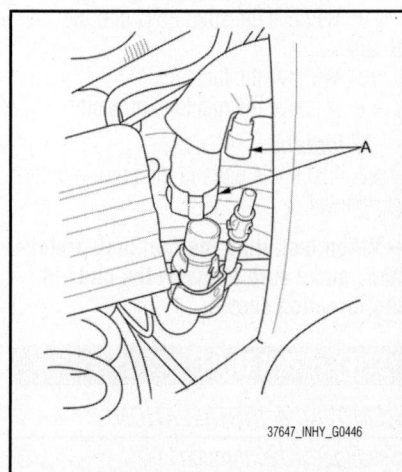

37647_INHY_G0446

Fig. 192 Disconnect the quick-connect fittings (A)

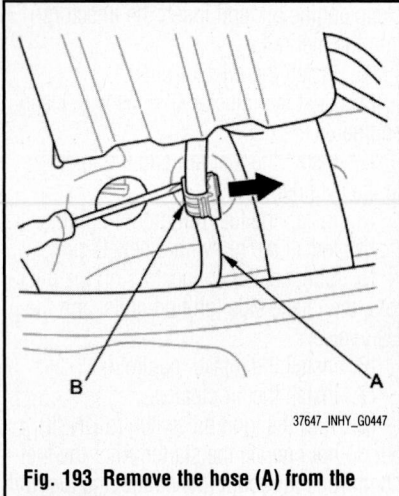

Fig. 193 Remove the hose (A) from the clamp (B)

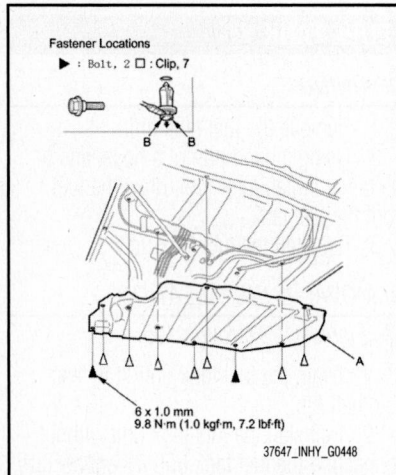

Fig. 194 Remove the middle floor under-cover (A)

6. Remove the middle floor undercover.
7. Remove the trailing arm braces.
8. Place a jack or other support under the fuel tank.
9. Remove the strap bolts and the straps.
10. Remove the fuel tank.
11. Remove the fuel tank protectors.

To install:

12. Install the parts in the reverse order of removal.

➡ When installing the fuel tank protectors, make sure to insert the clips in the direction shown.

FUEL TANK UNIT

REMOVAL & INSTALLATION

See Figures 196 through 200.

Special Tools Required: Fuel Pump Module Locknut Wrench 07AAA-SNAA100

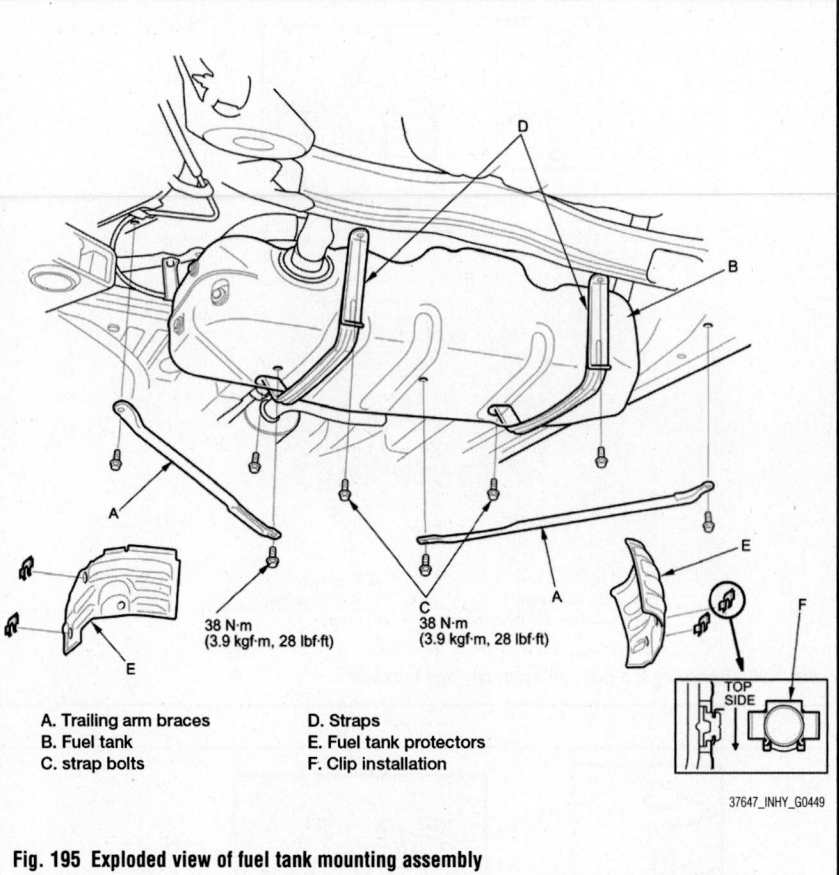

38 N·m
(3.9 kgf·m, 28 lbf·ft)

38 N·m
(3.9 kgf·m, 28 lbf·ft)

A. Trailing arm braces
B. Fuel tank
C. strap bolts
D. Straps
E. Fuel tank protectors
F. Clip installation

Fig. 195 Exploded view of fuel tank mounting assembly

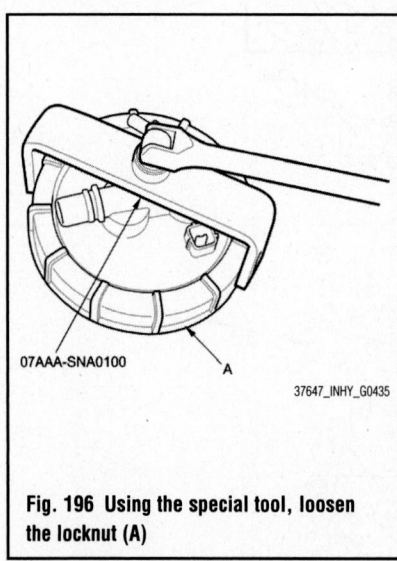

Fig. 196 Using the special tool, loosen the locknut (A)

1. Relieve the fuel pressure.
2. Remove the rear seat cushion.
   a. Remove the bolt from the slit in the seat cushion.
   b. Pull the seat hook knobs while pushing down on the seat cushion to release the hooks.
   c. Pull up the seat cushion to release the seat belt buckles from their slots, and remove it.

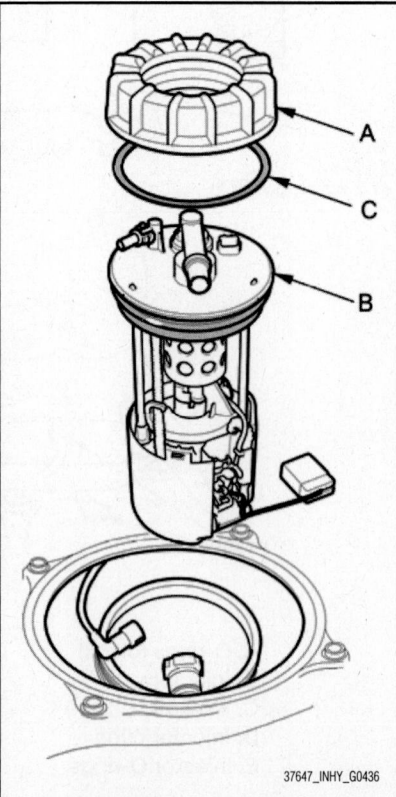

Fig. 197 Remove the locknut (A), the fuel tank unit (B), and the locknut plate (C)

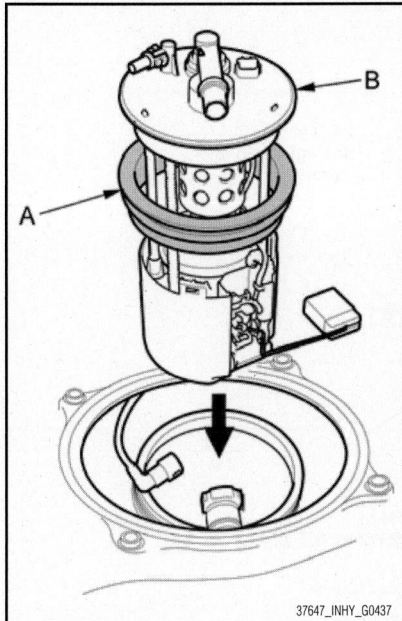

Fig. 198 Temporarily attach a new base gasket (A) to the fuel tank unit (B), then insert the fuel tank unit partially into the fuel tank

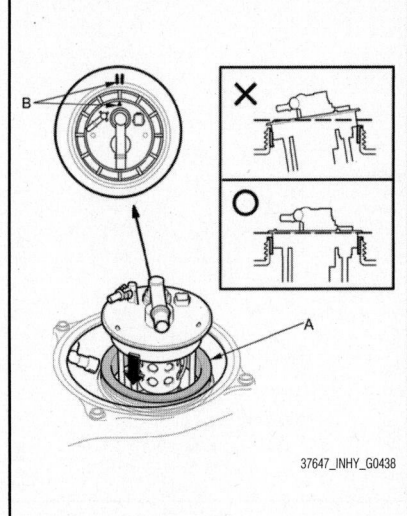

Fig. 199 Transfer the base gasket (A) and align the marks (B)

3. Remove the access panel from the floor.

4. Disconnect the fuel tank unit 4P connector.

5. Disconnect the quick-connect fittings from the fuel tank unit.

6. Using the special tool, loosen the locknut.

7. Remove the locknut, the fuel tank unit, and the locknut plate.

To install:

8. Temporarily attach a new base

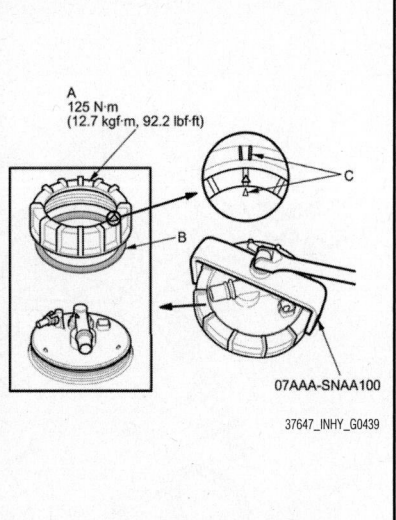

Fig. 200 Using the special tool, tighten a new locknut (A) with a new locknut plate (B)

gasket to the fuel tank unit, then insert the fuel tank unit partially into the fuel tank.

- Be careful not to damage the new base gasket.
- Be careful not to bend the fuel gauge sending unit.
- Do not coat the base gasket with oil.

9. Transfer the base gasket from the fuel tank unit to the fuel tank.

10. Align the marks on the fuel tank and fuel tank unit, then insert the fuel tank unit into the fuel tank until the fuel tank unit rests on top of the base gasket.

➡**To avoid a fuel leak, check the base gasket, visually or by hand, to make sure it is not pinched.**

11. Using the special tool, tighten a new locknut (A) with a new locknut plate (B) to 92 ft. lbs. (125 Nm).

   a. Before tightening , align the marks on the fuel tank and the locknut.

   b. After tightening, make sure the marks are still aligned.

   c. After installation, check the base gasket, visually or by hand, to make sure it is not pinched.

12. Connect the fuel tank unit 4P connector, then connect the quick-connect fitting, and the vent tube.

13. Reconnect the negative cable to the 12 volt battery, and turn the ignition switch to ON (II) (but do not operate the starter motor). The fuel pump will run for about 2 seconds, and fuel pressure rises. Repeat this two or three times, then check for fuel leaks.

14. Install the access panel.

15. Install the rear seat cushion.

THROTTLE BODY

REMOVAL & INSTALLATION

See Figure 201.

✳✳ CAUTION

Do not insert your fingers into the installed throttle body when you turn the ignition switch to ON (II) or while the ignition switch is in ON (II). If you do, you will seriously injure your fingers if the throttle valve is activated.

➡**If you are replacing or cleaning the throttle body, start at step 1. If you are removing the throttle body, start at step 4.**

1. Connect the HDS to the DLC while the engine is stopped.

2. Select the INSPECTION MENU on the HDS.

3. Do the TP POSITION CHECK in the ETCS TEST.

4. Turn the ignition switch to LOCK (0).

5. Remove the air cleaner.

6. Disconnect the MAP sensor 3P connector and the EVAP canister purge valve 2P connector, then remove the harness holder from the EVAP canister purge guard

7. Disconnect the throttle body connector, then disconnect the EVAP canister purge valve hose.

8. Disconnect the purge hoses.

9. Disconnect and plug the water bypass hoses.

10. Remove the throttle body.

11. Remove the purge pipe.

12. Install the parts in the reverse order of removal with a new gasket (M), then refill the radiator with engine coolant.

- If you replace or clean the throttle body, go to step 13.
- If you did not replace or clean the throttle body, this procedure is complete.
- Be careful not to drop or damage the plates.
- Align the marks on the hoses and the throttle body, then insert the hoses.
- Align the mark on the hoses, then insert the hoses.

13. Connect the HDS to the Data Link Connector (DLC) located under the driver's side of the dashboard.

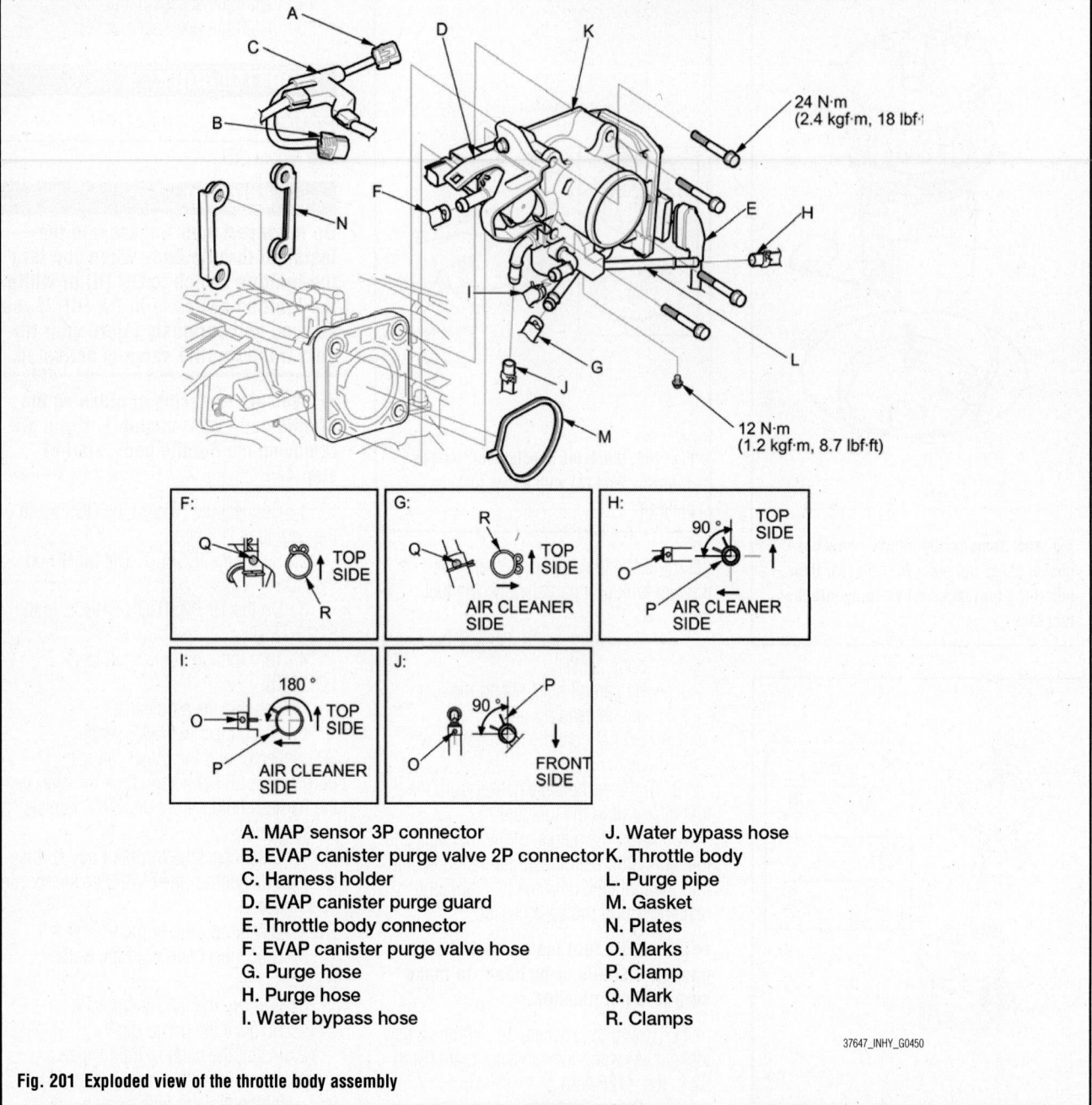

24 N·m
(2.4 kgf·m, 18 lbf·ft)

12 N·m
(1.2 kgf·m, 8.7 lbf·ft)

F:
Q
TOP SIDE
R

G:
R
Q
TOP SIDE
AIR CLEANER SIDE

H:
90°
TOP SIDE
O
P AIR CLEANER SIDE

I:
180°
O
TOP SIDE
P AIR CLEANER SIDE

J:
90°
P
O
FRONT SIDE

A. MAP sensor 3P connector
B. EVAP canister purge valve 2P connector
C. Harness holder
D. EVAP canister purge guard
E. Throttle body connector
F. EVAP canister purge valve hose
G. Purge hose
H. Purge hose
I. Water bypass hose
J. Water bypass hose
K. Throttle body
L. Purge pipe
M. Gasket
N. Plates
O. Marks
P. Clamp
Q. Mark
R. Clamps

37647_INHY_G0450

Fig. 201 Exploded view of the throttle body assembly

14. Turn the ignition switch to ON (II).

15. Reset the PCM with the HDS.

16. Select the ETCS TEST in the INSPECTION MENU with the HDS.

17. Select the TP POSITION CHECK, then clear the Throttle Position (TP) learned value.

18. Turn the ignition switch to LOCK (0).

19. Turn the ignition switch to ON (II), and wait 2 seconds without pressing the accelerator pedal.

20. Do the PCM idle learn procedure.

HEATING & AIR CONDITIONING SYSTEM

BLOWER MOTOR

REMOVAL & INSTALLATION

See Figures 202 through 206.

1. Remove the passenger's dashboard undercover and the glove box.
2. Cut the plastic cross brace in the glove box opening with diagonal cutters in the area, and discard it.
3. Remove the bolts and the glove box frame.
4. Disconnect the connectors from the A/C wire harness and the mode control motor and air mix control motor.
5. Remove the self-tapping screws and the passenger's heater duct.
6. Remove the self-tapping screws, the mounting nuts, and the blower unit.

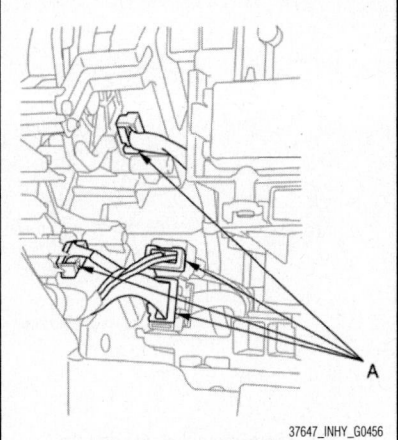

37647_INHY_G0456

Fig. 204 Disconnect the connectors (A) from the A/C wire harness and the mode control motor and air mix control motor

7. Install the unit in the reverse order of removal. Make sure that there is no air leakage.

HEATER CORE

REMOVAL & INSTALLATION

See Figures 267 through 212.

✳✳ WARNING

SRS components are located in this area. Review the SRS component locations, and the precautions and procedures before doing repairs or service.

1. Do the 12 volt battery terminal disconnection procedure.
2. Disconnect the A/C line from the evaporator core.

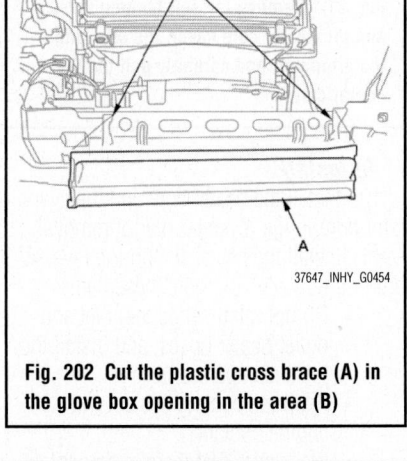

37647_INHY_G0454

Fig. 202 Cut the plastic cross brace (A) in the glove box opening in the area (B)

37647_INHY_G0457

Fig. 205 Remove the self-tapping screws and the passenger's heater duct (A)

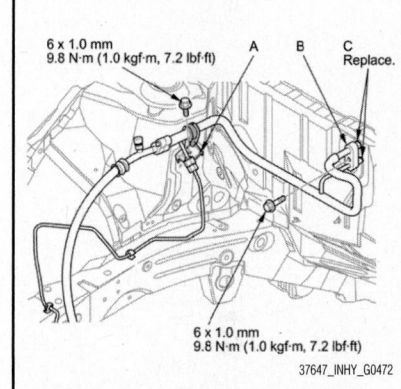

6 x 1.0 mm
9.8 N·m (1.0 kgf·m, 7.2 lbf·ft)

6 x 1.0 mm
9.8 N·m (1.0 kgf·m, 7.2 lbf·ft)

Replace.

37647_INHY_G0472

Fig. 207 Remove the bolts from the A/C line clamp (A); remove the bolt, then disconnect the A/C line (B) from the evaporator core and remove the O-rings (C)

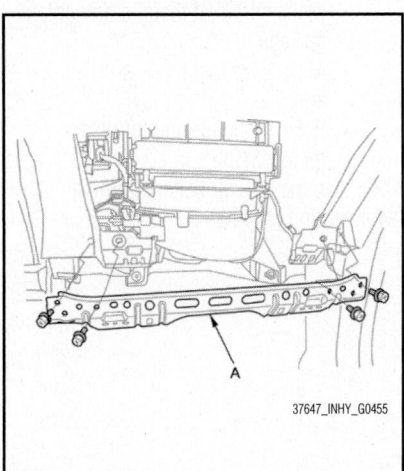

37647_INHY_G0455

Fig. 203 Remove the bolts and the glove box frame (A)

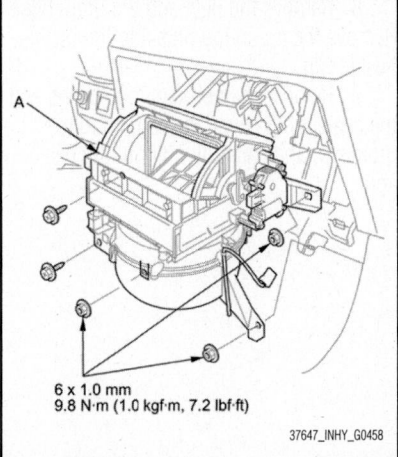

6 x 1.0 mm
9.8 N·m (1.0 kgf·m, 7.2 lbf·ft)

37647_INHY_G0458

Fig. 206 Remove the self-tapping screws, the mounting nuts, and the blower unit (A)

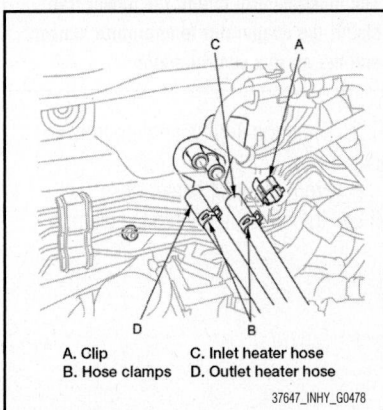

A. Clip
B. Hose clamps
C. Inlet heater hose
D. Outlet heater hose

37647_INHY_G0478

Fig. 208 Remove the clip; slide the hose clamps back, then disconnect the inlet heater hose and the outlet heater hose from the heater unit

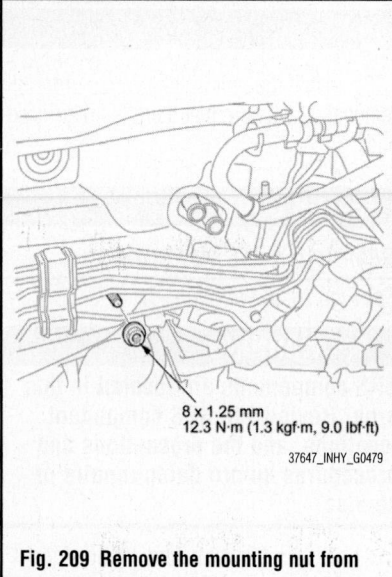

Fig. 209 Remove the mounting nut from the heater unit

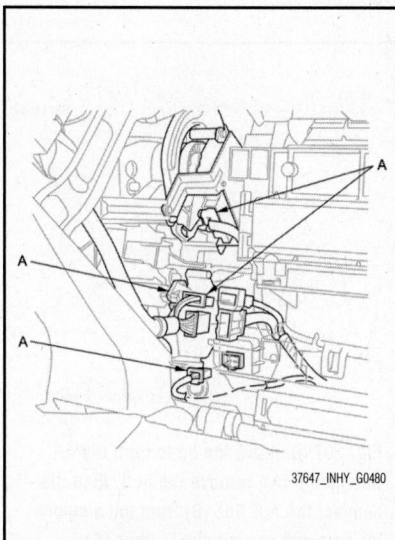

Fig. 210 Disconnect these connectors (A): The mode control motor, the power transistor, the evaporator temperature sensor, and the air mix control motor

3. Drain the engine coolant from the radiator.

4. From the inlet heater hose, remove the clip. Slide the hose clamps

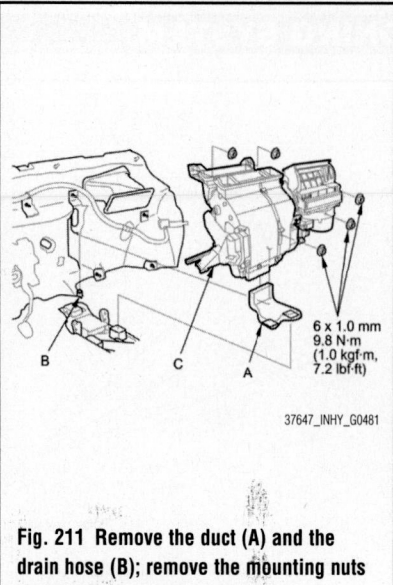

Fig. 211 Remove the duct (A) and the drain hose (B); remove the mounting nuts and the blower-heater unit (C)

back, then disconnect the inlet heater hose and the outlet heater hose from the heater unit. Note the layout of the hoses.

➡**Engine coolant will run out when the hoses are disconnected; drain it into a clean drip pan. Be sure not to let coolant spill on the electrical parts or the painted surfaces. If any coolant spills, rinse it off immediately.**

5. Remove the mounting nut from the heater unit. Take care not to damage or bend the fuel lines or the brake lines, etc.

6. Remove the dashboard.

7. Disconnect these connectors: The mode control motor, the power transistor, the evaporator temperature sensor, and the air mix control motor.

8. Remove the duct and the drain hose. Remove the mounting nuts and the blower-heater unit.

9. Remove the self-tapping screws and the heater core cover, the cap, and the grommet, and carefully pull out the heater core.

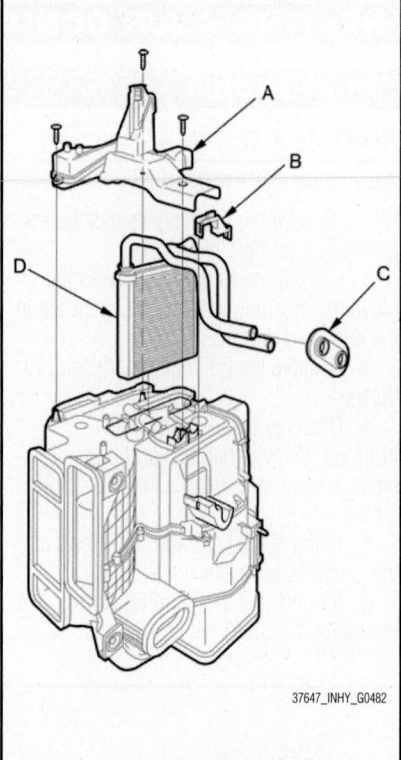

Fig. 212 Remove the self-tapping screws and the heater core cover, the cap, and the grommet, and carefully pull out the heater core

To install:

10. Install the heater core and the evaporator core in the reverse order of removal.

11. Install the heater unit in the reverse order of removal, and note these items:

- Do not interchange the inlet and outlet heater hoses, and install the hose clamps securely.
- Refill the cooling system with engine coolant.
- Make sure that there is no coolant leakage.
- Make sure that there is no air leakage.
- Refer to the evaporator core replacement.

12. Do the 12 volt battery terminal reconnection procedure.

STEERING

EPS CONTROL UNIT

REMOVAL & INSTALLATION
See Figure 213.

1. Do the 12 volt battery terminal disconnection procedure.
2. Remove the driver's dashboard undercover.
3. Disconnect EPS control unit connector A (2P), connector B (2P), and connector C (16P).
4. Remove the harness clip from EPS control unit bracket.
5. Remove the nuts from the EPS control unit bracket.
6. Remove the EPS control unit.

To install:

7. Install the EPS control unit in the reverse order of removal.
8. Do the 12 volt battery terminal reconnection procedure.
9. Do the memorizing of the torque sensor neutral position.
10. After installation, start the engine, allow it to idle, and turn the steering wheel to full left or full right. Check that the EPS indicator does not come on.

EPS MOTOR

REMOVAL & INSTALLATION
See Figures 214 and 215.

➡Do not allow dust, dirt, or other foreign materials to enter the steering gearbox.

1. Remove the steering gearbox.
2. Remove the torque sensor 4P connector from the connector bracket.
3. Remove the connector bracket mount bolt, then remove the EPS motor and the O-ring.

To install:

4. Clean the mating surfaces of the EPS motor and the steering gearbox.
5. Apply grease included in the motor set to the new O-ring and carefully fit it on the EPS motor.
6. Apply steering grease to the EPS motor shaft.
7. Install the EPS motor on the steering gearbox by engaging the EPS motor shaft and the worm shaft.
8. Before tightening the bolts, turn the

EPS motor two or three times to the right and left about 45 degrees.

➡Make sure the EPS motor is evenly seated on the steering gearbox, and that the O-ring is not pinched between the mating surfaces.

9. Tighten the EPS motor mounting bolts to 15 ft. lbs. (20 Nm).
10. Install the torque sensor 4P connector to the connector bracket.
11. Install the steering gearbox.

POWER RACK & PINION STEERING GEAR

REMOVAL & INSTALLATION
See Figures 216 through 227.

Special Tools Required:
• Ball Joint Thread Protector, 10 mm 07AAF-SECA120
• Ball Joint Remover, 28 mm 07MAC-SL0A202
• Ball Joint Thread Protector, 14 mm 071AF-S3VA000

➡Note these items during removal:

• Use solvent and a brush, wash any oil and dirt off the end of the steering gearbox, but avoid any electri-

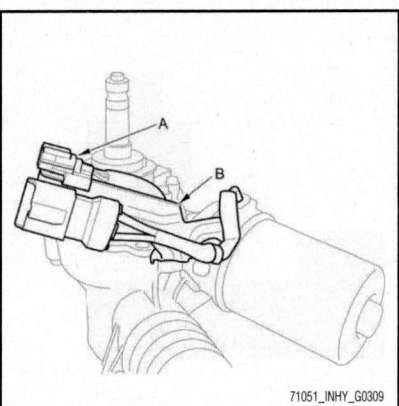

Fig. 214 Remove the torque sensor 4P connector (A) from the connector bracket (B).

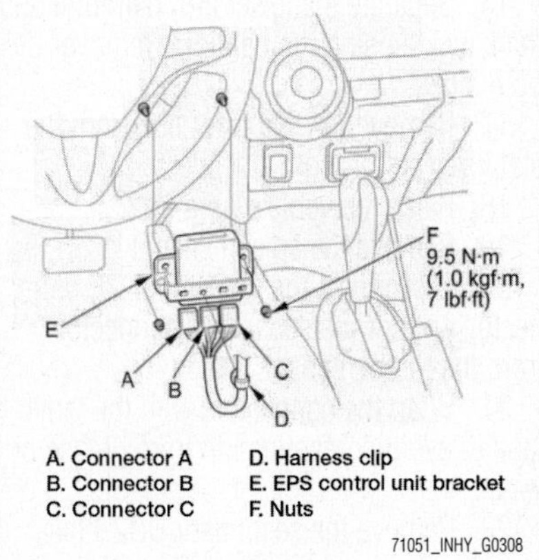

71051_INHY_G0309

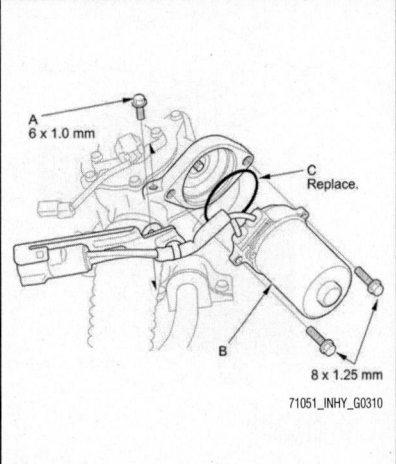

A
6 x 1.0 mm

C
Replace.

B

8 x 1.25 mm

71051_INHY_G0310

Fig. 215 Remove the connector bracket mount bolt (A), then remove the EPS motor (B) and the O-ring (C).

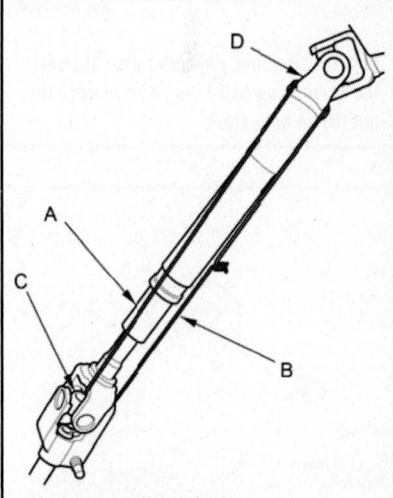

A. Lower slide shaft
B. Wire
C. Joint yoke of lower slide shaft
D. Joint yoke of upper shaft

37647_INHY_G0487

Fig. 216 Hold the lower slide shaft on the column with a piece of wire between the joint yoke of the lower slide shaft and joint yoke of the upper shaft to prevent the lower slide from pulling out

F
9.5 N·m
(1.0 kgf·m,
7 lbf·ft)

A. Connector A   D. Harness clip
B. Connector B   E. EPS control unit bracket
C. Connector C   F. Nuts

71051_INHY_G0308

Fig. 213 Removing the EPS control unit

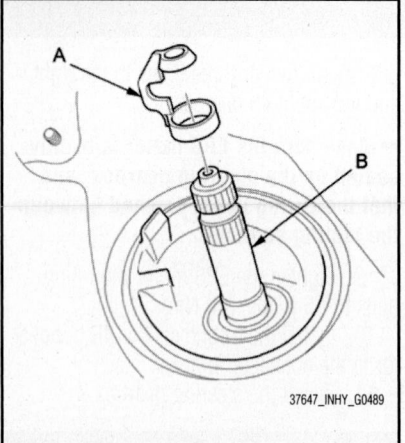

Fig. 217 Remove the center guide (A) (if equipped) from the top of the pinion shaft (B)

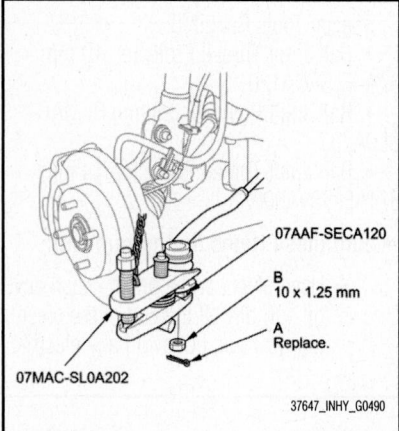

Fig. 218 Remove the cotter pin (A) from the tie-rod end ball joint, then remove the nut (B) on both sides

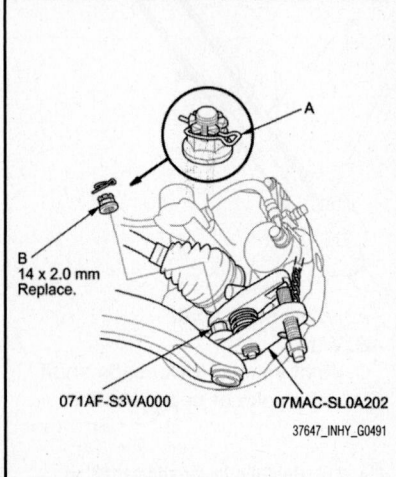

Fig. 219 Remove the clip (A) from the lower arm ball joint, then remove the castle nut (B) on both sides

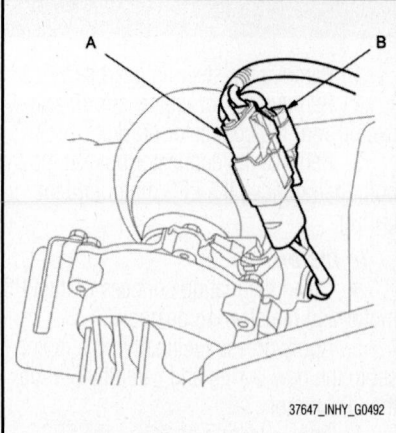

Fig. 220 Disconnect the EPS motor 2P connector (A), and torque sensor 4P connector (B) from the torque sensor

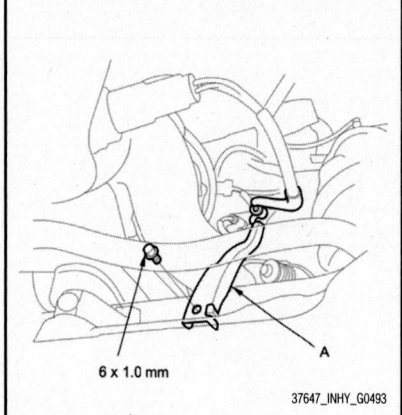

Fig. 221 Remove the secondary HO2S harness bracket (A) from the steering gearbox

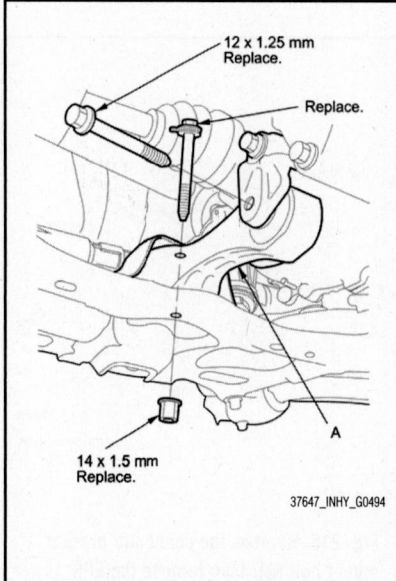

Fig. 222 Remove the torque rod (A)

cal parts. Blow dry with compressed air.
- Be sure to remove the steering wheel before disconnecting the steering joint or damage to the cable reel can occur.

1. Do the 12 volt battery terminal disconnection procedure.

2. Raise the vehicle on a lift.

3. Remove the front wheels.

4. Release the lock lever, and adjust the steering column to the full tilt down position, and to the full telescopic in position.

5. Tighten the lock lever.

6. Remove the steering joint cover.

7. Hold the lower slide shaft on the column with a piece of wire between the joint yoke of the lower slide shaft and joint yoke of the upper shaft to prevent the lower slide from pulling out.

8. Release the lock lever, and slide the steering column all the way out, then tighten the lock lever.

9. Remove the steering joint bolt from the steering joint.

10. Center the steering wheel spokes, and install a commercially available steering wheel holder tool.

11. Disconnect the steering joint by sliding the steering joint into the column.

12. Remove the center guide (if equipped) from the top of the pinion shaft and discard it. The center guide is for factory assembly use only.

13. Remove the cotter pin from the tie-rod end ball joint, then remove the nut on both sides.

14. Separate the tie-rod end ball joint and knuckle using the ball joint remover on both sides.

15. Remove the clip from the lower arm ball joint, then remove the castle nut on both sides.

16. Separate the lower arm ball joint and the knuckle using the ball joint remover on both sides.

17. Remove the stabilizer link from the stabilizer bar on both sides.

18. Raise the vehicle on a lift.

19. Remove the splash shield.

20. Disconnect the EPS motor 2P connector, and torque sensor 4P connector from the torque sensor.

21. Wrap the connectors with the vinyl tape to avoid contamination from grease or water.

22. Remove the secondary HO2S harness bracket from the steering gearbox. Do not disconnect secondary HO2S 4P connector and the secondary HO2S.

23. Place a jack under the engine oil pan.

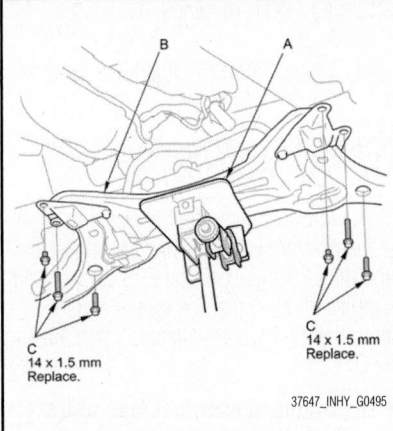

Fig. 223 Attach a transmission jack (A) to the center of the front subframe (B); remove the front subframe mounting bolts (C)

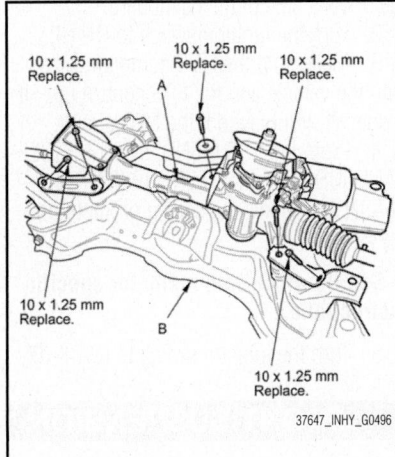

Fig. 224 Remove the steering gearbox (A) from the front subframe (B)

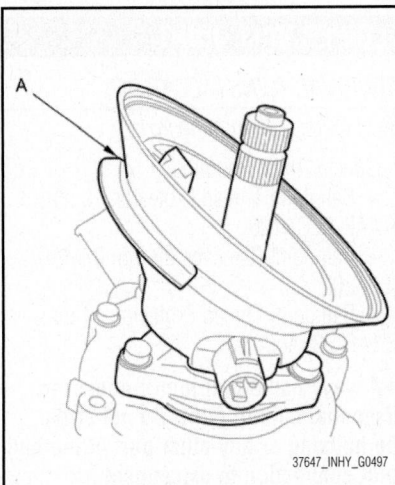

Fig. 225 Remove the pinion shaft grommet (A) from the top of the torque sensor

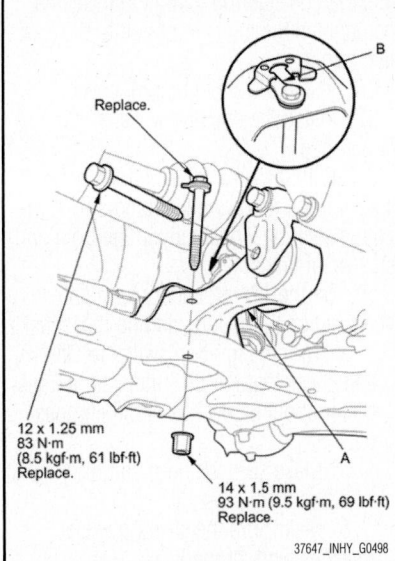

Fig. 226 Install the torque rod (A) with new mounting bolts and new mounting nut; make sure the guide plate (B) of the flange bolt is securely installed

24. Remove the torque rod.

25. Attach a transmission jack to the center of the front subframe, and support the front subframe securely by raising the transmission jack.

26. Remove the front subframe mounting bolts.

27. Lower the front subframe and the steering gearbox as an assembly by lowering the transmission jack slowly.

28. Remove the steering gearbox from the front subframe.

29. Remove the pinion shaft grommet from the top of the torque sensor.

To install:

30. Before installing the steering gearbox, make sure that no grease is on the mating surface of the steering gearbox and the front subframe. To prevent the gearbox mounting bolts from loosening after the installation, remove any grease from the bolt holes.

31. Install the pinion shaft grommet. Align the ledge portion in the pinion shaft grommet with the lug portion on the torque sensor 3P connector. The grommet must not have a gap at the mating surface of the grommet and torque sensor.

32. Place the steering gearbox on the front subframe.

33. Loosely install the stiffener plates, and new gearbox mounting bolts, then tighten the bolts.

34. Turn the lip of the pinion shaft grommet to ease installation.

35. Set the front subframe with the steer-ing gearbox on the transmission jack, and support it.

36. Carefully raise the front subframe with transmission jack, and pass the pinion shaft into the passenger's compartment.

➡ **Be sure that the pinion shaft grommet is securely in place. Make sure the lip of the pinion shaft grommet is not turned up. Incorrect installation can cause leakage of water or mud, or noise. Take care not to damage the lower arm ball joint boot with the edge of the knuckle, etc.**

37. Loosely install new front subframe mounting bolts, then tighten the bolts to 69 ft. lbs. (93 Nm).

38. Install the torque rod with new mounting bolts and new mounting nut, and tighten them to the specified torque.

➡ **Make sure the guide plate of the flange bolt is securely installed.**

39. Remove the jack.

40. Install the secondary HO2S harness bracket on the steering gearbox.

41. Remove the vinyl tape from the connectors.

42. Connect the EPS motor 2P connector and the torque sensor 4P connector to the torque sensor.

43. Install the splash shield.

44. Connect the stabilizer links.

45. Lower the vehicle.

46. Wipe off any grease contamination from the ball joint tapered section and threads. Reconnect the tie-rod end ball joints to the knuckles. Install the nuts, and tighten to 32 ft. lbs. (43 Nm) on both sides.

47. Install the new cotter pin.

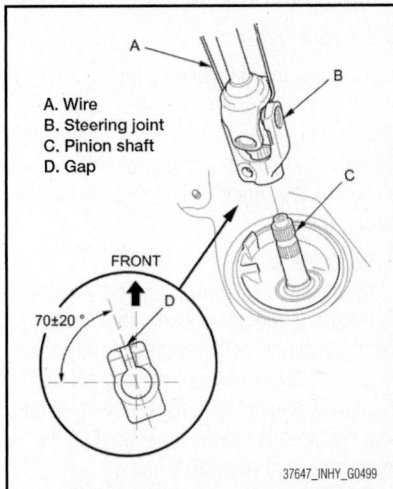

Fig. 227 Slip the lower end of the steering joint onto the pinion shaft taking care to align the gap within the angle shown

48. Wipe off any grease contamination from the lower arm ball joint tapered section and threads. Then reconnect the lower arm to the knuckle. Install the new castle nut and tighten it on both sides.

➡**Be careful not to damage the lower ball joint boot. Check the ball joint boot for deformation before connecting the knuckle. Torque the castle nut to the lower torque specification, then tighten it only far enough to align the slot with the joint pin clip hole. Do not align the castle nut by loosening it.**

49. Install the clip.
50. Install the front wheels, then set the wheels in the straight ahead position.

➡**Before installing the wheel, clean the mating surfaces of the brake disc and the inside of the wheel.**

51. Cut the wire holding the steering shaft.
52. Slip the lower end of the steering joint onto the pinion shaft taking care to align the gap within the angle shown.
53. Remove the steering wheel holder tool.
54. Align the bolt hole on the steering joint with the groove around the pinion shaft, then loosely install the lower steering joint bolt. Be sure that the joint bolt is securely in the groove in the pinion shaft.
55. Pull on the steering joint to make sure that the steering joint is fully seated, then tighten the lower joint bolt to the specified torque.
56. Install the steering joint cover.
57. Install the steering wheel and the driver's airbag.
58. With the tires raised off the ground, check for the following symptoms by turning the steering wheel fully to the right and left several times.
59. Do the 12 volt battery terminal reconnection procedure, and do these tasks:
   a. Turn the ignition switch to ON (II), and check that the SRS indicator comes on for about 6 seconds and then goes off.
   b. Make sure the horn and turn signal switches work properly.
   c. Make sure the steering wheel switches work properly.
60. Do the memorizing of the torque sensor neutral position.
61. After installation, do these checks:
   a. Start the engine, allow it to idle, and turn the steering wheel to full left or full right.
   b. Check that the EPS indicator does not come on.
   c. Check the steering wheel spoke angle. If steering spoke angles to the right and left are not equal (steering wheel and rack are not centered), correct the engagement of the joint/pinion shaft serrations, then adjust the front toe by turning the tie-rod ends, if necessary.
62. Check the wheel alignment, and adjust it if necessary.

TORQUE SENSOR NEUTRAL POSITION MEMORIZATION

The torque sensor neutral position must be memorized whenever the steering gearbox, the EPS motor, or the EPS control unit is replaced. Note that the torque sensor neutral position is not affected when erasing the DTC.

➡**The torque sensor is temperature sensitive. When memorizing the torque sensor neutral position, the ambient temperature must be above 68 °F (20 °C).**

1. With the ignition switch in LOCK (0), connect the HDS to the DLC located under the driver's side of the dashboard.
2. Turn the ignition switch to ON (II).
3. Make sure the HDS communicates with the vehicle and the EPS control unit. If it doesn't, troubleshoot the DLC circuit.
4. From the EPS MENU, select MISCELLANEOUS TEST then TORQUE SENSOR LEARN and follow the screen prompts on the HDS.

➡**See the HDS Help menu for specific instructions.**

5. Turn the ignition switch to LOCK (0).

SUSPENSION

CONTROL LINKS

REMOVAL & INSTALLATION

Front Stabilizer Link

See Figure 228.

1. Raise the vehicle on a lift.
2. Remove the front wheel.
3. Remove the self-locking nut and the flange nut while holding the respective joint pin with a hex wrench, then remove the stabilizer link.

To install:

4. Install the stabilizer link on the stabilizer bar and the damper with the joint pins set at the center of their range of movement.
5. Install the new self-locking nut and the new flange nut, and tighten them to the specified torque values while holding the respective joint pin with a hex wrench.
6. Clean the mating surfaces of the brake disc and the inside of the wheel, then install the front wheel.

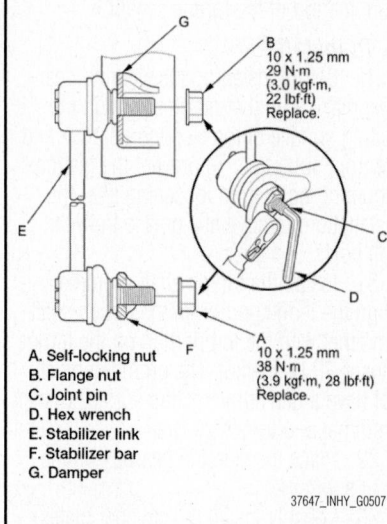

A. Self-locking nut
B. Flange nut
C. Joint pin
D. Hex wrench
E. Stabilizer link
F. Stabilizer bar
G. Damper

B
10 x 1.25 mm
29 N·m
(3.0 kgf·m,
22 lbf·ft)
Replace.

A
10 x 1.25 mm
38 N·m
(3.9 kgf·m, 28 lbf·ft)
Replace.

37647_INHY_G0507

Fig. 228 Remove the self-locking nut and the flange nut while holding the respective joint pin with a hex wrench, then remove the stabilizer link

---

FRONT SUSPENSION

7. Test-drive the vehicle.
8. After 5 minutes of driving, tighten the self-locking nut again to the specified torque.

LOWER BALL JOINT

REMOVAL & INSTALLATION

See Figures 229 through 231.

Special Tools Required:
• Ball Joint Thread Protector, 10 mm 07AAF-SECA120
• Ball Joint Remover, 28 mm 07MAC-SL0A202
• Ball Joint Thread Protector, 14 mm 071AF-S3VA000

➡**Always use a ball joint remover to disconnect a ball joint. Do not strike the housing or any other part of the ball joint connection to disconnect it.**

1. Install a hex nut or the ball joint thread protector onto the threads of the ball joint.

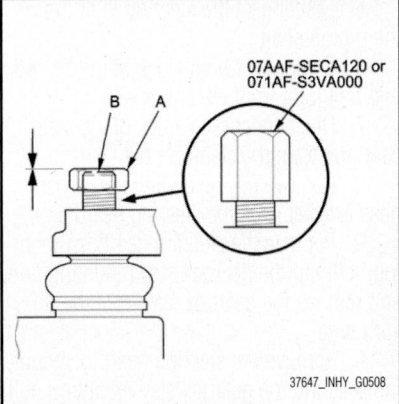

Fig. 229 Install a hex nut (A) or the ball joint thread protector onto the threads of the ball joint (B)

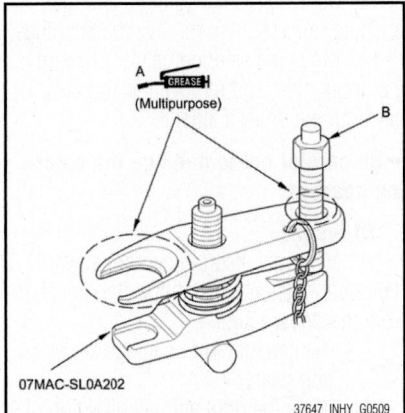

Fig. 230 Apply grease to the ball joint remover on the areas shown (A); this will ease installation of the tool and prevent damage to the pressure bolt (B) threads

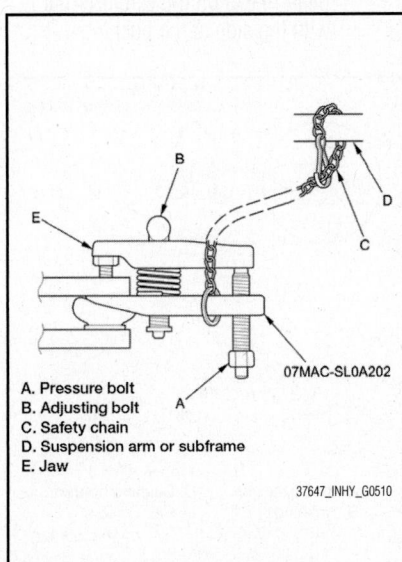

A. Pressure bolt
B. Adjusting bolt
C. Safety chain
D. Suspension arm or subframe
E. Jaw

Fig. 231 Loosen the pressure bolt, and install the ball joint remover as shown

➡️**Using a hex nut, make sure the nut is flush with the ball joint pin end to prevent damage to the threaded end of the ball joint pin.**

2. Apply grease to the ball joint remover on the areas shown. This will ease installation of the tool and prevent damage to the pressure bolt threads.

3. Loosen the pressure bolt, and install the ball joint remover as shown. Insert the jaws carefully, making sure not to damage the ball joint boot. Adjust the jaw spacing by turning the adjusting bolt.

➡️**Fasten the safety chain securely to a suspension arm or the subframe. Do not fasten it to a brake line or wire harness.**

4. After adjusting the adjusting bolt, make sure the head of the adjusting bolt is in the position shown to allow the jaw to pivot.

5. With a wrench, tighten the pressure bolt until the ball joint pin pops loose from the ball joint connecting hole. If necessary, apply penetrating type lubricant to loosen the ball joint pin.

➡️**Do not use pneumatic or electric tools on the pressure bolt.**

6. Remove the ball joint remover, then remove the nut from the end of the ball joint pin, and pull the ball joint out of the ball joint connecting hole. Inspect the ball joint boot, and replace it if damaged.

LOWER CONTROL ARM

REMOVAL & INSTALLATION

See Figures 232 and 233.

Special Tools Required:
- Ball Joint Remover, 28 mm 07MAC-SL0A202
- Ball Joint Thread Protector, 14 mm 071AF-S3VA000

➡️**Do not remove the lower arm from both sides at the same time. The lower arm mounting bolts also secure the subframe to the vehicle.**

1. Raise the vehicle on a lift.
2. Remove the front wheel.
3. Remove the splash shield.
4. Remove the lock pin from the lower arm ball joint, then remove the castle nut.

➡️**During installation, install the lock pin after tightening the new castle nut.**

5. Disconnect the lower ball joint from the knuckle using the ball joint remover.

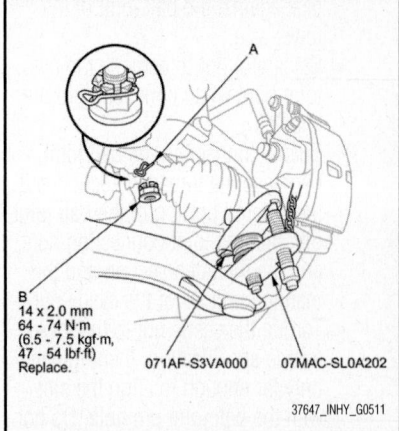

Fig. 232 Remove the lock pin (A) from the lower arm ball joint, then remove the castle nut (B)

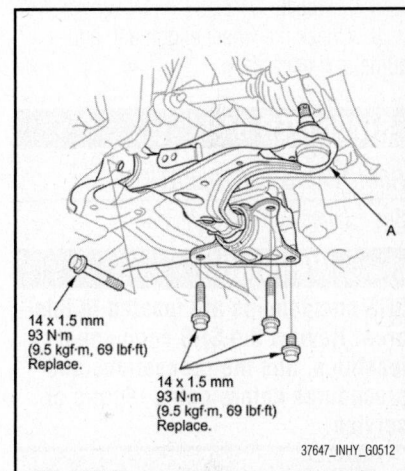

Fig. 233 Remove the lower arm mounting bolts, then remove the lower arm (A) from the front subframe

➡️**Be careful not to damage the ball joint boot when installing the remover. Do not force or hammer on the lower arm, or pry between the lower arm and the knuckle. You could damage the ball joint.**

6. Remove the lower arm mounting bolts, then remove the lower arm from the front subframe.

➡️**Use new lower arm mounting bolts during reassembly.**

To install:

7. Install the lower arm in the reverse order of removal, and note these items:

- First install all the components, and lightly tighten the bolts and the nuts, then raise the suspension to load it with the vehicle's weight before fully tightening to the specified torque values. Do not place the jack against the ball joint of the lower arm.

- Be careful not to damage the ball joint boot when connecting to the knuckle.
- Before connecting the ball joint, degrease the threaded section and the tapered portion of the ball joint pin, the ball joint connecting hole, and the threaded section and the mating surfaces of the castle nut.
- Torque the castle nut to the lower torque specification, then tighten it only far enough to align the slot with the ball joint pin hole. Do not align the castle nut by loosening it.
- Before installing the wheel, clean the mating surfaces of the brake disc and the inside of the wheel.

8. Check the wheel alignment, and adjust it if necessary.

STABILIZER BAR

REMOVAL & INSTALLATION

See Figures 234 through 238.

❊❊ WARNING

SRS components are located in this area. Review the SRS component locations, and the precautions and procedures before doing repairs or service.

1. Do the 12 volt battery terminal disconnection procedure.
2. Raise the vehicle on a lift.
3. Remove the front wheels.
4. Remove the steering wheel.

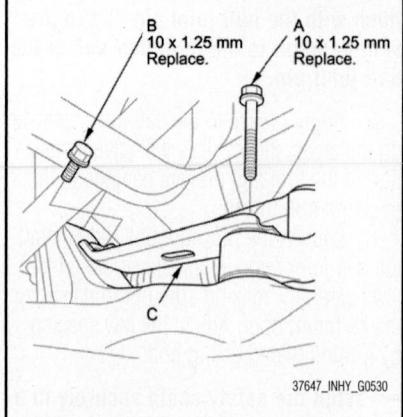

Fig. 235 Remove the steering gearbox mounting bolt (A) and the gearbox stay mounting bolt (B), and remove the gearbox stay (C) from the passenger's side

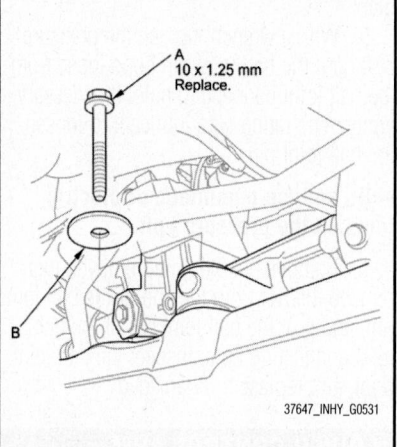

Fig. 236 Remove the steering gearbox mounting bolt (A) and the washer (B) from the rear side of the steering gearbox

5. Disconnect the steering joint from the pinion shaft.
6. Disconnect both sides of the tie-rod end ball joint from the knuckle.
7. Disconnect both sides of the stabilizer link from the stabilizer bar.
8. Remove the secondary HO2S harness bracket from the steering gearbox.
9. Remove the steering gearbox mounting bolt and the gearbox stay mounting bolt, and remove the gearbox stay from the driver's side.
10. Remove the steering gearbox mounting bolt and the gearbox stay mounting bolt, and remove the gearbox stay from the passenger's side.
11. Remove the steering gearbox mounting bolt and the washer from the rear side of the steering gearbox.
12. Remove the flange bolts and the bushing holders, then remove the bushings.
13. Move the steering gearbox toward the upper side, and remove the stabilizer bar from the driver's side.

➡**Be careful not to damage the steering gearbox.**

To install:

14. Move the steering gearbox toward the upper side, and install the stabilizer bar from the driver's side.

- Be careful not to damage the steering gearbox.
- Note the right and left direction of the stabilizer bar.
- Note the direction of installation for the bushings and the bushing holders.
- Align the stabilizer band or the paint marks on the stabilizer bar with the side of the bushings.

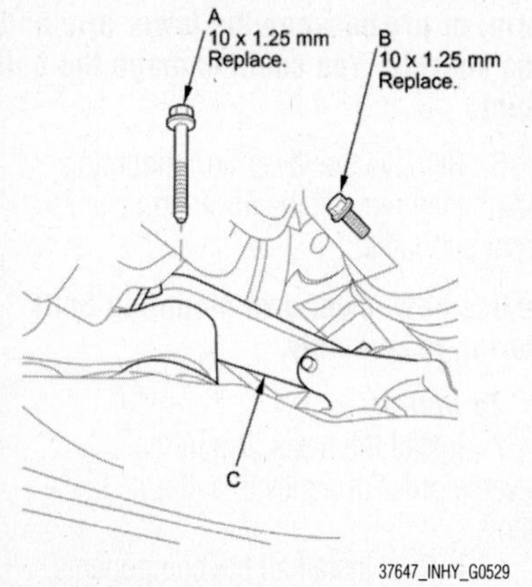

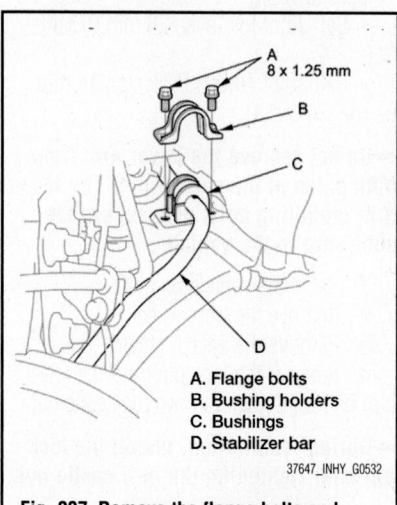

A. Flange bolts
B. Bushing holders
C. Bushings
D. Stabilizer bar

Fig. 237 Remove the flange bolts and the bushing holders, then remove the bushings

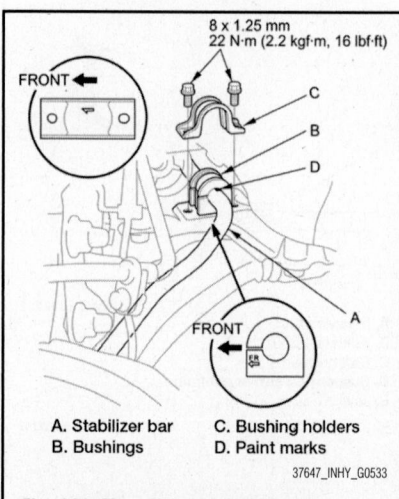

A. Stabilizer bar   C. Bushing holders
B. Bushings   D. Paint marks

Fig. 238 Move the steering gearbox toward the upper side, and install the stabilizer bar from the driver's side

Fig. 234 Remove the steering gearbox mounting bolt (A) and the gearbox stay mounting bolt (B), and remove the gearbox stay (C) from the driver's side

15. Install and loosely tighten the new steering gearbox mounting bolt and the washer from the rear side of the steering gearbox.

16. Install and loosely tighten the new steering gearbox mounting bolt and the new gearbox stay mounting bolt to the gearbox stay on the passenger's side.

17. Install and loosely tighten the new steering gearbox mounting bolt and the new gearbox stay mounting bolt to the gearbox stay on the driver's side.

18. Tighten the steering gearbox mounting bolts and the gearbox stay mounting bolts to 44 ft. lbs. (60 Nm).

19. Install the secondary HO2S harness bracket on the steering gearbox.

20. Connect both sides of the stabilizer link to the stabilizer bar.

21. Connect both sides of the tie-rod end ball joint to the knuckle.

22. Connect the steering joint.

23. Install the steering wheel.

24. Do the 12 volt battery terminal reconnection procedure, and do these tasks:

   a. Turn the ignition switch to ON (II), and check that the SRS indicator should come on for about 6 seconds and then go off.

   b. Make sure the horn and turn signal switches work properly.

   c. Make sure the steering wheel switches work properly.

25. Clean the mating surfaces of the brake disc and the inside of the wheel, then install the front wheels.

26. Check the wheel alignment, and adjust it if necessary.

STABILIZER LINK

REMOVAL & INSTALLATION

See Figure 239.

1. Raise and support the vehicle.
2. Remove the front wheel.
3. Remove the self-locking nut and the flange nut while holding the respective joint pin with a hex wrench, then remove the stabilizer link.

To install:

4. Install the stabilizer link on the stabilizer bar and the damper with the joint pins set at the center of their range of movement.

5. Install the new self-locking nut and the new flange nut, and tighten them to the specified torque while holding the respective joint pin with a hex wrench.

6. Clean the mating surfaces between the brake disc and the inside of the wheel, then install the front wheel.

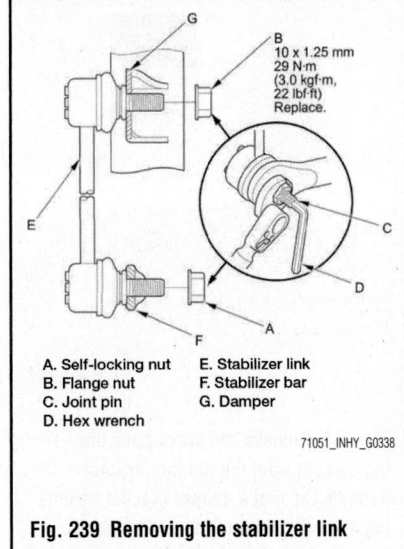

A. Self-locking nut
B. Flange nut
C. Joint pin
D. Hex wrench
E. Stabilizer link
F. Stabilizer bar
G. Damper

71051_INHY_G0338

Fig. 239 Removing the stabilizer link

7. Test-drive the vehicle.
8. After 5 minutes of driving, tighten the self-locking nut again to the specified torque.

STRUT (DAMPER/SPRING)

REMOVAL & INSTALLATION

See Figures 240 through 242.

1. Raise the vehicle on a lift.
2. Remove the front wheel.
3. Remove the wheel speed sensor from the knuckle. Do not disconnect the wheel speed sensor connector.
4. Disconnect the stabilizer link from the damper.
5. Remove the wheel speed sensor clip, the wire guide grommet, and the brake hose bracket from the damper. Do not disconnect the wheel speed sensor connector.
6. Remove the damper pinch bolts and the self-locking nuts from the damper.

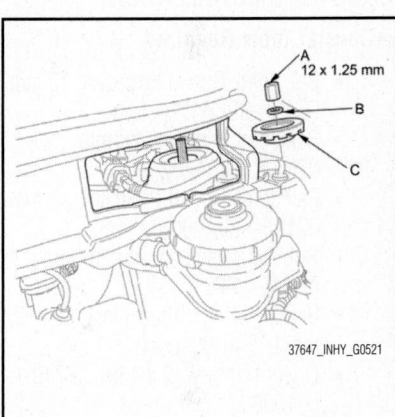

A. Damper cap
B. Damper shaft
C. Hex wrench
D. Damper mounting nut

37647_INHY_G0520

Fig. 240 Remove the damper cap from the top of the damper

➡**Do not allow the knuckle to rotate too far outward. This may allow the driveshaft inboard joint come apart. During installation, install new damper pinch bolts and new self-locking nuts, then tighten the damper pinch bolts to the specified torque value.**

7. Remove the cowl lid from the cowl cover.

8. Remove the damper cap from the top of the damper.

9. Hold the damper shaft using a hex wrench, and loosen the damper mounting nut.

10. Remove the damper mounting nut and the wave washer, then remove the damper mounting base from the top of the damper.

11. Remove the damper/spring and the damper rubber mount.

➡**Be careful not to damage the body.**

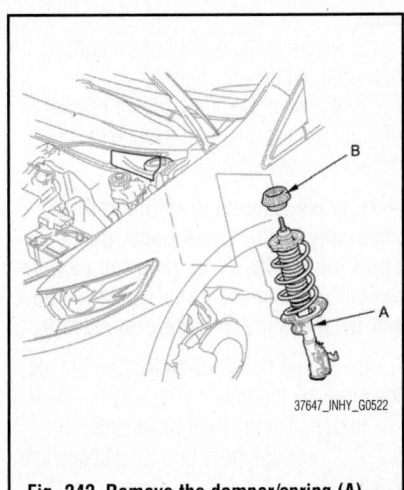

37647_INHY_G0521

Fig. 241 Remove the damper mounting nut (A) and the wave washer (B), then remove the damper mounting base (C) from the top of the damper

37647_INHY_G0522

Fig. 242 Remove the damper/spring (A) and the damper rubber mount (B)

To install:

12. Install all of the removed parts in the reverse order of removal, and note these items:

- First install all the components, and lightly tighten the bolts and the nuts, then raise the suspension to load it with the vehicle's weight before fully tightening to the specified torque values. Do not place the jack against the ball joint of the lower arm.
- Before installing the wheel, clean the mating surfaces of the brake disc and the inside of the wheel.

13. Check the wheel alignment, and adjust it if necessary.

STEERING KNUCKLE, WHEEL HUB & BEARING

REMOVAL & INSTALLATION

➡ **Special Tools Required:**

- Ball Joint Thread Protector, 10 mm 07AAF-SECA120
- Ball Joint Remover, 28 mm 07MAC-SL0A202
- Ball Joint Thread Protector, 14 mm 071AF-S3VA000
- Hub Dis/Assembly Pin, 40 mm 07GAF-SE00100
- Bearing Driver Attachment, 52 x 55 mm 07746-0010400
- Driver Handle, 15 x 135L 07749-0010000
- Support Base 07965-SD90100

Knuckle/Hub Replacement

See Figures 243 through 248.

1. Raise and support the vehicle.
2. Remove the wheel nuts and the front wheel.
3. Remove the brake hose mounting bolt from the damper bracket.
4. Remove the brake caliper bracket mounting bolts, then remove the caliper assembly from the knuckle.

➡ **To prevent damage to the caliper assembly or the brake hose, use a short piece of wire to hang the caliper assembly from the undercarriage. Do not twist the brake hose excessively.**

5. Pry up the stake on the spindle nut, then remove the nut.
6. Remove the front brake disc.
7. Check the front hub for damage and cracks.
8. Remove the wheel speed sensor

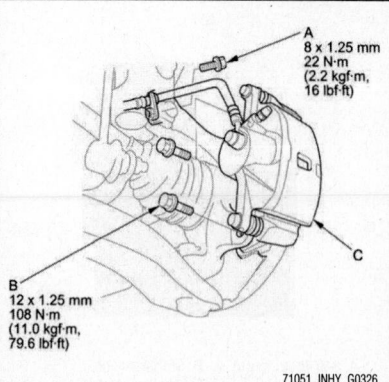

Fig. 243 Remove the brake hose mounting bolt (A) from the damper bracket. Remove the brake caliper bracket mounting bolts (B), then remove the caliper assembly (C) from the knuckle.

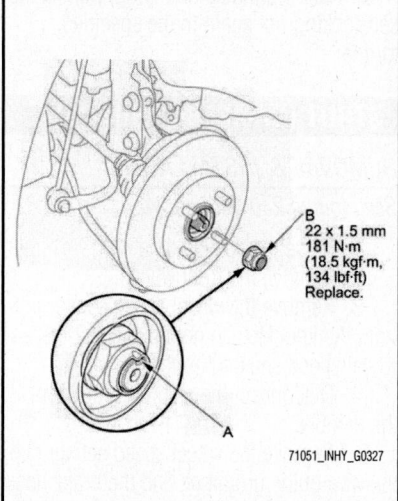

Fig. 244 Pry up the stake (A) on the spindle nut (B), then remove the nut.

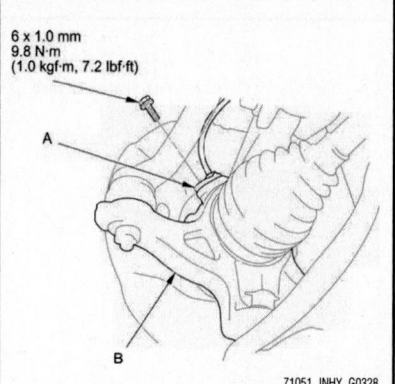

Fig. 245 Remove the wheel speed sensor (A) from the knuckle (B). Do not disconnect the wheel speed sensor 2P connector.

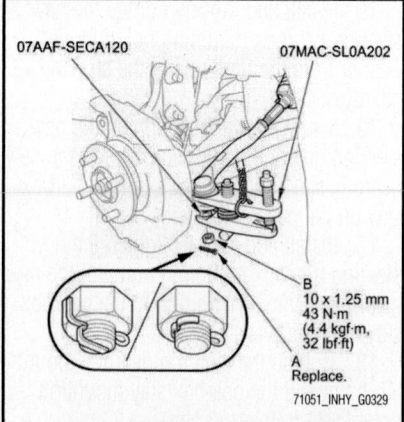

Fig. 246 Remove the cotter pin (A) from the tie-rod end ball joint, then remove the nut (B). During installation, install the new cotter pin after tightening the nut, and bend its end as shown.

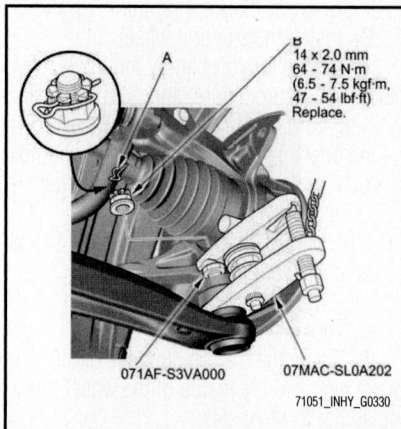

Fig. 247 Remove the lock pin (A) from the lower arm ball joint, then remove the castle nut (B). During installation, install the lock pin as shown after tightening the new castle nut.

from the knuckle. Do not disconnect the wheel speed sensor 2P connector.

9. Remove the cotter pin from the tie-rod end ball joint, then remove the nut.

➡ **During installation, install the new cotter pin after tightening the nut, and bend its end as shown.**

10. Disconnect the tie-rod end ball joint from the knuckle using the ball joint thread protector and the ball joint remover.
11. Remove the lock pin from the lower arm ball joint, then remove the castle nut.

➡ **During installation, install the lock pin as shown after tightening the new castle nut.**

12. Disconnect the lower ball joint from

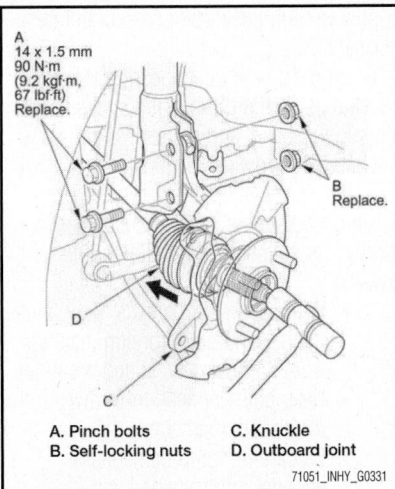

A. Pinch bolts
B. Self-locking nuts
C. Knuckle
D. Outboard joint

71051_INHY_G0331

Fig. 248 Remove the damper pinch bolts and the self-locking nuts from the damper. Use new damper pinch bolts and new self-locking nuts during reassembly. Pull the knuckle outward, and separate the outboard joint from the front hub using a soft face hammer.

the knuckle using the ball joint thread protector and the ball joint remover.

Note the following:

• Be careful not to damage the ball joint boot when installing the remover.

• Do not force or hammer on the lower arm, or pry between the lower arm and the knuckle. You could damage the ball joint.

13. Remove the damper pinch bolts and the self-locking nuts from the damper.

➡**Use new damper pinch bolts and new self-locking nuts during reassembly.**

14. Pull the knuckle outward, and separate the outboard joint from the front hub using a soft face hammer.

Note the following:

• Do not pull the driveshaft end outward. The driveshaft inboard joint may come apart.

• During installation, apply grease to the mating surfaces of the wheel bearing and the driveshaft outboard joint.

To install:

15. Install the knuckle/hub in the reverse order of removal, and note these items:

a. First install all the components, and lightly tighten the bolts and the nuts, then raise the suspension to load it with the vehicle's weight before fully tightening to the specified torque. Do not place the jack against the ball joint of the lower arm.

b. Be careful not to damage the ball joint boot when connecting the knuckle.

c. Before connecting the ball joint, degrease the threaded section and the

tapered portion of the ball joint pin, the ball joint connecting hole, and the threaded section and the mating surfaces of the castle nut.

d. Torque the castle nut to the lower torque specification, then tighten it only far enough to align the slot with the ball joint pin hole. Do not align the castle nut by loosening it.

e. Use a new spindle nut on reassembly.

f. Before installing the spindle nut, apply a small amount of engine oil to the seating surface of the nut. After tightening, use a drift to stake the spindle nut shoulder against the driveshaft.

g. Before installing the brake disc, clean the mating surfaces of the front hub and the inside of the brake disc.

h. Before installing the wheel, clean the mating surfaces between the brake disc and the inside of the wheel.

16. Check the wheel alignment, and adjust it if necessary.

Wheel Bearing Replacement

See Figures 249 through 254.

1. Separate the hub from the knuckle using the hub dis/assembly tool and a hydraulic press. Hold the knuckle with the attachment of the hydraulic press or equivalent tool. Be careful not to damage or deform the splash guard. Hold onto the hub to keep it from falling when pressed clear.

2. Press the wheel bearing inner race off

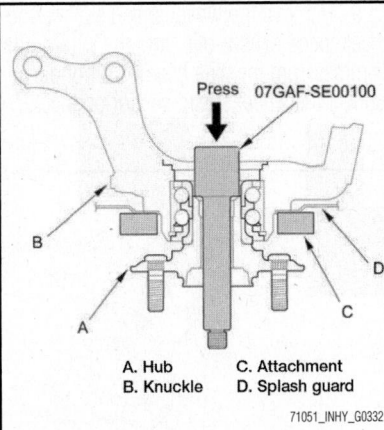

A. Hub
B. Knuckle
C. Attachment
D. Splash guard

71051_INHY_G0332

Fig. 249 Separate the hub from the knuckle using the hub dis/assembly tool and a hydraulic press. Hold the knuckle with the attachment of the hydraulic press or equivalent tool. Be careful not to damage or deform the splash guard. Hold onto the hub to keep it from falling when pressed clear.

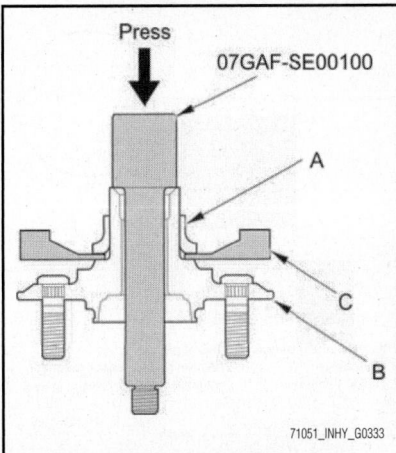

71051_INHY_G0333

Fig. 250 Press the wheel bearing inner race (A) off of the hub (B) using the hub dis/assembly tool, a commercially available bearing separator (C) and a press.

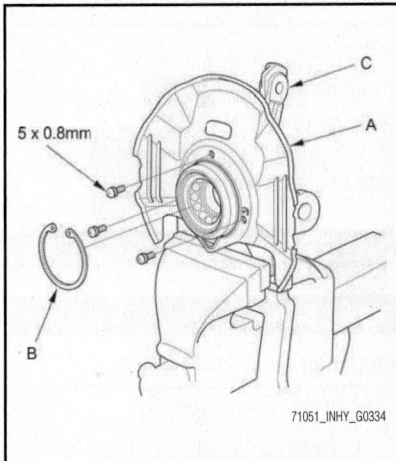

71051_INHY_G0334

Fig. 251 Remove the splash guard (A) and the snap ring (B) from the knuckle (C).

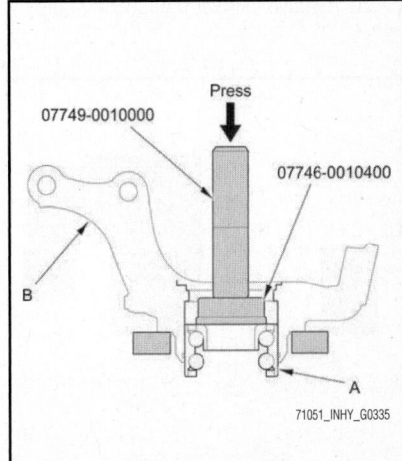

71051_INHY_G0335

Fig. 252 Press the wheel bearing (A) out of the knuckle (B) using the bearing driver attachment, the driver handle, and a press.

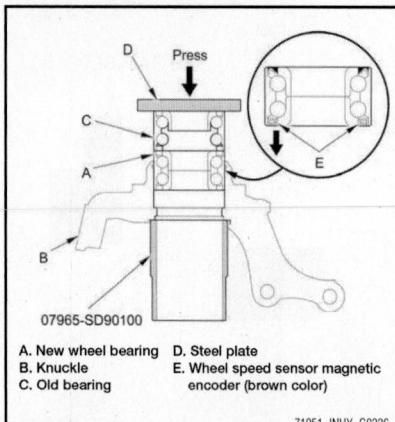

Fig. 253 Press a new wheel bearing into the knuckle using the old bearing, a steel plate the support base, and a press. Install the wheel bearing with the wheel speed sensor magnetic encoder (brown color) toward the inside of the knuckle.

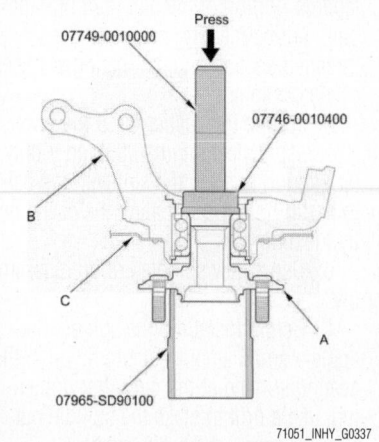

Fig. 254 Install the hub (A) onto the knuckle (B) using the bearing driver attachment, the driver handle, the support base, and a hydraulic press. Be careful not to damage the splash guard (C).

of the hub using the hub dis/assembly tool, a commercially available bearing separator and a press.

3. Remove the splash guard and the snap ring from the knuckle.

4. Press the wheel bearing out of the knuckle using the bearing driver attachment, the driver handle, and a press.

5. Wash the knuckle and the hub thor-oughly in high flash point solvent before reassembly.

6. Press a new wheel bearing into the knuckle using the old bearing, a steel plate the support base, and a press.

Note the following:

• Install the wheel bearing with the wheel speed sensor magnetic encoder (brown color) toward the inside of the knuckle.

  • Remove any oil, grease, dust, metal debris, and other foreign material from the magnetic encoder surface.

  • Keep any magnetic tools away from the encoder surface.

  • Be careful not to damage the encoder surface when you insert the wheel bearing.

7. Install the snap ring securely in the knuckle.

8. Install the splash guard and tighten the screws.

9. Install the hub onto the knuckle using the bearing driver attachment, the driver handle, the support base, and a hydraulic press. Be careful not to damage the splash guard.

SUSPENSION

REAR SUSPENSION

AXLE BEAM

REMOVAL & INSTALLATION

See Figures 255 through 259.

1. Raise the vehicle on a lift.
2. Remove the rear wheels.
3. Remove the rear suspension lower cover.
4. Remove the rear strake.
5. Remove the rear hub bearing unit.

6. Remove the parking brake cable from the axle beam.

7. Disconnect the brake line from the wheel cylinder, and plug the line with a shop towel.

8. Remove the backing plate with the brake shoes assembly from the spindle.

9. Remove the wheel speed sensor, the wheel speed sensor clip, and the wire guide grommet from the axle beam. Do not disconnect the wheel speed sensor connector.

10. Remove the rear spring.

11. Disconnect the brake hose from both sides of the brake line, then remove the brake line by removing the brake hose clip.

➡**Do not spill brake fluid on the vehicle; it may damage the paint; if brake fluid gets on the paint, wash it off immediately with water. Plug the end of a hose and joints to prevent spilling brake fluid. During installation, install new brake hose clips.**

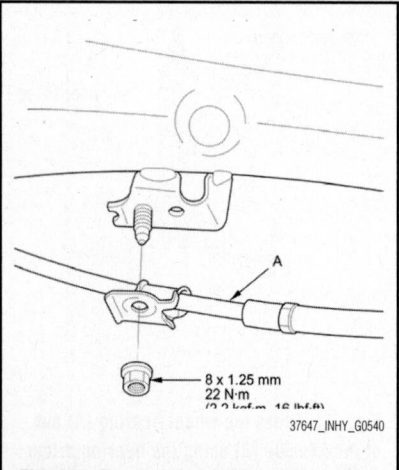

Fig. 255 Remove the parking brake cable (A) from the axle beam

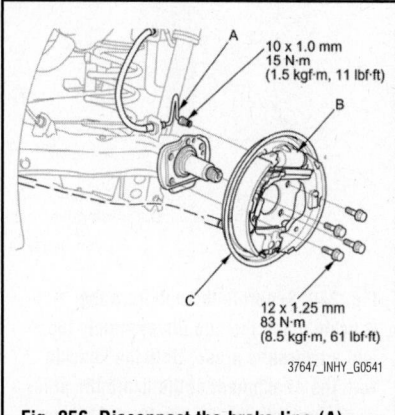

Fig. 256 Disconnect the brake line (A) from the wheel cylinder (B); remove the backing plate (C) with the brake shoes assembly

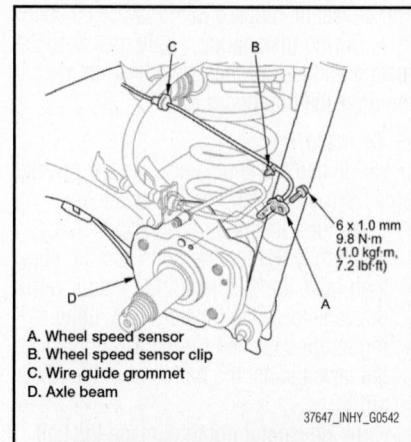

A. Wheel speed sensor
B. Wheel speed sensor clip
C. Wire guide grommet
D. Axle beam

Fig. 257 Remove the wheel speed sensor, the wheel speed sensor clip, and the wire guide grommet from the axle beam

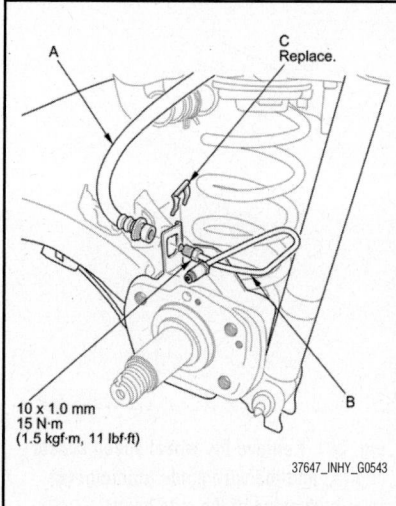

Fig. 258 Disconnect the brake hose (A) from both sides of the brake line (B), then remove the brake line by removing the brake hose clip (C)

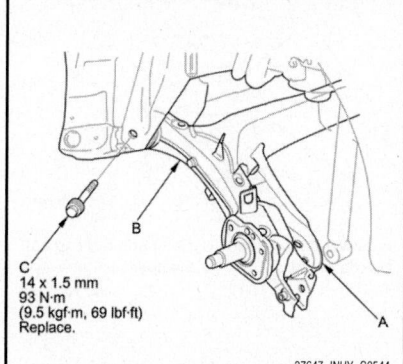

Fig. 259 Place a floor jack under the lower spring seat (A) on both sides of the axle beam (B), and support it by raising the floor jack; remove the axle beam mounting bolt (C) on both sides

12. Place a floor jack under the lower spring seat on both sides of the axle beam, and support it by raising the floor jack. Do not place the floor jack under the center of the axle beam.

13. Remove the axle beam mounting bolt on both sides.

➡**Use new axle beam mounting bolts during reassembly.**

14. Lower the jack slowly, then remove the axle beam.

To install:

15. Install the axle beam in the reverse order of removal, and note these items:
   • First install all the components, and lightly tighten the bolts, and place a jack under the lower spring

seat of the axle beam on both sides, then raise the suspension to load with the vehicle's weight before fully tightening to the specified torque values.
   • After installing the brake hose, the brake line, and the parking brake cable, check for interference and twisting of other parts.
   • Before installing the brake drum, clean the mating surfaces of the hub bearing unit and the inside of the brake drum.
   • After installing, fill the reservoir with new brake fluid, and bleed the brake system.
   • Check the brake hose and line joint for leaks, and tighten if necessary.
   • Before installing the wheel, clean the mating surfaces of the brake drum and the inside of the wheel.

16. Check the wheel alignment, and adjust it if necessary.

COIL SPRING

REMOVAL & INSTALLATION

See Figures 260 through 265.

1. Raise the vehicle on a lift.
2. Remove the rear wheels.
3. Remove the wheel speed sensor, the wheel speed sensor clip, and the wire guide grommet from both sides of the axle beam. Do not disconnect the wheel speed sensor connector.
4. Position the floor jack under spring seat on both sides of the axle beam. Raise the floor jack until the suspension begins to compress.
5. Remove the damper mounting bolt from both sides.
6. Lower the floor jack gradually.
7. Remove the spring, the upper rubber mount, and the lower rubber mount. Do not lower the jack more than necessary.

To install:

8. Install the upper rubber mount on the spring by aligning the upper end of the spring with the ledge portion of the upper rubber mount.
9. Install the lower rubber mount on the spring by aligning the lower end of the spring with the ledge portion of the lower rubber mount.
10. Install the tab of the lower rubber mount into the groove of the lower spring seat.

➡**Make sure that the tab of the lower rubber mount is properly installed into**

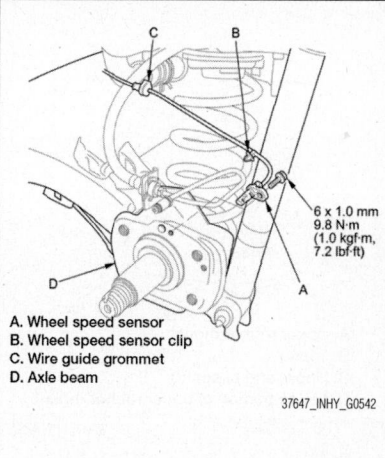

A. Wheel speed sensor
B. Wheel speed sensor clip
C. Wire guide grommet
D. Axle beam

Fig. 260 Remove the wheel speed sensor, the wheel speed sensor clip, and the wire guide grommet from both sides of the axle beam

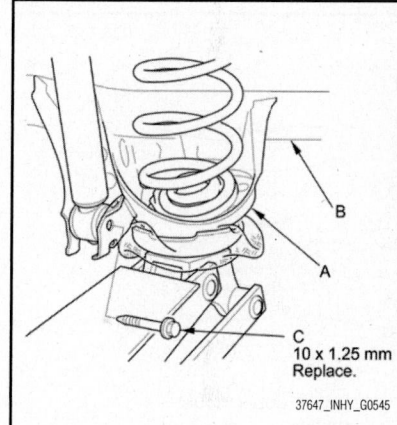

Fig. 261 Position the floor jack under spring seat (A) on both sides of the axle beam (B); remove the damper mounting bolt (C) from both sides

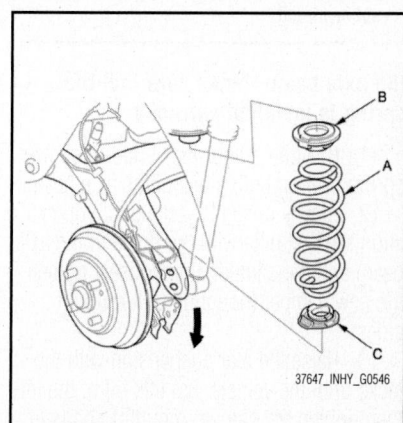

Fig. 262 Remove the spring (A), the upper rubber mount (B), and the lower rubber mount (C)

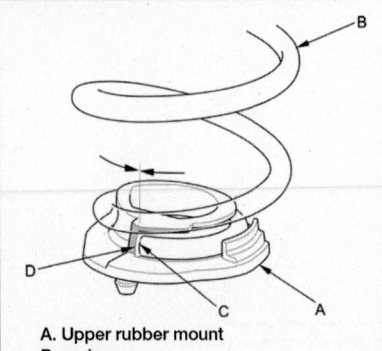

A. Upper rubber mount
B. spring
C. Upper end of spring
D. Ledge portion of upper rubber mount

37647_INHY_G0547

Fig. 263 Install the upper rubber mount on the spring by aligning the upper end of the spring with the ledge portion of the upper rubber mount

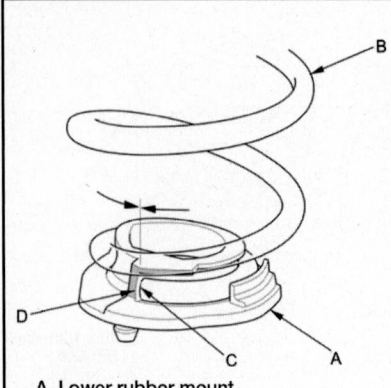

A. Lower rubber mount
B. Spring
C. Lower end of spring
D. Ledge portion of lower rubber mount

37647_INHY_G0548

Fig. 264 Install the lower rubber mount on the spring by aligning the lower end of the spring with the ledge portion of the lower rubber mount

the axle beam. Make sure that the spring is installed correctly.

11. Position a floor jack under the lower spring seat on both sides of the axle beam.

12. Slowly raise the jacks until you can align the bolt hole with the holes in the axle beam and the damper, then loosely tighten the new damper mounting bolt on both sides.

13. Raise the rear suspension with the jacks until the vehicle just lifts off of the lift, then tighten the damper mounting bolts to the specified torque.

14. Install the wheel speed sensor, the wheel speed sensor clip, and the wire guide grommet on both sides of the axle beam.

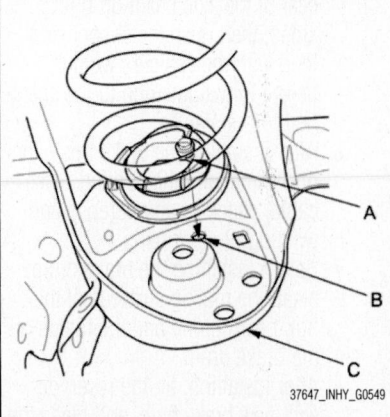

37647_INHY_G0549

Fig. 265 Install the tab (A) of the lower rubber mount into the groove (B) of the lower spring seat

15. Clean the mating surfaces of the brake drum and the inside of the wheel, then install the rear wheel.

DYNAMIC DAMPER

REMOVAL & INSTALLATION

See Figure 266.

1. Raise and support the vehicle.
2. Remove the dynamic damper from the axle beam.
3. Install the new dynamic damper in the reverse order of removal.
4. Use new self-locking nuts during reassembly.

SHOCK ABSORBER (DAMPER)

REMOVAL & INSTALLATION

See Figures 267 through 270.

1. Raise the vehicle on a lift.
2. Remove the rear wheels.

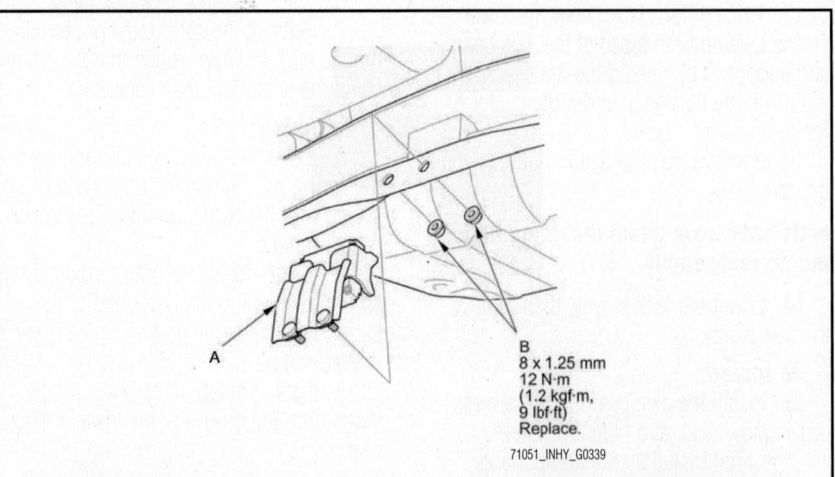

B
8 x 1.25 mm
12 N·m
(1.2 kgf·m,
9 lbf·ft)
Replace.

71051_INHY_G0339

Fig. 266 Remove the dynamic damper (A) from the axle beam. Use new self-locking nuts (B) during reassembly.

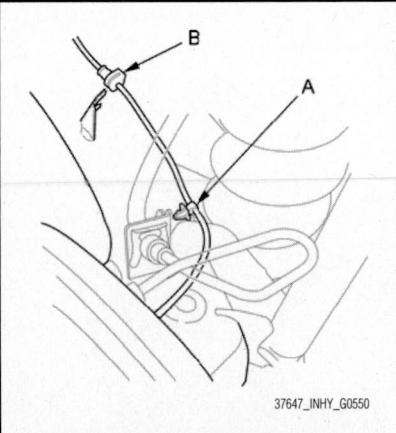

37647_INHY_G0550

Fig. 267 Remove the wheel speed sensor clip (A) and the wire guide grommet (B) from both sides of the axle beam

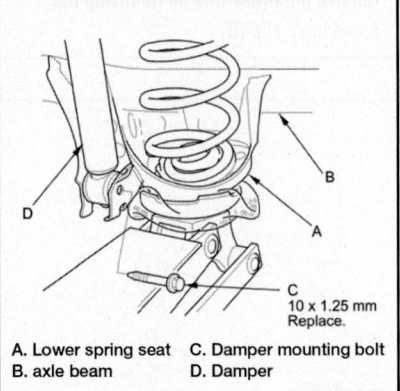

C
10 x 1.25 mm
Replace.

A. Lower spring seat   C. Damper mounting bolt
B. axle beam   D. Damper

37647_INHY_G0551

Fig. 268 Position a floor jack under lower spring seat on both sides of the axle beam

3. Remove the wheel speed sensor clip and the wire guide grommet from both sides of the axle beam.

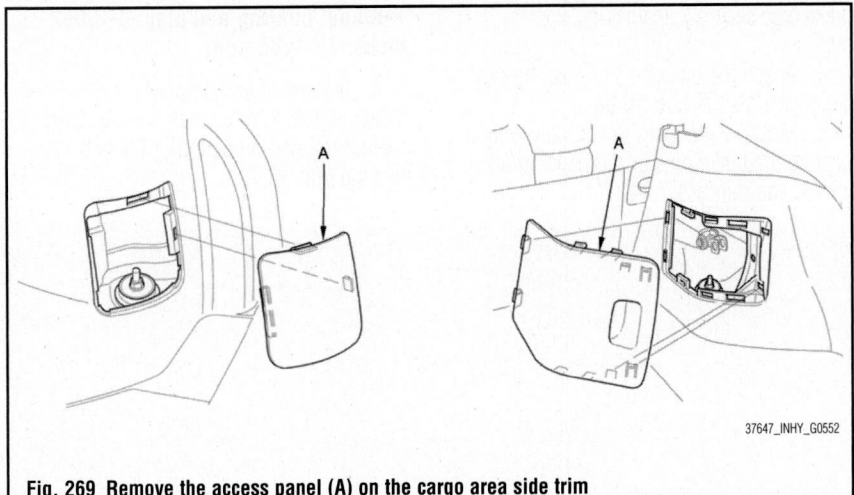

Fig. 269 Remove the access panel (A) on the cargo area side trim

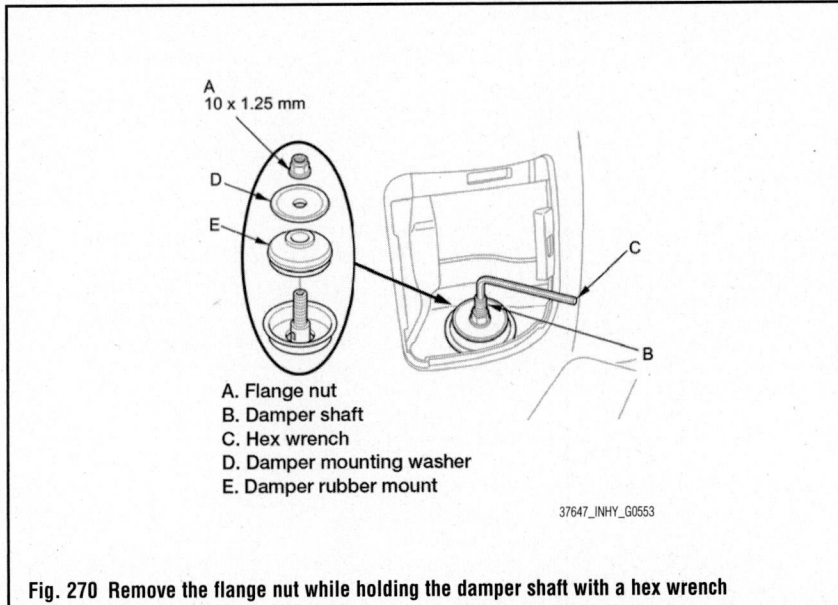

A. Flange nut
B. Damper shaft
C. Hex wrench
D. Damper mounting washer
E. Damper rubber mount

Fig. 270 Remove the flange nut while holding the damper shaft with a hex wrench

4. Position a floor jack under lower spring seat on both sides of the axle beam. Raise the floor jack until the suspension begins to compress.

5. Remove the damper mounting bolt that connects the axle beam and the damper.

6. Remove the access panel on the cargo area side trim.

7. Remove the flange nut while holding the damper shaft with a hex wrench.

8. Remove the damper mounting washer and the damper rubber mount from the top of the damper.

9. Compress the damper assembly by hand, and remove it from the vehicle.

➡Be careful not to damage the body.

10. Remove the damper rubber mount.

To install:

11. Install the damper rubber mount onto the damper unit. Position the damper

assembly between the body and the axle beam.

12. Position the floor jack under lower spring seat on both sides of the axle beam.

13. Slowly raise the jack until you can align the bolt hole with the holes in the axle beam and the damper, then loosely tighten the new damper mounting bolt on both sides.

14. Raise the rear suspension with the jack until the vehicle just lifts off of the lift, then tighten the damper mounting bolts to 40 ft. lbs. (54 Nm).

15. Install the damper rubber mount, the damper mounting washer, and the flange nut on the damper shaft.

➡During installation, note the direction of the damper rubber mount and the damper mounting washer.

16. Tighten the flange nut) to 22 ft. lbs. (29 Nm) while holding the damper shaft with a hex wrench.

17. Install the access panel on the cargo area side trim.

18. Install the wheel speed sensor clip and the wire guide grommet on both sides of the axle beam.

19. Clean the mating surfaces of the brake drum and the inside of the wheel, then install the rear wheel.

TESTING

1. Install the flange nut on the damper shaft end, and set the socket wrench and T-handle on the nut.

2. Compress the damper unit by hand, and check for smooth operation through a full stroke, both compression and extension. The damper should extend smoothly and constantly when compression is released. If it does not, the gas is leaking and the damper should be replaced.

3. Check for oil leaks, abnormal noises, and binding during these tests.

WHEEL HUB & BEARING

REMOVAL & INSTALLATION

1. Raise the vehicle on a lift.

2. Remove the wheel nuts and the rear wheel.

3. Remove the brake drum.

4. Remove the hub cap. Pry up the stake on the spindle nut, then remove the nut.

5. Remove the hub bearing unit from the spindle.

6. Check the hub bearing unit for damage and cracks.

To install:

7. Install the hub bearing unit in the reverse order of removal, and note these items:

- Tighten all mounting hardware to the specified torque values.
- Use a new spindle nut and hub cap on reassembly.
- Before installing the spindle nut, apply a small amount of engine oil to the seating surface of the nut. After tightening, use a drift to stake the spindle nut shoulder against the spindle.
- Before installing the brake drum, clean the mating surface of the hub bearing unit and the inside of the brake drum.
- Before installing the wheel, clean the mating surface of the brake drum and the inside of the wheel.

ADJUSTMENT

1. Raise the vehicle on a lift.
2. Remove the wheels.
3. Install suitable flat washers and the wheel nuts. Tighten the nuts to the specified torque to hold the brake disc securely against the hub.
4. Attach the dial gauge. Place the dial gauge against the hub flange.
5. Measure the bearing end play while moving the brake disc or the brake drum inward and outward.

→**Wheel bearing end play: 0–0.002 inches (0–0.05 mm)**

6. If the bearing end play measurement is more than the standard, replace the wheel bearing or the hub bearing unit.

HONDA

Odyssey

14

BRAKES **14-9**

ANTI-LOCK BRAKE SYSTEM..14-9
Wheel Speed Sensors 14-9
  Removal & Installation..........14-9
**BLEEDING THE BRAKE
SYSTEM** **14-10**
Bleeding Procedure..............14-10
  Bleeding the ABS System ...14-10
  Fluid Recommendations 14-10
  Level Check 14-10
  Manual 14-10
FRONT DISC BRAKES **14-11**
Brake Pads 14-11
  Removal & Installation.......14-11
PARKING BRAKE **14-13**
Adjustments 14-13
Parking Brake Shoes 14-13
  Removal & Installation.......14-13
REAR DISC BREAKS **14-12**
Break Pads 14-12
  Removal & Installation.......14-12

CHASSIS ELECTRICAL **14-15**

**AIR BAG (SUPPLEMENTAL
RESTRAINT SYSTEM)** **14-15**
General Information..................14-15
  Arming the System 14-16
  Disarming the System........14-16
  Service Precautions 14-15

DRIVE TRAIN **14-16**

Automatic Transaxle 14-16
  Drain & Refill 14-16
  Filter Replacement 14-17
  Fluid Level Check 14-17
Front Halfshafts 14-18
  Removal & Installation.......14-18

ENGINE COOLING **14-20**

Electric Engine Fan..................14-21
  Removal & Installation.......14-21
Engine Coolant........................14-20
  Bleeding 14-20
  Drain & Refill 14-20
  Level Check 14-21

Radiator................................14-22
  Removal & Installation........14-22
Thermostat 14-22
  Removal & Installation........14-22
Water Pump 14-22
  Removal & Installation........14-22

ENGINE ELECTRICAL **14-23**

BATTERY SYSTEM............ **14-23**
Battery................................14-23
  Battery
  Disconnection/Reconnection
  Procedure 14-23
  Removal & Installation........14-23
  Resetting the Power
  Tailgate Control Unit.........14-23
  Resetting the Power
  Sliding Door Control
  Unit 14-23
CHARGING SYSTEM **14-24**
Alternator 14-24
  Removal & Installation........14-24
IGNITION SYSTEM **14-24**
Firing Orders.......................14-24
Ignition Coils & Spark Plugs 14-24
  Removal & Installation........14-24
Ignition Timing......................14-24
  Inspection & Adjustment 14-24
STARTING SYSTEM **14-25**
Starter 14-25
  Removal & Installation........14-25

ENGINE MECHANICAL...... **14-25**

Accessory Drive Belt System...14-25
  Adjustment 14-25
  Belt Routings 14-25
  Inspection 14-25
  Removal & Installation........14-25
Air Cleaner 14-26
  Removal & Installation........14-26
Camshaft & Bearings 14-26
  Removal & Installation........14-26
Crankshaft Front Seal..............14-27
  Removal & Installation........14-27
Crankshaft Pulley 14-27
  Removal & Installation........14-27

Crankshaft Rear Seal 14-28
  Removal & Installation........14-28
Cylinder Head 14-28
  Removal & Installation........14-28
Cylinder Head Cover 14-30
  Removal & Installation........14-30
Engine Oil & Filter 14-32
  Oil & Filter Change.............14-32
  Oil Level Check.................14-32
Exhaust Manifold 14-33
  Removal & Installation........14-33
Intake Manifold 14-33
  Removal & Installation........14-33
Oil Pan 14-34
  Removal & Installation........14-34
Oil Pump 14-35
  Removal & Installation........14-35
Pistons & Rings 14-36
  Positioning 14-36
Rear Main Seal......................14-36
  Removal & Installation........14-36
Rocker Arms.........................14-37
  Removal & Installation........14-37
Timing Belt & Sprockets 14-38
  Removal & Installation........14-38
Timing Belt Front Cover 14-38
  Removal & Installation........14-38
Valve Lash (Clearance)
  Adjustment...........................14-42
  Adjustment 14-42

**ENGINE PERFORMANCE &
EMISSION CONTROLS** **14-44**

Camshaft Position Sensor........14-44
  Removal & Installation........14-44
Crankshaft Position Sensor.....14-44
  CKP Patteern Clear/CKP
  Pattern Learn
  Procedure 14-44
  Removal & Installation........14-44
EGR Valve 14-45
  Removal & Installation........14-45
Engine Coolant Temperature
Sensor 14-44
  Removal & Installation........14-44
EVAP Canister 14-45
  Removal & Installation........14-45

Heated Oxygen Sensor
(HO2S)......................14-46
    Removal & Installation........14-46
Knock Sensor (KS).................14-46
    Location.........................14-46
    Removal & Installation........14-46
Maintenance Reminder Lights ...14-47
    Reset Procedure.................14-47
Manifold Absolute Pressure
(MAP) Sensor 14-47
    Removal & Installation........14-47
Mass Airflow/Intake Air
Temperature Sensor 14-47
    Location.........................14-47
    Removal & Installation........14-47
PCV Valve...........................14-47
    Removal & Installation........14-47
Powertrain Control Module.....14-47
    CKP Pattern Clear/CKP
    Pattern Learn 14-49
    PCM Idle Learn Procedure....14-49
    PCM Update 14-48
    Removal & Installation........14-47

FUEL 14-50

GASOLINE FUEL INJECTION
SYSTEM 14-50
Fuel Filter 14-51
    Removal & Installation........14-51
Fuel Rail & Injectors 14-52
    Removal & Installation........14-52
Fuel Tank Unit 14-51
    Removal & Installation........14-51
Idle Speed 14-51
    Adjustment 14-51
Relieving Fuel System
    Pressure.......................14-50
Throttle Body.......................14-53
    Removal & Installation........14-53

HEATING & AIR CONDITIONING
SYSTEM 14-54
Blower Motor 14-54
    Removal & Installation........14-54
Heater Core 14-54
    Removal & Installation........14-54

PRECAUTIONS 14-9

SPECIFICATION AND
MAINTENANCE CHART 14-3

SPECIFICATIONS 14-3
Brake Specifications.................14-7
Camshaft & Bearings
    Specifications 14-5
Capacities 14-4
Crankshaft & Connecting Rod
    Specifications 14-5
Engine and Model Year
    Identification 14-3
Engine Tune-Up
    Specifications 14-3
Fluid Specifications.................14-4
General Engine
    Specifications 14-3
Piston & Ring
    Specifications 14-5
Scheduled Maintenance
    Intervals 14-8
Tire, Wheel, & Ball Joint
    Specifications 14-7
Torque Specifications................14-6
Valve Specifications 14-4
Wheel Alignment
    Specifications 14-6

STEERING 14-55
Power Rack & Pinion Steering
    Gear 14-55
    Removal & Installation........14-55
Power Steering Fluid...............14-58
    Fluid Fill Procedure...........14-58
    Fluid Level Check 14-58
Power Steering Pump..............14-57
    Removal & Installation........14-57

SUSPENSION 14-59
FRONT SUSPENSION 14-59
Knuckle & Hub Bearing Unit ...14-59
    Adjustment 14-59
    Removal & Installation........14-59
Lower Ball Joints...................14-60
    Removal & Installation........14-60
Lower Control Arms 14-61
    Removal & Installation........14-61
Stabilizer Bar & Links 14-61
    Removal & Installation........14-61
Struts (Damper/Spring)...........14-62
    Removal & Installation........14-62
REAR SUSPENSION 14-63
Coil Springs.......................14-63
    Removal & Installation........14-63
Knuckle/Hub Bearing Unit.......14-64
    Adjustment 14-64
    Removal & Installation........14-64
Lower Control Arms 14-65
    Removal & Installation........14-65
Shock Absorbers...................14-66
    Removal & Installation........14-66
Trailing Arms.......................14-67
    Removal & Installation........14-67
Upper Arms.......................14-68
    Removal & Installation........14-68
Upper Ball Joints...................14-67
    Removal & Installation........14-67

SPECIFICATIONS

ENGINE AND VEHICLE IDENTIFICATION

		Engine						Model Year	
Code/VIN ①	Liters (cc)	Cu. In.	Cyl.	Fuel Sys.	Engine Type	Eng. Mfg.		Code ②	Year
J35Z8/2	3.5 (3471)	211	6	MPFI	SOHC	Honda		B	2011
								C	2012

① 8th position of VIN

② 10th position of VIN

71051_HODY_C0001

GENERAL ENGINE SPECIFICATIONS

All measurements are given in inches.

Year	Model	Engine Displacement Liters (cc)	Engine ID/VIN	Fuel System Type	Net Horsepower @ rpm	Net Torque @ rpm (ft. lbs.)	Bore x Stroke (in.)	Com-pression Ratio	Oil Pressure @ rpm
2011	Odyssey	3.5 (3471)	J35Z8/2	MPFI	248@5700	250@4800	3.50x3.66	10.5:1	71@3000 10@Idle
2012	Odyssey	3.5 (3471)	J35Z8/2	MPFI	248@5700	250@4800	3.50x3.66	10.5:1	71@3000 10@Idle

71051_HODY_C0002

ENGINE TUNE-UP SPECIFICATIONS

Year	Engine Displacement Liters	Engine ID/VIN	Spark Plug Gap (in.)	Ignition Timing (deg.) AT	Fuel Pump (psi)	Idle Speed (RPM) AT	Valve Clearance Intake	Valve Clearance Exhaust
2011	3.5	J35Z8/2	0.039-0.043	10+/- BTDC	57-64	690-790	0.008-0.009	0.011-0.012
2012	3.5	J35Z8/2	0.039-0.043	10+/- BTDC	57-64	690-790	0.008-0.009	0.011-0.012

71051_HODY_C0003

CAPACITIES

Year	Model	Engine Displacement Liters	Engine ID/VIN	Engine Oil with Filter	Transaxle (pts.) 5-Speed	6-Speed	Fuel Tank (gal.)	Cooling System (qts.)
2011	Odyssey	3.5	J35Z8/2	4.5	①	②	21.0	10.6
2012	Odyssey	3.5	J35Z8/2	4.5	①	②	21.0	10.6

NOTE: All capacities are approximate. Add fluid gradually and ensure a proper fluid level is obtained.

① Fluid change: 3.3 qts.
   Overhaul: 8.0 qts.

② Fluid change: 3.6 qts.
   Overhaul: 7.4 qts.

71051_HODY_C0004

FLUID SPECIFICATIONS

Year	Model	Engine Disp. Liters	Engine Oil	Auto. Trans.	Power Steering Fluid	Brake Master Cylinder	Cooling System
2011	Odyssey	3.5	0W-20	①	Honda	Honda	②
2012	Odyssey	3.5	0W-20	①	Honda	Honda	②

DOT: Department Of Transpotation

① Honda Automatic Transmission Fluid (ATF DW-1)

② Honda Long Life Antifreeze/Coolant Type 2

71051_HODY_C0005

VALVE SPECIFICATIONS

Year	Engine Displacement Liters	Engine ID/VIN	Seat Angle (deg.)	Face Angle (deg.)	Spring Test Pressure (lbs. @ in.)	Spring Free-Length (in.)	Spring Installed Height (in.)	Stem-to-Guide Clearance (in.) Intake	Exhaust	Stem Diameter (in.) Intake	Exhaust
2011	3.5	J35Z8/2	NA	NA	NA	NA	NA	0.0008-0.0018	0.0022-0.0032	0.2159-0.2163	0.2146-0.2150
2012	3.5	J35Z8/2	NA	NA	NA	NA	NA	0.0008-0.0018	0.0022-0.0032	0.2159-0.2163	0.2146-0.2150

NA Not Available

71051_HODY_C0006

CAMSHAFT SPECIFICATIONS
All measurements in inches unless noted

Year	Engine Displacement Liters	Engine Code/VIN	Journal Diameter	Brg. Oil Clearance	Shaft End-play	Runout	Journal Bore	Lobe Height Intake	Lobe Height Exhaust
2011	3.5	J35Z8/2	NA	0.0020-0.0035	0.0020-0.0079	0.0012	NA	①	②
2012	3.5	J35Z8/2	NA	0.0020-0.0035	0.0020-0.0079	0.0012	NA	①	②

NA Not Available

① Intake Nos. 1, 2, 3, 4: 1.3843 inches

   Intake Nos. 5, 6: 1.3841 inches

② Exhaust No.s 1, 2, 3, 4: 1.4385 inches

   Exhaust Nos. 5, 6: 1.4375 inches

71051_HODY_C0007

CRANKSHAFT AND CONNECTING ROD SPECIFICATIONS
All measurements are given in inches.

Year	Engine Displacement Liters	Engine ID/VIN	Crankshaft Main Brg. Journal Dia.	Crankshaft Main Brg. Oi Clearance	Crankshaft Shaft End-play	Crankshaft Thrust on No.	Connecting Rod Journal Diameter	Connecting Rod Oil Clearance	Connecting Rod Side Clearance
2011	3.5	J35Z8/2	2.8337-2.8346	0.0008-0.0018	0.0039-0.0138	3	2.1644-2.1654	0.0008-0.0017	NA
2012	3.5	J35Z8/2	2.8337-2.8346	0.0008-0.0018	0.0039-0.0138	3	2.1644-2.1654	0.0008-0.0017	NA

NA Not Available

71051_HODY_C0008

PISTON AND RING SPECIFICATIONS
All measurements are given in inches.

Year	Engine Displacement Liters	Engine ID/VIN	Piston Clearance	Ring Gap Top Compression	Ring Gap Bottom Compression	Ring Gap Oil Control	Ring-To-Groove Clearance Top Compression	Ring-To-Groove Clearance Bottom Compression	Ring-To-Groove Clearance Oil Control
2011	3.5	J35Z8/2	0.0006-0.0016	0.008-0.013	0.016-0.021	0.008-0.027	0.003-0.003	0.002-0.002	NA
2012	3.5	J35Z8/2	0.0006-0.0016	0.008-0.013	0.016-0.021	0.008-0.027	0.003-0.003	0.002-0.002	NA

NA Not Available

25742_IMPA_C0009

TORQUE SPECIFICATIONS
All readings in ft. lbs.

Year	Engine Disp. Liters	Engine ID/VIN	Cylinder Head Bolts	Main Bearing Bolts	Rod Bearing Bolts	Crankshaft Pulley Bolt	Drive Plate Bolts	Intake Manifold	Spark Plugs	Oil Pan Drain Plug
2011	3.5	J35Z8/2	①	②	③	④	55	16	16	30
2012	3.5	J35Z8/2	①	②	③	④	55	16	16	30

① Step 1: 22 ft. lbs.

Step 2: Additional 90 degrees

Step 3: Additional 90 degrees

② Main bearing cap bolts: 55 ft. lbs.

Main bearing cap side bolts: 36 ft. lbs.

③ Step 1: 15 ft. lbs.

Step 2: Additional 90 degrees

④ Step 1: 48 ft. lbs.

Step 2: Additional 60 degrees

71051_HODY_C0010

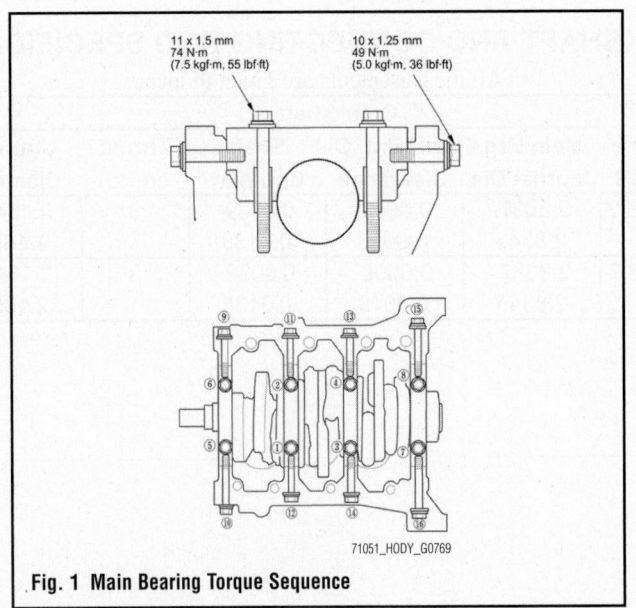

Fig. 1 Main Bearing Torque Sequence

WHEEL ALIGNMENT

Year	Model		Caster Range (+/-Deg.)	Caster Preferred Setting (Deg.)	Camber Range (+/-Deg.)	Camber Preferred Setting (Deg.)	Toe-in (in.)
2011	Odyssey	Front	0.36	2.36	0.45	0.00	0+/-0.08
		Rear	—	—	0.45	-0.30	0.08+/-0.08
2012	Odyssey	Front	0.36	2.36	0.45	0.00	0+/-0.08
		Rear	—	—	0.45	-0.30	0.08+/-0.08

71051_HODY_C0011

TIRE, WHEEL AND BALL JOINT SPECIFICATIONS

Year	Model	OEM Tires		Tire Pressures (psi)		Wheel Size	Ball Joint Inspection	Lug Nut (ft. lbs.)
		Standard	Optional	Front	Rear			
2011	Odyssey	235/65R17	235/60R18	①	②	17/18	NA	94
2012	Odyssey	235/65R17	235/60R18	①	②	17/18	NA	94

OEM: Original Equipment Manufacturer

PSI: Pounds Per Square Inch

NA: Information not available

① Standard tires: 33 psi
   Optional tires: 35 psi

② Standard tires: 35 psi
   Optional tires: 36 psi

71051_HODY_C0012

BRAKE SPECIFICATIONS

All measurements in inches unless noted

Year	Model		Brake Disc			Minimum Pad/Lining Thickness		Brake Caliper	
			Original Thickness	Minimum Thickness	Max. Runout	Front	Rear	Bracket Bolts (ft. lbs.)	Mounting Bolts (ft. lbs.)
2011	Odyssey	Front	1.102	1.024	0.0016	0.063	—	101	37
		Rear	0.433	0.354	0.0016	—	0.063	65	25
2012	Odyssey	Front	1.102	1.024	0.0016	0.063	—	101	37
		Rear	0.433	0.354	0.0016	—	0.063	65	25

71051_HODY_C0013

MAINTENANCE MINDER SCHEDULE
Honda Odyssey

The Odyssey displays engine oil life and maintenance service items in the information display to indicate when to perform maintenance service. If the engine oil life is 15% or less, based on the onboard computer's caluculations, you will see SERVICE DUE SOON in the information display every time the ignition key is turned to ON. The maintenance minder indicator will also come on and the maintenance code(s) for other scheduled maintenance items needing service will be displayed below the message.

Symbol	Item	Service
A	Engine oil ①	Change
B	Engine oil and filter	Change
	Tires	Rotate
	Brakes	Inspect
	Parking brake adjustment	Check
	Steering gear and linkage	Inspect
	Suspension components	Inspect
	Driveshaft boots	Inspect
	Brake hoses and lines	Inspect
	All fluid levels and condition	Inspect
	Exhaust system	Inspect
	Fuel lines and connections	Inspect
1	Tires	Rotate
2	Engine air filter ②	Replace
	Dust and pollen filter ③	Replace
	Accessory drive belt	Inspect
3	Transmission fluid ④	Replace
4	Spark plugs	Replace
	Timing belt ⑤	Replace
	Water pump	Inspect
	Valve clearance ⑥	Inspect
5	Engine coolant	Replace

① If the message SERVICE DUE NOW does not appear more than 12 months after the display is reset, change every year.

② If driven in dusty conditions, replace every 15,000 miles.

③ If driven in urban areas that have a high concentration of soot from industry and diesel, replace every 15,000 miles

④ If regularly driven in mountainous areas at very low speed or trailer towing, change the fluid every 30,000 miles.

⑤ If driven regularly in temperatures over 110 deg.F or below -20 deg.F, or towing a trailer, replace every 60,000 miles.

⑥ Adjust if necessary.

Additionally, replace the brake fluid every 3 years, and inspect the idle speed every 160,000 miles.

To reset the Engine Oil Life Display on LX, EX and EX-L models:

1. Turn the ignition switch to ON.
2. Press the SELECT/RESET button repeatedly until the engine oil life display or the service message is displayed.
3. Press the SELECT/RESET button for about 10 seconds. The engine oil life idicator and the codes will blink.
4. Press the SELECT/RESET knob for more than 5 seconds. The codes will disappear and the indicator will reset to 100.

To reset the Engine Oil Life Display on Touring models:

1. Turn the ignition switch to ON.
2. Press the SEL/RESET button on the steering wheel until the engine oil life is displayed.
3. Press the SEL/RESET button on the steering wheel for 10 seconds. The display will change to CUSTOM SETUP mode.
4. Press the SEL/RESET button on the steering wheel. The codes will disappear and the indicator will reset to 100.

PRECAUTIONS

Before servicing any vehicle, please be sure to read all of the following precautions, which deal with personal safety, prevention of component damage, and important points to take into consideration when servicing a motor vehicle:

• Never open, service or drain the radiator or cooling system when the engine is hot; serious burns can occur from the steam and hot coolant.

• Observe all applicable safety precautions when working around fuel. Whenever servicing the fuel system, always work in a well-ventilated area. Do not allow fuel spray or vapors to come in contact with a spark, open flame, or excessive heat (a hot drop light, for example). Keep a dry chemical fire extinguisher near the work area. Always keep fuel in a container specifically designed for fuel storage; also, always properly seal fuel containers to avoid the possibility of fire or explosion. Refer to the additional fuel system precautions later in this section.

• Fuel injection systems often remain pressurized, even after the engine has been turned **OFF**. The fuel system pressure must be relieved before disconnecting any fuel lines. Failure to do so may result in fire and/or personal injury.

• Brake fluid often contains polyglycol ethers and polyglycols. Avoid contact with the eyes and wash your hands thoroughly after handling brake fluid. If you do get brake fluid in your eyes, flush your eyes with clean, running water for 15 minutes. If eye irritation persists, or if you have taken

brake fluid internally, IMMEDIATELY seek medical assistance.

• The EPA warns that prolonged contact with used engine oil may cause a number of skin disorders, including cancer. You should make every effort to minimize your exposure to used engine oil. Protective gloves should be worn when changing oil. Wash your hands and any other exposed skin areas as soon as possible after exposure to used engine oil. Soap and water, or waterless hand cleaner should be used.

• All new vehicles are now equipped with an air bag system, often referred to as a Supplemental Restraint System (SRS) or Supplemental Inflatable Restraint (SIR) system. The system must be disabled before performing service on or around system components, steering column, instrument panel components, wiring and sensors. Failure to follow safety and disabling procedures could result in accidental air bag deployment, possible personal injury and unnecessary system repairs.

• Always wear safety goggles when working with, or around, the air bag system. When carrying a non-deployed air bag, be sure the bag and trim cover are pointed away from your body. When placing a non-deployed air bag on a work surface, always face the bag and trim cover upward, away from the surface. This will reduce the motion of the module if it is accidentally deployed. Refer to the additional air bag system precautions later in this section.

• Clean, high quality brake fluid from a sealed container is essential to the safe and

proper operation of the brake system. You should always buy the correct type of brake fluid for your vehicle. If the brake fluid becomes contaminated, completely flush the system with new fluid. Never reuse any brake fluid. Any brake fluid that is removed from the system should be discarded. Also, do not allow any brake fluid to come in contact with a painted surface; it will damage the paint.

• Never operate the engine without the proper amount and type of engine oil; doing so WILL result in severe engine damage.

• Timing belt maintenance is extremely important. Many models utilize an interference-type, non-freewheeling engine. If the timing belt breaks, the valves in the cylinder head may strike the pistons, causing potentially serious (also time-consuming and expensive) engine damage. Refer to the maintenance interval charts for the recommended replacement interval for the timing belt, and to the timing belt section for belt replacement and inspection.

• Disconnecting the negative battery cable on some vehicles may interfere with the functions of the on-board computer system(s) and may require the computer to undergo a relearning process once the negative battery cable is reconnected.

• When servicing drum brakes, only disassemble and assemble one side at a time, leaving the remaining side intact for reference.

• Only an MVAC-trained, EPA-certified automotive technician should service the air conditioning system or its components.

BRAKES   ANTI-LOCK BRAKE SYSTEM

WHEEL SPEED SENSORS

REMOVAL & INSTALLATION

Front

See Figure 2.

1. Turn the ignition switch to LOCK (0).
2. Remove the grommet, then disconnect the wheel speed sensor connector.
3. Remove the clamp, the clip, and the wire guide rubber.
4. Remove the bolt and the wheel speed sensor.

To install:

5. Install the wheel speed sensor in the reverse order of removal, and note these items:

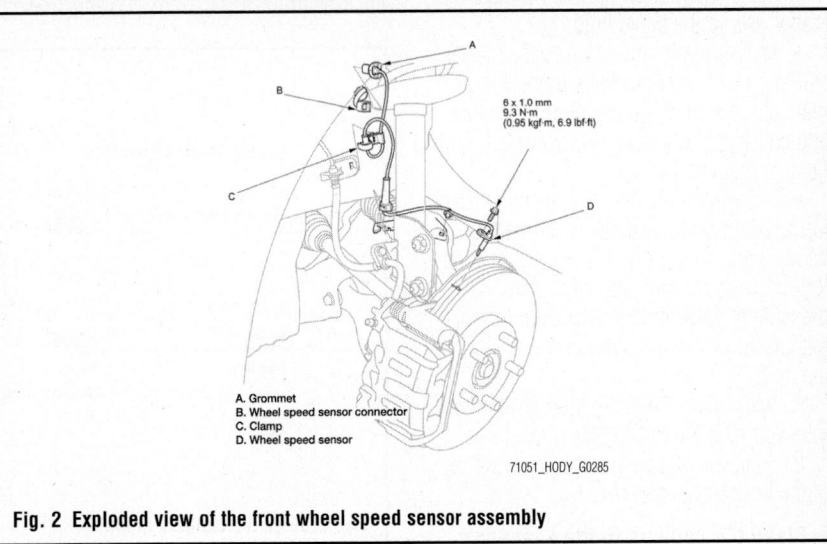

A. Grommet
B. Wheel speed sensor connector
C. Clamp
D. Wheel speed sensor

6 x 1.0 mm
9.3 N·m
(0.95 kgf·m, 6.9 lbf·ft)

71051_HODY_G0285

Fig. 2 Exploded view of the front wheel speed sensor assembly

a. Do not twist the sensor wires.

b. If the wheel speed sensor comes in contact with the hub bearing unit, it is faulty.

c. Make sure there is no debris in the sensor mounting hole.

Rear

See Figure 3.

1. Turn the ignition switch to LOCK (0).

2. Remove the clip, then disconnect the wheel speed sensor connector.

3. Remove the wire guide rubbers and the clips.

4. Remove the bolt and the wheel speed sensor.

5. Remove the wire guide rubber from the clamp, then pull out the wire harness from between the subframe and the body.

To install:

6. Install the wheel speed sensor in the reverse order of removal, and note these items:

a. Do not twist the sensor wires.

b. If the wheel speed sensor comes in contact with the hub bearing unit, it is faulty.

c. Make sure there is no debris in the sensor mounting hole.

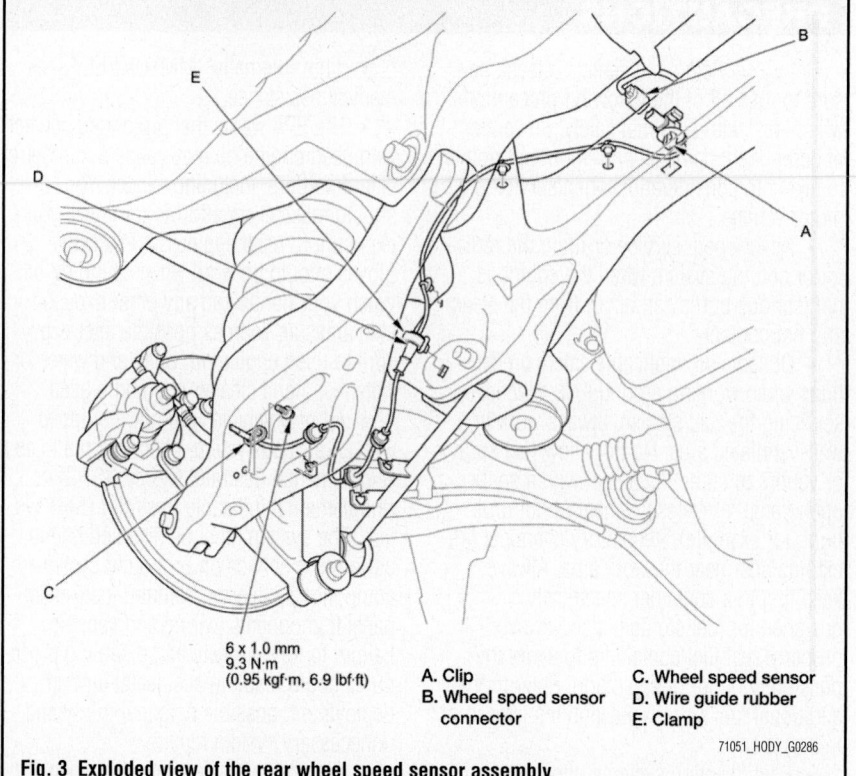

6 x 1.0 mm
9.3 N·m
(0.95 kgf·m, 6.9 lbf·ft)

A. Clip
B. Wheel speed sensor connector
C. Wheel speed sensor
D. Wire guide rubber
E. Clamp

71051_HODY_G0286

Fig. 3 Exploded view of the rear wheel speed sensor assembly

BRAKES

BLEEDING THE BRAKE SYSTEM

BLEEDING PROCEDURE

MANUAL

See Figure 4.

Note the following:

• Do not reuse the drained brake fluid. Use only new Honda DOT 3 Brake Fluid from an unopened container. Using a non-Honda brake fluid can cause corrosion and shorten the life of the system.

• Make sure no dirt or other foreign matter gets in the brake fluid.

• The reservoir connected to the master cylinder must be at the MAX (upper) level mark at the start of the bleeding procedure and checked after bleeding each wheel. Add fluid as required.

• There are three different methods used for bleeding brake systems. The method shown here is the preferred manual method for removing the air from the system. For pressure or vacuum bleeding, refer to tool manufacturer's instructions included with the tool.

1. Make sure the brake fluid level in the reservoir is at the MAX (upper) level line.

2. Start the bleeding at the driver's side of the front brake system.

➡**Bleed the calipers in the sequence shown.**

3. Attach a length of clear drain tube to the bleed screw. Submerge the other end of the drain tube into a clear plastic catch bottle of brake fluid.

4. Have an assistant slowly pump the brake pedal several times then apply steady continuous pressure.

5. Loosen the bleed screw slowly to bleed the fluid into the plastic catch bottle. The brake pedal will travel toward the floor as the fluid is bled from the system.

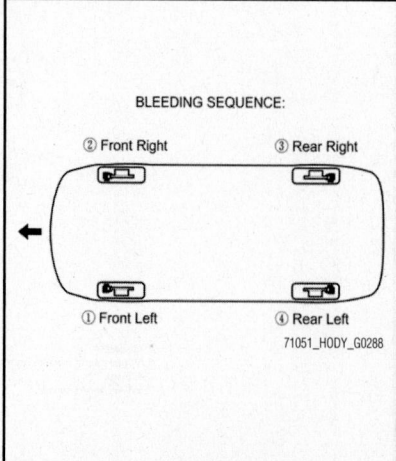

BLEEDING SEQUENCE:

② Front Right   ③ Rear Right

① Front Left   ④ Rear Left

71051_HODY_G0288

Fig. 4 Bleed the calipers in the sequence shown

6. When the brake pedal reaches the floor, have the assistant hold the pedal in that position, then tighten the bleed screw. The brake pedal can now be released.

7. Check and refill the master cylinder reservoir to the MAX (upper) level line. Be sure to reinstall the master cylinder reservoir cap.

8. Repeat steps 4 thru 7 until the brake fluid in the clear drain tube appears fresh and there are no air bubbles in the fluid.

9. Repeat this procedure for each brake in the bleeding sequence.

BLEEDING THE ABS SYSTEM

Bleeding the ABS brake system is the same procedure as bleeding conventional brake system.

FLUID RECOMMENDATIONS

Always use Honda DOT 3 Brake Fluid. Using a non-Honda brake fluid can cause corrosion and decrease the life of the system.

LEVEL CHECK

Check that the brake fluid level is between the "MIN" and "MAX" level marks on the master cylinder reservoir.

BRAKES   **FRONT DISC BRAKES**

BRAKE PADS

REMOVAL & INSTALLATION

See Figures 5 through 9.

Special Tools Required: Brake Caliper Piston Compressor 07AAE-SEPA101

1. Remove some brake fluid from the master cylinder.
2. Raise and support the vehicle.
3. Remove the front wheels.
4. Remove the brake hose mounting bolt.
5. Remove the flange bolt while holding the caliper pin with a wrench. Be careful not to damage the pin boot, and pivot the caliper body up out of the way. Check the hose and the pin boots for damage and deterioration.
6. Remove the pad return springs while holding the brake pads.

➡**The pad return springs are installed on the pads to prevent brake drag. Be careful when pivoting the caliper body up or the spring could pop out of position.**

7. Remove the pad shims and the brake pads.
8. Remove the pad retainers.
9. Clean the caliper bracket thoroughly; remove any rust, and check for grooves and cracks. Verify that the caliper pins move in and out smoothly. Clean and lube if needed.
10. Inspect the brake disc for runout, thickness, parallelism, and check for damage and cracks.

To install:

11. Apply a thin coat of M-77 assembly paste (P/N 08798-9010) to the retainer mat-

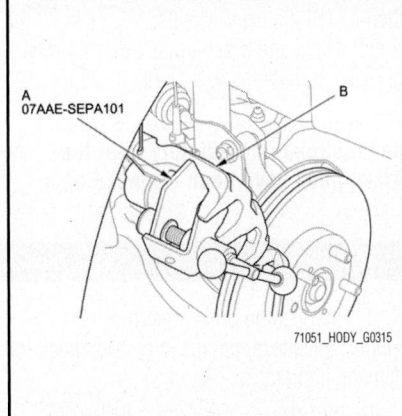

Fig. 6 Install the brake caliper piston compressor tool (A) on the caliper body (B)

ing surface of the caliper bracket (indicated by the arrows).

12. Install the pad retainers. Wipe off the excess assembly paste from the retainers. Keep the assembly paste away from the brake disc and the brake pads.
13. Install the brake caliper piston compressor tool on the caliper body.
14. Press in the piston with the brake caliper piston compressor tool so the caliper body will fit over the brake pads. Make sure the piston boot is in position to prevent damaging it when pivoting the caliper body down.

➡**Be careful when pressing in the piston; brake fluid might overflow from the master cylinder's reservoir. If brake fluid gets on any painted surface, wash it off immediately with water.**

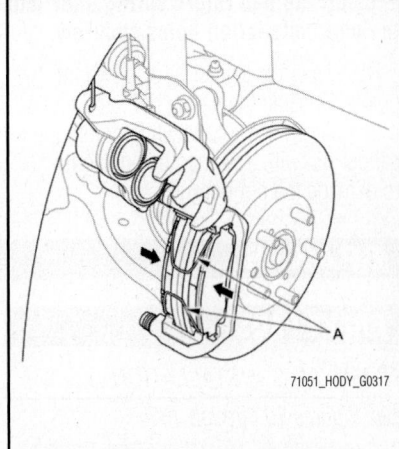

Fig. 8 Install the pad return springs (A) while holding the brake pads

15. Remove the brake caliper piston compressor tool.
16. Apply a thin coat of M-77 assembly paste (P/N 08798-9010) to the pad side of the shims, the back of the brake pads and the other areas indicated by the arrows. Wipe off the excess assembly paste from the pad shims and brake pads friction material. Keep grease and assembly paste away from the brake disc and brake pads. Contaminated brake disc or brake pads reduce stopping ability.
17. Install the brake pads and pad shims correctly. Install the brake pad with the wear indicator on the upper inside position. If you are reusing the brake pads, always reinstall the brake pads in their original positions to prevent a temporary loss of braking efficiency.

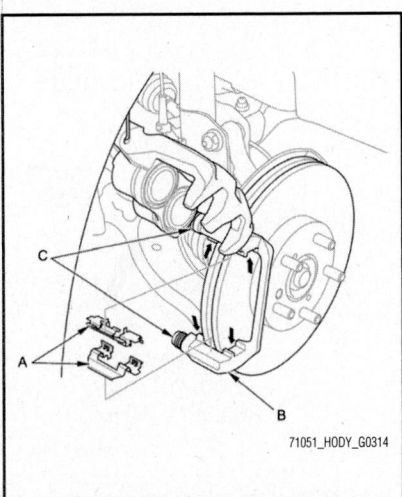

Fig. 5 Remove the pad retainers (A); caliper bracket (B), caliper pins (C)

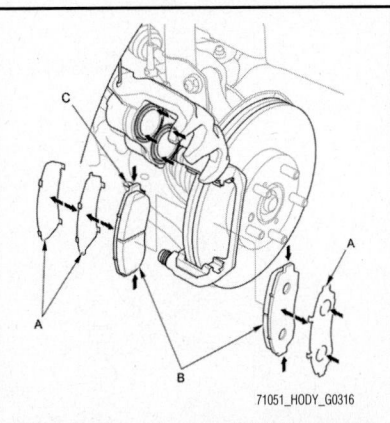

Fig. 7 Apply a thin coat of M-77 assembly paste to the pad side of the shims (A), the back of the brake pads (B), and the other areas indicated by the arrows; wear indicator (C)

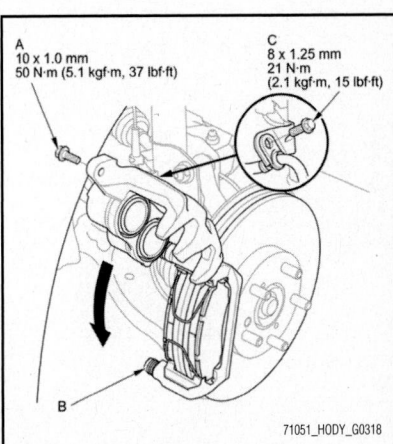

Fig. 9 Install the flange bolt (A), and tighten it to the specified torque while holding the caliper pin (B) with a wrench, then Install the brake hose mounting bolt (C)

18. Install the pad return springs while holding the brake pads.

→ **Insert the pad return spring ends into the pad installation holes securely.**

19. Pivot the caliper body down into position. Install the flange bolt, and tighten it to the specified torque while holding the caliper pin with a wrench being careful not to damage the pin boot.

20. Install the brake hose mounting bolt.

21. Clean the mating surfaces between the brake disc and the inside of the wheel, then install the front wheels.

22. Press the brake pedal several times to make sure the brakes work.

→ **Engagement may require a greater pedal stroke immediately after the brake pads have been replaced as a** set. **Several applications of the brake pedal will restore the normal pedal stroke.**

23. Add brake fluid as needed.

24. After installation, check for leaks at hose and line joints or connections, and retighten if necessary. Test-drive the vehicle, then recheck for leaks.

BRAKES | REAR DISC BRAKES

BRAKE PADS

REMOVAL & INSTALLATION

See Figures 10 through 13.

1. Remove some brake fluid from the master cylinder.

2. Raise and support the vehicle.

3. Remove the rear wheels.

4. Remove the brake hose mounting bolt.

5. Remove the flange bolt and pivot the caliper body down out of the way. Check the hose and pin boots for damage and deterioration.

6. Remove the pad shims and the brake pads.

7. Remove the upper and lower pad retainers.

→ **The upper and lower pad retainers are different. During installation, make sure the pad retainers are in the proper positions.**

8. Clean the caliper bracket thoroughly; remove any rust, and check for grooves and cracks. Verify that the caliper pins move in and out smoothly. Clean and lube if needed.

9. Inspect the brake disc/drum for runout, thickness, parallelism, and check for damage and cracks.

To install:

10. Apply a thin coat of M-77 assembly paste (P/N 08798-9010) to the retainer mating surface of the caliper bracket (indicated by the arrows).

11. Install the upper and lower pad retainers. Wipe off the excess assembly paste from the retainers. Keep the assembly paste away from the discs and pads.

12. Install the brake caliper piston compressor tool on the caliper body.

13. Press in the piston with the brake caliper piston compressor tool so the caliper body will fit over the brake pads. Make sure the piston boot is in position to prevent damaging it when pivoting the caliper body up.

→ **Be careful when pressing in the piston; brake fluid might overflow from the master cylinder's reservoir. If brake fluid gets on any painted surface, wash it off immediately with water.**

14. Remove the brake caliper piston compressor tool.

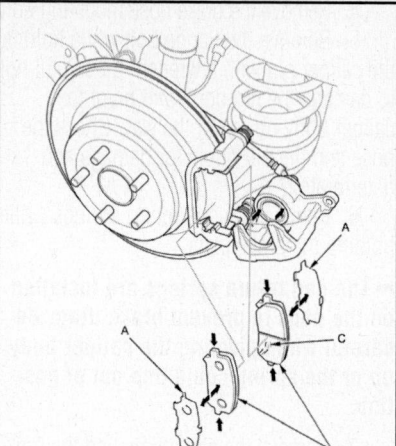

Fig. 12 Apply a thin coat of M-77 assembly paste to the pad side of the shims (A), the back of the brake pads (B), and the other areas indicated by the arrows; wear indicator (C)

15. Apply a thin coat of M-77 assembly paste (P/N 08798-9010) to the pad side of the shims, the back of the brake pads, and

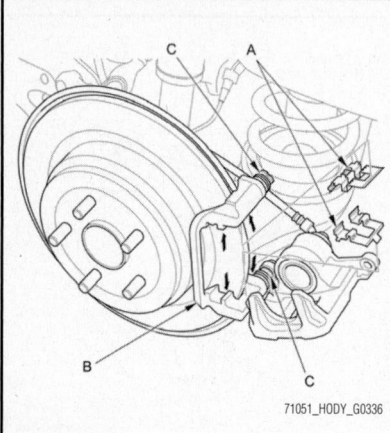

Fig. 10 Remove the upper and lower pad retainers (A), clean the caliper bracket (B), then verify that the caliper pins (C) move in and out smoothly

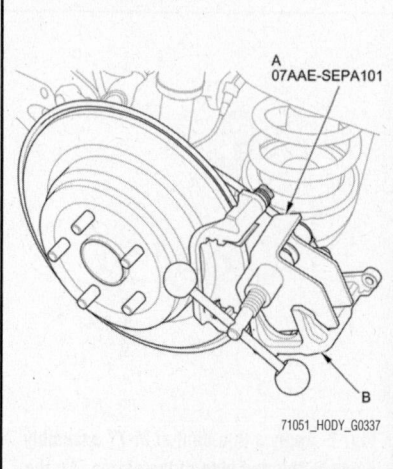

Fig. 11 Install the brake caliper piston compressor tool (A) on the caliper body (B)

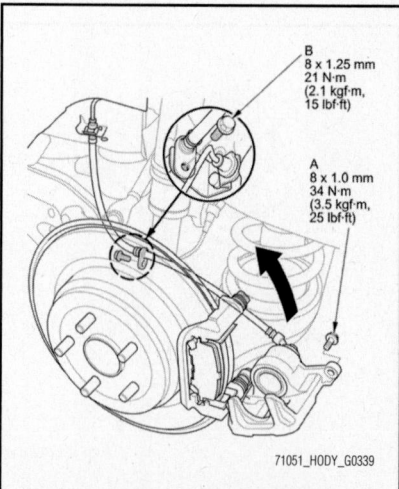

Fig. 13 Pivot the caliper body up into position, install the flange bolt (A), and install the brake hose mounting bolt (B)

the other areas indicated by the arrows. Wipe off the excess assembly paste from the pad shims and brake pads friction material. Keep grease and assembly paste away from the brake disc/drum and brake pads. Contaminated brake disc/drum or brake pads reduce stopping ability.

16. Install the brake pads and pad shims correctly. Install the brake pad with the wear indicator on the bottom inside position. If you are reusing the brake pads, always reinstall the brake pads in their original posi-

tions to prevent a temporary loss of braking efficiency.

17. Pivot the caliper body up into position. Install the flange bolt, and tighten it to the specified torque.

18. Install the brake hose mounting bolt.

19. Clean the mating surfaces between the brake disc/drum and the inside of the wheel, then install the rear wheels.

20. Press the brake pedal several times to make sure the brakes work.

➡Engagement may require a greater pedal stroke immediately after the brake pads have been replaced as a set. Several applications of the brake pedal will restore the normal pedal stroke.

21. Add brake fluid as needed.

22. After installation, check for leaks at hose and line joints or connections, and retighten if necessary.

23. Test-drive the vehicle, then recheck for leaks.

BRAKES

ADJUSTMENTS

Inspection

1. Press the parking brake pedal with 66 lbs. (294 N) of force to fully apply the parking brake. The parking brake pedal should be locked within the specified number of clicks. Pedal locked clicks: 8 to 10 clicks

2. Adjust the parking brake if the pedal clicks are not within the specification.

➡Minor parking brake pedal adjustments (1 to 2 clicks) can be made with the adjusting nut (see parking brake minor adjustment). If a larger adjustment is required, follow the major adjustment procedure using the adjuster nut at the parking brake drum (see parking brake major adjustment). After installing new parking brake shoes and/or new brake disc/drum, make sure you drive the vehicle for "break-in".

Minor Adjustment

See Figure 14.

1. Raise and support the vehicle.
2. Release the parking brake pedal fully.
3. Press the parking brake pedal 1 click.
4. Tighten the adjusting nut until the parking brakes drag slightly when the rear wheels are turned.
5. Release the parking brake pedal fully, and check that the parking brakes do not drag when the rear wheels are turned. Readjust if necessary.
6. Make sure the parking brake pedal is within the specified number of clicks (8 to 10 clicks).

Major Adjustment (to be done when replacing parking brake shoes and after lining surface break-in)

See Figure 15.

1. Raise and support the vehicle.
2. Release the parking brake pedal fully.
3. Loosen the adjusting nut.
4. Remove the rear wheels.
5. Remove the access plug.
6. Turn the ratchet teeth on the adjuster nut with a brake adjuster tool or a flat-tip screwdriver until the shoes lock against the

PARKING BRAKE

parking brake drum. Then back off the adjuster 15 clicks, and install the access plug.

7. Clean the mating surfaces between the brake disc/drum and the inside of the wheel, then install the rear wheels.

8. Do the minor adjustment procedure.

PARKING BRAKE SHOES

REMOVAL & INSTALLATION

See Figures 16 through 23.

1. Raise and support the vehicle.
2. Remove the rear wheels.
3. Release the parking brake, and remove the brake disc/drum.
4. Disconnect and remove the upper return springs.
5. Remove the tension pins by pushing the respective retainer springs and turning the pin.
6. Disconnect the rod spring from the connecting rod, and remove the rod spring and the connecting rod.
7. Lower the parking brake shoe assembly.

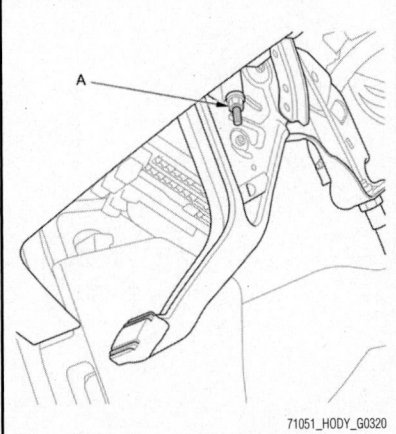

71051_HODY_G0320

Fig. 14 Tighten the adjusting nut (A) until the parking brakes drag slightly when the rear wheels are turned

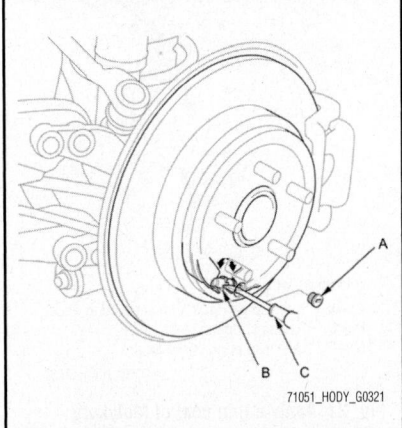

71051_HODY_G0321

Fig. 15 Remove the access plug (A); turn the ratchet teeth on the adjuster nut (B) with a brake adjuster tool or a flat-tip screwdriver (C)

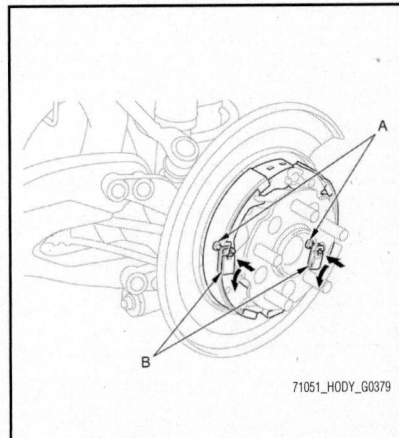

71051_HODY_G0379

Fig. 16 Remove the tension pins (A) by pushing the respective retainer springs (B) and turning the pin

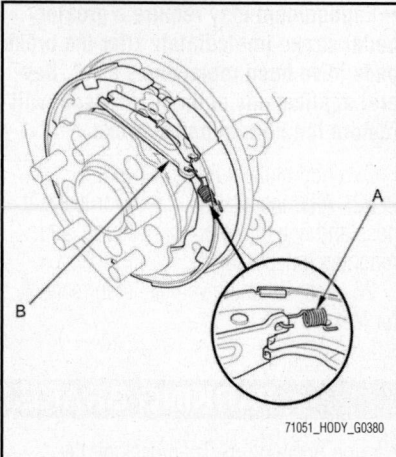

Fig. 17 Disconnect the rod spring (A) from the connecting rod (B), and remove the rod spring and the connecting rod

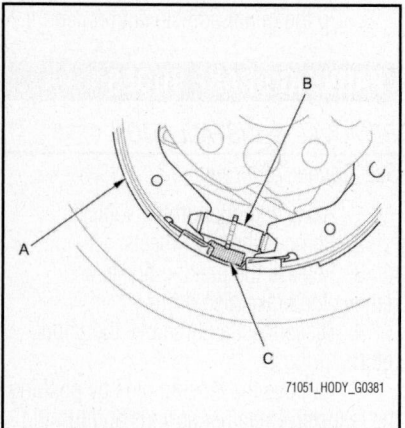

Fig. 18 Remove the forward brake shoe (A) and the adjuster assembly (B) by removing the lower return spring (C)

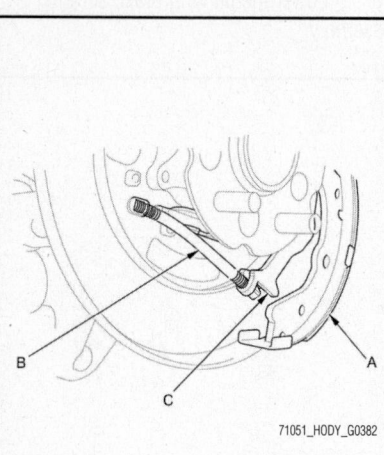

Fig. 19 Remove the rearward brake shoe (A) by disconnecting the parking brake cable (B) from the parking brake lever (C)

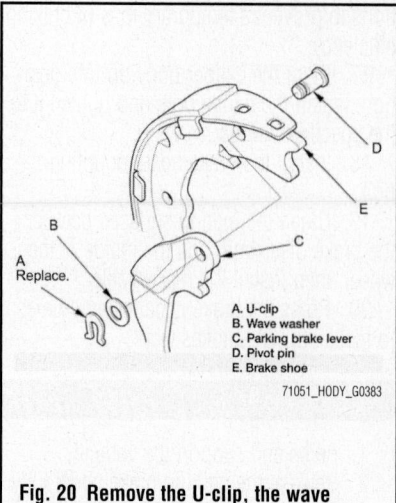

A. U-clip
B. Wave washer
C. Parking brake lever
D. Pivot pin
E. Brake shoe

Fig. 20 Remove the U-clip, the wave washer, the parking brake lever, and the pivot pin from the brake shoe

8. Remove the forward brake shoe and the adjuster assembly by removing the lower return spring.

9. Remove the rearward brake shoe by disconnecting the parking brake cable from the parking brake lever.

10. Remove the U-clip, the wave washer, the parking brake lever, and the pivot pin from the brake shoe.

To install:

11. Apply Molykote 44MA grease to the sliding surface of the pivot pin and insert the pin into the rearward brake shoe from the outside.

12. Install the parking brake lever and the wave washer on the pivot pin, and secure with a new U-clip.

   a. Install the wave washer with its convex side facing out.

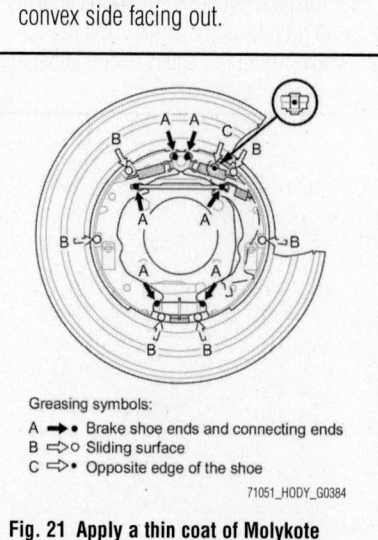

Greasing symbols:
A ➡• Brake shoe ends and connecting ends
B ⇨○ Sliding surface
C ⇨• Opposite edge of the shoe

Fig. 21 Apply a thin coat of Molykote 44MA grease to the shoe ends and connecting rod ends (A), sliding surfaces (B), and opposite edges of the parking brake shoe (C) as shown

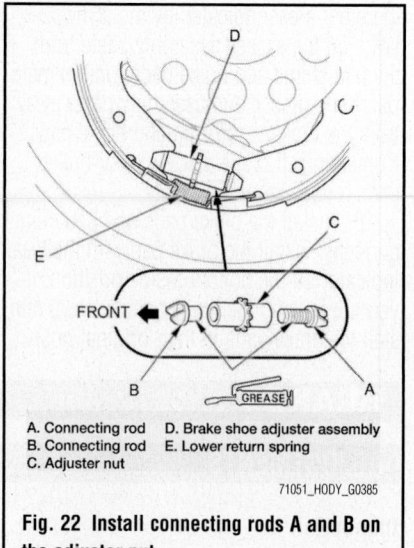

A. Connecting rod   D. Brake shoe adjuster assembly
B. Connecting rod   E. Lower return spring
C. Adjuster nut

Fig. 22 Install connecting rods A and B on the adjuster nut

   b. Pinch the U-clip securely to prevent the parking brake lever from coming off the brake shoe.

13. Connect the parking brake cable to the parking brake lever. Apply Molykote 44MA grease to the cable contact surface on the backing plate.

14. Apply a thin coat of Molykote 44MA grease to the shoe ends and connecting rod ends, sliding surfaces, and opposite edges of the parking brake shoe as shown. Wipe off any excess. Keep the grease away from the brake linings.

15. Install connecting rods A and B on the adjuster nut.

- Clean the threaded portions of connecting rod A and the sliding surface of connecting rod B, then coat them with Molykote 44MA grease.

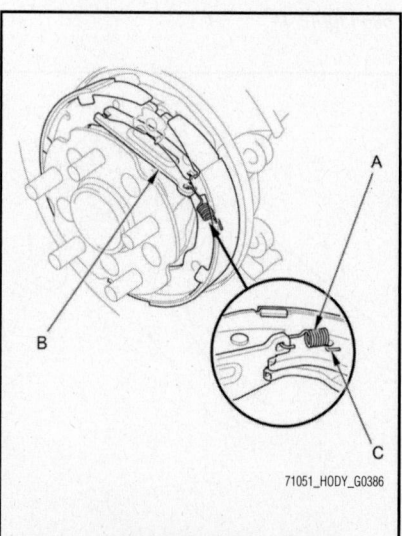

Fig. 23 Install the rod spring (A) to the connecting rod (B) first with the spring end (C) pointing downward

- Shorten connecting rod A by fully turning the adjuster nut.

16. Position the brake shoe adjuster assembly on the parking brake shoes.

➡**The rod B of the adjuster assembly shall point toward the front of the vehicle as shown.**

17. Hook the lower return spring on the parking brake shoes.

18. Install the rod spring to the connecting rod first with the spring end pointing downward. Then install the connecting rod on the parking brake shoes.

19. Install the tension pins and the retainer springs. Make sure the tension pin does not contact the parking brake lever.

20. Install the upper return springs.

21. Install the brake disc/drum and the brake caliper.

➡**Before installing the brake disc/drum, clean the mating surfaces between the hub bearing unit and the inside of the brake disc/drum.**

22. Do the major parking brake adjustment.

23. Clean the mating surfaces between the brake/drum and the inside of the wheel, then install the rear wheels.

CHASSIS ELECTRICAL

GENERAL INFORMATION

❋❋ CAUTION

These vehicles are equipped with an air bag system. The system must be disarmed before performing service on, or around, system components, the steering column, instrument panel components, wiring and sensors. Failure to follow the safety precautions and the disarming procedure could result in accidental air bag deployment, possible injury and unnecessary system repairs.

SERVICE PRECAUTIONS

General Precautions

➡**Some systems store data in memory that is lost when the battery is disconnected. Before disconnecting the battery, refer to Battery Terminal Disconnection and Reconnection.**

❋❋ CAUTION

Please read the following precautions carefully before servicing the SRS, or the airbags could accidentally deploy and cause damage or injuries.

- Except when doing electrical inspections that requires battery power, always turn the ignition switch to LOCK (0), disconnect the negative cable from the battery, then wait at least 3 minutes before starting work.
  NOTE: The SRS memory is not erased even if the ignition switch is turned to LOCK (0) or if the battery cables are disconnected from the battery.
- Use replacement parts which are manufactured to the same standards and quality as the original

AIR BAG (SUPPLEMENTAL RESTRAINT SYSTEM)

parts. Do not install used SRS parts. Use only new parts when making SRS repairs.
- Carefully inspect any SRS part before you install it.
Do not install any part that shows signs of being dropped or improperly handled, such as dents, cracks or deformation.
- Before disconnecting the SRS unit connectors, always disconnect the appropriate SRS parts connectors.
- Use only a digital multimeter to check the system. If it is not a Honda multimeter, make sure its output is 0.01 A (10 mA) or less when switched to the lowest value in the ohmmeter range. A tester with a higher output could cause accidental deployment and possible injury.
- Do not put objects on the front passenger's airbag.

Steering-Related Precautions

Cable Reel Alignment

❋❋ CAUTION

Misalignment of the cable reel could cause an open in the wiring, making the SRS, remote steering wheel controls, or the horn inoperative.

Center the cable reel whenever you do the following.
- Installation of the steering wheel.
- Installation of the cable reel.
- Installation of the steering column.
- Other steering-related adjustment or installation.
Do not disassemble the cable reel.
Do not apply grease to the cable reel.
If the cable reel shows any signs of damage, replace it with a new one. For example,

if the cable reel does not rotate smoothly, replace it.

Airbag Handling and Storage

Do not disassemble an airbag. It has no serviceable parts. Once an airbag has been deployed, it cannot be repaired or reused.

❋❋ CAUTION

For temporary storage of an airbag during service, observe the following precautions.

1. Store the removed airbag with the pad surface up. Never put anything on the airbag.

2. To prevent damage to the airbag, keep it away from any oil, grease, detergent, or water.

3. Store the removed airbag on a secure, flat surface away from any high heat source (exceeding 200°F (93°C)).

4. Never do electrical tests on the airbags, such as measuring resistance.

5. Do not position yourself in front of the airbag during removal, inspection, or replacement.

6. For proper disposal of a damaged airbag or tensioner, refer to Airbag and Tensioner Disposal.

7. The side curtain airbag module assembly is a long, jointed part containing two inflators, a flexible bag, and brackets. When removing or installing the side curtain airbag assembly, never:
- Handle the flexible bag.
- Drop the airbag assembly.
- Cut, tear, or unwrap the tape strips.

Wiring Precautions

Some of the SRS wiring can be identified by a special yellow outer covering, and the SRS connectors can be identified by their yellow color. Observe the following instructions.

1. Never attempt to modify, splice, or repair SRS wiring. If there is an open or damage to the SRS wiring, replace the harness.

2. Be sure to install the harness wires so they do not get pinched or interfere with other parts.

3. Make sure all SRS ground locations are clean, and the grounds are securely fastened for optimum metal-to-metal contact. Poor grounds can cause intermittent problems that are difficult to diagnose.

4. Do not use any silicone based cleaners or lubricants on any SRS connectors or terminals.

DISARMING THE SYSTEM

Do the battery terminal disconnection procedure, then wait at least 3 minutes before starting work.

ARMING THE SYSTEM

Do the battery terminal reconnection procedure.

DRIVE TRAIN

AUTOMATIC TRANSAXLE

DRAIN & REFILL

5-Speed Transaxle

See Figures 24 and 25.

✳✳ WARNING

Keep all foreign particles out of the transmission.

1. Warm up the engine to normal operating temperature (the radiator fan comes on), and turn the engine off.

2. Raise the vehicle on a lift, or apply the parking brake, block both rear wheels, and raise the front of the vehicle. Make sure it is securely supported.

3. Remove the front splash shield.

4. Remove the ATF filler bolt, the drain plug and the sealing washer, and drain the automatic transmission fluid (ATF).

5. Reinstall the drain plug with a new sealing washer.

6. Refill the transmission with the recommended fluid through the filler hole.

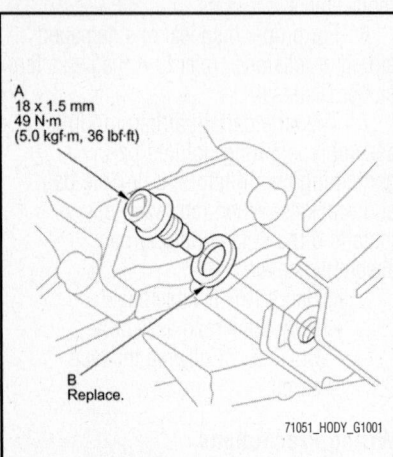

Fig. 24 Remove the ATF filler bolt, the drain plug (A) and the sealing washer (B), and drain the automatic transmission fluid (ATF)

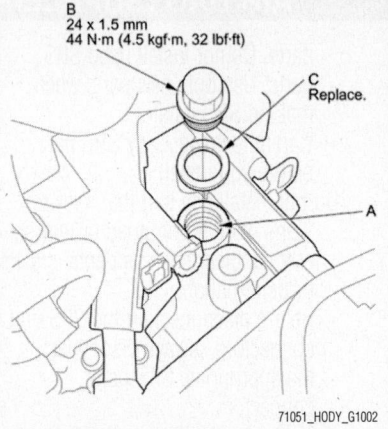

Fig. 25 Refill the transmission with the recommended fluid through the filler hole (A); ATF filler bolt (B) and sealing washer (C)

a. Always use Honda ATF DW-1 Automatic Transmission Fluid (ATF). Using a non-Honda ATF can affect shift quality.

b. Automatic Transmission Fluid Capacity:
- 3.3 qts. (3.1 L) at change
- 8.0 qts. (7.6 L) at overhaul

7. Install the ATF filler bolt with a new sealing washer.

8. Check the ATF level.

9. Install the front splash shield.

10. If the Maintenance Minder recommends replacing the ATF, reset the Maintenance Minder to complete this procedure. If the Maintenance Minder did not require you to replace the ATF, go to step 11.

11. Connect the HDS to the data link connector (DLC) located under the driver's side of the dashboard.

12. Turn the ignition switch to ON (II). Make sure the HDS communicates with the PCM. If it does not, go to the DLC circuit troubleshooting.

13. Select GAUGES in the BODY ELECTRICAL with the HDS.

14. Select ADJUSTMENT in the GAUGES with the HDS.

15. Select MAINTENANCE MINDER in the ADJUSTMENT with the HDS.

16. Select RESET in the MAINTENANCE MINDER with the HDS.

17. Select MAINTENANCE SUB ITEM 3 RESET, and reset the ATF life with the HDS.

6-Speed Transaxle

See Figures 26 and 27.

✳✳ WARNING

Keep all foreign particles out of the transmission.

1. Park the vehicle on the level ground.

2. Warm up the engine to normal operating temperature (the radiator fan comes on), and turn the engine off.

3. Remove the ATF filler bolt and the sealing washer.

4. Remove the drain plug and sealing washer, and drain the automatic transmission fluid (ATF).

➡ **If a cooler cleaning is necessary, refer to the ATF cooler cleaning.**

5. Reinstall the drain plug with a new sealing washer.

6. Refill the transmission with the recommended fluid into the filler hole.

a. Always use Honda ATF DW-1 Automatic Transmission Fluid (ATF). Using a non-Honda ATF can affect shift quality.

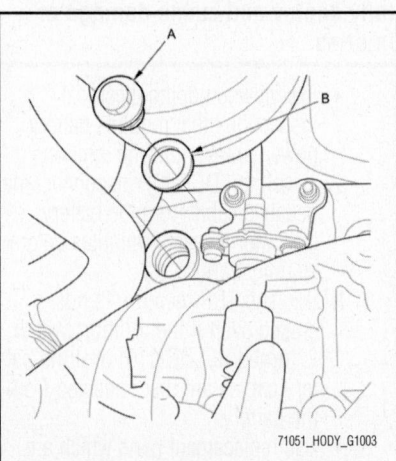

Fig. 26 Remove the ATF filler bolt (A) and the sealing washer (B)

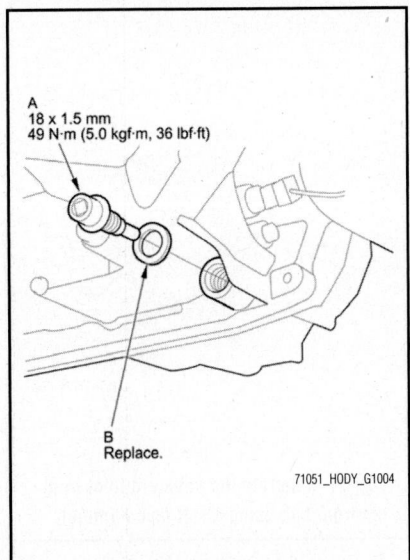

Fig. 27 Remove the drain plug (A) and sealing washer (B), and drain the automatic transmission fluid (ATF)

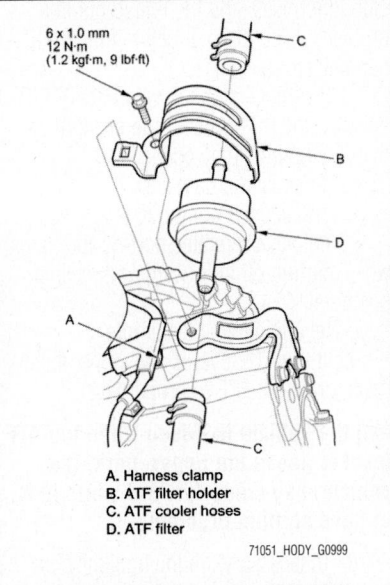

A. Harness clamp
B. ATF filter holder
C. ATF cooler hoses
D. ATF filter

Fig. 28 Exploded view of the inline ATF filter assembly

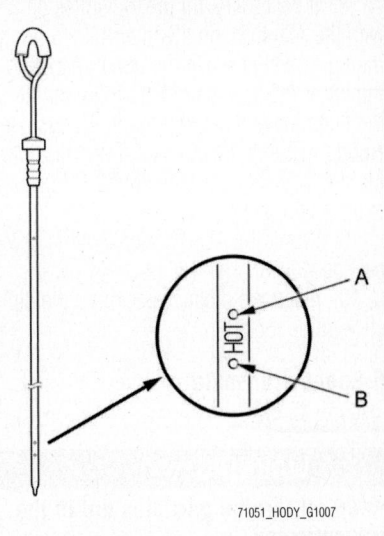

Fig. 29 Remove the dipstick, and check the ATF level; it should be between the upper mark (A) and the lower mark (B)

b. Automatic Transmission Fluid Capacity:
- 3.6 qts. (3.4 L) at change
- 7.4 qts. (7.0 L) at overhaul

7. Install the ATF filler bolt with a new sealing washer.

8. Remove the ATF dipstick.

9. Make sure the fluid level is between the upper mark and the lower mark on the dipstick, then insert the dipstick back into the dipstick guide tube.

10. If the Maintenance Minder required ATF replacement, reset the Maintenance Minder, and this procedure is complete. If the Maintenance Minder did not require you to replace the ATF, go to step 11.

11. Connect the HDS to the DLC located under the driver's side of the dashboard.

12. Turn the ignition switch to ON (II). Make sure the HDS communicates with the PCM. If it does not, go to the DLC circuit troubleshooting.

13. Select BODY ELECTRICAL with the HDS.

14. Select MAINTENANCE MINDER in the ADJUSTMENT with the HDS.

15. Select ADJUSTMENT in the GAUGES with the HDS.

16. Select RESET in the MAINTENANCE MINDER with the HDS.

17. Select MAINTENANCE SUB ITEM 3 RESET, and reset the ATF life with the HDS.

FILTER REPLACEMENT

6-Speed Transaxle

See Figure 28.

➡**The ATF filter is not a scheduled maintenance item. Replace the filter only if it is leaking, or contaminated, or when the transmission is being overhauled or replaced with a remanufactured unit.**

1. Raise the vehicle on a lift, and make sure it is securely supported.

2. Remove the front splash shield.

3. Remove the drain plug, sealing washer, and drain the ATF.

4. Reinstall the drain plug with a new sealing washer.

5. Remove the harness clamp from the ATF filter holder.

6. Disconnect the ATF cooler hoses from the ATF filter.

7. Remove the ATF filter holder, and remove the ATF filter.

To install:

8. Install the ATF filter, and secure it with its holder and bolt.

9. Connect the ATF cooler hose.

10. Install the harness clamp to the ATF filter holder.

11. Install the front splash shield.

12. Refill the transmission with ATF.

FLUID LEVEL CHECK

5-Speed Transaxle

See Figure 29.

※※ WARNING

Keep all foreign particles out of the transmission.

Note the following:
- Check the ATF level within 60–90 seconds after turning the engine off.
- Higher ATF level may be indicated if the radiator fan comes on twice or more.

1. Park the vehicle on the level ground.

2. Warm up the engine to normal operating temperature (the radiator fan comes on), and turn the engine off.

3. Remove the dipstick (yellow loop) from the dipstick guide tube, and wipe it with a clean cloth.

4. Insert the dipstick back into the dipstick guide tube.

5. Remove the dipstick, and check the ATF level. It should be between the upper mark and the lower mark.

6. If the ATF level is below the lower mark, check for fluid leaks at the transmission, the ATF cooler hoses, and the line joints. If a problem is found, fix it before filling the transmission with ATF.

If the vehicle is driven when the ATF level is below the lower mark, one or more of these symptoms may occur:
- Transmission damage.
- Vehicle does not move in any gear.
- Vehicle accelerates poorly, and flares when starting off in D, L, and R.
- The engine vibrates at idle.

7. If the level is above the upper mark, drain the ATF to the proper level.

➡**If the vehicle is driven when the ATF level is above the upper mark, the vehicle may creep forward while in N, or have shifting problems.**

8. If necessary, fill the transmission with the ATF through the filler hole to bring the fluid level between the upper mark and the lower mark of the dipstick. Do not fill the fluid above the upper mark. Always use Honda ATF DW-1 Automatic Transmission Fluid (ATF). Using a non-Honda ATF can affect shift quality.

9. Install the ATF filler bolt with a new sealing washer.

10. Insert the dipstick back into the dipstick guide tube.

6-Speed Transaxle

See Figure 30.

✳✳ WARNING

Keep all foreign particles out of the transmission.

Note the following:
- Check the fluid level within 60–90 seconds after turning the engine off.
- Higher fluid level may be indicated if the radiator fan comes on twice or more.

1. Park the vehicle on the level ground.

2. Warm up the engine to normal operating temperature (the radiator fan comes on), then turn the engine off.

3. Remove the ATF dipstick (yellow loop) from the dipstick guide tube, and wipe it with a clean cloth.

4. Insert the dipstick back into the dipstick guide tube aligning the notch with the guide tab.

5. Remove the dipstick and check the fluid level. It should be between the upper mark (A) and the lower mark (B).

6. If the ATF level is below the lower mark, check for fluid leaks at the transmis-

sion, the hoses, and the line joints. If a problem is found, fix it before filling the transmission with ATF.

If the vehicle is driven when the ATF level is below the lower mark, one or more of these symptoms may occur:
- Transmission damage.
- Vehicle does not move in any gear.
- Vehicle accelerates poorly, and flares when starting off in forward and reverse positions.
- The engine vibrates at idle.

7. If the ATF level is above the upper mark, drain the ATF to proper level.

➡ **If the vehicle is driven when the ATF level is above the upper mark, the vehicle may creep forward while in N, or have shifting problems.**

8. If necessary, fill the transmission with the recommended fluid through the filler hole to bring the fluid level between the upper mark and the lower mark of the dipstick. Do not fill past the upper mark. Always use Honda ATF DW-1 Automatic Transmission Fluid (ATF). Using a non-Honda ATF can affect shift quality.

9. Install the ATF filler bolt with a new sealing washer.

10. Insert the dipstick back into the dipstick guide tube.

FRONT HALFSHAFTS

REMOVAL & INSTALLATION

See Figures 31 through 41.

Special Tools Required:
- Ball Joint Thread Protector, 14 mm 071AF-SZNA100
- Ball Joint Remover, 32 mm 07MAC-SL0A102

1. Raise and support the vehicle.

2. Remove the front wheel.

3. Pry up the stake on the spindle nut, then remove the nut.

4. Drain the transmission fluid, then reinstall the drain plug with a new sealing washer.

5. With headlight leveling system: If removing the right driveshaft, remove the front suspension stroke sensor.

6. Remove the lock pin from the lower arm ball joint, then remove the castle nut. Separate the knuckle from the lower arm using the 14 mm ball joint thread protector and the 32 mm ball joint remover.
- Be careful not to damage the ball joint boot when installing the remover.
- Do not force or hammer on the lower arm, or pry between the

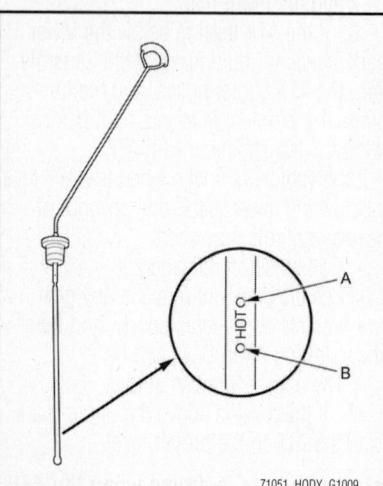

Fig. 30 Remove the dipstick and check the fluid level. It should be between the upper mark (A) and the lower mark (B)

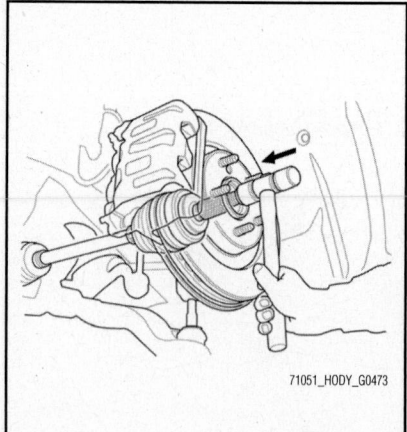

Fig. 31 Separate the outboard joint from the front hub using a soft face hammer

Fig. 32 Pull the knuckle outward, and separate the outboard joint from the front hub

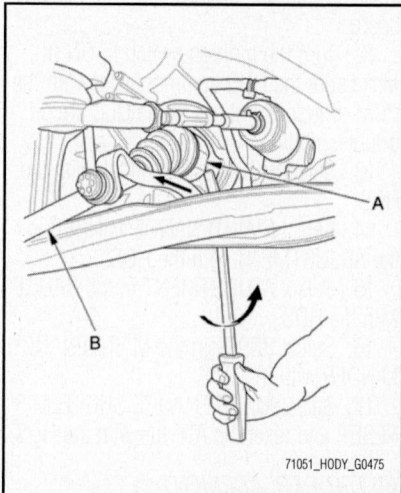

Fig. 33 Left driveshaft (B): Pry the inboard joint (A) from the differential using a pry bar

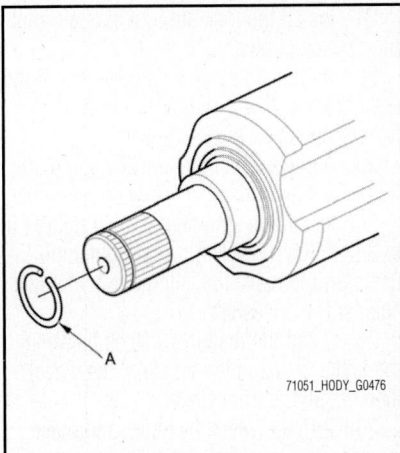

71051_HODY_G0476

Fig. 34 Left driveshaft: Remove the set ring (A) from the driveshaft inboard joint

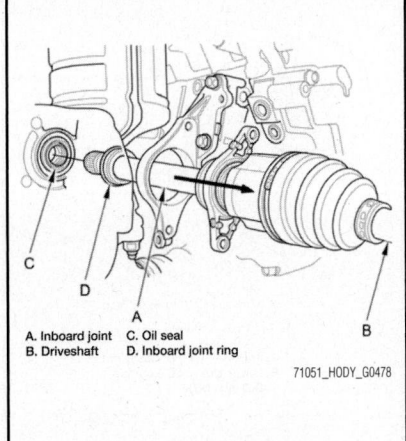

A. Inboard joint   C. Oil seal
B. Driveshaft   D. Inboard joint ring

71051_HODY_G0478

Fig. 36 Right driveshaft: Remove the inboard joint from the differential

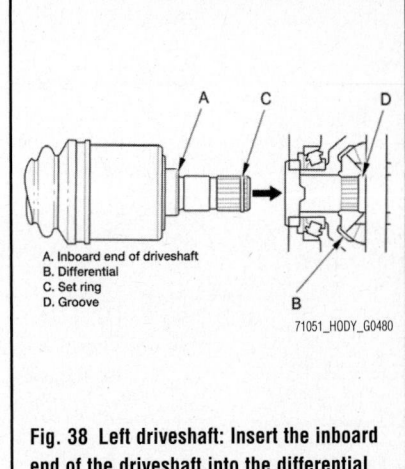

A. Inboard end of driveshaft
B. Differential
C. Set ring
D. Groove

71051_HODY_G0480

Fig. 38 Left driveshaft: Insert the inboard end of the driveshaft into the differential until the set ring locks in the groove

lower arm and the knuckle. You could damage the ball joint.

7. Separate the outboard joint from the front hub using a soft face hammer.

8. Pull the knuckle outward, and separate the outboard joint from the front hub.

9. Left driveshaft: Pry the inboard joint from the differential using a pry bar. Remove the driveshaft as an assembly.

- Do not pull on the driveshaft, or the inboard joint may come apart. Pull the inboard joint straight out to avoid damaging the oil seal.
- Be careful not to damage the oil seal or the end of the inboard joint with the pry bar.

10. Left driveshaft: Remove the set ring from the driveshaft inboard joint.

11. Right driveshaft: Remove the driveshaft bearing bracket bolts.

12. Right driveshaft: Remove the inboard joint from the differential.

- Do not pull on the driveshaft, or the inboard joint may come apart.
- Pull the inboard joint straight out to avoid damaging the oil seal.
- Be careful not to damage the inboard joint ring.

To install:

➡**Before starting installation, make sure the mating surfaces of the joint and the splined section are clean.**

13. Apply Moly 60 paste (P/N 08734-0001) to the contact area (A) of the outboard joint and the front wheel bearing.

➡**The paste helps prevent noise and vibration.**

14. Clean the areas where the driveshaft contacts the differential thoroughly with solvent, and dry them with compressed air.

➡**Do not wash the rubber parts with solvent.**

15. Left driveshaft: Install a new set ring into the set ring groove of the driveshaft inboard joint.

16. Left driveshaft: Insert the inboard end of the driveshaft into the differential until the set ring locks in the groove.

➡**Insert the driveshaft horizontally to prevent damaging the oil seal.**

17. Right driveshaft: Insert the inboard joint into the differential correctly.

- Insert the driveshaft horizontally to prevent damaging the oil seal.
- Be careful not to damage the inboard joint ring.
- Make sure to set the "TOP" mark of the driveshaft bearing bracket to upper side when you install it.

18. Right driveshaft: Install the driveshaft bearing bracket bolts.

19. Install the outboard joint into the front hub on the knuckle.

71051_HODY_G0477

Fig. 35 Right driveshaft: Remove the driveshaft bearing bracket bolts (A)

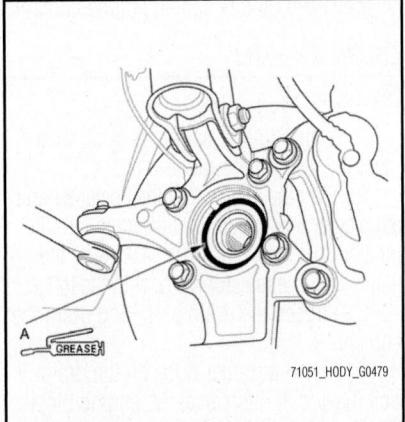

71051_HODY_G0479

Fig. 37 Apply Moly 60 paste to the contact area (A) of the outboard joint and the front wheel bearing

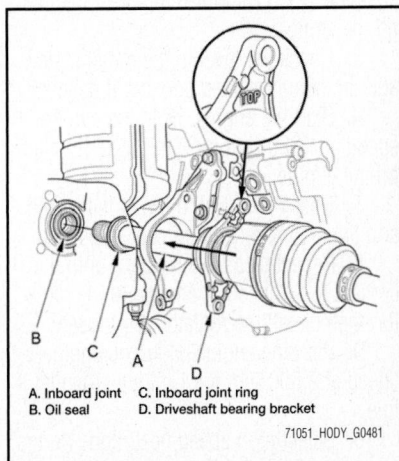

A. Inboard joint   C. Inboard joint ring
B. Oil seal   D. Driveshaft bearing bracket

71051_HODY_G0481

Fig. 39 Right driveshaft: Insert the inboard joint into the differential correctly

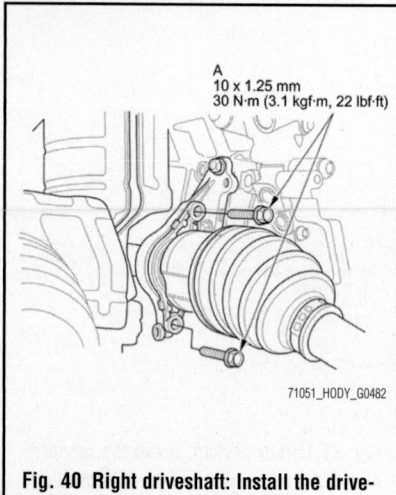

A
10 x 1.25 mm
30 N·m (3.1 kgf·m, 22 lbf·ft)

71051_HODY_G0482

Fig. 40 Right driveshaft: Install the drive-shaft bearing bracket bolts (A)

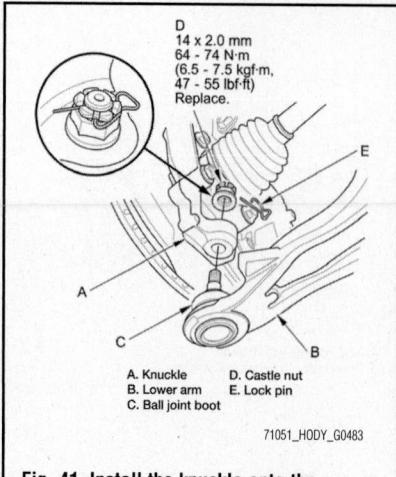

D
14 x 2.0 mm
64 - 74 N·m
(6.5 - 7.5 kgf·m,
47 - 55 lbf·ft)
Replace.

E

A

C

B

A. Knuckle   D. Castle nut
B. Lower arm   E. Lock pin
C. Ball joint boot

71051_HODY_G0483

Fig. 41 Install the knuckle onto the lower arm

20. Wipe off any grease contamination from the ball joint tapered section and threads, then install the knuckle onto the lower arm. Be careful not to damage the ball joint boot. Wipe off the grease before tightening the nut at the ball joint. Torque a new castle nut to the lower

torque specification, then tighten it only far enough to align the slot with the ball joint pin hole.

- Make sure the ball joint boot is not damaged or cracked.
- Do not align the nut by loosening it.

21. Install the lock pin into the ball joint pin hole as shown.

22. With headlight leveling system: If the right driveshaft was removed, install the front suspension stroke sensor.

23. Apply a small amount of engine oil to the seating surface of a new spindle nut.

24. Install the spindle nut, then tighten it to 242 ft. lbs. (328 Nm). After tightening, use a drift to stake the spindle nut shoulder against the driveshaft.

25. Clean the mating surfaces between the brake disc and the inside of the wheel, then install the front wheel.

26. Turn the wheel by hand, and make sure there is no interference between the driveshaft and surrounding parts.

27. Lower the vehicle.

28. Refill the transmission with the recommended transmission fluid:

29. Check the wheel alignment, and adjust it if necessary.

30. With headlight leveling system: If the right driveshaft was removed, do the headlight initial position learning procedure.

31. Test-drive the vehicle.

ENGINE COOLING

ENGINE COOLANT

BLEEDING

1. Pour Honda Long Life Antifreeze/Coolant Type 2 into the radiator up to the base of the filler neck.

2. Start the engine, and let it run until it warms up (the radiator fan comes on at least twice).

3. Turn off the engine. Check the level in the radiator, and add the recommended coolant, if needed.

4. Start the engine. Hold the engine speed at 1,500 rpm for 5 minutes, then turn off the engine.

5. Check the level in the radiator, and add the recommended coolant, if needed.

6. Start the engine. Hold the engine speed at 1,500 rpm for 3 minutes, then turn off the engine.

7. Check the level in the radiator, and add the recommended coolant, if needed.

8. Repeat steps 6 through 7 until the coolant level does not change in the radiator, then install the radiator cap loosely.

9. Start the engine. Hold the engine speed at 2,500 rpm for 1 minute, then let it idle.

10. Select high speed heat mode on the rear control panel. Measure the temperature of the air from the rear floor vent for 3 minutes. Make sure the temperature is above

the standard line of the graph as shown. If the temperature is below the standard line, repeat steps 6 through 4 several more times. Then loosely install the radiator cap and recheck.

11. With the engine idling, make sure you do not hear the sound of water flowing near the rear heater unit. If you hear water flowing, repeat steps 6 through 7 several more times.

12. Install the radiator cap on securely.

13. Fill the coolant reservoir with the recommended coolant up to the MAX mark, then add an extra 0.11 gal. (0.4 L) of coolant.

14. Clean up any spilled engine coolant.

DRAIN & REFILL

See Figure 42.

1. Wait until the engine is cool, then carefully remove the radiator cap.

2. Start the engine. Set the heater temperature control to maximum heat, then turn the ignition switch to LOCK (0). Make sure the engine and the radiator are cool to the touch.

3. Loosen the drain plug, and drain the coolant.

4. Install a rubber hose on the drain bolt located at the rear of the engine block, then loosen the drain bolt.

5. When the coolant stops draining, tighten the drain bolt. Remove the rubber hose.

6. Tighten the radiator drain plug securely.

7. Remove, drain, and reinstall the coolant reservoir.

8. Fill the coolant reservoir to the MAX mark with Honda Long Life Antifreeze/Coolant Type 2.

9. Pour Honda Long Life Antifreeze/Coolant Type 2 into the radiator up to the base of the filler neck.

10. Start the engine, and let it run until it warms up (the radiator fan comes on at least twice).

11. Turn off the engine. Check the level in the radiator, and add the recommended coolant, if needed.

12. Start the engine. Hold the engine speed at 1,500 rpm for 5 minutes, then turn off the engine.

13. Check the level in the radiator, and add the recommended coolant, if needed.

14. Start the engine. Hold the engine speed at 1,500 rpm for 3 minutes, then turn off the engine.

15. Check the level in the radiator, and add the recommended coolant, if needed.

16. Repeat steps 14 through 15 until the coolant level does not change in the radiator, then install the radiator cap loosely.

17. Start the engine. Hold the engine speed at 2,500 rpm for 1 minute, then let it idle.

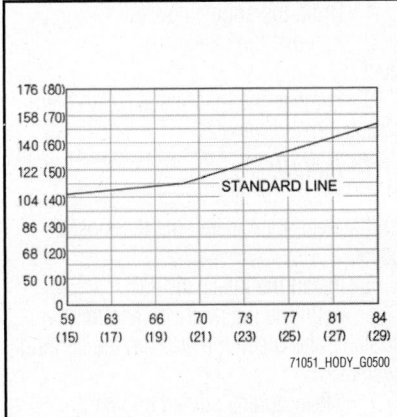

Fig. 42 Make sure the temperature is above the standard line of the graph as shown

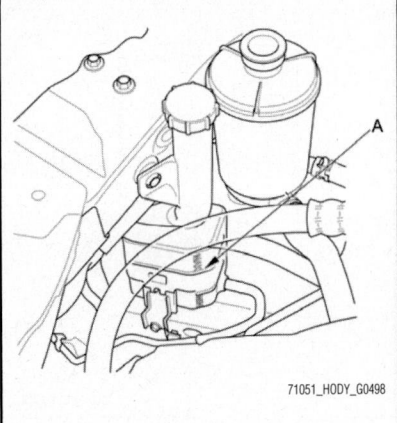

Fig. 43 Fill the coolant reservoir to the MAX mark (A)

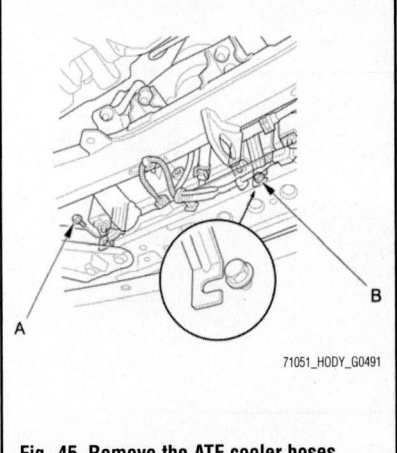

Fig. 45 Remove the ATF cooler hoses bracket bolt (A) and loosen the bolt (B)

18. Select high speed heat mode on the rear control panel. Measure the temperature of the air from the rear floor vent for 3 minutes. Make sure the temperature is above the standard line of the graph as shown. If the temperature is below the standard line, repeat steps 14 through 15 several more times. Then loosely install the radiator cap and recheck.

19. With the engine idling, make sure you do not hear the sound of water flowing near the rear heater unit. If you hear water flowing, repeat steps 14 through 15 several more times.

20. Install the radiator cap on securely.

21. Fill the coolant reservoir with the recommended coolant up to the MAX mark, then add an extra 0.11 gal. (0.4 L) of coolant.

22. Clean up any spilled engine coolant.

23. Inspect for engine coolant leaks.

24. If the Maintenance Minder required replacing the engine coolant, reset the Maintenance Minder; information display, multi information display, and this procedure is complete. If the Maintenance Minder did not require engine coolant replacement, go to step 25.

25. Turn the ignition switch to LOCK (0).

26. Connect the HDS to the DLC.

27. Turn the ignition switch to ON (II).

28. Make sure the HDS communicates with the vehicle and the PCM. If it does not communicate, troubleshoot the DLC circuit.

29. Select GAUGES in the BODY ELECTRICAL with the HDS.

30. Select ADJUSTMENT in the GAUGES with the HDS.

31. Select MAINTENANCE MINDER in the ADJUSTMENT with the HDS.

32. Select RESET in the MAINTENANCE MINDER with the HDS.

33. Select MAINTENANCE SUB ITEM 5 RESET with the HDS.

LEVEL CHECK
See Figure 43.

1. Fill the coolant reservoir to the MAX mark with Honda Long Life Antifreeze/Coolant Type 2.

ELECTRIC ENGINE FAN

REMOVAL & INSTALLATION
See Figures 44 through 47.

1. Remove the air intake scoop.

2. Remove the front bulkhead cover.

3. Disconnect the A/C condenser fan motor connector, and coolant reservoir hose, then remove the coolant reservoir hose from the clamps.

4. Remove the radiator brackets and the condenser brackets.

5. Remove the front grille bolts.

6. Remove the A/C condenser fan shroud assembly.

7. Disconnect the radiator fan motor connector and remove the harness clamps.

8. Raise the vehicle on the lift.

9. Remove the splash shield.

10. Remove the ATF cooler hoses bracket bolt and loosen the bolt.

11. Lower the vehicle on the lift.

12. With security system: Disconnect the hood latch switch connector.

13. Remove the hood latch with the wire.

14. Remove the radiator fan shroud assembly.

➡**Pull-up the radiator, then move the radiator fan shroud assembly toward the right side of the vehicle to allow for enough space to lift it up and away from the A/C condenser fan shroud assembly.**

15. Disassemble the fan shrouds.

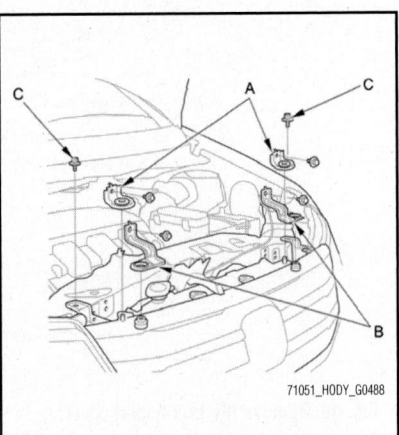

Fig. 44 Remove the radiator brackets (A) and the condenser brackets (B), then remove the front grille bolts(C)

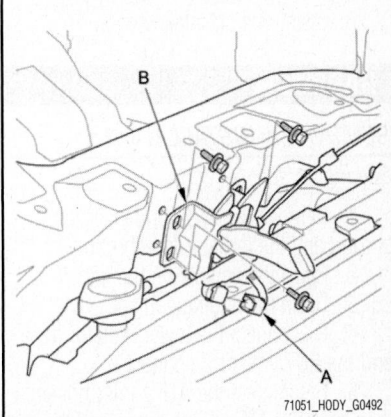

Fig. 46 With security system: Disconnect the hood latch switch connector (A); Remove the hood latch (B) with the wire

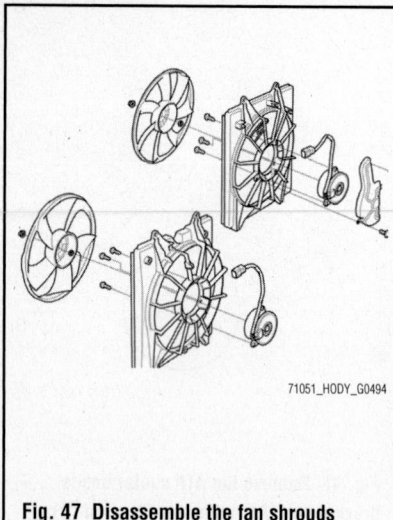

Fig. 47 Disassemble the fan shrouds

To install:

16. Assemble the fan shrouds.

17. Install the radiator fan shroud assembly, then tighten the bolts.

18. Install the hood latch.

19. With security system: Connect the hood latch switch connector, then adjust the hood latch.

20. Raise the vehicle on the lift.

21. Install the ATF cooler hoses bracket bolts.

22. Lower the vehicle on the lift.

23. Connect the radiator fan motor connector and Install the harness clamps.

24. Install the A/C condenser fan shroud assembly, then tighten the bolts.

25. Install the front grille bolts.

26. Install the radiator brackets and the A/C condenser brackets.

27. Connect the A/C condenser fan motor connector, and coolant reservoir hose, then install the coolant reservoir hose to the clamps.

28. Install the front bulkhead cover.

29. Install the air intake scoop.

RADIATOR

REMOVAL & INSTALLATION

See Figure 48.

1. Remove the A/C condenser fan shroud and the radiator fan shroud.

2. Raise the vehicle on the lift.

3. Drain the engine coolant.

4. Disconnect the lower radiator hose and the ECT sensor 2 connector.

5. Disconnect the ATF cooler hoses, then plug the ATF cooler hoses and the lines:

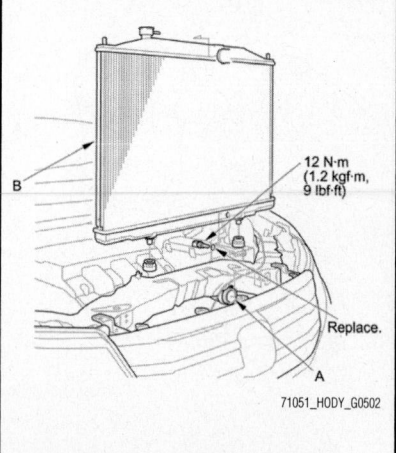

Fig. 48 Disconnect the upper radiator hose (A), then pull up the radiator (B)

6. Lower the vehicle on the lift.

7. Disconnect the upper radiator hose, then pull up the radiator.

8. Remove the other parts from the radiator.

To install:

9. Install radiator in the reverse order of removal. Make sure the upper and lower cushions are set securely.

10. Install the A/C condenser fan shroud and the radiator fan shroud.

11. Refill the transmission with the ATF:

12. Fill the radiator with engine coolant, and bleed the air from the cooling system.

13. Clean up any spilled engine coolant.

THERMOSTAT

REMOVAL & INSTALLATION

See Figure 49.

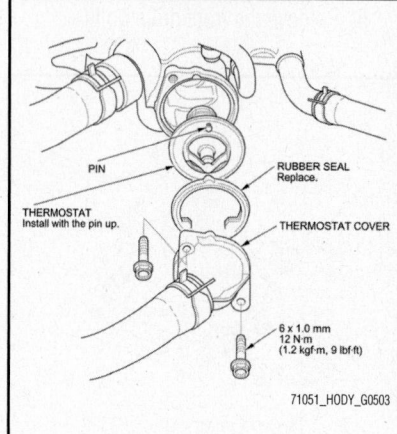

Fig. 49 Remove the thermostat cover, then remove the thermostat

1. Drain the engine coolant.

2. Remove the intake air duct.

3. Remove the thermostat cover, then remove the thermostat.

To install:

4. Install the new thermostat with a new rubber seal, then install the thermostat cover.

5. Install the intake air duct.

6. Refill the radiator with engine coolant, and bleed the air from the cooling system.

7. Clean up any spilled engine coolant.

WATER PUMP

REMOVAL & INSTALLATION

See Figure 50.

1. Drain the engine coolant.

2. Remove the timing belt.

3. Remove the timing belt adjuster.

4. Remove the five bolts securing the water pump, then remove the water pump and the O-ring.

To install:

5. Inspect and clean the mating surface of the engine block.

6. Install the water pump with a new O-ring.

7. Clean up any spilled engine coolant.

8. Install the timing belt adjuster.

9. Install the timing belt.

10. Refill the radiator with engine coolant, and bleed the air from the cooling system.

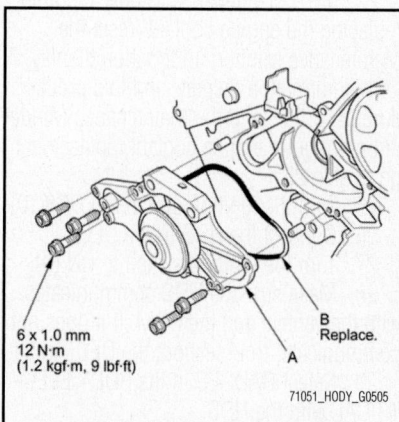

Fig. 50 Remove the five bolts securing the water pump (A), then remove the water pump and the O-ring (B)

BATTERY

REMOVAL & INSTALLATION

See Figure 51.

➡The battery terminal disconnection and reconnection procedure must be done before and after doing this procedure. Some systems store data in memory that is lost when the battery is disconnected.

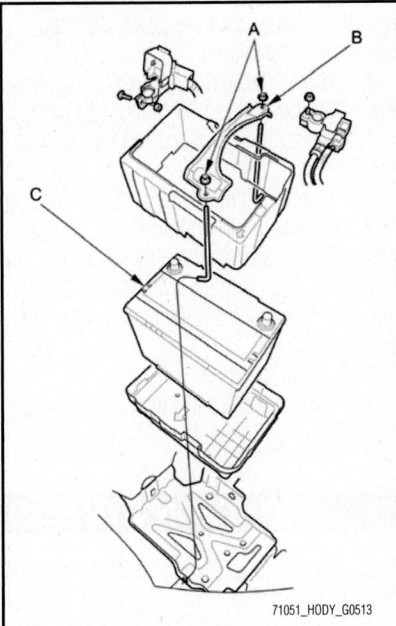

Fig. 51 Remove the two nuts (A) securing the battery setting plate (B), then remove the battery setting plate and the battery (C)

1. Do the battery terminal disconnection procedure.
2. Remove the two nuts securing the battery setting plate, then remove the battery setting plate and the battery.

To install:
3. Install the battery, then install the battery setting plate.
4. Tighten the two nuts equally until the battery is stable.

➡Do not deform the battery setting plate by over-tightening the nuts.

5. Do the battery terminal reconnection procedure.

➡Make sure the battery is correctly installed, and that the positive terminal and the negative terminal are not connected in reverse.

6. Do the resetting the power sliding door control unit procedure (if equipped).
7. Do the resetting the power tailgate control unit procedure (if equipped).

BATTERY DISCONNECTION/RECONNECTION PROCEDURE

Disconnection

➡Some systems store data in memory (including seat position, mirror position, etc.) that is lost when the battery is disconnected. Do the following procedures before disconnecting the battery.

1. Make sure to have the anti-theft code for the audio unit or the audio-navigation unit.
2. Make sure the ignition switch is in LOCK (0).
3. Close the power sliding doors (with power sliding door).
4. Close the power tailgate (with power tailgate).
5. Disconnect and isolate the negative cable from the battery.

➡Always disconnect the negative cable from the battery first.

6. Disconnect the positive cable from the battery.

Reconnection

See Figure 52.

➡Some systems store data in memory (including seat position, mirror position, etc.) that is lost when the battery is disconnected. Do the following procedures to restore the systems back to normal operation.

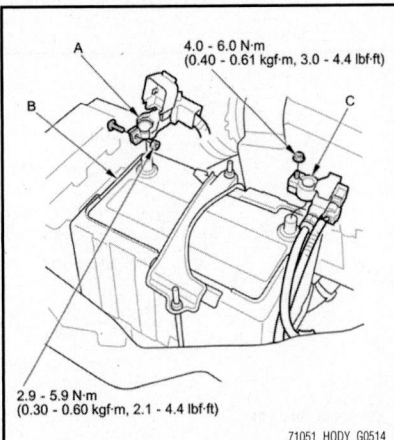

Fig. 52 Reconnect the positive cable (A) to the battery (B) first, then reconnect the negative cable (C) to the battery

1. Clean the battery terminals.
2. Test the battery.
3. Reconnect the positive cable to the battery first, then reconnect the negative cable to the battery.

➡Always connect the positive cable to the battery first.

4. Apply multipurpose grease to the terminals to prevent corrosion.
5. Enter the anti-theft code for the audio unit or the audio-navigation unit.
6. Enter the audio presets.
7. Set the clock (for vehicles without navigation).
8. If you disconnected the battery with the power sliding door open, or opened the door with the power disconnected, manually close the door once, then do the power sliding door rehoming procedure.
9. If you disconnected the battery with the power tailgate open, or opened the power tailgate with the power disconnected, manually close the door once, then do the power tailgate rehoming procedure.

RESETTING THE POWER SLIDING DOOR CONTROL UNIT

Rehome the power sliding doors whenever the power sliding door is manually opened after the battery power is disconnected from the power sliding door control unit (battery disconnected, control unit disconnected, etc.) or after replacing a power sliding door component. If the door was opened with the power disconnected, manually close the door once before doing the following rehoming procedure.

1. Clear the power sliding door DTCs by removing the No. B16 (10 A) fuse from the under-hood fuse/relay box.
2. Turn off the main switch for the doors. Make sure the ignition switch is in LOCK (0).
3. Fully close the power sliding door manually (the control unit must see the full latch switch closed and the ratchet switch open at the same time).
4. Turn the ignition switch to ON (II). Turn on the main switch.
5. Test the door operation with the power sliding door switch, the remote transmitter, and the door handles.

RESETTING THE POWER TAILGATE CONTROL UNIT

If the battery terminal is disconnected or the No. B16 (10 A) fuse in the under-hood fuse/relay box is removed while the power tailgate is operating, the power tailgate will not be able to open or close automatically until it is reset. To reset the system, fully close the tailgate manually. Once the battery terminals are reconnected or the No. B16 (10 A) fuse in the under-hood fuse/relay box is replaced, the power tailgate system automatically resets. Make sure the power tailgate operates properly.

ENGINE ELECTRICAL

ALTERNATOR

REMOVAL & INSTALLATION

See Figure 53.

1. Do the battery terminal disconnection procedure.
2. Remove the drive belt.
3. Disconnect the alternator connector and the positive alternator cable from the alternator.
4. Disconnect the A/C compressor clutch connector from the A/C compressor.
5. Remove the bolt securing the harness holder.
6. Remove the A/C compressor without disconnecting the A/C hoses. Do not bend the A/C hoses excessively.

➡Hang the A/C compressor with a wire tie.

7. Remove the power steering reservoir from the clamp, then remove the push pin from the coolant reservoir.

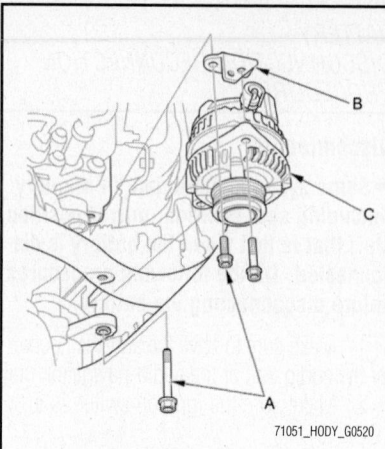

Fig. 53 Remove the mounting bolts (A) and the alternator bracket (B), then remove the alternator (C)

8. Remove the mounting bolts and the alternator bracket, then remove the alternator.

CHARGING SYSTEM

To install:

9. Install the alternator, then tighten the mounting bolts and the alternator bracket.
   a. Tighten the long bolt to 33 ft. lbs. (45 Nm).
   b. Tighten the short bolts to 16 ft. lbs. (22 Nm).
10. Install the push pin to the coolant reservoir, then install the power steering reservoir to the clamp.
11. Install the A/C compressor. Tighten the mounting bolts to 16 ft. lbs. (22 Nm).
12. Install the bolt securing the harness holder.
13. Connect the alternator connecter, the A/C compressor clutch connector, and the positive alternator cable to the alternator. Make sure the crimped side of the ring terminal faces away from the alternator when you connect it.
14. Install the drive belt.
15. Do the battery terminal reconnection procedure.

ENGINE ELECTRICAL

FIRING ORDERS

The firing order for the J35Z8 V6 engine is 1–4–2–5–3–6.

IGNITION COILS & SPARK PLUGS

REMOVAL & INSTALLATION

See Figures 54 and 55.

1. Remove the engine cover.
2. Disconnect the ignition coil connectors, then remove the ignition coils.
3. Remove the spark plugs and inspect them.
4. Apply a small amount of anti-seize compound to the plug threads, and screw the plugs into the cylinder head, finger-tight, then tighten the plugs to 16 ft. lbs. (22 Nm).

IGNITION SYSTEM

5. Install the ignition coils in the reverse order of removal.

IGNITION TIMING

INSPECTION & ADJUSTMENT

See Figure 56.

1. Connect the HDS to the DLC.
2. Turn the ignition switch to ON (II).

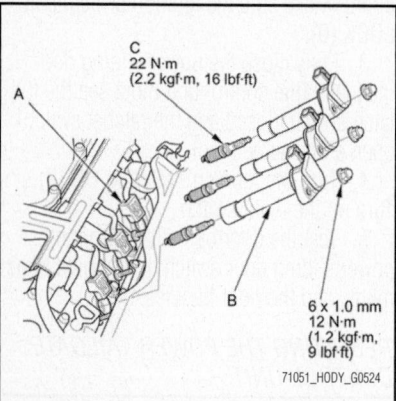

Fig. 54 Disconnect the ignition coil connectors (A), then remove the ignition coils (B) and spark plugs (C)

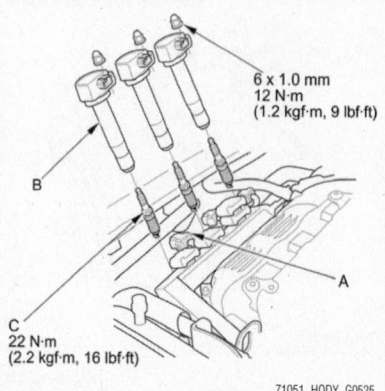

Fig. 55 Disconnect the ignition coil connectors (A), then remove the ignition coils (B) and spark plugs (C)

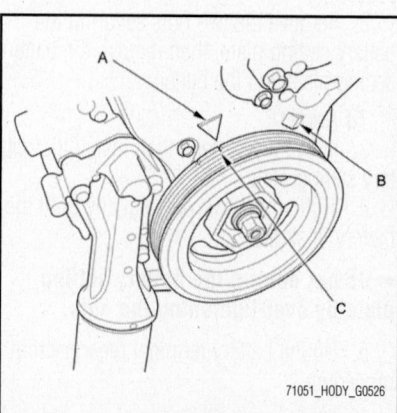

Fig. 56 Aim the light toward the pointer (A) on the timing belt lower cover; pointer not used (B); RED mark (C)

3. Make sure the HDS communicates with the vehicle and the PCM. If it does not communicate, troubleshoot the DLC circuit.

4. Check for DTCs. If a DTC is present, diagnose and repair the cause before continuing with this test.

5. Start the engine. Hold the engine speed at 3,000 rpm with no load (in P or N) until the radiator fan comes on, then let it idle.

6. Check the idle speed.

7. Jump the SCS line with the HDS.

8. Connect the timing light to the No.1 ignition coil harness.

9. Aim the light toward the pointer (A) on the timing belt lower cover. Check the ignition timing under a no load condition (headlights, blower fan, rear window defogger, and air conditioner are turned off).

➡**The other pointer (B) is not used.**

➡**Ignition Timing: 10±2 °BTDC (RED mark (C)) at idle in P or N**

10. If the ignition timing differs from the specification, check the camshaft timing. If the camshaft timing is OK, update the PCM if it does not have the latest software, or substitute a known-good PCM, then recheck. If the system works properly, and the PCM was substituted, replace the original PCM.

11. Disconnect the HDS and the timing light.

ENGINE ELECTRICAL

STARTER

REMOVAL & INSTALLATION

See Figure 57.

1. Do the battery removal procedure.
2. Remove the air cleaner.
3. Remove the battery base.
4. Disconnect the positive starter cable and the S terminal connector.
5. Remove the upper radiator hose bracket and the dipstick.
6. Remove the two bolts holding the starter, then remove the starter and the gasket.

To install:

7. Install the starter, then tighten the mounting bolts.

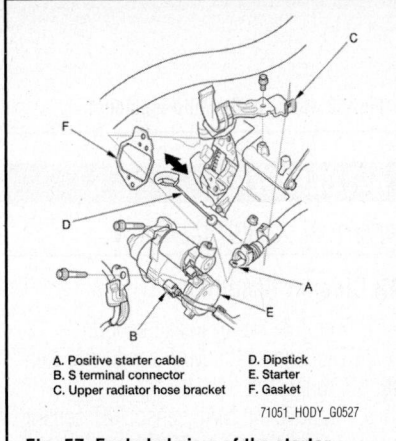

A. Positive starter cable
B. S terminal connector
C. Upper radiator hose bracket
D. Dipstick
E. Starter
F. Gasket

71051_HODY_G0527

Fig. 57 Exploded view of the starter assembly

STARTING SYSTEM

➡**Always use a new gasket.**

8. Install the upper radiator hose bracket and the dipstick.

9. Connect the positive starter cable and the S terminal connector. Make sure the crimped side of the ring terminal faces away from the starter when you connect it.

10. Install the battery base.

11. Install the air cleaner.

12. Do the battery installation procedure.

13. Start the engine to make sure the starter works properly.

ENGINE MECHANICAL

ACCESSORY DRIVE BELT SYSTEM

ADJUSTMENT

Tension for the drive belt is provided by the drive belt auto-tensioner. No adjustment is necessary.

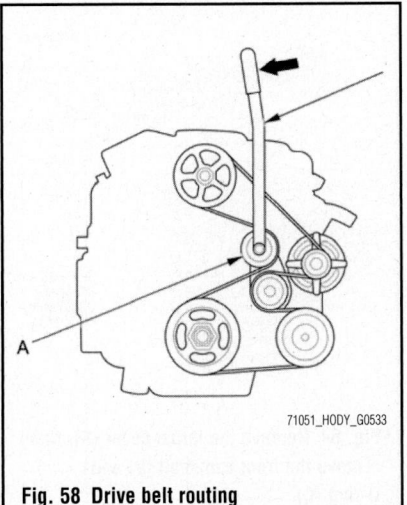

71051_HODY_G0533

Fig. 58 Drive belt routing

BELT ROUTINGS

See Figure 58.

INSPECTION

See Figure 59.

1. Inspect the belt for cracks or damage.

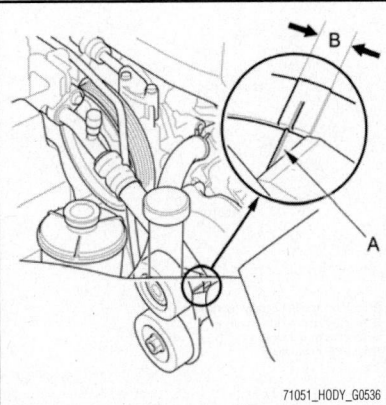

71051_HODY_G0536

Fig. 59 Check that the position of the auto-tensioner indicator (A) is within the standard range (B) as shown

If the belt is cracked or damaged, replace it.

2. Check that the position of the auto-tensioner indicator (A) is within the standard range (B) as shown. If it is out of the standard range, replace the drive belt.

REMOVAL & INSTALLATION

Drive Belt

See Figure 60.

Special Tools Required*: Belt Tension Release Tool Snap-On® TK9317 (*: Available through the Honda Tool and Equipment Program 888-424-6857).

1. Remove the power steering reservoir from the clamp.

2. Move the auto-tensioner using the belt tension release tool (TK9317) in the direction of the arrow to relieve tension from the drive belt, then remove the drive belt.

3. Install the new drive belt in the reverse order of removal.

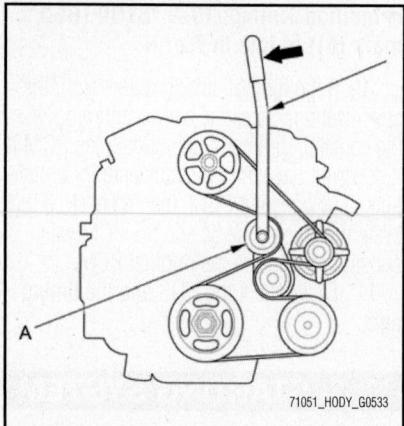

Fig. 60 Move the auto-tensioner (A) using the belt tension release tool in the direction of the arrow to relieve tension from the drive belt, then remove the drive belt

Drive Belt Tensioner Pulley

See Figure 61.

1. Remove the auto-tensioner.
2. Remove the pulley bolt (left-hand threads), and remove the tensioner pulley.
3. Install the tensioner pulley in the reverse order of removal.

Drive Belt Tensioner

See Figure 62.

1. Remove the drive belt.
2. Remove the splash shield.
3. Remove the auto-tensioner.
4. Install the auto-tensioner in the reverse order of removal.

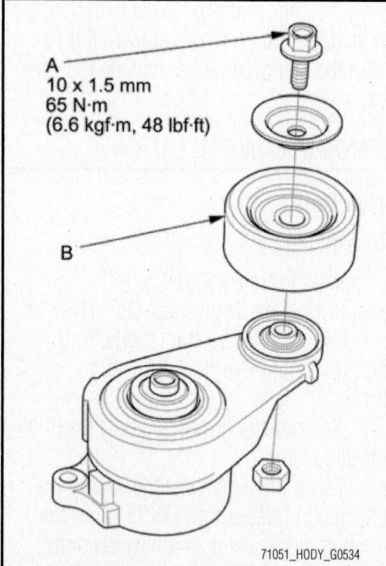

Fig. 61 Remove the pulley bolt (A) (left-hand threads), and remove the tensioner pulley (B)

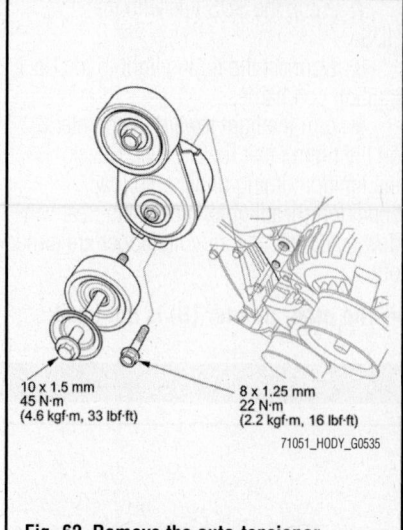

Fig. 62 Remove the auto-tensioner

AIR CLEANER

REMOVAL & INSTALLATION

Air Cleaner Assembly

1. Remove the air intake scoop.
2. Disconnect the MAF sensor/IAT sensor 5P connector.
3. Remove the MAF sensor/IAT sensor harness from the air cleaner housing holder and the bolts.
4. Loosen the clamp screw, then remove the air cleaner.
5. Install the parts in the reverse order of removal.

➡**When you tighten the clamp screw, refer to the throttle body removal/installation procedure.**

Air Filter Element

See Figure 63.

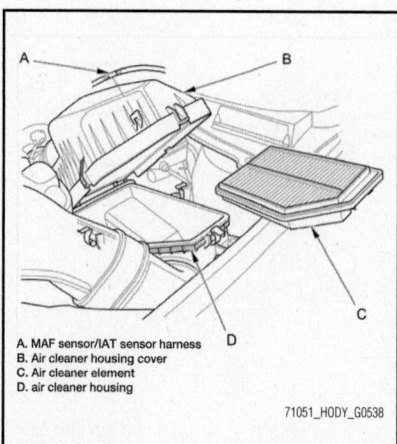

A. MAF sensor/IAT sensor harness
B. Air cleaner housing cover
C. Air cleaner element
D. air cleaner housing

Fig. 63 Remove the MAF sensor/IAT sensor harness from the air cleaner housing cover

1. Remove the MAF sensor/IAT sensor harness from the air cleaner housing cover.
2. Open the air cleaner housing cover.
3. Remove the air cleaner element from the air cleaner housing.
4. Check the air cleaner element for damage or clogging. If it is damaged or clogged, replace it.

➡**Do not use compressed air to clean the air cleaner element.**

5. Clean and remove any debris from inside the air cleaner housing.
6. Install the parts in the reverse order of removal.

➡**If the idle speed fluctuates, do the idle speed inspection.**

CAMSHAFT & BEARINGS

REMOVAL & INSTALLATION

Front

See Figure 64.

1. Do the battery removal procedure.
2. Remove the battery base.
3. Remove the EGR valve.
4. Remove the EGR valve stud bolts.
5. Remove the timing belt.
6. Remove the front rocker arm assembly.
7. Remove the front camshaft pulley.
8. Remove the thrust cover, then remove the front camshaft (B) and O-ring.

To install:

9. Install the front camshaft in the reverse order of removal. Always use a new O-ring. Apply new engine oil to the journals and the cam lobes.

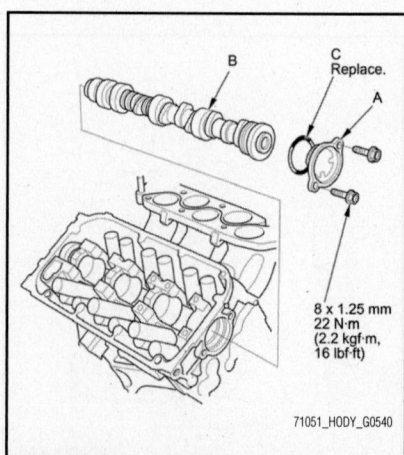

Fig. 64 Remove the thrust cover (A), then remove the front camshaft (B) and O-ring (C)

10. Apply new engine oil to the threads of the camshaft pulley mounting bolt, then install the front camshaft pulley.

11. Install the front rocker arm assembly, then tighten the mounting bolts to 18 ft. lbs.

12. Install the timing belt.

13. Adjust the valve clearance.

14. Install the EGR valve stud bolts, then install the EGR valve.

15. Install the battery base.

16. Do the battery installation procedure.

17. Do the CKP pattern clear/CKP pattern learn procedure.

Rear

See Figure 65.

1. Relieve the fuel pressure.
2. Drain the engine coolant.
3. Remove the intake air duct.
4. Remove the quick-connector fitting cover, then disconnect the fuel feed hose.
5. Disconnect the heater hose and remove the EVAP canister purge joint with the bracket.
6. Remove the timing belt.
7. Remove the rear rocker arm assembly.
8. Remove the rear camshaft pulley.
9. Remove the thrust cover, then remove the rear camshaft (B) and O-ring.

To install:

10. Install the rear camshaft in the reverse order of removal. Always use a new O-ring. Apply new engine oil to the journals and the cam lobes.

11. Apply new engine oil to the threads of the camshaft pulley mounting bolt, then install the rear camshaft pulley.

12. Install the rear rocker arm assembly, then tighten the mounting bolts to 18 ft. lbs..

13. Install the timing belt.

14. Adjust the valve clearance.

15. Connect the heater hoses and install the EVAP canister purge joint with the bracket.

16. Connect the fuel feed hose, then install the quick-connect fitting cover.

17. Inspect for fuel leaks. Turn the ignition switch to ON (II) (do not operate the starter) so the fuel pump runs for about 2 seconds and pressurizes the fuel line. Repeat this operation three times, then check for fuel leakage at any point in the fuel line.

18. Fill the radiator with engine coolant, and bleed the air from the cooling system.

19. Do the CKP pattern clear/CKP pattern learn procedure.

CRANKSHAFT FRONT SEAL

REMOVAL & INSTALLATION

See Figure 66.

Special Tools Required: Oil Seal Driver, 64 mm 07OAD-RCAA100

1. Remove the timing belt drive pulley.
2. Remove the pulley end crankshaft oil seal.

To install:

3. Clean and dry the crankshaft oil seal housing.

4. Apply a light coat of new engine oil to the lip of the crankshaft oil seal.

5. Using the oil seal driver, 64 mm, drive in the new crankshaft oil seal until the oil seal driver bottoms on the oil pump.

6. Clean the excess oil off the crankshaft, and check that the oil seal lip is not distorted.

7. Install the timing belt drive pulley.

CRANKSHAFT PULLEY

REMOVAL & INSTALLATION

See Figures 67 through 69.

Special Tools Required:
• Holder Handle 07JAB-001020B
• Holder Attachment, 50 mm, Offset 07MAB-PY3010A
• Socket, 19 mm 07JAA-001020A

1. Raise the vehicle on the lift.
2. Remove the right front wheel.
3. Remove the splash shield.
4. Remove the drive belt.
5. Hold the pulley with the holder handle and the holder attachment, 50 mm offset.
6. Remove the bolt with a heavy duty socket, 19 mm and a breaker bar, then remove the crankshaft pulley.

To install:

7. Remove any oil and clean the pulleys, the crankshaft, the bolt, and the

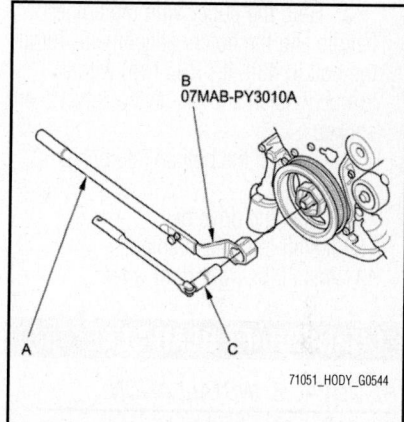

Fig. 67 Hold the pulley with the holder handle (A) and the holder attachment, 50 mm offset (B); 19mm socket (C)

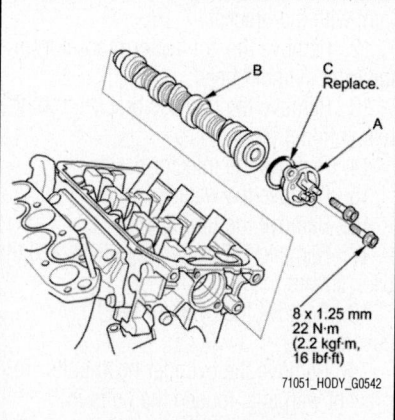

Fig. 65 Remove the thrust cover (A), then remove the rear camshaft (B) and O-ring (C)

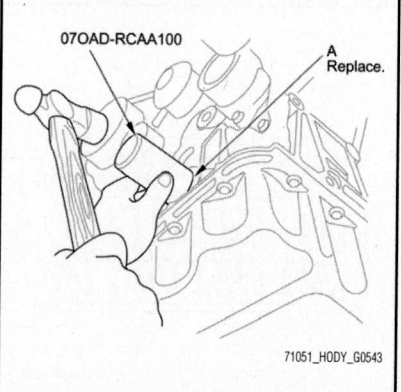

Fig. 66 Using the oil seal driver, 64 mm, drive in the new crankshaft oil seal (A) until the oil seal driver bottoms on the oil pump

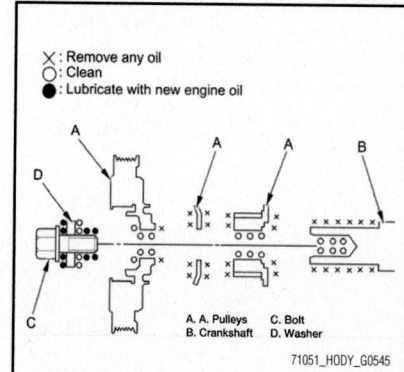

Fig. 68 Remove any oil and clean the pulleys, the crankshaft, the bolt, and the washer; lubricate with new engine oil as shown

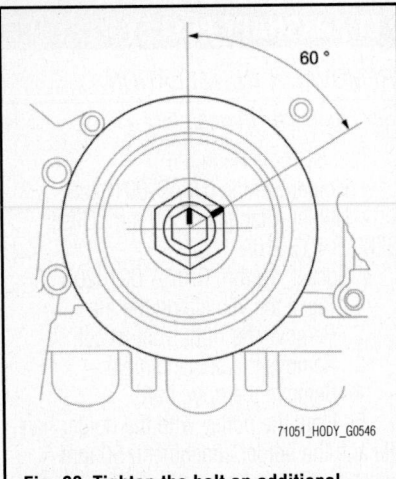

Fig. 69 Tighten the bolt an additional 60 degrees

washer. Lubricate with new engine oil as shown.

8. Install the crankshaft pulley, then tighten the bolt. Do not use an impact wrench.

   a. Hold the pulley with the holder handle and the holder attachment. Torque the bolt to 48ft. lbs. (65 Nm) with a torque wrench and the heavy duty 19 mm socket.

   b. Tighten the bolt an additional 60 degrees.

9. Install the drive belt.

10. Install the splash shield.

11. Install the right front wheel.

CRANKSHAFT REAR SEAL

REMOVAL & INSTALLATION

See Figure 70.

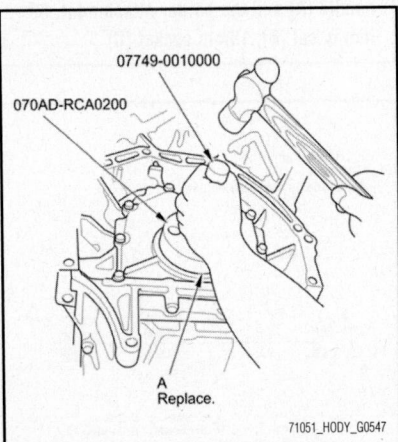

Fig. 70 Using the driver handle and the oil seal driver attachment, drive in the new crankshaft oil seal (A) until the oil seal driver attachment bottoms on the engine block end cover

Special Tools Required:
- Driver Handle, 15 x 135L 07749-0010000
- Oil Seal Driver Attachment, 106 mm 070AD-RCA0200

1. Remove the transmission:
2. Remove the drive plate:
3. Remove the transmission end crankshaft oil seal.

To install:

4. Clean and dry the crankshaft oil seal housing.

5. Apply a light coat of new engine oil to the lip of the crankshaft oil seal.

6. Using the driver handle, 15 x 135L and the oil seal driver attachment, 106 mm, drive in the new crankshaft oil seal until the oil seal driver attachment bottoms on the engine block end cover. Align the hole in the oil seal driver attachment with the pin on the crankshaft.

7. Clean the excess oil off the crankshaft, and check that the oil seal lip is not distorted.

8. Install the drive plate:

9. Install the transmission:

CYLINDER HEAD

REMOVAL & INSTALLATION

See Figures 71 through 79.

Note the following:
- To avoid damaging the wires and terminals, unplug the wiring connectors carefully while holding the connector portion.
- Connect the HDS to the DLC, and monitor ECT SENSOR 1. To avoid damaging the cylinder head, wait until the engine coolant temperature drops below

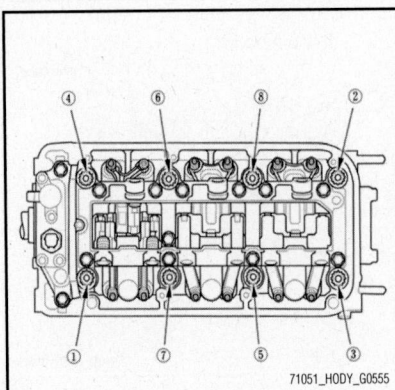

Fig. 71 Remove the cylinder head bolts; to prevent warpage, loosen the bolts in sequence—Front

100°F (38°C) before loosening the cylinder head bolts.
- Mark all wirings and hoses to avoid misconnection. Also, be sure that they do not contact any other wirings or hoses, or interfere with any other parts.

1. Relieve the fuel pressure.

2. Do the battery terminal disconnection procedure.

3. Drain the engine coolant.

4. Remove the six ignition coils.

5. Remove the alternator.

6. Remove the power steering pump and the power steering hose bracket with its hose connected.

7. Remove the intake manifold.

8. Disconnect the following engine wire harness connectors, and remove the wire harness clamps from the cylinder head:

- Six injector connectors
- Knock sensor connector
- ECT sensor 1 connector
- EGR valve connector
- Rocker arm oil pressure sensor connector
- Rocker arm oil control solenoid A (Bank 1) connector
- Rocker arm oil control solenoid A (Bank 2) connector
- Rocker arm oil control solenoid B (Bank 1) connector
- Front rocker arm oil pressure switch connector
- Rear rocker arm oil pressure switch connector
- CMP sensor connector
- Two A/F sensor connectors
- Two secondary HO2S connectors

9. Remove the front warm up TWC and the rear warm up TWC.

10. Remove the quick-connect fitting cover, then disconnect the fuel feed hose.

11. Remove the EVAP canister purge joint with the bracket.

12. Remove the connector bracket from the front cylinder head.

13. Remove the harness bracket from the rear cylinder head.

14. Remove the injector bases.

15. Remove the water passage.

16. Remove the timing belt.

17. Remove the camshaft pulleys and the back covers.

18. Remove the cylinder head covers.

19. Remove the cylinder head bolts. To prevent warpage, loosen the bolts in sequence 1/3 turn at a time; repeat the sequence until all bolts are loosened.

20. Remove the cylinder heads.

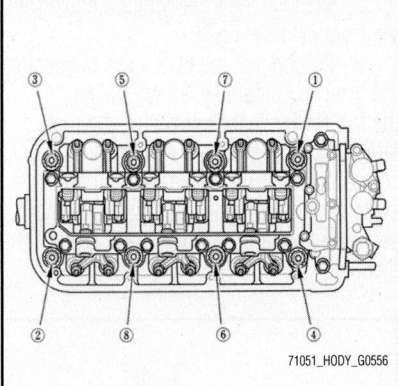

Fig. 72 Remove the cylinder head bolts; to prevent warpage, loosen the bolts in sequence—Rear

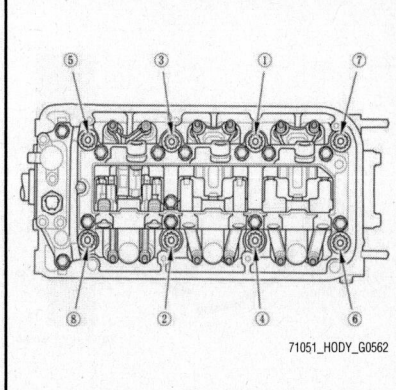

Fig. 74 Set the camshaft pulleys to TDC by aligning the TDC marks (A) on the camshaft pulleys with the pointers (B) on the back covers—Front

To install:

21. Clean the cylinder head and the engine block surface.

22. Clean and install the oil control orifices with new O-rings.

23. Install the dowel pins and the new cylinder head gaskets.

24. Clean the timing belt pulleys, the timing belt guide plate, and the upper and lower covers.

25. Set the timing belt drive pulley to top dead center (TDC) by aligning the TDC mark on the tooth of the timing belt drive pulley with the pointer on the oil pump.

26. Set the camshaft pulleys to TDC by aligning the TDC marks on the camshaft pulleys with the pointers on the back covers.

27. Install the cylinder heads on the engine block.

28. Measure the diameter of each cylinder head bolt at point A and point B.

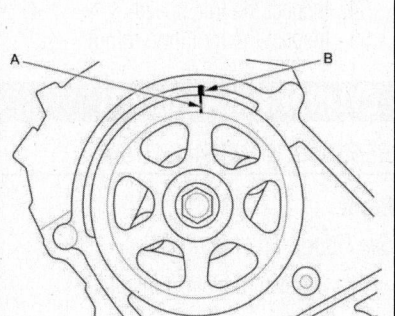

Fig. 75 Set the camshaft pulleys to TDC by aligning the TDC marks (A) on the camshaft pulleys with the pointers (B) on the back covers—Rear

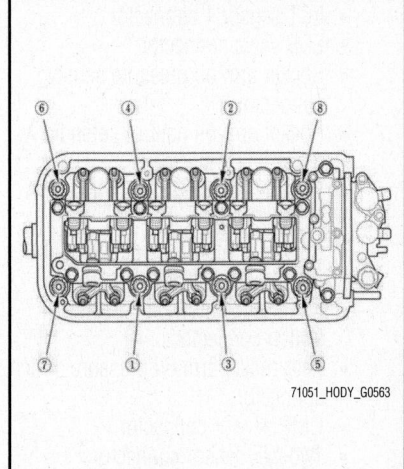

Fig. 77 Torque the cylinder head bolts in sequence to 22 ft. lbs. (30 Nm), using a beam-type torque wrench—Front

29. If either diameter is less than 0.445 inches (11.3 mm), replace the cylinder head bolt.

30. Apply new engine oil to the threads and under the bolt heads of all cylinder head bolts.

31. Torque the cylinder head bolts in sequence to 22 ft. lbs. (30 Nm), using a beam-type torque wrench. When using a preset click-type torque wrench, be sure to tighten slowly and do not overtighten. If a bolt makes any noise while you are torquing it, loosen the bolt and retighten it from the first step.

32. After torquing, tighten all cylinder head bolts in two steps (90 degrees per step) using the sequence shown in step 11. If you are using a new cylinder head bolt, tighten the bolt an extra 90 degrees.

➡**Remove the cylinder head bolt if you tightened it beyond the specified angle, and go back to step 8 of the procedure.**

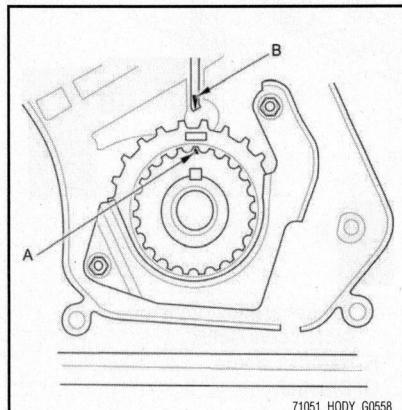

Fig. 73 Set the timing belt drive pulley to top dead center (TDC) by aligning the TDC mark (A) on the tooth of the timing belt drive pulley with the pointer (B) on the oil pump

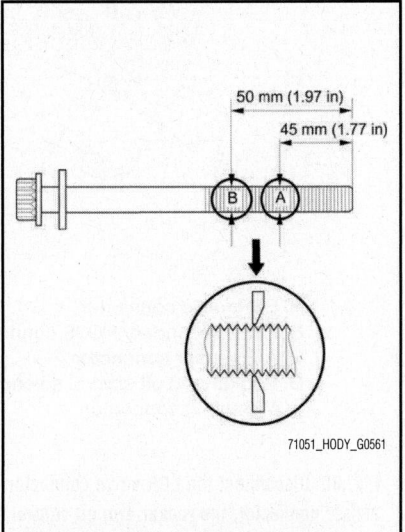

Fig. 76 Measure the diameter of each cylinder head bolt at point A and point B

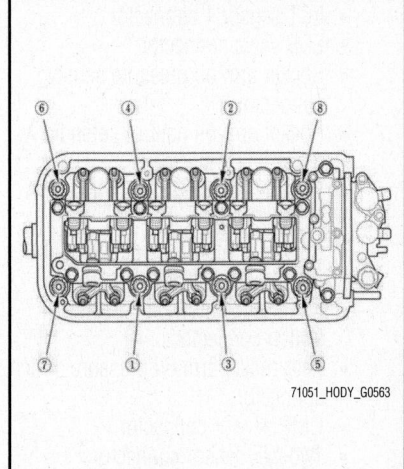

Fig. 78 Torque the cylinder head bolts in sequence to 22 ft. lbs. (30 Nm), using a beam-type torque wrench—Rear

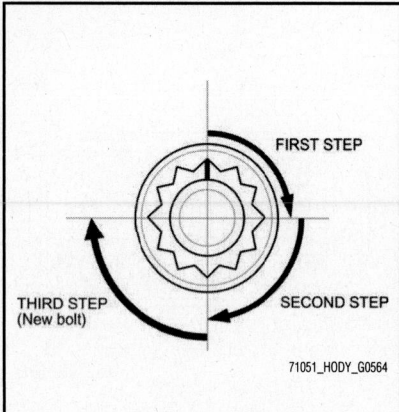

Fig. 79 After torquing, tighten all cylinder head bolts in two steps (90 degrees per step) using the sequence shown

Do not loosen it back to the specified angle.

33. Install the timing belt.
34. Adjust the valve clearance.
35. Install the cylinder head covers.
36. Install the water passage.
37. Install the injector bases.
38. Install the connector bracket to the front cylinder head.
39. Install the harness bracket to the rear cylinder head.
40. Install the EVAP canister purge joint with the bracket.
41. Connect the fuel feed hose, then install the quick-connect fitting cover.
42. Install the front warm up TWC and the rear warm up TWC.
43. Connect the following engine wire harness connectors, and install the wire harness clamps to the cylinder head:
   - Six injector connectors
   - Knock sensor connector
   - ECT sensor 1 connector
   - EGR valve connector
   - Rocker arm oil pressure sensor connector
   - Rocker arm oil control solenoid A (Bank 1) connector
   - Rocker arm oil control solenoid A (Bank 2) connector
   - Rocker arm oil control solenoid B (Bank 1) connector
   - Front rocker arm oil pressure switch connector
   - Rear rocker arm oil pressure switch connector
   - CMP sensor connector
   - Two A/F sensor connectors
   - Two secondary HO2S connectors
44. Install the intake manifold.
45. Install the power steering pump and the power steering hose bracket.

46. Install the alternator.
47. Install the six ignition coils.
48. Do the battery terminal reconnection procedure.
49. After installation, check that all tubes, hoses and connectors are installed correctly.
50. Inspect for fuel leaks. Turn the ignition switch to ON (II) (do not operate the starter) so the fuel pump runs for about 2 seconds and pressurizes the fuel line. Repeat this operation three times, then check for fuel leakage at any point in the fuel line.
51. Refill the radiator with engine coolant, and bleed the air from the cooling system.
52. Do the PCM idle learn procedure.
53. Do the CKP pattern clear/CKP pattern learn procedure.
54. Inspect the idle speed.
55. Inspect the ignition timing.

CYLINDER HEAD COVER

REMOVAL & INSTALLATION

Front

See Figures 80 through 83.

1. Remove the intake manifold.

2. Remove the three ignition coils from the front cylinder head.
3. Disconnect the EGR valve connector, the front secondary HO2S connector, the front A/F sensor connector, the rocker arm oil control solenoid A (Bank 2) connector, the front rocker arm oil pressure switch connector, and the harness clamp securing the harness holder, and remove the dipstick.

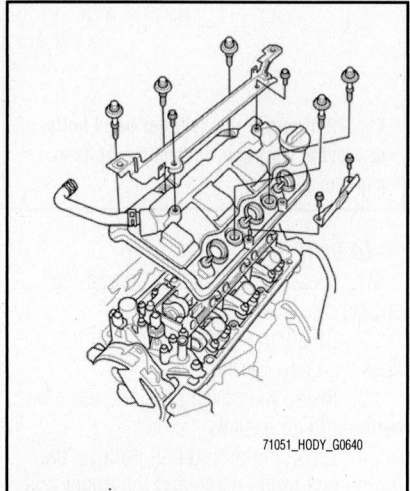

Fig. 81 Remove the front cylinder head cover

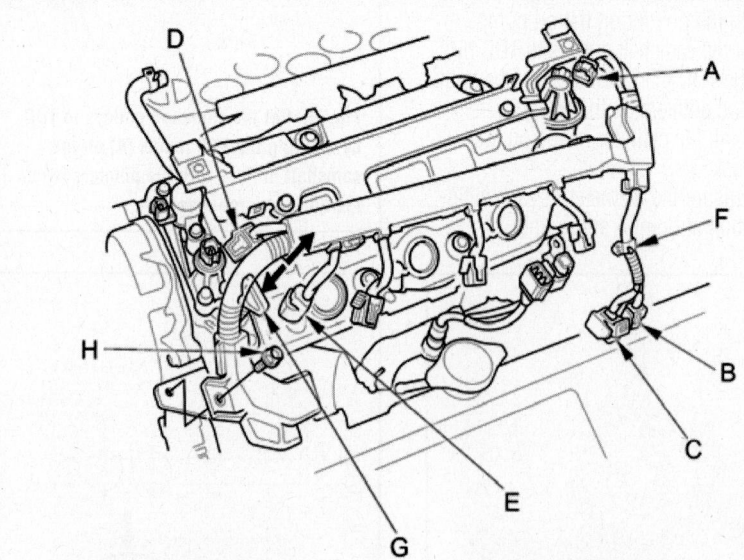

A. EGR valve connector
B. Front secondary HO2S connector
C. A/F sensor connector
D. Rocker arm oil control solenoid A (Bank 2) connector
E. Rocker arm oil pressure switch connector
F. Harness clamp
G. Dipstick
H. Bolt

71051_HODY_G0639

Fig. 80 Disconnect the EGR valve connector, the front secondary HO2S connector, the front A/F sensor connector, the rocker arm oil control solenoid A (Bank 2) connector, the front rocker arm oil pressure switch connector, and the harness clamp securing the harness holder, and remove the dipstick

4. Remove the bolt securing the harness holder.

5. Remove the front cylinder head cover.

To install:

6. Check the spark plug seals for damage. If any seals are damaged, replace it.

7. Thoroughly clean the head cover gasket and the groove of the cylinder head cover.

➡**Check and if necessary, replace the head cover gasket.**

8. Install the head cover gasket in the groove of the cylinder head cover. Make sure the head cover gasket is seated securely.

9. Remove all of the old liquid gasket from the front rocker arm oil control valve and the cylinder head.

10. Clean the head cover contacting surfaces with a shop towel.

11. Apply liquid gasket (P/N 08717-0004, 08718-0003, 08718-0004, or 08718-0009) to the rocker arm oil control valve mating surfaces. Install the component within 5 minutes of applying the liquid gasket.

- If you apply liquid gasket P/N 08718-0012, the component must be installed within 4 minutes.
- If too much time has passed after applying the liquid gasket, remove the old liquid gasket and residue, then reapply the new liquid gasket.

12. Set the spark plug seals on the spark plug tubes, and install the front cylinder head cover.

13. Inspect the spark plug seals for damage.

14. Inspect the cover washers. Replace any washer that is damaged or deteriorated.

15. Tighten the bolts in three steps. In the final step torque the bolts, in sequence, 9 ft. lbs. (12 Nm).

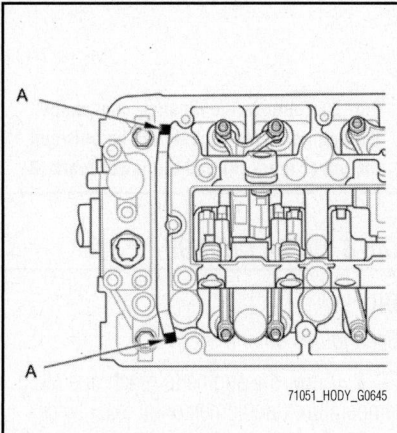

Fig. 82 Apply liquid gasket to the rocker arm oil control valve mating surfaces (A)

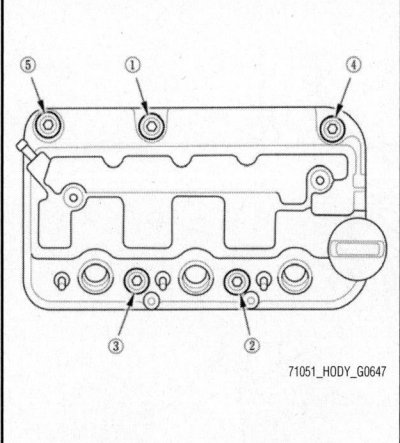

Fig. 83 Tighten the bolts in three steps in sequence

- Wait at least 30 minutes before filling the engine with oil.
- Do not run the engine for at least 3 hours after installing the cylinder head cover.

16. Tighten the bolt securing the harness holder.

17. Connect the EGR valve connector, the front secondary HO2S connector, the front A/F sensor connector, the rocker arm oil control solenoid A (Bank 2) connector, the front rocker arm oil pressure switch connector, and the harness clamp securing the harness holder, and install the dipstick.

18. Install the three ignition coils to the front cylinder head.

19. Install the intake manifold.

Rear

See Figures 82, 84 through 87.

1. Remove the intake manifold.

2. Remove the three ignition coils from the rear cylinder head.

3. Remove the drive belt.

4. Remove the power steering pump and the power steering hose bracket with its hoses connected.

5. Remove the harness holder mounting bolts.

6. Disconnect the three injector connectors and the two harness clamps.

7. Remove the harness from the upper cover.

8. Disconnect the rear rocker arm oil pressure switch connector, the rocker arm oil control solenoid B (Bank 1) connector, the rocker arm oil control solenoid A (Bank 1) connector, the rear A/F sensor connector, the rear secondary HO2S connector, and the harness clamps, then remove the harness holder.

9. Disconnect the breather hose.

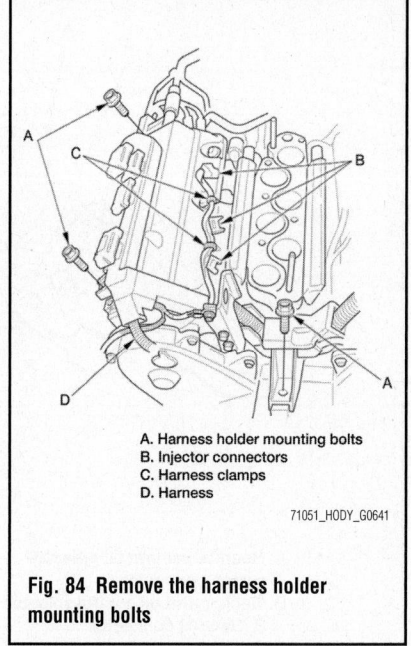

A. Harness holder mounting bolts
B. Injector connectors
C. Harness clamps
D. Harness

Fig. 84 Remove the harness holder mounting bolts

10. Remove the rear cylinder head cover.

To install:

11. Check the spark plug seals for damage. If any seals are damaged, replace it.

12. Thoroughly clean the head cover gasket and the groove of the cylinder head cover.

➡**Check and if necessary, replace the head cover gasket.**

13. Install the head cover gasket in the groove of the cylinder head cover. Make sure the head cover gasket is seated securely.

14. Remove all of the old liquid gasket from the rear rocker arm oil control valve and the cylinder head.

15. Clean the head cover contacting surfaces with a shop towel.

16. Apply liquid gasket (P/N 08717-0004, 08718-0003, 08718-0004, or 08718-0009) to the rocker arm oil control valve mating surfaces (A). Install the component within 5 minutes of applying the liquid gasket.

- If you apply liquid gasket P/N 08718-0012, the component must be installed within 4 minutes.
- If too much time has passed after applying the liquid gasket, remove the old liquid gasket and residue, then reapply the new liquid gasket.

17. Set the spark plug seals on the spark plug tubes, and install the rear cylinder head cover.

18. Inspect the spark plug seals for damage.

19. Inspect the cover washers. Replace any washer that is damaged or deteriorated.

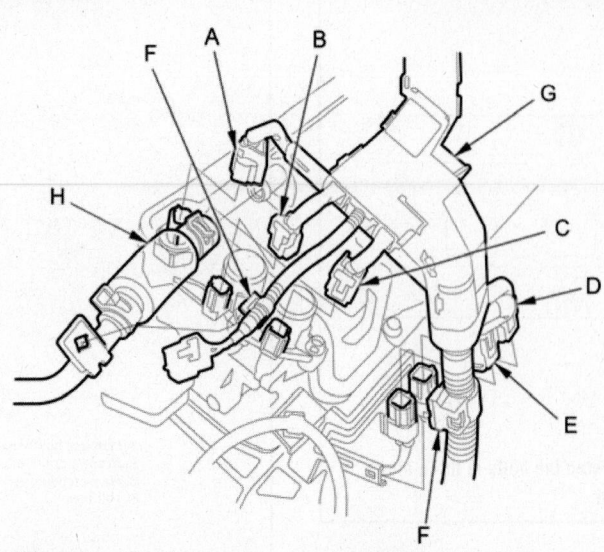

A. Rear rocker arm oil pressure switch connector
B. Rocker arm oil control solenoid B (Bank 1) connector
C. Rocker arm oil control solenoid A (Bank 1) connector
D. Rear A/F sensor connector
E. Rear secondary HO2S connector
F. Harness clamps
G. Harness holder
H. Breather hose

71051_HODY_G0642

Fig. 85 Disconnect the rear rocker arm oil pressure switch connector, the rocker arm oil control solenoid B (Bank 1) connector, the rocker arm oil control solenoid A (Bank 1) connector, the rear A/F sensor connector, the rear secondary HO2S connector, and the harness clamps, then remove the harness holder

connectors and the two harness clamps.

26. Install the power steering pump and the power steering hose bracket.

27. Install the drive belt.

28. Install the three ignition coils to the rear cylinder head.

29. Install the intake manifold.

ENGINE OIL & FILTER

OIL LEVEL CHECK

See Figure 88.

1. Park the vehicle on level ground, and start the engine. Hold the engine speed at 3,000 rpm with no load (in P or N) until the radiator fan comes on, then turn off the engine, and wait a few minute.

2. Remove the dipstick, and wipe off the dipstick, then reinstall the dipstick.

3. Remove the dipstick, and check the engine oil level. It should be between the upper mark and the lower mark.

4. If the engine oil level is near or below the lower mark, check for oil leakage, and add engine oil to bring it to the upper mark.

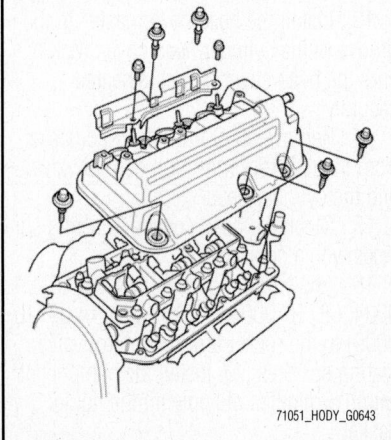

71051_HODY_G0643

Fig. 86 Remove the rear cylinder head cover

71051_HODY_G0648

Fig. 87 Tighten the bolts in three steps in sequence

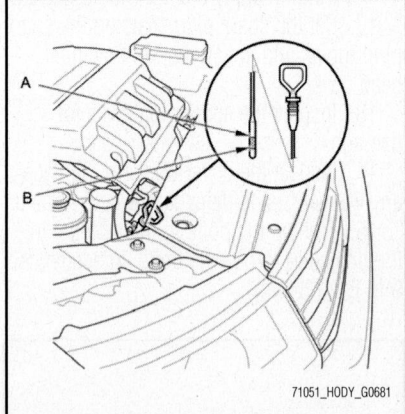

71051_HODY_G0681

Fig. 88 Remove the dipstick, and check the engine oil level; it should be between the upper mark (A) and the lower mark (B)

20. Tighten the bolts in three steps. In the final step torque the bolts, in sequence, 9 ft. lbs. (12 Nm).
  • Wait at least 30 minutes before filling the engine with oil.
  • Do not run the engine for at least 3 hours after installing the cylinder head cover.

21. Connect the rear rocker arm oil pressure switch connector, the rocker arm oil control solenoid B (Bank 1) connector, the

rocker arm oil control solenoid A (Bank 1) connector, the rear A/F sensor connector, the rear secondary HO2S connector, and the harness clamps, then install the harness holder.

22. Connect the breather hose.

23. Tighten the harness holder mounting bolts.

24. Install the harness to the upper cover.

25. Reconnect the three injector

OIL & FILTER CHANGE

Oil

See Figures 89 and 90.

1. Allow the engine to reach operating temperature (fan comes on at least twice).

2. Remove the drain bolt, and drain the engine oil.

3. Reinstall the drain bolt with a new washer and torque to specification.

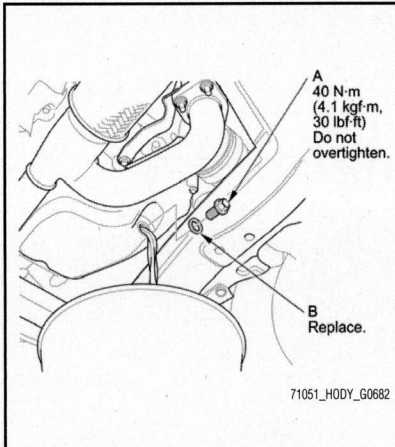

Fig. 89 Remove the drain bolt (A) and washer (B), and drain the engine oil

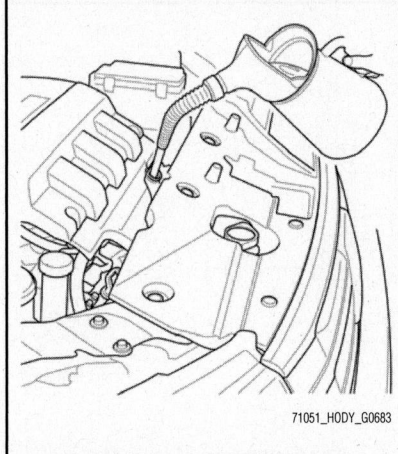

Fig. 90 Refill the engine with the recommended engine oil

4. Refill the engine with the recommended engine oil.

5. Run the engine for more than 3 minutes, then check the oil level and the oil leakage.

6. If the Maintenance Minder required replacing the engine oil, reset the Maintenance Minder; information display, multi information display, and this procedure is complete. If the Maintenance Minder did not require engine oil replacement, go to step 7.

7. Turn the ignition switch to LOCK (0).

8. Connect the HDS to the DLC.

9. Turn the ignition switch to ON (II).

10. Make sure the HDS communicates with the vehicle and the PCM. If it does not communicate, troubleshoot the DLC circuit.

11. Select GAUGES in the BODY ELECTRICAL with the HDS.

12. Select ADJUSTMENT in the GAUGES with the HDS.

13. Select MAINTENANCE MINDER in the ADJUSTMENT with the HDS.

14. Select RESET in the MAINTENANCE MINDER with the HDS.

15. Select RESETTING THE ENGINE OIL LIFE with the HDS.

Oil Filter

See Figures 91 and 92.

Special Tools Required: Oil Filter Wrench 07AAA-PLCA100

1. Remove the oil filter with the oil filter wrench.

2. Inspect the filter to make sure the rubber seal is not stuck to the oil filter seating surface of the engine.

3. Inspect the threads and the rubber seal on the new filter. Clean the seat on the oil filter base, then apply a light coat of new engine oil to the oil filter rubber seal. Use only filters with a built-in bypass system.

4. Install the oil filter by hand.

5. After the rubber seal seats, tighten the oil filter clockwise with the oil filter wrench.

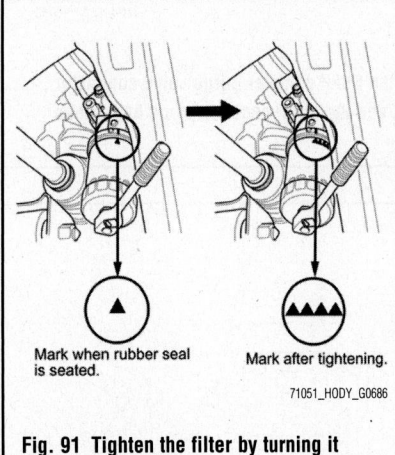

Fig. 91 Tighten the filter by turning it clockwise three numbers or marks from the one you noted

Number or Mark when rubber seal is seated	1 or ▼	2 or ▼▼	3 or ▼▼▼	4 or ▼▼▼▼
Number or Mark after tightening	4 or ▼▼▼▼	1 or ▼	2 or ▼▼	3 or ▼▼▼

71051_HODY_G0687

Fig. 92 Table for proper tightening of the oil filter

a. Tighten: ¾ Turn Clockwise

b. Tightening Torque: 9 ft. lbs. (12 Nm)

6. If four numbers or marks (1 to 4 or ▼ to ▼▼▼▼) are printed around the outside of the filter, you can use the following procedure to tighten the filter:

a. Spin the filter on until its seal lightly seats against the oil filter base, and note which number or mark is at the bottom.

b. Tighten the filter by turning it clockwise three numbers or marks from the one you noted. For example, if mark ▼ is at the bottom when the seal is lightly seated, tighten the filter until the mark ▼▼▼▼ comes around to the bottom.

7. After installation, fill the engine with engine oil to the specified level, run the engine for more than 3 minutes, then check for oil leakage.

EXHAUST MANIFOLD

REMOVAL & INSTALLATION

The exhaust manifold is manufactured as an integral part of the cylinder head.

INTAKE MANIFOLD

REMOVAL & INSTALLATION

See Figures 93 through 97.

1. Remove the engine cover.

2. Disconnect the breather pipe, then remove the intake air duct.

3. Disconnect the throttle actuator connector, the EVAP canister purge valve connector, the water bypass hoses, the MAP sensor connector, and the EVAP canister purge hose, then plug the water bypass hoses.

4. Remove the harness clamp.

5. Disconnect the PCV hose, the brake booster vacuum hose, and the IMT actuator connector.

6. Remove the upper cover mounting bolts and nuts sequentially in three steps, then remove the upper cover.

7. Remove the intake manifold mounting bolts and nuts sequentially in three steps.

8. Remove the intake manifold.

To install:

9. Install the intake manifold. Tighten the bolts and nuts sequentially in three steps. Always use a new intake manifold gasket. Specified Torque: 8 x 1.25 mm: 16 ft. lbs. (22 Nm)

10. Install the upper cover (A). Tighten the bolts and nuts sequentially in three

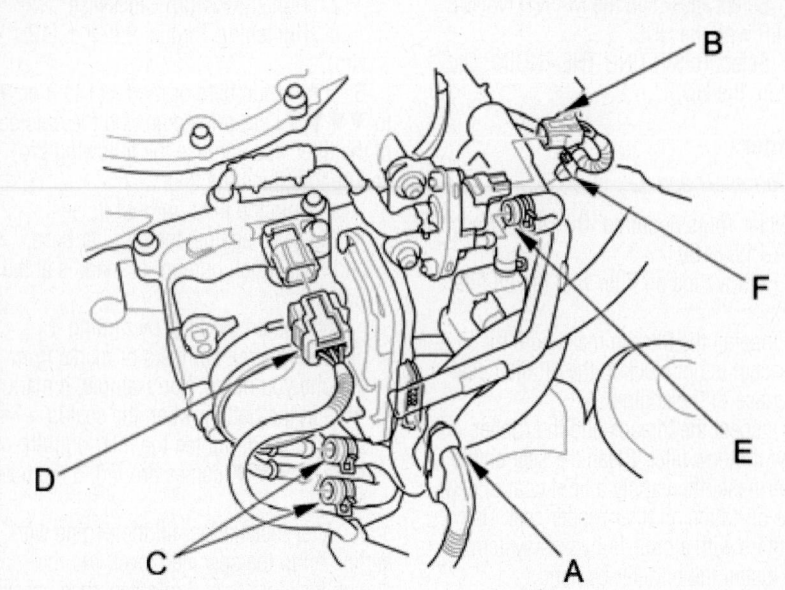

A. Throttle actuator connector
B. EVAP canister purge valve connector
C. Water bypass hoses
D. MAP sensor connector
E. EVAP canister purge hose
F. Harness clamp

71051_HODY_G0762

Fig. 93 Disconnect the throttle actuator connector, the EVAP canister purge valve connector, the water bypass hoses, the MAP sensor connector, and the EVAP canister purge hose, then plug the water bypass hoses

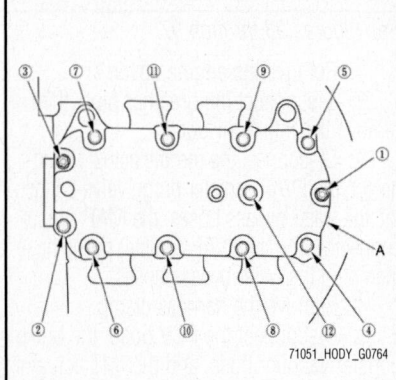

71051_HODY_G0764

Fig. 94 Remove the upper cover mounting bolts and nuts sequentially in three steps, then remove the upper cover (A)

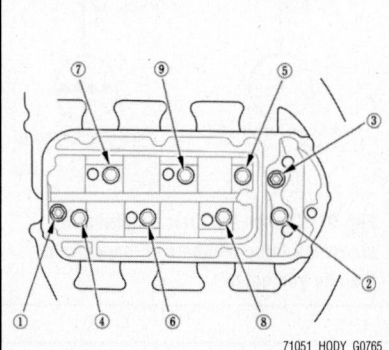

71051_HODY_G0765

Fig. 95 Remove the intake manifold mounting bolts and nuts sequentially in three steps

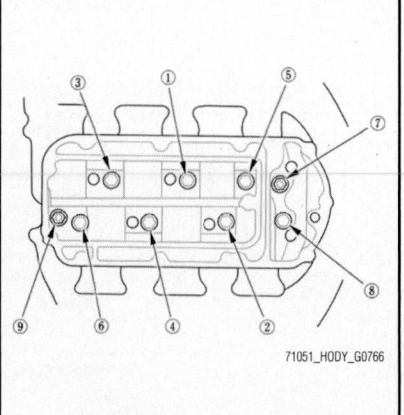

71051_HODY_G0766

Fig. 96 Tighten the bolts and nuts sequentially in three steps

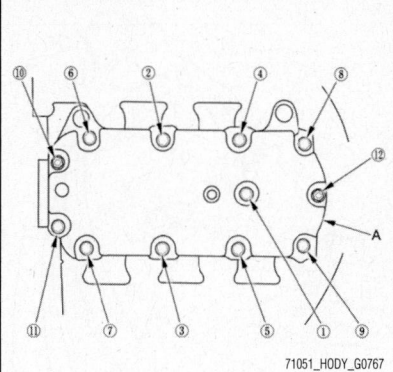

71051_HODY_G0767

Fig. 97 Install the upper cover (A), and tighten the bolts and nuts sequentially in three steps

steps. Always use a new gasket. Specified Torque: 6 x 1.0 mm: 9 ft. lbs. (12 Nm)

11. Connect the IMT actuator connector, the brake booster vacuum hose, and the PCV hose.

12. Connect the throttle actuator connector, the EVAP canister purge valve connector, the water bypass hoses, the MAP sensor connector, and the EVAP canister purge hose.

13. Install the harness clamp.

14. Install the intake air duct, then connect the breather pipe.

→When you tighten the band, refer to the throttle body removal/installation procedure.

15. Clean up any spilled engine coolant.

16. After installation, check that all tubes, hoses, and connectors are installed correctly.

17. Install the engine cover.

18. Refill the radiator with engine coolant, and bleed the air from the cooling system.

OIL PAN

REMOVAL & INSTALLATION

See Figures 98 through 101.

1. If the engine is already out of the vehicle, go to step 6.

2. Raise the vehicle on the lift.

3. Drain the engine oil.

4. Remove the inner fender.

5. Remove exhaust pipe A.

6. Remove the rear warm up TWC bracket.

7. Remove the CKP sensor cover and the bolt, then disconnect the CKP sensor connector.

8. Remove the torque converter case cover and the four bolts securing the transmission.

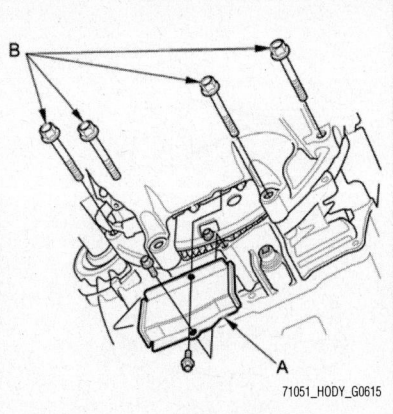

Fig. 98 Remove the torque converter case cover (A) and the four bolts (B) securing the transmission

9. Remove the bolts securing the oil pan.

10. Using a flat blade screwdriver, separate the oil pan from the engine block in the places shown.

11. Remove the oil pan.

To install:

12. Remove all of the old liquid gasket from the oil pan mating surfaces, the bolts, and the bolt holes.

13. Clean and dry the oil pan mating surfaces.

14. Apply liquid gasket (P/N 08717-0004, 08718-0003, 08718-0004, or 08718-0009) to the oil pan mating surface of the engine block and to the inside edge of the threaded bolt holes. Install the component within 5 minutes of applying the liquid gasket.

- Apply a bead of liquid gasket along the broken line.
- If you apply liquid gasket P/N 08718-0012, the component must be installed within 4 minutes.

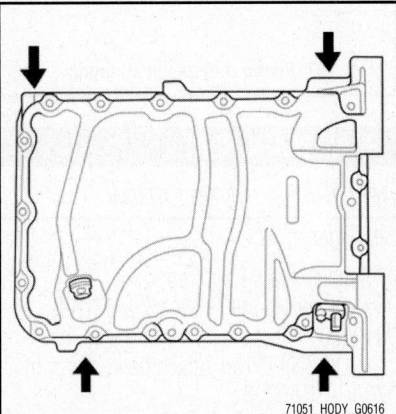

Fig. 99 Using a flat blade screwdriver, separate the oil pan from the engine block in the places shown

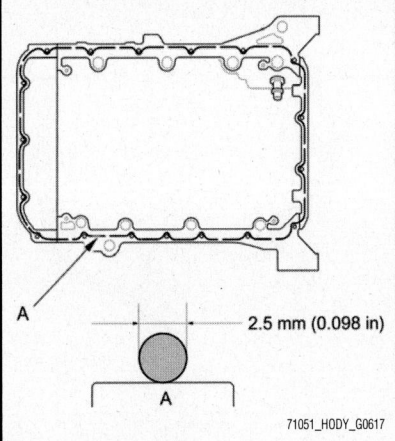

Fig. 100 Apply a bead of liquid gasket along the broken line (A)

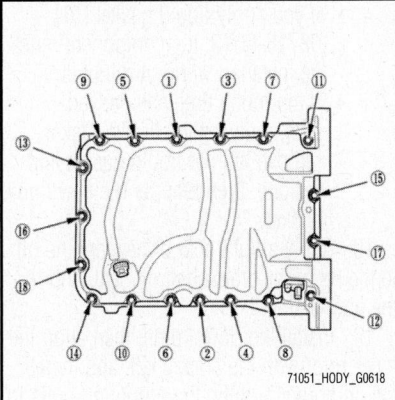

Fig. 101 Tighten the bolts in three steps in sequence

- If too much time has passed after applying the liquid gasket, remove the old liquid gasket and residue, then reapply the new liquid gasket.

15. Install the oil pan on the engine block.

16. Tighten the bolts in three steps. In the final step, torque the bolts, in sequence, to 9 ft. lbs. (12 Nm).

- Wait at least 30 minutes before filling the engine with oil.
- Do not run the engine for at least 3 hours after installing the oil pan.

17. Tighten the four bolts securing the transmission to 55 ft. lbs. (75 Nm), then install the torque converter case cover.

18. Connect the CKP sensor connector, then install the CKP sensor cover and the bolt.

19. Install the rear warm up TWC bracket.

20. If the engine is still in the vehicle, do the following steps.

21. Install exhaust pipe A using new gaskets and new self-locking nuts.

22. Install the splash shield.

23. Refill the engine with engine oil.

OIL PUMP

REMOVAL & INSTALLATION

See Figures 102 through 105.

Special Tools Required*:
- Oil Seal Driver, 64 mm 07OAD-RCAA100
- Engine Support Hanger, A and Reds AAR-T1256
- *: Available through the Honda Tool and Equipment Program 888-424-6857.

1. Open the hood, and secure it with the hood support rod in the wide-open position.

2. Drain the engine oil.

3. Remove the bulkhead cover.

4. Remove the cowl cover.

5. Remove the hood opener cable clip. Install the engine support hanger (AAR-T1256) onto the vehicle as shown, and attach the hook to the engine hook. Tighten the wing nut by hand, and lift and support the engine/transmission.

- Be careful not to damage the hood opener cable when installing the engine support hanger at the front bulkhead.
- AAR-T1256 two sets required for stacking additional cross section bar.

6. Remove the timing belt.

7. Remove the timing belt drive pulley from the crankshaft.

8. Remove the oil filter base/oil filter assembly .

9. Remove the oil pan.

10. Remove the oil strainer.

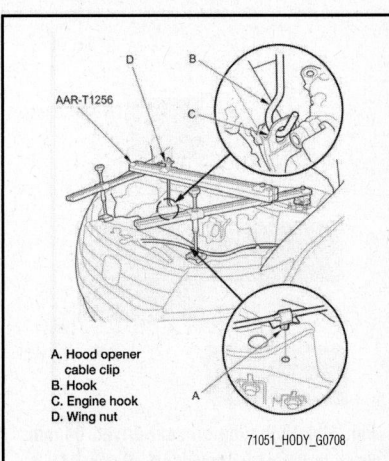

Fig. 102 Install the engine support hanger onto the vehicle as shown

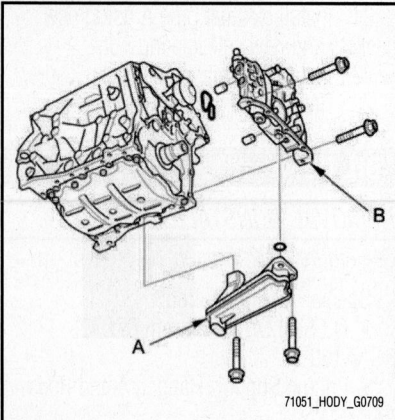

Fig. 103 Remove the oil strainer (A), remove the mounting bolts, then remove the oil pump assembly (B)

11. Remove the mounting bolts, then remove the oil pump assembly.

To install:

12. Remove the old oil seal from the oil pump.

13. Clean and dry the crankshaft oil seal housing.

14. Using the oil seal driver, 64 mm, drive in the new crankshaft oil seal until the oil seal driver bottoms on the pump.

15. Remove all of the old liquid gasket from the oil pump mating surfaces, the bolts, and the bolt holes.

16. Clean and dry the oil pump mating surfaces.

17. Apply liquid gasket (P/N 08717-0004, 08718-0003, 08718-0004, or 08718-0009) to the engine block mating surface of the oil pump and to the inside edge of the threaded bolt holes. Install the component within 5 minutes of applying the liquid gasket.

- Apply a bead of liquid gasket along the broken line.

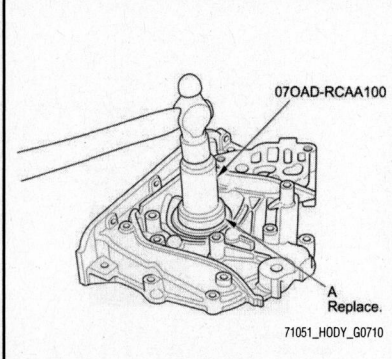

Fig. 104 Using the oil seal driver, 64 mm, drive in the new crankshaft oil seal (A) until the oil seal driver bottoms on the pump

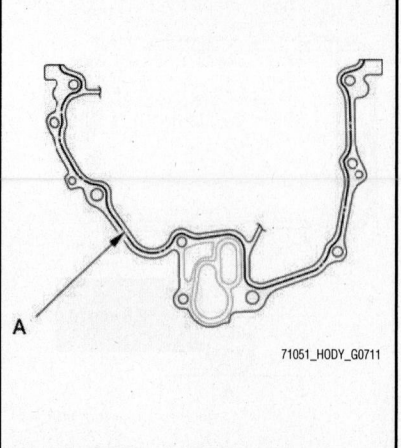

Fig. 105 Apply a bead of liquid gasket along the broken line (A)

- If you apply liquid gasket P/N 08718-0012, the component must be installed within 4 minutes.
- If too much time has passed after applying the liquid gasket, remove the old liquid gasket and residue, then reapply the new liquid gasket.

18. Apply a light coat of new engine oil to the lip of the crankshaft oil seal, and to the new O-ring.

19. Install the dowel pins, then align the inner rotor with the crankshaft, and install the oil pump. Tighten the mounting bolts to 9 ft. lbs. (12 Nm).

- Wait at least 30 minutes before filling the engine with oil.
- Do not run the engine for at least 3 hours after installing the oil pump.

20. Clean the excess oil off the crankshaft, and check that the oil seal lip is not distorted.

21. Install the oil strainer with a new O-ring.

22. Install the oil pan.

23. Install the oil filter base/oil filter assembly, with a new O-ring.

24. Install the timing belt drive pulley to the crankshaft.

25. Support the engine with a jack and a wood block under the oil pan.

26. Remove the engine support hanger.

27. Install the timing belt.

28. Refill the engine with engine oil.

29. Install the cowl cover.

30. Install the bulkhead cover.

PISTONS & RINGS

POSITIONING

See Figures 106 and 107.

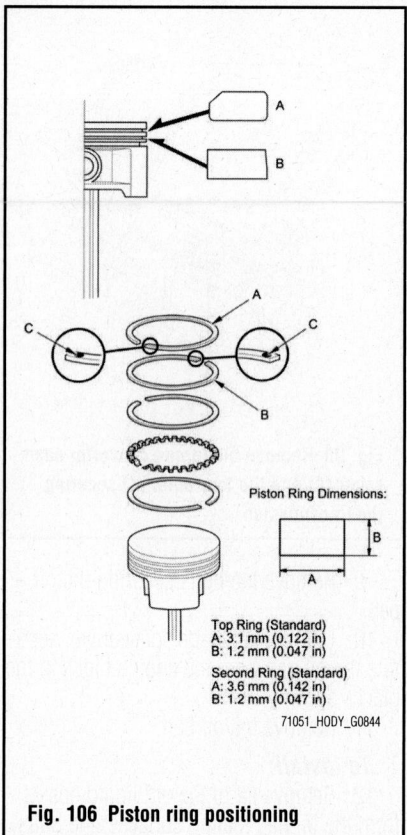

Fig. 106 Piston ring positioning

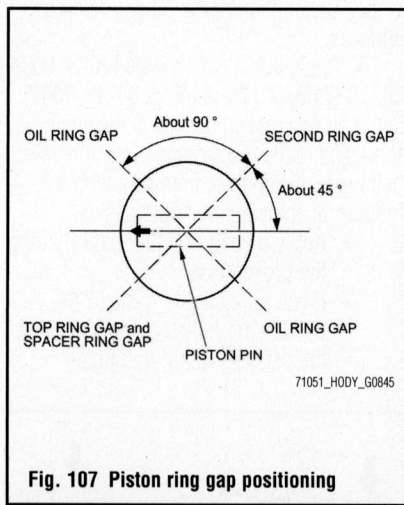

Fig. 107 Piston ring gap positioning

REAR MAIN SEAL

REMOVAL & INSTALLATION

See Figure 108.

Special Tools Required:
- Driver Handle, 15 x 135L 07749-0010000
- Oil Seal Driver Attachment, 106 mm 070AD-RCA0200

1. Remove the transmission.

2. Remove the drive plate.

3. Remove the transmission end crankshaft oil seal.

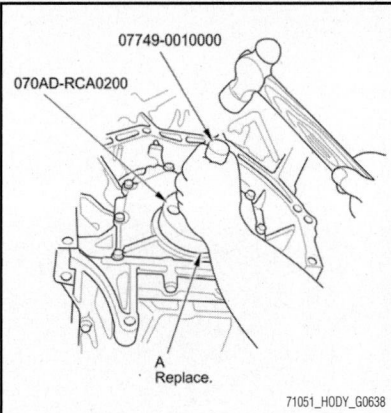

Fig. 108 Using the driver handle and the oil seal driver attachment, drive in the new crankshaft oil seal (A) until the oil seal driver attachment bottoms on the engine block end cover

To install:

4. Clean and dry the crankshaft oil seal housing.

5. Apply a light coat of new engine oil to the lip of the crankshaft oil seal.

6. Using the driver handle, 15 x 135L and the oil seal driver attachment, 106 mm, drive in the new crankshaft oil seal until the oil seal driver attachment bottoms on the engine block end cover. Align the hole in the oil seal driver attachment with the pin on the crankshaft.

7. Clean the excess oil off the crankshaft, and check that the oil seal lip is not distorted.

8. Install the drive plate.

9. Install the transmission.

ROCKER ARMS

REMOVAL & INSTALLATION

Front

See Figures 109 and 110.

1. Remove the cylinder head cover.
2. Loosen the locknuts and the adjusting screws.
3. Remove the rocker shaft bridge mounting bolts, the front rocker arm oil control valve mounting bolts, and the rocker arm assembly.

  a. Loosen the rocker shaft bridge mounting bolts and the front rocker arm oil control valve mounting bolts in sequence two turns at a time, to prevent damaging the valves or the rocker arm assembly.

  b. When removing the rocker arm assembly, do not remove the rocker shaft bridge mounting bolts and the front rocker arm oil control valve mounting

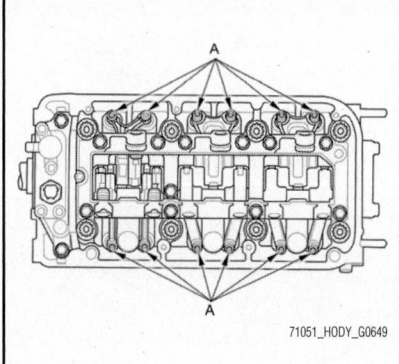

Fig. 109 Loosen the locknuts and the adjusting screws (A)

bolts. The bolts will keep the rocker arms on the shafts.

To install:

4. Remove all of the old liquid gasket from the front rocker arm oil control valve and the cylinder head.

5. Apply liquid gasket (P/N 08717-0004, 08718-0003, 08718-0004, or 08718-0009) to the rocker arm oil control valve mating surface of the cylinder head. Install the component within 5 minutes of applying the liquid gasket.

- Apply a bead of liquid gasket along the broken line (A).
- If you apply liquid gasket P/N 08718-0012, the component must be installed within 4 minutes.
- If too much time has passed after applying the liquid gasket, remove the old liquid gasket and residue, then reapply the new liquid gasket.

6. Set the rocker arm assembly in place, and loosely install the bolts. Make sure that

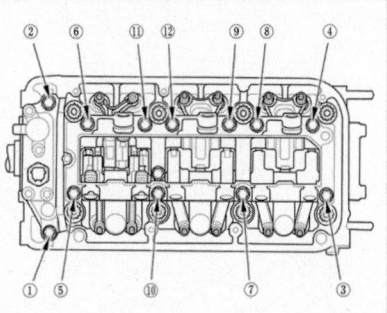

Fig. 110 Loosen the rocker shaft bridge mounting bolts and the front rocker arm oil control valve mounting bolts in sequence two turns at a time, to prevent damaging the valves or the rocker arm assembly

the rocker arms are properly positioned on the valve stems.

- Wait at least 30 minutes before filling the engine with oil.
- Do not run the engine for at least 3 hours after installing the rocker arm assembly.

7. Tighten each bolt two turns at a time in the sequence shown to ensure that the rockers do not bind on the valves.

  a. When the rocker shaft bridge is reused: Specified Torque: 18 ft. lbs. (24.5 Nm)

  b. When a new rocker shaft bridge is installed, torque as follows:

- 1st Step: 21 ft. lbs. (28 Nm), then loosen the bolts.
- 2nd Step: 18 ft. lbs. (24.5 Nm)

Rear

See Figures 111 and 112.

1. Remove the cylinder head cover.

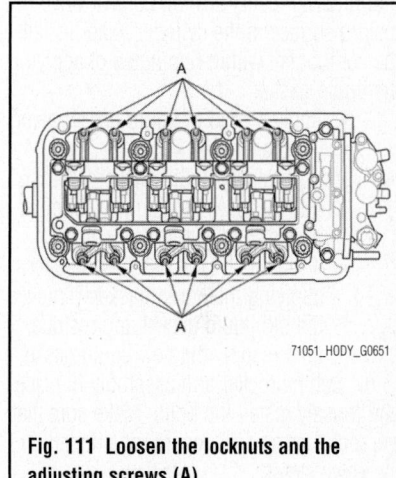

Fig. 111 Loosen the locknuts and the adjusting screws (A)

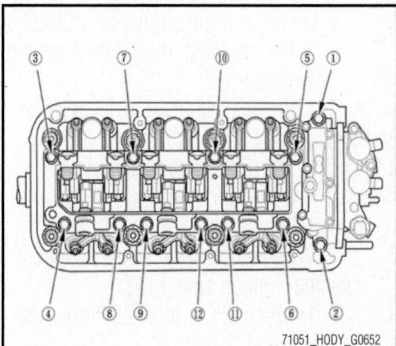

Fig. 112 Loosen the rocker shaft bridge mounting bolts and the rear rocker arm oil control valve mounting bolts in sequence two turns at a time, to prevent damaging the valves or the rocker arm assembly

2. Loosen the locknuts and the adjusting screws.

3. Remove the rocker shaft bridge mounting bolts, the rear rocker arm oil control valve mounting bolts, and the rocker arm assembly.

a. Loosen the rocker shaft bridge mounting bolts and the rear rocker arm oil control valve mounting bolts in sequence two turns at a time, to prevent damaging the valves or the rocker arm assembly.

b. When removing the rocker arm assembly, do not remove the rocker shaft bridge mounting bolts and the rear rocker arm oil control valve mounting bolts. The bolts will keep the rocker arms on the shafts.

To install:

4. Remove all of the old liquid gasket from the front rocker arm oil control valve and the cylinder head.

5. Apply liquid gasket (P/N 08717-0004, 08718-0003, 08718-0004, or 08718-0009) to the rocker arm oil control valve mating surface of the cylinder head. Install the component within 5 minutes of applying the liquid gasket.

- Apply a bead of liquid gasket along the broken line (A).
- If you apply liquid gasket P/N 08718-0012, the component must be installed within 4 minutes.
- If too much time has passed after applying the liquid gasket, remove the old liquid gasket and residue, then reapply the new liquid gasket.

6. Set the rocker arm assembly in place, and loosely install the bolts. Make sure that the rocker arms are properly positioned on the valve stems.

- Wait at least 30 minutes before filling the engine with oil.
- Do not run the engine for at least 3 hours after installing the rocker arm assembly.

7. Tighten each bolt two turns at a time in the sequence shown to ensure that the rockers do not bind on the valves.

a. When the rocker shaft bridge is reused, tighten to: 18 ft. lbs. (24.5 Nm)

b. When a new rocker shaft bridge is installed, tighten as follows:

- 1st Step: 21 ft. lbs. (28 Nm), then loosen the bolts.
- 2nd Step: 18 ft. lbs. (24.5 Nm)

TIMING BELT FRONT COVER

REMOVAL & INSTALLATION

See Figures 113 through 117.

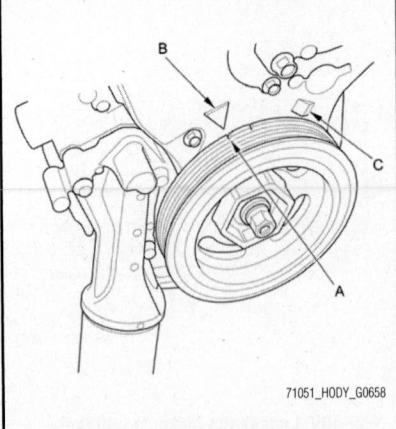

Fig. 113 Turn the crankshaft pulley so its white mark (A) on the crankshaft pulley lines up with the pointer (B), not other pointer (C)

1. Turn the crankshaft pulley so its white mark on the crankshaft pulley lines up with the pointer.

➡**The other pointer is not used.**

2. Check that the No. 1 piston top dead center (TDC) mark on the front camshaft pulley and the pointer on the front upper cover are aligned.

➡**If the marks are not aligned, rotate the crankshaft 360 degrees, and recheck the camshaft pulley mark.**

3. Raise the vehicle on the lift, then remove the right front wheel.

4. Remove the splash shield.

5. Remove the drive belt auto-tensioner.

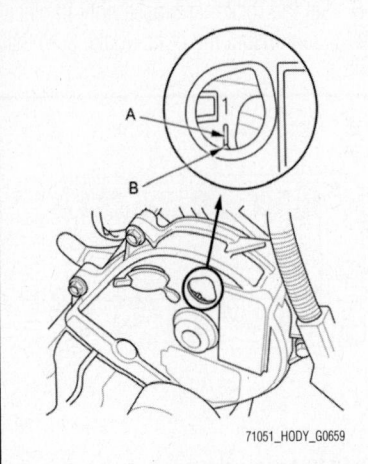

Fig. 114 Check that the No. 1 piston top dead center (TDC) mark (A) on the front camshaft pulley and the pointer (B) on the front upper cover are aligned

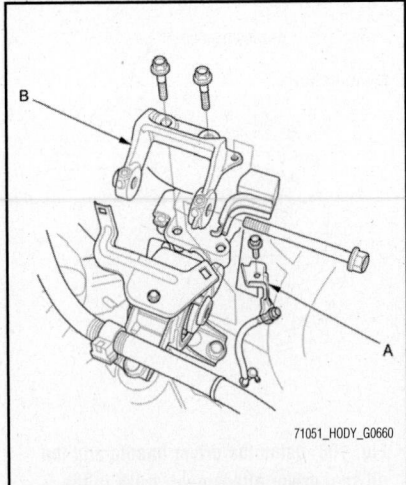

Fig. 115 Remove the ground cable bracket (A), then remove the upper half of the side engine mount bracket (B)

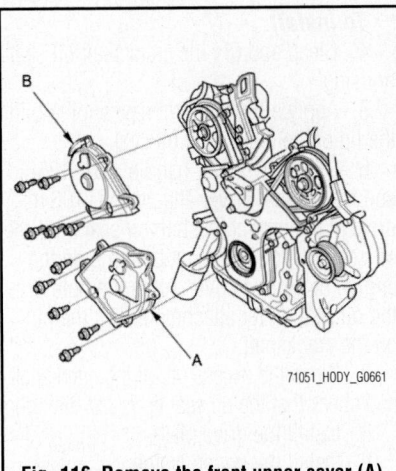

Fig. 116 Remove the front upper cover (A) and the rear upper cover (B)

6. Lift and support the engine with a jack and a wood block under the oil pan.

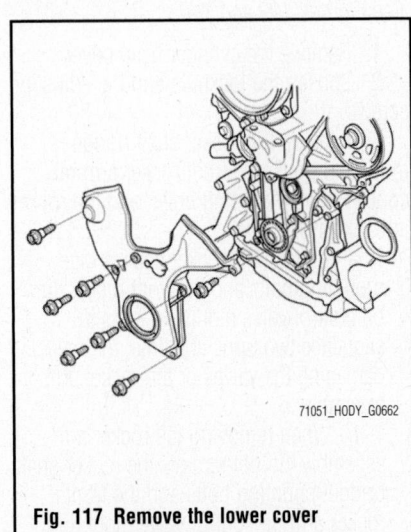

Fig. 117 Remove the lower cover

7. Remove the ground cable bracket, then remove the upper half of the side engine mount bracket.

8. Remove the crankshaft pulley.

9. Remove the front upper cover and the rear upper cover.

10. Remove the lower cover.

11. Installation is the reverse order of removal.

TIMING BELT & SPROCKETS

REMOVAL & INSTALLATION

See Figures 113 through 122.

1. Turn the crankshaft pulley so its white mark on the crankshaft pulley lines up with the pointer.

➡**The other pointer is not used.**

2. Check that the No. 1 piston top dead center (TDC) mark on the front camshaft pulley and the pointer on the front upper cover are aligned.

➡**If the marks are not aligned, rotate the crankshaft 360 degrees, and recheck the camshaft pulley mark.**

3. Raise the vehicle on the lift, then remove the right front wheel.

4. Remove the splash shield.

5. Remove the drive belt auto-tensioner.

6. Lift and support the engine with a jack and a wood block under the oil pan.

7. Remove the ground cable bracket, then remove the upper half of the side engine mount bracket.

8. Remove the crankshaft pulley.

9. Remove the front upper cover and the rear upper cover.

10. Remove the lower cover.

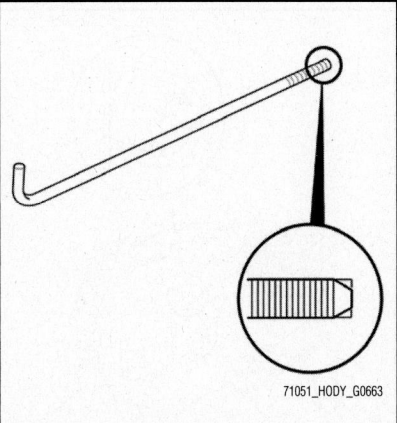

Fig. 118 Remove one of the battery clamp bolts from the battery tray, and grind the end of it as shown

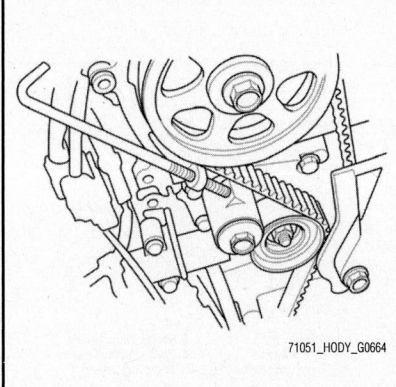

Fig. 119 Thread the battery clamp bolt in as shown to hold the timing belt adjuster in its current position

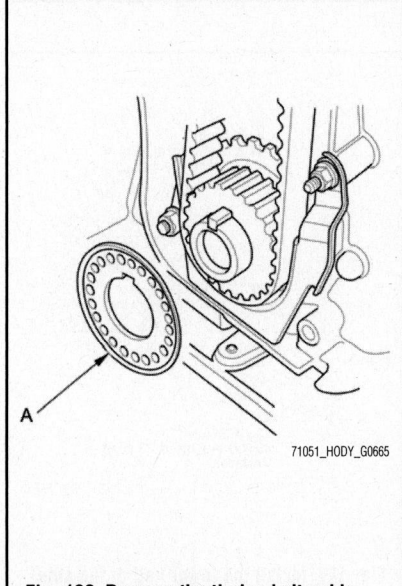

Fig. 120 Remove the timing belt guide plate (A)

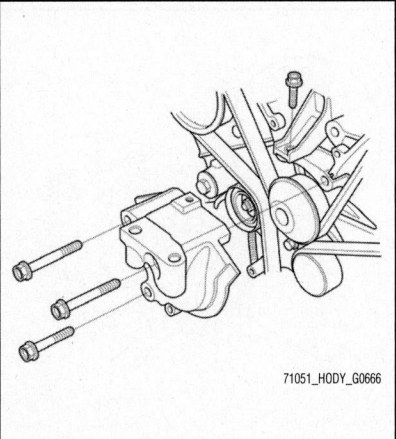

Fig. 121 Remove the lower half of the side engine mount bracket

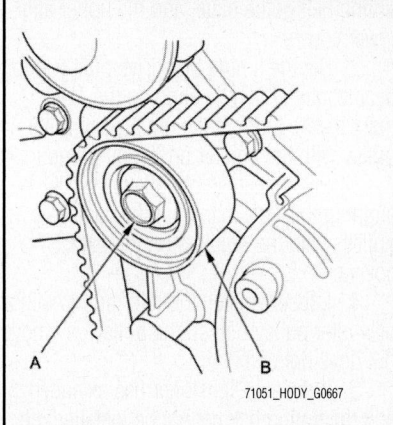

Fig. 122 Remove the idler pulley bolt (A) and the idler pulley (B), then remove the timing belt

11. Remove one of the battery clamp bolts from the battery tray, and grind the end of it as shown.

12. Thread the battery clamp bolt in as shown to hold the timing belt adjuster in its current position. Tighten it by hand, do not use a wrench.

13. Remove the timing belt guide plate.

14. Remove the lower half of the side engine mount bracket.

15. Remove the idler pulley bolt and the idler pulley, then remove the timing belt. Discard the idler pulley bolt.

To install:

Installing A Used Belt

See Figures 123 through 132.

➡**The following procedure is for installation of a used timing belt. If you are installing a new belt, refer to the timing belt replacement procedure.**

1. Clean the timing belt pulleys, the

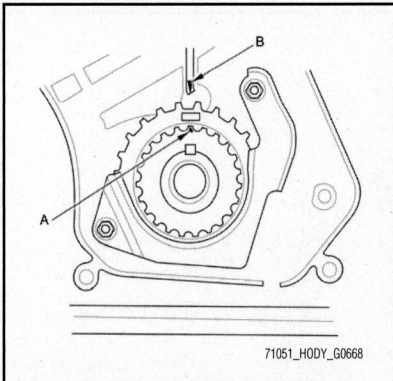

Fig. 123 Set the timing belt drive pulley to top dead center (TDC) by aligning the TDC mark (A) on the tooth of the timing belt drive pulley with the pointer (B) on the oil pump

timing belt guide plate, and the upper and lower covers.

2. Set the timing belt drive pulley to top dead center (TDC) by aligning the TDC mark on the tooth of the timing belt drive pulley with the pointer on the oil pump.

3. Set the camshaft pulleys to TDC by aligning the TDC marks on the camshaft pulleys with the pointers on the back covers.

4. Loosely install the idler pulley with a new idler pulley bolt so the pulley can move but does not come off.

5. If the auto-tensioner has extended and the timing belt cannot be installed, do the timing belt replacement procedure.

6. Install the timing belt in a counterclockwise sequence starting with the drive pulley. Take care not to damage the timing belt during installation:

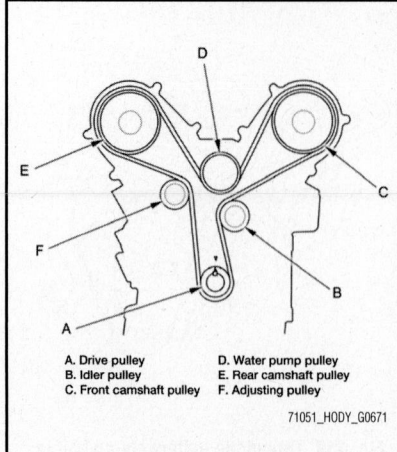

A. Drive pulley
B. Idler pulley
C. Front camshaft pulley
D. Water pump pulley
E. Rear camshaft pulley
F. Adjusting pulley

71051_HODY_G0671

Fig. 126 Install the timing belt in a counterclockwise sequence starting with the drive pulley

a. Drive pulley
b. Idler pulley
c. Front camshaft pulley
d. Water pump pulley
e. Rear camshaft pulley
f. Adjusting pulley

7. Tighten the idler pulley bolt to 33 ft. lbs. (45 Nm).

8. Remove the battery clamp bolt from the back cover.

9. Install the lower half of the side engine mount bracket using new bolts.

10. Install the timing belt guide plate as shown.

11. Install the lower cover. Tighten the bolts to 9 ft. lbs. (12 Nm).

12. Install the front upper cover and the rear upper cover. Tighten the bolts to 9 ft. lbs. (12 Nm).

13. Install the crankshaft pulley.

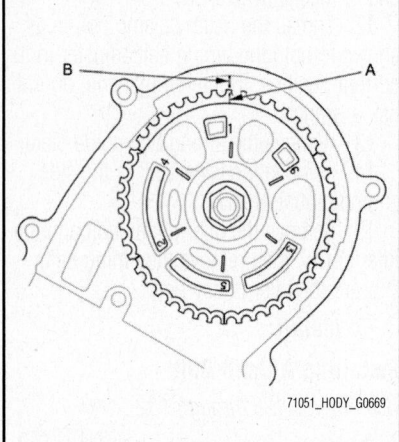

71051_HODY_G0669

Fig. 124 Set the camshaft pulleys to TDC by aligning the TDC marks (A) on the camshaft pulleys with the pointers (B) on the back covers—Front

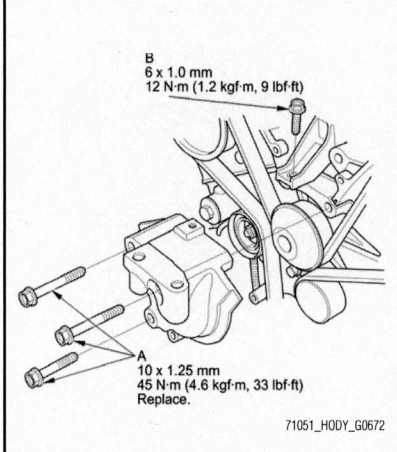

B
6 x 1.0 mm
12 N·m (1.2 kgf·m, 9 lbf·ft)

A
10 x 1.25 mm
45 N·m (4.6 kgf·m, 33 lbf·ft)
Replace.

71051_HODY_G0672

Fig. 127 Install the lower half of the side engine mount bracket using new bolts (A) and bolt (B)

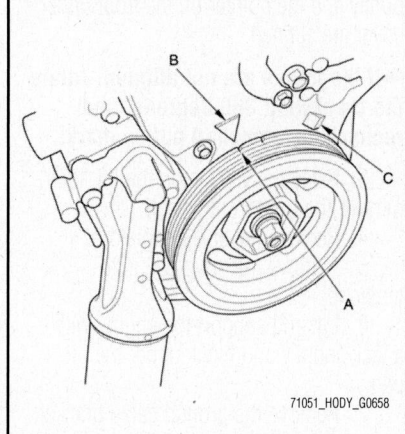

71051_HODY_G0658

Fig. 129 Turn the crankshaft pulley so its white mark (A) on the crankshaft pulley lines up with the pointer (B), not other pointer (C)

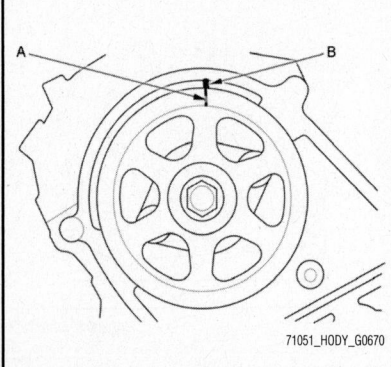

71051_HODY_G0670

Fig. 125 Set the camshaft pulleys to TDC by aligning the TDC marks (A) on the camshaft pulleys with the pointers (B) on the back covers—Rear

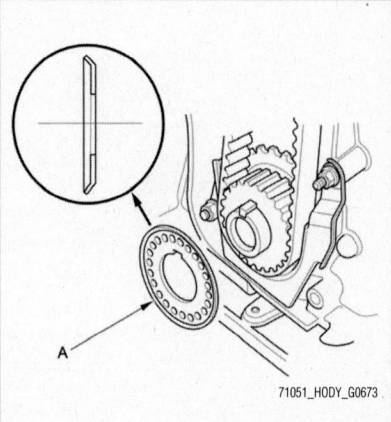

71051_HODY_G0673

Fig. 128 Install the timing belt guide plate (A) as shown

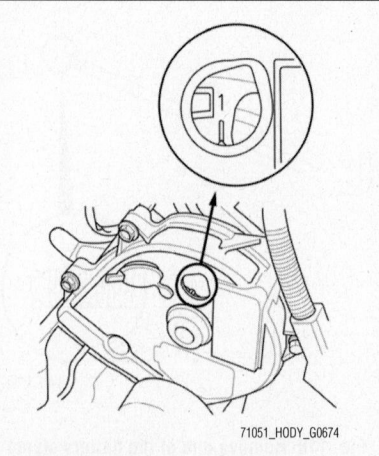

71051_HODY_G0674

Fig. 130 Check the camshaft pulley marks—Front

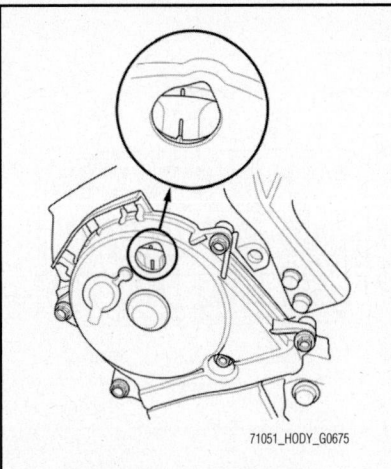

Fig. 131 Check the camshaft pulley marks—Rear

14. Rotate the crankshaft pulley about six turns clockwise so the timing belt positions itself on the pulleys.

15. Turn the crankshaft pulley so its white mark lines up with the pointer.

➡**The other pointer is not used.**

16. Check the camshaft pulley marks.

➡**If the marks are not aligned, rotate the crankshaft 360 degrees, and recheck the camshaft pulley mark:**

- If the camshaft pulley marks are at TDC, go to step 17.
- If the camshaft pulley marks are not at TDC, remove the timing belt and repeat steps 2 through 16.

17. Loosely install the upper half of the side engine mount bracket using new bolts, then install the ground cable bracket.

18. Remove the jack and the wood block.

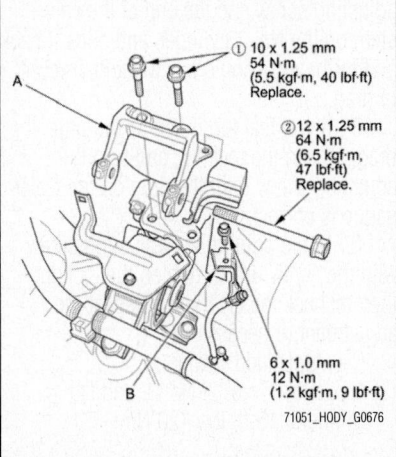

Fig. 132 Loosely install the upper half of the side engine mount bracket (A) using new bolts, then install the ground cable bracket (B)

19. Tighten the mounting bolts in the numbered sequence shown.

20. Install the drive belt auto-tensioner.

21. Install the splash shield.

22. Install the right front wheel.

23. Do the CKP pattern clear/CKP pattern learn procedure.

Installing A New Belt

See Figures 113, 123 through 128, 130 through 132, 133 through 137.

➡**The following procedure is for installation of a new timing belt. If you are installing a used belt, refer to the timing belt installation procedure.**

1. Clean the timing belt pulleys, the timing belt guide plate, and the upper and lower covers.

2. Set the timing belt drive pulley to top

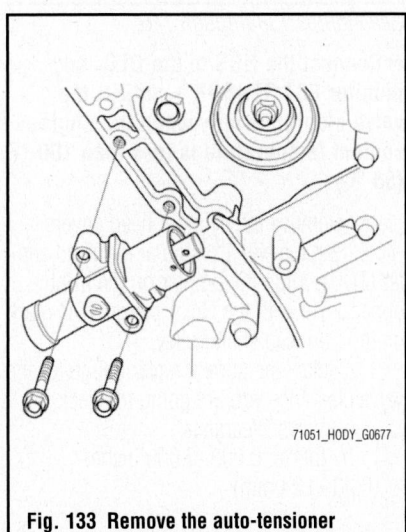

Fig. 133 Remove the auto-tensioner

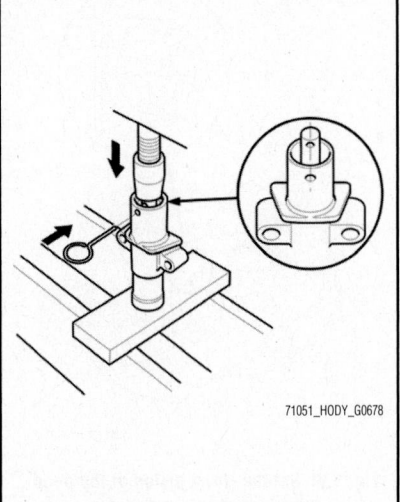

Fig. 134 Align the holes on the rod and the housing of the auto-tensioner

dead center (TDC) by aligning the TDC mark on the tooth of the timing belt drive pulley with the pointer on the oil pump.

3. Set the camshaft pulleys to TDC by aligning the TDC marks on the camshaft pulleys with the pointers on the back covers.

4. Remove the battery clamp bolt from the back cover.

5. Remove the auto-tensioner.

6. Align the holes on the rod and the housing of the auto-tensioner.

7. Use a hydraulic press to slowly compress the auto-tensioner. Insert a pin through the housing and the rod.

➡**The compression pressure should not exceed 2,203.1 lbs. (9,800 N).**

8. Install the auto-tensioner.

➡**Make sure the pin stays in place.**

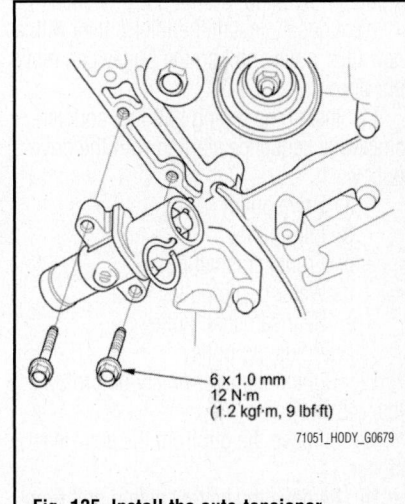

6 x 1.0 mm
12 N·m
(1.2 kgf·m, 9 lbf·ft)

Fig. 135 Install the auto-tensioner

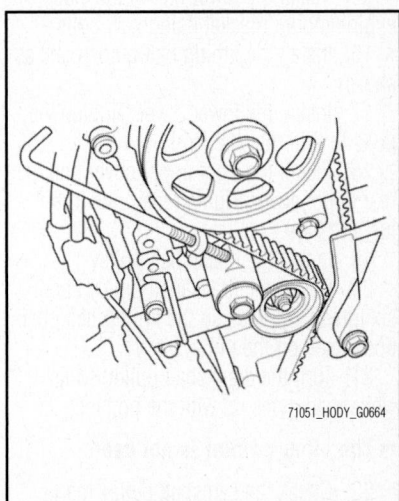

Fig. 136 Thread the battery clamp bolt in as shown to hold the timing belt adjuster in its current position

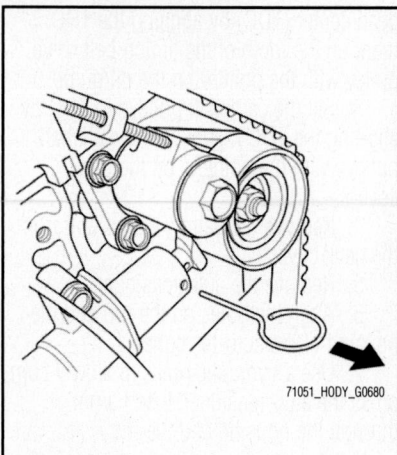

Fig. 137 Remove the pin from the auto-tensioner

9. Thread the battery clamp bolt in as shown to hold the timing belt adjuster. Tighten it by hand, do not use a wrench.

10. Loosely install the idler pulley with a new idler pulley bolt so the pulley can move but does not come off.

11. Install the timing belt in a counter-clockwise sequence starting with the drive pulley:

    a. Drive pulley
    b. Idler pulley
    c. Front camshaft pulley
    d. Water pump pulley
    e. Rear camshaft pulley
    f. Adjusting pulley

12. Tighten the idler pulley bolt to 33 ft. lbs. (45 Nm).

13. Remove the pin from the auto-tensioner.

14. Remove the battery clamp bolt from the back cover.

15. Install the lower half of the side engine mount bracket using new bolts.

16. Install the timing belt guide plate as shown.

17. Install the lower cover. Tighten the bolts to 9 ft. lbs. (12 Nm).

18. Install the front upper cover and the rear upper cover. Tighten the bolts to 9 ft. lbs. (12 Nm).

19. Install the crankshaft pulley.

20. Rotate the crankshaft pulley about six turns clockwise so the timing belt positions itself on the pulleys.

21. Turn the crankshaft pulley so its white mark lines up with the pointer.

➡The other pointer is not used.

22. Check the camshaft pulley marks.

➡If the marks are not aligned, rotate the crankshaft 360 degrees, and recheck the camshaft pulley mark:

- If the camshaft pulley marks are at TDC, go to step 24.
- If the camshaft pulley marks are not at TDC, remove the timing belt and repeat steps 3 through 23.

23. Loosely install the upper half of the side engine mount bracket using new bolts, then install the ground cable bracket.

24. Remove the jack and the wood block.

25. Tighten the mounting bolts in the numbered sequence shown.

26. Install the drive belt auto-tensioner.

27. Install the splash shield.

28. Install the right front wheel.

29. Do the CKP pattern clear/CKP pattern learn procedure.

VALVE LASH (CLEARANCE) ADJUSTMENT

ADJUSTMENT

See Figures 138 through 146.

➡**Connect the HDS to the DLC, and monitor ECT SENSOR 1. Adjust the valve clearance only when the engine coolant temperature is less than 100 °F (38 °C).**

1. Remove the cylinder head covers.

2. Set the No. 1 piston at top dead center (TDC). Align the pointer on the front upper cover with the No. 1 piston TDC mark on the front camshaft pulley.

3. Select the correct feeler gauge for the valve clearance you are going to check.

    a. Valve Clearance:
    b. Intake: 0.008–0.009 inches (0.20–0.24 mm)

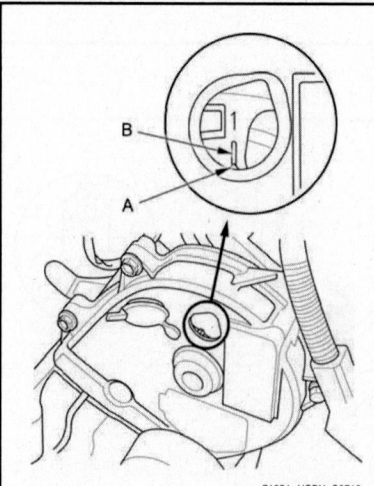

Fig. 138 Set the No. 1 piston at top dead center (TDC). Align the pointer (A) on the front upper cover with the No. 1 piston TDC mark (B) on the front camshaft pulley

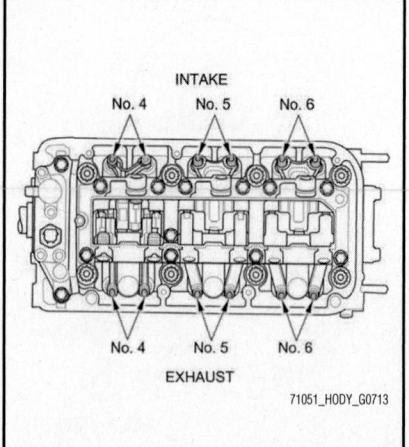

Fig. 139 Locations for checking valve clearance—Front

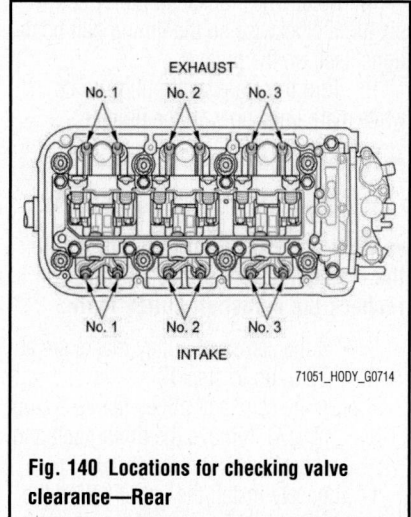

Fig. 140 Locations for checking valve clearance—Rear

    c. Exhaust: 0.011–0.012 inches (0.28–0.32 mm)

4. Insert the feeler gauge between the adjusting screw and the end of the valve stem on the No. 1 cylinder, and slide it back and forth; you should feel a slight amount of drag.

5. If you feel too much or too little drag, loosen the locknut, and turn the adjusting screw until the drag on the feeler gauge is correct.

6. While holding the adjusting screw with the screw driver, tighten the locknut, then recheck the clearance. Repeat the adjustment, if necessary.

    a. Specified Torque:
    b. No. 1, No. 2, No. 3, and No. 4 cylinders: 15 ft. lbs. (20 Nm)

➡**Apply new engine oil to the nut threads.**

    c. No. 5 and No. 6 cylinders: 10 ft. lbs. (14 Nm)

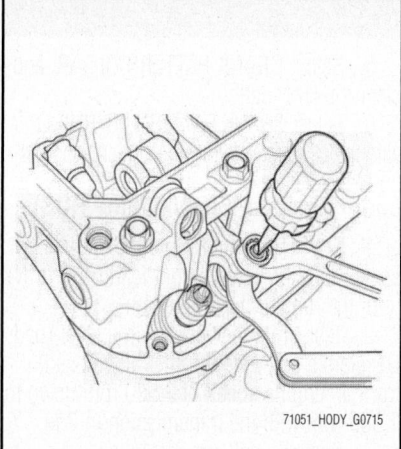

Fig. 141 If you feel too much or too little drag, loosen the locknut, and turn the adjusting screw until the drag on the feeler gauge is correct

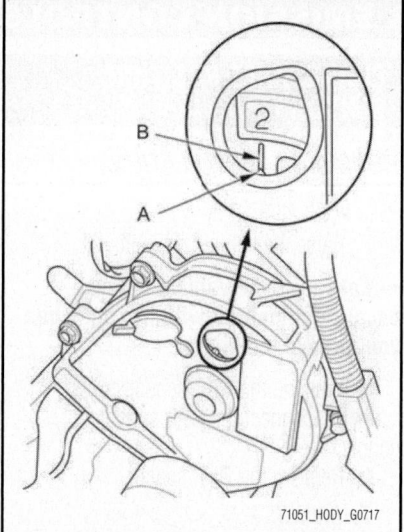

Fig. 143 Align the pointer (A) on the front upper cover with the No. 2 piston TDC mark (B) on the front camshaft pulley

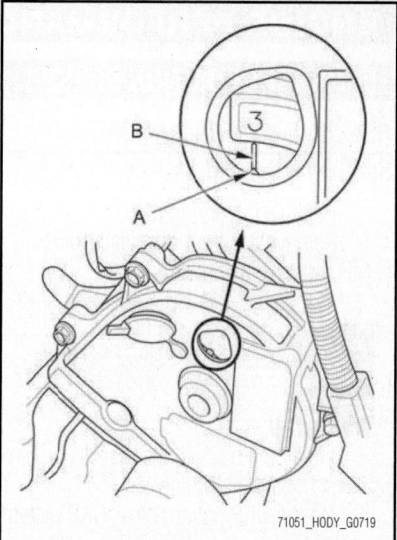

Fig. 145 Align the pointer (A) on the front upper cover with the No. 3 piston TDC mark (B) on the front camshaft pulley

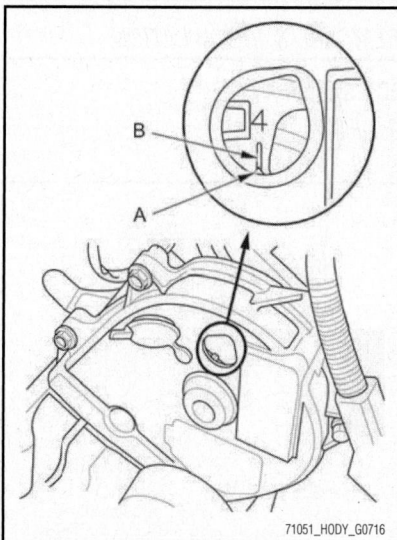

Fig. 142 Align the pointer (A) on the front upper cover with the No. 4 piston TDC mark (B) on the front camshaft pulley

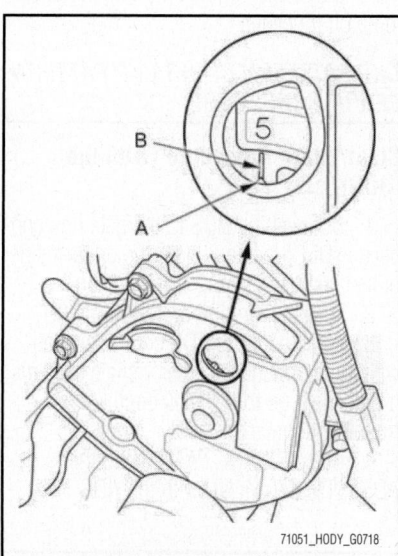

Fig. 144 Align the pointer (A) on the front upper cover with the No. 5 piston TDC mark (B) on the front camshaft pulley

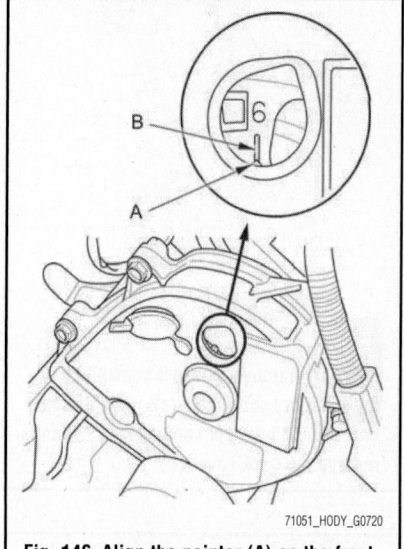

Fig. 146 Align the pointer (A) on the front upper cover with the No. 6 piston TDC mark (B) on the front camshaft pulley

➡**Apply new engine oil to the nut threads.**

7. Rotate the crankshaft clockwise. Align the pointer on the front upper cover with the No. 4 piston TDC mark on the front camshaft pulley.

8. Check and, if necessary, adjust the valve clearance on the No. 4 cylinder.

9. Rotate the crankshaft clockwise. Align the pointer on the front upper cover

with the No. 2 piston TDC mark on the front camshaft pulley.

10. Check and, if necessary, adjust the valve clearance on the No. 2 cylinder.

11. Rotate the crankshaft clockwise. Align the pointer on the front upper cover with the No. 5 piston TDC mark on the front camshaft pulley.

12. Check and, if necessary, adjust the valve clearance on the No. 5 cylinder.

13. Rotate the crankshaft clockwise. Align the pointer (A) on the front upper

cover with the No. 3 piston TDC mark (B) on the front camshaft pulley.

14. Check and, if necessary, adjust the valve clearance on the No. 3 cylinder.

15. Rotate the crankshaft clockwise. Align the pointer (A) on the front upper cover with the No. 6 piston TDC mark (B) on the front camshaft pulley.

16. Check and, if necessary, adjust the valve clearance on the No. 6 cylinder.

17. Install the cylinder head covers.

ENGINE PERFORMANCE & EMISSION CONTROLS

CAMSHAFT POSITION SENSOR

REMOVAL & INSTALLATION

See Figures 147 and 148.

1. Remove the timing belt.
2. Remove the front camshaft pulley (CMP sensor pulse plate).
3. Disconnect the CMP sensor connector, then remove the back cover.
4. Remove the CMP sensor from the back cover.

To install:

5. Install the parts in the reverse order of removal. Install the timing belt.
6. Do the CKP pattern clear/CKP pattern learn procedure.

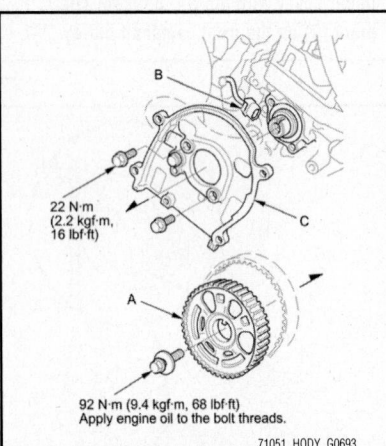

Fig. 147 Remove the front camshaft pulley (CMP sensor pulse plate) (A), disconnect the CMP sensor connector (B), then remove the back cover (C)

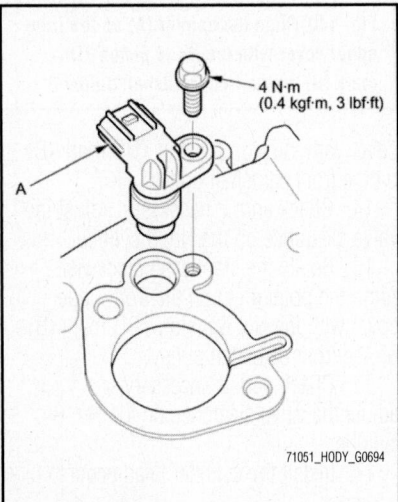

Fig. 148 Remove the CMP sensor (A) from the back cover

CRANKSHAFT POSITION SENSOR

REMOVAL & INSTALLATION

See Figure 149.

1. Raise the vehicle on a lift.

➡**Make sure the vehicle is level, because engine oil will drip out when you remove the sensor.**

2. Remove the CKP sensor cover.
3. Disconnect the CKP sensor connector.
4. Remove the CKP sensor.

To install:

5. Install the parts in the reverse order of removal with a new O-ring.
6. Do the CKP pattern clear/CKP pattern learn procedure.
7. Check the engine oil level, and add more oil if needed.

CKP PATTEERN CLEAR/CKP PATTERN LEARN PROCEDURE

Clear/Learn Procedure (with the HDS)

1. Connect the HDS to the data link connector (DLC) located under the driver's side of the dashboard.
2. Turn the ignition switch to ON (II).
3. Make sure the HDS communicates with the PCM and all other vehicle systems. If it doesn't, go to the DLC circuit troubleshooting.
4. Select CRANK PATTERN in the ADJUSTMENT MENU with the HDS.

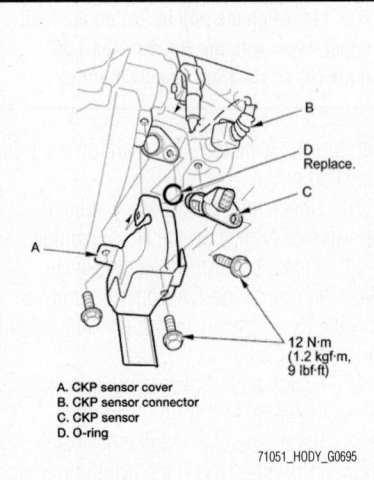

A. CKP sensor cover
B. CKP sensor connector
C. CKP sensor
D. O-ring

Fig. 149 Exploded view of the CKP sensor assembly

5. Select CRANK PATTERN CLEAR, and clear the CKP pattern.
6. Select CRANK PATTERN LEARNING with the HDS, and follow the screen prompts.

Learn Procedure (without the HDS)

1. Start the engine. Hold the engine speed at 3,000 rpm without load (in P or N) until the radiator fan comes on.
2. Test-drive the vehicle on a level road: Decelerate (with the throttle fully closed) from an engine speed of 2,500 rpm down to 1,000 rpm with the transmission in 2nd.
3. Repeat step 2 several times.
4. Turn the ignition switch to LOCK (0).
5. Turn the ignition switch to ON (II), and wait 30 seconds.

ENGINE COOLANT TEMPERATURE SENSOR

REMOVAL & INSTALLATION

ECT Sensor 1

See Figure 150.

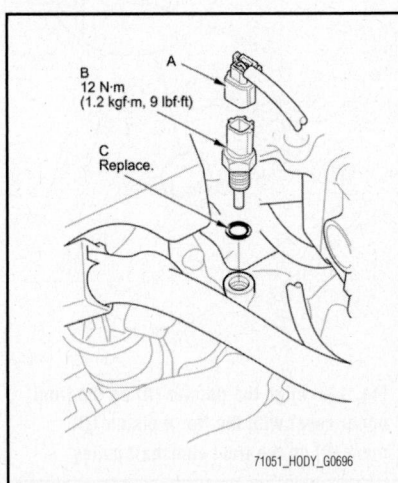

Fig. 150 Disconnect the ECT sensor 1 2P connector (A), then remove ECT sensor 1 (B) and O-ring (C)

1. Drain the engine coolant.
2. Remove the engine cover.
3. Disconnect the ECT sensor 1 2P connector.
4. Remove ECT sensor 1.
5. Install the parts in the reverse order of removal with a new O-ring, then refill the radiator with engine coolant.

ECT Sensor 2

See Figure 151.

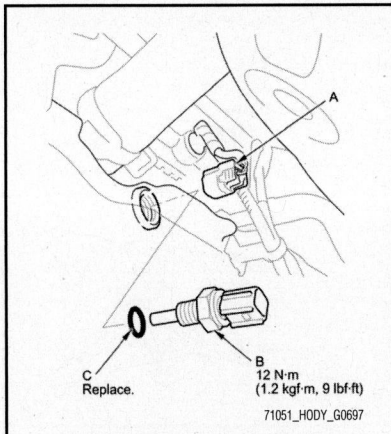

Fig. 151 Disconnect the ECT sensor 2 2P connector (A), then remove ECT sensor 1 (B) and O-ring (C)

1. Drain the engine coolant.
2. Disconnect the ECT sensor 2 2P connector.
3. Remove ECT sensor 2.
4. Install the parts in the reverse order of removal with a new O-ring, then refill the radiator with engine coolant.

EVAP CANISTER

REMOVAL & INSTALLATION

See Figures 152 through 155.

1. Raise the vehicle on a lift.
2. Remove the EVAP canister cover.
3. Disconnect the quick-connect fitting, the hoses, the EVAP canister vent shut valve

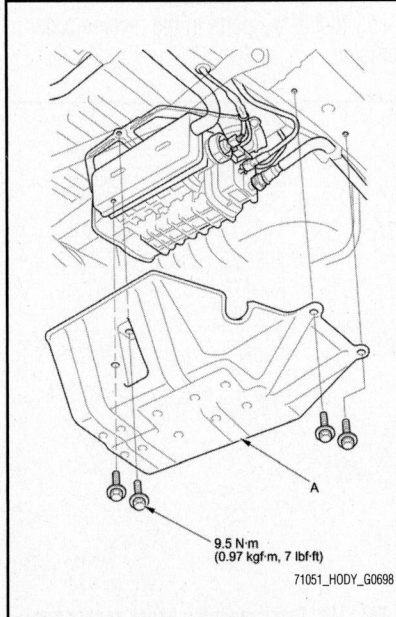

Fig. 152 Remove the EVAP canister cover (A)

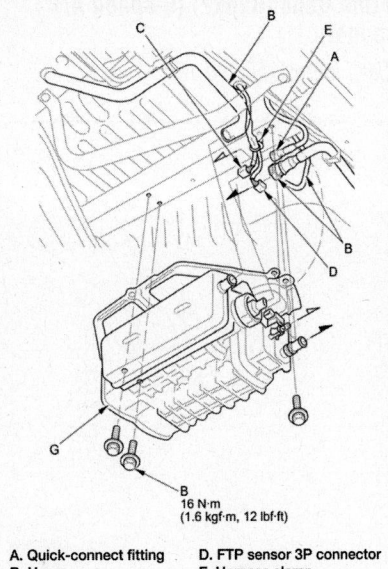

A. Quick-connect fitting
B. Hoses
C. EVAP canister vent shut valve 2P connector
D. FTP sensor 3P connector
E. Harness clamp
F. Bolts
G. EVAP canister assembly

Fig. 153 Disconnect the quick-connect fitting, the hoses, the EVAP canister vent shut valve 2P connector, and the FTP sensor 3P connector, and remove the harness clamp

2P connector, and the FTP sensor 3P connector, and remove the harness clamp.
4. Remove the bolts.
5. Remove the EVAP canister assembly.
6. Remove the EVAP canister from the bracket.
7. Remove the EVAP canister vent shut valve and the FTP sensor from the EVAP canister.

To install:

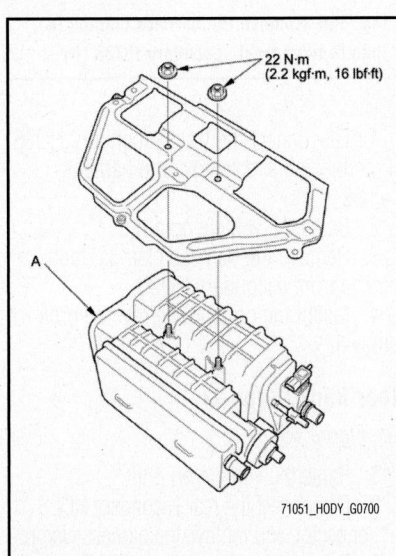

Fig. 154 Remove the EVAP canister (A) from the bracket

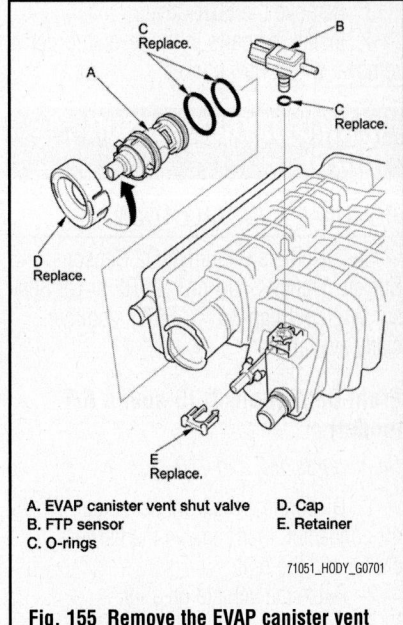

A. EVAP canister vent shut valve
B. FTP sensor
C. O-rings
D. Cap
E. Retainer

Fig. 155 Remove the EVAP canister vent shut valve and the FTP sensor from the EVAP canister

8. Reassemble the EVAP canister with new O-rings, a new cap, and a new retainer, then install the EVAP canister bracket.

➡**Do not coat the O-rings with oil.**

9. Install the parts in the reverse order of removal.

EGR VALVE

REMOVAL & INSTALLATION

See Figure 156.

1. Remove the engine cover.
2. Disconnect the EGR valve 5P connector.

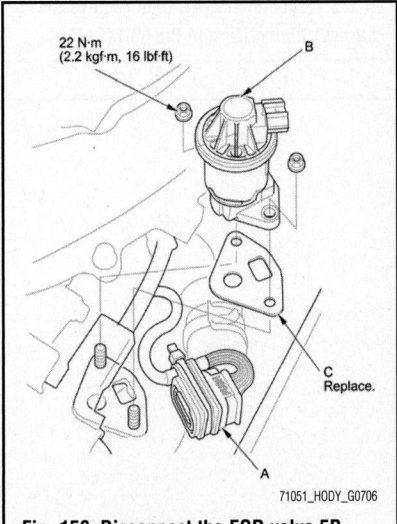

Fig. 156 Disconnect the EGR valve 5P connector (A), then remove the EGR valve (B) and gasket (C)

3. Remove the EGR valve.
4. Install the parts in the reverse order of removal with a new gasket.

HEATED OXYGEN SENSOR (HO2S)

REMOVAL & INSTALLATION

Special Tools Required: O2 Sensor Socket Wrench Snap-on® SWR2 or O2 Sensor Wrench Snap-on® YA8875, commercially available

Front Bank (Bank2) (5-speed A/T model)

See Figures 157 and 158.

1. Disconnect the front secondary HO2S 4P connector, then remove the harness clamp and the bolt.
2. Raise the vehicle on a lift.
3. Remove front secondary HO2S.
4. Install the parts in the reverse order of removal.

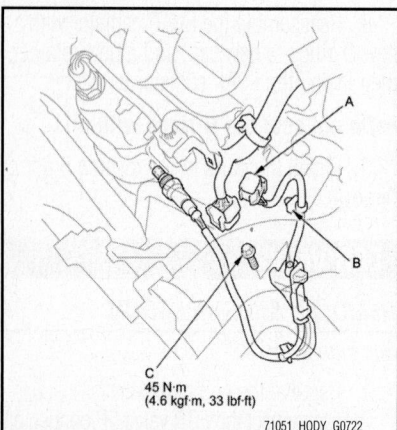

Fig. 157 Disconnect the front secondary HO2S 4P connector (A), then remove the harness clamp (B) and the bolt (C)

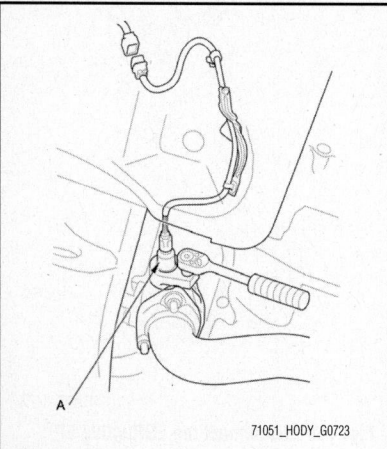

Fig. 158 Remove front secondary HO2S (A)

Front Bank (Bank2) (6-speed A/T model)

See Figures 159 and 160.

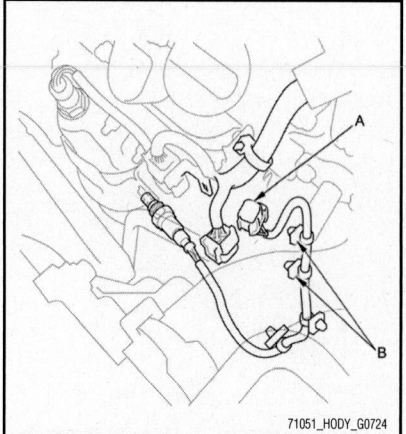

Fig. 159 Disconnect the front secondary HO2S 4P connector (A), and remove the harness clamps (B)

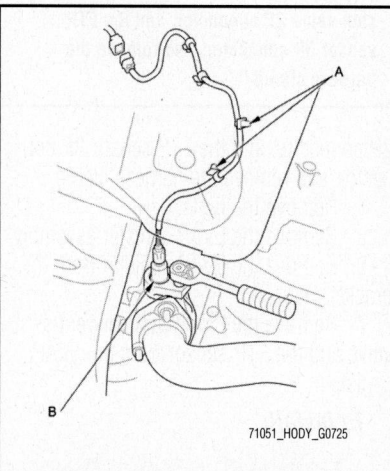

Fig. 160 Remove the harness clamps (A), then remove front secondary HO2S (B)

1. Disconnect the front secondary HO2S 4P connector, and remove the harness clamps.
2. Raise the vehicle on a lift.
3. Remove the harness clamps, then remove front secondary HO2S.
4. Install the parts in the reverse order of removal.

Rear Bank (Bank 1)

See Figure 161.

1. Raise the vehicle on a lift.
2. Disconnect the rear secondary HO2S 4P connector, and remove the harness clamps.
3. Remove rear secondary HO2S.
4. Install the parts in the reverse order of removal.

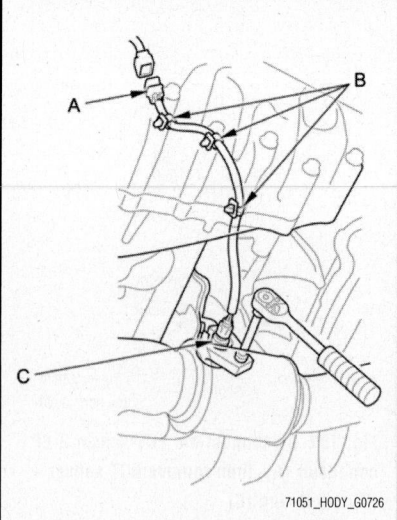

Fig. 161 Disconnect the rear secondary HO2S 4P connector (A), and remove the harness clamps (B), then remove rear secondary HO2S (C)

KNOCK SENSOR (KS)

LOCATION

The knock sensor is located in the engine block beneath the intake manifold.

REMOVAL & INSTALLATION

See Figure 162.

1. Remove the intake manifold.
2. Remove the injector base.
3. Disconnect the knock sensor connector.
4. Remove the knock sensor.
5. Install the parts in the reverse order of removal.

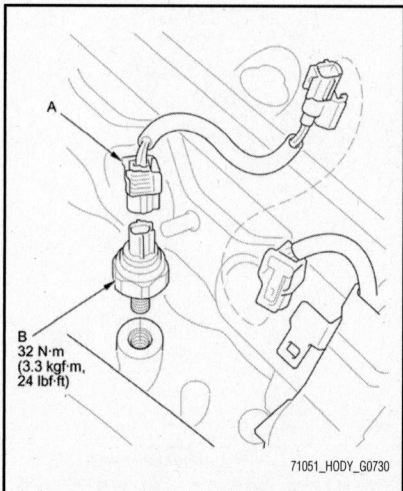

Fig. 162 Disconnect the knock sensor connector (A) and remove the knock sensor (B)

MAINTENANCE REMINDER LIGHTS

RESET PROCEDURE

Note the following:
• The vehicle must be stopped to reset the Maintenance Minder.
• If a required service is done and the Maintenance Minder is not reset, or if the Maintenance Minder is reset without doing the service, the system will not show the proper maintenance timing. This can lead to serious mechanical problems because there will be no accurate record of when the required maintenance is needed.
• The engine oil life and the maintenance item(s) can be independently reset with the HDS.
1. Turn the ignition switch to ON (II).
2. Push the SELECT/RESET knob repeatedly until the engine oil life indicator is displayed.
3. Press and hold the SELECT/RESET knob for about 10 seconds. The engine oil life indicator and the maintenance item code(s) will blink, then release the knob.

➡ **If you are resetting the Maintenance Minder when the engine oil life is more than 15 %, make sure any maintenance item(s) requiring service are done before resetting the display.**

4. Push the SELECT/RESET knob to select OIL LIFE, the display will begin to blink.
5. Push in on the SELECT/RESET knob to enter this selection, the OIL LIFE and the

maintenance item code(s) will begin to blink.
6. Press and hold the SELECT/RESET knob for another 5 seconds. The maintenance item code(s) will disappear, and the engine oil life will reset to "100."

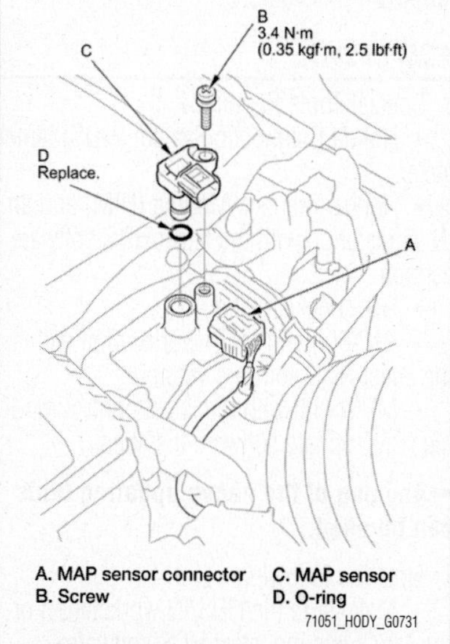

MANIFOLD ABSOLUTE PRESSURE (MAP) SENSOR

REMOVAL & INSTALLATION

1. Disconnect the MAP sensor connector.
2. Remove the screw.
3. Remove the MAP sensor.
4. Install the parts in the reverse order of removal with a new O-ring.

MASS AIRFLOW/INTAKE AIR TEMPERATURE SENSOR

LOCATION

The MAF sensor/IAT sensor is located on the air intake duct.

REMOVAL & INSTALLATION

See Figure 164.

1. Disconnect the MAF sensor/IAT sensor 5P connector.
2. Remove the screw.
3. Remove the MAF sensor/IAT sensor.
4. Install the parts in the reverse order of removal with a new O-ring.

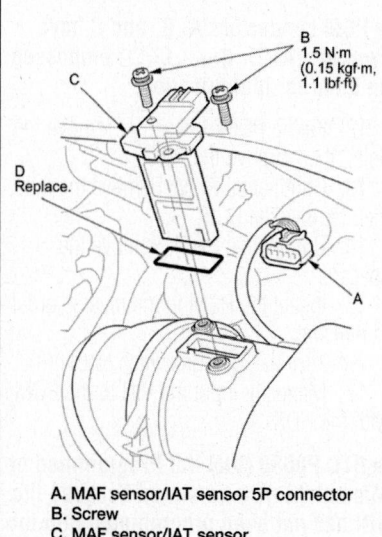

A. MAP sensor connector   C. MAP sensor
B. Screw   D. O-ring

71051_HODY_G0731

Fig. 163 Exploded view of the MAP sensor assembly

A. MAF sensor/IAT sensor 5P connector
B. Screw
C. MAF sensor/IAT sensor
D. O-ring

71051_HODY_G0729

Fig. 164 Exploded view of the MAF sensor/IAT sensor assembly

PCV VALVE

REMOVAL & INSTALLATION

See Figure 165.

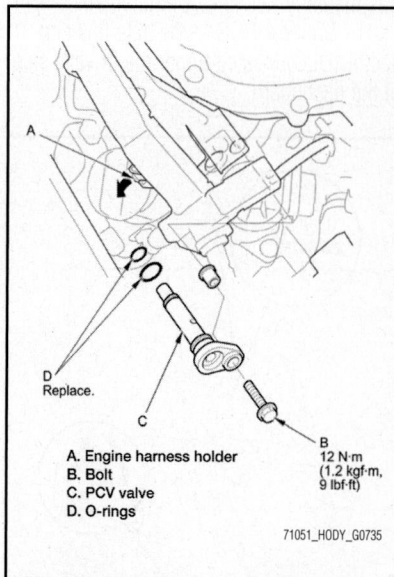

A. Engine harness holder
B. Bolt
C. PCV valve
D. O-rings
B. 12 N·m (1.2 kgf·m, 9 lbf·ft)

71051_HODY_G0735

Fig. 165 Exploded view of the PCV valve assembly

1. Remove the engine cover.
2. Lift the engine harness holder.
3. Remove the bolt.
4. Remove the PCV valve.

➡ **Take care not to spill oil on the hot exhaust manifold.**

5. Install the parts in the reverse order of removal.
• When installing a new PCV valve, make sure the O-rings are in place.
• When installing a used PCV valve, use new O-rings.

POWERTRAIN CONTROL MODULE

REMOVAL & INSTALLATION

See Figures 166 and 167.

Special Tools Required:
• Honda Diagnostic System (HDS) tablet tester
• Honda Interface Module (HIM) and an iN workstation with the latest HDS software version
• HDS pocket tester
• GNA600 and an iN workstation with the latest HDS software version
• MVCI unit with the latest control module (CM) update software installed

➡ **Any one of the above updating tools can be used.**

Note the following:

• Make sure the HDS/iN workstation or the MVCI has the latest HDS software version.

• Check the label on the PCM to see if it was made by Keihin or Continental.

1. Connect HDS to the data link connector (DLC) located under the driver's side of the dashboard.

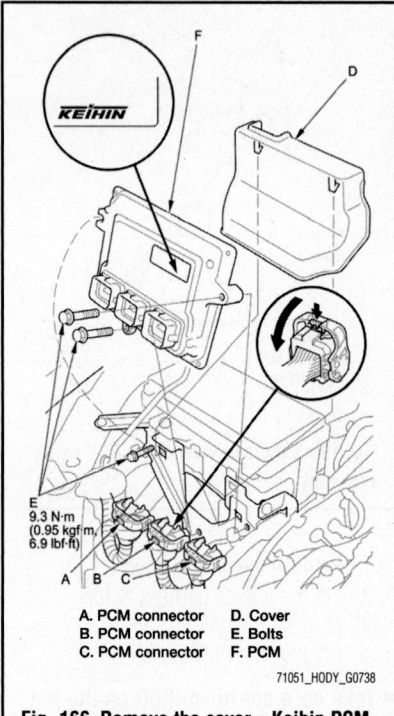

A. PCM connector   D. Cover
B. PCM connector   E. Bolts
C. PCM connector   F. PCM

71051_HODY_G0738

Fig. 166 Remove the cover—Keihin PCM

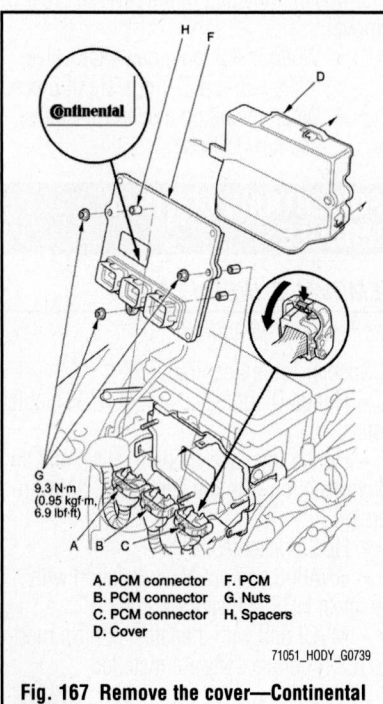

A. PCM connector   F. PCM
B. PCM connector   G. Nuts
C. PCM connector   H. Spacers
D. Cover

71051_HODY_G0739

Fig. 167 Remove the cover—Continental PCM

2. Turn the ignition switch to ON (II).

3. Make sure the HDS communicates with the PCM and all other vehicle systems. If it doesn't, go to the DLC circuit troubleshooting. If the result of the DLC circuit troubleshooting is to replace the PCM, skip steps 4 through 7, 18 through 23, and 26 through 27, and do the following procedures after replacing the PCM:

  a. Replace the engine oil and the engine oil filter.

  b. Replace the ATF.

4. Select the PGM-FI system with the HDS.

5. Select the REPLACE PCM MENU, then select READ DATA, and follow the screen prompts.

• Doing this step copies (READS) the engine oil life data from the original PCM so you can later download (WRITES) it into the new PCM.

• If the READ DATA indicates FAILED, continue with this procedure.

6. Select the A/T system with the HDS.

7. Select the REPLACE TCM/PCM MENU, then select READ DATA, and follow the screen prompts.

• Doing this step copies (READS) the ATF life data from the original PCM so you can later download (WRITES) it into the new PCM.

• If READ DATA indicates FAILED, continue with this procedure.

8. Turn the ignition switch to LOCK (0).

9. Jump the SCS line with the HDS.

10. Remove the cover.

11. Disconnect PCM connectors A, B, and C.

➡PCM connectors A, B, and C have symbols (A=□, B=△, C=○) embossed on them for identification.

12. Keihin PCM: Loosen or remove the bolts, then remove the PCM.

13. Continental PCM: Remove the nuts, then remove the PCM.

14. Continental PCM: Remove the spacers.

15. Install the PCM in the reverse order of removal.

16. Turn the ignition switch to ON (II).

17. Manually input the VIN to the PCM with the HDS.

➡DTC P0630 (VIN Not Programmed or Mismatch) may be stored because the VIN has not been programmed into the PCM; ignore it, and continue this procedure.

18. If the READ DATA (engine oil life) failed in step 5, go to step 21. Otherwise, go to step 19.

19. Select the PGM-FI system with the HDS.

20. Select the REPLACE PCM MENU, then select WRITE DATA, and follow the screen prompts.

➡If the WRITE DATA indicates FAILED, continue with this procedure.

If the READ DATA (ATF life) failed in step 7, go to step 24. Otherwise go to step 22.

21. Select the A/T SYSTEM with the HDS.

22. Select the REPLACE TCM/PCM MENU, then select WRITE DATA, and follow the screen prompts.

➡If the WRITE DATA indicates FAILED, continue with this procedure.

23. Select the IMMOBI system with the HDS.

24. Enter the immobilizer PCM code that you got from the iN, and use the PCM replacement procedure in the IMMOBI Menu of the HDS; it allows you to start the engine.

25. If the READ DATA failed in step 5 or the WRITE DATA failed in step 20, replace the engine oil and the engine oil filter, then go to step 27.

26. If the READ DATA failed in step 7 or the WRITE DATA failed in step 23, replace the ATF; then go to step 28.

27. Select the PGM-FI system, and reset the PCM with the HDS.

28. Clean the throttle body.

29. Update the PCM if it does not have the latest software.

30. Do the PCM idle learn procedure.

31. Do the CKP pattern clear/CKP pattern learn procedure.

PCM UPDATE

Special Tools Required:

• Honda Diagnostic System (HDS) tablet tester

• Honda Interface Module (HIM) and an iN workstation with the latest HDS software version

• HDS pocket tester

• GNA600 and an iN workstation with the latest HDS software version

• MVCI unit with the latest control module (CM) update software installed

➡Any one of the above updating tools can be used.

Note the following:

• Make sure the HDS/iN workstation or the MVCI has the latest HDS software version.

• Before you update the PCM, make sure

the battery in the vehicle is fully charged, and connect a jumper battery (not a battery charger) to maintain system voltage.

• Never turn the ignition switch to ACC (I) or LOCK (0) during the update. If there is a problem with the update, leave the ignition switch in ON (II).

• To prevent PCM damage, do not operate anything electrical (headlights, audio system, brakes, A/C, power windows, moonroof, door locks, etc.) during the update.

• To ensure the latest program is installed, do a PCM update whenever the PCM is substituted or replaced.

• You cannot update a PCM with a program it already has. It will only accept a new program.

• High temperature in the engine compartment might cause the PCM to become too hot to run the update. If the engine has been running before this procedure, open the hood and cool the engine compartment.

• If you need to diagnose the Honda interface module (HIM) because the HIM's red (#3) light came on or was flashing during the update, leave the ignition switch in ON (II) when you disconnect the HIM from the data link connector (DLC). This will prevent PCM damage.

1. Turn the ignition switch to ON (II), but do not start the engine.

2. Connect the HDS to the data link connector (DLC) located under the driver's side of the dashboard.

3. Make sure the HDS communicates with the PCM and other vehicle systems. If it doesn't, go to the DLC circuit troubleshooting.

4. Exit the HDS diagnostic system, then select the update mode, and follow the HDS prompts to update the PCM.

5. If the software in the PCM is the latest, disconnect the updating tool from the DLC, and go back to the procedure that you were doing. If the software in the PCM is not the latest, follow the instructions on the screen. If prompted to choose the PGM-FI

system or the A/T system, make sure you update both.

➡**If the PCM update system requires you to cool the PCM, follow the instructions on screen. If you run into a problem during the update procedure (programming takes over 15 minutes, status bar goes over 100 %, D or immobilizer indicator flashes, HDS tablet freezes, etc.), follow these steps to minimize the chance of damaging the PCM:**

    a. Leave the ignition switch in ON (II).

    b. Connect a jumper battery (do not connect a battery charger).

    c. Shut down the updating tool.

    d. Disconnect the updating tool from the DLC.

    e. Reboot the updating tool.

    f. Reconnect the updating tool to the DLC, and try the update procedure again.

6. Do the PCM idle learn procedure.

7. Do the CKP pattern clear/CKP pattern learn procedure.

PCM IDLE LEARN PROCEDURE

The idle learn procedure must be done so the PCM can learn the engine idle characteristics.

Do the idle learn procedure whenever you do any of these actions:

• Replace the PCM
• Reset the PCM
• Update the PCM
• Replace or clean the throttle body
• Disassemble the engine or the transmission

➡**Erasing DTCs with the HDS does not require you to do the idle learn procedure.**

1. Make sure all electrical items (A/C, audio, lights, etc.) are off.

2. Reset the PCM with the HDS.

3. Turn the ignition switch to ON (II), and wait 2 seconds.

4. Start the engine. Hold the engine speed at 3,000 rpm without load (in P or N) until the radiator fan comes on, or until the engine coolant temperature reaches 194 °F (90 °C).

5. Let the engine idle for about 5 minutes with the throttle fully closed.

➡**If the radiator fan comes on, do not include its running time in the 5 minutes.**

CKP PATTERN CLEAR/CKP PATTERN LEARN

Clear/Learn Procedure (with the HDS)

1. Connect the HDS to the data link connector (DLC) located under the driver's side of the dashboard.

2. Turn the ignition switch to ON (II).

3. Make sure the HDS communicates with the PCM and all other vehicle systems. If it doesn't, go to the DLC circuit troubleshooting.

4. Select CRANK PATTERN in the ADJUSTMENT MENU with the HDS.

5. Select CRANK PATTERN CLEAR, and clear the CKP pattern.

6. Select CRANK PATTERN LEARNING with the HDS, and follow the screen prompts.

Learn Procedure (without the HDS)

1. Start the engine. Hold the engine speed at 3,000 rpm without load (in P or N) until the radiator fan comes on.

2. Test-drive the vehicle on a level road: Decelerate (with the throttle fully closed) from an engine speed of 2,500 rpm down to 1,000 rpm with the transmission in 2nd.

3. Repeat step 2 several times.

4. Turn the ignition switch to LOCK (0).

5. Turn the ignition switch to ON (II), and wait 30 seconds.

FUEL SYSTEM SERVICE PRECAUTIONS

Safety is the most important factor when performing not only fuel system maintenance, but any type of maintenance. Failure to conduct maintenance and repairs in a safe manner may result in serious personal injury or death. Work on a vehicle's fuel system components can be accomplished safely and effectively by adhering to the following rules and guidelines.

• To avoid the possibility of fire and personal injury, always disconnect the negative battery cable unless the repair or test procedure requires that battery voltage be applied.

• Always relieve the fuel system pressure prior to disconnecting any fuel system component (injector, fuel rail, pressure regulator, etc.) fitting or fuel line connection. Exercise extreme caution whenever relieving fuel system pressure to avoid exposing skin, face and eyes to fuel spray. Please be advised that fuel under pressure may penetrate the skin or any part of the body that it contacts.

• Always place a shop towel or cloth around the fitting or connection prior to loosening to absorb any excess fuel due to spillage. Ensure that all fuel spillage is quickly removed from engine surfaces. Ensure that all fuel-soaked cloths or towels are deposited into a flame-proof waste container with a lid.

• Always keep a dry chemical (Class B) fire extinguisher near the work area.

• Do not allow fuel spray or fuel vapors to come into contact with a spark or open flame.

• Always use a second wrench when loosening or tightening fuel line connection fittings. This will prevent unnecessary stress and torsion on fuel piping. Always follow the proper torque specifications.

• Always replace worn fuel fitting O-rings with new ones. Do not substitute fuel hose where rigid pipe is installed.

RELIEVING FUEL SYSTEM PRESSURE

With the HDS

See Figures 168 and 169.

Before disconnecting fuel lines or hoses, relieve pressure from the system by disabling the fuel pump and disconnecting the fuel line/quick connect fitting in the engine compartment.

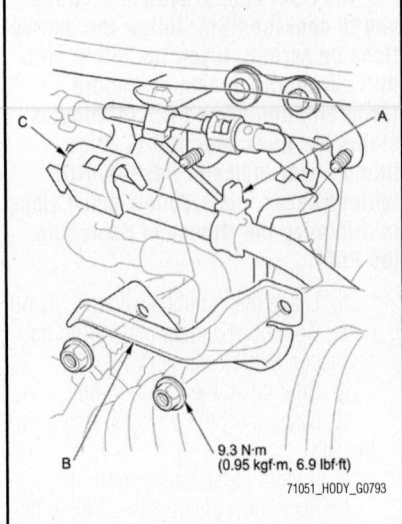

9.3 N·m
(0.95 kgf·m, 6.9 lbf·ft)
71051_HODY_G0793

Fig. 168 Remove the clamp (A) from the cover (B), then remove the cover and the quick-connect fitting cover (C)

1. Connect the HDS to the data link connector (DLC) located under the driver's side of the dashboard.
2. Turn the ignition switch to ON (II).
3. Make sure the HDS communicates with the PCM. If it doesn't, go to the DLC circuit troubleshooting.
4. Turn the ignition switch to LOCK (0).
5. Remove the fuel fill cap to relieve the pressure in the fuel tank.
6. Turn the ignition switch to ON (II).
7. From the INSPECTION MENU of the HDS, select Fuel Pump OFF, then start the engine, and let it idle until it stalls.
   • Do not allow the engine to idle above 1,000 rpm or the PCM will continue to operate the fuel pump.
   • A Pending or Confirmed DTC may be set during this procedure. Check for DTCs, and clear them as needed.
8. Turn the ignition switch to LOCK (0).
9. Do the battery terminal disconnection procedure.
10. Remove the clamp from the cover, then remove the cover.
11. Remove the quick-connect fitting cover.
12. Check the fuel quick-connect fitting for dirt, and clean it if needed.
13. Place a rag or shop towel over the quick-connect fitting.

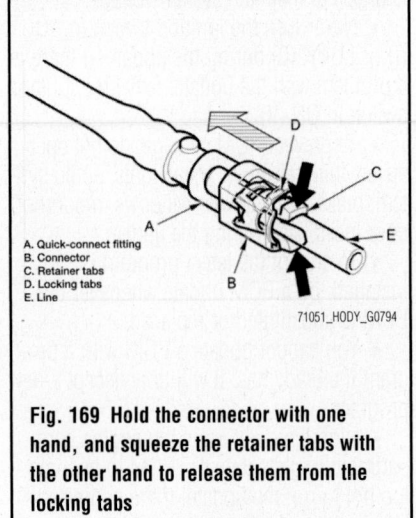

A. Quick-connect fitting
B. Connector
C. Retainer tabs
D. Locking tabs
E. Line

71051_HODY_G0794

Fig. 169 Hold the connector with one hand, and squeeze the retainer tabs with the other hand to release them from the locking tabs

14. Disconnect the quick-connect fitting: Hold the connector with one hand, and squeeze the retainer tabs with the other hand to release them from the locking tabs. Pull the connector off.
   • Be careful not to damage the line or other parts.
   • Do not use tools.
   • If the connector does not move, keep the retainer tabs pressed down, and alternately pull and push the connector until it comes off easily.
   • Do not remove the retainer from the line; once removed, the retainer must be replaced with a new one.
15. After disconnecting the quick-connect fitting, check it for dirt or damage.
16. Do the battery terminal reconnection procedure.

Without the HDS

See Figures 168 through 170

1. Remove PGM-FI main relay 2 from the driver's under-dash fuse/relay box.
2. Start the engine, and let it idle until it stalls.

➡**If any DTCs are stored, clear and ignore them.**

3. Turn the ignition switch to LOCK (0).
4. Remove the fuel fill cap to relieve the pressure in the fuel tank.
5. Do the battery terminal disconnection procedure.
6. Remove the clamp from the cover, then remove the cover.
7. Remove the quick-connect fitting cover.
8. Check the fuel quick-connect fitting for dirt, and clean it if needed.

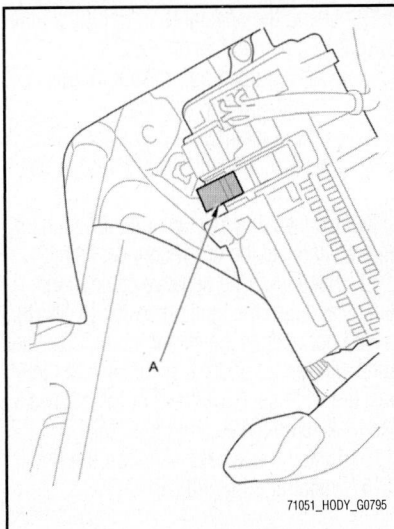

Fig. 170 Remove PGM-FI main relay 2 (A) from the driver's under-dash fuse/relay box

9. Place a rag or shop towel over the quick-connect fitting.

10. Disconnect the quick-connect fitting: Hold the connector with one hand, and squeeze the retainer tabs with the other hand to release them from the locking tabs. Pull the connector off.

- Be careful not to damage the line or other parts.
- Do not use tools.
- If the connector does not move, keep the retainer tabs pressed down, and alternately pull and push the connector until it comes off easily.
- Do not remove the retainer from the line; once removed, the retainer must be replaced with a new one.

11. After disconnecting the quick-connect fitting, check it for dirt or damage.

12. Do the battery terminal reconnection procedure.

IDLE SPEED

ADJUSTMENT

Before checking the idle speed, check these items:

- The malfunction indicator lamp (MIL) has not been reported on, and there are no DTCs.
- Ignition timing
- Spark plugs
- Air cleaner
- PCV system

1. Apply the parking brake, and make sure the headlights are off.

2. Disconnect the evaporative emission (EVAP) canister purge valve connector.

3. Connect the HDS to the data link connector (DLC) located under the driver's side of the dashboard.

4. Make sure the HDS communicates with the PCM. If it doesn't, go to the DLC circuit troubleshooting.

5. Start the engine. Hold the engine speed at 3,000 rpm without load (in P or N) until the radiator fan comes on, then let it idle.

6. Check the idle speed under no load conditions: headlights, blower fan, radiator fan, audio system, and air conditioner off. Idle speed should be: 740 ± 50 rpm (in P or N)

7. Let the engine idle for 1 minute with high electric load (A/C on, temperature set to max cool, blower fan on high, headlights on high beam). Idle speed should be: 740 ± 50 rpm (in P or N)

➡**If the idle speed is not within specification, do the PCM idle learn procedure. If the idle speed is still not within specification, go to the symptom troubleshooting.**

8. Reconnect the EVAP canister purge valve connector.

FUEL FILTER

REMOVAL & INSTALLATION

See Figure 171.

➡**The fuel filter should be replaced whenever the fuel pressure drops below the pressure of 57–64 psi**

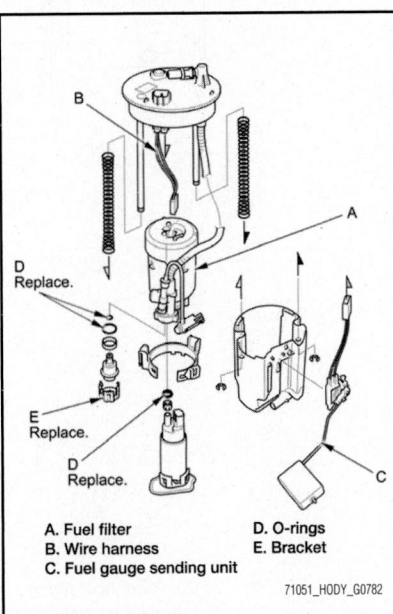

A. Fuel filter
B. Wire harness
C. Fuel gauge sending unit
D. O-rings
E. Bracket

71051_HODY_G0782

Fig. 171 Exploded view of the fuel tank unit showing the fuel filter location

(390–440 kPa), and after making sure that the fuel pump and fuel pressure regulator are OK.

1. Remove the fuel tank unit.

2. Remove the fuel filter set (A).

To install:

3. Check these items before installing the fuel tank unit:

   a. When connecting the wire harness (B), make sure the connection is secure and the connectors are firmly locked into place.

   b. When installing the fuel gauge sending unit (C), make sure the connection is secure and the connector is firmly locked into place. Be careful not to bend or twist it excessively.

4. Install the parts in the reverse order of removal with new O-rings (D) and new bracket (E). When installing the fuel tank unit, align the marks on the unit and the fuel tank.

- Coat the O-rings with clean engine oil; do not use any other oil or fluid.
- Do not pinch the O-rings during installation.
- Use all the new parts supplied in the fuel filter replacement kit.

FUEL TANK UNIT

REMOVAL & INSTALLATION

See Figures 172 through 174.

Special Tools Required: Fuel Sender Wrench 07AAA-S0XA100

1. Relieve the fuel pressure.

2. Move the second row seat forward.

3. Remove the access panel from the left side of the floor.

4. Disconnect the fuel tank unit 4P connector.

5. Disconnect the quick-connect fitting from the fuel tank unit.

6. Using the special tool, loosen the locknut.

7. Remove the locknut, the locknut plate, and the fuel tank unit.

To install:

8. Temporarily attach a new base gasket to the fuel tank unit, insert the fuel tank unit partially into the fuel tank.

- Be careful not to damage the new base gasket.
- Be careful not to bend the fuel gauge sending unit.
- Do not coat the base gasket with oil.

9. Transfer the base gasket from the fuel tank unit to the fuel tank.

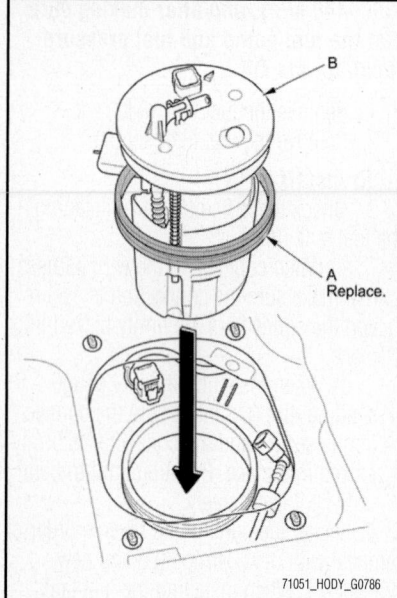

Fig. 172 Temporarily attach a new base gasket (A) to the fuel tank unit (B), insert the fuel tank unit partially into the fuel tank

71051_HODY_G0786

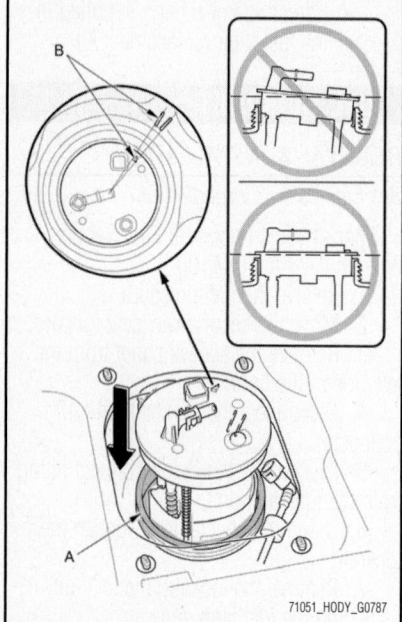

Fig. 173 Transfer the base gasket (A) from the fuel tank unit to the fuel tank and align the marks (B)

71051_HODY_G0787

10. Align the marks on the fuel tank and the fuel tank unit, then insert the fuel tank unit into the fuel tank until the fuel tank unit rests on top of the base gasket.

➡**To prevent a fuel leak, check the base gasket, visually or by hand, to make sure it is not pinched.**

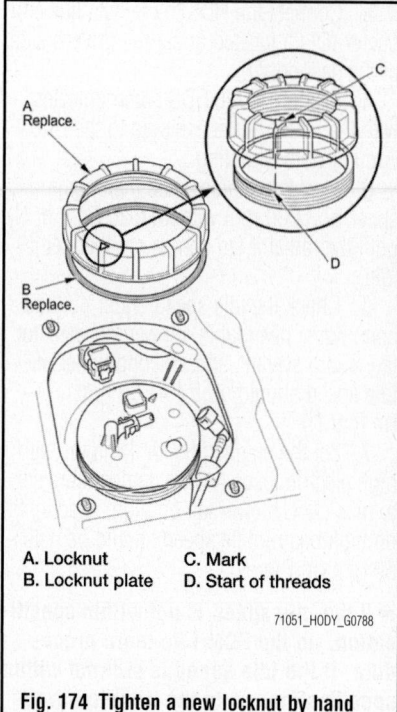

A. Locknut
B. Locknut plate
C. Mark
D. Start of threads

71051_HODY_G0788

Fig. 174 Tighten a new locknut by hand with a new locknut plate

11. Tighten a new locknut by hand with a new locknut plate.

➡**Before tightening, align the mark on the locknut with the start of the threads.**

12. Using the special tool, tighten a new locknut to the specified torque.
- After tightening, make sure the marks are still aligned.
- After installation, check the base gasket, visually or by hand, to be sure it is not pinched.

13. Connect the fuel tank unit 4P connector, then connect the quick-connect fitting.

14. Reconnect the negative cable to the battery, and turn the ignition switch to ON (II) (but do not operate the starter motor). The fuel pump will run for about 2 seconds, and fuel pressure will rise. Repeat two or three times, then make sure there are no fuel leaks.

15. Install the access panel on the floor.

16. Install the fuel fill cap.

FUEL RAIL & INJECTORS

REMOVAL & INSTALLATION

See Figures 175 and 176.

1. Relieve the fuel pressure.
2. Remove the intake manifold.
3. Disconnect the quick-connect fitting.
4. Remove the fuel joint hose mounting bolt.
5. Disconnect the connectors from the injectors.
6. Remove the fuel rail mounting bolts from the fuel rails.

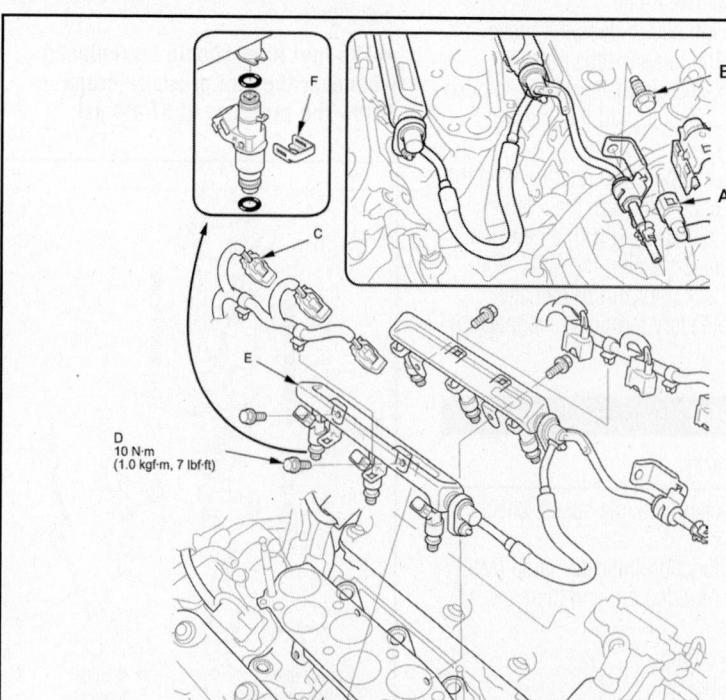

A. Quick-connect fitting
B. Fuel joint hose mounting bolt
C. Connectors
D. Fuel rail mounting bolts
E. Fuel rails
F. Injector clips

71051_HODY_G0842

Fig. 175 Exploded view of the fuel injectors and fuel rail assemblies

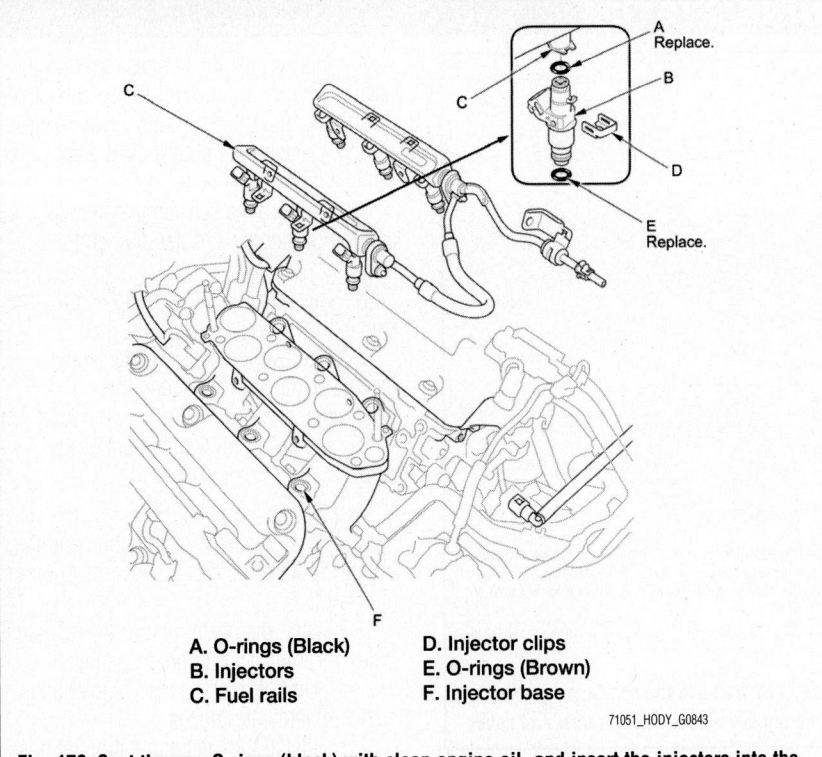

A. O-rings (Black)
B. Injectors
C. Fuel rails
D. Injector clips
E. O-rings (Brown)
F. Injector base

71051_HODY_G0843

Fig. 176 Coat the new O-rings (black) with clean engine oil, and insert the injectors into the fuel rails

7. Remove the fuel rails and the injectors from the injector base.

8. Remove the injector clips from the fuel rails.

9. Remove the injectors from the fuel rails.

To install:

10. Coat the new O-rings (black) with clean engine oil, and insert the injectors into the fuel rails.

11. Install the injector clips.

12. Coat the new O-rings (brown) with clean engine oil.

13. Install the fuel rails and the injectors in the injector base.

14. Reinstall the fuel rail mounting bolts, and connect the connectors on the injectors.

15. Reinstall the fuel joint hose mounting bolt.

16. Connect the quick-connect fitting.

17. Turn the ignition switch to ON (II), but do not operate the starter. After the fuel pump runs for about 2 seconds, the fuel rail is pressurized. Repeat this two or three times, then make sure there are no fuel leaks.

18. Reinstall the intake manifold with a new gasket.

THROTTLE BODY

REMOVAL & INSTALLATION

See Figure 177.

❊❊ CAUTION

Do not insert your fingers into the installed throttle body when you turn the ignition switch to ON (II) or while the ignition switch is in ON (II). If you

do, you will seriously injure your fingers if the throttle valve is activated.

1. Make sure the ignition switch turned LOCK (0).

2. Disconnect the MAP sensor connector.

3. Remove the intake air duct.

4. Disconnect the throttle body connector.

5. Disconnect and plug the water bypass hoses.

6. Remove the throttle body.

To install:

7. Install the parts in the reverse order of removal with a new gasket, then refill the radiator with engine coolant.

- After torquing the screw of the hose band, make sure the clearance is less than 0.04 inches (1.0 mm).
- If it is in case of replacing or cleaning the throttle body, go to step 8. If it isn't, this procedure is complete.

8. Connect the HDS to the data link connector (DLC) located under the driver's side of the dashboard.

9. Turn the ignition switch to ON (II).

10. Reset the PCM with the HDS.

11. Select the ETCS TEST in the INSPECTION MENU with the HDS.

12. Select TP POSITION CHECK and clear the throttle position (TP) learned value.

13. Turn the ignition switch to LOCK (0).

14. Turn the ignition switch to ON (II), and wait 2 seconds without pressing the accelerator pedal.

15. Do the PCM idle learn procedure.

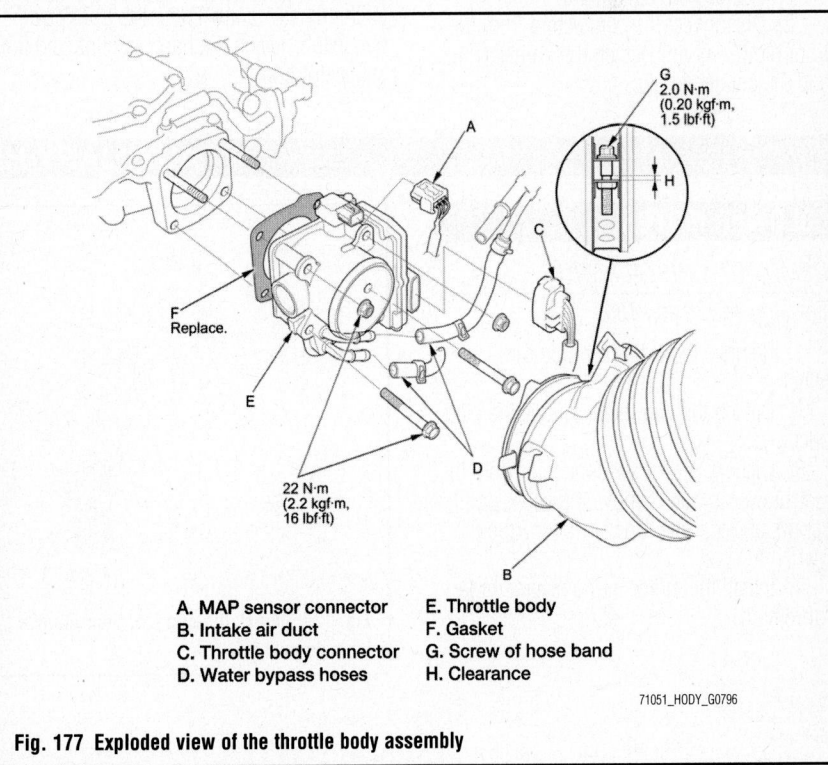

A. MAP sensor connector
B. Intake air duct
C. Throttle body connector
D. Water bypass hoses
E. Throttle body
F. Gasket
G. Screw of hose band
H. Clearance

71051_HODY_G0796

Fig. 177 Exploded view of the throttle body assembly

HEATING & AIR CONDITIONING SYSTEM

HEATER CORE

REMOVAL & INSTALLATION

See Figure 178.

✳✳ CAUTION

SRS components are located in this area. Review the SRS component locations and the precautions and procedures before doing repairs or service.

1. Do the battery terminal disconnection procedure.
2. Disconnect the A/C lines from the front evaporator core.
3. When the engine is cool, drain the engine coolant.
4. From under the hood, slide the hose clamps back.
5. Disconnect the inlet heater hose and the outlet heater hose from the heater unit. Note the layout of the hoses.

➡**Engine coolant will run out when the hoses are disconnected; drain it into a clean drip pan. Be sure not to let coolant spill on the electrical parts or the painted surfaces. If any coolant spills, rinse it off immediately.**

6. Remove the mounting nut from the heater unit. Take care not to damage or bend the fuel lines or the brake lines.
7. Remove the dashboard.
8. Disconnect the connectors from the front blower motor. Detach the harness clips and the connector clip.

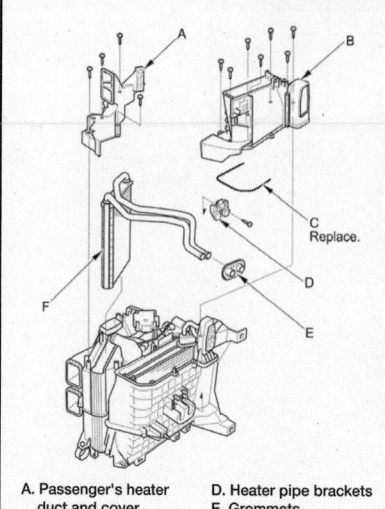

A. Passenger's heater duct and cover
B. Expansion valve cover
C. Seal
D. Heater pipe brackets
E. Grommets
F. Front heater core

71051_HODY_G0840

Fig. 178 Remove the self-tapping screws and the passenger's heater duct and cover

9. Disconnect the connector from the recirculation control motor. Remove the mounting nuts. Slide the blower-heater unit back, then remove the drain hose and the blower-heater unit.
10. Remove the blower unit.
11. Disconnect the connectors from the cool box control motor (with climate control), the passenger's air mix control motor (with climate control), and the front power transistor. Detach the harness clips and the connector clip.

12. Disconnect the connectors from the front mode control motor, the air mix control motor, and the evaporator temperature sensor. Detach the harness clips and remove the wire harness.
13. Remove the self-tapping screws and the passenger's heater duct and cover.
14. Remove the self-tapping screws, the expansion valve cover, and the seal.
15. Remove the self-tapping screws, the heater pipe brackets, and the grommets.
16. Carefully pull out the front heater core.

To install:

17. Install the front heater core, and the front evaporator core in the reverse order of removal.
18. Install the heater unit in the reverse order of removal, and note these items:
   a. Do not interchange the inlet and the outlet heater hoses.
   b. Install the inlet and the outlet hose clamps securely.
   c. Refill the cooling system with engine coolant.
   d. Make sure that there is no coolant leakage.
   e. Make sure that there is no air leakage.
   f. Make sure that there is no refrigerant leakage.
   g. Refer to the evaporator core replacement.
19. Do the battery terminal reconnection procedure.

AUXILIARY HEATING & AIR CONDITIONING

BLOWER MOTOR

REMOVAL & INSTALLATION

See Figures 179 and 180.

1. Remove the right rear side trim panel.
2. Detach the clip, then remove the side duct.
3. Disconnect the 2P connector from the rear blower motor. Remove the self-tapping screws and the rear blower motor from the rear HVAC unit.
4. Install the motor in the reverse order of removal.

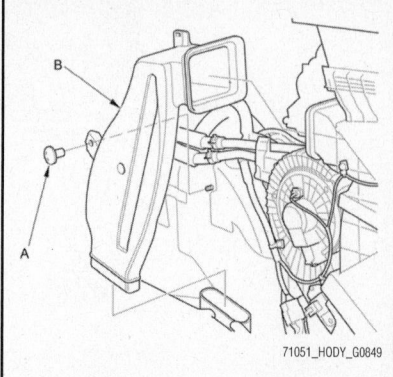

71051_HODY_G0849

Fig. 179 Detach the clip (A), then remove the side duct (B)

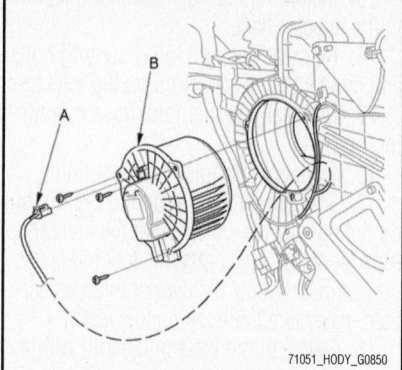

71051_HODY_G0850

Fig. 180 Disconnect the 2P connector (A) from the rear blower motor (B)

STEERING

POWER RACK & PINION STEERING GEAR

REMOVAL & INSTALLATION

See Figures 181 through 191.

Special Tools Required*:
• Ball Joint Thread Protector, 12 mm 07AAF-SDAA100
• Ball Joint Remover, 28 mm 07MAC-SL0A202
• Subframe Adapter VSB02C000016
• *Available through the Honda Tool and Equipment Program, 888-424-6857.

Note these items during removal:

a. Using clean solvent and a brush, wash any oil and dirt off the valve body unit, it's lines, and the end of the steering gearbox. Blow dry with compressed air.

b. Lower the front subframe from the body, and remove the steering gearbox through the gap produced by lowering the front subframe.

1. Place the hood support rod in the wide-open position.

2. Remove the engine cover.

3. Drain the power steering fluid.

4. Remove the air intake scoop.

5. Do the battery terminal disconnection procedure.

6. Remove the front bulkhead cover.

7. Raise and support the vehicle.

8. Remove the front wheels.

9. Remove the driver's heater panel.

10. Pull back the carpet.

11. Remove the steering joint cover.

12. Tilt the steering column at the way up, and move it all the way in, then tighten the lock lever.

13. Hold the lower slide shaft on the column with a piece of wire between the joint yoke of the lower slide shaft and joint yoke of the upper shaft to prevent the lower slide shaft from pulling out.

14. Remove the steering joint bolt.

15. Center the steering wheel spokes, and install a commercially available steering wheel holder tool.

16. Disconnect the steering joint by sliding the steering joint into the column shaft.

17. Remove the center guide (if equipped) from the top of the pinion shaft, and discard it.

18. Wrap vinyl tape over the splines on the pinion shaft.

19. Remove the power steering heat shield.

20. Remove the cotter pin from the tie-rod end ball joint, then remove the nut on both sides.

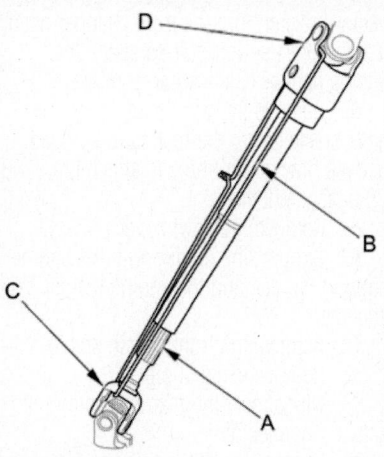

A. Lower slide shaft
B. Wire
C. Joint yoke of the lower slide shaft
D. Joint yoke of the upper shaft

71051_HODY_G0860

Fig. 181 Hold the lower slide shaft on the column with a piece of wire between the joint yoke of the lower slide shaft and joint yoke of the upper shaft to prevent the lower slide shaft from pulling out

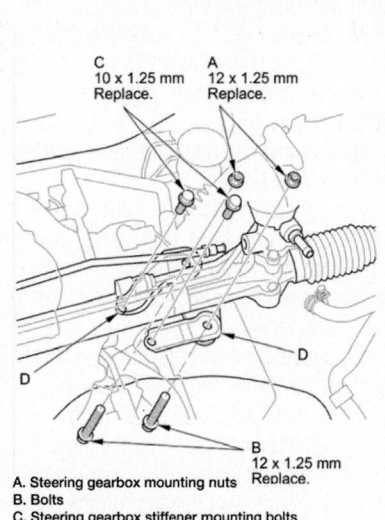

C 10 x 1.25 mm Replace.
A 12 x 1.25 mm Replace.

A. Steering gearbox mounting nuts
B. Bolts
C. Steering gearbox stiffener mounting bolts
D. Steering gearbox stiffeners

B 12 x 1.25 mm Replace.

71051_HODY_G0869

Fig. 182 Remove the steering gearbox mounting nuts and bolts, the steering gearbox stiffener mounting bolts, and the steering gearbox stiffeners from the driver's side of the steering gearbox

21. Disconnect the tie-rod end ball joint from the knuckle using the ball joint thread protector and the ball joint remover.

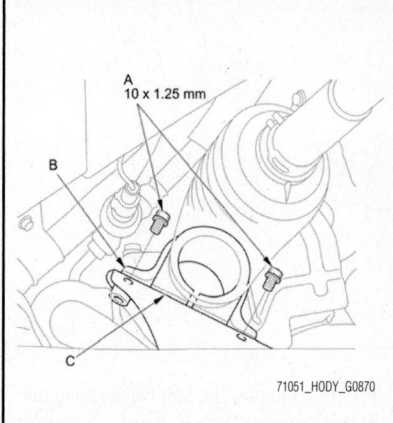

A 10 x 1.25 mm

71051_HODY_G0870

Fig. 183 Remove the steering gearbox mounting bolts (A) from the passenger's side of the steering gearbox, then remove the steering gearbox mounting bracket (B) and the mounting cushion (C)

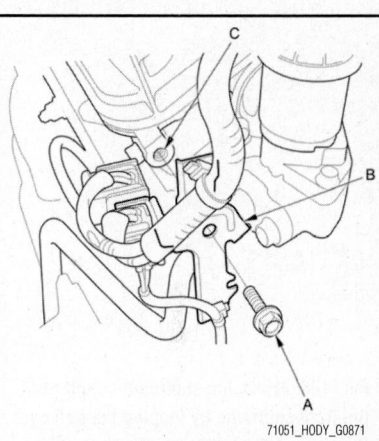

71051_HODY_G0871

Fig. 184 Remove the bolt (A) securing the connector bracket (B) from the front cylinder head; use the bracket bolt hole (C) to attach the engine hanger balance bar front arm

➡ Be careful not to damage the ball joint boot when installing the remover.

22. Loosen the adjustable hose clamp, and disconnect the return hose.

23. Remove the inlet line clamp bolt.

24. Loosen the flare nut, and disconnect the inlet line.

25. Remove the steering gearbox mounting nuts and bolts, the steering gearbox stiffener mounting bolts, and the steering gearbox stiffeners from the driver's side of the steering gearbox.

26. Remove the steering gearbox mounting bolts from the passenger's side of the steering gearbox, then remove the steering

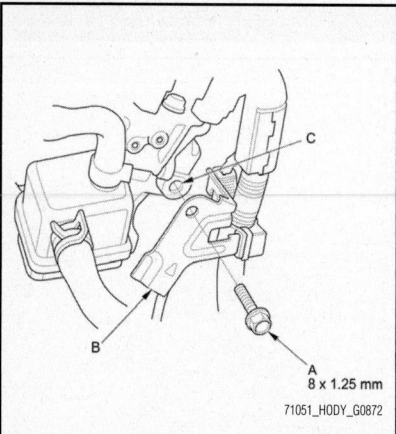

Fig. 185 Remove the bolt (A) securing the harness bracket (B) from the rear cylinder head; use the bracket bolt hole (C) to attach the 2008 V6 attachment arm

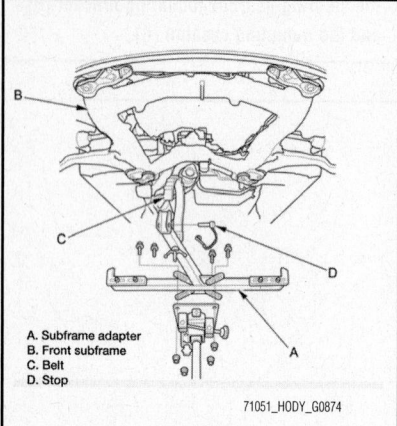

A. Subframe adapter
B. Front subframe
C. Belt
D. Stop

Fig. 186 Attach the subframe adapter to the front subframe by looping the belt over the front of the subframe, then secure the belt with its stop

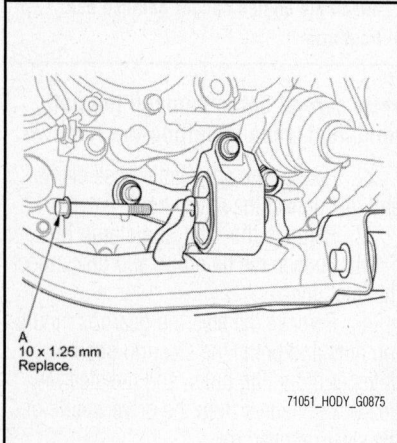

Fig. 187 Remove the lower transmission mount bolt (A)

gearbox mounting bracket and the mounting cushion.

27. Remove the bolt securing the connector bracket from the front cylinder head; use the bracket bolt hole to attach the engine hanger balance bar front arm.

28. Remove the bolt securing the harness bracket from the rear cylinder head; use the bracket bolt hole to attach the 2008 V6 attachment arm.

29. Remove the cowl cover.

30. Support the engine with the engine support hanger, and the engine hanger balance bar.

31. Remove the front splash shield.

32. Remove exhaust pipe A.

33. Disconnect the rear engine mount actuator connector.

34. Remove the rear engine mount mounting bolts and the rear engine mount base bracket mounting bolts.

35. Remove the rear engine mount and the rear engine mount base bracket.

36. Attach the subframe adapter to the front subframe by looping the belt over the front of the subframe, then secure the belt with its stop.

37. Raise the jack, and line up the slots in the subframe adapter arms with the bolt holes on the jack base, then securely attach them with bolts and nuts.

38. Support the front subframe securely by raising the transmission jack.

39. Remove the lower transmission mount bolt.

40. Remove the rear stiffener mounting bolts on both sides.

41. Loosen the front subframe mounting bolts so they are about 1.18 inches (30 mm) from the mounting surface. Do not loosen the subframe mounting bolts more than necessary.

42. Lower the transmission jack slowly until the front subframe rear side has dropped about 1.18 inches (30 mm).

43. Carefully move the steering gearbox toward the driver's side until the pinion shaft clears the fenderwell opening on the body.

44. Remove the steering gearbox through the fenderwell opening on the driver's side.

45. Remove the pinion shaft grommet from the top of the valve body unit (B).

46. Remove the return line joint if necessary.

To install:

47. Before installing the steering gearbox, make sure that no grease is on the mating surface of the steering gearbox and the front subframe. To prevent the gearbox mounting bolts from loosening after the

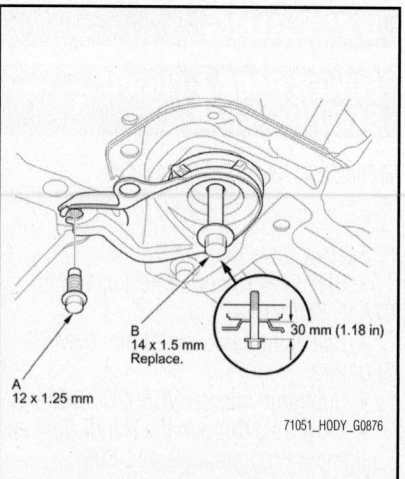

Fig. 188 Remove the rear stiffener mounting bolts (A) on both sides, then loosen the front subframe mounting bolts (B)

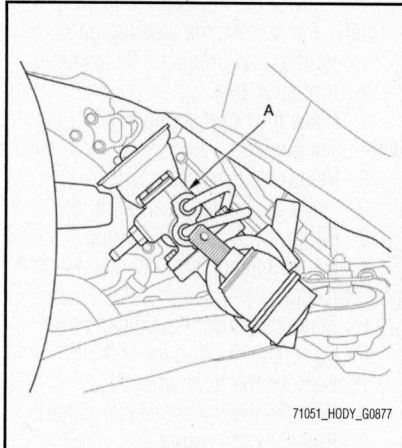

Fig. 189 Carefully move the steering gearbox (A) toward the driver's side until the pinion shaft clears the fenderwell opening on the body

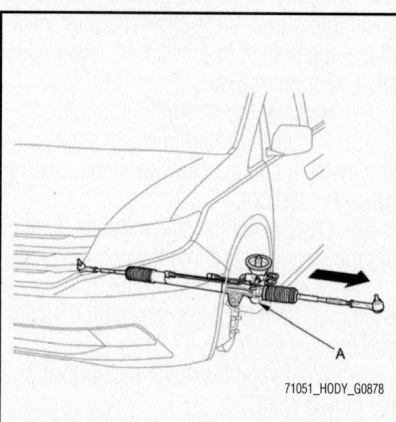

Fig. 190 Remove the steering gearbox (A) through the fenderwell opening on the driver's side

installation, remove any grease from the bolt holes.

48. Wrap vinyl tape over the splines on the pinion shaft.

49. Install the pinion shaft grommet. The grommet must not have a gap at the mating surface of the grommet and the valve body unit.

50. Install the return line joint if removed. Tighten to 21 ft. lbs. (29 Nm).

51. Slide the steering gearbox between the front subframe and the body from the driver's side.

52. Carefully move the steering gearbox toward the passenger's side until the pinion shaft clears the wheelwell opening on the body.

53. Continue moving the steering gearbox toward the passenger's side until the steering gearbox is in position.

54. Raise the front subframe up to the body. Loosely install the new subframe mounting bolts and the new rear stiffener mounting bolts.

55. Align the front subframe with the subframe alignment pin.

56. Install the lower transmission mount bolt, and tighten it to 40 ft. lbs. (54 Nm).

57. Lower the transmission jack supporting the front subframe.

58. Install the rear engine mount and the rear engine mount base bracket.

59. Install the new rear engine mount mounting bolts and the new rear engine mount base bracket mounting bolts.

   a. Tighten the rear engine mount mounting bolts to 40 ft. lbs. (54 Nm).

   b. Tighten the rear engine mount base bracket mounting bolts to 36 ft. lbs. (49 Nm).

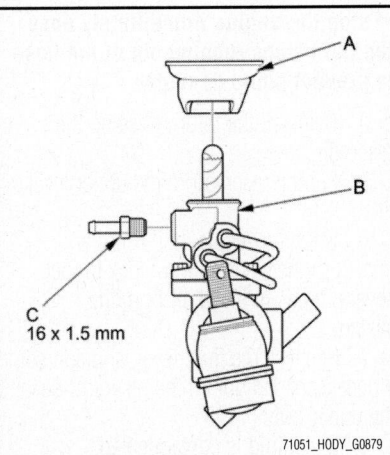

Fig. 191 Remove the pinion shaft grommet (A) from the top of the valve body unit (B); Remove the return line joint (C) if necessary

60. Connect the rear engine mount actuator connector.

61. Install exhaust pipe A.

62. Install the front splash shield.

63. Lower the vehicle.

64. Remove the engine support hanger, the engine hanger balance bar, and engine hanger adapter set.

65. Install the cowl cover.

66. Install the harness bracket to the rear cylinder head.

67. Install the connector bracket to the front cylinder head.

68. Position the cutout of the mounting cushion to the lower side, and install it on the passenger's side of the steering gearbox.

69. Install the steering gearbox mounting bracket over the mounting cushion, and loosely install the steering gearbox mounting bolts.

70. Install the steering gearbox stiffeners, the new steering gearbox mounting bolts and nuts, and the new steering gearbox stiffener mounting bolts on the driver's side of the steering gearbox, then loosely tighten the bolts and nuts.

71. Tighten the steering gearbox mounting bolts and nuts to 64 ft. lbs. (87 Nm), and steering gearbox stiffener mounting bolts to 44 ft. lbs. (59 Nm) alternately in two steps.

72. Connect the inlet line, and tighten the flare nut to 27 ft. lbs. (37 Nm).

73. Install the inlet line clamp bolt.

74. Connect the return hose securely, and tighten the adjustable hose clamp.

75. On both sides, wipe off any grease contamination from the ball joint tapered section and threads. Connect the tie-rod end ball joint to the knuckle. Install the new nut, and tighten it to 44 ft. lbs. (60 Nm).

76. Install a new cotter pin, and bend it.

77. Install the power steering heat shield with the flange bolts, and tighten them to 7 ft. lbs. (9.5 Nm).

78. Remove the vinyl tape around the splines on the pinion shaft.

79. Center the steering rack within its stroke.

80. Cut the wire.

81. Slip the lower end of the steering joint onto the pinion shaft taking care to align the gap within the angle.

82. Remove the steering wheel holder tool.

83. Align the bolt hole on the steering joint with the groove around the pinion shaft, then loosely install the steering joint bolt. Be sure that the joint bolt is securely in the groove in the pinion shaft.

84. Pull on the steering joint to make

sure that the steering joint is fully seated, then tighten the steering joint bolt to 21 ft. lbs. (29 Nm).

85. Install the steering joint cover.

86. Set the carpet in place.

87. Install the driver's heater panel.

88. Install the front wheels, then set the wheels in the straight ahead position.

➡**Before installing the wheel, clean the mating surfaces between the brake disc and the inside of the wheel.**

89. Install the front bulkhead cover.

90. Do the battery terminal reconnection procedure.

91. Install the air intake scoop.

92. Fill the system with power steering fluid, and bleed air from the system.

93. Install the engine cover.

94. After installation, check these items:

   a. Start the engine, allow it to idle, and turn the steering wheel from lock to lock several times to warm up the fluid. Check the steering gearbox for leaks.

   b. Check the steering wheel spoke angle. If steering spoke angles to the right and left are not equal (steering wheel and rack are not centered), correct the engagement of the joint/pinion shaft splines.

   c. Set the steering column to the center tilt position, and to the center telescopic position, then do the front toe inspection.

POWER STEERING PUMP

REMOVAL & INSTALLATION

See Figures 192 and 193.

1. Remove the engine cover.

2. Place a suitable container under the vehicle to catch any spilled fluid.

3. Drain the power steering fluid from the reservoir.

4. Release the front receiver line from the clips.

5. Disconnect the IMT actuator connector.

6. Remove the drive belt from the pump pulley.

7. Cover the auto-tensioner, the alternator, and A/C compressor with several shop towels to protect them from spilled power steering fluid. Disconnect the pump inlet hose and pump outlet hose from the pump, and plug them. Take care not to spill the fluid on the vehicle. Wipe off any spilled fluid at once. Do not turn the steering wheel with the pump removed.

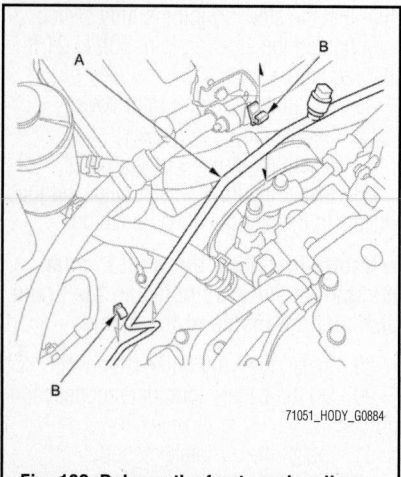

Fig. 192 Release the front receiver line (A) from the clips (B)

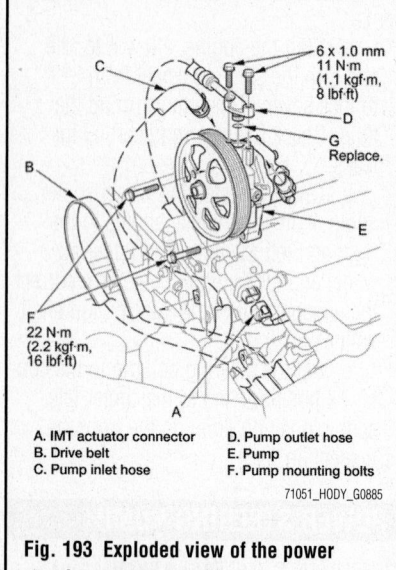

A. IMT actuator connector
B. Drive belt
C. Pump inlet hose
D. Pump outlet hose
E. Pump
F. Pump mounting bolts

6 x 1.0 mm
11 N·m
(1.1 kgf·m,
8 lbf·ft)

22 N·m
(2.2 kgf·m,
16 lbf·ft)

G
Replace.

Fig. 193 Exploded view of the power steering pump assembly

8. Remove the pump mounting bolts, then remove the pump.

9. Cover the opening of the pump to prevent foreign material from entering the pump.

To install:

10. Transfer the pump inlet hose and the pump outlet hose from the original pump onto the new pump with the new O-ring.

11. Loosely install the pump in the pump bracket with the mounting bolts, then tighten the pump fittings to the specified torque.

12. Tighten the pump mounting bolts to the specified torque.

13. Reconnect the IMT actuator connector.

14. Install the drive belt.

15. Note these items during drive belt installation:

a. Inspect the belt for wear and cracks. Replace the belt if necessary.

b. Make sure that the belt is properly positioned on the pulleys.

c. Do not get power steering fluid or grease on the auto-tensioner, alternator, A/C compressor, and drive belt or pulley faces. Clean off any fluid or grease before installation.

16. Install the front receiver line to the clip.

17. Fill the reservoir to the upper level line.

18. Install the engine cover.

19. Start the engine, and check for leaks.

POWER STEERING FLUID

FLUID LEVEL CHECK

See Figure 194.

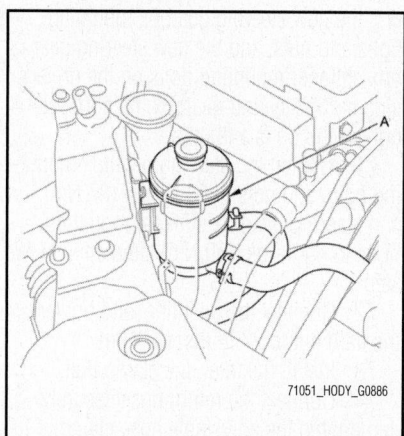

Fig. 194 Check the reservoir (A) at regular intervals, and add the recommended fluid as necessary

Check the reservoir (A) at regular intervals, and add the recommended fluid as necessary. Always use Honda Power Steering Fluid. Using any other type of power steering fluid or automatic transmission fluid can cause increased wear, fluid leaks, and poor steering in cold weather.

➡️**If the fluid is contaminated, the screen in the reservoir may be partially blocked. Inspect the reservoir screen for any debris. If the reservoir screen is clogged, replace the reservoir, and check for the source of the contamination.**

FLUID FILL PROCEDURE

See Figure 195.

Always use Honda Power Steering Fluid. Using any other type of power steering fluid

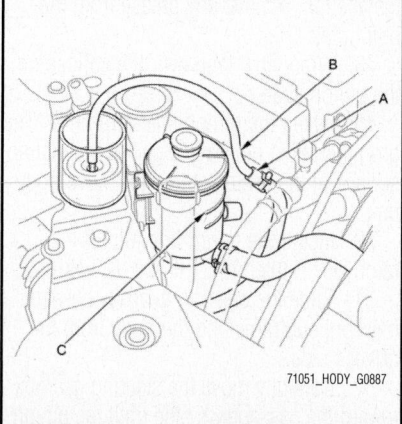

Fig. 195 Disconnect the return hose (A) to drain the reservoir; Connect a hose (B); Fill the reservoir to the upper level line (C)

or automatic transmission fluid can cause increased wear, fluid leaks, and poor steering in cold weather.

1. Disconnect the return hose to drain the reservoir. Take care not to spill the fluid on the vehicle. Wipe off any spilled fluid at once.

➡️**Inspect the reservoir screen for any debris. If the reservoir screen is clogged, replace the reservoir, and check for the source of the contamination.**

2. Connect a hose of suitable diameter to the disconnected return hose, and put the hose end in a suitable container.

3. Start the engine, let it run at idle, and turn the steering wheel from lock to lock several times. When fluid stops running out of the hose, shut off the engine. Discard the fluid.

➡️**Stop the engine immediately once the fluid stops running out of the hose to prevent pump damage.**

4. Reinstall the return hose on the reservoir.

5. Fill the reservoir to the upper level line.

6. Start the engine, and run it at idle. Turn the steering from lock to lock several times to bleed air from the system.

7. Recheck the fluid level, and add some if necessary. Do not fill the reservoir beyond the upper level line.

8. If the fluid is contaminated, dark, or discolored, repeat the procedure as necessary until the system is clean.

KNUCKLE & HUB BEARING UNIT

ADJUSTMENT

1. Raise and support the vehicle.
2. Remove the wheels.
3. Install suitable flat washers and the wheel nuts. Tighten the nuts to the specified torque to hold the brake disc or the brake disc/drum securely against the hub.
4. Attach the dial gauge. Place the dial gauge against the hub flange.
5. Measure the bearing end play while moving the brake disc or the brake disc/drum inward and outward.

➡**Wheel bearing end play: Front/Rear: 0–0.0020 inches (0–0.05 mm)**

6. If the bearing end play measurement is more than the standard, replace the hub bearing unit.

REMOVAL & INSTALLATION

Hub Bearing Unit Replacement

See Figures 196 and 197.

Special Tools Required:
• Ball Joint Thread Protector, 12 mm 07AAF-SDAA100
• Ball Joint Remover, 28 mm 07MAC-SL0A202
• Ball Joint Thread Protector, 14 mm 071AF-SZNA100
• Ball Joint Remover, 32 mm 07MAC-SL0A102

1. Raise and support the vehicle.
2. Remove the wheel nuts and the front wheel.

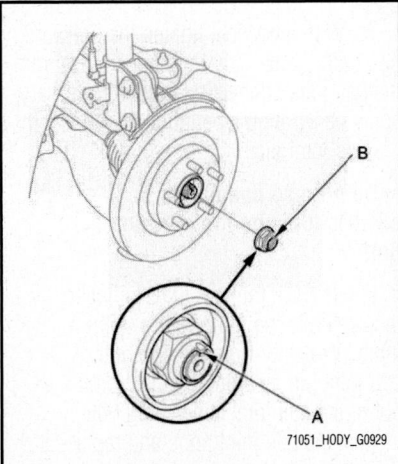

Fig. 196 Pry up the stake (A) on the spindle nut (B), then remove the nut

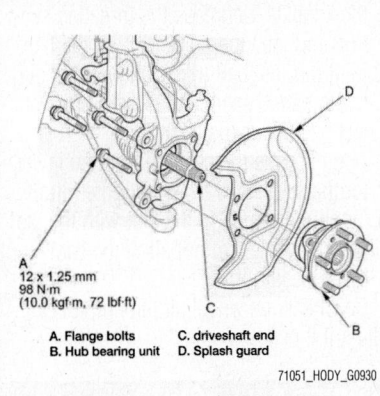

A. Flange bolts
B. Hub bearing unit
C. driveshaft end
D. Splash guard

12 x 1.25 mm
98 N·m
(10.0 kgf·m, 72 lbf·ft)

71051_HODY_G0930

Fig. 197 Remove the flange bolts, and remove the hub bearing unit by tapping the driveshaft end with a soft face hammer while drawing the hub bearing unit outward, then remove the splash guard

3. Remove the brake hose mounting bolt from the damper.
4. Remove the brake caliper bracket mounting bolts, then remove the caliper assembly from the knuckle. To prevent damage to the caliper assembly or brake hose, use a short piece of wire to hang the caliper assembly from the undercarriage. Do not twist the brake hose excessively.
5. Pry up the stake on the spindle nut, then remove the nut.
6. Remove the front brake disc.
7. Remove the flange bolts, and remove the hub bearing unit by tapping the driveshaft end with a soft face hammer while drawing the hub bearing unit outward, then remove the splash guard.
   • Do not pull the driveshaft end outward. The driveshaft inboard joint may come apart.
   • During installation, apply grease to the mating surfaces of the hub bearing unit and driveshaft outboard joint.
8. Check the hub bearing unit for damage and cracks.

To install:

9. Install the hub bearing unit in the reverse order of removal, and note these items:
   a. Use the new spindle nut during reassembly.
   b. Before installing the spindle nut, apply a small amount of engine oil to the seating surface of the nut. After tightening, use a drift to stake the spindle nut shoulder against the driveshaft.

c. Before installing the brake disc, clean the mating surfaces between the hub bearing unit and the inside of the brake disc.
d. Before installing the wheel, clean the mating surfaces between the brake disc and the inside of the wheel.
10. Check the wheel alignment, and adjust it if necessary.

Knuckle Replacement

See Figures 198 through 200.

Special Tools Required:
• Ball Joint Thread Protector, 12 mm 07AAF-SDAA100
• Ball Joint Remover, 28 mm 07MAC-SL0A202
• Ball Joint Thread Protector, 14 mm 071AF-SZNA100

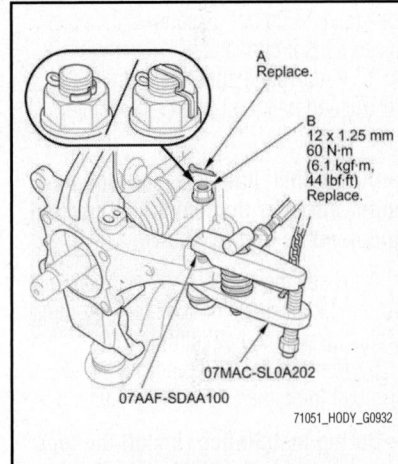

A
Replace.

B
12 x 1.25 mm
60 N·m
(6.1 kgf·m,
44 lbf·ft)
Replace.

07MAC-SL0A202
07AAF-SDAA100

71051_HODY_G0932

Fig. 198 Remove the cotter pin (A) from the tie-rod end ball joint, then remove the nut (B)

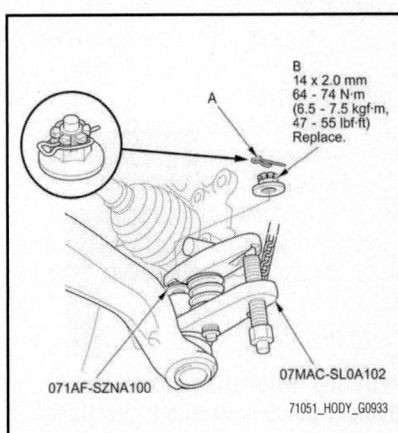

B
14 x 2.0 mm
64 - 74 N·m
(6.5 - 7.5 kgf·m,
47 - 55 lbf·ft)
Replace.

A

071AF-SZNA100
07MAC-SL0A102

71051_HODY_G0933

Fig. 199 Remove the lock pin (A) from the lower arm ball joint, then remove the castle nut (B)

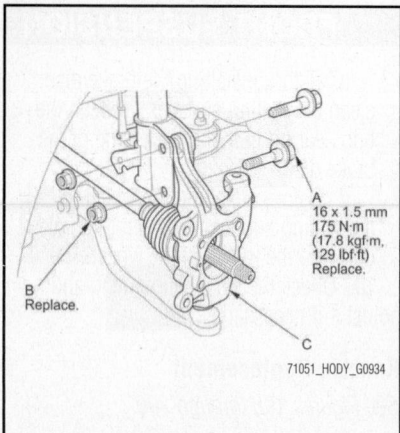

Fig. 200 Remove the damper pinch bolts (A) and flange nuts (B) from the damper, then remove the knuckle (C)

• Ball Joint Remover, 32 mm 07MAC-SL0A102

1. Remove the hub bearing unit.

2. Remove the wheel speed sensor from the knuckle. Do not disconnect the wheel speed sensor connector.

3. Remove the cotter pin from the tie-rod end ball joint, then remove the nut.

➡**During installation, install the new cotter pin after tightening the new nut, and bend its end as shown.**

4. Disconnect the tie-rod end ball joint from the knuckle using the ball joint thread protector and the ball joint remover.

5. Remove the lock pin from the lower arm ball joint, then remove the castle nut.

➡**During installation, install the lock pin as shown after tightening the new castle nut.**

6. Disconnect the lower arm ball joint from the knuckle using the ball joint thread protector and the ball joint remover.
   • Be careful not to damage the ball joint boot when installing the remover.
   • Do not force or hammer on the lower arm, or pry between the lower arm and the knuckle. You could damage the ball joint.

7. Remove the damper pinch bolts and flange nuts from the damper, then remove the knuckle.

To install:

➡**During installation, install new damper pinch bolts and a new flange nuts.**

8. Install the knuckle in the reverse order of removal, and note these items:

a. Be careful not to damage the ball joint boot when connecting the knuckle.

b. Before connecting the ball joint to the knuckle, degrease the threaded section and the tapered portion of the ball joint pin, the ball joint connecting hole, the threaded section and the mating surfaces of the castle nut.

c. Torque the castle nut to the lower torque specification, then tighten it only far enough to align the slot with the ball joint pin hole. Do not align the castle nut by loosening it.

9. Check the wheel alignment, and adjust it if necessary.

LOWER BALL JOINTS

REMOVAL & INSTALLATION

See Figures 201 through 203.

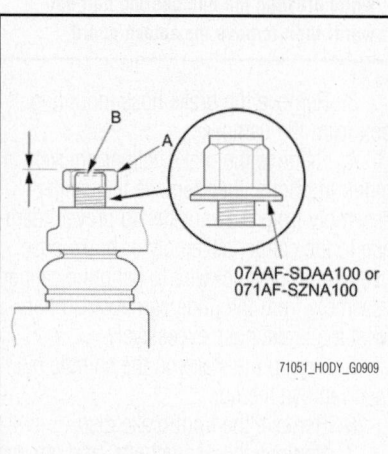

Fig. 201 Install a hex nut (A) or the ball joint thread protector onto the threads of the ball joint (B)

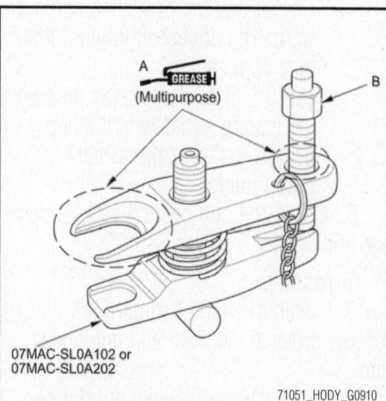

Fig. 202 Apply grease to the ball joint remover on the areas as shown (A); this will ease the installation of the tool, and prevent damage to the pressure bolt (B) threads

Special Tools Required:
• Ball Joint Thread Protector, 12 mm 07AAF-SDAA100
• Ball Joint Thread Protector, 14 mm 071AF-SZNA100
• Ball Joint Remover, 32 mm 07MAC-SL0A102
• Ball Joint Remover, 28 mm 07MAC-SL0A202

➡**Always use a ball joint remover to disconnect a ball joint. Do not strike the housing or any other part of the ball joint connection to disconnect it.**

1. Install a hex nut or the ball joint thread protector onto the threads of the ball joint.

➡**When using a hex nut, make sure the nut is flush with the ball joint pin end to prevent damage to the threaded end of the ball joint pin.**

2. Apply grease to the ball joint remover on the areas as shown (A). This will ease the installation of the tool, and prevent damage to the pressure bolt (B) threads.

3. Loosen the pressure bolt, and install the ball joint remover as shown. Insert the jaws carefully, making sure not to damage the ball joint boot. Adjust the jaw spacing by turning the adjusting bolt.

➡**Fasten the safety chain securely to a suspension arm or the subframe. Do not fasten it to a brake line or wire harness.**

4. After adjusting the adjusting bolt, make sure the head of the adjusting bolt is in the position shown to allow the jaw to pivot.

5. With a wrench, tighten the pressure bolt until the ball joint pin pops loose from the ball joint connecting hole. If necessary, apply penetrating type lubricant to loosen the ball joint pin.

➡**Do not use pneumatic or electric tools on the pressure bolt.**

6. Remove the ball joint remover, then remove the nut or the ball joint thread protector from the end of the ball joint pin, and pull the ball joint out of the ball joint connecting hole. Inspect the ball joint boot, and replace it if damaged.

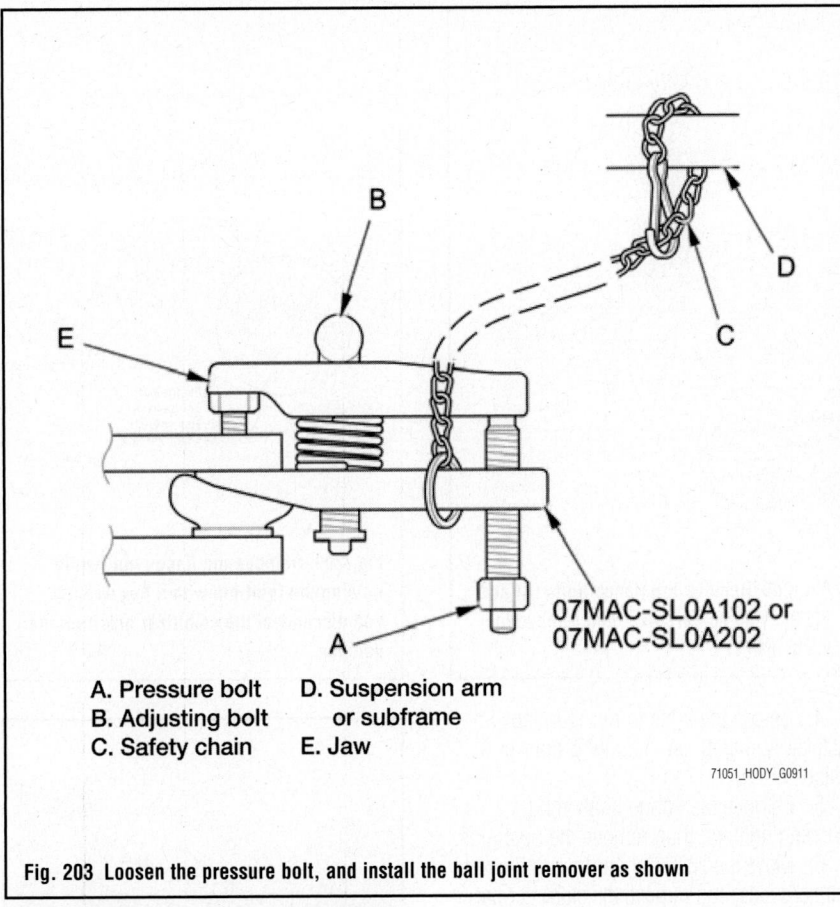

A. Pressure bolt
B. Adjusting bolt
C. Safety chain
D. Suspension arm or subframe
E. Jaw

07MAC-SL0A102 or 07MAC-SL0A202

71051_HODY_G0911

Fig. 203 Loosen the pressure bolt, and install the ball joint remover as shown

LOWER CONTROL ARMS

REMOVAL & INSTALLATION

See Figures 204 and 205.

Special Tools Required:
• Ball Joint Thread Protector, 14 mm 071AF-SZNA100
• Ball Joint Remover, 32 mm 07MAC-SL0A102

1. Raise and support the vehicle.

2. Remove the front wheel.
3. With headlight leveling system: If removing the right side lower arm, disconnect the suspension stroke sensor from the lower arm.
   a. Disconnect the 3P connector from the suspension stroke sensor.
   b. Remove the bolts and the nut, then remove the suspension stroke sensor.
4. Remove the lock pin from the lower arm ball joint, then remove the castle nut.

➡During installation, install the lock pin as shown after tightening the new castle nut.

5. Disconnect the lower arm ball joint from the knuckle using the ball joint thread protector and the ball joint remover.
   • Be careful not to damage the ball joint boot when installing the remover.
   • Do not force or hammer on the lower arm, or pry between the lower arm and the knuckle. You could damage the ball joint.
6. Remove the lower arm mounting bolts, then remove the lower arm from the front subframe.

To install:

➡Use new lower arm mounting bolts during reassembly.

7. Install the lower arm in the reverse order of removal, and note these items:
   a. First install all of the components, and lightly tighten the bolts and the nuts, then raise the suspension to load it with the vehicle's weight before fully tightening to the specified torque. Do not place the jack against the ball joint of the lower arm.
   b. Be careful not to damage the ball joint boot when connecting the knuckle.
   c. Before connecting the ball joint to the knuckle, degrease the threaded section and the tapered portion of the ball joint pin, the ball joint connecting hole, the threaded section and the mating surfaces of the castle nut.
   d. Torque the castle nut to the lower torque specification, then tighten it only far enough to align the slot with the ball joint pin hole. Do not align the castle nut by loosening it.
   e. Before installing the wheel, clean the mating surfaces between the brake disc and the inside of the wheel.
   f. With headlight leveling system: If the right side lower arm was removed, reconnect the suspension stroke sensor.
8. Check the wheel alignment, and adjust it if necessary.
9. With headlight leveling system: If the right side lower arm was removed, do the headlight initial position learning procedure.

STABILIZER BAR & LINKS

REMOVAL & INSTALLATION

Stabilizer Links

See Figure 206.

1. Raise and support the vehicle.
2. Remove the front wheel.

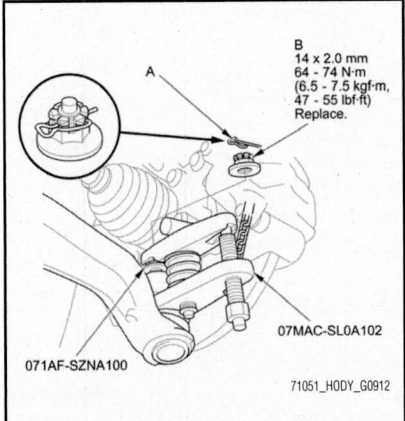

A

B
14 x 2.0 mm
64 - 74 N·m
(6.5 - 7.5 kgf·m, 47 - 55 lbf·ft)
Replace.

07MAC-SL0A102
071AF-SZNA100

71051_HODY_G0912

Fig. 204 Remove the lock pin (A) from the lower arm ball joint, then remove the castle nut (B)

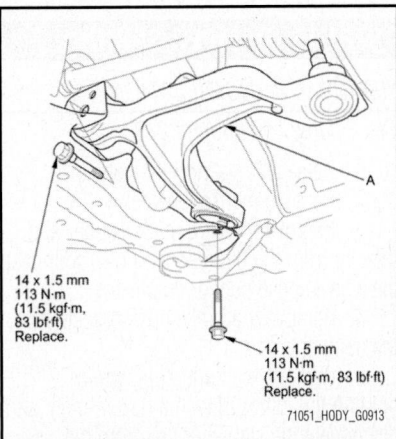

A

14 x 1.5 mm
113 N·m
(11.5 kgf·m, 83 lbf·ft)
Replace.

14 x 1.5 mm
113 N·m
(11.5 kgf·m, 83 lbf·ft)
Replace.

71051_HODY_G0913

Fig. 205 Remove the lower arm mounting bolts, then remove the lower arm (A) from the front subframe

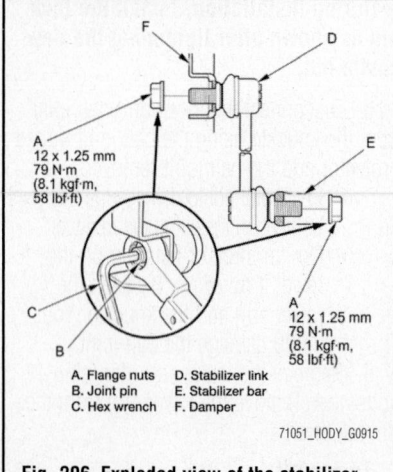

Fig. 206 Exploded view of the stabilizer link assembly

3. Remove the flange nuts while holding the respective joint pin with a hex wrench, then remove the stabilizer link.

To install:

4. Install the stabilizer link on the stabilizer bar and the damper with the joint pins set at the center of their range of movement.

5. Install the flange nuts, and tighten them to the specified torque while holding the respective joint pin with a hex wrench.

6. Clean the mating surfaces between the brake disc and the inside of the wheel, then install the front wheel.

Stabilizer Bar

See Figures 207 and 208.

1. Raise and support the vehicle.
2. Remove the front wheels.
3. Disconnect both sides of the stabilizer link from the stabilizer bar.

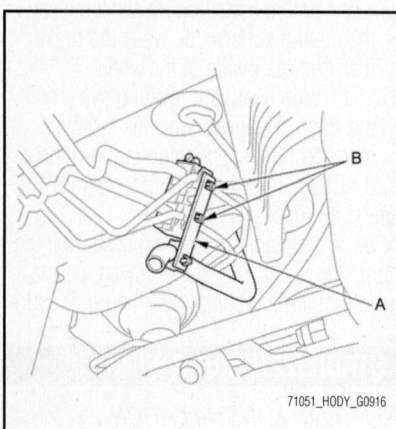

Fig. 207 Detach the clips (B) from mounting bracket A, then move mounting bracket A aside

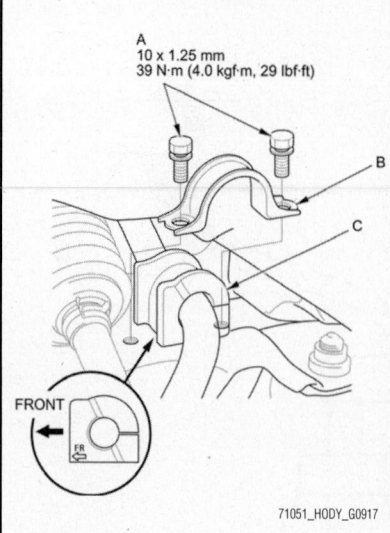

Fig. 208 Remove the flange bolts (A) and the bushing holders (B), then remove the bushings (C)

4. Detach the clips from mounting bracket A, then move mounting bracket A aside.

5. Remove the flange bolts and the bushing holders, then remove the bushings.

6. Move the stabilizer bar toward the passenger's side, and remove the stabilizer bar.

To install:

7. Install all of the removed parts in the reverse order of removal, and note these items:

   a. Note the direction of installation for the bushings.

   b. Refer to stabilizer link removal/installation to connect the stabilizer bar to the links.

   c. Before installing the wheel, clean the mating surfaces between the brake disc and the inside of the wheel.

STRUTS (DAMPER/SPRING)

REMOVAL & INSTALLATION

See Figures 209 through 213.

1. Raise and support the vehicle.
2. Remove the front wheel.
3. Remove the wire guide rubber and the clip from the damper. Do not disconnect the wheel speed sensor connector.

4. Remove the brake hose bracket from the damper.

5. Remove the flange nut, while holding the joint pin with a hex wrench, and disconnect the stabilizer link from the damper.

6. Remove the damper pinch bolts and flange nuts from the damper.

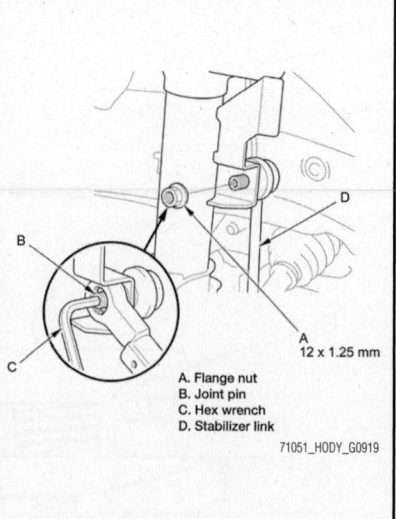

Fig. 209 Remove the flange nut, while holding the joint pin with a hex wrench, and disconnect the stabilizer link from the damper

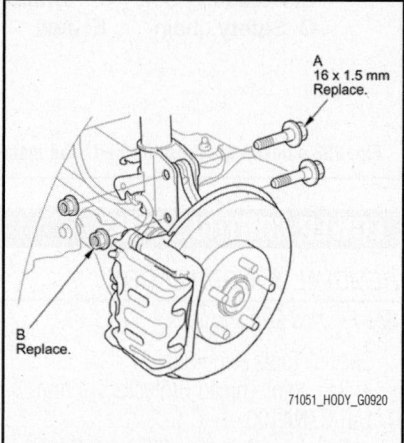

Fig. 210 Remove the damper pinch bolts (A) and flange nuts (B) from the damper

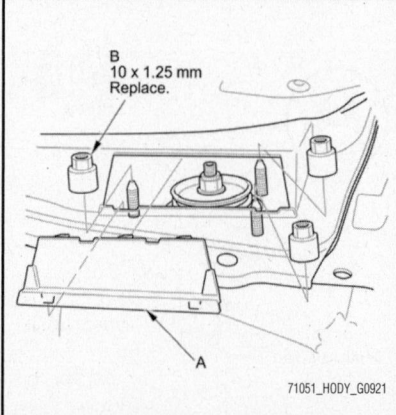

Fig. 211 Remove the cowl top wiper cover (A), then remove the flange nuts (B) from the top of the damper

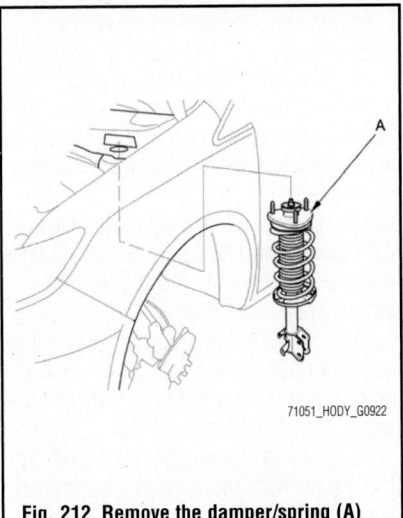

Fig. 212 Remove the damper/spring (A)

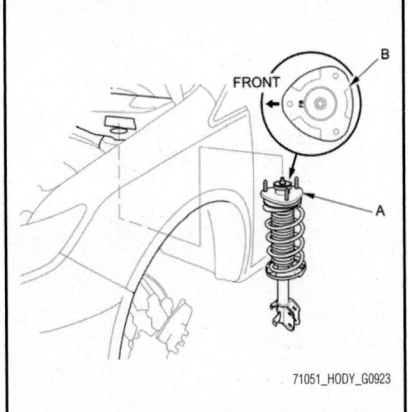

Fig. 213 Install the damper/spring (A) on to the frame; note the direction of the damper mounting base (B) as shown

➡**Do not allow the knuckle to rotate too far outward. This may allow the driveshaft inboard joint to come apart.**

7. Remove the cowl top wiper cover, then remove the flange nuts from the top of the damper. Do not let the damper/spring drop down under its own weight.

8. Remove the damper/spring.

➡**The left and right damper springs are different. Mark the springs L and R before you continue.**

To install:

9. Install the damper/spring on to the frame. Note the direction of the damper mounting base as shown.

10. Loosely install the new flange nuts to the top of the damper.

➡**Install the cowl top wiper cover after tightening the flange nuts to the specified torque.**

11. Loosely install new damper pinch bolts and new flange nuts to the damper.

12. Connect the stabilizer link to the damper, and loosely install the flange nut.

13. Tighten the flange nut to 58 ft. lbs. (79 Nm), while holding the joint pin with the hex wrench.

14. Tighten the flange nuts on top of the damper to 44 ft. lbs. (59 Nm).

15. Tighten the damper pinch bolts while holding the flange nuts to 129 ft. lbs. (175 Nm).

16. Install the brake hose bracket to the damper.

17. Install the wire guide rubber and the clip to the damper.

18. Clean the mating surfaces between the brake disc and the inside of the wheel, then install the front wheel.

19. Check the wheel alignment, and adjust it if necessary.

SUSPENSION

COIL SPRINGS

REMOVAL & INSTALLATION

See Figures 214 through 217.

1. Raise and support the vehicle.
2. Remove the rear wheel.
3. Position a floor jack under lower arm B. Raise the floor jack until the suspension begins to compress.
4. Remove the flange bolt from the knuckle.

5. Lower the floor jack gradually.
6. Remove the spring, the spring mounting cushion, and the lower spring seat.
7. Remove the bump stop from the body if necessary.

To install:

8. Install the bump stop to the body if removed. Tighten the bolt to 32 ft. lbs. (44 Nm).

9. Install the spring, spring mounting cushion, and the lower spring seat. Align the bottom of the spring with the stepped

REAR SUSPENSION

part of the lower spring seat and lower arm B as shown.

10. Position a floor jack under lower arm B. Raise the floor jack until the hole in lower arm B aligns with the hole in the knuckle, then loosely install the new flange bolt to the knuckle.

11. Raise the rear suspension with the floor jack to load the suspension with the vehicle's weight.

12. Tighten the flange bolt to 69 ft. lbs. (93 Nm).

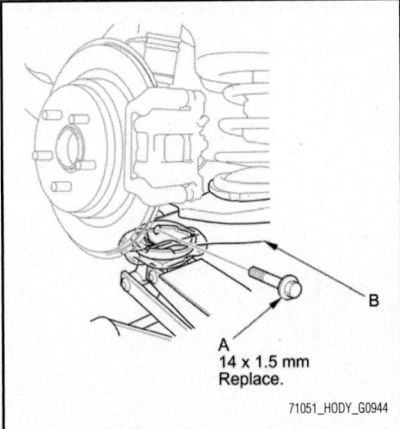

Fig. 214 Position a floor jack under lower arm B; remove the flange bolt (A) from the knuckle

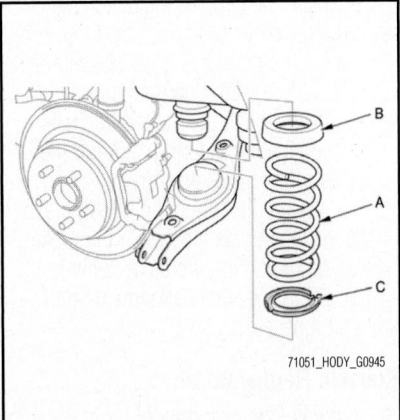

Fig. 215 Remove the spring (A), the spring mounting cushion (B), and the lower spring seat (C)

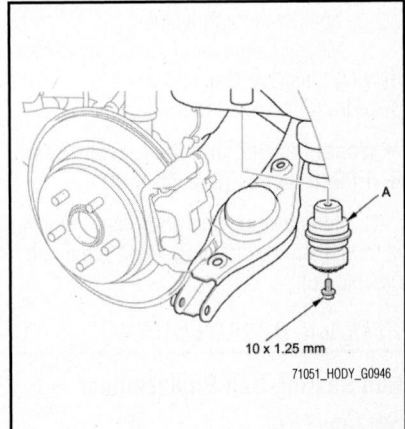

Fig. 216 Remove the bump stop (A) from the body if necessary

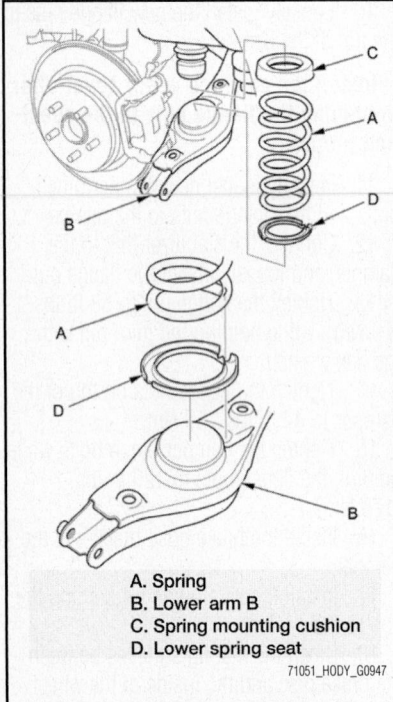

A. Spring
B. Lower arm B
C. Spring mounting cushion
D. Lower spring seat

71051_HODY_G0947

Fig. 217 Install the spring, spring mounting cushion, and the lower spring seat

13. Clean the mating surfaces between the brake disc/drum and the inside of the wheel, then install the rear wheel.

14. Check the wheel alignment, and adjust it if necessary.

KNUCKLE/HUB BEARING UNIT

ADJUSTMENT

1. Raise and support the vehicle.
2. Remove the wheels.
3. Install suitable flat washers and the wheel nuts. Tighten the nuts to the specified torque to hold the brake disc or the brake disc/drum securely against the hub.
4. Attach the dial gauge. Place the dial gauge against the hub flange.
5. Measure the bearing end play while moving the brake disc or the brake disc/drum inward and outward.

➡**Wheel bearing end play: Front/Rear: 0–0.0020 inches (0–0.05 mm)**

6. If the bearing end play measurement is more than the standard, replace the hub bearing unit.

REMOVAL & INSTALLATION

Hub Bearing Unit Replacement

See Figure 218.

Special Tools Required:
• Ball Joint Thread Protector, 12 mm 07AAF-SDAA100

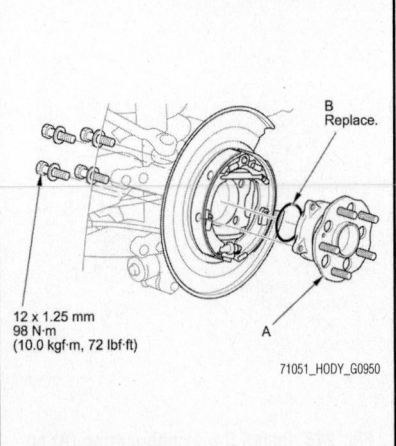

12 x 1.25 mm
98 N·m
(10.0 kgf·m, 72 lbf·ft)

71051_HODY_G0950

Fig. 218 Remove the hub bearing unit (A) and the O-ring (B)

• Ball Joint Remover, 28 mm 07MAC-SL0A202

1. Raise and support the vehicle.
2. Remove the wheel nuts and the wheel.
3. Remove the brake hose mounting bolt.
4. Remove the brake caliper bracket mounting bolts, then remove the caliper assembly from the knuckle. To prevent damage to the caliper assembly or the brake hose, use a short piece of wire to hang the caliper assembly from the undercarriage. Do not twist the brake hose excessively.
5. Remove the two washers.

➡**During installation, make sure the washers are installed between the brake caliper bracket and the knuckle.**

6. Release the parking brake lever fully, and remove the rear brake disc/drum.
7. Remove the hub bearing unit and the O-ring.

To install:

8. Check the hub bearing unit for damage and cracks.
9. Install the hub bearing unit in the reverse order of removal, and note these items:
    a. Use a new O-ring on reassembly.
    b. Before installing the brake disc/drum, clean the mating surfaces between the hub bearing unit and the inside of the brake disc/drum.
    c. Before installing the wheel, clean the mating surfaces between the brake disc/drum and the inside of the wheel.
10. Check the wheel alignment, and adjust it if necessary.

Knuckle Replacement

See Figures 219 through 224.

Special Tools Required:
• Ball Joint Thread Protector, 12 mm 07AAF-SDAA100

• Ball Joint Remover, 28 mm 07MAC-SL0A202

1. Remove the hub bearing unit.
2. Remove the flange nuts, then remove the backing plate with the parking brake shoe assembly. To prevent damage to the backing plate with the parking brake shoes assembly and parking brake cable, use a short piece of wire to hang the backing plate from the undercarriage. Do not twist the parking brake cable excessively.
3. Remove the wheel speed sensor and the brake hose bracket from the knuckle. Do not disconnect the wheel speed sensor connector.
4. Position a floor jack under lower arm B. Raise the floor jack until the suspension begins to compress.
5. Remove the lock pin from the upper arm ball joint, then remove the castle nut.

➡**During installation, install the lock pin as shown after tightening the new castle nut.**

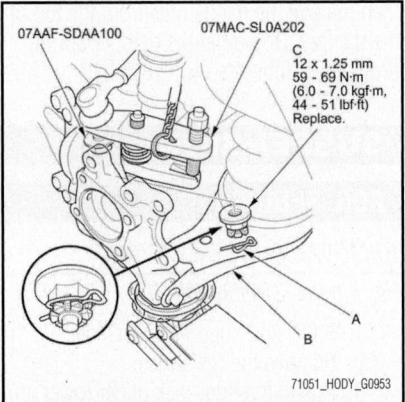

07AAF-SDAA100   07MAC-SL0A202

C
12 x 1.25 mm
59 - 69 N·m
(6.0 - 7.0 kgf·m,
44 - 51 lbf·ft)
Replace.

71051_HODY_G0953

Fig. 219 Position a floor jack under lower arm B; remove the lock pin (A) from the upper arm ball joint, then remove the castle nut (C)

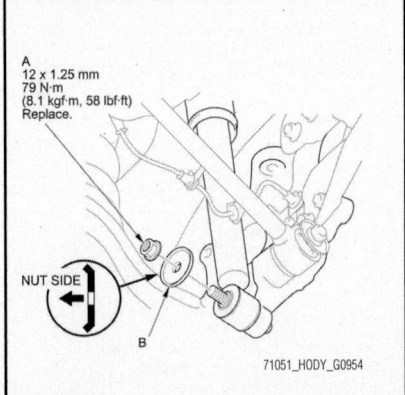

A
12 x 1.25 mm
79 N·m
(8.1 kgf·m, 58 lbf·ft)
Replace.

NUT SIDE

71051_HODY_G0954

Fig. 220 Remove the damper lower flange nut (A) and the washer (B) from the bottom of damper

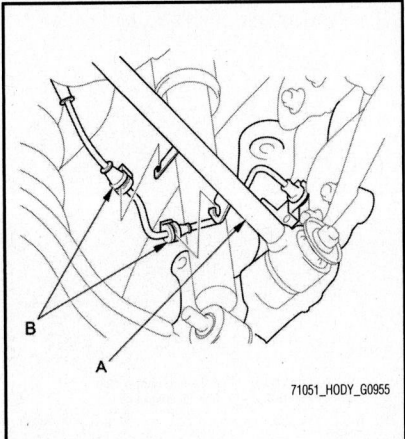

Fig. 221 Remove the wire guide rubbers (B) from lower arm A

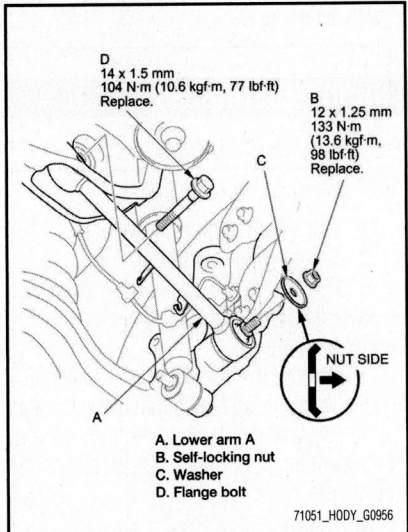

D
14 x 1.5 mm
104 N·m (10.6 kgf·m, 77 lbf·ft)
Replace.

B
12 x 1.25 mm
133 N·m
(13.6 kgf·m,
98 lbf·ft)
Replace.

C

NUT SIDE

A

A. Lower arm A
B. Self-locking nut
C. Washer
D. Flange bolt

Fig. 222 Remove the self-locking nut, the washer, and the flange bolt, then remove lower arm A

6. Disconnect the upper arm ball joint from the knuckle using the ball joint thread protector and the ball joint remover.

➡ **Be careful not to damage the ball joint boot when installing the remover.**

7. Remove the damper lower flange nut and the washer from the bottom of damper.

➡ **During installation, install a new damper lower flange nut.**

8. Remove the wire guide rubbers from lower arm A. Do not disconnect the wheel speed sensor connector.

9. Remove the self-locking nut, the washer, and the flange bolt, then remove lower arm A.

➡ **Use a new self-locking nut and the new flange bolt, during reassembly.**

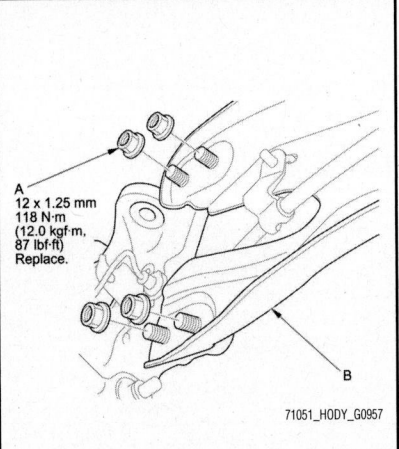

A
12 x 1.25 mm
118 N·m
(12.0 kgf·m,
87 lbf·ft)
Replace.

B

Fig. 223 Remove the flange nuts (A) from the trailing arm (B)

10. Remove the flange nuts from the trailing arm.

➡ **Use new flange nuts during reassembly.**

11. Remove the flange bolt, then remove the knuckle.

➡ **Use the new flange bolt during reassembly.**

To install:

12. Install the knuckle in the reverse order of removal, and note these items:

a. First install all of the components, and lightly tighten the bolts and the nuts, then raise the suspension to load it with the vehicle's weight before fully tightening to the specified torque.

b. Be careful not to damage the ball joint boot when connecting the knuckle.

c. Before connecting the ball joint, degrease the threaded section and the tapered portion of the ball joint pin, the ball joint connecting hole, and the

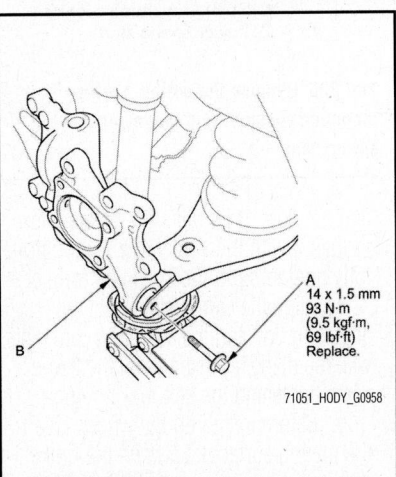

A
14 x 1.5 mm
93 N·m
(9.5 kgf·m,
69 lbf·ft)
Replace.

B

Fig. 224 Remove the flange bolt (A), then remove the knuckle (B)

threaded section and the mating surfaces of the castle nut.

d. Torque the castle nut to the lower torque specification, then tighten it only far enough to align the slot with the ball joint pin hole. Do not align the castle nut by loosening it.

e. Before installing the wheel, clean the mating surfaces between the brake disc/drum and the inside of the wheel.

13. Check the wheel alignment, and adjust it if necessary.

LOWER CONTROL ARMS

REMOVAL & INSTALLATION

Lower Arm A

See Figures 225 and 226.

1. Raise and support the vehicle.
2. Remove the rear wheel.
3. Position a floor jack under lower arm B. Raise the floor jack until the suspension begins to compress.
4. Remove the wire guide rubbers from lower arm A. Do not disconnect the wheel speed sensor connector.
5. Remove the self-locking nut, the washer, and the flange bolt, then remove lower arm A.

➡ **Use a new self-locking nut and the new flange bolt, during assembly.**

To install:

6. Install lower arm A in the reverse order of removal, and note these items:

a. First install all of the components, and lightly tighten the bolt and the nut, then raise the suspension to load it with the vehicle's weight before fully tightening to the specified torque.

b. Before installing the wheel, clean

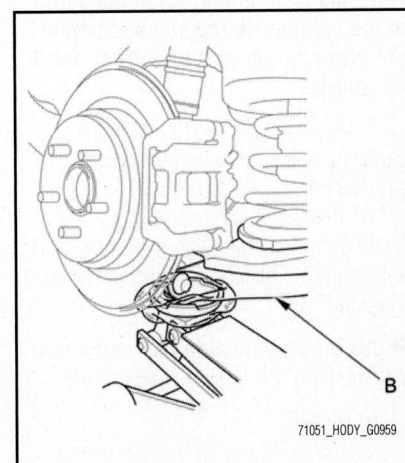

B

Fig. 225 Position a floor jack under lower arm B

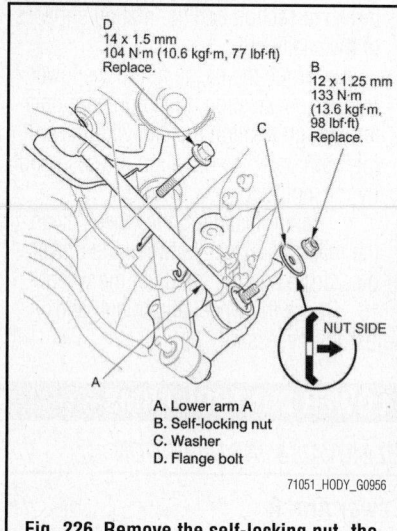

D
14 x 1.5 mm
104 N·m (10.6 kgf·m, 77 lbf·ft)
Replace.

B
12 x 1.25 mm
133 N·m
(13.6 kgf·m,
98 lbf·ft)
Replace.

NUT SIDE

A. Lower arm A
B. Self-locking nut
C. Washer
D. Flange bolt

71051_HODY_G0956

Fig. 226 Remove the self-locking nut, the washer, and the flange bolt, then remove lower arm A

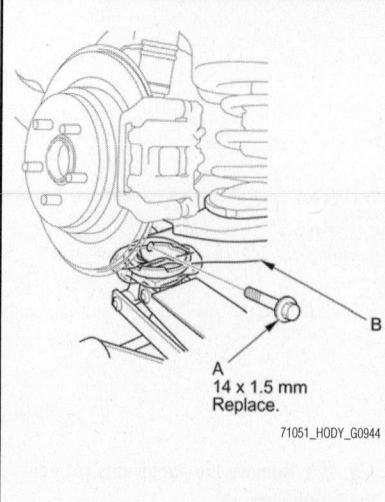

B

A
14 x 1.5 mm
Replace.

71051_HODY_G0944

Fig. 227 Position a floor jack under lower arm B; remove the flange bolt (A) from the knuckle

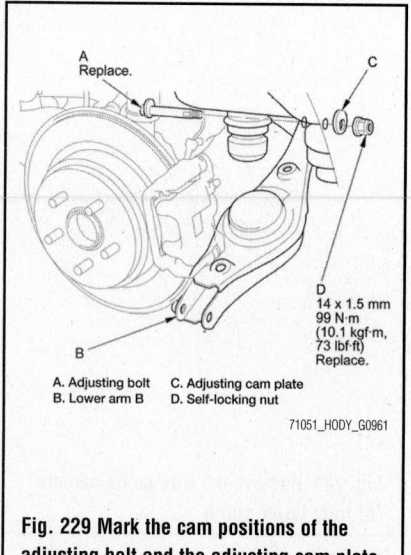

A
Replace.

C

D
14 x 1.5 mm
99 N·m
(10.1 kgf·m,
73 lbf·ft)
Replace.

B

A. Adjusting bolt
B. Lower arm B
C. Adjusting cam plate
D. Self-locking nut

71051_HODY_G0961

Fig. 229 Mark the cam positions of the adjusting bolt and the adjusting cam plate with the frame

the mating surfaces between the brake disc/drum and the inside of the wheel.

7. Check the wheel alignment, and adjust it if necessary.

Lower Arm B

See Figures 227 through 229.

1. Raise and support the vehicle.
2. Remove the rear wheel.
3. Position a floor jack under lower arm B. Raise the floor jack until the suspension begins to compress.
4. Remove the flange bolt from the knuckle.

➡Use the new flange bolt during reassembly.

5. Lower the floor jack gradually.
6. Remove the spring, the spring mounting cushion, and the lower spring seat.

➡During installation, align the bottom of the spring with the stepped part of the lower spring seat and lower arm B as shown.

7. Mark the cam positions of the adjusting bolt and the adjusting cam plate with the frame.

8. Remove the self-locking nut while holding the adjusting bolt, then remove the adjusting cam plate, the adjusting bolt, and lower arm B.

➡Use a new adjusting bolt and a new self-locking nut during reassembly.

To install:

9. Install lower arm B in the reverse order of removal, and note these items:
    a. First install all of the suspension components, and lightly tighten the bolts

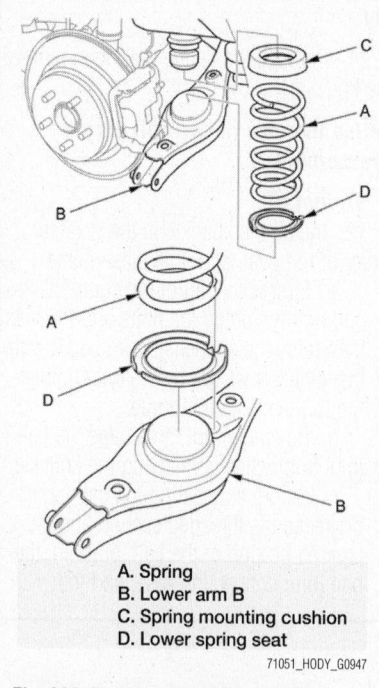

C
A

B

A

D

B

A. Spring
B. Lower arm B
C. Spring mounting cushion
D. Lower spring seat

71051_HODY_G0947

Fig. 228 Remove the spring, spring mounting cushion, and the lower spring seat

and the nuts, then raise the suspension to load it with the vehicle's weight before fully tightening to the specified torque.

    b. Align the cam positions of the adjusting bolt and the adjusting cam plate with the marked positions on the frame when tightening the self-locking nut.

    c. Before installing the wheel, clean the mating surfaces between the brake disc/drum and the inside of the wheel.

10. Check the wheel alignment, and adjust it if necessary.

SHOCK ABSORBERS

REMOVAL & INSTALLATION

See Figures 230 through 232.

1. Raise and support the vehicle.
2. Remove the rear wheel.
3. Position a floor jack under lower arm B. Raise the floor jack until the suspension begins to compress.
4. Remove the damper lower flange nut and the washer from the bottom of the damper.
5. Remove the flange nuts from the top of the damper, then remove the damper from the vehicle.

To install:

6. Position the damper between the body and knuckle.
7. Loosely install the flange nuts to the top of the damper.

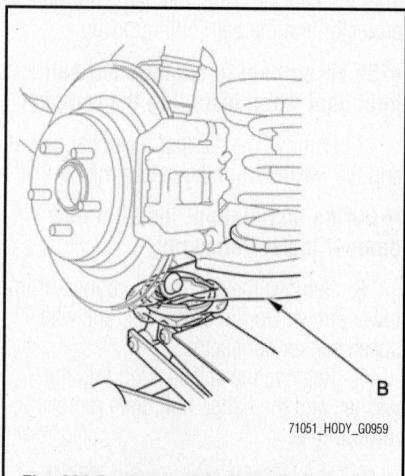

B

71051_HODY_G0959

Fig. 230 Position a floor jack under lower arm B

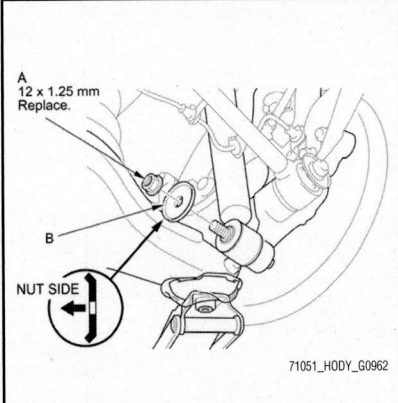

Fig. 231 Remove the damper lower flange nut (A) and the washer (B) from the bottom of the damper

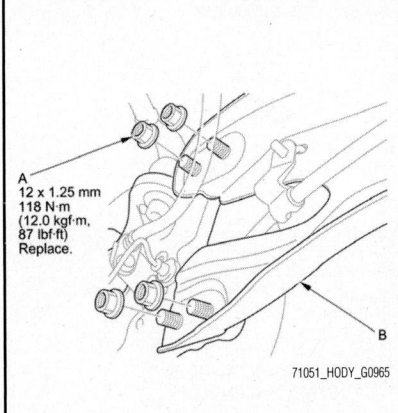

Fig. 233 Remove the flange nuts (A) from the trailing arm (B)

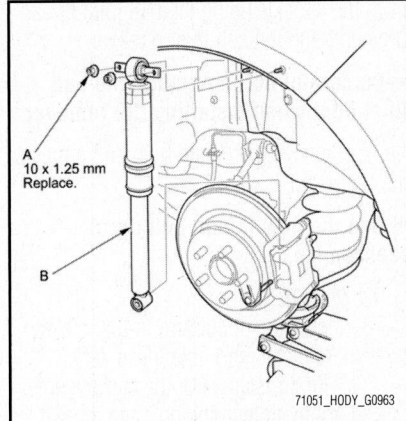

Fig. 232 Remove the flange nuts (A) from the top of the damper (B), then remove the damper from the vehicle

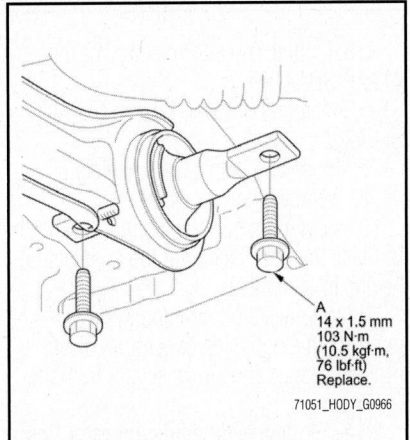

Fig. 234 Remove the flange bolts (A) from the trailing arm, then remove the trailing arm

8. Install the washer and loosely install the new flange nut to the bottom of the damper.

9. Raise the rear suspension with the floor jack to load the suspension with the vehicle's weight.

10. Tighten the flange nuts on the top and bottom of the damper to the specified torque.

    a. Tighten the top damper flange nuts to 32 ft. lbs. (44 Nm).

    b. Tighten the bottom flange nut to 58 ft. lbs. (79 Nm).

11. Clean the mating surfaces between the brake disc/drum and the inside of the wheel, then install the rear wheel.

12. Check the wheel alignment, and adjust it if necessary.

TRAILING ARMS

REMOVAL & INSTALLATION

See Figures 233 and 234.

1. Raise and support the vehicle.
2. Remove the rear wheel.

3. Remove the parking brake cable mounting nuts (A) from the trailing arm (B).

4. Remove the flange nuts from the trailing arm.

➡**Use new flange nuts during reassembly.**

5. Remove the flange bolts from the trailing arm, then remove the trailing arm.

➡**Use new flange bolts during reassembly.**

To install:

6. Install the trailing arm in the reverse order of removal, and note these items:

    a. First install all of the components, and lightly tighten the bolts and the nuts, then raise the suspension to load it with the vehicle's weight before fully tightening to the specified torque.

    b. Before installing the wheel, clean the mating surfaces between the brake disc/drum and the inside of the wheel.

7. Check the wheel alignment, and adjust it if necessary.

UPPER BALL JOINTS

REMOVAL & INSTALLATION

See Figures 235 through 237.

Special Tools Required:
- Ball Joint Thread Protector, 12 mm 07AAF-SDAA100
- Ball Joint Thread Protector, 14 mm 071AF-SZNA100
- Ball Joint Remover, 32 mm 07MAC-SL0A102
- Ball Joint Remover, 28 mm 07MAC-SL0A202

➡**Always use a ball joint remover to disconnect a ball joint. Do not strike the housing or any other part of the ball joint connection to disconnect it.**

1. Install a hex nut or the ball joint thread protector onto the threads of the ball joint.

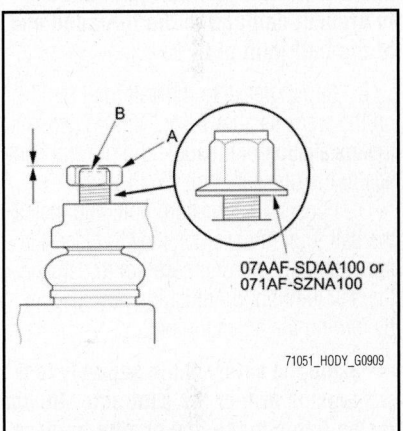

Fig. 235 Install a hex nut (A) or the ball joint thread protector onto the threads of the ball joint (B)

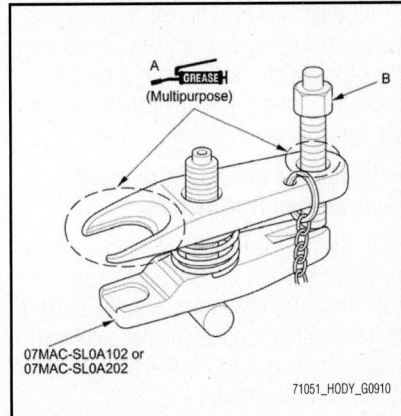

Fig. 236 Apply grease to the ball joint remover on the areas as shown (A); this will ease the installation of the tool, and prevent damage to the pressure bolt (B) threads

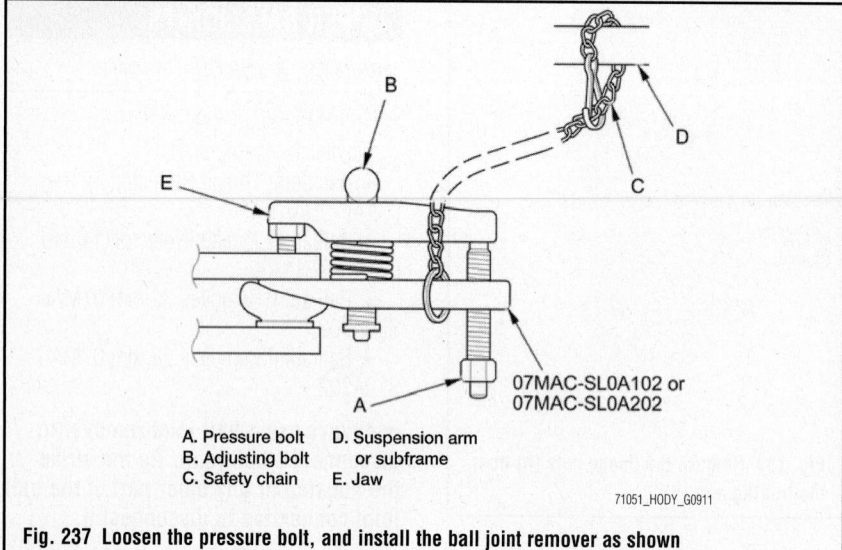

A. Pressure bolt
B. Adjusting bolt
C. Safety chain
D. Suspension arm or subframe
E. Jaw

07MAC-SL0A102 or 07MAC-SL0A202

71051_HODY_G0911

Fig. 237 Loosen the pressure bolt, and install the ball joint remover as shown

➡ **When using a hex nut, make sure the nut is flush with the ball joint pin end to prevent damage to the threaded end of the ball joint pin.**

2. Apply grease to the ball joint remover on the areas as shown (A). This will ease the installation of the tool, and prevent damage to the pressure bolt (B) threads.

3. Loosen the pressure bolt, and install the ball joint remover as shown. Insert the jaws carefully, making sure not to damage the ball joint boot. Adjust the jaw spacing by turning the adjusting bolt.

➡ **Fasten the safety chain securely to a suspension arm or the subframe. Do not fasten it to a brake line or wire harness.**

4. After adjusting the adjusting bolt, make sure the head of the adjusting bolt is in the position shown to allow the jaw to pivot.

5. With a wrench, tighten the pressure bolt until the ball joint pin pops loose from the ball joint connecting hole. If necessary, apply penetrating type lubricant to loosen the ball joint pin.

➡ **Do not use pneumatic or electric tools on the pressure bolt.**

6. Remove the ball joint remover, then remove the nut or the ball joint thread protector from the end of the ball joint pin, and pull the ball joint out of the ball joint connecting hole. Inspect the ball joint boot, and replace it if damaged.

UPPER ARMS

REMOVAL & INSTALLATION

See Figures 238 and 239.

Special Tools Required:

- Ball Joint Thread Protector, 12 mm 07AAF-SDAA100
- Ball Joint Remover, 28 mm 07MAC-SL0A202

1. Raise and support the vehicle.
2. Remove the rear wheel.
3. Position a floor jack under lower arm B. Raise the floor jack until the suspension begins to compress.
4. With headlight leveling system: If removing the right side upper arm, disconnect the suspension stroke sensor from the upper arm.

    a. Disconnect the 3P connector from the suspension stroke sensor.

    b. Remove the bolt and the nut, then remove the suspension stroke sensor.

5. Remove the lock pin from the upper arm ball joint, then remove the castle nut.

➡ **During installation, install the lock pin as shown after tightening the new castle nut.**

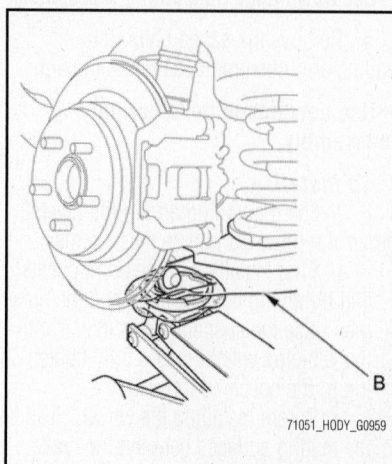

71051_HODY_G0959

Fig. 238 Position a floor jack under lower arm B

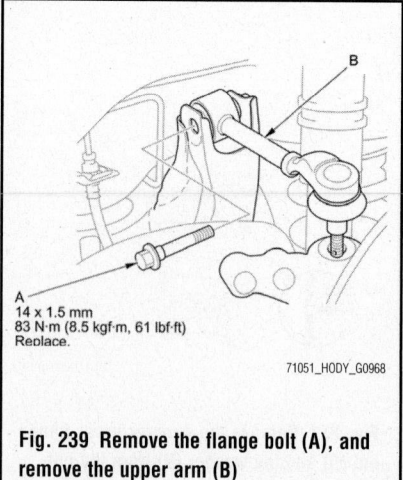

A
14 x 1.5 mm
83 N·m (8.5 kgf·m, 61 lbf·ft)
Replace.

71051_HODY_G0968

Fig. 239 Remove the flange bolt (A), and remove the upper arm (B)

6. Disconnect the upper arm ball joint from the knuckle using the ball joint thread protector and the ball joint remover.

➡ **Be careful not to damage the ball joint boot when installing the remover.**

7. Remove the flange bolt, and remove the upper arm.

➡ **Use the new flange bolt during reassembly.**

To install:

8. Install the upper arm in the reverse order of removal, and note these items:

    a. First install all of the components, and lightly tighten the bolt and the nut, then raise the suspension to load it with the vehicle's weight before fully tightening to the specified torque.

    b. Be careful not to damage the ball joint boot when connecting the knuckle.

    c. Before connecting the ball joint to the knuckle, degrease the threaded section and the tapered portion of the ball joint pin, the ball joint connecting hole, the threaded section and the mating surfaces of the castle nut.

    d. Torque the castle nut to the lower torque specification, then tighten it only far enough to align the slot with the ball joint pin hole. Do not align the castle nut by loosening it.

    e. Before installing the wheel, clean the mating surfaces between the brake disc/drum and the inside of the wheel.

    f. With headlight leveling system: If the right side upper arm was removed, reconnect the suspension stroke sensor.

9. Check the wheel alignment, and adjust it if necessary.

10. With headlight leveling system: If removing the right side upper arm, do the headlight initial position learning procedure.

HONDA

Pilot

15

BRAKES 15-9

ANTI-LOCK BRAKE SYSTEM..15-9
General Information..................15-9
Precautions...........................15-9
Wheel Speed Sensors 15-9
Removal & Installation....15-9
BLEEDING THE BRAKE
SYSTEM 15-11
Bleeding Procedure................15-11
Bleeding the ABS
System 15-11
Fluid Fill Procedure 15-11
Manual 15-11
FRONT DISC BRAKES 15-11
Brake Calipers.......................15-11
Removal & Installation........15-11
Brake Pads 15-11
Removal & Installation........15-11
PARKING BRAKE.............15-14
Parking Brake Cable...............15-14
Adjustments.......................15-14
Parking Brake Shoes 15-14
Adjustments.......................15-14
Removal & Installation........15-14
REAR DISC BRAKES 15-12
Brake Calipers.......................15-12
Removal & Installation........15-12
Brake Pads 15-12
Removal & Installation........15-12

CHASSIS ELECTRICAL 15-16

AIR BAGS (SUPPLEMENTAL
RESTRAINT SYSTEM) 15-16
Arming the System 15-16
Clockspring
Centering 15-16
Disarming the
System 15-16
Service Precautions 15-16

DRIVE TRAIN 15-17

Automatic Transmission
Fluid....................15-17
Drain & Refill...................15-17
Filter Replacement 15-17

Fluid Level Check...............15-17
Fluid Recommendations 15-18
Driveshaft (Propeller
Shaft) 15-18
Removal & Installation........15-18
Halfshafts 15-19
Removal & Installation........15-19
Transfer Case 15-21
Drain & Refill...................15-22
Removal & Installation........15-21

ENGINE COOLING 15-23

Electric Engine
Fan......................15-24
Removal & Installation.......15-24
Engine Coolant......................15-23
Drain & Refill...................15-23
Radiator...........................15-24
Removal & Installation........15-24
Thermostat 15-25
Removal & Installation........15-25
Water Pump 15-26
Removal & Installation........15-26

ENGINE ELECTRICAL 15-26

BATTERY SYSTEM...........15-26
Battery................................15-26
Battery Reconnect/Relearn
Procedure 15-26
Removal & Installation........15-26
CHARGING SYSTEM 15-27
Alternator 15-27
Removal & Installation........15-27
IGNITION SYSTEM 15-27
Adjustment.......................15-27
Firing Orders.....................15-27
Ignition Coil 15-27
Removal & Installation........15-27
Ignition Timing 15-27
Adjustment 15-27
Inspection 15-27
Spark Plugs.......................15-27
Removal & Installation........15-27
STARTING SYSTEM 15-28
Starter 15-28
Removal & Installation........15-28

ENGINE MECHANICAL......15-28

Accessory Drive Belt
System 15-28
Adjustment 15-28
Belt Routings 15-28
Inspection 15-28
Removal & Installation........15-28
Air Cleaner 15-29
Removal & Installation........15-29
Camshaft & Bearings 15-29
Removal & Installation........15-29
Crankshaft Front Seal.............15-31
Removal & Installation........15-31
Crankshaft Pulley 15-31
Removal & Installation........15-31
Cylinder Head 15-32
Removal & Installation.......15-32
Cylinder Head Covers 15-42
Removal & Installation........15-42
Engine Oil & Filter 15-33
OIL & Filter Change.............15-33
Exhaust Manifold 15-34
Component Locations.........15-34
Front Oil Seal 15-34
Removal & Installation.......15-34
Intake Manifold 15-34
Removal & Installation........15-34
Intake Manifold Cover 15-35
Component Locations.........15-35
Oil Pan...........................15-35
Removal & Installation........15-35
Oil Pump...........................15-36
Removal & Installation........15-36
Pistons & Rings 15-37
Positioning 15-37
Rear Main Seal 15-37
Removal & Installation........15-37
Rocker Arms.........................15-38
Removal & Installation........15-38
Timing Belt & Sprockets 15-39
Removal & Installation........15-39
Timing Belt Front Cover 15-38
Removal & Installation........15-38
Valve Lash (Clearance)
Adjustment..........................15-44
Adjustment 15-44

ENGINE PERFORMANCE & EMISSION CONTROLS 15-46

Accelerator Pedal Position
  (APP) Sensor 15-46
    Location 15-46
    Removal & Installation........15-46
Camshaft Position (CMP)
  Sensor 15-46
    Location 15-46
    Removal & Installation........15-46
Crankshaft Position (CKP)
  Sensor 15-46
    CKP Pattern Clear/CKP
      Pattern Learn Procedure ...15-46
    Removal & Installation........15-46
Engine Coolant Temperature
  (ECT) Sensor 15-47
    Removal & Installation........15-47
EVAP Canister 15-47
    Location 15-47
    Removal & Installation........15-47
Exhaust Gas Recirculation
  (EGR) Valve 15-47
    Location 15-47
    Removal & Installation........15-47
Heated Oxygen Sensor
  (HO2S) 15-47
    Location 15-47
    Removal & Installation........15-47
Intake Air Temperature (IAT)
  Sensor 15-48
    Location 15-48
    Removal & Installation........15-48
Knock Sensor (KS) 15-48
    Location 15-48
    Removal & Installation........15-48
Maintenance Reminder
  Lights 15-49
    Reset Procedure..................15-49
Manifold Absolute Pressure
  (MAP) Sensor 15-49
    Removal & Installation........15-49
Mass Air Flow (MAF) Sensor ...15-49
    Location 15-49
    Removal & Installation........15-49
Positive Crankcase Ventilation
  (PCV) Valve 15-49
    Location 15-49
    Removal & Installation........15-49

Powertrain Control Module
  (PCM) 15-50
    Location..............................15-50
    PCM Idle Learn
      Procedure 15-51
    PCM Update 15-51
    Removal & Installation........15-50

FUEL SYSTEM 15-52

GASOLINE FUEL INJECTION SYSTEM 15-52
  Fuel Filter 15-53
    Removal & Installation........15-53
  Fuel Pressure Regulator 15-54
    Removal & Installation........15-54
  Fuel Rail & Injectors 15-55
    Removal & Installation........15-55
  Fuel System Service
    Precautions 15-52
  Fuel Tank Unit 15-53
    Removal & Installation........15-53
  Idle Speed 15-53
    Adjustment 15-53
  Relieving Fuel System
    Pressure..............................15-52
  Throttle Body........................15-55
    Removal & Installation........15-55

HEATING & AIR CONDITIONING 15-56

  Blower Motor 15-56
    Removal & Installation........15-56
  Heater Core 15-57
    Removal & Installation........15-57

PRECAUTIONS 15-9

SCHEDULE MAINTENANCE MINDER 15-59

SPECIFICATIONS 15-3

  Brake Specifications..................15-7
  Camshaft & Bearings
    Specifications 15-5
  Capacities 15-4
  Crankshaft & Connecting Rod
    Specifications 15-5

Engine and Model Year
  Identification 15-3
Engine Tune-Up Specifications ...15-3
Fluid Specifications..................15-4
General Engine Specifications...15-3
Maintenance Intervals 15-8
Piston & Ring Specifications 15-5
Tire Wheel and Ball Joint
  Specifications 15-7
Torque Specifications 15-6
Valve Specifications 15-4
Wheel Alignment......................15-6

STEERING 15-59

Power Rack & Pinion Steering
  Gear 15-59
    Removal & Installation........15-59
Power Steering Pump..............15-63
    Fluid Level Check 15-63
    Fluid Replacement 15-64
    Removal & Installation........15-63

SUSPENSION 15-64

FRONT SUSPENSION 15-64
  Knuckle 15-64
    Removal & Installation........15-64
  Lower Ball Joints....................15-65
    Removal & Installation........15-65
  Lower Control Arms 15-65
    Removal & Installation........15-65
  Stabilizer Bar & Links 15-66
    Removal & Installation........15-66
  Struts...................................15-70
    Removal & Installation........15-70
  Wheel Hubs & Bearings 15-71
    Adjustment 15-71
    Removal & Installation........15-71
REAR SUSPENSION 15-71
  Coil Springs...........................15-71
    Removal & Installation........15-71
  Control Links 15-72
    Removal & Installation........15-72
  Shock Absorbers.....................15-74
    Removal & Installation........15-74
  Stabilizer Bar & Links 15-75
    Removal & Installation........15-75
  Wheel Hubs & Bearing............15-75
    Removal & Installation........15-75

ENGINE AND VEHICLE IDENTIFICATION CHART

Engine Code							Model Year	
Code	Liters (cc)	Cu. In.	Cyl.	Fuel Sys.	Engine Type	Eng. Mfg.	Code ①	Year
J35Z4	3.5 (3471)	212	6	SMFI	SOHC	Honda	B	2011
							C	2012

SOHC: Single Overhead Cam

SMFI: Sequential Multi-port Fuel Injection

① 10th position of VIN

71051_PILO_C0001

GENERAL ENGINE SPECIFICATIONS

Year	Model	Engine Displacement Liters (ID)	Net Horsepower @ rpm	Net Torque @ rpm (ft. lbs.)	Bore x Stroke (in.)	Com- pression Ratio	Oil Pressure @ rpm
2011	Pilot	3.5 (J35Z4)	250@5700	253@4800	3.50x3.66	10.0:1	71@3000
2012	Pilot	3.5 (J35Z4)	250@5700	253@4800	3.50x3.66	10.0:1	71@3000

NA: Not Available

71051_PILO_C0002

ENGINE TUNE-UP SPECIFICATIONS

Year	Engine Displacement Liters (ID)	Spark Plug Gap (in.)	Ignition Timing (deg.)		Fuel Pump (psi)	Idle Speed (rpm)		Valve Clearance (in.)	
			MT	AT		MT	AT	In.	Ex.
2011	3.5 (J35Z4)	0.039-0.043	NA	8-12B	57-64	NA	650-750	0.008-0.009	0.011-0.013
2012	3.5 (J35Z4)	0.039-0.043	NA	8-12B	57-64	NA	650-750	0.008-0.009	0.011-0.013

NOTE: The Vehicle Emission Control Information label often reflects changes made during production and must be used if they differ from this chart.

NOTE: Pressure with fuel pressure gauge connected

NA: Not Applicable

B: Before top dead center

71051_PILO_C0003

CAPACITIES

Year	Model	Engine Displacement Liters (ID)	Engine Oil with Filter (qts.)	Transmission (pts.) 5-Spd	Transmission (pts.) Auto.	Transfer Case (pts.)	Drive Axle Front (pts.)	Drive Axle Rear (pts.)	Fuel Tank (gal.)	Cooling System (qts.)
2011	Pilot	3.5 (J35Z4)	4.5	NA	3.6	0.9	NA	5.6	21.0	8.0
2012	Pilot	3.5 (J35Z4)	4.5	NA	3.6	0.9	NA	5.6	21.0	8.0

NOTE: All capacities are approximate. Add fluid gradually and check to be sure a proper fluid level is obtained.

NA: Not Applicable

71051_PILO_C0004

FLUID SPECIFICATIONS

Year	Model	Engine Displacement Liters (ID)	Engine Oil	Auto. Trans.	Transfer Case	Rear Differential	Power Steering Fluid	Brake Master Cylinder
2011	Pilot	3.5 (J35Z4)	①	②	③	④	⑤	⑥
2012	Pilot	3.5 (J35Z4)	①	②	③	④	⑤	⑥

DOT: Department Of Transportation

Note: If specification disagrees with specification in owners manual, use specification in owners manaual

① Honda Motor Oil: 0w-20

② Acura ATF-Z1 fluid

③ API classified GL4 or GL5 only. SAE 90 viscosity.

④ Honda VTM-4 differential fluid

⑤ Honda power steering fluid

⑥ Honda DOT 3 Brake Fluid

71051_PILO_C0012

VALVE SPECIFICATIONS

Year	Engine Displacement Liters (ID)	Seat Angle (deg.)	Face Angle (deg.)	Spring Test Pressure (lbs. @ in.)	Spring Installed Height (in.)	Stem-to-Guide Clearance (in.) Intake	Stem-to-Guide Clearance (in.) Exhaust	Stem Diameter (in.) Intake	Stem Diameter (in.) Exhaust
2011	3.5 (J35Z4)	NA	NA	NA	NA	0.0008-0.0018	0.0022-0.0032	0.2159-0.2163	0.2146-0.2150
2102	3.5 (J35Z4)	NA	NA	NA	NA	0.0008-0.0018	0.0022-0.0032	0.2159-0.2163	0.2146-0.2150

NA: Not Available

71051_PILO_C0005

CAMSHAFT SPECIFICATIONS
All measurements in inches unless noted

Year	Model	Engine Displacement Liters (ID)	Journal Dia.	Brg. Oil Clearance	Shaft End-play	Circle Runout	Lobe Height Intake	Lobe Height Exhaust
2011	Pilot	3.5 (J35Z4)	NA	0.00197-0.0035	0.0020-0.0079	NA	①	②
2012	Pilot	3.5 (J35Z4)	NA	0.00197-0.0035	0.0020-0.0079	NA	①	②

NA: Not Available
① Cylinders 1,2,3,4: 1.38433 inches
   Cylinders 5, 6: 1.38405 inches
② Cylinders 1,2,3,4: 1.43846 inches
   Cylinders 5, 6: 1.43748 inches

71051_PILO_C0006

CRANKSHAFT AND CONNECTING ROD SPECIFICATIONS
All measurements are given in inches

Year	Engine Displacement Liters (ID)	Crankshaft Main Brg. Journal Dia.	Crankshaft Main Brg. Oil Clearance	Crankshaft Shaft End-play	Crankshaft Thrust on No.	Connecting Rod Journal Diameter	Connecting Rod Oil Clearance	Connecting Rod Side Clearance
2011	3.5 (J35Z4)	2.8337-2.8346	0.0008-0.0018	0.0039-0.0138	3	2.283	0.0008-0.0017	0.0059-0.0138
2012	3.5 (J35Z4)	2.8337-2.8346	0.0008-0.0018	0.0039-0.0138	3	2.283	0.0008-0.0017	0.0059-0.0138

71051_PILO_C0007

PISTON AND RING SPECIFICATIONS
All measurements are given in inches

Year	Engine Displacement Liters (ID)	Piston Clearance	Ring Gap Top Compression	Ring Gap Bottom Compression	Ring Gap Oil Control	Ring Side Clearance Top Compression	Ring Side Clearance Bottom Compression	Ring Side Clearance Oil Control
2011	3.5 (J35Z4)	0.0006-0.0016	0.0079-0.0138	0.0157-0.0217	0.0079-0.0276	0.0022-0.0032	0.0012-0.0022	NA
2012	3.5 (J35Z4)	0.0006-0.0016	0.0079-0.0138	0.0157-0.0217	0.0079-0.0276	0.0022-0.0032	0.0012-0.0022	NA

NA: Not Available

71051_PILO_C0008

TORQUE SPECIFICATIONS
All readings in ft. lbs.

Year	Engine Displacement Liters (ID)	Cylinder Head Bolts	Main Bearing Bolts	Rod Bearing Bolts	Crankshaft Damper Bolts	Flywheel Bolts	Manifold Intake	Manifold Exhaust	Spark Plugs	Oil Pan Drain Plug
2011	3.5 (J35Z4)	①	②	③	④	54	16	23	16	29
2012	3.5 (J35Z4)	①	②	③	④	54	16	23	16	29

NOTE: Dip main bearing bolts and crankshaft damper bolt in clean engine oil prior to tightening.

① Step 1: 22 ft. lbs.
   Step 2: plus 90 degrees
   Step 3: plus 90 degrees
   Step 4: If using a new bolt, plus 90 degrees

② Cap bolts: 54 ft. lbs.
   Side bolts: 36 ft. lbs.

③ Step 1: 15 ft. lbs.
   Step 2: 90 degrees

④ Step 1: 47 ft. lbs.
   Step 2: plus 60 degrees

71051_PILO_C0009

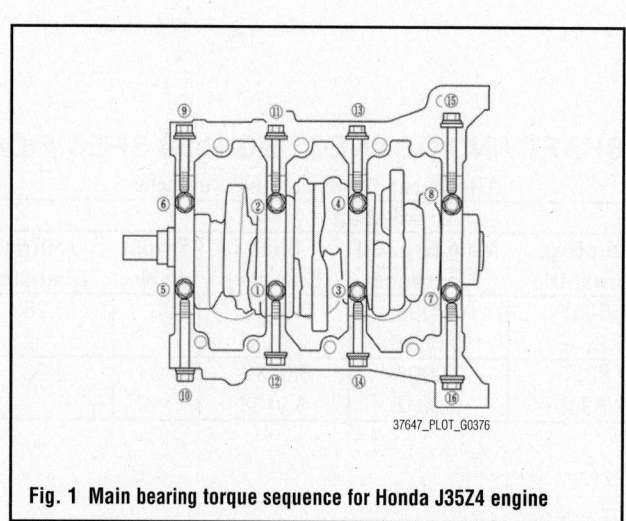

37647_PLOT_G0376

Fig. 1 Main bearing torque sequence for Honda J35Z4 engine

WHEEL ALIGNMENT

Year	Model	Position	Caster Range (+/-Deg.)	Caster Preferred Setting (Deg.)	Camber Range (+/-Deg.)	Camber Preferred Setting (Deg.)	Toe-in (in.)
2011	Pilot	F	0.35	4.11	0.45	-0.30	0+/-0.08
		R	—	—	0	-0.30	0.08+/-0.08
2012	Pilot	F	0.35	4.17	0.45	-0.30	0+/-0.08
		R	—	—	0	-0.30	0.08+/-0.08

71051_PILO_C0010

TIRE, WHEEL AND BALL JOINT SPECIFICATIONS

Year	Model	OEM Tires		Tire Pressures (psi)		Wheel Size	Ball Joint Inspection	Lug Nut (ft. lbs.)
		Standard	Optional	Front	Rear			
2011	Pilot	P245/65R17	None	33	33	R17	NS	80
2012	Pilot	P235/65R17	None	33	33	R17	NS	80

OEM: Original Equipment Manufacturer

PSI: Pounds Per Square Inch

NS: Not specified by manufacturer

71051_PILO_C0011

BRAKE SPECIFICATIONS

All measurements in inches unless noted

Year	Model	Position	Brake Disc			Minimum Lining Thickness		Brake Caliper	
			Original Thickness	Minimum Thickness	Maximum Runout	Front	Rear	Bracket Bolts (ft. lbs.)	Mounting Bolts (ft. lbs.)
2011	Pilot	F	1.100	1.024	0.0016	0.063	—	101	53
		R	0.430	0.354	0.0016	—	0.063	65	27
2012	Pilot	F	1.100	1.024	0.0016	0.063	—	101	53
		R	0.430	0.354	0.0016	—	0.063	65	27

F: Front

R: Rear

71051_PILO_C0013

MAINTENANCE MINDER SCHEDULE
Honda Pilot

All Honda's displays engine oil life and maintenance service items in the information display to indicate when to perform maintenance service. If the engine oil life is 15% or less, based on the onboard computer's caluculations, you will see SERVICE DUE SOON in the information display every time the ignition key is turned to ON. The maintenance minder indicator will also come on and the maintenance code(s) for other scheduled maintenance items needing service will be displayed below the message.

Symbol	Item	Service
A	Engine oil ①	Change
B	Engine oil and filter	Change
	Fluid levels	Inspect
	Brakes	Inspect
	Parking brake adjustment	Check
	Steering gear and linkage	Inspect
	Suspension components	Inspect
	Driveshaft boots	Inspect
	Brake hoses and lines	Inspect
	Exhaust system	Inspect
	Fuel lines and connections	Inspect
1	Tires	Rotate
2	Engine air filter ②	Replace
	Dust and pollen filter ③	Replace
	Accessory drive belt	Inspect
3	Transmission fluid ④	Replace
	Transfer case fluid ④	Replace
4	Spark plugs	Replace
	Timing belt ⑤	Replace
	Water pump	Inspect
	Valve clearance ⑥	Inspect
5	Engine coolant	Replace
6	VTM-4 rear differential fluid	Replace

① If the message SERVICE DUE NOW does not appear more than 12 months after the display is reset, change every year.

② If driven in dusty conditions, replace every 15,000 miles.

③ If driven in urban areas that have a high concentration of soot from industry and diesel, replace every 15,000 miles

④ If regularly driven in mountainous areas at very low speed or trailer towing, change the fluid every 30,000 miles.

⑤ If driven regularly in temperatures over 110 deg.F or below -20 deg.F, or towing a trailer, replace every 60,000 miles.

⑥ Adjust if necessary.

Additionally, replace the brake fluid every 3 years, and inspect the idle speed every 160,000 miles.
To reset the Engine Oil Life Display:
1. Turn the ignition switch to ON.
2. Press the SELECT button repeatedly until the engine oil life display or the service message is displayed.
3. Press the RESET button for about 10 seconds. You will see a MAINT RESET message.
4. Select the appropriate answer, MAINT RESET >N (NO) or MAINT RESET > y (YES) by pressing the SELECT button repeatedly.
    >N or >Y is displayed on the outside temperature >N or >Y is displayed on the outside temperature display.
5. Select the MAINT RESET > Y (YES), and press and hold the RESET button again to reset the engine oil life to 100%.

PRECAUTIONS

Before servicing any vehicle, please be sure to read all of the following precautions, which deal with personal safety, prevention of component damage, and important points to take into consideration when servicing a motor vehicle:

• Never open, service or drain the radiator or cooling system when the engine is hot; serious burns can occur from the steam and hot coolant.

• Observe all applicable safety precautions when working around fuel. Whenever servicing the fuel system, always work in a well-ventilated area. Do not allow fuel spray or vapors to come in contact with a spark, open flame, or excessive heat (a hot drop light, for example). Keep a dry chemical fire extinguisher near the work area. Always keep fuel in a container specifically designed for fuel storage; also, always properly seal fuel containers to avoid the possibility of fire or explosion. Refer to the additional fuel system precautions later in this section.

• Fuel injection systems often remain pressurized, even after the engine has been turned **OFF**. The fuel system pressure must be relieved before disconnecting any fuel lines. Failure to do so may result in fire and/or personal injury.

• Brake fluid often contains polyglycol ethers and polyglycols. Avoid contact with the eyes and wash your hands thoroughly after handling brake fluid. If you do get brake fluid in your eyes, flush your eyes with clean, running water for 15 minutes. If eye irritation persists, or if you have taken brake fluid internally, IMMEDIATELY seek medical assistance.

• The EPA warns that prolonged contact with used engine oil may cause a number of skin disorders, including cancer. You should make every effort to minimize your exposure to used engine oil. Protective gloves should be worn when changing oil. Wash your hands and any other exposed skin areas as soon as possible after exposure to used engine oil. Soap and water, or waterless hand cleaner should be used.

• All new vehicles are now equipped with an air bag system, often referred to as a Supplemental Restraint System (SRS) or Supplemental Inflatable Restraint (SIR) system. The system must be disabled before performing service on or around system components, steering column, instrument panel components, wiring and sensors. Failure to follow safety and disabling procedures could result in accidental air bag deployment, possible personal injury and unnecessary system repairs.

• Always wear safety goggles when working with, or around, the air bag system. When carrying a non-deployed air bag, be sure the bag and trim cover are pointed away from your body. When placing a non-deployed air bag on a work surface, always face the bag and trim cover upward, away from the surface. This will reduce the motion of the module if it is accidentally deployed. Refer to the additional air bag system precautions later in this section.

• Clean, high quality brake fluid from a sealed container is essential to the safe and proper operation of the brake system. You should always buy the correct type of brake fluid for your vehicle. If the brake fluid becomes contaminated, completely flush the system with new fluid. Never reuse any brake fluid. Any brake fluid that is removed from the system should be discarded. Also, do not allow any brake fluid to come in contact with a painted surface; it will damage the paint.

• Never operate the engine without the proper amount and type of engine oil; doing so WILL result in severe engine damage.

• Timing belt maintenance is extremely important. Many models utilize an interference-type, non-freewheeling engine. If the timing belt breaks, the valves in the cylinder head may strike the pistons, causing potentially serious (also time-consuming and expensive) engine damage. Refer to the maintenance interval charts for the recommended replacement interval for the timing belt, and to the timing belt section for belt replacement and inspection.

• Disconnecting the negative battery cable on some vehicles may interfere with the functions of the on-board computer system(s) and may require the computer to undergo a relearning process once the negative battery cable is reconnected.

• When servicing drum brakes, only disassemble and assemble one side at a time, leaving the remaining side intact for reference.

• Only an MVAC-trained, EPA-certified automotive technician should service the air conditioning system or its components.

BRAKES

GENERAL INFORMATION

PRECAUTIONS

• Certain components within the ABS system are not intended to be serviced or repaired individually.

• Do not use rubber hoses or other parts not specifically specified for and ABS system. When using repair kits, replace all parts included in the kit. Partial or incorrect repair may lead to functional problems and require the replacement of components.

• Lubricate rubber parts with clean, fresh brake fluid to ease assembly. Do not use shop air to clean parts; damage to rubber components may result.

• Use only DOT 3 brake fluid from an unopened container.

• If any hydraulic component or line is removed or replaced, it may be necessary to bleed the entire system.

• A clean repair area is essential. Always clean the reservoir and cap thoroughly before removing the cap. The slightest amount of dirt in the fluid may plug an orifice and impair the system function. Perform repairs after components have been thoroughly cleaned; use only denatured alcohol to clean components. Do not allow ABS components to come into contact with any substance containing mineral oil; this includes used shop rags.

• The Anti-Lock control unit is a microprocessor similar to other computer units in the vehicle. Ensure that the ignition switch is **OFF** before removing or installing controller harnesses. Avoid static electricity discharge at or near the controller.

ANTI-LOCK BRAKE SYSTEM

• If any arc welding is to be done on the vehicle, the control unit should be unplugged before welding operations begin.

WHEEL SPEED SENSORS

REMOVAL & INSTALLATION

Front

See Figure 2.

1. Turn the ignition switch to LOCK (0).
2. Release the clamp, then disconnect the wheel speed sensor connector.
3. Remove the bolts and the wheel speed sensor.

To install:

4. Install the wheel speed sensor in the reverse order of removal, and note these items:

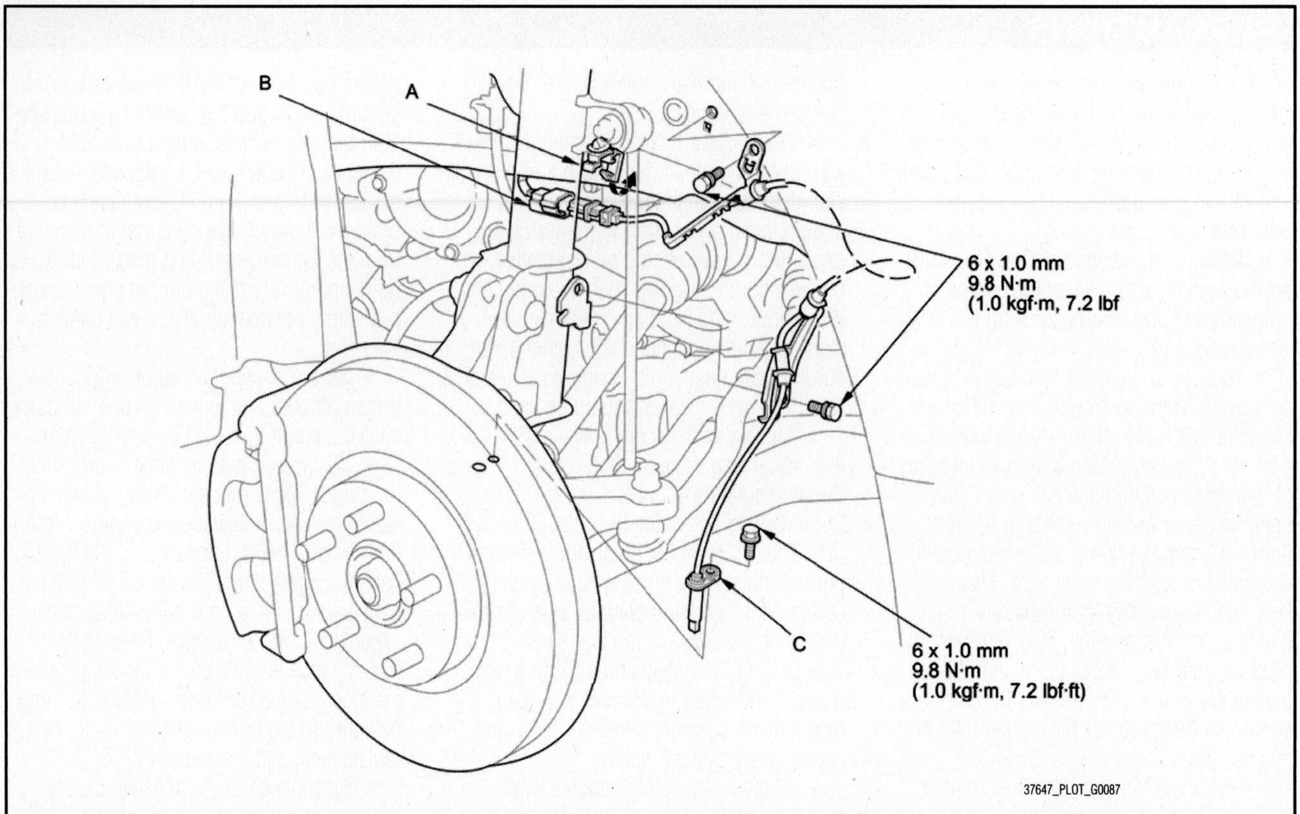

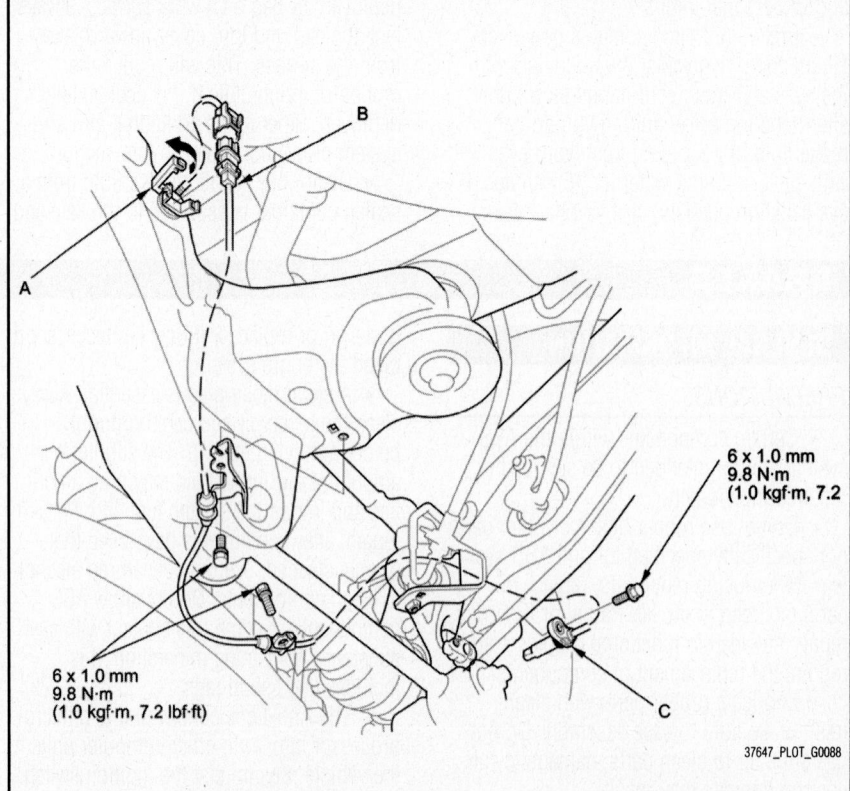

6 x 1.0 mm
9.8 N·m
(1.0 kgf·m, 7.2 lbf

6 x 1.0 mm
9.8 N·m
(1.0 kgf·m, 7.2 lbf·ft)

6 x 1.0 mm
9.8 N·m
(1.0 kgf·m, 7.2 lbf·ft)

37647_PLOT_G0087

Fig. 2 Release the clamp (A), then disconnect the wheel speed sensor connector (B); remove the bolts and the wheel speed sensor (C)

- Install the sensor carefully to avoid twisting the wires.
- If the wheel speed sensor comes in contact with the hub bearing unit, it is faulty.

5. Start the engine, and make sure the ABS and the VSA indicators go off.

6. Test-drive the vehicle, and make sure the ABS and the VSA indicators do not come on.

Rear

See Figure 3.

1. Turn the ignition switch to LOCK (0).

2. Release the clamp, then disconnect the wheel speed sensor connector.

3. Remove the bolts and the wheel speed sensor.

4. Install the wheel speed sensor in the reverse order of removal, and note these items:

- Install the sensor carefully to avoid twisting the wires.
- If the wheel speed sensor comes in contact with the hub bearing unit, it is faulty.

5. Start the engine, and make sure the ABS and the VSA indicators go off.

6. Test-drive the vehicle, and make sure the ABS and the VSA indicators do not come on.

37647_PLOT_G0088

Fig. 3 Release the clamp (A), then disconnect the wheel speed sensor connector (B); remove the bolts and the wheel speed sensor (C)

BRAKES

BLEEDING THE BRAKE SYSTEM

BLEEDING PROCEDURE

MANUAL

See Figure 4.

➡ **Do not spill brake fluid on the vehicle; it may damage the paint. If brake fluid gets on the paint, wash it off immediately with water.**

➡ **Note the following when bleeding the brake system:**

- Do not reuse the drained fluid. Use only new Honda DOT 3 Brake Fluid from an unopened container. Using a non-Honda brake fluid can cause corrosion and shorten the life of the system.
- Make sure no dirt or other foreign matter is allowed to contaminate the brake fluid.
- The reservoir connected to the master cylinder must be at the MAX (upper) level mark at the start of the bleeding procedure, and checked after bleeding each wheel location. Add fluid as required.

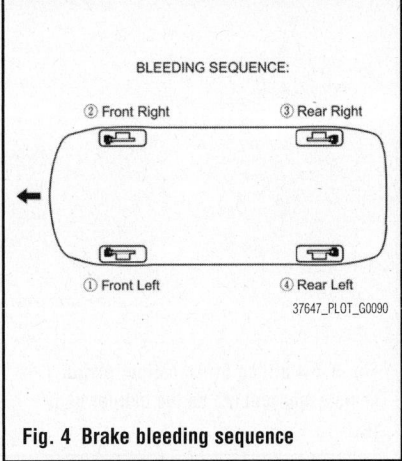

BLEEDING SEQUENCE:

② Front Right   ③ Rear Right

① Front Left   ④ Rear Left

37647_PLOT_G0090

Fig. 4 Brake bleeding sequence

1. Make sure the brake fluid level in the reservoir is at the MAX (upper) level line.
2. Have someone slowly pump the brake pedal several times, then apply steady pressure.
3. Start the bleeding at the driver's side of the front brake system.

➡ **Bleed the calipers in the sequence shown.**

4. Attach a length of clear drain tube to the bleed screw, then loosen the bleed screw to allow air to escape from the system. Then tighten the bleed screw securely.
5. Refill the master cylinder reservoir to the MAX (upper) level line.
6. Repeat the procedure for each brake circuit until there are no air bubbles in the fluid.

BLEEDING THE ABS SYSTEM

Bleeding the ABS brake system is done in the same manner as bleeding a conventional brake system.

FLUID FILL PROCEDURE

Honda DOT 3 Brake Fluid: Always use Honda DOT 3 Brake Fluid. Using a non-Honda brake fluid can cause corrosion and decrease the life of the system.

1. Unscrew the top from the master cylinder reservoir.
2. Carefully pour in the brake fluid until it reached the MAX (upper) line.
3. Replace the master cylinder reservoir cap.
4. Wipe up any spills.

BRAKES

FRONT DISC BRAKES

BRAKE CALIPERS

REMOVAL & INSTALLATION

See Figure 5.

➡ **Keep any grease off the brake disc and brake pads.**

1. Raise the front of the vehicle, and support it with safety stands in the proper locations.
2. Remove the front wheel.
3. Remove the brake hose mounting bolt.
4. Remove the brake caliper bracket mounting bolts, then remove the caliper assembly from the knuckle.
5. Disconnect the flexible brake line from the caliper.

To install:

6. Installation is the reverse of removal.
7. Tighten the caliper mounting bolts to 101 ft. lbs. (137 Nm).

BRAKE PADS

REMOVAL & INSTALLATION

See Figures 6 through 9.

1. Remove some brake fluid from the master cylinder.

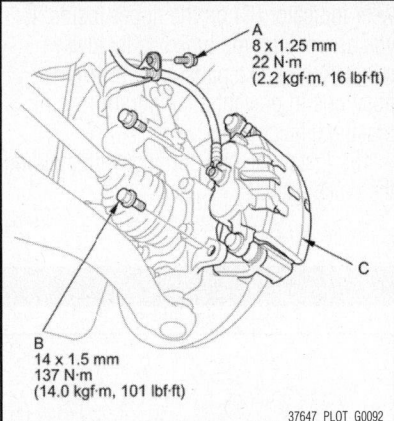

A
8 x 1.25 mm
22 N·m
(2.2 kgf·m, 16 lbf·ft)

B
14 x 1.5 mm
137 N·m
(14.0 kgf·m, 101 lbf·ft)

C

37647_PLOT_G0092

Fig. 5 Remove the brake hose mounting bolt (A), the brake caliper bracket mounting bolts (B), then remove the caliper assembly (C)

2. Raise and support the vehicle.
3. Remove the front wheels.
4. Remove the flange bolt, and pivot the caliper up out of the way. Check the hose and the pin boots for damage and deterioration.
5. Remove the pad springs while holding the brake pads.

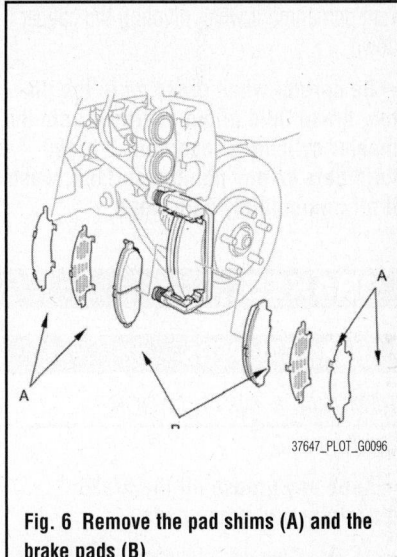

A

A

B

37647_PLOT_G0096

Fig. 6 Remove the pad shims (A) and the brake pads (B)

6. Remove the pad shims and the brake pads.
7. Remove the pad retainers.

➡ **The upper and lower pad retainers are different. During installation, make sure the pad retainers are in proper positions.**

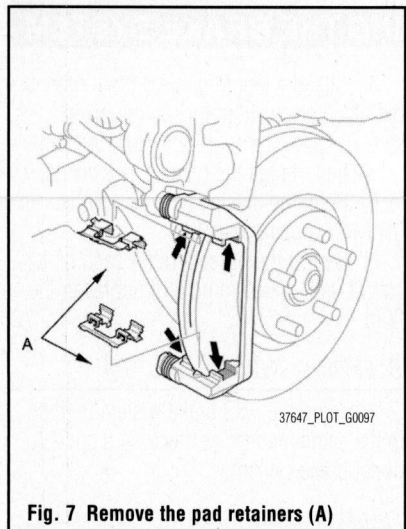

Fig. 7 Remove the pad retainers (A)

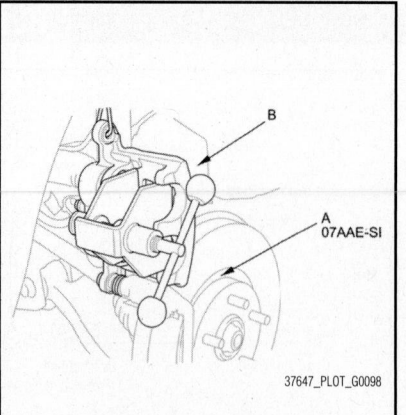

Fig. 8 Install the brake caliper piston compressor tool (A) on the caliper body (B)

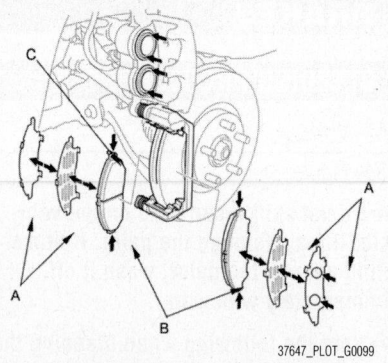

Fig. 9 Apply a thin coat of M-77 assembly paste (P/N 08798-9010) to the pad side of the shims (A), the back of the brake pads (B) and the other areas indicated by the arrows

To install:

8. Apply a thin coat of M-77 assembly paste (P/N 08798-9010) to the retainer mating surface of the caliper bracket.

9. Install the pad retainers. Wipe excess assembly paste off the retainers. Keep the assembly paste off the brake disc and brake pads.

10. Install the brake caliper piston compressor tool on the caliper body.

11. Press in the piston with the brake caliper piston compressor tool so the caliper will fit over the brake pads. Make sure the piston boot is in position to prevent damaging it when pivoting the caliper down.

➡Be careful when pressing in the piston; brake fluid might overflow from the master cylinder's reservoir. If brake fluid gets on any painted surface, wash it off immediately with water.

12. Remove the brake caliper piston compressor tool.

13. Apply a thin coat of M-77 assembly paste (P/N 08798-9010) to the pad side of the shims, the back of the brake pads and the other areas indicated by the arrows. Wipe excess assembly paste off the pad shims and the brake pads friction material. Keep grease and assembly paste off the brake disc and brake pads. Contaminated brake disc or brake pads reduce stopping ability.

14. Install the brake pads and pad shims correctly. Install the brake pad with the wear indicator (C) on the upper inside. If you are reusing the brake pads, always reinstall the brake pads in their original positions to prevent a temporary loss of braking efficiency.

15. Install the pad springs while holding the brake pads.

16. Pivot the caliper down into position. while holding the brake pads. Install the flange bolt, and tighten it to 53 ft. lbs. (72 Nm).

17. Clean the mating surfaces between the brake disc and the inside of the wheel, then install the front wheels.

18. Press the brake pedal several times to make sure the brakes work.

➡Engagement may require a greater pedal stroke immediately after the brake pads have been replaced as a set. Several applications of the brake pedal will restore the normal pedal stroke.

19. Add brake fluid as needed.

20. After installation, check for leaks at hose and line joints or connections, and retighten if necessary.

21. Test-drive the vehicle, then recheck for leaks.

BRAKES

BRAKE CALIPERS

REMOVAL & INSTALLATION

See Figure 10.

➡Keep any grease off the brake disc/drum and brake pads.

1. Raise the rear of the vehicle, and support it with safety stands in the proper locations.

2. Remove the rear wheel.

3. Release the parking brake pedal fully.

4. Remove the brake caliper bracket mounting bolts, and remove the caliper assembly from the knuckle.

5. Disconnect the flexible brake line from the caliper.

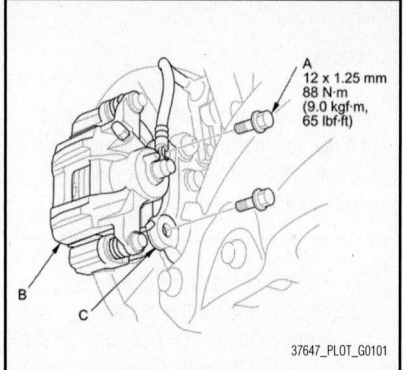

Fig. 10 Remove the brake caliper bracket mounting bolts (A), and remove the caliper assembly (B) from the knuckle

REAR DISC BRAKES

6. Installation is the reverse of removal.

7. Tighten the caliper mounting bolts to 65 ft. lbs. (88 Nm).

BRAKE PADS

REMOVAL & INSTALLATION

See Figures 11 through 13.

1. Remove some brake fluid from the master cylinder.

2. Raise and support the vehicle.

3. Remove the rear wheels.

4. Remove the flange bolt and pivot the caliper up out of the way. Check the hose and pin boots for damage and deterioration.

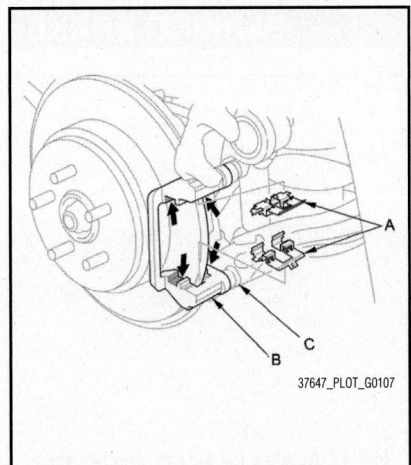

Fig. 11 Remove the upper and lower pad retainers (A)

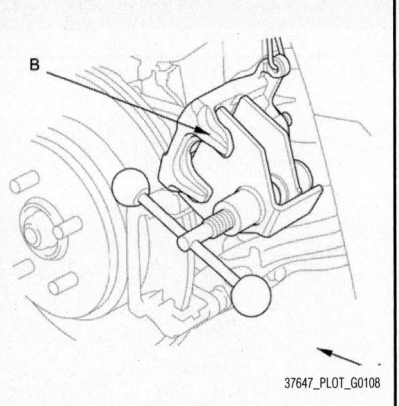

Fig. 12 Install the brake caliper piston compressor tool (A) on the caliper body (B)

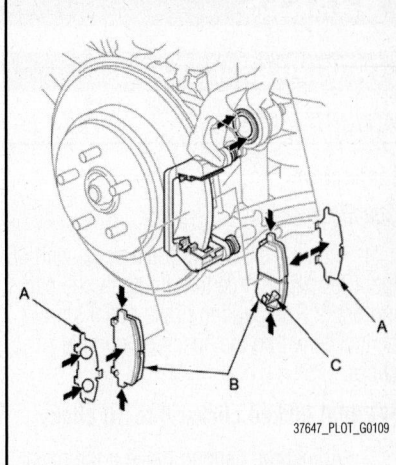

Fig. 13 Apply a thin coat of M-77 assembly paste to the pad side of the shims (A), the back of the brake pads (B), and the other areas indicated by the arrows

5. Remove the pad shims and the brake pads.

6. Remove the upper and lower pad retainers.

➡**The upper and lower pad retainers are different. During installation, make sure the pad retainers are in their proper positions.**

7. Clean the caliper bracket thoroughly; remove any rust, and check for grooves and cracks. Verify that the caliper pins move in and out smoothly. Clean and lube if needed.

8. Apply a thin coat of M-77 assembly paste (P/N 08798-9010) to the retainer mating surface of the caliper bracket (indicated by the arrows).

9. Install the upper and lower pad retainers. Wipe excess assembly paste off the retainers. Keep the assembly paste off the brake discs/drum and brake pads.

10. Install the brake caliper piston compressor tool on the caliper body.

11. Press in the piston with the brake caliper piston compressor tool so the

caliper will fit over the brake pads. Make sure the piston boot is in position to prevent damaging it when pivoting the caliper down.

➡**Be careful when pressing in the piston; brake fluid might overflow from the master cylinder's reservoir. If brake fluid gets on any painted surface, wash it off immediately with water.**

12. Remove the brake caliper piston compressor tool.

13. Apply a thin coat of M-77 assembly paste (P/N 08798-9010) to the pad side of the shims, the back of the brake pads, and the other areas indicated by the arrows. Wipe excess assembly paste from the pad shims and brake pads friction material. Keep grease and assembly paste away from the brake disc/drum and brake pads. Contaminated brake disc/drum or brake pads reduce stopping ability.

14. Install the brake pads and pad shims correctly. Install the brake pad with the wear indicator on the bottom inside. If you are reusing the brake pads, always

reinstall the brake pads in their original positions to prevent a temporary loss of braking efficiency.

15. Pivot the caliper down into position. Install the flange bolt, and tighten it to 27 ft. lbs. (37 Nm).

16. Clean the mating surfaces between the brake disc/drum and the inside of the wheel, then install the rear wheels.

17. Press the brake pedal several times to make sure the brakes work.

➡**Engagement may require a greater pedal stroke immediately after the brake pads have been replaced as a set. Several applications of the brake pedal will restore the normal pedal stroke.**

18. Add brake fluid as needed.

19. After installation, check for leaks at hose and line joints or connections, and retighten if necessary.

20. Test-drive the vehicle, then check for leaks.

PARKING BRAKE CABLE

ADJUSTMENTS

Inspection

See Figure 14.

1. Press the parking brake pedal with 66 lbs. (294 N) of force to fully apply the parking brake. The parking brake pedal should be locked within the specified number of clicks.

➡ **Pedal locked clicks: 8 to 10 clicks**

2. Adjust the parking brake if the pedal clicks are not within the specification.

➡ **Minor parking brake pedal adjustments (1 to 2 clicks) can be made with the adjusting nut (see parking brake minor adjustment). If a larger adjustment is required, follow the major adjustment procedure using the adjuster nut at the parking brake drum (see parking brake major adjustment). After installing new parking brake shoes and/or new brake disc/drum, make sure you drive the vehicle for "break-in".**

Minor Adjustment

See Figure 15.

1. Raise and support the vehicle.
2. Release the parking brake pedal fully.
3. Press the parking brake pedal 1 click.
4. Tighten the parking brake adjusting nut until the parking brakes drag slightly when the rear wheels are turned.
5. Release the parking brake pedal fully, and check that the parking brakes do not

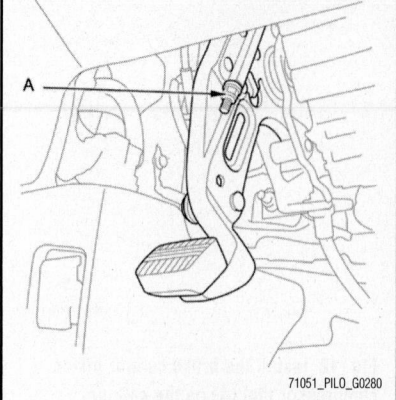

Fig. 15 Tighten the parking brake adjusting nut (A) until the parking brakes drag slightly when the rear wheels are turned

drag when the rear wheels are turned. Readjust if necessary.

6. Make sure the parking brake pedal is within the specified number of clicks (8 to 10 clicks).

Major Adjustment (to be done when replacing parking brake shoes and after lining surface break-in)

See Figures 16 and 17.

1. Raise and support the vehicle.
2. Release the parking brake pedal fully.
3. Loosen the parking brake adjusting nut (A).
4. Remove the rear wheels.
5. Remove the access plug.
6. Turn the ratchet teeth on the adjuster nut with a brake adjuster tool or a flat-tip screwdriver until the shoes lock against the parking brake drum. Then back off the

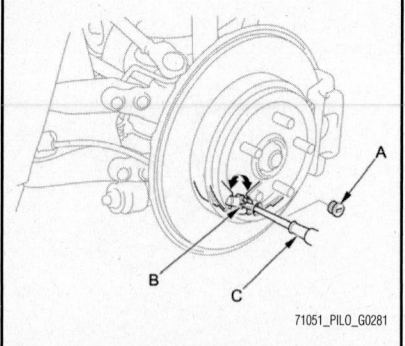

Fig. 17 Remove the access plug (A), then turn the ratchet teeth on the adjuster nut (B) with a brake adjuster tool or a flat-tip screwdriver (C) until the shoes lock against the parking brake drum

adjuster 10 clicks, and install the access plug.

7. Clean the mating surfaces between the brake disc/drum and the inside of the wheel, then install the rear wheels.

8. Do the minor adjustment procedure.

PARKING BRAKE SHOES

ADJUSTMENTS

Refer to Parking Brake Cable adjustment.

REMOVAL & INSTALLATION

See Figures 18 through 23.

1. Raise and support the vehicle.
2. Remove the rear wheels.
3. Release the parking brake, and remove the rear brake disc/drum.
4. Disconnect and remove the upper return springs, then remove the shoe guide plate.
5. Remove the tension pins by pushing the respective retainer, and turning the pin.
6. Remove the strut and the rod spring.
7. Lower the parking brake shoe assembly.
8. Remove the forward brake shoe and the adjuster assembly by removing the lower return spring.
9. Remove the rearward brake shoe by disconnecting the parking brake cable from the parking brake lever.
10. Remove the U-clip, wave washer and parking brake lever from the brake shoe.

To install:

11. Apply Molykote 44MA to the sliding surface of the pivot pin of the rearward brake shoe.

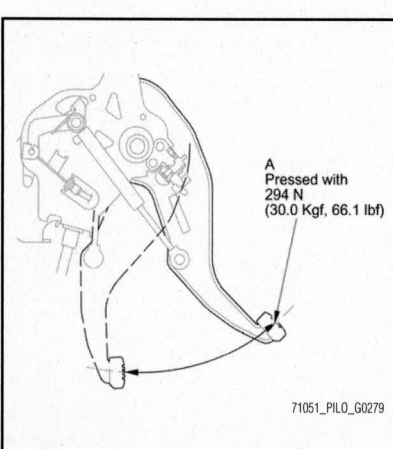

A
Pressed with
294 N
(30.0 Kgf, 66.1 lbf)

Fig. 14 Press the parking brake pedal (A) with 66 lbs. (294 N) of force to fully apply the parking brake

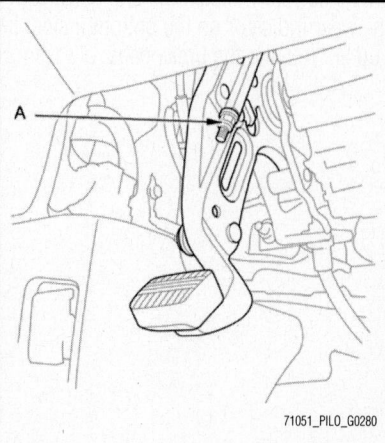

Fig. 16 Loosen the parking brake adjusting nut (A)

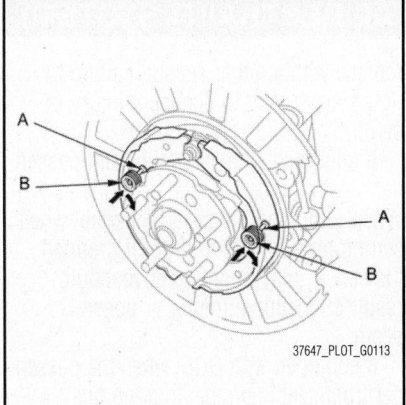

Fig. 18 Remove the tension pins (A) by pushing the respective retainer (B), and turning the pin

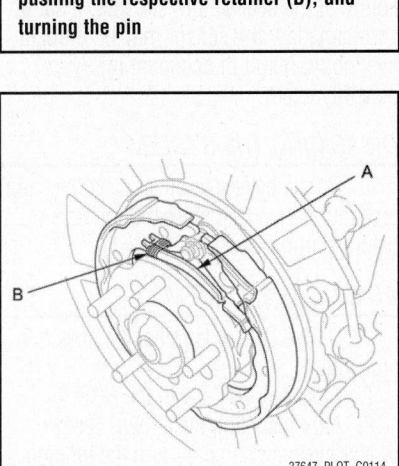

Fig. 19 Remove the strut (A) and the rod spring (B)

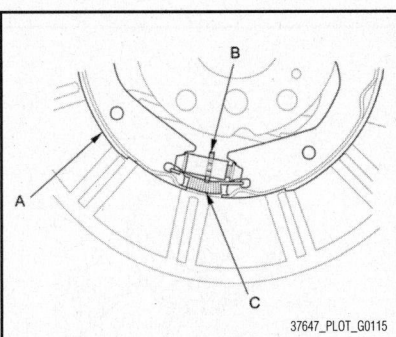

Fig. 20 Remove the forward brake shoe (A) and the adjuster assembly (B) by removing the lower return spring (C)

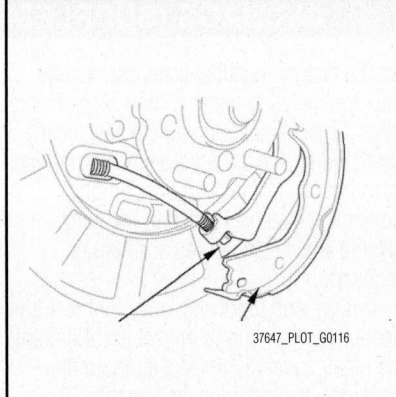

Fig. 21 Remove the rearward brake shoe by disconnecting the parking brake cable from the parking brake lever

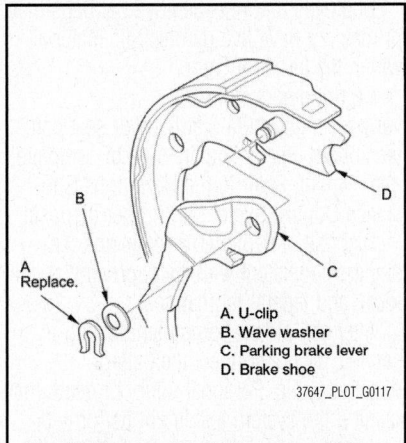

A. U-clip
B. Wave washer
C. Parking brake lever
D. Brake shoe

Fig. 22 Remove the U-clip, wave washer and parking brake lever from the brake shoe

12. Install the parking brake lever and the wave washer on the pivot pin, and secure with a new U-clip.

a. Install the wave washer with its convex side facing out.

b. Pinch the U-clip securely to prevent the parking brake lever from coming off the brake shoe.

13. Connect the parking brake cable to the parking brake lever. Apply Molykote 44MA to the cable contact surface on the backing plate.

14. Apply a thin coat of Molykote 44MA grease to the shoe ends and strut ends, sliding surfaces, and opposite edges of the parking brake shoe as shown. Wipe off any

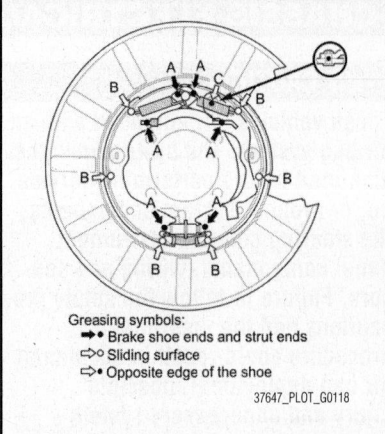

Greasing symbols:
➡● Brake shoe ends and strut ends
⇨○ Sliding surface
⇨● Opposite edge of the shoe

Fig. 23 Apply a thin coat of Molykote 44MA grease to the shoe ends and strut ends (A), sliding surfaces (B), and opposite edges of the parking brake shoe (C) as shown

excess. Keep the grease off the brake linings.

15. Install the tension pin, the retainer spring, and the retainer on the rearward brake shoe. Make sure the tension pin does not contact the parking brake lever.

16. Install connecting rods A and B on the adjuster nut.

➡**Clean the threaded portions of connecting rod A and the sliding surface of connecting rod B, then coat them with rubber grease. Shorten connecting rod A by fully turning the adjuster nut.**

17. Position the brake shoe adjuster assembly on the parking brake shoes.

18. Hook the lower return spring on the parking brake shoes.

19. Install the rod spring to the strut first. Then install the strut on the parking brake shoes.

20. Install the tension pin, the retainer spring, and the retainer on the forward brake shoe.

21. Install the shoe guide plate.

22. Install the upper return springs.

23. Install the brake disc/drum and the rear brake caliper.

24. Do the major parking brake adjustment.

25. Clean the mating surfaces between the brake/drum and the inside of the wheel, then install the rear wheels.

CHASSIS ELECTRICAL | AIR BAGS (SUPPLEMENTAL RESTRAINT SYSTEM)

✳✳ CAUTION

These vehicles are equipped with an air bag system. The system must be disarmed before performing service on, or around, system components, the steering column, instrument panel components, wiring and sensors. Failure to follow the safety precautions and the disarming procedure could result in accidental air bag deployment, possible injury and unnecessary system repairs.

SERVICE PRECAUTIONS

Disconnect and isolate the battery negative cable before beginning any airbag system component diagnosis, testing, removal, or installation procedures. Allow system capacitor to discharge for two minutes before beginning any component service. This will disable the airbag system. Failure to disable the airbag system may result in accidental airbag deployment, personal injury, or death.

Do not place an intact undeployed airbag face down on a solid surface. The airbag will propel into the air if accidentally deployed and may result in personal injury or death.

When carrying or handling an undeployed airbag, the trim side (face) of the airbag should be pointing towards the body to minimize possibility of injury if accidental deployment occurs. Failure to do this may result in personal injury or death.

Replace airbag system components with OEM replacement parts. Substitute parts may appear interchangeable, but internal differences may result in inferior occupant protection. Failure to do so may result in occupant personal injury or death.

Wear safety glasses, rubber gloves, and long sleeved clothing when cleaning powder residue from vehicle after an airbag deployment. Powder residue emitted from a deployed airbag can cause skin irritation. Flush affected area with cool water if irritation is experienced. If nasal or throat irritation is experienced, exit the vehicle for fresh air until the irritation ceases. If irritation continues, see a physician.

Do not use a replacement airbag that is not in the original packaging. This may result in improper deployment, personal injury, or death.

The factory installed fasteners, screws and bolts used to fasten airbag components have a special coating and are specifically designed for the airbag system. Do not use substitute fasteners. Use only original equipment fasteners listed in the parts catalog when fastener replacement is required.

During, and following, any child restraint anchor service, due to impact event or vehicle repair, carefully inspect all mounting hardware, tether straps, and anchors for proper installation, operation, or damage. If a child restraint anchor is found damaged in any way, the anchor must be replaced. Failure to do this may result in personal injury or death.

Deployed and non-deployed airbags may or may not have live pyrotechnic material within the airbag inflator.

Do not dispose of driver/passenger/curtain airbags or seat belt tensioners unless you are sure of complete deployment. Refer to the Hazardous Substance Control System for proper disposal.

Dispose of deployed airbags and tensioners consistent with state, provincial, local, and federal regulations.

After any airbag component testing or service, do not connect the battery negative cable. Personal injury or death may result if the system test is not performed first.

If the vehicle is equipped with the Occupant Classification System (OCS), do not connect the battery negative cable before performing the OCS Verification Test using the scan tool and the appropriate diagnostic information. Personal injury or death may result if the system test is not performed properly.

Never replace both the Occupant Restraint Controller (ORC) and the Occupant Classification Module (OCM) at the same time. If both require replacement, replace one, then perform the Airbag System test before replacing the other.

Both the ORC and the OCM store Occupant Classification System (OCS) calibration data, which they transfer to one another when one of them is replaced. If both are replaced at the same time, an irreversible fault will be set in both modules and the OCS may malfunction and cause personal injury or death.

If equipped with OCS, the Seat Weight Sensor is a sensitive, calibrated unit and must be handled carefully. Do not drop or handle roughly. If dropped or damaged,

replace with another sensor. Failure to do so may result in occupant injury or death.

If equipped with OCS, the front passenger seat must be handled carefully as well. When removing the seat, be careful when setting on floor not to drop. If dropped, the sensor may be inoperative, could result in occupant injury, or possibly death.

If equipped with OCS, when the passenger front seat is on the floor, no one should sit in the front passenger seat. This uneven force may damage the sensing ability of the seat weight sensors. If sat on and damaged, the sensor may be inoperative, could result in occupant injury, or possibly death.

DISARMING THE SYSTEM

1. Do the battery terminal disconnection procedure, then wait at least 3 minutes before starting work.

ARMING THE SYSTEM

1. Do the battery terminal reconnection procedure.
2. Clear any DTCs with the HDS.
3. After installing the airbag, confirm proper system operation, turn the ignition switch to ON (II); and check that the SRS indicator comes on for about 6 seconds and then goes off.

CLOCKSPRING CENTERING

See Figures 24 and 25.

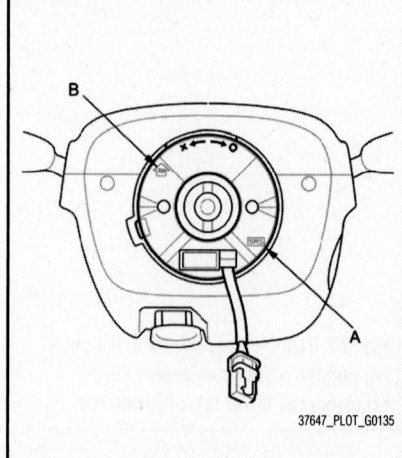

37647_PLOT_G0135

Fig. 24 Rotate the cable reel (A) clockwise until it stops, then rotate it counterclockwise (about three turns) until the arrow mark (B) on the cable reel label points straight up

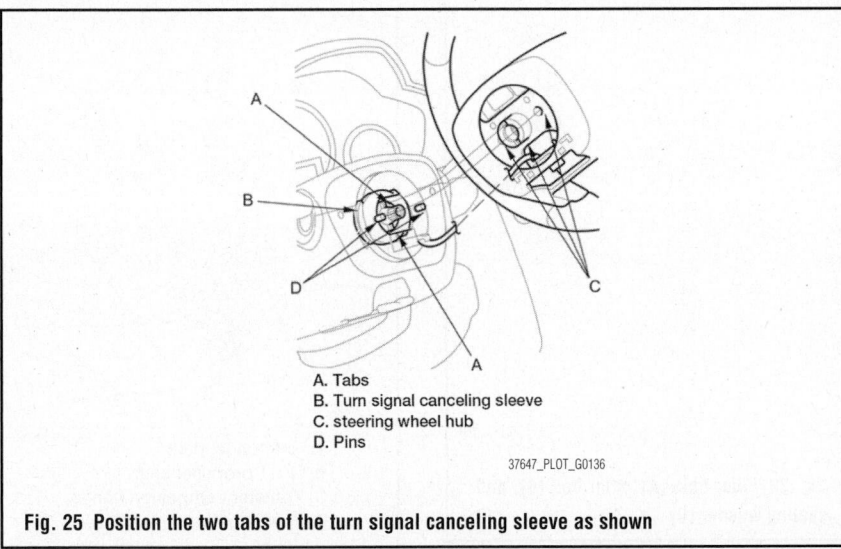

A. Tabs
B. Turn signal canceling sleeve
C. steering wheel hub
D. Pins

37647_PLOT_G0136

Fig. 25 Position the two tabs of the turn signal canceling sleeve as shown

1. Before installing the steering wheel, make sure the front wheels are pointing ahead, then center the cable reel. Do this by first rotating the cable reel clockwise until it stops. Then rotate it counterclockwise (about three turns) until the arrow mark on the cable reel label points straight up.

2. Position the two tabs of the turn signal canceling sleeve as shown, and install the steering wheel on to the steering column shaft, making sure the steering wheel hub engages the pins of the cable reel and tabs of the turn signal canceling sleeve. Do not tap on the steering wheel or steering column shaft when installing the steering wheel.

DRIVE TRAIN

AUTOMATIC TRANSMISSION FLUID

DRAIN & REFILL

➡**Keep all foreign particles out of the transmission.**

1. Park the vehicle on a level ground.

2. Warm up the engine to normal operating temperature (the radiator fan comes on), and turn the engine off.

3. Raise the vehicle on a lift, or apply the parking brake, block both rear wheels, and raise the front of the vehicle. Make sure it is securely supported.

4. Remove the ATF filler bolt and the drain plug, and drain the Automatic Transmission Fluid (ATF).

➡**If a ATF cooler cleaning is necessary, refer to ATF Cooler Cleaning.**

5. Reinstall the drain plug with a new sealing washer.

6. Refill the transmission with the recommended fluid through the filler hole. Always use Honda ATF-Z1 Automatic Transmission Fluid (ATF). Using a non-Honda ATF can affect shift quality.

7. Install the ATF filler bolt and a new sealing washer.

8. Remove the dipstick.

9. Make sure the ATF level is between the upper mark and the lower mark on the dipstick.

10. If the maintenance information recommends replacing the ATF, reset the maintenance information, and this procedure is complete. If the maintenance information did not require you to replace the ATF, go to step 11 and reset the ATF life with the HDS.

11. Connect the HDS to the DLC located under the driver's side of the dashboard.

12. Turn the ignition switch to ON (II). Make sure the HDS communicates with the PCM. If it does not, go to the DLC circuit troubleshooting.

13. Select BODY ELECTRICAL with the HDS.

14. Select ADJUSTMENT in the GAUGE MENU with the HDS.

15. Select RESET in the MAINTENANCE INFORMATION with the HDS.

16. Select MAINTENANCE SUB ITEM 3 RESET, and reset the ATF life with the HDS.

FILTER REPLACEMENT

See Figure 26.

1. Remove the three 6.0 mm bolts securing the ATF filter cover and the ATF pipe.

2. Remove the ATF pipe from the ATF filter cover, and remove the ATF filter from the cover.

3. Clean the ATF filter, then check that it is in good condition, and is not clogged. Replace the ATF filter if it is clogged or damaged.

4. Install the ATF filter with a new O-ring in the filter cover, and install the ATF pipe in the cover, then install them in the transmission housing.

5. Secure the ATF filter cover with the two bolts, then secure the ATF pipe with the bolt.

FLUID LEVEL CHECK

See Figures 27 through 29.

Note the following:
• Keep all foreign particles out of the transmission.

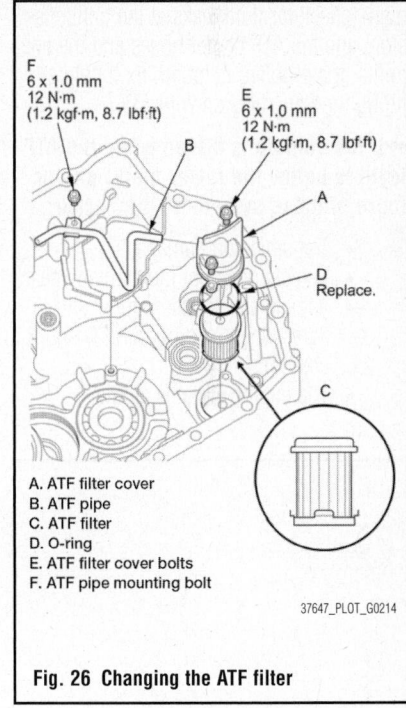

F
6 x 1.0 mm
12 N·m
(1.2 kgf·m, 8.7 lbf·ft)

E
6 x 1.0 mm
12 N·m
(1.2 kgf·m, 8.7 lbf·ft)

D
Replace.

A. ATF filter cover
B. ATF pipe
C. ATF filter
D. O-ring
E. ATF filter cover bolts
F. ATF pipe mounting bolt

37647_PLOT_G0214

Fig. 26 Changing the ATF filter

• Check the ATF level within 60–90 seconds after turning the engine off.
• Higher ATF level may be indicated if the radiator fan comes on twice or more.

1. Park the vehicle on the level ground.

2. Warm up the engine to normal operating temperature (the radiator fan comes on), then turn the engine off.

3. Remove the dipstick (yellow loop) from the dipstick guide tube, and wipe it with a clean cloth.

4. Insert the dipstick back into the dipstick guide tube.

5. Remove the dipstick, and check the

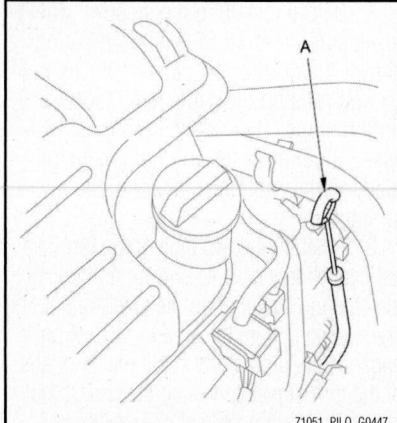

Fig. 27 Remove the dipstick (yellow loop) (A) from the dipstick guide tube, and wipe it with a clean cloth

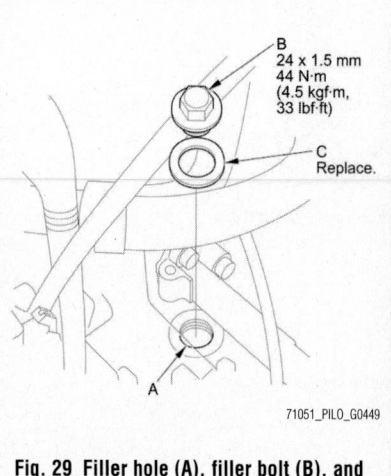

Fig. 29 Filler hole (A), filler bolt (B), and sealing washer (C)

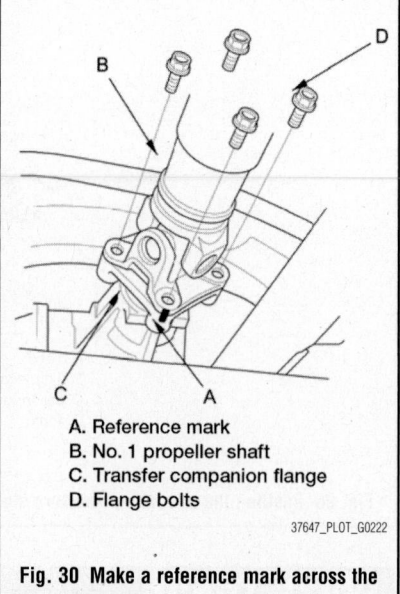

A. Reference mark
B. No. 1 propeller shaft
C. Transfer companion flange
D. Flange bolts

Fig. 30 Make a reference mark across the No. 1 propeller shaft and the transfer companion flange

ATF level. It should be between the upper mark and the lower mark.

6. If the ATF level is below the lower mark, check for fluid leaks at the transmission, and the ATF cooler hoses and the line joints. If a problem is found, fix it before filling the transmission with ATF.

➡**If the vehicle is driven when the ATF level is below the lower mark, one or more of these symptoms may occur:**

- Transmission damage.
- Vehicle does not move in any gear.
- Vehicle accelerates poorly, and flares when starting off in D and R positions.
- The engine vibrates at idle.

7. If the level is above the upper mark, drain the ATF to the proper level.

➡**If the vehicle is driven when the ATF level is above the upper mark, the vehicle may creep forward while in N, or have shifting problems.**

8. If necessary, fill the transmission with the ATF through the filler hole to bring the fluid level between the upper mark and the lower mark of the dipstick. Do not fill the fluid above the upper mark. Always use Honda ATF DW-1 Automatic Transmission Fluid (ATF). Using a non-Honda ATF can affect shift quality.

9. Install the ATF filler bolt and a new sealing washer.

10. Insert the dipstick back into the dipstick guide tube.

FLUID RECOMMENDATIONS

Honda Automatic Transmission Fluid (ATF DW-1): Always use Honda ATF DW-1. Using a non-Honda ATF can affect shift quality.

DRIVESHAFT (PROPELLER SHAFT)

REMOVAL & INSTALLATION

Rear

See Figures 30 through 32.

1. Raise and support the vehicle.
2. Remove the propeller shaft protector.
3. Make a reference mark across the No. 1 propeller shaft and the transfer companion flange.
4. Remove the flange bolts.
5. Remove the center support bearing mounting bolts.

6. Make a reference mark across the No. 2 propeller shaft and the rear differential companion flange.
7. Remove the flange bolts.

To install:

8. Set the No. 2 propeller shaft to the rear differential companion flange by aligning the reference mark you made during the removal procedure. Then install new flange bolts to 53 ft. lbs. (72 Nm).

➡ **When replacing the propeller shaft or the rear differential, align the factory reference marks.**

9. Install the center support bearing mounting bolts.
10. Set the No. 1 propeller shaft to the transfer companion flange by aligning the

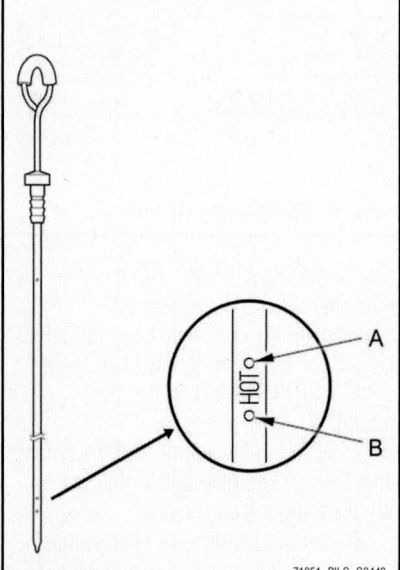

Fig. 28 Remove the dipstick, and check the ATF level. It should be between the upper mark (A) and the lower mark (B)

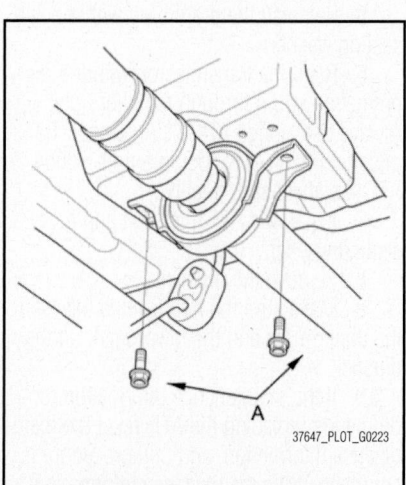

Fig. 31 Remove the center support bearing mounting bolts (A)

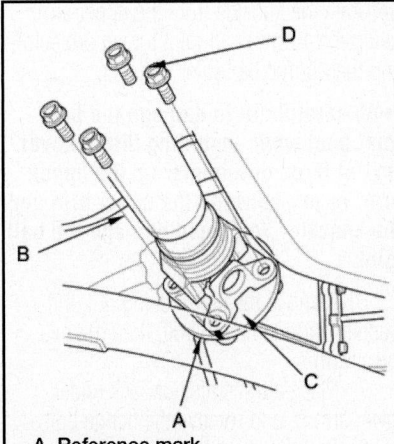

A. Reference mark
B. No. 2 propeller shaft
C. Rear differential companion flange
D. Flange bolts

37647_PLOT_G0224

Fig. 32 Make a reference mark across the No. 2 propeller shaft and the rear differential companion flange

reference mark you made during the removal procedure. Then install new flange bolts to 53 ft. lbs. (72 Nm).

11. Install the propeller shaft protector.

12. If you installed a new propeller shaft, test-drive the vehicle at 55 mph (88 km/h) and check for noise or vibration. If there is a noise or vibration, rotate the propeller shaft 180 degrees from its current alignment with the rear differential companion flange, and check.

HALFSHAFTS

REMOVAL & INSTALLATION

Front Halfshaft

See Figures 33 and 34.

Special Tools Required:
• Ball Joint Thread Protector, 14 mm 071AF-S3VA000
• Ball Joint Remover, 32 mm 07MAC-SL0A102

1. Raise and support the vehicle.
2. Remove the front wheels.
3. Pry up the stake on the spindle nut, then remove the nut.
4. Drain the transmission fluid, then reinstall the drain plug with a new sealing washer.
5. Remove the brake hose mounting bolt.
6. Remove the lock pin from the lower arm ball joint, then remove the castle nut. Separate the knuckle from the lower arm using the 14 mm ball joint thread protector and the 32 mm ball joint remover.

➡Be careful not to damage the ball joint boot when installing the remover. Do not force or hammer on the lower arm, or pry between the lower arm and the knuckle. You could damage the ball joint.

7. Pull the knuckle outward, and separate the outboard front driveshaft joint from the front hub using a soft face hammer.

8. Left driveshaft: Pry the inboard joint from the differential using a pry bar. Remove the driveshaft as an assembly.

➡Do not pull on the driveshaft, or the inboard joint may come apart. Pull the inboard joint straight out to avoid damaging the oil seal. Be careful not to damage the oil seal or the end of the inboard joint using the pry bar.

9. Right driveshaft: Drive the inboard joint off of the intermediate shaft using a drift and a hammer. Remove the driveshaft as an assembly.

➡Do not pull on the driveshaft, or the inboard joint may come apart.

10. Remove the set ring from the left driveshaft inboard joint.

11. Remove the set ring from the intermediate shaft.

To install:

➡Before starting installation, make sure the mating surfaces of the joint and the splined section are clean.

12. Apply about 0.18 oz. (5 g) of Moly 60 paste (P/N 08734-0001) to the contact area (A) of the outboard joint and the front wheel bearing.

➡The paste helps prevent noise and vibration.

13. Install a new set ring into the set ring groove of the left driveshaft inboard joint.

14. Install a new set ring into the set ring groove of the intermediate shaft.

15. Apply 2.0–3.0 g (0.07–0.10 oz.) of super high temp urea grease (P/N 08798-9002) to the whole splined surface of the right driveshaft. After applying grease, remove the grease from the splined grooves at intervals of 2–3 splines and from the set ring groove so that air can bleed from the intermediate shaft.

16. Clean the areas where the driveshaft contacts the differential thoroughly with solvent, and dry them with compressed air.

➡Do not wash the rubber parts with solvent.

17. Insert the inboard end of the drive-

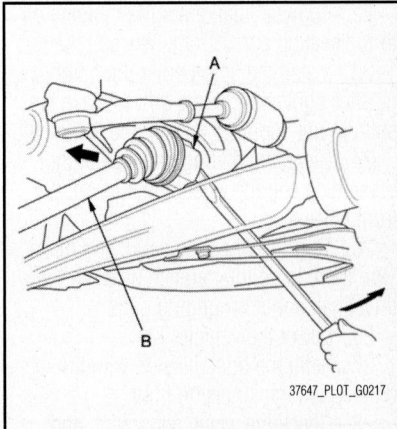

37647_PLOT_G0217

Fig. 33 Left driveshaft: Pry the inboard joint (A) from the differential; remove the driveshaft (B) as an assembly

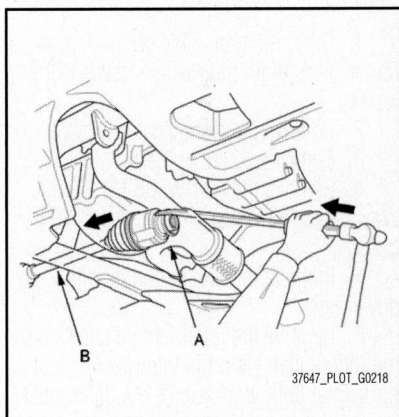

37647_PLOT_G0218

Fig. 34 Right driveshaft: Drive the inboard joint (A) off of the intermediate shaft; remove the driveshaft (B) as an assembly

shaft into the differential or the intermediate shaft until the set ring locks in the groove.

➡Insert the driveshaft horizontally to prevent damaging the oil seal.

18. Install the outboard joint into the front hub on the knuckle.

19. Wipe off any grease contamination from the ball joint tapered section and threads, then install the knuckle onto the lower arm. Be careful not to damage the ball joint boot. Wipe off the grease before tightening the nut at the ball joint. Torque a new castle nut to the lower torque specification, then tighten it only far enough to align the slot with the ball joint pin hole.

➡Make sure the ball joint boot is not damaged or cracked. Do not align the nut by loosening it.

20. Install the lock pin into the ball joint pin hole.

21. Install the brake hose mounting bolt.

22. Apply a small amount of engine oil to the seating surface of a new spindle nut.

23. Install the spindle nut, then tighten it. After tightening, use a drift to stake the spindle nut shoulder against the driveshaft.

24. Clean the mating surfaces of the brake disc and the wheel, then install the front wheels.

25. Turn the wheel by hand, and make sure there is no interference between the driveshaft and surrounding parts.

26. Lower the vehicle.

27. Refill the transmission with the recommended transmission fluid.

28. Check the wheel alignment, and adjust it if necessary.

29. Test-drive the vehicle.

Intermediate Shaft

See Figures 35 and 36.

1. Drain the transmission fluid, then reinstall the drain plug with a new sealing washer.

2. Remove the right front driveshaft.

3. Remove exhaust pipe A.

4. Remove the rear Warm Up Three Way Catalytic Converter (rear WU-TWC) bracket.

5. Remove the flange bolt, and the two dowel bolts.

6. Remove the intermediate shaft from the differential. Hold the intermediate shaft horizontal until it is clear of the differential to prevent damaging the oil seal, then remove the set ring from the intermediate shaft.

To install:

7. Clean the areas where the intermediate shaft contacts the differential thoroughly with solvent, and dry them with compressed air.

➡**Do not wash the rubber parts with solvent.**

8. Install a new set ring into the set ring groove of the intermediate shaft.

9. Insert the intermediate shaft into the differential correctly.

➡**Insert the intermediate shaft carefully to prevent damaging the oil seal.**

10. Install the flange bolt and the two dowel bolts.

11. Install the rear warm up three way catalytic converter (rear WU-TWC) bracket.

12. Install exhaust pipe A.

13. Install the right front driveshaft.

14. Refill the transmission with the recommended transmission fluid.

15. Check the wheel alignment, and adjust it if necessary.

16. Test-drive the vehicle.

Rear Halfshaft

See Figures 37 through 41.

Special Tools Required:
- Driveshaft Remover 07AAD-S9VA000
- Ball Joint Thread Protector, 12 mm 07AAF-SDAA100
- Ball Joint Remover, 28 mm 07MAC-SL0A202

➡**Be careful not to damage the brake hose, sensor, and harness.**

1. Raise and support the vehicle.

2. Remove the rear wheels.

3. Pry up the stake on the spindle nut, then remove the nut.

4. Remove the rear wheel speed sensor and harness stay.

5. Remove the lock pin from the upper arm ball joint, then remove the castle nut.

Separate the knuckle from the upper arm using the 12 mm ball joint thread protector and the 28 mm ball joint remover.

➡**Be careful not to damage the ball joint boot when installing the remover. Do not force or hammer on the upper arm, or pry between the upper arm and the knuckle. You could damage the ball joint.**

6. Remove the self-locking nut, the washer, and the flange bolt, then remove lower arm A.

7. Place a transmission jack under lower arm B, and remove the flange bolt.

8. Pull the knuckle outward, and separate the rear driveshaft outboard joint from the rear hub using a soft face hammer.

9. Using the driveshaft remover pry out the inboard joint from the rear differential.

➡**This is a prying tool, do not strike it with a hammer.**

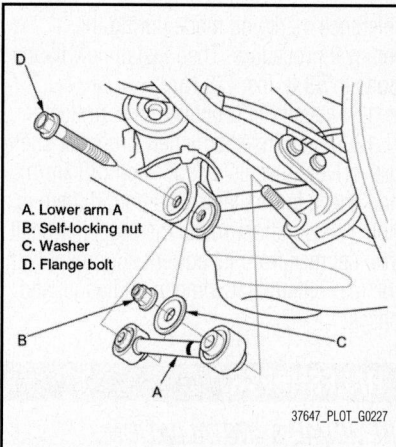

A. Lower arm A
B. Self-locking nut
C. Washer
D. Flange bolt

37647_PLOT_G0227

Fig. 37 Remove the self-locking nut, the washer, and the flange bolt, then remove lower arm A

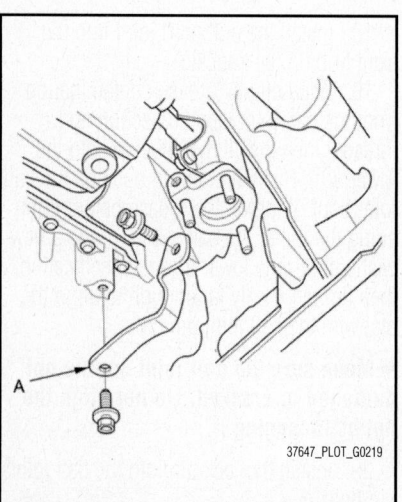

37647_PLOT_G0219

Fig. 35 Remove rear WU-TWC bracket (A)

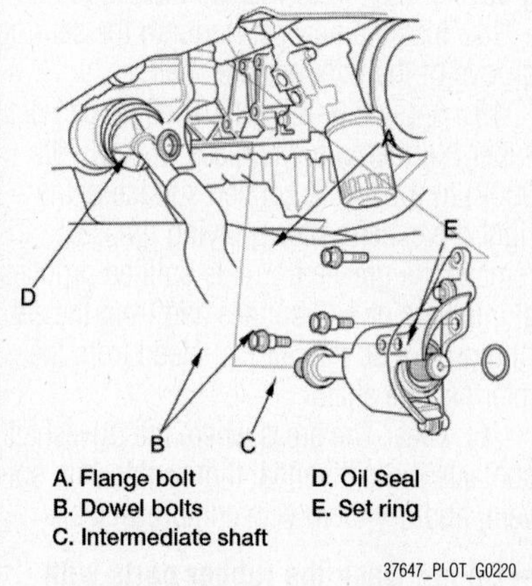

A. Flange bolt
B. Dowel bolts
C. Intermediate shaft
D. Oil Seal
E. Set ring

37647_PLOT_G0220

Fig. 36 Remove the flange bolt, and the two dowel bolts

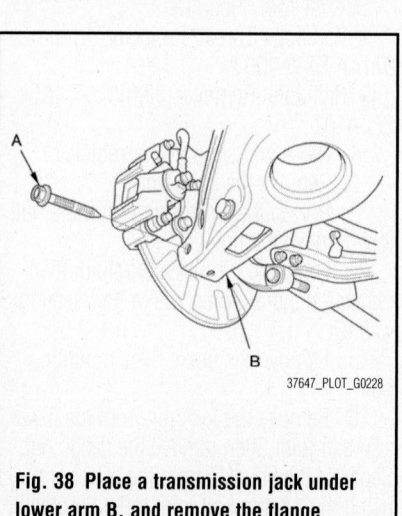

37647_PLOT_G0228

Fig. 38 Place a transmission jack under lower arm B, and remove the flange bolt (A)

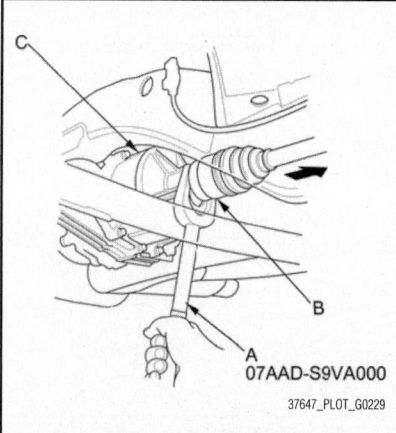

Fig. 39 Using the driveshaft remover (A) pry out the inboard joint (B) from the rear differential (C)

10. Remove the rear driveshaft.

➡**When removing the outboard joint, continue supporting both the knuckle and the lower arm with the transmission jack. Make sure not to over extend the brake hose.**

11. Remove the set ring from the rear differential.

To install:

➡**Before starting installation, make sure the mating surfaces of the joint and the splined section are clean.**

12. Apply 1.5–2.0 g (0.05–0.07 oz.) of super high temp urea grease (P/N 08798-9002) to the whole splined surface. After applying grease, remove the grease from the splined grooves at intervals of 2–3 splines and from the set ring groove so that air can bleed from the differential.

13. Install a new set ring into the set ring groove of the differential.

14. Clean the areas where the driveshaft contacts the differential thoroughly with solvent, and dry them with compressed air.

➡**Do not wash the rubber parts with solvent.**

15. Make sure the inboard joint is installed all the way into the rear differential and to ensure the set ring is properly seated.

16. Pull the knuckle outward, and insert the rear driveshaft outboard joint into the rear hub.

➡**When installing the outboard joint continue supporting both the knuckle and the lower arm with the transmission jack. Make sure not to over extend the brake hose.**

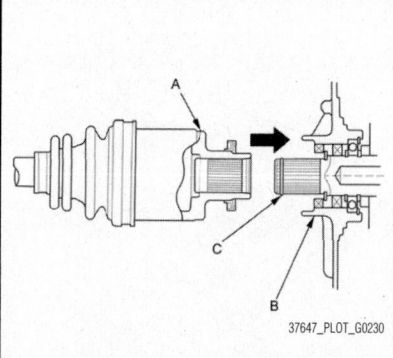

Fig. 40 Make sure the inboard joint (A) is installed all the way into the rear differential (B) and to ensure the set ring (C) is properly seated

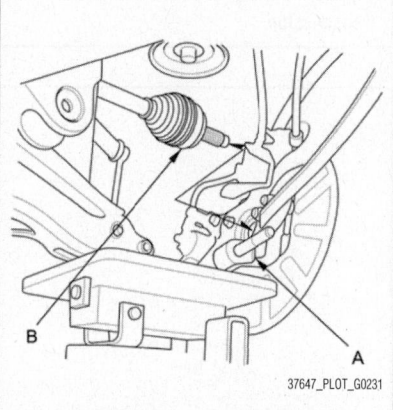

Fig. 41 Pull the knuckle (A) outward, and insert the rear driveshaft outboard joint (B) into the rear hub

17. Loosely install a new flange bolt onto lower arm B.

18. Install lower arm A, then loosely install a new flange bolt and a new self-locking nut with the washer.

➡**Position the paint mark on lower arm A toward the outside of the vehicle.**

19. Wipe off any grease contamination from the ball joint tapered section and threads, then install the upper arm. Be careful not to damage the ball joint boot. Wipe off the grease before tightening the nut at the ball joint. Torque the castle nut to the lower torque specification, then tighten it only far enough to align the slot with the ball joint pin hole.

20. Install the lock pin into the ball joint pin hole as shown.

21. Install the rear wheel speed sensor and the harness stay.

22. Place a floor jack under lower arm B,

and raise the suspension to load it with the vehicle's weight.

23. Tighten the lower arm B flange bolt to 83 ft. lbs. (113 Nm), the lower arm A self-locking nut to 83 ft. lbs. (113 Nm), and the lower arm A flange bolt to 69 ft. lbs. (93 Nm), then remove the floor jack.

24. Apply a small amount of engine oil to the seating surface of a new spindle nut.

25. Install the spindle nut, then tighten it to 181 ft. lbs. (245 Nm). After tightening, use a drift to stake the spindle nut shoulder against the driveshaft.

26. Clean the mating surfaces of the brake disc and the wheel, then install the rear wheels.

27. Turn the wheel by hand, and make sure there is no interference between the driveshaft and surrounding parts.

28. Lower the vehicle.

29. Check the wheel alignment, and adjust it if necessary.

30. Test-drive the vehicle.

TRANSFER CASE

REMOVAL & INSTALLATION

See Figures 42 through 45.

1. Raise the vehicle on a lift, and make sure it is supported securely.

2. Shift the shift lever to N.

3. Remove the drain plug, and drain the Automatic Transmission Fluid (ATF).

4. Reinstall the drain plug with a new sealing washer.

5. Remove the front subframe stiffener.

6. Remove the exhaust pipe A and the gaskets.

7. Remove the bolt securing the transfer breather hose bracket, and disconnect the breather hose from the breather pipe on the transfer assembly.

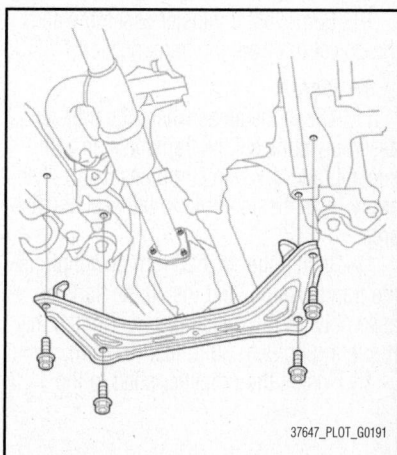

Fig. 42 Remove the front subframe stiffener

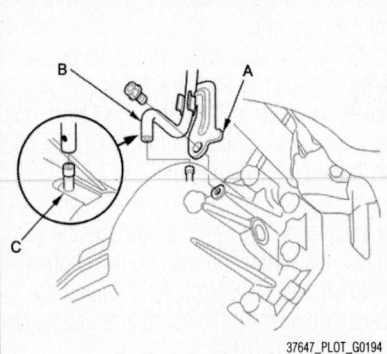

Fig. 43 Remove the bolt securing the transfer breather hose bracket (A), and disconnect the breather hose (B) from the breather pipe (C) on the transfer assembly

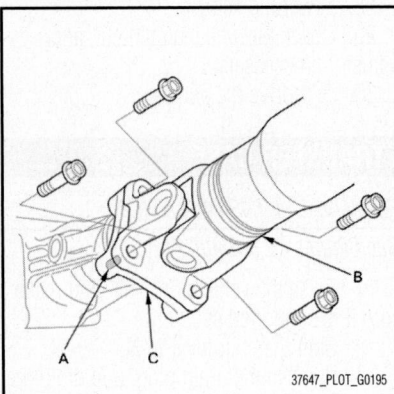

Fig. 44 Make a reference mark (A) across the propeller shaft (B) and the transfer companion flange (C)

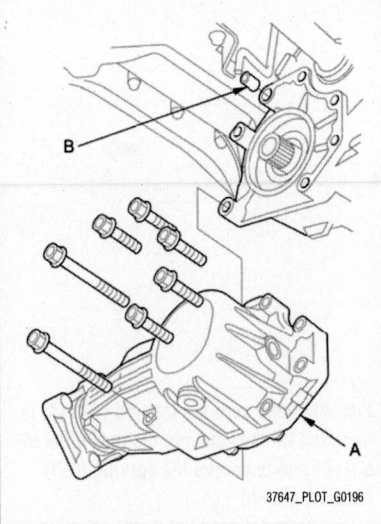

Fig. 45 Remove the transfer assembly (A) and the dowel pin (B) from the transmission

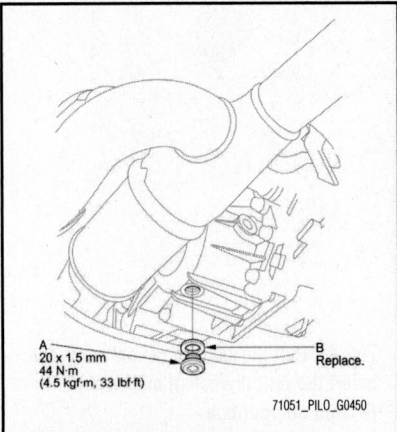

Fig. 46 Remove the drain plug (A) and sealing washer (B), and drain the transfer fluid (hypoid gear oil)

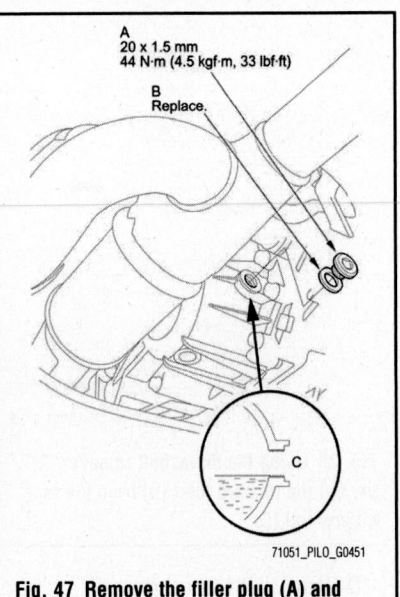

Fig. 47 Remove the filler plug (A) and the sealing washer (B); filler hole (C)

8. Make a reference mark across the propeller shaft and the transfer companion flange.

9. Separate the propeller shaft from the transfer companion flange.

10. Remove the transfer assembly and the dowel pin from the transmission.

To install:

11. Clean the areas where the transfer assembly contacts the transmission with solvent, and dry with compressed air. Then apply transmission fluid to the seal contact area.

12. Install the 14 x 20 mm dowel pin in the transmission, and install the transfer assembly on the transmission. Tighten the mounting bolts to 38 ft. lbs. (51 Nm).

13. Install the propeller shaft to the transfer companion flange by aligning the reference mark. Make sure you use new mounting bolts. Tighten the mounting bolts to 53 ft. lbs. (72 Nm).

14. Secure the transfer breather hose bracket on the transfer assembly with the bolt, and connect the breather hose over the breather pipe with the dot mark facing out.

15. Install the exhaust pipe A with new self-locking nuts, its mount, and new gaskets.

16. Install the front subframe stiffener with new mounting bolts. Tighten the mounting bolts to 43 ft. lbs. (59 Nm).

17. Refill the transfer assembly with the transfer fluid (hypoid gear oil), if necessary.

18. Refill the transmission with ATF.

DRAIN & REFILL

See Figures 46 and 47.

Use an SAE 90 viscosity hypoid gear oil, API classified GL4 or GL5 only.

1. Warm up the engine to normal operating temperature (the radiator fan comes on).

2. Park the vehicle on level ground, and turn the engine off.

3. Raise the vehicle on a lift, and make sure it is securely supported.

4. Remove the drain plug, and drain the transfer fluid (hypoid gear oil).

5. Reinstall the drain plug with a new sealing washer.

6. Remove the filler plug and the sealing washer.

7. Refill the transfer assembly with the recommended fluid (hypoid gear oil) through the filler hole until the fluid flows out. Use an SAE 90 viscosity hypoid gear oil, API classified GL4 or GL5 only.

   a. Viscosity: SAE 90: Above 0°F (−18°C)

   b. Transfer Fluid (Hypoid Gear Oil) Capacity:

     • 0.45 qt. (0.43 L) at fluid change

     • 0.48 qt. (0.45 L) at overhaul

8. Install the filler plug with a new sealing washer.

ENGINE COOLING

ENGINE COOLANT

DRAIN & REFILL

See Figures 48 through 52.

Always use Honda Long Life Antifreeze/Coolant Type 2. Using a non-Honda coolant can result in corrosion, causing the cooling system to malfunction or fail.

Honda Long Life Antifreeze/Coolant Type 2 is a mixture of 50% antifreeze and 50% water. Do not add water.

1. Remove the radiator cap.
2. Start the engine. Set the heater temperature climate control or HVAC control unit to maximum heat, then turn the ignition switch to LOCK (0). Make sure the engine and radiator are cool to the touch.
3. Open the splash shield cover.
4. Loosen the drain plug and drain the coolant.
5. Remove the rocker arm oil pressure sensor cover.
6. Install a rubber hose on the drain bolt located at the rear of the engine block, then loosen the drain bolt.
7. When the coolant stops draining, tighten the drain bolt. Remove the rubber hose, then install the rocker arm oil pressure sensor cover.
8. Tighten the radiator drain plug securely.
9. Close the splash shield cover.
10. Remove, drain, and reinstall the coolant reservoir.
11. Fill the coolant reservoir to the MAX mark with Honda Long Life Antifreeze/Coolant Type 2 (P/N OL999-9001).

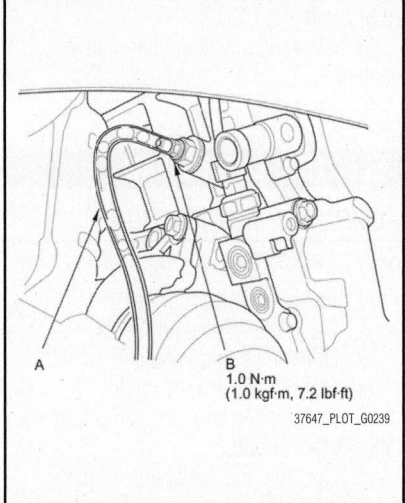

Fig. 49 Install a rubber hose (A) on the drain bolt (B), then loosen the drain bolt

➡Always use Honda Long Life Antifreeze/Coolant Type 2. Using a non-Honda coolant can result in corrosion, causing the cooling system to malfunction or fail. Honda Long Life Antifreeze/Coolant Type 2 is a mixture of 50% antifreeze and 50% water. Do not add water.

12. Pour Honda Long Life Antifreeze/Coolant Type 2 into the radiator up to the base of the filler neck.
13. Start the engine. Hold the engine speed at 1,500 rpm until it warms up (the radiator fan comes on at least twice). Make sure the thermostat is open.
14. Turn off the engine. Check the level in the radiator, and add Honda Long Life Antifreeze/Coolant Type 2, if needed.

15. Set the climate control or HVAC control unit to maximum cool. Start the engine. Hold the engine speed at 1,500 rpm for 5 minutes, then turn off the engine.
16. Check the level in the radiator, and add Honda Long Life Antifreeze/Coolant Type 2, if needed.
17. Set the climate control or HVAC control unit to maximum heat. Start the engine. Hold the engine speed at 1,500 rpm for 5 minutes, then turn off the engine.
18. Check the level in the radiator, and add Honda Long Life Antifreeze/Coolant Type 2, if needed.
19. Set the climate control or HVAC control unit to maximum cool. Start the engine. Hold the engine speed at 1,500 rpm for 3 minutes, then turn off the engine.
20. Check the level in the radiator, and add Honda Long Life Antifreeze/Coolant Type 2, if needed.
21. Set the climate control or HVAC control unit to maximum heat. Start the engine. Hold the engine speed at 1,500 rpm for 3 minutes, then turn off the engine.
22. Check the level in the radiator, and add Honda Long Life Antifreeze/Coolant Type 2, if needed.
23. Repeat steps 19 through 22 until the coolant level does not change in the radiator, then install the radiator cap loosely.
24. Set the climate control or HVAC control unit to maximum cool. Start the engine. Hold the engine speed at 2,500 rpm for 1 minute.
25. Set the climate control or HVAC control unit to maximum heat, and select

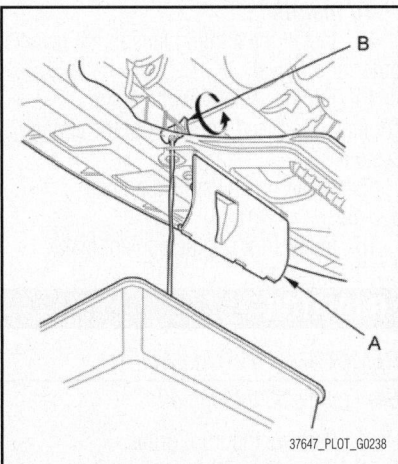

Fig. 48 Open the splash shield cover (A) and loosen the drain plug (B)

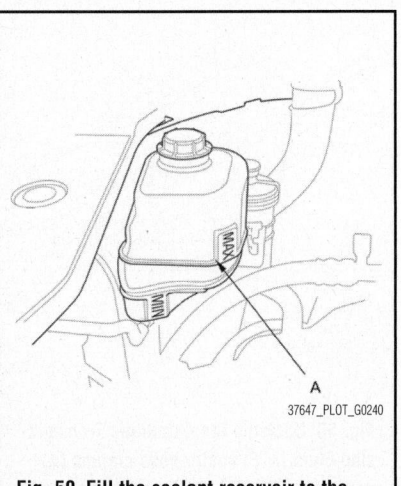

Fig. 50 Fill the coolant reservoir to the MAX mark (A)

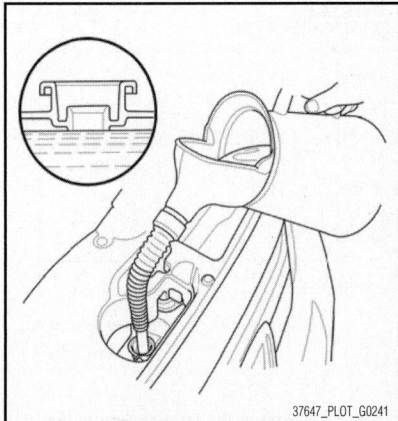

Fig. 51 Pour Honda Long Life Antifreeze/Coolant Type 2 into the radiator up to the base of the filler neck

high speed heat mode on the rear control panel. Measure the temperature of the air from the rear floor vent for 3 minutes. Make sure the temperature is above the standard line of the graph as shown. If the temperature is below the standard line, repeat steps 19 through 22 several more times. Then loosely install the radiator cap and recheck.

26. With the engine idling, make sure you do not hear the sound of water flowing near the rear heater unit. If you hear water flowing, repeat steps 19 through 22 several more times.

27. Install the radiator cap securely.

28. Fill the coolant reservoir with Honda Long Life Antifreeze/Coolant Type 2 up to the MAX mark on the coolant reservoir, then add an extra 0.11 gal. (0.4 L) of coolant.

29. If the maintenance minder required engine coolant replacement, reset the maintenance minder. If the maintenance minder did not require engine coolant replacement, go to step 30.

30. Turn the ignition switch to LOOK (0).

31. Connect the Honda Diagnostic System (HDS) to the Data Link Connector (DLC).

32. Turn the ignition switch to ON (II).

33. Make sure the HDS communicates with the vehicle and the Powertrain Control Module (PCM). If it does not communicate, troubleshoot the DLC circuit.

34. Select GAUGES in the BODY ELECTRICAL with the HDS.

35. Select ADJUSTMENT in the GAUGES with the HDS.

36. Select SERVICE REMINDER in the ADJUSTMENT with the HDS.

37. Select RESET in the SERVICE REMINDER with the HDS.

38. Select MAINTENANCE SUB ITEM 5 RESET with the HDS.

39. Clean up any spilled engine coolant.

40. Inspect for engine coolant leaks.

ELECTRIC ENGINE FAN

REMOVAL & INSTALLATION

See Figures 53 through 58.

1. Remove the front grille.
2. Raise the vehicle on the lift.
3. Drain the engine coolant.
4. Remove the splash shield.
5. Remove the harness clamp from the radiator fan shroud.
6. Unclamp the Automatic Transmission Fluid (ATF) cooler hose clamps and loosen the A/C condenser fan shroud mounting bolts.
7. Lower the vehicle on the lift.
8. Disconnect the radiator fan motor connector and the radiator upper hose, then remove the harness clamps.
9. Disconnect the A/C condenser fan motor connector and the coolant reservoir hose, then remove the harness clamps.
10. Remove the bulkhead bracket mounting bolt/nut and the upper brackets/cushions, then disconnect the coolant reservoir hose.
11. Remove the A/C condenser fan shroud assembly, then remove the radiator fan shroud assembly.

➡**Move the radiator fan shroud assembly toward the A/C compressor side of**

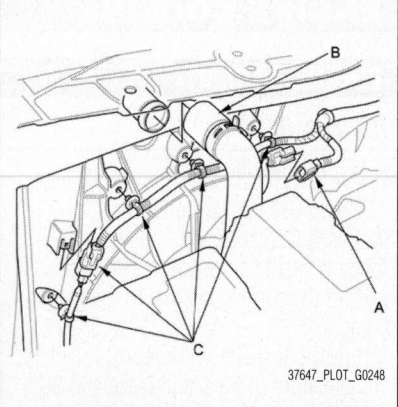

Fig. 54 Disconnect the radiator fan motor connector (A) and the radiator upper hose (B), then remove the harness clamps (C)

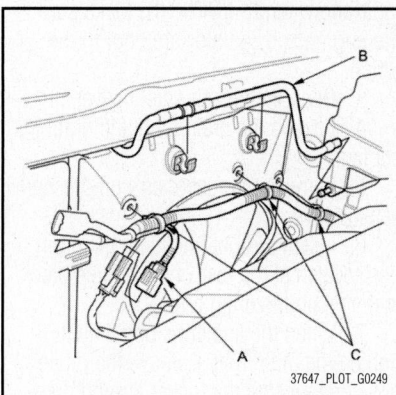

Fig. 55 Disconnect the A/C condenser fan motor connector (A) and the coolant reservoir hose (B), then remove the harness clamps (C)

the vehicle to allow for enough space to lift it up and away from the A/C condenser fan shroud assembly.

12. Disassemble the fan shrouds.

To install:

13. Install the cooling fans in the reverse order of removal.

14. Fill the radiator with engine coolant, and bleed the air from the cooling system.

15. Clean up any spilled engine coolant.

16. Inspect for engine coolant leaks.

RADIATOR

REMOVAL & INSTALLATION

See Figures 53 through 60.

1. Remove the front grille.
2. Raise the vehicle on the lift.
3. Drain the engine coolant.
4. Remove the splash shield.

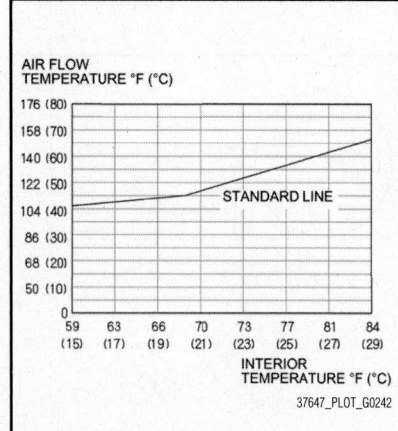

Fig. 52 Measure the temperature of the air from the rear floor vent for 3 minutes; make sure the temperature is above the standard line of the graph as shown

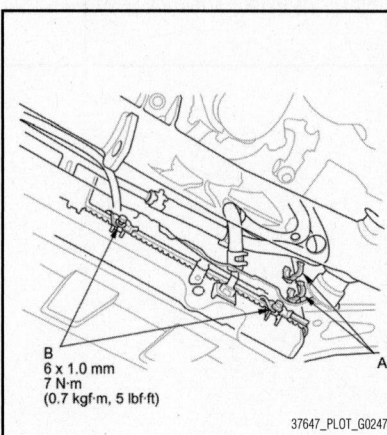

Fig. 53 Unclamp the Automatic Transmission Fluid (ATF) cooler hose clamps (A) and loosen the A/C condenser fan shroud mounting bolts (B)

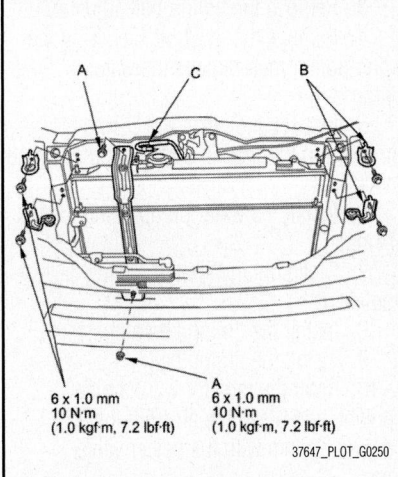

Fig. 56 Remove the bulkhead bracket mounting bolt/nut (A) and the upper brackets/cushions (B), then disconnect the coolant reservoir hose (C)

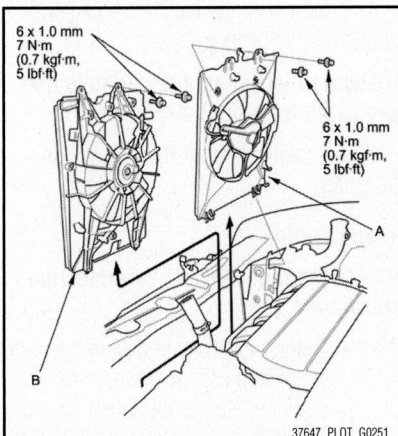

Fig. 57 Remove the A/C condenser fan shroud assembly (A), then remove the radiator fan shroud assembly (B)

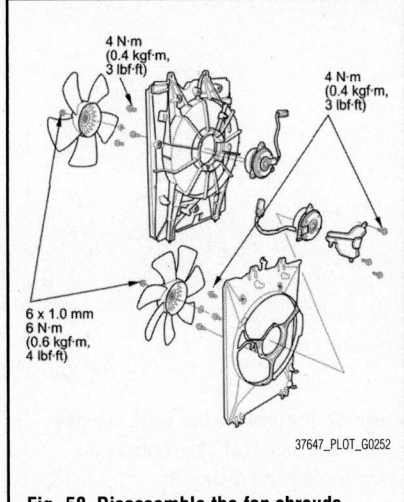

Fig. 58 Disassemble the fan shrouds

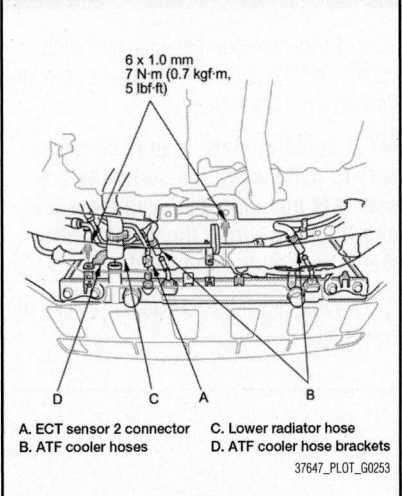

A. ECT sensor 2 connector
B. ATF cooler hoses
C. Lower radiator hose
D. ATF cooler hose brackets

Fig. 59 Disconnect the Engine Coolant Temperature (ECT) sensor 2 connector, the ATF cooler hoses, the lower radiator hose and remove the ATF cooler hose brackets

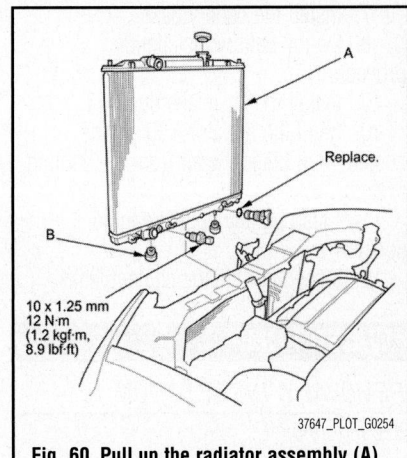

Fig. 60 Pull up the radiator assembly (A), then remove the lower mounting cushions (B)

5. Remove the harness clamp from the radiator fan shroud.

6. Unclamp the Automatic Transmission Fluid (ATF) cooler hose clamps and loosen the A/C condenser fan shroud mounting bolts.

7. Lower the vehicle on the lift.

8. Disconnect the radiator fan motor connector and the radiator upper hose, then remove the harness clamps.

9. Disconnect the A/C condenser fan motor connector and the coolant reservoir hose, then remove the harness clamps.

10. Remove the bulkhead bracket mounting bolt/nut and the upper brackets/cushions, then disconnect the coolant reservoir hose.

11. Remove the A/C condenser fan shroud assembly, then remove the radiator fan shroud assembly.

➡ Move the radiator fan shroud assembly toward the A/C compressor side of the vehicle to allow for enough space to lift it up and away from the A/C condenser fan shroud assembly.

12. Disassemble the fan shrouds.

13. Raise the vehicle on the lift.

14. Disconnect the Engine Coolant Temperature (ECT) sensor 2 connector, the ATF cooler hoses, the lower radiator hose and remove the ATF cooler hose brackets.

15. Lower the vehicle on the lift.

16. Pull up the radiator assembly, then remove the lower mounting cushions.

17. Remove the related parts from the radiator.

To install:

18. Install the radiator in the reverse order of removal. Make sure the upper and lower mounting cushions are set securely.

19. Fill the radiator with engine coolant, and bleed the air from the cooling system.

20. Clean up any spilled engine coolant.

21. Inspect for engine coolant leaks.

THERMOSTAT

REMOVAL & INSTALLATION

See Figure 61.

1. Drain the engine coolant.
2. Remove the air intake duct.
3. Remove the battery.
4. Remove the battery base.
5. Remove the thermostat cover, then remove the thermostat.
6. Install the new thermostat with a new rubber seal, then install the thermostat cover.

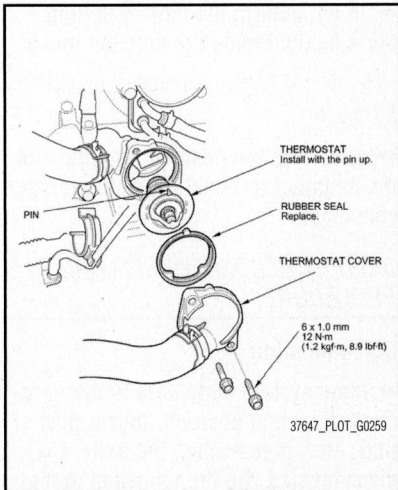

Fig. 61 Exploded view of thermostat assembly

7. Install the battery base.

8. Do the battery installation procedure.

9. Install the air intake duct.

10. Refill the radiator with engine coolant, and bleed the air from the cooling system.

11. Clean up any spilled engine coolant.

12. Inspect for engine coolant leaks.

WATER PUMP

REMOVAL & INSTALLATION

See Figure 62.

1. Drain the engine coolant.
2. Remove the timing belt.

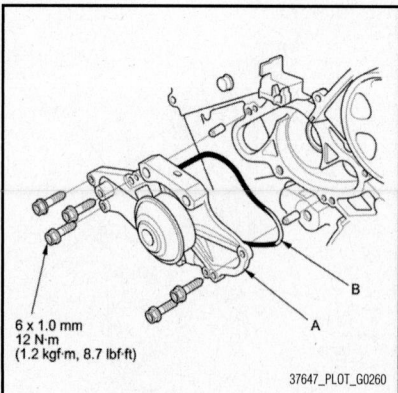

6 x 1.0 mm
12 N·m
(1.2 kgf·m, 8.7 lbf·ft)

37647_PLOT_G0260

Fig. 62 Remove the five bolts securing the water pump (A), then remove the water pump and O-ring (B)

3. Remove the timing belt adjuster.

4. Remove the five bolts securing the water pump, then remove the water pump.

5. Inspect and clean the O-ring groove and the mating surface of the engine block.

6. Install the water pump using a new O-ring.

7. Clean up any spilled engine coolant.

8. Install the timing belt adjuster.

9. Install the timing belt.

10. Refill the radiator with engine coolant, and bleed the air from the cooling system with the heater valve open.

ENGINE ELECTRICAL

BATTERY

REMOVAL & INSTALLATION

See Figure 63.

➡The battery terminal disconnection/reconnection procedure must be done before and after doing this procedure. Some system store data in memory is lost when the battery disconnected.

1. Do the battery terminal disconnection procedure.

2. Remove the two nuts securing the battery setting plate, then remove the battery setting plate and the battery.

To install:

3. Install the battery, then install the battery setting plate.

4. Tighten the two nuts equally until the battery is stable.

➡Do not deform the battery setting plate by tightening the nuts too much.

5. Do the battery terminal reconnection procedure.

➡Make sure the positive terminal and the negative terminal are not reverse-connected.

BATTERY RECONNECT/RELEARN PROCEDURE

Disconnection

➡Some system store data in memory (including seat position, mirror position, etc.) is lost when the battery is disconnected. Do the following procedures before disconnecting the battery.

1. Make sure you have the anti-theft code(s) for the audio system and/or navigation system (if equipped).

➡For some models, it may be necessary to write down the audio the presets (AM and FM), because the audio unit dose not retain the presets after the battery is disconnected.

2. Write down the XM audio presets (if equipped).

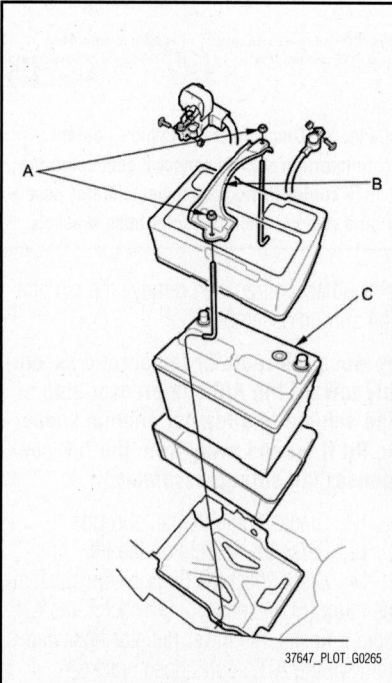

37647_PLOT_G0265

Fig. 63 Remove the two nuts (A) securing the battery setting plate, then remove the battery setting plate (B) and the battery (C)

BATTERY SYSTEM

3. Make sure the ignition switch is in LOCK (0).

4. Disconnect and isolate the negative cable from the battery.

➡Always disconnect the negative battery cable from the battery first.

5. Disconnect the positive cable from the battery.

Reconnection

➡Note the following for reconnection procedures:

- Some system store data in memory (including seat position, mirror position, etc.) is lost when the battery is disconnected. Do the following procedures to restore the system back to normal operation.

- Be sure to close the power tailgate before reconnecting the battery.

1. Clean the battery terminals.

2. Test the battery.

3. Reconnect the positive cable to the battery first, then reconnect the negative cable to the battery.

➡Always connect the positive cable to the battery first.

4. Apply multipurpose grease to the terminals to prevent corrosion.

5. Enter the anti-theft code(s) for the audio system and/or navigation system (if equipped).

6. Enter the audio presets (if applicable), and enter the XM audio presets (if equipped).

7. Set the clock (for vehicles without navigation).

ENGINE ELECTRICAL

ALTERNATOR

REMOVAL & INSTALLATION

See Figures 64 and 65.

1. Remove the air intake duct.
2. Do the battery terminal disconnection procedure.
3. Remove the engine cover.

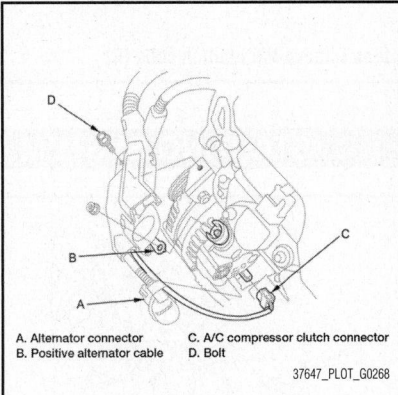

A. Alternator connector C. A/C compressor clutch connector
B. Positive alternator cable D. Bolt

37647_PLOT_G0268

Fig. 64 Disconnect the alternator connector and the positive alternator cable from the alternator

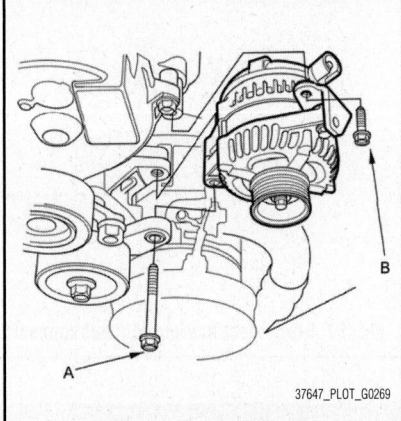

37647_PLOT_G0269

Fig. 65 Remove the mounting bolt (A) and the alternator bracket mounting bolt (B), then remove the alternator

4. Remove the coolant reservoir and the Power Steering (P/S) fluid reservoir, then remove the holder bracket.
5. Remove the accessory drive belt.
6. Disconnect the alternator connector and the positive alternator cable from the alternator.
7. Disconnect the A/C compressor clutch connector from the A/C compressor.

8. Remove the bolt securing the harness holder.
9. Remove the mounting bolt and the alternator bracket mounting bolt, then remove the alternator.

To install:

10. Install the alternator, then tighten the mounting bolt to 33 ft. lbs. (45 Nm), and the alternator bracket mounting bolt to 16 ft. lbs. (22 Nm).
11. Install the bolt securing the harness holder.
12. Connect the A/C compressor clutch connector to the A/C compressor.
13. Connect the alternator connector and the positive alternator cable to the alternator. Make sure the crimped it side of the ring terminal faces away from the alternator when you connect it.
14. Install the drive belt.
15. Install the holder bracket, then install the coolant reservoir and the Power Steering (P/S) fluid reservoir.
16. Install the engine cover.
17. Do the battery terminal reconnection procedure.
18. Install the air intake duct.

ENGINE ELECTRICAL

ADJUSTMENT

Ignition timing is controlled by the Powertrain Control Module (PCM). No adjustment is necessary or possible.

FIRING ORDERS

Firing order for the Honda J35Z4 engine is: 1–4–2–5–3–6.

IGNITION COIL

REMOVAL & INSTALLATION

See Figure 66.

1. Remove the engine cover.
2. Disconnect the ignition coil connectors, then remove the ignition coils.
3. Install the ignition coils in the reverse order of removal.

IGNITION TIMING

INSPECTION

1. Connect the Honda Diagnostic System (HDS) to the Data Link Connector (DLC).

2. Turn the ignition switch to ON (II).
3. Make sure the HDS communicates with the vehicle and the Powertrain Control Module (PCM). If it does not communicate, troubleshoot the DLC circuit.
4. Check for DTCs. If a DTC is present, diagnose and repair the cause before continuing with this test.
5. Start the engine. Hold the engine speed at 3,000 rpm with no load (in N or P) until the radiator fan comes on, then let it idle.
6. Check the idle speed.

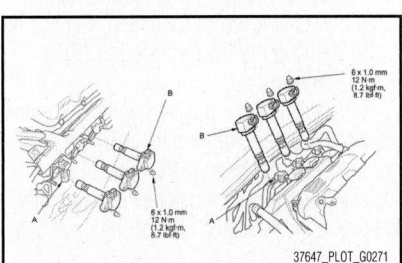

37647_PLOT_G0271

Fig. 66 Disconnect the ignition coil connectors (A), then remove the ignition coils (B)

7. Jump the SCS line with the HDS.
8. Connect the timing light to the No.1 ignition coil harness.
9. Aim the light toward the pointer on the timing belt lower cover. Check the ignition timing under a no load condition (headlights, blower fan, rear window defogger, and air conditioner are turned off).
10. If the ignition timing differs from the specification, check the cam timing. If the cam timing is OK, update the PCM if it does not have the latest software, or substitute a known-good PCM, then recheck. If the system works properly, and the PCM was substituted, replace the original PCM.
11. Disconnect the HDS and the timing light.

ADJUSTMENT

Ignition timing is controlled by the Powertrain Control Module (PCM). No adjustment is necessary or possible.

SPARK PLUGS

REMOVAL & INSTALLATION

See Figure 67.

→**Use NGK: ILZKR7B11 spark plugs.**

1. Remove the engine cover.
2. Disconnect the ignition coil connectors, then remove the ignition coils.
3. Remove the spark plugs and inspect them.
4. Check the spark plug gap:
   a. Standard (New): 0.039–0.043 inches (1.0–1.1 mm).
5. Apply a small amount of anti-seize compound to the plug threads, and screw the plugs into the cylinder head, finger-tight. Torque them to 16 ft. lbs. (22 Nm).
6. Install the ignition coils in the reverse order of removal.

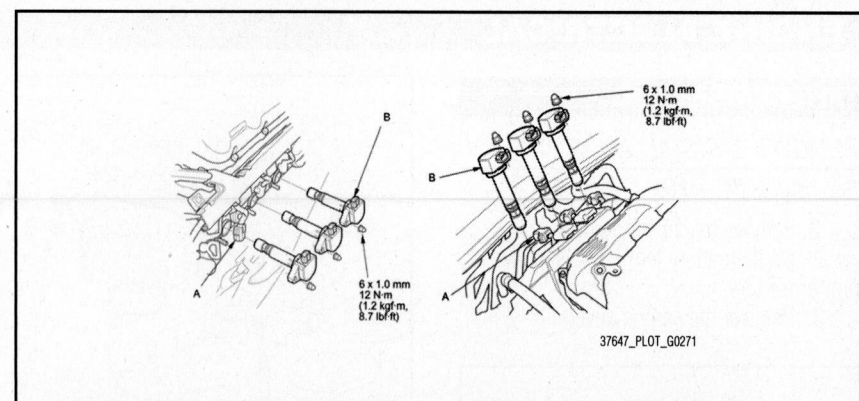

Fig. 67 Disconnect the ignition coil connectors (A), then remove the ignition coils (B)

ENGINE ELECTRICAL   STARTING SYSTEM

STARTER

REMOVAL & INSTALLATION

See Figures 68 and 69.

1. Remove the air intake duct.
2. Remove the battery.
3. Remove the battery base.
4. Remove the harness clamp, then disconnect the positive starter cable and the S terminal connector from the starter.
5. Remove the lower radiator hose bracket and the dipstick.
6. Remove the two bolts holding the starter, then remove the starter.

To install:

7. Install the starter, then tighten the mounting bolts to 33 ft. lbs. (44 Nm).

→**Always use a new gasket.**

8. Install the lower radiator hose bracket and the dipstick.
9. Connect the positive starter cable

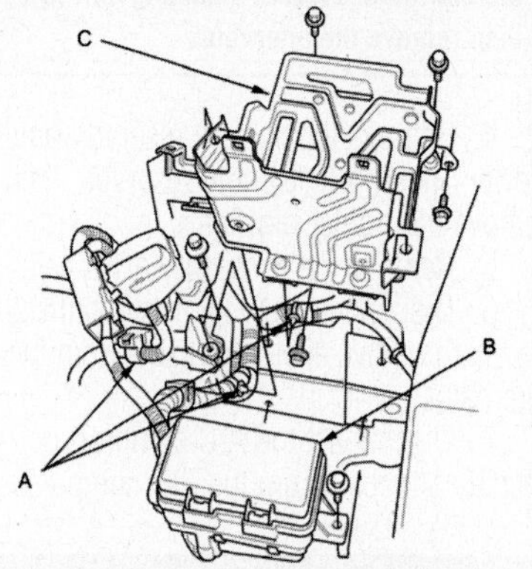

Fig. 68 Remove the harness clamps (A), and the auxiliary under-hood fuse/relay box (B), then remove the battery base (C)

and the S terminal connector to the starter, then install the harness clamp. Make sure the crimped side of the ring terminal faces away from the starter when you connect it.
10. Install the battery base.

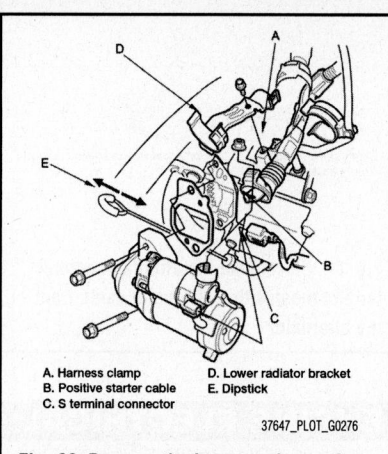

A. Harness clamp
B. Positive starter cable
C. S terminal connector
D. Lower radiator bracket
E. Dipstick

Fig. 69 Remove the harness clamp, then disconnect the positive starter cable and the S terminal connector from the starter

11. Do the battery installation procedure.
12. Install the air intake duct.
13. Start the engine to make sure the starter works properly.

ENGINE MECHANICAL

ACCESSORY DRIVE BELT SYSTEM

ADJUSTMENT

The accessory drive belt tension is set by the auto-tensioner. No adjustment is necessary or possible.

BELT ROUTINGS

Refer to the graphic in the Removal & Installation section for proper accessory drive belt routing.

INSPECTION

See Figure 70.

1. Inspect the belt for cracks or damage. If the belt is cracked or damaged, replace it.
2. Remove the Power Steering (P/S) fluid reservoir.
3. Check that the position of the auto-tensioner indicator pointer on the A/C compressor bracket is not beyond the edge of the indicator on the auto-tensioner. If the pointer is beyond the indicator, replace the drive belt.

REMOVAL & INSTALLATION

Drive Belt

See Figure 71.

1. Set a socket wrench on the drive belt

auto-tensioner, and slowly turn the wrench in the direction of the rotation arrow, then remove the drive belt.

→**This is a hydraulic type auto-tensioner, so you must turn the wrench slowly for at least 3 seconds.**

2. Install the new belt in the reverse order of removal.

Drive Belt Tensioner

See Figure 72.

1. Remove the drive belt.
2. Remove the splash shield.
3. Remove the auto-tensioner.

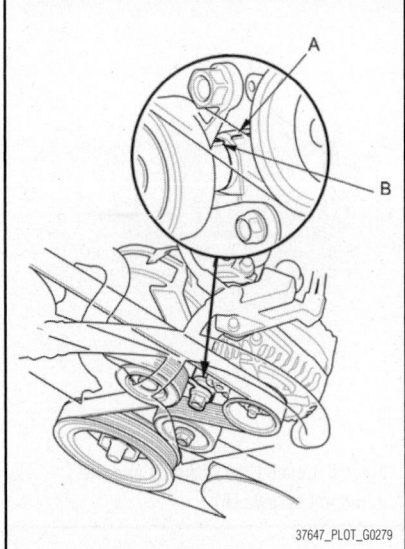

Fig. 70 Check that the position of the auto-tensioner indicator pointer (A) on the A/C compressor bracket is not beyond the edge of the indicator (B) on the auto-tensioner

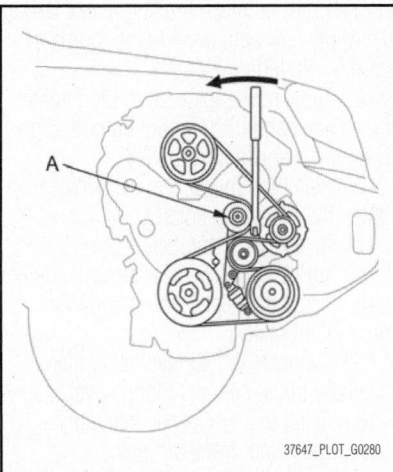

Fig. 71 Set a socket wrench on the drive belt auto-tensioner (A), and slowly turn the wrench in the direction of the rotation arrow, then remove the drive belt

4. Install the auto-tensioner in the reverse order of removal.

AIR CLEANER

REMOVAL & INSTALLATION

Air Cleaner Assembly

See Figure 73.

1. Disconnect the MAF sensor/IAT sensor connector.
2. Remove the harness clamps and the bolts.

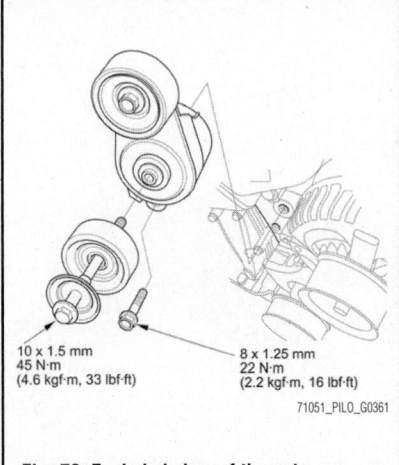

Fig. 72 Exploded view of the auto-tensioner

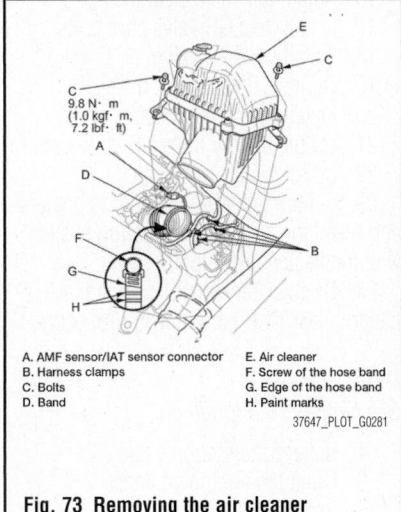

A. AMF sensor/IAT sensor connector
B. Harness clamps
C. Bolts
D. Band
E. Air cleaner
F. Screw of the hose band
G. Edge of the hose band
H. Paint marks

Fig. 73 Removing the air cleaner

3. Loosen the band, then remove the air cleaner.
4. Install the parts in the reverse order of removal.

➡**When tightening the screw of the hose band, align the edge of the hose band between the marks painted on the hose band.**

Air Filter Element

See Figure 74.

1. Open the air cleaner housing cover.
2. Remove the air cleaner element from the air cleaner housing.

➡**Be careful when removing the element, and do not turn it upside down. Debris from the element may fall into the MAF sensor/IAT sensor.**

3. Check the air cleaner element for damage or clogging. If it is damaged or clogged, replace it.

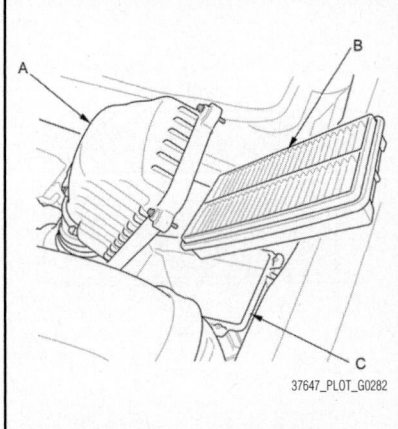

Fig. 74 Open the air cleaner housing cover (A); remove the air cleaner element (B) from the air cleaner housing (C)

➡**Do not use compressed air to clean the air cleaner element.**

4. Clean and remove any debris from inside the air cleaner housing.

To install:

5. Install the parts in the reverse order of removal.

a. If you did not replace the air cleaner element, this procedure is complete.

b. If the maintenance information required air cleaner element replacement, reset the maintenance information.

c. If the idle speed fluctuates, do the idle speed inspection procedure.

CAMSHAFT & BEARINGS

REMOVAL & INSTALLATION

Front

See Figures 75 through 77.

1. Remove the air intake duct.
2. Remove the battery.
3. Remove the battery base.
4. Drain the engine coolant.
5. Disconnect the radiator hoses.
6. Remove the Exhaust Gas Recirculation (EGR) valve.
7. Remove the EGR valve stud bolts.
8. Remove the timing belt.
9. Remove the rocker arm assembly.

a. Loosen the locknuts and adjusting screws.

b. Remove the rocker shaft bridge mounting bolts, the rocker shaft holder mounting bolts, and the rocker arm assembly.

c. Loosen the rocker shaft bridge mounting bolts and the rocker shaft holder mounting bolts in sequence two

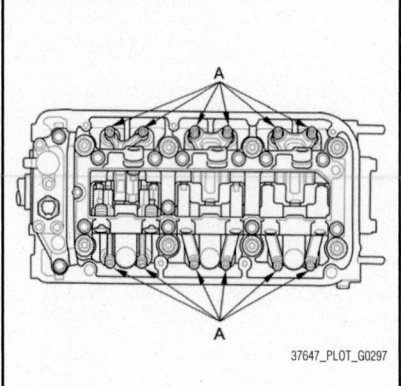

Fig. 75 Loosen the locknuts and adjusting screws (A)

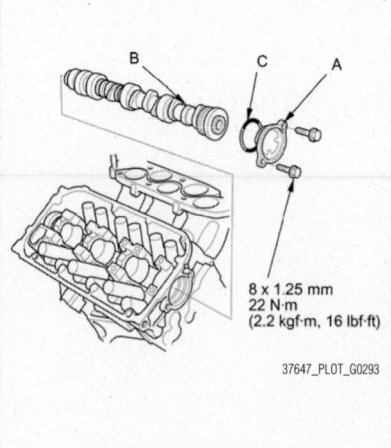

8 x 1.25 mm
22 N·m
(2.2 kgf·m, 16 lbf·ft)

Fig. 77 Remove the thrust cover (A), then remove the camshaft (B) and O-ring (C)

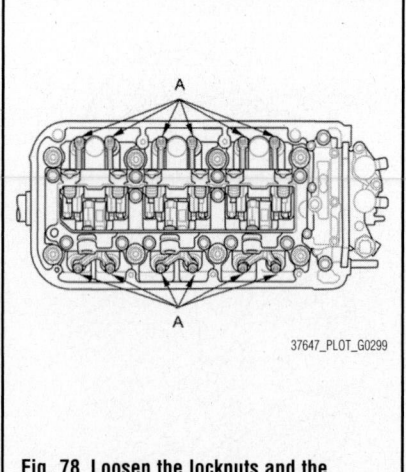

Fig. 78 Loosen the locknuts and the adjusting screws (A)

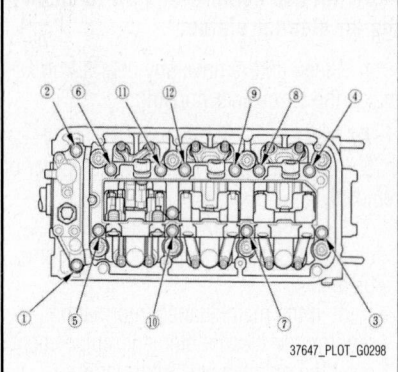

Fig. 76 Loosen the rocker shaft bridge mounting bolts and the rocker shaft holder mounting bolts in sequence

turns at a time, starting at the ends in a crisscross pattern working toward the middle to prevent damaging the valves or the rocker arm assembly.

d. When removing the rocker arm assembly, do not remove the rocker shaft bridge mounting bolts and the rocker shaft holder mounting bolts. The bolts will keep the rocker arms on the shafts.
10. Remove the front camshaft pulley.
11. Remove the thrust cover, then remove the camshaft.

To install:

12. Install the camshaft in the reverse order of removal using a new O-ring. Apply new engine oil to the journals and cam lobes.
13. Apply new engine oil to the threads of the camshaft pulley mounting bolt, then install the front camshaft pulley.
14. Install the rocker arm assembly, then tighten the mounting bolts.
15. Install the timing belt.

16. Adjust the valve clearance.
17. Install the EGR valve stud bolts.
18. Install the EGR valve.
19. Connect the radiator hoses.
20. Install the battery base.
21. Do the battery installation procedure.
22. Install the air intake duct.
23. Fill the radiator with engine coolant and bleed the air from the cooling system with the heater valve open.
24. Do the Crankshaft Position (CKP) pattern clear/CKP pattern lean procedure.

Rear

See Figures 78 through 80.

1. Relieve the fuel pressure.
2. Drain the engine coolant.
3. Remove the intake air duct.
4. Remove the air cleaner assembly.
5. Remove the quick-connect fitting cover, then disconnect the fuel feed hose.
6. Disconnect the heater hoses and the Evaporative Emission (EVAP) canister hose, then remove the purge joint bracket.
7. Remove the timing belt.
8. Remove the rocker arm assembly.
   a. Loosen the locknuts and the adjusting screws (A).
   b. Remove the rocker shaft bridge mounting bolts, the rocker shaft holder mounting bolts, and the rocker arm assembly.
   c. Loosen the rocker shaft bridge mounting bolts and the rocker shaft holder mounting bolts in sequence two turns at a time, starting at the ends in a crisscross pattern working toward the middle to prevent damaging the valves or the rocker arm assembly.
   d. When removing the rocker arm assembly, do not remove the rocker shaft bridge mounting bolts and the rocker

shaft holder mounting bolts. The bolts will keep the rocker arms on the shafts.
9. Remove the rear camshaft pulley.
10. Remove the thrust cover, then remove the rear camshaft.

To install:

11. Install the rear camshaft in the reverse order of removal using a new O-ring (C). Apply new engine oil to the journals and the cam lobes.
12. Apply new engine oil to the threads of the camshaft pulley mounting bolt, then install the rear camshaft pulley.
13. Install the rocker arm assembly, then tighten the mounting bolts.
14. Install the timing belt.
15. Adjust the valve clearance.
16. Install the heater hoses and the purge joint bracket.
17. Connect the fuel feed hose, then install the quick-connect fitting cover.
18. Install the air cleaner assembly.
19. Install the intake air duct.

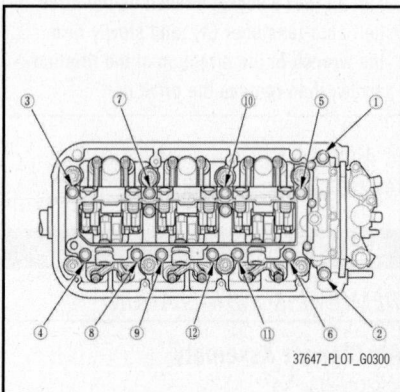

Fig. 79 Loosen the rocker shaft bridge mounting bolts and the rocker shaft holder mounting bolts in sequence

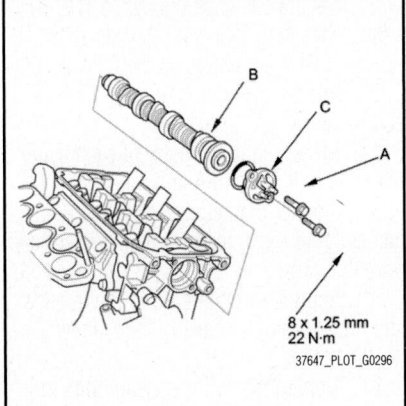

Fig. 80 Remove the thrust cover (A), then remove the rear camshaft (B)

20. Inspect for fuel leaks. Turn the ignition switch to ON (II) (do not operate the starter) so the fuel pump runs for about 2 seconds and pressurizes the fuel line. Repeat this operation three times, then check for fuel leakage at any point in the fuel line.

21. Fill the radiator with engine coolant, and bleed the air from the cooling system with the heater valve open.

22. Do the CKP pattern clear/CKP pattern learn procedure, as outlined under the Crankshaft Position (CKP) sensor in the Engine Performance & Emission Controls Section.

CRANKSHAFT FRONT SEAL

REMOVAL & INSTALLATION

Special Tools Required: Oil Seal Driver, 64 mm 070AD-RCAA100

1. Remove the timing belt, the timing belt stopper, and the timing belt drive pulley.

2. Remove the pulley end crankshaft oil seal.

3. Clean and dry the crankshaft oil seal housing.

4. Apply a light coat of new engine oil to the crankshaft and to the lip of the crankshaft oil seal.

5. Apply a light coat of new engine oil around the crankshaft oil seal, then using the oil seal driver, 64 mm, drive in the crankshaft oil seal until the driver bottoms against the oil pump. When the seal is in place, clean any excess grease off the crankshaft, and check that the oil seal lip is not distorted.

6. Install the timing belt drive pulley, the timing belt stopper, and the timing belt.

CRANKSHAFT PULLEY

REMOVAL & INSTALLATION

See Figures 81 through 83.

Special Tools Required:
- Handle, 6-25-660L 07JAB-001020B
- Holder Attachment, 50 mm, Offset 07MAB-PY3010A
- Socket Wrench, 19 x 90L 07JAA-001020A or equivalent

1. Raise the vehicle on the lift.
2. Remove the right front wheel.
3. Remove the splash shield.
4. Remove the drive belt.
5. Hold the pulley with the holder and the holder attachment, 50mm, offset.
6. Remove the bolt with a heavy duty socket, 19 mm and a breaker bar, then remove the crankshaft pulley.

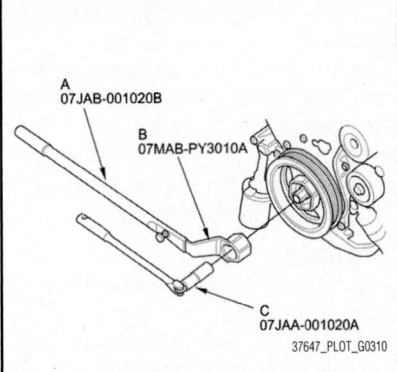

Fig. 81 Hold the pulley with the holder (A) and the holder attachment, 50mm, offset (B); remove the bolt with a heavy duty socket, 19 mm (C) and a breaker bar

To install:

7. Remove any oil and clean the pulleys, the crankshaft, the bolt, and the washer. Lubricate using new engine oil as shown.

8. Install the crankshaft pulley, and tighten the bolt. Do not use an impact wrench.

a. Hold the pulley with the handle and the holder attachment, 50mm, offset. Tighten the bolt to 47 ft. lbs. (64 Nm) with a torque wrench and a socket, 19 mm.

b. Mark the bolt head and the crankshaft pulley as shown, then tighten the bolt an additional 60 degrees (The mark on the bolt head lines up with the mark on the crankshaft pulley).

9. Install the drive belt.

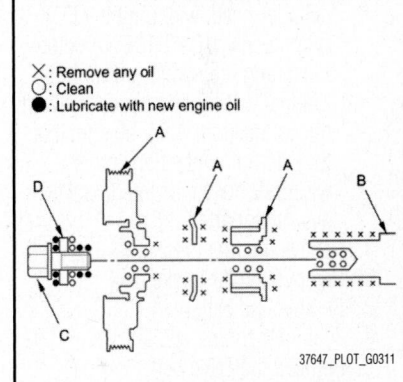

Fig. 82 Remove any oil and clean the pulleys, the crankshaft, the bolt, and the washer; lubricate using new engine oil as shown

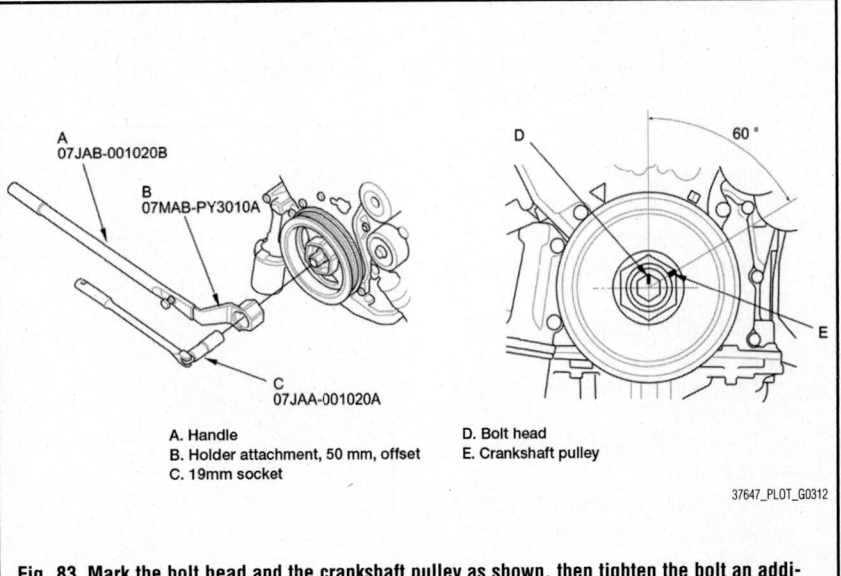

A. Handle
B. Holder attachment, 50 mm, offset
C. 19mm socket

D. Bolt head
E. Crankshaft pulley

Fig. 83 Mark the bolt head and the crankshaft pulley as shown, then tighten the bolt an additional 60 degrees

10. Install the splash shield.
11. Install the right front wheel.

CYLINDER HEAD

REMOVAL & INSTALLATION

See Figures 84 through 91.

➡️**Note the following when removing the cylinder heads:**

- Use fender covers to avoid damaging painted surfaces.
- To avoid damaging the wiring and terminals, unplug the wiring connectors carefully while holding the connector portion.
- Connect the Honda Diagnostic System (HDS) to the Data Link Connector (DLC), and monitor the Engine Coolant Temperature (ECT) sensor 1. To avoid damaging the cylinder head, wait until the ECT drops below 100°F (38°C) before loosening the cylinder head bolts.
- Mark all wiring and hoses to avoid misconnection. Also, be sure that they do not contact any other wiring or hoses, or interfere with any other parts.

1. Relieve the fuel pressure.
2. Remove the air intake duct.
3. Remove the battery.
4. Drain the engine coolant.
5. Remove the drive belt.
6. Remove the Power Steering (P/S) pump and the bolt securing the P/S hose bracket.
7. Remove the alternator.
8. Remove the timing belt.
9. Remove the intake manifold.
10. Remove the six ignition coils.
11. Disconnect the following engine wire harness connectors, and remove the wire harness clamps from the cylinder head:
   - Six injector connectors
   - Engine Coolant Temperature (ECT) sensor 1 connector
   - Front rocker arm oil pressure switch connector
   - Rear rocker arm oil pressure switch connector
   - Camshaft Position (CMP) sensor connector
   - Front Air Fuel Ratio (A/F) sensor 1 connector
   - Rear Air Fuel Ratio (A/F) sensor 1connector
   - Front secondary Heated Oxygen Sensor 2 (secondary HO2S) connector
   - Rear secondary Heated Oxygen

Sensor 2 (secondary HO2S) connector
   - Rocker arm oil control solenoid A (BANK 1) connector
   - Rocker arm oil control solenoid A (BANK 2) connector
   - Rocker arm oil control solenoid B (BANK 1) connector
   - Knock sensor connector

12. Remove the front warm up three way catalytic converter (front WU-TWC) and the rear warm up three way catalytic converter (rear WU-TWC).
13. Remove the quick-connect fitting cover, then disconnect the fuel feed hose.
14. Remove the connector bracket from the front cylinder head.
15. Remove the harness clamp bracket (A) from the rear cylinder head.
16. Remove the injector bases.
17. Remove the water passage.
18. Remove the camshaft pulleys and the back covers.
19. Remove the cylinder head covers.
20. Remove the cylinder head bolts. To prevent warping, loosen the bolts in sequence ⅓ turn at a time; repeat the sequence until all the bolts are loosened.
21. Remove the cylinder heads.

To install:

22. Clean the cylinder head and the engine block surface.
23. Clean and install the oil control orifices using the new O-rings.
24. Install the dowel pins and the new cylinder head gaskets.
25. Clean the timing belt pulleys, the timing belt guide plate, and the upper and lower covers.
26. Set the timing belt drive pulley to Top Dead Center (TDC) by aligning the TDC mark on the tooth of the timing belt drive pulley with the pointer on the oil pump.

27. Set the camshaft pulleys to TDC by aligning the TDC marks on the camshaft pulleys with the pointers on the back covers.
28. Put the cylinder head onto the engine block.
29. Measure the diameter of each cylinder head bolt at point A and point B.
30. If either diameter is less than 0.42 inches (10.6 mm), replace the cylinder head bolt.
31. Apply new engine oil to the threads and under the bolt heads of all cylinder head bolts.
32. Tighten the cylinder head bolts in sequence to 22 ft. lbs. (29 Nm) using a beam-type torque wrench. The tightening sequence starts in the middle and uses a crisscross pattern working out to the ends. When using a preset click-type torque wrench, be sure to torque slowly and do not over tighten. If a bolt makes any noise while you are torquing it, loosen the bolt and retighten it from the first step.
33. After torquing, tighten all cylinder head bolts in two steps (90° per step) using the sequence in step 11. If you are using a new cylinder head bolt, tighten the bolt an extra 90°.

➡️**Remove the cylinder head bolt if you tightened it beyond the specified angle, and go back to step 8 of the procedure. Do not loosen it back to the specified angle.**

34. Install the timing belt.
35. Adjust the valve clearance.
36. Install the cylinder head covers.
37. Install the water passage.
38. Install the injector bases.
39. Install the connector bracket to the front cylinder head.
40. Install the harness clamp bracket to the rear cylinder head.

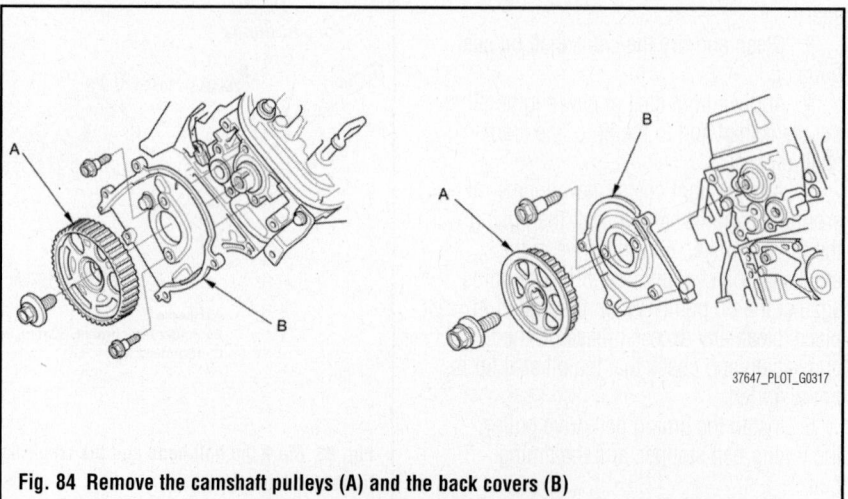

37647_PLOT_G0317

Fig. 84 Remove the camshaft pulleys (A) and the back covers (B)

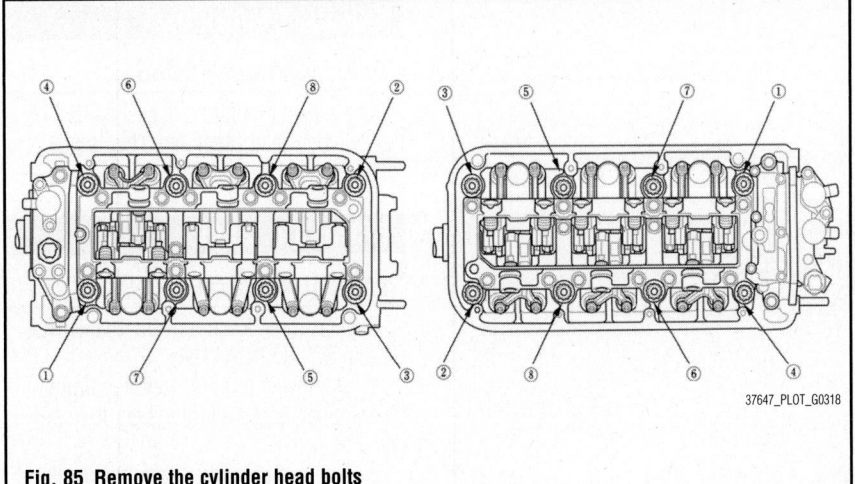

Fig. 85 Remove the cylinder head bolts

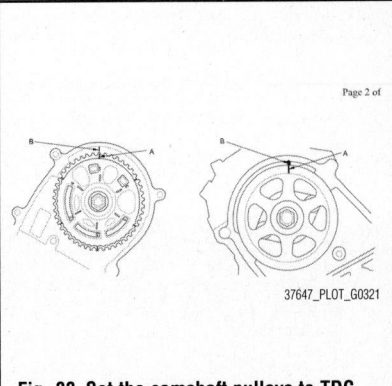

Fig. 88 Set the camshaft pulleys to TDC by aligning the TDC marks (A) on the camshaft pulleys with the pointers (B) on the back covers

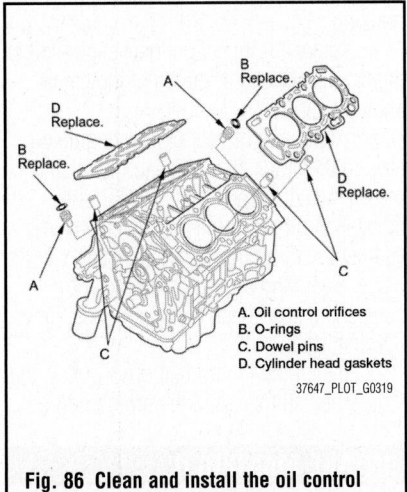

A. Oil control orifices
B. O-rings
C. Dowel pins
D. Cylinder head gaskets

Fig. 86 Clean and install the oil control orifices using the new O-rings

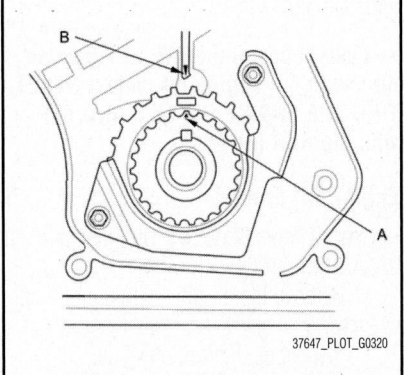

Fig. 87 Set the timing belt drive pulley to TDC by aligning the TDC mark (A) on the tooth of the timing belt drive pulley with the pointer (B) on the oil pump

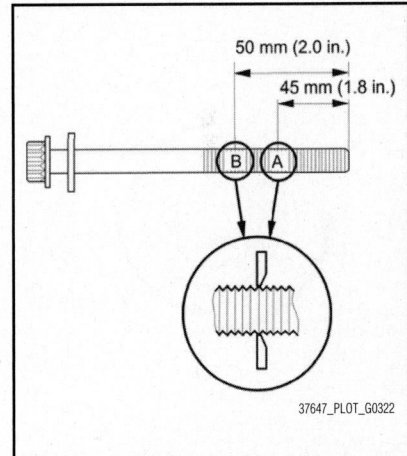

Fig. 89 Measure the diameter of each cylinder head bolt at point A and point B

41. Connect the fuel feed hose, then install the quick-connect fitting cover.

Install the and the rear Warm Up Three Way Catalytic Converter (rear WU-TWC).

42. Connect the following engine wire harness connectors, and install the wire harness clamps to the cylinder head:
- Six injector connectors
- Engine Coolant Temperature (ECT) sensor 1 connector
- Front rocker arm oil pressure switch connector
- Rear rocker arm oil pressure switch connector
- Camshaft Position (CMP) sensor connector
- Front Air Fuel Ratio (A/F) (sensor 1) connector
- Rear Air Fuel Ratio (A/F) (sensor 1) connector
- Front secondary Heated Oxygen Sensor 2 (secondary HO2S) connector
- Rear secondary Heated Oxygen

Sensor 2 (secondary HO2S) connector
- Rocker arm oil control solenoid A (BANK 1) connector
- Rocker arm oil control solenoid A (BANK 2) connector
- Rocker arm oil control solenoid B (BANK 1) connector
- Knock sensor connector

43. Install the six ignition coils.
44. Install the intake manifold.
45. Install the alternator.
46. Install the Power Steering (P/S) pump and tighten the bolt securing the P/S hose bracket.
47. Install the drive belt.
48. Do the battery installation procedure.
49. Install the air intake duct.
50. After installation, check that all tubes, hoses, and connectors are installed correctly.
51. Inspect for fuel leaks. Turn the ignition switch to ON (II) (do not operate the starter) so the fuel pump runs for about 2

seconds and pressurizes the fuel line. Repeat this operation three times, then check for fuel leakage at any point in the fuel line.

52. Refill the radiator with engine coolant, and bleed the air from the cooling system with the heater valve open.
53. Check for fluid leaks.
54. Do the Powertrain Control Module (PCM) idle learn procedure.
55. Do the CKP pattern clear/CKP pattern learn procedure, as outlined under the Crankshaft Position (CKP) sensor in the Engine Performance & Emission Controls Section.
56. Inspect the idle speed.
57. Inspect the ignition timing.

ENGINE OIL & FILTER

OIL & FILTER CHANGE

Engine Oil

1. Warm up the engine.
2. Remove the drain bolt and drain the engine oil.

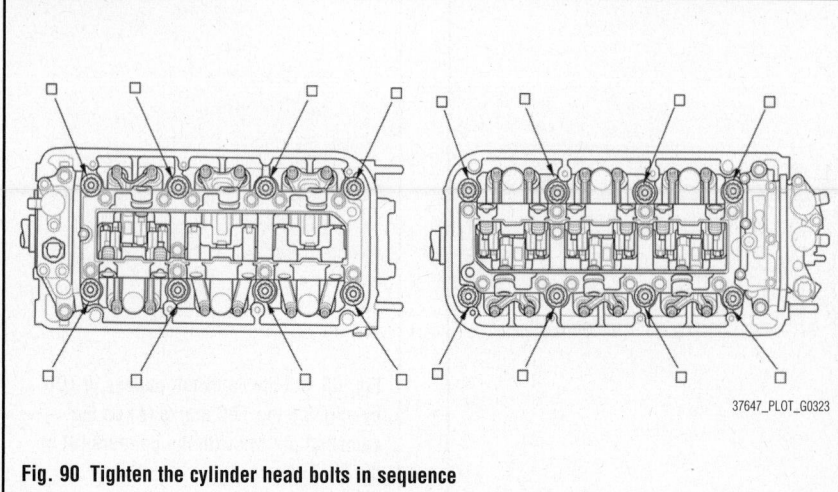

Fig. 90 Tighten the cylinder head bolts in sequence

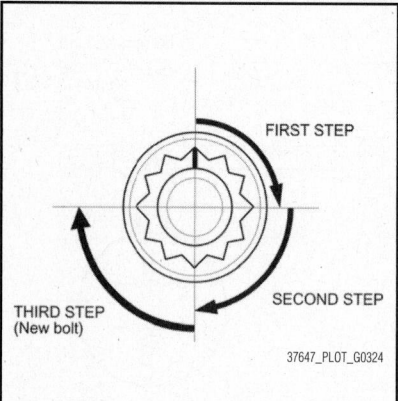

Fig. 91 After torquing, tighten all cylinder head bolts in two steps (90° per step) using the sequence

3. Reinstall the drain bolt with a new washer.

4. Refill the engine with the recommended engine oil.

5. Run the engine for more than 3 minutes, then check for oil leakage.

6. Turn the ignition switch to LOCK (0).

7. Connect the Honda Diagnostic System (HDS) to the Data Link Connector (DLC).

8. Turn the ignition switch to ON (II).

9. Make sure the HDS communicates with the vehicle and the Powertrain Control Module (PCM). If it does not communicate, troubleshoot the DLC circuit.

10. Select GAUGES in the BODY ELECTRICAL with the HDS.

11. Select ADJUSTMENT in the GAUGES with the HDS.

12. Select MAINTENANCE INFORMATION in the ADJUSTMENT with the HDS.

13. Select RESET in the MAINTENANCE INFORMATION with the HDS.

14. Select RESETTING THE ENGINE OIL LIFE with the HDS.

➡**If you changed the ATF at the same time with the engine oil, select RESETTING THE ENGINE OIL LIFE AND ATF with the HDS instead.**

Engine Oil Filter

Special Tools Required: Oil Filter Wrench 07AAA-PLCA100

1. Remove the oil filter with the oil filter wrench.

2. Inspect the filter to make sure the rubber seal is not stuck to the oil filter seating surface of the engine.

3. Inspect the threads and the rubber seal on the new filter. Clean the seat on the oil filter base, then apply a light coat of new engine oil to the filter rubber seal. Use only filters with a built-in bypass system.

4. Install the oil filter by hand.

5. After the rubber seal seats, tighten the oil filter ¾ turn clockwise with the oil filter wrench.

6. If four numbers or marks (1 to 4 or ▼ to ▼▼▼▼) are printed around the outside of the filter, you can use the following procedure to tighten the filter:

   a. Spin the filter on until its seal lightly seats against the oil filter base, and note which number or mark is at the bottom.

   b. Tighten the filter by turning it clockwise three numbers or marks from the one you noted. For example, if mark ▼ is at the bottom when the seal is lightly seated, tighten the filter until the mark ▼▼▼▼ comes around to the bottom.

7. After installation, fill the engine with oil up to the specified level, run the engine for more than 3 minutes, then check for oil leakage.

COMPONENT LOCATIONS

The exhaust manifold is cast as part of the cylinder head assembly. This is no longer a separate component.

FRONT OIL SEAL

REMOVAL & INSTALLATION

Special Tools Required: Oil Seal Driver, 64 mm 070AD-RCAA100

1. Remove the timing belt, the timing belt stopper, and the timing belt drive pulley.

2. Remove the pulley end crankshaft oil seal.

3. Clean and dry the crankshaft oil seal housing.

4. Apply a light coat of new engine oil to the crankshaft and to the lip of the crankshaft oil seal.

5. Apply a light coat of new engine oil around the crankshaft oil seal, then using the oil seal driver, 64 mm, drive in the crankshaft oil seal until the driver bottoms against the oil pump. When the seal is in place, clean any excess grease off the crankshaft, and check that the oil seal lip is not distorted.

6. Install the timing belt drive pulley, the timing belt stopper, and the timing belt.

INTAKE MANIFOLD

REMOVAL & INSTALLATION

See Figures 92 through 96.

1. Remove the engine cover.

2. Disconnect the breather pipe, then remove the intake air duct.

3. Disconnect the Positive Crankcase Ventilation (PCV) hose, the brake booster vacuum hose, and the Intake Manifold Tuning (IMT) actuator connector.

4. Disconnect the Evaporative Emission (EVAP) canister hose, the EVAP canister purge valve connector, the throttle actuator connector, and the Manifold Absolute Pressure (MAP) sensor connector.

5. Disconnect and plug the water bypass hoses.

6. Remove the upper cover mounting bolts and nuts sequentially in three steps, then remove the upper cover.

7. Remove the intake manifold mounting bolts and nuts sequentially in three steps, then remove the intake manifold.

To install:

8. Install the intake manifold. Tighten the bolts and nuts sequentially in three

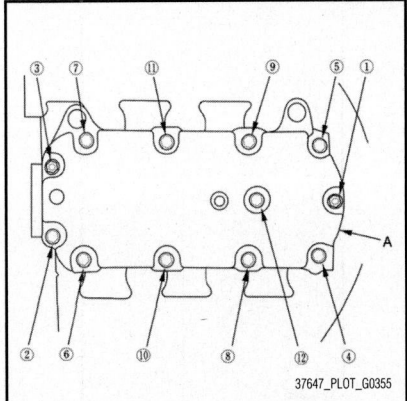

Fig. 92 Remove the upper cover mounting bolts and nuts sequentially in three steps, then remove the upper cover (A)

steps. Final torque should be 16 ft. lbs. (22 Nm). Always use a new intake manifold gasket.

9. Install the upper cover. Tighten the bolts and nuts sequentially in three steps. The final torque should be 9 ft. lbs. (12 Nm). Always use a new intake manifold gasket.

10. Connect the water bypass hoses.

11. Connect the MAP sensor connector, the throttle actuator connector, the EVAP canister purge valve connector, and the EVAP canister hoses.

12. Connect the IMT actuator connector, the brake booster vacuum hose, and the PCV hose.

13. Install the intake air duct, then connect the breather pipe.

➡**When tightening the screw of the hose band, align the edge of the hose band with the mark painted on the hose band. If you tighten the screw over the mark, replace the hose band.**

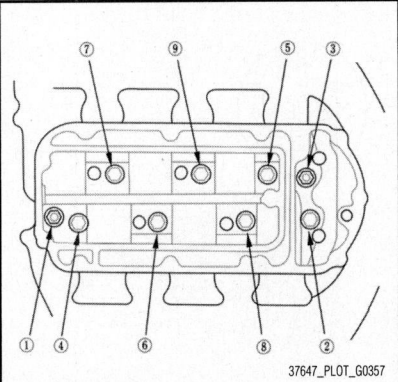

Fig. 93 Remove the intake manifold mounting bolts and nuts sequentially in three steps, then remove the intake manifold

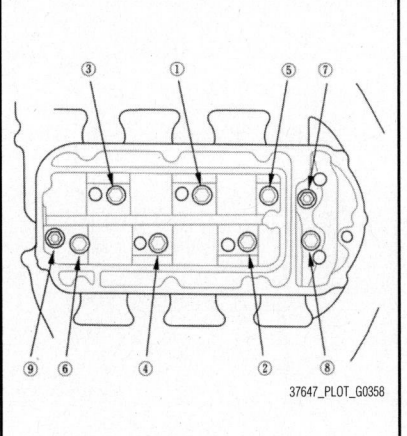

Fig. 94 Install the intake manifold

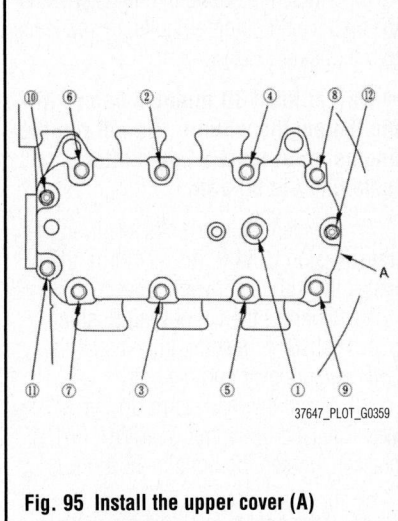

Fig. 95 Install the upper cover (A)

14. Install the engine cover.

15. Clean up any spilled engine coolant.

16. After installation, check that all

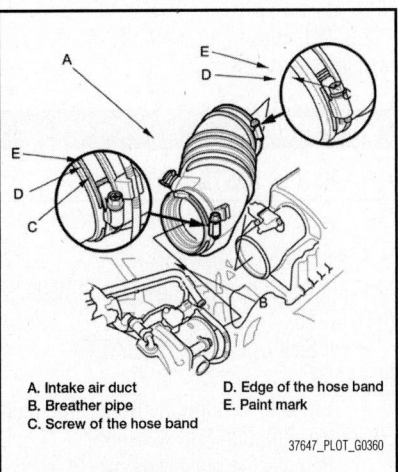

A. Intake air duct
B. Breather pipe
C. Screw of the hose band
D. Edge of the hose band
E. Paint mark

Fig. 96 Install the intake air duct, then connect the breather pipe

tubes, hoses, and connectors are installed correctly.

17. Refill the radiator with engine coolant, and bleed the air from the cooling system.

INTAKE MANIFOLD COVER

COMPONENT LOCATIONS

Refer to Intake Manifold Removal and Installation. This component is referenced during the procedure.

OIL PAN

REMOVAL & INSTALLATION
See Figures 97 through 100.

1. If the engine is already out of the vehicle, go to step 6.

2. Raise the vehicle on the lift.

3. Drain the engine oil.

4. Remove the front subframe stiffener.

5. Remove exhaust pipe A.

6. Remove the rear Warm Up Three Way Catalytic Converter (rear WU-TWC) bracket.

7. Remove the Crankshaft Position (CKP) sensor cover and the bolt, then disconnect the CKP sensor connector.

8. Remove the torque converter cover and the four bolts securing the transmission.

9. Remove the bolts securing the oil pan.

10. Using a flat blade screwdriver, separate the oil pan from the engine block in the places shown.

11. Remove the oil pan.

To install:

12. Remove all of the old liquid gasket from the oil pan mating surfaces, the bolts, and the bolt holes.

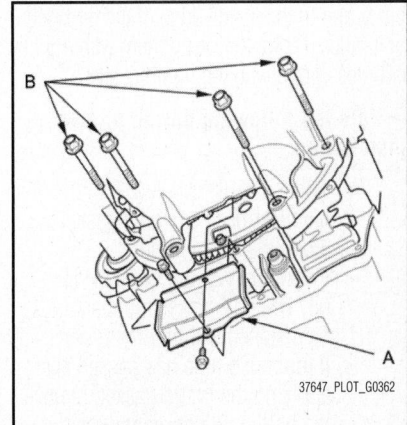

Fig. 97 Remove the torque converter cover (A) and the four bolts (B) securing the transmission

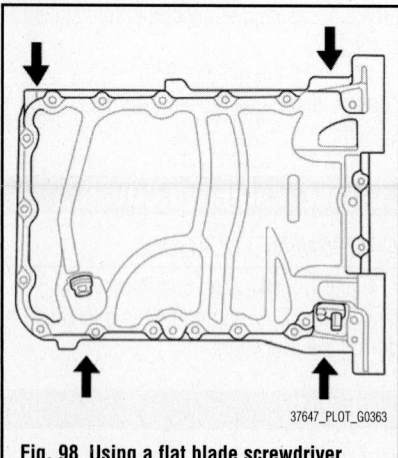

Fig. 98 Using a flat blade screwdriver, separate the oil pan from the engine block in the places shown

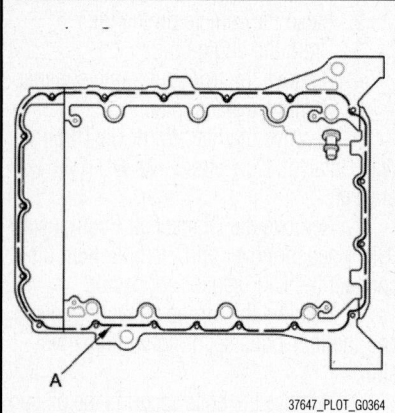

Fig. 99 Apply a 0.10 inches (2.5 mm) diameter bead of liquid gasket along the broken line (A)

13. Clean and dry the oil pan mating surfaces.

14. Apply liquid gasket (P/N 08717-0004, 08718-0003, 08718-0004, or 08718-0009) to the oil pan mating surface of the engine block and to the inside edge of the threaded bolt holes. Install the component within 5 minutes of applying the liquid gasket.

➡**Note the following during oil pan installation:**

- Apply a 0.10 inches (2.5 mm) diameter bead of liquid gasket along the broken line.
- If you apply liquid gasket P/N 08718-0012, the component must be installed within 4 minutes.
- If too much time has passed after applying the liquid gasket, remove the old liquid gasket and residue, then reapply the new liquid gasket.

15. Install the oil pan on the engine block.

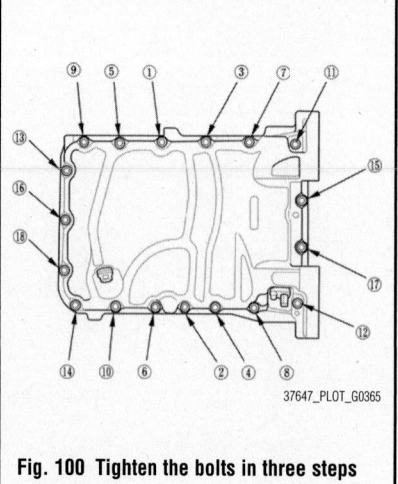

Fig. 100 Tighten the bolts in three steps in sequence

16. Tighten the bolts in three steps. In the final step, tighten all bolts, in sequence, to 9 ft. lbs. (12 Nm).

➡**Wait at least 30 minutes before filling the engine with oil. Do not run the engine for at least 3 hours after installing the oil pan.**

17. Tighten the four bolts securing the transmission to 54 ft. lbs. (74 Nm), then install the torque converter cover.

18. Connect the Crankshaft Position (CKP) sensor connector, then install the CKP sensor cover and the bolt.

19. Install the rear Warm Up Three Way Catalytic Converter (rear WU-TWC) bracket. Tighten the bolts to 16 ft. lbs. (22 Nm).

20. If the engine is still in the vehicle, do the following steps.

21. Install exhaust pipe A using new gaskets and new self-locking nuts.

22. Install the front subframe stiffener. Tighten the mounting bolts to 43 ft. lbs. (59 Nm).

23. Refill the engine with the recommended engine oil.

OIL PUMP

REMOVAL & INSTALLATION

See Figures 101 through 104.

Special Tools Required*:
- Engine Support Hanger, A and Reds AAR-T1256
- Oil Seal Driver, 64 mm 070AD-RCAA100
- *Available through the Honda Tool and Equipment Program 888-424-6857

1. Remove the bulkhead cover.

   a. Remove the clips, then remove the air intake duct.

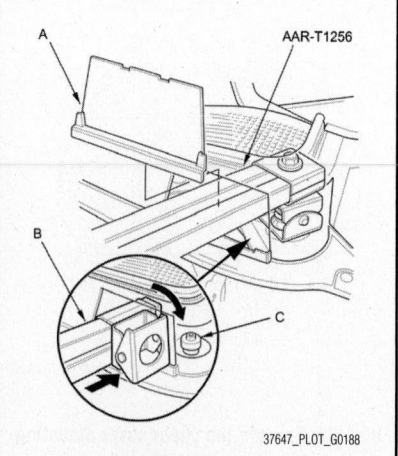

Fig. 101 Remove the cowl top side cover (A) from both sides; insert the hanger beam (B) through the opening then rotate it over the damper (C)

   b. Remove the clips, then remove the front bulkhead cover.

2. Remove the drive belt.

3. Remove the Power Steering (P/S) pump and the P/S line bracket.

4. Remove the cowl top side cover from both sides.

5. With the engine support hanger (AAR-T1256) on its side, insert the hanger beam through the opening, then rotate it over the damper.

6. Install the engine support hanger (AAR-T1256) onto the vehicle as shown, and attach the hook to the engine hanger. Tighten the wing nut by hand, and lift and support the engine/transmission.

Note the following:
- Be careful when working around the windshield.
- Be careful not to damage the hood

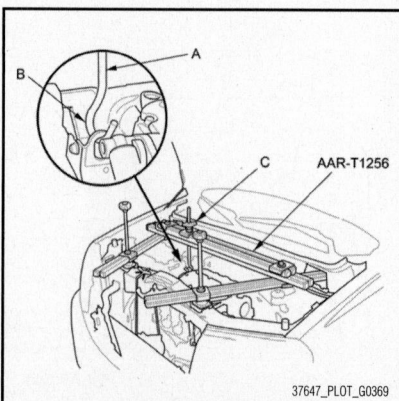

Fig. 102 Install the engine support hanger and attach the hook (A) to the engine hanger (B); tighten the wing nut (C) by hand

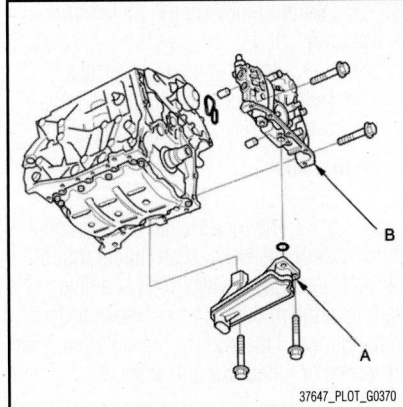

Fig. 103 Remove the oil strainer (A); remove the mounting bolts, then remove the oil pump assembly (B)

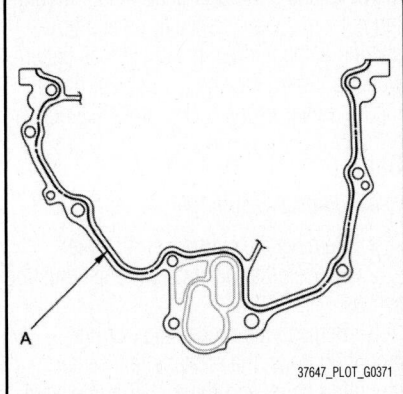

Fig. 104 Apply a 0.10 inches (2.5 mm) diameter bead of liquid gasket along the broken line (A)

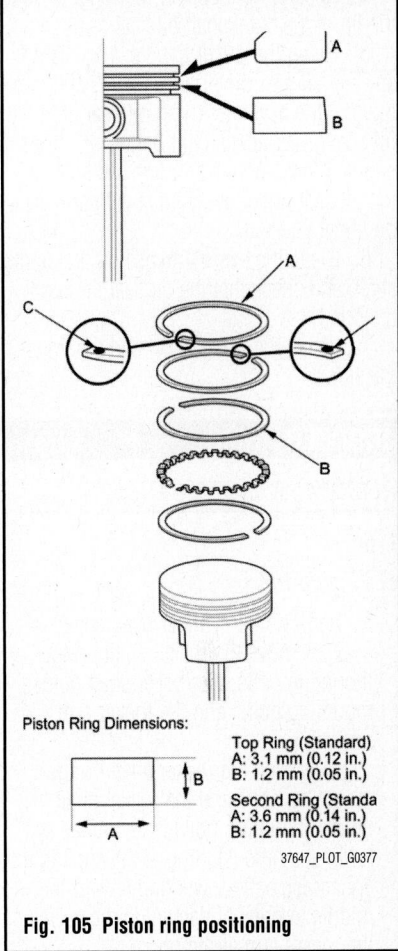

Fig. 105 Piston ring positioning

opener cable when installing the engine support hanger at the front bulkhead.

• AAR-T1256 two sets required for stacking additional cross section bar.

7. Remove the timing belt.

8. Remove the oil filter base/oil filter assembly.

9. Remove the oil pan.

10. Remove the oil strainer.

11. Remove the mounting bolts, then remove the oil pump assembly.

To install:

12. Remove the old oil seal from the oil pump.

13. Clean and dry the crankshaft oil seal housing.

14. Using the oil seal driver, 64 mm, drive in the new crankshaft oil seal until the oil seal driver bottoms on the pump.

15. Remove all of the old liquid gasket from the oil pump mating surfaces, the bolts, and the bolt holes.

16. Clean and dry the oil pump mating surfaces.

17. Apply liquid gasket (P/N 08717-0004, 08718-0003, 08718-0004, or 08718-0009) to the engine block mating surface of the oil pump and to the inside edge of the threaded bolt holes. Install the component within 5 minutes of applying the liquid gasket.

Note the following:

• Apply a 0.10 inches (2.5 mm) diameter bead of liquid gasket along the broken line.

- If you apply liquid gasket P/N 08718-0012, the component must be installed within 4 minutes.
- If too much time has passed after applying the liquid gasket, remove the old liquid gasket and residue, then reapply the new liquid gasket.

18. Apply a light coat of new engine oil to the lip of the crankshaft oil seal, and apply new engine oil to the new O-ring.

19. Install the dowel pins, then align the inner rotor with the crankshaft, and install the oil pump.

➡**Wait at least 30 minutes before filling the engine with oil. Do not run the engine for at least 3 hours after installing the oil pump.**

20. Clean the excess grease off the crankshaft, and check the seal for distortion.

21. Install the oil strainer with a new O-ring.

22. Install the oil pan.

23. Install the oil filter base/oil filter assembly with a new O-ring.

24. Install the timing belt.

25. Remove the engine support hanger from the vehicle.

26. Install the both sides cowl top side lid on the cowl cover.

27. Install the Power Steering (P/S) pump and the P/S line bracket.

28. Install the drive belt.

29. Install the bulkhead cover.

30. Refill the engine with the recommended engine oil.

PISTONS & RINGS

POSITIONING
See Figures 105 and 106.

REAR MAIN SEAL

REMOVAL & INSTALLATION

Special Tools Required:
• Driver Handle, 15 x 135L 07749-0010000

• Oil Seal Driver Attachment, 106 mm 070AD-RCA0200

1. Remove the transmission and the drive plate.

2. Remove the transmission end crankshaft oil seal.

3. Clean and dry the crankshaft oil seal housing.

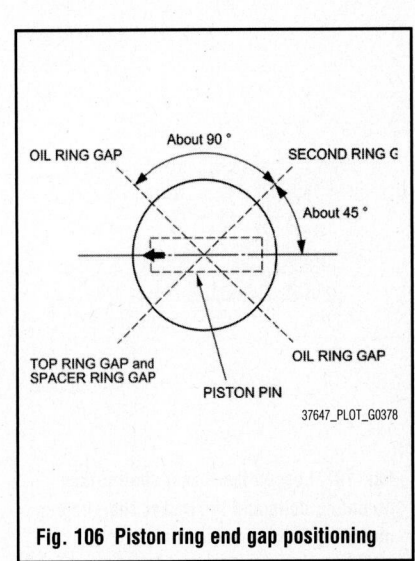

Fig. 106 Piston ring end gap positioning

4. Apply a light coat of new engine oil to the lip of the crankshaft oil seal.

5. Using the driver handle, 15 x 135L and the oil seal driver attachment, 106 mm, drive in the new crankshaft oil seal until the oil seal driver attachment bottoms on the engine block end cover. Align the hole in the oil seal driver attachment with the pin on the crankshaft.

6. Clean the excess grease off the crankshaft, and check that the oil seal lip is not distorted.

7. Install the drive plate, and the transmission.

ROCKER ARMS

REMOVAL & INSTALLATION

Front

See Figure 107.

1. Remove the cylinder head cover.
   a. Remove the rocker shaft bridge mounting bolts, the rocker shaft holder mounting bolts, and the rocker arm assembly.
   b. Loosen the rocker shaft bridge mounting bolts and the rocker shaft holder mounting bolts in sequence two turns at a time, starting at the ends in a crisscross pattern working toward the middle to prevent damaging the valves or the rocker arm assembly.
   c. When removing the rocker arm assembly, do not remove the rocker shaft bridge mounting bolts and the rocker shaft holder mounting bolts. The bolts will keep the rocker arms on the shafts.

To install:

2. Installation is the reverse of removal.
3. Tighten the rocker shaft bridge mounting bolts and the rocker shaft holder mounting bolts in sequence two turns at a time, starting at the middle in a crisscross pattern working toward the ends to prevent damaging the valves or the rocker arm assembly.

Rear

See Figures 108 and 109.

1. Remove the cylinder head cover.
2. Loosen the locknuts and the adjusting screws.
3. Remove the rocker shaft bridge mounting bolts, the rocker shaft holder mounting bolts, and the rocker arm assembly.
   a. Loosen the rocker shaft bridge mounting bolts and the rocker shaft holder mounting bolts in sequence two turns at a time, starting at the ends in a crisscross pattern working toward the middle to prevent damaging the valves or the rocker arm assembly.

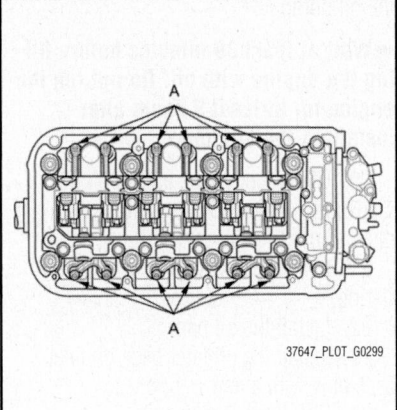

Fig. 108 Loosen the locknuts and the adjusting screws (A)

   b. When removing the rocker arm assembly, do not remove the rocker shaft bridge mounting bolts and the rocker shaft holder mounting bolts. The bolts will keep the rocker arms on the shafts.

To install:

4. Installation is the reverse of removal.
5. Tighten the rocker shaft bridge mounting bolts and the rocker shaft holder mounting bolts in sequence two turns at a time, starting at the middle in a crisscross pattern working toward the ends to prevent damaging the valves or the rocker arm assembly.

TIMING BELT FRONT COVER

REMOVAL & INSTALLATION

See Figures 110 through 115.

1. Remove the accessory drive belt.
2. Turn the crankshaft so the white mark lines up with the pointer.

➡ **The other pointer is not used.**

3. Check that the No. 1 piston Top Dead Center (TDC) mark on the front camshaft pulley and the pointer on the front upper cover are aligned.

➡ **If the marks are not aligned, rotate the crankshaft 360 degrees, and recheck the camshaft pulley mark.**

4. Raise the vehicle on the lift, then remove the right front wheel.
5. Remove the splash shield.
6. Remove the drive belt auto-tensioner.
7. Lift and support the engine with a jack and a wood block under the oil pan.
8. Remove the ground cable, then remove the upper half of the side engine mount bracket.

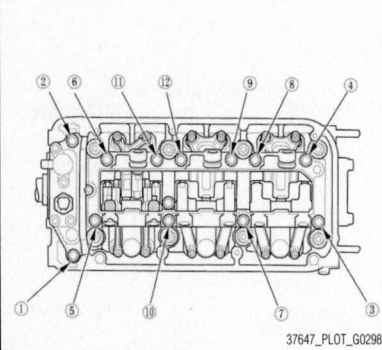

Fig. 107 Loosen the rocker shaft bridge mounting bolts and the rocker shaft holder mounting bolts in sequence

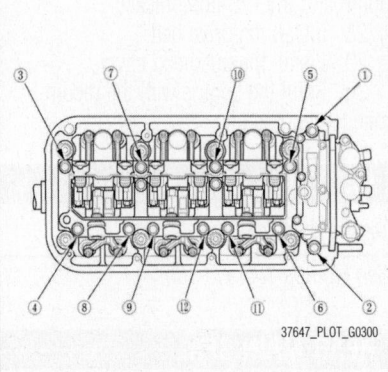

Fig. 109 Loosen the rocker shaft bridge mounting bolts and the rocker shaft holder mounting bolts in sequence

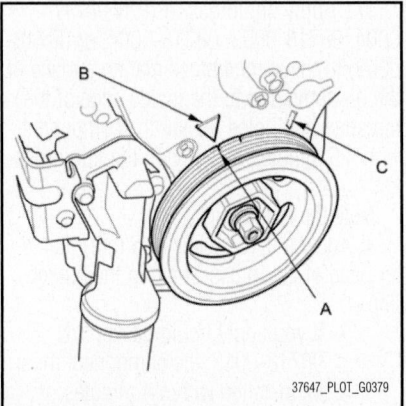

Fig. 110 Turn the crankshaft so the white mark (A) lines up with the pointer (B); pointer (C) is not used

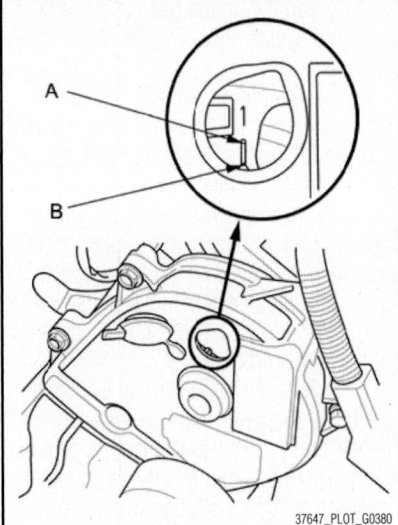

Fig. 111 Check that the No. 1 piston TDC mark (A) on the front camshaft pulley and the pointer (B) on the front upper cover are aligned

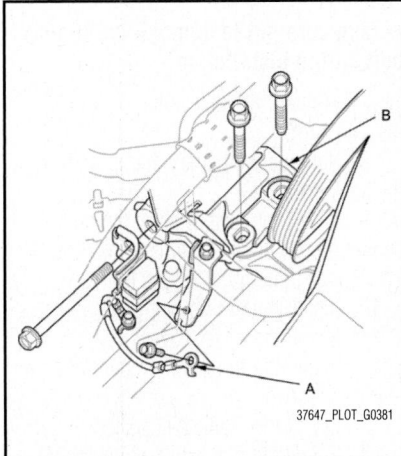

Fig. 112 Remove the ground cable (A), then remove the upper half of the side engine mount bracket (B)

9. Remove the crankshaft pulley.

10. Remove the front upper cover and the rear upper cover.

11. Remove the lower cover.

To install:

12. Install the lower cover.

13. Install the front upper cover and the rear upper cover.

14. Install the crankshaft pulley.

15. Rotate the crankshaft pulley about six turns clockwise so the timing belt positions itself on the pulleys.

16. Turn the crankshaft pulley so its white mark lines up with the pointer.

17. Check the camshaft pulley marks.

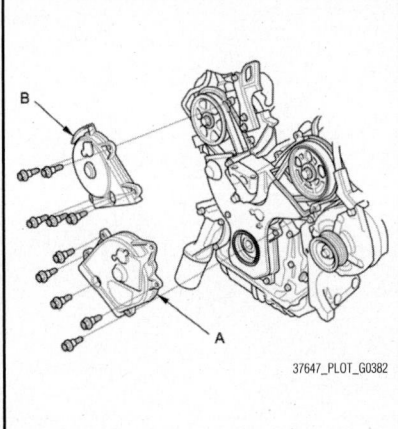

Fig. 113 Remove the front upper cover (A) and the rear upper cover (B)

➡ **If the marks are not aligned, rotate the crankshaft 360 degrees, and recheck the camshaft pulley mark.**

a. If the camshaft pulley marks are at TDC, go to step 17.

b. If the camshaft pulley marks are not at TDC, remove the timing belt and repeat steps 2 through 16.

18. Install the upper half of the side engine mount bracket, and tighten the new mounting bolts to 33 ft. lbs. (44 Nm), then tighten the mass damper mounting bolt to 40 ft. lbs. (54 Nm).

19. Install the ground cable.

20. Install the drive belt auto-tensioner.

21. Install the splash shield.

22. Install the right front wheel.

23. Do the Crankshaft Position (CKP) pattern clear/CKP pattern learn procedure.

TIMING BELT & SPROCKETS

REMOVAL & INSTALLATION

See Figures 110 through 114, 116 through 120.

1. Turn the crankshaft so the white mark lines up with the pointer.

➡ **The other pointer is not used.**

2. Check that the No. 1 piston Top Dead Center (TDC) mark on the front camshaft pulley and the pointer on the front upper cover are aligned.

➡ **If the marks are not aligned, rotate the crankshaft 360 degrees, and recheck the camshaft pulley mark.**

3. Raise the vehicle on the lift, then remove the right front wheel.

4. Remove the splash shield.

5. Remove the drive belt auto-tensioner.

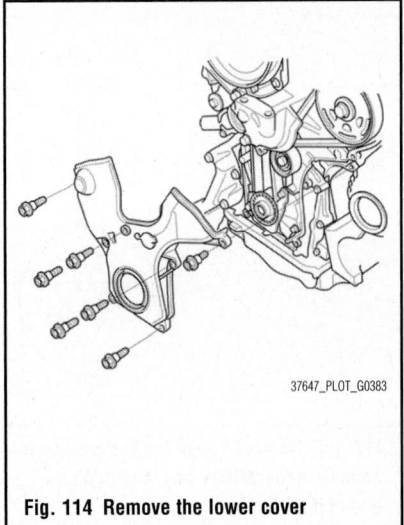

Fig. 114 Remove the lower cover

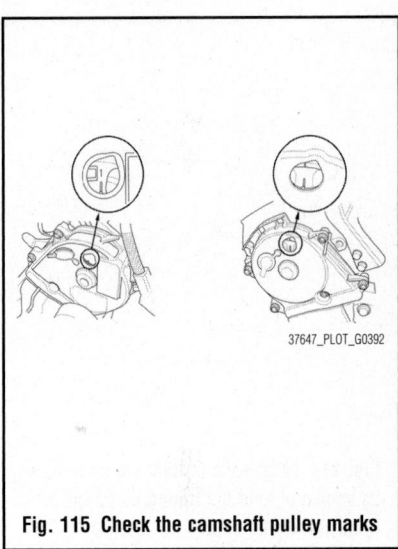

Fig. 115 Check the camshaft pulley marks

6. Lift and support the engine with a jack and a wood block under the oil pan.

7. Remove the ground cable, then remove the upper half of the side engine mount bracket.

8. Remove the crankshaft pulley.

9. Remove the front upper cover and the rear upper cover.

10. Remove the lower cover.

11. Remove one of the battery clamp bolts from the battery tray, and grind the end of it as shown.

12. Thread the battery clamp bolt in as shown to hold the timing belt adjuster in its current position. Tighten it by hand, do not use a wrench.

13. Remove the timing belt guide plate.

14. Remove the lower half of the side engine mount bracket.

15. Remove the idler pulley bolt and the idler pulley, then remove the timing belt. Discard the idler pulley bolt.

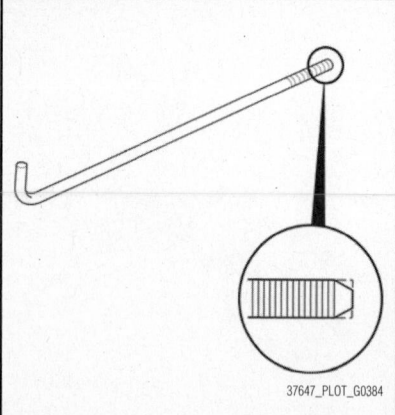

Fig. 116 Remove one of the battery clamp bolts from the battery tray, and grind the end of it as shown

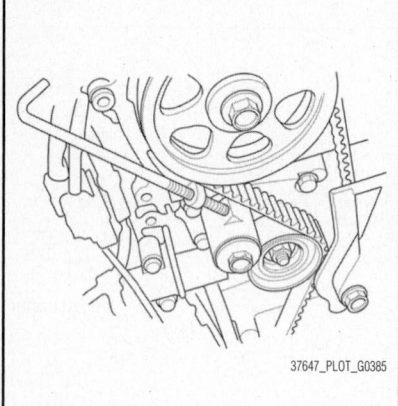

Fig. 117 Thread the battery clamp bolt in as shown to hold the timing belt adjuster in its current position

To install:

Installing a Used Timing Belt

See Figures 121 through 124.

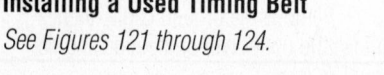

Fig. 118 Remove the timing belt guide plate (A)

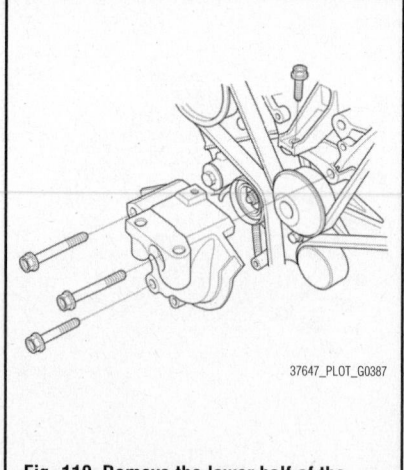

Fig. 119 Remove the lower half of the side engine mount bracket

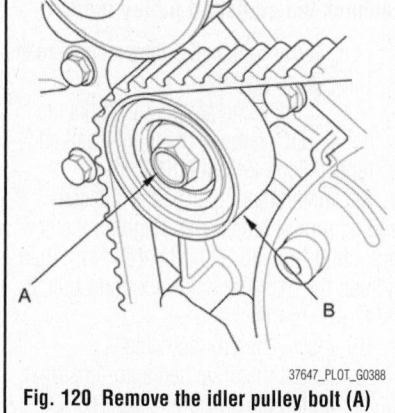

Fig. 120 Remove the idler pulley bolt (A) and the idler pulley (B), then remove the timing belt

➡ The following procedure is for installing a used timing belt. If you are installing a new belt, refer to the New Timing Belt replacement procedure.

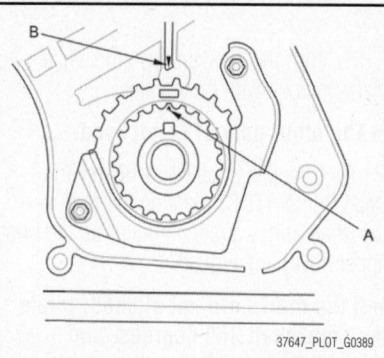

Fig. 121 Set the timing belt drive pulley to Top Dead Center (TDC) by aligning the TDC mark (A) on the tooth of the timing belt drive pulley with the pointer (B) on the oil pump

1. Clean the timing belt pulleys, the timing belt guide plate, and the upper and lower covers.

2. Set the timing belt drive pulley to Top Dead Center (TDC) by aligning the TDC mark on the tooth of the timing belt drive pulley with the pointer on the oil pump.

3. Set the camshaft pulleys to TDC by aligning the TDC marks on the camshaft pulleys with the pointers on the back covers.

4. Loosely install the idler pulley with a new idler pulley bolt so the pulley can move but does not come off.

5. If the auto-tensioner has extended and the timing belt cannot be installed, do the new timing belt replacement procedure.

6. Install the timing belt in a counter-clockwise sequence starting with the drive pulley in the following order:

- Drive pulley
- Idler pulley
- Front camshaft pulley
- Water pump pulley
- Rear camshaft pulley
- Adjusting pulley

➡ **Take care not to damage the timing belt during installation.**

7. Tighten the idler pulley bolt to 33 ft. lbs. (44 Nm).

8. Remove the battery clamp bolt from the back cover.

9. Install the lower half of the side engine mount bracket. Tighten the three 10 x 1.25 mm bolts to 33 ft. lbs. (44 Nm).

10. Install the timing belt guide plate.

11. Install the lower cover.

12. Install the front upper cover and the rear upper cover.

13. Install the crankshaft pulley.

14. Rotate the crankshaft pulley about six turns clockwise so the timing belt positions itself on the pulleys.

15. Turn the crankshaft pulley so its white mark lines up with the pointer.

➡ **The other pointer is not used.**

16. Check the camshaft pulley marks.

➡ **If the marks are not aligned, rotate the crankshaft 360 degrees, and recheck the camshaft pulley mark.**

a. If the camshaft pulley marks are at TDC, go to step 17.

b. If the camshaft pulley marks are not at TDC, remove the timing belt and repeat steps 2 through 16.

17. Install the upper half of the side engine mount bracket, and tighten the new mounting bolts to 33 ft. lbs. (44 Nm), then tighten the mass damper mounting bolt to 40 ft. lbs. (54 Nm).

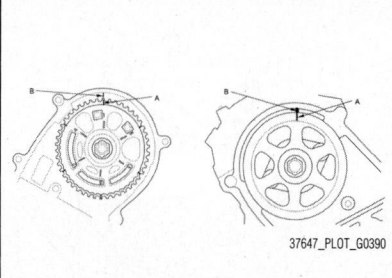

Fig. 122 Set the camshaft pulleys to TDC by aligning the TDC marks (A) on the camshaft pulleys with the pointers (B) on the back covers

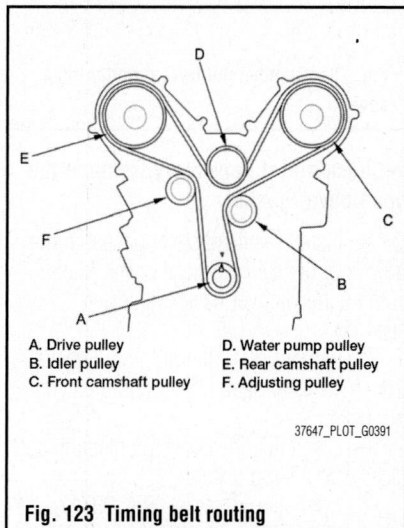

A. Drive pulley
B. Idler pulley
C. Front camshaft pulley
D. Water pump pulley
E. Rear camshaft pulley
F. Adjusting pulley

37647_PLOT_G0391

Fig. 123 Timing belt routing

18. Install the ground cable.
19. Install the drive belt auto-tensioner.
20. Install the splash shield.
21. Install the right front wheel.
22. Do the CKP pattern clear/CKP pattern learn procedure, as outlined under the Crankshaft Position (CKP) sensor in the Engine Performance & Emission Controls Section.

Installing a New Timing Belt

See Figures 115, 121 through 123, 125 through 129.

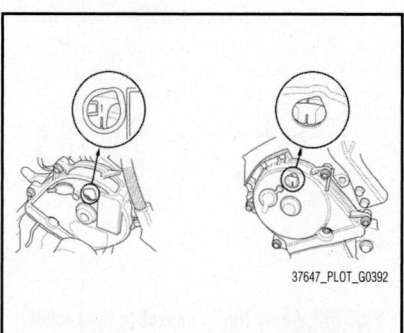

Fig. 124 Check the camshaft pulley marks

➡The following procedure is for installing a new timing belt. If you are installing a used belt, refer to the Used Timing Belt installation procedure.

1. Remove the timing belt.
2. Clean the timing belt pulleys, the timing belt guide plate, and the upper and lower covers.
3. Set the timing belt drive pulley to Top Dead Center (TDC) by aligning the TDC mark on the tooth of the timing belt drive pulley with the pointer on the oil pump.
4. Set the camshaft pulleys to TDC by aligning the TDC marks on the camshaft pulleys with the pointers on the back covers.
5. Remove the battery clamp bolt from the back cover.
6. Remove the auto-tensioner.
7. Align the holes on the rod and the housing of the auto-tensioner.
8. Use a hydraulic press to slowly

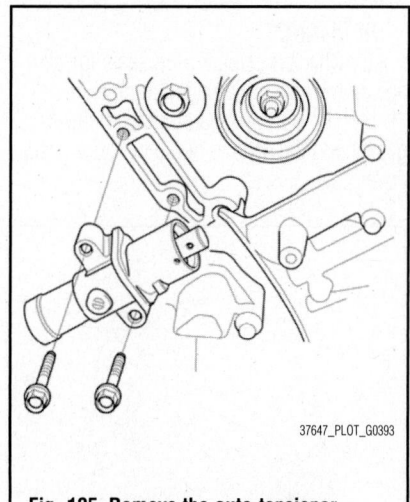

Fig. 125 Remove the auto-tensioner

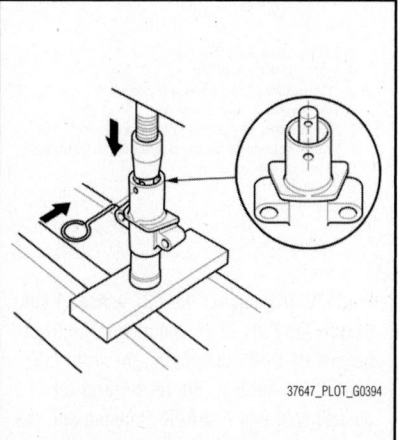

Fig. 126 Align the holes on the rod and the housing of the auto-tensioner

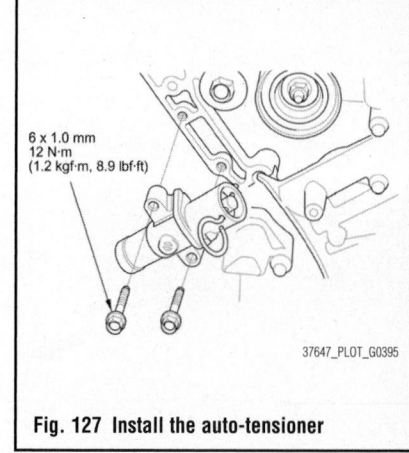

6 x 1.0 mm
12 N·m
(1.2 kgf·m, 8.9 lbf·ft)

37647_PLOT_G0395

Fig. 127 Install the auto-tensioner

compress the auto-tensioner. Insert a 0.08 inches (2.0 mm) pin through the housing and the rod.

➡**The compression pressure should not exceed 2,200 lbs. (9,800 N).**

9. Install the auto-tensioner.

➡**Make sure the pin stays in place.**

10. Thread the battery clamp bolt in as shown to hold the timing belt adjuster. Tighten it by hand, do not use a wrench.
11. Loosely install the idler pulley with a new idler pulley bolt so the pulley can move but does not come off.
12. Install the timing belt in a counter-clockwise sequence starting with the drive pulley in the following order:

• Drive pulley
• Idler pulley
• Front camshaft pulley
• Water pump pulley
• Rear camshaft pulley
• Adjusting pulley

13. Tighten the idler pulley bolt to 33 ft. lbs. (44 Nm).

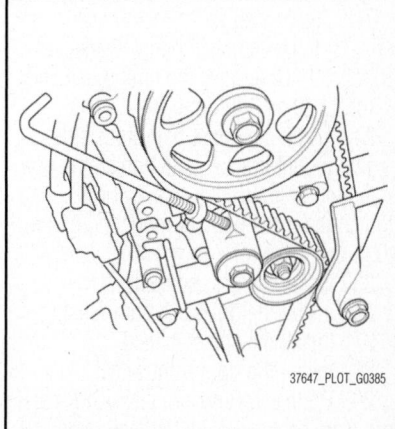

Fig. 128 Thread the battery clamp bolt in as shown to hold the timing belt adjuster

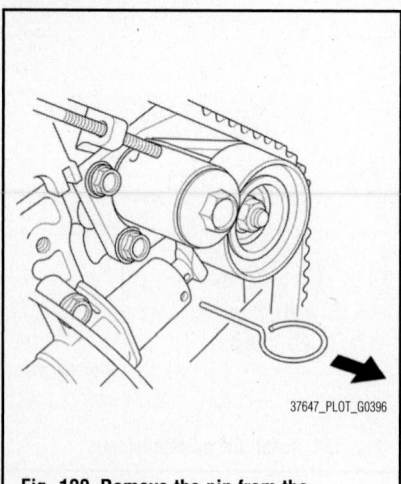

Fig. 129 Remove the pin from the auto-tensioner

14. Remove the pin from the auto-tensioner.

15. Remove the battery clamp bolt from the back cover.

16. Install the lower half of the side engine mount bracket. Tighten the three 10 x 1.25 mm bolts to 33 ft. lbs. (44 Nm).

17. Install the timing belt guide plate.

18. Install the lower cover.

19. Install the front upper cover and the rear upper cover.

20. Install the crankshaft pulley.

21. Rotate the crankshaft pulley about six turns clockwise so the timing belt positions itself on the pulleys.

22. Turn the crankshaft pulley so its white mark lines up with the pointer.

➡**The other pointer is not used.**

23. Check the camshaft pulley marks.

➡**If the marks are not aligned, rotate the crankshaft 360 degrees, and recheck the camshaft pulley mark.**

    a. If the camshaft pulley marks are at TDC, go to step 24.

    b. If the camshaft pulley marks are not at TDC, remove the timing belt and repeat steps 3 through 22.

24. Install the upper half of the side engine mount bracket, and tighten the new mounting bolts to 33 ft. lbs. (44 Nm), then tighten the mass damper mounting bolt to 40 ft. lbs. (54 Nm).

25. Install the ground cable.

26. Install the drive belt auto-tensioner.

27. Install the splash shield.

28. Install the right front wheel.

29. Do the CKP pattern clear/CKP pattern learn procedure, as outlined under the Crankshaft Position (CKP) sensor in the Engine Performance & Emission Controls Section.

CYLINDER HEAD COVERS

REMOVAL & INSTALLATION

Front

See Figures 130 through 133.

1. Remove the intake manifold.

2. Remove the three ignition coils from the front cylinder head.

3. Disconnect the Engine Coolant Temperature (ECT) sensor 1 connector, the Exhaust Gas Recirculation (EGR) valve connector, the front secondary Heated Oxygen Sensor 2 (secondary HO2S) connector, the front Air Fuel Ratio (A/F) sensor 1 connector, the rocker arm oil control solenoid A (BANK 2) connector, the front rocker arm oil pressure switch connector and the harness clamp securing the harness holder, and remove the dipstick.

4. Remove the bolt securing the harness holder.

5. Remove the front cylinder head cover.

To install:

6. Check the spark plug seals for damage. If any seal is damaged, replace it.

7. Thoroughly clean the head cover gasket and the groove of the cylinder head cover.

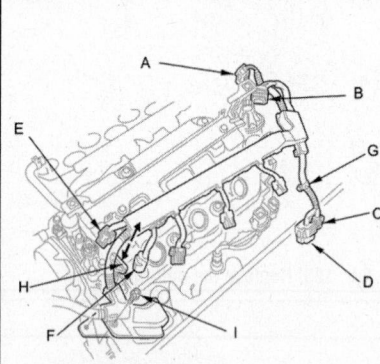

A. ECT sensor 1 connector
B. EGR valve connector
C. Secondary HO2S sensor 2 connector
D. A/F sensor 1 connector
E. Rocker arm oil control solenoid A connector
F. Front rocker arm oil pressure switch connector
G. Harness clamp
H. Dipstick
I. Bolt

37647_PLOT_G0397

Fig. 130 Disconnect the ECT sensor 1 connector, the EGR valve connector, the front secondary HO2S connector, the front A/F sensor 1 connector, the rocker arm oil control solenoid A (BANK 2) connector, the front rocker arm oil pressure switch connector and the harness clamp securing the harness holder, and remove the dipstick

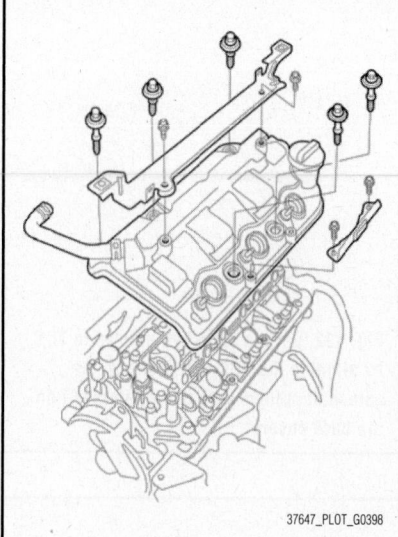

Fig. 131 Remove the front cylinder head cover

➡**Check and if necessary, replace the head cover gasket.**

8. Install the head cover gasket in the groove of the cylinder head cover. Make sure the head cover gasket is seated securely.

9. Remove all of the old liquid gasket from the rocker shaft holder and the cylinder head.

10. Clean the head cover contacting surfaces with a shop towel.

11. Apply liquid gasket (P/N 08717-0004, 08718-0003, 08718-0004, or 08718-0009) to the rocker shaft holder mating surface. Install the component within 5 minutes of applying the liquid gasket.

➡**If you apply liquid gasket P/N 08718-0012, the component must be installed within 4 minutes. If too much time has**

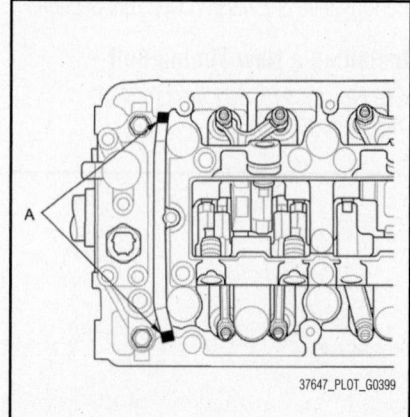

Fig. 132 Apply liquid gasket to the rocker shaft holder mating surface (A)

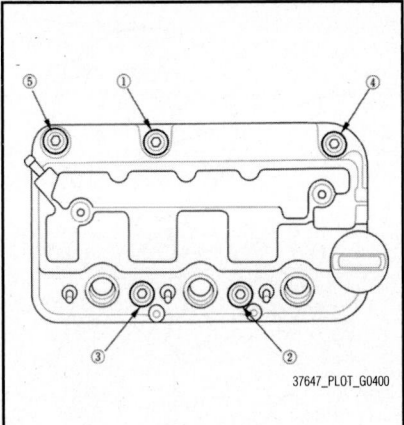

Fig. 133 Tighten the bolts in three steps, in sequence

passed after applying the liquid gasket, remove the old liquid gasket and residue, then reapply the new liquid gasket.

12. Set the spark plug seals on the spark plug tubes, and install the front cylinder head cover.

➡**Wait at least 30 minutes before filling the engine with oil. Do not run the engine for at least 3 hours after installing the cylinder head cover.**

13. Inspect the spark plug seals for damage.

14. Inspect the cover washers. Replace any washer that is damaged or deteriorated.

15. Tighten the bolts in three steps. In the final step torque the all bolts, in sequence, 9 ft. lbs. (12 Nm).

16. Install the bolt securing the harness holder.

17. Connect the Engine Coolant Temperature (ECT) sensor 1 connector, the Exhaust Gas Recirculation (EGR) valve connector, the front secondary Heated Oxygen Sensor 2 (secondary HO2S) connector, the front Air Fuel Ratio (A/F) sensor 1 connector, the rocker arm oil control solenoid A (BANK 2) connector, the front rocker arm oil pressure switch connector and the harness clamp securing the harness holder, and install the dipstick.

18. Install the three ignition coils to the front cylinder head.

19. Install the intake manifold.

Rear

See Figures 134 through 138.

1. Remove the intake manifold.
2. Remove the three ignition coils from the rear cylinder head.

3. Remove the drive belt.
4. Remove the power steering (P/S) pump and the P/S hose bracket with its hoses connected.
5. Remove the harness holder mounting bolts and the engine ground cable.
6. Disconnect the three injector connectors and the two harness clips.
7. Disconnect the breather hose.
8. Remove the harness from the upper cover.
9. Disconnect the rear rocker arm oil pressure switch connector, the rocker arm oil control solenoid B (BANK 1) connector, the rocker arm oil control solenoid A (BANK 1) connector, the rear Air Fuel Ratio (A/F) sensor 1 connector, the rear secondary Heated Oxygen Sensor 2 (secondary HO2S) connector, and the harness clamps, then remove the harness holder.
10. Disconnect the breather hose.
11. Remove the rear cylinder head cover.

To install:

12. Check the spark plug seals for damage. If any seal is damaged, replace it.
13. Thoroughly clean the head cover gasket and the groove of the cylinder head cover.

➡**Check and if necessary, replace the head cover gasket.**

14. Install the head cover gasket in the groove of the cylinder head cover. Make sure the head cover gasket is seated securely.

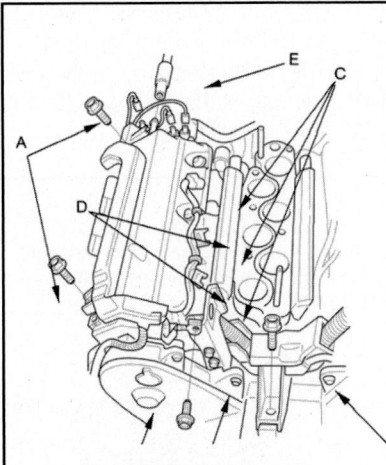

A. Harness holder mounting bolts
B. Engine ground cable
C. Injector connectors
D. Harness clips
E. Breather hose
F. Harness

Fig. 134 Remove the harness holder mounting bolts and the engine ground cable

15. Remove all of the old liquid gasket from the rocker shaft holder and the cylinder head.
16. Clean the head cover contacting surfaces with a shop towel.
17. Apply liquid gasket (P/N 08717-0004, 08718-0003, 08718-0004, or 08718-0009) to the rocker shaft holder mating surface. Install the component within 5 minutes of applying the liquid gasket.

➡**If you apply liquid gasket P/N 08718-0012, the component must be installed within 4 minutes. If too much time has passed after applying the liquid gasket, remove the old liquid gasket and residue, then reapply the new liquid gasket.**

18. Set the spark plug seals on the spark plug tubes, and install the rear cylinder head cover.

➡**Wait at least 30 minutes before filling the engine with oil. Do not run the engine for at least 3 hours after installing the cylinder head cover.**

19. Inspect the spark plug seals for damage.
20. Inspect the cover washers. Replace any washer that is damaged or deteriorated.

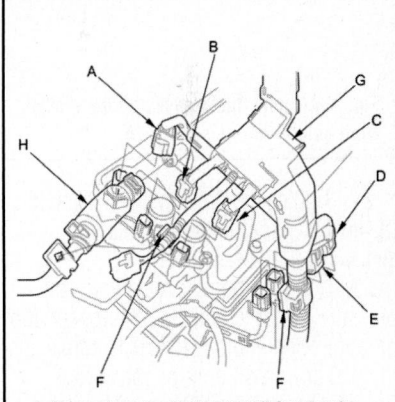

A. Rear rocker arm oil pressure switch connector
B. Rocker arm oil control solenoid B connector
C. Rocker arm oil control solenoid A connector
D. Rear A/F sensor 1 connector
E. Rear secondary HO2S connector
F. Harness clamps
G. Harness holder
H. Breather hose

Fig. 135 Disconnect the rear rocker arm oil pressure switch connector, the rocker arm oil control solenoid B connector, the rocker arm oil control solenoid A connector, the rear A/F sensor 1 connector, the rear secondary HO2S connector, and the harness clamps, then remove the harness holder

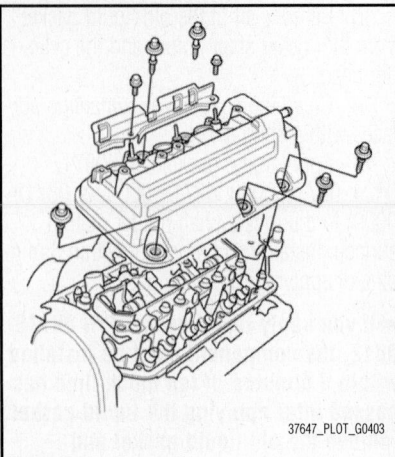

Fig. 136 Remove the rear cylinder head cover

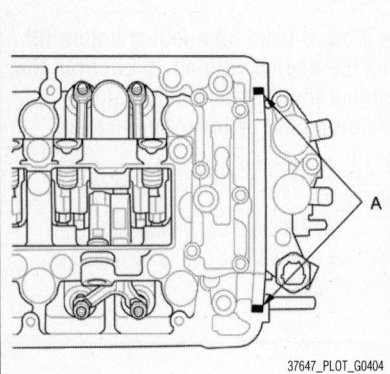

Fig. 137 Apply liquid gasket to the rocker shaft holder mating surface (A)

21. Tighten the bolts in three steps. In the final step torque the all bolts, in sequence, 9 ft. lbs. (12 Nm).

22. Connect the rear rocker arm oil pressure switch connector, the rocker arm oil control solenoid B (BANK1) connector, the rocker arm oil control solenoid A (BANK 1) connector, the rear Air Fuel Ratio (A/F) sensor 1 connector, the rear secondary Heated Oxygen Sensor 2 (secondary HO2S) connector, and the harness clamps, then install the harness holder.

23. Connect the breather hose.

24. Tighten the harness holder mounting bolt.

25. Install the engine ground cable and connect the breather hose, then install the harness to the upper cover.

26. Reconnect the three injector connectors and the two harness clips.

27. Install the Power Steering (P/S) pump and the P/S hose bracket.

28. Install the drive belt.

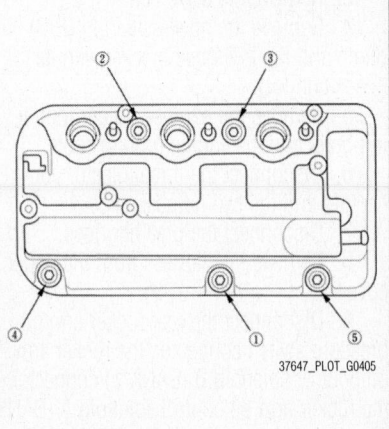

Fig. 138 Tighten the bolts in three steps, in sequence

29. Install the three ignition coils to the rear cylinder head.

30. Install the intake manifold.

VALVE LASH (CLEARANCE) ADJUSTMENT

ADJUSTMENT

See Figures 139 through 145.

➡**Connect the Honda Diagnostic System (HDS) to the Data Link Connector (DLC), and monitor the Engine Coolant Temperature (ECT) sensor 1 with the HDS. Adjust the valve clearance only when the engine coolant temperature is less than 100°F (38°C).**

1. Remove the cylinder head covers.

2. Set the No. 1 piston at Top Dead Center (TDC). Align the pointer on the front upper cover with the No. 1 piston TDC mark on the front camshaft pulley.

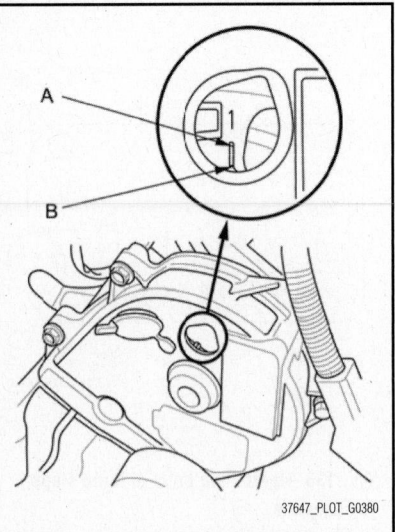

Fig. 139 Set the No. 1 piston at TDC; align the pointer (A) on the front upper cover with the No. 1 piston TDC mark (B) on the front camshaft pulley

3. Select the correct feeler gauge for the valve clearance you are going to check.

➡**Valve Clearance:**

- Intake: 0.008–0.009 inches (0.20–0.24 mm)
- Exhaust: 0.011–0.013 inches (0.28–0.32 mm)

4. Insert the feeler gauge between the adjusting screw and the end of the valve stem on the No. 1 cylinder and slide it back and forth; you should feel a slight amount of drag.

5. If you feel too much or too little drag, loosen the locknut, and turn the adjusting screw until the drag on the feeler gauge is correct.

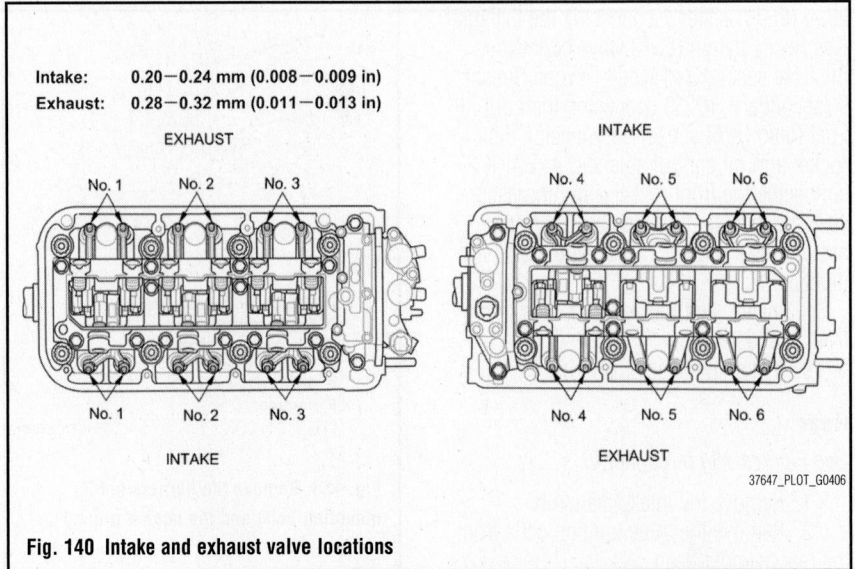

| Intake: | 0.20–0.24 mm (0.008–0.009 in) |
| Exhaust: | 0.28–0.32 mm (0.011–0.013 in) |

Fig. 140 Intake and exhaust valve locations

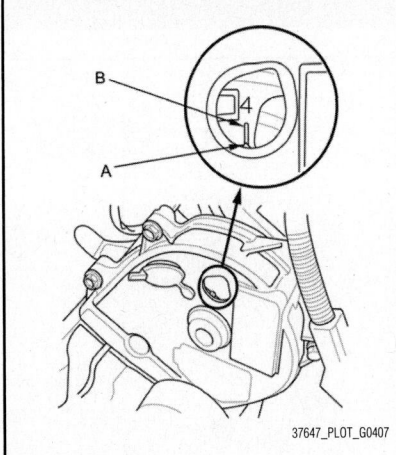

Fig. 141 Align the pointer (A) on the front upper cover with the No. 4 piston TDC mark (B) on the front camshaft pulley

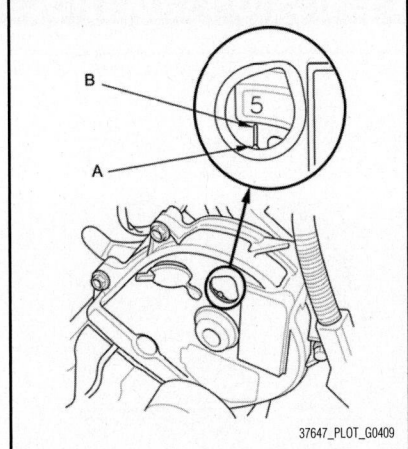

Fig. 143 Align the pointer (A) on the front upper cover with the No. 5 piston TDC mark (B) on the front camshaft pulley

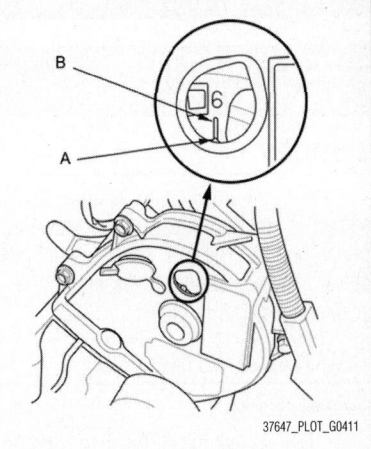

Fig. 145 Align the pointer (A) on the front upper cover with the No. 6 piston TDC mark (B) on the front camshaft pulley

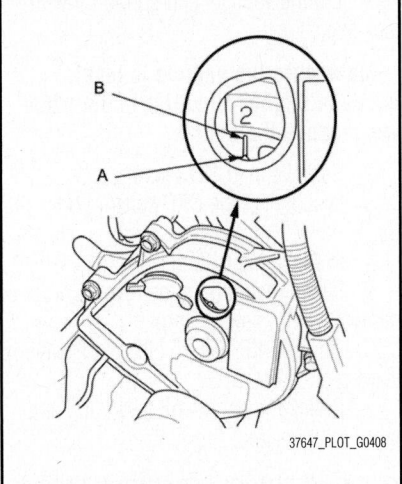

Fig. 142 Align the pointer (A) on the front upper cover with the No. 2 piston TDC mark (B) on the front camshaft pulley

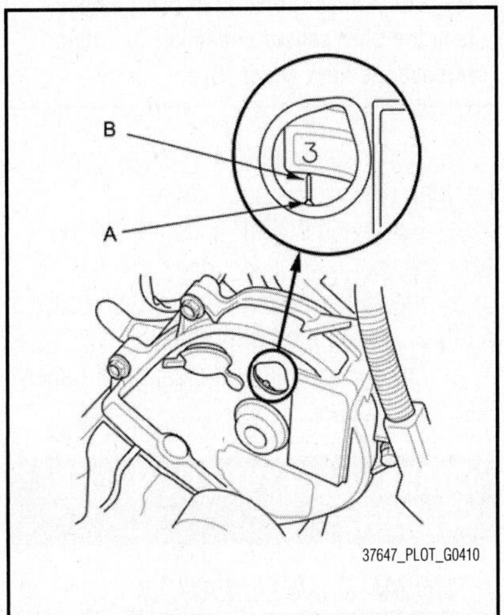

Fig. 144 Align the pointer (A) on the front upper cover with the No. 3 piston TDC mark (B) on the front camshaft pulley

6. While holding the adjusting screw with the screw driver, tighten the locknut, then recheck the clearance. Repeat the adjustment, if necessary.

➡**Specified Torque:**

- No. 1, No. 2,No. 3, and No. 4 cylinders: 14 ft. lbs. (20 Nm)

- No. 5 and No. 6 cylinders: 10 ft. lbs. (14 Nm)

7. Rotate the crankshaft clockwise. Align the pointer on the front upper cover with the No. 4 piston TDC mark on the front camshaft pulley.

8. Check and, if necessary, adjust the valve clearance on No. 4 cylinder.

Rotate the crankshaft clockwise. Align the pointer on the front upper cover with the No. 2 piston TDC mark on the front camshaft pulley.

9. Check and, if necessary, adjust the valve clearance on the No. 2 cylinder.

10. Rotate the crankshaft clockwise. Align the pointer on the front upper cover with the No. 5 piston TDC mark on the front camshaft pulley.

11. Check and, if necessary, adjust the valve clearance on No. 5 cylinder.

12. Rotate the crankshaft clockwise. Align the pointer on the front upper cover with the No. 3 piston TDC mark on the front camshaft pulley.

13. Check and, if necessary, adjust the valve clearance on No. 3 cylinder.

14. Rotate the crankshaft clockwise. Align the pointer on the front upper cover with the No. 6 piston TDC mark on the front camshaft pulley.

15. Check and, if necessary, adjust the valve clearance on the No. 6 cylinder.

16. Install the cylinder head covers.

ENGINE PERFORMANCE & EMISSION CONTROLS

ACCELERATOR PEDAL POSITION (APP) SENSOR

LOCATION

The Accelerator Pedal Position (APP) sensor is an integral part of the accelerator pedal module. The APP sensor is not available separately. Do not disassemble the accelerator pedal module.

REMOVAL & INSTALLATION

See Figure 146.

1. Disconnect the APP sensor 6P connector.
2. Remove the accelerator pedal module.

➡The APP sensor is not available separately. Do not disassemble the accelerator pedal module.

3. Install the parts in the reverse order of removal.

CAMSHAFT POSITION (CMP) SENSOR

LOCATION

The Camshaft Position (CMP) sensor is located on the back cover for the front camshaft.

REMOVAL & INSTALLATION

See Figures 147 and 148.

1. Remove the timing belt.
2. Remove the front camshaft pulley (CMP sensor pulse plate).

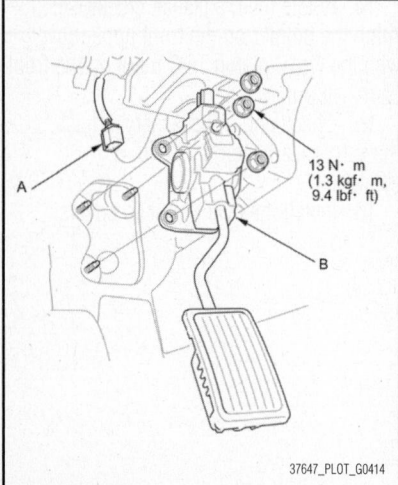

Fig. 146 Disconnect the APP sensor 6P connector (A); remove the accelerator pedal module (B)

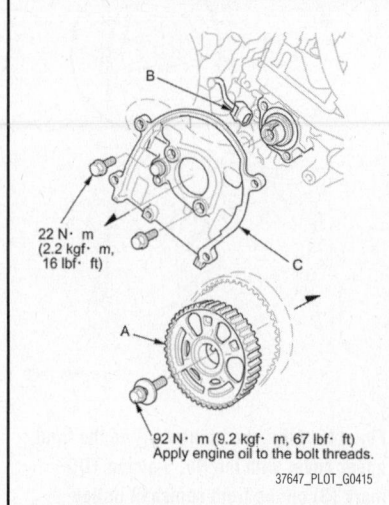

Fig. 147 Remove the front camshaft pulley (CMP sensor pulse plate) (A); disconnect the CMP sensor connector (B), then remove the back cover (C)

3. Disconnect the CMP sensor connector, then remove the back cover.
4. Remove the CMP sensor from the back cover.
5. Install the parts in the reverse order of removal. Install the timing belt.
6. Do the CKP pattern clear/CKP pattern learn procedure.

CRANKSHAFT POSITION (CKP) SENSOR

REMOVAL & INSTALLATION

See Figure 149.

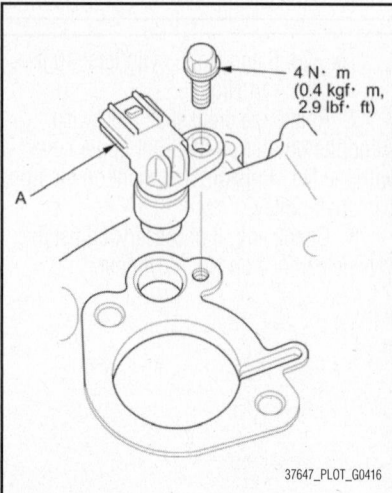

Fig. 148 Remove the CMP sensor (A) from the back cover

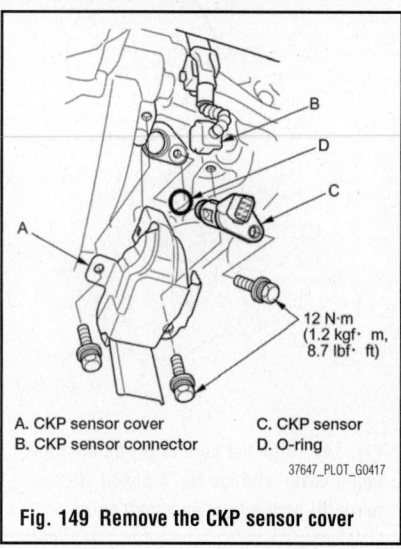

A. CKP sensor cover
B. CKP sensor connector
C. CKP sensor
D. O-ring

Fig. 149 Remove the CKP sensor cover

1. Lift the vehicle, and support it with jack stands.

➡Make sure the vehicle is level, because engine oil will drip out when you remove the sensor.

2. Remove the CKP sensor cover.
3. Disconnect the CKP sensor connector.
4. Remove the CKP sensor.
5. Install the parts in the reverse order of removal with a new O-ring.
6. Do the CKP pattern clear/CKP pattern learn procedure.
7. Check the engine oil level, and add more oil if needed.

CKP PATTERN CLEAR/CKP PATTERN LEARN PROCEDURE

Clear/Learn Procedure (with the HDS)

1. Connect the HDS to the Data Link Connector (DLC) located under the driver's side of the dashboard.
2. Turn the ignition switch to ON (II).
3. Make sure the HDS communicates with the PCM and all other vehicle systems. If it doesn't, go to the DLC circuit troubleshooting.
4. Select CRANK PATTERN in the ADJUSTMENT MENU with the HDS.
5. Select CRANK PATTERN LEARNING with the HDS, and follow the screen prompts.

Learn Procedure (without the HDS)

1. Start the engine. Hold the engine speed at 3,000 rpm without load (in P or N) until the radiator fan comes on.
2. Test-drive the vehicle on a level road:

Decelerate (with the throttle fully closed) from an engine speed of 2,500 rpm down to 1,000 rpm with the transmission in 2.

   3. Repeat step 2 several times.

   4. Turn the ignition switch to LOCK (0).

   5. Turn the ignition switch to ON (II), and wait for 30 seconds. The CKP pattern learn procedure is complete.

ENGINE COOLANT TEMPERATURE (ECT) SENSOR

REMOVAL & INSTALLATION

ECT Sensor 1

See Figure 150.

   1. Drain the engine coolant.

   2. Remove the engine cover.

   3. Disconnect the ECT sensor 1 2P connector.

   4. Remove ECT sensor 1.

   5. Install the parts in the reverse order of removal with a new O-ring, then refill the radiator with engine coolant.

ECT Sensor 2

See Figure 151.

   1. Remove the splash shield.

   2. Drain the engine coolant.

   3. Disconnect the ECT sensor 2 2P connector.

   4. Remove ECT sensor 2.

   5. Install the parts in the reverse order of removal with a new O-ring, then refill the radiator with engine coolant.

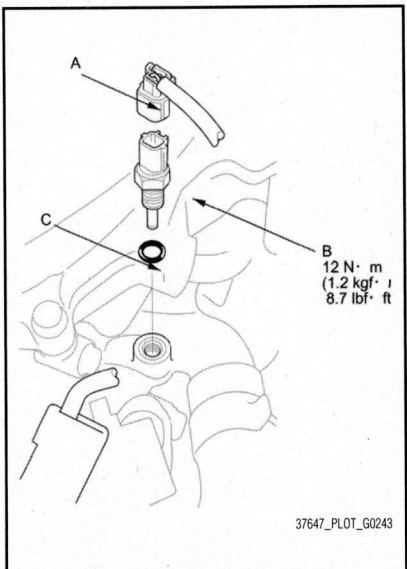

37647_PLOT_G0243

Fig. 150 Disconnect the ECT sensor 1 2P connector (A); remove ECT sensor 1 (B) and the O-ring (C)

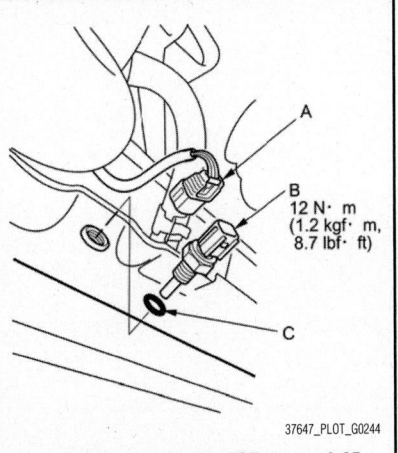

37647_PLOT_G0244

Fig. 151 Disconnect the ECT sensor 2 2P connector (A); remove ECT sensor 2 (B) and the O-ring (C)

EVAP CANISTER

LOCATION

The EVAP canister is mounted on the vehicle frame. Refer to the graphics in the Removal and Installation section for the location(s).

REMOVAL & INSTALLATION

See Figures 152 and 153.

   1. Raise the vehicle on a lift.

   2. Remove the EVAP canister cover.

   3. Disconnect the quick-connect fittings, the hoses, the EVAP canister vent shut valve connector, and the FTP sensor connector, then remove the harness clamps.

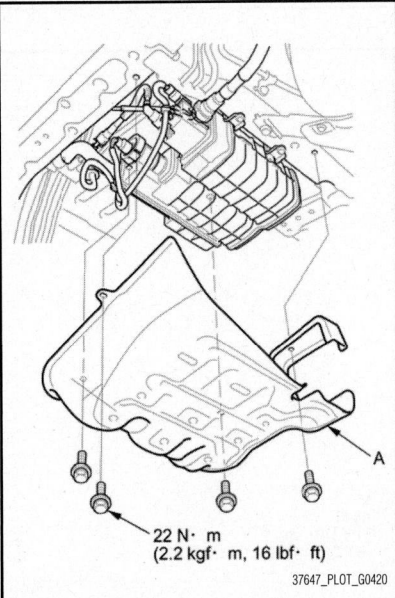

22 N· m
(2.2 kgf· m, 16 lbf· ft)

37647_PLOT_G0420

Fig. 152 Remove the EVAP canister cover (A)

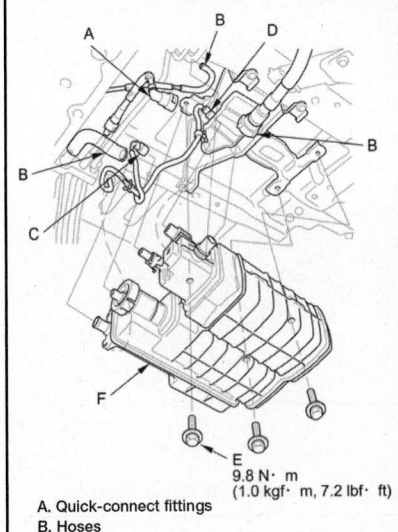

9.8 N· m
(1.0 kgf· m, 7.2 lbf· ft)

A. Quick-connect fittings
B. Hoses
C. EVAP canister vent shut valve connector
D. FTP sensor connector
E. Bolts
F. EVAP canister assembly

37647_PLOT_G0421

Fig. 153 Disconnect the quick-connect fittings, the hoses, the EVAP canister vent shut valve connector, and the FTP sensor connector, then remove the harness clamps

   4. Remove the bolts.

   5. Remove the EVAP canister assembly.

   6. Install the parts in the reverse order of removal.

EXHAUST GAS RECIRCULATION (EGR) VALVE

LOCATION

The EGR valve is located under the engine cover on the intake manifold.

REMOVAL & INSTALLATION

See Figures 154 and 155.

   1. Remove the engine cover.

   2. Disconnect the EGR 5P valve connector.

   3. Remove the EGR valve.

   4. Install the parts in the reverse order of removal with a new gasket.

HEATED OXYGEN SENSOR (HO2S)

LOCATION

See Figure 156.

REMOVAL & INSTALLATION

Front Bank (Bank2)

See Figures 157 and 158.

Fig. 154 Remove the engine cover

Special Tools Required: O2 Sensor Socket Wrench Snap-On® SWR2 or O2 Sensor Wrench Snap-On® YA8875, commercially available

1. Disconnect the front secondary HO2S 4P connector, and remove the harness clamps.
2. Raise the vehicle on a lift.
3. Remove the harness clamp, then remove the front secondary HO2S.
4. Install the parts in the reverse order of removal.

Rear Bank (Bank 1)

See Figure 159.

Special Tools Required: O2 Sensor Socket Wrench Snap-On® SWR2 or O2 Sensor Wrench Snap-On® YA8875, commercially available

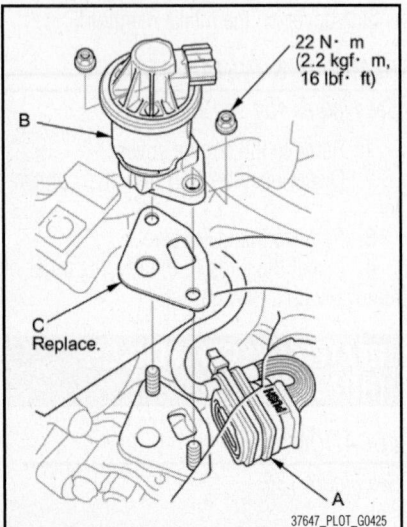

Fig. 155 Disconnect the EGR 5P valve connector (A); remove the EGR valve (B) and the gasket (C)

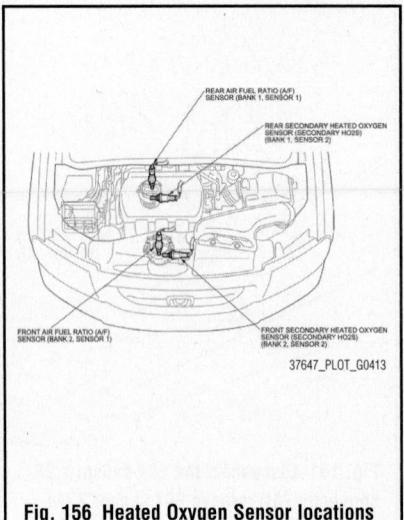

Fig. 156 Heated Oxygen Sensor locations

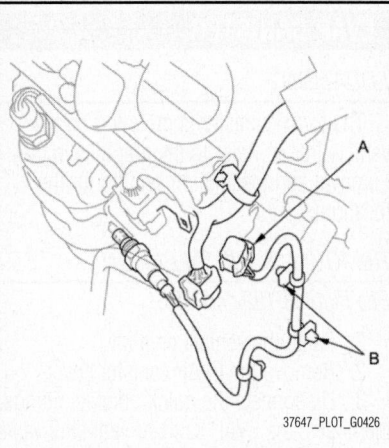

Fig. 157 Disconnect the front secondary HO2S 4P connector (A), and remove the harness clamps (B)

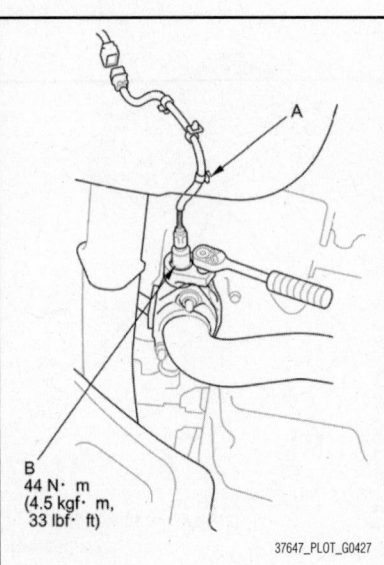

Fig. 158 Remove the harness clamp (A), then remove the front secondary HO2S (B)

1. Raise the vehicle on a lift.
2. Disconnect the rear secondary HO2S 4P connector, and remove the harness clamps.
3. Remove the rear secondary HO2S.
4. Install the parts in the reverse order of removal.

INTAKE AIR TEMPERATURE (IAT) SENSOR

LOCATION

The Intake Air Temperature (IAT) sensor is an integral part of the Mass Air Flow (MAF) sensor/Intake Air Temperature (IAT) sensor which is located on the engine air take duct. Refer to the Mass Air Flow (MAF) sensor/Intake Air Temperature (IAT) sensor section when servicing this component.

REMOVAL & INSTALLATION

Refer to the Mass Air Flow (MAF) sensor/Intake Air Temperature (IAT) sensor section when servicing this component.

KNOCK SENSOR (KS)

LOCATION

Refer to the graphics in the Removal and Installation section for the location(s).

REMOVAL & INSTALLATION

See Figures 160 and 161.

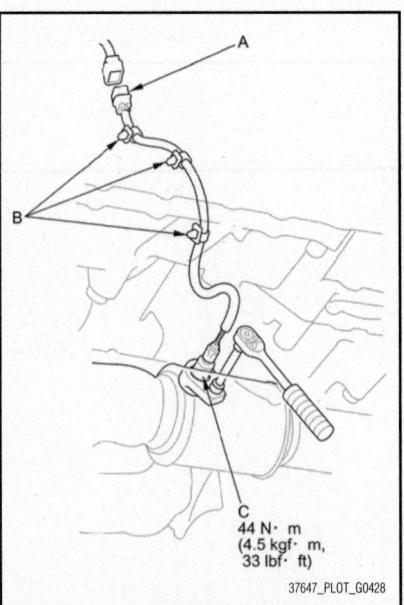

Fig. 159 Disconnect the rear secondary HO2S 4P connector (A), and remove the harness clamps (B); remove the rear secondary HO2S (C)

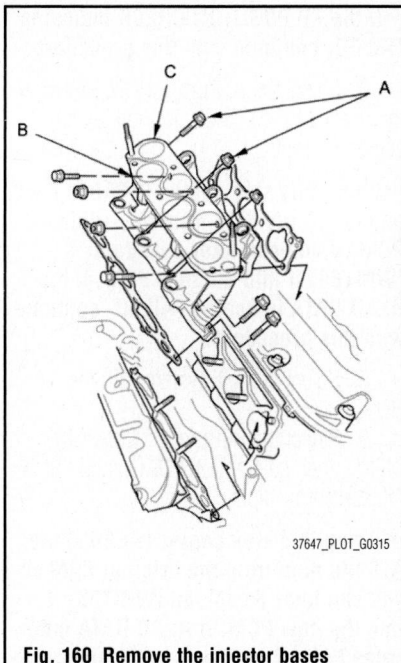

Fig. 160 Remove the injector bases

1. Remove the intake manifold.
2. Remove the injector base.
3. Disconnect the knock sensor connector.
4. Remove the knock sensor.
5. Install the parts in the reverse order of removal.

MAINTENANCE REMINDER LIGHTS

RESET PROCEDURE

Note the following:
• The vehicle must be stopped to reset the Maintenance Minder.

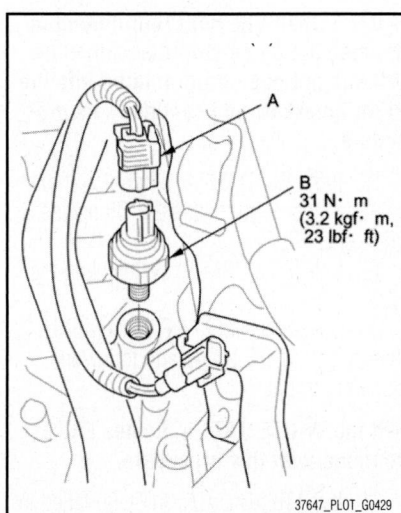

Fig. 161 Disconnect the knock sensor connector (A) and remove the knock sensor (B)

• If a required service is done and the Maintenance Minder is not reset, or if the Maintenance Minder is reset without doing the service, the system will not show the proper maintenance timing. This can lead to serious mechanical problems because there will be no accurate record of when the required maintenance is needed.
• The engine oil life and the maintenance item(s) can be independently reset with the HDS.

1. Turn the ignition switch to ON (II).
2. Push the SEL/RESET knob repeatedly until the engine oil life indicator is displayed.
3. Press and hold the SEL/RESET knob for about 10 seconds. The engine oil life indicator and the maintenance item code(s) will be blink, then release the knob.

➡**If you are resetting the Maintenance Minder when the engine oil life is more than 15 %, make sure any maintenance item(s) requiring service are done before resetting the display.**

4. Push the SEL/RESET knob to select OIL LIFE, the display will begin to blink.
5. Push in on the SEL/RESET knob to enter this selection, the OIL LIFE and the maintenance item code(s) will begin to blink.
6. Press and hold the SEL/RESET knob for another 5 seconds. The maintenance item code(s) will disappear, and the engine oil life will reset to "100."

Resetting Individual Maintenance Items

1. Connect the Honda Diagnostic System (HDS) to the data link connector (DLC).
2. Turn the ignition switch to ON (II).
3. Make sure the HDS communicates with the vehicle and the powertrain control module (PCM). If it doesn't communicate, troubleshoot the DLC circuit.
4. Select GAUGES in the BODY ELECTRICAL with the HDS.
5. Select ADJUSTMENT in the GAUGES with the HDS.
6. Select MAINTENANCE MINDER in the ADJUSTMENT with the HDS.
7. Select RESET in the MAINTENANCE MINDER with the HDS.
8. Select the individual maintenance item you wish to reset with the HDS.

MANIFOLD ABSOLUTE PRESSURE (MAP) SENSOR

REMOVAL & INSTALLATION

See Figure 162.

1. Disconnect the MAP sensor connector.
2. Remove the screw.
3. Remove the MAP sensor.
4. Install the parts in the reverse order of removal with a new O-ring.

MASS AIR FLOW (MAF) SENSOR

LOCATION

The MAF sensor/IAT sensor is located on the engine air intake duct.

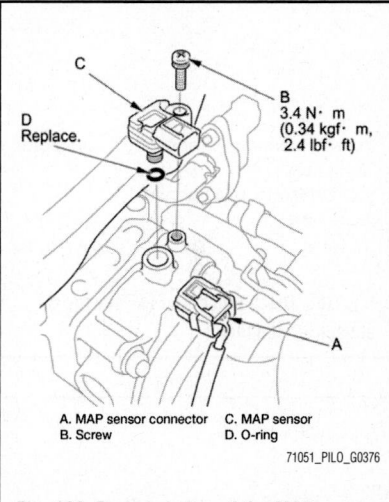

A. MAP sensor connector   C. MAP sensor
B. Screw   D. O-ring

Fig. 162 Exploded view of the MAP sensor assembly

REMOVAL & INSTALLATION

See Figure 163.

1. Disconnect the MAF sensor/IAT sensor 5P connector.
2. Remove the screw.
3. Remove the MAF sensor/IAT sensor.
4. Install the parts in the reverse order of removal with a new O-ring.

POSITIVE CRANKCASE VENTILATION (PCV) VALVE

LOCATION

Refer to the graphics in the Removal and Installation section for the location(s).

REMOVAL & INSTALLATION

See Figure 164.

1. Remove the engine cover.
2. Remove the bolt.
3. Remove the PCV valve.

➡**Do not to spill oil on the hot exhaust manifold.**

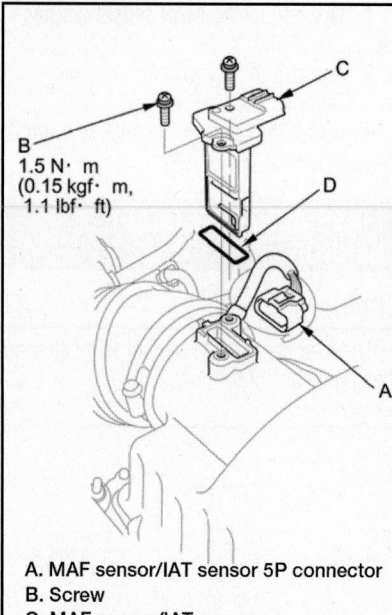

A. MAF sensor/IAT sensor 5P connector
B. Screw
C. MAF sensor/IAT sensor
D. O-ring

37647_PLOT_G0430

Fig. 163 Disconnect the MAF sensor/IAT sensor 5P connector

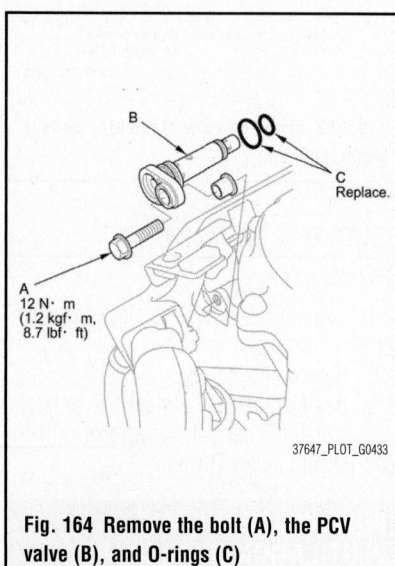

37647_PLOT_G0433

Fig. 164 Remove the bolt (A), the PCV valve (B), and O-rings (C)

4. Install the parts in the reverse order of removal.

➡ **When installing a new PCV valve, make sure the new O-rings are in place.**

POWERTRAIN CONTROL MODULE (PCM)

LOCATION

The Powertrain Control Module (PCM) is located in the engine compartment.

REMOVAL & INSTALLATION

See Figure 165.

Special Tools Required:
- Honda Diagnostic System (HDS) tablet tester
- Honda Interface Module (HIM) and an iN workstation with the latest HDS software version
- HDS pocket tester
- GNA600 and an iN workstation with the latest HDS software version

➡ **Any one of the above updating tools can be used.**

1. Connect HDS to the Data Link Connector (DLC) located under the driver's side of the dashboard.
2. Turn the ignition switch to ON (II).
3. Make sure the HDS communicates with the PCM and other vehicle systems. If it doesn't, go to the DLC circuit troubleshooting. If you are returning from the DLC circuit troubleshooting, skip steps 5 through 9, 18 through 23, and 26 through 28, and do this after replacing the PCM:
   - Replace the engine oil and the engine oil filter.
   - Replace the ATF.
   - Clean the throttle body.
4. Select the PGM-FI system with the HDS.
5. Select the INSPECTION MENU with the HDS.
6. Select the ETCS TEST, then select the TP POSITION CHECK, and follow the screen prompts.

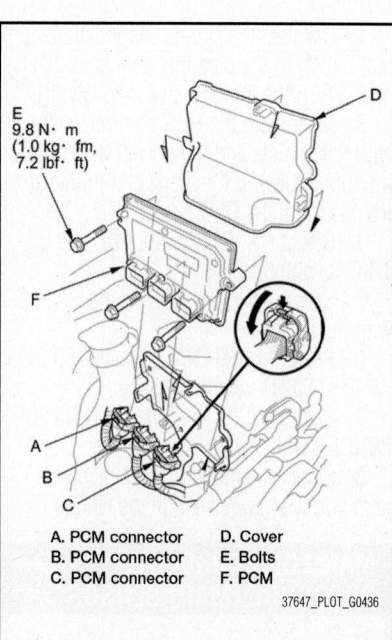

A. PCM connector
B. PCM connector
C. PCM connector
D. Cover
E. Bolts
F. PCM

37647_PLOT_G0436

Fig. 165 Exploded view of PCM assembly

➡ **If the TP POSITION CHECK indicates FAILED, continue with this procedure.**

7. Select the REPLACE PCM MENU, then select READ DATA, and follow the screen prompts.

➡ **Doing this step copies (READS) the engine oil life data from the original PCM so you can later download (WRITES) it into the new PCM. If the READ DATA indicates FAILED, continue with this procedure.**

8. Select the A/T system with the HDS.
9. Select the REPLACE TCM/PCM MENU, then select READ DATA, and follow the screen prompts.

➡ **Doing this step copies (READS) the ATF life data from the original PCM so you can later download (WRITES) it into the new PCM. If READ DATA indicates FAILED, continue with this procedure.**

10. Turn the ignition switch to LOCK (0).
11. Jump the SCS line with the HDS.
12. Remove the cover.
13. Disconnect PCM connectors A, B, and C.

➡ **PCM connectors A, B, and C have symbols (A=□, B=△, C=○) embossed on them for identification.**

14. Remove the bolts and remove the PCM.
15. Install the PCM in the reverse order of removal.
16. Turn the ignition switch to ON (II).
17. Manually input the VIN to the PCM with the HDS.

➡ **DTC P0630 VIN Not Programmed or Mismatch may be stored because the VIN has not been programmed into the PCM; ignore it, and continue this procedure.**

18. If the READ DATA (engine oil life) failed in step 7, go to step 21. Otherwise, go to step 19.
19. Select the PGM-FI system with the HDS.
20. Select the REPLACE PCM MENU, then select WRITE DATA, and follow the screen prompts.

➡ **If the WRITE DATA indicates FAILED, continue with this procedure.**

21. If the READ DATA (ATF life) failed in step 9, go to step 24. Otherwise go to step 22.
22. Select the A/T SYSTEM with the HDS.

23. Select the REPLACE TCM/PCM MENU, then select WRITE DATA, and follow the screen prompts.

➡**If the WRITE DATA indicates FAILED, continue with this procedure.**

24. Select the IMMOBI SYSTEM with the HDS.

25. Enter the immobilizer PCM code that you got from the iN, and use the PCM replacement procedure in the IMMOBI MENU of the HDS; it allows you to start the engine.

26. If the TP POSITION CHECK failed in step 6, clean the throttle body, then go to step 27.

27. If the READ DATA failed in step 7 or the WRITE DATA failed in step 20, replace the engine oil and engine oil filter, then go to step 28.

28. If the READ DATA failed in step 9 or the WRITE DATA failed in step 23, replace the ATF, then go to step 29.

29. Select the PGM-FI system, and reset the PCM with the HDS.

30. Update the PCM if it does not have the latest software.

31. Do the PCM idle learn procedure.

32. Do the CKP pattern clear/CKP pattern learn procedure.

PCM UPDATE

Special Tools Required:
• Honda Diagnostic System (HDS) tablet tester
• Honda Interface Module (HIM) and an iN workstation with the latest HDS software version
• HDS pocket tester
• GNA600 and an iN workstation with the latest HDS software version

➡**Any one of the above updating tools can be used.**

➡**Note the following for PCM update procedure:**

• Make sure the HDS/iN workstation has the latest HDS software version.
• Before you update the PCM, make sure the battery in the vehicle is fully charged, and connect a jumper battery (not a battery charger) to maintain system voltage.
• Never turn the ignition switch to ACC (I) or to LOCK (0) during the update. If there is a problem with the update, leave the ignition switch ON (II).

• To prevent PCM damage, do not operate anything electrical (headlights, audio system, brakes, A/C, power windows, moonroof (if equipped), door locks, etc.) during the update.
• To ensure the latest program is installed, do a PCM update whenever the PCM is substituted or replaced.
• You cannot update a PCM with a program it already has. It will only accept a new program.
• High temperature in the engine compartment might cause the PCM to become too hot to run the update. If the engine has been running before this procedure, open the hood and cool the engine compartment.
• If you need to diagnose the Honda Interface Module (HIM) because the HIM's red (#3) light came on or was flashing during the update, leave the ignition switch in ON (II) when you disconnect the HIM from the Data Link Connector (DLC). This will prevent PCM damage.

1. Turn the ignition switch to ON (II), but do not start the engine.

2. Connect the HDS to the Data Link Connector (DLC) located under the driver's side of the dashboard.

3. Make sure the HDS communicates with the PCM and other vehicle systems. If it doesn't, go to the DLC circuit troubleshooting. If you are returning from DLC circuit troubleshooting, skip steps 4 and 5, and clean the throttle body after updating the PCM.

4. Select the INSPECTION MENU with HDS.

5. Select the ETCS TEST, then select the TP POSITION CHECK, and follow the HDS screen prompts.

➡**If the TP POSITION CHECK indicates FAILED, continue this procedure.**

6. Exit the HDS diagnostic system, then select the update mode, and follow the screen prompts to update the PCM.

7. If the software in the PCM is the latest, disconnect the updating tool from the DLC, and go back to the procedure that you were doing. If the software in the PCM is not the latest, follow the instructions on the screen. If prompted to choose the PGM-FI system or the A/T system, make sure you update both.

➡**If the PCM update system requires you to cool the PCM, follow the instructions on screen. If you run in to a problem during the update procedure (programming takes over 15 minutes, status bar goes over 100%, D or immobilizer indicator flashes, HDS tablet freezes, etc.), follow these steps to minimize the chance of damaging the PCM.**

• Leave the ignition switch is in ON (II).
• Connect a jumper battery (do not connect a battery charger).
• Shut down the HDS.
• Disconnect the HDS from the DLC.
• Reboot the HDS.
• Reconnect the HDS to the DLC, and try the update procedure again.

8. If the TP POSITION CHECK failed in step 5, clean the throttle body.

9. Do the PCM idle learn procedure.

10. Do the CKP pattern clear/CKP pattern learn procedure.

PCM IDLE LEARN PROCEDURE

The idle learn procedure must be done so the PCM can learn the engine idle characteristics.

Do the idle learn procedure whenever you do any of these actions:
• Replace the PCM.
• Reset the PCM.
• Update the PCM.
• Replace or clean the throttle body.
• Disassemble the engine or transmission.

➡**Erasing DTCs with the HDS does not require you to do the idle learn procedure.**

1. Make sure all electrical items (the A/C, the audio, the lights, etc.) are off.

2. Reset the PCM with the HDS.

3. Turn the ignition switch to ON (II), and wait 2 seconds.

4. Start the engine. Hold the engine speed at 3,000 rpm without load (in P or N) until the radiator fan comes on, or until the engine coolant temperature reaches 194°F (90°C).

5. Let the engine idle for about 5 minutes with the throttle fully closed.

➡**If the radiator fan comes on, do not include its running time in the 5 minutes.**

FUEL SYSTEM   **GASOLINE FUEL INJECTION SYSTEM**

FUEL SYSTEM SERVICE PRECAUTIONS

Safety is the most important factor when performing not only fuel system mainte-nance, but any type of maintenance. Failure to conduct maintenance and repairs in a safe manner may result in serious personal injury or death. Work on a vehicle's fuel system components can be accomplished safely and effectively by adhering to the fol-lowing rules and guidelines.

• To avoid the possibility of fire and per-sonal injury, always disconnect the negative battery cable unless the repair or test proce-dure requires that battery voltage be applied.

• Always relieve the fuel system pres-sure prior to disconnecting any fuel system component (injector, fuel rail, pressure reg-ulator, etc.) fitting or fuel line connection. Exercise extreme caution whenever relieving fuel system pressure to avoid exposing skin, face and eyes to fuel spray. Please be advised that fuel under pressure may pene-trate the skin or any part of the body that it contacts.

• Always place a shop towel or cloth around the fitting or connection prior to loosening to absorb any excess fuel due to spillage. Ensure that all fuel spillage is quickly removed from engine surfaces. Ensure that all fuel-soaked cloths or towels are deposited into a flame-proof waste con-tainer with a lid.

• Always keep a dry chemical (Class B) fire extinguisher near the work area.

• Do not allow fuel spray or fuel vapors to come into contact with a spark or open flame.

• Always use a second wrench when loosening or tightening fuel line connection fittings. This will prevent unnecessary stress and torsion on fuel piping. Always follow the proper torque specifications.

• Always replace worn fuel fitting O-rings with new ones. Do not substitute fuel hose where rigid pipe is installed.

RELIEVING FUEL SYSTEM PRESSURE

With the HDS
See Figures 166 and 167.

Before disconnecting fuel lines or hoses, relieve pressure from the system by dis-abling the fuel pump and disconnecting the fuel line/quick connect fitting in the engine compartment.

1. Connect the HDS to the Data Link Connector (DLC) located under the driver's side of the dashboard.
2. Turn the ignition switch to ON (II).
3. Make sure the HDS communicates with the PCM. If it doesn't, go to the DLC circuit troubleshooting.
4. Turn the ignition switch to LOCK (0).
5. Remove the fuel fill cap to relieve the pressure in the fuel tank.
6. Turn the ignition switch to ON (II).
7. From the INSPECTION MENU of the HDS, select Fuel Pump OFF, then start the engine, and let it idle until it stalls.

➡ **Do not allow the engine to idle above 1,000 rpm or the PCM will continue to operate the fuel pump. Pending or Con-firmed DTC may be set during this pro-cedure. Check for DTCs, and clear them as needed.**

8. Turn the ignition switch to LOCK (0).
9. Do the battery terminal disconnec-tion procedure.
10. Remove the quick-connect fitting cover.
11. Check the fuel quick-connect fitting for dirt, and clean it if needed.
12. Place a rag or a shop towel over the quick-connect fitting.
13. Disconnect the quick-connect fitting: Hold the connector with one hand, and squeeze the retainer tabs with the other hand to release them from the locking tabs. Pull the connector off.

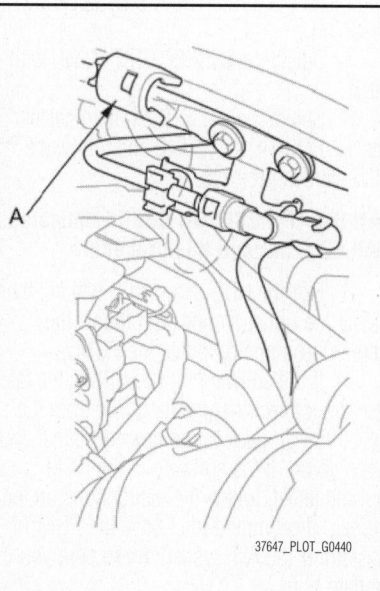

Fig. 166 Remove the quick-connect fitting cover (A)

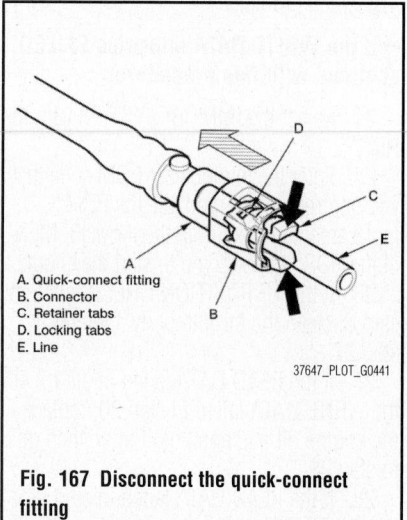

A. Quick-connect fitting
B. Connector
C. Retainer tabs
D. Locking tabs
E. Line

37647_PLOT_G0441

Fig. 167 Disconnect the quick-connect fitting

➡ **Be careful not to damage the line or other parts. Do not use tools. If the connector does not move, keep the retainer tabs pressed down, and alter-nately pull and push the connector until it comes off easily. Do not remove the retainer from the line; once removed, the retainer must be replaced with a new one.**

14. After disconnecting the quick-con-nect fitting, check it for dirt or damage.
15. Do the battery terminal reconnection procedure.

Without the HDS
See Figures 166 through 168.

Before disconnecting fuel lines or hoses, relieve pressure from the system by dis-abling the fuel pump and disconnecting the fuel line/quick connect fitting in the engine compartment.

1. Remove PGM-FI main relay 2 (FUEL PUMP) from the under-dash fuse/relay box.
2. Start the engine, and let it idle until it stalls.

➡ **If any DTCs are stored, clear and ignore them.**

3. Turn the ignition switch to LOCK (0).
4. Remove the fuel fill cap to relieve the pressure in the fuel tank.
5. Do the battery terminal disconnec-tion procedure.
6. Remove the quick-connect fitting cover.
7. Check the fuel quick-connect fitting for dirt, and clean it if needed.
8. Place a rag or a shop towel over the quick-connect fitting.

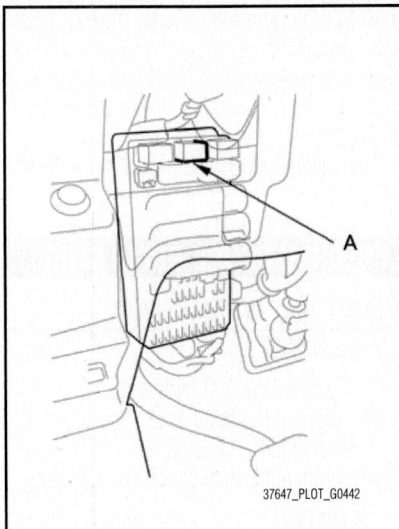

Fig. 168 Remove PGM-FI main relay 2 (FUEL PUMP) (A) from the under-dash fuse/relay box

9. Disconnect the quick-connect fitting: Hold the connector with one hand, and squeeze the retainer tabs with the other hand to release them from the locking tabs. Pull the connector off.

➡ **Be careful not to damage the line or other parts. Do not use tools. If the connector does not move, keep the retainer tabs pressed down, and alternately pull and push the connector until it comes off easily. Do not remove the retainer from the line; once removed, the retainer must be replaced with a new one.**

10. After disconnecting the quick-connect fitting, check it for dirt or damage.

11. Do the battery terminal reconnection procedure.

IDLE SPEED

ADJUSTMENT

Idle speed is controlled by the Powertrain Control Module (PCM). No adjustment is necessary or possible.

FUEL FILTER

REMOVAL & INSTALLATION

See Figure 169.

➡ **The fuel filter should be replaced whenever the fuel pressure drops below the specified value, after making sure that the fuel pump and the fuel pressure regulator are OK.**

1. Remove the fuel tank unit.

2. Remove the fuel filter set.

To install:

3. Check these items before installing the fuel tank unit:

a. When connecting the wire harness, make sure the connection is secure and the connectors are firmly locked into place.

b. When installing the fuel gauge sending unit, make sure the connection is secure and the connector is firmly locked into place. Be careful not to bend or twist it excessively.

4. Install the parts in the reverse order of removal with new O-rings and new bracket. When installing the fuel tank unit, align the marks on the fuel tank unit and the fuel tank.

- Do not coat the tank unit base gasket with engine oil.
- Coat the O-rings with clean engine oil; do not use any other type oils or fluids.
- Do not pinch the O-rings during installation.
- Use all the new parts supplied in the fuel filter replacement kit.

FUEL TANK UNIT

REMOVAL & INSTALLATION

See Figures 169 through 174.

Special Tools Required: Fuel Sender Wrench 07AAA-S0XA100

1. Relieve the fuel pressure.
2. Remove the second row seat.

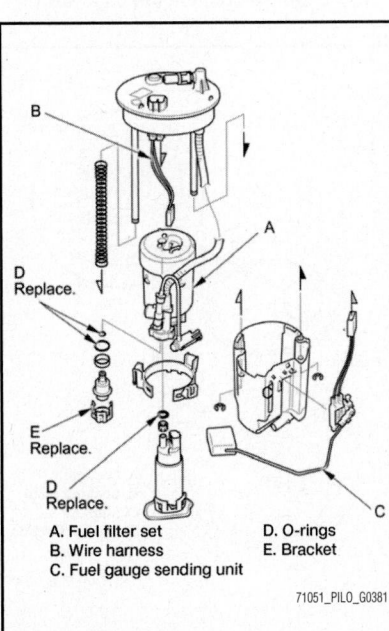

A. Fuel filter set   D. O-rings
B. Wire harness   E. Bracket
C. Fuel gauge sending unit

Fig. 169 Exploded view of the fuel tank unit assembly

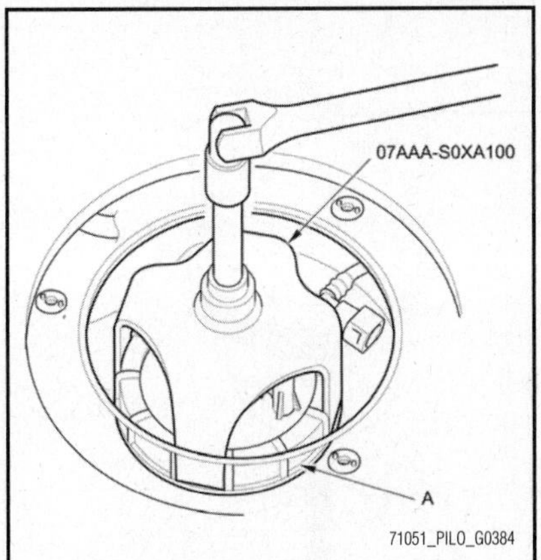

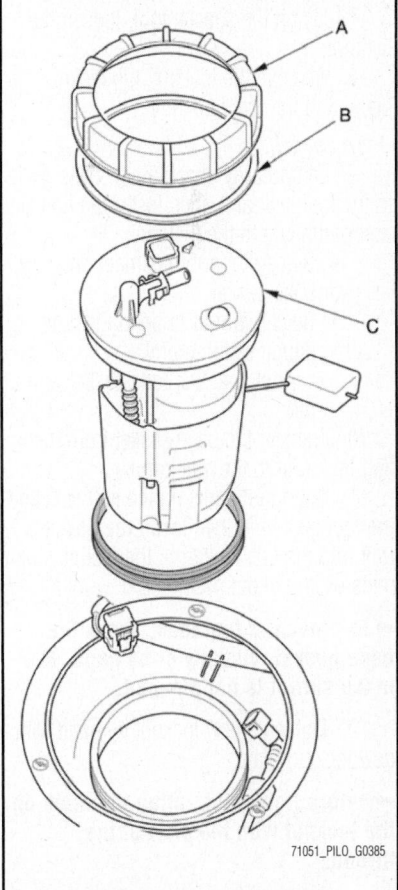

Fig. 170 Using the special tool, loosen the locknut (A)

3. Pull the carpet out of way.
4. Remove the access panel from the floor.
5. Disconnect the fuel tank unit connector.

- USA model: 5P connector
- Canada model: 4P connector

Fig. 171 Remove the locknut (A), the locknut plate (B), and the fuel tank unit (C)

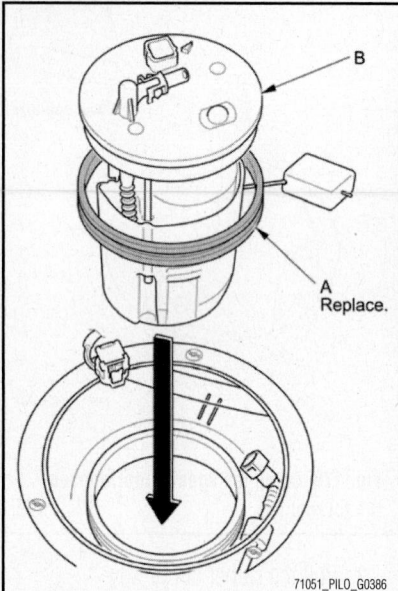

Fig. 172 Temporarily attach a new base gasket (A) to the fuel tank unit (B), then insert the fuel tank unit partially into the fuel tank

6. Disconnect the quick-connect fitting from the fuel tank unit.

7. Using the special tool, loosen the locknut.

8. Remove the locknut, the locknut plate, and the fuel tank unit.

To install:

9. Temporarily attach a new base gasket to the fuel tank unit, then insert the fuel tank unit partially into the fuel tank.

- Be careful not to damage the new base gasket.
- Be careful not to bend the fuel gauge sending unit.
- Do not coat the base gasket with oil.

10. Transfer the base gasket from the fuel tank unit to the fuel tank.

11. Align the marks on the fuel tank and the fuel tank unit, then insert the fuel tank unit into the fuel tank until the fuel tank unit rests on top of the base gasket.

➡**To prevent a fuel leak, check the base gasket, visually or by hand, to make sure it is not pinched.**

12. Tighten a new locknut by hand with a new locknut plate.

➡**Before tightening, align the mark on the locknut with the start of the threads.**

13. Using the special tool, tighten the locknut to the specified torque.

- After tightening, make sure the marks are still aligned.

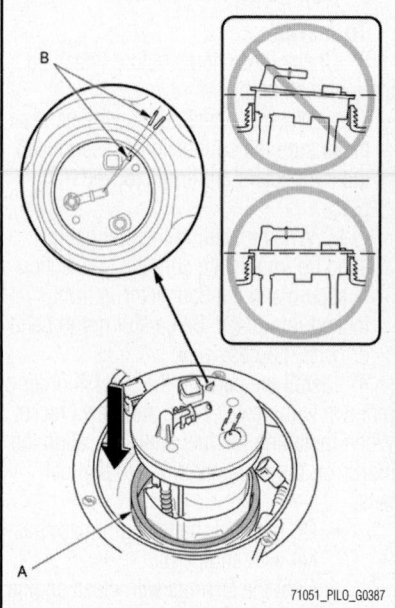

Fig. 173 Transfer the base gasket (A) from the fuel tank unit to the fuel tank, align the marks (B) on the fuel tank and the fuel tank unit, then insert the fuel tank unit into the fuel tank until the fuel tank unit rests on top of the base gasket

- After installation, check the base gasket, visually or by hand, to make sure it is not pinched.

14. Connect the fuel tank unit connector.

15. Reconnect the quick-connect fitting to the fuel tank unit.

Reconnect the negative cable to the battery, and turn the ignition switch to ON (II)

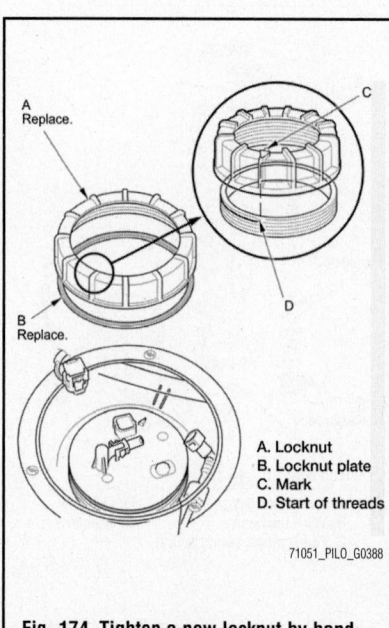

Fig. 174 Tighten a new locknut by hand with a new locknut plate

A. Locknut
B. Locknut plate
C. Mark
D. Start of threads

71051_PILO_G0388

(but do not operate the starter motor). The fuel pump runs for about 2 seconds, and the fuel pressure rises. Repeat this two or three times, then make sure there are no fuel leaks.

16. Install the access panel on the floor.

17. Install the second row seat.

FUEL PRESSURE REGULATOR

REMOVAL & INSTALLATION

See Figure 175.

1. Remove the fuel tank unit.
2. Remove the reservoir.
3. Remove the bracket.
4. Remove the fuel pressure regulator.

To install:

5. Install the parts in the reverse order of removal with new O-rings and new bracket. When installing the fuel tank unit, align the marks on the fuel tank unit and the fuel tank.

- Do not coat the tank unit base gasket with engine oil.
- Coat the O-rings with clean engine oil; do not use any other oils or fluids.
- Do not pinch the O-rings during installation.
- Use all the new parts supplied in the pressure regulator replacement kit.

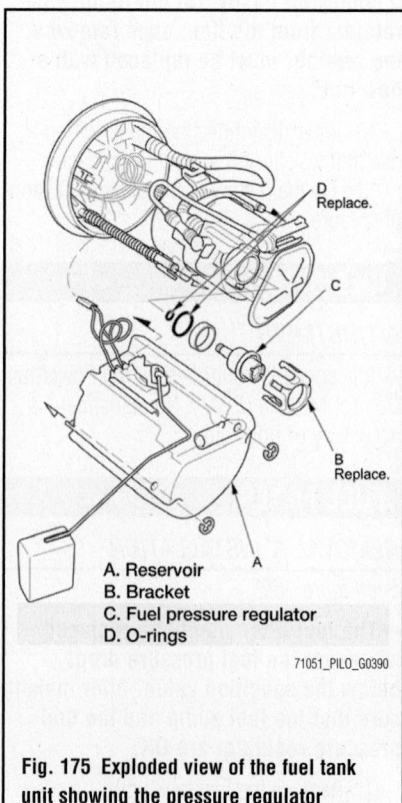

A. Reservoir
B. Bracket
C. Fuel pressure regulator
D. O-rings

71051_PILO_G0390

Fig. 175 Exploded view of the fuel tank unit showing the pressure regulator

FUEL RAIL & INJECTORS

REMOVAL & INSTALLATION

See Figure 176.

1. Relieve the fuel pressure.
2. Remove the intake manifold. Refer to the Engine Mechanical section for Intake Manifold Removal and Installation.
3. Disconnect the quick-connect fitting.
4. Remove the fuel joint hose mounting bolt.
5. Disconnect the connectors from the injectors.
6. Remove the fuel rail mounting bolts from the fuel rails.
7. Remove the fuel rail and the injectors from the injector base.
8. Remove the injector clips from the fuel rail.
9. Remove the injectors from the fuel rails.

To install:

10. Coat the new O-rings (black) with clean engine oil, and insert the injectors into the fuel rails.
11. Install the injector clips.
12. Coat the new injector O-rings (brown) with clean engine oil.
13. Install the fuel rails and injectors in the injector base.
14. Install the fuel rail mounting bolts, and connect the connectors on the injectors.
15. Install the fuel joint hose mounting bolt.
16. Connect the quick-connect fitting.
17. Turn the ignition switch to ON (II), but do not operate the starter. After the fuel pump runs for about 2 seconds, the fuel rail will be pressurized. Repeat this two or three times, then check the fuel leakage.
18. Install the intake manifold with a new gasket.

THROTTLE BODY

REMOVAL & INSTALLATION

See Figure 177.

❄❄ CAUTION

Do not insert your fingers into the installed throttle body when you turn the ignition switch to ON (II) or while the ignition switch is in ON (II). If you do, you will seriously injure your fingers if the throttle valve is activated.

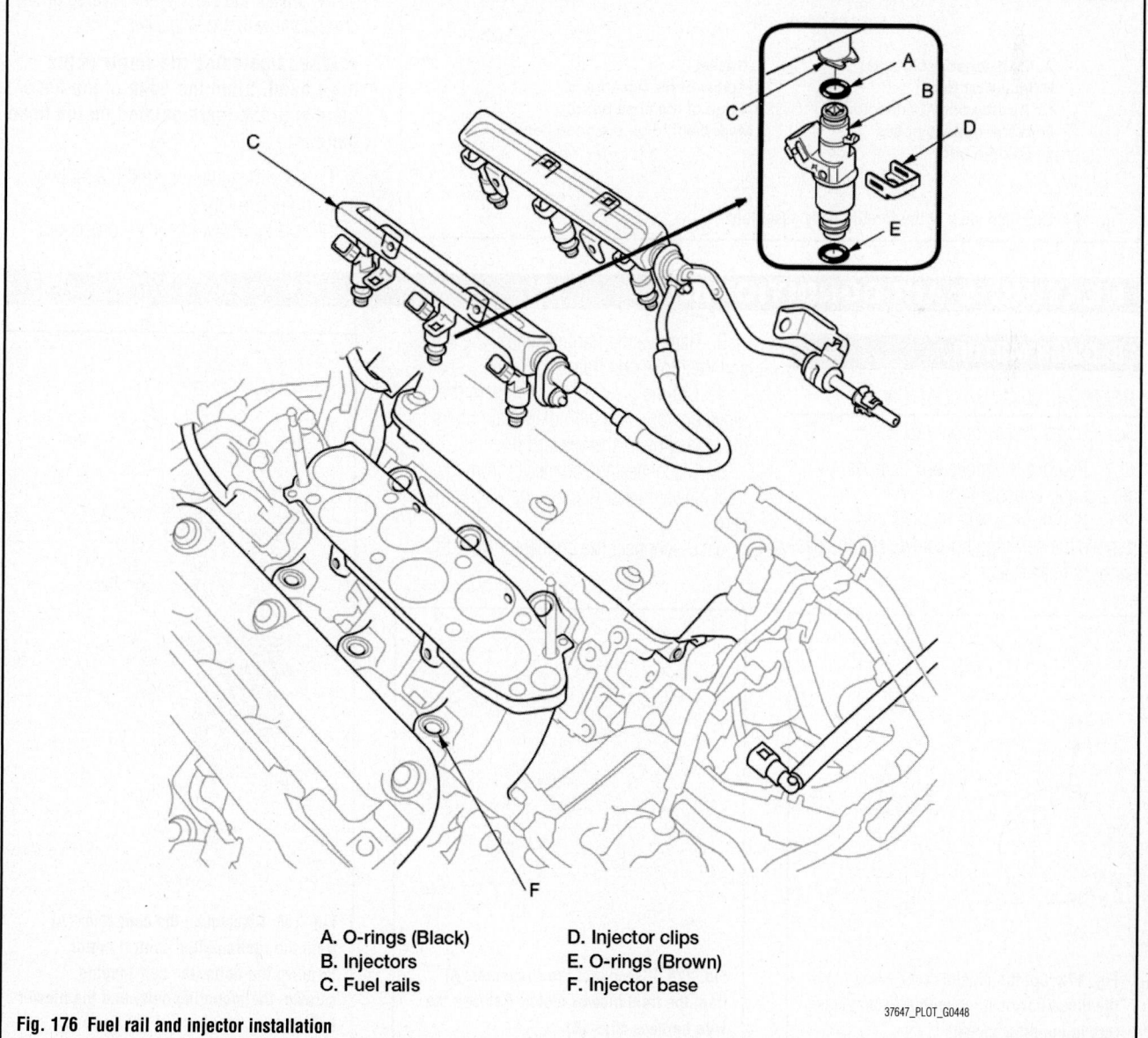

A. O-rings (Black)
B. Injectors
C. Fuel rails
D. Injector clips
E. O-rings (Brown)
F. Injector base

Fig. 176 Fuel rail and injector installation

37647_PLOT_G0448

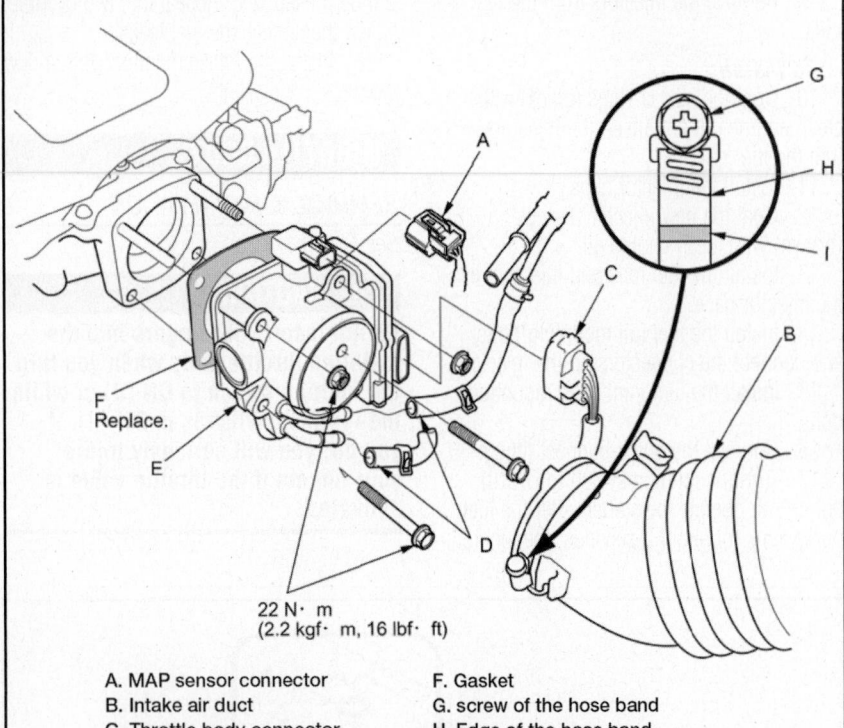

22 N· m
(2.2 kgf· m, 16 lbf· ft)

A. MAP sensor connector
B. Intake air duct
C. Throttle body connector
D. Water bypass hoses
E. Throttle body
F. Gasket
G. screw of the hose band
H. Edge of the hose band
I. Mark painted on the hose band

37647_PLOT_G0457

Fig. 177 Exploded view of the throttle body assembly

➡ **If you are replacing the throttle body, begin at step 1. If you are removing the throttle body temporarily, begin at step 4.**

1. Connect the HDS while the engine is stopped.
2. Select the INSPECTION MENU with the HDS.
3. Do the TP LEARNING CHECK in the ETCS TEST.
4. Turn the ignition switch to LOCK (0).
5. Disconnect the MAP sensor connector.
6. Remove the intake air duct.
7. Disconnect the throttle body connector.
8. Disconnect and plug the water bypass hoses.
9. Remove the throttle body.

To install:

10. Install the parts in the reverse order of removal with a new gasket.

➡ **When tightening the screw of the hose band, align the edge of the hose band with the mark painted on the hose band.**

11. After installation, refill the radiator with engine coolant.
12. Do the PCM idle learn procedure.

HEATING & AIR CONDITIONING

BLOWER MOTOR

REMOVAL & INSTALLATION

See Figures 178 through 180.

1. Remove the glove box and passenger's center console trim.
2. Remove the wire harness clip, then remove the self-tapping screws and the passenger's heater duct.
3. Remove the harness clips, the bolts, and the glove box frame.
4. Cut the plastic cross brace in the glove box opening with diagonal cutters in the area shown, and discard it.
5. Disconnect the connector from the front blower motor. Remove the wire harness clips.
6. Disconnect the connector from the

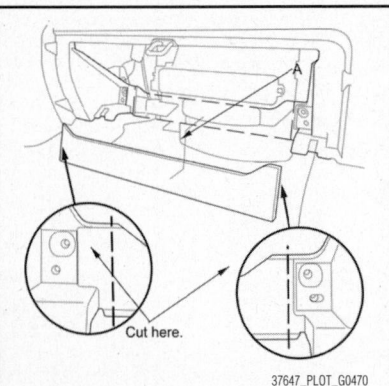

Cut here.

37647_PLOT_G0470

Fig. 178 Cut the plastic cross brace (A) in the glove box opening with diagonal cutters in the area shown

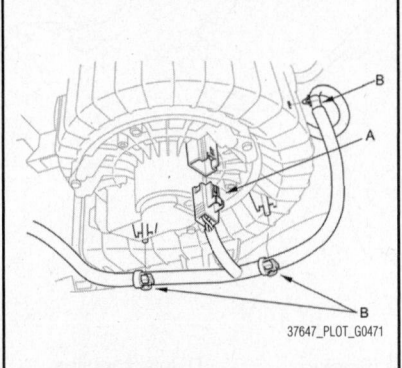

37647_PLOT_G0471

Fig. 179 Disconnect the connector (A) from the front blower motor. Remove the wire harness clips (B)

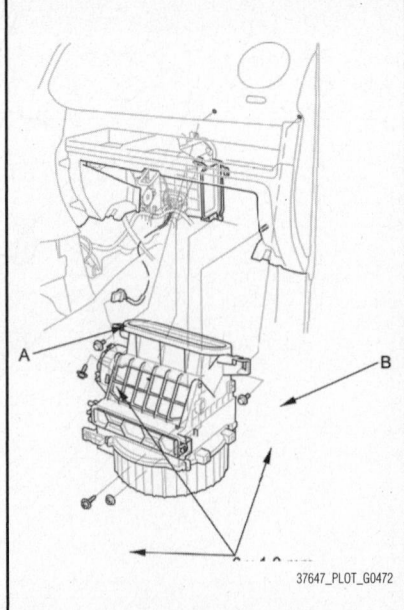

37647_PLOT_G0472

Fig. 180 Disconnect the connector (A) from the recirculation control motor. Remove the bolts, the self-tapping screws, the mounting nuts, and the blower unit (B)

recirculation control motor. Remove the bolts, the self-tapping screws, the mounting nuts, and the blower unit.

7. Install the unit in the reverse order of removal. Make sure that there is no air leakage.

HEATER CORE

REMOVAL & INSTALLATION

See Figures 181 through 188.

✳✳ WARNING

SRS components are located in this area. Review the SRS component locations and the precautions and procedures before doing repairs or service.

1. Do the battery terminal disconnection procedure.

2. Disconnect the front receiver line and front suction line from the front evaporator core.

 a. Remove the bolts.

 b. Remove the bolts, then disconnect the front receiver line and the front suction line from the front evaporator core. Remove the O-rings.

3. When the engine is cool, drain the engine coolant from the radiator.

4. From under the hood, slide the hose clamps back. Disconnect the inlet heater hose and the outlet heater hose from the heater unit. Note the layout of the hose.

➡**Engine coolant will run out when the hoses are disconnected; drain it into a clean drip pan. Be sure not to let coolant spill on the electrical parts or the painted surfaces. If any coolant spills, rinse it off immediately.**

5. Remove the mounting nuts from the heater unit. Take care not to damage or bend the fuel lines or the brake lines, etc.

6. Remove the dashboard.

7. Disconnect the connector from the front blower motor. Remove the wire harness clips.

8. Disconnect the connectors from the front mode control motor, the passenger's air mix control motor (with climate control), the recirculation control motor, and the front

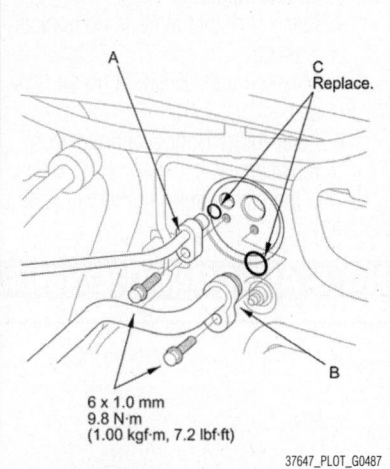

Fig. 181 Remove the bolts, then disconnect the front receiver line (A) and the front suction line (B) from the front evaporator core; remove the O-rings (C)

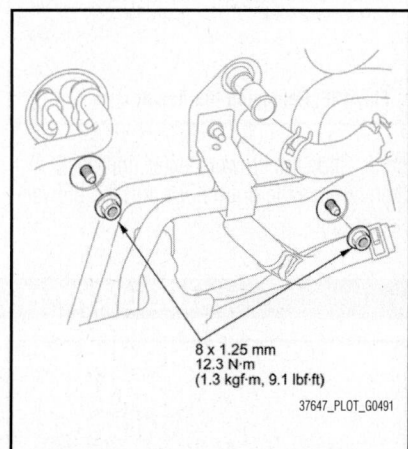

Fig. 183 Remove the mounting nuts from the heater unit

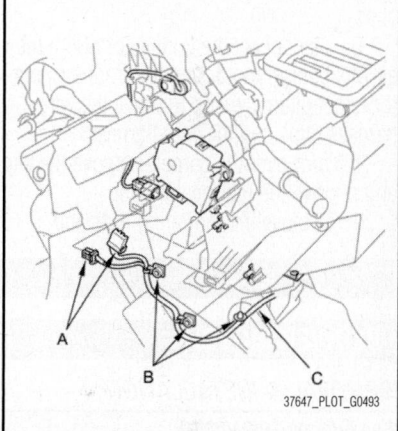

Fig. 185 Disconnect the connectors (A) from the air mix control motor and the front evaporator temperature sensor; remove the wire harness clips (B) and the wire harness (C)

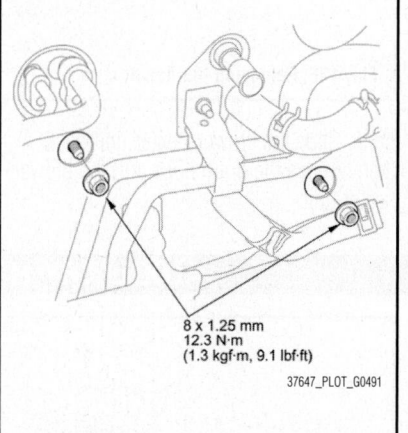

Fig. 182 Slide the hose clamps (A) back; disconnect the inlet heater hose (B) and the outlet heater hose (C) from the heater unit

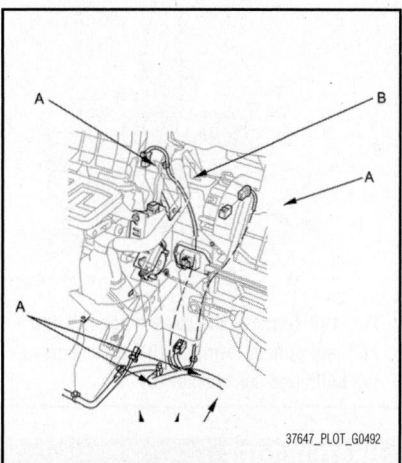

Fig. 184 Disconnect the connectors (A) from the front mode control motor, the passenger's air mix control motor, the recirculation control motor, and the front power transistor; remove the wire harness clips (B)

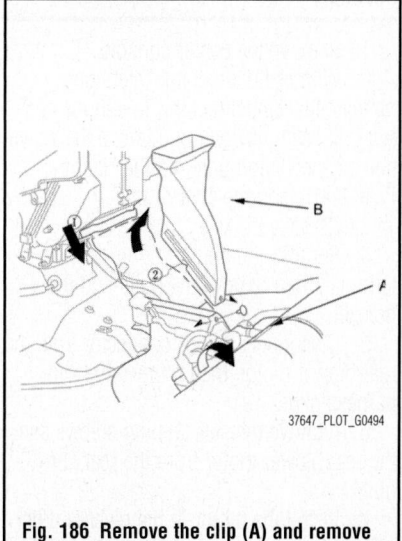

Fig. 186 Remove the clip (A) and remove the rear heater duct (B)

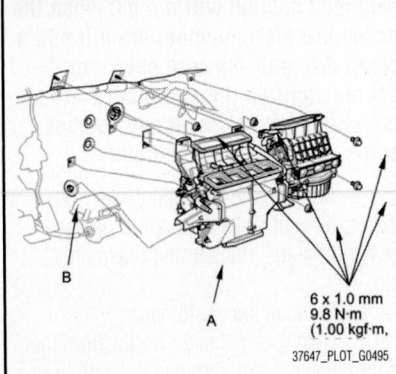

Fig. 187 Remove the mounting nuts and bolts; slide the blower-heater unit (A) back, then remove the drain hose (B) and the blower-heater unit

power transistor. Remove the wire harness clips.

9. Disconnect the connectors from the air mix control motor and the front evaporator temperature sensor. Remove the wire harness clips and the wire harness.

10. Turn over the carpet. Remove the clip and remove the rear heater duct.

11. Remove the mounting nuts and

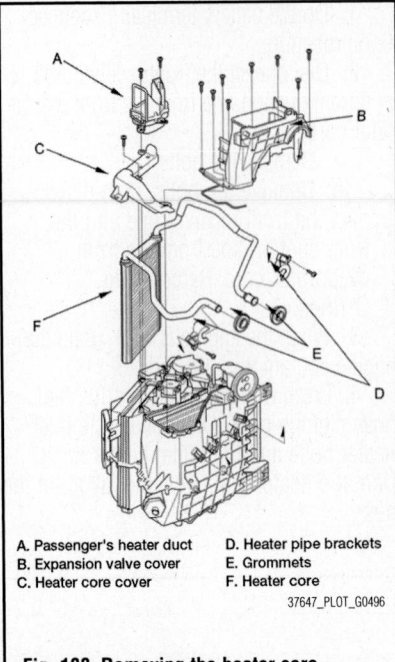

A. Passenger's heater duct
B. Expansion valve cover
C. Heater core cover
D. Heater pipe brackets
E. Grommets
F. Heater core

37647_PLOT_G0496

Fig. 188 Removing the heater core

bolts. Slide the blower-heater unit back, then remove the drain hose and the blower-heater unit.

12. Remove the self-tapping screws and the passenger's heater duct. Remove the self-tapping screws and the expansion valve cover. Remove the self-tapping screw and the front heater core cover. Remove the self-tapping screws, the heater pipe brackets, the grommets, and carefully pull out the front heater core.

To install:

13. Install the front heater core, and the front evaporator core in the reverse order of removal.

14. Install the heater unit in the reverse order of removal, and note these items:
- Do not interchange the inlet and the outlet heater hoses.
- Install the inlet and the outlet hose clamps securely.
- Refill the cooling system with engine coolant.
- Make sure that there is no coolant leakage.
- Make sure that there is no air leakage.
- Refer to the front evaporator core replacement.

15. Do the battery terminal reconnection procedure.

AUXILIARY HEATING & AIR CONDITIONING

BLOWER MOTOR

REMOVAL & INSTALLATION

See Figures 189 and 190.

❊❊ CAUTION

SRS components are located in this area. Review the SRS component locations and the precautions and procedures before doing repairs or service.

1. Remove the center console.
2. With HandsFreeLink Telephone, remove the HandsFreeLink Telephone control unit. With navigation, remove the active noise cancellation unit and the Hands-FreeLink Telephone control unit.
3. Remove the relay block.
4. Remove the bolt and ground terminal, then remove the bolts and the bracket.
5. Disconnect the 2P connector from the rear blower motor, then remove the wire harness clips.
6. Remove the self-tapping screws and the rear blower motor from the rear HVAC unit.
7. Install the motor in the reverse order of removal.

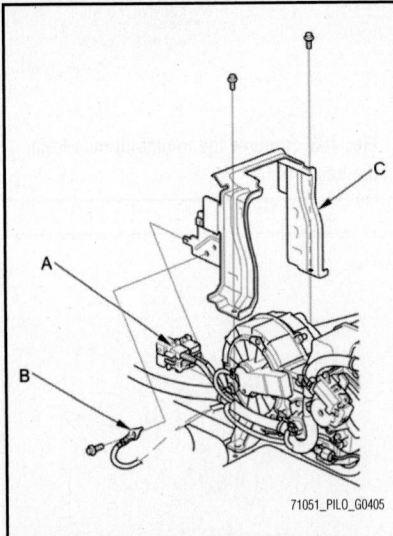

Fig. 189 Remove the relay block (A), the bolt and ground terminal (B), then remove the bolts and the bracket (C)

HEATER CORE

REMOVAL & INSTALLATION

See Figures 191 and 192.

❊❊ WARNING

SRS components are located in this area. Review the SRS component

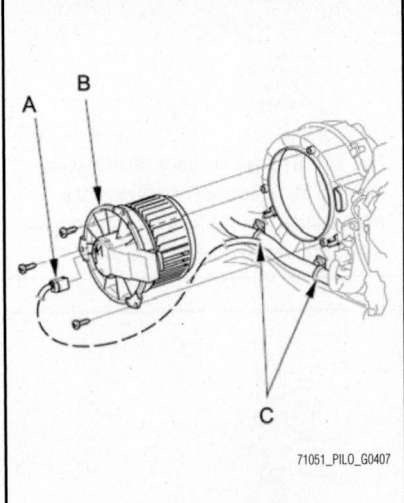

Fig. 190 Disconnect the 2P connector (A) from the rear blower motor (B), then remove the wire harness clips (C)

locations and the precautions and procedures before doing repairs or service.

1. When the engine is cool, drain the engine coolant from the radiator.
2. Remove the center console.
3. Slide the hose clamps back, then disconnect the rear inlet heater hose and the

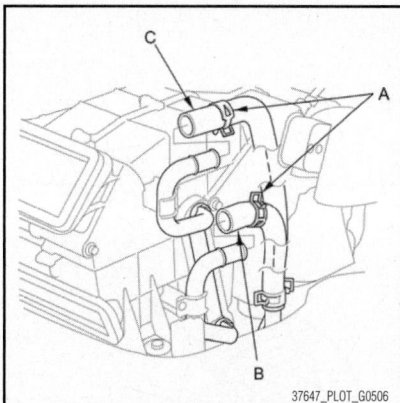

Fig. 191 Slide the hose clamps (A) back, then disconnect the rear inlet heater hose (B) and the rear outlet heater hose (C) from the rear heater core

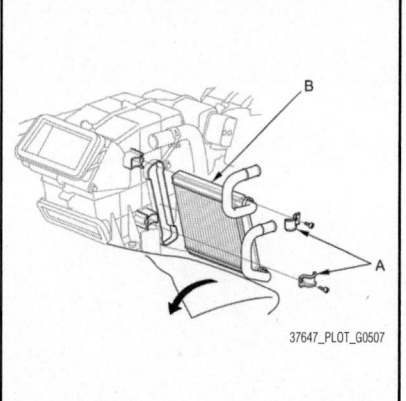

Fig. 192 Remove the self-tapping screws and the clamps (A); pull out the rear heater core (B) without bending lines

rear outlet heater hose from the rear heater core. Note the orientation of the hose.

➡**Engine coolant will run out when the hoses are disconnected; drain it into a** clean drip pan. Be sure not to let coolant spill on the electrical parts or the painted surfaces. If any coolant spills, rinse it off immediately.

4. Turn over the carpet. Remove the self-tapping screws and the clamps. Carefully pull out the rear heater core without bending lines.

To install:

5. Install the unit in the reverse order of removal, and note these items:
- Do not interchange the inlet and outlet heater hoses, and install the hose clamps securely.
- Refill the cooling system with engine coolant.
- Make sure that there is no coolant leakage.
- Make sure that there is no air leakage.

STEERING

POWER RACK & PINION STEERING GEAR

REMOVAL & INSTALLATION

See Figures 193 through 211.

Special Tools Required*:
- Ball Joint Thread Protector, 12 mm 07AAF-SDAA100
- Ball Joint Remover, 28 mm 07MAC-SL0A202
- Engine Support Hanger, A and Reds AAR-T1256
- 2008 V6 Attachment Arm SIL02C000033
- Subframe Adapter VSB02C000016
- Engine Hanger Balance Bar VSB02C000019
- *Available through the Honda Tool and Equipment Program 888-424-6857

✷✷ WARNING

SRS components are located in this area. Review the SRS component locations and the precautions and procedures before doing repairs or service.

➡**Note these items during removal:**
- Using clean solvent and a brush, wash any oil and dirt off the valve body unit, it's lines, and the end of the steering gearbox. Blow dry with compressed air.
- Be sure to remove the steering wheel before disconnecting the

steering joint, or damage to the cable reel can occur.
- Lower the front subframe from the body, and remove the steering gearbox through the gap produced by lowering the front subframe.

1. Remove the hood support rod, then use it to prop the hood in the wide-open position.
2. Remove the engine cover.
3. Drain the power steering fluid.
4. Do the battery terminal disconnection procedure.
5. Raise and support the vehicle.
6. Remove the front wheels.
7. Remove the driver's airbag, and the steering wheel.
8. Remove the driver's dashboard lower cover.
9. Remove the footrest.
   a. Detach the hooks, then remove the footrest plate.
   b. Remove the nut with a 6 mm hexagonal wrench, release the clip, then remove the footrest.
10. Remove the center console trim.
11. Pull back the carpet.
12. Remove steering joint cover.
13. Release the lock lever, and adjust the steering column to the full tilt up position, and to the full telescope in position.
14. Tighten the lock lever.
15. Hold the lower slide shaft on the column with a piece of wire between the joint yoke of the lower slide shaft and joint yoke of the upper shaft to prevent the lower slide shaft from pulling out.

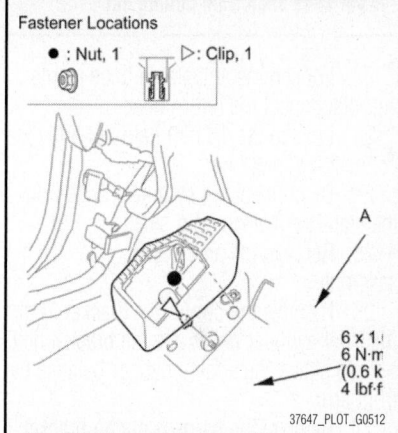

Fig. 193 Remove the nut with a 6 mm hexagonal wrench, release the clip, then remove the footrest (A)

16. Release the lock lever, and adjust the steering column to the full telescopic out position, then tighten the lock lever.
17. Remove the steering joint bolt, and disconnect the steering joint by sliding the steering joint into the column.
18. Remove the center guide (if equipped) from the top of the pinion shaft, and discard it. The center guide is for factory assembly use only.
19. Wrap vinyl tape over the splines on the pinion shaft.
20. Remove the cotter pin from the tie-rod end ball joint, then remove the nut on both sides.
21. Disconnect the tie-rod end ball joint from the knuckle using the ball joint thread protector and the ball joint remover.

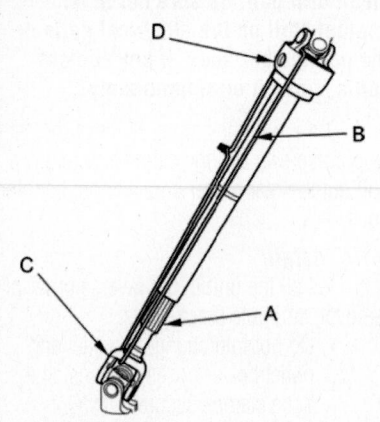

A. Lower slide shaft
B. Wire
C. Joint yoke of the lower slide shaft
D. Joint yoke of the upper shaft

37647_PLOT_G0514

Fig. 194 Hold the lower slide shaft on the column with a piece of wire between the joint yoke of the lower slide shaft and joint yoke of the upper shaft to prevent the lower slide shaft from pulling out

22. Loosen the adjustable hose clamp and disconnect the return hose.

23. Loosen the 16 mm flare nut and disconnect the inlet line.

24. Disconnect the stabilizer links from the stabilizer bar on both sides.

25. Remove the air cleaner assembly.

26. Remove the connector bracket from the front cylinder head; use the bracket bolt hole to attach the engine hanger balance bar front arm.

27. Remove the harness clamp bracket from the rear cylinder head; use the bracket

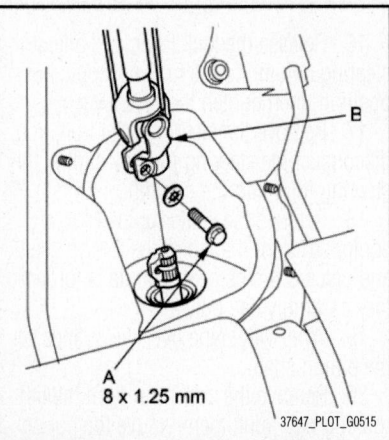

8 x 1.25 mm

37647_PLOT_G0515

Fig. 195 Remove the steering joint bolt (A), and disconnect the steering joint (B) by sliding the steering joint into the column

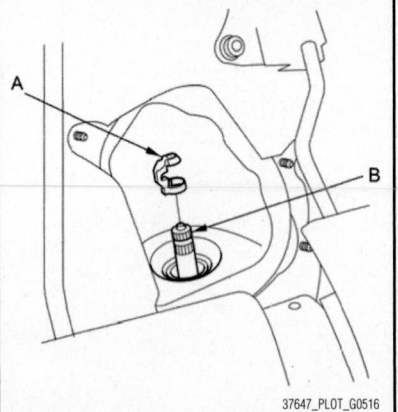

37647_PLOT_G0516

Fig. 196 Remove the center guide (A) (if equipped) from the top of the pinion shaft (B), and discard it

bolt hole to attach the attachment arm (SIL02C000033).

28. Remove the cowl top side cover from both sides, then install the engine support hanger (AAR-T1256). With the engine support hanger beam on its side, insert the hanger beam through the opening, then rotate it over the damper.

29. Install the engine hanger balance bar (VSB02C000019). Attach the front arm to the front cylinder head with several spacers and a 10 x 1.25 mm bolt. Remove the rear arm from the engine hanger balance bar, then install the 2008 V6 attachment arm (SIL02C000033). Attach the 2008 V6 attachment arm to the rear cylinder head with an 8 x 1.25 mm bolt.

➡**Be careful not to damage the hood opener cable when installing the engine support hanger (AAR-T1256) at the front bulkhead. AAR-T1256 two sets**

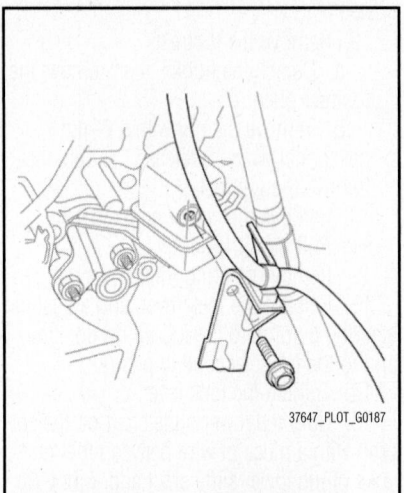

37647_PLOT_G0187

Fig. 197 Remove the harness clamp bracket (A) from the rear cylinder head

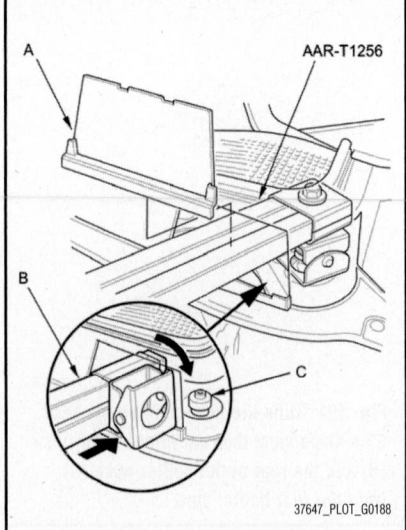

AAR-T1256

37647_PLOT_G0188

Fig. 198 Remove the cowl top side cover (A) from both sides, then install the engine support hanger; with the engine support hanger beam (B) on its side, insert the hanger beam through the opening, then rotate it over the damper (C)

required for stacking additional cross section bar.

30. Install the engine support hanger onto the vehicle, and attach the hook to the engine balancer bar slot. Tighten the wing nut by hand, and lift and support the engine.

31. Raise the vehicle.

32. Remove the splash shield.

33. Remove the front subframe stiffener.

34. Remove exhaust pipe.

35. 4WD: Remove the propeller shaft and propeller shaft protectors.

36. Remove the inlet line clamp bolts.

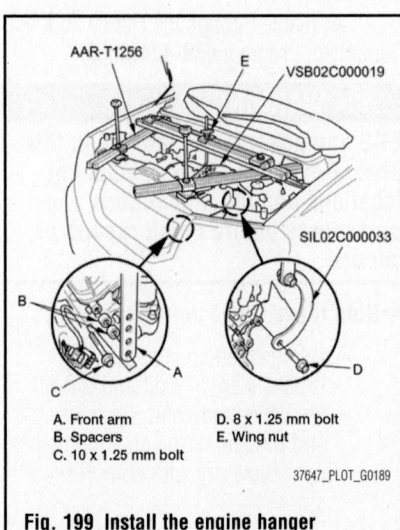

AAR-T1256   E   VSB02C000019

SIL02C000033

A. Front arm   D. 8 x 1.25 mm bolt
B. Spacers   E. Wing nut
C. 10 x 1.25 mm bolt

37647_PLOT_G0189

Fig. 199 Install the engine hanger balance bar

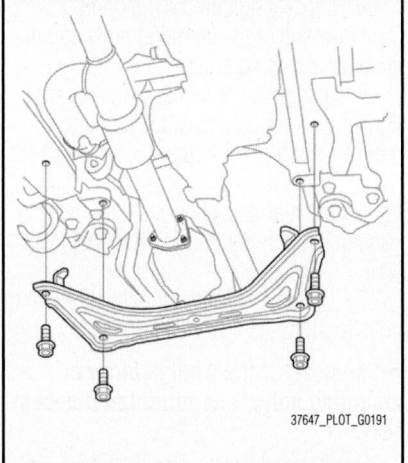

Fig. 200 Remove the front subframe stiff-ener (A)

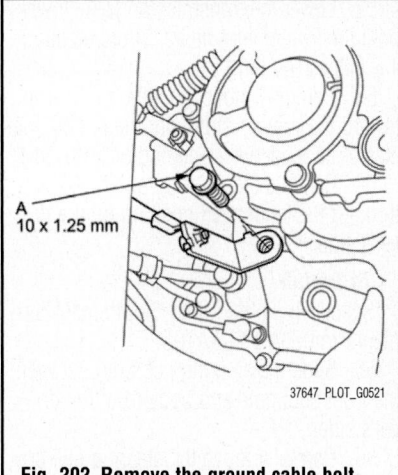

Fig. 202 Remove the ground cable bolt (A) from the transmission

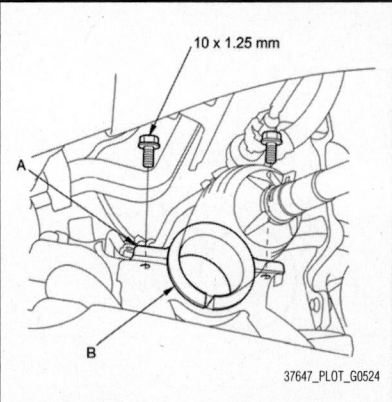

Fig. 205 Remove the mounting bolts from the passenger's side of the steering gear-box, then remove the mounting bracket (A) and cushion (B)

37. Disconnect the front engine mount actuator connector and the harness clamps. Remove the front engine mount stop, then remove the front engine mount bolt.

38. Remove the rear engine mount bolts, and disconnect the rear engine mount actuator connector.

39. Remove the rear engine mount from the rear engine mount base bracket, and move it aside.

40. Remove the ground cable bolt from the transmission.

41. Remove the mounting bolts from the driver's side of the steering gearbox.

42. Remove the gearbox stiffener bracket from the driver's side of the front subframe.

43. Remove the mounting bolts from the passenger's side of the steering gearbox, then remove the mounting bracket and cushion.

44. Remove the lower transmission mount bolts.

45. Attach the front subframe adapter

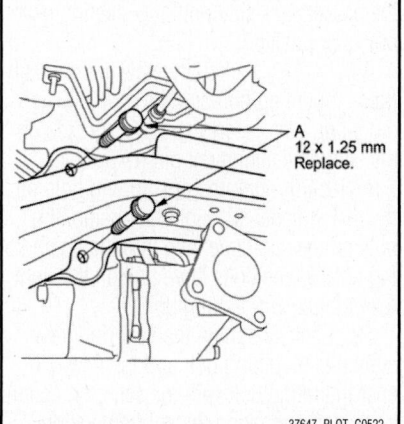

Fig. 203 Remove the mounting bolts (A) from the driver's side of the steering gear-box

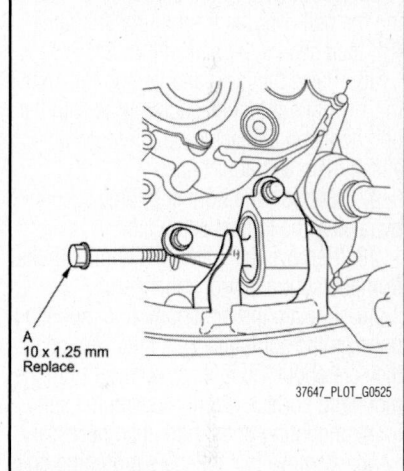

Fig. 206 Remove the lower transmission mount bolts (A)

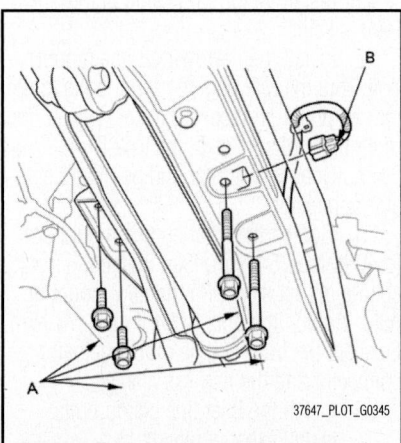

Fig. 201 Remove the rear engine mount bolts (A), and disconnect the rear engine mount actuator connector (B)

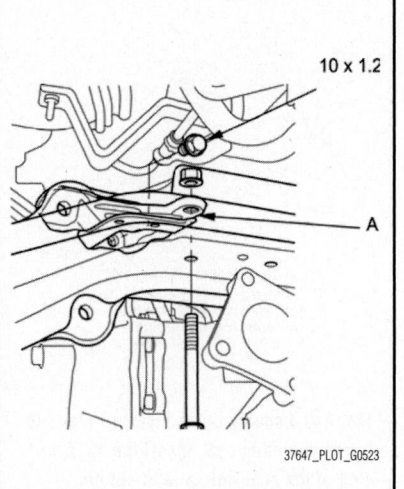

Fig. 204 Remove the gearbox stiffener bracket (A) from the driver's side of the front subframe

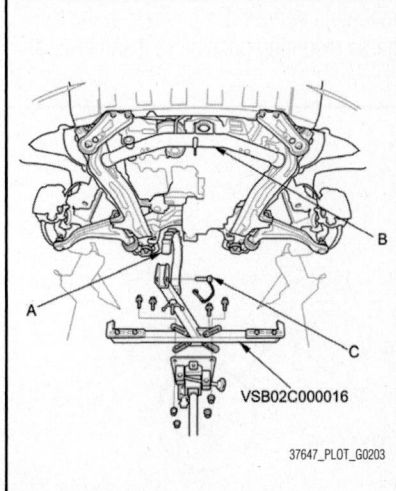

Fig. 207 Attach the front subframe adapter to the subframe by looping the belt (A) over the front of the subframe, then secure the belt with its stop

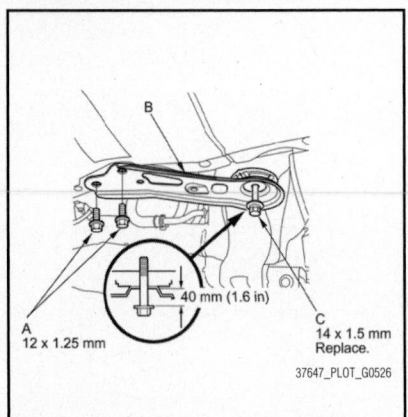

Fig. 208 Remove the stiffener mounting bolts (A) from the subframe front stiffeners (B)

(VSB02C000016) to the subframe by looping the belt over the front of the subframe (B), then secure the belt with its stop (C).

46. Raise the jack, and line up the slots in the front subframe adapter arms with the bolt holes on the jack base, then securely attach them with four bolts.

47. Support the front subframe securely by raising the transmission jack.

48. Remove the stiffener mounting bolts from the subframe front stiffeners.

49. Loosen the front subframe mounting bolts on the subframe front stiffeners so they are about 1.6 inches (40 mm) from the mounting surface. Do not loosen the subframe mounting bolts more than necessary.

50. Remove the stiffeners mounting bolts from the subframe rear stiffeners.

51. Loosen the front subframe mounting bolts on the subframe rear stiffeners so they are about 1.6 inches (40 mm) from the mounting surface. Do not loosen the subframe mounting bolts more than necessary.

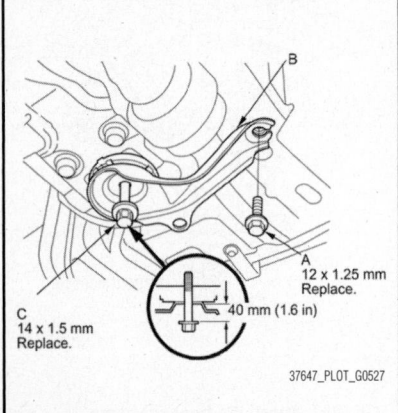

Fig. 209 Remove the stiffeners mounting bolts (A) from the subframe rear stiffeners (B)

52. Lower the transmission jack slowly until the front subframe has dropped about 1.6 inches (40 mm).

53. Carefully move the steering gearbox toward the driver's side until the pinion shaft clears the fenderwell opening on the body.

54. Remove the steering gearbox through the fenderwell opening on the driver's side.

To install:

Special Tools Required: Subframe Alignment Pin 070AG-SJAA10S

55. Slide the steering gearbox) between the front subframe and body from the driver's side.

56. Carefully move the steering gearbox toward the passenger's side until the pinion shaft clears the fenderwell opening on the body.

57. Continue moving the gearbox toward the passenger's side until the steering gearbox is in position.

58. Loosely tighten the new left rear subframe mounting bolt and the new stiffener mounting bolt; insert the 15.7 mm side of the subframe alignment pin (070AG-SJAA10S) through the positioning hole on the rear stiffener, through the positioning hole on the subframe, and into the positioning hole on the body, then tighten the right rear subframe mounting bolt.

59. Loosely tighten the new right rear subframe mounting bolt and the new stiffener mounting bolt with the same procedure as the left rear using the subframe alignment pin.

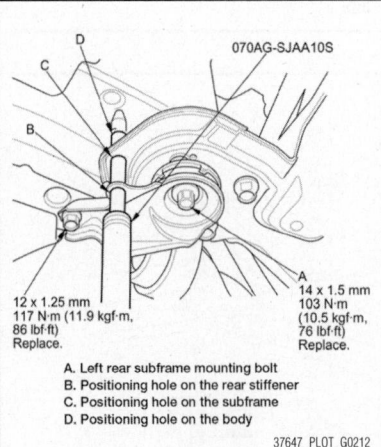

Fig. 210 Loosely tighten the left rear subframe mounting bolt; insert the 15.7 mm side of the subframe alignment pin through the positioning hole on the rear stiffener, through the positioning hole on the subframe, and into the positioning hole on the body

60. Loosely tighten the right and left front new subframe mounting bolts and the stiffener mounting bolts.

61. Using the subframe alignment pin, tighten the rear-driver's side subframe mounting bolt to 76 ft. lbs. (103 Nm).

62. Using the subframe alignment pin, tighten the rear-passenger's side subframe mounting bolt to 76 ft. lbs. (103 Nm).

63. Tighten the front subframe mounting bolts to 76 ft. lbs. (103 Nm).

➡**Check all of the front subframe mounting bolts, and retighten if necessary.**

64. Tighten the front and rear stiffener mounting bolts to 86 ft. lbs. (117 Nm).

➡**Before tightening the stiffener mounting bolts, check that the positioning holes and slot are aligned using the subframe alignment pin.**

65. Loosely tighten new lower transmission mount bolt.

66. Lower the transmission jack supporting the front subframe.

67. Install the gearbox stiffener bracket on the driver's side of the front subframe, and tighten the bolts and nut to 28 ft. lbs. (38 Nm).

68. Loosely install the new mounting bolts on the driver's side of the steering gearbox.

69. Position the cutout on the mounting cushion at the bottom, and install it on the passenger's side of the steering gearbox.

70. Install the gearbox mounting bracket over the mounting cushion, and loosely install the mounting bolts.

71. Tighten the mounting bolts on both sides of the steering gearbox to 37 ft. lbs. (50 Nm) alternately in two steps.

72. Install the ground cable bolt to the transmission.

73. Install the new rear engine mount bolts and the rear engine mount bolts to the rear engine mount base bracket. Tighten the mounting bolts to 31 ft. lbs. (42 Nm).

74. Connect the rear engine mount actuator connector.

75. Tighten the front engine mount bolt to 40 ft. lbs. (54 Nm), then install the front engine mount stop with the new mounting nuts. Tighten the nuts to 54 ft. lbs. (74 Nm). Connect the front engine mount actuator connector and the harness clamps.

76. Install the inlet line clamp bolts.

77. Install exhaust pipe A.

78. 4WD: Install the propeller shaft and propeller shaft protectors.

79. Install the front subframe stiffener

with new mounting bolts, and tighten to 43 ft. lbs. (59 Nm).

80. Install the front splash shield.

81. Lower the vehicle.

82. Remove the engine support hanger, then install the cowl top side lid on both sides.

83. Tighten the lower transmission mount bolt to the specified torque.

84. Connect the stabilizer links to the stabilizer bar. Tighten the mounting nuts to 58 ft. lbs. (78 Nm).

85. Connect the return hose securely, and tighten the adjustable hose clamp.

86. Connect the inlet line, and tighten the 16 mm flare nut to 31 ft. lbs. (42 Nm).

87. On both sides, wipe off any grease contamination from the ball joint tapered section and threads. Reconnect the tie-rod ball joint to the knuckle. Install the new nut, and tighten it to 44 ft. lbs. (60 Nm).

88. Install a new cotter pin, and bend it.

89. Install the front wheels, then set the wheels in the straight ahead position.

➡**Before installing the wheel, clean the mating surfaces of the brake disc and the inside of the wheel.**

90. Center the steering rack within its stroke.

91. Cut the wire from the steering joint.

92. Slip the lower end of the steering joint onto the pinion shaft taking care to align the gap within the angle.

93. Align the bolt hole on the steering joint with the groove around the pinion shaft, then loosely install the joint bolt. Be sure that the joint bolt is securely in the groove in the pinion shaft.

94. Pull on the steering joint to make sure that the steering joint is fully seated, then tighten the joint bolt to 21 ft. lbs. (28 Nm).

95. Install the steering joint cover.

96. Set the carpet.

97. Install the center console trim.

98. Install the footrest.

99. Install the dashboard driver's lower cover.

100. Install the steering wheel, and the driver's airbag.

101. Do the battery terminal reconnection procedure, and do these tasks:

a. Turn the ignition switch to ON (II).The SRS indicator should come on for about 6 seconds, and then go off.

b. Make sure the horn and turn signal switches work properly.

c. Make sure the steering wheel switches work properly.

102. Fill the system with power steering fluid, and bleed the air from the system.

103. Install the engine cover.

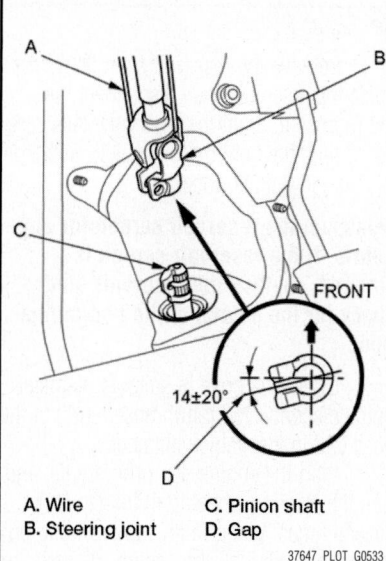

A. Wire   C. Pinion shaft
B. Steering joint   D. Gap

37647_PLOT_G0533

Fig. 211 Cut the wire, and slip the lower end of the steering joint onto the pinion shaft

104. Reinstall the support strut to the proper locations on the hood.

105. After installation, check these items.

a.Start the engine, allow it to idle, and turn the steering wheel from lock to lock several times to warm up the fluid. Check the gearbox for leaks.

b. Check the steering wheel spoke angle. If steering spoke angles to the right and left are not equal (steering wheel and rack are not centered), correct the engagement of the joint/pinion shaft serrations.

c. Set the steering column to the center tilt position, and to the center telescopic position, then do the front toe inspection.

POWER STEERING PUMP

REMOVAL & INSTALLATION

See Figure 212.

1. Remove the engine cover.

2. Place a suitable container under the vehicle to catch any spilled fluid.

3. Drain the power steering fluid from the reservoir.

4. Remove the drive belt from the pump pulley.

5. Cover the auto-tensioner, the alternator, and the A/C compressor with several shop towels to protect them from spilled power steering fluid. Disconnect the pump inlet hose and the pump outlet hose from the pump, and plug them. Take care not to

spill the fluid on the vehicle. Wipe off any spilled fluid at once. Do not turn the steering wheel with the pump removed.

6. Remove the pump mounting bolts, then remove the pump.

7. Cover the opening of the pump with a piece of tape to prevent foreign material from entering the pump.

8. Transfer the pump inlet hose and the pump outlet hose from the original pump onto the new pump with a new O-ring.

To install:

9. Loosely install the pump in the pump bracket with the mounting bolts, then tighten the pump fittings to the specified torque.

10. Tighten the pump mounting bolts to the specified torque.

11. Install the drive belt.

➡**Note these items during drive belt installation:**

a. Inspect the belt for wear and cracks. Replace the belt if necessary.

b. Make sure that the belt is properly positioned on the pulleys.

c. Do not get power steering fluid or grease on the auto-tensioner, alternator, A/C compressor, and drive belt or pulley faces. Clean off any fluid or grease before installation.

12. Install the engine cover.

13. Fill the reservoir to the upper level line.

14. Start the engine, and check for leaks.

FLUID LEVEL CHECK

See Figure 213.

Check the reservoir at regular intervals, and add the recommended fluid as necessary. Always use Honda Power Steering

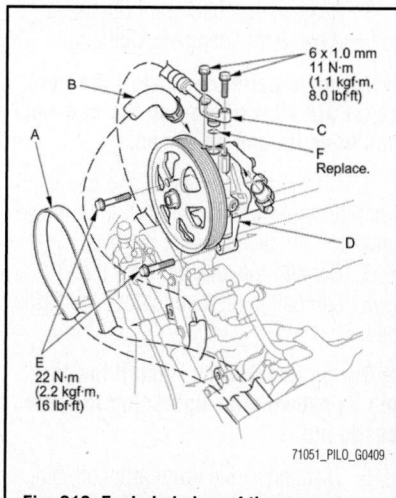

6 x 1.0 mm
11 N·m
(1.1 kgf·m,
8.0 lbf·ft)

F
Replace.

22 N·m
(2.2 kgf·m,
16 lbf·ft)

71051_PILO_G0409

Fig. 212 Exploded view of the power steering pump assembly

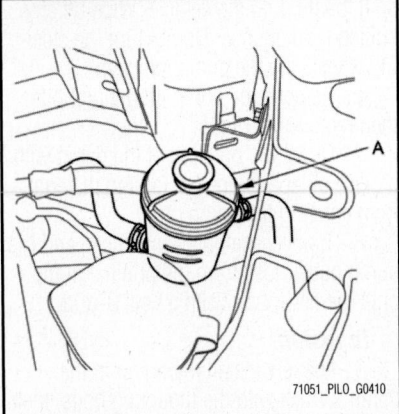

Fig. 213 Check the reservoir (A) at regular intervals, and add the recommended fluid as necessary

FLUID REPLACEMENT

See Figure 214.

1. Remove the reservoir from its holder. Raise the reservoir, then disconnect the return hose to drain the reservoir. Take care not to spill the fluid on the vehicle. Wipe off any spilled fluid at once.

→**Inspect the reservoir screen for any debris. If the reservoir screen is clogged, replace the reservoir, and check for the source of the contamination.**

2. Connect a hose of suitable diameter to the disconnected return hose, and put the hose end in a suitable container.
3. Start the engine, let it run at idle, and turn the steering wheel from lock to lock several times. When fluid stops running out of the hose, shut off the engine. Discard the fluid.

→**Stop the engine immediately once the fluid stops running out of the hose to prevent pump damage.**

4. Reinstall the return hose on the reservoir.
5. Fill the reservoir to the upper level line.

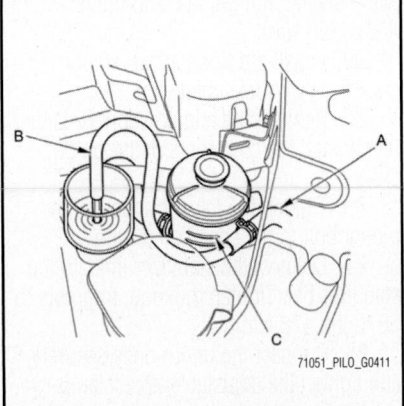

Fig. 214 Disconnect the return hose (A), connect a hose (B), remove the fluid, then fill the reservoir to the upper level line (C)

6. Start the engine, and run it at idle. Turn the steering from lock to lock several times to bleed air from the system.
7. Recheck the fluid level, and add some if necessary. Do not fill the reservoir beyond the upper level line.
8. If the fluid is contaminated, dark, or discolored, repeat the procedure as necessary until the system is clean.

Fluid. Using any other type of power steering fluid or automatic transmission fluid can cause increased wear, fluid leaks, and poor steering in cold weather.

→**If the fluid is contaminated, the screen in the reservoir may be partially blocked. Inspect the reservoir screen for any debris. If the reservoir screen is clogged, replace the reservoir, and check for the source of the contamination.**

SUSPENSION

FRONT SUSPENSION

KNUCKLE

REMOVAL & INSTALLATION

See Figures 215 and 216.

1. Remove the hub bearing unit.
2. Remove the wheel speed sensor from the knuckle. Do not disconnect the wheel speed sensor connector.
3. Remove the cotter pin from the tie-rod end ball joint, then remove the nut.

→**During installation, install the new cotter pin after tightening the new nut, and bend its end as shown.**

4. Disconnect the tie-rod end ball joint from the knuckle using the ball joint thread protector and the ball joint remover.
5. Remove the lock pin from the lower arm ball joint, then remove the castle nut.

→**During installation, install the lock pin as shown after tightening the new castle nut.**

6. Disconnect the lower arm ball joint from the knuckle using the ball joint thread protector and the ball joint remover.

→**Be careful not to damage the ball joint boot when installing the remover. Do not force or hammer on the lower arm, or pry between the lower arm and the knuckle. You could damage the ball joint.**

7. Remove the damper pinch bolts and

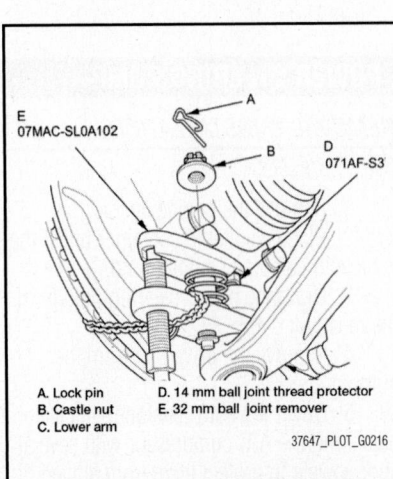

A. Lock pin
B. Castle nut
C. Lower arm
D. 14 mm ball joint thread protector
E. 32 mm ball joint remover

Fig. 215 Remove the lock pin from the lower arm ball joint, then remove the castle nut

the flange nuts from the damper, then remove the knuckle.

→**During installation, install new damper pinch bolts and new flange nuts.**

To install:

8. Install the knuckle in the reverse order of removal, and note these items:
   • First install all of the components, and lightly tighten the bolts and the

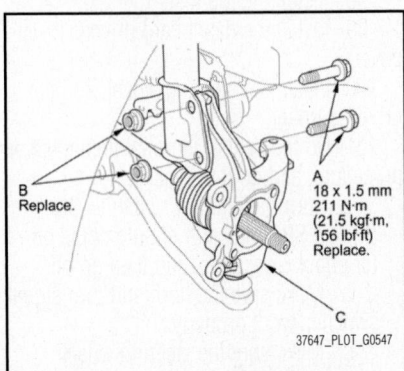

Fig. 216 Remove the damper pinch bolts (A) and the flange nuts (B) from the damper, then remove the knuckle (C)

nuts, then raise the suspension to load it with the vehicle's weight before fully tightening to the specified torque. Do not place the jack against the ball joint of the lower arm.
- Be careful not to damage the ball joint boot when connecting the knuckle.
- Before connecting the ball joint to the knuckle, degrease the threaded section and the tapered portion of the ball joint pin, the ball joint connecting hole, and the threaded section and the mating surfaces of the castle nut.
- Torque the castle nut to the lower torque specification, then tighten it only far enough to align the slot with the ball joint pin hole. Do not align the castle nut by loosening it.
- Before installing the wheel, clean the mating surfaces on the brake disc and the inside of the wheel.

9. Check the wheel alignment, and adjust it if necessary.

LOWER BALL JOINTS

REMOVAL & INSTALLATION

See Figures 217 through 219.

Special Tools Required:
- Ball Joint Thread Protector, 12 mm 07AAF-SDAA100
- Ball Joint Remover, 32 mm 07MAC-SL0A102
- Ball Joint Remover, 28 mm 07MAC-SL0A202
- Ball Joint Thread Protector, 14 mm 071AF-S3VA000

➡**Always use a ball joint remover to disconnect a ball joint. Do not strike the housing or any other part of the ball joint connection to disconnect it.**

1. Install a hex nut or the ball joint thread protector onto the threads of the ball joint.

➡**Using a hex nut, make sure the nut is flush with the ball joint pin end to prevent damage to the threaded end of the ball joint pin.**

2. Apply grease to the ball joint remover on the areas shown. This will ease the installation of the tool and prevent damage to the pressure bolt threads.

3. Loosen the pressure bolt, and install the ball joint remover as shown. Insert the jaws carefully, making sure not to damage

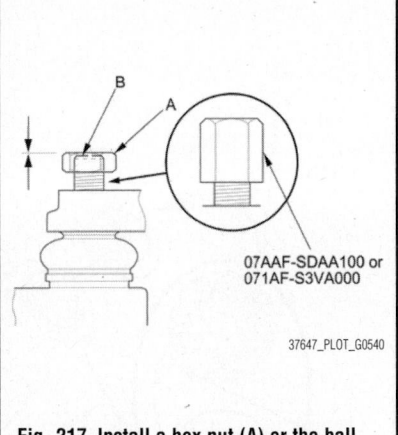

Fig. 217 Install a hex nut (A) or the ball joint thread protector onto the threads of the ball joint (B)

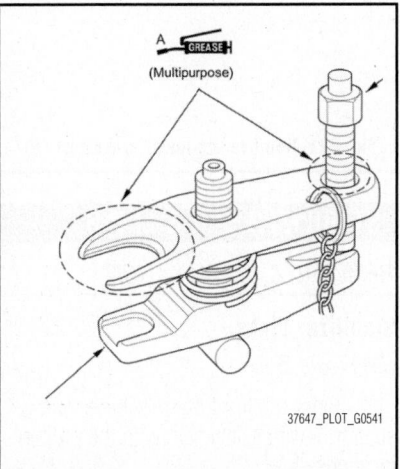

Fig. 218 Apply grease to the ball joint remover on the areas shown (A)

the ball joint boot. Adjust the jaw spacing by turning the adjusting bolt.

➡**Fasten the safety chain securely to a suspension arm or the subframe. Do not fasten it to a brake line or wire harness.**

4. After adjusting the adjusting bolt, make sure the head of the adjusting bolt is in the position shown to allow the jaw to pivot.

5. With a wrench, tighten the pressure bolt until the ball joint pin pops loose from the ball joint connecting hole. If necessary, apply penetrating type lubricant to loosen the ball joint pin.

➡**Do not use pneumatic or electric tools on the pressure bolt.**

6. Remove the ball joint remover, then remove the nut or the ball joint thread protector from the end of the ball joint pin, and pull the ball joint out of the ball joint con-

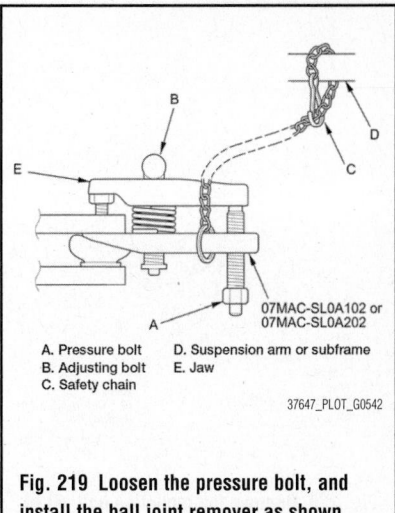

A. Pressure bolt
B. Adjusting bolt
C. Safety chain
D. Suspension arm or subframe
E. Jaw

Fig. 219 Loosen the pressure bolt, and install the ball joint remover as shown

necting hole. Inspect the ball joint boot, and replace it if damaged.

LOWER CONTROL ARMS

REMOVAL & INSTALLATION

See Figures 220 and 221.

Special Tools Required:
- Ball Joint Remover, 32 mm 07MAC-SL0A102
- Ball Joint Thread Protector, 14 mm 071AF-S3VA000

1. Raise the front of the vehicle, and support it with safety stands in the proper locations.

2. Remove the front wheel.

3. Remove the lock pin from the lower arm ball joint, then remove the castle nut.

➡**During installation, install the lock pin as shown after tightening the new castle nut.**

4. Disconnect the lower arm ball joint from the knuckle using the ball joint thread protector and the ball joint remover.

➡**Be careful not to damage the ball joint boot when installing the remover. Do not force or hammer on the lower arm, or pry between the lower arm and the knuckle. You could damage the ball joint.**

5. Remove the mounting bolt of the rear side on the stabilizer bar bushing holder.

➡**During installation, install the mounting bolt after tightening the lower arm mounting 14 mm bolts to the specified torque value.**

6. Remove the lower arm mounting 14 mm and 16 mm bolts, then remove the lower arm.

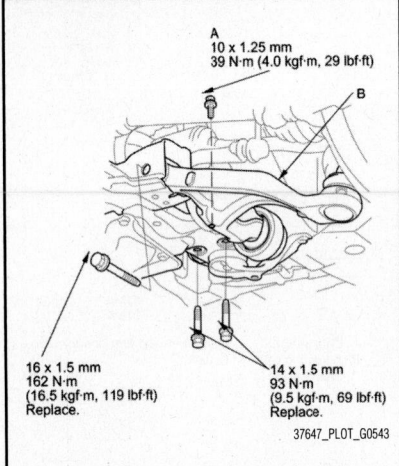

Fig. 220 Remove the mounting bolt (A) of the rear side on the stabilizer bar bushing holder; remove the lower arm mounting 14 mm and 16 mm bolts, then remove the lower arm (B)

➡ **Use new mounting bolts during reassembly.**

7. Remove the lower arm stops.

➡ **During installation, align the slot on the lower arm stop with the lug portion on the front side lower arm bushing.**

To install:

8. Install the lower arm in the reverse order of removal, and note these items:

- First install all of the components, and lightly tighten the bolts and the nuts, then raise the suspension to load it with the vehicle's weight before fully tightening to the specified torque values. Do not place the jack against the ball joint of the lower arm.
- Be careful not to damage the ball joint boot when connecting the knuckle.
- Before connecting the ball joint, degrease the threaded section and the tapered portion of the ball joint pin, the ball joint connecting hole, and the threaded section and the mating surfaces of the castle nut.
- Torque the castle nut to the lower torque specification, then tighten it only far enough to align the slot with the ball joint pin hole. Do not align the castle nut by loosening it.
- Before installing the wheel, clean the mating surfaces of the brake disc and the inside of the wheel.

9. Check the wheel alignment, and adjust it if necessary.

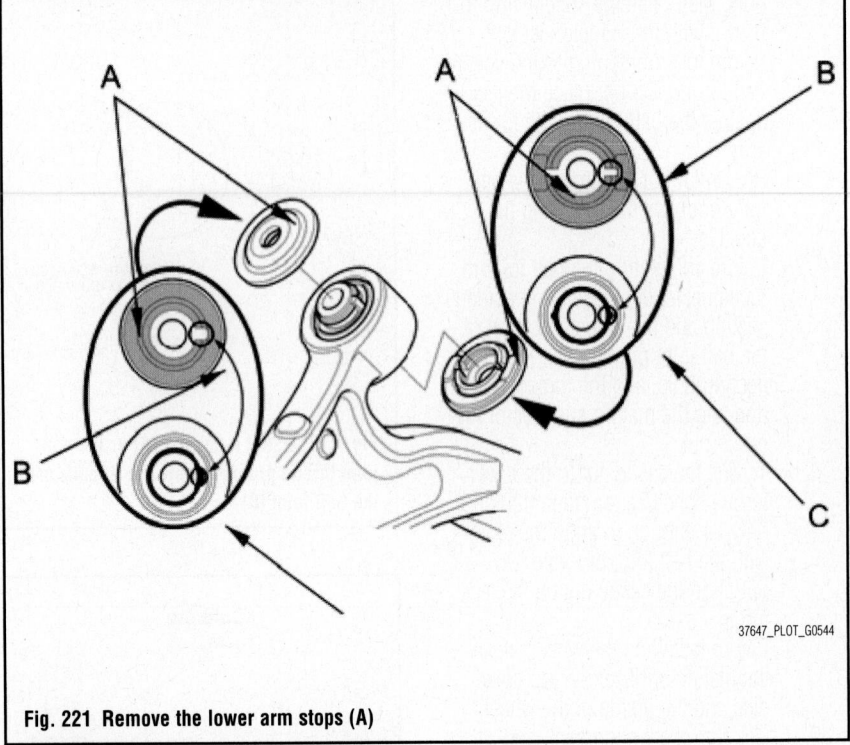

Fig. 221 Remove the lower arm stops (A)

STABILIZER BAR & LINKS

REMOVAL & INSTALLATION

Stabilizer Links

See Figure 222.

1. Raise the front of the vehicle, and support it with safety stands in the proper locations.

2. Remove the front wheel.

3. Remove the flange nuts while holding the respective joint pin with a hex wrench, then remove the stabilizer link.

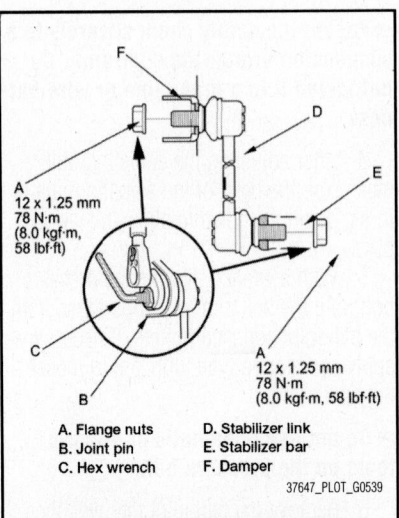

A. Flange nuts   D. Stabilizer link
B. Joint pin   E. Stabilizer bar
C. Hex wrench   F. Damper

Fig. 222 Remove the flange nuts while holding the respective joint pin with a hex wrench, then remove the stabilizer link

To install:

4. Install the stabilizer link on the stabilizer bar and the damper with the joint pins set at the center of their range of movement.

5. Install the flange nuts, and tighten them to the specified torque value while holding the respective joint pin with a hex wrench.

6. Clean the mating surfaces of the brake disc and the inside of the wheel, then install the front wheel.

Stabilizer Bar

See Figures 223 through 238.

Special Tools Required*:
- Ball Joint Thread Protector, 12 mm 07AAF-SDAA100
- Ball Joint Remover, 28 mm 07MAC-SL0A202
- Subframe Alignment Pin 070AG-SJAA10S
- 2008 V6 Attachment Arm SIL02C000033
- Engine Support Hanger, A and Reds AAR-T1256
- Engine Hanger Balance Bar VSB02C000019
- Subframe Adapter VSB02C000016
- *Available through the Honda Tool and Equipment Program, 888-424-6857

❋❋ WARNING

SRS components are located in this area. Review the SRS component

locations and the precautions and procedures before doing repairs or service.

1. Note these items during replacement:
   - Be sure to remove the steering wheel before disconnecting the steering joint. Damage to the cable reel can occur.
   - Lower the front subframe from the body, and replace the front stabilizer bar through the gap created by lowering the front subframe.

2. Remove the support struts from the engine hood. Move the engine hood to a vertical position, then reinstall the support strut.

3. Remove the engine cover.

4. Do the battery terminal disconnection procedure.

5. Raise and support the vehicle.

6. Remove the front wheels.

7. Remove the driver's airbag and the steering wheel.

8. Remove the footrest.
   a. Detach the hooks, then remove the footrest plate.
   b. Remove the nut with a 6 mm hexagonal wrench, release the clip, then remove the footrest.

9. Remove the center console trim.

10. Pull back the carpet, then remove the steering joint cover.

11. Release the lock lever, and adjust the steering column to the full tilt up position, and to the full telescope in position.

12. Tighten the lock lever.

13. Hold the lower slide shaft on the column with a piece of wire between the joint yoke of the lower slide shaft and joint yoke of the upper shaft to prevent the slider shaft from pulling out.

14. Release the lock lever, and adjust the steering column to the full telescopic out position, then tighten the lock lever.

15. Remove the steering joint bolt, disconnect the steering joint by moving the steering joint toward the column.

➡ If the center guide is in place and has not moved, leave it in place. If the center guide has moved or been removed, discard it.

16. Disconnect the Power Steering Pressure (PSP) switch connector).

17. Remove the cotter pin from the tie-rod end ball joint, then remove the nut.

➡ During installation, install the new cotter pin after tightening the new nut, and bend its end.

18. Disconnect the tie-rod end ball joint from the knuckle using the ball joint thread protector and the ball joint remover.

19. Disconnect both sides of the stabilizer link from the stabilizer bar.

20. Remove the power steering pump outlet hose mounting bolt.

21. Remove the air cleaner assembly.

22. Remove the connector bracket from the front cylinder head; use the bracket bolt hole to attach the engine hanger balance bar front arm.

23. Remove the harness clamp bracket from the engine rear cylinder head; use the

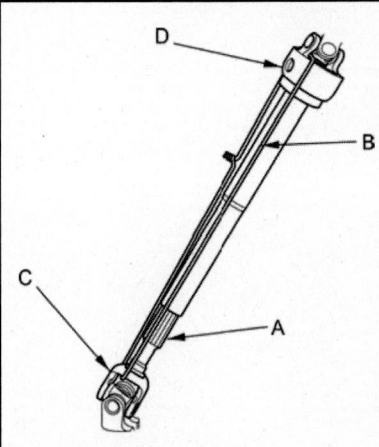

A. Lower slide shaft
B. Wire
C. Joint yoke of the lower slide shaft
D. Joint yoke of the upper shaft

37647_PLOT_G0514

Fig. 225 Hold the lower slide shaft on the column with a piece of wire between the joint yoke of the lower slide shaft and joint yoke of the upper shaft to prevent the lower slide shaft from pulling out

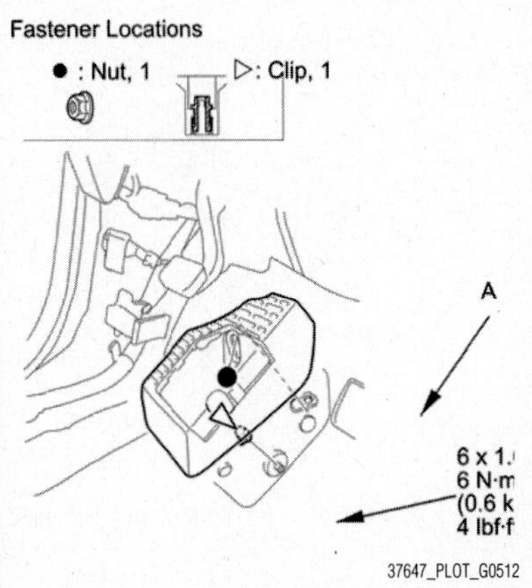

bracket bolt hole to attach the attachment arm.

24. Remove the cowl top side cover from both sides, then install the engine support hanger (AAR-T1256).With the engine support hanger beam on its side, insert the hanger beam through the opening, then rotate it over the damper.

25. Install the engine hanger balance bar (VSB02C000019). Attach the front arm to the front cylinder head with several spacers and a 10 x 1.25 mm bolt. Remove

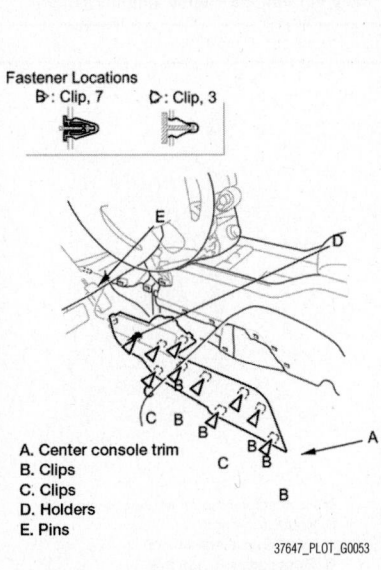

Fastener Locations
▷: Clip, 7   ▷: Clip, 3

A. Center console trim
B. Clips
C. Clips
D. Holders
E. Pins

37647_PLOT_G0053

Fig. 224 Gently pull out the center console trim to detach the clips and release the holders from the pins, then remove the trim

Fastener Locations
● : Nut, 1   ▷: Clip, 1

A

6 x 1.
6 N·m
(0.6 k
4 lbf·f

37647_PLOT_G0512

Fig. 223 Remove the nut with a 6 mm hexagonal wrench, release the clip, then remove the footrest (A)

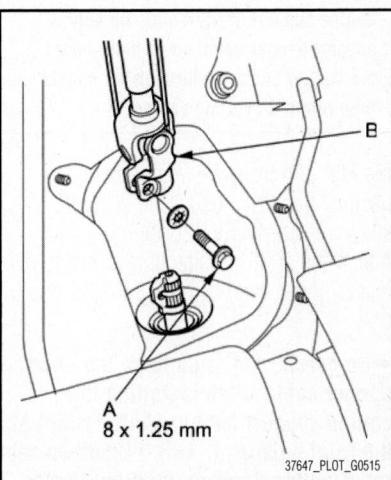

A
8 x 1.25 mm
B

37647_PLOT_G0515

Fig. 226 Remove the steering joint bolt (A), and disconnect the steering joint (B) by sliding the steering joint into the column

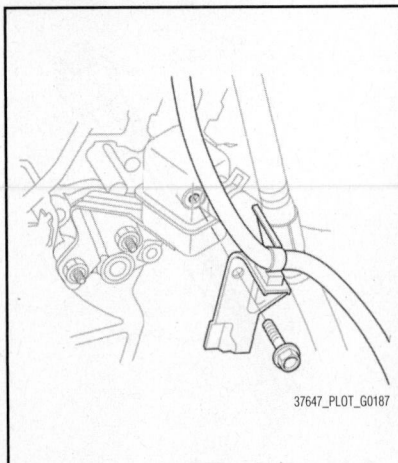

Fig. 227 Remove the harness clamp bracket from the engine rear cylinder head

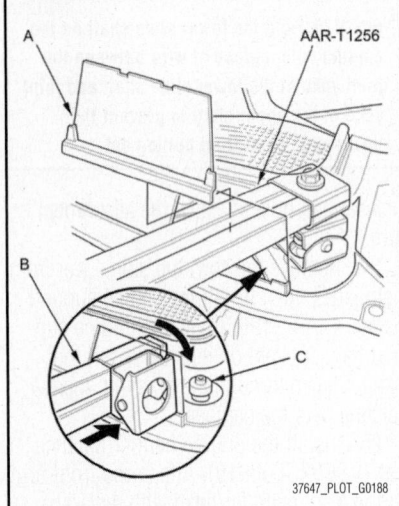

Fig. 228 Remove the cowl top side cover (A) from both sides, then install the engine support hanger; with the engine support hanger beam on its side, insert the hanger beam (B) through the opening, then rotate it over the damper (C)

the rear arm from the engine hanger balance bar, then install the 2008 V6 attachment arm (SIL02C000033). Attach the 2008 V6 attachment arm to the rear cylinder head with an 8 x 1.25 mm bolt.

➡**Be careful not to damage the hood opener cable when installing the engine support hanger (AAR-T1256) at the front bulkhead. AAR-T1256 two sets required for stacking additional cross section bar.**

26. Install the engine support hanger onto the vehicle, and attach the hook to the engine balancer bar slot. Tighten the wing

nut by hand, and lift and support the engine.

27. Raise the vehicle.

28. Remove the splash shield.

29. Remove the front subframe stiffener.

30. 4WD model: Remove the propeller shaft and propeller shaft protectors.

31. Disconnect the front engine mount actuator connector and the harness clamps. Remove the front engine mount stop, then remove the front engine mount bolt.

32. Remove the rear engine mount bolts and disconnect the rear engine mount actuator connector.

33. Remove the rear engine mount from the rear engine mount base bracket, and move it aside.

34. Remove the ground cable bolt from the transmission.

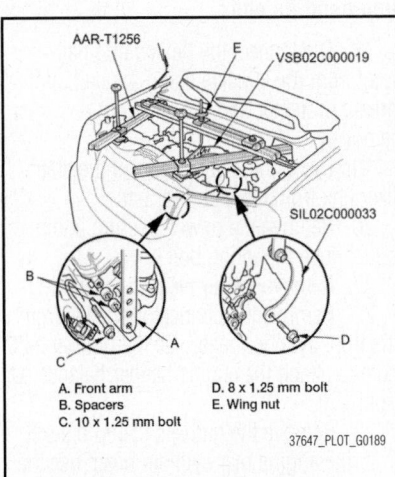

A. Front arm
B. Spacers
C. 10 x 1.25 mm bolt
D. 8 x 1.25 mm bolt
E. Wing nut

Fig. 229 Install the engine hanger balance bar and the engine support hanger

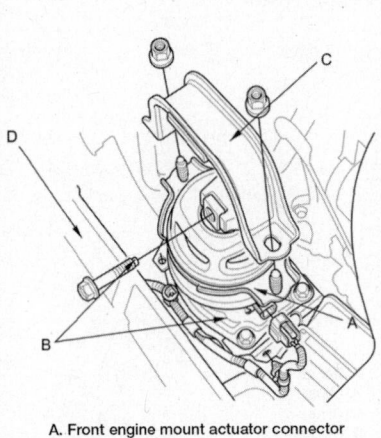

A. Front engine mount actuator connector
B. Harness clamps
C. Front engine mount stop
D. Front engine mount bolt

Fig. 230 Disconnect the front engine mount actuator connector and the harness clamps

35. Remove the lower transmission mount bolts.

36. Attach the front subframe adapter (VSB02C000016) to the subframe by looping the belt over the front of the subframe, then secure the belt with its stop.

37. Raise the jack and line up the slots in the front subframe adapter arms with the bolt holes on the jack base, then securely attach then with four bolts.

38. Remove the stiffener mounting bolts on both sides of the front subframe front stiffeners.

39. Loosen the front side subframe mounting bolts until there is 1.2 inches (30 mm) distance between the bolt seat and the mounting surface. Do not loosen the mounting bolts more than necessary.

40. Remove the stiffener mounting bolts

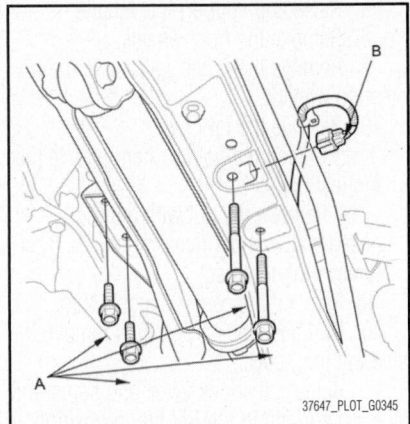

Fig. 231 Remove the rear engine mount bolts (A) and disconnect the rear engine mount actuator connector (B)

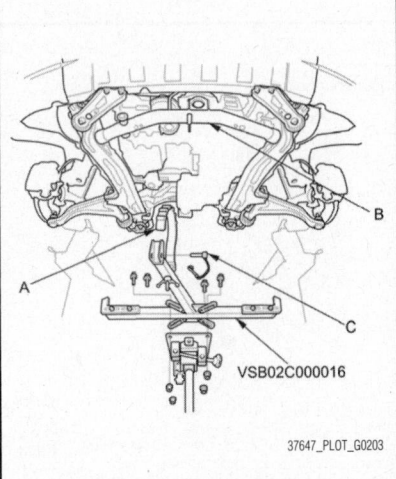

Fig. 232 Attach the front subframe adapter to the subframe by looping the belt (A) over the front of the subframe (B), then secure the belt with its stop (C)

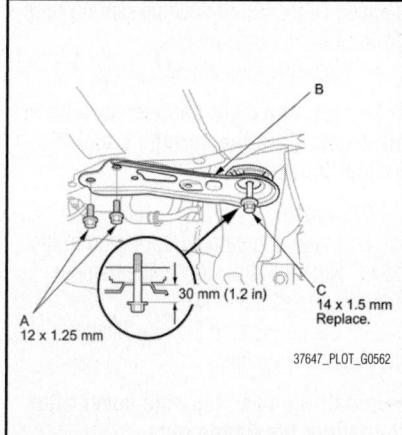

Fig. 233 Remove the stiffener mounting bolts (A) from the subframe front stiffeners (B)

on both sides of the front subframe rear bracket.

41. Loosen the rear side subframe mounting bolts until there is 1.2 inches (30 mm) distance between the bolt seat and the mounting surface. Do not loosen the mounting bolts more than necessary.

42. Lower the transmission jack slowly with the front subframe adapter until the front subframe has dropped about 1.2 inches (30 mm).

➡**Do not try to lower the front subframe beyond the loosened subframe mounting bolts clearance.**

43. Remove the flange bolts and the bushing holders, then remove the bushings.

➡**During installation, align the paint marks on the stabilizer bar with the side of the bushings.**

44. Move the stabilizer bar toward the

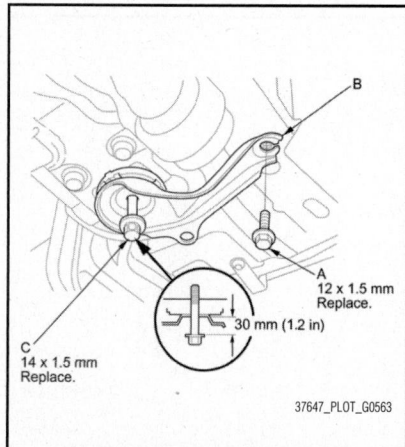

Fig. 234 Remove the stiffener mounting bolts (A) on both sides of the front subframe rear bracket (B)

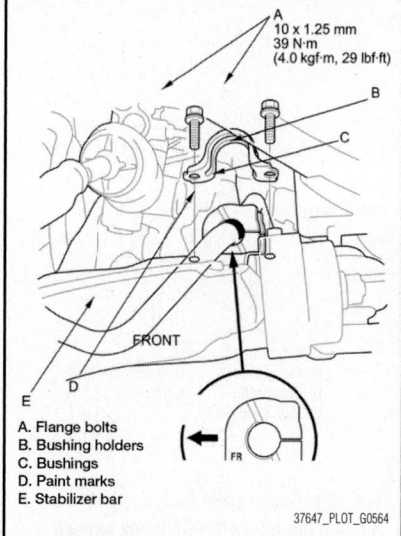

A. Flange bolts
B. Bushing holders
C. Bushings
D. Paint marks
E. Stabilizer bar

Fig. 235 Remove the flange bolts and the bushing holders, then remove the bushings

passenger's side, and remove the stabilizer bar.

To install:

45. Install the stabilizer bar.

➡**Note the right and left direction of the stabilizer bar. Note the direction of installation for the bushings.**

46. Loosely tighten the new left-rear subframe mounting bolt and the new stiffener mounting bolt; insert the subframe alignment pin (070AG-SJAA10S) through the positioning hole on the front subframe

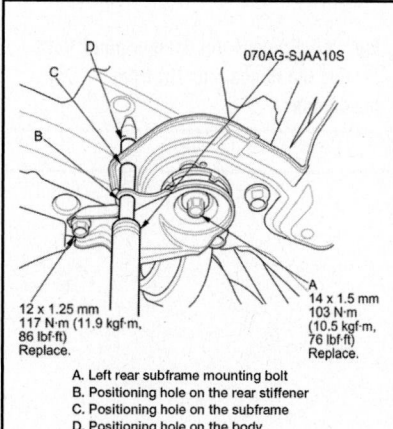

A. Left rear subframe mounting bolt
B. Positioning hole on the rear stiffener
C. Positioning hole on the subframe
D. Positioning hole on the body

Fig. 236 Loosely tighten the left rear subframe mounting bolt; insert the 15.7 mm side of the subframe alignment pin through the positioning hole on the rear stiffener, through the positioning hole on the subframe, and into the positioning hole on the body

rear stiffener, through the positioning hole on the subframe, and into the positioning hole on the body, then tighten the subframe mounting bolt.

47. Loosely tighten the new right-rear subframe mounting bolt using the same procedure as the left-rear with the subframe alignment pin.

48. Loosely tighten the new subframe mounting bolt and the stiffener mounting bolts on both sides to the front subframe front stiffener.

49. Using the subframe alignment pin, tighten the rear-driver's side subframe mounting bolt to the specified torque.

50. Using the subframe alignment pin, tighten the rear-passenger's side subframe mounting bolt to the specified torque.

51. Tighten the front subframe mounting bolts to the specified torque.

➡**Check all of the subframe mounting bolts, and retighten if necessary.**

52. Tighten the front and rear stiffener mounting bolts to the specified torque.

➡**Before tightening the stiffener mounting bolts, check that the positioning holes and slot are aligned using the subframe alignment pin.**

53. Install all of the removed parts in the reverse order of removal, and note these items:

- Refer to stabilizer link removal/installation to connect the stabilizer bar to the links.
- If the center guide is in place, use it to determine the steering joint installation angle.
- If the center guide is gone, check the steering joint installation angle.

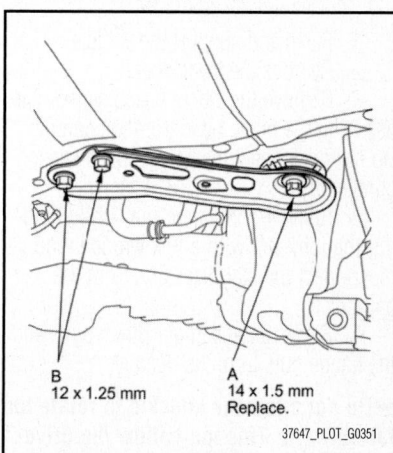

Fig. 237 Loosely install the right and left front subframe mounting bolts (A) and stiffener mounting bolts (B)

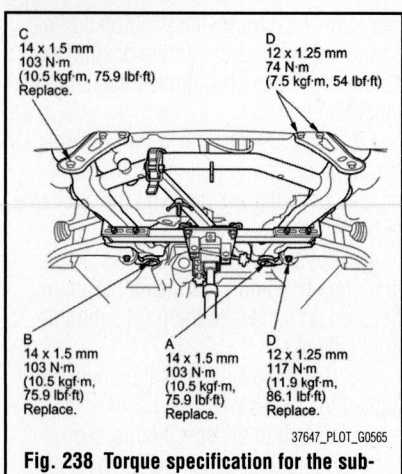

C
14 x 1.5 mm
103 N·m
(10.5 kgf·m, 75.9 lbf·ft)
Replace.

D
12 x 1.25 mm
74 N·m
(7.5 kgf·m, 54 lbf·ft)

B
14 x 1.5 mm
103 N·m
(10.5 kgf·m,
75.9 lbf·ft)
Replace.

A
14 x 1.5 mm
103 N·m
(10.5 kgf·m,
75.9 lbf·ft)
Replace.

D
12 x 1.25 mm
117 N·m
(11.9 kgf·m,
86.1 lbf·ft)
Replace.

37647_PLOT_G0565

Fig. 238 Torque specification for the sub-frame mounting bolts and the stiffener mounting bolts

- Check the steering wheel installation.
- When connecting the front and rear engine mount, first lightly tighten the mounting bolts, supporting the engine/transmission with a floor jack, then remove the jack, and tighten the bolts to the specified torque.
- Before installing the wheel, clean the mating surfaces on the brake disc and inside of the wheel.

54. Do the battery terminal reconnection procedure, then turn the ignition switch to ON (II) and check that the SRS indicator should come on for about 6 seconds and then go off.

55. Check the wheel alignment, and adjust it if necessary.

STRUTS

REMOVAL & INSTALLATION

See Figures 239 through 242.

1. Raise and support the vehicle.
2. Remove the front wheel.
3. Remove the wheel speed sensor harness and the brake hose from the damper. Do not disconnect the wheel speed sensor connector.
4. Remove the flange nut, while holding the joint pin with a hex wrench, and disconnect the stabilizer link from the damper.
5. Remove the damper pinch bolts and the flange nuts from the damper.

➡**Do not allow the knuckle to rotate too far outward. This may allow the driveshaft inboard joint to come apart.**

6. Remove the cowl top side cover, then remove the flange nuts from the top of the

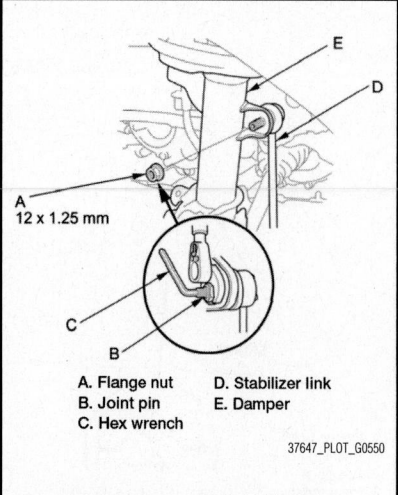

A
12 x 1.25 mm

A. Flange nut
B. Joint pin
C. Hex wrench
D. Stabilizer link
E. Damper

37647_PLOT_G0550

Fig. 239 Remove the flange nut, while holding the joint pin with a hex wrench, and disconnect the stabilizer link from the damper

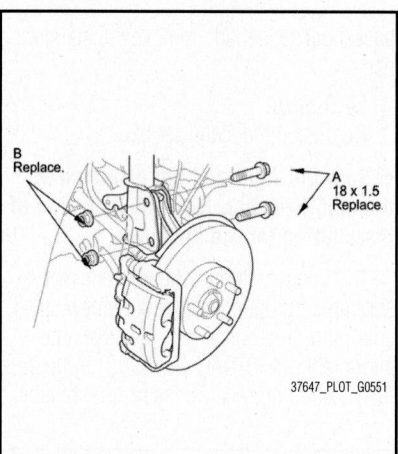

B
Replace.

A
18 x 1.5
Replace.

37647_PLOT_G0551

Fig. 240 Remove the damper pinch bolts (A) and the flange nuts (B) from the damper

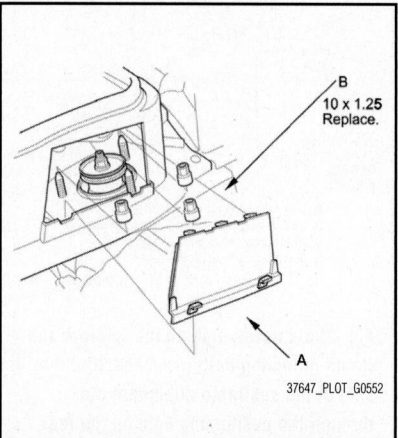

B
10 x 1.25
Replace.

A

37647_PLOT_G0552

Fig. 241 Remove the cowl top side cover (A), then remove the flange nuts (B) from the top of the damper

damper. Do not let the damper/spring drop down under its own weight.

7. Remove the damper/spring.

➡**The left and right damper springs are different. Mark the springs L and R before you continue.**

To install:

8. Install the damper/spring on to the frame. Note the direction of the damper mounting base as shown.

9. Loosely install the new flange nuts to the top of the damper.

➡**Install the cowl top side cover after tightening the flange nuts.**

10. Loosely install new damper pinch bolts and new flange nuts to the damper.

11. Connect the stabilizer link to the damper, and loosely install the flange nut.

12. Tighten the flange nut to 58 ft. lbs. (78 Nm), while holding the joint pin with the hex wrench.

13. Place a floor jack under the lower arm, and raise the suspension to load it with the vehicle's weight.

➡**Do not place the jack against the ball joint of the lower arm.**

14. Tighten the flange nuts on top of the damper to 43 ft. lbs. (59 Nm).

15. Tighten the damper pinch bolts and the flange nuts to 156 ft. lbs. (211 Nm).

16. Install the wheel speed sensor harness and the brake hose to the damper.

17. Clean the mating surfaces of the brake disc and the inside of the wheel, then install the front wheel.

18. Check the wheel alignment, and adjust it if necessary.

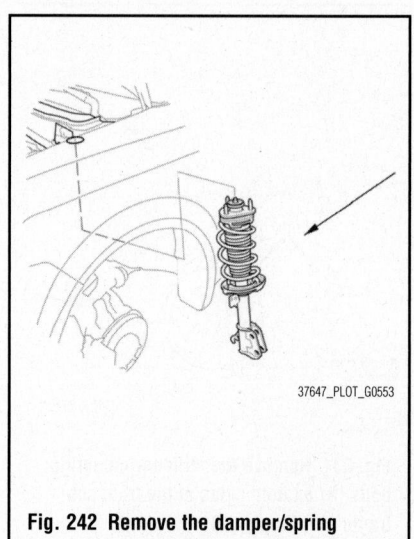

37647_PLOT_G0553

Fig. 242 Remove the damper/spring

WHEEL HUBS & BEARINGS

ADJUSTMENT

The hub bearing cannot be adjusted. If there is need for adjustment, replace the hub bearing.

REMOVAL & INSTALLATION

See Figure 243.

Special Tools Required:
- Ball Joint Thread Protector, 12 mm 07AAF-SDAA100
- Ball Joint Remover, 32 mm 07MAC-SL0A102
- Ball Joint Remover, 28 mm 07MAC-SL0A202
- Ball Joint Thread Protector, 14 mm 071AF-S3VA000

1. Raise and support the vehicle.
2. Remove the front wheel.
3. Remove the brake hose mounting bolt from the damper.
4. Remove the brake caliper bracket mounting bolts, then remove the caliper assembly from the knuckle. To prevent damage to the caliper assembly or the brake hose, use a short piece of wire to hang the caliper assembly from the undercarriage. Do not twist the brake hose excessively.

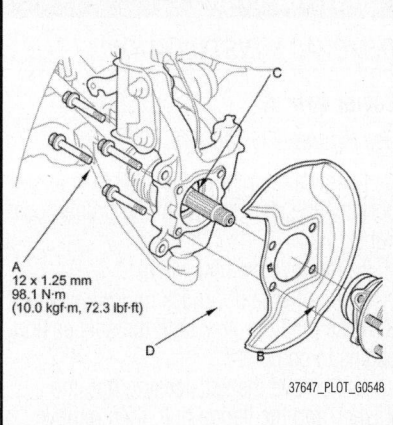

Fig. 243 Remove the flange bolts, and remove the hub bearing unit by tapping the driveshaft end with a soft face hammer while drawing the hub bearing unit outward, then remove the splash guard

5. Pry up the stake on the spindle nut, then remove the nut.
6. Remove the front brake disc.
7. Remove the flange bolts, and remove the hub bearing unit by tapping the driveshaft end with a soft face hammer while drawing the hub bearing unit outward, then remove the splash guard.

➡Do not pull the driveshaft end outward. The driveshaft inboard joint may come apart. During installation, apply grease to the mating surfaces of the hub bearing unit and the driveshaft outboard joint.

8. Check the hub bearing unit for damage and cracks.

To install:

9. Install the hub bearing unit in the reverse order of removal, and note these items:
- Use a new spindle nut on reassembly.
- Before installing the spindle nut, apply a small amount of engine oil to the seating surface of the nut. After tightening, use a drift to stake the spindle nut shoulder against the driveshaft.
- Before installing the brake disc, clean the mating surfaces on the hub bearing unit and the inside of the brake disc.
- Before installing the wheel, clean the mating surfaces on the brake disc and the inside of the wheel.

10. Check the wheel alignment, and adjust it if necessary.

SUSPENSION

COIL SPRINGS

REMOVAL & INSTALLATION

See Figures 244 through 247.

1. Raise the rear of the vehicle, and support it with safety stands in the proper locations.
2. Remove the rear wheel.
3. Remove the muffler from the muffler hangers.
4. Remove the flange nut while holding the joint pin with a hex wrench, and disconnect the stabilizer link from lower arm B.
5. Position a floor jack under lower arm B. Raise the floor jack until the suspension begins to compress.
6. Remove the flange bolt from the bottom of the damper.
7. Remove the flange bolt from the knuckle.
8. Lower the floor jack gradually.
9. Remove the spring and the lower spring seat.
10. Remove the flange bolt that connects the body, and remove the bump stop, the spring guide, and the spring mounting cushion if necessary.

To install:

11. Install the bump stop, the spring guide, and the spring mounting cushion, then tighten the flange bolt to 43 ft. lbs. (59 Nm) if removed.
12. Install the spring and the lower spring seat. Align the bottom of the spring

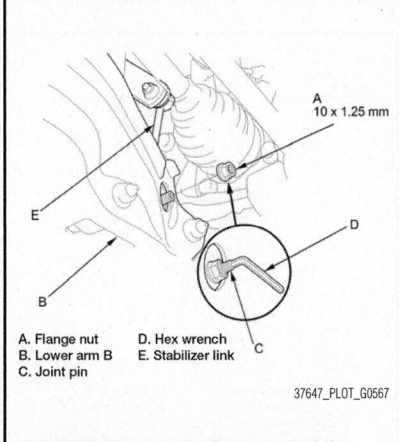

Fig. 244 Remove the flange nut while holding the joint pin with a hex wrench, and disconnect the stabilizer link from lower arm B

REAR SUSPENSION

with the stepped part of the lower spring seat and lower arm B.

13. Position the floor jack under lower arm B. Raise the floor jack until the hole in lower arm B aligns with the hole in the damper, then loosely install the new flange bolt to the bottom of the damper.

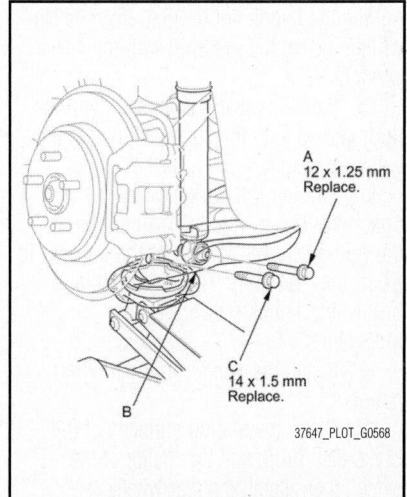

Fig. 245 Position a floor jack under lower arm B, remove the flange bolts (A, C)

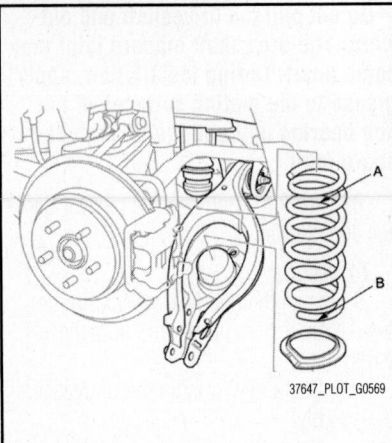

Fig. 246 Remove the spring (A) and the lower spring seat (B)

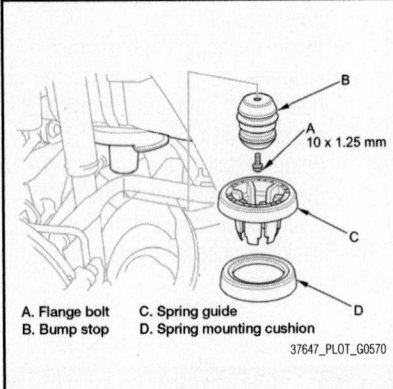

A. Flange bolt   C. Spring guide
B. Bump stop   D. Spring mounting cushion

37647_PLOT_G0570

Fig. 247 Remove the flange bolt that connects the body, and remove the bump stop, the spring guide, and the spring mounting cushion if necessary

14. Loosely install the new flange bolt to the knuckle.

15. Connect the stabilizer link to lower arm B, and install the new flange nut, and tighten the flange nut to 36 ft. lbs. (49 Nm), while holding the joint pin with the hex wrench.

16. Raise the rear suspension with the floor jack to load the suspension with the vehicle's weight.

17. Tighten the flange bolts to the specified torque values. Tighten the flange bolt in the bottom of the damper to 47 ft. lbs. (64 Nm). Tighten the flange bolt to the knuckle to 83 ft. lbs. (113 Nm).

18. Install the muffler to the muffler hangers.

19. Clean the mating surfaces of the brake disc/drum and the inside of the wheel, then install the rear wheel.

20. Check the wheel alignment, and adjust it if necessary.

CONTROL LINKS

REMOVAL & INSTALLATION

Lower Arm A

See Figures 248 and 249.

1. Raise the rear of the vehicle, and support it with safety stands in the proper locations.

2. Remove the rear wheel.

3. Position a floor jack under lower arm B. Raise the floor jack until the suspension begins to compress.

4. Remove the self-locking nut, the washer, and the flange bolt, then remove lower arm A.

To install:

➡ **During installation, position the paint mark on lower arm A toward outside of the vehicle. Use a new self-locking nut and the new flange bolt, during assembly.**

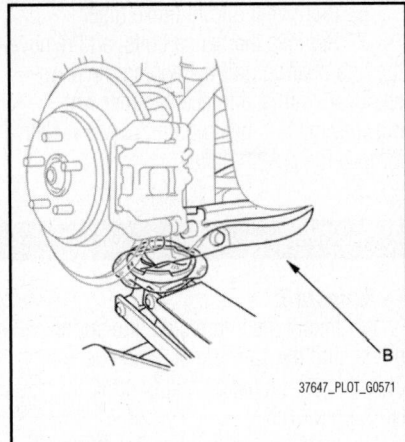

Fig. 248 Position a floor jack under lower arm B

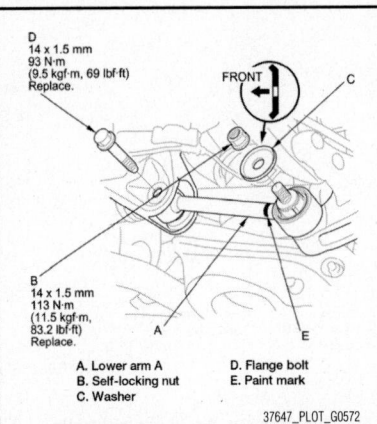

A. Lower arm A   D. Flange bolt
B. Self-locking nut   E. Paint mark
C. Washer

37647_PLOT_G0572

Fig. 249 Remove the self-locking nut, the washer, and the flange bolt, then remove lower arm A

5. Install lower arm A in the reverse order of removal, and note these items:
- First install all of the components, and lightly tighten the bolts and the nuts, then raise the suspension to load it with the vehicle's weight before fully tightening to the specified torque values.
- Before installing the wheel, clean the mating surfaces of the brake disc/drum and the inside of the wheel.

6. Check the wheel alignment, and adjust it if necessary.

Lower Arm B

See Figures 250 through 253.

1. Raise the rear of the vehicle, and support it with safety stands in the proper locations.

2. Remove the rear wheel.

3. Remove the muffler from the muffler hangers.

4. Remove the flange nut while holding the joint pin with a hex wrench, and disconnect the stabilizer link from lower arm B.

5. Position a floor jack under lower arm B. Raise the floor jack until the suspension begins to compress.

6. Remove the flange bolt from the bottom of the damper.

7. Remove the flange bolt from the knuckle.

8. Lower the floor jack gradually.

9. Remove the spring and the lower spring seat.

10. Mark the cam positions of the adjusting bolt and the adjusting cam plate with the frame.

11. Remove the self-locking nut while holding the adjusting bolt, then remove the adjusting cam plate, the adjusting bolt, and lower arm B.

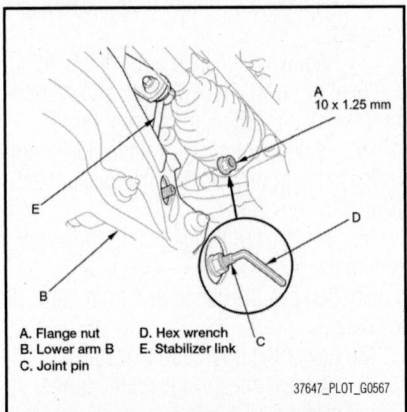

A. Flange nut   D. Hex wrench
B. Lower arm B   E. Stabilizer link
C. Joint pin

37647_PLOT_G0567

Fig. 250 Remove the flange nut while holding the joint pin with a hex wrench, and disconnect the stabilizer link from lower arm B

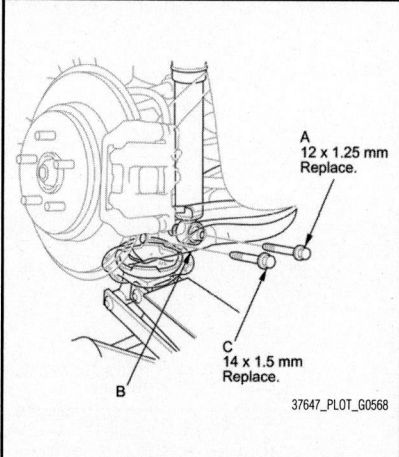

Fig. 251 Position a floor jack under lower arm B, remove the flange bolts (A, C)

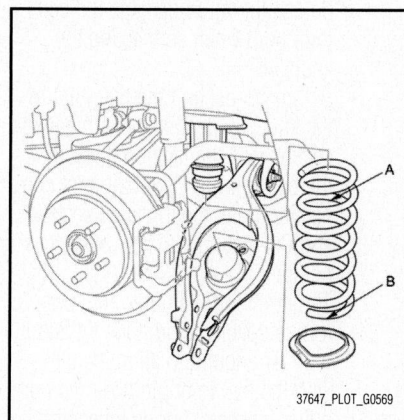

Fig. 252 Remove the spring (A) and the lower spring seat (B)

➡**Use a new adjusting bolt and a new self- locking nut during reassembly.**

To install:

12. Install lower arm B in the reverse order of removal, and note these items:
- First install all of the components, and lightly tighten the bolts and nuts, then raise the suspension to load it with the vehicle's weight before fully tightening to the specified torque.
- Align the cam positions of the adjusting bolt and the adjusting cam plate with the marked positions on the frame when tightening the self-locking nut.
- Before installing the wheel, clean the mating surfaces on the brake disc/drum and the inside of the wheel.

13. Check the wheel alignment, and adjust it if necessary.

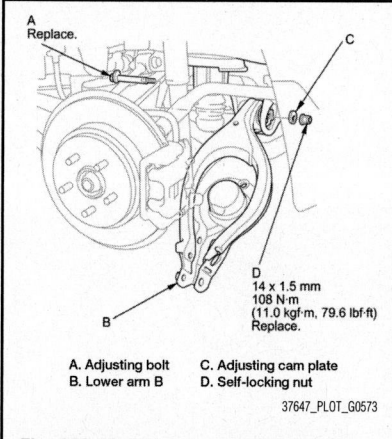

A. Adjusting bolt
B. Lower arm B
C. Adjusting cam plate
D. Self-locking nut

Fig. 253 Mark the cam positions of the adjusting bolt and the adjusting cam plate with the frame

Stabilizer Link

See Figure 254.

1. Raise the rear of the vehicle, and support it with safety stands in the proper locations.
2. Remove the rear wheel.
3. Remove the self-locking nut and the flange nut while holding the respective joint pin with a hex wrench, then remove the stabilizer link.

To install:

4. Install the stabilizer link on the stabilizer bar and lower arm B with the joint pins set at the center of their range of movement.

➡**The stabilizer link has a paint mark. The left stabilizer link is marked with**

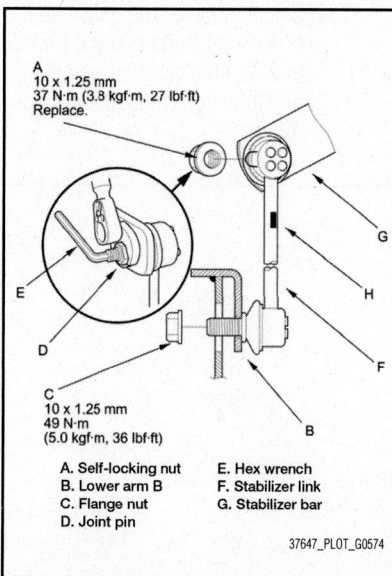

A. Self-locking nut
B. Lower arm B
C. Flange nut
D. Joint pin
E. Hex wrench
F. Stabilizer link
G. Stabilizer bar

Fig. 254 Remove the self-locking nut and the flange nut while holding the respective joint pin with a hex wrench, then remove the stabilizer link

yellow paint, and the right stabilizer link is marked with white paint. Install the end of the stabilizer link with the paint mark in the upper position.

5. Install a new self-locking nut and the new flange nut, and tighten them to the specified torque values while holding the respective joint pin with a hex wrench.
6. Clean the mating surfaces of the brake disc/drum and the inside of the wheel, then install the rear wheel.
7. Test-drive the vehicle.
8. After 5 minutes of driving, tighten the self-locking nut again to the specified torque value.

Trailing Arm

See Figures 255 and 256.

1. Raise the rear of the vehicle, and support it with safety stands in the proper locations.
2. Remove the rear wheel.
3. Remove the parking brake cable from the backing plate.
4. Remove the parking brake cable nut from the trailing arm.
5. Disconnect the brake line from the brake hose, then remove the brake hose clip.

➡**Plug the end of a hose and joint to prevent spilling brake fluid. Use the new brake hose clip during reassembly.**

6. Remove the wheel sensor harness bolt.
7. Remove the flange nuts from the trailing arm.

➡**Use new flange nuts during reassembly.**

8. Remove the flange bolts from the trailing arm, then remove the trailing arm.

➡**Use new flange bolts during reassembly.**

To install:

9. Install the trailing arm in the reverse order of removal, and note these items:
- First install all of the components, and lightly tighten the bolts and the nuts, then raise the suspension to load it with the vehicle's weight before fully tightening to the specified torque values.
- Do the parking brake adjustment.
- Fill the master cylinder reservoir to the MAX (upper) level line, and bleed the brake system. Check for a leak at the brake hose/line joint, and retighten it if necessary.

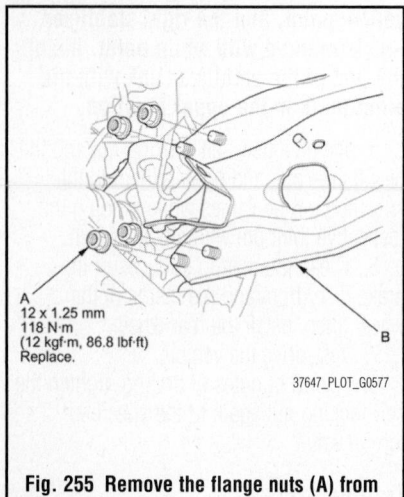

Fig. 255 Remove the flange nuts (A) from the trailing arm (B)

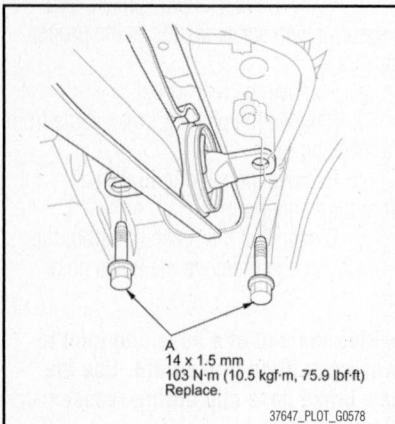

Fig. 256 Remove the flange bolts (A) from the trailing arm, then remove the trailing arm

• Before installing the wheel, clean the mating surfaces of the brake disc/drum and the inside of the wheel.

10. Check the wheel alignment, and adjust it if necessary.

Upper Arm

See Figures 257 through 259.

Special Tools Required:
• Ball Joint Thread Protector, 12 mm 07AAF-SDAA100
• Ball Joint Remover, 28 mm 07MAC-SL0A202

1. Raise and support the vehicle.
2. Remove the rear wheel.
3. Position a floor jack under lower arm B. Raise the floor jack until the suspension begins to compress.
4. Remove the lock pin from the upper arm ball joint, then remove the castle nut.

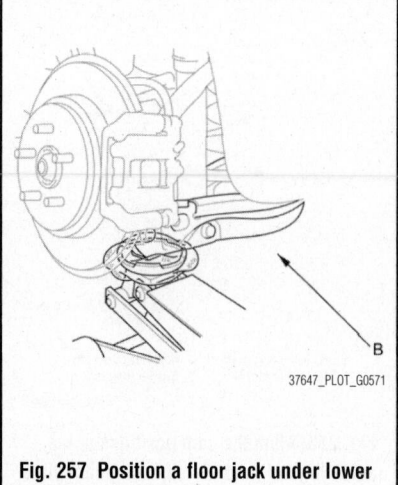

Fig. 257 Position a floor jack under lower arm B

➡**During installation, install the lock pin as shown after tightening the castle nut.**

5. Disconnect the upper arm ball joint from the knuckle using the ball joint thread protector and the ball joint remover.

➡**Be careful not to damage the ball joint boot when installing the remover.**

6. Remove the upper arm mounting bolt and remove the upper arm.

➡**Use the new mounting bolt during reassembly.**

To install:

7. Install the upper arm in the reverse order of removal, and note these items:
• First install all of the components, and lightly tighten the bolts and the nuts, then raise the suspension to load it with the vehicle's weight before fully tightening to the specified torque.

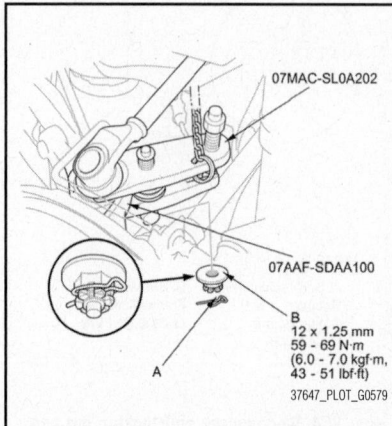

Fig. 258 Remove the lock pin (A) from the upper arm ball joint, then remove the castle nut (B)

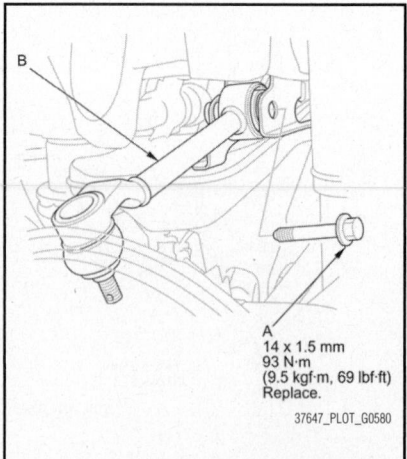

Fig. 259 Remove the upper arm mounting bolt (A), and remove the upper arm (B)

• Be careful not to damage the ball joint boot when connecting the knuckle.
• Before connecting the ball joint to the knuckle, degrease the threaded section and the tapered portion of the ball joint pin, the ball joint connecting hole, the threaded section and the mating surfaces of the castle nut.
• Torque the castle nut to the lower torque specification, then tighten it only far enough to align the slot with the ball joint pin hole. Do not align the castle nut by loosening it.
• Before installing the wheel, clean the mating surfaces on the brake disc/drum and the inside of the wheel.

8. Check the wheel alignment, and adjust it if necessary.

SHOCK ABSORBERS

REMOVAL & INSTALLATION

See Figures 260 through 262.

1. Raise the rear of the vehicle, and support it with safety stands in the proper locations.
2. Remove the rear wheel.
3. Remove the flange nut while holding the joint pin with a hex wrench, and disconnect the stabilizer link from lower arm B.
4. Position a floor jack under lower arm B. Raise the floor jack until the suspension begins to compress.
5. Remove the flange bolt from the bottom of the damper.
6. Remove the flange bolt and the nut from the top of the damper.
7. Compress the damper unit by hand, and remove it from the vehicle.

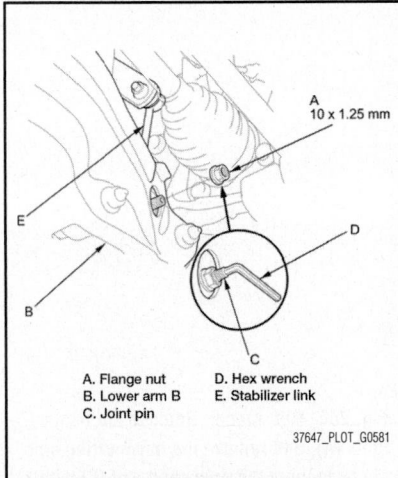

Fig. 260 Remove the flange nut while holding the joint pin with a hex wrench, and disconnect the stabilizer link from lower arm B

A. Flange nut   D. Hex wrench
B. Lower arm B   E. Stabilizer link
C. Joint pin

37647_PLOT_G0581

To install:

8. Position the damper unit between the body and lower arm B.

9. Loosely install the new flange bolt and the nut to the top of the damper.

10. Raise the floor jack until the hole in lower arm B aligns with the hole in the damper, then loosely install the new flange bolt to the bottom of the damper.

11. Connect the stabilizer link to lower arm B, and install the new flange nut, and tighten the flange nut to 36 ft. lbs. (49 Nm), while holding the joint pin with the hex wrench.

12. Raise the rear suspension with the floor jack to load the suspension with the vehicle's weight.

13. Tighten the flange bolt and the nut on top of the damper to 33 ft. lbs. (44 Nm).

14. Tighten the flange bolt on bottom of the damper to 47 ft. lbs. (64 Nm).

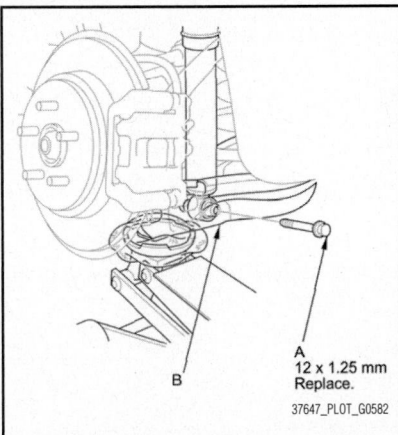

A. 12 x 1.25 mm Replace.
B.

37647_PLOT_G0582

Fig. 261 Position a floor jack under lower arm B; remove the flange bolt (A) from the bottom of the damper

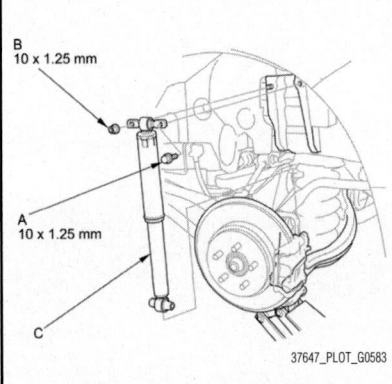

B. 10 x 1.25 mm
A. 10 x 1.25 mm
C.

37647_PLOT_G0583

Fig. 262 Remove the flange bolt (A) and the nut (B) from the top of the damper; compress the damper unit (C)

15. Clean the mating surfaces of the brake disc/drum and the inside of the wheel, then install the rear wheel.

16. Check the wheel alignment, and adjust it if necessary.

STABILIZER BAR & LINKS

REMOVAL & INSTALLATION

See Figure 265.

1. Raise and support the vehicle.
2. Remove the rear wheels.
3. Remove the spare tire from the vehicle.
4. Remove the spare tire hoist and the tire support bracket.
5. Remove the muffler from the muffler hangers.
6. Disconnect both sides of the stabilizer link from the stabilizer bar.
7. Remove the flange bolts and the bushing holders, then remove the bushings and the stabilizer bar.

➡**During installation, align the paint marks on the stabilizer bar with the side of the bushings.**

To install:

8. Install the stabilizer bar in the reverse order of removal, and note these items:
- Note the right and left direction of the stabilizer bar.
- Refer to the stabilizer link removal/installation to connect the stabilizer bar to the links.
- Before installing the wheel, clean the mating surfaces on the brake disc/drum and the inside of the wheel.
- Do not set the bushings on the bent or curved part of the stabilizer bar.

WHEEL HUBS & BEARING

REMOVAL & INSTALLATION

See Figures 264 through 266.

1. Raise and support the vehicle.
2. Remove the wheel nuts (A) and the rear wheel.
3. Release the parking brake lever fully.
4. Remove the brake caliper bracket mounting bolts (A), then remove the caliper assembly (B) from the knuckle. To prevent damage to the caliper assembly or the brake hose, use a short piece of wire to hang the caliper assembly from the undercarriage. Do not twist the brake hose excessively.
5. Remove the two washers.

➡**During installation, make sure the washers are installed between the brake caliper bracket and the knuckle.**

6. 4WD model: Pry up the stake on the spindle nut, then remove the nut.
7. Remove the rear brake disc/drum.
8. 2WD model: Remove the flange bolts and remove the hub bearing unit.
9. 4WD model: Remove the flange bolts, and remove the hub bearing unit by tapping the driveshaft end with a soft face hammer while drawing the hub bearing unit outward.

➡**Do not pull the driveshaft end outward. The driveshaft inboard joint may come apart.**

10. Check the hub bearing unit for damage and cracks.

To install:

11. Install the hub bearing unit in the reverse order of removal, and note these items:

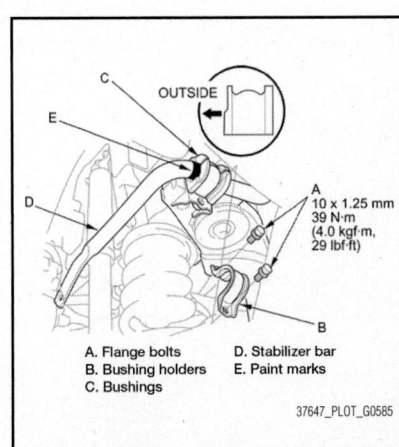

C. OUTSIDE
E.
D.
A. 10 x 1.25 mm
39 N·m
(4.0 kgf·m,
29 lbf·ft)
B.

A. Flange bolts   D. Stabilizer bar
B. Bushing holders   E. Paint marks
C. Bushings

37647_PLOT_G0585

Fig. 263 Remove the flange bolts and the bushing holders, then remove the bushings and the stabilizer bar

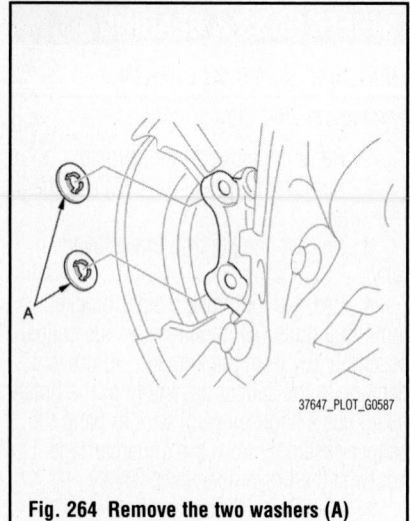

37647_PLOT_G0587

Fig. 264 Remove the two washers (A)

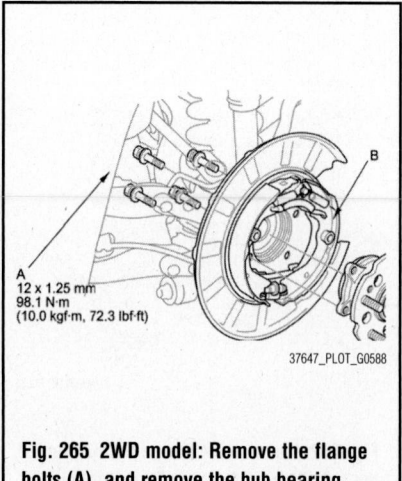

A
12 x 1.25 mm
98.1 N·m
(10.0 kgf·m, 72.3 lbf·ft)

37647_PLOT_G0588

Fig. 265 2WD model: Remove the flange bolts (A), and remove the hub bearing unit (B)

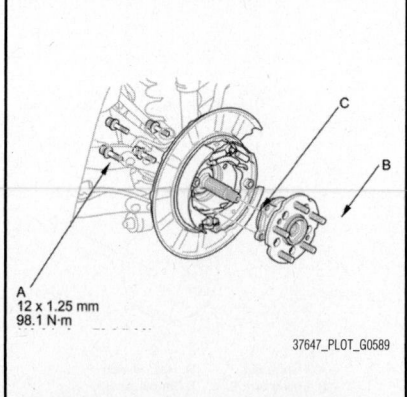

A
12 x 1.25 mm
98.1 N·m

37647_PLOT_G0589

Fig. 266 4WD model: Remove the flange bolts (A), and remove the hub bearing unit (B) by tapping the driveshaft end (C) with a soft face hammer

- 4WD model: Use a new spindle nut on reassembly.
- 4WD model: Before installing the spindle nut, apply a small amount of engine oil to the seating surface on the nut. After tightening, use a drift to stake the spindle nut shoulder against the driveshaft.
- Before installing the brake disc/drum, clean the mating surfaces of the hub bearing unit and the inside of the brake disc/drum.

- Before installing the wheel, clean the mating surfaces on the brake disc/drum and the inside of the wheel.

12. Check the wheel alignment, and adjust it if necessary.

HONDA

Ridgeline

16

BRAKES 16-9

ANTI-LOCK BRAKE SYSTEM..16-9
Precautions 16-9
Wheel Speed Sensors 16-9
Removal & Installation.............16-9
**BLEEDING THE BRAKE
SYSTEM 16-10**
Bleeding Procedure.................16-10
Bleeding the ABS
System 16-10
Manual 16-10
FRONT DISC BRAKES........16-10
Brake Pads 16-10
Removal & Installation...........16-10
PARKING BRAKE.............16-13
Adjustments 16-13
Major Adjustment 16-13
Minor Adjustment 16-13
Parking Brake Shoes 16-13
Removal & Installation...........16-13
REAR DISC BRAKES 16-11
Brake Caliper..........................16-11
Removal & Installation........16-11
Brake Pads 16-12
Removal & Installation........16-12

CHASSIS ELECTRICAL 16-15

**AIR BAGS (SUPPLEMENTAL
RESTRAINT SYSTEM) 16-15**
Precautions 16-15
Arming the System 16-15
Disarming the System.........16-15
General Precautions............16-15
Steering-related
Precautions 16-15

DRIVE TRAIN 16-15

Automatic Transmission 16-15
Drain & Refill......................16-15
Filter Replacement 16-16
Fluid Level Check 16-16
Driveshaft...............................16-16
Removal & Installation........16-16
Front Halfshafts 16-17
Removal & Installation........16-17

Rear Halfshafts 16-18
Removal & Installation........16-18
Transfer Case 16-19
Drain & Refill......................16-20
Fluid Level Check 16-20
Removal & Installation........16-19

ENGINE COOLING 16-21

Electric Engine Fan..................16-21
Removal & Installation........16-21
Engine Coolant........................16-21
Drain & Refill......................16-21
Radiator..................................16-21
Removal & Installation........16-21
Thermostat 16-22
Removal & Installation........16-22
Water Pump 16-22
Removal & Installation........16-22

ENGINE ELECTRICAL 16-23

BATTERY SYSTEM............16-23
Battery 16-23
Battery Reconnect/Relearn
Procedure 16-23
Removal & Installation........16-23
CHARGING SYSTEM 16-23
Alternator 16-23
Removal & Installation........16-23
IGNITION SYSTEM 16-24
Firing Orders...........................16-24
Ignition Coils & Spark Plugs....16-24
Removal & Installation........16-24
Ignition Timing........................16-24
Inspection & Adjustment 16-24
STARTING SYSTEM 16-25
Starter 16-25
Removal & Installation........16-25

ENGINE MECHANICAL......16-25

Accessory Drive Belt
System 16-25
Adjustment 16-25
Belt Routings 16-25
Inspection 16-25
Removal & Installation........16-25

Air Cleaner 16-26
Removal & Installation........16-26
Camshaft & Bearings 16-26
Removal & Installation........16-26
Crankshaft Front Seal..............16-27
Removal & Installation........16-27
Crankshaft Pulley 16-27
Removal & Installation........16-27
Crankshaft Rear Seal 16-27
Removal & Installation........16-27
Cylinder Head 16-27
Removal & Installation........16-27
Engine Oil & Filter 16-30
OIL & Filter Change............16-30
OIL Level Check..................16-30
Exhaust Manifold 16-31
Component Locations.........16-31
Intake Manifold 16-31
Removal & Installation........16-31
Oil Pan 16-31
Removal & Installation........16-31
Oil Pump.................................16-31
Removal & Installation........16-31
Pistons & Rings 16-33
Positioning 16-33
Rear Main Seal 16-33
Removal & Installation........16-33
Rocker Arm (Cylinder Head)
Cover 16-34
Removal & Installation........16-34
Rocker Arms & Shafts.............16-35
Removal & Installation........16-35
Timing Belt.............................16-36
Removal & Installation.......16-36
Valve Lash (Clearance)
Adjustment..........................16-39
Adjustment 16-39

ENGINE PERFORMANCE &
EMISSION CONTROLS 16-41

Accelerator Pedal Position
(APP) Sensor 16-41
Location.............................16-41
Removal & Installation........16-41
Camshaft Position (CMP)
Sensor 16-41
Removal & Installation........16-41

Crankshaft Position (CKP)
  Sensor 16-41
    CKP Pattern Clear/CKP
    Pattern Learn 16-42
    Removal & Installation........16-41
Engine Coolant Temperature
  (ECT) Sensor 16-42
    Location...............................16-42
    Removal & Installation........16-42
EVAP Canister 16-42
    Location...............................16-42
    Removal & Installation........16-42
Exhaust Gas Recirculation
  (EGR) Valve.........................16-42
    Removal & Installation........16-42
Heated Oxygen Sensor
  (HO2S) 16-43
    Removal & Installation........16-43
Knock Sensor (KS).................16-44
    Location...............................16-44
    Removal & Installation........16-44
Maintenance Reminder
  Lights..................................16-44
    Reset Procedure..................16-44
Malfunction Indicator
  Light....................................16-45
    Reset Procedure..................16-45
Manifold Absolute Pressure
  (MAP) Sensor 16-45
    Removal & Installation........16-45
Mass Air Flow/Intake Air
  Temperature Sensor 16-45
    Location...............................16-45
    Removal & Installation........16-45
Positive Crankcase Ventilation
  (PCV) Valve 16-45
    Location...............................16-45
    Removal & Installation........16-45
Powertrain Control Module
  (PCM)...................................16-45
    Location...............................16-45
    PCM Idle Learn
    Procedure..........................16-46
    PCM Update 16-46
    Removal & Installation........16-46
    Reset Procedure..................16-46

FUEL 16-47

GASOLINE FUEL INJECTION
SYSTEM 16-47
    Fuel Filter.............................16-48
    Removal & Installation........16-48
    Fuel Rail & Injectors 16-49
    Removal & Installation........16-49
    Fuel Tank Unit 16-49
    Removal & Installation........16-49
    Idle Speed 16-48
    Adjustment 16-48
    Relieving Fuel System
    Pressure............................16-47
    Throttle Body.......................16-50
    Removal & Installation........16-50

HEATING & AIR CONDITIONING
SYSTEM 16-51

    Blower Motor 16-51
    Removal & Installation........16-51
    Heater Core 16-51
    Removal & Installation........16-51

SPECIFICATIONS AND
MAINTENANCE CHARTS.....16-3

    Brake Specifications.................16-7
    Camshaft & Bearings
    Specifications 16-5
    Capacities 16-4
    Crankshaft & Connecting Rod
    Specifications 16-5
    Engine and Model Year
    Identification 16-3
    Engine Tune-Up
    Specifications 16-3
    Fluid Specifications...............16-4
    General Engine
    Specifications 16-3
    Maintenance Intervals 16-8
    Piston & Ring
    Specifications 16-5
    Tire, Wheel, & Ball Joint
    Specifications 16-7
    Torque Specifications 16-6

Valve Specifications 16-4
Wheel Alignment
    Specifications 16-6

STEERING 16-52

Power Rack & Pinion Steering
  Gear 16-52
    Removal & Installation........16-52
Power Steering Fluid..............16-56
    Fluid Fill Procedure...........16-56
    Fluid Level Check 16-56
Power Steering Pump.............16-56
    Removal & Installation........16-56

SUSPENSION 16-57

FRONT SUSPENSION 16-57
Knuckle 16-57
    Removal & Installation........16-57
Lower Ball Joints...................16-58
    Removal & Installation........16-58
Lower Control Arms 16-58
    Removal & Installation........16-58
Stabilizer Bar & Links 16-59
    Removal & Installation........16-59
Struts....................................16-61
    Removal & Installation........16-61
Wheel Hubs & Bearings..........16-62
    Removal & Installation........16-62
REAR SUSPENSION 16-63
Knuckle/Hub Assembly 16-63
    Removal & Installation........16-63
Rear Lower Arms....................16-64
    Removal & Installation........16-64
Stabilizer Bar & Links 16-64
    Removal & Installation........16-64
Struts (Damper/Spring)...........16-65
    Removal & Installation........16-65
Trailing Arms.........................16-65
    Removal & Installation........16-65
Upper Arms 16-66
    Removal & Installation........16-66
Upper Ball Joints....................16-66
    Removal & Installation........16-66
Wheel Bearing.......................16-67
    Removal & Installation........16-67

SPECIFICATIONS AND MAINTENANCE CHARTS

ENGINE AND VEHICLE IDENTIFICATION CHART

			Engine Code				Model Year	
Code	Liters (cc)	Cu. In.	Cyl.	Fuel Sys.	Engine Type	Eng. Mfg.	Code ①	Year
J35Z5	3.5 (3471)	222	6	SMFI	SOHC	Honda	B	2011
							C	2012

SOHC: Single Overhead Cam

SMFI: Sequential Multi-port Fuel Injection

① 10th position of VIN

71051_RIDG_C0001

GENERAL ENGINE SPECIFICATIONS

Year	Model	Engine Displacement Liters	Engine ID	Net Horsepower @ rpm	Net Torque @ rpm (ft. lbs.)	Bore x Stroke (in.)	Com-pression Ratio	Oil Pressure @ rpm
2011	Ridgeline	3.5	J35Z5	255@5750	252@4500	3.50x3.66	10:01	71@3000 ①
2012	Ridgeline	3.5	J35Z5	247@5750	245@4500	3.50x3.66	10:01	71@3000 ①

① At idle: 10 psi

71051_RIDG_C0002

ENGINE TUNE-UP SPECIFICATIONS

Year	Engine Displacement Liters	Engine ID	Spark Plug Gap (in.)	Ignition Timing (deg. BTDC)	Fuel Pump (psi)	Idle Speed (rpm)	Valve Clearance (in.) In.	Valve Clearance (in.) Ex.
2011	3.5	J35Z5	0.039-0.043	8-12	57-64	680-780	0.008-0.009	0.011-0.013
2012	3.5	J35Z5	0.039-0.043	8-12	57-64	680-780	0.008-0.009	0.011-0.013

NOTE: The Vehicle Emission Control Information label often reflects changes made during production and must be used if they differ from this chart.

BTDC: Before top dead center

71051_RIDG_C0003

CAPACITIES

Year	Model	Engine Displacement Liters	Engine ID	Engine Oil with Filter (qts.)	Transmission (pts.) *	Transfer Case (pts.)	Rear Drive Axle (pts.)	Fuel Tank (gal.)	Cooling System (qts.)
2011	Ridgeline	3.5	J35Z5	4.5	6.6	0.9	5.58	22.0	①
2012	Ridgeline	3.5	J35Z5	4.5	6.6	0.9	5.58	22.0	①

NOTE: All capacities are approximate. Add fluid gradually and check to be sure a proper fluid level is obtained.

* Fluid change

① Total capacity: 8.56 qts.
　At coolant change: 6.26 qts.

71051_RIDG_C0004

FLUID SPECIFICATIONS

Year	Model	Engine Displacement Liters	Engine Oil	Auto. Trans.	Drive Axle	Power Steering Fluid	Brake Master Cylinder
2011	Ridgeline	3.5	①	Honda ATF-Z1	Honda VTM-4	Honda P/S Fluid	DOT 3
2012	Ridgeline	3.5	①	Honda ATF-Z1	Honda VTM-4	Honda P/S Fluid	DOT 3

DOT: Department Of Transportation

① See oil filler cap.

71051_RIDG_C0013

VALVE SPECIFICATIONS

Year	Engine Displacement Liters	Engine ID	Seat Angle (deg.)	Face Angle (deg.)	Spring Test Pressure (lbs. @ in.)	Spring Free Length (in.)	Stem-to-Guide Clearance (in.) Intake	Stem-to-Guide Clearance (in.) Exhaust	Stem Diameter (in.) Intake	Stem Diameter (in.) Exhaust
2011	3.5	J35Z5	45	45	NA	①	0.0008-0.0018	0.0022-0.0031	0.2159-0.2163	0.2146-0.2150
2012	3.5	J35Z5	45	45	NA	①	0.0008-0.0018	0.0022-0.0031	0.2159-0.2163	0.2146-0.2150

NA: Information not available

① Intake: 1.9713 in.
　Exhaust: 2.1060 in.

71051_RIDG_C0005

CAMSHAFT AND BEARING SPECIFICATIONS

All measurements are given in inches.

Year	Engine Displacement Liters	Engine ID	Journal Diameter	Brg. Oil Clearance	Shaft End-play	Runout	Journal Bore	Lobe Lift Intake	Lobe Lift Exhaust
2011	3.5	J35Z5	NA	0.0002-0.0035	0.0002-0.0008	0.001	NA	①	1.4375
2012	3.5	J35Z5	NA	0.0002-0.0035	0.0002-0.0008	0.001	NA	①	1.4375

NA: Information not available
① Primary: 1.4024
   Secondary 1.3504

71051_RIDG_C0007

CRANKSHAFT AND CONNECTING ROD SPECIFICATIONS

All measurements are given in inches

Year	Engine Displacement Liters	Engine ID	Crankshaft Main Brg. Journal Dia.	Crankshaft Main Brg. Oil Clearance	Crankshaft Shaft End-play	Thrust on No.	Connecting Rod Journal Diameter	Connecting Rod Oil Clearance	Connecting Rod Side Clearance
2011	3.5	J35Z5	2.8337-2.8346	0.0007-0.0018	0.0040-0.0140	3	2.1644-2.1654	0.0002-0.0006	0.0060-0.0140
2012	3.5	J35Z5	2.8337-2.8346	0.0007-0.0018	0.0040-0.0140	3	2.1644-2.1654	0.0002-0.0006	0.0060-0.0140

71051_RIDG_C0006

PISTON AND RING SPECIFICATIONS

All measurements are given in inches

Year	Engine Disp. Liters	Engine ID	Piston Clearance	Ring Gap Top Compression	Ring Gap Bottom Compression	Ring Gap Oil Control	Ring Side Clearance Top Compression	Ring Side Clearance Bottom Compression	Ring Side Clearance Oil Control
2011	3.5	J35Z5	0.0006-0.0016	0.0080-0.0140	0.0160-0.0220	0.0080-0.0280	0.0022-0.0031	0.0012-0.0022	snug
2012	3.5	J35Z5	0.0006-0.0016	0.0080-0.0140	0.0160-0.0220	0.0080-0.0280	0.0022-0.0031	0.0012-0.0022	snug

71051_RIDG_C0008

TORQUE SPECIFICATIONS
All readings in ft. lbs.

Year	Engine Disp. Liters	Engine ID	Cylinder Head Bolts	Main Bearing Bolts	Rod Bearing Bolts	Crankshaft Damper Bolts	Flywheel Bolts	Manifold Intake	Manifold Exhaust	Spark Plugs	Oil Pan Drain Plug
2011	3.5	J35Z5	①	②	③	④	54	16	NA	13	29
2012	3.5	J35Z5	①	②	③	④	54	16	NA	13	29

NA: Information not available

① Step 1: 22 ft. lbs.
  Step 2: plus 90 degrees
  Step 3: plus 90 degrees
② Cap bolts: 54 ft. lbs.
  Side bolts: 36 ft. lbs.
③ Step 1: 12 ft. lbs.
  Step 2: plus 90 degrees
④ Step 1: 47 ft. lbs.
  Step 2: plus 60 degrees

71051_RIDG_C0009

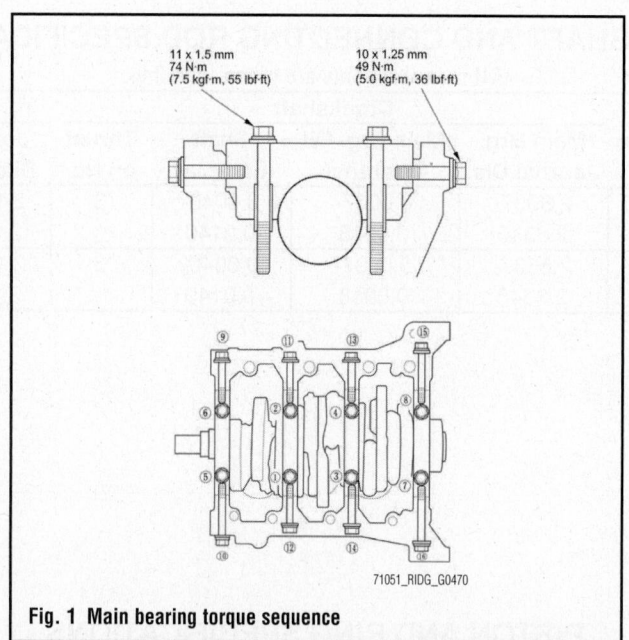

11 x 1.5 mm
74 N·m
(7.5 kgf·m, 55 lbf·ft)

10 x 1.25 mm
49 N·m
(5.0 kgf·m, 36 lbf·ft)

71051_RIDG_G0470

Fig. 1 Main bearing torque sequence

WHEEL ALIGNMENT

Year	Model		Caster Range (+/-Deg.)	Caster Preferred Setting (Deg.)	Camber Range (+/-Deg.)	Camber Preferred Setting (Deg.)	Toe-in (in.)
2011	Ridgeline	F	1.00	+1.70	1.00	-0.50	0 +/- 0.08
		R	—	—	0.75	-0.50	0 +/- 0.08
2012	Ridgeline	F	1.00	+1.70	1.00	-0.50	0 +/- 0.08
		R	—	—	0.75	-0.50	0 +/- 0.08

71051_RIDG_C0011

TIRE, WHEEL AND BALL JOINT SPECIFICATIONS

Year	Model	OEM Tires		Tire Pressures (psi)		Wheel Size	Ball Joint Inspection	Lug Nut Torque (ft. lbs.)
		Standard	Optional	Front	Rear			
2011	Ridgeline	①	None	32	②	7.5	NA	94
2012	Ridgeline	①	None	32	②	7.5	NA	94

NA: Information not available

OEM: Original Equipment Manufacturer

PSI: Pounds Per Square Inch

① P245/65R17 (except RTL model)

   P245/60R18 (RTL model)

② Consult door jamb label or Owner's Manual.

71051_RIDG_C0010

BRAKE SPECIFICATIONS
All measurements in inches unless noted

Year	Model		Brake Disc			Minimum Lining Thickness	Brake Caliper	
			Original Thickness	Minimum Thickness	Maximum Runout		Bracket Bolts (ft. lbs.)	Mounting Bolts (ft. lbs.)
2011	Ridgeline	F	1.10-1.11	1.020	0.0016	0.040	101	53
		R	0.43-0.44	0.350	0.0016	0.040	80	16
2012	Ridgeline	F	1.10-1.11	1.020	0.0016	0.040	101	53
		R	0.43-0.44	0.350	0.0016	0.040	80	16

F: Front

R: Rear

71051_RIDG_C0012

MAINTENANCE MINDER SCHEDULE
Honda Ridgeline

The Ridgeline displays engine oil life and maintenance service items in the information display to indicate when to perform maintenance service. If the engine oil life is 15% or less, based on the onboard computer's caluculations, you will see SERVICE DUE SOON in the information display every time the ignition key is turned to ON. The maintenance minder indicator will also come on and the maintenance code(s) for other scheduled maintenance items needing service will be displayed below the message.

Symbol	Item	Service
A	Engine oil ①	Change
B	Engine oil and filter	Change
	Brakes	Inspect
	Parking brake adjustment	Check
	Steering gear, boots and linkage	Inspect
	Suspension components	Inspect
	Driveshaft boots	Inspect
	Brake hoses and lines, including VSA	Inspect
	Exhaust system	Inspect
	All fluid levels and fluid condition	Inspect
	Exhaust system components	Inspect
	Fuel lines and connections	Inspect
1	Tires: condition and pressures	Rotate
2	Engine air filter ②	Replace
	Dust and pollen filter ③	Replace
	Accessory drive belt	Inspect
3	Transmission fluid	Replace
	Transfer case fluid	Replace
4	Spark plugs	Replace
	Timing belt ④	Replace
	Water pump	Inspect
	Valve clearance ⑤	Inspect
5	Engine coolant	Replace
6	Rear differential fluid ⑥	Replace

① If the message SERVICE DUE NOW does not appear more than 12 months after the display is reset, change every year.

② If driven in dusty conditions, replace every 15,000 miles.

③ If driven in urban areas that have a high concentration of soot from industry and diesel, replace every 15,000 miles

④ If driven regularly in temperatures over 110 deg.F or below -20 deg.F, or towing a trailer, replace every 60,000 miles.

⑤ Adjust if necessary.

⑥ Driving in mountainous areas at very low vehicle speeds or trailer towing results in higher level of mechanical (shear) stres to fluid. This requires differential fluid changes more frequently than recommended by the maintenance minder. If the vehicle regularly driven under these conditions, changed have the differential fluid at 7,500 miles, then every 15,000 miles.

Additionally, replace the brake fluid every 3 years, and inspect the idle speed every 160,000 miles.

To reset the Engine Oil Life Display:

1. Turn the ignition switch to ON.

2. Press the SELECT button repeatedly until the engine oil life display or the service message is displayed.

3. Press the TRIP/RESET button for about 10 seconds. The engine oil life indicator and maintenance item code(s) will blink.

NOTE: If you are resetting the display when the engine oil life is more than 15 %, make sure any maintenance item(s) requiring service are done before resetting the display.

4. Press TRIP/RESET button for another 5 seconds. The maintenance item code(s) will disappear, and the engine oil life will reset to "100".

BRAKES

PRECAUTIONS

❊❊ CAUTION

Dust and dirt accumulating on brake parts during normal use may contain asbestos fibers from production or aftermarket brake linings. Breathing excessive concentrations of asbestos fibers can cause serious bodily harm. Exercise care when servicing brake parts. Do not sand or grind brake lining unless equipment used is designed to contain the dust residue. Do not clean brake parts with compressed air or by dry brushing. Cleaning should be done by dampening the brake components with a fine mist of water, then wiping the brake components clean with a dampened cloth. Dispose of cloth and all residue containing asbestos fibers in an impermeable container with the appropriate label. Follow practices prescribed by the Occupational Safety and Health Administration (OSHA) and the Environmental Protection Agency (EPA) for the handling, processing, and disposing of dust or debris that may contain asbestos fibers.

This vehicle uses an Antilock Brake System (ABS). If the Antilock Brake Module (ABM) senses impending rear wheel lock-up, it will energize the isolation solenoid. This prevents a further increase of driver induced brake pressure to the rear wheels. If this initial action is not enough to prevent rear wheel lock-up, the ABM will momentarily energize a dump solenoid. This opens the dump valve to vent a small amount of isolated rear brake pressure to an accumulator. The action of fluid moving to the accumulator reduces the isolated brake pressure at the wheel cylinders. The dump (pressure venting) cycle is limited to very short time periods (milliseconds). The ABM

will pulse the dump valve until rear wheel deceleration reaches the desired slip rate programmed into the ABM. The system will switch to normal braking once wheel locking tendencies are no longer present.

As part of the anti-lock brake system, this vehicle is equipped with Vehicle Stability Assistance (VSA). VSA is a computerized technology that improves the safety of a vehicle's handling by detecting and preventing skids. When VSA detects loss of steering control, VSA automatically applies individual brakes to help "steer" the vehicle where the driver wants to go. Braking is automatically applied to individual wheels, such as the outer front wheel to counter oversteer, or the inner rear wheel to counter under steer. Some VSA systems also reduce engine power until control is regained.

VSA compares the driver's intended direction (by measuring steering angle) to the vehicle's actual direction (by measuring lateral acceleration, vehicle rotation (yaw) and individual road wheel speeds). If the vehicle is not going where the driver is steering, VSA then brakes individual front or rear wheels and/or reduces excess engine power as needed to help correct under steer (plowing) or oversteer (fishtailing).

VSA incorporates yaw rate control into the anti-lock braking system (ABS). Yaw is rotation around the vertical axis; i.e. spinning left or right. Anti-lock brakes enable VSA to brake individual wheels. The VSA used on this vehicle also incorporates a Traction Control System (TCS), which senses drive-wheel slip under acceleration and individually brakes the slipping wheel or wheels and/or reduces excess engine power until control is regained.

WHEEL SPEED SENSORS

REMOVAL & INSTALLATION

Front

1. Turn the ignition switch to LOCK (0).

2. Remove the three clips, then disconnect the wheel speed sensor connector.
3. Remove the bolts and the wheel speed sensor.

To install:

4. Install the wheel speed sensor in the reverse order of removal, and note these items:
   a. Do not twist the sensor wires.
   b. If the wheel speed sensor comes in contact with the wheel bearing, it is faulty.
   c. Make sure there is no debris in the sensor mounting hole.
5. Start the engine, and check that the ABS, the VSA indicators, and the trailer stability assist warning goes off.
6. Test-drive the vehicle, and check that the ABS, the VSA indicators, and the trailer stability assist warning do not come on.

Rear

1. Turn the ignition switch to LOCK (0).
2. Disconnect the wheel speed sensor connector, then release the clamp.
3. Remove the clip, the three bolts, and the wheel speed sensor.

To install:

4. Install the wheel speed sensor in the reverse order of removal, and note these items:
   a. Do not twist the sensor wires.
   b. If the wheel speed sensor comes in contact with the wheel bearing, it is faulty.
   c. Make sure there is no debris in the sensor mounting hole.
5. Start the engine, and check that the ABS, the VSA indicators, and the trailer stability assist warning goes off.
6. Test-drive the vehicle, and check that the ABS, the VSA indicators, and the trailer stability assist warning do not come on.

BRAKES

BLEEDING THE BRAKE SYSTEM

BLEEDING PROCEDURE

MANUAL

See Figure 2.

Always use Honda DOT 3 Brake Fluid. Using a non-Honda brake fluid can cause corrosion and decrease the life of the system.

> ❈❈ **WARNING**
>
> **Do not spill brake fluid on the vehicle; it may damage the paint. If brake fluid gets on the paint, wash it off immediately with water.**

Note the following:
- Do not reuse the drained fluid. Use only new Honda DOT 3 Brake Fluid from an unopened container. Using a non-Honda brake fluid can cause corrosion and shorten the life of the system.
- Make sure no dirt or other foreign matter gets in the brake fluid.
- The reservoir connected to the master

cylinder must be at the MAX (upper) level mark at the start of the bleeding procedure and check after bleeding each wheel location. Add fluid as required.

1. Make sure the brake fluid level in the reservoir is at the MAX (upper) level line.
2. Have someone slowly pump the brake pedal several times, then apply steady pressure.
3. Start the bleeding at the driver's side of the front brake system.

➡ **Bleed the calipers in the sequence shown.**

4. Attach a length of clear drain tube to the bleed screw, then loosen the bleed screw to allow air to escape from the system. Then tighten the bleed screw securely.
5. Refill the master cylinder reservoir to the MAX (upper) level line.
6. Repeat the procedure for each brake circuit until there are no air bubbles in the fluid.

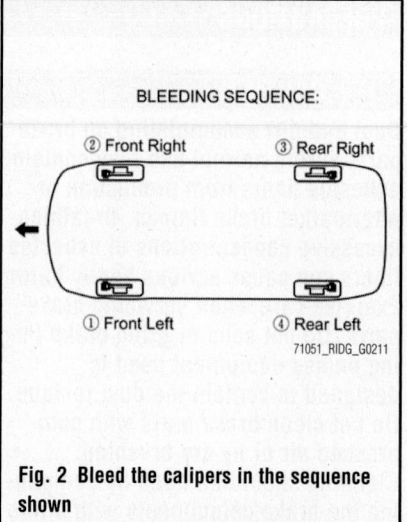

BLEEDING SEQUENCE:

② Front Right   ③ Rear Right

① Front Left   ④ Rear Left

71051_RIDG_G0211

Fig. 2 Bleed the calipers in the sequence shown

BLEEDING THE ABS SYSTEM

The bleeding procedure for the ABS System is the same as the conventional bleeding procedure. Refer to "BLEEDING THE BRAKE SYSTEM" in this section.

BRAKES

FRONT DISC BRAKES

BRAKE PADS

REMOVAL & INSTALLATION

See Figures 3 through 6.

> ❈❈ **CAUTION**
>
> **Frequent inhalation of brake pad dust, regardless of material composition, could be hazardous to your health. Avoid breathing dust particles. Never use an air hose or brush to clean brake assemblies. Use an OSHA-approved vacuum cleaner.**

1. Remove some of the brake fluid from the master cylinder.
2. Raise and support the vehicle.
3. Remove the front wheels.
4. Remove the flange bolt, and pivot the caliper up out of the way. Check the hose and the pin boots for damage and deterioration.
5. 2012 model: Remove the pad springs while holding the brake pads.
6. Remove the pad shims and the brake pads.
7. Remove the pad retainers.

➡ **The upper and lower pad retainers are different. During installation, make sure the pad retainers are in proper positions.**

8. Clean the caliper bracket thoroughly; remove any rust, and check for grooves and cracks. Verify that the caliper pins move in and out smoothly. Clean and lube if needed.
9. Inspect the brake disc for runout, thickness, parallelism Inspect for runout, thickness and parallelism, and check for damage and cracks.

To install:

10. Install the upper and lower pad retainers. Wipe off the excess assembly

paste from the retainers. Keep the assembly paste away from the brake disc and the brake pads.

11. Install the brake caliper piston compressor tool on the caliper body.
12. Press in the piston with the brake caliper piston compressor tool so the caliper will fit over the brake pads. Make sure the piston boot is in position to prevent damaging it when pivoting the caliper down.

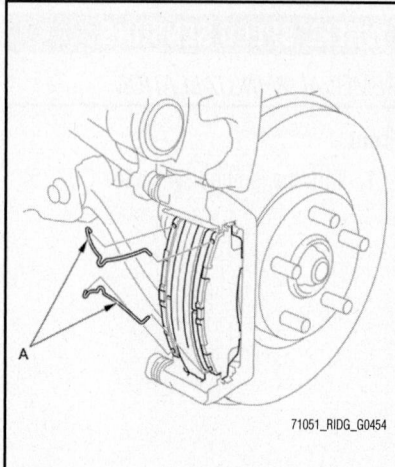

71051_RIDG_G0454

Fig. 3 2012 model: Remove the pad springs (A) while holding the brake pads

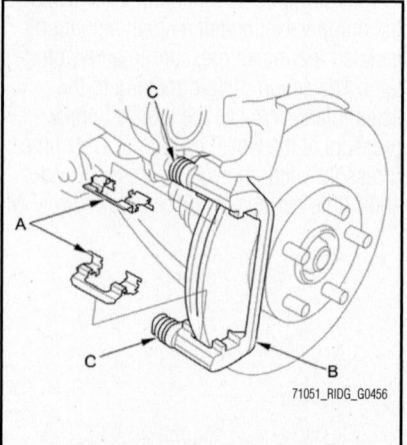

71051_RIDG_G0456

Fig. 4 Remove the pad retainers (A); showing the caliper bracket (B), and caliper pins (C)

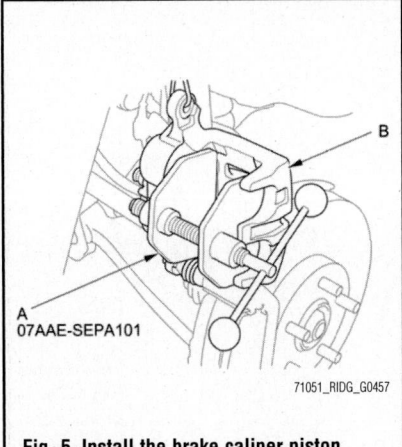

Fig. 5 Install the brake caliper piston compressor tool (A) on the caliper body (B)

➡**Be careful when pressing in the pistons; brake fluid might overflow from the master cylinder's reservoir. If brake fluid gets on any painted surface, wash it off immediately with water.**

13. Remove the brake caliper piston compressor tool.

14. Apply a thin coat of M-77 assembly paste (P/N 08798-9010) to the pad side of the shims, the back of the brake pads and

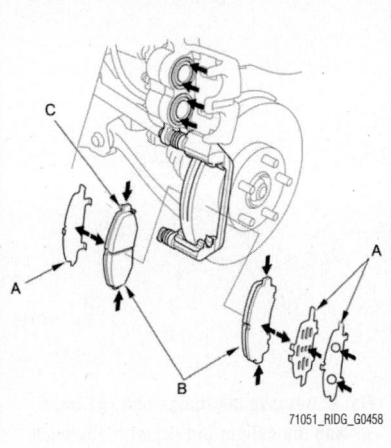

Fig. 6 Apply a thin coat of M-77 assembly paste (P/N 08798-9010) to the pad side of the shims (A), the back of the brake pads (B) and the other areas indicated by the arrows; wear indicator (C)

the other areas indicated by the arrows. Wipe excess assembly paste off the pad shims and the brake pads friction material. Keep grease and assembly paste off the brake disc and the brake pads. Contaminated brake disc or brake pads reduce stopping ability.

15. Install the brake pads and the pad shims correctly. Install the brake pad with the wear indicator on the upper inside. If you are reusing the brake pads, always reinstall the brake pads in their original positions to prevent a temporary loss of braking efficiency.

16. 2012 model: Install the pad springs while holding the brake pads.

17. Pivot the caliper down into position. Install the flange bolt, and tighten it to 53 ft. lbs. (72 Nm).

18. Clean the mating surface between the brake disc and the inside of the wheel, then install the front wheels.

19. Press the brake pedal several times to make sure the brakes work.

➡**Engagement may require a greater pedal stroke immediately after the brake pads have been replaced as a set. Several applications of the brake pedal will restore the normal pedal stroke.**

20. Add brake fluid as needed.

21. After installation, check for leaks at hose and line joints or connections, and retighten if necessary. Test-drive the vehicle, then recheck for leaks.

BRAKES   **REAR DISC BRAKES**

BRAKE CALIPER

REMOVAL & INSTALLATION

See Figures 7 and 8.

➡**Keep grease away from the brake disc/drum and the brake pads.**

1. Raise and support the vehicle.
2. Remove the rear wheels.
3. Release the parking brake pedal fully.
4. Remove and discard the brake hose clip from the brake hose.

➡**During installation, install a new brake hose clip.**

5. Remove the brake hose bracket mounting bolts from the knuckle.

6. Remove the brake caliper bracket mounting bolts, then remove the caliper assembly from the knuckle. To prevent damage to the caliper assembly or brake hose, use a short piece of wire to hang the caliper assembly from the undercarriage.

➡**Do not twist the brake hose excessively.**

7. Remove the brake disc/drum flathead screws.

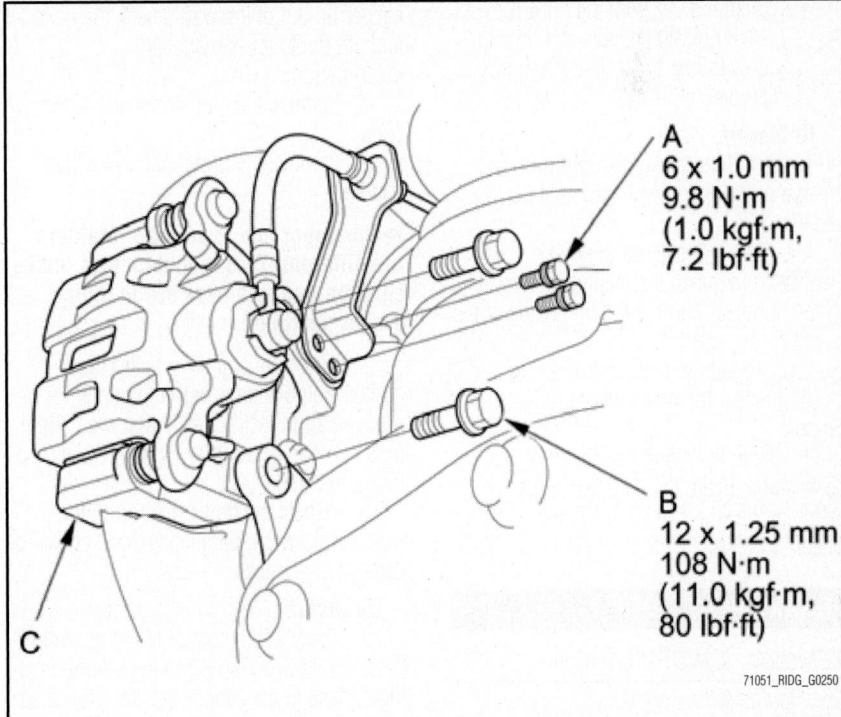

A
6 x 1.0 mm
9.8 N·m
(1.0 kgf·m,
7.2 lbf·ft)

B
12 x 1.25 mm
108 N·m
(11.0 kgf·m,
80 lbf·ft)

Fig. 7 Remove the brake hose bracket mounting bolts (A) from the knuckle; remove the brake caliper bracket mounting bolts (B), then remove the caliper assembly (C) from the knuckle

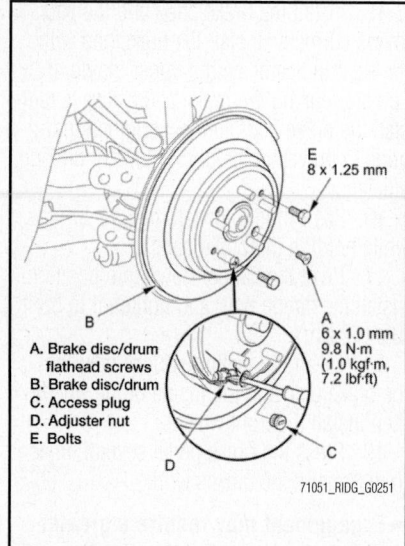

A. Brake disc/drum flathead screws
B. Brake disc/drum
C. Access plug
D. Adjuster nut
E. Bolts

Fig. 8 Remove the brake disc/drum flathead screws

8. Remove the brake disc/drum from the hub.

- Remove the access plug, then if necessary, turn the adjuster nut with a brake adjust tool or a flat-tip screwdriver until the shoes become loose.
- If the brake disc/drum is stuck to the hub, thread two 8 x 1.25 mm bolts into the brake disc/drum to push it away from the hub. Turn each bolt 90 degrees at a time to prevent the brake disc/drum from binding.

To install:

9. Install the brake disc/drum in the reverse order of removal, and note these items:

   a. Before installing the brake disc/drum, clean the mating surface of the hub bearing unit and the inside of the brake disc/drum.

   b. Adjust the parking brake.

10. Inspect the brake disc/drum runout.

11. Clean the mating surfaces between the brake disc/drum and the inside of the wheel, then install the rear wheel.

BRAKE PADS

REMOVAL & INSTALLATION

See Figures 9 through 12.

✳✳ CAUTION

Frequent inhalation of brake pad dust, regardless of material composi-

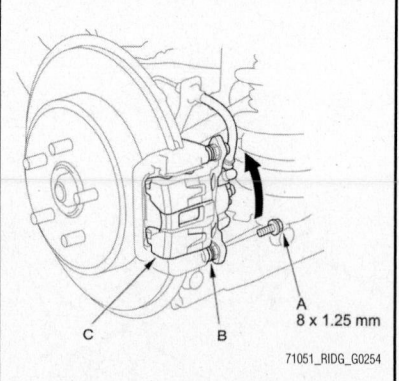

Fig. 9 Remove the flange bolt (A) while holding the caliper pin (B) with a wrench being careful not to damage the pin boot, and pivot the caliper (C) up out of the way

tion, could be hazardous to your health. Avoid breathing dust particles. Never use an air hose or brush to clean brake assemblies. Use an OSHA-approved vacuum cleaner.

1. Remove some brake fluid from the master cylinder.

2. Raise and support the vehicle.

3. Remove the rear wheels.

4. Remove the flange bolt while holding the caliper pin with a wrench being careful not to damage the pin boot, and pivot the caliper up out of the way. Check the hose and pin boots for damage and deterioration.

5. Remove the pad shims and brake pads.

6. Remove the upper and lower pad retainers.

➡ **The upper and lower pad retainers are different. During installation, make sure the pad retainers are in their proper positions.**

7. Clean the caliper bracket thoroughly; remove any rust, and check for grooves and cracks. Verify that the caliper pins move in and out smoothly. Clean and lube if needed.

8. Inspect the brake disc/drum for runout, thickness, parallelism, and check for damage and cracks.

To install:

9. Apply a thin coat of M-77 assembly paste (P/N 08798-9010) to the retainer mating surface of the caliper bracket (indicated by the arrows).

10. Install the upper and lower pad retainers. Wipe off the excess assembly paste from the retainers. Keep the

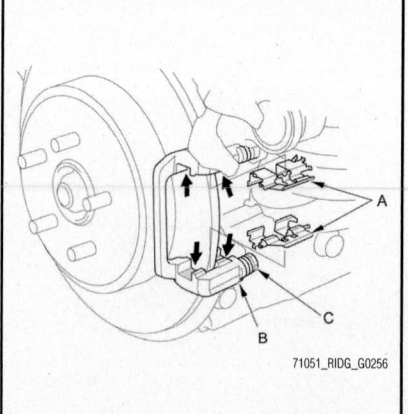

Fig. 10 Remove the upper and lower pad retainers (A); caliper bracket (B); caliper pins (C)

assembly paste away from the discs and pads.

11. Apply a thin coat of M-77 assembly paste (P/N 08798- 9010) to the pad side of the shims, the back of the brake pads, and the other areas indicated by the arrows. Wipe off the excess assembly paste from the pad shims and brake pads friction material. Keep grease and assembly paste away from the brake disc/drum and brake pads. Contaminated brake disc/drum or brake pads reduce stopping ability.

12. Install the brake pads and the pad shims as shown. Install the brake pad with the wear indicator on the bottom inside position. If you are reusing the brake pads, always reinstall the brake pads in their original positions to prevent a temporary loss of braking efficiency.

13. Install the brake caliper piston compressor tool on the caliper body.

14. Press in the piston with a brake caliper piston compressor tool so the caliper will fit over the brake pads. Make sure the piston boot is in position to prevent damaging it when pivoting the caliper down.

➡ **Be careful when pushing in the caliper piston; brake fluid might overflow from the master cylinder's reservoir. If brake fluid gets on any painted surface, wash it off immediately with water.**

15. Remove the brake caliper piston compressor tool.

16. Pivot the caliper down into position. Install the flange bolt, and tighten it to the specified torque while holding the caliper pin B with a wrench. Be careful not to damage the pin boot.

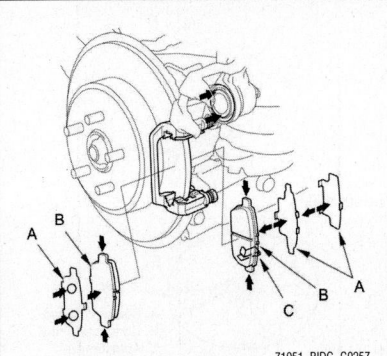

Fig. 11 Apply a thin coat of M-77 assembly paste to the pad side of the shims (A), the back of the brake pads (B), and the other areas indicated by the arrows; wear indicator (C)

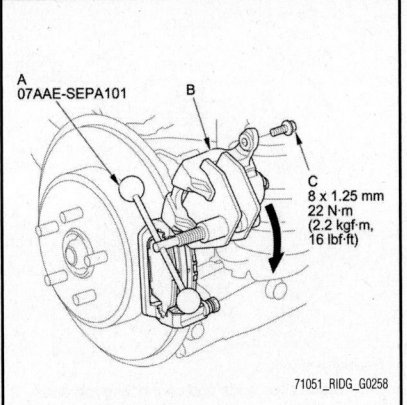

Fig. 12 Install the brake caliper piston compressor tool (A) on the caliper body (B); install the flange bolt (C)

17. Clean the mating surface between the brake disc/drum and the inside of the wheel, then install the front wheels.

18. Press the brake pedal several times to make sure the brakes work.

➡**Engagement may require a greater pedal stroke immediately after the brake pads have been replaced as a set. Several applications of the brake pedal will restore the normal pedal stroke.**

19. Add brake fluid as needed.

20. After installation, check for leaks at hose and line joints or connections, and retighten if necessary. Test-drive the vehicle, then recheck for leaks.

BRAKES

PARKING BRAKE

ADJUSTMENTS

MINOR ADJUSTMENT

See Figure 13.

1. Raise and support the vehicle.
2. Release the parking brake pedal fully.
3. Press the parking brake pedal 1 click.
4. Remove the adjusting nut cap.
5. Tighten the parking brake adjusting nut until the parking brakes drag slightly when the rear wheels are turned.
6. Release the parking brake pedal fully, and check that the parking brakes do not drag when the rear wheels are turned. Readjust if necessary.
7. Make sure the parking brake pedal is within the specified number of clicks (10 to 12 clicks).

MAJOR ADJUSTMENT

See Figures 13 and 14.

➡**To be done when replacing parking brake shoes and after lining surface break-in.**

1. Raise and support the vehicle.
2. Release the parking brake pedal fully.
3. Remove the adjusting nut cap, then loosen the parking brake adjusting nut.
4. Remove the rear wheels.
5. Remove the access plug.
6. Turn the ratchet teeth on the adjuster nut with a brake adjuster tool or a flat-tip screwdriver until the shoes lock against the parking brake drum. Then back off the adjuster 10 clicks, and install the access plug.
7. Clean the mating surface between the brake disc/drum and the inside of the wheel, then install the rear wheels.

8. Do the minor adjustment procedure.

PARKING BRAKE SHOES

REMOVAL & INSTALLATION

See Figures 15 through 20.

✳✳ CAUTION

Frequent inhalation of brake pad dust, regardless of material composition, could be hazardous to your health. Avoid breathing dust particles. Never use an air hose or brush to clean brake assemblies. Use an OSHA-approved vacuum cleaner.

1. Raise and support the vehicle.
2. Remove the rear wheels.
3. Release the parking brake, and remove the rear brake caliper and brake disc/drum.
4. Disconnect and remove the upper return springs, then remove the shoe guide plate.
5. Remove the tension pins by pushing the respective retainers and turning the pin.
6. Remove the connecting rod and the rod spring.
7. Lower the parking brake shoe assembly.
8. Remove the forward brake shoe and the adjuster assembly by removing the lower return spring.
9. Remove the rearward brake shoe by disconnecting the parking brake cable from the parking brake lever.
10. Remove the U-clip, wave washer, and parking brake lever from the brake shoe.

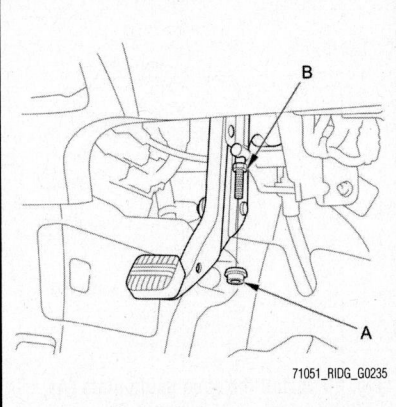

Fig. 13 Remove the adjusting nut cap (A); then tighten the parking brake adjusting nut (B)

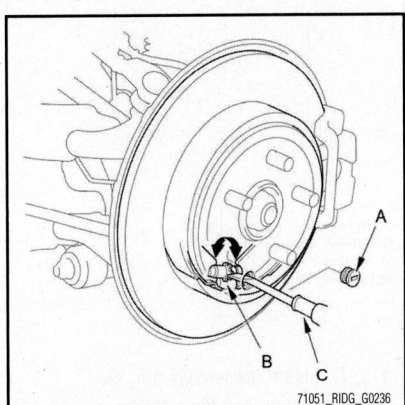

Fig. 14 Remove the access plug (A), turn the ratchet teeth on the adjuster nut (B) with a brake adjuster tool or a flat-tip screwdriver (C)

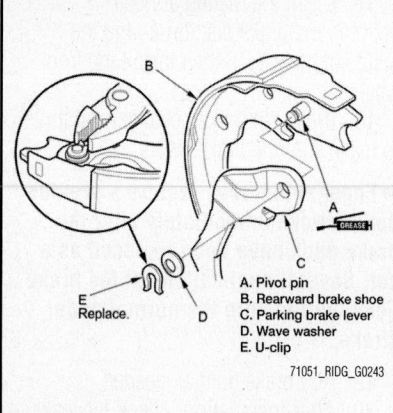

Fig. 15 Apply Molykote 44MA grease to the sliding surface of the pivot pin of the rearward brake shoe

A. Pivot pin
B. Rearward brake shoe
C. Parking brake lever
D. Wave washer
E. U-clip

E. Replace.

71051_RIDG_G0243

To install:

11. Apply Molykote 44MA grease to the sliding surface of the pivot pin of the rearward brake shoe.

12. Install the parking brake lever and the wave washer on the pivot pin, and secure with a new U-clip.

   a. Install the wave washer with its convex side facing out.

   b. Pinch the U-clip securely to prevent the parking brake lever from coming out from the brake shoe.

13. Connect the parking brake cable (A) to the parking brake lever (B).

14. Apply a thin coat of Molykote 44MA grease to the shoe ends and connecting rod ends, sliding surfaces, and opposite edges of the parking brake shoe as shown. Wipe off any excess. Keep grease off the brake linings.

15. Install the tension pin, the retainer spring, and the retainer on the rearward brake shoe. Make sure the tension pin does not contact the parking brake lever.

16. Install connecting rods on the adjuster nut.

- Clean the threaded portions of connecting rod A and the sliding surface of connecting rod B, then coat them with Molykote 44MA grease.
- Shorten connecting rod A by fully turning the adjuster nut.

17. Position the brake shoe adjuster assembly on the parking brake shoes.

18. Hook the lower return spring on the parking brake shoes.

19. Install the rod spring to the con-

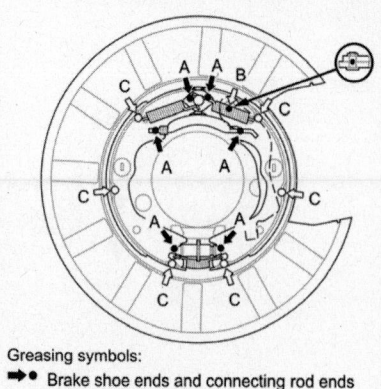

Greasing symbols:
➡️○ Brake shoe ends and connecting rod ends
⇨○ Opposite edge of the shoe
⇨● Sliding surface

71051_RIDG_G0244

Fig. 16 Apply a thin coat of Molykote 44MA grease to the shoe ends and connecting rod ends (A), sliding surfaces (B), and opposite edges of the parking brake shoe (C) as shown

necting rod first. Then install the connecting rod on the parking brake shoes.

20. Install the tension pin, the retainer spring, and the retainer on the forward brake shoe.

21. Install the shoe guide plate.

22. Install the upper return springs as shown.

23. Install the brake disc/drum and the rear brake caliper.

24. Do the major parking brake adjustment.

25. Clean the mating surfaces between the brake disc/drum and the inside of the wheel, then install the rear wheels.

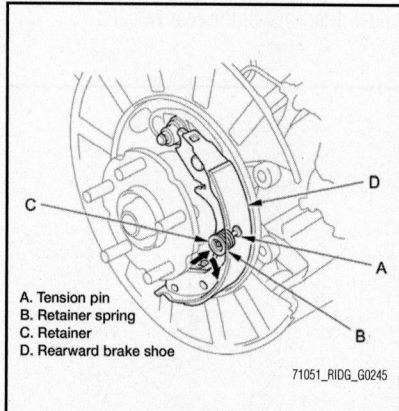

A. Tension pin
B. Retainer spring
C. Retainer
D. Rearward brake shoe

71051_RIDG_G0245

Fig. 17 Install the tension pin, the retainer spring, and the retainer on the rearward brake shoe

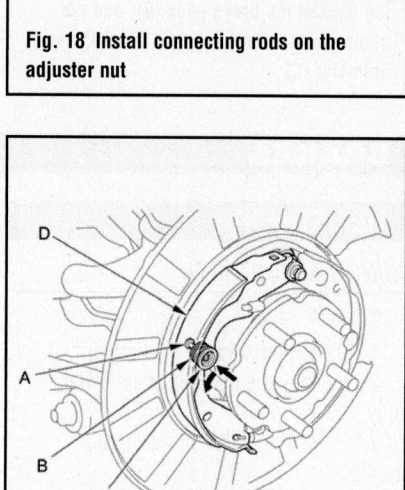

A. Connecting rod
B. Connecting rod
C. Adjuster nut
D. Brake shoe adjuster assembly
E. Lower return spring

71051_RIDG_G0246

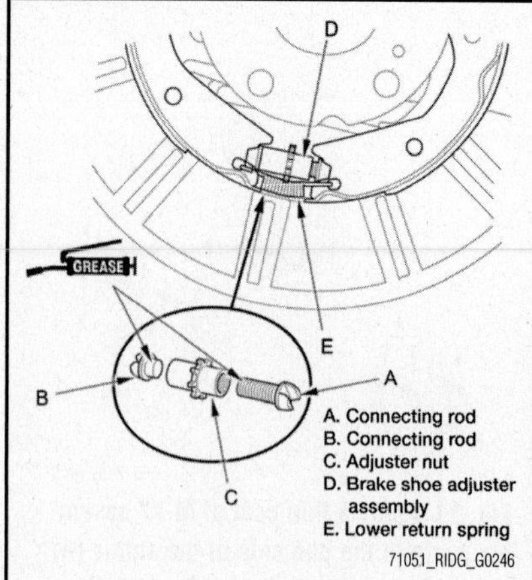

Fig. 18 Install connecting rods on the adjuster nut

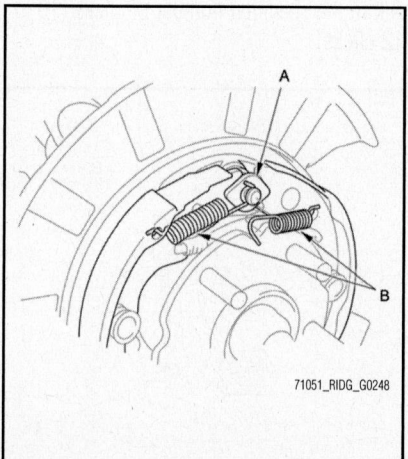

A. Tension pin
B. Retainer spring
C. Retainer
D. Forward brake shoe

71051_RIDG_G0247

Fig. 19 Install the tension pin, the retainer spring, and the retainer on the forward brake shoe

Fig. 20 Install the shoe guide plate (A) and the upper return springs (B) as shown

71051_RIDG_G0248

CHASSIS ELECTRICAL   AIR BAGS (SUPPLEMENTAL RESTRAINT SYSTEM)

PRECAUTIONS

GENERAL PRECAUTIONS

➡Some systems store data in memory that is lost when the battery is disconnected. Before disconnecting the battery, refer to Battery Terminal Disconnection and Reconnection.

Please read the following precautions carefully before servicing the SRS or the airbags could accidentally deploy and cause damage or injuries.

• Except when doing electrical inspections that requires battery power, always turn the ignition switch to LOCK (0), disconnect the negative cable from the battery, then wait at least 3 minutes before starting work.

NOTE: The SRS memory is not erased even if the ignition switch is turned to LOCK (0) or if the battery cables are disconnected from the battery.

- Use replacement parts which are manufactured to the same standards and quality as the original parts. Do not install used SRS parts. Use only new parts when making SRS repairs.
- Carefully inspect any SRS part before you install it.
Do not install any part that shows signs of being dropped or improperly handled, such as dents, cracks or deformation.
- Before disconnecting the SRS unit connectors, always disconnect the appropriate SRS parts connectors.

• Use only a digital multimeter to check the system. If it is not a Honda multimeter, make sure its output is 0.01 A (10 mA) or less when switched to the lowest value in the ohmmeter range. A tester with a higher output could cause accidental deployment and possible injury.
• Do not put objects on the front passenger's airbag.

STEERING-RELATED PRECAUTIONS

Cable Reel Alignment

✷✷ WARNING

Misalignment of the cable reel could cause an open in the wiring, making the SRS, remote steering wheel controls, or the horn inoperative. Center the cable reel whenever you do the following:

- Installation of the steering wheel.
- Installation of the cable reel.
- Installation of the steering column.
- Other steering-related adjustment or installation.
- Do not disassemble the cable reel.
- Do not apply grease to the cable reel.
- If the cable reel shows any signs of damage, replace it with a new one. For example, if the cable reel does not rotate smoothly, replace it.

Airbag Handling and Storage

✷✷ WARNING

Do not disassemble an airbag. It has no serviceable parts. Once an airbag has been deployed, it cannot be repaired or reused.

For temporary storage of an airbag during service, observe the following precautions:
• Store the removed airbag with the pad surface up. Never put anything on the airbag.
• To prevent damage to the airbag, keep it away from any oil, grease, detergent, or water.
• Store the removed airbag on a secure, flat surface away from any high heat source [exceeding 200 °F(93 °C)].
• Never do electrical tests on the airbags, such as measuring resistance.
• Do not position yourself in front of the airbag during removal, inspection, or replacement.
• For proper disposal of a damaged airbag or tensioner, refer to airbag disposal.

DISARMING THE SYSTEM

Disconnect and isolate the negative battery cable. Wait 3 minutes for the system capacitor to discharge before performing any service.

ARMING THE SYSTEM

To arm the system, connect the negative battery cable.

DRIVE TRAIN

AUTOMATIC TRANSMISSION

DRAIN & REFILL

See Figures 21 and 22.

Always use Honda ATF DW-1 Automatic Transmission Fluid (ATF). Using a non-Honda ATF can affect shift quality.

✷✷ WARNING

Keep all foreign particles out of the transmission.

1. Warm up the engine to normal operating temperature (the radiator fan comes on), and turn the engine off.

2. Park the vehicle on level ground.
3. Remove the ATF filler bolt and the drain plug, and drain the automatic transmission fluid (ATF).
4. Reinstall the drain plug with a new sealing washer.
5. Refill the transmission with the recommended fluid through the filler hole. Always use Honda ATF DW-1 Automatic Transmission Fluid (ATF). Using a non-Honda ATF can affect shift quality.
6. Install the ATF filler bolt and a new sealing washer.
7. Check that the ATF level is between the upper mark and the lower mark of the dipstick.

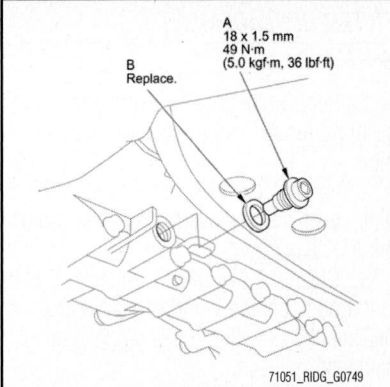

71051_RIDG_G0749

Fig. 21 Remove the ATF filler bolt, the drain plug (A), and the sealing washer (B), and drain the automatic transmission fluid (ATF)

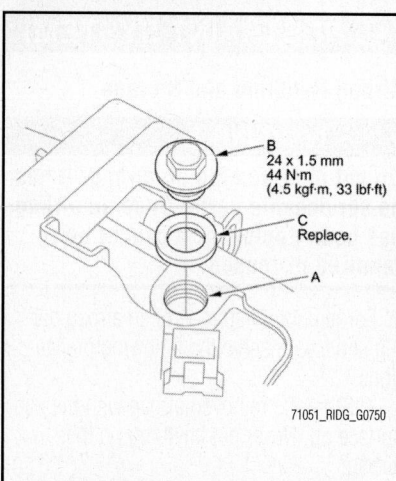

Fig. 22 Refill the transmission with the recommended fluid through the filler hole (A), then install the ATF filler bolt (B) and a new sealing washer (C)

8. If the maintenance minder recommends replacing the ATF, reset the maintenance minder, and this procedure is complete.

➡️**If the maintenance minder did not require you to replace the ATF, go to step 9.**

9. Connect the HDS to the DLC located behind the driver's dashboard lower cover.

10. Turn the ignition switch to ON (II). Make sure the HDS communicates with the PCM. If it does not, go to the DLC circuit troubleshooting.

11. Select BODY ELECTRICAL with the HDS.

12. Select ADJUSTMENT in the GAUGES MENU with the HDS.

13. Select RESET in the MAINTENANCE MINDER with the HDS.

14. Select MAINTENANCE SUB ITEM 3 RESET, and reset the ATF life with the HDS.

FILTER REPLACEMENT

See Figure 23.

1. Remove the three 6.0 mm bolts securing the ATF filter cover and the ATF pipe.

2. Remove the ATF pipe from the ATF filter cover, and remove the ATF filter from the ATF filter cover.

3. Clean the ATF filter, then check that it is in good condition, and is not clogged. Replace the ATF filter if it is clogged or damaged.

4. Install the ATF filter with a new O-ring in the ATF filter cover, and install the ATF pipe in the ATF filter cover, then install them in the transmission housing.

5. Secure the ATF filter cover with the

two bolts, then secure the ATF pipe with the bolt.

FLUID LEVEL CHECK

Always use Honda ATF DW-1 Automatic Transmission Fluid (ATF). Using a non-Honda ATF can affect shift quality.

✳✳ WARNING

Keep all foreign particles out of the transmission.

Note the following:
• Check the fluid level within 60–90 seconds after turning the engine off.
• Higher fluid level may be indicated if the radiator fan comes on twice or more.

1. Park the vehicle on the level ground.

2. Warm up the engine to normal operating temperature (the radiator fan comes on), and turn the engine off.

3. Remove the dipstick (yellow loop) from the dipstick guide tube, and wipe it with a clean cloth.

4. Insert the dipstick back into the tube.

5. Remove the dipstick and check the fluid level. It should be between the upper mark and the lower mark.

6. If the level is below the lower mark, check for fluid leaks at the transmission, the hose and the line joints, and the cooler lines. If a problem is found, fix it before filling the transmission.

➡️**If the vehicle is driven when the ATF level is below the lower mark, one or more of these symptoms may occur:**

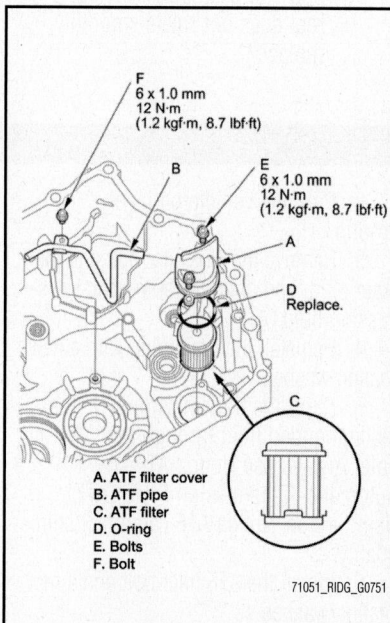

Fig. 23 Exploded view of the ATF filter assembly

• Transmission damage.
• Vehicle does not move in any gear.
• Vehicle accelerates poorly, and flares when starting off in D and R.
• The engine vibrates at idle.

7. If the level is above the upper mark, drain the ATF to the proper level.

➡️**If the vehicle is driven when the ATF level is above the upper mark, the vehicle may creep forward while in N, or have shifting problems.**

8. If necessary fill the transmission with the recommended fluid through the filler hole to bring the fluid level midway between the upper mark and the lower mark of the dipstick. Do not fill past the upper mark. Always use Honda ATF DW-1 Automatic Transmission Fluid (ATF). Using a non-Honda ATF can affect shift quality.

9. Install the ATF filler bolt and a new sealing washer.

10. Check the ATF level.

DRIVESHAFT

REMOVAL & INSTALLATION

Rear

See Figure 24.

1. Raise and support the vehicle.

2. Remove the No. 1 propeller shaft protector.

3. Remove the No. 2 propeller shaft protector.

4. Make reference marks across the No. 1 propeller shaft and the transfer companion flange.

5. Remove the flange bolts.

6. Remove the center support bearing mounting bolts.

7. Make reference marks across the No. 2 propeller shaft and rear differential companion flange.

8. Separate the propeller shaft from the rear differential, then remove the propeller shaft.

To install:

9. Set the No. 2 propeller shaft to the rear differential by aligning the reference marks you made during the removal procedure. Then install new flange bolts and torque to 53 ft. lbs. (72 Nm).

➡️**When replacing the propeller shaft or the rear differential, align the factory reference marks.**

10. Install the center support bearing with new bolts. tighten the bolts to 29 ft. lbs. (39 Nm).

11. Set the No. 1 propeller shaft to the

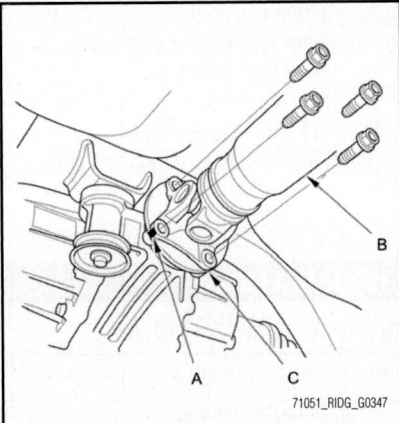

Fig. 24 Make reference marks (A) across the No. 2 propeller shaft (B) and rear differential companion flange (C)

transfer companion flange by aligning the reference marks you made during the removal procedure. Then install new flange bolts and torque to 53 ft. lbs. (72 Nm).

12. Install the No. 2 propeller shaft protector. Tighten the bolts to 16 ft. lbs. (22 Nm).

13. Install the No. 1 propeller shaft protector. Tighten the bolts to 16 ft. lbs. (22 Nm).

14. If you installed a new propeller shaft, test drive the vehicle at 55 mph (88 km/h), and check for noise or vibration. If there is a noise or vibration, rotate the propeller shaft 180 degrees from its current alignment with the rear differential companion flange, then recheck.

FRONT HALFSHAFTS

REMOVAL & INSTALLATION

See Figures 25 through 32.

1. Raise and support the vehicle.
2. Remove the front wheels.
3. Pry up the stake on the spindle nut, then remove the nut.
4. Drain the transmission fluid, then reinstall the drain plug with a new sealing washer.
5. Remove the lock pin from the lower arm ball joint, then remove the castle nut. Separate the ball joint from the knuckle, using the ball joint thread protector and the ball joint puller.

- To avoid damaging the ball joint, install the ball joint thread protector onto the threads of the ball joint. Be careful not to damage the ball joint boot when installing the remover.
- Do not force or hammer on the lower arm, or pry between the

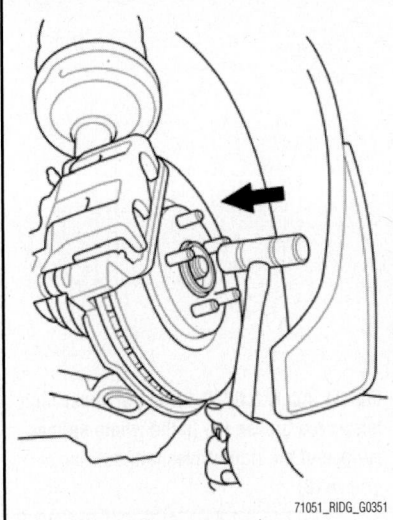

Fig. 25 Pull the knuckle outward, and separate the outboard joint from the front hub using a soft face hammer

lower arm and the knuckle. You could damage the ball joint.

6. Pull the knuckle outward, and separate the outboard joint from the front hub using a soft face hammer.

7. Left driveshaft: Pry the inboard joint from the differential using a prybar. Remove the driveshaft as an assembly.

- Do not pull on the driveshaft, or the inboard joint may come apart. Pull the inboard joint straight out to avoid damaging the oil seal.
- Be careful not to damage the oil seal or the end of the inboard joint with the prybar.

8. Right driveshaft: Drive the inboard joint off of the intermediate shaft using a drift punch and a hammer. Remove the driveshaft as an assembly.

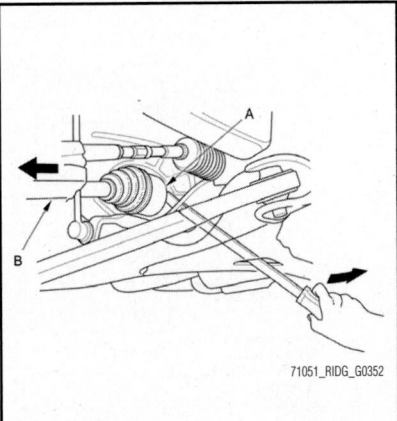

Fig. 26 Left driveshaft: Pry the inboard joint (A) from the differential using a prybar; driveshaft (B)

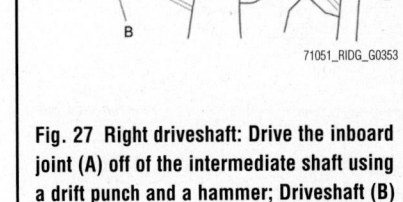

Fig. 27 Right driveshaft: Drive the inboard joint (A) off of the intermediate shaft using a drift punch and a hammer; Driveshaft (B)

- Do not pull on the driveshaft, or the inboard joint may come apart.
- Be careful not to damage the end of the inboard joint with the drift.

9. Remove the set ring from the left driveshaft inboard joint.

10. Remove the set ring from the intermediate shaft.

To install:

➡Before starting installation, make sure the mating surfaces of the joint and the splined section are clean.

11. Apply about 0.18 oz. Moly 60 paste, (P/N 08734-0001) to the contact area of the outboard joint and the front wheel bearing.

➡The paste helps to prevent noise and vibration.

12. Install a new set ring into the set ring groove of the left driveshaft inboard joint.

13. Install a new set ring into the set ring groove of the intermediate shaft.

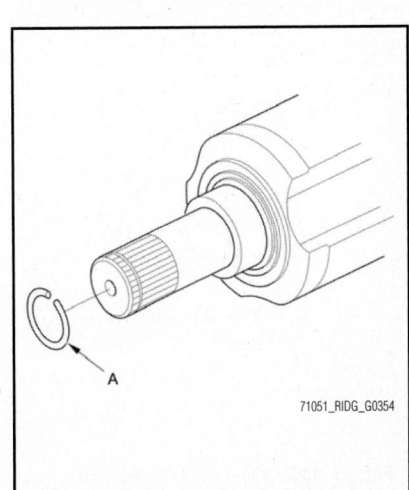

Fig. 28 Remove the set ring (A) from the left driveshaft inboard joint

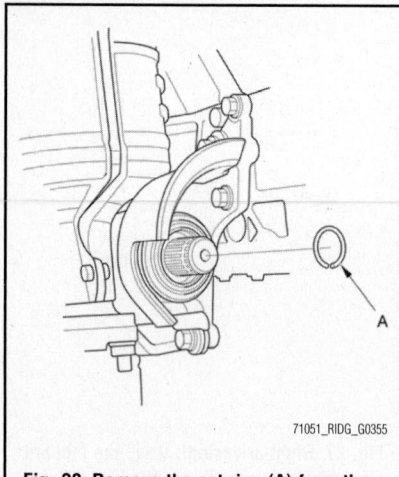

Fig. 29 Remove the set ring (A) from the intermediate shaft

14. Apply 0.07–0.11 oz.) of super high temp urea grease (P/N 08798-9002) to the whole splined surface of the right driveshaft. After applying grease, remove the grease from the splined grooves at intervals of 2–3 splines and from the set ring groove so that air can bleed from the intermediate shaft.

15. Clean the areas where the driveshaft contacts the differential thoroughly with solvent, and dry them with compressed air.

✳✳ CAUTION

Do not wash the rubber parts with solvent.

16. Insert the inboard end of the driveshaft into the differential or intermediate shaft until the set ring locks in the groove.

➡ **Insert the driveshaft horizontally to prevent damaging the oil seal.**

17. Install the outboard driveshaft CV joint end into the front hub on the knuckle.

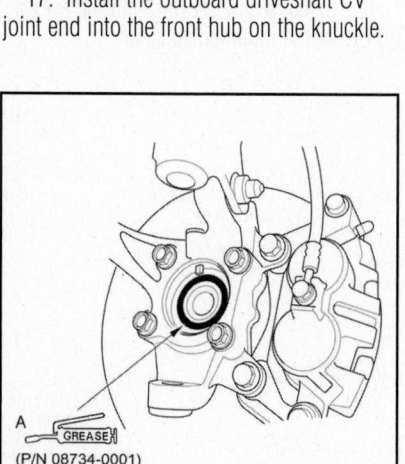

Fig. 30 Apply about 0.18 oz. Moly 60 paste (A), to the contact area of the outboard joint and the front wheel bearing

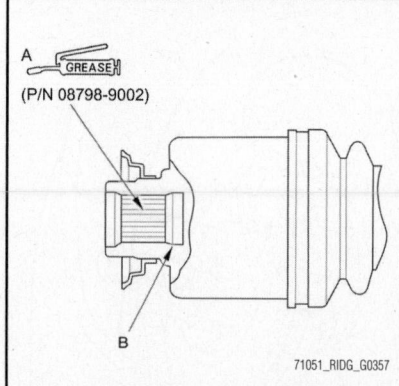

Fig. 31 Apply 0.07–0.11 oz.) of super high temp urea grease (A) to the whole splined surface of the right driveshaft; set ring groove (B)

18. Wipe off any grease contamination from the ball joint tapered section and threads, then install the knuckle onto the lower arm. Be careful not to damage the ball joint boot.

19. Install a new castle nut and torque the castle nut to the lower torque specification, then tighten it only far enough to align the slot with the ball joint pin hole. Tightening range is 65–72 ft. lbs. (88–98 Nm). Install the lock pin into the pin hole.

➡ **Do not align the nut by loosening it.**

20. Apply a small amount of engine oil to the seating surface of a new spindle nut.

21. Install the spindle nut, then tighten it to 242 ft. lbs. (328 Nm). After tightening, use a drift to stake the spindle nut shoulder against the driveshaft.

22. Clean the mating surfaces of the brake disc and the wheel, then install the front wheels.

23. Turn the front wheel by hand, and

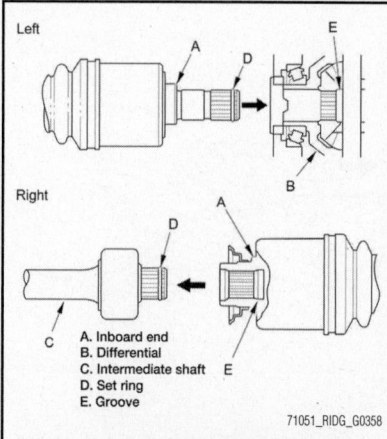

A. Inboard end
B. Differential
C. Intermediate shaft
D. Set ring
E. Groove

Fig. 32 Insert the inboard end of the driveshaft into the differential or intermediate shaft until the set ring locks in the groove

make sure there is no interference between the driveshaft and surrounding parts.

24. Lower the vehicle.

25. Refill the transmission with the recommended transmission fluid.

26. Check the wheel alignment, and adjust it if necessary.

27. Test-drive the vehicle.

REAR HALFSHAFTS

REMOVAL & INSTALLATION

See Figures 33 through 35.

Special Tools Required:
• Ball Joint Thread Protector, 12 mm 07AAF-SDAA100
• Driveshaft Remover 07AAD-S9VA000
• Ball Joint Remover, 28 mm 07MAC-SL0A202

✳✳ WARNING

Be careful not to damage the brake hose, sensor, and harness.

1. Raise and support the vehicle.
2. Remove the rear wheels.
3. Pry up the stake on the spindle nut, then remove the nut.
4. Remove the rear wheel speed sensor.
5. Remove the lock pin from the upper arm ball joint castle nut, and remove the nut.
6. Separate the ball joint from the upper arm using the 12 mm ball joint thread protector and the 28 mm ball joint remover.
7. Remove the flange nut, the washer, and the flange bolt, then remove the lower arm A.
8. Remove the flange bolt and separate the knuckle from the lower arm B.
9. Pull the knuckle outward, and separate the outboard joint from the rear hub using a soft face hammer.
10. Use the driveshaft remover, pry out the inboard joint from the differential.

➡ **This is a prying tool; do not strike it with a hammer.**

11. Remove the rear driveshaft.

➡ **When removing the outboard joint, the knuckle is supported with a floor jack. Make sure not to over extend the brake hose.**

12. Remove the set ring from the rear differential.

To install:

➡ **Before starting installation, make sure the mating surfaces of the joint and the splined section are clean.**

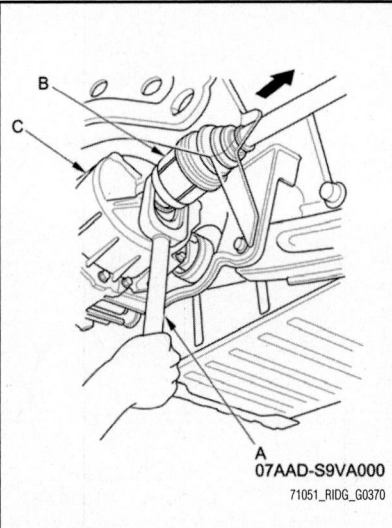

Fig. 33 Use the driveshaft remover (A), pry out the inboard joint (B) from the differential (C)

13. Apply super high temp urea grease (P/N 08798-9002) to the whole splined surface (A). After applying grease, remove the grease from the splined grooves at intervals of 2–3 splines and from the set ring groove (B) so that air can bleed from the differential.

14. Install a new set ring into the set ring groove of the rear differential.

15. Clean the areas where the driveshaft contacts the differential thoroughly with solvent, and dry them with compressed air.

➡**Do not wash the rubber parts with solvent.**

16. Install the outboard joint into the rear hub.

➡**When installing the outboard joint support, the knuckle with a floor jack.**

Make sure not to over extend the brake hose.

17. Make sure the inboard joint is installed all the way into the rear differential and to ensure the stop ring is properly seated.

18. Loosely install a new flange bolt onto lower arm B.

19. Install the lower arm A, then loosely install a new flange bolt and a new flange nut with the washer.

20. Wipe off any grease contamination from the ball joint tapered section and threads, then install the upper arm. Be careful not to damage the ball joint boot. Loosely install the castle nut.

➡**Make sure the ball joint boot is not damaged or cracked.**

21. Torque the castle nut to the lower torque specification [36–43 ft. lbs. (49–59 Nm)], then tighten it only far enough to align the slot with the ball joint pin hole and install the lock pin hole as shown.

➡**Do not align the nut by loosening it.**

22. Install the rear wheel speed sensor.
23. Place a floor jack under the lower arm B, and raise the suspension to load it with the vehicle's weight.

➡**Do not put the floor jack under the ball joint.**

24. Tighten the two of flange bolts and the flange nut to the specified torque values, then remove the floor jack.
   a. Tighten the flange bolt on lower B to 112 ft. lbs. (152 Nm).
   b. Tighten the flange bolt on lower arm A to 69 ft. lbs. (93 Nm).
   c. Tighten the flange nut on lower arm A to 75 ft. lbs. (101 Nm).

25. Apply a small amount of engine oil to the seating surface of a new spindle nut.
26. Install the spindle nut, then tighten it. After tightening, use a drift to stake the spindle nut shoulder against the driveshaft.
27. Clean the mating surfaces of the brake disc and the wheel, then install the rear wheels.
28. Turn the rear wheel by hand, and make sure there is no interference between the driveshaft and surrounding parts.
29. Lower the vehicle.
30. Check the wheel alignment, and adjust it if necessary.
31. Test-drive the vehicle.

TRANSFER CASE

REMOVAL & INSTALLATION
See Figures 36 through 38.

1. Raise the vehicle on a lift, and make sure it is supported securely.
2. Shift the shift lever into N.
3. Remove the drain plug, and drain the automatic transmission fluid (ATF).
4. Reinstall the drain plug with a new sealing washer.
5. Remove the front subframe stiffener (A).
6. Remove exhaust pipe A and its mount.
7. Remove the bolt securing the transfer breather hose bracket, and disconnect the breather hose from the breather pipe on the transfer assembly. Cap the breather pipe to prevent fluid from leaking out.
8. Make a reference mark across the propeller shaft and the transfer companion flange.
9. Separate the propeller shaft from the transfer companion flange.

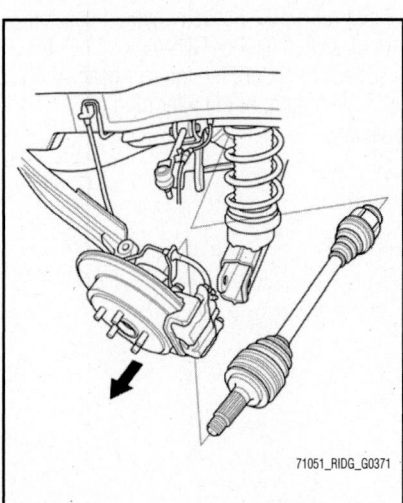

Fig. 34 Remove the rear driveshaft

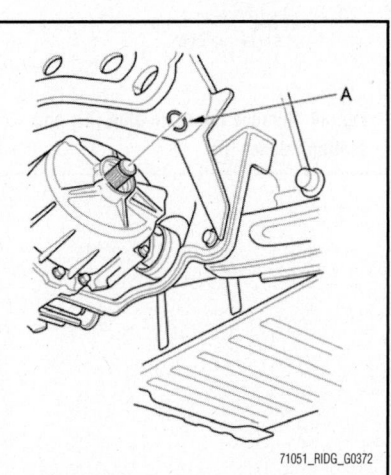

Fig. 35 Remove the set ring (A) from the rear differential

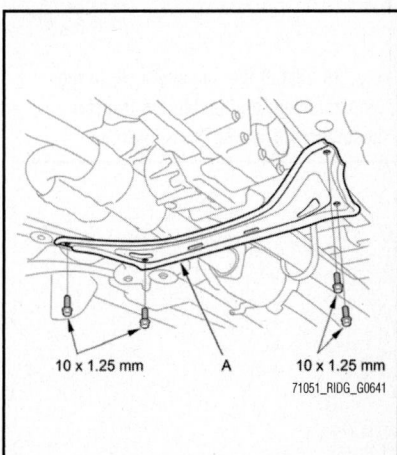

Fig. 36 Remove the front subframe stiffener (A)

10. Remove the transfer assembly from the transmission.

To install:

11. Clean the areas where the transfer

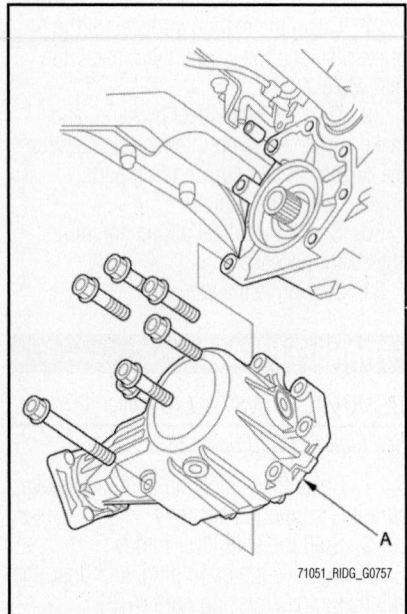

Fig. 37 Remove the transfer assembly (A) from the transmission

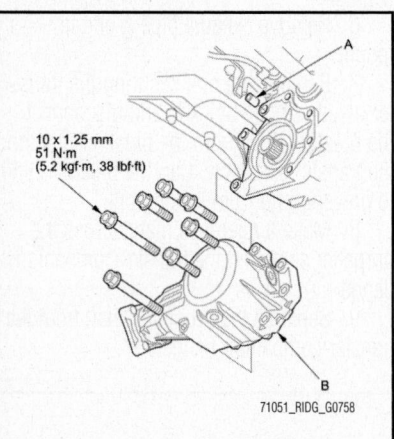

Fig. 38 Install the dowel pin (A) in the transmission, and install the transfer assembly (B) on the transmission

assembly contacts the transmission with solvent, and dry with compressed air. Then apply transmission fluid to the seal contact area.

12. Install the dowel pin in the transmission, and install the transfer assembly on the transmission.

13. Install the propeller shaft to the transfer companion flange by aligning the reference mark. Tighten the bolts to 53 ft. lbs. (72 Nm).

14. Secure the transfer breather hose bracket on the transfer assembly with the bolt, and install the breather hose over the breather pipe with the dot on the hose facing out.

15. Install exhaust pipe A with new self-locking nuts, its mount, and new gaskets.

16. Install the front subframe stiffener with new mounting bolts. Tighten the bolts to 40 ft. lbs. (54 Nm).

17. Refill the transfer assembly with transfer fluid (hypoid gear oil), if necessary.

18. Refill the transmission with ATF.

DRAIN & REFILL

See Figures 39 and 40.

Use a SAE 90 viscosity hypoid gear oil, API classified GL4 or GL5 only.

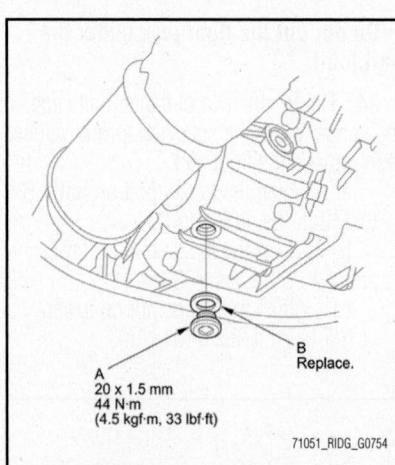

Fig. 39 Remove the drain plug (A), and sealing washer (B)

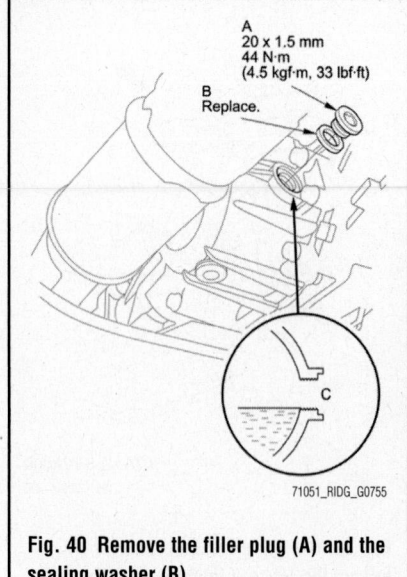

Fig. 40 Remove the filler plug (A) and the sealing washer (B)

1. Park the vehicle on level ground.

2. Warm up the engine to normal operating temperature (the radiator fan comes on), then turn the engine off.

3. Remove the drain plug, and drain the transfer fluid (hypoid gear oil).

4. Reinstall the drain plug with a new sealing washer.

5. Remove the filler plug and the sealing washer.

6. Refill the transfer assembly with the recommended fluid (hypoid gear oil) through the filler hole (C) until the fluid flows out. Use a SAE 90 viscosity hypoid gear oil, API classified GL4 or GL5 only.

7. Install the filler plug and a new sealing washer.

FLUID LEVEL CHECK

See Figure 70.

Use a SAE 90 viscosity hypoid gear oil, API classified GL4 or GL5 only.

Check to ensure the transfer gear oil is to the bottom of the filler plug opening.

ENGINE COOLING

ENGINE COOLANT

DRAIN & REFILL

See Figures 41 and 42.

1. Wait until the engine is cool, then carefully remove the radiator cap.
2. Remove the front splash shield.
   a. 2012 models: Remove the front strake.
   b. From under the front bumper, remove the bolts, then turn down the bumper, and remove the clip.
   c. Remove the clips that secure the front inner fender and the splash shield to the body.
   d. Pull out the splash shield.
3. Loosen the drain plug, and drain the coolant.
4. Install a rubber hose on the drain bolt located at the rear of the engine block, then loosen the drain bolt.
5. When the coolant stops draining, tighten the drain bolt. Remove the rubber hose.
6. Tighten the radiator drain plug securely.
7. Install the front splash shield.
8. Remove, drain, and reinstall the coolant reservoir.
9. Fill the coolant reservoir to the MAX mark with Honda Long Life Antifreeze/Coolant Type 2.
10. Pour Honda Long Life Antifreeze/Coolant Type 2 into the radiator up to the base of the filler neck.
    - Always use Honda Long Life Antifreeze/Coolant Type 2. Using a non-Honda coolant can result in corrosion, causing the cooling system to malfunction or fail.
    - Honda Long Life Antifreeze/Coolant Type 2 is a mixture of 50% antifreeze and 50% water. Do not add water.
11. Loosely install the radiator cap.
12. Start the engine and let it run until it warms up (the radiator fan comes on at least twice).
13. Turn off the engine. Check the level in the radiator and add Honda Long Life Antifreeze/Coolant Type 2, if needed.
14. Put the radiator cap on securely, then run the engine again, and check for leaks.
15. Clean up any spilled engine coolant.
16. If the maintenance minder required engine coolant replacement, reset the maintenance minder, and this procedure is complete.

➡**If the maintenance minder did not require engine coolant replacement, go to step 17.**

17. Turn the ignition switch to LOCK (0).
18. Connect the HDS to the DLC.
19. Turn the ignition switch to ON (II).
20. Make sure the HDS communicates with the vehicle and the PCM. If it does not communicate, troubleshoot the DLC circuit.
21. Select GAUGES in the BODY ELECTRICAL with the HDS.
22. Select ADJUSTMENT in the GAUGES with the HDS.
23. Select MAINTENANCE MINDER in the ADJUSTMENT with the HDS.
24. Select RESET in the MAINTENANCE MINDER with the HDS.
25. Select MAINTENANCE SUB ITEM 5 RESET with the HDS.

ELECTRIC ENGINE FAN

REMOVAL & INSTALLATION

See Figure 43.

1. Drain the engine coolant.
2. Remove the front bulkhead cover.
3. Disconnect the fan motor connectors and the engine coolant temperature sensor 2 connector, then remove the harness connector, the harness clamps, and the coolant reservoir hose.
4. Remove the upper radiator hose and the lower radiator hose.
5. Raise the vehicle on the lift to full height.
6. Remove the splash shield.
7. Remove the automatic transmission fluid (ATF) cooler hoses from the radiator, then plug the line and the hose.
8. Unclamp the ATF cooler hose clamp on the ATF cooler line.
9. Lower the vehicle on the lift.
10. Remove the upper bracket and cushions, then pull up the radiator.
11. Remove the fan shroud assemblies and other parts from the radiator.

To install:

12. Install the radiator in the reverse order of removal.
13. Make sure the upper and lower cushions are set securely.
14. Fill the radiator with engine coolant, then bleed the air from the cooling system.

RADIATOR

REMOVAL & INSTALLATION

See Figure 43.

1. Drain the engine coolant.
2. Remove the front bulkhead cover.
3. Disconnect the fan motor connectors and the engine coolant temperature sensor 2 connector, then remove the harness connector, the harness clamps, and the coolant reservoir hose.
4. Remove the upper radiator hose and the lower radiator hose.
5. Raise the vehicle on the lift to full height.
6. Remove the splash shield.
7. Remove the automatic transmission fluid (ATF) cooler hoses from the radiator, then plug the line and the hose.
8. Unclamp the ATF cooler hose clamp on the ATF cooler line.

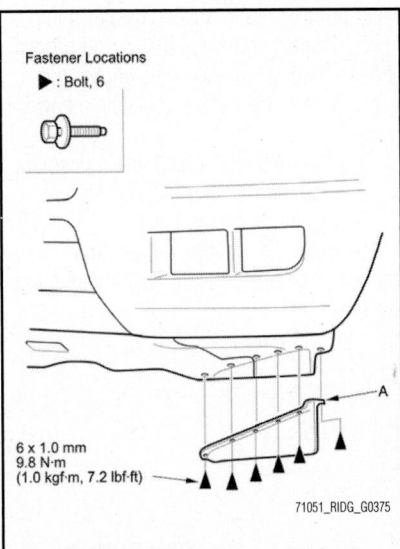

Fastener Locations
▶ : Bolt, 6

6 x 1.0 mm
9.8 N·m
(1.0 kgf·m, 7.2 lbf·ft)

71051_RIDG_G0375

Fig. 41 2012 models: Remove the front strake (A)

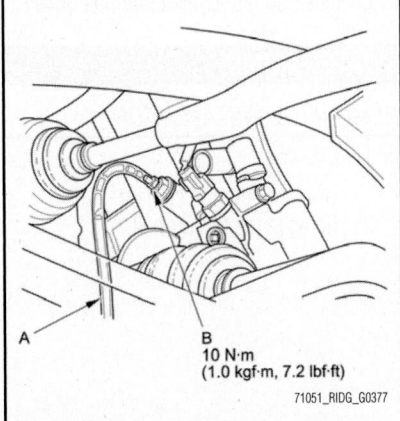

B
10 N·m
(1.0 kgf·m, 7.2 lbf·ft)

71051_RIDG_G0377

Fig. 42 Install a rubber hose (A) on the drain bolt (B) located at the rear of the engine block, then loosen the drain bolt

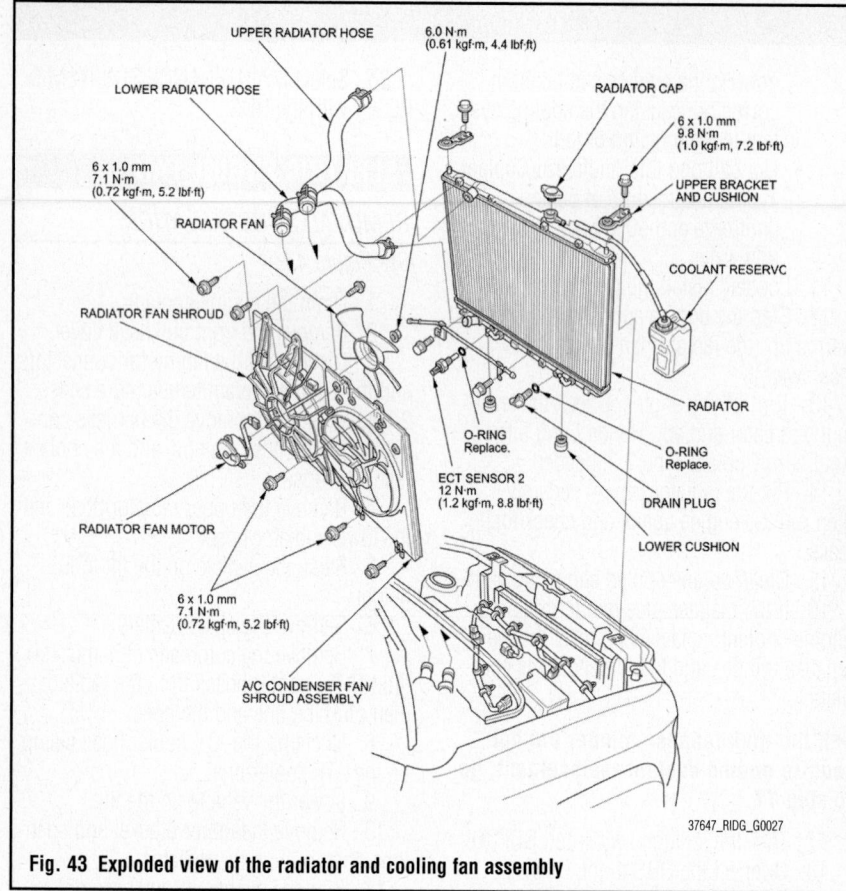

Fig. 43 Exploded view of the radiator and cooling fan assembly

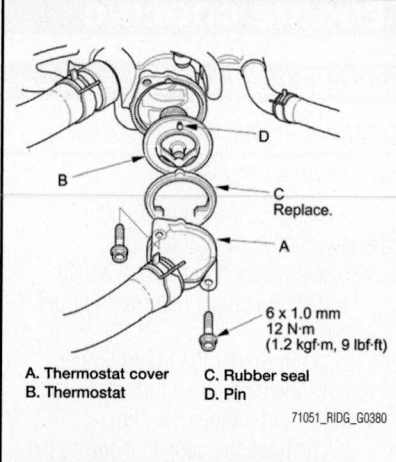

A. Thermostat cover   C. Rubber seal
B. Thermostat   D. Pin

71051_RIDG_G0380

Fig. 44 Exploded view of the thermostat assembly

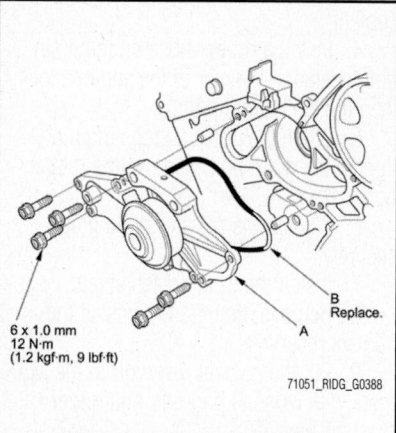

71051_RIDG_G0388

Fig. 45 Remove the five bolts securing the water pump (A), then remove the water pump and the O-ring (B)

9. Lower the vehicle on the lift.

10. Remove the upper bracket and cushions, then pull up the radiator.

11. Remove the fan shroud assemblies and other parts from the radiator.

To install:

12. Install the radiator in the reverse order of removal.

13. Make sure the upper and lower cushions are set securely.

14. Fill the radiator with engine coolant, then bleed the air from the cooling system.

THERMOSTAT

REMOVAL & INSTALLATION
See Figure 44.

1. Do the battery removal procedure.
2. Drain the engine coolant.

3. Remove the thermostat cover, then remove the thermostat.

To install:

4. Install the new thermostat with a new rubber seal, then install the thermostat cover.

➡**Install the thermostat with the pin as shown.**

5. Do the battery installation procedure.
6. Refill the radiator with engine coolant, and bleed the air from the cooling system.
7. Clean up any spilled engine coolant.

WATER PUMP

REMOVAL & INSTALLATION
See Figure 45.

1. Drain the engine coolant.
2. Remove the timing belt.

3. Remove the timing belt adjuster.
4. Remove the five bolts securing the water pump, then remove the water pump.
5. Inspect and clean the O-ring groove and the mating surface of the engine block.
6. Install the water pump with a new O-ring.
7. Clean up any spilled engine coolant.
8. Install the timing belt adjuster.
9. Install the timing belt.
10. Refill the radiator with engine coolant, then bleed the air from the cooling system.

ENGINE ELECTRICAL

BATTERY

REMOVAL & INSTALLATION

See Figure 46.

➡ **The battery terminal disconnection/ reconnection procedure must be done before and after doing this procedure. Some systems store data in memory that is lost when the battery is disconnected.**

1. Do the battery terminal disconnect procedure, then wait at least 3 minutes before beginning work.
2. Remove the two nuts securing the battery setting plate, then remove the battery setting plate and the battery.

To install:

3. Install the battery, then install the battery setting plate.
4. Tighten the two nuts equally until the battery is stable.

➡ **Do not deform the battery setting plate by tightening the nuts too much.**

5. Do the battery terminal reconnection procedure.

➡ **Make sure the battery is installed correctly, and the positive terminal and the negative terminal are not reversed.**

BATTERY RECONNECT/RELEARN PROCEDURE

Disconnection

➡ **Some systems store data in memory that is lost when the battery is discon-**

nected. **Do the following steps before disconnecting the battery.**

1. Make sure you have the anti-theft code(s) for the audio and/or the navigation system (if equipped).
2. If you are replacing the audio unit, write down the audio presets (AM and FM), and the XM radio presets (if equipped), because the audio unit does not retain the preset after the battery is disconnected.
3. Make sure the ignition switch is in LOCK (0).
4. Disconnect and isolate the negative cable from the battery.

✳✳ CAUTION

Always disconnect the negative cable from the battery first.

5. Disconnect the positive cable from the battery.

Reconnection

1. Clean the battery terminals.
2. Test the battery.
3. Reconnect the positive cable to the battery first, then reconnect the negative cable to the battery.

✳✳ CAUTION

Always connect the positive cable to the battery first.

4. Apply multipurpose grease to the terminals to prevent corrosion.
5. Enter the anti-theft code(s) for the audio system and/or the navigation system (if equipped).

6. Enter the audio presets (if applicable), and enter the XM radio presets (if equipped).
7. Set the clock (for vehicles without navigation).

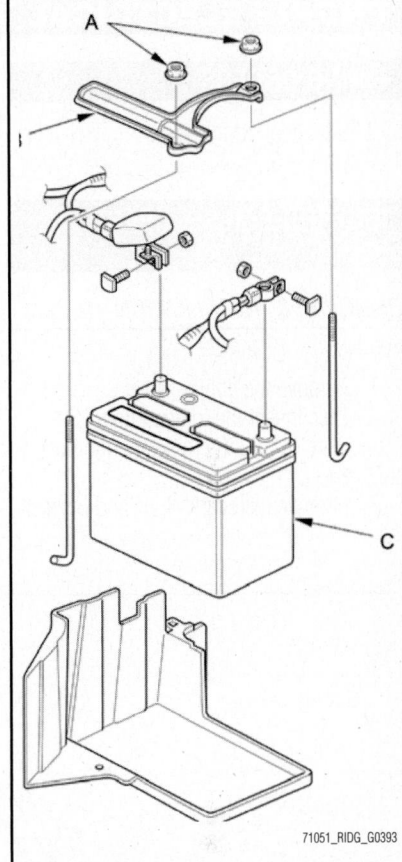

71051_RIDG_G0393

Fig. 46 Remove the two nuts (A) securing the battery setting plate, then remove the battery setting plate (B) and the battery (C)

ENGINE ELECTRICAL

ALTERNATOR

REMOVAL & INSTALLATION

See Figures 47 and 48.

1. Do the battery terminal disconnection procedure.
2. Remove the engine cover.
3. Remove the drive belt.
4. Disconnect the alternator connector, and the positive alternator cable from the alternator.
5. Remove the bolt securing the harness bracket.
6. Remove the mounting bolt and the alternator bracket mounting bolt, then remove the alternator.

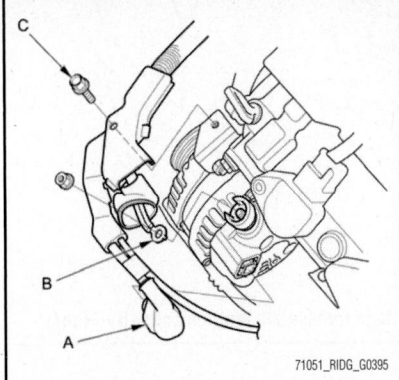

71051_RIDG_G0395

Fig. 47 Disconnect the alternator connector (A), and the positive alternator cable (B) from the alternator, then remove the bolt (C) securing the harness bracket

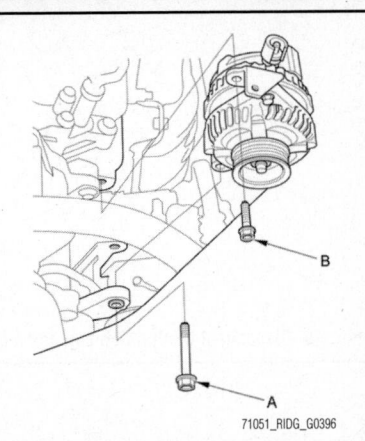

71051_RIDG_G0396

Fig. 48 Remove the mounting bolt (A) and the alternator bracket mounting bolt (B), then remove the alternator

To install:

7. Install the alternator, then tighten the mounting bolt to 33 ft. lbs. (45 Nm), and the alternator bracket mounting bolt to 16 ft. lbs. (22 Nm).

8. Connect the alternator connector and the positive alternator cable to the alternator. Make sure the crimped side of the ring terminal faces away from the alternator when you connect it.

9. Install the bolt securing the harness bracket.

10. Install the drive belt.

11. Install the engine cover.

12. Do the battery terminal reconnection procedure.

ENGINE ELECTRICAL

IGNITION SYSTEM

FIRING ORDERS

3.5L Engine Firing Order: 1–4–2–5–3–6.

IGNITION COILS & SPARK PLUGS

REMOVAL & INSTALLATION

See Figures 49 and 50.

1. Remove the engine cover.
2. Disconnect the ignition coil connectors (A), then remove the ignition coils (B).
3. Remove the spark plugs and inspect them.

4. Apply a small amount of anti-seize compound to the plug threads, and screw the plugs into the cylinder head, finger tight. Torque them to 16 ft. lbs. (22 Nm).

5. Install the ignition coils in the reverse order of removal.

IGNITION TIMING

INSPECTION & ADJUSTMENT

Ignition timing is controlled by the Powertrain Control Module (PCM). No adjustment is necessary or possible. Ignition timing is set at 10+/- 2° BTDC.

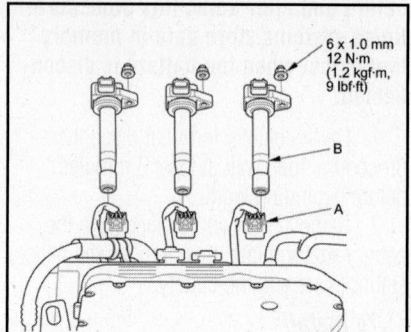

71051_RIDG_G0402

Fig. 50 Disconnect the ignition coil connectors (A), then remove the ignition coils (B)—Rear

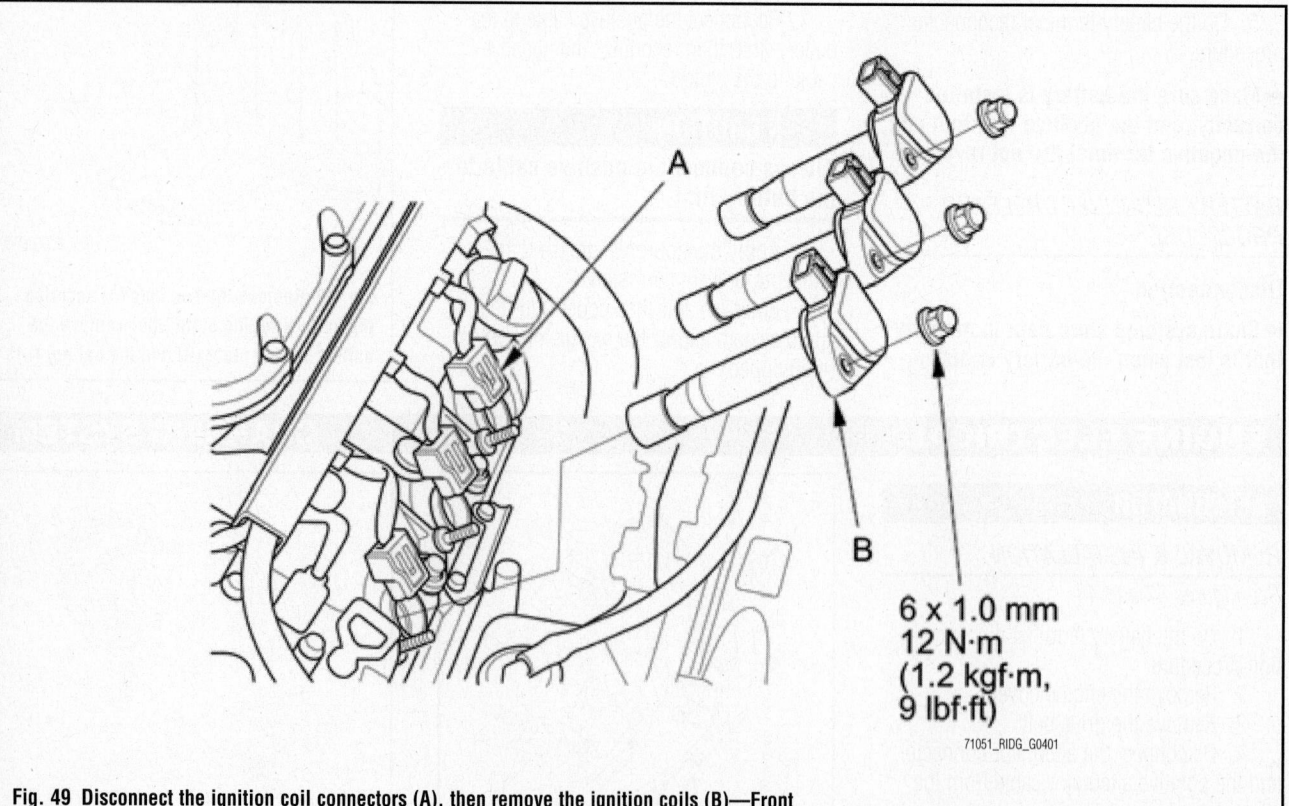

6 x 1.0 mm
12 N·m
(1.2 kgf·m,
9 lbf·ft)

71051_RIDG_G0401

Fig. 49 Disconnect the ignition coil connectors (A), then remove the ignition coils (B)—Front

ENGINE ELECTRICAL   **STARTING SYSTEM**

STARTER

REMOVAL & INSTALLATION

See Figure 51.

1. Do the battery terminal disconnection procedure.
2. Remove the ATF dipstick and the radiator hose bracket.
3. Remove the harness clamp.
4. Disconnect the positive starter cable and the S terminal connector.
5. Remove the two bolts holding the starter, then remove the starter.

To install:
6. Install the starter, then tighten

the mounting bolts to 33 ft. lbs. (44 Nm).

➡**Always use a new gasket.**

7. Install the radiator hose bracket and the ATF dipstick.
8. Connect the positive starter cable and the S terminal connector. Make sure the crimped side of the ring terminal faces away from the starter when you connect it.
9. Install the harness clamp.
10. Do the battery terminal reconnection procedure.
11. Start the engine to make sure the starter works properly.

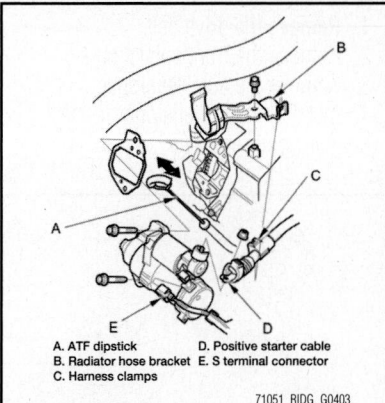

A. ATF dipstick
B. Radiator hose bracket
C. Harness clamps
D. Positive starter cable
E. S terminal connector

71051_RIDG_G0403

Fig. 51 Exploded view of the starter assembly

ENGINE MECHANICAL

ACCESSORY DRIVE BELT SYSTEM

ADJUSTMENT

Tension for the serpentine accessory drive belt is maintained by the belt tensioner. No adjustment is necessary.

BELT ROUTINGS

Refer to the graphic in the Removal and Installation section for the proper routing of the drive belt.

INSPECTION

See Figure 52.

1. Inspect the drive belt for cracks or damage. If the drive belt is cracked or damaged, replace the drive belt.
2. Check that the position of the auto-

tensioner indicator is within the standard range as shown. If it is out of the standard range, replace the drive belt.

REMOVAL & INSTALLATION

Drive Belt

See Figure 53.

Special Tools Required*: Belt Tension Release Tool Snap-on® YA9317, commercially available (*Available through the Honda Tool and Equipment Program, 888-424-6857).

1. Move the auto-tensioner using the belt tension release tool (YA9317) in the direction of the rotation arrow to relieve ten-

sion from the drive belt, then remove the drive belt.
2. Install the new belt in the reverse order of removal.

Drive Belt Tensioner Pulley

See Figure 54.

1. Remove the auto-tensioner.
2. Remove the pulley bolt (left-hand threads), and remove the tensioner pulley.
3. Install the tensioner pulley in the reverse order of removal.

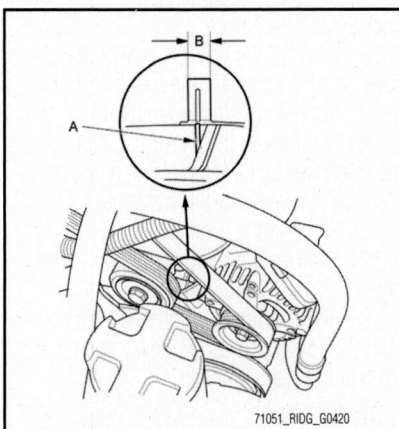

71051_RIDG_G0420

Fig. 52 Check that the position of the auto-tensioner indicator (A) is within the standard range (B) as shown

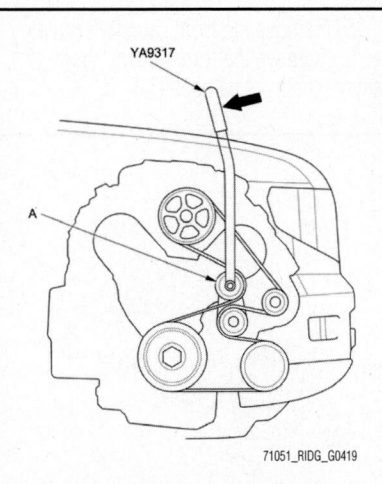

YA9317

71051_RIDG_G0419

Fig. 53 Move the auto-tensioner (A) using the belt tension release tool in the direction of the rotation arrow to relieve tension from the drive belt, then remove the drive belt

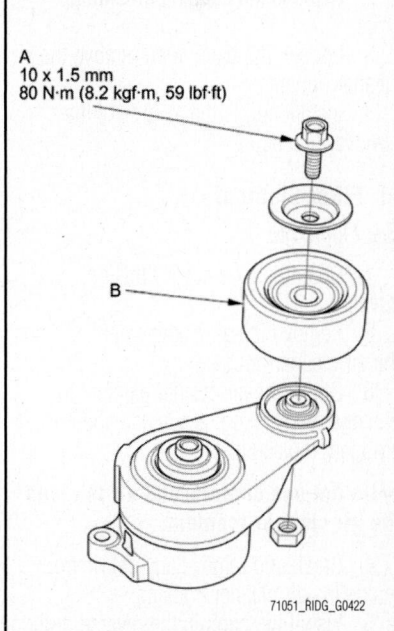

A
10 x 1.5 mm
80 N·m (8.2 kgf·m, 59 lbf·ft)

B

71051_RIDG_G0422

Fig. 54 Remove the pulley bolt (A) (left-hand threads), and remove the tensioner pulley (B)

Drive Belt Tensioner

See Figure 55.

1. Remove the drive belt.
2. Remove the front splash shield.
3. Remove the auto-tensioner.
4. Install the auto-tensioner in the reverse order of removal.

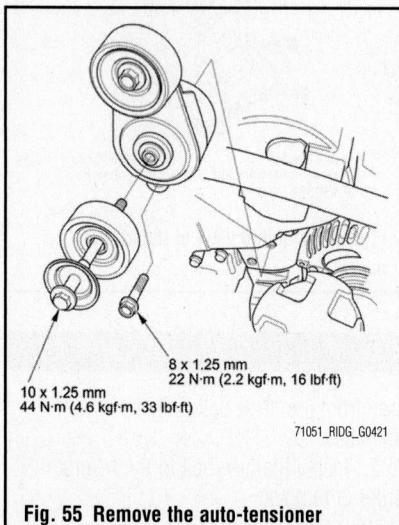

8 x 1.25 mm
22 N·m (2.2 kgf·m, 16 lbf·ft)

10 x 1.25 mm
44 N·m (4.6 kgf·m, 33 lbf·ft)

71051_RIDG_G0421

Fig. 55 Remove the auto-tensioner

AIR CLEANER

REMOVAL & INSTALLATION

Air Cleaner Assembly

1. Disconnect the MAF sensor/IAT sensor 5P connector.
2. Remove the housing mounting bolts.
3. Loosen the band, then remove the air cleaner housing.
4. Installation is the reverse of the removal procedure.

Air Filter Element

See Figure 56.

1. Open the air cleaner housing cover.
2. Remove the air cleaner element from the air cleaner housing.
3. Check the air cleaner element for damage or clogging. If it is damaged or clogged, replace it.

➡ **Do not use compressed air to clean the air cleaner element.**

4. Clean and remove any debris from inside the air cleaner housing.
5. Install the parts in the reverse order of removal.

  a. If you did not replace the air cleaner element, this procedure is complete.

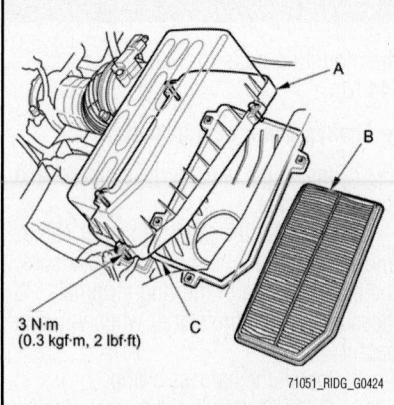

3 N·m
(0.3 kgf·m, 2 lbf·ft)

71051_RIDG_G0424

Fig. 56 Open the air cleaner housing cover, remove the air cleaner element from the air cleaner housing

  b. If the maintenance minder required air cleaner element replacement, reset the maintenance minder.

  c. If the idle speed fluctuates, do the idle speed inspection.

CAMSHAFT & BEARINGS

REMOVAL & INSTALLATION

Front

See Figures 57 and 58.

1. Do the battery removal procedure.
2. Drain the engine coolant.
3. Disconnect the radiator hoses.
4. Remove the EGR valve.
5. Remove the EGR valve stud bolts.
6. Remove the timing belt.
7. Remove the front rocker arm assembly.
8. Remove the front camshaft pulley.
9. Remove the thrust cover, then remove the front camshaft.

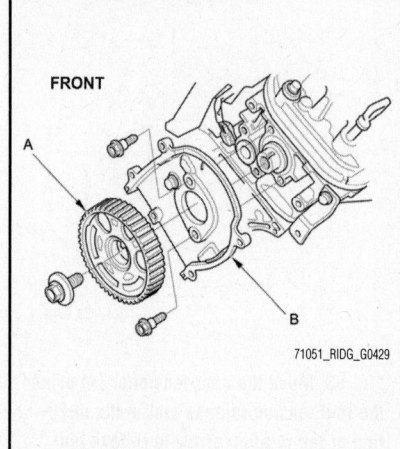

FRONT

71051_RIDG_G0429

Fig. 57 Remove the front camshaft pulley (A) and back cover (B)

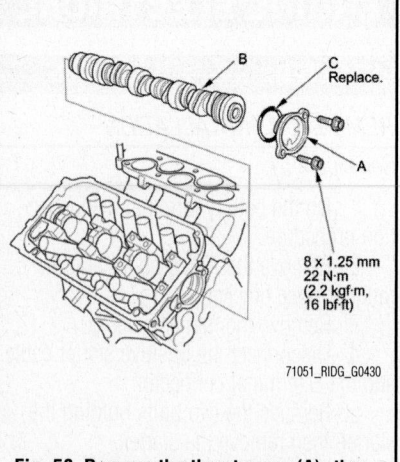

8 x 1.25 mm
22 N·m
(2.2 kgf·m,
16 lbf·ft)

71051_RIDG_G0430

Fig. 58 Remove the thrust cover (A), then remove the front camshaft (B); O-ring (C)

To install:

10. Install the front camshaft in the reverse order of removal. Always use a new O-ring. Apply new engine oil to the journals and the cam lobes.
11. Apply new engine oil to the threads of the camshaft pulley mounting bolt, then install the front camshaft pulley.
12. Install the front rocker arm assembly, then tighten the mounting bolts.
13. Install the timing belt.
14. Adjust the valve clearance.
15. Install the EGR valve stud bolts, then install the EGR valve.
16. Connect the radiator hoses (A).
17. Do the battery installation procedure.
18. Fill the radiator with engine coolant, and bleed the air from the cooling system.
19. Do the CKP pattern clear/CKP pattern learn procedure.

Rear

See Figures 59 and 60.

1. Relieve the fuel pressure. Refer to the Fuel System section.
2. Remove the quick-connect fitting cover, then disconnect the fuel feed hose.
3. Remove the intake manifold.
4. Disconnect the EVAP canister hose, then remove the EVAP canister purge joint.
5. Remove the water passage.
6. Remove the timing belt.
7. Remove the rear rocker arm assembly.
8. Remove the rear camshaft pulley.
9. Remove the thrust cover, then remove the rear camshaft.

To install:

10. Install the rear camshaft in the reverse order of removal. Always use a new O-ring (C). Apply new engine oil to the journals and the cam lobes.

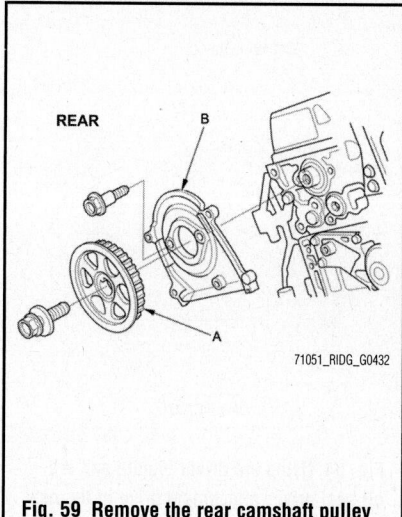

Fig. 59 Remove the rear camshaft pulley

11. Apply new engine oil to the threads of the camshaft pulley mounting bolt, then install the rear camshaft pulley.

12. Install the rear rocker arm assembly, then tighten the mounting bolts.

13. Install the timing belt.

14. Adjust the valve clearance.

15. Install the EVAP canister purge joint (A), then connect the EVAP canister hose (B).

16. Connect the fuel feed hose, then install the quick-connect fitting cover.

17. Inspect for fuel leaks. Turn the ignition switch to ON (II) (do not operate the starter) so the fuel pump runs for about 2 seconds and pressurizes the fuel line. Repeat this operation three times, then check for fuel leakage at any point in the fuel line.

18. Fill the radiator with engine coolant, and bleed the air from the cooling system.

19. Do the CKP pattern clear/CKP pattern learn procedure.

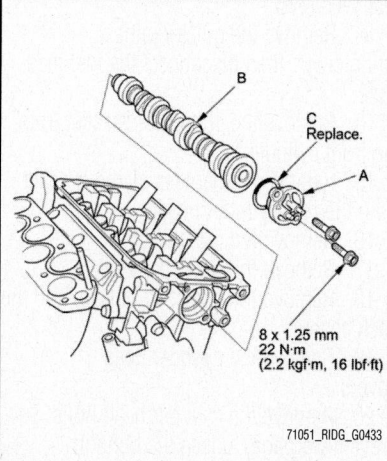

Fig. 60 Remove the thrust cover (A), then remove the rear camshaft (B); O-ring (C)

CRANKSHAFT FRONT SEAL

REMOVAL & INSTALLATION

See Figure 61.

Special Tools Required: Oil Seal Driver, 64 mm 07OAD-RCAA100

1. Remove the timing belt drive pulley.

2. Remove the pulley end crankshaft oil seal.

To install:

3. Clean and dry the crankshaft oil seal housing.

4. Apply a light coat of new engine oil to the lip of the crankshaft oil seal.

5. Using the oil seal driver, 64 mm, drive in the new crankshaft oil seal until the oil seal driver bottoms on the oil pump. When the seal is in place.

6. Clean the excess oil off the crankshaft, and check that the oil seal lip is not distorted.

7. Install the timing belt drive pulley.

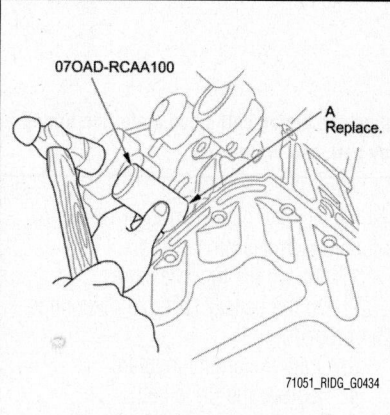

Fig. 61 Using the oil seal driver, 64 mm, drive in the new crankshaft oil seal (A) until the oil seal driver bottoms on the oil pump

CRANKSHAFT PULLEY

REMOVAL & INSTALLATION

See Figures 62 and 63.

Special Tools Required:
• Holder Handle 07JAB-001020B
• Holder Attachment, 50 mm, Offset 07MAB-PY3010A
• Socket, 19 mm 07JAA-001020A

1. Raise the vehicle on the lift.

2. Remove the right front wheel.

3. Remove the front splash shield.

4. Remove the drive belt.

5. Hold the pulley with the holder handle and the holder attachment, 50 mm, offset.

6. Remove the bolt with a heavy duty

socket, 19 mm and a breaker bar, then remove the crankshaft pulley.

To install:

7. Remove any oil and clean the crankshaft pulley, the timing belt guide plate, the timing belt drive pulley, the crankshaft, the crankshaft pulley bolt, and the washer. Lubricate with new engine oil as shown.

8. Install the crankshaft pulley, and tighten the bolt. Do not use an impact wrench.

a. Hold the pulley with the holder handle and the holder attachment, 50 mm, offset. Tighten the bolt to 48 ft. lbs. (65 Nm) with a torque wrench and a socket, 19 mm.

b. Mark the bolt head and the crankshaft pulley as shown, then tighten the bolt an additional 60 ° (The mark on the bolt head lines up with the mark on the crankshaft pulley).

9. Install the drive belt.

10. Install the front splash shield.

11. Install the right front wheel.

CRANKSHAFT REAR SEAL

REMOVAL & INSTALLATION

See Figure 64.

Special Tools Required:
• Driver Handle, 15 x 135L 07749-0010000
• Oil Seal Driver Attachment, 106 mm 070AD-RCA0200

1. Remove the transmission and the drive plate.

2. Remove the transmission end crankshaft oil seal.

To install:

3. Clean and dry the crankshaft oil seal housing.

4. Apply a light coat of new engine oil to the lip of the crankshaft seal.

5. Using the driver handle, 15 x 135L and the oil seal driver attachment, 106 mm, drive in the new transmission end crankshaft oil seal until the oil seal driver attachment bottoms on the engine block end cover. Align the hole in the oil seal driver attachment with the pin on the crankshaft.

6. Clean the excess oil off the crankshaft, and check that the oil seal lip is not distorted.

7. Install the drive plate, and the transmission.

CYLINDER HEAD

REMOVAL & INSTALLATION

See Figures 65 through 73.

Note the following:
• To avoid damaging the wiring and terminals, unplug the wiring connectors

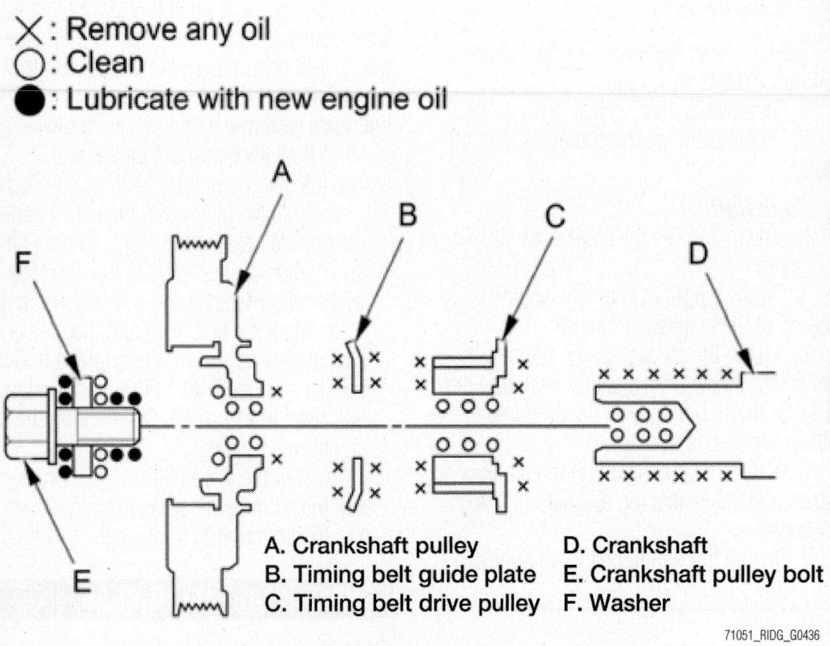

✕ : Remove any oil
◯ : Clean
● : Lubricate with new engine oil

A. Crankshaft pulley
B. Timing belt guide plate
C. Timing belt drive pulley
D. Crankshaft
E. Crankshaft pulley bolt
F. Washer

71051_RIDG_G0436

Fig. 62 Remove any oil and clean the crankshaft pulley, the timing belt guide plate, the timing belt drive pulley, the crankshaft, the crankshaft pulley bolt, and the washer

070AD-RCA0200
07749-0010000
A replace.
71051_RIDG_G0438

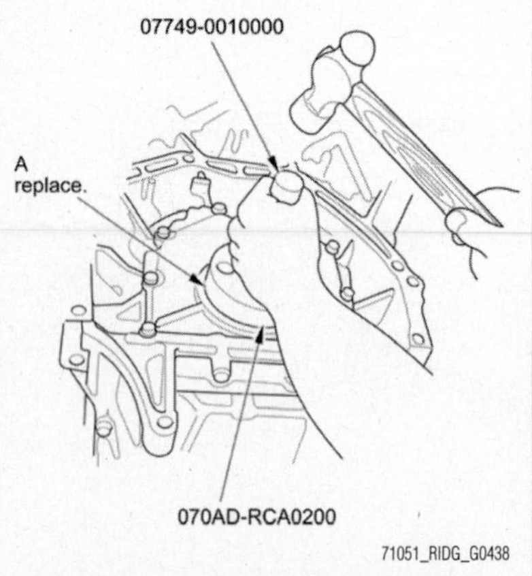

Fig. 64 Using the driver handle and the oil seal driver attachment drive in the new transmission end crankshaft oil seal until the oil seal driver attachment bottoms on the engine block end cover

carefully while holding the connector portion.

• Connect the HDS to the DLC, and monitor the ECT sensor 1. To avoid damaging the cylinder head, wait until the ECT drops below 100°F (38°C) before loosening the cylinder head bolts.

• Mark all wiring and hoses to avoid misconnection. Also, be sure that they do not contact any other wiring or hoses, or interfere with any other parts.

1. Relieve the fuel pressure.
2. Do the battery terminal disconnection procedure.
3. Drain the engine coolant.
4. Remove the drive belt.
5. Remove the alternator.
6. Remove the power steering pump

and the power steering hose bracket with its hoses connected.
7. Remove the intake manifold.
8. Remove the six ignition coils.
9. Remove the timing belt.
10. Remove the cylinder head covers.
11. Remove the following engine wire harness connectors and wire harness clamps from the cylinder head:
- Six injector connectors
- ECT sensor 1 connector
- Oil pressure switch connector
- CMP sensor connector
- Rocker arm oil control solenoid connector
- Rocker arm oil pressure switch connector
- Rear A/F sensor connector
- Rear secondary HO2S connector
12. Remove the front WU-TWC and the rear WU-TWC.
13. Remove the quick-connect fitting cover, then disconnect the fuel feed hose.
14. Remove the connector bracket from the front cylinder head.
15. Remove the harness clamp bracket from the rear cylinder head.
16. Remove the fuel rails.
17. Remove the water passage.
18. Remove the camshaft pulleys and the back covers.
19. Remove the cylinder head covers.
20. Remove the cylinder head bolts. To prevent warpage, loosen the bolts in sequence ⅓ turn at a time; repeat the sequence until all bolts are loosened.
21. Remove the cylinder heads.

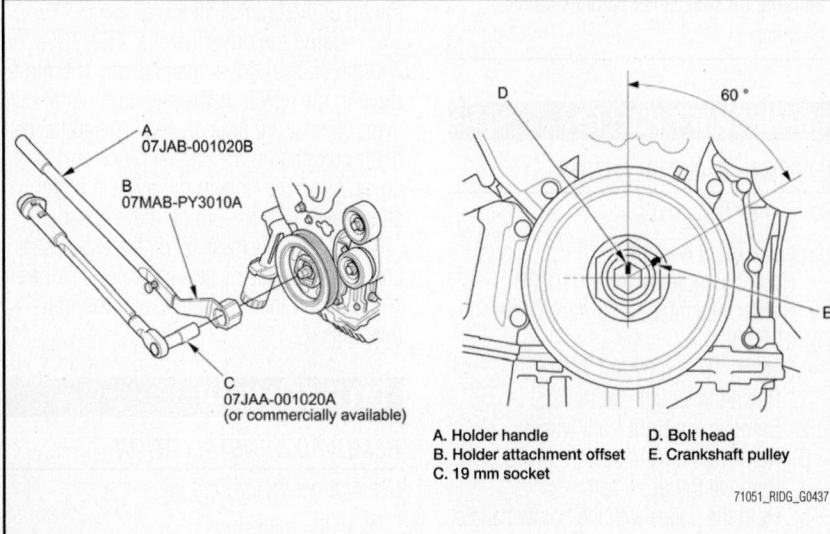

A. 07JAB-001020B
B. 07MAB-PY3010A
C. 07JAA-001020A (or commercially available)

A. Holder handle
B. Holder attachment offset
C. 19 mm socket
D. Bolt head
E. Crankshaft pulley

60°

71051_RIDG_G0437

Fig. 63 Mark the bolt head and the crankshaft pulley as shown

To install:

22. Clean the cylinder head and the engine block surface.

23. Clean and install the oil control orifices with new O-rings.

24. Install the dowel pins and the new cylinder head gaskets.

25. Clean the timing belt pulleys, the timing belt guide plate, and the upper and lower covers.

26. Set the timing belt drive pulley to top dead center (TDC) by aligning the TDC mark on the tooth of the timing belt drive pulley with the pointer on the oil pump.

27. Set the camshaft pulleys to TDC by aligning the TDC marks on the camshaft pulleys with the pointers on the back covers.

28. Install the cylinder heads on the engine block.

29. Measure the diameter of each cylinder head bolt at point A and point B.

30. If either diameter is less than 0.417 inches (10.6 mm), replace the cylinder head bolt.

31. Apply new engine oil to the threads and under the bolt heads of all cylinder head bolts.

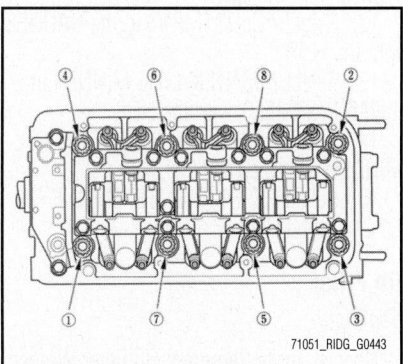

Fig. 65 Remove the cylinder heads—Front

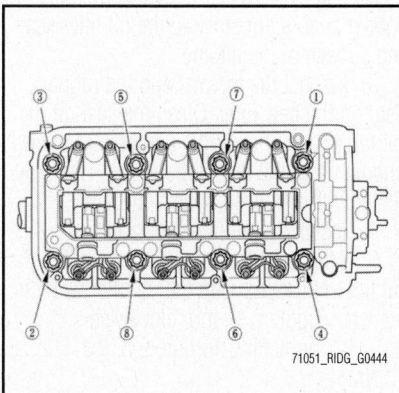

Fig. 66 Remove the cylinder heads—Rear

32. Torque the cylinder head bolts in sequence to 22 ft. lbs. (30 Nm) using a beam-type torque wrench. When using a preset click-type torque wrench, be sure to torque slowly and do not over tighten. If a bolt makes any noise while you are torquing it, loosen the bolt and retighten it from the first step.

33. After torquing, tighten all cylinder head bolts in two steps (90 degrees per step) using the sequence shown in step 11. If you are using a new cylinder head bolt, tighten the bolt an extra 90 degrees.

➡**Remove the cylinder head bolt if you tightened it beyond the specified angle, and go back to step 7 of the procedure. Do not loosen it back to the specified angle.**

34. Install the timing belt.

35. Adjust the valve clearance.

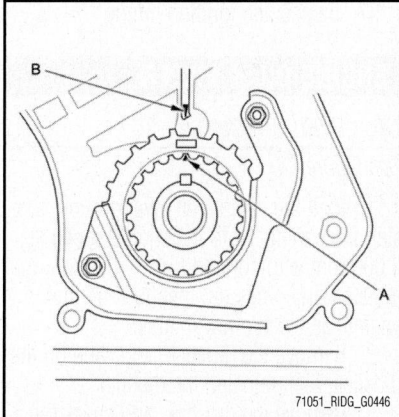

Fig. 67 Set the timing belt drive pulley to top dead center (TDC) by aligning the TDC mark (A) on the tooth of the timing belt drive pulley with the pointer (B) on the oil pump

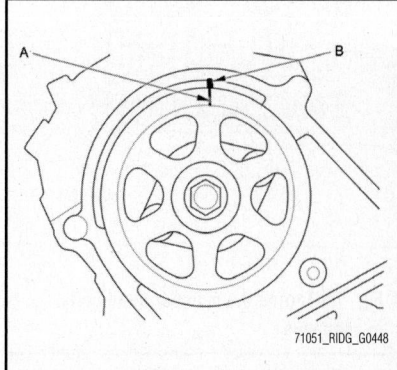

Fig. 68 Set the camshaft pulleys to TDC by aligning the TDC marks (A) on the camshaft pulleys with the pointers (B) on the back covers—Front

36. Install the cylinder head covers.

37. Install the water passage.

38. Install the fuel rails.

39. Install the connector bracket to the front cylinder head.

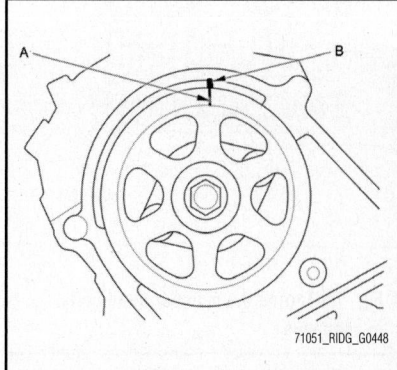

Fig. 69 Set the camshaft pulleys to TDC by aligning the TDC marks (A) on the camshaft pulleys with the pointers (B) on the back covers—Rear

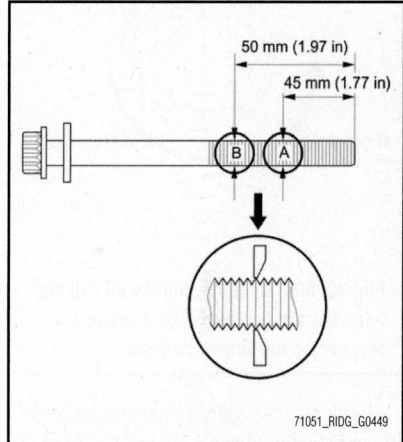

Fig. 70 Measure the diameter of each cylinder head bolt at point A and point B

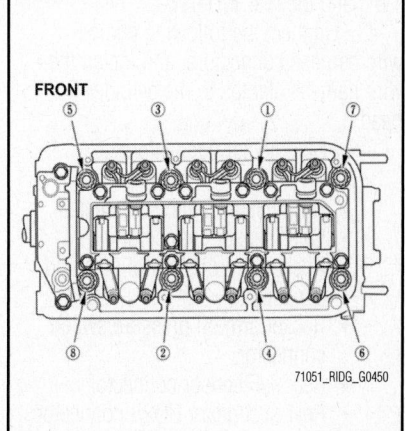

Fig. 71 Torque the cylinder head bolts in sequence

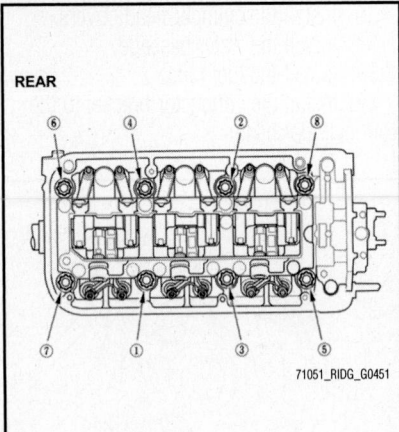

Fig. 72 Torque the cylinder head bolts in sequence

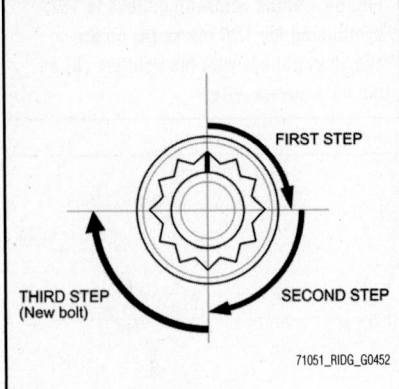

Fig. 73 After torquing, tighten all cylinder head bolts in two steps (90 degrees per step) using the sequence shown

40. Install the harness clamp bracket to the rear cylinder head.

41. Connect the fuel feed hose, then install the quick-connect fitting cover.

42. Install the front warm up front WU-TWC and the rear WU-TWC.

43. Connect the following engine wire harness connectors, and install the wire harness clamps to the cylinder head:

- Six injector connectors
- ECT sensor 1 connector
- Oil pressure switch connector
- CMP sensor connector
- Rocker arm oil control solenoid connector
- Rocker arm oil pressure switch connector
- Rear A/F sensor connector
- Rear secondary HO2S connector

44. Install the six ignition coils.

45. Install the intake manifold.

46. Install the alternator.

47. Install the power steering pump and the power steering hose bracket.

48. Install the drive belt.

49. Do the battery terminal reconnection procedure.

50. After installation, check that all tubes, hoses, and connectors are installed correctly.

51. Inspect for fuel leaks. Turn the ignition switch to ON (II) (do not operate the starter) so the fuel pump runs for about 2 seconds and pressurizes the fuel line. Repeat this operation three times, then check for fuel leakage at any point in the fuel line.

52. Refill the radiator with engine coolant, and bleed the air from the cooling system.

53. Check for fluid leaks.

54. Do the PCM idle lean procedure.

55. Do the CKP pattern clear/CKP pattern learn procedure.

56. Inspect the idle speed.

57. Inspect the ignition timing.

ENGINE OIL & FILTER

OIL LEVEL CHECK

See Figure 74.

1. Park the vehicle on level ground, and start the engine. Hold the engine speed at 3,000 rpm with no load (in P or N) until the radiator fan comes on, then turn off the engine, and wait a few minutes.

2. Remove the dipstick, and wipe off the dipstick, then reinstall the dipstick.

3. Remove the dipstick, and check the engine oil level. It should be between the upper mark and lower mark.

4. If the engine oil level is near or below the lower mark, check for oil leakage, and add engine oil to bring it between the upper mark.

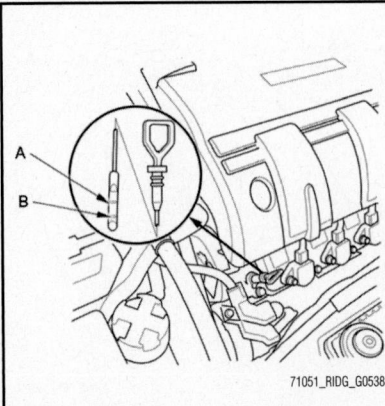

Fig. 74 Remove the dipstick, and check the engine oil level, it should be between the upper mark (A) and lower mark (B)

OIL & FILTER CHANGE

Engine Oil

1. Warm up the engine.

2. Remove the drain bolt, and drain the engine oil.

3. Reinstall the drain bolt with a new washer.

4. Refill the engine with the recommended oil.

5. Run the engine for more than 3 minutes, then check the oil level and oil leakage.

6. If the maintenance minder required engine coolant replacement, reset the maintenance minder, and this procedure is complete.

- Information Display.
- Multi-Information Display.

If the maintenance minder did not require engine coolant replacement, go to step 7.

7. Turn the ignition switch to LOCK (0).

8. Connect the HDS to the DLC.

9. Turn the ignition switch to ON (II).

10. Make sure the HDS communicates with the vehicle and the PCM. If it does not communicate, troubleshoot the DLC circuit.

11. Select GAUGES in the BODY ELECTRICAL with the HDS.

12. Select ADJUSTMENT in the GAUGES with the HDS.

13. Select MAINTENANCE MINDER in the ADJUSTMENT with the HDS.

14. Select RESET in the MAINTENANCE MINDER with the HDS.

15. Select RESETTING THE ENGINE OIL LIFE with the HDS.

Oil Filter

See Figure 75.

Special Tools Required: Oil Filter Wrench 07AAA-PLCA100

1. Remove the oil filter with the oil filter wrench.

2. Inspect the filter to make sure the rubber seal is not stuck to the oil filter seating surface of the engine.

3. Inspect the threads and the rubber seal on the new filter. Clean the seat on the oil filter base, then apply a light coat of new engine oil to the filter rubber seal. Use only filters with a built-in bypass system.

4. Install the oil filter by hand.

5. After the rubber seal seats, tighten the oil filter clockwise with the oil filter wrench.

   a. Tighten: ¾ turn clockwise

   b. Tightening Torque: 9 ft. lbs. (12 Nm)

6. If four numbers or marks (1 to 4 or ▼ to ▼▼▼▼) are printed around the outside

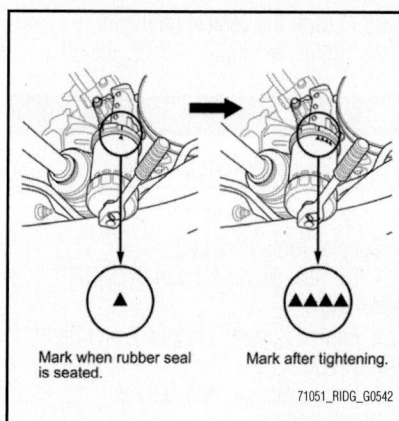

Fig. 75 If four numbers or marks are printed around the outside of the filter, you can use the following procedure

of the filter, you can use the following procedure to tighten the filter.

a. Spin the filter on until its seal lightly seats against the oil filter base, and note which number or mark is at the bottom.

b. Tighten the filter by turning it clockwise three numbers or marks from the one you noted. For example, if mark ▼ is at the bottom when the seal is lightly seated, tighten the filter until the mark ▼▼▼▼ comes around to the bottom.

7. After installation, fill the engine with the engine oil up to the specified level, run the engine for more than 3 minutes, then check for oil leakage.

EXHAUST MANIFOLD

COMPONENT LOCATIONS

Honda manufactures the cylinder head and exhaust manifold as one piece.

INTAKE MANIFOLD

REMOVAL & INSTALLATION

See Figures 76 through 79.

1. Remove the engine cover.
2. Remove the breather pipe, then remove the air intake duct.
3. Disconnect the engine wire harness connectors from the intake manifold:
   - Throttle body connector
   - MAP sensor connector
   - EVAP canister purge valve connector
   - IMT valve actuator connector
4. Disconnect the PCV hose, the brake booster vacuum hose, and the EVAP canister purge hose.

5. Disconnect the water bypass hoses, then plug the water bypass hoses.
6. Remove the upper cover mounting bolts and nuts sequentially, in three steps, then remove the upper cover.
7. Remove the intake manifold mounting bolts and nuts sequentially in three steps, then remove the intake manifold and the spacer.

To install:

8. Put a new gasket and the spacer on the injector base.
9. Install the intake manifold. Tighten the bolts and nuts sequentially in three steps. Always use a new intake manifold gasket. Specified Torque: 8 x 1.25 mm bolts: 16 ft. lbs. (22 Nm)
10. Install the upper cover. Tighten the bolts and nuts sequentially in three steps. Always use a new gasket. Specified Torque: 6 x 1.0 mm bolts: 9 ft. lbs. (12 Nm)
11. Connect the water bypass hoses.
12. Connect the PCV hose, the brake booster vacuum hose, and the EVAP canister purge hose.
13. Connect the engine wire harness connectors to the intake manifold:
    - Throttle body connector
    - MAP sensor connector
    - EVAP canister purge valve connector
    - IMT valve actuator connector
14. Install the air intake duct, then install the breather pipe.
15. Clean up any spilled engine coolant.
16. After installation, check that all tubes, hoses, and connectors are installed correctly.
17. Install the engine cover.
18. Refill the radiator with engine coolant, and bleed the air from the cooling system.

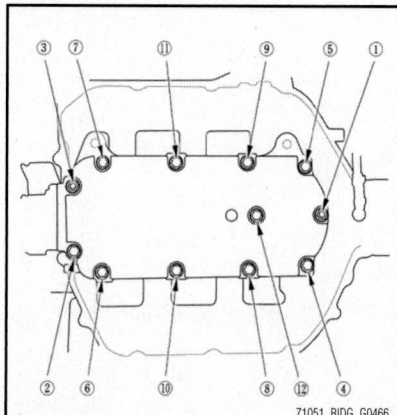

Fig. 76 Remove the upper cover mounting bolts and nuts sequentially, in three steps, then remove the upper cover

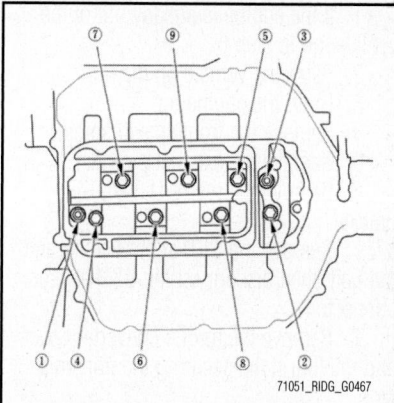

Fig. 77 Remove the intake manifold mounting bolts and nuts sequentially in three steps, then remove the intake manifold and the spacer

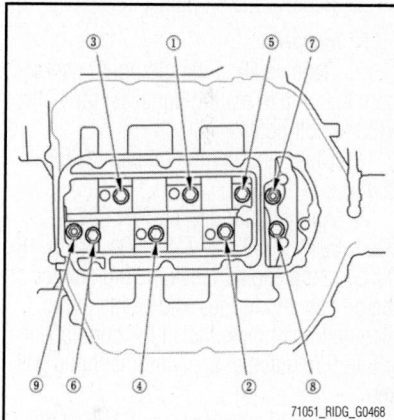

Fig. 78 Install the intake manifold, tighten the bolts and nuts sequentially in three steps

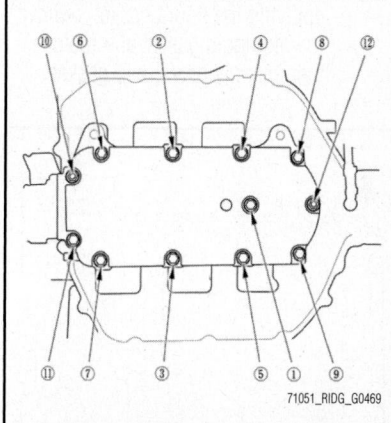

Fig. 79 Install the upper cover, tighten the bolts and nuts sequentially in three steps

OIL PAN

REMOVAL & INSTALLATION

See Figures 80 through 82.

1. If the engine is already out of the vehicle, go to step 6.

2. Raise the vehicle on the lift.

3. Drain the engine oil.

4. Remove the front splash shield.

5. Remove the exhaust pipe A.

6. Remove the rear WU-TWC bracket.

7. Remove the CKP sensor cover and the bolt, then disconnect the CKP sensor connector.

8. Remove the torque converter cover and the four bolts securing the transmission.

9. Remove the bolts securing the oil pan.

10. Using a flat blade screwdriver, separate the oil pan from the engine block in the places shown.

11. Remove the oil pan.

To install:

12. Remove all of the old liquid gasket from the oil pan mating surfaces, the bolts, and the bolt holes.

13. Clean and dry the oil pan mating surfaces.

14. Apply liquid gasket (P/N 08717-0004, 08718-0003, 08718-0004, or 08718-0009) to the engine block mating surface of the oil pan and to the inside edge of the threaded bolt holes. Install the component within 5 minutes of applying the liquid gasket.

- Apply a bead of liquid gasket along the broken line.
- If you apply liquid gasket P/N 08718-0012, the component must be installed within 4 minutes.
- If too much time has passed after applying the liquid gasket, remove the old liquid gasket and residue, then reapply new liquid gasket.

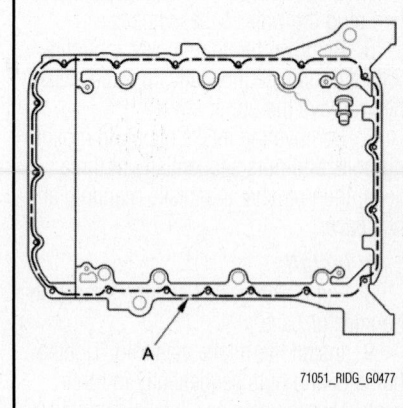

Fig. 81 Apply a bead of liquid gasket along the broken line (A)

15. Install the oil pan on the engine block.

16. Tighten the bolts in three steps. In the final step, tighten all bolts, in sequence, to 9 ft. lbs. (12 Nm).

➡**Wait at least 30 minutes before filling the engine with oil. Do not run the engine for at least 3 hours after installing the oil pan.**

17. Tighten the four bolts securing the transmission to 54 ft. lbs. (74 Nm), then install the torque converter cover.

18. Connect the CKP sensor connector, then install the CKP sensor cover and the bolt.

19. Install the rear WU-TWC bracket.

20. If the engine is still in the vehicle, do the following steps.

21. Install exhaust pipe A using new gaskets and new self-locking nuts.

22. Install the front splash shield.

23. Lower the vehicle on the lift.

24. Refill the engine with engine oil.

OIL PUMP

REMOVAL & INSTALLATION

See Figures 83 through 86.

Special Tools Required*:

- Oil Seal Driver, 64 mm 07OAD-RCAA100
- Engine Support Hanger, A and Reds AAR-T1256
- Engine Hanger Adapter Set VSB02C000024
- * Available through the Honda Tool and Equipment Program, 888-424-6857.

1. Drain the engine oil.

2. Open the hood, and secure it with the hood support rod in the wide-open position.

3. Remove the bulkhead cover.

4. Remove the intake manifold cover.

5. Remove the drive belt.

6. Remove the power steering pump and the power steering pump line bracket.

7. Remove the service caps (A) for the front damper flange nuts from the cowl cover (B). Position the engine hanger adapter set (VSB02C000024) with the "FRONT" mark facing forward over the damper flange nuts.

8. Install the A and Reds engine support hanger (AAR-T1256) to the vehicle, and attach the hook to the engine hanger. Tighten the wing nut by hand, and support the engine.

➡**AAR-T1256 two sets required for stacking additional cross section bar.**

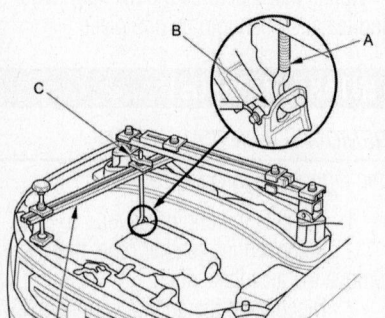

AAR-T1256

Fig. 83 Install the A and Reds engine support hanger (AAR-T1256) to the vehicle, and attach the hook (A) to the engine hanger (B), tighten the wing nut (C) by hand, and support the engine

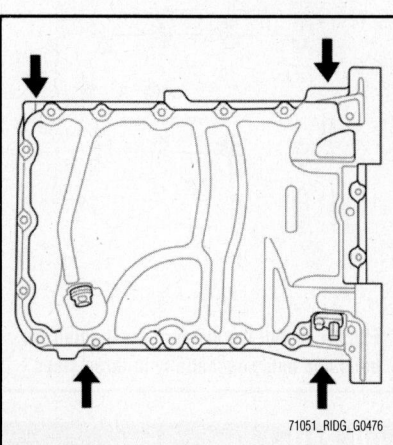

Fig. 80 Using a flat blade screwdriver, separate the oil pan from the engine block in the places shown

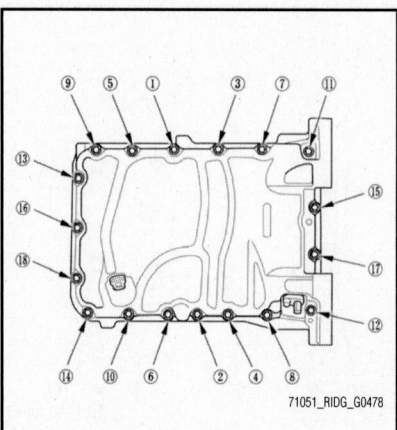

Fig. 82 Tighten the bolts in three steps in sequence

9. Remove the timing belt.

10. Remove the rocker arm oil control solenoid/oil filter assembly.

11. Remove the oil pan.

12. Remove the oil strainer, then remove the oil pump.

To install:

13. Remove the old oil seal from the oil pump.

14. Clean and dry the crankshaft oil seal housing.

15. Using the oil seal driver, 64 mm, drive in the new pulley end crankshaft oil seal until the oil seal driver bottoms on the pump.

16. Remove all of the old liquid gasket from the oil pump mating surfaces, the bolts, and the bolt holes.

17. Clean and dry the oil pump mating surfaces.

18. Apply liquid gasket (P/N 08717-0004, 08718-0003, 08718-0004, or 08718-0009) to the engine block mating surface of the oil pump and to the inside edge of the threaded bolt holes. Install the component within 5 minutes of applying the liquid gasket.

- Apply a bead of liquid gasket along the broken line.
- If you apply liquid gasket P/N 08718-0012, the component must be installed within 4 minutes.
- If too much time has passed after applying the liquid gasket, remove the old liquid gasket and residue, then reapply new liquid gasket.

19. Apply a light coat of new engine oil to the lip of the crankshaft oil seal, and apply new engine oil to the new O-ring.

20. Install the dowel pins, then align the

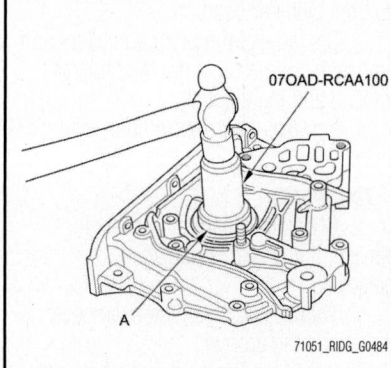

Fig. 85 Using the oil seal driver, 64 mm, drive in the new pulley end crankshaft oil seal (A) until the oil seal driver bottoms on the pump

inner rotor with the crankshaft, and install the oil pump.

➡ **Wait at least 30 minutes before filling the engine with oil. Do not run the engine for at least 3 hours after installing the oil pump.**

21. Clean the excess oil off the crankshaft, and check the seal for distortion.

22. Install the oil strainer with a new O-ring.

23. Install the oil pan.

24. Install the rocker arm oil control solenoid/oil filter assembly, with a new rocker arm oil control solenoid filter.

25. Install the timing belt.

26. Remove the engine support hanger.

27. Install the service caps on the cowl cover.

28. Install the bulkhead cover.

29. Refill the engine with engine oil.

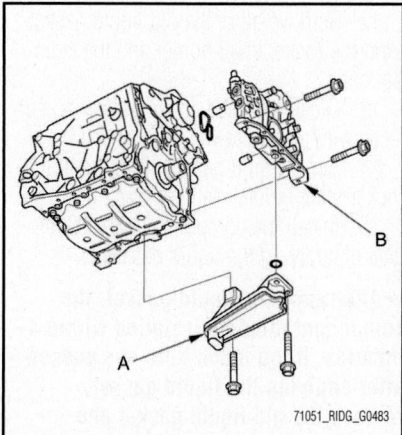

Fig. 84 Remove the oil strainer (A), then remove the oil pump (B)

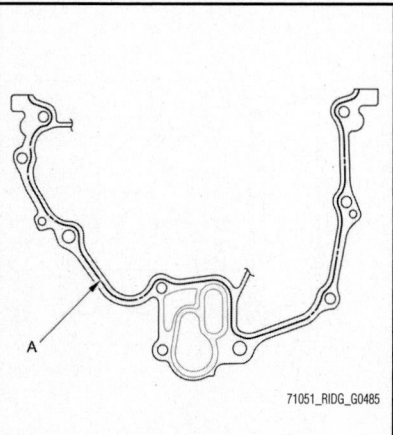

Fig. 86 Apply a bead of liquid gasket along the broken line (A)

PISTONS & RINGS

POSITIONING

See Figures 87 and 88.

1. Install the rings as shown. The top ring has a 1D or 1R mark and the second ring has a 2X or 2R mark. The manufacturing marks must be facing upward.

2. Position the ring end gaps as shown.

REAR MAIN SEAL

REMOVAL & INSTALLATION

See Figure 89.

Special Tools Required:
- Driver Handle, 15 x 135L 07749-0010000
- Oil Seal Driver Attachment, 106 mm 070AD-RCA0200

1. Remove the transmission and the drive plate.

2. Remove the transmission end crankshaft oil seal.

To install:

3. Clean and dry the crankshaft oil seal housing.

4. Apply a light coat of new engine oil to the lip of the crankshaft seal.

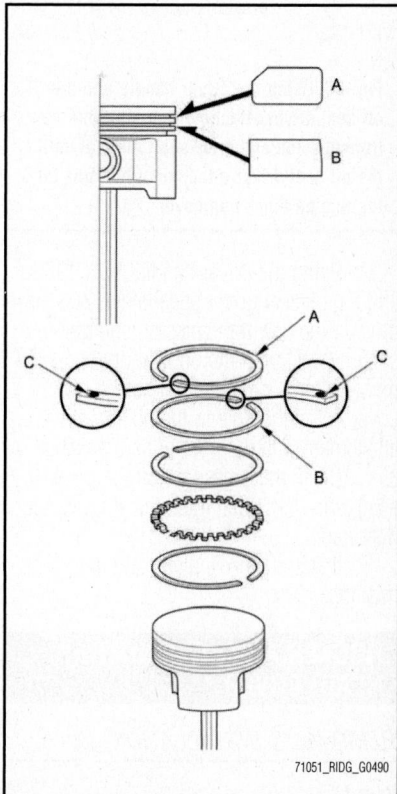

Fig. 87 Piston ring installation showing top ring (A), second ring (B) and manufacturing marks (C)

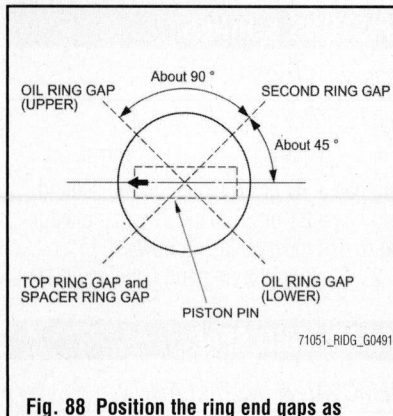

Fig. 88 Position the ring end gaps as shown

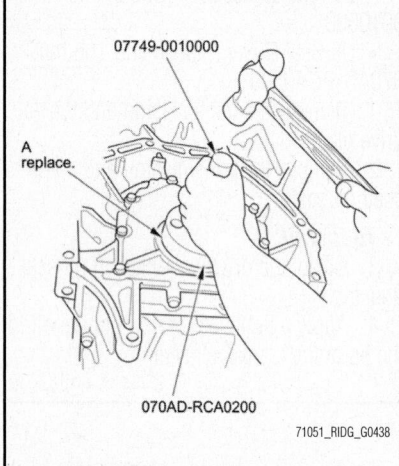

Fig. 89 Using the driver handle and the oil seal driver attachment drive in the new transmission end crankshaft oil seal until the oil seal driver attachment bottoms on the engine block end cover

5. Using the driver handle, 15 x 135L and the oil seal driver attachment, 106 mm, drive in the new transmission end crankshaft oil seal until the oil seal driver attachment bottoms on the engine block end cover. Align the hole in the oil seal driver attachment with the pin on the crankshaft.

6. Clean the excess oil off the crankshaft, and check that the oil seal lip is not distorted.

7. Install the drive plate, and the transmission.

ROCKER ARM (CYLINDER HEAD) COVER

REMOVAL & INSTALLATION

Front

See Figure 90.

1. Remove the intake manifold. See "Intake Manifold" in this section.

2. Remove the three ignition coils from the front cylinder head.

3. Disconnect the four connectors and remove the harness clamp, the harness holders, and the dipstick.

4. Remove the front cylinder head cover.

To install:

5. Check the spark plug seals for damage. If any seal is damaged, replace it.

6. Thoroughly clean the head cover gasket and the groove.

7. Install the head cover gasket in the groove of the cylinder head cover. Make sure the head cover gasket is seated securely.

8. Remove all of the old liquid gasket from the rocker shaft holder and the cylinder head.

9. Clean the head cover contacting surfaces with a shop towel.

10. Apply liquid gasket evenly to the rocker shaft holder mating areas.

11. Install the component within 5 minutes of applying the liquid gasket.

➡After applying liquid gasket, the component must be installed within 4 minutes. If too much time has passed after applying the liquid gasket, remove the old liquid gasket and residue, then reapply the new liquid gasket.

12. Set the spark plug seals on the spark plug tubes, and install the front cylinder head cover.

➡Wait at least 30 minutes before filling the engine with oil. Do not run the engine for at least 3 hours after installing the cylinder head cover.

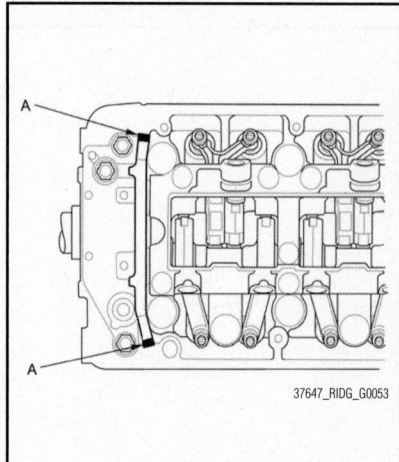

Fig. 90 Apply liquid gasket evenly to the rocker shaft holder mating areas (A).

13. Inspect the spark plug seals for damage. Inspect the cover washers. Replace any washer that is damaged or deteriorated.

14. Tighten the cover bolts, in an alternating pattern, in three steps. In the final step torque all bolts, in sequence to 8.8 ft. lbs. (12 Nm).

15. Connect the four connectors and install the harness clamp, the harness holders, and the dipstick.

16. Install the three ignition coils to the front cylinder head.

17. Install the intake manifold.

Rear

See Figure 90.

1. Remove the intake manifold. See "Intake Manifold" in this section.

2. Remove the three ignition coils from the rear cylinder head.

3. Remove the accessory drive belt.

4. Remove the bolt securing the power steering (P/S) hose bracket and the P/S pump without disconnecting the P/S hoses.

5. Remove the harness holder mounting bolts, the ground cable, and the breather hose.

6. Disconnect the three injector connectors and remove the harness clamps.

7. Remove the engine harness from the rear upper cover.

8. Remove the rear cylinder head cover.

To install:

9. Check the spark plug seals for damage. If any seal is damaged, replace it.

10. Thoroughly clean the head cover gasket and the groove.

11. Install the head cover gasket in the groove of the cylinder head cover. Make sure the head cover gasket is seated securely.

12. Remove all of the old liquid gasket from the rocker shaft holder and the cylinder head.

13. Clean the head cover contacting surfaces with a shop towel.

14. Apply liquid gasket evenly to the rocker shaft holder mating areas.

15. Install the component within 5 minutes of applying the liquid gasket.

➡After applying liquid gasket, the component must be installed within 4 minutes. If too much time has passed after applying the liquid gasket, remove the old liquid gasket and residue, then reapply the new liquid gasket.

16. Set the spark plug seals on the spark

plug tubes, and install the front cylinder head cover.

➡️**Wait at least 30 minutes before filling the engine with oil. Do not run the engine for at least 3 hours after installing the cylinder head cover.**

17. Inspect the spark plug seals for damage. Inspect the cover washers. Replace any washer that is damaged or deteriorated.

18. Tighten the cover bolts, in an alternating pattern, in three steps. In the final step torque all bolts, in sequence to 8.8 ft. lbs. (12 Nm).

19. Tighten the harness holder mounting bolts.

20. Install the ground cable and the breather hose, then install the engine wire harness to the rear upper cover.

21. Connect the three injector connectors, then install the harness clamps.

22. Install the bolt securing the power steering (P/S) pump and the P/S hose bracket.

23. Install the drive belt.

24. Install the three ignition coils form the rear cylinder head.

25. Install the intake manifold.

ROCKER ARMS & SHAFTS

REMOVAL & INSTALLATION

Front

See Figures 91 through 93.

1. Remove the cylinder head cover.
2. Loosen the locknuts and the adjusting screws.
3. Remove the rocker shaft bridge mounting bolts, the rocker shaft holder mounting bolts, and the rocker arm assembly.

   a. Loosen the rocker shaft bridge mounting bolts and the rocker shaft holder mounting bolts in sequence two turns at a time, to prevent damaging the valves or the rocker arm assembly.

   b. When removing the rocker arm assembly, do not remove the rocker shaft bridge mounting bolts and the rocker shaft holder mounting bolts. The bolts will keep the rocker arms on the shafts.

   To install:
4. Remove all of the old liquid gasket from the rocker shaft holder and the cylinder head.
5. Apply liquid gasket (P/N 08717-0004, 08718-0003, 08718-0004, or 08718-0009) to the rocker shaft holder mating surface of the cylinder head. Install the component within 5 minutes of applying the liquid gasket.

- Apply a bead of liquid gasket along the broken line.
- If you apply liquid gasket P/N 08718-0012, the component must be installed within 4 minutes.
- If too much time has passed after applying the liquid gasket, remove the old liquid gasket and residue, then reapply the new liquid gasket.

6. Set the rocker arm assembly in place, and loosely install the bolts. Make sure that the rocker arms are properly positioned on the valve stems.

➡️**Wait at least 30 minutes before filling the engine with oil. Do not run the engine for at least 3 hours after installing the rocker arm assembly.**

7. Tighten each bolt two turns at a time in the sequence shown to ensure that the rockers do not bind on the valves.

   a. When the rocker shaft bridge is reused: Specified Torque: 8 x 1.25 mm bolts: 18 ft. lbs. (24.5 Nm)

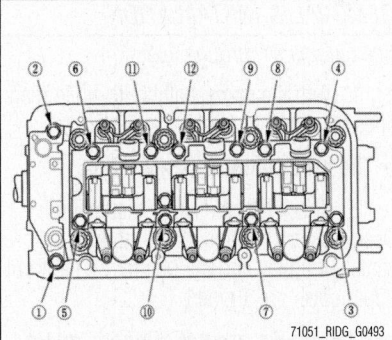

Fig. 91 Loosen the rocker shaft bridge mounting bolts and the rocker shaft holder mounting bolts in sequence two turns at a time, to prevent damaging the valves or the rocker arm assembly

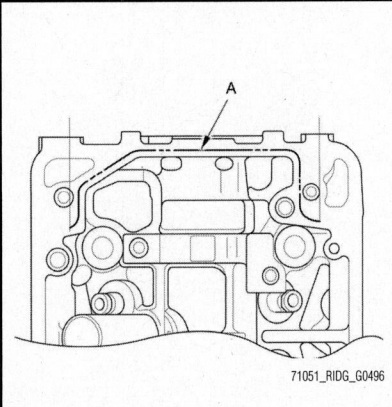

Fig. 92 Apply a bead of liquid gasket along the broken line (A)

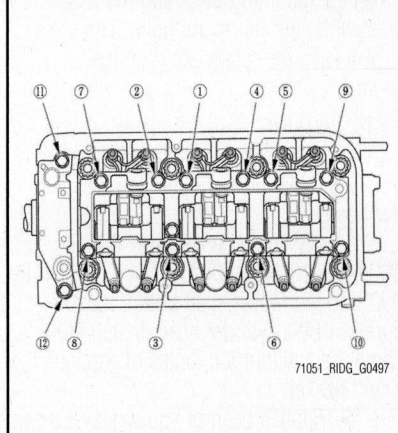

Fig. 93 Tighten each bolt two turns at a time in the sequence shown

   b. When a new rocker shaft bridge is installed: Specified Torque: 8 x 1.25 mm bolts:
- 1st: 21 ft. lbs. (28 Nm), then loosen the bolts.
- 2nd: 18 ft. lbs. (24.5 Nm)

Rear

See Figures 94 through 96.

1. Remove the cylinder head cover.
2. Loosen the locknuts and the adjusting screws.
3. Remove the rocker shaft bridge mounting bolts, the rocker shaft holder mounting bolts, and the rocker arm assembly.

   a. Loosen the rocker shaft bridge mounting bolts and the rocker shaft holder mounting bolts in sequence two turns at a time, to prevent damaging the valves or the rocker arm assembly.

   b. When removing the rocker arm assembly, do not remove the rocker shaft

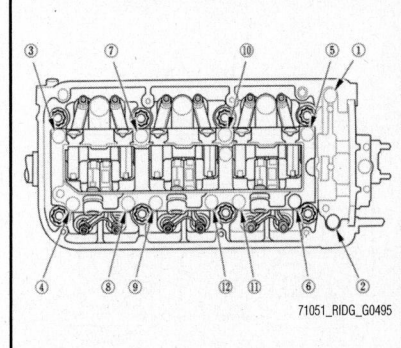

Fig. 94 Loosen the rocker shaft bridge mounting bolts and the rocker shaft holder mounting bolts in sequence two turns at a time, to prevent damaging the valves or the rocker arm assembly

bridge mounting bolts and the rocker shaft holder mounting bolts. The bolts will keep the rocker arms on the shafts.

To install:

4. Remove all of the old liquid gasket from the rocker shaft holder and the cylinder head.

5. Apply liquid gasket (P/N 08717-0004, 08718-0003, 08718-0004, or 08718-0009) to the rocker shaft holder mating surface of the cylinder head. Install the component within 5 minutes of applying the liquid gasket.

- Apply a bead of liquid gasket along the broken line.
- If you apply liquid gasket P/N 08718-0012, the component must be installed within 4 minutes.
- If too much time has passed after applying the liquid gasket, remove the old liquid gasket and residue, then reapply the new liquid gasket.

6. Set the rocker arm assembly in place, and loosely install the bolts. Make sure that the rocker arms are properly positioned on the valve stems.

➡**Wait at least 30 minutes before filling the engine with oil. Do not run the engine for at least 3 hours after installing the rocker arm assembly.**

7. Tighten each bolt two turns at a time in the sequence shown to ensure that the rockers do not bind on the valves.

a. When the rocker shaft bridge is reused: Specified Torque: 8 x 1.25 mm bolts: 18 ft. lbs. (24.5 Nm)

b. When a new rocker shaft bridge is installed: Specified Torque: 8 x 1.25 mm bolts:

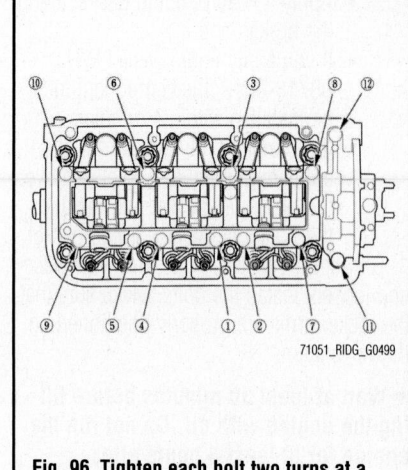

Fig. 96 Tighten each bolt two turns at a time in the sequence shown

- 1st: 21 ft. lbs. (28 Nm), then loosen the bolts.
- 2nd: 18 ft. lbs. (24.5 Nm)

TIMING BELT

REMOVAL & INSTALLATION

See Figures 97 through 102.

1. Turn the crankshaft so its white mark lines up with the pointer.

➡**The other pointer is not used.**

2. Check that the No. 1 piston Top Dead Center (TDC) mark on the front camshaft pulley and the pointer on the front upper cover are aligned.

➡**If the marks are not aligned, rotate the crankshaft 360 degrees, and recheck the camshaft pulley mark.**

3. Raise the vehicle on the lift, then remove the right front wheel.

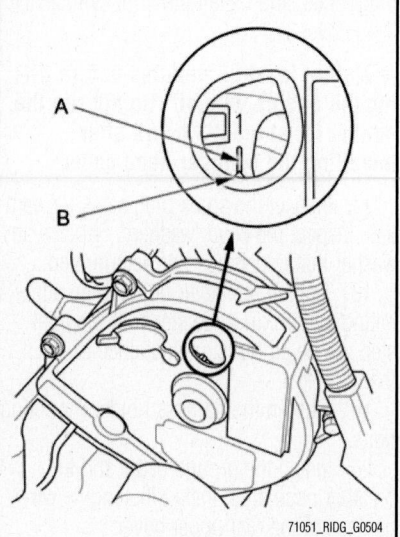

Fig. 98 Check that the No. 1 piston Top Dead Center (TDC) mark (A) on the front camshaft pulley and the pointer (B) on the front upper cover are aligned

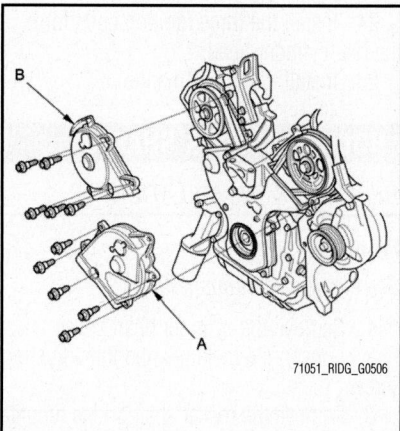

Fig. 99 Remove the front upper cover (A) and the rear upper cover (B)

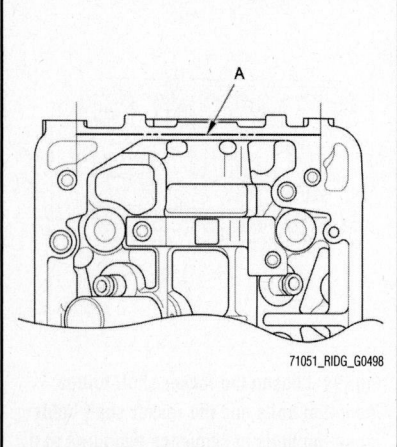

Fig. 95 Apply a bead of liquid gasket along the broken line (A)

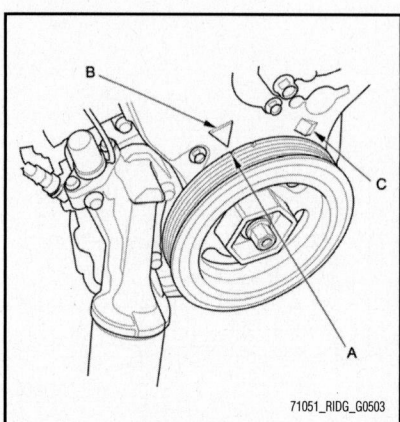

Fig. 97 Turn the crankshaft so its white mark (A) lines up with the pointer (B); other pointer (C) is not used

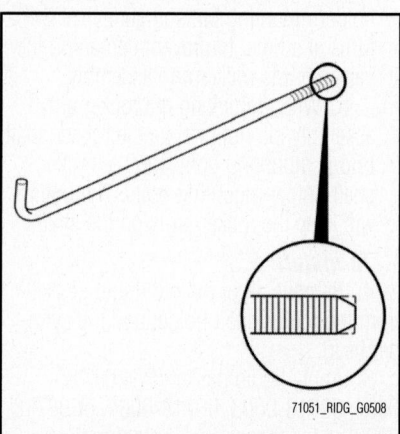

Fig. 100 Remove one of the battery clamp bolts from the battery tray, and grind the end of it as shown

4. Remove the front splash shield.

5. Remove the drive belt auto-tensioner.

6. Lift and support the engine with a jack and a wood block under the oil pan.

7. Remove the ground cable bracket, then remove the upper half of the side engine mount bracket.

8. Remove the crankshaft pulley.

9. Remove the front upper cover and the rear upper cover.

10. Remove the lower cover.

11. Remove one of the battery clamp bolts from the battery tray, and grind the end of it as shown.

12. Thread in the battery clamp bolt as shown to hold the timing belt adjuster in its current position. Tighten it by hand, do not use a wrench.

13. Remove the timing belt guide plate.

14. Remove the lower half of the side engine mount bracket.

15. Remove the idler pulley bolt and the

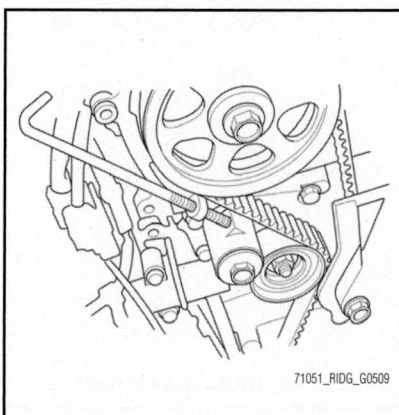

Fig. 101 Thread in the battery clamp bolt as shown to hold the timing belt adjuster in its current position

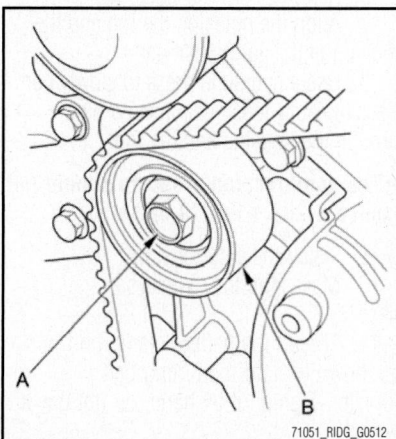

Fig. 102 Remove the idler pulley bolt (A) and the idler pulley (B), then remove the timing belt

idler pulley, then remove the timing belt. Discard the idler pulley bolt.

To install:

Installing Used Timing Belt

See Figures 103 through 110.

➡️**The following procedure is for installing a used timing belt. If you are installing a new belt, refer to the new timing belt replacement procedure.**

1. Clean the timing belt pulleys, the timing belt guide plate, and the upper and lower covers.

2. Set the timing belt drive pulley to top dead center (TDC) by aligning the TDC mark on the tooth of the timing belt drive pulley with the pointer on the oil pump.

3. Set the camshaft pulleys to TDC by aligning the TDC marks on the camshaft pulleys with the pointers on the back covers.

4. Loosely install the idler pulley with a new idler pulley bolt so the pulley can move but does not come off.

5. If the auto-tensioner has extended and the timing belt cannot be installed, do the timing belt replacement procedure.

6. Install the timing belt in a counter-clockwise sequence starting with the drive pulley. Take care not to damage the timing belt during installation.

- Drive pulley
- Idler pulley
- Front camshaft pulley
- Water pump pulley
- Rear camshaft pulley
- Adjusting pulley

7. Tighten the idler pulley bolt to 33 ft. lbs. (44 Nm).

8. Remove the battery clamp bolt from the back cover.

9. Install the lower half of the side engine mount bracket. Tighten the mounting bolts to 33 ft. lbs. (44 Nm).

10. Install the timing belt guide plate as shown.

11. Install the lower cover. Tighten the bolts to 9 ft. lbs. (12 Nm).

12. Install the front upper cover and the rear upper cover. Tighten the bolts to 9 ft. lbs. (12 Nm).

13. Install the crankshaft pulley.

14. Rotate the crankshaft pulley about six turns clockwise so the timing belt positions itself on the pulleys.

15. Turn the crankshaft pulley so its white mark lines up with the pointer.

➡️**The other pointer is not used.**

16. Check the camshaft pulley marks.

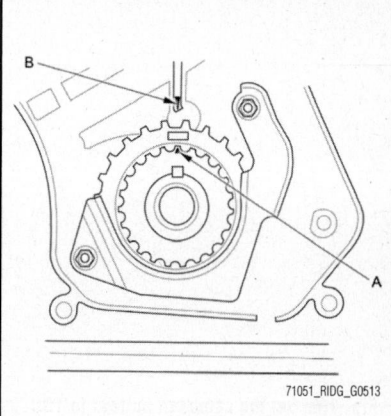

Fig. 103 Set the timing belt drive pulley to top dead center (TDC) by aligning the TDC mark (A) on the tooth of the timing belt drive pulley with the pointer (B) on the oil pump

➡️**If the marks are not aligned, rotate the crankshaft 360 degrees, and recheck the camshaft pulley mark.**

a. If the camshaft pulley marks are at TDC, go to step 17.

b. If the camshaft pulley marks are not at TDC, remove the timing belt and repeat steps 2 through 16.

17. Loosely install the upper half of the side engine mount bracket, then install the ground cable bracket.

18. Remove the jack and the wood block.

19. Tighten the mounting bolts in the numbered sequence shown.

20. Install the drive belt auto-tensioner.

21. Install the front splash shield.

22. Install the right front wheel.

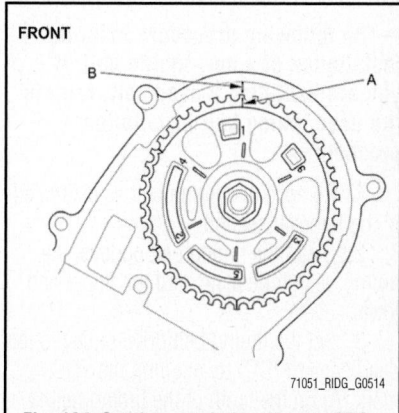

Fig. 104 Set the camshaft pulleys to TDC by aligning the TDC marks (A) on the camshaft pulleys with the pointers (B) on the back covers—Front

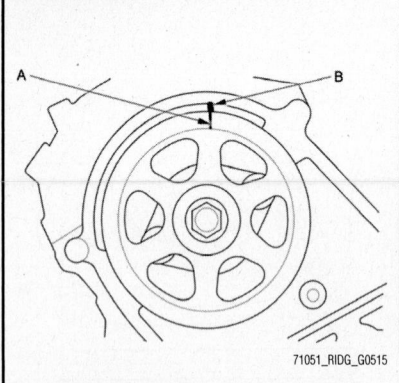

Fig. 105 Set the camshaft pulleys to TDC by aligning the TDC marks (A) on the camshaft pulleys with the pointers (B) on the back covers—Rear

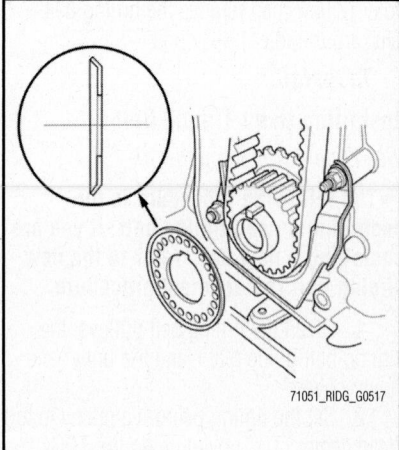

Fig. 107 Install the timing belt guide plate as shown

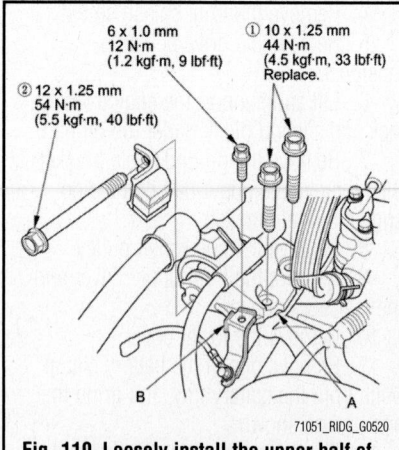

Fig. 110 Loosely install the upper half of the side engine mount bracket (A), then install the ground cable bracket (B)

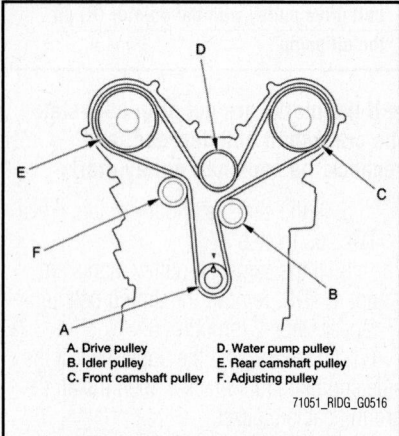

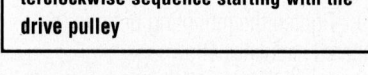

A. Drive pulley
B. Idler pulley
C. Front camshaft pulley
D. Water pump pulley
E. Rear camshaft pulley
F. Adjusting pulley

Fig. 106 Install the timing belt in a counterclockwise sequence starting with the drive pulley

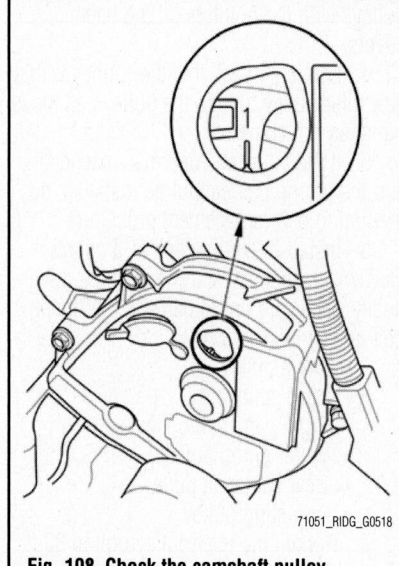

Fig. 108 Check the camshaft pulley marks—Front

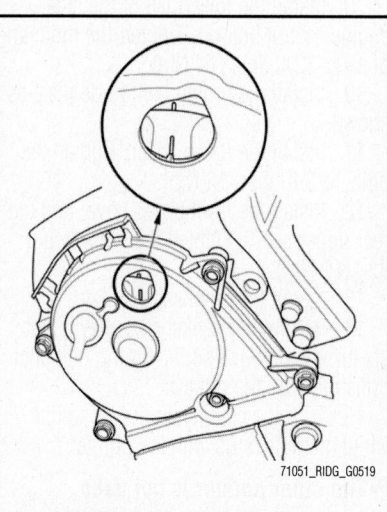

Fig. 109 Check the camshaft pulley marks—Rear

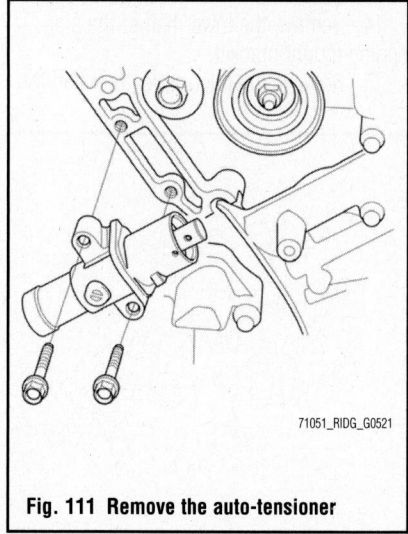

Fig. 111 Remove the auto-tensioner

23. Do the CKP pattern clear/CKP pattern learn procedure.

Installing New Timing Belt

See Figures 103 through 114.

➡The following procedure is for the installation of a new timing belt. If you are installing a used belt, refer to the used timing belt installation procedure.

1. Remove the timing belt as described in the removal section.
2. Clean the timing belt pulleys, the timing belt guide plate, and the upper and lower covers.
3. Set the timing belt drive pulley to top dead center (TDC) by aligning the TDC mark (A) on the tooth of the timing belt drive pulley with the pointer (B) on the oil pump.
4. Set the camshaft pulleys to TDC by aligning the TDC marks (A) on the camshaft

pulleys with the pointers (B) on the back covers.

5. Remove the battery clamp bolt from the back cover.
6. Remove the auto-tensioner.
7. Align the holes on the rod and the housing of the auto-tensioner.
8. Use a hydraulic press to slowly compress the auto-tensioner. Insert a pin through the housing and the rod.

➡The compression pressure should not exceed 2,203.1 lbs. (9,800 N).

9. Install the auto-tensioner.
10. Make sure the pin stays in place.
11. Thread the battery clamp bolt in as shown to hold the timing belt adjuster. Tighten it by hand, do not use a wrench.
12. Loosely install the idler pulley with a new idler pulley bolt so the pulley can move but does not come off.

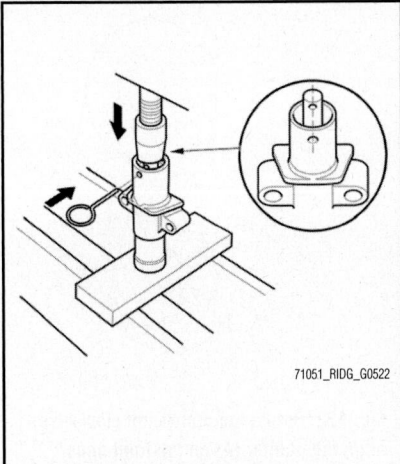

Fig. 112 Align the holes on the rod and the housing of the auto-tensioner

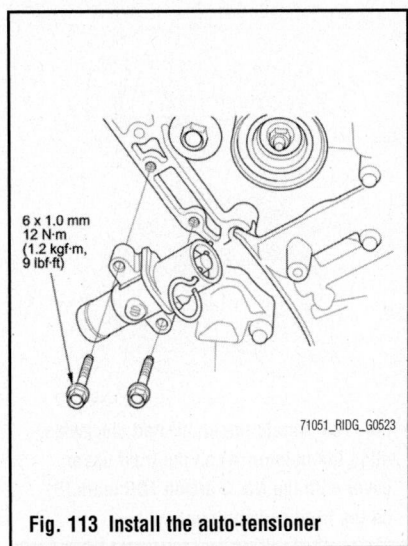

Fig. 113 Install the auto-tensioner

6 x 1.0 mm
12 N·m
(1.2 kgf·m,
9 lbf·ft)

13. Install the timing belt in a counter-clockwise sequence starting with the drive pulley.

- Drive pulley
- Idler pulley
- Front camshaft pulley
- Water pump pulley
- Rear camshaft pulley
- Adjusting pulley

14. Tighten the idler pulley bolt to 33 ft. lbs. (44 Nm).

15. Remove the pin from the auto-tensioner.

16. Remove the battery clamp bolt from the back cover.

17. Install the lower half of the side engine mount bracket. Tighten the mounting bolts to 33 ft. lbs. (44 Nm).

18. Install the timing belt guide plate as shown.

19. Install the lower cover. Tighten the bolts to 9 ft. lbs. (12 Nm).

20. Install the front upper cover and the

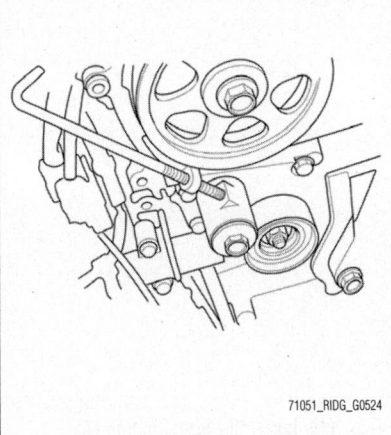

Fig. 114 Thread the battery clamp bolt in as shown to hold the timing belt adjuster

rear upper cover. Tighten the bolts to 9 ft. lbs. (12 Nm).

21. Install the crankshaft pulley.

22. Rotate the crankshaft pulley about six turns clockwise so the timing belt positions itself on the pulleys.

23. Turn the crankshaft pulley so its white mark lines up with the pointer.

➡ **The other pointer is not used.**

24. Check the camshaft pulley marks.

➡ **If the marks are not aligned, rotate the crankshaft 360 degrees, and recheck the camshaft pulley mark.**

a. If the camshaft pulley marks are at TDC, go to step 24.

b. If the camshaft pulley marks are not at TDC, remove the timing belt and repeat steps 3 through 23.

25. Loosely install the upper half of the side engine mount bracket, then install the ground cable bracket.

26. Remove the jack and the wood block.

27. Tighten the mounting bolts in the numbered sequence shown.

28. Install the drive belt auto-tensioner.

29. Install the front splash shield.

30. Install the right front wheel.

31. Do the CKP pattern clear/CKP pattern learn procedure.

VALVE LASH (CLEARANCE) ADJUSTMENT

ADJUSTMENT

See Figures 115 through 124.

➡ **Connect the HDS to the DLC, and monitor the ECT sensor 1. Adjust the valves clearance only when the ECT is less than 100°F (38°C).**

1. Remove the cylinder head covers.

2. Set the No. 1 piston at top dead center (TDC). Align the pointer on the front upper cover with the No. 1 piston TDC mark on the front camshaft pulley.

3. Select the correct feeler gauge for the valve clearance you are going to check.

a. Valve Clearance:

b. Intake: 0.008–0.009 inches (0.20–0.24 mm)

c. Exhaust: 0.011–0.012 inches (0.28–0.32 mm)

4. Insert the feeler gauge between the adjusting screw and the end of the valve stem on the No. 1 cylinder, and slide it back and forth; you should feel a slight amount of drag.

5. If you feel too much or too little

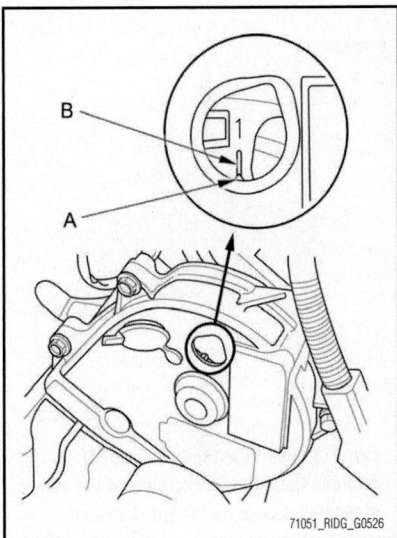

Fig. 115 Set the No. 1 piston at top dead center (TDC), align the pointer (A) on the front upper cover with the No. 1 piston TDC mark (B) on the front camshaft pulley

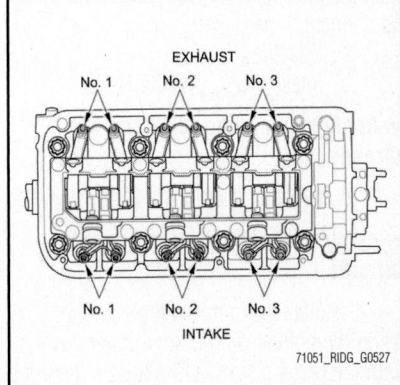

EXHAUST
No. 1   No. 2   No. 3

No. 1   No. 2   No. 3
INTAKE

Fig. 116 Select the correct feeler gauge for the valve clearance you are going to check—Rear

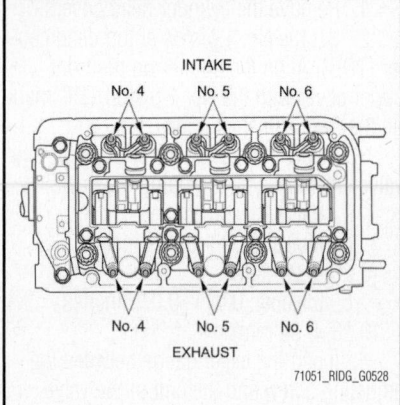

Fig. 117 Select the correct feeler gauge for the valve clearance you are going to check—Front

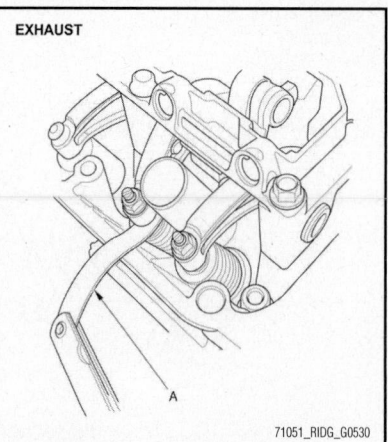

Fig. 119 Insert the feeler gauge (A) between the adjusting screw and the end of the valve stem on the No. 1 cylinder

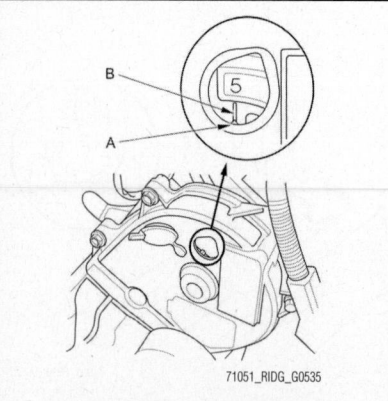

Fig. 122 Rotate the crankshaft clockwise, align the pointer (A) on the front upper cover with the No. 5 piston TDC mark (B) on the front camshaft pulley

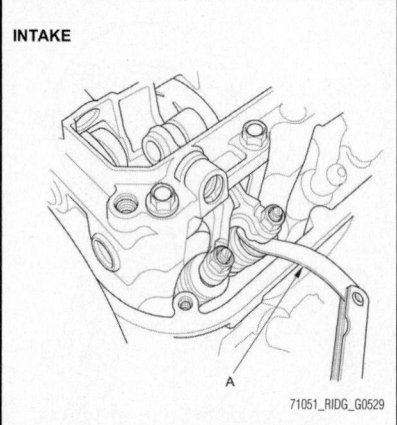

Fig. 118 Insert the feeler gauge (A) between the adjusting screw and the end of the valve stem on the No. 1 cylinder

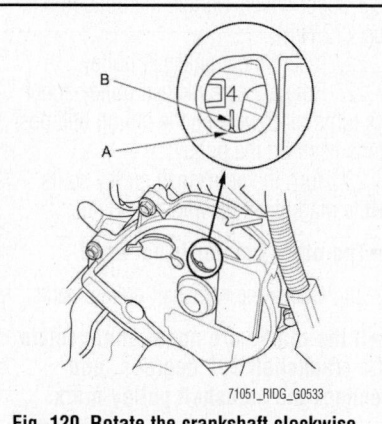

Fig. 120 Rotate the crankshaft clockwise, align the pointer (A) on the front upper cover with the No. 4 piston TDC mark (B) on the front camshaft pulley

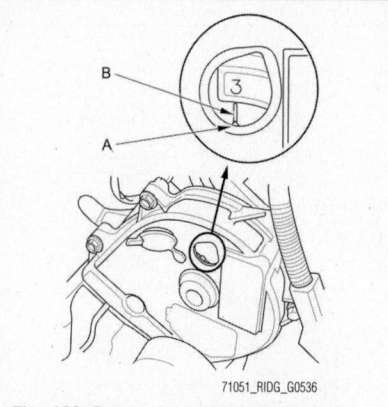

Fig. 123 Rotate the crankshaft clockwise, align the pointer (A) on the front upper cover with the No. 3 piston TDC mark (B) on the front camshaft pulley

drag, loosen the locknut, and turn the adjusting screw until the drag on the feeler gauge is correct.

6. While holding the adjusting screw with the screw driver, tighten the locknut, then recheck the clearance. Repeat the adjustment, if necessary.

   a. Specified Torque:

   b. Intake: 15 ft. lbs. (20 Nm)

➡**Apply new engine oil to the nut threads.**

   c. Exhaust: 10 ft. lbs. (14 Nm)

➡**Apply new engine oil to the nut threads.**

7. Rotate the crankshaft clockwise. Align the pointer on the front upper cover with the No. 4 piston TDC mark on the front camshaft pulley.

8. Check and, if necessary, adjust the valve clearance on the No. 4 cylinder.

9. Rotate the crankshaft clockwise.

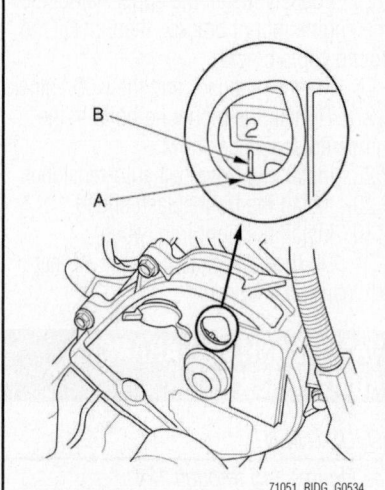

Fig. 121 Rotate the crankshaft clockwise, align the pointer (A) on the front upper cover with the No. 2 piston TDC mark (B) on the front camshaft pulley

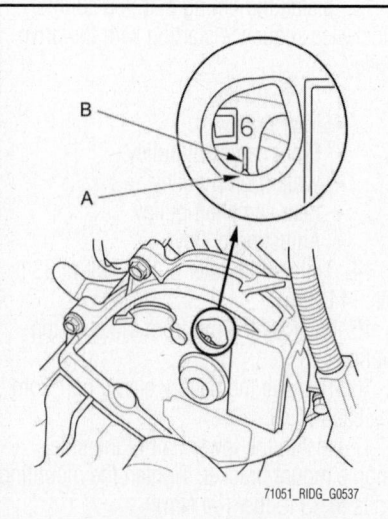

Fig. 124 Rotate the crankshaft clockwise, align the pointer (A) on the front upper cover with the No. 6 piston TDC mark (B) on the front camshaft pulley

Align the pointer on the front upper cover with the No. 2 piston TDC mark on the front camshaft pulley.

10. Check and, if necessary, adjust the valve clearance on the No. 2 cylinder.

11. Rotate the crankshaft clockwise. Align the pointer on the front upper cover with the No. 5 piston TDC mark on the front camshaft pulley.

12. Check and, if necessary, adjust the valve clearance on the No. 5 cylinder.

13. Rotate the crankshaft clockwise. Align the pointer on the front upper cover with the No. 3 piston TDC mark on the front camshaft pulley.

14. Check and, if necessary, adjust the valve clearance on the No. 3 cylinder.

15. Rotate the crankshaft clockwise. Align the pointer on the front upper cover with the No. 6 piston TDC mark on the front camshaft pulley.

16. Check and, if necessary, adjust the valve clearance on the No. 6 cylinder.

17. Install the cylinder head covers.

ENGINE PERFORMANCE & EMISSION CONTROLS

ACCELERATOR PEDAL POSITION (APP) SENSOR

LOCATION

See Figure 125.

REMOVAL & INSTALLATION

1. Remove the throttle cable.
2. Disconnect the Accelerator Pedal Position (APP) sensor 6P connector.
3. Remove the bolts and the APP sensor.
4. Install the parts in the reverse order of removal.

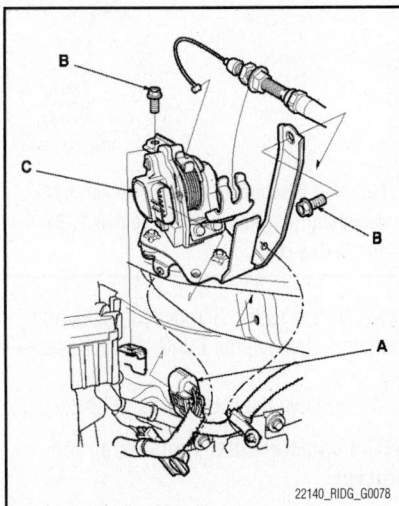

Fig. 125 APP sensor (C), attaching bolt (B) and connector (A)

CAMSHAFT POSITION (CMP) SENSOR

REMOVAL & INSTALLATION

See Figures 126 and 127.

1. Remove the timing belt.
2. Remove the front camshaft pulley (CMP sensor pulse plate).
3. Disconnect the CMP sensor connector, then remove the back cover.

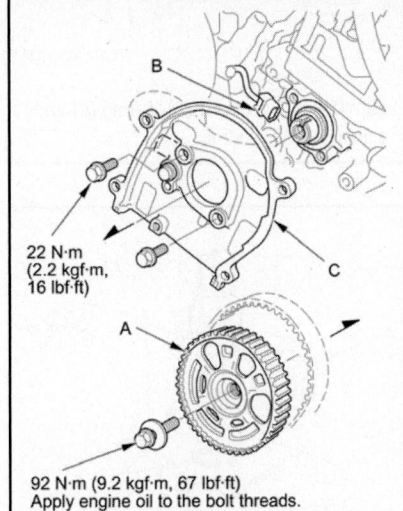

22 N·m
(2.2 kgf·m,
16 lbf·ft)

92 N·m (9.2 kgf·m, 67 lbf·ft)
Apply engine oil to the bolt threads.

71051_RIDG_G0545

Fig. 126 Remove the front camshaft pulley (CMP sensor pulse plate) (A), disconnect the CMP sensor connector (B), then remove the back cover (C)

4. Remove the CMP sensor from the back cover.

To install:

5. Install the parts in the reverse order of removal. Install the timing belt.

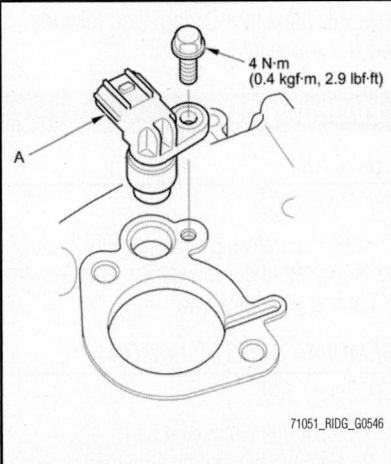

4 N·m
(0.4 kgf·m, 2.9 lbf·ft)

71051_RIDG_G0546

Fig. 127 Remove the CMP sensor (A) from the back cover

6. Do the CKP pattern clear/CKP pattern learn procedure.

CRANKSHAFT POSITION (CKP) SENSOR

REMOVAL & INSTALLATION

See Figure 128.

1. Raise the vehicle on a lift.

➡ **Make sure the vehicle is level, because engine oil will drip out when you remove the sensor.**

2. Remove the CKP sensor cover.
3. Disconnect the CKP sensor connector.
4. Remove the CKP sensor.

To install:

5. Install the parts in the reverse order of removal with a new O-ring.
6. Do the CKP pattern clear/CKP pattern learn procedure.

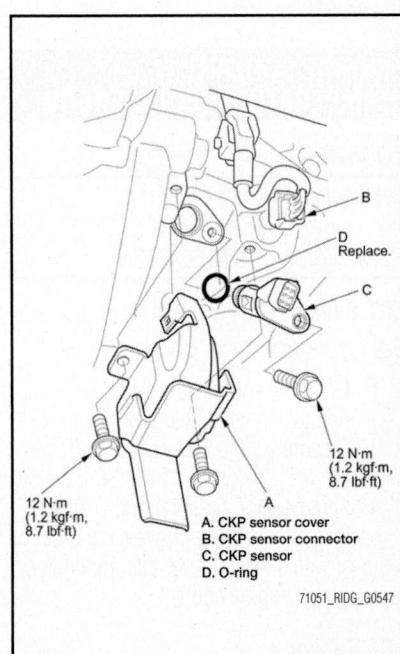

B

D
Replace.

C

12 N·m
(1.2 kgf·m,
8.7 lbf·ft)

12 N·m
(1.2 kgf·m,
8.7 lbf·ft)

A

A. CKP sensor cover
B. CKP sensor connector
C. CKP sensor
D. O-ring

71051_RIDG_G0547

Fig. 128 Exploded view of the CKP sensor assembly

CKP PATTERN CLEAR/CKP PATTERN LEARN

Clear/Learn Procedure (with the HDS)

1. Connect the HDS to the data link connector (DLC) located under the driver's side of the dashboard.

2. Turn the ignition switch to ON (II).

3. Make sure the HDS communicates with the PCM and other vehicle systems. If it does not, go to the DLC circuit troubleshooting.

4. Select CRANK PATTERN in the ADJUSTMENT MENU with the HDS.

5. Select CRANK PATTERN CLEAR, and clear the CKP pattern.

6. Select CRANK PATTERN LEARNING with the HDS, and follow the screen prompts.

Learn Procedure (without the HDS)

1. Start the engine. Hold the engine speed at 3,000 rpm without load (in P or N) until the radiator fan comes on.

2. Test-drive the vehicle on a level road: Decelerate (with the throttle fully closed) from an engine speed of 2,500 rpm down to 1,000 rpm with the transmission in 2.

3. Test-drive the vehicle on a level road: Decelerate (with the throttle fully closed) from an engine speed of 5,000 rpm down to 3,000 rpm with the transmission in 2.

4. Repeat steps 2 and 3 several times.

5. Turn the ignition switch to LOCK (0).

6. Turn the ignition switch to ON (II), and wait 30 seconds.

ENGINE COOLANT TEMPERATURE (ECT) SENSOR

LOCATION

See Figures 129 and 130.

REMOVAL & INSTALLATION

ECT Sensor 1

See Figure 131.

1. Drain the engine coolant.

2. Remove the engine cover.

3. Disconnect the ECT sensor 1 2P connector.

4. Remove ECT sensor 1 and O-ring.

5. Install the parts in the reverse order of removal with a new O-ring, then refill the radiator with engine coolant.

ECT Sensor 2

See Figure 132.

1. Remove the front splash shield.

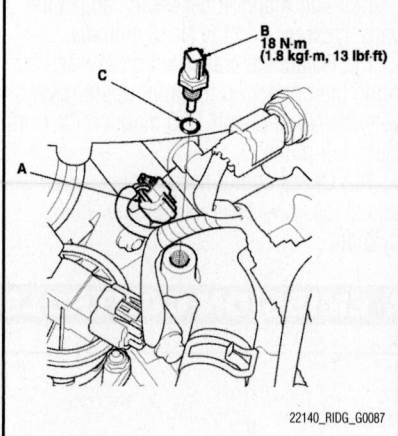

Fig. 129 ECT sensor 1 (B), O-ring (C) and connector (A) location

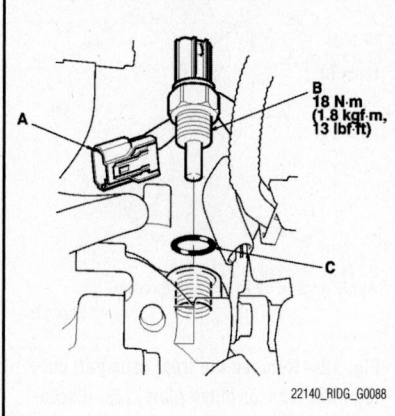

Fig. 130 ECT sensor 2 (B), O-ring (C) and connector (A) location

2. Drain the engine coolant.

3. Disconnect the ECT sensor 2 2P connector.

4. Remove ECT sensor 2 and O-ring.

5. Install the parts in the reverse order of removal with a new O-ring, then refill the radiator with engine coolant.

EVAP CANISTER

LOCATION

See Figure 133.

The Evaporative Emissions (EVAP) canister is located under the vehicle, just forward of the fuel tank, as shown.

REMOVAL & INSTALLATION

See Figure 134.

1. Raise the vehicle on a lift.

2. Remove the fuel tank guard.

3. Remove the hoses, the FTP sensor 3P connector, and the EVAP canister vent shut

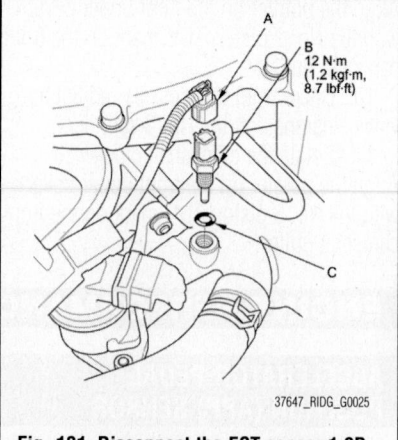

Fig. 131 Disconnect the ECT sensor 1 2P connector (A). Remove ECT sensor 1 (B) and O-ring (C).

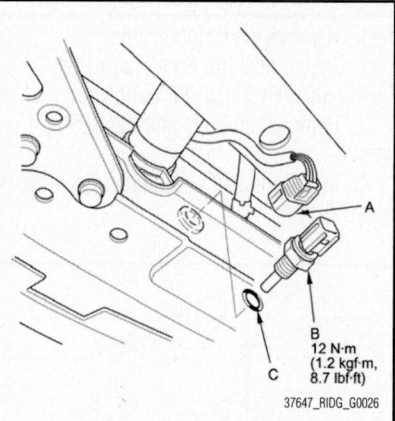

Fig. 132 Disconnect the ECT sensor 2 2P connector (A). Remove ECT sensor 1 (B) and O-ring (C).

valve 2P connector. Remove the bolts and the clips. Remove the EVAP canister assembly.

4. Remove the EVAP canister bracket.

➡**The canister filter remains on the bracket.**

5. Installation is the reverse of the removal procedure.

EXHAUST GAS RECIRCULATION (EGR) VALVE

REMOVAL & INSTALLATION

See Figure 135.

1. Remove the engine cover.

2. Disconnect the EGR valve 5P connector.

3. Remove the EGR valve.

4. Install the parts in the reverse order of removal with a new gasket.

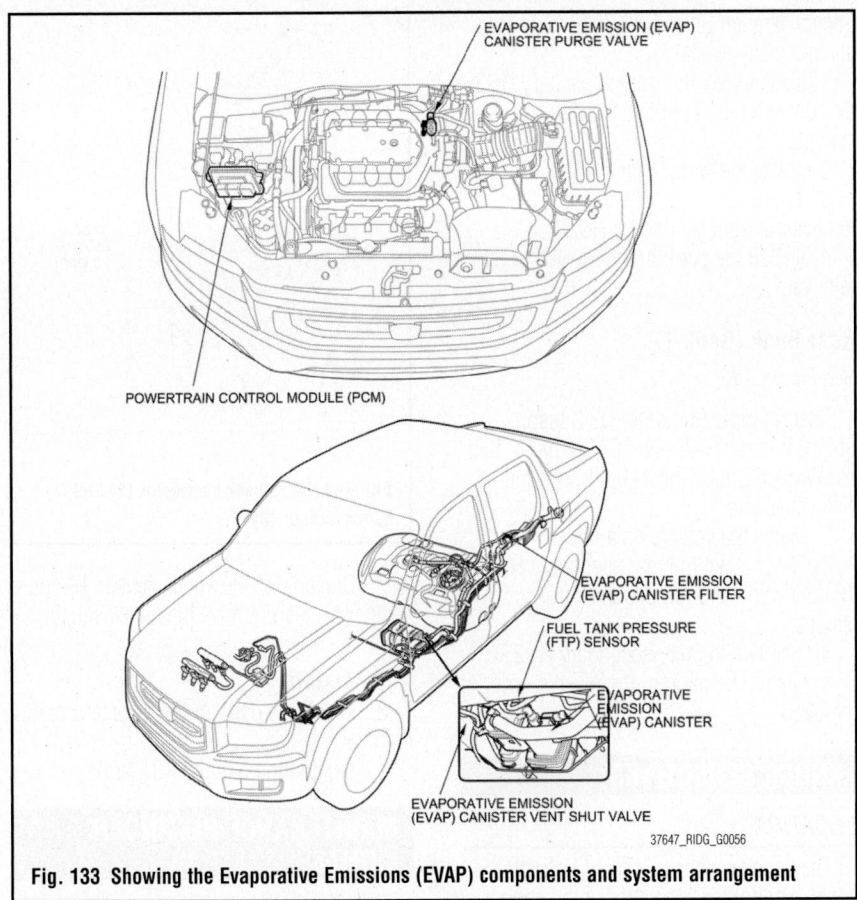

Fig. 133 Showing the Evaporative Emissions (EVAP) components and system arrangement

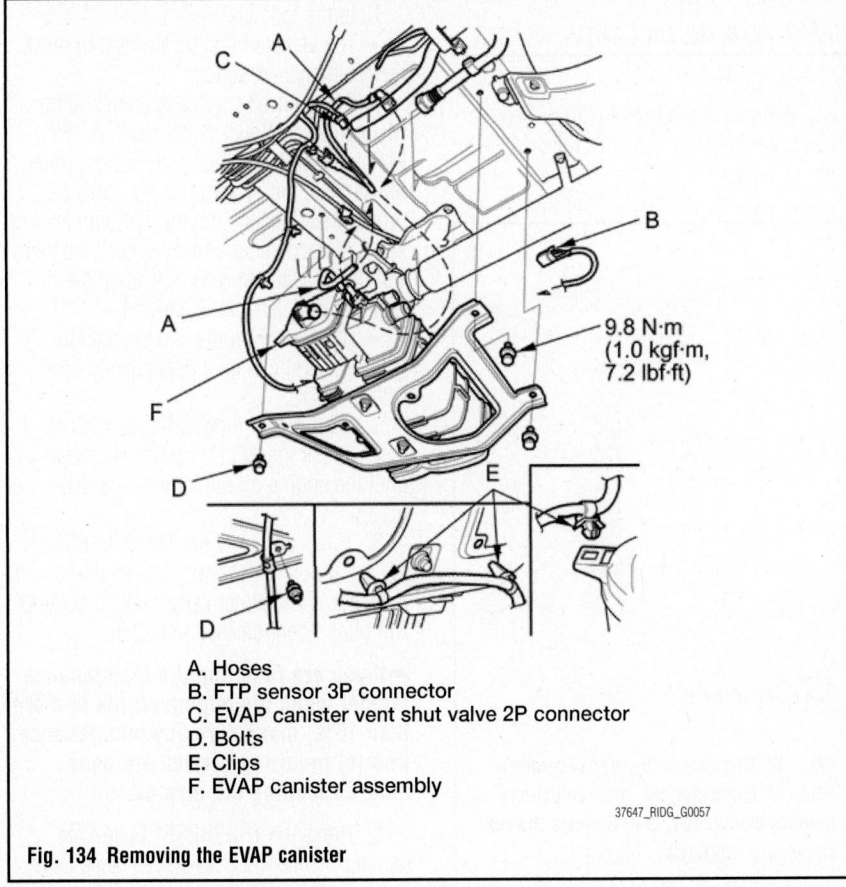

A. Hoses
B. FTP sensor 3P connector
C. EVAP canister vent shut valve 2P connector
D. Bolts
E. Clips
F. EVAP canister assembly

37647_RIDG_G0057

Fig. 134 Removing the EVAP canister

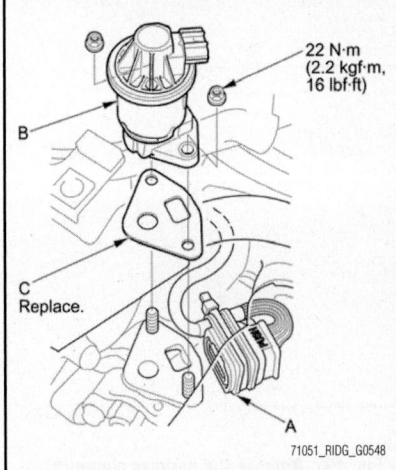

Fig. 135 Disconnect the EGR valve 5P connector (A), then remove the EGR valve (B) and gasket (C)

HEATED OXYGEN SENSOR (HO2S)

REMOVAL & INSTALLATION

Front Bank (Bank 2)

2011 Models

See Figures 136 and 137.

Special Tools Required: O2 Sensor Socket Wrench Snap-On SWR2 or O2 Sensor Wrench Snap-On YA8875, commercially available

1. Disconnect the front secondary HO2S 4P connector, and remove the harness clamps.
2. Raise the vehicle on a lift.
3. Remove the harness clamp, then remove the front secondary HO2S.
4. Install the parts in the reverse order of removal.

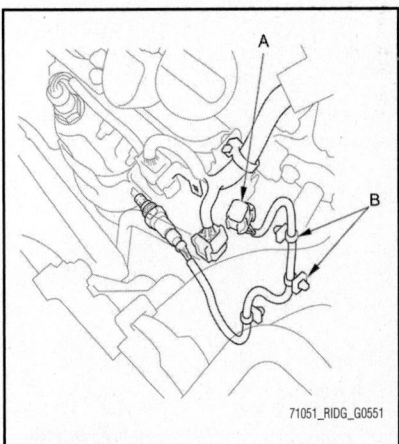

Fig. 136 Disconnect the front secondary HO2S 4P connector (A), and remove the harness clamps (B)

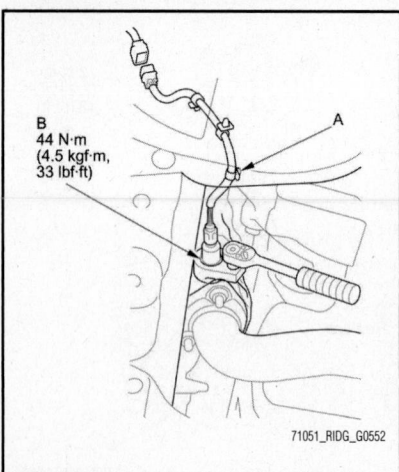

Fig. 137 Remove the harness clamp (A), then remove the front secondary HO2S (B)

2012 Models

See Figures 138 and 139.

Special Tools Required: O2 Sensor Socket Wrench Snap-On® SWR2 or O2

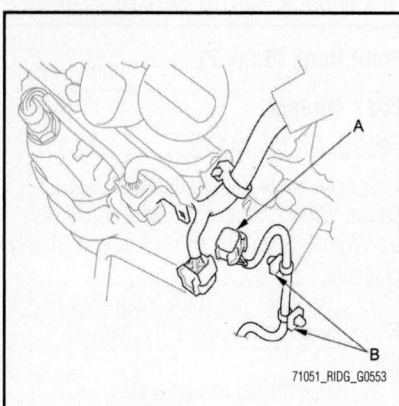

Fig. 138 Disconnect the front secondary HO2S 4P connector (A), and remove the harness clamps (B)

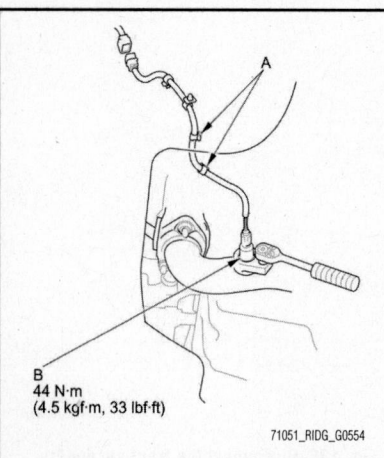

Fig. 139 Remove the harness clamps (A), then remove the front secondary HO2S (B)

Sensor Wrench Snap-On® YA8875, commercially available

1. Disconnect the front secondary HO2S 4P connector, and remove the harness clamps.
2. Raise the vehicle on a lift.
3. Remove the harness clamps, then remove the front secondary HO2S.
4. Install the parts in the reverse order of removal.

Rear Bank (Bank 1)

See Figure 140.

Special Tools Required: O2 Sensor Socket Wrench Snap-On® SWR2 or O2 Sensor Wrench Snap-On® YA8875, commercially available

1. Raise the vehicle on a lift.
2. Disconnect the rear secondary HO2S 4P connector, and remove the harness clamps.
3. Remove the rear secondary HO2S.
4. Install the parts in the reverse order of removal.

KNOCK SENSOR (KS)

LOCATION

The knock sensor is located beneath the intake manifold in the valley of the engine block.

REMOVAL & INSTALLATION

See Figure 141.

1. Remove the intake manifold. See "Intake Manifold" in "ENGINE MECHANICAL" section.

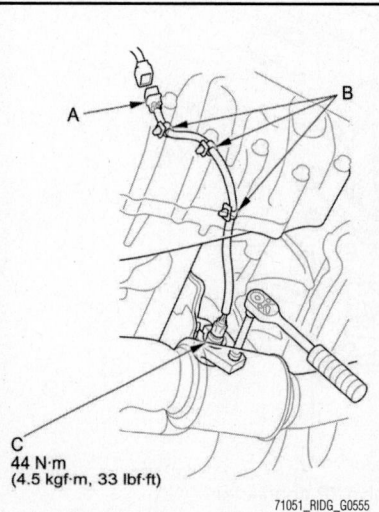

Fig. 140 Disconnect the rear secondary HO2S 4P connector (A), and remove the harness clamps (B), then Remove the rear secondary HO2S (C)

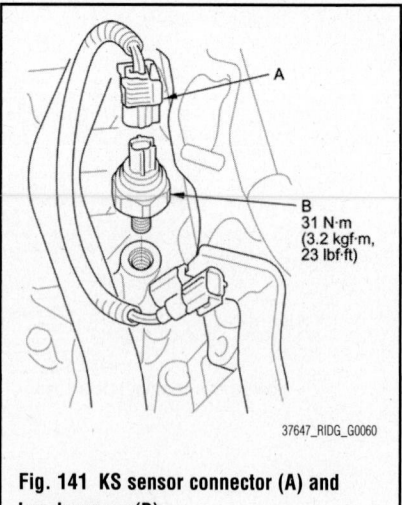

Fig. 141 KS sensor connector (A) and knock sensor (B)

2. Disconnect the knock sensor 1P connector (A), then remove the knock sensor (B).

To install:

3. Install the parts in the reverse order of removal.
4. Tighten to 23 ft. lbs. (31 Nm).

MAINTENANCE REMINDER LIGHTS

RESET PROCEDURE

Note the following:
• The vehicle must be stopped to reset the Maintenance Minder.
• If a required service is done and the Maintenance Minder is not reset, or if the Maintenance Minder is reset without doing the service, the system will not show the proper maintenance timing. This can lead to serious mechanical problems because there will be no accurate record of when the required maintenance is needed.
• The engine oil life and the maintenance item(s) can be independently reset with the HDS.

1. Turn the ignition switch to ON (II).
2. Press the SELECT button repeatedly until the engine oil life indicator is displayed.
3. Press and hold the TRIP/RESET button for about 10 seconds. The engine oil life indicator and the maintenance item code(s) will blink, then release the button.

➡**If you are resetting the Maintenance Minder when the engine oil life is more than 15%, make sure any maintenance item(s) requiring service are done before resetting the display.**

4. Press the TRIP/RESET button for another 5 seconds. The maintenance item

code(s) will disappear, and the engine oil life will reset to "100."

MALFUNCTION INDICATOR LIGHT

RESET PROCEDURE

1. Connect an HDS or equivalent OBD II scan tool to the diagnostic connector.
2. Turn the ignition switch to **ON**.
3. Turn the tester or scan tool **ON**.
4. Check whether any DTCs have been stored. Note them down if necessary.
5. Clear DTCs.
6. The MIL should turn **OFF**.

MANIFOLD ABSOLUTE PRESSURE (MAP) SENSOR

REMOVAL & INSTALLATION

See Figure 142.

1. Remove the engine cover.
2. Disconnect the MAP sensor 3P connector.
3. Remove the screw.
4. Remove the MAP sensor.
5. Install the parts in the reverse order of removal with a new O-ring.

MASS AIR FLOW/INTAKE AIR TEMPERATURE SENSOR

LOCATION

The Mass Air Flow/Intake Air Temperature (MAF/IAT) sensor is located on the intake air duct, near the throttle body.

REMOVAL & INSTALLATION

See Figure 143.

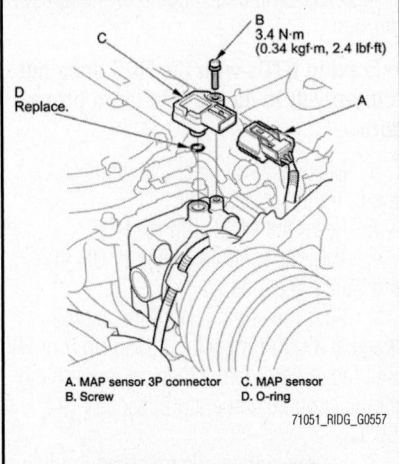

A. MAP sensor 3P connector   C. MAP sensor
B. Screw   D. O-ring

71051_RIDG_G0557

Fig. 142 Exploded view of the MAP sensor assembly

1. Disconnect the MAF/IAT sensor 5P connector.
2. Remove the screws.
3. Remove the MAF/IAT sensor and O-ring.
4. Install the parts in the reverse order of removal with a new O-ring.

POSITIVE CRANKCASE VENTILATION (PCV) VALVE

LOCATION

The Positive Crankcase Ventilation (PCV) valve is located at the left front corner of the engine, on the cylinder head.

REMOVAL & INSTALLATION

See Figure 144.

1. Remove the engine cover.
2. Remove the bolt.

➡**Do not spill oil on the hot exhaust manifold.**

3. Remove the PCV valve.

To install:

4. Install the parts in the reverse order of removal.

  a. When installing a new PCV valve, make sure the O-rings are in place.

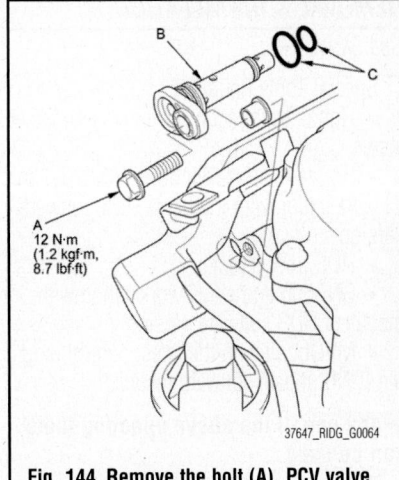

A
12 N·m
(1.2 kgf·m,
8.7 lbf·ft)

37647_RIDG_G0064

Fig. 144 Remove the bolt (A), PCV valve (B) and O-rings (C)

  b. When installing a used PCV valve, use new O-rings.

POWERTRAIN CONTROL MODULE (PCM)

LOCATION

The Powertrain Control Module (PCM) is located in the forward right corner of the engine compartment.

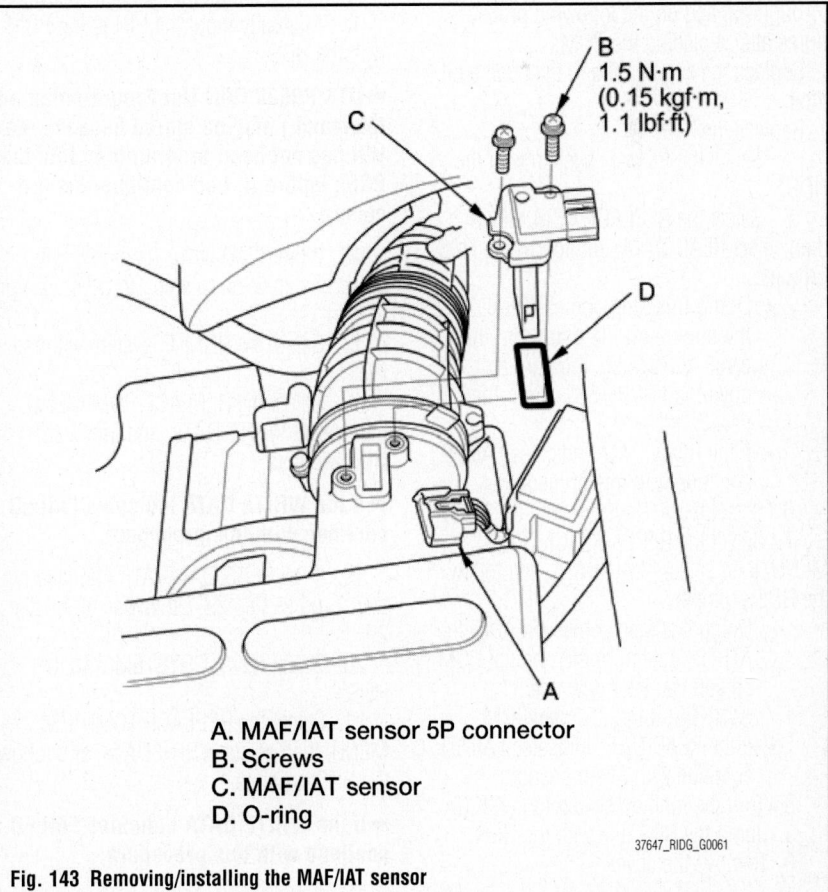

B
1.5 N·m
(0.15 kgf·m,
1.1 lbf·ft)

A. MAF/IAT sensor 5P connector
B. Screws
C. MAF/IAT sensor
D. O-ring

37647_RIDG_G0061

Fig. 143 Removing/installing the MAF/IAT sensor

REMOVAL & INSTALLATION

See Figure 145.

Special Tools Required:
- Honda Diagnostic System (HDS) tablet tester
- Honda Interface Module (HIM) and an iN workstation with the latest HDS software version
- HDS pocket tester
- GNA600 and an iN workstation with the latest HDS software version
- MVCI unit with the latest control module (CM) update software installed

➡**Any one of the above updating tools can be used.**

➡**Make sure the HDS/iN workstation or the MVCI unit has the latest HDS software version.**

1. Connect the HDS to the data link connector (DLC) located under the driver's side of the dashboard.

2. Turn the ignition switch to ON (II).

3. Make sure the HDS communicates with the PCM and other vehicle systems. If it does not, go to the DLC circuit troubleshooting. If the result of the DLC circuit troubleshooting is to replace the PCM, skip steps 4 through 7, 16 through 21, and 24 through 25, and do the following procedures after replacing the PCM:
Replace the engine oil and the engine oil filter.
Replace the ATF.

4. Select the PGM-FI system with the HDS.

5. Select the REPLACE PCM MENU, then select READ DATA, and follow the HDS prompts.
- Doing this step copies (READS) the engine oil life data from the original PCM so you can later download (WRITES) it into the new PCM.
- If the READ DATA indicates FAILED, continue with this procedure.

6. Select the A/T system with the HDS.

7. Select the REPLACE TCM/PCM MENU, then select READ DATA, and follow the HDS prompts.
- Doing this step copies (READS) the ATF life data from the original PCM so you can later download (WRITES) it into the new PCM.
- If the READ DATA indicates FAILED, continue with this procedure.

8. Turn the ignition switch to LOCK (0).

9. Jump the SCS line with the HDS.

10. Remove the cover.

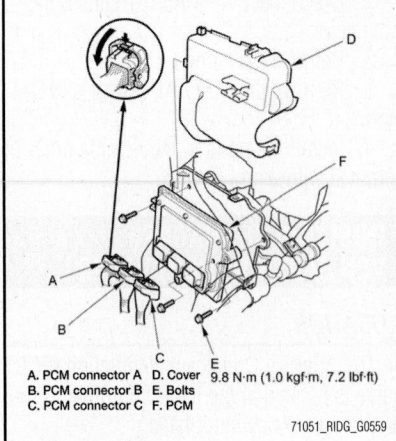

A. PCM connector A D. Cover
B. PCM connector B E. Bolts
C. PCM connector C F. PCM

9.8 N·m (1.0 kgf·m, 7.2 lbf·ft)

71051_RIDG_G0559

Fig. 145 Exploded view of the PCM assembly

11. Disconnect PCM connectors A, B, and C.

➡**PCM connectors A, B, and C have symbols (A=□, B=△, C=○) embossed on them for identification.**

12. Remove the bolts, then remove the PCM assembly.

To install:

13. Install the PCM in the reverse order of removal.

14. Turn the ignition switch to ON (II).

15. Manually input the VIN to the PCM with the HDS.

➡**DTC P0630 (VIN Not Programmed or Mismatch) may be stored because the VIN has not been programmed into the PCM; ignore it, and continue this procedure.**

16. If the READ DATA (engine oil life) failed in step 5, go to step 19. Otherwise go to step 17.

17. Select the PGM-FI system with the HDS.

18. Select the REPLACE PCM MENU, then select WRITE DATA, and follow the HDS prompts.

➡**If the WRITE DATA indicates FAILED, continue with this procedure.**

19. If the READ DATA (ATF life) failed in step 7, go to step 22. Otherwise go to step 20.

20. Select the A/T SYSTEM with the HDS.

21. Select the REPLACE TCM/PCM MENU, then select WRITE DATA, and follow the HDS prompts.

➡**If the WRITE DATA indicates FAILED, continue with this procedure.**

22. Select IMMOBI system with the HDS.

23. Enter the immobilizer PCM code that you got from the iN, and use the PCM replacement procedure in the IMMOBI MENU of the HDS; it allows you to start the engine.

24. If the READ DATA failed in step 5 or the WRITE DATA failed in step 18, replace the engine oil and the engine oil filter, then go to step 25.

25. If the READ DATA failed in step 7 or the WRITE DATA failed in step 21, replace the ATF, then go to step 26.

26. Clean the throttle body.

27. Select the PGM-FI system with the HDS, and reset the PCM with the HDS.

28. Update the PCM if it does not have the latest software.

29. Do the PCM idle learn procedure.

30. Do the CKP pattern clear/CKP pattern learn procedure.

RESET PROCEDURE

1. Reset the PCM with the HDS while the engine is stopped.

2. Turn the ignition to LOCK (0).

3. Turn the ignition switch to ON (II), and wait 30 seconds.

4. Turn the ignition to LOCK (0), and disconnect the HDS from the DLC.

5. Do the PCM idle learn procedure.

PCM IDLE LEARN PROCEDURE

The idle learn procedure must be done so the PCM can learn the engine idle characteristics.

Do the idle learn procedure whenever you do any of these actions:
- Replace PCM.
- Reset PCM.
- Update PCM.
- Clean or replace the throttle body.
- Disassemble the engine or the transmission.

➡**Erasing DTCs with the HDS does not require you to do the idle learn procedure.**

1. Make sure all electrical items (A/C, audio, lights, etc.) are off.

2. Reset the PCM with the HDS.

3. Turn the ignition switch to ON (II), and wait 2 seconds.

4. Start the engine. Hold the engine speed at 3,000 rpm without load (in P or N) until the radiator fan comes on, or until the engine coolant temperature reaches 194°F (90°C).

5. Let the engine idle for about 5 minutes with the throttle fully closed.

➡️**If the radiator fan comes on, do not include its running time in the 5 minutes.**

PCM UPDATE

Special Tools Required:
• Honda Diagnostic System (HDS) tablet tester
• Honda Interface Module (HIM) and an iN workstation with the latest HDS software version
• HDS pocket tester
• GNA600 and an iN workstation with the latest HDS software version
• MVCI unit with the latest control module (CM) update software installed

➡️**Any one of the above updating tools can be used.**

Note the following:
• Make sure the HDS/iN workstation or the MVCI unit has the latest HDS software version.
• Before you update the PCM, make sure the battery in the vehicle is fully charged, and connect a jumper battery (not a battery charger) to maintain system voltage.
• Never turn the ignition switch to ACC (I) or LOCK (0) during the update. If there is a problem with the update, leave the ignition switch in ON (II).
• To prevent PCM damage, do not operate anything electrical (headlights, audio system, brakes, A/C, power windows, moonroof (if equipped), door locks, etc.) during the update.
• To ensure the latest program is installed, do a PCM update whenever the PCM is substituted or replaced.
• You cannot update a PCM with a program it already has. It will only accept a new program.
• High temperature in the engine compartment might cause the PCM to become too hot to run the update. If the engine has been running before this procedure, open the hood and cool the engine compartment.
• If you need to diagnose the Honda interface module (HIM) because the HIM's red (#3) light came on or was flashing during the update, leave the ignition switch in ON (II) when you disconnect the HIM from the data link connector (DLC). This will prevent PCM damage.
1. Turn the ignition switch to ON (II), but do not start the engine.
2. Connect the HDS to the data link connector (DLC) located under the driver's side of the dashboard.
3. Make sure the HDS communicates with the PCM and other vehicle systems. If it does not, go to the DLC circuit troubleshooting.
4. Exit the HDS diagnostic system, then select the update mode, and follow the screen prompts to update the PCM.
5. If the software in the PCM is the latest, disconnect the updating tool from the DLC, and go back to the procedure that you were doing. If the software in the PCM is not the latest, follow the instructions on the screen. If prompted to choose the PGM-FI system or the A/T system, make sure you update both.

➡️**If the PCM update system requires you to cool the PCM, follow the instructions on screen. If you run into a problem during the update procedure (programming takes over 15 minutes, status bar goes over 100%, D or immobilizer indicator flashes, HDS tablet freezes, etc.), follow these steps to minimize the chance of damaging the PCM:**

a. Leave the ignition switch in the ON (II).
b. Connect a jumper battery (do not connect a battery charger).
c. Shut down the HDS.
d. Disconnect the HDS from the DLC.
e. Reboot the HDS.
f. Reconnect the HDS to the DLC, and try the update procedure again.
6. Do the PCM idle learn procedure.
7. Do the CKP pattern clear/CKP pattern learn procedure.

FUEL   GASOLINE FUEL INJECTION SYSTEM

FUEL SYSTEM SERVICE PRECAUTIONS

Safety is the most important factor when performing not only fuel system maintenance, but any type of maintenance. Failure to conduct maintenance and repairs in a safe manner may result in serious personal injury or death. Work on a vehicle's fuel system components can be accomplished safely and effectively by adhering to the following rules and guidelines.
• To avoid the possibility of fire and personal injury, always disconnect the negative battery cable unless the repair or test procedure requires that battery voltage be applied.
• Always relieve the fuel system pressure prior to disconnecting any fuel system component (injector, fuel rail, pressure regulator, etc.) fitting or fuel line connection. Exercise extreme caution whenever relieving fuel system pressure to avoid exposing skin, face and eyes to fuel spray. Please be advised that fuel under pressure may penetrate the skin or any part of the body that it contacts.
• Always place a shop towel or cloth around the fitting or connection prior to loosening to absorb any excess fuel due to spillage. Ensure that all fuel spillage is quickly removed from engine surfaces. Ensure that all fuel-soaked cloths or towels are deposited into a flame-proof waste container with a lid.
• Always keep a dry chemical (Class B) fire extinguisher near the work area.
• Do not allow fuel spray or fuel vapors to come into contact with a spark or open flame.
• Always use a second wrench when loosening or tightening fuel line connection fittings. This will prevent unnecessary stress and torsion on fuel piping. Always follow the proper torque specifications.
• Always replace worn fuel fitting O-rings with new ones. Do not substitute fuel hose where rigid pipe is installed.

RELIEVING FUEL SYSTEM PRESSURE

With the HDS

✳✳ CAUTION

Before disconnecting fuel lines or hoses, relieve pressure from the system by disabling the fuel pump, running the engine until it stalls, then disconnecting the fuel line/quick connect fitting in the engine compartment.

1. Connect the HDS to the data link connector (DLC) located under the driver's side of the dashboard.
2. Turn the ignition switch to ON (II).
3. Make sure the HDS communicates with the PCM. If it does not, go to the DLC circuit troubleshooting.

4. Turn the ignition switch to LOCK (0).

5. Remove the fuel fill cap to relieve the pressure in the fuel tank.

6. Turn the ignition switch to ON (II).

7. From the INSPECTION MENU of the HDS, select Fuel Pump OFF, then start the engine, and let it idle until it stalls.

- Do not allow the engine to idle above 1,000 rpm or the PCM will continue to operate the fuel pump.
- A DTC or a Temporary DTC may be set during this procedure. Check for DTCs, and clear them as needed.

8. Turn the ignition switch to LOCK (0).

9. Do the battery terminal disconnection procedure.

10. Remove the quick-connect fitting cover.

11. Check the fuel quick-connect fitting for dirt, and clean it if needed.

12. Place a rag or shop towel over the quick-connect fitting.

13. Disconnect the quick-connect fitting: Hold the connector with one hand, and squeeze the retainer tabs with the other hand to release them from the locking tabs. Pull the connector off.

- Be careful not to damage the line or other parts.
- Do not use tools.
- If the connector does not move, keep the retainer tabs pressed down, and alternately pull and push the connector until it comes off easily.
- Do not remove the retainer from the line; once removed, the retainer must be replaced with a new one.

14. After disconnecting the quick-connect fitting, check it for dirt or damage.

15. Do the battery terminal reconnection procedure.

Without the HDS

See Figure 146.

❈❈ CAUTION

Before disconnecting fuel lines or hoses, relieve pressure from the system by disabling the fuel pump, running the engine until it stalls, then disconnecting the fuel line/quick connect fitting in the engine compartment.

1. Remove the left kick panel, then remove PGM-FI main relay 2 from the under-dash fuse/relay box.

2. Start the engine, and let it idle until it stalls.

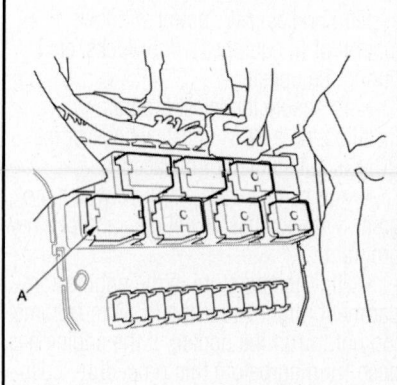

Fig. 146 Remove the left kick panel, then remove PGM-FI main relay 2 (A) from the under-dash fuse/relay box

➡**If any DTCs are stored, clear and ignore them.**

3. Turn the ignition switch to LOCK (0).

4. Remove the fuel fill cap to relieve the pressure in the fuel tank.

5. Do the battery terminal disconnection procedure.

6. Remove the quick-connect fitting cover.

7. Check the fuel quick-connect fitting for dirt, and clean it if needed.

8. Place a rag or shop towel over the quick-connect fitting.

9. Disconnect the quick-connect fitting: Hold the connector with one hand, and squeeze the retainer tabs with the other hand to release them from the locking tabs. Pull the connector off.

- Be careful not to damage the line or other parts.
- Do not use tools.
- If the connector does not move, keep the retainer tabs pressed down, and alternately pull and push the connector until it comes off easily.
- Do not remove the retainer from the line; once removed, the retainer must be replaced with a new one.

10. After disconnecting the quick-connect fitting, check it for dirt or damage.

11. Do the battery terminal reconnection procedure.

IDLE SPEED

ADJUSTMENT

Idle speed is controlled by the Powertrain Control Module (PCM). No adjustment is necessary or possible.

FUEL FILTER

REMOVAL & INSTALLATION

See Figure 147.

➡**The fuel filter should be replaced whenever the fuel pressure drops below the specified 57 psi, after making sure that the fuel pump and the fuel pressure regulator are OK.**

1. Remove the fuel tank unit.

2. Remove the fuel filter set.

To install:

3. Check these items before installing the fuel tank unit:

a. When connecting the wire harness, make sure the connection is secure and the connectors are firmly locked into place.

b. When installing the fuel gauge sending unit, make sure the connection is secure and the connector is firmly locked into place. Be careful not to bend or twist it excessively.

4. Install the parts in the reverse order of removal with new O-rings. When installing the fuel tank unit, align the marks on the unit and the fuel tank.

- Coat the O-rings with clean engine oil. Do not use any other oil or fluid.
- Do not pinch the O-rings during installation.
- Use all the new parts supplied in the pressure regulator replacement kit.

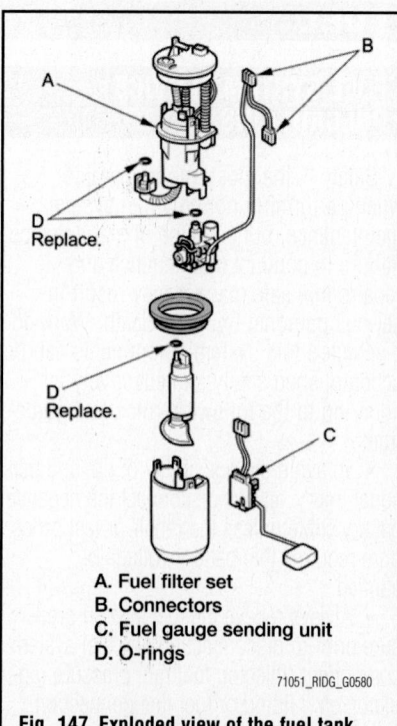

A. Fuel filter set
B. Connectors
C. Fuel gauge sending unit
D. O-rings

Fig. 147 Exploded view of the fuel tank unit showing the fuel filter set

FUEL TANK UNIT

REMOVAL & INSTALLATION

See Figure 148.

Special Tools Required: Fuel Sender Wrench 07AAA-S0XA100

1. Relieve the fuel pressure.
2. Remove the fuel fill cap.
3. Remove the rear seat.
4. Remove the access panel from the floor, then disconnect the fuel tank unit 5P connector.
5. Disconnect the quick-connect fitting from the fuel tank unit.
6. Using the special tool, loosen the fuel tank unit locknut.
7. Remove the locknut, the locknut plate, the fuel tank unit, and the base gasket.

To install:

8. Temporarily attach a new base gasket to the fuel tank unit, then insert the fuel tank unit partially into the fuel tank.
   - Be careful not to damage the new base gasket.
   - Be careful not to bend the fuel gauge sending unit.
   - Do not coat the base gasket with oil.
9. Transfer the base gasket from the fuel tank unit onto the fuel tank.
10. Align the marks on the fuel tank and the fuel tank unit, then insert the fuel tank unit into the fuel tank until the fuel tank unit rests on top of the base gasket.

➡**To prevent a fuel leak, check the base gasket, visually or by hand, to make sure it is not pinched.**

11. Tighten a new locknut by hand with a new locknut plate, then tighten the new locknut using the special tool.

- After tightening, make sure the marks are still aligned.
- After installation, check the base gasket, visually or by hand, to be sure it is not pinched.

12. Connect the fuel tank unit 5P connector, then connect the quick-connect fitting.
13. Reconnect the negative cable to the battery, and turn the ignition switch to ON (II) (but do not operate the starter motor). The fuel pump will run for about 2 seconds, and fuel pressure will rise. Repeat this two or three times, then check that there is no leakage in the fuel supply system.
14. Install the access panel.
15. Install the rear seat.
16. Install the fuel fill cap.

FUEL RAIL & INJECTORS

REMOVAL & INSTALLATION

See Figures 149 and 150.

1. Relieve the fuel pressure.
2. Remove the intake manifold.

3. Disconnect the quick-connect fitting.
4. Remove the fuel joint hose mounting bolt.
5. Disconnect the connectors from the injectors.
6. Remove the fuel rail mounting bolts from the fuel rails.
7. Remove the fuel rails and the injectors from the injector base.
8. Remove the injector clips from the fuel rails.
9. Remove the injectors from the fuel rails.

To install:

10. Coat the new O-rings (black) with clean engine oil, and insert the injectors into the fuel rails.
11. Install the injector clips.
12. Coat the new injector O-rings (green) with clean engine oil.
13. Install the fuel rails and the injectors in the injector base.
14. Install the fuel rail mounting bolts, and connect the connectors to the injectors.

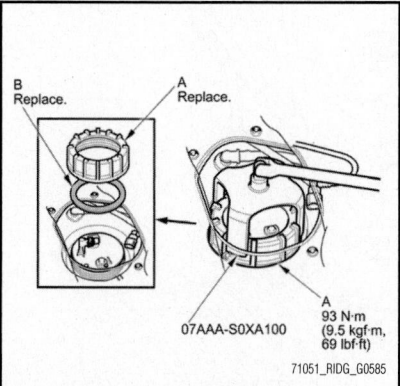

A
93 N·m
(9.5 kgf·m, 69 lbf·ft)

07AAA-S0XA100

71051_RIDG_G0585

Fig. 148 Tighten a new locknut (A) by hand with a new locknut plate (B), then tighten the new locknut using the special tool

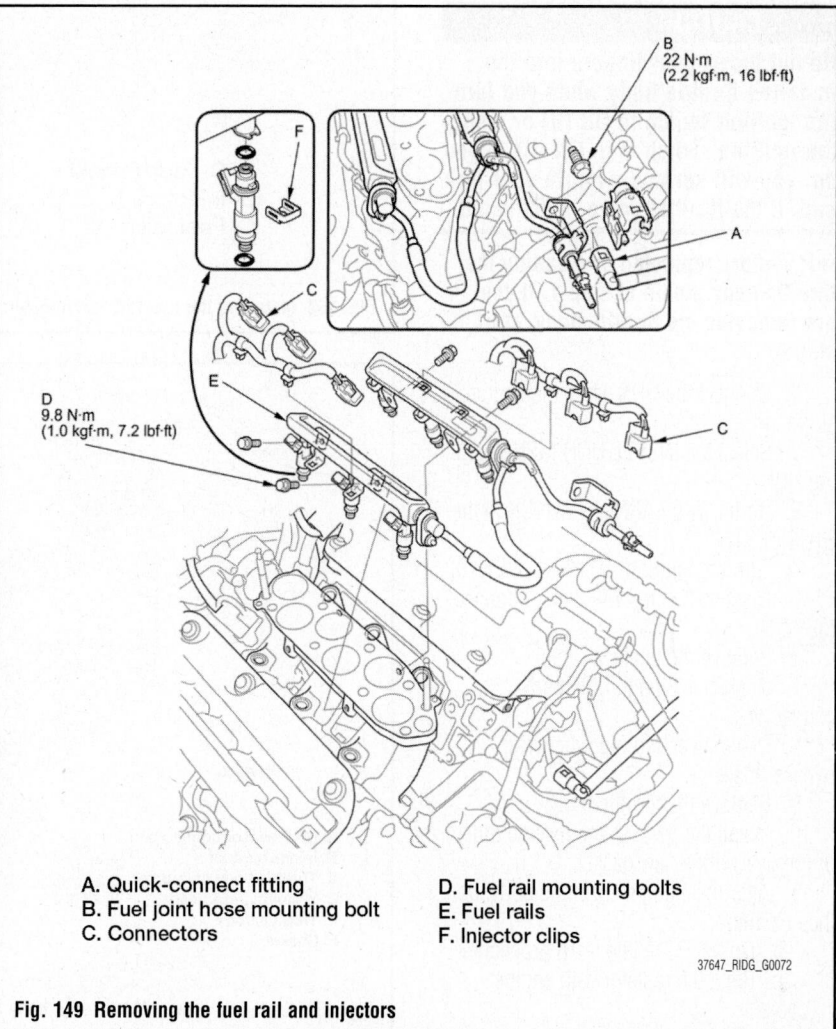

B
22 N·m
(2.2 kgf·m, 16 lbf·ft)

D
9.8 N·m
(1.0 kgf·m, 7.2 lbf·ft)

A. Quick-connect fitting
B. Fuel joint hose mounting bolt
C. Connectors
D. Fuel rail mounting bolts
E. Fuel rails
F. Injector clips

37647_RIDG_G0072

Fig. 149 Removing the fuel rail and injectors

15. Install the fuel joint hose mounting bolt.

16. Connect the quick-connect fitting.

17. Turn the ignition switch to ON (II), but do not operate the starter. After the fuel pump runs for about 2 seconds, the fuel rail will be pressurized. Repeat this two or three times, then check for fuel leaks.

18. Install the intake manifold with a new gasket.

THROTTLE BODY

REMOVAL & INSTALLATION

See Figure 151.

✳ CAUTION

Before disconnecting fuel lines or hoses, relieve pressure from the system by disabling the fuel pump, running the engine until it stalls, then disconnecting the fuel line/quick connect fitting in the engine compartment.

✳ CAUTION

Do not insert your fingers into the installed throttle body when you turn the ignition switch to ON (II) or while the ignition switch is in ON (II). If you do, you will seriously injure your fingers if the throttle valve is activated.

➡ **If you are replacing or cleaning the throttle body, begin at step 1. If you are removing the throttle body, start at step 4.**

1. Connect the HDS while the engine is stopped.

2. Select the INSPECTION MENU with the HDS.

3. Do the TP LEARNING CHECK in the ETCS TEST.

4. Turn the ignition switch to LOCK (0).

5. Disconnect the MAP sensor connector.

6. Remove the intake air duct.

7. Disconnect the throttle body connector.

8. Disconnect and plug the water bypass hoses.

9. Remove the throttle body.

10. Install the parts in the reverse order of removal with a new gasket.

11. After installing the throttle body, do these items:

   a. Do the PCM idle learn procedure.

   b. Refill the radiator with engine coolant.

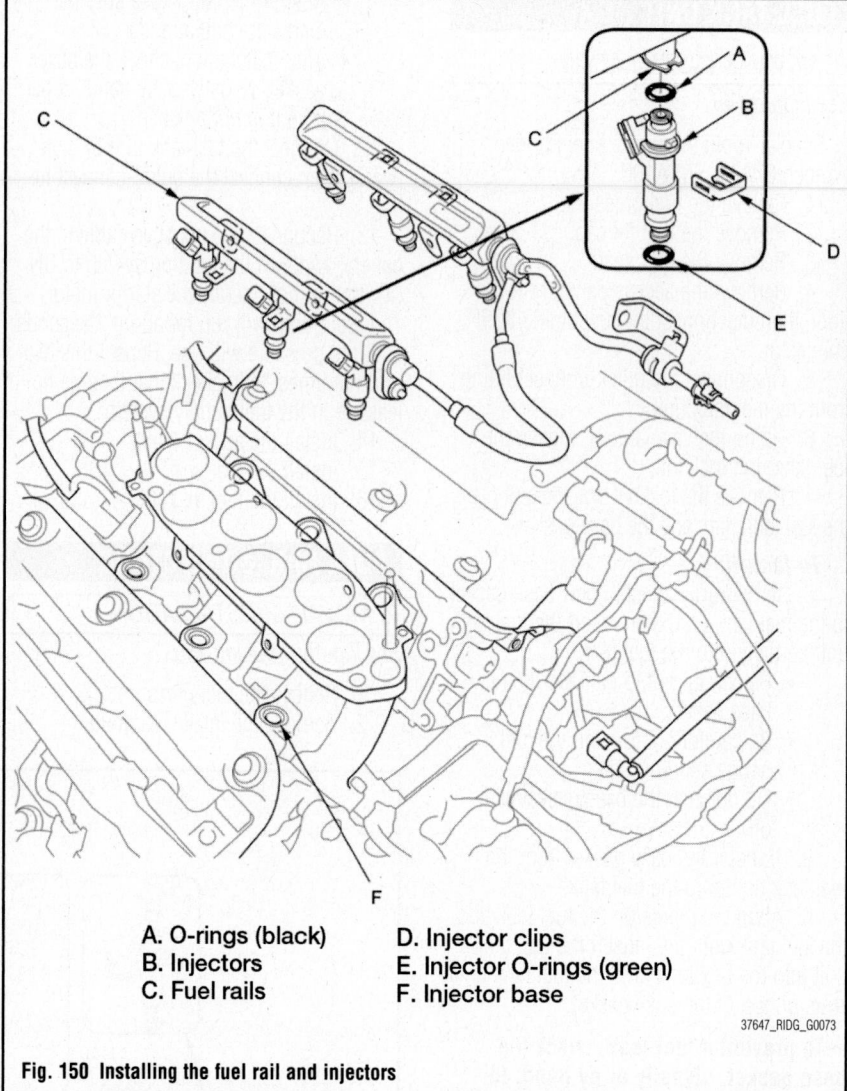

A. O-rings (black)
B. Injectors
C. Fuel rails
D. Injector clips
E. Injector O-rings (green)
F. Injector base

37647_RIDG_G0073

Fig. 150 Installing the fuel rail and injectors

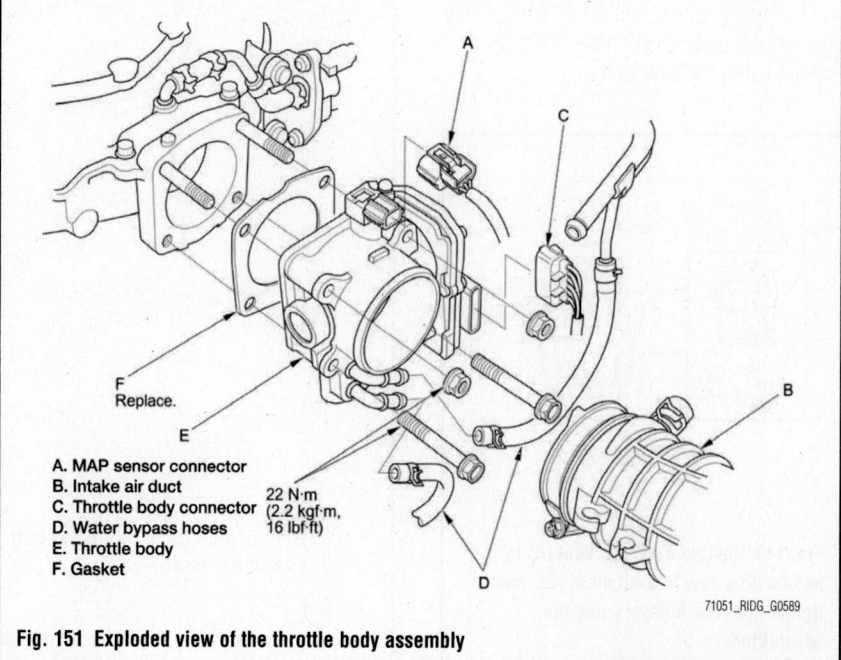

A. MAP sensor connector
B. Intake air duct
C. Throttle body connector
D. Water bypass hoses
E. Throttle body
F. Gasket

22 N·m
(2.2 kgf·m,
16 lbf·ft)

Replace.

71051_RIDG_G0589

Fig. 151 Exploded view of the throttle body assembly

HEATING & AIR CONDITIONING SYSTEM

BLOWER MOTOR

REMOVAL & INSTALLATION

See Figures 152 and 153.

1. Remove the glove box housing.
2. Remove the bolts and the glove box frame.
3. Cut the plastic cross brace in the glove box opening with diagonal cutters in the area shown, and discard it.
4. Disconnect these connectors from the blower motor and the power transistor, then detach the wire harness clips.
5. Disconnect the connector from the recirculation control motor, then remove the self-tapping screws, the mounting nuts, and the blower unit.

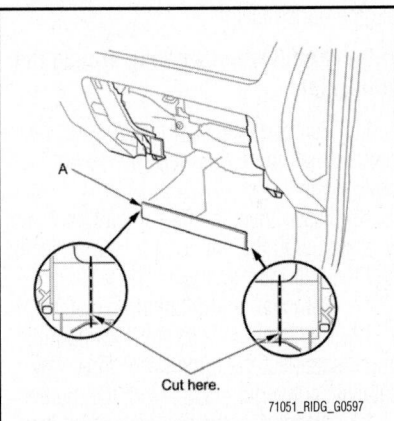

Fig. 152 Cut the plastic cross brace (A) in the glove box opening with diagonal cutters in the area shown, and discard it

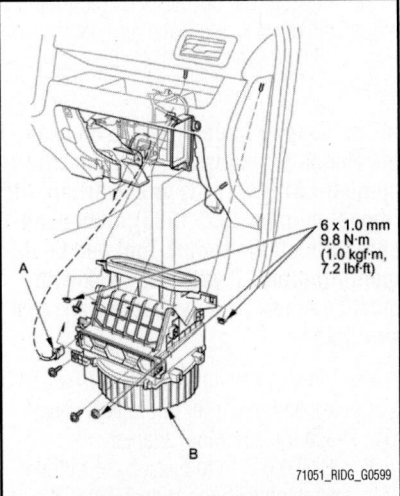

Fig. 153 Disconnect the connector (A) from the recirculation control motor, then remove the self-tapping screws, the mounting nuts, and the blower unit (B)

6. Install the unit in the reverse order of removal. Make sure that there is no air leakage.

HEATER CORE

REMOVAL & INSTALLATION

See Figures 154 through 156.

❊❊ CAUTION

SRS components are located in this area. Review the SRS component locations and the precautions and procedures before doing repairs or service.

1. Do the battery terminal disconnection procedure.
2. Recover the refrigerant with a recovery/recycling/ charging station
3. Disconnect the suction and the receiver lines from the evaporator core.
4. When the engine is cool, drain the engine coolant from the radiator.
5. From under the hood, slide the hose clamps back, then disconnect the inlet heater hose and the outlet heater hose from the heater unit. Note the layout of the hoses.

➡**Engine coolant will run out when the hoses are disconnected; drain it into a clean drip pan. Be sure not to let coolant spill on the electrical parts or the painted surfaces. If any coolant spills, rinse it off immediately.**

6. Remove the mounting nut from the

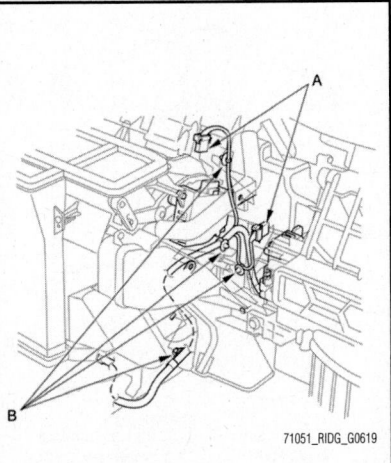

Fig. 154 Disconnect these connectors (A) from the mode control motor and the recirculation control motor, then detach the wire harness clips (B)

heater unit. Take care not to damage or bend the fuel lines or the brake lines.
7. Remove the dashboard.
8. Disconnect these connectors from the blower motor and the power transistor, then detach the wire harness clips.
9. Disconnect the connectors shown from the mode control motor and the recirculation control motor, then detach the wire harness clips.
10. Disconnect the connectors from the evaporator sensor and the air mix control motor, then detach the wire harness clips and the wire harness.

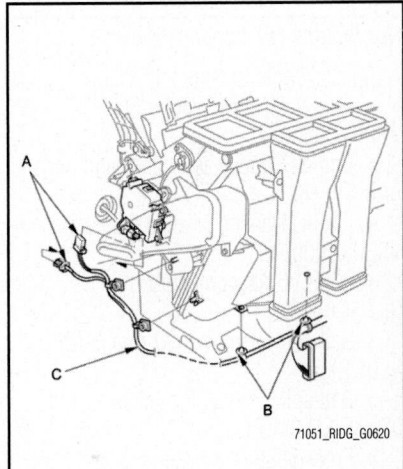

Fig. 155 Disconnect these connectors (A) from the evaporator sensor and the air mix control motor, then detach the wire harness clips (B) and the wire harness (C)

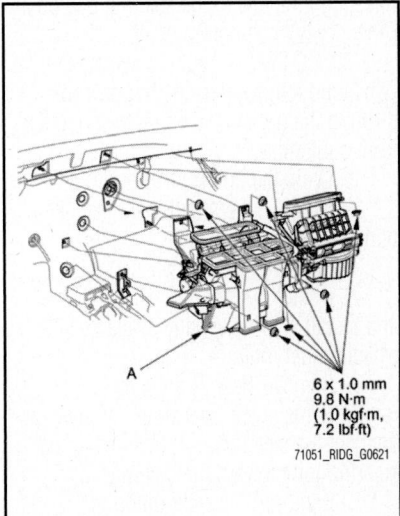

Fig. 156 Remove the mounting nuts and the blower-heater unit (A)

11. Remove the mounting nuts and the blower-heater unit.

12. Remove the self-tapping screws, the joint duct, and seal.

13. Remove the self-tapping screws, then remove the passenger's heater outlet, and the heater core cover.

14. Remove the self-tapping screws, the heater pipe brackets, the grommets, and carefully pull out the heater core so you don't bend the inlet and outlet pipes.

To install:

15. Install the heater core in the reverse order of removal.

16. Install the heater unit in the reverse order of removal, and note these items:

   a. Do not interchange the inlet and outlet heater hoses, and install the hose clamps securely.

   b. Refill the cooling system with engine coolant.

   c. Make sure that there is no coolant leakage.

   d. Make sure that there is no air leakage.

   e. Charge the system.

17. Do the battery terminal reconnection procedure.

18. After installation, operate the heater unit in various functions to confirm that it works properly.

STEERING

POWER RACK & PINION STEERING GEAR

REMOVAL & INSTALLATION

See Figures 157 through 173.

Special Tools Required*:
- Engine Hanger Adapter Set VSB02C000024
- Engine Support Hanger, A and Reds AAR-T1256
- Engine Hanger Balance Bar VSB02C000019
- 2008 V6 Attachment Arm SIL02C000033
- Ball Joint Remover, 32 mm 07MAC-SL0A102
- Ball Joint Thread Protector, 12 mm 07AAF-SDAA100
- Front Subframe Adapter EQS02BMDXSB0
- *Available through the Honda Tool and Equipment Program, 888-424-6857.

Note these items during removal:
- Using clean solvent and a brush, wash any oil and dirt off the valve body unit, it's lines, and the end of the steering gearbox. Blow dry with compressed air.
- Lower the front subframe from the body, and remove the steering gearbox through the gap produced by lowering the front subframe.

1. Drain the power steering fluid.

2. Do the battery terminal disconnection procedure.

3. Remove the front wheels.

4. Center the steering wheel spokes, and install a commercially available steering wheel holder tool.

5. Remove steering joint cover.

6. Remove the steering joint bolts, disconnect the steering joint by moving the steering joint toward the column.

7. Remove the center guide (if equipped) from the top of the pinion shaft, and discard it. The center guide is for factory assembly use only.

8. Disconnect the pump outlet hose from the power steering pump, and remove the clamp.

9. Remove the 10 mm flange bolts on the engine side mount bracket.

10. Remove the connector bracket from the front cylinder head; use the bracket bolt hole to attach the engine hanger balance bar front arm.

11. Remove the harness clamp bracket from the rear cylinder head; use the bracket bolt hole to attach the 2008 V6 attachment arm (SIL02C000033).

12. Remove the service caps for the front damper flange nuts from the cowl cover. Position the engine hanger adapters (VSB02C000024) with the "FRONT" mark facing forward over the damper flange nuts.

13. Install the engine hanger balance bar (VSB02C000019). Attach the front arm to the front cylinder head with a spacer and a 10 x 1.25 mm bolt. Remove the rear arm from the balance bar, then install the 2008 V6 attachment arm (SIL02C000033). Attach

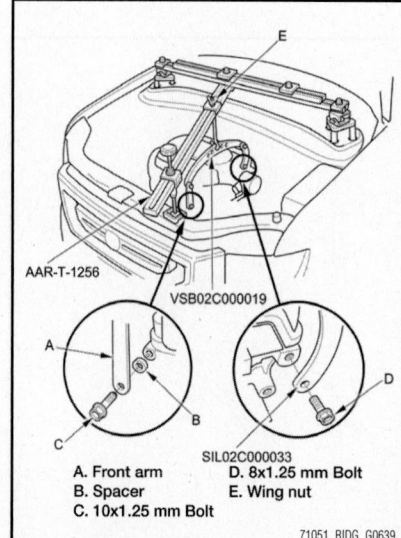

AAR-T-1256:

VSB02C000019

A. Front arm
B. Spacer
C. 10x1.25 mm Bolt
D. 8x1.25 mm Bolt
E. Wing nut

SIL02C000033

71051_RIDG_G0639

Fig. 157 Install the engine hanger balance bar

the 2008 V6 attachment arm to the rear cylinder head with an 8 x 1.25 mm bolt.

14. Install the engine support hanger (AAR-T1256) onto the vehicle as shown, and attach the hook to the slotted hole in the engine hanger balance bar. Tighten the wing nut by hand, and lift and support the engine/transmission.

➡ **Be careful when working around the windshield.**

15. Remove the cotter pin from the tie-rod end ball joint, then remove the nut on both sides.

16. Disconnect the tie-rod ball joint and knuckle using the ball joint thread protector and the ball joint remover.

17. Remove the front subframe stiffener.

18. Remove the three self-locking nuts, and disconnect the under-floor three way catalytic converter (TWC) from the muffler.

19. Disconnect the power steering pressure (PSP) switch connector.

20. Remove the No.1 propeller shaft protector.

21. Remove the front splash shield.

22. Loosen the four bolts holding the adjustable arms of the front subframe adapter (EQS02BMDXSB0) to its center plate.

➡ **The adapter is designed to be used the Honda transmission jack (model number LSL-W93714) or powertrain lift (model number OTC-1585), both available through the Honda Tool and Equipment Program. It will also work with most commercially available transmission jacks.**

23. Line up the slots in the arms with the bolt holes on the corner of the jack base, then attach the subframe adapter (EQS02BMDXB0) to the jack base with the bolts that came with the jack. Tighten the bolts securely.

24. Raise the jack to the vehicle height, then attach the front subframe adapter to the

front subframe using the subframe stiffener mounting bolts and bolt holes.

25. Remove the 12 mm flange bolts from both sides of the front suspension subframe front brackets.

26. Loosen the 14 mm flange bolts on the front suspension subframe so they are about 1.2 inches (30 mm) from both sides of the mounting surface. Do not loosen the 14 mm flange bolts more than necessary.

27. Support the front subframe securely by raising the transmission jack. Remove the 10 mm flange bolts, nuts, and the subframe bolt retainers.

28. Remove the 12 mm flange bolts, 14 mm special bolts from both sides of the front suspension subframe rear brackets.

29. Lower the transmission jack slowly until the front subframe has dropped about 2.0 inches (50 mm).

30. Remove the P/S line mounting brackets from the front subframe and the gearbox mounting bracket.

31. Loosen the adjustable hose clamp and disconnect the return hose.

32. Remove the two 10 mm flange bolts from the right side of the steering gearbox, then remove the mounting bracket and cushion.

33. Remove the two 10 mm flange bolts from the left side of the gearbox.

34. Lower the transmission jack slowly until the front subframe has dropped 3.9 inches (100 mm) total.

35. Remove the gearbox stiffener bracket from the left side of the front subframe.

36. Loosen the 16mm flare nut, and disconnect the inlet line.

37. Slide the steering gearbox between the body and front subframe toward the left, and remove the steering gearbox.

38. Remove the pinion shaft grommet from the frame.

39. After removing the steering gearbox, make sure that no power steering fluid gets on the gearbox mount cushions, the gearbox housing, the surface of the front subframe and the stiffener. Wipe off any spilled fluid at once.

To install:

40. Before installing the steering gearbox, make sure that no power steering fluid is on the mating surface of the steering gearbox and the front subframe. To prevent the gearbox mounting bolts from loosening after the installation, remove any power steering fluid from the mount cushions and bolt holes.

41. Install the pinion shaft grommet on the valve housing.

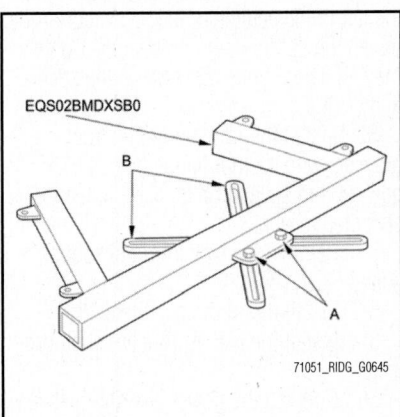

Fig. 158 Loosen the four bolts (A) holding the adjustable arms (B) of the front subframe adapter to its center plate

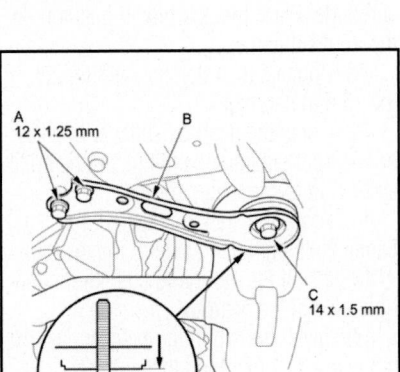

Fig. 160 Remove the 12 mm flange bolts (A) from both sides of the front suspension subframe front brackets (B); loosen the 14 mm flange bolts (C) on the front suspension subframe

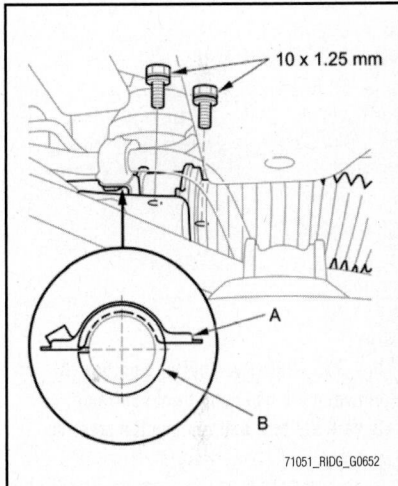

Fig. 162 Remove the two 10 mm flange bolts from the right side of the steering gearbox, then remove the mounting bracket (A) and cushion (B)

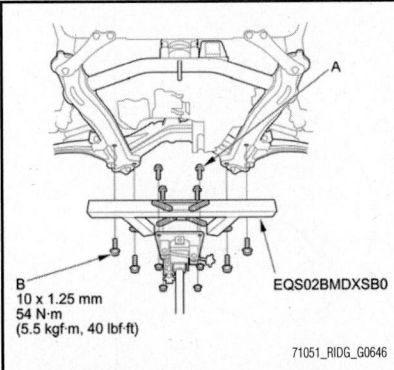

Fig. 159 Attach the subframe adapter to the jack base with bolts (A), then attach the front subframe adapter to the front subframe using the subframe stiffener mounting bolts (B) and bolt holes

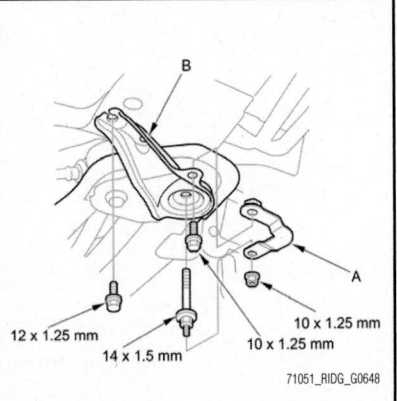

Fig. 161 Subframe bolt retainers (A), and front suspension subframe rear brackets (B)

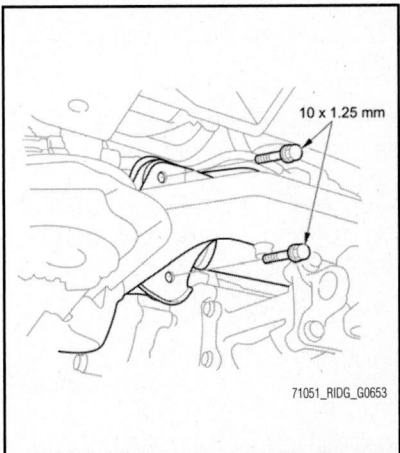

Fig. 163 Remove the two 10 mm flange bolts from the left side of the gearbox

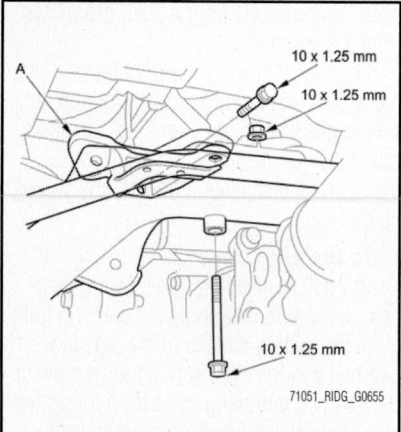

Fig. 164 Remove the gearbox stiffener bracket (A) from the left side of the front subframe

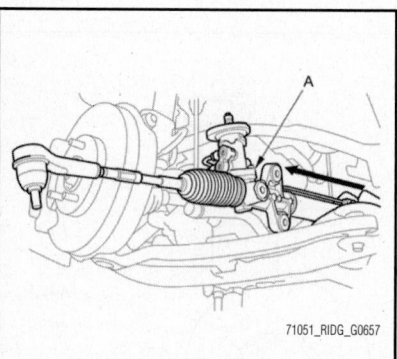

Fig. 165 Slide the steering gearbox (A) between the body and front subframe toward the left, and remove the steering gearbox

➡ **If the return line joint is removed, install it.**

42. Slide the steering gearbox between the front subframe and body from the dri-

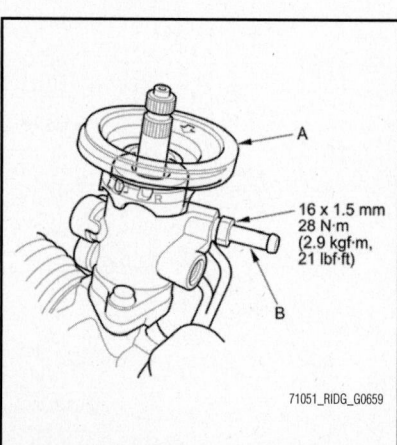

Fig. 166 Install the pinion shaft grommet (A) on the valve housing; return line joint (B)

ver's side. Place the gearbox in position on the front subframe.

43. Connect the inlet line, and tighten the 16 mm flare nut.

44. Install the gearbox stiffener bracket on the front subframe, and tighten the bolts and nut to the specified torque.

45. Loosely install the new 10 mm flange bolts on the left side of the gearbox.

46. Install the mounting cushion on the right side of the gearbox.

47. Install the mounting bracket over the mounting cushion, then install the two 10 mm flange bolts. Tighten the four 10 mm flange bolts for the gearbox to the specified torque.

48. Connect the return hose securely, and tighten the adjustable hose clamp.

49. Install the P/S line mounting brackets on the front suspension subframe and gearbox mounting bracket.

50. Carefully raise the front subframe with the front subframe adapter and the

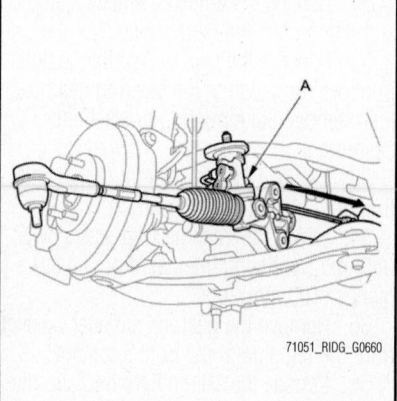

Fig. 167 Slide the steering gearbox (A) between the front subframe and body from the driver's side

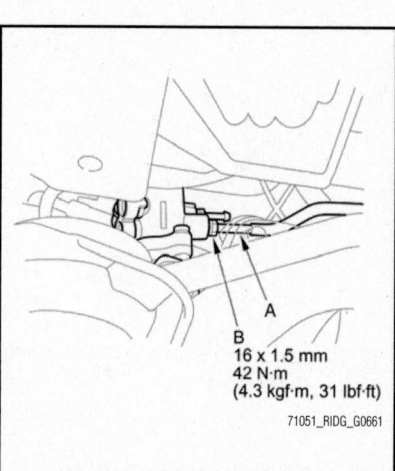

Fig. 168 Connect the inlet line (A), and tighten the 16 mm flare nut (B)

transmission jack or the power train lift until the subframe is in position.

51. Install the front subframe rear brackets. Tighten the 12 mm flange bolts and the new 14 mm special bolts on the right and the left of the vehicle to the specified torque. Install the subframe bolt retainers. Tighten the two flange bolts and nuts to the specified torque.

52. Install the front subframe front brackets with 12 mm flange bolts and the new 14 mm special bolts, and tighten to the specified torque.

53. Remove the jack from the subframe adapter.

54. Install the front splash shield.

55. Install the No. 1 propeller shaft protector.

56. Connect the power steering pressure (PSP) switch connector.

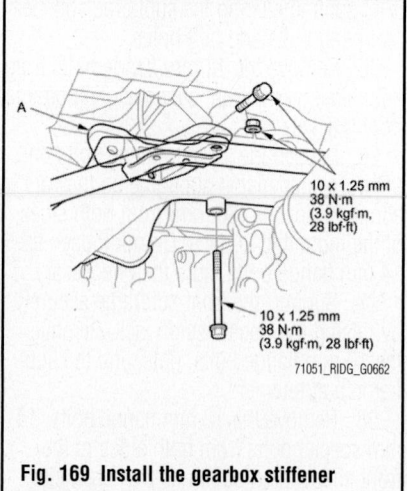

Fig. 169 Install the gearbox stiffener bracket (A) on the front subframe

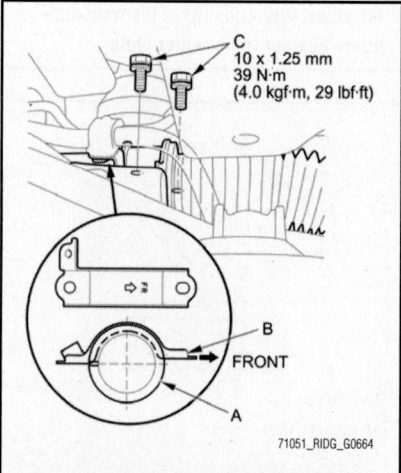

Fig. 170 Install the mounting cushion (A) on the right side of the gearbox, and the mounting bracket (B) over the mounting cushion, then install the two 10 mm flange bolts (C)

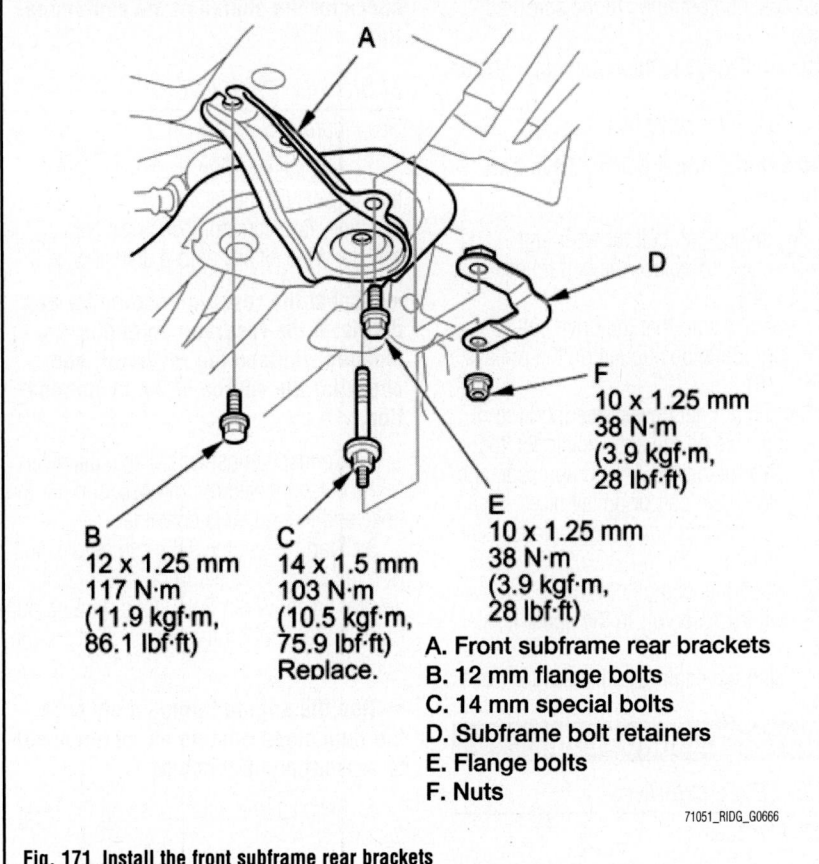

B
12 x 1.25 mm
117 N·m
(11.9 kgf·m,
86.1 lbf·ft)

C
14 x 1.5 mm
103 N·m
(10.5 kgf·m,
75.9 lbf·ft)
Replace.

F
10 x 1.25 mm
38 N·m
(3.9 kgf·m,
28 lbf·ft)

E
10 x 1.25 mm
38 N·m
(3.9 kgf·m,
28 lbf·ft)

A. Front subframe rear brackets
B. 12 mm flange bolts
C. 14 mm special bolts
D. Subframe bolt retainers
E. Flange bolts
F. Nuts

71051_RIDG_G0666

Fig. 171 Install the front subframe rear brackets

57. Connect the under-floor three way catalytic converter (TWC) to the muffler. Install the new 10 mm self-locking nuts and tighten them to 25 ft. lbs. (33 nm).

58. Install the front subframe stiffener and tighten to 40 ft. lbs. (54 Nm).

59. Install the new 10 mm flange bolts on the engine side mount bracket. Tighten to 33 ft. lbs. (44 Nm).

60. Remove the engine support hanger,

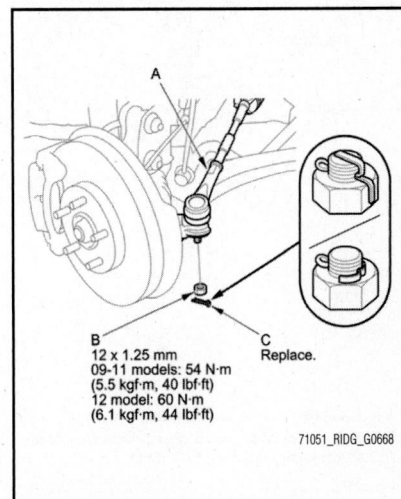

B
12 x 1.25 mm
74 N·m
(7.5 kgf·m, 54 lbf·ft)

C
14 x 1.5 mm
103 N·m
(10.5 kgf·m, 75.9 lbf·ft)
Replace.

71051_RIDG_G0667

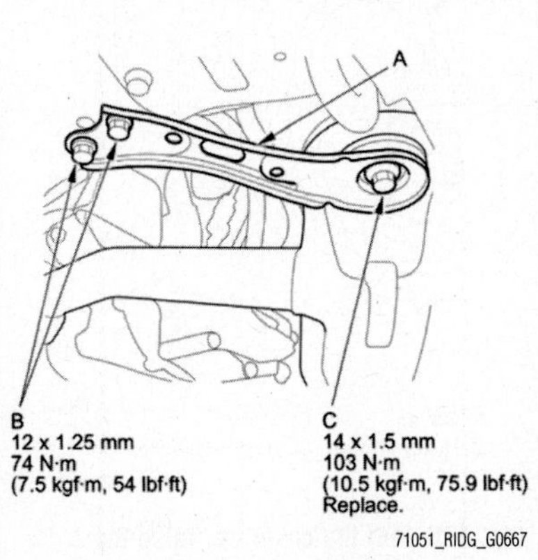

Fig. 172 Install the front subframe front brackets (A) with 12 mm flange bolts (B) and the new 14 mm special bolts (C)

the hanger balance bar, and the hanger adapter set.

61. Install the connector bracket to the front cylinder head. Tighten to 33 ft. lbs. (44 Nm).

62. Install the harness clamp bracket to the rear cylinder head. Tighten to 20 ft. lbs. (26 Nm).

63. Install the service caps on the cowl cover.

64. Install the power steering pump outlet line with a new O-ring to the pump, and secure the hose clamp with the bolt.

65. Wipe off any grease contamination from the ball joint tapered section and threads. Reconnect the tie-rod ends to the steering knuckles. Install the 12mm nut and tighten it on both sides.

66. Install a new cotter pin and bend it as shown on both sides.

67. Center the steering rack within its stroke.

68. Insert the upper end of the steering joint onto the steering shaft (line up the bolt hole with the flat portion on the shaft), and loosely install the upper joint bolt.

69. Slip the lower end of the steering joint onto the pinion shaft taking care to align the gap within the angle.

70. Align the bolt hole on the steering

joint with the groove around the pinion shaft, then loosely install the lower steering joint bolt. Be sure that the joint bolt is securely in the groove in the pinion shaft.

71. Pull on the steering joint to make sure that the steering joint is fully seated, then tighten the lower joint bolt to the specified torque.

72. Tighten the upper steering joint bolt to the specified torque.

73. Install steering joint cover.

74. Install the front wheel, then set the wheels in the straight ahead position.

➡**Before installing the wheel, clean the mating surfaces of the brake disc and the inside of the wheel.**

75. Remove the steering wheel holder tool.

76. Do the battery terminal reconnection procedure, and check these items:

a. Make sure the horn and turn signal switches work properly.

b. Make sure the steering wheel switches work properly.

77. Fill the system with power steering fluid, and bleed air from the system.

78. After installation, check these items:

a. Start the engine, allow it to idle, and turn the steering wheel from lock to lock several times to warm up the fluid. Check the gearbox for leaks.

b. Do the front toe inspection.

c. Check the steering wheel spoke angle.

If steering spoke angles to the right and left are not equal (steering wheel and rack are not centered), correct the engagement of the joint/pinion shaft ser-

B
12 x 1.25 mm
09-11 models: 54 N·m
(5.5 kgf·m, 40 lbf·ft)
12 model: 60 N·m
(6.1 kgf·m, 44 lbf·ft)

C
Replace.

71051_RIDG_G0668

Fig. 173 Reconnect the tie-rod ends (A) to the steering knuckles. Install the 12mm nut (B) and tighten it on both sides; cotter pin (C)

rations, then adjust the front toe by turning the tie-rod ends, if necessary.

POWER STEERING PUMP

REMOVAL & INSTALLATION

See Figure 174.

1. Place a suitable container under the vehicle to catch any spilled fluid.
2. Drain the power steering fluid from the reservoir.
3. Remove the engine cover.
4. Remove the drive belt from the pump pulley.
5. Cover the auto-tensioner, the alternator, and the A/C compressor with several shop towels to protect them from spilled power steering fluid. Disconnect the pump inlet hose and pump outlet hose from the pump, and plug them. Take care not to spill the fluid on the vehicle. Wipe off any spilled fluid at once. Do not turn the steering wheel with the pump removed.
6. Remove the pump mounting bolts then remove the pump.
7. Cover the opening of the pump with a piece of tape to prevent foreign material from entering the pump.

To install:

8. Transfer the pump inlet hose and the pump outlet hose from the original pump onto the new pump with a new O-ring.
9. Loosely install the pump in the pump bracket with the mounting bolts, then

tighten the pump fittings to the specified torque.
10. Tighten the pump mounting bolts to the specified torque.
11. Install the drive belt.

➡**Note these items during drive belt installation:**

- Inspect the belt for wear and cracks. Replace the belt if necessary.
- Make sure that the drive belt is properly positioned on the pulleys (B).
- Do not get power steering fluid or grease on the auto-tensioner, the alternator, the A/C compressor, and the drive belt or pulley faces. Clean off any fluid or grease before installation.

12. Install the engine cover.
13. Fill the reservoir to the upper level line.
14. Start the engine, and check for leaks.

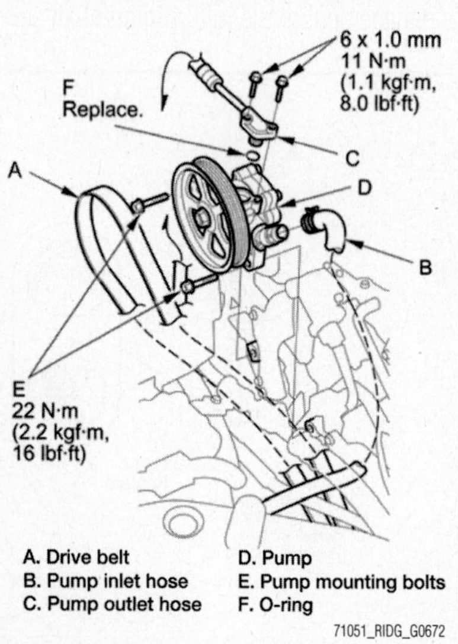

POWER STEERING FLUID

FLUID LEVEL CHECK

See Figure 175.

Check the reservoir at regular intervals, and add the recommended fluid as necessary. Always use Honda Power Steering Fluid. Using any other type of power steering fluid or automatic transmission fluid can cause increased wear, fluid leaks, and poor steering in cold weather.

➡**Inspect the reservoir screen for any debris. If the reservoir screen is clogged, replace the reservoir, and**

check for the source of the contamination.

FLUID FILL PROCEDURE

See Figure 176.

1. Pull the cover up, raise the reservoir, then disconnect the return hose to drain the reservoir. Take care not to spill the fluid on the vehicle. Wipe off any spilled fluid at once.

➡**Inspect the reservoir screen for any debris. If the reservoir screen is clogged, replace the reservoir, and check for the source of the contamination.**

2. Connect a hose of suitable diameter to the disconnected return hose, and put the hose end in a suitable container.
3. Start the engine, let it run at idle, and turn the steering wheel from lock to lock several times. When fluid stops running out of the hose, shut off the engine. Discard the fluid.

➡**Stop the engine immediately once the fluid stops running out of the hose to prevent pump damage.**

4. Reinstall the return hose on the reservoir.
5. Fill the reservoir to the upper level line.
6. Start the engine and run it at idle. Turn the steering wheel from lock to lock several times to bleed air from the system.
7. Recheck the fluid level, and add fluid if necessary. Do not fill the reservoir beyond the upper level line.
8. If the fluid is contaminated, dark, or discolored, repeat the procedure as necessary until the system is clean.

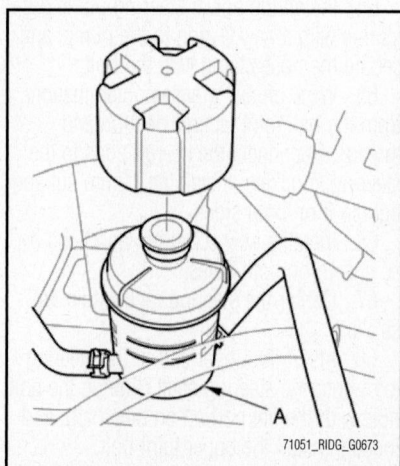

A. Drive belt
B. Pump inlet hose
C. Pump outlet hose
D. Pump
E. Pump mounting bolts
F. O-ring

F Replace.

6 x 1.0 mm
11 N·m
(1.1 kgf·m,
8.0 lbf·ft)

E
22 N·m
(2.2 kgf·m,
16 lbf·ft)

71051_RIDG_G0672

Fig. 174 Remove the drive belt from the pump pulley

Fig. 175 Check the reservoir (A) at regular intervals

71051_RIDG_G0673

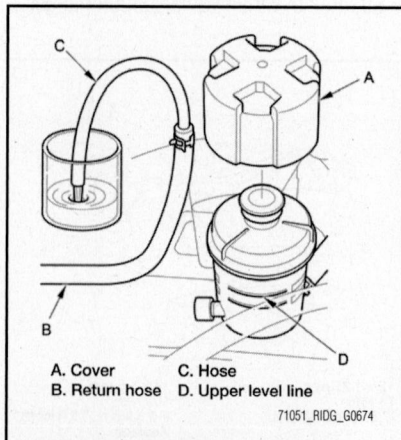

A. Cover
B. Return hose
C. Hose
D. Upper level line

71051_RIDG_G0674

Fig. 176 Pull the cover up, raise the reservoir, then disconnect the return hose to drain the reservoir

KNUCKLE

REMOVAL & INSTALLATION

See Figure 177.

Special Tools Required:
- Ball Joint Remover, 32 mm 07MAC-SL0A102
- Ball Joint Thread Protector, 12 mm 07AAF-SDAA100
- Ball Joint Puller 07AAC-SJCA120
- Ball Joint Thread Protector,14mm 07AAC-SJCA200
- Hub Dis/Assembly Pin, 42 mm 07GAF-SD4A100
- Driver Handle, 15 x 135L 07749-0010000
- Bearing Driver Attachment, 72 x 75 mm 07746-0010600
- Support Base 07965-SD90100
- Attachment, 96 mm 07948-SB00101

1. Raise and support the vehicle.
2. Remove the wheel nuts and the front wheel.
3. Remove the brake hose mounting bolt.
4. Remove the brake caliper bracket mounting bolts, then remove the caliper assembly from the knuckle. To prevent damage to the caliper assembly or brake hose, use a short piece of wire to hang the caliper assembly from the undercarriage. Do not twist the brake hose excessively.
5. Remove the wheel speed sensor from the knuckle. Do not disconnect the wheel speed sensor connector.
6. Pry up the stake on the spindle nut, then remove the nut.
7. Remove the front brake disc.
8. Check the front hub for damage and cracks.
9. Remove the cotter pin from the tie-rod end ball joint, then remove the nut.

➡**During installation, install the new cotter pin after tightening the nut, and bend its end as shown.**

10. Disconnect the tie-rod ball joint from the knuckle using the ball joint thread protector and the ball joint remover.
11. Remove the lock pin from the lower arm ball joint, then remove the nut.

➡**During installation, install the new lock pin after tightening the new castle nut.**

12. Disconnect the lower ball joint from the knuckle using the 14 mm ball joint thread protector and the ball joint puller.

13. Remove the damper pinch bolts and flange nuts from the damper.

➡**During installation, install the new damper pinch bolts and new flange nuts.**

14. Pull the knuckle outward, and separate the outboard joint from the front hub using a soft face hammer then remove the knuckle/hub.
- Do not pull the driveshaft end outward. The driveshaft inboard joint may come apart.
- During installation, apply grease to the mating surfaces of the wheel bearing and driveshaft outboard joint.

To install:
15. Install the knuckle/hub in the reverse order of removal, and note these items:
   a. Be careful not to damage the ball joint boot when installing the knuckle.
   b. Tighten all mounting hardware to the specified torque.
   c. Before connecting the lower ball joint to the knuckle, degrease the threaded section and tapered portion of the ball joint pin, the lower arm connecting hole, the threaded section and mating surface of the castle nut.
   d. First install all the components and lightly tighten the bolts and nuts, then raise the suspension to load it with the vehicle's weight before fully tightening to the specified torque.
   e. Torque the castle nut to the lower torque specification, then tighten it only far enough to align the slot with the ball joint pin hole. Do not align the castle nut by loosening it.
   f. Use a new spindle nut on reassembly.
   g. Before installing the spindle nut, apply a small amount of engine oil to the seating surface of the nut. After tightening, use a drift to stake the spindle nut shoulder against the driveshaft.
   h. Before installing the brake disc, clean the mating surface of the front hub and the inside of the brake disc.
   i. Before installing the wheel, clean the mating surface of the brake disc and the inside of the wheel.
16. Check the wheel alignment, and adjust it if necessary.

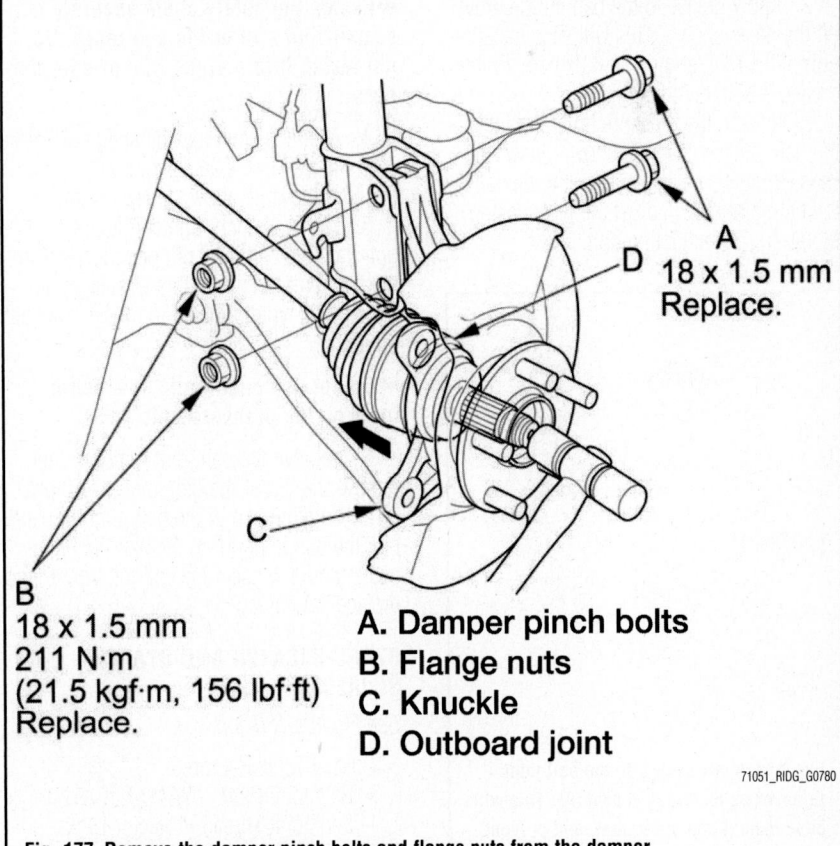

B
18 x 1.5 mm
211 N·m
(21.5 kgf·m, 156 lbf·ft)
Replace.

D 18 x 1.5 mm
Replace.

A. Damper pinch bolts
B. Flange nuts
C. Knuckle
D. Outboard joint

71051_RIDG_G0780

Fig. 177 Remove the damper pinch bolts and flange nuts from the damper

LOWER BALL JOINTS

REMOVAL & INSTALLATION

07MAC-SL0A102 OR 07MAC-SL0A202

See Figures 178 and 179.

Special Tools Required:
- Ball Joint Puller 07AAC-SJCA120
- Ball Joint Thread Protector,14mm 07AAC-SJCA200
- Ball Joint Thread Protector, 12 mm 07AAF-SDAA100
- Ball Joint Thread Protector, 10 mm 07AAF-SECA120
- Ball Joint Remover, 32 mm 07MAC-SL0A102
- Ball Joint Remover, 28 mm 07MAC-SL0A202

➡**Always use a ball joint remover to disconnect a ball joint. Do not strike the housing or any other part of the ball joint connection to disconnect it.**

1. Install a hex nut or the ball joint thread protector onto the threads of the ball joint.

➡**When using a hex nut, make sure the nut is flush with the ball joint pin end to prevent damage to the threaded end of the ball joint pin.**

2. Apply grease to the ball joint remover on the areas shown. This will ease installation of the tool, and prevent damage to the pressure bolt threads.

3. Loosen the pressure bolt, and install the ball joint remover as shown. Insert the jaws carefully, making sure not to damage the ball joint boot. Adjust the jaw spacing by turning the adjusting bolt.

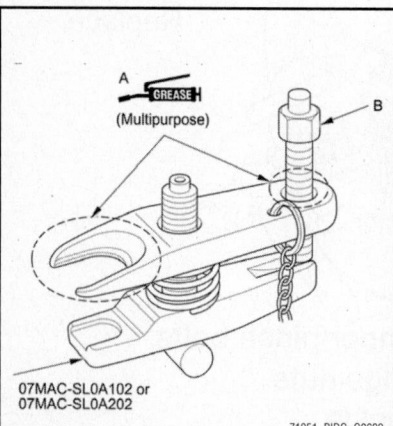

07MAC-SL0A102 or
07MAC-SL0A202

71051_RIDG_G0692

Fig. 178 Apply grease to the ball joint remover on the areas shown (A). This will ease installation of the tool, and prevent damage to the pressure bolt (B) threads

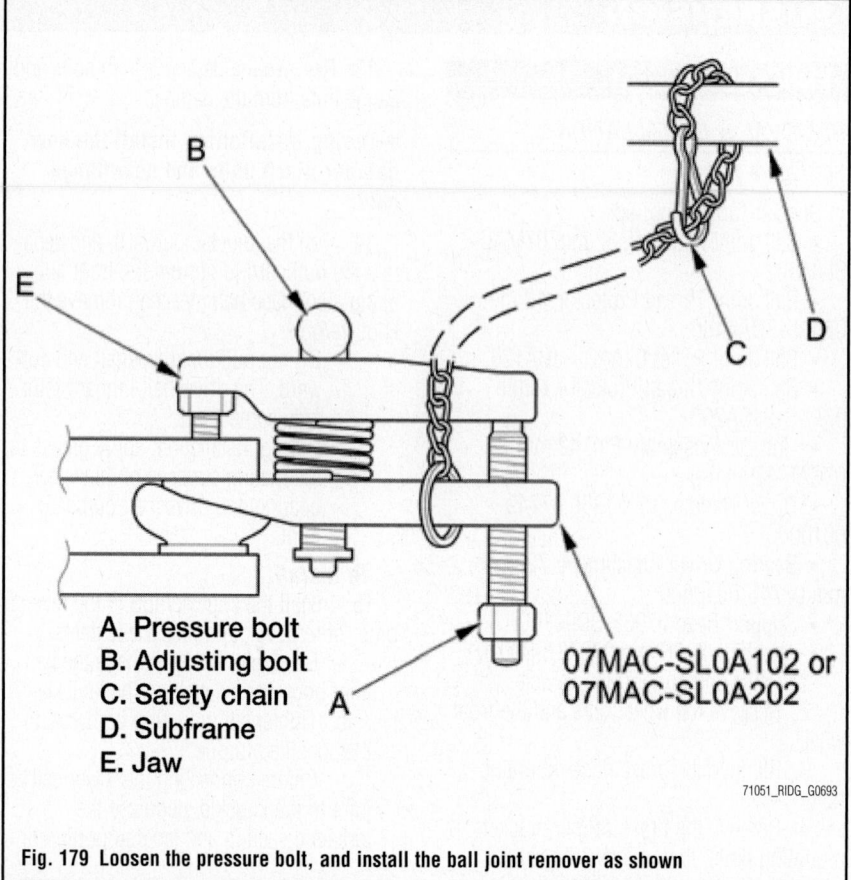

A. Pressure bolt
B. Adjusting bolt
C. Safety chain
D. Subframe
E. Jaw

07MAC-SL0A102 or
07MAC-SL0A202

71051_RIDG_G0693

Fig. 179 Loosen the pressure bolt, and install the ball joint remover as shown

➡**Fasten the safety chain securely to a suspension arm or the subframe. Do not fasten it to a brake line or wire harness.**

4. After adjusting the adjusting bolt, make sure the head of the adjusting bolt is in the position shown to allow the jaw to pivot.

5. With a wrench, tighten the pressure bolt until the ball joint pin pops loose from the ball joint pin hole. If necessary, apply penetrating type lubricant to loosen the ball joint pin.

➡**Do not use pneumatic or electric tools on the pressure bolt.**

6. Remove the ball joint remover, then remove the nut or the ball joint thread protector from the end of the ball joint pin, and pull the ball joint out of the ball joint connecting hole. Inspect the ball joint boot, and replace it if damaged.

07AAC-SJCA120 AND 07AAC-SCJA200

See Figures 180 and 181.

Special Tools Required:
- Ball Joint Puller 07AAC-SJCA120
- Ball Joint Thread Protector,14mm 07AAC-SJCA200

- Ball Joint Thread Protector, 12 mm 07AAF-SDAA100
- Ball Joint Thread Protector, 10 mm 07AAF-SECA120
- Ball Joint Remover, 32 mm 07MAC-SL0A102
- Ball Joint Remover, 28 mm 07MAC-SL0A202

➡**Always use a ball joint remover to disconnect a ball joint. Do not strike the housing or any other part of the ball joint connection to disconnect it.**

1. Apply grease to the 14 mm ball joint thread protector and the ball joint on the areas shown This will prevent damage to the ball joint puller and thread of ball joint.

2. Install the 14 mm ball joint thread protector by hand, all the way onto the threads of the ball joint.

3. Install the ball joint puller as shown.

4. With a wrench, unscrew the 14 mm ball joint thread protector to apply pressure against the ball joint puller until the ball joint pin pops loose from the knuckle.

5. Remove the ball joint puller and the ball joint thread protector from the knuckle and the end of the ball joint pin, and pull the ball joint out of the knuckle. Inspect the ball joint boot, and replace it if damaged.

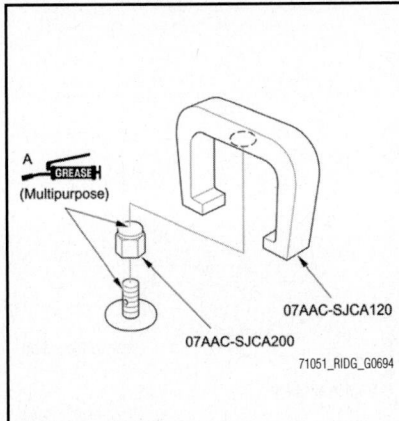

Fig. 180 Apply grease to the 14 mm ball joint thread protector and the ball joint on the areas shown

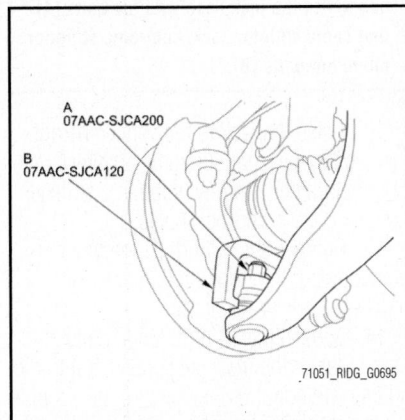

Fig. 181 Install the 14 mm ball joint thread protector (A), then install the ball joint puller (B) as shown

6. Remove any oil, grease, dust, metal debris, and other foreign material from the stud portion of the ball joint stud

LOWER CONTROL ARMS

REMOVAL & INSTALLATION

See Figure 182.

Special Tools Required:
• Ball Joint Puller 07AAC-SJCA120
• Ball Joint Thread Protector,14mm 07AAC-SJCA200

1. Raise and support the vehicle.
2. Remove the front wheels.
3. Remove the flange nut while holding the respective joint pin with a hex wrench, then disconnect the stabilizer links from the damper.
4. Turn the stabilizer bar backward to gain easier access to the front side of the lower arm mounting bolt.
5. Remove the lock pin from the

lower arm ball joint, then remove the nut.

➡**During installation, install the lock pin after tightening the new castle nut.**

6. Disconnect the lower ball joint from the knuckle using the 14 mm ball joint thread protector and the ball joint puller.
7. Remove the lower arm mounting bolt.

➡**Use the new mounting bolt during reassembly.**

8. Remove the lower arm mounting bolt, then remove the lower arm from the front suspension subframe.

➡**Use the new mounting bolt during reassembly.**

To install:

9. Install the lower arm in the reverse order of removal, and note these items:
    a. Be careful not to damage the ball joint boot when installing the knuckle.
    b. Tighten all mounting hardware to the specified torque.
    c. Before connecting the lower ball joint to the knuckle, degrease the threaded section and tapered portion of the ball joint pin, the lower arm connecting hole, the threaded section and mating surface of the castle nut.
    d. First install all the components and lightly tighten the bolts and nuts, then raise the suspension to load it with the vehicle's weight before fully tightening to the specified torque.
    e. Torque the castle nut to the lower torque specification, then tighten it only far enough to align the slot with the ball joint pin hole. Do not align the castle nut by loosening it.

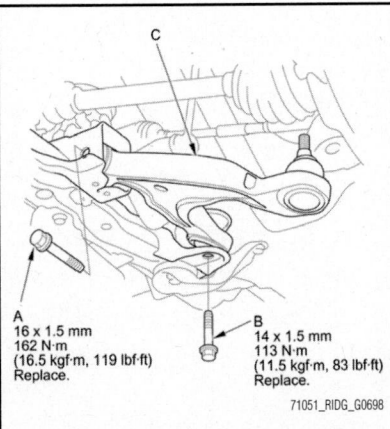

Fig. 182 Remove the lower arm mounting bolt (A), remove the lower arm mounting bolt (B), then remove the lower arm (C)

    f. Before installing the wheel, clean the mating surface of the brake disc and the inside of the wheel.
10. Check the wheel alignment, and adjust it if necessary.

STABILIZER BAR & LINKS

REMOVAL & INSTALLATION

Stabilizer Links

See Figure 183.

1. Raise and support the vehicle.
2. Remove the front wheel.
3. Remove the flange nuts while holding the respective joint pin with a hex wrench, then remove the stabilizer link.

To install:

4. Install the stabilizer link on the stabilizer bar and strut the joint pins set at the center of their range of movement.
5. Install the flange nuts, and lightly tighten them.
6. Clean the mating surface of the brake disc and the inside of the wheel, then install the front wheel.
7. Tighten the flange nuts to 58 ft. lbs. (78 Nm) while holding the respective joint pin with a hex wrench.
8. Test-drive the vehicle.

Stabilizer Bar

See Figures 184 through 189.

Special Tools Required*:
• Engine Hanger Adapter Set VSB02C000024
• Engine Hanger Balance Bar VSB02C000019
• Engine Support Hanger, A and Reds AAR-T1256
• Front Subframe Adapter EQS02BMDXSB0
• 2008 V6 Attachment Arm SIL02C000033
• *Available through Honda Tool and Equipment program 888-424-6857

1. Raise and support the vehicle.
2. Remove the front wheels.
3. Remove the connector bracket from the front cylinder head; use the bracket bolt hole to attach the engine hanger balance bar front arm.
4. Remove the harness clamp bracket from the rear cylinder head; use the bracket bolt hole to attach the 2008 V6 attachment arm (SIL02C000033).
5. Remove the service caps for the front damper flange nuts from the cowl cover. Position the engine hanger adapters (VSB02C000024) with the "FRONT" mark

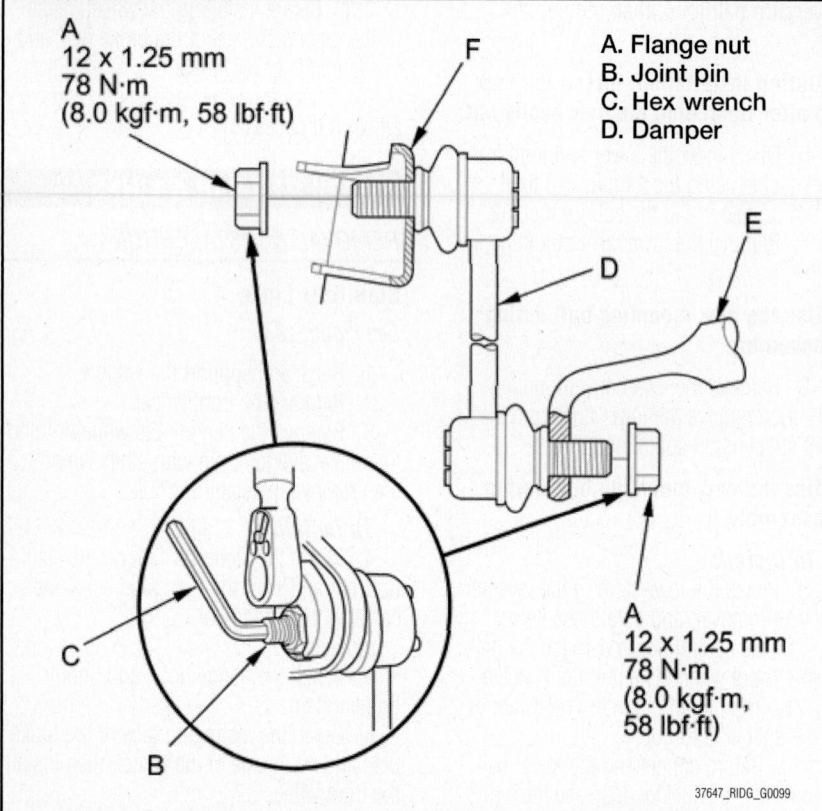

A
12 x 1.25 mm
78 N·m
(8.0 kgf·m, 58 lbf·ft)

A. Flange nut
B. Joint pin
C. Hex wrench
D. Damper

A
12 x 1.25 mm
78 N·m
(8.0 kgf·m, 58 lbf·ft)

37647_RIDG_G0099

Fig. 183 Remove the flange nut while holding the respective joint pin with a hex wrench, then disconnect the stabilizer links from the damper.

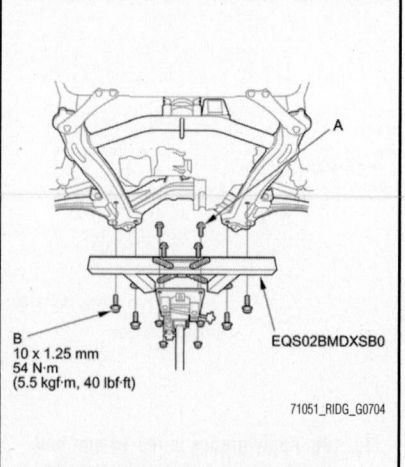

B
10 x 1.25 mm
54 N·m
(5.5 kgf·m, 40 lbf·ft)

EQS02BMDXSB0

71051_RIDG_G0704

Fig. 185 Line up the slots in the arms with the bolt holes on the corner of the jack base, then attach the subframe adapter to the jack base with the bolts (A) that came with the jack; subframe stiffener mounting bolts (B)

facing forward over the damper flange nuts.

6. Install the engine hanger balance bar (VSB02C000019). Attach the front arm to the front cylinder head with a spacer and a 10 x 1.25 mm bolt. Remove the rear arm from the balance bar, then install the 2008 V6 attachment arm (SIL02C000033). Attach the 2008 V6 attachment arm to the rear cylinder head with an 8 x 1.25 mm bolt.

7. Install the engine support hanger (AAR-T1256) onto the vehicle as shown, and attach the hook to the slotted hole in the engine hanger balance bar. Tighten the wing nut by hand, and lift and support the engine/transmission.

➡ **Be careful when working around the windshield.**

8. Disconnect both sides of the stabilizer link from the stabilizer bar.

9. Remove the front splash shield.

10. Loosen the four bolts holding the adjustable arms of the front subframe adapter (EQS02BMDXSB0) to its center plate.

➡ **The adapter is designed to be used the Honda transmission jack (model number LSL-W93714) or powertrain lift (model number OTC-1585), both avail-**

able through the Honda Tool and Equipment Program. It will also work with most commercially available transmission jacks.

11. Line up the slots in the arms with the bolt holes on the corner of the jack base, then attach the subframe adapter (EQS02BMDXB0) to the jack base with the bolts that came with the jack. Tighten the bolts securely.

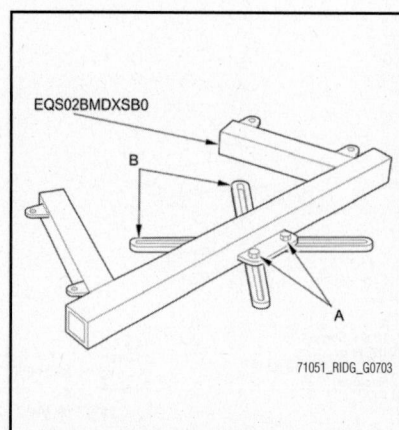

EQS02BMDXSB0

B

A

71051_RIDG_G0703

Fig. 184 Loosen the four bolts (A) holding the adjustable arms (B) of the front subframe adapter to its center plate

12. Raise the jack to the vehicle height, then attach the front subframe adapter to the front subframe using the subframe stiffener mounting bolts and bolt holes.

13. Remove the 12 mm flange bolts on both sides of the front subframe front bracket.

14. Loosen the 14 mm flange bolts on both sides of the front suspension subframe so they are about 0.6 inches (14 mm) from the mounting surface. Do not loosen the 14 mm flange bolts more than necessary.

15. Support the front suspension subframe securely by raising the transmission jack, then remove the 10 mm flange bolts, the nuts, and the subframe bolt retainers on both sides of the front suspension subframe rear brackets.

16. Remove the 12 mm flange bolts, 14

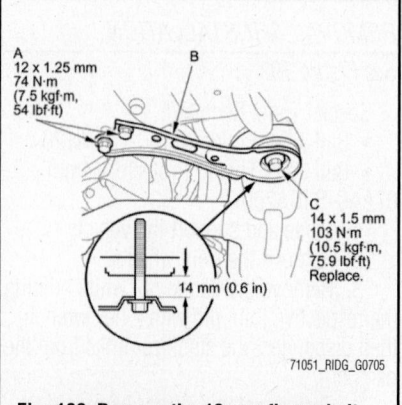

A
12 x 1.25 mm
74 N·m
(7.5 kgf·m,
54 lbf·ft)

B

C
14 x 1.5 mm
103 N·m
(10.5 kgf·m,
75.9 lbf·ft)
Replace.

14 mm (0.6 in)

71051_RIDG_G0705

Fig. 186 Remove the 12 mm flange bolts (A) on both sides of the front subframe front bracket (B), then loosen the 14 mm flange bolts (C)

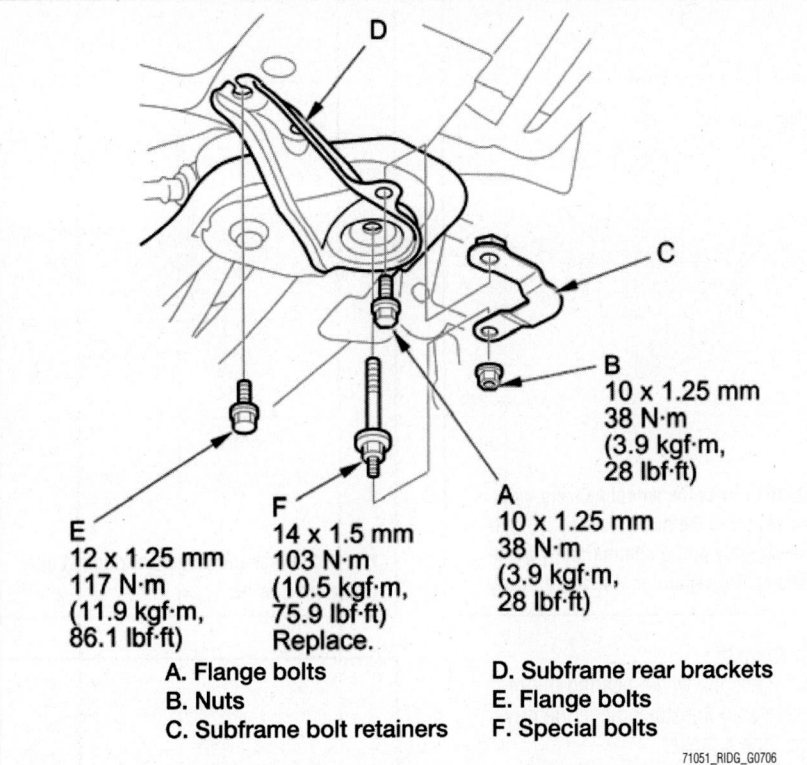

E
12 x 1.25 mm
117 N·m
(11.9 kgf·m,
86.1 lbf·ft)

F
14 x 1.5 mm
103 N·m
(10.5 kgf·m,
75.9 lbf·ft)
Replace.

B
10 x 1.25 mm
38 N·m
(3.9 kgf·m,
28 lbf·ft)

A
10 x 1.25 mm
38 N·m
(3.9 kgf·m,
28 lbf·ft)

A. Flange bolts
B. Nuts
C. Subframe bolt retainers
D. Subframe rear brackets
E. Flange bolts
F. Special bolts

71051_RIDG_G0706

Fig. 187 Support the front suspension subframe securely by raising the transmission jack, then remove the 10 mm flange bolts, the nuts, and the subframe bolt retainers on both sides of the front suspension subframe rear brackets

mm special bolts on both sides of the front suspension subframe rear brackets.

17. Lower the jack supporting the front suspension subframe with the special tool slowly until the front suspension subframe has dropped about 0.6 inches (14 mm).

18. Remove the flange bolts and the bushing holders, then remove the bushings and the stabilizer bar from the front suspension subframe.

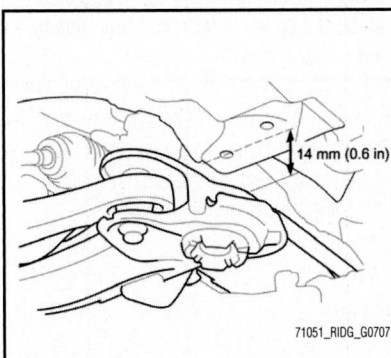

14 mm (0.6 in)

71051_RIDG_G0707

Fig. 188 Lower the jack supporting the front suspension subframe with the special tool slowly until the front suspension subframe has dropped about 0.6 inches (14 mm)

To install:

19. Install the stabilizer bar in the reverse order of removal, and note these items:

    a. Note the right and left direction of the stabilizer bar.

    b. Align the paint marks on the stabilizer bar with the sides of the bushings.

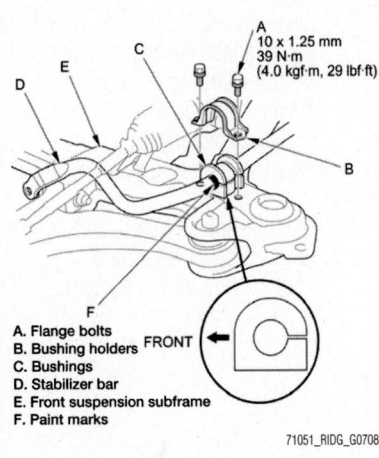

A
10 x 1.25 mm
39 N·m
(4.0 kgf·m, 29 lbf·ft)

A. Flange bolts
B. Bushing holders
C. Bushings
D. Stabilizer bar
E. Front suspension subframe
F. Paint marks

FRONT

71051_RIDG_G0708

Fig. 189 Remove the flange bolts and the bushing holders, then remove the bushings and the stabilizer bar from the front suspension subframe

c. Note the fore/aft direction of the bushing holders.

d. Raise the front suspension subframe up with the jack and special tool until it contacts the body frame, then tighten the mounting bolts to the specified torque.

e. Refer to stabilizer link removal/installation to connect the stabilizer bar to the links.

f. Do the subframe alignment.

g. Clean the mating surface of the brake disc and the inside of the wheel, then install the front wheel.

20. Check the wheel alignment, and adjust it if necessary.

STRUTS

REMOVAL & INSTALLATION

See Figure 190.

1. Raise and support the vehicle.
2. Remove the front wheel.
3. Disconnect the stabilizer link from the damper.
4. Remove the wheel speed sensor harness clips and the brake hose bracket from the damper. Do not disconnect the wheel speed sensor connector.
5. Remove the damper pinch bolts and the flange nuts from the damper.

➡**Do not allow the knuckle to rotate too far outward. This may allow the driveshaft inboard joint to come apart.**

6. Remove the service caps, and remove the three flange nuts from top of the damper, then remove the damper assembly.

➡**The damper springs are different, left and right. Mark the springs L and R before you continue the procedure.**

To install:

7. Install the damper assembly on to the frame. Note the direction of the damper mounting base as shown.
8. Loosely install new damper pinch bolts and new flange nuts to the damper.
9. Install the wheel sensor harness clips and the brake hose bracket to the damper.
10. Loosely install the stabilizer link to the damper.
11. Raise the front suspension with a floor jack to load the suspension with the vehicle's weight.
12. Tighten the damper pinch bolts and flange nuts. Tighten to 156 ft. lbs. (211 Nm).
13. Tighten the flange nuts on top of the

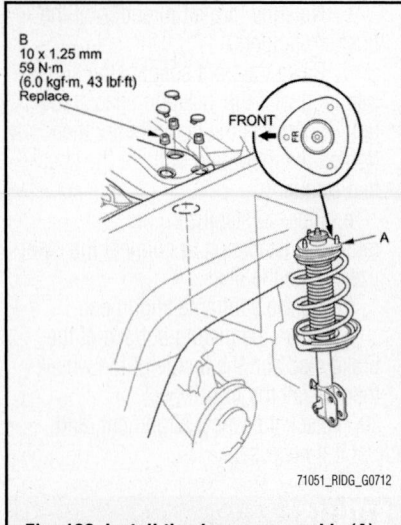

Fig. 190 Install the damper assembly (A) on to the frame, note the direction of the damper mounting base as shown

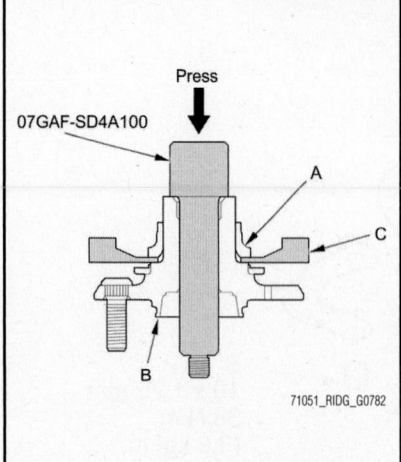

Fig. 191 Press the wheel bearing inner race (A) off of the hub (B) using the hub dis/assembly pin, a commercially available bearing separator (C), and a press

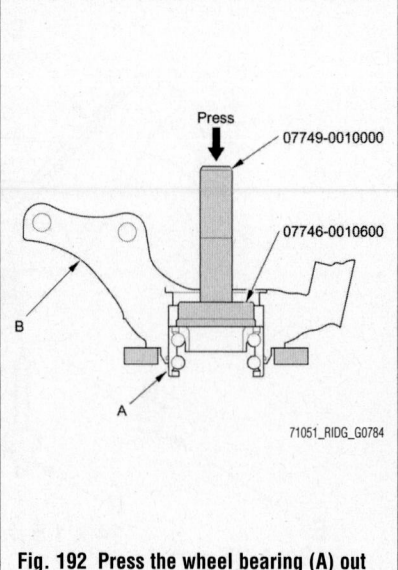

Fig. 192 Press the wheel bearing (A) out of the knuckle (B)

damper to 43 ft. lbs. (59 Nm), and the stabilizer link nuts to 58 ft. lbs. (78 Nm).

14. Install the service caps.

15. Clean the mating surface of the brake disc and the inside of the wheel, then install the front wheel.

16. Check the wheel alignment, and adjust it if necessary.

WHEEL HUBS & BEARINGS

REMOVAL & INSTALLATION

See Figures 191 through 193.

1. Remove the knuckle/hub.

2. Separate the hub from the knuckle using the hub dis/assembly pin and a hydraulic press. Hold the knuckle with the attachment of the hydraulic press or equivalent tool. Be careful not to damage or deform the splash guard. Hold onto the hub to keep it from falling when pressed clear.

3. Press the wheel bearing inner race off of the hub using the hub dis/assembly pin, a commercially available bearing separator, and a press.

4. Remove the splash guard and the snap ring from the knuckle.

To install:

5. Press the wheel bearing out of the knuckle using the attachment, the driver handle, and a press.

6. Wash the knuckle and hub thoroughly in high flash point solvent before reassembly.

7. Press a new wheel bearing into the knuckle using the old bearing, a steel plate, the 96 mm attachment, the support base, and a press.

- Install the wheel bearing with the wheel speed sensor magnetic encoder (brown color), toward the inside of the knuckle.

- Remove any oil, grease, dust, metal debris, and other foreign material from the magnetic encoder surface.

- Keep any magnetic tools away from the magnetic encoder surface.

- Be careful not to damage the encoder surface when you insert the wheel bearing.

8. Install the snap ring securely in the knuckle.

9. Install the splash guard, and tighten the screws to the specified torque.

10. Install the hub onto the knuckle

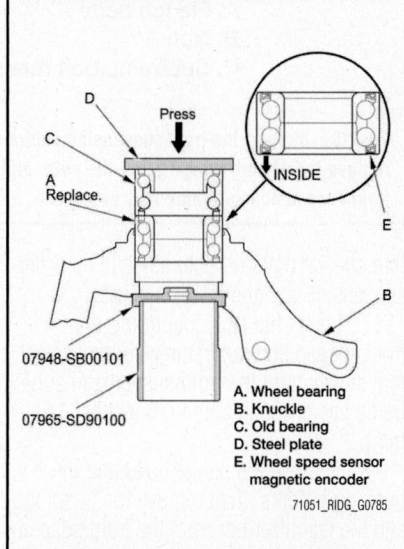

Fig. 193 Press a new wheel bearing into the knuckle using the old bearing, a steel plate, the 96 mm attachment, the support base, and a press

using the attachment, the driver handle, the support base, and a hydraulic press. Be careful not to damage the splash guard.

11. Install the knuckle/hub.

KNUCKLE/HUB ASSEMBLY

REMOVAL & INSTALLATION

See Figure 194.

Special Tools Required:
- Ball Joint Remover, 32 mm 07MAC-SL0A102
- Ball Joint Thread Protector, 12 mm 07AAF-SDAA100
- Hub Dis/Assembly Pin, 42 mm 07GAF-SD4A100
- Driver Handle, 15 x 135L 07749-0010000
- Bearing Driver Attachment, 62 x 68 mm 07746-0010500
- Oil Seal Driver Attachment, 96 mm 07ZAD-PNA0100
- Support Base 07965-SD90100

1. Raise and support the vehicle.
2. Remove the wheel nuts and rear wheel.
3. Release the parking brake lever fully.
4. Remove and discard the brake hose clip from the brake hose.

➡**Use the new brake hose clip during reassembly.**

5. Remove the brake hose bracket mounting bolts from the knuckle.
6. Remove the brake caliper bracket mounting bolts, and remove the caliper assembly from the knuckle. To prevent damage to the caliper assembly or brake hose, use a short piece of wire to hang the caliper assembly from the undercarriage. Do not twist the brake hose excessively.
7. Remove the wheel speed sensor from the knuckle. Do not disconnect the wheel speed sensor connector.
8. Pry up the stake on the spindle nut, then remove the nut.
9. Remove the brake disc/drum.
10. Check the rear hub for damage and cracks.
11. Remove the parking brake shoes assembly, and parking brake cable.
12. Remove the lock pin from the upper arm ball joint, then remove the castle nut.

➡**During installation, install the lock pin as shown after tightening the castle nut.**

13. Disconnect the upper arm ball joint from the knuckle using the ball joint thread protector and the ball joint remover.
14. Remove the self-locking nut, the washer, and the flange bolt, then remove lower arm A.

➡**Use the new self-locking nut, new washer and new flange bolt during reassembly.**

15. Remove the flange bolts, and separate the knuckle from the trailing arm.

➡**Use the new flange bolts during reassembly.**

16. Place a floor jack under lower arm B. Remove the flange bolt.

➡**Use the new flange bolts during reassembly.**

17. Remove the driveshaft outboard joint from the knuckle by tapping the driveshaft end with a soft face hammer while drawing the hub outward, then remove the knuckle.
- Do not pull the driveshaft end outward. The inner driveshaft joint may come apart.
- During installation, apply grease to the mating surfaces of the wheel bearing and driveshaft outboard joint.

To install:
18. Install the knuckle/hub in the reverse order of removal, and note these items:

a. Be careful not to damage the ball joint boot when installing the knuckle.
b. Tighten all mounting hardware to the specified torque.
c. First install all the components and lightly tighten the bolts and nuts, then raise the suspension to load it with the vehicle's weight before fully tightening to the specified torque.
d. Torque the castle nut to the lower torque specification, then tighten it only far enough to align the slot with the ball joint pin hole. Do not align the castle nut by loosening it.
e. Use a new spindle nut on reassembly.
f. Before installing the spindle nut, apply a small amount of engine oil to the seating surface of the nut. After tightening, use a drift to stake the spindle nut shoulder against the driveshaft.
g. Before installing the brake disc/drum, clean the mating surface of the rear hub and the inside of the brake disc/drum.
h. Before installing the wheel, clean the mating surface of the brake

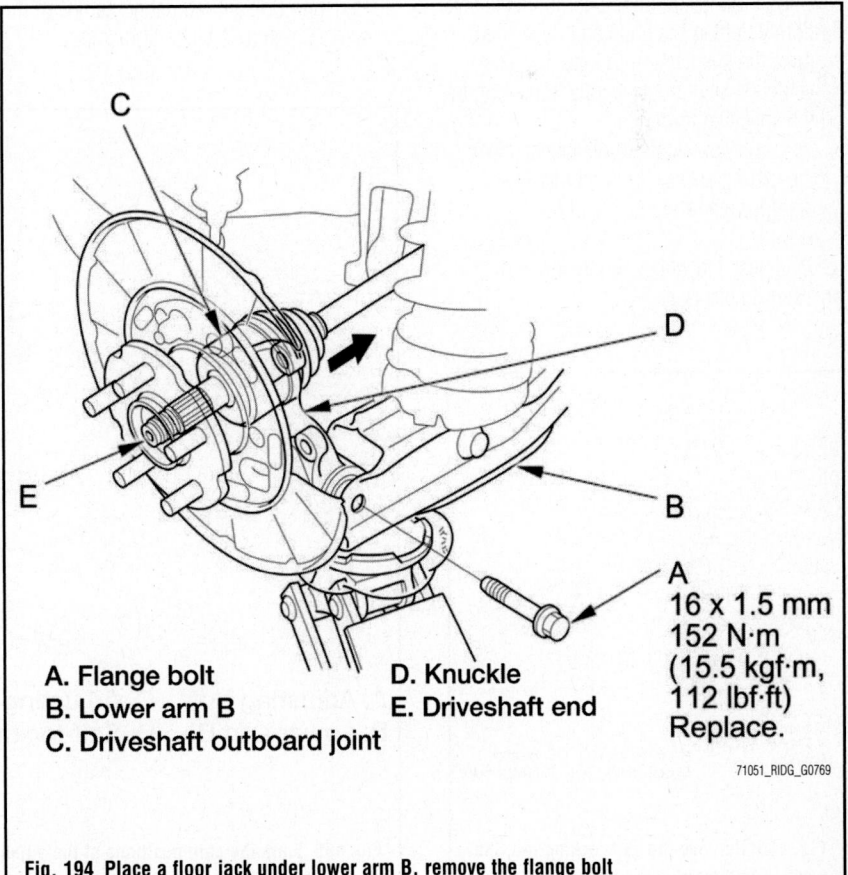

A. Flange bolt
B. Lower arm B
C. Driveshaft outboard joint
D. Knuckle
E. Driveshaft end

A
16 x 1.5 mm
152 N·m
(15.5 kgf·m,
112 lbf·ft)
Replace.

71051_RIDG_G0769

Fig. 194 Place a floor jack under lower arm B, remove the flange bolt

disc/drum and the inside of the wheel.

19. Check the wheel alignment, and adjust it if necessary.

REAR LOWER ARMS

REMOVAL & INSTALLATION

Lower Arm A

See Figure 195.

1. Raise and support the vehicle.
2. Remove the rear wheel.
3. Position a floor jack at the connecting point of the lower arm B and the knuckle. Raise the floor jack until the suspension begins to compress.
4. Remove the self-locking nut, the washer, and the flange bolt.

➡**Use a new self-locking nut and a new flange bolt during reassembly.**

5. Raise the rear suspension with the floor jack until you are finished removing lower arm A.

To install:

6. Install lower arm A in the reverse order of removal, and note these items:

   a. Tighten all mounting hardware to the specified torque.

   b. First install all the components and lightly tighten the bolts and nuts, then raise the suspension to load it with the vehicle's weight before fully tightening to the specified torque.

   c. Before installing the wheel, clean the mating surface on the brake disc/drum and the inside of the wheel.

7. Check the wheel alignment, and adjust it if necessary.

Lower Arm B

See Figure 196.

1. Raise and support the vehicle.
2. Remove the rear wheel.
3. Position a floor jack at the connecting point of lower arm B and the knuckle. Raise the floor jack until the suspension begins to compress.
4. Remove the flange nut, then remove flange bolt.

➡**Use the new flange and new flange bolt during reassembly.**

5. Remove the flange bolt, and lower the floor jack gradually.

➡**Use the new flange bolt during reassembly.**

6. Mark the cam positions of the adjusting bolt and the adjusting cam plate, then remove the self-locking nut, the adjusting cam plate, and the adjusting bolt.

➡**Use a new self-locking nut during reassembly.**

7. Remove the lower arm B.

To install:

8. Install the lower arm B in the reverse order of removal, and note these items:

   a. Align the cam positions of the adjusting bolt and adjusting cam with the marked positions when tightening.

   b. Tighten all mounting hardware to the specified torque.

   c. First install all the components and lightly tighten the bolts and nuts, then raise the suspension to load it with the vehicle's weight before fully tightening to the specified torque.

   d. Before installing the wheel, clean the mating surface on the brake disc/drum and the inside of the wheel.

9. Check the wheel alignment, and adjust it if necessary.

STABILIZER BAR & LINKS

REMOVAL & INSTALLATION

Stabilizer Bar

See Figure 197.

1. Raise and support the vehicle.
2. Remove the rear wheels.
3. Disconnect both sides of the stabilizer link from the stabilizer bar.
4. Remove the rear suspension subframe from the body.
5. Remove the flange bolts and bushing holders, then remove the bushings and the stabilizer bar.

To install:

6. Install the stabilizer bar in the reverse order of removal, and note these items:

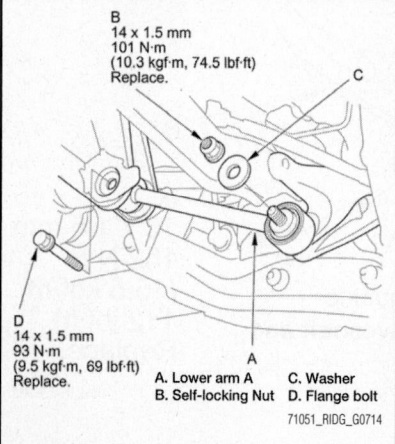

B
14 x 1.5 mm
101 N·m
(10.3 kgf·m, 74.5 lbf·ft)
Replace.

C

D
14 x 1.5 mm
93 N·m
(9.5 kgf·m, 69 lbf·ft)
Replace.

A. Lower arm A   C. Washer
B. Self-locking Nut   D. Flange bolt

71051_RIDG_G0714

Fig. 195 Remove the self-locking nut, the washer, and the flange bolt

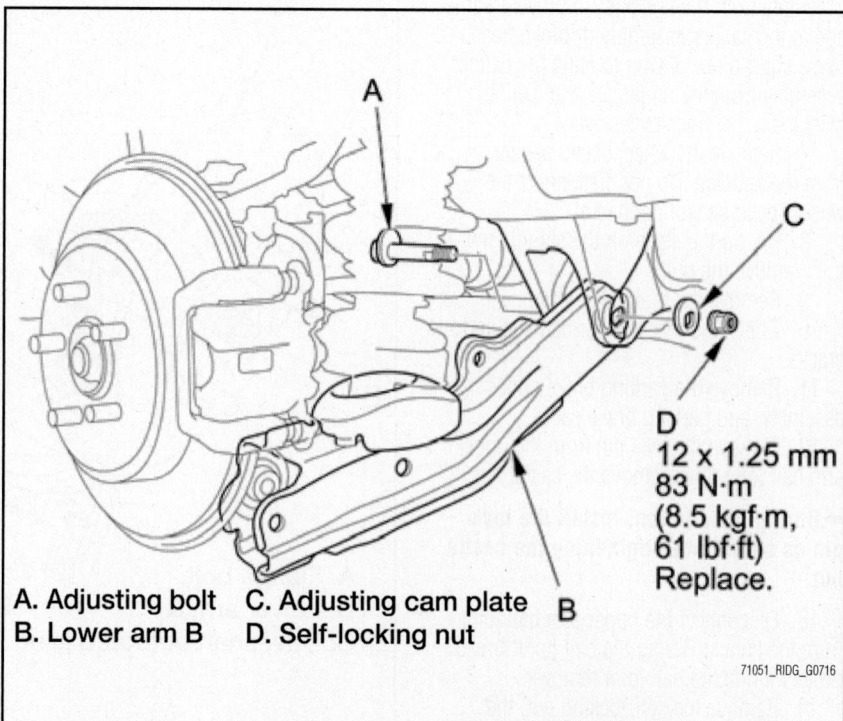

A

C

D
12 x 1.25 mm
83 N·m
(8.5 kgf·m,
61 lbf·ft)
Replace.

B

A. Adjusting bolt   C. Adjusting cam plate
B. Lower arm B   D. Self-locking nut

71051_RIDG_G0716

Fig. 196 Mark the cam positions of the adjusting bolt and the adjusting cam plate, then remove the self-locking nut, the adjusting cam plate, and the adjusting bolt

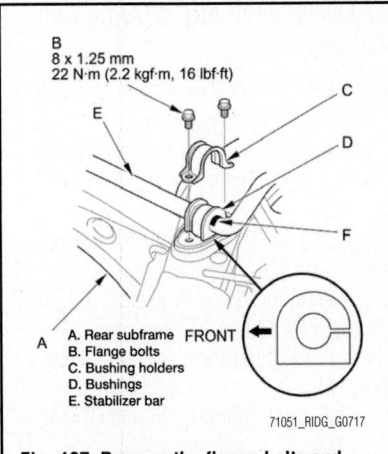

Fig. 197 Remove the flange bolts and bushing holders, then remove the bushings and the stabilizer bar

a. Note the right and left direction of the stabilizer bar.

b. Align the paint marks on the stabilizer bar with the sides of the bushings.

c. Note the fore/aft direction of the bushing holders.

d. Raise the rear suspension subframe up with the jack and special tool until it contacts the body frame, then tighten the mounting bolts to the specified torque.

e. Refer to stabilizer link removal/installation to connect the stabilizer bar to the links.

f. Do the subframe alignment.

g. Clean the mating surface of the brake disc/drum the inside of the wheel, then install the rear wheels.

7. Check the wheel alignment, and adjust it if necessary.

Stabilizer Link

See Figure 198.

1. Raise and support the vehicle.
2. Remove the rear wheel.
3. Remove the flange nut and self-locking nut while holding the respective joint pin with a hex wrench, then remove the stabilizer link.

To install:

4. Install the stabilizer link on the stabilizer bar and lower arm B with the joint pins set at the center of their range of movement.

➡**The left stabilizer link has a yellow paint mark, while the right stabilizer link has a white paint mark.**

5. Install the new self-locking nut and flange nut, and lightly tighten them.

6. Clean the mating surface of the brake disc/drum and the inside of the wheel, then install the rear wheel.

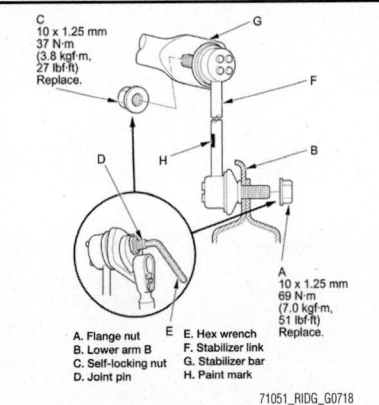

Fig. 198 Remove the flange nut and self-locking nut while holding the respective joint pin with a hex wrench, then remove the stabilizer link

7. Tighten the self-locking nut and the flange nuts to the specified torque while holding the respective joint pins with a hex wrench.

8. Test-drive the vehicle.

STRUTS (DAMPER/SPRING)

REMOVAL & INSTALLATION

See Figure 199.

1. Raise and support the vehicle.
2. Remove the rear wheel.
3. Remove the lower arm B.
4. Remove the three flange bolts from the top of the damper, then remove the damper assembly.

To install:

5. Install the damper assembly onto the body, then loosely install the new flange bolts.

6. Install the lower arm B.

7. Raise the rear suspension with a floor jack to load the vehicle weight.

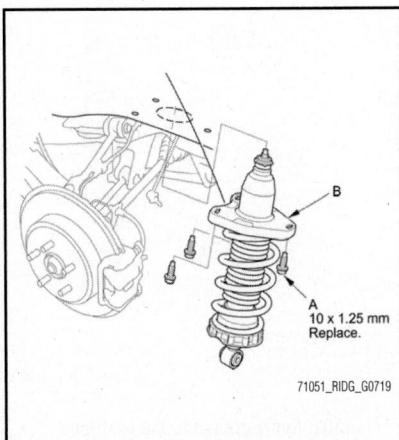

Fig. 199 Remove the three flange bolts (A) from the top of the damper, then remove the damper assembly (B)

8. Tighten all mounting hardware to the specified torque.

9. Clean the mating surface of the brake disc/drum and the inside of the wheel, then install the rear wheel.

10. Check the wheel alignment, and adjust it if necessary.

TRAILING ARMS

REMOVAL & INSTALLATION

See Figure 200.

1. Raise and support the vehicle.
2. Remove the rear wheel.
3. Position a floor jack at the connecting point of the lower arm B and the knuckle. Raise the floor jack until the suspension begins to compress.
4. Remove the brake disc/drum.
5. Remove the parking brake shoes assembly, and parking brake cable.
6. Disconnect the brake hose from the brake line.
7. Remove and discard the brake hose clip from the brake hose.

➡**Use the new clip during reassembly.**

8. Remove the parking brake cable from the trailing arm.

9. Remove the flange bolts, and separate the knuckle from the trailing arm.

➡**Use the new flange bolts during reassembly.**

10. Remove the trailing arm mounting bolts, and remove the trailing arm.

➡**Use the new mounting bolts during reassembly.**

To install:

11. Install the trailing arm in the reverse order of removal, and note these items:

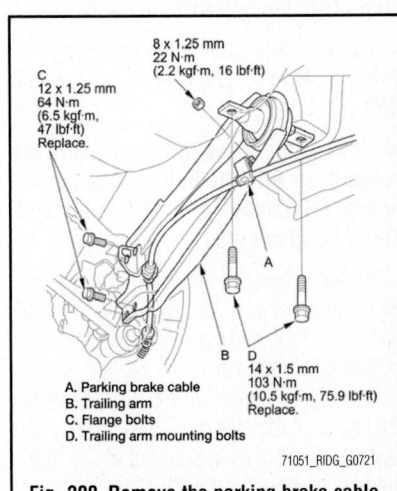

Fig. 200 Remove the parking brake cable from the trailing arm

a. Tighten all mounting hardware to the specified torque.

b. First install all the components and lightly tighten the bolts and nuts, then raise the suspension to load it with the vehicle's weight before fully tightening to the specified torque.

c. Check the brake hose for interference and twisting.

d. Before installing the wheel, clean the mating surfaces on the brake disc/drum and the inside of the wheel.

e. Fill up the brake reservoir, and bleed the brake system.

f. Inspect the parking brake, and adjust as needed.

g. After installation, check for leaks at the line joint, and retighten if necessary.

12. Check the wheel alignment, and adjust it if necessary.

UPPER BALL JOINTS

REMOVAL & INSTALLATION

07MAC-SL0A102 OR 07MAC-SL0A202

See Figures 201 through 203.

Special Tools Required:
- Ball Joint Puller 07AAC-SJCA120
- Ball Joint Thread Protector,14mm 07AAC-SJCA200
- Ball Joint Thread Protector, 12 mm 07AAF-SDAA100
- Ball Joint Thread Protector, 10 mm 07AAF-SECA120
- Ball Joint Remover, 32 mm 07MAC-SL0A102
- Ball Joint Remover, 28 mm 07MAC-SL0A202

➡**Always use a ball joint remover to disconnect a ball joint. Do not strike the housing or any other part of the ball joint connection to disconnect it.**

1. Install a hex nut or the ball joint thread protector onto the threads of the ball joint.

➡**When using a hex nut, make sure the nut is flush with the ball joint pin end to prevent damage to the threaded end of the ball joint pin.**

2. Apply grease to the ball joint remover on the areas shown. This will ease installation of the tool, and prevent damage to the pressure bolt threads.

3. Loosen the pressure bolt, and install the ball joint remover as shown. Insert the jaws carefully, making sure not to damage the ball joint boot. Adjust the jaw spacing by turning the adjusting bolt.

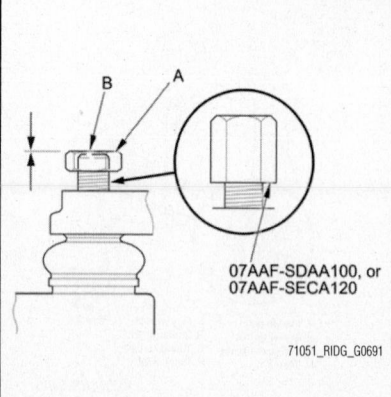

Fig. 201 Install a hex nut (A) or the ball joint thread protector onto the threads of the ball joint (B)

➡**Fasten the safety chain securely to a suspension arm or the subframe. Do not fasten it to a brake line or wire harness.**

4. After adjusting the adjusting bolt, make sure the head of the adjusting bolt is in the position shown to allow the jaw to pivot.

5. With a wrench, tighten the pressure bolt until the ball joint pin pops loose from the ball joint pin hole. If necessary, apply penetrating type lubricant to loosen the ball joint pin.

➡**Do not use pneumatic or electric tools on the pressure bolt.**

6. Remove the ball joint remover, then remove the nut or the ball joint thread protector from the end of the ball joint pin, and pull the ball joint out of the ball joint connecting hole. Inspect the ball joint boot, and replace it if damaged.

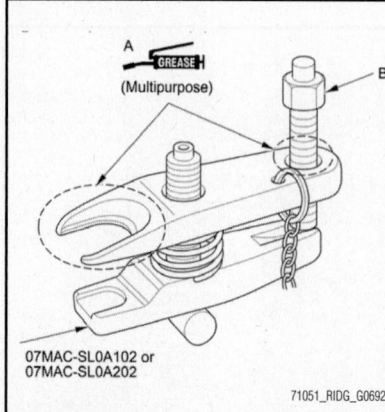

Fig. 202 Apply grease to the ball joint remover on the areas shown (A). This will ease installation of the tool, and prevent damage to the pressure bolt (B) threads

07AAC-SJCA120 AND 07AAC-SCJA200

See Figures 204 and 205.

Special Tools Required:
- Ball Joint Puller 07AAC-SJCA120
- Ball Joint Thread Protector,14mm 07AAC-SJCA200
- Ball Joint Thread Protector, 12 mm 07AAF-SDAA100
- Ball Joint Thread Protector, 10 mm 07AAF-SECA120
- Ball Joint Remover, 32 mm 07MAC-SL0A102
- Ball Joint Remover, 28 mm 07MAC-SL0A202

➡**Always use a ball joint remover to disconnect a ball joint. Do not strike the housing or any other part of the ball joint connection to disconnect it.**

1. Apply grease to the 14 mm ball joint thread protector and the ball joint on the areas shown This will prevent damage to the ball joint puller and thread of ball joint.

2. Install the 14 mm ball joint thread protector by hand, all the way onto the threads of the ball joint.

3. Install the ball joint puller as shown.

4. With a wrench, unscrew the 14 mm ball joint thread protector to apply pressure against the ball joint puller until the ball joint pin pops loose from the knuckle.

5. Remove the ball joint puller and the ball joint thread protector from the knuckle and the end of the ball joint pin, and pull the ball joint out of the knuckle. Inspect the ball joint boot, and replace it if damaged.

6. Remove any oil, grease, dust, metal debris, and other foreign material from the stud portion of the ball joint stud

UPPER ARMS

REMOVAL & INSTALLATION

See Figure 206.

Special Tools Required:
- Ball Joint Thread Protector, 12 mm 07AAF-SDAA100
- Ball Joint Remover, 32 mm 07MAC-SL0A102

1. Raise and support the vehicle.

2. Remove the rear wheel.

3. Remove the wheel speed sensor harness clip from the upper arm. Do not disconnect the wheel speed sensor connector.

4. Position a floor jack at the connecting point of the lower arm B and the knuckle.

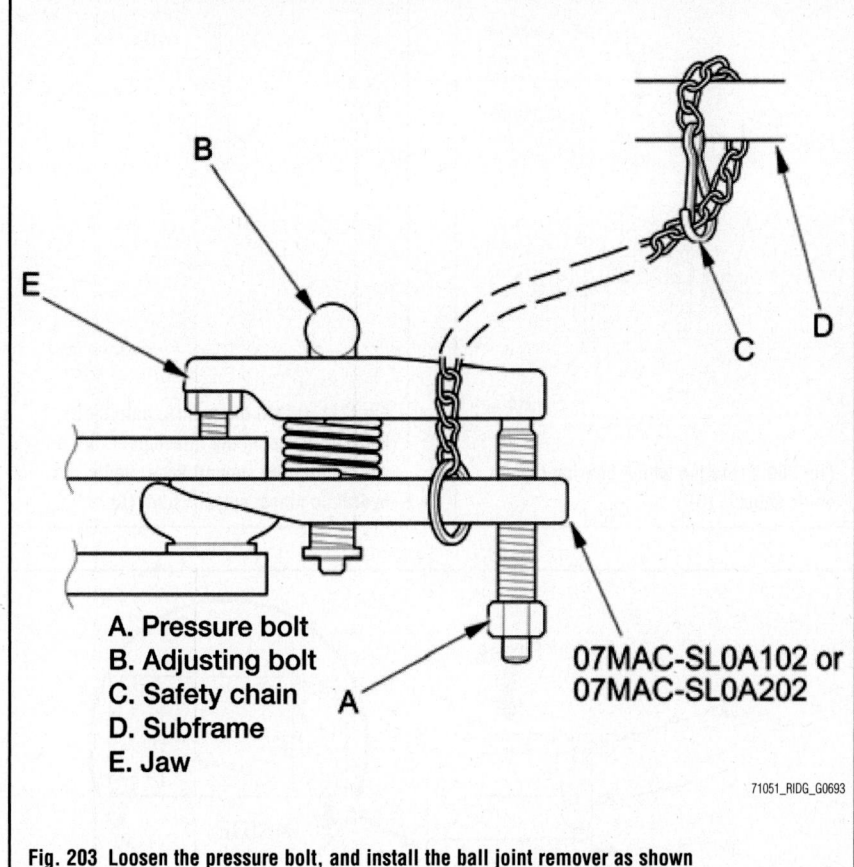

A. Pressure bolt
B. Adjusting bolt
C. Safety chain
D. Subframe
E. Jaw

07MAC-SL0A102 or
07MAC-SL0A202

71051_RIDG_G0693

Fig. 203 Loosen the pressure bolt, and install the ball joint remover as shown

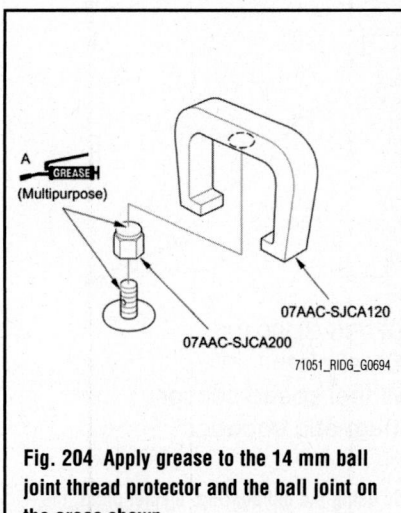

71051_RIDG_G0694

Fig. 204 Apply grease to the 14 mm ball joint thread protector and the ball joint on the areas shown

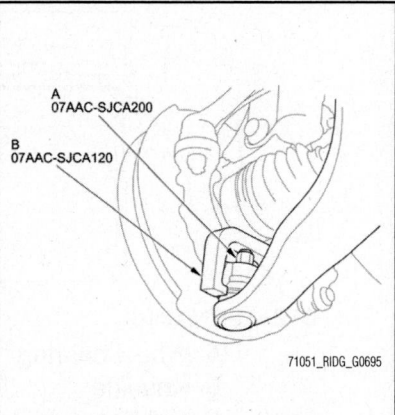

71051_RIDG_G0695

Fig. 205 Install the 14 mm ball joint thread protector (A), then install the ball joint puller (B) as shown

5. Raise the rear suspension with the floor jack until the suspension begins to compress.

6. Remove the lock pin from the upper arm ball joint, then remove the castle nut.

➡**During installation, install the lock pin as shown after tightening the castle nut.**

7. Disconnect the upper arm ball joint

from the knuckle using the ball joint thread protector and the ball joint remover.

8. Remove the upper arm mounting bolt and remove the upper arm from the vehicle.

➡**Use the new mounting bolt during reassembly.**

To install:

9. Install the upper arm in the reverse order of removal, and note these items:

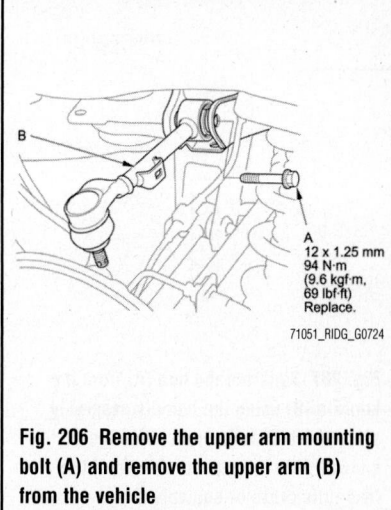

71051_RIDG_G0724

Fig. 206 Remove the upper arm mounting bolt (A) and remove the upper arm (B) from the vehicle

a. Be careful not to damage the ball joint boot when installing the knuckle.

b. Tighten all mounting hardware to the specified torque.

c. First install all the components and lightly tighten the bolts and nuts, then raise the suspension to load it with the vehicle's weight before fully tightening to the specified torque.

d. Torque the castle nut to the lower torque specification, then tighten it only far enough to align the slot with the ball joint pin hole. Do not align the castle nut by loosening it.

e. Before installing the wheel, clean the mating surface of the brake disc and the inside of the wheel.

10. Check the wheel alignment, and adjust it if necessary.

WHEEL BEARING

REMOVAL & INSTALLATION

See Figures 201 through 211.

1. Remove the knuckle/hub.

2. Separate the hub from the knuckle using the hub dis/assembly pin and a hydraulic press. Hold the knuckle with the attachment of the hydraulic press or equivalent tool. Be careful not to deform the backing plate. Hold onto the hub to keep it from falling when pressed clear.

3. Press the wheel bearing inner race off of the hub using the hub dis/assembly pin, a commercially available bearing separator, and a press.

4. Remove the flange nuts, the backing plate, and the snap ring from the knuckle.

5. Press the wheel bearing out of the knuckle using the 62 x 68 mm attachment, the driver handle, and a press.

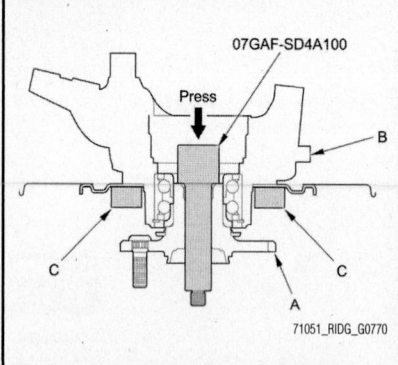

Fig. 207 Separate the hub (A) from the knuckle (B) using the hub dis/assembly pin and a hydraulic press; hold the knuckle with the attachment (C) of the hydraulic press or equivalent tool

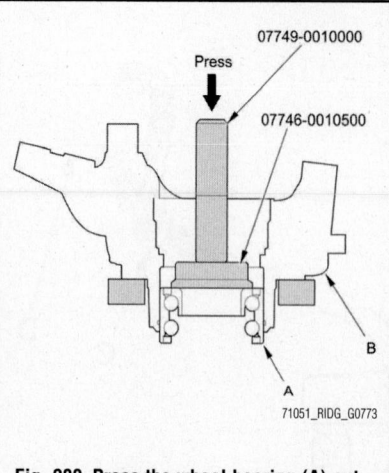

Fig. 209 Press the wheel bearing (A) out of the knuckle (B)

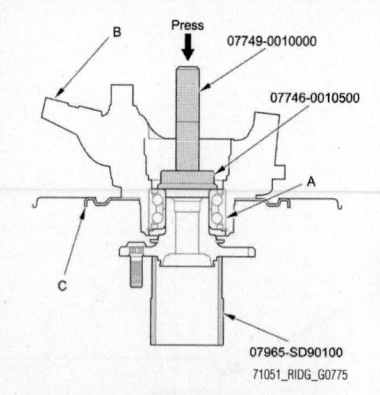

Fig. 211 Install the hub (A) onto the knuckle (B) using the attachment, the driver handle, the support base, and a hydraulic press; splash guard (C)

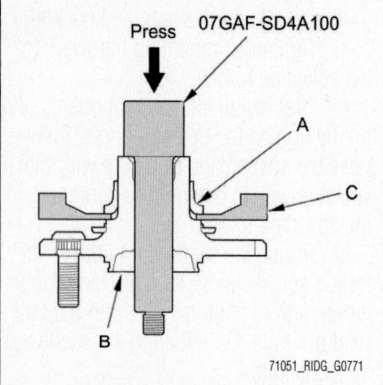

Fig. 208 Press the wheel bearing inner race (A) off of the hub (B) using the hub dis/assembly pin, a commercially available bearing separator (C), and a press

To install:

6. Wash the knuckle and hub thoroughly in high flash point solvent before reassembly.

7. Press a new wheel bearing into the knuckle using the old bearing, a steel plate, the 96 mm attachment, the support base, and a press.

- Install the wheel bearing with the wheel speed sensor magnetic encoder (brown color), toward the inside of the knuckle.
- Remove any oil, grease, dust, metal debris, and other foreign material from the magnetic encoder surface.
- Keep any magnetic tools away from the magnetic encoder surface.
- Be careful not to damage the mag-

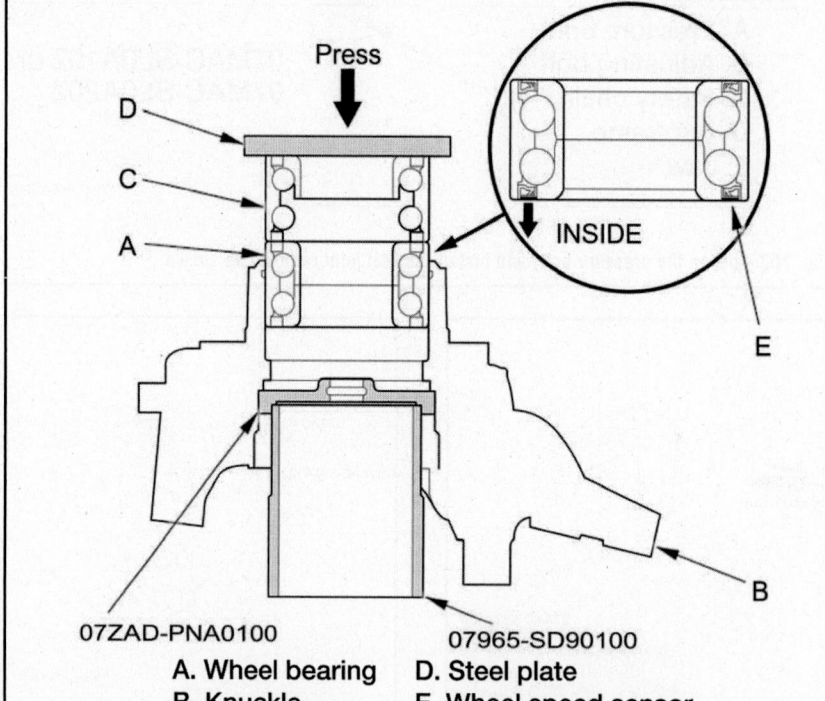

A. Wheel bearing
B. Knuckle
C. Old bearing
D. Steel plate
E. Wheel speed sensor magnetic encoder

07ZAD-PNA0100
07965-SD90100

Fig. 210 Press a new wheel bearing into the knuckle using the old bearing, a steel plate, the 96 mm attachment, the support base, and a press

netic encoder surface when you insert the wheel bearing.

8. Install the snap ring, backing plate, and flange nuts. Tighten the flange nuts to 54 ft. lbs. (74 Nm).

9. Install the hub onto the knuckle using the attachment, the driver handle, the support base, and a hydraulic press. Be careful not to damage the splash guard.

10. Install the knuckle/hub.

HONDA

Diagnostic Trouble Codes

DIAGNOSTIC TROUBLE CODES **DTC-1**

OBD II Trouble Code List (P0XXX Codes).. DTC-2
OBD II Trouble Code List (P1XXX Codes).. DTC-44
OBD II Trouble Code List (P2XXX Codes).. DTC-59
OBD II Vehicle Applications DTC-1
   Honda DTC-1

DIAGNOSTIC TROUBLE CODES

OBD II VEHICLE APPLICATIONS

HONDA

Accord
2011–2012
- 2.4L MPFI ..... Engine Code: K24Z2
- 2.4L MPFI ..... Engine Code: K24Z3
- 3.5L MPFI ..... Engine Code: J35Z2

Civic
2011–2012
- 1.3L PGM-FI .... Engine Code: LDA2
- 1.8L PGM-FI ... Engine Code: R18A1
- 1.8L CNG ...... Engine Code: R18A4
- 2.0L PGM-FI ... Engine Code: K20Z3

CrossTour
2011–2012
- 2.4L MPFI ..... Engine Code: K24Z2
- 2.4L MPFI ..... Engine Code: K24Z3
- 3.5L MPFI ..... Engine Code: J35Z2

CR-V
2011–2012
- 2.4L MPFI ..... Engine Code: K24Z1
- 2.4L MPFI ..... Engine Code: K24Z6

CR-Z
2011–2012
- 1.5L MPFI ...... Engine Code: LEA1

Element
2011
- 2.4L SMFI ..... Engine Code: K24A8

Fit
2011–2012
- 1.5L MPFI .... Engine Code: CD3

Insight
2011–2012
- 1.3L SMFI ...... Engine Code: LDA3

Odyssey
2011–2012
- 3.5L MPFI ..... Engine Code: J35Z8

Pilot
2011–2012
- 3.5L SMFI ..... Engine Code: J35Z4

Ridgeline
2011–2012
- 3.5L SMFI ..... Engine Code: J35Z5

OBD II Trouble Code List (P0XXX Codes)

DTC	Trouble Code Title and Conditions
DTC: P0010 **1T PCM, MIL: Yes** **Year:** 2011, 2012 **Model:** Accord, Civic, Crosstour, CR-V, Element **Engine:** 2.0L, 2.4L	**Variable Valve Timing Control (VTC) Oil Control Solenoid Valve Malfunction :** With the engine running one of the following conditions must be met: Condition 1: Output duty is 40% or more and the VTC current is 200 mA or less for at least 5 seconds. Condition 2: Output duty is 5% or less and the VTC current is 800 mA or more for at least 5 seconds. **NOTE: Before you troubleshoot, record all freeze data and any on-board snapshot, and review the general troubleshooting information.**
DTC: P0011 **2T PCM, MIL: Yes** **Year:** 2011, 2012 **Model:** Accord, Civic, Crosstour, CR-V, Element **Engine:** 2.0L, 2.4L, 3.5L	**Variable Valve Timing Control (VTC) System Malfunction :** With the engine running and at operating temperature. The difference between the timing control command and the actual timing of the camshaft is +/- 50 degrees or more for at least 15 seconds. **NOTE: Before you troubleshoot, record all freeze data and any on-board snapshot, and review the general troubleshooting information.**
DTC: P0101 **2T PCM, MIL: Yes** **Year:** 2011, 2012 **Model:** All **Engine:** All	**Mass Airflow (MAF) Sensor Circuit Range/Performance Problem:** Elapsed time after engine start 5 seconds, engine coolant temperature 156 ºF (69 ºC), engine speed 650-2,100 rpm. No active DTCs P0102, P0103, P0107, P0108, P0112, P0113, P0117, P0118, P0171, P0172, P0174, P0175, P0300, P0301, P0302, P0303, P0304, P0305, P0306, P0335, P0339, P0340, P0341, P0344, P0401, P0404, P0443, P0496, P0497, P0506, P0507, P1128, P1129, P145C, P2413, P2646, P2647, P2648, P2649. The difference between the amount of intake air measured by the MAF sensor and the amount of intake air calculated from the MAP sensor output is out of the normal area for at least 10 seconds.
DTC: P0102 **1T PCM, MIL: Yes** **Year:** 2011, 2012 **Model:** All **Engine:** All	**Mass Air Flow (MAF) Sensor Circuit Low Voltage:** The lower limit of the MAF sensor output is specified. If the output is below that limit, the PCM detects a malfunction and stores a DTC. Execution is continuous and the duration time is 2 seconds or more. The MAF sensor input voltage is 0.1 volt or less for at least 2 seconds. **NOTE: Before you troubleshoot, record all freeze data and any on-board snapshot.**
DTC: P0103 **1T PCM, MIL: Yes** **Year:** 2011, 2012 **Model:** Fit **Engine:** 1.5L	**Mass Airflow (MAF) Sensor Circuit High Voltage:** Ignition switch ON. The MAF sensor output voltage is 4.89 V or more for at least 2.0 seconds.
DTC: P0103 **1T PCM, MIL: Yes** **Year:** 2011, 2012 **Model:** Accord, Civic, Crosstour, CR-V, CR-Z, Element, Insight, Odyssey, Pilot, Ridgeline **Engine:** All	**Mass Airflow (MAF) Sensor Circuit High Voltage:** The upper limit of the MAF sensor output is specified. If the output is above that limit, the PCM detects a malfunction and stores a DTC. Execution is continuous and the duration time is 2 seconds or more. The MAF sensor input voltage is 4.89 V or more for at least 2 seconds. P0102 is not active. **NOTE: Before you troubleshoot, record all freeze data and any on-board snapshot.**
DTC: P0106 **1T PCM, MIL: Yes** **Year:** 2011, 2012 **Model:** Insight **Engine:** 1.3L	**Manifold Absolute Pressure (MAP) Sensor Circuit Range/Performance Problem:** The difference between the MAP sensor value measured before start-up (immediately after the ignition is turned ON) and the current value is 60 mV or less for at least 5 seconds. The execution time is once per driving cycle. **NOTE: Before you troubleshoot, record all freeze data and any on-board snapshot.**
DTC: P0107 **1T PCM, MIL: Yes** **Year:** 2011, 2012 **Model:** Accord, Civic, Crosstour, CR-V, CR-Z, Element, Insight **Engine:** All	**Manifold Absolute Pressure (MAP) Sensor Circuit Low Voltage:** With the engine running the Map sensor output voltage is 0.24 volts or less for at least 2 seconds.
DTC: P0107 **1T PCM, MIL: Yes** **Year:** 2011, 2012 **Model:** Crosstour, CR-V, Fit, Odyssey, Pilot, Ridgeline **Engine:** 1.5L, 2.4L, 3.5L	**Manifold Absolute Pressure (MAP) Sensor Circuit Low Voltage (PGM-FI System):** The MAP sensor outputs low signal voltage at high vacuum (throttle valve closed) and high signal voltage at low vacuum (throttle valve wide open). If a signal voltage from the MAP sensor is a set value or less, the powertrain Control Module (PCM) detects a malfunction and a DTC is stored. The execution time is continuous and the duration time is 2 seconds or more. The MAP sensor output voltage is 0.23 V or less for at least 2 seconds. DTC P0108 is not active. **NOTE: Before you troubleshoot, record all freeze data and any on-board snapshot.**

DTC	Trouble Code Title and Conditions
DTC: P0108 **1T PCM, MIL: Yes** **Year:** 2011, 2012 **Model:** Civic, Crosstour, CR-V, CR-Z, Element, Insight **Engine:** 1.3L, 1.5L, 1.8L, 2.0L, 2.4L	**Manifold Absolute Pressure (MAP) Sensor Circuit High Voltage:** With the engine running the Map sensor output voltage value is 4.49 V for at least 2 seconds. **NOTE: Before you troubleshoot, record all freeze data and any on-board snapshot, and review the general troubleshooting information.**
DTC: P0108 **1T CCM, MIL: Yes** **Year:** 2012 **Model:** Civic **Engine:** 2.4L	**Manifold Air Pressure Sensor Circuit High Input:** Engine started, engine running in closed loop, and the PCM detected the MAP sensor was near 29.9" Hg during the test period. **Note: The key on, engine off MAP sensor input should be near 2.9v.**
DTC: P0108 **1T PCM, MIL: Yes** **Year:** 2011, 2012 **Model:** Accord, Crosstour, Fit, Odyssey, Pilot, Ridgeline **Engine:** 1.5L, 2.4L, 3.5L	**MAP Sensor Circuit High Voltage (A/T/System) (With Navigation):** The MAP sensor outputs low signal voltage at high vacuum (throttle valve closed) and high signal voltage at low vacuum (throttle valve wide open). If a signal voltage from the MAP sensor is a set value or more, the powertrain Control Module (PCM) detects a malfunction and a DTC is stored. The execution time is continuous and the duration time is 2 seconds or more. DTC P0107 is not active. The MAP sensor output voltage is 4.49 V or more for at least 2 seconds. **NOTE: Before you troubleshoot, record all freeze data and any on-board snapshot.**
DTC: P0111 **2T PCM, MIL: Yes** **Year:** 2011, 2012 **Model:** All **Engine:** All	**Intake Air Temperature (IAT) Sensor Circuit Range/Performance:** The execution is once per driving cycle and the duration time is 10 seconds or more. Engine OFF time is 6 hours. DTCs P0112, P0113, P0116, P0117, P0118, P0125, P1116, P2183, P2184, P2185, P2610 are not active. A malfunction is detected if the following three conditions are present after the engine has stopped and the ignition switch has been turned to LOCK (0) for at least 6 hours before restarting the engine: * When the temperature difference between the IAT and ECT1 is 57 ºF (32 ºC) or more. * When the temperature difference between the IAT and ECT2 is 30 ºF (17 ºC) or more. * When the temperature difference between the ECT2 and ECT1 is 73 ºF (41 ºC) or more. **NOTE: Before you troubleshoot, record all freeze data and any on-board snapshot.**
DTC: P0111 **2T CCM, MIL: Yes** **Year:** 2012 **Model:** Civic **Engine:** 2.4L	**Intake Air Temperature Sensor Range/Performance:** Engine started, engine runtime over 10 minutes, and the PCM detected the Intake Air Temperature (IAT) sensor signal changed too much in too short a period of time during the CCM test.
DTC: P0112 **1T PCM, MIL: Yes** **Year:** 2011, 2012 **Model:** Accord, Crosstour, Fit, Pilot **Engine:** 1.5L, 2.4L, 3.5L	**Intake Air Temperature (IAT) Sensor Circuit Low Voltage:** If the IAT sensor output voltage is excessively low, the powertrain Control Module (PCM) detects a malfunction and a DTC is stored. The execution time is continuous and the duration time is 2 seconds or more. The IAT sensor output voltage is 0.08 V or less for at least 2 seconds. **NOTE: Before you troubleshoot, record all freeze data and any on-board snapshot.**
DTC: P0112 **1T PCM, MIL: Yes** **Year:** 2011, 2012 **Model:** Ridgeline **Engine:** 3.5L	**Intake Air Temperature (IAT) Sensor 1 Circuit Low Voltage:** With the ignition ON the IAT sensor 1 output voltage is 0.07 V or less for at least 5 seconds. **NOTE: Before you troubleshoot, record all freeze data and any on-board snapshot, and review the general troubleshooting information.**
DTC: P0112 **1T PCM, MIL: Yes** **Year:** 2011, 2012 **Model:** Civic, Crosstour, CR-V, CR-Z, Element, Insight, Odyssey **Engine:** All	**Intake Air Temperature (IAT) Sensor Circuit Low Voltage:** With the ignition ON the IAT sensor output voltage is 0.08 V or less for at least 2 seconds.
DTC: P0112 **1T CCM, MIL: Yes** **Year:** 2012 **Model:** Civic **Engine:** 2.4L	**Intake Air Temperature Sensor Circuit Low Input:** Key on or engine running, and the PCM detected the IAT sensor signal indicated less than 0.1v (Scan Tool reads over 302ºF).

DTC	Trouble Code Title and Conditions
DTC: P0113 **1T PCM, MIL: Yes** **Year:** 2011, 2012 **Model:** Civic, Crosstour, CR-V, CR-Z, Insight **Engine:** 1.3L, 1.5L, 1.8L, 2.0L, 2.4L	**Intake Air Temperature (IAT) Sensor Circuit High Voltage:** With ignition ON, the IAT sensor output voltage is 4.92 V or more for at least 2 seconds.
DTC: P0113 **1T CCM, MIL: Yes** **Year:** 2012 **Model:** Civic **Engine:** 2.4L	**Intake Air Temperature Sensor Circuit High Input:** Key on or engine running, and the PCM detected the IAT sensor signal indicated more than 4.90 (Scan Tool reads less than -4ºF).
DTC: P0113 **2T PCM, MIL: Yes** **Year:** 2011, 2012 **Model:** Accord, Crosstour **Engine:** 2.4L, 3.5L	**IAT Sensor Circuit High Voltage:** * The ignition is ON or the engine is running. * DTC P0113 runs continuously once the above condition has been met. • ECM/PCM
DTC: P0113 **1T PCM, MIL: Yes** **Year:** 2011, 2012 **Model:** CR-V, Element, Fit, Odyssey, Pilot **Engine:** 1.5L, 2.4L, 3.5L	**Intake Air Temperature (IAT) Sensor Circuit High Voltage:** The IAT sensor resistance varies depending on temperature. The output voltage and the sensor resistance increase as the intake air temperature decreases. Conversely, the output voltage and the sensor resistance decrease as the intake air temperature increases. If the IAT sensor output voltage is excessively high, the powertrain Control Module (PCM) detects a malfunction and a DTC is stored. The execution time is continuous and the duration time is 2 seconds or more. P0112 is not active. The IAT sensor output voltage is 4.92 V or more for at least 2 seconds. **NOTE: Before you troubleshoot, record all freeze data and any on-board snapshot.**
DTC: P0113 **1T PCM, MIL: Yes** **Year:** 2011, 2012 **Model:** Ridgeline **Engine:** 3.5L	**Intake Air Temperature (IAT) Sensor 1 Circuit High Voltage:** With the ignition ON the IAT sensor 1 output voltage is 4.93 V or less for at least 5 seconds. **NOTE: Before you troubleshoot, record all freeze data and any on-board snapshot, and review the general troubleshooting information.**
DTC: P0116 **2T PCM, MIL: Yes** **Year:** 2011, 2012 **Model:** Civic, CR-Z **Engine:** 1.5L	**Engine Coolant Temperature (ECT) Sensor 1 Range/Performance Problem:** Malfunction Threshold: * Malfunction determination 2011 model: With a completely cooled engine (one that has been off for at least 6 hours): When the change in coolant temperature after 36 minutes or more of running time or 4 miles (6 km) or more of driving distance is 18 ºF (10 ºC) or less, a malfunction is detected. * Malfunction determination 2012 model on and after: With a partially cooled engine (one that has been off for less than 6 hours): When the difference between the coolant temperature after 36 minutes or more of running time or 4 miles (6 km) or more of driving distance the coolant temperature after the engine has been off for 4 hours, 30 minutes and then ran for 10 seconds is 18 ºF (10 ºC) or less, a malfunction is detected.
DTC: P0116 **2T PCM, MIL: Yes** **Year:** 2011, 2012 **Model:** Accord, Civic, Crosstour, CR-V, Element, Fit, Odyssey, Pilot, Ridgeline **Engine:** All	**Engine Coolant Temperature (ECT) Sensor 1 Circuit Range/Performance Problem :** The execution time is once per driving cycle and the duration time is 10 seconds or more. The following DTCs are not active P0101, P0102, P0103, P0107, P0108, P0117, P0118, P0134, P0135, P0154, P0155, P0171, P0172, P0174, P0175, P0300, P0301, P0302, P0303, P0304, P0305, P0306, P0335, P0339, P0340, P0344, P0401, P0404, P0443, P0496, P0497, P0506, P0507, P0627, P1077, P1078, P1128, P1129, P1172, P1174, P145C, P2195, P2197, P2227, P2228, P2229, P2237, P2238, P2240, P2241, P2243, P2245, P2247, P2249, P2251, P2252, P2254, P2255, P2413, P2610, P2646, P2647, P2648, P2649. Malfunction determination 1: With a completely cooled engine (one that has been off for at least 6 hours): When the change in coolant temperature after 42 minutes or more of running time or drive 5 miles (7 km) or more of driving time is 50 ºF (10 ºC) or less, a malfunction is detected. Malfunction determination 2: With a partially cooled engine (one that has been off for less than 6 hours): When the difference between the coolant temperature after 42 minutes or more of running time or drive 5 miles (7 km) or more of driving time the coolant temperature after the engine has been off for 150 minutes and then run for 10 seconds is 50ºF (10 ºC) or less, a malfunction is detected. **NOTE: Before you troubleshoot, record all freeze data and any on-board snapshot.**

DTC	Trouble Code Title and Conditions
DTC: P0116 **2T PCM, MIL: Yes** **Year:** 2011, 2012 **Model:** Insight **Engine:** 1.3L	**Engine Coolant Temperature (ECT) Sensor 1 Circuit Range/Performance Problem:** A malfunction is detected if the following three conditions are present after the engine has stopped and the ignition switch has been turned to LOCK (0) for at least 6 hours before restarting the engine: (1) When the temperature difference between the IAT and ECT1 is 57 ºF (32 ºC) or more. (2) When the temperature difference between the IAT and ECT2 is 30 ºF (17 ºC) or more. (3) When the temperature difference between the ECT2 and ECT1 is 73 ºF (41 ºC) or more. **NOTE: If DTC P0111 is stored at the same time as DTC P1116, troubleshoot DTC P0111 first, then recheck for DTC P1116.**
DTC: P0116 **1T CCM, MIL: Yes** **Year:** 2012 **Model:** Civic **Engine:** 2.4L	**Engine Coolant Temperature Sensor Range/Performance:** Engine running, and the PCM detected the Engine Coolant Temperature (ECT) sensor signal changed too much too quickly. **Note: The ECT sensor should read 0.47v-0.78v at hot idle speed.**
DTC: P0117 **1T CCM, MIL: Yes** **Year:** 2012 **Model:** Civic **Engine:** 2.4L	**Engine Coolant Temperature Sensor Circuit Low Input:** Key on or engine running, and the PCM detected the Engine Coolant Temperature (ECT) sensor signal indicated more than 302ºF (0.1v). **Note: The normal range of the ECT sensor is from 0.47v to 0.78v.**
DTC: P0117 **1T PCM, MIL: Yes** **Year:** 2011, 2012 **Model:** All **Engine:** All	**Engine Coolant Temperature (ECT) Sensor 1 Circuit Low Voltage:** If the ECT sensor 1 output voltage is less than a set value when the engine coolant temperature is high, the PCM detects a malfunction and a DTC is stored. The execution time is continuous and the duration time is 2 seconds or more. P0118 is not active. The ECT sensor 1 output voltage is 0.08 V or less for at least 2 seconds. **NOTE: Before you troubleshoot, record all freeze data and any on-board snapshot.**
DTC: P0118 **1T PCM, MIL: Yes** **Year:** 2011, 2012 **Model:** All **Engine:** All	**Engine Coolant Temperature (ECT) Sensor 1 Circuit High Voltage:** If the ECT sensor 1 output voltage is less than the set value when the engine coolant temperature is low, the PCM detects a malfunction and a DTC is stored. The execution time is continuous and the duration time is 2 seconds or more. DTC P0117 is not active. The ECT sensor 1 output voltage is 4.92 V or more for at least 2 seconds. **NOTE: Before you troubleshoot, record all freeze data and any on-board snapshot.**
DTC: P0118 **1T CCM, MIL: Yes** **Year:** 2012 **Model:** Civic **Engine:** 2.4L	**Engine Coolant Temperature Sensor Circuit High Input:** Key on or engine running, and the PCM detected the Engine Coolant Temperature (ECT) sensor signal indicated less than -4ºF (4.9v). **Note: The normal range of the ECT sensor is from 0.47v to 0.78v.**
DTC: P0122 **1T PCM, MIL: Yes** **Year:** 2012 **Model:** Crosstour **Engine:** 2.4L	**Throttle Position (TP) Sensor A Circuit Low Voltage :** With the ignition ON the TP sensor 1 output voltage value is 0.20 V or less for at least 0.1 second. **NOTE: Before you troubleshoot, record all freeze data and any on-board snapshot, and review the general troubleshooting information.**
DTC: P0122 **1T CCM, MIL: Yes** **Year:** 2012 **Model:** Civic **Engine:** 2.4L	**Throttle Position Sensor Circuit Low Input:** Engine running in closed loop conditions, and the PCM detected the closed throttle TP signal was less than 0.16v (less than 10% open).
DTC: P0122 **1T PCM, MIL: Yes** **Year:** 2011, 2012 **Model:** Accord, Crosstour, Pilot, Ridgeline **Engine:** 2.4L, 3.5L	**Throttle Position (TP) Sensor A Circuit Low Voltage:** With the ignition switch ON (II), the TP sensor output voltage is 0.3 V or less for at least 200 milliseconds. **NOTE: Before you troubleshoot, record all freeze data and any on-board snapshot, and review the general troubleshooting information.**
DTC: P0122 **1T PCM, MIL: Yes** **Year:** 2011, 2012 **Model:** Civic, Crosstour, CR-V, CR-Z, Element, Fit, Insight, Odyssey **Engine:** All	**Throttle Position (TP) Sensor A Circuit Low Voltage:** With the ignition switch ON (II), the TP sensor output voltage is 0.3 V or less for at least 200 milliseconds. **NOTE: Before you troubleshoot, record all freeze data and any on-board snapshot, and review the general troubleshooting information.**

DTC	Trouble Code Title and Conditions
DTC: P0123 1T PCM, MIL: Yes **Year:** 2011, 2012 **Model:** All **Engine:** All	**Throttle Position (TP) Sensor A Circuit High Voltage:** If the signal from TP sensor A is more than a fixed value for a set time, the PCM detects a TP sensor A malfunction and stores a DTC. The execution time is continuous and the duration time is 200 milliseconds or more. Ignition is running and DTCs P0122, P2101, P2118, P2135, P2176 are not active. The TP sensor A output voltage is 4.8 V or more for at least 200 milliseconds. **NOTE: Before you troubleshoot, record all freeze data and any on-board snapshot.**
DTC: P0123 1T CCM, MIL: Yes **Year:** 2012 **Model:** Civic **Engine:** 2.4L	**Throttle Position Sensor Circuit High Input:** Engine running in closed loop conditions, and the PCM detected the wide-open-throttle TP signal was more than 4.60v (more than 90%).
DTC: P0125 2T PCM, MIL: Yes **Year:** 2011, 2012 **Model:** All **Engine:** All	**Engine Coolant Temperature (ECT) Sensor 1 Malfunction/Slow Response:** As the engine coolant warms, the ECT sensor 1 resistance decreases, and the PCM detects a low signal voltage. If the ECT sensor 1 output voltage does not reach a specified temperature at which closed-loop control for stoichiometric air/fuel ratio starts within a set time, depending on the initial coolant temperature after starting the engine, the PCM detects a malfunction. **NOTE: Before you troubleshoot, record all freeze data and any on-board snapshot.**
DTC: P0128 2T ECT, MIL: Yes **Year:** 2012 **Model:** Civic **Engine:** 2.4L	**Thermostat Range/Performance:** DTC P0107, P0108, P0112, P0113, P0116-118, P0335, P0336, P0300-P0306, P0401, P0505, P1106-P1108, P1259 and P1519, engine running at road load for 10 minutes, and the PCM detected the ECT sensor input did not reach the correct closed loop value. **Note: It is possible for this code to set if the engine is left running while the hood is open for an extended period in a warm climate.**
DTC: P0128 2T PCM, MIL: Yes **Year:** 2011, 2012 **Model:** Civic, Crosstour, CR-V, CR-Z, Element, Fit, Insight, Odyssey, Pilot, Ridgeline **Engine:** All	**Cooling System Malfunction :** With the engine OFF for at least 6 hours, coolant temperature between 20 to 123 ºF (-7 to 50 ºC), the malfunction threshold are as follows. Malfunction determination 1: If the difference between the current measured coolant temperature at the radiator (ECT 2) and the initial coolant temperature at the radiator is at least 13 ºF (7 ºC) when the estimated coolant temperature at the engine (ECT 1) reaches 164 ºF (73 ºC), a malfunction is detected (Thermostat Stuck Open); or if the coolant temperature at the radiator (ECT 2) only reaches 68 ºF (20 ºC), a malfunction is detected. Malfunction determination 2: When the calculated engine coolant temperature (ECT 1) reaches 158 ºF (70 ºC) before the measured engine coolant temperature (ECT 1) reaches 158 ºF (70 ºC), a malfunction is detected.
DTC: P0128 2T PCM, MIL: Yes **Year:** 2011, 2012 **Model:** Accord, Accord Crosstour, Crosstour **Engine:** 2.4L, 3.5L	**Cooling System Malfunction:** With engine running the ECT sensor output (70 C) 158 F or less, an estimated engine coolant temperature is (75 C) 168 F or more. The difference between the estimated engine coolant temperature and the ECT sensor output is (15 C) 27 F or more. **NOTE: If the DTCs listed below are stored at the same time as DTC P0128, troubleshoot those DTCs first:** P0107, P0108, P1128, P1129, P1106, P1107, P1108, P1259, P0401, P0116, P0117, P0118, P0112, P0113, P0335, P0336, P0300-P0306, P0505, P1519, then recheck for P0128.
DTC: P0131 1T CCM, MIL: Yes **Year:** 2012 **Model:** Civic **Engine:** 2.4L	**HO2S-11 (Bank 1 Sensor 1) Circuit Low Input:** Engine running in closed loop in 4th or 6th gear at cruise speed, and the PCM detected the HO2S signal was fixed at less than 0.50v. **Note: The actual value to set the code is stored in the PCM memory.**
DTC: P0132 1T CCM, MIL: Yes **Year:** 2012 **Model:** Civic **Engine:** 2.4L	**HO2S-11 (Bank 1 Sensor 1) Circuit High Input:** Engine running in closed loop in 4th or 6th gear at cruise speed, and the PCM detected the HO2S signal was fixed at more than 0.90v. **Note: The actual value to set the code is stored in the PCM memory.**
DTC: P0133 1T O2S1, MIL: Yes **Year:** 2012 **Model:** Civic **Engine:** 2.4L	**HO2S-11 (Bank 1 Sensor 1) Circuit Slow Response:** Engine running in closed loop in 4th or 6th gear at over 55 mph at steady speed, and the PCM detected the HO2S response time to switch between 300-600 mv was too slow, or that the rich to lean or lean to rich switch time was too slow.

DTC	Trouble Code Title and Conditions
DTC: P0133 **2T PCM, MIL: Yes** **Year:** 2011, 2012 **Model:** Accord, Civic, Crosstour, CR-V, CR-Z, Element, Fit, Insight **Engine:** 1.3L, 1.5L, 1.8L, 2.0L, 2.4L	**Air/Fuel Ratio (A/F) Sensor (Sensor 1) Malfunction/Slow Response:** With the engine running and in closed loop, and the vehicle speed at least 33 mph. The total A/F sensor (SENSOR 1) output value for A/T models is 20 and for M/T models 38 or less in 0.8 seconds. **NOTE: Before you troubleshoot, record all freeze data and any on-board snapshot, and review the general troubleshooting information. If DTC P0139 is stored at the same time as DTC P0133, troubleshoot DTC P0139 first, then recheck for DTC P0133.**
DTC: P0133 **2T O2S1, MIL: Yes** **Year:** 2011, 2012 **Model:** Accord, Crosstour, Odyssey, Pilot, Ridgeline **Engine:** 3.5L	**Rear Air/Fuel Ratio (A/F) Sensor (Bank 1, Sensor 1) Malfunction/Slow Response:** The execution time is once per drive cycle and the duration time is 6.8 seconds or more. Engine coolant temperature 156 ºF (69 ºC), Intake air temperature, Fuel trim, 0.73-1.47, Engine speed 1,250-2,200 rpm, Vehicle speed 33 mph or more. The total rear A/F sensor output value is 11 or less in 6.5 seconds. **NOTE: Before you troubleshoot, record all freeze data and any on-board snapshot.** Conditions for illuminating the MIL When a malfunction is detected during the first drive cycle, a Temporary DTC is stored in the PCM memory. If the malfunction recurs during the next (second) drive cycle, the MIL comes on and the DTC and the freeze frame data are stored.
DTC: P0134 **2T PCM, MIL: Yes** **Year:** 2011, 2012 **Model:** Accord, Civic, Crosstour, CR-V, CR-Z, Element, Fit, Insight **Engine:** 1.3L, 1.5L, 1.8L, 2.0L, 2.4L	**Air/Fuel Ratio (A/F) Sensor (Sensor 1) Heater System Malfunction :** With the engine running for 40 seconds or more and battery voltage above 10.5 V. Malfunction 1: From start up the A/F sensor (SENSOR 1) internal resistance value is 90 ohms or more for at least 40 seconds right after the engine starts. Malfunction 2: With the engine hot and at operating temperature the A/F sensor (SENSOR 1) internal resistance value is 250 ohms or more for at least 1.0 second. **NOTE: Before you troubleshoot, record all freeze data and any on-board snapshot, and review the general troubleshooting information. If DTC P0135 is stored at the same time as DTC P0134, troubleshoot DTC P0135 first, then recheck for DTC P0134.**
DTC: P0134 **2T PCM, MIL: Yes** **Year:** 2011, 2012 **Model:** Accord, Crosstour, Odyssey, Pilot, Ridgeline **Engine:** 3.5L	**Rear Air/Fuel Ratio (A/F) Sensor (Bank 1, Sensor 1) Heater System Malfunction:** The execution time is continuous and the duration time is 40 seconds or more. Battery voltage minimum 10.5 volts. Malfunction determination 1: The rear A/F sensor (bank 1, sensor 1) internal resistance value is 110 ohms or more for at least 40 seconds right after the engine starts. Malfunction determination 2: The rear A/F sensor (bank 1, sensor 1) internal resistance value is 110 ohms or more for at least 15 seconds. **NOTE: Before you troubleshoot, record all freeze data and any on-board snapshot.** The rear A/F sensor (bank 1, sensor 1) internal resistance value is 200 ohms or more for at least 1 second.
DTC: P0135 **1T PCM, MIL: Yes** **Year:** 2011, 2012 **Model:** Accord, Crosstour, Odyssey, Pilot, Ridgeline **Engine:** 3.5L	**Rear Air/Fuel Ratio (A/F) Sensor (Bank 1, Sensor 1) Heater Circuit Malfunction:** The execution time is continuous and the duration time is 2 seconds or more. Engine running and battery at 10 volts or more. No return signal HIGH is detected when the PCM output duty is less than 2%. Return signal does not change when the PCM output duty is more than 20% and less than 80%. No return signal LOW is detected when the PCM output duty is more than 8%. **NOTE: Before you troubleshoot, record all freeze data and any on-board snapshot.**
DTC: P0135 **2T PCM, MIL: Yes** **Year:** 2011, 2012 **Model:** Accord, Civic, Crosstour, CR-V, CR-Z, Element, Fit, Insight **Engine:** 1.3L, 1.5L, 1.8L, 2.0L, 2.4L	**Air/Fuel Ratio (A/F) Sensor (Sensor 1) Heater Circuit Malfunction :** With the engine running and the engine temperature at least 69 degrees the following malfunctions are detected. Malfunction 1: The heater current is 0.8 A or less for at least 4 seconds while the heater is active, and the heater current is 0.8 A or more for at least 4 seconds while the heater is not active. Malfunction 2: The heater current is 15.2 A or more for at least 0.6 second. **NOTE: Before you troubleshoot, record all freeze data and any on-board snapshot, and review the general troubleshooting information.**
DTC: P0135 **1T O2S HTR1, MIL: Yes** **Year:** 2012 **Model:** Civic **Engine:** 2.4L	**HO2S-11 (Bank 1 Sensor 1) Heater Circuit Malfunction:** Engine runtime over 80 seconds, and the PCM detected an incorrect signal value at the HO2S heater circuit during the test period.
DTC: P0137 **1T PCM, MIL: Yes** **Year:** 2011, 2012 **Model:** Accord, Crosstour, Odyssey, Pilot, Ridgeline **Engine:** 3.5L	**Rear Secondary Heated Oxygen Sensor (Secondary HO2S (Bank 1, Sensor 2) Circuit Low Voltage:** The execution time is continuous and the duration time is 40 seconds or more. The system is in closed loop with the engine coolant temperature at 156 ºF (69 ºC) and the intake air temperature at -13 ºF (-25 ºC). The rear secondary HO2S output voltage is 0.05 V or less for at least 40 seconds. After current is applied to the rear secondary HO2S heater, if the rear secondary HO2S output continues low (lean) during feedback control, a malfunction is detected and a DTC is stored. **NOTE: Before you troubleshoot, record all freeze data and any on-board snapshot.**

DTC	Trouble Code Title and Conditions
DTC: P0137 **1T PCM, MIL: Yes** **Year:** 2011, 2012 **Model:** Accord, Civic, Crosstour, CR-V, CR-Z, Element, Fit, Insight **Engine:** 1.3L, 1.5L, 1.8L, 2.0L, 2.4L	**Secondary HO2S (Sensor 2) Circuit Low Voltage :** With engine running the secondary HO2S (Sensor 2) output voltage remains in Zone 1 (0.29 V or below) for at least 72 seconds.
DTC: P0137 **1T CCM, MIL: Yes** **Year:** 2012 **Model:** Civic **Engine:** 2.4L	**HO2S-12 (Bank 1 Sensor 2) Circuit Low Input:** Engine running in closed loop in 4th or 6th gear at cruise speed, and the PCM detected the HO2S signal was fixed at less than 0.30v. **Note: The actual value where the code sets is in the PCM memory.**
DTC: P0138 **1T CCM, MIL: Yes** **Year:** 2012 **Model:** Civic **Engine:** 2.4L	**HO2S-12 (Bank 1 Sensor 2) Circuit High Input:** Engine running in closed loop in 4th or 6th gear at cruise speed, and the PCM detected the HO2S signal was fixed at more than 0.60v. **Note: The actual value where the code sets is in the PCM memory.**
DTC: P0138 **2T PCM, MIL: Yes** **Year:** 2011, 2012 **Model:** Accord, Civic, Crosstour, CR-V, CR-Z, Element, Fit, Insight **Engine:** 1.3L, 1.5L, 1.8L, 2.0L, 2.4L	**Secondary HO2S (Sensor 2) Circuit High Voltage :** With engine running the Secondary HO2S (Sensor 2) remains in Zone 4 (0.80 V or above) for at least 72 seconds.
DTC: P0138 **1T PCM, MIL: Yes** **Year:** 2011, 2012 **Model:** Accord, Crosstour, Odyssey, Pilot, Ridgeline **Engine:** 3.5L	**Rear Secondary Heated Oxygen Sensor (Secondary HO2S (Bank 1, Sensor 2) Circuit High Voltage:** The execution time is continuous and the duration time is 2 seconds or more. The rear secondary HO2S output voltage is 1.270 V or more for at least 5 seconds. After current is applied to the rear secondary HO2S heater, if the rear secondary HO2S output continues to exceed the upper limit used during feedback control, a malfunction is detected and a DTC is stored. **NOTE: Before you troubleshoot, record all freeze data and any on-board snapshot.**
DTC: P0139 **2T PCM, MIL: Yes** **Year:** 2011, 2012 **Model:** Odyssey, Pilot **Engine:** 3.5L	**Rear Secondary Heated Oxygen Sensor (Secondary HO2S (Bank 1, Sensor 2) Circuit Slow Response:** If the response time of the rear secondary HO2S becomes longer than the specified time after current to the secondary HO2S heater is applied, a malfunction is detected and a DTC is stored. The execution time is once per driving cycle and the duration time is 20.7 seconds or less. When a malfunction is detected during the first drive cycle, a Temporary DTC is stored in the PCM memory. If the malfunction recurs during the next (second) drive cycle, the MIL comes on and the DTC and the freeze frame data are stored. **NOTE: Before you troubleshoot, record all freeze data and any on-board snapshot.**
DTC: P0139 **2T PCM, MIL: Yes** **Year:** 2011, 2012 **Model:** Crosstour, CR-V, Element, Insight **Engine:** 1.3L, 2.4L	**Secondary HO2S (Sensor 2) Slow Response:** With the engine running for at least 2 minutes, coolant temperature at a minimum of 156 F (69 C) and the engine speed a maximum of 4,000 rpm, the secondary HO2S output voltage is 0.05 V or more for at least 10 seconds.
DTC: P0139 **2T PCM, MIL: Yes** **Year:** 2011, 2012 **Model:** Accord, Civic, Crosstour, CR-Z, Fit **Engine:** 1.5L, 2.4L	**Secondary HO2S (Sensor 2) Slow Response :** Engine started, vehicle driven in closed loop in 4th or 6th gear at over 55 mph at steady speed, and the PCM detected the HO2S response time to switch between 300-600 mv was too slow, or that the rich to lean or lean to rich switch time was too slow.
DTC: P0139 **2T PCM, MIL: Yes** **Year:** 2011, 2012 **Model:** Accord, Crosstour, Ridgeline **Engine:** 3.5L	**Rear Secondary HO2S (Bank 1, Sensor 2) Slow Response:** With the engine running in closed loop and driven at a steady speed of 35 mph (57 km/h) for at least 22.7 seconds. The rear secondary HO2S output drops to the response deterioration judgment threshold value and the response characteristics measurement is finished. OR. The voltage does not drop to the response deterioration judgment threshold value after the predetermined time has elapsed. (Elapsed time Minimum 0.51 seconds–Maximum 4.35 seconds) Before you troubleshoot, record all freeze data and any on-board snapshot, and review the general troubleshooting information.

DTC	Trouble Code Title and Conditions
DTC: P0139 **2T O2S1, MIL: Yes** **Year:** 2012 **Model:** Civic **Engine:** 2.4L	**HO2S-12 (Bank 1 Sensor 2) Circuit Slow Response:** Engine running in closed loop in 4th or 6th gear at over 55 mph at steady speed, and the PCM detected the HO2S response time to switch between 300-600 mv was too slow, or that the rich to lean or lean to rich switch time was too slow.
DTC: P0141 **1T PCM, MIL: Yes** **Year:** 2011, 2012 **Model:** Accord, Civic, Crosstour, CR-V, CR-Z, Element, Fit, Insight **Engine:** 1.3L, 1.5L, 1.8L, 2.0L, 2.4L	**Secondary HO2S (Sensor 2) Heater Circuit Malfunction :** With engine running the current is 0.5 A or less, or 3.6 A or more for at least 5 seconds when the heater is on.
DTC: P0141 **1T PCM, MIL: Yes** **Year:** 2011, 2012 **Model:** Accord, Crosstour, Odyssey, Pilot, Ridgeline **Engine:** 3.5L	**Rear Secondary Heated Oxygen Sensor (Bank 1 Sensor 2) Heater Circuit Malfunction:** The rear secondary HO2S heater output is 0.38 A or less, or 3.33 A or more, for at least 5 seconds when the heater is on. Engine running the battery voltage (IGP terminal of PCM) is 10.5-16 V, DTCs P0117, P0118 are not active. If the rear secondary HO2S heater draws more or less than a specified amperage, the PCM detects a malfunction and a DTC is stored. **NOTE: Before you troubleshoot, record all freeze data and any on-board snapshot.**
DTC: P0141 **1T ECM, MIL: Yes** **Year:** 2012 **Model:** Civic **Engine:** 2.4L	**HO2S-12 (Bank 1 Sensor 2) Heater Circuit Malfunction:** Engine runtime over 80 seconds, and the PCM detected an incorrect signal value at the HO2S heater circuit during the test period.
DTC: P0153 **2T PCM, MIL: Yes** **Year:** 2011, 2012 **Model:** Accord, Crosstour, Odyssey, Pilot, Ridgeline **Engine:** 3.5L	**Front Air/Fuel Ratio (A/F) Sensor (Bank 2, Sensor 1) Malfunction/Slow Response:** The execution time is once per drive cycle and the duration time is 6.8 seconds or more. Engine coolant temperature 156 ºF (69 ºC), Intake air temperature, Fuel trim, 0.73-1.47, Engine speed 1,250-2,200 rpm, Vehicle speed 33 mph or more. The total front A/F sensor (bank 2, sensor 1) output value is 34 or less in 6.8 seconds.
DTC: P0154 **2T PCM, MIL: Yes** **Year:** 2011, 2012 **Model:** Odyssey, Pilot, Ridgeline **Engine:** 3.5L	**Front Air Fuel Ratio (A/F) Sensor (Bank 2, Sensor 1) Heater System Malfunction :** The execution time is continuous and the duration time is 40 seconds or more. Battery voltage minimum 10.5 volts. Malfunction determination 1: The front A/F sensor (bank 2, sensor 1) internal resistance value is 200 or more for at least 40 seconds right after the engine starts. Malfunction determination 2: The front A/F sensor (bank 2, sensor 1) internal resistance value is 270 or more for at least 1.0 second. **NOTE: Before you troubleshoot, record all freeze data and any on-board snapshot.**
DTC: P0155 **2T PCM, MIL: Yes** **Year:** 2011, 2012 **Model:** Odyssey, Pilot, Ridgeline **Engine:** 3.5L	**Front Air/Fuel Ratio (A/F) Sensor (Bank 2, Sensor 1) Heater Circuit Malfunction:** The execution time is continuous and the duration time is 2 seconds or more. Engine running and battery voltage at 10 volts or more. One of these 2 conditions must be met for at least 2 seconds: (1) No return signal HIGH is detected when the PCM output duty is less than 20%. (2) Return signal does not change when the PCM output duty is more than 20% and less than 80%. **NOTE: Before you troubleshoot, record all freeze data and any on-board snapshot.**
DTC: P0157 **1T O2S HTR2, MIL: Yes** **Year:** 2011, 2012 **Model:** Accord, Crosstour, Odyssey, Pilot, Ridgeline **Engine:** 3.5L	**Front Secondary HO2S (Bank 2, Sensor 2) Circuit Low Voltage :** The execution time is continuous and the duration time is 40 seconds or more. The system is in closed loop with the engine coolant temperature at 156 ºF (69 ºC) and the intake air temperature at -13 ºF (-25 ºC). The rear secondary HO2S output voltage is 0.05 V or less for at least 40 seconds. After current is applied to the rear secondary HO2S heater, if the rear secondary HO2S output continues low (lean) during feedback control, a malfunction is detected and a DTC is stored. **NOTE: Before you troubleshoot, record all freeze data and any on-board snapshot.**
DTC: P0158 **2T PCM, MIL: Yes** **Year:** 2011, 2012 **Model:** Accord, Crosstour, Odyssey, Pilot, Ridgeline **Engine:** 3.5L	**Front Secondary HO2S (Bank 2, Sensor 2) Circuit High Voltage :** With the vehicle at operating temperature and driven between 1,000-3,000 rpm for at least 1 minute, the rear secondary HO2S output voltage is 1.270 V or more for at least 5 seconds. After current is applied to the rear secondary HO2S heater, if the rear secondary HO2S output continues to exceed the upper limit used during feedback control, a malfunction is detected and a DTC is stored. **NOTE: Before you troubleshoot, record all freeze data and any on-board snapshot.**

DTC	Trouble Code Title and Conditions
DTC: P0159 **2T PCM, MIL: Yes** **Year:** 2011, 2012 **Model:** Accord, Crosstour, Odyssey, Pilot, Ridgeline **Engine:** 3.5L	**Front Secondary HO2S (Bank 2, Sensor 2) Slow Response :** The execution time is once per driving cycle and the duration time is 20.7 seconds or less. Engine coolant temperature is 156 ºF (69 ºC), intake air temperature -13 ºF (-25 ºC), engine speed 1,180-2,000 rpms and vehicle speed is 30 mph. When the front secondary HO2S output drops to the response deterioration judgment threshold value and the response characteristics measurement is finished. (0.77-4.58 seconds)
DTC: P0161 **1T PCM, MIL: Yes** **Year:** 2011, 2012 **Model:** Accord, Crosstour, Odyssey, Pilot, Ridgeline **Engine:** 3.5L	**Front Secondary HO2S (Bank 2, Sensor 2) Heater Circuit Malfunction :** The front secondary HO2S heater output is 0.38 A or less, or 3.33 A or more, for at least 5 seconds when the heater is on. Engine running the battery voltage (IGP terminal of PCM) is 10.5-16 V, DTCs P0117, P0118 are not active. If the front secondary HO2S heater draws more or less than a specified amperage, the PCM detects a malfunction and a DTC is stored. **NOTE: Before you troubleshoot, record all freeze data and any on-board snapshot.**
DTC: P0171 **2T , MIL: Yes** **Year:** 2012 **Model:** Civic **Engine:** 2.4L	**Fuel System Too Lean (Bank 1):** DTC P0107, P0108, P0135, P0137, P0138, P0141, P1128, P1129 and P1259 not set, engine running in closed loop, and the PCM detected the LONGFT value exceeded the calibrated lean limit value.
DTC: P0171 **2T PCM, MIL: Yes** **Year:** 2011, 2012 **Model:** Accord, Civic, Crosstour, CR-V, CR-Z, Element, Fit, Insight **Engine:** 1.3L, 1.5L, 1.8L, 2.0L, 2.4L	**Fuel System Too Lean:** With the the engine at operating temperature in closed loop, and the engine speed between 550-650 rpm. The Long term fuel trim is higher than 1.33 (+33%). If any related DTCs are listed first troubleshoot those DTCs first, then recheck for P0171. **NOTE: Before you troubleshoot, record all freeze data and any on-board snapshot, and review the general troubleshooting information.**
DTC: P0171 **2T PCM, MIL: Yes** **Year:** 2011, 2012 **Model:** Accord, Crosstour, Odyssey, Pilot, Ridgeline **Engine:** 3.5L	**Rear Bank (Bank 1) Fuel System Too Lean:** If long term fuel trim is higher than normal (too lean), a malfunction in the fuel metering components is detected and a DTC is stored. Long term fuel trim is higher than 1.188 (+18.8 %). Engine is at running temperature, vehicle speed at 550-4,000 rpms and the system is in closed loop. **NOTE: Before you troubleshoot, record all freeze data and any on-board snapshot.**
DTC: P0172 **2T PCM, MIL: Yes** **Year:** 2011, 2012 **Model:** Accord, Crosstour, Odyssey, Pilot, Ridgeline **Engine:** 3.5L	**Rear Bank (Bank 1) Fuel System Too Rich:** The execution time is once per driving cycle and the duration time is 13 seconds or more. If long term fuel trim is lower than normal (too rich), a malfunction in the fuel metering components is detected and a DTC is stored. Engine is at running temperature, vehicle speed at 550-4,000 rpm and the system is in closed loop the long term fuel trim is lower than 0.820 (-18.0 %). **NOTE: Before you troubleshoot, record all freeze data and any on-board snapshot.**
DTC: P0172 **2T PCM, MIL: Yes** **Year:** 2011, 2012 **Model:** Accord, Civic, Crosstour, CR-V, CR-Z, Element, Fit, Insight **Engine:** 1.3L, 1.5L, 1.8L, 2.0L, 2.4L	**Fuel System Too Rich:** With the vehicle running, engine rpm between 550-650 and in closed loop. The long term fuel trim is lower than 0.785 (-215%). If any related DTCs are present, troubleshoot those DTCs first, then recheck P0172. **NOTE: Before you troubleshoot, record all freeze data and any on-board snapshot, and review the general troubleshooting information.**
DTC: P0172 **2T Fuel, MIL: Yes** **Year:** 2012 **Model:** Civic **Engine:** 2.4L	**Fuel System Too Rich (Bank 1):** DTC P0107, P0108, P0135, P0137, P0138, P0141, P1128, P1129 and P1259 not set, engine running in closed loop, and the PCM detected the LONGFT value exceeded the calibrated rich limit. **Note: A high MAP sensor signal at idle can cause this code to set.**
DTC: P0174 **1T PCM, MIL: Yes** **Year:** 2011, 2012 **Model:** Accord, Crosstour, Odyssey, Pilot, Ridgeline **Engine:** 3.5L	**Front Bank (Bank 2) Fuel System Too Lean :** If long term fuel trim is higher than normal (too lean), a malfunction in the fuel metering components is detected and a DTC is stored. Long term fuel trim is higher than 1.188 (+18.8 %). Engine is at running temperature, vehicle speed at 550-4,000 rpm and the system is in closed loop. **NOTE: Before you troubleshoot, record all freeze data and any on-board snapshot**

DTC	Trouble Code Title and Conditions
DTC: P0175 **2T PCM, MIL: Yes** **Year:** 2011, 2012 **Model:** Accord, Crosstour, Odyssey, Pilot, Ridgeline **Engine:** 3.5L	**Front Bank (Bank 2) Fuel System Too Rich :** The execution time is once per driving cycle and the duration time is 13 seconds or more. If long term fuel trim is lower than normal (too rich), a malfunction in the fuel metering components is detected and a DTC is stored. Engine is at running temperature, vehicle speed at 550-4,000 rpm and the system is in closed loop the long term fuel trim is lower than 0.820 (-18.0 %). **NOTE: Before you troubleshoot, record all freeze data and any on-board snapshot.**
DTC: P0181 **2T PCM, MIL: Yes** **Year:** 2011, 2012 **Model:** Civic **Engine:** 1.8L	**Fuel Temperature Sensor Circuit Range/Performance Problem :** With the engine OFF for a minimum of 6 hours, then start the engine and run for at least 10 seconds. When the difference between the fuel temperature sensor and the IAT is 45 ºF (25 ºC) or more a malfunction is detected.
DTC: P0182 **2T PCM, MIL: Yes** **Year:** 2011, 2012 **Model:** Civic **Engine:** 1.8L	**Fuel Temperature Sensor Circuit Low Voltage :** With the ignition switch ON and P0183 not active, the fuel temperature output voltage is 0.08 V or less for at least 2 seconds.
DTC: P0183 **2T PCM, MIL: Yes** **Year:** 2011, 2012 **Model:** Civic **Engine:** 1.8L	**Fuel Temperature Sensor Circuit High Voltage :** With the ignition switch ON and the Intake Air Temperature (IAT) a minimum of 22 ºF (30 ºC), the fuel temperature ouput voltage is 4.92 V or more for at least 2 seconds. **NOTE: Before you troubleshoot, record all freeze data and any on-board snapshot, and review the general troubleshooting information.**
DTC: P0191 **2T PCM, MIL: Yes** **Year:** 2011, 2012 **Model:** Civic **Engine:** 1.8L	**Fuel Rail Pressure Sensor Circuit Range/Performance Problem :** With the engine running and the coolant temperature a minimum of 156 ºF (69 ºC), the 2 malfunction threshold modes are as follows. Mode 1: The difference between PF2A vaules when the MAP is maximum and when MAP is minimum, is 4 kPa (0.04 kgf/cm2, 0.5 psi) or less. Mode 2: PF2A is not between 204-326 kPa (2.1-3.3 kgf/cm2, 30-47 psi) for at least 3 seconds. (PF2S=PF2A-Map)
DTC: P0192 **1T PCM, MIL: Yes** **Year:** 2011, 2012 **Model:** Civic **Engine:** 1.8L	**Fuel Rail Pressure Sensor Circuit Low Voltage :** With the ignition switch ON and P0193 not active, the fuel rail pressure sensor output voltage is 0.23 V or less for at least 2 seconds.
DTC: P0196 **T BCM** **Year:** 2011 **Model:** Civic **Engine:** 1.3L	**Engine Oil Temperature Sensor Circuit Range/Performance:** Malfunction Threshold: The engine oil temperature sensor voltage change is 0.06 V or less.
DTC: P0197 **T PCM** **Year:** 2011 **Model:** Civic **Engine:** 1.3L	**Engine Oil Temperature Sensor Circuit Low Voltage:** Malfunction Threshold: The engine oil temperature sensor output voltage is 0.08 V or less for at least 2 seconds when the engine oil temperature is 158 °F (70 °C) or less. When a malfunction is detected during the first drive cycle, a Pending DTC is stored in the PCM memory. If the malfunction returns in the next (second) drive cycle, a Confirmed DTC and the freeze data are stored. The MIL does not come on.
DTC: P0198 **2T PCM** **Year:** 2011 **Model:** Civic **Engine:** 1.3L	**Engine Oil Temperature Sensor Circuit High Voltage:** Malfunction Threshold: The engine oil temperature sensor output voltage is 4.92 V or less for at least 2 seconds.
DTC: P0201 **1T PCM, MIL: Yes** **Year:** 2011, 2012 **Model:** Civic, CR-Z, Fit, Insight **Engine:** 1.3L, 1.5L, 2.0L	**No. 1 Cylinder Injector Circuit Malfunction:** The return signal from the injector does not change at least 30 times (counts once per engine cycle).
DTC: P0202 **1T PCM, MIL: Yes** **Year:** 2011, 2012 **Model:** Civic, CR-Z, Fit, Insight **Engine:** 1.3L, 1.5L, 2.0L	**No. 2 Cylinder Injector Circuit Malfunction:** The return signal from the injector does not change at least 30 times (counts once per engine cycle).

DTC	Trouble Code Title and Conditions
DTC: P0203 **1T PCM, MIL: Yes** **Year:** 2011, 2012 **Model:** Civic, CR-Z, Insight **Engine:** 1.3L, 1.5L, 2.0L	**No. 3 Cylinder Injector Circuit Malfunction:** The return signal from the injector does not change at least 30 times (counts once per engine cycle).
DTC: P0204 **1T PCM, MIL: Yes** **Year:** 2011, 2012 **Model:** Civic, CR-Z, Fit, Insight **Engine:** 1.3L, 1.5L, 2.0L	**No. 4 Cylinder Injector Circuit Malfunction:** The return signal from the injector does not change at least 30 times (counts once per engine cycle).
DTC: P0222 **1T PCM, MIL: Yes** **Year:** 2011, 2012 **Model:** All **Engine:** All	**Throttle Position (TP) Sensor B Circuit Low Voltage:** The execution time is continuous and the duration time is 200 milliseconds or more. Ignition on and DTCs P0223, P2101, P2118, P2135, P2176 are not active. The TP sensor B output voltage is 0.3 V or less for at least 200 milliseconds. If the signal from TP sensor B is less than a fixed value for a set time, the PCM detects a TP sensor B malfunction and stores a DTC. **NOTE: Before you troubleshoot, record all freeze data and any on-board snapshot.**
DTC: P0223 **1T PCM, MIL: Yes** **Year:** 2011, 2012 **Model:** All **Engine:** All	**Throttle Position (TP) Sensor B Circuit High Voltage :** The execution time is continuous and the duration time is 200 milliseconds or more. Ignition ON and DTCs P0222, P2101, P2118, P2135, P2176 are not active. The TP sensor B output voltage is 4.8 V or more for at least 200 milliseconds. If the signal from TP sensor B is more than a fixed value for a set time, the PCM detects a TP sensor B malfunction and stores a DTC. **NOTE: Before you troubleshoot, record all freeze data and any on-board snapshot.**
DTC: P0300 **2T MISFIRE, MIL: Yes** **Year:** 2012 **Model:** Civic **Engine:** 2.4L	**Multiple Misfire Detected:** DTC P0107, P0108, P0131, P0132, P0171, P0172, P1128, P0335, P0336, P0505, P1128, P1129, P1259, P1361, P1362, P1366, P1367 and P1519 not set, engine running under positive torque conditions, and the PCM detected a misfire in 2 or more cylinders. **Note: If the misfire is severe, the MIL will flash on/off on the 1st trip!**
DTC: P0300 **2T PCM, MIL: Yes** **Year:** 2011, 2012 **Model:** All **Engine:** All	**Random Misfire Detected:** The execution time is once per driving cycle and the duration time is 200 revolutions for Type 1 and 1,000 revolutions for Type 2. Conditions for illuminating the MIL Type 1 and Type 2. * Misfire Type 1 (Severe) Per 200 revolutions 27-90 times. If a type 1 misfire (catalyst damaging) occurs once, the MIL blinks once per second, a Temporary DTC is stored, and the high rpm fuel injection stop system activates. The fuel injection stops, at high rpm only, on the cylinder that has the highest misfire counts. the MIL continues to blink, and the fuel injection stays off at high rpm, until the drive is completed. If a type 1 misfire occurs during a second drive cycle, the MIL and fuel injection behave the same and a DTC is stored. After a type 1 misfire has been detected during two drive cycles, the MIL comes on and stays on beginning with the third drive cycle, unless the Temporary DTC has been cleared by the PCM. Even if the MIL is on, it will start blinking if a type 1 misfire occurs. If the malfunction recurs during the next (second) drive cycle, the MIL comes on and the DTC and the freeze frame data are stored. * Misfire Type 2 (Light) Per 1,000 revolutions 55 times. If a type 2 misfire (emission-related but not severe enough to immediately damage the catalyst) occurs, a Temporary DTC is stored, but the MIL does come on or blink. If a type 2 misfire occurs during a second drive cycle, the MIL comes on and stays on unless the Temporary DTC has been cleared by the PCM. **NOTE: Before you troubleshoot, record all freeze data and any on-board snapshot.**
DTC: P0301 **1T MISFIRE, MIL: Yes** **Year:** 2012 **Model:** Civic **Engine:** 2.4L	**Cylinder 1 Misfire Detected:** DTC P0107, P0108, P0131, P0132, P0171, P0172, P1128, P0335, P0336, P0505, P1128, P1129, P1259, P1361, P1362, P1366, P1367 and P1519 not set, engine running under positive torque conditions, and the PCM detected a misfire condition in one cylinder. **Note: If the misfire is severe, the MIL will flash on/off on the 1st trip!**

DTC	Trouble Code Title and Conditions
DTC: P0301 **2T PCM, MIL: Yes** **Year:** 2011, 2012 **Model:** All **Engine:** All	**Cylinder 1 Misfire Detected :** The execution time is once per driving cycle and the duration time is 200 revolutions for Type 1 and 1,000 revolutions for Type 2. Conditions for illuminating the MIL Type 1 and Type 2. * Misfire Type 1 (Severe) Per 200 revolutions 27-90 times. If a type 1 misfire (catalyst damaging) occurs once, the MIL blinks once per second, a Temporary DTC is stored, and the high rpm fuel injection stop system activates. The fuel injection stops, at high rpm only, on the cylinder that has the highest misfire counts. the MIL continues to blink, and the fuel injection stays off at high rpm, until the drive is completed. If a type 1 misfire occurs during a second drive cycle, the MIL and fuel injection behave the same and a DTC is stored. After a type 1 misfire has been detected during two drive cycles, the MIL comes on and stays on beginning with the third drive cycle, unless the Temporary DTC has been cleared by the PCM. Even if the MIL is on, it will start blinking if a type 1 misfire occurs. If the malfunction recurs during the next (second) drive cycle, the MIL comes on and the DTC and the freeze frame data are stored. * Misfire Type 2 (Light) Per 1,000 revolutions 55 times. If a type 2 misfire (emission-related but not severe enough to immediately damage the catalyst) occurs, a Temporary DTC is stored, but the MIL does come on or blink. If a type 2 misfire occurs during a second drive cycle, the MIL comes on and stays on unless the Temporary DTC has been cleared by the PCM. **NOTE: Before you troubleshoot, record all freeze data and any on-board snapshot.**
DTC: P0302 **2T PCM, MIL: Yes** **Year:** 2011, 2012 **Model:** All **Engine:** All	**Cylinder 2 Misfire Detected :** The execution time is once per driving cycle and the duration time is 200 revolutions for Type 1 and 1,000 revolutions for Type 2. Conditions for illuminating the MIL Type 1 and Type 2. * Misfire Type 1 (Severe) Per 200 revolutions 27-90 times. If a type 1 misfire (catalyst damaging) occurs once, the MIL blinks once per second, a Temporary DTC is stored, and the high rpm fuel injection stop system activates. The fuel injection stops, at high rpm only, on the cylinder that has the highest misfire counts. the MIL continues to blink, and the fuel injection stays off at high rpm, until the drive is completed. If a type 1 misfire occurs during a second drive cycle, the MIL and fuel injection behave the same and a DTC is stored. After a type 1 misfire has been detected during two drive cycles, the MIL comes on and stays on beginning with the third drive cycle, unless the Temporary DTC has been cleared by the PCM. Even if the MIL is on, it will start blinking if a type 1 misfire occurs. If the malfunction recurs during the next (second) drive cycle, the MIL comes on and the DTC and the freeze frame data are stored. * Misfire Type 2 (Light) Per 1,000 revolutions 55 times. If a type 2 misfire (emission-related but not severe enough to immediately damage the catalyst) occurs, a Temporary DTC is stored, but the MIL does come on or blink. If a type 2 misfire occurs during a second drive cycle, the MIL comes on and stays on unless the Temporary DTC has been cleared by the PCM. **NOTE: Before you troubleshoot, record all freeze data and any on-board snapshot.**
DTC: P0302 **1T MISFIRE, MIL: Yes** **Year:** 2012 **Model:** Civic **Engine:** 2.4L	**Cylinder 2 Misfire Detected:** DTC P0107, P0108, P0131, P0132, P0171, P0172, P1128, P0335, P0336, P0505, P1128, P1129, P1259, P1361, P1362, P1366, P1367 and P1519 not set, engine running under positive torque conditions, and the PCM detected a misfire condition in one cylinder. **Note: If the misfire is severe, the MIL will flash on/off on the 1st trip!**
DTC: P0303 **2T PCM, MIL: Yes** **Year:** 2011, 2012 **Model:** All **Engine:** All	**Cylinder 3 Misfire Detected :** The execution time is once per driving cycle and the duration time is 200 revolutions for Type 1 and 1,000 revolutions for Type 2. Conditions for illuminating the MIL Type 1 and Type 2. * Misfire Type 1 (Severe) Per 200 revolutions 27-90 times. If a type 1 misfire (catalyst damaging) occurs once, the MIL blinks once per second, a Temporary DTC is stored, and the high rpm fuel injection stop system activates. The fuel injection stops, at high rpm only, on the cylinder that has the highest misfire counts. the MIL continues to blink, and the fuel injection stays off at high rpm, until the drive is completed. If a type 1 misfire occurs during a second drive cycle, the MIL and fuel injection behave the same and a DTC is stored. After a type 1 misfire has been detected during two drive cycles, the MIL comes on and stays on beginning with the third drive cycle, unless the Temporary DTC has been cleared by the PCM. Even if the MIL is on, it will start blinking if a type 1 misfire occurs. If the malfunction recurs during the next (second) drive cycle, the MIL comes on and the DTC and the freeze frame data are stored. * Misfire Type 2 (Light) Per 1,000 revolutions 55 times. If a type 2 misfire (emission-related but not severe enough to immediately damage the catalyst) occurs, a Temporary DTC is stored, but the MIL does come on or blink. If a type 2 misfire occurs during a second drive cycle, the MIL comes on and stays on unless the Temporary DTC has been cleared by the PCM. **NOTE: Before you troubleshoot, record all freeze data and any on-board snapshot.**
DTC: P0303 **1T MISFIRE, MIL: Yes** **Year:** 2012 **Model:** Civic **Engine:** 2.4L	**Cylinder 3 Misfire Detected:** DTC P0107, P0108, P0131, P0132, P0171, P0172, P1128, P0335, P0336, P0505, P1128, P1129, P1259, P1361, P1362, P1366, P1367 and P1519 not set, engine running under positive torque conditions, and the PCM detected a misfire condition in one cylinder. **Note: If the misfire is severe, the MIL will flash on/off on the 1st trip!**

DTC	Trouble Code Title and Conditions
DTC: P0304 **2T PCM, MIL: Yes** **Year:** 2011, 2012 **Model:** All **Engine:** All	**Cylinder 4 Misfire Detected :** The execution time is once per driving cycle and the duration time is 200 revolutions for Type 1 and 1,000 revolutions for Type 2. Conditions for illuminating the MIL Type 1 and Type 2. * Misfire Type 1 (Severe) Per 200 revolutions 27-90 times. If a type 1 misfire (catalyst damaging) occurs once, the MIL blinks once per second, a Temporary DTC is stored, and the high rpm fuel injection stop system activates. The fuel injection stops, at high rpm only, on the cylinder that has the highest misfire counts. the MIL continues to blink, and the fuel injection stays off at high rpm, until the drive is completed. If a type 1 misfire occurs during a second drive cycle, the MIL and fuel injection behave the same and a DTC is stored. After a type 1 misfire has been detected during two drive cycles, the MIL comes on and stays on beginning with the third drive cycle, unless the Temporary DTC has been cleared by the PCM. Even if the MIL is on, it will start blinking if a type 1 misfire occurs. If the malfunction recurs during the next (second) drive cycle, the MIL comes on and the DTC and the freeze frame data are stored. * Misfire Type 2 (Light) Per 1,000 revolutions 55 times. If a type 2 misfire (emission-related but not severe enough to immediately damage the catalyst) occurs, a Temporary DTC is stored, but the MIL does come on or blink. If a type 2 misfire occurs during a second drive cycle, the MIL comes on and stays on unless the Temporary DTC has been cleared by the PCM. **NOTE: Before you troubleshoot, record all freeze data and any on-board snapshot.**
DTC: P0304 **1T MISFIRE, MIL: Yes** **Year:** 2012 **Model:** Civic **Engine:** 2.4L	**Cylinder 4 Misfire Detected:** DTC P0107, P0108, P0131, P0132, P0171, P0172, P1128, P0335, P0336, P0505, P1128, P1129, P1259, P1361, P1362, P1366, P1367 and P1519 not set, engine running under positive torque conditions, and the PCM detected a misfire condition in one cylinder. **Note: If the misfire is severe, the MIL will flash on/off on the 1st trip!**
DTC: P0305 **2T PCM, MIL: Yes** **Year:** 2011, 2012 **Model:** Accord, Crosstour, Odyssey, Pilot, Ridgeline **Engine:** 3.5L	**Cylinder 5 Misfire Detected :** The execution time is once per driving cycle and the duration time is 200 revolutions for Type 1 and 1,000 revolutions for Type 2. Conditions for illuminating the MIL Type 1 and Type 2. * Misfire Type 1 (Severe) Per 200 revolutions 27-90 times. If a type 1 misfire (catalyst damaging) occurs once, the MIL blinks once per second, a Temporary DTC is stored, and the high rpm fuel injection stop system activates. The fuel injection stops, at high rpm only, on the cylinder that has the highest misfire counts. the MIL continues to blink, and the fuel injection stays off at high rpm, until the drive is completed. If a type 1 misfire occurs during a second drive cycle, the MIL and fuel injection behave the same and a DTC is stored. After a type 1 misfire has been detected during two drive cycles, the MIL comes on and stays on beginning with the third drive cycle, unless the Temporary DTC has been cleared by the PCM. Even if the MIL is on, it will start blinking if a type 1 misfire occurs. If the malfunction recurs during the next (second) drive cycle, the MIL comes on and the DTC and the freeze frame data are stored. * Misfire Type 2 (Light) Per 1,000 revolutions 55 times. If a type 2 misfire (emission-related but not severe enough to immediately damage the catalyst) occurs, a Temporary DTC is stored, but the MIL does come on or blink. If a type 2 misfire occurs during a second drive cycle, the MIL comes on and stays on unless the Temporary DTC has been cleared by the PCM. **NOTE: Before you troubleshoot, record all freeze data and any on-board snapshot.**
DTC: P0306 **1T PCM, MIL: Yes** **Year:** 2011, 2012 **Model:** Accord, Crosstour, Odyssey, Pilot, Ridgeline **Engine:** 3.5L	**Cylinder 6 Misfire Detected:** The execution time is once per driving cycle and the duration time is 200 revolutions for Type 1 and 1,000 revolutions for Type 2. Conditions for illuminating the MIL Type 1 and Type 2. * Misfire Type 1 (Severe) Per 200 revolutions 27-90 times. If a type 1 misfire (catalyst damaging) occurs once, the MIL blinks once per second, a Temporary DTC is stored, and the high rpm fuel injection stop system activates. The fuel injection stops, at high rpm only, on the cylinder that has the highest misfire counts. the MIL continues to blink, and the fuel injection stays off at high rpm, until the drive is completed. If a type 1 misfire occurs during a second drive cycle, the MIL and fuel injection behave the same and a DTC is stored. After a type 1 misfire has been detected during two drive cycles, the MIL comes on and stays on beginning with the third drive cycle, unless the Temporary DTC has been cleared by the PCM. Even if the MIL is on, it will start blinking if a type 1 misfire occurs. If the malfunction recurs during the next (second) drive cycle, the MIL comes on and the DTC and the freeze frame data are stored. * Misfire Type 2 (Light) Per 1,000 revolutions 55 times. If a type 2 misfire (emission-related but not severe enough to immediately damage the catalyst) occurs, a Temporary DTC is stored, but the MIL does come on or blink. If a type 2 misfire occurs during a second drive cycle, the MIL comes on and stays on unless the Temporary DTC has been cleared by the PCM. **NOTE: Before you troubleshoot, record all freeze data and any on-board snapshot**
DTC: P0325 **1T PCM, MIL: Yes** **Year:** 2011, 2012 **Model:** All **Engine:** All	**Knock Sensor Circuit Malfunction :** The execution time is continuous and the duration time is 5 seconds or more. Engine running, speed 2,000 rpm or more, coolant temerature at 140 °F (60 °C) or more. No signals from the knock sensor are detected for at least 5 seconds. If the signals from the knock sensor do not vary for a set time, the PCM detects a malfunction and stores a DTC. **NOTE: Before you troubleshoot, record all freeze data and any on-board snapshot.**

DTC	Trouble Code Title and Conditions
DTC: P0325 **2T ECM, MIL: Yes** **Year:** 2012 **Model:** Civic **Engine:** 2.4L	**Knock Sensor Circuit Malfunction:** Engine running for over 1 minute, and the PCM detected an incorrect signal at the rear Knock Sensor (KS) circuit during the test.
DTC: P0326 **1T PCM, MIL: Yes** **Year:** 2011, 2012 **Model:** Civic, Insight **Engine:** 1.3L, 1.8L, 2.0L	**Knock Sensor Malfunction:** The knock sensor output signal strength is 445 or less at the engine speed [ENGINE SPEED] of 3,000 rpm for at least 5.0 seconds. Engine running, speed 2,000 rpm or more, coolant temperature at 140 ºF (60 ºC) or more.
DTC: P0327 **1T PCM, MIL: Yes** **Year:** 2011, 2012 **Model:** Civic, Insight **Engine:** 1.3L, 1.8L, 2.0L	**Knock Sensor Circuit Low Voltage:** The knock sensor output voltage [KNOCK SENSOR] is 0.20 V or less for at least 2 seconds. Engine running, speed 2,000 rpm or more, coolant temerature at 140 ºF (60 ºC) or more.
DTC: P0328 **1T PCM, MIL: Yes** **Year:** 2011, 2012 **Model:** Civic **Engine:** 1.8L, 2.0L	**Knock Sensor Circuit High Voltage:** The knock sensor output voltage [KNOCK SENSOR] is 4.50 V or more for at least 2 seconds. Engine running, speed 2,000 rpm or more, coolant temerature at 140 ºF (60 ºC) or more.
DTC: P0330 **2T CCM, MIL: Yes** **Year:** 2012 **Model:** Civic **Engine:** 2.4L	**Knock Sensor (Front) Circuit Malfunction:** Engine running for over 1 minute, and the PCM detected an incorrect signal at the front Knock Sensor (KS) circuit during the test.
DTC: P0335 **1T PCM, MIL: Yes** **Year:** 2011, 2012 **Model:** Pilot, Ridgeline **Engine:** 3.5L	**Crankshaft Position (CKP) Sensor No Signal:** With the engine running for 2 seconds or more at 750 rpm, no input signals from the CKP sensor are detected while the signals from the CMP sensor are detected for at least 125 times.
DTC: P0335 **1T PCM, MIL: Yes** **Year:** 2011, 2012 **Model:** Accord, Civic, Crosstour, CR-Z, Insight **Engine:** 1.3L, 1.5L, 1.8L, 2.0L, 2.4L	**Crankshaft Position (CKP) Sensor No Signal:** With the engine running the PCM has detected no CKP pulses over 75 times.
DTC: P0335 **1T PCM, TCIL: Yes** **Year:** 2011, 2012 **Model:** Accord, Crosstour, CR-V, Element, Fit **Engine:** 1.5L, 2.4L, 3.5L	**Crankshaft Position (CKP) Sensor A No Signal :** The execution is continuous and the duration is for 2 seconds or more (when the engine speed is 750 rpm). MIL OFF, D indicator blinks. No signals from the CKP sensor are input at least 63 times. If no pulsing signals from the CKP sensor are detected, a malfunction is detected and a DTC is stored.
DTC: P0335 **1T PCM, MIL: Yes** **Year:** 2011, 2012 **Model:** Odyssey **Engine:** 3.5L	**Crankshaft Position (CKP) Sensor No Signal :** With the engine running with the speed over 750 rpm, and DTCs P0365, P0369 are not active. No input signals from the CKP sensor are detected while the signals from the CMP sensor are detected at least 125 times. **NOTE: Before you troubleshoot, record all freeze data and any on-board snapshot, and review the general troubleshooting information.**
DTC: P0335 **1T CCM, MIL: Yes** **Year:** 2012 **Model:** Civic **Engine:** 2.4L	**CKP Sensor 'A' Circuit Malfunction (No Signal):** Engine running, and the PCM did not detect any signals from the Crankshaft Position (CKP) Sensor 'A' during the test period. **Note: The engine will crank for a longer period of time, may buck or jerk, but it will start and run without the CKP sensor signal present.**
DTC: P0336 **1T CCM, MIL: Yes** **Year:** 2012 **Model:** Civic **Engine:** 2.4L	**CKP Sensor 'A' Circuit Range/Performance:** Engine running, and the PCM detected the Crankshaft Position (CKP) Sensor 'A' signal was missing for a short period of time. **Note: This trouble code is usually caused by an intermittent fault.**

DTC	Trouble Code Title and Conditions
DTC: P0336 **2T PCM, MIL: Yes** **Year:** 2011, 2012 **Model:** Civic, CR-Z **Engine:** 1.5L	**Crankshaft Position (CKP) Sensor Circuit Range/Performance:** The engine speed measured by the CKP sensor is less than the drive pulley speed measured by the CVT input shaft (drive pulley) speed sensor by a factor of 0.56 for at least 12 seconds.
DTC: P0336 **1T PCM, MIL: Yes** **Year:** 2011, 2012 **Model:** Insight **Engine:** 1.3L	**CKP Sensor Intermittent Interruption :** With the engine running the PCM has detected no CKP pulses over 75 times.
DTC: P0339 **1T PCM, MIL: Yes** **Year:** 2011, 2012 **Model:** Ridgeline **Engine:** 3.5L	**Crankshaft Position (CKP) Sensor A Intermittent Interruption:** With the engine running, other than 22 pulses are detected during intervals between reference pulses for each crankshaft revolution. This condition has been detected at least 30 times. DTCs P0335, P0340, P0385 and P0389 are not active. **Note: Before you troubleshoot, record all freeze data and any on-board snapshot, and review the general troubleshooting information.**
DTC: P0339 **1T PCM, MIL: Yes** **Year:** 2011, 2012 **Model:** Accord, Civic, Crosstour, CR-V, CR-Z, Element, Fit, Insight, Odyssey, Pilot **Engine:** All	**Crankshaft Position (CKP) Sensor Circuit Intermittent Interruption :** With the engine running othet than 58 pulses are detected during intervals between reference pulses for each crank revolution. This condition has been detected 30 times. **NOTE: Before you troubleshoot, record all freeze data and any on-board snapshot, and review the general troubleshooting information.**
DTC: P0340 **1T PCM, MIL: Yes** **Year:** 2011, 2012 **Model:** Accord, Crosstour, Odyssey **Engine:** 3.5L	**Camshaft Position (CMP) Sensor No Signal:** With the engine running no signals from CMP sensor A are detected while signals from the CKP sensor are detected 352 times in succession. The execution time is continuous and DTCs P0335, P0339, P0344 are not active. **NOTE: Before you troubleshoot, record all freeze data and any on-board snapshot.**
DTC: P0340 **1T PCM, MIL: Yes** **Year:** 2011, 2012 **Model:** Accord, Civic, Crosstour, CR-V, Element **Engine:** 2.0L, 2.4L	**Camshaft Position (CMP) Sensor A No Signal :** Engine running, NO signals from the CMP sensor A are detected 25 times or more in succession. **NOTE: Before you troubleshoot, record all freeze data and any on-board snapshot, and review the general troubleshooting information.**
DTC: P0341 **2T PCM, MIL: Yes** **Year:** 2011, 2012 **Model:** Accord, Civic, Crosstour, CR-V, Element **Engine:** 2.0L, 2.4L	**Camshaft Position (CMP) Sensor A and CKP Sensor Incorrect Phase Detected :** With the engine running and at operating temperature, and the vehicle speed between 19-38 mph (30-60 km/h) for at least 5 seconds. Malfunctions are as follows: Malfunction with VTC OFF: The gap between the CMP sensor A pulse and the middle of the CMP sensor A assembly is 10 degrees or more for at least 5 seconds. Malfunction with VTC ACTIVE: The timing of the camshaft is out of specified range (Other than when BTDC is between 10-100 degrees) for at least 5 seconds. **NOTE: Before you troubleshoot, record all freeze data and any on-board snapshot, and review the general troubleshooting information.**
DTC: P0344 **1T PCM, MIL: Yes** **Year:** 2011, 2012 **Model:** Accord, Accord Crosstour, Crosstour **Engine:** 3.5L	**Camshaft Position (CMP) Sensor Circuit Intermittent Interruption:** Engine running, more or less that two CMP pulses are detected during intervals between the standard CKP pulses. This condition occurs at least 30 times. **NOTE: Before you troubleshoot, record all freeze data and any on-board snapshot, and review the general troubleshooting information.**
DTC: P0344 **1T PCM, MIL: Yes** **Year:** 2011, 2012 **Model:** Odyssey **Engine:** 3.5L	**Camshaft Position (CMP) Sensor Intermittent Interruption:** With the engine running more or less than two CMP sensor pulses are detected during intervals between the CKP standard pulses. This condition occurs at least 30 times. The execution time is continuous and DTCs P0335, P0339, P0340 are not active. **NOTE: Before you troubleshoot, record all freeze data and any on-board snapshot.**

DTC	Trouble Code Title and Conditions
DTC: P0344 **2T PCM, MIL: Yes** **Year:** 2011, 2012 **Model:** Accord, Crosstour, CR-V, Element **Engine:** 2.4L	**Camshaft Position (CMP) Sensor A Circuit Intermittent Interruption:** With the engine running, more or less than two CMP sensor A pulses are detected during intervals between the CKP standard pulses. This condition occurs at least 30 times. **NOTE: Before you troubleshoot, record all freeze data and any on-board snapshot, and review the general troubleshooting information.**
DTC: P0351 **1T PCM, MIL: Yes** **Year:** 2011, 2012 **Model:** Civic, Crosstour, Fit, Odyssey, Ridgeline **Engine:** 1.5L, 1.8L, 2.4L, 3.5L	**Cylinder 1 Ignition Coil Circuit Malfunction:** With the engine running, the execution time is continuous and the duration time is 5 seconds or more. (when the engine speed is 700 rpm) The return signal does not change for at least 5 seconds when the ignition coil is triggered. **NOTE: Before you troubleshoot, record all freeze data and any on-board snapshot.**
DTC: P0351 **1T PCM, MIL: Yes** **Year:** 2011, 2012 **Model:** Accord, Civic, Crosstour, CR-Z **Engine:** 1.5L, 2.4L, 3.5L	**1 Cylinder Ignition Coil Circuit Malfunction:** No IGF signal to ECM while engine is running.
DTC: P0352 **1T PCM, MIL: Yes** **Year:** 2011, 2012 **Model:** Accord, Civic, Crosstour, CR-Z **Engine:** 1.5L, 2.4L, 3.5L	**2 Cylinder Ignition Coil Circuit Malfunction:** No IGF signal to ECM while engine is running.
DTC: P0352 **1T PCM, MIL: Yes** **Year:** 2011 **Model:** Civic **Engine:** 1.3L	**Exhaust Side Ignition Coil Power Circuit Malfunction:** Malfunction Threshold: The IG RTNe terminal voltage is 4.0 V or less for at least 5 seconds.
DTC: P0352 **1T PCM** **Year:** 2011, 2012 **Model:** Crosstour, Fit, Odyssey, Ridgeline **Engine:** 1.5L, 2.4L, 3.5L	**Cylinder 2 Ignition Coil Circuit Malfunction:** With the engine running, the execution time is continuous and the duration time is 5 seconds or more. (when the engine speed is 700 rpm) The return signal does not change for at least 5 seconds when the ignition coil is triggered.
DTC: P0353 **1T PCM, MIL: Yes** **Year:** 2011, 2012 **Model:** Fit, Odyssey, Ridgeline **Engine:** 1.5L, 3.5L	**Cylinder 3 Ignition Coil Circuit Malfunction:** With the engine running, the execution time is continuous and the duration time is 5 seconds or more. (when the engine speed is 700 rpm) The return signal does not change for at least 5 seconds when the ignition coil is triggered. **NOTE: Before you troubleshoot, record all freeze data and any on-board snapshot**
DTC: P0353 **1T PCM, MIL: Yes** **Year:** 2011, 2012 **Model:** Accord, Civic, Crosstour, CR-Z **Engine:** 1.5L, 2.4L, 3.5L	**3 Cylinder Ignition Coil Circuit Malfunction:** No IGF signal to ECM while engine is running.
DTC: P0354 **1T PCM, MIL: Yes** **Year:** 2011, 2012 **Model:** Accord, Civic, Crosstour, CR-Z **Engine:** 1.5L, 2.4L, 3.5L	**4 Cylinder Ignition Coil Circuit Malfunction:** No IGF signal to ECM while engine is running.
DTC: P0354 **1T PCM, MIL: Yes** **Year:** 2011, 2012 **Model:** Fit, Odyssey, Ridgeline **Engine:** 1.5L, 3.5L	**Cylinder 4 Ignition Coil Circuit Malfunction:** With the engine running, the execution time is continuous and the duration time is 5 seconds or more. (when the engine speed is 700 rpm) The return signal does not change for at least 5 seconds when the ignition coil is triggered. **NOTE: Before you troubleshoot, record all freeze data and any on-board snapshot**

DTC	Trouble Code Title and Conditions
DTC: P0355 **1T PCM, MIL: Yes** **Year:** 2011, 2012 **Model:** Accord, Crosstour **Engine:** 3.5L	**5 Cylinder Ignition Coil Circuit Malfunction:** No IGF signal to ECM while engine is running.
DTC: P0355 **1T PCM, MIL: Yes** **Year:** 2011, 2012 **Model:** Odyssey, Ridgeline **Engine:** 3.5L	**Cylinder 5 Ignition Coil Circuit Malfunction:** With the engine running, the execution time is continuous and the duration time is 5 seconds or more. (when the engine speed is 700 rpm) The return signal does not change for at least 5 seconds when the ignition coil is triggered. **NOTE: Before you troubleshoot, record all freeze data and any on-board snapshot**
DTC: P0356 **1T PCM, MIL: Yes** **Year:** 2011, 2012 **Model:** Odyssey, Ridgeline **Engine:** 3.5L	**Cylinder 6 Ignition Coil Circuit Malfunction:** With the engine running, the execution time is continuous and the duration time is 5 seconds or more. (when the engine speed is 700 rpm) The return signal does not change for at least 5 seconds when the ignition coil is triggered. **NOTE: Before you troubleshoot, record all freeze data and any on-board snapshot**
DTC: P0356 **1T PCM, MIL: Yes** **Year:** 2011, 2012 **Model:** Accord, Crosstour **Engine:** 2.4L, 3.5L	**6 Cylinder Ignition Coil Circuit Malfunction:** No IGF signal to ECM while engine is running.
DTC: P0365 **1T PCM, MIL: Yes** **Year:** 2011, 2012 **Model:** Fit, Odyssey, Pilot, Ridgeline **Engine:** 1.5L, 3.5L	**Camshaft Position (CMP) Sensor Circuit No Signal:** With the engine running at 750 rpm, DTCs P0335, P0339 and P0369 are not active. No input signals from the CMP sensor are detected while the signals from the CKP sensor are detected at least 300 times. **NOTE: Before you troubleshoot, record all freeze data and any on-board snapshot**
DTC: P0365 **1T PCM, MIL: Yes** **Year:** 2011, 2012 **Model:** Accord, Civic, Crosstour, CR-V, CR-Z, Element **Engine:** 1.3L, 1.5L, 1.8L, 2.0L, 2.4L	**Camshaft Position (CMP) Sensor B No Signal:** With the engine running, NO CMP sensor B pulsing signals are detected for at least 50 times in succession.
DTC: P0369 **1T PCM, MIL: Yes** **Year:** 2011, 2012 **Model:** Civic, CR-Z, Fit, Odyssey, Pilot, Ridgeline **Engine:** 1.3L, 1.5L, 1.8L, 3.5L	**Camshaft Position (CMP) Sensor Circuit Intermittent Interruption:** Other than 30 times of CMP pulses are detected during two cycles of CKP. This condition has been detected at least 10 times. **NOTE: Before you troubleshoot, record all freeze data and any on-board snapshot.**
DTC: P0369 **1T PCM, MIL: Yes** **Year:** 2011, 2012 **Model:** Accord, Civic, Crosstour, CR-V, Element **Engine:** 2.0L, 2.4L	**Camshaft Position (CMP) Sensor B Intermittent Interruption:** With the engine running, more or less than four CMP B pulsing signals are detected during intervals between standard pulses at least 30 times in succession. **NOTE: Before you troubleshoot, record all freeze data and any on-board snapshot, and review the general troubleshooting information.**
DTC: P0385 **1T PCM, MIL: Yes** **Year:** 2011, 2012 **Model:** Odyssey **Engine:** 3.5L	**Crankshaft Position (CKP) Sensor B No Signal :** With the engine running the execution is continuous and P0385, P0389 are not active. No signals from CKP sensor A are detected while signals from CKP sensor B are detected 127 times in succession. If no pulsing signals are detected from CKP sensor A, a malfunction is detected and a DTC is stored. **NOTE: Before you troubleshoot, record all freeze data and any on-board snapshot.**
DTC: P0389 **1T PCM, MIL: Yes** **Year:** 2011, 2012 **Model:** Odyssey **Engine:** 3.5L	**Crankshaft Position (CKP) Sensor B Intermittent Interruption :** Engine running, other than 22 pulses are detected during intervals between reference pulses for each crankshaft revolution. This condition has been detected at least 30 times. An abnormal amount of pulsing signals are detected from CKP sensor B. **NOTE: Before you troubleshoot, record all freeze data and any on-board snapshot.**

DTC	Trouble Code Title and Conditions
DTC: P0400 **2T PCM, MIL: Yes** **Year:** 2011, 2012 **Model:** Accord, Civic, Crosstour, CR-Z, Fit, Odyssey, Pilot, Ridgeline **Engine:** 1.3L, 1.5L, 1.8L, 2.0L, 3.5L	**Exhaust Gas Recirculation (EGR) System Leak Detected:** With engine running, rpm 1,500-3,500 and system in closed loop the A/F sensor response parameter according to the opening and closing of the EGR is 14,746 (LT80A) or less. **NOTE: Before you troubleshoot, record all freeze data and any on-board snapshot.**
DTC: P0401 **2T EGR1, MIL: Yes** **Year:** 2012 **Model:** Civic **Engine:** 2.4L	**EGR System Insufficient Flow Detected:** Cold engine startup requirement met (ECT) sensor less than 76ºF at startup), engine running in closed loop at 40-55 mph for 2 minutes in Drive (D4), followed by a deceleration period back to 35 mph with the throttle closed, and the PCM detected a signal from the EGR position sensor that indicated insufficient EGR flow during the test.
DTC: P0401 **2T PCM, MIL: Yes** **Year:** 2011, 2012 **Model:** Accord, Civic, Crosstour, CR-Z, Fit, Insight, Odyssey, Pilot, Ridgeline **Engine:** 1.3L, 1.5L, 1.8L, 2.0L, 3.5L	**Exhaust Gas Recirculation (EGR) Insufficient Flow:** Engine temperature is at a minimum of 156 ºF (69 ºC), engine speed is between 1,100-2,400 rpm with the map value of 14 kPa (4.0 in.Hg, 100 mmHg), throttle fully closed and battery 10.5 volt minimum. The execution time is once per driving cycle and the duration time is 3 seconds or more. The ratio of the current EGR flow to the normal EGR flow is 15% or less for at least 3 seconds. **NOTE: Before you troubleshoot, record all freeze data and any on-board snapshot.**
DTC: P0404 **2T PCM, MIL: Yes** **Year:** 2011, 2012 **Model:** Accord, Civic, Crosstour, CR-V, CR-Z, Fit, Insight, Odyssey, Pilot, Ridgeline **Engine:** All	**Exhaust Gas Recirculation (EGR) Control Circuit Range/Performance Problem :** Vehicle at a speed between 15-75 mph, actual valve lift command at 0.012 in. (0.3 mm). The execution time is once per driving cycle and the duration time is 5 seconds or more. The difference between the command value of the amount of EGR valve lift in the PCM and the actual amount of valve lift is 0.041 in. (1.020 mm) or more for at least 5 seconds. **NOTE: Before you troubleshoot, record all freeze data and any on-board snapshot.**
DTC: P0406 **1T PCM, MIL: Yes** **Year:** 2011, 2012 **Model:** Accord, Civic, Crosstour, CR-Z, Fit, Insight, Odyssey, Pilot, Ridgeline **Engine:** 1.3L, 1.5L, 1.8L, 2.0L, 3.5L	**Exhaust Gas Recirculation (EGR) Valve Position Sensor Circuit High Voltage:** Wth the engine running the execution time is continuous and the duration time is 2 seconds or more. The EGR valve position sensor output voltage is 4.88 V or more for at least 2 seconds. **NOTE: Before you troubleshoot, record all freeze data and any on-board snapshot.**
DTC: P0420 **2T PCM, MIL: Yes** **Year:** 2011, 2012 **Model:** Accord, Crosstour, Odyssey, Pilot, Ridgeline **Engine:** 3.5L	**Rear Bank Catalyst System Efficiency Below Threshold (Bank 1):** The execution time is once per driving cycle and the duration time is 102 seconds or more. The number of detections is 752 (CTAGLT67) or more. **NOTE: Before you troubleshoot, record all freeze data and any on-board snapshot.**
DTC: P0420 **2T CAT1, MIL: Yes** **Year:** 2012 **Model:** Civic **Engine:** 2.4L	**Catalyst Efficiency Below Threshold (Bank 1):** DTC P0137, P0138 and P0141 not set, engine running in closed loop at 40-55 mph for 2 minutes, followed by a deceleration period to 35 mph at closed throttle, and the PCM detected excessive activity in the Catalyst oxygen sensor (rear HO2S) during the test period.
DTC: P0420 **2T PCM, MIL: Yes** **Year:** 2011, 2012 **Model:** Accord, Civic, Crosstour, CR-V, CR-Z, Element, Fit, Insight **Engine:** 1.3L, 1.5L, 1.8L, 2.0L, 2.4L	**Catalytic System Efficiency Below Threshold :** With the vehicle driven to a speed of 16-75 mph at less than 3500 rpm in closed loop for 2-5 minutes, ECT sensor more than 140ºF, MAF sensor from 8-50 g/sec, engine load less than 99% (±8%), predicted Catalyst temperature over 750ºF, and the PCM detected the catalyst oxygen storage capacity was below an acceptable threshold during the test. **NOTE: If some of the DTCs listed below are stored at the same time as DTC P0420, troubleshoot those DTCs first, then recheck for DTC P0420.**
DTC: P0430 **2T PCM, MIL: Yes** **Year:** 2011, 2012 **Model:** Accord, Crosstour, Odyssey, Pilot, Ridgeline **Engine:** 3.5L	**Front Bank Catalyst System Efficiency Below Threshold (Bank 2) :** The execution time is once per driving cycle and the duration time is 102 seconds or more. The number of detections is 720 (CTAGLT68) or more. **NOTE: Before you troubleshoot, record all freeze data and any on-board snapshot.**

DTC	Trouble Code Title and Conditions
DTC: P0441 T **Year:** 2012 **Model:** Civic **Engine:** 2.4L	**EVAP System Incorrect Purge Flow:** Cold engine startup (ECT sensor signal less than 154ºF and IAT sensor signal more than 14ºF at startup), engine runtime from 3-5 minutes, followed by an acceleration period to 50-60 mph at steady throttle, and the PCM did not detect enough purge flow through the EVAP system during the EVAP Monitor flow test.
DTC: P0443 **1T PCM, MIL: Yes** **Year:** 2011, 2012 **Model:** All **Engine:** All	**Evaporative Emission (EVAP) Canister Purge Valve Circuit Malfunction:** With the engine running, EVAP canister purge valve output duty at 2% to 98% the execution time is continuous and the duration time is 5 seconds or more. When the return signal does not change according to the EVAP canister purge valve duty cycle for a set time, the PCM detects a malfunction. The return signal does not change according to the EVAP canister purge valve output for at least 5 seconds. **NOTE: Before you troubleshoot, record all freeze data and any on-board snapshot.**
DTC: P0451 **2T PCM, MIL: Yes** **Year:** 2011, 2012 **Model:** All **Engine:** All	**Fuel Tank Pressure (FTP) Sensor Circuit Range/Performance Problem:** Elapsed time after starting the engine is 2 seconds with the throttle fully closed. The execution time is once per driving cycle and the duration time is 20 seconds or more. The FTP sensor output fluctuates by 0.3 kPa (0.1 in.Hg, 2 mmHg) or more at least five times within 3 seconds. **NOTE: Before you troubleshoot, record all freeze data and any on-board snapshot.**
DTC: P0451 **2T CCM, MIL: Yes** **Year:** 2012 **Model:** Civic **Engine:** 2.4L	**Fuel Tank Pressure (FTP) Sensor Range/Performance:** Engine running, and the PCM detected the fuel tank pressure (FTP) sensor signal was less than the allowable range stored in the PCM memory (a calibrated range adjusted to current conditions).
DTC: P0452 **1T CCM, MIL: Yes** **Year:** 2012 **Model:** Civic **Engine:** 2.4L	**Fuel Tank Pressure (FTP) Sensor Circuit Low Input:** Key on or engine running, and the PCM detected the fuel tank pressure (FTP) sensor signal was less than 0.16v during the test. **Note: The FTP sensor PID should be near 2.5v with the fuel cap off.**
DTC: P0452 **2T PCM, MIL: Yes** **Year:** 2011, 2012 **Model:** All **Engine:** All	**Fuel Tank Pressure (FTP) Sensor Circuit Low Voltage:** Elapsed time after starting the engine is 2 seconds at idle. The execution time is once per driving cycle and the duration time is 3 seconds or more. The output from the fuel tank pressure sensor is less than -7 kPa (-2.1 in.Hg, -55 mmHg) for at least 3 seconds. **NOTE: Before you troubleshoot, record all freeze data and any on-board snapshot.**
DTC: P0453 **1T PCM, MIL: Yes** **Year:** 2011, 2012 **Model:** Accord, Crosstour **Engine:** 2.4L, 3.5L	**FTP Sensor Circuit High Voltage (with paddle shifter):** The EVAP pressure is higher than 123.8 kPa for 0.5 seconds. Enabling conditions are as follows: Battery voltage 8 V or more, Starter OFF, Either of following conditions met (a) Engine switch On (IG) (b) Soak timer ON
DTC: P0453 **2T PCM, MIL: Yes** **Year:** 2011, 2012 **Model:** Accord Crosstour, Civic, Crosstour, CR-V, CR-Z, Element, Fit, Insight, Odyssey, Pilot, Ridgeline **Engine:** All	**Fuel Tank Pressure (FTP) Sensor Circuit High Voltage:** Elapsed time after starting the engine is 2 seconds at idle. The execution time is once per driving cycle and the duration time is 3 seconds or more. If the FTP sensor output voltage is higher than a target value within a set time after starting the engine in a cold condition. The output from the fuel tank pressure sensor is more than 8 kPa (2.2 in.Hg, 55 mmHg) for at least 3 seconds. **NOTE: Before you troubleshoot, record all freeze data and any on-board snapshot.**
DTC: P0453 **1T CCM, MIL: Yes** **Year:** 2012 **Model:** Civic **Engine:** 2.4L	**Fuel Tank Pressure Sensor Circuit High Input:** Key on or engine running, and the PCM detected the fuel tank pressure (FTP) sensor signal was more than 4.90v during the test. **Note: The FTP sensor PID should be near 2.5v with the fuel cap off.**

DTC	Trouble Code Title and Conditions
DTC: P0455 **2T PCM, MIL: Yes** **Year:** 2011, 2012 **Model:** Accord, Crosstour **Engine:** 2.4L, 3.5L	**Evaporative Emission Control System Leak Detected (Gross Leak):** Atmospheric pressure: 70 to 110 kPa-a (525 to 825 mmHg-a) Battery voltage: 10.5 V or higher Vehicle speed: Less than 4 km/h (2.5 mph) Ignition switch: OFF Time after key-off: 5, 7 or 9.5 hours Purge VSV: Not operated by scan tool Vent valve: Not operated by scan tool Leak detection pump: Not operated by scan tool Both of following conditions met before key-off: Conditions 1 and 2 1. Duration that vehicle driven: 5 minutes or more 2. EVAP purge operation: Performed Engine coolant temperature: 4.4 to 35ºC (40 to 95ºF) Intake air temperature: 4.4 to 35ºC (40 to 95ºF)
DTC: P0455 **2T PCM, MIL: Yes** **Year:** 2011, 2012 **Model:** Accord Crosstour, Civic, Crosstour, CR-V, CR-Z, Element, Fit, Insight, Odyssey, Pilot **Engine:** All	**Evaporative Emission (EVAP) System Large Leak Detected:** The execution time is once per driving cycle and the duration time is 36 minutes and 37 seconds, or less. Here is an overview of the malfunction detection for the EONV method: 1: Judgement of detection of 0.09 inch leak as normal operation 2: Judgement of detection of 0.02 inch leak as normal operation 3: Detection of 0.02 inch leak 4: Detection of atmospheric pressure failure 5: Flickering of the FTP sensor The execution time is once per driving cycle and the duration time is 36 minutes and 37 seconds, or less. The variation of pressure inside the fuel tank is 0.03 kPa (0.009 in.Hg, 0.24 mmHg) or more. **NOTE: Before you troubleshoot, record all freeze data and any on-board snapshot.**
DTC: P0455 **2T PCM, MIL: Yes** **Year:** 2011, 2012 **Model:** Ridgeline **Engine:** 3.5L	**EVAP System Large Leak Detected:** After 6 hours, with initial engine coolant temperature between 40-95 ºF (5-35 ºC), Barometric pressure 76 kPa (569 mmHg, 22.5 in. Hg), battery voltage a minimum of 10.5 V, fuel trim 0.73-1.47 and fuel level is not full. The variation of pressure inside the fuel tank is -1 kPa (-7 mmHg, -0.3 in. Hg) or less.
DTC: P0456 **2T PCM, MIL: Yes** **Year:** 2011, 2012 **Model:** All **Engine:** All	**EVAP System Very Small Leak Detected :** The execution time is once per driving cycle and the duration time is at least 16 minutes and 37 seconds, but not more than 36 minutes and 37 seconds. Here is an overview of the malfunction detection for the EONV method: 1: Judgement of detection of 0.09 inch leak as normal operation 2: Judgement of detection of 0.02 inch leak as normal operation 3: Detection of 0.02 inch leak 4: Detection of atmospheric pressure failure 5: Flickering of the FTP sensor The execution time is once per driving cycle and the duration time is 36 minutes and 37 seconds, or less. The variation of pressure inside the fuel tank is 0.03 kPa (0.009 in.Hg, 0.24 mmHg) or more. **NOTE: Before you troubleshoot, record all freeze data and any on-board snapshot.**
DTC: P0457 **3T PCM, MIL: Yes** **Year:** 2011, 2012 **Model:** Accord, Civic, Crosstour, CR-V, Element, Odyssey, Pilot, Ridgeline **Engine:** 1.3L, 1.8L, 2.0L, 2.4L, 3.5L	**Evaporative Emission (EVAP) System Leak Detected/Fuel Fill Cap Loose or Missing:** Engine coolant temperature before EVAP purge control starts 140 ºF (60 ºC), 2 mph (2 km/h), Barometric pressure, 76 kPa (22.5 in.Hg, 569 mmHg), Battery voltage 10.5 volts or more, fuel trim 0.73-1.47, system is in close loop. The execution time is continuous and the duration time is 12 seconds or more. The output from the fuel cap monitor is 0.053 or less for at least 12 seconds (when there is no NG judgment history in this drive cycle). P0455 or P0456 are judged as NG. **NOTE: Before you troubleshoot, record all freeze data and any on-board snapshot.**
DTC: P0461 **1T PCM** **Year:** 2011, 2012 **Model:** Accord, Crosstour, CR-V, Element, Insight, Odyssey, Pilot, Ridgeline **Engine:** 1.3L, 2.4L, 3.5L	**Fuel Level Sensor (Fuel Gauge Sending Unit) Range/Performance Problem :** The execution time is every 125 miles (200 km), DTCs P0462, P0463, U0028, U0155 are not active. If the powertrain control module (PCM) receives no change in the fuel level sensor output after driving for a specified number of miles, it detects a malfunction. The change in the fuel level sensor output is 3.5 % or less.

DTC	Trouble Code Title and Conditions
DTC: P0461 **1T PCM** **Year:** 2011, 2012 **Model:** Civic, CR-Z, Fit **Engine:** 1.3L, 1.5L, 1.8L, 2.0L	**Fuel Level Sensor (Fuel Gauge Sending Unit) Circuit Range/Performance Problem :** The execution time is every 125 miles (200 km).The change in the fuel level sensor output is 3.5 % or less. **NOTE: Because it requires 162 miles (260 km) of driving without refueling to complete this diagnosis, DTC P0461 cannot be duplicated during this troubleshooting.**
DTC: P0462 **1T PCM** **Year:** 2011, 2012 **Model:** All **Engine:** All	**Fuel Level Sensor (Fuel Gauge Sending Unit) Circuit Low Voltage:** The fuel level sensor (fuel gauge sending unit) output voltage is 0.10 V or less for at least 5 seconds. The execution time is continuous DTCs P0463, U0155 are not active. **NOTE: Before you troubleshoot, record all freeze data and any on-board snapshot.**
DTC: P0463 **1T PCM** **Year:** 2011, 2012 **Model:** All **Engine:** All	**Fuel Level Sensor (Fuel Gauge Sending Unit) Circuit High Voltage :** The execution time is every 125 miles (200 km), DTCs P0462, P0463, U0028, U0155 are not active. If the powertrain Control Module (PCM) receives no change in the fuel level sensor output after driving for a specified number of miles, it detects a malfunction. The fuel level sensor output voltage is 4.92 V or more for at least 5 seconds.
DTC: P0496 **2T PCM, MIL: Yes** **Year:** 2011, 2012 **Model:** All **Engine:** All	**Evaporative Emission (EVAP) System High Purge Flow Detected:** Monitor execution is once per driving cycle. The output from the EVAP canister purge valve is 0.2 kPa (0.07 in.Hg, 2 mmHg) or more for at least 10 seconds. **NOTE: Before you troubleshoot, record all freeze data and any on-board snapshot.**
DTC: P0497 **2T PCM, MIL: Yes** **Year:** 2011, 2012 **Model:** All **Engine:** All	**Evaporative Emission (EVAP) System Low Purge Flow Detected:** The execution time is once per driving cycle and P145C is judged as NG. Enable conditions are low load duration time 10 seconds, Wait for 10 seconds after the ignition switch is turned to LOCK (0) and Engine coolant temperature before EVAP purge control starts is 140 ºF (60 ºC). The output from the fuel tank pressure sensor is 0.2 kPa (0.07 in.Hg, 2 mmHg) or less for at least 10 seconds. **NOTE: Before you troubleshoot, record all freeze data and any on-board snapshot.**
DTC: P0498 **1T PCM, MIL: Yes** **Year:** 2011, 2012 **Model:** Crosstour, CR-V, Element, Fit, Odyssey, Pilot, Ridgeline **Engine:** 1.5L, 2.4L, 3.5L	**Evaporative Emission (EVAP) Canister Vent Shut Valve Control Circuit Low Voltage:** The execution time is continuous and the duration time is 5 seconds or more. If the return signal is OFF (Low) when the powertrain Control Module (PCM) outputs the ON signal to the EVAP canister vent shut valve, the PCM detects a malfunction. The return signal is Low for at least 5 seconds when the PCM outputs the ON signal to the EVAP canister vent shut valve.
DTC: P0498 **1T PCM, MIL: Yes** **Year:** 2011, 2012 **Model:** Accord, Civic, Crosstour, CR-Z, Insight **Engine:** All	**Evaporative Emission (EVAP) Canister Vent Shut Valve Control Circuit Low Voltage:** The execution time is continuous and the duration time is 5 seconds or more. If the return signal is OFF when the powertrain control module (PCM) outputs the ON signal to the EVAP canister vent shut valve, the PCM detects a malfunction. DTC P0499 is not active. The return signal is Low for at least 5 seconds when the PCM outputs the ON signal to the EVAP canister vent shut valve.
DTC: P0499 **1T PCM, MIL: Yes** **Year:** 2011, 2012 **Model:** All **Engine:** All	**Evaporative Emission (EVAP) Canister Vent Shut Valve Control Circuit High Voltage:** If the return signal is ON when the powertrain Control Module (PCM) outputs the OFF signal to the EVAP canister vent shut valve, the PCM detects a malfunction.The execution time is continuous and the duration time is 5 seconds or more. DTC P0498 is not active. The return signal is ON for at least 5 seconds when the PCM outputs the Low signal to the EVAP canister vent shut valve. **NOTE: Before you troubleshoot, record all freeze data and any on-board snapshot.**
DTC: P0501 **1T PCM, MIL: Yes** **Year:** 2011, 2012 **Model:** Civic, CR-Z **Engine:** 1.5L	**CVT Speed Sensor Circuit Range/Performance:** The driven pulley speed measured by the vehicle speed sensor is less than the driven pulley speed measured by the CVT output shaft (driven pulley) speed sensor by a factor of 0.53 for at least 4 seconds. **NOTE: Before you troubleshoot, record all freeze data and any on-board snapshot.** This code is caused by an electrical circuit problem and cannot be caused by a mechanical problem in the transmission.
DTC: P0501 **T PCM** **Year:** 2011, 2012 **Model:** Insight **Engine:** 1.3L	**Range/Performance Problem in CVT Speed Sensor Circuit:** With the vehicle speed at a minimum or 25 mph (40km/h) during deceleration, the engine speed measured by the Vehicle Speed Sensor (VSS) is less than engine speed measured by the CVT output shaft speed sensor by a factor of 0.53 for at least 4 seconds. **Note: This code is caused by an electrical circuit problem and cannot be caused by a mechanical problem in the transmission.**
DTC: P0502 **1T PCM, MIL: Yes** **Year:** 2011, 2012 **Model:** Civic, CR-Z **Engine:** 1.5L	**CVT Speed Sensor Circuit Open/Short:** The driven pulley speed measured by the vehicle speed sensor is 100 rpm or less for at least 4 seconds when the vehicle speed measured by the CVT output shaft (driven pulley) speed sensor is 8 mph (14 km/h) or more.

DTC	Trouble Code Title and Conditions
DTC: P0502 **1T PCM, MIL: Yes** **Year:** 2011, 2012 **Model:** Insight **Engine:** 1.3L	**Problem in CVT Speed Sensor Circuit:** With the vehicle speed at a minimum of 25 mph (40 km/h) during deceleration, the engine speed measured by the CVT output shaft speed sensor is 100 rpm or less for at least 4 seconds when the vehicle speed measured by the vehicle speed sensor is 9 mph (14 km/h) or more. **NOTE: This code is caused by an electrical circuit problem and cannot be caused by a mechanical problem in the transmission.**
DTC: P0506 **2T PCM, MIL: Yes** **Year:** 2011, 2012 **Model:** All **Engine:** All	**Idle Control System RPM Lower Than Expected:** The execution time is once per driving cycle and the duration time is 20 seconds or more. Throttle fully closed, fuel trim 0.75-1.47, intake air temperature 19 ºF (-7 ºC) and battery voltage 10.5 or more. If the actual idle speed varies beyond a specified value from the target speed over a certain period of time, the PCM detects a malfunction. The actual idle speed is at least 100 rpm less than the target idle speed for at least 20 seconds. **NOTE: Before you troubleshoot, record all freeze data and any on-board snapshot.**
DTC: P0507 **2T PCM, MIL: Yes** **Year:** 2011, 2012 **Model:** All **Engine:** All	**Idle Control System RPM Higher Than Expected:** Enable conditions are as follows: coolant temperature 156 ºF (69 ºC) minimum, intake air temperature 19 ºF (-7 ºC), fuel trim 0.73-1.47, throttle closed and battery voltage at least 10.5 volts. The execution time is once per driving cycle and the duration time is 20 seconds or more. If the actual idle speed varies beyond a specified value from the target speed over a certain period of time, the PCM detects a malfunction. The actual idle speed is at least 200 rpm greater than the target idle speed for at least 20 seconds. **NOTE: Before you troubleshoot, record all freeze data and any on-board snapshot.**
DTC: P050A **2T PCM, MIL: Yes** **Year:** 2011, 2012 **Model:** All **Engine:** All	**Cold Start Idle Air Control System Performance Problem:** The execution time is once per driving cycle and the duration time is 10 seconds or more. When the actual amount of air is less than the target amount, a malfunction is detected. The total airflow is decreased by a factor of 0.693 for at least 10 seconds. **NOTE: Before you troubleshoot, record all freeze data and any on-board snapshot.**
DTC: P050B **2T PCM, MIL: Yes** **Year:** 2011, 2012 **Model:** Crosstour, Odyssey **Engine:** 2.4L, 3.5L	**Cold Start Ignition Timing Performance Problem:** The execution time is once per driving cycle and the duration time is 3.5 seconds or more and the engine speed is 2,100 rpm or more for at least 3.5 seconds. When the actual engine speed is a specified value or more, and it continues for a specified time, a malfunction is detected. **NOTE: Before you troubleshoot, record all freeze data and any on-board snapshot.**
DTC: P050B **2T PCM, MIL: Yes** **Year:** 2011, 2012 **Model:** Accord, Civic, Crosstour, CR-V, CR-Z, Element, Fit, Insight, Pilot, Ridgeline **Engine:** All	**Cold Start Ignition Timing Control System Performance Problem:** With the vehicle at idle and throttle position fully closed, the engine speed is 2,150 rpm or more for at least 3.5 seconds. **NOTE: Before you troubleshoot, record all freeze data and any on-board snapshot, and review the general troubleshooting information.**
DTC: P0522 **2T PCM, MIL: Yes** **Year:** 2011, 2012 **Model:** Civic, Insight **Engine:** 1.3L, 1.8L, 2.0L	**Rocker Arm Oil Pressure Sensor A Low Voltage:** With the ignition switch ON and the barometric pressure a minimum of 56 Kpa (17.3 in. Hg. 439 mmHg), the rocker arm oil pressure sensor A output voltage is 0.18 V or less for at least 2 seconds.
DTC: P0522 **2T CAT, MIL: Yes** **Year:** 2011, 2012 **Model:** Accord, Crosstour **Engine:** 2.4L, 3.5L	**Rocker Arm Oil Pressure Sensor Circuit Low Voltage:** The barometric pressure (BARO) sensor is 69 kPa (20.5 inHg, 520 mmHg) or more and the rocker arm oil pressure sensor voltage is 0.18 V or less for at least 2 seconds.
DTC: P0522 **1T PCM, MIL: Yes** **Year:** 2011, 2012 **Model:** Odyssey, Pilot **Engine:** 3.5L	**Rocker Arm Oil Pressure Sensor Circuit Low Voltage :** With the engine running and the barometric pressure a minimum of 69 Kpa (20.5 in. Hg. 520 mmHg) or more, the rocker arm oil pressure sensor output voltage is 0.18 V or less for at least 2 seconds.
DTC: P0523 **2T PCM, MIL: Yes** **Year:** 2011, 2012 **Model:** Accord, Crosstour **Engine:** 2.4L, 3.5L	**Rocker Arm Oil Pressure Sensor Circuit High Voltage:** The rocker arm oil pressure sensor output voltage is 4.79 V or more for at least 2.0 seconds.

DTC	Trouble Code Title and Conditions
DTC: P0523 **1T PCM, MIL: Yes** **Year:** 2011, 2012 **Model:** Odyssey, Pilot **Engine:** 3.5L	**Rocker Arm Oil Pressure Sensor Circuit High Voltage:** With the engine speed at a maximum of 6,000 rpm, engine coolant temperature a minimum of 104 °F (40 °C), the EOP sensor output voltage is 4.79 V or more for at least 2 seconds.
DTC: P0523 **2T PCM, MIL: Yes** **Year:** 2011, 2012 **Model:** Civic, Insight **Engine:** 1.3L, 1.8L, 2.0L	**Rocker Arm Oil Pressure Sensor A High Voltage:** With the engine speed a maximum of 4,000 rpm and engine coolant a minimum of 140 °F (60 °C), the rocker arm oil pressure sensor A output voltage is 4.79 volts or more for at least 2 seconds.
DTC: P0532 **2T PCM** **Year:** 2011, 2012 **Model:** Civic, Crosstour, CR-V, CR-Z, Insight **Engine:** 1.3L, 1.5L, 1.8L, 2.0L, 2.4L	**A/C Pressure Sensor Circuit Low Voltage:** Ignition switch ON (II), DTC P0533 is not active. The A/C pressure sensor output voltage is 0.24 V for at least 10 seconds. **NOTE: Before you troubleshoot, record all freeze data and any on-board snapshot, and review the general troubleshooting information.**
DTC: P0533 **2T PCM** **Year:** 2011, 2012 **Model:** Accord, Crosstour **Engine:** 2.4L, 3.5L	**A/C Pressure Sensor Circuit High Voltage (with paddle shifter):** The A/C pressure sensor output voltage is 4.76 V or more for at least 10 seconds.
DTC: P0533 **1T PCM** **Year:** 2011, 2012 **Model:** Civic, Crosstour, CR-V, CR-Z, Insight **Engine:** 1.3L, 1.5L, 1.8L, 2.0L, 2.4L	**A/C Pressure Sensor Circuit High Voltage:** With the ignition switch on, the A/C pressure sensor output voltage is 4.74 V or more for at least 10 seconds. DTC P0532 is not active. **NOTE: Before you troubleshoot, record all freeze data and any on-board snapshot, and review the general troubleshooting information.**
DTC: P0555 **1T PCM** **Year:** 2011, 2012 **Model:** Civic, CR-Z **Engine:** 1.5L	**Brake Booster Pressure Sensor A Circuit High Voltage:** High voltage detected at pressure sensor A.
DTC: P0555 **2T PCM, MIL: Yes** **Year:** 2011, 2012 **Model:** Insight **Engine:** 1.3L	**Brake Booster Pressure Sensor A Circuit high Voltage:** With the ignition switch ON (II), a malfunction has been detected in the brake booster pressure sensor A circuit.
DTC: P0557 **1T PCM** **Year:** 2011, 2012 **Model:** Civic, CR-Z **Engine:** 1.5L	**Brake Booster Pressure Sensor A Circuit Low Voltage:** Low voltage detected at the brake booster sensor.
DTC: P0557 **2T PCM, MIL: Yes** **Year:** 2011, 2012 **Model:** Insight **Engine:** 1.3L	**Brake Booster Pressure Sensor A Circuit Low Voltage:** With the ignition switch ON (II), a malfution has been detected in the brake booster pressure sensor A circuit.
DTC: P055B **1T PCM, MIL: Yes** **Year:** 2011, 2012 **Model:** Insight **Engine:** 1.3L	**Rocker Arm Oil Pressure Sensor A Stuck High:** With the engine running at operating temperature, and the vehicle speed is a minimum of 12 mph (18 km/h) or more with the engine speed between 900-3,000 rpm the following conditions occur for 2 seconds or more. Condition 1: rocker arm oil pressure sensor A is ON Condition 2: rocker arm oil pressure sensor B is ON Condition 3: rocker arm oil pressure sensor A output HIGH Condition 4: rocker arm oil pressure sensor B output LOW B

DTC	Trouble Code Title and Conditions
DTC: P055B **2T PCM, MIL: Yes** **Year:** 2011, 2012 **Model:** Civic **Engine:** 1.3L, 1.8L, 2.0L	**Rocker Arm Oil Pressure Sensor A Stuck High:** Malfunction Threshold, The following conditions occur for at least 2.0 seconds in the valve pause mode: 1). Rocker Arm Oil Control Solenoid A ON 2). Rocker Arm Oil Control Solenoid B ON 3). Rocker Arm Oil Pressure Sensor A output High (39 kPa (11.6 inHg, 293 mmHg) or more) 4). Rocker Arm Oil Pressure Sensor B output Low (39 kPa (11.5 inHg, 292 mmHg) or less)
DTC: P055C **1T PCM, MIL: Yes** **Year:** 2011, 2012 **Model:** Insight **Engine:** 1.3L	**Rocker Arm Oil Pressure Sensor B Low Voltage:** With the ignition switch ON (II) and the barometric pressure at 56 kPa (13.3 in.Hg, 439 mmHg), the rocker arm oil presseure sensor B output voltage is 0.18 V or less for at least 2 seconds.
DTC: P055C **2T PCM, MIL: Yes** **Year:** 2011, 2012 **Model:** Civic **Engine:** 1.3L, 1.8L, 2.0L	**Rocker Arm Oil Pressure Sensor B Low Voltage:** Malfunction Threshold: The rocker arm oil pressure sensor B output voltage is 0.18 V or less for at least 2 seconds.
DTC: P055D **2T PCM, MIL: Yes** **Year:** 2011, 2012 **Model:** Civic **Engine:** 1.3L, 1.8L, 2.0L	**Rocker Arm Oil Pressure Sensor B High Voltage:** Malfunction Threshold: The rocker arm oil pressure sensor B output voltage is 4.79 V or more for at least 2.0 seconds.
DTC: P055D **1T PCM, MIL: Yes** **Year:** 2011, 2012 **Model:** Insight **Engine:** 1.3L	**Rocker Arm Oil Pressure Sensor B High Voltage:** With the vehicle running and the engine speed 4,000 rpm the rocker arm oi pressure sensor B output voltage is or more than 4.79 V for at least 2 seconds.
DTC: P0560 **1T CCM, MIL: Yes** **Year:** 2012 **Model:** Civic **Engine:** 2.4L	**PCM Backup Circuit Low Voltage:** Key on or engine running, and the PCM detected a low voltage condition on the PCM Backup circuit. **Note: This circuit is connected to the Backup/Radio 7.5 amp fuse.**
DTC: P0560 **1T PCM, MIL: Yes** **Year:** 2011, 2012 **Model:** Accord, Crosstour **Engine:** 3.5L	**ECM/PCM Power Source Circuit Unexpected Voltage:** The back-up power supply voltage (at the VBUM terminal of the ECM/PCM) is 6.0 V or less for at least 5 seconds.
DTC: P0562 **1T PCM** **Year:** 2011, 2012 **Model:** Accord, Civic, Crosstour, CR-V, Element, Fit, Odyssey, Pilot, Ridgeline **Engine:** All	**Charging System Low Voltage:** The execution time is continuous and the duration time is 60 seconds or more, engine speed 550 rpm. When the IGP (power source) terminal voltage is a set value or less for a set time, the PCM detects a malfunction. The IGP terminal voltage is 11.0 V or less for at least 60 seconds. **NOTE: Before you troubleshoot, record all freeze data and any on-board snapshot. If any high current load accessories are installed, this DTC can be set.**
DTC: P0562 **1T PCM, MIL: Yes** **Year:** 2011, 2012 **Model:** Civic, CR-Z **Engine:** 1.5L	**Motor Control Module (MCM) Power Source Circuit Unexpected Voltage (Battery):** The +B BACK UP terminal voltage of the MCM is 3.1 V or less for at least 2.0 seconds, IMA system indicator on
DTC: P0562/15 **1T PCM, MIL: Yes** **Year:** 2011, 2012 **Model:** Civic, Insight **Engine:** 1.3L	**Motor Control Module (MCM) Power Source Circuit Unexpected Voltage (HYBRID MODEL):** With the ignition switch ON and the MCM power supply a minimum of 9.0 V, the VBU terminal voltage of the MCM is 4.0 V or less for at least 2 seconds. (IMA system indicator ON) WARNING: The IMA system uses high voltage (144 V), before servicing turn OFF the battery module switch.
DTC: P0562/94 **1T ECM, MIL: Yes** **Year:** 2011, 2012 **Model:** Insight **Engine:** 1.3L	**Motor Control Module (MCM) System Low Voltage (HYBRID MODELS):** With the ignition ON and the MCM power-supply voltage a minimum of 10.5 V, a system low voltage malfunction has been detected. (IMA system indicator is ON) WARNING: The IMA system uses high voltage (144 V), before servicing turn OFF the battery module switch.

DTC	Trouble Code Title and Conditions
DTC: P0563 **1T PCM, MIL: Yes** **Year:** 2011, 2012 **Model:** Accord, Civic, Crosstour, CR-V, CR-Z, Element, Fit **Engine:** All	**Engine Control Module (ECM) Powertrain Control Module (PCM) Power Source Circuit Unexpected Voltage:** The ECM/PCM operates for at least 5 seconds after the ignition switch is turned to LOCK (0). Battery voltage 10.1 volts. (IGP terminal of PCM) **NOTE: Before you troubleshoot, record all freeze data and any on-board snapshot.**
DTC: P0563 **1T PCM, MIL: Yes** **Year:** 2011, 2012 **Model:** Insight, Odyssey, Pilot, Ridgeline **Engine:** 1.3L, 3.5L	**Powertrain Control Module (PCM) Power Source Circuit Unexpected Voltage:** The PCM operates for at least 5 seconds after the ignition switch is turned to LOCK (0). Battery voltage 10.1 volts. (IGP terminal of PCM) **NOTE: Before you troubleshoot, record all freeze data and any on-board snapshot.**
DTC: P0602 **1T PCM, MIL: Yes** **Year:** 2011, 2012 **Model:** Civic, CR-Z **Engine:** 1.5L	**Motor Control Module (MCM) Program Not Installed (Battery Condition Monitor Program):** **NOTE: Before you troubleshoot, record all freeze data, and any on-board snapshot, and review the general troubleshooting information. If the DTC cannot be cleared, do the troubleshooting for DTC U1204.** This DTC is indicated when an MCM update is not completed. WARNING: Do not turn the ignition switch to ACCESSORY (I) or to LOCK (0) while updating the MCM. If you do, the MCM can be damaged.
DTC: P0602 **1T PCM, MIL: Yes** **Year:** 2011, 2012 **Model:** Accord, Civic, Crosstour, CR-V, CR-Z, Element, Fit, Odyssey, Pilot, Ridgeline **Engine:** 1.5L, 2.4L, 3.5L	**ECM/PCM Programming Error:** With ignition on the execution time is continuous and the duration time is 1 second or less. The ECM/PCM program update stops 1 second before it is finished. **NOTE: This DTC is indicated when a PCM update is not completed.** WARNING: Do not turn the ignition switch to LOCK (0) or ACC (I) while updating the PCM. If you turn the ignition switch to LOCK (0) before completion, the ECM/PCM can be damaged.
DTC: P0602 **1T , MIL: Yes** **Year:** 2011, 2012 **Model:** Civic, Insight **Engine:** 1.3L, 1.8L, 2.0L	**PCM Programming Error/ Hybrid BCM Programming Error:** With the ignition ON (II), a programing error was detected for at least 2 seconds. **NOTE: Do not turn the ignition switch to ACC (I) or to LOCK (0) while updating the PCM. If you turn the ignition switch to ACC (I) or to LOCK (0) before completion, the PCM can be damaged.**
DTC: P0602/91 **1T PCM, MIL: Yes** **Year:** 2011, 2012 **Model:** Civic, Insight **Engine:** 1.3L	**Motor Control Module (MCM) Programming Error:** With the ignition switch ON (II), the power is interrupted during reprogramming, and the MCM is not updated and programmed normally. (IMA system indicator is ON) **NOTE: Do not turn the ignition switch to ACC (I) or to LOCK (0) while updating the MCM. If you do, the MCM can be damaged.**
DTC: P0602/92 **1T ECM, MIL: Yes** **Year:** 2011, 2012 **Model:** Insight **Engine:** 1.3L	**Motor Control Module (MCM) Program Not Installed (Battery Condition Monitor Program) (HYBRID MODELS):** With the ignition switch ON (II), a program malfunction has been detected. WARNING: Do not turn the ignition switch to ACC (I) or to LOCK (0) while updating the MCM. If you do, the MCM can be damaged.
DTC: P0603/60 **1T PCM, MIL: Yes** **Year:** 2011 **Model:** Civic **Engine:** 1.3L	**Motor Control Module (MCM) Internal Circuit KAM Error (HYBRID MODELS):** Immediately after the ignition is turned ON (II), one of the following conditions occurs. Condition 1: The data input to or read out from the KAM is bad. Condition 2: The check sum of the data readout is abnormal three times.
DTC: P0606 **1T PCM, MIL: Yes** **Year:** 2011, 2012 **Model:** Accord, Civic, Crosstour, CR-V, CR-Z, Element, Fit **Engine:** 1.3L, 1.5L, 1.8L, 2.0L, 2.4L	**ECM/PCM Processor Malfunction:** After 30 seconds have elapsed since start-up, or after the engine speed exceeds 1,000 rpm once. No signal from the DKS CPU is detected or is abnormal for at least 5 seconds. **NOTE: Before you troubleshoot, record all freeze data and any on-board snapshot, and review the general troubleshooting information.**
DTC: P060A **1T PCM, MIL: Yes** **Year:** 2011, 2012 **Model:** Crosstour, Odyssey, Ridgeline **Engine:** 2.4L, 3.5L	**Powertrain Control Module (PCM) (A/T System) Internal Control Module Malfunction:** (Symptom 1) The execution time is continuous and the duration time 500 milliseconds or more The serial communication between the FI CPU and A/T CPU is abnormal or the serial communication is interrupted for at least 500 milliseconds. (Symptom 2) The execution time is continuous and the duration time , 30 milliseconds or more The watchdog timer that monitors the A/T CPU detects an abnormality, and the FI CPU receives the A/T CPU check signal for at least 30 milliseconds. **NOTE: Before you troubleshoot, record all freeze data and any on-board snapshot.**

DTC	Trouble Code Title and Conditions
DTC: P060A **1T PCM, MIL: Yes** **Year:** 2011, 2012 **Model:** Accord, Civic, Crosstour, CR-V, CR-Z, Fit, Insight **Engine:** 1.3L, 1.5L, 2.4L, 3.5L	**Powertrain Control Module (PCM) Internal Control Module Malfunction:** With the ignition ON and battery voltage at a minimum of 10. 0 volts. The execution time is continuous and the duration time is 200 milliseconds or more. With keyless access system 110 milliseconds or more. One of these 2 symptoms occurs: (1) Symptom 1: The internal communication between the FI CPU and the A/T CPU is abnormal or the internal communication is interrupted for at least 200 milliseconds. (2) Symptom 2: The watchdog timer that monitors the A/T CPU detects an abnormality, and the FI CPU receives the A/T CPU check signal for at least 110 milliseconds.
DTC: P060A **1T PCM, MIL: Yes** **Year:** 2011, 2012 **Model:** Civic **Engine:** 1.3L, 1.8L, 2.0L	**Powertrain Control Module (PCM) (A/T System) Internal Control Module Malfunction:** Malfunction Threshold One of these symptoms occurs for at least 0.5 second: 1). No sub CPU monitor signals are detected. 2). Sub CPU monitor signals are abnormal.
DTC: P060F **1T PCM, MIL: Yes** **Year:** 2012 **Model:** Crosstour, CR-V **Engine:** 2.4L	**Powertrain Control Module (PCM) Internal Control Module Keep Alive Memory (KAM) Error :** With the ignition switch ON a malfunction is detected whenever the keep alive data retrieval and writing process is not completed normally. **NOTE: Before you troubleshoot, record all freeze data and any on-board snapshot with the HDS, and review General Troubleshooting Information.**
DTC: P0615 **1T PCM** **Year:** 2011, 2012 **Model:** Pilot **Engine:** 3.5L	**Starter Cut Relay STRLD Circuit Malfunction:** Withe the ignition switch ON (II). The diagnosis line (STRLD) input voltage is between 2.4 V to 2.6 V for at least 1 second. The execution time is continuous and the duration time is 1 seconds or more. **NOTE: Before you troubleshoot, record all freeze data and any on-board snapshot.**
DTC: P062F **1T PCM, TCIL: Yes** **Year:** 2011, 2012 **Model:** All **Engine:** All	**ECM/PCM Internal Control Module Keep Alive Memory (KAM) Error :** A malfunction is detected whenever the keep alive data retrieval and writing process is not completed normally. **NOTE: Before you troubleshoot, record all freeze data and any on-board snapshot.**
DTC: P062F/60 **1T ECM, MIL: Yes** **Year:** 2011, 2012 **Model:** Civic, CR-Z, Insight **Engine:** 1.3L, 1.5L	**Motor Control Module (MCM) Internal Circuit EEPROM Error (HYBRID MODELS):** With the ignition switch ON (II), the data input to or read out from the EEPROM does not match.
DTC: P062F/61 **1T PCM, MIL: Yes** **Year:** 2011, 2012 **Model:** Civic, CR-Z **Engine:** 1.5L	**Battery Control Module (BCM) Internal Circuit EEPROM Error:** IMA system indicator on.
DTC: P0630 **1T PCM, MIL: Yes** **Year:** 2011, 2012 **Model:** All **Engine:** All	**VIN Not Programmed or Mismatch:** The VIN is not registered in the keep-alive memory in the PCM. **NOTE: Before you troubleshoot, record all freeze data and any on-board snapshot.**
DTC: P0641 **1T PCM, MIL: Yes** **Year:** 2011, 2012 **Model:** Accord, Crosstour, Insight, Odyssey, Pilot **Engine:** 1.3L, 3.5L	**Sensor Reference Voltage A Malfunction:** With the ignition ON the execution time is continuous and the sensor power voltage is 5.2 V or more, or 4.5 V or less, for at least 2.0 seconds. **NOTE: Before you troubleshoot, record all freeze data and any on-board snapshot. It may be possible to locate the fault by disconnecting one component at a time from the 5-volt reference circuit while viewing the 5-Volt Reference circuit parameter on the scan tool. The scan tool parameter would change from Fault to OK when the source of the fault is disconnected. If all 5-volt reference components have been disconnected and a Fault is still indicated, the fault may exist in the wiring harness.**
DTC: P0651 **1T PCM, MIL: Yes** **Year:** 2011, 2012 **Model:** Civic, Insight, Odyssey **Engine:** 1.3L, 1.8L, 2.0L, 3.5L	**Sensor Reference Voltage B Malfunction:** The sensor power voltage is 5.2 V or more, or 4.5 V or less, for at least 2.0 seconds.

DTC	Trouble Code Title and Conditions
DTC: P065A **1T PCM, MIL: Yes** **Year:** 2011, 2012 **Model:** Civic **Engine:** 1.3L, 1.8L, 2.0L	**ACG No Charging Malfunction:** Either one of the conditions is met for at least 5 seconds: * B terminal voltage is regulated voltage or more, and F terminal voltage is HIGH. * B terminal voltage is 10 V or less (when the alternator speed is 3,000 rpm or less), or 11 V or less (when the alternator speed is 3,000 rpm or more). * P terminal voltage is 2 V or less (when the alternator speed is 3,000 rpm or less), or 6 V or less (when the alternator speed is 3,000 rpm or more). * Alternator speed is 400 rpm or less.
DTC: P065E **1T PCM, MIL: Yes** **Year:** 2011, 2012 **Model:** Civic **Engine:** 1.3L, 1.8L, 2.0L	**Intake Manifold Tuning (IMT) Actuator Circuit Malfunction:** Either one of the conditions is met: * Open circuit failure The return signal from the IMT actuator is HIGH for at least 2 seconds while the current is not applied to the IMT actuator. * Short circuit failure Over current from the IMT actuator continues for at least 0.4 second while the current is applied to the IMT actuator.
DTC: P0685 **2T PCM, MIL: Yes** **Year:** 2011, 2012 **Model:** Accord Crosstour, Insight, Odyssey, Pilot, Ridgeline **Engine:** 1.3L, 3.5L	**Powertrain Control Module (PCM) Power Control Circuit/Internal Circuit Malfunction:** When the voltage to the PCM is turned off and the PCM shuts down without the normal shut down procedure, a malfunction in the PGM-FI main relay 1 control circuit is detected. **NOTE: Before you troubleshoot, record all freeze data and any on-board snapshot.**
DTC: P0685 **2T PCM, MIL: Yes** **Year:** 2011, 2012 **Model:** Accord, Civic, Crosstour, CR-V, CR-Z, Element, Fit **Engine:** All	**ECM/PCM Power Control Circuit Malfunction:** When the voltage to the ECM/PCM is turned off and the ECM/PCM shuts down without the normal shut down procedure, a malfunction in the PGM-FI main relay 1 control circuit is detected. **NOTE: Before you troubleshoot, record all freeze data and any on-board snapshot.**
DTC: P06A8 **1T PCM, MIL: Yes** **Year:** 2011, 2012 **Model:** Civic, Insight **Engine:** 1.3L, 1.8L, 2.0L	**Internal VCC Power Malfunction:** The ECM/PCM internal comparison power supply voltage is 3.13 V or less, 3.84 V or more for at least 2.0 seconds.
DTC: P06B1/79 **1T ECM, MIL: Yes** **Year:** 2011, 2012 **Model:** Civic, CR-Z, Insight **Engine:** 1.3L, 1.5L	**Battery Current Sensor Circuit Malfunction (HYBRID MODELS):** With the ignition switch ON (II), the battery current sensor power-supply voltage is 4.8 V or less for at least 2 seconds. (IMA system indicator is ON) WARNING: The IMA system uses high voltage (144 V), before servicing turn OFF the battery module switch.
DTC: P0705 **1T PCM, MIL: Yes, TCIL: Yes** **Year:** 2011, 2012 **Model:** Accord, Civic, Crosstour, CR-V, CR-Z **Engine:** 1.3L, 1.5L, 1.8L, 2.4L, 3.5L	**Transmission Range Switch Circuit (Multiple Shift-position Input):** Malfunction 1: The PCM detects the selected range switch input and another range switch (except L2 switch) input simultaneously for at least 1 second. Malfunction 2: The PCM detects the P,R or N range switch input and the L2 switch input simultaneously for at least 1 second. **NOTE: Record all freeze data and review General Troubleshooting Information before you troubleshoot. This code is caused by an electrical circuit problem and cannot be caused by a mechanical problem in the transmission.**
DTC: P0705 **1T PCM, MIL: Yes, TCIL: Yes** **Year:** 2011, 2012 **Model:** Crosstour, CR-V, Element, Fit, Insight, Odyssey, Pilot, Ridgeline **Engine:** 1.3L, 1.5L, 2.4L, 3.5L	**Short in Transmission Range Switch Circuit (Multiple Shift-position Input):** One of 3 conditions occurs: (1) The PCM detects the selected range switch input and another range switch input simultaneously for at least 1 seconds. (2) The PCM detects the P, R, or N range switch input and the FWD switch input simultaneously for at least 1 seconds (3) The PCM detects the D or D3 range switch input and the RVS switch input simultaneously for at least 1 second. **NOTE: Before you troubleshoot, record all freeze data and any on-board snapshot. This code is caused by an electrical circuit problem and cannot be caused by a mechanical problem in the transmission.**
DTC: P0706 **2T PCM, MIL: Yes** **Year:** 2011, 2012 **Model:** All **Engine:** 1.3L, 1.5L, 1.8L, 2.4L, 3.5L	**Open in Transmission Range Switch Circuit:** No FWD position signal is detected when the vehicle speed changes from 6 mph (10 km/h) 25 mph (40 km/h) 6 mph (10 km/h) in D or D3. DTCs P0705, P0721, P0722 are not active. **NOTE: This code is caused by an electrical circuit problem and cannot be caused by a mechanical problem in the transmission.**

DTC	Trouble Code Title and Conditions
DTC: P0711 **1T PCM, TCIL: Yes** **Year:** 2011, 2012 **Model:** Accord, Civic, Crosstour, CR-V, Element, Fit, Odyssey, Pilot, Ridgeline **Engine:** 1.3L, 1.5L, 1.8L, 2.4L, 3.5L	**Problem in ATF Temperature Sensor Circuit:** The ATF temperature sensor signal does not change. Stuck at low temperature or stuck at high temperature is detected. **NOTE: This code is caused by an electrical circuit problem and cannot be caused by a mechanical problem in the transmission.**
DTC: P0712 **1T PCM, TCIL: Yes** **Year:** 2011, 2012 **Model:** Accord, Civic, Crosstour, CR-V, Element, Fit, Odyssey, Pilot, Ridgeline **Engine:** 1.3L, 1.5L, 1.8L, 2.4L, 3.5L	**Short in ATF Temperature Sensor Circuit:** When the ATF temperature sensor signal voltage to the PCM is under the specification, indicating that the temperature is above the specification (a short to ground), a malfunction is detected. The ATF temperature sensor output voltage is less than 0.07 V for at least 10 seconds. **NOTE: This code is caused by an electrical circuit problem and cannot be caused by a mechanical problem in the transmission.**
DTC: P0712 **2T PCM** **Year:** 2011, 2012 **Model:** Insight **Engine:** 1.3L	**CVTF Temperature Sensor (Short):** The CVTF temperature sensor output voltage is less than 0.07 V for at least 10 seconds. (D indicator blinks) **NOTE: Before you troubleshoot, record all freeze data and any on-board snapshot with the HDS, and review the General Troubleshooting Information.** This code is caused by an electrical circuit problem and cannot be caused by a mechanical problem in the transmission.
DTC: P0713 **2T PCM** **Year:** 2011, 2012 **Model:** Insight **Engine:** 1.3L	**CVTF Temperature Sensor (Open):** The CVTF temperature sensor output voltage is more than 4.93 V for at least 10 seconds. (D indicator blinks)
DTC: P0713 **1T PCM, TCIL: Yes** **Year:** 2011, 2012 **Model:** Accord, Civic, Crosstour, CR-V, Element, Fit, Odyssey, Pilot, Ridgeline **Engine:** All	**Open in ATF Temperature Sensor Circuit:** When the ATF temperature sensor signal voltage to the PCM is above the specification, indicating that the temperature is under the specification (open), a malfunction is detected. The ATF temperature sensor output voltage is 4.93 V or more for at least 10 seconds. **NOTE: This code is caused by an electrical circuit problem and cannot be caused by a mechanical problem in the transmission.**
DTC: P0714 **2T PCM, TCIL: Yes** **Year:** 2011, 2012 **Model:** Accord, Crosstour, Odyssey **Engine:** 2.4L, 3.5L	**ATF Temperature Sensor Circuit Intermittent Failure:** The temperature in the ATF temperature sensor deviates from the temperature in ECT sensor 2, less than -43 ºF (-24 ºC) or greater than 57 ºF (32 ºC). **NOTE: Greater than 108 ºF (60 ºC) if engine block heater is equipped.**
DTC: P0714 **2T PCM, TCIL: Yes** **Year:** 2011 **Model:** Accord Crosstour **Engine:** 3.5L	**ATF Sensor Intermittent Failure:** With the ignition OFF for at least 6 hours and batter voltage a minimum of 11 V, the temperature in the ATF temperature sensor deviates from the temperature in the ECT 2 sensor, lees than 11 ºF (-24 ºC) or greater than 9 ºF (32 ºC). Greater than 140 ºF (60 ºC) if equipped with a block heater. **NOTE: This code is caused by an electrical circuit problem and cannot be caused by a mechanical problem in the transmission.**
DTC: P0715 **1T ECM** **Year:** 2011, 2012 **Model:** Accord **Engine:** 3.5L	**Input Shaft (Mainshaft) Speed Sensor Circuit Malfunction (M/T Model):** With the engine running and engine speed at a minimum of 1,000 rpm, No signal from the input shaft (mainshaft) speed sensor is detected for at least 30 seconds. **NOTE: Before you troubleshoot, record all freeze data and any on-board snapshot, and review the general troubleshooting information.**
DTC: P0716 **1T PCM, MIL: Yes** **Year:** 2011, 2012 **Model:** Civic, CR-Z **Engine:** 1.5L	**CVT Input Shaft (Drive Pulley) Speed Sensor "A" Circuit Range/Performance:** The drive pulley speed measured by the CVT input shaft (drive pulley) speed sensor is less than the engine speed measured by the CKP sensor by a factor of 0.53 for at least 12 seconds. **NOTE: Before you troubleshoot, record all freeze data and any on-board snapshots.** This code is sometimes caused by an electrical circuit problem and sometimes caused by a worn or defective flywheel drive plate in the transmission.

DTC	Trouble Code Title and Conditions
DTC: P0716 **1T PCM, MIL: Yes, TCIL: Yes** **Year:** 2011, 2012 **Model:** All **Engine:** 1.3L, 1.5L, 1.8L, 2.4L, 3.5L	**Problem in Input Shaft (Mainshaft) Speed Sensor Circuit:** If no pulses occur with the input shaft (mainshaft) rotating, the PCM detects a malfunction that may be caused by an open, a temporary open, or a short to ground. The vehicle speed measured by the input shaft (mainshaft) speed sensor/(divided by) the vehicle speed measured by the output shaft (countershaft) speed sensor is less than 0.156 for at least 10 seconds. **NOTE: This code is caused by an electrical circuit problem and cannot be caused by a mechanical problem in the transmission.**
DTC: P0717 **1T PCM, MIL: Yes** **Year:** 2011, 2012 **Model:** Civic, CR-Z **Engine:** 1.5L	**CVT Input Shaft (Drive Pulley) Speed Sensor "A" Circuit No Signal:** The drive pulley speed measured by the CVT input shaft (drive pulley) speed sensor is 100 rpm or less for at least 3 seconds when the engine speed measured by the CKP sensor is 600 rpm or more.
DTC: P0717 **1T PCM, MIL: Yes, TCIL: Yes** **Year:** 2011, 2012 **Model:** Accord, Civic, Crosstour, CR-V, Element, Fit, Insight, Odyssey, Pilot, Ridgeline **Engine:** 1.3L, 1.5L, 1.8L, 2.4L, 3.5L	**Problem in Input Shaft (Mainshaft) Speed Sensor Circuit (No Signal Input):** If no pulses occur with the input shaft (mainshaft) rotating, the PCM detects a malfunction that may be caused by an open, a temporary open, or a short to ground. When the vehicle speed measured by the output shaft (countershaft) speed sensor is 13 mph (20 km/h) or more, the vehicle speed measured by the input shaft (mainshaft) speed sensor is 1 mph (2 km/h) or less for at least 10 seconds. **NOTE: This code is caused by an electrical circuit problem and cannot be caused by a mechanical problem in the transmission.**
DTC: P0718 **2T PCM, MIL: Yes, TCIL: Yes** **Year:** 2011, 2012 **Model:** Accord, Civic, Crosstour, CR-V, Element, Fit, Odyssey, Pilot, Ridgeline **Engine:** 1.3L, 1.5L, 1.8L, 2.4L, 3.5L	**Input Shaft (Mainshaft) Speed Sensor Intermittent Failure:** If no pulses occur with the input shaft (mainshaft) rotating, the PCM detects a malfunction that may be caused by an open, a temporary open, or a short to ground. The fluctuation of the vehicle speed measured by the input shaft (mainshaft) speed sensor in 10 milliseconds is 4 mph (6km/h) or more, and it fluctuates at least six times within 500 milliseconds. **NOTE: This code is caused by an electrical circuit problem and cannot be caused by a mechanical problem in the transmission.**
DTC: P0720 **2T PCM, MIL: Yes** **Year:** 2011, 2012 **Model:** Accord, Crosstour, CR-V, CR-Z **Engine:** 1.5L, 2.4L, 3.5L	**Output Shaft (Countershaft) Speed Sensor Circuit Malfunction :** With the engine speed at a minimum of 4,000 rpm and during fuel cut-off operation for deceleration. NO signal from the output shaft (Countershaft) speed sensor is detected for at least 5 seconds. **NOTE: Before you troubleshoot, record all freeze data and any on-board snapshot, and review the general troubleshooting information.**
DTC: P0720 **2T PCM, MIL: Yes** **Year:** 2011, 2012 **Model:** Fit **Engine:** 1.5L	**Output Shaft (Countershaft) Speed Sensor Circuit Malfunction (M/T model):** No signal from the output shaft (countershaft) speed sensor is detected for at least 5 seconds during deceleration with the engine speed at 4,000 rpm or less.
DTC: P0721 **1T PCM, MIL: Yes** **Year:** 2011, 2012 **Model:** Civic, CR-Z **Engine:** 1.5L	**CVT Output Shaft (Driven Pulley) Speed Sensor Range/Performance:** The driven pulley speed measured by the CVT output shaft (driven pulley) speed sensor is less than the driven pulley speed measured by the vehicle speed sensor by a factor of 0.53 for at least 15 seconds. **NOTE: Before you troubleshoot, record all freeze data and any on-board snapshot with the HDS, and review the General Troubleshooting Information.** This code is sometimes caused by an electrical circuit problem and sometimes caused by a broken steel belt in the transmission.
DTC: P0721 **1T PCM, MIL: Yes, TCIL: Yes** **Year:** 2011, 2012 **Model:** Accord, Civic, Crosstour, CR-V, Element, Fit, Insight, Odyssey, Pilot, Ridgeline **Engine:** 1.3L, 1.5L, 1.8L, 2.4L, 3.5L	**Problem in Output Shaft (Countershaft) Speed Sensor Circuit:** If pulse dropouts occur with the output shaft (countershaft) rotating, the PCM detects a malfunction that may be caused by an open, a temporary open, or a short to ground. The vehicle speed measured by the input shaft (mainshaft) speed sensor/(divided by) the vehicle speed measured by the output shaft (countershaft) speed sensor is greater than 6.0 for at least 10 seconds. **NOTE: This code is caused by an electrical circuit problem and cannot be caused by a mechanical problem in the transmission.**
DTC: P0722 **1T PCM, MIL: Yes, TCIL: Yes** **Year:** 2011, 2012 **Model:** Accord, Civic, Crosstour, CR-V, Element, Fit, Insight, Odyssey, Pilot, Ridgeline **Engine:** 1.3L, 1.5L, 1.8L, 2.4L, 3.5L	**Problem in Output Shaft (Countershaft) Speed Sensor Circuit (No Signal Input):** If pulse dropouts occur with the output shaft (countershaft) rotating, the PCM detects a malfunction that may be caused by an open, a temporary open, or a short to ground. When the vehicle speed measured by the input shaft (mainshaft) speed sensor is 13 mph (20 km/h) or more, the vehicle speed measured by the output shaft (countershaft) speed sensor is 1 mph (2 km/h) or less for at least 10 seconds. **NOTE: This code is caused by an electrical circuit problem and cannot be caused by a mechanical problem in the transmission.**

DTC	Trouble Code Title and Conditions
DTC: P0722 **1T PCM, MIL: Yes** **Year:** 2011, 2012 **Model:** Civic, CR-Z **Engine:** 1.5L	**CVT Output Shaft (Driven Pulley) Speed Sensor No Signal Input:** The driven pulley speed measured by the CVT output shaft (driven pulley) speed sensor is 100 rpm or less for at least 15 seconds.
DTC: P0723 **2T PCM, MIL: Yes, TCIL: Yes** **Year:** 2011, 2012 **Model:** Accord, Civic, Crosstour, CR-V, Element, Fit, Odyssey, Pilot, Ridgeline **Engine:** 1.3L, 1.5L, 1.8L, 2.4L, 3.5L	**Output Shaft (Countershaft) Speed Sensor Intermittent Failure:** If pulse dropouts occur with the output shaft (countershaft) rotating, the PCM detects a malfunction that may be caused by an open, a temporary open, or a short to ground. Based on the fluctuation of the vehicle speed measured by the output shaft (countershaft) speed sensor, a malfunction is detected. The fluctuation of the vehicle speed measured by the output shaft (countershaft) speed sensor in 10 milliseconds is 4 mph (6 km/h) or more, and it fluctuates at least six times within 500 milliseconds. **NOTE: This code is caused by an electrical circuit problem and cannot be caused by a mechanical problem in the transmission.**
DTC: P0729 **2T PCM, TCIL: Yes** **Year:** 2011, 2012 **Model:** Civic, Odyssey **Engine:** 1.3L, 1.8L, 3.5L	**Problem in 6th Clutch and 6th Clutch Hydraulic Circuit:** With the vehicle running in drive and engine speed at a minimum of 7 mph, battery voltage 11 volts. The actual gear ratio must match one of these conditions for at least 12 seconds with the 6th gear shift command: Actual gear ratio is greater than the 6th gear ratio by a factor of 1.25. Actual gear ratio is less than the 6th gear ratio by a factor of 0.8. **NOTE: Before you troubleshoot, record all freeze data and any on-board snapshots.**
DTC: P0731 **2T PCM, TCIL: Yes** **Year:** 2011, 2012 **Model:** Accord, Civic, Crosstour, CR-V, Element, Fit, Odyssey, Ridgeline **Engine:** 1.5L, 1.8L, 2.4L, 3.5L	**Problem in 1st Clutch and 1st Clutch Hydraulic Circuit (1st gear incorrect ratio):** The Powertrain Control Module (PCM) computes the ratio of the input shaft (mainshaft) speed to the output shaft (countershaft) speed. When the ratio is not the 1st gear ratio, it is detected as a malfunction of the hydraulic circuit or the 1st clutch. (Symptom 1) The actual gear ratio must match one of these conditions for at least 12 seconds with the 1st gear command: * Actual gear ratio is greater than the 1st gear ratio by a factor of 1.2 * Actual gear ratio is less than the 1st gear ratio by a factor of 0.75 (Symptom 2) The actual gear position is neutral for at least 3 seconds and then the gear up-shifted from 2nd to 3rd, even though 1st gear shift is commanded. **NOTE: Before you troubleshoot, record all freeze data and any on-board. snapshot,**
DTC: P0732 **2T PCM, TCIL: Yes** **Year:** 2011, 2012 **Model:** Accord, Civic, Crosstour, CR-V, Element, Fit, Odyssey, Pilot, Ridgeline **Engine:** 1.5L, 1.8L, 2.4L, 3.5L	**Problem in 2nd Clutch and 2nd Clutch Hydraulic Circuit (2nd gear incorrect ratio):** The Powertrain Control Module (PCM) computes the ratio of the input shaft (mainshaft) speed to the output shaft (countershaft) speed. When the ratio is not the 2nd gear ratio, it is detected as a malfunction of the hydraulic circuit or the 2nd clutch. The actual gear ratio must match one of these conditions for at least 12 seconds with the 2nd gear command. * Actual gear ratio is greater than the 2nd gear ratio by a factor of 1.2 * Actual gear ratio is less than the 2nd gear ratio by a factor of 0.75. **NOTE: Before you troubleshoot, record all freeze data and any on-board snapshot.**
DTC: P0733 **2T PCM, TCIL: Yes** **Year:** 2011, 2012 **Model:** Accord, Civic, Crosstour, CR-V, Element, Fit, Odyssey, Pilot, Ridgeline **Engine:** 1.5L, 1.8L, 2.4L, 3.5L	**Problem in 3rd Clutch and 3rd Clutch Hydraulic Circuit (3rd gear incorrect ratio):** The powertrain control module (PCM) computes the ratio of the input shaft (mainshaft) speed to the output shaft (countershaft) speed. When the ratio is not the 3rd gear ratio, it is detected as a malfunction of the hydraulic circuit or the 3rd clutch. The actual gear ratio must match one of these conditions for at least 12 seconds with the 3rd gear command. * Actual gear ratio is greater than the 3rd gear ratio by a factor of 1.2. * Actual gear ratio is less than the 3rd gear ratio by a factor of 0.75. **NOTE: Before you troubleshoot, record all freeze data and any on-board snapshot.**
DTC: P0734 **2T PCM, TCIL: Yes** **Year:** 2011, 2012 **Model:** Accord, Civic, Crosstour, CR-V, Element, Fit, Odyssey, Pilot, Ridgeline **Engine:** 1.5L, 1.8L, 2.4L, 3.5L	**Problem in 4th Clutch and 4th Clutch Hydraulic Circuit (4th gear incorrect ratio):** The Powertrain Control Module (PCM) computes the ratio of the input shaft (mainshaft) speed to the output shaft (countershaft) speed. When the ratio is not the 4th gear ratio, it is detected as a malfunction of the hydraulic circuit or the 4th clutch. The actual gear ratio must match one of these conditions for at least 12 seconds with the 4th gear command. * Actual gear ratio is greater than the 4th gear ratio by a factor of 1.2. * Actual gear ratio is less than the 4th gear ratio by a factor of 0.75. **NOTE: Before you troubleshoot, record all freeze data and any on-board snapshot.**
DTC: P0735 **2T PCM, TCIL: Yes** **Year:** 2011, 2012 **Model:** Accord, Civic, Crosstour, CR-V, Element, Fit, Odyssey, Pilot, Ridgeline **Engine:** 1.5L, 1.8L, 2.4L, 3.5L	**Problem in 5th Clutch and 5th Clutch Hydraulic Circuit (5th gear incorrect ratio):** The Powertrain Control Module (PCM) computes the ratio of the input shaft (mainshaft) speed to the output shaft (countershaft) speed. When the ratio is not the 5th gear ratio, it is detected as a malfunction of the hydraulic circuit or the 5th clutch. The actual gear ratio must match one of these conditions for at least 12 seconds with the 5th gear command. * Actual gear ratio is greater than the 5th gear ratio by a factor of 1.2 * Actual gear ratio is less than the 5th gear ratio by a factor of 0.75 **NOTE: Before you troubleshoot, record all freeze data and any on-board snapshot.**

DTC	Trouble Code Title and Conditions
DTC: P0741 **2T PCM, MIL: Yes** **Year:** 2011, 2012 **Model:** Accord, Civic, Crosstour, CR-V, Element, Fit, Odyssey, Pilot, Ridgeline **Engine:** 1.5L, 1.8L, 2.0L, 2.4L, 3.5L	**Torque Converter Clutch Hydraulic Circuit Stuck OFF:** If the ratio of engine speed and input shaft (mainshaft) speed is not about 1:1 while the PCM is issuing the command to turn shift solenoid valve D and A/T clutch pressure control solenoid valve C ON, the PCM detects a faulty lock-up control system. The ratio of the engine revolutions to the transmission input pulses does not reach about 100 % for at least 22 seconds. **NOTE: Before you troubleshoot, record all freeze data and any on-board snapshot.**
DTC: P0746 **2T PCM, MIL: Yes, TCIL: Yes** **Year:** 2011, 2012 **Model:** Accord, Civic, Crosstour, Fit, Odyssey, Pilot, Ridgeline **Engine:** 1.5L, 1.8L, 3.5L	**A/T Clutch Pressure Control Solenoid Valve A Stuck OFF:** When an improper gear ratio is output compared to the predetermined gear ratio, an A/T clutch pressure control solenoid valve A OFF failure is detected. One of these symptoms occur: * Transmission is held in 1st gear. * The engine speed flares when upshifting to 2nd-3rd. * The engine speed flares when upshifting to 3rd-4th or 5th if applicable. **NOTE: Before you troubleshoot, record all freeze data and any on-board snapshot.**
DTC: P0746 **2T PCM, MIL: Yes, TCIL: Yes** **Year:** 2011, 2012 **Model:** Civic, CR-Z, Insight **Engine:** 1.3L, 1.5L	**CVT Drive Pulley Pressure Control Valve Stuck OFF:** With the vehicle at operating temperature, and driven for 15 seconds or more, both of the following judgement conditions must be met. Judgement A: The pulley ratio is 0.3-0.8 for at least 2.5 seconds. Judgement B: The pulley ratio is less than 3 for at least 10 seconds.
DTC: P0747 **2T PCM, MIL: Yes, TCIL: Yes** **Year:** 2011, 2012 **Model:** Accord, Civic, Crosstour, CR-V, Element, Fit, Odyssey, Pilot, Ridgeline **Engine:** 1.5L, 1.8L, 2.4L, 3.5L	**A/T Clutch Pressure Control Solenoid Valve A Stuck ON:** When an improper gear ratio is output compared to the predetermined gear ratio, an A/T clutch pressure control solenoid valve A ON failure is detected. The execution time is continuous and the duration time is 20 seconds. One of these conditions occur: * The transmission is held in 2nd gear against the 2nd-3rd gear upshift command as long as 20 seconds though there is no record of being neutral when the 1st gear shift is commanded. * The transmission is held in 4th gear against the 4th-5th gear upshift command as long as 20 seconds though there is no record of being neutral when the 1st gear shift is commanded. **NOTE: Before you troubleshoot, record all freeze data and any on-board snapshot.**
DTC: P0751 **2T PCM, MIL: Yes, TCIL: Yes** **Year:** 2011, 2012 **Model:** Accord, Civic, Crosstour, Fit, Odyssey, Pilot, Ridgeline **Engine:** 1.5L, 1.8L, 2.4L, 3.5L	**Shift Solenoid Valve A Stuck OFF:** When an improper gear ratio is output compared to the predetermined gear change mode, a shift solenoid valve A OFF failure is detected and a DTC is stored. The execution time is continuous and the duration time is 2 seconds or more. The transmission is held in 5th gear against the 3rd gear command for at least 2 seconds. **NOTE: Before you troubleshoot, record all freeze data and any on-board snapshot.**
DTC: P0752 **2T PCM, MIL: Yes, TCIL: Yes** **Year:** 2011, 2012 **Model:** Accord, Civic, Crosstour, CR-V, Element, Fit, Odyssey, Pilot, Ridgeline **Engine:** 1.5L, 1.8L, 2.4L, 3.5L	**Shift Solenoid Valve A Stuck ON:** When the wrong transmission fluid switch is turned on for a given speed change mode, a shift solenoid valve turn-on malfunction is detected. The execution time is continuous depending on the driving pattern. The 3rd clutch transmission fluid switch is ON against the 4th-5th gear upshift command for at least 11 seconds. **NOTE: Before you troubleshoot, record all freeze data and any on-board snapshot.**
DTC: P0756 **2T PCM, MIL: Yes, TCIL: Yes** **Year:** 2011, 2012 **Model:** Accord, Civic, Crosstour, CR-V, Element, Fit, Odyssey, Pilot, Ridgeline **Engine:** 1.5L, 1.8L, 2.4L, 3.5L	**Shift Solenoid Valve B Stuck OFF:** When an improper gear ratio is output compared to the predetermined gear change mode, a shift solenoid valve B OFF failure is detected. The transmission is held in 4th gear against the 2nd gear command for at least 2 seconds. The execution time is continuous and the duration time is 2 seconds or more. **NOTE: Before you troubleshoot, record all freeze data and any on-board snapshot.**
DTC: P0757 **2T PCM, MIL: Yes, TCIL: Yes** **Year:** 2011, 2012 **Model:** Accord, Civic, Crosstour, CR-V, Element, Fit, Odyssey, Pilot, Ridgeline **Engine:** 1.5L, 1.8L, 2.4L, 3.5L	**Shift Solenoid Valve B Stuck ON:** When the wrong gear ratio is output for a given speed change mode, or when the wrong transmission fluid pressure switch is turned-on, a shift solenoid valve turn-on malfunction is detected. The execution time is continuous depending on the driving pattern. One of these conditions occur: * The 2nd clutch transmission fluid switch is ON against the 3rd-4th gear upshift command for at least 11 seconds. * After the 3rd-4th gear upshift command is output, it is neutral for at least 2 seconds when the 5th gear shift command is output, though there is no history of being neutral. **NOTE: Before you troubleshoot, record all freeze data and any on-board snapshot.**

DTC	Trouble Code Title and Conditions
DTC: P0761 **2T PCM, MIL: Yes, TCIL: Yes** **Year:** 2011, 2012 **Model:** Accord, Civic, Crosstour, CR-V, Element, Fit, Odyssey, Pilot, Ridgeline **Engine:** 1.5L, 1.8L, 2.4L, 3.5L	**Shift Solenoid Valve C Stuck OFF:** The execution time is continuous and the duration time is 20 seconds. When an improper gear ratio is output compared to the predetermined gear change mode, a shift solenoid valve C OFF failure is detected. One of these symptoms occurred when the actual gear position was neutral when the 1st gear shift is commanded: * The transmission is held in 2nd gear against the 2nd-3rd gear upshift command for at least 17 seconds. * The transmission is held in 4th gear against the 4th-5th gear upshift command for at least 17 seconds. **NOTE: Before you troubleshoot, record all freeze data and any on-board snapshot.**
DTC: P0762 **2T PCM, MIL: Yes, TCIL: Yes** **Year:** 2011, 2012 **Model:** Accord, Civic, Crosstour, Fit, Odyssey, Pilot, Ridgeline **Engine:** 1.5L, 1.8L, 3.5L	**Shift Solenoid Valve C Stuck ON:** When an improper gear ratio is output compared to the predetermined gear change mode, a shift solenoid valve C ON failure is detected. The execution time is continuous and the duration time is 20 seconds. The transmission is held in 3rd gear against the 3rd-4th gear upshift command for as long as 20 seconds, without records that the gear change time was short when the 2nd-3rd gear upshift were commanded. **NOTE: Before you troubleshoot, record all freeze data and any on-board snapshot.**
DTC: P0766 **2T PCM, MIL: Yes, TCIL: Yes** **Year:** 2011, 2012 **Model:** Civic, Fit, Odyssey **Engine:** 1.5L, 1.8L, 3.5L	**Shift Solenoid Valve D Stuck OFF:** When an improper gear ratio is output compared to the predetermined gear change mode, a shift solenoid valve D OFF failure is detected. **NOTE: Before you troubleshoot, record all freeze data and any on-board snapshot.**
DTC: P0766 **2T PCM, MIL: Yes, TCIL: Yes** **Year:** 2011, 2012 **Model:** Accord, Crosstour **Engine:** 2.4L, 3.5L	**Shift Solenoid Valve D Stuck OFF:** The PCM detects a shift solenoid valve D OFF failure, 3-11 seconds or more.
DTC: P0767 **2T PCM, MIL: Yes, TCIL: Yes** **Year:** 2011, 2012 **Model:** Civic, Fit **Engine:** 1.5L, 1.8L	**Shift Solenoid Valve D Stuck ON:** When an improper gear ratio is output compared to the predetermined gear change mode, a shift solenoid valve D ON failure is detected. One of these conditions occur: * The actual gear position is neutral for at least 3 seconds when 1st gear in-gear is commanded, though there is no history of being neutral when reverse gear in-gear is commanded. * The actual gear position is neutral for at least 3 seconds, though 1st gear in-gear is commanded and reverse drive occurred during this driving cycle. **NOTE: Before you troubleshoot, record all freeze data and any on-board snapshot.**
DTC: P0767 **2T PCM, MIL: Yes, TCIL: Yes** **Year:** 2011, 2012 **Model:** Accord, Crosstour **Engine:** 2.4L, 3.5L	**Shift Solenoid Valve D Stuck ON:** One of these conditions occurs: * The actual gear position is neutral for at least 3.0 seconds when 1st gear in-gear is commanded with no record of being neutral when reverse gear in-gear is commanded. * The actual gear position is neutral for at least 3.0 seconds though 1st gear in-gear is commanded and reverse drive occurred during this driving cycle.
DTC: P0771 **2T PCM, MIL: Yes, TCIL: Yes** **Year:** 2011, 2012 **Model:** Accord, Crosstour, CR-V, Element **Engine:** 2.4L	**Shift Solenoid Valve E Stuck OFF:** While driving the vehicle with the torque converter lock-up ON a maalfunction is detected when shifting from 3rd gear into 4th gear. **NOTE: Before you troubleshoot, record all freeze data and any on-board snapshot with the HDS, and review General Troubleshooting Information.**
DTC: P0776 **2T PCM, MIL: Yes, TCIL: Yes** **Year:** 2011, 2012 **Model:** Accord, Civic, Crosstour, CR-V, Element, Fit, Odyssey, Pilot, Ridgeline **Engine:** 1.5L, 1.8L, 2.4L, 3.5L	**A/T Clutch Pressure Control Solenoid Valve B Stuck OFF:** When an improper gear ratio is output compared to the predetermined gear change mode, an A/T clutch pressure control solenoid valve B OFF failure is detected. The transmission is held in 3rd gear against the 3rd-4th gear upshift command for as long as 20 seconds, with records that the gear change time was short when the 2nd-3rd gear upshift was commanded. **NOTE: Before you troubleshoot, record all freeze data and any on-board snapshot.**
DTC: P0777 **2T PCM, MIL: Yes, TCIL: Yes** **Year:** 2011, 2012 **Model:** Accord, Civic, Crosstour, CR-V, Element, Fit, Odyssey, Pilot, Ridgeline **Engine:** 1.5L, 1.8L, 2.4L, 3.5L	**A/T Clutch Pressure Control Solenoid Valve B Stuck ON:** When an improper gear ratio is output compared to the predetermined gear change mode, an A/T clutch pressure control solenoid valve B ON failure is detected. The engine speed flares during 2nd-3rd and 3rd-4th upshifts for at least 1 second. The execution time is continuous depending on the driving pattern. **NOTE: Before you troubleshoot, record all freeze data and any on-board snapshot.**

DTC	Trouble Code Title and Conditions
DTC: P0777 **2T PCM, MIL: Yes, TCIL: Yes** **Year:** 2011, 2012 **Model:** Civic, CR-Z, Insight **Engine:** 1.3L, 1.5L	**CVT Driven Pulley Pressure Control Valve Stuck ON:** With the vehicle at operating temperature, and driven for 15 seconds or more, both of the following judgement conditions must be met. Judgement A: The pulley ratio is 0.3-0.8 for at least 2.5 seconds. Judgement B: The pulley ratio is less than 3 for at least 0.2 seconds.
DTC: P0780 **2T PCM, MIL: Yes, TCIL: Yes** **Year:** 2011, 2012 **Model:** Insight **Engine:** 1.3L	**CVT Drive Pulley Pressure Control Valve Stuck ON or CVT Driven Pulley Pressure Control Valve Stuck OFF:** With the vehicle driven between 40-50 mph (65-80 km/h) on a flat road for at least 15 seconds, then stop the vehicle pressing the brake pedal for 5 seconds. Both of the following judgement malfunctions must be met. Judgement A: The pulley ratio is 2.2-2.7 for at least 0.4 second. Judgement B: The pulley ratio is 1.8 to 2.7 for at least 5 seconds.
DTC: P0780 **2T PCM, MIL: Yes, TCIL: Yes** **Year:** 2011, 2012 **Model:** Accord, Crosstour, CR-V, Element **Engine:** 2.4L, 3.5L	**Shift Control System:** This code is stored whenever DTCs P1730, P1731, P1732, P1733, and P1734 are detected. Refer to specific DTC information. Before you troubleshoot, record all freeze data and any on-board snapshot with the HDS, and review General Troubleshooting Information.
DTC: P0780/100 **2T PCM, MIL: Yes** **Year:** 2011, 2012 **Model:** Civic, CR-Z **Engine:** 1.5L	**Shift Error:** With the vehicle driven between 40-50 mph (65-80 km/h) on a flat road for at least 15 seconds, then stop the vehicle pressing the brake pedal for 5 seconds. Both of the following judgement malfunctions must be met. Judgement A: The pulley ratio is 2.2-2.7 for at least 0.4 second. Judgement B: The pulley ratio is 1.8 to 2.7 for at least 5 seconds.
DTC: P0796 **2T PCM, MIL: Yes, TCIL: Yes** **Year:** 2011, 2012 **Model:** Accord, Civic, Crosstour, CR-V, Element, Fit, Odyssey, Pilot, Ridgeline **Engine:** 1.5L, 1.8L, 2.4L, 3.5L	**A/T Clutch Pressure Control Solenoid Valve C Stuck OFF:** When an improper gear ratio is output compared to the predetermined gear ratio, an A/T clutch pressure control solenoid valve C OFF failure is detected. The execution time is continuous depending on the driving pattern. **NOTE: Before you troubleshoot, record all freeze data and any on-board snapshot.**
DTC: P0796 **2T PCM, MIL: Yes, TCIL: Yes** **Year:** 2011, 2012 **Model:** Civic, CR-Z, Insight **Engine:** 1.3L, 1.5L	**CVT Start Clutch Pressure Control Valve Stuck OFF:** With the vehicle at operating temperature, and driven for 15 seconds or more, the oil pressure of the start clutch is at specified value or more for at least 2 seconds.
DTC: P0797 **2T PCM, MIL: Yes, TCIL: Yes** **Year:** 2011, 2012 **Model:** Accord, Civic, Crosstour, CR-V, Element, Fit, Odyssey, Pilot, Ridgeline **Engine:** 1.5L, 1.8L, 2.4L, 3.5L	**A/T Clutch Pressure Control Solenoid Valve C Stuck ON:** When the wrong transmission fluid pressure switch is turned on, an A/T clutch pressure control solenoid valve C turn-on malfunction is detected. The 2nd clutch transmission fluid switch is ON against the 3rd-4th gear upshift command for at least 11 seconds. The execution time is continuous and the duration time is 20 seconds. **NOTE: Before you troubleshoot, record all freeze data and any on-board snapshot,.**
DTC: P0797 **T TCM, TCIL: Yes** **Year:** 2012 **Model:** Crosstour, CR-V **Engine:** 2.4L	**Pressure Control Solenoid Valve C (Stuck On):** **NOTE: If DTC U1000 is displayed with this DTC, first perform the trouble diagnosis for DTC U1000.** This malfunction will not be detected while the O/D OFF indicator lamp is indicating another self-diagnosis malfunction. This is not caused by electrical malfunction (circuits open or shorted) but by mechanical malfunction such as control valve sticking, improper solenoid valve operation, etc. The pressure control solenoid valve C is normally low, 3-port linear pressure control solenoid. The pressure control solenoid valve C is activated to control the apply and release of the 2nd brake and 1st and reverse brake, and torque converter clutch. Lock-up operation, however, is prohibited when A/T fluid temperature is too low. When the accelerator pedal is depressed (less than 1/8) in lock-up condition, the engine speed should not change abruptly. If there is a big jump in engine speed, there is no lock-up. This is an OBD-II self-diagnostic item. Diagnostic trouble code "PC SOL C STC ON" with CONSULT-III or P0797 without CONSULT-III is detected when condition of pressure control solenoid valve C is different from monitor value, and relation between gear position and actual gear ratio or lock-up status is irregular.

DTC	Trouble Code Title and Conditions
DTC: P0812 **2T PCM, MIL: Yes, TCIL: Yes** **Year:** 2011, 2012 **Model:** Accord, Crosstour **Engine:** 2.4L, 3.5L	**Open in Transmission Range Switch ATPR Switch Circuit:** The RVS position signal input is detected but the R position signal input is not detected.
DTC: P0812 **2T PCM, MIL: Yes, TCIL: Yes** **Year:** 2011, 2012 **Model:** Civic, Crosstour, Fit, Odyssey, Pilot, Ridgeline **Engine:** 1.5L, 1.8L, 2.4L, 3.5L	**Open in Transmission Range Switch ATP RVS Switch Circuit:** If the R switch is OPEN with the shift lever in R, the PCM detects a switch OPEN failure. The RVS signal is detected but the R switch signal is not detected for at least 2 seconds. The execution time is continuous depending on the driving pattern. **NOTE: This code is caused by an electrical circuit problem and cannot be caused by a mechanical problem in the transmission.**
DTC: P0842 **1T PCM, MIL: Yes, TCIL: Yes** **Year:** 2011, 2012 **Model:** Accord, Civic, Crosstour, CR-V, Element, Fit, Odyssey, Pilot, Ridgeline **Engine:** 1.5L, 1.8L, 2.4L, 3.5L	**Short in 2nd Clutch Transmission Fluid Pressure Switch Circuit, or 2nd Clutch Transmission Fluid Pressure Switch Stuck ON:** The input signal from the 2nd clutch transmission fluid pressure switch to the PCM is low when driving in 1st gear, 3rd gear, or 5th gear. The execution time is continuous and the duration time is 2 seconds or more. **NOTE: This code is caused by an electrical circuit problem and cannot be caused by a mechanical problem in the transmission.**
DTC: P0842 **2T PCM** **Year:** 2011, 2012 **Model:** Insight **Engine:** 1.3L	**CVT Driven Pulley Pressure Sensor (Short):** **NOTE: Before you troubleshoot, record all freeze data and any on-board snapshot with the HDS, and review the General Troubleshooting Information.** This code is caused by an electrical circuit problem and cannot be caused by a mechanical problem in the transmission. (D indicator blinks)
DTC: P0843 **2T PCM** **Year:** 2011, 2012 **Model:** Insight **Engine:** 1.3L	**CVT Driven Pulley Pressure Sensor (Open):** The CVT driven pulley pressure sensor output voltage is more than 4.84 V for at least 10 seconds. (D indicator blinks)
DTC: P0843 **1T PCM, MIL: Yes, TCIL: Yes** **Year:** 2011, 2012 **Model:** Accord, Civic, Crosstour, CR-V, Element, Fit, Odyssey, Pilot, Ridgeline **Engine:** 1.5L, 1.8L, 2.4L, 3.5L	**Open in 2nd Clutch Transmission Fluid Pressure Switch Circuit, or 2nd Clutch Transmission Fluid Pressure Switch Stuck OFF :** The input signal from the 2nd clutch transmission fluid pressure switch to the PCM is high when driving in 2nd gear. The execution time is continuous and the duration time is 2 seconds or more. **NOTE: Before you troubleshoot, record all freeze data and any on-board snapshot.**
DTC: P0847 **1T PCM, MIL: Yes, TCIL: Yes** **Year:** 2011, 2012 **Model:** Accord, Civic, Crosstour, CR-V, Element, Fit, Odyssey, Pilot, Ridgeline **Engine:** 1.5L, 1.8L, 2.4L, 3.5L	**Short in 3rd Clutch Transmission Fluid Pressure Switch Circuit, or 3rd Clutch Transmission Fluid Pressure Switch Stuck ON:** The input signal from the 3rd clutch transmission fluid pressure switch to the PCM is low when driving in 1st gear, 2nd gear, 4th gear, or 5th gear. The execution time is continuous and the duration time is 2 seconds or more. **NOTE: This code is caused by an electrical circuit problem and cannot be caused by a mechanical problem in the transmission.**
DTC: P0848 **1T PCM, MIL: Yes, TCIL: Yes** **Year:** 2011, 2012 **Model:** Accord, Civic, Crosstour, CR-V, Element, Fit, Odyssey, Pilot, Ridgeline **Engine:** 1.5L, 1.8L, 2.4L, 3.5L	**Open in 3rd Clutch Transmission Fluid Pressure Switch Circuit, or 3rd Clutch Transmission Fluid Pressure Switch Stuck OFF :** The input signal from the 3rd clutch transmission fluid pressure switch to the PCM is high when driving in 3rd gear. The execution time is continuous and the duration time is 2 seconds or more. **NOTE: Before you troubleshoot, record all freeze data and any on-board snapshot.**
DTC: P0872 **1T PCM, MIL: Yes, TCIL: Yes** **Year:** 2011, 2012 **Model:** Accord, Crosstour, Odyssey, Pilot, Ridgeline **Engine:** 2.4L, 3.5L	**Short in 4th Clutch Transmission Fluid Pressure Switch Circuit, or 4th Clutch Transmission Fluid Pressure Switch Stuck ON :** The input signal from the 4th clutch transmission fluid pressure switch to the PCM is low when driving in 5th gear. The execution time is continuous and the duration time is 2 seconds or more. **NOTE: This code is caused by an electrical circuit problem and cannot be caused by a mechanical problem in the transmission.**

DTC	Trouble Code Title and Conditions
DTC: P0873 **1T PCM, MIL: Yes, TCIL: Yes** **Year:** 2011, 2012 **Model:** Accord, Crosstour, Odyssey, Pilot, Ridgeline **Engine:** 2.4L, 3.5L	**Open in 4th Clutch Transmission Fluid Pressure Switch Circuit, or 4th Clutch Transmission Fluid Pressure Switch Stuck OFF:** The input signal from the 4th clutch transmission fluid pressure switch to the PCM is high when driving in 4th gear. The execution time is continuous and the duration time is 2 seconds or more. **NOTE: Before you troubleshoot, record all freeze data and any on-board snapshot.**
DTC: P0877 **2T PCM, MIL: Yes, TCIL: Yes** **Year:** 2011, 2012 **Model:** Odyssey **Engine:** 3.5L	**Short in Transmission Fluid Pressure Switch D (5th Clutch) Circuit, or Transmission Fluid Pressure Switch D (5th Clutch) Stuck ON:** One of these 3 conditions occurs: (1) The input signal from transmission fluid pressure switch D (5th clutch) to the PCM is low for at least 5 seconds when the shift lever is in P, R, or N. (2) The input signal from transmission fluid pressure switch D (5th clutch) to the PCM is low for at least 5 seconds when driving in 1st or 2nd gear. (3) The input signal from transmission fluid pressure switch D (5th clutch) to the PCM is low for at least 5 seconds when the 2nd gear shift is commanded or when the shift lever is in P, R, or N, after the signal is low for at least 3 seconds when driving in 3rd, 4th, or 6th gear. **NOTE: Before you troubleshoot, record all freeze data and any on-board snapshot. This code is caused by an electrical circuit problem and cannot be caused by a mechanical problem in the transmission.**
DTC: P0878 **2T PCM, MIL: Yes, TCIL: Yes** **Year:** 2011, 2012 **Model:** Odyssey **Engine:** 3.5L	**Open in Transmission Fluid Pressure Switch D (5th Clutch) Circuit, or Transmission Fluid Pressure Switch D (5th Clutch) Stuck OFF:** The input signal from transmission fluid pressure switch D (5th clutch) to the PCM is high when driving in 5th gear. The execution time is continuous and the duration time is 4 seconds or more. **NOTE: Before you troubleshoot, record all freeze data and any on-board snapshot. If DTC P0878 is stored in the PCM, the transmission does not shift to any gear other than 2nd through 5th gear because of the fail-safe function**
DTC: P0962 **1T PCM, MIL: Yes, TCIL: Yes** **Year:** 2011, 2012 **Model:** Accord, Civic, Crosstour, CR-V, Element, Fit, Odyssey, Pilot, Ridgeline **Engine:** 1.5L, 1.8L, 2.4L, 3.5L	**Problem in A/T Clutch Pressure Control Solenoid Valve A Circuit:** If the measured current for the PCM output duty cycle is not within a specified range (open or short), a malfunction's detected. The execution time is continuous and the duration time is 2 seconds or more. The measured current for the PCM command value is Duty 57-89 %, current less than 0.2, low input. **NOTE: This code is caused by an electrical circuit problem and cannot be caused by a mechanical problem in the transmission.**
DTC: P0962 **1T PCM, MIL: Yes, TCIL: Yes** **Year:** 2011, 2012 **Model:** Civic, CR-Z, Insight **Engine:** 1.3L, 1.5L	**CVT Drive Pulley Pressure Control Valve Circuit Low Voltage (Hybrid Models):** With the the engine running for at least 1 second and battery voltage a minimum of 10.0 V, the measured current for the PCM's command value is as follows. Duty % 56.5-90.2, Current A less than 0.27, Failure mode (Low Input)
DTC: P0963 **1T PCM, MIL: Yes, TCIL: Yes** **Year:** 2011, 2012 **Model:** Civic, CR-Z, Insight **Engine:** 1.3L, 1.5L	**CVT Drive Pulley Pressure Control Valve Circuit High Voltage:** With the engine running, battery voltage a minimum of 10.0 V, the measured current for the PCM's command is less than 0.57-0.89 (High input) for at least 1 second. **NOTE: This code is caused by an electrical circuit problem and cannot be caused by a mechanical problem in the transmission.**
DTC: P0963 **1T PCM, MIL: Yes, TCIL: Yes** **Year:** 2011, 2012 **Model:** Accord, Civic, Crosstour, CR-V, Element, Fit, Odyssey, Pilot, Ridgeline **Engine:** 1.5L, 1.8L, 2.4L, 3.5L	**Problem in A/T Clutch Pressure Control Solenoid Valve A:** Duty cycle is less than 13-27 %, current (A) 0.6-0.9, high input failure. The execution time is continuous and the duration time is 2 seconds or more. DTCs P0962, P0966, P0967, P0970, P0971 are not active. **Note: This code is caused by an electrical circuit problem and cannot be caused by a mechanical problem in the transmission.**
DTC: P0966 **1T PCM, MIL: Yes, TCIL: Yes** **Year:** 2011, 2012 **Model:** Civic, CR-Z, Insight **Engine:** 1.3L, 1.5L, 1.8L	**CVT Driven Pulley Pressure Control Valve Circuit Low Voltage:** With the engine running, battery voltage a minimum of 10.0 V, the measured current for the PCM's command is less than 0.19-0.27 (Low Input) for at least 1 second. **NOTE: This code is caused by an electrical circuit problem and cannot be caused by a mechanical problem in the transmission.**
DTC: P0966 **1T PCM, MIL: Yes, TCIL: Yes** **Year:** 2011, 2012 **Model:** Accord, Civic, Crosstour, CR-V, Element, Fit, Odyssey, Pilot, Ridgeline **Engine:** 1.5L, 1.8L, 2.4L, 3.5L	**Problem in A/T Clutch Pressure Control Solenoid Valve B Circuit:** The measured current is less than 0.2, duty cycle 57-89 %, low input. The execution time is continuous and the duration time is 1 seconds or more. **Note: This code is caused by an electrical circuit problem and cannot be caused by a mechanical problem in the transmission.**

DTC	Trouble Code Title and Conditions
DTC: P0967 **1T PCM, MIL: Yes, TCIL: Yes** **Year:** 2011, 2012 **Model:** Civic, CR-Z, Insight **Engine:** 1.3L, 1.5L	**CVT Driven Pulley Pressure Control Valve Circuit High Voltage:** With the engine running, battery voltage a minimum of 10.0 V, the measured current for the PCM's command is less than 0.57-0.89 (High Input) for at least 1 second. **NOTE: This code is caused by an electrical circuit problem and cannot be caused by a mechanical problem in the transmission.**
DTC: P0967 **1T PCM, MIL: Yes, TCIL: Yes** **Year:** 2011, 2012 **Model:** Accord, Civic, Crosstour, CR-V, Element, Fit, Odyssey, Pilot, Ridgeline **Engine:** 1.5L, 1.8L, 2.4L, 3.5L	**Problem in A/T Clutch Pressure Control Solenoid Valve B:** The measured current for the PCMs command is 0.6-0.9, duty cycle 13-27%, high input failure. Engine running, DTCs P0962, P0963, P0966, P0970, P0971 are not active. The execution time is continuous and the duration time is 1 seconds or more. **NOTE: This code is caused by an electrical circuit problem and cannot be caused by a mechanical problem in the transmission.**
DTC: P0970 **1T PCM, MIL: Yes, TCIL: Yes** **Year:** 2011, 2012 **Model:** Civic, CR-Z, Insight **Engine:** 1.3L, 1.5L	**CVT Start Clutch Pressure Control Valve Circuit Low Voltage:** With the engine running, battery voltage a minimum of 10.0 V, the measured current for the PCM's command is less than 0.19-0.27 (Low Input) for at least 1 second. **NOTE: This code is caused by an electrical circuit problem and cannot be caused by a mechanical problem in the transmission.**
DTC: P0970 **1T PCM, MIL: Yes, TCIL: Yes** **Year:** 2011, 2012 **Model:** Accord, Civic, Crosstour, CR-V, Element, Fit, Odyssey, Pilot, Ridgeline **Engine:** 1.5L, 1.8L, 2.4L, 3.5L	**Problem in A/T Clutch Pressure Control Solenoid Valve C Circuit:** The measured current for the PCMs command value is 0.2-0.4, duty cycle 57-89%, low imput failure. Engine is running with battery voltage at 11 volts, DTCs P0962, P0963, P0966, P0967, P0971 are not active. The execution time is continuous and the duration time is 1 seconds or more. **NOTE: This code is caused by an electrical circuit problem and cannot be caused by a mechanical problem in the transmission.**
DTC: P0971 **1T PCM, MIL: Yes, TCIL: Yes** **Year:** 2011, 2012 **Model:** Accord, Civic, Crosstour, CR-V, Element, Fit, Odyssey, Pilot, Ridgeline **Engine:** 1.5L, 1.8L, 2.4L, 3.5L	**Problem in A/T Clutch Pressure Control Solenoid Valve C:** The measured current for the PCMs command value is 0.6-0.9, duty cycle 13-27, high imput failure. The execution time is continuous and the duration time is 1 seconds or more. **NOTE: This code is caused by an electrical circuit problem and cannot be caused by a mechanical problem in the transmission.**
DTC: P0971 **1T PCM, MIL: Yes, TCIL: Yes** **Year:** 2011, 2012 **Model:** Civic, CR-Z, Insight **Engine:** 1.3L, 1.5L	**CVT Start Clutch Pressure Control Valve Circuit High Voltage:** With the engine running, battery voltage a minimum of 10.0 V, the measured current for the PCM's command is less than 0.57-0.89 (High Input) for at least 1 second. **NOTE: This code is caused by an electrical circuit problem and cannot be caused by a mechanical problem in the transmission.**
DTC: P0973 **1T PCM, MIL: Yes, TCIL: Yes** **Year:** 2011, 2012 **Model:** Accord, Civic, Crosstour, CR-V, Element, Fit, Odyssey, Pilot, Ridgeline **Engine:** 1.5L, 1.8L, 2.4L, 3.5L	**Short in Shift Solenoid Valve A Circuit:** The return signal does not match the command to turn ON shift solenoid valve A for at least 1 second. The execution time is continuous and the duration time is 2 seconds or more. DTCs P0974, P0982, P0983 are not active. **NOTE: This code is caused by an electrical circuit problem and cannot be caused by a mechanical problem in the transmission.**
DTC: P0974 **1T PCM, MIL: Yes, TCIL: Yes** **Year:** 2011, 2012 **Model:** Accord, Civic, Crosstour, CR-V, Element, Fit, Odyssey, Pilot, Ridgeline **Engine:** 1.5L, 1.8L, 2.4L, 3.5L	**Open in Shift Solenoid Valve A Circuit:** The return signal does not match the command to turn OFF shift solenoid valve A for at least 1 second. The execution time is continuous and the duration time is 1 seconds or more. DTCs P0973, P0982, P0983 are not active. **NOTE: This code is caused by an electrical circuit problem and cannot be caused by a mechanical problem in the transmission.**
DTC: P0976 **1T PCM, MIL: Yes, TCIL: Yes** **Year:** 2011, 2012 **Model:** Accord, Civic, Crosstour, CR-V, Element, Fit, Odyssey, Pilot, Ridgeline **Engine:** 1.5L, 1.8L, 2.4L, 3.5L	**Short in Shift Solenoid Valve B Circuit:** The return signal does not match the command to turn ON shift solenoid valve B for at least 1 second. The execution time is continuous and the duration time is 2 seconds or more. DTCs P0977, P0979, P0980 are not active. **NOTE: This code is caused by an electrical circuit problem and cannot be caused by a mechanical problem in the transmission.**

DTC	Trouble Code Title and Conditions
DTC: P0977 **1T PCM, MIL: Yes, TCIL: Yes** **Year:** 2011, 2012 **Model:** Accord, Civic, Crosstour, CR-V, Element, Fit, Odyssey, Pilot, Ridgeline **Engine:** 1.5L, 1.8L, 2.4L, 3.5L	**Open in Shift Solenoid Valve B Circuit:** The return signal does not match the command to turn OFF shift solenoid valve B for at least 1 second. The execution time is continuous and the duration time is 1 seconds or more. Battery voltage minimum 11 volts, and DTCs P0976, P0979, P0980 are not active. **NOTE: This code is caused by an electrical circuit problem and cannot be caused by a mechanical problem in the transmission.**
DTC: P0979 **1T PCM, MIL: Yes, TCIL: Yes** **Year:** 2011, 2012 **Model:** Accord, Civic, Crosstour, CR-V, Element, Fit, Odyssey, Pilot, Ridgeline **Engine:** 1.5L, 1.8L, 2.4L, 3.5L	**Short in Shift Solenoid Valve C Circuit:** The return signal does not match the command to turn ON shift solenoid valve C for at least 1 second. The execution time is continuous and the duration time is 1 seconds or more. Battery voltage is 11 volts and DTCs P0976, P0977, P0980 are not active. **NOTE: This code is caused by an electrical circuit problem and cannot be caused by a mechanical problem in the transmission.**
DTC: P0980 **1T PCM, MIL: Yes, TCIL: Yes** **Year:** 2011, 2012 **Model:** Accord, Civic, Crosstour, CR-V, Element, Fit, Odyssey, Pilot, Ridgeline **Engine:** 1.5L, 1.8L, 2.4L, 3.5L	**Open in Shift Solenoid Valve C Circuit:** The return signal does not match the command to turn OFF shift solenoid valve C for at least 1 second. The execution time is continuous and the duration time is 1 seconds or more. Battery voltage is 11 V and DTCs P0976, P0977, P0979 are not active. **NOTE: This code is caused by an electrical circuit problem and cannot be caused by a mechanical problem in the transmission.**
DTC: P0982 **1T PCM, MIL: Yes, TCIL: Yes** **Year:** 2011, 2012 **Model:** Accord, Civic, Crosstour, CR-V, Element, Fit, Odyssey, Pilot, Ridgeline **Engine:** 1.5L, 1.8L, 2.4L, 3.5L	**Short in Shift Solenoid Valve D Circuit:** The return signal does not match the command to turn ON shift solenoid valve D for at least 1 second. The execution time is continuous and the duration time is 1 seconds or more. Battery voltage is 11v, and DTCs P0973, P0974, P0983 are not active. **NOTE: This code is caused by an electrical circuit problem and cannot be caused by a mechanical problem in the transmission.**
DTC: P0983 **1T PCM, MIL: Yes, TCIL: Yes** **Year:** 2011, 2012 **Model:** Accord, Civic, Crosstour, CR-V, Element, Fit, Odyssey, Pilot, Ridgeline **Engine:** 1.5L, 1.8L, 2.4L, 3.5L	**Open in Shift Solenoid Valve D Circuit:** The return signal does not match the command to turn OFF shift solenoid valve D for at least 1 second. The execution time is continuous and the duration time is seconds or more. Battery voltage is 11v and DTCs P0973, P0974, P0982 are not active. **NOTE: This code is caused by an electrical circuit problem and cannot be caused by a mechanical problem in the transmission.**
DTC: P0985 **1T PCM, MIL: Yes, TCIL: Yes** **Year:** 2011, 2012 **Model:** Accord, Crosstour, CR-V, Element **Engine:** 2.4L	**Short in Shift Solenoid Valve E Circuit:** With the engine started and in park the return signal does not match the command to turn ON shift solenoid valve E for at least 1 second. **NOTE: Before you troubleshoot, record all freeze data and any on-board snapshot with the HDS, and review General Troubleshooting Information. This code is caused by an electrical circuit problem and cannot be caused by a mechanical problem in the transmission.**
DTC: P0986 **1T PCM, MIL: Yes, TCIL: Yes** **Year:** 2011, 2012 **Model:** Accord, Crosstour **Engine:** 2.4L, 3.5L	**Open in Shift Solenoid Valve E Circuit:** The return signal does not match the command to turn OFF shift solenoid valve E for at least 1 second.
DTC: P0986 **1T PCM, MIL: Yes, TCIL: Yes** **Year:** 2011, 2012 **Model:** Crosstour, CR-V, Element **Engine:** 2.4L	**Open in Shift Solenoid Valve E Circuit:** With the engine running and driven in 2nd gear in the D position, the return signal does not match the command to turn OFF shift solenoid E for at least 1 second. **NOTE: Before you troubleshoot, record all freeze data and any on-board snapshot with the HDS, and review General Troubleshooting Information. This code is caused by an electrical circuit problem and cannot be caused by a mechanical problem in the transmission.**

DTC	Trouble Code Title and Conditions
DTC: P0989 **2T PCM, MIL: Yes, TCIL: Yes** **Year:** 2011, 2012 **Model:** Odyssey **Engine:** 3.5L	**Short in Transmission Fluid Pressure Switch E (6th Clutch) Circuit, or Transmission Fluid Pressure Switch E (6th Clutch) Stuck ON:** One of these 3 conditions occurs: (1) The input signal from transmission fluid pressure switch E (6th clutch) to the PCM is low for at least 4 seconds when the shift lever is in P, R, or N. (2) The input signal from transmission fluid pressure switch E (6th clutch) to the PCM is low for at least 4 seconds when driving in 1st, 2nd, or 3rd gear. (3) The input signal from transmission fluid pressure switch E (6th clutch) to the PCM is low for at least 4 seconds when the 3rd gear shift is commanded or when the shift lever is in P, R, or N, after the signal is low for at least 2 seconds when driving in 4th or 5th gear. **NOTE: Before you troubleshoot, record all freeze data and any on-board snapshot. This code is caused by an electrical circuit problem and cannot be caused by a mechanical problem in the transmission.**
DTC: P0990 **2T PCM, MIL: Yes, TCIL: Yes** **Year:** 2011, 2012 **Model:** Odyssey **Engine:** 3.5L	**Open in Transmission Fluid Pressure Switch E (6th Clutch) Circuit, or Transmission Fluid Pressure Switch E (6th Clutch) Stuck OFF:** The input signal from transmission fluid pressure switch E (6th clutch) to the PCM is high when driving in 6th gear. **NOTE: Before you troubleshoot, record all freeze data and any on-board snapshot. If DTC P0990 is stored in the PCM, the transmission does not shift to any gear other than 3rd or 6th gear because of the fail-safe function.**
DTC: P0A14 **1T PCM** **Year:** 2011, 2012 **Model:** Accord, Crosstour, Odyssey, Pilot **Engine:** 3.5L	**Front Engine Mount Actuator Circuit Malfunction:** Engine running, and DTCs P0A15, P0AB6, P0AB7, P16C4 and P16C6 are not active. The current valve during one cycle of the front engine mount actuator, which outputs in 1 TDC duration as one cycle, is 7.5 A or more as a maximum current or 5.3 A or more as a minimum current, or the engine mounts cycle during start up and continue to cycle for at least 20 seconds. **NOTE: When testing the engine mount actuator, be sure to use a VOM capable of reading 0.01 ohms change of resistance. If the vehicle has been driven, allow the engine mount actuator to cool down for about 2 hours before testing.**
DTC: P0A15 **1T PCM** **Year:** 2011, 2012 **Model:** Accord, Crosstour, Odyssey, Pilot **Engine:** 3.5L	**Front Engine Mount Actuator Control Circuit Low Current:** With the engine running and the DTCs P0A14, P0AB6, P0AB7, P15AB, P16C4 and P16C6 are not active. The duty cycle of the front engine mount actuator, which outputs in 1 TDC duration as one cycle, reads 100% all the time during engine the mount operation and continues at least 20 seconds. **NOTE: When testing the engine mount actuator, be sure to use a VOM capable of reading 0.01 ohms change of resistance. If the vehicle has been driven, allow the engine mount actuator to cool down for about 2 hours before testing.**
DTC: P0A16 **1T PCM** **Year:** 2011, 2012 **Model:** Pilot **Engine:** 3.5L	**Front Engine Mount Actuator Control Circuit High Current:** With the ignition switch ON and DTCs P15AF, P15B0, P15B1, P16C5, P16C9 not active. The minimum front ACM actuator current in one cycle is 6.6 A or more for at least 2.6 seconds, the front ACM actuator output cycle is one rotation of the crankshaft.
DTC: P0A1B/84 **1T PCM, MIL: Yes** **Year:** 2011, 2012 **Model:** Civic, CR-Z, Insight **Engine:** 1.3L, 1.5L	**Motor Control Module (MCM) Internal Circuit Malfunction (HYBRID MODELS):** With the ignition switch ON, the communication information can not be received, or the received data malfunction for at least 0.5 seconds. P16C1 and P16C2 are not active. WARNING: The IMA system uses high voltage (144 V), before servicing turn OFF the battery module switch.
DTC: P0A1F/112 **1T BCM, MIL: Yes** **Year:** 2011 **Model:** Civic **Engine:** 1.3L	**F-CAN Malfunction (BCM Module-MCM) (HYBRID MODELS):** With the ignition ON, MCM supply-power voltage a minimum of 10.0 V, and the elapsed time after the ignition switch is tuned for 3 seconds. One of the following conditions occurs. (IMA system indicator is ON) Condition 1: No information is sent from the BCM for at least 0.5 seconds. Condition 2: The information sent from the BCM is abnormal at least 10 times. WARNING: The IMA system uses high voltage (144 V), before servicing turn OFF the battery module switch.
DTC: P0A27/46 **1T** **Year:** 2011, 2012 **Model:** Civic, CR-Z, Insight **Engine:** 1.3L, 1.5L	**High Voltage Contactor/Bypass Contactor Stays Activated (HYBRID MODELS):** When the ignition switch is turned OFF and MCM power-supply voltage is a minimum of 10.5 V, IMA motor speed 300 rpm, MPI module voltage is between 90–240 V. The difference of the MPI module voltage before the high voltage contactor opening process and the MPI voltage after the high voltage contactor opening process is 3 V or less after at least 10 seconds. WARNING: The IMA system uses high voltage (144 V), before servicing turn OFF the battery module switch.
DTC: P0A3C/39 **1T , MIL: Yes** **Year:** 2011, 2012 **Model:** Civic, CR-Z **Engine:** 1.3L, 1.5L	**Motor Control Module (MCM) Overheating (HYBRID MODELS):** After the ignition switch has been turned ON (II) for 2 seconds, the following conditions occurs. (IMA system indicator is ON) Condition 1: The MPI module temperature is 230 °F (110 °C) or more. Condition 2: The MCM internal temperature is 257 °F (125 °C) or more. WARNING: The IMA system uses high voltage (144 V), before servicing turn OFF the battery module switch.

DTC	Trouble Code Title and Conditions
DTC: P0A3C/39 **1T PCM, MIL: Yes** **Year:** 2011, 2012 **Model:** Insight **Engine:** 1.3L	**Motor Power Inverter (MPI) Module Overheating (HYBRID MODEL):** With the ignition switch ON and MCM power-supply voltage a minimum of 10.5 V, the overheating information from the MPI module is transmitted for at least 0.5 seconds. (IMA system indicator is ON) WARNING: The IMA system uses high voltage (144 V), before servicing turn OFF the battery module switch.
DTC: P0A3F/89 **1T PCM, MIL: Yes** **Year:** 2011, 2012 **Model:** Civic, CR-Z, Insight **Engine:** 1.3L, 1.5L	**Motor Rotor Position Sensor Circuit Malfunction (HYBRID MODELS):** With the ignition switch ON and battery voltage at a minimum of 10.0 V, one of the following symptoms can occur. Symptom 1: An abnormal signal is output from the R/D converter for at least 0.4 seconds within 0.5 seconds. Symptom 2: The output voltage to the motor rotor position sensor is 0.1 V or less for at least 0.5 seconds. Symptom 3: The input voltage to the motor rotor position sensor is 1.2 V or less for at least 0.5 seconds. Symptom 4: The deviation of the IMA motor speed and the engine speed is 500 rpm or more for at least 2 seconds. (IMA system indicator is ON) WARNING: The IMA system uses high voltage (144 V), before servicing turn OFF the battery module switch.
DTC: P0A5E/24 **1T PCM, MIL: Yes** **Year:** 2011, 2012 **Model:** Civic, CR-Z, Insight **Engine:** 1.3L, 1.5L	**U Phase Motor Current Sensor Circuit Low Voltage (HYBRID MODELS):** With the ignition ON and the MCM power-supply voltage a minimum of 9.0 V, the MCM input voltage from the U phase motor current sensor is 0.156 V or less for at least o.5 seconds. P0A5F, P16C1, P16C2 and U1203 are not active. (IMA system indicator is ON) WARNING: The IMA system uses high voltage (144 V), before servicing turn OFF the battery module switch.
DTC: P0A5F/25 **1T PCM, MIL: Yes** **Year:** 2011, 2012 **Model:** Civic, CR-Z, Insight **Engine:** 1.3L, 1.5L	**U Phase Motor Current Sensor Circuit High Voltage (HYBRID MODELS):** With the ignition ON and MCM power-supply voltage a minimum of 9.0 V, the MCM input voltage from the U phase motor current sensor is 4.848 V or more for at least 0.5 seconds. P0A5E, P16C1, P16C2 and U1203 are not active. (IMA system indicator is ON) WARNING: The IMA system uses high voltage (144 V), before servicing turn OFF the battery module switch.
DTC: P0A61/26 **1T PCM, MIL: Yes** **Year:** 2011, 2012 **Model:** Civic, CR-Z, Insight **Engine:** 1.3L, 1.5L	**V Phase Motor Current Sensor Circuit Low Voltage (HYBRID MODELS):** With the ignition switch ON and MCM power-supply voltage is a minimum of 9.0 V, the MCM input voltage from the V phase motor current sensor is 0.156 V or less for at least 0.5 seconds. (IMA system indicator is ON) WARNING: The IMA system uses high voltage (144 V), before servicing turn OFF the battery module switch.
DTC: P0A62/27 **1T PCM, MIL: Yes** **Year:** 2011, 2012 **Model:** Civic, CR-Z, Insight **Engine:** 1.3L, 1.5L	**V Phase Motor Current Sensor Circuit High Voltage (HYBRID MODELS):** With the ignition ON and the MCM power-supply voltage a minimum of 9.0 V, the MCM input voltage from the V phase motor current sensor is 4.848 V or more for at least 0.5 seconds. (IMA system indicator is ON) WARNING: The IMA system uses high voltage (144 V), before servicing turn OFF the battery module switch.
DTC: P0A64/28 **1T PCM, MIL: Yes** **Year:** 2011, 2012 **Model:** Civic, CR-Z, Insight **Engine:** 1.3L, 1.5L	**W Phase Motor Current Sensor Circuit Low Voltage (HYBRID MODELS):** With the ignition ON and the MCM power-supply voltage a minimum of 9.0 V, the MCM input voltage from the W phase motor current sensor is 0.156 V or less for at least 0.5 seconds. (IMA system indicator is ON) WARNING: The IMA system uses high voltage (144 V), before servicing turn OFF the battery module switch.
DTC: P0A65/29 **1T PCM, MIL: Yes** **Year:** 2011, 2012 **Model:** Civic, CR-Z, Insight **Engine:** 1.3L, 1.5L	**W Phase Motor Current Sensor Circuit High Voltage (HYBRID MODELS):** With the ignition ON and the MCM supply-power voltage a minimum of 9.0 V, the MCM input voltage from the W phase motor current sensor is 4.848 V or more for at least 0.5 seconds. (IMA system indicator is ON) WARNING: The IMA system uses high voltage (144 V), before servicing turn OFF the battery module switch.

DTC	Trouble Code Title and Conditions
DTC: P0A78/32 **1T , MIL: Yes** **Year:** 2011 **Model:** Civic **Engine:** 1.3L	**Motor Control Module (MCM) Internal Circuit Malfunction (HYBRID MODELS):** With the ignition ON, battery voltage a minimum of 7.1 V. One of the following symtoms occurs. (IMA system indicator is ON) Symptom 1: Information is not sent from the motor control sub processor for at least 500 milliseconds, or information sent from the motor control subprocessor includes abnormal condition information for at least 0.5 second. Symptom 2: (one of the following conditions occurs) * The power supply circuit voltage (L=8V) is 4.30 V or more, or 2.47 V or less, for at least 40 milliseconds, or its deviation is 1.07 V or more for at least 40 milliseconds. * The constant curreny circuit detection voltage is 0.91 V or more, or 0.53 V or less, for at least 40 milliseconds, or its deviation is 1.07 V or more for at least 40 milliseconds. * The constant current circuit reference voltage is 0.84 V or more, or 0.6 V or less, for at least 40 milliseconds, or its deviation is 1.07 V or more for at least 40 milliseconds. * The platinum sensor disconnection threshold value is 4.66 V or more, or 1.8 V or less, for at least 0.8 second. Symptom 3: (one of the following conditions occur) * The power suppl circuit voltage (+8V0 is 3.78 V or more, or 2.2 V or less, for at least 40 milliseconds, or its deviation is 1.07 V or more for at least 40 milliseconds. * The over current upper limit threshold voltage is 0.5 V or more, or 3.47 V or less, for at least 40 milliseconds, or its deviation is 1.07 V or more for at least 40 milliseconds. Symptom 4: The signal to the CAN controller and the return signal differ at least three times. Symptom 5: The over voltage notification is input three times in 25 seconds even though the control voltage is 235 V or less.
DTC: P0A78/32 **1T ECM, MIL: Yes** **Year:** 2011, 2012 **Model:** Civic, CR-Z, Insight **Engine:** 1.3L, 1.5L	**Motor Power Inverter (MPI) Module Internal Circuit Malfunction:** With the ignition ON and the MCM power-supply voltage a minimum of 7.5 V, the MCM receives the nonvolatile memory failure, the internal temperature sensor failure, or the IGBT gate over voltage failure for at least 0.5 second. (IMA system indicator is ON) WARNING: The IMA system uses high voltage (144 V), before servicing turn OFF the battery module switch.
DTC: P0A7E/72 **1T PCM, MIL: Yes** **Year:** 2011, 2012 **Model:** Civic, CR-Z, Insight **Engine:** 1.3L, 1.5L	**Battery Module Overheating:** With the ignition ON and the MCM supply-power voltage a minimum of 10.6 V, the battery module temperature is 167 ºF (75 ºC) or more for at least 2 seconds. (IMA system indicator is ON) WARNING: The IMA system uses high voltage (144 V), before servicing turn OFF the
DTC: P0A7F/78 **1T PCM, MIL: Yes** **Year:** 2011, 2012 **Model:** Civic, CR-Z, Insight **Engine:** 1.3L, 1.5L	**Battery Module Deterioration (HYBRID MODELS):** With the ignition ON, battery temperature between 32-113 ºF (O-45 ºC) and the MCM power-supply a minimum of 10.5 V, the battery capacity is less than 10% of 6.5 Ah. (IMA system indicator is ON) WARNING: The IMA system uses high voltage (144 V), before servicing turn OFF the battery module switch.
DTC: P0A94 **T PCM** **Year:** 2011, 2012 **Model:** Civic, CR-Z **Engine:** 1.5L	**DC-DC Converter Output Low Voltage:** Before you troubleshoot, record all freeze data, and any on-board snapshot, and review the general troubleshooting information. **NOTE: If the DTC cannot be cleared, do the troubleshooting for DTC U1204.**
DTC: P0A94/48 **1T** **Year:** 2011, 2012 **Model:** Insight **Engine:** 1.3L	**DC-DC Converter Output Low Voltage :** With the ignition ON, the MCM supply-power voltage a minimum of 10.5 V and the MPI module voltage is between 135-192 V, the difference between the DC-DC converter output voltage and the 12 V battery voltage is 2.1 V or more for at least 60 seconds. (IMA system indicator is OFF) WARNING: The IMA system uses high voltage (144 V), before servicing turn OFF the battery module switch.
DTC: P0A9D/49 **1T PCM, MIL: Yes** **Year:** 2011, 2012 **Model:** Civic, CR-Z, Insight **Engine:** 1.3L, 1.5L	**Battery Module Temperature Sensor 1 Circuit Low Voltage :** With the ignition ON and the MCM supply-power voltage a minimum of 9.0 V, the battery module temperature sensor is less than 0.05 V at least 2 seconds. (IMA system indicator is ON) WARNING: The IMA system uses high voltage (144 V), before servicing turn OFF the battery module switch.
DTC: P0A9E/50 **1T PCM, MIL: Yes** **Year:** 2011, 2012 **Model:** Civic, CR-Z, Insight **Engine:** 1.3L, 1.5L	**Battery Module Temperature Sensor 1 Circuit High Voltage (HYBRID MODELS):** With the ignition ON and the MCM supply-power voltage a minimum of 9.0 V, the battery module temperature sensor 1 output voltage is more than 4.95 V for at least 2 seconds. (IMA system indicator is ON) WARNING: The IMA system uses high voltage (144 V), before servicing turn OFF the battery module switch.

DTC	Trouble Code Title and Conditions
DTC: P0AA6/59 **1T PCM** **Year:** 2011, 2012 **Model:** Civic, CR-Z, Insight **Engine:** 1.3L, 1.5L	**High Voltage Short Circuit :** With the ignition ON, MCM supply-power voltage a minimum of 10.5 V, and the MPI module voltage 120-192 V the resistance value between the high voltage circuit resistance and the body ground is 100 k ohms or less for at least 20 seconds. (IMA system indicator is ON) WARNING: The IMA system uses high voltage (144 V), before servicing turn OFF the battery module switch. **NOTE: If the wrong A/C compressor oil was used in the A/C system, P0AA6 may be stored.**
DTC: P0AA7/76 **1T PCM** **Year:** 2011, 2012 **Model:** Civic, CR-Z, Insight **Engine:** 1.3L, 1.5L	**Motor Control Module (MCM) Internal Circuit Malfunction (HYBRID MODELS):** With the ignition ON and the MCM supply-power voltage a minimum of 9.0 V, the photoMOS relay check is not finished normally. (IMA system indicator is ON) WARNING: The IMA system uses high voltage (144 V), before servicing turn OFF the battery module switch.
DTC: P0AB6 **1T PCM** **Year:** 2011, 2012 **Model:** Accord, Crosstour, Odyssey, Pilot **Engine:** 3.5L	**Rear Engine Mount Actuator Circuit Malfunction:** Engine running, and DTCs P0A15, P0AB6. P0AB7, P16C4 and P16C6 are not active. The current valve during one cycle of the front engine mount actuator, which outputs in 1 TDC duration as one cycle, is 7.5 A or more as a maximum current or 5.3 A or more as a minimum current, or the engine mounts cycle during start up and continue to cycle for at least 20 seconds. **NOTE: When testing the engine mount actuator, be sure to use a VOM capable of reading 0.01 ohms change of resistance. If the vehicle has been driven, allow the engine mount actuator to cool down for about 2 hours before testing.**
DTC: P0AB7 **1T PCM** **Year:** 2011, 2012 **Model:** Accord, Crosstour, Odyssey, Pilot **Engine:** 3.5L	**Rear Engine Mount Actuator Control Circuit Low Current:** With the engine running and the DTCs P0A14, P0AB6, P0AB7, P15AB, P16C4 and P16C6 are not active. The duty cycle of the rear engine mount actuator, which outputs in 1 TDC duration as one cycle, reads 100% all the time during engine the mount operation and continues at least 20 seconds. **NOTE: When testing the engine mount actuator, be sure to use a VOM capable of reading 0.01 ohms change of resistance. If the vehicle has been driven, allow the engine mount actuator to cool down for about 2 hours before testing.**
DTC: P0AB8 **1T PCM** **Year:** 2011, 2012 **Model:** Odyssey, Pilot **Engine:** 3.5L	**Rear Engine Mount Actuator Control Circuit High Current:** With the ignition switch ON and DTCs P15AF, P15B0, P15B1, P16C5, P16C9 not active. The minimum front ACM actuator current in one cycle is 6.6 A or more for at least 2.6 seconds, the front ACM actuator output cycle is one rotation of the crankshaft.
DTC: P0AC0/65 **1T ECM, MIL: Yes** **Year:** 2011, 2012 **Model:** Civic, CR-Z, Insight **Engine:** 1.3L, 1.5L	**Battery Current Sensor 1 Circuit Malfunction (HYBRID MODELS):** With the ignition switch ON (II), the average value of the battery current sensor input deviates from the range of 2.24 V to 2.76 V. (IMA system indicator is ON) WARNING: The IMA system uses high voltage (144 V), before servicing turn OFF the battery module switch.
DTC: P0AC1/115 **1T PCM, MIL: Yes** **Year:** 2011, 2012 **Model:** Insight **Engine:** 1.3L	**Battery Current Sensor 1 Circuit Low Voltage:** The battery current sensor ISOC signal voltage is 0.5 V or less for at least 0.5 second. (IMA system indicator on) **NOTE: If the DTC cannot be cleared, do the troubleshooting for DTC U1204.**
DTC: P0AC2/114 **1T ECM, MIL: Yes** **Year:** 2011, 2012 **Model:** Civic, CR-Z, Insight **Engine:** 1.3L, 1.5L	**Battery Current Sensor 1 Circuit High Voltage (HYBRID MODELS):** With the ignition switch ON (II), the battery current sensor signal is 4.5 V or more for at least 0.5 seconds. (IMA system indicator is ON) WARNING: The IMA system uses high voltage (144 V), before servicing turn OFF the battery module switch.
DTC: P0AC7/51 **1T PCM, MIL: Yes** **Year:** 2011, 2012 **Model:** Civic, CR-Z, Insight **Engine:** 1.3L, 1.5L	**Battery Module Temperature Sensor 2 Circuit Low Voltage :** With the ignition ON and the MCM supply-power voltage a minimum of 9.0 V, the battery module temperature sensor 2 output voltage is less than 0.05 V for at lest 2 seconds. (IMA system indicator is ON) WARNING: The IMA system uses high voltage (144 V), before servicing turn OFF the battery module switch.
DTC: P0AC8/52 **1T PCM, MIL: Yes** **Year:** 2011, 2012 **Model:** Civic, CR-Z, Insight **Engine:** 1.3L, 1.5L	**Battery Module Temperature Sensor 2 Circuit High Voltage (HYBRID MODELS):** With the ignition ON and the MCM supply-power voltage a minimum of 9.0 V, the battery module temperature sensor 2 output voltage is more than 4.95 V for at least 2 seconds. (IMA system indicator is ON) WARNING: The IMA system uses high voltage (144 V), before servicing turn OFF the battery module switch.
DTC: P0ACC/53 **1T PCM, MIL: Yes** **Year:** 2011, 2012 **Model:** Civic, CR-Z, Insight **Engine:** 1.3L, 1.5L	**Battery Module Temperature Sensor 3 Circuit Low Voltage (HYBRID MODELS):** With the ignition ON and the MCM supply-power voltage a minimum of 9.0 V, the battery module temperature sensor 3 output voltage is less than 0.05 V for at least 2 seconds. (IMA system indicator is ON) WARNING: The IMA system uses high voltage (144 V), before servicing turn OFF the battery module switch.

DTC	Trouble Code Title and Conditions
DTC: P0ACD/54 **1T PCM, MIL: Yes** **Year:** 2011, 2012 **Model:** Civic, CR-Z, Insight **Engine:** 1.3L, 1.5L	**Battery Module Temperature Sensor 3 Circuit High Voltage (HYBRID MODELS):** With the ignition ON and the MCM supply-power voltage a minimum of 10.5 V, the battery module temperature sensor 3 output voltage is more than 4.95 V or at least 2 seconds. (IMA system indicator is ON) WARNING: The IMA system uses high voltage (144 V), before servicing turn OFF the battery module switch.
DTC: P0AE1/62 **1T ECM, MIL: Yes** **Year:** 2011, 2012 **Model:** Civic, CR-Z, Insight **Engine:** 1.3L, 1.5L	**Bypass Contactor Malfunction (HYBRID MODELS):** With the ignition ON and the MCM power-supply voltage a minimum of 7.5 V, the difference between the voltage on the MPI module side and the battery side is more than 8.01 V for at least 210 milliseconds after the bypass contactor is turned on. **NOTE: When the ignition switch is turned to ON (II) while the battery module switch is OFF, DTC P0AE1/62 is detected. (IMA system indicator is ON)** WARNING: The IMA system uses high voltage (144 V), before servicing turn OFF the battery module switch.
DTC: P0AEE/109 **1T , MIL: Yes** **Year:** 2011, 2012 **Model:** Civic, CR-Z, Insight **Engine:** 1.3L, 1.5L	**Motor Control Module (MCM) Internal Circuit Malfunction:** With the ignition ON, MCM supply-power voltage a minimum of 9 V, and the module temperature detection value at a minimum of 221 ºF (105 ºC), the MPI module temperatue change is less than 39 ºF (4 ºC) within 10 minutes. (IMA system indicator is ON)
DTC: P0AEF/110 **1T , MIL: Yes** **Year:** 2011, 2012 **Model:** Civic, CR-Z **Engine:** 1.3L, 1.5L	**Motor Control Module (MCM) Internal Temperature Sensor Circuit Low Voltage (HYBRID MODELS):** With the ignition ON (II), the voltage change value of the MPI module temperature less than 1.79 V for at least 0.8 second. (IMA system indicator is ON)
DTC: P0AF0/111 **1T , MIL: Yes** **Year:** 2011, 2012 **Model:** Civic, CR-Z **Engine:** 1.3L, 1.5L	**Motor Control Module (MCM) Internal Temperature Sensor Circuit High Voltage (HYBRID MODELS):** With the ignition ON (II), the voltage change value of the MPI module temperature is more than 4.63 V for at least 0.8 second. (IMA system indicator is ON)
DTC: P0B0F/113 **1T ECM, MIL: Yes** **Year:** 2011, 2012 **Model:** Civic, CR-Z, Insight **Engine:** 1.3L, 1.5L	**Battery Current Sensor 2 Circuit Malfunction (HYBRID MODELS):** With the ignition switch ON (II), one of the following 2 symptoms can occur. Symptom 1: The average value of the battery current sensor input deviates from the range of 2.24 V to 2.67 V. Symptom 2: The absolute current value difference between the ISOC signal and the ISOCF signal is 10 A or more for at least 2 seconds when the battery current sensor value is in the range of -30 A to 30 A. (IMA system indicator is ON) WARNING: The IMA system uses high voltage (144 V), before servicing turn OFF the battery module switch.
DTC: P0B10/117 **1T ECM, MIL: Yes** **Year:** 2011, 2012 **Model:** Civic, CR-Z, Insight **Engine:** 1.3L, 1.5L	**Battery Current Sensor 2 Circuit Low Voltage (HYBRID MODELS):** With the ignition switch ON (II), the battery current sensor ISOCF signal voltage is 0.05 V or less for at least 0.5 seconds. (IMA system indicator is ON) WARNING: The IMA system uses high voltage (144 V), before servicing turn OFF the battery module switch.
DTC: P0B11/116 **1T ECM, MIL: Yes** **Year:** 2011, 2012 **Model:** Civic, CR-Z, Insight **Engine:** 1.3L, 1.5L	**Battery Current Sensor 2 Circuit High Voltage (HYBRID MODELS):** With the ignition switch ON (II), the battery current sensor ISOCF signal voltage is 4.95 V or more for at least 0.5 seconds. (IMA system indicator is ON) WARNING: The IMA system uses high voltage (144 V), before servicing turn OFF the battery module switch.
DTC: P0BE6/86 **1T ECM, MIL: Yes** **Year:** 2011, 2012 **Model:** Civic, CR-Z, Insight **Engine:** 1.3L, 1.5L	**U Phase Motor Current Sensor Circuit Malfunction (HYBRID MODELS):** With the ignition switch ON (II) or during idle stop, the average signal voltage of sixteen acquired U phase motor current voltage is 2.4 V or less, or 2.6 V or more. (IMA system indicator is ON) WARNING: The IMA system uses high voltage (144 V), before servicing turn OFF the battery module switch.
DTC: P0BEA/87 **1T ECM, MIL: Yes** **Year:** 2011, 2012 **Model:** Civic, CR-Z, Insight **Engine:** 1.3L, 1.5L	**V Phase Motor Current Sensor Circuit Malfunction (HYBRID MODELS):** With the ignition switch ON (II) or during idle stop, the average signal voltage of sixteen acquired V phase motor current signal voltage is 2.4 V or less, or 2.6 V or more. (IMA system indicator is ON) WARNING: The IMA system uses high voltage (144 V), before servicing turn OFF the battery module switch.

DTC	Trouble Code Title and Conditions
DTC: P0BEE/88 **1T ECM, MIL: Yes** **Year:** 2011, 2012 **Model:** Civic, CR-Z, Insight **Engine:** 1.3L, 1.5L	**W Phase Motor Current Sensor Circuit Malfunction:** With the ignition ON (II) or during idle stop, the average signal voltage of sixteen acquired W phase motor current signal voltage is 2.4 V or less, or 2.6 V or more. (IMA system indicator is ON) WARNING: The IMA system uses high voltage (144 V), before servicing turn OFF the battery module switch.
DTC: P0BFD/30 **1T ECM, MIL: Yes** **Year:** 2011, 2012 **Model:** Civic, CR-Z, Insight **Engine:** 1.3L, 1.5L	**Motor Current Sensor Circuit Malfunction (HYBRID MODELS):** With the engine running at a minimum speed of 500 rpm and the battery voltage a minimum of 6.5 V, the sum of the counter value is 1,000 or more. (IMA system indicator is ON) WARNING: The IMA system uses high voltage (144 V), before servicing turn OFF the battery module switch.

OBD II Trouble Code List (P1XXX Codes)

DTC	Trouble Code Title and Conditions
DTC: P1009 **1T PCM, MIL: Yes** **Year:** 2011, 2012 **Model:** Civic, Crosstour, CR-V, Element **Engine:** 2.0L, 2.4L	**Variable Valve Timing Control (VTC) Advance Malfunction:** With the engine idling at operating temperature, the camshaft phase value is not 24.0 degrees or less within the monitored area (camshaft phase control directed value plus failure judgment value) after 3 seconds have passed, or when the camshaft phase value is 5.0 degrees for M/T and 20.0 degrees for A/T or less and continues for 0.5 seconds or more, even when the VTC does not actuate. **NOTE: Before you troubleshoot, record all freeze data and any on-board snapshot, and review the general troubleshooting information. If DTC P0341 is stored at the same time as DTC P1009, troubleshoot DTC P1009 first, then recheck for DTC P0341.**
DTC: P1077 **2T PCM, MIL: Yes** **Year:** 2011, 2012 **Model:** Odyssey **Engine:** 3.5L	**Intake Manifold Tuning (IMT) Valve Stuck in High RPM Position:** When the PCM sends a close (long runner) command, no long runner return signal is received for at least 7 seconds. The execution time is once per driving cycle. **NOTE: Before you troubleshoot, record all freeze data and any on-board snapshot.**
DTC: P1077 **2T PCM, MIL: Yes** **Year:** 2011, 2012 **Model:** Accord, Accord Crosstour, Crosstour **Engine:** 3.5L	**Intake Manifold Runner Control (IMRC) Valve Stuck Open: Short Runner Position:** With the engine running the PCM sends a close (Long Runner) command, no long runner command is received for at least 1 second. **NOTE: If DTC P0651 is indicated at the same time as DTC P1077, troubleshoot DTC P0651 first, then recheck for P1077.**
DTC: P1077 **2T PCM, MIL: Yes** **Year:** 2011, 2012 **Model:** Civic, Pilot, Ridgeline **Engine:** 1.8L, 2.0L, 3.5L	**Intake Manifold Tuning (IMT) Valve Stuck in High RPM Position:** When the PCM sends a close (long runner) command, no long runner return signal is received for at least 7 seconds. Execution is once per drive cycle, engine speed, 3,600 rpm and intake air temperature a minimum of 5 ºF (-15 ºC). **NOTE: Before you troubleshoot, record all freeze data and any on-board snapshot.**
DTC: P1078 **2T PCM, MIL: Yes** **Year:** 2011, 2012 **Model:** Accord, Accord Crosstour, Crosstour **Engine:** 3.5L	**Intake Manifold Runner Control (IMRC) Valve Stuck Closed: Long Runner Position:** When the PCM sends an open (short runner) command, no short runner return signal is received for at least 1 second. The execution time is once per driving cycle. **NOTE: Before you troubleshoot, record all freeze data and any on-board snapshot.**
DTC: P1078 **2T PCM, MIL: Yes** **Year:** 2011, 2012 **Model:** Civic, Odyssey, Pilot, Ridgeline **Engine:** 1.8L, 2.0L, 3.5L	**Intake Manifold Tuning (IMT) Valve Stuck in Low RPM Position:** When the PCM sends an open (short runner) command, no short runner return signal is received for at least 3 seconds. Intake air temperature 5 ºF (-15 ºC), engine speed is 3,800 rpm and battery voltage minimum of 10.5v. DTCs P0107, P0108, P0112, P0113, P0117, P0118, P0563, P1128, P1129, P2227, P2228, P2229 are not active. **NOTE: Before you troubleshoot, record all freeze data and any on-board snapshot.**
DTC: P1106 **1T CCM, MIL: Yes** **Year:** 2012 **Model:** Civic **Engine:** 2.4L	**BARO Pressure Sensor Performance:** Engine running in 4th gear, then accelerated to WOT, and the PCM detected the BARO sensor input did not change sufficiently within a specified period of time.

DTC	Trouble Code Title and Conditions
DTC: P1107 **1T CCM, MIL: Yes** **Year:** 2012 **Model:** Civic **Engine:** 2.4L	**BARO Pressure Sensor Circuit Low Input:** Key on or engine running, and the PCM detected the BARO sensor signal was less than a value in stored in backup memory.
DTC: P1108 **1T CCM, MIL: Yes** **Year:** 2012 **Model:** Civic **Engine:** 2.4L	**BARO Pressure Sensor Circuit High Input:** Key on or engine running, and the PCM detected the BARO sensor signal was more than a value in stored in backup memory.
DTC: P1109 **1T PCM, MIL: Yes** **Year:** 2011, 2012 **Model:** Accord, Civic, Crosstour, CR-V, CR-Z, Fit, Insight, Odyssey, Pilot, Ridgeline **Engine:** All	**Barometric Pressure (BARO) Sensor Circuit Out of Range High:** The BARO sensor output voltage is between 3.59 V to 4.49 V for at least 2 seconds. The execution time is continuous and DTCs P2228, P2229 are not active. **NOTE: Before you troubleshoot, record all freeze data and any on-board snapshot.**
DTC: P1116 **2T PCM, MIL: Yes** **Year:** 2011, 2012 **Model:** All **Engine:** All	**Engine Coolant Temperature (ECT) Sensor 1 Circuit Range/Performance Problem:** A malfunction is detected if the following 3 conditions are present after the engine has stopped and the ignition switch has been turned to LOCK (0) for at least 6 hours before restarting the engine: (1) When the temperature difference between the IAT and ECT1 is 57 ºF (32 ºC) or more. (2) When the temperature difference between the IAT and ECT2 is 30 ºF (17 ºC) or more. (3) When the temperature difference between the ECT2 and ECT1 is 73 ºF (41 ºC) or more. **NOTE: If DTC P0111 is stored at the same time as DTC P1116, troubleshoot DTC P0111 first, then recheck for DTC P1116.**
DTC: P1121 **1T CCM, MIL: Yes** **Year:** 2012 **Model:** Civic **Engine:** 2.4L	**TP Sensor Input Lower Than Expected:** Engine running, and the PCM detected the TP sensor input was lower than an expected value with the throttle wide open (<13.7%). **Note: This trouble code sets if this circuit fails the rationality test.**
DTC: P1122 **1T CCM, MIL: Yes** **Year:** 2012 **Model:** Civic **Engine:** 2.4L	**TP Sensor Input Higher Than Expected:** Engine running, and the PCM detected that the TP sensor input higher than the expected value with the throttle closed (>16.9%). **Note: This trouble code sets if this circuit fails the rationality test.**
DTC: P1128 **1T CCM, MIL: Yes** **Year:** 2012 **Model:** Civic **Engine:** 2.4L	**MAP Sensor Value Less Than Expected:** Engine running at cruise speed, then back to idle speed, and the PCM detected a MAP sensor signal lower than the expected value. **Note: This trouble code sets if this circuit fails the rationality test.**
DTC: P1128 **2T PCM, MIL: Yes** **Year:** 2011, 2012 **Model:** Accord, Civic, Crosstour, CR-V, CR-Z, Element, Fit, Odyssey, Pilot, Ridgeline **Engine:** 1.5L, 1.8L, 2.0L, 2.4L, 3.5L	**Manifold Absolute Pressure (MAP) Sensor Signal Lower Than Expected:** The MAP sensor output is 33 kPa (9.8 in.Hg, 249 mmHg) or less for at least 2 seconds when atmospheric pressure is 52 kPa (15.4 in.Hg, 392 mmHg). The execution time is once per driving cycle. **NOTE: Before you troubleshoot, record all freeze data and any on-board snapshot.**
DTC: P1129 **1T CCM, MIL: Yes** **Year:** 2012 **Model:** Civic **Engine:** 2.4L	**MAP Sensor Value Higher Than Expected:** Engine running at cruise speed, then back to idle speed, and the PCM detected a MAP sensor signal higher than the expected value. **Note: This trouble code sets if this circuit fails the rationality test.**
DTC: P1129 **2T PCM, MIL: Yes** **Year:** 2011, 2012 **Model:** Civic, Crosstour, CR-V, Element, Fit, Odyssey, Pilot, Ridgeline **Engine:** 1.5L, 1.8L, 2.0L, 2.4L, 3.5L	**Manifold Absolute Pressure (MAP) Sensor Signal Higher Than Expected:** The MAP sensor output is 36 kPa (10.9 in.Hg, 277 mmHg) or more for at least 2 seconds. The execution time is once per driving cycle. **NOTE: Before you troubleshoot, record all freeze data and any on-board snapshot.**

DTC	Trouble Code Title and Conditions
DTC: P1129 **2T PCM, MIL: Yes** **Year:** 2011, 2012 **Model:** Accord, Civic, Crosstour, CR-Z **Engine:** 1.5L, 2.4L, 3.5L	**MAP Sensor Signal Higher Than Expected:** With the vehicle speed 1,750 or more and the throttle position is at 22° or more. The difference between the barometric pressure sensor measured during boost conditions and negative pressure (NO boost) is a specified output or less, an upward shift in the MAP sensor signal is detected. **NOTE: Before you troubleshoot, record all freeze data and any on-board snapshot, and review the general troubleshooting information.**
DTC: P1149 **1T O2S1, MIL: Yes** **Year:** 2012 **Model:** Civic **Engine:** 2.4L	**HO2S-11 (Bank 1 Sensor 1) Performance:** Vehicle driven at 55 mph in closed loop at a steady throttle in Drive, then back to idle speed, and the PCM detected the front HO2S response time was slow between 300-600 mv, or it detected the rich to lean or lean to rich switch rate was too slow.
DTC: P1157 **1T PCM, MIL: Yes** **Year:** 2011, 2012 **Model:** Accord, Civic, Crosstour, CR-V, CR-Z, Element, Fit, Insight **Engine:** 1.3L, 1.5L, 1.8L, 2.0L, 2.4L	**Air Fuel Ratio (A/F) Sensor (Sensor 1) AFS Line High Voltage:** With the engine running and battery voltage between 10.5-16.0 V, the A/F sensor (SENSOR 1) heater element resistance is 250 ohms or more for at least 5 seconds. **NOTE: Before you troubleshoot, record all freeze data and any on-board snapshot, and review the general troubleshooting information.**
DTC: P1163 **2T PCM, MIL: Yes** **Year:** 2012 **Model:** Civic **Engine:** 2.4L	**HO2S (Bank 1 Sensor 1) Signal Slow Response:** Engine at idle speed, then accelerated to over 55 mph for 5 seconds at a steady throttle, then back to idle speed for 5 seconds, and the PCM detected the front HO2S response time was too slow, or that the R-L or L-R switch rate was too slow during the HO2S test period.
DTC: P1163 **2T O2S1, MIL: Yes** **Year:** 2012 **Model:** Civic **Engine:** 2.4L	**Lean A/F Sensor (Bank 1 Sensor 1) Slow Response:** Engine at idle speed, then accelerated to 55 mph for 5 seconds, then back to idle speed for 5 seconds, and the PCM detected the LAF sensor response time was too slow, or that the R-L or L-R switch rate was too slow during the HO2S Monitor test period.
DTC: P1164 **2T O2S1, MIL: Yes** **Year:** 2012 **Model:** Civic **Engine:** 2.4L	**Lean A/F Sensor (Bank 1 Sensor 1) Range/Performance:** Engine speed over 1500 rpm in 4th gear in closed loop, then a quick acceleration to WOT, followed by a 5 second deceleration period with the throttle closed, and the PCM detected a problem in the LAF sensor response time, or that the rich-to lean or lean-to rich switch rate was too slow during the test.
DTC: P1164 **2T O2S1, MIL: Yes** **Year:** 2012 **Model:** Civic **Engine:** 2.4L	**HO2S-11 (Bank 1 Sensor 1) Signal Range/Performance:** Vehicle driven to over 1500 rpm in high gear in closed loop, then a quick acceleration to WOT, followed by a 5 second deceleration period with the throttle closed, and the PCM detected the front HO2S response time or the R-L or L-R switch rate was too slow.
DTC: P1165 **2T CCM, MIL: Yes** **Year:** 2012 **Model:** Civic **Engine:** 2.4L	**Lean A/F-11 (Bank 1 Sensor 1) Range/Performance:** Vehicle driven at an engine speed of 1500-2500 rpm at 45-60 mph for 2-3 minutes, followed by a deceleration period of 3 seconds back to idle speed with the throttle closed, and the PCM detected the LAF sensor signal was too high or too low during the HO2S Monitor test.
DTC: P1172 **2T PCM, MIL: Yes** **Year:** 2011, 2012 **Model:** Accord, Crosstour, Odyssey, Pilot, Ridgeline **Engine:** 3.5L	**Rear Air/Fuel Ratio (A/F) Sensor (Bank 1, Sensor 1) Circuit Out of Range High:** A malfunction is detected when the rear A/F sensor (bank 1, sensor 1) output voltage is 4.7 V or more. The execution time is continuous and the duration time is 7 seconds or more. **NOTE: Before you troubleshoot, record all freeze data and any on-board snapshot.**
DTC: P1172 **1T PCM, MIL: Yes** **Year:** 2011, 2012 **Model:** Accord, Civic, Crosstour, CR-V, CR-Z, Element, Fit, Insight **Engine:** 1.3L, 1.5L, 1.8L, 2.0L, 2.4L	**Air/Fuel Ratio (A/F) Sensor (Sensor 1) Circuit Out of Range High:** A malfunction is detected when the rear A/F sensor (Sensor 1) output voltage is 4.9 V or more. The execution time is continuous and the duration time is 7 seconds or more. **NOTE: Before you troubleshoot, record all freeze data and any on-board snapshot.**

DTC	Trouble Code Title and Conditions
DTC: P1174 **2T PCM, MIL: Yes** **Year:** 2011, 2012 **Model:** Accord, Crosstour, Odyssey, Pilot, Ridgeline **Engine:** 3.5L	**Front Air/Fuel Ratio (A/F) Sensor (Bank 2, Sensor 1) Circuit Out of Range High:** A malfunction is detected when the front A/F sensor (bank 2, sensor 1) output voltage is 4.7 V or more. The execution time is continuous and the duration time is 7 seconds or more. **NOTE: Before you troubleshoot, record all freeze data and any on-board snapshot.**
DTC: P1187 **1T PCM** **Year:** 2011, 2012 **Model:** Civic **Engine:** 1.8L, 2.0L	**Fuel Tank Temperature (FTT) Sensor Circuit Low Voltage :** With the ignition switch ON, the FTT output voltage is 0.08 V or less for at least 2 seconds. P01188 is not active.
DTC: P1188 **1T PCM** **Year:** 2011, 2012 **Model:** Civic **Engine:** 1.8L, 2.0L	**Fuel Tank Temperature (FTT) Sensor Circuit High Voltage :** With the ignition switch ON, the FTT output voltage is 4.92 V or more for at least 2 seconds. P01187 is not active.
DTC: P1192 **1T PCM** **Year:** 2011, 2012 **Model:** Civic **Engine:** 1.8L, 2.0L	**Fuel Tank Pressure (FTP) Sensor Circuit Low Voltage :** With the ignition switch ON, the FTP sensor output voltage is 0.23 V or less for at least 2 seconds. P1193 is not active. **NOTE: Before you troubleshoot, record all freeze data and any on-board snapshot, and review the general troubleshooting information.**
DTC: P1193 **1T PCM** **Year:** 2011, 2012 **Model:** Civic **Engine:** 1.8L, 2.0L	**Fuel Tank Pressure (FTP) Sensor Circuit High Voltage:** With the ignition switch ON, the FTP sensor output voltage is 4.61 V or more for at least 2 seconds. P1193 is not active. **NOTE: Before you troubleshoot, record all freeze data and any on-board snapshot, and review the general troubleshooting information.**
DTC: P1220/34 **1T O2S4, MIL: Yes** **Year:** 2011 **Model:** Civic **Engine:** 1.3L	**DC-DC Converter Lost Communication with Battery Condition Monitor (BCM) Module (Hybrid Models):** With the ignition switch ON and the BCM power supply votage a minimum of 10.5 V, The serial communication data is not updated for at least 2 seconds. (IMA system indicator is ON) WARNING: The IMA system uses high voltage (144 V), before servicing turn OFF the battery module switch.
DTC: P1221/35 **1T PCM, MIL: Yes** **Year:** 2011 **Model:** Civic **Engine:** 1.3L	**Battery Condition Monitor (BCM) Module Lost Communication with DC-DC Converter (Hybrid Models):** With the ignition switch ON and the battery voltage a minimum of 8.4 V, the communication information from the DC-DC converter is not received or the received data is abnormal for at least 2 seconds. (IMA system indicator is ON) WARNING: The IMA system uses high voltage (144 V), before servicing turn OFF the battery module switch.
DTC: P1286 **1T PCM, MIL: Yes** **Year:** 2011, 2012 **Model:** Civic, Insight **Engine:** 1.3L	**Rocker Arm Oil Pressure Sensor B Stuck Low:** With the engine running at operating temperature, and the vehicle speed is a minimum of 12 mph (18 km/h) or more with the engine speed between 900-3,000 rpm the following conditions occur for 2 seconds or more. Condition 1: rocker arm oil pressure sensor A is ON Condition 2: rocker arm oil pressure sensor B is ON Condition 3: rocker arm oil pressure sensor A output HIGH Condition 4: rocker arm oil pressure sensor B output LOW
DTC: P1287 **2T PCM, MIL: Yes** **Year:** 2011 **Model:** Civic **Engine:** 1.3L	**Rocker Arm Oil Pressure Switch Circuit High Voltage:** With the engine running at a speed between 2,000-4,600 rpm in (P or N) for at least 5 seconds, the following conditions occur. * Rocker Arm Oil Control Solenoid 1 is OFF * Rocker Arm Oil Control Solenoid 2 is OFF * Rocker Arm Oil Pressure Switch is OFF * Rocker Arm Oil pressure Sensor Output HIGH 39 kPa or more.
DTC: P1288 **2T PCM, MIL: Yes** **Year:** 2011 **Model:** Civic **Engine:** 1.3L	**Rocker Arm Oil Pressure Switch Circuit Low Voltage:** With the engine running at a speed between 2,000-4,600 rpm in (P or N) for at least 5 seconds, the following conditions occur. * Rocker Arm Oil Control Solenoid 1 is ON. * Rocker Arm Oil Control Solenoid 2 is OFF. * Rocker Arm Oil Pressure Switch is ON. * Rocker Arm Oil pressure Sensor Output HIGH 39 kPa or less.

DTC	Trouble Code Title and Conditions
DTC: P1289 **2T PCM, MIL: Yes** **Year:** 2011 **Model:** Civic **Engine:** 1.3L	**Rocker Arm Oil Pressure Sensor Stuck High:** With the engine running at a speed between 3,000-4,800 rpm in (P or N) for at least 5 seconds in high valve mode, the following conditions occur. * Rocker Arm Oil Control Solenoid 1 is ON * Rocker Arm Oil Control Solenoid 2 is OFF * Rocker Arm Oil Pressure Switch is OFF * Rocker Arm Oil pressure Sensor Output HIGH 39 kPa or more
DTC: P128A **1T PCM, MIL: Yes** **Year:** 2011, 2012 **Model:** Insight **Engine:** 1.3L	**Rocker Arm Oil Pressure Sensor B Stuck High:** With the engine running at operating temperature, and the vehicle speed is a minimum of 12 mph (18 km/h) or more with the engine speed between 900-3,000 rpm the following conditions occur for 2 seconds or more. Condition 1: rocker arm oil pressure sensor A is ON Condition 2: rocker arm oil pressure sensor B is ON Condition 3: rocker arm oil pressure sensor A output HIGH Condition 4: rocker arm oil pressure sensor B output LOW
DTC: P128A **2T PCM, MIL: Yes** **Year:** 2011 **Model:** Civic **Engine:** 1.3L	**Valve Pause System (VPS) Stuck OFF:** With the engine running at a speed between 1,000-3,000 rpm in (P or N) for at least 5 seconds in the valve pause mode, the following conditions occur. * Rocker Arm Oil Control Solenoid 1 is ON * Rocker Arm Oil Control Solenoid 2 is ON * Rocker Arm Oil Pressure Switch is ON * Rocker Arm Oil pressure Sensor Output HIGH 39 kPa or more
DTC: P128B **2T PCM, MIL: Yes** **Year:** 2011 **Model:** Civic **Engine:** 1.3L	**Valve Pause System (VPS) Malfunction:** With the engine running at a speed between 1,000-3,000 rpm in (P or N) for at least 5 seconds in the valve pause mode, the following conditions occur. * Rocker Arm Oil Control Solenoid 1 is ON * Rocker Arm Oil Control Solenoid 2 is ON * Rocker Arm Oil Pressure Switch is OFF * Rocker Arm Oil pressure Sensor Output HIGH 39 kPa or less
DTC: P128C **1T PCM, MIL: Yes** **Year:** 2011, 2012 **Model:** Insight **Engine:** 1.3L	**Rocker Arm Oil Pressure Sensor A Stuck Low:** With the engine running at operating temperature, and the vehicle speed is a minimum of 12 mph (18 km/h) or more with the engine speed between 900-3,000 rpm the following conditions occur for 2 seconds or more. Condition 1: rocker arm oil pressure sensor A is ON Condition 2: rocker arm oil pressure sensor B is ON Condition 3: rocker arm oil pressure sensor A output HIGH Condition 4: rocker arm oil pressure sensor B output LOW
DTC: P128C **2T PCM, MIL: Yes** **Year:** 2011, 2012 **Model:** Civic **Engine:** 1.3L, 1.8L, 2.0L	**Rocker Arm Oil Pressure Sensor A Stuck Low:** Malfunction Threshold, The following conditions occur for at least 2.0 seconds in the normal valve lift mode 2. 1). Rocker Arm Oil Control Solenoid A OFF 2). Rocker Arm Oil Control Solenoid B ON 3). Rocker Arm Oil Pressure Sensor A output Low (39 kPa (11.5 inHg, 292 mmHg) or less) 4). Rocker Arm Oil Pressure Sensor B output High (39 kPa (11.6 inHg, 293 mmHg) or more)
DTC: P128D **2T PCM, MIL: Yes** **Year:** 2011, 2012 **Model:** Civic **Engine:** 1.3L, 1.8L, 2.0L	**Rocker Arm Oil Pressure Sensor A Stuck High:** Malfunction Threshold, The following conditions occur for at least 2.0 seconds in the normal valve lift mode 1. 1). Rocker Arm Oil Control Solenoid A OFF 2). Rocker Arm Oil Control Solenoid B OFF 3). Rocker Arm Oil Pressure Sensor A output High (39 kPa (11.6 inHg, 293 mmHg) or more) 4). Rocker Arm Oil Pressure Sensor B output High (39 kPa (11.6 inHg, 293 mmHg) or more)
DTC: P128D **1T PCM, MIL: Yes** **Year:** 2011, 2012 **Model:** Insight **Engine:** 1.3L	**Rocker Arm Oil Pressure Sensor A Stuck High:** With the engine running at operating temperature, and the vehicle speed is a minimum of 12 mph (18 km/h) or more with the engine speed between 900-3,000 rpm the following conditions occur for 2 seconds or more. Condition 1: rocker arm oil pressure sensor A is ON Condition 2: rocker arm oil pressure sensor B is ON Condition 3: rocker arm oil pressure sensor A output HIGH Condition 4: rocker arm oil pressure sensor B output LOW

DTC	Trouble Code Title and Conditions
DTC: P1297 **2T PCM** **Year:** 2011, 2012 **Model:** All **Engine:** All	**Electrical Load Detector (ELD) Circuit Low Voltage:** The ELD output voltage is 0.27 V or less for at least 5 seconds. The execution time is continuous and DTC P1298 is not active. **NOTE: Before you troubleshoot, record all freeze data and any on-board snapshot.**
DTC: P1298 **2T PCM** **Year:** 2011, 2012 **Model:** All **Engine:** All	**Electrical Load Detector (ELD) Circuit High Voltage:** With the ignition switch ON (II) the ELD output voltage is 4.57 V or more for at least 5 seconds. The execution time is continuous and DTC P1297 is not active. **NOTE: Before you troubleshoot, record all freeze data and any on-board snapshot.**
DTC: P1336 **1T CCM, MIL: Yes** **Year:** 2012 **Model:** Civic **Engine:** 2.4L	**Engine Speed Fluctuation Sensor Circuit Malfunction:** Engine running, and PCM detected an unexpected or intermittent interruption of the engine speed fluctuation (ESF) sensor signal.
DTC: P1359 **1T CCM, MIL: Yes** **Year:** 2012 **Model:** Civic **Engine:** 2.4L	**CKP/TDC Sensor Circuit Malfunction:** Engine running, and the PCM detected an unexpected voltage condition on the CKP/TDC sensor circuit during the CCM test.
DTC: P1361 **1T CCM, MIL: Yes** **Year:** 2012 **Model:** Civic **Engine:** 2.4L	**Top Dead Center Sensor 1 Circuit Intermittent Signal:** Engine running, and the PCM detected an unexpected or intermittent interruption of the Top Dead Center 1 (TDC1) sensor signal.
DTC: P1362 **1T CCM, MIL: Yes** **Year:** 2012 **Model:** Civic **Engine:** 2.4L	**Top Dead Center Sensor 1 No Signal:** Engine cranking or running, and the PCM did not receive any signals from the Top Dead Center 1 (TDC1) sensor during the CCM test. **Note: The engine will start and run without the TDC sensor 1 signal.**
DTC: P1381 **1T CCM, MIL: Yes** **Year:** 2012 **Model:** Civic **Engine:** 2.4L	**Camshaft Position Sensor 1 Circuit Malfunction:** Engine running and the PCM detected an unexpected or intermittent interruption of the Camshaft Position (CMP) sensor 1 signal.
DTC: P1382 **2T CCM, MIL: Yes** **Year:** 2012 **Model:** Civic **Engine:** 2.4L	**Camshaft Position Sensor 1 No Signal:** Engine cranking or running, and the PCM did not detect any signals from the Camshaft Position (CMP) sensor 1 during the CCM test. **Note: The engine will start and run without the CMP sensor 1 signal.**
DTC: P1437/41 **1T PCM, MIL: Yes** **Year:** 2011, 2012 **Model:** Civic, CR-Z, Insight **Engine:** 1.3L, 1.5L	**Motor Power Inverter (MPI) Module Short Circuit (HYBRID MODELS):** With the ignition ON and the MCM supply-power voltage a minimum of 10.5 V, the MPI module transmits the shorting information for at least 2 seconds. (IMA system indicator is ON) WARNING: The IMA system uses high voltage (144 V), before servicing turn OFF the battery module switch.
DTC: P1440/57 **1T PCM, MIL: Yes** **Year:** 2011, 2012 **Model:** Civic, CR-Z, Insight **Engine:** 1.3L, 1.5L	**Motor Power Inverter (MPI) Module Output Circuit Malfunction (HYBRID MODELS):** With the ignition ON, MCM supply-power voltage a minimum of 7.0 V, IMA motor speed between 500-1500 rpm and IMA driving torque a minimum of 74 ft lbs. (10 Nm). The following malfunctions are detected: Malfunction 1: The ratio of maximum to minimum for the three motor current sensor fluctuation values become 70% or less for at least 0.5 seconds. Malfunction 2: The deviation of the target torque and the actual torque for the motor control is at least 0.5 seconds and 30 ft. lbs. (40 Nm). (IMA system indicator is ON) WARNING: The IMA system uses high voltage (144 V), before servicing turn OFF the battery module switch.

DTC	Trouble Code Title and Conditions
DTC: P1446/74 **1T PCM, MIL: Yes** **Year:** 2011, 2012 **Model:** Civic, CR-Z, Insight **Engine:** 1.3L, 1.5L	**Battery Module Individual Voltage Input Deviation (HYBRID MODELS):** With the ignition ON, MCM supply-power voltage a minimum of 10.5 V, the battery module temperature a minimum of 32 ºF, detected individual voltage channel 1-10 is a minimum of 14.5 V, the voltage difference between channels is 1.2 V or more for at least 25.4 seconds. (IMA system indicator is ON) WARNING: The IMA system uses high voltage (144 V), before servicing turn OFF the battery module switch.
DTC: P1448/63 **1T PCM, MIL: Yes** **Year:** 2011, 2012 **Model:** Civic, CR-Z, Insight **Engine:** 1.3L, 1.5L	**Intelligent Power Unit (IPU) Module Fan Problem (HYBRID MODELS):** With the ignition ON, MCM supply-power voltage a minimum of 10.5 V, the IPU module fan speed is 2,500 rpm, the actual IPU fan speed is 1,500 or less for at least 30 seconds. (IMA system indicator is ON) WARNING: The IMA system uses high voltage (144 V), before servicing turn OFF the battery module switch
DTC: P1454 **2T PCM, MIL: Yes** **Year:** 2011, 2012 **Model:** All **Engine:** 1.3L, 1.5L, 2.4L, 3.5L	**Fuel Tank Pressure (FTP) Sensor Circuit Range/Performance Problem:** One of these 2 conditions is met. (1) The FTP sensor output fluctuates by 0.6 kPa (0.1 in.Hg, 5 mmHg) or more, or -0.6 kPa (-0.1 in.Hg, -5 mmHg) or less for at least 3 seconds. (2) The FTP sensor output value is -1.3 kPa (-0.3 in.Hg, -10 mmHg) or less for at least 3 seconds. DTCs P0452, P0453 are judged as OK. **NOTE: Before you troubleshoot, record all freeze data and any on-board snapshot.**
DTC: P1454 **2T PCM, MIL: Yes** **Year:** 2011, 2012 **Model:** Civic **Engine:** 1.3L, 1.8L, 2.0L	**Fuel Tank Pressure (FTP) Sensor Range/Performance Problem:** Malfunction Threshold: 1). The FTP sensor output fluctuates by 0.7 kPa (0.2 inHg, 5 mmHg) or more, or -0.7 kPa (-0.2 inHg, -5 mmHg) or less, for at east 8.0 seconds. 2). The FTP sensor output [FTP SENSOR] value is -1.4 kPa (-0.4 inHg, -10 mmHg) or less for at least 8.0 seconds.
DTC: P1456 **T** **Year:** 2012 **Model:** Civic **Engine:** 2.4L	**EVAP System Leak Detected (Fuel Tank Area):** Cold startup completed (IAT sensor signal from 32-86ºF at engine startup), vehicle driven at over 5 mph for over 2 minutes, then with ECT sensor signal more than 154ºF and the EVAP Control and Vent solenoids enabled, the PCM detected the fuel tank pressure was incorrect due to a leak in the fuel tank area during the Leak Test.
DTC: P1457 **T** **Year:** 2012 **Model:** Civic **Engine:** 2.4L	**EVAP System Leak Detected (Canister Area):** Cold startup completed (IAT sensor signal from 32-86ºF at engine startup), vehicle driven at over 5 mph for over 2 minutes, then with ECT sensor signal more than 154ºF and the EVAP Control and Vent solenoids enabled, the PCM detected the fuel tank pressure was incorrect due to a leak in the canister area during the Leak Test.
DTC: P1458 **2T PCM, MIL: Yes** **Year:** 2011, 2012 **Model:** Civic, Odyssey **Engine:** 1.3L, 1.8L, 2.0L, 3.5L	**Fuel Tank Pressure (FTP) Sensor Circuit Range/Performance Problem:** Malfunction Threshold: The misalignment of zero point pressure inside the fuel tank is 0.6 kPa (0.1 inHg, 5 mmHg) or more. The output from the FTP sensor is flickering 3.1 seconds or more.
DTC: P1458 **2T PCM, MIL: Yes** **Year:** 2011, 2012 **Model:** Accord, Civic, Crosstour, CR-Z **Engine:** 1.5L, 2.4L, 3.5L	**FTP Sensor Circuit Range/Performance Problem:** **NOTE: If any of DTCs listed below are indicated at the same time as DTC P1458, troubleshoot those DTCs first, then recheck for P1458. P0451, P0452, P0453, P1454: FTP sensor.** The misalignment of zero point pressure inside the fuel tank is 0.6 kPa (0.1 inHg, 5 mmHg) or more. The output from the FTP sensor is flickering 3.1 seconds or more.
DTC: P1458 **2T PCM, MIL: Yes** **Year:** 2011, 2012 **Model:** Fit **Engine:** 1.5L	**Fuel Tank Pressure (FTP) Sensor Circuit Range/Performance:** The misalignment of zero point pressure inside the fuel tank is 0.6 kPa (0.1 inHg, 5 mmHg) or more. The output from the FTP sensor is flickering 3.1 seconds or more.
DTC: P145C **2T PCM, MIL: Yes** **Year:** 2011, 2012 **Model:** Civic, CR-Z **Engine:** 1.3L, 1.5L, 1.8L, 2.0L	**Evaporative Emission (EVAP) System Purge Flow Malfunction:** Malfunction Threshold: The pulses detected by the FTP sensor are 50 % or less for at least 18 seconds.

DTC	Trouble Code Title and Conditions
DTC: P145C **2T PCM, MIL: Yes** **Year:** 2011, 2012 **Model:** Accord, Crosstour, CR-V, Element, Fit, Insight, Odyssey, Ridgeline **Engine:** 1.3L, 1.5L, 2.4L, 3.5L	**Evaporative Emission (EVAP) System Purge Flow Malfunction:** The pulses detected by the fuel tank pressure sensor are 1.0 % of the duty cycle or less for at least 31 seconds. The execution time is continuous. **NOTE: This DTC is representative of an EVAP system purge flow problem. If DTC P145C is indicated alone, troubleshoot P0496 and P0497 using the freeze data for P145C.** If any of the DTCs listed below are indicated at the same time as DTC P145C, troubleshoot those DTC first, then recheck for P145C: * P0496, P0497: EVAP system purge flow.
DTC: P1508 **1T CCM, MIL: Yes** **Year:** 2012 **Model:** Civic **Engine:** 2.4L	**Idle Air Control System Low RPM:** Engine running at idle speed while in closed loop, and the PCM detected the Actual idle speed was more than 100 rpm below the Target idle speed during the CCM test.
DTC: P1509 **1T CCM, MIL: Yes** **Year:** 2012 **Model:** Civic **Engine:** 2.4L	**Idle Air Control System High RPM:** Engine running at idle speed while in closed loop, and the PCM detected the Actual idle speed was more than 100 rpm below the Target idle speed during the CCM test.
DTC: P1509 **1T CCM, MIL: Yes** **Year:** 2012 **Model:** Civic **Engine:** 2.4L	**Idle Air Control System High RPM:** Engine running at idle speed while in closed loop, and the PCM detected the Actual idle speed was more than 100 rpm above the Target idle speed during the CCM test.
DTC: P1549 **1T PCM** **Year:** 2011, 2012 **Model:** Accord, Civic, Crosstour, CR-V, Element, Fit, Odyssey, Pilot, Ridgeline **Engine:** All	**Charging System High Voltage:** The IGP terminal voltage is 16.0 volts or more for at least 60 seconds. The execution time is continuous. **NOTE: If a high voltage battery (24 V, etc.) is connected to the vehicle, this DTC can be stored.**
DTC: P154A **1T PCM** **Year:** 2011, 2012 **Model:** Civic **Engine:** 1.3L, 1.8L, 2.0L	**Battery Sensor Internal Failure:** Either one of the conditions is met for at least 5 seconds: * Internal current detection circuit open or A/D conversion circuit abnormal * Internal voltage detection circuit open or A/D conversion circuit abnormal * Internal temperature detection circuit open or A/D conversion circuit abnormal * Calibration data abnormal
DTC: P1570/66 **1T PCM, MIL: Yes** **Year:** 2011, 2012 **Model:** Civic, CR-Z, Insight **Engine:** 1.3L, 1.5L	**Battery Module Individual Voltage Problem (HYBRID MODELS):** With the ignition ON, MCM supply-power voltage a minimum of 9.0 V, and the detected individual voltage (one or more channels between channel 1-10) is at a minimum of 15 V. The difference between the maximum and minimum voltage is 3 V or more for at least 1.5 seconds. (IMA system indicator is ON) WARNING: The IMA system uses high voltage (144 V), before servicing turn OFF the battery module switch.
DTC: P1574/68 **2T PCM, MIL: Yes** **Year:** 2011, 2012 **Model:** Civic, CR-Z, Insight **Engine:** 1.3L, 1.5L	**Battery Module Temperature Signal Circuit Malfunction (HYBRID MODELS):** With the ignition ON, MCM supply-power voltage a minimum of 10.5 V, the battery module temperature is 131 ºF (55 ºC) with IPU module fan on high. The battery module temperature change is less than 1 ºF (0.5 ºC) within 10 minutes. (IMA system indicator is ON) WARNING: The IMA system uses high voltage (144 V), before servicing turn OFF the battery module switch.
DTC: P1575/12 **1T PCM, MIL: Yes** **Year:** 2011, 2012 **Model:** Civic, CR-Z, Insight **Engine:** 1.3L, 1.5L	**Motor Power Inverter (MPI) Module Voltage Malfunction (HYBRID MODELS):** With the ignition ON, MCM supply-power voltage a minimum of 10.5 V, MCM module voltage 10-250 V, battery module voltage 20-200 V. The difference between the MPI module voltage and the battery module voltage is more than 10 V for at least 2 seconds. (IMA system indicator is ON) **NOTE: If the battery module switch is turned OFF while the ignition switch is ON (II) , DTC P1575 may be detected.** WARNING: The IMA system uses high voltage (144 V), before servicing turn OFF the battery module switch.
DTC: P1585/30 **1T PCM, MIL: Yes** **Year:** 2011 **Model:** Civic **Engine:** 1.3L	**Motor Current Sensor Circuit Malfunction (HYBRID MODELS):** With the ignition ON and the MCM supply-power voltage a minimum of 9.0 V, the output electrical current value of the MCM internal summation circuit is -50 A or less or 50 A or more, for at least 0.5 seconds. (IMA system indicator is ON) WARNING: The IMA system uses high voltage (144 V), before servicing turn OFF the battery module switch.

DTC	Trouble Code Title and Conditions
DTC: P1586/23 **1T PCM, MIL: Yes** **Year:** 2011, 2012 **Model:** Insight **Engine:** 1.3L	**Battery Current Sensor Signal Circuit Problem (HYBRID MODELS):** With the ignition ON and the MCM supply-power voltage a minimum of 10.5 V. One of the following conditions are met: Condition 1: The difference between the MPI module current and the battery module current is more than 44 A for at least 1.0 seconds. Condition 2: When the driving torque of the IMA motor is 74 ft lbs. (10 Nm) or more, the battery module current drops 3 A for at least 0.5 seconds. (IMA system indicator is ON) WARNING: The IMA system uses high voltage (144 V), before servicing turn OFF the battery module switch.
DTC: P15A5/85 **1T PCM, MIL: Yes** **Year:** 2011, 2012 **Model:** Civic, CR-Z, Insight **Engine:** 1.3L, 1.5L	**Motor Current Sensor Circuit Malfunction (HYBRID MODELS):** With the ignition ON and the MCM supply-power voltage a minimum of 10.0 V, the motor phase current is more than 270 A. (IMA system indicator is ON) WARNING: The IMA system uses high voltage (144 V), before servicing turn OFF the battery module switch.
DTC: P15A6/86 **1T PCM, MIL: Yes** **Year:** 2011, 2012 **Model:** Civic, CR-Z **Engine:** 1.3L, 1.5L	**U Phase Motor Current Sensor Circuit Malfunction (HYBRID MODELS):** With the ignition ON and the battery voltage a minimum of 6.0 V, the sensor output voltage when the U phase motor current sensor is 0 A, is 2.32 V or less, or 2.68 V or more for at least 0.5 seconds. (IMA system indicator is ON) WARNING: The IMA system uses high voltage (144 V), before servicing turn OFF the battery module switch.
DTC: P15A7/87 **1T PCM, MIL: Yes** **Year:** 2011 **Model:** Civic **Engine:** 1.3L	**V Phase Motor Current Sensor Circuit Malfunction (HYBRID MODELS):** With the ignition ON, and the battery voltage a minimum of 6.0 V, the sensor output voltage when the V phase motor current sensor is 0 A, is 2.32 V or less or 2.68 V or more for at least 0.5 seconds. (IMA system indicator is ON) WARNING: The IMA system uses high voltage (144 V), before servicing turn OFF the battery module switch.
DTC: P15A8/88 **1T PCM, MIL: Yes** **Year:** 2011 **Model:** Civic **Engine:** 1.3L	**W Phase Motor Current Sensor Circuit Malfunction (HYBRID MODELS):** With the ignition ON and the battery voltage a minimum of 6.0 V, the sensor output voltage when the W phase motor sensor is 0 A, is 2.32 V or less, or 2.68 V or more for at least 0.5 seconds. (IMA system indicator is ON) WARNING: The IMA system uses high voltage (144 V), before servicing turn OFF the battery module switch.
DTC: P15AA/93 **1T** **Year:** 2011, 2012 **Model:** Civic, CR-Z, Insight **Engine:** 1.3L, 1.5L	**Motor Rotor Position Not Learned (HYBRID MODELS):** With the engine running between 500-1500 rpm, battery voltage a minimum of 10.0 V and the battery module voltage between 120.0-245.0 V. One of the following conditions occurs: Condition 1: Angle detection value is -30 degrees or less. Condition 2: Angle detection value is +30 degrees or more. Condition 3: Angle detection value is not stable for at lest 15 seconds. (IMA system indicator is ON) WARNING: The IMA system uses high voltage (144 V), before servicing turn OFF the battery module switch.
DTC: P15AB **1T PCM** **Year:** 2011, 2012 **Model:** Accord, Crosstour, Odyssey, Pilot **Engine:** 3.5L	**Engine Mount Control Unit Power Source Circuit Low Voltage:** With the ignition on (II), the IG1 terminal voltage is 4.0 V or less for at least 2.6 seconds. **NOTE: Before you troubleshoot, record all freeze data and any on-board snapshots, and review the general troubleshooting information.**
DTC: P15AC **1T PCM** **Year:** 2011, 2012 **Model:** Pilot **Engine:** 3.5L	**Engine Mount Control Unit Internal Circuit Malfunction:** With the ignition switch ON, P16C4 and P16C9 are not active. The ACM actuator driving power voltage is 24 V or less for at least 2.6 seconds. **NOTE: Before you troubleshoot, record all freeze data and any on-board snapshots, and review the general troubleshooting information.**
DTC: P15AE **1T PCM** **Year:** 2011, 2012 **Model:** Pilot **Engine:** 3.5L	**Cylinder Pause Signal Malfunction:** With the ignition switch ON, P16C9 not active, the SCP signal input value differs from the PCM out put value (received by the CAN communication) for at least 2.6 seconds. **NOTE: Before you troubleshoot, record all freeze data and any on-board snapshots, and review the general troubleshooting information.**
DTC: P15AE **1T PCM** **Year:** 2011, 2012 **Model:** Accord, Crosstour, Odyssey **Engine:** 3.5L	**Cylinder Pause Signal 1 Malfunction:** With the engine running, the SCPA signal input value differs from the PCM output value (Received by CAN communication) for at least 2.6 seconds. **NOTE: Select the VTEC TEST in the PGM-FI INSPECTION menu, and do the VPS TEST of the 3 CYLINDER ACTIVATION TEST with the HDS.**

DTC	Trouble Code Title and Conditions
DTC: P15B0 **1T PCM** **Year:** 2011, 2012 **Model:** Accord, Crosstour, Odyssey, Pilot **Engine:** 3.5L	**Crankshaft Position (CKP) Sensor Signal Malfunction:** With the engine running at idle, the CKP signal is not input to the engine mount control unit for at least 20 seconds. **NOTE: If PGM-FI system's DTC P0335 and/or P0339 is stored at the same time as DTC P15B0 troubleshoot P0335 and/or P0339 first, then recheck for P15B0.**
DTC: P15BD **1T PCM** **Year:** 2011, 2012 **Model:** Accord, Crosstour, Odyssey, Pilot **Engine:** 3.5L	**Cylinder Pause Signal 2 Malfunction:** With the engine running, the SCPA signal input value differs from the PCM output value (Received by CAN communication) for at least 2.6 seconds. **NOTE: Select the VTEC TEST in the PGM-FI INSPECTION menu, and do the VPS TEST.**
DTC: P15BE **1T PCM** **Year:** 2011, 2012 **Model:** Accord, Crosstour, Odyssey, Pilot **Engine:** 2.4L, 3.5L	**Camshaft Position (CMP) Sensor Signal Malfunction:** With the engine running at idle, the CMP signal is not input to the engine mount control unit for at least 20 seconds. **NOTE: If PGM-FI system's DTC P0365 and/or P0369 is stored at the same time as DTC P15BE, troubleshoot P0365 and/or P0369 first, then recheck for P15BE.**
DTC: P15BF **1T PCM** **Year:** 2011, 2012 **Model:** Accord, Crosstour, Odyssey, Pilot **Engine:** 3.5L	**Camshaft Position (CMP) Sensor Signal Intermittent Interruption:** With the engine running at idle, other than 10 CMP pulses are detected during two revolutions of the crankshaft. This condition has been detected for at least 127 times. **NOTE: If PGM-FI system's DTC P0365 and/or P0369 is stored at the same time as DTC P15BF, troubleshoot P0365 and/or P0369 first, then recheck for P15BF.**
DTC: P15C0 **1T PCM** **Year:** 2011, 2012 **Model:** Accord, Crosstour, Odyssey, Pilot **Engine:** 3.5L	**Crankshaft Position (CKP) Sensor Signal Intermittent Interruption:** With the engine running at idle, The number of CKP pulse is not 58 during intervals between reference pulses for each crank revolution. OR. No interval is detected while the CKP sensor pulse inputs 82.
DTC: P15C1 **1T PCM, MIL: Yes** **Year:** 2011, 2012 **Model:** Insight **Engine:** 1.3L	**Brake Booster Pressure Sensor B Circuit High Voltage:** Withe the ignition ON, an open condition was detected in the brake booster pressure sensor B circuit. **NOTE: Before you troubleshoot, record all freeze data and any on-board snapshot, and review the general troubleshooting information.**
DTC: P15C2 **1T PCM, MIL: Yes** **Year:** 2011, 2012 **Model:** Insight **Engine:** 1.3L	**Brake Booster Pressure Sensor B Circuit Low Voltage:** With the ignition ON, a malfunction in the brake booster pressure sensor B circuit was detected. **NOTE: Before you troubleshoot, record all freeze data and any on-board snapshot, and review the general troubleshooting information.**
DTC: P15C3 **1T PCM, MIL: Yes** **Year:** 2011, 2012 **Model:** Insight **Engine:** 1.3L	**Brake Booster Pressure sensor A/B Incorrect Voltage Correlation :** With the ignition ON and engine running, a malfuntion was detected between sensor A and sensor B. **NOTE: Before you troubleshoot, record all freeze data and any on-board snapshot, and review the general troubleshooting information.** B
DTC: P15C4/118 **2T PCM, MIL: Yes** **Year:** 2011, 2012 **Model:** Civic, CR-Z, Insight **Engine:** 1.3L, 1.5L	**Auto Idle Stop System Malfunction (HYBRID MODELS):** With the ignition ON and the MCM power-supply voltage a minimum of 10.2 V, the MCM detects a failure related to the auto idle stop prohibition for at least 2 seconds.
DTC: P1607 **1T CCM, MIL: Yes** **Year:** 2012 **Model:** Civic **Engine:** 2.4L	**PCM Internal Circuit 'A' Malfunction:** Key on, and the PCM detected an Internal Fault 'A' condition. **Note: This trouble code indicates an internal failure in the PCM. The OEM repair procedure recommends replacing the original PCM with a "known good" PCM and then verify the code does not reset.**

DTC	Trouble Code Title and Conditions
DTC: P1634/47 **1T PCM, MIL: Yes** **Year:** 2011, 2012 **Model:** Civic, CR-Z, Insight **Engine:** 1.3L, 1.5L	**Motor Power Inverter (MPI) Module Signal Circuit Malfunction (HYBRID MODELS):** With the ignition ON and the MCM supply-power voltage a minimum of 9.0 V. One of the following conditions occurs: Condition 1: The number of abnormal MCM reception data or checksum abnormalities is at least three. Condition 2: Communication interruption does not occur on the MCM for at least 0.5 seconds. (IMA system indicator is ON) WARNING: The IMA system uses high voltage (144 V), before servicing turn OFF the battery module switch.
DTC: P1658 **1T PCM, MIL: Yes** **Year:** 2011, 2012 **Model:** Accord, Civic, Crosstour, CR-V, CR-Z, Fit, Insight, Odyssey, Pilot, Ridgeline **Engine:** All	**Electronic Throttle Control System (ETCS) Control Relay ON Malfunction:** The communication signal is input from the throttle actuator control module for at least 2 seconds, after the throttle actuator control module relay is turned off. The execution time is once per driving cycle. DTCs P0122, P0123, P0222, P0223, P1659, P2101, P2118, P2122, P2123, P2127, P2128, P2135, P2138, P2176 are not active. **NOTE: Before you troubleshoot, record all freeze data and any on-board snapshot.**
DTC: P1659 **1T PCM, MIL: Yes** **Year:** 2011, 2012 **Model:** Accord, Civic, Crosstour, CR-V, CR-Z, Fit, Insight, Odyssey, Pilot, Ridgeline **Engine:** All	**Electronic Throttle Control System (ETCS) Control Relay OFF Malfunction:** Ignition switch ON (II), battery voltage 10 volts minimum and DTCs P0122, P0123, P0222, P0223, P2101, P2118, P2122, P2123, P2127, P2128, P2135, P2138, P2176 are not active. The voltage is not applied from the throttle actuator controller relay for at least 200 milliseconds. The execution time is once per driving cycle. **NOTE: Before you troubleshoot, record all freeze data and any on-board snapshot.**
DTC: P1673/22 **1T PCM** **Year:** 2011, 2012 **Model:** Civic, CR-Z, Insight **Engine:** 1.3L, 1.5L	**Motor Control Module (MCM) Relay Stays Activated (HYBRID MODELS):** With the ignition ON and the MCM supply-power voltage a minimum of 9.0 V, the MCM power is not turned off for at least 5 seconds after the MCM relay 1 is directed to turn off. (IMA system indicator is ON) WARNING: The IMA system uses high voltage (144 V), before servicing turn OFF the battery module switch.
DTC: P1683 **1T PCM, MIL: Yes** **Year:** 2011, 2012 **Model:** All **Engine:** All	**Throttle Valve Default Position Spring Performance Problem:** Ignition switch LOCK (0), battery voltage a minimum of 6 volts, engine coolant minimum 158 ºF (70 ºC) and DTCs P0117, P0118, P0122, P0123, P0222, P0223, P2101, P2118, P2122, P2123, P2127, P2128, P2135, P2138, P2176 not active. The throttle valve position is more than +5 º from fully closed, or less than +3 º from fully closed for at least 3 seconds. **NOTE: Before you troubleshoot, record all freeze data and any on-board snapshot.**
DTC: P1684 **1T PCM, MIL: Yes** **Year:** 2011, 2012 **Model:** All **Engine:** All	**Throttle Valve Return Spring Performance Problem:** Ignition switch LOCK (0), coolant temperature minimum 158 ºF (70 ºC), battery voltage minimum 6 volts and DTCs P0117, P0118, P0122, P0123, P0222, P0223, P2101, P2118, P2122, P2123, P2127, P2128, P2135, P2138, P2176 not active. The throttle valve opening angle is 17 º or more, or 11 º or less, for at least 3 seconds. Before you troubleshoot, record all freeze data and any on-board snapshot.
DTC: P16BB **1T PCM** **Year:** 2011, 2012 **Model:** Accord, Civic, Crosstour, CR-V, Element, Fit, Odyssey, Pilot, Ridgeline **Engine:** 1.5L, 1.8L, 2.0L, 2.4L, 3.5L	**Alternator B Terminal Circuit Low Voltage:** Engine speed 500-3,000 rpm, alternator control mode 14.5 volts. The IGP terminal voltage is 12.0 V or less, and the alternator power generation amount is within 1.0 % to 50.0 %, for at least 60 seconds. The execution time is continuous. **NOTE: Before you troubleshoot, record all freeze data and any on-board snapshot.**
DTC: P16BC **1T PCM** **Year:** 2011, 2012 **Model:** Accord, Crosstour, CR-V, Element, Fit, Odyssey, Pilot, Ridgeline **Engine:** 1.5L, 2.4L, 3.5L	**Alternator FR Terminal Circuit/IGP Circuit Low Voltage:** Engine speed 500-3,000 rpm, alternator control mode 14.5 volts minimum. The IGP terminal voltage is 12.0 V or less, and the alternator power generation amount is 0.5 % or less, for at least 60 seconds. The execution time is continuous. **NOTE: Before you troubleshoot, record all freeze data and any on-board snapshot.**
DTC: P16BD **1T PCM** **Year:** 2011, 2012 **Model:** Odyssey, Pilot **Engine:** 3.5L	**Starter Cut Relay 2 Malfunction:** Engine running, starter switch OFF, DTC P16BE not active. The terminal voltage of the STRLD drops to 2.2 V for at least 5 seconds when the starter cut relay output is turned off. **NOTE: Before you troubleshoot, record all freeze data and any on-board snapshot.**

DTC	Trouble Code Title and Conditions
DTC: P16BE **1T PCM** **Year:** 2011, 2012 **Model:** Odyssey, Pilot **Engine:** 3.5L	**Starter Cut Relay 1 Malfunction:** Engine running, Starter switch OFF cut relay turn on switching failure, ON STRLD open circuit failure. One of these conditions occurs. (1) The terminal voltage of the STRLD exceeds 3.0 volts for at least 5 seconds when the starter cut relay output is turned off. (2) The terminal voltage of the STRLD is 2.4-2.6 volts for at least 0.8 second when the starter cut relay output is turned on. **NOTE: Before you troubleshoot, record all freeze data and any on-board snapshot.**
DTC: P16BF **1T PCM** **Year:** 2011, 2012 **Model:** Odyssey, Pilot **Engine:** 3.5L	**Starter Cut Relay STRLY Circuit Malfunction:** With ignition switch ON (II) the signal of the on command from the PCM to the starter and the return signal from the starter cut relay do not coincide for at least 5 seconds. The execution time is continuous. **NOTE: Before you troubleshoot, record all freeze data and any on-board snapshot.**
DTC: P16C0 **1T PCM, MIL: Yes** **Year:** 2011, 2012 **Model:** Civic, CR-Z **Engine:** 1.5L	**PCM CVT Control System Incomplete Update:** The program rewriting process does not finish normally due to an irregular process, such as turning off the PCM power during the CVT CPU rewriting process, and the ignition switch is turned to ON (II) again. **NOTE: Before you troubleshoot, record all freeze data and any on-board snapshots.** This code is indicated when PCM updating is incomplete. WARNING: Do not turn the ignition switch to LOCK (0) or ACCESSORY (I) while updating the PCM. If you turn the ignition switch to LOCK (0) or ACCESSORY (I) before completion, the PCM can be damaged.
DTC: P16C0 **1T PCM, MIL: Yes** **Year:** 2011, 2012 **Model:** Accord, Civic, Crosstour, CR-V, Odyssey, Pilot, Ridgeline **Engine:** 1.8L, 2.0L, 2.4L, 3.5L	**PCM A/T Control System Incomplete Update:** The program rewriting process does not finish normally due to an irregular process, such as turning off the PCM power during the A/T CPU rewriting process, and then turning the ignition switch to ON (II) again. The execution time is continuous and the duration time is 1 seconds or more. **NOTE: This code is indicated when PCM updating is incomplete.**
DTC: P16C0 **1T PCM, MIL: Yes** **Year:** 2011, 2012 **Model:** Fit **Engine:** 1.5L	**PCM A/T Control System Incomplete Update:** The program rewriting process does not finish normally due to an irregular process, such as turning off the PCM power during the A/T CPU rewriting process, and the ignition switch is turned to ON (II) again.
DTC: P16C0/99 **1T PCM, MIL: Yes, TCIL: Yes** **Year:** 2011, 2012 **Model:** Civic, Insight **Engine:** 1.3L	**Powertrain Control Module (PCM) CVT Control System Incomplete Update (Hybrid Models):** With the ignition switch ON, the program rewritiing process does not finish normally due to an irregular process, such as turning off the power during the CVT CPU rewriting process, and the ignition switch is turned to ON, again. This code is indicated when PCM updating is incomplete. **NOTE: Do not turn the ignition switch to LOCK (0) or ACCESSORY (I) while updating the PCM. If you turn the ignition switch to LOCK (0) or ACCESSORY (I) before completion, the PCM can be damaged.**
DTC: P16C3/31 **1T** **Year:** 2011, 2012 **Model:** Civic, CR-Z, Insight **Engine:** 1.3L, 1.5L	**DC-DC Converter Temperature Sensor Circuit Malfunction (HYBRID MODELS):** With the ignition ON and the MCM supply-power voltage a minimum of 10.5 V, the abnormal temperature sensor information continues for at least 2 seconds. (IMA system indicator is ON) WARNING: The IMA system uses high voltage (144 V), before servicing turn OFF the battery module switch.
DTC: P16C4 **1T PCM** **Year:** 2011, 2012 **Model:** Accord, Crosstour, Odyssey, Pilot **Engine:** 3.5L	**Engine Mount Actuator Control Power Circuit (Stuck Off):** With the ignition switch ON (II), the IGSOL voltage is 4.0 V or less for at least 2.6 seconds during the active engine mount (ACM) control relay is on. **NOTE: Before you troubleshoot, record all freeze data and any on-board snapshots, and review the general troubleshooting information.**
DTC: P16C5 **1T PCM** **Year:** 2011, 2012 **Model:** Accord, Crosstour, Odyssey, Pilot **Engine:** 3.5L	**Engine Mount Actuator Control Power Circuit (Stuck ON):** With the ignition ON (II), the IGSOL voltage is 8.0 V or more for at least 2.6 seconds during the active control engine mount (ACM) control relay is off. **NOTE: Before you troubleshoot, record all freeze data and any on-board snapshots, and review the general troubleshooting information.**

DTC	Trouble Code Title and Conditions
DTC: P16C6 **1T PCM** **Year:** 2011, 2012 **Model:** Accord, Crosstour, Odyssey, Pilot **Engine:** 3.5L	**Engine Mount Actuator High Voltage During Function Test:** With the ignition switch ON (II), one of these 3 conditions occur: 1). The value during one cycle of the front engine mount actuator current is 7.5 A or more as a maximum current or 5.3 A or more as a minimum current. 2). The engine mount cycle occurs during current start up and continues. 3). The duty cycle of the front engine mount actuator eads 100% all the time during engine mount operation and continues at least 20 seconds. **NOTE: Before you troubleshoot, record all freeze data and any on-board snapshots, and review the general troubleshooting information.**
DTC: P16D5 **1T PCM, MIL: Yes** **Year:** 2011, 2012 **Model:** Civic, CR-Z, Insight **Engine:** 1.3L, 1.5L	**F-CAN Malfunction (Internal Malfunction):** With the ignition switch ON, for 3 seconds and battery voltage at a minimum of 10.0 V, one of the following malfunction conditions exist. Condition 1: The PCM does not receive any signals via the F-CAN lines for at least 1 second. Condition 2: The information sent from the CVT CPU is abnormal at least 20 times.
DTC: P16D6 **1T PCM, MIL: Yes** **Year:** 2011, 2012 **Model:** Civic, CR-Z, Insight **Engine:** 1.3L, 1.5L	**IMA-CAN Malfunction (Internal Malfunction) (HYBRID MODELS):** With the Ignition ON an minimum of 3 seconds, and battery voltage minimum of 10.0 V, one of the following conditions exist, Condition 1: The PCM does not receive any signals via the IMA-CAN lines. Condition 2: The information sent from the CVT CPU is abnormal at least 20 times.
DTC: P16D7 **1T PCM** **Year:** 2011, 2012 **Model:** Civic, CR-Z **Engine:** 1.5L	**PCM Internal F-CAN Communication Circuit Malfunction:** No signals from the FI CPU via the F-CAN are received for at least 1.5 seconds.
DTC: P16D7/107 **1T PCM, TCIL: Yes** **Year:** 2011, 2012 **Model:** Civic, Insight **Engine:** 1.3L	**Powertrain Control Module (PCM) Internal F-CAN Communication Circuit Malfunction:** With the ignition switch ON for at least 5 seconds and the battery voltage at a minimum of 10.0 V, no signal from the FI CPU via the F-CAN are received for at least 1.5 seconds.
DTC: P16D8 **1T PCM, MIL: Yes** **Year:** 2011, 2012 **Model:** Civic, CR-Z **Engine:** 1.5L	**PCM Internal IMA-CAN Communication Circuit Malfunction :** No signals from the FI CPU via the IMA-CAN are received for at least 1.5 seconds.
DTC: P16D8/120 **1T PCM, MIL: Yes, TCIL: Yes** **Year:** 2011, 2012 **Model:** Civic, Insight **Engine:** 1.3L	**PCM Internal IMA-CAN Communication Circuit Malfunction (HYBRID MODELS):** With the ignition switch ON for at least 5 seconds and the battery voltage a minimum of 10.0 V, no signals from the FI CPU via the IMA-CAN are received for at least 1.5 seconds.
DTC: P16E2 **1T PCM** **Year:** 2011, 2012 **Model:** Civic **Engine:** 1.3L, 1.8L, 2.0L	**PGM-FI-ACG LIN Communication Error:** Either one of the conditions is met for at least 5 seconds: * The ECM/PCM cannot receive any information via the LIN line. * The information sent from the alternator is abnormal.
DTC: P16E3 **1T PCM** **Year:** 2011, 2012 **Model:** Civic **Engine:** 1.3L, 1.8L, 2.0L	**PGM-FI-Battery Sensor LIN Communication Error:** Either one of the conditions is met for at least 5 seconds: * The ECM/PCM cannot receive any information via the LIN line. * The information sent from the battery sensor is abnormal.
DTC: P16E4 **1T PCM** **Year:** 2011, 2012 **Model:** Civic **Engine:** 1.3L, 1.8L, 2.0L	**ACG High-temperature:** The alternator internal temperature is 190 °C (374 °F) or more for at least 60 seconds.

DTC	Trouble Code Title and Conditions
DTC: P1717 **2T PCM, MIL: Yes, TCIL: Yes** **Year:** 2011, 2012 **Model:** Civic, Crosstour, CR-V, Element, Fit, Odyssey, Ridgeline **Engine:** 1.5L, 1.8L, 2.4L, 3.5L	**Open in Transmission Range Switch ATPRVS Switch Circuit :** With the vehicle driven in reverse at a speed of 3 (5 km/h) mph or less for at least 2 seconds, no RVS signal is detected.
DTC: P1717 **2T PCM, MIL: Yes** **Year:** 2011, 2012 **Model:** Accord, Crosstour **Engine:** 2.4L, 3.5L	**Open in Transmission Range Switch ATPRVS Switch Circuit:** No RVS signal is detected with the shift lever in R while shifting between P, R, and N.
DTC: P1730 **2T PCM, MIL: Yes, TCIL: Yes** **Year:** 2011, 2012 **Model:** Accord, Crosstour, CR-V **Engine:** 2.4L	**Problem in Shift Control System:** With the engine running and in Drive position allow trans mission to shift to 5th gear. Shift solenoid A or D stuck OFF, Shift solenoid B stuck ON, Shift Valves A, B or D stuck.
DTC: P1731 **2T PCM, MIL: Yes, TCIL: Yes** **Year:** 2011, 2012 **Model:** Accord, Crosstour, CR-V, Element **Engine:** 2.4L	**Problem in Shift Control System:** With the engine running and in Drive position allow transmission to shift to 5th gear. Shift solenoid E stuck ON, Shift solenoid E stuck, A/T Clutch pressure control solenoid A stuck OFF. **NOTE: Before you troubleshoot, record all freeze data and any on-board snapshot with the HDS, and review General Troubleshooting Information.**
DTC: P1732 **2T PCM, MIL: Yes, TCIL: Yes** **Year:** 2011, 2012 **Model:** Accord, Crosstour, CR-V **Engine:** 2.4L	**Problem in Shift Control System: Shift Solenoid B or C Stuck ON:** With the engine running and in Drive position allow transmission to shift to 5th gear. Shift solenoid B or C stuck ON, Shift solenoid B or C Stuck. **NOTE: Before you troubleshoot, record all freeze data and any on-board snapshot with the HDS, and review General Troubleshooting Information.**
DTC: P1733 **2T PCM, MIL: Yes, TCIL: Yes** **Year:** 2011, 2012 **Model:** Accord, Crosstour, CR-V, Element **Engine:** 2.4L	**Problem in Shift Control System:** With the engine running and in Drive position allow transmission to shift to 5th gear. Shift solenoid D stuck ON, Shift solenoid D Stuck. A/T clutch pressure control switch valve C Stuck OFF. **NOTE: Before you troubleshoot, record all freeze data and any on-board snapshot with the HDS, and review General Troubleshooting Information.**
DTC: P1734 **2T PCM, MIL: Yes, TCIL: Yes** **Year:** 2011, 2012 **Model:** Accord, Crosstour, CR-V, Element **Engine:** 2.4L	**Problem in Shift Control System: :** With the engine running and in Drive position allow transmission to shift to 5th gear. Shift solenoid B or C stuck OFF, Shift solenoid B or C Stuck. **NOTE: Before you troubleshoot, record all freeze data and any on-board snapshot with the HDS, and review General Troubleshooting Information.**
DTC: P1743 **2T PCM, TCIL: Yes** **Year:** 2011, 2012 **Model:** Accord, Crosstour, Odyssey, Ridgeline **Engine:** 3.5L	**Problem in Shift Control System; Shift Valve E Stuck OFF:** Vehicle speed is 5 mph (9 km/h), torque converter slip ratio 96-110 %. battery minimun of 11 volts. The transmission is neutral against the 5th gear shift command for at least 2 seconds, without records that the engine speed flares when upshifting to 3rd-4th. The execution time is continuous depending on the driving pattern. **NOTE: Before you troubleshoot, record all freeze data and any on-board snapshot,.**
DTC: P1744 **2T PCM, TCIL: Yes** **Year:** 2011, 2012 **Model:** Accord, Crosstour, Odyssey, Ridgeline **Engine:** 3.5L	**Problem in Shift Control System; Shift Valve E Stuck ON:** The transmission is held in 5th gear against the 1st gear shift command for at least 2 seconds. After driving the vehicle in 3rd gear in D for at least 2 seconds. The execution time is continuous depending on the driving pattern. **NOTE: Before you troubleshoot, record all freeze data and any on-board snapshot.**
DTC: P1745 **2T PCM, TCIL: Yes** **Year:** 2011, 2012 **Model:** Accord, Crosstour, Odyssey, Ridgeline **Engine:** 3.5L	**Problem in Shift Control System; Servo Control Valve Stuck OFF or Servo Valve Stuck OFF:** The engine speed flares when upshifting to 2nd-3rd, the engine speed flares when upshifting to 3rd-4th, driving in 5th gear neutral condition. The execution time is continuous and the duration time is 1-2 seconds or more. **NOTE: Before you troubleshoot, record all freeze data and any on-board snapshot.**

DTC	Trouble Code Title and Conditions
DTC: P1746 **2T PCM, TCIL: Yes** **Year:** 2011, 2012 **Model:** Fit **Engine:** 1.5L	**Hydraulic Control System (Cut Valve A Stuck OFF or Cut Valve B Stuck ON):** With the vehicle driven at a minimum of 5 mph (9km/h) and the accelerator position pedal a minimum of 2.0%, an improper gear ratio output compared to the predetermined gear change mode, the cut valve A OFF failure or the cut valve B ON failure is detected. (Poor shift quality) **NOTE: This code is stored simultaneously with DTC P1780, which is caused by a hydraulic control system problem.**
DTC: P1746 **1T PCM** **Year:** 2011, 2012 **Model:** Civic **Engine:** 1.3L, 1.8L	**Hydraulic Control System (Cut A Valve Stuck OFF or Cut B Valve Stuck ON):** Transmission shift malfunction.
DTC: P1747 **2T PCM, TCIL: Yes** **Year:** 2011, 2012 **Model:** Fit **Engine:** 1.5L	**Hydraulic Control System (Cut Valve A Stuck ON or Cut Valve B Stuck OFF):** With the vehicle driven at a minimum of 5 mph (9km/h) and the accelerator position pedal a minimum of 2.0%, an improper gear ratio output compared to the predetermined gear change mode, the cut valve A ON failure or the cut valve B OFF failure is detected. (Poor shift quality) **NOTE: This code is stored simultaneously with DTC P1780, which is caused by a hydraulic control system problem.**
DTC: P1747 **1T PCM** **Year:** 2011, 2012 **Model:** Civic **Engine:** 1.3L, 1.8L	**Hydraulic Control System (Cut A Valve Stuck ON or Cut B Valve Stuck OFF):** One of these symptoms occurs. Symptom 1 (Symptom 1-B occurs after Symptom 1-A occurs) Symptom 2 Symptom Gear position commanded by the PCM Actual gear position 1-A N → D Driving in 1st gear 1-B Driving in 1st gear Neutral Symptom Gear position commanded by the PCM Actual gear position 2 Driving in 1st gear Neutral
DTC: P177A **2T PCM, TCIL: Yes** **Year:** 2011, 2012 **Model:** Odyssey **Engine:** 3.5L	**Line Pressure Solenoid Valve A Stuck OFF, or Line Pressure Switch Stuck OFF:** The input signal from the line pressure switch to the PCM is high when the low line pressure mode is commanded. The execution time is continuous and the duration time is 4 seconds or more.
DTC: P1780 **2T PCM, MIL: Yes, TCIL: Yes** **Year:** 2011, 2012 **Model:** Accord, Crosstour, Fit, Odyssey, Ridgeline **Engine:** 1.5L, 3.5L	**Problem in Shift Control System (Transmission is in Default Mode):** The A/T control switches to the default mode due to a mechanical malfunction. The execution time is continuous depending on the driving pattern. **NOTE: DTC P1780 means there is one or more A/T DTCs about the shift control system. Before you troubleshoot, record all freeze data and any on-board snapshot.**
DTC: P1860/33 **1T PCM, TCIL: Yes** **Year:** 2011, 2012 **Model:** Civic, CR-Z, Insight **Engine:** 1.3L, 1.5L	**Inhibitor Solenoid Circuit Low Voltage (Hybrid Models):** When the transmission gear selector is shifted into the reverse position by mistake during the forward operation, the inhibitor solenoid intercepts the activation of the reverse brake operation to prevent transmission damage. The return signal is inappropriate for the turn on command to the inhibitor solenoid. **NOTE: This code is caused by an electrical circuit problem and cannot be caused by a mechanical problem in the transmission.**
DTC: P1861/33 **1T PCM, TCIL: Yes** **Year:** 2011, 2012 **Model:** Civic, CR-Z, Insight **Engine:** 1.3L, 1.5L	**Inhibitor Solenoid Circuit High Voltage:** When the transmission gear selector is shifted into the reverse position by mistake during the forward operation, the inhibitor solenoid intercepts the activation of the reverse brake operation to prevent transmission damage. The return signal is inappropriate for the turn on command to the inhibitor solenoid. **NOTE: This code is caused by an electrical circuit problem and cannot be caused by a mechanical problem in the transmission.**
DTC: P1890/42 **1T PCM, TCIL: Yes** **Year:** 2011, 2012 **Model:** Civic, CR-Z, Insight **Engine:** 1.3L, 1.5L	**Problem in CVT Speed Control System:** With the vehicle driven at 38 mph (60 km/h) for at least 1 minute, the difference between the target drive pulley speed and the actual drive pulley speed is 500 rpm or more for at least 18 seconds. **NOTE: Before you troubleshoot, record all freeze data and any on-board snapshot with the HDS, and review General Troubleshooting Information.**
DTC: P1891/43 **1T PCM, TCIL: Yes** **Year:** 2011, 2012 **Model:** Civic, CR-Z, Insight **Engine:** 1.3L, 1.5L	**Problem in Start Clutch Control System:** With the vehicle driven at 38 mph (60 km/h), the deviation in the vehicle speed measured by the vehicle speed sensor and vehicle speed measured by the driven pulley speed sensor is in correct.

DTC	Trouble Code Title and Conditions
DTC: P1898/100 **2T PCM, MIL: Yes, TCIL: Yes** **Year:** 2011, 2012 **Model:** Civic, CR-Z, Insight **Engine:** 1.3L, 1.5L	**CVT Drive Pulley Pressure Control Valve Stuck ON or CVT Driven Pulley Pressure Control Valve Stuck OFF:** With the vehicle driven between 40-50 mph (65-80 km/h) on a flat road for at least 15 seconds, then stop the vehicle pressing the brake pedal for 5 seconds. Both of the following judgment malfunctions must be met. Judgment A: The pulley ratio is 2.2-2.7 for at least 0.4 second. Judgment B: The pulley ratio is 1.8 to 2.7 for at least 5 seconds.
DTC: P1899/100 **2T PCM, MIL: Yes, TCIL: Yes** **Year:** 2011, 2012 **Model:** Civic, Insight **Engine:** 1.3L	**CVT Drive Pulley Pressure Control Valve Stuck OFF or CVT Driven Pulley Pressure Control Valve Stuck ON:** With the vehicle driven between 40-50 mph (65-80 km/h) on a flat road for at least 15 seconds, then stop the vehicle pressing the brake pedal for 5 seconds. Both of the following judgment malfunctions must be met. Judgment A: The pulley ratio is 0.3-0.8 for at least 2.5 seconds. Judgment B: The pulley ratio is 3 for at least 12 seconds.

OBD II Trouble Code List (P2XXX Codes)

DTC	Trouble Code Title and Conditions
DTC: P2101 **1T PCM, MIL: Yes** **Year:** 2011, 2012 **Model:** All **Engine:** All	**Electronic Throttle Control System (ETCS) Malfunction (Includes Hybrid Models):** Ignition switch ON (II), battery voltage minimum of 6 volts and DTCs P0122, P0123, P0222, P0223, P2118, P2122, P2123, P2127, P2128,P2135, P2138, P2176 are not active. Difference between throttle valve target position and actual throttle valve position, 4-6 º or more. The execution time is continuous and the duration time is 250-500 milliseconds or more. **NOTE: Before you troubleshoot, record all freeze data and any on-board snapshot.**
DTC: P2108 **1T PCM, MIL: Yes** **Year:** 2011, 2012 **Model:** Accord, Crosstour, CR-V, Element **Engine:** 2.4L, 3.5L	**Throttle Actuator Control Module Problem:** Withe the ignition ON (II) one of these 4 conditions must be met for at least 200 milliseconds. (1) Data read from the ROM is abnormal. (2) Data read from the RAM is abnormal. (3) The A/D converter standard voltage is out of specified value. (4) The serial signals between the PCM and the throttle actuator control module do not agree. **NOTE: Before you troubleshoot, record all freeze data and any on-board snapshot.**
DTC: P2118 **1T PCM, MIL: Yes** **Year:** 2011, 2012 **Model:** Accord, Civic, Crosstour, CR-V, CR-Z, Fit, Odyssey, Pilot, Ridgeline **Engine:** All	**Throttle Actuator Current Range/Performance Problem (Includes Hybrid Models):** With ignition ON (II), battery voltage minimum of 6 volts and DTCs P0122, P0123, P0222, P0223, P2101, P2122, P2123, P2127, P2128, P2135, P2138, P2176 are not active. The current flow to the throttle actuator is 11 A or more for at least 200 milliseconds. The execution time is continuous. **NOTE: Before you troubleshoot, record all freeze data and any on-board snapshot.**
DTC: P2122 **1T PCM, MIL: Yes** **Year:** 2011, 2012 **Model:** All **Engine:** All	**APP Sensor A or 1 (TP Sensor D) Circuit Low Voltage (Includes Hybrid Models):** With the ignition switch ON (II) the APP sensor A output voltage is 0.2 V or less for at least 200 milliseconds. The execution time is continuous. DTC P2123 is not active. **NOTE: Before you troubleshoot, record all freeze data and any on-board snapshot.**
DTC: P2123 **1T PCM, MIL: Yes** **Year:** 2011, 2012 **Model:** All **Engine:** All	**APP Sensor A or 1 (TP Sensor D) Circuit High Voltage (Includes Hybrid Models):** With engine running the APP sensor A output voltage is 4.9 V or more for at least 200 milliseconds. The execution time is continuous. DTC P2122 is not active.
DTC: P2127 **1T PCM, MIL: Yes** **Year:** 2011, 2012 **Model:** All **Engine:** All	**APP Sensor B or 2 (Throttle Position (TP) Sensor E) Circuit Low Voltage (Includes Hybrid Models):** With the ignition ON (II) the APP sensor B output voltage is 0.2 V or less for at least 200 milliseconds. The execution time is continuous. DTC P2128 is not active. **NOTE: Before you troubleshoot, record all freeze data and any on-board snapshot.**
DTC: P2128 **1T PCM, MIL: Yes** **Year:** 2011, 2012 **Model:** All **Engine:** All	**APP Sensor B or 2 (Throttle Position (TP) Sensor E) Circuit High Voltage (Includes Hybrid Models):** With engine running the APP sensor B output voltage is 4 V or more for at least 200 milliseconds. The execution time is continuous. DTC P2127 is not active. **NOTE: Before you troubleshoot, record all freeze data and any on-board snapshot.**

DTC	Trouble Code Title and Conditions
DTC: P2129 **2T PCM, MIL: Yes** **Year:** 2011, 2012 **Model:** Civic **Engine:** 1.3L, 1.8L, 2.0L	**Barometric Pressure (BARO) Sensor Circuit High Voltage:** The BARO sensor output voltage [BARO SENSOR] is 4.49 V or more for at least 2.0 seconds.
DTC: P2135 **1T PCM, MIL: Yes** **Year:** 2011, 2012 **Model:** All **Engine:** All	**Throttle Position (TP) Sensor A/B or 1/2 Incorrect Voltage Correlation (Includes Hybrid Models):** The difference between the throttle valve positions indicated by TP sensor A and TP sensor B exceeds the value as follows, for at least 200 milliseconds. Difference between TP sensor A and TP sensor B 1.8-14.7º or more, or 5º or less. **NOTE: Before you troubleshoot, record all freeze data and any on-board snapshot.**
DTC: P2138 **1T PCM, MIL: Yes** **Year:** 2011, 2012 **Model:** All **Engine:** All	**APP Sensor A/B or 1/2 (Throttle Position (TP) Sensor D/E) Incorrect Voltage Correlation (Includes Hybrid Models):** With the ignition ON (II) one of these conditions must be met for at least 300 milliseconds: (1) APP sensor B voltage exceeds the range from 0 V or less to 0.37 V or more when the APP sensor A voltage is 0.37 V. (2) APP sensor B voltage exceeds the range from 2.31 V or less to 2.69 V or more when the APP sensor A voltage is 5 V. The execution time is continuous and DTCs P2122, P2123, P2127, P2128 are not active. **NOTE: Before you troubleshoot, record all freeze data and any on-board snapshot.**
DTC: P2176 **1T PCM, MIL: Yes** **Year:** 2011, 2012 **Model:** All **Engine:** All	**Throttle Actuator Control System Idle Position Not Learned (Includes Hybrid Models):** One of these condition must be met for at least 0.7 seconds. (1) The registration of the throttle valve fully closed position is not completed within a predetermined time after the ignition switch is turned ON. (2) The registered value of the throttle valve fully closed position is 0.74 volts TP1, 1.61 volts TP2 or more, or 0.49 volts TP1, 1.37 volts TP2 or less. The execution time is once per driving cycle and DTCs P0122, P0123, P0222, P0223, P2101, P2118, P2135 are not active. **NOTE: If DTC P2135 is stored at the same time as DTC P2176, troubleshoot DTC P2135 first, then recheck for DTC P2176.**
DTC: P2183 **2T PCM, MIL: Yes** **Year:** 2011, 2012 **Model:** All **Engine:** All	**Engine Coolant Temperature (ECT) Sensor 2 Circuit Range/Performance Problem (Includes Hybrid Models):** A malfunction is detected if the following three conditions are present after the engine has stopped and the ignition switch has been turned to LOCK (0) for at least 6 hours before restarting the engine: (1) When the temperature difference between the IAT and ECT1 is 57 ºF (32 ºC) or more. (2) When the temperature difference between the IAT and ECT2 is 30 ºF (17 ºC) or more. (3) When the temperature difference between the ECT2 and ECT1 is 73 ºF (41 ºC) or more. The execution time is once per driving cycle and the duration time is 10 seconds or more. **NOTE: If DTC P0111 is stored at the same time as DTC P2183, troubleshoot DTC P0111 first, then recheck for DTC P2183.**
DTC: P2184 **1T PCM, MIL: Yes** **Year:** 2011, 2012 **Model:** Accord, Civic, Crosstour, CR-V, CR-Z, Fit, Insight, Odyssey, Pilot, Ridgeline **Engine:** All	**Engine Coolant Temperature (ECT) Sensor 2 Circuit Low Voltage (Includes Hybrid Models):** With ignition switch ON (II) the ECT sensor 2 output voltage is 0.08 V or less for at least 2 seconds. The execution time is continuous. DTC P2185 is not active. **NOTE: Before you troubleshoot, record all freeze data and any on-board snapshot.**
DTC: P2185 **1T PCM, MIL: Yes** **Year:** 2011, 2012 **Model:** All **Engine:** All	**Engine Coolant Temperature (ECT) Sensor 2 Circuit High Voltage (Includes Hybrid Models):** With ignition ON (II) the ECT sensor 2 output voltage is 4.92 volts or more for at least 2 seconds. The execution time is continuous and DTC P2184 is not active. **NOTE: Before you troubleshoot, record all freeze data and any on-board snapshot.**
DTC: P2195 **1T PCM, MIL: Yes** **Year:** 2011, 2012 **Model:** Accord, Civic, Crosstour, CR-V, CR-Z, Element, Fit, Insight **Engine:** 1.3L, 1.5L, 1.8L, 2.0L, 2.4L	**A/F Sensor (Sensor 1) Signal Stuck Lean (Includes Hybrid Models):** The A/F sensor (Sensor 1) output voltage is 2.5 V or more for at least 7 seconds. The execution time is once per driving cycle and DTCs P0134, P0135, P0171, P0300, P0301, P0302, P0303, P0304, P0305, P0306, P0627, P1172, P2237, P2238, P2243, P2245, P2251, P2252 are not active. **NOTE: If DTC P2101, P2118, P2135, P2138, P2176, or a combination of P2122 and P2127, P2122 and P2138 or P2127 and P2138 is stored at the same time as DTC P2195, troubleshoot those DTCs first, then recheck for DTC P2195.**
DTC: P2195 **2T PCM, MIL: Yes** **Year:** 2011, 2012 **Model:** Accord, Crosstour, Odyssey, Pilot, Ridgeline **Engine:** 3.5L	**Rear Air/Fuel Ratio (A/F) Sensor (Bank 1, Sensor 1) Signal Stuck Lean:** The rear A/F sensor (bank 1, sensor 1) output voltage is 3.48 V or more for at least 7 seconds. The execution time is once per driving cycle and DTCs P0134, P0135, P0171, P0300, P0301, P0302, P0303, P0304, P0305, P0306, P0627, P1172, P2237, P2238, P2243, P2245, P2251, P2252 are not active. **NOTE: If DTC P2101, P2118, P2135, P2138, P2176, or a combination of P2122 and P2127, P2122 and P2138 or P2127 and P2138 is stored at the same time as DTC P2195, troubleshoot those DTCs first, then recheck for DTC P2195.**

DTC	Trouble Code Title and Conditions
DTC: P2197 **2T PCM, MIL: Yes** **Year:** 2011, 2012 **Model:** Accord, Crosstour, Odyssey, Pilot, Ridgeline **Engine:** 3.5L	**Front Air/Fuel Ratio (A/F) Sensor (Bank 2, Sensor 1) Signal Stuck Lean:** The front A/F sensor (bank 2, sensor 1) output voltage is 3.48 V or more for at least 7 seconds. The execution time is once per driving cycle and DTCs P0154, P0155, P0174, P0300, P0301, P0302, P0303, P0304, P0305, P0306, P0627, P1174, P2240, P2241, P2247, P2249, P2254, P2255 are not active. **NOTE: If DTC P2101, P2118, P2135, P2138, P2176, or a combination of P2122 and P2127, P2122 and P2138 or P2127 and P2138 is stored at the same time as DTC P2195, P2197, troubleshoot those DTCs first, then recheck for DTC P2197.**
DTC: P219A **2T PCM, MIL: Yes** **Year:** 2011, 2012 **Model:** Civic, CR-Z **Engine:** 1.3L, 1.5L, 1.8L, 2.0L	**Air/Fuel Ratio Variation of Cylinders:** The integrated A/F sensor (sensor 1) output variation value is 56.67 or more in 4.4 seconds.
DTC: P2227 **2T PCM, MIL: Yes** **Year:** 2011, 2012 **Model:** All **Engine:** All	**Barometric Pressure (BARO) Sensor Circuit Range/Performance Problem:** Throttle position 17.0-33.0 degrees at 1,100-3,000 rpm. The difference between the BARO sensor output and the MAP sensor output is 26 kPa (7.5 in.Hg, 190 mmHg) or more for at least 2.5 seconds. The execution time is once per driving cycle and the duration time is 2.5 seconds or more. **NOTE: If DTC P0107, P0108, P1128, and/or P1129 are stored at the same time as DTC P2227, troubleshoot those DTCs first, then recheck for DTC P2227.**
DTC: P2228 **1T PCM, MIL: Yes** **Year:** 2011, 2012 **Model:** All **Engine:** All	**Barometric Pressure (BARO) Sensor Circuit Low Voltage:** With ignition ON (II) the BARO sensor output voltage is 1.31 V or less for at least 2 seconds. The execution time is continuous and DTCs P1109, P2229 are not active. **NOTE: Before you troubleshoot, record all freeze data and any on-board snapshot.**
DTC: P2229 **1T PCM, MIL: Yes** **Year:** 2011, 2012 **Model:** All **Engine:** All	**Barometric Pressure (BARO) Sensor Circuit High Voltage:** With ignition switch ON (II) the BARO sensor output voltage is 4.49 V or more for at least 2 seconds. The execution time is continuous and DTCs P1109, P2228 are not active. **NOTE: Before you troubleshoot, record all freeze data and any on-board snapshot.**
DTC: P2237 **1T PCM, MIL: Yes** **Year:** 2011, 2012 **Model:** Accord, Crosstour, Odyssey, Pilot, Ridgeline **Engine:** 3.5L	**Rear Air/Fuel Ratio (A/F) Sensor (Bank 1, Sensor 1) IP Circuit High Voltage:** With the engine running the IPB1 terminal voltage is 2 volts or less, or 5.6 volts or more, for at least 15 seconds. The execution time is continuous and DTCs P0135, P2195, P2238, P2243, P2245, P2251, P2252, P2627, P2628 are not active. **NOTE: Before you troubleshoot, record all freeze data and any on-board snapshot.**
DTC: P2238 **1T PCM, MIL: Yes** **Year:** 2011, 2012 **Model:** Accord, Crosstour, Odyssey, Pilot, Ridgeline **Engine:** 3.5L	**Rear Air/Fuel Ratio (A/F) Sensor (Bank 1, Sensor 1) IP Circuit Low Voltage:** With engine running the IPB1 input terminal voltage is 1 volts or less for at least 5 seconds after the sensor becomes active and 85 seconds before the sensor becomes active. The execution time is continuous, DTCs P0135, P2195, P2237, P2243, P2245, P2251, P2252, P2627, P2628 are not active. **NOTE: Before you troubleshoot, record all freeze data and any on-board snapshot.**
DTC: P2238 **1T PCM, MIL: Yes** **Year:** 2011, 2012 **Model:** Accord, Civic, Crosstour, CR-V, CR-Z, Element, Fit, Insight **Engine:** 1.3L, 1.5L, 1.8L, 2.0L, 2.4L	**Air/Fuel Ratio (A/F) Sensor (Sensor 1) AFS+ Circuit Low Voltage:** With the engine running and battery voltage at a minimum of 10.5 V, the AFS+ voltage is 0.4 V or less for at least 4 seconds. **NOTE: Before you troubleshoot, record all freeze data and any on-board snapshot, and review the general troubleshooting information.**
DTC: P2240 **1T PCM, MIL: Yes** **Year:** 2011, 2012 **Model:** Accord, Crosstour, Odyssey, Pilot, Ridgeline **Engine:** 3.5L	**Front Air/Fuel Ratio (A/F) Sensor (Bank 2, Sensor 1) IP Circuit High Voltage:** With the engine running the IPB2 terminal voltage is 5.6 volts or more, or 2 volts or less, for at least 15 seconds. The execution time is continuous and DTCs P0155, P2197, P2240, P2241, P2247, P2249, P2254, P2255, P2630, P2631 are not active. **NOTE: Before you troubleshoot, record all freeze data and any on-board snapshot.**
DTC: P2241 **1T PCM, MIL: Yes** **Year:** 2011, 2012 **Model:** Accord, Crosstour, Odyssey, Pilot, Ridgeline **Engine:** 3.5L	**Front Air/Fuel Ratio (A/F) Sensor (Bank 2, Sensor 1) IP Circuit Low Voltage:** With engine running the IPB1 input terminal voltage is 1 volts or less for at least 5 seconds after the sensor becomes active and 85 seconds before the sensor becomes active. The execution time is continuous DTCs P0135, P2195, P2237, P2243, P2245, P2251, P2252, P2627, P2628 are not active. **NOTE: Before you troubleshoot, record all freeze data and any on-board snapshot.**

DTC	Trouble Code Title and Conditions
DTC: P2243 **1T PCM, MIL: Yes** **Year:** 2011, 2012 **Model:** Accord, Crosstour, Odyssey, Pilot, Ridgeline **Engine:** 3.5L	**Rear Air/Fuel Ratio (A/F) Sensor (Bank 1, Sensor 1) VCENT Circuit High Voltage:** With engine running the VSB1 terminal voltage repeatedly fluctuates from a value above 4.8 volts to a value below 3.4 volts, at least 150 times. The execution time is continuous and DTCs P0135, P2195, P2237, P2238, P2245, P2251, P2252, P2627, P2628 are not active. **NOTE: Before you troubleshoot, record all freeze data and any on-board snapshot.**
DTC: P2245 **1T PCM, MIL: Yes** **Year:** 2011, 2012 **Model:** Accord, Crosstour, Odyssey, Pilot, Ridgeline **Engine:** 3.5L	**Rear Air/Fuel Ratio (A/F) Sensor (Bank 1, Sensor 1) VCENT Circuit Low Voltage:** With engine running the IPB1 input terminal voltage is 1 volts or less for at least 5 seconds after the sensor becomes active, 85 seconds before the sensor becomes active. The execution time is continuous and DTCs P0135, P2195, P2237, P2238, P2243, P2251, P2252, P2627, P2628 are not active.
DTC: P2247 **1T PCM, MIL: Yes** **Year:** 2011, 2012 **Model:** Accord, Crosstour, Odyssey, Pilot, Ridgeline **Engine:** 3.5L	**Front A/F Sensor (Bank 2, Sensor 1) VCENT Circuit High Voltage :** With engine running the VSB1 terminal voltage repeatedly fluctuates from a value above 4.8 volts to a value below 3.4 volts, at least 150 times. The execution time is continuous and DTCs P0135, P2195, P2237, P2238, P2245, P2251, P2252, P2627, P2628 are not active. **NOTE: Before you troubleshoot, record all freeze data and any on-board snapshot.**
DTC: P2249 **1T PCM, MIL: Yes** **Year:** 2011, 2012 **Model:** Accord, Crosstour, Odyssey, Pilot, Ridgeline **Engine:** 3.5L	**Front A/F Sensor (Bank 2, Sensor 1) VCENT Circuit Low Voltage :** With the engine running the VSB2 terminal voltage between reads 0.3-1.5 volts.The execution time is continuous and DTCs P0155, P2197, P2240, P2241, P2247, P2254, P2255 are not active. The IPB2 input terminal voltage is 1.0 volts or less for at least 5 seconds, after the sensor becomes active, 85 seconds before the sensor becomes active. **NOTE: Before you troubleshoot, record all freeze data and any on-board snapshot.**
DTC: P2251 **1T PCM, MIL: Yes** **Year:** 2011, 2012 **Model:** Accord, Crosstour, Odyssey, Pilot, Ridgeline **Engine:** 3.5L	**Rear Air/Fuel Ratio (A/F) Sensor (Bank 1, Sensor 1) VS Circuit High Voltage:** With engine running the VSB1 terminal voltage is 6 V or more, and the PCM internal signal voltage is 4.6 V or more, for at least 5 seconds. The execution time is continuous and DTCs P0135, P2195, P2237, P2238, P2243, P2245, P2252, P2627, P2628 are not active. **NOTE: If DTC P2251 is stored at the same time as DTC P0134, troubleshoot DTC P2251 first, then recheck for P0134.**
DTC: P2252 **1T PCM, MIL: Yes** **Year:** 2011, 2012 **Model:** Accord, Civic, Crosstour, CR-V, CR-Z, Element, Fit, Insight **Engine:** 1.3L, 1.5L, 1.8L, 2.0L, 2.4L	**Air/Fuel Ratio (A/F) Sensor (Sensor 1) AFS- Circuit Low Voltage :** With the engine running the AFS-terminal voltage is 0.4 V or less for at least 5 seconds. **NOTE: Before you troubleshoot, record all freeze data and any on-board snapshot, and review the general troubleshooting information.**
DTC: P2252 **1T PCM, MIL: Yes** **Year:** 2011, 2012 **Model:** Accord, Crosstour, Odyssey, Pilot, Ridgeline **Engine:** 3.5L	**Rear Air/Fuel Ratio (A/F) Sensor (Bank 1, Sensor 1) VS Circuit Low Voltage:** With engine running the VSB1 terminal voltage is 0.3 V or less, and the IPB1 terminal voltage is 1 V or more, for at least 5 seconds after the sensor becomes active, 85 seconds before the sensor becomes active.
DTC: P2254 **1T PCM, MIL: Yes** **Year:** 2011, 2012 **Model:** Accord, Crosstour, Odyssey, Pilot, Ridgeline **Engine:** 3.5L	**Front A/F Sensor (Bank 2, Sensor 1) VS Circuit High Voltage:** With engine running the VSB2 terminal voltage is 6 V or more, and the PCM internal signal voltage is 4.6 V or more, for at least 5 seconds. The execution time is continuous and DTCs P0155, P2197, P2240, P2241, P2247, P2249, P2255, P2630, P2631 are not active. **NOTE: If DTC P2254 is stored at the same time as DTC P0154, troubleshoot DTC P2254 first, then recheck for P0154.**
DTC: P2255 **1T PCM, MIL: Yes** **Year:** 2011, 2012 **Model:** Accord, Crosstour, Odyssey, Pilot, Ridgeline **Engine:** 3.5L	**Front A/F Sensor (Bank 2, Sensor 1) VS Circuit Low Voltage :** With engine running the VSB2 terminal voltage is 0.3 V or less, and the IPB2 terminal voltage is 1 V or more, for at least 5 seconds after the sensor is active, 85 seconds before the sensor becomes active. **NOTE: Before you troubleshoot, record all freeze data and any on-board snapshot.**

DTC	Trouble Code Title and Conditions
DTC: P2270 **2T PCM, MIL: Yes** **Year:** 2011, 2012 **Model:** Accord, Civic, Crosstour, CR-V, Element **Engine:** 1.3L, 1.8L, 2.0L, 2.4L	**Secondary HO2S (Sensor 2) Circuit Signal Stuck Lean:** With the engine running, and in closed loop with a speed of 30 mph or more, the secondary HO2S output voltage is 0.650 V or less. **NOTE: Before you troubleshoot, record all freeze data and any on-board snapshot, and review the general troubleshooting information.**
DTC: P2270 **2T PCM, MIL: Yes** **Year:** 2011, 2012 **Model:** Accord, Crosstour, Odyssey, Pilot, Ridgeline **Engine:** 3.5L	**Rear Secondary Heated Oxygen Sensor (Secondary HO2S (Bank 1, Sensor 2) Circuit Signal Stuck Lean:** The rear secondary HO2S (bank 1, sensor 2) output voltage is 0.650 V or less. The execution time is once per driving cycle and the duration time is 36.3 seconds or less. **NOTE: Before you troubleshoot, record all freeze data and any on-board snapshot.**
DTC: P2270 **2T PCM, MIL: Yes** **Year:** 2011, 2012 **Model:** Fit **Engine:** 1.5L	**Secondary Heated Oxygen Sensor (Secondary HO2S (Sensor 2)) Circuit Signal Stuck Lean:** The secondary HO2S output voltage is 0.05 V or less for at least 30 seconds.
DTC: P2271 **2T PCM, MIL: Yes** **Year:** 2011, 2012 **Model:** Accord, Civic, Crosstour, CR-V, Element, Fit **Engine:** 1.3L, 1.5L, 1.8L, 2.0L, 2.4L	**Secondary HO2S (Sensor 2) Circuit Signal Stuck Rich :** The secondary HO2S (Sensor 2) output voltage is 0.293 V or more. The execution time is once per driving cycle and the duration time is 24.2 seconds or more. **NOTE: Before you troubleshoot, record all freeze data and any on-board snapshot.**
DTC: P2271 **2T PCM, MIL: Yes** **Year:** 2011, 2012 **Model:** Accord, Crosstour, Odyssey, Pilot, Ridgeline **Engine:** 3.5L	**Rear Secondary HO2S (Bank 1, Sensor 2) Circuit Signal Stuck Rich :** If, after current is applied to the rear secondary HO2S heater, the rear secondary HO2S does not fluctuate and the output is stuck within the specified area, a malfunction is detected. The rear secondary HO2S (bank 1, sensor 2) output voltage is 0.293 V or more. The execution time is once per driving cycle and the duration time is 36.3 seconds or more. **NOTE: Before you troubleshoot, record all freeze data and any on-board snapshot.**
DTC: P2272 **2T PCM, MIL: Yes** **Year:** 2011, 2012 **Model:** Accord, Crosstour, Odyssey, Pilot, Ridgeline **Engine:** 3.5L	**Front Secondary HO2S (Bank 2, Sensor 2) Circuit Signal Stuck Lean :** If, after current is applied to the front secondary HO2S heater, the front secondary HO2S does not fluctuate and the output is stuck within the specified area, a malfunction is detected. The front secondary HO2S (bank 2, sensor 2) output voltage is 0.650 V or less.
DTC: P2273 **2T PCM, MIL: Yes** **Year:** 2011, 2012 **Model:** Odyssey, Pilot, Ridgeline **Engine:** 3.5L	**Front Secondary HO2S (Bank 2, Sensor 2) Circuit Signal Stuck Rich :** If, after current is applied to the front secondary HO2S heater, the front secondary HO2S does not fluctuate and the output is stuck within the specified area, a malfunction is detected. The front secondary HO2S (bank 2, sensor 2) output voltage is 0.293 V or more. **NOTE: Before you troubleshoot, record all freeze data and any on-board snapshot.**
DTC: P2279 **2T PCM, MIL: Yes** **Year:** 2011, 2012 **Model:** Accord, Civic, Crosstour, CR-V, Odyssey, Pilot, Ridgeline **Engine:** 1.3L, 2.4L, 3.5L	**Intake Air System Leak:** With the engine running in closed loop for a minimum of 15 seconds. Either of these 2 conditions is met: (1) The estimated volume of intake air is 310 l/min (327.6 US qt/min, 272.8 Imp qt/min) or more when the MAP value is 35 kPa (10.3 in. Hg, 260 mmHg). (2) The estimated volume of intake air is 302 l/min (319.2 US qt/min, 265.8 Imp qt/min) or more when the MAP value is 62 kPa (18.2 in. Hg, 460 mmHg). The execution time is once per driving cycle and the duration time is 22 seconds or more. **NOTE: If DTC P0443 is stored at the same time as DTC P2279, troubleshoot DTC P0443 first, then recheck for DTC P2279.**
DTC: P2413 **2T PCM, MIL: Yes** **Year:** 2011, 2012 **Model:** Accord, Civic, Crosstour, CR-V, CR-Z, Fit, Insight, Odyssey, Pilot, Ridgeline **Engine:** 1.3L, 1.5L, 1.8L, 2.4L, 3.5L	**Exhaust Gas Recirculation (EGR) System Malfunction:** With the engine running at 4,400 rpm, commanded EGR valve lift is 0.040 in. (1 mm). If the actual valve lift is 0.006 in. (0.15 mm) or less for at least 5 seconds, the valve is considered stuck closed. **NOTE: Before you troubleshoot, record all freeze data and any on-board snapshot.**

DTC	Trouble Code Title and Conditions
DTC: P2422 **2T PCM, MIL: Yes** **Year:** 2011, 2012 **Model:** Civic **Engine:** 1.3L, 1.8L, 2.0L	**Evaporative Emission (EVAP) Canister Vent Shut Valve Stuck:** Malfunction Threshold: The output from the FTP sensor [FTP SENSOR] is -4 kPa (-1.0 inHg, -25 mmHg)*, -2 kPa (-0.4 inHg, -10 mmHg)** or less for at least 1.04*, 8.0** seconds.
DTC: P2422 **2T PCM, MIL: Yes** **Year:** 2011, 2012 **Model:** All **Engine:** 1.3L, 1.5L, 2.4L, 3.5L	**EVAP Canister Vent Shut Valve Close Malfunction :** The output from the fuel tank pressure sensor is -4 kPa (-1.0 in. Hg, -25 mmHg), -2 kPa (-0.4 in. Hg, -10 mmHg) or less for at least 1.04-8 seconds. Elapsed time after the FTP sensor output exceeds the malfunction threshold 1.04 seconds. Excessive negative pressure is detected 8 seconds.
DTC: P2446 **1T PCM, MIL: Yes** **Year:** 2011, 2012 **Model:** Odyssey **Engine:** 3.5L	**Rocker Arm Oil Pressure Switch Circuit Low Voltage (J35A6 Engine):** With the engine running, engine speed a minimum of 4,400 rpm (Variable that is depending on the engine load). The rocker arm oil control solenoid is ON, the rocker arm oil pressure switch remains ON. The execution time is once per driving cycle and DTCs P0522, P0523, P2648, P2649 are not active. **NOTE: Before you troubleshoot, record all freeze data and any on-board snapshot.**
DTC: P2552 **1T PCM, MIL: Yes** **Year:** 2011, 2012 **Model:** CR-V, Element **Engine:** 2.4L	**Throttle Actuator Control Module Relay Malfunction :** The serial signal is input from the throttle actuator control module for at least 2 seconds after the throttle actuator control module relay is turned OFF. **NOTE: Before you troubleshoot, record all freeze data and any on-board snapshot.**
DTC: P2610 **1T PCM, MIL: Yes** **Year:** 2011, 2012 **Model:** All **Engine:** All	**ECM/PCM Ignition Off Internal Timer Malfunction:** The access process to the ignition off timer fails, or a malfunction is found in the read data for at least 10 seconds. Ignition switch ON (II) when a battery is disconnected and connected again is excluded. **NOTE: Before you troubleshoot, record all freeze data and any on-board snapshot.**
DTC: P2627 **1T PCM, MIL: Yes** **Year:** 2011, 2012 **Model:** Pilot, Ridgeline **Engine:** 3.5L	**Rear Air/Fuel Ratio (A/F) Sensor (Bank 1, Sensor 1) LABEL Circuit Low Voltage:** With the engine running at 3,000 rpm for 2 minutes the VLBLB1 is 0.4 V or less for at least 5 seconds.
DTC: P2627 **1T PCM, MIL: Yes** **Year:** 2011, 2012 **Model:** Odyssey **Engine:** 3.5L	**VTEC System Stuck ON (J35A7 Engine):** With the engine running, engine speed between 1,500-3,000 rpm, the rocker arm oil control solenoid is ON, the EOP sensor output remains HIGH (39 kPa 11.6 in. Hg, 293 mmHg) or more. The execution time is once per driving cycle and DTCs P0522, P0523, P2227, P2228, P2229 are not active. **NOTE: Before you troubleshoot, record all freeze data and any on-board snapshot.**
DTC: P2628 **1T PCM, MIL: Yes** **Year:** 2011, 2012 **Model:** Pilot, Ridgeline **Engine:** 3.5L	**Rear Air/Fuel Ratio (A/F) Sensor (Bank 1, Sensor 1) LABEL Circuit High Voltage:** With the engine running at 3,000 rpm for 2 minutes the VLBLB1 is 4.69 V or more for at least 5 seconds.
DTC: P2630 **1T PCM, MIL: Yes** **Year:** 2011, 2012 **Model:** Pilot, Ridgeline **Engine:** 3.5L	**Front Air/Fuel Ratio (A/F) Sensor (Bank 2, Sensor 1) LABEL Circuit Low Voltage:** With the engine running at 3,000 rpm for 2 minutes the VLBLB2 is 0.3 V or less for at least 5 seconds.
DTC: P2631 **1T PCM, MIL: Yes** **Year:** 2011, 2012 **Model:** Pilot, Ridgeline **Engine:** 3.5L	**Front Air/Fuel Ratio (A/F) Sensor (Bank 2, Sensor 1) LABEL Circuit High Voltage:** With the engine running at 3,000 rpm for 2 minutes the VLBLB1 is 4.69 V or more for at least 5 seconds.
DTC: P2646 **1T PCM, MIL: Yes** **Year:** 2011, 2012 **Model:** Accord, Crosstour, CR-V, Element, Fit, Odyssey, Pilot, Ridgeline **Engine:** 1.5L, 2.4L, 3.5L	**Rocker Arm Oil Pressure Switch (VTEC Oil Pressure Switch) Circuit Low Voltage :** With the engine running, engine speed a minimum of 4,800 rpm (Variable that is depending on the engine load). The rocker arm oil control solenoid is ON, the rocker arm oil pressure switch remains ON. The execution time is once per driving cycle and DTCs P2648, P2649 are not active. **NOTE: Before you troubleshoot, record all freeze data and any on-board snapshot.**

DTC	Trouble Code Title and Conditions
DTC: P2646 **1T PCM, MIL: Yes** **Year:** 2011, 2012 **Model:** Insight **Engine:** 1.3L	**Rocker Arm Oil Control Valve A Stuck ON/OFF:** With the engine running at operating temperature, and the vehicle speed is a minimum of 12 mph (18 km/h) or more with the engine speed between 900-3,000 rpm the following conditions occur for 2 seconds or more. Condition 1: Rocker arm oil pressure sensor A is ON Condition 2: Rocker arm oil pressure sensor B is ON Condition 3: Rocker arm oil pressure sensor A output HIGH Condition 4: Rocker arm oil pressure sensor B output LOW
DTC: P2646 **1T PCM, MIL: Yes** **Year:** 2011, 2012 **Model:** Civic **Engine:** 1.3L, 1.8L, 2.0L	**VTEC System Stuck OFF :** With the engine running and at operating temperature, vehicle speed a minimum of 9 mph (14 km/h), the throttle plate anything but wide open, vehicle speed between 1,100-3,300 rpm, the rocker arm oil control solenoid is on and the EOP sensor remains (High).
DTC: P2646 **2T PCM, MIL: Yes** **Year:** 2011, 2012 **Model:** Civic, CR-Z **Engine:** 1.5L	**Rocker Arm Oil Pressure Switch Circuit Low Voltage:** When the rocker arm oil control solenoid is ON, the rocker arm oil pressure switch remains ON.
DTC: P2646 **1T PCM, MIL: Yes** **Year:** 2011, 2012 **Model:** Odyssey **Engine:** 3.5L	**VTEC System Stuck OFF (J35A7 Engine):** With the engine running at idle, battery voltage a minimum of 10.5 V and DTCs P0522, P0523, P2227, P2228 and P2229 not active, the rocker arm oil control solenoid is OFF, the EOP sensor output remains LOW (39 kPa, 11.5 in. Hg, 292 mmHg) or less.
DTC: P2647 **2T PCM, MIL: Yes** **Year:** 2011, 2012 **Model:** Civic **Engine:** 1.3L, 1.8L, 2.0L	**VTEC System Stuck ON :** With the engine running for at least 2.5 seconds and battery voltage a minimum of 10.5 V, the rocker oil control solenoid is OFF and the EOP sensor remains (LOW).
DTC: P2647 **2T PCM, MIL: Yes** **Year:** 2011, 2012 **Model:** Civic, CR-Z **Engine:** 1.5L	**Rocker Arm Oil Pressure Switch Circuit High Voltage:** When the rocker arm oil control solenoid is OFF, the rocker arm oil pressure switch remains OFF.
DTC: P2647 **1T PCM, MIL: Yes** **Year:** 2011, 2012 **Model:** Odyssey **Engine:** 3.5L	**Rocker Arm Oil Pressure Switch Circuit High Voltage (J35A6 Engine):** With the engine running at idle, and battery voltage a minimum of 10.5 V. The rocker arm oil control solenoid is OFF, the rocker arm oil pressure switch remains OFF. The execution time is once per driving cycle and DTCs P0522, P0523, P2648, P2649 are not active. **NOTE: Before you troubleshoot, record all freeze data and any on-board snapshot.**
DTC: P2647 **1T PCM, MIL: Yes** **Year:** 2011, 2012 **Model:** Accord, Crosstour, CR-V, Element, Fit, Pilot, Ridgeline **Engine:** 1.5L, 2.4L, 3.5L	**Rocker Arm Oil Pressure Switch (VTEC Oil Pressure Switch) Circuit High Voltage :** When the rocker arm oil control solenoid is OFF, the rocker arm oil pressure switch remains OFF. (Low lift cam operation) **NOTE: Before you troubleshoot, record all freeze data and any on-board snapshot.**
DTC: P2648 **1T PCM, MIL: Yes** **Year:** 2011, 2012 **Model:** Odyssey **Engine:** 3.5L	**Rocker Arm Oil Control Solenoid Circuit Low Voltage (J35A6 Engine):** With the vehicle driven in a lower gear at 4,400 rpm or more and battery voltage a minimum of 10.5 V, the return signal is OFF (LOW) for at least 2 seconds when the PCM outputs the ON (HIGH) signal to the rocker arm oil control solenoid.
DTC: P2648 **1T PCM, MIL: Yes** **Year:** 2011, 2012 **Model:** Accord, Civic, Crosstour, CR-V, CR-Z, Element, Fit, Insight, Pilot, Ridgeline **Engine:** All	**Rocker Arm Oil Control Solenoid (VTEC Solenoid Valve) Circuit Low Voltage :** The return signal is OFF (low) for at least 2 seconds when the PCM outputs the ON (high) signal to the rocker arm oil control solenoid. DTC P2649 is not active. **NOTE: Before you troubleshoot, record all freeze data and any on-board snapshot.**

DTC	Trouble Code Title and Conditions
DTC: P2648 **1T PCM, MIL: Yes** **Year:** 2011, 2012 **Model:** Accord, Crosstour **Engine:** 2.4L, 3.5L	**Rocker Arm Oil Control Solenoid Circuit Low Voltage:** The return signal is OFF (low) for at least 2.0 seconds when the ECM/PCM outputs the ON (high) signal to the rocker arm oil control solenoid. .
DTC: P2648 **1T PCM, MIL: Yes** **Year:** 2011, 2012 **Model:** Odyssey **Engine:** 3.5L	**Rocker Arm Oil Control Solenoid A (Bank 1) Circuit Low Voltage (J35A7 Engine):** With the vehicle driven in a lower gear and engine speed at 3,300 rpm, battery voltage a minimum of 10.5 V, the return signal is OFF (LOW) for at least 1 second when the PCM outputs the ON (HIGH) signal to the rocker arm oil control solenoid A (Bank 1).
DTC: P2649 **1T PCM, MIL: Yes** **Year:** 2011, 2012 **Model:** Accord, Civic, Crosstour, CR-V, CR-Z, Element, Fit, Insight, Pilot, Ridgeline **Engine:** All	**Rocker Arm Oil Control Solenoid Circuit High Voltage :** The return signal is ON (High) for at least 2 seconds when the PCM outputs the OFF (Low) signal to the rocker arm oil control solenoid. With the engine running the execution time is continuous. DTC P2648 is not active. **NOTE: Before you troubleshoot, record all freeze data and any on-board snapshot.**
DTC: P2649 **1T PCM, MIL: Yes** **Year:** 2011, 2012 **Model:** Odyssey **Engine:** 3.5L	**Rocker Arm Oil Control Solenoid Circuit High Voltage (J35A6 Engine):** With the engine running at idle for at least 7 seconds and battery voltage a minimum of 10.0 V. The return signal is ON (HIGH) for at least 2 seconds when the PCM outputs the OFF (LOW) signal to the rocker arm oil control solenoid. DTC P2648 is not active. **NOTE: Before you troubleshoot, record all freeze data and any on-board snapshot.**
DTC: P2649 **1T PCM, MIL: Yes** **Year:** 2011, 2012 **Model:** Odyssey **Engine:** 3.5L	**Rocker Arm Oil Control Solenoid A (Bank 1) Circuit High Voltage (J35A7 Engine):** The return signal is ON (High) for at least 2 seconds when the PCM outputs the OFF (Low) signal to the rocker arm oil control solenoid. With the engine running the execution time is continuous. DTC P2648 is not active. **NOTE: Before you troubleshoot, record all freeze data and any on-board snapshot.**
DTC: P2651 **1T PCM, MIL: Yes** **Year:** 2011 **Model:** Civic **Engine:** 1.3L	**Valve Pause System (VPS) Stuck OFF:** With the vehicle driven in a low gear at 4,800 rpm or more at around 7 mph for at least 5 seconds, The following conditions occur for at least 5 seconds. * Rocker arm oil control solenoid 1, ON * Rocker arm oil control solenoid 2, OFF * Rocker arm oil pressure switch, ON * Rocker arm oil pressure sensor, High 39 kPa or more
DTC: P2652 **1T PCM, MIL: Yes** **Year:** 2011 **Model:** Civic **Engine:** 1.3L	**Valve Pause System (VPS) Stuck ON:** With the engine running at a speed between 2,000-4,600 rpm in (P or N) for at least 5 seconds, the following conditions occur. * Rocker Arm Oil Control Solenoid 1 is OFF * Rocker Arm Oil Control Solenoid 2 is OFF * Rocker Arm Oil Pressure Switch is OFF * Rocker Arm Oil pressure Sensor Output HIGH 39 kPa or more
DTC: P2653 **2T PCM, MIL: Yes** **Year:** 2011, 2012 **Model:** Accord, Crosstour, Odyssey, Pilot **Engine:** 3.5L	**Rocker Arm Oil Control Solenoid B (Bank 1) Circuit Low Voltage:** With the engine running at about 3,000 rpm for 10 seconds, the return signal is OFF (Low) for at least 1.0 second when the PCM outputs the ON (High) signal to the rocker arm oil control solenoid B (Bank 1). **NOTE: Before you troubleshoot, record all freeze data and any on-board snapshot, and review the general troubleshooting information.**
DTC: P2653 **2T PCM, MIL: Yes** **Year:** 2011, 2012 **Model:** Civic **Engine:** 1.3L, 1.8L, 2.0L	**Rocker Arm Oil Control Solenoid B Circuit Low Voltage:** Malfunction Threshold: The return signal is OFF (low) for at least 2 seconds when the PCM outputs the ON (high) signal to rocker arm oil control solenoid B.
DTC: P2654 **2T PCM, MIL: Yes** **Year:** 2011, 2012 **Model:** Accord, Crosstour, Insight **Engine:** 1.3L, 2.4L	**Rocker Arm Oil Control Solenoid B (Exhaust Valve Side) Circuit High Voltage:** With the engine running at idle for at least 10 seconds, the return signal is ON (High) for at least 2.0 seconds when the PCM outputs the OFF (Low) signal to the rocker arm oil control solenoid B. **NOTE: Before you troubleshoot, record all freeze data and any on-board snapshot, and review the general troubleshooting information.**

DTC	Trouble Code Title and Conditions
DTC: P2654 **1T PCM, MIL: Yes** **Year:** 2011 **Model:** Civic **Engine:** 1.3L	**Rocker Arm Oil Control Solenoid 1 Circuit High Voltage:** With the ignition ON and the battery voltage a minimum of 10.0 V, the return signal is ON (HIGH) for at least 1.1 seconds when the PCM outputs the OFF (LOW) signal to the rocker arm oil control solenoid 1.
DTC: P2654 **2T PCM, MIL: Yes** **Year:** 2011, 2012 **Model:** Accord, Crosstour, Odyssey, Pilot **Engine:** 3.5L	**Rocker Arm Oil Control Solenoid B (Bank 1) Circuit High Voltage:** With the engine running at idle for at least 10 seconds, the return signal is ON (High) for at least 2.0 seconds when the PCM outputs the OFF (Low) signal to the rocker arm oil control solenoid B (Bank 1). **NOTE: Before you troubleshoot, record all freeze data and any on-board snapshot, and review the general troubleshooting information.**
DTC: P2658 **2T PCM, MIL: Yes** **Year:** 2011, 2012 **Model:** Accord, Crosstour, Odyssey, Pilot **Engine:** 3.5L	**Rocker Arm Oil Control Solenoid A (Bank 2) Circuit Low Voltage:** With the engine running at about 3,000 rpm for 10 seconds, the return signal is OFF (Low) for at least 1.0 second when the PCM outputs the ON (High) signal to the rocker arm oil control solenoid A (Bank 2). **NOTE: Before you troubleshoot, record all freeze data and any on-board snapshot, and review the general troubleshooting information.**
DTC: P2659 **2T PCM, MIL: Yes** **Year:** 2011, 2012 **Model:** Accord, Crosstour, Odyssey, Pilot **Engine:** 3.5L	**Rocker Arm Oil Control Solenoid A (Bank 2) Circuit High Voltage:** With the engine running at idle for at least 10 seconds, the return signal is ON (High) for at least 2.0 seconds when the PCM outputs the OFF (Low) signal to the rocker arm oil control solenoid A (Bank 2). **NOTE: Before you troubleshoot, record all freeze data and any on-board snapshot, and review the general troubleshooting information**
DTC: P2714 **2T PCM, MIL: Yes, TCIL: Yes** **Year:** 2011, 2012 **Model:** Odyssey **Engine:** 3.5L	**A/T Clutch Pressure Control Solenoid Valve D Stuck OFF:** Shift lever position D, S, R, with ATF the temperature between -4 ºF (-20 ºC), 248 ºF (120 ºC) improper gear ratio output has been detected for the shift schedule. The execution time is continuous and the duration time is between 3-10 seconds or more.
DTC: P2720 **2T PCM, MIL: Yes, TCIL: Yes** **Year:** 2011, 2012 **Model:** Odyssey **Engine:** 3.5L	**Problem in A/T Clutch Pressure Control Solenoid Valve D Circuit:** The measured current for the PCM's command value is less than 0.19 with duty cylce at 56.5-89%, with duty cycle at 89% or more it is Less than 0.27. **NOTE: Before you troubleshoot, record all freeze data and any on-board snapshot.This code is caused by an electrical circuit problem and cannot be caused by a mechanical problem in the transmission.**
DTC: P2721 **2T PCM, MIL: Yes, TCIL: Yes** **Year:** 2011, 2012 **Model:** Odyssey **Engine:** 3.5L	**Problem in A/T Clutch Pressure Control Solenoid Valve D:** The measured current for the PCM's command value is more than 0.6 duty cycle less than 13.7%. More than 0.9 with duty cycle at 13.7-27%. (High Imput) **NOTE: Before you troubleshoot, record all freeze data and any on-board snapshot. This code is caused by an electrical circuit problem and cannot be caused by a mechanical problem in the transmission.**
DTC: P2821 **2T PCM, MIL: Yes, TCIL: Yes** **Year:** 2011, 2012 **Model:** Odyssey **Engine:** 3.5L	**Line Pressure Solenoid Valve A Stuck ON:** The input signal from the line pressure switch to the PCM is high during the self shut down mode after the signal is low when the normal line pressure mode was commanded. **NOTE: Before you troubleshoot, record all freeze data and any on-board snapshot. When DTC P2821 is stored in the PCM, engine torque is restricted because of the fail-safe function.**
DTC: P2821 **2T PCM, MIL: Yes** **Year:** 2011, 2012 **Model:** Civic **Engine:** 1.3L, 1.8L	**Line Pressure Solenoid Valve A Stuck ON:** The clutch speed difference between the speed input to the clutch and the speed output from the clutch is 5 rpm or more for at least 2 seconds.
DTC: P2826 **2T PCM, MIL: Yes** **Year:** 2011, 2012 **Model:** Civic **Engine:** 1.3L, 1.8L	**Line Pressure Solenoid Valve A Low Input:** The return signal does not match the command to turn on line pressure solenoid valve A for at least 1 second.
DTC: P2826 **2T PCM, TCIL: Yes** **Year:** 2011, 2012 **Model:** Odyssey **Engine:** 3.5L	**Short in Line Pressure Solenoid Valve A Circuit:** The return signal does not match the command to turn on line pressure solenoid valve A for at least 1 second. The circuit detects malfunctions such as a circuit short or open. **NOTE: Before you troubleshoot, record all freeze data and any on-board snapshot. This code is caused by an electrical circuit problem and cannot be caused by a mechanical problem in the transmission.**

DTC	Trouble Code Title and Conditions
DTC: P2827 **2T PCM, MIL: Yes** **Year:** 2011, 2012 **Model:** Civic **Engine:** 1.3L, 1.8L, 2.0L	**Line Pressure Solenoid Valve A High Input:** The return signal does not match the command to turn off line pressure solenoid valve A for at least 1 second.
DTC: P2827 **2T PCM, TCIL: Yes** **Year:** 2011, 2012 **Model:** Odyssey **Engine:** 3.5L	**Open in Line Pressure Solenoid Valve A Circuit:** The return signal does not match the command to turn off line pressure solenoid valve A for at least 1 second. **NOTE: Before you troubleshoot, record all freeze data and any on-board snapshot. This code is caused by an electrical circuit problem and cannot be caused by a mechanical problem in the transmission.**
DTC: P2A00 **2T PCM, MIL: Yes** **Year:** 2011, 2012 **Model:** Accord, Civic, Crosstour, CR-V, CR-Z, Element, Fit, Insight **Engine:** 1.3L, 1.5L, 1.8L, 2.0L, 2.4L	**Air/Fuel Ratio (A/F) Sensor (Sensor 1) Circuit Range/Performance Problem:** The A/F sensor (Sensor 1) output voltage is 3.0 V or less, or 4.8 V or more. The execution time is once per driving cycle and the duration time is 3.5 seconds or more. **NOTE: Before you troubleshoot, record all freeze data and any on-board snapshot.**
DTC: P2A00 **2T PCM, MIL: Yes** **Year:** 2011, 2012 **Model:** Accord, Crosstour, Odyssey, Pilot, Ridgeline **Engine:** 3.5L	**Rear Air/Fuel Ratio (A/F) Sensor (Bank 1, Sensor 1) Circuit Range/Performance Problem:** The rear A/F sensor (bank 1, sensor 1) output voltage is 2.55 V or less, or 4.50 V or more. The execution time is once per driving cycle and the duration time is 5.3 seconds or more. **NOTE: Before you troubleshoot, record all freeze data and any on-board snapshot.**
DTC: P2A03 **2T PCM, MIL: Yes** **Year:** 2011, 2012 **Model:** Accord, Crosstour, Odyssey, Pilot, Ridgeline **Engine:** 3.5L	**Front Air/Fuel Ratio (A/F) Sensor (Bank 2, Sensor 1) Circuit Range/Performance Problem:** The front A/F sensor (bank 2, sensor 1) output voltage is 2.55 V or less, or 4.50 V or more. The execution time is once per driving cycle and the duration time is 5.3 seconds or more.

Commonly Used Abbreviations

2

2WD	Two Wheel Drive

4

4WD	Four Wheel Drive

A

A/C	Air Conditioning
ABDC	After Bottom Dead Center
ABS	Anti-lock Brakes
AC	Alternating Current
ACL	Air cleaner
ACT	Air Charge Temperature
AIR	Secondary Air Injection
ALCL	Assembly Line Communications Link
ALDL	Assembly Line Diagnostic Link
AT	Automatic Transaxle/Transmission
ATDC	After Top Dead Center
ATF	Automatic Transmission Fluid
ATS	Air Temperature Sensor
AWD	All Wheel Drive

B

BAP	Barometric Absolute Pressure
BARO	Barometric Pressure
BBDC	Before Bottom Dead Center
BCM	Body Control Module
BDC	Bottom Dead Center
BPT	Backpressure Transducer
BTDC	Before Top Dead Center
BVSV	Bimetallic Vacuum Switching Valve

C

CAC	Charge Air Cooler
CARB	California Air Resources Board
CAT	Catalytic Converter
CCC	Computer Command Control
CCCC	Computer Controlled Catalytic Converter
CCCI	Computer Controlled Coil Ignition
CCD	Computer Controlled Dwell
CDI	Capacitor Discharge Ignition
CEC	Computerized Engine Control
CFI	Continuous Fuel Injection
CIS	Continuous Injection System
CIS-E	Continuous Injection System - Electronic
CKP	Crankshaft Position
CL	Closed Loop
CMP	Camshaft Position
CPP	Clutch Pedal Position
CTOX	Continuous Trap Oxidizer System
CTP	Closed Throttle Position
CVC	Constant Vacuum Control
CYL	Cylinder

D

DBC	Dual Bed Catalyst
DC	Direct Current
DFI	Direct Fuel Injection
DIS	Distributorless Ignition System
DLC	Data Link Connector
DMM	Digital Multimeter
DOHC	Double Overhead Camshaft
DRB	Diagnostic Readout Box
DTC	Diagnostic Trouble Code
DTM	Diagnostic Test Mode
DVOM	Digital Volt/Ohmmeter

E

EBCM	Electronic Brake Control Module
ECM	Engine Control Module
ECT	Engine Coolant Temperature
ECU	Engine Control Unit or Electronic Control Unit
EDIS	Electronic Distributorless Ignition System
EEC	Electronic Engine Control
EEPROM	Electrically Erasable Programmable Read Only Memory
EFE	Early Fuel Evaporation
EGR	Exhaust Gas Recirculation
EGRT	Exhaust Gas Recirculation Temperature
EGRVC	EGR Valve Control
EPROM	Erasable Programmable Read Only Memory
EVAP	Evaporative Emissions
EVP	EGR Valve Position

F

FBC	Feedback Carburetor
FEEPROM	Flash Electrically Erasable Programmable Read Only Memory
FF	Flexible Fuel
FI	Fuel Injection
FT	Fuel Trim
FWD	Front Wheel Drive

G

GND	Ground

H

HAC	High Altitude Compensation
HEGO	Heated Exhaust Gas Oxygen sensor
HEI	High Energy Ignition
HO2 Sensor	Heated Oxygen Sensor

I

IAC	Idle Air Control
IAT	Intake Air Temperature
ICM	Ignition Control Module
IFI	Indirect Fuel Injection
IFS	Inertia Fuel Shutoff
ISC	Idle Speed Control
IVSV	Idle Vacuum Switching Valve

Commonly Used Abbreviations

K

KOEO	Key On, Engine Off
KOER	Key ON, Engine Running
KS	Knock Sensor

M

MAF	Mass Air Flow
MAP	Manifold Absolute Pressure
MAT	Manifold Air Temperature
MC	Mixture Control
MDP	Manifold Differential Pressure
MFI	Multiport Fuel Injection
MIL	Malfunction Indicator Lamp or Maintenance
MST	Manifold Surface Temperature
MVZ	Manifold Vacuum Zone

N

NVRAM	Nonvolatile Random Access Memory

O

O2 Sensor	Oxygen Sensor
OBD	On-Board Diagnostic
OC	Oxidation Catalyst
OHC	Overhead Camshaft
OL	Open Loop

P

P/S	Power Steering
PAIR	Pulsed Secondary Air Injection
PCM	Powertrain Control Module
PCS	Purge Control Solenoid
PCV	Positive Crankcase Ventilation
PIP	Profile Ignition Pick-up
PNP	Park/Neutral Position
PROM	Programmable Read Only Memory
PSP	Power Steering Pressure
PTO	Power Take-Off
PTOX	Periodic Trap Oxidizer System

R

RABS	Rear Anti-lock Brake System
RAM	Random Access Memory
ROM	Read Only Memory
RPM	Revolutions Per Minute
RWAL	Rear Wheel Anti-lock Brakes
RWD	Rear Wheel Drive

S

SBC	Single Bed Converter
SBEC	Single Board Engine Controller
SC	Supercharger
SCB	Supercharger Bypass
SFI	Sequential Multiport Fuel Injection
SIR	Supplemental Inflatable Restraint
SOHC	Single Overhead Camshaft
SPL	Smoke Puff Limiter
SPOUT	Spark Output
SRI	Service Reminder Indicator
SRS	Supplemental Restraint System
SRT	System Readiness Test
SSI	Solid State Ignition
ST	Scan Tool
STO	Self-Test Output

T

TAC	Thermostatic Air Cleaner
TBI	Throttle Body Fuel Injection
TC	Turbocharger
TCC	Torque Converter Clutch
TCM	Transmission Control Module
TDC	Top Dead Center
TFI	Thick Film Ignition
TP	Throttle Position
TR Sensor	Transaxle/Transmission Range Sensor
TVV	Thermal Vacuum Valve
TWC	Three-way Catalytic Converter

V

VAF	Volume Air Flow, or Vane Air Flow
VAPS	Variable Assist Power Steering
VRV	Vacuum Regulator Valve
VSS	Vehicle Speed Sensor
VSV	Vacuum Switching Valve

W

WOT	Wide Open Throttle
WU-TWC	Warm Up Three-way Catalytic Converter

ENGLISH TO METRIC CONVERSION: TORQUE

To convert foot-pounds (ft. lbs.) to Newton-meters (Nm), multiply the number of ft. lbs. by 1.36
To convert Newton-meters (Nm) to foot-pounds (ft. lbs.), multiply the number of Nm by 0.7376

ft. lbs.	Nm	ft. lbs.	Nm	ft. lbs.	Nm	ft. lbs.	Nm
0.1	0.1	34	46.2	76	103.4	118	160.5
0.2	0.3	35	47.6	77	104.7	119	161.8
0.3	0.4	36	49.0	78	106.1	120	163.2
0.4	0.5	37	50.3	79	107.4	121	164.6
0.5	0.7	38	51.7	80	108.8	122	165.9
0.6	0.8	39	53.0	81	110.2	123	167.3
0.7	1.0	40	54.4	82	111.5	124	168.6
0.8	1.1	41	55.8	83	112.9	125	170.0
0.9	1.2	42	57.1	84	114.2	126	171.4
1	1.4	43	58.5	85	115.6	127	172.7
2	2.7	44	59.8	86	117.0	128	174.1
3	4.1	45	61.2	87	118.3	129	175.4
4	5.4	46	62.6	88	119.7	130	176.8
5	6.8	47	63.9	89	121.0	131	178.2
6	8.2	48	65.3	90	122.4	132	179.5
7	9.5	49	66.6	91	123.8	133	180.9
8	10.9	50	68.0	92	125.1	134	182.2
9	12.2	51	69.4	93	126.5	135	183.6
10	13.6	52	70.7	94	127.8	136	185.0
11	15.0	53	72.1	95	129.2	137	186.3
12	16.3	54	73.4	96	130.6	138	187.7
13	17.7	55	74.8	97	131.9	139	189.0
14	19.0	56	76.2	98	133.3	140	190.4
15	20.4	57	77.5	99	134.6	141	191.8
16	21.8	58	78.9	100	136.0	142	193.1
17	23.1	59	80.2	101	137.4	143	194.5
18	24.5	60	81.6	102	138.7	144	195.8
19	25.8	61	83.0	103	140.1	145	197.2
20	27.2	62	84.3	104	141.4	146	198.6
21	28.6	63	85.7	105	142.8	147	199.9
22	29.9	64	87.0	106	144.2	148	201.3
23	31.3	65	88.4	107	145.5	149	202.6
24	32.6	66	89.8	108	146.9	150	204.0
25	34.0	67	91.1	109	148.2	151	205.4
26	35.4	68	92.5	110	149.6	152	206.7
27	36.7	69	93.8	111	151.0	153	208.1
28	38.1	70	95.2	112	152.3	154	209.4
29	39.4	71	96.6	113	153.7	155	210.8
30	40.8	72	97.9	114	155.0	156	212.2
31	42.2	73	99.3	115	156.4	157	213.5
32	43.5	74	100.6	116	157.8	158	214.9
33	44.9	75	102.0	117	159.1	159	216.2

METRIC TO ENGLISH CONVERSION: TORQUE

To convert foot-pounds (ft. lbs.) to Newton-meters (Nm), multiply the number of ft. lbs. by 1.36
To convert Newton-meters (Nm) to foot-pounds (ft. lbs.), multiply the number of Nm by 0.7376

Nm	ft. lbs.	Nm	ft. lbs.	Nm	ft. lbs.	Nm	ft. lbs.	Nm	ft. lbs.
0.1	0.1	34	25.0	76	55.9	118	86.8	160	117.6
0.2	0.1	35	25.7	77	56.6	119	87.5	161	118.4
0.3	0.2	36	26.5	78	57.4	120	88.2	162	119.1
0.4	0.3	37	27.2	79	58.1	121	89.0	163	119.9
0.5	0.4	38	27.9	80	58.8	122	89.7	164	120.6
0.6	0.4	39	28.7	81	59.6	123	90.4	165	121.3
0.7	0.5	40	29.4	82	60.3	124	91.2	166	122.1
0.8	0.6	41	30.1	83	61.0	125	91.9	167	122.8
0.9	0.7	42	30.9	84	61.8	126	92.6	168	123.5
1	0.7	43	31.6	85	62.5	127	93.4	169	124.3
2	1.5	44	32.4	86	63.2	128	94.1	170	125.0
3	2.2	45	33.1	87	64.0	129	94.9	171	125.7
4	2.9	46	33.8	88	64.7	130	95.6	172	126.5
5	3.7	47	34.6	89	65.4	131	96.3	173	127.2
6	4.4	48	35.3	90	66.2	132	97.1	174	127.9
7	5.1	49	36.0	91	66.9	133	97.8	175	128.7
8	5.9	50	36.8	92	67.6	134	98.5	176	129.4
9	6.6	51	37.5	93	68.4	135	99.3	177	130.1
10	7.4	52	38.2	94	69.1	136	100.0	178	130.9
11	8.1	53	39.0	95	69.9	137	100.7	179	131.6
12	8.8	54	39.7	96	70.6	138	101.5	180	132.4
13	9.6	55	40.4	97	71.3	139	102.2	181	133.1
14	10.3	56	41.2	98	72.1	140	102.9	182	133.8
15	11.0	57	41.9	99	72.8	141	103.7	183	134.6
16	11.8	58	42.6	100	73.5	142	104.4	184	135.3
17	12.5	59	43.4	101	74.3	143	105.1	185	136.0
18	13.2	60	44.1	102	75.0	144	105.9	186	136.8
19	14.0	61	44.9	103	75.7	145	106.6	187	137.5
20	14.7	62	45.6	104	76.5	146	107.4	188	138.2
21	15.4	63	46.3	105	77.2	147	108.1	189	139.0
22	16.2	64	47.1	106	77.9	148	108.8	190	139.7
23	16.9	65	47.8	107	78.7	149	109.6	191	140.4
24	17.6	66	48.5	108	79.4	150	110.3	192	141.2
25	18.4	67	49.3	109	80.1	151	111.0	193	141.9
26	19.1	68	50.0	110	80.9	152	111.8	194	142.6
27	19.9	69	50.7	111	81.6	153	112.5	195	143.4
28	20.6	70	51.5	112	82.4	154	113.2	196	144.1
29	21.3	71	52.2	113	83.1	155	114.0	197	144.9
30	22.1	72	52.9	114	83.8	156	114.7	198	145.6
31	22.8	73	53.7	115	84.6	157	115.4	199	146.3
32	23.5	74	54.4	116	85.3	158	116.2	200	147.1
33	24.3	75	55.1	117	86.0	159	116.9	201	147.8

ENGLISH/METRIC CONVERSION: TEMPERATURE

To convert Fahrenheit (F°) to Celsius (C°), take F° temperature and subtract 32, multiply the result by 5 and divide the result by 9
To convert Celsius (C°) to Fahrenheit (F°), take C° temperature and multiply it by 9, divide the result by 5 and add 32

F°	C°	F°	C°	C°	F°	C°	F°
-40	-40.0	150	65.6	-38	-36.4	46	114.8
-35	-37.2	155	68.3	-36	-32.8	48	118.4
-30	-34.4	160	71.1	-34	-29.2	50	122
-25	-31.7	165	73.9	-32	-25.6	52	125.6
-20	-28.9	170	76.7	-30	-22	54	129.2
-15	-26.1	175	79.4	-28	-18.4	56	132.8
-10	-23.3	180	82.2	-26	-14.8	58	136.4
-5	-20.6	185	85.0	-24	-11.2	60	140
0	-17.8	190	87.8	-22	-7.6	62	143.6
1	-17.2	195	90.6	-20	-4	64	147.2
2	-16.7	200	93.3	-18	-0.4	66	150.8
3	-16.1	205	96.1	-16	3.2	68	154.4
4	-15.6	210	98.9	-14	6.8	70	158
5	-15.0	212	100.0	-12	10.4	72	161.6
10	-12.2	215	101.7	-10	14	74	165.2
15	-9.4	220	104.4	-8	17.6	76	168.8
20	-6.7	225	107.2	-6	21.2	78	172.4
25	-3.9	230	110.0	-4	24.8	80	176
30	-1.1	235	112.8	-2	28.4	82	179.6
35	1.7	240	115.6	0	32	84	183.2
40	4.4	245	118.3	2	35.6	86	186.8
45	7.2	250	121.1	4	39.2	88	190.4
50	10.0	255	123.9	6	42.8	90	194
55	12.8	260	126.7	8	46.4	92	197.6
60	15.6	265	129.4	10	50	94	201.2
65	18.3	270	132.2	12	53.6	96	204.8
70	21.1	275	135.0	14	57.2	98	208.4
75	23.9	280	137.8	16	60.8	100	212
80	26.7	285	140.6	18	64.4	102	215.6
85	29.4	290	143.3	20	68	104	219.2
90	32.2	295	146.1	22	71.6	106	222.8
95	35.0	300	148.9	24	75.2	108	226.4
100	37.8	305	151.7	26	78.8	110	230
105	40.6	310	154.4	28	82.4	112	233.6
110	43.3	315	157.2	30	86	114	237.2
115	46.1	320	160.0	32	89.6	116	240.8
120	48.9	325	162.8	34	93.2	118	244.4
125	51.7	330	165.6	36	96.8	120	248
130	54.4	335	168.3	38	100.4	122	251.6
135	57.2	340	171.1	40	104	124	255.2
140	60.0	345	173.9	42	107.6	126	258.8
145	62.8	350	176.7	44	111.2	128	262.4

LENGTH CONVERSION

To convert inches (in.) to millimeters (mm), multiply the number of inches by 25.4
To convert millimeters (mm) to inches (in.), multiply the number of millimeters by 0.04

Inches	Millimeters	Inches	Millimeters	Inches	Millimeters	Inches	Millimeters
0.0001	0.00254	0.005	0.1270	0.09	2.286	4	101.6
0.0002	0.00508	0.006	0.1524	0.1	2.54	5	127.0
0.0003	0.00762	0.007	0.1778	0.2	5.08	6	152.4
0.0004	0.01016	0.008	0.2032	0.3	7.62	7	177.8
0.0005	0.01270	0.009	0.2286	0.4	10.16	8	203.2
0.0006	0.01524	0.01	0.254	0.5	12.70	9	228.6
0.0007	0.01778	0.02	0.508	0.6	15.24	10	254.0
0.0008	0.02032	0.03	0.762	0.7	17.78	11	279.4
0.0009	0.02286	0.04	1.016	0.8	20.32	12	304.8
0.001	0.0254	0.05	1.270	0.9	22.86	13	330.2
0.002	0.0508	0.06	1.524	1	25.4	14	355.6
0.003	0.0762	0.07	1.778	2	50.8	15	381.0
0.004	0.1016	0.08	2.032	3	76.2	16	406.4

ENGLISH/METRIC CONVERSION: LENGTH

To convert inches (in.) to millimeters (mm), multiply the number of inches by 25.4
To convert millimeters (mm) to inches (in.), multiply the number of millimeters by 0.04

Inches Fraction	Inches Decimal	Millimeters Decimal	Inches Fraction	Inches Decimal	Millimeters Decimal	Inches Fraction	Inches Decimal	Millimeters Decimal
1/64	0.016	0.397	11/32	0.344	8.731	11/16	0.688	17.463
1/32	0.031	0.794	23/64	0.359	9.128	45/64	0.703	17.859
3/64	0.047	1.191	3/8	0.375	9.525	23/32	0.719	18.256
1/16	0.063	1.588	25/64	0.391	9.922	47/64	0.734	18.653
5/64	0.078	1.984	13/32	0.406	10.319	3/4	0.750	19.050
3/32	0.094	2.381	27/64	0.422	10.716	49/64	0.766	19.447
7/64	0.109	2.778	7/16	0.438	11.113	25/32	0.781	19.844
1/8	0.125	3.175	29/64	0.453	11.509	51/64	0.797	20.241
9/64	0.141	3.572	15/32	0.469	11.906	13/16	0.813	20.638
5/32	0.156	3.969	31/64	0.484	12.303	53/64	0.828	21.034
11/64	0.172	4.366	1/2	0.500	12.700	27/32	0.844	21.431
3/16	0.188	4.763	33/64	0.516	13.097	55/64	0.859	21.828
13/64	0.203	5.159	17/32	0.531	13.494	7/8	0.875	22.225
7/32	0.219	5.556	35/64	0.547	13.891	57/64	0.891	22.622
15/64	0.234	5.953	9/16	0.563	14.288	29/32	0.906	23.019
1/4	0.250	6.350	37/64	0.578	14.684	59/64	0.922	23.416
17/64	0.266	6.747	19/32	0.594	15.081	15/16	0.938	23.813
9/32	0.281	7.144	39/64	0.609	15.478	61/64	0.953	24.209
19/64	0.297	7.541	5/8	0.625	15.875	31/32	0.969	24.606
5/16	0.313	7.938	41/64	0.641	16.272	63/64	0.984	25.003
21/64	0.328	8.334	21/32	0.656	16.669	1/1	1.000	25.400
			43/64	0.672	17.066			

CHILTON® LABOR GUIDE

Chilton's labor times are so trusted, even a competing publisher uses them!

The *Chilton 2013 Labor Guide* features new models and new labor operations in order to stay current with new technologies. Labor times have also been refined for normal and severe maintenance schedules, if applicable. The 2013 edition Labor Guide Manuals provide repair times for 1990-current import and domestic vehicles and the 2013 edition Labor Guide CD-ROM provides repair times for 1981-current import and domestic vehicles. Chilton's editors consider warranty times, component locations, component type, the environment in which technicians work, the training they receive, and the tools they use when calculating a labor time. To allow for vehicle age, operating conditions, and type of service, the *Chilton 2013 Labor Guide* provides standard and severe service times, plus OEM warranty times. Vehicle makes and models conform to current Automotive Aftermarket Industry Association (AAIA) standards.

Chilton 2013 Labor Guide Manual Set (Domestic & Import) ISBN 978-1-2851-9293-2, Part No. 209293

Chilton 2013 Labor Guide CD-ROM (Domestic & Import) ISBN 978-1-2851-9296-3, Part No. 209296

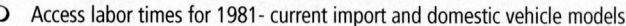

CD-ROM FEATURES

- ○ Access labor times for 1981- current import and domestic vehicle models.
- ○ Save time with automatically calculated labor charges, taxes, & parts as total job is estimated.
- ○ Create professional estimates for your customer and worksheets for your technicians, printing them whenever needed.
- ○ Keep track of customers, prior estimates, and your own parts or package jobs with less paper.
- ○ Choose part names for estimates from an industry standard database to reduce typing.
- ○ Estimate and track your work status with improved forms.
- ○ Communicate easily with customers using re-designed printouts which show all labor and parts in an easy-to-read format.
- ○ Simplify adding parts to your estimate or work order with a helpful parts list.
- ○ Locate information quick with a keyword search engine.
- ○ Quickly locate work requests by day, week and month using the calendar feature.

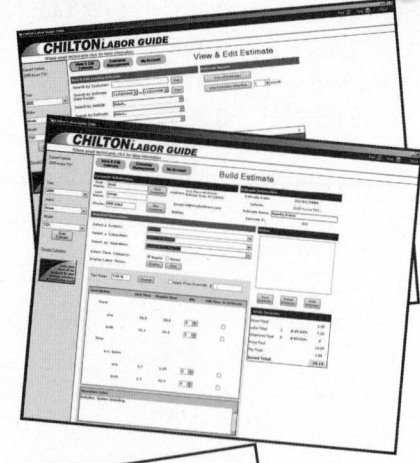

Manual FEATURES

- ○ Nearly 4,000 pages of updated Chilton labor times split into two volumes includes vehicle information from 1990 to current models.
- ○ Trusted by more service professionals than any other labor guide.
- ○ Less flipping though pages with separate domestic and imported vehicle manuals and more specific vehicle groups.
- ○ Convenient tabs display contents by manufacturer and model.
- ○ Easy-to-find manufacturers are arranged alphabetically within each volume.
- ○ Search using two-indexes - labor operations and systems - in each model group.
- ○ Page numbers include manufacturer code so you know where you are in the book.

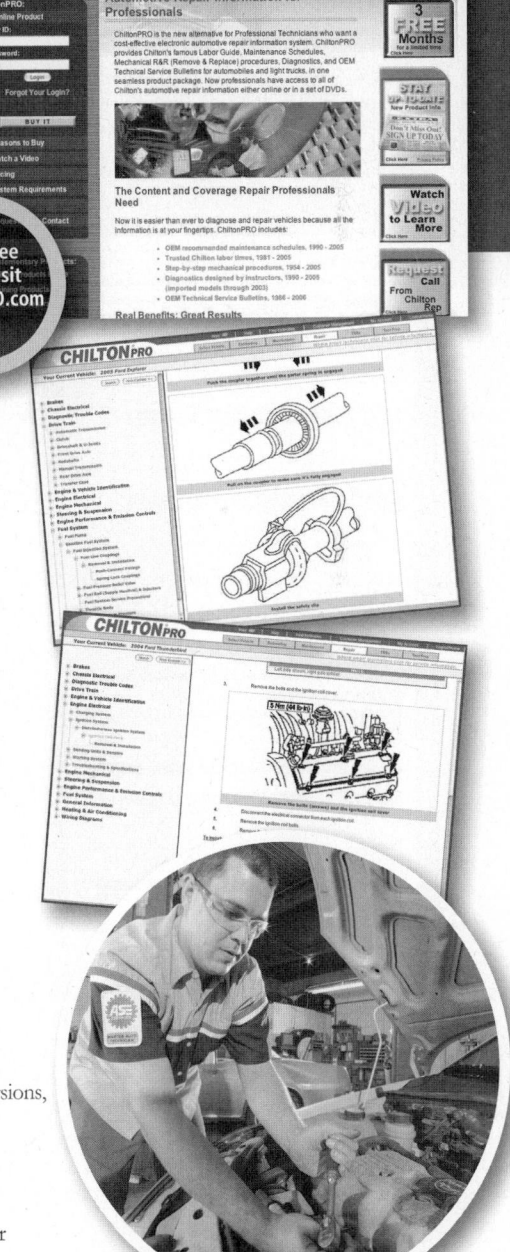

Proudly Partnered With:

CHILTON® ESTIMATING

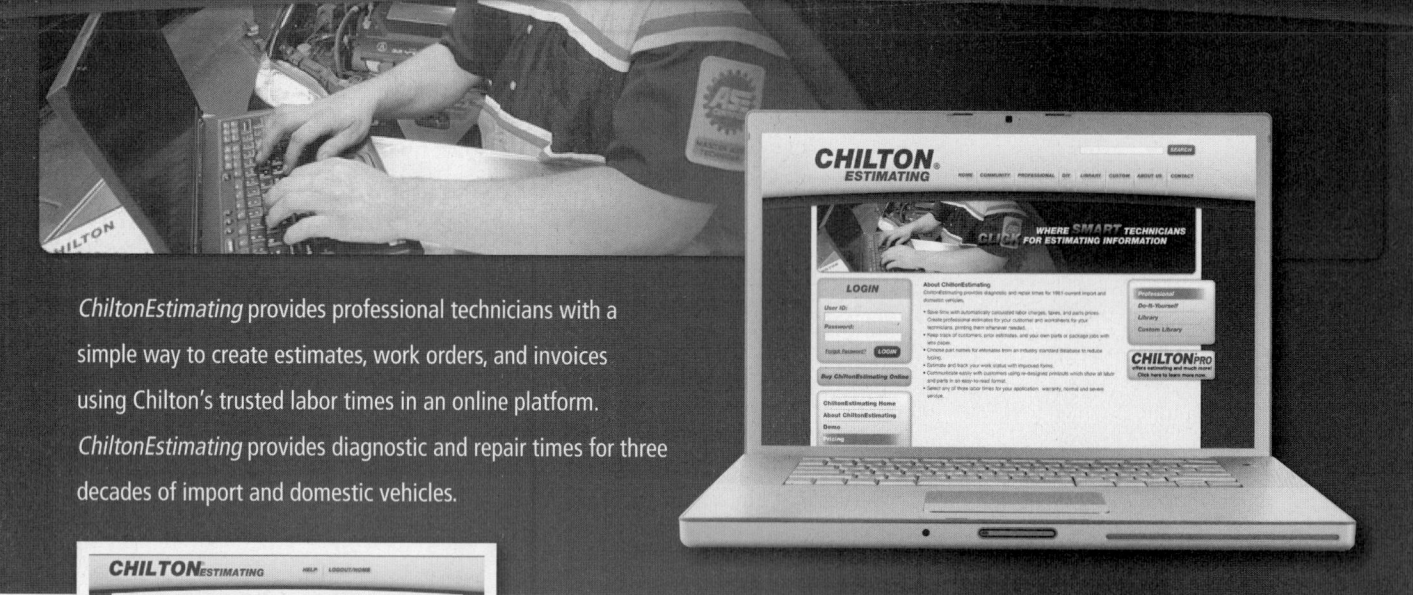

ChiltonEstimating provides professional technicians with a simple way to create estimates, work orders, and invoices using Chilton's trusted labor times in an online platform. *ChiltonEstimating* provides diagnostic and repair times for three decades of import and domestic vehicles.

○ Access up-to-date information immediately. *ChiltonEstimating* is continuously updated!

○ Enjoy a hassle-free product with nothing to download and nothing to install.

○ Never fret over lost or damaged software or books again.

○ Secure your valuable customer data on our server, which won't be lost if your computer crashes.

○ Easily access the program from any web-enabled computer.

○ Work on more than one job at a time using *ChiltonEstimating's* two shop-user accounts.

○ Download all customer contact information easily for marketing purposes.

○ Cancel your subscription at any time by going to the "My Account" tab. No contract or obligation required. Customer data will be available to download for up to six months after a subscription has expired.

○ Save time with automatically calculated labor charges, taxes, and parts prices. Create professional estimates for your customer and worksheets for your technicians, printing them whenever needed.

○ Keep track of customers, prior estimates, and your own parts or package jobs with less paper.

○ Choose part names for estimates from an industry standard database to reduce typing.

○ Estimate and track your work status with improved forms.

○ Communicate easily with customers using re-designed printouts which show all labor and parts in an easy-to-read format.

○ Select any of three labor times for your application: warranty, normal and severe service.

System Requirements:
Web browser
- Internet Explorer 7.0 or above (recommended)
- Firefox 3.6 or 4 or Safari
- High-speed internet connection
- Adobe Flash Player
- Adobe Reader
- Windows XP or above